CASSELL
POCKET
ENGLISH
DICTIONARY

CASSELL POCKET ENGLISH DICTIONARY

CASSELL

Cassell
Wellington House, 125 Strand
London WC2R 0BB, England

Cassell's English Dictionary first published 1891
Last complete revision 1962
Cassell Pocket English Dictionary first published 1990
This edition first published 1995

© Cassell 1995
Gazetteer and atlas © Geddes & Grosset Ltd 1995

All rights reserved.
No part of this publication may be reproduced
or transmitted in any form or by any means,
electronic or mechanical including photocopying,
recording or any information storage or retrieval system,
without prior permission in writing
from the publishers.

British Library Cataloguing in Publication Data
A catalogue entry for this book is
available from the British Library.

ISBN 0-304-34639-X

Printed and bound in Finland

CONTENTS

vi	Acknowledgments
vii	How to use the *Cassell Pocket English Dictionary*
viii	Chief Abbreviations
xii	Guide to Pronunciation
xiii	Symbols
xv	**Introduction**
xv	The History of the English Language
xix	A Summary of Spelling Rules
xxi	Words Commonly Misspelt
xxiii	Common Prefixes and Suffixes
1	**The Pocket English Dictionary**
969	Appendix I: REGISTER OF NEW WORDS
985	Appendix II: GAZETTEER AND INDEX TO MAPS
	Atlas

ACKNOWLEDGMENTS

Managing Editor (*First edition*)	Betty Kirkpatrick
Lexicographers	Callum Brines
	Steve Curtis
	Carolyn Donaldson
	Elaine Higgleton
	Pandora Kerr Frost
	Fergus McGauran
	Margaret McPhee
	Katherine Seed
Pronunciations Consultant	Jock Graham
New Words Editor	Jane Garden
New Words Consultants	Pandora Kerr Frost
	Jonathon Green
	Fiona McKenzie
	David Pickering
	Adrian Stenton

HOW TO USE THE
CASSELL POCKET ENGLISH DICTIONARY

The entry

Each entry in the dictionary begins with an entry-word or headword in bold type. This is immediately followed by the pronunciation, the relevant part of speech and the meaning/meanings.

Arrangement of entries

By no means all words defined in the dictionary are headwords. Many words and expressions which are derived from the same root have been grouped or 'nested' together, e.g. **execrable** is under **execrate**. This has the great merit not only of demonstrating at a glance the relationship of the words but of acting as a significant space-saving device. The system allows many more words to be included in the dictionary than would otherwise be the case.

The majority of such words are easy to find since their positions in the dictionary are very close alphabetically to what they would have been if they had been entered as separate headwords. Where this is not the case cross-references have been added for facility of use, e.g. **elision** is cross-referred to **elide**.

Organization of entries

Most headwords have more than one meaning and more than one word derived from them. The words and expressions derived from headwords fall into three categories – idioms/phrases, compounds and direct derivatives.

Idioms consist of phrases including the headword, e.g. **to gain on** or compound words not beginning with the headword, e.g. **old gold**. They are placed immediately after the last meaning of the last part of speech of the headword.

Compounds, which consist of two elements beginning with the headword, are placed immediately after the last meaning of the last idiom. The compound word may be hyphenated, e.g. **cross-bow**, two words, e.g. **emergency landing** or one word, e.g. **eyesight**, according to convention.

Direct derivatives are words formed from the root of the headword or its stem by adding a suffix, e.g. *-ness*, *-ly* etc., e.g. **gauntness**, **ghostly**. These are placed after the last meaning of the last compound.

Derivatives which themselves are derived from derivatives of the headword follow on from the words in the entry from which they are derived. Thus **endless band** follows the direct derivative **endless**.

Labels

Labels in round brackets have been added where necessary. They are divided into two categories – stylistic labels, such as (*offensive*), (*sl.*), (*coll.*) etc., and field labels, such as (*Med.*), (*Comput.*) etc. A list of abbreviations of labels appears under *Chief Abbreviations* (p.viii).

Cross-references

The word cross-referred to appears in small caps, e.g. **enure** INURE.

CHIEF ABBREVIATIONS

All are given here in roman, though most of them may also appear in italics as labels.

a.	adjective	Boh.	Bohemian
abbr.	abbreviation	Bot.	Botany
abl.	ablative	Braz.	Brazilian
Abor.	Aboriginal, Aborigines	Bret.	Breton
		Build.	Building
acc.	accusative; according	Bulg.	Bulgarian
		Byz.	Byzantine
adapt.	adaptation		
adv.	adverb	c.	circa, about
A-F	Anglo-French	Camb.	Cambridge
Afr.	African	Campan.	Campanology
aft.	afterwards	Can.	Canada, Canadian
Agric.	Agriculture	Carib.	Caribbean
Alch.	Alchemy	Carp.	Carpentry
Alg.	Algebra	Cat.	Catalan
alln.	allusion	Celt.	Celtic
alt.	alternative	Ceram.	Ceramics
Am. Ind.	American Indian	Ch.	Church
anal.	analogous	Chem.	Chemistry
Anat.	Anatomy	Chin.	Chinese
Ang.-Ind.	Anglo-Indian	Civ. Eng.	Civil Engineering
Ang.-Ir.	Anglo-Irish	Class.	Classical
Ang.-Lat.	Anglo-Latin	Coal-min.	Coal-mining
appar.	apparently	cogn.	cognate
Arab.	Arabic	coll.	colloquial; collateral
Aram.	Aramaic	collect.	collective
Arch.	Architecture	comb.	combination
Archaeol.	Archaeology	comb. form.	combining form
Arith.	Arithmetic	Comm.	Commerce
Art.	Artistic	comp.	comparative
Artill.	Artillery	Comput.	Computing
assim.	assimilated, assimilation	Conch.	Conchology
		cond.	conditional
Assyr.	Assyrian	conf.	confusion
Astrol.	Astrology	conj.	conjunction
Astron.	Astronomy	conn.	connected
attrib.	attribute, attributive	contr.	contraction
augm.	augmentative	Cook.	Cooking
Austral.	Australian	Copt.	Coptic
Austr.-Hung.	Austro-Hungarian	Corn.	Cornish
aux.v.	auxiliary verb	corr.	corruption; corresponding
Aviat.	Aviation		
		Cosmog.	Cosmogony
b.	born	cp.	compare
Bibl.	Bible, biblical	Cryst.	Crystallography
Bibliog.	Bibliography		
Biol.	Biology	d.	died

Dan.	Danish	Geog.	Geography
dat.	dative	Geol.	Geology
def.	definition	Geom.	Geometry
deriv.	derivation	ger.	gerund, gerundive
derog.	derogatory	Goth.	Gothic
dial.	dialect	Gr.	Greek
dim.	diminutive	grad.	gradually
Diplom.	Diplomatics	Gram.	Grammar
dist.	distinct, distinguished		
Dut.	Dutch	Heb.	Hebrew
Dynam.	Dynamics	Her.	Heraldry
		Hind.	Hindi
E	East, Eastern	Hist.	History
Eccles.	Ecclesiastical	Hort.	Horticulture
Econ.	Economics	Hung.	Hungarian
EFris	East Frisian	Hydrostat.	Hydrostatics
e.g.	exempli gratia, for example	Hyg.	Hygiene
Egypt.	Egyptian	Icel.	Icelandic
Egyptol.	Egyptology	Ichthyol.	Ichthyology
EInd.	East Indian	ident.	identical; identified
Elec.	Electricity	i.e.	id est, that is
ellipt.	elliptical, elliptically	imag.	imaginary
Embryol.	Embryology	imit.	imitative
emphat.	emphatic	imper.	imperative
Eng.	English; Engineering	impers.	impersonal
Ent.	Entomology	incept.	inceptive
erron.	erroneously	incorr.	incorrectly
esp.	especially	Ind.	India, Indian
Ethn.	Ethnology	ind.	indicative
euphem.	euphemistic	indef. art.	indefinite article
Eur.	European	Indo-Port.	Indo-Portuguese
Exam.	Examination	inf.	infinitive
exc.	except	influ.	influenced
		inst.	instinctive
F	French	instr.	instrumental
f.	feminine	int.	interjection
facet.	facetiously	intens.	intensive
fem.	feminine	Internat.	International
Feud.	Feudal	interrog.	interrogative
fig.	figuratively	intr.	intransitive
fl.	floruit, flourished	Ir.	Irish
Flem.	Flemish	iron.	ironical
foll.	the following	irreg.	irregular
For.	Foreign	It.	Italian
Fort.	Fortification		
freq.	frequentative	Jap.	Japanese
Fris.	Frisian	Jav.	Javanese
fut.	future	Jewel.	Jewellery
G	German	L	Latin
Gael.	Gaelic	lat.	latitude
gen.	genitive	LG	Low German
Geneal.	Genealogy	Lit.	Literature, literary

lit.	literal, literally	OFris.	Old Frisian
Lit. crit.	Literary criticism	OHG	Old High German
Lith.	Lithuanian	OLG	Old Low German
loc.	locative	ON	Old Norse
Log.	Logic	ONF	Old Norman French
		onomat.	onomatopoeic
m.	masculine	OPers.	Old Persian
Mach.	Machinery	opp.	opposed, opposition
Manufact.	manufacturing	Opt.	Optics
Math.	Mathematics	orig.	origin, originally
MDan.	Middle Danish	Ornith.	Ornithology
MDut.	Middle Dutch	OS	Old Saxon
ME	Middle English	o.s.	old style
Mech.	Mechanics	OSlav.	Old Slavonic
Med.	Medicine	OSp.	Old Spanish
med.	mediaeval	OTeut.	Old Teutonic
Merc.	Mercian		
Metal.	Metallurgy	Palaeont.	Palaeontology
Metaph.	Metaphysics	paral.	parallel
Meteor.	Meteorology	Parl.	Parliamentary
Mex.	Mexican	part.	participle, participial
MF	Middle French	pass.	passive
MG	Middle German	Path.	Pathology
Microsc.	Microscopy	perf.	perfect
Mil.	Military	perh.	perhaps
Min.	Mineralogy	Pers.	Persian
mistrans.	mistranslation	pers.	person; personal
mod.	modern	Peruv.	Peruvian
Mus.	Music	Petrol.	Petrology
Myth.	Mythology	Phil.	Philosophy
		Philol.	Philology
N	North	Phoen.	Phoenician
n.	noun	phon.	phonetics; phonology
N Am.	North American	Phot.	Photography
Nat. Hist.	Natural History	phr.	phrase
Naut.	Nautical	Phys.	Physics
Nav.	Naval	Phys. Sci.	Physical Science
neg.	negative	pl.	plural
neol.	neologism	poet.	poetry, poetical
neut.	neuter	Pol.	Polish
Newsp.	Newspaper	Polit.	Political
nom.	nominative	pop.	popular, popularly
Norm.	Norman	Port.	Portuguese
North.	Northern	poss.	possessive
Northum.	Northumbrian	p.p.	past participle
Norw.	Norwegian	prec.	the preceding
NT	New Testament	pred.	predicative
Numis.	Numismatics	pref.	prefix
		prep.	preposition
obj.	objective	pres.	present
obs.	obsolete	pres.p.	present participle
OED	the Oxford English Dictionary	pret.	preterite
		prev.	previously
OF	Old French	Print.	Printing

priv.	privative	Surg.	Surgery
prob.	probably	Swed.	Swedish
pron.	pronoun; pronounced	syl.	syllable
prop.	proper, properly	Syr.	Syriac
Pros.	Prosody		
Prov.	Provencal	Teleg.	Telegraphy
prov.	provincial	Teut.	Teutonic
Psych.	Psychology	Theat.	Theatre
pubd.	published	Theol.	Theology
		Therap.	Therapeutics
Radiol.	Radiology	Therm.	Thermionics
redupl.	reduplicate	tr.	transitive
ref.	referring, reference	trans.	translation
reflex.	reflexive	Trig.	Trigonometry
rel.	related	Turk.	Turkish
Relig.	Religion	TV	Television
rel. pron.	relative pronoun		
remonstr.	remonstrative	ult.	ultimately
Rhet.	Rhetoric	Univ.	University
Rom.	Roman; Romance	US	United States of America
Rus.	Russian		
		usu.	usually
S	South		
Sansk.	Sanskrit	v.	verb
Sc.	Scottish	var.	variant
Scand.	Scandinavian	Venet.	Venetian
Sci.	Science	verb.a.	verbal adjective
Sculp.	Sculpture	Vet.	Veterinary Surgery
Semit.	Semitic	v.i.	verb intransitive
Serb.	Serbian	viz.	videlicet, namely
Shak.	Shakespeare	voc.	vocative
Sic.	Sicilian	v.r.	verb reflexive
sing.	singular	v.t.	verb transitive
sl.	slang		
Slav.	Slavonic	W	West; Welsh
Sp.	Spanish	WG	West German
Spens.	Spenser	WInd.	West Indian
Stock. Exch.	Stock Exchange	wr.	written
subj.	subjunctive		
suf.	suffix	Zool.	Zoology
superl.	superlative		

GUIDE TO PRONUNCIATION

Introduction

The revised respelling scheme used for pronunciations in this edition of the dictionary has been designed to provide as good a compromise as possible between accuracy and understanding by the majority of users. Therefore, as few specialized phonetic symbols and additional accents or marks on letters have been used as will fulfil this aim. A full list of symbols/letters and their equivalents follows below, with transcriptions given alongside the words used as examples.

As in the previous edition, the particular variety of pronunciation aimed for is that of the 'ordinary educated English speaker', which some readers will no doubt recognize under the labels of 'Oxford' or 'BBC' English, or 'Received Pronunciation'.

Where sub-headwords differ in pronunciation (and this includes stress) from the headword, partial or full pronunciations are also given for these; where partials appear, it should be assumed that the remaining (untranscribed) part of the word concerned is pronounced as before.

In longer entries, there may be more than one variety of difference in pronunciation from the headword. In such cases, any subhead *not* given a transcription should be assumed to revert to the pronunciation pattern of the headword. The exception to this is derivatives of the subhead which closely follow the subhead and usually have minimal difference from it in form; for example, under **drama** (drah′mə), the subhead **dramatic, -ical** has the partial (-mat′-), and is followed by **dramatically**, in which case **dramatically** follows **dramatic, -ical** in pattern and *not* **drama**.

It can also be seen from this illustration that derivatives formed by adding suffixes which are consistently pronounced, are assumed to be known by the reader (eg **-ly, -ness** etc.) and the pronunciation of such suffixes is only given in rare instances of possible confusion.

Further, cases where the only change in the subhead is one of stress position, *and where this change is consistently predictable*, are not given pronunciations; an example would be the suffix **-ation**, where the sound and stress pattern are always (-ā′shən).

Stress

Stress (′) is shown in pronunciations immediately *after* the syllable which is stressed, e.g. (tī′gə) = **tiger**. Stress is *not* given on compounds composed of two or more separate words, nor on idioms.

American English

In a *very* limited number of cases where a North American English pronunciation of a word has become widespread also in British English (e.g. **schedule**), the variant is given with the label *esp. N Am*.

SYMBOLS

Vowel sounds

ah	far	(fah)	o	not	(not)
a	fat	(fat)	ō	note	(nōt)
ā	fate	(fāt)		sower	(sō'ə)
aw	fall	(fawl)	oo	blue	(bloo)
	north	(nawth)	ŭ	sun	(sŭn)
	paw	(paw)	u	foot	(fut)
	soar	(saw)		bull	(bul)
e	bell	(bel)	ū	muse	(mūz)
ē	beef	(bēf)	ə	again	(əgen')
œ	her	(hœ)		current	(kŭ'rənt)
	fur	(fœ)		sailor	(sā'lə)
i	bit	(bit)		publicity	(pəblis'iti)
ī	bite	(bīt)			

Note: the neutral sound of many unstressed vowels is represented, as shown above, by the symbol ə; some unstressed vowels in this dictionary are (more correctly) transcribed as (-i-), as in (ilek'trik).

Consonants

p	pit	(pit)	s	sit	(sit)
b	bit	(bit)	v	van	(van)
t	tin	(tin)	w	win	(win)
d	dance	(dahns)	y	yet	(yet)
k	kit	(kit)	z	haze	(hāz)
m	man	(man)	ng	sing	(sing)
n	nut	(nŭt)	th	thin	(thin)
l	lid	(lid)	dh	this	(dhis)
f	fit	(fit)	sh	ship	(ship)
h	hit	(hit)	zh	measure	(mezh'ə)
g	get	(get)	kh	loch	(lokh)
j	just	(jŭst)	ch	church	(chœch)
r	run	(rŭn)			

Note: where a sound represented by two consonants, e.g. (-ng-) is followed by another syllable which begins with the second consonant (-g-) and where the stress mark falls elsewhere, a centred dot is used to show where the syllable break occurs, e.g. as in (ling·gwis'tiks).

Foreign words

r' macabre (məkahbr″)
l' honorable (onorahbl″)
y' merveille (mervāy″)

Diphthongs

(i) Vowel sounds incorporating the final unpronounced 'r' of standard British English:

eə	fair	(feə)
	mare	(meə)
	mayor	(meə)
iə	fear	(fıə)
	seer	(siə)
īə	fire	(fīə)
ūə	pure	(pūə)
uə	poor	(puə)

(ii) Others:

ow	bout	(bowt)
	cow	(kow)
oi	join	(join)

Foreign vowels not dealt with by the main system

(i) Nasalized:

ã	(ãsyen′)	a̱ncienne
ẽ	(ẽfã′)	e̱nfant
ĩ	(lĩfam′)	(écraser) l'i̱nfâme
õ	(kõ′zhã)	(co̱ngé)
ũ	(verdũ′)	(Verdu̱n)

(ii) Other:

ü	(ẽtẽdü′))	entendu̱
	(ü′bə)	ü̱ber

Proprietary terms

This book includes some words which are or are asserted to be proprietary names. The presence or absence of such assertions should not be regarded as affecting the legal status of any proprietary name or trade mark.

Introduction
THE HISTORY OF THE ENGLISH LANGUAGE

1. **An examination of the European** and some Asian languages has shown that they can be divided into several groups, the members of which resemble one another because they were derived from the same original tongue. Thus English, with German, Dutch, Norwegian, Danish etc., belongs to the *Germanic* group of languages. All of these tongues were developed from a primitive language spoken in prehistoric times by the early Germanic tribes. Similarly, French, Italian, Spanish, Portuguese etc. (called *Romance* languages, because they are derived from the speech of the Romans) are the offspring of Latin, which was once of the Italic family; Irish, Welsh, Scots, Gaelic, Manx and Breton belong to the Celtic group; while Russian, Polish, Serbo-Croat etc. belong to the Slavonic group. Not only do the various members of any *one* of these groups of languages exhibit strong resemblances one to another, but members of *different* groups also show signs of kinship: European languages even show likenesses to the languages of India and Persia. These facts have been accounted for by assuming that there existed thousands of years ago a primitive language called *Indo-European*, which was the common origin of the various groups described above.

2. **The relationship of the various languages** belonging to the Indo-European family is shown by the following genealogical tree:

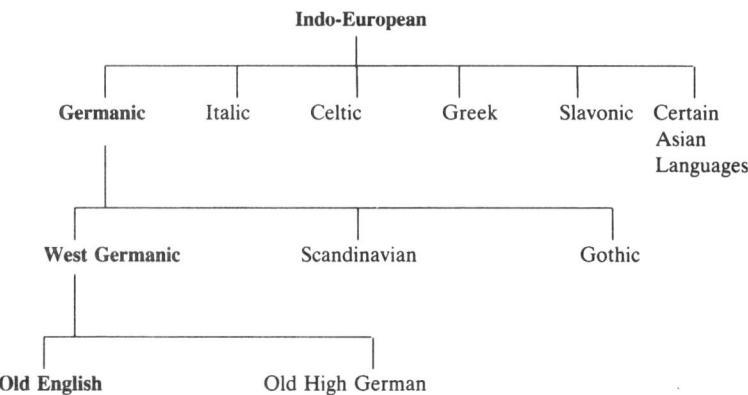

3. **The English language** was brought to the British Isles by the Germanic tribes (Angles, Saxons and Jutes) who settled here in the fifth and sixth centuries. Old English (i.e. the language spoken in England before the Norman Conquest) differed greatly from our present language in pronunciation, vocabulary and grammar: in fact, a passage of English written in the time of King Alfred is unintelligible to a modern reader. Old English, like Latin and Greek, had a complicated system of *inflections*: that is to say, nouns, pronouns, adjectives, verbs etc. had many different forms according to their grammatical relationship. The vocabulary of the language was almost purely Germanic: very little borrowing from other languages had taken place.

4. **During the last thousand years** English has undergone great changes. Large additions to vocabulary have been made by groups of new settlers (esp. Danes and Normans) and through literary and scientific borrowing from Latin and Greek. The development of English vocabulary is looked at in more detail below.

5. **Latin influence** on English vocabulary.

(a) Latin began to influence our language before the Anglo-Saxons arrived in England. The Germanic tribes were in contact with the outposts of Roman civilization, and borrowed a few words, which are still to be found in the different branches of the Germanic groups of languages: e.g.

street (Lat. *strata via*): cheese (Lat. *caseum*); *mint* (O.E. *mynet*, Lat. *moneta*).

(b) When the English tribes came to settle in this country, they came in contact with a people (the Britons) that had for long been part of the Roman Empire. It is probable that the educated population of the British towns spoke a form of Latin. Certainly a large number of Latin words were in use among them, and some of these words passed into the language of the new conquerors. Latin borrowings of this period are distinguished by their form, because the Latin spoken in Britain had undergone considerable modification: e.g.

Chester (Lat. *castra*); *cowl* (O.E. *cugela*, Lat. *cucullus*); *provost* (O.E. *prafost*, Lat. *praepositus*).

(c) In the sixth century, Christianity was reintroduced into this country by Roman missionaries. As the new religion spread, the English language adopted a large number of Latin words to express new ideas connected with the faith: e.g.

Pope (Lat. *papa*), *martyr, mass, monk*.

(d) In later times, especially since the Revival of Learning in the sixteenth century (which led to wide study of Latin and Greek literature), Latin words have frequently been borrowed from the literary language. Such words have undergone little change in form in passing from the one language to the other.

6. **Scandinavian influence** on English vocabulary. From the end of the eighth century to the time of Alfred, the Danes made continual raids upon the English coasts. Mercia and the southern part of Northumbria were invaded and settled by them, and the invaders, who, after overrunning Wessex, were defeated by Alfred, retired to East Anglia and settled there. A century later the Danish king Svein invaded England, and eventually his son Canute (Cnut) became King of the English. These extensive Danish settlements were not without their influence on the language. Many words were borrowed by the English, such as *skin, skill, ill, get, leg, Thursday*, and the forms of the third person pronouns beginning with *th* (*they, their, them*).

7. **French influence** on English vocabulary. It should be noted that French influence is indirectly Latin influence, since the French language is derived from Latin, and so a very large proportion of English vocabulary is, either directly or indirectly, of Latin origin.

(a) After the Norman Conquest the French language, as spoken by the Normans, was the tongue of the ruling classes in this country, and was also used largely by Englishmen. Moreover, from the thirteenth to the fifteenth centuries much French literature was translated into English. A large number of French words were thus incorporated in our language. Classified examples are:

i. Words for the flesh of animals used for food – *beef, mutton, veal, pork.* (The names of the living animals – *ox, sheep, calf, pig* – are English.)
ii. Words connected with the household – *master, servant, dinner, banquet.*
iii. Words connected with law, government and property – *court, assize, prison, custom, rent, price.*
iv. Names of titles – *duke, marquis, viscount, baron.*
v. Military terms – *battle, siege, standard, fortress.*
vi. Words for the remoter relationships – *uncle, aunt, nephew, niece, cousin.*

(b) The Norman conquerors of English spoke the French dialect of Normandy and Picardy. When the Angevin dynasty came to the throne of England in the twelfth century, the dialect of central France became the language of the Court, and the incorporation of French words in English continued. Some words were borrowed twice, first from one dialect and then from the other:

catch – chase; warden – guardian; wage – gage.

(c) In the reign of Charles II there was close intimacy between the English and the French courts, and a knowledge of French language and literature was fashionable in England. Many French words thus passed into English, and the process has continued ever since that time:

campaign, memoir, prestige.

8. **Greek influence** on English vocabulary. The Greek element in the English language is chiefly of modern origin, and is used mainly to express scientific ideas. New words from this source are constantly being introduced because it is very easy to coin words from Greek roots:

telephone, ballistic, paediatrics, macrobiotic, biotechnology.

9. **The effect of the mixed nature of the English vocabulary.** As an instrument of expression, the English language has been enormously improved by its borrowings from other tongues. It surpasses most other languages in its wealth of synonymous words and hence in its power of drawing precise and subtle distinctions. Very often we have a choice between a native English word and a synonym of Latin or French origin:

almighty – omnipotent; blessing – benediction; bloom – flower; calling – vocation; manly – virile; womanly – feminine.

In the course of centuries, many of these originally equivalent terms have acquired slightly divergent meanings, and our means of expression have consequently been increased.

10. **The alphabet**. The sounds of spoken language are represented in writing by means of symbols known as the letters of the alphabet. In a perfect alphabet, every letter would be a phonetic symbol representing one sound and one only, and each sound would have its appropriate symbol. Judged by this standard, the English alphabet is obviously defective. We have not enough symbols to represent all the sounds, and hence:

(a) The same symbol may represent many different sounds: e.g.

'a' in *rat, tall, many, mane, want, bare.*
'o' in *hot, woman, whose, hero, son.*

(b) The same sound may be represented by various symbols: e.g.

hit, nymph, busy, women, sieve. (All these words contain the vowel sound of *hit.*)

fate, champagne, pail, vein, they, reign, gauge, dahlia, steak. (All these words contain the diphthongal sound of *fate.*)

On the other hand, some letters are superfluous: e.g.

'q' (*qu* might equally well be written *kw*), 'x' (=*ks* or *gz*), 'c' (=*k* or *s*).

In pointing out the deficiencies of the English alphabet, we are really calling attention to the fact that modern English spelling is not phonetic; that is, it does not accurately and consistently represent the sounds of speech. The spelling of Old English was very nearly phonetic. How is it, then, that the spelling of today is so defective as a symbolic representation of the spoken language?

11. **The history of English spelling.** The answer is, briefly, that modern spelling was fixed in the fifteenth century, and, so far as it represents any pronunciation at all, it represents the pronunciation of that century. Before that time the scribes had observed no uniformity in the matter of spelling, but when printing was invented and books began to multiply, it was found necessary to adhere to some definite system. Thus the early printers produced a system of spelling which has persisted, with few changes, ever since. When it is added that English pronunciation has undergone many and far-reaching changes since Caxton's time, one reason for the lack of correspondence between the written word and the spoken sound will become clear. The symbol *gh* for instance, which is now silent in *sought, bought* etc., and which, in *laugh, enough* etc., has the sound of 'f', originally represented one and the same sound in all these words. Again, the symbols *ee* in *seed* or *ei* in *receive* in the fifteenth century represented a different sound from that of *ea* in *bead*, though these three symbols now all represent the same sound.

A further reason for the chaotic state of modern English spelling is to be found in the fact that even as early as the fifteenth century there were many anomalies, due largely to French scribes who had introduced symbols from their own language to represent English sounds. This explains the use of *c* for *s* in *city, mice* etc., *gu* for *g* in *guest, guess* etc., *qu* for OE *cw* in *quick, queen* etc., and *ou* or *ow* for the diphthongal sound in *house, cow* etc.

Further confusion resulted from attempts to make the spelling of certain words indicate their etymology. The Norman-French words *dette* and *doute*, for example, retained this spelling when they were first introduced. They were later written *debt* and *doubt* in order to show their connection with the Latin *debitum* and *dubitum*. The *b* has never been pronounced.

A SUMMARY OF SPELLING RULES

1. Plurals of nouns:

(a) In most words add -s: cat, cats.
(b) In words ending in s, x, z, ch, sh (i.e. the sibilants) add -es: bosses, taxes, topazes, churches, bushes, brushes.
(c) After f preceded by a long vowel change f to v: e.g. thief, thieves; wolf, wolves. But *not* after oo: e.g. roofs, hoofs, proofs.
(d) y becomes ies when preceded by a consonant: e.g. lady, ladies; penny, pennies. But y becomes ys when preceded by a vowel: e.g. boy, boys; donkey, donkeys; holiday, holidays.
(e) Words ending in o take -es: e.g. potatoes, tomatoes, heroes, echoes.
But a few take just -s: e.g. pianos, dynamos, cantos, sopranos.
(f) Some words have the same form in the singular and plural: sheep, deer, grouse, swine, salmon, trout etc.
(g) Some words have a plural form but take a singular verb: news, mathematics, statistics, means, politics, innings. e.g. The news is bad; he played a good innings; politics is in his blood.

2. The doubling of the final consonant of some words can be decided by the position of the stress:

(a) Words ending in VOWEL plus CONSONANT double the final consonant when the stress falls on the end of the word
e.g. transfér, transférred or transférring; begín, begínning, begínner; occúr, occúrring, occúrred.
(b) All words of one syllable ending in VOWEL plus CONSONANT observe this rule, since the stress is bound to fall on them:
e.g. rot, rotten, rotter, rotting; stop, stopper, stopping, stopped; sob, sobbing, sobbed.
(c) In words where the stress does not fall on the final VOWEL PLUS CONSONANT, do not double the final consonant:
e.g. rívet, ríveted, ríveting; díffer, díffered, díffering.
Note, however, the exception: worship worshipped, worshipping.

3. When the last syllable ends in a VOWEL PLUS L the final 'l' is doubled, wherever the stress:
travel, traveller, travelling, travelled; level, levelling, levelled; control, controlling, controlled, controller.
Note the exception: paralleled.

4. Words ending in mute (silent) e:

(a) In words ending in e, drop the e coming before a vowel:
e.g. give, giving; love, lovable.
But retain the e before a consonant:
e.g. care, careful; love, lovely.
Note exceptions, true, truly; whole, wholly; nine, ninth; argue, argument.

(b) Retain the e in words ending in -ce or -ge before a, o, u:
e.g. change, changeable; courage, courageous;
peace, peaceable; notice, noticeable.

5. Words beginning or ending in -all, -full, -fill, or beginning in well-, drop one -l when combined with another word:

e.g. almost, already, faithful, doubtful, skilful, fulfil, welcome.

6. Words spelt with -ie with the sound *ee* (*see*). In words spelt with *ie* sounded as *ee* in 'see', put i before e except after c:

e.g. belief, chief, *but* receive, ceiling.

Note, however: seize, weird, weir, counterfeit. In words like *ancient, either, their, foreign*, the sound is not like *ee*.

WORDS COMMONLY MISSPELT

absence
accident
accommodation
achieve
acknowledgment (or -ledgement)
acquiesce
address
adviser
advisory
aggravate
agreeable
albeit
all right
already
altogether (*totally*)
ancillary
anxiety
apologize (*or* -ise)
argument
ascend
ascendancy (*or* -ency)
ascension
atrophy
auxiliary
awful

bachelor
beginning
beige
believe
benefit, benefited
bureau
bureaucracy
business

calendar
cancel, cancelled
cannon (*gun*)
canon (*law, church*)
category
ceiling
chaos
college
commitment
committee

comparison
competence (*competency*)
complement (*completion*)
compliment (*flattery*)
conscientious
consciousness
correspondence
courteous
courtesy
credible
currant (*fruit*)
current (*electric*)

deceit
defer, deferred
definite
desperate
differ, differed
difficult
dignitary
disastrous
discipline
dispatch (despatch)
disseminate
doubt
draft (*rough note*)
draught (*drink*)

eighth
elixir
embarrass
equip, equipped
especially
essential
exhaustion
existence
extrovert (extravert)

familiar
favour
favourable
February
fictitious
financial

financier
forty
fulfil, fulfilled

gauge
grammar
gramophone
guardian
guerrilla
guild

harass
height
honorary
humorous
humour
hungrily
hurriedly
hygiene
hypothesis

immediately
immigrant
independent
influential
install
instil, instilled
intelligible
interfere
introvert

jeopardy
jewel, jewelled
jewellery (*or* jewelry)
judgment (*or* judgement)

kaleidoscope
knack
knobbly
knowledgeable

laborious
labour
laundry
legible

leisurely
level, levelled
livelihood
luscious

manoeuvre
marriage
medicine
Mediterranean
meridional
miniature
minutia (*pl* -iae)
moustache
movable
movement
myxomatosis

necessary
niece
nought
nutritious

occasionally
occur, occurred
offer, offered
omit, omitted
omission

parallel
paralleled
parliament
planning
pleasant
practice (*noun*)
(to) practise (*verb*)

preceding
principal (*chief*)
principle (*rule*)
privilege
proceeding
professor
proficient
pronunciation
proprietary
proprietor
psychiatrist
pusillanimous

queue
quiet
quotient

recommend
refer, referred
repetition
replaceable
replacement
resuscitation
rheumatism
rhinoceros
rhododendron

sanctuary
scarcely
scholastic
seize
separate
siege
signatory
sincerely
skilful

successful
summary
supersede
synonym

tendentious
tenterhooks
thoroughfare
transfer, transferred
truly
twelfth

unconscionable
unctuous
unmistakable
unparalleled
until
unwieldy

vaccination
valuable
vigour
vigorous

Wednesday
wholly
woollen
worship, worshipped
wreathe (*verb*)
wring

yeoman

zoological
zoology

COMMON PREFIXES AND SUFFIXES

Prefixes

a-, an- not, without *asexual*.
aero- aircraft *aerodrome*; air *aerobic*.
agro- agriculture *agrobiology*.
ambi- two, both *ambidextrous*.
Anglo- English, British *Anglo-Catholic*.
ante- before *antenatal*.
anthropo- human being *anthropology*.
anti- against, opposite, opposing *anti-establishment*.
arch- chief, highest *archbishop*.
astro- star *astrophysics*.
audio- sound, hearing *audiovisual*.
auto- self, oneself *autobiography*; automatic *autopilot*; automotive, automobile *autocross*.
bi- two, twice *bicycle*.
biblio- book *bibliophile*.
bio- life, living material *biology*.
by(e)- secondary *by-product*.
cardi(o)- heart *cardiovascular*.
centi- hundred *centipede*; hundredth part *centimetre*.
chron(o)- time *chronology*.
circum- round *circumnavigate*.
co- together *cooperate*.
con-, col-, com-, cor- with, together *conjoin*.
contra- against *contraception*.
counter- against, opposite *counter-clockwise*; matching *counterpart*.
crypto- secret, hidden *cryptofascist*.
cyclo- circle *cyclorama*.
de- to do the opposite *decentralize*; to remove *deseed*; to make less *devalue*.
deca- ten *decathlon*.
deci- tenth part *decilitre*.
demi- half, part *demigod*.
derm- skin *dermatitis*.
di- two, double *dioxide*.
dis- opposite, not *dishonest*; to remove *dismember*.
electro- electricity, electrical *electromagnetic*.
en-, em- to cause to be *enlarge*; to put in or on *endanger*.
equi- equal *equidistant*.
Eur(o)- European *Eurasian*; European Community *Eurocurrency*.
ex- former *ex-president*.
extra- beyond, outside *extra-terrestrial*.
fore- before *foretell*; front, front part *foreleg*.
geo- earth *geography*.
gyn(o)-, gynaec(o)- female, woman *gynaecology*.
haem(o)- blood *haemorrhage*.
hemi- half *hemisphere*.
hetero- different, other *heterosexual*.
hex(a)- six *hexagram*.
hol(o)- complete, whole *holistic*.
homo- same, alike *homogeneous*.
hydro- water *hydroelectric*; hydrogen *hydrochloric*.
hyper- more than normal, excessive *hyperactive*.
hypo- less than normal, low, too low *hypothermia*.
in-, il-, im-, ir- not *insensitive*.
infra- below, beneath *infrared*.
inter- between *intercity*.
intra- in, within *intravenous*.
iso- equal, uniform, the same *isobar*.
kilo- thousand *kilogram*.
macro- large, large-scale *macroclimate*.
mal- bad, badly *malnutrition*.
matri- mother *matriarch*.
mega- million *megaton*; large, extremely large *megalith*.
micro- small, small-scale *microcomputer*.
milli- thousandth part *millilitre*.
mini- small *mini-skirt*.
mis- bad, badly *misbehave*; not *mistrust*.

mono- single, one *monoplane*.
multi- many, several *multiracial*.
neo- new, recent *neo-Georgian*.
neuro- nerve, nervous system *neuroscience*.
non- not *nonsmoker*.
oct-, octa-, octo- eight *octopus*.
omni- all *omnivore*.
osteo- bone *osteoarthritis*.
out- beyond, exceeding, surpassing *outlive*; forth *outpouring*; outside, external *outlying*.
over- above *overlord*; outer *overcoat*; too much *overeat*.
paed(i)(o)- child, *paediatrician*.
palaeo- old, archaic, early *Palaeolithic*.
pan- all *pan-American*.
para- beside *parallel*; beyond *paranormal*; abnormal *paranoia*; resembling *paratyphoid*; associated, supplementary *paramedical*.
patri- father *patriarch*.
pent(a)- five *pentagon*.
photo- light *photosensitive*; photography *photocopy*.
physi(o)- nature, living things *physiology*; physical *physiotherapy*.
poly- many *polyglot*.
post- after *postdate*.
pre- before *prehistoric*.
pro- favouring *pro-American*; substitute for *pro-consul*.
prot(o)- first, original *prototype*.
pseud(o)- false *pseudonym*.
psych(o)- mind *psychoanalysis*.
quadr(i)- four *quadruped*.
quasi- partly, seemingly *quasi-judicial*.
radio- radiation *radiology*; radioactive, radioactivity *radioisotope*.
re- again *rewrite*.
retro- back, backwards *retrogress*.
self- oneself, itself *self-discipline*.
semi- half *semicircle*; partly *semi-conscious*.
socio- social, society *sociology*.
step- related by remarriage *stepsister*.
sub- below, under *subsoil*; less than, incompletely *subhuman*; subordinate, subdivision *subcontinent*.

super- above, greater, exceeding, more, superior *superpower*.
syn-, sym- together, with *synthesis*.
techno- technology, technical *technocracy*.
tele- over a distance *telecommunications*.
tetra- four *tetrahedron*.
theo- gods, God *theology*.
thermo- heat *thermometer*.
trans- across *transatlantic*.
tri- three *triangle*.
ultra- above, beyond *ultraviolet*.
un- not *unhappy*; to do the opposite *unknot*.
under- below, underneath *underpass*; insufficient *underfunding*; less important *undersecretary*.
uni- one *unicycle*.
vice- one next below *vice-president*.

Suffixes

-able, -ible that can be *washable*; having the quality of *comfortable*.
-ability, -ibility.
-ade fruit drink *lemonade*.
-aholic, -oholic (one) addicted to *workaholic*.
-ana, -iana objects etc. belonging to *Victoriana*.
-arch ruler, leader, governor *monarch*. **-archy.**
-arian believer in, supporter of *vegetarian*; one connected with *librarian*.
-athon, -thon large-scale contest or event *swimathon*.
-cide killing, killer *fungicide*. **-cidal.**
-cracy government, rule *democracy*; dominant or ruling class *aristocracy*.
-crat supporter of (the type of government) *democrat*; member of (a dominant class) *aristocrat*.
-cratic.
-dom state or condition of being *boredom*; realm, domain *kingdom*.
-ectomy surgical removal *mastectomy*.
-ee one to whom something is done *payee*; one who is *absentee*; small version of *bootee*.
-eer one engaged in *profiteer*.

-er one who or that which does or is *employer*; one engaged in *lawyer*; one coming from *Londoner*.
-ese (people or language) of or from *Chinese*; language associated with *journalese*.
-esque in the style of *statuesque*.
-ess female *lioness*.
-ette small version of *kitchenette*; female *majorette*; imitation *satinette*.
-fold times *two-fold*.
-form having the form of *cruciform*.
-free without *lead-free*.
-friendly helpful to, supporting *user-friendly*.
-ful amount that fills something *bucketful*; full of *colourful*; having or causing *peaceful*.
-fy -IFY.
-gram drawn or written record *cardiogram*; message, greeting *kissogram*.
-graph instrument that records *seismograph*; something recorded or represented *autograph*. **-graphic, -graphy**.
-hood time or condition of being *childhood*.
-iana -ANA.
-ible, -ibility -ABLE.
-ics science, study *electronics*.
-ify to make or become *purify*; to fill with *terrify*.
-ise -IZE.
-ish like, similar to *childish*; somewhat *baldish*.
-ism state, condition *heroism*; doctrine, movement, system, theory *Buddhism*; discrimination on grounds of *racism*.
-ist follower or practitioner of a doctrine, science etc. *botanist*.
-ite (person) of or from, or that adheres to or supports *Israelite*; mineral *calcite*.
-itis disease *bronchitis*.
-ize, -ise to make or become *neutralize*.
-kin small version of *lambkin*.
-latry worship *idolatry*.
-less without *harmless*.

-let small version of *piglet*.
-like resembling *ladylike*.
-ling small, young, or lesser version of *duckling*.
-logy science, theory *biology*; writing, treatise *trilogy*. **-logical, -logist**.
-meter instrument for measuring *barometer*.
-monger dealer in *fishmonger*.
-nik one connected with *beatnik*.
-oholic -AHOLIC.
-oid like, resembling *planetoid*.
-ology, -ologist -LOGY.
-or -ER.
-osis action, process *metamorphosis*; diseased condition *thrombosis*.
-ped foot *quadruped*. **-pedal**.
-phile, -phil lover of *Anglophile*.
-philia love of, attraction or tendency towards *necrophilia*. **-philiac**.
-phobe one who hates or fears *xenophobe*. **-phobia** hatred or fear of *claustrophobia*. **-phobic**.
-phone speaker of (a language) *Francophone*; sound *xylophone*.
-proof resisting, protecting against *heatproof*.
-scape view, scene *landscape*.
-scope instrument for viewing *microscope*. **-scopy**.
-ship condition, state, position *membership*; skill *craftsmanship*.
-some characterized by, full of *troublesome*; group of so many *foursome*.
-speak language, jargon *computer-speak*.
-thon -ATHON.
-tomy surgical incision *lobotomy*.
-ward, -wards towards *upward*.
-ware articles *silverware*.
-ways in the direction, manner or position of *sideways*.
-wise in the direction, manner or position of *clockwise*; concerning *money-wise*.
-y having, full of, covered with *dirty*; inclined to *sleepy*; like *wintry*; affectionate term used esp. with children *doggy*.

A

A¹, a¹, the first letter in the English alphabet. **A** is used as a symbol to denote the first of a series; the first known quantity in an algebraic expression; (the scale of a composition in which the keynote is) the sixth note of the diatonic scale of C major, corresponding to *la* in tonic sol-fa notation; in Britain formerly, a film certified as suitable for all but requiring parental consent for children under 14; one of the human blood types. **from A to B**, from one point or position to another. **from A to Z**, from beginning to end. **A1**, *a*. first class. **A-bomb**, *n*. an atomic bomb. **A-level**, *n*. (a pass in) an examination in a subject at the Advanced level of the General Certificate of Education. **A-road**, *n*. a trunk road or a main road. **A-team**, *n*. the first or best team in a sport.
A², (*abbr.*) academy, academician; ampere; Ångström unit; Associate.
a² (ə; *when stressed* ā), **an** (ən; *when stressed* an), *a.* a weakened form of *one*, the indefinite article, used before singular nouns to denote an individual of a class. *A* is used before words beginning with (the sound of) a consonant, e.g., *Europe*, *one*, *usual*. *An* is used before vowels and sometimes before *h* in an unaccented syllable, e.g. *an historian*. In such phrases as *50 pence a pound*, *twice a week*, it equates with *each* or *every*. Also used before collective phrases like *a hundred men*, *a dozen eggs*, *a few*.
a³, (*abbr.*) acre; alto; ante (before).
a-, *pref.* (1) on, as in *aboard*; (2) away, out, as in *awake*; (3) of, from, as in *akin*; (4) from, as in *avert*; (5) not, without, as in *amoral*.
AA, (*abbr.*) Alcoholics Anonymous; anti-aircraft; Automobile Association.
AAA, (*abbr.*) Amateur Athletic Association; American Automobile Association.
A and M, (*abbr.*) Ancient and Modern (hymns).
aardvark (ahd'vahk), *n*. the African ant-eater. **aardwolf** (-wulf), *n*. a hyena-like carnivorous mammal of southern Africa.
Aaron's beard (eə'ronz), *n*. pop. name for large-flowered St-John's wort, and for a Chinese herb with hanging stems bearing clusters of hairy leaves. **Aaron's rod**, *n*. pop. name for certain plants that flower on long stems.
AB, (*abbr.*) able-bodied seaman; (US) Bachelor of Arts.
ab-¹, *pref.* off, from, away, apart, as in *abrogate*, *absent*.
ab-², *pref.* to, as in *abbreviate*.
aback (əbak'), *adv.* backwards; behind; by surprise. **taken aback** TAKE.
abacus (ab'əkəs), *n*. (*pl.* **-ci** (-sī), **-cuses**) a counting-frame; an apparatus made of beads sliding on wires for arithmetical calculations; a flat stone crowning the capital of a column and supporting the architrave.
abaft (əbahft'), *adv.*, *prep.* in, on or towards the back part of a ship.
abalone (abəlō'ni), *n*. an edible gasteropod mollusc.
abandon (əban'dən), *v.t.* to give up, yield; to desert or forsake; to surrender (oneself) unreservedly, e.g. to indolence or vice. *n*. freedom from conventional restraint, careless or joyful freedom of manner. **abandoned**, *a*. deserted; wholly given up to enjoyment, wickedness, etc. **abandonment**, *n*. the act of abandoning; self-surrender to a cause, passion or vice; relinquishment of property, desertion (of a relation, friend, servant).

abase (əbās'), *v.t.* to lower, humble, degrade. **abased**, *a*. **abasement**, *n*. the act of abasing, a state of humiliation, degradation.
abash (əbash'), *v.t.* to embarrass or shame by exciting a sense of guilt, mistake or inferiority. **abashed**, *a*. **abashment**, *n*.
abasia (əbā'siə), *n*. lack of power to coordinate the movements of the muscles in walking.
abate (əbāt'), *v.t.*, *v.i.* to diminish, reduce, lessen. **abatable**, *a*. **abatement**, *n*. **abater**, *n*.
abatis, abattis (a'batis, -tē), *n*. a defence made of felled trees with their boughs directed outwards.
abattoir (ab'ətwah), *n*. a public slaughter-house.
abaxial (abak'siəl), *a*. (*Bot.*) facing away from the stem.
abba (ab'ə), *n*. father (in the invocation *Abba, father*); an episcopal title in the Syriac and Gothic churches.
abbacy (ab'əsi), *n*. the office and jurisdiction of an abbot. **abbat**, *n*. ABBOT. **abbatial** (-bā'-), *a*. pertaining to an abbey or an abbot.
abbé (ab'ā), *n*. an ecclesiastic without a benefice; a cleric in minor orders; generally a mere title without any definite office or responsibility.
abbess (ab'is, -es), *n*. the female superior of an abbey.
abbey (ab'i), *n*. a monastic community governed by an abbot or abbess; a building either now or formerly inhabited by a body of monks or nuns; a church attached to an abbey.
abbot (ab'ət), *n*. a monk; the superior of a monastery or an abbey.
abbreviate (əbrē'viāt), *v.t.* to shorten, abridge, reduce. **abbreviate** (-ət), *a*. shortened, cut short. **abbreviation**, *n*. the act of abridging or contracting; an abridged or shortened form, e.g. of a word. **abbreviator**, *n*. **abbreviatory**, *a*. abbreviating or tending to abbreviate. **abbreviature**, *n*. an abbreviation.
ABC¹, *n*. the alphabet; rudiments, first principles. [the first letters of the alphabet]
ABC², (*abbr.*) American Broadcasting Company; Associated British Cinemas; Australian Broadcasting Commission.
abdicate (ab'dikāt), *v.t.* to resign, to formally renounce, to give up. *v.i.* to abandon or relinquish a throne, or other dignity or privilege. **abdicable**, *a*. **abdicant**, *a*. abdicating, renouncing. *n*. one who abdicates, an abdicator. **abdication**, *n*. the act of abdicating. **abdicator**, *n*.
abdomen (ab'dəmən), *n*. that portion of the trunk which lies between the thorax and the pelvis; the belly; the posterior division of the body in the higher arthropods. **abdominal** (-dom'-), *a*. pertaining to the abdomen. **abdominal regions**, *n.pl.* certain portions of the body near to or including the belly. **abdominally**, *adv.*
abduce (abdūs'), *v.t.* to draw or pull from one part to another by an abductor; to lead away. **abducent**, *a*. having the property of drawing back or away.
abduct (abdŭkt'), *v.t.* to take away (esp. a woman or child) by guile or force; to kidnap. **abduction**, *n*. a leading or drawing away; separation of parts or a bone after a fracture, or of sides of a wound; the illegal taking away of a person, esp. a child or a woman, by fraud or force. **abductor**, *n*.
abeam (əbēm'), *adv.* on a line at right angles to the keel of

a ship.
abecedarian (ābəsədēə'riən), *a.* alphabetical; having verses distinguished by letters alphabetically arranged like the 119th Psalm.
abed (əbed'), *adv.* in bed, gone to bed.
Aberdeen (abədēn'), *n.* a rough-haired Scotch terrier. **Aberdonian** (-dō'-), *n.* (*sometimes derog.*) a native or inhabitant of Aberdeen, supposedly noted for thrift. *a.* belonging to Aberdeen.
aberrance (abe'rəns), **-cy**, *n.* a wandering from the right way. **aberrant**, *a.* wandering from the right way; deviating from the normal type. **aberration**, *n.* deviation from the normal course or standard; departure from rule; deviation from type; the difference between the true and observed position of a heavenly body; deviation of focused rays preventing them from uniting in a point.
abet (əbet'), *v.t.* (*past, p.p.* **abetted**) to encourage or aid (a person or cause) by word or deed, esp. in wrongdoing; to countenance, stimulate or instigate (chiefly in a bad sense). **abetment**, *n.* **abetter**, (*Law*) **abettor**, *n.* one who encourages or instigates another; an accessory.
abeyance (əbā'əns), *n.* the state of being held back, suspended; dormancy, quiescence. **in abeyance**, being suspended or set aside temporarily; (*Law*) waiting for an occupant or owner.
abhor (əbhaw', əbaw'), *v.t.* (*past, p.p.* **abhorred**) to hate extremely, detest; to shrink from with horror. **abhorrence** (-ho'-), **-ency**, *n.* **abhorrent**, *a.* exciting repugnance, hatred. **abhorrently**, *adv.* **abhorrer**, *n.*
Abib (ā'bib), *n.* the first month of the ancient Hebrew calendar, corresponding to Nisan.
abide (əbīd'), *v.i.* to dwell or live in a place; to stay, wait; to continue, remain firm. *v.t.* to await; to submit to; to endure, tolerate. **to abide by**, to remain beside, adhere to (rules, wishes etc.). **abidance**, *n.* continuance. **abider**, *n.* **abiding**, *a.* continuing, permanent, durable. *n.* continuance, residence. **abidingly**, *adv.*
Abies (ab'iēz), *n.* a genus of conifers, containing the silver firs, spruces, larches and cedars. **abietic** (-et'-), *a.* pertaining to or derived from trees of this genus. **abiet-**, *comb. form.* stem of various chemical terms relating to substances so derived.
ability (əbil'iti), *n.* physical, mental or moral power; capacity, competence; wealth, means; (*pl.*) intellectual gifts.
ab initio (ab inish'iō), *adv.* from the beginning. [L]
abiogenesis (ābīōjen'əsis), *n.* the theory that living matter can be produced from that which has no life; spontaneous generation. **abiogenetic** (-net'-), *a.* **abiogenetically**, *adv.* **abiogenist** (-oj'-), *n.* one who believes in abiogenesis. **abiogenous** (-oj'-), *a.* produced by abiogenesis.
abiotic (ābīot'ik), *a.* not living, not produced by living organisms.
abject (ab'jekt), *a.* cast away; sunk to a low condition; servile, degraded, morally debased; mean, low. **abjectedness** (-jek'-), **abjection**, *n.* the act of casting away; the state of being cast away; abasement. **abjectly**, *adv.* **abjectness**, *n.*
abjure (əbjooə'), *v.t., v.i.* to renounce or retract (anything) upon oath. **abjuration** (ab-), *n.* the act of abjuring on oath; a denial or renunciation on oath. **abjuratory** (-rā'-), *a.* **abjurement**, *n.* **abjurer**, *n.*
ablactate (əblak'tāt), *v.t.* to wean from the breast. **ablactation** (ab-), *n.* the weaning of a child from the breast; grafting by inarching.
ablation (əblā'shən), *n.* removal, carrying away; wearing away. **ablate**, *v.t.* **ablative** (ab'lə-), *a.* taking away, separating, subtractive. *n.* the grammatical case in Latin and other languages expressing separation, instrumentality, and other relations expressed in English by the prepositions from, by, with etc. **ablative absolute**, *n.* in Latin grammar, a construction with noun and participle, noun and adjective, in the ablative case, expressing time or circumstances. **ablatival** (ablətī'-), *a.*
ablator (əblā'tə), *n.* an instrument for excising diseased parts; an instrument for removing the tails of sheep.
ablaut (ab'lowt), *n.* a vowel change in the middle of a word to indicate modification in meaning, as *sit, set*; *rise, raise*; *ring, rang, rung*.
ablaze (əblāz'), *adv.*, *a.* on fire, in a blaze; brilliant; excited.
able (ā'bl), *a.* having sufficient physical, mental, moral or spiritual power, or acquired skill, or financial or other resources (to do something); gifted, vigorous, active. **ablebodied**, *a.* having a sound, strong body; experienced, skilled (applied to a sailor who is classed as AB, and called an **able-seaman**). **ableism**, *n.* discrimination in favour of able-bodied people. **ableist**, *a.* **ably** (āb'li), *adv.* in an able manner; with ability.
-able (-əbl), *suf.* able, or likely, to; fit, suitable for, that may be, full of, as in *likeable, eatable, saleable, reasonable.* **-ably** (-əbli), *suf.* **-ability** (-əbiliti), *suf.*
ablegate (ab'ligət), *n.* a papal envoy sent with insignia to new cardinals etc.
abloom (əbloom'), *a.*, *adv.* blooming, in a state of bloom.
abluent (ab'luənt), *a.* cleansing, washing away. *n.* that which washes off or carries off impurities.
ablution (əbloo'shən), *n.* (*often pl.*) the act of washing, cleansing or purifying by means of water or other liquids; a ceremonial or symbolical washing or cleansing; (*coll.*) a building containing washing facilities, as in a military camp. **ablutionary**, *a.*
ABM, (*abbr.*) antiballistic missile.
abnegate (ab'nigāt), *v.t.* to deny, to refuse, to renounce. **abnegation**, *n.* **abnegative** (-gā'-), *a.* implying denial, negative. **abnegator**, *n.*
abnormal (abnaw'ml), *a.* not according to rule, departing from the ordinary type. **abnormality** (-mal'-), *n.* irregularity, deformity. **abnormally**, *adv.*
Abo (ab'ō), *n.* (*offensive*) an aboriginal native of Australia.
aboard (əbawd'), *adv.* on board, on a ship or boat. *prep.* into a ship.
abode (əbōd'), past of ABIDE, dwelt, stayed. *n.* continuance for a longer or shorter period in any place; residence; a habitation. **to make (one's) abode**, to dwell, reside.
abolish (əbol'ish), *v.t.* to do away with, put an end to, destroy; to annul, cancel or revoke (used of laws, customs, institutions or offices). **abolishable**, *a.* **abolisher**, *n.* **abolishment**, *n.* **abolition**, *n.* the act of abolishing or doing away with; the state of being abolished. **abolitionism**, *n.* **abolitionist**, *n.* one who holds views in favour of abolition, esp. one who favoured the abolition of slavery during the 18th and 19th cents.
abomasum (abəmā'səm), **-masus** (-səs), *n.* (*pl.* **-sa** (-sə), **-si** (-sī)) the fourth stomach in a ruminating animal.
abominate (əbom'ināt), *v.t.* to loathe, to hate exceedingly. **abominable**, *a.* very hateful or odious, physically or morally. **abominable snowman**, *n.* the yeti. **abominableness**, *n.* **abominably**, *adv.* **abomination**, *n.* the act of doing something hateful; the state of being greatly hated or loathed; an object of extreme hatred or loathing.
aborigine (abərij'ini), *n.* an indigenous or original inhabitant of a continent, country or district; (**Aborigine**) a member of a dark-skinned people indigenous to Australia. **aboriginal**, *a.* original, indigenous, inhabiting a place from the earliest times. *n.* an original inhabitant (esp. of Australia); a member of the original fauna or flora. **aboriginally**, *adv.* from the beginning, from the first; originally.
abort (əbawt'), *v.i.* to miscarry, bring forth prematurely; to undergo partial or entire arrest of development. *v.t.* to give birth to before the proper time; to induce the abortion of; to terminate prematurely or in the early stages. **abortifacient** (əbawtifā'shənt), *n.* a device or drug to induce abortion. **abortion**, *n.* the act of miscarrying; the

production of a foetus before the proper time; a procedure to induce the premature production of a foetus; the product of a miscarriage; anything which fails instead of coming to maturity; a monster, a misshapen creature. **contagious abortion**, a contagious or infectious disease, esp. brucellosis, which causes abortion in some farm animals. **abortionist**, *n*. one who performs abortions. **abortive** (-baw'-), *a*. brought forth in an immature state; imperfectly formed; procuring or intended to procure abortion; ineffectual, failing in its effect. *n*. a drug causing or intended to cause abortion. **abortively**, *adv*. **abortiveness**, *n*.
ABO system, *n*. a system for typing human blood according to the presence or absence of certain antigens.
aboulia (aboo'liə), **abulia** (-ū'-), *n*. abnormal loss of willpower.
abound (əbownd'), *v.i*. to be rich (in), to be copiously supplied (with); to be present in great plenty. **abounding**, *a*. plentiful, copious. *n*. abundance.
about (əbowt'), *prep*. around, surrounding, on the outside or surface of; near in time, space, number, quantity or quality; concerning, in connection with. *adv*. around, circuitously, nearly; here and there. **about face, (right-) about turn**, turn right round, face the opposite way. **be about to**, be going to (do something) imminently. **to bring about**, to cause to happen. **to come about**, to happen. **to go about**, to prepare to do (something); to change the course of a ship or boat.
above (əbŭv'), *prep*. over, at or to a higher point than; in excess of, superior to, more important than, beyond; in writing, previous, preceding. *adv*. overhead; in a higher place or position; previously; in heaven. *n*. the aforesaid; heaven. **above all**, before everything else. **above-board**, *adv*., *pred.a*. openly, without trickery. **above par**, *a*. at a premium; of superior quality.
Abp., (*abbr*.) archbishop.
abracadabra (abrəkədab'rə), *n*. a cabbalistic word used as a word-charm, a jingle or nonsensical phrase.
abrade (əbrād'), *v.t*. to rub or wear away by friction. **abradant**, *a*. abrasive.
Abrahamic (ābrəham'ik), **-mitic** (-mit'-), **-mitical**, *a*. pertaining to the patriarch Abraham.
abranchial (əbrang'kiəl), **abranchiate** (-ət), *a*. without gills. *n*. an animal that at no period possesses gills.
abrasion (əbrā'zhən), *n*. the act of rubbing away or wearing down; the state of being rubbed away or worn down; a superficial lesion of the skin; the substance worn or rubbed off. **abrasive** (-siv), *a*. tending to rub or wear away; of a person's manner, causing friction or irritation. *n*. a substance, as emery, used for grinding or rubbing down.
abraum (ab'rowm), *n*. a red clay used to deepen the colour of mahogany.
abreaction (abriak'shən), *n*. the ridding oneself of a complex by reliving in feeling or action repressed fantasies or experiences. **abreact**, *v.i*.
abreast (əbrest'), *adv*. side by side with the fronts in line; up to the standard (of); up-to-date, aware (of).
abreption (əbrep'shən), *n*. complete severance.
abridge (əbrij'), *v.t*. to shorten, epitomize; to deprive (a person of). **abridger**, *n*. **abridgment**, *n*. the act of abridging; the state or process of being abridged; a condensed form, an abstract, a summary.
abroad (əbrawd'), *adv*. widely, at large; beyond the bounds of a house or country; in or to a foreign country; before the public generally.
abrogate (ab'rəgāt), *v.t*. to annul by an authoritative act; to repeal, make void. **abrogation**, *n*. the act of abrogating; repeal. **abrogative**, *a*. tending to abrogate. **abrogator**, *n*.
abrupt (əbrŭpt'), *a*. broken, very steep, precipitous; sudden, disconnected; brusque, curt. **abruptly**, *adv*. **abruptness**, *n*.

ABS, (*abbr*.) anti-lock braking system.
abs-, *pref*. away, off, from, as in *abstain, abstruse*.
abscess (ab'ses), *n*. a gathering of pus in any tissue or organ, accompanied by pain and heat. **abscessed**, *a*.
abscind (əbsind'), *v.t*. to cut off. **abscission** (-sish'ən), *n*.
abscissa (əbsis'ə), *n*. (*pl*. **-ssae** (-ē), **-ssas**) the x-coordinates in a system of fixed rectilinear axes.
abscond (əbskond'), *v.i*. to go away secretly, esp. out of the jurisdiction of a court, or hide oneself to avoid legal proceedings. **abscondence**, *n*. **absconder**, *n*.
abseil (ab'sāl), *v.i*. to descend a vertical or steeply sloping surface, using a rope attached at the top and wound round the body.
absent[1] (ab'sənt), *a*. away from or not present in a place; wanting, not existing. **absent-minded**, *a*. inattentive, abstracted in mind. **absent-mindedly**, *adv*. **absent-mindedness**, *n*. **absence**, *n*. the state of, or an act of, being absent. **absentee** (-tē'), *n*., *a*. (of) one who is habitually absent; (of) a landlord who lives away from his/her estate. **absenteeism**, *n*. unjustified failure of workers to report for work. **absently**, *adv*.
absent[2] (əbsent'), *v.refl*. to keep (oneself) away.
absidal (ab'sidəl), *a*. apsidal.
absinthe (ab'sinth, absit'), *n*. (a liqueur flavoured with) wormwood. **absinthian, absinthic** (-sin'-), *a*. pertaining to or derived from wormwood; hence, bitter.
absolute (ab'səloot), *a*. independent, unlimited, under no restraint; self-existent; arbitrary, despotic; highly accomplished, perfect; (*Phil*.) existing independently of any other cause; measured from vacuum, as 'the absolute pressure of steam'. **absolute magnitude**, *n*. the magnitude of a star at a distance of 32·6 light years (10 parsecs) from the Earth. **absolute majority**, *n*. a number of votes polled which exceeds the combined total for all other candidates. **absolute temperature**, *n*. temperature measured from the absolute zero. **absolute zero**, *n*. the zero of the absolute scale of temperature, equal to −273·1° C. **absolutely**, *adv*. totally, unconditionally; used to express agreement or assent. **absoluteness**, *n*. **absolutism**, *n*. despotic government; the theological doctrine of absolute predestination. **absolutist**, *n*. one who is in favour of arbitrary government. *a*. pertaining to absolutism or despotism. **absolutistic** (-tis'-), *a*. absolutist.
absolution (absəloo'shən), *n*. acquittal, remission, forgiveness; the declaration of pardon of sins by a priest to an individual or a congregation after private or general public confession.
absolve (əbzolv'), *v.t*. to set free, pardon, acquit; to forgive the sins of (a penitent). **absolver**, *n*. **absolvitor** (-taw), *n*. (*Sc. Law*) a favourable verdict; an acquittal.
absonant (ab'sənənt), *a*. discordant, inharmonious, unreasonable.
absorb (absawb', -z-), *v.t*. to suck up, drink in; to soak up by capillary action; to incorporate; to fully occupy the attention of, to engross; to take in and transform (radiant energy) without transmission or reflection. **absorbable**, *a*. **absorbability** (-bil'-), *n*. **absorbed**, *a*. **absorbent**, *a*. capable of or tending to absorb. *n*. a vessel in an organism which takes nutritive matter into the system; a substance which has the power of absorbing gases or liquids. **absorbent cotton**, *n*. (*N Am*.) cotton wool. **absorber**, *n*. that which absorbs. **absorbing**, *a*. **absorption** (-sawp'-, -z-), *n*. the act of absorbing; the process of being absorbed. **absorption spectrum**, *n*. the spectrum produced when electromagnetic radiation passes through a selectively absorbing medium. **absorptive** (-sawp'-, -z-), *a*. absorbent.
abstain (əbstān'), *v.i*. to keep oneself away, refrain (from); to refrain from intoxicating liquors voluntarily; to refrain from voting. **abstainer**, *n*.
abstemious (əbstē'miəs), *a*. sparing, not self-indulgent, esp. in eating, and drinking strong liquors; moderate,

temperate, inclined to abstinence. **abstemiously,** *adv.* **abstemiousness,** *n.*
abstention (əbsten'shən), *n.* the act of abstaining or refraining, esp. from exercising one's right to vote.
absterge (əbstœj'), *v.t.* to wipe clean, to cleanse; to purge by medicine. **abstergent,** *n., a.* (something) that cleanses, esp. a medicine which cleanses or purges. **abstersion** (-shən), *n.* the act of cleansing or purgation. **abstersive,** *n., a.* (something) which has cleansing, purifying qualities. **abstersiveness,** *n.*
abstinence (ab'stinəns), *n.* the act or practice of refraining from some indulgence. **abstinent,** *n., a.* (one) practising abstinence. **abstinently,** *adv.*
abstract[1] (əbstrakt'), *v.t.* to draw or take away, to remove; (*euphem.*) to steal; to separate mentally, to consider separately from other things; to summarize; to separate by chemical distillation; to extract.
abstract[2] (ab'strakt), *a.* separated from particular things, ideal; existing in the mind only; abstruse; theoretical; geometrical or non-naturalistic in design. *n.* an abstract term; a summary; an abstract work of art. **abstract of title,** a summary of the evidences of ownership. **in the abstract,** without reference to individual cases, ideally, theoretically. **abstract expressionism,** *n.* a 20th-cent. artistic movement in which the artist's emotions and feelings are expressed by non-naturalistic forms. **abstract nouns** or **terms,** *n.pl.* names of qualities, in contradistinction to concrete nouns or terms which are names of things. **abstracted** (-strak'-), *a.* inattentive, withdrawn in thought. **abstractedly,** *adv.* absent-mindedly; in the abstract, separately. **abstractedness,** *n.* **abstractly,** *adv.* **abstractness,** *n.*
abstraction (əbstrak'shən), *n.* the act of abstracting or separating; taking away; (*euphem.*) stealing; the state of being engrossed in thought; the process of considering separately the quality of an object; a mental conception so formed; an abstract idea; the process by which people form abstract ideas. **abstractive,** *a.* possessing the power or quality of abstracting; tending to abstraction. **abstractively,** *adv.*
abstruse (əbstroos'), *a.* hidden from observation or knowledge; off the beaten track of human thought. **abstrusely,** *adv.* **abstruseness,** *n.*
absurd (əbsœd'), *a.* incongruous; contrary to or inconsistent with reason; nonsensical, logically contradictory; ridiculous. **absurdity,** *n.* the quality or state of being absurd; an absurd notion, statement or action. **absurdly,** *adv.* **absurdness,** *n.*
ABTA (ab'tə), (*acronym*) Association of British Travel Agents.
abulia ABOULIA.
abundance (əbūn'dəns), *n.* fullness; (a) great quantity or number (of). **abundant,** *a.* in great supply; plentiful, fully sufficient, ample. **abundantly,** *adv.*
abuse[1] (əbūz'), *v.t.* to misuse; to use in an illegitimate sense, to pervert the meaning of; to maltreat, act or speak cruelly to; to violate. **abusable,** *a.* **abuser,** *n.*
abuse[2] (əbūs'), *n.* improper treatment, misuse; a corrupt practice or custom; grossly insulting language; perversion from the proper meaning; physical maltreatment; violation. **abusive,** *a.* given to the use of harsh language or ill-treatment; grossly insulting. **abusively,** *adv.* **abusiveness,** *n.*
abut (əbūt'), *v.i.* (*past, p.p.* **abutted**) to be contiguous; to border (on or upon); to form a point or line of contact; of a building, to lean (on or upon). *v.t.* to border on. **abutment,** *n.* the state of abutting; that which abuts or borders; a pier or wall, or the part of a pier or wall, against which an arch rests. **abuttal,** *n.* abutment, esp. the abutting part of a piece of land. **abutter,** *n.* one who or that which abuts; the owner of property that abuts.
abysmal (əbiz'məl), *a.* profound, immeasurable; (*coll.*) extremely bad. **abysmally,** *adv.*
abyss (əbis'), *n.* a vast geographical depth, chasm or cavity; primeval chaos; anything profound and unfathomable, as ignorance or degradation. **abyssal,** *a.* pertaining to an abyss; pertaining to the lowest depths of the sea beyond 300 fathoms (about 550 m).
Abyssinian (abəsin'iən), *a.* pertaining to Abyssinia or its inhabitants. *n.* an inhabitant of Abyssinia or Ethiopia.
ac-, *pref.* AD-, assim. to *c, k, qu,* e.g. *accommodate, accord, acquire.*
-ac (-ak), *suf.* pertaining to, e.g. *cardiac, demoniac.* (Adjectives so formed are often used as nouns.)
AC, (*abbr.*) alternating current; ante Christum (before Christ); of wine, appellation contrôlée.
Ac, (*chem. symbol*) actinium.
a/c, (*abbr.*) account.
ACA, (*abbr.*) Associate of the Institute of Chartered Accountants.
Acacia (əkā'shə), *n.* an extensive genus of trees with small flowers in balls or spikes; (**acacia**) any plant of this genus. **acacia-tree,** *n.* the N American locust-tree or false acacia.
academic (akədem'ik), *a.* pertaining to an academy, college or university; scholarly, theoretic. *n.* a staff member of an academy, college or university. **academical,** *a.* academic; unpractical. *n.pl.* scholarly dress, cap and gown. **academically,** *adv.*
academy (əkad'əmi), *n.* a place of study, a high school; a seminary for higher education; a society or association for promoting literature, science or art; (**Academy**) the Royal Academy. **academe** (ak'ədēm, -dēm'), *n.* (*poet.*) an academy; the world of academics. **academician** (-mish'ən), *n.* a person belonging to an academy or association for the promotion of science, literature or art; (**Academician**) a Royal Academician. **academicism** (akədem'-), *n.* the system of teaching in an academy or high school; academical mannerism. **academist,** *n.* a member of an academy.
Acadian (əkā'diən), *a., n.* (one) belonging to Nova Scotia.
acajou (ak'əzhoo), *n.* a gummy substance derived from the cashew-nut tree; a wood resembling mahogany.
-acal (-əkəl), *suf.* adjectives ending in *-ac* being often used as nouns, *-al* was added to distinguish the adjective, e.g. *demoniacal, maniacal; -al* is also added to adjectives to show a less intimate connection with the original noun, e.g. *cardiacal.*
acaleph (ak'əlef), *n.* any of a class of marine animals containing the sea-nettles, jelly-fish etc. **acalephan,** *n., a.* **acalephoid** (-oid), *a.*
acalycine (əkal'isin), *a.* without calyx or flower-cup. **acalycinous** (akəlis'in-), *a.*
Acanthus (əkan'thəs), *n.* (any of) a genus of prickly-leaved plants; an ornament resembling the foliage of the acanthus, used to decorate the capitals of Corinthian columns. **acanthaceous** (-thā'shəs), *a.* armed with spines or prickles. **acanthine** (-thin), *a.* pertaining to or resembling the acanthus; prickly; ornamented with the acanthus leaf. **acanthoid** (-thoid), **acanthous,** *a.* prickly, spinous. **acantho-,** *comb.form* (*Bot.*) spiny, having thorns or thorn-like processes.
a cappella (a, ah kəpel'ə), *a., adv.* without instrumental accompaniment. [It. *a cappella,* in chapel style]
acardiac (əkah'diak), *a.* without a heart. *n.* a foetus destitute of a heart.
acarpous (əkah'pəs), *a.* producing no fruit; sterile, barren.
Acarus (ak'ərəs), *n.* a genus of Arachnida, comprising the mites and ticks. **acaricide** (əka'risid), *n.* a substance that kills mites, a remedy for the itch. **Acarida** (əka'ridə), **Acarina** (-nə), *n.pl.* the order including the mites and ticks. **acarid, acaridan** (-kə'-), **acaridean** (-rid'i-), *n., a.* (one) of the Acarida. **acaroid** (-roid), *a.*
ACAS (ā'kas), (*acronym*) Advisory, Conciliation and Arbitration Service.

acatalectic (əkatəlek'tik), *a.* not breaking off short; complete; having the full number of metrical feet. *n.* a verse having the complete number of feet.
acatalepsy (əkat'əlepsi), *n.* incomprehensibleness; the sceptical doctrine that things are unknowable; mental confusion. **acataleptic** (-lep'-), *a.* incomprehensible; not to be known with certainty.
acauline (əkaw'lin), **-lose**, **-lous**, *a.* without apparent stem, stemless. **acaulescence** (akawles'-), *n.* the occasional apparent suppression of the stem. **acaulescent**, *a.*
acc., (*abbr.*) according; account; accusative.
Accadian (əkā'diən), *n.* a member of one of the races of ancient Babylonia; the language of this race. *a.* pertaining to this race or its Semitic language.
accede (əksēd'), *v.i.* to come (to a certain view), to agree, to assent; to come (to an office or dignity). **accedance**, *n.*
accelerando (əchelərən'dō), *a.*, *adv.* (*Mus.*) with increasing speed. [It.]
accelerate (əksel'ərāt), *v.t.* to hasten; to increase the rate of progress or velocity of; to bring (e.g. events) nearer in point of time. *v.i.* to increase in velocity or rate of progress, to move faster. **acceleratedly**, *adv.* **acceleration**, *n.* the act of accelerating, or the state of being accelerated; (rate of) increase of velocity or progress. **accelerative**, *a.* **accelerator**, *n.* that which accelerates; the pedal in a vehicle which is used to cause the engine to run at increased speed; any chemical or apparatus for speeding up the development of a photographic picture, plate or print; an electrical appliance for accelerating charged particles such as electrons or protons to high velocities or energies. **acceleratory**, *a.* **accelerometer** (-rom'itə), *n.* an instrument for measuring acceleration.
accensor (aksen'sə), *n.* one who lights and trims the tapers in the Roman Catholic Church.
accent[1] (ak'sənt), *n.* a particular prominence given to a syllable by means of stress or higher musical pitch; manner of speaking or pronunciation expressive of feeling, or peculiar to an individual, a locality or a nation; a mark used in writing or printing to direct the stress of the voice or alter pronunciation; musical, metrical or rhythmical stress; distinctive emphasis or intensity.
accent[2] (əksent'), *v.t.* to lay stress upon (a syllable or word, or a note or passage of music); to mark with emphasis, make conspicuous; to mark with an accent. **accentual**, *a.* pertaining to accent; rhythmical. **accentuate**, *v.t.* to pronounce or mark with an accent; to lay stress on, to emphasize. **accentuation**, *n.*
accept (əksept'), *v.t.* to consent to take (something offered); to view with favour; to admit the truth of, acknowledge; to agree to, to admit, to take responsibility for; to promise to pay (a bill of exchange) when due. **acceptable**, *a.* **acceptability** (-bil'-), **acceptableness**, *n.* **acceptably**, *adv.* **acceptance**, *n.* the act of receiving; favourable reception; agreement to terms or proposals; admission to favour; an accepted bill of exchange; the act of subscribing, or the subscription to, a bill of exchange. **acceptancy**, *n.* acceptance; willingness to accept. **acceptant**, *a.* willingly receiving. *n.* one who accepts. **accepter**, *n.* one who accepts. **acceptor**, *n.* one who accepts a bill of exchange; an impurity added to a semiconductor to increase its conductivity.
access (ak'ses), *n.* admission to a place or person; freedom to obtain or use something; the means of (approach), passage, channel; increase, addition; attack of (disease or emotion). *v.t.* to gain access to, esp. to retrieve (data) from (computer storage); to place (data) in (computer storage). *a.* designating or pertaining to radio and television programmes made by the general public. **access time**, *n.* the time interval between requesting data and its delivery from computer storage. **accessary** (-ses'-), *n.*, *a.* (an) accessory. **accessible** (-ses'-), *a.* capable of being approached or reached; easy of access; approachable, attainable. **accessibility** (-bil'-), *n.* **accessibly**, *adv.* **accession** (-sesh'ən), *n.* the act of going or coming to; agreeing or consenting to; coming to (the throne, an office or a dignity); an increase, addition; an improvement or addition to property by growth or labour expended.
accessory (-ses'-), *a.* contributive, helpful to some effect, aiding, or acting in subordination to a principal; accompanying, additional; guilty, not as the chief actor, but *before the fact*, by counselling or commanding the act, or *after the fact*, by assisting or concealing the offender. *n.* one who abets anything that is wrong; an accomplice; something added merely for ornament.
acciaccatura (achəkətoo'rə), *n.* (*pl.* **-ras**, **-re** (-rā)) a short grace note played rapidly.
accidence (ak'sidəns), *n.* that part of grammar which deals with the inflections of words.
accident (ak'sidənt), *n.* an event proceeding from an unknown cause; the unforeseen effect of a known cause; something unexpected; a casualty, a mishap (e.g. of birth). **accidental** (-den'-), *a.* occurring by chance, unexpectedly; not according to the usual order of things. *n.* a sharp, flat or natural sign occurring in music before a particular note, not in the key-signature. **accidental colours**, *n.pl.* the complementary colours seen after looking fixedly at a bright-coloured object, and then at a white or light-coloured surface. **accidentalism**, *n.* accidental character or effect. **accidentality** (-tal'-), *n.* **accidentally**, *adv.* **accidentalness**, *n.*
Accipiter (əksip'itə), *n.* a genus of birds of prey, containing the hawks. **accipitral**, *a.* **accipitrine** (-trīn, -trin), *a.* belonging to or resembling hawks; rapacious, predatory; keen-sighted.
acclaim (əklām'), *v.t.* to applaud loudly; to welcome or announce with enthusiasm. *v.i.* to shout applause. *n.* acclamation. **acclamation** (aklə-), *n.* a demonstration of joy or applause made by a meeting or gathering. **acclamatory** (-klam'-), *a.*
acclimatize, -ise (əklī'mətiz), (*N Am.*) **acclimate** (-māt), *v.t.*, *v.i.* to habituate to a new climate or environment; to adapt for existence and propagation in a new climate. **acclimatization**, *n.* the act or process of acclimatizing; the state of being acclimatized. **acclimatation, acclimation** (akli-), *n.* in nature, spontaneous accommodation to new conditions as distinguished from acclimatization by humans.
acclivity (əkliv'iti), *n.* an upward slope. **acclivitous**, *a.*
accolade (ak'əlād), *n.* the ceremony of conferring knighthood by an embrace, putting hand on neck, or a gentle stroke with the flat of a sword; an award or honour, (an expression of) praise and approval.
accommodate (əkom'ədāt), *v.t.* to fit, adapt to, make room for; to oblige (someone) with a want; to provide lodging for. **accommodating**, *a.* obliging, complying, yielding to others' desires. **accommodatingly**, *adv.* **accommodation**, *n.* the act of accommodating; adjustment, compromise; the state of being accommodated; anything that supplies a want in respect of convenience, lodging etc. **accommodation address**, *n.* an address to which mail may be sent, used instead of a permanent address. **accommodation ladder**, *n.* a light ladder fixed outside a vessel at the gangway. **accommodation land**, *n.* land bought by a speculator to be leased out for building purposes. **accommodation unit**, *n.* a dwelling. **accommodative**, *a.* **accommodativeness**, *n.*
accompany (əkŭm'pəni), *v.t.* to go with as a companion; to exist along with, to characterize; to play the instrumental accompaniment for. *v.i.* to play the accompaniment. **accompanier**, *n.* **accompaniment**, *n.* something added to or attendant upon another thing; something which gives greater completeness (to); the part or parts performed by instruments accompanying the soloist. **accompanist**, *n.* the performer who plays the

accomplice   6   **acetic**

instrumental accompaniment.
accomplice (əkŭm′plĭs), *n.* a partner, esp. in crime. **accompliceship**, *n.* **accomplicity** (akəmplĭs′-), *n.* assistance in crime.
accomplish (əkŭm′plĭsh), *v.t.* to carry out, achieve, finish. **accomplishable**, *a.* **accomplished**, *a.* complete, finished; highly skilled; having the graces and social skills perfecting one for society. **accomplisher**, *n.* **accomplishment**, *n.* the act of accomplishing; the state of being accomplished; something achieved; an acquirement, attainment, esp. a social skill.
accompt (əkownt′), *v.*, *n.* (*dated*) ACCOUNT. **accomptant** (*dated*) ACCOUNTANT.
accord (əkawd′), *v.t.* to cause to agree; to make consistent, to grant. *v.i.* to agree, to be in correspondence or harmony. *n.* agreement, harmony, assent; a treaty. **of one's own accord**, voluntarily. **with one accord**, unanimously. **accordance**, **-ancy**, *n.* **accordant**, *a.* agreeing, harmonious. **accordantly**, *adv.* **accorder**, *n.* **according**, *part.a.*, *adv.* agreeing, corresponding (to); harmonious. **according as**, in proportion to. **according to**, in relation to; as stated by; depending on. **accordingly**, *adv.* suitably; therefore, consequently.
accordion (əkaw′dĭən), *n.* a small portable keyed musical instrument in which the notes are produced by bellows action on metallic reeds. **accordion-pleating**, *n.* pleats with very narrow folds resembling the bellows of an accordion. **accordionist**, *n.* a player on the accordion.
accost (əkost′), *v.t.* to approach and speak to; of a prostitute, to solicit.
accouchement (akooshmē′), *n.* confinement at childbirth. **accoucheur** (-shœr′), *n.* a person who assists women at childbirth. **accoucheuse** (-shœz′), *n.fem.*
account (əkownt′), *v.t.* to reckon, compute, count; to regard as, to deem. *v.i.* to give a reckoning or explanation. *n.* reckoning, computation; a description, explanation; a statement of receipts and expenditure showing the balance; a statement of goods or services supplied with calculation of money due; on the Stock Exchange, the fortnightly period from one settlement to another; a business arrangement involving the establishing of a record of debit and credit; **of no account**, valueless, negligible. **on account**, as an interim payment. **on account of**, for the sake of, because of. **on no account**, by no means. **on one's account**, for one's sake, on one's behalf. **to account for**, to render an account of; to give an explanation of. **to bring** or **call to account**, to require an explanation from; to reprimand. **to give a good account of**, to be successful, do (oneself) credit. **to (good) account**, to one's advantage or profit. **to hold to account**, to hold responsible. **to take account of**, to take into account TAKE. **account-book**, *n.* a register of business transactions. **account day**, *n.* a day of financial reckoning. **accountable**, *a.* responsible. **accountability** (-bĭl′-), *n.* **accountableness**, *n.*
accountant (əkown′tənt), *n.* one whose occupation is the keeping of accounts; a public officer charged with the duty of keeping and inspecting accounts. **accountancy**, *n.* **accountantship**, *n.*
accoutre (əkoo′tə), *v.t.* to dress, to equip, esp. for military service. **accoutrement** (-trəmənt), *n.* a soldier's equipment, excepting arms and dress; (*usu. pl.*) dress, outfit, equipment.
accredit (əkred′ĭt), *v.t.* to confer official recognition on, vouch for; to send with credentials (as an ambassador). **accreditation**, *n.* **accredited**, *a.* recognized officially, generally accepted as conforming to an official standard of quality.
accrete (əkrēt′), *v.i.* to grow together; to combine round a nucleus. **accretion**, *n.* increase by organic growth, or growth by external additions; the result of such growth.
accrue (əkroo′), *v.i.* to increase; to fall, come to (as a natural growth). **accrual**, *n.*

acct., (*abbr.*) account; accountant.
acculturate (əkŭl′chərāt), *v.t.*, *v.i.* to adopt the values and traits of another culture. **acculturation**, *n.*
accumulate (əkū′mūlāt), *v.t.* to gather together by degrees, to amass. *v.i.* to grow in size, number or quantity, by repeated additions. **accumulation**, *n.* the act of accumulating or amassing; that which is accumulated; a mass. **accumulative**, *a.* accumulating. **accumulatively**, *adv.* **accumulator**, *n.* one who or that which accumulates; an apparatus for storing hydraulic or electric energy, esp. a rechargeable electric cell or battery; a bet in which the winnings from previous races are staked on the next.
accurate (ak′ūrət), *a.* exact, in precise accordance with rule or standard of truth; without error or defect. **accuracy**, *n.* **accurately**, *adv.* **accurateness**, *n.*
accursed (əkœ′sĭd), **accurst**, *a.* under a curse; execrable, detestable.
accusative (əkū′zətĭv), *a.* of or relating to the objective case of declinable words in inflected languages; also applied to the word that represents the object in uninflected languages. *n.* the grammatical case defined above. **accusatival** (-tī′-), *a.* pertaining to the accusative. **accusatively**, *adv.*
accuse (əkūz′), *v.t.* to charge with a crime, offence or fault, to indict; to lay the blame formally on (a person or thing). **accusal**, *n.* **accusation** (akū-), *n.* the act of accusing; a charge brought against one. **accusatory**, *a.* containing or involving an accusation. **accusatorial** (-taw′-), *a.* involving accusation or indictment in a case in which judge and prosecutor are distinct (contrasted with inquisitorial). **accusatorially**, *adv.* **accused**, *a.* **the accused**, the defendant(s) in a criminal case. **accuser**, *n.* **accusingly**, *adv.*
accustom (əkŭs′təm), *v.t.* to habituate (usually in *pass.* or *reflex.*, oneself to, or to do), to make familiar by use. **accustomed**, *a.* often practised, familiar, habitual. **accustomedness**, *n.*
AC/DC, *a.* (*sl.*) bisexual. [alternating current, direct current]
ace (ās), *n.* a card, die or domino with but one mark upon it; a very small amount, a hair's-breadth; a fighter-pilot who has brought down ten or more hostile aircraft; a person of first rank in sport etc. *a.* (*coll.*) excellent.
-acea, *suf.* used to form names of classes or orders of animals, e.g. *Crustacea*.
-aceae, *suf.* used to form names of orders or families of plants, e.g. *Rosaceae*.
-acean, *suf.* forms singular nouns or adjectives corresponding to collective nouns in *-acea*, e.g. (a) *crustacean*.
acentric (āsen′trĭk), *a.* without centre; not about a centre.
-aceous (-āshəs, -āsĭəs), *suf.* of the nature of, like; forming adjectives from nouns in natural science, e.g. *crustaceous*, *farinaceous*.
acephalous (əsef′ələs), *a.* without a head; (*Zool.*) with no distinct head. **acephalia**, *a.*, *n.*
Acer (ā′sə), *n.* a genus of plants comprising over 100 species including the sycamore and maples.
acerbic (əsœ′bĭk), *a.* sour, astringent; bitter or harsh in speech or manner. **acerbically**, *adv.*
acerbity (əsœ′bĭtĭ), *n.* sourness, with roughness or astringency, as of unripe fruit; bitterness of suffering; harshness of speech, action or temper.
acet-, *comb. form.* of the nature of vinegar.
acetabulum (əsətab′ūləm), *n.* (*pl.* **-la** (-lə), **-lums**) an ancient Roman vessel for holding vinegar; a cavity in any bone designed to receive the protuberant head of another bone, e.g. the socket of the hip-joint in man; the socket in which the leg of an insect is inserted; one of the suckers on the arms of a cuttlefish; the cup-shaped fructification of many lichens; the receptacle of certain fungi.
acetaldehyde (əsətal′dəhīd), *n.* a volatile liquid aldehyde used in the manufacture of organic compounds.
acetic (əsē′tĭk, -set′-), *n.* pertaining or akin to vinegar;

sour. **acetic acid**, *n.* the acid which imparts sourness to vinegar. **acetate** (as'ətāt), *n.* a salt of acetic acid; cellulose acetate; a photographic film or an audio disc made from cellulose acetate. **acetated**, *a.* treated with acetic acid. **acetify** (əset'ifī), *v.t.* to convert into vinegar; to render sour. *v.i.* to become sour. **acetification** (-setifī-), *n.* **acetous** (as'itəs), **-ose** (-ōs), *a.* having the character of vinegar, sour; causing acetification.
acetone (as'itōn), *n.* an inflammable liquid obtained by distilling acetated or organic substances and used in the manufacture of chloroform and as a solvent; a ketone. **acetyl** (as'itil, əsē'-), *n.* the radical of acetic acid. **acetylsalicylic acid** (-salisil'ik), *n.* aspirin.
acetylene (əset'əlēn), *n.* a gas composed of carbon and hydrogen, which burns with an intensely brilliant flame.
ache (āk), *v.i.* to suffer mental or physical pain or distress. *n.* continuous pain (in contradistinction to a twinge); distress.
Acheulian (əshoo'liən), *a.* of or pertaining to the period of Lower Palaeolithic culture, typified by remains discovered in St Acheul, and placed by archaeologists between the Chellean and the Mousterian epochs. [St *Acheul*, France]
achieve (əchēv'), *v.t.* to perform, accomplish, finish; to attain or bring about by an effort. **achievable**, *a.* **achievement**, *n.* the act of accomplishing; the thing achieved; a heroic deed, an exploit **achiever**, *n.*
Achillean (akilē'ən), *a.* like Achilles; heroic, invulnerable; invincible. **Achilles heel** (əkil'iz), *n.* a person's vulnerable point, the heel being the part where Achilles was said to be vulnerable, his mother Thetis holding him by the heel when she dipped him in the river Styx to make him invulnerable. **Achilles' tendon**, *n.* the tendon or ligature connecting the muscles of the calf to the heel-bone.
achromatic (akrəmat'ik), *a.* colourless; transmitting light without decomposing it into its primary colours; (*Mus.*) having no sharps or flats. **achromatically**, *adv.* **achromatism** (əkrō'-), **achromaticity** (əkrōmətis'-), *n.* the quality or state of being achromatic. **achromatize** (əkrō'-), **-ise**, *v.t.* to deprive of colour. **achromatopsy** (əkrō'mətopsi), *n.* colour blindness.
acicular (əsik'ūlə), *a.* resembling a needle in shape or sharpness. **acicularly**, *adv.* **aciculate** (-lət), **-ated** (-ātid), *a.* having needle-like bristles; marked with fine, irregular streaks.
acid (as'id), *a.* sour, tart, sharp to the taste; sharp or sour in manner or speech; having the properties of an acid, reacting as an acid; of rocks which have a large proportion of silica. *n.* a sour substance; a compound of hydrogen in which the hydrogen can be replaced by a metal, or with a basic metallic oxide, to form a salt of that metal and water; (*sl.*) LSD. **acid cloud**, *n.* an area of mist or low cloud containing high concentrations of pollutant acids harmful to crops etc. **acid house, Acid House**, *n.* (a youth cult concerned with) highly electronically synthesized disco or pop music (and associated with the taking of psychedelic drugs). **acid rain**, *n.* precipitation made acidic and thus harmful to crops etc. by the release of (industrial) pollutants into the atmosphere. **acid rock**, *n.* rock music featuring bizarre amplified instrumental effects. **acid test**, *n.* (*coll.*) an absolute and definite test; a critical ordeal.g **acidic** (-sid'-), *a.* acid. **acidify** (əsid'ifī), *v.t.* to render acid or sour; to convert into an acid. *v.i.* to become acid. **acidifiable** (əsidifī'-), *a.* **acidification** (-fī-), *n.* **acidimeter** (asidim'itə), *n.* instrument for measuring the strength of acids. **acidimetry** (-dim'itri), *n.* **acidity** (-sid'-), **acidness**, *n.*
acidosis (əsidō'sis), *n.* condition characterized by abnormal quantites of acids in the urine and bloodstream. **acidotic** (-dot'-) *a.*
acidulous (əsid'ūləs), *a.* a little sour or acid, moderately sharp to the taste. **acidulate**, *v.t.* to render slightly acid; to flavour with an acid. **acidulated**, *a.* rendered slightly acid; flavoured with acid; soured, embittered in temper.
aciniform (əsin'fawm), *a.* clustered like grapes.
-acious (-āshəs), *suf.* abounding in, characterized by, inclined to: added to verbal stems to form adjectives, e.g. *loquacious, tenacious.*
-acity (-asiti), *suf.* the quality of: forms nouns of quality from adjectives in -ACIOUS.
ack-ack (akak'), *n.* anti-aircraft.
ack-emma (ak em'ə), *n.* (*dated*) morning, a.m.
acknowledge (əknol'ij), *v.t.* to admit (the truth of), to recognize formally, to confess; to recognize (the authority of); to give a reply to or receipt for; to express appreciation or gratitude for. **acknowledgeable**, *a.* **acknowledgment**, *n.* the act of acknowledging; recognition, confession, admission; receipt for money or goods; reply to a communication; an expression of gratitude.
acme (ak'mi), *n.* the top or highest point, the culmination.
acne (ak'ni), *n.* a skin disease characterized by pimples or tubercles.
acolyte (ak'əlit), *n.* an assisting officer in the Roman Catholic Church; an attendant, ministrant; a faithful follower.
aconite (ak'ənit), *n.* a plant of the genus *Aconitum,* esp. the common monk's-hood or wolf's-bane; a poison drug used medicinally, obtained from the root of this plant. **aconitic** (-nit'-), *a.*
acorn (ā'kawn), *n.* the fruit of the oak. **acorned**, *a.* (*Her.*) bearing acorns.
acotyledon (əkotilē'dən), *n.* any plant of the class Acotyledones; a plant which does not possess distinct seed-lobes. **acotyledonous**, *a.*
acoustic (əkoos'tik), **-ical**, *a.* pertaining to the ear, constituting part of the physical apparatus for hearing; pertaining to hearing, sound or acoustics; pertaining to musical instruments whose sound is not electronically amplified. *n.* the acoustics of a room or building. **acoustically**, *adv.* **acoustician** (-tish'ən), *n.* one skilled in acoustics. **acoustics**, *n.* the science of sound and its phenomena, and of the phenomena of hearing; (*pl.*) the properties of a room or building that determine sound quality. **New Acoustics**, *n.pl.* popular musicians who employ acoustic instruments and intimate venues as a reaction against over-use of electronics and mass audiences.
acquaint (əkwānt'), *v.t.* to make aware of, inform. *v.reflex.* to give (oneself) knowledge of or acquaintance with. **acquaintance**, *n.* knowledge of any person or thing; the state of knowing, or becoming known to, a person; a person, or the persons collectively, whom one knows, but with whom one is not intimate. **acquaintanceship**, *n.* the state of being acquainted; the relation of mutual acquaintance. **acquainted**, *a.* known to another or each other; familiar (with).
acquiesce (akwies'), *v.i.* to submit or remain passive; to assent tacitly (to), to concur (in). **acquiescence**, *n.* **acquiescent**, *a.* **acquiescently**, *adv.* **acquiescingly**, *adv.*
acquire (əkwiə'), *v.t.* to gain, or obtain possession of, by one's own exertions or abilities; to come into possession of, often by purchasing. **acquired characteristic**, *n.* a characteristic of an organism that is attained through environmental influences rather than genetically. **acquired immune deficiency syndrome** AIDS. **acquired taste**, *n.* something which one learns to like. **acquirable**, *a.* **acquirability** (-bil'-), *n.* **acquirement**, *n.* the act of acquiring; the object gained; a personal attainment of body or mind.
acquisition (akwizish'ən), *n.* the act of acquiring; the object acquired. **acquisitive** (əkwiz'-), *a.* enthusiastic in making acquisitions. making, or disposed to make acquisitions. **acquisitively**, *adv.* **acquisitiveness**, *n.* the quality of being acquisitive.
acquit (əkwit'), *v.t.* (*past, p.p.* **acquitted**) to release from an obligation, suspicion or charge; to declare not guilty. *v.reflex.* to discharge (oneself) of (the duties of one's

acre position); to conduct (oneself) in a particular way.
acquittal, *n.* discharge or release from a promise, debt or other obligation; performance; a decision of not guilty.
acre (ā'kə), *n.* a measure of land containing 4840 sq. yd. (0·4 ha); a piece of tilled or enclosed land. **acreage,** *n.* the area of any piece of land in acres; acres taken collectively or in the abstract.
acrid (ak'rid), *a.* pungent, biting to the taste; irritating, esp. of temper and manners; bitter. **acridness, acridity** (əkrid'-), *n.* sharpness, pungency, bitterness of manner or speech.
Acrilan® (ak'rilan), *n.* a type of acrylic fibre or fabric used for clothing, carpets etc.
acrimony (ak'riməni), *n.* sharpness, bitterness of temper, manner or speech. **acrimonious** (-mō'-), *a.* bitter and irritating in temper or manner. **acrimoniously,** *adv.* **acrimoniousness,** *n.*
acro-, *comb. form* situated on the outside, beginning, termination, extremity, point or top, e.g. *acrobat, acronym.*
acrobat (ak'rəbat), *n.* a performer of daring gymnastic feats, as a tumbler or a tight-rope walker. **acrobatic** (-bat'-), *a.* pertaining to an acrobat or the performances of an acrobat. **acrobatically,** *adv.* **acrobatics,** *n.pl.* the feats performed by an acrobat.
acromegaly (akrōmeg'əli), *n.* a disease the chief feature of which is the enlargement of the face and extremities of the limbs. **acromegalic** (-gal'-), *a.*
acronym (ak'rənim), *n.* a word formed from initial letters of words, e.g. *NATO, laser.* **acronymic** (-nim'-), *a.*
acrophobia (akrəfō'biə), *n.* a morbid fear of high places.
acropolis (əkrop'əlis), *n.* the citadel of a Greek town, esp. that of Athens.
across (əkros'), *adv.,g prep.* transversely, from side to side, crosswise; upon (e.g. *come across,* to come upon accidentally); over (e.g. *across the Channel*). **across-the-board,** *a.* affecting or applying in all cases.
acrostic (əkros'tik), *n.* a composition in which the lines are laid out so that their initial letters taken in order constitute a word or short sentence; an abecedarian poem. *a.* relating to or containing an acrostic. **acrostical,** *a.* **acrostically,** *adv.*
acroterium, acroterion (akrətēriəm, -on), **acroter** (əkrō'tə, ak'-), *n.* (*pl.* **-teria, -ters**) a pedestal on a pediment, for carrying a figure. **acroterial,** *a.*
acrylic (əkril'ik), *n.* a synthetic textile fibre; paint containing an acrylic resin; a painting executed in acrylic paint. **acrylic acid,** *n.* an acid used in the manufacture of acrylic resins. **acrylic resin,** *n.* a resin consisting of a polymer of acrylic acid or one of its derivatives.
act (akt), *n.* that which is done or being done, a deed; the exertion of physical, mental or moral power; one of the principal divisions of a play, usually subdivided into smaller portions called scenes; a statute or edict of a legislative or judicial body; *v.t.* to perform; to play the part of; to impersonate. *v.i.* to exert power, to produce an effect; to be in action or motion; to carry out a purpose or determination; to behave (in a stated manner); to perform as (if) an actor. **in the (very) act,** in the actual commission (of some deed). **to act up,** (*coll.*) to behave badly; to function badly, to give trouble. **to get in on the act,** (*coll.*) to become involved in an undertaking, esp. so as to benefit. **act of God,** the operation of uncontrollable natural forces in causing an event; such an event. **actable,** *a.* capable of being performed on the stage; practically possible. **acting,** *a.* doing temporary duty. *n.* performance, action; dramatic performance. **actor,** *n.* a performer; one (male or female) who represents a character on the stage; a doer, an agent. **actress** (-tris), *n. fem.*
ACTH ADRENOCORTICOTROPHIC HORMONE.
actinic (aktin'ik), *a.* pertaining to rays; pertaining to the chemical rays of the sun. **actinic rays,** *n.pl.* electromagnetic rays capable of affecting photographic emulsions, including X-rays, ultraviolet, infrared rays etc. **actinism** (ak'-), *n.* the property in rays of light by which chemical changes are produced; the radiation of light or heat. **actinograph** (-graf), *n.* an instrument which registers the variations of chemical influence in solar rays. **actinometer** (aktinom'itə), *n.* an instrument for measuring the heating power of the sun's rays. **actinometric** (-met'-), *a.* **actinometry** (-nom'-), *n.* **actinotherapy** (-the'-), *n.* the treatment of disease by exposure to actinic radiation.
actinium (aktin'iəm), *n.* a radioactive metallic element, at. no. 89; chem. symbol Ac, found in pitchblende. **actinide** (ak'tinīd), **actinoid** (-oid), *n.* any of a series of radioactive elements beginning with actinium and ending with lawrencium.
action (ak'shən), *n.* the state or condition of acting or doing; activity; anything done or performed; a battle, an engagement; combat; the mechanism or movement of a compound instrument e.g. a clock, a piano; gesture, gesticulation; the trained motion of a horse; the events constituting the main subjects of a play or other work of fiction; a legal suit; (*coll.*) lively activity. **action committee,** *n.* a committee formed to take positive action to achieve an end. **action painting,** *n.* abstract expression using spontaneous actions of smearing, throwing etc. of paint. **action replay,** *n.* the repetition, often in slow motion, of a small piece of film showing an important or decisive (sporting) incident. **action-taking,** *a.* litigious. **actionable,** *a.* furnishing grounds for an action at law. **actionably,** *adv.*
active (ak'tiv), *a.* possessed of or exerting the power of acting; communicating action or motion; exerting influence; quick in movement, agile; continually employed, busy (opposed to idle or indolent); characterized by action, work or the performance of business; in actual operation; of a volcano, still liable to erupt; applied to intransitive verbs, or transitive verbs that attribute the action expressed to the subject of the sentence. **active suspension,** *n.* a system in which a vehicle's suspension is continuously monitored and adjusted by computer. **activate,** *v.t.* to make active; to make radioactive. **activation,** *n.* **activator,** *n.* **actively,** *adv.* **activism,** *n.* the policy of decisive action. **activist,** *n.* one who takes decisive, sometimes militant, action in support of a (political or social) cause. **activity** (-tiv'-), *n.* the quality or state of being active; exertion of energy; busy movement; a pursuit, occupation. **activity centre,** *n.* a toy for amusing and providing sensory stimuli for babies or toddlers.
actor, actress ACT.
actual (ak'chuəl), *a.* existing in reality; real, current. **actuality** (-al'-), *n.* the state of being actual; (a) reality. **actualize, -ise,** *v.t.* to make actual; to describe realistically. **actualization,** *n.* a making real or actual; realization. **actually,** *adv.*
actuary (ak'chuəri), *n.* an officer of a mercantile or insurance company, skilled in the statistics of risks. **actuarial** (-eə'ri-), *a.*
actuate (ak'chuāt), *v.t.* to excite to action, to put in action. **actuation,** *n.* **actuator,** *n.*
acuity (əkū'iti), *n.* sharpness, acuteness (of a point, an acid, disease or wit).
aculeus (əkū'liəs), *n.* (*Zool.*) a sting; (*Bot.*) a prickle. **aculeate** (-ət), **-ated** (-ātid), *a.*
acumen (ak'ūmən), *n.* acuteness of mind, shrewdness, keen perception.
acuminate (əkū'minət), *a.* tapering to a point. *v.t.* (-nāt), to sharpen, to give a point to. **acuminated,** *part.a.* **acumination,** *n.*
acupressure (ak'ūpreshə), *n.* massage using the fingertips applied to the points of the body used in acupuncture.
acupuncture (ak'ūpŭngkchə), *n.* a system of medical treatment in which the body surface is punctured by needles

at specific points to relieve pain, cure disease or produce anaesthesia. **acupuncturist** (ak'-, -püngk'-), *n.* a practitioner in acupuncture.

acute (əkūt'), *a.* terminating in a sharp point; sharp, penetrating, perceptive; shrill, high in pitch; applied also to the accent (') marking such sounds; of an illness, violent and severe; of an angle, less than 90 degrees. *n.* an acute accent. **acutely,** *adv.* **acuteness,** *n.*

-acy (-əsi), *suf.* forming nouns of quality, state, condition etc.; e.g. *fallacy, magistracy, piracy.*

ad (ad), *n. (coll.)* short for ADVERTISEMENT.

AD, *(abbr.) anno Domini,* in the year of our Lord.

ad-, *pref.* to, at, into; signifying motion towards, direction to, adherence etc. e.g. *adduce, adhere, adjacent.*

adage (ad'ij), *n.* a proverb; a pithy maxim handed down from old time.

adagio (ədah'jiō, *esp. Am.* -zhiō), *adv. (Mus.)* slowly, gracefully. *a.* slow, graceful. *n.* a slow movement of a soft, tender, elegiac character. **adagietto** (-et'ō), *n.* a slow, graceful movement but somewhat quicker than adagio. **adagissimo** (-jis'imō), *adv.* very slowly.

Adam[1] (ad'əm), *n.* the name of the first man; the fundamentally unchanged state of man. **Adam's ale,** *n.* water. **Adam's apple,** *n.* a protuberance on the forepart of the throat formed by the thyroid cartilage. **to not know someone from Adam** KNOW.

Adam[2] (ad'əm), *a.* pertaining to a decorative style of architecture and furniture designed by the Scottish brothers Robert and James Adam in the 18th cent.

adamant (ad'əmənt), *n.* a stone of impenetrable hardness; the lodestone, the diamond. *a.* made of or pertaining to adamant; hard, devoid, or incapable of feeling; resolute. **adamancy,** *n.* **adamantine** (-man'tin), *a.* made of adamant; incapable of being broken. **adamantly,** *adv.*

adapt (ədapt'), *v.t.* to fit, to adjust, to make suitable, to remodel. *v.i.* to become fit or suitable. **adaptability** (-bil'-), *n.* **adaptable,** *a.* esp. through change. **adaptation** (adəp-), *n.* the act of adapting; the state of being adapted; that which is adapted. **adaptedness,** *n.* **adapter, -or,** *a.* one who adapts; an accessory for connecting a plug etc., fitted with terminals of one type to a supply point fitted with terminals of another type, or for connecting several appliances to a single supply point. **adaptive,** *a.* tending to adapt. **adaptively,** *adv.*

adaxial (adak'siəl), *a. (Bot.)* facing the stem.

ADC, *(abbr.)* aide-de-camp.

add (ad), *v.t.* to set or put together; to join, to unite; to put into one total; to annex, to subjoin. *v.i.* to serve as an increment (to); to perform the operation of addition. **to add up,** to produce a correct total when added; to amount (to); to make sense, show a consistent pattern. **add-on,** *n.* something supplementary; a computer peripheral. **addability** (-bil'-), *n.* **addible,** *a.* **addition,** *n.* the act of adding; the thing added; the process of collecting two or more numbers or quantities into one sum; a title or designation given to a person beyond his name and surname; **in addition,** as well as, also. **additional,** *a.* added; supplementary. **additionally,** *adv.* **additive,** *a.* that may be or is to be added. **additively,** *adv.*

addax (ad'aks), *n.* a species of antelope.

addendum (əden'dəm), *n. (pl.* **-da** (-də)) a thing to be added, an addition; an appendix.

adder (ad), *n.* the common viper; applied, with epithet, to some of the foreign Viperidae, as puff-adder, death-adder.

addict[1] (ədikt'), *v.t.* (usu. *pass.*) to habituate; to make a slave to (a vice or drug). **addicted,** *part.a.* wholly devoted; slavishly dependent. **addictedness,** *n.* **addiction,** *n.* the act or state of being addicted or devoted; the thing addicted to.

addict[2] (ad'ikt), *n.* one who is addicted to some habit, esp. the taking of drugs.

Addison's disease (ad'isənz), *n.* a disease characterized by undersecretion of steroid hormones from the adrenal cortex, causing weakness, weight loss and browning of the skin.

additament (ad'itəmənt), *n.* something added.

addition, additive ADD.

addle (ad'l), *a.* putrid, bad, as an egg; empty, muddled. *v.t.* to make addled. *v.i.* to grow putrid (as an egg). **addle-headed, addle-brained, addle-pated,** *a.* terms applied to one whose brain seems muddled. **addled,** *a.* **addlement,** *n.*

address (ədres'), *v.t.* to direct (an oral or written communication) to (a person or location); to accost or speak to; to write the address or direction on. *v.reflex.* apply (oneself) to. *n.* (ədres', *esp. Amer.* ad'res) the act of addressing oneself to a person or persons; any speech or writing in which one person or body makes a communication to another person or body; the direction of a letter; the name of the place where one lives; a number that identifies a location in a computer memory where a particular piece of data is stored. **to address oneself to,** to speak to. **addressable,** *a. (Comput.)* able to be accessed by means of an address. **addressee,** *n.* one to whom a communication is addressed. **addresser, -or,** *n.* one who addresses. **Addressograph**® (-graf), *n.* a machine for addressing envelopes, wrappers etc.

adduce (ədūs'), *v.t.* to bring forward as a proof or illustration, to cite, to quote. **adducent,** *a.* bringing or drawing to a given point (used of the adductor muscles). **adducer,** *n.* **adducible,** *a.* **adduct** (ədŭkt'), *v.t.* to draw (a body part) inwards or towards another part. **adduction,** *n.* the act of adducing or adducting. **adductive,** *a.* tending to adduct. **adductor,** *n.* a muscle which brings one part of the body towards or in contact with another part.

-ade, *suf.* forms nouns denoting action, e.g. *cannonade, ambuscade;* person or body involved in action, e.g. *brigade, cavalcade;* product of action, e.g. *masquerade;* sweet drink, e.g. *lemonade.*

aden-, adeni-, adeno- *comb. form.* connected with a gland or glands; glandular; in medical terms, e.g. *adenitis, adenoid.*

adenine (ad'ənēn, -nin), *n.* one of the four purine bases in DNA and RNA.

adenitis (adəni'tis), *n.* inflammation of the lymphatic glands.

adenoid (ad'ənoid), *a.* having the form of a gland, glandular. *n.pl.* adenoid tissue; a spongy growth at the back of the nose and throat, impeding respiration and speech when enlarged. **adenoidal** (-noi'-), *a.* pertaining to the adenoids; (sounding as if) suffering from enlarged adenoids.

adenoma (adənō'mə), *n. (pl* **-mas, -mata** (-tə)) a benign tumour formed of glandular tissue.

adenosine (əden'əsēn, -sin), *n.* a compound of adenine and the sugar ribose that forms part of RNA and various compounds that provide energy in cells.

adept (ad'ept), *n.* one highly knowledgeable in any science or art. *a.* (ədept', ad'ept), thoroughly versed, well skilled. **adeption,** *a.,* *n.* **adeptly,** *adv.*

adequate (ad'ikwət), *a.* equal to a requirement, sufficient. **adequately,** *adv.* **adequateness, adequacy,** *n.*

à deux (a dœ'), *a., adv.* of or between two (people). [F]

adhere (ədhiə', ədiə'), *v.i.* to stick (to); to remain firmly attached (to); to continue to support or to follow.

adherence (ədhiə'rəns, ədiə'-, -he'-), *n.* the state or quality of adhering; firm attachment. **adherent,** *a.* sticking; tenaciously attached. *n.* one who adheres; a supporter, a follower.

adhesion (ədhē'zhən, ədē'-), *n.* the act or state of sticking, attaching oneself to, or joining; the union of structures or tissues that are normally separate; the fusion of two surfaces, as the two opposing surfaces of a wound in healing.

adhesive (-siv, -ziv), *a.* having the power of adhering; sticky, clinging. *n.* a substance used for sticking things together. **adhesively**, *adv.* **adhesiveness**, *n.*

ad hoc (ad hok, hōk), *a.*, *adv.* for one particular purpose, specially. [L]

ad hominem (ad hom'inem), *a.*, *adv.* of arguments, directed to or against the person, not objective. [L, to the man]

adiabatic (adiəbat'ik), *a.* without loss or gain of heat. **adiabatically**, *adv.*

adieu (ədū', ədyœ'), *int.*, *n.* (*pl.* **adieux** (-z), **adieus**) God be with you, farewell.

ad infinitum (ad infini'təm), *adv.* to infinity, without end. [L]

adipocere (ad'ipəsiə), *n.* a greyish-white fatty or soapy substance, into which the flesh of dead bodies buried in moist places is converted. **adipocerous** (-pos'ərəs), *a.*

adipose (ad'ipōs), *a.* pertaining to animal fat, fatty. *n.* animal fat, esp. the fat on the kidneys. **adipose tissue**, *n.* the vesicular structure in which fat is deposited. **adiposity** (-pos'-), *n.*

adit (ad'it), *n.* an entrance, passage, esp. to a mine.

Adj., Adjt, (*abbr.*) adjutant.

adj., (*abbr.*) adjective; adjust(ment).

adjacent (əjā'sənt), *a.* lying near (to); contiguous; neighbouring. **adjacency**, *n.* **adjacently**, *adv.*

adjective (aj'iktiv), *n.* a part of speech joined to a substantive to qualify and describe it. *a.* dependent. **adjectival** (-tī'-), *a.* **adjectivally**, *adv.*

adjoin (ajoin'), *v.t.* to join or add, to unite; to be contiguous to. *v.i.* to be contiguous. **adjoining**, *a.* adjacent, contiguous; neighbouring.

adjourn (əjœn'), *v.t.* to put off or defer till a later period; to suspend (a meeting) in order to meet at a later period or elsewhere; to postpone (a topic) till a future meeting. *v.i.* to cease proceedings till a later period; to move elsewhere. **adjournment**, *n.* the act of adjourning; the time during which or until which business or a meeting (esp. of a public body) is postponed.

adjudge (əjūj'), *v.t.* to make a judicial decision, to decide, pronounce, condemn. **adjudg(e)ment**, *n.* the act of adjudging; the judgment or verdict given.

adjudicate (əjoo'dikāt), *v.t.* to judge, decide. *v.i.* to act as a judge in a competition; to give one's judgment. **adjudication**, *n.* the act of adjudicating; the decision or judgment of a judge or court. **adjudicator**, *n.*

adjunct (aj'ūngkt), *n.* any thing joined to another without being an essential part of it; (*Gram.*) an extension of the subject or predicate; an associate. *a.* added to, or conjoined with any person or thing. **adjunction**, *n.* **adjunctive** (əjūngk'-), *a.* (having the quality of) joining. *n.* anything joined to another. **adjunctively**, **adjunctly**, *adv.* by way of adjunct, in connection with.

adjure (əjooə'), *v.t.* to order (someone to do something) upon oath, or upon pain of the divine displeasure; to entreat with great earnestness. **adjuration**, *n.* the act of adjuring; an order under penalty of a curse; a solemn entreaty. **adjuratory**, *a.*

adjust (əjŭst'), *v.t.* to put together, to order, to arrange; to fit, adapt; to settle, to assess (an insurance claim). *v.i.* to adapt or conform (to a new situation, environment etc.). **adjustable**, *a.* **adjusted**, *a.* (with adverbs) psychologically fit, as *well-adjusted*. **adjuster**, *n.* **loss adjuster** LOSS. **adjustment**, *n.* the act of adjusting; the state of being adjusted; a settlement of claims, liabilities, etc.

adjutant (aj'ətənt), *n.* an officer in each regiment who assists the commanding officer in matters of business, duty and discipline. **adjutant bird**, *n.* a large wading bird of the stork family. **adjutant-general**, *n.* an officer assisting a general. **adjutancy**, *n.* the office of adjutant.

ad lib (ad'lib), *adv.* at will, as one pleases. **ad-lib**, *v.t.*, *v.i.* (*past*, *p.p.* **ad-libbed**) to deliver (e.g. a speech) without notes or preparation. *a.* improvised, extempore. *n.* an improvised speech, line etc. [L *ad libitum*, at pleasure]

admass (ad'mas), *n.* the mass audience to whom television and radio advertising is directed.

admeasure (admezh'ə), *v.t.* to measure out, to apportion. **admeasurement**, *n.*

admin (ad'min), *n.* (*coll.*) short for ADMINISTRATION, administrative work.

administer (ədmin'istə), *v.t.* to manage or conduct as chief agent; to superintend the execution of (as laws) or the giving of (an oath); to manage and dispose of (the estate of a deceased person); to dispense (as medicine). *v.i.* to act as administrator. **administrable**, *a.* **administrant**, *a.*, *n.* **administration**, *n.* the act of administering; the executive functions of government; the executive officers; (*N Am.*) government. **administrative** (-trə-), *a.* **administratively**, *adv.* **administrator** (-trā-), *n.* one who administers or manages; (*Law*) one who administers the estate of an intestate. **administratorship**, *n.* **administratrix** (-triks), *n. fem.* (*pl.* **-trices** (-sēz)) a female administrator.

admirable (ad'mirəbl), *a.* worthy of admiration; excellent, highly satisfactory. **admirability** (-bil'-), **admirableness**, *n.* **admirably**, *adv.*

admiral (ad'mirəl), *n.* the commander of a fleet or a division of a fleet. This rank in Britain has four grades: Admiral of the Fleet, Admiral, Vice-Admiral and Rear-Admiral; **red admiral**, *n.* a common species of brightly-coloured butterfly. **admiralship**, *n.*

admiralty (ad'mirəlti), *n.* the office of admiral. **the Admiralty**, *n.* the government department that deals with the British navy; the Lords Commissioners who administer naval affairs in Great Britain; the building where they transact business.

admire (ədmīə'), *v.t.* to regard with wonder and pleasure; to have a high estimation of. **admiration** (-mi-), *n.* wonder excited by anything pleasing or excellent; esteem. **admirer**, *n.* one who feels admiration; a suitor, lover. **admiringly**, *adv.*

admissible (ədmis'ibl), *a.* fit to be considered as an opinion or as evidence; (*Law*) allowable as evidence. **admissibility** (-bil'-), *n.* **admissibly**, *adv.* **admission** (-shən), *n.* the act of admitting; the state of being admitted; permission to enter; concession in argument; acknowledgment, confession; a person allowed to enter (an institution, membership, etc.). **admissive**, *a.* tending to admit, implying admission.

admit (ədmit'), *v.t.* (*past*, *p.p.* **admitted**) to let in; to permit to enter, as a place, a post, or the mind; to accept as valid; to concede, to acknowledge. **admittable**, *a.* admissible. **admittance**, *n.* the act of admitting; entrance given or permitted. **admittedly**, *adv.*

admix (admiks'), *v.t.* to mix, to mingle. **admixture** (admiks'chə, ad'-), *n.* the act of mixing; something, esp. a foreign element, added to something else; an alloy, mixture.

admonish (ədmon'ish), *v.t.* to reprove gently; to caution. **admonisher**, *n.* **admonishment**, *n.* the act of admonishing; an admonition. **admonition** (ad-), *n.* (a gentle) reproof, friendly caution. **admonitive**, *a.* **admonitorily**, *adv.* **admonitor**, *n.* **admonitory**, *a.*

ad nauseam (ad naw'ziam, -si-), *adv.* to the point of producing disgust or nausea, interminably.

ado (ədoo'), *n.* business, activity; trouble, difficulty, fuss.

adobe (ədō'bi), *n.* a sun-dried brick; a clay used in making such bricks; a building made of adobe bricks.

adolescent (adəles'ənt), *a.* growing up; advancing to maturity but not yet fully mature. *n.* a person in the age of adolescence. **adolescence**, *n.* the period between childhood and adulthood; collectively, those passing through this period.

Adonai (ədō'nī, adōnī'), *n.* the Lord.

Adonis (ədō'nis), *n.* a beau, a dandy; a very handsome

young man.
adopt (ədopt'), *v.t.* to take into any relationship, as child, heir, citizen, candidate etc.; to take (a child) as one's own; to embrace, to espouse (as a principle, cause etc.); to choose for or take (e.g. citizenship, (a name) as one's own. **adopted**, *part.a.* taken as someone's own, accepted into some intimate relation such as that of child. **adopter**, *n.* **adoption**, *n.* (an) act of adopting. **adoptional**, *a.* **adoptive**, *a.* due to or by reason of adoption. **adoptively**, *adv.*
adore (ədaw'), *v.t.* to worship; to love greatly; (*coll.*) to like very much. **adorable**, *a.* worthy of worship; worthy of the utmost love and respect; (*coll.*) charming, delightful, fascinating. **adorableness**, *n.* **adorably**, *adv.* (*coll.*) delightfully. **adoration** (ad-), *n.* divine worship. **adorer**, *n.* **adoringly**, *adv.*
adorn (ədawn'), *v.t.* to decorate, ornament, embellish; to add attractiveness to. **adorner**, *n.* **adorning**, *n.* adornment. **adorningly**, *adv.* **adornment**, *n.* (a) decoration, ornament, (an) embellishment.
adrenal (ədrē'nəl), *a.* near the kidneys. **adrenal gland**, *n.* a small gland adjacent to each kidney that secretes adrenalin and steroid hormones. **adren(o)-**, *comb. form.* adrenal; adrenalin.
adrenalin, adrenaline (ədren'əlin), *n.* a hormone, secreted by the adrenal glands, which accelerates the heart; (**Adrenalin**®) synthetic adrenalin; **adrenocorticotrophic hormone** (ədrēnokawtikōtrof'ik), *n.* a hormone produced by the pituitary gland that stimulates the adrenal cortex.
adrift (ədrift'), *adv.* in a drifting condition; at the mercy of the wind and waves; (*coll.*) off course, in the wrong direction.
adroit (ədroit'), *a.* dexterous, clever, ready in mental or bodily resource. **adroitly**, *adv.* **adroitness**, *n.*
adscititious (adsitish'əs), *a.* assumed, supplemental. **adscititiously**, *adv.*
adsorb (ədsawb', -z-), *v.t.* (of a substance) to take up and cause to adhere in a thin film on the surface, in contradistinction to *absorb. v.i.* to concentrate and adhere to the surface of a solid. **adsorbent**, *n., a.* (pertaining to) a solid substance that adsorbs gases, vapours or liquids that contact it. **adsorption** (-sawp'-, -z-), *n.* concentration of a substance on a surface. **adsorptive**, *a.*
aduki bean (ədoo'ki), **adsuki** (-soo'-), **adzuki** (-zoo'-), *n.* a bean, with small reddish-brown edible seeds, grown esp. in China and Japan.
adulate (ad'ūlāt), *v.t.* to admire in the extreme, to flatter servilely. **adulation**, *n.* servile flattery. **adulator**, *n.* **adulatory**, *a.*
adult (ad'ŭlt, *esp. N Am.* ədŭlt'), *a.* (grown to be) physically and/or mentally mature; of or for adults; (*euphem.*) containing sexually explicit material, pornographic. *n.* one grown to maturity. **adult education**, *n.* part-time, usu. non-vocational courses for people over school-leaving age. **adulthood** (-hud), **adultness**, *n.*
adulterate[1] (ədŭl'tərāt), *v.t.* to corrupt or debase (anything) by mixing with it something of less worth. *a.* (-rət) adulterated, debased by admixture. **adulterant**, *n., a.* that which adulterates or is used to adulterate; adulterating. **adulteration**, *n.* the act or result of adulterating; the state of being adulterated; an adulterated substance. **adulterator**, *n.* one who adulterates.
adultery (ədŭl'təri), *n.* sexual intercourse on the part of a married person with someone not his/her spouse. **adulterate**[2] (-rət), *a.* adulterous; born of an adulterous union. **adulterer**, *n.* (*fem.* **adulteress**) a person guilty of adultery. **adulterous**, *a.* pertaining to or guilty of adultery. **adulterously**, *adv.*
adumbrate (ad'əmbrāt), *v.t.* foreshadow; to overshadow. **adumbrant**, *a.* **adumbration**, *n.* **adumbrative**, *a.*
adv., (*abbr.*) adverb(ial); advocate.
ad valorem (ad vəlaw'rem), *a., adv.* of a tax, in proportion to the value of the goods.

advance (ədvahns'), *v.t.* to bring or move forward or upwards; to promote, benefit; to supply before or on credit; to put forward for attention. *v.i.* to move forward; to progress, to rise (e.g. in one's profession, in price). *n.* the act or process of moving forward; promotion, improvement; a rise (in price); a movement towards; (*pl.*) amorous overtures; part-payment beforehand, a loan. *a.* being before in time or place; beforehand. **in advance**, beforehand; in front. **advance(d) guard**, *n.* a detachment which precedes the main body of an army. **advance man, woman**, *n.* one who visits in advance of a VIP to ensure that all arrangements are in order. **advanced**, *a.* in the front rank; far on; mature beyond one's age; (technically) highly developed. **advanced gas-cooled reactor**, a nuclear reactor in which the uranium fuel core is cooled by circulating carbon dioxide. **Advanced level** A-LEVEL. **advancer**, *n.*
advancement (ədvahns'mənt), *n.* the act of advancing; the state of being advanced; preferment, promotion; furtherance, improvement.
advantage (ədvahn'tij), *n.* favourable condition or circumstance; gain, profit, superiority of any kind; in tennis, the next point won after deuce points. **to advantage**, so as to display the best points. **to one's advantage**, in one's favour. **advantageous** (advəntā'jəs), *a.* conferring advantage; profitable, beneficial. **advantageously**, *adv.* **advantageousness**, *n.*
advent (ad'vent), *n.* the Incarnation of Christ; the Second Coming; (**Advent**) the season including the four Sundays before Christmas; any important arrival. **Adventist** (-vən-), *n.* one who believes that the Second Coming of Christ is imminent.
adventitious (advəntish'əs), *a.* extraneous; foreign; accidental. **adventitiously**, *adv.* **adventitiousness**, *n.*
adventure (ədven'chə), *n.* (a) hazard, risk; an enterprise in which hazard or risk is incurred; any novel or unexpected event. **adventure holiday**, *n.* a holiday centred round outdoor activities having an element of danger, as rock-climbing and canoeing. **adventure playground**, *n.* a children's playground containing (often old or recycled) objects that can be used in creative play. **adventurer**, *n.* (*fem.* **adventuress**) one who seeks adventures; one who seeks to gain money or social position by false pretences. **adventuresome** (-səm), *a.* adventurous. **adventuresomeness**, *n.* **adventurism**, *n.* hasty, opportunistic action, esp. in politics. **adventurous**, *a.* fond of adventure; venturesome, daring not shirking risk. **adventurously**, *adv.* **adventurousness**, *n.*
adverb (ad'vœb), *n.* a word or phrase qualifying a verb, an adjective or another adverb. **adverbial** (-vœ'-), *a.* **adverbially**, *adv.*
adversary (ad'vəsəri), *n.* an opponent, an enemy. *a.* (*Law*) opposed, hostile. **adversarious** (-seə'-), *a.* denoting opposition.
adverse (ad'vœs), *a.* acting in a contrary direction; hostile, inimical, unpropitious. **adversely**, *adv.* **adverseness**, *n.* **adversity** (-vœ'-), *n.* adverse circumstances; misfortune, trouble.
advert[1] (ədvœt'), *v.t.* to turn (one's attention to).
advert[2] (ad'vœt), *n.* (*coll.*) short for ADVERTISEMENT.
advertise (ad'vətiz), *v.t.* to give public notice of; to make publicly known; publicly to describe (a product or service) in order to promote awareness or increase sales. *v.i.* to give public notice; to issue advertisements. **advertisement** (-vœ'tiz-), *n.* the act of advertising; a public notice; a paid announcement by journal, radio, television etc. **advertiser**, *n.* one who or a journal which advertises. **advertising**, *n.* publicity; advertisements; the business of publicizing products or services.
advice (ədvīs'), *n.* counsel, opinion as to a course of action; a formal or official notice. **advice note**, *n.* a notice accompanying a delivery of goods stating what has been

advise — aestival

delivered.
advise (ədvīz'), *v.t.* to counsel; to communicate intelligence to; to inform, to notify. *v.i.* to give advice; **advisable**, *a.* capable of being advised; right, proper, wise. **advisability** (-bil'-), *n.* **advisableness**, *n.* **advisably**, *adv.*
advised, *a.* acting with deliberation; well considered, deliberate. **ill-advised** ILL. **well-advised** WELL [1]. **advisedly** (-zid-), *adv.* with careful thought; intentionally. **adviser**, **-or**, *n.* one who advises, esp. in a professional capacity. **advisership**, *n.* **advising**, *n.* advice, counsel. **advisorate**, *n.* a body of professional advisors. **advisory**, *a.* having power to advise; containing advice.
advocaat (ad'vəkah), *n.* a sweet thick liqueur containing raw egg and brandy.
advocate (ad'vəkət), *n.* one who defends or recommends a course of action, a cause; one who pleads a case in a civil or criminal court. *v.t.* (-kāt) to plead in favour of, recommend. **advocacy**, *n.* a pleading for; judicial pleading; the office of advocate; recommendation. **advocateship**, *n.* the office of an advocate; advocacy. **advocatory** (-kā'-), *a.* pertaining to advocacy.
advowson (ədvow'zən), *n.* the right of putting (a clergyman) forward to a vacant benefice in the Church of England.
advt., (*abbr.*) advertisement.
adytum (ad'itəm), *n.* (*pl.* **-ta** (-tə)) a shrine; the innermost and most sacred part of a temple.
adze (adz), *n.* a wood-cutting tool with an arched blade at right angles to the handle.
adzuki bean ADUKI BEAN.
ae (ā), *adj.* (*Sc.*) one, a.
-ae, *suf.* forming plural of unnaturalized Latin words, e.g. *areolae, rebulae*. [L]
aegis (ē'jis), *n.* protection, protective influence.
aegrotat (ī'grətat), *n.* a note certifying that a student is sick; a degree awarded to a student unable to sit the relevant examinations because of illness.
-aemia, (*esp. N Am.*) **-emia** (ē'miə), *comb. form.* pertaining to or denoting blood, esp. a specified condition of the blood.
Aeneid (ənē'id), *n.* the great epic poem of Virgil which has Aeneas for its hero.
Aeolian harp (ēō'liən), *n.* a stringed instrument played by a current of air.
aeon (ē'ən, -on), *n.* an age of the universe; a period of immense duration.
aerate (eə'rāt), *v.t.* to subject to the action of atmospheric air; to charge with carbon dioxide; to oxygenate (the blood) by respiration. **aerated**, *part.a.* exposed to the action of the air, charged with air; charged with carbon dioxide gas; effervescent. **to get aerated**, (*often mispronounced* (eə'riātid), (*sl.*) to become over-excited or angry. **aeration**, *n.* the act of aerating. **aerator**, *n.*
AERE, (*abbr.*) Atomic Energy Research Establishment.
aerial (eə'riəl), *a.* belonging to the air; resembling, produced by, operating in or inhabiting the air; growing in the air; atmospheric; high, elevated; imaginary, nonmaterial; of, for or using aircraft; effected by or operating from or against aircraft. *n.* a collector or radiator of electromagnetic waves for radio, television etc. **aerial perspective**, *n.* the representation of distance and space on a plane surface. **aerial photograph**, *n.* a photograph made from an aeroplane. **aerial surveying**, *n.* a method of surveying by the use of aerial photographs. **aerially**, *adv.*
aerie, aery (eə'ri, iə'-), **eyrie, eyry** (īə'ri, iə'-, eə'-), *n.* the nest of any bird of prey, esp. of an eagle; a human dwelling or retreat perched on a high, precipitous site.
aer(o)- *comb. form.* pertaining to the air or atmosphere; aerial, atmospheric; e.g. *aerodynamics, aeronaut, aeroplane.*
aerobatics (eərəbat'iks), *n.pl.* (the performing of) aerial acrobatics, esp. in an aeroplane.
aerobe (eə'rōb), *n.* an organism that requires oxygen for life. **aerobic** (-rō'-), *a.* using or requiring oxygen; occurring in the presence of oxygen; pertaining to aerobics. **aerobically**, *adv.* **aerobics**, *n.pl.* physical exercises designed to improve heart and lung function, esp. (*often sing. in constr.*) a system of exercises consisting of routines of rhythmic dance-like movements and stretches, usu. performed to music.
aerobiology (eərōbiol'əji), *n.* the study of air-borne microorganisms.
aerocamera (eərōkam'ərə), *n.* a special form of camera used vertically for photographing the ground from an aeroplane.
aerodensimeter (eərōdensim'itə), *n.* a pressure gauge for gases.
aerodrome (eə'rədrōm), *n.* an area, with any buildings attached, for the operation of (esp. small or military) aircraft.
aerodynamics (eərōdīnam'iks), *n.* the science which deals with the forces exerted by gases in motion. **aerodynamic**, *a.* of or involving aerodynamics; designed so as to minimize wind resistance. **aerodynamically**, *adv.*
aeroembolism (eərōem'bəlizm), *n.* the formation of nitrogen bubbles in the blood and tissues, caused by too rapid a reduction in atmospheric pressure (see CAISSON DISEASE).
aerofoil (eə'rəfoil), *n.* a wing-like structure constructed to obtain lift on its surfaces from the air, with little drag.
aerogram (eə'rəgram), *n.* **aerogramme**, an air letter.
aerography (eərog'rəfi), *n.* the description of the properties etc. of the air.
aerolite (eə'rəlit), **-lith** (-lith), *n.* a meteoric stone. **aerolitic** (-lit'-), *a.*
aerology (eərol'əji), *n.* the department of science that deals with the atmosphere. **aerological** (-loj'-), *a.* **aerologist**, *n.*
aerometer (eərom'itə), *n.* an instrument for measuring the weight and density of air and gases.
aeronaut (eə'rənawt), *n.* one concerned with the navigation of balloons or airships. **aeronautic, -ical** (-naw'-), *a.* pertaining to aerial navigation. **aeronautics**, *n.* the science or art which deals with aerial navigation.
aerophyte (eə'rəfit), *n.* a plant which grows entirely in the air, as distinguished from one growing in the ground.
aeroplane (eə'rəplān), (*N Am.*) **airplane** (eə'plān), *n.* a mechanically-driven heavier-than-air flying-machine with wings as lifting surfaces.
aerosol (eə'rəsol), *n.* a suspension of fine particles in air or gas, as in smoke or mist; a substance dispersed as such a suspension from a pressurized metal container; the container from which it is dispersed.
aerospace (eə'rəspās), *n.* the earth's atmosphere and the space beyond; the science or industry concerned with aerospace. *a.* pertaining to aerospace, to travel or operation in aerospace, or to vehicles used in aerospace.
aerostatic (eərəstat'ik), *a.* pertaining to aerostatics. **aerostatics**, *n.* the science which deals with the equilibrium and pressure of air and gases.
aerotherapy (eərōthe'rəpi), *n.* the treatment of disease by fresh or suitably medicated air.
aery (eə'ri), *a.* aerial, ethereal, unsubstantial.
Aesculapius (ēskūlā'piəs), *n.* the Greek god of medicine. **Aesculapian**, *a.*
aesthete (ēs'thēt, *esp. Amer.* es'-, (*esp. N Am.*) **esthete** (es'-), *n.* one who professes a special appreciation of the beautiful, and endeavours to carry his or her ideas into practice. **aesthetic** (əsthet'-), **-ical**, *a.* pertaining to aesthetics; appreciating the beautiful in nature and art; in accord with the laws of the beautiful, or with principles of taste. **aesthetically**, *adv.* **aestheticism**, *n.* the quality of being aesthetic; devotion to the study of the beautiful. **aestheticize, -ise**, *v.t.* **aesthetics**, *n.* the theory or philosophy of the perception of the beautiful.
aestival (ēstī'vəl), (*esp. N Am.*) **estival** (es-), *a.* of or

belonging to the summer; produced in the summer. **aestivate** (ēs'-, es'-), *v.i.* to remain (in a place) during the summer; of an animal, to fall into a summer sleep or torpor. **aestivation**, *n.* the act of remaining torpid in the summer.
aether (ē'thə) etc. ETHER etc.
aetiology (ētiol'əji), (*esp. N Am.*) **etiology**, *n.* an account of the cause of anything, assignment of a cause; the philosophy of causation; the science of the causes of disease. **aetiological** (-loj'-), *a.* pertaining to aetiology. **aetiologically**, *adv.*
AF, (*abbr.*) audio frequency.
afar (əfah'), *adv.* from, at or to a (great) distance.
AFC, (*abbr.*) Air Force Cross; Association Football Club.
affable (af'əbl), *a.* (of a person) easily approached; friendly, pleasant. **affability** (-bil'-), *n.* **affableness**, *n.* **affably**, *adv.*
affair (əfeə'), *n.* any kind of business; that which is to be done; a thing; a concern, matter or object of slight importance; a love intrigue; (*pl.*) public or private business; (*pl.*) finances; (*pl.*) circumstances.
affect (əfekt'), *v.t.* to practise, use, assume; to feign, to make a pretence of; to act upon, exert an influence upon; to attack; to impress, move emotionally. **affectation** (af-), *n.* (an) artificial appearance, assumption, pretence. **affected**, *a.* given to false show; pretending to what is not natural or real. **affectedly**, *adv.* **affectedness**, *n.* **affectible**, *a.* **affectibility** (-bil'-), *n.* **affecting**, *a.* touching, moving to emotion. **affectingly**, *adv.*
affection (əfek'shən), *n.* the state of being affected, esp. in the emotions; fondness, love; a state of the body due to any malady, disease. **affectionate** (-nət), *a.* of a loving disposition; tenderly disposed; indicating or expressing love. **affectionately**, *adv.* **affectionateness**, *n.*
afferent (af'ərənt), *a.* bringing or conducting inwards or towards, esp. conducting nerve impulses towards the brain or spinal cord.
affettuoso (afetuō'sō), *adv.* (*Mus.*) with feeling. [It.]
affiance (əfī'əns), *v.t.* (usu. *pass.*) to promise solemnly in marriage. *n.* contract of marriage, betrothal. **affianced**, *a.* promised in marriage, betrothed.
affidavit (afidā'vit), *n.* a voluntary statement of facts sworn before a person qualified to administer an oath.
affiliate (əfil'iāt), *v.t.* to receive as a member or branch. *v.i.* to become connected or associated, combine. *n.*, *a.* (-ət) (of) a person, organization etc. affiliated. **affiliable**, *a.* **affiliation**, *n.* the act or an instance of affiliating; adoption. **affiliation order**, *n.* a legal order requiring the father of an illegitimate child to make maintenance payments.
affinity (əfin'iti), *n.* connection; resemblance due to common origin; physical or spiritual attraction; chemical attraction, the property by which elements unite to form new compounds. **affinity card**, *n.* a credit card whose sponsoring bank etc makes a donation to a charity each time the holder uses it. **affinity marketing**, *n.* the concept of associating credit cards and charities by means of affinity cards.
affirm (əfœm'), *v.t.* to assert the truth of positively or solemnly; to confirm, reinforce. **affirmable**, *a.* **affirmance**, *n.* **affirmation** (af-), *n.* the act of affirming anything; that which is affirmed; a solemn declaration made under penalties. **affirmative**, *a.* relating to or containing an affirmation; confirmatory, positive. *n.* that which affirms a 'yes'. **in the affirmative**, yes. **affirmatively**, *adv.*
affix (əfiks'), *v.t.* to fix, fasten, attach. **affix** (af'iks), *n.* an addition; a word or syllable added to the beginning or the end of, or inserted in a word or root to produce a derived word or inflection: a prefix, suffix or infix.
afflatus (əflā'təs), *n.* inspiration; poetic impulse.
afflict (əflikt'), *v.t.* to inflict bodily or mental pain on; to trouble greatly. **afflicted**, *a.* **afflictingly**, *adv.* **affliction**, *n.* the state of being afflicted; trouble, misery, distress;

mental or bodily ailment. **afflictive**, *a.* causing affliction; distressing. **afflictively**, *adv.*
affluent (af'luənt), *a.* wealthy. **affluently**, *adv.* **affluence**, *n.* prosperity, wealth.
afford (əfawd'), *v.t.* to provide the means of, to supply; to be able to bear the expense of. **affordable**, *a.* **affordability** (-bil'-), *n.*
afforest (əfo'rist), *v.t.* to convert into forest, to plant trees on. **afforestation**, *n.* the act of converting waste or other land into forest.
affranchise (əfran'chiz), *v.t.* to make free; to set at liberty physically or morally. **affranchisement** (-chiz-), *n.* the act of making free; emancipation.
affray (əfrā'), *n.* commotion, tumult; a fight between two or more persons in a public place.
affront (əfrŭnt'), *v.t.* to make ashamed; to offend. *n.* an insult. **affrontingly**, *adv.*
Afghan (af'gan), *a.* belonging to Afghanistan. *n.* a native of Afghanistan; the language of Afghanistan, Pushtoo; an Afghan hound; an Afghan coat. **Afghan coat**, *n.* a type of sheepskin coat. **Afghan hound**, *n.* a tall slim hunting dog with long silky hair.
aficionado (əfishənah'do), *n.* (*pl.* **-dos**) a keen, knowledgeable follower or fan. **aficionada** (-də), *n.fem.* (*pl.* **-das**).
afield, FAR AFIELD under FAR.
aflame (əflām'), *adv.*, *pred.a.* flaming; in or into flame.
aflatoxin (af'lətoksin), *n.* a carcinogenic toxin produced in badly stored peanuts, maize etc. by the mould *Aspergillus flavus*. [*Aspergillus flavus*, TOXIN]
afloat (əflōt'), *a.*, *adv.* floating; at sea, aboard ship; out of debt, financially solvent.
AFM, (*abbr.*) Air Force Medal.
afoot (əfut'), *adv.* in motion, in action.
afore (əfaw'), *adv.*, *prep.* before, in front of. **afore-cited**, *a.* already cited. **afore-going**, *n.*, *a.* (the) preceding. **aforementioned**, *a.* previously-mentioned. **aforenamed**, *a.* previously-named. **aforesaid**, *a.* said or mentioned before. **aforethought**, *pred. a.* premeditated.
a fortiori (ā fawtiaw'ri, ah), *adv.* with still more reason; still more conclusively. [L]
afraid (əfrād'), *a.* filled with fear, terrified; apprehensive; regretfully admitting or of the opinion.
afresh (əfresh'), *adv.* again, anew.
African (af'rikən), *a.* pertaining to Africa. *n.* a native of Africa; a person, wherever born, who belongs ethnologically to one of the African races. **Afric**, *a.* (*poet.*) African. **African oak**, *n.* a wood from Sierra Leone resembling oak or mahogany, sometimes called African teak or African mahogany. **African violet**, *n.* a tropical African plant with velvety leaves and pink, white or violet flowers. **Africanism**, *n.* a characteristic of Africa or of the African culture or language. **Africanize**, **-ise**, *v.t.* to make African; to bring under Black African influence or control. **Africanization**, *n.*
Afrikaans (afrikahnz', -s), the South African Dutch language, an official language of South Africa.
Afrikander (afrikan'də), *n.*, *a.* a South African breed of sheep and of cattle.
Afrikaner (afrikah'nə), *n.* a person born in South Africa of white parents, whose mother tongue is Afrikaans.
Afro-, *comb. form.* pertaining to Africa or Africans.
Afro (af'rō), *n.*, *a.* (of) a hairstyle characterized by thick, bushy, curly hair.
Afro-American (afrōəme'rikən), *n.*, *a.* (of) an American of African descent.
afrormosia (afrəmō'ziə), *n.* any tree of the African genus *Afrormosia*, with dark hard wood used for furniture; this wood.
aft (ahft), *a.*, *adv.* towards or at the stern of a vessel; abaft.
after (ahf'tə), *adv.*, *prep.*, *conj.* in the rear; behind; in pursuit of; in imitation of, according to; following (in time),

aftermath / **aggression**

next to; subsequently to, at the time subsequent. *a.* later, subsequent; located (further) towards the rear or stern, posterior. **after all** ALL. **after-birth**, *n.* the placenta. **after-burner**, *n.* a device for burning fuel unspent in the main burner of a jet engine. **after-care**, *n.* care or supervision following a person's discharge from hospital, prison etc. **after-damp**, *n.* carbon dioxide gas resulting from the combustion of fire-damp in coal-mines. **after-effect**, *n.* an effect that follows some time after the cause. **after-glow**, *n.* glow in the western sky after sunset; (*coll.*) a feeling of pleasure after an enjoyable experience. **after-image**, *n.* the image that remains for a moment after looking away from an object at which one has been gazing steadily. **afterlife**, *n.* life after death. **afternoon**, *n.* the time between midday and evening. **afternoon-buyer**, *n.* a purchaser who waits till after the market dinner, in the hope of cheaper prices. **after-pains**, *n.pl.* the pains which follow childbirth, and by which the after-birth is expelled. **after-piece**, *n.* a short piece acted after a more important play. **afters**, *n.pl.* (*coll.*) what follows the main course at a meal, dessert. **aftershave**, *n.* a cosmetic lotion applied to the face after shaving. **after-taste**, *n.* a (usu. unpleasant) taste that persists after eating or drinking; an impression or feeling that remains. **afterthought**, *n.* reflection after the act; something done belatedly, after the main action. **afterward**, **afterwards**, *adv.* subsequently; at a later period. **afterword**, *n.* an epilogue in a piece of writing, in contradistinction to *foreword*. **afterworld**, *n.* the world supposedly inhabited by dead souls.
aftermath (ahf'təmahth), *n.* consequences.
aftermost (ahf'təməst), *a.* nearest the stern.
AG, (*abbr.*) adjutant-general; attorney general; joint-stock company (from G *Aktiengesellschaft*).
Ag, (*chem. symbol*) silver.
aga, **agha** (ah'gə), *n.* a Turkish civil or military officer of high rank. **Aga Khan**, *n.* the nominated hereditary spiritual head of the Ismaili sect of Muslims.
again (əgen', əgān'), *adv.* a second time, once more; moreover; in addition; on the other hand. **again and again**, *adv.* with frequent repetition, repeatedly.
against (əgenst', əgānst'), *prep.* in opposition to; opposite to, in contrast to; in contact with, on; in preparation or provision for. **against the grain**, contrary to inclination, reluctantly, with aversion. **up against** UP.
Agapanthus (agəpan'thəs), *n.* a genus of ornamental plants with bright blue flowers.
agape (əgāp'), *adv.*, *pred.a.* (of the mouth) wide open, in an attitude of amazement or wondering expectation.
agar-agar (āgah-ā'gah), **agar**, *n.* a gelatinous substance obtained from seaweeds and used for the artificial cultivation of bacteria and in cooking.
agaric (əga'rik), *n.* a mushroom; the name of several species of fungus. *a.* fungoid.
agate (ag'ət, -āt), *n.* any semi-pellucid variety of chalcedony, marked with bands or clouds, or infiltrated by other minerals, and used for seals, jewellery etc.
Agave (əgā'vi), *n.* a genus of spiny-leaved plants found in Central America.
age (āj), *n.* a period of existence; duration to date of the existence of a person or thing; a period or stage of life; the latter portion of life; senility; (*legal*) maturity, majority; an epoch, a generation; an aeon. (*usu. pl.*) (*coll.*) a long time. *v.t.*, *v.i.* to (cause to) grow old or to show signs of age. **age of consent**, the age at which a person's consent is legally valid, esp. a female's consent to sexual intercourse (16 in English and Scottish law). **age of discretion**, the age when one is judged able to use one's discretion, in English law, 14. **of age**, having reached the age of 18. **aged** (ājd), *a.* of a certain age; (ā'jid) old. **the aged** (ā'jid), old people. **agedness** (-jid-), *n.* the state of being old, the state of having attained a certain age. **ageism**, *n.* unfair discrimination on grounds of age. **ageist**, *a.* supporting or influenced by ageism. **ageless** (-lis), *a.* never growing aged, never coming to an end; perenially tashionable.
-age (-ij), *suf.* appertaining to, aggregate of: forms abstract or collective nouns, e.g. *baronage*, *courage*, *foliage*; notes act of doing or thing done, e.g. *passage*, *voyage*.
agency AGENT.
agenda (əjen'də), *n.* a memorandum-book; a list of the pieces of business to be transacted; (*pl.*) things to be done, engagements to be kept; (*coll.*) a timetable of events. [L, pl. of *agendum*, neut. ger. of *agere*, to do]
agent (ā'jənt), *n.* one who or that which exerts power; something or someone that produces an effect, the material cause or instrument; one who transacts business on behalf of another; a spy. **agent orange**, *n.* a herbicide containing dioxin as an impurity, used as a defoliant in the Vietnam war (from the orange marking on the container). **agency**, *n.* the office or business of an agent; causative action, instrumentality; (place of business, office of) a service organization, e.g. *employment agency*. **agential** (əjen'-), *a.* of or pertaining to an agent or agency.
agent provocateur (a'zhē provokatœ'), *n.* (*pl.* **agents provocateurs**) a person employed to detect suspected political offenders by leading them on to some overt action.
Ageratum (ajərā'təm), *n.* a genus of low plants of the aster family.
agglomerate (əglom'ərāt), *v.t.*, *v.i.* to heap up or collect into a ball or mass. *v.i.* to gather in a mass. **agglomerate** (-rət), *a.* heaped up. *n.* a mass; a mass of volcanic fragments united by heat. **agglomeration**, *n.* the act of agglomerating; a mass, a heap. **agglomerative**, *a.*
agglutinate (əgloo'tināt), *v.t.* to glue together; to turn into glue; to cause to adhere; to compound, e.g. simple words into compounds; to cause (bacteria, blood cells etc.) to collect into clumps. *v.i.* to unite, cohere; to form compound words. **agglutinate** (-nət), *a.* glued together; (*Philol.*) consisting of simple words, or roots, combined into compounds by straightforward suffixation. **agglutination**, *n.* the act of gluing or cementing; the formation of simple words or roots into compound terms by mere suffixation; the clumping together of red blood cells, bacteria etc. **agglutinative** (-nə-), *a.*
aggrandize, **-ise** (əgran'dīz), *v.t.* to enlarge; to make great(er) in power, wealth, rank or reputation. **aggrandization**, **aggrandizement** (-diz-), *n.* the act of aggrandizing; the state of being aggrandized.
aggravate (ag'rəvāt), *v.t.* to render less tolerable; to make worse or more severe; (*coll.*) to exasperate, to irritate. **aggravated**, *a.* (*Law*) more serious than normal, often due to the use of violence. **aggravating**, *a.* **aggravatingly**, *adv.* **aggravation**, *n.* the act of aggravating; the state of being aggravated; that which aggravates; an addition to a crime or charge; (*coll.*) trouble, often involving physical violence.
aggregate[1] (ag'rigāt), *v.t.*, *v.i.* to collect together into a mass or whole; *v.i.* to unite.
aggregate[2] (ag'rigət), *a.* collected together; formed of separate parts combined into a mass or whole; consisting of individuals united in a compound organism; composed of distinct minerals. *n.* a mass formed by the union of individual particles; the total, the whole; particles to be bonded together to form concrete. **in (the) aggregate**, collectively, as a whole. **aggregation**, *n.* the act of collecting together; the state of being aggregated; an aggregate. **aggregative**, *a.*
aggression (əgresh'ən), *n.* (an) unprovoked attack or injury; violation of a country's territorial integrity or sovereignty by another country; hostile attitude, outlook or behaviour. **aggressive**, *a.* involving an act of aggression; making the first attack; hostile, pugnacious. **aggressively**, *adv.* **aggressiveness**, *n.* **aggressor**, *n.* one who begins a quarrel or performs a hostile act.

aggrieved (əgrēvd'), *a.* caused grief, annoyance or pain; perpetrated injustice against.
aggro (ag'rō), *n.* (*coll.*) aggressive, annoying behaviour; trouble(-making).
agha AGA.
aghast (əgahst'), *a.* appalled, horrified.
agile (aj'īl), *a.* having the faculty of moving quickly and gracefully; mentally quick. **agilely,** *adv.* **agility** (əjil'-), *n.*
agitate (aj'itāt), *v.t.* to shake or move briskly; to excite, to perturb. *v.i.* to arouse public feeling or opinion for or against something. **agitation,** *n.* **agitator,** *n.* one who or that which agitates; one who excites or keeps up (esp. political) agitation; a mechanical contrivance for stirring and mixing.
agitato (ajitah'tō), *adv.* (*Mus.*) in an agitated manner.
agitpop (aj'itpop), *n.* political propaganda communicated via pop music.
agitprop (aj'itprop), *n.* political propaganda, esp. pro-Communist.
agley (əglā'), *adv.* (*Sc.*) astray, awry.
aglow (əglō'), *adv.* in a glow.
AGM, (*abbr.*) annual general meeting.
agnail (ag'nāl), **hang-nail** (hang'-), *n.* a sore at the root of toe- or fingernail, a whitlow.
agnostic (əgnos'tik), *n.* one who is uncertain about the existence of a God; one who denies that humans have any knowledge except of material phenomena. *a.* pertaining to agnostics or their teachings. **agnostically,** *adv.* **agnosticism** (-sizm), *n.* the teachings of the agnostics; the state of being agnostic.
Agnus Dei (ag'nus dā'ē, an'yus), *n.* a figure of a lamb bearing a flag or cross; (a musical setting of) the part of the Mass beginning with the words *Agnus Dei.*
ago (əgō'), *pred.a.*, *adv.* gone by, bygone, passed, passed away; since.
agog (əgog'), *adv.* in a state of eager expectation and rapt attention.
-agogue *comb. form.* a leader, a leading, as in *demagogue, pedagogue, synagogue.*
agonize, -ise (ag'əniz), *v.t.* to subject to extreme pain; to torture. *v.i.* to suffer agony; to make desperate or convulsive efforts; to put oneself through great mental distress (esp. over the rights and wrongs of a course of action). **agonized,** *a..* **agonizing,** *a.* causing or suffering agony. **agonizingly,** *adv.*
agony (ag'əni), *n.* anguish of mind; extreme physical pain; a paroxysm of pain or pleasure; the death struggle; a painful struggle or contest; the mental anguish of Christ in Gethsemane. **to pile on the agony,** (*coll.*) to exaggerate, describe in the most sensational terms. **agony aunt,** *n.* the person in charge of an agony column; a woman who gives sympathetic advice. **agony column,** *n.* the column in a newspaper devoted to readers' problems.
agora (ag'ərə), *n.* the public square, forum or marketplace of an ancient Greek town. **agoraphobia,** *n.* abnormal dread of open spaces. **agoraphobic,** *a.*, *n.*
agouti, agouty (əgoo'ti), *n.* (*pl.* **-tis, -ties**) a small W Indian and S American rodent.
AGR, (*abbr.*) advanced gas-cooled reactor.
agr., agric., (*abbr.*) agriculture, agricultural.
agraphia (əgraf'iə), *n.* loss of the cerebral power of expressing one's ideas in writing. **agraphic,** *a.*
agrarian (əgreə'riən), *a.* pertaining to landed property or cultivated land. *n.* a person in favour of the redistribution of landed property. **agrarian crime, agrarian outrage,** *n.* crime or outrage arising out of disputes about land. **agrarianism,** *n.* a redistribution of land; political agitation concerning land or land-tenure. **agrarianize, -ise,** *v.t.* to apportion (land) by an agrarian law; to convert to agrarianism. **agrarianization,** *n.*
agree (əgrē'), *v.i.* to be of the same opinion, intention etc.; to live in concord; to consent, to accede; to settle by stipulation; to harmonize with, to coincide; to be of the same grammatical number, gender or case. *v.t.* to make agreeable or harmonious; to settle, to reconcile; to arrange, to render consistent.
agreeable (əgrē'əbl), *a.* affording pleasure, pleasing, pleasant; favourable, disposed to agree (to). **agreeability** (-bil'-), **agreeableness,** *n.* **agreeably,** *adv.*
agreement (əgrē'mənt), *n.* a coming into accord; mutual conformity, understanding, accordance; grammatical similarity of number, gender or case; a mutually accepted decision; a contract duly executed and legally binding.
agribusiness (ag'ribizniz), *n.* the businesses involved in farming and marketing farm produce taken as a whole.
agriculture (ag'rikŭlchə), *n.* the science and practice of cultivating the soil, growing crops and rearing livestock. **agricultural** (-kŭl'-), *a.* pertaining to the culture of the soil. **agriculturalist, agriculturist** (-kŭl'-), *n.* one engaged in agriculture.
agrimotor (ag'rimōtə), *n.* a motor tractor used in agriculture.
agro-, *comb. form.* pertaining to fields, soil or agriculture.
agrobiology (agrōbiol'əji), *n.* the study of plant nutrition etc., in relation to soil management. **agrobiological** (-loj'-), *a.* **agrobiologist,** *n.*
agrochemical (agrōkem'ikl), *n.* a chemical for use on the land or in farming. *a.* of or producing agrochemicals.
agronomy (əgron'əmi), *n.* the management of cultivated land. **agronomic** (-nom'-), **-ical,** *a.* of or pertaining to agronomy. **agronomics,** *n.* science of cultivated land management as a branch of economics. **agronomist,** *n.* a rural economist; one skilled in agronomy.
aground (əgrownd'), *adv., pred.a.* on the ground; on the shallow bottom of any water.
agt., (*abbr.*) agent.
agterskot (ahkh'təskot), *n.* (*S Afr.*) a percentage paid to (fruit) farmers after the first payment has been made.
ague (ā'gū), *n.* a malarial fever, marked by successive hot and cold paroxysms, the latter attended with shivering; any fit of shivering or shaking. **agued,** *part.a.* affected with ague; shaking with fear. **aguish,** *a.*
ah (ah), *int.* an exclamation expressive of various emotions, according to the manner in which it is uttered, e.g. sorrow, regret, fatigue, relief, surprise, admiration, appeal, remonstrance, aversion, contempt, mockery.
aha (əhah'), *int.* an exclamation expressive of surprise, triumph or mockery.
ahead (əhed'), *adv., pred.a.* in advance; forward, onward; at the head, in front. **to get ahead** GET[1]. **to go ahead** GO[1].
ahem (əhem'), *int.* an exclamation used to attract attention or merely to gain time.
ahimsa (əhim'sə), *n.* the Hindu and Buddhist doctrine of non-violence towards all living things.
ahoy (əhoi'), *int.* (*Naut.*) a word used in hailing to attract attention.
AI, (*abbr.*) artificial insemination; (*Comput.*) artificial intelligence.
AID, (*abbr.*) Agency for International Development (US); artificial insemination by donor.
aid (ād), *v.t.* to assist, to help. *n.* help, assistance, succour, relief; anything, as an apparatus, by which assistance is rendered; money etc. given to poorer countries or disaster areas by the government or by charitable organizations. **in aid of,** in support of, so as to help; (*coll.*) intended for.
aide (ād), *n.* an assistant, a help esp. in the army or government. **aide-de-camp** (ād-dəkamp', -kā', ed-), *n.* (*pl.* **aides-de-camp**) an officer who receives and transmits the orders of a general. **aide-memoire** (ādmemwah'), *n.* an aid to memory, a memorandum(-book).
AIDS (ādz), *n.* acronym for *acquired immune deficiency syndrome,* a condition in which the body's immune system is attacked by a virus, leaving the body

aiguille (āgwēl', ā'-), *n.* a slender, needle-shaped peak of rock; an instrument used in boring holes in rock for blasting.
AIH, (*abbr.*) artificial insemination by husband.
aikido (īkē'dō), *n.* a Japanese martial art using throws, locks and twisting techniques to turn an opponent's momentum against him- or herself.
ail (āl), *v.t.* to trouble; to cause pain or uneasiness of body or mind to. **ailing**, *part.a.* affected with illness, sick, physically or mentally suffering. **ailment**, *n.* a (slight) disorder or illness, sickness.
aileron (ā'lǝron), *n.* the hinged portion on the rear edge of the wing-tip of an aeroplane for purposes of control.
aim (ām), *v.t.* to point at a target with (a missile or weapon); to direct (a blow). *v.i.* to point a missile or weapon (at a target); to form plans, intend (to do something). *n.* the act of aiming; skill in aiming; the point or object aimed at; direction of a missile; purpose, intention, design. **aimer**, *n.* **aimless** (-lis), *a.* without direction or (specific) purpose. **aimlessly**, *adv.* **aimlessness**, *n.*
ain't (ānt), (*coll.*) are not; is not, am not; have not, has not.
air (eǝ), *n.* the mixture of oxygen and nitrogen enveloping the earth, the atmosphere; open space; a light wind, a breeze; manner, appearance, mien; (*usu. pl.*) affectation, haughtiness; (the operation of, or transportation by) aircraft; the medium of broadcasting, airwaves; a tune, melody, either solo or in harmony. *v.t.* to expose to open or fresh air; to ventilate, to dry or warm (as clothes) by exposing to heat; to show off, to parade. *v.i.* to become aired; to be broadcast (at a given time). **castles in the air** CASTLE. **in the air**, projected, dreamed of, anticipated; rumoured. **off (the) air**, not broadcasting. **on (the) air**, broadcasting. **to take the air**, to go for a walk outdoors. **up in the air** UP. **air bag**, *n.* a safety device in a car consisting of a bag that inflates automatically in a collision, cushioning the passengers against the impact. **air base**, *n.* a place used as a base for operations or the housing of aircraft. **air-bed**, *n.* a (*usu.* temporary) bed or mattress inflated with air. **airborne**, *a.* carried by air. **air-brake**, *n.* a brake worked by compressed air. **air-brick**, *n.* a perforated brick or iron grating for admitting air through a wall. **airbrush**, *n.* a device for spraying paint by compressed air. *v.t.* to paint with an airbrush. **airbus**, *n.* a large passenger jet aeroplane used for short intercity flights. **Air Chief Marshal**, *n.* an officer in the RAF corresponding in rank to general in the Army. **Air Commodore**, *n.* an officer in RAF corresponding in rank to brigadier in the army. **air-condition**, *v.t.* to equip (as a building) with an air-conditioning system. **air-conditioned**, *a.* **air-conditioner**, *n.* **air-conditioning**, *n.* an apparatus for, or the process of, purifying the air circulating in a room or building and controlling its temperature and humidity. **air-cool**, *v.t.* to cool (an engine) by circulating air. **air-cooled**, *a.* **air-corridor**, *n.* a path for air traffic in an area where flying is restricted. **aircraft**, *n. pl.* collective term for all types of flying-machines, both heavier and lighter than air. (*sing.*) one of these machines. **aircraft-carrier**, *n.* a ship designed for the housing and servicing of aircraft, with a deck where they can take off and land. **aircraftman**, *n.*, **aircraftwoman**, *n.fem.* a person of lowest rank in the RAF. **aircrew**, *n.* the crew of a aircraft. **air cushion**, *n.* a cushion or pillow inflated to make it resilient; the body of air supporting a hovercraft. **air drop**, *n.* a delivery of supplies or troops by parachute from an aircraft. **air-drop**, *v.t.* **airfield**, *n.* a field specially prepared for the landing and taking-off of aircraft. **air force**, *n.* the branch of a country's armed forces organized for warfare in the air. **airframe**, *n.* the structure and external surfaces of an aircraft or rocket, excluding the engines. **airgun**, *n.* a gun from which missiles are projected by compressed air. **airhead**,
n. (*derog. sl.*) a person of minimal mental capacity. **airhole**, *n.* an opening to admit air. **air hostess**, *n.* a woman employed to attend to the comfort of passengers on aeroplanes. **air-lane**, *n.* a path regularly used by aircraft. **air letter**, *n.* (a letter on) a single sheet of lightweight paper to be folded and sent by airmail, an aerogram. **air-lift**, *n.* an operation mounted for the purpose of transporting supplies, goods etc. by air, esp. to an area under attack, a disaster area, etc. **airline**, *n.* a commercial organization operating regular transport by air. **airliner**, *n.* a passenger-carrying aeroplane flying along a regular air route. **air-lock**, *n.* a pneumatic chamber allowing entrance to or exit from a caisson without loss of air-pressure; an obstruction in a pipe caused by a bubble of air. **airmail**, *n.* mail conveyed by aircraft; the postal system of conveying mail by air. **airman**, *n.*, **airwoman**, *n.fem.* aviator, pilot of an aeroplane or airship. **Air Marshal**, *n.* an officer in the RAF corresponding in rank to lieutenant-general in the army. **air-miss**, *n.* a near-collision of aircraft. **airplane** AEROPLANE. **air-plant**, *n.* an epiphyte. **air pocket**, *n.* an area of rarefied atmosphere where an aircraft drops unexpectedly. **airport**, *n.* a facility for the take-off and landing, maintenance and fuelling of passenger aircraft. **air power**, *n.* the strength of a country in military aircraft. **air-pump**, *n.* an instrument for pumping air in and/or out. **air-raid**, *n.* an attack by hostile aircraft. **Air Raid Precautions**, *n.pl.* official regulations for the prevention of air raids, or for minimizing the damage caused by them. **air rifle**, *n.* a rifle from which missiles are projected by compressed air. **air sac**, *n.* a tiny air cell in the lungs, an alveolus; an air-filled space connecting with the lungs in birds. **airscrew**, *n.* the propeller of an aircraft. **air-shaft**, *n.* a vertical passage into a mine for the purpose of ventilation. **airship**, *n.* a lighter-than-air flying-machine driven by an engine. **air-sickness**, *n.* nausea caused by the motion of aircraft. **air-sick**, *a.* **air side**, *n.* the part of an airport complex beyond the passport control, towards the tarmac. **airspace**, *n.* the atmosphere above (a certain part of) the earth, esp. above a particular country and regarded as part of its territory. **air speed**, *n.* the speed of an aeroplane or airship relative to the air, as distinct from its speed relative to the ground. **airstrip**, *n.* a (*usu.* temporary or makeshift) strip of even ground for taking-off and landing of aircraft. **air terminal**, *n.* a building where passengers assemble to be taken to an airport. **airtight**, *a.* so tight as to prevent the passage of air. **airtime**, *n.* broadcasting time, esp. on radio, allotted to a particular topic, political party or record etc. **air-to-air**, *a.* between aircraft in flight; launched from one aircraft at another. **air-to-surface**, *a.* launched from an aircraft at a target on the ground. **air-traffic control**, *n.* the ground-based organization which determines the altitudes, routes etc. to be used by aircraft in a particular area. **Air Vice-Marshal**, *n.* an officer in the RAF corresponding in rank to major-general in the army. **airways**, *n.pl.* (radio) broadcasting channels. **air-way**, *n.* a tunnel in a mine, fitted with valve-like doors, for the passage of air in one direction; a passage cleared in the throat to enable or aid breathing; a tube employed thus. **airer**, *n.* a clothes-horse. **airing**, *n.* exposure to the free action of the air, or to a fire or heat; a walk or ride in the open air. **to get, give an airing**, to (cause to) be heard or publicized. **airing cupboard**, *n.* a heated cupboard fitted with racks for airing esp. sheets and other household linen. **airless** (-lis), *a.* not open to the air; close, musty; calm, still. **airlessness**, *n.* **airworthy**, *a.* of an aeroplane, examined and passed as fit for flying. **airworthiness**, *n.* **airy**, *a.* of or belonging to the air; consisting of or open to the air; of a room etc., having a great deal of space and light; light, unsubstantial as air. **airy-fairy**, *a.* (*coll.*) fanciful, unrealistic. **airily**, *adv.* **airiness**, *n.*
airedale (eǝ'dāl), *n.* the largest type of terrier, often used

airt (eət), *n.* (Sc.) a point of the compass; a direction.
aisle (īl), *n.* a wing or lateral division of a church; hence, a passage between the seats in a place of worship; an avenue; a corridor, gangway, e.g. in a theatre. **aisled**, *a.* furnished with aisles.
ait (āt), *n.* a small island, esp. one in a river or lake.
aitch (āch), *n.* the letter *h*.
aitchbone (āch'bōn), *n.* the rump bone; the cut of beef over this bone.
ajar¹ (əjah'), *a., adv.* partly open, as a door or (less usually) a window.
akimbo (əkim'bō), *adv.* with the hands resting on the hips and the elbows turned outwards.
akin (əkin'), *a.* allied by blood relationship; similar in properties or character.
Al, (*chem. symbol*) aluminium.
-al (-əl), *suf.* belonging to, capable of, like, e.g. *annual, equal, mortal*; forming substantives, e.g. *animal, canal, hospital*.
Ala., (*abbr.*) Alabama.
à la (a la), *prep.* in the fashion of, after the manner of; of food, prepared with (an ingredient) or in the style of (a person or country). **à la française, grècque, à l'anglaise** etc., in the French, Greek, English etc. style. [F]
alabaster (al'əbastə), *n.* massive gypsum, and other kinds of sulphate or carbonate of lime, either white or delicately shaded. *a.* made of alabaster; white and translucent like alabaster. **alabastrine** (-bas'trin, -trīn), *a.* made of or resembling alabaster.
à la carte (a la kaht'), *a., adv.* (according to a menu) having each dish priced separately, in contradistinction to *table d'hôte*.
alack (əlak'), *int.* (*dated*) an exclamation of sorrow.
alacrity (əlak'riti), *n.* briskness, eagerness; speed.
à la mode, alamode (a la mōd'), *adv., a.* fashionable.
alar (ā'lə), **alary**, *a.* pertaining to a wing; wing-like, wing-shaped.
Alar® (ē'lah), *n.* trade name of a chemical sprayed on apples to enhance their appearance and reduce their rate of decay.
alarm (əlahm'), *n.* a summons to arms; (*poet.* **alarum** (əla'rəm)) (a device for giving) warning of approaching danger; terror mingled with surprise; a device for waking persons from sleep or arousing attention. *v.t.* to rouse to a sense of danger; to inspire with apprehension of coming evil. **alarm-bell**, *n.* a bell rung to sound an alarm. **alarm-clock**, *n.* a clock that can be set to sound an alarm at a particular hour. **alarming**, *part.a.* exciting apprehension or fear. **alarmingly**, *adv.* **alarmist**, *n.* one who needlessly raises doubts; a panic-monger. *a.* provoking needless panic.
alas (əlas'), *int.* (*dated*) an exclamation of sorrow, grief, pity or concern.
alb (alb), *n.* a kind of surplice with close sleeves worn by priests when celebrating Mass.
Albanian (albā'niən), *a.* pertaining to Albania or its inhabitants. *n.* a native of Albania; the language of Albania.
albatross (al'bətros), *n.* the English name of a genus of petrels; the largest known sea-bird, the great albatross.
albeit (awlbē'it), *conj.* although, even though, notwithstanding.
albert (al'bət), *n.* a short kind of watch-chain, fastened to a waistcoat buttonhole.
albescent (albes'ənt), *a.* becoming or passing into white; whitish. **albescence**, *n.*
albino (albē'nō), *n.* a human being, or animal, having the colour pigment absent from the skin, the hair and the eyes, so as to be abnormally light in colour; a plant in which little or no chlorophyll is developed. **albinism**, (al'bi-), *n.*
Albion (al'biən), *n.* an old name for Britain or Scotland, still retained in poetry.
album (al'bəm), *n.* a blank book for the insertion of photographs, autographs, drawings or the like; a collection of pieces of recorded music issued on one or more long-playing records, cassettes etc; such a record.
albumen (al'būmin), *n.* the white of an egg; albumin; the substance interposed between the skin and embryo of many seeds. **albumenize, -ise** (-bū'-), *v.t.* to coat (a photographic paper or plate) with an albuminous solution.
albumin (al'būmin), *n.* any of several water-soluble proteins existing in animals, in the white of egg, in blood serum and in plants. **albuminize, -ise** (-bū'-), *v.t.* to convert into albumen. **albuminoid** (-bū'minoid), **-oidal** (-oi'-), *a.* resembling or of the nature of albumen. *n.* a protein; a fibrous protein, as collagen or keratin. **albuminous** (-bū'-), **-ose** (-nōs), *a.* consisting of, resembling or containing albumen. **albuminuria** (-nū'riə), *n.* the presence of albumen in the urine; the morbid condition causing this.
alchemy (al'kəmi), *n.* the chemistry of the Middle Ages, the search for the philosophers' stone and the elixir of life; the power of transmutation into gold. **alchemic, -ical** (-kem'-), *a.* of or pertaining to alchemy. **alchemically**, *adv.* **alchemist**, *n.* one who studies or practises alchemy. **alchemistic** (-mis'-), *a.* **alchemize, -ise**, *v.t.* to transmute into gold.
alcohol (al'kəhol), *n.* a colourless liquid produced by fermenting sugars and constituting the intoxicating agent in various drinks; any of a class of compounds analogous to common alcohol that contain one or more hydroxyl groups; intoxicating drink containing alcohol. **alcohol abuse**, *n.* excessive use of alcohol. **alcoholic** (-hol'-), *n.* an addict to alcohol, *a.* of or pertaining to alcohol or alcoholics. **alcoholism**, *n.* the action of (excessive) alcohol on the human system; the state of being affected by alcohol; addiction to or excessive use of alcohol. **alcoholize, -ise**, *v.t.* to mix, saturate with alcohol. **alcoholization**, *n.* mixing or saturation with alcohol; alcoholism. **alcoholometer** (-om'itə), **alcoometer** (-kōom'-), *n.* an instrument for measuring the proportion of pure alcohol in a liquor. **alcoholometry, alcoometry** (-om'itri), *n.*
alcoran (alkərahn'), *n.* the Koran. **alcoranist**, *n.* one who adheres very strictly to the commands of the Koran.
alcove (al'kōv), *n.* a vaulted recess; a recess in a wall.
Ald., (*abbr.*) alderman.
aldehyde (al'dihīd), *n.* a volatile liquid that can be obtained from alcohol by oxidation, acetaldehyde, ethanol; any of an extensive class of organic compounds of the same type.
al dente (al den'ti, -tā), *a.* esp. of cooked pasta, firm when chewed. [It., to the teeth]
alder (awl'də), *n.* a well-known English tree growing in moist places; applied also, with distinguishing epithet, to many plants whose leaves more or less resemble those of the alder.
alderman (awl'dəmən), *n.* a civic dignitary next in rank below the mayor. **aldermanic** (-man'-), *a.* pertaining to an alderman. **aldermanlike, aldermanly**, *a.* like or befitting an alderman. **aldermanry**, *n.* a district having its own alderman, a ward; the dignity or office of alderman. **aldermanship**, *n.* the office or dignity of alderman.
Alderney (awl'dəni), *n.* an individual of that breed of cattle originating on Alderney Island.
aldosterone (aldos'tərōn), *n.* a steroid hormone produced by the adrenal glands that regulates salt levels.
ale (āl), *n.* (a variety of) intoxicating drink made from malt by fermentation. **ale-house**, *n.* (*dated*) a tavern licensed to sell ale. **ale-taster**, *n.* an ale-conner. **ale-wife**, *n.* a woman who keeps an ale-house.
aleatory (ā'liətəri), *a.* depending upon an uncertain event. **aleatoric** (-to'-), *a.*
Alemmanic (aləman'ik), *n.* the group of High German dialects of Alsace, Switzerland, and SW Germany.

aleph (al'ef), *n.* the first letter of the Hebrew alphabet. **aleph-null**, *n.* the cardinal number indicating the number of elements in the set of all positive integers; the smallest infinite cardinal number.

alert (əlœt'), *a.* watchful, vigilant; brisk, sprightly. *n.* an alarm; a warning by siren or otherwise (e.g. of a threatened air-raid). *v.t.* to warn; to put on guard; to arouse (suspicion). **on the alert**, on one's guard; ready, prepared. **alertly**, *adv.* **alertness**, *n.*

A-level A [1].

alexandrine (aligzan'drīn), *n.* an iambic verse with six feet.

alexandrite (aligzan'drīt), *n.* a dark green chrysoberyl.

alexia (əlek'siə), *n.* the loss of power to understand written or printed words; word-blindness.

alfalfa (alfal'fə), *n.* general term for lucerne. [Sp.]

alfresco (alfres'kō), *adv., a.* in the open air; open-air. [It. *al fresco*, in the fresh]

alg., (*abbr.*) algebra.

alga (al'gə), *n.* (*pl.* **algae** (-gē)) a seaweed or other plant belonging to the Algae.

Algae (al'gē), *n.pl.* a major group of simple aquatic or sub-aquatic plants, including the seaweeds, that lack differentiation into stems, roots and leaves. **algal**, *a.* pertaining to seaweeds or other Algae. **algin** (-j-), *n.* a jelly-like substance obtained from seaweed. **alginate** (-jināt), *n.* a salt of alginic acid used as a stabilizing and thickening agent in pharmaceuticals, food and plastics. **alginic** (-jin'-), *a.* pertaining to or obtained from seaweed. **algoid** (-goid), *a.* of the nature of or like Algae.

algebra (al'jibrə), *n.* universal arithmetic in which letters are used as symbols for quantities, and signs represent arithmetical processes. **algebraic, -ical** (-brā'-), *a.* of or relating to algebra; involving or employing algebra. **algebraically**, *adv.* **algebraist** (-brāist), **algebrist**, *n.* one who is versed in algebra.

-algia, *comb. form.* denoting pain (in a particular place), e.g. *neuralgia*.

algin, algoid, etc. ALGA.

Algol, ALGOL (al'gol), *n.* acronym for *algo*rithmic *l*anguage, a computer language used chiefly for mathematical purposes.

Algonquian (algong'kwiən), **Algonkian** (-kiən), *n.* a family of New American Indian languages; a member of a tribe speaking an Algonquian language.

algorithm (al'gəridhm), *n.* a rule or set procedure for solving a mathematical problem.

Alhambra (alham'brə), *n.* the Moorish palace and citadel at Granada in Spain. **alhambraesque** (-esk'), *a.* resembling the Alhambra or its style of architecture.

ali-, *comb. form.* pertaining to a wing.

alia INTER ALIA.

alias (āl'iəs), *adv.* otherwise (named or called). *n.* a second name, an assumed name.

alibi (al'ibī), *n.* the plea (of a person accused) of having been elsewhere when the offence was committed; the evidence to support such a plea; (*coll.*) an excuse (for failing to do something).

alicante (alikan'ti), *n.* a red, sweet wine from the Spanish town of this name.

alien (ā'liən), *a.* belonging to another culture, planet etc.; of foreign extraction; averse, repugnant (to); incongruous. *n.* a foreign-born non-naturalized resident; a being from another planet. **alienability** (-bil'-), *n.* **alienable**, *a.* that may be alienated.

alienate (ā'liənāt), *v.t.* to turn away the affections of; to cause (someone) to become hostile to oneself. **alienation**, *n.* act of alienating; state of being alienated. **alienator**, *n.*

aliform (ā'lifawm), *a.* shaped like a wing. [ALI-]

alight[1] (əlīt'), *v.i.* to get down, descend, dismount; to reach the ground, to settle; to light on, happen on something.

alight[2] (əlīt'), *pred.a.* on fire; illuminated.

align (əlīn'), *v.t.* to range or place in a line; to place (oneself, one's views) in a position of agreement with others. *v.i.* to fall into line. **alignment**, *n.* the act of ranging in line or being ranged; objects arranged in a line or lines; agreement or alliance with others; the act of taking a side or associating with a party, cause, country etc.

alike (əlīk'), *a.* similar. *adv.* equally, in the same manner, similarly.

aliment (al'imənt), *n.* nutriment, food; support, sustenance. **alimental** (-men'-), *a.* pertaining to aliment; nutritive. **alimentally**, *adv.* **alimentary**, *a.* pertaining to aliment or nutrition; nutritious, nourishing. **alimentary canal**, *n.* the great tube or duct from mouth to anus conveying food to the stomach and carrying off solid excreta from the system. **alimentation**, *n.* the act or quality of affording nourishment; the state of being nourished. **alimentative** (-men'-), *a.* connected with the function of nutrition.

alimony (al'imənī), *n.* maintenance; payment of part of a person's income allowed for the support of a spouse on legal separation, or for other causes.

aliphatic (alifat'ik), *a.* fatty; belonging or pertaining to a class of organic compounds containing open chains of carbon atoms in the molecular structure, not aromatic.

aliquot (al'ikwot), *a.* pertaining to a number that is contained an integral number of times by a given number. *n.* an integral factor, an aliquot part. **aliquot part**, *n.* a part that is a division of the whole without remainder, as 50p of £1, 10 g of 1 kg.

alive (əlīv'), *pred.a., adv.* living, existent; in force or operation; lively. **alive and kicking**, in a very lively state. **alive to**, fully aware of (possibilities etc.). **alive with**, full of, crawling etc. with; infested by. **look alive!** look sharp, make haste!

alkali (al'kəlī), *n.* (*pl.* **-lis, -lies**) a compound of hydrogen and oxygen with sodium, potassium or other substances, which is soluble in water, and produces caustic and corrosive solutions capable of neutralizing acids and changing the colour of vegetable substances; any water-soluble chemical base; alkaline products, such as caustic potash and caustic soda. **alkali-metals**, *n.pl.* metals, the hydroxides of which are alkalis: these are potassium, sodium, caesium, lithium, rubidium, francium. **alkalescence, -cy** (-les'-), *n.* the state or condition of becoming alkaline, tendency to become alkaline. **alkalescent**, *a.* becoming or tending to become alkaline; slightly alkaline. **n.** an alkalescent substance. **alkalifiable** (-lifī'-), *a.* capable of being converted into an alkali. **alkalify** (-li-), *v.t., v.i.* to (cause to) be converted into an alkali. **alkaline** (-līn), *a.* having the properties of an alkali. **alkalize, -ise**, *v.t.* to render alkaline. **alkalization, alkaloid** (-loid), *a.* resembling an alkali in properties. *n.* any of a large group of natural organic nitrogenous bases derived from plants, some of which are used as medicinal drugs.

alkoran (alkərahn') ALCORAN, KORAN.

alkyl (al'kil), *n.* general name for a monovalent hydrocarbon radical of the alkane series, e.g. methyl, ethyl, butyl. **alkylation**, *a.* the introduction of an alkyl into a compound.

all (awl), *a.* the whole (quantity, duration, extent, amount, quality or degree) of; every one of the greatest possible; any whatever. *n.* the whole, everything, every one. *adv.* wholly, entirely, completely; each, apiece. **after all**, after everything has been taken into consideration. **all aboard**, take your seats. **all about it**, the whole of the matter. **all along**, throughout, from start to finish. **all and sundry**, (often *derog.*) everyone; all (taken distributively). **all at once** ONCE. **all but**, almost. **all for**, completely in favour of. **all found** FOUND[3]. **all in**, everything included; (*coll.*) exhausted, very tired. **all in all**, everything considered. **all of**, as much, far etc. as, no less than (followed by a quantity). **all one**, the same in all respects; immaterial. **all out**, with maximum effort. **all over**, (*coll.*) completely,

Allah everywhere; finished (in the phrases *all over with*, *is all over*); typically. **all right** RIGHT. **all the best** BEST. **all the better**, so much the better. **all the same**, nevertheless; in spite of what has been said. **all there**, (*coll.*) sharp in intellect, alert, quick. **all told** TELL. **all up with**, (*coll.*) no more hope for. **and all**, (*coll.*) too, withal. **and all that**, with all the rest of it. **at all**, in any respect; to any extent; in any degree; of any kind; whatever. **in all**, in total. **on all fours** FOUR. **all-American**, *a*. typifying US ideals, as *all-American boy*. **All Blacks**, *n.pl.* the New Zealand international Rugby Union team. **all clear**, *n*. a signal indicating that danger has passed or that one can proceed safely. **All Fools' Day** APRIL. **all-hail**, *int*. a phrase expressive of respect or welcome. **All-Hallows**, *n*. All Saints' Day. **All-Hallows' Eve**, *n*. Hallowe'en. **all-important**, *a*. of utmost importance. **all-in**, *a*. including everything. **all-in wrestling**, *n*. a form of wrestling with almost no restrictions. **all-or-nothing**, *a*. allowing no compromise; with an outright win or loss in prospect. **all-out**, *a*., with maximum effort. **all-purpose**, *a*. suitable for all purposes. **all-round**, *a*. good in all respects. **all-rounder**, *n.*, *adv.* one whose merits, acquirements or skill are not limited to one or two pursuits; one widely competent or versatile. **All Saints' Day**, *n*. a church festival (1 Nov.) in honour of the saints collectively. **all-singing all-dancing**, *a*. (*coll.*) having every available feature; performing any function required. **All Souls' Day**, *n*. the day (2 Nov.) on which the Roman Catholic Church commemorates all the faithful departed. **all-star**, *a*. composed of star performers. **all-time**, *a*. exceeding all others, as yet unsurpassed. **all-time high**, or **low**, *n*. a record high (or low) level. **all together**, *adv*. in a body, altogether. **all-up**, *a*. expressing the total weight of an aircraft with its load when in the air.

Allah (al'ə), *n*. the name of God among the Muslims.

allay (əlā'), *v.t.* to quiet, to still; to abate, to alleviate. **allaying**, *n*.

allegation (aligā'shən), *n*. the act of alleging; an assertion without proof; a specific charge, a statement of what one undertakes to prove.

allege (əlej'), *v.t.* to affirm positively but without or before proof. **alleged**, *a*. stated but not yet proven. **allegedly** (-jid-), *adv*. according to a statement which is yet to be proved.

allegiance (əlē'jəns), *n*. the obligation of a subject to his sovereign or government; respect, devotion, loyalty.

allegory (al'igəri), *n*. a description of one thing under the image of another; an instance of such description, an extended metaphor, an emblem, an allegorical representation. **allegoric, -ical** (-go'-), *a*. pertaining to an allegory; resembling an allegory. **allegorically**, *adv*. **allegorist**, *n*. a writer of allegories. **allegorize, -ise**, *v.t.* to convert into an allegory; to interpret allegorically. *v.i.* to use allegory, to speak or write in a figurative manner.

allegro (əleg'rō), *a*. (*Mus.*) briskly, lively, merry. *adv*. briskly, quickly. *n*. a movement in allegro time or manner. **allegretto** (aləgret'ō), *adv*. somewhat briskly but less so than allegro.

allele (əlēl', a'-), *n*. an allelomorph. **allelic**, *a*.

allelomorph (əlē'ləmawf, əlel'-), *n*. any of two or more contrasted characteristics, inherited as alternatives, and assumed to depend on genes in homologous chromosomes; any of two or more genes determining such alternative characteristics. **allelomorphic** (-maw'-), *a*.

alleluia (aliloo'yə) HALLELUJAH.

allemande (al'əmahnd), *n*. any of various German dances of the 17th and 18th cents.; the music for or suitable for this (occurring as a movement in a suite).

allergy (al'əji), *n*. an abnormal response or reaction to some food or substance innocuous to most people; hypersensitiveness to certain substances inhaled or touched; (*coll.*) an aversion, antipathy. **allergen** (-jən), *n*. a substance that induces allergy. **allergenic** (-jen'-), *a*. **allergic** (-lœ'-), *a*. caused by allergy; having an allergic response (to); (*coll.*) averse (to). **allergist**, *n*. a specialist in the treatment of allergy.

alleviate (əlē'viāt), *v.t.* to lessen, mitigate. **alleviation**, *n*. the act of alleviating; relief, mitigation. **alleviative**, *n*. that which alleviates. **alleviator**, *n*. **alleviatory**, *a*.

alley[1] (al'i), *n*. a narrow street, passage or lane; a narrow enclosure for playing at skittles etc.

All-Hallows ALL.

alliance (əlī'əns), *n*. the state of being allied; union by marriage, affinity; (union by) a treaty or league; union or connection of interests; the parties allied.

allied, Allies etc. ALLY[1,2].

alligator (al'igātə), *n*. an animal of the genus *Alligator*, found in America and China, that differs from the crocodiles esp. in having a broader snout.

alliterate (əlit'ərāt), *v.i.* to commence with the same letter or sound; to practise alliteration. **alliteration**, *n*. commencement of two or more words or accented syllables, in close sequence, with the same letter or sound. **alliterative**, *a*. pertaining to alliteration; alliterating. **alliteratively**, *adv*.

allo-, *comb. form.* different, other, as in *allomorph, allopathy*.

allocate (al'əkāt), *v.t.* to assign, allot, apportion. **allocation**, *n*. the act of allocating; a quantity allocated.

allogamy (əlog'əmi), *n*. (*Bot.*) cross-fertilization. **allogamous**, *a*. reproducing by cross-fertilization.

allomorph (al'əmawf), *n*. any of the two or more forms of a morpheme; any of two or more crystalline forms of a substance. **allomorphic** (-maw'-), *a*.

allopathy (əlop'əthi), *n*. the treatment of disease by inducing effects of a different kind from those produced by the disease; ordinary medical practice, as opposed to homoeopathy. **allopathic** (aləpath'-), *a*. pertaining to or practising allopathy. **allopathically**, *adv*. **allopathist**, *n*. one who practises allopathy.

allophone (al'əfōn), *n*. any of the two or more forms of a phoneme. **allophonic** (-fon'-), *a*.

allot (əlot'), *v.t.* (*past, pp.* **allotted**) to distribute, to grant, to assign as a share. **allotment**, *n*. the act of allotting; the share assigned; a small plot of land let for cultivation. **allottee** (-ē'), *n*. the person to whom allotment is made.

allotropy (əlot'rəpi), *n*. variation of physical properties without change of substance; thus, diamond, graphite and charcoal are allotropic forms of carbon. **allotrope** (al'ətrōp), *n*. one of the forms in which a substance exhibiting allotropy exists. **allotropic** (alətrop'ik), *a*. pertaining to allotropy; existing in diverse states. **allotropically**, *adv*. **allotropism**, *n*. allotropy.

allow (əlow'), *v.t.* to admit, permit; to assign, set aside (for a purpose); to take into account, give credit for. **to allow for**, to make concession or deduction for. **to allow of**, to accept, to admit. **allowable**, *a*. **allowableness**, *n*. **allowably**, *adv*. **allowance**, *n*. the act of allowing; a deduction; a fixed quantity or sum (esp. of money) regularly allowed. **to make allowance** or **allowances for**, to take (mitigating circumstances) into account when considering (something or someone).

alloy (al'oi), *n*. an inferior metal mixed with one of greater value; a mixture of metals; an amalgam; any base admixture. *v.t.* (əloi') to mix with a baser metal; to mix metals; to mix with anything base or inferior.

allspice (awl'spīs), *n*. (a spice prepared from) the berry of the pimento, said to combine the flavour of cinnamon, cloves and nutmeg.

allude (əlood', əlūd'), *v.i.* to make indirect reference (to), to hint; (*loosely*) to mention, to refer (to). **allusion** (-zhən), *n*. a reference to anything not directly mentioned; a hint. **allusive**, *a*. containing an allusion; hinting at an implied meaning, characterized by allusion.

allusively, *adv*. **allusiveness**, *n*.
allure (əlūə'), *v.t.* to attract or tempt by the offer of some real or apparent good; to entice; to fascinate, to charm. *n.* charm, sex appeal. **allurement**, *n.* the act of alluring or enticing; that which allures; a bait, an enticement. **alluring**, *a.* enticing, attractive. **alluringly**, *adv.*
allusion, etc. ALLUDE.
alluvium (əloo'viəm), *n.* (*pl.* **alluvia**) earth, sand, gravel, stones or other transported matter which has been washed away and thrown down by rivers, floods or similar causes. **alluvial**, *a.* pertaining to alluvium; deposited from flowing water.
ally[1] (əlī'), *v.t.* to unite by treaty, confederation, marriage or friendship. **allied**, *a.* united, associated; of the same type, related. **Allied**, *a.* of the Allies.
ally[2] (al'ī), *n.* one united with others by treaty, confederation, marriage or friendship; something akin to another in structure or properties. **the Allies** (al'īz), in World War I, the nations united against the central European powers; in World War II, the nations united against the Axis powers.
alma, almah (al'mə), *n.* an Egyptian dancing-girl.
alma mater (al'mə mä'tə, mah'-), *n.* name used by an ex-student for his college or (usu. public) school.
almanac, (*dated*) **almanack** (awl'mənak), *n.* a register of the days of the year, with astronomical data and calculations, civil and ecclesiastical festivals etc.
almighty (awlmī'ti), *a.* omnipotent; possessed of unlimited ability, strength or power. *a.*, *adv.* (*coll.*) mighty, great, exceedingly. **the Almighty**, God. **almighty dollar**, *n.* (*coll.*) money; feverish love of money. **almightiness**, *n.*
almond (ah'mənd), *n.* a small widely cultivated tree of the rose family; the edible kernel of the fruit of the almond-tree. **almond-eyed**, *a.* having eyes the shape of the almond kernel, i.e., elliptical with pointed ends.
almoner (al'mənə, ah'-), *n.* an official distributor of alms or bounty; a former name for a medical social worker attached to a hospital. **almonry** (ah'mənri), *n.* a place where alms are distributed; the residence of an almoner.
almost (awl'məst), *adv.* very nearly.
alms (ahmz), *n.pl.* anything given out of charity to the poor; charity. **alms-giving**, *n.* **almshouse**, *n.* a house where poor persons are lodged and provided for by charitable endowment.
Aloe (al'ō), *n.* a genus of succulent plants, with bitter juice; (**aloe**) any of various other plants, e.g. the American aloe; (**aloe**) the condensed juice of the plants of the genus *Aloe*, a purgative drug. **Aloe vera** (veə'rə), *n.* a Mediterranean aloe whose juice is used in various medical and cosmetic preparations.
aloft (əloft'), *adv.* on high; above the ground; in the rigging, at the mast-head.
alone (əlōn'), *pred.a.* single, solitary, by oneself or itself; without equal, unique. *adv.* only, merely, simply; solely.
along (əlong'), *adv.* lengthwise, in a line with the length, in progressive motion; onward. *prep.* by the side of, from end to end, over or through lengthwise. **all along**, *adv.* throughout, all the time. **along with**, in the company of, together with. **alongside**, *prep.*, *adv.* beside, by the side (of); up to the side (of); side by side. **alongside of**, side by side with.
aloof (əloof'), *adv.* away at a distance from; apart. *a.* distant or unsympathetic in manner. **to stand, keep aloof**, to take no part in, keep away; to remain by oneself, remain unsympathetic. **aloofness**, *n.* the state of keeping aloof.
alopecia (aləpē'shə), *n.* baldness due to (possibly nervous) illness.
aloud (əlowd'), *adv.* loudly; with a loud voice; audibly.
alp (alp), *n.* a high mountain; pasture ground on the side of mountain. **Alps**, *n.pl.* the chain of mountains separating France from Italy etc. **alphorn, alpenhorn**, *n.* a very long wooden horn used by herdsmen in the Swiss Alps.

Alpine (-pīn), *a.* pertaining to the Alps or to any high mountains; growing on the Alps or on any high mountain; growing above the tree line; pertaining to ski events such as slalom and downhill racing. **alpinism** (-pin-), *n.* mountain-climbing. **alpinist** (-pin-), *n.* a person expert in Alpine climbing.
alpaca (alpak'ə), *n.* the domesticated llama of Peru; the wool of the domesticated llama; cloth made from this wool.
alpenhorn ALP.
alpenstock (al'pənstok), *n.* a long stick shod with iron, used in mountaineering.
alpha (al'fə), *n.* the first letter of the Greek alphabet, used to designate numerical sequence; the chief or brightest star in a constellation. **alpha and omega**, the beginning and the end. **alpha particle**, *n.* a positively-charged particle emitted by certain radioactive substances, e.g. radium. **alpha plus**, *a.* superlatively good. **alpha-rays**, *n.pl.* rays consisting of streams of alpha particles. **alpha rhythm, alpha wave**, *n.* the pattern of electrical activity of the brain associated with a person awake but at rest.
alphabet (al'fəbet), *n.* the letters used to represent a language, arranged in customary order. **alphabet soup**, *n.* (*coll. derog.*) a confusing mass of letters, esp. initials. **alphabetic, -ical** (-bet'-), *a.* pertaining to the alphabet, arranged in the customary order of the alphabet. **alphabetically**, *adv.* **alphabetize, -ise**, *v.t.* to arrange alphabetically.
alphanumeric, -ical (alfənūme'rik), **alphameric, -ical** (-me'rik), *a.* consisting of or using both letters and numbers.
alphasort (al'fəsawt), *v.t.*, *v.i.* to sort (computer data) into alphabetical order. *n.* an act of alphasorting.
Alpine etc. ALP.
already (awlred'i), *adv.* beforehand, before some specified time, in anticipation.
alright (awlrīt'), a nonstandard spelling of *all right* (see RIGHT).
Alsatian (alsā'shən), *a.* belonging to Alsace. *n.* a native of Alsace; a large, intelligent, German wolf-like dog, also known as a German Shepherd.
also (awl'sō), *adv.*, *conj.* likewise, in like manner; in addition, as well. **also-ran**, *n.* an unplaced horse in a race; (*coll.*) an unimportant person, a failure.
alt., (*abbr.*) alteration; alternate; altitude; alto.
Alta, (*abbr.*) Alberta (Canada).
Altaian (altā'ən), *a.* a term applied to the peoples, and to the languages of the peoples (Turanian or Ural-Altaic), lying near the Altai mountains and the Arctic Ocean. *n.* a member of this group. **Altaic**, *a.* Altaian.
altar (awl'tə), *n.* a sacrificial block; a place of sacrifice, commemoration or devotion; the communion-table. **to lead to the altar**, to marry. **altar-bread**, *n.* wafer bread used in the celebration of the Eucharist. **altar-cloth**, *n.* the linen cloth which covers an altar. **altar-piece**, *n.* a picture or ornamental sculpture over the altar (or communion-table) in a church. **altar-plate**, *n.* the plate used in the celebration of the Eucharist. **altar-rails**, *n.pl.* the low railings separating the altar from the main body of the church. **altar-screen**, *n.* the reredos wall or screen at the back of an altar. **altar-stone**, *n.* a portable altar on which Mass is said.
altazimuth (altaz'iməth), *n.* an instrument for measuring altitude and azimuth.
alter (awl'tə), *v.t.*, *v.i.* to (cause to) vary or change in some degree; to modify. **alterable**, *a.* **alterability** (-bil'-), *n.* **alteration**, *n.* the act of altering; the change made. **alterative**, *a.* tending to produce alteration.
altercation (awltəkā'shən), *n.* wrangling; a vehement dispute.
alter ego (awltə ē'gō, al'-, e'-), *n.* (*Psych.*) a second self; a very close friend.

alternate¹ (awl'tənāt), *v.t.* to arrange or perform by turns; to cause to succeed by turns or reciprocally; to interchange. *v.i.* to happen by turns; to change repeatedly from one condition or state to another; (*Elec.*) to change from positive to negative and back again in turns. **alternating current**, *n.* an electric current that changes from positive to negative regularly and frequently. **alternation**, *n.* the act of alternating; the state of being alternate.

alternate² (awltœ'nət), *a.* done or happening by turns, first one and then the other; reciprocal; every other, every second; of plant parts, placed on opposite sides of an axis at successive levels; of angles, succeeding regularly on opposite sides of a straight line. **alternately**, *adv.*

alternative (awltœ'nətiv), *a.* offering a choice of two (or more) things; being the other of two (or more) things open to choice; denoting or pertaining to a life-style, practice, art form etc. which functions outside, and constitutes an option other than, conventional or institutionalized methods or systems. *n.* the permission or opportunity to choose between two (or more) things; either of two (or more) courses which may be chosen. **alternative medicine**, *n.* any system of medicine or medical treatment, as homoeopathy or osteopathy, that does not use orthodox practices or substances. **alternatively**, *adv.*

alternator (awl'tənātə), *n.* a dynamo for generating an alternating electric current.

although (awldhō'), *conj.* though, notwithstanding, despite the fact that.

alti- *comb. form.* high, highly, height.

altimeter (altim'itə), *n.* an instrument that indicates height above a given datum, usu. sea-level.

altissimo (altis'imō), *adv.* (*Mus.*) very high; in the second octave above the treble stave.

altitude (al'titūd), *n.* vertical height; elevation of an object above its base; height above sea-level; the elevation of a heavenly body above the horizon.

alto (al'tō), *n.* the highest adult male voice, countertenor; the lowest female voice, contralto; a singer possessing such a voice; the part of a piece of vocal music sung by persons possessing the alto voice. **alto clarinet, alto viola**, *n.* musical instruments of alto pitch.

altogether (awltəge'dhə), *adv.* wholly, completely, entirely; inclusive of everything; on the whole, considering all things. **the altogether**, (*coll. euphem.*) the nude.

alto-relievo (altōrəlyā'vō), *n.* high relief, standing out from the background by more than half the true proportions of the figures carved.

altruism (al'trooizm), *n.* devotion to the good of others. **altruist**, *n.* one who practises altruism. **altruistic** (-is'-), *a.* **altruistically**, *adv.*

alum (al'əm), *n.* a double sulphate salt of aluminium and potassium; any of a series of double salts including this; a family of analogous compounds; (*Min.*) name of various minerals, alums or pseudo-alums.

alumina (əloo'minə), *n.* the oxide of aluminium.

aluminium (alūmin'iəm), *n.* a white, ductile metallic element, at.no.13; chem. symbol Al, with good resistance to corrosion, used as a basis for many light alloys. **aluminize, -ise** (-loo'-), *v.t.* to coat with aluminium.

aluminous (əloo'minəs), *a.* composed of or pertaining to alum or alumina.

aluminum (əloo'minəm), *n.* (*N Am.*) aluminium.

alumnus (əlŭm'nəs), *n.* (*pl.* **-ni** (-nī)) a former pupil or student (of a stated place of education); (*N Am.*) a graduate, old scholar. **alumna** (-nə), *n. fem.* (*pl.* **-nae** (-nē)).

alveolus (alvē'ələs), *n.* (*pl.* **-li** (-lī)) a little cavity; a tooth socket; an air sac in the lungs. **alveolar**, *a.* pertaining to or having alveoli or an alveolus; pertaining to the sockets of the teeth; socket-shaped; (*phon.*) produced with the tip of the tongue touching the roof of the mouth behind the front teeth. *n.* an alveolar sound.

always (awl'wāz), *adv.* continuously or continually; on all occasions; for ever; while one lives; in all cases; in any event.

Alyssum (əlis'əm), *n.* a genus of cruciferous plants; a related plant, sweet alyssum.

Alzheimer's disease (alts'hīməz), *n.* a degenerative disease of the central nervous system characterized by a deterioration of mental faculties resembling premature mental senility or dementia.

AM, (*abbr.*) amplitude modulation; associate member; (*chiefly N Am.*) Master of Arts.

Am.¹, (*abbr.*) America, American.

Am², (*chem. symbol*) americium.

am¹ (am), *1st pers. sing. pres. ind. of the v.* to BE.

am², **a.m.** (*abbr.*) ante meridiem (before noon). **the a.m.**, (*coll.*) the morning.

amabile (amah'bili), *adv.* (*Mus.*) amiably, tenderly, sweetly.

amah (ah'mə), *n.* in the East, a nanny.

amalgam (əmal'gəm), *n.* a mixture of any other metal with mercury; a compound of different things. **amalgamate**, *v.t.* to mix, unite, combine into one mixture; to combine into a single entity; to combine (another metal) with mercury. *v.i.* to combine, to blend, to merge into one. **amalgamation**, *n.* the act of amalgamating; the result of amalgamating; the blending or uniting of different things. **amalgamative**, *a.* tending to combine.

amanuensis (əmanüen'sis), *n.* (*pl.* **-ses** (-sēz)) a person employed to write what another dictates, or to copy manuscripts.

amaranth (am'əranth), *n.* an imaginary flower supposed never to fade; any of a genus of plants that includes love-lies-bleeding and prince's feather. **amaranthine** (-ran'thin), **amarantine** (-ran'tīn), *a.* pertaining to amaranth; unfading.

Amaryllis (aməril'is), *n.* a genus of autumn-flowering bulbous plants.

amass (əmas'), *v.t.* to make or gather into a heap; to collect together, to accumulate.

amateur (am'ətə, -chə), *n.* one who cultivates anything as a pastime, as distinguished from one who does so professionally; one who competes in a sport for enjoyment rather than payment; (*derog.*) one who dabbles or is unskilled in a subject. *a.* pertaining to an amateur, as *amateur gardener*; not professional; not receiving payment, as *amateur boxer.* **amateurish**, *a.* (*derog.*) not up to the professional standard. **amateurishness**, *n.* the quality of being amateurish; inferior standard of execution. **amateurism**, *n.* state, condition or practice of an amateur. (*derog.*) dilettantism.

amatol (am'ətol), *n.* an explosive consisting of a mixture of ammonium nitrate and trinitrotoluene.

amatory (am'ətəri), *a.* pertaining to love, amorous; causing or designed to cause love.

amaze (əmāz'), *v.t.* to astound, to overwhelm with wonder, to bewilder. **amazedly** (-zid-), *adv.* **amazedness** (-zid-), *n.* **amazement**, *n.* overwhelming surprise; the state of being amazed. **amazing**, *a.* **amazingly**, *adv.*

Amazon (am'əzən), *n.* one of a fabled race of female warriors; (**amazon**) a female warrior; (**amazon**) a tall, strong woman, a virago. **amazonian, Amazonian** (-zō'-), *a.* of or pertaining to the fabled Amazon women, hence, war-like, strong; pertaining to the river Amazon, named from the female warriors recorded there by the early Spaniards.

ambassador (ambas'ədə), *n.* a minister of high rank, representing his or her country at a foreign court or government, being styled **ordinary** when resident, and **extraordinary** when sent on a special mission; a representative, messenger. **ambassador-at-large**, *n.* an ambassador not accredited to any particular foreign government. **ambassador plenipotentiary**, *n.* an ambassador armed with full powers to sign treaties etc. **ambassadorial**

amber   22   **amine**

(-daw'-), *a.* pertaining to an ambassador. **ambassadorship**, *n.* **ambassadress** (-dris), *n. fem.* a female ambassador; the wife of an ambassador.

amber (am'bə), *n.* a yellowish translucent fossil resin, found chiefly on the southern shores of the Baltic, used for ornaments, mouthpieces of pipes, and in the manufacture of some varnishes; a traffic or railway signal of this colour, advising caution. *a.* made of or coloured like amber.

ambergris (am'bəgrēs), *n.* a light, fatty, inflammable substance, ashy in colour, found floating in tropical seas, a secretion from the intestines of the cachalot or spermaceti whale, and used in perfumery, formerly in cookery and medicine.

ambidextrous, -terous (ambidek'strəs), *a.* using both hands with equal facility. **ambidexterity** (-ste'-), *n.* **ambidextrously**, *adv.* **ambidextrousness**, *n.*

ambient (am'biənt), *a.* surrounding, encompassing on all sides. **ambient light**, *n.* (esp. *Phot.*) light occurring naturally in a place, not added to or produced artificially. **ambience, ambiance** (am'biəns, äbiäs'), *n.* the surrounding atmosphere or influence, environment.

ambiguous (əmbig'ūəs), *a.* susceptible of two or more meanings, having the potential for being interpreted in more than one way; of doubtful meaning, equivocal, obscure. **ambiguously**, *adv.* **ambiguousness**, *n.* **ambiguity** (-bigū'-), *n.* the state or an instance of being ambiguous; uncertainty of meaning.

ambit (am'bit), *n.* bounds, precincts, scope.

ambition (ambi'shən), *n.* a desire for power, success, superiority or excellence; strong desire to achieve anything (advantageous or creditable); the object of such desire. **ambitious**, *a.* actuated by or indicating ambition; full of or displaying (sometimes too much) ambition. **ambitiously**, *adv.* **ambitiousness**, *n.*

ambivalence, -cy (ambiv'ələns, -si), *n.* the simultaneous existence in the mind of two incompatible feelings or wishes. **ambivalent**, *a.*

amble (am'bl), *v.i.* to walk at an easy pace, to stroll. *n.* an easy pace; a leisurely walk, a stroll. **ambling**, *a.*

amblyopia (amblio'piə), *n.* dimness of vision. **amblyopic** (-op'-), *a.* affected with or pertaining to amblyopia.

ambrosia (ambrō'ziə), *n.* the fabled food of the gods; anything very pleasant to the taste or the smell; the pollen collected by bees for food. **ambrosial**, *a.* containing the qualities of ambrosia; delicious, fragrant; ethereal, divine. **ambrosially**, *adv.*

ambulance (am'būləns), *n.* a moving hospital which follows an army in the field; a vehicle for the transport of wounded, injured, disabled or sick people. **ambulance-chaser**, *n.* (*coll.*) one who seeks to make profit from another's tragedy, grief etc., esp. a lawyer who offers to pursue a claim for damages on behalf of accident victims; (*coll.*) one who seeks profit (esp. politically) from another's mistakes.

ambulate (am'būlāt), *v.i.* to walk about. **ambulation**, *n.* the act of walking. **ambulant**, *a.* walking or moving about; able to walk.

ambulatory (am'būlətəri), *a.* pertaining to walking; fitted for walking; not confined or confining to bed.

ambuscade (ambəskād'), *n.* an ambush, a lying in wait to attack an enemy.

ambush (am'bush), *n.* the concealment of forces to entrap an enemy; the locality chosen; the force employed; any (*lit.* or *fig.*) lying in wait or entrapment. *v.t.* to lie in wait for; to attack from ambush.

ameba (əmē'bə) AMOEBA.

ameer, amir (əmiə'), *n.* the title of several Muslim rulers in India and Afghanistan.

ameliorate (əmē'liərāt), *v.t.* to make better; to improve. *v.i.* to grow better. **amelioration**, *n.* the act of making better; the state of being made better; improvement. **ameliorative**, *a.* **ameliorator**, *n.*

amen (ahmen', *esp. Amer.* ā-), *int.* truly, verily; so be it; may it be as has been asked, said or promised. *n.* the word 'Amen', an expression of assent; a concluding word.

amenable (əmēn'əbl), *a.* tractable, responsive; cooperative. **amenability** (-bil'-), **amenableness**, *n.* the quality or state of being amenable; tractableness. **amenably**, *adv.*

amend (əmend'), *v.t.* to alter (a person or thing) for the better, to improve; to reform, to correct; to formally alter (a bill or resolution). **amendable**, *a.* **amendment**, *n.* (a) change for the better; something added to a bill or motion; a correction of error in a writ or process. **to make amends**, to make reparation (for), to give compensation (for).

amenity (əmē'niti, -men'-), *n.* the quality of being pleasant or agreeable; a feature or facility conducive to the attractiveness of something; (*pl.*) facilities, attractions, charms.

amenorrhoea (əmenərē'ə), *n.* the abnormal cessation of menstruation.

Amer., (*abbr.*) America, American.

Amerasian (amərā'zhən), *n.*, *a.* (a person) of mixed US and Asian parents.

amerce (əmœs'), *v.t.* to punish by fine; to exact something from; to punish. **amercement**, *n.* the infliction of an arbitrary fine; the fine inflicted. **amerciable**, *a.* liable to amercement.

American (əme'rikən), *a.* pertaining to the continent of America, esp. to the US. *n.* a native or inhabitant of N, S or Central America; a citizen of the US. **American football**, *n.* a football game somewhat resembling rugby, played with an oval ball and teams of 11 players. **American Indian**, *n.* a member of any of the indigenous peoples of N, S or Central America, usu. with the exception of the Eskimos. **American plan**, *n.* inclusive terms at a hotel etc. **Americanism**, *n.* attachment to or political sympathy with the US; anything characteristic of the US, esp. a word or phrase peculiar to or borrowed from the US. **Americanize, -ise**, *v.t.* to naturalize as an American; to assimilate (esp. political) customs or institutions, or citizens from other countries, to those of the US. *v.i.* to become American in character, manners or speech.

americium (aməris'iəm), *n.* an artificially-created, metallic radioactive element, at. no. 95; chem. symbol Am.

Amerind (am'ərind), **Amerindian** (-rin'-), *n.*, *a.* (an) American Indian.

amethyst (am'əthist), *n.* a violet-blue variety of crystalline quartz, used in jewellery. **amethystine** (-tin), *a.* composed of, containing or resembling amethyst.

Amharic (amha'rik), *n.* the official language of Ethiopia.

amiable (ā'miəbl), *a.* friendly, kindly-disposed, lovable; possessed of qualities fitted to evoke friendly feeling. **amiably**, *adv.* **amiability** (-bil'-), *n.*

amicable (am'ikəbl), *a.* friendly; designed to be friendly; resulting from friendliness. **amicable suit**, *n.* (*Law*) a suit promoted by arrangement in order to obtain an authoritative decision on some point of law. **amicability** (-bil'-), *n.* **amicableness**, *n.* **amicably**, *adv.*

amicus curiae (əmēkəs kū'riē), *n.* (*pl.* **-ci** (-kī) **curiae**) a disinterested counsellor. [L, friend of the court]

amid (əmid'), **amidst** (əmidst'), *prep.* in the midst or middle; among. **amidships**, *adv.* in the middle part of a ship.

amide (am'īd), *n.* any of various organic compounds constituted as if obtained from ammonia by the substitution of one or more univalent organic acid radicals for one or more atoms of hydrogen; any of various compounds formed by substitution of another element or radical for an atom of hydrogen in ammonia. **amidic** (əmid'-), *a.* **amido-**, *comb. form.* containing the characteristic amide group of ammonia with one hydrogen atom replaced by an acid radical.

amidst AMID.

amine (am'īn, -ēn, ā'-), *n.* any of various organic

compounds derived from ammonia by the substitution of one or more univalent hydrocarbon radicals for one or more atoms of hydrogen. **aminic** (əmin'-), *a.* **amino** (əmē'nō), *a.* containing the characteristic amine group of ammonia with one hydrogen atom replaced by a hydrocarbon radical. **amino acid,** *n.* an organic acid containing one or more amino groups, esp. any of those that occur as the constituents of proteins.
amir AMEER.
amiss (əmis'), *a.* faulty, beside the mark, unsatisfactory; wrong. *adv.* wrongly; in a faulty manner.
amity (am'iti), *n.* friendship, mutual good feeling, friendly relations.
ammeter (am'ətə), *n.* an instrument for measuring the strength of the electric current (amps) in a circuit.
ammo (am'ō), *n.* (*coll.*) short for AMMUNITION.
ammonia (əmō'niə), *n.* a pungent volatile gas, powerfully alkaline, a compound of nitrogen and hydrogen first obtained from sal ammoniac; ammonium hydroxide. **ammonia water, aqueous ammonia, liquid ammonia,** *n.* ammonium hydroxide. **ammoniac** (əməni'ak), **-acal,** *a.* pertaining to or possessing the properties of ammonia. **ammoniated** (-ātid), *a.* combined with ammonia. **ammonium** (-əm), *n.* the ion or radical derived from ammonia by addition of a hydrogen ion or atom. **ammonium hydroxide,** *n.* a solution of ammonia in water.
ammonite (am'ənīt), *n.* the shell of a genus of fossil cephalopods, curved like a ram's horn.
ammonium AMMONIA.
ammunition (amūnish'ən), *n.* powder, shot, shell etc.; offensive missiles generally; arguments for use in a dispute. *v.t.* to supply with ammunition. **ammunitioned,** *a.* provided with ammunition.
amnesia (amnē'ziə, -zhə), *n.* loss of memory. **amnesiac** (-ziak), **amnesic** (-zik), *n.* a person suffering from amnesia. *a.* pertaining to amnesia.
amnesty (am'nəsti), *n.* a period during which laws or rules are relaxed; a general overlooking or pardon of misdeeds, prisoners etc.
amniocentesis (amniōsentē'sis), *n.* the removal of a sample of amniotic fluid from the womb, by insertion of a hollow needle, in order to test for chromosomal abnormalities in the foetus.
amnion (am'niən), *n.* (*pl.* **-ions, -nia**) the innermost membrane with which the foetus in the womb is surrounded. **amniotic** (-ot'-), *a.* pertaining to, characterized by, contained in or of the nature of an amnion. **amniotic fluid,** *n.* the fluid contained by the amnion in which the foetus is suspended.
amoeba (əmē'bə), (*N Am.*) **ameba,** *n.* (*pl.* **-bas, -bae**) a microscopic organism of the simplest structure, consisting of a single protoplasmic cell, which is extensile and contractile, so that the shape is continually changing. **amoeban, amoebic,** *a.* **amoebiform** (-ifawm), **amoeboid** (-oid), *a.* amoeba-like.
amok AMUCK.
among (əmŭng'), **amongst,** *prep.* mingled with, in the number of, included in; in the midst of; surrounded by.
amontillado (əmontilah'dō), *n.* a kind of medium dry sherry.
amoral (āmo'rəl, ə-), *a.* not concerned with morals, non-moral. **amoralism,** *n.* **amoralist,** *n.* a non-moral person. **amorality** (-ral'-), *n.*
amorous (am'ərəs), *a.* naturally inclined to love; in love; lecherous; relating to, or belonging to, love. **amorously,** *adv.* **amorousness,** *n.*
amorphous (āmaw'fəs, ə-), *a.* shapeless; irregularly shaped; (*Biol.*) not conforming to a normal standard; not crystalline; ill-arranged, unsystematic, unorganized. **amorphism,** *n.* lack of regular form; absence of crystallization. **amorphousness,** *n.* the quality of being amorphous.
amortize, -ise (əmaw'tīz), *v.t.* to extinguish (a debt) by payments to a sinking fund. **amortization,** *n.*
amount (əmownt'), *v.i.* to mount up (to), to add up (to); to be equivalent (to). *n.* the total sum, effect, substance, result, significance; a (numerical) quantity.
amour (əmuə'), *n.* (one's partner in) a love affair.
amour propre (amuə propr"), *n.* self-esteem. [F]
amp (amp), *n.* short for AMPERE, AMPLIFIER.
ampere (am'peə), *n.* a unit by which an electric current is measured, specifically, the current sent by 1 volt through a resistance of 1 ohm. **ampere-hour,** *n.* the quantity of electricity delivered in 1 hour by a 1-ampere strength current. **amperage** (-rij), *n.* the strength of an electric current measured in amperes.
ampersand (am'pəsand), *n.* the sign '&'.
amphetamine (amfet'əmēn, -min), *n.* (a derivative of) a synthetic drug (*alpha,* **m**ethyl, **ph**enyl, **a**mine) which has a stimulant action on the brain.
Amphibia (amfib'iə), *n.pl.* a class of vertebrate animals, between reptiles and fishes, which in their early stage breathe by gills; (**amphibia**) animals which can live either on land or water. **amphibian,** *n.* any amphibious animal; an aircraft, tank or other vehicle adapted for both land and water. *a.* pertaining to the Amphibia; amphibious. **amphibious,** *a.* capable of living both on land and in water; designed for operation on land and in water; pertaining to or trained for the invasion of foreign shores via the sea. **amphibiousness,** *n.*
amphibiology (amfibiol'əji), *n.* the department of science which treats of the Amphibia. **amphibiological** (-loj'-), *a.*
amphibrach (am'fibrak), *n.* a metrical foot of three syllables, the middle one long and the first and third short, as in *in-hū-man.*
amphigam (am'figam), *n.* one of the lower cryptogams, having no distinct sexual organs.
Amphipoda (amfip'ədə), *n.pl.* an order of crustaceans having two kinds of feet, one for walking and one for swimming. **amphipod** (am'fipod), *n.,* *a.* (of) any animal of the Amphipoda. **amphipodous,** *a.*
amphitheatre (am'fithiətə), *n.* an oval or circular building with rows of seats rising one above another round an open space; a place of public contest; a valley surrounded by hills. **amphitheatrical** (-at'-), *a.*
amphitryon (amfit'riən), *n.* a host; the giver of a banquet.
amphora (am'fərə), *n.* an ancient two-handled vessel for holding wine, oil etc.; an ancient liquid measure containing about 6 gallons (27·3 litres) among the Romans, and about 9 gallons (41 litres) among the Greeks.
ample (am'pl), *a.* of large dimensions; wide, great, fully sufficient, liberal. **ampleness,** *n.* **amply,** *adv.*
amplify (am'plifī), *v.t.* to enlarge upon, to give more details of; to increase, make greater; to increase the strength of (a signal), esp. the loudness of (sound). *v.i.* to speak or write diffusely; to give further details. **amplification** (-fi-), *n.* enlargement or extension; further detail; diffuseness; increase in strength of a signal or sound; an enlarged representation. **amplifier,** *n.* a complete unit which performs amplification of signals; an electrical or electronic circuit or system to amplify signals.
amplitude (am'plitūd), *n.* extent, size, bulk; greatness, abundance; the angular distance of a heavenly body, at its rising or setting, from the east or the west point of the horizon; the magnitude of the variation from a mean position or value of a vibration or oscillation, or of an alternating current or wave. **amplitude modulation,** *n.* (transmission of a signal by) modulation of the amplitude of a radio carrier wave in accordance with the characteristics of the signal carried; broadcasting using this system of transmission.
ampoule (am'pool), *n.* a sealed phial containing one dose of a drug.
ampulla (ampul'ə), *n.* (*pl.* **-llae** (-ē)) a nearly globular flask with two handles, used by the ancient Romans; a vessel

amputate | 24 | analytic

for holding consecrated oil, wine etc. **ampullaceous** (-ā'shəs), *a.* resembling a globular flask; bottle-shaped.
amputate (am'pūtāt), *v.t.* to cut off from an animal body. **amputation**, *n.* the act of amputating. **amputator**, *n.* **amputee** (-tē'), *n.* a person or animal having had an amputation.
amt, (*abbr.*) amount.
AMU, (*abbr.*) atomic mass unit.
amuck (əmŭk'), **amok** (əmok'), *adv.* in **to run amuck**, to attack indiscriminately, actuated by a frenzied desire for blood; hence to run wild or headlong, to go crazy.
amulet (am'ūlit), *n.* anything worn about the person as an imagined preservative against sickness, witchcraft etc.
amuse (əmūz'), *v.t.* to divert attention from serious business by anything entertaining; to please with anything light and cheerful; to entertain. **amusement**, *n.* that which amuses; play, diversion; excitement of laughter; the state of being amused. **amusement arcade**, *n.* a covered space containing coin-operated game and gambling machines. **amusing**, *a.* entertaining, diverting, provoking laughter. **amusingly**, *adv.*
amyl (am'il), *n.* a monovalent alcohol radical. **amylaceous** (-ā'shəs), *a.* pertaining to or of the nature of starch. **amylase** (-ās), *n.* any of various enzymes that break down starch and glycogen. **amylene** (-ēn), *n.* a hydrocarbon with anaesthetic properties. **amyloid** (-oid), *a.* resembling or containing starch; starchy. *n.* a non-nitrogenous starchy substance.
an, (an, *unstressed* ən), see a [2].
an., (*abbr.*) in the year.
an-, *pref.* (1) on, as in *anent*, *anon;* (2) AD- before *n*, as in *annex*, *announce;* (3) see ANA-; (4) A-, not, as in *anaesthetic*, *anarchy*.
-an (-ən), *suf.* of, belonging to, pertaining to, e.g. *human*, *pagan*, *publican*, *Christian*, *Unitarian*, *European* etc.
ana-, **an-**, *pref.* up, back, backwards; as in *anachronism*, *anagram*, *analogy*, *aneurism;* again, as in *anabaptism*.
-ana, -iana, *suf.* things about, sayings of, anecdotes concerning, objects relating to, as in *Johnsoniana*, *Shakespeareana*, *Virgiliana*, *railwayana*.
anabaptism (ənəbap'tizm), *n.* a second baptism; the doctrine of the anabaptists. **anabaptist**, *n.* one who rebaptizes; a member of a German sect which arose in the 16th cent. and advocated baptism only of adults; applied (as a term of reproach) to the modern Baptists who adhere to this doctrine. *a.* of or pertaining to anabaptism. **anabaptistical** (-tis'-), *a.*
anabatic (ənəbat'ik), *a.* of wind or air currents, moving upwards.
anabolic (ənəbol'ik), *a.* in **anabolic steroids**, *n.pl.* synthetic steroid hormones that cause rapid growth in body tissues, esp. skeletal muscle, and are sometimes (illegally) taken by athletes.
anabranch (an'əbrahnch), *n.* (*Austral.*) a tributary rejoining the main stream of a river and thus forming an island.
anachronism (anak'rənizm), *n.* the reference of an event, custom or circumstance to a wrong period or date; anything out of date or incongruous with the present, or with the time portrayed (in a film, play etc.). **anachronic** (anəkron'ik), *a.* **anachronically**, *adv.* **anachronistic** (-nis'-), *a.* pertaining to or involving an anachronism. **anachronistically**, *adv.*
anaconda (anəkon'də), *n.* a python from Sri Lanka; a large S American boa; any large snake which kills its prey by constriction.
anacrusis (anəkroo'sis), *n.* (*pl.* **-ses** (-sēz)) an upward beat at the beginning of a verse, consisting of an unstressed syllable or syllables.
anadromous (anad'rəməs), *a.* of fish, ascending rivers to deposit spawn.
anaemia (ənē'miə), (*esp. N Am.*) **anemia**, *n.* want of blood, deficiency of blood; lack of haemoglobin or of red corpuscles. **anaemic**, *a.* of, relating to or suffering from anaemia; lacking vitality; pale.
anaerobe (ənēə'rōb, an'ə-), *n.* an organism that thrives best, or only, in the absence of oxygen. **anaerobic** (anərō'-), *a.*
anaesthesia (ənəsthē'ziə), (*esp. N Am.*) **anesthesia** (anəsthē'ziə), *n.* loss of feeling and consciousness, produced intentionally by drugs; insensibility. **anaesthesis**, *n.* anaesthesia. **anaesthetic** (-thet'-), *a.* producing anaesthesia. *n.* a substance which produces anaesthesia (during surgical operations). **anaesthetically**, *adv.* by way of an anaesthetic; so as to cause anaesthesia. **anaesthetist** (-nēs'-), *n.* one who administers an anaesthetic. **anaesthetize, -ise** (-nēs'-), *v.t.* to administer an anaesthetic to. **anaesthetization**, *n.* the process of administering an anaesthetic.
anaglyph (an'əglif), *n.* a figure cut or embossed in low relief. **anaglyphic** (-glif'-), **anaglyptic** (-glip'-), *a.* of or pertaining to an anaglyph; wrought in low relief. **anaglypta** (-glip'-), *n.* a type of thick white wallpaper with a heavily embossed pattern. **anaglyptics**, *n.pl.* the art of working in low relief.
anagram (an'əgram), *n.* a word or sentence formed by transposing the letters of another word or sentence. **anagrammatical** (-mat'-), *a.* of, pertaining to or containing an anagram. **anagrammatically**, *adv.* **anagrammatism** (-gram'-), *n.* the art or practice of making anagrams. **anagrammatist**, *n.* one who makes anagrams. **anagrammatize, -ise**, *v.t.* to transpose so as to form into an anagram.
anal (ā'nəl), *a.* pertaining to or situated near the anus.
anal., (*abbr.*) analogous, analogy; analyse, analysis.
analeptic (anəlep'tik), *a.* restorative, increasing the strength. *n.* a restorative medicine.
analgesia (anəljē'ziə), *n.* loss of sensibility to pain. **analgesic**, *n.* a drug that relieves pain. *a.* insensible to pain.
analogy (ənal'əji), *n.* similarity; reasoning from a parallel case; comparative example of similar but apparently different things; imitation of existing words or linguistic patterns in forming new words, inflectional forms etc. **analogic** (-loj'-), **-ical**, *a.* of, pertaining to, or involving analogy. **analogically**, *adv.* **analogist**, *n.* one who is occupied with analogy. **analogize, -ise**, *v.t.* to represent or explain by analogy. *v.i.* to reason from analogy.
analogous (ənal'gəs), *a.* presenting some analogy or resemblance. **analogously**, *adv.*
analogue, (*esp. N Am.*) **analog** (an'əlog), *n.* an analogous word or thing; a parallel. *a.* pertaining to information having a continuous range of values; measuring or displaying information on a continuous scale (in contradistinction to *digital*). **analogue computer**, *n.* a computer in which directly measurable, varying physical quantities, such as current or voltage, represent the numbers on which arithmetical operations are to be performed.
analyse (an'əliz), (*esp. N Am.*) **analyze**, *v.t.* to take to pieces, resolve into the constituent elements; to examine minutely; to determine the elements of a chemical compound; (*Lit.*) to examine critically; to resolve a sentence into its grammatical elements; to psychoanalyse. **analysable**, *a.* **analysand** (-and'), *n.* a person undergoing psychoanalysis. **analyser**, *n.* one who or that which analyses.
analysis (ənal'isis), *n.* (*pl.* **-ses** (-sēz)) the process of analysing; separation into constituent elements; resolution of a chemical compound into its elements to ascertain composition, purity etc.; resolution of mathematical problems by reducing them to equations; psychoanalysis.
analyst (an'əlist), *n.* one who analyses; a psychoanalyst.
analytic (anəlit'ik), **-ical**, *a.* pertaining to analysis; resolving anything into its constituent parts; (*Philol.*) using separate words instead of inflections capable of, given to, precise critical examination. **analytical geometry**, *n.* geometry that uses coordinates to determine the position of a point. **analytically**, *adv.* **analytics**, *n.* the science of

analysis.
anamorphosis (anəmaw'fəsis), *n.* a distorted projection of any object so contrived that if looked at from one point of view, or reflected from a suitable mirror, it will appear properly proportioned; (*Bot.*) degeneration causing change of appearance; abnormal alteration of form.
ananas (ənah'nəs, an'ənas), *n.* the pineapple plant or its fruit.
anapaest (an'əpest, -pēst), (*esp. N Am.*) **anapest**, *n.* a metrical foot consisting of three syllables, the first two short and the third long, a reversed dactyl. **anapaestic** (-pes'-), *a.* composed of anapaests. *n.* an anapaestic line or verse.
anaphora (ənaf'ərə), *n.* the commencement of successive sentences or clauses with the same word or words; use of a word, such as a pronoun, to refer to a preceding word or phrase without repetition. **anaphoric** (anəfo'-), *a.* pertaining to anaphora; referring to a preceding word or phrase. **anaphorically,** *adv.*
anaphrodisiac (ənafrədiz'iak), *n.* an agent destroying or decreasing sexual desire.
anaphylaxis (anəfilak'sis), *n.* a condition of increased or extreme sensitivity to a foreign substance introduced into the body following previous contact. **anaphylactic,** *a.*
anaplasty (an'əplasti), *n.* plastic surgery. **anaplastic** (-plas'-), *a.*
anaptotic (anəptot'ik), *a.* becoming uninflected again (a term applied to languages, English for example, in which the inflections have been replaced by particles).
anarchy (an'əki), *n.* absence of government; want of settled government; disorder, lawlessness; political anarchism or the utopian society resulting from this. **anarchic, -ical** (-nah'-), *a.* **anarchically,** *adv.* **anarchism,** *n.* the principles of anarchy; a theory of government based on the free agreement of individuals rather than on submission to law and authority. **anarchist,** *n.* one who aims at producing anarchy; one opposed to all forms of government, a supporter of anarchism.
anarthrous (ənah'thrəs), *a.* (*Physiol.*) without joints.
anastigmat, anastigmat lens (anəstig'mat), *n.* a lens free from astigmatism, which refers every point on the scene accurately to a corresponding point on the lens. **anastigmatic** (-mat'-), *a.* free from astigmatism.
anastrophe (ənas'trəfi), *n.* inversion of the natural order of the words in a sentence or clause.
anat., (*abbr.*) anatomical, anatomy.
anathema (ənath'əmə), *n.* (*pl.* **-mas, -mata** (-them'ətə)) the formal act by which a person or thing is cursed, excommunication; the person or thing cursed; a curse, denunciation; an object of loathing. **anathematize, -ise,** *v.t.* to excommunicate, to curse, to put under a ban. *v.i.* to curse.
anatomy (ənat'əmi), *n.* the science of dissecting an organized body so as to discover its structure, and the makeup, arrangement and inter-relation of its parts; a treatise on the science of anatomy; the act of dissecting; a minute examination, reduction to parts or elements, analysis. **anatomic** (anətom'-), **-ical,** *a.* pertaining to or connected with anatomy. **anatomically,** *adv.* **anatomize, -ise,** *v.t.* to dissect; to make a dissection of; to examine minutely, analyse. **anatomist,** *n.* one who practises or is skilled in anatomy.
anatta (ənat'ə), **anatto, annatto** (-tō), *n.* an orange-red dye from Central America, used to colour cheese.
anc., (*abbr.*) ancient.
-ance (-əns), *suf.* noting state or action, as *distance, fragrance, parlance, riddance*.
ancestor (an'sistə), *n.* one from whom a person is descended; a progenitor; an organism of low type from which others of higher type have been developed. **ancestral** (-ses'-), *a.* pertaining to ancestors; derived from or owned by ancestors. **ancestry** (-tri), *n.* a line of ancestors; (high) birth, (honourable) lineage; ancient descent.
anchor (ang'kə), *n.* a heavy hooked iron instrument dropped from a ship to embed itself in the bottom and prevent drifting; anything shaped like an anchor; something that holds an object in place; a source of security or confidence. *v.t.* to secure by means of an anchor; to fix firmly. *v.i.* to take up a position at anchor; to settle, rest, to sit down. **at anchor,** held by an anchor; at rest. **to cast, drop anchor,** to drop the anchor into the sea; to settle permanently. **to weigh anchor,** to raise the anchor preparatory to sailing. **anchor-hold,** *n.* the hold which an anchor takes; a secure grip. **anchorman, -woman,** *n.* in sport, the last team member to compete, esp. in a relay race; a television or radio broadcaster who introduces and links the various reports etc. making up a (news) programme. **anchor-watch,** *n.* a watch set on board ship whilst at anchor; the people composing such a watch. **anchored,** *a.* held by an anchor; firmly fixed. **anchorless** (-lis), *a.* without an anchor or firm hold; drifting.
anchorage (ang'kərij), *n.* a place suitable or designed for anchoring in; the hold on the sea-bottom by an anchor; a source of security.
anchovy (an'chəvi), *n.* a small, strongly-flavoured fish of the herring family, caught in the Mediterranean, pickled for exportation, and used in sauces etc.
anchylose, ankylose (ang'kilōz), *v.t.* to stiffen (a joint) by anchylosis; to consolidate (two separate bones). *v.i.* to become stiff; to grow together. **anchylosis, ankylosis** (-lō'-), *n.* the formation of a stiff joint by the union of bones or fibrous tissue; the coalescence of two bones.
ancient (än'shənt), *a.* of or belonging to long past time; past, former, esp. of the times before the Middle Ages, that is before the end of the Western Empire (AD 476); (*coll.*) very old; (*coll.*) antiquated. *n.pl.* those who lived in former (esp. Classical) times. **the Ancient of Days,** the Deity. **ancient history,** *n.* history of ancient times, esp. to the end of the Western Empire, AD 476; (*coll.*) information, gossip etc. that is widely known. **ancient lights,** *n.pl.* windows that have acquired by long usage (not less than 20 years) the right to light from outside. **anciently,** *adv.* **ancientness,** *n.*
ancillary (ansil'əri), *a.* subservient, subordinate; auxiliary, supplementary. *n.* one who assists or supplements others.
-ancy (-ənsi), *suf.* expressing quality or state, e.g. *constancy, elegancy, infancy, vacancy.*
and (and. *unstressed* ənd), *conj.* the copulative which joins words and sentences; plus.
andante (andan'ti), *adv., a.* (*Mus.*) moderately slow(ly). *n.* a moderately slow movement or piece. **andante affettuoso,** *adv.* (*Mus.*) slowly and tenderly. **andante con moto** (kon mō'tō), *adv.* slowly but with movement. **andante grazioso,** *adv.* slowly and gracefully. **andante maestoso,** *adv.* slowly and majestically. **andante sostenuto,** *adv.* slow, but sustained. **andantino** (-tē'nō), *adv., a.* slightly less slow(ly) than andante. *n.* a movement or piece of this character. [It.]
Anderson shelter (an'dəsən), *n.* an air-raid shelter formed of arched corrugated steel. [Sir J. *Anderson,* 1882–1958, Home Secretary, 1939–40]
andiron (an'diən), *n.* a horizontal bar raised on short legs, with an ornamental upright in front, placed on each side of a hearth to support logs in a wood fire; a fire-dog.
andr(o)-, *comb. form.* pertaining to the male sex, or to male flowers.
androgen (an'drəjən), *n.* a male sex hormone; any substance with male sex hormone activity. **androgenic** (-jen'-), *a.*
androgynous (androj'inəs), *a.* presenting the characteristics of both sexes in the same individual; bearing both stamens and pistils in the same flower or on the same plant. **androgyny** (-droj'ini), *n.* hermaphroditism, the presence of male and female organs in one individual.

android (an'droid), *n.* a robot having human form.
andropetalous (andrəpet'ələs), *a.* a term applied to flowers made double by the conversion of stamens into petals.
androphagous (androf'əgəs), *a.* man-eating, cannibal.
androsterone (andros'tərōn), *n.* a male sex hormone occurring in the testes and in urine.
-androus, *suf.* having male organs or stamens, e.g. *diandrous, monandrous.*
ane (yin, än), *n., a., pron.* (*Sc.*) one, a, an.
-ane, *suf.* forming adjectives, e.g. *humane, mundane, urbane;* forming names of hydrocarbons, e.g. *methane, pentane, hexane.*
anecdote (an'ikdōt), *n.* the relation of an isolated fact or incident; a short, pithy, often humorous narrative. **anecdotal** (-dō'-), *a.* pertaining to or consisting of anecdotes. **anecdotally**, *adv.*
anechoic (anikō'ik), *a.* free from echoes.
anemia ANAEMIA.
anemometer (anəmom'itə), *n.* an instrument for measuring the velocity of wind, a wind gauge. **anemometric** (-met'-), *a.* **anemometry** (-tri), *n.*
Anemone (ənem'əni), *n.* a genus of plants with brilliantly-coloured flowers common in Britain; (**anemone**) a plant or flower of this genus. See also SEA-ANEMONE under SEA.
anencephaly (anənsef'əli, -kef'-), *n.* a congenital defect in which part or all of the brain is missing. **anencephalic** (anensəfal'ik, -kə-), *a.*
-aneous (-äniəs), *suf.* belonging to, e.g. *extraneous, instantaneous.*
aneroid (an'əroid), *a.* operating without liquid. **aneroid barometer**, *n.* a barometer which measures the pressure of air by its action on a springy metallic box from which the air has been partially exhausted.
anesthesia, anesthetic etc. ANAESTHESIA.
aneurism, aneurysm (an'ūrizm), *n.* an abnormal dilatation in an artery, particularly of the aorta. **aneurysmal, -ismal** (-riz'-), *a.*
anew (ənū'), *adv.* (once) again; afresh.
angary (ang'gəri), **angaria** (-gah'riə), *n.* (*Law*) the confiscation or destruction by a belligerent of neutral property, esp. shipping, subject to claim for compensation.
angel (än'jəl), *n.* a messenger from God; a ministering spirit; a guardian or attendant spirit; (*sl.*) in popular journalism, a nurse; a benign, innocent or adorable creature; an old English gold coin, orig. the *angel-noble,* varying in value from 33p upwards, and bearing the figure of the archangel Michael; a conventional representation of the heavenly messenger; (*coll.*) a financial backer, esp. of a theatrical production. **angels on horseback,** oysters rolled in bacon. **on the side of the angels,** of a person, basically good, well-intentioned, supporting the morally right. **angel dust**, *n.* (*coll.*) the hallucinogenic drug phencyclidine. **angel-fish**, *n.* a fish allied to the rays and sharks, named from the wing-like expansion of the pectoral fins; any of several brightly-coloured tropical fishes with laterally compressed bodies; a small tropical American fish with black and silver stripes, often kept in aquariums. **angel(-food) cake**, *n.* a light sponge cake made with egg whites. **angelhood**, *n.* **angelic** (anjel'ik), **-ical**, *a.* resembling, or of the nature of, an angel. **angelically**, *adv.*
angelolatry (-ol'ətri), *n.* angel-worship. **angelology** (-ol'-), *n.* the doctrine of angelic beings.
Angelica (anjel'ikə), *n.* a genus of umbelliferous plants, including *arch-angelica,* used in medicine, and as a preserve or sweetmeat; (**angelica**) candied angelica root.
angelus (an'jələs), *n.* a short devotional exercise in the Roman Catholic Church in honour of the Incarnation. **angelus(-bell)**, *n.* a bell rung early in the morning, at noon, and in the evening, as a signal to say the angelus.
anger (ang'gə), *n.* rage, fierce displeasure, passion excited by a sense of wrong. *v.t.* to make angry, to excite to wrath; to enrage. *v.i.* to become angry. **angry** (ang'gri), *a.* wrathful, expressing anger; inflamed, painful; suggesting anger, threatening. **angrily**, *adv.*
angina (anji'nə), *n.* quinsy; angina pectoris. **anginal**, *a.*
angina pectoris (anji'nə pek'təris), *n.* a heart condition marked by paroxysms of intense pain due to over-exertion when the heart is weak or diseased.
angi(o)-, *comb. form.* vascular; pertaining to the vessels of organisms.
angiography (anjiog'rəfi), *n.* X-ray photography of the blood vessels. **angiograph** (an'-), **angiogram**, *n.* a photograph made by angiography.
Angle (ang'gl), *n.* a member of one of the Low German tribes that settled in Northumbria, Mercia and East Anglia. **Anglian** (-gli-), *a.* of or pertaining to the Angles. *n.* an Angle.
angle [1] (ang'gl), *v.i.* to go angling (*fig.*) to fish (for), to try to elicit, as a compliment; to get something by craft. **angler**, *n.* one who fishes with a rod. **angler-fish**, *n.* a small British fish which attracts its prey by filaments attached to its head. **angling**, *n.* the art or practice of fishing with a rod and line; trying to find out by craft. **to go angling,** to fish with rod and line.
angle [2] (ang'gl), *n.* a corner; the inclination (in degrees) of two lines towards each other; an angular projection; a point of view from which something is considered, an approach. *v.t.* to move, place, turn, direct at an angle; to present (a report, news story etc.) in a particular way or from a particular point of view; to bias. *v.i.* to proceed or turn at an angle. **angle of incidence,** the angle between a ray of light meeting a surface and a line perpendicular to the surface. **angle of inclination,** the angle between an inclined plane and the horizon. **angle of refraction,** the angle at which a ray of light is turned from its direct course in passing through a given medium. **angle of vision,** the angle at which objects are seen and which determines their apparent magnitudes. **angle-iron**, *n.* an angular piece of iron used to strengthen framework of any kind. **angled**, *a.* having an angle or angles; biased.
angler ANGLE [1].
Anglican (ang'glikən), *a.* of or belonging to the Church of England or any church in communion with it. *n.* a member of the Anglican church. **Anglicanism**, *n.* the teachings and practices of the Anglican church.
anglice (ang'glisē, -si), *adv.* in English.
Anglicism (ang'glisizm), *n.* an English idiom; an English custom or characteristic; attachment to what is English. **Anglicist**, *n.* a student or specialist in English language, literature or culture. **Anglicize, -ise**, *v.t.* to make (similar to) English; to give an English form to; to turn into English (language).
Anglo-, *comb. form.* English; of or belonging to England or the English; partially English (the meaning completed by another word), as *Anglo-Irish.*
Anglo-American (ang·glōəme'rikən), *n.* an American of English or British parentage or descent. *a.* of or belonging to such Americans, or to England (or Britain) and the US.
Anglo-French (ang·glōfrensh'), *a.* pertaining to England or Britain and France pertaining to the Anglo-French language. *n.* the French language of mediaeval England.
Anglo-Indian (ang·glōin'diən), *n.* an English person born, or long resident, in India; a person of mixed English or British and Indian blood. *a.* of or belonging to such people, or to England (or Britain) and India.
Anglo-Irish (ang·glōi'rish), *a.* of or between England (or Britain) and N or S Ireland; of the Anglo-Irish people. *n.pl.* Irish people of English Protestant descent.
Anglomania (ang·glōmā'niə), *n.* excessive fondness for English manners and customs. **Anglomaniac** (-ak), *n.*
Anglophile (ang'glōfīl), **-phil** (-fil), *n.* an admirer of

England or of the English. **Anglophilia** (-fil′iə), *n.* **Anglophilic** (-fil′ik), *a.*
Anglophobe (ang′glōfōb), *n.* a hater of England or of the English. **Anglophobia** (-fō′-), *n.* fear or distrust of England. **Anglophobic**, *a.*
Anglophone (ang′glōfōn), *n.* a person who speaks English. *a.* of or belonging to an English-speaking nation.
Anglo-Saxon (ang·glōsak′sən), *a.* of or belonging to the English race or language as distinct from Continental Saxons; of the whole English people before the Norman Conquest; of English people of Teutonic descent; of English-speaking people generally. *n.* a member of the Anglo-Saxon race; the Old English language.
angora (ang·gaw′rə), *n.* a goat with long silky hair; the hair itself, or a fabric made therefrom; a long-haired variety of the domestic cat; a breed of rabbit with long, fine fur; a yarn or fabric made from angora rabbit hair.
angostura, angustura (ang·gəstū′rə), *n.* the bark of certain trees, used originally for treating fever, and currently in the preparation of bitters.
angry, angrily ANGER.
angst (angst), *n.* a nonspecific feeling of anxiety and guilt produced esp. by considering the imperfect human condition.
Ångström unit (ang′strəm), *n.* a unit of length used to express the wavelengths of different kinds of radiations, equivalent to 1/254,000,000 in. (10^{-10}m). [A. J. *Ångström*, 1814–74]
anguish (ang′gwish), *n.* excessive pain or distress of body or mind. *v.t.* to afflict with extreme pain or grief.
angular (ang′gūlə), *a.* having angles or sharp corners; forming an angle; in an angle; measured by an angle; bony, lacking in plumpness or smoothness. **angularity** (-la′-), *n.* **angularly**, *adv.* **angulate** (-lāt), *a.* angular, formed with angles or corners. **angulose, -lous**, *a.* angular, having angles or corners.
anharmonic (anhahmon′ik), *a.* (*Math.*) not harmonic.
anhydride (anhī′drīd), *n.* a chemical substance formed from another, esp. an acid, by removing the elements of water.
anhydrite (anhī′drīt), *n.* a colourless, orthorhombic mineral, calcium sulphate, anhydrous gypsum.
anhydrous (anhī′drəs), *a.* having no water in the composition.
anil (an′il), *n.* the indigo-plant; indigo (dye).
aniline (an′ilīn), *n.* a chemical base used in the production of many beautiful dyes, and originally obtained from indigo, now chiefly from nitrobenzene. *a.* produced using aniline.
anima (an′imə), *n.* (*Psych.*) a person's true inner self; the feminine aspect of the male personality.
animadvert (animədvœt′), *v.i.* to direct the attention to; to criticize or censure (with *on* or *upon*). **animadversion** (-shən), *n.* criticism, censure, reproof.
animal (an′iməl), *n.* any organized being possessing life, sensation and the power of voluntary motion, including humans; one of the lower animals as distinct from humans, esp. a mammal or quadruped; a human being whose animal nature is abnormally strong, a brute; (*coll.*) a (variety of) person, thing or organization, as in *there's no such animal*. *a.* of, belonging to or derived from animals, their nature or functions; carnal; pertaining to animals as distinguished from vegetables or minerals. **animal-free**, *a.* not containing or using animal products. **animal heat**, **animal warmth**, *n.* the warm temperature characterizing the bodies of living animals. **animal husbandry**, *n.* the breeding and care of farm animals. **animal kingdom**, *n.* animals generally, viewed as one of the three great divisions of natural objects. **animal liberation**, *n.* a movement aimed at securing animal rights, protecting them from laboratory experiments etc. **animal magnetism**, *n.* the quality of being powerfully attractive, esp. to members of the opposite sex. **animal rights**, *n.* the bestowing on animals of various rights usu. attributed only to humans. **animalism**, *n.* the theory which views mankind as merely animal; sensuality. **animalist**, *n.* a believer in animalism; a supporter of animal rights. **animality** (-mal′-), *n.* animal nature. **animalize, -ise**, *v.t.* to make into an animal; to brutalize. **animalization, animally**, *adv.* physically, as opposed to intellectually.
animalcule (animal′kūl), **animalculum** (-ləm), *n.* (*pl.* **-cules, -cula** (-lə)) an animal so small as to be invisible to the naked eye. **animalcular**, *a.*
animalism, animality ANIMAL.
animate (an′imāt), *v.t.* to give life or spirit to; to vivify, to inspire; to stir up; to give the appearance of movement to; to produce as an animated cartoon. *a.* (-mət), living, endowed with life; lively. **animated**, *part.a.* possessing life; full of life or spirits; vivacious, lively; moving as if alive. **animated cartoon**, *n.* a film produced by photographing a series of drawings or objects, each varying slightly in position from the preceding one, to give the illusion of movement. **animated graphics**, *n.pl.* computer graphics featuring moving pictures or shapes. **animatedly**, *adv.* **animating**, *a.* **animatingly**, *adv.* **animation**, *n.* the act of animating; state of being animated, vitality; life, vivacity; (the techniques used in the production of) an animated cartoon. **animative**, *a.* having the power to impart life or spirit. **animator**, *n.* an artist who prepares material for animated cartoons.
animism (an′imizm), *n.* the doctrine that vital phenomena are produced by an immaterial soul distinct from matter; the attribution of a living soul to inanimate objects and to natural phenomena. **animist**, *n.* a believer in animism. **animistic** (-mis′-), *a.*
animosity (animos′əti), *n.* enmity tending to show itself in action.
animoso (animō′sō), *adv.* (*Mus.*) with spirit.
animus (an′iməs), *n.* spirit actuating feeling, usu. of a hostile character; animosity; the masculine part of the female personality.
anion (an′īən), *n.* an ion that moves towards the anode; a negatively charged ion.
anise (an′is), *n.* an umbelliferous plant, cultivated for its aromatic seeds, which are carminative. **aniseed** (-sēd), *n.* the seed of the anise, used as a flavouring. **anisette** (-zet′), *n.* a liqueur made from aniseed.
aniso-, *comb. form.* odd, unequal, unsymmetrical.
anisotropy (anīsot′rəpi), *n.* aeolotropy. **anisotropic** (-trop′-), *a.*
ankh (angk), *n.* a keylike cross being the emblem of life, or male symbol of generation.
ankle (ang′kl), *n.* the joint by which the foot is united to the leg; the part of the leg between foot and calf. **anklet** (-lit), *n.* a fetter, strap or band for the ankle; an ornamental chain or band worn round the ankle.
ankylose, ankylosis ANCHYLOSE.
ann., (*abbr.*) annals; annual.
anna (an′ə), *n.* a former monetary unit and coin of India, Burma and Pakistan, equal to one-sixteenth of a rupee.
annals (an′əlz), *n.pl.* a narrative of events arranged in years; historical records; in the Roman Catholic Church, Masses said for the space of a year. **annalist**, *n.* one who writes annals. **annalistic** (-lis′-), *a.*
annates (an′āts), *n.pl.* the first year's revenue of Roman Catholic ecclesiastics on their appointment to a benefice, paid to the Pope.
annatto ANATTA.
anneal (ənēl′), *v.t.* to burn metallic colours into (glass etc.); to temper, as glass or metals, by subjecting them to intense heat, and then allowing them to cool slowly; (*fig.*) to temper; to render tough.
annectent (ənek′tənt), *a.* connecting, linking.
Annelida (ənel′idə), *n.pl.* a class of invertebrate animals,

including the earthworm, with elongated bodies composed of annular segments. **annelid** (an'-), *n.* one of the Annelida. **annelidan,** *n.*, *a.*

annex (əneks'), *v.t.* to unite to, add on to; to take possession of (as territory); to append as a condition, qualification or consequence; (*coll.*) to steal. **annexable,** *a.* **annexation** (an-), *n.* the act of annexing; something annexed (often with the idea of unlawful acquisition). **annexe, annex** (an'eks), *n.* an appendix; a supplementary or subsidiary building.

annihilate (əni'əlāt), *v.t.* to reduce to nothing; to blot out of existence; to destroy the organized existence of; to reduce to constituent elements or parts; (*coll.*) to defeat overwhelmingly. *v.i.* (*Phys.*) to undergo annihilation. **annihilation,** *n.* the act of annihilating; the state of being annihilated; complete destruction of soul and body; the combining of an elementary particle and its antiparticle with the spontaneous transformation into energy.

anniversary (anivœ'səri), *n.* the annual return of any remarkable date; the celebration of such annually recurring date.

Anno Domini (anō dom'inī), *phr.* in the year of our Lord (abbr. AD), reckoned from the Christian era. *n.* (*coll.*) old age.

annotate (an'ətāt), *v.t.*, *v.i.* to make notes or comments (upon). **annotation,** *n.* the act of annotating; an explanatory note. **annotator,** *n.*

announce (ənowns'), *v.t.* to make known, to proclaim; to declare officially, or with authority; to make known the approach or arrival of. **announcement,** *n.* **announcer,** *n.* the person who announces the items of a broadcasting programme, reads news summaries etc.

annoy (ənoi'), *v.t.* to tease, to molest, to put to inconvenience by repeated or continued acts. *v.i.* to cause annoyance. **annoyance,** *n.* the act of annoying; the state of being annoyed; that which annoys. **annoyingly,** *adv.*

annual (an'ūəl), *a.* returning or happening every year; reckoned by, or done or performed in a year; (*Bot.*) lasting but a single year or season. *n.* a book published every year, a year-book; a plant which lives for a year only. **annual ring,** *n.* a ring of wood seen in the cross-section of a plant stem or root, indicating one year's growth. **annualize, -ise,** *v.t.* to adjust or calculate according to a yearly rate. **annually,** *adv.* year by year, yearly.

annuity (ənū'iti), *n.* a sum of money payable annually; an investment insuring fixed annual payments. **annuitant,** *n.* one who receives an annuity.

annul (ənūl'), *v.t.* (*past*, *p.p.* **annulled**) to render void, cancel, abolish; to destroy the validity of. **annulment,** *n.*

annular (an'ūlə), *a.* ring-shaped, ringed. **annular eclipse,** *n.* an eclipse of the sun in which the silhouette of the moon obscures only the central portion of the sun's surface and leaves a ring of light showing round the moon. **annularly,** *adv.* **annulate** (-āt), **-lated,** *a.* having or marked with rings; composed of rings or ring-like segments. **annulation,** *n.* the state of being annulate; ring-like structure or markings. **annulet** (-lit), *n.* a little ring; (*Arch.*) a small fillet encircling a column. **annulus** (-ləs), *n.* (*pl.* -**li** (-lī)) a ring-shaped structure or part.

annunciate (ənūn'siāt), *v.t.* to announce. **annunciation,** *n.* the act of announcing; the announcement of the Incarnation made by the angel Gabriel to the Virgin Mary; the church festival (Lady Day, 25 March) in honour of that event. **annunciator,** *n.* one who, or that which, announces; an indicator for electric bells or telephones to show who has rung or spoken.

annus mirabilis (an'əs, -us mirah'bilis), *n.* (*pl.* **anni** (-nē) **mirabiles** (-lēz)) a remarkable year (usu. applied in English history to 1666, year of the Great Fire of London etc.).

anode (an'ōd), *n.* the positive electrode or pole in an electrolytic cell; the negative electrode of a primary cell delivering current; the positive electrode which collects electrons in an electronic valve. **anodal, anodic** (-nod'-), *a.* **anodize, -ise,** *v.t.* to give a protective surface coating of an oxide to (a metal) by making it the anode of an electrolytic cell.

anodyne (an'ədin), *a.* assuaging pain; alleviating distress of mind, soothing. *n.* a medicine which assuages pain; anything which alleviates distress of mind or soothes the feelings.

anoint (ənoint'), *v.t.* to smear with oil or an unguent; esp. to pour oil on as a religious ceremony; to consecrate with oil. **anointed,** *n.*, *a.*

anomaly (ənom'əli), *n.* (an) irregularity; (a) deviation from the common or established order, abnormality; the angular distance of a planet or satellite from its last perihelion or perigee. **anomalistic** (-lis'-), *a.* irregular, abnormal. **anomalistic month,** *n.* the time in which the moon passes from perigee to perigee. **anomalistic year,** *n.* the time occupied by the earth (or other planet) in passing from perihelion to perihelion: slightly longer than a tropical or sidereal year. **anomalous,** *a.* deviating from rule; irregular, abnormal. **anomalously,** *adv.* **anomalousness,** *n.*

anomie, anomy (an'omi), *n.* the breakdown or absence of moral and social standards in an individual or society.

anon (ənon'), *adv.* immediately, thereupon; soon after; in a little while.

anon. (ənon'), (*abbr.*) anonymous.

anonymous (ənon'iməs), *a.* nameless; having no name attached; of unknown or unavowed authorship or origin; lacking distinctive characteristics, nondescript. **anonym** (an'ənim), *n.* a person whose name is not made known; a pseudonym. **anonymity** (anənim'iti), *n.* **anonymized, -ised,** *a.* made anonymous. **anonymized screening,** *n.* the testing of unidentified (blood) samples, as for the presence of the HIV virus, without the patient's knowledge. **anonymously,** *adv.* **anonymousness,** *n.*

Anopheles (ənof'əlēz), *n.* a genus of mosquitoes including the malarial mosquito *Anopheles maculipennis*.

Anoplura (anəploo'rə), *n.* an order of parasitic insects including the human louse.

anorak (an'ərak), *n.* a warm waterproof jacket, usu. with a hood.

anorexia (anərek'siə), *n.* loss of appetite; anorexia nervosa. **anorexia nervosa** (nœvō'sə), *n.* a psychological disorder characterized by an aversion to eating and fear of gaining weight. **anorectic** (-rek'-), *a.* suffering from anorexia (nervosa); causing loss of appetite. *n.* a substance that causes loss of appetite. **anorexic,** *a.* anorectic. *n.* a person suffering from anorexia (nervosa).

anosmia (ənoz'miə), *n.* absence of the sense of smell. **anosmatic** (anozmat'-), *a.*

another (ənū'dhə), *pron.*, *a.* an other one, one more; one of the same kind; a different one; any other. **another place,** *n.* the other House (of Parliament).

anoxia (ənok'siə), *n.* deficiency of oxygen to the tissues. **anoxic,** *a.*

Anschluss (an'shlus), *n.* the forced union of Austria with Germany in 1938.

anserine (an'sərin), *a.* of or belonging to the goose; gooselike, stupid, silly.

answer (an'sə, ahn'-), *n.* a reply to a charge, objection, appeal or question; a solution of a problem; something done in return; a practical reply; (*Law*) a counter-statement to a bill of charges. *v.t.* to reply or respond to; to be sufficient for or suitable to; to be opposite to; to solve. *v.i.* to reply, to respond; to suit, to correspond. **to answer back,** to reply rudely or cheekily. **to answer for,** to be responsible or answerable for. **answerable,** *a.* liable to be called to account; capable of being answered. **answerer,** *n.* one who answers (to a question etc.).

ant (ant), *n.* a small, social, hymenopterous insect of the

family Formicidae. **ant-bear**, *n.* the giant ant-eater of S America; an aardvark. **ant-eater**, *n.* a genus of edentate mammals, with long extensile tongues, which they thrust into ant-hills and withdraw covered with ants; an echidna; an aardvark. **ant-eggs**, *n.pl.* the popular name for the pupae of ants. **ant-hill**, *n.* the mound or hillock raised by a community of ants. **ant-lion**, *n.* a neuropterous insect, the larvae of which construct a kind of pitfall for ants and other insects. **anting**, *n.* the placing by birds of ants in their plumage.
ant., *(abbr.)* antonym.
an't (ant) AIN'T.
ant-, *pref.* against, as in *antagonist, Antarctic*.
-ant (-ənt), *suf.* forming adjectives, as *distant, elegant, trenchant;* denoting an agent, one who or thing which produces effect, as in *accountant, merchant*.
antacid (antas'id), *a.* counteracting acidity. *n.* a medicine that counteracts acidity of the stomach.
antae (an'tē), *n.pl.* square pilasters on each side of a door, or at the angles of a building.
antagonist (antag'ənist), *n.* an opponent; one who contends or strives with another; a muscle which counteracts another, and is in turn counteracted by it; a drug that counteracts the action of another or of a substance occurring naturally in the body. **antagonism**, *n.* opposition; conflict, active disagreement; (an) opposing force, action or principle. **antagonistic** (-nis'-), *a.* **antagonistically**, *adv.* **antagonize, -ise**, *v.t.* to counteract, put in active opposition; to arouse hostility or opposition in. *v.i.* to act in opposition.
antalkali (antal'kəlī), *n.* something that neutralizes an alkali. **antalkaline** (-līn), *n.*, *a.* (a medicine) that counteracts the effect of an alkali.
antaphrodisiac (antafrədiz'iak), *n.*, *a.* (a medicine or agent) that allays or prevents sexual desire.
Antarctic (antahk'tik), *a.* opposite to the Arctic; southern; of or belonging to the S Pole or the region within the Antarctic Circle. *n.* the Antarctic regions. **Antarctic Circle**, *n.* a parallel of the globe, 23° 28' distant from the S Pole, which is its centre.
antarthritic (antahthrit'ik), *n.*, *a.* (a medicine) that prevents or relieves gout.
antasthmatic (antəsmat'ik), *n.*, *a.* (a medicine) that prevents or relieves asthma.
ante (an'ti), *n.* the stake which a poker-player puts down after looking at the cards, but before drawing; *(coll.)* amount paid, price. *v.t.* to stake; to pay. **to up the ante**, to increase the (asking) price or cost.
ante-, *pref.* before.
antebellum (antibel'əm), *a.* existing before the war, esp. the American Civil War.
antecede (antisēd'), *v.t.* to precede; to go before or in front of. **antecedence**, *n.* **antecedent**, *a.* going before in time; prior; anterior. *n.* that which goes before; the word to which a relative pronoun refers; the conditional clause of a hypothetical proposition; the first term of a mathematical ratio; *(pl.)* past circumstances. **antecedently**, *adv.*
antechamber (an'tichāmbə), *n.* an anteroom.
antechapel (an'tichapəl), *n.* the part of a chapel between the western wall and the choir screen.
antedate (an'tidāt), *n.* a date preceding the actual date. *v.t.* to date before the true date; to cause to happen prematurely; to happen earlier, precede; to anticipate.
antediluvian (antidiloo'viən), *a.* of or pertaining to the period before the biblical Flood; old-fashioned, antiquated. *n.* one who lived before the Flood; a very old or old-fashioned person.
antelope (an'təlōp), *n.* a animal ruminant, genus *Antilope*, akin to the deer and the goat.
ante meridiem (anti mərid'iem), *phr.* before noon (abbr. a.m.).
antenatal (antinā'təl), *a.* happening or existing before birth; dealing with pregnancy or pregnant women.
antenna (anten'ə), *n.* *(pl.* **-nnae** (-nē), **-nnas**) a sensory organ composed of segments in pairs on the heads of insects and crustaceans; a palp, a feeler; a filament in the male flowers of orchids that ejects the pollen when touched; an aerial. **antennal, antennary,** *a.* pertaining to the antennae. **antenniferous** (-nif'-), *a.* bearing antennae. **antenniform** (-fawm), *a.* shaped like an antenna.
antenuptial (antinŭp'shəl, -chəl), *a.* happening before marriage.
antependium (antipen'diəm), *n.* a covering for the front of an altar; a frontal.
antepenult (antipənŭlt'), **antepenultimate** (-mət), *a.* pertaining to the last syllable but two; last but two. *n.* the last (syllable) but two.
anterior (antiə'riə), *a.* going before, preceding in time or place. **anteriority** (-o'-), *n.* **anteriorly**, *adv.*
anteroom (an'tirum), *n.* a room leading into or forming an entrance to another.
anth-, *pref.* against, opposite to; used before aspirates, e.g. *anthelion, anthelmintic*.
anthelion (anthē'liən), *n.* *(pl.* **-lia** (ə)) a mock sun, a luminous ring projected on a cloud or fog-bank opposite the sun.
anthelmintic (anthəlmin'tik), *a.* destroying or remedial against parasitic, esp. intestinal, worms. *n.* a remedy for intestinal worms.
anthem (an'thəm), *n.* a hymn in alternate parts; a portion of Scripture or of the Liturgy set to music; a song of gladness or triumph.
Anthemis (an'thəmis), *n.* a genus of composite plants including the camomile.
anther (an'thə), *n.* the pollen-bearing organ of flowering plants.
antheridium (anthərid'iəm), *n.* *(pl.* **-dia** (-diə)) the male spore-bearing organ, analogous to an anther, of cryptogams. **antheridial,** *a.*
anthocyanin (anthōsī'ənin), *n.* any of a class of scarlet to blue plant pigments.
anthology (anthol'əji), *n.* collection of selected poems or other literary pieces; a collection of songs, paintings etc. **anthological** (-loj'-), *a.* **anthologist**, *n.* the compiler of an anthology. **anthologize, -ise,** *v.t.* to compile or put into an anthology.
Anthozoa (anthəzō'ə), *n.pl.* the Actinozoa. **anthozoan**, *n.*
anthracene (an'thrəsēn), *n.* a crystalline substance with blue fluorescence obtained from tar, used in the manufacture of dyes.
anthracite (an'thrəsīt), *n.* a non-bituminous coal, burning with intense heat, without smoke, and with little flame. **anthracitic** (-sit'-), **-citous** (an'thrəsitəs), *a.* bearing anthracite; composed of anthracite.
anthrax (an'thraks), *n.* an infectious, often fatal bacterial disease of sheep and cattle transmissible to humans; a malignant pustule in humans derived from animals suffering from this.
anthrop(o)-, *comb. form.* human; pertaining to mankind.
anthropocentric (anthrəpəsen'trik), *a.* centring in human beings; regarding mankind as the measure and aim of the universe.
anthropogeny (anthrəpoj'əni), **anthropogenesis** (-jen'əsis), *n.* the science or study of the origin of human beings.
anthropogenic, -genetic (-net'-), *a.*
anthropogeography (anthrəpōjiog'rəfi), *n.* the geography of the distribution of the races of mankind.
anthropography (anthrəpog'rəfi), *n.* the science which investigates the geographical distribution of human beings; ethnography.
anthropoid (an'thrəpoid), *a.* resembling human beings, of human form; of a person, apelike. *n.* a creature, esp. one of the higher apes, resembling a human being in form.
anthropology (anthrəpol'əji), *n.* the science of human

anthropometry   30   **antiphon**

beings; esp. the study of mankind as to body, mind, evolution, race and environment. **anthropological** (-loj'-), *a*.
anthropologically, *adv*. **anthropologist**, *n*. one versed in anthropology.
anthropometry (anthrəpom'ətri), *n*. the scientific measurement of the human body. **anthropometric, -ical** (-met'-), *a*.
anthropomorphous (anthrəpəmaw'fəs), **-morphic** (-fik), *a*. possessed of a form resembling that of a human being; pertaining to anthropomorphism. **anthropomorphism**, *n*. the attribution of a human form or character to the Deity, or of human characteristics to the lower animals. **anthropomorphist**, *n*. **anthropomorphise, -ize**, *v.t*. to give a human shape or attribute human characteristics to.
anthropophagous (anthrəpof'əgəs), *a*. feeding on human flesh, cannibal. **anthropophagy**, *n*. cannibalism.
Anthropopithecus (anthrəpōpith'ikəs), *n*. the genus of apes, including the chimpanzee and the gorilla, which most resembles man.
anti (an'ti), *prep*. opposed to. *n*. an opponent of a policy, political party etc.
anti-, *pref*. opposite, opposed to, against, instead of, in exchange, acting against, counteracting, as in *anti-bilious, anti-phlogistic, antiseptic, anti-social;* the opposite of, an opponent of, one of a contrary kind, the reverse of.
anti-abolitionist (antiabəlish'ənist), *n*. one opposed to the abolition of slavery.
anti-aircraft (antieə'krahft), *a*. employed against hostile aircraft.
antiballistic missile (antibəlis'tik), *n*. a missile designed to intercept and destroy a ballistic missile in flight.
antibiosis (antibīō'sis), *n*. antagonistic association between two organisms or between one organism and a substance produced by the other. **antibiotic** (-ot'-), *a*. inimical to life, esp. bacteria. *n*. a substance produced by a microorganism which inhibits the growth of or kills another microorganism.
antibody (an'tibodi), *n*. a substance produced in the blood in response to the presence of an antigen and capable of counteracting toxins.
antic (an'tik), *n*. (*usu. pl.*) a ludicrous act, trick or gesture.
anticancer (antikan'sə), *a*. used in the treatment of cancer.
Antichrist (an'tikrīst), *n*. a personal antagonist of Christ spoken of in the NT; an opponent of Christ. **antichristian** (-kris'-), *a*. opposed to Christ or to Christianity; pertaining to Antichrist. *n*. one opposed to Christ or to Christianity; an adherent of Antichrist. **antichristianism**, *n*.
anticipate (antis'ipāt), *v.t*. to use in advance; to deal with or be before (another); to forestall; to cause to happen earlier; to lock forward to, consider or deal with anything before the proper time. *v.i*. to speak, write or do something in expectation of something occurring later. **anticipant**, *a*. anticipating, expecting. *n*. one who anticipates.
anticipation, *n*. the act of anticipating; preconception, expectation; the introduction of a note before the chord about to be played. **anticipative, -patory**, *a*. **anticipatively**, *adv*. **anticipator**, *n*.
anticlerical (antikle'rikəl), *a*. opposed to (the political influence of) the clergy. **anticlericalism**, *n*.
anticlimax (antiklī'maks), *n*. the opposite of climax; a descent or decrease in impressiveness; bathos. **anticlimactic** (-mak'-), *a*.
anticlinal (antiklī'nəl), *a*. (*Geol.*) forming a ridge so that the strata lean against each other and in opposite directions. **anticline** (an'-), *n*. an anticlinal fold.
anticlockwise (antiklok'wīz), *a., adv*. in the reverse direction from that taken by the hands of a clock.
anticoagulant (antikōag'ūlənt), *n., a*. (a drug) that hinders blood clotting.
anticonvulsant (antikənvŭl'sənt), *n., a*. (a drug) used in treating or controlling (epileptic) convulsions.
anticyclone (antisī'klōn), *n*. the rotary outward flow of air from an atmospheric region of high pressure. **anticyclonic** (-klon'-), *a*.
antidepressant (antidipres'ənt), *n., a*. (a drug) used in treating or preventing mental depression.
antidote (an'tidōt), *n*. a medicine designed to counteract poison or disease; anything intended to counteract something harmful or unpleasant. **antidotal** (-dō'-), *a*.
antifreeze (an'tifrēz), *n*. a substance added to the water in car radiators to lower the freezing point.
antigen (an'tijən), *n*. a substance introduced into the body which stimulates the production of antibodies.
antihero (an'tihiərō), *n*. (*pl*. **-roes**) a principal character in a play, novel etc. who lacks noble or traditional heroic qualities. **antiheroic** (-hirō'-), *a*.
antihistamine (antihis'təmēn), *n*. a drug that counteracts the effects of histamine in allergic reactions.
antiknock (antinok'), *n*. a compound which is added to petrol to prevent knocking.
anti-life (antilīf'), *a*. not pro-life, opposing the views of pro-lifers.
antilogarithm (antilog'əridhm), **antilog** (an'-), *n*. the number represented by a logarithm. **antilogarithmic** (-ridh'-), *a*.
antilogy (antil'əji), *n*. contradiction in terms or in ideas.
antimacassar (antimakəs'ə), *n*. a covering for chairs, sofas etc. to prevent their being soiled by (macassar) oil on the hair, or as an ornament.
anti-marketeer (antimahkitiə'), *n*. an opponent of Britain's membership of the European Economic Community.
antimatter (antimat'ə), *n*. hypothetical matter composed of antiparticles.
antimony (an'timəni), *n*. a bright bluish-white brittle metallic element, at. no. 51; chem. symbol Sb, occurring native, and of great use in the arts and medicine. **antimonial** (-mō'-), *a*. of or containing antimony. **antimonic** (-mon'-), *a*. of or containing pentavalent antimony. **antimonious** (-mō'-), *a*. of or containing trivalent antimony.
anti-national (antinash'ənəl), *a*. opposed to the interest of one's country, or the national party.
anting ANT.
antinode (an'tinōd), *n*. (*Phys*.) a region of maximum vibration between two nodes.
antinomian (antinō'miən), *a*. opposed to the moral law; of or pertaining to antinomians. *n*. one who holds that the moral law is not binding on Christians. **antinomianism**, *n*.
antinomy (antin'əmi), *n*. a contradiction between two laws; a conflict of authority; intellectual contradiction, opposition between laws or principles that appear to be equally founded in reason.
anti-novel (an'tinovəl), *n*. a novel in which the conventional features of plot, characterization etc. are lacking.
anti-nuclear (antinū'kliə), *a*. opposed to nuclear power or weapons.
antipapal (antipā'pəl), *a*. opposed to the pope or to papal doctrine.
antiparticle (an'tipahtikl), *n*. an elementary particle with the same mass as but opposite charge to another particle. Collision of a particle with its antiparticle produces mutual annihilation.
antipasto (antipas'tō), *n*. (*pl*. **-tos**) hors d'oeuvre.
antipathy (antip'əthi), *n*. contrariety of nature or disposition; hostile feeling towards; aversion, dislike. **antipathetic** (-thet'-), **-ical**, *a*. having an antipathy or contrariety to. **antipathetically**, *adv*.
antipersonnel (antipœsənel'), *a*. of a weapon etc., designed to kill or injure people.
antiperspirant (antipœ'spirənt), *n., a*. (a substance) used to reduce perspiration.
antiphlogistic (antiflǝjis'tik), *n., a*. a remedy that allays inflammation.
antiphon (an'tifon), *n*. a sentence sung by one choir in response to another; a series of such responsive sentences or

antiphrasis   31   **apathy**

versicles; a short sentence said or sung before the psalms, canticles etc., in the Roman Catholic Church. **antiphonal** (-tif'-), *a.* consisting of antiphons; sung alternately. *n.* an antiphonary. **antiphonally**, *adv.* **antiphonary** (-tif'-), *n.* a book containing a collection of antiphons. **antiphony** (-tif'-), *n.* opposition of sound; alternate chanting or singing by a choir divided into two parts; an antiphon.
antiphrasis (antif'rəsis), *n.* the use of words in a sense contrary to their ordinary meaning. **antiphrastic** (-fras'-), *a.*
antipodes (antip'ədēz), *n.pl.* those who dwell directly opposite to each other on the globe; a place on the surface of the globe diametrically opposite to another, esp. Australasia; a pair of places diametrically opposite; the direct opposite of some other person or thing. **antipodal**, *a.* **antipodean** (-dē'-), *a.*
antipope (an'tipōp), *n.* a pope elected in opposition to the one canonically chosen.
antipyretic (antipīret'ik), *n., a.* (a medicine) that prevents or allays fever.
antiquary (an'tikwəri), *n.* a student, investigator, collector or seller of antiquities or antiques; a student of ancient times. **antiquarian** (-kweə'ri-), *a.* pertaining to the study of antiquities. *n.* an antiquary. **antiquarianism**, *n.* **antiquarianize, -ise**, *v.i.*
antiquated (an'tikwātid), *a.* old-fashioned, out of date.
antique (antēk'), *a.* ancient, old, that has long existed; old-fashioned, antiquated. *n.* a relic of antiquity; a piece of furniture, ornament etc., made in an earlier period and valued by collectors. *v.t.* to give the appearance of an antique to. **the antique**, the ancient style in art. **antiquity** (antik'witi), *n.* the state of having existed long ago; the state of being ancient; great age; ancient times; the ancients; manners, customs, events etc. of ancient times; (*usu. pl.*) a relic of ancient times.
Antirrhinum (antirī'nəm), *n.* a genus of plants that includes the snapdragon.
anti-Semite (antisem'īt), *n.* one hostile towards Jews. **anti-Semitic** (-mit'-), *a.* **anti-Semitism** (-sem'i-), *n.*
antiseptic (antisep'tik), *a.* counteracting sepsis, or putrefaction; free from contamination; lacking interest, warmth or excitement, sterile. *n.* a substance which inhibits the growth of microorganisms. **antisepsis**, *n.* the principle of antiseptic treatment.
antiserum (antisiə'rəm), *n.* (*pl.* **-rums, -ra** (-ə)) serum containing antibodies.
anti-social (antisō'shəl), *a.* opposed to the interest of society, or to the principles on which society is constituted; unsociable.
antistatic (antistat'ik), *n., a.* (an agent) that counteracts the effects of static electricity.
antistrophe (antis'trəfi), *n.* the returning of the Greek chorus, exactly answering to, but reversing the movement of, a previous strophe; the poem or choral song recited during this movement; the rhetorical figure of retort; an inverted grammatical construction. **antistrophic** (-strof'-), *a.*
antitank (antitangk'), *a.* used against armoured vehicles.
antitetanus (antitet'ənəs), *a.* preventing tetanus. **antitetanic** (-tan'-), *a.* **antitetanin** (-ənin), *n.* an antitoxin used for curing or preventing tetanus.
antithesis (antith'əsis), *n.* (*pl.* **-ses** (-sēz)) sharp opposition or contrast between words, clauses, sentences or ideas; a counter proposition; opposition, contrast; the direct opposite. **antithetic** (-thet'-), **-ical**, *a.* pertaining to or marked by antithesis; sharply opposed. **antithetically**, *adv.*
antitoxin (antitok'sin), *n.* an antibody or antiserum formed in the body which neutralizes the action of toxins.
antitrade (an'titrād), *n., a.* (a wind) blowing in an opposite direction to that of the trade-winds.
anti-trust (antitrŭst'), *a.* (*N Am.*) opposing trusts or monopolies which adversely affect trade.
antitype (an'titīp), *n.* that which is represented by a type or symbol; an opposite type. **antitypal** (-tī'-), *a.* **antitypical** (-tip'-).
antivenin (antiven'in), *n.* serum obtained from animals immunized against snake venom, used as an antidote against snake-bite.
antivirus (an'tivīrəs), *n.* software designed to protect a computer system from a virus. **antiviral**, *a.*
anti-vivisection (antivivisek'shən), *n.* (active) opposition to vivisection. **anti-vivisectionist**, *n., a.*
antler (ant'lə), *n.* a branch of the horns of a stag or other deer; either of the branched horns of a deer. **antlered**, *part.a.*
antonomasia (antənəmā'ziə), *n.* the substitution of an epithet for a proper name, as *the Corsican* for Napoleon; the use of a proper name to describe one of a class, as a *Cicero* for an orator. **antonomastic** (-mas'-), *a.*
antonym (an'tənim), *n.* a term expressing the reverse of some other term, as 'good' to 'bad'.
antrum (antrəm), *n.* a natural anatomical cavity, particularly one in bone.
anus (ā'nəs), *n.* the lower, excretory opening of the intestinal tube.
anvil (an'vil), *n.* the iron block on which smiths hammer and shape their work; anything resembling a smith's anvil in shape or use; esp. the incus of the ear.
anxious (angk'shəs), *a.* troubled or solicitous about some uncertain or future event; distressing, worrying; eagerly desirous (to do something). **anxiety** (angzī'əti), *n.* the state of being anxious; trouble or mental distress. **anxiously**, *adv.*
any (en'i), *a., pron.* one indefinitely; some or any number indefinitely; whichever, whatever; (*coll.*) anything, in *I'm not taking any.* **anybody**, *n., pron.* any person, any one; a person of little importance; (*pl.*) persons of no importance. **anyhow**, *adv., conj.* at any rate; in any way; in any case; imperfectly, haphazardly. **anyone**, *n., pron.* any person, anybody. **anyplace**, *adv.* (*N Am.*) anywhere. **anything**, *n., pron.* any thing (in its widest sense) as distinguished from any person. **anyway**, *adv., conj.* anyhow. **anywhere**, *adv.* in any place. **anywise**, *adv.* in any manner, case, or degree; anyhow.
Anzac (an'zak), *n.* a soldier in the Australian or New Zealand forces, in World War 1. [acronym for *Australian (and) New Zealand Army Corps*]
Anzus (an'zəs), *n.* a pact for the security of the Pacific, formed in 1952 by Australia, New Zealand and the US. [acronym for *Australia, New Zealand, US*]
a/o, (*abbr.*) account of.
AOB, (*abbr.*) any other business.
AOC, (*abbr.*) Air Officer Commanding.
aorist (ā'ərist), *n.* a Greek tense expressing indefinite past time. *a.* aoristic. **aoristic** (-ris'-), *a.* indefinite in point of time; pertaining to an aorist tense.
aorta (āaw'tə), *n.* the main trunk of the arterial system proceeding from the left ventricle of the heart. **aortic**, *a.*
AP, (*abbr.*) Associated Press.
ap-, *pref.* AD-, assim. to *p* e.g. *appear*, *approve*.
apace (əpās'), *adv.* at a quick pace; speedily, fast.
Apache (əpach'i), *n.* a member of a N American Indian people of the SW US and N Mexico; a (Parisian) ruffian who robs and maltreats people, a hooligan.
apanage, appanage (ap'ənij), *n.* lands or office assigned for the maintenance of a royal house; a dependency; a perquisite; a necessary adjunct or attribute.
apart (əpaht'), *adv.* to one side; separately with regard to place, purpose or things; independently; parted, at a distance; separate; into two or more pieces or parts.
apartheid (əpah'tāt, -tit), *n.* (a policy of) racial segregation.
apartment (əpaht'mənt), *n.* a portion of a house; a single room in a house; (*pl.*) a suite of rooms, lodgings; (*chiefly N Am.*) a flat.
apathy (ap'əthi), *n.* absence of feeling or passion;

apatite | 32 | **aposiopesis**

indifference; mental indolence. **apathetic** (thet'-), *a.* characterized by apathy; unemotional, indifferent. **apathetically,** *adv.*
apatite (ap'ətīt), *n.* a common mineral, calcium phosphate.
APB, (*abbr.*) all-points bulletin.
ape (āp), *n.* a tailless monkey; a gorilla, chimpanzee, orang-outan or gibbon; a mimic, a servile imitator. *v.t.* to imitate or mimic. **apeman,** *n.* an extinct primate intermediate between the higher apes and humans. **apery** (ā'pəri), *n.* mimicry; apish behaviour. **apish,** *a.* of the nature of or befitting an ape. **apishly,** *adv.* **apishness,** *n.*
apepsy (əpep'si), **apepsia** (-ə), *n.* indigestion, dyspepsia.
aperçu (apœsoo'), *n.* a concise exposition, an outline; an insight.
aperient (əpiə'riənt), **aperitive** (əpe'ritiv), *n., a.* (a) laxative.
aperiodic (āpiəriod'ik), *a.* not occurring regularly; (*Phys.*) not having a periodic motion, not oscillatory. **aperiodically,** *adv.* **aperiodicity** (-dis'-), *n.*
aperitif (əperitēf'), *n.* a short drink, usu. alcoholic, taken as an appetizer.
aperitive APERIENT.
aperture (ap'əchə), *n.* an opening, a hole; (the diameter of) the space through which light passes in an optical instrument; the diameter of a lens.
apery APE.
apetalous (əpet'ələs), *a.* without petals.
APEX (ā'peks), *n.* a discounted fare on some air or sea journeys paid for at least 28 days before departure. [acronym for *A*dvance *P*urchase *Ex*cursion]
apex (ā'peks), *n.* (*pl.* **apices** (ā'pisēz), **apexes**) the tip, top, vertex or summit of anything; the culmination, climax. **apical** (-pi-), *a.* pertaining to an apex; placed at the summit. **apically,** *adv.*
aphaeresis, apheresis (əfiə'rəsis), *n.* the taking away of a letter or syllable at the commencement of a word.
aphasia (əfā'ziə), *n.* (partial) loss of the power of articulate speech. **aphasic,** *a.*
aphelion (əfē'liən), *n.* (*pl.* **-lia** (-ə)) the point most distant from the sun in the orbit of a planet or a comet.
aphesis (af'əsis), *n.* a form of aphaeresis, in which an unaccented vowel at the beginning of a word is gradually lost. **aphetic** (əfet'-), *a.* **aphetize, -ise,** *v.t.* to shorten by aphesis.
aphid (ā'fid), *n.* any of a group of minute insects very destructive to vegetation, including the green-fly, black fly, etc.
aphis (ā'fis, af'-), *n.* (*pl.* **aphides** (-dēz)) an aphid, esp. of the genus *Aphis.*
aphonia (əfō'niə), **aphony** (af'əni), *n.* inability to speak; loss of voice. **aphonic** (əfon'-), *a.*
aphorism (af'ərizm), *n.* a detached, pithy sentence, containing a maxim or wise precept. **aphorismic** (-riz'-), *a.* **aphorist,** *n.* one who writes or utters aphorisms. **aphoristic** (-ris'-), *a.* **aphoristically,** *adv.* **aphorize, -ise,** *v.i.* to utter or write aphorisms.
aphrodisiac (afrədiz'iak), *n., a.* (a substance) that excites sexual desire. **aphrodisian,** *a.*
aphthae (af'thē), *n.pl.* the minute specks seen in the mouth and tongue in thrush.
aphyllous (əfil'əs), *a.* without leaves.
apian (ā'piən), *a.* pertaining to bees. **apiarian** (āpieə'riən), *a.* relating to bees or bee-keeping. *n.* an apiarist. **apiarist** (ā'-), *n.* one who rears or keeps bees. **apiary** (ā'-), *n.* a place where bees are kept.
apical etc. APEX.
apiculture (ā'pikŭlchə), *n.* bee-keeping; bee-rearing.
apiece (əpēs'), *adv.* for or to each, severally.
apish etc. APE.
APL, *n.* a computer programming language designed for mathematical applications. [*a programming language*]
aplanatic (aplənat'ik), *a.* of a lens etc., free from spherical aberration.
aplasia (əplā'ziə), *n.* defective or arrested development in a body tissue or organ. **aplastic** (-plas'-), *a.*
aplenty (əplen'ti), *adv.* in plenty, in abundance.
aplomb (əplom'), *n.* self-possession, coolness.
apnoea (apnē'ə), (*N Am.*) **apnea,** *n.* a cessation of respiration.
apo-, *pref.* away, detached, separate; as in *apology, apostrophe.*
Apoc., (*abbr.*) Apocalypse; Apocrypha, apocryphal.
Apocalypse (əpok'əlips), *n.* the revelation granted to St John the Divine; the book of the New Testament in which this is recorded; (**apocalypse**) any revelation or prophetic disclosure; (**apocalypse**) a vast decisive event or confrontation. **apocalyptic, -ical** (-lip'-), *a.* pertaining to the revelation of St John; of the nature of a revelation or apocalypse; prophesying disaster or doom. **apocalyptically,** *adv.*
apocope (əpok'əpi), *n.* a cutting off or dropping of the last letter or syllable of a word.
apocrypha (əpok'rifə), *n.* writings or statements of doubtful authority; (**Apocrypha**) a collection of 14 books in the Old Testament, included in the Septuagint and the Vulgate, but not written in Hebrew originally, nor reckoned genuine by the Jews, nor inserted in the Authorized Version of the Bible. **apocryphal,** *a.* pertaining to the apocrypha; spurious, fabulous. **apocryphally,** *adv.*
apodal (ap'ədl), *a.* footless; having no ventral fin.
apodictic (apədik'tik), **apodeictic** (-dīk'-), *a.* clearly demonstrative; established on uncontrovertible evidence. **apodictically,** *adv.*
apodosis (əpod'əsis), *n.* the consequent clause in a conditional sentence, answering to the protasis.
apogamy (əpog'əmi), *n.* (*Bot.*) the absence of sexual reproductive power, the plant perpetuating itself from an unfertilized female cell. **apogamous,** *a.* **apogamously,** *adv.*
apogee (ap'əjē), *n.* the point in the orbit of the moon or any planet or satellite which is at the greatest distance from the earth or the planet round which it revolves; the furthest or highest point, the culmination. **apogean,** *a.*
apolitical (āpəlit'ikəl), *a.* uninterested in political affairs, politically neutral; without political significance.
Apollyon (əpol'iən), *n.* the destroyer; the Devil.
apologetic, apologia, apologize etc. APOLOGY.
apologue (ap'əlog), *n.* a fable designed to impress some moral truth upon the mind; esp. a beast-fable or a fable of inanimate things.
apology (əpol'əji), *n.* a defence, vindication; an explanation, excuse; a regretful acknowledgment of offence; a wretched substitute for the real thing. **apologetic** (-jet'-), **-ical,** *a.* excusing, explanatory, vindicatory. **apologetically,** *adv.* **apologetics,** *n.sing.* defensive argument; esp. the argumentative defence of Christianity. **apologia** (apəlō'jiə) *n.* a vindication, formal defence, excuse. **apologist,** *n.* one who defends or apologizes by speech or writing; a professed defender of Christianity. **apologize, -ise,** *v.i.* to make an apology or excuse for.
apomixis (apəmik'sis), *n.* reproduction without fertilization. **apomictic,** *a.* **apomictically,** *adv.*
apophthegm, apothegm (ap'əthem), *n.* a terse pointed saying, a maxim expressed in few but weighty words. **apophthegmatic** (-theg-), *a.* pertaining to, or using apophthegms; sententious, pithy. **apophthegmatically,** *adv.*
apoplexy (ap'əpleksi), *n.* a sudden loss of sensation and of power of motion, generally caused by rupture or obstruction of a blood vessel in the brain. **apoplectic** (-plek'-), *a.* pertaining to or tending to cause apoplexy; predisposed to apoplexy; violently angry.
aport (əpawt'), *adv.* on or towards the port side of a ship.
aposiopesis (əpəsiəpē'sis), *n.* a stopping short for rhetorical effect.

apostasy (əpos'təsi), *n.* renunciation of religious faith, moral allegiance or political principles; in the Roman Catholic Church, renunciation of religious vows.
apostate (əpos'tāt), *n.* one who apostatizes. *a.* unfaithful to creed or principles; rebel, rebellious. **apostatic, -ical** (apəstat'-), *a.* **apostatize, -ise** (-tə-), *v.i.* to abandon one's creed, principles or party; to commit apostasy.
a posteriori (a postiəriaw'ri, ah-), *a., adv.* (applied to or derived from reasoning) from consequences, effects, things observed to causes; inductive, as opposed to *a priori* or deductive.
apostle (əpos'l), *n.* one of the 12 men appointed by Christ to preach the gospel; a first Christian missionary to any region, or one who has pre-eminent success; the leader of a reform; a supporter. **Apostles' Creed,** *n.* a Christian creed, each clause of which is said to have been contributed by one of the Apostles. **apostle-spoons,** *n.pl.* (tea)spoons, with handles ending in figures of the Apostles. **apostleship,** *n.* **apostolate** (-təlāt), *n.* the office of apostle; leadership; propagation of a doctrine. **apostolic** (apəstol'-), **-ical,** *a.* pertaining to the Apostles or their doctrine or practice; of the character or nature of an apostle; pertaining to the Pope as St Peter's successor. **Apostolic Fathers,** *n.pl.* those Christian Fathers or writers contemporaneous with the Apostles or their immediate disciples. **Apostolic See,** *n.* the Papacy. **apostolic succession,** *n.* uninterrupted transmission of spiritual authority through bishops, from the Apostles. **apostolically,** *adv.*
apostrophe¹ (əpos'trəfi), *n.* a rhetorical figure in which the speaker addresses one person in particular, or turns away from those present to address the absent or dead. **apostrophic** (apəstrof'ik), *a.* **apostrophize, -ise,** *v.t.* to address in or with apostrophe.
apostrophe² (əpos'trəfi), *n.* the sign (') used to denote the omission of a letter or letters, and as the sign of the English possessive case. **apostrophize, -ise,** *v.t.* to mark an omission of a letter or letters from a word by inserting an apostrophe.
apothecary (əpoth'ikəri), *n. (formerly)* a druggist or pharmaceutical chemist; a licentiate of the Apothecaries' Society. **apothecaries' measure,** *n.* a system of liquid capacity measure formerly used in pharmacy, based on the minim, fluid drachm and fluid ounce. **apothecaries' weight,** *n.* a system of weights formerly used in pharmacy, based on the grain, scruple, drachm, and troy ounce.
apothegm APOPHTHEGM.
apotheosis (əpothiō'sis), *n. (pl.* **-ses** (-sēz)) deification; transformation into a god; a deified ideal. **apotheosize, -ise** (-poth'-), *v.t.* to deify, to exalt.
app., *(abbr.)* apparent, apparently; appendix; appointed; apprentice.
appal (əpawl'), *(esp. N Am.)* **appall,** *v.i. (past, p.p.* **appalled)** *v.t.* to horrify, to terrify; to dismay. **appalling,** *a.* **appallingly,** *adv.*
Appaloosa (apəloo'sə), *n.* a N American breed of horse with a spotted coat.
appanage APANAGE.
apparat (apərat', a'-), *n.* the apparatus of the Soviet Communist party. **apparatchik** (-chik), *n.* a member of the Soviet apparat.
apparatus (apərā'təs, -rah'-), *n. (pl.* **apparatuses, apparatus)** equipment or arrangements generally; the instruments employed in scientific or other research; the organs by which any natural process is carried on; the administrative workings of a (political) system or organization. **apparatus criticus** (krit'ikəs), *n.* critical equipment, the materials employed, as variant readings etc., in literary criticism and investigation.
apparel (əpa'rəl), *n.* dress, attire, clothes. *v.t. (past, p.p.* **apparelled)** to dress, to clothe; to adorn, to ornament.
apparent (əpa'rənt, -peə'-), *a.* to be seen, visible; plain, obvious, indubitable; appearing (in a certain way), seeming; *(Phys.)* as observed without adjustment or correction. **apparent horizon** HORIZON. **apparently,** *adv.* to external appearances; seemingly, as distinguished from actually.
apparition (apərish'ən), *n.* an act or the state of becoming visible; a strange appearance; a spectre, phantom, ghost. **apparitional,** *a.*
appeal (əpēl'), *v.t.* to refer (a case) to a higher court or authority. *v.i.* to refer to a superior judge, court or authority; to refer to some person or thing as corroboration; to invoke aid, pity, mercy etc.; to have recourse (to); to apply (to); to attract or interest. *n.* the act of appealing; the right of appeal; reference or recourse to another; entreaty; a request for aid, esp. for money for charitable purposes; power of attracting or interesting. **to appeal to the country** COUNTRY. **appealable,** *a.* that may be appealed against; that can be appealed to. **appealing,** *a.* of the nature of an appeal, suppliant; having appeal, arousing interest. **appealingly,** *adv.* **appealingness,** *n.*
appear (əpiə'), *v.i.* to become or be visible; to present oneself; to come before the public; to be manifest; to seem. **appearance,** *n.* the act of appearing; the thing seen; the act of appearing formally or publicly; a phenomenon; a phantom; mien, aspect; external show, pretence; *(pl.)* the aspect of circumstances. **to keep up appearances,** to keep up an outward show; to conceal the absence of something desirable. **appearance money,** *n.* money paid to a celebrity or top performer, e.g. an athlete, to take part in an event.
appease (əpēz'), *v.t.* to quiet, to pacify, to assuage, to allay; to conciliate by acceding to demands. **appeasable,** *a.* **appeasement,** *n.* the act of appeasing; the endeavour to preserve peace by giving way to the demands of an aggressor power.
appellant (əpel'ənt), *a.* appellate. *n.* one who appeals to a higher tribunal or authority; one who makes an appeal. **appellate** (-ət), *a.* pertaining to or dealing with appeals. **appellation** (ap-), *n.* a name, designation; naming. **appellative,** *a.* common as opposed to proper; designating a class. *n.* an appellation, a name; a common as opposed to a proper noun. **appellatively,** *adv.* **appellation contrôlée,** *n.* a certification guaranteeing the origin and quality of a French wine.
append (əpend'), *v.t.* to hang to or upon; to add or subjoin. **appendage** (-dij), *n.* something added or appended; a subordinate or subsidiary organ or process, as a limb or branch. **appendant,** *a.* attached, annexed, joined on. *n.* that which is attached or annexed; an appendix, a corollary.
appendicectomy (əpendisek'təmi), **appendectomy** (əpəndek'-), *n.* the excision of the vermiform appendix.
appendicitis (əpendisī'tis), *n.* inflammation of the vermiform appendix.
appendix (əpen'diks), *n. (pl.* **-dixes, -dices** (-sēz)) something appended; an adjunct or concomitant; a supplement to a book or document containing useful material; a small process or prolongation of an organ, esp. the vermiform appendix of the intestine.
apperception (apəsep'shən), *n.* perception of one's own mental processes; consciousness of one's self; understanding and assimilation of a new perception in terms of previous experiences. **apperceive** (-sēv), *v.t.* **apperceptive,** *a.*
appertain (apətān'), *v.i.* to belong (as a part to a whole, as a possession, or as a right or privilege); to relate (to); to be suitable or appropriate.
appetence (ap'itəns), **-tency** (-si), *n.* instinctive desire, craving, appetite; natural propensity; affinity. **appetent,** *a.*
appetite (ap'ətīt), *n.* inclination, disposition; the desire to satisfy a natural function; desire; relish for food. **appetitive** (əpet'i-), *a.* **appetizer, -iser,** *n.* a whet; stimulant to appetite, esp. food or drink served before or at the

applaud   34   **approximate**

beginning of a meal. **appetizing**, *a.* stimulating appetite or hunger.
applaud (əplawd'), *v.i.* to express approbation, esp. by clapping the hands. *v.t.* to approve or praise in an audible and significant manner. **applause** (əplawz') *n.* the act of applauding; praise loudly expressed. **applausive**, *a.* **applausively**, *adv.*
apple (ap'l), *n.* the round, firm, fleshy fruit of the apple-tree; any similar fruit; a tree that bears apples. **apple of discord**, the golden apple contended for as prize of beauty by Juno, Minerva and Venus; a cause of contention. **apple of one's eye**, the pupil, formerly supposed to be a solid body; anything very dear or precious. **apple-pie bed**, a bed whose sheets are so doubled as to prevent one stretching one's full length. **apple-pie order**, perfect order. **to upset the apple-cart**, to disrupt plans or arrangements. **apple-faced**, **apple-cheeked**, *a.* having a chubby face or cheeks. **apple-green**, *n.*, *a.* (of) a light, yellowish-green colour. **apple-jack**, *n.* (*N Am.*) spirit made from apples. **apple-sauce**, *n.* sauce made from apples; (*N Am., coll.*) insincere praise, nonsense. **apple-wife**, **apple-woman**, *n.* a woman who sells apples in the street.
Appleton layer (ap'ltən), *n.* an ionized layer in the upper atmosphere, above the Heaviside layer, which reflects radio waves. [the British physicist, Sir E. *Appleton*, 1892–1965]
appliance, **applicant** etc. APPLY.
appliqué (ap'likā), *n.* ornamental work laid on some other material. **appliquéd**, *a.* treated with work of this kind.
apply (əplī'), *v.t.* to put or lay on; to put close to; to administer (as an external remedy); to employ, to devote; to make suitable, adapt, conform to. *v.i.* to harmonize, to be relevant; to have recourse (to); to study; to offer oneself (for a job, position etc.). **appliance** (-plī'-), *n.* the act of applying; anything applied as a means to an end; an apparatus, device or contrivance. **applicable** (ap'li-, əplik'-), *a.* capable of being applied; fit, suitable. **applicability** (-bil'-), *n.* **applicant** (ap'li-), *n.* one who applies; a petitioner. **applicate** (ap'likət), *a.* applied to practical use. **application** (apli-), *n.* the act of applying; the thing applied; the use to which something is put; (a) petition, request; close attention; study. **applications program**, *n.* a computer program that performs a specific task for a user. **applicative** (ap'likə-), *a.* **applicator** (ap'li-), *n.* a tool for applying medicine, make-up etc. **applied**, *a.* practical; put to practical use. **applied science**, *n.* science of which the abstract principles are put to practical use in solving problems.
appoggiatura (əpojətoo'rə), *n.* a grace-note before a significant note.
appoint (əpoint'), *v.t.* to decree, ordain, fix; to nominate, designate; to assign, to grant (a thing to a person). *v.i.* to decree, ordain. **appointed**, *a.* furnished, equipped. **appointee** (-tē'), *n.* one who receives an appointment; (*Law*) one in whose favour an appointment is executed. **appointment**, *n.* the act of appointing; the office or situation assigned; that which is appointed or fixed; an engagement or assignation; (*pl.*) equipment, fittings (of a ship); (*Law*) the official declaration of the destination of any specific property.
apportion (əpaw'shən), *v.t.* to mete out in just proportions; to divide in suitable proportion. **apportionment**, *n.*
apposite (ap'əzit), *a.* fit, apt, appropriate. **appositely**, *adv.* **appositeness**, *n.* **apposition** (-zish'-), *n.* the act of putting together or side by side; juxtaposition, addition; (*Gram.*) the placing together of two words, esp. of two substantives, one being a complement to the other. **appositional**, *a.*
appraise (əprāz'), *v.t.* to set a price on; to value; to estimate the worth of. **appraisable**, *a.* **appraisal**, *n.* an authoritative valuation; an estimate of worth. **appraisement**, *n.* the act of appraising; estimated value or worth.

appraiser, *n.* one who appraises; a person authorized to fix the value of property.
appreciate (əprē'shiāt, -si-), *v.t.* to perceive the value, merit, or quality of; to estimate aright; to be sensible of (delicate impressions); to esteem highly; to raise in value. *v.i.* to rise in value. **appreciable** (-shə-), *a.* capable of being appreciated; quite large. **appreciably**, *adv.* **appreciation**, *n.* the act of appreciating; a (favourable) estimate; a critical study; adequate recognition; a rise in value. **appreciative** (-shə-), *a.* capable of, expressing appreciation; esteeming favourably. **appreciatively**, *adv.* **appreciator**, *n.* **appreciatory** (-shə-), *a.*
apprehend (aprihend'), *v.t.* to take hold of; to arrest; to seize or grasp mentally; to anticipate, esp. with fear. *v.i.* to understand. **apprehensible**, *a.* **apprehension** (-shən), *n.* the act of seizing or arresting; the mental faculty which apprehends; conception, idea; fear, dread of what may happen. **apprehensive**, *a.* anticipative of something unpleasant or harmful, fearful, anxious. **apprehensively**, *adv.* **apprehensiveness**, *n.*
apprentice (əpren'tis), *n.* one bound by indentures to serve an employer for a term of years in order to learn some trade or craft which the employer agrees to teach; a learner, a novice. *v.t.* to bind as an apprentice. **apprenticeship**, *n.* the state or position of an apprentice; the term for which an apprentice is bound to serve.
apprise (əpriz'), *v.t.* to inform, to make aware.
appro (ap'rō), *n.* short for APPROVAL.
approach (əprōch'), *v.i.* to come, go or draw near or nearer. *v.t.* to come near to; to resemble, to approximate; (*coll.*) to come near or address with a view to securing something, as a favour or intimate relations. *n.* the act of drawing near; approximation, resemblance; avenue, entrance, access; (*usu. pl.*) works thrown up by a besieging force to protect it in its advance; (*Golf*) a stroke that should take the ball on to the green; manner of or attitude towards dealing with a job, problem etc.; (*usu. pl.*) advances towards establishing a relationship. **approachable**, *a.* capable of being approached; easy to deal with; friendly. **approachability** (-bil'-), *n.*
approbation (aprəbā'shən), *n.* (formal) approval, commendation. **approbatory** (aprəbā'təri, ap'-), *a.*
appropriate¹ (əprō'priāt), *v.t.* to take as one's own; to take possession of; to devote to or set apart for a special purpose or use. **appropriable**, *a.* **appropriation**, *n.* the act of appropriating; a sum of money or a portion of revenue appropriated to a specific end. **appropriative**, *a.* **appropriator**, *n.*
appropriate² (əprō'priət), *a.* suitable, fit, becoming. **appropriately**, *adv.* **appropriateness**, *n.*
approve (əproov'), *v.t.* to esteem, accept or pronounce as good; to commend, sanction, confirm. *v.i.* to express or to feel approval. **approvable**, *a.* **approval**, *n.* favourable judgement; commendation, sanction. **on approval**, **on appro**, on trial to ascertain if suitable; of goods, to be returned if not suitable. **approved**, *a.* tried, proved, tested; regarded with approval; officially sanctioned. **approved school**, *n.* formerly, a state boarding school for juvenile offenders (boys under 15, girls under 17). **approver**, *n.* **approving**, *a.* **approvingly**, *adv.*
approx. (əproks'), (*abbr.*) approximate, approximately.
approximate¹ (əprok'simāt), *v.t.* to draw or bring near, to cause to approach. *v.i.* to draw near, to approach. **approximation**, *n.* approach, proximity; (*Math.*) a coming or getting nearer to a quantity sought, when no process exists for ascertaining it exactly; something approximate; a mathematical value that is sufficiently accurate for a purpose though not exact. **approximative**, *a.* approaching. **approximatively**, *adv.*
approximate² (əprok'simət), *a.* very close to; closely resembling; (nearly) approaching accuracy. **approximately**, *adv.*

appurtenance (əpœ'tinəns), *n.* that which belongs to something else; an adjunct, an accessory. **appurtenant**, *a.* pertaining to, belonging to, pertinent. *n.* an appurtenance.
APR, *(abbr.)* annual percentage rate (of credit etc.).
Apr., *(abbr.)* April.
apraxia (əprak'siə), *n.* (partial) loss of the ability to execute voluntary movements.
après-ski (apreske'), *n.*, *a.* (of or intended for) the social time following a day's skiing.
apricot (ā'prikot, ap'-), *n.* a pinkish-yellow stone-fruit allied to the plum; the tree on which it grows; its colour.
April (ā'prəl), *n.* the fourth month of the year. **April-fool**, *n.* a victim of a practical joke on 1 April. **April-fool day, April-Fools' day**, *n.* 1 April, from the custom of playing practical jokes on people on that day. **April shower**, *n.* a sudden, brief shower of rain (common in the month of April).
a priori (ā priaw'ri, -rī, ah), *a.*, *adv.* (applied to or derived from reasoning) from cause to effect; from abstract ideas to consequences; prior to or independently of experience; deductive as opposed to *a posteriori* or inductive. **apriority** (āprio'rəti), *n.*
apron (ā'prən), *n.* a garment worn in front of the body to protect the clothes, or as part of a distinctive dress, e.g. of bishops, Freemasons; anything resembling an apron in shape or function; the fat skin covering the belly of a roast goose or duck; the extension of the stage in some theatres beyond the proscenium; a platform of planks at the entrance to a dock; the surfaced area on an airfield; an extensive deposit of sand, gravel etc. **tied to the apron-strings**, unduly controlled (by a wife, mother etc.). **aproned**, *a.* wearing an apron.
apropos (aprəpō'), *adv.* opportunely; appropriately; by the way. *a.* opportune; appropriate; to the point. **apropos of**, *prep.* as bearing upon the subject; as suggested by.
apse (aps), **apsis** (ap'sis), *n.* (*pl.* **apses** (-sēz) **apsides** (-sidēz, -si'-) (*usu.* **apse**) a semicircular or polygonal, and generally dome-roofed, recess in a building, esp. in a church; (*usu.* **apsis**) one of two points at which a planet or satellite is at its greatest or least distance from the body round which it revolves. **apsidal** (-si-), *a.* pertaining to or of the shape of an apse or apsis.
apt (apt), *a.* fit, suitable, relevant; having a tendency (to), likely; quick, ready; quick to learn. **aptly**, *adv.* **aptness**, *n.*
apterous (ap'tərəs), *a.* wingless; having only rudimentary wings; (*Bot.*) without membranous wing-like expansions.
apteryx (ap'təriks), *n.* the kiwi, a bird from New Zealand, about the size of a goose, with rudimentary wings.
aptitude (ap'titūd), *n.* fitness, suitability, adaptation; natural ability or talent, esp. for learning.
aqua (ak'wə), *n.* water, liquid, solution.
aqu(a)-, aqu(i)-, *comb. form.* pertaining to water.
aquaculture (ak'wəkülchə), **aquiculture** (ak'wi-), *n.* hydroponics, the cultivation of aquatic organisms for human use.
aqua fortis, aquafortis (akwəfaw'tis), *n.* nitric acid. **aquafortist**, *n.* (*Art*) one who etches or engraves with aquafortis.
aqualung (ak'wəlŭng), *n.* a portable diving apparatus, strapped on the back and feeding air to the diver as required.
aquamarine (akwəmərēn'), *n.* a bluish-green variety of beryl, named from its colour. *a.* bluish-green.
aquanaut (ak'wənawt), *n.* a deep-sea diver or explorer.
aquaplane (ak'wəplān), *n.* a board on which one is towed, standing, behind a motor-boat. *v.i.* to ride on an aquaplane; of a car etc., to slide on a film of water on a road surface.
aqua regia (akwə rē'jə), *n.* a mixture of nitric and hydrochloric acids, capable of dissolving gold and platinum.

aquarelle (akwərel'), *n.* a kind of painting in Chinese ink and very thin transparent water-colours; the design so produced. **aquarellist**, *n.* one who paints in aquarelle.
aquarist (ak'wərist), *n.* the keeper of an aquarium.
aquarium (əkweə'riəm), *n.* (*pl.* **-riums, -ria**) an artificial tank, pond or vessel in which aquatic animals and plants are kept alive; a place in which such tanks are exhibited.
Aquarius (əkweə'riəs), *n.* a zodiacal constellation, the water carrier; the 11th sign of the zodiac.
aquatic (əkwat'ik), *a.* of or pertaining to water; living or growing in or near water. *n.* an aquatic animal or plant. **aquatics**, *n.pl.* sports or athletic exercises on or in the water.
aquatint (ak'wətint), *n.* a method of etching on copper; a design so produced.
aquavit (ak'wəvēt, -vit), *n.* an alcoholic spirit flavoured with caraway seeds.
aqua vitae (akwə vē'tī), *n.* unrectified alcohol; strong spirits, brandy etc.
aqueduct (ak'widŭkt), *n.* an artificial channel, esp. one raised on pillars or arches, for the conveyance of (drinking) water from place to place; a small canal, in the body.
aqueous (ak'wiəs, ā'-), *a.* consisting of, containing, formed in or deposited from water; watery. **aqueous humour**, *n.* the watery fluid in the eye between the cornea and the lens. **aqueous rocks**, *n.pl.* rocks built up from deposits in water.
aquiculture AQUACULTURE.
aquifer (ak'wifə), *n.* a water-bearing layer of rock, gravel etc. **aquiferous** (əkwif'-), *a.* conveying, bearing or yielding water.
Aquilegia (akwilē'jə), *n.* a genus of acrid plants, order Ranunculaceae, commonly known as columbine.
aquiline (ak'wilin), *a.* of, pertaining to or like an eagle; esp. of noses, hooked, curved.
Ar, *(chem. symbol)* argon.
ar.[1], *(abbr.)* in the year of the reign. [L *anno regni*]
ar.[2], *(abbr.)* arrival, arrive(s).
ar-, *pref.* AD-, assim, to r, e.g. *arrest, arrogate*.
-ar[1] (-ə), *suf.* belonging to, of the nature of, e.g. *angular, linear, lunar, regular*; thing pertaining to, e.g. *altar, exemplar, pillar*.
-ar[2] (-ə), *suf.* ER[1,2], e.g. *beggar, liar*.
ARA, *(abbr.)* Associate of the Royal Academy.
Arab (a'rəb), *n.* a member of a Semitic people orig. inhabiting Arabia and now much of the Middle East; an Arabian horse; (also **street arab**) an outcast or vagrant child. *a.* Arabian. **Arabian** (ərā'biən), *a.* of or pertaining to Arabia or to Arabs. *n.* a native of Arabia. **Arabian Nights**, *n. sing.* a famous collection of stories; a collection of fantastic stories. **Arabic**, *a.* pertaining to Arabia, the Arabs, or to Arabic. *n.* the language of the Arabs. **Arabic numerals**, *n.pl.* the figures, 1,2,3 etc. **arabist**, *n.* a student of the Arabic language or culture.
Arab., *(abbr.)* Arabia, Arabian, Arabic.
arabesque (arəbesk'), *a.* Arabian in design; in the style of arabesque. *n.* surface decoration composed of flowing line fancifully intermingled, usu. representing foliage in a conventional manner, without animal forms; a posture in ballet-dancing with one leg raised behind and the arms extended.
arabica (ərab'ikə), *n.* a coffee from the seeds of the coffee tree *Coffea arabica*.
Arabis (a'rəbis), *n.* a genus of cruciferous plants largely grown on rockwork, also called rock-cress.
arable (a'rəbl), *a.* capable of being ploughed; fit for tillage.
Araby (a'rəbi), *n.* (*poet.*) Arabia.
araceous (ərā'shəs), *a.* belonging to the *Arum* genus of plants.
Arachis (a'rəkis), *n.* a small genus of low Brazilian leguminous herbs including the peanut. **arachis oil**, *n.* peanut oil.

arachnid (ərak'nid), *n.* any individual of the class Arachnida, which contains the spiders, scorpions and mites. **arachnidan, -dean** (-nid'i-), *n.*, *a.* **arachnoid** (-noid), *a.* (*Bot.*) cobweb-like, covered with long, filamentous hairs; of or resembling the Arachnida. *n.* the transparent membrane lying between the pia mater and the dura mater, that is the middle of the three membranes enveloping the brain and spinal cord. **arachnology** (araknol'-), *n.* the scientific study of spiders or of the Arachnida generally. **arachnologist,** *n.*

aragonite (a'rəgənit), *n.* a carbonate of lime, first found in Aragon, Spain.

arak ARRACK.

ARAM, (*abbr.*) Associate of the Royal Academy of Music.

Aramaean (arəmē'ən), *a.* pertaining to ancient Aram, or Syria, or its language. *n.* a Syrian; the Syrian language.

Aramaic (-mā'ik), *a.* of or belonging to Aram; applied to the ancient northern branch of the Semitic family of languages, including Syriac and Chaldean. *n.* Syriac.

Aran (a'rən), *a.* knitted in a style that originated in the Aran Islands off the W coast of Ireland, typically with a thick cream-coloured wool.

Araucaria (arawkeə'riə), *n.* a genus of coniferous plants, one species of which (*A. imbricata*), the monkey-puzzle, is common in England as an ornamental tree. **araucarian,** *a.*

arb (ahb), *n.* short for ARBITRAGEUR under ARBITRAGE.

arbalest (ah'bəlest), **arbalist** (-list), *n.* a steel crossbow for throwing arrows and other missiles. **arbalester,** *n.* a man armed with an arbalest.

arbiter (ah'bitə), *n.* a judge; a person appointed to decide between contending parties; an umpire; one who has absolute power to decide. **arbitress** (-tris), *n.fem.*

arbitrage (ah'bitrij), *n.* traffic in bills of exchange or stocks so as to take advantage of rates of exchange in different markets. **arbitrageur** (-trahzhœ'), *n.*

arbitrament (ahbit'rəmənt), *n.* power or liberty of deciding; decision by authority; the award given by an arbitrator.

arbitrary (ah'bitrəri), *a.* determined by one's own will or caprice; (apparently) random, irrational; subject to the will or control of no other; of a fine etc., determined by the court. **arbitrarily** (ah'-, treə'rə-), *adv.* **arbitrariness,** *n.*

arbitrate (ah'bitrāt), *v.t.* to hear and judge as an arbitrator; to decide, to settle. *v.i.* to act as arbitrator or umpire. **arbitral,** *a.* of or pertaining to arbitration. **arbitration,** *n.* the hearing or determining of a dispute by means of an arbitrator. **arbitrator,** *n.* an umpire, an arbiter; a person chosen or appointed to settle a dispute. **arbitratorship,** *n.* **arbitress** ARBITER.

arbor[1] (ah'bə), *n.* the main support or chief axis of a piece of mechanism; a spindle. **arbor-vitae** (-vē'tī), *n.* the popular name of several evergreens of the genus *Thuja*.

arbor[2] ARBOUR.

arboraceous (ahbərā'shəs), *a.* resembling a tree; woody, wooded. **arboreal** (-baw'ri-), *a.* pertaining to trees; connected with or living in trees. **arboreous** (-baw'ri-), *a.* wooded; arboreal; arborescent. **arborescent** (-res'ənt), *a.* having tree-like characteristics; branching like a tree. **arborescence,** *n.* **arborescently,** *adv.* **arboretum** (-rē'təm), *n.* a botanical garden for the rearing and exhibition of rare trees. **arboriculture** (-baw'ri-), *n.* the culture of trees and shrubs. **arboricultural,** *a.* **arboriculturist,** *n.* **arborization, -isation,** *n.* tree-like appearance or markings. **arborous** (ah'-), *a.* of, belonging to or formed by trees.

arbour (ah'bə), (*esp. N Am.*) **arbor,** *n.* a bower formed by trees or shrubs closely planted or trained on lattice-work; a shady retreat.

Arbutus (ah'būtəs), *n.* a genus of evergreen shrubs and trees, of which *A. unedo*, the strawberry-tree, is cultivated as an ornamental tree in Britain.

ARC[1], (*abbr.*) Agricultural Research Council.

ARC[2] (ahk), (*acronym*) AIDS-related complex or condition, a set of symptoms (e.g. weight loss, herpes infection) that may be manifested by someone infected with HIV before AIDS develops.

arc (ahk), *n.* a portion of the circumference of a circle or other curve; something curved in shape; that part of a circle which a heavenly body appears to pass through above or below the horizon; the luminous arc or bridge across a gap between two electrodes when an electric current is sent through them. *v.i.* (*past, p.p.* **arced, arcked**) to form an (electric) arc. **arc-lamp,** *n.* an electric lamp in which such an arc or bridge is the source of illumination. **arc-weld,** *v.t.* to weld (metal) by means of an electric arc. **arc welding,** *n.*

arcade (ahkād'), *n.* a series of arches sustained by columns or piers; a walk arched over; a covered passage with shops on each side. **arcaded,** *a.* furnished with or formed like an arcade.

Arcadian (ahkā'diən), *a.* of or pertaining to Arcadia, a district of the Peloponnesus, the ideal region of rural happiness; hence, ideally rustic or pastoral. *n.* an inhabitant of Arcadia; an ideal rustic. **arcadianism,** *n.*

arcane (ahkān'), *a.* secret, esoteric.

arcanum (ahkā'nəm), *n.* (*pl.* **-na** (-nə)) a mystery, a secret; esp. one of the supposed secrets of the alchemists; an elixir, a miraculous remedy.

arch[1] (ahch), *n.* a curved structure so arranged that the parts support each other by mutual pressure; anything resembling this, a vault, a curve; a curved anatomical structure, as of the bony part of the foot; an archway. *v.t.* to cover with or form into an arch or arches; to overarch, to span. *v.i.* to assume an arched form. **arch-stone,** *n.* a wedge-shaped stone used in building arches; a key-stone. **archway,** *n.* an arched entrance or vaulted passage. **archwise,** *adv.* in the shape of an arch or vault. **arching,** *a.* forming an arch, curved. *n.* arched structure.

arch[2] (ahch), *a.* chief, pre-eminent, principal (in this sense generally in composition with a hyphen). **archenemy,** *n.* a principal enemy; esp. Satan, the devil. **archfiend,** *n.* the chief fiend; Satan. **arch-heresy,** *n.* extreme heresy. **arch-heretic,** *n.* **arch-priest,** *n.* a chief priest; a kind of dean or vicar to a bishop; a rural dean.

arch[3] (ahch), *a.* clever, cunning, roguish, sly. **archly,** *adv.* **archness,** *n.*

arch., (*abbr.*) archaic; architect, architecture.

arch- ARCH(1)-.

-arch, *comb. form.* ruler, leader, e.g. *monarch, matriarch*.

-archy, *comb. form.* government, rule, e.g. *oligarchy*.

Archaean, (*esp. N Am.*) **Archean** (ahkē'ən), *a.* pertaining or belonging to the earliest geological period or the rocks formed in this time.

archaeo-, (*esp. N Am.*) **archeo-,** *pref.* pertaining to past time; primitive.

archaeol., (*abbr.*) archaeology.

archaeology (ahkiol'əji), *n.* the science or special study of antiquities, esp. of prehistoric remains. **archaeologic** (-loj'-), **-ical,** *a.* **archaeolog'ically,** *adv.* **archaeologist,** *n.*

Archaeopteryx (ahkiop'təriks), *n.* (a fossil genus containing) the oldest known bird.

Archaeozoic (arkiəzō'ik), *a.* pertaining to the earliest geological era, the dawn of life on the earth.

archaic (ahkā'ik), *a.* pertaining to antiquity; belonging to an earlier period, no longer in general use; old-fashioned, antiquated. **archaism** (ah'-), *n.* an old-fashioned habit or custom; an archaic word or expression; affectation or imitation of ancient style or idioms. **archaist** (ah'-), *n.* one who affects the archaic, an imitator of ancient style; an antiquary. **archaistic** (-is'-), *a.* imitating or affecting the archaic; tending to archaism. **archaize, -ise** (ah'kāīz), *v.i.* to imitate or affect ancient manners, language or style. *v.t.* to make archaic.

archangel (ahk'ānjəl), *n.* an angel of the highest rank; a kind of dead-nettle; a kind of fancy pigeon.

archbishop   37   **argand**

archangelic (-anjel'-), *a.*
archbishop (ahchbish'əp), *n.* a chief bishop; a metropolitan; the spiritual head of an archiepiscopal province.
archbishopric, *n.* the office of archbishop; the district under the jurisdiction of an archbishop.
archdeacon (ahchdē'kən), *n.* a chief deacon; a church dignitary next below a bishop in the care of the diocese.
archdeaconry, *n.* the portion of a diocese over which an archdeacon exercises jurisdiction; the rank or office of an archdeacon; an archdeacon's residence. **archdeaconship**, *n.*
archdiocese (ahchdī'əsis, -sēs), *n.* the see of an archbishop.
archduke (ahch'dūk), *n.* a chief duke, esp. a son of an Emperor of Austria. **archducal** (-dū'-), *a.* of or pertaining to an archduke. **archduchess**, *n.* the wife of an archduke; a daughter of an Emperor of Austria. **archduchy** (-dūch'i), *n.* the territory ruled over by an archduke.
Archean ARCHAEAN.
archegonium (ahkigō'niəm), *n.* (*pl.* -**nia**) the female sex organ in mosses, ferns and some conifers.
archeo- ARCHAEO-.
archer (ah'chə), *n.* one who uses the bow and arrow; the constellation of Sagittarius. **archer-fish**, *n.* a fish, *Toxotes jaculator*, from the E Indies, that has the power of projecting water from its mouth to a considerable distance.
archeress (-ris), *n. fem.* **archery**, *n.* the sport or art of shooting with bow and arrow.
archetype (ah'kitīp), *n.* the primitive or original type, model or pattern on which anything is (assumed to be) formed. **archetypal** (ah'-, -tī'-), **-typical** (-tip'-), *a.* pertaining to an archetype; primitive, original. **archetypally** (ah'-, -tī'-), **-typically**, *adv.*
arch(i)-, *pref.* chief, principal; leading, pre-eminent; first; e.g. *archangel, archbishop.*
archidiaconal (ahkidīak'ənəl), *a.* of, or pertaining to, or holding the office of an archdeacon. **archidiaconate** (-nət), *n.* the office or territory of an archdeacon.
archiepiscopal (ahkiəpis'kəpəl), *a.* of or pertaining to an archbishop or an archbishopric. **archiepiscopate** (-pət), *n.* the office, dignity, or jurisdiction of an archbishop; an archbishop's tenure of office.
archil (ah'chil, -kil), *n.* any of certain lichens of the genus *Roccella*; a purple or violet dye prepared from these lichens.
archimandrite (ahkiman'drīt), *n.* the superior of a monastery or convent in the Greek Church, corresponding to an abbot in the Roman Catholic Church.
Archimedean (ahkimē'diən), *a.* of, pertaining to, or invented by Archimedes, a Greek mathematician (*c.* 287–212 BC). **Archimedean screw, Archimedes' screw**, *n.* an instrument for raising water, formed by winding a tube into the form of a screw round a long cylinder.
archipelago (ahkipel'əgō), *n.* (*pl.* **-goes, -gos**) any sea or water studded with islands; these islands collectively.
archipelagic (-laj'-), **-gian** (-lā'ji-), *a.*
architect (ah'kitekt), *n.* one who plans and draws the designs of buildings and superintends their erection; a designer of some complex work; a contriver, planner. **architecture**, *n.* the art of building edifices or constructions of any kind; the art or profession of designing buildings; architectural work; style of building; construction; the design and structural arrangement of the hardware components of a computer. **architectural** (-tek'-), *a.* **architecturally**, *adv.*
architectonic (ahkitekton'ik), **-ical**, *a.* of or pertaining to architecture; constructive; directive; controlling; pertaining to the organization of knowledge. **architectonics**, *n. sing.* the science of architecture; the systematization of knowledge; construction or systematic design in a literary or other artistic work.
architrave (ah'kitrāv), *n.* the lowest portion of the entablature of a column, immediately resting on the column itself; the ornamental moulding round a door or window.
archive (ah'kīv), *n.* (*usu. pl.*) a place in which (historical) records are kept; (historical) records officially preserved. **archival** (-kī'-), *a.* **archivist** (-ki-), *n.* one who has charge of archives; a keeper of records.
archivolt (ah'kivōlt), *n.* (the mouldings and ornaments on) the inner contour of an arch.
archly etc. ARCH[3].
archon (ah'kon), *n.* (*Hist.*) a chief magistrate of Athens.
archway ARCH[1].
-archy -ARCH.
Arctic (ahk'tik), *a.* of or pertaining to the north, the North Pole, or the region within the Arctic Circle. *n.* the North Pole; Arctic regions. **Arctic Circle**, *n.* a parallel of the globe, 23° 28' distant from the North Pole, which is its centre. **Arctic fox**, *n.* a small species of fox, with beautiful fur, found in N America within the Arctic Circle.
Arcturus (ahktū'rəs), *n.* the bright star in the constellation Boötes.
arcuate (ah'kūāt, -tid), *a.* curved like a bow. **arcuately**, *adv.*
-ard, *suf.* noting disposition or character, with augmentative force; e.g. *drunkard, sluggard.*
Ardea (ah'diə), *n.* a genus of birds including herons, bitterns and egrets.
ardent (ah'dənt), *a.* glowing; intense, eager, zealous. **ardent-spirits**, *n.pl.* alcoholic spirits (orig. meaning inflammable, combustible spirits). **ardency**, *n.* **ardently**, *adv.*
ardour (ah'də), *n.* heat of passion; warmth of emotion.
arduous (ah'dūəs), *a.* steep and lofty, hard to climb; involving much labour, strenuous; laborious, difficult. **arduously**, *adv.* **arduousness**, *n.*
are[1] (ah), *n.* a metric unit of area equal to 100 square metres (1076·44 sq. ft.).
are[2] (ah, *unstressed* ə), *pl. pres. ind. of the verb* to be.
area (eə'riə), *n.* any clear or open space; the sunken court, partly enclosed by railings, giving access to the basement of some dwelling-houses; space left open round a basement to obviate damp; the extent of a surface; a particular extent of surface, a region, a tract of country; a section of a larger space or surface or of a building etc.; a limited extent of the surface of any organism, distinguished from that which surrounds it; a sphere of interest or study.
Areca (a'rikə, ərē'-), *n.* a genus of palms, esp. *A. catechu*, which yields the betel-nut.
arefy (a'rəfī), *v.t.* to make dry, to dry up, to parch. **arefaction** (-fak'-), *n.*
arena (ərē'nə), *n.* the floor of an amphitheatre where combats took place, orig. strewn with sand to absorb the blood; an amphitheatre; a field of conflict; a sphere of action. **arenaceous** (arənā'shəs), *a.* sandy; growing in sand; composed partly or entirely of sand. **arenose** (a'rənōs), *a.* full of grit or sand.
aren't (ahnt), contr. form of *are not; am not* (in questions).
areocentric (ariəsen'trik), *a.* centring in the planet Mars.
areola (ərē'ələ), *n.* (*pl.* **-lae** (-lē)) a very small defined area; one of the interstices in organized tissue; any minute space enclosed by lines or markings; a slightly depressed spot; a dark circle round the human nipple; a similar circle round a pustule. **areolar**, *a.* of, pertaining to, or consisting of areolae. **areolate** (-lət), *a.* marked or marked off by intersecting lines. **areola'tion** (a-), *n.* **areole** (a'-), *n.* an areola.
arête (ərət'), *n.* a sharp ascending ridge of a mountain.
argala (ah'gələ), *n.* the adjutant-bird, a gigantic stork from India.
argali (ah'gəli), *n.* (*pl.* **-lis, -li**) the wild rock-sheep of Asia.
argand, argand lamp (ah'gand), *n.* a lamp having a circular hollow wick or gas-burner, which admits air so as to

secure more complete combustion and brighter light. [Aimé *Argand*, 1755–1803, Swiss inventor]

argent (ah'jənt), *n., a. (poet.)* silver; (*Her.*) (of) the white colour representing silver. **argentiferous** (-tif'-), *a.* producing silver.

Argentine (ah'jəntīn), **Argentinian** (-tin'iən), *n., a.* (a native) of Argentina.

argentine (ah'jəntīn), *a.* of or containing silver; silvery. *n.* electro-plate; a small fish with silvery scales.

argie-bargie ARGY-BARGY.

argil (ah'jil), *n.* white clay, potter's earth. **argillaceous** (-lā'shəs), *a.* of the nature of clay; containing a large amount of clay. **argilliferous** (-lif'-), *a.* producing or yielding clay.

Argive (ah'gīv, -jīv), *a.* of or pertaining to the ancient Greek city of Argos; hence, Greek. *n.* a native of Argos; a Greek.

argol (ah'gol), *n.* an impure acid potassium tartrate deposited from wines; crude cream of tartar.

argon (ah'gon), *n.* an inert gas, at. no. 18; chem. symbol Ar, one of the gaseous constituents of the atmosphere, discovered in 1894.

Argonaut (ah'gənawt), *n.* one of the legendary heroes who accompanied Jason in the ship *Argo* to seek the Golden Fleece; (**argonaut**) (the popular name of a genus of cephalopod molluscs containing) the paper-nautilus. **argonautic** (-naw'-), *a.*

argosy (ah'gəsi), *n.* a large vessel for carrying merchandise; (*fig.*) a richly-laden ship.

argot (ah'gō), *n.* thieves' slang; the phraseology of a class; slang generally. **argotic** (-got'-), *a.* slangy.

argue (ah'gū), *v.t.* to prove, to show, to evince; to (try to) exhibit or prove by reasoning; to convince by logical methods; to discuss, debate. *v.i.* to bring forward reasons, to discuss; to reason in opposition, to dispute. **arguable**, *a.* capable of being argued. **argufy**, *v.i.* (*coll.*) to argue. **argufier**, *n.* **argument** (ah'gūmənt), *n.* proof; (a) reason, series of reasons or demonstration put forward; process of reasoning; (a) debate, discussion; an abstract or summary of a book; the subject of a discourse; a mathematical variable whose value determines that of a dependent function. **argumentation**, *n.* the act or process of reasoning; methodical reasoning; a systematic argument. **argumentative** (-men'-), *a.* consisting of or pertaining to argument; controversial; having a natural tendency to argue. **argumentatively**, *adv.* **argumentativeness**, *n.*

Argus (ah'gəs), *n.* in Greek mythology, the guardian of Io, fabled to have a hundred eyes; a vigilant watcher or guardian; a pheasant from the E Indies, having the plumage marked with eye-like spots; a butterfly of the genus *Polyommatus*, which has eye-like spots on the wings. **argus-eyed**, *a.* very observant, sharp-sighted.

argy-bargy, argie-bargie (ahjibah'ji), *n.* (*coll.*) (a) dispute, argument.

aria (ah'riə), *n.* an air; a song for one voice supported by instruments.

Arian[1] (eə'riən), *a.* pertaining to Arius or his doctrine. *n.* a follower of Arius of Alexandria (*c.* 250–336) who denied that Christ was consubstantial with the Father. **Arianism**, *n.* **Arianize, -ise**, *v.t., v.i.* to convert to or propagate Arianism.

Arian[2] ARYAN.

-arian, *suf.* belonging to, believing in; one who belongs to, believes in, or is associated with; e.g. *humanitarian, sabbatarian, sexagenarian, trinitarian.*

arid (a'rid), *a.* dry, parched; barren, bare; dry, uninteresting. **aridity** (ərid'-), **aridness**, *n.*

ariel (eə'riəl), *n.* a W Asiatic and African gazelle.

Aries (eə'rēz), *n.* the Ram, the first of the zodiacal constellations, which the sun enters in March.

arietta (ariet'ə), **ariette** (ariet'), *n.* a short lively air, tune or song.

aright (ərīt'), *adv.* right, rightly, properly, becomingly; without failure or mistake.

aril (a'ril), *n.* an accessory seed-covering formed by a growth near the hilum. **arillate** (əril'āt), **arilled**, *a.* furnished with an aril.

-arious, *suf.* connected with, belonging to; forming adjectives, e.g. *gregarious, vicarious.*

arise (əriz'), *v.i.* (*past,* **arose** (əroz'), *p.p.* **arisen** (əriz'n)) to assume an upright position from an attitude of repose, to get up; to rise from the dead; to appear, to come into being, notoriety etc.; to originate; to take place, occur (as a result).

arista (əris'tə), *n.* (*pl.* **-tae** (-tē), **-tas**) an awn; a bristle or bristle-like process. **aristate** (-tāt), *a.* furnished with an arista.

aristo (a'ristō), *n.* (*pl.* **-tos**) short for ARISTOCRAT.

aristocracy (aristok'rəsi), *n.* government by the best citizens or by the nobles; a state so governed; a ruling body of nobles; the nobility; the best of any class or group. **aristocrat** (a'ristəkrat), *n.* a noble; a member of an aristocracy; (*rare*) one who favours aristocratic government. **aristocratic, -ical** (-krat'-), *a.* pertaining or relating to an aristocracy or an aristocrat, grand, stylish. **aristocratically**, *adv.*

Aristotelian, -lean (aristətē'liən), *a.* of or pertaining to the Greek philosopher Aristotle (384–322 BC), or to his philosophy. *n.* one who adheres to or is learned in the philosophy of Aristotle. **Aristotelianism**, *n.*

arith., (*abbr.*) arithmetic, arithmetical.

arithmetic (ərith'mətik), *n.* the science of numbers; computation by figures; arithmetical knowledge. **arithmetic** (arithmet'ik), **arithmetical**, *a.* of or pertaining to arithmetic. **arithmetic and logic unit**, (*Comput.*) the section of a central processing unit where arithmetic operations are carried out. **arithmetic mean**, *n.* the average value of a set of numbers or terms, found by dividing the sum of the terms by the number. **arithmetic progression**, *n.* a series of numbers that increase or decrease consecutively by a constant quantity. **arithmetically**, *adv.* **arithmetician** (ərithmətish'ən), *n.* one skilled in arithmetic.

-arium, *suf.* thing connected with or used for; place for; *aquarium, herbarium, sacrarium.*

a rivederci, arrivederci (ərēvədœ'chi), *int.* (*It.*) goodbye, to our next meeting.

Ariz., (*abbr.*) Arizona.

ark (ahk), *n.* (also **Noah's ark**) the vessel in which Noah and his family were saved from the Deluge; a chest, a box; a sacred repository; esp. one for the scrolls of the Torah; a refuge; (also **Ark of the Covenant**) the wooden coffer containing the tables of the Law etc. in the Jewish tabernacle; a boat held to resemble Noah's ark, a model of Noah's ark with toy animals. **arkite**, *a.* pertaining to Noah's ark. *n.* an inmate of the ark.

Ark., (*abbr.*) Arkansas.

arm[1] (ahm), *n.* the upper limb of the human body on either side, from the shoulder to the hand; anything resembling the human arm; a sleeve; a projecting branch, as of the sea, a mountain, machine, instrument or the like; the fore-limb of any of the lower mammals; a flexible appendage, with arm-like functions, in invertebrates; the part of a chair etc. on which the arm rests; a division of a service or organization; power, authority. **an arm and a leg**, (*coll.*) a great amount of money. **arm in arm, arm-in-arm**, with the arms interlinked. **at arm's length**, at a distance. **with open arms**, enthusiastically. **arm-band**, *n.* a band of material encircling the coat-sleeve, usu. black to indicate mourning. **armchair**, *n.* a chair with arms to support the elbows. **armhole**, *n.* the hole in a garment to admit the arm. **armlet** (-lit), *n.* a small (ornamental) band worn on the arm; a small arm of the sea. **armpit**, *n.* the hollow under the arm at the shoulder. **armrest**, *n.* the arm of a chair etc. **arm wrestling**, *n.* a contest of strength in

which two people, with elbows resting on a surface, lock right or left hands and attempt to force the other's arm down on to the surface. **armful**, *n*. **armless** (-lis), *a*.
arm[2] (ahm), *n*. a weapon; any branch of the military service; (*pl.*) war; (*pl.*) the military profession; (*pl.*) heraldic bearings. *v.t.* to furnish or equip with offensive or defensive weapons; something that protects, strengthens or fortifies; to prepare for war; to equip with tools or other appliances; to make ready (a bomb etc.) for explosion. *v.i.* to prepare for war. **to arms!** take your weapons; prepare for battle. **under arms**, bearing arms; ready for service. **up in arms**, in revolt; on the aggressive defensive. **arms race**, *n*. rivalry between nations, esp. the US and USSR, in building up stocks of (nuclear) weapons. **armed**, *a*. **armed forces**, *n.pl.* the army, navy and air force of a nation. **armless** (-lis), *a*.
armada (ahmah'də), *n*. an armed fleet, esp. the fleet sent by Philip II of Spain against England in 1588; any large (armed) force.
Armadillo (ahmədil'ō), *n*. (*pl.* **-llos**) the name of several small burrowing edentate animals, native to S America, encased in bony armour, and capable of rolling themselves into a ball.
Armageddon (ahməged'ən), *n*. the final battle between good and evil; a great and destructive battle.
Armagnac (ah'mənyak), *n*. a dry brandy from S W France.
armament (ah'məmənt), *n*. the act of arming a fleet or army for war; the munitions of war, esp. the guns of a warship; an armed force; the total military capacity of a nation.
armature (ah'məchə), *n*. armour; means of defence in general; the supportive framework for a model in clay etc.; a piece of soft iron placed in contact with the poles of a magnet to preserve and increase its power; the revolving part of an electric motor or dynamo; the moving part of an electromagnetic device.
Armenian (ahmē'niən), *a*. of or pertaining to Armenia. *n*. a native of Armenia; the language spoken by the Armenians.
armiger (ah'mijə), *n*. an esquire; one entitled to heraldic bearings. **armigerous** (-mij'-), *a*. entitled to heraldic bearings.
armilla (ahmil'ə), *n*. a bracelet, an armlet. **armillary** (ah'-, -mil'-), *a*. pertaining to bracelets; consisting of parts resembling bracelets. **armillary sphere**, *n*. a skeleton celestial globe or sphere consisting of metallic circles mechanically fixed to represent the celestial equator, the ecliptic, the colures etc.
Arminian (ahmin'iən), *a*. of or pertaining to Arminius, i.e. James Harmensen, the Dutch theologian (1560–1609) who maintained the doctrine of free-will against Calvin. *n*. a follower of Arminius. **Arminianism**, *n*.
armistice (ah'mistis), *n*. a cessation of arms for a stipulated time during war; a truce. **Armistice Day**, *n*. 11 Nov., the day on which an armistice was signed in 1918. Since the 1939–45 war, Remembrance Day (in Britain, the Sunday nearest to 11 Nov.) is solemnly observed to commemorate the fallen in both wars.
armlet ARM[1].
armoire (ahmwah'), *n*. a chest, a cupboard.
Armoric (ahmo'rik), *a*. of or pertaining to Brittany, the ancient Armorica. *n*. the language of Armorica. **Armorican**, *a*. Armoric. *n*. a Breton.
armorial (ahmaw'riəl), *a*. pertaining or relating to heraldic arms. *n*. a book containing coats of arms. **armorist**, *n*. one learned in heraldry; one skilled in blazoning arms.
armour (ah'mə), (*esp*. *N Am*.) **armor**, *n*. a defensive covering worn by a person in combat, esp. a mediaeval warrior; a protective covering of animals or plants; the protective metal plating of a warship, tank, aircraft etc.; the watertight dress of a diver; heraldic bearings; collectively tanks and other armoured vehicles. *v.t.* to furnish with protective covering, esp. armour or armour-plating. **armour-bearer**, *n*. one who carried the weapons of a warrior; an esquire. **armour-clad**, *a*. ironclad. **armour-plate**, *n*. a plate of iron or steel for covering the sides of warships, tanks etc.; armour-plating. **armour-plated**, *a*. **armour-plating**, *n*. (a defensive covering of) iron or steel plates. **armoured**, *a*. clad in armour; protected; of a ship, ironclad. **armoured car, train**, *n*. a motor-car or train protected by steel plates. **armourer**, *n*. one who made armour; a manufacturer of arms; a non-commissioned officer in charge of the arms of a regiment, ship etc. **armoury**, *n*. armour or arms; a place for keeping arms, an arsenal; a store of resources.
armpit ARM[1].
army (ah'mi), *n*. a body of people organized for land warfare; a multitude, a host; an organized body (e.g. the *Salvation Army*). **army ant**, *n*. any of various ants which travel in vast numbers destroying animals and plants. **army-corps**, *n*. a main division of an army. **army list**, *n*. an official list of the officers of an army.
arnica (ah'nikə), *n*. a tincture prepared from *Arnica montana*, mountain tobacco, and used as an application for bruises, sprains etc.; any of a genus of compositous plants including this.
aroint, aroynt (əroint'), *int.*, *v*. avaunt! begone!
aroma (ərō'mə), *n*. the fragrance in a plant, spice, fruit, wine etc.; an agreeable odour or smell; a subtle pervasive quality. **aromatherapy**, *n*. the use of (massage with) essential plant oils to promote physical and mental well-being and in healing. **aromatherapist**, *n*.
aromatic (ərəmat'ik), *a*. of or pertaining to an aroma; fragrant, spicy; belonging or pertaining to a class of organic compounds containing a benzene ring in the molecular structure. *n*. a fragrant drug, spice etc. **aromatically**, *adv*. **aromaticity** (-rōmətis'-), *n*. **aromatize**, **-ise** (ərō'-), *v.t.* to render aromatic or fragrant; to perfume, to scent. **aromatization**, *n*.
arose (ərōz'), *pret*. ARISE.
around (ərownd'), *prep*. surrounding; round about; on all sides of; along the circuit of. *adv*. all round; in a circle; about, here and there, in all directions; in existence, in evidence. **been around**, (*coll.*) to have gained experience, esp. become worldly-wise.
arouse (ərowz'), *v.t.* to raise, stir up, awaken; to excite, stimulate. **arousal**, *n*. **arouser**, *n*.
ARP, (*abbr*.) air raid precautions.
arpeggio (ahpej'iō), *n*. (*pl.* **-ggios**) a method of playing a chord on a keyed instrument by striking the notes in rapid succession instead of simultaneously; a chord so played.
arquebus (ah'kwibəs) HARQUEBUS.
arrack, arak (a'rək), *n*. a distilled spirit from the East; esp. one distilled from coconut or rice.
arraign (ərān'), *v.t.* to cite before a tribunal to answer a criminal charge; to accuse; to charge with fault. **arraigner**, *n*. **arraignment**, *n*.
arrange (ərānj'), *v.t.* to draw up in rank or ranks; to adjust, to put in proper order; to adapt (a musical composition) for other instruments or voices; to plan or settle circumstances in readiness for. *v.i.* to come to agreement; to prepare. **arrangement**, *n*. the act of arranging, the state of being arranged; the manner in which things are arranged; settlement, agreement; a grouping or combination of things in a particular way; (*often pl.*) preparation; the adaptation of a musical composition for instruments or voices for which it was not written.
arrant (a'rənt), *a*. notorious, downright, unmitigated; complete, thorough. **arrantly**, *adv*. shamelessly, infamously.
arras (a'rəs), *n*. a kind of tapestry made at Arras in Artois; a rich fabric of coloured tapestry; a wall hanging.
array (ərā'), *v.t.* to put in readiness (as troops), to marshal;

to arrange, order; to dress up, to deck, to equip; (*Law*) to set (a jury) in order for a trial. *n.* order, esp. of battle; an imposing or orderly arrangement or disposition; (*poet.*) (fine) dress, attire; (*Law*) the order of empanelling a jury; the panel of jurors; an arrangement of numbers or mathematical symbols in rows and columns; a collection of elements that form a unit.
arrear (əriə'), *n.* (*usu. pl.*) that which is behindhand, unpaid, or unsatisfied. **in arrears**, behindhand, esp. in discharging a debt; unpaid, unsatisfied. **arrearage** (-rij), *n.* the state of being in arrears; that which is in arrears or outstanding, items overdue.
arrest (ərest'), *v.t.* to stop, check; to seize and fix (the sight, mind etc.); to stay (legal proceedings etc.); to apprehend, esp. to apprehend and take into legal custody; to seize by legal authority. *n.* a stoppage, stay, check; seizure, detention, esp. by legal authority. **arrest of judgment**, staying of a judgment after a verdict. **under arrest**, in legal custody. **arrested development**, *n.* development arrested at some stage of its progress. **arrestable**, *a.* **arrestee**, *n.* (*N Am.*) a person who is arrested. **arrester**, *n.* one who, or that which, arrests; a contrivance for cutting off a force (e.g. lightning); (usu. **arrestor**) (*Sc. Law*) the person who arrests a debt or property in another's hands. **arresting**, *a.* striking, catching the attention. **arrestingly**, *adv.* **arrestive**, *a.* tending to arrest (e.g. the conj. *but*). **arrestment**, *n.* the act of arresting; stop, stay, check; (*Law*) seizure of property by legal authority, esp. (*Sc. Law*) the process by which a creditor detains the effects of a debtor, which are in the hands of third parties, till the money owing is paid. **arrestor** ARRESTER. **arrestor-hook**, *n.* a device that enables an aircraft landing on a carrier-ship to check speed by catching on a cable.
arrhythmia (əridh'miə, ā-), *n.* an irregularity or alteration in the rhythm of the heartbeat. **arrhythmic**, *a.*
arrière-pensée (arieə'pē'sā), *n.* (*F*) a mental reservation; an unrevealed intention.
arris (a'ris), *n.* (*pl.* **arris**, **arrises**) the line in which two straight or curved surfaces forming an exterior angle meet each other.
arrive (əriv'), *v.i.* to come to, reach a place, position, state of mind etc.; to gain, reach to or attain an object; to come about, to occur; (*coll.*) to attain notoriety, become eminent, make one's fortune. **arrival**, *n.* the act of coming to a journey's end or destination; the coming to a position, state of mind etc.; one who, or that which, has arrived; (*coll.*) a new-born child.
arrivederci A RIVEDERCI.
arriviste (arēvēst'), *n.* a social climber, a parvenu; a self-seeker, esp. in politics.
arrogant (a'rəgənt), *a.* claiming or assuming too much; insolent, assuming, overbearing, haughty. **arrogantly**, *adv.* **arrogance, -ancy**, *n.* the act or quality of being arrogant; undue assumption.
arrogate (a'rəgāt), *v.t.* to make unduly exalted claims or baseless pretensions to (a thing) for oneself or for someone else. **arrogation**, *n.*
arrondissement (arŏdēs'mē), *n.* a territorial division of a French department; a ward in Paris.
arrow (a'rō), *n.* a slender, straight missile shot from a bow; anything resembling an arrow in shape or function. **arrowhead**, *n.* the pointed head of an arrow; a mark shaped like an arrowhead, indicating direction; a plant of the genus *Sagittaria*, the leaves of which resemble arrowheads. **arrow-headed**, *a.* **arrowroot**, *n.* a nutritious starch extracted from the tubers of several species of *Maranta*; a plant of the genus *Maranta*, which includes *M. arundinacea*, the tubers of which were used to absorb poison from wounds, esp. those made by poisoned arrows. **arrowy**, *a.* consisting of arrows; resembling an arrow or arrows in form or motion.
arroyo (ərō'yō), *n.* (*pl.* **-yos**) (*N Am.*) a dried-up watercourse, a rocky ravine. [Sp.]
arse (ahs), (*esp. N Am.*) **ass** (as), *n.* (*taboo*) the buttocks, the rump, the hind parts; (*sl.*) the fag-end; (*taboo*) the anus. **to arse about** or **around**, (*taboo sl.*) to mess around, act in a stupid or irritating manner. **arsehole**, (*esp. N Am.*) **asshole**, *n.* (*taboo*) the anus; (*sl.*) a stupid or worthless person. **arselicker**, *n.* (*sl.*) a sycophant, toady. **arselicking**, *n.*, *a.*
arsenal (ah'sənəl), *n.* a place for the (manufacture and) storage of naval and military weapons and ammunition; a store of spiritual, literary or other weapons.
arsenic[1] (ahs'nik), *n.* a brittle, semi-metallic steel-grey element, at.no. 33; chem. symbol As; the trioxide of this element, a virulent poison.
arsenic[2] (ahsen'ik), *a.* of or containing (pentavalent) arsenic. **arsenical**, *a.* pertaining to arsenic; having arsenic in the composition. **arsenious** (-sē'-), *a.* of or containing arsenic; esp. applied to compounds in which (frivalent) arsenic. **arsine** (ah'sēn), *n.* hydrogen arsenide, a very poisonous gas.
arsis (ah'sis), *n.* (*pl.* **-ses** (-sez)) the stressed syllable in metre; the stressed note in barred music.
arson (ah'sən), *n.* the wilful setting on fire of another's house or other property, or to one's own with intent to defraud the insurers.
art[1] (aht), *v.* the archaic second pers. sing. pres. ind. of the verb TO BE.
art[2] (aht), *n.* skill, human skill or workmanship, as opposed to nature; skill applied to the creation of (visual) beauty and aesthetic objects; (any of) the fine arts, esp. the arts of representation and design; (visual) works of art; perfection of workmanship for its own sake; a craft, a profession; acquired skill; a knack; craft, cunning, artifice; (*pl.*) the humanities or liberal arts, the learning of the schools; (*pl.*) the subjects studied in a nonscientific or nontechnical university course. *a.* decorative; consciously displaying artistic qualities. **Bachelor of Arts, Master of Arts**, titles conferred on those who have attained certain degrees of proficiency in the humanities. **Art Deco** (dek'ō), *n.* a style of decorative art of the 1920s and 1930s characterized by bold geometrical forms. **art form**, *n.* an established form in music or literature; a medium of artistic expression. **art-house**, *a.* pertaining to films produced for aesthetic or artistic purposes rather than commercial success. **Art Nouveau** (noovō'), *n.* a style of decorative art of the late 19th and early 20th cents. characterized by sinuous curving forms. **art paper**, *n.* paper coated with a composition of china clay, making it suitable for fine printing. **art union**, *n.* an association for the promotion and the encouragement of artists; (*Austral.*) a lottery. **artwork**, *n.* the illustrative material in a magazine, book etc. **artful**, *a.* crafty, cunning; characterized by art or skill; artificial, unreal. **artfully**, *adv.* **artfulness**, *n.* **artless** (-lis), *a.* guileless, simple, unaffected; without art; unskilful, clumsy; uncultured, natural. **artlessly**, *adv.* **artlessness**, *n.* **arty** *a.* (*coll.*) self-consciously or pretentiously aping the artistic. **arty crafty**, *a.* more showily artistic than functional.
art., (*abbr.*) article; artificial; artillery.
artefact, artifact (ah'tifakt), *n.* a product of human skill or workmanship; esp. a simple object of archaeological importance or interest.
Artemisia (artimiz'iə), *n.* a genus of composite plants, containing wormwood, southern-wood etc.
artery (ah'təri), *n.* any of the membranous pulsating vessels, conveying blood from the heart to all parts of the body; a main channel of communication or transport. **arterial** (-tiə'-), *a.* pertaining to or affecting an artery or arteries; pertaining to the oxygenated blood that circulates in the arteries. **arterial road**, *n.* a main road for swift, long-distance traffic between the chief industrial centres. **arterialize, -ise**, *v.t.* to convert venous into arterial blood

by exposing to the action of oxygen in the lungs; to endow with arteries. **arterialization**, *n.* **arteriole** (ahtiə′riōl), *n.* a small branch of an artery. **arteriosclerosis** (ahtiəriōskləro′sis), *n.* thickening and loss of elasticity in the walls of the arteries. **arteriosclerotic** (-rot′-), *a.* **arteritis** (ahtərī′tis), *n.* inflammation occurring in the arteries.

artesian well, *n.* a well in which water is obtained by boring through an upper retentive stratum to a subjacent water-bearing stratum, the water being forced to the surface by natural pressure.

Artex® (ah′teks), *n.* a textured paint covering for ceilings and walls. *v.t.* to give a textured surface to.

artful etc. ART².

arthralgia (ahthral′jə), *n.* pain in a joint. **arthralgic**, *a.*

arthritic (ahthrit′ik), *a.* pertaining to or affecting the joints; of or suffering from arthritis. *n.* a person with arthritis.

arthritis (-thrī′-), *n.* (painful) inflammation of one or more joints causing stiffness.

arthr(o)-, *comb. form.* pertaining to joints; characterized by joints.

arthropathy (ahthrop′əthi), *n.* (a) disease of the joints.

Arthropoda (ahthrop′ədə), *n.pl.* a phylum of invertebrate animals with segmented bodies and jointed limbs, including the insects, arachnids and crustaceans. **arthropodal**, **-podous**, *a.* **arthropod** (ah′thrəpod), *n.* a member of the Arthropoda.

arthrosis (ahthrō′sis), *n.* (*pl.* **-ses** (-sēz)) a joint uniting two bones, an articulation.

Arthurian (ahthū′riən), *a.* pertaining to Arthur, a legendary King of Britain, or his knights.

artic, *n.* short for ARTICULATED LORRY under ARTICULATE¹.

artichoke (ah′tichōk), *n.* (also **globe artichoke**) a composite plant, *Cynara scolymus*, somewhat like a large thistle: the receptacle and fleshy bases of the scales are eaten as a vegetable; JERUSALEM ARTICHOKE.

article (ah′tikl), *n.* a distinct member or portion; a prose composition, complete in itself, in a newspaper, magazine, encyclopaedia etc.; an item, a piece, a distinct detail; a distinct statement, clause or provision in an agreement, statute, code or other document; an item of trade, use or property; a thing, an object; a name for the adjectives, *a*, *an*, *the*, when these are considered to form a separate part of speech; (*pl.*) a formal agreement; (*pl.*) terms, conditions. *v.t.* to draw up in the form of articles; to bind (an apprentice), indenture; to indict. **articles of association**, the statutes of a limited liability company. **Articles of War**, a code of discipline for the British Army. **Thirty-nine Articles**, the thirty-nine statements subscribed to by the clergy of the Church of England. **articled**, *a.* bound under article of apprenticeship, esp. of a lawyer's clerk.

articulate¹ (ahtik′ūlāt), *v.t.* to connect by means of a joint; to utter distinctly; to express clearly and coherently; to article. *v.i.* to form a joint (with); to speak distinctly. **articular** (ahtik′ūlə), *a.* pertaining to the joints. **articulated**, *a.* **articulated lorry**, *n.* a long lorry with separate tractor and trailer sections connected so as to allow the tractor to turn at an angle to the remainder. **articulation**, *n.* the process or method of jointing; the act or process of speaking; articulate sound, utterance; a consonant; a joint; a jointed structure; (*Bot.*) the point at which a deciduous member separates from the plant. **articulator**, *n.* **articulatory**, *a.*

articulate² (ahtik′ūlāt), *a.* jointed; distinguished by distinct and intelligent movements of the organs of speech; able to express oneself clearly and coherently; expressed in this manner; (*Biol.*) composed of segments. **articulately**, *adv.* **articulateness**, *n.*

artifact ARTEFACT.

artifice (ah′tifis), *n.* anything contrived by art; human skill; cunning, trickery; a contrivance; a trick. **artificer** (-tif′-), *n.* one who practises an art; a craftsman; a maker, a contriver; a mechanic employed to make and repair military stores.

artificial (ahtifish′əl), *a.* made or produced by human skill; not natural, not real; affected in manner; factitious, feigned, fictitious. **artificial horion**, *n.* a small trough containing mercury, the surface of which affords a reflected image of a heavenly body, used in taking altitudes etc. in places where there is no visible horizon; an instrument that indicates an aircraft's position with respect to the horizontal. **artificial insemination**, *n.* artificial injection of semen into a female. In human beings from the husband (AIH), or from an anonymous donor (AID). **artificial intelligence**, *n.* the ability of a computer, robot etc. to perform as an intelligent being; the area of study dealing with the development of machines capable of imitating intelligent human-like mental processes. **artificial respiration**, *n.* a method of reviving a person who has lost consciousness through drowning etc. **artificially**, *adv.* **artificiality** (-al′-), *n.* **artificialize**, **-ise**, *v.t.* to render artificial. **artificialness**, *n.*

artillery (ahtil′əri), *n.* implements of war; large-calibre guns, cannons, ordnance; the science and practice of gunnery; the branch of the military service in charge of the ordnance; any immaterial weapon. **artilleryman**, **artillerist**, *n.* an artillery soldier.

artiodactyl (ahtiōdak′til), **-tyle** (-til), *n.*, *a.* (an ungulate) having an even number of toes. **Artiodactyla** (-tilə), *n.pl.* a division of the Ungulata, containing those with an even number of toes.

artisan (ah′tizan, -zan′), *n.* one trained to practise a manual art; a handicraftsman, a mechanic.

artist (ah′tist), *n.* one who practises any of the fine arts, esp. that of painting; a craftsman who applies the principles of taste and imagination; any artistic performer, an artiste; (*sl.*) one who frequently practises, or is proficient in, a particular, esp. dubious, activity. **artistic**, **-ical** (-tis′-), *a.* of or pertaining to art or artists; proficient in the fine arts. **artistically**, *adv.* **artistry** (ah′-), *n.* artistic quality or workmanship; skill, proficiency.

artiste (ahtēst′), *n.* a public performer, an actor, dancer, musician, acrobat etc.; a highly proficient cook, hairdresser etc.

artless, **arty** etc. ART.

arty., (*abbr.*) artillery.

Arum (eə′rəm), *n.* a genus of plants, containing the wakerobin or cuckoo-pint. **arum lily**, *n.* an ornamental plant of the same family.

arundinaceous (ərundinā′shəs), *a.* resembling a reed; reedy. **arundineous** (ərəndin′iəs), *a.* abounding in reeds; reedy.

arvo (ah′vo), *n.* (*Austral. coll.*) afternoon.

-ary¹ (-əri), *suf.* pertaining to, connected with; belonging to, engaged in; thing connected with, used in; a place for; e.g. *elementary*, *necessary*, *voluntary*; *antiquary*, *statuary*; *aviary*, *granary*.

-ary² (-əri), *suf.* equivalent to -AR and sometimes to -ARY¹; e.g. *exemplary*, *military*, *contrary*.

Aryan, **Arian** (eə′riən), *a.* of or belonging to an ancient race of Europe or Central Asia, from whom many of the Indian and most of the European races are descended; of the Indo-European or Indo-Germanic race; of or speaking the Indo-European or Indo-Iranian languages; in Nazi terminology, non-Semitic. *n.* the old Aryan language; a member of the Aryan race; a non-semitic Caucasian, esp. a Nordic. **Aryanize**, **-ise**, *v.t.* to imbue with Aryan characteristics.

aryl (a′ril), *n.* a general name for a monovalent aromatic hydrocarbon radical, e.g. phenyl.

AS, (*abbr.*) Anglo-Saxon; antisubmarine.

As, (*chem. symbol*) arsenic.

as¹ (az), *adv.*, *conj.* in the same manner; in or to the same degree; equally with; thus; for instance; while, whilst;

as since, because, that. *rel. pron.* that, who, which; in the role, position, or state of. **as for, as regards, as to**, regarding, concerning. **as from, as of**, from (the specified time or date). **as if, as though**, as it would be if. **as it is**, in the present state, actually. **as it were**, in a certain way, to some extent, so to speak. **as was**, (*coll.*) in a previous state.

as² (as), *n.* a Roman copper coin, originally of 12 oz (340 g) but frequently reduced.

as- *pref.* AD-, assim. to *s*, as *assimilate, assume*.

ASA¹, (*abbr.*) Advertising Standards Authority; Amateur Swimming Association; American Standards Association.

ASA², **ASA speed**, *n.* a rating of the speed of a photographic film according to standards set by the American Standards Association.

asafoetida, asafetida (asəfet'idə, -fē'-), *n.* a gum, with a strong smell of garlic, obtained from *Narthex asafetida* and allied plants, used in medicine and cookery.

asap, (*abbr.*) as soon as possible.

asbestos (asbes'təs, az-), *n.* a mineral, esp. a variety of amphibole, of flax-like fibrous structure, practically incombustible, and resistant to chemicals. **asbestic**, *a.* **asbestine** (-tīn), *a.* **asbestosis** (-tō'sis), *n.* a lung disease caused by breathing in asbestos particles.

Ascaris (as'kəris), *n.* a genus of intestinal nematode worms, parasitic in man and the lower animals. **ascarid** (-rid), *n.* a worm of the genus *Ascaris*.

ascend (əsend'), *v.i.* to go or come from a lower to a higher place or position; to rise, to be raised; to slope upwards; to proceed from a lower to a higher plane of thought, quality, degree, rank; to go back in order of time; (*Astron.*) to move towards the zenith; (*Mus.*) to rise in pitch. *v.t.* to climb or go up, to go to a higher position upon; to go to the top, summit or source of; to mount. **ascendable**, *a.* **ascendancy**, **-ency**, **-ance**, **-ence**, *n.* controlling influence, governing power. **ascendant**, **-ent**, *a.* moving upwards, rising; predominating, ruling; (*Astrol.*) just above the eastern horizon; (*Astron.*) moving towards the zenith. *n.* ascent, slope, acclivity; an ancestor; a position of superiority, supremacy; (*Astrol.*) (the sign of the zodiac containing) the point of the ecliptic which is rising in the eastern point of the horizon at the moment of a person's birth. **house of the ascendant**, the space from 5° of the zodiac above to 25° below the ascendant. **in the ascendant**, dominant, predominant, supreme. **ascender**, *n.* that part of a letter (e.g. b, t) which is above the line of type.

ascension (əsen'shən), *n.* the act of ascending; (**Ascension**) the ascent of Christ to Heaven; Ascension Day. **right ascension** RIGHT. **Ascension Day**, *n.* the day on which the Ascension of Jesus Christ is commemorated – the Thursday but one before Whitsuntide, Holy Thursday. **ascensional**, *a.*

ascent (əsent'), *n.* the act or process of ascending, upward motion; an eminence; a slope; a way by which one may ascend; a movement back in time or ancestry; advancement, rise.

ascertain (asərtān'), *v.t.* to find out or learn by investigation, examination or experiment; to make sure of; to find out. **ascertainable**, *a.* **ascertainment**, *n.*

ascetic (əset'ik), *a.* of or pertaining to the ascetics or their mode of life; severely abstinent, austere; practising rigorous self-discipline. *n.* one of the early hermits who practised rigorous self-denial and mortification; hence, any person given to rigorous self-denial and mortification; (*pl.*) asceticism. **ascetical**, *a.* **ascetically**, *adv.* **asceticism**, *n.* the mode of life of an ascetic.

asci, pl. of ASCUS.

ascidian (əsid'iən), *n.* a tunicate of the order Ascidiacea.

ascidium (əsid'iəm), *n.* (*pl.* **-dia** (-ə)) a pitcher- or flask-shaped plant part, as the leaf of the pitcher plant.

ASCII (as'ki), *n.* a standard system for representing alphanumeric symbols as binary numbers, used in data processing. [acronym for American Standard Code for Information Interchange]

ascites (əsī'tēz), *n.* dropsy of the belly or abdomen. **ascitic** (-sit'-), **-ical**, *a.* suffering from this.

Ascomycetes ASCUS.

ascorbic acid (əskaw'bik), *n.* vitamin C, occurring in vegetables, fruits etc.

ascribe (əskrīb'), *v.t.* to attribute, to assign, to claim (something) for (someone). **ascribable**, *a.* **ascription** (əskrip'shən), *n.* the act of attributing; an expression, statement etc. ascribing something to someone. **ascriptitious** (askriptish'-), *a.*

ascus (as'kəs), *n.* (*pl.* **asci** (-kī)) a cell in which spores are formed in an ascomycete. **ascomycete** (askōmī'sēt), *n.* a fungus of the Ascomycetes. **Ascomycetes**, *n.pl.* a large class of fungi, including *Penicillium* and yeasts, having spores formed in asci.

asdic (az'dik), *n.* instruments and apparatus for detecting the presence and position of submarines. [acronym for Allied Submarine Detection *I*nvestigation *C*ommittee]

ASEAN (as'ian), (*abbr.*) Association of South-East Asian Nations.

aseismic (asīz'mik), *a.* free from or proof against earthquake shocks.

asepsis (āsep'sis), *n.* the condition of being aseptic; the process of making aseptic. **aseptic**, *a.* not liable to or free from putrefaction or pathogenic micro-organisms; preventing putrefaction or infection. *n.* an aseptic substance. **asepticism**, *n.* treatment by aseptic or antiseptic principles. **asepticize, -ise,** *v.t.*

asexual (asek'sūəl, -shəl), *a.* without sex, sexual organs or sexual functions; of reproduction, without union of gametes; without sexual content or interest. **asexuality** (-sūal'-, -shual'-), *n.* **asexually**, *adv.*

ASH (ash), (*abbr.*) Action on Smoking and Health.

ash¹ (ash), *n.* the residuum left after the burning of anything combustible; powdery mineral matter ejected from volcanoes; (*pl.*) the remains of anything burnt; (*pl.*) the remains of a cremated or disintegrated dead body. **The Ashes**, *n.pl.* a term used by the *Sporting Times* in 1882 in a mock In Memoriam to the demise of English cricket after the successful visit of the Australians. Since then English and Australian teams visiting one another have endeavoured to 'bring back the ashes'. **ash-bin**, *n.* a dustbin. **ash-blond**, *n.* a very pale blond colour; a person with hair of this colour. *a.* of or having hair of this colour. **ash-blonde**, *n.fem.* **ash-can**, *n.* (*N Am.*) a dustbin. **ash-coloured**, *a.* of a colour between brown and grey. **ash-pan**, *n.* a pan beneath a furnace or grate for the reception of ashes. **ashtray**, *n.* a small container for tobacco ash, cigarette butts etc. **Ash Wednesday**, *n.* the first day of Lent, so called from the Roman Catholic practice of sprinkling the foreheads of the people with ashes on that day. **ashy**, *a.* of or composed of ashes; covered with ashes; pale, ashen.

ash² (ash), *n.* a forest tree with grey bark, pinnate leaves and tough, close-grained wood; the wood of the ash-tree. *a.* made from ash.

ashamed (əshāmd'), *a.* affected with shame; abashed by consciousness of error or guilt. **ashamedly** (-mid-), *adv.*

ashen¹ (ash'ən), *a.* ash-coloured; very pale, whitish-grey.

ashen² (ash'ən), *a.* of or pertaining to the ash-tree; made of ash.

Ashkenazi (ash'kənahzi), *n.* (*pl.* **-zim** (-zim)) an E European or German Jew.

ashlar, ashler (ash'lə), *n.* square-hewn stone used in building; masonry built of this; thin masonry built as a facing to rubble or brick work. **ashlared**, *a.* covered with ashlar.

ashore (əshaw'), *adv.* to the shore; on the shore; on land.

ashram (ash'rəm), in India, a hermitage, place for reflection; a religious retreat.

ashy ASH¹.
Asian (ā'shən, -zhən), *a.* of, pertaining to or belonging to Asia or its people. *n.* (a descendant of) a native or inhabitant of Asia, esp. (in Britain) of the Indian sub-continent.
Asian flu, *n.* a severe type of influenza caused by a virus isolated during an epidemic in Asia in 1957.
Asiatic (āshiat'ik, -zhi-), *a.* Asian. **Asiatic cholera** CHOLERA.
aside (əsīd'), *adv.* at, to or towards one side; away. *n.* something spoken aside so as to be audible only to the person addressed, esp. by an actor, which the others on the stage are not supposed to hear; an indirect effort, a digression.
asinine (as'inīn), *a.* of, pertaining to or resembling asses; stupid, obstinate. **asininely**, *adv.* **asininity** (-nin'-), *n.* asinine behaviour, obstinate stupidity.
ASIO, (*abbr.*) Australian Security Intelligence Organization.
asitia (əsish'iə), *n.* pathological distaste for food, want of appetite.
ask (ahsk), *v.t.* to request; to seek to obtain by words; to solicit, to demand, to state (a price required); to question, to inquire of; to inquire concerning; to request to be informed about; to invite. *v.i.* to make a request, petition or demand; to inquire. **to ask for**, to behave in such a way as to invite (trouble etc.). **asking price**, *n.* the price set by a seller. **asker**, *n.*
askance (əskans'), **askant** (əskant'), *adv.* obliquely, sideways, askew; with mistrust, suspicion or disapproval.
askari (askah'ri), *n.* an E African native soldier.
askew (əskū'), *adv.* askance, asquint; in an oblique direction. *a.* oblique, awry, skew.
aslant (əslahnt'), *adv., a.* in a slanting or oblique direction. *prep.* across in a slanting direction.
asleep (əslēp'), *adv., pred.a.* in or into a state of sleep; (*euphem.*) dead.
ASLEF (az'lef), (*abbr.*) Associated Society of Locomotive Engineers and Firemen.
A/S level, *n.* an examination equivalent to half an A level. [Advanced Supplementary]
aslope (əslōp'), *a., adv.* in a sloping or oblique direction.
ASM, (*abbr.*) air-to-surface missile.
asocial (āsō'shəl), *a.* not social; antisocial.
asp¹ (asp), **aspic** (as'pik), *n.* a small venomous hooded serpent, *Naja haje*, the Egyptian cobra; a European viper, *Vipera aspis*; any venomous serpent.
asp² ASPEN.
asparagus (əspa'rəgəs), *n.* a culinary plant of the lily family, the tender shoots of which are eaten.
aspartame (əspah'tām), *n.* an artificial sweetener.
aspect (as'pekt), *n.* look, view; looking, way of looking; (*Astrol.*) the situation of one planet with respect to another; the direction in which something is turned, phase; appearance, expression. **aspect ratio**, *n.* the ratio of the width to the height of a television or cinema screen image; the ratio of the span of an aerofoil to its mean chord.
aspen (as'pən), **asp**, *n. Populus tremula*, the trembling poplar, remarkable for its quivering leaves.
asperge (əspœj'), *v.t.* to besprinkle, esp. with holy water. *n.* an aspergillum. **asperges** (-jēz), *n.* the sprinkling of the congregation with holy water by the celebrant of High Mass. **aspergillum** (aspəjil'əm), *n.* (*pl.* **-lla, -llums**) the brush used to sprinkle holy water.
Aspergillus (-jil'əs), *n.* a genus of fungi including many moulds that grow on decaying organic matter, named from their resemblance to the aspergillum.
asperity (əspe'riti), *n.* roughness of surface; a rugged excrescence; harshness of sound; severity, bleakness; harshness of manner, acrimony.
aspermia (āspœ'miə, ə-), *n.* total absence of semen.
aspermous, *a.* (*Bot.*) without seed; destitute of seed.
asperse (əspœs'), *v.t.* to scatter or strew upon, to besprinkle; to spread disparaging reports about, to defame. **aspersion** (-shən), *n.* calumny, slander, a false report or insinuation; the act of sprinkling. **aspersive, aspersory**, *a.* **aspersively**, *adv.* **aspersorium** (aspəsaw'riəm), *n.* the vessel from which holy water is sprinkled.
asphalt, asphalte (as'falt), *n.* mineral pitch, a dark brown or black form of bitumen; bituminous limestone, or an artificial substitute (often made with tar), used for roofing, road surfacing etc. *v.t.* to cover, surface or line with asphalt. **asphaltic** (-fal'-), *a.* pertaining to asphalt; consisting of or containing asphalt.
asphodel (as'fədel), *n.* a mythical undying flower, said to bloom in the Elysian fields; a plant of the liliaceous genus *Asphodelus*.
asphyxia (əsfik'siə, ās-), **asphyxy** (əsfik'si), *n.* lack of oxygen in the blood resulting in loss of consciousness or death; suffocation. **asphyxial, asphyxiant**, *a.* **asphyxiate**, *v.t., v.i.* to affect with or undergo asphyxia; to suffocate. **asphyxiation**, *n.* **asphyxiator**, *n.*
aspic¹ ASP¹.
aspic² (as'pik), *n.* a savoury jelly used as a garnish or in which game, hard-boiled eggs, fish etc., may be embedded; a dish of meat etc. moulded in aspic.
Aspidistra (aspidis'trə), *n.* a liliaceous genus of plants often grown as house plants.
aspirant ASPIRE.
aspirate (as'pirāt), *v.t.* to pronounce with a full breath; to prefix the letter *h* or its equivalent; to draw out (gas) from a vessel; to remove (blood etc.) by suction. *a.* aspirated; pronounced with audible breathing. *n.* a letter pronounced with the sound of *h*. **aspiration**, *n.* the act of breathing; the act of aspirating; an aspirated sound. **aspirator**, *n.* one who or that which aspirates; an instrument for drawing air or gas through a tube; (*Med.*) an instrument for evacuating a cavity by means of an exhausted receiver.
aspire (əspīə'), *v.i.* to long, desire eagerly; to seek to attain; (*fig.*) to rise, to mount up. **aspirant** (as'pi-), *a.* aspiring, aiming at a higher position. *n.* one who aspires; a candidate. **aspiration** (aspi-), *n.* the act of aspiring; steadfast desire; a seeking for better things. **aspiring**, *a.* eagerly desirous of some high object, ambitious. **aspiringly**, *adv.* **aspiringness**, *n.*
aspirin (as'prin), *n.* (a tablet containing) acetylsalicylic acid, used as a pain-killer.
asplenium (əsplē'niəm) SPLEENWORT under SPLEEN.
ass¹ (as), *n.* (*pl.* **asses**) a quadruped, *Equus asinus*, allied to the horse, but of smaller size, with long ears and a tufted tail; (also *pron.* ahs) a stupid, obstinate person. **to make an ass of**, to render ridiculous, make appear foolish.
ass² ARSE.
assafetida (asəfet'idə, -fē'-), ASAFOETIDA.
assagai, assegai (as'əgī), *v.t., n.* to wound or kill with a slender lance of hard wood, chiefly applied to the missile lances of the southern African tribes.
assai (əsī'), *adv.* (*Mus.*) very; as *largo assai*, very slow.
assail (əsāl'), *v.t.* to attack violently by physical means or with argument, abuse, entreaty, temptation, snares and the like; to dash against; to approach with intent to overcome; **assailable**, *a.* **assailant**, *a.* assailing, attacking. *n.* one who assails.
assassin (əsas'in), *n.* one of a body of Muslim fanatics, in the time of the Crusades, who were sent forth to murder secretly the Christian leaders; one who kills by surprise or secret assault (generally for money or for fanatical, political etc. motives). **assassinate**, *v.t.* to kill by surprise or secret assault; to murder (as a political or religious leader) by sudden violence; to injure or destroy (a person's character or reputation). **assassination**, *n.* **assassinator**, *n.*
assault (əsawlt'), *n.* the act of assailing; a violent attack with material or immaterial weapons; the charge of an

assay 44 associate

attacking body on a fortified post; (*Law*) a threatening word or act; an attempt at rape. *v.t.* to make an assault on. **assault and battery,** *n.* (*Law*) an assault with action as well as words. **assault course,** *n.* an obstacle course used for training soldiers. **assaultable,** *a.*

assay (as'ā, əsā'), *n.* a trial, examination; the scientific determination of the quantity of metal in an ore, alloy, bullion or coin; the chemical analysis of a substance to determine its content; a metal or other substance analysed. *v.t.* (as'ā), to try, to test; to determine the amount of metal in (an ore, alloy, bullion or coin); to subject to chemical analysis; to try to do, attempt. **assayer,** *n.*

assegai ASSAGAI.

assemblance (əsem'bləns), *n.* appearance, show.

assemble (əsem'bl), *v.t.* to call together; to bring together into one mass or heap; to fit together the component parts of. *v.i.* to meet or come together. **assemblage** (-blij), *n.* a gathering, assembling; a collection. **assembler,** *n.* one who, or that which, assembles; a computer program that automatically translates assembly language into machine code; assembly language. **assembly,** *n.* the act of assembling; the state of being assembled; a body of people met together for some common purpose; a deliberative, legislative or religious body; a lower house in some legislatures; (*Mil.*) the second beat of the drum summoning soldiers to prepare to march; the conversion of assembly language into machine code. **assembly language,** *n.* a low-level computer language in which instructions written in mnemonics correspond directly to instructions in machine code. **assembly line,** *n.* a serial collection of workers and machines operating from stage to stage in assembling some product. **assemblyman,** *n.* a member of a legislative assembly. **assembly-room,** *n.* a room in which public assemblies, balls, concerts etc. are held.

assent (əsent'), *v.i.* to agree to or sanction something proposed. *n.* sanction; agreement, acquiescence. **assentation** (as-), *n.* assenting, esp. with obsequiousness or servility. **assentient** (-shənt), *a.* assenting to. *n.* one who assents or agrees. **assentingly,** *adv.* **assentor,** *n.* one who gives assent, esp. one who signs the nomination of a Parliamentary candidate after the proposer and seconder.

assert (əsœt'), *v.t.* to affirm, to declare positively; to maintain; to insist on (a claim, right etc.); to put (oneself) forward, insist on one's rights etc. **assertable,** *a.* **assertion,** *n.* the act of asserting; a positive statement, an affirmation. **assertive,** *a.* characterized by assertion, dogmatic. **assertively,** *adv.* **assertiveness,** *n.* **assertor,** *n.*

assess (əses'), *v.t.* to fix by authority the amount of (a tax, fine etc. for a person or community); to value (property, income etc.) for the purpose of taxation; to value; to evaluate, to judge. **assessable,** *a.* capable of being assessed; liable to be assessed. **assessably,** *adv.* **assessment,** *n.* the act of assessing; the amount assessed; a scheme of rating or taxation; an official valuation for those purposes; (an) evaluation, appraisal (e.g. of a person's work). **assessment centre,** *n.* a centre where juvenile offenders are sent whilst their situation is assessed. **assessor,** *n.* one who makes an assessment; one who sits near and advises a judge or magistrate on technical points, commercial usage, navigation etc.; one who evaluates insurance claims. **assessorship,** *n.*

assets (as'ets), *n.pl.* goods sufficient to satisfy a testator's debts and legacies; property or effects that may be applied for this purpose; the effects of an insolvent debtor; all the property of a person or company which may be liable for outstanding debts; property in general. **asset,** *n.* a useful or valuable resource. **asset-stripping,** *n.* the practice of buying a company and selling off its assets to make a profit.

asseverate (əsev'ərāt), *v.t.* to affirm with solemnity.

assiduous (əsid'ūəs), *a.* constant in application; diligent.

assiduously, *adv.* **assiduousness,** *n.* **assiduity** (asidū'-), *n.* constant or close application to the matter in hand, diligence; (*pl.*) persistent endeavours to please.

assign (əsin'), *v.t.* to allot, to apportion; to designate for a specific purpose; to name, to fix; to point out, to ascribe, to attribute; (*Law*) to transfer, to surrender. *n.* one to whom a property or right is transferred. **assignable,** *a.* **assignation** (asig-), *n.* (the appointment of a particular time or place for) a meeting, esp. an illicit one between lovers; the act of assigning. **assignee** (-nē'), *n.* an agent, a representative; (*Law*) one to whom a right or property is transferred. **assignment,** *n.* the act of assigning; something assigned; a specific task or mission; a position or job to which one is assigned; a legal transference of right or property; the instrument by which such transference is effected; the right or property transferred. **assignor** (-naw'), *n.* one who transfers a right or property.

assimilate (əsim'əlāt), *v.t.* to make similar or alike; to take as nutriment and convert into living tissue, to incorporate in the substance of an organism; to take in (information) and comprehend; to absorb into a population or group; to incorporate; to adapt (a speech sound) so as to resemble an adjacent sound. *v.i.* to become similar; to be incorporated in the substance of a living organism; to become absorbed or incorporated. **assimilable,** *a.* capable of being assimilated. **assimilability** (-bil'-), *n.* **assimilation,** *n.* **assimilative** (-lā-), *a.* having the power of assimilating. **assimilator,** *n.* **assimilatory** (-lə-), *a.* tending to assimilate.

assist (əsist'), *v.t.* to help, to give support or succour to; to act as a subordinate to. *v.i.* to give help or aid; to be present (at). **assisted reproduction,** *n.* a collective term for all methods of human fertilization and reproduction involving medical intervention, e.g. gamete intra-fallopian transfer, artificial insemination. **assistance,** *n.* help, support. **assistant,** *a.* aiding, helping, auxiliary. *n.* one who assists another; an auxiliary. **assistor,** *n.* (*Law*) an assistant.

assize (əsīz'), *n.* a trial in which sworn assessors decide questions of fact; an action so tried or decided; (*pl.*) from 1815 to 1971, the sessions held periodically by the judges of the Supreme Court in each county in England for the administration of civil and criminal justice. **assizer,** *n.* (*Sc. Law*) a jury member.

assoc., (*abbr.*) associate, associated, association.

associate (əsō'shiət, -sō'si-), *v.t.* to join, to unite, to combine, to connect; to bring together in the mind; to connect (oneself) as a partner, supporter, friend, companion etc. *v.i.* to unite or combine for a common purpose; to keep company or mix (with). *a.* (-ət), connected, joined; confederate, allied; in the same group or category; having subordinate status. *n.* (-ət), a confederate, an ally; a partner, colleague, coadjutor; a person having partial membership or a subordinate status in an association or institution; something generally found with another. **associateship,** *n.* **associable,** *a.* capable of being (mentally) associated. **associability** (-bil'-), *n.* liable. **association,** *n.* the act of combining for a common purpose; a society formed for the promotion of some common object; fellowship, intimacy, connection; mental connection of ideas, memories, sensations etc.; a memory, thought or feeling connected with some object, place etc. and recalled to the mind in connection therewith; a plant community growing in a uniform habitat and forming part of a larger ecological unit; the formation of loosely held aggregates of molecules, ions etc. **deed of association,** a legal instrument in which the particulars of a limited liability company must be recorded on its formation. **Association Football,** *n.* football played between two teams of eleven players, with a round ball which may not be touched with the hands except by the goalkeepers. **associational,** *a.* **associationism,** *n.* (*Psych.*) the theory which accounts for mental and moral phenomena by association of ideas. **associationist**, *n*. **associative** (-ə-), *a*. tending to associate.

assonant (as'ənənt), *a*. corresponding in sound; rhyming in the accented vowels, but not in the consonants; also, correspondence of consonant sounds with different vowels. *n*. an assonant word. **assonance**, *n*.

assort (əsawt'), *v.t*. to arrange or dispose in sorts or lots; to arrange into different classes; to furnish with articles so arranged. *v.i*. to suit, to agree, to match; to be in congruity or harmony. **assorted**, *a*. arranged in sorts; of various sorts. **assortment**, *n*. a collection of things assorted; a collection of things of various kinds; the act of assorting; the state of being assorted.

asst., (*abbr*.) assistant.

assuage (əswāj), *v.t*. to sweeten, allay, mitigate; to soothe, to lessen the violence of; to appease, satisfy. **assuagement**, *n*. **assuasive**, *a*. assuaging, mitigating, soothing.

assume (əsūm'), *v.t*. to take to oneself; to receive, adopt; to take upon oneself, to put on, to undertake; to arrogate, appropriate; to take for granted; to pretend, feign. *v.i*. to be arrogant or pretentious; to claim more than is one's due. **assumed**, *a*. usurped, pretended; feigned, false; taken for granted. **assumedly** (-mid-), *adv*. **assuming**, *a*. arrogant, haughty.

assumption (əsŭmp'shən), *n*. the act of assuming; the thing assumed; a supposition, a postulate; arrogance; (**Assumption**) ascent to heaven, esp. the reception of the Virgin Mary into heaven; the feast (15 Aug.) in honour of this event. **assumptive**, *a*. assumed; taken to oneself; taken for granted.

assure (əshuə'), *v.t*. to make safe, secure or certain; to give confidence to, to encourage; to tell positively; to ensure, guarantee; to insure the payment of compensation in case of loss of (esp. life), to insure. **assurance**, *n*. the act of assuring; positive declaration; certainty, security; self-reliance, intrepidity; audacity, impudence; insurance, esp. a contract to pay a given sum on a person's death in return for an annual premium; (*Law*) evidence of the conveyance of property. **assured**, *a*. safe; made certain; confident, convinced; self-confident, full of assurance; insured. *n*. one whose life is insured; the beneficiary of an assurance policy. **assuredly** (-rid-), *adv*. **assuredness** (-rid-), *n*. **assurer**, *n*. one who or that which gives assurance; an insurer, an underwriter; one who takes out a policy of assurance. **assuror**, *n*. (*Law*) an assurer, an underwriter.

assurgent (əsœ'jənt), *a*. rising, ascending; rising aggressively; (*Bot*.) pointing upwards; rising upwards in a curve.

Assyrian (əsi'riən), *a*. of or pertaining to the ancient kingdom of Assyria. *n*. a native of Assyria; the language of Assyria.

AST, (*abbr*.) Atlantic Standard Time.

astable (āstā'bl), *a*. not stable; of an electrical circuit, switching between two states.

astatic (əstat'ik, ā-), *a*. not remaining fixed; not influenced by the earth's magnetism.

astatine (as'tətēn), *n*. a radioactive element, at.no. 85; chem. symbol At, formed in minute amounts by radioactive decay or made artificially.

Aster (as'tə), *n*. a genus of compositous plants with showy, daisy-like heads; a star-shaped figure seen in a cell during mitosis. **China aster** CHINA.

-aster, *suf*. one who is (somewhat) after the manner of; e.g. *criticaster, poetaster*.

asteriated (əstiə'riātid), *a*. exhibiting asterism.

asterisk (as'tərisk), *n*. a mark (*) used in printing to call attention to a note, to mark omission etc. *v.t*. to mark with an asterisk.

asterism (as'tərizm), *n*. a constellation; a small cluster of stars; three asterisks placed thus (*⁎*) to draw attention to something important; the star-like figure visible in some mineral crystals, as in the asteriated sapphire.

astern (əstœn'), *adv*., *a*. in, at or towards the stern of a ship, behind a ship; in the rear, behind.

asteroid (as'təroid), *a*. having the figure or appearance of a star; resembling a starfish. *n*. any of the small celestial bodies that orbit the sun, esp. between the orbits of Mars and Jupiter, a planetoid, a minor planet. **asteroidal** (-roi'-), *a*.

asthenia (əsthē'niə), *n*. absence of strength; debility, diminution or loss of vital power. **asthenic** (-then'-), *a*.

asthma (as'mə), *n*. a disorder of respiration characterized by wheezing, constriction of the chest, and usu. coughing. **asthmatic** (-mat'-), *a*. of, pertaining to, affected with or good for asthma. *n*. a person affected with asthma. **asthmatical**, *a*. **asthmatically**, *adv*.

astigmatism (əstig'mətizm), *n*. a defect of the eye or of a lens as a result of which a point source of light tends to be focused as a line. **astigmatic** (astigmat'ik), *a*. of, pertaining to, characterized by or correcting for astigmatism.

Astilbe (əstil'bi), *n*. a genus of perennial plants with pink or white flower spikes.

astir (əstœ'), *a*. in motion; in commotion, in excitement; out of bed.

Asti spumante (as'tē spooman'tē, spū-), *n*. a sparkling white wine made at Asti, NW Italy.

astonish (əston'ish), *v.t*. to strike with sudden surprise or wonder, to amaze. **astonishing**, *a*. **astonishingly**, *adv*. **astonishment**, *n*. amazement; a cause of amazement.

astound (əstownd'), *v.t*. to strike with amazement; to shock with alarm, wonder, or surprise. **astounding**, *a*. **astoundingly**, *adv*. **astoundment**, *n*.

astraddle (əstrad'l), *adv*. in a straddling position; astride.

astragal (as'trəgəl), *n*. the astragalus; (*Arch*.) a small semicircular moulding or bead, as round the top or the bottom of a column. **astragalus** (əstrag'-), *n*. the ball of the ankle-joint; the bone which the tibia articulates below.

astrakhan (astrəkan', -kahn'), *n*. the tightly curled, usu. black or grey fleece obtained from lambs orig. from Astrakhan; a fabric with a pile in imitation of this.

astral (as'trəl), *a*. of or pertaining to the stars; starry; star-shaped; pertaining to one's astral body or to the material of which the astral body is composed. *n*. an astral-lamp. **astral-body**, *n*. a kind of spiritual body which occultists claim to be able to project to a distance, and so to exercise the power of bilocation; the ethereal or spiritual body round which the physical body is built up, the spirit.

astray (əstrā'), *adv*., *pred. a*. out of or away from the right way.

astride (əstrīd'), *adv*., *pred. a*. in a striding position; with legs on either side. *prep*. astride of.

astringent (əstrinj'ənt), *a*. causing contraction of body tissues; styptic; stern, severe, harsh; sharp, acid, pungent. *n*. an astringent substance. **astringency**, *n*. **astringently**, *adv*.

astro-, *comb. form*. pertaining to the heavenly bodies, planets or stars; e.g. *astrology*, *astronomy*.

astrobiology (astrōbīol'əji), *n*. a branch of biology concerned with the search for life beyond the earth.

astrodome (as'trədōm), *n*. a dome window in an aircraft to enable astronomical observations to be made; a large sports stadium covered by a translucent domed roof.

astrol., (*abbr*.) astrologer, astrology.

astrolabe (as'trəlāb), *n*. an instrument formerly used in astrology and astronomical observations for taking altitudes.

astrology (əstrol'əji), *n*. a spurious science that professes to establish a connection between the changing aspects of the heavenly bodies and the changing course of human life, thence claiming to predict events and to be competent to advise on life's conduct. **astrologer**, *n*. one versed

in astrology. **astrological** (astrəloj′-), **-logic**, *a.* **astrologically**, *adv.*
astron., *(abbr.)* astronomer, astronomy.
astronautics (astrənaw′tiks), *n. sing.* the science of aerial navigation. **astronaut** (as′-), *n.* one who travels into space beyond the earth's atmosphere in a suitable projectile. **astronautical**, **-nautic**, *a.*
astronomy (əstron′əmi), *n.* the science which studies all the phenomena of the heavenly bodies. **astronomer**, *n.* one who studies or is versed in astronomy. **astronomer-royal**, *n.* the officer in charge of a royal or national observatory in Great Britain. **astronomic** (astrənom′-), **-ical**, *a.* of or pertaining to astronomy; enormously large or great. **astronomical clock**, *n.* a pendulum clock which gives sidereal time. **astronomical unit**, *n.* a unit of length equal to the mean distance of the earth from the sun, about 93 million miles (150 million km), used for measuring distances in the solar system. **astronomical-year**, *n.* a year determined by astronomical observations, as opposed to a civil year. **astronomically**, *adv.*
astrophotometer (aströfətom′itə), *n.* an instrument for measuring the intensity of sidereal light.
astrophysics (astrōfiz′iks), *n. sing.* the study of stellar physics. **astrophysical**, *a.* **astrophysicist**, *n.*
Astroturf® (as′trōtœf), *n.* an artificial grass surface, esp. for sports fields.
astute (əstūt′), *a.* acute, discerning, shrewd; clever, wily, cunning. **astutely**, *adv.* **astuteness**, *n.*
astylar (əstī′lə), *a.* without columns or pilasters.
asunder (əsun′də), *adv.* apart, separately, in different pieces or places.
asylum (əsī′ləm), *n.* a place of refuge for criminals and debtors, a sanctuary; an institution affording relief and shelter to the afflicted, unfortunate or destitute, esp. (formerly) an institution for the treatment of the mentally ill; protection from extradition given by one country to a person, esp. a political refugee, from another; (a) shelter, (a) refuge.
asymmetry (əsim′ətri), *n.* want of symmetry, or of proportion; *(Math.)* incommensurability. **asymmetric** (asimet′-, ā-), **-ical**, *a.* **asymmetrically**, *adv.*
asymptomatic (əsimptəmat′ik, ā-), *a.* not exhibiting symptoms of disease.
asymptote (as′imtōt), *n.* a straight mathematical line continually approaching some curve but never meeting it within a finite distance. **asymptotic** (-tot′-), **-ical**, *a.*
asynchronous (əsing′krənəs, ā-), *a.* not coincident in point of time. **asynchronism**, **-chrony**, *n.*
asyndeton (əsin′ditən), *n.* a rhetorical figure by which the conjunction is omitted, as 'I came, I saw, I conquered'.
At, *(chem. symbol)* astatine.
at (at), *prep.* denoting nearness or precise position in time or space; denoting position or situation as regards occupation, condition, quality or degree, effect, relation, value; denoting direction to or towards. **at it**, at work, engaged, busy. **at that**, moreover.
at., *(abbr.)* atomic.
at-, *pref.* AD-, assim. to *t*, e.g. *attain, attend.*
ataraxia (atərak′siə), **ataraxy** (at′-), *n.* impassiveness, calmness, indifference, stoicism. **ataractic** (-rak′-), *a.* calming, tranquillizing. *n.* a tranquillizing drug.
atavism (at′əvizm), *n.* recurrence of some characteristic of a more or less remote ancestor; recurrence of a disease after the lapse of some generations; reversion to a primitive or ancestral form. **atavistic** (-vis′-), *a.*
ataxia (ətak′siə), **ataxy** (ətak′si, at′-), *n.* loss of the power of co-ordination of the muscles, resulting in irregular, jerky movements; disorder, confusion. **locomotor ataxia** LOCOMOTION. **ataxic**, *a.*
ATB[1], *n.* a bicycle with wide tyres and a large range of gears, designed for riding off the road. [*all-terrain bicycle*]
ATB[2] *(abbr.)* advanced technology bomber (aircraft).

ATC, *(abbr.)* air-traffic control; Air Training Corps.
ate (et, āt), *pret.* EAT.
-ate (-āt, -ət), *suf.* (1) forming nouns of office or function, e.g. *curate, episcopate, aldermanate;* participial nouns, e.g. *delegate, mandate;* chemical names for salts of acids, e.g. *acetate, carbonate;* (2) forming participial adjectives, e.g. *desolate, situate* (cp. *desolated, situated,* in which the p.p. gives rise to a causative verb); and other adjectives formed by analogy, e.g. *roseate, ovate;* (3) (-āt) forming verbs, e.g. *desolate, separate,* corresponding to adjectives in same form, or others produced on the same model, e.g. *fascinate, isolate, felicitate.*
ateleo-, atelo-, *comb. form* indicating incomplete development or imperfection of structure.
atelier (ətel′yā, at′-), *n.* a workshop, an artist's studio.
a tempo (a tem′pō), *adv., a.* in the original tempo or time.
Athanasian (athənā′shən, -zhən), *a.* of or pertaining to Athanasius, bishop of Alexandria AD 326. *n.* a follower of Athanasius; one holding his views with respect to the Trinity. **Athanasian creed**, *n.* a creed stating the doctrine of the Trinity and the Incarnation, with damnatory clauses, formerly attributed to Athanasius.
atheism (ā′thiizm), *n.* disbelief in the existence of a God or gods. **atheist**, *n., a.* **atheistic** (-is′-), **-ical**, *a.* **atheistically**, *adv.*
Athenaeum (athənē′əm), *n.* the temple of Athene in ancient Athens, where professors taught and orators and poets declaimed; hence (**athenaeum**) a literary or scientific club or institution.
Athenian (əthē′niən), *n., a.* (a native or inhabitant) of Athens.
athermancy (athœ′mənsi), *n.* the power of stopping radiant heat. **athermanous**, *a.* impermeable by radiant heat.
atheroma (athərō′mə), *n.* the deposition of fatty material on the inner coat of the arteries. **atheromatosis** (-tō′sis), *n.* **atheromatous**, *a.*
atherosclerosis (athərōsklərō′sis), *n.* arteriosclerosis characterized by deposits of fatty material in the arteries. **atherosclerotic** (-rot′-), *a.*
athetosis (athətōsis), *n.* a type of cerebral palsy characterized by constant involuntary movement. **athetoid**, *a.* pertaining to or affected by athetosis. *n.* an athetoid spastic.
athlete (ath′lēt), *n.* one trained to perform feats of strength and activity; esp. one trained to compete in events, as running, weight-throwing and jumping, requiring strength, agility, speed or stamina; a powerful, vigorous person. **athlete's foot**, *n.* a fungal infection of the foot. **athletic** (-let′-), *a.* of or for athletes or athletics; physically strong and active; muscular, robust. **athletic support**, *n.* a jockstrap. **athletically**, *adv.* **athleticism**, *n.* the practice of athletics; devotion (esp. excessive) to athletics; the state of being athletic. **athletics**, *n. sing.* the practice of physical exercises by which muscular strength is developed; the type of competitive sporting events engaged in by athletes.
-athon, *suf.* denoting an event or contest that continues for a long time, e.g. *talkathon, danceathon.*
athwart (əthwawt′), *prep.* from side to side of, across; against, opposing. *adv.* transversely, from side to side; so as to thwart. **athwart-ships**, *adv.* from side to side of the ship.
-atic, *suf.* forming adjectives, e.g. *aquatic, fanatic, lunatic.*
-ation (-ā′shən), *suf.* forming abstract nouns from verbs, e.g. *agitation, appreciation, ovation.*
-ative (-ətiv, -ā-), *suf.* forming adjectives, e.g. *demonstrative, representative, talkative.*
Atlantean[1] (ətlan′tiən), *a.* of or like the Titan Atlas who held up the heavens, very strong.
Atlantean[2] (ətlan′tiən), *a.* of Atlantis.
atlantes ATLAS.
Atlantic (ətlan′tik), *n.* the ocean between Europe and

Africa in the E and America in the W. *a*. of or pertaining to the Atlas mountains in N Africa; of or occurring in or near the Atlantic Ocean. **Atlantic Charter**, *n*. a joint declaration by Great Britain and the US, signed in 1941, laying down 'certain principles as a basis for a better future for the world'. **Atlantis** (ətlan'tis), *n*. the legendary island in the West whose site is occupied by the Atlantic Ocean.
atlas (at'ləs), *n*. a collection of maps in a volume; a collection of charts or plates in a volume; a large size of drawing paper; (*pl.* **atlantes** (-lan'tēz)) a colossal statue of a man used to support an entablature; the first cervical vertebra, on which the skull is supported. **atlas-beetle**, *n*. a large lamellicorn beetle (*Chalcosoma atlas*) from the East. **atlas-moth**, *n*. *Attacus atlas*, a large moth from China. [Gr. *Atlas -antos*, a Titan, fabled to hold up the pillars of the universe]
ATM, (*abbr*.) automated teller machine.
atm., (*abbr*.) atmosphere; atmospheric.
atman (aht'mən), *n*. in Hinduism, the innermost self, the soul.
atm(o)-, *comb. form.* pertaining to vapour or to the atmosphere.
atmolysis (atmol'əsis), *n*. the separation of gases in combination.
atmosphere (at'məsfiə), *n*. the gaseous envelope of any of the celestial bodies; that surrounding the earth; a gaseous envelope surrounding anything; the air in any given place; a unit of pressure corresponding to the average pressure of the earth's atmosphere at sea level and equal to a pressure of about 15 lb/sq. in. (101,325 N/m^2); mental or moral environment; a prevailing emotional etc. mood. **atmospheric** (-fe'-), **-ical**, *a*. of or pertaining to the atmosphere; of the nature of air; existing in the atmosphere, or produced by the atmosphere. **atmospherically**, *adv.* **atmospherics**, *n.pl.* (audible radio interference produced by) electromagnetic waves generated by an electric discharge between two clouds or from a cloud to earth.
at. no., (*abbr*.) atomic number.
atoll (at'ol), *n*. a coral island, consisting of an annular reef surrounding a lagoon.
atom (at'əm), *n*. the smallest conceivable portion of anything; a body or particle of matter originally thought to be incapable of further division; the smallest particle taking part in chemical action, the smallest particle of matter possessing the properties of an element. **atom bomb**, *n*. a bomb in which the explosion is due to atomic energy released when atoms of uranium, plutonium etc. undergo nuclear fission. **atom smasher**, *n*. (*coll*.) an accelerator for increasing the energy of charged particles. **atomic** (ətom'-), *a*. consisting of separate atoms; pertaining to an atom or atoms; pertaining to or using atomic energy or atom bombs. **atomic bomb** ATOM BOMB. **atomic clock**, *n*. an electronic apparatus which makes use of molecular or atomic resonances to generate precise intervals of time. **atomic energy**, *n*. nuclear energy. **atomic mass unit**, a unit of mass equal to 1/12 of the mass of an atom of carbon-12. **atomic number**, *n*. the number of protons in the nucleus of an atom. The atomic number determines the chemical properties of an atom. **atomic pile**, *n*. a nuclear reactor. **atomic theory**, *n*. the theory that all matter is composed of atoms. **atomic warfare**, *n*. nuclear warfare. **atomic weight**, *n*. the weight of an atom of an element expressed on a scale in which the weight of an atom of carbon-12 is 12. **atomically**, *adv.* **atomicity** (-is'-), *n*. the number of atoms in a molecule of an element or of a compound; the combining capacity of an element or radical, valency. **atomize**, **-ise**, *v.t.* to reduce to atoms; to reduce to fine particles or to a spray. **atomization**, *n*. **atomizer**, *n*. an instrument for reducing a liquid, as a disinfectant or perfume, into spray.
atonal (ātō'nəl), *a*. (*Mus*.) without a fixed key. **atonality** (-nal'-), *n*.
at one (ət wŭn'), *adv*. in harmony, at one, in a state of reconciliation; of the same opinion.
atone (ətōn'), *v.i.* to make expiation or satisfaction for some crime, sin or fault. **atonable**, *a*. **atonement**, *n*. the act of atoning; reparation, expiation, amends, reconciliation; the propitiation of God by the expiation of sin; the Redemption. **atoningly**, *adv*.
atonic (əton'ik), *a*. without an accent, unaccented; lacking physiological or muscular tone. *n*. an unaccented word in Greek. **atony** (at'-), *n*. lack of physiological, esp. muscular, tone; enervation; lack of intellectual energy.
atop (ətop'), *adv*. on or at the top. **atop of**, *prep.* on or at the top of.
-ator (-ātə), *suf.* -OR, e.g. *commentator*.
-atory (-ətəri, -ātəri), *suf.* -ORY, forming adjectives, e.g. *commendatory*.
ATP, (*abbr*.) adenosine triphosphate.
atrabilious (atrəbil'yəs), *a*. (*formerly*) of or affected by black bile; melancholic, hypochondriacal; splenetic, bitter-tempered. **atrabiliousness**, *n*.
atrip (ətrip'), *pred. a.* of an anchor, just drawn out of the ground at right angles to it; of the top sails, hoisted as high as possible on the masts.
atrium (at'riəm, ā'-), *n*. (*pl.* **atria**, **atriums**) the court or portico in an ancient Roman house; a covered court or portico; a body cavity; esp. either of the two upper chambers of the heart into which the veins pour the blood. **atrial**, *a*.
atrocious (ətrō'shəs), *a*. savagely and wantonly cruel, characterized by heinous wickedness; appalling, shocking; very bad, execrable. **atrociously**, *adv*. **atrocity** (ətros'-), *n*. excessive cruelty or other flagrant wickedness; an atrocious act; a bad blunder; a barbarism.
atrophy (at'rəfi), *n*. a wasting of the body, or (one of) its organs, through want of nourishment or disease; mental or spiritual starvation. *v.t.* to affect with atrophy, to cause to waste away. *v.i.* to waste away. **atrophied**, *a*.
atropine (at'rəpēn, -pin), *n*. an organic base obtained from deadly nightshade. **atropism**, *n*. atropine poisoning.
att., (*abbr*.) attorney.
attaboy (at'əboi), *int*. (*chiefly N Am.*, *coll.*) an exclamation of encouragement.
attach (ətach'), *v.t.* to fasten on, connect; to affix; to lay hold on, arrest, indict, esp. to seize (a person or goods) by a writ of attachment; to appoint to an organization, military, police etc. unit, temporarily; to join to in sympathy or feelings; to attract and cause to adhere to oneself; to attribute. *v.i.* to adhere; (*Law*) to apply. **attachable**, *a*. **attached**, *a*. **attachment**, *n*. the act of attaching; the means by which anything is attached; connection; fidelity, affection, devotion; the thing attached; (*Law*) apprehension, esp. for contempt of court; the seizure of goods or estate to secure a debt or demand; the writ or precept by which such apprehension or seizure is effected.
attaché (ətash'ā), *n*. one attached to the suite of an ambassador. **attaché case**, *n*. a leather case for carrying papers etc.
attack (ətak'), *v.t.* to fall upon with force; to assault; to assail by hostile words, writings etc.; to begin (a work) with determination; of a physical agent, disease etc., to exert a destructive influence on; to take offensive action against. *v.i.* to make an attack; to take offensive action in a game or sport. *n*. the act of attacking; an onset, an assault; violent abuse or injury; the beginning of active work on something; a fit of illness; the commencement of destructive action; a (crisp and decisive) manner of beginning a musical piece or passage; an offensive or scoring move in a game or sport; the players in a team who attack. **attackable**, *a*.
attain (ətān'), *v.i.* to arrive at some object. *v.t.* to reach,

attainder / **au courant**

gain; to 'arrive at; to accomplish. **attainable**, *a.* **attainability** (-bil'-), *n.* **attainableness**, *n.* **attainment**, *n.* the act of attaining; that which is attained; a personal acquirement.

attainder (ətān'də), *n.* the forfeiture of civil rights as the legal consequence of a sentence of death or outlawry for treason or felony; an act or bill of attainder; condemnation.

attar (at'ə), *n.* a fragrant essence, or essential oil, esp. of roses.

attempt (ətempt', ətemt'), *v.t.* to try, endeavour to do, achieve, effect etc. *n.* an endeavour, effort, undertaking; an effort as contrasted with attainment; an assault (on life, honour etc.). **attemptable**, *a.* **attemptability** (-bil'-), *n.*

attend (ətend'), *v.t.* to apply the mind to; to accompany, escort; to look after, wait upon; to be present at; to go regularly to (church, a school etc.). *v.i.* to pay attention, apply the mind; to apply one's efforts; to be present; to be in attendance; to wait upon or for a person. **attendance**, *n.* the act of attending; service, presence; (the number of) persons attending. **in attendance**, waiting, attendant on. **attendance centre**, *n.* a centre at which young offenders attend regularly, as an alternative to a prison sentence. **attendant**, *a.* accompanying, waiting on, ministering to; following as a consequence; present. *n.* one who, or that which, attends or accompanies; a servant.

attention (əten'shən), *n.* the act or state of attending; the mental faculty of attending; (*usu. pl.*) an act of courtesy, kindness or love; watchful care, close observation, notice; a military attitude of readiness.

attentive (əten'tiv), *a.* heedful, intent, regardful; polite, courteous. **attentively**, *adv.* **attentiveness**, *n.*

attenuate (əten'ūāt), *v.t.* to make thin or slender; to dilute, diminish the density of; to reduce the strength, intensity or force of; to weaken; to extenuate. *v.i.* to become thin or weak. **attenuate** (-ət), *a.* slender; tapering; thin in consistency. **attenuated**, *a.* **attenuation**, *n.* the act of attenuating; diminution of thickness, density, strength or force; reduction in strength of radiation as it passes through the medium between the source and destination. **attenuator**, *n.* a circuit to provide attenuation of the current, voltage or power of a signal.

attest (ətest'), *v.t.* to testify, esp. in a formal manner; to vouch for; to affirm to be true or valid; to put (a person) on oath or solemn declaration. *v.i.* to bear witness. *n.* evidence, attestation. **attestable**, *a.* **attestation** (at-), *n.* the act of attesting; evidence, proof; formal confirmation or verification; the administration of an oath, esp. of allegiance. **attested**, *a.* certified as being free from the tuberculosis bacillus. **attestor**, *n.* one who attests or vouches for.

Attic (at'ik), *a.* of or belonging to Attica or its capital, Athens; classical, refined. *n.* a native of Attica; an Athenian; the Attic dialect. **Attic salt** or **wit**, *n.* refined, delicate wit, for which the Athenians were famous. **Atticism** (-sizm), *n.* attachment to Athens; idiom and style characteristic of Attic Greek; concise and elegant expression.

attic (at'ik), *n.* a low storey placed above an entablature or cornice; (a room in) the top storey of a house; (*sl.*) the head, the brain.

attire (ətīə'), *v.t.* to dress; to array in apparel. *n.* dress, clothes. **attiring**, *n.* dress, apparel, trappings.

attitude (at'itūd), *n.* the posture in which a figure is represented in painting or sculpture; bearing or gesture, expressing action or emotion; a mental position or mood as regards someone or something; posture or disposition of a person, animal or object; behaviour indicating opinion and sentiment; the position of an aircraft or spacecraft in relation to a plane of reference. **to strike an attitude**, to assume an exaggerated or theatrical attitude. **attitude of mind**, habitual mode of thinking and feeling. **attitudinal** (-tū'din-), *a.* **attitudinize** (-tū'din-), **-ise**, *v.i.* to practise or assume attitudes; to pose; to behave or act affectedly.

attn., (*abbr.*) attention, for the attention of.

atto- (atō-), *pref.* a million million millionth part of, 10^{-18}.

attorney (ətœ'ni), *n.* a legally authorized agent or deputy; formerly, a qualified practitioner in the Common Law courts, who prepared the case for the barristers or counsel, as distinguished from a solicitor who practised in a court of equity, the title is now Solicitor of the Supreme Court; (*N Am.*) a lawyer, a barrister, a solicitor, esp. one qualified to act for another in legal proceedings. **District Attorney** DISTRICT. **power, letter, warrant of attorney**, a written authority by which one person authorizes another to act in his or her stead. **Attorney-General**, *n.* (*pl.* **Attorneys-General, Attorney-Generals**) the functionary whose duty it is to transact all legal business in which the State is a party. **Attorney-Generalship**, *n.* **attorneyship**, *n.*

attract (ətrakt'), *v.t.* to draw to or cause to approach (in a material or immaterial sense); to cause to approach by some influence; to entice, to allure; to draw the notice of. *v.i.* to exert the power of attraction, to be attractive. **attractable**, *a.* **attractability** (-bil'-), *n.* **attractingly**, *adv.* **attractor**, *n.* **attraction**, *n.* the action or power of attracting; an attracting quality or characteristic; a force causing two objects, molecules etc. to be drawn together or to resist separation; that which attracts. **attractive**, *a.* having the power of attracting; alluring. **attractively**, *adv.* **attractiveness**, *n.*

attrib., (*abbr.*) attribute, attributed (to); attributive, attributively.

attribute (at'ribūt), *n.* a quality ascribed or imputed to any person or thing, as an essential characteristic; a characteristic; a symbol or other object recognized as peculiar or characteristic; an attributive word; (*Log.*) that which may be predicated of any subject. **attribute** (ətrib'-), *v.t.* to ascribe; to impute as belonging or due to; to ascribe as consequence. **attributable** (ətrib'-), *a.* **attribution** (at-), *n.* the act of attributing; that which is ascribed; function, authority etc. formally assigned; (*Log.*) predication. **attributive** (ətrib'-), *a.* characterized by attributing; (*Log.*) assigning an attribute to a subject; (*Gram.*) expressing an attribute without actual predication. *n.* a word denoting an attribute, now generally restricted to adjectives. **attributively**, *adv.*

attrited (ətrīt'id), *a.* worn down by friction. **attriteness**, *n.* **attrition** (-tri'-), *n.* the act or process of rubbing down or away; abrasion; wearing away by friction; (*Theol.*) sorrow for sin on account of the punishment due to it; a constant wearing down or weakening, as of an adversary.

attune (ətūn'), *v.t.* to bring to the right pitch; to make tuneful; to bring into accord; to accustom, acclimatize.

atty, (*abbr.*) attorney.

ATV, (*abbr.*) all-terrain vehicle; Associated Television.

at. wt., (*abbr.*) atomic weight.

atypical (ātip'ikəl), *a.* not typical, not conforming to type. **atypically**, *adv.*

AU, (*abbr.*) Ångström unit; astronomical unit.

Au, (*chem. symbol*) gold.

aubade (ōbahd'), *n.* a poem or musical piece announcing or greeting dawn.

auberge (ōbœzh'), *n.* (*F*) an inn. **aubergiste**, *n.* a keeper of an auberge.

aubergine (ō'bəzhēn), *n.* the egg-plant; its ovoid, characteristically dark purple fruit used as a vegetable and in stews; a dark purple colour.

Aubrietia (awbrē'shə), *n.* a genus of rock plants of the family Cruciferae.

auburn (aw'bən), *a.* reddish-brown.

au courant (ō koorā'), *a.* fully informed, up-to-date with

the situation. [F]
auction (awk'shən), *n.* a public sale by a person licensed for the purpose, in which each bidder offers a higher price than the preceding. *v.t.* to sell by auction. **auction bridge**, *n.* a development of bridge in which the players bid for the advantage of choosing trump suit. **auctioneer** (-niə'), *n.* a person who sells goods by auction, one licensed to conduct auctions. *v.t.* to sell by auction.
auctorial (awktaw'riəl), *a.* pertaining to an author or this occupation.
audacious (awdā'shəs), *a.* bold, daring, spirited; impudent, shameless. **audaciously**, *adv.* **audaciousness, audacity** (-das'-), *n.*
audible (aw'dibl), *a.* capable of being heard; clear or loud enough to be heard. **audibility** (-bil'-), **audibleness**, *n.* **audibly**, *adv.*
audience (aw'diəns), *n.* the act of hearing, attention; reception at a formal interview granted by a superior to an inferior; an assemblage of hearers or spectators; the readers of a book; the people who regularly watch or listen to a particular television or radio programme, performer etc.
audile (aw'dīl), *a.* pertaining to sound or hearing; characterized by mental pictures of sounds. *n.* a person whose recollection is based mainly on terms of sounds.
audio (aw'diō), *a.* of or pertaining to sound or its reproduction, transmission or broadcasting; pertaining to or using audio-frequencies. *n.* the (electronic) reproduction and transmission of sound.
audio-, *comb. form.* pertaining to hearing; pertaining to sound or sound reproduction.
audio-frequency (aw'diōfrēkwənsi), *n.* a frequency in the range corresponding to that of audible sound waves.
audiology (awdiol'əji), *n.* the science of hearing. **audiological** (-loj'-), *a.* **audiologist**, *n.*
audiometer (awdiom'itə), *n.* an application of the telephone for testing the sense of hearing. **audiometric** (-met'-), *a.* **audiometry** (-tri), *n.*
audiophile (aw'diōfīl), *n.* an enthusiast of high-fidelity sound reproduction.
audiotypist (aw'diōtīpist), *n.* a typist trained to type directly from material on a dictating machine. **audiotyping**, *n.*
audiovisual (awdiōvizh'uəl), *a.* directed at or involving hearing and sight, as in *audiovisual aids.* **audiovisually**, *adv.*
audit (aw'dit), *n.* an official examination of accounts; any formal review or solemn rendering of accounts; a check, inspection. *v.t.* to examine officially and pronounce as to the accuracy of (accounts).
audition (awdish'ən), *n.* the act or faculty of hearing; a trial performance by a singer, musician, actor etc. applying for a position or role. *v.t.* to test by an audition. *v.i.* to give a trial performance; to hold auditions. **auditive** (aw'-), *a.* pertaining to hearing.
auditor (aw'ditə), *n.* a hearer, one of an audience; one appointed to audit accounts. **auditorship**, *n.* **auditorial** (-taw'-), *a.* auditory; of or pertaining to an audit of accounts. **auditorially**, *adv.*
auditorium (awditaw'riəm), *n.* (*pl.* **-riums, -ria** (-ə)) the part of a building occupied by the audience; the reception-room in a monastic building.
auditory (aw'ditəri), *a.* of or pertaining to the organs or sense of hearing, perceived by the ear.
au fait (ō fā), *a.* familiar, well-acquainted with; up to the mark. [F, to the point]
auf Wiedersehen (owf vē'dəzān), *int.* (G) farewell, goodbye.
Aug., (*abbr.*) August.
Augean (awjē'ən), *a.* filthy. [from *Augeas*, mythical king of Elis, whose stable had not been cleaned for 30 years, till Hercules, by turning the river Alpheus through it, did so in a day]

auger (aw'gə), *n.* a carpenter's tool, somewhat resembling a very large gimlet, for boring holes in wood; a similar instrument for boring into soil or rock.
aught (awt), *n.* anything whatever; a whit, a jot or tittle; (*erroneously*) the figure 0, a naught. *adv.* in any respect.
augite (aw'gīt), *n.* a greenish, brownish-black or black variety of aluminous pyroxene. **augitic** (-git'-), *a.*
augment[1] (awgment'), *v.t.* to increase, to make larger or greater in number, degree, intensity etc.; to prefix a grammatical augment to. *v.i.* to increase, to become greater in size, number, degree etc. **augmentation**, *n.* the act of augmenting; the state of being augmented; the thing added; increase, addition; the reproduction of a melody or passage in notes of greater length than those in which it was first treated. **augmentative** (-men'-), *a.* having the power or quality of augmenting; of an affix, increasing the force of a word; of a word, extending the force of an idea. *n.* an augmentative element or word.
augment[2] (awg'mənt), *n.* a grammatical prefix (*a*) used in the older Aryan languages to denote past time.
au gratin (ō grati), *a.* cooked with a covering of breadcrumbs or grated cheese or both. [F]
augur (aw'gə), *n.* a religious official among the Romans who professed to foretell future events from omens derived chiefly from the actions of birds, inspection of the entrails of slaughtered victims etc.; a soothsayer, a diviner. *v.t.* to foretell from signs or omens; to betoken, portend. *v.i.* to make predictions of future events from signs or omens; to be a sign or foreboding. **augural** (-gū-), *a.* **augurship**, *n.* **augury** (-gū-), *n.* the art or practice of the augur; divination from the actions of birds; an augural ceremony; an omen.
August[1] (aw'gəst), *n.* the eighth month of the year, named in honour of Augustus Caesar.
august[1] (awgŭst'), *a.* majestic, stately, inspiring reverence and admiration; dignified, worshipful. **augustly**, *adv.* **augustness**, *n.*
august[2] AUGUSTE.
Augustan (awgŭs'tən), *a.* of or belonging to Augustus Caesar (63 BC–AD 14), or his age in which Latin literature reached its highest development; hence, classical, refined, distinguished by correct literary taste. *n.* a writer of any Augustan literary period.
auguste, august (owgoost'), *n.* a clown with maladroit antics.
Augustinian (awgŭstin'iən, awgəs-), *a.* of or pertaining to St Augustine (354–430), Bishop of Hippo (396–430), or to his doctrine of grace and predestination. *n.* an adherent of these doctrines; one of an order of friars named after him.
auk (awk), *n.* a northern sea-bird with rudimentary wings, esp. the great auk (now extinct), the little auk and the razor-bill. **auklet**, *n.* any of several small auks.
auld (awld), *a.* (*Sc., North.*) old. **auld lang syne**, old long since, long ago.
au naturel (ō natürel'), *a., adv.* in the natural state; uncooked or plainly cooked; (*coll. euphem.*) naked. [F]
aunt (ahnt), *n.* the sister of one's father or mother; one's uncle's wife; (*coll.*) a woman friend of a child, esp. a benevolent, practical woman. **Aunt Sally**, *n.* a game at fairs, in which a figure with a pipe in its mouth is set up, and the players endeavour to break the pipe by throwing sticks at it; an object of ridicule. **aunthood** (-hud), *n.* **auntie, aunty**, *n.* a familiar form of AUNT; a familiar term for an elderly woman. **Auntie**, *n.* (*coll.*) the BBC. **auntship**, *n.*
au pair (ō peə'), *n.* a person, esp. a girl, from a foreign country who performs domestic tasks in exchange for board and lodging. *v.i.* to work as an au pair.
aura (aw'rə), *n.* (*pl.* **-ras, -rae** (-rē) a subtle emanation from any body; a distinctive atmosphere or quality; a sensation (as of a current of cold air rising to the head) that precedes an attack in epilepsy, hysteria etc. **aural**[1], *a.*

aural — autism

aural² (aw'rəl), *a.* of or pertaining to the ear; received by the ear. **aurally,** *adv.* **aurist,** *n.* a specialist in ear diseases.

aureate (aw'riət), *a.* gold-coloured; covered with gold, gilded; brilliant, splendid; of language or literary style, over-elaborate and embellished.

Aurelia (awrē'liə), *n.* a genus of phosphorescent marine jellyfish. **aurelian,** *a.*

aureole (aw'riōl), **aureola** (-rē'ələ), *n.* (the glory attaching to) the crown which is the special reward of virgins, martyrs and doctors; the gold disc surrounding the head in early pictures, and denoting glory, a nimbus; a luminous envelope surrounding the body, a vesica piscis; a glorifying halo, glory; the halo round the moon in total eclipses of the sun, a corona; a halo of radiating light round the sun or moon.

au revoir (ō rəvwah'), *int.* farewell, goodbye. [F]

auric (aw'rik), *a.* of or containing (trivalent) gold.

auricle (aw'rikəl), *n.* the external ear, that part which projects from the head; any process shaped like the lobe of the ear; an atrium of the heart. **auricled,** *a.* having an auricle or auricles.

auricula (awrik'ūlə), *n.* a garden flower, *Primula auricula*, sometimes called bear's ear, from the shape of its leaves.

auricular (awrik'ūlə), *a.* of, pertaining to, using or known by the sense of hearing; whispered in the ear, hence secret; shaped like an auricle; of or pertaining to an atrium of the heart. **auricularly,** *adv.* **auriculate** (-lət), *a.* having ears, or appendages resembling ears.

auriferous (awrif'ərəs), *a.* yielding or producing gold.

Aurignacian (awrignā'shən), *a.* pertaining to the period of Upper Palaeolithic culture typified by human remains and implements etc. of stone, horn and bone found in the cave of Aurignac, Haute-Garonne.

aurist AURAL².

aurochs (aw'roks), *n.* (*pl.* **aurochs**) the extinct wild ox, *Bos urus* or *primigenius*, of Central Europe; erroneously applied to the European bison, *Bos bonasus*, strictly preserved in Lithuania.

aurora (awraw'rə), *n.* (*pl.* **-ras, -rae** (-rē)) morning twilight, dawn; the colour of the sky at sunrise; (**Aurora**) the Roman goddess of the dawn; a peculiar illumination of the night sky common within the polar circles, consisting of streams of light ascending towards the zenith, called **aurora borealis** (bawriah'lis, -ā'-), or **aurora australis** (ostrah'lis, -ā'-) according to whether it is seen in the northern or southern hemisphere. **auroral,** *a.* of or pertaining to the dawn; rosy, roseate; of or pertaining to an aurora.

aurous (aw'rəs), *a.* of or containing (univalent) gold.

auscultation (awskəltā'shən), *n.* the act of listening; listening with the ear or stethoscope to the sounds made by the internal organs, to judge their condition. **auscultator** (aws'-), *n.* **auscultatory** (-kŭl'-), *a.* **auscultate** (aws'-), *v.t.* to examine by auscultation.

Auslese (ows'lāzə), *n.* a usu. sweetish white wine from Germany or Austria made from selected ripe grapes.

auspice (aw'spis), *n.* an omen drawn from the actions of birds; (*often pl.*) a (favourable) portent, sign or omen; (*usu. pl.*) patronage, protection. **under the auspices of,** under the leadership, encouragement or patronage of. **auspicious** (-spish'-), *a.* having favourable omens; auguring good fortune; conducive to prosperity or success. **auspiciously,** *adv.* **auspiciousness,** *n.*

Aussie (oz'i), *n.*, *a.* (*coll.*) (an) Australian.

Auster (aws'tə), *n.* (*poet.*) the south wind.

austere (ostiə'), *a.* severe, stern, rigorous; ascetic, abstemious; sober, simple, unadorned. **austerely,** *adv.* **austereness,** *n.* **austerity** (-te'-), *n.* sternness, severity; self-denial, asceticism; lack of adornment; (*pl.*) ascetic or penitential practices.

Austin (os'tin), *n.*, *a.* (an) Augustinian.

austral (os'trəl), *a.* of or pertaining to, or situated in or towards the south; southern.

Australasian (ostrəlā'zhən), *a.* of or pertaining to Australasia, a general name for Australia, New Zealand, Tasmania and the surrounding islands. *n.* a native or inhabitant of Australasia.

Australian (ostrāl'yən), *a.* of or belonging to Australia. *n.* a native or inhabitant of Australia. **Australian rules,** *n. sing.* a variety of rugby football played in Australia. **Australianism,** *n.* an Australian idiom or characteristic. **Australoid** (os'-), *a.* denoting or resembling the racial type that includes the Aborigines of Australia. *n.* a member of this group.

Australopithecus (ostrələpith'ikəs), *n.* a genus of fossil primates whose remains have been found in Southern Africa. **australopithecine** (-sēn), *n.*, *a.* (an individual) of the genus *Australopithecus* or a related genus.

Australorp (os'trəlawp), *n.* an Australian utility type of Black Orpington fowl.

Austrian (os'triən), *n.*, *a.* (a native or inhabitant) of Austria.

Austro-¹, *comb. form.* southern.

Austro-², *comb. form.* Austrian.

AUT, (*abbr.*) Association of University Teachers.

autacoid (aw'təkoid), *n.* (*Physiol.*) an internal secretion, a hormone or chalone.

autarch (aw'tahk), *n.* an absolute sovereign, an autocrat. **autarchy,** *n.* absolute sovereignty, autocracy. **autarchic, -ical** (-tah'-), *a.*

autarky (aw'tahki), *n.* self-sufficiency, esp. national economic self-sufficiency.

auteur (ōtœ'), *n.* the director of a film, esp. when viewed as the film's creative force. **auteurist,** *a.* supporting the view of director as creator of a film or film genre.

authentic (awthen'tik), *a.* entitled to acceptance or belief; trustworthy, credible; of undisputed origin, genuine; (*Mus.*) having the notes between the keynote or tonic and the octave above. **authentically,** *adv.* **authenticate,** *v.t.* to render authentic or valid; to establish the truth or credibility of; to verify the authorship of. **authentication,** *n.* **authenticator,** *n.* **authenticity** (-tis'-), *n.*

author (aw'thə), *n.* the originator, producer or efficient cause of anything; the composer of a literary work; one whose profession is writing, esp. books; the books written by an author. **authoress** (-ris), *n. fem.* **authorial** (-thaw'-), *a.* **authorless** (-lis), *a.* without an acknowledged author; anonymous. **authorship,** *n.* the profession of a writer of books; origin of a literary work.

authority (awthŏ'rəti), *n.* legitimate power to command or act; (*often pl.*) a person or body exercising this power; power, weight or influence, derived from character, station, mental superiority and the like; weight of testimony, credibility; delegated power or right to act; the author or the source of a statement; the standard book or work of reference on any subject; an expert, one entitled to speak with authority on any subject. **authoritarian** (-teə'-), *n.* one who places obedience to authority above personal liberty. *a.* obedience to authority above liberty; dictatorial. **authoritative,** *a.* imperative, commanding; possessed of authority, founded on sufficient authority; (entitled to be) accepted as true. **authoritatively,** *adv.* **authoritativeness,** *n.*

authorize, -ise (aw'thəriz), *v.t.* to give authority to, to empower; to establish by authority; to sanction; to warrant legally; to justify, afford just ground for; to make or prove legitimate; to vouch for, to confirm. **authorized,** *a.* **Authorized Version,** *n.* the English translation of the Bible published in 1611. **authorizable,** *a.* **authorization,** *n.* the act of authorizing; establishment by authority; a document etc. that authorizes something.

autism (aw'tizm), *n.* abnormal absorption in fantasy, delusions etc., accompanied by withdrawal from reality; a disorder of mental development marked by lack of social

communication and inability to form relationships. **autistic** (-tis'-), *a.*
auto (aw'tō), *n.* (*chiefly N Am.*) short for AUTOMOBILE.
auto., (*abbr.*) automatic.
auto-, *comb. form.* self, from within or by oneself; one's own, independently, e.g. *automatic, automotive.*
Autobahn (aw'təbahn), *n.* a motorway in Germany.
autobiography (awtəbiog'rəfi), *n.* a memoir of one's life, written by oneself. **autobiographer**, *n.* **autobiographic, -ical** (-graf'-), *a.* **autobiographically**, *adv.*
autocephalous (awtōsef'ələs), *a.* having an independent head or chief; esp., of a church, having its own bishop. **autocephaly**, *n.*
autochthon (awtok'thən), *n.* (*pl.* **-thons, -thones**) one of the original or earliest known inhabitants; an aboriginal animal or plant. **autochthonous**, *a.* native, indigenous, occurring, formed or originating in the place where found. **autochthonism, autochthony**, *n.*
autoclave (aw'təklāv), *n.* a sealed vessel used for chemical reactions at high temperature and pressure; an apparatus using super-heated steam for sterilizing, cooking etc. *v.t.* to sterilize etc. in an autoclave.
autocracy (awtok'rəsi), *n.* absolute government; controlling power. **autocrat** (aw'təkrat), *n.* a sovereign of uncontrolled authority; a dictatorial person. **autocratic, -ical** (-krat'-), *a.* pertaining to autocracy; absolute, despotic. **autocratically**, *adv.* **autocratrix** (-tok'rətriks), *n. fem.*
autocross (aw'tōkros), *n.* the sport of motor racing on grass.
Autocue® (aw'təkū), *n.* a device that displays the text to be spoken by a person on television.
auto-da-fé (awtōdafā'), *n.* (*pl.* **autos-da-fé** (awtō-)) a sentence pronounced by the Inquisition; the execution of this judgment; the burning of a heretic.
autodidact (awtōdi'dakt), *n.* a self-taught person. **autodidactic** (-dak'-), *a.*
autoerotism (awtōe'rətizm), **-ticism** (-rot'-), *n.* self-produced sexual pleasure or emotion, e.g. masturbation. **autoerotic** (-rot'-), *a.*
autofocus (aw'təfōkəs), *n.* a facility in some cameras for automatically focusing the lens. *v.i.* to focus automatically by means of this facility.
autogamy (awtog'əmi), *n.* self-fertilization. **autogamous**, *a.*
autogenous (awtoj'ənəs), **-genic** (-jen'-), *a.* self-engendered, self-produced; originating from sources within the same body. **autogeny** (-toj'-), *n.* (*Biol.*) a kind of spontaneous generation.
autogiro, -gyro (aw'təjirō), *n.* an aircraft in which the lifting surfaces are the freely-rotating blades of a large horizontal air-screw.
autograph (aw'təgrahf, -graf), *n.* a person's own handwriting, esp. his or her signature; a manuscript in an author's own handwriting. *v.t.* to write with one's own hand; to sign. **autographic, -ical** (-graf'-), *a.* **autographically**, *adv.* **autography** (-tog'-), *n.*
Autoharp® (aw'tōhahp), *n.* a zither-like instrument having dampers which stop selected strings from sounding and allow chords to be played.
auto-immune (awtōimūn'), *a.* of or caused by antibodies that attack the normally present molecules, cells etc. of the organism producing them. **auto-immunity, -isation, -ization**, *n.*
auto-intoxication (awtōintoksikā'shən), *n.* reabsorption of toxic matter produced by the body.
autolysis (awtol'isis), *n.* the breakdown of cells by the action of enzymes produced in the cells themselves. **autolyse** (aw'təliz), (*chiefly N Am.*) **-lyze**, *v.t.* to cause autolysis in. *v.i.* to undergo autolysis. **autolytic** (-lit'-), *a.*
automat (aw'təmat), *n.* (*N Am.*) a restaurant equipped with automatic machines for supplying food etc.; a vending machine.
automate AUTOMATION.

automatic (awtəmat'ik), *a.* self-acting, self-regulating; acting as an automaton, having the power of movement within itself; of a firearm, repeatedly ejecting the empty shell, introducing a new one and firing, until the trigger is released; carried on unconsciously; involuntary, reflex; merely mechanical, occurring as a normal or habitual consequence. *n.* an automatic firearm; a motor vehicle with automatic transmission. **automatic gain control**, a control for maintaining the output of an amplifier at a constant level despite varying input signals. **automatic pilot**, *n.* a device which automatically maintains an aircraft or spacecraft on a predetermined course. **automatic transmission**, *n.* power transmission in a motor vehicle in which the gears change automatically. **automatic writing**, *n.* writing performed without the consciousness of the writer. **automatically**, *adv.* **automaticity** (-tis'-), *n.*
automation (awtəmā'shən), *n.* the use of self-regulating or automatically programmed machines in the manufacture of goods. **automate** (aw'-), *v.t.* to make automatic. *v.i.* to apply automation. **automated teller machine**, an electronic machine, operated by a bank or building society, from which a customer can obtain cash or account details at any time by inserting a special card.
automatism (awtom'ətizm), *n.* the quality of being automatic; involuntary action; the theory that animals are automatons performing their functions as mere machines without conscious control, as the result of natural laws; unconscious action, automatic routine. **automatist**, *n.* **automatize, -ise**, *v.t.* to reduce to the condition of an automaton; to make automatic.
automaton (awtom'ətən), *n.* (*pl.* **-tons, -ta** (-tə)) a mechanism that moves under its own (hidden) power; a piece of machinery simulating human or animal action; a person whose actions are merely mechanical. **automatous**, *a.*
automobile[1] (awtəməbēl'), *a.* self-moving. **automobilism**, *n.*
automobile[2] (aw'təməbēl), *n.* (*chiefly N Am.*) a motor-car. **automobilist** (-bēl'-), *n.*
automotive (awtəmō'tiv), *a.* self-propelling; pertaining to motor vehicles.
autonomy (awton'əmi), *n.* the right of self-government; an independent state or community; freedom to act as one pleases; in Kantian philosophy, freedom of the will; organic independence. **autonomous**, *a.* of or possessing autonomy; self-governing; independent; having organic independence. **autonomously**, *adv.* **autonomic** (-nom'-), *a.* autonomous; independent; (*Biol.*) occurring involuntarily, spontaneous; pertaining to or mediated by the autonomic nervous system. **autonomic nervous system**, the part of the vertebrate nervous system that regulates the involuntary actions of the heart, glands and some muscles. **autonomist**, *n.* an advocate of autonomy. **autonomize, -ise**, *v.t.* to render independent; to make self-governing.
autopilot AUTOMATIC PILOT under AUTOMATIC.
autopsy (aw'topsi), *n.* personal observation; dissection; a post-mortem examination; a critical examination. *v.t.* to perform a post-mortem examination on. **autoptic, -ical** (-top'-), *a.* seen by one's own eyes; based on personal examination. **autoptically**, *adv.*
autoroute (aw'tōroot), *n.* a motorway in France.
autosome (aw'təsōm), *n.* a chromosome other than a sex chromosome. **autosomal** (-sō'-), *a.*
autostrada (aw'təstrahdə), *n.* a motorway in Italy.
autosuggestion (awtōsəjes'chən), *n.* suggestion arising from oneself, esp. the unconscious influencing of one's own beliefs, physical condition etc. **autosuggestive**, *a.*
autoteller (aw'tətelə), AUTOMATED TELLER MACHINE under AUTOMATION.
autotimer (aw'tətīmə), *n.* a control on a cooker etc. that can be preset to switch the device on and off.
autotomy (awtot'əmi), *n.* voluntary separation of a part of the body, e.g. the tail, as in certain lizards.

autotrophic (awtətrof'ik), *a.* self-nourishing; of or pertaining to organisms capable of manufacturing organic foods from inorganic sources, as by photosynthesis. **autotroph**, *n.* **autotrophically**, *adv.*

autumn (aw'təm), *n.* the season of the year between summer and winter (astronomically, it extends from the autumnal equinox, 21 Sept., to the winter solstice, 21 Dec.; popularly, it comprises September, October and often November); the early stages of decline of human life; the fruits of harvest. **autumn crocus**, *n.* meadow-saffron. **autumnal** (-tŭm'nəl), *a.* of or pertaining to, characteristic of or produced in autumn; pertaining to the declining period of life. *n.* a plant which flowers in autumn.

aux., (*abbr.*) auxiliary.

auxiliary (awgzil'yəri), *a.* helping, aiding; subsidiary; applied to verbs used in the conjugation of other verbs. *n.* one who, or that which, helps or assists; an auxiliary verb; (*pl.*) foreign or allied troops in the service of a nation at war.

auxin (awk'sin), *n.* a growth-promoting plant hormone.

AV, (*abbr.*) audio-visual; Authorized Version.

av., (*abbr.*) average; avoirdupois; (**av., Av.**) avenue.

avail (əvāl'), *v.i.* to be of value, use, profit or advantage; to be helpful, or effectual or sufficient. *v.t.* to be of use or advantage to. *n.* worth, value, profit, advantage, use. **of no avail, without avail,** ineffectual. **to avail oneself of,** (*N Am.*) **to avail of,** to take advantage of, make use of. **to little avail,** ineffectually. **available,** *a.* capable of being employed; at one's disposal; at hand, valid. **availability** (-bil'-), *n.* the quality of being available; (*N Am.*) a qualification in a candidate which implies strong probability of his or her success. **availableness,** *n.* **availably,** *adv.*

avalanche (av'əlahnh), *n.* a mass of snow, ice and debris falling or sliding from the upper parts of a mountain; a sudden inundation; the cumulative production of charged particles resulting from the collisions of a single charged particle with matter to produce further particles which in turn collide etc. *v.t., v.i.* to descend as an avalanche (on).

avant-garde (avāgahd'), *n.* a. as a forerunner in music, art etc.; in advance of contemporary artistic taste or trend. *n.* the people who create or take up avant-garde or experimental ideas, esp. in the arts. **avant-gardism,** *n.* **avant-gardist,** *n.*

avarice (av'əris), *n.* an excessive craving after wealth; eager desire to get and keep. **avaricious** (-rish'əs), *a.* **avariciously,** *adv.*

avast (əvahst'), *int.* (*Naut.*) stay! stop! desist!

avatar (av'ətah), *n.* the descent of a Hindu deity to the earth; the incarnation of a deity; a manifestation, phase.

avdp (*abbr.*) avoirdupois.

avaunt (əvawnt'), *int.* be off! away with you! begone!

ave (ah'vā), *int.* hail! welcome! farewell! (in allusion to the classical custom of greeting the dead). **Ave,** *n.* an Ave Maria. **Ave Maria, Ave Mary,** *n.* the Hail Mary; the angelical salutation (Luke i.28) with that of St Elisabeth (i.42), to which a prayer is added, the whole being used as a form of devotion.

ave., Ave., (*abbr.*) avenue.

avenge (əvenj'), *v.t.* to vindicate by punishing a wrongdoer; to exact satisfaction for (an injury etc.); to inflict punishment on account of. *v.i.* to execute vengeance. **avengeful,** *a.* **avengement,** *n.* **avenger,** *n.*

avens (av'əns), *n.* a plant of the genus *Geum*, as the wood avens or herb bennet, *G. urbanum*, and the water avens, *G. rivale*; the mountain avens, *Dryas octopetala*.

aventurine, aventurin (əven'churin), *n.* a gold-spangled glass made first at Murano (the process was accidentally discovered, whence the name); a quartz spangled with scales of mica or some other mineral.

avenue (av'inū), *n.* a way or means of access or approach; an approach to a country house or similar building; a broad alley bordered with trees; the rows of trees bordering such an alley; a fine wide thoroughfare.

aver (əvœ'), *v.t.* (*past, p.p.* **averred**) to assert or declare positively; (*Law*) to prove; to allege, declare. **averrable,** *a.* **averment,** *n.* the act of averring; affirmation, positive assertion; (*Law*) an affirmation alleged to be true, and followed by an offer to verify.

average (av'ərij), *n.* loss arising from damage to ship or cargo at sea; apportionment of such loss among the parties interested; a number or quantity intermediate to several different numbers or quantities; a mean; the rate, proportion, degree, quantity etc. generally prevailing. *v.t.* to calculate the average of; to take the ordinary standard of; to divide proportionately to the number involved; to be or consist of on an average; to do, have or take as a mean rate or value. *v.i.* to be or amount to as an average. *a.* ascertained by taking a mean proportion between given quantities; medium, ordinary. **on an average,** taking the mean deduced from a number of examples; typically. **average adjuster,** *n.* an assessor who deals with claims for losses at sea. **averagely,** *adv.*

averse (əvœs'), *a.* turned away mentally; feeling repugnance or dislike; unwilling, disinclined, reluctant (to). **aversely,** *adv.* **averseness,** *n.* **aversion** (-shən), *n.* an averted state of feeling or mind; disinclination, dislike, repugnance; an object of dislike. **aversion therapy,** *n.* therapy designed to stop undesirable behaviour by associating it with an unpleasant sensation (as an electric shock).

avert (əvœt'), *v.t.* to turn away; to ward off. **avertible,** *a.*

Avesta (əves'tə), *n.* the sacred scriptures of Zoroastrianism. **Avestan,** *a.* of the Avesta. *n.* the Iranian language of the Avesta, Zend.

avian (ā'viən), *a.* of or pertaining to birds. **aviary,** *n.* a large cage or building in which birds are kept. **aviarist,** *n.*

aviation (āviā'shən), *n.* the art of flying or travelling in the air; all matters to do with aircraft or flying in an aircraft; the design and manufacture of aircraft. **aviate** (ā'-), *v.i.* to fly, to travel in an aircraft. **aviator,** *n.* a person who flies an aircraft. **aviatrix** (āviā'triks), *n. fem.*

aviculture (ā'vikŭlchə), *n.* the breeding and rearing of birds.

avid (av'id), *a.* greedy, covetous; ardently desirous; extremely eager, hungry. **avidly,** *adv.* **avidity** (-vid'-), *n.*

avifauna (ā'vifawnə), *n.* the birds in any district taken collectively. **avifaunal,** *a.*

avionics (āvion'iks), *n. sing.* (the science concerned with) the development and use of electronic and electric equipment in aircraft and spacecraft.

avitaminosis (āvitəminō'sis), *n.* (*pl.* **-ses** (-sēz)) disease resulting from vitamin deficiency.

AVM, (*abbr.*) Air Vice-Marshal.

avocado (avəkah'dō), *n.* (*pl.* **-dos**) (also **avocado pear**) the pear-shaped fruit of a West Indian tree, *Persea gratissima*; this tree; a dull green colour.

avocation (avəkā'shən), *n.* a minor employment or occupation, a hobby; ordinary employment, vocation.

avocet, avoset (av'əset), *n.* a wading bird allied to the snipes and stilts, having a long slender bill curved upwards.

avoid (əvoid'), *v.t.* to shun; to keep away from; to escape, evade; (*Law*) to defeat, to invalidate, to quash. **avoidable,** *a.* **avoidably,** *adv.* **avoidability** (-bil'-), *n.* **avoidance,** *n.* **avoidless** (-lis), *a.* (*poet.*) incapable of being avoided; inevitable.

avoirdupois (avwahdoopwah', avədəpoiz'), *n.* a system of weights based on the unit of a pound of 16 ounces, equal to 7000 grains (0·4536 kg); (*chiefly N Am.*) weight, heaviness.

avoset AVOCET.

avouch (əvowch'), *v.t.* to affirm, vouch for, guarantee as certain; acknowledge, avow; to maintain, to justify. *v.i.* to

vouch, give assurance or guarantee. **avouchable**, *a*.
avow (əvou'), *v.t.* to acknowledge, to admit (of one's free will); to state, allege, declare. **avowable**, *a*. **avowal**, *n*. an open declaration, a free admission. **avowed**, *a*. acknowledged; self-acknowledged. **avowedly** (-id-), *adv*.
avulsion (əvŭl'shən), *n*. the act of tearing away or violently separating; (*Law*) sudden removal of land (without change of ownership) by flood, alteration in the course of a river or the like. **avulsive**, *a*.
avuncular (əvŭng'kūlə), *a*. of, pertaining to or resembling an uncle; friendly, genial.
AWACS (ā'waks), (*abbr.*) Airborne Warning and Control System.
await (əwāt'), *v.t.*, *v.i.* to wait (for), look out (for), expect; to be in store (for).
awake (əwāk'), *v.i.* (*past* **awoke** (əwōk'), *awaked*, *p.p.* **awoken**) to wake from sleep, cease sleeping; to become conscious of or alive to something; to become active or alert. *v.t.* to arouse from sleep, or from lethargy or inaction; to stir up, excite. *a*. not asleep; roused from sleep; active, vigilant, aware, alive (to). **awakable**, *a*. **awaken**, *v.t.* to arouse, awake; to arouse to a sense of sin. *v.i.* to awake. **awakenable**, *a*. **awakening, awakenment**, *n*.
award (əwawd'), *v.t.* to adjudge, to assign by judicial sentence; to grant or confer, esp. as a prize for merit or as something needed. *n*. the decision of judge, arbitrator or umpire; that which is awarded.
aware (əweə'), *a*. apprised, cognizant, conscious; watchful, vigilant. **awareness**, *n*.
awash (əwosh'), *adv*. on a level with the water; at the mercy of the waves. *a*. covered with water. **awash with**, full of, having an abundance of.
away (əwā'), *adv*. implying motion from a place, person, cause or condition; absent, in the other direction, at another place; at a distance; out of existence or consciousness continuously, constantly; straightaway, directly. Used elliptically as a verb, be off! begone! (to) go away. *a*. absent; distant; played on an opponent's ground. *n*. a (football) match played or won at an opponent's ground. **away back**, long ago. **away with**, take away. **to do away with** DO. **to make away with** MAKE. **far and away, out and away**, beyond comparison. **awayday**, *n*. a day trip using public rail or bus services.
awe (aw), *n*. dread mingled with veneration; solemn, reverential wonder. *v.t.* to inspire with solemn fear or reverence; to restrain by profound respect or reverential fear. **aweless** (-lis), *a*. **awelessness**, *n*. **awesome** (-səm), *a*. full of or displaying awe; inspiring awe. **awesomely**, *adv*. **awesomeness**, *n*. **awe-stricken, awe-struck**, *a*. overwhelmed with awe.
aweigh (əwā'), *adv*. of an anchor, raised vertically just off the bottom.
awful (aw'fəl), *a*. Inspiring or worthy of awe; dreadful, fearful, appalling; filled with awe; extremely disagreeable, frightful, terrible; (*coll.*) often used as an intensive. **awfully**, *adv*. in an awful manner; (*coll.*) exceedingly, very. **awfulness**, *n*.
awheto, aweto (əwä'tō), *n*. (*N. Zealand*) a vegetable-eating caterpillar that, when dried, yields a tattoo dye.
awhile (əwīl'), *adv*. for some time; for a little; (*loosely*) a while.
awkward (awk'wəd), *a*. unhandy, ill-adapted for use; lacking dexterity, bungling, clumsy; ungraceful, ungainly; embarrassed, ill at ease; embarrassing; not easy to manage or deal with. **awkwardly**, *adv*. **awkwardness**, *n*.
awl (awl), *n*. a tool with a cylindrical tapering blade, sharpened at the end, for making holes for stitches in leather.
awn (awn), *n*. the beard of corn and grasses, one of the bristles springing from a bract in the inflorescence of grasses. **awned**, *a*. **awnless** (-lis), *a*.

awning (aw'ning), *n*. a covering of tarpaulin, canvas or other material used as a protection from sun or rain, as above the deck of a ship; the part of the poop-deck which is continued forward beyond the bulk-head of the cabin; a shelter. **awned**, *part.a.* fitted with an awning.
awoke (əwōk'), *past* AWAKE.
AWOL (ā'wol), *a.*, *adv*. absent without authorization from one's post or position of duty. *n*. a member of the armed forces who is absent without authorization. [*absent without leave*]
awry (əri'), *adv*. obliquely, crookedly; erroneously, amiss. *a*. crooked, distorted, oblique; wrong.
axe, (*esp. N Am.*) **ax** (aks), *n*. (*pl*. **axes**) an instrument for cutting or chopping consisting of an iron head with a sharp edge, fitted to a wooden handle or helve; a celt probably used as an axe. *v.t.* to chop or cut with an axe; to dismiss (staff) for reasons of economy; to make drastic reductions in (expenditure, services etc.). **the axe**, *n*. dismissal from employment; drastic reduction in expenditure. **to have an axe to grind**, to have an ulterior motive; to have a grievance to air. **axe-man**, *n*. a woodman; a warrior armed with a battle-axe; a psychopath who kills with an axe.
axel (ak'sl), *n*. a jump in ice-skating incorporating one and a half turns. [*Axel* Paulsen, 1855-1938, Norw. skater]
axial, axile, etc. AXIS.
axil (ak'sl), *n*. the hollow where the base of a leaf joins the stem or where a branch leaves the trunk. **axilla** (-sil'ə), *n*. (*pl*. **-llas, -llae** (-lē)) the armpit. **axillar, axillary**, *a*. pertaining to the armpit; pertaining to or arising from the axil.
axiom (ak'siəm), *n*. a self-evident or generally accepted truth; (*Math.*) a self-evident proposition, assented to as soon as enunciated. **axiomatic, -ical** (-mat'-), *a*. self-evident, containing an axiom or axioms; full of maxims. **axiomatically**, *adv*.
axis[1] (ak'sis), *n*. (*pl*. **axes** (-sēz)) a real or imaginary straight line round which a body or geometric figure revolves, or is conceived to revolve, or round which its parts are arranged, or to which they have a symmetrical relation; a fixed reference line used, as on a graph, in locating a point; the second cervical vertebra; the central stem, core or main skeletal support of (a part of) an organism; the central shaft of growth of a plant; an imaginary line round which a crystal can be symmetrically built up; (*Geol.*) a central ridge; a ray of light passing through the centre of or falling perpendicularly on a lens; the straight line from the lens of the eye to the object seen. **axial**, *a*. pertaining to an axis; forming an axis. **The Axis**, the term used to describe the political collaboration (Rome – Berlin axis) between Nazi Germany and Fascist Italy from 1935-43. **axiality** (-al'-), *n*. the quality of being axial. **axially**, *adv*. in the direction of the axis. **axile** (-īl), *a*. situated in the axis of anything.
axis[2] (ak'sis), **axis deer**, *n*. a S Asian deer with a white-spotted coat.
axle (ak'sl), *n*. the pin or bar on which a wheel revolves or which revolves with the wheel; (the thin ends of) the axle-tree. **axle-box**, *n*. a case in which the ends of axles revolve; a metal cover for the hub. **axle-tree**, *n*. the beam or bar connecting wheels, on the ends of which the wheels revolve.
Axminster (aks'minstə), *n*. a variously coloured and patterned woven carpet with a tufted pile. [*Axminster*, town in Devon where a type of patterned carpet was originally woven]
axolotl (ak'səlotl), *n*. a salamander of the genus *Ambystoma* that retains the larval form when fully grown; esp. a small Mexican salamander, *Ambystoma mexicanum*.
axon (ak'son), *n*. the projection from a nerve cell that typically conducts impulses away from the cell.
ay, aye[1] (ī), *adv.*, *int.* yes. *n*. an affirmative vote in the House of Commons; (*pl.*) those who vote in the affirma-

tive. **aye, aye, sir,** (*Naut.* or *facet.*) yes, sir; very well, sir.
ayah (i′ə), *n.* a Hindu nurse for children; a lady's maid.
ayatollah (iətol′ə), *n.* a leader of the Shiite Muslims in Iran.
aye[2] (ā), *adv.* always, ever; in all cases, on all occasions. **for aye, for ever and aye,** for ever, to all eternity.
aye-aye (i′ī), *n.* a small lemur found in Madagascar.
Aylesbury (ālz′bəri), *n.* a breed of table ducks.
Ayrshire (eə′shə), *n.* a breed of cattle named after the former county of Ayrshire, Scotland, and highly prized for dairy purposes.
Ayurveda (ahyuvā′də), *n.* the ancient Hindu writings on medicine, health and healing. **Ayurvedic,** *a.* of or pertaining to the Hindu philosophy of medicine and healing as set down in the Ayurveda.
az., (*abbr.*) azimuth.
Azalea (əzāl′yə), *n.* a genus of shrubby plants with showy and occasionally fragrant flowers.
azan (ahzahn′), *n.* the Muslim call to prayer.
azarole (az′ərōl), *n.* the Neapolitan medlar, *Crataegus azarolus,* or its fruit.
azidothymidine AZT.
Azilian (əzil′iən), *a.* pertaining to the period of culture typified by the remains found in the cavern of Mas-d'*Azil* in the Pyrenees.
azimuth (az′iməth), *n.* horizontal angle or direction, point of the compass, bearing; (**azimuth, true azimuth**) the arc of the horizon intercepted between the north (or, in the southern hemisphere, the south) point of the horizon and the point where the vertical circle passing through a heavenly body cuts the horizon; (**azimuth, magnetic azimuth**) the arc intercepted between the true azimuth and the magnetic meridian. **azimuth circle,** *n.* a circle passing through the zenith and cutting the horizon perpendicularly. **azimuth compass,** *n.* an instrument for finding the magnetic azimuth.
azo-, *comb. form.* nitrogen.
azoic (əzō′ik), *a.* having no trace of life; (*Geol.*) destitute of organic remains, of the time that antedates life.
azonal (əzōn′əl), *a.* not divided into separate zones or regions.
azote (əzōt′), *n.* an old name for nitrogen, from its fatal effects upon animal life. **azotic** (əzot′-), *a.*
AZT, *n.* an antiviral drug that suppresses the activity of the virus that causes AIDS and is used to alleviate some of the AIDS symptoms. [*az*ido*t*hymidine]
Aztec (az′tek), *a.* of, pertaining to or naming the leading Mexican tribe at the time of the Spanish invasion (1519); loosely applied to Mexican antiquities generally. *n.* a member of the Aztec tribe; its language.
azure (azh′ə, ā′-), *n.* lapis-lazuli; the deep blue of the sky; the vault of heaven; a bright blue pigment or dye; (*Her.*) the blue of coats of arms, represented in engraving by horizontal lines. *a.* resembling the clear bright blue of the sky; clear, unclouded; (*Her.*) blue. **azure-spar, -stone,** *n.* lazulite. **azurite** (-rīt), *n.* blue carbonate of copper. **azurn** (-az′œn), *a.* (*poet.*) azure.
azygous (az′igəs), *a.* (*Biol.*) unpaired, occurring singly, not as one of a pair. **azygously,** *adv.*
azyme (az′im, -im), *n.* the Passover cake of unleavened bread. **Azymite** (az′imit), *n.* one who uses unleavened bread in the Eucharist.

B

B¹, b¹, the second letter in the English and other alphabets, representing a flat labial mute; (*pl.* **Bs, B's, Bees**) used as a symbol to denote the second of a series; the second known quantity in an algebraic expression; (*Mus.*) the seventh note of the diatonic scale of C major; one of a second class or order; one of the human blood groups. **B road**, *n.* a road of secondary importance.
B², (*abbr.*) Bachelor; Baron; bel; Belgium; of pencil lead, black; British.
B³, (*chem. symbol*) boron.
b¹, (*abbr.*) barrel; billion; book; born; bottle; bowled.
BA, (*abbr.*) Bachelor of Arts; British Academy; British Airways; British America; British Association (for the Advancement of Science); Buenos Aires.
Ba, (*chem. symbol*) barium.
BAA, (*abbr.*) British Airports Authority.
baa (bah), *n.* the cry or bleat of a sheep. *v.t.* to make this noise.
Baal (bä′əl, bahl), *n.* (*pl.* **Baalim**, (-lim)) the chief male divinity among the Phoenicians; a false god. **Baalism** (-izm), *n.* idolatry. **Baalist, Baalite** (-it), *n.*
baas (bahs), *n.* boss, overseer.
baba (bah′bah), *n.* a small cake soaked in rum (**rum baba**).
babacoote (ba′bəkoot), *n.* the indri, a short-tailed woolly lemur from Madagascar.
Babbit (bab′it), *n.* a dull, complacent man with orthodox views and little interest in cultural values. **Babbitry,** *n.* [after the character in the novel (1922) by Sinclair Lewis]
babbit metal, babbit's metal (bab′it). *n.* an alloy of tin, antimony and copper, used in bearings to diminish friction. [Isaac *Babbit*, 1799–1862, American inventor]
babble (ba′bl), *v.i.* to talk childishly or inopportunely; to prattle; of streams, birds etc, to make inarticulate sounds. *v.t.* to prate; to utter. *n.* prattle; shallow, foolish talk; confused murmur; water. **babblement,** *n.* **babbler,** *n.* one who babbles; **babbling,** *n., a.*
babe (bāb), *n.* a baby; a foolish or childish person; (*sl., often derog.*) a girl, woman.
Babel (bā′bl), *n.* the city and tower described in Gen. xi, where the confusion of tongues is said to have occurred; (**babel**) noisy tumult, disorder.
baboo, babu (bah′boo), *n.* in Bengal, a term corresponding to English Mr; an Indian clerk who writes English.
baboon (bəboon′), *n.* the popular name of a large division of monkeys, with dog-like snout, callosities on the buttocks, and capacious cheek-pouches; an epithet of abuse. **baboonery,** *n.*
babouche, babuche (bəboosh′), *n.* a Turkish heelless slipper.
babushka (bəboosh′kə), *n.* in Russia, an old woman; a headscarf tied under the chin.
baby (bā′bi), *n.* a child in arms; a foolish, childish person; (*coll.*) a girl; a pet project. *v.t.* to treat like a baby. **to hold the baby,** (*coll.*) to be left to bear the brunt of something. **baby boomer,** *n.* (*coll.*) one born during an upswing in the birth-rate, esp. in the years following World War II. **baby-carriage,** *n.* (*N Am.*) a pram. **baby grand,** *n.* a small grand piano. **Babygro**® (-grō), *n.* an all-in-one baby garment made of a stretch fabric. **baby-minder,** *n.* one who looks after infants when their parents are at work. **baby-sitter,** *n.* one who looks after a child while the parents are out. **baby-sit,** *v.i.* **baby-snatcher,** *n.* one who abducts an infant; a person who marries or goes out with someone much younger. **baby-walker,** *n.* a frame on wheels for supporting a baby learning to walk. **babyhood,** *n.* **babyish,** *a.* **babyishness,** *n.* **babyism,** *n.*
Babylon (bab′ilon), *n.* the âncient capital of the Chaldaean empire; Rome; the papacy; a great and dissolute city. **Babylonian** (-lō′-), *a.* **Babylonic** (-lon′-), **Babylonish,** *a.*
baccalaureate (bakəlaw′riət), *n.* the university degree of bachelor.
baccara, baccarat (bak′ərah), *n.* a gambling card game between banker and punters.
baccate (bak′āt), *a.* berried, bearing berries; berry-like.
bacchanal (bak′ənəl), *a.* of or pertaining to Bacchus, the god of wine, or his festivities; hence characterized by drunken revelry. *n.* a follower of Bacchus; hence, a drunken reveller; a song or dance, or (*pl.*) a festival, in honour of Bacchus; an orgy. **bacchanalia** (-nä′liə), *n.pl.* the festival of Bacchus; drunken revelry. **bacchanalian,** *n., a.* bacchanal. **bacchanalianism,** *n.* **bacchant,** *n.* a follower of Bacchus; hence, a drunken reveller. *a.* worshipping Bacchus; fond of drinking. **bacchante** (-kan′ti), *n.* a priestess of Bacchus. **bacchantic** (-kan′-), *a.* **bacchic,** *a.* pertaining or relating to Bacchus or his worship; hence, frenzied; riotously festive.
bacci-, *comb. form* pertaining to a berry or berries. **bacciferous** (baksif′ərəs), *a.* **bacciform** (bak′sifawm), *a.* **baccivorous** (baksiv′ərəs), *a.*
baccy, *n.* short for TOBACCO.
bach BACHELOR.
bachelor (bach′ələ), *n.* an unmarried man; (also **bachelor girl**) an unmarried woman; a man or woman who has taken the first degree of a university below master or doctor. **bachelor flat,** *n.* a small flat suitable for a single person. **bachelor's buttons,** *n.pl.* a name for several plants with button-like flowers. **bachelorhood** (-hud), **bachelorship,** *n.* **bachelorism,** *n.* a peculiarity of a bachelor. **bach¹,** *n.* (*coll.*) a bachelor. *v.i.* to live as a bachelor.
bacillus (bəsil′əs), *n.* (*pl.* **-lli,** (-ī)) a microscopic, rod-like (disease-causing) bacterium. **bacillar, bacillary,** *a.* of, pertaining to or consisting of little rods; of, pertaining to or caused by bacilli. **bacilliform** (-ifawm), *a.* rod-shaped.
back¹ (bak), *n.* the hinder part of the human body, from the neck to the lower end of the spine; the corresponding portion in other animals; the surface of any object opposite to the face or front; the hinder part, the part away from the actor or speaker; the keel of a ship; one of the players whose duty it is to defend the goal in football (**half** and **three-quarter backs** are stationed nearer the front). *a.* situated behind or in the rear; coming back, turned back; behind in time; distant, inferior. *adv.* in a direction to the rear; to the place from which one came; to a former state, position or condition; behind, not advancing, behindhand; in return, in retaliation; in a position behind; in a state of check; in time past; again; in returning. *v.t.* to furnish with or constitute a back or backing; to support materially or morally; to bet in favour of; to mount or get on the back of; to countersign, to endorse; to cause to move back; to reverse the action of. *v.i.* to retreat; to move in a reversed direction. **at the back of**

one's mind, not consciously thought of. **back and forth**, backwards and forwards, up and down. **back of**, (*N Am.*) behind. **back to front**, the wrong way round. **behind one's back**, surreptitiously. **on the back burner**, not of immediate importance. **on the back of**, weighing as a heavy burden on; in addition to. **the back of beyond**, an extremely remote place. **to back down, out**, to move backwards; to retreat from a difficult situation. **to back the field**, to bet against all the horses except one. **to back the wrong horse**, to make a bad choice. **to back up**, to support; to duplicate a computer data file as security against damage to the original. **to back water**, to reverse the motion of the oars when rowing. **to break the back of**, to perform the greater part of (a piece of work). **to put one's back into**, to make a strenuous effort to perform (a task). **to put one's back up**, to cause resentment. **to see the back of**, to get rid of. **to turn the back (up)on**, to abandon, to forsake. **with one's back to the wall**, in a critical position. **back-bencher**, *n.* a member of Parliament without portfolio. **backbite**, *v.t., v.i.* to slander, censure or speak ill of. **back-biter**, *n.* **back-blocks**, *n.pl.* (*Austral.*) the interior parts of the continent or a station, esp. those far from a river. **back-blocker**, *n.* **back-board**, *n.* a board forming the back of anything; a board attached to the rim of a water-wheel to prevent the water running off the floats; a board strapped across the back to prevent stooping. **backbone**, *n.* the bony framework of the back, the spine; the spinal column; a main support or axis; strength of character. **to the backbone**, thoroughly. **backboned**, *a.* **back-breaking**, *a.* physically exhausting. **back-chat**, *n.* (*coll.*) flippant or impudent answering back. **back-cloth**, *n.* the curtain at the back of a stage; background. **back-comb**, *v.t.* to comb backwards with short, sharp strokes, making (the hair) fuzzy. **back-country**, *n.* thinly populated districts. **backdate**, *v.t.* to apply retrospectively from a particular date (e.g. a pay rise). **back-door**, *n.* a back or private entrance; an indirect or circuitous way. *a.* clandestine. **back-draught**, *n.* a backward draught of air; a hood for producing back-draught in a fire. **back-fire**, *n.* (*Motor*) premature combustion in the cylinder. *v.i.* to emit a loud noise as a result of premature combustion in the cylinder; (*coll.*) to fail and have the opposite effect. **back-formation**, *n.* the formation of a new word as if, e.g. by contraction, from an existing one (as *burgle* from *burglar*). **background**, *n.* the ground or surface behind the chief objects of a picture, stage-scene etc.; the setting; (*fig.*) obscurity; a person's upbringing, education and history. **background radiation**, *n.* low-level radiation present in the soil and atmosphere. **back hair**, *n.* the long hair at the back of a woman's head. **back-hand**, *n.* handwriting sloped backwards; the hand turned backwards (as at tennis) to take a ball at the left. **back-handed**, *a.* with the back of the hand; directed backwards; indirect. **back-hander**, *n.* a blow with the back of the hand; a bribe. **backlash**, *n.* jarring reaction in a piece of mechanism; an excessive reaction (against). **back-light**, *n.* (*Cinema*) a light projected on a subject from a source behind the camera. **backlog**, *n.* reserves or arrears of unfulfilled orders; an accumulation of business. **back-number**, *n.* a past issue of a newspaper or magazine; an out-of-date person or thing. **back pack**, *n.* a rucksack; the oxygen supply etc. carried by an astronaut. *v.i.* to hike with a rucksack. **back-packer**, *n.* **back passage**, *n.* the rectum. **back-pay**, *n.* arrears of pay. **back-pedal**, *v.i.* to press back the pedals of a cycle; to reverse a course of action; to restrain one's enthusiasm. **back-piece**, *n.* a piece forming the back of anything. **back-plate**, *n.* the piece forming the back of anything. **back-pressure**, *n.* resistance to the working of the piston, caused by waste steam or atmospheric pressure. **back-projection**, *n.* the projection of film on to a translucent screen so that the picture is visible from the other side. **backroom boys**, *n.pl.* (*coll.*) scientists and others who work in the background unrecognized. **back-scratcher**, *n.* an appliance for scratching the back; a flatterer. **back-scratching**, *n.* flattery; toadyism. **back seat**, *n.* the seat at the back of anything, such as a car or theatre; a position of less importance. **to take a back seat**, *n.* to accept an inferior role. **back-seat driver**, *n.* a passenger in a car who offers unwanted advice; one who offers advice on matters which do not concern him or her. **backside**, *n.* the back or hinder portion of anything; (*coll.*) the buttocks. **back-sight**, *n.* a sight taken backwards in land surveying; the sight of a rifle near the stock. **back-slang**, *n.* a slang in which ordinary words are pronounced backwards. **back-slide**, *v.i.* to fall into wrongdoing or false opinions; to relapse. **backslider**, *n.* **backspacer**, *n.* a typewriter key for moving the carriage backwards. **backspace**, *v.i.* **back spin**, *n.* in tennis, golf etc., a backward spin of a ball imparted to dull the bounce. **backstage**, *a., adv.* behind the scenes; out of public view. **backstairs**, *n.pl.* stairs in a house for the use of servants etc. *a.* clandestine, underhand. **back-stitch**, *n.* a method of sewing with stitches that are made to overlap. *v.t., v.i.* to sew in this manner. **back street**, *n.* a street away from the centre of the town; (*pl.*) the poorer streets of a town. **back-street abortion**, *n.* one performed by an unqualified person. **back-stroke**, *n.* a swimming stroke performed on the back. **backtrack**, *v.i.* to retrace one's steps; to reverse an opinion, attitude etc. **backup**, *n.* support; reinforcements. **backveld**, *n.* (*S Afr.*) country far removed from towns. **backwash**, *n.* the wash from the oars of a boat; eddy or swirl caused by a ship's propeller; reaction; aftermath; the rush of air from an aircraft engine. **backwater**, *n.* still water away from the main current; a condition or place of social etc. stagnation. **back-way**, *n.* a way leading to the back; a roundabout way. **backwoods**, *n.pl.* remote, uncleared forest land; (*derog.*) a remote, uncultured area. **backwoodsman**, *n.* **back yard**, *n.* a yard or garden at the back of a house. **in one's back yard**, (too) close to home; closely concerning one. **backed**, *a.* (*chiefly in comb.*) provided with a back; supported, betted on; accepted. **backer**, *n.* one who backs or supports; one who bets on a horse etc.; a book-maker. **backing**, *n.* supporting, seconding; one who or that which supports; money supplied for a project by an investor; musical accompaniment, esp. for a popular song; a piece forming the back or lining the back; backward motion, esp. of the wind in the opposite direction to that of the sun. **backing group**, *n.* one that provides a musical backing.

back[2] (bak), *n.* a large tub used in brewing, distilling, dyeing etc.

backgammon (bak'gamən), *n.* a game played by two persons on a table with draughtsmen, the moves being determined by throwing dice; the highest win in backgammon. *v.t.* to defeat at backgammon.

backing BACK[1].

backsheesh (bak'sheesh), BAKSHEESH.

backwards, backward (bak'wəd), *adv.* with the back foremost; towards the back or rear; behind, towards the starting-point; towards past time; towards a worse state or condition, in reverse order. **backward**, *a.* directed to the back or backwards; reversed, reluctant; esp. of the season, crops etc., behind in time, late; behind in progress; towards or into past time. **backward(s) and forward(s)**, to and fro; vacillating. **to bend, lean, fall over backwards**, to go to extreme lengths to please or accommodate someone. **backwardation**, *n.* (*Stock Exchange*) a consideration paid by a seller of stock for the privilege of delaying its delivery. **backwardly**, *adv.* **backwardness**, *n.*

baclava (bəklah'və), BAKLAVA.

bacon (bā'kən), *n.* the back and sides of a pig, cured by salting and drying with or without wood-smoke. **to bring home the bacon**, (*coll.*) to succeed; to provide a living. **to save one's bacon**, to escape from injury or loss. **bacony**, *a.*

Baconian (bəkō'niən), *a.* of or pertaining to Francis *Bacon* (1561-1626) or his inductive philosophy. *n.* a follower of Bacon; (*pop.*) a believer in the conceit that Bacon was really the author of Shakespeare's works.
bacteria, bacterial BACTERIUM.
bactericide (baktiə'risid), *n.* an agent that destroys bacteria. **bactericidal**, *a.*
bacteri(o)-, *comb. form* pertaining to bacteria.
bacteriology (baktiəriol'əji), *n.* the scientific study of bacteria. **bacteriological** (-loj'-), *a.* **bacteriologist**, *n.*
bacteriolysis (baktiəriol'isis), *n.* the destruction of bacteria. **bacteriolytic** (-lit'-), *a.*
bacteriophage (baktiə'riəfāj), *n.* a virus which destroys bacteria. **bacteriophagic** (-faj'-), *a.*
bacteriostasis (baktiəriəstā'sis), *n.* inhibition of the growth of bacterial cells. **bacteriostatic** (-stat'-), *a.*
bacterium (baktiə'riəm), *n.* (*pl.* -**ria** (-riə)) a member of a class (Schizomycetes) of microscopic unicellular organisms, often causing disease. **bacterial**, *a.*
bacteroid (bak'təroid), *a.* of the nature of or resembling a bacterium.
Bactrian (bak'triən), *a.* descriptive of a camel with two humps.
bad (bad), *a.* (*comp.* **worse** (wœs), *superl.* **worst**, (wœst)) not good, worthless; defective, incorrect; evil, hurtful, wicked, painful, dangerous, pernicious; in ill-health; injured, diseased; (*Law*) invalid; (*sl.*) very good. *n.* that which is bad; a bad state or condition. **to go bad**, to decay. **to go to the bad**, to go to ruin. **bad blood**, *n.* enmity. **bad debt**, *n.* a debt that cannot be recovered. **bad egg, lot, penny**, *n.* a bad speculation; a ne'er-do-well. **bad form**, *n.* bad manners. **bad grace**, *n.* unwillingness, reluctance. **bad hat**, *n.* a rogue; **bad lands**, *n.pl.* tracts of arid country in the western States of America; unsafe parts of a country. **bad mouth**, *v.t.* (*coll.*) to abuse. **bad word**, *n.* criticism of someone; (*coll.*) a swear-word.
baddie, baddy, *n.* (*coll.*) a criminal or wrong-doer, esp. an evil character in fiction, cinema etc. **baddish**, *a.* rather bad. **badly**, *adv.* (*comp.* **worse**, *superl.* **worst**) in a bad manner; improperly, wickedly, unskilfully, imperfectly; defectively; faultily; dangerously, disastrously; (*coll.*) very much, by much. **badness**, *n.* the quality of being bad; inferiority; incorrectness, faultiness; wickedness; worthlessness.
bade (bad. bād), *past* BID.
badge (baj), *n.* a distinctive mark, sign or token; an emblem sewn on clothing; a feature or quality that characterizes. *v.t.* to mark with or as with a badge.
badger (baj'ə), *n.* a plantigrade animal about the size of a fox, with thick body and short legs, having a black and white striped coat. *v.t.* to worry, to tease, to annoy. **badger-baiting**, *n.* the setting of dogs to draw a badger from its earth or from a barrel. **badger-dog**, *n.* the German dachshund, with long body and short legs, used to draw badgers.
badinage (bad'inahzh, -nij), *n.* light good-humoured, playful talk; banter.
badminton (bad'mintən), *n.* a game resembling lawn-tennis, but played, usu. indoors, with shuttlecocks instead of balls; a kind of claret-cup.
baffle (baf'l), *v.t.* to frustrate, elude, escape, circumvent; to thwart, defeat; to confound. *v.i.* to struggle ineffectually. *n.* (*Acoustics*) a rigid appliance that regulates the distribution of sound-waves from a producer. **baffle-board**, *n.* a device to prevent the carrying of noise. **baffle-plate**, *n.* (*Eng.*) a plate used to direct the flow of fluid. **bafflement**, *n.* **baffler**, *n.* **baffling**, *a.* bewildering; thwarting; of winds, variable. **bafflingly**, *adv.*
baft (bahft), *adv.* abaft, astern.
BAFTA (baf'tə), (*abbr.*) British Academy of Film and Television Arts.
bag (bag), *n.* a pouch, small sack or other flexible receptacle; a measure of quantity, varying with different commodities; the contents of such a measure; the result of a day's sport or of a hunting expedition; a purse; an udder; a sac in animal bodies containing some secretion; (*pl.*) loose clothes, esp. trousers; (*pl.*) (*coll.*) quantities; (*sl.*) a slovenly, bad tempered or ugly woman, often in *old bag*. *v.t.* to put into a bag; to shoot, to catch; (*coll.*) to take, seize, appropriate (also **bags**). *v.i.* to swell as a bag; to hang loosely. **bag and baggage**, with all belongings; entirely, completely. **bag of bones**, someone very thin. **in the bag**, (*sl.*) certain to happen, succeed etc. **the whole bag of tricks**, everything; all means or expedients. **to let the cat out of the bag**, to reveal a secret. **bag lady**, *n.* (*coll.*) a female vagrant. **bagman**, *n.* (*coll.*) a travelling salesman; a vagrant; (*N Am. sl.*) one who collects and transports money for gangsters. **bag-swinger**, *n.* (*Austral.*) a book-maker. **bag-wig**, *n.* a wig fashionable in the 18th cent. in which the back hair was enclosed in a bag. **bagful**, *n.* as much as a bag will hold. **bagging**, *n.* material for making bags. **baggy**, *a.* loose; bulging out like a bag; of trousers, sweater etc., stretched by wear. **baggily**, *adv.* **bagginess**, *n.*
bagasse (bəgas'), *n.* the refuse products in sugar-making; cane-trash.
bagatelle (bagətel'), *n.* a trifle, a trumpery amount; a game played on a nine-holed board, with nine balls and a cue; a light piece of music.
bagel (bā'gl), *n.* a doughnut-shaped bread roll.
baggage (bag'ij), *n.* luggage; portable belongings of an army; a woman of loose character; a playful arch young woman. *a.* used for carrying or looking after or conveying baggage. **baggage-car**, *n.* (*N Am.*) a railway luggage-van. **baggage-man, -master**, *n.* (*N Am.*) a guard in charge of passengers' luggage. **baggage-train**, *n.* the part of an army that convoys the baggage.
bagman BAG.
bagnio (ban'yō), *n.* (*pl.* -**nios**) a bathing-house, a bath; a brothel.
bagpipe (bag'pīp), *n.* a musical instrument consisting of a wind-bag and several reed-pipes into which the air is pressed by the player.
baguette, -guet (baget'), *n.* a precious stone cut into a rectangular shape; a narrow stick of French bread.
bah (bah), *int.* an expression of contempt.
Bahai (bəhah'i), *n.* a follower of a religious movement originating in Iran in the 19th cent., which stresses the validity of all world religions and the spiritual unity of all humanity. **Bahaism**, *n.* **Bahaist**, *n.* [Pers. *baha'i*, lit. of glory]
bail[1] (bāl), *n.* the temporary release of a prisoner from custody on security given for his or her due surrender when required; the money security, or the person or persons giving this security. *v.t.* to obtain the liberation of by giving sureties; to admit to or release on bail; to deliver (goods) in trust on an expressed or implied contract. **to bail out**, to procure release on bail from prison. **to stand bail**, to secure freedom until trial for an accused on payment of surety. **bail-bond**, *n.* a bond entered into by a prisoner upon release on bail, and his sureties. **bailsman**, *n.* one who gives bail. **bailable**, *a.* entitled to be admitted to bail; admitting of bail.
bail[2] (bāl), *n.* a hoop or ring; the handle of a kettle.
bail[3] (bāl), *n.* a division between the stalls of a stable; (*Austral.*) a framework for securing the head of a cow while she is being milked; (*pl.*) (*Cricket*) the crosspieces laid on the top of the wicket. *v.i.* to surrender by throwing up the arms. **bailer**, *n.* (*Cricket*) a ball that hits off the bails.
bail[4], **bale**[3] (bāl), *v.t.* to throw (water) out of a boat with a shallow vessel; to empty a boat of water. **to bail out** BALE[3]. **bailer, baler**, *n.* one who or that which bails water out of a boat etc.

bailey (bā'li), *n.* the wall enclosing the outer court of a feudal castle; the outer court itself; any other courts or enclosures of courts, the *outer bailey* or the *inner bailey*. **Old Bailey**, the Central Criminal Court standing at the outer boundary of the old wall of London.
Bailey bridge (bā'li), *n.* (*Mil.*) a bridge of lattice steel construction made of standard parts for rapid erection and transport. [Sir Donald *Bailey*, 1901-85, its inventor]
bailie (bā'li), *n.* a Scottish municipal magistrate corresponding to an English alderman.
bailiff (bā'lif), *n.* an officer appointed for the administration of justice in a certain bailiwick or district; a sheriff's officer who executes writs and distrains; an agent or steward to a land-owner.
bailiwick (bā'liwik), *n.* the district within which a bailie or bailiff possesses jurisdiction; one's field of interest or expertise.
bailment BAIL [1].
bain-marie (bimərē'), *n.* a vessel of boiling water into which saucepans are put to warm; a double saucepan.
Bairam (birəm'), *n.* the name of two Muslim festivals following the Ramadan, the *Lesser* lasting three days, the *Greater*, which falls seventy days later, lasting four days.
bairn (beən), *n.* (*Sc.*) a child of either sex.
bait (bāt), *v.t.* to furnish (a hook, gin, snare etc.) with real or sham food; to tempt, allure; to give food to (a horse) on a journey; to set dogs to worry (an animal); to worry, harass, torment. *n.* a worm, insect etc. put on a hook, gin, snare etc., to attract fish or animals; food, refreshment on a journey; a temptation, allurement. **live bait**, small fish used alive for bait. **baiting**, *n.* (*usu. in comb.*) worrying with dogs, as *badger-baiting*, *bear-baiting*, *bull-baiting*.
baize (bāz), *n.* a woollen-like flannel fabric.
bake (bāk), *v.t.* to cook by dry conducted heat in an oven or on a heated surface; to dry and harden by means of fire or by the sun's rays. *v.i.* to undergo the process of baking; to become dry and hard by heat. **to bake blind** BLIND. **bakehouse**, *n.* a building in which baking is carried on. **baked**, *a.* **half-baked**, *a.* (*coll.*) raw, uncouth, half-witted, soft. **baked Alaska**, *n.* a dessert of ice-cream covered with meringue baked in an oven. **baked beans**, *n.pl.* haricot beans baked and usu. tinned in tomato sauce. **baker**, *n.* one whose occupation is to bake bread, biscuits etc. **baker's dozen**, *n.* thirteen. **bakery**, *n.* the trade or calling of a baker; a baker's establishment. **baking**, *n.* the action of the verb **to bake**; the quantity baked at one operation. **baking-powder**, *n.* a powder of bicarbonate of soda and tartaric acid used as a raising agent.
Bakelite® (bā'kəlīt), *n.* a synthetic resin used for insulation and in the manufacture of plastics, paints and varnishes.
baklava (bəklah'və), *n.* a cake made from layered pastry strips with nuts and honey.
baksheesh, bakhshish (bak'shēsh), *n.* a gratuity, a tip.
balaclava helmet (baləklah'və), *n.* a woollen hood covering the ears, chin and neck.
balalaika (baləlī'kə), *n.* a Russian three-stringed triangular musical instrument.
balance (bal'əns), *n.* (*often pl.*) a pair of scales or other instrument used for weighing; equality of weight or power; the amount necessary to make two unequal amounts equal; an impartial state of mind; that which renders weight or authority equal; the difference between the debtor and creditor side of the account; harmony of design; a zodiacal constellation, Libra; the seventh sign of the zodiac; (*coll.*) the remainder. *v.t.* to weigh; to compare by weighing; to compare; to equalize, to steady; to adjust an account, to make two amounts equal. *v.i.* to have equal weight or force; to oscillate. **balance of payments**, the difference over a period of time between the total payments (for goods and services) to, and total receipts from, abroad. **balance of power**, a condition of equilibrium among sovereign states, supposed to be a guarantee of peace. **balance of trade**, the difference between the imports and exports of a country. **in the balance**, in an uncertain or undecided state. **on balance**, taking all factors into consideration. **to hold the balance**, to have the power of deciding. **to lose one's balance**, to tumble; to be upset mentally. **to strike a balance**, to reckon up the balance on a statement of credit and indebtedness. **balance-sheet**, *n.* a tabular statement of accounts, showing receipts and expenditure. **balance-wheel**, *n.* the wheel regulating the beat in watches. **balanceable**, *a.* **balanced**, *a.* having good balance; sane, sensible (*often in comb.*, as *well-balanced*). **balancer**, *n.* one who or that which balances; an acrobat.
balanid (bal'ənid), *n.* (*Zool.*) a member of the Balanidae, or acorn shells. **balaniferous** (-nif'-), *a.* acorn-bearing. **balanite**, *n.* a precious stone. **Balanoglossus** (-ōglos'əs), *n.* a genus of worm-like animals. **balanoid** (-oid), *a.* acorn-shaped.
balas (bal'əs), *n.* a rose-red variety of the spinel ruby.
balcony (bal'kəni), *n.* a gallery or platform projecting from a house or other building; in theatres, a tier of seats between the dress-circle and the gallery; (*N Am.*) dress-circle. **balconied** (-nid), *a.*
bald (bawld), *a.* without hair upon the crown of the head; bare, treeless, leafless; meagre; destitute of ornament or grace; undisguised, shameless. **bald-coot, baldicoot**, *n.* the coot, from its broad white frontal plate; bald-head. **bald-eagle**, *n.* an American eagle with a white head, used as a national emblem. **bald-faced**, *a.* of horses, having the face marked with white. **bald-head**, *n.* one who is bald; a variety of pigeon. **bald-headed**, *a.* **to go at it bald-headed**, to attack or undertake something boldly, regardless of consequences. **bald-pate**, *n.* one who is bald; a variety of duck and pigeon. *a.* bald. **bald-pated**, *a.* **bald-rib**, *n.* a joint of pork cut from nearer the rump than the spare-rib; a lean person. **balding**, *a.* **baldish**, *a.* **baldly**, *adv.* in a bald manner; nakedly, shamelessly, inelegantly; plainly, directly. **baldness**, *n.*
baldachin, -quin (bawl'dəkin), **baldachino** (-kē'nō), *n.* a canopy over an altar, throne or doorway, generally supported by pillars, but sometimes suspended from above.
balderdash (bawl'dədash), *n.* confused speech or writing; rubbish, nonsense.
baldric (bawl'drik), *n.* a richly ornamented girdle or belt, passing over one shoulder and under the opposite, to support dagger, sword, bugle etc. **baldric-wise**, *adv.*
bale[1] (bāl), *n.* evil, mischief, calamity; pain, sorrow. **baleful**, *a.* full of evil; pernicious, harmful, deadly. **balefully**, *adv.* **balefulness**, *n.*
bale[2] (bāl), *n.* a package, a certain quantity of goods or merchandise, wrapped and corded for transportation. *v.t.* to pack in a bale or bales. **bale-goods**, *n.pl.* goods done up in bales. **baling**, *n.* the process of putting goods into bales. **baling-paper**, *n.* (*N Am.*) **baling-press**, *n.* one used to compress goods before putting them into bales.
bale[3], **bail**[4] (bāl), **to bale out**, to abandon an aeroplane in the air and descend by parachute; to help out of a difficulty.
baleen (bəlēn'), *n., a.* (of) whalebone.
balefire (bāl'fīə), *n.* a beacon-fire; a bonfire.
balistite (bal'istīt, -lis'-), *n.* ballistite.
balk, baulk (bawlk, bawk), *n.* a beam of timber; the headline of a fishing-net; the part of a billiard table behind a transverse line; a hindrance, a check; a disappointment. *v.t.* to refuse; to avoid, let slip; to check, hinder; to disappoint; to evade, frustrate; to dispute, argue contentiously. *v.i.* to turn aside, to swerve, to refuse a leap. **to make a balk**, (*Billiards*) to leave one's own ball and the red inside the balk when the opponent's is in hand. **balked**, *a.* **balky**, *a.* of a horse, prone to balk or swerve.

Balkanize, -ise (bawl'kəniz), *v.t.* to split (a region) into a number of smaller and often mutually hostile states, as occurred in the *Balkan* peninsula, SE Europe, during the 19th and early 20th cents. **Balkanization,** *n.*
ball¹ (bawl), *n.* a spherical object, a globe; a round or ovall object used in games; a game with a ball; a throw or cast of the ball in games; a bullet (not now usually spherical) or larger round projectile for ordnance, esp. a solid projectile; things or parts of things with spherical or rounded outlines; (*pl, sl.*) testicles. *v.t.* to clog (as a horse's foot with a collection of snow). *v.i.* to gather into a ball; to become clogged. **ball and socket joint,** one (e.g. the hip joint) in which a round head fits in a round cavity and moves freely in any direction. **ball of fire,** (*coll.*) a dynamic or lively individual. **ball of the foot,** the rounded part of the base of the great toe. **ball of the thumb,** the corresponding part of the hand. **on the ball,** alert; in control. **the ball's in your court,** it's your turn to act. **three balls,** a pawnbroker's sign. **to balls up, to make a balls of,** (*sl.*) to make a mess of; to botch. **to keep the ball rolling,** to keep the conversation etc. from flagging. **to play ball,** (*coll.*) to cooperate. **ball-bearing,** *n.* (*usu. pl.*) a bearing containing loose metallic balls for lessening friction; one of these metal balls. **ball-cartridge,** *n.* a cartridge containing a bullet. **ball-cock, -tap,** *n.* a self-acting tap which is turned off or on by the rising or falling of a hollow ball on the surface of the water in a cistern, boiler etc. **ballgame,** *n.* a game played with a ball; (*N Am.*) baseball. **a different ballgame,** (*coll.*) something quite different. **ball lightning,** *n.* floating luminous balls sometimes seen during thunderstorms. **ballpark,** *n.* (*N Am.*) a baseball field; a sphere of activity. **ball-point pen,** *n.* a fountain pen with a tiny ball in place of a nib. **ball-proof,** *a.* impenetrable by bullets. **ball-valve,** *n.* a valve opened or closed by the rising of a ball. **balled,** *a.* formed into a ball. **balls!** *int.* (*sl.*) nonsense.
ball² (bawl), *n.* a social assembly for dancing. **to have a ball,** (*coll.*) to have a good time. **ball-room,** *n.*
ballad (bal'əd), *n.* light simple song; a popular song, generally of a personal or political character; a simple narrative poem. *v.t.* to make (someone) the subject of a ballad or ballads; to satirize ballad-wise. **ballad-maker,** *n.* a writer of ballads. **ballad-monger,** *n.* one who sells ballads; a contemptuous epithet for a composer of ballads. **ballad-singer, balladeer,** *n.* one who sings ballads, esp. in the streets. **balladry,** *n.* the ballad style of composition; ballads collectively. **ballad-wise,** *adv.* in the form of a ballad.
ballade (bəlahd'), *n.* a poem consisting of three eight-lined stanzas, each having the same line as a refrain, and with an envoy of four lines.
ballast (bal'əst), *n.* heavy substances placed in the bottom of a ship to lower the centre of gravity and steady it; gravel or other material laid as foundation for a railway or road; that which tends to give intellectual or moral stability. *v.t.* to furnish with ballast; to steady. **in ballast,** having only ballast in the hold; used for ballasting. **ballastage** (-ij), *n.* a toll paid for the privilege of taking ballast. **ballasting,** *n.*
ballerina (baləre'nə), *n.* (*pl.* **-rine, -rinas**) a female ballet dancer; a female dancer taking a leading part in a ballet.
ballet (bal'ā), *n.* a dramatic representation, consisting of dancing and pantomime; an artistic exhibition of dancing. **ballet-girl,** *n.* **ballet-master, -mistress,** *n.* the director of a ballet. **balletic,** *a.* **balletomane** (bəlet'əmān), *n.* an enthusiast for the ballet. **balletomania,** *n.*
ballista (bəlis'tə), *n.* (*pl.* **-ae** (-ē), **-as**) a military engine used in ancient times for hurling stones etc.
ballistic (bəlis'tik), *a.* of or pertaining to the hurling and flight of projectiles; (*sl.*) extremely angry. **ballistic missile,** *n.* (*Mil.*) one guided over the first part of its course but then descending according to the laws of ballistics. **ballistics,** *n.sing.* the science of the flight of projectiles.
ballistite (-it), *n.* a propellant explosive based on nitroglycerine and nitrocellulose.
ballocks (bol'əks) BOLLOCKS.
balloon (bəloon'), *n.* a bag of rubber etc. filled with hot air or a light gas so as to float in the air with a basket carrying passengers or instruments; an inflatable rubber bag used as a child's toy; a large brandy glass; a line enclosing the words or thoughts of a cartoon character; anything inflated or hollow. *v.i.* to go up in a balloon; to swell out. **captive balloon,** a balloon held by a rope. **like a lead balloon,** a complete failure. **pilot balloon,** a small balloon sent up in advance to show the direction and strength of the wind. **when the balloon goes up,** when the troubles start. **balloon angioplasty,** *n.* a technique for treating blocked arteries, in which a tiny balloon is inserted into the blockage and inflated. **balloon barrage** BARRAGE². **balloon-fish,** *n.* popular name for fishes belonging to the genus *Diodon*, which are able to distend their bodies with air. **balloon tyre,** *n.* a low-pressure tyre, large in section. **ballooner,** *n.* **ballooning,** *n.* the practice of making balloon ascents; (*N Am.*) the practice of running up stock above its value. **balloonist,** *n.*
ballot (bal'ət), *n.* paper or other instrument used to give a secret vote; the method or system of secret voting; the total votes recorded. *v.t.* to ask to vote secretly. *v.i.* to vote secretly. **to ballot for,** to choose by secret voting. **ballot-box,** *n.* a box into which ballot-papers are put. **ballot-paper,** *n.* the voting-paper used in voting by ballot.
bally (bal'i), *a.* (*sl., euphem.*) bloody.
ballyhoo (balihoo'), *n.* noisy and unprincipled propaganda; a great fuss about nothing.
ballyrag (bal'irag), *v.t.* to victimize with abuse or practical jokes. *v.i.* to engage in horseplay.
balm (bahm), *n.* the fragrant juice, sap or gum of certain trees or plants; fragrant ointment or oil; anything which soothes or heals; the popular name of several fragrant garden herbs. *v.t.* to anoint or impregnate with balm; to soothe, to assuage. **Balm of Gilead** (gil'iad), the gum of *Balsamodendron gileadense*, used as antiseptic and vulnerary; an imitation of this. **balm-cricket,** *n.* the cicada. **balmy,** *a.* producing balm; impregnated with or having the qualities of balm; soft, soothing, healing; mild; (*sl.*) barmy. **balmily,** *adv.* **balminess,** *n.*
Balmoral (balmor'rəl), *n.* a kind of Scottish cap; (*pl.*) ankle boots laced in front.
balneology (balniol'əji), *n.* the science of treating diseases by bathing and medicinal springs.
baloney, boloney (bəlō'ni), *n.* (*sl.*) nonsense.
BALPA (bal'pə), (*acronym*) British Airline Pilots' Association.
balsa (bawl'sə), *n.* an American tropical tree with light, strong wood used for rafts, model aircraft etc.
balsam (bawl'səm), *n.* a fragrant resin produced by any of several trees, used medicinally and in perfumes; a tree yielding balsam; popular name of the genus *Impatiens*; CANADA BALSAM; anything that possesses healing or soothing qualities. *v.t.* to impregnate or perfume with balsam; to heal, soothe. **balsam-apple,** *n.* a tropical plant of the gourd family bearing a highly coloured fruit; (*erroneously*) the common garden balsam. **balsam-fir,** *n.* a N American fir which yields Canada balsam. **balsamic** (-sam'-), **balsamous,** *a.* having the qualities of balsam; assuaging pain, soothing. **balsamically,** *adv.* **balsamiferous** (-mif'-), *a.* **Balsamodendron,** *n.* a genus of trees which exude balsam. **balsamy,** *a.* balsam-like; balmy.
balsamine (bal'səmēn), *n.* the English name of *Impatiens balsamina*; (*erroneously*) the balsam-apple.
Baltic (bawl'tik), *a.* pertaining to the Baltic sea in N Europe or its bordering provinces; of, or denoting Baltic as a group of languages. *n.* a branch of the Indo-European languages comprising Latvian, Lithuanian and

Old Prussian. Baltoslav, *n.* **Baltoslavic, -slavonic,** *a.* a subfamily of Indo-European languages containing Baltic and Slavonic.
Baltimore, Baltimore bird (bawl'timaw), *n.* a N American bird of the starling family with black head and orange plumage.
baluster (bal'əstə), *n.* a small column forming part of a series called a balustrade; a post supporting a hand-rail, a banister. **balustered,** *a.* **balustrade** (-strād), *n.* a range of balusters, resting on a plinth, supporting a coping or rail.
bambino (bambē'nō), *n.* (*pl.* **-nos, -ni**) a child, a baby; esp. an image of the infant Jesus in the crib.
bamboo (bamboo'), *n.* (*pl.* **-boos**) any of a genus, *Bambusa,* of giant tropical grasses; the stem of such grass used as a stick, thatch, building material etc. **bamboo curtain,** *n.* the barrier set up between Communist China and the rest of the world.
bamboozle (bamboo'zl), *v.t.* to mystify for purposes of fraud; to swindle; to bewilder. *v.i.* to practise trickery. *n.* bamboozlement. **bamboozlement,** *n.* the act or process of bamboozling; a hoax.
ban (ban), *v.t.* (*past, p.p.* **banned**) to forbid; to proscribe. *n.* an edict of excommunication, an interdict; a formal or official prohibition; a proclamation of outlawry. (*pl.*) BANNS.
banal (bənahl'), *a.* commonplace, trite, petty. **banality** (-nal'-), *n.* a trite remark; commonplaceness.
banana (bənah'nə), *n.* a tropical and subtropical tree, *Musa sapientum,* closely allied to the plantain; its long, yellow, crescent-shaped fruit. **to be, go bananas,** (*sl.*) to be or go insane. **Bananaland,** *n.* (*coll.*) Queensland. **banana republic,** *n.* (*offensive*) a small tropical country, politically unstable and dominated by foreign capital. **banana skin,** *n.* any episode or occurrence which leads to humiliation or embarrassment. **banana split,** *n.* a dessert consisting of a banana sliced lengthwise and filled with ice-cream etc.
banausic (bənaw'sik), *a.* suitable for a mechanic.
Banbury-cake (ban'bəri), *n.* a kind of cake filled with mincemeat, supposed to be made at Banbury in Oxfordshire. **Banbury-man,** *n.* an overzealous Puritan; a puritanical hypocrite.
banc (bangk), **banco**[1] (-ō), *n.* the Bench. **in banc, in banco,** in full court.
banco[2] (bang'kō), *a.* a term applied to bank money of account, as distinguished from ordinary currency.
band[1] (band), *n.* that which binds or restrains; a fillet, a tie, a chain; one of the cords on which a book is sewn; a bond, a tie, a uniting influence; (*pl.*) fetters.
band[2] (band), *n.* a flat slip or band (BAND [1]), used to bind together, encircle or confine, or as part of an article of apparel; (*pl.*) a pair of linen strips hanging down in front from the collar and forming part of clerical, legal or academical dress; a bandage; a broad, endless strap for communicating motion; a specific range of frequencies or wavelengths; a track of a record or magnetic tape; a division of pupils according to ability. **Band-Aid**[※], *n.* a small adhesive plaster with a medicated gauze pad. **band-aid,** *a.* of measures etc., temporary. **bandbox,** *n.* a box for holding collars, hats, millinery etc.; a flimsy affair. **band brake,** *n.* (*Mech.*) a flexible band that grips the periphery of a drum or wheel. **band-fish,** *n.* a Mediterranean fish of the genus *Cepola,* from their ribbon-like shape. **band-saw,** *n.* an endless saw running over wheels. **band-wheel,** *n.* a wheel worked by means of an endless strap. **bandwidth,** *n.* the range of frequencies used for a particular radio transmission. **banding,** *n.* grouping schoolchildren according to ability.
band[3] (band), *n.* an organized company; a confederation; a company of musicians trained to play together. **Band of Hope,** a name given about 1850 to any association of children pledged to total abstinence. **band-master,** *n.* the leader of a band of musicians. **bandsman,** *n.* a member of a band of musicians. **band-stand,** *n.* a platform for the use of a band of musicians. **band-wagon,** *n.* the musicians' wagon in a circus parade. **to climb on the band-wagon,** to try to be on the winning side.
band[4] (band), *v.t.* to bind or fasten with a band; to mark with a band; to form into a band, troop or society. *v.i.* to unite, to assemble.
band[5] BANDY [1].
bandage (ban'dij), *n.* a strip of flexible material used to bind up wounds, fractures etc.; *v.t.* the operation of bandaging; a similar strip used to cover up something; (*Arch.*) a tie or bond. *v.t.* to bind up with a bandage.
bandanna, bandana (bandan'ə), *n.* a silk or cotton handkerchief with a spotted pattern.
b. and b., (*abbr.*) bed and breakfast.
bandeau (ban'dō), *n.* (*pl.* **-deaux** (-dōz) a narrow band or ribbon for the head.
bandelet (ban'dələt), *n.* a small stripe or band; a small flat moulding round a column.
banderilla (bandəre'yə, -rēl'-), *n.* a little dart ornamented with ribbons, which bull-fighters stick in the neck of the bull.
banderol, banderole (ban'dərōl), *n.* a long narrow flag with a cleft end; any small ornamental streamer; the small square of silk hanging from a trumpet.
bandicoot (ban'dikoot), *n.* a large Indian rat (*Mus giganteus*); the marsupial genus *Perameles,* which has some resemblance to this.
bandit (ban'dit), *n.* (*pl.* **-ditti** (-dit'ē), **-dits**) an outlaw; a brigand. **banditti,** *n.sing.* a company of bandits.
bandog (ban'dog), *n.* a large fierce dog, kept chained, a mastiff, a bloodhound.
bandoleer, bandolier (bandəliə'), *n.* a belt worn over the shoulder with little leather loops to receive cartridges; (*usu. pl.*) the cases or boxes containing charges.
bandore (bandaw'. ban'-), *n.* an old musical instrument somewhat resembling a lute.
bandy[1] (ban'di), *v.t.* to toss to and fro or toss about like a ball; to exchange (esp. blows, arguments etc.); to band together, make into a faction. **to bandy words,** to wrangle. **bandy ball,** *n.* hockey.
bandy[2] (ban'di), *a.* crooked, bent. **bandy-legged,** *a.*
bane (bān), *n.* poison (*chiefly in comb.*, as *henbane, rat's bane* etc.); that which causes ruin; destruction, woe. **baneberry,** *n.* herb Christopher, a plant with black poisonous berries. **banewort,** *n.* the lesser spearwort; the deadly nightshade. **baneful,** *a.* poisonous, harmful. **banefully,** *adv.* **banefulness,** *n.*
bang[1] (bang), *v.t.* to beat with loud blows; to thrash; to handle roughly; to slam (a door), fire (a gun), beat (a musical instrument) with a loud noise; to cut (the front hair) square across; (*sl.*) to have sexual intercourse with. *v.i.* to resound with a loud noise; to jump or bounce up noisily. *n.* a resounding blow; a sudden explosive noise; (*sl.*) an act of sexual intercourse; the front hair cut straight across. *adv.* with a violent blow or noise; suddenly, abruptly, all at once. **bang on,** (*sl.*) exactly (right); to talk at great length. **to bang away at,** to do something violently or noisily. **to bang up,** (*sl.*) to shut up in a (prison) cell. **with a bang,** very well, successfully. **bang-up,** *a.* (*sl.*) fine, first-rate. **banger,** *n.* (*sl.*) a sausage; (*coll.*) a decrepit old car; a small explosive firework.
bang[2] BHANG.
bangle (bang'gl), *n.* a ring-bracelet or anklet. **bangled,** *a.* adorned with bangles.
banian BANYAN.
banish (ban'ish), *v.t.* to condemn to exile; to drive out or away; to expel. **banishment,** *n..*
banister (ban'istə), *n.* a shaft or upright supporting a hand-rail at the side of a staircase; (*pl.*) the whole railing protecting the outer side of a staircase.

banjo   61   baptize

banjo (ban'jō), *n.* (*pl.* **-jos, -joes**) a stringed musical instrument, having a head and neck like those of a guitar and a body like that of a tambourine, and played with the fingers. **banjoist,** *n.*

bank[1] (bangk), *n.* a raised shelf or ridge of ground; a slope; a mound of sand etc. in the sea etc.; (the ground near) the edge of a river; an embankment; the sides of a road, cutting or any hollow; an incline on a railway; a bed of shellfish; a long flat-topped mass, as of ice, snow, cloud or the like. *v.t.* to form a bank to; to confine within a bank or banks; to embank; to bring to land; to fortify with earthworks; (*Aviat.*) to incline inwards at a high angle in turning. *v.i.* to rise into banks. **to bank up,** to make up (a fire) by putting on and pressing down fuel. **bank engine,** *n.* a locomotive employed to assist trains up inclines. **bank-martin, -swallow,** *n.* the sand-martin. **bank-side,** *n.* the sloping side of a bank; (**Bankside**) the district bordering the Thames at Southwark. **banksman,** *n.* a workman who superintends unloading at a pit-mouth. **banker,** *n.* a horse good at jumping on and off high banks; (*Austral.*) a swollen river. **bankless,** *a.* not defined or limited by a bank.

bank[2] (bangk), *n.* an establishment which deals in money, receiving it on deposit from customers and investing it; (*Gaming*) the money which the dealer in some gambling games has before him or her; any store or reserve of material or information, as in *blood bank. v.i.* to keep a bank; to act as a banker; to be a depositor in a bank; (*Gaming*) to form a bank, to challenge all comers; (*coll.*) to count or depend (on). *v.t.* to deposit in a bank; convert into money. **Bank of England,** the central bank of England and Wales which manages the monetary systems on behalf of the government. **to break the bank,** to win the limit set by the management of a gambling house for a particular period. **bank-bill,** *n.* (*N Am.*) a bank note. **bankbook,** *n.* a pass-book in which the cashier enters the debits and credits of a customer. **bank credit,** *n.* permission to draw on a bank to a certain amount. **bank holiday,** *n.* a day on which all banks are legally closed, observed as a national holiday. **bank note,** *n.* a note issued by a bank and payable on demand. **bank rate,** *n.* (formerly) the rate at which the Bank of England was prepared to discount bills of exchange (SEE MINIMUM LENDING RATE). **bankable,** *a.* capable of being banked; guaranteed to produce a profit. **banker,** *n.* a proprietor of a bank; one involved in banking; one who keeps the bank at a gaming-table; the dealer in certain card games. **banker's card,** *n.* a card issued by a bank guaranteeing payment of cheques up to a certain limit.

bank[3] (bangk), *n.* the bench for rowers, or a tier of oars, in a galley; a range of instruments on a panel or keys on a typewriter; a bench or table used in various trades; (*Print.*) the table on which sheets are laid; the raised floor of a glass-furnace; a row of organ keys. **banker,** *n.* a sculptor's revolving table; a bench used by bricklayers or stonemasons.

banket (bang'kit, -ket'), *n.* a gold-bearing conglomerate.

bankrupt (bangk'rŭpt), *n.* one who cannot pay his/her debts and whose assets are distributed among the creditors. *a.* judicially declared bankrupt; insolvent; (*fig.*) without credit; at the end of one's resources. *v.t.* to render (a person) bankrupt; to render insolvent; to reduce to beggary, or to discredit. **bankruptcy** (-si), *n.*

banksia (bangk'siə), *n.* an Australian flowering shrub or tree of the family Proteaceae.

banlieue (bä'lyœ), *n.* (*F*) suburbs, precincts.

banner (ban'ə), *n.* the standard of a feudal lord, used as a rallying-point in battle; hence (*fig.*) **to join, follow, fight under the banner of;** an ensign or flag painted with some device or emblem; a flag, generally square, painted or embroidered with the arms of the person in whose honour it is borne; an ensign or symbol of principles or fellowship. **banner headline,** *n.* a headline in heavy type running across the entire page of a newspaper. **banner-screen,** *n.* a fire-screen suspended from a pole or mantelpiece by its upper edge. **bannered,** *a.* furnished with banners.

banneret (banərit), *n.* a title borne by certain officers in Switzerland and in some of the old Italian republics.

bannerol (ban'ərōl), *n.* a banderol.

bannock (ban'ək), *n.* a flat round cake baked on an iron plate over the fire.

banns (banz), *n.pl.* proclamation in church of an intended marriage, so that any impediment thereto may be made known and inquired into. **to forbid the banns,** to allege an impediment to an intended marriage.

banquet (bang'kwit), *n.* a sumptuous feast, usu. of a ceremonial character, followed by speeches. *v.t.* to entertain at a sumptuous feast. *v.i.* to take part in a banquet, to feast luxuriously. **banqueter,** *n.*

banquette (bäket'), *n.* built-in cushioned seating along a wall.

banshee (ban'shē), *n.* a supernatural being, supposed to wail round a house when one of the inmates is about to die.

bant BANTING.

bantam (ban'təm), *n.* a small domestic fowl, of which the cocks are very pugnacious; a small and conceited or very pugnacious person. **bantam-weight,** *n.* a boxer not exceeding 8 st. 6 lb. (53·5 kg) in weight if professional, or between 8 st. and 8 st. 7 lb. (51–54 kg) if amateur.

banter (ban'tə), *v.t.* to ridicule good-humouredly. *v.i.* to indulge in good-natured teasing or pleasantries. *n.* good-natured teasing or chat, chaff. **banterer,** *n.*

banting (ban'ting), *n.* the reduction of obesity by abstinence from fat, starch and sugar. **bant,** *v.i.* (*coll.*) to practise this method. [W. Banting, 1797–1878, inventor]

bantling (bant'ling), *n.* a little child, a brat.

Bantu (ban'too), *n.* a group of languages of S and Central Africa; a member of the peoples inhabiting these areas; an official name for Black S Africans; (*pl.* **-tu, -tus**) (*offensive*) a Bantu speaker. *a.* relating to these languages or peoples. **Bantustan** (-stahn), *n.* (*coll.*) a name applied to semi-autonomous regions of S Africa reserved for Black people.

banxring (bangks'ring), *n.* a Javanese squirrel-like tree-shrew.

banyan (ban'yan), **banian** (-yən), *n.* a Hindu merchant or shopkeeper, esp. in Bengal; a loose morning-gown or jacket; the banian-tree. **banian-day,** *n.* a day when sailors have no meat (in allusion to the vegetarian diet of Hindus). **banian-, banyan-tree,** *n.* an Indian fig tree with many aerial roots.

banzai (ban'zī, -zī'), *int.* Japanese battle-cry, patriotic salute or cheer.

baobab (bā'əbab), *n.* an African tree, *Adansonia digitata*, called also monkey-bread.

BAOR, (*abbr.*) British Army of the Rhine.

bap (bap), *n.* a large soft roll.

baptize, -ise (baptīz'). *v.t.* to sprinkle with or immerse in water as a sign of purification and consecration, esp. into the Christian Church; to consecrate, initiate; to christen, to give a name or nickname to; to initiate into or to introduce to for the first time. *v.i.* to administer baptism. **baptism** (bap'tizm), *n.* the act of baptizing. **baptism of blood,** martyrdom before baptism. **baptism of fire,** a soldier's first experience of actual war. **baptismal** (-tiz'-), *a.* conferred at baptism. **baptismally,** *adv.* **baptist** (bap'tist), *n.* one who baptizes; (**Baptist**) a special title of St John, the forerunner of Christ; (**Baptist**) a member of a Christian church which holds that baptism should be administered only to adult believers, and by immersion. **baptistery** (-təri), **baptistry** (-tri), *n.* the place where baptism is administered; the tank used for baptism in Baptist churches.

bar¹ (bah), *n.* a piece of wood, iron or other solid material, long in proportion to breadth; a transverse piece in a gate, window etc.; a connecting piece in various structures; a straight stripe, any thing that constitutes a hindrance or obstruction; a bank of silt, sand or gravel deposited at the mouth of a river or harbour; (a space marked off by) a rail or barrier; (*Law Courts*) the barrier at which prisoners stand during trial; the profession of a barrister; barristers collectively; any tribunal; the barrierr cutting off a space near the door in both Houses of Parliament, to which non-members are admitted; a counter in a public house etc. across which liquors etc. are sold; the room containing this; (*Mus.*) a vertical line drawn across the stave to divide a composition into parts of equal duration; the portion contained between two such lines; a metal strip attached to a medal, indicative of an additional award; any physical or moral barrier or obstacle; a counter or place where foods, goods or services are sold or provided. *v.t.* (*past, p.p.* **barred**) to fasten with a bar or bars; to obstruct, to exclude; to take exception to; to prevent; to mark with or form into bars; (*sl.*) to object to, dislike. **to be called within the bar**, to be made a Queen's Counsel. **to call to the bar**, to admit as a barrister. **trial at bar**, a trial in the Queen's Bench division. **bar-bell**, *n.* a metal bar with heavy disks at each end used for weightlifting and exercising. **bar chart, -graph,** *n.* a graph containing vertical or horizontal bars representing comparative quantities. **bar code,** *n.* an arrangement of lines of varied lengths and thicknesses which is machine-readable, e.g. printed on goods, giving coded details of price, quantity etc. **bar-iron,** *n.* iron in malleable bars. **bar-keeper,** *n.* a bartender; a toll-bar keeper. **barmaid,** *n. fem.* **barman,** *n.* a bartender. **bar-parlour,** *n.* a small room adjoining or containing a bar in a public house. **bar-posts,** *n.pl.* posts sunk in the ground to admit movable bars serving the purpose of a gate. **bar-room,** *n.* the room in a public house in which the bar is situated. **bar-sinister** BEND SINISTER. **bartender,** *n.* one who serves at the bar of a public house etc. **bar-tracery,** *n.* window tracery in which the stonework resembles a twisted bar. **barred,** *a.* furnished or secured with a bar or bars; obstructed by a bar; striped, streaked. **barring,** *n., a.*
bar² (bah), *n.* the maigre, a large European fish.
bar³ (bah), *n.* a unit of atmospheric pressure equivalent to 10^6 dynes per square centimetre (10^5 newtons per square metre).
bar⁴ (bah), *prep.* except, apart from. **bar none**, without exception. **bar one**, except one. **barring accidents,** apart from accidents.
bar., (*abbr.*) baritone; barometric; barrel; barrister.
bar- BAR(O)-.
barathea (barəthiə'), *n.* a light soft fabric made of a wool mixture.
barb¹ (bahb), *n.* a recurved point, as in a fish-hook or arrow; a point, a sting; a biting or pointed remark or comment; one of the lateral filaments from the shaft of a feather; (*Bot.*) a hooked hair. *v.t.* to furnish (fish-hooks, arrows etc.) with barbs. **barbed wire,** *n.* a wire armed with sharp points, used for fences etc.
barb² (bahb), *n.* a fine breed of horse; a fancy breed of pigeons (both orig. from Barbary).
barbarian (bahbeə'riən), *n.* a savage, an uncivilized person; one destitute of pity or humanity; one outside the pale of Christian civilization; a foreigner with outlandish manners and language. *a.* rude, uncivilized; cruel, inhuman. **barbaric** (-ba'-), *a.* of or pertaining to barbarians; rude, uncouth, uncivilized. **barbarism** (bah'bə-), *n.* an impropriety of speech, a foreign idiom; absence of civilization or culture; a concrete instance of this defect. **barbarity** (-ba'-), *n.* brutality, inhumanity, cruelty; an act of brutality or cruelty; the state or quality of being barbaric; a barbarism. **barbarize, -ise** (bah'bə-), *v.t.* to render barbarous; to corrupt (a language). *v.i.* to utter a barbarism in speech; to grow barbarous. **barbarization,** *n.* **barbarous** (bah'bə-), *a.* foreign in speech, harsh-sounding; uncivilized; cruel. **barbarously,** *adv.* **barbarousness,** *n.*
Barbary (bah'bəri), *n.* an extensive region in the north of Africa. **Barbary ape,** *n.* a tailless ape found in the north of Africa, with a colony on the rock of Gibraltar. **Barbary gum,** *n.* a gum obtained from *Acacia gummifera*. **Barbary hen** GUINEA-HEN. **Barbary-horse** BARB².
barbate (bah'bāt), *n.* (*Bot., Zool.*) bearded; having small tufts of hair.
barbecue (bah'bikū), *n.* a grill on which meat is smoked; an animal broiled or roasted whole; a party at which food is prepared outdoors over a charcoal fire; food so cooked. *v.t.* to smoke or dry (meat etc.) on a framework over a fire; to broil or roast whole.
barbel (bah'bl), *n.* a European freshwater fish allied to the carp.
barber (bah'bə), *n.* one who shaves and cuts beards and hair. *v.t.* to shave or dress the hair of. **barber's block,** *n.* a round block on which wigs were made up and displayed. **barbershop,** *n., a.* (of) a type of close harmony singing for male voices, usu. quartets. **barber's itch, rash,** *n.* sycosis, an inflammation of the roots of the hair. **barber's pole,** *n.* a pole usu. striped spirally, exhibited as a sign in front of a barber's shop.
barberry (bah'bəri), **berberry** (bœ'-), *n.* a shrub of the genus *Berberis*, esp. *B. vulgaris*; its red acid berry.
barbet (bah'bit), *n.* a tropical bird having tufts of hair at the base of its bill.
barbette (bahbet'), *n.* a mound of earth in a fortification on which guns are mounted to be fired over the parapet; a platform for a similar purpose on a warship. **barbette-cruiser,** *n.* a cruiser equipped with barbettes.
barbican (bah'bikən), *n.* an outer fortification to a city or castle, esp. over a gate or bridge and serving as a watchtower.
barbituric (bahbitū'rik), *a.* term applied to an acid obtained from malonic and uric acids. **barbitone** (bah'bitōn), (*N Am.*) **barbital,** *n.* a derivative of this used as a sedative, veronal. **barbiturates** (-bit'ūrits), *n.pl.* compounds with hypnotic and sedative properties derived from barbituric acid.
barbola (bahbō'lə), *n.* the attachment of small flowers etc. in paste to embellish vases etc.
barbule (bah'būl), *n.* a hooked or serrated filament given off from the barb of a feather.
barcarole, -rolle (bahkəról', -rol'), *n.* a song sung by Venetian gondoliers; a composition of a similar kind.
bard (bahd), *n.* a Celtic minstrel; hence, a poet generally; (*Welsh*) a poet honoured at the Eisteddfod. **Bard of Avon,** Shakespeare. **bardic,** *a.* **bardish,** *a.* **bardism,** *n.* the sentiments, maxims or system of the bards. **bardling** (-ling), *n.* a young bard, a tyro; a poetaster. **bardolatry** (-dol'ətri), *n.* the worship of Shakespeare.
bare (beə), *a.* unclothed, naked, destitute of natural covering, as hair, leaves etc.; defenceless; unsheathed; ill-furnished, empty; mere, unsupported, undisguised; bald, meagre; unadorned. *v.t.* to strip, to make bare; to uncover, unsheathe; to make manifest. **bareback, -backed,** *a., adv.* without a saddle. **barefaced,** *a.* having the face bare or uncovered; impudent, shameless; beardless. **barefacedly,** *adv.* **barefacedness,** *n.* **barefoot,** *a., adv.* with the feet naked. **barefoot doctor,** *n.* a villager, esp. in Asia, who has been trained in basic health care to meet the simple medical needs of the community. **bare-footed,** *a.* **bare-headed,** *a.* **bare poles,** *n.pl.* masts with no sails set. **barely,** *adv.* nakedly, poorly; scarcely; openly, explicitly. **bareness,** *n.* **barish,** *a.*
baresark (beə'sahk), BERSERK.
bargain (bah'gin), *n.* haggling, discussions as to terms; an

agreement generally concerning a sale; the thing bought or sold; an advantageous purchase. *v.i.* to haggle over terms; to make a contract or agreement for purchase or sale. **a bad bargain,** a purchase or sale adverse to the party under consideration. **Dutch bargain, wet bargain,** a bargain concluded over a glass of liquor. **into the bargain,** over and above what is stipulated. **to bargain for,** to count on, to expect. **to be off one's bargain,** to be released from a purchase or engagement. **to make the best of a bad bargain,** to do the best one can in adverse circumstances. **to strike a bargain,** to come to terms. **bargain-basement, -counter,** *n.* a part of a store where goods are sold cheaply. **bargainer,** *n.* a trafficker, a haggler.
bargan (bah'gən), **barragan** (ba'rə-), *n.* (*Austral.*) a boomerang.
barge (bahj), *n.* a flat-bottomed freight-boat used principally on canals or rivers; the second boat of a man-of-war; a large ornamental state or pleasure boat, an ornamental houseboat. *v.i.* to lurch (into), rush (against). **bargeman,** *n.* **barge-master,** *n.* **barge-pole,** *n.* the pole with which a barge is propelled or kept clear of banks etc. **not fit to be touched with a barge-pole,** not fit to come near on account of dirt, disease or ill temper. **bargee** (-jē'), *n.* a bargeman.
barge- (bahj-), *comb. form.* **barge-board,** *n.* a projecting horizontal board at the gable-end of a building, warding off the rain. **barge-course,** *n.* the tiling projecting beyond the principal rafters in a building; a wall-coping formed of bricks set on edge.
baric (beər'ik), BARIUM.
barilla (bəril'ə), *n.* an impure alkali obtained from the ash of certain seaside plants or from kelp.
baritone (ba'ritōn), *n.* (a male singer with a) voice intermediate between a bass and a tenor; the smaller bass saxhorn in B flat or C. *a.* having a compass between tenor and bass. **baritone-clef,** *n.* the F clef on the middle line of the bass stave.
barium (beə'riəm), *n.* a metallic divalent element, at. no. 56; chem. symbol Ba, the metallic base of baryta. **barium meal,** *n.* a mixture of barium-sulphate, administered to allow X-ray examination of a patient's stomach or intestines. **baric** (beə'-), *a.* containing barium.
bark[1] (bahk), *v.i.* to utter a sharp, harsh cry, like that of a dog; to speak in a sharp manner; to cough. *n.* a sharp, harsh cry of dogs or other animals; the report of a firearm; a cough. **to bark up the wrong tree,** to be on a false trail. **barker,** *n.* one who or that which barks; a dog; a vocal advertiser for a circus, fun-fair etc.; (*sl.*) a pistol, a cannon. **barking,** *n.*, *a.* **barking-bird,** *n.* a S American bird named from its cry. **barking-iron,** *n.* (*sl.*) a pistol.
bark[2] (bahk), *n.* the exterior covering of a tree, formed of tissues parallel to the wood; spent bark, tan; an outer covering. *v.t.* to strip the bark from or cut a ring in the bark of (a tree); to steep in a solution of bark, to tan; to graze (the shins, elbows etc.); to cover with or as with bark, to encrust; to strip or scrape off. **bark-bed,** *n.* a hot-bed formed of spent bark. **bark-bound,** *a.* having the bark so close as to hinder the growth. **bark-mill,** *n.* a mill for crushing bark. **bark-pit,** *n.* a pit in which hides are tanned. **bark-tree,** *n.* the popular name of the genus *Cinchona.* **barker,** *n.* one who strips the bark from a tree. **barky,** *a.*
bark[3], **barque** (bahk), *n.* (*Poet.*) a ship or boat, esp. a small sailing vessel; (*usu.* **barque**) a type of sailing-vessel with three or more masts. **barkentine** BARQUENTINE. **barque-rigged,** *a.* rigged like a barque.
barley (bah'li), *n.* the grain or the plant of the genus *Hordeum,* a hardy, awned cereal, used for soups, malt liquors and spirits, animal feeds etc. **pearl-barley,** barley stripped of the husk and ground to a small white lump. **barley-broth,** *n.* broth made with barley; strong beer. **barley-corn,** *n.* a grain of barley; a measure, the third part of an inch (about 0·8 cm) **John Barleycorn,** barley personified as the grain from which malt liquor is made; malt liquor. **barley-mow,** *n.* a stack of barley. **barley-sugar,** *n.* a hard confection, prepared by boiling down sugar, formerly with a decoction of barley. **barley-water,** *n.* a soothing drink made from pearl-barley. **barley-wine,** *n.* a strong kind of ale.
barm (bahm), *n.* the frothy scum which rises to the surface of malt liquor in fermentation, used as a leaven; yeast. **barmy,** *a.* of or full of barm or yeast; frothing, fermenting; crazy, cracked, silly.
barmecide (bah'misid), *n.* one who gives illusory benefits. *a.* barmecidal. **Barmecide feast,** *n.* short commons. **barmecidal** (-si'-), *a.* unreal, unsatisfying, illusory.
bar mitzvah (bah mits'və), *n.* a Jewish boy who has reached the age of religious responsibility, usu. on his 13th birthday; the ceremony and celebration marking this event.
barn (bahn), *n.* a covered farm building for storage; a large (uncomfortable) building; (*N Am.*) a stable, a cowshed. **barn dance,** *n.* a kind of country-dance; a party at which such dances are performed. **barn door,** *n.* the large door of a barn; a target too big to be easily missed. *a.* of fowls, reared at the barn door. **barn-owl,** *n.* the white, church and screech owl, *Strix flammea*. **barnstorm,** *v.i.* to tour the country giving theatrical performances; (*N Am.*) to tour rural areas giving political speeches at election time. **barnstormer,** *n.* **barnstorming,** *n.*, *a.* **barnyard,** *n.* the yard adjoining a barn; a farmyard.
Barnaby (bah'nəbi), *n.* Barnabas. **Barnaby-Bright,** *n.* St Barnabas's Day, 11 June; according to the Old Style, the longest day.
barnacle (bah'nəkl), *n.* (also **bernacle**) the barnacle-goose; a small marine crustacean that lives attached to rocks, ship bottoms etc.; a person who is difficult to get rid of. **barnacle-goose,** *n.* a kind of black-and-white wild goose.
barney (bah'ni), *n.* a noisy lark or quarrel.
bar(o)-, *comb. form* weight, pressure.
barograph (ba'rəgrahf), *n.* an aneroid barometer recording variations of atmospheric pressure. **barogram** (-gram), *n.* the record produced by a barograph. **barographic,** *a.*
barology (bərol'əji), *n.* the science of weight.
barometer (bərom'itə), *n.* an instrument used for measuring atmospheric pressure, thus indicating weather change, and also altitude; any indicator of change (e.g. in public opinion). **barometric** (barəmet'-), **-ical,** *a.* **barometrically,** *adv.* **barometry,** *n.*
barometrography (barōmitrog'rəfi), *n.* the branch of meteorology which deals with the measurement of atmospheric pressure.
baron (ba'rən), *n.* one who held land by military service from the king; a member of the lowest rank of nobility; a title of the judges of the Court of Exchequer; a powerful head of a business or financial organization. **baron of beef,** a joint consisting of the two sirloins. **baronage** (-nij), *n.* the whole body of barons; the dignity of a baron; a published list of barons. **baroness** (-nis), *n. fem.* **baronial** (-rō'-), *a.* **barony,** *n.* the rank or dignity of a baron.
baronet (ba'rənit), *n.* a hereditary titled order of commoners ranking next below barons. *v.t.* to confer a baronetcy on. **baronetage** (-netij), *n.* baronets collectively; a list of the baronets. **baronetcy** (-si), *n.* the title or rank of a baronet.
baroque (bərōk', -rok'), *n.* an artistic and architectural style prevalent esp. in 17th-cent. Europe, characterized by extravagant ornamentation; a similar style in music or literature. *a.* baroque in style; grotesque; flamboyant.
baroscope (ba'rəskōp), *n.* a weather glass.
barothermograph (barōthœ'məgrahf), *n.* an instrument combining a barometer and a thermometer.
barouche (bəroosh'), *n.* a double-seated four-wheeled

barque 64 base

horse-drawn carriage, with a movable top, and a seat outside for the driver.
barque BARK [3].
barquentine, barkentine (bah'kənten), *n.* a three-masted vessel, with the foremast square-rigged, and the main and mizzen fore-and-aft rigged.
barrack [1] (ba'rək), *n.* a temporary hut; (*pl.*) buildings used to house troops; any large building resembling barracks. *v.t.* to provide with barracks; to put in barracks. *v.i.* to lodge in barracks. **barrack-master**, *n.* an officer in charge of barracks.
barrack [2] (ba'rək), *v.i.* to jeer; (*Austral.*) to cheer (for). *v.t.* to shout or cheer derisively at (e.g. a sports side); (*Austral.*) to shout support for (a team).
barracoon (barəkoon'), *n.* a fortified African slave-house.
barracouta (barəkoo'tə), *n.* a large edible fish of the Pacific.
barracuda (barəkū'də), *n.* (*pl.* **-da, -das**) a predatory tropical fish, dangerous to man.
barrage [1] (ba'rahzh), *n.* (the formation of) an artificial bar or dam to raise the water in a river.
barrage [2] (ba'rahzh), *n.* (*Mil.*) a screen of artillery fire to protect troops; heavy or continuous questioning or criticism. **box barrage**, a barrage surrounding a particular area. **creeping barrage**, a barrage that can move forward or backward. **balloon barrage**, disposition of anchored balloons to prevent hostile aircraft making machine-gun attacks.
barramundi, burramundi (barəmŭn'di), *n.* a variety of perch found in Queensland rivers.
barranca (bərang'kə), *n.* (*N Am.*) a deep gorge, with steep sides.
barrator, -er (ba'rətə), *n.* one who out of malice or for his own purposes stirs up litigation or discord. **barratry** (-tri), *n.* the offence of vexatiously exciting or maintaining lawsuits; traffic in church or public offices. **barratrous**, *a.*
barre (bah), *n.* a wall-mounted horizontal rail used for ballet exercises.
barrel (ba'rəl), *n.* a cylindrical wooden vessel bulging in the middle, formed of staves held together by hoops, and with flat ends; the capacity or contents of such a vessel; anything resembling such a vessel, as the tube of a firearm; the belly and loins of a horse, ox etc.; a measure of capacity for liquid and dry goods, varying with the commodity; a revolving cylinder or drum round which a chain or rope is wound; (*N Am. sl.*) money to be used for political campaigning. *v.t.* to draw off into, or put or stow in barrels. *v.i.* (*N Am.*) to drive fast. **to have someone over a barrel**, to have power over someone. **to scrape the barrel**, to get the last remaining bit. **barrel-bulk**, *n.* (*Naut.*) a measure of 5 cu. ft. used in estimating the capacity of a vessel for freight. **barrel campaign**, *n.* (*N Am.*) an election fought by means of bribery. **barrel-chested**, *a.* having a wide, strong chest. **barrel-drain**, *n.* a cylindrical drain. **barrel-organ**, *n.* a musical instrument in which the keys are mechanically acted on by a revolving cylinder (barrel) studded with pins. **barrel roll**, *n.* a manoeuvre in aerobatics in which an aircraft rolls about its longitudinal axis. **barrel-vault**, *n.* (*Arch.*) a semi-cylindrical vault. **barrelled**, *a.* packed in barrels; barrel-shaped; having a barrel or barrels.
barren (ba'rən), *a.* incapable of producing offspring; bearing no fruit; producing no vegetation; unprofitable; uninventive, dull. *n.* a tract of barren land, esp. in the US. **barren-wort**, *n.* the English name of the genus *Epimedium*. **barrenly**, *adv.* **barrenness**, *n.*
barret (ba'rit), *n.* a little flat cap; a biretta.
barette (bəret'), *n.* (*N Am.*) a hair-clasp.
barretter (bəret'ə), *n.* an appliance for keeping current in a circuit at constant strength.
barricade (barikād', ba'-), **barricado** (-kā'dō), *n.* a hastily-formed rampart erected across a street or passage; any bar or obstruction. *v.t.* to block with a barricade or other barrier.
barrico (bərē'kō), *n.* (*pl.* **-coes**) a small cask.
barrier (ba'riə), *n.* that which hinders approach or attack; an enclosing fence; a limit, a boundary; (*fig.*) any material or immaterial obstruction. *v.t.* to close (in) or shut (off) with a barrier. *a.* pertaining to an obstruction or separating agent, often protective, as in *barrier contraceptive*. **barrier cream**, *n.* a cream used to protect the skin from dirt etc. **barrier ice**, *n.* ice-floe, ice-pack. **barrier-pillar**, *n.* a large pillar of coal supporting the roof of a mine. **barrier-reef**, *n.* a coral reef running nearly parallel to the land, with a lagoon between.
barring (bah'ring), *prep.* (*coll.*) except, omitting.
barrio (ba'riō), *n.* (*pl.* **-rios**) (*N. Am.*) a Spanish-speaking community or district in a city.
barrister (ba'ristə), *n.* a lawyer who has been admitted to practise as an advocate at the bar. **revising barrister**, one formerly appointed to hold an annual court for the revision of the register of Parliamentary voters. **barristership**, *n.*
barrow [1] (ba'rō), *n.* a hill; a prehistoric grave-mound.
barrow [2] (ba'rō), *n.* a shallow cart with two wheels pushed by hand. **barrow-boy**, *n.* a street trader who sells from a barrow. **barrowful**, *n.*
Bart., (*abbr.*) Baronet.
barter (bah'tə), *v.t.* to give (anything except money) in exchange for some other commodity; to exchange. *v.i.* to trade by exchanging one thing for another. *n.* trade by exchanging one commodity for another. **to barter away**, to dispose of by barter; to part with for a consideration (usually an inadequate one). **barterer**, *n.*
Bartholomew (bahthol'əmū), *n.* one of the twelve Apostles. **Massacre of St Bartholomew**, the slaughter of some 30,000 French Huguenots on St Bartholomew's Day, 24 August, 1572. **Bartholomew-day, -tide**, *n.* his feast day, 24 August. **Bartholomew Fair**, *n.* a fair formerly held annually about this date at Smithfield, notorious for its roughness and licence.
bartizan (bah'tizan, -zan'), *n.* a battlement on top of a house or castle; a small overhanging turret projecting from the angle on the top of a tower. **bartizaned**, *a.*
barton [1] (bah'tən), *n.* a farm-yard.
barton [2] (bah'tən), BURTON.
barwood (bah'wud), *n.* a red wood from W Africa used for dyeing.
baryon (ba'rion), *n.* any member of the heavier class of subatomic particles that have a mass equal to or greater than that of the proton.
barysphere (ba'risfiə), *n.* the solid, heavy core of the earth.
baryta (bəri'tə), *n.* the monoxide of barium. **barytes** (-tēz), *n.* native sulphate of barium, heavy spar (used as white paint). **barytic** (-rit'-), *a.*
barytone (ba'ritōn), *n.* BARITONE.
basal BASE [2].
basalt (bas'awlt), *n.* a dark igneous rock of a uniform and compact texture; a black stone-ware first used by Wedgwood. **basaltic** (-sawl'-), *a.* **basaltiform** (-sawl'tifawm), *a.*
basan, bazan (baz'ən), *n.* a tanned sheepskin for bookbinding.
basanite (bas'ənit), *n.* a velvet-black variety of quartz; Lydian-stone, touchstone.
bas-bleu (bah blœ), *n.* BLUESTOCKING.
bascinet (bas'inet, -net'), BASINET.
bascule (bas'kūl), *n.* an apparatus on the principle of the lever, in which the depression of one end raises the other; a bascule-bridge. **bascule-bridge**, *n.* one balanced by a counterpoise which falls or rises as the bridge is raised or lowered.
base [1] (bās), *a.* of little height; occupying a low position; unworthy, despicable; inferior in quality; alloyed,

debased, counterfeit. **base-born**, *a.* of humble birth; of base origin or nature. **base-court**, *n.* the outer court of a mansion, the farm-yard. **base-hearted**, *a.* **base-heartedness**, *n.* **base metals**, *n.pl.* those which are not precious metals. **base-tenant**, *n.* one holding land as a villein. **basely**, *adv.* **baseness**, *n.*
base² (bās), *n.* the lowest part on which anything rests; fundamental principle; a pedestal; the bottom of anything; the extremity of a part by which it is attached to the trunk; the side on which a plane figure stands or is supposed to stand; that line or place from which a combatant draws reinforcements of men, ammunition etc.; that with which an acid combines to form a salt; the place from which a commencement is made in some ball-games; the starting-post; any substance used in dyeing as a mordant; the original stem of a word; the line from which trigonometrical measurements are calculated; the number on which a system of calculations depends; the datum or basis for any process of reckoning measurement or argument; an old English game, still played by boys, and often called 'prisoner's base'. *v.t.* to make a foundation for; to lay on a foundation. **to make first base**, (*N Am. coll.*) to complete the initial stage in a process; to seduce. **baseball**, *n.* the national ballgame of America, akin to English rounders, also called 'ball-game'; the ball used in this. **base-burner**, *n.* an iron stove fed at the top. **base-line**, *n.* the base; the back line at each end of a tennis court. **base-plate**, *n.* a foundation-plate. **base rate**, *n.* the rate of interest on which a bank bases its lending rates. **basal**, *a.* pertaining to, situated at or constituting the base of anything; fundamental. *n.* a basal part. **basal metabolism**, *n.* the amount of energy consumed by an individual in a resting state for functions such as respiration and blood circulation. **baseless**, *a.* **baselessness**, *n.* **basement**, *n.* the lowest or fundamental portion of a structure; the lowest inhabited storey of a building, esp. when below the ground level. **basic**, *a.* of, pertaining to or constituting a base, fundamental; without luxury, extras etc.; (*Chem.*) having the base in excess; (of igneous rock) with little silica present in its composition. **BASIC**, *n.* a computer programming language using simple English terms (acronym of *Beginners All-purpose Symbolic Instruction Code*). **basically**, *adv.* **basics**, *n.pl.* fundamental principles. **Basic English**, *n.* a fundamental selection of 850 English words used as a first step in English teaching and as an auxiliary language. **basicity** (-sis′-), *n.* the combining power of an acid. **basilar** (-lə), *a.* (*Bot., Zool.*) growing from, or situated near, the base.
bash¹ (bash), *v.t.* to strike, so as to smash. *v.i.* to strike violently. *n.* a heavy blow, a bang; a social entertainment. **to have a bash at**, to attempt. **basher**, *n.* a rough, a hooligan. **-bashing**, *n.* (*in comb.*) physical or verbal attack on a disliked or despised group of people. **-basher**, *n.*
bash² (bash), *v.t.* abash. **bashful**, *a.* shy; characterized by excessive modesty. **bashfully**, *adv.* **bashfulness**, *n.*
bashaw (bəshaw′), PASHA.
bashi-bazouk (bash′ibəzook′), *n.* (*Hist.*) a Turkish irregular soldier.
basi-, *comb. form.* pertaining to or forming the base, or at the base of.
basic BASE².
basicranial (bāsikrā′niəl), *a.* of or at the base of the cranium.
basidium (bəsid′iəm), *n.* (*pl.* **-dia**) (*Bot.*) a mother-cell carried on a stalk and bearing spores characteristic of various fungi. **Basidiomycetes** (-ōmisēts), *n.* a group of fungi (including many toadstools and mushrooms) in which the spores are borne on basidia.
basifugal (bāsifū′gl), *a.* growing away from the base.
basil¹ (baz′l), *n.* the popular name of the genus *Ocymum*, species of which are used as culinary herbs, e.g. the sweet basil, *O. basilicum.*

basil² (baz′l), BASAN.
basilar BASE².
basilateral (bāsilat′ərəl), *a.* at the side of a base.
basilica (bəsil′ikə), *n.* (*pl.* **-cas**) a large oblong building with double colonnades and an apse, used as a court of justice and an exchange; such a building used as a Christian church. **basilical**, *a.* **basilican**, *a.* **basilicum** (-kəm), *n.* a name given to several ointments from their reputed sovereign virtues.
basilisk (baz′ilisk, bas′-), *n.* a fabulous reptile whose look and breath were reputed fatal; a tropical American lizard named from its inflatable crest.
basin (bā′sn), *n.* a hollow vessel for holding water, esp. for washing; a bowl; the quantity contained by such a vessel, a reservoir; a land-locked harbour; the scale-dish of a balance; the tract of country drained by a river and its tributaries; a hollow. **basinful**, *n.* (*coll.*) as much work or trouble as one can cope with.
basinet (bas′inet, -net′), **basnet** (basə-), *n.* a light helmet.
basiophil (bās′iəfil), *a.* having an affinity for basic stains.
basipetal (bāsip′itəl), *a.* proceeding in the direction of the base.
basis (bā′sis), *n.* (*pl.* **-ses** (-sēz)) the base or foundation; the fundamental principle, ingredient or support.
basitemporal (bāsitem′pərəl), *a.* of or pertaining to the base of the temples.
bask (bahsk), *v.i.* to expose oneself to the heat of the sun, good fortune etc. **basking-shark**, *n.* the largest species of shark.
basket (bahs′kit), *n.* a wickerwork vessel of plaited osiers, twigs or similar flexible material; as much as will fill a basket; a basket-hilt; the net or hoop used as a goal in basketball. *v.t.* to put in a basket. **the pick of the basket**, the best of the lot. **basketball**, *n.* a game in which opposing teams try to drop a large ball into suspended nets or hoops. **basket case**, *n.* (*sl., offensive*) an unbalanced or frenzied person. **basket-chair**, *n.* a wickerwork chair. **basket-fish**, *n.* a starfish of the genus *Astrophyton*. **basket-hilt**, *n.* a basket-shaped hilt of a sword which protects the swordsman's hand. **basket-hilted**, *a.* **basket-stitch**, *n.* in knitting, alternate purl and plain stitches which create a basketwork pattern. **basket-stones**, *n.pl.* fragments of the stems of the fossil Crinoidea. **basket weave**, *n.* a form of textile weave resembling chequered basketwork. **basketwork**, *n.* wickerwork. **basketful**, *n.* **basketry**, *n.* BASKETWORK.
basmati rice (basmah′ti), *n.* a type of rice with a slender grain.
bason¹ (bā′sn), BASIN.
bason² (bā′sn), *n.* a bench with a slab or iron plate and a fire underneath for felting hats. *v.t.* to harden the felt in hat-making.
basophil (bā′səfil), **-phile** (-fil), *n.* a white blood cell with basophilic contents. **basophilic** (-fil′-), *a.* (of cells) readily stained with basic dyes.
Basque (bahsk), *n.* a member of a people occupying both slopes of the western Pyrenees; the language spoken by this people. *a.* of this people or language. **basque**, *n.* a woman's jacket or bodice, extended slightly below the waist. **basqued**, *a.* furnished with a basque or short skirt.
bas-relief (bahrəlēf′, bas-), *n.* low relief; a sculpture in which the figures project less than one-half of their true proportions above the plane forming the background.
bass¹ (bas), *n.* bast. **bass-wood**, *n.* the American lime-tree, *Tilia americana;* its wood.
bass², **basse** (bas), *n.* a sea-fish, *Labrax lupus*, common in European waters. **black bass,** *Perca huro*, from Lake Huron. **sea bass,** (*N Am.*) a serranoid food-fish, *Centropristis striatus*, common on the Atlantic shores of the US. **striped bass,** (*N Am.*) the rockfish, *Roccus lineatus* and *R. sexatilis.*
bass³ (bās), *n.* the lowest part in harmonized musical

compositions; (a singer with) the deepest male voice; the lowest tones of an instrument; a bass string; a bass instrument, esp. bass guitar or double-bass. *a.* of or pertaining to a low voice, sound or instrument. **bass-bar**, *n.* a bar of wood fixed lengthwise in the belly of stringed instruments to enable them to resist pressure. **bass clef**, *n.* (*Mus.*) the F clef on the fourth line. **bass drum**, *n.* a large drum with a low pitch played in an orchestra or band. **bass-viol**, *n.* a stringed instrument for playing bass; a violoncello. **bassist**, *n.* **bassy**, *a.*
basset ¹ (bas'it), *n.* an obsolete game of cards.
basset ² (bas'it), **basset hound**, *n.* a short-legged smooth-haired dog.
basset ³ (bas'it), *n.* an outcrop of strata at the surface of the ground.
basset-horn (bas'it), *n.* a tenor clarinet with a recurved mouth.
bassinet (basinet'), *n.* a wicker basket with a hood at the end used as a cradle; a pram of similar shape.
basso (bas'ō), *n.* bass ³. **basso-continuo** (-kontin'ūō), *n.* thoroughbass. **basso-profundo** (-profun'dō), *n.* the lowest male voice. **basso-ripieno** (-ripyä'nō), *n.* the bass of the grand chorus, which comes in only occasionally. [It.]
bassoon (bəsoon'), *n.* a wooden double-reed instrument, the bass to the clarinet and oboe; an organ-stop of similar tone. **bassoonist**, *n.*
basso-rilievo (basōrilyä'vō), *n.* (*pl.* **-vos**) bas-relief.
bast (bast), *n.* the inner bark of the lime or linden-tree; any similar fibrous bark; a rope, mat etc. made from this fibre.
bastard (bahs'təd), *n.* an illegitimate child or person; anything spurious, counterfeit or false; (*sl.*, *often considered taboo*) an obnoxious or disagreeable person, hence any person in general; (*sl.*, *often considered taboo*) something annoying or unpleasant. *a.* illegitimate; spurious, not genuine; resembling something of a higher quality or kind; of abnormal shape or size. **bastard title**, *n.* a half-title. **bastard type**, *n.* (*Print.*) a fount of type with a face too large or too small in proportion to its body. **bastard-wing**, *n.* three or four quill-like feathers placed at a small joint in the middle of a bird's wing. **bastardize, -ise**, *v.t.* to declare one a bastard; to debase. **bastardization**, *n.* **bastardy**, *n.*
baste ¹ (bāst), *v.t.* to moisten (a roasting joint etc.) with liquid fat, gravy etc.
baste ² (bāst), *v.t.* to beat with a stick, to thrash.
baste ³ (bāst), *v.t.* to tack, to sew together with long stitches.
Bastille (bastēl'), *n.* the State prison in Paris, destroyed in 1789; (**bastille**) a fortified tower; a prison.
bastinado (bastinā'dō), *n.* (*pl.* **-dos, -does**) a method of torture, beating with a stick on the soles of the feet; a rod, a stick, a cudgel. *v.t.* to beat thus.
bastion (bas'tyən), *n.* a projecting tower in a fortification; (*fig.*) a rampart, a defence. **bastioned**, *a.* **last bastion**, one of a small set of people or things left defending a principle, way of life etc.
bat ¹ (bat), *n.* a wooden instrument with a cylindrical handle and broad blade used to strike the ball at cricket or similar games; a blow with a bat or club; a batsman. *v.t.* (*past, p.p.* **batted**) to strike with a bat. *v.i.* to take an innings as batsman. **off his/her own bat**, by his/her own exertions. **bat-fowling**, *n.* catching birds by holding a light before a net, and beating their roosting-places with bats. **batsman**, *n.* one who uses the bat at cricket and other ball games; (*Aviat.*) the person on an airfield or aircraft carrier who guides landing aircraft by waving a round, plainly visible bat in each hand. **batlet** (-lit), *n.* a small bat.
bat ² (bat), *n.* a small nocturnal mouse-like mammal, having the digits extended to support a wing-membrane stretching from the neck to the tail, by means of which it flies. **blind as a bat**, having very poor eyesight. **like a bat out of hell**, (*coll.*) extremely quickly. **to have bats in the belfry**, (*coll.*) to be crazy. **bats**, *a.* (*coll.*) batty. **batty**, *a.* batlike; (*coll.*) mentally unstable; crazy.
bat ³ (bat), *n.* (*only in comb.*) a pack-saddle. **batman**, *n.* the military servant of an officer. **bat-money, bat-pay**, *n.* an allowance for carrying baggage in the field. **bat-needle**, *n.* a packing-needle.
bat ⁴ (hat), *v.t.* to blink. **not bat an eyelid, eyelash**, not to blink; to show no surprise or emotion.
batata (bətah'tə), *n.* a plant with a tuberous root, from the West Indies, the sweet potato.
Batavian (bətā'viən), *a.* of or pertaining to the ancient Batavians or the modern Dutch. *n.* one of the ancient Batavi; a Dutchman.
batch ¹ (bach), *n.* as much bread as is produced at one baking; hence, any quantity produced at one operation; a group of people. *v.t.* to collect into batches; to group (items) for computer processing. **batch processing**, *n.* a system by which a number of jobs submitted by users are run through a computer as a single batch.
batch ² (bach), (*Austral.*) BACH ².
bate ¹ (bāt), *v.t.* to abate, diminish; to reduce, restrain; to deduct; to deprive, remove. *v.i.* to abate. **with bated breath**, in suspense, anxiously.
bate ² (bāt), *n.* alkaline lye used in tanning.
bateau (bat'ō), *n.* (*pl.* **bateaux** (-z)) a long, light, flat-bottomed river-boat, tapering at both ends, used by French-Canadians. **bateau-bridge**, *n.* a floating bridge supported by bateaux.
Bath ¹ (bahth), *n.* a city in Somerset, famous for its hot springs. **Bath bun**, *n.* a rich bun, generally with currants. **bath chair**, *n.* a wheeled chair for invalids. **Bath chap**, *n.* a small pig's cheek cured for the table. **Bath Oliver**, *n.* an unsweetened biscuit invented by Dr W. Oliver (1695–1764), of Bath. **Bath stone**, *n.* a white building-stone quarried from the oolite near Bath.
bath ² (bahth), *n.* (*pl.* **badhz, baths**) the act of washing or immersing the body in water or other fluid; (a vessel containing) water or other fluid for bathing; a wash, a lotion; (*often pl.*) a room or building for bathing in; a hydropathic establishment; a town having medicinal springs; immersion of any substance in a solution for scientific, art or trade purposes; (the vessel containing) such solution; the Order of the Bath. *v.t.* to wash or put (usu. a child) in a bath. **Order of the Bath**, a British order of knighthood. **bath-house**, *n.* **bath-oil**, *n.* (perfumed) oil for use in bath-water. **bath robe**, *n.* (*esp. N Am.*) a dressing-gown. **bathroom**, *n.* an apartment containing a bath or shower; (*esp. N Am.*) a lavatory. **bath salts**, *n.pl.* perfumed crystals used for softening bath water. **bath-tub**, (*esp. N Am.*) tub, a vessel for containing water for bathing.
bathe (bādh), *v.t.* to immerse in or as in a bath; to moisten, to wet copiously; to cleanse by applying water. *v.i.* to swim in water for pleasure; (*esp. N Am.*) to take a bath. *n.* the act of taking a bath (esp. in the sea, a river etc.). **bathing-costume, -dress, -suit**, *n.* **bathing hut**, *n.* a hut for bathers to undress and dress in. **bathing-machine**, *n.* (*formerly*) a kind of covered carriage to bathe from. **bather**, *n.* **bathers**, *n.* (*Austral.*) a swimming costume or swimming trunks.
bath mitzvah (bahth mits'və), (also **bas** (bahs), **bat** (baht)), *n.* a Jewish girl who has reached the age (usu. 13 years) of religious responsibility; the celebrations marking this event.
batho-, bathy-, *comb. form.* deep, depth.
batholite (bath'əlit), **-lith** (-lith), *n.* a great mass of intrusive igneous rock, esp. granite. **batholitic** (-lit'), **-lithic** (-lith'-), *a.*
bathometer (bəthom'itə), *n.* an instrument used to ascertain the depths reached in soundings. **bathometric**, *a.* **bathometry**, *n.*
bathos (bā'thos), *n.* ridiculous descent from the sublime to

the commonplace in writing or speech; anticlimax. **bathetic** (bəthet'-), *a.*
bathymetry (bəthim'itri), *n.* the art of taking deep soundings. **bathymetric, -metrical** (bathimet'-), *a.* **bathymetrically**, *adv.*
bathyscaph (bath'iskaf), **-scaphe** (skāf), **-scape** (-skāp), *n.* a submersible vessel for deep-sea observation and exploration.
bathysphere (bath'isfiə), *n.* a strong steel deep-sea observation chamber.
batik (bat'ik), *n.* a method of printing designs on fabric by masking areas to be left undyed with wax; fabric produced by this method.
bating (bā'ting), *prep.* leaving out of the question; excepting.
batiste (bətēst'), *n.* a fine cotton or linen fabric.
batlet BAT[1].
batman BAT[3].
baton (bat'on, bat'n), *n.* a staff or club; a stick used as a badge or symbol of authority; a short stick transferred between successive team-mates in a relay-race; a knobbed staff carried and swung into the air at the head of a parade or twirled by majorettes etc.; the thin stick used by a conductor to beat time. *v.t.* to strike with a policeman's baton or truncheon. **baton charge**, *n.* a charge by police or troops with batons. **baton gun**, *n.* a gun which fires rubber or plastic bullets to control rioters.
Batrachia (bətrā'kiə), *n.pl.* an order of reptiles including those breathing by gills; an order of Amphibia containing frogs and toads. **batrachian**, *n.*, *a.* **batrachoid** (bat'rəkoid), *a.*
battalion (bətal'yən), *n.* a main division of an army, several companies of infantry, about 1000 strong on a war footing. *v.t.* to form into battalions.
battels (bat'lz), *n.pl.* (*Univ. of Oxford*) provisions from the college buttery; the account for these; college accounts generally. **battel**, *v.i.* to have an account for battels; to get one's provisions at the college buttery.
batten[1] (bat'n), *n.* a strip of sawn wood used for flooring, clamping together boards etc.; a thin piece of wood nailed on masts etc. to prevent chafing, or to fasten down the edges of tarpaulins over the hatches. *v.t.* to fasten or strengthen with battens. **to batten down the hatches**, to secure the hatches of a ship; to prepare for action, trouble, danger etc. **battening**, *n.*
batten[2] (bat'n), *v.i.* to grow fat; to thrive (at the expense of).
batten[3] (bat'n), *n.* the movable bar of a loom which strikes the weft in.
batter[1] (bat'ə), *v.t.* to strike with successive blows so as to bruise, shake, demolish; to wear or impair by beating or rough usage; to subject to hard, crushing attack; to bombard. *v.i.* to hammer (at) a door. *n.* in cooking, a mixture of several ingredients, esp. eggs, flour and milk, well beaten together; a blow. **battered baby**, *n.* an infant who has suffered violent injury at the hands of an adult. **battering-charge**, *n.* the heaviest charge for a siege-gun. **battering-engine, -machine**, *n.* an engine used for battering down walls or ramparts. **battering-gun, -piece**, *n.* **battering-ram**, *n.* an ancient military engine used for battering down walls, a heavy beam with a metal head, orig. in the form of a ram's head. **battering-train**, *n.* a train of artillery for siege purposes.
batter[2] (bat'ə), *v.i.* to incline (as a wall etc.) from the perpendicular with a receding slope. *n.* a receding slope. **battering**, *a.*
battery (bat'əri), *n.* an assailing by blows; (*Law*) an unlawful attack by beating, or even touching in a hostile manner; the tactical unit of artillery; a ship's armament; a place in which artillery is mounted; a connected series of electric cells etc. forming a source of electric energy; any apparatus for providing voltaic electricity; a combined series of lenses, instruments or general apparatus for use in various arts or sciences; an embankment; a series of nesting-boxes in which hens are confined to increase laying. **cross batteries**, two batteries commanding the same point from different directions. **enfilading battery**, a battery that rakes a whole line with its fire. **floating battery**, an armoured vessel for bombarding fortresses. **masked battery**, one concealed from the enemy's observation. **battery-piece**, *n.* a siege-gun. **battery-wagon**, *n.* a vehicle used for transporting tools and material for a battery.
batting (bat'ing), *n.* cotton fibre prepared for quilting.
battle (bat'l), *n.* a fight or hostile engagement between opposing armies etc.; fighting, hostilities, war. *v.i.* to fight, to contend (with or against); (*esp. Austral.*) to struggle for a living. **half the battle**, an immense advantage. **line of battle**, the arrangement of troops or warships in readiness for a general engagement. **pitched battle**, a general engagement the time and place of which have been settled beforehand. **to have the battle**, to be victorious. **to join battle**, to commence a general combat. **battle-array**, *n.* the order of troops prepared for engagement. **battle-axe**, *n.* an axe formerly used in battle; (*coll.*) a formidable woman. **battle-cry**, *n.* **battle-cruiser**, *n.* a large, heavily-armed cruiser. **battle-dress**, *n.* comfortable, loose-fitting uniform worn by soldiers in battle. **battle fatigue** SHELL SHOCK. **battle-field**, *n.* **battle-piece**, *n.* a pictorial etc. description of a battle. **battle-plane**, *n.* a large, fighting aircraft. **battle-royal**, *n.* a general engagement; a free fight. **battle-ship**, *n.* a warship; a ship adapted by armament for line of battle as opposed to a cruiser. **battled**, *a.* drawn up in line of battle; contested; protected with battlements.
battledore (bat'ldaw), *n.* the light racket used to strike a shuttlecock; the game in which this is used. **battledore and shuttlecock**, *n.* the game of battledore.
battlement (bat'lmənt), *n.* a parapet with openings or embrasures, on the top of a building, originally for defensive purposes, afterwards used as an ornament; a roof having a battlement. **battlemented**, *a.*
battue (batoo'), *n.* driving game from cover by beating the bushes; a shoot on this plan; wholesale slaughter.
battuta (batoo'tə), *n.* a bar; the beating of time.
baubee (bawbē'), BAWBEE.
bauble (baw'bl), *n.* a short stick carried by the fools or jesters of former times; a showy trinket; a piece of childish folly; a mere toy; a thing of no value.
baud (bawd), *n.* (*pl.* **baud, -s**) a unit which measures the rate of telegraphic or electronic transmission: one equals one bit of data per second.
baudekin (baw'dikin), **baudkin** (bawd'kin), BALDACHIN.
baudric (bawd'rik), BALDRIC.
Bauhaus (bow'hows), *n.* a radical German school of architecture and the arts dedicated to achieving a functional synthesis of art, design and technology. [G, lit. building house]
baulk BALK.
bauxite (bawk'sīt), *n.* a clay which is the principal source of aluminium.
bavin (bav'in), *n.* a bundle of brushwood.
bawd[1] (bawd), *n.* a procuress, a brothel-keeper; a prostitute. **bawdy**, *a.* dirty; of or befitting a bawd; obscene. *n.* bawdiness. **bawdy-house**, *n.* a brothel. **bawdily**, *adv.* **bawdiness**, *n.* **bawdry**, *n.* fornication; obscene talk.
bawd[2] (bawd), *n.* a hare.
bawl (bawl), *v.i.* to howl, bellow; to shout at the top of one's voice. *v.t.* to utter with bawling. *n.* a loud, prolonged shout or cry. **to bawl out**, (*coll.*) to reprove fiercely.
bay[1] (bā), *n.* an arm or inlet of the sea extending into the land with a wide mouth; a recess or cirque in a range of hills; (*N Am.*) an arm of a prairie extending into woods.

bay-floe, bay-ice, *n.* new ice formed in bays or sheltered waters. **bay-salt,** *n.* coarse-grained crystals of salt originally of sea-water, now of a saturated solution of chloride of sodium. **Bay State,** *n.* (*N Am.*) Massachusetts. **baywood,** *n.* a coarse mahogany.

bay[2] (bā), *n.* an opening or recess in a wall; a main compartment or division, like the interval between two pillars; a division of a barn or other building; (*Railway*) a platform with a cul-de-sac, forming the terminus of a side-line; a compartment or division in a ship or in the fuselage of an aircraft. **sick bay,** a ship's hospital. **bay-window,** *n.* an angular window structure forming a recess in a room.

bay[3] (bā), *n.* a dam or embankment retaining water. *v.t.* to dam, hold (back) water.

bay[4] (bā), *n.* barking; the prolonged hoarse bark of a dog; the barking of a pack that has tracked down its prey; hence, the final encounter between hounds and their prey. *v.i.* to bark hoarsely, as a hound at its prey. *v.t.* to bark at; to bring to bay; to express by barking. **at bay,** in a position of defence, in the last extremity. **to stand at bay, hold (hounds) at bay,** to keep back the assailing dogs or other form of attack. **to bring** or **drive to bay,** to come to close quarters with the animal hunted; to reduce to extremities.

bay[5] (bā), *n.* the bay-tree or bay-laurel (*Laurus nobilis*); (*N Am.*) a place covered with bay trees; (*pl.*) leaves or twigs of laurel woven into a garland as a reward for a conqueror or poet; fame, renown. **bayberry,** *n.* the berry of the bay; (*N Am.*) the fruit of *Myrica cerifera* or wax myrtle; the plant itself. **bayberry tallow,** *n.* a kind of tallow obtained from the berries of the wax myrtle. **bay-cherry,** *n.* the cherry laurel, *Cerasus laurocerasus*. **bay-leaf,** *n.* a leaf from the bay-tree, dried and used in cooking. **bay-rum,** *n.* an aromatic, spirituous liquid, used in medicines and cosmetics, and prepared by distilling rum in which bay-leaves have been steeped. **bay-tree,** *n. Laurus nobilis*. **baywood,** *n.* wood of the mahogany tree, *Swietenia mahogani*.

bay[6] (bā), *a.* reddish-brown in colour, approaching chestnut. *n.* a bay horse.

bay[7] (bā), *n.* the second branch of a stag's horn, the next to the brow antler.

bayadère (bayadeə′), *n.* a Hindu dancing-girl.

bayonet (bā′ənit), *n.* a weapon for stabbing or thrusting, attached to the muzzle of a rifle; (*pl.*) infantry; a type of connection used to secure light-bulbs, camera lenses etc. *v.t.* to stab with a bayonet; to compel by military force. **Spanish bayonet,** a species of yucca with lanceolate leaves. **bayonet catch, joint,** *n.* device for securing in place two cylindrical parts by means of a turn.

bayou (bah′yoo), *n.* (*N Am.*) the sluggish outlet of a lake or river.

bazaar (bəzah′), *n.* an Eastern market-place; a sale of useful or ornamental articles in aid of charity; a shop where a variety of (ornamental) goods are sold.

bazooka (bəzoo′kə), *n.* an anti-tank or rocket-firing gun.

BB, (*abbr.*) Boys' Brigade; (on lead pencils) double black.

BBC, (*abbr.*) British Broadcasting Corporation.

BC, (*abbr.*) before Christ; British Columbia; British Council.

BCE, (*abbr.*) before the Common Era.

BCG, (*abbr.*) Bacillus Calmette-Guérin, used in antituberculosis vaccine.

BD, (*abbr.*) Bachelor of Divinity.

bdellium (del′iəm), *n.* (gum-resin produced by any of) several species of *Balsamodendron*.

BDS, (*abbr.*) Bachelor of Dental Surgery.

be (bē), *inf., pres. subj.* and *imper. v.* to exist, to live, to have a real state or existence, physical or mental; to become, to remain; to happen, come to pass; to have come or gone to or to occupy a certain place; to have a certain state or quality; used to assert a connection between the subject and the predicate. **the be-all and end-all,** the sole object or idea in view.

Be, (*chem. symbol*) beryllium.

be-, *pref.* (1) about, by; e.g. *besmear, bedaub, before, below, besiege;* (2) making intransitive verbs transitive or reflective; e.g. *bemoan, bespeak, bethink;* (3) forming verbs from nouns or adjectives, as *befool, befriend, benumb;* (4) having a privative force, as in *behead, bereave;* (5) compounded with nouns, signifying to call this or that, as *bedevil, belady, bemadam;* (6) intensive, e.g. *becrowd, bedrug, bescorch;* (7) making adjectives, e.g. *bejewelled, bewigged.*

BE, (*abbr.*) bill of exchange; Bachelor of Engineering; Board of Education.

beach (bēch), *n.* shingle; a sandy or pebbly seashore; the strand on which the waves break. *v.t.* to haul or run (a ship or boat) on a beach. **raised beach,** an ancient beach or shore left dry by change of sea level. **beach-comber,** *n.* a long wave rolling in from the ocean; one who lives by fishing and gathering jetsam on beaches. **beach-grass,** *n.* a coarse grass, growing on the sea-shore. **beachhead,** *n.* a position held on the beach of a hostile coast. **beachmaster,** *n.* an officer who directs the process of disembarking of troops. **beached,** *a.* **beachy,** *a.*

beacon (bē′kən), *n.* (a signal-fire on) a hill; a watch-tower; a lighthouse; a signalling buoy; a transmitter concentrating its radiation in a narrow beam, to act as a guide to aircraft; anything which gives notice of danger. *v.t.* to light up with beacon-fires; to mark with beacons; to lead, to guide. *v.i.* to shine like a beacon. **beaconage** (-ij), *n.*

bead (bēd), *n.* a small perforated drop of glass, coral, metal etc. threaded on a string to form a rosary or an ornamental necklace; a bead-like drop of a liquid, a bubble; the front sight of a gun; a narrow semicircular moulding; an ornament resembling a string of beads; (*pl.*) a necklace; a rosary. *v.t.* to ornament with beads or beading; to thread beads. *v.i.* to form beads. **Baily's beads,** a phenomenon resembling a string of beads observed on the sun in total eclipses, first described by the astronomer Francis *Baily* in 1836. **to draw a bead on,** to aim at. **to tell** or **say one's beads,** to count the rosary, to say one's prayers. **bead-frame,** *n.* an abacus. **bead-house,** *n.* an almshouse. **bead-roll,** *n.* a list of names (originally of benefactors) to be prayed for. **beadsman, -woman, bedesman, -woman,** *n.* one appointed to pray for another; an almsman or almswoman. **bead-tree,** *n.* the pride of India, *Melia azedirach*, and other trees, the seeds of which are used as rosary beads. **beadwork,** *n.* ornamental work in beads. **beaded,** *a.* **beading,** *n.* the formation of beads; beadwork; a bead-moulding. **beady,** *a.* of eyes, small and bright like beads; covered with beads or bubbles, foaming.

beadle (bē′dl), *n.* a petty officer of a church, parish, college, city company etc. **beadledom** (-dəm), *n.* **beadleship,** *n.*

beadsman, beadswoman BEAD.

beagle (bē′gl), *n.* a small dog originally bred for hunting hares; one who scents out or hunts down; an officer of the law.

beak (bēk), *n.* the pointed bill of a bird; anything similarly pointed, as the mandibles of a turtle or an octopus; the prow of an ancient war-galley, often sheathed with brass, and used as a ram; a promontory of land etc.; a spout; any beak-like process; (*sl.*) a magistrate, a headmaster or headmistress. *v.t.* to seize or strike with the beak (esp. in cock-fighting). **beaked,** *a.*

beaker (bē′kə), *n.* a large wide-mouthed drinking-vessel; the contents of a beaker; an open-mouthed glass vessel with a lip, used in scientific experiments. **Beaker Folk,** *n.pl.* a prehistoric people inhabiting Britain and Europe during the Bronze Age, named from the beakers found in their burial sites.

beal (bēl), **bull** (bul), *n.* (*Austral.*) a sweet honey drink.

beam (bēm), *n.* a large, long piece of timber squared on its sides, esp. one supporting rafters in a building; the part of a balance from which the scales are suspended; the part of a loom on which the warp is wound; a cylinder on which cloth is wound as it is woven; the main piece of a plough to which the handles are fixed; the main trunk of a stag's horn; a ray or collection of rays of light or radiation; a bright smile; the width of a ship or boat; the shank of an anchor. *v.t.* to send forth, to radiate, to emit in rays; to send forth by beam-transmission. *v.i.* to send forth rays of light; to shine radiantly; to smile brightly. **off** or **on the beam**, off or on the course indicated by a radio beam; off or on the mark. **on the beam**, (*Naut.*) at right-angles to the keel. **on one's beam-ends**, quite destitute. **beam-engine**, *n.* an engine with a beam connecting piston-rod and crank. **beam transmission**, *n.* short-wave radio transmission in which the energy radiated is concentrated for reception in a particular zone. **beam-tree**, *n.* the white-beam, whose timber is used for axle-trees. **beamy**, *a.* massive, shining, radiant, brilliant; antlered; broad in the beam (of ships). **beaming**, *a.*

bean (bēn), *n.* the kidney-shaped seed in long pods of *Faba vulgaris* and allied plants; other seeds resembling those of the common bean. **full of beans**, energetic, vigorous. **old bean**, (*dated sl.*) old fellow, old chap. **to give someone beans**, (*sl.*) to punish; to scold. **beanbag**, *n.* a small cloth bag filled with dried beans used in games; a large cushion filled with foam or polystyrene beads, used as a seat. **bean-feast**, *n.* a celebration. **bean-fed**, *a.* in good condition. **bean-fly**, *n.* an insect of purple colour found on beans. **bean-goose**, *n.* a migratory goose, *Anser segetum*. **beanpole**, *n.* a tall, thin pole used to support bean plants; (*coll.*) a tall thin person. **bean sprouts**, *n.pl.* the young shoots of mung beans used as a vegetable. **bean-stalk**, *n.* stem of the bean. **bean-straw**, *n.* the haulm of bean plants. **bean-tree**, *n.* a popular name for several trees bearing seeds in pod, esp. the carob-tree. **beantrefoil**, *n.* a popular name for the leguminous genus *Anagyris*, the laburnum, *Cytisus laburnum*, and the buck-bean or bog-bean, *Menyanthes trifoliata*. **beano** (-ō), *n.* (*coll.*) a treat, a spree, a bean-feast.

bear[1] (beə), *n.* a large plantigrade mammal with long shaggy hair and a stumpy tail; a rough unmannerly man; either of the northern constellations, the Great or the Little Bear; one who sells stock for future delivery in the expectation that prices will fall. *v.i.* to speculate for a fall in stocks. *v.t.* to produce a fall in the price of (stock etc.). **bear-baiting**, *n.* BAITING under BAIT. **bear-berry**, *n.* the genus *Arctostaphylos*, a procumbent heath; (*erroneously*) the barberry. **bear-garden**, *n.* a noisy, disorderly scene. **bear-leader**, *n.* (*coll.*) a travelling tutor. **bear's-breech**, *n.* the genus *Acanthus*. **bear's ear**, *n.* the common auricula. **bear's foot**, *n.* stinking hellebore. **bearskin**, *n.* the skin of a bear; a shaggy woollen cloth, used for overcoats; the tall fur cap worn by the Foot Guards and some other regiments in the British Army. **bearish**, *a.* bear-like; rough, rude, uncouth; in the stock market, characterized by a fall in prices.

bear[2] (beə), *v.t.* (*past* **bore** (baw), *p.p.* **borne** (bawn),) to carry, to wear, to show or display; to bring; to support the burden of; to be responsible for, to wield, to endure; to thrust, to press; to give birth to; to produce, to yield. *v.i.* to behave; to suffer, to be patient; to imply, to take effect; to have relation to; to incline, take a certain direction (as to the point of the compass) with respect to something else. **to bear against**, to rest upon; to be in contact with. **to bear arms**, to be a soldier; (*Her.*) to be entitled to a coat of arms. **to bear a hand**, to lend assistance. **to bear away**, to carry off; to win. **to bear down**, to overwhelm, to crush, to subdue; to use the abdominal muscles to assist in giving birth. **to bear down on**, to sail in the direction of; to approach purposefully. **to bear hard**, to press, to urge. **to bear in mind**, to remember. **to bear on**, to press against. **to bear out**, to confirm, to justify. **to bear up**, to endure cheerfully. **to bear upon**, to be relevant to. **to bring to bear**, to apply, bring into operation. **to bear with**, to put up with. **borne in upon one**, become one's firm conviction, realized by one. **bearable**, *a.* **bearably**, *adv.* **bearer**, *n.* one who or that which bears, carries or supports; one who assists to carry a corpse to the grave or to hold the pall; a porter; one who holds or presents a cheque; a bringer of anything; the holder of any rank or office; a support; in India, Africa etc. a personal or domestic servant. **bearing**, *n.* endurance, toleration; deportment, manner, behaviour; connection; a carrier or support for moving parts of any machine; any part of a machine that bears the friction; (*Her.*) a charge, a device; relation, relevance, aspect; the direction in which an object lies from a ship; (*pl.*) relative position. **to lose one's bearings**, to be uncertain of one's position. **bearing-rein**, *n.* a fixed rein for holding a horse's head up.

beard (biəd), *n.* the hair on the lower part of a man's face, esp. on the chin; analogous hairy appendage in animals; the barb of an arrow; a byssus; the bristles of a feather; the awn of grasses; hairs occurring in tufts. *v.t.* to oppose with resolute effrontery; to defy. **old man's beard**, the wild clematis or traveller's joy. **bearded**, *a.* **beardie** (-i), *n.* (*Austral.*) a variety of cod-fish. **beardless**, *a.* **beardlessness**, *n.*

bearskin BEAR[1].

beast (bēst), *n.* any of the animals other than man; a quadruped, esp. a large one; an animal to ride or drive; a brutal or objectionable person or thing. **the Beast**, Antichrist (Rev. xiii.1); (*fig.*) man's carnal instincts. **beast-like**, *a.* **beastly**, *a.* like a beast in form or nature; brutal, filthy; disgusting; disagreeable. **beastliness**, *n.*

beastings (bēs'tingz), BEEST.

beat (bēt), *v.t.* (*past* **beat**, *p.p.* **beaten**) to strike with repeated blows, to thrash; to bruise or break by striking or pounding; to work (metal etc.) by striking; to strike, as bushes, in order to rouse game; to mix or agitate by beating; to dash against (of water, wind etc.); to overcome, master; to tread, as a path; to play (an instrument or tune) by striking; to indicate time with a baton. *v.i.* to strike against some obstacle; to pulsate, to knock; to mark time in music; (*Naut.*) to make way against the wind. *n.* a stroke or blow; a pulsation; a certain assigned space regularly traversed at intervals by patrols, police etc.; hence, sphere, department, range; the rise or fall of the hand or foot in regulating time; a periodic variation in amplitude caused by the combination of oscillations of different frequencies. *a.* a shortened form of **beaten**. **dead-beat**, overcome, worn out. **to beat about**, to tack. **to beat about the bush**, to approach a matter in a roundabout way; to shilly-shally. **to beat a retreat**, to retire. **to beat back**, to compel to retire. **to beat down**, to force down (price) by haggling. **to beat hollow**, to excel or surpass in a great degree. **to beat in**, to crush. **to beat into**, to knock into by dint of blows: to instil. **to beat it**, (*sl.*) to go away. **to beat off**, to drive away by blows. **to beat one's brains**, to puzzle, to ponder laboriously. **to beat out**, to extend by beating, to hammer out; to extinguish by beating. **to beat the clock**, to complete a task within the allotted time. **to beat up**, to bring to a fluid or semi-fluid mass by beating; to make way against wind or tide; to injure seriously by beating. **beat generation**, *n.* (*orig. N Am.*) a bohemian movement of poets, writers etc. of the 1950s who rejected prevailing social and cultural values; young people of that time characterized by unconventional attitudes and self-conscious bohemianism. **beatnik** (-nik), *n.* (*often derog.*) a member of this movement. **beat music**, *n.* popular music characterized by a pulsating rhythm. **beaten**, *a.* subjected to

repeated blows; defeated, weary; trodden smooth or bare. **beater**, *n.* one who beats; a man employed to rouse game, esp. grouse or pheasant; an instrument for beating or pounding. **beating**, *n.* the action of striking repeated blows; punishment or chastisement by blows; throbbing; defeat; sailing against the wind. **to take a beating**, to suffer verbal or physical punishment. **to take some, a lot of beating**, to be difficult to improve upon.
beatify (biat'ifī), *v.t.* to render supremely blessed or happy; in the Roman Catholic Church, to declare (deceased person) blessed. **beatific** (bēətif'-), *a.* **beatific vision, beatific vision,** *n.* the vision of the glories of heaven. **beatifically**, *adv.* **beatification**, *n.* **beatitude** (-tūd), *n.* supreme felicity; heavenly bliss; esp. the special blessedness announced in the Sermon on the Mount.
beatnik BEAT.
beau (bō), *n.* (*pl.* **beaus, beaux** (bōz)) a fop, a dandy; a suitor, sweetheart. **beau geste** (zhest), *n.* (F) gracious gesture. **beau-ideal**, *n.* the highest conceivable type of excellence. **beau-monde** (mōd'), *n.* the fashionable world. **beauish**, *a.*
Beaufort scale (bō'fət), *n.* a scale of wind velocity devised by the English admiral and hydrographer Sir Francis Beaufort, 1774-1857, ranging from 0 = calm to 12 = hurricane.
Beaujolais (bō'zhəlā), *n.* a usu. red, light Burgundy wine.
Beaune (bōn), *n.* a usu. red Burgundy wine.
beauty (bū'ti), *n.* that quality or assemblage of qualities which gives the eye or the other senses intense pleasure; a beautiful person, esp. a woman; beautiful women generally; a beautiful feature or characteristic; a particular aspect that gives satisfaction or (*iron.*) the reverse; a very fine example of its kind; (*coll.*) an egregious person; a scamp. **beauty parlour**, *n.* a shop specializing in beauty treatments. **beauty queen**, *n.* a woman picked as the most attractive in a beauty contest. **beauty-sleep**, *n.* sleep before midnight. **beauty-spot**, *n.* a patch or spot placed upon the face to heighten some beauty; a foil; (*coll.*) a beautiful place or landscape. **beauty treatment**, *n.* improvement of women's appearance by artificial means. **beaut**, *n.* (*sl.*, *esp. Austral.*) something or someone outstanding. *int.* great, excellent. **beauteous**, *a.* (*poet.*) beautiful. **beauteously**, *adv.* **beauteousness**, *n.* **beautician**, *n.* one who administers beauty treatment. **beautiful**, *a.* full of beauty; possessing the attributes that constitute beauty; satisfactory, delicious; (*iron.*) egregious. **beautifully**, *adv.* **beautifulness**, *n.* **beautify** (-fī), *v.t.* to make beautiful. **beautification**, *n.* **beautifier**, *n.*
beaux arts (bōz ah'), *n.pl.* (F) fine arts.
beaux esprits BEL ESPRIT.
beauxite (bō'zīt), BAUXITE.
beaver (bē'və), *n.* an amphibious rodent mammal, *Castor fiber*, with broad tail, soft fur and habits of building huts and dams; the fur of this animal; a hat made of such fur; (*sl.*) a man with a beard. **to beaver away**, to work hard. **beaver board**, *n.* a building board of wood-fibre material. **beaver-dam**, *n.* an obstruction placed across a stream by beavers. **beaver-rat**, *n.* the musquash or musk-rat. **beaver-tree, beaver-wood**, *n.* (*N Am.*) the sweet-bay or laurel-magnolia, *Magnolia glauca*.
bebop (bē'bop), *n.* a variety of jazz music distinguished by its harsh melodies, dissonant harmonies and faster tempos.
becalm (bikahm'), *v.t.* to render calm or still; to quiet, to soothe; to deprive (a ship) of wind.
became (bikām'), *past* BECOME.
because (bikoz'), *conj.* by reason of, on account of, inasmuch as.
beccafico (bekəfē'kō), *n.* a small migratory songbird of the genus *Sylvia*.
béchamel (bā'shəmel), *n.* a white sauce made with cream or milk and flavoured with onions and herbs. [After Louis de *Béchamel*, d. 1704, its F inventor]
bêche-de-mer (beshdəmeə'), *n.* the sea-slug or trepang, *Holothuria edulis*, an echinoderm eaten by the Chinese.
beck[1] (bek), *n.* a bow or curtsy; a nod, a gesture of the finger or hand; the slightest indication of will. *v.i.* to make a mute signal; to curtsy. *v.t.* to call by a beck. **beck and call**, absolute control.
beck[2] (bek), *n.* a brook, a rivulet; esp. a stream.
becket (bek'it), *n.* (*Naut.*) anything used to confine loose ropes, tackle or spars, as a large hook, a rope with an eye at one end; a bracket, pocket, loop etc.
beckon (bek'ən), *v.t., v.i.* to make a signal by a gesture of the hand or a finger or by a nod (to).
become (bikŭm'), *v.i.* (*past* **became** (-kām'), *p.p.* **become**) to pass from one state or condition into another; to come into existence; to come to be. *v.t.* to be suitable to, to befit, to be proper to or for; to be in harmony with; to look well upon. **becoming**, *a.* suitable, proper; in harmony or keeping with; graceful in conduct, attire etc. **becomingly**, *adv.* **becomingness**, *n.*
becquerel (bek'ərel), *n.* a unit which measures the activity of a radioactive source. [A. H. *Becquerel*, 1852-1908, French physicist]
bed (bed), *n.* an article of domestic furniture to sleep upon; hence, marriage, conjugal rights, childbirth and, with qualifying adjective, the grave; the resting-place of an animal; the flat surface on which anything rests; a plot of ground in a garden; the channel of a river; the bottom of the sea; a horizontal course in a wall; a layer or stratum of rock etc.; the foundation of a road, street or railway. *v.t.* (*past, p.p.* **bedded**) to put in bed; to plant in a bed or beds; to have sexual intercourse with; to fix in a stratum or course; to place in a matrix of any kind; to embed. *v.i.* to go to bed. **bed and board**, lodgings and food. **bed and breakfast**, in a hotel etc., overnight accommodation with breakfast. **bed of roses**, a comfortable place. **to be brought to bed**, to be delivered of a child. **to bed out**, to plant out in beds. **to get out of bed on the wrong side**, to begin the day in a foul mood. **to keep one's bed**, to remain in bed. **to lie in the bed one has made**, to suffer for one's own misdeeds or mistakes. **to make a bed**, to put a bed in order after it has been used. **to take to one's bed**, to be confined to bed (from sickness etc.). **bedbug**, *n.* a bloodsucking insect which infests filthy bedding. **bed-chamber**, *n.* a bedroom. **lords, gentlemen (ladies, women) of the bed-chamber**, officers of the Royal Household who wait upon a male (female) sovereign. **bed-clothes**, *n.pl.* sheets, blankets and coverlets for a bed. **bed-fellow**, *n.* one who sleeps in the same bed with another. **bed-hangings**, *n.pl.* hangings or curtains for a bed. **bed-linen**, *n.* sheets and pillow for a bed. **bed-mate**, *n.* a bed-fellow. **bed-pan**, *n.* a chamber-utensil for use in bed. **bed-plate**, *n.* (*Eng.*) the cast-iron or steel plate used as the base plate of an engine or machine. **bedpost**, *n.* one of the upright supports of a bedstead. **bed-ridden**, *a.* confined to bed through age or sickness. **bedrock**, *n.* the rock underlying superficial formations; hence, fundamental principles; the lowest possible state. **bed-roll**, *n.* bedding rolled up so as to be carried by a camper etc. **bedroom**, *n.* a sleeping apartment. **bedside**, *n.* place by, or companionship by a bed. **bedside manner**, *n.* suave manner in attending a patient. **bed-sitting-room, bed-sitter, -sit**, *n.* bedroom and sitting-room combined. **bedsore**, *n.* a sore produced by long confinement to bed. **bedspread**, *n.* a counterpane. **bedstead** (-sted), *n.* the framework on which a mattress is placed. **bedstraw**, *n.* English name of the genus *Galium*. **bedtime**, *n.* the usual hour for going to bed. **bedward**, *adv.* in the direction of bed. **bedder**, *n.* a plant for bedding-out. **bedding**, *n.* a bed with the clothes upon it; bed-clothes; litter for domestic animals; a bottom layer or foundation; stratification; the line or plane of stratification. **bedding-plants,**

bedding-out plants, *n.pl.* plants intended to be set in beds.
BEd., (*abbr.*) Bachelor of Education.
bedad (bidad'), *int.* (*Ir.*) begad.
bedaub (bidawb'), *v.t.* to daub over, to bedizen.
bedazzle (bidaz'l), *v.t.* to confuse by dazzling. **bedazzlingly**, *adv.* **bedazzlement**, *n.*
bede, bedesman etc. BEAD.
bedeck (bidek'), *v.t.* to deck out, to adorn.
bedeguar (bed'igah), *n.* a mossy growth on rose-briers.
bedel, bedell (bē'dl), *n.* an officer at the Universities of Oxford and Cambridge who performs ceremonial functions.
bedevil (bidev'l), *v.t.* to bewitch; to torment; to confuse; to obstruct. **bedevilment**, *n.*
bedew (bidū'), *v.t.* to moisten or sprinkle with dew-like drops.
bedim (bidim'), *v.t.* to render dim; to obscure.
bedizen (bidīz'n, -diz'n), *v.t.* to deck out in gaudy vestments or with tinsel finery. **bedizenment**, *n.*
bedlam (bed'ləm), *n.* a lunatic asylum; a scene of wild uproar; madness, lunacy. *a.* mad, foolish, lunatic. **bedlamite** (-it), *n.* a madman, a lunatic.
Bedlington (bed'lingtən), *n.* a grey, crisp-haired terrier. [town in Northumberland]
Bedouin, -duin (bed'uin), *n.* a nomadic Arab, as distinguished from one living in a town. *a.* nomadic.
bedraggle (bidrag'l), *v.t.* to soil by trailing in the wet or mire. **bedraggled**, *a.*
Beds. (bedz), (*abbr.*) Bedfordshire.
bee (bē), *n.* a four-winged insect which collects nectar and pollen and is often kept in hives for the honey and wax it produces; any closely allied insect, e.g. *carpenter-bee*; a busy worker; (*N Am.*) a social meeting for work usually on behalf of a neighbour. **spelling-bee**, a social contest in spelling. **the bee's knees**, (*coll.*) someone or something wonderful, admirable. **to have a bee in one's bonnet**, to be cranky on some point. **bee-bird**, *n.* the spotted flycatcher, *Muscicapa grisola;* (*N Am.*) the king-bird, *Tyrannus tyrannus.* **bee-bread**, *n.* a mixture of honey and pollen, on which bees feed their larvae. **bee-cuckoo**, *n.* an African bird, called also the honey-guide. **bee-eater**, *n.* a tropical Old World bird of the genus *Merops.* **bee-fold**, *n.* an enclosure for beehives. **bee-glue**, *n.* the substance with which bees fill up crevices in their hives. **beehive**, *n.* a receptacle (usually of wood or straw) for bees. *a.* dome-shaped. **bee-line**, *n.* the shortest route between two places. **bee-master, -mistress**, *n.* one who keeps bees. **bee-moth**, *n.* the wax-moth, which lays its eggs in hives. **bee-orchis**, *n.* a British orchid whose flower resembles a bee. **bee-skep**, *n.* a straw beehive. **beeswax**, *n.* the wax secreted by bees for their cells, used to make polishes. *v.t.* to rub or polish with beeswax. **beeswing**, *n.* a fine filmy deposit in an old port wine; old port.
Beeb (bēb), *n.* (*coll.*) an informal, humorous name for the BBC.
beech (bēch), *n.* a forest tree of the genus *Fagus;* esp. *F. sylvatica,* the common beech. the wood of this tree. **beech-drops**, *n.pl.* (*N Am.*) any of several plants parasitic on the roots of the beech. **beech-fern**, *n.* popular name of *Polypodium phegopteris,* **beech-mast**, *n.* the fruit of the beech-tree. **beech-nut**, *n.* **beech-oil**, *n.* **beech-wheat** BUCK WHEAT. **beechen**, *a.* **beechy**, *a.*
beef (bēf), *n.* the flesh of the ox, cow or bull, used as food; an ox (*usu.* in *pl.,* **beeves** (bēvz)), esp. one fatted for the market; flesh, muscle; (*sl. pl.* **beefs**) a complaint. *v.i.* to grumble. **to beef up**, (*coll.*) to strengthen, reinforce. **beefburger**, *n.* a hamburger. **beefcake**, *n.* (*sl.*) men with muscular physiques, esp. as displayed in photographs. **Beefeater**, *n.* a Yeoman of the Guard, instituted in 1485; a warder of the Tower of London. **beefsteak**, *n.* a slice of meat from the hindquarters of an ox. **beef-tea**,

n. the juice extracted from beef by simmering. **beef-wood**, *n.* the timber of the *Casuarina* and other trees. **beefy**, *a.* like beef; fleshy; muscular. **beefiness**, *n.*
beehive BEE.
Beelzebub (biel'zibūb), *n.* a god worshipped in Ekron (II Kings i.2); the prince of evil spirits; an evil spirit.
been (bēn), *p.p.* BE.
beep (bēp), *n.* a short sound as made by a car horn or an electronic device. *v.i.* to make such a sound. *v.t., v.i.* to cause to make such a sound. **beeper**, *n.*
beer (biə), *n.* an alcoholic drink brewed from fermented malt, hops, water and sugar; any malt liquor prepared by brewing; other fermented liquors, as *ginger-beer* etc. **small beer**, things of no account. **beer and skittles**, enjoyment or pleasure. **beer-barrel**, *n.* **beer-engine, -pump**, *n.* a machine for pumping up beer from the cellar to the bar. **beer-garden**, *n.* an outdoor area with tables where beer and other refreshments may be consumed. **beer-money**, *n.* a money allowance in lieu of beer. **beery**, *a.* abounding in beer; like beer; under the influence of beer; fuddled. **beerily**, *adv.* **beeriness**, *n.*
beest, beestings (bēst), *n.* the first milk drawn from a cow after calving.
beeswax, beeswing BEE.
beet (bēt), *n.* a genus of plants, comprising red beet, used as a salad, and white beet, used in sugar-making, cultivated for its esculent root. **beet-radish, -rave** (-rāv), *n.* the common beet, *Beta vulgaris,* when raised for salad. **beetroot**, *n.* the root of this used as a salad; the red colour of beetroot.
beetle[1] (bē'tl), *n.* a heavy wooden mallet for driving stakes etc. into the ground, hammering down paving-stones etc. *v.t.* to beat with a beetle. **beetlebrain**, *n.* (*coll.*) an idiot. **beetlebrained**, *a.*
beetle[2] (bē'tl), *n.* an insect of the order Coleoptera, the upper wings of which have been converted into hard wing-cases, the under ones being used for flight; also applied to other similar insects; a game in which the players attempt to complete a beetle-shaped drawing according to the throw of a dice. *v.i.* to jut out, to hang over. *a.* overhanging, scowling. **to beetle along, off** etc., to hurry, scuttle along. **beetle-browed**, *a.* having projecting or overhanging brows. **beetle-crusher**, *n.* (*sl.*) a large foot; a heavy boot; a policeman; a soldier. **beetling**, *a.*
beetroot BEET.
beeves BEEF.
BEF, (*abbr.*) British Expeditionary Force.
befall (bifawl'), *v.t., v.i.* (*past* **befell** (-fel'), *p.p.* **befallen**) to happen (to).
befit (bifit'), *v.t.* to be suitable to or for. **befitting**, *a.* **befittingly**, *adv.*
befog (bifog'), *v.t.* to involve in a fog; to obscure.
before (bifaw'), *prep.* in front of, in time, space, rank or degree; in presence of; under the cognizance of; under the influence or impulsion of; in preference to. *adv.* in front; previously. *conj.* earlier than; rather than. **before the mast**, in the fo'c'sle; applied to common sailors who live in the fo'c'sle in front of the foremast. **before the wind**, with the wind right aft. **before Christ**, before the birth of Christ. **before God**, *a.* with the knowledge or in the sight of God. **before-going**, *a.* preceding. **beforehand**, *adv.* in anticipation, in advance. **to be beforehand**, to forestall; to be earlier than expected. **before-mentioned**, *a.* mentioned before. **before-time**, *adv.* formerly.
befoul (bifowl'), *v.t.* to render dirty, to soil.
befriend (bifrend'), *v.t.* to favour, help.
befuddle (bifūd'l), *v.t.* to confuse; to stupefy with drink. **befuddlement**, *n.*
beg[1] (beg), *v.i.* (*past, p.p.* **begged**) to ask for alms, to live by asking alms; (of a dog) to sit up on the hind quarters expectantly. *v.t.* to ask earnestly, to crave, entreat. **to beg off**, to seek to be released from some obligation. **to beg**

the question, to assume the thing to be proved. **to go a-begging**, to be left without a taker. **beggar**, *n.* one who begs; one who lives by asking alms; one in indigent circumstances; *(coll.)* a fellow. *v.t.* to impoverish; to outdo. **to beggar description**, to go beyond one's power of expression. **beggar-my-neighbour**, *n.* a game of cards. **beggarly**, *a.* like a beggar; poverty-stricken; poor, contemptible. *adv.* in the manner of a beggar. **beggarliness**, *n.* **beggary**, *n.* extreme indigence.
beg² BEY.
begad (bigad'), *int.* by God.
began (bigan'), *past* BEGIN.
begem (bijem'), *v.t.* to cover or set as with gems.
beget (biget'), *v.t.* *(past* **begot**, *p.p.* **begotten**) to procreate; to cause to come into existence. **begetter**, *n.*
beggar etc. BEG¹.
Beghard (beg'əd), *n.* a lay brother, belonging to a 13th-cent. Flemish religious order like the Beguines.
begin (bigin'), *v.i.* *(past* **began** (-gan'), *p.p.* **begun** (-gŭn')) to come into existence; to commence. *v.t.* to be the first to do, to do the first act of, to enter on, to commence. **to begin with**, to take first; firstly. **beginner**, *n.* one who originates anything; one who is the first to do anything; a learner; the actor or actors who appear first on the stage at the start of a play. **beginning**, *n.* the first cause, the origin; the first state or commencement; rudiments.
begird (bigœd'), *v.t.* to encircle with or as with a girdle.
begone (bigon'), *imper. v.* go away, depart.
begonia (bigō'niə), *n.* a genus of tropical plants cultivated chiefly for their ornamental foliage.
begot (bigot'), *past*, **begotten**, *p.p.* BEGET.
begrime (bigrīm'), *v.t.* to blacken or soil with grime.
begrudge (bigrŭj'), *v.t.* to grudge; to envy (a person) the possession of.
beguile (bigīl'), *v.t.* to deceive, cheat; to deprive of or lead into by fraud; to charm away tedium or weariness, to amuse; to bewitch. **beguilement**, *n.* **beguiler**, *n.* **beguiling**, *a.* **beguilingly**, *adv.*
Beguine (bigēn'. beg'in), *n.* a member of certain sisterhoods in the Netherlands in the 12th cent. **beguinage** (bāgēnahzh', beg'inij), *n.* a house or establishment for Beguines.
beguine (bigēn'), *n.* music or dance in bolero rhythm, of S American or W Indian origin.
begum (bā'gəm), *n.* a lady of high rank in some Muslim countries.
begun (bigŭn'), *p.p.* BEGIN.
behalf (bihahf'), *n.* interest, lieu, stead. **on behalf of**, on account of, for the sake of; representing.
behave (bihāv'), *v.r.* to conduct, to demean. *v.i.* to conduct oneself or itself; to display good manners. **well-behaved**, having good manners. **behaviour**, *(esp. N Am.)* **behavior**, *n.* manners, conduct, demeanour; the manner in which a thing acts. **behaviour therapy**, *n.* a method of treating neurotic disorders by gradually conditioning the patient to react normally. **behavioural**, *a.* **behavioural science**, *n.* the scientific study of the behaviour of human beings and other organisms. **behaviourism**, *n.* the guiding principle of certain psychologists who hold that the proper basis of psychological science is the objective study of behaviour under stimuli. **behaviourist**, *n.* **behaviouristic**, *a.*
behead (bihed'), *v.t.* to cut the head off, to kill by decapitation.
beheld (biheld'), *past*, *p.p.* BEHOLD.
behemoth (bē'əmoth, bihē'-), *n.* a huge person or thing.
behest (bihest'), *n.* a command; an injunction.
behind (bihīnd'), *prep.* at the back of; after; in the rear of. *adv.* at the back, in the rear; towards the rear; in the past; backwards, out of sight, on the farther side of; in reserve; in arrears. *n.* the back part of a person or garment; the buttocks. **behind one's back**, without one's knowledge. **behind the scenes**, out of sight, without being obvious, private, secret. **behindhand**, *a.* dilatory, tardy; backward, unfinished; in arrear.
behold (bihōld'), *v.t.* *(past, p.p.* **beheld** (-held)) to look attentively at; to see; view; to consider. *v.i.* to look. *int.* lo!, see. **beholder**, *n.* **beholden**, (orig. *p.p.* of BEHOLD) *a.* obliged, indebted (with *to*).
behoof (bihoof'), *n.* advantage, use, profit, benefit.
behove (bihōv'), *v.t.* to befit, to be due to, to suit. *v.i.* to be needful to; due to; to be incumbent.
beige (bāzh), *n.* a fabric made of undyed and unbleached wool. *a.* pale brownish yellow.
being (bē'ing), *n.* the state of existing; lifetime; existence; nature, essence; a thing or person existing. *conj.* seeing that, since. *a.* existing, present. **Supreme Being**, God. **the time being**, the present.
bekko-ware (bek'ō), *n.* Chinese pottery veined with colour like tortoise-shell.
bel (bel), *n.* a measure for comparing the intensity of noises, currents etc., the logarithm to the base 10 of the ratio of one to the other.
belabour (bilā'bə), *v.t.* to beat, to thrash; to assault verbally.
belar, belah (bē'lah), *n.* a variety of casuarina tree.
belated (bilā'tid), *a.* very late; too late; benighted. **belatedly**, *adv.* **belatedness**, *n.*
belaud (bilawd'), *v.t.* to praise excessively.
belay (bilā'), *v.t.* to fasten a running rope by winding it round a cleat or belaying-pin; to turn a rope round an object; to secure a climber to a rope. *n.* a turn of a rope round an object; that around which a climber's rope is belayed. *int.* stop, enough. **belaying-pin**, *n.*
bel canto (bel kan'tō), (It.) *n.* a style of operatic singing characterized by purity of tone and exact phrasing.
belch (belch), *v.t., v.i.* to expel (wind) from the mouth; to throw or pour out copiously or violently. *n.* an eructation; an eruption, a burst (of smoke or fire).
belcher (bel'chə), *n.* a blue and white spotted neckerchief.
beldam (bel'dəm), *n.* a grandmother; an old woman; a witch.
beleaguer (bilē'gə), *v.t.* to besiege; to harass. **beleaguerment**, *n.*
belemnite (bel'əmnīt), *n.* a conical, sharply pointed fossil shell of a genus of cephalopods, allied to the cuttle-fish. **belemnitic** (-nit'-), *a.*
bel esprit (bel esprē'), *n.* *(pl.* **beaux esprits** (bōz)) a person of genius; a wit.
belfry (bel'fri), *n.* a bell-tower attached to or separate from a church or other building; the chamber for the bells in a church-tower.
Belg., *(abbr.)* Belgian; Belgium.
Belgian (bel'jən), *a.* of or pertaining to Belgium or the Belgians. *n.* a native of Belgium; a kind of canary. **Belgian hare**, *n.* a large breed of domestic rabbit, dark-red in colouring. **Belgic**, *n.*, *a.*
Belial (bē'liəl), *n.* the Devil, Satan; one of the fallen angels. **son, man of Belial**, *n.* a worthless, wicked man.
belie (bilī'), *(pres.p.* **belying**, *past, p.p.* **belied**) *v.t.* to tell lies about, to slander; to misrepresent; to be faithless to; to fail to perform or justify. **belier**, *n.*
belief (bilēf'), *n.* reliance; the mental act or operation of accepting a fact or proposition as true; the thing so believed; opinion, persuasion; religion.
believe (bilēv'), *v.t.* to have confidence in or reliance on; to give credence to; to accept as true; be of opinion that. *v.i.* to think; to have faith. **believe it or not**, although it may seem incredible, the statement is true. **to believe in**, to trust in, to rely on. **to make believe**, to pretend. **believable**, *a.* **believableness**, *n.* **believer**, *n.* **believing**, *a.* **believingly**, *adv.*
Belisha beacon (bilish'ə), *n.* a flashing orange globe on a post to indicate a street-crossing for pedestrians.

belittle (bilit'l), *v.t.* to dwarf; to depreciate verbally. **belittlement**, *n.*

bell (bel), *n.* a hollow body of cast metal, usually in the shape of an inverted cup with recurved edge, so formed as to emit a clear musical sound when struck by a hammer; applied to many objects of a similar form; the body of a Corinthian capital; the cry of a stag at rutting time; a bell-shaped corolla; the catkin containing the female flowers of the hop; the bell struck on board ship every half-hour to indicate the time; a space of half an hour. *v.i.* of stags at rutting time, to bellow; to be in flower (of hops). *v.t.* to furnish with a bell; to utter loudly. **sound, clear as a bell**, sound or clear, free from any flaw. **to bear away the bell**, to carry off the prize. **to bear the bell**, to be first. **to bell the cat**, to be a ringleader in a dangerous enterprise. **to curse by bell, book and candle**, to excommunicate solemnly. **bell-animalcules**, *n.pl.* the infusorial family Vorticellidae. **bell-bird**, *n.* a S American or Austral. bird with a note like a bell. **bell-bottomed**, *a.* (of trousers) with wide, bell-shaped bottoms. **bell-boy, bell-hop**, *n.* (*N Am.*) hotel page-boy. **bell-buoy**, *n.* a buoy to which a bell is attached, rung by the motion of the waves. **bell-cot, bell-cote**, *n.* a small turret for a bell or bells. **bell-crank**, *n.* a crank adapted to communicate motion from one bell-wire to another at right angles to it. **bell-faced**, *a.* having a convex face (as a hammer). **bell-flower**, *n.* **bell-founder**, *n.* one who casts bells. **bell-founding, -foundry**, *n.* **bell-gable, -turret**, *n.* a gable or turret in which bells are hung. **bell-glass**, *n.* a bell-shaped glass for protecting plants. **belljar**, *n.* a bell-shaped glass cover used in laboratories to protect apparatus or contain gases in experiments etc. **bellman**, *n.* a public crier who attracts attention by ringing a bell. **bell-metal**, *n.* an alloy of copper and tin, used for bells. **bell-pull**, *n.* a cord or handle by which a bell is rung. **bell-punch**, *n.* a ticket punch in which a bell is rung each time it is used. **bell push**, *n.* a button which operates an electric bell. **bell-ringer**, *n.* **bell-rope**, *n.* the rope by which a bell is rung. **bell-shaped**, *a.* **bell-telegraph**, *n.* a telegraph instrument in which two bells are used to represent Morse dots and dashes. **bell-tent**, *n.* a conical tent. **bell-turret**, BELL-GABLE. **bell-wether**, *n.* the sheep that wears a bell and leads a flock; (*fig.*) a leader. **bell-wort**, *n.* any plant of the Campanulaceae.

belladonna (beladon'a), *n.* deadly nightshade *Atropa belladonna*; a drug prepared from it.

belle[1] (bel), *n.* a beautiful woman.

belle[2] (bel), fem. of BEAU. **belle amie** (amē'), *n.* a female friend; mistress. **belle époque** (äpok'), *n.* the period of security and comfort before the outbreak of World War I.

belle vue (vü), *n.* a fine sight. [F]

belles-lettres (bel·let'r'), (F) *n.pl.* polite literature, the humanities, pure literature. **belletrist**, *n.* **belletristic**, *a.*

bellicose (bel'ikōs), *a.* warlike; inclined to war or fighting. **bellicosity** (-kos'-), *n.*

bellied BELLY.

belligerent (bilij'arant), *a.* (of or pertaining to persons or nations) carrying on war; aggressive. *n.* a nation, party or individual engaged in war. **belligerence, belligerency**, *n.*

bellow (bel'ō), *v.i.* to emit a loud hollow sound (as a bull); to raise an outcry or clamour. *v.t.* to utter with a loud hollow voice. *n.* the roar of a bull, or any similar sound.

bellows (bel'ōz), *n.pl.* or *n.sing.* an instrument or machine for supplying a strong blast of air to a fire or organ pipes; the expansible portion of a camera; (*fig.*) the lungs. **pair of bellows**, a two-handled bellows for fanning fire. **bellows-fish**, *n.* the trumpet-fish.

belly (bel'i), *n.* that part of the body in front which extends from the breast to the insertion of the lower limbs; the part containing the stomach and bowels; the stomach, the womb; that part of the body which demands food; hence, appetite, gluttony; the front or lower surface of an object; anything swelling out or protuberant; a cavity, a hollow surface; the interior; the bulging part of a violin or a similar instrument. *v.t.* to cause to swell out. *v.i.* to swell or bulge out. **belly-ache**, *n.* a pain in the stomach. *v.i.* (*coll.*) to express discontent, to whine. **bellyband**, *n.* a band passing under the belly of a beast of burden to keep the saddle in place. **bellybutton**, *n.* (*coll.*) the navel. **belly dance**, *n.* a solo dance involving undulating movements of the abdomen. **belly flop**, *n.* an awkward dive into the water flat against the surface. *v.t.* to perform a belly flop. **belly landing**, *n.* landing without using the landing-wheels. **belly-laugh**, *n.* a deep, hearty laugh. **bellied**, *a.* having a belly (*in comb.*); corpulent. **bellyful**, *n.* as much as fills the belly, as much food as satisfies the appetite; (*coll.*) more than enough. **bellying**, *a.*

belomancy (bel'ōmansi), *n.* divination by means of arrows.

belong (bilong'), *v.i.* to be appropriate, to pertain; to be the property, attribute, appendage, member, right, duty, concern or business of; to be connected with; to be a native or resident of. **belonger**, *n.* **belongings**, *n. pl.* one's possessions.

Belorussian (belōrūsh'an), **Byelorussian** (byelō-), *a.* of Belorussia, a region of the W Soviet Union, of its language or people. *n.* the Slavonic language of Belorussia; a native or citizen of Belorussia.

beloved (bilūvd'. -lŭv'id), *n., a.* (one) loved greatly.

below (bilō'), *prep.* under in place, rank, excellence etc.; down stream from; unworthy of. *adv.* in or to a lower place, rank or station; on earth (as opp. to heaven); in hell (as opp. to earth); downstairs; down stream; lower on the same page, or on a following page. **below one's breath**, in a whisper.

bel paese (bel paä'zi), *n.* a mild Italian cream cheese.

belt (belt), *n.* a broad, flat strip of leather etc. worn around the waist or over the shoulder, esp. as a badge of rank or distinction; anything resembling such a belt in shape; a broad strip or stripe; a strait; a zone or region; a flat endless strap passing round two wheels and communicating motion from one to the other; (*coll.*) a blow. *v.t.* to encircle with or as with a belt; to fasten on with a belt; to invest with a belt; to thrash. **below the belt**, unfairly. **to belt out**, to sing or emit a sound vigorously or loudly. **to belt up**, (*sl.*) to stop talking (often *imp.*); to fasten with a belt. **to tighten one's belt**, to make economies, to reduce expenditure. **under one's belt**, secured in one's possession. **belted**, *a.* wearing a belt, esp. as a mark of rank or distinction; affixed by a belt; surrounded as with a belt. **belted earl**, *n.* an earl wearing (or entitled to wear) his distinctive cincture. **belting**, *n.* belts collectively; material for belts; (*sl.*) a beating.

Beltane (bel'tān), *n.* May-day (o.s.); a Celtic festival with bonfires on May-day.

beluga (biloo'ga), *n.* the great white sturgeon, *Acipenser huso*, from the Black and Caspian Seas; the white whale.

belvedere (belvidia'), *n.* a turret, lantern or cupola, raised above the roof of a building to command a view; a summer-house built on an eminence.

belying BELIE.

BEM, (*abbr.*) British Empire Medal.

bemire (bimia'), *v.t.* to cover or soil with mire. **bemired**, *a.*

bemoan (bimōn'), *v.t.* to moan over, to deplore. *v.i.* to lament.

bemuse (bimūz'), *v.t.* to make utterly confused or dazed.

bench (bench), *n.* a long seat or form; a seat where judges and magistrates sit in court; hence judges or magistrates collectively; the office of judge; (*pl.*) groups of seats in the Houses of Parliament; (*N Am.*) a level tract between a river and neighbouring hills; a terrace or ledge in masonry, quarrying, mining, earthwork etc.; a work-bench; a platform for exhibiting dogs. *v.t.* to furnish with benches; to exhibit (dogs) at a show; (*N Am.*) to remove a player from a game. **Queen's** (or **King's**) **Bench**, the court for-

merly presided over by the Sovereign; now one of the divisions of the Supreme Court. **the Bench of Bishops,** the Episcopate collectively. **to be raised to the bench,** to be made a judge. **treasury bench, front bench,** etc., seats appropriated to certain officers, parties or groups in Parliament. **bench-mark,** *n.* a mark cut in rock etc. in a line of survey for reference at a future time; anything that serves as a standard of comparison or point of reference. **bench-plane,** *n.* the jack-plane, the trying-plane or the smoothing plane. **bench-show,** *n.* a dog-show, in which the dogs are exhibited on benches. **bench-table,** *n.* a low seat of stone in churches and cloisters. **bench-warrant,** *n.* a warrant issued by a judge, as distinct from a magistrate's warrant. **bencher,** *n.* one of the senior members of an Inn of Court.
bend (bend), *v.t.* (*past, p.p.* **bent** (bent), *exc.* in **bended knees**) to bring into a curved shape (as a bow); to deflect; to direct to a certain point; to incline from the vertical; to subdue; to make fast; to tie into a knot; *v.i.* to assume the form of a curve or angle; to incline from an erect position, to bow, stoop; to surrender, submit; to turn in a new direction. *n.* a bending curve or flexure; incurvation; a sudden turn in a road or river; an inclination; (*Her.*) an ordinary formed by two parallel lines drawn across from the dexter chief to the sinister base point of an escutcheon; a similar ordinary from the sinister chief to the dexter base point is a mark of bastardy, and is called **bend sinister,** a knot; (*pl.*) caisson disease. **on bended knees,** with the knees bent; as a suppliant. **round the bend,** (*coll.*) crazy, insane. **to bend a sail,** to extend or make it fast to its proper yard or stay. **to bend the brows,** to frown. **to bend the elbow,** (*coll.*) to be fond of drinking alcohol. **bender,** *n.* (*sl.*) a bout of heavy drinking. **bendy,** *a.*
beneath (bineth'), *prep.* under, in point of place or position; unworthy of. *adv.* in a lower place, below.
benedicite (benidi'siti), *int.* bless you. *n.* the invocation of a blessing; grace before meat.
benedick (ben'idik), **benedict** (-dikt), *n.* a newly married man (from *Benedick,* a character in *Much Ado About Nothing*).
Benedictine (benidik'tēn), *a.* of or pertaining to St Benedict, 480–543, or to the order of monks founded by him. *n.* a monk or nun of his order; a liqueur first made by Benedictine monks.
benediction (benidik'shən), *n.* the act of blessing or invoking a blessing; grace before or after meals; blessedness; a Roman Catholic devotion including a blessing with the Host. **benedictory,** *a.* **benedictus** (-tus), *n.* the hymn of Zacharias (Luke i.68), used in the Church of England and the Roman Catholic Church.
benefaction (benifak'shən), *n.* the conferring of a benefit; a benefit conferred; a gift or endowment for charitable purposes. **benefactor** (be'nifaktər), *n.* one who gives another help or friendly service; one who gives to a religious or charitable institution. **benefactress** (-tris), *n. fem.*
benefice (ben'ifis), *n.* an ecclesiastical living. **beneficed,** *a.*
beneficent (binef'isənt), *a.* generous; characterized by benevolence. **beneficently,** *adv.* **beneficence,** *n.* the habitual practice of doing good; charity. **beneficial** (benifish'əl), *a.* advantageous; remedial; (*Law*) of or belonging to usufruct; enjoying the usufruct of. **beneficially,** *adv.* **beneficiary** (-shiəri), *n.* one who receives a favour; a feudatory; the holder of a benefice; one who benefits under a trust.
benefit (ben'ifit), *n.* profit, advantage, gain; a theatrical, music-hall or other performance, the receipts from which, with certain deductions, are given to some person or charity; money or services provided under government social security or private pension schemes etc. *v.t.* to do good to; to be of advantage to. *v.i.* to derive advantage. **benefit of clergy** CLERGY. **benefit of the doubt,** assumption of innocence in the absence of clear evidence of guilt. **benefit club, society,** *n.* a friendly society.
Benelux (ben'ilūks), *n.* name given to Belgium, the Netherlands and Luxembourg.
benevolent (binev'ələnt), *a.* disposed to do good; charitable, generous. **benevolently,** *adv.* **benevolence,** *n.*
BEng., (*abbr.*) Bachelor of Engineering.
Bengali (bengaw'li, beng-), *a.* of or pertaining to Bengal, its people or language. *n.* a native of Bengal; the language of Bengal. **Bengal light,** *n.* a firework giving a vivid and sustained light.
benighted (bini'tid), *a.* overtaken by night; morally or intellectually enlightened; uncivilized.
benign (binīn'), *a.* kind-hearted; propitious; genial; not malignant; mild. **benignly,** *adv.* **benignant** (-nig'nənt), *a.* benevolent; propitious. **benignantly,** *adv.* **benignity** (-nig'ni-), *n.* kindly feeling.
benison (ben'izən), *n.* a blessing.
Benjamin (ben'jəmin), *n.* the youngest son of a family. **benjamin** (ben'jəmin), *n.* benzoin. **benjamin-tree,** *n.* a tree yielding benzoin; the spice-bush, also *Ficus benjamina.*
bennet[1] (ben'it), *n.* herb bennet.
bennet[2] (ben'it), BENT[2].
bent[1] (bent), *n.* inclination, bias; propensity; tension; extent, capacity. *a. p.p.* of BEND; curved; intent (on); (*sl.*) crooked; (*sl.*) stolen; (*sl.*) homosexual. **to the top of one's bent,** to one's utmost capacity. **bentwood,** *n.* wood steamed and curved in moulds for making furniture. *a.* made using bentwood.
bent[2] (bent), *n.* stiff, rush-like grass; old grass-stalks; unenclosed pasture. **bent-brass,** *n.* the genus *Agrostis.*
benthal (ben'thəl), *a.* of or pertaining to the depths of the ocean beyond 1000 fathoms.
Benthamism (ben'thəmizm), *n.* the Utilitarian philosophy based on the principle of the greatest happiness of the greatest number. **Benthamite** (-it), *n.* [Jeremy *Bentham,* 1748–1832]
benthos (ben'thos), *n.* animal and plant life on the ocean bed. **benthoscope** (-skōp), *n.* a submersible sphere for studying deep-sea life. **benthic,** *a.*
bentonite (ben'tənīt), *n.* an absorbent clay used as a filler, bonding agent etc.
benumb (binŭm'), *v.t.* to render torpid or numb; to deaden, to paralyse. **benumbment,** *n.*
Benzedrine® (ben'zidrēn), *n.* amphetamine.
benzene (ben'zēn), *n.* an aromatic hydrocarbon used in the synthesis of organic chemical compounds, as a solvent and insecticide. **benzene ring,** *n.* a closed chain of six carbon atoms each bound to a hydrogen atom in the benzene molecule. **benzine** (-zēn), *n.* a mixture of liquid hydrocarbons used esp. as a solvent and motor fuel. **benzocaine** (-zō-), *n.* a local anaesthetic. **benzodiazepine** (-zōdīaz'əpīn), *n.* a sedative and tranquillizing drug. **benzol** (-zol), **-zole** (-zōl), *n.* unrefined benzene used as a fuel; (*erroneously*) benzene. **benzoline** (-zəlēn), *n.* impure benzene; benzine. **benzyl** (-zil), *n.* an organic radical derived from benzene.
benzoin (ben'zoin, -zōin), *n.* a resin obtained from *Styrax benzoin,* used in medicine and in perfumery, called also gum benzoin. **benzoic** (-zō'-), *a.* **benzoic acid,** *n.* an acid present in benzoin etc., used in medicines, dyes, as a food preservative and in organic synthesis.
bequeath (bikwēdh'), *v.t.* to leave by will; to transmit to future generations. **bequeathable,** *a.* **bequeathment,** *a.* **bequest** (-kwest'), *n.* the act of bequeathing; that which is bequeathed; a legacy.
berate (birāt'), *v.t.* to scold vehemently.
Berber (bœ'bə), *n.* a member of the Hamitic peoples of N Africa; their language. *a.* of or belonging to this people or their language.
berberis (bœ'bəris), *n.* any of the barberry genus of

shrubs. **berberine** (-rēn), n. an alkaloid obtained from barberry roots.
berberry BARBERRY.
berceuse (beəsœz'), n. a cradle-song; lulling music.
bereave (birēv'), v.t. (past, p.p. **bereaved, bereft** (-reft')) to deprive, rob or spoil of anything; to render desolate (usu. in p.p. **bereaved**, of loss by death). **bereavement**, n.
Berenice's hair (berəni'siz), n. a small northern constellation, near the tail of Leo.
beret (be'rā), **berret** (-rit), n. a round, brimless flat cap fitting the head fairly closely.
berg[1] (bœg), ICEBERG.
berg[2] (bœg), (S Afr.) n. mountain. **berg wind**, n. a hot dry wind blowing from the north to the coast.
bergamask (bœg'əmahsk), n. a rustic dance associated with the people of Bergamo in Italy.
bergamot[1] (bœ'gəmot), n. (the bergamot orange, Citrus bergamia, which yields) a fragrant essential oil used in perfumery; a kind of mint which yields a similar oil.
bergamot[2] (bœ'gəmot), n. a juicy kind of pear.
bergschrund (bœg'shroont), n. a crevasse between the base of a steep slope and a glacier or nevé.
Bergsonian (bœgsō'niən), a. pertaining to the French philosopher Henry Bergson's (1859-1941) theory of creative evolution and the life force. n. a follower of Bergson.
beriberi (ber'ibe'ri, be'-), n. a degenerative disease prevalent in S and E Asia due to a deficiency of vitamin B_1.
berk, burk (bœk), n. (sl.) an idiot.
Berkeleian (bahklē'ən), n., a. (an adherent) of Bishop Berkeley (1685-1753) or his philosophy, which denied that the mind could know the external world objectively.
berkelium (bœkē'liəm), n. an artificially produced radioactive element, at. no. 97; chem. symbol Bk.
Berks. (bahks), (abbr.) Berkshire.
berm (bœm), n. the bank of a canal opposite the towing-path.
Bermuda shorts (bəmü'də), n. tight-fitting knee-length shorts.
bernacle (bœ'nəkəl), BARNACLE.
Bernardine (bœ'nədēn), n., a. Cistercian.
berretta (biret'ə), BIRETTA.
berry (be'ri), n. any smallish fleshy fruit; (Bot.) a many-seeded, indehiscent, pulpy fruit; one of the eggs of a fish or lobster; a coffee bean; cereal grain. v.i. to bear or produce berries; to swell; to go berry-gathering. **berried**, a. having or bearing berries; (of a hen lobster) bearing eggs.
bersagliere (beəsalyeə'rā), n. (pl. **-ri** (-rē)) a sharpshooter.
berserk, berserker (bəsœk'), **baresark** (beə'sahk), n. a Norse warrior possessed of preternatural strength and fighting with desperate fury and courage; a bravo. a., adv. frenzied; filled with furious rage. **to go berserk**, to lose control of one's actions in violent rage.
berth (bœth), n. sea-room; a convenient place for mooring; a place for a ship at a wharf; a situation on board ship; a permanent job or situation of any kind; a sleeping-place on board ship or in a train. v.t. to moor; to furnish with a berth. **to give a wide berth to**, to keep away from; to steer clear of. **berthage** (-ij), n. room or accommodation for mooring ships; dock dues.
Berthon boat (bœ'thən), n. a collapsible, canvas lifeboat.
beryl (be'ril), n. a gem nearly identical with the emerald, but varying in colour from pale green to yellow or white; a silicate of aluminium and beryllium. **berylline** (-in), a.
beryllium (biril'iəm), n. a light metallic element, at. no. 4; chem. symbol Be, used as a component in nuclear reactors and to harden alloys etc.
bescreen (biskrēn'), v.t. to hide from view; to envelop in shadow.
beseech (bisēch'), v.t. (past, p.p. **besought** (-sawt)), to ask earnestly, implore, entreat. **beseeching**, a. **beseechingly**, adv.
beseem (bisēm'), v.t. to be fit, suitable, proper for or be-coming to. v.i. to be seemly or proper. (usu. impersonal in either voice). **beseeming**, a. **beseemingly**, adv. **beseemingness**, n.
beseen (bisēn'), a. seen, appearing; dressed, furnished, accomplished.
beset (biset'), v.t. (past, p.p. **beset**) to set or surround (with); to besiege, occupy; to set upon, to fall upon; to encompass, to assail. **beset**, a. set or encumbered (with difficulties, snares etc.). **besetting**, a. assailing.
beshrew (bishroo'), v.t. to curse.
beside (bisīd'), prep. by the side of, side by side with; in comparison with; close to; away from, wide of. adv. besides. **beside oneself**, out of one's wits. **beside the point, question**, irrelevant. **besides**, prep. in addition to, over and above; other than, except. adv. moreover, in addition; otherwise.
besiege (bisēj'), v.t. to surround a place with intent to capture it by military force; to invest; to crowd round; to assail importunately. **besieger**, n. **besiegingly**, adv.
besmear (bismiə'), v.t. to smear or daub; to soil.
besmirch (bismœch'), v.t. to soil; to sully.
besom (bē'zəm), n. a broom made of twigs bound round a handle; (Sc., derog.) a woman.
besot (bisot'), v.t. to make sottish; to stupefy; to cause to dote upon. **besotted**, a. infatuated. **besottedly**, adv.
besought (bisawt'), p.p. BESEECH.
bespangle (bispang'gl), v.t. to cover over with or as with spangles.
bespatter (bispat'ə), v.t. to spatter over or about.
bespeak (bispēk'), v.t. (past, **-spoke** (-spōk'), p.p. **-spoken**) to arrange for, to order beforehand; to request; to betoken. **bespoke** (for BESPOKEN), p.p. ordered beforehand. a. of a suit etc., made to a customer's specific requirements.
bespeckle (bispek'l), v.t. to speckle over, to variegate.
bespectacled (bispek'təkld), a. wearing spectacles.
bespoke (bispōk'), **bespoken** BESPEAK.
bespread (bispred'), v.t. to spread over; to adorn.
besprinkle (bispring'kl), v.t. to sprinkle or scatter over.
Bessemer process (bes'əmə), n. a process invented in 1856 for the elimination of carbon and silicon from cast iron. **Bessemer iron** or **steel**, n. **Bessemerize, -ise**, v.t.
best (best), a. of the highest excellence; surpassing all others; most desirable. v.t. to get the better of; to cheat, outwit. adv. superlative of WELL; in the highest degree; to the most advantage; with most ease; most intimately. n. the best thing; the utmost; (collect.) the best people. **at best**, as far as can be expected. **Sunday-best**, best clothes. **the best part**, the largest part. **to get, have the best of**, to get or have the advantage. **to make the best of**, to make the most of; to be content with. **to the best of**, to the utmost extent of. **best man**, n. a groomsman. **bestseller**, n. a popular book which has sold in large numbers; a writer of such a book. **bestsell**, v.i.
bestead[1] (bisted'), v.t. to help; to profit; to be of service to. v.i. to avail.
bested, bestead[2] (bisted'), a. situated, circumstanced (usu. with adv. ill, hard, hardly, sore etc.).
bestial (bes'tiəl), a. resembling a beast; brutish, obscene, sexually depraved. **bestiality** (-tial'-), n. bestial behaviour; sexual relations between a person and an animal. **bestialize, -ise**, v.t. **bestially**, adv. **bestiary** (bes'tiəri), n. a moralized natural history of animals.
bestick (bistik'), v.t. to bedeck; to transfix. **bestuck** (-stŭk'), a.
bestir (bistœ'), v.t. to rouse into activity.
bestow (bistō'), v.t. to stow away, to provide with quarters; to expend, to lay out; to give as a present. **bestowal**, n. **bestower**, n. **bestowment**, n.
bestrew (bistroo'), v.t. to strew over; to lie scattered over.
bestride (bistrīd'), v.t. to sit upon with the legs astride; to span, overarch.

bet (bet), *n.* a wager; a sum wagered. *v.t.* (*past, p.p.* **bet, betted**) to wager; to stake upon a contingency. *v.i.* to lay a wager. **you bet**, (*sl.*) certainly. **better²**, *n.*

beta (bē'tə), *n.* the second letter of the Greek alphabet; the second star in a constellation; the second of a series or group. **beta-blocker**, *n.* a drug that reduces the heart-rate, esp. used to treat high blood-pressure. **beta particle**, *n.* an electron or positron emitted by certain radioactive substances. **beta rays**, *n.pl.* rays consisting of a stream of beta particles. **beta rhythm, wave**, *n.* the normal electrical activity of the brain. **betatron** (-tron), *n.* an electrical apparatus for accelerating electrons to high energies.

betake (bitāk'), *v.r.* (*past* **betook** (-tuk'), *p.p.* **betaken**) to take oneself to; to have recourse to.

bête noire (bet'nwah'), *n.* a bugbear, pet aversion.

betel (bē'tl), *n. Piper betle*, a shrubby evergreen plant, called also **betel-pepper** and **betel-vine**; its leaf, chewed with areca nut and shell lime by the peoples of SE Asia. **betel-nut**, *n.* the nut of the areca palm. **betel-tree**, *n. Areca catechu*, nut is chewed with betel-leaves.

beth (beth), *n.* the second letter of the Hebrew alphabet.

bethel (beth'əl), *n.* a Nonconformist chapel; a missionroom; a seamen's church, esp. afloat.

bethesda (bithez'də), *n.* a Nonconformist chapel.

bethink (bithingk'), *v.r.* (*past, p.p.* **bethought**) to consider, think; to meditate.

bethrall (bithrawl'), *v.t.* to enslave.

betide (bitīd'), *v.t.* to happen to; (*erron.*) to betoken. *v.i.* to happen, to come to pass.

betimes (bitīmz'), *adv.* early; in good time; soon.

betitle (bitī'tl), *v.t.* to give a name or title to.

betoken (bitō'kn), *v.t.* to be a type of; to foreshow, be an omen of, to indicate.

betony (bet'əni), *n.* a labiate plant, *Stachys betonica*, with purple flowers.

betook (bituk'), *past.* BETAKE.

betray (bitrā'), *v.t.* to deliver up treacherously; to be false to; to lead astray; to disclose treacherously; to reveal incidentally. **betrayal**, *n.* **betrayer**, *n.*

betroth (bitrōdh'), *v.t.* to contract in an engagement to marry; to affiance. **betrothal, betrothment**, *n.* **betrothed**, *n., a.* (a person) engaged to be married.

better¹ (bet'ə), *a.* superior, more excellent; more desirable; greater in degree; improved in health. *v.t.* to make better; to surpass, to improve on. *v.i.* to become better; *adv.* comp. of WELL; in a superior, more excellent or more desirable manner or degree; more correctly or fully; with greater profit; more. *n.pl.* social superiors. **better off**, in better circumstances. **for better (or) for worse**, whatever the circumstances. **for the better**, in the way of improvement. **the better part of**, the most. **to get the better of**, to defeat, to outwit. **to think better of**, to reconsider. **better half**, *n.* (*facet.*) a spouse, esp. a wife. **betterment**, *n.* amelioration; an improvement of property. **bettermost**, *a.* best.

better² BET.

between (bitwēn'), *prep.* in, on, into, along or across the place, space or interval of any kind separating two points, lines, places or objects; intermediate in relation to; related to both of; related so as to separate; related so as to connect, from one to another; among; in shares among, so as to affect all. *n.* an interval of time; (*pl.*) an intermediate size and quality of sewing-needles. *adv.* intermediately; in an intervening space or time; in relation to both of; to and fro; during or in an interval. **go-between**, an intermediary. **between-decks**, *n.* the space between two decks. **between-maid** TWEENY under TWEEN. **betweenwhiles, betweentimes**, *adv.* now and then; at intervals. **between ourselves**, in confidence. **betwixt and between**, neither one thing nor the other.

betwixt (bitwikst'), *prep., adv.* between.

BeV, (*abbr.*) (*N Am.*) billion electron-volt(s): equivalent to gigaelectronvolt(s), GeV.

bevatron (bev'ətron), *n.* an electrical apparatus for accelerating protons to high energies.

bevel (bev'l), *a.* oblique, slanting; at more than a right angle. *n.* a tool for measuring angles; a slope from the right angle, an inclination of two planes, except one of 90°. *v.t.* to give a bevel angle to. *v.i.* to slant. **bevel-edge**, *n.* the oblique edge of a chisel or similar cutting tool. **bevel-gear, -gearing**, *n.* gear for transmitting motion by means of bevel-wheels. **bevel-wheels**, *n.pl.* cogged wheels whose axes form an angle (usually 90°) with each other. **bevelling**, *n., a.* **bevelment**, *n.* the process of bevelling.

beverage (bev'ərij), *n.* any drink other than water. **bevvied** (-id), *a.* (*sl.*) drunk. **bevvy**, *n.* (*sl.*) alcoholic drink.

bevy (bev'i), *n.* a flock of larks or quails; a herd of roes; a company of women.

bewail (biwāl'), *v.t.* to wail over, to lament for. *v.i.* to express grief. **bewailing**, *n., a.* **bewailingly**, *adv.* **bewailment**, *n.*

beware (biweə'), *v.i.* to be on one's guard; to take care. *v.t.* to be wary of; to look out for.

beweep (biwēp'), *v.t.* to weep over or for; to moisten with tears.

bewet (biwet'), *v.t.* to wet profusely.

bewig (biwig'), *v.t.* to adorn with a wig. **bewigged**, *a.*

bewilder (biwil'də), *v.t.* to perplex, confuse, lead astray. **bewildering**, *a.* **bewilderingly**, *adv.* **bewilderment**, *n.*

bewitch (biwich'), *v.t.* to practise witchcraft against; to charm, to allure. **bewitching**, *a.* **bewitchingly**, *adv.* **bewitchment**, *n.*

bewray (birā'), *v.t.* to reveal, to disclose. **bewrayingly**, *adv.*

bey (bā), *n.* a governor of a Turkish town, province or district. **beylic** (-lik), *n.* the district governed by a bey.

beyond (biyond'), *prep.* on, to or towards the farther side of; past, later than; more than; surpassing in quality or degree, outside the limit of; in addition to. *adv.* at a greater distance than; farther away. *n.* that which lies beyond human experience or after death. **the back of beyond**, an out-of-the-way place.

bezant (bizant'. bez'-), *n.* a gold coin struck at Constantinople; (*Her.*) a gold roundel borne as a charge.

bezel (bez'l), *n.* one of the oblique sides of a cut gem; the groove by which a watch-glass or a jewel is held.

bezique (bizēk'), *n.* a game of cards of French origin.

bezoar (bē'zaw), *n.* a calculous concretion found in the stomach of certain animals and supposed to be an antidote to poisons. **bezoar-goat**, *n.* the Persian wild goat. **bezoar-stone**, *n.*

bf, (*abbr.*) bloody fool; bold face (print); brought forward.

bhang, bang (bang), *n.* an intoxicating or stupefying preparation made from the dried leaves of hemp.

bhangra (bang'grə), *n., a.* (of) a type of popular music combining Western and Asian forms.

bhp, (*abbr.*) brake horsepower.

Bi (*chem. symbol*) bismuth.

bi-, *pref.* two, double, twice.

biacuminate (bīəkū'minət), *a.* having two tapering points.

biangular (biang'gūlə), *a.* having two angles.

biannual (biə'nūəl), *a.* half-yearly, twice a year. **biannually**, *adv.*

biarticulate (bīahtik'ūlət), *a.* (*Zool.*) two-jointed.

bias (bī'əs), *n.* an oblique motion; hence, inclination, prejudice, prepossession. *a.* slanting, oblique. *adv.* obliquely, to prepossess. **bias binding**, *n.* a strip of material cut slantwise used for binding hems in sewing. **bias(s)ing**, *pres.p.* **bias(s)ed**, *past, p.p.*

biathlon (bīath'lon), *n.* an athletic event combining crosscountry skiing and rifle shooting.

biaxial (bīak'siəl), **biaxal** (-səl), *a.* having two (optical) axes.

bib (bib), *v.t., v.i.* to drink; to drink frequently; to tipple.

n. a cloth or piece of shaped plastic put under a child's chin to keep the front of the clothes clean; the front section of a garment (e.g. an apron, dungarees) above the waist; the whiting-pout, a food fish with a chin barbel. **bibcock**, *n.* a tap with the nozzle bent downwards. **bibber**, *n.* **bibbing**, *n.*
Bib., (*abbr.*) Bible; Biblical.
bibacious (bibā'shəs), *a.* addicted to drinking.
bibelot (bib'əlō), *n.* a knick-knack.
Bibl., bibl, (*abbr.*) Biblical.
Bible (bī'bl), *n.* the sacred writings of the Christian religion comprising the Old and New Testament; a copy of the Scriptures, a particular edition; a text-book, an authority. **Bible-basher, -thumper**, *n.* an aggressive preacher; an ardent exponent of the Bible. **Bible belt**, *n.* those regions of the southern US characterized by fervent religious fundamentalism. **Bible-class**, *n.* a class for studying the Bible. **Bible-reader, Bible-woman**, *n.* one employed as a lay missioner. **Bible Society**, *n.* a society for the distribution of the Bible. **biblical** (bib'li-), *a.* **biblically**, *adv.* **biblicism** (-sizm), *n.* strict adherence to the letter of the Bible. **biblicist**, *n.* **biblism** (bib'-), *n.* adherence to the Bible as the only rule of faith. **biblist**, *n.*
biblio-, *comb. form.* pertaining to books.
bibliography (bibliog'rəfi), *n.* the methodical study of books, printing, editions, etc.; a systematic list of books of any author, printer or country, or on any subject. **bibliographer**, *n.* **bibliographical** (-graf'-), *a.*
bibliolatry (bibliol'ətri), *n.* excessive admiration of a book or books; excessive reverence for the letter of the Bible. **bibliolater**, *n.* **bibliolatrous**, *a.*
bibliology (bibliol'əji), *n.* scientific study of books; bibliography; biblical study. **bibliological** (-loj'-), *a.*
bibliomancy (bib'liōmansi), *n.* divination by books or verses of the Bible.
bibliomania (bibliōmā'niə), *n.* a mania for collecting and possessing books. **bibliomaniac** (-ak), *n.*
bibliopegy (bibliop'əji), *n.* the art of binding books. **bibliopegic** (-pej'-), *a.* **bibliopegist**, *n.* one who collects bindings; a bookbinder.
bibliophile (bib'liōfīl), *n.* a lover of books. **bibliophilism** (-of'i-), *n.* **bibliophilist** (-of'i-), *n.*
bibliophobia (bibliōfō'biə), *n.* a dread or hatred of books.
bibliopole (bib'liōpōl), *n.* a bookseller. **bibliopolic, bibliopolical** (-pol'-), *a.* **bibliopolist** (-op'-), *n.* **bibliopoly** (-op'-), *n.* bookselling.
bibliotheca (bibliōthē'kə), *n.* a library; a bibliography. **bibliothecal**, *a.*
bibulous (bib'ūləs), *a.* addicted to alcohol. **bibulously**, *adv.*
bicameral (bikam'ərəl), *a.* having two legislative chambers.
bicarbonate (bikah'bənət), *a.* a salt of carbonic acid. **bicarbonate of soda**, *n.* (*coll. contr.* **bicarb** (bī'-)) sodium bicarbonate.
bice (bīs), *n.* a pale blue or green pigment.
bicentenary (bisəntē'nəri), *a.* consisting of or pertaining to 200 years. *n.* the 200th anniversary. **bicentennial** (-ten'-), *a.* occurring every 200 years; lasting 200 years. *n.* a bicentenary.
bicephalous (bīsef'ələs), *a.* having two heads.
biceps (bī'seps), *n., a.* (of) a muscle having two attachments, esp. the large muscles in front of the upper arms and the thigh. **bicipital** (-sip'itəl), *a.*
bichromate (bīkrō'māt), *n.* a salt containing dichromic acid.
bicker (bik'ə), *v.i.* to squabble over petty issues; to quiver, flicker. *n.* an altercation.
biconcave (bīkon'kāv), *a.* concave on both sides.
biconvex (bīkon'veks), *a.* convex on both sides.
bicorporal (bīkaw'pərəl), *a.* having two bodies, biporporate.
bicorporate (bīkaw'pərət), *a.* double-headed; (*Her.*) having two bodies with a single head.

bicuspid (bīkūs'pid), *a.* having two cusps. *n.* a premolar tooth in man. **bicuspidate** (-dāt), *a.* two-pointed.
bicycle (bī'sikl), *n.* a two-wheeled pedal-driven vehicle, with the wheels one behind the other and usually with a saddle for the rider mounted on a metal frame. *v.i.* to ride on a bicycle. **tandem bicycle** TANDEM. **bicycle clip**, *n.* a metal clip worn around the ankle to prevent trousers from catching on a bicycle chain. **bicycle pump**, *n.* a hand pump for filling bicycle tyres with air. **bicyclist**, *n.*
bid (bid), *v.t.* (*past* **bid**, *p.p.* **bid, bidden** (-n)) to command; to invite, to announce; to make a tender of (a price, esp. at an auction). *v.i.* to make an offer at an auction. *n.* an offer of a price, esp. at an auction; the call at bridge whereby a player contracts to make as many tricks as he names. **to bid farewell, welcome**, to salute at parting or arrival. **to bid fair**, to seem likely, to promise well. **to bid up**, to raise the price of a commodity by a succession of overbids. **biddable**, *a.* obedient, willing. **bidder**, *n.* **bidding**, *n.* invitation, command; a bid at an auction. **bidding-prayer**, *n.* prayer in which the congregation is exhorted to pray for certain objects.
biddy (bid'i), *n.* (*derog.*) an old woman.
bide (bīd), *v.t.* (*past* **bided, bode** (bōd), *p.p.* **bided**) to abide, await; (*arch. exc. in* **bide one's time**, await an opportunity). *v.i.* to stay; to remain. **biding**, *n.*
bident (bī'dent), *n.* a two-pronged fork. **bidentate, bidentated** (-tātid), *a.* having two teeth or toothlike processes.
bidet (bē'dā), *n.* a low basin for bathing the genital and anal area.
biennial (bien'iəl), *a.* happening every two years; lasting two years; having a two-year life cycle. *n.* a biennial plant. **biennially**, *adv.* **biennium** (-əm), *n.* a period of two years.
bier (biə), *n.* a stand on which a corpse or coffin is placed.
bifacial (bīfā'shəl), *a.* having two faces.
bifarious (bifeə'riəs), *a.* double; ranged in two rows.
biff (bif), *v.t.* (*coll.*) to strike, to cuff. *n.* a blow.
biffin (bif'in), *n.* a deep-red cooking-apple much cultivated in Norfolk.
bifid (bī'fid), *a.* split into two lobes by a central cleft. **bifidity** (bifid'-), **bifidly**, *adv.*
bifocal (bīfō'kəl), *a.* with two foci. **bifocal lenses**, *n.pl.* spectacle lenses divided for near and distant vision.
bifold (bī'fōld), *a.* twofold, double.
bifoliate (bīfō'liət), *a.* having two leaves.
biform (bī'fawm), having two forms.
bifurcate (bī'fəkāt), *v.i.* to divide into two branches or forks. **bifurcate** (-kāt, -fœ'kət), *a.* divided into two forks or branches. **bifurcation**, *n.*
big (big), *a.* large or great in bulk; grown up; advanced in pregnancy; important; magnanimous; pretentious. *adv.* (*coll.*) boastfully; pretentiously. **too big for one's boots**, unduly self-important. **to talk big**, to boast. **Big Bang**, *n.* the deregulation of the London Stock Exchange in 1986; (*fig.*) any fundamental change in organization. **big bang theory**, the theory that the universe evolved from a cataclysmic explosion of superdense matter and is still expanding. **big-bellied**, *a.* corpulent; heavily pregnant. **Big Ben**, *n.* the great bell and clock in the Houses of Parliament in Westminster. **Big Brother**, *n.* a sinister and ruthless person or organization that exercises totalitarian control. **big business**, *n.* large corporations and enterprises, used collectively esp. when regarded as exploitative. **big deal**, *int.* (*sl.*) a derisory exclamation or response. **big dipper**, *n.* ROLLER COASTER; (**Big Dipper**, *N Am.*) the constellation Great Bear. **big end**, *n.* the crankpin end of the connecting-rod in an internal-combustion engine. **big game**, *n.* large animals hunted or fished for sport. **big gun**, *n.* (*sl.*) an important person. **big-head**, *n.* (*coll.*) a conceited individual. **big-horn**, *n.* the Rocky Mountain sheep. **big-mouth**, *n.* (*sl.*) an indiscreet, boastful person.

big noise, shot, *n. (coll.)* a person of importance. **big stick,** *n. (coll.)* brutal force. **big talk,** *n.* boasting, bragging. **big time,** *n. (coll.)* the highest rank in a profession. **big top,** *n.* a large circus tent. **big-wig,** *n. (coll.)* a man of importance. **bigness,** *n.*
bigamy (big'əmi), *n.* marriage with a second person while a legal spouse is living. **bigamist,** *n.* **bigamous,** *a.* **bigamously,** *adv.*
bight (bīt), *n.* a wide bay; a bend or loop of a rope.
Bignonia (bignō'niə), *n.* a genus of plants, containing the trumpet flower.
bigot (big'ət), *n.* a person unreasonably and intolerantly devoted to a particular creed, system or party. **bigoted,** *a.* **bigotedly,** *adv.* **bigotry** (-ri), *n.*
bijou (bē'zhoo), *n. (pl.* **bijoux**) a jewel, a trinket; anything that is small, pretty or valuable. *a.* small and pretty.
bike[1] (bīk), *n. (coll.)* a bicycle; a motorcycle. *v.i.* to ride a bicycle. **biker,** *n. (coll.)* a motorbike enthusiast.
bike[2] (bīk), *n. (Austral. sl.)* a prostitute.
bikini (bikē'ni), *n.* a brief, two-piece swimming costume.
bilabial (bīlā'biəl), *a.* of or denoting a consonant produced with two lips, e.g. b, p, w. *n.* a bilabial consonant. **bilabiate** (-ət), *a. (Bot.)* having two lips.
bilateral (bīlat'ərəl), *a.* having, arranged on, or pertaining to two sides; affecting two parties. **bilateralism,** *n.* **bilaterally,** *adv.*
bilberry (bil'bəri), *n.* (the fruit of) a dwarf moorland shrub, *Vaccinium myrtillus;* other species of *Vaccinium.*
Bildungsroman (bil'dungzrōmahn), *n.* (G) a novel dealing with the emotional and spiritual education of its central figure.
bile (bīl), *n.* a bitter yellowish fluid secreted by the liver to aid digestion; *(fig.)* anger, irritability. **bile-pigment,** *n.* colouring matter existing in bile. **biliary** (bil'i-), *a.* of or pertaining to the bile, the ducts which convey the bile, the small intestine or the gall-bladder. **bilious** (bil'yəs), *a.* biliary; produced or affected by bile; *(fig.)* peevish, ill-tempered. **biliously,** *adv.* **biliousness,** *n.*
bilge (bilj), *n.* the bottom of a ship's floor; dirty water which collects there; *(sl.)* worthless nonsense. *v.i.* to spring a leak. **bilge-pump,** *n.* a pump to carry off bilge-water.
Bilharzia (bilhah'ziə), *n.* a genus of trematode worms parasitic in the blood of birds, humans and other mammals. **bilharzia, -iasis** (-ī'əsis), **-iosis** (-iō'sis), *n.* a disease caused by the worm, characterized by blood loss and tissue damage.
bilingual (biling'gwəl), *a.* speaking or written in two languages. **bilingually,** *adv.* **bilinguist,** *n.*
bilious, biliary, etc. BILE.
bilirubin (biliroo'bin), *n.* the chief pigment of the bile.
bilk (bilk), *v.t.* to cheat; to evade payment of; to escape from, to elude. *n.* a swindle; *(sl.)* a swindler.
Bill[1], **the** *(sl.)* THE OLD BILL under BILL.
bill[2] (bil), *n.* the horny beak of birds or of the platypus; a beak-like projection or promontory; the point of the fluke of an anchor. *v.i.* to exhibit affection. **to bill and coo,** to kiss and fondle. **billed,** *a.* having a beak or bill (usually in comb., as *hard-billed* etc.).
bill[3] (bil), *n.* an obsolete weapon resembling a halberd; a bill-hook. **bill-hook,** *n.* a thick, heavy knife with a hooked end, used for chopping brushwood etc.
bill[4] (bil), *n.* a statement of particulars of goods delivered or services rendered; a promissory note; a draft of a proposed law; an advertisement or public announcement distributed or posted up; *(Law)* a written statement of a case; *(N Am.)* a bank-note. *v.t.* to announce by bills or placards; to put into a programme; to present an account for payment (to). **bill of exchange,** a written order from one person to another to pay a sum on a given date to a designated person. **bill of fare,** a menu. **bill of health,** a document certifying the health of a ship's company (hence *(fig.)* a **clean bill of health**). **bill of lading,** a list of goods to be shipped. **bill of rights,** a summary of rights and liberties claimed by a people and guaranteed by the state, esp. the English statute of 1689 and the first ten amendments to the US Constitution. **bill of sale,** a legal document for the transfer of personal property. **to fill the bill,** to prove satisfactory. **to head, top the bill,** to be the star attraction. **billboard,** *n. (N Am.)* a street hoarding. **bill-broker, -discounter,** *n.* one who deals in bills of exchange and promissory notes. **bill-fold,** *n. (N Am.)* a wallet for notes. **billhead,** *n.* a business form with the name and address of the firm etc. at the top. **billposter, -sticker,** *n.* a person who sticks bills on walls, etc. **billed,** *a.* named in a programme or advertisement. **billing,** *n.* sending out invoices; the relative position of a performer or act in a programme or advertisement.
billabong (bil'əbong), *n.* a branch from a river; a creek that fills seasonally.
billet[1] (bil'it), *n.* a ticket requisitioning food and lodgings for a soldier or others; the quarters so assigned; *(coll.)* a situation, an appointment. *v.t.* to quarter soldiers or others.
billet[2] (bil'it), *n.* a small log for firing; a bar of gold or silver.
billet-doux (bilādoo'), *n. (pl.* **billets-doux**) a love-letter.
billiards (bil'yədz), *n.pl.* a game with balls, which are driven about on a cloth-lined table with a cue. **billiard-cue,** *n.* a tapering stick used to drive the balls. **billiard-marker,** *n.* one who marks the points made by players; an apparatus for registering these.
billingsgate (bil'ingzgāt), *n.* scurrilous abuse, foul language.
billion (bil'yən), *n.* formerly in Britain, one million million, i.e. 1,000,000,000,000 or 10^{12}; in the US (and now in Britain and elsewhere this usage is the more common) one thousand million, i.e. 1,000,000,000 or 10^9; *(pl.)* any very large number. **billionaire** (-neə'), *n.* **billionth,** *a.*
billon (bil'ən), *n.* base metal, esp. silver alloyed with copper.
billow (bil'ō), *n.* a great swelling wave of the sea; *(fig.)* the sea; anything sweeping on or surging. *v.i.* to surge. **billowy,** *a.*
billy, billie (bil'i), *n. (esp. Austral.)* a metal can or pot for boiling water etc. over a campfire; *(N Am.)* a policeman's club. **billy-can,** *n. (esp. Austral.)* a billy. **billy goat,** *n.* a male goat.
bilobed, bilobate (bīlōbd', -lō'bāt), *a.* having or divided into two lobes.
bilocation (bīlōkā'shən), *n.* the state or faculty of being in two places at once.
bilocular (bīlok'ūlə), *a.* having two cells or compartments.
biltong (bil'tong), *(S Afr.) n.* strips of lean meat dried in the sun.
bimanal (bim'ənəl), **bimanous** (-ənəs), *a.* having two hands.
BIM, *(abbr.)* British Institute of Management.
bimbashi (bimbash'i), *n.* a Turkish or Egyptian army officer.
bimbo (bim'bō), *n. (sl.)* an attractive woman, naive or of limited intelligence.
bimensal (bīmen'səl), **bimestrial** (-mes'tri-), *a.* continuing for two months; occurring every two months.
bimeridian (bīmərid'iən), *a.* pertaining to or recurring at midday and midnight.
bimetallism (bīmet'əlizm), *n.* the use of two metals (gold and silver) in the currency of a country, at a fixed ratio to each other. **bimetallist,** *n.* **bimetallic** (-tal'-), *a.*
bimillenary (bīmilen'əri), *n.* a period of two thousand years.
bimonthly (bīmŭnth'li), *a.* occurring every two months; lasting two months or twice a month.
bin (bin), *n. (pres.p.* **binning,** *past, p.p.* **binned**) a box or

other receptacle for corn, bread, wine etc.; wine from a particular bin; a container for rubbish; (*sl.*, *derog.*) a lunatic asylum. *v.t.* to stow in a bin. **bin-end**, *n.* a bottle of wine sold off cheaply because there are so few left of the bin. **bin-liner**, *n.* a plastic bag used to line a rubbish bin.

binary (bī'nəri), *a.* consisting of a pair or pairs; double, dual. **binary compound**, *n.* a chemical compound of two elements. **binary fission**, *n.* the division of a cell into two parts. **binary form**, *n.* a musical composition having two themes or sections. **binary notation**, *n.* a number system using the base two (instead of base ten), numbers being represented as combinations of one and zero (or on and off etc.). **binary star, system**, *n.* a system of two stars revolving around a common centre of gravity. **binary weapon**, *n.* a chemical weapon consisting of two substances, each harmless, which become toxic when mixed.

binaural (bīnaw'rəl), *a.* relating to, having or using two ears; employing two channels in recording or transmitting sound.

bind (bīnd), *v.t.* (*past, p.p.* **bound** (bownd)) to tie, or fasten together, to or on something; to wrap with a cover or bandage; to form a border to; to cover, secure or strengthen, by means of a band; to sew (a book) and put into a cover; to tie up; to cause to cohere; to make constipated; to oblige to do something by contract; to compel; to confirm or ratify. *v.i.* to cohere; to grow stiff and hard; (*sl.*) to complain. *n.* a band or tie; a bine; a tie or brace; (*coll.*) an annoying or frustrating predicament. **to bind down**, to restrain by formal stipulations. **to bind over**, to place under legal obligation. **bind-weed**, *n.* a plant of the genus *Convolvulus;* several other climbing plants. **binder**, *n.* one who or that which binds; a book-binder; a person or machine which binds sheaves; a cover or folder for loose papers, correspondence etc.; a clip; a tie-beam; a bandage; a cementing agent. **bindery**, *n.* a book-binder's workshop. **binding**, *a.* obligatory. *n.* the act of binding; that which binds; the state of being bound; the act, art or particular style of bookbinding; a book-cover, braid or other edging. **bindingly**, *adv.* **bindingness**, *n.*

bine (bīn), *n.* a flexible shoot or stem, esp. of the hop.

binervate (bīnœ'vət), *a.* having two nerves or leaf-ribs.

Binet-Simon scale (benăsēmō', -sī'mən), *n.* an intelligence test employing graded tasks for subjects (usually children) according to age. [after Alfred *Binet*, 1857–1911, and Theodore *Simon*, 1873–1961, F psychologists]

bing (bing), *n.* a heap, a pile; (*Mining*) a heap of alum or of metallic ore.

binge (binj), *n.* (*coll.*) a drinking spree; overindulgence in anything. *v.i.* to overeat.

bingo[1] (bing'gō), *n.* (*sl.*) brandy.

bingo[2] (bing'gō), *n.* a game in which random numbers are called out and then marked off by players on a card; an exclamation expressing suddenness.

binnacle (bin'əkl), *n.* the case in which the ship's compass is kept.

binocular (binok'ūlə), *a.* having two eyes; suited for use by both eyes. *n.* a binocular microscope; (*pl.*) a field or opera glass with tubes for both eyes.

binomial (bīnō'miəl), *n.*, *a.* (a mathematical expression) consisting of two terms united by the signs + or −. **binomial theorem**, *n.* a formula by which a binomial quantity can be raised to any power without actual multiplication.

binominal (-nom'inəl), *a.* having two names, the first denoting the genus, the second the species.

bint (bint), *n.* (*sl. derog.*) a girl or woman.

bio-, *comb. form.* pertaining to life or living things.

bioastronautics (bīōastrənaw'tiks), *n.sing.* the study of the effects of space travel on living organisms.

biochemistry (bīōkem'istri), *n.* the chemistry of physiological processes occurring in living organisms. **biochemical**, *a.* **biochemist**, *n.*

biocide (bī'ōsīd), *n.* a chemical which kills living organisms. **biocidal** (-sī'-), *a.*

biocoenosis (bīōsēnō'sis), *n.* the relationship between plants and animals that are ecologically interdependent.

biodegradable (bīōdigrā'dəbl), *a.* capable of being broken down by bacteria.

bioengineering (bīōenjiniə'ring), *n.* the provision of aids such as artificial limbs to restore body functions; the design, construction and maintenance of equipment used in biosynthesis. **bioengineer**, *n.*

bioethics (bīōeth'iks), *n.sing.* the study of ethical issues arising from advances in medicine and science.

biofeedback (bīōfēd'bak), *n.* a method of regulating involuntary body functions, e.g. heartbeat, by conscious mental control.

biogenesis (bīōjen'isis), *n.* the doctrine that living matter originates only from living matter. **biogenetic** (-jinet'-), **biogenic**, *a.*

biogeography (bīōjiog'rəfi), *n.* the study of the distribution of plant and animal life over the globe.

biograph (bī'ōgraf), *n.* an early name for the cinematograph.

biography (bīog'rəfi), *n.* the history of the life of a person; literature dealing with personal history. **biographer**, *n.* **biographic, -ical**, (-graf'-), *a.* **biographically**, *adv.*

biology (bīol'əji), *n.* the science of physical life or living matter in all its phases. **biologic, -ical** (-loj'-), *a.* **biological clock**, *n.* the inherent mechanism that regulates cyclic physiological processes in living organisms. **biological control**, *n.* the control of pests etc. by using other organisms that destroy them. **biological warfare**, *n.* warfare involving the use of disease germs. **biologist**, *n.*

bioluminescence (bīōloomines'əns), *n.* the production of light by living organisms.

biomass (bī'ōmas), *n.* the total weight of living organisms in a unit of area.

biomedicine (bīōmed'sin, -isin), *n.* the study of the medical and biological effects of stressful environments, esp. space travel.

biometry (bīom'itri), *n.* the statistical measurement of biological data. **biometrics** (-met'-), *n.sing.* **biometric**, *a.* **-ical**, *a.*

bionics (bīon'iks), *n.sing.* the science of applying knowledge of biological systems to the development of electronic equipment; the replacement of parts of the body or enhancement of physiological functions by electrical or mechanical equipment. **bionic**, *a.* of or pertaining to bionics; (*coll.*) having exceptional powers through the electronic augmentation of physical processes.

bionomics (bīōnom'iks), *n.sing.* ecology. **bionomic**, *a.*

biont (bī'ont), *n.* a living organism. **biontic** (-on'-), *a.* **-biont** (-biont), *n. comb. form* belonging to a specific environment. **-biontic** (-bion'-), *a. comb. form.*

biophysics (bīōfiz'iks), *n.sing.* the application of physics to living things.

biopic (bī'ōpik), *n.* a biographical film.

bioplasm (bī'ōplazm), *n.* protoplasm. **bioplast** (-plast), *n.* a nucleus of germinal matter.

biopsy (bī'opsi), *n.* (*pl.* **-sies**) the removal and diagnostic examination of tissue or fluids from a living body.

biorhythms (bī'ōridhmz), *n.pl.* supposed biological cycles governing physical, emotional and intellectual moods and performance.

bioscope (bī'ōskōp), *n.* a cinematograph; (*S Afr.*) cinema.

-biosis (-bīō'sis), *comb. form* a specific mode of life. **-biotic** (-biot'ik), *a. comb. form.*

biosphere (bī'ōsfiə), *n.* the portion of the earth's surface and atmosphere which is inhabited by living things.

biosynthesis (bīōsin'thəsis), *n.* the production of chemical compounds by living organisms. **biosynthetic** (-thet'-), *a.*

biota (bīō'tə), *n.* the flora and fauna of a region.

biotechnology (bīōteknol'əji), *n.* the use of microorganisms and biological processes in industry.

biotin (bī′ōtin), *n.* a vitamin of the B complex (also known as vitamin H) found esp. in liver and egg yolk.
biparous (bip′ərəs), *a.* bringing forth two at a birth.
bipartite (bīpah′tīt), *a.* comprising or having two parts; affecting or corresponding to two parties (e.g. as an agreement). **bipartisan** (-tīzan′, -pah′-), *a.* involving or supported by two (political) parties. **bipartisanship** (-zan′-), **bipartition** (-tish′ən), *n.* division into two.
biped (bī′ped), *a.* having two feet. *n.* an animal having two feet. **bipedal** (-pē′-), *a.*
bipetalous (bīpet′ələs), *a.* having two petals in a flower.
bipinnaria (bīpinəə′riə), *n.* a starfish larva with two bands of cilia.
bipinnate (bīpin′ət), **bipinnated**, *a.* applied to pinnated leaflets of a pinnate leaf.
biplane (bī′plān), *n.* an aircraft with two pairs of wings one above the other.
bipolar (bīpō′lə), *a.* having two poles or opposite extremities. **bipolarity**, *n.*
biquadratic (bikwodrat′ik), *a.* of or pertaining to the fourth power. *n.* the fourth power, the square of a square. **biquadratic equation**, *n.* one containing the fourth power of the unknown quantity.
birch (bœch), *n.* a genus of northern forest trees, *Betula*, with slender limbs and thin, tough bark; the wood of any of these trees; a rod made from birch twigs for flogging. *a.* birchen. *v.t.* to flog. **birchen**, *a.* composed of birch. **birching**, *n.* a flogging.
bird (bœd), *n.* any feathered vertebrate animal; (*sl.*) a girl, young woman; (*sl.*) a prison term; (*coll.*) a person. **bird of paradise**, any of the New Guinea Paradisidea which have brilliantly coloured plumage. **bird of passage**, a migratory bird; a person who rarely stays long in one place. **bird of peace**, (*fig.*) the dove. **bird of prey**, one which feeds on carrion or hunts other animals for food. **birds of a feather**, persons of similar tastes. **strictly for the birds**, unimportant. **to get the bird**, to be hissed, hence dismissed. **birdbath**, *n.* a small basin for birds to bathe in. **bird-brained**, *a.* (*coll.*) stupid, silly. **birdcage**, *n.* a wire or wicker cage for holding birds. **birdcall**, *n.* the cry of a bird; an instrument for imitating the cry of birds. **bird-catcher**, *n.* **bird-catching**, *n.*, *a.* **bird-fancier**, *n.* one who collects, breeds or rears birds. **bird-lime**, *n.* a sticky substance used to snare birds. **bird's-eye**, *a.* resembling a bird's eye; having eye-like marking; seen from above, as by the eye of a bird (esp. in **bird's-eye view**). **bird's-foot**, **bird-foot**, *n.* a popular name for certain plants e.g. *Cheilanthes radiata*, a small fern widely distributed. **bird's-foot trefoil**, *n.* a British wild flower, *Lotus corniculatus*. **bird's-nest**, *n.* the nest of a bird; a cask or other shelter for the lookout man at the masthead. *v.i.* to search for birds'-nests. **bird-nesting**, *n.* seeking birds' nests to steal the eggs. **bird strike**, *n.* a collision of a bird with an aircraft. **bird table**, *n.* a small elevated platform for wild birds to feed from. **bird-watcher**, *n.* one who observes wild birds in their natural habitat. **bird-watching**, *n.* **birdie** (-ī), *n.* a little bird; a hole in golf made in one under par. **birding**, *n.* bird-catching; bird-watching. *a.* pertaining to or used in fowling, bird-catching or bird-watching.
birefringence (bīrifrin′jəns), *n.* the formation of two unequally refracted rays of light from a single unpolarized ray. **birefringent**, *a.*
biretta (biret′ə), *n.* a square cap worn by clerics of the Roman Catholic and Anglican Churches.
birk BIRCH.
biriani, biryani (biryah′nī), *n.* an Indian dish of spiced rice mixed with meat or fish.
birl (bœl), *v.i.* to spin round, to rotate noisily. *v.t.* to spin; to throw, toss.
Biro® (bī′rō), *n.* (*pl.* **-ros**) a type of ballpoint pen. [*Biró*, Hungarian inventor]
birostrate (bīros′trāt), **birostrated** (bīros′trətəd), *a.* having two beaks or beak-like processes.
birth (bœth), *n.* the act of bringing forth; the bearing of offspring; the act of coming into life or being born; that 'which is brought forth; parentage; condition resulting from birth; origin, beginning. **birth certificate**, *n.* an official document giving particulars of one's birth. **birth control**, *n.* the artificial control of reproduction, esp. by means of contraceptives. **birthday**, *n.*, *a.* (pertaining to) the day on which one was born, or its anniversary. **birthday honours**, *n.pl.* knighthoods etc. conferred on the sovereign's birthday. **birthday-suit**, (*coll.*) *n.* bare skin; nudity. **birthmark**, *n.* a mark or blemish formed on the body of a child at or before birth. **birthplace**, *n.* the place at which someone or something was born. **birthrate**, *n.* the percentage of births to the population. **birthright**, *n.* rights belonging to an eldest son, to a member of a family, order or people, or to a person as a human being.
bis (bis), *adv.* again; twice (indicating that something occurs twice).
biscuit (bis′kit), *n.* a thin flour-cake baked until it is highly dried; pottery moulded and baked in an oven, but not glazed. *a.* light brown in colour. **to take the biscuit**, (*coll.*) to be the best of the lot.
bise (bēz), *n.* a keen, dry, northerly wind prevalent in Switzerland and adjacent countries.
bisect (bīsekt′), *v.t.* to divide into two (equal) parts. *v.i.* to fork. **bisection**, *n.* **bisector**, *n.*
biserial (bisiə′riəl), **biseriate** (-ət), *a.* arranged in two rows.
bisexual (bīsek′shuəl), *a.* hermaphrodite; attracted sexually to both sexes; of or relating to both sexes. **bisexual**, *n.* **bisexuality** (-shual′-), *n.*
bishop [1] (bish′əp), *n.* a senior clergyman presiding over a diocese; a beverage composed of wine, oranges and sugar; a piece in chess, having the upper part shaped like a mitre. **bishop's-cap**, *n.* the genus *Mitella*, or mitre-wort. **bishop's weed**, *n.* *Aegopodium podagraria;* the umbelliferous genus *Ammi*. **bishopric** (-rik), *n.* the diocese, jurisdiction or office of a bishop.
bishop [2] (bish′əp), *v.t.* to tamper with the teeth (of a horse) so as to conceal its age.
bisk BISQUE [1].
bismillah (bismil′ə), *int.* in the name of Allah.
bismuth (biz′məth), *n.* a reddish white crystalline metallic element, at. no 83; chem. symbol Bi, used in alloys and in medicine.
bison (bī′sən), *n.* a large bovine mammal with a shaggy coat and a large hump; the European bison, now very rare; the American bison, commonly called buffalo.
bisque [1] (bisk), *n.* a rich soup made by boiling down shellfish.
bisque [2] (bisk), *n.* in tennis, golf etc. a stroke allowed at any time to the weaker party to equalize the players.
bisque [3] (bisk), *n.* a kind of unglazed white porcelain.
bissextile (bisek′stīl), *a.* of or pertaining to leap-years. *n.* leap-year.
bistort (bis′tawt), *n.* a plant with a twisted root, and spike of flesh-coloured flowers, *Polygonum bistorta*.
bistoury (bis′təri), *n.* a small surgical knife.
bistre (bis′tə), *n.* a transparent brownish yellow pigment prepared from soot. *a.* coloured like this pigment. **bistred**, *a.*
bistro (bēs′trō), *n.* (*pl.* -**tros**) a small bar or restaurant.
bisulcate (bīsŭl′kāt), *a.* (*Zool.*) cloven-hoofed.
bit [1] (bit), *n.* a small portion; a fragment; the smallest quantity, a jot; a brief period of time; a small coin (usually with the value expressed, as a **threepenny-bit**); (*N Am.*) an eighth of a dollar; (*coll.*) a poor little thing; somewhat or something of. **a bit**, a little; somewhat. **a bit of muslin**, **bit of stuff**, (*sl.*) a young woman. **bit by bit**, gradually. **every bit**, quite, entirely. **to do one's bit**, to do one's share. **bit (-part)**, *n.* a small role in a play. **bit-player**, *n.* an actor who plays small parts. **bitty**, *a.* scrappy, dis-

jointed, piecemeal; lacking unity.
bit² (bit), *n.* a bite, the act of biting; the iron part of the bridle inserted in the mouth of a horse; the cutting part of a tool; the movable boring-piece in a drill; the part of the key at right angles to the shank; short sliding piece of tube in a cornet for modifying the tone etc. *v.t.* to furnish with, or accustom (a horse) to, a bit; to restrain. **to take the bit in one's teeth**, to become unmanageable.
bit³ (bit), *n.* in binary notation, either of two digits, one or zero; a unit of information in computers and information theory representing either of two states, such as *on* and *off*.
bitch (bich), *n.* the female of the dog, fox or wolf; (*sl., derog.*) an offensive, malicious or spiteful woman; (*sl.*) a complaint; (*sl.*) an awkward problem. *v.i.* (*sl.*) to complain. *v.t.* (*sl.*) to mess up, botch. **bitchy**, *a.* **bitchily**, *adv.* **bitchiness**, *n.*
bite (bīt), *v.t.* (*past* **bit** (bit), *p.p.* **bitten** (bit'n)) to seize, cut, pierce or crush anything with the teeth; to wound; to affect with severe cold; to inflict sharp physical or mental pain on; to hold fast, as an anchor or screw; to corrode; to cheat, to trick. *v.i.* to have a habit, or exercise the power, of biting; to sting, to be pungent; to take a bait; to act upon something (of weapons, tools etc.). *n.* the act of biting; a wound made by the teeth; a mouthful, a small quantity; a piece seized or detached by biting; a hold, a grip. **to bite in**, to corrode or eat into by means of acid. **to bite off**, to seize with the teeth and detach. **to bite someone's head off**, (*coll.*) to snap at someone. **to bite the bullet**, to submit to an unpleasant situation. **to bite the dust**, to be slain in battle, to die. **to bite the hand that feeds one**, to be ungrateful. **biter**, *n.* **the biter bit**, the cheater cheated. **biting**, *a.* **bitingly**, *adv.* **bitten**, *a.* **bitten with**, infected by (a passion, mania etc.).
bitt (bit), *n.* a strong post on the deck of a ship for fastening cables etc. *v.t.* to put around a bitt. **bitter end**, *n.* the loose end of a belayed rope; the last extremity.
bitten (bit'n), *p.p.* BITE.
bitter (bit'ə), *a.* sharp or biting to the taste; acrid; piercingly cold; painful, distressing. *n.* anything bitter; bitterness; (*coll.*) bitter beer; (*pl.*) liquors flavoured with bitter herbs etc., used as appetizers or stomachics. *v.t.* to make bitter. **to the bitter end** BITT. **bitter-almond**, *n.* a bitter variety of the common almond. **bitter-sweet**, *a.* sweet with a bitter after-taste; pleasant with admixture of unpleasantness or sadness. *n.* a kind of apple; woody nightshade. **bitter-sweeting**, *n.* the bitter-sweet apple. **bittervetch**, *n.* a popular name for some species of the genus *Vicia*. **bitterish**, *a.* **bitterly**, *adv.* **bitterness**, *n.*
bittern¹ (bit'ən), *n.* the liquid obtained when sea-water is evaporated to extract the salt.
bittern² (bit'ən), *n.* a wading bird smaller than a heron.
bitumen (bit'ūmin), *n.* any of various solid or sticky mixtures of hydrocarbons that occur naturally or as a residue from petroleum distillation, e.g. tar, asphalt. **bituminiferous** (-nif'-), *a.* yielding bitumen. **bituminize, -ise** (-tū'-), *v.t.* to impregnate with or convert into, bitumen. **bituminization**, *n.* **bituminous** (-tū'-), *a.*
bivalent (bīvā'lənt, biv'ə-), *a.* having a valency of two; (of homologous chromosomes) associated in pairs.
bivalve (bī'valv), *a.* having two shells or valves which open and shut. *n.* a mollusc with such shells, as the oyster. **bivalved, bivalvular** (-val'vū-), *a.*
bivious (biv'iəs), *a.* leading two different ways.
bivouac (biv'uak), *n.* a temporary encampment in the field without tents etc. *v.i.* (*past, p.p.* **bivouacked**) to camp thus.
biweekly (bīwēk'li), *n., a., adv.* (a periodical) occurring once a fortnight; occurring twice a week.
biz (biz), *n.* (*coll.*) business, work, employment.
bizarre (bizah'), *a.* odd, fantastic, eccentric; of mixed or discordant style; irregular, in bad taste. **bizarrely**, *adv.*
bizarreness, *n.*
Bk, (*chem. symbol*) berkelium.
bk, (*abbr.*) bank; book.
BL, (*abbr.*) Bachelor of Law; Bachelor of Letters; Barrister-at-Law; British Legion; British Library.
blab (blab), *v.t.* to tell or reveal indiscreetly. *v.i.* to talk indiscreetly. *n.* a tell-tale; babbling. **blabber**, *n.*
black (blak), *a.* intensely dark in colour (the opposite of white); destitute of light; obscure; dirty; angry; dark-skinned, of or pertaining to the Negro race (often offensive but accepted in many countries); wearing black clothes, uniform or armour; gloomy, dirty; denoting total absence of colour due to absence or entire absorption of light; atrociously wicked; disastrous, dismal, mournful; subject to a trade-union ban. *n.* the darkest of all colours (the opposite of white); a black pigment or dye; a member of a dark-skinned race, a Negro, W Indian, Austral. Aborigine etc. (often offensive but accepted in many countries esp. **Black**); mourning garments; a minute particle of soot or dirt. *v.t.* to blacken; to soil; to place under a trade-union ban. **black-and-blue**, *a.* discoloured by beating; livid. **Black and Tan**, *n.* a member of the auxiliary police force employed in Ireland in 1919-20 (from the colour of their uniforms). **black-and-white**, *n. n., a.* (material) using or reproduced in only black and white, not coloured; printed; divided into two extremes, not admitting of compromise. **black art**, *n.* magic. **blackball**, *n.* vote of rejection in a ballot. *v.t.* to vote against; to exclude. **black-beetle**, *n.* a cockroach. **black belt**, *n.* a belt awarded for highest proficiency in judo, karate etc.; one entitled to wear this. **blackberry**, *n.* the common bramble, *Rubus fruticosus* or *discolor;* its fruit. **blackberrying**, *n.* gathering blackberries. **blackbird**, *n.* a species of European thrush, the male of which has black plumage and an orange beak; any of several dark plumaged American birds. **blackboard**, *n.* a board painted black for writing and drawing on. **black body**, *n.* a hypothetical body which absorbs all radiation falling upon it, and reflects none. **black book**, *n.* a book recording the names of persons liable to censure or punishment. **black box**, *n.* a closed unit in an electronic system whose circuitry remains hidden and is irrelevant to understanding its function; (*coll.*) FLIGHT RECORDER. **black bread**, *n.* rye bread. **black-browed**, *a.* gloomy; threatening. **black buck**, *n.* a common Indian antelope. **black cap**, *n.* formerly a cap put on by judges when pronouncing sentence of death; **(black-cap warbler)** an English bird having a black crown. **black-cock**, *n.* the male of the black grouse; the heathcock, *Tetrao tetrix*. **black coffee**, *n.* coffee without milk or cream. **black comedy**, *n.* a play or film in which grotesque humour or farce serves to underline and expose true reality. **Black Country**, *n.* a term applied to the heavily industrialized Midlands of England. **black-currant**, *n.* a garden bush, *Ribes nigrum*, and its small black fruit. **black death**, *n.* a form of bubonic plague which ravaged Europe and Asia during the 14th cent. **black earth**, *n.* a fertile soil covering regions in southern USSR. **black economy**, *n.* illegal and undeclared economic activity. **black eye**, *n.* discoloration produced by a blow upon the parts round the eye. **black-face**, *n.* a black-faced sheep or other animal. **black-fish**, *n.* a salmon just after spawning; in Australia, a small species of whale. **black flag**, *n.* one used as a sign that no quarter will be given or taken, as an ensign by pirates, and as the signal for an execution. **blackfly**, *n.* a black aphid that infests beans and other plants. **Blackfoot**, *n.* one of a tribe of North American Indians, so called from their dark moccasins. **black friar**, *n.* a Dominican friar. **black grouse**, *n.* a large N European grouse with dark plumage. **blackguard** (blag'ahd), *n., a.* (of) a scoundrel. *v.t.* to revile. *v.i.* to act the part of a blackguard. **blackguardism**, *n.* **blackguardly** (-gəd-), *a., adv.* **black-head**, *n.* various birds with dark plumage on

bladder   82   **blast**

the head; a pimple with a black head. **black-hearted**, *a.* wicked. **black-hole**, *n.* a punishment cell; a hypothetical celestial region formed from a collapsed star, surrounded by a strong gravitational field from which no matter or energy can escape. **black ice**, *n.* a thin layer of transparent ice on roads. **black-jack**, *n.* (*N Am.*) a loaded stick, a bludgeon; pontoon or a similar card game. **blacklead** (-led'), *n.* plumbago or graphite, made into pencils, also used to polish ironwork. **blackleg**, *n.* a swindler, esp. on the turf; a workman who works while his comrades are on strike. **black-letter**, *n.* the 𝔒𝔩𝔡 𝔈𝔫𝔤𝔩𝔦𝔰𝔥 or 𝔊𝔬𝔱𝔥𝔦𝔠 as distinguished from the Roman character. **black light**, *n.* invisible infrared or ultraviolet light. **blacklist**, *n.* a list of persons in disgrace, or who have incurred censure or punishment. *v.t.* to ban or prohibit. **black magic** BLACK ART. **blackmail**, *n.* any payment extorted by intimidation or pressure. *v.t.* to levy blackmail on. **blackmailer**, *n.* **Black Maria** (məri'ə), *n.* a prison van. **black mark**, *n.* a note of disgrace put against one's name. **black market**, *n.* illegal buying and selling of rationed goods. **black mass**, *n.* a travesty of the Mass performed by diabolists. **black monks**, *n.pl.* the Benedictines. **blackout**, *n.* the extinguishing or concealment of lights; a temporary loss of consciousness, sight or memory; an electrical power failure or cut; an interruption of broadcasting, communications etc. *v.t.* to cause a blackout. *v.i.* to suffer a blackout. **black, Black power**, *n.* the use of political and economic pressure by Blacks in the US to achieve equality with Whites. **black-pudding**, *n.* a kind of sausage made with blood, rice and chopped fat. **Black Rod**, *n.* the chief usher of the House of Lords and of the Garter. **black sheep**, *n.* an unsatisfactory member of a group. **Blackshirt**, *n.* a member of a Fascist organization in Europe before and during World War II, esp. in Italy. **blacksmith**, *n.* a smith who works in iron. **black-snake**, *n.* (*N Am.*) a large non-poisonous snake; any of several Old World venomous snakes. **black spot**, *n.* an area of a road where accidents are common; any dangerous area. **blackthorn**, *n.* the sloe, *Prunus spinosa*; a walking-stick of its wood. **black tie**, *n.* denoting an occasion when a dinner jacket and black bow tie should be worn. **black velvet**, *n.* a mixture of stout and champagne or cider. **Black Watch**, *n.* the 42nd Highland Regiment, from the colour of their tartan. **blackwater fever**, *n.* a form of malaria disease in which the urine is very dark in colour. **black widow**, *n.* a venomous American and Far Eastern spider, of which the female has a black body. **blacken**, *v.t.* to make black; to defame. *v.i.* to become black. **blacking**, *n.* the action of making black; a composition for giving a shining black polish to boots etc. **blackish**, *a.* **blackness**, *n.*

bladder (blad'ə), *n.* a membranous bag in the animal body which receives the urine; any similar membranous bag as the gall-, swim-bladder etc.; an inflated pericarp; a vesicle; the prepared (urinary) bladder of an animal; the membrane of this bladder used for airtight coverings; anything inflated and hollow, esp. the air-filled bag inside a football. **bladder-fern**, *n.* the genus *Cystopteris*. **bladderwort**, *n.* the genus *Utricularia*. **bladder-wrack**, *n.* the seaweed, *Fucus vesiculosus*, which has air-bladders in its fronds. **bladdered**, *a.* **bladdery**, *a.*

blade (blād), *n.* a leaf of a plant; the culm and leaves of a grass or cereal; the expanded part of a leaf or petal as distinguished from the petiole; any broad, flattened part, as of a paddle, bat, oar etc.; the thin cutting part of a knife, sword etc.; the front part of the tongue; a sword; a dashing, reckless fellow. **blade-bone**, *n.* the shoulder-blade in man and the lower mammals. **bladed**, *a.*

blaes (blāz), *n.* a hardened shale used in powdered form to make surfaces for tennis courts.

blague (blahg), *n.* pretentiousness, humbug. [F]

blah, blah blah (blah), *n.* foolish talk, chatter, exaggeration.

blain (blān), *n.* a pustule, blister or sore.

blame (blām), *v.t.* to censure, to find fault with, to reproach; to hold responsible. *n.* the act of censuring; responsibility, accountability. **to be to blame**, to be culpable. **blamable**, *a.* deserving blame; culpable. **blamableness**, *n.* **blamably**, *a.* **blamelessly**, *adv.* **blamelessness**, *n.* **blameworthy**, *a.* **blameworthiness**, *n.*

blanch (blahnch), *v.t.* to whiten by taking out the colour; to make pale; to take off the outward covering of (as of almonds, walnuts etc.); to whiten (as plants) by the deprivation of light; to plunge (vegetables, fruit, meat etc.) briefly into boiling water. *v.i.* to lose colour; to become white.

blancmange (bləmonzh'), *n.* milk (usu. sweetened) thickened with cornflour or gelatine to form a jelly-like dessert.

blanco (blang'kō), *n.* a substance used by the armed forces to whiten or colour uniform belts, webbing etc.

bland (bland), *a.* mild, soft, gentle; genial, balmy; dull, insipid. **blandly**, *adv.* **blandness**, *n.*

blandish (blan'dish), *v.t.* to flatter gently; to coax, to cajole. **blandishment**, *n.* flattering speech or action; cajolery, charm.

blank (blangk), *a.* empty, void; not written or printed on; not filled up; confused, nonplussed; pure, downright, sheer. *n.* the white point in the centre of a target; a blank space in a written or printed document; a blank form; a lottery ticket that draws no prize; a piece of metal before stamping; (*fig.*) a vacant space; an uneventful space of time; a meaningless thing. *v.t.* to render blank; to nonplus, confuse; to block out; (*int.*) a mild execration. **blank-cartridge**, *n.* cartridge containing no bullet. **blank cheque**, *n.* a cheque with the amount left for the payee to insert; complete freedom of action. **blank verse**, *n.* unrhymed verse, esp. iambic pentameter. **blankly**, *adv.* **blankness**, *n.*

blanket (blang'kət), *n.* a coarse, loosely-woven woollen material; a bed-covering or covering for an animal made of this; a concealing cover, as of smoke or leaves; *a.* covering all cases, as in *blanket medical screening*. *v.t.* to cover with or as with a blanket; to bring under one coverage; to apply over a wide area. **born on the wrong side of the blanket**, illegitimate. **blanket bath**, *n.* a wash given to a bedridden person. **blanket stitch**, *n.* a reinforcing stitch for the edge of blankets and other thick material. **wet blanket**, *n.* a person who is a damper to conversation or enjoyment. **blanketing**, *n.* material for blankets.

blankety (blang'kəti), *a.* (*euph.*) used in place of any swear-word.

blare (bleə), *v.t. v.i.* to roar, bellow; to sound loudly and harshly *n.* sound as of a trumpet; roar, noise, bellowing.

blarney (blah'ni), *n.* smooth, flattering speech; cajolery. *v.t., v.i.* to wheedle, to cajole. **Blarney Stone**, *n.* an inscribed stone in the wall of Blarney Castle, near Cork, Ireland, said to give a cajoling tongue to anyone who kisses it.

blasé (blah'zā, blah'-), *a.* indifferent or unenthusiastic esp. through over-familiarity.

blaspheme (blasfēm'), *v.t.* to utter profane language against (God or anything sacred); to abuse. *v.i.* to utter blasphemy, to rail. **blasphemous** (blas'fə-), *a.* uttering or containing blasphemy; grossly irreverent or impious. **blasphemously**, *adv.* **blasphemy** (blas'fəmi), *n.* profane language towards God or about sacred things; impious irreverence; irreverent or abusive speaking about any person or thing held in high esteem.

blast (blahst), *n.* a violent gust of wind; the sound of a trumpet or the like; any destructive influence on animals or plants; the strong current of air used in iron-smelting; an explosion, a detonation; the charge of explosive used; a violent gust of air caused by an explosion. *v.t.* to blow or breathe on so as to wither; to blight, to ruin; to make, re-

-blast   83   **Blighty**

move or destroy with explosive; to shoot at; to criticize severely; to curse. *v.i.* to use explosives; to make a loud noise. *int.* used to express annoyance. **in, at full blast,** at full speed, power, volume etc. **blast-furnace,** *n.* a furnace into which a current of air is introduced to assist combustion. **blast-off,** *n.* the launch of a rocket-propelled missile or space vehicle; (*coll.*) the start of something. **blast off,** *v.i.* **blast-pipe,** *n.* a pipe conveying steam from the cylinders to the funnel of a locomotive to aid the draught. **blasted,** *a.* blighted; confounded, cursed. **blaster,** *n.* **blasting,** *n.*, *a.*
-blast (blast), *comb. form.* used in biological terms indicating an embryonic cell or cell layer; e.g. *mesoblast, statoblast.*
blasto- *comb. form.* pertaining to germs or buds; germinal.
blastocyst (blas′tōsist), *n.* the modified blastula in mammals.
blastoderm (blas′tōdœm), *n.* the germinal membrane enclosing the yolk of an impregnated ovum.
blastomere (blastōmiə), *n.* one of the cells formed during the primary divisions of an egg.
blastula (blastūlə), *n.* (*pl.* **-las, -lae** (-lē),) a hollow sphere composed of a single layer of cells, produced by the cleavage of an ovum. **blastular,** *a.* **blastulation,** *n.*
blatant (blā′tənt), *a.* loud, clamorous; very obvious, palpable. **blatancy,** *n.* **blatantly,** *adv.*
blather, blatherskite BLETHER.
blaze[1] (blāz), *n.* a bright glowing flame; a bright light; a brillian display; an outburst of passion; (*pl.*) the flames of hell. *v.i.* to burn with a bright flame; to shine, to glitter; to be bright with colour; to be eminent or conspicuous from character, talents etc. **like blazes,** furiously. **to blazes,** to the devil. **to blaze away,** to burn brightly and strongly; to fire continuously (with guns); to work continuously and enthusiastically. **to blaze up,** suddenly to burst into flames; to burst into anger. **blazer,** *n.* a usu. lightweight, coloured jacket often worn as part of school uniform or to show membership of a team or club. **blazing,** *a.* emitting flame or light; radiant, lustrous; very angry.
blaze[2] (blāz), *n.* a white mark on the face of a horse or other animal; a white mark made on a tree by chipping off bark; (*N Am.*) the path or boundary indicated by a line of such marks. *v.t.* to mark (a tree); to indicate a path or boundary by such marks.
blaze[3] (blāz), *v.t.* to proclaim; to blazon.
blazer BLAZE[1].
blazon (blā′zən), *n.* a coat of arms; a banner bearing a coat of arms; the art of describing and explaining coats of arms; renown, reputation *v.t.* to describe or depict according to the rules of heraldry; to decorate with heraldic devices; to proclaim, to trumpet. **blazonment,** *n.* **blazonry,** *n.* the art of depicting or describing a coat of arms; armorial bearings; brilliant display.
-ble (-bl), *suf.* tending to, able to, fit to (forming verbal adjectives); e.g. *conformable, durable, flexible, suitable, visible.*
bldg., (*abbr.*) building.
bleaberry BLAEBERRY.
bleach (blēch), *v.t.* to make white by exposure to the sun or by chemical agents. *v.i.* to grow white; to become pale or colourless. **bleach-field,** *n.* a field in which bleaching is carried on. **bleaching-clay,** *n.* kaolin, used for sizing cotton goods. **bleaching-powder,** *n.* chloride of lime, a whitish powder consisting of chlorinated calcium hydroxide. **bleachers,** *n. pl.* (*N Am.*) uncovered seating at a baseball ground. **bleachery,** *n.*
bleak[1] (blēk), *a.* bare of vegetation; cold, chilly, desolate, cheerless. **bleakish,** *a.* **bleakly,** *adv.* **bleakness,** *n.*
bleak[2] (blēk), *n.* a small European river fish, *Leuciscus alburnus*, with silvery scales.
blear (bliə), *a.* dim, indistinct. *v.t.* to make (the eyes) dim; to blur with or as with tears. **blearedness** (-rid-) *n.* dim-

ness, dullness; haziness; indistinctness. **bleary,** *a.* **bleary-eyed,** *a.*
bleat (blēt), *v.i.* to cry like a sheep, goat or calf. *v.t.* to utter in a bleating tone; to say feebly and foolishly. *n.* the cry of a sheep, goat or calf; a complaint, whine.
bleb (bleb), *n.* a small blister or bladder; a bubble in glass or anything similar.
bleed (blēd), *v.i.* (*past, p.p.* **bled** (-bled),) to emit or run with blood; to emit sap, resin or juice from a cut or wound; to be wounded; to die from a wound; to lose money; to have money extorted; (*coll.*) to feel acute mental pain. *v.t.* to draw blood from; (*coll.*) to extort money from; (*Bookbinding*) to cut margins too much and trench on the print; to extract liquid, air or gas from a container or closed system (such as hydraulic brakes). **bled,** *a.* **bled-off,** (*Print*) illustration pages so arranged that the outside edges of the illustration are cut off in trimming when binding. **bleeding heart,** *n.* a plant of the genus Dicentra, characterized by heart shaped flowers; an excessively soft-hearted person. **bleeder,** *n.* a haemophiliac; (*sl.*) a contemptible person. **bleeding,** *n.* haemorrhage; the operation of letting blood, or of drawing sap from a tree. *a.* running with blood; (*sl.*) bloody.
bleep (blēp), *n.* a high-pitched sound from an electronic device. *v.i.* to emit this sound. **to bleep out,** (*coll.*) to replace offensive words with bleeps. **bleeper,** *n.* a small radio receiver which bleeps allowing the user to be contacted.
blemish (blem′ish), *v.t.* to impair, tarnish. *n.* a physical or moral defect; an imperfection, a flaw.
blench[1] (blench), *v.i.* to shrink back; to flinch.
blench[2] (blench), *v.i.* to become pale. *v.t.* to make pale.
blend (blend), *v.t.* to mix, to mingle (esp. teas, wines, spirits, tobacco etc.). *v.i.* to become mingled or indistinguishably mixed; to form a harmonious whole; to pass imperceptibly into each other. *n.* a mixture. **blender,** *n.* a type of electric liquidizer esp. for mixing and puréeing.
blende (blend), *n.* a native sulphide of zinc.
blenny (blen′i), *n.* a genus of small, spiny-finned seafishes.
blent (blent), *a.* mingled.
blesbok (bles′bok, -būk), *n.* a S African white-faced antelope.
bless (bles), *v.t.* to consecrate; to invoke God's favour on; to render happy or prosperous; to wish happiness to; to extol, worship. **bless me** *or* **bless my soul,!** an ejaculation of surprise etc. **bless you,** a phrase used to someone who has sneezed; an expression of gratitude. **to bless oneself,** to make the sign of the cross (as a defence against evil spirits). **without a penny to bless oneself with,** penniless. **blessed** (bles′d, blest), **blest** (blest), *a.* consecrated; worthy of veneration; fortunate; beatified, enjoying the bliss of heaven; joyful, blissful; (*euphem.*) cursed. *n.* (*collect.*) the saints in heaven. **blessedly** (-id-), *adv.* **blessedness** (-id-), *n.* **blessing,** *n.* consecration; divine favour; an invocation of divine favour or happiness; a cause of happiness; a gift; grace before or after a meal. **to ask a blessing,** to say grace before a meal.
blether (bledh′ə), **blather** (bladh′ə), *v.i.* (*coll.*, *dial.*) to talk nonsense volubly. *n.* voluble nonsense. **bletherskate** (-skāt), **blatherskite** (-skīt), *n.* one who talks blatant nonsense.
blew (bloo), *past* BLOW[1].
blight (blīt), *n.* any baleful atmospheric influence affecting the growth of plants; disease caused in plants by fungoid parasites and various insects; any obscure malignant influence; a condition of urban decay. *v.t.* to affect with blight; to exert a baleful influence; to mar, frustrate. **blighter,** *n.* (*sl.*) a nasty fellow, a blackguard. **blightingly,** *adv.*
Blighty (blī′ti), *n.* (*sl.*) soldier's name for Britain, home;

(*Mil.*) a wound that invalids one home (*also* **Blighty one**).
blimey (bli'mi), *int.* exclamation of astonishment.
blimp (blimp), *n.* a small airship; (*Cinema*) a sound-proof covering for the camera mechanism; (*coll.*) someone narrow-minded and conservative; a die hard army officer.
blimpish, *a.* [from Colonel *Blimp*, the cartoon character created by David Low, 1891-1963]
blind (blind), *a.* unseeing; destitute of sight either naturally or by deprivation; dark, admitting no light; obstructing the vision; having no outlet; of, pertaining to or for the use or benefit of, the sightless; destitute of understanding, judgment or foresight; reckless, heedless, drunk; purposeless, random; (*Bot.*) having no buds, eyes or terminal flower; abortive (of a bud). *n.* anything which obstructs the light or sight; a blinker for a horse; (*coll.*) a pretence, a pretext; a window-screen or shade, esp. one on rollers for coiling up, or of slats on strips of webbing; (*sl.*) a drunken fit. *pl.* blind persons collectively; *v.t.* to deprive of sight (permanently or temporarily); to darken, make dim; (*coll.*) to deceive; to make unaware, heedless. *v.i.* (*sl.*) to swear. **to bake blind**, to bake pastry intended for a pie or flan before adding the filling. **to blind with science**, to confuse or intimidate with a display of technical knowledge or jargon. **to fly blind**, (*Aviat.*) to fly by the use of instruments only. **blind alley**, *n.* a street, road or alley walled-up at the end. **blind-blocking**, **-tooling**, *n.* (*Bookbinding*) ornamentation done by impressing hot tools without gold-leaf. **blind-coal**, *n.* a flameless anthracite. **blind date**, *n.* a social engagement arranged between two people previously unknown to one another. **blind drunk**, *adv.* too drunk to be able to see straight. **blind-fish**, *n.* a fish without functional eyes found in underground streams. **blindfold**, *v.t.* to cover the eyes, esp. with a bandage. *a.* having the eyes bandaged; devoid of foresight. **blind-man's-buff**, *n.* a game in which a player has his eyes bandaged, and has to catch and identify one of the others. **blind screening**, (also **anonymized screening**), *n.* the testing of unidentified samples (of blood) without the patient's knowledge. **blind side**, *n.* the direction in which one is most easily assailed; a weakness. **blind spot**, *n.* a part of the retina insensitive to light; (*Radio.*) a point where signals are received very faintly; a tendency to overlook faults etc.; a failure of understanding or judgment; a weakness. **blind-stitch**, *n.* sewing that does not show, or that shows at the back only. *v.t.*, *v.i.* to sew in this manner. **blind-worm**, *n.* = SLOW-WORM. **blinder**, *n.* one who or that which blinds; (*N Am.*) a horse's blinker (*sl.*) an outstanding piece of play, performance etc.; (*sl.*) a lengthy bout of drinking. **blindly**, *adv.* **blindness**, *n.* sightlessness; lack of intellectual or moral perception; ignorance, folly, recklessness.
blink (blingk), *v.i.* to move the eyelids; to open and shut the eyes; to look unsteadily; to shine fitfully, to wink at; to be surprised. *v.t.* to shut the eyes to; evade, to shirk. *n.* a gleam, a glimmer, a twinkle; a glance, a twinkling (cp. ICE-BLINK). **on the blink**, (*coll.*) not functioning properly (of a machine). **blinker**, *n.* one who blinks; (*pl.*) spectacles to cure squinting, or to protect the eyes from cold, dust etc.; leather screens to prevent a horse from seeing sideways. **blinkered**, *a.* wearing blinkers; not understanding what is going on around one; having a distorted or biased view or opinion. **to wear blinkers**, (*fig.*) not to see or understand what is going on around one.
blinking, *a.* (*coll.*) a euphemism for BLOODY used for emphasis.
blintz(e) (blints), *n.* a thin, stuffed pancake.
blip (blip), *n.* (*coll.*) an irregularity in the linear trace on a radar screen indicating the presence of an aircraft, vessel etc.; any small or temporary irregularity in an established pattern; a bleep.
bliss (blis), *n.* supreme happiness; the perfect joy of heaven; heaven. **blissful**, *a.* very happy; causing bliss. **blissfully**, *adv.* **blissfully ignorant of**, quite unaware of. **blissfulness**, *n.*
blister (blis'tər), *n.* a pustule or thin vesicle raised on the skin and containing a watery fluid or serum; any similar swelling on a plant, metal, a painted surface etc.; anything applied to raise a blister. *v.i.* to rise in blisters; to be covered with blisters; (*Austral. coll.*) to overcharge, to demand an exorbitant sum. *v.t.* to raise blisters on, esp. by a vesicatory; to criticize spitefully. **blister pack**, *n.* a type of clear plastic and cardboard packaging for small products. **blistered**, *a.* **blistery**, *a.*
blithe (blidh), *a.* gay, cheerful, joyous; merry, sprightly. **blithely**, *adv.* **blitheness**, *n.* **blithesome** (-səm), *a.* blithe; cheery. **blithesomeness**, *n.*
blithering (blidh'əring), *a.* (*coll., dial.*) nonsensical, contemptible.
blitz (blits), *n.* (*coll.*) an intense enemy onslaught, esp. an air raid; an intensive campaign; intensive activity or action. *v.t.* to attack suddenly and ferociously; to mount an intensive campaign; to subject to intensive activity.
blizzard (bliz'əd), *n.* a snow-squall; a furious storm of snow and wind.
bloat (blōt), *v.t.* to cause to swell; to puff up; to make vain or conceited. *v.i.* to swell; to grow turgid. **bloated**, *a.* swollen, inflated, pampered, puffed up with pride. **bloatedness**, *n.*
bloater (blō'tə), *n.* a herring partially cured by steeping in dry salt and smoking.
blob (blob), *n.* a globular drop of liquid; a spot of colour; any vague, soft form.
bloc (blok), *n.* a combination of parties, or of nations. [F]
block (blok), *n.* a solid mass of wood or stone; a log, a tree-stump; a building block; the piece of wood on which criminals were beheaded; death by beheading; a compact or connected group of buildings, esp. when bounded by intersecting streets, regarded in the US as a method of measuring distances; a mould on which a thing is shaped; a piece of wood or metal on which figures are engraved for printing from; a cliché taken from such a block; a solid unshaped mass of any material; a starting block; a pulley, or system of pulleys, mounted in a frame or shell; (*Parl.*) a notice of opposition to a Bill (*see below*); an obstruction, a hindrance; a blocking manoeuvre; a psychological barrier; a stupid or impassive person; (*sl.*) the head. *v.t.* to enclose, to shut up; to stop up, to obstruct; to impede progress or advance; to shape a hat on the block; (*Bookbinding*) to emboss a cover by impressing a device; (*Cricket*) to stop a ball dead; (*Parl.*) to give notice of opposition to a Bill, thus preventing its being proceeded with at certain times. **block and tackle**, an arrangement of pulley-blocks and ropes for hoisting. **to block in**, to sketch roughly the broad masses of a picture or drawing. **to block out**, to mark out work roughly; to prevent from entering. **to block up**, to obstruct, to stop up. **barber's block**, BARBER. **blockboard**, *n* a board made of strips of wood sandwiched between two strips of veneer. **block-booking**, *n.* the reserving of a number of seats at a single booking. **block-buster**, *n.* (*coll.*) a very heavy aerial bomb; a particularly effective or successful thing or person, esp. a film or book. **block-busting**, *a.* **block diagram**, *n.* a diagram using rectangles and connecting lines to represent a process or system. **blockhead**, *n.* a stupid, dull person. **block-hole**, *n.* (*Cricket*) a mark made a yard in front of the wicket. **blockhouse**, *n.* a detached fort covering some strategical point; a building, esp. one with loop-holes; a house of squared timber. **block-letters**, *n.pl.* imitation in handwriting of printed capital letters. **block release**, *n.* the short-term release of employees for formal study or training. **block-signal**, *n.* a signal to stop a train when the next section of the line is not clear. **block-system**, *n.* a system by which a railway line is divided into sections, governed by block-signals. **blockage**, *n.*, an obstruction.

blocker, *n.* **blockish**, *a.* stupid; dull; rough, clumsy. **blockishly**, *adv.* **blockishness**, *n.*
blockade (blokād'), *n.* the prevention of any movement of people or supplies into or out of an enemy area. *v.t.* to block up, esp. by troops or ships. **blockade-runner**, *n.* (the owner, captain or a sailor of) a vessel that runs or attempts to run into a blockaded port.
bloke (blōk), *n.* (*coll.*) a man, a fellow.
blond, blonde (used with fem. substantives), (blond), *a.* fair or light in colour; having light hair and a fair complexion. *n.* a person with light hair and a fair complexion (the form **blonde** is used of women).
blood (blŭd), *n.* the red fluid circulating through the bodies of humans and other vertebrates; any analogous fluid in the invertebrates; lineage, descent; honourable or high birth; family relationship, kinship; murder, bloodshed; the guilt of murder; temperament, passion; vitality, mettle; a dashing young man. *v.t.* to stain with blood, esp. as a form of initiation; to inure to blood (as a hound); to initiate. **bad blood**, resentment, ill-feeling. **blood and thunder**, sensational literature. **flesh and blood**, the carnal nature of man; human nature. **half-blood**, HALF. **in cold blood**, not in anger; deliberately. **new blood**, new entrants to a community or group who add freshness or vigour. **the blood**, Royal blood; the royal family. **blood bank**, *n.* the place where blood for transfusion is stored. **blood-brother**, *n.* a brother by both parents; a man who has sworn fidelity to another, usu. by a ceremony involving the mingling of their blood. **blood count**, *n.* a calculation of the number of red and white corpuscles in a sample of blood. **blood-curdling**, *a.* harrowing, horrifying. **blood donor**, *n.* one who gives blood for transfusion. **blood-feud**, *n.* a feud arising out of murder or homicide; a vendetta. **blood groups**, *n.pl.* the four groups into which human beings have been classified for purposes of blood-transfusion. **blood-heat**, *n.* the ordinary temperature of blood in a healthy human body (about 98° F or 37°C). **bloodhound**, *n.* a variety of hound remarkable for keenness of scent, used for tracking fugitives; one who relentlessly pursues an opponent; a detective, a spy. **blood-letting**, *n.* taking blood from the body; phlebotomy; bloodshed; (*coll.*) excessive financial demands. **blood-money**, *n.* money paid for information leading to a conviction on a capital charge; compensation paid to the next of kin for the murder of a relative. **blood-orange**, *n.* an orange having pulp and juice of a reddish hue. **blood plasma**, *n.* blood from which all red corpuscles have been removed. **blood poisoning**, *n.* a diseased condition caused by septic matter in the blood. **blood pressure**, *n.* pressure of the blood on the walls of the arteries. **blood-relation**, *n.* a relation by descent, not merely by marriage. **blood-shed**, *n.* the act of shedding blood; murder; slaughter in war. **bloodshot**, *a.* red and inflamed; (of the eye) suffused with blood. **blood sports**, *n.pl.* sports entailing the killing of animals, such as fox-hunting. **blood-stain**, *n.* a stain produced by blood. **blood-stained**, *a.* **blood stock**, *n.* collective term for thoroughbred horses. **blood-stone**, *n.* heliotrope, a variety of quartz with blood-like spots of jasper. **blood stream**, *n.* the circulatory movement of the blood in the body. **blood-sucker**, *n.* any animal which sucks blood, esp. the leech; an extortioner. **blood test**, *n.* the medical examination of a blood sample. **bloodthirsty**, *a.* eager to shed blood; delighting in sanguinary deeds. **bloodthirstiness**, *n.* **blood-transfusion**, *n.* transference of blood from a healthy person to one whose blood is deficient in quantity or quality. **blood-vessel**, *n.* a vessel in which blood circulates in the animal body; an artery or a vein. **bloodwood**, **blood-tree**, *n.* a term applied to several varieties of trees that exude a bright red gum. **blood-worm**, *n.* a small red earth-worm used by anglers. **bloodied** (-id), *a.* stained with blood. **bloodless**, *a.* without blood; without shedding blood; spiritless; unfeeling. **bloodlessly**, *adv.* **bloody**, *a.* of or pertaining to blood; stained or running with blood; involving bloodshed; cruel, murderous; (*sl.*) damned; very, exceedingly; (*sl.*) annoying, wretched etc. **Bloody Mary**, *n.* a cocktail made of vodka and tomato juice. **bloody-minded**, *a.* obstinate; unhelpful. **bloodily**, *adv.* **bloodiness**, *n.*
bloom (bloom), *n.* a blossom, a flower; the delicate dust on newly gathered plums, grapes etc.; the yellow sheen on well-tanned leather; lustre, efflorescence; prime, perfection. *v.i.* to blossom, to come into flower; to be at the highest point of perfection or beauty. **bloomer**, *n.* a plant that blooms (esp. in *comb.*, as an *early-bloomer*); (*sl.*) a mistake, a foolish blunder. **in bloom**, flowering, blossoming. **blooming**, *a.* in a state of bloom, flourishing; bright, lustrous; (*sl.*, *euph.*) bloody. **bloomingly**, *adv.* **bloomless**, *a.* **bloomy**, *a.*
bloomers (bloo'məz), *n.* baggy knickers; loose trousers for women gathered round the ankles, formerly worn esp. for cycling. [from the American Mrs *Bloomer*, who introduced them *c.* 1850]
blossom (blos'əm), *n.* the flower of a plant, esp. considered as giving promise of fruit; the mass of flowers on a fruit-tree. *v.i.* to put forth flowers; to bloom; to flourish. **blossomless**, *a.* **blossomy**, *a.*
blot[1] (blot), *n.* a spot or stain of ink or other discolouring matter; a blotting out by way of correction; a dark patch; a blemish; a disgrace. *v.t.* (*past, p.p.* **blotted**) to spot or stain with ink or other discolouring matter; to obliterate; to dry with blotting-paper; to disfigure, to sully. *v.i.* to make blots, to become blotted. **to blot one's copybook**, (*coll.*) to commit an indiscretion; to spoil one's good record. **to blot out**, to obliterate, to efface. **blotter**, *n.* a blotting-pad. **blotting**, *n.*, *a.* **blotting-paper**, *n.* absorbent paper for drying up ink. **blotting-pad**, *n.* a pad made up of this.
blot[2] (blot), *n.* an exposed piece at backgammon. **to hit a blot**, to take an exposed piece at backgammon; to detect a fault.
blotch (bloch), *n.* an irregular spot or discoulouration. *v.t.* to mark with blotches. **blotched**, *a.* **blotchy**, *a.*
blotto (blot'ō), *a.* (*sl.*) unconscious with drink.
blouse (blowz), *n.* a light, loose, upper garment. [F (*etym.* unknown)]
blouson (bloo'zon), *n.* a short, loose jacket fitted or belted in at the waist. [F]
blow[1] (blō), *v.i.* (*past* **blew**, *p.p.* **blown**) to move as a current of air; to send a current of air from the mouth; to pant; to sound (as a horn); to eject water and air from the spiracles; of a fuse, to burn out; of a tyre, to burst; (*sl.*) to explode. *v.t.* to drive a current of air upon; to shape by blowing; to drive by a current of air; to put out of breath; to sound a wind instrument or a note on it; to taint by depositing eggs upon (as flies); to shatter by explosives; to cause (a fuse) to burn out; to burst by excess pressure; (*coll.*) to lavish, to squander; (*sl.*) to lose or make a mess of through a blunder; (*sl.*) to leave hurriedly. *int.* (*sl.*) curse, confound. *n.* a blowing, a blast of air; (a walk to take) a breath of fresh air. **blow it!** Confound it. **I'll be blowed!** *int.* (*sl.*) I'll be damned etc. **to blow hot and cold**, to vacillate; to do one thing at one time, and its opposite at another. **to blow in**, to make an unexpected visit. **to blow it**, (*coll.*) to lose a chance or advantage by committing a blunder. **to blow off**, to escape with a blowing noise, as steam; to discharge (steam, energy, anger etc.). **to blow one's own trumpet**, to boast, to sing one's own praises. **to blow out**, to extinguish by blowing; to clear by means of blowing. **to blow over**, to pass away, to subside. **to blow the gaff**, (*sl.*) to let out a secret. **to blow up**, to inflate; to destroy with explosives; to explode; to lose one's temper; to criticize severely; to enlarge (a photograph), to increase or exaggerate the importance of. **blowdry**, *n.* a

method of styling hair while drying it with a small hair-dryer. **blow-fly**, *n*. the meat-fly. **blowhard**, *n*. a braggart. **blow-hole**, *n*. an air-hole; a breathing hole in the ice for seals and whales; (*pl.*) the spiracles of a cetacean. **blow-job**, *n*. (*sl.*) fellatio. **blow-lamp, -torch**, *n*. lamp used in soldering, brazing etc.; burner used to remove paint. **blow-line**, *n*. (*Angling*) a light line with real or artificial bait at the end, allowed to float over the surface of water with the wind. **blow moulding**, *v.i.* a method of manufacturing plastic goods. **blow-out**, *n*. (*sl.*) a hearty meal; a celebration; an explosion of oil and gas from an oil well; the puncturing of a tyre; the burning out of an electrical fuse or a valve. **blow-pipe**, *n*. a tube used for increasing combustion by directing a current of air into a flame; a pipe used in glass-blowing; a tube used by American Indians for shooting darts. **blow-torch**, BLOW-LAMP. **blow up**, *n*. the enlargement of part or whole of a photograph; (*coll.*) a burst of anger, a heated argument. **blower**, *n*. a contrivance for creating an artificial current of air; (*coll.*) a telephone. **blowing**, *n*, *a*. **blowy**, *a*. windy; exposed to the wind.

blow[2] (blō), *v.i.* to blossom; to open out. *n*. the state of blossoming; bloom; a display of blossoms.

blow[3] (blō), *n*. a stroke with the fist or any weapon or instrument; an act of hostility; a severe shock; a sudden and painful calamity. **blow-by-blow**, minutely detailed. **to come to blows**, to fight.

blowzy, blowsy, *a.* having a coarse, reddish complexion; of a woman, untidy, sluttish.

blub (blŭb), *v.i.* (*coll.*) to weep, shed tears.

blubber (blŭb′ə), *n*. the fat underlying the skin in whales and other cetaceans, from which train-oil is prepared. *v.i.* (*coll.*) to weep noisily. *v.t.* to wet and disfigure with weeping; to utter with sobs and tears.

bludgeon (blŭj′ən), *n*. a short, thick stick, sometimes loaded; a black-jack. *v.t.* to strike with this; to coerce verbally, or by physical force.

blue (bloo), *a*. of the colour of the cloudless sky or deep sea; livid; dressed in blue; belonging to the political party which adopts blue for its colour (in Britain, usually the Conservative); (*coll.*) miserable, low-spirited; (*sl.*) obscene, smutty. *n*. a blue colour; a blue pigment; a blue powder used by laundresses; a blue substance, object or animal (as explained by context); the sky; the sea; a man who plays for his university in sport or athletics; (*Austral.*) a summons. *v.t.* to make blue; to treat with laundress's blue; (*sl.*) to squander money. **light-blue, dark-blue**, the respective colours of Eton and Harrow schools, and of Cambridge and Oxford Universities in their athletic contests. **old blue**, a former University athlete. **out of the blue**, unexpected, unpredicted. **the Blues**, the Royal Horse Guards. **the blues**, a form of melancholy, Black American folk-song originating in the deep south, usu. consisting of 3, 4-bar phrases in 4/4 time; low spirits, depression. **true blue**, staunch, faithful, genuine. **blue baby**, *n*. a baby with a bluish discolouration of the skin due to a shortage of oxygen in the blood. **bluebell**, *n*. the blue bell of Scotland, *Campanula rotundifolia;* the wild hyacinth of England, *Scilla nutans*. **blueberry**, *n*. (*N Am.*) the genus *Vaccinium;* (*Austral.*) the native currant. **blue-black**, *a.* of a blue colour that is almost black; black with a tinge of blue. **blue-bird**, *n*. a small American bird, *Sylvia sialis;* a symbol of happiness. **blue blood**, *n*. aristocratic descent. **blue blooded**, *a.* **Blue book**, *n*. an official report of Parliament (bound in volumes which have blue covers); (*N Am.*) a list of Government officials with their salaries etc. **blue-bottle**, *n*. the blue cornflower, *Centaurea cyanus;* applied also loosely to other blue flowers; the meat-fly or blow fly, *Musca vomitoria;* (*coll.*) a policeman. **blue cat**, *n*. a Siberian cat, valued for its slaty-blue fur. **blue cheese**, *n*. a cheese threaded by blue veins of mould induced by the insertion of copper wires during its making. **blue chip**, *n*. an issue of stocks or shares believed to be dependable in maintaining or increasing its value; hence, anything of worth and stability. **blue-chip**, *a*. **blue-coat boy**, *n*. a boy wearing the blue coat of a charity school, esp. a scholar of Christ's Hospital. **blue-collar**, *a*. pertaining to manual work and manual workers in contrast to desk work and office employees (see WHITE-COLLAR). **blue-devils**, *n.pl.* low spirits, depression (the blues); the illusions of delirium tremens. **blue-eyed**, *a.* having an eye with a blue iris. **blue-eyed boy, girl**, someone especially favoured by a person or group. **blue film, movie**, *n*. a sexually explicit or pornographic film. **blue funk**, *n.* (*sl.*) abject terror. **blue-grass**, *n.* (*N Am.*) the rich grass of the limestone lands of Kentucky and Tennessee (blue-grass country); a kind of folk music originating from these regions. **Blue-grass State**, *n.* Kentucky. **blue-gum tree**, *n.* an Australian tree, *Eucalyptus globulus*. **bluejacket**, *n.* a sailor in the British Navy. **Blue Mantle**, *n.* one of the four pursuivants in the College of Arms. **blue moon**, *n.* a very rare or unknown occurrence, never. **blue mould**, *n.* a blue coloured fungus which grows on rotting food and other vegetable matter, and is induced in blue cheese. **blue movie**, BLUE FILM. **blue pencil**, *v.t.* (*coll.*) to censor, edit or mark with corrections (trad. using a blue pencil). **Blue Peter**, *n.* a small blue flag, with a white square in the centre used as a signal for sailing. **blue pointer**, *n.* (*Austral.*) a voracious shark with a blue back. **blueprint**, *n.* a plan or drawing printed on specially sensitized paper: the print is composed of white lines on a blue background, and is much used for scale and working drawings of engineering designs, electrical circuits etc.; any original plan or guideline for future work; a prototype. **blue ribbon**, *n.* the ribbon of the Garter; hence, the greatest distinction; the first prize. **blue-sky**, *a.* (*coll.*) purely theoretical speculative or experimental; lacking specific goals (as of a research project). **blue-sky laws**, *n.pl.* American legislation against a form of fraud involving stocks and shares. **bluestocking**, *n.*, *a.* (*often derog.*) (a woman) with intellectual or literary interests. [from a literary society that met at Montagu House, London, the latter part of the 18th cent.] **bluetit**, *n.* a common European tit with a blue crown and yellow underparts. **blue whale**, *n.* a blueish-grey whale: the largest living animal. **blueing**, *n.* (*N Am.*) laundress's blue. **bluely**, *adv.* **blueness**, *n.* **bluey**, **bluish**, *a.* **bluishly**, *adv.* **bluishness**, *n.*

bluff[1] (blŭf), *a.* having a broad, flattened face or front; abrupt, blunt, frank, outspoken. *n.* a cliff or headland with a steep, broad front. **bluffly**, *adv.* **bluffness**, *n.* **bluffy**, *a.* having bold headlands; blunt, off-handed.

bluff[2] (blŭf), *v.t.* to make one's adversary (at cards) believe one's hand is stronger than it is, and induce him to throw up the game; to deceive or intimidate by a show of confidence and/or a pretence of strength. *n.* an act or instance of bluffing. **to call somebody's bluff**, to challenge a person to reveal his or her true strength, intentions, etc.

blunder (blŭn′də), *v.i.* to err grossly; to act blindly or stupidly; to flounder, to stumble. *n.* a gross mistake, a stupid error. **blunderer**, *n.* **blundering**, *a.* **blunderingly**, *adv.*

blunderbuss (blŭn′dəbŭs), *n.* a short gun, with a wide muzzle.

blunt (blŭnt), *a.* without edge or point; abrupt, unceremonious; rough, unpolished. *v.t.* to make less sharp, keen, or acute; to deaden, to dull. *v.i.* to become blunt. **bluntish**, *a.* **bluntly**, *adv.* **bluntness**, *n.*

blur (blœ), *n.* a smear, a smudge; a dim, misty effect. *v.t.* to smear, to smudge; to render misty and indistinct. **blurriness**, *n.* **blurry**, *a.*

blurb (blœb), *n.* a promotional description of a book, usually printed on the dust-jacket.

blurt (blœt), *v.t.* to utter abruptly (*usu.* with *out*).

blush (blŭsh), *v.i.* to become red in the face from shame or other emotion; to be ashamed. *n.* the reddening of the face produced by shame, modesty or any similar cause; a crimson or roseate hue; a flush of light; rosé wine. **at the first blush, at first blush,** at the first glance; at first sight. **blush-rose,** *n.* a white rose with pink tinge. **blusher,** *n.* a cosmetic for reddening the cheeks. **blushful,** *a.* suffused with blushes; modest, self-conscious. **blushfully,** *adv.* **blushing,** *a.* modest; ruddy, roseate. **blushingly,** *adv.* **blushless,** *a.*

bluster (blus'tə), *v.i.* to blow boisterously; to make a loud boisterous noise; to boast. *n.* boisterous, inflated talk; empty threats. **blusterer,** *n.* **blustering, blustery,** *a.* blowing boisterously; tempestuous; hectoring, boastful. **blusteringly,** *adv.*

BM, (*abbr.*) Bachelor of Medicine; (surveying) bench mark; British Museum.

BMA, (*abbr.*) British Medical Association.

BMC, (*abbr.*) British Medical Council.

B Mus (bē mūs'), Bachelor of Music.

BMW, (*abbr.*) Bayerische Motoren Werke, a make of car. [G, Bavarian motor works]

BMX, (*abbr.*) bicycle motocross, bicycle stunt riding over an obstacle course; (**BMX**®) a bicycle designed for this.

Bn., (*abbr.*) Baron; battalion; billion.

bo, boh (boo), *int.* an exclamation intended to surprise or frighten. **to say bo to a goose,** to open one's mouth, to speak. **bo-beep,** *n.* a children's game in which a player suddenly looks out from a hiding-place and cries 'bo!' to startle the others.

BO, (*abbr.*) body odour; box office.

Boa (bō'ə), *n.* a genus of large S American snakes which kill their prey by crushing; a long fur or feather tippet worn round the neck. **boa-constrictor,** *n.* a Brazilian snake, the best-known species of the genus Boa.

boar (bôr), *n.* the uncastrated male of the domesticated or the wild pig. **wild boar,** *n.* the male of *Sus scrofa,* wild in Europe, Asia and Africa. **boarish,** *a.* swinish, brutal; sensual, cruel.

board (bawd), *n.* a piece of timber of considerable length, and of moderate breadth and thickness; a flat slab of wood, used for displaying notices, and other purposes; a specially marked surface on which games (as chess, draughts etc.) are played; a thick substance formed of layers of paper etc., pasted or squeezed together; a piece of stout pasteboard or millboard used as one of the sides of a bound book; a table, esp. for meals; food served at table; daily provisions; one's keep, or money in lieu of keep; the members of a council; the governors or directors of a public trust or a business concern; the side of a ship; (*pl.*) the stage. *v.t.* to furnish or cover with boards; to provide with daily meals (and now *usu.* with lodging); to board out; to attack and enter (a ship) by force; to embark; *v.i.* to have one's meals (and *usu.* lodging) at another person's house. **above board,** open, unconcealed, openly. **across the board,** inclusive of all categories or types. **bed and board,** conjugal relations. **board and lodging,** meals and sleeping-quarters. **by the board,** overboard, by the ship's side; ignored, rejected or disused. **on board,** in or into a ship, train, bus or aeroplane. **to board out,** to arrange board and lodging for away from home. **to sweep the board,** to win a total victory, as in a game. **board game,** *n.* a game, such as chess, which is played with pieces or counters on a special board. **board-room,** *n.* the meeting place of a company's board of directors. **board sailing,** *n.* windsurfing. **board-sailing,** *v.i.* **board-school,** *n.* a school managed by a Board, as established by the Elementary Education Act, 1870. **board-walk,** *n.* a seaside promenade made of planks. **boarder,** *n.* a lodger; a schoolchild who is boarded and lodged at a school; (*Naut.*) one who boards an enemy's ship. **boarding,** *n.* the action of the verb TO BOARD; a structure of boards. **boarding-clerk,** *n.* a clerk in the Customs or in a mercantile firm, who communicates with the masters of ships on their arrival in port. **boarding-house,** *n.* a house which provides accommodation and meals. **boarding officer,** *n.* officer who boards a ship to examine bill of health etc. **boarding pass,** *n.* a ticket authorising one to board an aeroplane, ship etc. **boarding-school,** *n.* a school in which pupils are boarded as well as taught.

boast (bōst), *n.* proud, vainglorious assertion, a brag; a cause of pride. *v.i.* to brag, to praise oneself, to speak ostentatiously or vaingloriously. *v.t.* to extol, to speak of with pride; to have as worthy of pride. **boaster,** *n.* **boastful,** *a.* **boastfully,** *adv.* **boastfulness,** *n.* **boastingly,** *adv.*

boat (bōt), *n.* a small vessel, generally undecked and propelled by oars or sails; applied also to fishing vessels, packets, and passenger steamers; a vessel or utensil resembling a boat, a sauce-boat. *v.i.* to take boat, to row in a boat. **in the same boat,** in the same circumstances or position. **to rock the boat,** to disrupt existing conditions, to cause trouble. **boat-bill,** *n.* a S American genus of birds, allied to the herons. **boat-fly,** *n.* a boat-shaped water-bug, *Notonecta glauca*. **boat-hook,** *n.* a pole with an iron point and hook, used to push or pull a boat. **boat-house,** *n.* a house by the water in which boats are kept. **boatman,** *n.* a man who lets out boats on hire; a man who rows or sails a boat for hire. **boat people,** *n.pl.* refugees (*usu.* Vietnamese) who flee from their countries in small boats. **boat-race,** *n.* a race between rowing-boats. **boat-train,** *n.* a train conveying passengers to or from a ship. **boatable,** *a.* (*N Am.*) that may be traversed by boat; navigable. **boatage** (-ij), *n.* charges for carriage by boat. **boater,** *n.* a man's stiff straw hat. **boatful, boat-load,** *n.* as much or as many as a boat will hold.

boatel, botel, *n.* a floating hotel, a moored ship functioning as a hotel; a water-front hotel accommodating boaters.

boatswain, bos'n (bō'sn), *n.* the foreman of the crew (in the RN a warrant officer) who looks after the ship's equipment.

bob[1] (bob), *n.* a weight or pendant at the end of a cord, chain, plumb-line, pendulum etc.; a knot or bunch of hair, a short curl; the docked tail of a horse; a shake, a jog; a short jerking action, a curtsy; (*sl.*) a shilling or 5 pence; a peal of courses or set of changes in bell-ringing. *v.t.* to move with a short jerking motion; to cut short (as a horse's tail); to rap, to strike lightly; *v.i.* to have a short jerking motion; to move to and fro or up and down; to dance, to curtsy. **to bob up,** to emerge suddenly. **treble bob, bob major, bob minor,** peals in which the bells have a jerking or dodging action: in the first the treble bell is dominant, the others are rung on eight and six bells respectively. **bobcat,** *n.* a N American mammal similar to but smaller than a lynx. **bob-sled,** *n.* a conveyance formed of two sleds or sleighs coupled together, used to transport large timber. **bob-sleigh,** *n.* a sleigh with two pairs of runners, one behind the other often used for racing. **bob-tail,** *n.* a tail (of a horse) cut short; a horse or dog with its tail cut short. **bob-tail, bob-tailed,** *a.* **bob-wig,** *n.* a wig having the bottom turned up in bobs or curls.

bobbie (-by) pin (bob'i), *n.* (*esp. N Am.*) a hair-grip.

bobbin (bob'in), *n.* a wooden pin with a head on which thread for making lace, cotton, yarn, wire etc., is wound and drawn off as required; a piece of wood with a string for actuating a door-latch; a reel, spool. **bobbin-lace, -work,** *n.* work woven with bobbins. **bobbinet** (-net'), *n.* machine-made cotton net, orig. imitated from bobbin-lace.

bobble (bob'l), *n.* a fabric or wool ball used as decorative trimming, a pom-pom.

bobby (bob'i), *n.* (*coll.*) a policeman. [from Sir *Robert* Peel, who introduced the new police, 1828]

bobby sox, socks (bob'i), *n.* (*N Am.*) ankle socks *usu.* worn by young girls. **bobby soxer,** *n.* an adolescent girl.

bobcat BOB.

bobolink (bob'əlink), *n.* an American song-bird, *Dolichonyx oryzivorus*.
bobstay (bob'stā), *n.* a chain or rope for drawing the bowsprit downward and keeping it steady.
bob-tail etc. BOB[1].
Boche (bosh), *n.* (*offensive*) a German; (collect.) the Germans. a German. [F *sl.*]
bock (bok), *n.* a large beer-glass; a large glass of beer.
bod (bod), *n.* (*coll.*) a person.
bode (bōd), *v.t.* to foretell, to forebode. *v.i.* to portend (well or ill). **bodement,** *n.* an omen. **boding,** *a.* presaging, ominous. *n.* an omen, presentiment. **bodingly,** *adv.*
bodega (bədē'gə), *n.* a wine-shop.
bodge (boj), *v.t.* to botch; to mend in a clumsy fashion; to construct clumsily. **bodger,** *n.*
bodice (bod'is), *n.* a tight-fitting outer vest for women; the upper part of a woman's dress. **bodice-ripper,** *n.* a sexy and sensational historical novel.
bodikin (bod'ikin), *n.* a little body. **od's bodikins,** by God's dear body.
bodkin (bod'kin), *n.* a large-eyed and blunt-pointed needle for leading a tape or cord through a hem, loop etc.; a pin for fastening up women's hair; an instrument for piercing holes in cloth; (*hist.*) a small dagger.
body (bod'i), *n.* the material frame of man or the lower animals; the trunk; the upper part of a dress BODICE; a corpse; the main or central part of a building, ship, document, book etc.; the part of a motor-car in which the driver and passengers sit; a collective mass of persons, things or doctrine, precepts etc.; matter, substance, as opposed to spirit; (*coll.*) a person, an individual; a society, a corporate body, a military force; (*Phil.*) matter, substance; (*Geom.*) a figure of three dimensions; strength, substantial quality. **heavenly body,** HEAVENLY. **to body forth,** to give mental shape to; to exhibit, to typify. **body blow,** *n.* in boxing, a punch landing between the breast bone and navel; a harsh disappointment or set-back, a severe shock. **body builder,** *n.* one who develops his/her muscles through exercise, an exercising machine and/or eating high-protein food. **body building,** *n.*, *a.* **body-check,** *v.t.* (*Ice Hockey etc.*) to stop or block an opponent with one's body. **body-colour,** *n.* a pigment having a certain degree of consistence and tingeing power as distinct from a wash; a colour rendered opaque by the addition of white. **bodyguard,** *n.* personal escort protecting a dignitary. **body language,** *n.* a form of non-verbal communication by means of conscious or unconscious gestures, postures and facial expressions. **body-line,** *a.*, *n.* (*Cricket*) (of) bowling aimed at the batsman's body. **body politic,** *n.* organized society; the State. **body-servant,** *n.* a valet. **body-snatcher,** *n.* one who steals a body from a grave for the purpose of dissection. **body shop,** *n.* a workshop where vehicle bodies are made or repaired. **body stocking,** *n.* a clinging all-in-one undergarment often of a sheer material. **body warmer,** *n.* a usu. quilted jerkin. **bodywork,** *n.* the metal shell of a motor vehicle. **bodiless,** *a.* **bodily,** *a.* of, pertaining to or affecting the body or the physical nature; corporeal. *adv.* corporeally, united with matter; wholly, completely.
Boeotian (biō'-shən), *a.* stupid, dull.
Boer (buə, bōə, baw), *n.* a S African of Dutch birth or extraction.
boffin (bof'in), *n.* a scientist, esp. one employed by the Services or the government.
boffo *a.* (*sl.*) indicating enthusiastic approval.
Bofors gun (bō'fəz), *n.* an automatic anti-aircraft gun [from *Bofors* the Swedish munition works]
bog (bog), *n.* a marsh, a morass; wet, spongy soil, a quagmire; (*sl.*) a lavatory. *v.t.* to sink or submerge in a bog. **to bog down,** to overwhelm, as with work; to hinder. **bog-oak,** *n.* oak found preserved in bogs, black from impregnation with iron. **bog-trotter,** *n.* a person used to traversing boggy country; (*offensive*) an Irishman.
boggy, *a.* of or characterized by bogs; swampy. **bogginess,** *n.* **boglet** (-lit), *n.* a little bog.
bogey[1] (bō'gi), *n.* a fair score for a good player, orig. **Colonel Bogey,** an ideal opponent against whom a solitary player could pit himself; one stroke over par on a hole. [imag. person]
bogey[2] (bō'gi), BOGIE.
bogey[3] (bō'gi) BOGY[2].
boggle (bog'l), *v.i.* to shrink back, to hesitate; to start with fright; (*coll.*) to be astounded; (*coll.*) to be unable to imagine or understand.
bogie, bogy[1] (bō'gi), *n.* a swivelling undercarriage having one or more pairs of wheels. **bogie-car, -engine,** *n.* a railway-carriage or locomotive mounted on these.
bogle (bō'gl), *n.* a hobgoblin, a spectre; a scarecrow.
bogus (bō'gəs), *a.* sham, counterfeit.
bogy[1] (bō'gi) BOGIE.
bogy[2], **bogey**[3] (bō'gi), *n.* a spectre, a bugbear; (*sl.*) a piece of dried snot. **bogy(-gey,) man,** *n.* an evil person or spirit, used to menace children.
bohea (bōhē'), *n.* an 18th cent. name for the finest kind of black tea.
Bohemian[1] (bəhē'miən), *a.* of or pertaining to Bohemia, its people or their language. *n.* a native or inhabitant of Bohemia.
bohemian[2] (bəhē'miən), *n.* a gipsy; one who leads a free, irregular life, despising social convention. *a.* of or characteristic of the gipsies or of social bohemians. **bohemianism,** *n.* **bohemianize, -ise,** *v.i.* to live in an unconventional way.
boil[1] (boil), *v.i.* to be agitated by the action of heat, as water or other fluids; to reach the temperature at which these are converted into gas; to be subjected to the action of boiling, as meat etc., in cooking; to bubble or seethe like boiling water (also of the containing vessel); to be agitated with passion. *v.t.* to cause a liquid to bubble with heat; to bring to the boiling point; to cook by heat in boiling water; to prepare in a boiling liquid; *n.* an act of boiling; the state of boiling; boiling-point. **to boil away,** to evaporate in boiling. **to boil down,** to lessen the bulk of by boiling; to condense. **to boil over,** to bubble up, so as to run over the sides of the vessel; to be effusive. **boiled,** *a.* **boiled shirt,** *n.* (*coll.*) a dress shirt. **boiler,** *n.* a vessel in which anything is boiled; the large vessel in a steam-engine in which water is converted into steam; a tank in which water is heated for domestic use. **boiler-suit,** *n.* a combined overall garment, esp. for dirty work. **boiler-tube,** *n.* one of a system of tubes by which heat is transmitted to the water in a boiler. **boiling,** *a.* in a state of ebullition by heat; inflamed, greatly agitated. *n.* the action of boiling. **boiling-point,** *n.* the temperature at which a fluid is converted into the gaseous state; esp. the boiling-point of water at sea-level (212°F, 100°C).
boil[2] (boil), *n.* a hard, inflamed, suppurating tumour.
boisterous (boi'stərəs), *a.* wild, unruly, intractable; stormy, roaring, noisy; tumultuous, rudely violent. **boisterously,** *adv.* **boisterousness,** *n.*
bolas (bō'ləs), *n.* a missile, used by the S American Indians, formed of balls or stones strung together and flung round the legs of the animal aimed at.
bold (bōld), *a.* courageous, daring, planned or executed with courage; vigorous, striking; audacious, forward, presumptuous; steep, prominent, projecting (of a cliff or headland). **bold as brass,** wholly impudent or audacious. **to make or be so bold,** to venture, to presume. **bold-face,** *a.* of type, heavy, conspicuous. **boldly,** *adv.* **boldness,** *n.*
bole (bōl), *n.* the stem or trunk of a tree.
bolero (bələro'rō), *n.* (the music for) a lively Spanish dance; (bol'ərō), a short jacket worn over a bodice.
Boletus (bəlē'təs), *n.* a genus of fungi having the under surface of the pileus full of pores instead of gills. **boletic**

(-let'-), a.
bolide (bō'lid), n. a large meteor; usually one that explodes and falls in the form of aerolites.
bolivar (bol'ivah), n. (pl. **-vars, -vares**) the standard unit of currency in Venezuela.
boll (bōl), n. a rounded seed-vessel or pod. **boll-weevil**, n. a weevil (*Anthonomus grandis*) that infests the flowers and bolls of the cotton plant.
bollard (bol'əd, -ahd), n. (*Naut.*) a large post on a wharf, dock or on ship-board for securing ropes or cables; a short post preventing motor-vehicle access.
bollocks (bol'əks), n.pl. (sl.) testicles; (often int.) rubbish, nonsense, a mess. v.i. to make a mess of. **bollocking**, n. (sl.) a strong rebuke.
bolometer (bəlom'itə), n. an extremely sensitive instrument for measuring radiant heat. **bolometric** (boləmet'-), a.
boloney (bəlō'ni), BALONEY.
Bolshevik (bol'shəvik), **-vist**, n. a member of the Russian majority Socialist party which came to power under Lenin in 1917; a revolutionary; (*often offensive*) a political troublemaker. a. **Bolshevism**, n. **bolshie, -shy** (-shi), n. (*coll.*) a Bolshevik; (*often offensive*) a political agitator. a. (sl.) stubborn and argumentative.
bolster (bōl'stə), n. a long under-pillow, used to support the pillows in a bed; a pad, cushion or anything resembling a pad or cushion, in an instrument, machine, ship, architecture or engineering; a punching-tool. v.t. (also **bolster up**) to support with or as with a bolster; to strengthen, to reinforce; to prevent from falling or collapsing. **bolstering**, n.
bolt[1] (bōlt), n. a short thick arrow with a blunt or thick head; a discharge of lightning; the act of gulping food without chewing; a measured roll of woven fabric, esp. canvas; a sliding piece of iron for fastening a door, window etc.; a metal pin for holding objects together, frequently screw-headed at one end to receive a nut; that portion of a lock which engages with the keeper to form a fastening; a sudden start, a sudden flight or desertion; (*N Am.*) sudden desertion from a political party. v.t. to shut or fasten by means of a bolt; to fasten together with a bolt or bolts; to gulp, to swallow hastily and without chewing; (*N Am.*) to desert (a political party). v.i. to start suddenly forward or aside; to run away (as a horse); (*N Am.*) to break away from a political party. **a bolt from the blue**, lightning from a cloudless sky, an unexpected sudden event. **bolt-head**, n. the head of a bolt; a globular flask with a long, cylindrical neck, used in distilling. **bolt-hole**, n. a hole by which or into which one escapes; a means of escape. **bolt upright**, a. straight upright. **bolter**,[1] n. one that bolts or runs; a horse given to bolting; (*N Am.*) one who suddenly breaks away from his party; (*Austral. Hist.*) a runaway convict. **bolting**, n.
bolt[2], **boult** (bōlt), n. a sieve for separating bran from flour. v.t. to pass through a bolt or bolting cloth; to examine, to try. **bolter**, n. a sieve; a bolting-cloth; a sifting-machine. **bolting**, n.
bolus (bō'ləs), n. medicine in a round mass larger than a pill; a round lump of anything, esp. unswallowed food.
bomb (bom), n. an explosive device triggered by impact or a timer usu. dropped from the air, thrown or placed by hand; (*coll.*) a great success; (*coll.*) a large amount of money; (*coll.*) of a play etc., utter failure, a flop. v.t. to attack, destroy or harm with bombs. v.i. to throw, drop or detonate bombs; (*coll.*) to fail utterly, to flop. **the bomb**, the atom or hydrogen bomb; nuclear arms. **volcanic bomb**, a roundish solid mass of lava ejected from a volcano. **bomb crater**, n. crater caused by the explosion of a bomb. **bomb disposal**, n. the detonation or diffusing of an unexploded bomb rendering it harmless. **bomb-proof**, a. affording safety from the explosion of a bomb. **bomb-shell**, n. a bomb thrown by artillery; a total (often unpleasant) surprise. **bomb-sight**, n. device for aiming a bomb from an aircraft. **bomb site**, n. a piece of ground on which the buildings have been destroyed by bombs. **bomber**, n. one who throws, drops, places or triggers bombs; an aircraft used for bombing. **bomber jacket**, n. a waist-length jacket elasticated at the wrists and waist.
bombard (bəmbahd', bom-), v.t. to attack with artillery fire; to assail with arguments or invective; to subject atoms to a stream of high-speed particles. n. (bom'-), the earliest form of cannon. **bombardier** (bombədiə'), n. in the British army, a non-commissioned artillery officer ranking as corporal. **bombardier-beetle**, n. the genus *Brachinus*, which, when disturbed, emits fluid from the abdomen, with blue vapour and a perceptible report. **bombardment**, n. an attack by artillery. **bombardon, bombardone** (bom'bədən, -bah'-), n. a brass instrument related to the tuba; a bass-reed stop on the organ.
bombasine, bombazine (bom'bəzēn, -zēn'), n. a twilled dress fabric of silk and worsted cotton and worsted or of worsted alone.
bombast (bom'bast), n. padding; padding, stuffing; inflated speech, fustian; high-sounding words. **bombastic** (-bas'-), a. inflated, turgid; given to inflated language. **bombastically**, a.
Bombay duck (bombā), n. a small E Indian fish, *Harpodon nehereus*, when salted and dried eaten as a relish; called also bummalo.
bomber BOMB.
bombe (bom, bō), n. an ice-cream dessert moulded into a rounded, bomb shape.
bombé (bom'bā, bō'-), a. protruding or round-fronted, as of furniture.
Bombyx (bom'biks), n. a genus of moths, containing the silk-worm, *Bombyx mori*. **bombycid** (-sid), a.
bona fide (bō'nə fi'di), adv. in good faith. a. genuine.
bona fides (-dēz), n. good faith, sincerity. [L]
bonanza (bənan'zə), n. a rich mine; a successful enterprise; a run of luck. a. very successful; highly profitable.
Bonapartism (bō'nəpahtizm), n. attachment to the dynasty founded in France by Napoleon Bonaparte. **Bonapartist**, n., a.
bonbon (bon'bon, bō'bō), n. a sweet esp. of fondant; a Christmas cracker.
bonce (bons), n. a large playing-marble; ((sl.)) the head.
bond[1] (bond), n. a thing which binds or confines, as a cord or band; (pl.) chains, imprisonment; a thing which joins or unites; a close mutual attachment; a binding agreement or engagement; a mode of overlapping bricks in a wall so as to tie the courses together (as with English bond and Flemish bond); (*Law*) a deed by which one person (the obligor) binds himself to pay a certain sum to another person (the obligee); a document by which a government or a public company undertakes to repay borrowed money. v.t. to put into a bonded warehouse; to mortgage; to bind or connect (as bricks or stones) by overlapping or by clamps; to cause to adhere tightly. v.i. to adhere tightly; to become emotionally attached. **in bond**, in a bonded warehouse and liable to customs duty. **bond-holder**, n. a person holding a bond or bonds granted by a private person or by a government. **bond paper**, n. a good quality paper. **bonded**, a. bound by a bond; put in bond; secured by bonds; composed of several layers held together by adhesive. **bonded goods**, n.pl. goods stored, under the care of customs officers, in warehouses until the duties are paid. **bonded warehouse**, n. see BONDED GOODS. **bonder**, n. one who puts or holds goods in bond; a stone or brick reaching a considerable distance through a wall so as to bind it together. **bonding**, n. storing goods in bond; strengthening by bonders; the adherence of two surfaces glued together; any union or attachment, esp. the emotional one formed between a parent and his/her new-born child.
bond[2] (bond), a. in serfdom or slavery. **bond-maid**, n. a

slave-girl. **bond-servant**, *n.* a slave. **bond-service**, *n.* villainage. **bond-slave**, *n.* an emphatic term for a slave. **bondsman, bondman**, *n.* a slave; a surety. **bondswoman, bondwoman**, *n. fem.* a female slave. **bondage** (-dij), *n.* slavery, captivity; subjection, restraint, obligation.
bone (bōn), *n.* the hard material of the skeleton; any separate and distinct part of such a skeleton; the substance of which the skeleton consists; a stiffening material for garments; (*pl.*) dice; a domino; (*pl.*) two pieces of bone held between the fingers of each hand, and used as a musical accompaniment; the performer on these; the body; mortal remains. *a.* of or pertaining to bone; made of bone. *v.t.* to take out the bones of (for cooking); (*sl.*) to steal; to stiffen a garment. **a bone of contention**, a subject of dispute. **close to** or **near the bone**, improper, risqué. **the bare bones**, the essentials. **to bone up**, (*sl.*) to study hard, to swot. **to have a bone to pick with someone**, to have a cause of quarrel with or complaint against someone. **to make no bones**, to act or speak without hesitation or scruple; to present no difficulty or opposition. **to point a bone**, (*Austral.*) in aboriginal magic, to will the death of an enemy; to put a jinx on someone. **to the bone**, to the inmost part; drastically, leaving a bare minimum. **bone-ash**, *n.* the mineral residue of bones burnt in the air. **bone china**, *a.* porcelain made with china clay (kaolin) and bone-ash (calcium phosphate). **bone-dry**, *a.* quite dry. **bone-dust**, *n.* bones ground for manure. **bone-grafting**, *n.* introduction of a piece of bone obtained elsewhere to replace bone lost by injury or disease. **bonehead**, *n.* (*sl.*) a dolt. **bone idle**, extremely lazy. **bone-lace**, *n.* a kind of thread-lace originally made with bone bobbins. **bone meal**, *n.* bone-dust used as animal feed or fertilizer. **bone-setter**, *n.* a non-qualified practitioner who sets fractured and dislocated bones. **bone-shaker**, *n.* an old-fashioned bicycle without india-rubber tyres; any dilapidated or old-fashioned vehicle. **boned**, *a.* possessed of bones (*in comb.*); deprived of bones (for cooking). **big-boned**, *a.* of large and massive build. **boneless**, *a.* without bones; without backbone, having no stamina. **bonelessness**, *n.* **boner**, *n.* (*N Am.*) gross mistake, a howler. **boning**, *n.* removing bones from poultry, fish etc. **bony**, *a.* of, pertaining to or of the nature of bone or bones; big-boned; skinny.
bonfire (bon'fīə), *n.* a large fire lit in the open air on occasion of some public rejoicing; a fire for burning up garden rubbish.
bongo (bong'gō), *n.* (*pl.* -**gos**, -**goes**), **bongo drum**, (*pl.* **drums**) a small Cuban hand drum often played in pairs.
bonhom(m)ie (bonəmē'), *n.* good-nature, geniality. [F *bon*, good, *homme*, man]
boning, etc. BONE.
bonito (bonē'tō), *n.* the striped tunny, *Thynnus pelamys*; some other species of the mackerel family.
bonjour (bōzhuə'), good day. [F]
bonk (bongk), *v.t.* (*coll.*) to hit; (*sl.*) to have sexual intercourse with. *v.i.* to have sexual intercourse. *n.* an act of sexual intercourse. **bonking**, *n.*
bonkers (bong'kəz), *a.* crazy.
bon mot (bō mō'), *n.* (*pl.* **bon mots**) a witticism. [F, lit. good word]
bonnet (bon'it), *n.* a head-covering tied beneath the chin, of various shapes and materials, formerly worn by women out of doors and now usu. by babies; (*esp. Sc.*) a flat cap; a chimney-cowl; the front part of a motor-car covering the engine; an additional piece of canvas laced to the bottom of a sail to enlarge it. **Scotch** or **Lowland bonnet**, a round, flat, woollen cap, like a beret, with a tassel in the middle. **bonneted**, *a.*
bonny (bon'i), *a.* beautiful, handsome, pretty; healthy-looking. **bonnily**, *adv.* **bonniness**, *n.*
bonsai (bon'sī), *n.* (*pl.* -**sai**) a potted tree or shrub cultivated into a dwarf variety by skilful pruning of its roots; the art of cultivating trees and shrubs in this manner. [Jap. *bon*, bowl, *sai*, to grow]
bonsella (bonsel'ə), *n.* (*S Afr.*) a tip, a present.
bon ton (bōtō'), *n.* fashion, good style. [F]
bonus (bō'nəs), *n.* something over and above what is due e.g. on top of a fixed salary; a premium given in addition to interest for a loan; an extra dividend; a distribution of profits to policy-holders in an insurance company. **bonus share, bonus issue**, *n.* a share or number of shares issued free to the holder of a paid-up share in a joint-stock company.
bon vivant (bō vēvä'), (*fem.* **bonne vivante** (bon vēvät'), *n.* one fond of good living. [F]
bon voyage (bon voiahzh, bō vwayahzh'), a pleasant journey, farewell. [F]
bony, etc. BONE.
bonza, bonzer (bon'zə), *a.* (*Austral. sl.*) excellent.
bonze (bonz), *n.* a Buddhist priest in Japan, China and adjacent regions.
boo, booh (boo), *int.* and *n.* a sound used as an expression of contempt, aversion and the like. *v.i., v.t.* to show one's displeasure (at) by shouting 'boo'. **boo-hoo** (hoo'), *n.* the sound of noisy weeping. *v.i.* to weep noisily.
boob[1] (boob), **boo boo** (boo'boo), *n.* (*coll.*) an error, a blunder. *v.t.* to err, commit a blunder.
boob[2] (boob), *n.* (*usu. in pl., sl.*) a woman's breast. **boob tube**, *n.* (*sl.*) a woman's elasticated, strapless top.
booby (boo'bi), *n.* a dull, stupid person; a dunce; a gannet, esp. *Sula fusca*. **booby hatch**, *n.* (*N Am., coll.*) madhouse. **booby-prize**, *n.* the prize, usu. a worthless one, given to the player who makes the lowest score, esp. in whist-drives. **booby-trap**, *n.* a practical joke arranged to catch an unsuspecting victim; a bomb made to explode when some object is touched. *v.t.* to set a booby-trap in or on (some object). **boobyish**, *a.*
boodle (boo'dl), *n.* (*sl.*) money, esp. when acquired dishonestly.
boogie-woogie (boogiwoo'gi), (*contr.*) **boogie**, *n.* a jazz piano style of a rhythmic and percussive nature based on 12-bar blues.
book (buk), *n.* a collection of sheets printed, written on or blank, bound in a volume; a literary composition of considerable extent; one of its principal divisions; a libretto; a set of tickets, cheques, forms of receipt, stamps or the like, fastened together; (*Turf*) bets on a race or at a meeting taken collectively; anything that can be read or that conveys instruction; (*Cards*) the first six tricks gained by a side at whist etc. (*pl.*) a set of accounts. *v.t.* to enter in a book; to reserve by payment in advance; to hand in or to receive for transmission (as a parcel, goods etc.); (of police) to take name etc. prior to making a charge; (of referee) to take a player's name for possible disciplinary action after a foul etc. **by the book**, with strict observance of regulations, proper forms etc. **in someone's good** or **bad books**, regarded with favour/disfavour by someone. **like a book**, formally, pedantically, as if one were reciting from a book. **The Book, The Book of God**, the Bible. **to be on the books**, to have one's name on the official list. **to bring to book**, to convict, call to account. **to throw the book at**, to charge with every relevant offence; to reprimand most severely. **without book**, from memory; without authority. **bookbinder**, *n.* one who binds books. **bookbindery**, *n.* a place for binding books. **bookbinding**, *n.* **book-case**, *n.* a case with shelves for books. **book-club**, *n.* a business which sells to its members a choice of books at below publishers' prices. **book-cover**, *n.* a pair of boards (usu. cloth- or leather-covered) for binding a book; case for periodicals, music etc. **book-ends**, *n.pl.* props placed at the ends of a row of books to keep them upright. **book-keeper**, *n.* one who keeps the accounts in a merchant's office etc. **book-keeping**, *n.* **book-learned**, *a.* **book-learning**, *n.* learning derived from books; theory,

not practical knowledge or experience. **book-maker,** *n.* one who takes bets, principally in relation to horse races. **book-making,** *n.* **bookman,** *n.* a literary man; a bookseller. **book-mark(er),** *n.* a piece of ribbon, paper, leather etc. put in a book to mark a place. **book-plate,** *n.* a label pasted in a book to show the ownership. **book-post,** *n.* the postal system for conveying books. **book-rest, -stand,** *n.* a support for a book. **bookseller,** *n.* one who sells books. **bookshop,** *n.* a shop where books are sold. **bookstall, -stand,** *n.* a stall or stand at which books and periodicals are sold. **book-store,** *n.* a bookshop. **book token,** *n.* a gift token exchangeable for books. **book value,** *n.* the value of an asset, commodity or enterprise as it is recorded on paper (not always the same as its market value). **bookwork,** *n.* study of text-books, as opposed to practice and experiment. **bookworm,** *n.* any worm or insect which eats holes in books; an avid reader. **bookable,** *a.* **booked,** *a.* registered; entered in a book; *(coll.)* caught, arrested. **booker,** *n.* **bookful,** *n.* all that a book contains. **bookie** (-i), *n.* *(coll.)* short for bookmaker. **booking,** *n.* a reservation. **booking-clerk,** *n.* one who issues tickets or takes bookings. **booking-office,** *n.* an office where tickets are issued or bookings made. **bookish,** *a.* learned, studious; acquainted with books only. **bookishly,** *adv.* **bookishness,** *n.* **booklet** (-lit), *n.* a little book, a pamphlet.
Boolean (boo'liən), *a.* being or pertaining to a logical system using symbols to represent relationships between entities. **Boolean algebra,** *n.* a branch of symbolic logic used in computers.
boom[1] (boom), *n.* a loud, deep, resonant sound; a sudden rapid increase in e.g. demand, prices; a burst of commercial activity and prosperity. *v.i.* to make a loud, deep, resonant sound; to rush with violence; to become very important, prosperous or active. *v.t.* to utter with a booming sound. **boom town,** *n.* a town undergoing rapid expansion or enjoying sudden commercial prosperity. **booming,** *n., a.*
boom[2] (boom), *n.* a long spar to extend the foot of a particular sail; a bar, chain etc. forming an obstruction to the mouth of a harbour; a barrier enclosing floating logs; an inflatable barrier to confine an oil spill; a movable overhead pole carrying a microphone.
boomer (boo'mə), *n.* *(Austral.)* a large kangaroo; *(coll.)* anything of a large size.
boomerang (boo'mərang), *n.* an Aborigine weapon, consisting of a curved flat stick that returns to the thrower; an action, speech or argument that recoils on the person who makes it. *v.i.* to recoil on the doer.
boon (boon), *n.* a favour, a gift; a benefit, a blessing. **boon companion** *n.* one who is convivial or congenial; a close or special friend.
boondocks (boon'doks), *n.pl.* *(N Am.)* remote or uncultivated country; *(sl.)* a provincial area.
boor (booə), *n.* a rude, awkward or insensitive person. **boorish,** *a.* clumsy, insensitive, unmannerly. **boorishly,** *adv.* **boorishness,** *n.*
boost (boost), *vt.,* *vi.* to push or shove upwards; to advertise widely; to promote or encourage; to enlarge or increase, e.g. the voltage in an electric circuit; to elevate or raise, e.g. the pressure of an internal combustion engine. *n.* an upward push; something that promotes, encourages or energizes; an increase or rise. **booster,** *n.* a contrivance for intensifying the strength of an alternating current; an auxiliary motor in a rocket that usu. breaks away when exhausted; any thing or person which boosts; a supplementary vaccination. **booster cushion** or **seat,** *n.* padded support used to raise a child while sitting e.g. in a car or at table.
boot[1] (boot), *n.* a covering for the foot and part of the leg; a heavy sports shoe; an instrument of torture applied to the leg and foot; *(pl.)* a hotel servant who cleans boots, runs errands etc.; a luggage compartment in a motor car; *(sl.)* summary dismissal, e.g. from employment; *(sl.)* a kick; *v.t.* to equip with boots; to kick; to start a computer program running. **the boot is on the other foot, leg,** the situation is reversed. **to bet one's boots,** to be absolutely certain. **to boot out,** *(sl.)* to eject, dismiss, sack. **to get the boot,** to be dismissed. **to put, stick the boot in,** *(sl.)* to kick someone hard esp. when they are already down; to deal with or finish off brutally. **boot-black,** *n.* a person who cleans and polishes shoes. **boot boy,** *n.* a hooligan, a bovver boy. **boot-jack,** *n.* a device for removing boots. **bootlace,** *n.* a string for fastening boots. **bootleg,** *a.* illicit, smuggled, e.g. of alcohol. *n.* an illicit or smuggled commodity. *v.i.* **bootlegger,** *n.* **bootlegging,** *n.* **bootlicker,** *n.* a sycophant. **boot-maker,** *n.* **boot strap,** *n.* a looped strap on a boot-top enabling it to be pulled up. **to pull oneself up by the bootstraps,** to improve one's situation by one's own efforts. **boot-tree,** *n.* a block inserted into a boot to stretch it or keep it in shape. **booted,** *a.* **bootee** (-tē'), *n.* a short boot; a knitted boot for infants. **bootless** (-lis), *a.*
boot[2] (boot), *n.* *(hist.)* profit, advantage. *v.t.* to benefit. **to boot,** into the bargain, besides. **bootless** (-lis), *a.* profitless, unavailing.
Bootes (bōō'tēz), *n.* a northern constellation including Arcturus, situated near the tail of the Great Bear.
booth (boodh), *n.* a stall, tent or other temporary erection at a fair, in a market etc.; a compartment or structure affording some privacy and containing a telephone, a table in a restaurant etc. **polling-booth,** *n.* POLL.
booty (boo'ti), *n.* spoil taken in war; property carried off by thieves; gain, a prize.
booze (booz), *n.* *(coll.)* drink; a drinking bout. *v.i.* to drink to excess, to tipple. **booze-up,** *n.* a drinking session. **boozer,** *n.* a heavy drinker; *(sl.)* a public house. **boozy, boosy,** *a.* drunk, tipsy; addicted to boozing.
bop (bop), *n.* an innovative style of jazz music dating from the 1940s, *v.i.* to dance to pop music. **bopper,** *n.* a fan of bop; any follower of popular music, one who dances to it.
bo-peep BO.
boracic BORAX.
borage (bŭr'ij, bo'-), *n.* a hairy, blue-flowered plant.
borax (baw'raks), *n.* a native salt used as a flux and a solder, and as a detergent. **boracic** (-ras'ik), **boric,** *a.* of, pertaining to or derived from borax or boron. **boracic, boric acid,** *n.* an acid obtained from borax. **borate** (-rāt), *n.* a salt of boric acid.
borazon BORON.
Bordeaux (bawdō'), *n.* wine from the region around Bordeaux in SW France; claret. **Bordeaux mixture,** *n.* a preparation of sulphate of copper and lime for destroying fungi and other garden pests.
bordello (bawdel'ō), *n.* a brothel.
border (baw'də), *n.* brim, edge, margin; boundary line or region; frontier, frontier region, esp. the one straddling the boundary between England and Scotland; an edging designed as an ornament; an edging to a plot or flowerbed. *v.t.* to put a border or edging to; to form a boundary to. *v.i.* to lie on the border; to be contiguous. **Border collie,** *n.* a rough-haired, usu. black and white, British sheepdog. **borderland,** *n.* land near the border between two countries or districts. **border-line,** *n.* a line of demarcation. **border-line case,** *n.* any case which is on the edge of one category and might easily fall into another. **border-plant,** *n.* a decorative plant for flower borders. **Border terrier,** *n.* a type of small rough-haired terrier. **bordered,** *a.* **borderer,** *n.* one who dwells on a border or frontier, esp. on that between England and Scotland. **bordering,** *n.* an ornamental border. **bordering upon,** adjoining; resembling. **borderless,** *a.*
bore[1] (baw), *v.t.* to perforate or make a hole through; to hollow out. *v.i.* to make a hole; to push forward persistently; to thrust the head straight forward (of a

bore horse); to push a horse, boat or other competitor out of the course; to drive a boxing adversary on to the ropes by sheer weight. *n.* a hole made by boring; the diameter of a tube; the cavity of a gun-barrel. **bore-hole,** *n.* a shaft or pit cut by means of a special tool. **borer,** *n.* a person, tool or machine that bores or pierces; a horse that bores; popular name for *Myxine glutinosa*, the glutinous hag or blind fish, the genus *Teredo* or shipworm, the annelid genus *Terebella*, and some insects that bore holes in wood. **boring,** *n.*

bore² (baw), *n.* a tidal wave of great height and velocity, caused by the meeting of two tides or the rush of the tide up a narrowing estuary.

bore³ (baw), *n.* a tedious or tiresome person, or thing. *v.t.* to weary with dullness, monotony, repetition. **boredom** (-dəm), *n.* the condition of being bored. **bored,** *a.* **boring,** *a.* **boringly,** *adv.* [etym. doubtful]

bore⁴ *past* BEAR².

Boreas (bo'rias), *n.* the god of the north wind; (*Poet.*) the north wind. **boreal,** *a.* pertaining to the north or the north wind; northern; living near the north; sub-arctic.

borer BORE ¹.

boring BORE ¹·³.

born (bawn), *p.p.*, *a.* brought into the world; brought forth, produced; having certain characteristics from birth. **born again,** regenerate; having been converted, esp. to evangelical Christianity. **born to,** destined to. **born with a silver spoon in one's mouth,** born in luxury.

borne (bawn), *p.p.* BEAR².

boron (baw'ron), *n.* the element, at. no. 5; chem. symbol B, present in borax and boracic acid. **borazon** (-əzon), *n.* a substance compounded of boron and nitrogen that for industrial use is harder than a diamond. **boride,** *n.* a compound containing boron.

borough (bŭ'rə), *n.* a town or urban district with its own elected council, esp. one which sends a representative to Parliament. **close** or **pocket borough,** formerly a borough owned by a person or persons. **rotten borough,** a borough (before 1832) in which there were few or no voters.

borrow (bo'rō), *v.t.* to obtain and make temporary use of; to obtain under a promise or understanding to return; to adopt, to assume, to derive from other people; to copy, imitate, feign; (*Golf*) to play a ball uphill in order that it may roll back. **borrowed,** *a.* obtained on loan; not genuine; hypocritical.

borsch (bawsh), **-t, bortsch** (bawch), **-t,** *a.* Russian beetroot soup.

borstal (baw'stəl), *n.* a place of detention and corrective training for juvenile offenders in Britain, officially called **youth custody centre. Borstal system,** *n.* a system of treating juvenile offenders by education and technical instruction. [named from the first system of its kind at the institute in *Borstal*, near Rochester in Kent]

bort, boart (bawt), *n.* small fragments split from diamonds in roughly reducing them to shape, used to make diamond powder.

borzoi (baw'zoi), *n.* a Russian wolfhound.

boscage, boskage (bos'kij), *n.* wood, woodland; underwood or ground covered with it; thick foliage; wooded landscape.

bosh (bosh), *n.* empty talk, nonsense. *int.* rubbish!

bosk (bosk), *n.* a thicket, a small forest. **bosky,** *a.* bushy, woody; covered with boscage; **boskiness,** *n.*

bosket, bosquet (bos'kit), *n.* a grove; a plantation of small trees and underwood in a garden or park.

bos'n (bō'sn) BOATSWAIN.

bosom (buz'm), *n.* the breast of a human being esp. of a woman; that part of the dress which covers this; the breast as the seat of emotions or the repository of secrets; secret counsel or intention; the interior of anything. **bosom of one's family,** midst of one's family. **in one's bosom,** clasped in one's embrace; in one's inmost feelings. **bosom friend,** dearest and most intimate friend.

boson (bō'son, -zon), *n.* a particle, with an integral or zero spin, which behaves in accordance with the statistical relations laid down by Bose and Einstein.

boss¹ (bos), *n.* a protuberant part; an ornamental stud; the knob in the centre of a shield; (*Arch.*) an ornamental projection at the intersection of the ribs in vaulting. **bossed,** *a.* **bossy,** *a.* having a boss or bosses, studded with bosses.

boss² (bos), *n.* a foreman, manager; a chief, leader or master; the manager or dictator of a party machine. *a.* chief, best; first-rate, excellent. *v.t.* to manage, to direct, to control. **bossy,** *a.* managing; domineering.

boss³ (bos), *n.* a miss, a bad shot. **boss-eyed,** *a.* (*coll.*) having only one eye; having one eye injured; squinting. **boss shot,** *n.* (*coll.*) a clumsy or unsuccessful attempt at something. **bosser,** *n.*

bossa nova (bos'ə nō'və), *n.* a Brazilian dance resembling the samba; the music for such a dance.

bot, bott (bot), *n.* a parasitic worm, the larva of the genus *Oestrus*. **the bots, botts,** a disease caused by these in horses; an analogous disease in cattle and in sheep. **botfly,** *n.* a fly of the genus *Oestrus*; a gadfly.

bot., (*abbr.*) botany, botanical, botanist; bought; bottle.

botany (bot'əni), *n.* the science which deals with plants and plant-life. **Botany Bay** (in New South Wales, named by Capt. Cook after the abundance of botanical specimens found there), *n.* a convict settlement established there; (*fig.*) transportation. **botanic, botanical** (-tan'-), *a.* of or pertaining to botany. **botanic garden,** *n.* (*often in pl.*) a garden laid out for the scientific culture and study of plants. **botanically,** *adv.* **botanist,** *n.* **botanize, -ise,** *v.i.* to collect plants for scientific study; to study plants. *v.t.* to explore botanically.

botch¹ (boch), *n.* a clumsy patch; a bungled piece of work. *v.t.* to mend or patch clumsily; to put together in an unsuitable or unskilful manner; to ruin, to bungle. **botcher,** *n.* **botchery,** *n.* the results of botching; clumsy workmanship. **botchy,** *a.* characterized by botching or bungling.

botch² (boch), *n.* an ulcerous swelling. **botchy,** *a.* marked with botches or excrescences.

botel (bōtel'), BOATEL.

both (bōth), *a.*, *pron.* the one and also the other, the two. *adv.* as well the one thing as the other; equally in the two cases.

bother (bodh'ə), *v.t.* to tease, to vex; to annoy, to pester. *v.i.* to make a fuss, to be troublesome; to worry oneself; to take trouble. *int.* an exclamation of annoyance. *n.* worry, disturbance, fuss. **botheration,** *n.* the act of bothering; bother. **bothersome** (-səm), *a.* troublesome, annoying.

bothy, bothie (both'i), *n.* (*esp. Sc.*) a rough cottage; a hut, a hovel; esp. a lodging place for unmarried labourers on a Scottish farm.

bo-tree (bō'trē), **bodhi tree** (bō'di), *n.* the peepul or peepla tree; the tree, *Ficus religiosa*, sacred to Buddhists, under which Gautama is said to have received the enlightenment which constituted him the Buddha.

botryoid (bot'rioid), **botryoidal** (-oi'-), *a.* in form resembling a bunch of grapes.

bots, bott BOT.

bottine (botēn'), *n.* a light kind of boot for women and children.

bottle¹ (bot'l), *n.* a vessel with a narrow neck for holding liquids (usu. of glass); the quantity in a bottle. (*sl.*) temerity, courage, strength of will. *v.t.* to put into bottles. **the bottle,** drinking. **to hit the bottle,** (*sl.*) to drink a great deal of alcoholic drinks. **to bottle-feed,** to feed a baby from a bottle instead of the breast. **to bottle up,** to conceal; to restrain, repress (one's emotions). **bottle bank,** *n.* a public repository for empty glass jars and bottles which

bottle   93   **boutique**

are to be recycled. **bottle-brush**, *n.* a brush for cleaning bottles; a plant or tree bearing brush-like flowers. **bottle gas**, *n.* butane gas in liquid form supplied in containers for use in heating, cooking etc. **bottle-glass**, *n.* coarse green glass for making bottles. **bottle-green**, *n.* dark green, like bottle-glass. **bottle-nosed**, *a.* having a large thick nose. **bottle-nosed dolphin**, *n.* a dolphin with a prominent beak, esp. *Tursiops truncatus*. **bottle-party**, *n.* a party to which each person brings his/her own alcoholic drink. **bottle-washer**, *n.* a person or machine that washes bottles; a general factotum, an understrapper. **bottled**, *a.* stored in jars or bottles; (*sl.*) drunk. **bottler**, *n.*
bottle² (bot'l), *n.* a bundle of hay or straw. [OF *botel*, dim. of *botte*]
bottom (bot'əm), *n.* the lowest part or point of anything, the part on which anything rests; the buttocks; the seat of a chair; the bed or channel of any body of water; an alluvial hollow; low-lying land; a deep cavity, an abyss; the inmost part, the furthest point of a recess, gulf or inland sea; the end of a table remote from a host, chairman etc.; the lowest rank; the keel of a ship and the underwater part of the keel, the hull; a ship as receptacle for cargo; foundation, basis; stamina, power of endurance. *a.* of or pertaining to the bottom; lowest; fundamental. **at bottom**, in reality; at heart. **to bottom out**, to drop to, and level out at, the lowest point, as of prices. **to get to the bottom of**, to fathom out, to find the cause of. **at the bottom of**, the cause or origin of. **bottom dollar**, *n.* one's last coin. **bottom drawer**, *n.* a drawer in which a woman keeps her new clothes etc. before marriage. **bottom-lands**, *n.pl.* (*N Am.*) rich flat lands on the banks of rivers in the western states. **bottom line**, *n.* the concluding line in a statement of accounts, giving net profit or loss figures; the final word on; the crux of a matter. **bottom-up**, *a.*, *adv.* upside-down. **bottomed**, *a.* (*usu. in comb.*) having a bottom, as *flat-bottomed*. **bottomless**, *a.* without a bottom; having no seat; fathomless, unfathomable. **bottomless pit**, *n.* hell; (*coll.*) a very hungry or greedy person. **bottommost**, *a.* lowest of all.
botulism (bot'ūlizm), *n.* a form of food-poisoning caused by eating preserved food infected by *Bacillus botulinus*.
bouclé (boo'klā), *n.* a looped yarn; the thick, curly material woven from such yarn.
boudoir (boo'dwah), *n.* a lady's small, elegantly furnished private dressing or sitting room.
bouffant (boo'fõ), *a.* full, puffed out, as a hairstyle.
bouffe OPERA BOUFFE.
bougainvillaea, -vilia (boogən vil'iə), *n.* a genus of tropical plants belonging to the Nyctaginaceae, the red or purple bracts of which almost conceal the flowers.
bough (bow), *n.* a large arm or branch of a tree.
bought (bawt), *p.p.* BUY.
bougie (boo'zhē), *n.* a smooth, flexible, slender cylinder used for exploring or dilating passages in the human body.
bouillabaisse (booyəbes'), *n.* a rich fish stew or chowder, popular in the south of France.
bouillon (boo'yõ), *n.* broth, soup.
boulder (bōl'də), *n.* a water-worn, rounded stone; a large rounded block of stone transported some distance from its parent rock; a large detached piece of ore. **boulder-clay, -drift**, *n.* a clayey deposit of the glacial period.
boules (bool), *n.pl.* a French game resembling bowls, played with metal balls.
boulevard (boo'ləvahd), *n.* a broad street planted with trees; (*esp. N Am.*) an arterial road, trunk road. **boulevardier** (boolvah'dyä), **-dist**, *n.* a (Parisian) man-about-town.
bounce (bowns), *v.i.* to rebound; to bound like a ball; to move suddenly, excitedly or unceremoniously; (of a cheque) to be returned to drawer. *v.t.* to throw or drop (e.g. a ball) so that it rebounds; to cause to move up and down springily; to return (a cheque) to drawer; (*N Am.*) to discharge suddenly from employment; (*N Am.*) to throw or turn out. *n.* rebound; a leap, a spring; swagger, self-assertion; impudence; (*N Am.*) dismissal from employment. **bouncer**, *n.* someone employed to eject (undesirable) people from a public place. **bouncing**, *a.* big, heavy; stout, strong; bustling, noisy. **bouncing chair, cradle**, *n.* seat or cradle which enables a baby to rock itself lightly up and down. **bouncingly**, *adv.* **bouncy**, *a.* vivacious, bouncing.
bound¹ (bownd), *n.* a leap, a spring, a rebound. *v.i.* to leap, to spring; to rebound, to bounce. **by leaps and bounds**, with astonishing speed. **bounder**, *n.* (*dated*) a scoundrel.
bound² (bownd), *n.* a limit, a boundary; limitation, restriction. *v.t.* to set bounds to; to confine; to form the boundary of. **out of bounds**, of an area, topic or person, forbidden, prohibited. **boundary**, *n.* (a mark indicating) a limit; (*Cricket*) (a stroke that sends the ball beyond) the limit of the playing area. **boundary-rider**, *n.* (*Austral.*) a man who keeps the boundary fences of a station in repair. **bounded**, *a.* **boundless**, *a.* limitless. **boundlessly**, *adv.* **boundlessness**, *n.*
bound³ (bownd), *a.* under obligation; compelled, obliged, certain (*with inf.*); in a cover, esp. of leather or other permanent material as distinguished from paper. **bound up with**, intimately associated with; having identical aims or interests with. **bounden**, *a.* bound; obliged. **bounden duty**, *n.* obligatory duty.
bound⁴ (bownd), *a.* prepared, ready; starting, destined; directing one's course. **homeward bound**, on the way home.
bound⁵ (bownd), *past*, *p.p.* BIND.
bounty (bown'ti), *n.* an act of generosity, a gift; a financial inducement or reward, offered usu. by the government, for e.g. killing animals regarded as vermin or (in the US) capturing a criminal. **bounty hunter**, *n.* one who hunts vermin or tracks down criminals for the bounty. **bounteous** (-tiəs), *a.* liberal, beneficent; generously given. **bounteously**, *adv.* **bounteousness**, *n.* **bountiful**, *a.* liberal, munificent; plenteous, abundant. **Lady Bountiful**, a wealthy woman charitable in her neighbourhood. **bountifully**, *adv.*
bouquet (bukā', bō-), *n.* a bunch of flowers; the perfume exhaled by wine. **bouquet garni** (boo'kā gah'nē), *n.* a bunch (trad. five sprigs) of herbs for flavouring meat dishes and soups.
bourbon (bœ'bən), *n.* (*N Am.*), a kind of whisky made of wheat or Indian corn. **Bourbon biscuit** (buə'-), *n.* one consisting of two chocolate-flavoured pieces with chocolate cream between.
bourgeois (buə'zhwah), *n.* (*sometimes derog.*) one of the mercantile, shop-keeping or middle class; a small-minded, unintellectual or materialistic person. *a.* of or pertaining to the bourgeoisie; middle-class as distinguished from working-class; humdrum, unintellectual; materialistic, middle-class in outlook. **bourgeoisie** (-zē'), *n.* (*sometimes derog.*) the mercantile or shop-keeping class; the middle class as opposed to the proletariat.
bourgeon BURGEON.
bourguignon (buə'gēnyõ), *a.* of meat dishes, stewed with (Burgundy) wine.
bourn¹ (bawn), *n.* a small stream.
bourne, bourn² (bon), *n.* a limit, a goal.
bourree (boo'rä), *n.* a folk-dance from the Auvergne and Basque provinces; a musical composition in this rhythm.
bourse (buəs), *n.* a stock exchange esp. that of Paris.
bout (bowt), *n.* a turn, a round, a set-to; trial, attempt; a spell of work; a fit of drunkenness or of illness.
boutique (bootēk'), *n.* a fashionable clothes shop; any small specialist shop; a shop within a department store, hotel, airport lounge etc.

boutonnière (booto̅nyeə'), *n.* a flower for the button hole.
bouzouki (boozoo'ki), *n.* a Greek stringed instrument similar to the mandolin.
bovine (bō'vīn), *a.* of or resembling oxen; sluggish; dull, stupid.
Bovrilʳ (bov'ril), *n.* a concentrated beef extract used for flavouring stews etc.
bovver (bov'ə), *n.* (*sl.*) a boisterous or violent commotion, a street fight. **bovver boots,** *n.pl.* (*sl.*) heavy workboots worn as weapons by teenage thugs. **bovver boy,** *n.* (*sl.*) a member of a violent teenage gang; a hooligan.
bow[1] (bō), *n.* a curve, a rainbow; a stringed weapon for discharging arrows; the doubling of a string in a slip-knot; a single-looped bow; an ornamental knot in which neckties, ribbons etc. are tied; a necktie, ribbon or the like, tied in such a knot; a thing shaped like a bow; the appliance with which instruments of the violin family are played; a single stroke of such an appliance. *v.t.* to play with or use the bow on (a violin etc.). **to draw the long bow,** to exaggerate; to tell lies. **to have two strings to one's bow,** to have more resources, plans or opportunities than one. **bow and string beam, bridge** or **girder,** a structure in the form of a bent bow, with a horizontal beam or girder in the position of the string. **bow-compasses,** *n.pl.* compasses with the legs jointed, so that the points can be turned inwards. **bow-fronted,** *a.* having a convex front. **bow-hand,** *n.* the hand that holds the bow in archery or in playing a stringed instrument. **bow-head,** *n.* the Greenland right whale. **bow-legged,** *a.* having the legs bowed or bent. **bowman,** *n.* one who shoots with the bow, an archer. **bow-saw,** *n.* a saw fitted in a frame like a bowstring in a bow. **bowshot,** *n.* the distance to which an arrow can be shot. **bowstring,** *n.* the string by which a bow is stretched. **bow tie,** *n.* short necktie tied in a bow; piece of pasta in this shape. **bow-window,** *n.* a bay-window segmentally curved.
bow[2] (bow), *v.i.* to bend forward as a sign of assent, submission or salutation; to incline the head; to submit, to yield (to). *v.t.* to cause to bend; to crush; to express by bowing; to usher (in or out). *n.* an inclination of the body or head, as a salute or token of respect. **to bow out,** to withdraw gracefully. **to take a bow,** to acknowledge praise, applause. **bowed,** *a.* bent, crooked; bent down.
bow[3] (bow), *n.* (*often in pl.*) the rounded fore-end of a ship or boat; the rower nearest this. **on the bow,** (*Naut.*) within 45° of the point right ahead. **bowline** (bō'lin, -lin), *n.* a rope fastened to the middle part of the weather side of a sail to make it stand close to the wind. **bowline knot,** *n.* a safe kind of knot. **on a bowline,** close-hauled; sailing close to the wind. **bow-oar,** *n.* the rower nearest the bow; his oar. **bowsprit** (bō'sprit), *n.* (*Naut.*) a spar running out from the bows of a vessel to support sails and stays. **bower,** *n.* either of two anchors (**best bower** and **small bower**) carried in the bows; the cable attached to either. **Bow bells** (bō), *n.pl.* the bells of St. Mary le Bow, Cheapside. **to be born within the sound of Bow bells,** to be born in the City of London; to be a true Cockney.
bowdlerize, -ise (bowd'ləriz), *v.t.* to expurgate (a book). **bowdlerism, bowdlerization, -isation,** *n.* the act or practice of expurgating.
bowel (bow'əl), *n.* one of the intestines; (*pl.*) the entrails, the intestines; (*pl.*) the seat of pity, compassion; the interior, the centre. **bowel movement,** *n.* (a discharge of) faeces.
bower[1] (bow'ə), *n.* (*poet.*) an arbour, a shady retreat, a summer-house; the run of a bower-bird. **bower-bird,** *n.* the name given to several Australian birds of the starling family, which build bowers or runs, adorning them with feathers, shells etc. **bowery,** *a.* **The Bowery,** *n.* a district in New York formerly notorious for political graft, now for its numerous bars, shops and cheap hotels.
bower[2] BOW[3].

bowie-knife (bō'i. *Amer. usu.* boo'i), *n.* a long knife, double-edged towards the point, used as a weapon in the south and south-west of US.
bowl[1] (bōl), *n.* a hollow (usually hemispherical) vessel for holding liquids; a basin; the contents of such a vessel; a drinking-vessel; a basin-shaped part or concavity.
bowl[2] (bōl), *n.* a solid ball, generally made of wood and slightly biased or one sided; (*pl.*) a game with bowls; a large, heavy ball with finger-holes used in ten-pin bowling; a spell of bowling in cricket. *v.i.* to play at bowls; to roll a bowl along the ground; to deliver the ball at cricket; to move rapidly and smoothly (usu. with *along*). *v.t.* to cause to roll or run along the ground; to deliver (as a ball at cricket); (also **bowl out**) to strike the wicket and put a man out. **to bowl over,** to knock over; to throw into a helpless condition. **bowler**[1], *n.* one who plays at bowls; the player who delivers the ball at cricket. **bowling,** *n.* playing at bowls, skittles or ten-pin bowling; delivering the ball in cricket. **bowling-alley,** *n.* a covered space for playing skittles or ten-pin bowls. **bowling-crease,** *n.* the line from behind which the bowler delivers the ball at cricket. **bowling-green,** *n.* a level green on which bowls are played.
bowler[2], **bowler hat** (bō'lə), *n.* an almost-hemispherical stiff felt hat.
bowline BOW[3].
bowman[1] BOW[1].
bowsprit BOW[3].
Bow-street (bō'-), *n.* a street in London where the principal police-court is situated. **Bow-street runner,** *n.* old name for a detective police officer.
bow-window BOW[1].
bow-wow (bow wow'), *int.,* *n.* (an exclamation imitating) the bark of a dog. (*childish*) (bow'-), a dog.
box[1] (boks), *n.* a genus of small evergreen shrubs, *Buxus,* esp. the common box-tree; box-wood. **box-tree,** *n.* the common box, *Buxus sempervirens.* **box-wood,** *n.*
box[2] (boks), *n.* a case or receptacle usually with a lid and rectangular or cylindrical, adapted for holding solids, not liquids; the contents of such a case; a compartment partitioned off in a theatre, court-room, or for animals in a stable, railway-truck etc.; the driver's seat on a horse-drawn coach; a hut, a small house; one of the compartments into which a type-case is divided; a post-office box; a case for the protection of some piece of mechanism from injury; a protective pad for the genitals worn by cricketers. *v.t.* to enclose in or furnish with a box. **in the wrong box,** mistaken, out of place. **the box,** television; a television set. **to box in,** to enclose, to restrict the freedom of movement. **to box off,** to partition off. **to box the compass,** to name the points of the compass in proper order; to go right round (in direction, political views etc.) and end at the starting point. **to box up,** to shut in; to squeeze together. **box camera,** *n.* a very simple kind of camera. **box-car,** *n.* (*N Am.*) a goods van. **box girder,** *n.* a rectangular or square hollow girder. **box junction,** *n.* a road junction with a box-shaped area painted with criss-crossed yellow lines into which traffic is prohibited from entering until there is a clear exit. **box-key,** *n.* a T-shaped implement for turning a water cock. **box kite,** *n.* a kite composed of open-ended connected cubes. **box mattress, box spring mattress,** *n.* a mattress consisting of spiral springs contained in a wooden frame. **box number,** *n.* a number in a newspaper office to which replies to advertisements may be sent; the number of a post-office box. **box-office,** *n.* an office in a theatre or concert-hall for booking seats. **box-pleat,** *n.* a double fold or pleat. **box-room,** *n.* a room for storing. **box-spanner,** *n.* a tubular spanner with the ends shaped to fit the nuts. **boxer**[1], *n.* one who puts or packs things up in boxes. **Boxing Day,** *n.* the first week-day after Christmas, when Christmas-boxes are given. **boxful,** *n.*

box³ (boks), *n.* a punch or slap esp. on the ear or side of the head. *v.t.* to strike (on the ear etc.). *v.i.* to fight or spar with fists or with gloves. **boxer²**, *n.* one who boxes; a pugilist; (*Boxer*) a member of a late 19th cent. secret society in China, which took the leading part in the movement for the expulsion of foreigners. a large, smooth-haired mastiff derived from the German bulldog. **boxer shorts**, *n.pl.* men's baggy underpants. **boxing**, *n.* the sport of fist-fighting with gloves. **boxing gloves**, *n.pl.* a pair of protective leather mittens worn by boxers. **Box and Cox** (boks and koks), *n.*, *adv.* taking turns. [From two fictional men thus named who rented the same room by night and day respectively and never met.]

boy (boi), *n.* a male child; a lad, a son; (*offensive*) a native, a native servant or labourer; (*offensive*) any male servant; (*pl.*) grown up sons; (*pl. coll.*) a group of male friends. **oh boy**, (*int.*) an exclamation of surprise, appreciation, delight or derision. **old boy** OLD. **boy friend**, *n.* (*coll.*) a man or boy in whom a girl is especially interested. **Boys' Brigade**, *n.* an organization founded in Britain in 1883 for the training and welfare of boys. **boyhood** (-hud), *n.* the state of being a boy; the time of life at which one is a boy. **Boy Scout**, *n.* former name for a member of the scout movement. [*See* SCOUT]. **boyish**, *a.* characteristic of or suitable to a boy; puerile. **boyishly**, *adv.* **boyishness**, *n.*

boyar (boi'ah, boyah'), **boyard** (boi'əd), *n.* a member of the old Russian nobility.

boycott (boi'kot), *v.t.* to combine to ostracize (a person) or refuse to have dealings with (a company etc.) or refuse to buy (goods). *n.* the action of boycotting. **boycottee** (-tē'), *n.* **boycotter**, *n.* **boycottism**, *n.*

Boyle's law (boilz), *n.* the principle that the pressure of a gas varies inversely with its volume at constant temperature.

boysenberry (boi'znberi), *n.* an edible hybrid fruit related to the loganberry and the raspberry.

BP, (*abbr.*) British Petroleum; British Pharmacopeia.

bp, (*abbr.*) of alcoholic density, below proof; bills payable; bishop; boiling point.

bpi, (*abbr.*) of computer tape, bits per inch.

Bq (*chem.* symbol) becquerel.

BR (*abbr.*) British Rail.

Br (*chem. symbol*) bromine.

Br. (*abbr.*) British; Brother.

bra (brah), short for BRASSIÈRE. **bra-burning**, *a.* exceptionally feministic, from the practice of burning brassieres as a token of women's independence. **braless**, *a.* not wearing a bra.

brace (brās), *n.* something which clasps, tightens, connects or supports; (*pl.*) straps to support the trousers; a sign in writing, printing or music uniting two or more words, lines, staves etc.; a couple, a pair; a timber or scantling to strengthen the framework of a building; a rope attached to a yard for trimming the sail; an appliance for supporting the trunk or a limb; a wire dental appliance for straightening crooked teeth. *v.t.* to bind or tie close; to tighten or make tense; to strengthen, to fill with energy or firmness. **brace and bit**, a tool used by carpenters for boring, consisting of a kind of crank in which a bit or drill is fixed. **to splice the main brace**, (*Naut. sl.*) to serve an extra rum ration. **bracer**, *n.* **bracing**, *a.* invigorating.

bracelet (brās'lit), *n.* an ornamental ring or band for the wrist or arm; (*pl.*) (*sl.*) handcuffs.

brachial (brā'kiəl, brak'-), *a.* of or belonging to the arm; resembling an arm. **brachiate** (-ət), *v.i.* of various arboreal mammals, to move along by swinging from each arm alternately. **brachiation**, *n.*

brachio-, *comb. form.* having arms or arm-like processes.

brachiopod (brak'iəpod), *n.* (*pl.* **-pods, -poda** (-op'ədə), a bivalve mollusc with tentacles on each side of the mouth. **brachiopodous** (-op'-), *a.*

brachiosaurus (brakiəsaw'rəs), *n.* a large herbivorous dinosaur with long front legs.

brachy-, *comb. form.* short.

brachycephalic (brakisəfal'ik), *a.* having a skull in which the breadth is at least four-fifths of the length. **brachycephaly**, **brachycephalism** (-sef'-), *n.*

brack (brak), *n.* a flaw or tear in a cloth or dress.

bracken (brak'n), *n.* a fern, esp. the brake-fern, *Pteris aquilina.*

bracket (brak'it), *n.* a projection with horizontal top fixed to a wall, a shelf with a stay underneath for hanging against a wall; an angular support; a gas pipe projecting from a wall; a mark used in printing to enclose words or mathematical symbols; a particular category or classification within a series, as *tax bracket.* *v.t.* to furnish with a bracket or brackets; to place within brackets; to connect (names of equal merit) in honour list; to associate, categorize or group like things together; (*Artill.*) to find the range of a target by dropping shots alternately short of and over it. **bracketing**, *n.*

brackish (brak'ish), *a.* partly fresh, partly salt; of a saline taste. **brackishness**, *n.*

bract (brakt), *n.* a small modified leaf or scale on the flower-stalk. **bracteal** (-tiəl), *a.* of the nature of a bract. **bracteate** (-tiət), *a.* formed of metal beaten thin; furnished with bracts. **bracteole** (-tiōl), *n.* a small bract. **bracteolate** (-tiəlāt), *a.* furnished with bracteoles. **bractless**, *a.*

brad (brad), *n.* a thin, flattish nail, with a small lip or projection on one side instead of a head.

bradawl (brad'awl), *n.* a small boring-tool.

brady-, *comb. form.* slow.

bradycardia (bradikah'diə), *n.* a slow heartbeat.

brae (brā), *n.* a slope bounding a river valley; a hill.

brag (brag), *v.i.* to boast. *v.t.* to boast; to challenge; to bully. *n.* a boast; boasting; a game of cards. *adv.* proudly, conceitedly. **braggadocio** (bragədō'chiō), *n.* an empty boaster; empty boasting. **braggart** (-ət), *n.* a boastful person. *a.* given to bragging; boastful. **bragging**, *n.*, *a.* **braggingly**, *adv.*

Brahma (brah'mə), *n.* the chief Hindu divinity, the Creator God.

Brahman (brah'mən), *n.* a member of the highest Hindu caste, the priestly order; a breed of Indian cattle. **Brahmanic, -ical** (-min'-), *a.* of or pertaining to Brahmans or to Brahmanism. **Brahmanee** (1) (-nē'), *n.* a female Brahman. (brah'minē), pertaining to the Brahman caste. **Brahmanism** *n.*

Brahmin (brah'min), *n.* Brahman; (*N Am.*) a person of superior intellectual or social status, a highbrow. **Brahminic, -ical**, *a.* **Brahminism**, *n.*

braid (brād), *n.* anything plaited or interwoven; a narrow band; a woven fabric for trimming or binding. *v.t.* to intertwine, to plait; to dress the hair in plaits or bands; to tie the hair with ribbon or bands; to trim or bind with braid. **braided**, *a.* **braiding**, *n.*

brail (brāl), *n.* (*pl.*) ropes used to gather up the foot and leeches of a sail before furling. *v.t.* to haul up by means of the brails.

Braille (brāl), *n.* a system of writing or printing for the blind, by means of combinations of points stamped in relief. **Braille writer**, *n.* an instrument for stamping paper with these.

brain (brān), *n.* the soft, whitish, convoluted mass of nervous substance contained in the skull of vertebrates; any analogous organ in the invertebrates (*sing.* the organ, *pl.* the substance); the seat of intellect, thought etc.; the centre of sensation; intellectual power; (*coll.*) an intelligent person. *v.t.* to dash out the brains of; to kill in this way. **to have something on the brain**, to be obsessed with it. **brain child**, *n.* a plan or project which is the product of creative thought. **brain-coral**, *n.* coral resembling the con-

volutions of the brain. **brain death,** *n.* the cessation of brain function, taken as an indication of death. **brain drain,** *n.* the loss of highly qualified people by emigration. **brain-fever,** *n.* inflammation of the brain; fever with brain complications. **brain-pan,** *n.* the skull. **brain stem,** *n.* the stalk-shaped part of the brain which connects it to the spinal cord. **brain storm,** *n.* a sudden, violent mental disturbance. **brainstorming,** *n.* (*esp. N Am.*) intensive discussion, e.g. to generate ideas. **brains trust,** *n.* a panel of experts answering questions from an audience. **brain teaser,** *n.* a perplexing problem or puzzle. **brain-washing,** *n.* the subjection of a victim to sustained mental pressure, or to indoctrination, in order to extort a confession or to induce him to change his views. **brain-wash,** *v.t.* **brainwave,** *n.* (*coll.*) a (sudden) brilliant idea. **brainless,** *a.* destitute of brain; silly, witless. **brainy,** *a.* having brains; acute, clever. **braininess,** *n.*
braise (brāz), *v.t.* to cook slowly in little liquid in a tightly closed pan.
brake[1] (brāk), *n.* bracken.
brake[2] (brāk), *n.* an appliance to a wheel to check or stop motion; anything which slows down an activity. *v.t.* to retard by means of a brake. **brake-block,** *n.* a block applied to a wheel as a brake. **brakeman,** (*N Am.*) **brakesman,** *n.* a man in charge of a brake, a railway guard. **brake horsepower,** *n.* the measurement of an engine's power calculated from its resistance to a brake. **brake light,** *n.* the red light on the rear of a vehicle which indicates braking. **brake-van,** *n.* a railway carriage containing a brake; a guard's van. **brakeless,** *a.*
brake[3] (brāk), *n.* a mass of brushwood, a thicket.
bramble (bram'bl), *n.* the blackberry, or any allied thorny shrub. **bramble finch, brambling** (-bling), *n.* the mountain finch, *Fringilla montifringilla.* **brambled,** *a.* **brambly,** *a.*
bran (bran), *n.* the husks of ground corn separated from the flour by bolting. **bran-mash,** *n.* bran soaked in water.
branch (brahnch), *n.* a shoot or limb of a tree or shrub, esp. one from a bough; any offshoot, member, part or subdivision of an analogous kind; anything considered as a subdivision or extension of a main trunk, as of a mountain-range, river, road, railway, family, subject, commercial organization etc. *v.i.* to shoot out into branches or subdivisions; to diverge from a main direction; to divide, to ramify. **to branch out,** to broaden one's interests or activities. **branched,** *a.* **brancher,** *n.* **branchless,** *a.* **branchlet** (-lit), *n.* **branchy,** *a.*
branchia (brang'kiə), **-chiae** (-ē), *n.* (*pl.*) a gill of fishes and some amphibia. **branchial,** *a.* pertaining to or of the nature of gills. **branchiate** (-ət), *a.* characterized by gills. **branchiform** (-fawm), *a.* shaped like gills.
branchio-, *comb. form.* pertaining to gills.
branchiopod (brang'kiəpod), *n.* (*pl.* **branchiopoda** (-op'ədə)), an individual of a group of molluscoid animals with gills on the feet. **branchiopodous,** *a.*
brand (brand), *n.* a piece of burning wood; a piece of wood partially burnt; a torch; a mark made by or with a hot iron, an instrument for stamping a mark; a trademark, hence a particular kind of manufactured article; a sword; a stigma; class, quality. *v.t.* to mark with a brand; to imprint on the memory; to stigmatize. **a brand from the burning,** a person rescued or converted from sin or irreligion. **brand name,** *n.* a trade name for the commodities of a particular manufacturer. **brand-new,** *a.* as if just from the furnace, quite new. **branding-iron,** *n.* an iron to brand with. **branded,** *a.* **brander,** *n.*
brandish (bran'dish), *v.t.* to wave or flourish about (as a weapon etc.). *n.* a flourish; waving.
brand-, bran-new BRAND.
brandy (bran'di), *n.* a spirit distilled from wine. *v.t.* to mix with brandy; to furnish or refresh with brandy. **brandyball,** *n.* a kind of sweet. **brandy glass,** *n.* a balloon-shaped glass with a short stem. **brandy-snap,** *n.* thin, crisp,

wafer-like gingerbread, usually scroll-shaped.
brant BRENT[2].
brash[1] (brash), *n.* loose, disintegrated rock or rubble. *a.* (*N Am.*) tender, brittle. **brash-ice,** *n.* broken ice. **brashy,** *a.* crumbly, rubbly.
brash[2] (brash), *n.* a slight indisposition arising from disorder of the alimentary canal. **water-brash,** a belching of water from the stomach; heartburn; a dash of rain.
brash[3] (brash), *a.* impertinent, cheeky; vulgarly assertive or pushy; impudent.
brasier (brā'ziə), BRAZIER[2].
brass (brahs), *n.* a yellow alloy of copper and zinc; anything made of this alloy; a brazen vessel; an engraved sepulchral tablet of this metal; musical wind-instruments of brass; (*also pl.*) the section in an orchestra composed of brass instruments; (*sl.*) money; effrontery, impudence. *a.* made of brass. **the brass,** those in authority; officers, brass hats. **top brass,** those in highest authority; the highest-ranking officers. **brass band,** *n.* a band performing chiefly on brass instruments. **brass farthing,** *n.* (*coll.*) the lowest measure of value. **brass hat,** *n.* (*coll.*) a staff officer. **brass neck,** *n.* (*coll.*) impudence, audacity. **brass plate,** *n.* a plate of brass engraved with name, trade or profession etc. fixed at doors etc. **brass rubbing,** *n.* the transfer of an image from a brass tablet to paper by placing the paper over the original and rubbing it with crayon or chalk; the image copied by this method. **brass tacks,** (*coll.*) details; the essential facts of a matter. **brassy,** *a.* resembling brass; unfeeling, impudent, shameless; debased, cheap, pretentious. *n.* a wooden golf club faced with brass. **brassily,** *adv.* **brassiness,** *n.*
brassard (bras'ahd), *n.* a badge worn on the arm, an armband, armlet.
brasserie (bras'əri), *n.* a (usu. small) restaurant which serves beer as well as wine etc.
brassica (bras'ikə), *n.* any plant belonging to the genus *Brassica* (turnip, cabbage etc.).
brassière (braz'iə, bras'-, *N Am. usu.* brəziə'), *n.* an undergarment for supporting the breasts.
brassy BRASS.
brat (brat), *n.* a child, usu. one who is badly behaved or ragged and dirty. **Brat Pack, brat pack,** *n.* a group of precociously successful young practitioners in any field, esp a group of young American film actors and directors in the 1980s. **Brat Packer, brat packer,** *n.* a member of a Brat Pack.
Bratwurst (brat'vuəst), *n.* a kind of German sausage.
bravado (brəvah'dō), *n.* (*pl.* **-oes**) ostentatious defiance; swaggering behaviour.
brave (brāv), *a.* daring, courageous; gallant, noble; excellent, fine. *n.* a N Am. Indian warrior. *v.t.* to defy, to challenge; to meet with courage.
bravo[1] (brahvō'), *n.* (*pl.* **-oes**), a hired assassin; a bandit, a desperado.
bravo[2] (brahvō'), *int.* (*pl.* **-voes, vos**; *fem.* **-va**; *superl.* **-vissimo** (-vis'imō), **-ma** (-mə),). Well done! *n.* a cry of approval; a cheer.
bravura (brəvüə'rə), *n.* (*Mus.*) brilliance of execution; a display of daring and skill in artistic execution; a brilliant and demanding piece of music displaying or requiring virtuosity. [It., bravery]
braw (braw), (*Sc.*) BRAVE.
brawl (brawl), *v.i.* to quarrel noisily; to babble (as running water); *n.* a noisy quarrel, disturbance, a tumult. **brawler,** *n.* **brawling,** *a.* **brawlingly,** *adv.*
brawn (brawn), *n.* muscle, flesh; a potted meat dish usu. made from pig's head. **brawny,** *a.* muscular, strong, hardy. **brawniness,** *n.*
bray[1] (brā), *v.t.* to pound or grind small, esp. with pestle and mortar.
bray[2] (brā), *v.i.* to make a harsh, discordant noise, like an ass. *v.t.* to utter harshly or loudly (often with *out*). *n.* a

loud cry; the cry of the ass; a harsh, grating sound.
braze[1] (brāz), *v.t.* to solder with an alloy of brass and zinc.
braze[2] (brāz), *v.t.* to cover or ornament with brass; to colour like brass. **brazen** *a.* made of brass; resembling brass; shameless, impudent. *v.t.* to face impudently (often with *out*); to harden, make shameless. **brazen-face,** *n.* an impudent person. **brazen-faced,** *a.* **brazenly,** *adv.* **brazenness, brazenry,** *n.* **brazier**[1] (-ziə), *n.* a worker in brass. **braziery,** *n.*
brazier[2] (brā'ziə), *n.* a large pan to hold lighted charcoal.
brazil (brəzil'), **brazil wood,** *n.* a red dyewood produced by the genus *Caesalpinia*. **brazil-nut,** *n.* the triangular, edible seed of *Bertholletia excelsa*.
BRCS, (*abbr.*) British Red Cross Society.
breach (brēch), *n.* the act of breaking; a break, a gap, esp. one made by guns in a fortification; violation, by act or by omission, of a law, duty, right, contract or engagement; a rupture, a quarrel; a whale's leap from the water. *v.t.* to make a breach or gap in. *v.i.* to leap from the water (as a whale). **breach of promise,** (formerly) failure to keep a promise to marry. **breach of the peace,** violation of the public peace; a riot, an affray. **to step, throw oneself, into the breach,** to act as a substitute in an emergency.
bread (bred), *n.* a food, made of flour or other meal kneaded into dough, generally with yeast, made into loaves and baked; (*sl.*) money; livelihood. *v.t.* to dress with breadcrumbs before cooking. **bread and circuses,** free food and entertainment, esp. to placate the population. **bread and wine,** the Holy Communion; the eucharistic elements. **bread buttered on both sides,** very fortunate circumstances. **on the bread line,** in extreme poverty. **to break bread** BREAK. **to take the bread out of someone's mouth,** to deny a person the means of living. **bread and butter,** *n.* a slice of buttered bread; livelihood. *a.* (**bread-and-butter**) plain, practical; routine, basic; giving thanks for hospitality (of a letter). **bread-basket,** *n.* a basket for holding bread; (*sl.*) the stomach; rich grain lands. **bread-board,** *n.* **bread-crumb,** *n.* **bread-fruit,** *n.* the farinaceous fruit of a S Sea tree, *Artocarpus incisa*. **bread-sauce,** *n.* a sauce made with bread-crumbs, milk and onions. **bread-winner,** *n.* the member of a family who supports it with his or her earnings. **breaded,** *a.* **breadless,** *a.*
breadth (bredth), *n.* measure from side to side; a piece of material of full breadth; width, extent, largeness; broad effect; liberality, catholicity, tolerance. **breadthways, -wise,** *adv.* by way of the breadth, across.
break (brāk), *v.t.* (*past* **broke** (brōk), *earlier* **brake** (brāk), *p.p.* **broken** (brō'kən), **broke**,)) to part by violence; to rend apart, to shatter, to rupture; to damage; to destroy the completeness or continuity of; to split into smaller units; to subdue, to tame, to ruin financially; to cashier, to reduce to the ranks; to disable, to wear out, to exhaust the strength or resources of; to disconnect, to interrupt; to cause to discontinue (a habit); to intercept, to lessen the force of; to infringe, to violate; to improve on (a record etc.); to make known; to decipher; to violate; *v.i.* to separate into two or more portions; of boxers, to separate after a clinch; to burst, to burst forth; to appear suddenly; to become known; to become bankrupt; to give way physically or mentally; to change direction; to twist, as a ball at cricket; to make the first stroke at billiards or snooker; to alter the pace (as a horse); to alter (as a boy's voice at puberty); to come to an end (as a spell) of weather; to pause, to take a rest. *n.* the act of breaking; an opening, gap, breach; interruption of continuity in time or space; a pause, a rest; a line in writing or printing noting suspension of the sense; irregularity; the twist of a ball at cricket; a number of points scored continuously in billiards or snooker; the point where one musical register changes to another; (*coll.*) a lucky opportunity. **break of day, dawn. to break away,** to remove by breaking; to start away; to secede. **to break bread with,** to be entertained at table by; to take Communion with. **to break camp,** to take down one's tent in preparation for leaving. **to break cover,** to dart out from a hiding-place. **to break down,** to destroy, to overcome; to cease to function; to suffer a mental or physical collapse; to analyse costs etc. into component parts. **to break even,** to emerge without gaining or losing. **to break free or loose,** to escape from captivity; to shake off restraint. **to break in,** to tame, to train to something; to wear in (e.g. shoes). **to break into,** to enter by force; to interrupt. **to break new ground,** to be a pioneer or innovator. **to break off,** to detach from; to cease, to desist. **to break open,** to force a door or cover; to penetrate by violence. **to break out,** to escape; to begin (as a war); to appear (as an eruption on the skin). **to break service, to break someone's serve,** to win a game of tennis in which the opposing player served. **to break the back,** to break the keel of a ship; (*fig.*) to get through the greater part of. **to break the heart,** to overwhelm with grief. **to break the ice,** to take the first steps, esp. in establishing social contacts. **to break the mould,** to make unique; to effect a fundamental change. **to break up,** to disintegrate; to lay open (as ground); to dissolve into laughter; to disband, to separate; to start school holidays. **to break wind** WIND. **to break with,** to cease to be friends with; to quarrel with. **break-away,** *n.* (*Austral.*) a stampede of cattle or sheep; an animal that breaks away from the herd; a secession, withdrawal; any person, thing or group which breaks away from a main body. *a.* which withdraws or secedes; made to break or shatter easily (as stage properties). **break dancing,** *n.* an energetic type of modern dancing characterized by spinning on various parts of the body. **break-down,** *n.* a mechanical failure; total failure resulting in stoppage; a nervous breakdown; an analysis. **break-neck,** *a.* hazardous; of speed, very fast. **break-through,** *n.* penetration of enemy lines; an advance, a discovery. **break-up,** *n.* disruption, disintegration, dissolution; dispersal. **breakwater** *n.* a pier, mole or anything similar, to break the force of the waves. **breakable,** *a.* **breakage** (-ij), *n.* the act of breaking; being broken; (compensation for) loss or damage from breaking. **breaker**[1], *n.* one or that which breaks; a heavy wave breaking against the rocks or shore; (*sl.*) a CB user. **breaking,** *n.*, *a.* **breaking and entering,** illegal forced entry into premises for criminal purposes. **breaking point,** *n.* the limit of endurance.
breaker[2] (brā'kə), *n.* a keg, a water-cask.
breakfast (brek'fəst), *n.* the first meal of the day. *v.i.* to take breakfast.
bream (brēm), *n.* a freshwater fish of the genus *Abramis*.
breast (brest), *n.* one of the organs for the secretion of milk in women; the rudimentary part corresponding to this in men; the fore-part of the human body between the neck and the abdomen; the analogous part in the lower animals; the upper fore-part of a coat or other piece of clothing; source of nourishment; the seat of the emotions; the front, the fore-part. *v.t.* to apply or oppose the breast to; to stem, to face; to reach the summit of. **to beat one's breast,** to show remorse publicly or ostentatiously. **to breast-feed,** to feed a baby from the breast instead of the bottle. **to make a clean breast,** to confess all that one knows. **breast-bone,** *n.* the flat bone in front of the chest to which certain ribs are attached, the sternum. **breast-high,** *a.*, *adv.* **breast-plate,** *n.* armour worn upon the breast. **breast-pocket,** *n.* inside pocket on the breast of a man's jacket. **breast stroke,** *n.* a swimming stroke involving wide circling motions of the arms and legs while facing forward on one's breast. **breast-work,** *n.* a hastily constructed parapet thrown up breast-high for defence. **breasted,** *a.*
breath (breth), *n.* the air drawn in and expelled by the

breathe / bride

lungs in respiration; the act or power of breathing, life; a single respiration; phonetically, the expulsion of air without vibrating the vocal cords; a very slight breeze; respite; an instant; a whiff, an exhalation; a rumour, a whisper, a murmur. **below, under one's breath**, in a whisper. **out of breath**, panting from exertion. **to catch one's breath**, to cease breathing momentarily; to regain even breathing after exertion or a shock. **to save one's breath**, (*coll.*) to avoid talking to no purpose. **to take breath**, to pause. **to take one's breath away**, to astonish. **breathtaking**, *a.* astonishing, marvellous. **breath test**, *n.* a test to determine the amount of alcohol in the breath. **breathalyse, -yze** (-əlīz), *v.i.* to test for the level of alcohol in a driver's breath. **breathalyser**, *n.* an instrument containing crystals for measuring the level of alcohol in the breath. **breathful**, *a.* **breathless**, *a.* out of breath; dead, lifeless; without a movement of the air; excited, eager. **breathlessly**, *adv.* **breathlessness**, *n.* **breathy**, *a.* aspirated; giving the sound of breathing. **breathiness**, *n.*

breathe (brēdh), *v.i.* to inhale or exhale air; to live; to take breath; to move or sound like breath. *v.t.* to inhale or exhale (as air); to emit, to send out, by means of the lungs; to utter; to utter softly; to express, to manifest; to instil (into); to allow breathing space to. **to breathe again, breathe freely, easily**, to be relieved from fear or anxiety. **to breathe down someone's neck**, to cause someone discomfort with one's close supervision or constant attention. **to breathe one's last**, to die. **breathable**, *a.* **breathableness**, *n.* **breather**, *n.* a rest in order to gain breath; a vent in an airtight container. **breathing**, *a.* living; lifelike. *n.* the action of breathing; an aspirate; (*Gr. Gram.*) either of the two signs ['] or ['] used to mark the presence or absence of the aspirate. **breathing-place, -space**, *n.* a pause, place or opening for breathing. **breathing-time**, *n.* time for recovering one's breath; a pause.

breccia (brech'iə), *n.* a rock composed of angular fragments cemented together in a matrix. **brecciated** (-ātid), *a.*

bred (bred), *p.p.* BREED.

breech (brēch), *n.* (*pl.* **breeches** (brich'iz)), the buttocks; the hinder part of anything; the portion of a gun behind the bore; (*pl.*) trousers reaching to just below the knees. **to wear the breeches**, to head the household. **breech birth, delivery**, *n.* a birth in which the baby's buttocks or feet emerge first. **breech-block**, *n.* a movable piece to close the breech of a gun. **breeches-buoy** (-brē), *n.* a lifebuoy, with a support in the form of canvas breeches, run on a rope and used for ship-to-ship transfers and rescues at sea. **breech-loader**, *n.* a fire-arm loaded at the breech. **breech-loading**, *a.*

breed (brēd), *v.t.* (*past, p.p.* **bred** (bred),) to give birth to; to raise (cattle etc.), to rear; to give rise to, to engender, to cause to develop; to train up, to educate, to bring up. *v.i.* to be pregnant; to produce offspring; to come into being, to spread; to be produced or engendered. *n.* a line of descendants from the same parents or stock; family, race, offspring. **breeder**, *n.* one who breeds, esp. one who breeds cattle and other animals. **breeder reactor**, *n.* a nuclear reactor which produces more plutonium than it consumes. **breeding**, *n.* the act of giving birth to; the raising of a breed; bringing-up, nurture, rearing, education, deportment, good manners. **breeding ground**, *n.* a favourable environment or atmosphere for generating or nurturing ideas, bacteria etc.

breeks (brēks), (*Sc.*) BREECHES, see BREECH.

breeze[1] (brēz), *n.* a light wind; a disturbance, a row; a whisper, rumour; (*coll.*) something which can be done or got with ease. *v.i.* to blow gently or moderately; to move in a lively way; (*coll.*) to do or achieve something easily. **to breeze up**, to approach in a carefree or lively manner. **breezeless**, *a.* **breezy**, *a.* open, exposed to breezes, windy; lively, brisk, jovial. **breezily**, *adv.* **breeziness**, *n.*

breeze[2] (brēz), *n.* small cinders and cinder-dust; small coke, siftings of coke. **breeze block, brick**, *n.* a brick or block made of breeze and cement.

bremsstrahlung (brem'shtrahlung), *n.* the electromagnetic radiation caused by an electron colliding with or slowed down by the electric field of a positively charged nucleus. [G *bremsen*, to brake, *Strahlung*, radiation]

Bren gun (bren), *n.* a type of light machine-gun. [first letters of *Br*no (Czechoslovakia) and *En*field]

brent-goose (brent), *n.* the smallest of the wild geese, *Bernicla brenta*, which visits Britain in the winter.

brer (breə), *n.* brother. [Black Am. contr. of BROTHER]

brethren (bredh'rən), *n.pl.* BROTHER.

Breton (bret'ən), *n.* a native of, or the language of, Brittany; *a.* belonging to Brittany.

breve (brēv), *n.* a sign (˘) used in printing to mark a short vowel; a musical note equal to two semibreves.

brevet (brev'it), *n.* a warrant conferring nominally higher rank on an officer without the pay. *v.t.* to confer (a certain rank) by brevet.

breviary (brē'viəri), *n.* in the Roman Catholic Church, a book containing the divine office.

brevity (brev'iti), *n.* briefness, shortness; conciseness.

brew (broo), *v.t.* to make (beer, ale etc.) by boiling, steeping and fermenting; to convert into (beer, ale etc.) by such processes; to prepare other beverages by mixing or infusion; to concoct; to contrive, to plot. *v.i.* to make beer etc. by boiling, fermenting etc.; to undergo these or similar processes; to be in preparation or impending. *n.* the action, process or product of brewing; the quantity brewed at one process; the quality of the thing brewed. **to brew up**, (*coll.*) to make tea. **brewage** (-ij), *n.* a mixture; a concocted beverage; the process of brewing. **brewer**, *n.* one whose trade is to brew malt liquors. **brewer's droop**, *n.* (*sl.*) temporary sexual impotence in men due to overindulgence in alcohol. **brewer's yeast**, *n.* a yeast used in brewing and as a source of vitamin B. **brewery** (-əri), *n.* **brewhouse**, *n.* a place where beer is brewed.

briar BRIER.

bribe (brīb), *n.* a gift or consideration of any kind offered to anyone to influence his or her judgment or conduct; an inducement. *v.t.* to influence action or opinion by means of a gift or other inducement. *v.i.* to practise bribery. **bribable**, *a.* **bribability** (-bil'-), *n.* **bribee** (-ē'), *n.* **bribeless**, *a.* incapable of being bribed. **briber**, *n.* **bribery** (-əri), *n.* the act of giving or receiving bribes.

bric-a-brac (brik'əbrak), *n.* fancy ware, curiosities, knickknacks.

brick (brik), *n.* a block of clay and sand, usually oblong, moulded and baked, used in building; a brick-shaped block of any material; a child's block for toy building; a brick-shaped loaf; (*sl.*) a good person. *a.* made of brick. *v.t.* to lay or construct with bricks; to imitate brickwork in plaster. **like a ton of bricks**, with great force. **to brick up**, to block up with brickwork. **to drop a brick**, to say the wrong thing, to commit a blunder. **to make bricks without straw**, to perform the impossible. **brickbat**, *n.* a broken piece of brick, esp. for use as a missile; (*coll.*) a critical remark. **brick-clay, brick-earth**, *n.* clay used for making brick; a clayey earth in the London basin. **brick-dust**, *n.* powdered brick. **brick-field**, *n.* a field in which brick-making is carried on. **brick-kiln**, *n.* a kiln for baking bricks. **bricklayer**, *n.* one who lays or sets bricks. **brick-nogging**, *n.* brickwork built into a timber framework. **brick-red**, *a.* **brickwork**, *n.* builder's work in brick; bricklaying. **brickie** (-i), *n.* (*coll.*) a bricklayer. **bricking**, *n.* brickwork. **bricky**, *a.*

bride[1] (brīd), *n.* a woman newly married or on the point of being married. **bride-cake**, *n.* wedding cake. **bridegroom**, *n.* a man about to be married or recently married. **bride(s)maid, bride(s)man**, *n.* the unmarried friends of the bride and groom who attend them at their wedding. **bridal** *n.* the nuptial ceremony or festival; a wedding; marri-

age. *a.* of or pertaining to a bride or a wedding.
bride² (brīd), *n.* the foundation net-work of lace; a bonnet-string.
bridewell (brīd'wel), *n.* a house of correction, a prison.
bridge¹ (brij), *n.* a structure thrown over a body of water, a ravine, another road etc. to carry a road, railway or path etc. across; anything more or less resembling a bridge in form or function; the upper bony part of the nose; the thin wooden bar over which the strings are stretched in a violin or similar instrument; a support for a billiard cue in an awkward stroke; the raised deck or cabin from which a ship is controlled; an electrical circuit used for the accurate measurement of electrical quantities, e.g. resistance; a partial denture. *v.t.* to span or cross with or as with a bridge. **to cross a bridge when one comes to it**, to cope with a difficulty only when it occurs. **bridge camera**, *n.* a type of camera intermediate between a compact and a single-lens reflex, which has a built-in zoom facility. **bridgehead**, *n.* a fortification protecting the end of a bridge nearest the enemy; an advanced position seized in enemy territory, usu. as a basis for a further advance. **bridge-of-boats**, *n.* a bridge supported on a number of boats moored abreast. **bridge roll**, *n.* a small, finger-shaped roll of soft bread. **bridgework**, *n.* a dental bridge. **bridgeable**, *a.* **bridgeless**, *a.* **bridging**, *n.* the structure of a bridge; the act of making or forming a bridge. **bridging loan**, *n.* a short-term loan with a high interest rate which covers a financial transaction until a long-term loan is arranged.
bridge² (brij), *n.* a card game resembling whist. **auction, contract, bridge** AUCTION, CONTRACT. **bridge-marker**, *n.* a device for registering the points made at bridge. **bridge-scorer**, *n.*
bridle (brī'dl), *n.* a head-stall, bit and rein, forming the head-gear of a horse or other beast of burden; a check, a restraint. *v.t.* to put a bridle on; to control with a bridle; to hold in, to check, to control. *v.i.* to draw back the head as an expression of scorn or resentment. **bridle-hand**, *n.* the left hand. **bridle-path, -road, -way**, *n.* a path for horsemen. **bridled**, *a.*
Brie (brē), *n.* a soft white cheese orig. produced in France.
brief (brēf), *n.* short in duration or extent; concise; curt. *n.* instructions; a short statement; a summary of facts and points of law given to counsel in charge of a case; (*N Am.*) pleadings; (*sl.*) counsel, a lawyer; (*pl.*) close fitting pants, underpants or knickers without legs. *v.t.* to instruct or retain a barrister by brief; to give detailed instructions. **in brief**, briefly. **brief-case**, *n.* a small usu. leather case for carrying papers. **briefing**, *n.* (a meeting for) giving instructions or information. **briefless**, *a.* **briefly**, *adv.* **briefness**, *n.*
brier, briar¹ (brī'ə), *n.* a thorny or prickly shrub, esp. a wild rose; the stem of a wild rose on which a garden rose is grafted. **sweet brier**, *n.* a wild rose with fragrant leaves. **brier-rose**, *n.* the dog-rose or the field-rose. **briery, briary**, *a.*
brier, briar² (brī'ə), *n.* the white or tree heath, *Erica arborea;* a tobacco-pipe made from the root of this. **brier-root**, *n.*
brig¹ (brig), (*Sc., North.*) BRIDGE¹.
brig² (brig), *n.* a square-rigged vessel with two masts; a US Navy prison; (*sl.*) any prison.
Brig., (*abbr.*) Brigade; Brigadier.
brigade (brigād'), *n.* a large unit within an army division; an organized body of workers, often wearing a uniform. *v.t.* to form into a brigade. **Boys' Brigade** BOY. **Fire Brigade** FIRE. **Household Brigade** HOUSEHOLD. **brigade-major**, *n.* a staff officer who assists a brigadier. **brigadier** (-ədiə'), *n.* the officer in command of a brigade; the rank below that of major-general.
brigand (brig'ənd), *n.* a bandit, an outlaw. **brigandage** (-dij), *n.* **brigandry**, *n.* the practices of brigands; highway robbery. **brigandish**, *a.* **brigandism**, *n.*
brigantine (brig'əntēn), *n.* a two-masted vessel square-rigged on both masts but with a fore-and-aft mainsail, and mainmast much longer than the foremast.
bright (brīt), *a.* lighted up, emitting or reflecting abundance of light; shining; unclouded; cheerful, happy; clever, intelligent. *adv.* brightly. **bright and early**, very early in the morning. **bright-eyed and bushy-tailed**, radiant with health and vigour. **the bright lights**, the area of a city where places of entertainment are concentrated; the city. **brightwork**, *n.* shiny metal fittings on cars etc. **brighten**, *v.t.* to make bright; to make happy, hopeful etc. *v.i.* to become bright; of the weather, to clear up. **brightly**, *adv.* **brightness**, *n.*
Bright's disease (brīts), *n.* a term including several forms of kidney disease, associated with albuminuria.
brill¹ (bril), *n.* a flat seafish, *Rhombus vulgaris*, similar to the turbot.
brill² (bril), *a.* (*coll.*) short for BRILLIANT, used as a general term of approbation.
brilliant (bril'yənt), *a.* shining, sparkling; lustrous; illustrious, distinguished; extremely clever and successful. *n.* a diamond or other gem of the finest cut, consisting of lozenge-shaped facets alternating with triangles. **brilliance, brilliancy**, *n.* **brilliantine** (-tēn), *n.* a cosmetic for rendering the hair glossy. **brilliantly**, *adv.*
brim (brim), *n.* the upper edge, margin or brink of a vessel, hollow or body of water; the rim of a hat. *v.t.* to fill to the brim. *v.i.* to be full to the brim. **to brim over**, to overflow. **brimful**, *a.* **brimless**, *a.* **brimmed**, *a.* having a brim; brimful. **brimming**, *a.*
brimstone (brim'stōn), *n.* sulphur, esp. in the Bible; the sulphur butterfly. **brimstone butterfly**, *n.* an early sulphur, *Gonepteryx rhamni.* **brimstone moth**, *n.* a sulphur-coloured moth, *Rumia crataegata.*
brindle (brind'l), **brindled** *a.* tawny, with bars of darker hue; streaked, spotted.
brine (brīn), *n.* water strongly impregnated with salt; the sea; (*poet.*) tears. *v.t.* to treat with brine, to pickle. **brinepan**, *n.* a shallow vessel or pit in which brine is evaporated in salt-making. **briny**, *a.* **the briny**, (*coll.*) the sea.
bring (bring), *v.t.* (*past, p.p.* **brought**) to cause to come along with oneself, to lead; to bear, to carry, to convey; to induce, to prevail upon, to produce, to yield result in. **to bring about**, to cause, to bring to pass; to reverse the ship. **to bring back**, to recall to memory. **to bring down**, to humble, to abase, to overthrow; to shoot, to kill; to lower (a price); to carry on (a history) to a certain date. **to bring down the house**, to create tumultuous applause. **to bring forth**, to produce, to give birth to; to cause. **to bring forward**, to produce, to adduce; to carry on a sum from the bottom of one folio to the top of the next (in book-keeping). **to bring home to**, to prove conclusively; to convince. **to bring in**, to produce, to yield; to introduce (a legislative bill); to return (a verdict). **to bring off**, to accomplish. **to bring on**, to cause to begin; to introduce for discussion; to cause to develop (more quickly). **to bring out**, to make plain or prominent; to introduce to society; to launch (as a product); to publish; to expose; to utter. **to bring over**, to convert; to cause to change sides. **to bring round**, to revive; to convert. **to bring to**, to restore to health or consciousness; to halt (a ship). **to bring to pass**, to cause to happen. **to bring up**, to educate, to rear; to lay before a meeting; to vomit; to come to a stop. **to bring up the rear**, to come last.
brinjal (brin'jəl), *n.* (*esp. in India and Africa*) an aubergine.
brink (bringk), *n.* the edge of a precipice, pit, body of water etc.; the verge. **brinkmanship** *n.* the art of maintaining one's position on the brink of a decision or crisis.
brio (brē'ō), *n.* liveliness. **brioso** (-ō'sō), *adv.* vigorously.
brioche (briosh'), *n.* a light, slightly sweet bread roll.
briony (brī'əni), BRYONY.

briquette, briquet (brikėt'), *n.* a block of compressed coal-dust; a slab of artificial stone.
brisk (brisk), *a.* lively, animated; stimulating, bracing. **briskly,** *adv.* **briskness,** *n.*
brisket (bris'kit), *n.* that part of the breast of an animal next to the ribs; this joint of meat. **brisket-bone,** *n.* the breast-bone.
brisling (briz'ling, bris'-), **bristling,** (bris'-), *n.* a sprat.
bristle (bris'l), *n.* a short, stiff, coarse hair, particularly on the back and sides of pigs; (*pl.*) a beard cropped short; stiff hairs on plants. *v.i.* to stand erect (as hair); to show indignation or defiance; to be thickly beset (with difficulties, dangers etc.). **bristling,** *a.* bristling with, full of, with many of. **bristled,** *a.* **bristly,** *a.* thickly covered with or as with bristles, (*coll.*) quick to anger, touchy. **bristliness,** *n.*
Bristol (bris'təl), an English city on the lower Avon; (*pl., sl.*) breasts. **Bristol-board,** *n.* a thick smooth white cardboard. **Bristol fashion,** in good order.
Brit[1] (brit), *n.* (*coll. abbr., some. derog.*) a Briton.
Brit[2], (*abbr.,*) Britain; British.
Britain (brit'ən), **Great Britain** *n.* England, Wales and Scotland. **Britannia** (-tan'yə), *n.* Britain personified as a female figure. **Britannic** (-tan'-), *a.* British. **British,** *a.* of or pertaining to ancient Britain, Great Britain or its inhabitants or the British Commonwealth. **British Summer Time** SUMMER TIME under SUMMER. **British Thermal Unit** THERM. **the British,** British people or soldiers. **Britisher,** *n.* a British subject or a native of Britain. **Britishism, Briticism** (-sizm), *n.* an idiom employed in Britain and not elsewhere. **Briton** (-ən), *n.* a member of the people inhabiting S Britain at the Roman invasion; a native of Britain.
brittle (brit'l), *a.* liable to break or be broken, fragile; not malleable. *n.* a brittle sweet (e.g. peanut brittle). **brittle-bone disease,** *n.* a disease which causes the bones to break easily. **brittle star,** *n.* a type of starfish with long flexible arms. **brittleness,** *n.*
bro., Bro., (*abbr.*) brother; Brother.
broach (brōch), *n.* a tapering iron tool; a roasting-spit. *v.t.* to pierce (as a cask), so as to allow liquor to flow; to open, and begin to use; to bring up for discussion. **to broach to,** to veer to windward so as to present a ship's broadside to the sea.
broad (brawd), *a.* wide, large; extensive, spacious; of wide range, general; open, clear (as *broad daylight*); obvious; tolerant, liberal; rough, strong, rustic; coarse, obscene; bold, vigorous, free in style or effect. *n.* a large, freshwater lake formed by the broadening of a river; (*N Am. sl.*) a woman; a prostitute; *adv.* in breadth; broadly, widely. **as broad as it's long,** equal upon the whole; the same either way. **broad arrow,** *n.* a mark cut or stamped on British Government property. **broad band,** *a.* receiving, transmitting or involving a wide range of frequencies. **broad bean,** *n.* a leguminous plant with edible seeds in a pod, *Faba vulgaris*. **broadcast,** *a.* scattered by the hand (as seed); widely disseminated; transmitted by radio or television. *n.* broadcast sowing; anything transmitted to the public by radio or television. *adv.* by scattering widely. *v.t.* to sow by scattering with the hand; to transmit by radio or television; to disseminate widely. **broadcaster,** *n.* **Broad Church,** *a.*, *n.* (of) a party in the Church of England interpreting doctrine in a liberal sense. **broadcloth,** *n.* a fine, wide, dressed black cloth, used for men's coats etc. **broadleaf, broad-leaved,** *a.* having a broad leaf. **broadloom,** *n.*, *a.* (carpet) woven on a wide loom. **broad-minded,** *a.* tolerant, having an open mind. **broadsheet,** *n.* a large sheet printed on one side only; a large format newspaper. **broadside,** *n.* the side of a ship above the water; a volley from all the guns on one side of a ship of war; an broadsheet; a attack on a person, policy, etc. **broad-spectrum,** *a.* of antibiotics etc., wide-ranging. **broadsword,** *n.* a sword with a broad blade. **broad-spoken,** *a.* plain spoken; using a dialect or coarse language. **Broadway,** New York's theatre and restaurant district. **broadways, -wise,** *adv.* **broaden,** *v.t.* to become broader, to spread. *v.t.* to make broader. **broadly,** *adv.* **broadness,** *n.*
Brobdingnagian (brobdingnag'iən), *a.* gigantic, huge. *n.* a giant.
brocade (brəkād'), *n.* silken stuff with raised figures. *v.t.* to weave or work with raised patterns; to decorate with brocade. **brocaded,** *a.*
broccoli, brocoli (brok'əli), *n.* a variety of cauliflower with green flowerheads.
brochette (broshet'), *n.* a skewer; small pieces of food grilled together on a skewer (like a kebab).
brochure (brō'shə), *n.* a small pamphlet.
brock (brok), *n.* a badger.
broderie anglaise (brō'dəri āglez'), *n.* open embroidery on cambric or linen. [F]
brogue (brōg), *n.* a sturdy shoe; dialectal pronunciation, esp. Irish.
broil[1] (broil), *n.* a tumult, disturbance.
broil[2] (broil), *v.t.* to cook on a gridiron; to grill. *v.i.* to be very hot; to be subjected to heat. **broiler,** *n.* one who or that which broils; a gridiron: a chicken 8–10 weeks old for broiling or roasting.
broke[1] (brōk), *part.a.* BROKEN.
broke[2] (brōk), *a.* (*sl.*) ruined, penniless. **to go for broke,** (*sl.*) to risk or gamble everything on a venture.
broken *a.* in pieces; not whole or continuous; interrupted, intermittent; weakened, infirm; crushed, humbled; transgressed, violated; bankrupt, ruined; (*Painting*) reduced by the addition of some other colour; of language, incoherent, defective in grammar or pronunciation. **broken-backed,** *a.* having the back broken; drooping at stem and stern from injury to the keel. **broken-down,** *a.* not in working order; dilapidated; worn-out; ruined in health, in character or financially. **broken-hearted,** *a.* crushed by grief or disappointment. **broken home,** *n.* the home of children with separated or divorced parents. **broken water,** *n.* choppy water. **broken-winded,** *a.* having defective respiratory organs; habitually short of breath. **brokenly,** *adv.* with breaks, jerkily, spasmodically.
broker (brō'kə), *n.* an agent, a middleman; one who buys and sells for others; a dealer in second-hand goods; a stockbroker. **brokerage** (-rij), *n.* the business or commission of a broker; a broker's commission on sales etc. **broking,** *n.*
brolly (brol'i), *n.* (*coll.*) an umbrella.
bromine (brō'mēn, -min), *n.* a non-metallic, dark red, liquid element; at. no. 35; chem. symbol Br. with a strong, irritating odour. **bromal,** *n.* a liquid like chloral produced by the action of bromine upon alcohol. **bromate** (-māt), *n.* a salt of bromic acid. **bromic** *a.* of or pertaining to bromine; having bromine in its composition. **bromide** (-mīd), *n.* a combination of bromine with a metal or a radical, esp. bromide of potassium, which is used as a sedative; (*coll.*) a commonplace remark, a platitude. **bromide paper,** *n.* a sensitized paper used in printing a photograph from a negative. **bromide process,** *n.* in photography, printing or enlarging on paper coated with silver bromide emulsion. **bromidic** (-mid'-), *a.* (*coll.*) dull, commonplace. **bromism** *n.* the condition produced by long treatment with bromide of potassium. **bromize, -mise,** *v.t.*
bronchi (brong'ki), **bronchia** (-kiə), *n.pl.* the main divisions of the windpipe; the ramifications into which these divide within the lungs. **bronchial,** *a.* **bronchial tubes,** *n.pl.* the bronchia.
bronchiectasis (brongkiek'təsis), *n.* abnormal dilation of the bronchial tubes.
bronchio- (brong'kio), **broncho-** (brong'kō), *comb. forms.* pertaining to the windpipe or the tubes into which it divides beneath.

bronchiole (brongk'iōl), *n.* any of the tiny branches of the bronchi.
bronchitis (brongkī'tis), *n.* inflammation of the bronchia. **bronchitic** (-kit'-), *a.*
bronchopneumonia (brongkōnūmō'niə), *n.* pneumonia originating in the bronchial tubes.
bronchoscope (brong'kəskōp), *n.* an instrument inserted in the bronchial tubes for the purpose of examination or extraction. **bronchoscopic** (-skop'-), *a.* **bronchoscopically**, *adv.* **bronchoscopy** (-kos'kəpi), *n.*
bronco (brong'kō), *n.* (*pl.* **-cos**) a native half-tamed horse of California or New Mexico. **bronco-buster**, *n.* a breaker-in of broncos.
Brontosaurus (brontəsaw'rəs), *n.* a genus of huge fossil dinosaurian reptiles, notable for their small head and diminutive brain-cavity.
bronze (bronz), *n.* a brown alloy of copper and tin; a brown colour, like that of bronze; a work of art in bronze; a bronze medal. *a.* made of or the colour of bronze. *v.t.* to give a bronze-like appearance to (wood, metal, plaster etc.); to brown, to tan. *v.i.* to become brown or tanned. **Bronze Age, Period,** *n.* a period after the Stone and before the Iron Age when weapons and implements were made of bronze. **bronze medal,** *n.* a medal made of bronze awarded for third place in a contest. **bronzed,** *a.* overlaid with bronze; coloured like bronze, sun-tanned. **bronzing,** *n.* **bronzy,** *a.*
brooch (brōch), *n.* an ornamental clasp with a pin, for fastening some part of a woman's dress.
brood (brood), *n.* a family of birds hatched at once; offspring, progeny; (*facet.*) children; a group of a particular kind. *v.i.* to sit on eggs; to hang close over (as clouds); to meditate moodily. *v.t.* to sit upon eggs to hatch them; to cherish under the wings; to prepare by long meditation; to cherish moodily. **brood-hen, -mare,** *n.* a hen or a mare kept for breeding. **brooder,** *n.* a cover for sheltering young chickens. **broodiness,** *n.* **broody,** *a.* inclined to sit on eggs; sullen, morose; inclined to brood over matters; of a woman, feeling a strong urge to become a mother.
brook[1] (bruk), *n.* a small stream, a rivulet. **brooklime,** *n.* a kind of speedwell. **brooklet** (-lit), *n.* **brooky,** *a.*
brook[2] (bruk), *v.t.* to endure, to put up with.
broom (broom), *n.* a shrub with yellow flowers of the genus *Sarothamnus* or *Cytisus*; the allied genus *Genista;* a besom for sweeping, orig. made of broom; a long-handled brush. **new broom,** a newly appointed person who sweeps away old practices or attitudes. **broom-rape,** *n.* the parasitic genus *Orobanche*. **broom-stick,** *n.* the handle of a broom. **broomy,** *a.*
bros., Bros. (bros), (*abbr.*) brothers, Brothers.
brose (brōz), *n.* a kind of porridge. **Athol brose,** (ath'l), *n.* a mixture of whisky and honey.
broth (broth), *n.* the liquor in which anything, esp. meat, has been boiled; thin soup; a medium for growing cultures (e.g. of bacteria). **a broth of a boy,** a high-spirited fellow.
brothel (broth'l), *n.* premises where prostitutes sell their services.
brother (brŭdh'ə), *n.*, (*pl.* **brothers,** and in more solemn senses **brethren** (bredh'rən)), a son of the same parents or parent; one closely connected with another; an associate; one of the same community, country, city, church, order, profession or society; a member of a male religious order. **half-brother,** brother on one side only. **brother-in-arms,** *n.* fellow-soldier. **brother-in-law,** *n.* (*pl.* **brothers-in-law**) the brother of one's husband or wife, one's sister's husband. **brotherhood** (-hud), *n.* the relationship of a brother; a fraternity, an association for mutual service; brotherly affection or feeling. **brotherless,** *a.* **brotherlike,** *a.* **brotherly,** *a. adv.* **brotherliness,** *n.*
brougham (broom. broo'əm, brō'əm), *n.* a close, four-wheeled carriage drawn by one horse.

brought (brawt), *past, p.p.* BRING.
brouhaha (broo'hahhah), *n.* a tumult, a row.
brow (brow), *n.* an eyebrow; the forehead; the countenance generally; aspect, appearance; the projecting edge of a cliff or hill; the top of a hill. **to knit the brows,** to frown. **browbeat,** *v.t.* to bear down arrogantly; to bully. **browbeaten,** *a.* intimidated.
brown (brown), *a.* of the colour of scorched wood or paper; dusky, dark, sun-tanned. *v.t.* to make brown; to give a brown lustre to (gun-barrels etc.). *v.i.* to become brown; to get sunburnt. *n.* a brown colour; a compound colour produced by a mixture of red, black and yellow; pigment of this colour; a brown butterfly; brown clothes. **brown bear,** *n.* a brown-coloured bear common to Europe, Asia and N America. **brown bread,** *n.* bread made from whole-meal; bread in which bran is mixed with the flour. **brown coal,** *n.* lignite. **brown fat,** *n.* a dark fatty tissue which generates body heat. **brown goods,** *n.pl.* household appliances, such as TV sets, record players etc. as opposed to *white goods*. **brown paper,** *n.* coarse, unbleached paper for packing parcels etc. **brown rice,** *n.* husked rice left unpolished. **brown-shirt,** *n.* a uniformed member of the Nazi party. **brownstone,** *n.* (*N Am.*) a dark-brown sandstone; a building of this material. **brown study,** *n.* reverie, day-dream. **brown sugar,** *n.* coarse, half-refined sugar. **brown trout,** *n.* a common European trout with a dark spotted back. **browned,** *a.* **browned off,** *a.* bored, fed up. **Brownie, Brownie Guide,** *n.* a junior Girl Guide from 8 to 11 years of age. **Brownie point,** (sometimes *derog.*). *n.* a supposed mark to one's credit for some achievement. **brownie,** (-ni), *n.* (*orig. N Am.*) a kind of nutty, dark chocolate cake cut into flat squares; a kindly domestic elf. **browning,** *n.* colouring material for gravy. **brownish,** *a.* **brownness,** *n.*
Brownian movement (brow'nion), *n.* random agitation of particles suspended in a fluid caused by bombardment of the particles by molecules of the fluid.
browse (browz), *v.t.* to nibble and eat off (twigs, young shoots etc.). *v.i.* to feed on twigs, young shoots etc.; to graze; to read in a desultory way, to leaf through; to look through, among articles in an idle manner. *n.* the tender shoots of trees and shrubs fit for cattle to feed on; the act of browsing. **browser,** *n.*
BRS, (*abbr.*) British Road Services.
brucellosis (broosəlō'sis), *n.* an infectious bacterial disease in animals which is also contagious to man.
Bruin (broo'in), *n.* familiar name for the brown bear.
bruise (brooz), *v.t.* to crush, indent or discolour, by a blow from something blunt and heavy; to injure without breaking skin or bone; to batter, pound, grind up; to injure psychologically. *v.t.* to become bruised. *n.* an injury caused by something blunt and heavy; a contusion. **bruised,** *a.* **bruiser,** *n.* one who or that which bruises; (*coll.*) a large strong man, a prize-fighter.
bruit (broot), *n.* rumour, report; an abnormal sound heard in auscultation. *v.t.* to rumour; to noise abroad.
brûlé (broo'lā), *a.* cooked with brown sugar. [F]
Brum (brŭm), (*abbr.*) Birmingham. **Brummy, -mmie,** (-i), *n.* (*coll.*) a person from Birmingham.
brume (broom), *n.* mist, fog.
brunch (brŭnch), *n.* (*coll.*) a meal which combines a late breakfast with an early lunch.
brunette (brunet'), *n.* a girl or woman of dark hair and complexion. *a.* brown-haired; of dark complexion.
brunt (brŭnt), *n.* the shock, impetus or stress of an attack, danger or crisis. **to bear the brunt,** to take the main force (e.g. of an attack).
brush (brŭsh), *n.* an instrument for sweeping or scrubbing, generally made of bristles, twigs or feathers; an instrument consisting of hair or bristle attached to a handle, for colouring, white-washing, painting etc.; a hair-pencil; a brushing; an encounter, a skirmish; a bushy tail, as of a

brusque fox; a piece of metal or carbon or bundle of wires or plates, forming a good electrical conductor; a brush-like discharge of electric sparks; a brush-like appearance produced by polarized light; brushwood, underwood, a thicket of small trees; loppings, faggots of brushwood. *v.t.* to sweep or scrub with a brush; to remove by brushing; to touch lightly, as in passing. *v.i.* to move with a sweeping motion; to pass lightly over. **to brush aside, off,** to dismiss curtly. **to brush up,** to clean by brushing; to tidy one's appearance; to refresh one's memory. **brush fire,** *n.* a fast spreading fire which consumes dry brush and scrub. **brush-off,** *n.* (*coll.*) a brusque rebuff. **brush-up,** *n.* a brushing. **brush-wheel,** *n.* a circular revolving brush. **brushwood,** *n.* a thicket, underwood; low scrubby thicket; loppings. **brushwork,** *n.* a painter's manipulation of the brush; style of manipulation of the brush. **brushed,** *a.* cleaned or smoothed with a brush; having a raised nap (of a fabric). **brusher,** *n.* **brushing,** *n.* **brushy,** *a.* resembling a brush; rough, shaggy; covered with brushwood.
brusque (brŭsk, broosk), *a.* rough, blunt, unceremonious. **brusquely,** *adv.* **brusqueness,** *n.* **brusquerie** (-kəri), *n.*
Brussels (brŭs'əlz), *a.* made at or derived from Brussels. *n.* a Brussels carpet. **Brussels carpet,** *n.* a kind of carpet with a backing of linen and wool face. **Brussels lace,** *n.* a kind of pillow-lace. **brussels sprouts,** *n.pl.* the small sprouts springing from the stalks of a variety of cabbage, and used as a vegetable.
brut (broot), *a.* of wine, dry, unsweetened. [F]
brute (broot), *a.* stupid, irrational; beastlike, sensual; purely physical. *n.* an irrational animal; a beast; the animal nature in man; a cruel or stupid person. **brutal,** *a.* resembling a brute; savage, cruel; coarse, unrefined, sensual. **brutally,** *adv.* **brutality** (-tal'-), **brutalism,** *n.* the qual- ity of being brutal; a brutal action. **brutalize, -ise,** *v.t.* to render brutal; *v.i.* to become brutal. **brutalization,** *n.* **brutify** (-fi), *v.t.* **brutification** (-fi-), *n.* **brutish,** *a.* like a brute; animal, bestial. **brutishly,** *adv.* **brutishness,** *n.*
bryology (briol'əji), *n.* the science of mosses; mosses collectively. **bryologist,** *n.*
bryony (bri'əni), *n.* a genus of climbing plants of the gourd family, esp. *Bryonia dioica,* white or common bryony; a similar plant, black bryony, *Tamus communis.*
Bryophyta (briöfi'tə), *n.* a division of the higher cryptogams consisting of the liverworts and mosses. **bryophyte** (bri'-), *n.* a member of this group.
Brython *n.* a Celt of the S British group. **Brythonic,** *n.* a group of Celtic languages comprising Welsh, Cornish and Briton. *a.* relating to Brythons or Brythonic.
BS, (*abbr.*) British Standard(s).
b.s., (*abbr.*) balance sheet; bill of sale.
B.Sc., (*abbr.*) Bachelor of Science.
BSI, (*abbr.*) British Standards Institution.
BST (*abbr.*) British Standard Time; British Summer Time.
Bt., (*abbr.*) Baronet; bought.
btu, BTU, B.Th.U., (*abbr.*) British Thermal Unit.
bu., (*abbr.*) bushel.
bubble (bŭb'l), *n.* a vesicle of water or other liquid filled with air or other gas; a cavity in a solidified material, such as ice, amber, glass etc.; a bubbling sound; a transparent dome; anything unsubstantial or unreal; (*hist.*) a swindling project. *v.i.* to rise up in or as in bubbles; to make a noise like bubbling water. **to bubble over,** to overflow with laughter, anger, etc. **bubble and squeak,** leftover potato and cabbage fried together. **bubble bath,** *n.* a foaming bath preparation. **bubble-car,** *n.* a midget motorcar with rounded line and transparent top. **bubble chamber,** *n.* an apparatus for tracking the path of a charged particle by the stream of bubbles left in its wake. **bubblegum,** *n.* a kind of chewing-gum that can be blown up into a bubble. *a.* appealing mainly to adolescents. **bubble memory,** *n.* a data-storage system in computers composed of tiny areas of bubbles of magnetism. **bubble pack** BLISTER PACK. **bubbly,** *a.* full of bubbles; excited, vivacious. *n.* (*coll.*) champagne.
bubo (bū'bō), *n.* (*pl.* **-boes**) an inflamed swelling of the lymphatic glands, esp. in the groin or armpit. **bubonic** (-bon'-), *a.* **bubonic plague,** *n.* a type of plague characterized by buboes. **bubonocele** (-bon'əsēl), *n.* hernia of the groin.
buccal (bŭk'l), *a.* pertaining to the cheek or the mouth.
buccaneer (bŭkənia'), *n.* one of the piratical rovers who formerly infested the Spanish Main; an adventurous, opportunistic and unscrupulous businessman, politician etc. *v.i.* to act the part of a buccaneer. **buccaneering,** *a.*
buccinator (bŭk'sinātə), *n.* the flat, thin muscle forming the wall of the cheek, used in blowing.
Buchmanism (bŭk'mənizm, buk'-), *n.* Moral Rearmament. **Buchmanite** (-nit), *n.*
buck (bŭk), *n.* the male of the fallow-deer, reindeer, goat, hare and rabbit; (*S. Afr.*) an antelope (of either sex); a dashing young fellow; (*offensive*) a male Indian or Negro; (*N Am. sl.*) a dollar; a marker in poker which indicates the next dealer; the responsibility; a jump by a vicious or unbroken horse, with the feet drawn together and the back arched, to unseat the rider. *v.i.* of a horse, to jump in the air to try to unseat the rider. *v.t.* to throw by bucking; (*coll.*) to resist, to refuse to go along with; (*sl.*) to cheer up, to invigorate. **to buck up,** to hurry; to improve; to beecome cheerful or lively. **to make a fast buck,** to make money quickly and easily, but not always strictly legally. **to pass the buck,** (*sl.*) to shift responsibility to someone else. **buck-eye,** *n.* the horse-chestnut of the US; (*offensive*) a native of Ohio. **buckhorn,** *n.* the horn of a buck; the material of a buck's horn used for knife-handles etc. **buck-hound,** *n.* a small variety of the stag-hound. **buck-passer,** *n.* a person who avoids responsibility. **buck-passing,** *n.* **buck-shot,** *n.* a kind of shot larger than swan-shot. **buck-skin,** *n.* the skin of a buck; a soft yellowish leather made from deer and sheepskins; (*pl.*) buckskin breeches; *a.* made of buckskin. **buckthorn,** *n.* the genus *Rhamnus,* esp. *R. catharticus,* berries of which yield sapgreen. **bucktooth,** *n.* a large, protruding tooth. **bucked,** *a.* invigorated; pleased. **bucker,** *n.*
buckbean (bŭk'bēn), *n.* a water-plant having pinkish-white flowers, of the genus *Menyanthes,* esp. *M. trifoliata.*
buckboard, *n.* a projecting board or ledge over the wheels of a cart; (*N Am.*) a light four-wheeled vehicle with a sprung platform.
bucket (bŭk'it), *n.* a vessel with a handle, for drawing or carrying water; a scoop or receptacle for lifting mud, gravel, coal, grain etc. in a dredger, excavator or elevator; as much as a bucket will hold (*pl., coll.*) large quantities. *v.t.* to lift or draw in buckets. *v.i.* to travel fast and bumpily; (also **bucket down**) (*coll.*) to rain heavily. **to kick the bucket,** (*sl.*) to die. **bucket-seat,** *n.* a round-backed seat for one person in a motor or aeroplane. **bucket-shop,** *n.* the office of unofficial brokers who deal in trashy stock; a place where cheap airline tickets are sold. **bucketful,** *n.*
buckle (bŭk'l), *n.* a link of metal, with a tongue or catch, for fastening straps etc.; the state of being crisped, curled or twisted. *v.t.* to fasten with or as with a buckle; to bend, to twist, to warp; to prepare (oneself) resolutely. *v.i.* to bend, to be put out of shape. **to buckle to, down,** to set to work, to set about energetically. **to buckle under,** to give way under stress. **buckler,** *n.* a small round shield.
Buckley's chance (bŭk'liz), *n.* (*esp. Austral. coll.*) no chance at all.
bucko (bŭk'ō), *n.* (*pl.* **-koes**) a young fellow.
buckram (bŭk'rəm), *n.* a strong coarse kind of linen cloth, stiffened with gum. *a.* made of buckram.
Bucks. (bŭks), (*abbr.*) Buckinghamshire.
buckshee (bŭkshē'), *a.* (*sl.*) free, gratuitous.
buckthorn BUCK [1].

buckwheat — Bulgarian

buckwheat (bŭk'wĕt), *n.* a cereal plant, *Polygonum fagopyrum*, the three-cornered seeds of which are used for animal fodder and in the US for cakes.

bucolic (būkol'ĭk), *a.* (often *derog.*) pastoral, rustic. *n.* a pastoral poem; a pastoral poet. **bucolically**, *adv.*

bud[1] (bŭd), *n.* the germ of a branch, cluster of leaves or flower; an unexpanded leaf or flower; a gemmule which develops into a complete animal; something undeveloped. *v.i.* to put forth buds; to begin to grow; to develop. *v.t.* to graft (on) by inserting a bud under the bark; to produce by germination. **in bud**, about to flower or put forth leaves. **to nip in the bud**, to put a stop to at the outset. **budded**, *a.* in bud. **budding**, *n.* grafting with a bud; asexual reproduction from a parent cell, as in yeast; (*Zool.*) gemmation. *a.* having buds; beginning to develop; promising; aspiring. **budless**, *a.* **budlet** (-lit), *n.*

bud[2] (bŭd), (*N Am.*, *sl.*) short for BUDDY, used as a form of address.

Buddha (bud'ə), *n.* the title, meaning 'the enlightened', given to Gautama, the founder of Buddhism, by his disciples. **Buddhism**, *n.* the religious system founded in India in the 5th cent. BC by Sakyamuni, Gautama or Siddartha. **esoteric Buddhism** THEOSOPHY. **Buddhist**, *n.* a follower of Buddha. *a.* of or connected with Buddhism. **Buddhistic**, **-ical** (-is'-), *a.*

Buddleia (bŭd'lia), *n.* a genus of shrubs of the family Loganiaceae.

buddy (bŭd'i), *n.* (*pl.* **-ddies**) (*coll.*) close friend, pal esp. a man's close male friend or partner; (*N Am.*, *coll.*) used as a not always friendly form of address; one who visits and counsels (in a voluntary capacity) someone suffering from AIDS. *a.* of a film, story, dealing with adventures of and relationship between usu. two male partners. **buddy-buddy**, *a.* (*coll.*) friendly, comradely; involving or dealing with buddies.

budge (bŭj), *v.i.* to stir; to move from one's place. *v.t.* to cause (something heavy or obstinate) to move.

budgerigar (bŭj'ərigah), *n.* the Australian green parrakeet, *Melopsittacus undulatus*; a small cage-bird bred in many different colours.

budget (bŭj'it), *n.* an estimate of receipts and expenditure, esp. the annual financial statement of the Chancellor of the Exchequer in the House of Commons; a plan for expenditure over a specified period; the amount of money available for a project. *a.* low-cost, economical. *v.i.* to prepare a budget or estimate (for). *v.t.* to make provision for in a budget. **budget account**, *n.* an account which allows one to regularize payments as prescribed in a budget. **budgetary**, *a.*

budgie (bŭj'i), short for BUDGERIGAR.

buff (bŭf), *n.* soft, stout leather prepared from the skin of the buffalo; the skins of other animals similarly prepared; the colour of buff leather, light yellow; (*coll.*) the bare skin; an instrument for polishing with; an expert on or devotee of a subject. *v.t.* to polish with a buff; to give a velvety surface to leather. **in the buff**, (*coll.*) naked. **buffstick**, **buff-wheel**, *n.* a stick or wheel used for polishing metals. **buffy**, *a.*

buffalo (bŭf'əlō), *n.* (*pl.* **-loes**) the name of various kinds of ox, *Bos bubalus*, *B caffer*, and the American bison. **buffalo-grass**, *n.* prairie grass of various kinds.

buffer, *n.* a mechanical apparatus for deadening or sustaining the force of a concussion; an apparatus fixed to railway carriages for this purpose; the fender of a ship; a fellow; a chemical compound which maintains the balance of acidity/alkalinity in a solution; a short-term storage unit in a computer. *v.t.* to add or treat with a buffer; to protect with a buffer. **old buffer**, a doddering old man. **buffer state**, *n.* a small neutral state separating two larger rival states and tending to prevent hostilities. **buffered**, *a.*

buffet[1] (bŭf'it), *n.* a blow with the hand or fist. *v.t.* to strike with the hand; to thump; to cuff; to beat back, to contend with. **buffeting**, *n.* repeated blows.

buffet[2] (bŭf'it), *n.* a cupboard or sideboard for the display of plate, china etc.; (buf'ā), a refreshment bar; (buf'ā), dishes of food set out on a table from which diners help themselves.

buffo (buf'ō), *n.* a singer in a comic opera. *a.* burlesque, comic.

buffoon (bə'foon'), *n.* one who clowns around; a vulgar, clowning fool. **buffoonery**, *n.*

bug (bŭg), *n.* any insect esp. a blood-sucking, evil-smelling insect e.g. a bedbug; a virus; a viral infection; a concealed microphone; a technical hitch, a flaw; an obsession, a temporary craze or fashion. *v.t.* (*past, p.p.* **bugged**) to plant a hidden microphone; to eavesdrop on using a bug; to pester or irritate. **big bug**, an important, aristocratic or wealthy person. **bugged**, *a.* **bugging**, *n.* **buggy**,[1] *a.* infested with bugs.

bugaboo (bŭgəboo), *n.* a bogy; a source of worry.

bugbear (bŭgbeə), *n.* a hobgoblin; a source of terror; a nuisance.

bugger (bŭg'ə), *n.* a sodomite; (*taboo*) a contemptible person; (*sl.*) something difficult, disliked, unwanted etc., a nuisance; (*esp. N Am.*) term of affection used to a child etc. *int.* (*often considered taboo*) used to indicate annoyance, frustration etc. *v.t.* to have anal intercourse with; (*sl.*) to exhaust; (*sl.*) to destroy or spoil. **bugger all**, (*sl.*) nothing. **to bugger about**, (*sl.*) to muddle about, to interfere with a thing. **to bugger off**, (*sl.*) to leave. **buggery**, *n.* anal intercourse.

buggy[2] (bŭg'i), *n.* a light, four-wheeled or two-wheeled vehicle with single seat; a pushchair, a baby buggy; any small, light vehicle (e.g. beach buggy).

bugle (bū'gl), *n.* a small military trumpet used to sound signals. **bugler**, *n.* one who plays a bugle; a soldier who transmits signals on a bugle.

bugloss (bū'glos), *n.* name of plants of the borage family with rough, hairy leaves.

build (bild), *v.t.* (*past, p.p.* **built** (bilt),) to construct, to erect, to make by putting together parts and materials; to put (into a structure); to establish. *v.i.* to erect a building or buildings; to make a nest; to increase in intensity. *n.* form, style or mode of construction; shape, proportions, figure. **to build on, upon**, to found or rely on (as a basis). **to build up**, to establish or strengthen by degrees; to erect many buildings in an area. **build-up**, *n.* a progressive increase; advance publicity; the leading to the climax in a speech etc. **builder**, *n.* one who builds esp. a masterbuilder or contractor who erects or repairs buildings. **builder's merchant**, *n.* a trades person who supplies building materials to builders. **building**, *n.* the act of constructing or erecting; an edifice. **building block**, *n.* any part or unit from which a construction is made; *pl.* small blocks of wood etc. used as toys. **building society**, *n.* an organization lending money to contributors enabling them to purchase houses. **built**, *a.* constructed, erected, fashioned, formed (in *comb.* as *well-built*). **built-in**, *a.* part of the main structure, e.g. cupboards, wardrobe; integral to, included. **built-up**, *a.* of an urban area, having many buildings.

bulb (bŭlb), *n.* a subterranean stem or bud sending off roots below and leaves above, as in the onion or lily; a bulbil; any bulb-shaped thing or part, as e.g. on a thermometer; an electric-light globe. **bulbar**, **bulbed**, *a.* having the form of a bulb. **bulbiferous** (-bif'-), *a.* producing bulbs. **bulbiform** (-bifawm), *a.* **bulbil** (-bil), *n.* a small bulb developed at the side of a larger one, or in an axil. **bulbous**, **bulbose** (-bos), *a.* of or pertaining to a bulb; bulb-shaped.

bulbul (bul'bul), *n.* an Eastern bird belonging to the thrush family.

Bulgarian (bŭlgeə'riən), *n.* the language or people of Bulgaria. *a.* belonging to Bulgaria.

bulge (bŭlj), *n.* a convex protuberance or swelling; a temporary increase in volume or numbers; *v.i.* to swell irregularly; to be protuberant. *v.t.* to swell out (a bag); to push out of shape. **bulginess,** *n.* **bulging,** *a.* protuberant.
bulgy, *a.* swollen so as to be clumsy.
bulimia (nervosa) (bŭlim'iə nœvō'sə), **bulimy** (bū'limi), *n.* a medical condition characterized by abnormal hunger.
bulk (bŭlk), *n.* size, great size, mass; the greater portion, the main mass; anything of great size; the trunk of the body, esp. if large; roughage. *v.i.* to appear relatively big or important; to amount. *v.t.* to pile in heaps; to pack in bulk. **in bulk,** cargo loose in the hold; in large quantities. **bulk buying,** *n.* the purchase of goods in large quantities to obtain cheaper prices; the purchase by one customer of the whole of a producer's output. **bulk carrier,** *n.* a vessel or vehicle which carries a large, undivided cargo. **bulkily,** *adv.* **bulkiness,** *n.* **bulky,** *a.* of great bulk or dimensions; large and unwieldy.
bulkhead (bŭlk'hed), *n.* an upright partition dividing a ship or aircraft into compartments.
bull [1] (bul), *n.* the uncastrated male of any bovine mammal; the male of some other large animals, e.g. the elk, the elephant, the whale; one who speculates for a rise in stocks (see also BEAR); the constellation and sign Taurus; (a hit in) the bull's eye; (*sl.*) in the armed forces, unnecessary and tiresome discipline and fatigues; (*sl.*) rubbish, nonsense. *a.* of large size; thickset; coarse; male. *v.i.* to speculate for a rise (in stocks); of a cow, to low when in season. *v.t.* to produce a rise in (stocks etc.). **a bull in a china shop,** an indelicate or tactless person, a blunderer. **John Bull,** the English people personified; an Englishman. **to take the bull by the horns,** to grapple with a difficulty boldly. **bull-baiting,** *n.* the baiting of a bull with dogs. **bulldog** *n.* a powerful breed of dogs formerly used to bait bulls; one who possesses obstinate courage; one of the proctor's attendants at Oxford and Cambridge. **Bulldog clip,**® *n.* a metal spring clip for fastening papers together or onto a board. **bullfight,** *n.* a Spanish sport in which a bull is baited and then killed. **bull-fighter,** *n.* **bull-fighting,** *n.* **bullfinch** (bul'finch), *n.* an English song-bird with handsome plumage, belonging to the genus *Pyrrhula*; a high, quick-set hedge with a ditch on one side. **bull-frog,** *n.* a large American frog with a deep voice, *Ranma pipiens*. **bullhead,** *n.* the miller's thumb, a small river-fish *Cottus gobio*, with a big head. **bull-headed,** *a.* with a massive head; stupid; obstinate; impetuous. **bull horn,** *n.* a loudspeaker. **bull-puncher,** *n.* (*Austral.*) a cattle-driver. **bull-pup,** *n.* a young bulldog. **bull-ring,** *n.* an arena for a bullfight; a place where bulls used to be baited. **bull-roarer,** *n.* a thin slat of wood that produces a formidable noise when swung rapidly with a string, now a plaything orig. used in religious rites of e.g. Australian aborigines, N Am. Indians. **bull's-eye,** *n.* a boss of glass in the middle of a blown sheet; a peppermint-flavoured sweet; a hemispherical disk of glass in the side or deck of a ship to give light below; a hemispherical lens in a lantern; a lantern with such a lens; a small round window; (a hit in) the centre of a target; something that achieves its aim. **bullshit** (bul'shit), *n.* (*sl.*) rubbish, deceptive nonsense. *v.i.* to talk rubbish, to attempt to deceive with nonsense. **bull-terrier,** *n.* a cross between a bulldog and a terrier, but now an acknowledged breed. **bullish,** *a.* resembling a bull; obstinate; on the stockmarket, a tendency towards rising share prices. **bullishly,** *adv.* **bullishness,** *n.* **bullock** (-ək), *n.* a castrated bull; an ox.
bull [2] (bul), *n.* a Papal edict.
bullace (bul'əs), *n.* a wild plum, *Prunus insititia*, having two varieties, one with white, the other with dark fruit.
bulldoze (bul'dōz), *v.t.* to level ground using a bulldozer; to force or bully. **bulldozer,** *n.* a power-operated machine with a large blade, employed for removing obstacles,

levelling ground and spreading material; a person who bulldozes, a bully.
bullet (bul'it), *n.* a metal ball or cone used in fire-arms of small calibre; a small round ball; a round missile; a fisherman's sinker. **to get the bullet,** (*sl.*) to be dismissed, get the sack. **bullet-head,** *n.* a round-shaped head; (*esp. N Am.*) an obstinate fellow. **bullet-headed,** *a.* **bullet-proof,** *a.* impenetrable to bullets. **bullet train,** *n.* a high-speed train, esp. in Japan.
bulletin (bul'ətin), *n.* an official report of some matter of public interest, e.g. of the condition of an invalid; a brief news item on radio or television, a news bulletin. **bulletin board,** *n.* a notice-board; a store of computerized information or computer files accessible via modem to users. **bulletinist,** *n.*
bullion (bul'yən), *n.* uncoined gold and silver in the mass; solid gold or silver; fringe made of gold or silver wire. *a.* made of solid gold or silver.
bullock (bul'ək) BULL [1].
bully [1] (bul'i), *n.* a blustering, overbearing person; a cowardly tyrant; a bravo, a swashbuckler, a hired ruffian. *a.* jolly, first-rate, capital. *v.t.* to treat in a tyrannical manner; to tease, oppress, terrorize. *v.i.* to act as a bully. **bully for you,** (*sometimes iron.*) well done! bravo! **bully boy,** *n.* a thug; a hired ruffian.
bully [2], **bully-off** (bul'i), *n.* in hockey, the starting of a game. *v.t.* to start a game of hockey.
bully [3] (bul'i), *n.* tinned beef, also called **bully beef.**
bullyrag (bul'irag), BALLYRAG.
bulrush (bul'rŭsh), *n.* a tall rush growing in water, *Scirpus lacustris*, or *Typha latifolia*, the reed-mace or cat's-tail; (*Bibl.*) the papyrus. **bulrushy,** *a.*
bulwark (bul'wək), *n.* a rampart or fortification; a mole, a breakwater; any shelter, protection, screen; that part of the sides of a ship which rises above the upper deck. *v.t.* to furnish with or protect as with bulwarks.
bum [1] (bŭm), *n.* the buttocks; the anus.
bum [2] (bŭm), **bummer** *n.* (*N Am.*) an idler, a loafer; (*coll.*) a tramp; (*coll.*) a scrounger; (*coll.*) a devotee of a particular form of recreation, e.g. a *beach-bum*. *a.* (*coll.*) useless; broken; worthless. *v.i.* (*coll.*) to live like a tramp; to idle; to scrounge. *v.t.* (*coll.*) to acquire by scrounging.
bumble (bŭm'bl), *v.i.* too speak, act or move in a clumsy, muddled manner. **bumble-bee,** *n.* a large bee belonging to the genus *Bombus*; a humble-bee. **bumbler,** *n.*
bumboat (bŭm'bōt), *n.* a boat used to carry provisions to vessels.
bumf, bumph (bŭmf), *n.* (*sl.*) toilet paper; (*derog.*) official documents; any unwanted paperwork.
bum(m)alo (bŭm'əlō), *n.* Bombay duck; a small Asiatic fish, dried and used as a relish.
bummer BUM [2].
bump (bŭmp), *n.* a thump, a dull, heavy blow, an impact or collision; a swelling; a protuberance on the skull, supposed to indicate distinct character traits; a sudden movement of an aircraft caused by turbulence. *v.t.* to cause to strike forcibly against anything hard or solid; to hurt by striking against something; to hit (against). *v.i.* to strike heavily; to collide; to move along with a bump or succession of bumps. *adv.* with a bump; with a sudden shock. **to bump into,** to meet unexpectedly; to encounter accidentally. **to bump off,** (*coll.*) *v.t.* to murder. **to bump start,** to start a motor vehicle by pushing it while engaging the gears; to jump start. **to bump up,** to increase (prices); to raise. **bumper,** *n.* a glass filled to the brim, esp. for drinking a toast; the fender of a motor-car; a buffer; (*coll.*) anything very large and wonderful or full. *a.* (*coll.*) extraordinary, startling, fine; full to the brim. **bumpy,** *a.* full of bumps, uneven, jolty. **bumpily,** *adv.* **bumpiness,** *n.*
bumph BUMF.
bumpkin (bŭmp'kin), *n.* a clumsy, unsophisticated country

bumptious — **burger**

person.
bumptious (bŭmp'shəs), *a.* disagreeably self-assertive. **bumptiously,** *adv.* **bumptiousness,** *n.*
bun[1] (bŭn), *n.* a small sweet cake; a compact ball of hair worn at the back of the head. **hot cross bun** HOT. **bunfight,** *n.* a crowded tea-party; a disturbance at an assembly.
bunch (bŭnch), *n.* a cluster of several things of the same kind growing or tied together; a tuft, a knot, a bow; a lot, a collection, a group. *v.t.* to tie up or form into a bunch; to gather into folds. *v.i.* to come or grow into a cluster or bunch. **to bunch up,** to form into a compact group. **bunching,** *n.* forming too tight or close a group. **bunchy,** *a.* forming a bunch; growing in bunches. **bunchiness,** *n.*
buncombe BUNKUM.
Bundestag (bun'dəstahg), *n.* the parliament of the Federal Republic of Germany.
bundle (bŭn'dl), *n.* a number of things or a quantity of anything bound together loosely; a package, a parcel; a set of rods, wires, fibres, nerves etc., bound together; a group of characteristics; (*sl.*) a large amount of money. *v.t.* to tie up in a bundle; to throw hurriedly together; to put or hustle unceremoniously. *v.i.* to start hurriedly (in, off, away or out). **to bundle off,** to send away hurriedly or unceremoniously; to dismiss. **to bundle up,** to gather into a bundle; to clothe warmly. **to go a bundle on,** to like enormously, to be enthusiastic for.
bundu (bun'doo), *n.* (*S Afr.*) the back of beyond, the far interior.
bung (bŭng), *n.* a large cork stopper for a bung-hole; a bung-hole. *v.t.* to stop with a bung; to close, to shut up; (*coll.*) to throw, to sling; (*coll.*) to put or shove carelessly. **to go bung,** (*Austral. coll.*) to go bankrupt. **bunged-up,** *a.* (*coll.*) congested, blocked; suffering from congestion of the nose or sinuses. **bung-hole,** *n.* the hole in the bulge of a cask through which it is filled.
bungalow (bŭng'gəlō), *n.* a one-storied house.
bungle (bŭng'gl), *v.t.* to botch; to manage clumsily or awkwardly. *v.i.* to act clumsily or awkwardly; to fail in a task. *n.* a mistake, botch. **bungler,** *n.* **bungling,** *a.* clumsy, awkward, unskilful. **bunglingly,** *adv.*
bunion (bŭn'yən), *n.* a swelling on the foot, esp. of the joint of the great toe.
bunk[1] (bŭngk), *n.* a box or recess serving for a bed; a sleeping-berth; (*N Am.*) a piece of timber on a sled to support heavy timber. *v.i.* to sleep in a bunk. **bunk bed,** *n.* one of a pair of narrow beds built one above the other. **bunker** *n.* a sandy hollow or other obstruction on a golf course; a container or bin usu. for coal or fuel, e.g. on a ship; a fortified underground shelter. **bunkered,** *a.* in golf, having hit one's ball into a bunker.
bunk[2] (bŭngk), *v.i.* (*sl.*) to make off, to bolt. *n.* a bolt; a making off, an escape. **to bunk off,** (*sl.*) to play truant. **to do a bunk,** (*sl.*) to run away.
bunk[3] BUNKUM.
bunkum, buncombe (bŭng'kəm), **bunk,** *n.* political claptrap; tall talk, humbug.
bunny (bŭn'i), *n.* a childish name for a rabbit. **bunny girl,** *n.* a waitress in a night-club who wears a sexually provocative costume including rabbit ears and tail. **bunny-hug, bunny-hugging,** *n.* a romping kind of dance in which the partners closely embrace each other.
Bunsen burner, lamp (bŭn'sən), *n.* a burner or lamp in which air is mingled with gas to produce an intense flame.
bunt[1] (bŭnt), *n.* the middle part of a sail, formed into a cavity to hold the wind; the baggy part of a fishing-net. **buntline,** *n.* a rope passing from the foot-rope of a square sail and in front of the canvas to prevent bellying.
bunt[2] (bŭnt), *n.* a fungus, *Tilletia caries,* which attacks wheat. **bunted, bunty,** *a.*

bunt[3] (bŭnt), *v.t., v.i.* to hit, push, butt.
bunting[1] (bŭn'ting), *n.* a group of birds, the Emberizinae, allied to the larks.
bunting[2] (bŭn'ting), *n.* a thin fabric of which flags are made; a flag; flags collectively (e.g. strung up as decoration).
bunya-bunya, bunya (bŭnyəbŭn'yə), *n.* a large conifer with edible seeds.
bunyip (bŭn'yip), *n.* (*Austral.*) the fabulous rainbow-serpent that lives in pools.
buoy (boi), *n.* an anchored float indicating a fairway, reef, shoal etc. *v.t.* to place a buoy upon, to mark with a buoy. **life-buoy,** *n.* a float to sustain a person in the water. **to buoy up,** to keep afloat, to bear up, bring to the surface. **buoyage** (-ij), *n.* the act of providing with buoys. **buoyancy** (-ənsi), *n.* ability to float; loss of weight due to immersion in a liquid; ability to recover from setbacks, resilience; lightheartedness; tendency to rise (of stocks, prices etc.). **buoyant,** *a.* tending to float; tending to keep up; elastic, light; resilient; cheerful. **buoyantly,** *adv.*
bur[1]**, burr**[2] (bœ), *n.* any prickly or spinous head or seed-case of a plant; any plant which produces these; knot of excrescence on a tree; hence the series of markings left in the timber, which are valuable for the effect in polished veneer etc.; someone or thing hard to get rid of; a lump in the throat; a small drill used by dentists and surgeons. **burdock** (-dək), *n.* a coarse plant with prickly flower-heads, of the genus *Arctium,* esp. *A. lappa.* **bur-thistle,** *n.* the spear-thistle, *Carduus lanceolatus.*
bur[2] (bœ), BURR[1].
Burberry® (bœ'bəri), *n.* a type of weatherproof cloth or clothing; a raincoat.
burble (bœ'bl), *v.i.* to bubble, gurgle, to talk inconsequently or excitedly.
burbot (bœ'bət), *n.* the eel-pout, *Lota vulgaris,* a flat-headed freshwater fish.
burden (bœ'dən), *n.* something borne or carried; a load; a load of labour, sin, sorrow, care, obligation, duty, taxation, expense, fate etc.; the carrying capacity of a vessel; tonnage; the principal theme, the gist of a composition of any kind; a refrain, a chorus. *v.t.* to load; to lay a burden on; to oppress, to encumber. **the burden of proof,** the obligation of proving a contention or assertion. **burdensome** (-səm), *a.* hard to bear; grievous, oppressive. **burdensomely,** *adv.* **burdensomeness,** *n.*
burdock BUR[1].
bureau (bū'rō), *n.* (*pl.* **bureaux** (-z),) a writing table with drawers for papers; a chest of drawers; an office; a public office; a Government department. **bureau de change** (də rshahzh'), *n.* an office or kiosk (e.g. in an airport, railway station) for exchanging currencies.
bureaucracy (bŭrok'rəsi), *n.* government by departments of state; centralization of government; officials as a body; officialism. **bureaucrat** (bū'rəkrat), *n.* a government official; a bureaucratist. **bureaucratic** (-krat'-), *a.* pertaining to or constituting a bureaucracy; tending towards bureaucracy. **bureaucratically,** *adv.* **bureaucratism** (-rok'rə-), *n.* **bureaucratist** (-rok'rə-), *n.* one who advocates or supports bureaucracy. **bureaucratize, -ise,** *v.t.* to make into a bureaucracy. **bureaucratization,** (-rok'rə-), *n.*
burette, buret (būret'), *n.* a graduated glass tube for measuring small quantities of liquid.
burg (bœg), *n.* a fortress; a walled town; (*N Am. coll.*) a town or city.
burgee (bœ'jē), *n.* a triangular or swallow-tailed flag.
burgeon, bourgeon (bœ'jən), *v.t.* to sprout, to bud; to begin to grow.
burger (bœ'gə), *n.* a flat round cake of minced meat or vegetables, e.g. *hamburger, beefburger,* which is grilled or fried; a burger served in a bread roll or bun often with a topping, e.g. *cheese burger, chilli burger.* **burger bar,** *n.* a small restaurant or take-away which mainly sells burgers.

burgess (bŭr'jis), *n.* an inhabitant of a borough possessing full municipal rights, a citizen; a freeman of a borough. **burgess-ship,** *n.*

burgh (bŭ'rə), *n.* a Scottish town holding a charter; a borough. **burghal** (bŭr'gəl), *a.* pertaining to a burgh. **burgher** (bŭr'gə), *n.* a citizen or inhabitant of a burgh, borough or corporate town, esp. of a Continental town. **burghership,** *n.*

burglar (bŭr'glə), *n.* one who breaks into premises with intent to commit a felony, esp. theft. **burglarious** (-gleə'ri-), *a.* **burglariously,** *adv.* **burglarize, -ise,** *v.t.* (*N Am.*) to enter or rob burglariously. **burglary,** *n.* **burgle,** (-gl), *v.t.* to commit burglary against.

burgomaster (bŭr'gəmahstə), *n.* the mayor of a municipal town in Austria, Germany, Holland or Flanders.

Burgundy (bŭr'gəndi), *n.* an old province in France; red or white wine made in Burgundy. **Burgundy mixture,** *n.* a fungicide used for spraying potatoes.

burial (bĕ'riəl), *n.* the act of burying, esp. of a dead body; interment; a funeral. **burial-ground, -place** *n.* **burial-mound,** *n.* a tumulus. **burial-service,** *n.* a religious service (esp. of the Church of England) for the burial of the dead.

burin (bū'rin), *n.* the cutting-tool of an engraver on copper; a triangular steel tool used by marble-workers; an early Stone Age flint tool.

burk (bŭk), BERK.

burk(h)ah (bŭr'kə), *n.* the long veil or veiled, loose overgarment worn by Muslim women.

burl[1] (bŭl), *n.* a knot or lump in wool or cloth; a knot in wood. *v.t.* to dress (cloth) by removing knots or lumps.

burl[2], **birl**[2] (bŭl), *n.* (*Sc., New Zealand, Austral., coll.*) a spin (in a motor vehicle); an attempt, a try.

burlap (bŭr'lap), *n.* a coarse kind of canvas used for sacking, upholstering etc.

burlesque (bŭlesk'), *a.* drolly or absurdly imitative; mock-serious or mock-heroic. *n.* mockery, grotesque imitation; literary or dramatic representation caricaturing other work; (*N Am.*) a form of theatrical variety show characterized by lewd humour, singing and dancing and strip-tease. *v.t.* to produce a grotesque imitation of; to travesty. **burlesquely,** *adv.*

burly (bŭr'li), *a.* stout. **burliness,** *n.* sturdy.

Burmese (bœmēz'), **Burman** (bœ'mən), *n.* the language, people of Burma. *a.* belonging to Burma.

burn[1] (bŭrn), *v.t.* (*past, p.p.* **burnt** (-t), sometimes **burned** (-d),), to consume, destroy, scorch or injure by fire; to subject to the action of fire; to make by, or as if by, the action of fire; to produce an effect (on anything) similar to the action of fire; to treat with heat for some purpose of manufacture etc.; to corrode, eat into; to use as fuel; to put to death by fire; to cauterize. *v.i.* to be on fire; to be or become intensely hot; to emit light, to shine; to act with destructive effect; to be bright, to glow with light or colour; to be passionately eager; to experience violent emotion; to travel at high speed. *n.* the effect of burning; a burnt place; a firing of a space-rocket engine to obtain thrust. **to burn a hole in one's pocket,** describing money one is keen to spend immediately. **to burn away,** to consume entirely by fire. **to burn down,** to reduce to ashes. **to burn in,** to render indelible by or as by burning. **to burn off,** to remove by burning. **to burn one's boats, bridges,** to commit oneself to something without possibility of retreat. **to burn one's fingers,** to hurt oneself by meddling. **to burn out,** to consume the inside or contents of; (*coll.*) to exhaust or render inoperative through overwork or overheating; to eradicate or expel by burning. **to burn up,** to destroy, to get rid of, by fire; to blaze (up). **to go for the burn,** (*coll.*) to try to achieve the burning sensation in the muscles, produced by strenuous exercise, to exercise hard. **burn-up,** *n.* (*coll.*) a fast drive in a motor vehicle; the consumption of nuclear fuel in a reactor.

burned up, *a.* (*N Am. sl.*) angry. **burnable,** *a.* **burner,** *n.* that part of a lamp or gas-jet from which the flame issues. **burning,** *a.* on fire; very hot; ardent, passionate; crucial. **burning bush,** *n.* the bush that burned and was not consumed (Exod. iii.2); any of various shrubs with vivid foliage, fruit etc. **burning-glass,** *n.* a convex lens used for causing intense heat by concentrating the sun's rays. **burning-mirror, -reflector,** *n.* a concave mirror, or a combination of plane-mirrors arranged to act as a burning-glass. **burning question, issue** *n.* one that excites heated discussion or that demands immediate solution. **burnt offering,** *n.* an offering or sacrifice to a deity by fire, esp. one offered to God by the Jews. **burnt-sienna** SIENNA. **burnt umber** UMBER.

burn[2] (bŭrn), *n.* (*chiefly Sc.*) a small stream, a brook.

burnet (bŭr'nit), *n.* brown-flowered plants of the genera *Poterium* and *Sanguisorba,* **burnet-fly, -moth,** *n.* a crimson-spotted, greenish-black moth, *Zygaena filipendulae.*

burnish (bŭr'nish), *v.t.* to polish, esp. by rubbing. *n.* polish, gloss. **burnisher,** *n.*

burnous, -nouse (bənoos'), *n. sing.* a mantle or cloak with a hood, worn by Arabs.

burnt (bŭrnt), *past, p.p.* BURN[1].

burp (bŭp), *n.* a belch. *v.i.* to belch. *v.t.* to make a baby burp by massaging or patting on the back.

burr[1] (bŭr), *n.* the round, knobby base of a deer's horn; a rough ridge or edge left on metal or other substance after cutting, punching etc.; the roughness made by the graver on a copper plate; a triangular hollow chisel; a mass of semi-vitrified brick; a rough sounding of the letter *r*; a whirring noise; an electric rotary filing tool. *v.i., v.t.* to speak with a rough sounding of the *r*. **burry** (bŭr'ri), *a.* rough, prickly.

burr[2] BUR[1].

burro (bu'rō), *n.* (*pl.* **-rros**) (*mainly N Am.*) a donkey.

burrow (bŭ'rō), *n.* a hole in the ground lived in by rabbits, foxes etc. *v.i.* to excavate a burrow for shelter or concealment; to live in a burrow; to bore or excavate; to nestle into; to dig deep while searching (e.g. in a pocket). *v.t.* to make by excavating. **burrower,** *n.*

bursa (bŭr'sə), *n.* (*pl.* **-sas, -sae**) a synovial sac found among tendons in the body and serving to reduce friction. **bursal,** *a.* **bursar** *n.* a treasurer, esp. of a college; one who holds a bursary. **bursarial,** *a.* **bursarship,** *n.* **bursary,** *n.* a bursar's office; a scholarship. **bursitis** (-sitis), *n.* inflammation of a bursa.

burst (bŭst), *v.t.* (*past, p.p.* **burst**) to break, split or rend asunder with suddenness and violence. *v.i.* to be broken suddenly from within; to fly open; to rush forth with suddenness and energy or force. *n.* a sudden, violent breaking forth; a sudden explosion; an outbreak; a spurt, a vigorous fit of activity; a volley of bullets. **to burst in,** to enter suddenly; to interrupt. **to burst out,** to break out; to exclaim. **bursting,** *n., a.* **bursting at the seams,** being too full for comfort.

burthen BURDEN.

burton (bŭr'tən), *n.* **gone for a burton,** (*sl.*) dead; absent, missing; out of action, useless.

bury (bĕ'ri), *v.t.* to inter, to consign to the grave (whether earth or sea); to perform funeral rites for; to put under ground; to consign to obscurity, oblivion etc.; to hide, to cover up, to embed; to occupy deeply, engross, absorb. **to bury the hatchet,** to forget and forgive, to effect a reconciliation (in allusion to an American Indian custom of burying a tomahawk when peace was concluded). **burying,** *n.* burial. **burying-ground, -place,** *n.*

bus (bŭs), *n.* (*pl.* **buses**) an omnibus; (*sl.*) an aeroplane, car etc; a series of conductors in a computer which carry information or power. *v.i.* to go by omnibus. *v.t.* to transport by bus. **to miss the bus,** (*coll.*) to miss an opportunity, to be too late. **busbar,** *n.* in an electric system, a conductor or series of conductors connecting several circuits; in computers, a bus. **busboy, -girl,** *n.* (*N Am.*) a restaurant employee who assists the waiters or waitresses. **bus fare,** *n.* the payment for a bus journey. **bus lane,** *n.* a traffic lane restricted to the use of buses. **busman's holiday,** *n.* (*coll.*) holiday spent doing one's everyday work. **bus shelter,** *n.* a shelter erected at a bus stop to protect waiting passengers against the weather. **bus stop,** *n.* a place marked by a sign at which buses stop to pick up or let off passengers. **busing, bussing,** *n. sing.* (*N Am.*) transporting children by bus to schools outside their areas to achieve evenly balanced racial numbers in classrooms.

busby (búz'bi), *n.* the tall fur cap worn by hussars; a bearskin hat worn by the Guards.

bush[1] (bush), *n.* a thick shrub; a clump of shrubs; a thicket; anything resembling a bush; uncleared land, more or less covered with wood, esp. in Australasia; the interior, the wild; a thick growth of hair; *v.i.* to grow bushy. **to beat about the bush,** to take circuitous methods. **bush-baby,** *n.* a small nocturnal African primate, *Galago maholi.* **bushbuck,** *n.* a small bush-dwelling African antelope. **bush-cat,** *n.* the servil. **bush fire,** *n.* a usu. fast spreading fire in the bush. **bush jacket, shirt,** *n.* a belted upper garment of a light-weight material equipped with large pockets. **bush-lawyer,** *n.* (*Austral.*) an irregular legal practitioner. **bushman** *n.* one who lives in the Australian bush; (**Bushman**) a member of a primitive nomadic tribe in S Africa. **bushmanship,** *n.* **bushmaster** *n.* a large and poisonous rattlesnake in S America. **bushranger,** *n.* an outlaw who has taken to the Australian bush. **bush-telegraph,** *n.* the rapid dissemination of rumours, information etc. **bushveld, bosveld,** *n.* wooded S African grasslands. **bushwhack,** *v.i.* to clear a way in the bush; to live as a bushwhacker. *v.t.* to ambush. **bushwhacker,** *n.* guerilla; (*N Am.*) a backwoodsman; (*Austral.*) an inhabitant of the outback, a country bumpkin. **bushwhacking,** *n.* **bushed,** *a.* (*Austral.*) lost in the bush; (*sl.*) confused; (*sl.*) exhausted. **bushiness,** *n.* **bushy,** *a.* abounding with bushes; shrubby, thick; growing like a bush.

bush[2] (bush), **bushing** *n.* the metal lining of an axle-hole or similar orifice. *v.t.* to furnish with a bush; to line with metal. **bush-metal,** *n.* an alloy of copper and tin used for bearings etc; gunmetal.

bushel[1] (bush'l), *n.* a dry measure of 8 gal. (36·37 litres). **to hide one's light under a bushel,** to conceal one's skills or talents. **bushelful,** *n.*

bushido (bushē'dō), *n.* the code of honour of the Japanese Samurai.

busily, *adv.* BUSY.

business (biz'nis), *n.* employment, occupation, trade, profession; serious occupation, work; concern, province; commercial, industrial or professional affairs; commercial activity a particular matter demanding attention; a commercial establishment; a shop, with stock, fixtures etc.; (*Theat.*) action, as distinct from speech; (*coll.*) an affair, a matter. **like nobody's business,** vigorously; zealously. **to mean business,** to be in earnest. **to mind one's own business,** to attend to one's own affairs; to refrain from meddling. **to send someone about his/her business,** to send someone off brusquely or summarily. **business card,** *n.* a card printed with a company's name, address and phone number, and the identity of the employee or executive who carries it. **business end,** *n.* the point (of a tool or weapon). **business hours,** *n.* fixed hours of work or for transaction of business in a shop, office etc. (esp. 9 am to 5 pm). **businessman, -woman,** *n.* one who deals with matters of commerce etc. **business studies,** *n. pl.* a college or university course relating to business. **business suit,** *n.* a lounge suit. **businesslike,** *a.* suitable for or befitting business; methodical, practical;

prompt, punctual; energetic.

busk (búsk), *v.i.* to perform in the street or in a public place, esp. beside a queue, in order to collect money. **busker,** *n.* **busking,** *n.*

buskin (bús'kin), *n.* a kind of high-boot reaching to the calf or knee; the thick-soled boot worn by actors in Athenian tragedy. **buskined,** *a.*

bust[1] (bust), *n.* a sculpture of the head, shoulders and breast of a person; the breast, the bosom, esp. of a woman. **busted,** *a.* having breasts. **busty,** *a.* (*coll.*) having ample breasts.

bust[2] (bust), *v.i.* (*coll.*) to break or burst. *v.t.* (*coll.*) to break or burst; (*sl.*) to raid or arrest, esp for a drug offence. *n.* (*sl.*) a drinking spree; (*sl.*) a raid or arrest; (*N Am.*) a punch. *a.* broken; bankrupt. **a bust-up,** a quarrel. **to go bust,** to go bankrupt. **buster,** *n.* (*Austral.*) a gale; (*coll., sometimes derog.*) a form of address to a boy or man; (*sl.*) something or person that breaks or destroys (esp. in combs. e.g. *crimebuster*)

bustard (bús'təd), *n.* a large bird allied to the plovers and the cranes, belonging to the genus *Otis.*

bustle[1] (bús'l), *n.* activity with noise and excitement. *v.i.* to be active, esp. with excessive fuss and noise; to make a show of activity. **bustler,** *n.*

bustle[2] (bús'l), *n.* a pad, cushion or framework, worn under a woman's dress to expand the skirts behind.

busy (biz'i), *a.* fully occupied; actively employed; characterized by activity, unresting, always at work; fussy, officious, meddlesome; overcrowded with detail. *v.t.* †to make or keep busy. **busy Lizzie** (liz'i), *n.* a popular flowering house plant belonging to the genus *Impatiens.* **busybody,** *n.* an officious person; a meddler; a mischiefmaker. **busily,** *adv.* **busyness,** *n.* the state of being busy.

but[1] (but), *prep.* except, barring; (*Sc.*) apart from, outside of; *conj.* yet still; notwithstanding which; except that; otherwise than, not that; on the contrary, nevertheless, however. *n.* a verbal objection; (*Sc.*) an outer room. *adv.* only; (*Sc.*) outwards. **a but and ben,** (*Sc.*) a two-roomed cottage. **all but,** almost, very nearly. **anything but,** definitely not. **but and ben,** (*Sc.*) out and in. **but for, that,** were it not for, not that.

butadiene (būtədī'ēn), *n.* the gas used in making synthetic rubber.

butane (bū'tān), *n.* an inflammable gaseous hydrocarbon of the paraffin series found in petroleum.

butch (buch), *a.* (*sl.*) aggressively masculine in manner or appearance; *n.* (*derog. sl.*) a lesbian with masculine manners or appearance; the dominant or masculine partner in a lesbian relationship; a tough, aggressive man.

butcher (buch'ə), *n.* a tradesman who slaughters domestic animals for food; one who sells meat; one who delights in killing. *v.t.* to slaughter animals for food; to put to death in a wanton or sanguinary fashion; to spoil by bad workmanship or performance; to criticize savagely. **butcherbird,** *n.* a shrike. **butcher's,** *n.* (*sl.*) a look. **butcher's knife,** *n.* a carving-knife. **butcher's-broom,** *n.* a prickly, evergreen British shrub. **butcher('s) meat,** *n.* meat sold fresh by butchers. **butcherly,** *adv.* **butchery,** *n.* the business of a butcher; a slaugher-house; cruel and remorseless slaughter, carnage.

butene (bū'tēn), *n.* BUTYLENE under BUTYL.

butler (but'lə), *n.* a servant in charge of the wine, plate etc.; a head servant. **butlership,** *n.* **butlery,** *n.* a butler's pantry; a buttery.

butment (but'mənt), ABUTMENT.

butt[1] (but), *n.* the hinder, larger or blunter end of anything, esp. of a tool, weapon and the like; the square end of a piece of timber coming against another piece; the joint so formed; the bole of a tree; the base of a leaf-stalk. *v.i.* to abut; to meet with the end against (of timber, planks etc.); to meet end to end. **butt-end,** *n.* the thick and heavy end; the remnant. **butt-hinge,** *n.* a kind of

hinge screwed to the edge of the door and the abutting edge of the casing. **butt-joint**, *n.* a joint in which the pieces come square against each other. **butt-weld**, *n.* a weld made without overlapping.
butt² (büt), *n.* a large cask.
butt³ (büt), *n.* a target, a mark for shooting; hence the mound behind targets, the shelter for the marker, and (*pl.*) the shooting-range; a target for ridicule, criticism or abuse.
butt⁴ (büt), *v.i., v.t.* to strike, thrust or push with the head or as with the head. *n.* a blow with the head or horns. **to butt in**, to interfere, interrupt. **butter**¹, *n.* an animal which butts.
butte (büt), *n.* (*N Am.*) an abrupt, isolated hill or peak.
butter² (büt'ə), *n.* the fatty portion of milk or cream solidified by churning; applied also to various substances of the consistency or appearance of butter. *v.t.* to spread or cook with butter. **to butter up**, (*coll.*) to flatter. **butter bean**, *n.* a variety of lima bean. **butter-bur**, **-dock**, *n.* the sweet coltsfoot. **buttercup**, *n.* popular name for the genus *Ranunculus*, esp. those species with yellow cup-shaped flowers. **butterfat**, *n.* the fat in milk from which butter is made. **butter-fingered**, *a.* apt to drop things. **butterfingers**, *n.sing.* one who is butter-fingered. **butterfly**, *n.* (*pl.* **-flies**) an insect with erect wings and knobbed antennae belonging to the diurnal Lepidoptera; a showily dressed, vain or fickle person; a swimming stroke performed on the front and characterized by simultaneous wide, upward strokes of the arms. (*pl., coll.*) nervous tremors. **butterflies in the stomach**, (*coll.*) nervous tremors in the stomach. **butterfly-nut**, **-screw**, *n.* a screw with a thumb-piece, a wing nut. **buttermilk**, *n.* that part of the milk which remains when the butter is extracted. **butter-muslin**, *n.* a fine loosely woven, cotton material used for protecting food from insects. **butter-nut**, *n.* the N American white walnut-tree, *Juglans cinerea*, and its fruit. **butter-scotch**, *n.* a kind of toffee. **butterwort**, *n.* a British bog-plant belonging to the genus *Pinguicula*. **buttered**, *a.* **butteriness**, *n.* **buttery**¹, *a.* having the qualities or appearance of butter.
buttercup BUTTER.
buttery² (büt'əri), *n.* a room in which food and drink are stored and/or, esp. in a college, sold.
buttock (büt'ək), *n.*, (*usu. in pl.*) one of the protuberant parts of the rump, a manoeuvre in wrestling.
button (büt'n), *n.* a knob or disk used for fastening or ornamenting garments; a small bud; a small handle, knob, etc. actuating electrical apparatus etc.; the knob on a foil. *a.* of mushrooms, blooms etc., having a small round shape. *v.t.* to fasten or furnish with buttons. **not to care a button**, to be quite indifferent about something. **not worth a button**, of no value. **the button**, a button which, when pushed, puts the apparatus for nuclear war into operation. **to button one's lip**, (*sl.*) to shut up. **to button up**, to arrange, to settle satisfactorily; to keep silent; to silence. **buttonhole**, *n.* a hole, slit or loop to admit a button; a small bouquet for the buttonhole of a coat. *v.t.* to hold by the buttonhole; to detain in conversation; to make buttonholes. **buttonholer**, *n.* (*coll.*) one who detains in conversation. **buttonhook**, *n.* a hook for drawing buttons through buttonholes. **button-mould**, *n.* a disk of metal or other substance to be covered with cloth, so as to form a button. **button-through**, *a.* of a garment, having button fastenings from top to bottom. **buttoned**, *a.* **buttonless**, *a.* **buttonlessness**, *n.* **buttons**, *n.sing.* (*coll.*) a page in buttoned livery. **buttony**, *a.*
buttress (büt'ris), *n.* a structure built against a wall to strengthen it; a prop, support; a spur or supporting ridge of a hill. *v.t.* to support by or as by a buttress.
butty¹ (büt'i), *n.* (*dial.*) a companion, a mate. **butty-system**, *n.* the letting of work to a body of men who divide the proceeds.

butty² (büt'i), *n.* (*dial.*) a sandwich, a snack.
butyl (bū'til, -til), *n.* any of four isomeric forms of the chemical group C_4H_9. **butylene**, *n.* a colourless gas, formula C_4H_8.
butyraceous (būtirā'shəs), *a.* of the nature or consistency of butter. **butyrate** (bū'tirət), *n.* a salt of butyric acid. **butyric** (-ti'-), *a.* of or pertaining to butter. **butyric acid**, *n.* a colourless acid occurring in butter and other fats.
buxom (būk'səm), *a.* plump, big-bosomed (of women). **buxomly**, *adv.* **buxomness**, *n.*
buy (bī), *v.t.* (*past, p.p.* **bought** (bawt),) to purchase; to procure by means of money or something equivalent as a price; to gain by bribery; (*sl.*) to believe. **a good buy**, (*coll.*) a bargain, a good thing to have bought. **to buy in**, to buy back for the owner (at an auction); to obtain a stock of anything by purchase; (*Stock Exch.*) to purchase stock and charge the extra cost to the person who had undertaken to deliver it. **to buy into**, to purchase a share of or interest in (e.g. a company). **to buy off**, to pay a price for release or non-opposition; to get rid of by a payment. **to buy out**, to purchase the release of a member of the forces from service; to buy a majority share in or complete control over (e.g. a property, a company), thereby dispossessing the original owner(s). **to buy up**, to purchase all the available stock of. **buy-out**, *n.* the purchase of a company, esp. by former employees. **buyable**, *a.* **buyer**, *n.* one who buys; esp. one who buys stock for a mercantile house. **buyer's market**, *n.* one favourable to buyers, i.e. when supply exceeds demand.
buzz (bŭz), *n.* a sibilant hum, like that of a bee; a confused, mingled noise; stir, bustle, movement; (*coll.*) report, rumour; (*coll.*) a telephone call; (*coll.*) a euphoric feeling, a boost. *v.i.* to make a noise like humming or whirring; to make this noise from excitement or busyness; to signal by electric buzzer. *v.t.* (*Aviat.*) to interfere with by flying very near to; (*coll.*) to make a telephone call to; signal to by buzzer; (*sl.*) to throw with some violence. **buzz-bomb**, *n.* a flying bomb. **buzz off**, *int.* go away! **buzz-saw**, *n.* a circular saw. **buzzword**, *n.* a vogue word adopted from the jargon of a particular subject or discipline. **buzzer**, *n.* a buzzing insect; an electric warning apparatus that makes a buzzing sound. **buzzing**, *a.* **buzzingly**, *adv.*
buzzard (bŭz'əd), *n.* a kind of falcon, esp. *Buteo vulgaris*.
BVM., (*abbr.*) *Beta Virgo Maria*, Blessed Virgin Mary.
bwana (bwah'nə), *n.* sir, master.
by (bī), *prep.* near, at, in the neighbourhood of, beside; along, through, via; with, through (as author, maker, means, cause); according to, by direction, authority or example of; in the ratio of; to the amount of; during; not later than; concerning, with regard to; sired by. *adv.* near at hand; in the same place; aside, in reserve; past. *a.* side, subordinate, secondary, of minor importance; private, secret, clandestine, sly. *n.* BYE. **by and by**, soon, presently; later on; the future; time to come. **by and large**, on the whole. **by oneself**, alone, without help; of one's own initiative. **by the by(e)**, **by the way**, casually, apart from the main subject. **by-blow**, *n.* a side-blow; a bastard. **by-election**, *n.* an election caused by the death or resignation of a member. **bygone**, *a.* past. *n.* a past event; (*pl.*) the past; past injuries. **let bygones be bygones**, think no more of past injuries. **bylaw** BYLAW. **byline**, *n.* a sideline; the name of the author of a newspaper or magazine article printed beside it. **bypass**, *n.* an alternative route or passage which goes around an obstruction in the normal route; a road for the purpose of diverting traffic from crowded areas; a cutting-out of undesirable radio frequencies. *v.t.* to avoid, evade, to circumvent; to go around; to cause to use a bypass; to supply a bypass. **bypass surgery**, *n.* an operation performed to by-pass blocked or damaged arteries as a cure for certain heart conditions. **by-play**, *n.* action carried on aside while the

main action is proceeding. **by-product,** *n.* a secondary product. **byroad,** *n.* a road little frequented. **bystander,** *n.* one standing near; an onlooker, an eye-witness. **byway,** *n.* a bypath; a secret or obscure way; a short cut; an out-of-the-way side of a subject. **byword,** *n.* a common saying; a proverb; an object of general contempt.

bye (bī), *n.* a subsidiary object; something of an incidental or secondary kind; in cricket, a run scored from a ball not struck by the batsman; in golf, holes left over after end of contest and played as a new game; free passage to the next round in a contest for a competitor left without an opponent.

bye[1] (bī), **bye-bye** (-bī), *int.* (*coll.*) good-bye.
bye-byes[2] (bī'bī', -bīz), *n.* a childish word for sleep, bedtime, bed.
bylaw, byelaw *n.* a private statute made by a corporation or local authority; rules adopted by an incorporated or other society.
Byelorussian (byelorüsh'ən), BELORUSSIAN.
byre (bīə), *n.* a cow-house.

Byronic (bīron'ik), *a.* like Lord Byron or his poetry; theatrical, moody; affecting volcanic passion, gloom or remorse. **Byronically,** *adv.* **Byronism** (bī'-), *n.*
byssus (bis'əs), *n.* (*hist.*) a textile fabric, the fine linen of the Scriptures; the tuft of fibres by which molluscs of the genus *Pinna* attach themselves to other bodies. **byssinosis** (-inō'sis), *n.* a lung disease contracted by cotton workers.
byte (bīt), *n.* in a computer, a series of usu. eight binary digits treated as a unit.
Byzantine (bī'zan'tin, biz'əntīn, -tēn), *a.* of or pertaining to Byzantium or Istanbul (formerly Constantinople); hierarchical, inflexible; convoluted, complex; in the style of the art or architecture developed in the Eastern Empire. *n.* an inhabitant of Byzantium. **Byzantine Church,** *n.* the Greek or E Church. **Byzantine Empire,** *n.* the E or Greek Empire (AD 395–1453). **Byzantinesque** (-esk'), *a.* **Byzantinism,** *n.* **Byzantinist,** *n.* a specialist in Byzantine history, arts etc.
bz., bz, (*abbr.*) benzene.

C

C¹, c, the third letter and the second consonant of the English alphabet, is borrowed in shape from the Latin. Before *a, o, u, l* and *r* it is sounded like guttural mute *k*, and before *e, i* and *y* like the soft sibilant *s* (when it has this sound before other letters it is marked ç). C is used as a symbol to denote the third serial order; (*Alg.*) the third quantity known; (*Mus.*) the first note of the diatonic scale, corresponding to the Italian *do;* the natural major mode; common time; (*Roman numeral*) 100. **C₃,** *n.* (*Mil.*) lowest category of a medical board. *a.* of a person, of low physique.
C², (*abbr.*) capacitance; catholic; Celsius; century; Conservative; coulomb.
C³, (*chem. symbol*) carbon.
c, (*abbr.*) caught; cent; centi-; chapter; circa; copyright; cubic.
CA, (*abbr.*) chartered accountant; Consumers' Association.
Ca, (*chem. symbol*) calcium.
ca, (*abbr.*) circa.
CAA, (*abbr.*) Civil Aviation Authority.
Caaba (kah'bə), KAABA.
CAB, (*abbr.*) Citizens' Advice Bureau.
cab (kab), *n.* a public covered carriage with two or four wheels; a taxi; the guard, or covered part, of a locomotive which protects the driver and fireman from the weather; the driver's compartment in a lorry, crane etc. **cabman,** *n.* a cab-driver. **cab-rank, cab-stand,** *n.* a place where cabs are authorized to stand for hire. **cabbie, cabby,** *n.* (*coll.*) a cab-driver. **cabless,** *a.*
cabal (kəbal'), *n.* a small body of persons closely united for some secret purpose; a plot, conspiracy. *v.i.* (*past, p.p.* **caballed**) to intrigue secretly with others for some private end. **caballer,** *n.*
cabala (kəbah'lə), CABBALA.
caballero (kabəlyeə'rō), *n.* (*pl.* **-ros**) a Spanish gentleman; a stately kind of Spanish dance.
cabana (kəban'ə), *n.* (*chiefly N Am.*) a cabin used as a changing room at the beach.
cabaret (kab'ərā), *n.* an entertainment or floor show consisting of singing, dancing etc.; (*N Am.*) a restaurant or nightclub where such entertainment is provided.
cabbage (kab'ij), *n.* the plain-leaved, hearted varieties of *Brassica oleracea;* the terminal bud of palm-trees; (*coll.*) an inert or apathetic person; (*coll.*) a person with severely limited or damaged mental faculties who is totally dependent on others. **cabbage-butterfly,** *n.* two kinds of butterfly the larvae of which cause injury to cabbages, *Pieris brassicae, P. rapae.* **cabbage-moth,** *n.* a nocturnal moth, whose larvae feed on the cabbage. **cabbage-palm** CABBAGE-TREE. **cabbage-rose,** *n.* a double red rose *Rosa centifolia,* with large, compact flowers. **cabbage-tree,** *n.* a palm with an edible terminal bud. **cabbage white** CABBAGE BUTTERFLY. **cabbage-worm,** *n.* the larva of the cabbage-moth and other insects. **cabbagy,** *a.*
cabbala, cabala (kəbah'lə), *n.* a traditional exposition of the Pentateuch attributed to Moses; mystic or esoteric doctrine. **cabbalism,** *n.* the system of the cabbala; occult doctrine. **cabbalist,** *n.* one skilled in the Jewish cabbala, or in mystic learning. **cabbalistic, -ical** (-bəlist'-), *a.* **cabbalistically,** *adv.*
caber (kā'bə), *n.* a pole, the roughly-trimmed stem of a young tree, used in the Highland sport of tossing the caber.
cabin (kab'in), *n.* a small hut or house; a temporary shelter; a little room; a room or compartment in a ship or aircraft for officers or passengers; a driver's cab. *v.t.* to confine (as) in a cabin, to coop in. **cabin-boy,** *n.* a boy who waits on the officers of a ship or passengers in the cabin. **cabin class,** *n.* in a passenger ship, a class between tourist and first. **cabin crew,** *n.* the crew in an aircraft responsible for looking after passengers. **cabin cruiser,** *n.* a motor-boat with living accommodation.
cabinet (kab'init), *n.* a piece of furniture with drawers, shelves etc., in which to keep curiosities or articles of value; an outer case for a television set etc.; a cabinet photograph; the secret council of a sovereign; a kind of deliberative committee of the principal members of the British Government; a meeting of such a committee. **cabinet-maker,** *n.* one who makes the finer kinds of household furniture. **cabinet-making,** *n.* **Cabinet Minister,** *n.* a member of the Cabinet. **cabinet photograph,** *n.* a photographic print measuring about 6 × 4 in. (about 10 × 15 cm). **cabinet pudding,** *n.* a sort of bread-and-butter pudding, with dried fruit. **cabinet-work,** *n.* cabinet-making; a piece of such furniture.
cable (kā'bl), *n.* a strong rope, more than 10 in. (25.4 cm) round; one-tenth of a nautical mile (202 yds./185 m); the rope or chain to which an anchor is fastened; a nautical unit of length equal to 120 fathoms (240 yds./219 m); a wire rope; an electrical circuit of one or more conductors insulated and in a sheath; a cablegram; cable television. *v.t.* to fasten with a cable; to send (a message) by cable; to inform by cablegram; to connect to a cable television service. **cable car,** *n.* a passenger cabin suspended from an overhead cable and moved by it; a carriage on a cable railway. **cablegram,** *n.* a telegraphic message by submarine cable, communications satellite etc. **cablegrammic** (-gram'-), **cablegraphic** (-graf'-), *a.* **cable-laid,** *a.* twisted like a cable. **cable railway,** *n.* a funicular railway. **cable stitch,** *n.* a plaited stitch in knitting. **cable television,** *n.* a television service transmitted by an underground cable connected to subscribers' television sets. **cableway,** *n.* a transport system for freight or passengers using containers or cable cars suspended from overhead cables.
cabochon (kab'əshon), *n.* a precious stone polished, and having the rough parts removed, but without facets. **en cabochon,** polished, but without facets.
caboodle (kəboo'dl), *n.* (*coll.*) crowd, lot. **the whole caboodle,** all the lot.
caboose (kəboos'), *n.* the cook's house or galley of a ship; (*N Am.*) a carriage for the use of workmen or train crew on a goods train.
cabotage (kab'ətij), *n.* trade between ports of the same country; the restriction of a country's internal air traffic to carriers belonging to that country.
cabriole (kab'riōl), *a.* of table and chair legs, shaped in a reflex curve.
cabriolet (kabriōlā', kab'-), *n.* (also erron. **cabriole** (kab'riōl)) a covered carriage drawn by two horses; a type of motor-car with a folding top.
cacao (kəkah'ō, -kā'ō), *n.* a tropical American tree, *Theobroma cacao,* from the seeds of which chocolate and cocoa

are prepared. **cacao-butter** COCOA BUTTER.
cachalot (kash'əlot, -lō), *n*. a member of a genus of whales having teeth in the lower jaw, esp. the sperm whale.
cache (kash), *n*. a hole in the ground or other place in which provisions, goods or ammunition are hidden; the stores hidden. *v.t.* to hide or conceal in a cache. **cachepot** (kash'pō), *n*. an ornamental holder for a plant-pot.
cachectic CACHEXIA.
cachet (kash'ā), *n*. a paper capsule in which nauseous or other drugs can be administered; a seal; a characteristic mark; a sign of authenticity; a mark of excellence; prestige.
cachexia, cachexy (kəkek'siə, -si), *n*. loss of weight, weakness etc. of body resulting from chronic disease. **cachectic** (-kek'-), *a*.
cachinnate (kak'ināt), *v.i.* to laugh immoderately. **cachinnation**, *n*. loud or immoderate laughter. **cachinnatory**, *a*.
cachou (ka'shoo, -shoo'), *n*. a small pill-like sweetmeat for perfuming the breath.
cachucha (kəchoo'chə), *n*. a lively Spanish dance in triple time.
cacique, cazique (kəsēk'), *n*. a chief of the aborigines of the W Indies or the neighbouring parts of America; a local political leader in this area. **caciquism**, *n*.
cack-handed (kak-han'did), *a*. *(sl.)* left-handed; inept.
cackle (kak'l), *n*. the cackling of a hen; silly chatter. *v.i.* to make a noise like a hen after laying an egg; to chatter in a silly manner; to giggle. **to cut the cackle**, to get down to business. **cackler**, *n*. **cackling**, *n*.
caco-, *comb. form.* bad, malformed, evil to the senses.
cacodemon (kakədē'mən), *n*. an evil spirit; a nightmare; an evil person.
cacodyl (kak'ədil), *n*. a stinking organic compound of arsenic and methyl. **cacodylic** (-dil'-), *a*.
cacoethes (kakōē'thēz), *n*. a bad habit; an irresistible propensity.
cacography (kəkog'rəfi), *n*. bad spelling; bad writing.
cacophony (kəkof'əni), *n*. rough, discordant sound. **cacophonous**, *a*. harsh-sounding, discordant.
cactus (kak'təs), *n*. *(pl.* **-ti** (-tī), **-tuses)** any of a family (Cactaceae) of succulent spiny plants. **cactaceous**, *a*. **cactal**, *a*. allied to the cactuses. **cactoid**, *a*.
CAD, *(abbr.)* computer-aided design; compact audio disc.
cad (kad), *n*. a low, vulgar fellow; an ill-mannered person, a person guilty of ungentlemanly conduct. **caddish**, *a*.
cadaver (kədav'ə), *n*. a corpse, dead body. **cadaveric**, *a*. **cadaverous**, *a*. corpse-like; deathly pale. **cadaverously**, *adv*. **cadaverousness**, *n*.
caddice (kad'is), **CADDIS**.
caddie, caddy (kad'i), *n*. one who attends on a golfer; *v.i.* to act as a caddie. **caddie car, cart**, *n*. a two-wheeled cart for carrying golf clubs.
caddis (kad'is), *n*. the larva of any species of *Phryganea*, esp. of the may-fly. **caddis-fly**, *n*. **caddis-worm**, *n*.
caddy[1] (kad'i), *n*. a small box in which tea is kept.
caddy[2] CADDIE.
cadence (kā'dəns), **cadency** (-si), *n*. the sinking of the voice, esp. at the end of a sentence; modulation of the voice, intonation; rhythm, poetical rhythm or measure; rhythmical beat or movement; (*Mus.*) the close of a movement or phrase; a cadenza. **cadenced**, *a*. **cadent**, *a*. having rhythmical cadence. **cadenza** (-den'zə), *n*. a vocal or instrumental flourish of indefinite form at the close of a movement.
cadet (kədet'), *n*. a younger son; the younger branch of a family; formerly, a volunteer who served in hope to gain a commission; one undergoing training for the armed services or police force. **cadetship**, *n*.
cadge (kaj), *v.t.* to get by begging. *v.i.* to beg. **cadger**, *n*.
cadi (kah'di, kā'-), *n*. the judge in a Muslim town or village.

Cadmean (kadmē'ən), **Cadmian** (kad'-), *a*. of or belonging to Cadmus, the mythical founder of Thebes, and inventor of letters; Theban. **Cadmean victory**, *n*. a victory that ruins the victor; a moral victory.
cadmium (kad'miəm), *n*. a bluish-white metallic element, at. no. 51; chem. symbol Cd. **cadmium-yellow**, *n*. a pigment prepared from cadmium sulphide. **cadmic**, *a*. **cadmiferous** (-mif'-), *a*.
cadre (kah'də, kah'dri), *n*. a framework, a scheme; the permanent establishment or nucleus of a regiment; a group of usu. Communist activists; a member of such a group.
caduceus (kədū'siəs), *n*. *(pl.* **-cei** (-siī)) (an emblem resembling) the winged staff of Mercury, borne by him as messenger of the gods. **caducean**, *a*.
caduciary (kədū'shiəri), *a*. (*Law*) heritable; subject to forfeiture.
caducous (kədū'kəs), *a*. *(Bot.)* falling off quickly or prematurely. **caducity**, *n*.
caecum (sē'kəm), *(esp. N Am.)* **cecum**, *n*. *(pl.* **-ca**) the blind gut, the first part of the large intestine which is prolonged into a blind pouch; any blind tube. **caecal**, *a*. pertaining to the caecum; having a blind end. **caecally**, *adv*. **caeciform** (-si-), *a*. **caecitis** (-sī'-), *n*. inflammation of the caecum.
Caenozoic (sēnəzō'ik), CAINOZOIC.
Caerphilly (keəfil'i), *n*. a mild crumbly moist white cheese.
Caesar (sē'zə), *n*. the title of the Roman emperors down to Hadrian, and of the heirs presumptive of later emperors; the Emperor (i.e. of the Holy Roman Empire), the German Kaiser; an autocrat; *(coll.)* a Caesarean section. **Caesarian, -rean** (sizeə'-), *a*. of or belonging to Caesar; imperial. *n*. a follower of Caesar; a supporter of autocratic government; *(esp. N Am.* **Cesarian, -rean**) a Caesarean section. **Caesarian** (esp. N. Am. **Cesarian**) **section, birth**, *n*. the delivery of a child through the walls of the abdomen (as Julius Caesar is said to have been brought into the world). **Caesarism**, *n*. absolute government; imperialism. **Caesarist**, *n*.
caesium, *(esp. N Am.)* **cesium** (sē'ziəm), *n*. a highly-reactive, silvery-white metallic element. at. no. 55; chem. symbol Cs, similar to sodium in many properties.
caespitose, cespitose (ses'pitōs), *a*. growing in tufts; matted; turfy.
caesura, cesura (sizū'rə), *n*. *(Classic pros.)* the division of a metrical foot between two words, esp. in the middle of a line; *(Eng. pros.)* a pause about the middle of a line. **caesural**, *a*.
cafard (kafah', kaf'ah), *n*. depression, low spirits.
café, cafe (kaf'ā, *coll.* käf, kaf), *n*. a coffee-house; a restaurant; coffee. **café au lait** (ō lā), *n*. coffee with milk. **café noir** (nwah), *n*. coffee without milk. **cafeteria** (kafitiə'riə), *n*. a restaurant in which customers fetch their own food from the counter. **cafetière** (kafet'ieə), *n*. a coffee-making device consisting of a jug with built-in filter which holds the grounds in place when the coffee is poured.
caff (kaf), *n*. *(coll.)* a café.
caffeine (kaf'ēn), *n*. a vegetable alkaloid derived from the coffee and tea plants.
caftan, kaftan (kaftan), *n*. a kind of long belted tunic worn in the East; a woman's long loose dress resembling this.
cage (kāj), *n*. a box or enclosure wholly or partly of wire, wicker-work or iron bars, in which birds or other animals are kept; an open framework resembling this; a prison, a lock-up; the cabin of a lift; *(Mining)* an iron structure used as a lift in a shaft; an outer work of timber enclosing another. *v.t.* to shut up (as) in a cage. **cagebird**, *n*. a cage-ling; a type of bird normally kept in a cage. **cageling**, *n*. a bird kept in a cage.
cagey, cagy (kā'ji), *a*. *(coll.)* wary, shrewdly knowing; secretive, not frank. **cagily**, *adv*. **caginess, cageyness**, *n*.

cagoule 112 Caledonian

cagoule (kəgool'), *n.* a lightweight long anorak, usu. hooded.
cahier (ka'yā), *n.* a number of sheets of paper loosely put together; the report of a committee, esp. concerning policy.
cahoots (kəhoots'), *n.pl.* (*coll.*) partnership, collusion.
caiman CAYMAN.
Cain (kān), *n.* the brother and murderer of Abel (Gen. iv); hence, a murderer. **to raise Cain**, (*sl.*) to make a disturbance, to make trouble. **Cainite**, *n.* a son of Cain; one of a heretical sect (2nd cent.) who reverenced Cain and other bad Scriptural characters.
Cainozoic (kīnəzō'ik), **Caenozoic**, (*esp. N Am.*) **Cenozoic** (sēn-), *n., a.* (*Geol.*) Tertiary.
caïque (kaēk'), *n.* a light boat used on the Bosporus; a small Levantine sailing vessel.
cairn (keən), *n.* a pyramidal heap of stones, esp. one raised over a grave or to mark a summit, track or boundary; a cairn terrier. **cairn terrier**, *n.* a small rough-haired terrier orig. from Scotland.
cairngorm (keəngawm'), *n.* a yellow or brown variety of rock crystal, found in the Cairngorm mountains in Scotland.
caisson (kā'sən), *n.* an ammunition-chest or wagon; a large, watertight case or chamber used in laying foundations under water; a similar apparatus used for raising sunken vessels; a floating vessel used as a dock-gate; a sunken panel in ceilings etc. **caisson disease**, *n.* symptoms resulting from a sudden return from high air pressure to normal pressure conditions, the bends.
caitiff (kā'tif), *n.* a despicable wretch; a cowardly fellow. *a.* cowardly, base, despicable.
cajole (kəjōl'), *v.t.* to persuade, beguile, or deceive by flattery or fair speech; to wheedle, to coax; to beguile (into or out of something). *v.i.* to use artful flattery. **cajoler**, *n.* **cajolement, cajolery** (-lə-), *n.* **cajolingly**, *adv.*
Cajun (kā'jun), *n.* a descendant of French-speaking Acadians who settled in Louisiana; the music of the Cajuns. *a.* pertaining to the Cajuns or their music; of a style of cooking characterized by hot spicy sauces.
cake (kāk), *n.* a small mass of dough baked; a composition of flour, butter, sugar and other ingredients, baked usu. in a tin; (*Sc.*) oatcake; a flat shaped mass of food; a mass or crust of a solidified or compressed substance. *v.t.* (*usu. pass.*) to encrust. *v.i.* to form a hardened mass or crust. **cakes and ale**, a good time. **like hot cakes**, with great speed; with energy. **piece of cake**, (*coll.*) something achieved without effort. **slice of the cake**, a share in the benefits. **to take the cake**, (*iron., sl.*) to come out first; to take first prize. **cake-walk**, *n.* a form of dance using high marching steps; (*sl.*) something easily accomplished.
Cal., (*abbr.*) California; Calorie (kilocalorie).
cal., (*abbr.*) calendar; calibre; (small) calorie.
Calabar bean (kal'əbah), *n.* Physostigma.
calabash (kal'əbash), *n.* a kind of gourd or pumpkin; the calabash-tree, *Crescentia cujete;* the shell enclosing the fruit of this, used for drinking-vessels and other domestic utensils, and tobacco-pipes. **calabash-pipe**, *n.*
caboose (kaləboos', kal'-), *n.* (*N Am. coll.*) a prison.
calabrese (kaləbrā'zi), *n.* a form of green broccoli.
Caladium (kəlā'diəm), *n.* a genus of plants belonging to the arum family, with starchy tuberous roots used in the tropics for food.
calamander (kaləman'də), *n.* a hard wood, beautifully marked, from India and Sri Lanka.
calamary (kal'əməri), *n.* a cuttle-fish of the genus *Loligo* or the family *Teuthidae;* a squid, a pen-fish.
calamine (kal'əmin), *n.* (*formerly*) native zinc carbonate; a pinkish powder of this or zinc oxide used in a lotion to soothe the skin.
calamint (kal'əmint), *n.* a plant of the genus *Calamintha;* esp. an aromatic herb, *Calamintha officinalis.*

calamity (kəlam'iti), *n.* extreme misfortune, adversity, disaster; distress. **calamity Jane**, *n.* (*coll.*) a person who heralds or brings disaster. **calamitous**, *a.* causing or characterized by great or widespread distress or unhappiness. **calamitously**, *adv.* **calamitousness**, *n.*
calamus (kal'əməs), *n.* (*pl.* **-mi**) the sweet flag, *Acorus calamus;* any of a genus (*Calamus*) of palm trees producing enormously long canes; the quill of a feather.
calandria (kəlan'driə), *n.* a sealed cylindrical vessel with tubes passing through it, used as a heat-exchanger, e.g. in nuclear reactors.
calash (kəlash'), *n.* a light pleasure-carriage, with low wheels and removable top; a woman's silk hood supported by a framework of whalebone.
calc- (kalk), *comb. form.* lime. **calc-spar**, *n.* calcite.
calcaneum (kalkā'niəm), *n.* (*pl.* **-nea**) the bone of the heel. **calcaneal**, *a.*
calcareo- (kalkeəriō-), *comb. form.* (*Geol.*) calcareous.
calcareous, -ious (kalkeə'riəs), *a.* of the nature of lime or limestone. **calcareous-spar**, *n.* calcite. **calcareously**, *adv.* **calcareousness**, *n.*
Calceolaria (kalsiəleə'riə), *n.* a genus of plants with slipper-like flowers. **calceolate** (-lāt), *a.* (*Bot.*) shaped like a slipper.
calces, *pl.* CALX.
calcic (kal'sik), *a.* pertaining to or composed in whole or in part of lime. **calciferous** (-sif'-), *a.* (*Chem.*) yielding or containing calcium salts. **calcific** (-sif'-), *a.* **calciform**, *a.* **calcify**, *v.t.* to convert into lime; to harden by deposition of calcium salts. *v.i.* to become calcified. **calcification**, *n.*
calciferol (kalsif'ərol), *n.* vitamin D_2.
calcine (kal'sin, -sīn), *v.t.* to reduce to powder, expel water and other volatile matter, or cause to combine with oxygen by heating. *v.i.* to undergo calcination. **calcination** (-sinā'-), *n.* **calciner**, *n.*
calcite (kal'sīt), *n.* native crystallized carbonate of lime. **calcitic** (-sit'-), *a.*
calcium (kal'siəm), *n.* a silver-white metallic element, at. no. 20; chem. symbol Ca, usually met with in the form of its oxide, lime. **calcium carbide** CARBIDE. **calcium carbonate**, *n.* a white crystalline compound occurring in limestone, chalk, marble etc. **calcium chloride**, *n.* chloride of lime, bleaching-powder.
calc-spar CALC-.
calculate (kal'kūlāt), *v.t.* to compute, to reckon up; to estimate; to ascertain by mathematical process; to plan beforehand; to adjust, to arrange. *v.i.* to reckon, to form an estimate; to rely (upon); (*N Am.*) to think, to suppose. **calculable**, *a.* that may be calculated. **calculated**, *a.* prearranged, intended; cold-blooded; suitable, well-adapted (to). **calculating**, *a.* that calculates; shrewd, acting with forethought. **calculating machine**, *n.* a mechanical device which performs one or more of the fundamental arithmetical operations. **calculation**, *n.* the act of reckoning or computing in numbers; the result of such process; estimate, opinion, inference; careful planning, esp. selfish. **calculative**, *a.* **calculator**, *n.* one who calculates; a series of tables for use in calculating; an electronic device, usu. small and portable, which can carry out mathematical calculations.
calculus (kal'kūləs), *n.* (*pl.* **-li** (-lī)) a stony concretion formed in various organs of the body; (*pl.* **-luses**) (*Math.*) a method of calculation. **differential calculus** DIFFERENTIAL. **integral calculus** INTEGRAL. **calculous**, *a.* affected with or of the nature of a calculus.
caldarium (kaldeər'iəm), *n.* a Roman hot bath or hot bath-room.
caldera (kaldeə'rə), *n.* a large, deep volcanic crater.
caldron CAULDRON.
Caledonian (kalidō'niən), *a.* (*poet.*) of or pertaining to Scotland; Scottish; denoting a mountain-building movement in the Palaeozoic era. *n.* a Scotsman.

calefacient (kalifā'shənt), *a.* (*Med.*) causing or exciting heat or warmth. *n.* a medicine for increasing the heat of the body. **calefaction,** *n.* **calefactive,** *a.* **calefactory,** *a.* producing or communicating heat. *n.* a room in which monks used to warm themselves.

calendar (kal'ində), *n.* a register or list of the months, weeks, and days of the year, often with civil and ecclesiastical holidays, festivals and other dates; the system by which the beginning, length and subdivisions of the civil year are defined; a list or register, esp. a schedule of events, appointments etc. in chronological order; a list of courses etc. offered by a university. *v.t.* to insert in a calendar or list. **calendar month, year,** *n.* a month, or year, according to the calendar, as distinct from *lunar month* etc. **calendric, -ical** (-len'-), *a.*

calender (kal'ində), *n.* a press or machine in which cloth or paper is passed between rollers to make it glossy. *v.t.* to glaze by passing between rollers. **calenderer, calendrer,** *n.* **calendry,** *n.*

calends, kalends (kal'əndz), *n.pl.* the first day of any month in the old Roman calendar.

Calendula (kələn'dūlə), *n.* the genus of plants to which the marigold belongs; a marigold.

calescence (kəles'əns), *n.* increasing warmth or heat.

calf[1] (kahf), *n.* (*pl.* **calves** (kahvz)) the young of any bovine animal, esp. of the domestic cow; leather made from calfskin; the young of some large animals, as of the elephant, rhinoceros, whale etc; a stupid, childish fellow; a small iceberg broken off from a larger ice mass. **Golden Calf,** the idol set up by the Israelites (Ex. xxxii); Mammon-worship, the pursuit of riches. **in calf, with calf,** of the above animals, pregnant. **calf-bound,** *a.* bound in calfskin. **calf-love,** *n.* attachment between a boy and a girl. **calfskin,** *n.* calf-leather used in bookbinding and for boots and shoes.

calf[2] (kahf), *n.* (*pl.* **calves** (kahvz)) the thick fleshy part of the leg below the knee. **-calved,** *a.* (*in comb.*), as *thick-calved*).

calibre, (*esp. N Am.*) **caliber** (kal'ibə), *n.* the internal diameter of the bore of a gun or any tube; quality, capacity, compass; ability, character, standing. **calibrate,** *v.t.* to ascertain the calibre of; to test the accuracy of (an instrument) against a standard; to graduate (as a gauge). **calibration,** *n.* **calibrator,** *n.* **-calibred,** *a.* (*in comb.*).

calices, *pl.* CALIX.

caliciform (kəlis'ifawm), CALYCIFORM under CALYX.

calico (kal'ikō), *n.* (*pl.* **-coes, -cos**) cotton cloth formerly imported from the East; white or unbleached cotton cloth; printed cotton cloth.

calif, califate CALIPH.

California poppy ESCHSCHOLTZIA.

californium (kalifaw'niəm), *n.* an artificially-produced radioactive element, at. no. 98; chem. symbol Cf.

calipash (kal'ipash), *n.* that part of a turtle next to the upper shell, containing a dull green gelatinous substance.

calipee (kal'ipē), *n.* that part of a turtle next to the lower shell, containing a light yellow substance.

calipers CALLIPERS.

caliph, calif (kā'lif, kal'-), *n.* the chief ruler in certain Muslim countries, who is regarded as the successor of Mohammed. **caliphate, califate** (-fāt), *n.* the office or dignity of a caliph; his term of office; the dominion of a caliph.

calix (kā'liks), *n.* (*pl.* **-lices** (-sēz)) a cup-like body cavity or organ.

calk[1] CAULK.

calk[2] (kawk), *v.t.* to copy (a drawing etc.) by rubbing the back with colouring matter, and tracing the lines with a style on to paper beneath.

calk[3] (kawk), *n.* a calkin. *v.t.* to furnish with a calkin.

calkin (kaw'kin, kal'-), *n.* a sharp projection on a horseshoe to prevent slipping.

call (kawl), *v.t.* to utter in a loud voice; to name, to designate; to describe as; to regard or consider as; to summon; to cite; to invite; to command; to invoke; to appeal to; to rouse from sleep; to nominate; to read (a register etc.) aloud; to lure (as birds), to attract by imitating their cry; to ring up on the telephone; (*Comput.*) to transfer control to (a subroutine) by means of a code (**calling sequence**). *v.i.* to speak in a loud voice; to cry aloud, to shout; to pay a short visit; in bridge, to make a bid; in poker, to ask an opponent to show his or her cards; in whist, to show by special play that trumps are wanted; to telephone. *n.* a loud cry; a vocal address or supplication; the cry of an animal, esp. of a bird; a whistle to imitate the cry of an animal; the act of calling at a house or office on one's way; a short, formal visit; a summons, an invitation; an invitation to become minister to a congregation; a summons or signal on a bugle, whistle or telephone; appeal, allure; a requirement of duty; duty, necessity, justification, occasion; a demand for payment of instalments due (of shares etc.); (also **call option**) the option of claiming stock at a certain time at a price agreed on. **at call, on call,** at command; available at once. **call of nature,** a need to urinate or defecate. **to call back,** to revoke, to withdraw; to visit again; to call later by telephone. **to call down,** to invoke. **to call for,** to desire the attendance of; to appeal, demand; to signal for (trumps); to visit any place to bring (some person or thing) away; to require, necessitate. **to call forth,** to elicit; to summon to action. **to call in,** to summon to one's aid; to withdraw (money) from circulation; to order the return of; to pay a short visit (on, upon, at etc.). **to call in question,** to dispute. **to call into being,** to give existence to, create. **to call into play,** to put in operation. **to call names,** to abuse. **to call off,** to summon away, to divert; to cancel. **to call on upon,** to invoke, to appeal to; to pay a short visit to; to demand explanation, payment etc. **to call one's own,** to regard as one's possession; to own. **to call out,** to bawl; to challenge to a duel; to summon (as troops etc.) to service; to elicit; to order (workers) to strike. **to call the tune** TUNE. **to call to mind,** to recall. **to call to the Bar,** to admit as a barrister. **to call up,** to bring into view or remembrance; to rouse to activity; to telephone; (*Mil., Nav.*) to mobilize; to summon to appear (before). **to call upon** CALL ON. **within call,** within hearing. **call-box,** *n.* a public telephone booth. **call-boy,** *n.* a boy who calls actors when they are wanted on the stage; one who transmits the orders of the captain of a (river) steamer to the engineer. **call-girl,** *n.* a prostitute who makes appointments by telephone. **call-loan, -money,** *n.* money lent on condition that repayment may be demanded without notice. **call number,** *n.* a set of numbers and/or letters identifying the position of a book in a library. **call sign,** *n.* a set of numbers and/or letters identifying a radio transmitter or station. **callable,** *a.* **caller,** *n.* one who calls, esp. one who pays a call or visit. **calling,** *n.* the action of the verb TO CALL; habitual occupation, trade, profession; a vocation; a solemn summons to duty, renunciation, faith etc.; duty; the body of persons employed in a particular occupation, business or vocation. **calling card,** *n.* (*N Am.*) a visiting card.

calla, calla lily (kal'ə), *n.* the arum lily, *Richardia* (or *Calla*) *aethiopica*.

callanetics (kalənet'iks), *n.sing.* a system of exercise using small precise movements to increase muscle tone. [*Callan* Pinckney, its deviser]

calligraphy (kəlig'rəfi), *n.* the art of beautiful handwriting; (*coll.*) handwriting. **calligraph** (kal'igraf), *n.*, *v.t.* **calligrapher, -phist,** *n.* **calligraphic** (-graf'-), *a.*

calling CALL.

Calliope (kəli'əpi), *n.* the ninth Muse, of eloquence and heroic poetry; (**calliope,** *N Am.*) a series of steam-whistles toned to produce musical notes, and played by a keyboard.

callipers, calipers (kal'ipəz), *n.pl.* compasses with bow legs for measuring convex bodies, or with points turned out for measuring calibres. *v.t.* to measure by means of callipers. **calliper (splint)**, *n.* a form of splint for the leg which takes pressure off the foot when walking. **calliper rule**, **calliper-square**, *n.* a rule for measuring diameters, internal or external.

callisthenics (kalisthen'iks), *n.pl.* rhythmic exercises for promoting strength and gracefulness. **callisthenic**, *a.*

callous (kal'əs), *a.* hardened, indurated; unfeeling, unsympathetic. **callously**, *adv.* **callousness**, *n.* **callosity** (-los'-), *n.* hardened or thick skin, caused by friction, pressure, disease or other injury; a callus; insensibility, want of feeling.

callow (kal'ō), *a.* of a bird, unfledged, downy; youthful, inexperienced. **callowness**, *n.*

calluna (kəloo'nə), *n.* the ling, *Calluna vulgaris*.

callus (kal'əs), *n.* a hardening of the skin from pressure or friction; (*Med.*) a bony formation serving to unite a fracture; (*Bot.*) a hard formation. *v.t., v.i.* to make or form a callus.

calm (kahm), *a.* still, quiet, serene; tranquil, undisturbed. *n.* the state of being calm; (*Naut.*) entire absence of wind. *v.t., v.i.* to still, to quiet, to soothe (often with *down*). **calmative**, *a.* tending to calm. *n.* a sedative medicine. **calmed**, *a.* rendered calm; becalmed. **calmly**, *adv.* **calmness**, *n.*

calomel (kal'əmel), *n.* mercurous chloride, an active purgative.

Calor gas® (kal'ə), *n.* a type of bottled gas for cooking etc.

caloric (kəlo'rik), *a.* pertaining to heat or calories. **calorificient** (kalərifā'shənt), *a.* esp. of foods, heat-producing.

calorie, calory (kal'əri), *n.* a unit of heat. The **(small) calorie** is the quantity of heat required to raise the temperature of 1 gram of water by 1° C; now officially superseded by the *joule* (1 joule = 4.1868 calories). The **kilocalorie** or **Calorie**, equalling 1000 calories, is used in measuring the energy content of food. **calorific**, *a.* producing heat, thermal. **calorific value**, *n.* the amount of heat produced by the complete combustion of a given amount of fuel or food. **calorifically**, *adv.* **calorimeter** (-rim'itə), *n.* an instrument for measuring quantities of heat, or the specific heat of a body. **calorimetric** (-met'-), *a.* **calorimetry** (kalərim'-), *n.*

calotte (kal'ot), *n.* a small skull-cap (worn by Roman Catholic ecclesiastics).

calotype (kal'ətīp), *n.* a photographic process invented by Fox Talbot and now disused; a Talbotype.

calque (kalk), *n.* a loan translation, a literal translation into English of a foreign idiom.

Caltha (kal'thə), *n.* a genus of ranunculaceous marsh plants containing the marsh marigold, *Caltha palustris*.

caltrop (kal'trəp), *n.* an instrument formed of four iron spikes joined at the bases, thrown on the ground to impede the advance of cavalry; a name for several trailing plants, with spiny fruit, esp. the water chestnut *Trapa natans*.

calumet (kal'ūmet), *n.* the tobacco-pipe of the N American Indians, used as a symbol of peace and friendship.

calumniate (kəlŭm'niāt), *v.t.* to slander; to charge falsely with something criminal or disreputable. *v.i.* to utter calumnies. **calumniation**, *n.* **calumniator**, *n.* **calumniatory**, *a.* **calumnious**, *a.* **calumniously**, *adv.* **calumniousness**, *n.*

calumny (kal'-), *n.* a malicious misrepresentation of the words or actions of another; slander; a false charge.

calvados (kal'vədos), *n.* apple brandy made in Normandy.

Calvary (kal'vəri), *n.* the place where Christ was crucified; a representation of the Crucifixion, or of the successive scenes of the passion.

calve (kahv), *v.i.* to bring forth a calf; to bring forth young; of icebergs, to detach and cast off a mass of ice. *v.t.* to bear, bring forth.

calves, *pl.* CALF.

Calvinism (kal'vinizm), *n.* the tenets of the Swiss theologian John Calvin (1509–64), esp. his doctrine of predestination and election. **Calvinist**, *n.* **Calvinistic** (-nist'-), **-ical**, *a.* pertaining to Calvin.

calx (kalks), *n.* (*pl.* **calces** (kal'sēz)) ashes or fine powder remaining from metals, minerals etc. after they have undergone calcination.

calyc(i)-, *comb. form* calyx.

calycine (kal'isīn), **calycinal** (-lis'-), *a.* of, belonging to, or in the form of a calyx.

calycle (kal'ikl), **calyculus** (kalik'ūləs), *n.* a small cup-shaped animal structure; a row of small bracts at the base of the calyx on the outside. **calycular** (-lik'-), **calyculate** (-lik'ūlət), *a.*

calypso (kəlip'sō), *n.* a W Indian narrative song made up as the singer goes on.

calyptra (kəlip'trə), *n.* (*Bot.*) a hood or cover. **calyptrate** (-trāt), *a.*

calyx (kā'liks, kal'-), *n.* (*pl.* **calyces** (-sēz)), the whorl of sepals (usu. green) forming the outer integument of a flower; a calix. **calycled**, *a.* having a calyx.

CAM, (*abbr.*) computer-aided manufacturing; content-addressable memory.

cam (kam), *n.* an eccentric projection attached to a revolving shaft for the purpose of giving linear motion to another part or follower. **camshaft**, *n.* a shaft bearing cams which operate the valves of internal-combustion engines.

camaraderie (kamərah'dəri), *n.* comradeship; good fellowship and loyalty among intimate friends.

camarilla (kaməril'ə), *n.* a band or company of intriguers; a private cabinet; a cabal.

camber (kam'bə), *n.* the condition of a piece of timber, ship's deck etc. of being slightly convex above; the curvature given to a road surface to make water run off it. *v.t., v.i.* to bend, to arch. **cambered**, *a.*

Camberwell beauty (kam'bəwel), *n.* a butterfly, *Vanessa antiopa*.

cambist (kam'bist), *n.* one skilled in the science of financial exchange. **cambism**, *n.*

cambium (kam'biəm), *n.* a cellular tissue formed between the xylem and phloem of dicotyledenous plants, which divides to increase the diameter of stems and roots.

Cambodian (kambōd'iən), *n., a.* (a native or inhabitant) of Cambodia (previously Kampuchea) SE Asia.

Cambrian (kam'briən), *a.* of or belonging to Wales; the name given to the system of Palaeozoic strata lying below the Silurian. *n.* a Welshman.

cambric (kam'brik), *n.* a very fine white linen.

Cambridge blue (kām'brij), *n., a.* pale blue.

Cambs., (*abbr.*) Cambridgeshire.

camcorder (kam'kawdə), *n.* a video camera and recorder combined in one unit.

came[1] (kām), *n.* a strip of lead used in framing glass in lattice windows.

came[2] (kām), *past* COME.

camel (kam'l), *n.* a large, hornless, humpbacked ruminant with long neck and padded feet, used in Africa and the East as a beast of burden; two species, the Arabian camel, *Camelus dromedarius*, with one hump, and the Bactrian, *C. bactrianus*, with two; a watertight float attached to a boat to raise it in the water; a pale brownish-yellow colour. *a.* of this colour; made of camel-hair fabric. **camel corps**, *n.* troops mounted on camels. **camel-hair**, *n.* camel's hair used as a material for various fabrics; a painter's brush made of hairs from squirrels' tails. **cameleer** (-liə'), *n.* a camel-driver. **camelish**, *a.* obstinate.

cameleon (kəmēl'yən), CHAMELEON.

Camellia (kəmē'liə), *n.* a genus of evergreen shrubs with beautiful flowers. [G.J. *Kamel*, a Moravian Jesuit, and Eastern traveller]

camelopard (kəmel'əpahd), *n*. (*formerly*) the giraffe.
Camembert (kam'əmbeə), *n*. a soft rich cheese from Normandy.
cameo (kam'iō), *n*. (*pl*. **cameos**) a precious stone with two layers of colours, the upper being carved in relief, the lower serving as background; a piece of jewellery using such carving; a design based on such carving, esp. a profiled head; a short literary piece; a small part in a play or film which allows an actor to display his or her skill. *a*. of a cameo or cameos; small and perfect.
camera (kam'ərə), *n*. the private chamber of a judge; an apparatus for taking photographs which records an image (or a series of images in a movie camera) on a light-sensitive surface; an apparatus which records (moving) images and converts them to electrical signals for TV transmission. **in camera** in private, the public being excluded from the court. **off camera**, not being filmed. **on camera**, being recorded on film. **camera crew**, *n*. a group of people, including cameraman, sound recordist etc., needed to make a television film (usu. on location). **cameraman**, *n*. a person who operates a movie or television camera. **camera obscura** (əbskū'rə), *n*. a dark box, or chamber, admitting light through a pinhole or a double-convex lens, at the focus of which an image is formed of external objects on paper, glass etc. **camera-shy**, *a*. unwilling to be photographed or filmed.
camerlengo (kamərleng'gō), **-lingo** (-ling'-), *n*. (*pl*. **-gos**) a papal treasurer.
camisole (kam'isōl), *n*. an under-bodice. **camiknickers**, *n.pl*. camisole and knickers in one piece.
camlet (kam'lit), *n*. a fabric orig. of camel's hair, now a mixture of silk, wool and hair (applied at different times to various substances).
camomile, chamomile (kam'əmil), *n*. an aromatic creeping plant belonging to the genera *Anthemis* or *Matricaria*, esp. *A. nobilis;* applied also to some other plants. **camomile tea**, *n*.
Camorra (kəmo'rə), *n*. a lawless secret society in S Italy, dating from the old kingdom of Naples; any similar group. **Camorist**, *n*.
camouflage (kam'əflahzh), *n*. disguise, esp. the concealment of guns, camps, vehicles etc., from the enemy by means of deceptive painting, foliage etc.; concealment of an animal from predators by blending with the background etc.; concealment of one's actions. *v.t*. to disguise.
camp[1] (kamp), *n*. the place where an army is lodged in tents or other temporary structures; a station for training troops; military life; temporary quarters of gipsies, holidaymakers, refugees etc.; the occupants of such quarters; (*Austral*.) a halting-place for cattle; a body of adherents; a side; a ruined prehistoric fort. *v.i*. to encamp; to live temporarily in makeshift conditions. **to camp out**, to lodge in a camp in the open; to sleep outdoors. **camp-bed**, *n*. a light folding bedstead. **camp-chair**, *n*. a folding chair. **camp-fire**, *n*. a fire lighted at the centre of a camp. **camp-follower**, *n*. a civilian who follows an army in the field; a hanger-on. **camp-meeting**, *n*. a religious meeting in the open air or in a tent, often prolonged for days. **camp-site**, *n*. a place set aside, or suitable, for camping. **camp-stool**, *n*. a folding stool. **camper**, *n*. one who camps; a vehicle having living accommodation in the back.
camp[2] (kamp), *a*.affectedly homosexual; effeminate; bizarre. *v.i*.to behave in a camp manner. *n*. camp behaviour. **high camp**, deliberately exaggerated camp. **to camp it up**, to act in an exaggeratedly camp manner. **campy**, *a*.
Campagna (kampah'nyə), *n*. the flat country around Rome.
campaign (kampān'), *n*. the operations and continuance of an army in the field; any analogous operations or course of action, esp. a course of political propaganda. *v.i*. to conduct or serve on a campaign. **campaigner**, *n*.
campanile (kampənē'li), *n*. (*pl*. **-niles**, **-nili** (-li)) a bell-tower, esp. a detached one.
campanology (kampənol'əji), *n*. the principles of bell-ringing etc. **campanologer**, **-gist**, *n*. **campanological** (-loj'-), *a*.
Campanula (kampan'ūlə), *n*. a genus of plants with bell-shaped flowers, containing the bluebell of Scotland, the Canterbury bell etc. **campanulaceous** (-lā'-), *a*. **campanular**, **-ulate** (-lət), *a*. (*Bot., Zool.*) bell-shaped.
Campari® (kampah'ri), *n*. a bitter-tasting aperitif.
Campeachy wood (kampē'chi), *n*. logwood.
campestral (kampes'trəl), *a*. pertaining to or growing in the fields or open country.
camphene (kam'fēn), **camphine** (-fēn, -fin), *n*. an oily compound distilled from turpentine.
camphor (kam'fə), *n*. a whitish, translucent, volatile, crystalline substance with a pungent odour, obtained from *Camphora officinarum*, *Dryobalanops aromatica* and other trees, used as an insect repellent, in liniment (**camphorated oil**), and in the manufacture of celluloid. **camphor-laurel**, **-tree**, *n*. *Cinnamomum camphora*. **camphor-wood**, *n*. the wood of this or of an Australian timber-tree, *Callitris robusta*. **camphoraceous** (-ā'-), *a*. **camphorate** (-rət), *n*. **camphorate**, *v.t*. to wash or impregnate with camphor. **camphoric** (kamfo'rik), *a*.
campion (kam'piən), *n*. any of various British flowering plants of the genera *Lychnis* and *Silene*.
campus (kam'pəs), *n*. the buildings and grounds of a university or college, or (*N Am*.) a school; (*N Am*.) a geographically separate part of a university; the academic world in general.
CAMRA (kam'rə), (*abbr*.) Campaign for Real Ale.
camshaft CAM.
Camu-Camu, camu-camu (kam'oo kam'oo), *n*. a S American shrub with a fruit rich in vitamin C.
camwood (kam'wud), *n*. barwood, a hard red wood from W Africa.
can[1] (kan), *n*. a metal vessel for holding liquid; a vessel of tinned iron or aluminium in which meat, fruit, fish etc. are hermetically sealed up for preservation; a canful; (*coll*.) a shallow metal container for film; (*sl*.) prison; (*sl*.) a lavatory; (*pl*., *sl*.) headphones. *v.t*. (*past*, *p.p*. **canned**) to put in cans for preservation. **can it!**, (*sl*.) stop doing that! **in the can**, of film, processed and ready for showing; (*fig*.) agreed, arranged. **to carry the can**, to take responsibility, accept blame. **can-opener**, *n*. a tin-opener. **canful**, *n*. **canned**, *a*. preserved in a can; (*sl*.) drunk; of music, recorded in advance; (*sl*.) of laughter, not spontaneous. **canner**, *n*. **cannery**, *n*. a factory where foods are canned.
can[2] (kan), *aux. v*. (*pres* **can**, *neg*. **cannot** (kan'ət, -not'), *past* **could** (kud)) to be able to; to be allowed to; to be possible to. **can-do**, *a*. showing a positive attitude to achieving a desired objective.
Can., (*abbr*.) Canada; Canadian.
Canaan (kā'nən), *n*. an ancient region of western Palestine; (*fig*.) the land of promise. **Canaanite**, *n*. an inhabitant of the land of Canaan; a descendant of Canaan, the son of Ham. **Canaanitic** (-nit'-), **Canaanitish** (-ni'-), *a*.
Canada balsam (kan'ədə), *n*. a pale resin obtained from *Abies balsamea* and *A. canadensis*, used in medicine and to mount microscopic objects.
Canada goose (kan'ədə), *n*. a large N American wild goose, *Branta canadensis*, grey and brown in colour.
Canadian (kənā'diən), *n*., *a*. (a native or inhabitant) of Canada.
canaille (kanī'), *n*. the dregs of the people; the rabble, the mob.
canakin (kan'əkin), CANNIKIN.
canal (kənal'), *n*. an artificial watercourse, esp. one used for navigation; (*Physiol., Bot*.) a duct; (*Zool*.) a siphonal groove; (*Arch*.) a fluting, a groove. *v.t*. (*past*, *p.p*. **cannalled**) to make a canal across or through. **canals of**

Mars, linear markings on the surface of the planet Mars, supposed by some astronomers to be waterways, or zones of vegetation produced by periodical diffusion of moisture. **canal boat**, *n*. a long narrow boat used on canals. **canal rays**, *n*. a steady flow of positively electrified particles which take part in the electrical discharge in a rarefied gas. **canalize, -ise** (kan'-), *v.t.* to make a canal across or through; to convert into a navigable waterway; to give a desired direction to; to channel. **canalization, -isation**, *n*.

canaliculate, -ated (kanəlik'ūlət, -lātid), *a*. (*Physiol*.) minutely grooved; striated. **canaliculus**, *n*. (*pl*. **-li**) a small anatomical groove.

canape, canapé (kan'əpā), *n*. a thin piece of bread or toast spread with cheese, fish etc.

canard (kanahd'), *n*. a hoax, a false report; an aircraft having a tailplane mounted in front of the wings.

canary (kəneə'ri), *n*. a light sweet wine made in the Canary Islands; a well-known cage-bird *Fringilla canaria*. *a*. bright yellow. **canary-coloured**, *a*. **canary-creeper, canariensis** (-en'sis), *n*. a climbing-plant with yellow flowers. **canary-seed**, *n*. the seed of *Phalaris canariensis*, the **canary-grass**, used as food for canaries.

canasta (kənas'tə), *n*. a card game similar to rummy, played by two to six players.

canaster (kənas'tə), *n*. a coarse kind of tobacco.

cancan (kan'kan), *n*. a French stage dance performed by female dancers, involving high kicking of the legs.

cancel (kan'sl), *v.t.* (*past, p.p.* **cancelled**) to obliterate by drawing lines across; to annul, countermand, revoke, neutralize; to suppress; (*Math*.) to strike out common factors; to mark (a stamp, ticket) to prevent reuse. *n*. a cancelling, countermanding; the deletion and reprinting of a part of a book; a page or sheet substituted for a cancelled one. **to cancel out**, to cancel one another; to make up for. **cancellate** (-lət), **-llated** (-lātid), **-llous**, *a*. (*Bot., Zool*.) reticulated; of bones, having a spongy internal structure. **cancellation**, *n*. **canceller**, *n*.

Cancer (kan'sə), *n*. the fourth of the 12 signs of the zodiac, the Crab; one born under this sign; (**cancer**) a malignant spreading growth affecting different parts of the human body; (**cancer**) a vice or other evil of an inveterate spreading kind. **Tropic of Cancer** TROPIC [1]. **cancer stick**, *n*. (*sl*.) a cigarette. **cancerous**, *a*. **cancriform** (kang'krifawm), *a*. crab-like; of the form of a cancer. **cancroid**, *a*. crab-like; having some of the qualities of cancer. *n*. a disease resembling cancer.

candela (kandel'ə, -dē'-), *n*. a unit of luminous intensity. **candelabrum, -bra** (kandəlah'brəm, -brə), *n*. (*pl*. **-bra, -bras**) a tall lamp-stand; a high, ornamental candlestick, usually branched.

candescent (kandes'ənt), *a*. glowing (as) with white heat. **candescence**, *n*.

C and G, (*abbr*.) City and Guilds.

candid (kan'did), *a*. frank, sincere, open, ingenuous; unbiased; outspoken, freely critical; pertaining to photographs of people taken informally or without their knowledge. **candidly**, *adv*. **candidness**, *n*.

Candida (kan'didə), *n*. a genus of yeastlike fungi, esp. *Candida albicans*, which causes thrush.

candidate (kan'didət, -dāt), *n*. one who seeks or is proposed for some office or appointment; a person considered suitable or worthy for an office or dignity; a person taking an examination. **candidacy**, *n*. **candidature** (-chə), **candidateship**, *n*.

candied CANDY.

candle (kan'dl), *n*. a cylindrical body of tallow, wax etc. with a wick in the middle, used as an illuminant; candlepower. *v.t.* to test (eggs) by holding before a candle. **not fit to hold a candle to**, not to be named in comparison with. **not worth the candle**, not worth the trouble. **Roman candle**, a firework consisting of a tube from which coloured fireballs are discharged. **to burn the candle at both ends**, to expend one's energies excessively or waste resources. **candleberry(-myrtle)**, *n*. a N American shrub, *Myrica cerifera*, yielding wax used for candle-making. **candle-holder** CANDLESTICK. **candlelight**, *n*. the light of a candle; evening. **Candlemas** (-məs), *n*. the feast of the Purification of the Virgin (2 Feb.), when candles are blessed and carried in procession. **candle-nut**, *n*. the fruit of *Aleurites triloba*, which furnishes a kind of wax; this tree. **candle-power**, *n*. intensity of light emitted, expressed in candelas. **candlestick**, *n*. a utensil for holding a candle. **candlewick**, *n*. a cotton fabric with a pattern of raised tufts.

candour, (*esp. N Am*.) **candor** (kan'də), *n*. candidness, sincerity, openness; freedom from malice or bias.

C and W, (*abbr*.) Country and Western.

candy (kan'di), *n*. sugar crystallized by boiling and evaporation; (*N Am*.) sweetmeats. *v.t.* to preserve with sugar, to coat with crystallized sugar; to crystallize. *v.i.* to become candied. **candy-floss**, *n*. coloured spun sugar on a stick. **candy-store**, *n*. (*N Am*.) a sweet-shop. **candystripe**, *n*. a pattern of alternate stripes of white and a colour. **candied**, *a*. preserved in or coated with sugar; crystalline; flattering, honeyed.

candytuft (kan'dititft), *n*. a plant of the genus *Iberis*, esp. *I. sempervivum*, the perennial candytuft, or *Iberis umbellata*.

cane (kān), *n*. a slender, hollow, jointed stem of the bamboo, sugar-cane or other reeds or grasses; the thin stem of the rattan or other palms; (such a stem or similar used as) a walking-stick or an instrument of punishment; the stem of a raspberry and other plants. *v.t.* to beat with a cane; to thrash (a lesson, with *into*); to put a cane bottom to or repair (as a chair); to weave with cane; to defeat decisively. **cane-brake**, *n*. (*N Am*.) a thicket of canes. **cane-sugar**, *n*. sugar made from canes as distinguished from beet-sugar; sucrose. **cany**, *a*. **caning**, *n*. a beating with a cane; a thorough defeat.

canescent (kənes'ənt), *a*. hoary, approaching to white. **canescence**, *n*.

cangue, cang (kang), *n*. a heavy wooden collar or yoke, formerly fixed round the neck of criminals in China.

canikin CANNIKIN.

canine (kā'nīn), *a*. of or pertaining to dogs; dog-like. *n*. an animal of the family Canidae, including the dogs and wolves; a canine tooth. **canine teeth**, *n.pl*. two pointed teeth in each jaw, one on each side, between the incisors and the molars.

canister (kan'istə), *n*. a metal case or box for holding tea, coffee etc.; canister-shot; in the Roman Catholic Church, the box in which the eucharistic wafers are kept before consecration. **canister-shot**, *n*. bullets packed in metal cases which burst when fired, called also *case-shot*.

canker (kang'kə), *n*. a corroding ulceration, esp. in the human mouth; an inflammatory condition of the ear in cats and dogs; a fungous excrescence in a horse's foot; a fungus growing on and injuring fruit trees; anything which corrupts or consumes. *v.t.* to infect or rot with canker; to infect, corrode. *v.i.* to become cankered, infected or corrupt. **canker-worm**, *n*. a caterpillar that feeds on buds and leaves; (*chiefly N Am*.) the larva of the geometer moths. **cankerous**, *a*. corroding, destroying.

Canna[1] (kan'ə), *n*. a genus of ornamental plants with bright coloured flowers.

canna[2] (kan'ə), (Sc.) CANNOT under CAN [2].

cannabis (kan'əbis), *n*. any of a genus, *Cannabis*, of plants containing the Indian hemp; a narcotic drug obtained from the leaves and flowers of plants of the genus, esp. *C. sativa* and *C. indica*. **cannabis resin**, *n*. cannabin. **cannabic**, *a*. **cannabin**, *n*. a sticky resin, the active principle of the drug cannabis.

canned, cannery, *a*. CAN [1].

cannel (kan'əl), **cannel-coal**, *n.* a hard, bituminous coal, burning with a bright flame.
cannelloni (kanəlō'ni), *n.* an Italian dish, rolls of sheet pasta filled with meat etc. and baked.
cannibal (kan'ibəl), *n.* a human being that feeds on human flesh; an animal that feeds on its own kind. *a.* pertaining to cannibalism; ravenous, bloodthirsty. **cannibalism**, *n.* the act or practice of feeding on one's own kind; barbarity, atrocity. **cannibalistic** (-lis'-), *a.* **cannibalize, -ise**, *v.t.* to dismantle (a machine) for spare parts or use (such spare parts) to build into a similar machine.
cannikin, canikin (kan'ikin), **canakin** (-əkin), *n.* a little can or cup.
cannily, canniness CANNY.
cannon[1] (kan'ən), *n.* a piece of ordnance; a heavy mounted gun; artillery, ordnance; a hollow sleeve or cylinder revolving independently on a shaft. **cannon-ball**, *n.* a solid shot fired from a cannon. **cannon-bone**, *n.* the metacarpal or metatarsal bone of a horse, ox etc. **cannon-fodder**, *n.* (*iron.*) soldiers, esp. infantrymen. **cannon-proof**, *a.* proof against artillery. **cannonade** (-nād'), *n.* a continued attack with artillery against a town, fortress etc. *v.t.* to attack or batter with cannon. **cannoneer** (-niə'), *n.* (*esp. formerly*) a gunner, an artilleryman. **cannonry**, *n.* cannon collectively; cannonading.
cannon[2] (kan'ən), *n.* (*Billiards*) a stroke by which two balls are hit successively. *v.i.* to make a cannon; to come into violent contact.
cannot CAN[2].
cannula (kan'ūlə), *n.* (*pl.* **-las, -lae**) a small tube introduced into a body cavity to withdraw a fluid. **cannular**, *a.* **cannulate**, *v.t.* to insert a cannula into.
canny (kan'i), *a.* (*chiefly Sc., North.*) knowing, shrewd, wise; quiet, gentle; comely, good; prudent, cautious; frugal, thrifty. **cannily**, *adv.* **canniness**, *n.*
canoe (kənoo'), *n.* a kind of light boat formed of the trunk of a tree hollowed out, or of bark or hide, and propelled by paddles; a light narrow boat propelled by paddles. *v.i.* to go in a canoe. **to paddle one's own canoe**, to be independent. **canoeist**, *n.*
canon (kan'ən), *n.* a rule, a regulation, a general law or principle; a standard, test or criterion; a decree of the Church; the catalogue of canonized saints; the portion of the Mass in which the words of consecration are spoken; (a list of) the books of Scriptures received as inspired; the list of an author's recognized works; a resident member of a cathedral chapter; a member of a religious body, from the fact that some cathedral canons lived in community; a musical composition in which the several parts take up the same subject in succession. **canon law**, *n.* ecclesiastical law as laid down by popes and councils. **canoness**, *n.* a member of a female community living by rule but not bound by vows. **canonic, -ical** (-non'-), *a.* pertaining to or according to canon law; included in the canon of Scripture; authoritative, accepted, approved; belonging or pertaining to a cathedral chapter; (*Mus.*) in canon form. **canonical hours**, *n.* 8 am to 6 pm during which marriages may legally be celebrated. **canonically**, *adv.* **canonicals**, *n.pl.* the full robes of an officiating clergyman as appointed by the canons. **canonicity** (-nis'-), *n.* the quality of being canonical, esp. the authority of a canonical book. **canonist**, *n.* one versed in canon law. **canonistic, -ical** (-nis'-), *a.* **canonize, -ise**, *v.t.* to enrol in the canon or list of saints; to recognize officially as a saint; to recognize as canonical; to sanction as conforming to the canons of the Church. **canonization**, *n.* **canonry**, *n.* the dignity, position or benefice of a canon.
cañon CANYON.
canoodle (kənoo'dl), *v.i.* (*coll.*) to embrace, to cuddle.
Canopic vase, jar (kənō'pik), *n.* an Egyptian vase with a lid shaped like a god's head, for holding the viscera of embalmed bodies.
canopy (kan'əpi), *n.* a rich covering of state suspended over an altar, throne, bed etc.; or borne over some person, relics, or the Host, in procession; a shelter, a covering, the sky; an ornamental projection over a niche or doorway; the transparent roof of an aircraft's cockpit; the fabric portion of a parachute; in a forest, the topmost layer of branches and leaves. *v.t.* to cover with or as with a canopy. **canopied**, *a.*
canst (kanst), archaic 2nd pers. sing. CAN[2].
cant[1] (kant), *n.* the peculiar dialect or jargon of beggars, thieves etc.; slang; (*often derog.*) a method of speech or phraseology peculiar to any sect or party; hypocritical or sanctimoniousness talk. *a.* pertaining to or of the nature of cant. *v.i.* to speak whiningly or insincerely; to talk cant. **canter**[1], *n.* **canting**, *n.* cant. *a.* whining, hypocritical. **canting crew**, *n.* the brotherhood or fraternity of vagabonds, thieves, sharpers etc. **cantingly**, *adv.*
cant[2] (kant), *n.* a slope, a slant; an inclination; a jerk producing a slant or upset; an external angle; a bevel; a slanting position. *v.t.* to tip, to tilt; to throw with a jerk; to bevel, give a bevel to. *v.i.* to tilt or slant over. **cant dog, hook**, *n.* a metal hook on a pole, used esp. for handling logs. **canted**, *a.*
can't (kahnt), contr. form of *cannot*.
Cantab. (kan'tab), (*abbr.*) of Cambridge. [L *Cantabrigiensis*]
cantabile (kantab'bilā), *n.*, *a.* (*Mus.*) (a piece performed) in an easy, flowing style.
Cantabrigian (kantəbrij'iən), *a.* of or relating to the town or Univ. of Cambridge, England or to the city of Cambridge, Massachusetts. *n.* a member of Cambridge or Harvard Univ.
cantal (kan'tal), *n.* a hard strong-flavoured French cheese.
cantaloup, cantaloupe (kan'təloop), *n.* a small, round, ribbed musk-melon, first raised at Cantalupo near Rome.
cantankerous (kantang'kərəs), *a.* quarrelsome, crotchety. **cantankerously**, *adv.* **cantankerousness**, *n.*
cantata (kantah'tə), *n.* a poem, short lyrical drama or (usu.) a biblical text, set to music, with solos and choruses.
cantatore (kantətaw'ri), *n.* a male professional singer. **cantatrice** (kantətrē'che, kätatrēs'), *n.* a female professional singer.
canteen (kantēn'), *n.* a place in a barracks, factory, office etc. where refreshments are sold at low prices; a soldier's mess tin; a chest for cutlery; a water-bottle.
canter[1] CANT[1].
canter[2] (kan'tə), *n.* an easy gallop. *v.t.* to cause (a horse) to go at this pace. *v.i.* to ride at a canter; to move at this pace. **in, at a canter**, easily.
canterbury (kan'təbəri), *n.* a light stand with divisions for music portfolios etc. **Canterbury bell**, *n.* name of plants belonging to the genus *Campanula*, esp. the exotic *C. medium*.
cantharis (kan'tharis), *n.* (*pl.* **cantharides** (kantha'ridēz)) Spanish fly, a coleopterous insect having vesicatory properties; applied to similar beetles; (*pl.*) a preparation of Spanish flies dried and used as a blister-producing counter-irritant and as an aphrodisiac.
canthus (kan'thəs), *n.* (*pl.* **-thi**) the angle made by the meeting of the eyelids.
canticle (kan'tikl), *n.* a brief song, a chant; applied to certain portions of Scripture appointed in the Prayer Book to be said or sung in churches. **Canticles**, *n.sing.* the Song of Solomon.
cantilena (kantilā'nə), *n.* a ballad; plain-song.
cantilever (kan'tilēvə), *n.* a projecting beam, girder or bracket for supporting a balcony or other structure. **cantilever bridge**, *n.* a bridge formed with cantilevers, resting in pairs on piers of masonry or ironwork, the ends meeting or connected by girders.
cantillate (kan'tilāt), *v.t.* to chant; to intone, as in Jewish

synagogues. **cantillation**, *n.*
cantina (kantē'nə), *n.* a bar, wine shop, esp. in Spanish-speaking countries.
cantle (kan'tl), *n.* a fragment, a piece; the projection at the rear of a saddle.
canto (kan'tō), *n.* (*pl.* **-tos**) one of the principal divisions of a poem; the upper voice part in concerted music. **canto fermo** (fer'mō), *n.* plainsong; the main theme, which is treated contrapuntally.
canton[1] (kan'ton, -ton'), *n.* a division of a country, a small district; a political division of Switzerland; (*Her.*) a small division in the corner of a shield. **canton** (-ton'), *v.t.* to divide into cantons. **cantonal** (kan'-), *a.*
canton[2] (kantoon'), *v.t.* to provide (troops) with quarters. **cantonment**, *n.* temporary or winter quarters for troops; a permanent military station in British India.
Cantonese (kantənēz'), *a.* of the city of Canton in S China, or its inhabitants, or the dialect of Chinese spoken there; of a highly-spiced style of cookery originating there. *n.* a native or inhabitant of Canton; the Cantonese dialect.
cantor (kan'taw), *n.* a precentor; the Jewish religious official who sings the liturgy. **cantorial** (-taw'-), *a.* pertaining to the precentor or to the north side of the choir. **cantoris** (-taw'ris), *a.* (sung) by the cantorial side of the choir.
cantus firmus (kantəs fœ'məs), CANTO FERMO under CANTO.
Canuck (kənŭk'), *n.* (*chiefly N Am., coll.*) a Canadian; (*Canada*) a French Canadian.
canvas (kan'vəs), *n.* a coarse unbleached cloth, made of hemp or flax, formerly used for sifting, now for sails, tents, paintings, embroidery etc.; the sails of a ship; a tent or tents; a sheet of canvas for oil-painting; a picture, esp. in oils; a covering for the ends of a racing-boat; this part of a boat; (*fig.*) background. *a.* made of canvas. **under canvas**, in a tent or tents; with sails set. **canvasback**, *n.* a N American sea-duck.
canvass (kan'vəs), *n.* close examination, discussion; the act of soliciting votes. *v.t.* to examine thoroughly, to discuss; to solicit votes, interest, orders etc. from. *v.i.* to solicit votes etc. **canvasser**, *n.*
canyon, cañon (kan'yən), *n.* a deep gorge or ravine with precipitous sides, esp. of the type of those formed by erosion in the western plateaux of the US.
canzone (kantsō'nā), *n.* a Provençal or Italian song. **canzonet** (kanzənet'), **-netta** (-tə), *n.* a short air or song; a light air in an opera.
caoutchouc (kow'chook), *n.* India-rubber, the coagulated juice of certain tropical trees, which is elastic and waterproof.
CAP, (*abbr.*) Common Agricultural Policy (of the EEC).
cap[1] (kap), *n.* a covering for the head, esp. a brimless head-covering for a man or a boy; a woman's head-dress, usu. for indoor wear; a natural or artificial covering resembling this in form or protective or ornamental function; a special form of head-dress distinguishing the holder of an office or occupation, membership of a sports team etc.; a percussion cap; the top part of anything; the pileus of a mushroom; (also **Dutch cap**) a form of contraceptive device; the crown of a tooth or an artificial replacement. *v.t.* (*past, p.p.* **capped**) to cover the top of with a cap; to put a cap on; to protect or cover with or as with a cap; to be on the top of; to complete, to surpass; to set an upper limit on. **cap and bells**, the insignia of a jester. **cap and gown**, full academic dress. **cap in hand**, in a humble or servile manner. **if the cap fits**, if the general remark applies to you, take it to yourself. **percussion cap** PERCUSSION. **to cap it all**, (*coll.*) as a finishing touch. **to set one's cap at**, of a woman, to endeavour to captivate. **cap-paper**, *n.* coarse paper used by grocers for wrapping up sugar etc. **cap sleeve**, *n.* a short sleeve just covering the shoulder. **cap-stone**, *n.* the top stone; a coping. **capper**, *n.*

cap[2] (kap), (*abbr.*) capital (letter); chapter.
capable (kā'pəbl), *a.* susceptible (of); competent, able, skilful, qualified, fitted. **capably**, *adv.* **capability** (-bil'-), *n.* the quality of being capable; capacity; (*pl.*) resources, abilities, intellectual attainments.
capacious (kəpā'shəs), *a.* able to contain much; wide, large, extensive; comprehensive, liberal. **capaciously**, *adv.* **capaciousness**, *n.*
capacity (kəpas'iti), *n.* power of containing or receiving; the amount that can be contained, held, stored etc.; cubic extent; power to absorb; capability, ability; opportunity, scope; relative position, character or office; legal qualification; a term used to denote the output of a piece of electrical apparatus. **to capacity**, (full) to the limit. **capacitance**, *n.* the ability of a conductor, system etc. to store electric charge; the amount stored, measured in farads. **capacitor**, *n.* a device for storing electric charge in a circuit.
cap-à-pie (kap'əpē'), *adv.* from head to foot (armed or accoutred).
caparison (kəpa'risən), *n.* (*often pl.*) housings, often ornamental, for a horse or other beast of burden; outfit, equipment. *v.t.* to furnish with trappings; to deck out.
cape[1] (kāp), *n.* a covering for the shoulders, attached to another garment or separate; a cloak. **caped**, *a.*
cape[2] (kāp), *n.* a headland projecting into the sea. **the Cape**, the Cape of Good Hope; the province of South Africa containing it. **Cape Coloured**, *n.* in South Africa, a person of mixed race. **Cape doctor**, *n.* a south easterly wind in the Cape. **Cape Dutch**, *n.* an architectural style characterized by high ornamental front gables, typical of early buildings at the Cape; (*formerly*) Afrikaans. **Cape gooseberry**, *n.* a tropical plant with a small yellow edible fruit; this fruit.
capelin, caplin (kap'əlin, kap'lin), *n.* a small Newfoundland fish like a smelt, used as bait for cod.
capellmeister (kəpel'mistə), KAPELLMEISTER.
caper[1] (kā'pə), *n.* a frolicsome leap; an escapade; a questionable act or enterprise. *v.i.* to leap; to skip about. **to cut capers**, to caper; to act in a ridiculous manner. **caperer**, *n.*
caper[2] (kā'pə), *n.* a prickly shrub, *Capparis spinosa*; (*pl.*) the flower-buds of this, pickled. **caper sauce**, *n.* a white sauce flavoured with capers, eaten with boiled mutton etc.
capercailzie, -caillie (kapəkā'li), *n.* the wood-grouse, *Tetrao urogallus*, called also the mountain cock or cock of the wood.
capias (kā'pias, kap'-), *n.* a judicial writ ordering an officer to arrest.
capillary (kəpil'əri), *a.* resembling a hair in slenderness; having a minute bore; pertaining to the capillary vessels or to capillary attraction etc. *n.* one of the minute blood-vessels in which the arterial circulation ends and the venous begins. **capillary attraction, repulsion**, *n.* capillarity.
capillarity (-la'-), *n.* the cause which determines the ascent or descent of the surface of a liquid in contact with a solid. **capillose** (kap'ilōs), *a.* hairy.
capital[1] (kap'itl), *n.* the head of a pillar.
capital[2] (kap'itl), *a.* principal, chief, most important; excellent, first-rate; involving or affecting the head or the life; punishable by death; fatal, injurious to life; of letters, initial; hence, of a larger size and a shape distinguishing chief letters; relating to the main fund or stock of a corporation or business firm. *n.* a capital letter; a head city or town; a metropolis; wealth appropriated to reproductive employment; a principal or fund employed in earning interest or profits; net worth. **capital transfer tax**, a tax levied on the transfer of money or property, either by gift or inheritance. **circulating capital** FLOATING CAPITAL. **fixed capital**, buildings, machinery, tools etc. used in industry. **floating capital**, raw material, money,

capitate goods etc. **to make capital out of,** to make profit from, turn to one's advantage. **capital assets** FIXED CAPITAL. **capital expenditure,** *n.* expenditure on buildings, equipment etc. **capital gain,** *n.* profit made from the sale of shares or other property. **capital goods,** *n.pl.* raw materials and tools used in the production of consumers' goods. **capital levy,** *n.* (*Fin.*) a levy on capital. **capital punishment,** *n.* the death penalty. **capital ship,** *n.* a warship of the most powerful kind. **capitalism,** *n.* the economic system under which individuals employ capital and employees to produce wealth. **capitalist,** *n.* one who possesses capital; one who supports capitalism. **capitalistic** (-list'tik), *a.* **capitalize, -ise,** *v.t.* to convert into capital; to use as capital; to provide with capital; to calculate or realize the present value of (periodical payments); to write or print with a capital letter. *v.i.* to take advantage. **capitalization,** *n.* **capitally,** *adv.* excellently.

capitate (kap'itāt), **-ated** (-tātid), *a.* having a head; having the inflorescence in a head-like cluster.

capitation (kapitā'shən), *n.* a tax, fee or grant per head. **capitation grant, allowance,** *n.* a subsidy or allowance calculated on the number of persons passing an examination or fulfilling specified conditions.

Capitol (kap'itol), *n.* the great national temple of ancient Rome, situated on the *Capitoline* Hill, dedicated to Jupiter; the building in which the US Congress meets; (*N Am.*) the senate-house of a state. **Capitolian** (-tō'-), **Capitoline** (-pit'əlin), *a.* of or pertaining to the Roman Capitol.

capitular (kəpit'ūlə), *a.* of or pertaining to an ecclesiastical chapter. *n.* a member of a chapter; a statute passed by a chapter. **capitularly,** *adv.*

capitulary (kəpit'ūləri), *n.* a collection of ordinances, esp. those of the Frankish kings.

capitulate (kəpit'ūlāt), *v.i.* to surrender on stipulated terms. **capitulation,** *n.* the act of capitulating; a document containing the terms of surrender. **capitulator,** *n.*

capitulum (kəpit'ūləm), *n.* (*pl.* **-la** (-lə)) a close cluster or head of sessile flowers; a head-shaped anatomical part.

caplin CAPELIN.

capo (kap'ō), *n.* (*pl.* **-pos**) a capotasto.

capon (kā'pən), *n.* a castrated cock fowl, esp. fattened for cooking.

capotasto (kap'ōtastō), *n.* a bar fitted across the fingerboard of a guitar, to alter the pitch of all the strings simultaneously.

capote (kəpōt'), *n.* a long cloak or overcoat, usu. with a hood.

cappuccino (kapuchē'nō), *n.* (*pl.* **-nos**) white coffee, esp. from an espresso machine, often with whipped cream or powdered chocolate.

capric (kap'rik), *a.* pertaining to a goat. **capric acid,** *n.* an acid, having a slight goat-like smell, contained in butter, coconut oil, and other compounds.

capriccio (kəprē'chō), *n.* (*pl.* **-ccios, -cci**) (*Mus.*) a lively composition more or less free in form. **capriccioso** (-sō), *adv.* (*Mus.*) in a free, fantastic style.

caprice (kəprēs'), *n.* a sudden impulsive change of opinion or humour; a whim, a freak; disposition to this kind of behaviour; a freakish or playful work of art; a capriccio. **capricious** (-pri'-), *a.* influenced by caprice; whimsical, uncertain, fickle, given to unexpected and incalculable changes. **capriciously,** *adv.* **capriciousness,** *n.*

Capricorn (kap'rikawn), *n.* the zodiacal constellation of the Goat; the 10th sign of the zodiac; one born under this sign. **Tropic of Capricorn** TROPIC [1].

caprification (kaprifikā'shən), *n.* the practice of suspending branches of the wild fig on the cultivated fig, that the (female) flowers of the latter may be pollinated by wasps parasitic on the flowers of the former.

caprifig (kap'rifig), *n.* the wild fig of S Europe and Asia Minor, used in caprification.

caprine (kap'rin), *a.* like a goat.

capriole (kap'riōl), *n.* a leap made by a horse without advancing. *v.i.* to perform a capriole.

Capri pants (kaprē'), **Capris,** *n.pl.* women's tight-fitting trousers, reaching to above the ankle.

caproic (kəprō'ik), *a.* pertaining to a goat. **caproic acid,** *n.* an acid contained, like capric and butyric acids, in butter etc.

capsicum (kap'sikəm), *n.* any of a genus of plants of the potato family, with mild or pungent fruit and seeds; the fruit of a capsicum used as a vegetable or ground as the condiments chilli, cayenne etc.

capsid[1] (kap'sid), *n.* any bug of the family Miridae, feeding on plants.

capsid[2] (kap'sid), *n.* the outer protein casing of some viruses.

capsize (kapsīz'), *v.t., v.i.* to upset, to overturn. *n.* an overturn. **capsizal,** *n.*

capstan (kap'stən, -stan), *n.* a revolving pulley or drum with a belt or cable running over it, used to increase the force exerted by the cable or belt; a revolving shaft that drives the tape on in a tape recorder. **capstan lathe,** *n.* one with a revolving turret, so that several different tools can be used in rotation.

cap-stone CAP [1].

capsule (kap'sūl), *n.* a metallic cover for a bottle; (*Physiol.*) an envelope or sac; (*Bot.*) a dry dehiscent seed-vessel; a spore-containing sac; (*Chem.*) a shallow saucer; a small envelope of gelatine containing medicine; a self-contained detachable part of a spacecraft or aircraft; a spacecraft. **capsular,** *a.* **capsuliform,** *a.* **capsulize, -ise,** *v.t.* to enclose in a capsule; to put (information) into a very condensed form.

capt., (*abbr.*) captain.

captain (kap'tin), *n.* a leader, a commander; in the army, a rank between major and lieutenant; in the navy, a rank between commodore and commander; (*N Am.*) a head-waiter or the supervisor of bell boys in a hotel; (*N Am.*) a police officer in charge of a precinct; the master of a merchant ship; the person in command of a civil aircraft; the head of a gang, side or team; the chief boy or girl in a school; a great soldier, a veteran commander. *v.t.* to act as captain to; to lead, to head. **captain's chair,** *n.* a wooden chair with back and arms in one semicircular piece, supported on wooden shafts. **captaincy,** *n.* **captainship,** *n.* **captainless,** *a.*

caption (kap'shən), *n.* the heading or descriptive preamble of a legal document; the wording under an illustration, the legend; the heading of a chapter, section or newspaper article; a subtitle or other printed or graphic material in a television or cinematograph film.

captious (kap'shəs), *a.* fault-finding, carping, cavilling. **captiously,** *adv.* **captiousness,** *n.*

captivate (kap'tivāt), *v.t.* to fascinate, to charm. **captivating,** *a.* **captivation,** *n.*

captive (kap'tiv), *n.* one taken prisoner, held in confinement or bondage, or fascinated. *a.* taken prisoner; held in bondage; held in control; captivated, fascinated; unable to move away or otherwise exercise choice, as in a *captive audience, market* etc. **captive balloon,** *n.* a balloon held by a rope from the ground. **captive time,** *n.* time during which a person is not working but must be available if needed. **captivity** (-tiv'-), *n.* **captor,** *n.* **captress,** *n. fem.*

capture (-chə), *n.* the act of seizing as a prisoner or a prize; the person or thing so taken. *v.t.* to make a capture of; to seize as a prize; to succeed in describing in words or by drawing (a likeness etc.). **capturer,** *n.*

capuche (kəpoosh'), *n.* a hood, esp. the long pointed hood of the Capuchins.

Capuchin (kap'əchin, -ū-), *n.* a Franciscan friar of the reform of 1528; a hooded cloak, like the habit of the Capuchins, worn by women. **capuchin monkey,** *n.* an

American monkey, *Cebus capucinus*.
capybara (kəpibah'rə), *n*. a S American mammal, *Hydrochaerus capybara*, the largest living rodent, allied to the guinea-pig.
CAR, (*abbr*.) Central African Republic.
Car., (*abbr*.) Carolina; Charles. [L *Carolus*]
car (kah), *n*. a motor-car; (*poet*.) a wheeled vehicle, a chariot; (*Ir*.) a jaunting-car; (*N Am*.) any railway coach or wagon; (*N Am*.) a lift cage; the pendent carriage of an airship; in Britain, used of certain types of passenger railway carriages, as in *dining-car*, *Pullman-car*. **car-bomb**, *n*. an explosive device hidden in a parked car, which destroys the car and anything nearby. **car-boot sale**, *n*. a sale of second-hand goods, from the boot of a car. **car-coat**, *n*. a short coat which can be worn comfortably in a car. **car ferry**, *n*. a ferry-boat built so that motor vehicles can be driven on and off it. **car-park**, *n*. a place where cars may be left for a limited period. **car port**, *n*. an open-sided shelter for a car beside a house. **car-sick**, *a*. affected with motion sickness when travelling in a car, bus etc. **car-wash**, *n*. an establishment with equipment for the automatic washing of cars. **carful**, *n*.
carabine, carabineer etc. CARBINE.
caracal (ka'rəkal), *n*. the Persian lynx, *Felis caracal*.
caracol, -cole (ka'rəkol, -kōl), *n*. a half turn or wheel made by a horse or horseman. *v.i.* to perform a caracol.
carafe (kəraf', -rahf, ka'rəf), *n*. a wide-mouthed glass jar for holding wine or water at table; as much wine or water as a carafe will hold.
carambola (karəmbō'lə), STAR-FRUIT.
carambole (ka'rəmbōl), *n.*, *v.i.* (*Billiards*) (a) cannon.
caramel (ka'rəmel, -məl), *n*. burnt sugar used as a food colouring; a kind of chewy sweetmeat; the colour of caramel, a pale brown. *v.t.*, *v.i.* **caramelize, -ise**, to turn into caramel.
carapace (ka'rəpās), *n*. the upper body-shell of the tortoise family, crab etc.
carat (ka'rət), *n*. a weight (standardized as the International Carat of 0.200 g) used for precious stones, esp. diamonds; a proportional measure of one 24th part, used to describe the fineness of gold.
caravan (ka'rəvan), *n*. a company of merchants or pilgrims, travelling together (esp. in desert regions) for mutual security; a travelling house, a carriage for living in drawn by horse or motor-car; a showman's covered wagon. *v.i.* to live, esp. temporarily in a caravan. **caravaneer** (-niə'), *n*. the leader of an Eastern caravan. **caravanner**, *n*. **caravanning**, *n*. travelling or holidaying in a caravan.
caravanserai, -sera, -sary (karəvan'səri, -rə, -ri), *n*. an Oriental inn with a large courtyard for the accommodation of caravans.
caravel (ka'rəvel), **carvel** (kah'vəl), *n*. a name applied at different times to various kinds of ships; e.g. a swift Spanish or Portuguese merchant vessel; a Turkish frigate.
caraway (ka'rəwā), *n*. a European umbelliferous plant, *Carum carvi*. **caraway-seeds**, *n.pl.* the small dried fruit of this, used as a flavouring.
carbide (kah'bīd), *n*. a compound of carbon with a metal, esp. **calcium carbide**, used for generating acetylene.
carbine (kah'bīn), **carabine** (ka'rəbin), *n*. a short rifle formerly used by cavalry; an automatic or semi-automatic rifle. **carbineer** (-biniə'), **carabinier** (-biniə'), *n*. a soldier armed with a carbine.
carb(o)-, *comb. form*. of, with, containing, or pertaining to carbon.
carbohydrate (kahbəhī'drāt), *n*. an organic compound of carbon, hydrogen and oxygen. Usually there are two atoms of hydrogen to every one of oxygen as in starch, glucose etc.
carbolic (kahbol'ik), *a*. derived from coal or coal-tar. **carbolic acid**, *n*. an antiseptic and disinfectant acid.

carbon (kah'bən), *n*. a non-metallic element, at. no.6; chem. symbol C, found in nearly all organic substances, in carbon dioxide, and the carbonates, and uncombined in diamond, graphite and charcoal; a pencil of fine charcoal used in arc-lamps; carbon-paper; a carbon-copy. **carbon-copy**, *n*. a typewritten duplicate. **carbon dating**, *n*. a method of calculating the age of organic material (wood, bones etc.) by measuring the decay of the isotope carbon-14. **carbon dioxide**, *n*. a gaseous combination of one atom of carbon with two of oxygen, a normal constituent of the atmosphere and of expired breath. **carbon fibre**, *n*. a very strong thread of pure carbon, used for reinforcing plastics, metals etc. **carbon monoxide**, *n*. a poisonous gas containing one atom of oxygen for each atom of carbon. **carbon-paper**, *n*. a dark-coated paper for taking impressions of writing, drawing etc. **carbon steel**, *n*. any of several steels containing carbon in varying amounts. **carbon tetrachloride**, *n*. a colourless toxic liquid, used as a dry-cleaning solvent. **carbonaceous** (-nā'-), *a*. like coal or charcoal; like or containing carbon. **carbonate** (-nət), *n*. a salt of carbonic acid. *v.t.* to impregnate with carbonic acid; to aerate (water etc.); to form into a carbonate. **carbonic** (-bon'-), *a*. pertaining to or containing carbon. **carbonic acid**, *n*. a weak acid, the compound formed by carbon dioxide and water. **carbonic-acid gas**, carbon dioxide. **carboniferous** (-nif'-), *a*. producing coal or carbon; applied to the strata between the Old Red Sandstone (below) and the Permian (above) or the geological epoch during which these strata were deposited. **carbonize, -ise**, *v.t.* to convert into carbon, esp. by the action of fire; to cover with carbon. **carbonization**, *n*.
carbonade, -nnade (kahbənād', -nahd), *n*. a beef stew made with beer. **carbonado**[1] (-nā'dō, -nah'-), *n*. (*pl.* **-dos, -does**) a carbonade.
carbonado[2] (kahbənā'dō, -nah'-), *n*. a black, opaque diamond of poor quality, used industrially in drills etc.
carburet, carbonic CARBON.
Carborundum® (kahbərŭn'dəm), *n*. a silicon carbide used for grinding-wheels etc.
carboxyl (kahboks'il), *n*. the chemical group characteristic of organic acids. **carboxylic** (-sil'ik), *a*.
carboy (kah'boi), *n*. a large glass bottle, usu. protected with wickerwork, used for holding corrosive liquids.
carbuncle (kah'bŭngkl), *n*. a precious stone of a red or fiery colour; a garnet cut in a concave cabochon; a hard, painful boil without a core, caused by bacterial infection; (*coll*.) an unattractive building etc. which defaces the appearance of the landscape. **carbuncled**, *a*. **carbuncular** (-bung'kū-), *a*.
carburet (kah'būret, -ret'), *v.t.* (*past*, *p.p.* **carburetted**) to combine (another element) with carbon. **carburation, -retion** (-re'-), *n*. **carburettor**, *n*. an apparatus in an internal combustion engine designed to vaporize fuel and to mix it intimately with air in proportions to ensure ready ignition and complete combustion. **carburize, -ise**, *v.t.* to carburet; to mix carbon with (esp. iron). **carburization, -isation**, *n*.
carcajou (kah'kəzhoo), *n*. the glutton or wolverine.
carcass, -case (kah'kəs), *n*. the trunk of a slaughtered beast without the head and offal; the dead body of a beast; (*derog.* or *facet.*) the human body dead or alive; the framework of a building, ship etc.; a mere body, mere shell, or husk. **carcass meat**, *n*. raw meat as sold in a butcher's shop.
carcinogen (kahsin'əjən), *n*. a substance that can give rise to cancer. **carcinogenic** (-jen'-), *a*.
carcinoma (kahsinō'mə), *n*. (*pl.* **-mata** (-tə), **mas**) the disease cancer; a malignant tumour. **carcinomatous**, *a*. **carcinosis** (-sis), **carcinomatosis** (-ō'sis), *n*. the spread of cancer through the body.
card[1] (kahd), *n*. one of a pack of oblong pieces of pasteboard, marked with pips and pictures, used in play-

card 121 **carina**

ing games of chance or skill; a flat, rectangular piece of stiff pasteboard or paper for writing or drawing on or the like; one used for sending greetings; a visiting-card, a ticket of admission, an invitation; a programme, a menu, a list of events at races, regattas etc., and various other senses denoted by a prefixed substantive; a compass card; a small rectangle of plastic issued by a bank, department store etc. entitling the holder to obtain credit, use a cash dispenser etc. (*pl.*) a game or games with cards; card-playing; (*sl.*) a character, an eccentric; (*pl., coll.*) a worker's employment documents; (*Comput.*) a punched card. **compass card** COMPASS. **house of cards** HOUSE. **on the cards**, possible; not improbable. **to get one's cards**, to be dismissed or made redundant. **to lay, put, one's cards on the table**, to disclose one's situation, plans etc. **to play one's cards well**, to be a good strategist. **to play with one's cards close to one's chest** PLAY. **to show one's cards**, to reveal one's plan. **visiting card** VISIT. **cardboard**, *n*. pasteboard for making boxes and other articles. *a*. without substance or reality; pertaining to the homeless who live and sleep on the streets. **card-carrying**, *a*. being a full member of (a political party etc.). **card catalogue, index**, *n*. a catalogue or index in which each item is entered on a separate card. **cardphone**, *n*. a public telephone where a special card (*phonecard*) is inserted rather than coins. **card punch**, *n*. (*Comput.*) a device which can take data from a store or processor and transfer it to punched cards. **card reader**, *n*. (*Comput.*) a device which can read the data on punched cards and convert it to a form in which it can be stored or processed. **card-sharp, -sharper**, *n*. one who swindles by means of card games or tricks with cards. **card-table**, *n*. a table to play cards on. **card vote**, *n*. a ballot where the vote of each delegate counts for the number of his constituents.
card² (kahd), *n*. an instrument for combing wool or flax. *v.t.* to comb (wool, flax or hemp) with a card; to raise a nap on. **carder, carding,** *a*. **carding-machine**, *n*. a machine for combing out and cleaning wool, cotton etc.
cardamom (kah'dəməm), *n*. a spice obtained from the seed capsules of various species of *Amomum* and other genera.
cardiac (kah'diak), *a*. of or pertaining to the heart; of or pertaining to the upper orifice of the stomach. *n*. a person suffering from heart disease. **cardialgy** (-al'ji), **-algia** (-al'jiə), *n*. an affection of the heart; heartburn. **cardialgic** (-al'-), *a*.
cardi(e) (kah'di), *n*. short for CARDIGAN.
cardigan (kah'digən), *n*. a knitted jacket buttoned up the front. [7th Earl of *Cardigan* 1797-1868]
cardinal (kah'dinəl), *a*. fundamental, chief, principal; of the colour of a cardinal's cassock, deep scarlet. *n*. one of the ecclesiastical princes of the Roman Church who elect a new pope, usu. from among their own number; a short cloak (orig. of scarlet) for women; a cardinal-bird; (*coll.*) mulled red wine. **cardinal-bird**, *n*. a N American red-plumaged song-bird. **cardinal bishop**, *n*. the highest rank of cardinal. **cardinal deacon**, *n*. the third rank of cardinal. **cardinal numbers**, *n.pl.* the simple numbers 1, 2, 3 etc., as distinguished from 1st, 2nd, 3rd etc. **cardinal points**, *n.pl.* the four points of the compass: north, south, east and west. **cardinal priest**, *n*. the second rank of cardinal. **cardinal virtues**, *n.pl.* (*Phil.*) Prudence, Temperance, Justice and Fortitude; (*Theol.*) Faith, Hope and Charity. **cardinalate** (-āt), **cardinalship**, *n*. the office or dignity of a cardinal. **cardinally**, *adv*. fundamentally.
cardio-, *comb. form.* pertaining to the heart.
cardiograph (kah'diəgrahf), *n*. an instrument for registering the movement of the heart. **cardiogram** (-gram), *n*. a reading from a cardiograph. **cardiography** (-og'-), *n*.
cardioid (kah'dioid), *n*. (*Math.*) a heart-shaped curve.
cardiology (kahdiol'əji), *n*. the science of the heart and heart disease.
cardiovascular (kahdiōvas'kūlə), *a*. relating to the heart and blood-vessels.
carditis (kahdī'tis), *n*. inflammation of the heart.
cardoon (kahdoon'), *n*. a kitchen-garden plant, *Cynara cardunculus*, allied to the artichoke.
CARE (keə), (*abbr.*) Cooperative for American Relief Everywhere.
care (keə), *n*. solicitude, anxiety, concern; a cause of these; caution, serious attention, heed; oversight, protection; an object of regard or solicitude. *v.i.* to be anxious or solicitous; to be concerned (about); to provide (for), attend (upon); to have affection, respect or liking (for); to be desirous, willing or inclined (to). **in care**, in the guardianship of the local authority. **who cares?** (*coll.*) I don't care. **carefree**, *a*. free from responsibility, light-hearted. **care label**, *n*. a label on a garment giving cleaning instructions. **care-laden**, *a*. **caretaker**, *n*. a person in charge of something, esp. a building, *a*. interim. **care-worn**, *a*. showing the effects of stress or anxiety. **carer**, *n*. one who cares; one who looks after someone, e.g. an invalid, dependent relative etc. **caring**, *a*. showing care or concern; providing medical care or social services, as in *caring professions*. **careful**, *a*. solicitous; watchful, cautious, circumspect; provident, painstaking, attentive, exact. **carefully**, *adv*. **carefulness**, *n*. **careless**, *a*. free from care, without anxiety, unconcerned; heedless, thoughtless, inaccurate; inattentive, negligent (of); negligently done. **carelessly**, *adv*. **carelessness,** *n*.
careen (kərēn'), *v.t.* to turn (a ship) on one side in order to clean or caulk her. *v.i.* to heel over. **careenage**, *n*. the act of, a place for, or the expense of careening.
career (kəriə'), *n*. a running, a swift course; course or progress through life; the progress and development of a nation, party etc; a way of making a living in business, professional or artistic fields etc. *a*. having a specified career, professional as *career diplomat*. *v.i.* to move in a swift, head-long course; to gallop at full speed. **career break**, *n*. an extended period of leave from one's job, sometimes for several years, in order to rear children. **careers master, mistress, teacher**, *n*. a teacher who gives advice on careers. **career girl, woman**, *n*. one who pursues a full-time career. **careerism**, *n*. **careerist**, *n*. one who makes personal advancement his or her main objective.
caress (kəres'), *n*. an embrace, a kiss; an act of endearment. *v.t.* to fondle, to stroke affectionately; to pet, to court, to flatter. **caressingly**, *adv*.
caret (ka'rət), *n*. a mark (∧) used to show that something, which may be read above or in the margin, has been left out.
carfuffle (kəfŭf'l), *n*. commotion, disorder.
cargo (kah'gō), *n*. (*pl.* **-goes**) the freight of a ship or aircraft; a load. **cargo cult**, *n*. a religion popular in some Polynesian islands, according to which the ancestors will come back bringing wealth for the islanders.
Carib (ka'rib), *n*. one of the aboriginal race in the southern islands of the W Indies; the language spoken by these people. **Caribbean** (-bē'ən), *a*. of or pertaining to the W Indies or their inhabitants.
caribou (ka'riboo), *n*. the N American reindeer.
caricature (ka'rikəchə, -tūə), *n*. a representation of a person or thing exaggerating characteristic traits in a ludicrous way; a burlesque, a parody; a laughably inadequate person or thing. *v.t.* to represent in this way; to burlesque. **caricaturable** (-tūrə'-), *a*. **caricatural**, *a*. **caricaturist** (-tūər'-), *n*.
caries (keə'riēz), *n*. decay of the bones or teeth; decay of vegetable tissue. **carious,** *a*.
carillon (kəril'yən, ka'rilən), *n*. a set of bells so arranged as to be played by the hand or by machinery; an air played on such bells; a musical instrument (or part of one) to imitate such bells.
carina (kərē'nə, -rī'-), *n*. (*pl.* **-nas, -nae**) (*Zool., Bot.*) a ridge-like structure; keel-shaped. **carinal,** *a*. **carinate**

caring

(ka'rənăt), **-ated,** *a.*
caring CARE.
carioca (kariŏk'ə), *n.* a S American dance like the samba; music for this dance; (*coll.*) a native or inhabitant of Rio de Janeiro.
cariole (kā'riōl), CARRIOLE.
carl, carle (kahl), *n.* (*Sc.*) a countryman; a man of low birth; a strong, sturdy fellow.
carline[2] (kah'lin), *n.* any of a genus of plants allied to the thistle, commonest species *Carlina vulgaris.*
Carlovingian CAROLINGIAN under CAROLINE.
Carmelite (kah'məlīt), *n.* one of an order of mendicant friars, founded in the 12th cent. on Mount Carmel (also *White Friars*); a nun of this order. *a.* belonging or pertaining to this order.
carminative (kah'minətiv, -min'-), *n., a.* (a medicine) that expels flatulence.
carmine (kah'min, -min), *n.* a beautiful red or crimson pigment obtained from cochineal. *a.* of this colour.
carnage (kah'nij), *n.* butchery, slaughter, esp. of people.
carnal (kah'nəl), *a.* fleshly, sensual; sexual, unregenerate, as opp. to *spiritual;* temporal, secular. **carnal knowledge,** *n.* sexual intercourse. **carnalism,** *n.* sensualism. **carnalist,** *n.* **carnality** (-nal'-), *n.* the state of being carnal. **carnalize, -ise,** *v.t.*
carnassial (kahnas'iəl), *n.* in the Carnivora, a large tooth adapted for tearing flesh. *a.* relating to such a tooth.
carnation (kahnā'shən), *n.* a light rose-pink; the cultivated clove-pink, *Dianthus caryophyllus;* of this colour.
carnauba (kahnow'bə), *n.* a Brazilian palm; its yellow wax, used in polishes.
carnelian (kahněl'yən), CORNELIAN.
carnet (kah'nā), *n.* a document allowing the transport of vehicles or goods across a frontier; a book of vouchers, tickets etc.
carnival (kah'nivəl), *n.* the season immediately before Lent, in many Roman Catholic countries devoted to pageantry and riotous amusement; a festival; (a period or instance of) riotous amusement, revelry; a fun-fair.
Carnivora (kahniv'ərə), *n.pl.* a large order of mammals subsisting on flesh. **carnivore** (kah'nivaw), *n.* a carnivorous animal or plant. **carnivorous,** *a.* feeding on flesh; also applied to insectivorous plants.
carob (ka'rəb), *n.* the Mediterranean locust-tree, *Ceratonia siliqua;* its fruit, with an edible pulp, used as a substitute for chocolate.
carol (ka'rəl), *n.* a joyous hymn, esp. in honour of the Nativity; joyous warbling of birds. *v.i.* (*past, p.p.* **carolled**) to sing carols; to warble. *v.t.* to celebrate in songs. **caroller,** *n.*
Caroline (ka'rəlīn), **Carolean** (-lē'-), **Carolinian** (-lī'-), *a.* pertaining to the reigns of Charles I and II of Britain; pertaining to Charlemagne; of a script (**Caroline minuscule**) developed during the reign of Charlemagne. **Carolingian** (-lin'jiən), **Carlovingian** (kahləvin'jiən), *n., a.* (a member) of the dynasty of French kings founded by Charlemagne.
carom (ka'rəm), CANNON[2].
carotene (ka'rətēn), *n.* any of a group of orange-red plant pigments that are convertible to vitamin A in the body. **carotenoid, -tinoid** (-rot'inoid), *n.* any of various yellow to red pigments including the carotenes.
carotid (kərot'id), *n., a.* (of or related to) either of the arteries (one on each side of the neck) supplying blood to the head.
carouse (kərowz'), *n.* a carousal. *v.i.* to drink freely; to participate in a carousal. **carouser,** *n.* **carousingly,** *adv.*
carousel (*esp. N Am.*) **carrousel** (karəsel'), *n.* (*N Am.*) a merry-go-round; a rotating conveyor belt for luggage at an airport; a rotating container which delivers slides to a projector.

carp[1] (kahp), *v.i.* to find fault, to cavil. **carping,** *n., a.* **carpingly,** *adv.*
carp[2] (kahp), *n.* a freshwater fish of the genus *Cyprinus,* esp. *C. cyprio,* the common carp, a pond-fish.
carpaccio (kahpahch'ēō, -pahch'ō), *n.* raw beef sliced thinly and served with a sauce. [from the Italian painter Vittore *Carpaccio,* c.1465–c.1522, alluding to his use of red]
carpal CARPUS.
carpel (kah'pəl), *n.* the female reproductive organ of a flower, comprising ovary, style and stigma. **carpellary,** *a.*
carpenter (kah'pintə), *n.* an artificer who prepares and fixes the wood-work of houses, ships etc; a wood-worker; (*chiefly N Am.*) a joiner. *v.i.* to do carpenter's work. *v.t.* to make by carpentry. **carpenter-ant, -bee, -bird, -moth,** *n.* insects and birds that bore into wood. **carpentry,** *n.* the trade of a carpenter; carpenter's work, esp. the kind of wood-work prepared at the carpenter's bench.
carpet (kah'pit), *n.* a woollen or other thick fabric, usu. with a pattern, for covering floors and stairs. *v.t.* to cover with or as with a carpet; (*coll.*) to reprimand. **on the carpet,** under consideration; (*coll.*) being reprimanded. **to sweep under the carpet,** to conceal or ignore deliberately (an object, happening etc.). **carpet-bag,** *n.* a travelling-bag orig. made with sides of carpet. **carpet-bagger,** *n.* (*chiefly N Am.*) an adventurer, esp. political. **carpet-baggery,** *n.* **carpet-beater,** *n.* a racket-shaped cane utensil for beating carpets. **carpet-bombing,** *n.* bombing of a whole area, rather than of selected targets. **carpet-rod,** *n.* a rod for holding down stair-carpet. **carpet-shark,** *n.* a shark of the genus *Orectolobus,* with two dorsal fins and the back patterned like a carpet. **carpet-slippers,** *n.pl.* comfortable slippers made of tapestry. **carpet-snake,** *n.* an Australian snake, *Morelia variegata.* **carpet-sweeper,** *n.* an apparatus equipped with revolving brushes and dustpans, used for sweeping carpets. **carpet tiles,** *n.pl.* small squares of carpeting which can be laid like tiles to cover a floor. **carpeting,** *n.* the stuff of which carpets are made; (*coll.*) a dressing-down. **carpetless,** *a.*
carpo-[1], *comb. form.* pertaining to the wrist.
carpo-[2], *comb. form.* pertaining to fruit.
carpus (kah'pəs), *n.* (*pl.* **-pi**) the wrist, the part of the human skeleton joining the hand to the forearm; the corresponding part in animals, in horses the knee. **carpal,** *a.* of the wrist. *n.* a wrist bone.
carrack (ka'rək), *n.* a large merchant ship; a galleon.
carrageen, -gheen (ka'rəgēn), *n.* Irish moss, a nutritious seaweed found on N Atlantic shores; carrageenan. **carrageenan, -g(h)eenin,** *n.* an extract of carrageen used in food-processing. [*Carragheen,* Co. Waterford, Ireland, where it is particularly plentiful]
carrel, carrell (ka'rəl), *n.* a cubicle for private study in a library.
carriage (ka'rij), *n.* carrying, transporting, conveyance, esp. of merchandise; the cost of conveying; manner of carrying; mien, bearing; carrying (of a motion, Bill etc.); a conveyance, a wheeled vehicle, esp. a horse-drawn vehicle kept for pleasure; the sliding or wheeled portion of machinery carrying another part; a passenger vehicle in a train. **carriage and pair,** a four-wheeled private vehicle drawn by two horses. **carriage clock,** *n.* a portable clock in an oblong metal case with a handle on top. **carriage dog,** *n.* (*dated*) a Dalmatian. **carriage forward,** *adv.* the cost of carriage to be paid by the receiver. **carriage rug,** *n.* a rug to cover the knees. **carriage trade,** *n.* trade from well-off customers. **carriageway,** *n.* that part of a road used for vehicular traffic. **carriageful,** *n.* **carriageless,** *a.*
carrick bend (ka'rik), *n.* a particular knot for splicing two ropes together.
carrier (ka'riə), *n.* one who carries, esp. one who conveys goods and merchandise for hire; a framework on a bicycle for holding luggage; applied also to various parts of

carriole | 123 | **carvel**

machines or instruments which act as transmitters or bearers; an organism that transmits a disease-causing agent, esp. a person who transmits an infectious disease without personally suffering from the disease; a carrier bag; (*Chem.*) an inert substance that acts as a vehicle for another; an electron or hole that carries charge in a semi-conductor; a carrier wave; an aircraft carrier. **common carrier**, (*Law*) a person or company transporting goods or merchandise for hire. **carrier bag**, *n.* a strong paper or plastic bag with handles. **carrier pigeon**, *n.* a breed of pigeons trained to carry communications. **carrier rocket**, *n.* one which carries, e.g. a satellite into orbit. **carrier wave**, *n.* an electromagnetic wave which is modulated for the radio transmission etc. of a signal.

carriole, cariole (ka'riōl), *n.* a small open carriage; a light, covered cart.

carrion (ka'riən), *n.* dead, putrefying flesh; garbage, filth. *a.* feeding on carrion; putrid; loathsome. **carrion-crow**, *n.* a species of crow, *Corvus corone*, that feeds on small animals and carrion.

carronade (karənād'), *n.* a short naval cannon of large bore, orig. made at Carron, near Falkirk, Scotland.

carrot (ka'rət), *n.* a plant with an orange-coloured tapering root used as a vegetable; (*pl., coll.*) (a person with) red hair; an incentive. **carroty**, *a.* of the colour of a carrot; red, red-haired.

carrousel CAROUSEL.

carry (ka'ri), *v.t.* to convey, to bear, to transport from one place to another by lifting and moving with the thing carried; to transfer, as from one book, page or column to another; to convey or take with one; to transmit; to conduct; to bring, to enable to go or come; to support; to effect, to accomplish; to win (as an election); to bear, to stand (as sail); to wear (as clothes); to bear or hold in a distinctive way; to extend in any direction in time or space (back, up etc.); to imply, to import, to contain, to include; to keep in stock for sale; to have in or on (esp. as armament); to take by assault; to be pregnant with. *v.i.* to act as bearer; to be transmitted; of a firearm etc., to propel a projectile to a distance; to be propelled, as a missile. *n.* the act of carrying; the range of a firearm; (*N Am.*) a portage. **to carry all before one**, to bear off all the honours; to succeed. **to carry away**, to excite, to deprive of self-control. **to carry it off**, to brave it out. **to carry off**, to remove; to win; to do successfully; to deprive of life. **to carry on**, to manage; to continue; to behave in a particular way, esp. to flirt outrageously; to make a fuss. **to carry oneself**, to behave (in a particular way). **to carry out**, to perform; to accomplish. **to carry over, forward**, to transfer to another page or column, or to a future occasion. **to carry through**, to accomplish; to bring to a conclusion in spite of obstacles. **to carry weight**, to be handicapped; of an argument etc., to be cogent. **to carry with one**, to bear in mind, to convince. **carry-all**, *n.* (*N Am.*) a hold-all. **carrycot**, *n.* a light portable cot for a baby. **carryings-on**, *n.pl.* course of behaviour (usu. of a questionable kind).

carse (kahs), *n.* (*Sc.*) low fertile land, usu. near a river. **carse-land**, *n.*

cart (kaht), *n.* a strong two-wheeled vehicle for heavy goods etc.; a light two-wheeled vehicle (usu. with attrib., as **dog-cart, hand-cart** etc.). *v.t.* to carry or convey in a cart; (*sl.*) to defeat badly; (*coll.*) to carry or pull, esp. with difficulty. *v.i.* to use carts for cartage. **in the cart**, (*sl.*) in a predicament. **cart-horse**, *n.* one of a breed of horses for drawing heavy carts. **cart-load**, *n.* **cart-road, -way**, *n.* a rough road on a farm etc. **cartwheel**, *n.* the wheel of a cart; a somersault taken sideways. **cartwright**, *n.* one whose trade is to make carts. **cartage**, *n.* the act of carting; the price paid for carting. **carter**, *n.*

carte (kaht), a bill of fare; a carte-de-visite. **carte-blanche** (-blāsh'), *n.* unlimited power to act. **carte-de-visite**

(-dəvēzēt'), *n.* a visiting card; a photographic likeness on a small card.

cartel (kahtel'), *n.* an agreement (often international) among manufacturers to keep prices up; in politics, an alliance between two parties to further common policies. **cartelize, -ise**, *v.t., v.i.*

Cartesian (kahtē'ziən, -zhən), *a.* of or pertaining to the French philosopher Descartes (1596–1650), or his philosophy or mathematical methods. *n.* an adherent of his philosophy. **Cartesian coordinates**, *n.pl.* numbers that determine the position of a point on a plane (or in space) with reference to two (or three) mutually perpendicular lines. **Cartesianism**, *n.*

Carthusian (kahthū'ziən), *a.* of or belonging to a religious order, orig. of monks, founded by St Bruno in 1086; of or belonging to Charterhouse School. *n.* a scholar of Charterhouse; a Carthusian monk or nun.

cartilage (kah'tilij), *n.* an elastic, pearly-white animal tissue, gristle; a cartilaginous structure. **cartilaginoid** (-laj'in-), *a.* **cartilaginous** (-laj'in-), *a.* of, like or pertaining to cartilage. **cartilaginous fishes**, *n.pl.* fishes with a cartilaginous skeleton, as sharks and rays.

cartogram (kah'təgram), *n.* a map showing statistical information in diagrammatic form. **cartography** (kahtog'rəfi), *n.* the art or business of making maps and charts. **cartographer**, *n.* **cartographic** (-graf'-), *a.* **cartology**, *n.* the science of maps and charts.

cartomancy (kah'təmansi), *n.* divination or fortune-telling by cards.

carton (kah'tən), *n.* a cardboard box; a box made of waxed paper for holding liquids. *v.t.* to put into a carton.

cartoon (kahtoon'), *n.* a design on strong paper for painting tapestry, mosaic, stained-glass etc.; an illustration, esp. comic, dealing with a social or political subject; a comic strip; an animated film. **cartoonist**, *n.*

cartouche (kahtoosh'), *n.* an ornamental tablet in the form of a scroll, for inscriptions etc.; an elliptical figure containing the hieroglyphics of Egyptian royal or divine names or titles.

cartridge (kah'trij), *n.* a case of paper, pasteboard, metal etc., holding the exact charge of a gun; a removable, sealed container holding photographic film or magnetic tape; a removable part of the pick-up arm of a record player, containing the stylus etc.; a replaceable container holding ink for a pen. **cartridge-belt**, *n.* a belt with pockets for cartridges. **cartridge-clip**, *n.* a removable container for cartridges in an automatic firearm. **cartridge-paper**, *n.* a stout, rough-surfaced paper, orig. used for cartridge-making, now for drawing, strong envelopes etc.

cartulary (kah'tūləri), *n.* the register, or collection of documents, relating to a monastery or church; the place where this is kept.

caruncle (ka'rŭngkl, -rŭng'), *n.* a small, morbid, fleshy excrescence; a wattle or the like; (*Bot.*) a protuberance round or near the hilum. **caruncular** (-rŭng'-), *a.* **carunculate, -ated** (-rŭng'-), *a.*

carve (kahv), *v.t.* (p.p. **carved** or **carven**) to cut; to cut into slices, as meat at table; to apportion; to make or shape by cutting; to cut or hew (some solid material) into the resemblance of some object; to adorn by cutting. *v.i.* to exercise the profession of a sculptor or carver; to carve meat. **to carve out**, to take (a piece) from something larger; to create by one's own effort. **to carve up**, to divide by, or as if by, carving. **carver**, *n.* one who carves; a large table-knife for carving; (*pl.*) a carving-knife and fork; a dining-chair with arms. **carvery**, *n.* a restaurant in which, for a set price, the customer can take as much as he or she wishes from a variety of dishes. **carving**, *n.* the action of the verb TO CARVE; carved work. **carving-knife**, *n.* a knife to carve meat at table.

carvel (kah'vl), *n.* a caravel. **carvel-built**, *a.* (*Naut.*) (of a vessel) having the planks flush at the edges, as distinct

caryatid (kariat'id), *n.* (*pl.* **-tids, -tides** (-dēz)) a figure of a woman in long robes, serving to support an entablature. **caryatic,** *a.*

caryo-, *comb. form.* nut, kernel.

caryopsis (kariop'sis), *n.* (*pl.* **-ses** (-sēz), **-sides** (-dēz)) a fruit with a single seed, to which the pericarp adheres throughout, as in grasses.

carz(e)y (kah'zi), **kars(e)y, kazi,** *n.* (*sl.*) a lavatory.

casbah (kaz'bah), KASBAH.

cascade (kəskād'), *n.* a small waterfall; anything resembling a cascade, as a loose, wavy fall of lace, a firework imitating a waterfall; a sequence of actions, processes or pieces of apparatus, each triggered, fuelled or acted on by the previous one. *v.i.* to fall in or like a cascade.

cascara (kəskah'rə), *n.* (the bark, used as an aperient, of) the Californian buckthorn *Cascara sagrada.*

cascarilla (kaskəril'ə), *n.* the aromatic bark of *Croton eleutheria.* [Sp., dim. of prec.]

case[1] (kās), *n.* that which contains or encloses something else; a box, covering or sheath; an oblong frame, with divisions, for type (see also LOWER CASE under LOW [1], UPPER-CASE under UPPER); a stiff cover for a book; a glass box for exhibits; the outer cover of an instrument, seed-vessel, pupa, projectile etc.; the casing of a door or window. *v.t.* to cover with or put into a case. **case-bound,** *a.* of a book, hardback. **case-harden,** *v.t.* to harden the outside surface, esp. of iron; to make callous. **case-knife,** *n.* a knife carried in a sheath. **case-shot,** *n.* small projectiles put in cases to be discharged from cannon shrapnel. **casing,** *n.* something that encases; an outside covering; the frame of a door or window.

case[2] (kās), *n.* that which happens or befalls; an event, a condition of things, position, state, circumstances; an instance; a question at issue; change in the termination of a declinable word to express relation to some other word in the sentence; used also of such relation in uninflected languages; (*sl.*) an eccentric or difficult character; a cause or suit in court; a statement of facts or evidence for submission to a court; the evidence and arguments supporting a claim, action etc. considered collectively; a cause that has been decided and may be quoted as a precedent; a doctor's patient; a particular instance of any disease; a solicitor's or social worker's client. *v.t.* (*sl.*) to reconnoitre with a view to burglary. **in any case,** in any event, whatever may happen. **in case,** if, supposing that, lest. **in case of,** in the event of. **in that case,** if that should happen. **case-book,** *n.* a book describing (medical or legal) cases for record or for instruction. **case-history,** *n.* a record of a patient's ancestry and personal history. **case law,** *n.* (*Law*) law as settled by precedent. **case-load,** *n.* the number of cases assigned to a medical or social worker. **case study,** *n.* a case-history; study or analysis of a case-history. **case-work,** *n.* medical or social work concentrating on individual cases. **case-worker,** *n.*

casein (kā'siin, -sēn), *n.* a protein in milk, forming the basis of cheese. **caseous** (-siəs), *a.* of or like cheese.

casemate (kās'māt), *n.* a bomb-proof vault or chamber in a fortress or ship, containing an embrasure. **casemated,** *a.*

casement (kās'mənt), *n.* a window or part of a window opening on hinges; (*poet.*) a window. **casemented,** *a.*

cash (kash), *n.* ready money; coin, specie, bank-notes. *a.* involving cash, paid for, or paying in cash. *v.t.* to turn into or exchange for cash. **cash down,** money paid on the spot. **cash on delivery,** a system by which goods are paid for on delivery. **hard cash,** actual coin; ready money. **to cash in, to cash in one's checks,** to hand over in exchange for money; (*sl.*) to die. **to cash in on,** (*coll.*) to seize a chance to profit from. **to cash up,** to add up the money taken (in a shop etc.) at the end of the day. **cash-and-carry,** *a., adv.* sold for cash, without a delivery service. *n.* a shop which trades in this way. **cash-book,** *n.* a book in which money transactions are entered. **cash card,** *n.* a card which operates a cash dispenser. **cash cow,** *n.* a productive source of liquid assets; (*loosely*) a source of wealth or profit. **cash crop,** *n.* one grown for sale, not for consumption. **cash desk,** *n.* the desk in a shop where payments are made by customers. **cash dispenser,** *n.* an electronic machine operated by a bank or building society, which dispenses cash on insertion of a special card. **cash flow,** *n.* (the balance of) the flow of money into and out from a business in the course of trading. **cash limit,** *n.* an upper limit imposed on the amount of money available for a particular use or by an individual, council etc. *v.t.* to put a cash limit on. **cash point** CASH DESK; CASH DISPENSER. **cash register,** *n.* a calculating till used in a retail shop. **cashless,** *a.* without (ready) money; of financial transactions, made without using cash, e.g. by computer transfer etc.

cashew (kash'oo, -shoo'), *n.* the kidney-shaped fruit of a tropical tree, *Anacardium occidentale;* this tree. **cashew-nut,** *n.* **cashew-tree,** *n.*

cashier[1] (kashiə'), *n.* one who has charge of the cash or of money transactions.

cashier[2] (kashiə'), *v.t.* to dismiss, esp. dishonourably from (military) service.

cashmere (kash'miə), *n.* (material made from) the hair of the Kashmir goat; a fine woollen fabric.

casing CASE [1].

casino[1] (kəsē'nō), *n.* (*pl.* **-nos**) an establishment, or part of one, used for gambling.

casino[2] CASSINO.

cask (kahsk), *n.* a barrel; the quantity contained in a cask.

casket (kahs'kit), *n.* a small case for jewels etc.; (*chiefly N Am.*) a coffin. *v.t.* to enclose in a casket.

casque (kask), *n.* a helmet; a horny cap or protuberance on the head of some birds.

Cassandra (kəsan'drə), *n.* one who prophesies evil or who takes gloomy views of the future; a prophet who is not listened to. [daughter of Priam, king of Troy, who had the gift of prophecy but was not believed]

cassareep (kas'ərēp), *n.* the boiled down juice of the cassava, used as a condiment.

cassata (kəsah'tə), *n.* a type of ice-cream containing nuts and candied fruit.

cassava (kəsah'və), *n.* a W Indian plant, the manioc, of the genus *Manihot;* a nutritious starch obtained from its roots.

casserole (kas'ərōl), *n.* a stew-pan; an earthenware etc. cooking-pot with a lid; the food cooked in such a pot. *v.t.* to cook in such a pot.

cassette (kəset'), *n.* a small plastic container with magnetic tape wound on two spools, for use with a tape deck or computer; a lightproof container of film, for loading into a camera.

cassia (kas'iə, kash'ə), *n.* a coarse kind of cinnamon, esp. the bark of *Cinnamomum cassia;* any of a genus of leguminous plants, including the senna. **cassia-bark,** *n.*

cassingle (kəsing'gl), *n.* a music cassette with one song etc. recorded on each side.

cassino, casino[2] (kəsē'nō), *n.* a game at cards for two or four players.

cassis (kas'ēs), *n.* a cordial made from blackcurrants.

cassiterite (kəsit'ərīt), *n.* native stannic dioxide, common tin-ore.

cassock (kas'ək), *n.* a long, close-fitting garment worn by clerics, choristers, vergers etc. **cassocked,** *a.*

cassoulet (kas'əlā), *n.* a dish of haricot beans stewed with bacon, pork etc.

cassowary (kas'əwəri), *n.* any of a genus (*Casuarius*) of large cursorial birds of New Guinea and Australia.

cast[1] (kahst), *v.t.* (*past, p.p.* **cast**) to throw, fling, hurl (now chiefly poet. or archaic except in certain uses); to drive, to toss; to throw off, to shed; to throw by reflec-

cast — tion; to direct; to consider, to contrive, to record or formally give (a vote); to allot, to assign (as the parts in a play); to select for a part; to condemn, to reject; to drop (as young) prematurely; to add up, compute, calculate; to found, to mould. *v.i.* to throw a fishing-line; to reckon accounts; to take form or shape (in a mould); to warp. **to cast about**, to look hither and thither for something; to search for mentally. **to cast aside**, to reject; to give up. **to cast away**, to reject; (*usu. pass.*) to shipwreck. **to cast back**, to turn (one's mind) back to the past. **to cast down**, to deject, to depress. **to cast off**, to discard; to estimate the number of words in a manuscript; to untie, to unmoor (a boat); in knitting, to finish by closing loops and making a selvedge. **to cast on**, in knitting, to form new stitches. **to cast out**, to expel. **casting**, *n.* the action of the verb TO CAST; anything formed by casting or founding; esp. a metal object as distinguished from a plaster cast. **casting voice, vote**, *n.* the deciding vote of a president when the votes are equal.

cast² (kahst), *n.* the act of casting or throwing; a throw; the thing thrown; the distance thrown; the allotment of parts in a play; the set of actors allotted; a throw of dice; the number thrown; chance; that which is shed or egested; feathers, fur etc. ejected from the stomach by a bird of prey; the end portion of a fishing line, usu. of gut or gimp, carrying hooks etc.; a motion or turn of the eye; a twist, a squint; a tinge, trace; characteristic quality; a plaster cast; shape, appearance. **cast-iron**, *n.* iron melted and run in moulds; *a.* made of cast-iron; rigid, unyielding, unadaptable; hard, indefatigable. **cast-off**, *a.* laid aside, rejected. **cast-steel**, *n.* steel melted and run into moulds. **caster**, *n.* one who or that which casts; CASTOR². **caster-, castor-sugar**, *n.* white powdered sugar for table use.

castanet (kastənet'), *n.* (*usu. pl.*) a small spoon-shaped concave instrument of ivory or hard wood, a pair of which is fastened to each thumb and rattled as an accompaniment to music.

castaway (kahs'təwā), *a.* rejected, useless; shipwrecked. *n.* an outcast; a shipwrecked person.

caste (kahst), *n.* one of the hereditary classes of society in India; any hereditary, exclusive class; the class system; the dignity or social influence due to position; a term used to describe specialized individuals among insects, e.g. queen bee, worker bee etc. **to lose caste**, to descend in the social scale. **caste mark**, *n.* a red mark on the forehead showing one's caste. **casteless**, *a.*

castellan (kas'tələn), *n.* the governor of a castle.

castellated (kas'təlātid), *a.* having turrets and battlements.

caster CAST².

castigate (kas'tigāt), *v.t.* to chastise, to punish, to rebuke. **castigation**, *n.* **castigator**, *n.* **castigatory**, *a.*

Castile soap (kastēl'), *n.* a fine, hard soap, whose main constituents are olive oil and soda.

Castilian (kastil'iən), *n.* a native of Castile; the dialect of Castile, the official language of Spain. *a.* of Castile or the Spanish language.

casting CAST¹.

castle (kah'sl), *n.* a fortified building, a fortress; a mansion that was formerly a fortress; the mansion of a noble or prince; a piece at chess in the shape of a tower, a rook. *v.i.* in chess, to move the king two squares to the right or left and bring up the castle to the square the king has passed over. *v.t.* to treat (the king) thus. **castles in the air** or **in Spain**, visionary projects. **castled**, *a.* having a castle.

castor¹ (kahs'tər), **castoreum** (-taw'riəm), *n.* an oily compound secreted by the beaver, used in medicine and perfumery.

castor², **caster** (kahs'tər), *n.* a small vessel with a perforated top for sprinkling sugar etc.; a small swivelled wheel attached to the leg or base of a chair, sofa etc.

castoreum CASTOR¹.

castor-oil (kahstəroil'), *n.* an oil, used as a cathartic and lubricant, obtained from the seeds of Palma Christi or **castor-oil plant** (*Ricinus communis*, also grown as a house plant).

castrate (kəstrāt'), *v.t.* to cut away the testicles of, to geld; to deprive of generative power; to emasculate, to deprive of force or vigour. **castration**, *n.*

castrato (kəstrah'tō), *n.* (*pl.* **-ti** (-tē)), a male emasculated for the purpose of retaining the pitch of his voice.

casual (kazh'uəl, -zhəl), *a.* happening by chance; offhand, trivial; occasional; unmethodical, careless; unconcerned, apathetic; informal. *n.* (*pl.*) flat-heeled shoes that slip on without lacing; (*pl.*) informal clothes; an occasional worker; (*coll.*) a (football) hooligan. **casual labour**, *n.* workers employed irregularly. **casually**, *adv.* **casualness**, *n.*

casualty (kazh'əlti), *n.* an accident, esp. one attended with personal injury or loss of life; one who is killed or injured in war or an accident; one who, or that which, is damaged, lost etc. **casualty ward, department**, *n.* the ward in a hospital for receiving the victims of accidents.

Casuarina (kasūəri'nə), *n.* a genus of trees of the E Indies with jointed leafless branches.

casuist (kaz'ūist), *n.* one who studies doubtful questions of conduct, esp. one who discovers exceptions; a sophist, a hair-splitter. **casuistic, -ical** (-is'-), *a.* **casuistically**, *adv.* **casuistry**, *n.* that part of ethics or theology which deals with cases of conscience.

CAT (kat), (*abbr.*) college of advanced technology; computer-aided typesetting; computerized axial tomography.

cat (kat), *n.* any species of the family Felidae, comprising the lion, tiger, leopard etc., esp. the domestic cat; a strong tackle used to hoist the anchor to the cat-heads; various parts of this tackle; a cat-o'-nine-tails; (*coll.*) a spiteful woman; (*sl.*) a man, esp. fashionable. *v.t.* (*past, p.p.* **catted**) to draw to the cat-head; (*coll.*) to vomit. **cat-and-dog**, *a.* quarrelsome. **cat-and-mouse**, *a.* involving stealthy observation and waiting in order to choose the best moment to close in and defeat an opponent. **to let the cat out of the bag**, to give away a secret, to be indiscreet. **to rain cats and dogs**, to pour. **cat-bird**, *n.* an American songbird of the thrush family. **cat burglar**, *n.* a thief who enters a house by climbing up the outside. **catcall**, *n.* a raucous shout or whistle expressing disapproval. *v.i., v.t.* to utter or deride with a catcall. **cat-door, -flap**, *n.* a small flap set into a door to allow a cat to pass through. **cat-fish**, *n.* any of various fishes with barbels round the mouth; a N Atlantic food fish. **cat-head**, *n.* a beam projecting from a ship's bows to which the anchor is secured. **cat-house**, *n.* (*sl.*) a brothel. **cat litter**, *n.* an absorbent material spread on a tray for a cat to urinate or defecate on. **cat-mint**, *n.* a European labiate plant, *Nepeta cataria*. **cat-nap**, *n.* a short sleep. **cat-nip**, *n.* cat-mint. **cat-o'-nine-tails**, *n.* a whip or scourge with nine lashes, formerly used as an instrument of punishment in the Army and Navy. **cat's cradle**, *n.* a children's game with string. **cat's-eye**, *n.* a precious stone, from Sri Lanka, Malabar etc., a vitreous variety of quartz; (**Cat's eye**®) a reflector stud on a road. **cat's-foot**, *n.* the ground-ivy, *Nepeta glechoma*; the mountain cudweed, *Antennaria dioica*. **cat's meat**, *n.* horse-flesh, used as food for cats. **cat's paw**, *n.* a dupe used as a tool (in allusion to the fable of the monkey who used the cat's paw to pick chestnuts out of the fire); a light wind which just ripples the surface of the water. **cat's-tail**, *n.* any of several species of *Typha*, the bulrush; a catkin. **cat-suit**, *n.* a one-piece trouser-suit. **cat's whisker**, *n.* (*Radio.*) a very fine wire in contact with a crystal receiver to rectify current and cause audibility. **cat's whiskers**, *n.sing.* (*coll.*) one who, or that which, is excellent or the best. **cat-walk**, *n.* a narrow walkway high above the ground, as above the stage in a theatre. **cathood**, *n.* **catlike**, *a.* **cattery**, *n.* a place where cats are bred or boarded. **cattish**, *a.* **catty**, *a.* spiteful, malicious.

cat., (*abbr.***)** catalogue; catamaran; catechism.

cat(a)-, cath-, *pref.* down; against; away; wrongly; entirely; thoroughly; according to.

catabolism (kətab'əlizm), KATABOLISM.

catachresis (katəkrē'sis), *n.* the wrong use of one word for another. **catachrestic** (-kres'-), *a.* **catachrestically**, *adv.*

cataclysm (kat'əklizm), *n.* a deluge, esp. the Noachian Flood; a terrestrial catastrophe; a vast and sudden social or political change. **cataclysmal, -mic,** *a.*

catacomb (kat'əkoom), *n.* a subterranean burying-place, with niches for the dead; (*pl.*) the subterranean galleries at Rome; an underground series of passages.

catadioptric (katədiop'trik), *a.* reflecting and refracting light.

catadromous (kətad'rəməs), *n.* of fish, descending periodically to spawn (in the sea or the lower waters of a river).

catafalque (kat'əfalk), *n.* a temporary stage or tomb-like structure for the coffin during a state funeral service.

Catalan (kat'əlan), *a.* of or pertaining to Catalonia. *n.* a native, or the language, of Catalonia.

catalectic (katəlek'tik), *a.* having the metrical foot at the end of a line incomplete.

catalepsy (kat'əlepsi), *n.* a trance or suspension of voluntary sensation. **cataleptic** (-lep'-), *n., a.* (a person) subject to attacks of catalepsy.

catalogue, (*esp. N Am.***)** **catalog** (kat'əlog), *n.* a methodical list, arranged alphabetically or under class-headings; a book or pamphlet containing details of items for sale. *v.t.* to enter in a list; to make a complete list of. **cataloguer**, *n.*

Catalpa (kətal'pə), *n.* a genus of trees, chiefly N American, with long, thin seed-pods.

catalysis (kətal'isis), *n.* change, esp. increase, in the rate of a chemical reaction, brought about by a catalyst. **catalyse** (kat'əliz), (*esp. N Am.*) **-yze**, *v.t.* to subject to catalysis; to spark off, bring about. **catalyst** (kat'əlist), *n.* any substance that changes, esp. increases, the speed of a chemical reaction without itself being changed; one who, or that which, sparks off a wide-ranging change, series of events etc. **catalytic** (katəlit'-), *a.* relating to or effected by catalysis or a catalyst. **catalytic converter**, *n.* a device fitted to the exhaust pipe of a motor vehicle to remove toxic impurities from the exhaust gases. **catalytic cracker**, *n.* an industrial apparatus used to break down the heavy hydrocarbons of crude oil and yield petrol, fuel oils etc.

catamaran (kat'əmərən), *n.* a primitive raft or float; a double-hulled boat; (*coll.*) a vixenish woman.

catamite (kat'əmīt), *n.* a boy kept for homosexual purposes.

catamount (kat'əmownt), **catamountain, cat-o'-mountain** (-mown'tən), *n.* a wild cat, as the puma, panther etc.

cataplexy (kat'əpleksi), *n.* temporary paralysis; a hypnotic condition affecting animals supposedly shamming dead. **cataplectic** (-plek'-), *a.*

catapult (kat'əpŭlt), *n.* an ancient military engine for hurling darts or stones; hence, a toy for propelling small stones; a device for launching aircraft. *v.t.* to throw, shoot or launch with or as with a catapult.

cataract (kat'ərakt), *n.* a large, rushing waterfall; a deluge of rain; a violent rush of water; a condition of the eye in which the crystalline lens or its envelope becomes opaque and vision is impaired or destroyed; such an opaque area.

catarrh (kətah'), *n.* a running or discharge of an inflamed mucous membrane, esp. from the nose. **catarrhal, -rhous,** *a.*

catarrhine (kat'ərīn), *a.* a term applied to the Old World monkeys, from the close, oblique position of their nostrils. *n.* a monkey of the Old World.

catastrophe (kətas'trəfi), *n.* the change which brings about the conclusion of a dramatic piece; a final event; a great misfortune; a violent convulsion of the globe, producing changes in the relative extent of land or water. **catastrophic** (katəstrof'-), *a.* **catastrophism**, *a.* the view that geological changes have been produced by the action of catastrophes. **catastrophist**, *n.*

catatonia (katətō'niə, kata-), *n.* a syndrome often associated with schizophrenia, marked by periods of catalepsy; (*loosely*) catalepsy, or a state of apathy or stupor. **catatonic** (-ton'-), *a.*

catawba (kətaw'bə), *n.* a grape-vine, *Vitis abrusca;* wine made therefrom.

catch (kach), *v.t.* (*past, p.p.* **caught** (kawt)) to grasp, to seize, esp. in pursuit; to take in a snare, to entrap; to take by angling or in a net; to intercept (as a ball) when falling; to dismiss (a batsman) by this; to check, to interrupt, to come upon suddenly; to surprise; to detect: to take hold of (as fire); to reproduce exactly; to receive by infection or contagion; to be in time for; to grasp, perceive, comprehend; to attract, gain over, fascinate. *v.i.* to become fastened or attached suddenly; to communicate; to ignite; to spread epidemically; to take hold; to become entangled. *n.* the act of seizing or grasping; anything that seizes, takes hold, or checks; the amount of fish caught; seizing and holding the ball at cricket; an acquisition; an opportunity; an advantage seized; (*coll.*) a person with capturing matrimonially; profit; a concealed difficulty, a trap, a snare; a part-song in which each singer in turn catches up, as it were, the words of his or her predecessor. **to catch at**, to attempt to seize. **to catch it**, to get a scolding. **to catch napping** NAP. **to catch on**, to hit the public taste; to grasp, to understand. **to catch one's eye**, to attract attention. **to catch out**, to discover (someone) in error or wrong-doing; in cricket, to dismiss (a batsman) by catching a ball. **to catch up**, to overtake; to make up arrears. **catch-all**, *a.* of a rule etc., which covers all situations, or any not previously covered. **catch-crop**, *n.* a quick-growing green crop sown between main crops; a crop which springs up on fallow land from seed dropped from the previous year's crop. **catch-drain**, *n.* an open drain along the side of a hill or canal to catch the surplus water. **catch-penny**, *a.* worthless, made only to sell. **catch-phrase**, *n.* a phrase which comes into fashion for a time and is much (or over-) used. **catch-22**, *n.* a situation from which escape is impossible because rules or circumstances frustrate effort in any direction. **catchword**, *n.* a popular cry; an actor's cue; a word printed under the last line of a page, being the first word of the next; the first word in a dictionary entry. **catchable**, *a.* **catcher**, *n.* **catching**, *a.* that catches; infectious; taking, attractive. **catchment**, *n.* a surface on which water may be caught and collected. **catchment-area**, *n.* (also **-basin**) an area the rainfall in which feeds a river-system; the area from which a particular school, hospital etc. officially takes its pupils, patients etc. **catchy**, *a.* catching; easy to remember (as a tune); tricky, deceptive; irregular, fitful.

catchup (kach'əp), **catsup** (kat'səp), KETCHUP.

catechize, -ise (kat'əkīz), *v.t.* to instruct by means of questions and answers; to instruct in the Church catechism; to question closely. **catechizer**, *n.* **catechetic, -ical**, *a.* consisting of questions and answers, pertaining to catechism. **catechetically**, *adv.* **catechism**, *n.* a form of instruction by means of question and answer; esp. the authorized manuals of Christian religious doctrine; a series of interrogations. **catechismal** (-kiz'-), *a.* **catechist**, *n.* one who teaches by catechizing; one who imparts elementary instruction, esp. in the principles of religion. **catechistic, -ical** (-kis'-), *a.* **catechistically**, *adv.*

catechu (kat'əchoo), *n.* a brown astringent gum, furnished chiefly by *Acacia catechu*. **catechuic** (-choo'-), *a.*

catechumen (katikū'mən), *n.* one who is under Christian instruction preparatory to receiving baptism.

category (kat'igəri), *n.* an order, a class, a division; one of the ten predicaments or classes of Aristotle, to which all

catena objects of thought or knowledge can be reduced; one of Kant's twelve primitive forms of thought, contributed by the understanding, apart from experience. **categorial** (-gaw'-), *a.* **categorical** (-go'-), *a.* pertaining to a category or the categories; absolute, unconditional. **categorical imperative**, *n.* in Kantian ethics, the absolute command of the reason as interpreter of the moral law. **categorically**, *adv.* **categorize, -ise**, *v.t.* to place in a category.

catena (kətē'nə), *n.* (*pl.* **-nae** (-nē)) a connected series. **Catena Patrum** (pah'trəm), *n.* a series of extracts from the writings of the Fathers. **catenate** (kat'ənāt), *v.t.* to chain, to link together. **catenation**, *n.*

catenary (kətē'nəri, kat'ə-), *n.* a curve formed by a chain or rope of uniform density hanging from two points of suspension. *a.* relating to a chain; relating to a catena.

cater (kā'tə), *v.i.* to supply food, amusement etc. (for). *v.t.* to provide food etc. for (a party etc.). **caterer**, *n.* one who provides food etc. for a social function. **cateress**, *n. fem.* **catering**, *n.* the trade of a caterer; the provisions etc. for a social function.

cater-cornered (kā'təkawnəd), *a.* (*N Am.*) not square (applied to a house built at a corner, and therefore more or less oblique in plan; and to a sheet of paper not cut square).

caterpillar (kat'əpilə), *n.* the larva of a lepidopterous insect; (**Caterpillar**®) a device whereby motor vehicles are fitted with articulated belts (**Caterpillar tracks**®) in lieu of wheels for operation on difficult ground; (**Caterpillar**®, **Caterpillar tractor**®), *n.* a tractor fitted with an articulated belt.

caterwaul (katəwawl), *v.i.* to make a noise as cats in the rutting season. *n.* such a noise.

catgut (kat'gŭt), *n.* cord made from the intestines of animals and used for strings of musical instruments, and for surgical sutures.

cath- CAT(A)-.

Cathar (kath'ə), *n.* a member of a mediaeval Manichaean sect in S France. **Catharism**, *n.* **Catharist**, *n.*

catharsis (kəthah'sis), *n.* purgation of the body; the purging of the emotions by tragedy; (*Psych.*) the bringing out and expression of repressed ideas and emotions. **cathartic**, *a.* cleansing the bowels; purgative; causing or resulting in catharsis. *n.* a purgative medicine. **cathartical**, *a.* **cathartically**, *adv.*

cathedral (kəthē'drəl), *n.* the principal church in a diocese, containing the bishop's throne. **cathedral church**, *n.* a cathedral.

catherine wheel (kath'ərin wēl), *n.* a firework that rotates like a wheel; an ornamental circular window with spoke-like mullions or shafts; a cartwheel somersault. [referring to the martyrdom of St Catherine]

catheter (kath'itə), *n.* a tube used to introduce fluids to, or withdraw them from, the body, esp. to withdraw urine from the bladder. **catheterize, -ise**, to introduce a catheter into.

cathexis (kəthek'sis), *n.* concentration of mental or emotional energy on a single object. **cathectic**, *a.*

cathiodermie (kathiōdœmi), *n.* a deep cleansing treatment for the face using an electric current passed through a gel spread on the skin.

cathode (kath'ōd), *n.* the negative electrode in an electrolytic cell; the positive electrode of a primary cell; the source of electrons in an electronic valve. **cathode ray**, *n.* a stream of electrons emitted from the surface of a cathode during an electrical discharge. **cathode ray tube**, *n.* a vacuum tube in which a beam of electrons, which can be controlled in direction and intensity, is projected on to a fluorescent screen thus producing a point of light. **cathodic, -dal** (-thod'-), *a.*

catholic (kath'əlik), *a.* universal, general, comprehensive; liberal, large-hearted, tolerant. **Catholic**, *a.* of or pertaining to the whole Christian church; not heretical; in the Middle Ages, of the Western or Latin Church; since the Reformation, of the Roman Church, as opposed to the Protestant churches; occasionally used of the Anglican Church, as claiming continuity from the old, undivided Christian church. *n.* a Roman Catholic; an Anglo-Catholic. **Old Catholics**, the German Catholics who separated from the Roman Communion in 1870. **Roman Catholic**, a member of the Roman Church. **Catholic Emancipation**, *n.* the removal of restrictions and penal laws from Roman Catholics in the United Kingdom. **Catholic Epistles**, *n.pl.* certain epistles addressed to the Church at large, including those of Peter, James, Jude and the 1st of John (sometimes also the 2nd and 3rd). **catholicly, -cally** (-thol'-), *adv.* **catholicism**, *n.* (Roman) Catholic christianity. **catholicity** (-lis'-), *n.* the quality of being catholic or Catholic. **catholicize, -ise** (-thol'-), *v.t.* to make Catholic. *v.i.* to become Catholic. **catholico-**, *comb. form.*

cation (kat'iən), *n.* the positive ion which in electrolysis is attracted towards the cathode. **cationic** (-on'-), *a.*

catkin (kat'kin), *n.* the pendulous unisexual inflorescence of the willow, birch, poplar etc.

catoptric (kətop'trik), *a.* pertaining to a mirror or reflector, or to reflection.

Cat scanner (kat skan'ə), *n.* a machine which produces X-ray photographs of sections of the body with the assistance of a computer. **Cat scan**, *n.* [Computerized axial tomography]

catsup (kat'səp), KETCHUP.

cattalo (kat'əlō), *n.* (*pl.* **-loes, -los**) a cross between domestic cattle and American bison, very hardy. [from *catt*le + *buffalo*]

cattle (kat'l), *n.* domesticated animals, esp. oxen and cows. **cattle-cake**, *n.* a concentrated processed food for cattle. **cattle grid**, *n.* a trench in a road, covered by a grid which hinders cattle from passing over it but leaves the road free for traffic. **cattle-guard**, *n.* (*N Am.*) a cattle grid. **cattleman**, *n.* one who looks after cattle; (*N Am.*) one who breeds and rears cattle, a ranch-owner. **cattle-show**, *n.* an exhibition of cattle at which prizes are given.

Cattleya (kat'liə), *n.* a genus of beautifully-coloured epiphytic orchids. [Wm. *Cattley*, English horticulturist]

catty CAT.

Caucasian (kawkā'zhən), *a.* of or pertaining to Mount Caucasus or the district adjoining; belonging to one of the main ethnological divisions of mankind, native to Europe, W Asia, and N Africa, comprising peoples with pale skin. *n.* a member of this race.

caucus (kaw'kəs), *n.* (*N Am.*) a preparatory meeting of representatives of a political party to decide upon a course of action; a small group or committee within a larger group or political party, deciding tactics, policy etc. *v.i.* to hold a caucus. *v.t.* to control by means of a caucus. **caucusdom**, *n.* **caucuser**, *n.*

caudal (kaw'dəl), *a.* pertaining to the tail or the posterior part of the body. **caudally**, *adv.* **caudate** (-dāt), *a.* having a tail or tail-like process.

caudex (kaw'deks), *n.* (*pl.* **-dices**) the stem and root of a plant, esp. of a palm or tree-fern.

caudillo (kowdē'lyō, kaw-), *n.* (*pl.* **-llos**) in Spanish-speaking countries, a military leader or head of state.

caudle (kaw'dl), *n.* a warm drink of wine and eggs formerly given to invalids.

caught CATCH.

caul (kawl), *n.* the rear part of a woman's cap; a net for the hair; a membrane enveloping the intestines, the omentum; a part of the amnion, sometimes enclosing the head of a child when born.

cauldron, caldron (kawl'drən), *n.* a large kettle or deep, bowl-shaped vessel with handles, for boiling.

cauliflower (kol'iflowə), *n.* a variety of cabbage with an edible white flowering head. **cauliflower ear**, *n.* a

caulk (kawk), *v.t.* to stuff the seams of (a ship) with a waterproofing material; to fill (cracks etc.). **caulking-iron**, *n.* a blunt chisel used by caulkers. **caulker**, *n.*

permanently swollen or misshapen ear, usu. caused by boxing injuries.

cause (kawz), *n.* that which produces or contributes to an effect; the person or other agent bringing about something; the reason or motive that justifies some act or mental state; a ground of action; a side or party; a movement, agitation, principle or propaganda; (*Law*) the grounds for an action; a suit, an action. *v.t.* to effect; to produce; to make or induce (to do). **to make common cause**, to unite for a definite purpose. **cause célèbre** (kōz sălebr''), *n.* a famous or notorious law-suit. **causal**, *a.* relating to or expressing cause; being a cause. **causally**, *adv.* **causality** (-zal'-), *n.* the operation of a cause; relation of cause and effect; the theory of causation. **causation** (-zā'-), *n.* the act of causing; connection between cause and effect; (*Phil.*) the theory that there is a cause for everything. **causationism** (-zā'-), *n.* the doctrine that all things are due to the agency of a causal force. **causationist**, *n.* **causative**, *a.* that causes; effective as a cause; (*Gram.*) expressing cause. **causatively**, *adv.* **causeless**, *a.* **causelessly**, *adv.*

'cause, (*coll.*) BECAUSE.

causerie (kōzərē'), *n.* a chatty kind of essay or article.

causeway (kawz'wā), **causey** (kaw'zi), *n.* a raised road across marshy ground or shallow water; a raised footway beside a road; a path or road of any kind.

caustic (kaws'tik), *a.* burning, hot, corrosive; bitter, sarcastic. *n.* a substance that burns or corrodes organic matter. **caustic curve**, *n.* (*Math.*) a curve to which the rays of light reflected or refracted by another curve are tangents. **caustic potash**, *n.* potassium hydroxide, an alkaline solid used in the manufacture of soap, detergents etc. **caustic soda**, *n.* sodium hydroxide, an alkaline solid used in the manufacture of rayon, paper, soap etc. **caustically**, *adv.* **causticity** (-tis'-), *n.*

cauterize, -ise (kaw'təriz), *v.t.* to burn or sear (a wound etc.) with a hot iron or caustic; (*fig.*) to sear. **cauterization**, *n.* **cautery, cauter**, *n.* burning with a hot iron, electricity or a caustic; an instrument or agent for effecting such burning.

caution (kaw'shən), *n.* wariness, prudence; care to avoid injury or misfortune; advice to be prudent, a warning; (*sl.*) something or someone extraordinary, or amusing; a formal warning to a person under arrest that what is said may be taken down and used in evidence. *v.t.* to warn; to administer a caution to. **cautionary**, *a.* containing, or serving as, a caution. **cautious**, *a.* heedful, careful, wary. **cautiously**, *adv.* **cautiousness**, *n.*

cavalcade (kavəlkād'), *n.* a company or train of riders on horseback; or (loosely) motor-cars; a procession.

cavalier (kavəliə'), *n.* a horseman, a knight; a gallant, a lady's man; a lover; (**Cavalier**) a partisan of Charles I, a Royalist. *a.* knightly, warlike, gallant; off-hand, haughty, supercilious. **cavalierish**, *a.* **cavalierly**, *adv.* in a haughty or off-hand manner.

cavalry (kav'əlri), *n.* horse soldiers trained to act as a body; one of the arms of the service. **cavalry twill**, *n.* a strong woollen twill fabric, used esp. for trousers.

cavass (kəvas'), KAVASS.

cavatina (kavətē'nə), *n.* a short, simple and smooth song; a similar instrumental composition.

cave[1] (kāv), *n.* a hollow place in the earth; a den; (*Hist.*) the secession of a discontented faction from their party; the body of seceders (see ADULLAMITE). *v.t.* to hollow out. *v.i.* to explore caves as a sport. **to cave in**, *v.i.* to fall in; to give in, to yield. **cave-man, -dweller**, *n.* a prehistoric man who dwelt in caves; (*facet.*) a man of primitive instincts. **caver**, *n.* **caving**, *n.* the sport of exploring caves.

cave[2] (kā'vi), *int.* Look out!

caveat (ka'viat), *n.* (*Law*) a process to stop procedure; a warning, a caution. **caveat emptor** (emp'taw), let the purchaser beware.

cavendish (kav'əndish), *n.* a kind of tobacco softened and pressed into cakes.

cavern (kav'ən), *n.* a cave, esp. a deep hollow cave. *v.t.* to shut or enclose in a cavern; to hollow out. **caverned**, *a.* **cavernous**, *a.* hollow or huge, like a cavern; full of caverns.

caviar, -are (kav'iah, -ah'), *n.* the salted roes of various fish, esp. the sturgeon.

cavil (kav'il), *n.* a frivolous objection. *v.i.* (*past, p.p.* **cavilled**) to argue captiously; to quibble. **caviller**, *n.* **cavillingly**, *adv.*

cavitation (kavitā'shən), *n.* the formation of a cavity or partial vacuum between a solid and a liquid in rapid relative motion, e.g. on a propeller.

cavity (kav'iti), *n.* a hollow place or part; a decayed hole in a tooth. **cavity wall**, *n.* one consisting of two rows of bricks with a space between.

cavort (kəvawt'), *v.i.* to prance about.

cavy (kā'vi), *n.* any of the S American genus *Cavia*, esp. *C. cobaya*, the guinea-pig.

caw (kaw), *v.i.* to cry like a rook. *n.* the cry of a rook.

cawker CAULKER.

CAWU, (*abbr.*) Clerical and Administrative Workers' Union.

cay (kā), **key** (kē), *n.* a reef, a shoal.

cayenne (kāen'), **cayenne pepper**, *n.* the powdered fruit of various species of capsicum, a very hot, red condiment.

cayman, caiman (kā'mən), *n.* a tropical American alligator.

cayuse (kiūs'), *n.* a small Indian horse.

cazique CACIQUE.

CB, (*abbr.*) citizen's band; Companion of the Order of the Bath; confined to barracks; county borough.

Cb, (*chem. symbol*) columbium.

CBC, (*abbr.*) Canadian Broadcasting Corporation.

CBE, (*abbr.*) Commander of the Order of the British Empire.

CBI, (*abbr.*) Confederation of British Industry.

CBS, (*abbr.*) Columbia Broadcasting System.

CC, (*abbr.*) chamber of commerce; county council(lor); cricket club.

cc, (*abbr.*) carbon copy; chapters; cubic centimetre.

CCD, (*abbr.*) charge coupled device.

CCTV, (*abbr.*) closed-circuit television.

CD, (*abbr.*) civil defence; compact disc; corps diplomatique.

Cd, (*chem. symbol*) cadmium.

cd, (*abbr.*) candela.

Cdr, (*abbr.*) commander.

CDV, (*abbr.*) compact disc video.

CE, (*abbr.*) chief engineer; Church of England; civil engineer; Common (or Christian) Era; Council of Europe.

Ce, (*chem. symbol*) cerium.

Ceanothus (seənō'thəs), *n.* a genus of ornamental flowering N American shrubs of the buckthorn family.

cease (sēs), *v.i.* to come to an end; to desist (from). *v.t.* to put a stop to; to discontinue. **without cease**, without intermission. **cease fire**, *int.* a command to stop firing. *n.* an agreement to stop fighting. **ceaseless**, *a.* incessant, unceasing. **ceaselessly**, *adv.* **ceaselessness**, *n.*

cecity (sē'siti), *n.* blindness (physical or mental).

cecum CAECUM.

cedar (sē'də), *n.* any of a genus, *Cedrus*, of evergreen coniferous trees with durable and fragrant wood, including the **cedar of Lebanon**, *Cedrus libani*, and many others; the wood of any of these trees. **cedared**, *a.* covered with cedars. **cedarn** (-dən), *a.* (*poet.*) made of cedar-wood; consisting of cedars.

cede (sēd), *v.t.* to give up, to surrender; to yield, grant.

cedilla (sədil'ə), *n.* a mark (¸) placed under a *c* in French, Spanish etc., to show that it has the sound of *s*.

Ceefax® (sē'faks), *n.* a teletext service operated by the BBC.
CEGB (*abbr.*) Central Electricity Generating Board.
ceil (sēl), *v.t.* to line the roof of (a room), esp. with plaster. **ceiling,** *n.* the inner, upper surface of a room; the plaster or other lining of this; the maximum height to which an aircraft can climb; the upper limit of prices, wages etc. **ceilinged,** *a.* having a ceiling.
ceilidh (kā'li), *n.* an informal gathering, esp. in Scotland or Ireland, for music, dancing etc.
ceinture (sān'chuə), *n.* the belt of leather, rope etc. worn round the waist outside the cassock.
celadon (sel'ədon), *n., a.* a soft, pale green colour; (a type of porcelain with) a glaze of this colour.
celandine (sel'əndīn), *n.* the name of two plants with yellow flowers, the **greater celandine,** *Chelidonium majus,* related to the poppy, and the **lesser celandine,** *Ranunculus ficaria,* also called the pile-wort or figwort.
-cele, *comb. form.* a tumour or hernia.
celebrate (sel'ibrāt), *v.t.* to praise, extol; to commemorate; to observe; to perform, to say or sing (as Mass), to administer (as Communion). *v.i.* to officiate, esp. at the Eucharist; to mark an occasion with festivities. **celebrated,** *a.* famous, renowned. **celebration,** *n.* **celebrator,** *n.* **celebratory,** *a.* **celebrant,** *n.* the priest who officiates, esp. at the Eucharist. **celebrity** (-leb'-), *n.* fame, renown; a celebrated person.
celeriac (səle'riak), *n.* a turnip-rooted variety of celery.
celerity (səle'riti), *n.* speed, swiftness, promptness.
celery (sel'əri), *n.* a plant, *Apium graveolens,* the blanched stems of which are eaten cooked or as a salad vegetable.
celesta (səles'tə), *n.* a keyboard instrument in which steel plates are struck by hammers.
celestial (səles'tiəl), *a.* pertaining to heaven; spiritual, angelic, divine; pertaining to the heavens or sky. *n.* an inhabitant of heaven; a native of China. **celestial sphere,** *n.* an imaginary sphere with the observer at its centre and all heavenly objects on its surface. **celestially,** *adv.*
celiac COELIAC.
celibate (sel'ibət), *n.* an unmarried or sexually inactive person. *a.* unmarried; devoted or vowed to a single life; abstaining from sexual activity. **celibatarian** (-teə'-), *a.* **celibacy,** *n.* the unmarried state; abstention from sexual activity.
cell (sel), *n.* a small room, esp. one in a monastery or prison; a small religious house dependent on a larger one; the retreat of a hermit; (*Biol.*) a small cavity or compartment; a compartment in a comb made by bees; the unit-mass of living matter in animals or plants; a subsidiary unit of a political organization, esp. a proscribed or revolutionary one; a division of a galvanic battery, or a battery having only one pair of metallic plates; a subdivision of a computer memory that stores one unit of data; see CELLULAR RADIO under CELLULE. **cellphone,** *n.* a telephone apparatus suitable for use with the cellular radio system. **celled,** *a.* **celliferous** (-lif'-), *a.* **celliform,** *a.*
cellar (sel'ə), *n.* a place for storing wine; a stock of wine; an underground vault, chamber beneath a building, used for storage. *v.t.* to put or store in a cellar. **cellarage,** *n.* cellars collectively; space for, or charge for storage in, cellars. **cellarer,** *n.* a monk in charge of the stores; an officer of a chapter in charge of the provisions. **cellaret** (-ret'), *n.* a small case, or a sideboard, with compartments for holding bottles. **cellarman,** *n.* one employed in a wine or beer cellar.
'cello (chel'ō), VIOLONCELLO.
Cellophane® (sel'əfān), *n.* a transparent material made of viscose, chiefly used for wrapping.
cellule (sel'ūl), *n.* a little cell or cavity. **cellular,** *a.* of, pertaining to, or resembling a cell or cells; (*Physiol.*) composed of cells; of textiles, woven with a very open texture. **cellular radio,** *n.* a type of radio communication, used esp. for car telephones, which connects directly to the public telephone network and uses a series of transmitting stations, each covering a small area or cell. **cellular telephone,** *n.* **cellulate** (-lət), **-lated** (-lātid), *a.* formed of cells. **cellulation,** *n.* **celluliferous** (-lif'-), *a.* **cellulite** (-līt), *n.* subcutaneous fat which gives the skin a dimpled appearance. **cellulitis** (-lī'tis), *n.* inflammation of subcutaneous tissue, caused by bacterial infection.
Celluloid® (sel'ūloid), *n.* a flammable thermoplastic made from cellulose nitrate, camphor and alcohol, used e.g. in cinema film; cinema film.
cellulose (sel'ūlōs), *n.* a carbohydrate of a starchy nature that forms the cell walls of all plants. **cellulose acetate,** *n.* any of several chemical compounds formed e.g. by the action of acetic acid on cellulose, used in the manufacture of photographic film, varnish, some textile fibres etc. **cellulose nitrate,** *n.* cellulose treated with nitric acid, used in making plastics, explosives etc. **cellulosity** (-los'-), *n.*
Celsius (sel'siəs), *a.* denoting a temperature scale in which the freezing point of water is designated 0° and the boiling point 100°. [Anders *Celsius,* 1701–44, Swed. astronomer, who invented it]
Celt (kelt, selt), **Kelt,** *n.* a member or descendant of an ancient race comprising the Welsh, Cornish, Manx, Irish, Gaels and Bretons, whose descendants are still found in the Highlands of Scotland, Ireland, Wales and the north of France. **Celtic,** *a.* pertaining to the Celts. *n.* the language of the Celts. **Celtic cross,** *n.* a Latin cross with a circle round the intersection of the arms. **Celticism,** *n.* a custom peculiar to the Celts. **Celticize, -ise,** *v.i., v.t.* to make or become Celtic.
celt (selt), *n.* a prehistoric cutting or cleaving implement of stone or bronze.
cembalo (chem'balō), *n.* (*pl.* **-li, -los**) a harpsichord.
cement (siment'), *n.* an adhesive substance, esp. one used in building for binding mortar and concrete and hardening like stone; any analogous material, paste, gum or mucilage for sticking things together; a substance for stopping teeth; (also **cementum** (-təm)) the bony substance forming the outer layer of the root of a tooth; (*fig.*) a bond of union. *v.t.* to unite with or as with cement; to line or coat with cement. *v.i.* to cohere. **cementation** (sē-), *n.* the act of cementing; the conversion of iron into steel by heating in a mass of charcoal.
cemetery (sem'ətri), *n.* a public burial-ground; esp. one that is not a churchyard.
cenobite (sē'nəbīt), COENOBITE.
cenotaph (sen'ətahf), *n.* a sepulchral monument raised to a person buried elsewhere; an empty tomb.
Cenozoic (sēnəzō'ik), CAINOZOIC.
cense (sens), *v.t.* to perfume with incense. **censer** (sen'sə), *n.* a vessel for burning incense.
censor (sen'sə), *n.* a Roman officer who registered the property of the citizens, imposed the taxes, and watched over manners and morals; anyone who watches over behaviour and morals; esp. a public officer appointed to examine books, plays etc., before they are published, to see that they contain nothing immoral, seditious or offensive; a public servant whose duty it is in war-time to see that nothing is published, or passes through the post, that might give information to the enemy; an unconscious mechanism in the mind that excludes disturbing factors from the conscious. *v.t.* to subject to censorial control; to expurgate or delete objectionable matter from. **censorial** (-saw'-), *a.* **censorious** (-saw'-), *a.* expressing or addicted to criticism or censure. **censoriously,** *adv.* **censoriousness,** *n.* **censorship,** *n.*
censure (sen'shə), *n.* disapproval, condemnation; an expression of this; reproach. *v.t.* to blame; to find fault with. **censurable,** *a.* **censurableness,** *n.* **censurably,** *adv.*
census (sen'səs), *n.* an official enumeration of the inhabitants of a country; the statistical result of such enumera-

tion; any similar official enumeration, as a *traffic census*.
cent (sent), *n*. (a coin of the value of) a hundredth part of the basic unit of many currencies (e.g. of the American dollar). **cental**, *n*. a weight of 100 lb (45·4 kg) used for grain.
centaur (sen'taw), *n*. a Greek mythological figure, half man, half horse; any incongruous union of diverse natures.
centaury (sen'tawri), *n*. the name of various plants (esp. genus *Centaurium*) once used medically; the lesser centaury, *Erythraeum centaurium*.
centenarian (sentənee'riən), *n*. a person who has reached the age of 100 years. **centenary** (sǝntē'nǝri, -ten'-), *a*. relating to 100 or a period of 100 years; recurring once in a 100 years. *n*. 100 years; the 100th anniversary of any event, or the celebration of this. **centennial** (-ten'-), *a*. pertaining to a 100th anniversary; 100 or more years old; completing 100 years. *n*. a centenary.
center CENTRE.
centering (sen'tǝring), *n*. the woodwork or framing on which an arch or vault is constructed.
centesimal (sǝntes'imǝl), *a*. hundredth; by fractions of a hundred. *n*. a hundredth part. **centesimally**, *adv*.
centi-, *comb. form*. a hundred; a hundredth part; esp. denoting a hundredth part of a metric unit, as in *centigram, centilitre, centimetre*.
centiare (sent'ieǝ), *n*. a metric unit of area equal to 1/100 are or 1 square metre (107.6 sq. ft.).
centigrade (sen'tigrād), *a*. divided into 100 degrees; applied esp. to the Celsius scale of temperature.
centime (sě'těm), *n*. a French, Belgian etc. coin worth a hundredth part of a franc.
centipede (sen'tipěd), *n*. an animal of the Arthropoda with many segments, each with a pair of legs.
CENTO (sen'tō), (*abbr*.) Central Treaty Organization.
cento (sen'tō), *n*. (*pl*. **-tos**) a composition of verses from different authors, arranged in a new order; a string of quotations, scraps and tags.
centr- CENTRO-.
central (sen'trǝl), *a*. relating to, containing, proceeding from, situated in or forming the centre; principal, of chief importance. **central nervous system**, that part of the nervous system of vertebrates consisting of the brain and spinal cord. **central processing unit**, central processor. **central bank**, *n*. the principal bank of a country that acts as banker to the government, regulates credit etc. **central heating**, *n*. a system of warming buildings from one furnace by steam or hot-water pipes or other devices. **central processor**, *n*. the part of a computer which performs arithmetical and logical operations on data. **central reservation**, *n*. the strip of ground that separates the carriageways of a motorway. **centralism**, *n*. a system or policy of centralization. **centralist**, *n*. **centrality** (-tral'-), *n*. the quality of being central. **centralize, -ise**, *v.t.* to bring to a centre; to concentrate; to bring under central control. *v.i.* to come to a centre. **centralization**, *n*. the act of centralizing; the system or policy of carrying on the government or any administrative organization at one central spot. **centrally**, *adv*. **centralness**, *n*.
centre, (*esp. N Am*.) **center** (sen'tǝ), *n*. the middle of anything; the middle or central object; the point round which anything revolves, the pivot or axis; the principal point; a point or place at which an activity is concentrated; the nucleus, the source from which anything radiates or emanates; the head or leader of an organization; a political party or group occupying a place between two extremes (**left centre**, the more radical portion, and **right centre**, the more conservative of this); a player in the middle of a forward line; the centering. *v.t.* to place on or at a centre; to collect to a point; to find the centre of. *v.i.* to be fixed on a centre; to have as a centre or focal point; to be collected at a centre or at one point. *a*. at or of the centre. **centre of attraction**, one who draws general attention; (*Phys*.) the point towards which bodies gravitate. **centre of gravity**, the point about which all the parts of a body exactly balance each other. **centre of inertia, mass**, a point through which a body's inertial force acts (coincident with the centre of gravity). **centre-bit**, *n*. a carpenter's tool with a central guiding point for boring large round holes. **centre-board**, *n*. a sliding keel which can be raised or lowered; a boat fitted with this. **centrefold**, *n*. (an illustration or article occupying) the two facing pages at the centre of a newspaper or magazine. **centre-forward**, *n*. in soccer and hockey, a player occupying the middle of the front line. **centre-half**, *n*. a player occupying the middle of the half-back line. **centre-piece**, *n*. an ornament for the middle of a table, ceiling etc. **centre spread** CENTREFOLD. **centric, -ical**, *a*. central. **centrically**, *adv*. **centricity** (-tris'-), *n*. **centrism**, *n*. **centrist**, *n*. one holding moderate political opinions.
centri- CENTRO-.
-centric, *comb. form*. having a specified centre, as *heliocentric*.
centrifugal (sentrif'ūgǝl, sen'-), *a*. tending to fly or recede from the centre; operating by or using centrifugal force. **centrifugal force**, *n*. the tendency of a revolving body to fly off from the centre. **centrifuge** (sen'trifūj), *n*. a machine using centrifugal force for separating liquids of different density, simulating rapid accelerations etc. **centrifugally**, *adv*.
centripetal (sentrip'ǝtǝl), *a*. tending to approach the centre; operating by or using centripetal force. **centripetal force**, *n*. the force which draws a revolving body towards the centre. **centripetally**, *adv*.
centro-, centr(i)-, *comb. form*. central, centrally.
centrosome (sen'trǝsōm), *n*. a small body of protoplasm near a cell nucleus.
centumvir (sentūm'vir), *n*. (*pl*. **-viri** (-rī)) one of the judges appointed by the praetor to decide common causes among the Romans. **centumviral**, *a*. **centumvirate** (-rǝt), *n*.
centuple (sen'tūpl), *n., a*. (a) hundredfold. *v.t.* to multiply a hundredfold. **centuplicate** (-tū'plikāt), *v.t.* to multiply a hundredfold. **centuplicate** (-tū'plikǝt), *n., a*. (a) centuple. **centuplication**, *n*.
centurion (sentū'riǝn, sen-), *n*. a Roman military officer commanding a company of a hundred men.
century (sen'chǝri), *n*. an aggregate of 100 things; a hundred; a period of 100 years; a division of the Roman people for the election of magistrates etc.; a division of a legion, consisting originally of 100 men; 100 runs in cricket. **century plant**, *n*. the American aloe, *Agave americanus*, erroneously supposed to flower only once in 100 years. **centurial** (-tū'-), *a*.
cep (sep), *n*. a type of edible mushroom with a brown shiny cap.
cephal- CEPHAL(O)-.
cephalic (sifal'ik), *a*. pertaining to the head. **cephalic index**, *n*. the ratio of a transverse to the longitudinal diameter of the skull.
-cephalic, -cephalous, *comb. forms*. headed, see HYDROCEPHALOUS, MICROCEPHALOUS, BRACHYCEPHALIC, ORTHOCEPHALIC.
cephal(o)-, *comb. form*. pertaining to the head.
cephalopod (sef'ǝlǝpod), *n*. a mollusc having a distinct head with prehensile and locomotive organs attached, e.g. the squids and octopuses.
cephalothorax (sefǝlōthaw'raks), *n*. the anterior division of the body, consisting of the coalescence of head and thorax in spiders, crabs and other arthropods.
Cepheid (variable) (sēfiid), *n*. any of a class of stars, whose brightness varies regularly.
ceramic (sǝram'ik), **keramic** (kǝ-), *a*. of or pertaining to pottery; applied to any material made by applying great heat to clay or another non-metallic mineral. *n*. such a

substance. **ceramics,** *n.sing.* the art of pottery. **ceramist** (se'-, kc'-), *n.*
cerastes (səras'tēz), *n.* a horned viper.
cerat(o)-, *comb. form.* horned; horny; having processes like horns; cornea.
ceratoid (ser'ətoid), *a.* horny; horn-like.
cercaria (sœkeə'riə), *n.* a trematode worm or fluke in its second larval stage.
cere (siə), *n.* the naked, wax-like skin at the base of the bill in many birds. *v.t.* to cover with wax. **cerecloth,** *n.* a cloth dipped in melted wax, formerly used to wrap embalmed bodies in. **cerement,** *n.* a cerecloth; (*pl.*) grave-clothes. **cereous,** *a.* waxen, waxy; like wax.
cereal (siə'riəl), *a.* pertaining to wheat or other grain. *n.* a plant producing an edible grain; any edible grain; (breakfast) food made from a cereal. **cerealian** (-ā'li-), *a.* **cerealin** (-lin), *n.* a nitrogenous substance found in bran. [from *Ceres*, the goddess of corn]
cerebellum (serəbel'əm), *n.* (*pl.* **-llums, -lla** (-lə)) a portion of the brain situated beneath the posterior lobes of the cerebrum, responsible for balance and muscular coordination. **cerebellar,** *a.* **cerebellar syndrome,** *n.* impairment of muscular coordination, balance, speech etc. resulting from a diseased condition of the cerebellum. **cerebellous,** *a.*
cerebro-, *comb. form.* relating to the brain.
cerebrospinal (serəbrōspī'nəl), *a.* pertaining to the brain and to the spinal cord. **cerebrospinal meningitis,** *n.* inflammation of the brain and spinal cord, spotted fever.
cerebrovascular (serəbrōvas'kūlə), *a.* pertaining to the brain and its blood-vessels. **cerebrovascular accident,** *n.* a paralytic stroke.
cerebrum (se'rəbrəm), *n.* (*pl.* **-brums, -bra** (-brə)) the chief portion of the brain, filling the upper cavity of the skull. **cerebral,** *a.* of or pertaining to the brain or the intellect; intellectual rather than emotional; of sounds, made by touching the roof of the mouth with the tip of the tongue. **cerebral cortex,** *n.* the much-folded mass of grey matter forming the outer layer of the cerebrum and responsible for intelligent behaviour. **cerebral haemorrhage,** *n.* bleeding into brain tissue from a cerebral artery. **cerebral hemisphere,** *n.* one of the two great divisions of the cerebrum. **cerebral palsy,** *n.* a disability caused by brain damage before or during birth, characterized by lack of balance and muscular coordination, often with speech impairment. **cerebrate,** *v.i.* to think. **cerebration,** *n.* the action of the brain, whether conscious or unconscious.
cerement CERE.
ceremony (se'rəməni), *n.* a prescribed rite or formality; a usage of politeness; formality, punctilio. **master of ceremonies,** one whose duty it is to see that due formalities are observed on public or state occasions; the person responsible for the running of a dance etc. **to stand on ceremony,** to be rigidly conventional. **ceremonial** (-mō'-), *a.* relating to or performed with ceremonies or rites. *n.* the prescribed order for a ceremony or function; observance of etiquette; in the Roman Catholic Church, the rules for rites and ceremonies; the book containing these. **ceremonialism,** *n.* fondness for or adherence to ceremony. **ceremonialist,** *n.* **ceremonially,** *adv.* **ceremonious** (-mō'-), *a.* involving or befitting a ceremony; punctiliously observant of ceremony according to prescribed form. **ceremoniously,** *adv.* **ceremoniousness,** *n.*
cereous CERE.
cerise (sərēs', -rēz'), *n., a.* (of) a cherry colour.
cerium (siə'riəm), *n.* a malleable grey metallic element of the rare earth group, at. no. 58; chem. symbol Ce, found in cerite. **cerite** (-rīt), *n.* a siliceous oxide of cerium. [after the planetoid *Ceres*]
cermet (sœ'mit), *n.* an alloy of a heat-resistant ceramic and a metal. [*ceramic, metal*]
CERN, (*abbr.*) European organization for nuclear research. [F, *Conseil européen pour la recherche nucléaire*]

cerography (siərog'rəfi), *n.* the art of writing or engraving on wax. **cerographic, -ical** (-graf'-), *a.* **cerographist,** *n.*
ceroplastic (siərəplas'tik), *a.* modelled in wax; pertaining to modelling in wax. **ceroplastics,** *n.sing.*
cert CERTAIN.
cert., (*abbr.*) certificate; certified.
certain (sœ'tən), *a.* sure, convinced, absolutely confident; established beyond a doubt; absolutely determined, fixed; sure to happen, inevitable; sure to do, reliable, unerring; not particularized, indefinite. *pron.* an indefinite number or quantity, some. **for certain,** assuredly. **cert,** *n.* (*sl.*) a certainty. **certainly,** *adv.* assuredly; beyond doubt; without fail; admittedly, yes. **certainty,** *n.* that which is certain; absolute assurance.
certificate[1] (sətif'ikət), *n.* a written testimony or voucher, esp. of character or ability.
certificate[2] (sətif'ikāt), *v.t.* to give a certificate to; to license by certificate. **certificated** (-kā-), *a.* possessing a certificate from some examining body. **certification,** *n.*
certify (sœ'tifī), *v.t.* to assure, to testify to in writing; to guarantee the standard of; to give certain information of or to; to certify as insane. **certified milk,** *n.* milk guaranteed free from tubercle bacillus. **certifiable,** *a.* **certifier,** *n.*
certiorari (sœtiəreə'rī, -rah'rē), *n.* a writ issuing from a superior court calling for the records of, or removing a case from, a court below.
certitude (sœ'titūd), *n.* certainty, conviction.
cerulean (səroo'liən), *a.* of a sky-blue colour; sky-coloured.
cerumen (səroo'men), *n.* the wax-like secretion of the ear. **ceruminous,** *a.*
ceruse (siə'roos, siroos'), *n.* white lead; a pigment made from this. **cerusite, cerussite** (-sit), *n.* a native carbonate of lead.
cervelat (sœ'vəlaht, -lah), *n.* a kind of smoked sausage made from pork or beef.
cervical (sœ'vikl, -vī'-), *a.* of or pertaining to the neck or cervix. **cervical smear,** *n.* a specimen of cells taken from the cervix of the uterus to test for the presence of cancer. **cervico-,** *comb. form.* pertaining to or connected with the neck or cervix. **cervicography,** *n.* photography of the cervix of the uterus, used to detect the early stages of cancer.
cervine (sœ'vīn), *a.* pertaining to or like deer.
cervix (sœ'viks), *n.* a necklike part of the body, esp. the passage between the uterus and the vagina.
Cesarian, -rean CAESARIAN under CAESAR.
cesium CAESIUM.
cespitose CAESPITOSE.
cess (ses), *n.* (*Ir., sl.*) luck. **bad cess to you,** ill luck befall you.
cessation (səsā'shən), *n.* the act of ceasing; pause, rest.
cession (sesh'ən), *n.* a yielding, a surrender, a ceding of territory, or of rights or property. **cessionary,** *n.* one who is the recipient of a legal transfer of rights or property.
cesspit (ses'pit), *n.* a pit for sewage; a filthy, squalid or corrupt place or situation.
cesspool (ses'pool), *n.* a deep hole in the ground for sewage to drain into.
cestoid (ses'toid), *a.* ribbon-like. **cestode** (-tōd), **-toid,** *n.* an intestinal worm of the group Cestoidea, a tape-worm.
Cestr., (*abbr.*) of Chester (bishop of Chester's signature).
cesura (sizū'rə), CAESURA.
Cetacea (sətā'shə), *n.pl.* a group of marine mammals, containing the whales, manatees etc. **cetacean,** *a.* of or pertaining to the Cetacea. *n.* any individual of the Cetacea. **cetaceous,** *a.*
cetane (sē'tān), *n.* an oily, colourless hydrocarbon found in petroleum. **cetane number,** *n.* a measure of the ignition quality of diesel fuel.
Cf, (*chem. symbol*) californium.
cf., (*abbr.*) compare. [L *confer*]

cfc (*abbr.*) chlorofluorocarbon.
c.f.i., (*abbr.*) cost, freight and insurance.
cg, (*abbr.*) centigram.
CGS, (*abbr.*) centimetre-gram-second; chief of general staff.
CGT, (*abbr.*) capital gains tax; *Confédération générale du travail* (the French TUC).
CH, (*abbr.*) Companion of Honour; Confederatio Helvetica (L, Switzerland).
ch, (*abbr.*) central heating; chain (unit of length in knitting); champion (of dogs); chapter; check (in chess); chief; child; church.
cha (chah), *n*. (*Anglo-Indian, sl.*) tea.
chablis (shab'lē), *n*. a white wine made at Chablis, in central France.
cha-cha (chah'chah), **cha-cha-cha**, *n*. a ballroom dance of W Indian origin.
chacma (chak'mə), *n*. a S African baboon.
chaconne (shəkon'), *n*. an old Spanish dance in triple time; the music for this.
chad (chad), *n*. (*pl*. **chad**) the small piece of paper removed when a hole is punched in a punched card, paper tape etc.
chador (chŭd'ə), *n*. a large veil, worn over the head and body by Muslim women.
chaeta (kē'tə), *n*. (*pl*. **-tae** (tē)) a bristle or bristle-like animal part, esp. on the body of a worm.
chafe (chāf), *v.t*. to make warm by rubbing; to rub so as to make sore, to fret; to gall, to irritate. *v.i*. to be worn or sore by rubbing; to fret. *n*. a sore caused by rubbing; irritation, a fit of rage. **chafing-dish**, *n*. a vessel for keeping food warm or cooking at table.
chafer (chāf'ə), *n*. a beetle, a cockchafer.
chaff¹ (chaf, chahf), *n*. the husks of grain; winnowings; hay or straw cut fine for fodder; the scales and bracts of grain and other flowers; anything worthless; thin strips of metal foil thrown from an aeroplane to confuse enemy radar. **chaffy**, *a*. like or full of chaff; light, worthless.
chaff² (chaf), *n*. banter; teasing. *v.t*. to tease. *v.i*. to indulge in banter or teasing.
chaffer (chaf'ə), *v.i*. to dispute about price, to haggle, to bargain; to chatter. *n*. bargaining, haggling. **chafferer**, *n*.
chaffinch (chaf'inch), *n*. a common British small bird, *Fringilla coelebs*.
chagrin (shəgrin'), *n*. vexation, disappointment, mortification; ill-humour. *v.t*. to vex; to put out of humour.
chain (chān), *n*. a series of links or rings fitted into or connected with each other, for binding, connecting, holding, hauling or ornamenting; a measure of 100 links, or 66 ft. (20·12 m); (*pl*.) bonds, fetters, bondage, restraint; a connected series, a sequence, a range; a series of atoms linked together in a molecule; (*pl*.) strong plates of iron bolted to a ship's sides and used to secure the shrouds; a group of shops, hotels etc. under the same ownership and run in a similar style. *v.t*. to fasten or bind with or as with a chain or chains. **chain armour**, *n*. chain mail. **chain-bridge**, *n*. a suspension bridge. **chain-gang**, *n*. a gang of convicts working in chains. **chain letter**, *n*. a circular letter each recipient of which forwards a copy to friends and others. **chain-mail**, *n*. armour of interwoven links. **chain reaction**, *n*. a self-perpetuating chemical or nuclear reaction in which the products of one step initiates or takes part in a further reaction and so on; a series of events, each precipitating the next. **chainring** CHAIN-WHEEL. **chainsaw**, *n*. a power saw whose teeth are in a continuous revolving chain. **chain-smoke**, *v.i*. (*coll*.) to smoke continuously, lighting one cigarette from another. **chain-smoker**, *n*. **chain-stitch**, *n*. an ornamental stitch resembling a chain; a loop-stitch (made by a sewing-machine). **chain-store**, *n*. one of a series of retail stores under the same ownership and selling the same kind of wares. **chainwheel**, *n*. a toothed wheel which receives or transmits power by means of an endless chain. **chainless**,

a. **chainlet**, *n*.
chair (cheə), *n*. a movable seat with a back, for one person; a seat of authority or office; a professorship; a chairmanship or mayoralty; (the seat of) the person presiding at a meeting; a sedan; an iron socket to support and secure the rails in a railway. *v.t*. to carry publicly in a chair in triumph; to install as president of a meeting or society; to act as chairperson at. **to take the chair**, to preside at a meeting. **chair-lift**, *n*. a series of seats suspended from a cable for conveying people up a mountain. **chairman**, *n*. a chairperson; (*Hist*.) one of a pair of men who carried a sedan. **chairmanship**, *n*. **chairperson**, *n*. the president of a meeting or the permanent president of a society, committee etc. **chairwoman**, *n. fem.*
chaise (shāz), *n*. a light travelling or pleasure carriage of various patterns. **chaise longue** (lŏg), *n*. a chair with support for the legs.
chalaza (kəlah'zə), *n*. (*pl*. **-zae** (zē), **-zas**) one of the two twisted albuminous threads holding the yolk in position in an egg; an analogous part of a plant ovule.
chalcedony, **calcedony** (kalsed'əni), *n*. a cryptocrystalline variety of quartz. **chalcedonic** (-don'-), *a*.
chalcography (kalkog'rəfi), *n*. the art or process of engraving on brass or copper. **chalcographer**, **-ist**, *n*. **chalcographic** (-graf'-), *a*.
chalcopyrite (kalkōpī'rīt), *n*. yellow or copper pyrites, a copper ore.
Chaldean, -dee (kaldē'ən, -dē'), *a*. of or belonging to ancient Chaldea in Babylonia or its language. *n*. the language of Chaldea; a native of Chaldea. **chaldaic** (-dā'-), *n., a.*
chaldron (chawl'drən), *n*. a measure (36 bushels; about 1·3 cu. m) for coals.
chalet (shal'ā), *n*. a small house or villa on a mountainside; a Swiss cottage; a small flimsy dwelling used esp. for holiday accommodation.
chalice (chal'is), *n*. a cup or drinking vessel; the cup used in the Eucharist; (*poet*.) a flower-cup.
chalk (chawk), *n*. soft white limestone or massive carbonate of lime, chiefly composed of marine shells; a piece of this or of a coloured composition prepared from it, used for writing and drawing. *v.t*. to rub, mark or write with chalk. **by a long chalk**, by a great deal. **as different, as like as chalk and cheese**, totally different. **to chalk it up**, to give or take credit for something. **to chalk out**, to sketch out, to plan. **chalk-stripe**, *n*. a pattern of narrow white stripes on a dark-coloured background. **chalky**, *a*. **chalkiness**, *n*.
challenge (chal'inj), *n*. a summons or defiance to fight a duel; an invitation to a contest of any kind; the cry of hounds on finding scent; a calling in question; exception taken to a juror or voter; the call of a sentry in demanding the counter-sign; a difficult task which stretches one's abilities. *v.t*. to invite or defy to a duel; to invite to a contest of any kind; to call on to answer; to demand, to invite, to claim; to object to, to dispute, contest; to stimulate, stretch. **challenge cup**, *n*. a cup competed for annually by football teams, yacht clubs etc. **challengeable**, *a*. **challenged**, *a*. handicapped as in *visually challenged*. **challenger**, *n*.
challis (shal'is, -li, chal'is), *n*. a light woollen fabric; formerly a fabric of silk and wool, for ladies' dresses.
chalybeate (kəlib'iət), *a*. impregnated with iron. *n*. a mineral water or spring so impregnated.
chamber (chām'bə), *n*. (*chiefly poet*.) a room, esp. a sleeping room; the place where a legislative assembly meets; the assembly itself; a hall of justice; an association of persons for the promotion of some common object; a hollow cavity or enclosed space; the space between the gates of a canal lock; that part of the bore of a gun or other firearm where the charge lies; a judge's private room in a court; (*pl*.) the office or apartments of a

barrister in an Inn of Court; a suite of apartments; a chamber-pot. **Chamber of Commerce,** an organization to promote the interests of business in a district. **Chamber of Horrors,** a room at Madame Tussaud's waxwork exhibition devoted to famous criminals; (*fig.*) a place full of horrifying objects. **chamber concert,** *n.* one where chamber music is given. **chambermaid,** *n.* a woman who cleans the bedrooms at a hotel. **chamber music,** *n.* music adapted for performance in a room, as distinguished from that intended for theatres, churches etc. **chamber orchestra,** *n.* a small orchestra suitable for playing chamber music. **chamber-pot, -utensil,** *n.* a bedroom receptacle for slops and urine. **chambered,** *a.* enclosed; divided into compartments or sections.

chamberlain (chām'bəlin), *n.* an officer in charge of the household of a sovereign or nobleman; the treasurer of a city or corporation. **Lord Chamberlain (of the Household),** one of the principal British officers of State, controlling the servants of the royal household above stairs, and the licensing of theatres and plays. **Lord Great Chamberlain of England,** a British hereditary officer of State in charge of the Palace of Westminster and performing ceremonial functions. **chamberlainship,** *n.*

Chambertin (shä'bətī), *n.* a dry red Burgundy wine. [a vineyard near Dijon]

chambré (sham'brä), *a.* of wine, warmed to room temperature.

chameleon (kəmēl'yən), *n.* a lizard having the power of changing colour and formerly fabled to live on air; a changeable person. **chameleonic** (-lion'-), *a.* **chameleon-like,** *a.*, *adv.*

chamfer (cham'fə), *n.* in carpentry, an angle slightly pared off; a bevel, a groove, a fluting. *v.t.* to groove; to bevel off.

chamlet (cham'lit), CAMLET.

chamois (sham'wah), *n.* (*pl.* **chamois**) a goat-like European antelope, *Antilope rupicapra*. **chamois-leather, chamois** (sham'i), *n.* a soft, pliable leather, orig. prepared from the skin of the chamois; a piece of this used for polishing.

chamomile CAMOMILE.

champ[1] (champ), *v.t.*, *v.i.* to bite with a grinding action or noise; to chew, to crunch. *n.* champing; the noise of champing. **to champ at the bit,** to be impatient.

champ[2] (champ), *n.* (*coll.*) short for CHAMPION.

champagne (shampān'), *n.* a light sparkling wine made in the province of Champagne, France; a pale yellow colour. **fine champagne** (fēn shäpahny''), liqueur brandy. **champenize, -ise,** *v.t.* to convert (still wine) into a champagne-type sparkling wine.

champaign (shampān'. sham'-), *n.* flat, open country.

champers (sham'pəz), (*coll.*) CHAMPAGNE.

champerty (cham'pəti), *n.* maintenance of a party in a suit on condition of sharing the property at issue if recovered.

champignon (shä'pinyō), *n.* a mushroom, esp. the fairy-ring agaric.

champion (cham'piən), *n.* one who engages in single combat on behalf of another; one who argues on behalf of or defends a person or a cause; the acknowledged superior in any athletic exercise or trial of skill; the person, animal or exhibit that defeats all competitors. *v.t.* to defend as a champion; to support (a cause). *a.* superior to all competitors; (*dial.*) first-class, supremely excellent. **championless,** *a.* **championship,** *n.* the fact of being a champion; the act of championing or defending; a contest to find a champion.

champlevé (shä'ləvā), *n.* enamelling by the process of inlaying vitreous powders into channels cut in the metal base.

Chanc., (*abbr.*) chancellor; chancery.

chance (chahns), *n.* fortune, luck, the course of events; event, issue, result; undesigned result or occurrence; accident, risk; possibility, opportunity; (*usu. pl.*) likelihood, probability; fate, the indeterminable course of events, fortuity. *v.t.* (*coll.*) to risk. *v.i.* to happen, to come to pass. *a.* fortuitous, unforeseen. **by chance,** as things fall out; accidentally; undesignedly. **on the (off-) chance,** on the possibility; in case. **the main chance,** the most important issue; gain; self-interest. **to chance upon,** to come upon accidentally. **to stand** (or **have**) **a good chance,** to have a reasonable prospect of success. **chanceful,** *a.* fortuitous, accidental; (*poet.*) eventful. **chancer,** *n.* (*sl.*) a person who takes risks in order to make a profit. **chancy,** *a.* risky, doubtful.

chancel (chahn'sl), *n.* the eastern part of a church, formerly cut off from the nave by a screen.

chancellery, -ory (chahn'sələri), *n.* a chancellor's court or council and official establishment; the building or room in which a chancellor has an office; the office or department attached to an embassy or consulate.

chancellor (chahn'sələ), *n.* the president of a court or public department; the titular head of a British university; the head of government in some European countries; a bishop's law-officer or a vicar-general. **Chancellor of the Duchy of Lancaster,** the representative of the Crown as holder of the Duchy of Lancaster. **Chancellor of the Exchequer,** the principal finance minister of the British Government. **Lord (High) Chancellor,** the highest officer of the British Crown, the president of the Chancery division of the Supreme Court (formerly the High Court of Chancery) and Speaker of the House of Lords **chancellorship,** *n.* **chancellory** CHANCELLERY.

chance-medley (chahnsmed'li), *n.* (*Law*) homicide by misadventure, as accidental homicide in repelling an unprovoked attack; inadvertency.

chancery (chahn'səri), *n.* the court of the Lord Chancellor, before 1873; the highest English court of justice next to the House of Lords, now a division of the High Court of Justice; (*N Am.*) a court of equity; a court or office for the deposit of records; a chancellery.

chancre (shang'kə), *n.* a hard syphilitic lesion. **chancroid,** *n.* a soft ulcer caused by venereal infection. **chancroid, chancrous,** *a.*

chancy CHANCE.

chandelier (shandəliə'), *n.* a hanging branched frame for a number of lights.

chandler (chahnd'lə), *n.* one who makes or sells candles; a retail dealer, esp. of a specified kind, as *ship's chandler*. **chandlery,** *n.* the establishment or the stock in trade of a chandler.

change (chānj), *v.t.* to make different, to alter; to give up or substitute for something else; to give or take an equivalent for in other coin; to exchange; to replace the coverings or clothes of (as a bed or baby), to transfer from one to another. *v.i.* to become different; to be altered in appearance; to pass from one state or phase to another; to put on different clothes; to transfer from one vehicle to another. *n.* alteration, variation; shifting, transition; the passing of the moon from one phase to another; alteration in order, esp. of ringing a peal of bells; substitution of one thing for another; small coin or foreign money given in return for other coins; balance of money paid beyond the value of goods purchased; (an) exchange; novelty, variety. **change of front,** (*Mil.*) a wheeling movement; a change of attitude, a reversal of policy. **change of life,** the menopause. **to change colour,** to turn pale; to blush. **to change down,** in driving etc., to change to a lower gear. **to change hands,** to pass from one person's ownership to another's. **to change one's mind,** to form a new plan or opinion. **to change one's tune,** to adopt a humble attitude; to become sad or vexed. **to change up,** in driving, etc., to change to a higher gear. **to get no change out of,** not to be able to take any advantage of. **to ring the changes,** to try all ways of doing something. **changeover,**

n. an alteration or reversal from one state to another; in a relay race, the passing of the baton from one runner to the next. **change-ringing**, *n.* a form of bell-ringing in which a set of bells is rung repeatedly but in slightly varying order. **changeable**, *a.* liable to change; inconstant, fickle, variable. **changeability**, *n.* **changeableness**, *n.* **changeably**, *adv.* **changeful**, *a.* **changefully**, *adv.* **changefulness**, *n.* **changeless**, *a.* **changeling** (-ling), *n.* anything substituted for another; a child substituted for another, esp. an elf-child. **changer**, *n.* **changing**, *a.* **changing bag**, **mat**, **table**, *n.* used in changing a baby's nappy.

channel (chan'əl), *n.* the bed of a stream or an artificial watercourse; the deep part of an estuary; a narrow piece of water joining two seas; a tube or duct, natural or artificial, for the passage of liquids or gases; means of passing, conveying or transmitting; a furrow, a groove, a fluting; a gutter; a course, line, or direction; a band of frequencies on which radio and television signals can be transmitted without interference from other channels; a path for an electrical signal; in a computer, a route along which data can be transmitted. *v.t.* (*past*, *p.p.* **channelled**) to cut a channel or channels in; to cut (a way) out; to groove. **The Channel**, the English Channel. **channelize**, **-ise**, *v.t.* to channel. **channelization**, *n.*

chanson (shä'sö), *n.* a song. [F]

chant (chahnt), *v.t.*, *v.i.* to recite to music or musically, to intone. *n.* a composition consisting of a long reciting note and a melodic phrase; a psalm, canticle or other piece sung in this manner; a musical recitation or monotonous song. **chanter**, *n.* one who chants; the pipe on a bagpipe that plays the melody.

chanterelle (shahntərel'), *n.* a type of edible fungus. [F]

chanteuse (shätœz'), *n.* a female nightclub singer.

chanticleer (chan'tikliə), *n.* a name for a cock, esp. as the herald of day.

chantry (chahn'tri), *n.* an endowment for a priest or priests to say mass daily for some person or persons deceased; the chapel or the part of a church used for this purpose; the body of priests who perform this duty. **chantry-priest**, *n.*

Chanukah HANUKKAH.

chaos (kā'os), *n.* the void, the confusion of matter said to have existed at the Creation; confusion, disorder. **chaotic** (-ot'-), *a.* **chaotically**, *adv.*

chap[1] (chap), *v.t.*, *v.i.* (*past*, *p.p.* **chapped**) to (cause to) crack or open in long slits. *n.* (*usu.* pl.) a longitudinal crack, cleft or seam on the surface of the skin. **chapped**, *a.*

chap[2] (chap), **chop** (chop), *n.* (*pl.*) the jaws (usu. of animals), the mouth and cheeks; the lower part of the cheek. **to lick one's chops**, to relish in anticipation. **chap-fallen**, *a.* having the lower jaw depressed; downcast, dejected, dispirited.

chap[3] (chap), *n.* (*coll.*) a man, a fellow.

chaparral (shap'əral, -ral'), *n.* a thicket of low evergreen oaks, or of thick bramble-bushes and thorny shrubs.

chapati, **-tti** (chəpat'i), *n.* in Indian cookery, a round, thin piece of unleavened bread.

chap-book (chap'buk), *n.* a small book, usually of wonderful tales, ballads or the like, formerly hawked by chapmen.

chapeau (shap'ō), *n.* (*pl.* **chapeaux** (-z)) esp. in heraldry, a hat.

chapel (chap'l), *n.* a place of worship connected with and subsidiary to a church; an area, room etc. containing an altar in a church; a place of worship other than a church or cathedral, esp. one in a palace, mansion, airport etc.; a Nonconformist place of worship; a service, or the sort of service, at a chapel; a printing-office (from the legend that Caxton set up his printing press in Westminster Abbey); a printers' or journalists' trade union, or a branch of it. *a.* belonging to a Nonconformist church. **chapel of ease**, a subordinate church in a parish. **chapel of rest**, room at a graveyard where bodies lie for viewing prior to burial. **father**, **mother of chapel**, the president of a branch of a printers' or journalists' trade union.

chaperon(e) (shap'ərōn), *n.* a person (esp. a married or elderly woman) who supervises a young unmarried lady in public places. *v.t.* to act as chaperone to. **chaperonage** (-nij), *n.* the duties or position of a chaperone.

chaplain (chap'lin), *n.* a clergyman who officiates at court, in the house of a person of rank, or in a regiment, ship or public institution. **chaplaincy**, *n.* **chaplainship**, *n.*

chaplet (chap'lit), *n.* a wreath or garland for the head; a string of beads one-third the number of a rosary; a necklace; a bird's crest.

chapman (chap'mən), *n.* (*dated*) one who buys and sells; an itinerant merchant, a pedlar, a hawker.

chappie (chap'i), (*coll.*) CHAP[3].

chaps (chaps), *n.pl.* leather leggings worn by cowboys.

chapter (chap'tə), *n.* a division of a book; a part of a subject; a piece of narrative, an episode; a division of Acts of Parliament arranged in chronological order for reference; the general membership of certain orders and societies; the council of a bishop, consisting of the clergy attached to a cathedral or collegiate church; a meeting of the members of a religious order; a chapter-house. **chapter and verse**, full and precise reference (in order to verify a fact or quotation). **chapter of accidents**, a series of accidents. **chapterhouse**, *n.* the place in which a religious chapter meets.

char[1] (chah), *n.* a small fish of the salmon family, found in the Lake District and N Wales.

char[2] (chah), *n.* a charwoman. *v.i.* (*past*, *p.p.* **charred**) to work as a charwoman. **charwoman**, **-lady**, *n.* a woman employed to do cleaning.

char[3] (chah), *v.t.* (*past*, *p.p.* **charred**) to reduce to charcoal; to burn slightly, to blacken with fire. *v.i.* to become blackened with fire.

char[4] (chah), CHA.

charabanc (sha'rəbang), *n.* (*dated*) a coach for day-trippers.

character (ka'riktə), *n.* a mark made by cutting, engraving or writing; a letter, a sign; (*pl.*) letters distinctive of a particular language; peculiar distinctive qualities or traits; the sum of a person's mental and moral qualities; moral excellence, moral strength; reputation, standing; (of a person, thing, place etc.) individuality; a person, a personage; a personality created by a novelist, poet or dramatist; a part in a play; an actor's part; (*coll.*) an eccentric person; a characteristic (of a species etc.); (*Comput.*) a symbol, e.g. a letter, punctuation mark etc., that can be used in representing data. **in**, **out of character**, similar or dissimilar to what one would expect of a person. **character actor**, *n.* one who specializes in portraying eccentric or complicated characters. **character assassination**, *n.* the destruction of a person's good reputation by, e.g. the spreading of malicious rumour. **character part**, *n.* the role of an eccentric or complicated person. **character reference**, *n.* a (usu. written) testament to a person's moral qualities. **characteristic** (-ris'-), *n.* that which marks or constitutes part of the character. **characteristic**, **-ical** (-ris'-), *a.* constituting or exhibiting typical qualities; typical (of the person etc. in question). **characteristically**, *adv.* **characterize**, **-ise**, *v.t.* to give character to, to stamp, to distinguish; to describe; to be characteristic of. **characterization**, *n.* **characterless**, *a.* without definite character; ordinary, commonplace.

charade (shərahd'), *n.* a kind of riddle based upon a word the key to which is given by description or action representing each syllable and the whole word; (*pl.*) a game based on such riddles; a ridiculous pretence, a travesty.

charcoal (chah'kōl), *n.* wood partially burnt under turf; an impure form of carbon prepared from vegetable or animal substances; a stick of charcoal used for drawing; a drawing made with such a stick; a dark grey colour.

charge (chahj), *v.t.* to put the proper load or quantity of material into (any apparatus), as to load (a gun), to fill (a glass), to accumulate electricity in (a battery) etc.; to rush on and attack; to put (weapons) in an attacking position; to lay on or impose a duty on; to enjoin, to command, to exhort; to accuse; to allege; to debit to; to ask a price for; to give directions to, as a judge to a jury etc., or a bishop to his clergy. *v.i.* to make an attack or onset; (*coll.*) to demand high prices or payments. *n.* a load, a burden; an office, duty or obligation; care, custody; the thing or person under one's care, a minister's flock; command, commission; an entry on the debit side of an account; price demanded, cost; accusation, allegation; attack, onset; the quantity with which any apparatus, esp. a firearm, is loaded; instructions, directions, esp. those of a judge to a jury, or of a bishop to his clergy; the electrical property of matter, negative or positive; the amount or accumulation of electricity, e.g. in a battery. **charge coupled device**, (*Comput.*) a storage device built into a chip which can be used only so long as it has an electric charge. **in charge**, in authority, control (of). **to give in charge**, to commit to the care of another. **charge account**, *n.* a credit account at a shop. **charge card**, *n.* a card issued by a finance company or similar organization, the balance on which must be totally paid off at the end of each month (in contradistinction to *credit card*). **charge hand**, *n.* a workman in charge of several other men and responsible to a foreman. **charge nurse**, *n.* a nurse in charge of a ward; male equivalent of (ward) *sister*. **charge-sheet**, *n.* a list of offenders taken into custody, with their offences, for the use of a police-magistrate. **chargeable**, *a.* liable to be charged or accused; liable to a monetary demand; liable to be an expense (to); capable of being properly charged (to); rateable. **chargeability**, *n.* **chargeless**, *a.* **charger**, *n.* one who charges; a war-horse; a cavalry horse; a large dish; a battery charger.
chargé d'affaires (shah′zhä dəfeə), **chargé**, *n.* (*pl.* **chargés d'affaires**) a diplomatic agent acting as deputy to an ambassador; an ambassador to a court of minor importance. [F, charged with affairs]
charily, etc. CHARY.
chariot (cha′riət), *n.* (*Hist.*) a carriage used in war, public triumphs and racing. **charioteer** (-tiə′), *n.* a chariot-driver. **charioteering** (-tiə′-), *n.* the act, art or practice of driving a chariot.
charisma (kəriz′mə), *n.* personal magnetism or charm enabling one to inspire or influence other people. **charismatic** (karizmat′-), *a.*
charity (cha′riti), *n.* love of one's fellow, one of the theological virtues; liberality to the poor; alms-giving; alms; an act of kindness; kindness, goodwill; leniency, tolerance of faults and offences; a foundation or organization for assisting the poor, the sick, or the helpless. **cold as charity**, cold-hearted, unsympathetic. **Charity Commissioners**, *n.pl.* members of a board instituted in 1853 for the control of charitable foundations. **charitable**, *a.* full of, pertaining to, or supported by charity; liberal to the poor; benevolent, kindly, lenient, large-hearted; dictated by kindness. **charitableness**, *n.* **charitably**, *adv.*
charivari (shahrivah′ri), *n.* a mock serenade of discordant music, intended to insult and annoy; a satirical journal.
charlady CHAR [2].
charlatan (shah′lətən), *n.* an empty pretender to skill or knowledge, a quack; an impostor. **charlatanism, -tanry**, *n.*
Charles's wain (chahl′ziz), *n.* seven stars in the constellation the Great Bear, also called the Plough.
Charleston (chahl′stən), *n.* a lively dance in 4/4 time with characteristic kicking outwards of the lower part of the legs.
Charlie (chah′li), *n.* (dated *sl.*) an utterly foolish person. often in *a proper Charlie*.
charlotte (shah′lət), *n.* a kind of pudding made of fruit and bread or cake crumbs, as in *apple charlotte*. **charlotte russe** (roos), *n.* custard or whipped cream enclosed in sponge cake.
charm (chahm), *n.* a spell, a thing, act or formula having magical power; an article worn to avert evil or ensure good luck, an amulet; the power or gift of alluring, pleasing or exciting love or desire; a pleasing or attractive feature; a trinket worn on a bracelet. *v.t.* to enchant, to fascinate, to delight; (*coll.*) to please; (*usu. pass.*) to protect with occult power; to remove by charms (with *away*). **like a charm**, perfectly (as planned). **charmer**, *n.* one who uses charms; one who fascinates. **charming**, *a.* highly pleasing; delightful. **charmingly**, *adv.* **charmingness**, *n.* **charmless**, *a.* having no attractive features.
charnel-house (chah′nəlhows), *n.* a place where dead bodies or the bones of the dead are deposited.
Charon (keə′rən), *n.* the ferryman who took departed spirits across the Styx into Hades; a ferryman.
charpoy (chah′poi), *n.* a light Indian bedstead.
charr (chah), CHAR [1].
chart (chaht), *n.* a map of some part of the sea or air, with coasts, islands, rocks, shoals etc.; a statement of facts in tabular form; a projection of relative facts, statistics or observations in the form of a graphic curve; a skeleton map for special purposes, e.g. weather *chart*; (*usu. pl.*) a weekly list of best-selling records. *v.t.* to make a chart of. **chartless**, *a.*
charter (chah′tə), *n.* an instrument in writing granted by the sovereign or Parliament, incorporating a borough, company or institution, or conferring certain rights and privileges; a chartered aeroplane, boat, vehicle etc., *a.* of an aircraft, boat, vehicle etc., hired; of a journey, made in a hired conveyance. *v.t.* to establish by charter; to license by charter; to hire or let. **charter-member**, *n.* a founder-member of a chartered body. **chartered**, *a.* professionally qualified to the standards set by a professional body, as in *chartered accountant*.
Chartism (chah′tizm), *n.* the principles of the Chartists, an English democratic party (1838–48), briefly, universal suffrage, vote by ballot, annual parliaments, payment of members, equal electoral districts and the abolition of property qualifications for members. **Chartist**, *n., a.*
chartreuse (shahtrœz′), *n.* a pale green or yellow liqueur made by the monks at la Grande Chartreuse, near Grenoble, France.
charwoman CHAR [2].
chary (cheə′ri), *a.* wary, prudent, cautious; frugal, sparing. **charily**, *adv.* **chariness**, *n.*
Charybdis (kərib′dis), *n.* a dangerous monster or whirlpool off the coast of Sicily, opposite Scylla, a monster or rock on the Italian shore; one of a pair of alternative risks.
Chas., (*abbr.*) Charles.
chase[1] (chās), *v.t.* to pursue; to hunt; to drive (away); to put to flight. *v.i.* to ride or run rapidly (after). *n.* (a) fast pursuit; the hunting of wild animals; an open (former) hunting-ground or preserve for game. **to chase up**, (*coll.*) to pursue or investigate in order to obtain information etc. **to give chase** GIVE [1]. **wild-goose chase** WILD. **chaser**, *n.* a horse used for steeplechasing; a drink of one kind (e.g. beer) taken after another kind (e.g. whisky).
chase[2] (chās), *v.t.* to engrave, to emboss. **chaser**, *n.* a tool for chasing; one skilled in chasing. **chasing**, *n.* the art of engraving or embossing metals; the pattern embossed.
chase[3] (chās), *n.* a rectangular iron frame in which metal type is locked for printing.
chasm (kaz′m), *n.* a large geographical cleft, fissure, a yawning gulf; a breach or division between persons or parties; a gap or void. **chasmed**, *a.* having chasms.
chasseur (shasœ′), *n.* a huntsman. *pred. a.* cooked in a sauce of white wine and mushrooms, as in *chicken chasseur*.
chassis (shas′i), *n.* (*pl.* **chassis**) the framework of a

chaste (chāst), *a.* abstaining from all sexual intercourse, or from sex outside marriage; modest, innocent, virginal; free from obscenity; pure in style, thought, deed etc. **chastely**, *adv.* **chastity** (chas'-), *n.* the state of being chaste; virginity; celibacy.
motor-car, aeroplane etc.; a framework supporting the components of a piece of electronic equipment.
chasten (chā'sn), *v.t.* to punish with a view to reformation; to subdue. **chastener**, *n.*
chastise (chastīz'), *v.t.* to punish, esp. physically; to chasten. **chastisement** (chas'tiz-. -tīz'-), *n.* **chastiser**, *n.*
chastity CHASTE.
chasuble (chaz'ūbl), *n.* a sleeveless vestment worn by a priest over the alb while celebrating Mass.
chat[1] (chat), *v.i.* (*past, p.p.* **chatted**) to talk easily and familiarly; to gossip. *n.* easy, familiar talk; gossip. **to chat up**, (*sl.*) to talk to in order to establish a (sexual) relationship. **chatline**, *n.* a telephone service where callers are connected with each other for informal conversation. **chat show**, *n.* a television show in which a host(ess) informally interviews invited celebrities. **chatty**, *a.* **chattiness**, *n.*
chat[2] (chat), *n.* the name of various birds, mostly warblers, e.g. *whinchat.*
chateau, château (shat'ō), *n.* (*pl.* **-teaux** (-tōz)) a castle; a country house in French-speaking countries.
chatline CHAT[1].
chattel (chat'əl), *n.* (*usu. pl.*) moveable property; (*Law*) any article of property except those which are freehold.
chatter (chat'ə), *v.i.* to utter rapid, inharmonious sounds like a magpie, jay etc.; to make a noise by or like the rattling together of the teeth; to talk idly and thoughtlessly. *n.* sounds like those of a magpie, jay etc.; idle talk. **chatterbox**, *n.* an incessant talker. **chatterer**, *n.* **the chattering classes**, (*derog.*) a set of people who spend much time in discussion and conversation, rather than in action; people who regard themselves as having informed opinions and ideas, e.g. about the arts or politics.
chatty CHAT[1].
Chaucerian (chawsiə'riən), *a.* pertaining or relating to the poet Chaucer; resembling his style. *n.* a student of Chaucer. **Chaucerism** (-sə-), *n.* something characteristic of Chaucer.
chauffeur (shō'fə, -fœ'), *n.* a person employed to drive a motor-car. *v.t.* to act as driver for. **chauffeuse** (shōfœz'), *n. fem.*
chauvinism (shō'vənizm), *n.* exaggerated patriotism of an aggressive kind; jingoism, an exaggerated and excessive attachment to any cause, such as sexism. **chauvinist**, *n.* one who believes his own race, sex etc. to be superior and despises all others, as in *male chauvinist.* **chauvinistic** (-nis'-), *a.*
ChB, (*abbr.*) Bachelor of Surgery.
cheap (chēp), *a.* low in price; worth more than its price or cost; easy to get; easily got; of low esteem. *adv.* cheaply. **cheap and nasty**, low in price and quality. **on the cheap**, cheaply; in a miserly way. **to hold cheap**, to despise. **to make oneself cheap**, to behave with undignified familiarity. **cheap-jack**, *a.* cheap; inferior. **cheap-skate**, *n.* (*coll.*) a miserly person. **dirt-cheap** DIRT. **cheapen**, *v.t., v.i.* **cheapish**, *a.* **cheapness**, *n.*
cheat (chēt), *n.* a fraud, a swindle; a trickster, a swindler. *v.t.* to defraud, to deprive (of); to deceive. *v.i.* to act as a cheat. **cheater**, *n.*
check[1] (chek), *n.* a sudden stoppage, arrest or restraint of motion; the person, thing or means of arrest; a reverse, a repulse; a mark put against names or items in going over a list; a test by which correctness, authenticity or accuracy may be ascertained; a term in chess when one player obliges the other to move or guard his king; the situation of such a king; (*esp. N Am.*) a bill at a restaurant etc. *v.t.* to cause to stop; to repress, to curb; to rebuke; to test or confirm the accuracy of, esp. by comparison with a list; in chess, to put (an opponent's king) in check. *v.i.* to pause, to halt; to agree, correspond (with); to confirm the accuracy of something. *int.* a call when an opponent's king is exposed (see CHECKMATE); a term expressing agreement or identicalness. **to check in**, to register on arrival at a hotel, airport etc. **to check out**, to depart from a hotel etc. after a stay; to test for accuracy, to investigate. **to check up**, to investigate, to test (often with *on*). **check-action**, *n.* a device for preventing the hammer in a piano from striking twice. **check bit, digit**, *n.* in computing, a bit or digit used to detect error. **check-in**, *n.* a place where one's arrival is registered. **check-list**, *n.* one used in checking for accuracy, completeness etc.; an inventory. **check-out**, *n.* a cash desk at a supermarket. **checkpoint**, *n.* a place (as at a frontier) where documents etc. are checked. **check-rein**, *n.* a branch rein coupling horses in a team. **check-up**, *n.* a general examination (esp. medical). **check-valve**, *n.* a valve that allows a flow in only one direction. **checker**[1], *n.*
check[2] (chek), *n.* a chequered pattern, a cross-lined pattern; a chequered fabric. **checked**, *a.*
check[3] (chek), (*N Am.*) CHEQUE. **checking account**, *n.* (*N Am.*) a current account.
checker[2] CHEQUER.
checkmate (chek'māt), *n.* in chess, the winning movement when one king is in check and cannot escape from that position; a complete defeat; a position from which there is no escape. *int.* the call when an opponent's king is put into this position. *v.t.* to put in checkmate; to defeat utterly, to frustrate.
Cheddar (ched'ə), *n.* a hard, strong-flavoured yellow cheese.
cheek (chēk), *n.* the side of the face below the eye; (*coll.*) impudence, sauciness; effrontery, (self-)assurance; (*sl.*) a buttock. *v.t.* to be impudent to. **cheek by jowl**, side by side; in the closest proximity. **cheek-bone**, *n.* the bone above the cheek. **cheeky**, *a.* impudent, saucy. **cheekily**, *adv.* **cheekiness**, *n.*
cheep (chēp), *v.i.* to chirp feebly (as a young bird). *n.* the feeble cry of a young bird.
cheer (chiə), *n.* a shout of joy or applause. *v.t.* to make glad or cheerful; to applaud, to encourage. *v.i.* to utter cheers. **to be of good cheer**, to be in a good, happy etc. frame of mind. **to cheer up**, to make or become (more) cheerful. **cheer-leader**, *n.* (*N Am.*) a girl who leads organized cheering at a rally, football game etc. **cheerful**, *a.* full of good spirits; lively, animated. **cheerfully**, *adv.* **cheerfulness**, *n.* **cheering**, *a.* **cheeringly**, *adv.* **cheerio** (-riō'), *int.* good-bye, au revoir; a drinking toast. **cheerless**, *a.* dull, gloomy, dispiriting. **cheerlessness**, *n.* **cheers**, *int.* a drinking toast; thank-you; good-bye. **cheery**, *a.* lively, sprightly, full of good spirits. **cheerily**, *adv.* **cheeriness**, *n.*
cheese[1] (chēz), *n.* the curd of milk pressed into a solid mass and ripened by keeping; a block of this; anything of cheese-like form. **hard cheese**, (*coll.*) hard luck. **say cheese**, used by photographers to encourage people to smile. **cheese-board**, *n.* a board on which cheese is served at table; the variety of cheeses on such a board. **cheeseburger**, *n.* a hamburger with a slice of cheese on top. **cheesecake**, *n.* a kind of tart made of pastry or crumbs with a filling of cream cheese, sugar etc.; (*coll.*) (pictures of) young and shapely women, esp. scantily clad or nude. **cheesecloth**, *n.* thin cotton cloth loosely woven; butter muslin. **cheese-paring**, *a.* niggardly, mean, miserly. *n.* meanness, stinginess. **cheese-plate**, *n.* a small plate used for cheese at the end of a meal. **cheese straw**, *n.* a long, thin, cheese-flavoured biscuit. **cheese-taster**, *n.* a gouge-like knife for scooping pieces of cheese as samples. **cheese-wire**, *n.* a thin wire used for cutting cheese. **cheesy**, *a.* resembling or tasting like cheese. **cheesiness**, *n.*
cheese[2] (chēz): **big cheese**, *n.* (*sl.*) an important person.

cheese — chevrotain

cheese³ (chēz): **cheesed off,** (*coll.*) bored, annoyed.
cheetah (chē'tə), *n.* the hunting leopard, renowned for its great speed.
chef (shef), *n.* a head or professional cook. **chef de cuisine** (də kwēzēn'), a head cook.
chef-d'oeuvre (shādœvr'ʼ), *n.* (*pl.* **chefs-**) a masterpiece.
cheiroptera (kīrop'tərə), *n.pl.* a group of mammals with membranes connecting their fingers and used as wings, consisting of the bats. **cheiropteran,** *n.* **cheiropterous,** *a.*
Cheka (chek'ə), *n.* name for the secret police in Soviet Russia (1917–22).
chela (kē'lə), *n.* (*pl.* **-lae** (-lē)) a claw (as of a lobster or crab), a modified thoracic limb. **chelate** (-lāt), *a.*
cheliform (kel'ifawm), *a.* like a claw in form or shape.
cheliped (kel'iped), *n.* one of the pair of legs carrying chelae.
Chelonia (kilō'niə), *n.pl.* an order of reptiles containing the turtles and tortoises. **chelonian,** *n.*, *a.*
Chelsea bun (chel'si bun'), *n.* a bun made of a roll of sweet dough with raisins. **Chelsea pensioner,** *n.* a retired or invalid old soldier living in the Chelsea Royal Hospital in London. **Chelsea ware,** *n.* a type of 18th-cent. china.
chem(i)-, chemic(o)-, chem(o)- *comb. forms.* chemical.
chemical (kem'ikəl), *a.* pertaining to chemistry, its laws, or phenomena; of or produced by chemical process. *n.* a substance or agent produced by or used in chemical processes. **chemical change,** *n.* a change involving the formation of a new substance. **chemical engineering,** *n.* the branch of engineering concerned with the design and building of industrial chemical plants. **chemical reaction,** *n.* (an instance of) the process of changing one substance into another. **chemical symbol,** *n.* a letter or letters used to represent (an atom of) a chemical element. **chemical toilet,** *n.* a toilet using chemicals rather than running water, the latter being unavailable. **chemical warfare,** *n.* war waged using poisonous chemicals (gases, sprays etc.). **chemically,** *adv.*
chemico-electric (kemikōilek'trik), *a.* pertaining to or produced by chemistry in conjunction with electricity.
chemiluminescence (kemiloomines'əns), *n.* luminescence occurring as a result of a chemical reaction, without production of heat.
chemin de fer (shimi də feə'), *n.* a variety of baccarat.
chemise (shəmēz'), *n.* a body garment of linen or cotton worn next to the skin by women.
chemist (kem'ist), *n.* one versed in chemistry; one qualified to dispense drugs, a pharmacist. **analytical chemist,** a chemist who carries out the process of analysis by chemical means.
chemistry (kem'istri), *n.* (the practical application of) the science which investigates the elements of which bodies are composed, the combination of these elements, and the reaction of these chemical compounds on each other (**inorganic chemistry** deals with mineral substances, **organic chemistry** with animal and vegetable substances); any process or change conceived as analogous to chemical action, esp. emotional attraction.
chemo- (kēm'ō), *comb. form.* CHEM(I)-.
chemoreceptor (kēmōrisep'tə, kem-), *n.* a sensory nerve-ending which responds to a chemical stimulus.
chemotherapy (kēmōthe'rəpi, kem-), *n.* treatment of disease, esp. cancer, by drugs.
chemurgy (kem'əji), *n.* that branch of chemistry which is devoted to the industrial utilization of organic raw material, esp. farm products.
chenille (shənēl'), *n.* round tufted or fluffy cord of silk or worsted; a pile fabric made with similar yarn.
cheongsam (chongsam'), *n.* a Chinese woman's long, tight-fitting dress with slit sides.
cheque (chek), *n.* a draft on a banker for money payable to bearer or to a third party; the form on which such a draft is written. **crossed cheque,** a cheque marked as negotiable only through a banker. **cheque-book,** *n.* a book containing cheques. **cheque-book journalism,** *n.* sensational journalism, using stories bought at high prices. **cheque (guarantee) card,** *n.* a card issued by a bank, guaranteeing payment of cheques up to a specified limit.
chequer, (*N Am.*) **checker** (chck'ə), *n.* (*usu. pl.*) a pattern made of squares in alternating colours, like a chess-board; (*pl.*, *N Am.*) the game of draughts. *v.t.* to form into a pattern of little squares; to variegate. **checker-board,** *n.* (*N Am.*) a draughts-board; a chess-board. **chequered flag,** *n.* a flag with black and white squares used to signal the winner in a motor race. **chequer-work,** *n.* work executed in diaper pattern or chequers. **chequered,** *a.* having periods of good and bad luck, events etc. (as a career).
cherish (che'rish), *v.t.* to hold dear, to treat with affection, to hold closely to, to cling to (a hope etc.). **cherishable,** *a.*
cheroot (shəroot'), *n.* a cigar with both ends cut square off.
cherry (che'ri), *n.* a small stone-fruit of the plum family; the tree on which it grows; the wood of this tree. *a.* of the colour of a red cherry; ruddy. **two bites at the cherry,** a second chance. **cherry brandy,** *n.* brandy in which cherries have been steeped. **cherry-picker,** *n.* a long extending mechanical arm with a platform at its upper end for standing on. **cherry-stone,** *n.* the endocarp of the cherry. **cherry-tree,** *n.* the tree on which the cherry grows. **cherry-wood,** *n.* the wood of the cherry-tree.
chert (chœt), *n.* (an) impure flinty rock. **cherty,** *a.* resembling or containing chert.
cherub (che'rəb), *n.* (*pl.* **-s, -bim, -bims** (-im, -imz)) a celestial spirit next in order to the seraphim; a beautiful child; in art, a winged child. **cherubic** (-roo'-), *a.* of or pertaining to cherubs; angelic; full-cheeked and ruddy. **cherubically,** *adv.*
chervil (chœ'vəl), *n.* a garden pot-herb and salad-herb.
Cheshire (chesh'ə): **to grin like a Cheshire cat,** to grin very widely. **Cheshire cheese,** *n.* a reddish cheese made in Cheshire.
chess (ches), *n.* a game played by two persons with 16 pieces each on a board divided into 64 squares. **chess-board,** *n.* the board on which chess is played. **chess-man,** *n.* one of the pieces used in chess.
chest (chest), *n.* a large box; a case for holding particular commodities (as *tea-chest*); the quantity such a case holds; the coffer, treasury or funds of an institution; the fore part of the human body from the neck to the belly. **chest of drawers,** a movable piece of furniture containing only drawers. **to get something off one's chest,** to unburden oneself (of a secret etc.). **chest-note,** *n.* a deep note sounded from the chest. **-chested,** *comb. form.* having a chest of a specified kind. **chesty,** *a.* (*coll.*) occurring in the chest; suffering from, or subject to, bronchitis etc.
chesterfield (ches'təfēld), *n.* a deeply upholstered sofa.
chestnut (ches'nŭt), *n.* a tree, esp. the Spanish or sweet chestnut, having edible fruit of a reddish-brown colour; this fruit; hence, a reddish-brown colour; a horse of this colour; (*coll.*) a stale joke or anecdote. *a.* reddish-brown.
cheval-glass (shəval'), *n.* a large swing mirror mounted on a frame.
chevalier (shevəliə'), *n.* a member of some foreign orders of knighthood or of the French Legion of Honour.
cheverel (chev'ərəl), *n.*, *a.* (leather) made from kidskin.
cheviot (chē'viət, chev'-), *n.* a sheep bred on the *Cheviot* Hills; rough cloth made from the wool of such sheep.
chevrette (shəvret'), *n.* a thin goatskin leather used for gloves.
chevron (shev'rən), *n.* an inverted V, the distinguishing mark on the coat-sleeves of non-commissioned officers; (*pl.*) a road-sign consisting of a black and white V-shaped pattern giving warning of a bend.
chevrotain, -tin (shev'rətān, -tin), *n.* a small animal allied to the musk-deer.

chew (choo), *v.t.* to masticate, to grind with the teeth; to ruminate on, to digest mentally. *v.i.* to masticate food, tobacco or gum; to meditate. *n.* something, esp. a sweet, which is chewed in the mouth; a quid of tobacco. **to chew over**, to discuss. **to chew the cud** CUD. **to chew the rag**, **fat**, *(sl.)* to grumble, to complain. **chewing-gum**, *n.* a preparation of flavoured insoluble gum for chewing. **chewy**, *a. (coll.)* firm-textured, suitable for chewing.
chez (shā), *prep.* at the house of.
Chianti (kian'ti), *n.* a dry red or white wine from Tuscany.
chiaroscuro (kiahrəskoo'rō), *n.* the treatment of light and shade; effects of light and shade; a drawing in black and white; relief, contrast (in a literary work etc.). *a.* obscure; half-revealed.
chiasmus (kiaz'məs), *n.* inversion of order in parallel phrases, as *I don't live to work, I work to live.*
chibouk, chibouque (chibook'), *n.* a long Turkish pipe for smoking.
chic (shēk), *n.* smartness, style; the best fashion or taste. *a.* stylish; fashionable.
chica[1] (chē'kə), *n.* a red colouring-matter used by S American Indians to stain the skin.
chica[2] (chē'kə), *n.* an old Spanish dance of an erotic character, forerunner of the fandango and bolero.
chicane (shikān'), *n.* an artificial obstacle on a motor-racing track. **chicanery** (-nəri), *n.* (the employment of) mean, petty subterfuge, esp. legal trickery.
chicano (chikah'nō. -kānō), *n. (pl.* **-nos**), *a.* (of, pertaining to) a person of Mexican origin living in the US.
chick (chik), *n.* a young bird about to be hatched or newly hatched; *(coll.)* a little child; *(dated coll., sometimes derog.)* a young woman. **chickabiddy** (-əbidi), *n.* a term of endearment for a child. **chickweed**, *n.* a small weed with white flowers, often fed to cagebirds.
chicken (chik'ən), *n.* the young of the domestic fowl; (the flesh of) the domestic fowl as eaten; *(coll.)* a coward. *a. (coll.)* cowardly. **chicken-and-egg situation**, one in which it is unclear what is cause and what effect. **no (spring) chicken**, older than he or she appears. **to chicken out**, *(coll.)* to lose one's nerve. **to count one's chickens before they are hatched**, to make plans which depend on something uncertain. **chicken-feed**, *n.* feed for chickens; *(coll.)* a trifling matter; *(derog.)* an insignificant sum of money. **chicken-hearted**, *a.* timid, cowardly. **chickenpox**, *n.* a pustulous, contagious disease, usually occurring in childhood. **chicken-wire**, *n.* wire netting with a small hexagonal mesh.
chick-pea (chik'pē), *n.* a dwarf species of pea.
chickweed CHICK.
chicle (chik'l), *n.* the juice of the sapodilla, used in the making of chewing-gum.
chicory (chik'əri), *n.* the succory, a blue-flowered plant, or its root, which, when roasted and ground, is used as a coffee additive; endive.
chide (chīd), *v.t. (past* **chided, chid** (chid), *p.p.* **chided, chid, chidden** (chid'n)) to find fault with, to reprove, to blame. *v.i.* to scold, to fret, to make complaints. **chider**, *n.* **chidingly**, *adv.*
chief (chēf), *a.* principal, first; highest in authority, most senior; most important, leading, main. *n.* (the title given to) a leader or commander, esp. the leader of a tribe or clan; the prime mover, the principal agent; the head of a department. **Chief Constable**, *n.* the highest-ranking police officer in a given force. **chief justice**, *n.* in several countries, the judge presiding over the highest court. **chief of staff**, *n.* the senior officer of a division of the armed forces. **chiefdom**, *n.* **chiefless**, *a.* **chiefly**, *adv.* principally, especially; for the most part. **chiefship**, *n.* **-in-chief**, *comb. form.* leading, most important, as **commander-in-chief**.
chieftain (chēf'tən), *n.* a general, a leader; the head of a tribe or a Highland clan. **chieftainess**, *n. fem.* **chieftaincy**,

-ry, -ship, *n.*
chiff-chaff (chif'chaf), *n.* a variety of European warbler.
chiffon (shif'on), *n.* a gauzy, semi-transparent silk or nylon fabric. *a.* made of chiffon; of puddings, having a fine, light consistency. **chiffonier** (-niə'), *n.* a movable piece of furniture serving as a cupboard and sideboard.
chignon (shēn'yō), *n.* a coil or knot of long hair at the back of the head.
chihuahua (chiwah'wah), *n.* a very small dog with large eyes and pointed ears.
chilblain (chil'blān), *n. (usu. pl.)* an inflamed swelling of the hands or feet caused by bad circulation and cold. **chilblained**, *a.* **chilblainy**, *a.*
child (chīld), *n. (pl.* **children** (chil'drən)) a descendant in the first degree; a boy; a girl; an infant; a young person; a son or daughter; one young in experience, judgment or attainments; *(pl.)* descendants; a person whose character is the result (of a specified environment etc.). **second childhood**, dotage. **with child**, *(dated)* pregnant. **child abuse**, *n.* (esp. physical) maltreatment of a child. **child-bearing**, *a.* (suitable for) bringing forth children. *n.* the act of bearing children. **child benefit**, *n.* a sum of money paid regularly by government to the parent of a child under a certain age. **childbirth**, *n.* the time or act of bringing forth a child. **child-minder**, *n.* a person whose profession is looking after other (working) persons' children. **child-proof**, *a.* of bottle tops, locks etc., designed to be impossible for a child to operate or open. **child's-play**, *n.* easy work. **childhood**, *n.* the state of being a child; the period from birth till puberty. **childish**, *a.* of or befitting a child; silly, puerile. **childish-minded**, *a.* **childishly**, *adv.* **childishness**, *n.* **childless**, *a.* without child or offspring. **childlessness**, *n.* **childlike**, *a.* resembling or befitting a child; docile, simple, innocent.
chili CHILLI.
chiliad (kil'iad), *n.* a thousand; a thousand years.
chill (chil), *n.* coldness, a fall in bodily temperature; a cold; a cold, shivering sensation preceding fever or ague; (a) discouragement. *v.t.* to make cold; to preserve (meat etc.) by cold; to cool (metal) suddenly so as to harden; to depress, to dispirit, to discourage. *v.i.* to become cold. *a.* cold; causing a sensation of coolness; coldly formal. **to chill out**, *(sl.)* to relax to a marked degree, sometimes under the influence of drugs; to hang around aimlessly, as on street corners. **to take the chill off**, to warm slightly. **chiller**, *n.* a chilled container for food or drink; *(coll.)* a frightening novel, film etc. **chilling**, *a.* making cold; depressing, distant in manner. **chillingly**, *adv.* **chillness**, *n.* **chilly**, *a.* rather cold; susceptible of cold; cold or distant in manner. **chilliness**, *n.*
chilli, chili (chil'i), *n. (pl.* **-(l)ies**) (also **chilli pepper**) the dried ripe pod of red pepper. **chilli con carne** (kon kah'ni-), *n.* a Mexican dish of minced meat with beans in a chilli sauce.
Chiltern Hundreds (chil'tən), *n.pl.* certain Crown lands in Buckinghamshire and Oxfordshire, the nominal stewardship of which is granted to a Member of Parliament who wishes to vacate his or her seat. **to apply for the Chiltern Hundreds**, to resign membership of the House of Commons.
chimaera CHIMERA.
chime (chīm), *n.* the harmonic or consonant sounds of musical instruments or bells; a number of bells tuned in diatonic succession; the sounds so produced. *v.i.* to sound in harmony or accord; of bells, to ring; to strike the hour etc.; to accord, to agree (with). *v.t.* to ring (a series of bells); to ring a chime on (bells); to cause to sound in harmony; to recite musically or rhythmically. **to chime in**, to join in; to express agreement.
chimera, chimaera (kimiə'rə, ki-), *n.* a fabulous fire-eating monster, with a lion's head, a serpent's tail, and the body of a goat; any incongruous conception of the fancy; an

chimichanga — chivalry

imaginary terror. **chimerical** (-me'-), *a*. purely imaginary. **chimerically**, *adv*.

chimichanga (chimichang'gə), *n*. a deep-fried flour tortilla, typically filled with beef, pork or chicken.

chimney (chim'ni), *n*. the flue, vent or passage through which smoke escapes from a fire or engine into the open air; a glass tube placed over the flame of a lamp to intensify combustion; a vent from a volcano; a vertical or nearly vertical fissure in rock. **chimney-breast**, *n*. the projecting part of the wall of a room containing the fireplace and lower chimney. **chimney-corner**, *n*. a nook or seat beside the fire, esp. beside a wide, old-fashioned fireplace. **chimney-piece**, *n*. a mantelpiece. **chimney-pot**, *n*. a tube of pottery or sheet-metal carried up above the chimney-shaft to prevent smoking; (*dated coll.*) a tall silk hat. **chimney-stack**, *n*. a number of chimney-stalks united in a block of masonry or brickwork; a tall factory chimney. **chimney-stalk, -top,** *n*. the part of the chimney-stack carried up above the roof. **chimney-sweep**, *n*. one whose business is to sweep chimneys.

chimpanzee (chimpanzē'), **chimp** *n*. an intelligent African anthropoid ape.

chin (chin), *n*. the front part of the lower jaw. **to keep one's chin up,** (*coll.*) to remain cheerful in adversity. **chinwag,** *n*. (*coll.*) chat, talk. *v.i.* (*coll.*) to chat. **chinless,** *a*. (*coll.*) having a receding chin; weak-spirited, ineffectual. **chinless wonder,** *n*. (*derog. sl.*) an upper-class idiot.

China (chī'nə), *a*. of or belonging to China. **China ink,** *n*. a black solid which, when mixed with water, yields a black indelible ink. **Chinaman,** *n*. (*derog.*) a native of China, or one of Chinese blood. **China tea,** *n*. a smoky-flavoured tea from China. **Chinatown,** *n*. the Chinese quarter of a town. **Chinese** (-nēz'), *n*. a native of China, or one of Chinese blood; the language of the Chinese. *a*. of or belonging to China. **Chinese restaurant syndrome,** a disorder caused by eating too much monosodium glutamate, common in Chinese food. **Chinese cabbage, leaves,** *n*. a vegetable with crisp leaves, like a cabbage. **Chinese chequers,** *n*. a board game like draughts. **Chinese gooseberry** KIWI FRUIT. **Chinese lantern,** *n*. a collapsible lantern made of thin paper.

china (chī'nə), *n*. porcelain, first brought from China; porcelain ware. *a*. made of porcelain. **china-clay**, *n*. kaolin. **china-closet,** *n*. a cupboard for storing china-ware. **china-ware,** *n*. articles made of china.

chinchilla (chinchil'ə), *n*. (the fur of) a genus of S American rodents; a fur coat made of this.

chin-chin (chinchin'), *n*. (*dated coll.*) a familiar form of salutation or health-drinking.

chinchona (chingkō'nə), CINCHONA.

chindit (chin'dit), *n*. a commando in Burma during World War II.

chine¹ (chīn), *n*. the backbone or spine of any animal; a ridge (e.g. of hills) resembling this. *v.t.* to cut or break the backbone of. **chined,** *a*. (*usu. in comb.*) having a backbone.

chine² (chīn), *n*. (*S Eng. dial.*) a deep and narrow ravine.

Chinese CHINA.

chink¹ (chingk), **Chinkie** (ching'ki), *n., a*. (*offensive*) (a) Chinese.

chink² (chingk), *n*. a narrow cleft or crevice; a small longitudinal opening; a slit.

chink³ (chingk), *n*. a jingling sound, as of coins. *v.t.* to cause to jingle. *v.i.* to emit a jingling sound.

chino (chē'nō), *n*. (*pl.* **-nos**) a tough, twilled cotton fabric; (*pl.*) trousers, often off-white, made of this fabric.

Chinook (chinook'), *n*. a jargon of American Indian and European words used in intercourse between traders and Indians in the region of the Columbia River; (**chinook**) a warm west wind from the Pacific Ocean occurring in the Rocky Mountains.

chintz (chints), *n*. printed cotton cloth with floral devices etc., usu. glazed. **chintzy,** *a*. like, or covered with, chintz; (*coll.*) gourdy, esp. with flowery patterns.

chip (chip), *n*. a small piece of wood, stone etc. detached or chopped off; a thin strip of wood; a thin fragment; a potato chip; a playing-counter used in gambling games; a very small piece of semiconducting material, esp. silicon, with an integrated circuit on it; a chip-shot. *v.t.* (*past, p.p.* **chipped**) to cut into chips; to cut or break chips off; to crack; to perform a chip-shot on. *v.i.* to break or fly off in chips; to play a chip-shot. **to have had one's chips,** (*coll.*) to have run out of luck, be about to fail, die etc. **when the chips are down,** when things are at their most desperate, their lowest ebb. **chip off the old block,** a son resembling his father. **to chip in,** (*coll.*) to cut into a conversation; to contribute (money). **to have a chip on one's shoulder,** to nourish a grievance. **chipboard,** *n*. a thin board made of compressed wood fragments. **chip-shot,** *n*. in football or golf, a shot played so that the ball travels high but not far. **chippy,** *n*. (*coll.*) a fish-and-chip shop; (*coll.*) a carpenter. **chippiness,** *n*.

chipmunk (chip'mŭngk), *n*. a N American rodent like the squirrel.

chipolata (chipəlah'tə), *n*. a small sausage.

Chippendale (chip'əndāl), *a*. applied to furniture of the style introduced by Chippendale about the middle of the 18th cent.

chipper (chip'ə), *a*. (*coll.*) energetic and cheerful; smart.

chirognomy (kīrog'nəmi), *n*. judgment of character from the lines in the hand.

chirography (kīərog'rəfi), *n*. character and style in handwriting. **chirographic, -ical** (-graf'-), *a*.

chirology (kīrol'əji), *n*. the art or practice of conversing by signs made with the hands or fingers; finger-speech. **chirologist,** *n*.

chiromancy (kī'rəmansi), *n*. divination by means of the hand; palmistry. **chiromancer,** *n*. **chiromantic** (-man'-), *a*.

chiropodist (kirop'ədist, shi-), *n*. one skilled in the care of the hands and feet, esp. in the removal of corns etc. **chiropody,** *n*.

chiropractic (kīrəprak'tik), *n*. spinal manipulation as a method of curing disease. **chiropractor** (kī'-), *n*.

Chiroptera (kīrop'tərə), CHEIROPTERA.

chirp (chœp), *v.i.* to make a quick, sharp sound (as birds and their young, insects etc.); to talk cheerfully. *v.t.* to utter or sing with a sharp, quick sound. *n*. a sharp, quick sound of a bird; a sound resembling this. **chirpingly,** *adv*. **chirpy,** *a*. cheerful; vivacious. **chirpiness,** *n*.

chirr (chœ), *v.i.* to make a trilling monotone sound like that of the grasshopper.

chirrup (chi'rəp), *v.i.* to chirp, to make a twittering sound. **chirruper,** *n*. **chirrupy,** *a*. cheerful, chatty.

chisel (chiz'əl), *n*. an edged tool for cutting wood, iron or stone, operated by pressure or striking. *v.t.* to cut, pare or grave with a chisel; (*sl.*) to take advantage of, to cheat. **chiselled,** *a*. cut with or as with a chisel; clear-cut. **chiseller,** *n*.

chit¹ (chit), *n*. (*coll.*) a child; a young thing; (*derog.*) a young girl.

chit² (chit), **chitty** (chit'i), *n*. a voucher; a receipt.

chit-chat (chit'chat), *n*. trifling talk; chat, gossip. *v.i.* to chat, gossip.

chitin (kī'tin), *n*. the horny substance that gives firmness to the integuments of crustaceans, arachnidans and insects. **chitinous,** *a*.

chitterlings (chit'əlingz), *n.pl.* the smaller intestines of animals, esp. as prepared for food.

chivalry (shiv'əlri), *n*. the knightly system of the Middle Ages; the ideal qualities which inspired it, nobleness and gallantry of spirit, courtesy, respect for and defence of the weak, gallantry, devotion to the service of women. **chivalric** (-val'-), *a*. pertaining to chivalry; gallant. **chivalrous,** *a*. gallant, noble; courteous. **chivalrously,** *adv*.

chive (chīv), *n.* a small onion-like herb, the leaves of which are eaten.
chivvy (chiv′i), *v.t.* (*coll.*) (esp. with *along* or *up*) to chase, to hurry.
chlor- CHLOR(O)-.
chloral hydrate (klaw′rəl), *n.* a white crystalline substance obtained from chloral, used as a hypnotic and anaesthetic.
chlorate (klaw′rāt), *n.* a salt of chloric acid.
chloric (klaw′rik), *a.* pertaining to pentavalent chlorine. **chloric acid**, *n.* an acid containing hydrogen, chlorine and oxygen.
chloride (klaw′rīd), *n.* a compound of chlorine with another element. **chloride of lime**, a compound of chlorine with lime, used as a disinfectant and for bleaching. **chloridate, -dize, -dise**, *v.t.* to treat or prepare (as a photographic plate) with a chloride.
chlorine (klaw′rēn), *n.* a yellow-green, poisonous, gaseous element, at. no. 17; chem. symbol Cl, obtained from common salt, used as a disinfectant and for bleaching. **chlorinate** (-ri-), *v.t.* **chlorination**, *n.* the extraction of gold by exposure of ore to chlorine gas; the sterilization of water with chlorine.
chlorite (klaw′rīt), *n.* a green silicate mineral. **chloritic** (-rit′-), *a.*
chlor(o)-, *comb. form.* of a green colour; denoting a chemical compound in which chlorine has replaced some other element.
chloroform (klo′rəfawm), *n.* a volatile fluid formerly used in surgery as an anaesthetic. *v.t.* to administer chloroform to; to render insensible with chloroform.
chlorophyll (klo′rəfil), *n.* the green colouring-matter of plants which absorbs the energy from sunlight, used in producing carbohydrates from water and carbon dioxide.
chloroplast (klor′əplast), *n.* a plastid containing chlorophyll.
ChM, (*abbr.*) Master of Surgery.
choc (chok), *n.* (*coll.*) short for CHOCOLATE. **choc-ice**, *n.* a bar of (vanilla) ice-cream coated with chocolate.
chock (chok), *n.* a wood block, esp. a wedge-shaped block used to prevent a cask or other body from shifting. *v.t.* to wedge, support, make fast, with a chock or chocks; to place a boat on chocks. **chock-a-block**, chock-full. **chock-full**, *adv.* full to overflowing. **chocker**, *a.* (*dated coll.*) full up; crammed.
chocoholic (chokōhol′ik), *n.* (*facet.*) a person obsessively fond of eating chocolate.
chocolate (chok′lit), *n.* an edible paste made from the roasted, ground seeds of the cacao tree; a sweetmeat made of or coated with this paste; (also **hot chocolate, drinking chocolate**) a drink made of this paste dissolved in hot water or milk; a dark brown colour. *a.* made of or flavoured with chocolate; of the colour of chocolate. **milk chocolate**, chocolate prepared with a great deal of milk. **plain chocolate**, chocolate that is less creamy and sweet than milk chocolate. **chocolate-box**, *a.* sentimentally pretty.
choice (chois), *n.* the power or act of choosing; the person or thing chosen; the things to be selected from; selection, preference; the best and preferable part. *a.* selected, picked, chosen with care; of great value. **for choice**, preferably. **Hobson's choice** (hob′sənz), no alternative. **to have no choice**, to have no option or alternative. **choicely**, *adv.* **choiceness**, *n.*
choir (kwīə), *n.* an organized body of singers, esp. in a church or chapel; the part of the church or chapel allotted to the singers; the part of a cathedral or large church where service is performed, the chancel. **choirboy**, *n.* a boy singer in a church choir. **choir-screen**, *n.* a screen of lattice-work, wood or other open work separating the choir from the nave.
choke (chōk), *v.t.* to block or compress the windpipe (of), so as to prevent breathing; to suffocate (as by gas, water etc.); to repress, to silence (often with *back, down*); to stop up, to obstruct, to clog. *v.i.* to have the windpipe stopped; to be wholly or partially suffocated; to be blocked up. *n.* a noise of suffocation in the throat; an inductance coil constructed to prevent high-frequency currents from passing; a device to prevent the passage of too much air to a carburettor. **to choke off**, to discourage, to suppress. **to choke up**, to fill up until blocked. **choked**, *a.* (*coll.*) disappointed; angry. **choker**, *n.* one who or that which chokes; a necklace that fits closely round the neck.
choky[1], *a.* having a sensation of choking.
choky[2], **chokey** (chō′ki), *n.* (*coll.*) a lock-up, police-station; a prison.
choler (kol′ə), *n.* bile, the humour supposed to cause irascibility of temper; (tendency to) anger. **choleric**, *a.* full of choler; irascible, passionately angry.
cholera, Asiatic cholera (kol′ərə), *n.* an acute, often fatal, bacterial infection, spread by contaminated water supplies in which severe vomiting and diarrhoea cause dehydration. **choleraic** (-rā′-), *a.* **choleroid**, *a.*
choleric CHOLER.
cholesterol (kəles′tərol), (*formerly*) **cholesterin**, *n.* a white solid alcohol occurring in gall-stones, nerves, blood etc., thought to be a cause of arteriosclerosis. **cholesteric** (koləste′-), *a.*
choli (chō′li), *n.* an Indian woman's garment, a short tight-fitting bodice worn under a sari.
chomp (chomp), CHAMP[1].
chondr(i)-, chondro- *comb. form* composed of or pertaining to cartilage.
chondrify (kon′drifī), *v.t.* to be converted into cartilage. **chondrification** (-fi-), *n.*
chondrine (kon′drin), *n.* gelatine from the cartilage of the ribs, joints etc.
chondrite (kon′drīt), *n.* a meteorite containing stony granules.
chondritis (kondrī′tis), *n.* inflammation of cartilage.
chondro- CHONDRI-.
chondroid (kon′droid), *a.* like cartilage.
choose (chooz), *v.t.* (*past* **chose** (chōz), *p.p.* **chosen** (chō′zn)) to take by preference, to select (from a number or from options). *v.i.* to feel inclined, to prefer (to do something rather than something else); to decide willingly (to do); to make one's choice; to have the power of choice. **cannot choose but**, has/have no alternative but. **to pick and choose**, to make a careful choice, to be overparticular. **chooser**, *n.* **choosy, -sey**, *adv.* (*coll.*) hard to please, particular. **choosingly**, *adv.*
chop[1] (chop), *v.t.* (*past, p.p.* **chopped**) to cut off suddenly; to strike off; to cut short or into parts; to strike (a ball) with backspin; (*coll.*) to reduce or abolish. *v.i.* to do anything with a quick motion like that of a blow. *n.* the act of chopping; a cutting stroke; a piece chopped off; a rib (of a sheep or pig) chopped off and cooked separately. **the chop**, (*sl.*) dismissal (from a job etc.). **to chop up**, to cut into small pieces, to mince. **chop-house**, *n.* (*dated*) a restaurant specializing in chops and steaks. **chopper**, *n.* one who or that which chops; a butcher's cleaver; an axe; (*sl.*) a helicopter; (*sl.*) a motorcycle or bicycle with very high handlebars. **chopping-block**, *n.* a wooden block on which anything is chopped. **choppy**, *a.* of the sea, rough, with short quick waves.
chop[2] (chop), *v.i.* to shift suddenly, as the wind. **to chop and change**, to vary continuously; to fluctuate. **to chop logic**, to wrangle pedantically. **chop logic**, *n.* pedantic and complicated argument(s). **choppy**, *a.* variable, continually changing.
chop[3] CHAP[2].
chop-chop (chopchop′), (*coll.*) at once, quickly.
chopsticks (chop′stiks), *n.pl.* two small sticks of wood or ivory used by the Chinese and Japanese to eat with.
chop suey (chop soo′i), *n.* a Chinese dish of shredded

choragus (kərā'gəs), *n.* the leader or director of the chorus in the ancient Greek theatrical performances. **choragic** (-raj'-), *a.*
choral[1] (kaw'rəl), *a.* belonging to or sung by a choir or chorus; chanted or sung. **chorally,** *adv.*
choral[2], **chorale** (kərahl'), *n.* a simple choral hymn or song, usually of slow rhythm and sung in unison.
chord[1] (kawd), *n.* the string of a musical instrument; a straight line joining the extremities of an arc or two points in a curve; (*Anat.*) CORD. **chordal,** *a.* **Chordata** (-dā'tə), *n.* a phylum of the animal kingdom, animals with a backbone or notochord. **chordate** (-dāt), *n., a.* (a member) of the Chordata.
chord[2] (kawd), *n.* the simultaneous and harmonious sounding of notes of different pitch. **chordal,** *a.*
chore (chaw), *n.* a small, regular (often boring) task; a daily or other household job.
chorea (kərē'ə), *n.* a nervous disorder characterized by irregular convulsive movements of an involuntary kind, St Vitus's dance.
choree (korē'), *n.* a trochee, a metrical foot consisting of a long syllable followed by a short one.
choreograph (koriəgrahf, -af), *v.t.* to compose or arrange the steps of (a stage dance or ballet). **choreographer** (-og'-), *n.* the composer or designer of a stage dance or ballet. **choreographic** (-graf'-), *a.* **choreography** (-og'-), *n.* (an example of) the art of dance design.
choriamb (ko'riamb), **-ambus,** *n.* (*pl.* **-ambs, -ambi** (-bī)) a metrical foot of four syllables, of which the first and fourth are long, and the second and third short. **choriambic** (-am'-), *n., a.* (pertaining to or of the nature of) a choriamb.
choric CHORUS.
chorion (kaw'rin), *n.* the outer membrane which envelops the foetus in the womb; the external membrane of a seed. **chorionic** (-on'-), *a.* **chorionic villus sampling,** a diagnostic test for detecting abnormalities in a foetus, whereby small pieces of the chorion are removed and examined.
chorister (ko'ristə), *n.* a singer; one who sings in a choir, esp. a choirboy.
chorography (kərog'rəfi), *n.* the art or practice of describing and making maps of particular regions or districts. **chorographer,** *n.* **chorographic, -ical** (-graf'-), *a.* **chorographically,** *adv.*
chorology (kərol'əji), *n.* the science of the geographical distribution of plants and animals. **chorological** (-loj'-), *a.*
chortle (chaw'tl), *v.i.* to make a loud chuckle. *v.t.* to utter with a loud chuckle. *n.* a loud chuckle.
chorus (kaw'rəs), *n.* a band of dancers and singers in the ancient Greek drama; the song or recitative between the acts of a Greek tragedy; (the speaker of) the prologue and epilogue in an Elizabethan play; a band of persons singing or dancing in concert; a piece of vocal music for such a band of singers; the refrain of a song in which the company joins the singer. **chorus girl,** *n.* a young woman who sings or dances in the chorus in a musical comedy etc. **chorus master,** *n.* the director of a band of singers. **choric** (ko'-), *a.* pertaining to a chorus; like the chorus in a Greek play.
chose, *past,* **chosen,** *p.p.* CHOOSE.
chough (chŭf), *n.* a large black bird of the crow family, with a red bill.
choux pastry (shoo pās'tri), *n.* a rich light pastry made with eggs.
chow[1] (chow), *n.* (*coll.*) food.
chow[2] (chow), **chow-chow** (chow'chow), *n.* an orig. Chinese breed of large dog with thick coat and curled tail.
chowder (chow'də), *n.* a thick soup or stew made of fish, bacon etc.
chow mein (chow mān'), *n.* a Chinese dish of meat and vegetables served with fried noodles.

Chr., (*abbr.*) Christ(ian).
chrematistic (krēmətis'tik), *a.* concerning money-making. **chrematistics,** *n.sing.* political economy so far as it relates to the production of wealth.
chrestomathy (krestom'əthi), *n.* a selection of passages with notes etc., to be used in learning a foreign language. **chrestomathic** (-math'-), *a.*
chrism (kriz'm), *n.* consecrated oil, used in the Roman and Greek Churches in administering baptism, confirmation, ordination and extreme unction. **chrismal,** *a.* **chrismatory,** *n.* a vessel for holding chrism.
Christ (krist), *n.* the Anointed One, a title given to Jesus the Saviour, and synonymous with the Hebrew Messiah. *int.* (*taboo*) expressing anger, annoyance etc. **Christhood,** *n.* **Christless,** *a.* without faith in or without the spirit of Christ. **Christlessness,** *n.* **Christlike,** *a.* **Christlikeness,** *n.* **Christly,** *a.* **Christward, -wards,** *adv.* [OE *Crist,* L *Christus,* Gr. *Christos* (*chriein,* to anoint)]
Christadelphian (kristədel'fiən), *n., a.* (a member) of a millenarian Christian sect, calling themselves the brethren of Christ, and claiming apostolic origin.
christen (kris'n), *v.t.* to receive into the Christian Church by baptism; to baptize; to name; to nickname. *v.i.* to administer baptism. **christening,** *n.* an act, or the ceremony, of baptism.
Christendom (kris'ndəm), *n.* that portion of the world in which Christianity is the prevailing religion; Christians collectively.
Christian (kris'chən), *n.* one who believes in or professes the religion of Christ; one belonging to a nation or country of which Christianity is the prevailing religion; one whose character is consistent with the teaching of Christ; (*facet.*) a civilized person as distinguished from a savage. *a.* pertaining to Christ or Christianity; professing the religion of Christ; Christlike; civilized. **Christian Democrat,** *n.* a member of a moderate Roman Catholic Church political party in Belgium, France, Italy, Germany etc. **Christian era,** *n.* the chronological period since the birth of Christ. **Christian name,** *n.* a name given in baptism, a first name, as distinct from a surname. **Christian Science,** *n.* a religion based on the belief that diseases are the result of wrong thinking and can be healed without medical treatment. **Christian Scientist,** *n.* **Christianity** (-tian'-), *n.* (faith in) the doctrines and precepts taught by Christ; Christian character and conduct; the state of being a Christian. **christianize, -ise,** *v.t.* to convert to Christianity. *v.i.* to be converted to Christianity. **christianization,** *n.* **christianlike,** *a.* **christianly,** *a., adv.* **Christiano-,** *comb. form.*
Christiania (kristiah'niə), **Christie** (kris'ti), *n.* a turn in skiing in which the skis are kept parallel, used esp. for stopping or turning sharply.
Christmas (kris'məs), *n.* the festival of the nativity of Jesus Christ celebrated on 25 Dec.; Christmastide. *a.* pertaining or appropriate to Christmas or its festivities. **Christmasbox,** *n.* a present or tip given at Christmas, esp. to tradesmen. **Christmas bush,** *n.* (*Austral.*) a tree that comes into flower about Christmastime, with bright red blooms. **Christmas cactus,** *n.* a S American branching cactus which produces red flowers in winter. **Christmas cake,** *n.* a type of heavy, iced fruit cake made at Christmastime. **Christmas card,** *n.* an ornamental card sent as a Christmas greeting. **Christmas carol,** *n.* a song of praise sung at Christmas. **Christmas Day,** *n.* the festival of Christmas, 25 Dec. **Christmas Eve,** *n.* the day before Christmas Day, 24 Dec. **Christmas number,** *n.* a special issue of a magazine or other periodical at Christmastime. **Christmas pudding,** *n.* a rich pudding made at Christmastime, usu. containing fruit, spices and suet. **Christmas rose,** *n.* a white-flowered hellebore flowering in winter. **Christmas stocking,** *n.* a stocking hung up by children at Christmas as a container for small presents. **Christmastide, -time,** *n.* the season of Christmas. **Christmas tree,** *n.* an evergreen or artificial

tree kept indoors and decorated at Christmastime. **Christmasy, -massy**, *a.*
Christo-, *comb. form.* pertaining to Christ.
chrom- CHROM(O)-.
chromat- CHROMAT(O)-.
chromate (krō'māt), *n.* a salt of chromic acid.
chromatic (krəmat'ik), *a.* relating to colour; coloured; (*Mus.*) including notes not belonging to the diatonic scale; increasing by semitones. **chromatic printing**, *n.* colour printing. **chromatic scale**, *n.* a succession of notes a semitone apart. **chromatic semitone**, *n.* the interval between a note and its flat or sharp. **chromatically**, *adv.* **chromaticity** (-tis'-), *n.* **chromatics**, *n. sing.* the science of colour.
chromat(o)-, *comb. form.* pertaining to colour.
chromatography (krōmətog'rəfi), *n.* a technique for separating or analysing the components of a mixture which relies on the differing capacity for adsorption of the components (in a column of powder, strip of paper etc.). **chromatogram** (-mat'əgram), *n.* the visual record produced by separating the components of a mixture by chromatography.
chromatrope (krō'mətrōp), *n.* a rotating magic-lantern slide for producing a kaleidoscopic effect.
chrome (krōm), *n.* chromium; a pigment containing chromium. *v.t.* to plate with chromium; to treat with a chromium compound. **chrome-green**, *n.* a dark green pigment obtained from oxide of chromium. **chrome steel**, *n.* a kind of steel containing chromium. **chrome-yellow**, *n.* chromate of lead; a brilliant yellow pigment.
chromite (krō'mīt), *n.* a mineral consisting of chromium and iron oxide.
chromium (krō'miəm), *n.* a bright steel-grey metallic element, at. no. 24; chem. symbol Cr, remarkable for the brilliance of colour of its compounds, used as a protective plating. **chromium-plated**, *a.* electroplated with chromium to give a shiny appearance. **chromic**, *a.* pertaining to or containing chromium.
chrom(o)- (krō'mō), *comb. form.* pertaining to colour.
chromograph (krō'məgrahf), *n.* an apparatus for reproducing writing or drawing in colours by lithography from an impression on gelatine. *v.t.* to make copies in this way.
chromolithograph (krōməlith'əgrahf), *n.* a picture printed in colours by lithography. **chromolithographic** (-graf'-), *a.* **chromolithographer** (-og'-), *n.* **chromolithography**, *n.*
chromosome (krō'məsōm), *n.* any of the microscopic rod-shaped structures in a cell nucleus that carry the genes which transmit hereditary characteristics. **chromosomal** (-sō'-), *a.*
chromosphere (krō'məsfiə), *n.* the gaseous envelope of the sun through which light passes from the photosphere.
chromotypography (krōmətīpog'rəfi), *n.* colour printing.
chronic (kron'ik), *a.* applied to (persons suffering from) diseases of long duration, or apt to recur; (*coll.*) very bad, severe. **chronically**, *adv.* **chronicity** (-nis'-), *n.*
chronicle (kron'ikl), *n.* a register or history of events in order of time. *v.t.* to record in a chronicle. **Chronicles**, *n.pl.* the two books of the Old Testament immediately following I and II Kings. **chronicler**, *n.*
chrono-, *comb. form.* pertaining to time or dates.
chronogram (kron'nəgram), *n.* a device by which a date is given by taking the letters of an inscription which coincide with Roman numerals and printing them larger than the rest; thus, GEORGIVs DVX BVCKINGAMMIAE (1 + 5 + 500 + 5 + 10 + 5 + 100 + 1 + 1000 + 1) = 1628 when the Duke was murdered. **chronogrammatic** (-mat'-), *a.*
chronograph (krō'nəgrahf), *n.* an instrument for measuring and registering minute portions of time with great precision; a stop-watch. **chronographer** (-nog'-), *n.* a chronicler; a chronologer. **chronography**, *n.* a description of past events. **chronographic** (-graf'-), *a.* pertaining to a chronograph.

chronology (krənol'əji), *n.* the science of computing time; an arrangement of dates of historical events; a tabular list of dates. **chronologer, -gist**, *n.* **chronological** (kronəloj'-), *a.* arranged in time sequence; pertaining to time. **chronological age**, *n.* age in years, as opposed to mental etc. age. **chronologically**, *adv.*
chronometer (krənom'itə), *n.* an instrument such as a sundial, clock or watch that measures time, esp. one that measures time with great exactness, such as is used to determine the longitude at sea by the difference between its time and solar time. **chronometric, -ical** (kronəmet'-), *a.* **chronometrically**, *adv.* **chronometry**, *n.*
chronoscope (kron'əskōp), *n.* an instrument for measuring the velocity of projectiles.
chrys- CHRYSO-.
chrysalis, -alid (kris'əlis, -əlid), *n.* (*pl.* **-lises, -lides** (-dēz), **-lids**) the last stage through which a lepidopterous insect passes before becoming a perfect insect; the pupa, the shell or case containing the imago.
Chrysanthemum (krisan'thəməm, -zan'-), *n.* a genus of composite plants containing the ox-eye daisy, the corn-marigold and the garden chrysanthemum; (**chrysanthemum**; often shortened to **chrysanth**) any of the cultivated varieties of the garden variety of this genus.
chryselephantine (kriselətan'tīn), *a.* made partly of gold and partly of ivory; overlaid with gold and ivory.
chrys(o)-, *comb. form* golden; of a bright yellow colour.
chrysoberyl (kris'əberəl, -be'-), *n.* a gem of a yellowish-green colour, composed of beryllium aluminate.
chrysolite (kris'əlīt), *n.* a green-coloured translucent orthorhombic mineral; olivine.
chub (chŭb), *n.* (*pl.* in general **chub**; in particular **chubs**) a coarse river-fish, also called the chevin; applied to various American fishes.
chubby (chŭb'i), *a.* fat, plump (esp. in the face). **chubbiness**, *n.*
Chubb® (chŭb), *n.* name of a tumbler-lock. [from the inventor, Charles *Chubb*, 1772–1846]
chuck[1] (chŭk), *n.* the call of a hen to her chickens. *v.i.* to make such a noise. *v.t.* to call, as a hen does her chickens.
chuck[2] (chŭk), *n.* a slight, friendly tap under the chin; a toss or throw. *v.t.* to tap gently and affectionately under the chin; to fling, to throw; (*coll.*) to resign (a job etc.); (*coll.*) to sack (from a team etc.). **to chuck away**, (*coll.*) to discard. **to chuck out**, (*sl.*) to eject forcibly (from a public meeting, licensed premises etc.); to throw out (rubbish etc.). **to chuck up**, (*sl.*) to abandon. **chucker-out**, *n.* (*coll.*) a bouncer.
chuck[3] (chŭk), *n.* an appendage to a lathe for holding the work to be turned, or to a drill for holding the bit; (also **chuck steak**) a cut of beef from the neck and shoulder; (*coll.*, esp. N Am.) food. **chuck key**, *n.* an instrument for tightening or loosening a chuck. **chuck wagon**, *n.* (*N Am.*) a wagon carrying food, cooking utensils etc.
chuck[4] (chŭk), *n.* darling, dear.
chuckle (chŭk'l), *v.i.* to laugh to oneself; to make a half-suppressed sound of laughter. *n.* such a laugh or call.
chuff (chŭf), *v.i.* to make a short puffing sound, as of a steam locomotive; to move while making such sounds. *n.* such a sound. **chuff-chuff**, *n.* a child's word for a steam locomotive.
chuffed (chŭft), *a.* (*coll.*) pleased, happy.
chug (chŭg), *n.* a short dull explosive sound, as of an internal-combustion engine. *v.i.* (*past, p.p.* **chugged**) to make such a noise; to move while making such a noise.
chukka, chukker (chŭk'ə), *n.* name of each of the periods into which a polo game is divided.
chum (chŭm), *n.* a comrade and close companion. **to chum up** (*past, p.p.* **chummed**), to become close friends (with). **chummy**, *a.*
chump (chŭmp), *n.* (*dated sl.*) a silly fellow; a cut of meat

chunder

from the loin and hindleg. **off one's chump,** crazy.
chump chop, *n.* a thick chop from the chump.
chunder (chŭn'də), *v.i.* (*esp. Austral., sl.*) to vomit.
chunk (chŭngk), *n.* a short, thick lump of anything; a (large) portion. **chunky,** *a.* containing or consisting of chunks; (*coll.*) small and sturdy.
Chunnel (chŭn'əl), *n.* (*coll.*) a tunnel under construction between England and France beneath the English Channel. [*Channel tunnel*]
chunter (chŭn'tə), *v.i.* (*coll.*; esp. with *on*) to talk at length and irrelevantly.
church (chœch), *n.* a building set apart and consecrated for Christian worship; a body of Christian believers worshipping in one place, with the same ritual and doctrines; Christians collectively; a section of Christians organized for worship under a certain form; the whole organization of a religious body or association; the clergy as distinct from the laity; divine service; ecclesiastical authority or influence; *a.* of or pertaining to church; ecclesiastical. **Church of England,** the English established or Anglican Church. **Church of Scotland,** the established Church of Scotland. **Free Church of Scotland,** the church formed at the disruption of the Scottish Church in 1843. **poor as a church mouse** POOR. **to go into, enter the church,** to take Holy Orders. **Church Army,** *n.* an organization in the Church of England based on the Salvation Army. **church-burial,** *n.* burial according to the rites of the church. **Church Commissioners,** *n.pl.* a body of administrators who manage the finances and property of the Church of England. **church-goer,** *n.* a regular attendant at church. **church-going,** *n.* the practice of regularly attending divine service. *a.* habitually attending divine service. **church-living,** *n.* a benefice. **churchman,** *n.* a cleric, an ecclesiastic. **churchwoman,** *n. fem.* **church member,** *n.* one eligible for communion in a church. **church membership,** *n.* **Church militant,** *n.* Christians (collectively) on earth, regarded as warring against evil. **church music,** *n.* sacred music, such as is used in church services. **church-owl,** *n.* the barn-owl. **church-rate,** *n.* a rate (now voluntary) for the support of a parish church. **church service,** *n.* service in a church; the Book of Common Prayer with the daily lessons added. **Church triumphant,** *n.* Christians (collectively) in heaven. **church-warden,** *n.* one of two officers, chosen annually at Easter, to protect church property, to superintend the performance of divine worship etc. and to act as the legal representatives of the parish generally; a long clay pipe with a large bowl. **churchward, -wards,** *adv.* **churchyard,** *n.* the ground adjoining the church consecrated for the burial of the dead. **churchism,** *n.* preference for and adherence to the principles of a church, esp. of the establishment. **churchless,** *a.* without a church. **churchlike,** *a.* befitting the church or clerics. **churchly,** *a.* **churchy,** *a.* making a hobby of church-work and church matters; aggressively devoted to the church and intolerant of dissenters. **churchify** (-fī), *v.t.* **churchiness,** *n.*
churl (chœl), *n.* a man of low birth; a peasant; a surly person. **churlish,** *a.* **churlishly,** *adv.* **churlishness,** *n.*
churn (chœn), *n.* a vessel in which milk or cream is agitated or beaten in order to produce butter; a large can for carrying milk long distances. *v.t.* to agitate in a churn for the purpose of making butter; to agitate with violence or continued motion (often with *up*). *v.i.* to perform the operation of churning; of waves, to foam, to swirl about. **to churn out,** to produce rapidly and prolifically, usu. without concern for excellence. **churning,** *n.* the action of the verb TO CHURN; the butter made at one operation.
churr (chœ), *n.* the deep, trilling cry of the night-jar. *v.i.* to make this cry.
chute (shoot), *n.* an inclined trough for conveying water, timber, grain etc. to a lower level; an inclined watercourse; a toboggan-slide; (*coll.*) short for PARACHUTE.
chutney (chŭt'ni), *n.* a hot seasoned condiment or pickle.

chutzpah (khuts'pə), *n.* barefaced audacity.
chyle (kīl), *n.* the milky fluid separated from the chyme by the action of the pancreatic juice and the bile, absorbed by the lacteal vessels, and assimilated with the blood. **chylaceous** (-lā'-), *a.* **chylify,** *v.t.,* *v.i.* **chylification** (-fi'-), *n.*
chyme (kīm), *n.* the pulpy mass of digested food before the chyle is separated from it. **chymify,** *v.t.,* *v.i.* **chymification** (-fi'-), *n.*
chypre (shēpr¨), *n.* a strong sandalwood perfume.
CI, (*abbr.*) Channel Islands.
Ci, (*abbr.*) curie.
CIA, (*abbr.*) Central Intelligence Agency (US).
ciao (cha'ō), *int.* (*coll.*) expressing greeting or leave-taking.
cicada (sikah'də), *n.* (*pl.* **dae** (-dē)) also **cicala** (sikah'lə), **cigala** (-gah'lə), (*pl.* **-le** (-le)) (an individual of) a genus of insects which rub their legs together with a creaking sound.
cicatrice CICATRIX.
cicatrix (sik'ətriks), **cicatrice** (-tris), *n.* (*pl.* **-trices**) the mark or scar left after a wound or ulcer has healed; a mark on a stem or branch of a plant where a leaf was attached. **cicatrize, -ise,** *v.t.* to heal (a wound or ulcer) by inducing the formation of a cicatrix. *v.i.* to skin over a wound or ulcer. **cicatrization,** *n.* **cicatrose** (-trōs), *a.* full of scars; scarry.
cicely (sis'əli), *n.* the name of several plants of the parsley family.
cicerone (chichərō'ni, sisə-), *n.* (*pl.* **-roni**) a guide; one who explains the curiosities and interesting features of a place to strangers. *v.t.* to conduct (visitors) in this manner. [It., from *Cicero ōnem,* the Roman orator, 106–43 BC]
Ciceronian (sisərō'niən), *a.* resembling the style of Cicero; easy, flowing. *n.* an admirer or imitator of the style of Cicero. **Ciceronianism,** *n.*
Cicestr., (*abbr.*) of Chichester, used as the Bishop of Chichester's signature.
CID, (*abbr.*) Criminal Investigation Department.
cid (sid), *n.* a prince or commander, esp. the Spanish hero Ruy Diaz, Count of Bivar (*c.* 1043–99), champion against the Moors and theme of several epic poems (*El Cid*).
-cide, *comb. form.* a person or substance that kills, as *fratricide* (= a person who kills their brother), *insecticide;* (a) killing, as *homicide.*
cider (sī'də), *n.* the fermented juice of pressed apples. **cider-brandy,** *n.* apple-brandy. **cider-mill,** *n.* a mill in which cider is made; a machine for grinding or crushing apples. **cider-press,** *n.* a press for squeezing the juice from crushed apples.
ci-devant (sēdəvā'), *a.* former, of a past time. **ci-gît** (sēzhē'), here lies (inscribed on gravestones).
Cie., (*abbr., F*) Company, Co.
cierge (sœj), *n.* a wax candle used in religious processions in the Roman Catholic Church.
c.i.f., (*abbr.*) cost, insurance, freight.
cig (sig), *n.* (*coll.*) short for CIGARETTE.
cigala CICADA.
cigar (sigah'), *n.* a roll of tobacco leaf for smoking. **cigar-holder,** *n.* a mouthpiece for holding a cigar. **cigar-shaped,** *a.* cylindrical, with tapering ends. **cigar-store,** *n.* (*N Am.*) a tobacconist's shop. **cigarette** (sigəret'), *n.* cut tobacco or aromatic herbs rolled in paper for smoking. **cigarette-card,** *n.* a picture card enclosed in cigarette packets. **cigarette-holder,** *n.* a mouthpiece for holding a cigarette. **cigarette machine,** *n.* a machine for making cigarettes; a vending machine for cigarettes. **cigarette-paper,** *n.* thin paper, usu. rice-paper, for wrapping the tobacco in cigarettes. **cigarillo** (sigəril'ō), *n.* a very small cigar.
cilium (sil'iəm), *n.* (*pl.* **-lia** (-ə)) an eyelash; a whip-like appendage in a unicellular organism. **ciliary,** *a.* **ciliate** (-ət), **-ated** (-ātid), *a.* **ciliation,** *n.* **ciliform** (-fawm), *a.*
cimex (sī'meks), *n.* (*pl.* **cimices** (sim'isēz)) any of a genus

of insects containing the bed-bug.

C.-in-C., (*abbr.*) Commander-in-Chief.

cinch (sinch), *n*. a broad kind of saddle-girth; a firm grip or hold; (*sl.*) a certainty, an easy task. *v.t.* to furnish or fasten with a cinch; to hold firmly. **to cinch up**, to tighten a cinch.

cinchona (singkō'nə), *n*. a genus of S American trees whose bark (Peruvian bark) yields quinine. **cinchonaceous** (-nā'-), *a*. **cinchenic** (-ken-'), *a*. **cinchonine** (sing'kōnīn), *n*. an organic alkaloid contained in Peruvian bark. **cinchonism** (sing'-), *n*. the disturbed condition of the body caused by overdoses of quinine. **cinchonize, -ise** (sing'-), *v.t.* to treat with quinine.

Cincinnatus (sinsinah'təs), *n*. a great man summoned from retirement to save the state in a crisis.

cincture (singk'chə), *n*. a belt, a girdle, a band; the fillet at the top and bottom of a column. *v.t.* to gird, to encircle.

cinder (sin'də), *n*. a coal that has ceased to burn but retains heat; a partly-burnt coal or other combustible; light slag; (*pl.*) the refuse of burnt coal or wood; the remains of anything that has been subject to combustion; (*pl.*) ashes ejected from a volcano. **cinder-path, -track**, *n*. a racecourse or footpath made up with the refuse of burnt coal or wood. **cindery**, *a*.

Cinderella (sindərel'ə), *n*. a person whose merits are unrecognized; one who is made use of, left behind when others enjoy themselves, etc. [scullery-maid who marries a prince in the fairy tale]

cine-, ciné-, *comb. forms.* cinema; cinematographic, as *cine-projector.*

cineaste (sin'iast), *n*. a cinema enthusiast; a person who makes films.

cine-camera, ciné- (sin'ikamərə), *n*. a camera for taking motion pictures.

cine-film (sin'ifilm), *n*. film suitable for use in a cine-camera.

cinema (sin'əmə), *n*. a theatre where cinematographic films are shown; films collectively; the making of films as artform or industry. **cinema-goer**, *n*. one who goes regularly to the cinema. **Cinemascope®**, *n*. a method of film-projection on a wide screen to give a three-dimensional effect. **cinematic** (-mat'-), *a*. pertaining to, or similar in effect to, the cinema.

cinematograph (sinəmat'əgrahf), *n*. an apparatus for projecting very rapidly on to a screen a series of photographs, so as to create the illusion of continuous motion. **cinematographer** (-tog'rəfə), *n*. **cinematographic** (-graf'-), *a*. **cinematography**, *n*. the art of filming motion pictures.

cinéma vérité (sē'nāmə vā'rētā), *n*. cinema which approaches a documentary style by using realistic settings, characters etc.

Cinerama® (sinərah'mə), *n*. a method of film-projection on a wide screen to give a three-dimensional effect.

cineraria (sinərea'riə), *n*. a variety of garden or hot-house plants with bright flowers and greyish leaves.

cinerary (sin'ərəri), *a*. pertaining to ashes. **cinerarium** (-reə'riəm), *n*. a place for the deposit of human ashes after cremation. **cineration**, *n*. reduction to ashes. **cinerator**, *n*. a furnace for cremating corpses.

Cingalese (sing-gəlēz'), *n*. SINHALESE.

cingulum (sing'gūləm), *n*. (*pl.* **-la** (-lə)) (*Anat., Zool.*) a band of various kinds; the girdle of a priest's alb.

cinnabar (sin'əbah), *n*. a native mercuric sulphide; vermilion; a large moth with red and black markings. *a*. vermilion in colour. **cinnabaric** (-ba'-), *a*.

cinnamon (sin'əmən), *n*. the aromatic inner bark of an E Indian tree, used as a spice; applied also to other trees and their bark; a light brownish-yellow colour. *a*. of the colour (of) cinnamon. **cinnamon toast**, *n*. toast spread with cinnamon and sugar. **cinnamic** (-nam'-), *a*. **cinnamomic** (-mō'mik), **-monic** (-mon'-), *a*.

cinque, cinq (singk), *n*. five; the five at cards or dice.

Cinque Ports, *n.pl.* the five English ports: Dover, Sandwich, Hastings, Hythe and Romney (to which Winchelsea and Rye were afterwards added), which enjoyed special privileges from the fact that they offered a defence against invasion.

cinquecento (chingkwichen'tō), *n., a*. (of) the revived classical style of art and literature that characterized the 16th cent., esp. in Italy. **cinquecentist**, *n*.

cinquefoil (singk'foil), *n*. a five-leaved variety of clover; (*Arch.*) an ornamental foliation in five compartments, used in tracery etc. **cinque-foiled**, *a*.

CIO, (*abbr.*) Congress of Industrial Organizations (US).

cipher, cypher (sī'fə), *n*. the arithmetical symbol 0; a character of any kind used in writing or printing; a monogram, a device; a code or alphabet used to carry on secret correspondence, designed to be intelligible only to the persons concerned; anything written in this; a key to it; (*derog.*) a person or thing of no importance. *v.i.* to do arithmetic. *v.t.* to express in code; to work by means of arithmetic. **cipher-key**, *n*. a key for reading writing in code.

circ (sœk), *n*. (*Archaeol.*) a stone circle.

circa (sœ'kə), *prep*. about, around. *adv*. about, nearly, often used with dates.

circadian (sœkā'diən), *a*. recurring or repeated (approximately) every 24 hours, as some biological cycles.

circassian (sœkas'iən), *n., a*. (of) a type of light cashmere of silk and mohair.

Circe (sœ'si), *n*. an enchantress; a woman who seduces. **Circean** (-sē'ən), *a*. [mythical enchantress, fabled to have turned the companions of Ulysses into swine]

circle (sœ'kl), *n*. a ring, a round figure; (*loosely*) a round body, a sphere; a round enclosure; a number of persons gathered in a ring; any series ending as it begins, and perpetually repeated; a number of persons or things considered as bound together by some bond; a class, a set, a coterie, an association of persons having common interests; a sphere of action or influence; a tier of seats at a theatre; a plane figure bounded by a curved line, called the circumference, every point in which is equidistant from the centre of the figure. *v.t.* to move round; to surround. *v.i.* to form a circle; to revolve. **dress circle**, the principal tier of seats in a theatre, in which evening dress is optional. **great circle**, a circle dividing a sphere into two equal parts. **lesser, small circle**, a circle dividing a sphere into two unequal parts. **Polar circles**, the Arctic and Antarctic parallels of latitude. **stone circle**, a ring of prehistoric monoliths. **to circle in**, to confine. **to come full circle**, to come round (esp. of events) to where one started. **to go, run round in circles**, to be very active in a confused manner or without achieving much. **to square the circle**, to undertake an impossible task; to construct geometrically a square of an area equal to that of a given circle. **circled**, *a*. having the form of a circle; encircled; marked with a circle or circles. **circler, n. circlet** (-lit), *n*. a little circle; a ring or circular band worn on the finger, head etc. **circlewise**, *adv*.

circs (sœks), *n*. (*coll.*) short for CIRCUMSTANCES.

circuit (sœ'kit), *n*. an act of revolving or moving round, a revolution; the distance round about; the space enclosed in a circle or within certain limits; formerly the periodical visitation of judges for holding assizes; the district thus visited; the barristers making the circuit; among Methodists, a group of churches associated together for purposes of government and organization of the ministry; a continuous electrical communication between the poles of a battery; (a diagram showing) a series of conductors, including the lamps, motors etc., through which a current passes; a motor-racing track; a series of sporting tournaments visited regularly by competitors; a group of theatres or cinemas under the same ownership, putting on the same entertainment in turn. **open circuit** OPEN. **short circuit** SHORT. **circuit board**, *n*. a board on which an elec-

circulable 145 **cismontane**

tronic circuit is built, with a connector to plug into a piece of equipment. **circuit-breaker**, *n.* a device which stops the electric current in the event of e.g. a short circuit; a device which artificially prevents share or currency values dropping below a certain level. **circuit training**, *n.* a form of athletic training consisting of repeated cycles of exercises. **circuitous** (-kū'i-), *a.* indirect, roundabout. **circuitously**, *adv.* **circuitousness**, *n.* **circuitry**, *n.* electric or electronic circuits collectively; the design of such a circuit. **circuity** (-kū'i-), *n.* indirect procedure.
circulable CIRCULATE.
circular (sœ'kūlə), *a.* in the shape of a circle; round; pertaining to a circle; (*loosely*) spherical; moving in a circle; cyclic; consisting of an argument which ends where it began. *n.* a letter or printed notice of which a copy is sent to many persons. **circular letter**, *n.* a notice, advertisement or appeal printed or duplicated for sending to a number of persons. **circular numbers**, *n.pl.* those whose powers terminate in the same digits as their roots. **circular saw**, *n.* a rotating disk notched with teeth for cutting timber etc. **circular tour**, *n.* a journey to a number of places ending at the starting-point. **circularity** (-la'-), *n.* the state of being circular. **circularize, -ise**, *v.t.* to send circulars to. **circularly**, *adv.*
circulate (sœ'kūlāt), *v.i.* to move round; to pass through certain channels, as blood in the body, sap in plants etc.; to pass from point to point or hand to hand, as money; (*coll.*) to attend to all one's guests, or to move round meeting people, at a social gathering. *v.t.* to cause to pass from point to point or hand to hand; to spread, to diffuse. **circulating**, *a.* that circulates; current; (*Math.*) recurring. **circulating decimal**, *n.* a decimal which cannot be expressed finitely in figures, and in which one or more figures recur continually in the same order. **circulating library**, *n.* (*esp. N Am.*) a lending library. **circulable**, *a.* **circulation**, *n.* the act of circulating; the state of being circulated; the motion of the blood in a living animal, by which it is propelled by the heart through the arteries to all parts of the body, and returned to the heart through the veins; the analogous motion of sap in plants; free movement (of water, air etc.); distribution of books, newspapers, news etc.; the amount of distribution, the number of copies sold. **circulative**, *a.* tending to circulate; promoting circulation. **circulator**, *n.* one who or that which circulates; a circulating decimal. **circulatory** (-lə-), *a.* circular, circulating.
circum-, *pref.* round, round about; surrounding; indirect(ly); pertaining to the circumference.
circumambient (sœkəmam'biənt), *a.* going round about; surrounding. **circumambiency**, *n.*
circumambulate (sœkəmam'būlāt), *v.t., v.i.* to walk or go round about. **circumambulation**, *n.* **circumambulatory**, *a.*
circumcise (sœ'kəmsīz), *v.t.* to remove surgically, sometimes as part of a religious ritual, the prepuce or foreskin in the male, or the clitoris in the female. **circumcision** (-sizh'ən), *n.* the operation of circumcising.
circumference (səkūm'fərəns), *n.* the line that bounds a circle; a periphery; the distance round a space or a body. **circumferential** (-ren'-), *a.*
circumflex (sœ'kəmfleks), *n.* a mark (ˆ or in Gr. ˜) placed above a vowel to indicate accent, quality, or contraction. *a.* (*pred.*) marked with such accent; (*Anat.*) bent, turning, or curving round something. *v.t.* to mark or pronounce with a circumflex. **circumflexion, -flection** (-flek'-), *n.*
circumfluent (səkūm'fluənt), *a.* flowing round on all sides. **circumfluence**, *n.*
circumfuse (sœkəmfūz'), *v.t.* to pour round, as a fluid; to surround, to bathe in or with. **circumfusion**, *n.*
circumgyrate (sœkəmjīrāt', -jī'-), *v.i.* to turn, roll or spin round. **circumgyration**, *n.* **circumgyratory** (-rā'-), *a.*
circumjacent (sœkəmjā'sənt), *a.* lying round; bordering.
circumlittoral (sœkəmlit'ərəl), *a.* pertaining to the shore; pertaining to the zone immediately outside of the littoral.
circumlocution (sœkəmləkū'shən), *n.* the use of roundabout, indirect or evasive language; the use of many words where few would suffice. **circumlocutional**, *a.* **circumlocutionary**, *a.* **circumlocutionist**, *n.* **circumlocutory** (-lok'ū-), *a.*
circumlunar (sœkəmloo'nə), *a.* revolving round or situated near the moon.
circumnavigate (sœkəmnav'igāt), *v.t.* to sail completely round. **circumnavigation**, *n.* **circumnavigator**, *n.*
circumoral (sœkəmaw'rəl), *a.* surrounding the mouth.
circumpolar (sœkəmpō'lə), *a.* (*Geog.*) situated round or near the pole; (*Astron.*) revolving about the pole (never setting).
circumscribe (sœ'kəmskrīb), *v.t.* to write or draw around; to limit, to define by bounds, to restrict; (*Log.*) to define; (*Geom.*) to surround with a figure that touches at every possible point. **circumscriber**, *n.* **circumscription** (-skrip'-), *n.* **circumscriptive** (-skrip'-), *a.* **circumscriptively**, *adv.*
circumsolar (sœkəmsō'lə), *a.* revolving round or situated near the sun.
circumspect (sœ'kəmspekt), *a.* looking on all sides; cautious, wary. **circumspection** (-spek'-), **circumspectness**, *n.* **circumspective** (-spek'-), *a.* **circumspectly**, *adv.*
circumstance (sœ'kəmstahns, -stans, -stəns), *n.* something attending or relative to a fact or case; an incident, an event; a concomitant; abundance of detail (in a narrative), circumstantiality; ceremony, pomp, fuss; (*pl.*) the facts, relations, influences and other conditions that affect an act or an event; (*pl.*) the facts, conditions etc. that affect one's living. **in, under, the circumstances**, in the particular situation for which allowance should be made. **straitened circumstances**, (esp. financial) hardship. **circumstanced**, *a.* situated.
circumstantial (sœkəmstan'shəl), *a.* depending on circumstances; incidental, not essential. **circumstantial evidence**, *n.* evidence inferred from circumstances which usually attend facts of a particular nature. **circumstantiality** (-shial'-), *n.* **circumstantially**, *adv.* **circumstantiate** (-shiāt), *v.t.* to provide evidence for.
circumterrestrial (sœkəmtəres'triəl), *a.* revolving round, or situated near, Earth.
circumvallate (sœkəmval'āt), *v.t.* to surround or enclose with a rampart. **circumvallation**, *n.*
circumvent (sœkəmvent', sœ'-), *v.t.* to go round, to avoid; to deceive, to outwit, to cheat, to get the best of. **circumvention** (-ven'-), *n.*
circus (sœ'kəs), *n.* the Circus Maximus in ancient Rome; any similar building; a circle of buildings at the intersection of streets; a travelling company of clowns, acrobats, trained animals etc.; the place, usu. a circular tent, where they perform; (*coll.*) a scene of noisy, disorganized activity.
ciré (sē'rā), *n.* satin with a waxed surface.
cirque (sœk), *n.* a circular recess among hills.
cirrhosis (sirō'sis), *n.* a disease of the liver in which it becomes yellowish and nodular because of the death of liver cells and the growth of fibrous tissue. **cirrhotic** (-rot'-), *a.*
cirrocumulus (sirōkū'myələs), *n.* a type of cloud broken up into small fleecy masses.
cirrostratus (sirōstrah'təs), *n.* a type of horizontal or slightly inclined sheet of fleecy cloud.
cirrus (si'rəs), *n.* a lofty type of feathery cloud.
cis-, *pref.* on this side of.
cisalpine (sisal'pīn), *a.* on the Roman side of the Alps; south of the Alps.
cisatlantic (sisətlan'tik), *a.* on the user's side of the Atlantic, as distinct from *transatlantic*.
cislunar (sisloo'nə), *a.* between the moon and the earth.
cismontane (sismon'tān), *a.* on the north side of the mountains (as regards France and Germany).

cispontine (sispon'tin), *a.* on the north side of the Thames, in London.
cissy, sissy (sis'i), *a.* (*derog.*) effeminate. *n.* an effeminate person.
cist (sist), *n.* a tomb consisting of a kind of stone chest formed of rows of stones, with a flat stone for cover; a casket or chest, esp. one used for carrying the sacred utensils in the Greek mysteries.
Cistercian (sistœ'shən), *n.* a member of a monastic order founded in 1098 in France. *a.* pertaining to the Cistercians.
cistern (sis'tən), *n.* a tank for storing water; a water-tank for a water-closet.
citable CITE.
citadel (sit'ədəl), *n.* a castle or fortified place in a city; a stronghold; a Salvation Army hall.
cite (sīt), *v.t.* to quote, to allege as an authority; to quote as an instance; to summon to appear in court. **citable**, *a.*
citation, *n.* (a) summons; (a) mention in dispatches etc.; (a) quotation.
cither (sith'ə), **cithern** (sith'ən), **cittern** (sit'ən), *n.* a mediaeval kind of guitar with wire strings; a zither.
citified CITY.
citizen (sit'izən), *n.* a member of a state in the enjoyment of political rights; a burgess or freeman of a city or town; a dweller in a town, a civilian. **citizen's arrest**, *n.* one made by a member of the public. **citizens' band**, *n.* a band of radio frequencies designated for use by private citizens for communication between individuals. **citizenhood**, *n.* **citizenry**, *n.* citizens collectively. **citizenship**, *n.* the state of being a citizen; the right of being a citizen.
citole (si'tōl, -tōl'), *n.* a type of stringed musical instrument.
citr- CITR(O)-.
citrate (sit'rāt, sī'-), *n.* a salt of citric acid. **citric** (sit'-), *a.* derived from the citron. **citric acid**, *n.* the acid found in lemons, citrons, limes, oranges etc.
citrine (sit'rin), *a.* like a citron; greenish-yellow. *n.* (*Min.*) a yellow, pellucid variety of quartz. **citrinous**, *a.* lemon-coloured.
citr(o)-, *comb. form.* citric.
citron (sit'rən), *n.* a tree bearing large lemon-like fruit.
citronella (sitrənel'ə), *n.* a fragrant oil used to drive away insects.
citrus (sit'rəs), *n.* any of a genus of trees and shrubs containing the orange, lemon, citron etc. **citrous**, *a.*
cittern CITHER.
city (sit'i), *n.* a town incorporated by a charter; (*pop.*) a large and important town; a cathedral town. *a.* pertaining to a city; characteristic of a city. **the Celestial City**, Heaven. **the City**, the part of London governed by the Lord Mayor and Corporation; the banks and financial institutions located there. **the Eternal City**, Rome. **the Holy City**, Jerusalem. **city article**, *n.* an article in a newspaper dealing with commerce or finance. **city desk**, *n.* the editorial department of a newspaper dealing with financial news; (*N Am.*) the department of a newspaper dealing with local news. **city editor**, *n.* one in charge of a city desk. **city fathers**, *n.pl.* men in charge of the administration of a city. **city technology college**, *n.* a secondary school providing technical and scientific education in city areas. **citied**, *a.* containing cities (*usu. in comb.*, as *many-citied*). **citified**, *a.* townish; having the peculiarities of dwellers in cities.
cityless, *a.* **cityward, -wards**, *adv.*
civet (siv'it), *n.* a resinous musky substance obtained from the anal pouch of the civet-cat, and used as a perfume. *v.t.* to perfume with civet. **civet-cat**, *n.* a carnivorous quadruped from Asia and Africa.
civic (siv'ik), *a.* pertaining to a city or citizens; urban; municipal; civil. **civic centre**, *n.* a group of buildings including the town hall and local administrative offices. **civically**, *adv.* **civics**, *n. sing.* the study of citizenship and municipal government.
civil (siv'l), *a.* pertaining to citizens; domestic, not foreign; municipal, commercial, legislative; civilized, polite, courteous; pertaining to social, commercial and administrative affairs, not warlike, not military or naval; (*Law*) pertaining to private matters, not criminal. **civil action, process**, *n.* an action or process in civil law. **civil architecture**, *n.* the construction of buildings for the purposes of civil life. **civil aviation**, *n.* civilian, non-military airlines and their operations. **civil defence**, *n.* a civilian service for the protection of lives and property in the event of enemy attack. **civil disobedience**, *n.* a concerted plan in a political campaign taking the form of refusal to pay taxes or perform civil duties. **civil engineer**, *n.* **civil engineering**, *n.* the science of constructing docks, railways, roads etc. **civil law**, *n.* the law dealing with private rights, not criminal matters. **civil liberties**, *n.pl.* personal freedoms, e.g. freedom of speech, within the framework of the state. **civil list**, *n.* the yearly sum granted for the support of a sovereign or ruler; the officers of a government who are paid from the public treasury. **civil list pension**, a small pension granted by the state to selected artists, writers, musicians etc. **civil magistrate**, *n.* a magistrate not dealing with ecclesiastical matters. **civil marriage**, *n.* one performed by a civil official, not by a clergyman. **civil rights**, *n.pl.* the rights of an individual or group within a state to certain freedoms, e.g. from discrimination. **civil servant**, *n.* a member of the civil service. **civil service**, *n.* the non-military branch of the public service, dealing with public administration. **civil state**, *n.* the entire body of the citizens, as distinct from the military, ecclesiastical and naval establishments. **civil suit**, *n.* (*Law*) a suit for a private claim or injury. **civil war**, *n.* a war between citizens of the same country. **civilian** (-vil'yən), *n.*, *a.* (a person) engaged in civil life, not belonging to the army or navy. **civilianize, -ise**, *v.t.* **civility** (-vil'-), *n.* the quality of being civil; politeness, courtesy. **civilly**, *adv.*
civilize, -ise (siv'əliz), *v.t.* to reclaim from barbarism; to instruct in the arts and refinements of civilized society. **civilizable**, *a.* **civilization**, *n.* the act or process of civilizing; the state of being civilized; refinement, social development; (a) civilized society. **civilized**, *a.* **civilizer**, *n.*
civvies (siv'iz), *n.pl.* (*Army coll.*) civilian clothes. **Civvy Street**, *n.* civilian life.
Cl, (*chem. symbol*) chlorine.
cl., (*abbr.*) centilitre; class.
clachan (klakh'ən), *n.* a small village or hamlet in the Highlands.
clack (klak), *v.i.* to make a sharp, sudden noise like a clap or crack; (*coll.*) to chatter rapidly and noisily. *v.t.* to cause to emit a sudden, sharp noise; to knock (together) with this sound. *n.* a sudden, sharp sound frequently repeated; rapid and noisy chattering. **clack-valve**, *n.* a valve hinged by one edge.
clad, *p.p.* CLOTHE.
cladding (klad'ing), *n.* a protective coating, e.g. of stone on a building or insulating material on a hot-water pipe.
cladistics (klədis'tiks), *n.sing.* a method of classifying organisms based on common characteristics.
claim (klām), *v.t.* to demand, or request, as a right; to assert (that one has or is something or has done something); to affirm, to maintain; to be deserving of. *n.* a real or supposed right; a title; the thing asserted, affirmed or maintained; a piece of land marked out by a settler or miner with the intention of buying it when it is offered for sale. **claim-jumper**, *n.* one who seizes on land claimed by another. **claim-jumping**, *n.* **claimable**, *a.* **claimant**, *n.* one who makes a claim.
clairaudience (kleəraw'diəns), *n.* the faculty of hearing voices and other sounds not perceptible to the senses. **clairaudient**, *n.*, *a.*
clair-obscure (kleərəbskūə'), CHIAROSCURO.

clairvoyance (kleəvoi'əns), *n.* the power of perceiving objects not present to the senses; unusual sensitivity or insight. **clairvoyant,** *n.* one having the power of clairvoyance. *a.* pertaining to or having the power of clairvoyance.

clam (klam), *n.* a name for several edible types of shellfish. **to clam up,** to become silent. **clambake,** *n.* (*N Am.*) a beach party at which clams are cooked and eaten; any noisy gathering.

clamant (klā'mənt), *a.* crying or begging earnestly.

clamber (klam'bə), *v.i.* to climb (up, across etc.) any steep place with hands and feet, esp. with difficulty.

clammy (klam'i), *a.* moist, damp and sticky; of weather, humid. **clammily,** *adv.* **clamminess,** *n.*

clamour (klam'ə), *n.* a loud and continuous shouting or calling out; a continued and loud expression of complaint, demand or appeal; popular outcry. *v.t.* to utter or express with loud noise. *v.i.* to cry out loudly and earnestly; to demand or complain importunately. **clamorous,** *a.* **clamorously,** *adv.* **clamorousness,** *n.*

clamp[1] (klamp), *n.* anything rigid which strengthens, fastens or binds; a piece of timber or iron used to fasten work together; a frame with two tightening screws to hold pieces of wood together; a wheel-clamp. *v.t.* to unite, fasten or strengthen with a clamp or clamps; to immobilize with a wheel-clamp. **to clamp down (on),** to impose (heavier) restrictions (on). **clamp-down,** *n.*

clamp[2] (klamp), *n.* a pile of bricks for burning; a heap, mound or stack of turf, rubbish, potatoes etc. *v.t.* to pile into a heap; to store in a clamp.

clamp[3] (klamp), *n.* a heavy footstep or tread. *v.i.* to tread heavily and noisily.

clan (klan), *n.* a tribe or number of families bearing the same name, descended from a common ancestor, and united under a chieftain representing that ancestor; a large extended family; a clique, a set. **clannish,** *a.* united closely together, as the members of a clan; of or pertaining to a clan; cliquish. **clannishly,** *adv.* **clannishness,** *n.* **clanship,** *n.* the system or state of clans. **clansman,** *n.* a member of a clan. **clanswoman,** *n.fem.*

clandestine (klandes'tin), *a.* secret, surreptitious, underhand. **clandestinely,** *adv.* **clandestineness,** *n.*

clang (klang), *v.t.*, *v.i.* to strike together, so as to cause a sharp, ringing sound. *n.* a sharp, ringing noise, as of two pieces of metal struck together. **clanger,** *n.* (*coll.*) a foolish mistake; a social blunder. **clangour** (-gə), *n.* a sharp, ringing sound or series of sounds. **clangorous,** *a.* **clangorously,** *adv.*

clank (klangk), *v.t.* to strike together so as to make a heavy, metallic rattling sound. *v.i.* to make such a sound. *n.* a sound as of solid metallic bodies struck together (usu. denotes a deeper sound than *clink*, and a less resounding one than *clang*).

clannish, etc. CLAN.

clap[1] (klap), *v.t.* (*past, p.p.* **clapped**) to strike together noisily; to strike quickly or slap with something flat; to put or place suddenly or hastily; to applaud, by striking the hands together. *v.i.* to strike the hands together in applause. *n.* the noise made by the collision of flat surfaces; a peal (of thunder); applause shown by clapping; a heavy slap. **to clap eyes on,** (*coll.*) to catch sight of. **to clap shut,** to close hastily. **clapped-out,** *a.* (*sl.*) finished, exhausted; (*sl.*) of no more use. **claptrap,** *n.* showy words or deeds designed to win applause or public favour; pretentious or insincere nonsense. **clapper,** *n.* one who or that which claps; the tongue of a bell; a noisy rattle for scaring birds. **like the clappers,** (*sl.*) extremely fast. **clapper-board,** *n.* a pair of hinged boards clapped together at the start of a take during film shooting to help in synchronizing sound and vision.

clap[2] (klap), *n.* (*sl.*) gonorrhoea.

clap-board (klap'bawd), *n.* a feather-edged board used to cover the roofs and sides of houses. *v.t.* to cover with clap-board. **clap-boarding,** *n.*

claptrap CLAP[1].

claque (klak), *n.* a body of hired applauders; the system of engaging applauders. **claquer, claqueur** (-kœ'), *n.* a hired applauder.

clarence (kla'rəns), *n.* a closed four-wheeled carriage for four passengers, with a seat for the driver.

clarendon (kla'rəndən), *n., a.* (*Print.*) a condensed type with heavy face.

clare-obscure (kleərəbskūə'), CHIAROSCURO.

claret (kla'rit), *n.* a light red Bordeaux wine; any light red wine resembling Bordeaux. **claret-coloured,** *a.* reddish-violet. **claret-cup,** *n.* a beverage composed of iced claret, brandy, lemon, borage etc.

clarify (kla'rifi), *v.t.* to clear from visible impurities; to make transparent; to make lucid or clear to the intellect. *v.i.* to become transparent. **clarification** (-fi'-), *n.* **clarifier,** *n.* one who or that which clarifies; a vessel in which sugar is clarified.

clarinet (klarinet'), *n.* a keyed single-reed musical instrument. **clarinettist,** *n.*

clarion (kla'riən), *n.* a kind of trumpet, with a narrow tube, and loud and clear note; a sound (as) of a clarion; an organ stop giving a similar tone. *a.* loud and clear, as *clarion call*.

clarity (kla'rəti), *n.* clearness.

clash (klash), *v.i.* to make a loud noise by striking against something; to come into collision; to disagree; to conflict; of colours, to lack harmony. *v.t.* to cause (one thing) to strike against another so as to produce a noise. *n.* the noise produced by the violent collision of two bodies; opposition, contradiction; conflict, esp. military or anti-authority; disharmony of colours.

clasp (klahsp), *n.* a catch, hook or interlocking device for fastening; a buckle or brooch; a close embrace; a grasp; a metal bar attached to a ribbon carrying a medal commemorating a battle or other exploit. *v.t.* to fasten or shut with or as with a clasp or buckle; to fasten (a clasp); to embrace; to grasp. **clasp-knife,** *n.* a pocket-knife in which the blade shuts into the hollow part of the handle. **clasper,** *n.* one who or that which clasps; one of a pair of organs in some insects and fishes by which the male holds the female.

class (klahs), *n.* a number of persons or things ranked together; (*a*) social rank; a number of scholars or students taught together; (*N Am.*) the students taken collectively who expect to graduate at the same time; a division according to quality; (*coll.*) high quality; a number of individuals having the same essential or accidental qualities; a division of animals or plants next above an order. *v.t.* to arrange in a class or classes. (*coll.*) of good quality. **-class,** *a.* (*in comb.*) of a particular quality, e.g. *first-class, second-class* etc. **in a class of one's, its, own,** of matchless excellence. **class conscious,** *a.* over-sensitive to social differences. **class-consciousness,** *n.* **class-list,** *n.* a classified list of candidates issued by examiners. **class-man,** *n.* one who takes honours at an examination, as opposed to a *passman.* **class-mate, -fellow,** *n.* one who is or has been in the same academic class. **class war, warfare,** *n.* overt antagonism between the social classes in a community. **classable,** *a.* capable of being classed. **classism,** *n.* discrimination on the ground of social class. **classist,** *n., a.* **classless,** *a.* not divided into classes; not belonging to any one class. **classy,** *a.* genteel; of superior quality.

classic (klas'ik), *n.* a Greek or Latin author of the first rank; an author of the first rank; a literary work by any of these; a recognized masterpiece; (*pl.*) ancient Greek and Latin literature; (*pl.* in form, *sing.* in construction) the study of these. *a.* pertaining to the literature of the ancient Greeks and Romans; in the style of these; of the first rank in literature or art; harmonious, well-

classical 148 **clearing**

proportioned; pure, refined, restrained; of standard authority; of the features, clear-cut, regular; of clothes, simple and well cut, that will not become dated. **classic orders**, *n.pl.* (*Arch.*) Doric, Ionic, Corinthian, Tuscan and Composite. **classic races**, *n.pl.* the five principal horse-races in England, being the 2000 Guineas, 1000 Guineas, Derby, Oaks and St Leger.

classical (klas'ikəl), *a.* belonging to or characteristic of the ancient Greeks and Romans or their civilization or literature; of education, based on a study of Latin and Greek; of any of the arts, influenced by Roman or Greek models, restrained, simple and pure in form; of music, composed esp. in the 18th and 19th cents., simple and restrained in style; (*loosely*) of orchestral music, opera etc. rather than pop, jazz etc.; of physics, not involving relativity or quantum mechanics. **classicism**, *n.* **classicality** (-kal'-), *n.* **classically**, *adv.* **classicism** (-sizm), *n.* a classic style or idiom; devotion to or imitation of the classics; classical scholarship. **classicist**, *n.* a scholar in classics. **classico-**, *comb. form.*

classify (klas'ifi), *v.t.* to distribute into classes or divisions; to assign to a class. **classifiable**, *a.* **classification** (-fi-), *n.* **classificatory** (-kā'-), *a.* **classified**, *a.* arranged in classes; of information, of restricted availability, esp. for security reasons; of printed advertisements, arranged according to the type of goods or services offered or required. **classifier**, *n.*

clatter (klat'ə), *v.i.* to make a sharp rattling noise; to fall or move with such a noise. *v.t.* to cause to make a rattling sound. *n.* a continuous rattling noise; loud, tumultuous noise.

Claudian (klaw'diən), *a.* pertaining to or of the period of the Roman emperors of the name *Claudius.*

clause (klawz), *n.* a complete simple grammatical sentence; a subdivision of a compound or complex sentence; a separate and distinct portion of a document; a particular stipulation. **clausal**, *a.*

claustrophobia (klawstrəfō'biə), *n.* a morbid dread of being in a confined space. **claustrophobic**, *n.*, *a.*

clavate (klā'vāt), *a.* (*Biol.*) club-shaped.

clavichord (klav'ikawd), *n.* one of the first stringed instruments with a keyboard, a predecessor of the pianoforte.

clavicle (klav'ikl), *n.* the collar-bone. **clavicular** (-vik'ū-), *a.*

clavier (klav'iə, klā'-), *n.* the keyboard of an organ, pianoforte etc.; a keyboard instrument.

claw (klaw), *n.* the sharp hooked nail of a bird or beast; the foot of any animal armed with such nails; the pincer of a crab, lobster or crayfish; anything resembling the claw of one of the lower animals; an implement for grappling or holding; a grasp, a clutch. *v.t.* to tear or scratch with the claws; to clutch or drag with or as with claws. to **claw back**, to get back by clawing or with difficulty; to take back part (as of a benefit or allowance) by extra taxation etc. **clawback**, *n.* **claw hammer**, *n.* a hammer furnished at the back with claws to extract nails. **clawed**, *a.* furnished with claws; damaged by clawing. **clawless**, *a.*

clay (klā), *n.* (*poet.*) the human body; a hydrous silicate of aluminium, with a mixture of other substances pliable when wet, but hard when dry, and used in the manufacture of china, pottery, bricks etc. (*coll.*) a clay pipe. **clay-cold**, *a.* cold and lifeless as clay. **clay-pigeon**, *n.* (*Sport*) a clay disk thrown into the air as a target. **clay pipe**, *n.* a pipe made of baked clay, usu. long; a churchwarden. **clay-pit**, *n.* a pit from which clay is dug. **clayey**, *a.* **clayish**, *a.*

claymore (klā'maw), *n.* a two-edged sword used by the Scottish Highlanders; (*incorrectly*) a basket-hilted broadsword.

clean (klēn), *a.* free from dirt, stain, alloy, blemish, imperfection, disease, ceremonial defilement, awkwardness or defect; pure, holy, guiltless; free from evidence of criminal activity; of a driving licence, free from endorsements or penalty points; free from sexual references, smut, innuendo etc.; producing relatively little radioactive fallout; (*sl.*) not carrying or containing a gun, drugs, illegal or incriminating articles etc.; (*Print.*) needing no correction, as a proof; smart, dexterous, unerring. *v.t.* to make clean; to cleanse, to purify. *adv.* quite, completely. **clean as a whistle**, completely empty; without any hitch. **to clean bowl**, to bowl out (a batsman at cricket) without the ball touching bat or body before hitting the wicket. **to clean out**, to clean by clearing the inside of; (*sl.*) to deprive of all money. **to clean up**, to put tidy; (*sl.*) to collect all the money, profits etc. **to come clean** (*coll.*), to confess. **clean bill of health** BILL OF HEALTH, under BILL^2. **clean-cut**, *a.* sharply defined; smart, with short hair, tidy appearance etc. **clean-handed**, *a.* free from blame in any matter. **clean-limbed**, *a.* having well-proportioned limbs. **clean-shaven**, *a.* without beard or moustache. **clean sheet, slate**, *n.* a new start, all debts etc. written off. **cleanable**, *a.* **cleaner**, *n.* one who or that which cleans; (*pl.*) a dry-cleaners' shop. **to take to the cleaners**, (*sl.*) to deprive of all one's money, goods etc. **cleanly**^{1}, *adv.* in a clean manner. **cleanness**, *n.*

cleanly^2 (klen'li), *a.* clean in person and habits. **cleanlily**, *adv.* **cleanliness**, *n.*

cleanse (klenz), *v.t.* to make clean, to purge, to purify; (*Bibl.*) to cure. **cleanser**, *n.*

clear (kliə), *a.* free from darkness, dullness or opacity; luminous, bright; transparent, translucent; unclouded; brightly intelligent; evident; indisputable, perspicuous, easily apprehended; irreproachable; free; unobstructed; distinctly audible; certain, unmistaken; free from deduction, net. *adv.* clearly, completely; quite, entirely. *v.t.* to make clear; to free from darkness, dimness, opacity, ambiguity, obstruction, imputation or encumbrance; to empty; to remove, to disengage; to acquit, to exonerate; to pay off all charges or debts; to gain, to realize as profit; to pass or leap over without touching; to sail out of, away from (a harbour etc.); to obtain authorization for; to pass (a cheque or bill) through a clearing house. *v.i.* to become clear, bright or serene; (*Naut.*) to sail. **a clear day**, a complete day. **in the clear**, (*coll.*) freed from suspicion. **to clear a ship**, to pay the charges at the customhouse and receive permission to sail. **to clear a ship for action**, to remove all encumbrances from the deck ready for an engagement. **to clear away**, to remove; to remove plates etc., after a meal; to disappear; to melt away. **to clear land**, to remove trees, rubble etc. in order to cultivate or build. **to clear off**, to remove; (*coll.*) to go away. **to clear out**, (*coll.*) to eject; (*coll.*) to depart; to melt away. **to clear the air**, to remove misunderstandings or suspicion. **to clear the decks**, (*coll.*) to tidy up (in preparation for something). **to clear up**, to become bright and clear; to remove (misunderstanding); to tidy up. **to get clear** GET. **clear-cut**, *a.* without possibility for confusion or misinterpretation. **clear days**, *n.pl.* time reckoned apart from the first day and the last. **clear-headed**, *a.* acute, sharp, intelligent. **clear-sighted**, *a.* acute, discerning, farseeing. **clear-sightedness**, *n.* **clearway**, *n.* a non-motorway road on which parking is forbidden. **clearer**, *n.* **clearly**, *adv.* in a clear manner; distinctly, audibly, plainly, evidently, certainly, undoubtedly. **clearness**, *n.* the state of being clear; perspicuity, distinctness to or of apprehension.

clearance (kliə'rəns), *n.* the act of clearing; the state of being cleared; a clearance sale; (*Sport*) an act of potting all the (remaining) balls on a snooker table; the removal of people, buildings etc. from an area; (*Banking*) the clearing of cheques or bills; a document giving (official permission to do something); the distance between a moving and a stationary part of a machine. **clearance sale**, *n.* a sale of stock at reduced prices to make room for new stock.

clearing (kliə'ring), *n.* the act of making clear, freeing or justifying; a tract of land in a wood, free of trees and large plants or undergrowth; (*Banking*) the passing of cheques etc. through a clearing house. **clearing bank,** *n.* one which is a member of a clearing house. **clearing house,** *n.* a financial establishment where cheques, transfers, bills etc. are exchanged between member banks, so that only outstanding balances have to be paid; a person or agency acting as a centre for the exchange of information etc.

cleat (klēt), *n.* a strip fastened on the soles of shoes to obviate slipping; (*Naut.*) a piece of wood or iron for fastening ropes upon. *v.t.* to fasten or fit with a cleat.

cleave[1] (klēv), *v.i.* to stick, to adhere; to be attached closely; to be faithful (to).

cleave[2] (klēv), *v.t.* (*past* **clove** (klōv), **cleft** (kleft), *p.p.* **cloven** (-vən), **cleft**) to split asunder with violence, to cut through, to divide forcibly; to make one's way through. *v.i.* to part asunder violently; to split, to crack. **cleavable,** *a.* **cleavage,** *n.* the act of cleaving; the particular manner in which a mineral with a regular structure may be cleft or split; the way in which a group etc. splits up; the hollow between a woman's breasts, esp. as revealed by a low-cut dress or top. **line, plane of cleavage,** the line or plane of weakness along which a mineral or a rock tends to split. **cleaver,** *n.* one who or that which cleaves; a butcher's instrument for cutting meat into joints.

cleavers (klē'vəz), **clivers** (kliv'-), *n.* a loose-growing plant with hooked prickles that catch in clothes.

clef (klef), *n.* a character at the beginning of a stave denoting the pitch and determining the names of the notes according to their position on the stave; a stave so indicated, as *bass clef*, *treble clef*.

cleft[1] (kleft), *past*, *p.p.* of **cleave**[2]. **cleft-footed,** *a.* having the hoof divided. **cleft palate,** *n.* congenital fissure of the hard palate. **cleft stick,** *n.* a stick split at the end. **in a cleft stick,** in a situation where going forward or back is impossible or very difficult; in a tight place, a fix.

cleft[2] (kleft), *n.* a split, a crack, a fissure.

cleg (kleg), *n.* a gadfly, a horsefly.

cleistogamic (klīstəgam'ik), *a.* having flowers that never open and are self-fertilized.

clematis (klem'ətis, -mā'-), *n.* a climbing plant cultivated for its large, colourful flowers.

clement (klem'ənt), *a.* gentle; merciful; of weather, mild. **clemency,** *n.*

clementine (klem'əntīn), *n.* a small, bright orange citrus fruit with a sweet flavour.

clench (klench), *v.t.* to rivet; to fasten firmly by bending the point of (with a hammer); to grasp firmly; to close or fix firmly (as the hands or teeth).

clepsydra (klep'sidrə), *n.* an instrument used by the ancients to measure time by the dropping of water from a graduated vessel through a small opening; a water-clock.

clerestory (kliə'stawri), *n.* the upper part of the nave, choir or transept of a large church containing windows above the roofs of the aisles; a railway carriage having windows above the roof-line.

clergy (klœ'ji), *n.* the body of men set apart by ordination for the service of the Christian Church; the ecclesiastics collectively of a church, district or country. **clergyman,** *n.* a male member of the clergy; an ordained male Christian minister, esp. of the Established Church.

cleric (kle'rik), *n.* a member of the clergy. **clerico-,** *comb. form.*

clerical (kle'rikəl), *a.* relating to the clergy; relating to (the work of) a clerk, copyist or writer. *n. pl.* the dress of a clergyman. **clerical collar** DOG-COLLAR. **clerical error,** *n.* an error in copying.

clerihew (klc'rihū), *n.* a satirical or humorous poem, usu. biographical, consisting of four rhymed lines of uneven length. [E. *Clerihew* Bentley, 1875–1956]

clerk (klahk, *Amer* klœk), *n.* a cleric, a clergyman; the lay officer of a parish church; one employed in an office, bank, shop etc. to assist in correspondence, book-keeping etc.; (*N Am.*) a shop assistant; (*N Am.*) a hotel receptionist. *v.i.* to work as a clerk. **clerk in holy orders,** an ordained clergyman. **clerk of the course,** an official in charge of administration of a motor- or horse-racing course. **clerk of works,** *n.* a surveyor appointed to watch over the performance of a building contract and test the quality of materials etc. **town clerk,** the chief officer of a corporation, usu. a solicitor. **clerkdom** (-dəm), *n.* **clerkess,** *n. fem.* (*not N Am.*). **clerkish,** *a.* **clerkly,** *a.* **clerkship,** *n.* scholarship; the office or position of a clerk.

cleromancy (kliə'rəmansi, kle'-), *n.* divination by casting lots with dice.

clever (klev'ə), *a.* dexterous, skilful; talented; very intelligent; expert, ingenious. **clever dick,** *n.* (*sl.*) one who shows off his or her own cleverness. **cleverish,** *a.* somewhat clever. **cleverly,** *adv.* **cleverness,** *n.*

clevis (klev'is), *n.* a forked iron at the end of a shaft or beam, or an iron loop, for fastening tackle to.

clew (kloo), *n.* the lower corner of a square sail; the aftermost corner of a staysail.

cliché (klē'shä), *n.* a hackneyed phrase, a tag; anything hackneyed or overused. **clichéd, cliché'd,** *a.*

click (klik), *v.i.* to make a slight, sharp noise, as small hard bodies knocking together; (*coll.*) to fall into place, to make sense; to be successful; (*coll.*) to become friendly with someone, esp. of the opposite sex. *v.t.* to cause to click. *n.* a slight sharp sound; a sharp clicking sound used in some languages of southern Africa.

client (klī'ənt), *n.* one who employs a lawyer as his agent or to conduct a case; one who entrusts any business to a professional man; a customer; a person who is receiving help from a social work or charitable agency. **clientage, clientelage** (-tē'lij), *n.* one's clients collectively; the system of patron and client; the condition of a client. **clientless,** *a.* **clientship,** *n.*

clientele (klēontel'), *n.* clients collectively; customers, patients, frequenters etc.; clientship.

cliff (klif), *n.* a steep, precipitous rock; a precipice. **cliff face,** *n.* the sheer drop of a cliff. **cliff-hanger,** *n.* a story, film etc. that keeps one in suspense till the end; a highly dramatic, unresolved ending to an instalment of a serial. **cliff-hanging,** *a.* **cliffy,** *a.* having cliffs; craggy.

climacteric (klīmak'tərik, -te'-), *n.* a critical period in human life; a period in which some great change is supposed to take place in the human constitution, or in the fortune of an individual; the menopause. *a.* critical. **climacterical** (-te'-), *a.* climacteric.

climactic CLIMAX.

climate (klī'mət), *n.* a region considered with reference to its weather; the temperature of a place, and its meteorological conditions generally, with regard to their influence on animal and vegetable life; a currently prevailing character, as of political conditions. **climatic** (-mat'-), *a.* **climatically,** *adv.* **climatology** (-tol'-), *n.* the investigation of climatic phenomena and their causes. **climatological** (-loj'-), *a.*

climax (klī'maks), *n.* a rhetorical figure in which the sense rises gradually in a series of images, each exceeding its predecessor in force or dignity; the highest point, culmination; (an) orgasm. *v.i.* to reach a climax. *v.t.* to bring to a culminating point. **climactic** (-mak'-), *a.*

climb (klīm), *v.t.* to ascend, esp. by means of the hands and feet; of a plant, to ascend by means of tendrils. *v.i.* to ascend; to slope upwards; to rise in rank or prosperity; to partake in the sport of climbing. *n.* an ascent. **to climb down,** to descend (using hands and feet); to abandon one's claims, withdraw from a position, opinion etc. **climbable,** *a.* **climber,** *n.* one who or that which climbs, esp. for sport; a creeper or climbing plant. **climbing,** *n.* mountaineering. *a.* that climbs. **climbing-frame,** *n.* a framework of bars for children to climb on.

climbing-irons, *n.pl.* a set of spikes fastened to the feet to assist in climbing.
clime (klīm), *n.* (*poet.*) a region, a country.
clinch (klinch), *v.t.* to secure a nail by hammering down the point; to drive home or establish (an argument etc.). *v.i.* to hold an opponent by the arms in boxing etc. *n.* an act of clinching; a grip, a hold-fast; (*coll.*) a lovers' embrace. **clinch-nail**, *n.* a nail with a malleable end adapted for clinching. **clincher**, *n.* one who or that which clinches; a conclusive argument or statement. **clincher-built** CLINKER-BUILT.
cline (klīn), *n.* a gradation of forms seen in a single species over a given area.
cling (kling), *v.i.* (*past, p.p.* **clung** (klŭng)) to adhere closely and tenaciously, esp. by twining, grasping or embracing; to be faithful (to). **to cling together**, to form one mass; to resist separation. **Clingfilm®**, *n.* a kind of thin polythene film which clings to itself or other plastics, glass etc., used for airtight wrapping. **clingy**, *a.* (*derog.*) clinging; showing great emotional dependence. **clinginess**, *n.*
clinic (klin'ik), *n.* medical and surgical instruction, esp. in hospitals; a private hospital, or one specializing in one type of ailment or treatment; a specialist department in a general hospital, esp. for out-patients; a session in which advice and instruction are given on any topic. **clinical**, *a.* pertaining to a patient in bed, or to instruction given to students in a hospital ward; detached, unemotional. **clinical thermometer**, *n.* one for observing the temperature of a patient. **clinically**, *adv.* **clinician** (-ni'shən), *n.* a doctor who works with patients, as opposed to a teacher or researcher.
clink[1] (klingk), *n.* a sharp, tinkling sound, as when two metallic bodies are struck lightly together. *v.t., v.i.* to (cause to) make this sound.
clink[2] (klingk), *n.* (*sl.*) (a) gaol.
clinker[1] (kling'kə), *n.* vitrified slag; fused cinders; bricks run together in a mass by heat.
clinker-built (klingkəbilt'), *a.* (*Naut.*) built with overlapping planks fastened with clinched nails.
clinometer (klīnom'itə, kli-), *n.* an instrument for measuring angles of inclination. **clinometric, -ical** (-met'-), *a.*
Clio (klī'ō), *n.* the muse of epic poetry and history.
clip[1] (klip), *v.t.* (*past, p.p.* **clipped**) to cut with shears or scissors; to trim; to cut out or away; to pare the edges of (as coin); to cut short (speech) by omitting (letters, syllables etc.); to cancel (a ticket) by snipping a piece out; to hit sharply. *v.i.* to run or go swiftly (along). *n.* a shearing or trimming; the whole wool of a season; a blow; an extract from a film; (*sl.*) a (fast) rate. **to clip the wings of**, to put a check on the ambitions of. **clip joint**, *n.* (*sl.*) a night-club etc. which overcharges. **clipper**, *n.* one who or that which clips; a fast sailing-vessel with a long sharp bow and raking masts; (*pl.*) a tool for clipping hair, nails etc. **clippie**, *n.* (*coll., dated*) a bus conductress. **clipping**, *n.* a piece clipped off; (*esp. N Am.*) a press-cutting; the action of the verb.
clip[2] (klip), *n.* a small appliance for gripping, holding or attaching. *v.t.* to attach with a chip. **clipboard**, *n.* a flat board with a spring clip at one end, to hold paper for writing.
clip-clop CLOP.
clique (klēk), *n.* an exclusive set; a coterie of snobs. **cliquish**, *a.* **cliquishness**, *n.* **cliquism**, *n.* **cliquy**, *a.*
clitoris (klit'əris), *n.* (*pl.* **-ides** (-dēz)) a small erectile body situated at the apex of the vulva in the female and corresponding to the penis in the male. **clitoral**, *a.*
Cllr, (*abbr.*) councillor.
cloaca (klōā'kə), *n.* the excretory cavity in certain animals, birds, insects etc. **cloacal**, *a.*
cloak (klōk), *n.* a loose, wide, outer garment; a covering; a disguise, a blind, a pretext. *v.t.* to cover with or as with a cloak; to disguise; to hide. **cloak-and-dagger**, *a.* of a story etc., involving mystery and intrigue. **cloakroom**, *n.* a room where coats, small parcels etc. can be deposited; (*euphem.*) a lavatory. **cloaking**, *n.* disguise, concealment; a rough, woollen material for cloaks.
clobber (klob'ə), *n.* a kind of coarse paste used by cobblers to conceal cracks in leather; (*sl.*) clothes; (*sl.*) belongings, equipment. *v.t.* (*sl.*) to beat; to criticize harshly. **clobberer**, *n.*
cloche (klosh), *n.* an orig. bell-shaped glass cover put over young or tender plants to preserve them from frost; a close-fitting hat shaped like a cloche.
clock[1] (klok), *n.* an instrument for measuring time, consisting of a mechanism actuated by a spring, weight or electricity; (*coll.*) a taximeter; (*coll.*) a speedometer. *v.t.* to time using a clock; to achieve (a stated speed or time) in a race; (*sl.*) to hit; (*sl.*) to see, notice. **against the clock**, of a task etc., (requiring) to be finished by a certain time. **round the clock**, continuously through the day and night. **to clock in, on, out, off**, to register on a specially constructed clock the times of arrival at, and departure from, work. **to clock up**, to register (a specified time, speed, amout of time, distance etc.). **to put the clock back**, to revert (as if) to an earlier era. **clock golf**, *n.* a putting game played on lawns. **clock-maker**, *n.* one who makes clocks. **clock radio**, *n.* an alarm clock combined with a radio, which can use the radio as an alarm instead of a bell. **clock-watcher**, *n.* one who is careful not to work any longer than necessary. **clock-watching**, *n.* **clockwise**, *adv.* as the hands of a clock, from north to east to south to west. **clockwork**, *n.* the movements of a mechanical clock; a train of wheels producing motion in a similar fashion. **like clockwork**, with unfailing regularity: mechanically, automatically.
clock[2] (klok), *n.* an ornamental pattern on the side of the leg of a sock. **clocked**, *a.*
clod (klod), *n.* a lump of earth or clay; a mass of earth and turf; a clod-hopper. **clod-hopper**, *n.* an awkward rustic; a bumpkin; a clumsy shoe. **clod-pate, -poll**, *n.* a stupid, thick-headed fellow. **clod-pated**, *a.* **cloddish**, *a.* loutish, coarse, clumsy. **cloddishness**, *n.*
clog (klog), *n.* a kind of shoe with a wooden sole; a wooden shoe, a sabot. *v.t.* (*past, p.p.* **clogged**) to hinder; to obstruct; to choke up. *v.i.* to be obstructed or encumbered with anything heavy or adhesive. **clog-dance**, *n.* a dance in which the performer wears clogs in order to produce a loud accompaniment to the music. **clogged**, *a.* **cloggy**, *a.* clogging; adhesive, sticky. **clogginess**, *n.*
cloisonné (klwazənā', klwa'-), *a.* partitioned, divided into compartments. *n.* (also **cloisonné enamel**) enamelwork in which the coloured parts are separated by metallic partitions.
cloister (klois'tə), *n.* a place of religious seclusion; a religious house or convent; a series of covered passages usu. arranged along the sides of a quadrangle in monastic, cathedral or collegiate buildings. *v.t.* to shut up in or as in a cloister or convent. †**cloister-garth**, *n.* a yard or grass-plot surrounded with cloisters, often used as a burial-ground. **cloistered**, *a.* (*fig.*) out of things; sheltered from the world, reality etc. **cloistral**, *a.*
clone (klōn), *n.* a number of organisms produced asexually from a single progenitor; any such organism; (*coll.*) an exact copy. *v.t.* to produce a clone of. **clonal**, *a.*
clonk (klongk), *v.i.* to make a short dull sound, as of two solid objects striking each other. *v.t.* (*coll.*) to hit. *n.* such a dull sound.
clop (klop), **clip-clop** (klip-), *n.* the sound of a horse's hoof striking the ground. *v.i.* (*past, p.p.* **clopped**) to make such a sound; to move while making such a sound.
close[1] (klōz), *v.t.* to shut to; to fill (up) an opening in; to enclose, to shut in; to bring together or unite together; to be the end of, conclude; to complete, to settle. *v.i.* to

shut; to coalesce; to come to an end, to cease; to agree, to come to terms; to come (nearer) together; to grapple, to come to hand-to-hand fighting. *n.* the act of closing; an end, a conclusion; a grapple, a hand-to-hand struggle. **to close down**, of factories, works etc., to (cause to) shut, cease work; (*Radio, TV*) to go off the air. **to close in**, to shut in, to enclose; to come nearer; (of nights) to get longer. **to close on, over, upon**, to shut over; to grasp. **to close up**, to block up, fill in; to bring or come together. **to close with**, to accede to, to agree or consent to; to grapple with. **a closed book**, something about which one is completely ignorant. **closed circuit**, *n.* a circuit with a complete, unbroken path for the current to flow through. **closed-circuit television**, a television system for a restricted number of viewers in which the signal is transmitted to the receiver by cable. **closed shop**, *n.* (a workplace or trade having) an arrangement whereby an employer may hire only members of a specified trades union. **closing time**, *n.* the hour at which a shop, office, pub or other establishment is declared closed for work or business. **closer**, *n.* one who or that which closes or concludes; a cut or moulded brick used in some bonds to complete the pattern at the corner of a wall, etc.

close[2] (klōs), *a.* closed, shut fast; pronounced with the lips or mouth partly shut; solid, dense, compact; near together in time or space; intimate, familiar; of a family, mutually caring and loyal; concise, compressed, coherent; nearly alike; attentive; following the original closely; precise, minute; without ventilation, oppressive, stifling; of the weather, warm, still and damp; restricted, limited; difficult to obtain, scarce, as money; secret, reticent; parsimonious, penurious. *adv.* nearby, near; tightly, securely; thickly or compactly. *n.* the precincts of a cathedral or abbey; a small enclosed field; (often used in street-names for) a narrow passage or street; (*Scot.*) an entry from the street to a tenement. **at close quarters**, (fighting) hand to hand; close together. **close by, to, upon**, within a short distance (from); very near. **close breeding**, *n.* breeding between animals closely akin. **close call, shave, thing**, a narrow escape or failure. **close file**, *n.* a row of people standing or moving one immediately behind the other. **close-fisted**, *a.* niggardly, miserly, penurious. **close-fistedness**, *n.* **close-fitting**, *a.* of clothes, fitting tightly to the outline of the body. **close harmony**, *n.* a kind of singing in which all the parts lie close together. **close-hauled**, *a.* (*Naut.*) kept as near as possible to the point from which the wind blows. **close-knit**, *a.* of a family or community, tightly bound together. **close season**, *n.* the season during which it is illegal to kill certain fish or game. **close-tongued**, *a.* reticent, silent. **close-up**, *n.* a view taken with the camera near to the subject. **close vowel**, *n.* one pronounced with a small opening of the lips, or with the mouth-cavity contracted. **closely**, *adv.* **closeness**, *n.*

closet (kloz'it), *n.* a small room for privacy and retirement; a water-closet; (*esp. N Am.*) a cupboard. *a.* (*attrib.*) secret; private. **to be closeted with**, to hold a confidential conversation with. **to come out of the closet**, to declare or make public one's inclinations, intentions etc., esp. to declare one's homosexuality.

closure (klō'zhə), *n.* the act of shutting; the state of being closed; the power of terminating debate in a legislative or deliberative assembly; a device which closes e.g. a paper bag, or fastens e.g. a handbag, brooch etc.

clot (klot), *n.* a small coagulated mass of soft or fluid matter, esp. of blood; (*sl.*) a silly person. *v.t.* (*past, p.p.* **clotted**) to make into clots. *v.i.* to become clotted. **clotted cream**, *n.* cream produced in clots on new milk when it is simmered, orig. made in Devonshire.

cloth (kloth), *n.* a woven fabric of wool, hemp, flax, silk or cotton, used for garments or other coverings (the name of the material is expressed except in the case of wool); any textile fabric, material; a piece of this; a tablecloth. **cloth of gold, silver**, *n.* a fabric of gold or silver threads interwoven with silk or wool. **man of the cloth**, a clergyman. **the cloth**, the dress of a profession, esp. the clerical, from their usu. wearing black cloth. **cloth binding**, *n.* book covers in linen or cotton cloth. **cloth-cap**, *a.* (*sometimes derog.*) belonging to or characteristic of the working class. **cloth-eared**, *a.* (*coll., derog.*) deaf; inattentive. **cloth hall**, *n.* a cloth exchange. **cloth-worker**, *n.* a maker of cloth.

clothe (klōdh). *v.t.* (*past, p.p.* **clothed, clad** (klad)) to furnish, invest or cover with or as with clothes. **clothes** (klōdhz, klōz), *n.pl.* garments; dress; bed-clothes. **clothes-basket**, *n.* a basket for soiled clothes. **clothes-brush**, *n.* a brush for removing dust etc. from clothes. **clothes-horse**, *n.* a frame for drying clothes on; (*coll.*) a fashionably-dressed person. **clothes-line**, *n.* a line for drying clothes on. **clothes-moth**, *n.* a type of moth, the larvae of which are destructive to cloth. **clothes-peg, -pin**, *n.* a peg used to fasten clothes on a line. **clothes-prop**, *n.* a pole for supporting a clothes-line. **clothing**, *n.* clothes, dress, apparel.

clothier (klō'dhiə), *n.* a manufacturer of cloth; one who deals in cloth or clothing.

clotted CLOT.

cloud (klowd), *n.* a mass of visible vapour condensed into minute drops or vesicles, and floating in the upper regions of the atmosphere; a volume of smoke or dust resembling a cloud; the dusky veins or markings in marble, precious stones etc.; a dimness or patchiness in liquid; obscurity, bewilderment, confusion of ideas; suspicion, trouble; any temporary depression; a great number, a multitude (of living creatures, or snow, arrows etc.) moving in a body. *v.t.* to overspread with clouds, to darken; to make (a liquid) less transparent; to confuse or adversely affect (one's judgement); to make gloomy or sullen. *v.i.* to grow cloudy. **in the clouds**, absent (minded). **on cloud nine**, very happy, elated. **under a cloud**, in temporary disgrace or misfortune. **cloud-burst**, *n.* a sudden and heavy fall of rain. **cloud-capped**, *a.* of mountains, with summit or summits veiled with clouds; very lofty. **cloud chamber**, *n.* an apparatus in which high-energy particles are tracked as they pass through a vapour. **cloud-cuckoo-land**, *n.* an unreal world. **cloudscape**, *n.* a view or picture of clouds; **cloudless**, *a.* unclouded; clear, bright. **cloudlessly**, *adv.* **cloudlessness**, *n.* **cloudlet** (-lit), *n.* a little cloud. **cloudwards**, *adv.* **cloudy**, *a.* consisting of or overspread with clouds; marked with veins or spots; obscure, confused; wanting in clearness. **cloudily**, *adv.* **cloudiness**, *n.*

clough (klŭf), *n.* a ravine; a narrow valley.

clout (klowt), *n.* a piece of cloth, rag etc., used to patch or mend; (*coll.*) a blow with the open hand, esp. on the head; (*coll.*) power, influence. *v.t.* to stud or fasten with clout-nails; (*coll.*) to strike with the open hand. **clout-nail**, *n.* a short nail with a large head for the soles of heavy boots and shoes.

clove[1] (klōv), *n.* one of the dried, unexpanded flower-buds of the **clove-tree**, used as a spice; (*pl.*) a spirituous cordial flavoured with this. **oil of cloves**, a medicinal oil obtained from cloves.

clove[2] (klōv), *n.* a small bulb forming one part of a compound bulb, as in garlic, the shallot etc.

clove[3] (klōv), *past,* CLEAVE. **clove-hitch**, *n.* a safe kind of rope-fastening round a spar or another rope. **cloven** (-vən), *a.* divided into two parts; cleft. **cloven-footed, -hoofed**, *a.* having the hoof divided in the centre, as have the ruminants. **cloven hoof**, *n.* an emblem of Pan or the Devil, an indication of guile or devilish design.

clover (klō'və), *n.* a (usually) three-leaved plant used for fodder. **to be, live in clover**, to be in greatly enjoyable circumstances; to live luxuriously. **cloverleaf**, *n.* a traffic device in which one crossing road passes over the other, and the connecting carriageways, having no abrupt turns, make the shape of a four-leaved clover.

clown (klown), *n.* a clumsy, awkward lout; a rough, ill-bred person; a buffoon in a circus or pantomime. *v.i.* to play silly jokes, to act the buffoon. **clownish,** *a.* **clownishly,** *adv.* **clownishness,** *n.*

cloy (kloi), *v.t.* to satiate, to glut. *v.i.* to tire with sweetness, richness or excess. **cloyless,** *a.* that does not or cannot cloy.

club¹ (klŭb), *n.* a piece of wood with one end thicker and heavier than the other, used as a weapon; a stick bent and (usually) weighted at the end for driving a ball in e.g. golf; a card of the suit clubs. *v.t. (past, p.p.* **clubbed)** to beat with a club. **club-foot,** *n.* a short deformed foot. **club-footed,** *a.* **club-headed,** *a.* having a club-shaped head or top. **club root,** *n.* a disease of plants of the *Brassica* (cabbage) genus in which the lower part of the stem becomes swollen and misshapen owing to the attacks of larvae. **clubs,** *n.sing.* or *pl.* one of the four suits of playing-cards, symbolized by the trefoil. **clubbed,** *a.* club-shaped.

club² (klŭb), *n.* an association of persons combined for some common object, as social intercourse, literature, politics, sport, hobbies etc., governed by self-imposed regulations; the house or building in which such an association meets; the body of members collectively. *v.t. (past, p.p.* **clubbed)** to contribute for a common object. *v.i.* to join (together) for a common object. **in the club,** *(sl.)* pregnant. **club car,** *n.* (*N Am.*) a railway-coach designed like a lounge, usu. with a bar. **clubhouse,** *n.* the building occupied by a club, or in which it holds its meetings; the establishment maintained by the members of a social or sports club, at which they meet, drink, dine or lodge temporarily. **clubland,** *n.* the district round St James's and Pall Mall where the principal London clubs are situated. **clubman, -woman,** *n.* a member of a club. **clubroom,** *n.* a room in which a club or society meets. **club sandwich,** *n.* one made with three slices of bread and two different fillings. **club soda,** *n.* (a) soda water. **clubbable,** *a.* having the qualities necessary for club life; sociable. **clubber,** *n.* a member of a club; one who uses a club. **clubdom** (-dəm), *n.*

cluck (klŭk), *n.* the guttural call of a hen; any similar sound. *v.i.* to utter the cry of a hen to her chickens. *v.t.* to call, as a hen does her chickens.

clue (kloo), *n.* anything of a material or mental nature that serves as guide, direction or hint for the solution of a problem or mystery. **I haven't a clue,** I have no idea whatever. **to clue in, up,** *(sl.)* to inform. **clued up,** well-informed. **clueless,** *a.* ignorant; stupid.

clumber (klŭm'bə), *n.* a variety of spaniel.

clump (klŭmp), *n.* a thick cluster (of trees, shrubs or flowers); a thick mass (of small objects or organisms); a heavy blow. *v.i.* to tread in a heavy and clumsy fashion; to form or gather into a clump or clumps. *v.t.* to make a thick cluster of. **clumpy,** *a.*

clumsy (klŭm'zi), *a.* awkward, ungainly, ill-constructed; rough, rude, tactless. **clumsily,** *adv.* **clumsiness,** *n.*

clung, *past, p.p.* CLING.

Cluniac (kloo'niak), *n.* one of a reformed branch of Benedictines founded at Cluny, Saône-et-Loire, France, in the 10th cent. *a.* pertaining to this order.

clunk (klŭngk), *v.i., v.t.* to (cause to) make a short, dull sound, as of metal striking a hard surface. *n.* such a sound. **clunky,** *a.* (*coll.*) heavy, unwieldy.

cluster (klŭs'tə), *n.* a number of things of the same kind growing or joined together; a bunch; a number of persons or things gathered into or situated in a close body; a group, a crowd. *v.i.* to come or to grow into clusters. *v.t.* to bring or cause to come into a cluster or clusters. **cluster bomb,** *n.* a bomb which explodes to scatter a number of smaller bombs.

clutch¹ (klŭch), *n.* a snatch, a grip, a grasp; the hands; a device for connecting and disconnecting two revolving shafts in an engine; a gripping device; (*pl.*) claws, tyrannical power. *v.t., v.i.* to seize, clasp or grip with the hand; to snatch. **disc, plate clutch,** a clutch which operates as a result of friction between the surfaces of discs. **clutch bag,** *n.* a woman's handbag, without a handle, carried in the hand.

clutch² (klŭch), *n.* a batch (of eggs in a nest); a brood (of chickens); a group, set (of people or things).

clutter (klŭt'ə), *n.* a mess, confusion; irrelevant echoes on a radar screen from sources other than the target. **to clutter up,** to fill or cover untidily.

Cm, (*chem. symbol*) curium.

cm, (*abbr.*) centimetre(s).

Cmd., (*abbr.*) command paper (before 1956).

CMG, (*abbr.*) Companion of (the Order of) St Michael and St George.

Cmnd., (*abbr.*) command paper (since 1956).

CND, (*abbr.*) Campaign for Nuclear Disarmament.

CNS, (*abbr.*) central nervous system.

CO, (*abbr.*) Colonial Office (before 1966); commanding officer; Commonwealth Office (since 1966); conscientious objector.

Co, (*chem. symbol*) cobalt.

Co., (*abbr.*) company; county.

c/o, (*abbr.*) care of.

co-, *pref.* with, together, jointly, mutually; joint, mutual; as in *coalesce, cooperate; coeternal, coefficient, coequal; co-partner.*

coach (kōch), *n.* a large, closed, four-wheeled horse-drawn vehicle, used for purposes of state, for pleasure, or (with regular fares) for travelling; a railway carriage; a long-distance, usu. single-deck, bus; a tutor who prepares for examinations; one who trains sports players; (*N Am.*) in an aircraft, second class. *v.t.* to prepare for an examination; to train; to instruct or advise in preparation for any event. *v.i.* to work as a tutor. **coach-box,** *n.* the seat on which the driver of a horse-drawn coach sits. **coach-builder,** *n.* one who builds or repairs the bodywork of road or rail vehicles. **coachbuilt,** *a.* of vehicles, built individually by craftsmen. **coach-driver,** *n.* the driver of a long-distance bus. **coach-house,** *n.* an outhouse to keep a coach or carriage in. **coachline,** *n.* a decorative paint line on the side of a motor vehicle. **coachman,** *n.* the driver of a horse-drawn coach; a livery servant who drives a carriage. **coachmanship,** *n.* **coach station,** *n.* a building where long-distance buses arrive and depart. **coach-whip,** *n.* a whip used by a driver of a horse-drawn coach. **coach-work,** *n.* the bodywork of a road or rail vehicle. **coachful,** *n.* as many as will fill a coach.

coact (kōakt'), *v.i.* to act in concert. **coaction,** *n.* **coactive,** *a.* **coactively,** *adv.*

coadapted (kōədap'tid), *a.* mutually adapted or suited. **coadaptation** (-adəptā'-), *n.*

coadjacent (kōəjā'sənt), *a.* mutually near, contiguous. **coadjacence, -ency,** *n.*

coadjutor (kōaj'ətə), *n.* an assistant, a helper, esp. to a bishop; a colleague. **coadjutorship,** *n.* **coadjutrix** (-triks), *n. fem.*

coagent (kōā'jənt), *n.* one who or that which acts with another. *a.* acting together (with). **coagency,** *n.*

coagulate (kōag'ūlāt), *v.t.* to cause to curdle; to convert from a fluid into a curd-like mass. *v.i.* to become curdled. **coagulable,** *a.* **coagulant,** *n.* a substance which causes coagulation. **coagulation,** *n.* **coagulative,** *a.* **coagulator,** *n.* **coagulum** (-ləm), *n.* (*pl.* **-la** (-lə)) a coagulated mass; a coagulant; a blood-clot.

coal (kōl), *n.* a black solid opaque carbonaceous substance of vegetable origin, obtained from the strata usu. below the surface of the earth, and used for fuel; a piece of this or other combustible substance, ignited, burning or charred. *v.t.* to supply with coal. *v.i.* to take in a supply of coal. **to carry coals to Newcastle**, to do anything superfluous or unnecessary, esp. taking something to where plenty more of the same is already present. **to haul over the coals**, to reprimand. **coal-bed, -seam**, *n.* a stratum of or containing coal. **coal-black**, *a.* as black as coal; jet-black. **coal-bunker**, *n.* a receptacle for coal. **coal-cellar**, *n.* a basement for storing coal. **coal-dust**, *n.* powdered coal. **coal face**, *n.* the exposed surface of a coal-seam. **coalfield**, *n.* a district where coal abounds. **coal-fired**, *a.* of a furnace, heating system etc., fuelled by coal. **coal-fish**, *n.* the black cod. **coal-flap, -plate**, *n.* an iron cover for the opening in a pavement etc. for putting coal into a cellar. **coal gas**, *n.* impure carburetted hydrogen obtained from coal and used for lighting and heating. **coal-hole**, *n.* a small cellar for keeping coals. **coal-measures**, *n.pl.* the upper division of the carboniferous system. **coal-merchant**, *n.* a retail seller of coal. **coalmine**, *n.* a mine from which coal is obtained. **coalminer**, *n.* **coal-owner**, *n.* (*formerly*) the owner of a colliery. **coal-pit**, *n.* a coalmine. **coal-screen**, *n.* a large screen or sifting-frame for separating large and small coals. **coal-scuttle**, *n.* a utensil for holding coals for immediate use. **coal-seam** COAL-BED. **coal-tar**, *n.* tar produced in the destructive distillation of bituminous coal. **coal-tit**, *a* dark species of the tit family. **coaling station**, *n.* a port where steamships may obtain coal, esp. (formerly) one established by a government for the supply of coal to warships. **coaler**, *n.* a ship that transports coal. **coalless**, *a.* **coaly**, *a.*

coalesce (kōəles'), *v.i.* to grow together; to unite into masses or groups spontaneously; to combine; to form a coalition. **coalescence**, *n.* **coalescent**, *a.*

coalition (kōəlish'ən), *n.* a union of separate bodies into one body or mass; a combination (often temporary) of persons, parties or states, having different interests. **coalition government**, *n.* a government in which two or more parties of varying politics unite in order between them to obtain a majority for the implementation of policies. **coalitionist**, *n.*

coamings (kō'mingz), *n.pl.* (*Naut.*) the raised borders round hatches etc. for keeping water from pouring into the hold.

coarse (kaws), *a.* of inferior quality; large in size or rough in texture; rude, rough, vulgar; unpolished, unrefined; indecent. **coarse fish**, *n.* any freshwater fish not of the salmon family. **coarse fishing**, *n.* **coarsely**, *adv.* **coarsen**, *v.t.* to make coarse. *v.i.* to grow or become coarse. **coarseness**, *n.* **coarsish**, *a.*

coast (kōst), *n.* that part of the border of a country which is washed by the sea; the seashore; a swift rush downhill on cycle or motor-car, without using motive power or applying brakes. *v.i.* to sail near or in sight of the shore; to descend an incline on a cycle or a mechanically propelled vehicle without applying motive power or brakes; to proceed without much effort (e.g. to victory). **the coast is clear**, the road is free; the danger (esp. of being caught) is over. **coast-to-coast**, *a.* from coast to coast, across a whole continent or island. **coastguard**, *n.* (one of) a body of people who watch the coast to save those in danger, give warning of wrecks, and prevent the illegal landing of persons and goods. **coastline**, *n.* the exact outline of a coast. **coastal**, *a.* of, pertaining to or bordering on a coastline. **coaster**, *n.* a ship which never ventures far from land, but goes from port to port; a small tray for a bottle or decanter on a table; a small mat under a glass to protect a surface from spills. **coasting-trade**, *n.* trade between the ports of the same country. **coastward, -wards**, *adv.* **coastwise**, *adv.*

coat (kōt), *n.* an upper outer garment with sleeves; (*N Am.*) a suit jacket or sports jacket; the hair or fur of any beast; the natural external covering of an animal; a layer of any substance covering or protecting something. *v.t.* to cover; to overspread with a layer of anything. **coat of arms**, an escutcheon or shield of arms; armorial bearings. **coat of mail**, armour worn on the upper part of the body, consisting of iron rings or scales fastened on a stout linen or leather jacket. **greatcoat** GREAT. **redcoat** RED. **to trail one's coat, coat-tails**, to invite attack. **to turn one's coat**, to change sides, hence **turn-coat** (see TURN). **coat-hanger**, *n.* a shaped piece of plastic, wood or metal on which to hang up coats, dresses etc. **coated**, *a.* **coatee** (-tē', *n.* a short coat for a baby. **coating**, *n.* a covering, layer or integument; the act of covering; a substance spread over as a cover or defence; cloth for coats. **coatless**, *a.*

coati (kōah'ti), **coatimundi** (-mŭn'di), *n.* a raccoon-like carnivorous animal with a long, flexible snout, from S America, Central America and Mexico.

co-author (kōaw'thə), *n.* one who writes a book together with someone else.

coax (kōks), *v.t.* to persuade by fondling or flattery; to wheedle, to cajole. *v.i.* to practise cajolery in order to persuade. **coaxer**, *n.* **coaxingly**, *adv.*

coaxial, -ial (kōak'səl, -siəl), *a.* having a common axis. **coaxial cable, coax** (kō'aks), *n.* a cable with both a central conductor and an outer conductor.

cob (kob), *n.* a short stout horse for riding; a kind of wicker basket; a cobnut; (*also* **cob-swan**) a male swan; the spike of Indian corn; a mixture of clay and straw used for building walls in the west of England; a small round loaf. **cobnut**, *n.* a variety of the cultivated hazel. **cob-wall**, *n.* a wall built of mud or clay, mixed with straw. **cobby**, *a.* [etym. doubtful]

cobalt (kō'bawlt), *n.* a reddish-grey, or greyish-white, brittle, hard metallic element, at. no. 27; chem. symbol Co. **cobalt-blue**, *n.* a deep blue pigment of alumina and cobalt. **cobaltic** (-bawl'-), *a.* **cobaltiferous** (-tif'-), *a.* **cobaltous**, *a.* **cobalto-**, *comb. form.*

cobber (kob'ə), *n.* (*Austral., sl.*) a pal, a chum.

cobble[1] (kob'l), *v.t.* to mend or patch (as shoes); to put (together), make or do clumsily. **cobbler**, *n.* one who mends shoes; a mender or patcher; a clumsy workman; (*N Am.*) a cooling drink of wine, sugar, lemon and ice; a type of (fruit) pudding with a crunchy topping. **cobbler's wax**, *n.* a resinous substance used for waxing thread. **cobblers**, *n.pl., inter. (sl.)* nonsense.

cobble[2] (kob'l), *n.* a rounded stone or pebble used for paving. *v.t.* to pave with cobbles.

Cobdenism (kob'dənizm), *n.* the doctrines of Richard Cobden (1804–65), esp. Free Trade, pacifism and non-intervention. **Cobdenite**, *n.* an adherent of Cobdenism.

cobelligerent (kōbəlij'ərənt), *a.* waging war jointly with another. *n.* one who joins another in waging war.

coble (kō'bl), *n.* a flat, square-sterned fishing-boat with a lug-sail and six oars.

Cobol (kō'bol), *n.* a high-level computer language for commercial use.

cobra (kō'brə), *n.* a viperine snake from Africa and tropical Asia, which distends the skin of the neck into a kind of hood when excited.

cobweb (kob'web), *n.* the web or net spun by a spider for its prey; the material or a thread of this. **to blow away the cobwebs**, to refresh oneself in the open air. **cobwebbed**, *a.* covered with or full of cobwebs. **cobwebby**, *a.*

coca (kō'kə), *n.* the dried leaf of a Peruvian plant chewed as a narcotic stimulant; the plant itself. **cocaine** (-kān', *n.* an alkaloid contained in coca leaves, used as a narcotic and medicinally as a local anaesthetic. **cocaine baby**, *n.* a baby (likely to be) born addicted to cocaine owing to the addiction of the mother during pregnancy.

Coca-Cola® (kōkəkō'lə), *n.* a brown-coloured soft drink flavoured with coca leaves etc.

coccus (kok'əs), *n.* (*pl.* **-cci** (kok'sī, kok'ī)) one of the dry one-seeded carpels into which a fruit breaks up; a spherical bacterium. **coccal**, *a.* **coccoid**, *a.*

coccyx (kok'siks), *n.* (*pl.* **-xes, -ges** (-jēz)) the lower solid portion of the vertebral column, the homologue in man of the tail of the lower vertebrates. **coccygeal** (koksij'iəl), *a.* **coccyg(eo)-,** *comb. form.*

cochineal (kochinēl'), *n.* a dye-stuff made from the dried bodies of the female cochineal insect, used in dyeing, as a food colouring and in the manufacture of scarlet and carmine pigments.

cochlea (kok'liə), *n.* (*pl.* **-leae** (-liē)) the anterior spiral division of the internal ear. **cochlean,** *a.* **cochleate** (-ət), **-ated** (-ātid), *a.* circular, spiral; (*Bot.*) twisted like a snail-shell.

cock[1] (kok), *n.* the male of birds, particularly of domestic fowls; a male salmon; a vane in the form of a cock, a weathercock; (used as a form of address) good fellow; a tap or valve for regulating the flow through a spout or pipe; (*taboo sl.*) the penis; the hammer of a gun or pistol, which, striking against a piece of flint or a percussion-cap, produces a spark and explodes the charge; (*sl.*) nonsense. **cock-a-doodle-doo** (-ədoodəldoo'), the crow of the domestic cock; a nursery name for the bird. **cock-a-hoop,** strutting like a cock; triumphant, exultant. *adv.* exultantly, with crowing and boastfulness. **cock-and-bull,** applied to silly, exaggerated stories or canards. **cock of the walk,** a masterful person; a leader, a chief. **cock-crow, -crowing,** *n.* the crow of a cock; early dawn. **cock-eye,** *n.* (*sl.*) an eye that squints. **cock-eyed,** *a.* (*sl.*) having squinting eyes; irregular, ill-arranged; askew; eccentric, ludicrous. **cock-fight, -fighting,** *n.* a battle or match of gamecocks. **cock-horse,** *n.* a stick with a horse's head at the end, on which children ride. **cockloft,** *n.* an upper loft, a garret. **cock-match,** *n.* a cock-fight. **cockpit,** *n.* a pit or area where game-cocks fight; a part of the lower deck of a man-of-war, used as a hospital in action; that portion of the fuselage of an aircraft where the pilot and crew (if any) are accommodated; the driver's compartment of a racing car. **cock-robin,** *n.* a male robin. **cockscomb,** *n.* the comb of a cock; a fool's cap. **cock-sparrow,** *n.* a male sparrow. **cocksure,** *a.* self-confident, arrogantly (sometimes too) certain. **cocksurely,** *adv.* **cocksureness,** *n.*

cock[2] (kok), *n.* the act of turning or sticking anything upward; the turn so given, as of a hat, a nose, an inclination of the head etc. *v.t.* to set erect; to set (the hat) jauntily on one side; to turn up (the nose), to turn (the eye) in an impudent or knowing fashion; to raise the hammer of (a gun). **at half cock,** before being fully prepared. **to cock a snook,** to put the thumb to the nose with the fingers spread out. **to cock up,** (*sl.*) to ruin, spoil, by incompetence. **cocked hat,** *n.* a pointed triangular hat. **knocked into a cocked hat,** (*sl.*) completely defeated or out-performed. **cock-up,** *n.* (*sl.*) a bungled failure, a mess.

cockade (kəkād'), *n.* a knot of ribbons or feathers, or a rosette, worn in the hat as a badge. **cockaded,** *a.*

cock-a-leekie (kokəlē'ki), *n.* a soup made from a fowl boiled with leeks.

cockatoo (kokətoo'), *n.* a large crested parrot, usu. white, from the Indian Archipelago and Australia. **cockatiel** (-tēl'), *n.* a small cockatoo or parrot.

cockatrice (kok'ətris, -tris), *n.* the fabled basilisk; (*Her.*) a cock with a serpent's tail.

cockchafer (kok'chāfə), *n.* a large brown beetle that makes a whirring noise in flying.

cockerel (kok'ərəl), *n.* a young make domestic fowl.

cocker spaniel (kokə span'yəl), *n.* a small breed of spaniel often kept as a pet.

cocket (kok'it), *n.* (formerly) a custom-house seal; a customs' receipt for duty on exported goods.

cockle[1] (kok'l), *n.* a bivalve belonging to a genus of molluscs; its ribbed shell; a shallow skiff. **to warm the cockles of one's heart,** to make one feel emotionally lifted. **cockle-boat,** *n.* a small and shallow skiff. **cockle-shell,** *n.* the shell of any species of cockle.

cockle[2] (kok'l), *v.i.* to crease. *v.t.* to curl, pucker up, crease or make to bulge. *n.* a crease or wrinkle (on paper). **cockled,** *a.*

cockney (kok'ni), *n.* a native of London (traditionally, a person born within sound of the bells of St-Mary-le-Bow, Cheapside); (*loosely*) the East London accent; one who speaks with it; a city resident. *a.* pertaining to (a) cockney. **cockneydom** (-dəm), *n.* **cockneyfy,** *v.t.* **cockneyish,** *a.* **cockneyism,** *n.* **cockneyize, -ise,** *v.t., v.i.*

cockroach (kok'rōch), *n.* an insect resembling a beetle, and a pest in kitchens etc.

cockswain (kok'sn, -swān), COXSWAIN.

cocktail (kok'tāl), *n.* a drink taken before a meal, usu. gin or other spirit with bitters and flavourings; an appetizer consisting of a mixture of cold foods, as *prawn cocktail*; a dessert of a like mixture, as *fruit cocktail*; any (often dangerous) mixture of assorted ingredients, e.g. drinks, drugs. **cocktail dress,** *n.* a dress suitable for semi-formal occasions. **cocktail stick,** *n.* a thin pointed stick for eating snack foods with.

cock-up COCK[2].

cocky (kok'i), *a.* impudent, overly self-confident. **cockily,** *adv.* **cockiness,** *n.*

cocoa[1] (kō'kō), **coconut palm,** *n.* a tropical palm tree. **coconut,** *n.* the fruit of this, a large, rough, hard-shelled nut with a white edible lining containing a sweet liquid known as **coconut milk. coconut butter,** *n.* the solid oil obtained from the lining of the coconut. **coconut matting,** *n.* coarse matting made from the fibrous husk of the nut. **coconut shy,** *n.* a kind of skittles in which the aim is to knock coconuts off sticks.

cocoa[2] (kō'kō), *n.* a powdered preparation from the seeds of the cacao tree; a drink made from this. **cocoa-bean,** *n.* the cacao seed. **cocoa butter,** *n.* a buttery substance extracted from the cacao nut in the manufacture of cocoa.

cocoon (kəkoon'), *n.* a silky covering spun by the larvae of certain insects in the chrysalis state; any analogous case made by other animals; any protective covering; a preservative coating sprayed onto machinery etc. *v.t.* to wrap in, or as if in, a natural or synthetic cocoon.

cocotte (kəkot'), *n.* a small dish in which food is cooked and served.

COD, (*abbr.*) cash on delivery.

cod[1] (kod), *n.* a large deep-sea food-fish. **codling** (-ling), *n.* a young cod. **codfish,** *n.* **cod-liver oil,** *n.* oil from the liver of the cod, rich in vitamins A and D. **cod-piece** (kod'pēs), *n.* a baggy appendage in the front of breeches or of the tight hose worn in the 15th and 16th cents to cover male genitals.

cod[2] (kod), *v.t.* (*past, p.p.* **codded**) (*sl.*) to hoax.

coda (kō'də), *n.* (*Mus.*) an adjunct to the close of a composition to enforce the final character of the movement. **codetta,** *n.* a short coda.

coddle (kod'l), *v.t.* to treat as an invalid or baby, to pamper; to cook (esp. eggs) gently in water. **coddled,** *a.*

code (kōd), *n.* a collection of statutes; a digest of law; a body of laws or regulations systematically arranged; (*Mil., Nav.*) a system of signals; a series of characters, letters or words used for the sake of brevity or secrecy; the principles accepted in any sphere of art, taste, conduct etc. *v.t.* to put into a code. **code name, number,** *n.* a short name or number used for convenience or secrecy. **codify,** *v.t.* to reduce to a code. **codification** (-fi-), *n.* **codifier,** *n.*

codeine (kō'dēn), *n.* an alkaloid obtained from opium and used as a narcotic and analgesic.

codex (cō'deks), *n.* (*pl.* **codices** (-disēz)) a manuscript volume, esp. of the Bible or of texts of classics; (*Med.*) a list of prescriptions.

codger (koj'ə), *n.* (*coll.*) an odd old person.

codicil (kod'isil), *n.* an appendix to a will, treaty etc.

codicillary (-sil'-), *a.*
codify etc. CODE.
codling COD[1].
codswallop (kodz'wolǝp), *n.* (*sl.*) nonsense.
coed, co-ed (kō'ed), *n.* (*chiefly N Am.*) a girl being educated in a coeducational establishment. *a.* coeducational.
coeducation (kōedūkā'shǝn), *n.* education of the two sexes together. **coeducational,** *a.*
coefficient (kōifish'ǝnt), *n.* anything cooperating; the cofactor of an algebraical number; in 4*ab*, 4 is the **numerical** and *ab* the **literal coefficient;** (*Phys.*) a number denoting the degree of a quality. **differential coefficient,** the ratio of the change of a function of a variable to the change in that variable.
coel- COEL(O)-.
coelacanth (sē'lǝkanth), *n.* the only known living representative of the fossil group of fish Crossopterygii, first captured off S Africa in 1953.
coelenterate (sēlen'tǝrāt, -rǝt), *n.*, *a.* (of or belonging to) any individual of the suborder of creatures containing the sponges, jellyfish, etc.
coeliac, (*esp. N Am.*) **celiac** (sē'liak), *a.* pertaining to the abdomen. **coeliac disease,** *n.* a condition involving defective digestion of fats.
coel(o)-, (*esp. N Am.*) **cel(o)-,** *comb. form.* (*Biol.*) hollow.
coen(o)-, cen(o)-, *comb. form.* common.
coenobite, cenobite (sē'nǝbīt), *n.* a monk living in a religious community. **coenobitic, -ical** (-bit'-), *a.* **coenobitism,** *n.*
coequal (kōē'kwǝl), *a.* equal with another; of the same rank, dignity etc. *n.* one of the same rank. **coequality** (-kwol'-), *n.* **coequally,** *adv.*
coerce (kōœs'), *v.t.* to restrain by force; to enforce by compulsion. *v.i.* to employ coercion (in government). **coercible,** *a.* **coercibleness,** *n.* **coercion** (-shǝn), *n.* compulsion of a free agent; government by force. **coercionary,** *a.* **coercionist,** *n.* **coercive,** *a.* having power or authority to coerce. **coercively,** *adv.*
coessential (kōisen'shǝl), *a.* of the same essence or substance, consubstantial. **coessentiality** (-shial'-), *n.* **coessentially,** *adv.*
coeternal (kōitœ'nǝl), *a.* equally eternal with another. **coeternally,** *adv.* **coeternity,** *n.*
coeval (kōē'vǝl), *a.* of the same age, date of birth or origin; existing at or for the same period. *n.* a contemporary. **coevality** (-val'-), *n.* **coevally,** *adv.*
coexecutor (kōigzek'ūtǝ), *n.* a joint executor. **coexecutrix** (-triks), *n. fem.* (*pl.* **-trices** (-sēz)).
coexist (kōigzist'), *v.i.* to exist together with and tolerate mutually. **coexistence,** *n.* mutual toleration by regimes with differing ideologies or systems of government. **coexistent,** *a.*
coextension (kōiksten'shǝn), *n.* equal extension. **coextensive,** *a.* **coextensively,** *adv.*
C of E, (*abbr.*) Church of England.
coffee (kof'i), *n.* (a cup of) a beverage made from the ground roasted seeds of a tropical Asiatic and African shrub; the last course at dinner consisting of coffee; the seeds of the tree; the tree itself; a pale brown colour, like milky coffee. **coffee bar,** *n.* a cafe where coffee, snacks etc. are served. **coffee-bean, -berry,** *n.* a coffee seed. **coffee-grounds,** *n.pl.* the sediment or lees of coffee-berries after infusion. **coffee-house,** *n.* a house where coffee and other refreshments are sold, esp. in 18th cent. London. **coffee-mill,** *n.* a machine for grinding coffee-beans. **coffee morning,** *n.* a party, held at mid-morning, where coffee is served. **coffee shop** COFFEE BAR. **coffee-stall,** *n.* a street stall where non-alcoholic beverages and snacks are sold throughout the night. **coffee table,** *n.* a low table in a sitting room. *a.* suitable for display on a coffee table, esp. of large and expensively produced illustrated books.

coffer (kof'ǝ), *n.* a chest or box for holding valuables; (*pl.*) a treasury, funds, financial resources. **coffer-dam,** *n.* a watertight enclosure which exposes a river bed etc., used in laying foundations of piers, bridges etc. **coffered,** *a.* enclosed in a coffer.
coffin (kof'in), *n.* the box in which a corpse is enclosed for burial or cremation. *v.t.* to put into a coffin. **coffin-nail,** *n.* (*sl.*) a cigarette. **coffin-plate,** *n.* a metal plate recording name etc., fastened on the lid of a coffin.
cog (kog), *n.* a tooth or projection in the rim of a wheel or other gear for transmitting motion to another part; a person playing a small and unimportant part in any enterprise. *v.t.* to furnish with cogs. **cog-wheel,** *n.* a wheel furnished with cogs. **cogged,** *a.*
cog., (*abbr.*) cognate.
cogent (kō'jǝnt), *a.* powerful, convincing, well-argued. **cogently,** *adv.* **cogency,** *n.*
cogitate (koj'itāt), *v.i., v.t.* reflect, to meditate (on). **cogitable,** *a.* capable of being thought; conceivable by the reason. **cogitation,** *n.* **cogitative,** *a.* meditative. **cogitatively,** *adv.* **cogitativeness,** *n.*
cognac (kon'yak), *n.* French brandy of fine quality, esp. that distilled in the neighbourhood of *Cognac*, in SW France.
cognate (kog'nāt), *a.* akin, related; of common origin; of the same kind or nature; (*Philol.*) derived from the same linguistic family or from the same word or root. *n.* (*Law*) a blood relation, as distinct from *agnate*, which is through the father only; a cognate word. **cognately,** *adv.* **cognateness,** *n.*
cognition (kognish'ǝn), *n.* the act of coming to know something; the faculty of perceiving, conceiving, and knowing, as distinguished from the feelings and the will; a perception, intuition, or conception; (*Law*) cognizance. **cognitive** (kog'-), *a.*
cognizance, -isance (kog'nizǝns), *n.* knowledge; recognition; (*Law*) judicial notice; knowledge not requiring proof; acknowledgment. **cognizable,** *a.* knowable. **cognizably,** *adv.* **cognizant,** *a.* having cognizance or knowledge (of); (*Law*) competent to take judicial notice of.
cognomen (kognō'mǝn), *n.* a surname; the last of the three names of an ancient Roman citizen; a nickname.
cognoscente (konyǝshen'ti, -sen'-), *n.* (*pl.* **-ti** (-ti)) a connoisseur, one who is knowledgeable on a subject.
cohabit (kōhab'it), *v.i.* to live together, esp. as husband and wife (without being legally married). **cohabitant, cohabiter,** *n.* **cohabitation,** *n.*
coheir (kōeǝ'), *n.* a joint heir. **coheiress,** *n. fem.*
cohere (kǝhiǝ'), *v.i.* to stick together; to hold together, remain united; to be logically consistent. **coherence, -ency,** *n.* **coherent,** *a.* that coheres; remaining united; logically connected, consistent. **coherently,** *adv.*
coheritor (kōhe'ritǝ), *n.* a coheir.
cohesion (kǝhē'zhǝn), *n.* coherence; consistency; the force uniting molecules of the same nature. **cohesive** (-siv), *a.* **cohesively,** *adv.* **cohesiveness,** *n.*
coho (kō'hō), *n.* a Pacific variety of salmon.
cohort (kō'hawt), *n.* the tenth part of a Roman legion; a body of soldiers; (*coll.*) a colleague or accomplice; a set of people in a population sharing a common attribute, e.g. age or class.
COHSE (kō'zi), (*abbr.*) Confederation of Health Service Employees.
coif (koif), *n.* a close-fitting cap worn in the Middle Ages. *v.t.* to cover with a coif. **coifed,** *a.*
coiffeur (kwafœ'), *n.* a hairdresser. **coiffeuse** (-fœz'), *n.fem.* **coiffure** (-fū ǝ'), *n.* a hairstyle.
coign (koin), *n.* a quoin. **coign of vantage,** a projecting corner affording a good view.
coil (koil), *v.t.* to wind (e.g. a rope) into rings; to twist. *v.i.* to wind itself, as a snake or creeping plant. *n.* a series of concentric rings into which anything is coiled up, a

length of anything coiled up; a single turn of anything coiled up; a coiled lock of hair; a wire wound in rings to form a resistance or an inductance; a metal or plastic device inserted in the uterus as a contraceptive. **to coil up**, to twist or be twisted into rings or a spiral shape.
coin (koin), *n.* a piece of metal stamped and current as money; money, esp. metal money. *v.t.* to mint or stamp (as money); to invent. *v.i.* to make counterfeit money. **false coin**, an imitation of money in base metal; a spurious fabrication, untruths, deceits etc. **to coin a phrase**, to use a supposed new expression (usu. a cliché or a pun). **to coin money, to coin it in**, *(sl.)* to make money rapidly. **coin-box**, *n.* a coin-operated telephone. **coin-op, -operated**, *a.* of a machine, operated by inserting a coin. **coin-op**, *n.* a launderette with coin-operated machines. **coiner**, *n.* one who coins money, esp. one who makes counterfeit coin.
coinage (koi'nij), *n.* the act of coining; the pieces coined (collectively); the monetary system in use; something (esp. a verbal usage) invented.
coincide (kōinsīd'), *v.i.* to correspond in time, place, relations etc.; to happen at the same time; (*Geom.*) to occupy the same position in space; to agree, to concur. **coincidence** (-in'si-), *n.* the act, fact, or condition of coinciding; a remarkable instance of apparently fortuitous concurrence. **coincident** (-in'si-), *a.* that coincides. **coincidently**, *adv.* **coincidental** (-siden'təl), *a.* coincident; characterized by or of the nature of coincidence. **coincidentally**, *adv.*
coinheritance (kōinhe'ritəns), *n.* a joint inheritance. **coinheritor**, *n.* a coheir.
Cointreau® (kwan'trō, kwon'-), *n.* a colourless orange-flavoured liqueur. [F]
coir (koiə), *n.* coconut fibre; ropes or matting manufactured therefrom.
coition (kōish'ən), *n.* copulation. **coitus** (ko'itəs, koi'təs), *n.* the act of copulation. **coitus interruptus** (intərŭp'təs), *n.* coitus interrupted before ejaculation into the vagina.
Coke® (kōk), *n.* short for COCA-COLA.
coke[1] (kōk), *n.* (*sl.*) short for COCAINE.
coke[2] (kōk), *n.* coal from which gas has been extracted. *v.t.* to convert into coke.
Col., (*abbr.*) colonel; Colorado; Colossians.
col., (*abbr.*) column.
col (kol), *n.* a depression in a mountain ridge; a saddle or elevated pass; an area of low pressure between two anticyclones.
col- *pref.* form of COM- before *l*.
cola, kola (kō'lə), *n.* a tropical African tree bearing a nut which is used as a condiment, digestive and tonic; a soft drink flavoured with cola-nuts. **cola-nut, -seed**, *n.* the fruit of this.
colander (kol'əndə, kŭl'-), **cullender** (kŭl'-), *n.* a culinary strainer having the bottom perforated with small holes; a similar contrivance used in casting small shot.
cold (kōld), *a.* low in temperature, esp. in relation to normal temperature; lacking heat or warmth; causing or suffering a sensation of loss of heat; without intensity, indifferent; unconcerned; sad, dispirited; not hasty or violent, spiritless; (*Hunting*) not affecting the scent strongly; bluish in tone, as opposed to warm tones such as red, yellow etc.; frigid; dead; (*sl.*) unconscious. *adv.* finally, absolutely; without rehearsal. *n.* (the sensation produced by) absence of warmth; COMMON COLD. **in cold blood**, without passion or excitement. **to catch (a) cold**, to contract a cold; (*coll.*) to run into difficulties. **to have cold feet**, (*sl.*) to be afraid. **to leave cold**, (*coll.*) to fail to excite or interest. **to throw cold water on**, to discourage. **cold-blooded**, *a.* having a body temperature which varies with that of the environment; unfeeling, unimpassioned. **cold-bloodedly**, *adv.* **cold-bloodedness**, *n.* **cold-calling**, *n.* the practice of sales representatives etc. of making unsolicited and unexpected calls in order to sell products or services. **cold cathode**, *n.* one which emits electrons at normal temperatures. **cold chisel**, *n.* a chisel for cutting cold metals. **cold comfort**, *n.* poor consolation, depressing reassurance. **cold cream**, *n.* a cooling ointment of oil and wax used as a cosmetic. **cold cuts**, *n.pl.* cold sliced meat. **cold-drawn**, *a.* of wire etc., drawn in a cold state. **cold feet**, *n.pl.* (*coll.*) a feeling of apprehension or doubt. **cold fish**, *n.* (*coll.*) an unemotional person. **cold frame**, *n.* a glass frame to protect seedlings etc., without actual heat. **cold front**, *n.* (*Meteor.*) the front edge of an advancing mass of cold air. **cold hammer**, *v.t.* to hammer (metals) in a cold state. **cold-hearted**, *a.* unfeeling, indifferent. **cold-heartedly**, *adv.* **cold-heartedness**, *n.* **cold shoulder**, *n.* a rebuff; studied indifference. *v.t.* to treat with studied coolness. **cold sore**, HERPES SIMPLEX. **cold steel**, *n.* cutting weapons, such as sword and bayonet, as opposed to firearms. **cold-storage**, *n.* preservation of perishable foodstuffs by refrigeration; abeyance. **cold sweat**, *n.* sweating accompanied by chill, caused esp. by fear. **cold turkey**, *n.* (*coll.*) symptoms caused by sudden withdrawal of drugs from an addict. **cold war**, *n.* a state of psychological tension between two countries without actual hostilities. **coldish**, *a.* **coldly**, *adv.* **coldness**, *n.*
cold-short (kōld'shawt), *a.* of metals, brittle when cold.
cole (kōl), *n.* the cabbage; the rape. **cole-rape**, *n.* a turnip. **cole-seed**, *n.* rape-seed. **coleslaw**, *n.* a salad made of shredded raw cabbage. **colewort**, *n.* the common cabbage.
colemouse COALMOUSE.
coleopter, -teran (koliop'tə, -rən), *n.* any individual of the Coleoptera or beetles, an order of insects having the fore wings converted into sheaths for the hinder wings. **coleopterist**, *n.* **coleopterous**, *a.*
cole-tit (kōl'tit), COALMOUSE.
coley (kō'li), COAL-FISH.
col(l)-, colo-, *comb. form.* colon.
colibri (kol'ibri), *n.* a kind of humming-bird.
colic (kol'ik), *n.* acute pains in the bowels. **colicky**, *a.*
coliseum (kolǝsē'əm), COLOSSEUM.
colitis (kəlī'tis), *n.* inflammation of the colon.
collaborate (kəlab'ərāt), *v.t.* to work jointly with another; to cooperate with an enemy in occupation of one's own country. **collaboration**, *n.* **collaborator**, *n.*
collage (kolahzh'), *n.* a picture made of pieces of paper, fabric etc., glued on to a surface; any collection of diverse things or ideas. **collagist**, *n.*
collagen (kol'əjən), *n.* a fibrous protein that yields gelatin.
collapse (kəlaps'), *v.i.* to fall in, as the sides of a hollow vessel; to shrink together; to break down, to suffer from physical or nervous prostration; to come to nothing. *n.* a falling in; complete failure; general prostration. **collapsed**, *a.* **collapsible**, *a.*
collar (kol'ə), *n.* something worn round the neck, either as a separate article of dress, or as forming part of some garment; a leather etc. loop round the neck of a horse, dog etc.; meat pickled and rolled; anything shaped like a collar or ring; the chain or other ornament for the neck worn by the knights of an order; a ring or round flange; an astragal, a cincture. *v.t.* to seize by the collar; to put a collar on; to capture; to pickle and roll (as meat); (*coll.*) to seize; (*sl.*) to steal. **collar of SS, esses**, a chain worn as badge by adherents of the House of Lancaster, and still a part of certain official costumes. **to slip the collar**, to free oneself. **collar-beam**, *n.* a tie-beam. **collar-bone**, *n.* the clavicle. **collar-harness**, *n.* harness attached to the collar. **collar-stud**, *n.* a stud to hold a collar to a shirt etc. **collar-work**, *n.* uphill work for a horse or (*fig.*) for a person; drudgery. **collared**, *a.* **collarless**, *a.*
collard (kol'əd), *n.* a variety of cabbage.
collarette (kolǝret'), *n.* a small collar.
collate (kəlāt'), *v.t.* to bring together in order to compare; to examine critically (esp. old books and manuscripts in

order to ascertain by comparison points of agreement and difference); to place in order (as printed sheets for binding); to present to a benefice (used when a bishop presents to a living in his own diocese). **collation**, *n.* the act of collating; a light meal. **collator**, *n.*

collateral (kəlat'ərəl), *a.* side by side; parallel; concurrent, subordinate; having the same common ancestor but not lineally related (as the children of brothers). *n.* a collateral relation; collateral security. **collateral security**, *n.* security for the performance of any contract over and above the main security. **collaterally**, *adv.*

collation COLLATE.

colleague[1] (kol'ēg), *n.* one associated with another in any office or employment. **colleagueship**, *n.*

colleague[2] (kəlēg'), *v.t.* to join as ally.

collect[1] (kol'ekt), *n.* a brief comprehensive form of prayer, adapted for a particular day or occasion.

collect[2] (kəlekt'), *v.t.* to gather together into one body, mass or place; to gather (money, taxes, subscriptions, books, curiosities etc.) from a number of sources; to concentrate, to bring under control; to infer; to fetch. *v.i.* to come together; to meet together. *a., adv.* (*N Am.*) of a telephone call, paid for by the recipient. **to collect on**, (*sl.*) to make money out of. **to collect oneself**, to recover one's self-possession. **collectable**, **-ible**, *a.* **collected**, *a.* gathered, brought together; cool, self-possessed. **collectedly**, *adv.* **collectedness**, *n.* **collector**, *n.* **collectorate** (-rət), **-ship**, *n.*

collectanea (kolektā'niə), *n.pl.* a collection of passages; a miscellany. **collectaneous**, *a.*

collection (kəlek'shən), *n.* the act of collecting; that which is collected; an assemblage of natural objects, works of art etc.; money collected for a charitable etc. purpose; an accumulation.

collective (kəlek'tiv), *a.* tending to collect; collected, formed by gathering; collectivized. *n.* a cooperative or collectivized organization or enterprise. **collective bargaining**, *n.* the method whereby employer and employees determine the conditions of employment. **collective noun**, *n.* a noun in the singular number expressing an aggregate of individuals. **collective ownership**, *n.* ownership of land, capital, and other means of production by those engaged in the production. **collective unconscious**, *n.* in Jungian theory, the part of the unconscious mind which is inherited and contains universal thought patterns and memories. **collectively**, *adv.* **collectivism**, *n.* the principle of public ownership of land, industry etc. **collectivist**, *n.*, *a.* **collectivity** (kolektiv'-), *n.* **collectivize**, **-ise**, *v.t.* to organize on collectivist lines. **collectivization**, *n.*

colleen (kolēn'), *n.* a girl. [Ir]

college (kol'ij), *n.* a body or community of persons, having certain rights and privileges, and devoted to common pursuits; an independent corporation of scholars, teachers, and fellows forming one of the constituent bodies of a university; a similar foundation independent of a university; an institution for higher education; a large secondary school. **College of Arms, Herald's College**, a corporation presided over by the Earl Marshal, for granting armorial bearings etc. **College of Cardinals, Sacred College**, in the Roman Catholic Church, the papal council of cardinals. **college of education**, a training college for teachers. **College of Justice**, (*Sc.*) the supreme civil courts. **college pudding**, *n.* a small baked pudding for one person. **colleger**, *n.* a pupil on the foundation of a school, esp. at Eton. **collegial** (-lē'jəl), *a.* constituted as a college. **collegian** (-lē'jən), *n.* a member of a college; a student at a university.

collegiate (kəlē'jət), *a.* pertaining to a college; containing a college; instituted or regulated as a college. **collegiate church**, *n.* a church which, though not a cathedral, has an endowed chapter of canons; (*Sc., N Am.*) a Presbyterian church under a joint pastorate. **collegiate school**, *n.* a school organized to resemble a college.

collegium (kəlē'jiəm), *n.* (*pl.* **-gia**) an ecclesiastical body not under state control.

collet (kol'ət), *n.* a band or ring; a flange or socket; the part of a ring in which a stone is set.

collide (kəlīd'), *v.i.* to come into collision or conflict.

collie (kol'i), *n.* a Scottish sheep-dog; a breed of showdogs.

collier (kol'yə), *n.* one who works in a coal-mine; a ship employed in the coal trade; one of her crew. **colliery**, *n.* a coal-mine.

colligate (kol'igāt), *v.t.* to bind together; to bring into connection. **colligation**, *n.* **colligative** (kəlig'ətiv), *a.* (*Chem.*) of a physical property, dependent on the concentration of particles present rather than their nature.

collimate (kol'imāt), *v.t.* to adjust the line of sight in a telescope; to make the axes of lenses or telescopes collinear. **collimation**, *n.* **collimator**, *n.* an instrument which fixes the line of sight of a telescope.

collinear (kəlin'iə), *a.* (*Geom.*) in the same straight line.

collingual (kəling'gwəl), *a.* having the same language.

Collins[1] (kol'inz), *n.* a letter of thanks after a visit.

collins[2] (kol'inz), *n.* a drink made of spirits mixed with soda water, fruit juice, ice etc.

collision (kəlizh'ən), *n.* the act of striking violently together; the state of being dashed or struck violently together; conflict; clashing of interests. **collision course**, *n.* a course which will result inevitably in a collision.

collocate (kol'əkāt), *v.t.* to place together; to arrange; to station in a particular place. **collocation**, *n.*

collocutor (kəlok'utə, kol'-), *n.* one who takes part in a conversation or conference.

collodion (kəlō'diən), *n.* a gummy solution of pyroxylin in ether and spirit, formerly used in photography and medicine. **collodioned**, *a.* **collodionize**, **-ise**, *v.t.*

collogue (kəlōg'), *v.i.* to talk confidentially or plot together.

colloid (kol'oid), *n.* an uncrystallizable, semisolid substance, capable of only very slow diffusion or penetration. *a.* like a colloid; like glue. **colloidal**, *a.* **colloidize**, **-ise**, *v.t.*

collop (kol'əp), *n.* a slice of meat; a small piece or slice of anything.

colloquium (kəlō'kwiəm), *n.* (*pl.* **-quia** (-ə)) an academic conference; a seminar.

colloquy (kol'əkwi), *n.* a conference, conversation or dialogue; a court or presbytery in the Presbyterian Churches. **colloquial** (-lō'-), *a.* pertaining to or used in common or familiar conversation; not used in correct writing or in literature. **colloquialism** (-lō'-), *n.* **colloquially**, *adv.* **colloquist**, *n.* a collocutor.

collotype (kol'ətīp), *n.* a method of lithographic printing using a film of gelatine as the negative; a print obtained in this way.

collude (kəlood'), *v.i.* to act in concert, to conspire. **colluder**, *n.*

collusion (kəloo'zhən), *n.* secret agreement for a fraudulent or deceitful purpose, esp. to defeat the course of law. **collusive**, *a.* **collusively**, *adv.*

colluvies (kəloo'viēz), *n.* filth; a mixed mass of refuse.

collyrium (kəli'riəm), *n.* (*pl.* **-ria** (-ə)) an eye-salve, an eyewash.

collywobbles (kol'iwoblz), *n.* (*coll.*) a stomach-ache; extreme nervousness.

Colo., (*abbr.*) Colorado.

colo-, COL(1)-.

colobus (kol'əbəs), *n.* (a member of) a genus of African monkeys, long-haired and lacking thumbs.

Colocasia (koləkā'ziə), *n.* a genus of plants of the arum family.

colocynth (kol'əsinth), *n.* the bitter cucumber or bitter apple, *Citrullus colocynthis*, or its fruit; a purgative drug

Cologne 158 **column**

obtained from it.
Cologne (kəlōn'), *n.* a city of Germany; eau-de-Cologne.
Cologne water EAU-DE-COLOGNE.
colon¹ (kō'lon), *n.* a punctuation mark (:) used to mark the start of a list or long quotation etc.; also used in expressing an arithmetical ratio.
colon² (kō'lon), *n.* the largest division of the intestinal canal, from the caecum to the rectum. **colonic**, *a.*
colonel (kœ'nəl), *n.* the commander of a regiment or of a battalion. **colonelcy, -ship**, *n.*
colonial (kəlō'niəl), *a.* of or pertaining to a colony, esp. to those of the British Empire or to those in America that became the US in 1776. *n.* an inhabitant of a colony. **colonial goose**, *n.* (*Austral., coll.*) baked leg of mutton boned and stuffed. **colonialism**, *n.* a policy of tight control over, or exploitation of, colonies. **colonialist**, *n.* **colonially**, *adv.*
colonize, -ise (kol'əniz), *v.t.* to found a colony in; to settle in; to people with colonists; of animals and plants, to establish a population in (a new environment). *v.i.* to found a colony or colonies. **colonist**, *n.* **colonization**, *n.* **colonizer**, *n.*
colonnade (kolənād'), *n.* a series or range of columns at certain intervals.
colony (kol'əni), *n.* a settlement founded by emigrants in a foreign country, and remaining subject to the parent state; a group of people of the same nationality or occupation in an area of a town; a body of organisms living or growing together. **crown colony** CROWN. **the Colonies**, (*Hist.*) those constituting the British Empire; those in America which became the United States.
colophon (kol'əfon), *n.* a publisher's identifying symbol on a book.
colophony (kəlof'əni), *n.* a dark-coloured resin obtained from turpentine. **colophonate** (-nət), *n.* **colophonic** (koləfon'-), *a.*
Colorado (kolərah'dō), *n.* one of the States of the American Union. **Colorado beetle**, *n.* a small yellow and black-striped beetle, very destructive to the potato.
coloration (kŭlərā'shən), *n.* the act of colouring; method of putting on or arranging colours; (*Biol.*) particular marking, arrangement of colours. **colorant** (kŭl'-), *n.* a substance used to impart colour, a pigment etc.
coloratura (kolərətoo'rə), *n.* (*Mus.*) the ornamental use of variation, trills etc.; a singer, esp. a soprano, capable of singing such ornamented music.
colorific (kŭlərif'ik), *a.* imparting colour; highly-coloured.
colorimeter (kŭlərim'itə), *n.* an instrument for measuring the brightness etc. of colours. **colorimetry**, *n.*
colosseum (koləsē'əm), *n.* a large amphitheatre or place of entertainment.
colossus (kəlos'əs), *n.* (*pl.* **-ssi** (-sī), **-ssuses**) a statue of gigantic size; a person or thing of great power or influence. **colossal**, *a.* **colossally**, *adv.*
colostomy (kəlos'təmi), *n.* the surgical formation of an artificial anus by an incision made into the colon.
colostrum (kəlos'trəm), *n.* the first milk secreted after parturition.
colotomy (kəlot'əmi), *n.* surgical incision into the colon.
colour¹ (kŭl'ə), (*esp. N Am.*) **color**, *n.* the sensation produced by waves of resolved light upon the optic nerve; that property of bodies by which rays of light are resolved so as to produce certain effects upon the eye; any one of the hues into which light can be resolved; a pigment; effect of colour, and of light and shade in drawings and engravings; a (healthy) complexion; (dark) pigmentation of the skin; any tint or hue, as distinguished from black or white; (*pl.*) a flag, standard, or ensign; (*pl.*) a badge of party, membership of a league, society, club etc.; (*fig.*) appearance, esp. false appearance; pretext; (*Mus.*) timbre; emotional quality; vividness, animation. **off-colour**, *a.* faulty; out of sorts; slightly obscene. **primary colours**, the fundamental colours from which others can be obtained by mixing (for paints red, blue and yellow; for transmitted light red, blue, green). **prismatic colours**, those into which pure white light is resolved when dispersed. **secondary colours**, colours produced by combinations of two primary colours. **to change colour**, to turn pale; to blush. **to join the colours**, to enlist. **to show one's colours**, to reveal one's opinions, feelings or designs. **water-colour** WATER. **with flying colours**, brilliantly, successfully; with signal credit. **colour bar**, *n.* social etc. discrimination against non-Caucasian people. **colour-blind**, *a.* **colour-blindness**, *n.* total or partial inability to distinguish different colours, esp. the primary colours; DALTONISM. **colour-code**, *n.* a system of marking different things, e.g. electric wires, in different colours for ease of identification. **colour-fast**, *a.* dyed with fast colours. **colour printing**, *n.* reproduction in two or more colours. **colour-scheme**, *n.* a set of colours used together in decorating. **colour-sergeant**, *n.* a non-commissioned officer in the infantry ranking above an ordinary sergeant. **colour supplement**, *n.* a (usu. weekly) supplement to a newspaper printed in colour and containing articles on lifestyle, entertainment etc. **colourway**, *n.* a particular colour scheme, e.g. in a fabric. **colourable**, *a.* plausible; feigned. **colourableness**, *n.* **colourably**, *adv.* **colouration** COLORATION. **Coloured**, *a.* (sometimes *derog.* or *offensive*) of other than Caucasian race; in S Africa, of mixed race. *n.* a person of non-Caucasian or (in S Africa) of mixed race. **coloured**, *a.* having a colour; esp. marked by any colour except black or white; having a specious appearance; (*Bot.*) of any colour except green. **colourful**, *a.* having bright colour(s); exotic. **colouring**, *n.* the act of giving a colour to; the colour applied; the art or style of using colour; a false appearance; the colour of a person's skin, hair etc. **colourist**, *n.* one who colours; a painter distinguished for management of colour. **colouristic** (-ris'-), *a.* **colourize, -ise**, *v.t.* to put (e.g. a black-and-white film) into colour, esp. using computers. **colourization**, *n.* **colourless**, *a.* without colour; pale; neutral-tinted, subdued, dull; lacking vigour. **colourlessly**, *adv.* **colourlessness**, *n.*
colour² (kŭl'ə), *v.t.* to give colour to; to paint, to dye; to put in a false light, to misrepresent. *v.i.* to become coloured; to turn red, to blush.
-colous, *comb. form.* (*Biol.*) inhabiting (a certain environment).
colposcope (kol'pəskōp), *n.* an instrument for examining the cervix and upper vagina. **colposcopy** (-pos'kəpi), *n.*
Colt®¹ (kōlt), *n.* an early type of American revolver. [S. *Colt*, 1814–62]
colt² (kōlt), *n.* a young male horse, till about the age of four; a young, inexperienced fellow; in sport, a member of a junior team. **colt's-foot**, *n.* a coarse-leaved, yellow-flowered weed, *Tussilago farfara.* **colthood**, *n.* **coltish**, *a.*
colter (kōl'tə), COULTER.
coluber (kol'ūbə), *n.* a genus of innocuous snakes. **colubriform** (-lū'brifawm), *a.* shaped like the genus *Coluber;* belonging to the group Colubriformes, which contains the innocuous snakes. **colubrine** (-brīn), *a.* resembling snakes.
columbarium (koləmbeə'riəm), *n.* (*pl.* **-ria** (-ə)) a pigeon house.
Columbian (kəlŭm'biən), *a.* pertaining to the United States of America.
Columbine¹ (kol'əmbīn), *n.* the female dancer in a pantomime, the sweetheart of Harlequin.
columbine² (kol'əmbīn), *a.* pertaining to or resembling a dove. *n.* a plant with five-spurred flowers, constituting the genus *Aquilegia.*
columbium (kəlŭm'biəm), NIOBIUM.
column (kol'əm), *n.* an upright pillar usu. of stone or wood, supporting a structure or ornamental; anything re-

colza (kŏl′zə), *n.* rape; rape-seed. **colza-oil,** *n.* oil expressed from this.
Com.¹, (*abbr.*) commander; commission(er); committee; Commonwealth.
com.², (*abbr.*) commerce; committee; common; commune.
com-, *pref.* with; together; in combination; completely.
coma¹ (kō′mə), *n.* a state of absolute unconsciousness, characterized by the absence of any response to external stimuli or inner need. **comatose** (-tōs), *a.* in a coma; (*coll.*) sleepy, sluggish.
coma² (kō′mə), *n.* (*pl.* **-mae** (-mē)), the nebulous covering of the nucleus of a comet; the assemblage of branches constituting the head of a forest tree; the tuft of hairs terminating certain seeds.
comate, comose (kō′māt, -mōs), *a.* (*Bot.*) bearing a tuft of hair at the end.
comb¹ (kōm), *n.* a toothed instrument for separating and dressing the hair; an ornamental toothed contrivance for fastening women's hair when dressed; a rake-shaped instrument with a short handle for cleaning wool or flax; the red, fleshy tuft on the head of a fowl, esp. the cock; the cellular substance in which bees deposit their honey; the crest of a wave; a ridge. *v.t.* to separate, dress, or arrange with a comb; to curry a horse; to make a thorough search of. *v.i.* to form a crest and roll over as waves. **to comb out,** to remove with a comb; to find and remove; to search thoroughly. **comb-out,** *n.* **-combed,** *comb. form.* **comber¹,** *n.* one who or that which combs; a wave that forms a long crest and rolls over. **combing,** *n.* a cleaning or dressing with a comb; (*pl.*) hair removed by a comb.
comb.², (*abbr.*) combined; combination.
combat (kom′bat), *v.i.* (*past, p.p.* **combated**) to contend, to fight, to struggle. *v.t.* to oppose, to contend against, to fight with. *n.* a fight, a battle. **single combat,** *n.* a duel. **combat fatigue,** *n.* nervous disturbance occurring in a very stressful situation, such as on the battlefield. **combatable,** *a.* **combatant** (-bə-), *a.* engaged in combat; bearing arms; antagonistic. *n.* one who fights or contends with another. **combative** (-bə-), *a.* inclined to combat. **combatively,** *adv.* **combativeness,** *n.*
combe (koom), *n.* a valley on the side of hills or mountains; a valley running up from the sea.
comber¹ COMB¹.
comber² (kom′bə), *n.* the wrasse, *Serranus cabrilla*, and the gaper, *Labrus maculatus*, var. *comber*; both British fish.
combination (kombinā′shən), *n.* the act or process of combining; the state of being combined; a combined body or mass; a union or association of people; combined action; chemical union; (*pl. Math.*) the different collections which may be made of certain given quantities in groups of a given number; a motor-cycle and sidecar; (*pl.*) vest and knickers combined in one garment; a sequence of chess moves; the sequence of numbers that will open a combination lock. **combination lock,** *n.* a lock which opens only when a set of dials is turned to show a particular combination of numbers. **combinative,** *a.* **combinatory,** *a.*
combine¹ (kəmbīn′), *v.t.* to cause to unite or coalesce; to bring together; to have at the same time (properties or attributes usu. separate). *v.i.* to unite, to coalesce; to be joined or united in friendship or plans; (*Chem.*) to unite by chemical affinity. **combined operations,** *n.pl.* operations in which sea, air and land forces work together under a single command. **combining form,** *n.* a form of a word not used alone but only in making up compound words.
combine² (kom′bīn), *n.* a combination, esp. of persons or companies to further their own commercial interests; a ring; (also **combine harvester**) a combined reaping and threshing machine.
combo (kom′bō), *n.* a combination of instruments, a small band in jazz and popular music.
combust (kəmbŭst′), *v.t.* to consume with fire. **combustible,** *a.* flammable; irascible, hot-tempered. *n.* a flammable substance. **combustibility** (-bil′-), *n.* **combustibleness,** *n.* **combustion** (-chən), *n.* the act or process of burning; (*Chem.*) the oxidation of a substance accompanied by light and heat; oxidation of the tissue of organisms or of decomposing organic matter. **spontaneous combustion,** the ignition of a body by the development of heat within itself. **combustive,** *a.*
Comdr, (*abbr.*) Commander.
Comdt, (*abbr.*) Commandant.
come (kŭm), *v.i.* (*past* **came** (kām), p.p. **come**) to move from a distance to a place nearer to the speaker; to approach; to be brought to or towards; to arrive; to advance; to appear; to arrive at some state or condition; to happen; to result, to originate (from); to become; (*sl.*) to experience orgasm. *v.t.* (*sl.*) to act the part of, to produce. *int.* used to excite attention or rouse to action (when repeated it expresses remonstrance or rebuke). **as it comes,** without additions or alterations. **come again,** say that again. **come along,** make haste. **come off it,** (*coll.*) stop behaving (or talking) so stupidly or pretentiously. **come to that,** in that case. **come what may,** whatever happens. **how come?** how does this happen? **to come,** in the future. **to come about,** to result, to happen; to recover; to change direction. **to come across,** to meet with acccidentally. **to come and go,** to appear and disappear; to pay a short call. **to come at,** to reach; to gain access to. **to come away,** to become parted or separated. **to come back,** to return; to recur to memory; to retort. **to come between,** to damage a relationship between (two people). **to come by,** to pass near; to obtain. **to come down,** to descend (to); to be humbled; to decide. **to come down to,** to amount to. **to come down (up)on,** to reprimand; to pay out. **to come down with,** to pay over (money); to contract (an ailment). **to come easy, expensive** etc., to prove easy, costly etc. **to come forward,** to make oneself known. **to come home,** to return home; to affect nearly; to be fully comprehended. **to come in,** to enter; to become fashionable; to yield; to become (useful etc.); to assume power. **to come in for,** to obtain, to get (a share of). **to come into,** to acquire, to inherit. **to come into the world,** to be born. **to come it strong,** (*sl.*) to exaggerate. **to come near,** to approach; nearly to succeed. **to come of,** to be descended from; to result from. **to come off,** to part from; to fall off; to get off free; to take place successfully; to appear. **to come on,** to advance; to prosper; to happen; to arise; (*imper.*) approach; proceed. **to come out,** to be revealed, become public; to be introduced into society; to be published; to declare something openly, esp. one's homosexuality; to be covered (in); to engage in a strike. **to come out of,** to proceed from. **to come out with,** to utter. **to come over,** to change sides; to prevail upon; (*coll.*) to become; to be perceived (as). **to come round,** to change; to recover. **to come short,** to fail. **to come through,** to survive. **to come to,** to amount to; to recover from faintness. **to come to an end,** to cease. **to come to a point,** to taper; to culminate; to reach a crisis. **to come to blows,** to begin fighting. **to come to harm,** to be injured. **to come**

to oneself, to recover one's senses. **to come to pass**, to happen. **to come to stay**, to have qualities of a permanent nature. **to come under**, to be classed as; to be subjected to (authority, influence etc.). **to come up**, to become public or fashionable; to be introduced as a topic. **to come up against**, to encounter (some difficulty). **to come upon**, to attack; to meet with unexpectedly. **to come up to**, to amount to; to be equal to. **to come up with**, to overtake; to produce. **where one came in**, back at the beginning. **come-back**, *n*. a retort; a return to popular favour. **come-down**, *n*. a fall or abasement. **come-hither**, *a*. sexually alluring. **come-on**, *n*. an invitation, encouragement, esp. sexual.

comeatable (kŭmat'əbl), *a*. easy to come at, accessible.

Comecon (kom'ikon), *n*. an economic organization of E European states, founded in 1949.

comedian (kəmē'diən), *n*. an actor or writer of comedy. **comedienne** (-dien'), *n*. *fem*.

comedietta (kəmēdiet'ə), *n*. a slight or brief comedy.

comedo (kom'idō), *n*. (*pl*. **-dos**, **-dones** (-dō'nēz) a blackhead.

comedy (kom'ədi), *n*. a dramatic composition of a light and entertaining character; life or any incident or situation regarded as an amusing spectacle. **comedic** (-mē'-), *a*. **comedist**, *n*.

comely (kŭm'li), *a*. pleasing in person, or in behaviour. **comeliness**, *n*.

comer (kŭm'ə), *n*. one who comes or arrives; a visitor. **all comers**, any one who accepts a challenge.

comestible (kəmes'tibl), *n*. (*usu. pl.*) food.

comet (kom'it), *n*. a heavenly body having a frozen nucleus and a luminous tail, revolving round the sun in a very eccentric orbit. **cometary**, *a*. **cometic** (-met'-), *a*. **cometology** (-tol'-), *n*. the science dealing with comets.

comeuppance (kŭmŭp'əns), *n*. retribution for past misdeeds.

comfit (kŭm'fit), *n*. a dry sweetmeat; a nut or seed coated with sugar.

comfort (kŭm'fət), *v.t*. to cheer, to encourage, to console; to make comfortable. *n*. support or assistance in time of weakness; (that which affords) consolation or encouragement; quiet enjoyment; general well-being, absence of trouble or anxiety; (*pl*.) the material things that contribute to bodily satisfaction; a comforter. **cold comfort** COLD. **comfort station**, *n*. (*N Am*.) a public convenience. **comfortable**, *a*. at ease, in good circumstances, free from hardship, trouble or pain; quietly happy, contented; providing comfort or security. **comfortableness**, *n*. **comfortably**, *adv*. **comforter**, *n*. one who or that which comforts; a woollen scarf; a baby's dummy; (*N Am*.) a quilted coverlet. **Job's comforter**, one who makes a show of comforting but does exactly the opposite. **comforting**, *a*. **comfortless**, *a*. without comfort; cheerless.

comfrey (kŭm'fri), *n*. a tall wild plant with rough leaves and yellowish or purplish flowers, formerly used for healing wounds.

comfy (kŭm'fi), *a*. (*coll.*) short for COMFORTABLE under COMFORT.

comic (kom'ik), *a*. pertaining to comedy, laughable, absurd, provoking mirth. *n*. a comedian; an amusing person; a magazine containing comic strips; the comic aspect of things. **comic opera**, *n*. a type of opera with humorous episodes and usu. some spoken dialogue. **comic strip**, *n*. a usu. comic narrative told in a series of pictures. **comical**, *a*. ludicrous, laughable; exciting mirth. **comicality** (-kal'-), *n*. **comically**, *adv*.

Cominform (kom'infawm), *n*. the Information Bureau of the Communist Parties, founded in 1947, orig. including Yugoslavia.

coming (kŭm'ing), *a*. approaching; future; promising. *n*. the act of approaching or arriving, arrival. **to have it coming**, (*coll.*) to deserve what (unpleasant thing) is about to happen.

comingle (kəming'gl), COMMINGLE.

comity (kom'iti), *n*. affability, courtesy. **comity of nations**, the courtesy by which a nation allows another's laws to be recognized within its territory, so far as is practicable.

Comm.[1], (*abbr.*) Commodore.

comm.[2], (*abbr.*) commentary; commerce; commercial; commonwealth.

comma (kom'ə), *n*. a punctuation mark (,), denoting the shortest pause in reading; (*Mus.*) a minute difference of tone; a butterfly with a white comma-shaped mark beneath the hind-wing. **inverted commas**, raised commas as thus: ' - '; " - " used to indicate quotations. **comma bacillus**, *n*. a comma-shaped bacillus which causes cholera.

command (kəmahnd'), *v.t.* to order, to enforce; to exercise authority over; to dominate; to control; to master. *v.i.* to give orders; to exercise supreme authority. *n*. an order, a mandate; power, authority; the power of dominating or overlooking; a naval or military force under the command of a particular officer; (*Comput.*) an instruction; a working knowledge (of). **at command**, ready for orders. **command-in-chief**, *n*. the supreme command. **to command-in-chief**, to be commander-in-chief (of an army etc.). **command performance**, *n*. a theatrical performance given by royal command. **command paper**, *n*. a government report presented to Parliament. **command post**, *n*. a place used as temporary headquarters by a military commander. **command sequence**, *n*. (*Comput.*) a series of commands for a specific task. **commandant** (koməndant'), *n*. the governor or commanding officer of a place. **commandantship** (-dant'-), *n*. **commanding**, *a*. **commandingly**, *adv*.

commandeer (koməndiə'), *v.t.* to seize for military purposes.

commander (kəmahn'də), *n*. one who commands or is in authority; a general or leader of a body of men; a member of one of the higher grades in some orders of knighthood; a naval officer between a lieutenant and a captain; a police officer in London in charge of a district. **commander-in-chief**, *n*. the officer in supreme command of military forces in an area. **commander-in-chiefship**, *n*. **commandership**, *n*. **commandery**, **-dry**, *n*. in military orders of knighthood, a district or manor administered by a commander.

commandment (kəmahnd'mənt), *n*. a command; a Divine command esp. one of the decalogue. **the Ten Commandments**, the decalogue.

commando (kəmahn'dō), *n*. (*pl*. **-dos**) a body of men called out for military service; a body of men selected and trained to undertake a specially hazardous raid on or behind the enemy lines; a man thus selected; a mobile amphibious force.

commedia dell'arte (kəmā'diə delah'ti), *n*. Italian comedy of the 16th-18th cents., using improvisation and stock characters.

comme il faut (kom ēl fō), *a*. as it should be, correct.

commemorate (kəmem'ərāt), *v.t.* to keep in remembrance (by some solemn act); to be a memorial of. **commemorable**, *a*. **commemoration**, *n*. the act of commemorating; a service, ceremony or festival in memory of some person, deed or event. **commemorative**, *a*. **commemoratively**, *adv*.

commence (kəmens'), *v.i.* to start, to begin; to begin (to do or be). *v.t.* to enter upon; to perform the first act of. **commencement**, *n*. beginning, origin, rise; first instance, first existence; in N American schools, speech day.

commend (kəmend'), *v.t.* to commit to the charge of; to recommend as worthy of notice, regard or favour; to praise. **commend me to**, remember me to. **commendable**, *a*. **commendableness**, *n*. **commendably**, *adv*. **commendation** (kom-), *n*. the act of commending; recommendation of a person to the consideration or favour of

commensal   161   common

another. **commendatory**, *a.* that serves to commend. *n.* commendation, eulogy.
commensal (kəmen'səl), *a.* eating at the same table. *n.* an animal that lives in intimate association with another, without being parasitic. **commensalism**, *n.* **commensality** (komənsal'-), *n.*
commensurable (kəmen'shərəbl), *a.* measurable by a common unit; proportionate (to). **commensurability** (-bil'-), **-ableness**, *n.* **commensurably**, *adv.*
commensurate (kəmen'shərət), *a.* having the same measure or extent; proportional. **commensurately**, *adv.* **commensurateness**, *n.*
comment (kom'ent), *n.* a remark; a criticism; a note interpreting or illustrating a work or portion of a work. *v.i.* to make explanatory or critical remarks or notes (on a book or writing). *v.t.* to remark. **no comment**, I refuse to answer or comment. **commentary**, *n.* a comment; a series of explanatory notes on a whole work; a broadcast description of an event as it takes place. **commentate**, *v.t., v.i.* to act as commentator (of). **commentation**, *n.* **commentator**, *n.* the author of a commentary; an annotator; the broadcaster of a commentary.
commerce (kom'œs), *n.* trade, traffic; the interchange of commodities between nations or individuals.
commercial (kəmœ'shəl), *a.* pertaining to or connected with commerce; done for profit; of chemicals, of poor quality and produced in bulk for industry. *n.* an advertisement broadcast on radio or television. **commercial art**, *n.* graphic art used in advertising etc. **commercial broadcasting**, *n.* broadcasting paid for by advertising or sponsorship. **commercial traveller**, *n.* a company's travelling representative. **commercial vehicle**, *n.* one used for the transport of goods or passengers. **commercialism**, *n.* a trading spirit; commercial practices. **commercialist**, *n.* **commerciality** (-shial'-), *n.* **commercialize**, **-ise**, *v.t.* **commercially**, *adv.*
commerge (kəmœj'), *v.t.* to merge together.
commie (kom'i), *n., a.* (*coll.* often *derog.*) short for COMMUNIST.
comminate (kom'ināt), *v.t.* to threaten vengeance, to denounce. **commination**, *n.* a threat, a denunciation. **comminatory**, *a.*
commingle (kəming'gl), *v.t.* to mingle or mix together; to blend.
comminute (kom'inūt), *v.t.* to reduce to minute particles or to powder; to divide into small portions. **comminuted fracture**, *n.* one in which the bone is broken into small pieces. **comminution** (-nū'-).
commis (kom'i), *n.* an apprentice or assistant waiter or chef.
commiserate (kəmiz'ərāt), *v.t.* to pity; to express pity or compassion for. **commiseration**, *n.* **commiserative**, *a.* **commiseratively**, *adv.*
commissar (kom'isah), *n.* formerly, the head of a department of government in the USSR; a party official responsible for political education.
commissariat (komisea'riət), *n.* that department of the army charged with supplying provisions and stores; formerly, a government department in the USSR.
commissary (kom'isəri), *n.* a commissioner; a deputy; (*Mil.*) an officer in charge of the commissariat; the deputy who supplies a bishop's place in the remote parts of his diocese. **commissary-general**, *n.* (*Mil.*) the head of the commissariat. **commissarial** (-seə'-), *a.* **commissaryship**, *n.*
commission (kəmish'ən), *n.* the act of doing or committing; entrusting a duty to another; hence, trust, command; delegation of authority; a number of persons entrusted with authority; the document conferring authority, esp. that of military and naval officers; a body of commissioners; an allowance made to a factor or agent; a percentage. *v.t.* to authorize, to appoint to an office; to put (a ship) in commission; to order (the painting of a picture, writing of a book etc.). **in commission**, entrusted with authority; (*Nav.*) prepared for active service. **on commission**, a percentage of the proceeds of goods sold being paid to the agent or retailer. **Royal Commission**, a commission of enquiry ordered by Parliament. **commission agent, merchant**, *n.* one who acts as agent for others, and is paid by a percentage. **commissionaire** (-neə'), *n.* a uniformed doorman at a hotel, theatre etc. **commissional**, *a.* **commissioned**, *a.* holding a commission, esp. from the Crown. **commissioner**, *n.* one empowered to act by a commission or warrant; a member of a commission or government board; the head of some department of the public service. **High Commissioner**, *n.* the chief representative of a Commonwealth country in another Commonwealth country. **Lord High Commissioner**, *n.* the sovereign's representative in the Church of Scotland. **commissionership**, *n.*
commissure (kom'isūə), *n.* a joint, a seam; the point of junction of two sides of anything separated, or of two similar organs; a suture; a line of closure; (*Arch.*) the joint of two stones; the application of one surface to another. **commissural** (-sūə'-), *a.*
commit (kəmit'), *v.t.* (*past, p.p.* **committed**) to entrust, to deposit, to consign, to perpetrate; to refer (as a Bill) to a Parliamentary committee; (*Law*) to send for trial or to prison; to assign, to pledge. **to commit oneself**, to pledge oneself. **to commit to memory**, to learn by heart. **commitment**, *n.* the action of the verb TO COMMIT; the state of being committed; the delivery of a prisoner to the charge of the prison authorities; an engagement to carry out certain duties or meet certain expenses. **committable**, *a.* **committal**, *n.* a sending for trial, to prison or to the grave. **committer**, **committor**, *n.*
committee (kəmit'i), *n.* a board elected or appointed to examine, consider, and report on any business referred to them; (*Law*) (kəmitē') the person to whom the care of a mentally incompetent person etc. is committed. **Committee of the Whole House**, the House of Commons sitting informally as a committee to discuss a bill. **committee-man, -woman**, *n.* a member of a committee.
commix (kəmiks'), *v.t., v.i.* to mix together, to blend. **commixtion** (-chən), *n.* **commixture** (-chə), *n.*
commode (kəmōd'), *n.* a bureau; a night-stool.
commodious (kəmō'diəs), *a.* roomy; convenient, suited to its purpose. **commodiously**, *adv.* **commodiousness**, *n.*
commodity (kəmod'iti), *n.* an article of commerce.
commodore (kom'ədaw), *n.* an officer ranking above captain or below rear-admiral; the president of a yacht-club; the leading ship or the senior captain of a fleet of merchantmen.
common (kom'ən), *a.* belonging equally to more than one; open or free to all; pertaining to or affecting the public; ordinary, usual; (*derog.*) vulgar; (*Math.*) belonging to several quantities; (*Gram.*) applicable to a whole class; (*Pros.*) variable in quantity. *n.* a tract of open ground, the common property of all members of a community; (*Law*) conjoint possession. **in common**, equally with another or others. **out of the common**, extraordinary, unusual. **common carrier** CARRIER. **common chord**, *n.* a note accompanied by its third and fifth. **common cold**, *n.* a viral infection of the mucous membranes of the respiratory tract, accompanied by sneezing and coughing. **common era**, *n.* the Christian era. **common gender**, *n.* applied to a word used both for the masculine and the feminine. **common ground**, *n.* matter in a discussion accepted by both sides. **common law**, *n.* the unwritten law, based on immemorial usage. **common-law husband, wife**, *n.* a person recognized as a husband or wife after long cohabitation. **common lawyer**, *n.* **common market**, *n.* the European Economic Community. **common measure**, *n.* (*Math.*) a number which will divide two or more numbers

exactly; (*Mus.*) common time, two or four beats to the bar, esp. four crotchets to the bar. **common multiple**, *n.* any number containing two or more numbers an exact number of times without a remainder. **common noun**, *n.* the name of any one of a class of objects. **common or garden**, *a.* (*coll.*) ordinary. **Common Prayer**, *n.* the liturgy of the Church of England. **common room**, *n.* a communal sitting room in a college or school. **common sense**, *n.* sound practical judgment. **common-sense, -sensical**, *a.* **Common Serjeant**, *n.* the judge of the City of London ranking next to the Recorder. **common time**, *n.* (*Mus.*) time with two beats, or any multiple of two beats, in a bar. **commonweal**, *n.* the welfare of the community. **commonish**, *a.* **commonly**, *adv.* usually, frequently; in an ordinary manner. **commonness**, *n.*
commonable (kom'ənəbl), *a.* held in common. **commonage**, *n.* the right of using anything in common; common property in land; common land; commonalty.
commonalty (kom'ənəlti), *n.* the common people; a corporation.
commoner (kom'ənə), *n.* one of the commonalty, below the rank of a peer; a member of the House of Commons; a student at Oxford or Winchester not on the foundation; one having a joint right in common ground.
commonplace (kom'ənplās), *a.* common, trivial, trite, unoriginal. *n.* a general idea; a trite remark; anything occurring frequently or habitually. **commonplace-book**, *n.* a book in which thoughts, extracts from books etc. are entered for future use. **commonplaceness**, *n.*
commons (kom'ənz), *n.pl.* the common people; the House of Commons; food provided at a common table; a ration or allowance of food. **House of Commons**, the lower House of Parliament in the British and some other constitutions, the third estate of the realm. **short commons**, a scanty allowance of food.
commonweal COMMON.
commonwealth (kom'ənwelth), *n.* the whole body of citizens; a republic; (*Hist.*) the form of government in England from 1649 to 1659; the federation of Australian States; the Commonwealth of Nations. **(British) Commonwealth of Nations**, a loose association of states that have been, or are, ruled by Britain.
commotion (kəmō'shən), *n.* violent agitation, excitement; a popular tumult.
commune[1] (kom'ūn), *n.* a small territorial district in France etc. governed by a mayor and council; the inhabitants or members of the council of a commune; a group of people, not related, living together and sharing property and responsibilities; the house used by such a group. **communal**, *a.* pertaining to a commune or a community; for the common use or benefit; shared. **communalism**, *n.* the theory of government by communes of towns and districts; the theory or practice of communal living. **communalist**, *n.* **communalistic** (-lis'-), *a.* **communalize, -ise**, *v.t.* **communally**, *adv.* **communard** (-nahd), *n.* one who lives in a commune.
commune[2] (kəmūn'), *v.i.* to talk together familiarly; to be in close touch (with); (*N Am.*) to receive Holy Communion. *n.* (kom'-) communion; intimate converse. **communer**, *n.*
communicate (kəmū'nikāt), *v.t.* to impart, to give a share of, to transmit; to reveal; to give Holy Communion to. *v.i.* to share; to confer by speech or writing; to be connected; to partake of the Holy Communion; to establish mutual understanding (with). **communicable**, *a.* **communicability** (-bil'-), **-ableness**, *n.* **communicably**, *adv.* **communicant**, *n.* one who communicates (information etc.); one who partakes of Holy Communion. **communication**, *n.* the act of communicating; that which is communicated; news; intercourse; means of passing from one place to another; (*pl.*) (the science of) means of communicating considered collectively (e.g. telecommunications, the press etc.); (*pl.*) (*Mil.*) a system of routes and vehicles for transport. **communication cord**, *n.* device whereby a passenger can stop a train in an emergency. **communication lines**, *n.pl.* (*Mil.*) communications. **communications satellite**, *n.* an artificial satellite orbiting round the earth and relaying television, telephone etc. signals. **communicative**, *a.* inclined to communicate; not reserved. **communicatively**, *adv.* **communicativeness**, *n.* **communicator**, *n.* **communicatory**, *a.*
communion (kəmūn'yən), *n.* the act of communicating or communing; sharing; fellowship, intercourse; union in religious faith; the act of partaking of the Eucharist; a religious body. **Holy Communion**, the celebration of the Eucharist. **communion table**, *n.* the table used in the celebration of the Eucharist.
communiqué (kəmū'nikā), *n.* an official announcement.
communism (kom'ūnizm), *n.* a theory of government based on common ownership of all property and means of production; the system of government, based on Marxist socialism, practised in the USSR etc. **communist**, *n.*, *a.* an adherent of, or pertaining to, communism. **communistic** (-nis'-), *a.* **communize, -ise**, *v.t.* to make communal or communistic.
community (kəmū'niti), *n.* a body of people having common rights, interests etc.; an organized body, municipal, national, social or political; society at large, the public; a body of individuals living in a common home; common possession or enjoyment; fellowship; identity of nature or character; (*Ecol.*) a set of interdependent plants and animals inhabiting an area. **community centre**, *n.* a building where all residents in the locality can come to enjoy social etc. activities. **community charge**, *n.* a flat-rate tax levied on all adults to raise money for local government, a poll tax. **community home**, *n.* a boarding school for young offenders. **community policing**, *n.* (the policy of) assigning a police officer to a particular area, to become familiar with that area and its inhabitants. **community radio**, *n.* radio broadcasting to a smaller audience than local radio, to a town or part of a city. **community service order**, *n.* a form of sentence ordering a convicted person to work for a specified time for the benefit of the community.
commute (kəmūt'), *v.t.* to substitute one (payment, punishment etc.) for another; to reduce the severity of (a punishment); to commutate. *v.i.* to travel (a considerable distance) daily to and from one's place of work. **commutable**, *a.* **commutability** (-bil'-), *n.* **commutate** (kom'ūtāt), *v.t.* to reverse the direction of (an electric current); to convert (an alternating current) to a direct current. **commutation**, *n.* the act of commuting or commutating; change, exchange. **commutation ticket**, *n.* (*N Am.*) a season ticket. **commutative** (kəmū'mū'-), *a.* **commutatively**, *adv.* **commutator** (kom'-), *n.* an instrument which reverses an electric current. **commutator transformer**, *n.* a device for converting from low to high voltage direct current and vice versa. **commuter**, *n.* one who commutes to and from work.
comose COMATE.
comp.[1] (komp), *n.* short for COMPOSITOR, ACCOMPANIMENT.
comp.[2] (*abbr.*) company; comparative; comparison; competition.
compact[1] (kom'pakt), *n.* an agreement, a bargain, a covenant.
compact[2] (kəmpakt'), *a.* closely packed or joined together; solid, succinct. *n.* (kom'-) a small box with face-powder, puff and mirror; (*N Am.*) a middle-sized motor car. *v.t.* to consolidate; to join closely and firmly together; to compose. **compact disc**, *n.* a small audio disc, read by laser beam, on which sound or data are stored digitally as microscopic pits. **compacted**, *a.* **compactedly**, *adv.* **compactedness**, *n.* **compaction**, *n.* **compactly**, *adv.* **compactness**, *n.*

compages (kəmpā'jēz), *n.* (*pl.* **compages**) a structure or system of many parts united.

companion[1] (kəmpan'yən), *n.* one who associates or keeps company with another; a comrade; a member of the lowest grade in some orders of knighthood; a person employed to live with another; a handbook. *a.* accompanying; going along with or matching something. *v.t.* to accompany. **companionable**, *a.* sociable. **companionableness**, *n.* **companionably**, *adv.* **companionless**, *a.* **companionship**, *n.* fellowship, association, company.

companion[2] (kəmpan'yən), *n.* (*Naut.*) the raised window-frame upon the quarter-deck through which light passes to the cabins and decks below. **companion-ladder, -way**, *n.* the ladder or staircase from the cabins to the quarter-deck.

company (kŭm'pəni), *n.* companionship, fellowship; a number of persons associated together by interest or for carrying on business; a corporation; associates, guests; a subdivision of an infantry regiment. **ship's company**, the crew of a ship. **to keep company (with)**, to associate (with).

comparative (kəmpa'rətiv), *a.* involving or expressing comparison; estimated by comparison; grounded on comparison; expressing a higher or lower degree of a quality. *n.* (*Gram.*) the comparative degree or the word or inflection expressing it. **comparatively**, *adv.* **comparator**, *n.* an apparatus for comparing.

compare (kəmpeə'), *v.i.* to show how one thing is like (or unlike) another; to liken one thing to another; (*Gram.*) to inflect according to degrees of comparison. *v.i.* to bear comparison. **beyond compare**, unequalled. **to compare notes**, to exchange opinions. **comparable** (kom'pə-), *a.* capable of being compared (with); worthy of being compared (to). **comparability** (kompərəbil'-, kəmparəbil'-), *n.*

comparison (kəmpar'isən), *n.* the act of comparing; a comparative estimate; a simile; (*Gram.*) the inflection of an adjective or adverb.

compartment (kəmpaht'mənt), *n.* a division; a portion of an enclosed space separated from the other parts. **compartmental** (kompahtmen'-), *a.* **compartmentalize, -ise,** *v.t.* to divide into separate units or categories.

compass (kŭm'pəs), *n.* a circumference; area, extent; (*fig.*) reach, capacity; the range or power of the voice or a musical instrument; an instrument indicating the magnetic meridian, used to ascertain direction; (*pl.*) an instrument with two legs connected by a joint for describing circles, measuring distances etc. *v.t.* to go round; to beleaguer, surround; to comprehend; to accomplish. **to fetch a compass**, to make a circuit. **compass-card**, *n.* the card or dial of a mariner's compass on which the points are drawn. **compass-plane**, *n.* (*Carp.*) a plane convex underneath for planing concave surfaces. **compass-saw**, *n.* a saw which cuts circularly. **compass-signal**, *n.* (*Naut.*) a flag indicating a point of the compass. **compass-window**, *n.* a semicircular window. **compassable**, *a.*

compassion (kəmpash'ən), *n.* pity, sympathy for the sufferings and sorrows of others. **compassion fatigue**, *n.* the state of being unwilling to contribute to charities because of the apathy etc. induced by the sheer number of charities and the promotion of these. **compassionable**, *a.* **compassionate** (-nət), *a.* merciful, inclined to pity; sympathetic. **compassionate leave**, *n.* leave granted on account of domestic difficulties. **compassionately**, *adv.* **compassionateness**, *n.*

compatible (kəmpat'ibl), *a.* that may coexist; consistent, harmonious; of electronic machinery of different types or by different manufacturers, able to work together without modification. **compatibly**, *adv.* **compatibility** (-bil'-), *n.*

compatriot (kəmpat'riət), *n.* a fellow-countryman. **compatriotic** (-ot'-), *a.* **compatriotism**, *n.*

compeer (kom'piə), *n.* an equal, peer.

compel (kəmpel'), *v.t.* (*past, p.p.* **compelled**) to force, to oblige; to cause by force; to drive with force. **compellable**, *a.* **compelling**, *a.* very interesting. **compellingly**, *adv.*

compendium (kəmpen'diəm), *n.* (*pl.* **-diums, -dia** (-ə)) an abridgment; a brief compilation; a collection of board or card games in one box. **compendious**, *a.* abridged; succinct; comprehensive. **compendiously**, *adv.* **compendiousness**, *n.*

compensate (kom'pənsāt), *v.t.* to counterbalance; to make amends for; to recompense. *v.i.* to supply an equivalent; (*Psych.*) to make up for a perceived or imagined deficiency by developing another aspect of the personality. **compensation**, *n.* the act of compensating; recompense, amends; that which balances or is an equivalent for something else; payment of a debt by an equal credit. **compensational**, *a.* **compensative** (kom'-, -pen'-), *a.* **compensator**, *n.* **compensatory** (kom'-, -pen'-), *a.*

compere (kom'peə), *n.* one who introduces the items in a (stage or broadcast) entertainment. *v.t., v.i.* to act as compere (of).

compete (kəmpēt'), *v.i.* to contend as a rival; to strive.

competent (kom'pitənt), *a.* qualified, sufficient; suitable, adequate; legally qualified; (*coll.*) admissible, permissible. **competence, -ency**, *n.* the state of being competent; sufficiency; adequate pecuniary support; legal capacity or qualification; ability (for or to do some task). **competently**, *adv.*

competition (kompətish'ən), *n.* the act of competing; rivalry; the struggle for existence or gain in industrial and mercantile pursuits; a competitive game or match; people or organizations competing against one. **competitive** (-pet'-), *a.* pertaining to or involving competition; liking competition; of prices etc., such as to give one an advantage against competitors. **competitively**, *adv.* **competitiveness**, *n.* **competitor** (-pet'-), *n.* one who competes; a rival.

compile (kəmpīl'), *v.t.* to compose out of materials from various authors; to assemble various items as in an index or dictionary; to gather such materials into a volume; (*Comput.*) to put (a program or instruction written in a high-level language) into machine code. **compilation** (kompilā'-), *n.* the act of compiling; that which is compiled. **compiler**, *n.* one who compiles; (*Comput.*) a program which compiles.

complacent (kəmplā'sənt), *a.* self-satisfied. **complacently**, *adv.* **complacence, -ency**, *n.* a feeling of inward satisfaction; smugness.

complain (kəmplān'), *v.i.* to express dissatisfaction or objection; to state a grievance; to make a charge; to express grief or pain. **complainant**, *n.* a plaintiff. **complainer**, *n.* **complaining**, *n., a.* **complainingly**, *adv.* **complaint**, *n.* an act of complaining; a reason for complaining; an accusation; a malady; (*Law*) a formal allegation or charge.

complaisant (kəmplā'sənt), *a.* courteous, obsequious, obliging. **complaisantly**, *adv.* **complaisance**, *n.*

compleat (kəmplēt'), COMPLETE (skilled).

complement (kom'plimənt), *n.* full quantity; the full number required to man a vessel; that which is necessary to make complete; a word or phrase required to complete the sense, the predicate. *v.t.* to supply a deficiency; to complete. **complemental** (-men'-), *a.* **complementally**, *adv.* **complementary** (-men'-), *a.* that complements. **complementary colour**, *n.* a colour which produces white when mixed with another to which it is complementary. **complementary medicine**, *n.* alternative medicine.

complete (kəmplēt'), *a.* fulfilled, finished; free from deficiency; absolute; skilled, highly accomplished. *v.t.* to bring to a state of perfection; to finish; to make whole. **completely**, *adv.* **completeness**, *n.* **completion**, *n.*

complex (kom'pleks), *a.* composed of several parts; composite; complicated. *n.* a complicated whole; a collec-

complexion | 164 | **compress**

tion; a complicated system; a group of emotions, ideas etc., partly or wholly repressed, which can influence personality or behaviour; (*loosely*) an obsession; a set of interconnected buildings for related purposes, forming a whole. **complex number**, *n.* one consisting of a real and an imaginary component. **complex sentence**, *n.* one consisting of a principal clause and at least one subordinate clause. **complexity** (-plek'-), *n.* **complexly**, *adv.*

complexion (kəmplek'shən), *n.* colour and appearance of the skin, esp. of the face; nature, character, aspect. **complexioned**, *a. usu. in comb.* **complexionless**, *a.*

complexity, complexus COMPLEX.

compliance (kəmpli'əns), *n.* the act of complying; submission, agreement, consent. **compliable, compliant**, *a.* yielding; tending to comply. **compliantly**, *adv.*

complicate (kom'plikāt), *v.t.* to make complex or intricate; to involve. **complicacy**, *n.* the state of being complicated. **complicated**, *a.* **complicatedly**, *adv.* **complication**, *n.* the act of complicating; the state of being complicated; a complicated or complicating matter or circumstance; a disease or morbid condition arising in the course of another disease.

complicity (kəmplis'əti), *n.* participation, partnership, esp. in wrong-doing.

complier COMPLY.

compliment (kom'plimənt), *n.* an expression or act of courtesy, approbation or respect; delicate flattery; (*pl.*) ceremonious greetings; courtesies, respects. *v.t.* (-ment), to pay compliments to; to congratulate, to praise, to flatter courteously. **complimental** (-men'-), *a.* **complimentary** (-men'-), *a.* **complimentary ticket**, *n.* a free ticket.

compline (kom'plin), *n.* in the Roman Catholic Church, the last part of the divine office of the breviary, sung after vespers.

comply (kəmpli'), *v.i.* to assent, to agree; to act in accordance with the wishes of another. **complier**, *n.*

compo[1] (kom'pō), *n.* (*pl.* **-pos**), applied to different compounds in various trades, as a kind of stucco etc. **compo rations**, *n.pl.* (*Mil.*) rations for several days for use in the field.

compo[2] (kom'pō), *n.* (*pl.* **-pos**) (*Austral.*) short for COMPENSATION (for injury etc.).

component (kəmpō'nənt), *a.* serving to make up a compound. *n.* a constituent part. **componential** (kompənen'-), *a.*

comport (kəmpawt'), *v.t.* to conduct, to behave (oneself). *v.i.* to suit, to agree. **comportment**, *n.*

compose (kəmpōz'), *v.t.* to make, arrange or construct, by putting together several parts; to constitute, to make up by combination; to write (a literary or musical work); to soothe; to settle, to adjust; to arrange in proper order; to set up (type). *v.i.* to practise composition. **composed**, *a.* calm, tranquil. **composedly** (-zid-), *adv.* **composedness** (-zid-), *n.* **composer**, *n.* one who composes, esp. one who writes music. **composing**, *n.*, *a.* **composing-machine**, *n.* a machine for setting type. **composing-room**, *n.* the room in a printing-office where the compositors work.

composite (kom'pəzit), *a.* made up of distinct parts or elements; compound; pertaining to the Compositae, a large family of plants, so called because the heads are made up of many small flowers. *n.* a composite substance, plant or thing; (-zit) a composited motion. *v.t.* (-zit) to merge related motions from different branches of e.g. a trade union, political party, for presentation to e.g. a national conference. **composite number**, *n.* a number which is the product of two other numbers greater than unity. **composite order**, *n.* (*Arch.*) the last of the five orders, which partakes of the characters of the Corinthian and Ionic. **composite resolution** (-zit), *n.* **compositely**, *adv.* **compositeness**, *n.* **compositive** (-poz'-), *a.*

composition (kompəzish'ən), *n.* the act of composing or putting together to form a whole; the thing composed (esp. used of literary and musical productions); orderly disposition of parts, structural arrangement; a combination of several parts or ingredients, a compound; compensation in lieu of that demanded; the amount so accepted; settlement by compromise; the process of setting type; the act of forming sentences; a piece written for the sake of practice in literary expression; the arrangement of columns, piers, doors etc. in a building; the arrangement of different figures in a picture. **compositional**, *a.*

compositor (kəmpoz'itə), *n.* one who sets type.

compos mentis (kompəs men'tis), *a.* in one's right mind. **non compos** (non), (*coll.*) not in one's right mind.

compost (kom'post), *n.* a fertilizing mixture of decomposed vegetable matter etc. *v.t.* to make into or manure with compost. **compost heap**, *n.* a heap of waste plant material decomposing into compost.

composure (kəmpō'zhə), *n.* calmness, tranquillity, a calm frame of mind.

compote (kom'pōt), *n.* fruit stewed or preserved in syrup.

compound[1] (kəmpownd'), *v.t.* to make into one mass by the combination of several constituent parts; to mix, to make up, to form a composite; to settle amicably; to adjust by agreement; to compromise; to pay a lump sum instead of a periodical subscription. *v.i.* to settle with creditors by agreement; to come to terms by abating something of the first demand. **to compound a felony**, (*Law*) to forbear to prosecute a felony for some valuable consideration. **compoundable**, *a.* **compounder**, *n.*

compound[2] (kom'pownd), *a.* composed of two or more parts, ingredients or elements; collective, combined, composite. *n.* a combination, a mixture; something made from two or more parts, elements etc.; a combination of two or more elements by chemical action. **compound addition, subtraction**, *n.* processes dealing with numbers of different denominations. **compound animal**, *n.* one consisting of a combination of organisms. **compound eye**, *n.* one made up of many separate light sensitive units (as in insects). **compound flower**, *n.* an inflorescence consisting of numerous florets surrounded by an involucre; one of the flower-heads of any of the Compositae. **compound fracture**, *n.* one in which the broken bone pierces the skin. **compound interest**, *n.* interest added to the principal and bearing interest; the method of computing such interest. **compound interval**, *n.* (*Mus.*) an interval greater than the octave. **compound leaf**, *n.* one with branched petioles. **compound microscope**, *n.* one with a combination of lenses. **compound quantity**, *n.* an arithmetical quantity of more than one denomination; an algebraic quantity, consisting of two or more terms connected by the signs + (plus), or − (minus), or expressed by more letters than one. **compound raceme**, *n.* one composed of several small ones. **compound sentence**, *n.* one consisting of two or more principal clauses.

compound[3] (kom'pownd), *n.* the yard surrounding a dwelling-house in India, China etc.; any similar walled or fenced space, as in a prison; living quarters for Black workers in S African mines etc.

comprehend (komprihend'), *v.t.* to grasp mentally; to understand; to comprise, to include. **comprehensible**, *a.* clear, intelligible. **comprehensibility** (-bil'-), *n.* **comprehensibly**, *adv.*

comprehension (komprihen'shən), *n.* the act or power of comprehending or comprising; the faculty by which ideas are comprehended by the intellect; (also **comprehension test**) a school exercise to test a pupil's understanding of a given passage. **comprehensive**, *a.* extending widely; including much or many things. **comprehensive school**, *n.* a secondary school serving all children of all abilities in an area. **comprehensively**, *adv.* **comprehensiveness**, *n.*

compress (kəmpres'), *v.t.* to squeeze or press together; to bring into narrower limits; to condense. *n.* (kom'-) a pad

pressed to the skin to reduce bleeding, inflammation etc. **compressible,** *a.* **compressibility** (-bil'-), *n.* **compression,** *n.* the act of compressing; the state of being compressed; condensation. **compressive,** *a.* **compressor,** *n.* a machine which compresses a gas.
comprise (kəmprīz'), *v.t.* to contain, to comprehend; to consist of. **comprisable,** *a.*
compromise (kom'prəmīz), *n.* a settlement (of controversy etc.) by mutual concession; a mean between two different qualities, courses of action etc. *v.t.* to settle by mutual concession; to place in a position of difficulty or danger; to expose to risk of disgrace. *v.i.* to make a compromise. **compromising,** *a.*
Comptometer® (komptom'itə), *n.* a type of calculating machine.
comptroller (kəntrō'lə, komp-), CONTROLLER.
compulsion (kəmpŭl'shən), *n.* the act of compelling or being compelled; (*Psych.*) an irresistible impulse to perform actions against one's will. **compulsive,** *a.* involving compulsion; tending to compel. **compulsively,** *adv.* **compulsiveness,** *n.* **compulsory,** *a.* exercising compulsion; enforced, necessitated. **compulsory purchase,** *n.* purchase of a property against the owner's wishes, for a public development. **compulsorily,** *adv.*
compunction (kəmpŭngk'shən), *n.* remorse, contrition; regret. **compunctionless,** *a.* **compunctious,** *a.* **compunctiously,** *adv.*
compute (kəmpūt'), *v.t.* to determine by calculation; to number, to estimate; to calculate using a computer. *v.i.* to calculate; to use a computer. **computable,** *a.* **computative,** *a.* **computation** (kompūtā'-), *n.* **computer,** *n.* an electronic device which does complex calculations or processes data according to the instructions contained in a program. **computer game,** *n.* a game of skill in which the player uses a computer keyboard to react to graphics on the screen. **computer graphics,** *n.pl.* visual images produced by a computer program on a screen, which can be manipulated and developed very rapidly, used in computer games and for simulators etc.; (*sing. in constr.*) the design of programs to generate such images. **computer-language,** *n.* a programming language. **computer literacy,** *n.* ability to understand computers, their uses and working. **computer-literate,** *a.* **computer science,** *n.* the sciences connected with the construction and operation of computers. **computer system,** *n.* a self-contained unit consisting of items of hardware and the necessary software to carry out a particular range of tasks. **computer virus,** *n.* a self-replicating computer program which damages or destroys the memory or other programs of the host computer. **computerize, -ise,** *v.t.* to perform or control by means of computer; to install computers in (a business, etc.). *v.i.* to install computers. **computerization,** *n.*
comrade (kom'rəd), *n.* a companion; an intimate associate. **comradely,** *a.* **comradeship,** *n.*
comsat® (kom'sat), *n.* short for COMMUNICATIONS SATELLITE.
Con.[1], (*abbr.*) Conservative.
con.[2], (*abbr.*) conclusion; convenience; conversation.
con[3] (kon), *v.t.* (*past, p.p.* **conned**) to read carefully; to learn.
con[4] (kon), *v.t.* (*past, p.p.* **conned**) to direct the steering of (a ship). **conning-tower,** *n.* the armoured shelter in a warship or submarine from which the vessel is steered. **conner,** *n.*
con[5] (kon), *n., prep.* short for CONTRA. **pro and con,** for and against.
con[6] (kon), *n.* (*sl.*) a confidence trick; a fraud, swindle. *v.t.* to deceive; to swindle. **con-man,** *n.* **con trick,** *n.*
con[7] (kon), *n.* (*sl.*) short for CONVICT[2].
con- COM-.
concamerate (kənkam'ərāt), *v.t.* to divide into chambers (as a shell). **concameration,** *n.*

concatenate (kənkat'ənāt), *v.t.* to join or link together in a successive series. **concatenation,** *n.*
concave (kon'kāv), *a.* having a hollow curve like the inner side of a circle or globe. *n.* such a hollow curve; an arch. **concavely,** *adv.* **concavity** (-kav'-), *n.* the state of being concave; a concave surface. **concavous,** *a.*
concavo-, *comb. form.* (*Opt.*) concave; concavely. **concavo-concave** (konkā'vō-), *a.* concave on both sides. **concavo-convex,** *a.* concave on one side and convex on the other.
conceal (kənsēl'), *v.t.* to hide from sight or observation; to keep secret or private. **concealable,** *a.* **concealment,** *n.* the act of concealing; the state of being concealed; a hiding-place; (*Law*) a suppression of material matters.
concede (kənsēd'), *v.t.* to give up, to surrender; to admit to be true; to allow to pass unchallenged. *v.i.* to yield; to make concessions.
conceit (kənsēt'), *n.* a vain opinion of oneself, overweening self-esteem; a fanciful idea; in literature, an elaborate or far-fetched image. **conceited,** *a.* inordinately vain; egotistical. **conceitedly,** *adv.* **conceitedness,** *n.*
conceive (kənsēv'), *v.t.* to receive into and form in the womb; to form, as an idea or concept, in the mind; to imagine or suppose as possible; to think. *v.i.* to become pregnant; to form an idea or concept in the mind. **conceivable,** *a.* **conceivability** (-bil'-), *n.* **-ableness,** *n.* **conceivably,** *adv.*
concelebrate (kənsel'əbrāt), *v.i.* to celebrate (Mass or the Eucharist) along with another priest. **concelebrant,** *n.* **concelebration,** *n.*
concentrate (kon'səntrāt), *v.t.* to bring to a common focus, centre, or point; to reduce to a greater density by removing water, etc. *v.i.* to come to a common focus or centre; to direct all one's thoughts or efforts to one end. *a.* concentrated. *n.* a concentrated or condensed substance. **concentration,** *n.* **concentration camp,** *n.* a camp for housing political prisoners and interned persons. **concentrative,** *a.* **concentrativeness,** *n.* **concentrator,** *n.* an apparatus for concentrating solutions; a pneumatic apparatus for separating dry comminuted ores.
concentre (kənsen'tə), *v.t.* to draw or direct to a common centre. *v.i.* to have a common centre; to combine for a common object. **concentric,** *a.* having a common centre. **concentrically,** *adv.* **concentricity** (konsəntris'-), *n.*
concept (kon'sept), *n.* a general notion; an idea. *a.* designed round a single central concept or idea. **conception** (-sep'-), *n.* the act of conceiving; the impregnation of the ovum; concept. **to have no conception of,** to be unable to imagine. **conceptional,** *a.* **conceptive** (-sep'-), *a.*
conceptacle (kənsep'təkl), *n.* that in which anything is contained; (*Biol.*) a cavity used in reproduction.
conceptual (kənsep'tūəl), *a.* of conception or a concept. **conceptualism,** *n.* the doctrine that universals exist only in the mind of the thinking subject. **conceptualist,** *n.* **conceptualize, -ise,** *v.t.* to form a concept (of).
concern (kənsœn'), *v.t.* to relate or belong to; to affect; to be of interest to; to disturb. *n.* that which affects or is of interest or importance to a person; interest, regard; solicitude; a business establishment; a matter of personal importance; (*pl.*) affairs; (*coll.*) an affair, a thing. **concerned,** *a.* interested, involved, engaged (with); anxious (about). **concernedly** (-nid-), *adv.* **concerning,** *prep.* with respect to. **concernment,** *n.* an affair, business; importance.
concert[1] (kənsœt'), *v.t.* to plan, to arrange mutually; to contrive. **concerted,** *a.* mutually planned or devised; (*Mus.*) arranged in parts.
concert[2] (kon'sət), *n.* harmony, accordance of plan or ideas; a public musical entertainment. **in concert,** acting together; of musicians, performing live on stage. **concert grand,** *n.* a powerful grand piano for use at concerts. **concert party,** *n.* a group of companies or financiers engaged together in a (shady) project. **concert pitch,** *n.* (*Mus.*) the

concertina pitch used at concerts, slightly higher than the ordinary; a high degree of readiness.
concertina (konsətē'nə), *n.* a portable instrument having a keyboard at each end, with bellows between. *v.i.* to collapse, fold up, like a concertina.
concertino (konchətē'nō), *n.* a short concerto.
concerto (kənchœ'tō), *n.* a composition for a solo instrument or instruments with orchestral accompaniment. **concerto grosso** (gros'ō), *n.* a composition for an orchestra and a group of soloists.
concession (kənsesh'ən), *n.* the act of conceding; the thing conceded; esp. a privilege or right granted by a government; the (exclusive) right to market a particular product or service in a particular area; a reduction in price for a special group of people; a subdivision of a township in Canada. **concessionnaire** (-neə'), *n.* one who holds a concession. **concessionary**, *a.* **concessive** (-siv), *a.*
conch (konch), *n.* a marine shell of a spiral form; such a shell used as a trumpet. **conchiferous** (kongkif'-), *a.* shell-bearing.
conch(o)- *comb. form.* shell.
conchology (kongkol'əji), *n.* the study of shells and the animals inhabiting them. **conchological** (-loj'-), *a.* **conchologist**, *n.*
oonohio, oonohy (kon'shi), *n. (coll. derog.)* short for CONSCIENTIOUS (objector).
concierge (konsiœzh'), *n.* a door-keeper, a porter, a janitor.
conciliar (kənsil'iə), *a.* pertaining to a council, esp. an ecclesiastical council.
conciliate (kənsil'iāt), *v.t.* to win the regard or goodwill of; to appease; to reconcile (conflicting views or parties). **conciliation**, *n.* **conciliative**, *a.* **conciliator**, *n.* **conciliatory**, *a.* **conciliatoriness**, *n.*
concinnous (kənsin'əs), *a.* harmonious; elegant. **concinnity**, *n.*
concise (kənsīs'), *a.* condensed, brief, terse. **concisely**, *adv.* **conciseness**, *n.* **concision** (-sizh'ən), *n.*
conclamation (konkləmā'shən), *n.* a united or general outcry.
conclave (kon'klāv), *n.* the assembly of cardinals met for the election of a pope; a secret assembly.
conclude (kənklood'), *v.t.* to bring to an end, to finish; to determine, to settle; to infer. *v.i.* to make an end; to come to a decision; to draw an inference. **in conclusion**, to conclude. **to conclude**, in short, in fine. **to try conclusions**, to contest; to try which is superior. **concluding**, *a.* **concludingly**, *adv.* **conclusion**, *n.* the end, the finish; the result; an inference; settlement (of terms etc.); a final decision. **conclusive**, *a.* that puts an end to argument, final. **conclusively**, *adv.* **conclusiveness**, *n.* **conclusory**, *a.*
concoct (kənkokt'), *v.t.* to prepare by mixing together different ingredients; to plot, to devise. **concoction**, *n.* **concoctive**, *a.* **concoctor**, *n.*
concolorous (konkūl'ərəs), *a. (Biol.)* uniform in colour.
concomitant (kənkom'itənt), *a.* accompanying; existing in conjunction with. *n.* one who or that which accompanies. **concomitantly**, *adv.* **concomitance, -ancy**, *n.* the state of being concomitant.
concord (kong'kawd, kon'-), *n.* agreement; union in opinions or interests; the agreement of one word with another in number, gender etc.; a combination of notes satisfactory to the ear. **concordance** (kənkaw'-), *n.* the state of being concordant; agreement; a list of the words in a book with exact references to the places where they occur. **concordant** (kənkaw'-), *a.* in harmony or accord; correspondent. **concordantly**, *adv.*
concordat (kənkaw'dat), *n.* a convention between a pope and a secular government.
concourse (kong'kaws, kon'-), *n.* a confluence, a gathering together; an assembly; a main hall or open space at an airport, railway station etc.
concremation (konkrimā'shən), *n.* cremation at the same time; consumption by fire.
concrescence (kəngkres'əns, kən-), *n.* a growing together, coalescence; union of parts, organs or organisms.
concrete[1] (kong'krēt, kon'-), *a.* formed by the union of many particles in one mass; material, solid, not abstract; specific, not general; made of concrete. *n.* cement, coarse gravel and sand mixed with water, used in building. *v.t.* to treat with concrete. *v.i.* to apply concrete. **reinforced concrete**, *n.* REINFORCE. **concrete music**, *n.* music consisting of pre-recorded music or other sound put together and electronically modified. **concrete poetry**, *n.* poetry which uses the visual shape of the poem to help convey meaning. **concretely**, *adv.* **concreteness**, *n.* **concretize, -ise** (-krə-), *v.t.* to render concrete, solid or specific.
concrete[2] (kəngkrēt', kən-), *v.i.* to coalesce; to grow together. *v.t.* to form into a solid mass. **concreter**, *n.*
concretion (kəngkrē'shən, kən-), *n.* the act of concreting; the mass thus formed; *(Geol.)* an aggregation of particles into a more or less regular ball; *(Med.)* a calculus. **concretionary**, *a.*
concubine (kong'kūbīn), *n.* a woman who cohabits with a man without being married to him; a wife of inferior rank. **concubinage** (konkū'bi-), *n.* **concubinary** (konkū'bi-), *a.*
concupiscence (kənkū'pisəns), *n.* strong or excessive (sexual) lust. **concupiscent**, *a.*
concur (kənkœ'), *v.i. (past, p.p.* **concurred**) to coincide; to agree; to act in conjunction (with). **concurrence** (-kū'-), *n.* **concurrent** (-kū'-), *a.* that concurs; happening or existing at the same time; acting in union or conjunction; consistent, harmonious; contributing to the same effect or result. *n.* a concurrent person or thing. **concurrently**, *adv.*
concuss (kənkūs'), *v.t.* to subject to concussion of the brain. **concussion** (-shən), *n.* shaking by sudden impact; a shock; a state of unconsciousness suddenly produced by a blow to the skull, usu. followed by amnesia. **concussive**, *a.*
concyclic (kənsī'klik), *a.* of points, lying upon the circumference of one circle.
condemn (kəndem'), *v.t.* to pronounce guilty; to give judgment against; to pass sentence on; to pronounce incurable or unfit for use; to censure, to blame. **condemned cell**, *n.* the cell in which prisoners condemned to death are confined before execution. **condemnable** (-dem'ə-, -dem'nə-), *a.* **condemnation** (kondemnā'-), *n.* **condemnatory** (-dem'nə-), *a.*
condense (kəndens'), *v.t.* to make more dense or compact; to concentrate; to reduce into another and denser form (as a gas into a liquid). *v.i.* to become dense or denser. **condensed milk**, *n.* a thickened and usu. sweetened form of preserved milk. **condensable, condensability** (-bil'-), *n.* **condensate** (-den'sāt), *n.* something made by condensation. **condensation**, *n.* the act of condensing; the state of being condensed; a liquid condensed from a vapour. **condensation trail**, *n.* a vapour trail. **condensational**, *a.*
condenser (kənden'sə), *n.* one who or that which condenses; a lens for concentrating light on an object; an apparatus for reducing steam to a liquid form; *(Elec.)* a capacitor. **condensity**, *n.*
condescend (kondisend'), *v.i.* to lower oneself voluntarily to an inferior position; to deign. **condescendence**, *n.* **condescending**, *a.* patronizing. **condescendingly**, *adv.* **condescension**, *n.* the act of condescending; patronizing behaviour.
condign (kəndīn'), *a.* of a punishment, well-deserved. **condignly**, *adv.*
condiment (kon'dimənt), *n.* anything used to give relish to food.

condition (kəndish'ən), *n.* a stipulation; a term of a contract; that on which anything depends; (*pl.*) circumstances or external characteristics; state or mode of existence; a (good) state of health or fitness; a (long-standing) ailment. *v.t.* to stipulate; to impose conditions on; to test; to accustom; to establish a conditioned reflex in; to put in a certain condition; to put in a good or healthy condition. **in, out of, condition,** in good, or bad, condition. **conditional,** *a.* containing, implying or depending on certain conditions; made with limitations or reservations; (*Gram.*) expressing condition. *n.* a conditional clause, mood etc. **conditionality** (-nal'-), *n.* **conditionally,** *adv.* **conditioned,** *a.* limited by certain conditions; (*usu. in comb.*) having a certain disposition, as *ill-conditioned, well-conditioned.* **conditioned by,** depending on; limited by. **conditioned reflex, response,** *n.* (*Psych.*) a natural response to a stimulus which, by much repetition, becomes attached to a different stimulus. **conditioner,** *n.* a lotion applied to improve the condition (of e.g. hair). **conditioning,** *n.*
condo (kon'dō), *n.* (*N Am., coll.*) short for CONDOMINIUM.
condole (kəndōl'), *v.i.* to sympathize (with). **condolatory,** *a.* **condolence,** *n.* **condolent,** *a.*
condom (kon'dom), *n.* a contraceptive device, a rubber sheath worn over the penis during sexual intercourse.
condominium (kondəmin'iəm), *n.* joint sovereignty over a state; (*N Am.*) a block of flats of which each unit is separately owned; any such flat.
condone (kəndōn'), *v.t.* to forgive; to overlook. **condonation,** *n.*
condor (kon'daw), *n.* a large S American vulture.
condottiere (kondotyeə'ri), *n.* (*pl.* **-ri** (-rē)) an Italian captain of mercenaries.
conduce (kəndūs'), *v.i.* to contribute (to a result). **conducement,** *n.* **conducive,** *a.* **conduciveness,** *n.*
conduct[1] (kon'dŭkt), *n.* the act of leading or guiding; the way in which anyone acts or lives, behaviour; management, control.
conduct[2] (kəndŭkt'), *v.t.* to lead, to guide; to manage, to direct; (*Phys.*) to transmit (as heat etc.); to direct (as an orchestra); (*reflex.*) to behave. *v.i.* to act as a conductor. **conductance,** *n.* the ability of a substance to allow electricity to flow through it. **conductible,** *a.* **conductibility** (-bil'-), *n.* **conduction,** *n.* transmission by a conductor; conveyance (of liquids, etc.). **conductive,** *a.* **conductively,** *adv.* **conductivity** (konduktiv'-), *n.* a measure of the ease with which a substance transmits electricity. **conductor,** *n.* a leader, a guide; the director of an orchestra; (*N Am.*) the guard of a train; the person in charge of a bus or tramcar; a body capable of transmitting heat, electricity, etc. **conductorship,** *n.* **conductress,** *n. fem.*
conduit (kon'dit, -dūit), *n.* a channel or pipe to convey water; a pipe surrounding and protecting electric cables.
condyle (kon'dil), *n.* an eminence with a flattened articular surface on a bone. **condylar, -loid,** *a.*
cone (kōn), *n.* a solid figure described by the revolution of a right-angled triangle about the side containing the right-angle; a solid pointed figure with straight sides and circular or otherwise curved base; anything cone-shaped, as a wafer holder for ice-cream, a temporary marker for traffic on roads etc.; the fruit of a conifer, having overlapping scales; a marine shell of the genus *Conus. v.i.* to bear cones. *v.t.* to mark (off) with cones. **conoid,** *n., a.* (a) cone-shaped (object). **conoidal** (-noi-), *a.* conoid.
coney CONY.
conf., (*abbr.*) compare.
confab (kon'fab), *n., v.i.* (*coll.*) short for CONFABULATION, CONFABULATE.
confabulate (kənfab'ūlāt), *v.i.* to talk familiarly; to chat. **confabulation,** *n.* **confabulatory,** *a.*
confect (kənfekt'), *v.t.* to make (by combining ingredients); to construct. **confection,** *n.* the act of compounding; a compound, esp. a sweetmeat, a preserve; an elaborate piece of clothing. *v.t.* to make confectionery; to make a confection. **confectioner,** *n.* one who prepares or sells confections, sweetmeats etc. **confectionery,** *n.* sweetmeats or candies generally; a confectioner's shop.
confederate (kənfed'ərət), *a.* united in a league; allied by treaty; (*Hist.*) (**Confederate**) applied to the Southerners in the American Civil War (1861–65). *n.* a member of a confederation; an ally, esp. an accomplice; (*Hist.*) a Southerner. *v.t., v.i.* (-rāt), to unite in a league. **confederacy,** *n.* a number of persons, parties or states united for mutual aid and support; conspiracy, collusion. **confederal,** *a.* **confederalist,** *n.* **confederation,** *n.* **confederatism,** *n.* **confederative,** *a.*
confer (kənfœ'), *v.t.* (*past, p.p.* **conferred**) to bestow, to grant. *v.i.* to consult together. **conferee** (-rē'), *n.* one who is conferred with; one on whom something is conferred. **conference** (kon'fə-), *n.* the act of conferring; a meeting for consultation or deliberation; the annual meeting of some Protestant churches to transact church business. **in conference,** at a meeting. **conferencing,** *n.* holding a conference between geographically separated people using telephone or computer links. **conferential** (konfəren'-), *a.* **conferment,** *n.* **conferrable,** *a.* **conferrer,** *n.*
confess (kənfes'), *v.t.* to acknowledge, to admit; to declare one's adherence to or belief in; to hear the confession of. *v.i.* to make confession, esp. to a priest. **confessant,** *n.* one who confesses to a priest. **confessedly** (-sid-), *adv.* avowedly. **confession** (-shən), *n.* the act of confessing; avowal, declaration; formal acknowledgment of sins to a priest in order to receive absolution. **confession of faith,** a formulary containing the creed of a Church. **confessional,** *n.* the place where a priest hears confessions; the practice of confession. *a.* pertaining to confession. **confessionary,** *a.* **confessor,** *n.* one who confesses; a priest who hears confessions.
confetti (kənfet'i), *n.pl.* bits of coloured paper thrown at weddings etc. **confetti money,** *n.* paper money made almost valueless by inflation.
confidant (konfidant'), *n.* one entrusted with secrets; a bosom friend. **confidante,** *n. fem.*
confide (kənfid'), *v.i.* to have trust or confidence (in); to talk confidentially (to). *v.t.* to entrust (to); to reveal in confidence (to). **confidence** (kon'fi-), *n.* trust, belief; self-reliance, boldness; something told in confidence. **in confidence,** as a secret. **confidence man,** *n.* one who practises confidence tricks. **confidence trick,** *n.* a trick by which one is induced to part with valuable property for something worthless. **confident** (kon'fi-), *a.* full of confidence; assured; self-reliant, bold. **confidential** (konfiden'-), *a.* entrusted with the private concerns of another; told or carried on in confidence. **confidentiality** (-shial'-), *n.* **confidentially,** *adv.* **confidentialness,** *n.* **confidently** (kon'fi-), *adv.* **confider,** *n.* **confiding,** *a.* trusting. **confidingness,** *n.*
configure (kənfig'yə), *v.t.* to give shape or form to. **configuration,** *n.* form; structural arrangement; contour or outline; (*Psych.*) a gestalt; (the layout of) the several items of hardware making up a computing or word-processing system.
confine[1] (kon'fin), *n.* (*usu. pl.*) boundaries, limits.
confine[2] (kənfin'), *v.t.* to imprison, to keep within bounds; to limit in application. **to be confined,** to be in child-bed. **confinement,** *n.* the act of confining; the state of being confined, esp. in child-bed; restraint, seclusion. **confiner,** *n.*
confirm (kənfœm'), *v.t.* to establish; to ratify; to bear witness to; to strengthen (in a course or opinion); to administer confirmation to. **confirmand** (kon'fœmand), *n.* one being prepared for the rite of confirmation. **confirmation** (kon-), *n.* the act of confirming; corroborative testimony; the rite of admitting into full communion with an episcopal church by the laying on of hands. **confirmative,** *a.*

confirmatively, *adv.* **confirmatory**, *a.* **confirmed**, *a.* established; beyond hope of recovery or help. **confirmedly** (-mid-), *adv.* **confirmedness** (-mid-, -fœmd'-), *n.* **confirmee** (konfœmē'), *n.* one who has received confirmation.

confiscate (kon'fiskāt), *v.t.* to seize as forfeited to the public treasury; to remove (property) as punishment etc. *a.* (-kət), confiscated. **confiscation**, *n.* **confiscable**, **confiscatable** (-kā'-), *a.* **confiscator**, *n.* **confiscatory** (-fis'kə-), *a.*

confiteor (kənfit'iaw), *n.* a Roman Catholic formula of confession. [L,]

confiture (kon'fichə), COMFITURE.

conflagration (konfləgrā'shən), *n.* a large and destructive fire.

conflate (kənflāt'), *v.t.* to fuse together; to blend (two variant texts) into one. **conflation**, *n.*

conflict[1] (kon'flikt), *n.* a struggle, a contest; opposition of interest, opinions, or purposes.

conflict[2] (kənflikt'), *v.i.* to come into collision; to struggle; to differ; to be discrepant. **conflicting**, *a.* **confliction**, *n.* **conflictive**, *a.*

confluent (kon'fluənt), *a.* flowing together; uniting in a single stream. *n.* a stream which unites with another. **confluence**, *n.* a flowing together; the point of junction of two or more streams.

conflux (kon'flūks), *n.* confluence.

conform (kənfawm'), *v.t.* to make like in form. *v.i.* to comply (with accepted standards); to be in harmony or agreement. **conformable**, *a.* having the same shape or form; compliant; (*Geol.*) arranged (as strata) in parallel planes. **conformability** (-bil'-), *n.* **conformably**, *adv.* **conformal**, *a.* of maps, showing small areas in their true shape. **conformation**, *n.* the manner in which a body is formed; form, shape, structure. **conformer**, *n.* **conformism**, *n.* **conformist**, *n.* one who conforms, esp. to the worship of the Church of England. **conformist**, *a.* **conformity**, *n.* resemblance; agreement, compliance, congruity; conforming to the worship of the Established Church.

confound (kənfownd'), *v.t.* to throw into confusion; to perplex, to terrify; to put to shame; to destroy; to overthrow; to confuse; to curse (used as a mild oath). **confounded**, *a.* **confoundedly**, *adv.*

confraternity (konfrətœ'niti), *n.* a brotherhood associated esp. for religious or charitable purposes.

confrère (kō'freə), *n.* a fellow-member of a profession or association. [F]

confront (kənfrŭnt'), *v.t.* to face; to bring face to face; to face defiantly or in hostility; to compare (with). **confrontation** (kon-), *n.*

Confucian (kənfū'shən), *a.* pertaining to *Confucius* (d. 479 BC), the Chinese philosopher, or his philosophical system. *n.* a follower of Confucius. **Confucianism**, *n.* **Confucianist**, *n.*

confuse (kənfūz'), *v.t.* to mix or mingle so as to render indistinguishable; to perplex; to disconcert. **confusedly** (-zid-, -zd-), *adv.* **confusedness** (-zid-, -zd-), *n.* **confusible**, **-able**, *a.* **confusion**, *n.* the act of confusing; the state of being confused; disorder; perplexity.

confute (kənfūt'), *v.t.* to overcome in argument; to prove to be false. **confutable**, *a.* **confutation** (kon-), *n.*

cong., (*abbr.*) congregation(al); Congregationalist; congress; congressional.

conga (kong'gə), *n.* a Latin American dance performed by several people in single file; music for this dance. *v.i.* (*pres. p.* **congaing**, *past*, *p.p.* **congaed** (-gəd)) to perform this dance. **conga drum**, *n.* a narrow bass drum beaten by the hand.

congé (kō'zhā), *n.* a courtesy before taking leave; departure, farewell; dismissal.

congeal (kənjēl'), *v.t.* to convert from the liquid to the solid state by cold; to coagulate. *v.i.* to coagulate. **congealable**, *a.* **congealment**, *n.*

congelation (konjəlā'shən), *n.* the act of congealing; the state of being congealed; a congealed mass.

congener (kon'jənər, -jē'-), *n.* one of the same kind or class; an organism of the same stock or family. *a.* akin, closely allied (to). **congeneric** (-ne'-), *a.* of the same race or genus. **congenerous** (-jen'-), *a.*

congenial (kənjē'nyəl), *a.* partaking of the same natural characteristics; sympathetic; suitable; pleasant. **congeniality** (-al'-), *n.* **congenially**, *adv.*

congenital (kənjen'itəl), *a.* existing from birth; constitutional. **congenitally**, *adv.*

conger (kong'gə), *n.* one of a genus (*Conger*) of large marine eels; the conger-eel.

congeries (kənjiə'rēz, -je'riēz), *n.* (*pl.* **congeries**) a collection or heap of things or bodies.

congest (kənjest'), *v.i.* to become congested. *v.t.* to overcharge (with blood). **congested**, *a.* closely crowded; unduly distended with an accumulation of blood; of the nose, blocked with mucus. **congestion** (-chən), *n.* an abnormal accumulation of blood in the capillaries; abnormal accumulation (of inhabitants, traffic etc.). **congestive**, *a.*

conglobate (kon'glōbāt), *v.t.*, *v.i.* to form into a ball. *a.* (-bət), formed into a ball. **conglobation**, *n.*

conglomerate (kənglom'ərət), *a.* gathered into a round body. *n.* a rock composed of pebbles cemented together; a large firm formed by the merger of several smaller firms with diverse interests. *v.t.*, *v.i.* (-rāt), to collect into a mass. **conglomeration**, *n.* a miscellaneous collection.

conglutinate (kəngloo'tināt), *v.t.* to glue together. *v.i.* to stick together, to adhere. **conglutination**, *n.*

Congolese (kong-gəlēz'), *n.*, *a.* a native of, or pertaining to, the Congo.

congou (kong'goo), *n.* a kind of Chinese black tea.

congratulate (kəngra'chəlāt), *v.t.* to express pleasure or joy to, on account of some event; to compliment upon, rejoice with, felicitate. **congratulant**, *a.* congratulating. **congratulation**, (*often pl.*) *n.* **congratulative**, *a.* **congratulator**, *n.* **congratulatory**, *a.*

congregate (kong'grigāt), *v.t.* to gather or collect together into a crowd. *v.i.* to come together, to assemble. **congregant**, *n.* a member of a congregation. **congregation**, *n.* the act of gathering together; an assembly of persons, esp. for religious worship; such an assembly habitually meeting in the same place; an administrative department at the Vatican; the assembly of qualified members of a university. **congregational** (-gā'-), *a.* pertaining to a congregation, or to Congregationalism. **Congregationalism**, *n.* that form of church government in which each church is self-governed, and independent of any other authority. **Congregationalist**, *n.*, *a.* **congregationalize**, **-ise**, *v.t.*

congress (kong'gres), *n.* a conference; a formal meeting of delegates for the settlement of international affairs; (**Congress**) the legislature of the US, consisting of a Senate and a House of Representatives; the legislative body in other countries. **Congressman**, **-woman**, *n.* a member of the US Congress. **Congress Party**, *n.* one of the political parties in India. **congressional** (-gresh'-), *a.*

congruent (kong'-), *a.* agreeing, suitable, correspondent; of geometrical figures, having the same shape. **congruence**, **-ency**, *n.* **congruous** (kong'-), *a.* appropriate, fitting. **congruously**, *adv.* **congruousness**, **congruity**, *n.*

conic (kon'ik), *a.* pertaining to or having the form of a cone. **conic sections**, *n.pl.* curves formed by the intersection of a cone and a plane – the parabola, the hyperbola, and the ellipse. **conical**, *a.* **conically**, *adv.* **conicalness**, *n.*

conico-, *comb. form.* conical, or tending to be conical.

conidium (kənid'iəm), *n.* (*pl.* **-dia** (-ə)) an asexual reproductive cell or spore in certain fungi. **conidial**, **-ioid**, *a.* **conidiiferous** (-if'-), *a.*

conifer (kon'ifə), *n.* a cone-bearing plant or tree; any tree

or shrub of the Coniferae. **Coniferae** (kənif'ərē), *n.pl.* (*Bot.*) an order of resinous trees, as the fir, pine and cedar, bearing a cone-shaped fruit. **coniferous** (nif'-), *a.*
coniine (kō'niēn), *n.* a poisonous alkaloid found in hemlock. **Conium** (-əm, kəni'-), *n.* the genus of Umbelliferae containing the hemlock; the fruit of the hemlock or the drug extracted therefrom.
conj., (*abbr.*) conjugation; conjunction.
conjecture (kənjek'chə), *n.* opinion based on inadequate evidence; the formation of such an opinion. *v.t.*, *v.i.* to guess, to surmise. **conjecturable**, *a.* **conjecturably**, *adv.* **conjectural**, *a.* **conjecturally**, *adv.* **conjectured**, *a.*
conjoin (kənjoin'), *v.t.*, *v.i.* **conjoiner**, *n.* to (cause to) unite. **conjoint**, *a.* united, associated, cooperating. **conjointly**, *adv.*
conjugal (kon'jəgəl), *a.* of or pertaining to matrimony or to married life. **conjugality** (-gal'-), *n.* **conjugally**, *adv.*
conjugate (kon'jəgət), *v.t.* to inflect (a verb) by going through the voices, moods, tenses etc. *v.i.* of a verb, to be inflected; (*Biol.*) to unite sexually; to become fused. *a.* (-gət), joined in pairs, coupled; of words, agreeing in derivation. *n.* a conjugate thing, substance, quantity etc. **conjugation**, *n.* the act or process of conjugating; the inflection of a verb; a class of verbs conjugated alike; the fusion of two or more cells or distinct organisms into a single mass. **conjugational**, *a.* **conjugative**, *a.*
conjunct (kənjŭngkt'), *a.* conjoined; closely connected; in union; conjoint. **conjunction**, *n.* union, association, connection; a word connecting sentences or clauses or coordinating words in the same clause. **conjunctional**, *a.* **conjunctionally**, *adv.* **conjunctive**, *a.* serving to unite; (*Gram.*) used as a conjunction, copulative. *n.* a conjunctive word. **conjunctively**, *adv.* **conjunctly**, *adv.* **conjuncture** (-chə), *n.* a combination of circumstances or events; a crisis.
conjunctiva (konjŭngktī'və), *n.* (*pl.* **-vas, -vae**) the mucous membrane lining the inner surface of the eyelids and the front of the eyeball. **conjunctival**, *a.* **conjunctivitis** (kənjŭngktivi'tis), *n.* inflammation of the conjunctiva.
conjure (kənjuə'), *v.t.* to appeal to solemnly; to bind by an oath. **conjuration** (kon-), *n.* the act of conjuring or invoking; a magic spell, a charm; a solemn appeal. **conjurator** (kon'jərā-), *n.* a conspirator. **conjurement**, *n.* a solemn adjuration. **conjuror**, *n.* one bound with others by a common oath.
conjure[2] (kŭn'jə, kon'-), *v.t.* to effect by magical influence; to raise up by or as by magic; to effect by jugglery. *v.i.* to practise the arts of a conjurer. **to conjure with**, of great influence. **to conjure up**, to arouse the imagination about. **conjurer**, **-or**, *n.* one who performs tricks by sleight of hand.
conk (kongk), *n.* (*sl.*) the head; the nose; a punch on the nose. *v.t.* (*sl.*) to hit (someone) on the nose. **to conk out**, (*sl.*) to give out, to fail; to die.
conker (kong'kə), *n.* a horse-chestnut; (*pl.*) a game played with conkers threaded on strings.
Conn., (*abbr.*) Connecticut.
connate (kon'āt), *a.* innate, born with one, congenital.
connect (kənekt'), *v.t.* to link or fasten together; to associate (in one's mind); to associate (with) as a cause or a result; to establish telephone communication between. *v.i.* to be or become connected; (*coll.*) to punch, kick etc.; of a train etc., to have its arrival and departure times arranged to be convenient for those of other trains etc. **connecting rod**, *n.* one that transmits power from one part of a machine to another, esp. from the piston to the crankshaft in an internal-combustion engine. **connected**, *a.* united, esp. by marriage. **well connected**, related to rich or socially powerful people. **connectedly**, *adv.* **connectedness**, *n.* **connecter**, **-or**, *n.* **connectible**, **-able**, *a.* **connective**, *a.* that connects. *n.* a connecting word. **connective tissue**, *n.* the fibrous tissue supporting and connecting the various parts throughout the body. **connectively**, *adv.*
connection, connexion (kənek'shən), *n.* the act of connecting; the state of being connected; relationship (esp. by marriage); one so connected; sexual intercourse; a connecting part; acquaintanceship; the fitting of the departure and arrival of trains, aeroplanes etc.; vehicle whose timetable is so fitted; the apparatus used in linking up electric current by contact; a telephone link; (*sl.*) a supplier of illegal drugs. **in connection with**, connected with (esp. of trains, steam-packets etc.). **in this connection**, in relation to this matter. **connectional**, *a.*
conniption (kənip'shən), *n.* (*N Am.*, *sl.*) (a fit of) rage or hysteria.
connive (kəniv'), *v.i.* to wink (at); voluntarily to omit or neglect to see or prevent any wrong or fault. **connivance**, *n.* passive cooperation in a fault or crime; tacit consent. **conniver**, *n.*
connoisseur (konəsœ'), *n.* one skilled in judging of the fine arts; a person of taste. **connoisseurship**, *n.* [F]
connote (kənōt'), *v.t.* to imply, to betoken indirectly; to involve. **connotation** (kon-), *n.* **connotative** (kon'-, nō'-), *a.* **connotatively**, *adv.*
connubial (kənū'biəl), *a.* relating to marriage. **connubiality** (-al'-), *n.* **connubially**, *adv.*
conoid, conoidal CONE.
conquer (kong'kə), *v.t.* to win by conquest; to overcome; to gain sovereignty over; to subdue. *v.i.* to be victorious. **conquerable**, *a.* **conqueringly**, *adv.* **conqueror**, *n.* **The Conqueror**, William of Normandy, who conquered England in 1066.
conquest (kong'kwest), *n.* the act of conquering; that which is conquered; (*coll.*) a person whose affection has been gained. **The (Norman) Conquest**, the conquest of England by William of Normandy in 1066.
conquistador (konkwis'tədaw, -kēstədaw'), *n.* (*pl.* **-res** (-res)) one of the Spanish conquerors of America in the 16th cent. [Sp.]
Cons., (*abbr.*) Conservative.
cons., (*abbr.*) consecrated; consignment; consolidated; consonant; construction; consultant.
consanguine (kənsang'gwin), **-guineous** (-gwin'-), *a.* of the same blood; related by birth. **consanguinity** (-gwin'-), *n.*
conscience (kon'shəns), *n.* the sense of right and wrong; consciousness; a feeling of guilt. **in conscience**, in truth; assuredly. **in all conscience**, (*coll.*) in all reason or fairness. **to have on one's conscience**, to feel guilt or remorse about. **conscience clause**, *n.* a clause in an Act of Parliament to relieve persons with conscientious scruples from certain requirements. **conscience money**, *n.* money paid voluntarily (and often anonymously), as compensation for evasion of commitments. **conscience-smitten**, **-stricken**, *a.* stung by conscience on account of some misdeed. **conscienceless**, *a.*
conscientious (konshien'shəs), *a.* acting according to conscience; scrupulous; painstaking. **conscientious objector**, *n.* one who refuses on principle to take part, or help in any way in war or in activities connected with it. **conscientiously**, *adv.* **conscientiousness**, *n.*
conscionable (kon'shənəbl), *a.* regulated by conscience; scrupulous, just. **conscionableness**, *n.* **conscionably**, *adv.*
conscious (kon'shəs), *a.* aware of one's own existence; self-conscious; aware, awake; present to consciousness, deliberate. *n.* the conscious mind. **-conscious**, *comb. form.* very aware of; attaching importance to. **consciously**, *adv.* **consciousness**, *n.* the state of being conscious; immediate knowledge, sense, perception.
conscript[1] (kon'skript), *a.* enlisted by conscription. *n.* one compelled to serve as a soldier. **conscription**, *n.* compulsory enrolment for military service.

conscript² (kənskript'), v.t. to enlist compulsorily.
consecrate (kon'sikrāt), v.t. to set apart as sacred; to devote to the service of; to dedicate, to hallow. a. consecrated. **consecration**, n. **consecrator**, n. **consecratory** (-krā'-), a.
consecution (konsikū'shən), n. the state of being consecutive; a succession or series; logical or grammatical sequence.
consecutive (konsek'ūtiv), a. following without interval or break; expressing logical or chronological sequence. **consecutively**, adv. **consecutiveness**, n.
conseil d'état (kōsāy' dātā'), n. a council of state. [F]
consenescence (konsənes'əns), n. general decay with age.
consensus (kənsen'səs), n. a general agreement, unanimity. **consensual**, a. (Physiol.) happening by sympathetic action, as opp. to volition; (Law) existing by consent.
consent (kənsent'), v.i. to assent, to agree, to yield. n. acquiescence in feeling, thought or action; compliance; permission; agreement. **with one consent**, unanimously. **consenter**, n. **consenting**, a. **consenting adult**, n. a person over the age of consent, esp. legally able to enter into a homosexual relationship. **consentingly**, adv.
consentaneous (konsəntā'niəs), a. mutually consenting, unanimous; accordant; simultaneous, concurrent. **consentaneously**, adv. **consentaneity** (-sentənē'i-), n. **consentaneousness**, n.
consentient (kənsen'shənt), a. of one mind, unanimous; consenting.
consequent (kon'sikwənt), a. following as a natural or logical result; consistent. n. the correlative to an antecedent; that which follows as a natural and logical result; (Math.) the second term in a ratio. **consequence**, n. a result or effect; inference; importance; social importance; (pl.) a parlour game. **in consequence**, as a result. **consequential** (-kwen'-), a. following as a result; resulting indirectly; important; self-important. **consequentiality** (-al'-), n. **consequentially**, adv. **consequently**, adv. as a consequence; therefore.
conservancy (kənsœ'vənsi), n. (a commission or court for) official preservation of forests, fisheries, rivers etc. **conservant**, a.
conservation (konsəvā'shən), n. the act of conserving; protection of natural resources and the environment, esp. from destruction by human activity. **conservation of energy** etc., the theory that the sum of energy etc. in the universe remains the same although particular forces are continually being transformed. **conservational**, a. **conservationist**, n.
conservative (kənsœ'vətiv), a. tending or inclined to conserve established values and institutions; opposed to change; (**Conservative**) pertaining to the Conservative Party; moderate, as a *conservative estimate;* relating to conservatism; conventional. n. a conservative person; (**Conservative**) a member or supporter of the Conservative Party. **Conservative Party**, n. a political party that supports private ownership and free enterprise. **conservatively**, adv. **conservatism**, n. conservative character; dislike of change; the political principles of the Conservative Party.
conservatoire (kənsœ'vətwah'), **conservatorium** (-taw'riəm), n. a public school of music.
conservator (kon'səvātə, -sœ'və-), n. a member of a conservancy; a custodian, curator.
conservatory (kənsœ'vətri), n. a glass-house for exotic plants; a conservatoire.
conserve (kənsœv'), v.t. to preserve from injury, decay or loss; to preserve (as fruit), to candy. n. anything preserved, as candied fruit. **conserver**, n.
consider (kənsid'ə), v.t. to think on, to contemplate; to observe and examine; to estimate; to have regard for; to bear in mind; to discuss. v.i. to reflect, to deliberate. **in consideration of**, as a payment for; because of. **to take into consideration**, to consider, to bear in mind. **under consideration**, being considered; under discussion. **considerable**, a. worth consideration or regard; important; moderately large or great. **considerably**, adv. **considerate** (-rət), a. characterized by consideration for others. **considerately**, adv. **considerateness**, n. **consideration**, n. the act of considering; reflection, thought; regard for others; a motive or ground for action; importance, worth; a reward; (Law) the material equivalent given in exchange for something and forming the basis of a contract. **considered**, a. carefully thought out. **considering**, prep. taking into consideration; in view of.
consign (kənsīn'), v.t. to commit to the keeping or trust of another; to send (as goods); to relegate. **consignable**, a. **consignee** (-nē'), n. **consignment**, n. the act of consigning; goods consigned. **consigner, -or**, n.
consilient (kənsil'iənt), a. concurring, agreeing. **consilience**, n.
consist (kənsist'), v.i. to be composed (of); to be founded or constituted (in). **consistence, -ency**, n. the state of being consistent; degree of density; coherence; firmness, solidity; **consistent**, a. harmonious; not self-contradictory; compatible. **consistently**, adv.
consistory (kənsis'təri, kon'-), n. the court of a bishop for dealing with ecclesiastical causes; the college of cardinals at Rome; a local governing body in the Lutheran and Calvinist Churches. **consistorial** (-taw'-), a.
consociate¹ (kənsō'shiət), a. associated together. n. an associate; an accomplice.
consociate² (kənsō'shiāt), v.t. to unite. v.i. to associate. **consociation**, n. association, fellowship.
console¹ (kənsōl'), v.t. to comfort or cheer in trouble or distress. **consolable**, a. **consolation** (kon-), n. that which consoles; alleviation of misery or mental distress. **consolation prize**, n. one awarded to a runner-up. **consolatory** (-sol'-, -sō'-), a. **consolatorily**, adv. **consoler**, n.
console² (kon'sōl), n. a bracket or corbel to support a cornice, table etc.; the frame enclosing the manuals, draw-knobs etc., of an organ; a free-standing cabinet for a television set etc.; (the desk or cabinet holding) the control panel of an electric or electronic system.
consolidate (kənsol'idāt), v.t. to form into a solid and compact mass; to strengthen; to combine. v.i. to become solid. a. (-dət), solidified, combined, hardened. **consolidated annuities, consols** (kon'solz), n.pl. the British Government securities, consolidated into a single stock in 1751. **consolidated fund**, n. a national fund for the payment of certain public charges. **consolidation**, n. **consolidator**, n. **consolidatory**, a.
consols CONSOLIDATE.
consommé (kənsom'ā), n. a soup made by boiling meat and vegetables to a jelly. [F]
consonant (kon'sənənt), a. agreeing or according, esp. in sound; in harmony. n. any letter of the alphabet other than a vowel. **consonance, -ancy**, n. accord or agreement of sound; agreement, harmony; assonance; pleasing agreement of sounds, concord. **consonantal** (-nan'-), a. **consonantly**, adv.
consort¹ (kon'sawt), n. a companion; a husband, a wife; a vessel accompanying another. **queen consort**, the wife of a king. **king, prince consort**, the husband of a queen. **consortism**, n. symbiosis. **consortship**, n.
consort² (kənsawt'), v.i. to keep company with; to be in harmony (with). v.t. to associate; to unite in harmony.
consort³ (kon'sawt), n. a group of musical instruments of the same type playing together.
consortium (kənsaw'tiəm), n. (pl. **-tia** (-ə)) temporary association of states or of companies or financial interests.
conspectus (kənspek'təs), n. a general sketch or survey; a synopsis.
conspicuous (kənspik'ūəs), a. attracting the eye; prominent, obvious. **conspicuous consumption**, n. lavish

spending as a display of wealth. **conspicuously,** *adv.* **conspicuousness,** *n.*
conspire (kənspīə'), *v.i.*, to plot together; to combine secretly to do any unlawful act; to unite. *v.t.* to plot, to concert. **conspiracy** (-spi'-), *n.* the act of conspiring; a secret agreement or combination between two or more persons to commit an unlawful act. **conspiracy of silence,** an agreement not to talk about a particular subject. **conspirator** (-spi'-), *n.* **conspiratorial** (-spirətaw'-), *a.* **conspiratress** (-spi'-), *n. fem.* **conspiringly,** *adv.*
constable (kon'stəbl, kūn'-), *n.* a policeman of the lowest rank; an officer charged with the preservation of the peace; a warden, a governor. **chief constable,** an officer in charge of a police force in an area (as a county). **special constable,** a citizen sworn in to aid the police force in times of emergency. **constableship,** *n.* **constabulary** (-stab'ū-), *n.* a body of police under one authority. *a.* pertaining to the police.
constant (kon'stənt), *a.* firm, unshaken; unmoved in purpose or opinion; unchanging, steadfast, faithful; unceasing. *n.* anything unchanging or unvarying; any property or relation, expressed by a number, that remains unchanged under the same conditions; a quantity that does not vary. **constancy,** *n.* **constantly,** *adv.*
constantan (kon'stəntan, -tən), *n.* an alloy of copper and nickel used for electrical components because of its high electrical resistance at any temperature.
Constantia (kənstan'shiə), *n.* a S African wine from Constantia, near Cape Town.
constellate (kon'stəlāt), *v.i.* to cluster. *v.t.* to set or adorn with or as with stars; to combine into a constellation. **constellation,** *n.* a number of fixed stars grouped within the outlines of an imaginary figure in the sky; an assemblage of similar people or things; a grouping of related ideas etc. **constellatory** (-stel'ə-), *a.*
consternate (kon'stənāt), *v.t.* to affright, to dismay. **consternation,** *n.*
constipate (kon'stipāt), *v.t.* to affect with constipation. **constipation,** *n.* an undue retention or imperfect evacuation of the faeces.
constituent (kənsti'chuənt), *a.* composing a whole; having power to elect or appoint, or to construct or modify a political constitution. *n.* a component part; one of a body which elects a representative. **constituency,** *n.* a body of electors; the place or body of persons represented by a member of Parliament; (*coll.*) a body of clients, customers, supporters etc.
constitute (kon'stitūt), *v.t.* to establish; to enact; to give legal form to; to give a definite nature or character to; to make up or compose; to elect or appoint to an office or employment. **constitutor,** *n.*
constitution (konstitū'shən), *n.* the act of constituting; the nature, form or structure of a system or body; natural strength of the body; mental qualities; a system of fundamental rules or principles for the government of a state; a statute embodying this. **constitutional,** *a.* inherent in the bodily or mental constitution; pertaining to or in accordance with an established form of government; legal. *n.* a walk or other exercise for the benefit of one's health. **constitutional government, monarchy,** *n.* a government or monarchy in which the head of the state is subject to a written or unwritten constitution. **constitutionalism,** *n.* government based on a constitution; adherence to constitutional government. **constitutionalist,** *n.* **constitutionality** (-nal'-), *n.* **constitutionalize, -ise,** *v.t.* to render constitutional. **constitutionally,** *adv.*
constitutive (kon'stitūtiv), *a.* that constitutes or composes; that enacts or establishes. **constitutively,** *adv.*
constr., (*abbr.*) construction.
constrain (kənstrān'), *v.t.* to compel (to do or not to do); to restrain; to confine, repress. **constrained,** *a.* acting under compulsion; forced; embarrassed. **constrainedly**

(-nid-), *adv.* **constraint,** *n.* restraint, compulsion, necessity; a compelling force; self-control.
constrict (kənstrikt'), *v.t.* to compress; to cause to contract; to keep within limits. **constriction,** *n.* **constrictive,** *a.* **constrictor,** *n.* that which constricts; a muscle which contracts or draws together; BOA-CONSTRICTOR.
construct[1] (kənstrŭkt'), *v.t.* to build up, frame; to put together in proper order; to combine words in clauses and sentences; to form by drawing; to form mentally. **constructor,** *n.* **constructorship,** *n.*
construct[2] (kon'strŭkt), *n.* something constructed; (*Psych.*) a concept or idea built up from sense-impressions etc.
construction (kənstrŭk'shən), *n.* the act or art of constructing; the thing constructed; style or form of structure; the syntactical arrangement and connection of words in a sentence; explanation, interpretation (of words, conduct etc.). **constructional,** *a.* **constructive,** *a.* having ability or power to construct; tending to construct, as opposed to *destructive.* **constructively,** *adv.* **constructivism,** *n.* an abstract style of art using geometric shapes and man-made materials. **constructivist,** *n.*
construe (kənstroo'), *v.t.* to combine syntactically; to arrange (as words) in order, so as to show the meaning; to translate; to explain, interpret. *v.i.* to apply the rules of syntax; to translate.
consubstantial (konsəbstan'shəl), *a.* having the same substance or essence, esp. of the three persons of the Trinity. **consubstantiality** (-shial'-), *n.*
consubstantiate (konsəbstan'shiāt), *v.t., v.i.* to unite in one substance. **consubstantiation,** *n.* the Lutheran doctrine that the body and blood of Christ are present along with the eucharistic elements after consecration, as distinct from *transubstantiation.*
consuetude (kon'switūd), *n.* custom, habit; familiarity. **consuetudinary** (-tū'din-), *a.*
consul (kon'səl), *n.* one of the two supreme magistrates of ancient Rome; an officer appointed by a state to reside in a foreign country to promote its mercantile interests and protect its citizens. **consul general,** *n.* the chief consul of a state. **consular** (-sū-), *a.* **consulate** (-sūlət), *n.* the official residence, jurisdiction, office or term of office of a consul. **consulship,** *n.*
consult (kənsŭlt'), *v.i.* to take counsel together; to deliberate. *v.t.* to ask advice or counsel from; to refer to (a book) for information; to have regard to. **consultable,** *a.* **consultancy,** *n.* **consultant,** *n.* one who consults; a person who is consulted, esp. an expert called on for advice and information; a doctor holding the most senior appointment in a branch of medicine in a hospital. **consultation** (kon-), *n.* a meeting for deliberation or to seek advice. **consultative, consultatory, consultive,** *a.* **consultee** (kon-), *n.* a person consulted. **consulter,** *n.* **consulting,** *a.* giving advice; called in for consultation; used for consultation.
consume (kənsūm', *esp. N Am.* -soom'), *v.t.* to destroy by fire, waste or decomposition; to use up; to eat or drink; to dissipate, squander. **consumable,** *a.* that may be consumed. *n.* something that may be consumed, esp. (*pl.*) food. **consumer,** *n.* one who or that which consumes; a person who purchases goods and services for his or her own use. **consumer goods,** *n.pl.* manufactured goods destined for purchase by consumers. **consumerism,** *n.* protection of the interests of consumers; the economic theory that increased consumption of goods and services is desirable. **consumerist,** *n.*
consummate[1] (kon'səmət, kənsūm'ət), *a.* complete, perfect; of the highest quality or degree. **consummately,** *adv.*
consummate[2] (kon'səmāt), *v.t.* to bring to completion, finish; to complete (a marriage) by sexual intercourse. **consummation,** *n.* the act of consummating; the end or

completion of something already begun; perfection, perfect development. **consummative**, *a.* **consummator**, *n.*
consumption (kənsŭmp'shən), *n.* the act of consuming; the state or process of being consumed; the purchase and use by individuals of goods and services; the amount consumed; pulmonary tuberculosis. **consumptive**, *a.* consuming, destructive; disposed to or affected with tuberculosis. *n.* a person suffering from tuberculosis. **consumptively**, *adv.* **consumptiveness**, *n.*
cont., (*abbr.*) contents; continent(al); continued.
contact (kŏn'tăkt), *n.* touch, meeting, the relation of touching; a person who has been exposed to an illness and is likely to carry contagion; a person who can provide one with introductions etc.; (the touching of) conductors allowing electric current to flow. *a.* caused or effected by contact. *v.t.* to establish contact or communication with. **in contact with**, in touch, close proximity or association with. **to make contact**, to complete an electric circuit; to get in touch with. **contact lens**, *n.* one worn in contact with the eyeball in place of spectacles. **contact print**, *n.* one made by placing a negative directly on to photographic paper. **contactable** (-tăk'-), *a.* **contactual** (-tăk'-), *a.*
contadino (kontədē'nō), *n.* (*pl.* **-ni** (-nē)) an Italian peasant. **contadina** (-nə), *n. fem.* (*pl.* **-ne** (-nā)). [It.]
contagion (kəntā'jən), *n.* communication of disease by contact with an infected person; contagious disease; a harmful influence. **contagious**, *a.* communicable by contact, communicating disease by contact; (*loosely*) infectious. **contagious abortion**, *n.* brucellosis in cattle. **contagiously**, *adv.* **contagiousness**, *n.*
contain (kəntān'), *v.t.* to hold within fixed limits, as a vessel; to be capable of holding; to comprise, include; (*Mil.*) to hem in; to restrain. **containable**, *a.* **container**, *n.* that which contains or encloses; a large rigid box of standard size and shape used for bulk transport and storage of goods. **container lorry, ship**, *n.* one designed for the transport of containers. **containerize, -ise**, *v.t.* to put into containers; to convert (e.g. a transportation system) to the use of containers. **containerization**, *n.* **containment**, *n.* the act of containing or restraining, esp. hostilities to a small area, or radioactive emission to a permitted zone in a nuclear reactor.
contaminate (kəntăm'ināt), *v.t.* to pollute, esp. food with disease-causing bacteria or anything with radioactivity. **contaminable**, *a.* **contaminant**, *n.* **contamination**, *n.* **contaminative**, *a.*
contango (kəntang'gō), *n.* the commission paid by a buyer for the postponement of transactions on the Stock Exchange.
contd., (*abbr.*) continued.
conte (kŏt), *n.* a tale, esp. a short amusing story in prose. [F]
contemn (kəntem'), *v.t.* to despise; to slight. **contemner** (-mə, -mnə), *n.*
contemplate (kŏn'təmplāt), *v.t.* to look at, study; to meditate and reflect on; to intend; to regard as possible or likely. *v.i.* to meditate. **contemplation**, *n.* **contemplative** (-tem'-), *a.* given to contemplation; thoughtful, studious. *n.* one who practises religious meditation. **contemplatively**, *adv.* **contemplativeness**, *n.* **contemplator**, *n.*
contemporaneous (kəntempərā'niəs), *a.* existing, living or happening at the same time; lasting the same period. **contemporaneously**, *adv.* **contemporaneousness**, *n.* **contemporaneity** (-nē'ə-), *n.*
contemporary (kəntem'pərəri), *a.* living or existing at the same time; of the same age; up-to-date, modern. *n.* a contemporary person or thing. **contemporize, -ise**, *v.t.* to make contemporary. **contemporization**, *n.*
contempt (kəntempt'), *n.* the act of contemning; scorn, disdain; shame, disgrace; (*Law*) disobedience to the authority of a court or a legislative body. **contemptible**, *a.* worthy of contempt, despicable. **contemptibility** (-bĭl'-),

n. **contemptibleness**, *n.* **contemptibly**, *adv.* **contemptuous** (-tū-), *a.* expressive of contempt; disdainful, scornful. **contemptuously**, *adv.* **contemptuousness**, *n.*
contend (kəntend'), *v.i.* to struggle (to defend or oppose); to strive to obtain or keep; to compete; to dispute. *v.t.* to maintain by argument. **contender**, *n.*
content[1] (kəntent'), *a.* satisfied, willing. *v.t.* to satisfy, appease; to make easy in any situation. *n.* satisfaction, ease of mind. **contented**, *a.* satisfied with what one has. **contentedly**, *adv.* **contentedness**, *n.* **contentment**, *n.*
content[2] (kŏn'tent), *n.* capacity or power of containing; volume; the meaning (of an utterance etc.) as opposed to the form; (*pl.*) that which is contained; the amount (of one substance) contained in a mixture, alloy etc.; (*pl.*) a table or summary of subject-matter. **contentless**, *a.* without any content or meaning.
contention (kəntĕn'shən), *n.* the act of contending; quarrel, controversy; a point contended for. **contentious**, *a.* disposed to or characterized by contention. **contentiously**, *adv.* **contentiousness**, *n.*
conterminous (kontœ'minəs), *a.* having a common boundary-line; having the same limits. **conterminal**, *a.* bordering, neighbouring, contiguous. **conterminously**, *adv.*
contest[1] (kəntest'), *v.t.* to contend for or about; call in question, oppose. *v.i.* to contend, vie. **contestable**, *a.* **contestant**, *n.* one who contests. **contestation** (kon-), *n.* **contester**, *n.*
contest[2] (kŏn'test), *n.* a struggle for victory or superiority; a dispute; competition, rivalry.
context (kŏn'tekst), *n.* the parts of a discourse or book immediately connected with a sentence or passage quoted; the setting, surroundings. **contextual** (-teks'tū-), *a.* **contextually**, *adv.*
contexture (kənteks'chə), *n.* a weaving together; the disposition and relation of parts in a compound body or a literary composition; structure. *v.t.* to give contexture to.
contiguous (kəntig'ūəs), *a.* touching; adjoining, neighbouring. **contiguity** (kontigū'-), *n.* **contiguously**, *adv.*
continent[1] (kŏn'tinənt), *a.* practising self-restraint, esp. sexual; able to control the evacuations of the body. **continence, -ency**, *n.* **continently**, *adv.*
continent[2] (kŏn'tinənt), *n.* a large tract of land not disjoined or interrupted by a sea; one of the great geographical divisions of land; (**Continent**) the mainland of Europe. **continental**, *a.* **continental breakfast**, *n.* a light breakfast of rolls and coffee. **continental climate**, *n.* one characteristic of the interior of a continent, with hot summers, cold winters and low rainfall. **continental drift**, *n.* the theory that the continents were orig. one landmass and have drifted apart slowly to their present positions. **continental quilt**, *n.* a duvet. **continental shelf**, *n.* an area of shallow water round a landmass before the ground begins to slope sharply down to the ocean depths. **continentalism** (-nen'-), *n.* **continentalist**, *n.* **continentalize, -ise**, (-nen'-), *v.t.* **continentally**, *adv.*
contingent (kəntin'jənt), *a.* dependent on an uncertain issue; uncertain; accidental. *n.* a fortuitous event; a military force, or any group of people, forming part of a larger force or group. **contingency**, *n.* the state of being contingent; a chance or possible occurrence; an accident; something dependent on an uncertain issue; (*pl.*) incidental expenses. **contingency fee**, *n.* (*Law*) one (including a percentage of the damages) charged only if the suit is successful. **contingency fund, plan**, *n.* a sum of money or plan of action kept in reserve in case some situation should arise. **contingently**, *adv.*
continual (kəntin'ūəl), *a.* incessant; without interruption; (*coll.*) very frequent. **continually**, *adv.*
continuance (kəntin'ūəns), *n.* the act of continuing; duration; (*Law*) adjournment. **continuant**, *a.* continuing. *n.* a

consonant whose sound can be prolonged, as *f, v, s, r*.
continuate (kəntin′ūət), *a.* continuous, uninterrupted; long-continued. **continuation**, *n.* the act of continuing; extension or prolongation in a series; a sequel; in the Stock Exchange, the carrying over of accounts for stock (see CONTANGO). **continuative**, *a.* **continuator**, *n.*
continue (kəntin′ū), *v.t.* to carry on without interruption; to keep up; to resume; to extend; (*Law*) to adjourn. *v.i.* to remain, to stay; to last, abide; to remain in existence; to persevere. **continuable**, *a.*
continuity (kontinū′iti), *n.* uninterrupted connection; union without a break or interval; the detailed description of a film in accordance with which the production is carried out. **continuity girl, man**, *n.* the person responsible for seeing that there are no discrepancies between the scenes of a film.
continuo (kəntin′ūō), *n.* (*pl.* **-nuos**) thoroughbass.
continuous (kəntin′ūəs), *a.* connected without a break in space or time; uninterrupted, unceasing. **continuous assessment**, *n.* assessment of the progress of a pupil by means of checks carried out at intervals throughout the course of study. **continuous creation**, *n.* the theory that the creation of the universe is a continuous process. **continuous stationery**, *n.* (*Comput.*) paper in a long strip with regular perforations, which can be fed through a printer. **continuously**, *adv.* **continuousness**, *n.*
continuum (kəntin′ūəm), *n.* (*pl.* **-nua** (-ə)) (*Phys.*) an unbroken mass, series or course of events; a continuous series of component parts that pass into each other. **space-time continuum** SPACE.
contort (kəntawt′), *v.t.* to twist with violence, to distort. **contortion**, *n.* the act of twisting; a writhing movement. **contortionist**, *n.* an acrobat who bends his or her body into various shapes; one who twists the sense of words. **contortive**, *a.*
contour (kon′tuə), *n.* the defining line of any figure or body; outline; outline of coast or other geographical feature. *v.t.* to make an outline of; to mark with contour lines; to carry (a road) round a valley or hill. **contour line**, *n.* a line on a map marking a particular level. **contour map**, *n.* one using contour lines. **contour ploughing**, *n.* ploughing round sloping ground on a level instead of up and down.
contr., (*abbr.*) contracted; contraction.
contra (kon′trə), *prep.* against, opposite. *n.* the opposite; the other side. **pro and contra**, for and against.
contra-, *pref.* against; denoting opposition, resistance or contrariety; in music, signifying extreme.
contraband (kon′trəband), *a.* prohibited; smuggled. *v.i.* to deal in contraband goods. *n.* prohibited traffic; articles forbidden to be exported or imported; smuggled articles. **contrabandist**, *n.* a smuggler.
contra-bass (kon′trəbās), *n.* double-bass.
contrabassoon (kon′trəbəsoon), *n.* the double bassoon.
contraception (kontrəsep′shən), *n.* the taking of measures to prevent conception. **contraceptive**, *n., a.* (a device or drug) for preventing conception.
contract[1] (kəntrakt′), *v.t.* to draw together; to make smaller or narrower; (*Gram.*) to abbreviate, shorten; to become liable for; to be attacked by (disease); to agree to (take part in); to establish. *v.i.* to shrink; to agree (to do any act or supply certain articles for a settled price). **to contract in, out**, to agree (not) to participate in some scheme, esp. a pension scheme. **contractable**, *a.* **contracted**, *a.* drawn together; betrothed; mean, narrow, selfish. **contractible**, *a.* **contractibility** (-bil′-), *n.* **contractile** (-tīl), *a.* tending to contract; having the power to shorten itself. **contractility** (-til′-), *n.* **contraction**, *n.* the act of contracting, the state of being contracted; the shortening of a word by the omission of a letter or syllable. **contractive**, *a.*
contract[2] (kon′trakt), *n.* a formal agreement; the document by which this is entered into; a formal betrothal; an undertaking to do certain work or supply certain articles for a specified consideration; (*Law*) an agreement recognized as a legal obligation. **contract bridge**, *n.* a form of auction bridge in which points are gained only for tricks made as well as bid. **contractual** (-trak′chəl), *a.* implying or relating to a contract.
contractor (kəntrak′tə), *n.* one who contracts to do or supply anything for a stipulated consideration, esp. to supply building materials and labour; a muscle that contracts an organ or other part of the body.
contra-dance (kon′trədahns), COUNTRY-DANCE.
contradict (kontrədikt′), *v.t.* to deny the truth of; to assert the opposite of; to oppose; to be inconsistent with. *v.i.* to deny the truth of a statement. **contradictable**, *a.* **contradiction**, *n.* **contradiction in terms**, a statement that is obviously self-contradictory or inconsistent. **contradictive**, *a.* **contradictively**, *adv.* **contradictiveness**, *n.* **contradictor**, *n.* **contradictory**, *a.* affirming the contrary; inconsistent; disputatious. *n.* (*Log.*) a contradictory proposition; the contrary. **contradictorily**, *adv.* **contradictoriness**, *n.*
contradistinguish (kontrədisting′gwish), *v.t.* to distinguish by contrasting opposite qualities. **contradistinction** (-tingk′-), *n.*
contraflow (kon′trəflō), *n.* two-way traffic on one carriage-way of a motorway so that the other may be closed.
contrail (kon′trāl), *n.* a condensation trail.
contraindicant (kontrəin′dikənt), *n.* (*Med.*) a symptom which indicates that a particular treatment or drug would be unsuitable. **contraindicate**, *v.t.* **contraindication**, *n.*
contralto (kəntral′tō), *n.* the lowest variety of the female voice; one who has such a voice. *a.* singing or arranged for contralto.
contraposition (kontrəpəzish′ən), *n.* a placing opposite to, or in contrast. **contrapositive** (-poz′ə-), *a.*
contraption (kəntrap′shən), *n.* a contrivance.
contrapuntal (kontrəpŭn′təl), *a.* (*Mus.*) pertaining or according to counterpoint. **contrapuntist**, *n.*
contrary (kon′trəri), *a.* opposite; diametrically different; contradictory; adverse; (*coll.*) (kəntreə′ri), antagonistic, wayward, perverse. *n.* a thing of opposite qualities; the opposite; a thing that contradicts. *adv.* contrarily; adversely; in an opposite manner or direction. **by contraries**, by way of contrast; by negation instead of affirmation, and vice versa. **on the contrary**, on the other hand; quite the reverse. **to the contrary**, to the opposite effect. **contrariety** (-rī′ə-), *n.* the state of being contrary; disagreement; inconsistency. **contrarily** (-treə′-), *adv.* **contrariness** (-treə′-), *n.* **contrariwise** (-treə′riwīz), *adv.* on the other hand, conversely; perversely.
contrast[1] (kəntrahst′), *v.t.* to set in opposition, so as to show the difference between. *v.i.* to stand in contrast or opposition.
contrast[2] (kon′trahst), *n.* opposition or unlikeness of things or qualities; the presentation of opposite things with a view to comparison; the degree of difference in tone between the light and dark parts of a photograph or TV picture. **contrastive** (-trahs′-), *a.* **contrasty**, *a.* showing great contrast between light and dark tones.
contrasuggestible (kontrəsəjes′tibl), *a.* reacting to a suggestion by doing the opposite.
contra-tenor (kon′trətenə), COUNTERTENOR.
contrat social (kōtra sōsyahl′), *n.* a social contract. [F]
contravene (kontrəvēn′), *v.t.* to violate, to transgress; to be inconsistent with. **contravention** (-ven′-), *n.*
contretemps (kō′trətō), *n.* an embarrassing event; a disagreement, confrontation. [F]
contribute (kəntrib′ūt), *v.t.* to give for a common purpose; to pay as one's share; to write (an article or chapter) for a publication. *v.i.* to give a part; to have a share in any act or effect; to write for a newspaper etc. **contributable**, *a.* **contribution** (kontribū′-), *n.* the act of contributing; that

contrite

which is contributed; a subscription. **contributive,** *a.* **contributiveness,** *n.* **contributor,** *n.* **contributory,** *a.* contributing to the same fund, stock or result; promoting the same end. **contributory negligence,** *n.* partial responsibility for injury etc., by reason of failure to take adequate precautions.

contrite (kəntrīt', kon'-), *a.* deeply sorry for sin; thoroughly penitent; characterized by penitence. **contritely,** *adv.* **contrition** (-trish'-), *n.*

contrive (kəntriv'), *v.t.* to devise, invent; to bring to pass, manage. *v.i.* to form designs, to scheme (against); to manage (successfully). **contrivable,** *a.* **contrivance,** *n.* **contrived,** *a.* forced, artificial. **contriver,** *n.*

control (kəntrōl'), *n.* check, restraint; restraining, directing and regulating power; a person who controls, esp. a spirit controlling a medium; a standard of comparison for checking the results of experiment; (*pl.*) the mechanisms which govern the operation of a vehicle or machine. *v.t.* (*past, p.p.* **controlled**) to govern, command; to regulate, to hold in check; to verify or check. **control board, panel,** *n.* one containing the switches etc. for operating an electrical or mechanical system. **control character,** *n.* (*Comput.*) one which functions as a signal to control some operation, e.g. start, print etc. **control column,** *n.* the lever by which the elevators and ailerons of an aircraft are operated. **control experiment,** *n.* one carried out on two objects so as to have a means of checking and confirming the inferences deduced. **controlling interest,** *n.* a shareholding sufficiently large to ensure some control over the running of a company. **control panel** CONTROL BOARD. **control room,** *n.* a room from which a large electric or other installation is controlled. **control tower,** *n.* a tower at an airport from which traffic in and out is controlled. **control unit,** *n.* (*Comput.*) the part of a central processor which controls the execution of a program. **controllable,** *a.* **controller,** *n.* one who controls; a ruler, director; a person in charge of financial planning, accounts etc. **controllership,** *n.* **controlment,** *n.* control, regulation; the power or act of controlling.

controversy (kontrov'əsi, kon'trəvœ-), *n.* disputation, esp. a dispute carried on in writing. **controversial** (kontrəvœ'shəl), *a.* inclined, pertaining to or arousing controversy. **controversialism** (-vœ'-), *n.* **controversialist** (-vœ'-), *n.* **controversially,** *adv.*

controvert (kon'trəvœt, -vœt'), *v.t.* to call in question; to oppose or refute by argument. **controvertible,** *a.* **controvertist** (kon'-), *n.*

contumacious (kontūmā'shəs), *a.* perverse, stubborn; stubbornly opposing lawful authority. **contumaciously,** *adv.* **contumaciousness,** *n.* **contumacy** (kon'tūməsi), *n.*

contumely (kon'tūmli), *n.* rude, scornful abuse or reproach; insolence, contempt. **contumelious** (-mē'-), *a.* **contumeliously,** *adv.* **contumeliousness,** *n.*

contuse (kəntūz'), *v.t.* to bruise without breaking the skin. **contusion** (-zhən), *n.*

conundrum (kənŭn'drəm), *n.* a riddle; a puzzling question.

conurbation (konəbā'shən), *n.* the aggregation of urban districts.

conv., (*abbr.*) convent; convention(al); conversation.

convalesce (konvəles'), *v.i.* to recover health. **convalescence,** *n.* **convalescent,** *n., a.* (one) recovering from illness. **convalescent hospital, home,** *n.*

convallaria (konvəleə'riə), *n.* a Liliaceous genus, lily of the valley.

convection (kənvek'shən), *n.* the act of conveying; the propagation of heat through liquids and gases by the movement of the heated particles. **convectional,** *a.* **convective,** *a.* **convector,** *n.* a heater which works by the circulation of currents of heated air.

convenance (kō'vənās), *n.* (*usu. pl.*) conventional usages, the proprieties. [F]

convene (kənvēn'), *v.t.* to call together; to summon to appear. *v.i.* to meet together. **convenable,** *a.* **convener, -nor,** *n.* one who calls a committee etc. together; (*Sc.*) the chairman of a public body or committee.

convenient (kənvēn'yənt), *a.* suitable; commodious; useful, handy; close by. **convenience,** *n.* the quality or state of being convenient; comfort, a cause or source of comfort; advantage; a thing that is useful or saves trouble; a water-closet or urinal. **convenience food,** *n.* food bought already prepared and (almost) ready to eat. **convenience store,** *n.* a shop which sells a wide range of useful articles as well as food, and is open at times convenient to the public. **conveniently,** *adv.*

convent (kon'vənt), *n.* a community of religious persons (now usu. for women); the building occupied by such a community. **conventual** (-ven'tū-), *a.* belonging to a convent. *n.* a member of a convent. **conventually,** *adv.*

conventicle (kənven'tikl), *n.* a clandestine or irregular gathering, esp. religious.

convention (kənven'shən), *n.* the act of coming together; a meeting; the persons assembled; an assembly of representatives; an agreement, treaty; an accepted usage. **conventional,** *a.* agreed on by compact; founded on custom or use; slavishly observant of the customs of society; of energy sources, warfare etc., not nuclear. **conventionalism,** *n.* **conventionalist,** *n.* **conventionality** (-al'-), *n.* **conventionalize, -ise,** *v.t.* **conventionally,** *adv.*

converge (kənvœj'), *v.i.* to tend towards one point; (*Math.*) to approach a definite limit by an indefinite number of steps. *v.t.* to cause to converge. **convergence, -ency,** *n.* **convergent,** *a.* tending to meet in one point; (*Biol.*) developing similar characteristics in a similar environment; (*Psych.*) referring to thinking which produces a logical or conventional result.

conversant (kənvœ'sənt), *a.* having knowledge acquired by study or familiarity; proficient; familiar (with). **conversance, -ancy,** *n.*

conversation (konvəsā'shən), *n.* the act of conversing; familiar talk. **conversation piece,** *n.* representation of figures in familiar groupings; something that provides a topic of conversation. **conversational,** *a.* **conversationalist,** *n.* **conversationally,** *adv.*

converse[1] (kənvœs'), *v.i.* to discourse easily and familiarly (with). **conversable,** *a.* inclined to conversation; sociable. **conversableness,** *n.* **conversably,** *adv.* **converser,** *n.*

converse[2] (kon'vœs), *n.* conversation; opposite, counterpart or complement; (*Math.*) an inverted proposition; (*Log.*) a converted proposition. *a.* opposite, reciprocal, contrary. **conversely** (-vœs'-), *adv.* in a contrary order; reciprocally.

conversion (kənvœ'shən), *n.* change from one state to another; change to a new mode of life, religion, morals or politics; (*Math.*) change to a different system of units in the expression of a quantity; (*Log.*) transposition of the terms of a proposition; (*Stock Exch.*) change of one kind of securities into another kind; a change in the structure or use of a building; a building so changed; the transformation of fertile to fissile material in a nuclear reactor.

convert[1] (kənvœt'), *v.t.* to change from one physical state to another; to cause to turn from one religion or party to another; to change (one kind of securities) into another kind; (*Log.*) to transpose the terms of; in Rugby football, to complete (a try) by kicking a goal. *v.i.* to be converted or changed; to undergo a change. **converter,** *n.* one who converts; a device for changing alternating current to direct current or vice versa; (*Comput.*) a device that converts data from one format to another; a reactor that converts fertile to fissile nuclear material. **convertible,** *a.* that may be converted or changed; exchangeable for another kind of thing (as paper money for coin); of a car, having a roof that folds back. *n.* such a car. **convertibility** (-bil'-), *n.* **convertibly,** *adv.*

convert² (kən'vœt), *n.* one who is converted from one religion or party to another.
convex (kon'veks), *a.* having a rounded form on the exterior surface. *n.* a convex body. **convexity** (-veks'-), *n.* **convexly**, *adv.* **convexo-**, *comb. form.* **convexo-concave** (konvek'sō-), *a.* convex on one side and concave on the other. **convexo-convex**, *a.* convex on both sides.
convey (kənvā'), *v.t.* to carry, to transport, transmit; to impart; (*Law*) to transfer (property). **conveyable**, *a.* **conveyance**, *n.* the act of conveying; a vehicle; (*Law*) the act of transferring real property from one person to another; the document by which it is transferred. **conveyancer**, *n.* (*Law*). **conveyancing**, *n.* **conveyer, -or**, *n.* one who or that which conveys; (also **conveyor belt**) an endless mechanical belt or moving platform which carries objects.
convict¹ (kənvikt'), *v.t.* to prove guilty; to return a verdict of guilty against. **conviction**, *n.* the act of convicting; the state of being convicted or convinced; strong belief, persuasion. *a.* actuated by strong belief. **to carry conviction**, to be convincing. **convictive**, *a.*
convict² (kon'vikt), *n.* a criminal sentenced to a term in prison.
convince (kənvins'), *v.t.* to make (someone) agree or believe; to persuade. **convincible**, *a.* capable of conviction or refutation. **convincing**, *a.* **convincingly**, *adv.* **convincingness**, *n.*
convivial (kənviv'iəl), *a.* festive, social, jovial. **convivialist** (-viv'-), *n.* **conviviality** (-al'-), *n.* **convivially**, *adv.*
convocation (konvəkā'shən), *n.* the act of calling together; an assembly, esp. of qualified graduates of certain universities or the clergy of a province. **convocational**, *a.*
convoke (kənvōk'), *v.t.* to call or summon together; to convene.
convolute (kon'vəloot), **-luted** (-loo'tid), *a.* rolled together; of petals, leaves etc., rolled up in another of the same kind; intricate, complex. **convolution** (-loo'-), *n.* a twisting or winding together; a fold, esp. of brain matter; intricacy. **convolve** (-volv'), *v.t.* to roll or wind together.
convolvulus (-vol'vūləs), *n.* a genus of climbing plants, containing the bindweed.
convoy (kon'voi), *v.t.* to accompany, by land or sea, for the sake of protection, esp. with a warship. *n.* the act of convoying; a group of ships with a protecting force; an escort, guard.
convulse (kənvŭls'), *v.t.* to agitate violently; to affect with convulsions; to excite spasms of laughter in. **convulsant**, *n.* a drug that induces convulsions. **convulsion** (-shən), *n.* (*usu. pl.*) an involuntary contraction of the muscular tissues of the body; hence, a violent agitation, disturbance or commotion. **convulsionary**, *a.* **convulsive**, *a.* producing or attended with convulsions. **convulsively**, *adv.* **convulsiveness**, *n.*
cony, coney (kō'ni), *n.* a rabbit; rabbit fur; (*Bibl.*) the hyrax.
coo (koo), *v.i.* (*past, p.p.* **cooed**) to make a soft low sound, like a dove; to speak lovingly. *v.t.* to say in cooing fashion. *n.* the characteristic note of a dove. *int.* expressing astonishment. **to bill and coo** BILL¹.
cooee (koo'ē), *n.* a call used to attract attention. *v.i.* (*past, p.p.* **cooeed**) to make this call.
cook (kuk), *n.* one who dresses or prepares food for the table. *v.t.* to prepare (as food) for the table by applying heat; to garble, falsify; to concoct (often with *up*). *v.i.* to act as a cook; to undergo the process of cooking. **to cook one's goose**, (*sl.*) to stop one's game. **to cook the books**, (*coll.*) to falsify the accounts. **what's cooking?** (*coll.*) what's happening? **cook book**, *n.* one containing recipes and advice on preparing food. **cook-chill**, *v.t.* to cook and chill (convenience foods). *a.* of foods prepared in this way. **cook general**, *n.* a person employed to do cooking and housework. **cookhouse**, *n.* (*Naut.*) a galley; also a detached kitchen in warm countries. **cookout**, *n.* (*N Am.*) a party at which food is cooked out of doors. **cookable**, *a.* **cooker**, *n.* a stove or other apparatus for cooking; a cooking apple. **cookery**, *n.* the act or art of cooking; the occupation of a cook. **cookery book** COOK BOOK. **cooking**, *a.* used in cooking; suitable for cooking rather than eating raw.
cookie (kuk'i), *n.* (*N Am.*) a small sweet cake, a biscuit; (*coll.*) a person. **the way the cookie crumbles**, the way things are, an unalterable state of affairs.
cool (kool), *a.* slightly or moderately cold; not retaining or causing heat; not ardent or zealous; indifferent; aloof; dispassionate; deliberate; impudent; (*coll.*) sophisticated; (*coll.*) relaxed; (*coll.*) without exaggeration. *n.* coolness. *v.t.* to make cool; to calm, allay. *v.i.* to become cool. **to cool one's heels**, (*coll.*) to be kept waiting. **to keep, lose one's cool**, (*coll.*) to remain (stop being) calm. **cool bag, box**, *n.* an insulated container in which food is kept cold. **cool cupboard**, *n.* a refrigerated storage cupboard for food or drink, esp. in a shop. **cool-headed**, *a.* dispassionate, self-possessed. **cooling tower**, *n.* a tower in which water is cooled for industrial reuse. **coolant**, *n.* a liquid used for cooling or lubricating. **cooler**, *n.* that which cools; a vessel in which liquors are set to cool; (*sl.*) prison; a drink consisting of wine and fruit juice. **coolish**, *a.* **coolly**, *adv.* **coolness**, *n.*
coolabah (koo'ləbah), *n.* (*Austral.*) name given to several species of eucalyptus trees.
coolie (koo'li), *n.* a hired labourer in or from any part of the East.
coomb (koom), COMBE.
coon (koon), *n.* (*coll.*) short for RACOON; (*offensive*) a Negro. **gone coon**, (*N Am., sl.*) one hopelessly ruined. **coonskin**, *n.* a hat made of the skin and tail of a racoon.
coon-can (koon'kan), *n.* a card game like rummy.
co-op (kō'op), *n., a.* short for COOPERATIVE (society or shop).
coop (koop), *n.* a box for confining domestic birds or small animals; a confined space. *v.t.* to confine in or as in a coop.
cooper (koo'pə), *n.* one whose trade is to make and repair barrels, tubs etc. *v.t.* to make or repair (casks etc.); (*coll.*) to furnish, to rig (up). **cooperage**, *n.* the trade or workshop of a cooper; the price paid for cooper's work. **coopery**, *n.*
cooperate (kōop'ərāt), *v.i.* to work or act with another or others for a common end; to be helpful. **cooperant**, *a.* **cooperation**, *n.* the act of cooperating; a form of partnership or association for the production or distribution of goods. **cooperative**, *a.* working with others for a common end; helpful; of a business venture, owned jointly by the workers etc., for the economic benefit of them all. *n.* a cooperative business, shop etc. **cooperative shop, store**, *n.* the shop of a **cooperative society** for the production or distribution of goods and the division of profits among the members. **cooperatively**, *adv.* **cooperator**, *n.*
co-opt (kōopt'), *v.t.* to elect into a body by the votes of the members. **co-optation, co-option**, *n.*
coordinate (kōaw'dinət), *a.* of the same order, rank or authority; of terms or clauses in a sentence, of equal order, as distinct from *subordinate*. *n.pl.* (*Math.*) lines used as elements of reference to determine the position of any point; clothes in harmonizing colours and patterns, designed to be worn together. *v.t.* (-nāt), to make coordinate; to bring into orderly relation of parts and whole. **coordinately**, *adv.* **coordination**, *n.* **coordinative**, *a.* **coordinator**, *n.*
coot (koot), *n.* a small black British aquatic bird, *Fulica atra;* a stupid person. **bald as a coot**, quite bald.
cootie (koo'ti), *n.* (*N Am., sl.*) a body louse.
cop¹ (kop), *n.* the top of a hill; a conical roll of thread on a spindle.
cop² (kop), *v.t.* (*past, p.p.* **copped**) (*sl.*) to seize; to arrest;

to catch or get (something unpleasant). *n.* (*coll.*) a policeman; (*sl.*) an arrest. **cop this**, (*sl.*) look at this. **not much cop**, (*sl.*) worthless. **to cop it**, (*sl.*) to be caught or punished. **to cop out**, (*sl.*) to refuse responsibility, or to do something. **cop-out**, *n.* **cop-shop**, *n.* (*sl.*) a police station. **copper**, *n.* (*coll.*) one who cops or seizes; a policeman.
copal (kō'pəl), *n.* (varnish made from) a resin from a Mexican plant.
copartner (kōpaht'nə), *n.* a partner, an associate. **copartnership**, *n.*
copatriot (kōpat'riət), COMPATRIOT.
cope[1] (kōp), *n.* an ecclesiastical sleeveless vestment worn at solemn ceremonies; anything spread overhead, as the sky. *v.t.* to cover with or as with a cope or coping. **copestone** COPING-STONE. **coping**, *n.* the course projecting horizontally on the top of a wall. **coping-stone**, *n.* the topmost stone of a building; a stone forming part of the coping.
cope[2] (kōp), *v.i.* to contend successfully (with); to deal (with) successfully.
cope[3] (kōp), *v.t.* to buy; to barter. *v.i.* to make a bargain, to deal. **coper**, *n.* a dealer, esp. in horses. **horse-coper**, *n.*
copeck (kō'pek), KOPECK.
copepod (kō'pipod), *n.* one of the Copepoda (-pep'-), a class of very small marine crustaceans, found in plankton.
coper COPE[3].
Copernican (kəpœ'nikən), *a.* pertaining to the system of the Polish astronomer *Copernicus* (1472-1543), which has the sun as its centre.
copier COPY.
copilot (ko'pīlət), *n.* a second or assistant pilot of an aircraft.
copious (kō'piəs), *a.* plentiful, abundant; prolific. **copiously**, *adv.* **copiousness**, *n.*
copita (kəpē'tə), *n.* a tulip-shaped sherry glass. [Sp.]
copper[1] (kop'ə), *n.* a red malleable, ductile metallic element, at. no. 29; chem. symbol Cu; a vessel, esp. a cooking or laundry boiler (formerly of copper); a copper (or bronze) coin; a reddish-brown colour. *a.* made of or resembling copper. *v.t.* to deposit a coating of copper on. **copper beech**, *n.* a variety of beech with copper-coloured leaves. **copper-bottomed**, *a.* (*Naut.*) sheathed with copper; (financially) reliable. **copper-butterfly**, *n.* the popular name for the genus *Lycaena*. **copperhead**, *n.* a highly venomous N American snake allied to the rattlesnake; its counterpart in Tasmania. **copper-nose**, *n.* a red nose. **copperplate**, *n.* a polished plate of copper on which something is engraved for printing; an impression from such a plate; a neat, sloping handwriting. **copper pyrites**, *n.* (*Min.*) a compound of copper and sulphur. **coppersmith**, *n.* a worker in copper. **copper-work**, *n.* **coppery**, *a.* made of, containing or resembling copper.
copper[2] COP[2].
copperas (kop'ərəs), *n.* a green sulphate of iron.
coppice (kop'is), *n.* a small wood of small trees and underwood, cut periodically for firewood. *v.t.* to cut (trees and bushes) to make a coppice. **coppicewood**, *n.*
copra (kop'rə), *n.* the dried kernel of the coconut, yielding coconut oil.
copro-, *comb. form.* pertaining to or living on or among dung.
coprolite (kop'rəlit), *n.* the fossil dung of various extinct animals. **coprolitic** (-lit'-), *a.*
coprology (kəprol'əji), *n.* filth in literature or art.
coprophagous (koprof'əgəs), *a.* feeding on dung. **coprophagy** (-ji), *n.*
coprophilia (koprəfil'iə), *n.* morbid, esp. sexual, interest in excrement. **coprophiliac** (-ak), *n.* **coprophilous** (-prof'iləs), *a.* growing in dung.
copse (kops), *n., v.t.* coppice. **copsewood**, *n.* underwood, brushwood. **copsy**, *a.*
Copt (kopt), *n.* one of the old Egyptian race; a Coptic Christian. **Coptic**, *n., a.*
copula (kop'ūlə), *n.* (*pl.* **-lae** (-lē)) that which couples; the word linking a subject and predicate together. **copular**, *a.*
copulate (kop'ūlāt), *v.i.* to have sexual intercourse. *a.* (-lət), joined, connected. **copulation**, *n.* the act of coupling; sexual intercourse; (*Log.*, *Gram.*) connection. **copulative**, *a.* serving to unite; connecting subject and predicate; pertaining to sexual intercourse. *n.* a copulative conjunction. **copulatively**, *adv.* **copulatory**, *a.*
copy (kop'i), *n.* a transcript or imitation of an original; a thing made in imitation of or exactly like another; an original to be imitated; manuscript ready for setting; something that will make a good (newspaper) story; an example of a particular work or book; the words, as opposed to the pictures or graphic material, in an advertisement etc. *v.t.* to imitate, make a copy of; (*fig.*) to follow as pattern or model. *v.i.* to make a copy. **copybook**, *n.* a book giving written material to be copied by children learning to write. *a.* correct, conventional; perfectly executed. **copycat**, *n.*, *a.* (*coll.*) (of) one who imitates someone else. **copyhold**, *n.* (*Law*) a tenure evidenced by a copy of the manorial rolls; property held by such tenure. **copyholder**, *n.* **copyright**, *n.* the exclusive right to publish or sell reproductions of a literary or artistic work. *a.* protected by copyright. *v.t.* to secure copyright for. **copy typist**, *n.* one who types from written copy, rather than from shorthand or tape. **copy-writer**, *n.* one who writes advertisements. **copier**, *n.* one who copies; a transcriber; a photocopier. **copying**, *a.* pertaining to or used for copying. **copying-ink**, *n.* a viscid ink allowing copies to be taken from documents written with it. **copyist**, *n.*
coq au vin (kok ō vī), *n.* a stew of chicken in wine. [F]
coquelicot (kok'likō), *n.* poppy-colour, bright red.
coquet, -ette (kəket'), *v.i.* (*past, p.p.* **coquetted**) to flirt (with); to trifle. **coquetry** (kō'-), *n.* **coquette**, *n.* a woman who flirts. **coquettish**, *a.* **coquettishly**, *adv.* **coquettishness**, *n.*
coquilla (kəkil'yə), *n.* the nut of *Attaka funifera*, a Brazilian palm, used in turnery.
coquito (kəkē'tō), *n.* (*pl.* **-tos**) a Chilean nut-bearing palm-tree.
cor[1] (kaw), *n.* (*Mus.*) a horn. **cor anglais** (kawrong'glā), *n.* the English horn, the tenor oboe.
cor[2] (kaw), *int.* (*sl.*) expressing surprise. **cor blimey**, *int.*
Cor. (*abbr.*) (*Bibl.*) Corinthians; coroner.
cor. (*abbr.*) (*Mus.*) cornet; correction; corrective; correlative.
cor- COM- (used before *r*).
coracle (ko'rəkl), *n.* a light, round boat used in Wales and Ireland, made of wickerwork covered with leather or oiled cloth.
coracoid (ko'rəkoid), *n.* a hook-like process of the scapula in mammals.
coraggio (korah'jō), *int.* courage! bravo! [It.]
coral (ko'rəl), *n.* the hard, calcareous skeletal structure secreted by certain polyps or zoophytes, esp. those of the genus *Corallium*, and deposited in masses on the bottom of the sea; the animal or colony of animals forming these structures; a deep orange-pink colour; an object made of coral; the unimpregnated eggs of a lobster (from their colour). *a.* made of or resembling coral. **coral rag**, *n.* a coralliferous limestone of the Middle Oolite. **coral reef**, *n.* a ridge or series of ridges of coral, tending to form a **coral island**. **coral snake**, *n.* any of the genus *Elaps*. **coral tree**, *n.* a tropical tree of the genus *Erythrina*, bearing blood-red flowers. **coralliferous** (-lif'-), *a.* **coralliform** (-fawm), *a.* (*Bot.*) branching, like coral. **coralligenous** (-lij'in-), *a.* producing coral. **coralline** (-lin), *a.* of the nature of coral; containing or resembling coral. **corallite** (-lit), *n.* a fossil coral; the skeleton or case of a polyp; coralline marble. **corallitic** (-lit'-), *a.* **coralloid**, *n.*, *a.* (an organism) re-

sembling coral.
coranto (kəran'tō), *n.* a rapid kind of dance.
corbeil (kaw'bəl, -bā'), *n.* (*Arch.*) an ornamental sculptured basket; (*Fort.*) a basket filled with earth, and set upon parapets as a protection.
corbel (kaw'bəl), *n.* a bracket or projection of stone, wood or iron projecting from a wall. *v.t.* to support by means of corbels. **corbel-table**, *n.* a projecting course, parapet etc. supported by corbels. **corbel-wood**, *n.* wood piled up to be sold by the cord.
corbie (kaw'bi), *n.* (*Sc.*) a raven, a crow. **corbie-steps**, *n.pl.* the stepped slopes of gables.
cord (kawd), *n.* thick string or thin rope composed of several strands; an electric flex; a raised rib in woven cloth; ribbed cloth, esp. corduroy; (*pl.*) corduroy trousers; a measure for cut wood, 128 cu. ft. (approx. 3·6 m³); (*Anat.*) a cord-like structure; anything which binds or draws. *v.t.* to bind with a cord. **corded**, *a.* bound or fastened with cords; made with cords; ribbed or twilled (like corduroy). **cordless**, *a.* of an electrical appliance, operated by stored electricity, e.g. batteries.
cordate (kaw'dāt), *a.* heart-shaped.
cordelier (kawdəliə'), *n.* a Franciscan friar of the strictest rule (from the knotted rope worn round the waist).
cordial (kaw'diəl), *a.* proceeding from the heart; sincere, warm-hearted; cheering or stimulating. *n.* anything which cheers or comforts; a sweetened drink made with fruit juice; a liqueur. **cordiality** (-al'-), *n.* **cordialize, -ise**, *v.i.* to become cordial; to have the warmest relations (with). **cordially**, *adv.*
cordiform (kaw'difawm), *a.* heart-shaped.
cordillera (kawdilyeə'rə), *n.* a ridge or chain of mountains, esp. used in *pl.*) in America.
cordite (kaw'dīt), *n.* a smokeless explosive, prepared in string-like grains.
cordon (kaw'dən), *n.* a ribbon or cord worn as a mark of rank or a badge; a line or series of soldiers, ships etc. placed so as to guard or blockade an area; a string-course; a fruit-tree trained and closely pruned to grow as a single stem. **sanitary cordon**, (F) **cordon sanitaire** (kawdō sanēteə'), a line of military posts to isolate an infected district. **to cordon off**, to protect by surrounding with a cordon. **cordon bleu** (kawdō blœ'), *n.* a trained cook of the highest calibre. *a.* of food or cookery, of the highest standard. [F]
cordovan (kaw'dəvən), *n.* fine, soft leather, orig. made at Cordova in Spain.
corduroy (kaw'dəroi, -roi'), *n.* a stout-ribbed cotton cloth made with a pile; (*pl.*) corduroy trousers.
core (kaw), *n.* the heart or inner part of anything; the hard middle of an apple, pear or similar fruit, containing the seeds; the central strand of a rope; the insulated conducting wires of a cable; the cylindrical mass of rock brought up by an annular drill; the gist, essence; the central part of the earth; (*Archaeol.*) the central portion of a flint left after flakes have been struck off; the essential part of a school curriculum, studied by all pupils; a piece of magnetic material, such as soft iron, inside an induction coil; the part of a nuclear reactor containing the fissile material; (*Comput.*) a small ferromagnetic ring formerly used in a memory to store one bit; (also **core memory**) a memory which uses cores. *v.t.* to remove the core from. **core time**, *n.* in a flexitime system, the central part of the day when everyone is at work. **coreless**, *a.* **corer**, *n.*
corelation (kōrəlā'shən), CORRELATION.
coreless CORE.
coreligionist (kōrəlij'ənist), *n.* one of the same religion.
coreopsis (koriop'sis), *n.* a genus of yellow garden plants.
co-respondent (kōrispon'dənt), *n.* a joint respondent in a suit, esp. a divorce suit.

corf (kawf), *n.* (*pl.* **corves** (-vz)) a basket or wagon for carrying ore or coal in mines.
corgi (kaw'gi), *n.* a small, smooth-haired, short-legged, Welsh dog.
coriaceous (koriā'shəs), *a.* made of or resembling leather.
coriander (korian'də, *esp. N Am.* ko'-), *n.* an umbellifer, *Coriandrum sativum*, with aromatic seeds used as a spice.
Corinthian (kərin'thiən), *a.* of or pertaining to Corinth, a city of Greece. *n.* a native of Corinth; a dandy. **Corinthian order**, *n.* (*Arch.*) the most ornate of the three Grecian orders, the capital having graceful foliated forms added to the volutes of the Ionic capital.
corium (kaw'riəm), *n.* the innermost layer of the skin in mammals.
cork (kawk), *n.* the very light outer layer of bark of the cork-tree, from which stoppers, floats etc. are made; a stopper for a bottle or cask. *a.* made of cork. *v.t.* to stop with a cork; to blacken with burnt cork. **cork-oak**, CORK-TREE. **corkscrew**, *n.* a screw for drawing corks. *v.t.* to direct or push forward in a wriggling fashion. *a.* twisted to resemble a corkscrew, spiral. **cork-tree**, *n.* a Mediterranean oak, *Quercus suber*, grown for its bark. **corkwood**, *n.* any light porous wood. **corkage**, *n.* a charge levied at hotels on wines consumed by guests but not supplied by the hotel. **corked**, *a.* stopped with cork; blackened with burnt cork; of wine, tasting of the cork. **corker**, *n.* (*coll.*) something or somebody astounding; a statement that puts an end to the discussion. **corky**, *a.*
corm (kawm), *n.* a bulb-like, fleshy subterranean stem.
cormorant (kaw'mərənt), *n.* a species of voracious British sea-bird, dark-coloured with a long neck; a glutton.
Corn., (*abbr.*) Cornish; Cornwall.
corn[1] (kawn), *n.* grain; the seed of cereals; wheat; (*Sc.*) oats; (*chiefly N Am.*) maize; something corny, as a song, joke etc.; a single seed or grain of certain plants. *v.t.* to preserve and season with salt. **corn on the cob**, maize boiled and eaten direct from the cob. **corn-ball**, *n.* (*N Am.*) a sweetmeat composed of popped corn and white of egg; (*N Am., coll.*) a rustic person. **corn-brash**, *n.* a calcareous sandstone belonging to the Inferior Oolite. **corn-bread**, *n.* (*N Am.*) maize bread. **corn-chandler**, *n.* a retail dealer in corn etc. **corn-cob**, *n.* a spike of maize. **corn-cockle**, *n.* a purple flower of the campion tribe, growing in cornfields. **corn-crake**, *n.* the landrail, *Crex pratensis*. **corn dolly**, *n.* a decorative figure made of plaited straw. **corned beef**, *n.* tinned seasoned and cooked beef. **corn exchange**, *n.* a market where corn is sold from samples. **cornfield**, *n.* **corn-flag**, *n.* a gladiolus. **cornflakes**, *n.pl.* a breakfast cereal made from toasted flakes of maize. **cornflour**, *n.* finely-ground meal of maize or rice, used to thicken sauces. **cornflower**, *n.* any of several plants that grow amongst corn, esp. the common bluebottle, *Centaurea cyanus*. **corn land**, *n.* land suitable for or devoted to growing corn. **corn-loft**, *n.* a store for corn. **corn-marigold**, *n.* a yellow-flowered composite plant, *Chrysanthemum segetum*. **corn meal**, *n.* (*N Am.*) meal of maize. **corn pone**, *n.* (*N Am.*) corn-bread baked or fried. **corn-rent**, *n.* rent paid in corn at the market-price. **corn-shuck**, *n.* (*N Am.*) the husk of maize. **cornstarch**, *n.* (*N Am.*) cornflour. **corny**, *a.* trite; old-fashioned and sentimental; unsophisticated.
corn[2] (kawn), *n.* a horny excrescence on the foot or hand, produced by pressure over a bone. **to tread on someone's corns**, to upset a person's feelings. **corn-plaster**, *n.* **corny**, *a.*
cornea (kaw'niə), *n.* (*pl.* **-neas, -neae** (-niē)) the transparent forepart of the external coat of the eye, through which the rays of light pass. **corneal**, *a.*
cornel (kaw'nəl), *n.* the genus *Cornus*, which includes the cornelian cherry-tree and the dogwood.
cornelian[1] (kawnēl'yən), *n.* a variety of semi-transparent chalcedony.

cornelian² (kawnēl'yən), *n.* the wild cornel or dogwood, or the cherry-tree, *Cornus mascula*, or its fruit.

corneous (kaw'niəs), *a.* horny.

corner (kaw'nə), *n.* the place where two converging lines or surfaces meet; the space included between such lines or surfaces; an angle; a place where two streets meet; either of two opposite angles of a boxing ring; a remote or secluded place; a position of difficulty or embarrassment; control over the available supply of any commodity, in order to raise the price, a ring; in football, a free kick from a corner. *v.t.* to drive into a corner, or into a position of difficulty; to furnish with corners; to buy up (a commodity) so as to raise the price. *v.i.* esp. of vehicles, to turn a corner. **to cut (off) a corner,** to take a short cut; to sacrifice quality to speed. **to turn the corner,** to pass the crisis of an illness; to get past a difficulty. **corner shop,** *n.* a small neighbourhood shop, often on a street corner, selling a variety of goods. **cornerstone,** *n.* the stone which unites two walls of a building; the principal stone; the foundation; something of the first importance. **cornerwise,** *adv.* diagonally, with the corner in front. **cornered,** *a.* having corners or angles (*usu. in comb.*); (*fig.*) placed in a difficult position. **cornerer,** *n.* a member of a corner or ring.

cornet¹ (kaw'nit), *n.* a metallic wind-instrument of the trumpet class, but furnished with valves and stoppers; a conical paper bag; a piece of paper twisted into a conical receptacle for small wares; the lower part of a horse's pastern; an ice-cream cone. **cornetist, cornettist** (-net'-), *n.*

cornet² (kaw'nit), *n.* (*formerly*) the lowest commissioned officer in a cavalry regiment. **cornetcy,** *n.*

cornflower CORN¹.

cornice (kaw'nis), *n.* a moulded horizontal projection crowning a wall or other part of a building; an ornamental band of plaster between a wall and ceiling; a projecting mass of snow. **cornice-pole,** *n.* a pole carried along the tops of windows to support curtains. **corniced,** *a.*

corniche (kawnēsh'), *n.* a coast road, esp. one along the face of a cliff.

cornific (kawnif'ik), *a.* producing horns or horny matter. **corniform** (kaw'nifawm), *a.* horn-shaped. **cornigerous** (-nij'ə-), *a.* bearing horns; horned.

Cornish (kaw'nish), *a.* of or pertaining to Cornwall. *n.* the ancient Celtic language of Cornwall. **Cornish engine,** *n.* a single-acting steam pumping-engine. **Cornish pasty,** *n.* a half-moon shaped pasty filled with seasoned meat and vegetables.

corno (kaw'nō), *n.* (*Mus.*) a horn. [It.]

cornu (kaw'nū), *n.* (*pl.* **-nua** (-ə)) (*Anat.*) a horn-like process. **cornual,** *a.* **cornuate** (-ət), *a.* [L]

cornucopia (kawnūkō'piə), *n.* (*pl.* **-pias**) the horn of plenty; a goat's horn wreathed and filled to overflowing with flowers, fruit, corn etc., the symbol of plenty and peace; a representation of a cornucopia; an abundant stock. **cornucopian,** *a.*

cornuted (kawnū'tid), *a.* horned or having horn-like projections; horn-shaped.

corolla (kərol'ə), *n.* the petals of a flower. **corollaceous** (korəlā'-), *a.* **corollate** (ko'rələt), **-lated** (-lātid), *a.* **corolline** (kor'əlin, -līn), *a.* pertaining to a corolla.

corollary (kərol'əri), *n.* (*Log.*) an additional inference from a proposition; a natural consequence; something appended.

corona (kərō'nə), *n.* (*pl.* **-nas, -nae** (-nē)) a broad projecting face forming the principal member of a cornice; a circular chandelier; the trumpet of a daffodil or similar flower; a disk or halo round the sun or the moon; an anthelion; the zone of radiance round the moon in a total eclipse of the sun; (*Anat.*) any structure which is like a crown in shape; a glowing electrical discharge round a charged conductor; a kind of long cigar with straight sides. **coronal**¹, *a.*

coronach (ko'rənakh), *n.* a dirge, a funeral lamentation, in the Scottish Highlands and Ireland.

coronal² (kərō'nəl), *a.* pertaining to a crown or the crown of the head; (*Bot.*) pertaining to a corona. *n.* (ko'rə-), a coronet; a wreath, garland. **coronal suture,** *n.* the suture extending over the crown of the skull. **coronally,** *adv.*

coronary (ko'rənəri), *a.* resembling a crown; placed as a crown. *n.* a small bone in a horse's foot; a coronary thrombosis. **coronary arteries,** *n.pl.* those supplying blood to the heart. **coronary thrombosis,** *n.* the formation of a clot in one of the arteries of the heart.

coronate, -nated (ko'rənət, -nātid), *a.* (*Bot., Zool.*) having a crown, arranged like a crown.

coronation (korənā'shən), *n.* the act or ceremony of crowning a sovereign. **coronation oath,** *n.* that taken by a sovereign at the coronation. **coronation stone,** *n.* the stone in the seat of the chair in Westminster Abbey in which British sovereigns are crowned, taken from the Scots in 1296.

coroner (ko'rənə), *n.* an official who presides over inquests into sudden or suspicious deaths. **coronership,** *n.*

coronet (ko'rənit), *n.* a woman's ornamental head-dress; a small crown worn by princes and noblemen; the part of a horse's pastern where the skin turns to horn. **coroneted,** *a.* entitled to wear a coronet.

coronoid (ko'rənoid), *a.* (*Anat.*) hooked at the tip.

corozo (kərō'zō), *n.* (*pl.* **-zos**) a S American ivory-nut tree, *Phytelephas macrocarpa,* the source of vegetable ivory.

corp., (*abbr.*) corporal; corporation.

corpora CORPUS.

corporal¹ (kaw'pərəl, -prəl), *n.* an army non-commissioned officer of the lowest grade; a petty officer who serves under the master-at-arms. **corporalship,** *n.*

corporal² (kaw'pərəl), *a.* relating to the body; material, corporeal. *n.* the white linen cloth on which the elements are consecrated in the Eucharist. **corporal punishment,** *n.* punishment inflicted on the body. **corporality** (-ral'-), *n.* **corporally,** *adv.*

corporate (kaw'pərət), *a.* united in a body and acting as an individual; pertaining to a corporation. **corporate body,** *n.* a corporation. **corporate hospitality,** *n.* (lavish) entertainment by a company of (potential) clients or customers. **corporate raider,** *n.* one who clandestinely builds up a shareholding in a company in order to gain some control over it. **corporate state,** *n.* a system of government based on trade and professional corporations. **corporately,** *adv.*

corporation (kawpərā'shən), *n.* (*Law*) a corporate body empowered to act as an individual; (*loosely*) a company or association for commercial or other purposes; an elected body charged with the conduct of civic business; (*coll.*) a prominent abdomen. **corporation tax,** *n.* one levied on the profits of companies. **corporatism** (-rə-), *n.*, CORPORATE STATE. **corporative** (kaw'pərə-), *a.* **corporator** (kaw'-), *n.* a member of a corporation.

corporeal (kawpaw'riəl), *a.* having a body; pertaining to the body; material, physical; (*Law*) tangible, visible. **corporeality** (-al'-), **corporeity** (-rē'əti), *n.* **corporeally,** *adv.*

corposant (kaw'pəzənt), *n.* a luminous electric discharge often seen on masts and rigging on stormy nights.

corps (kaw), *n.* (*pl.* **corps** (kawz)) a body of troops having a specific function. **corps de ballet** (də bal'ā), *n.* a body of dancers in a ballet. **corps diplomatique** (diplōmatēk'), *n.* diplomatic corps. [F]

corpse (kawps), *n.* a dead human body. *v.i.* of an actor, to laugh uncontrollably or forget one's lines on stage. **corpse-candle, -light,** *n.* an ignis fatuus seen in churchyards and regarded as an omen of death.

corpulent (kaw'pūlənt), *a.* excessively fat or fleshy. **corpulence, -ency,** *n.* **corpulently,** *adv.*

corpus (kaw'pəs), *n.* (*pl.* **corpora** (-pərə)) the body or

corpuscle — **corvette**

mass of anything; a collection of writings or of literature; (*Physiol.*) the main body of an organ or part. **Corpus Christi** (kris'tī), *n.* a festival in honour of the real presence in the Eucharist, on the Thursday after Trinity Sunday. **corpus delicti** (dilik'tī), *n.* (*Law*) the aggregation of facts which constitute a breach of the law. **corpus luteum** (loo'tiəm), *n.* (*pl.* **corpora lutea** (-iə)) a mass of tissue which develops in the ovary after the discharge of an ovum. [L]

corpuscle (kaw'pəsl), **corpuscule** (-pŭs'kŭl), *n.* (*Physiol.*) a minute particle or cell; a cell, esp. a *white* or *red corpuscle*, suspended in the blood. **corpuscular** (-pŭs'-), *a.*

corr., (*abbr.*) correspond(ing); correspondence.

corrade (kərād'), *v.t.* (*Geol.*) to wear down by wind- or water-borne rock fragments. **corrasion** (-zhən), *n.* **corrasive**, *a.*

corral (kərahl'), *n.* an enclosure for cattle or horses; a defensive circle of wagons. *v.t.* (*past, p.p.* **corralled**) to pen up; to form into a corral.

correct (kərekt'), *v.t.* to set right; to remove faults or errors from; to mark errors for rectification; to admonish, to punish; to obviate, to counteract; to eliminate an aberration. *a.* free from fault or imperfection; conforming to a fixed standard or rule; proper, decorous; true, exact, accurate. **to stand corrected**, to acknowledge a mistake. **correctly**, *adv.* **correctness**, *n.* **corrector**, *n.* **correction**, *n.* the act of correcting; something substituted for what is wrong; improvement; punishment; a compensatory adjustment. **house of correction**, a gaol. **under correction**, as liable to correction; perhaps in error. **correctional**, *a.* **corrective**, (*n., a.*) (something) having power or tending to correct.

corregidor (kərekh'idaw), *n.* the chief magistrate of a Spanish town.

correlate (ko'rəlāt), *v.i.* to be reciprocally related. *v.t.* to bring into mutual relation. *n.* a correlative. **correlation**, *n.* reciprocal relation; act of bringing into correspondence or interaction; (*Phys.*) interdependence of forces and phenomena; the mutual relation of structure, functions etc. in an organism. **correlational** (-lā'-), *a.* **correlative** (-rel'ə-), *a.* reciprocally connected or related; (*Gram.*) corresponding to each other, as *either* and *or*, *neither* and *nor*. *n.* one who or that which is correlated with another. **correlatively**, *adv.* **correlativity** (-relətiv'-), *n.*

correspond (korəspond'), *v.i.* to be congruous; to fit; to suit, to agree; to communicate by letters sent and received. **correspondence**, *n.* mutual adaptation; congruity; intercourse by means of letters; the letters which pass between correspondents. **correspondence college, school**, *n.* one whose courses (**correspondence courses**) are conducted by post. **correspondent**, *a.* agreeing or congruous with; answering. *n.* a person with whom intercourse is kept up by letters; a person or firm having business relations with another; one who sends news from a particular place or on a particular subject, to a newspaper, radio or TV station etc. **correspondently**, *adv.* **corresponding**, *a.* suiting; communicating by correspondence. **correspondingly**, *adv.*

corrida (kərē'də), *n.* a bull-fight.

corridor (ko'ridaw), *n.* a passage communicating with the rooms or apartments of a building; a narrow strip of territory belonging to one state, which passes through the territory of another; a restricted path for air traffic; a passageway along the side of a railway carriage (**corridor carriage**). **corridors of power**, the higher ranks in any organization. **corridor train**, *n.* one with corridors, allowing passage between carriages.

corrie (ko'ri), *n.* a semi-circular hollow or cirque in a mountain side.

corrigendum (korijen'dəm), *n.* (*pl.* **-da** (-ə)) an error needing correction, esp. in a book.

corrigible (ko'rijibl), *a.* capable of being corrected; submissive, docile. **corrigibly**, *adv.*

corroborate (kərob'ərāt), *v.t.* to strengthen, to confirm; to provide additional evidence for. **corroboration**, *n.* the act of strengthening or confirming; confirmation by additional evidence. **corroborative**, *a.* **corroborator**, *n.* **corroboratory**, *a.*

corroboree (kərob'ərē), *n.* a festive or warlike dance of the Australian Aborigines; (*Austral.*) any noisy party or gathering. [Abor.]

corrode (kərōd'), *v.t.* to wear away by degrees; to consume gradually, to rust. *v.i.* to be eaten away gradually. **corrodible**, *a.* **corrosion** (-zhən), *n.* **corrosive**, *a.* tending to corrode; fretting, vexing, virulent. *n.* anything which corrodes. **corrosively**, *adv.* **corrosiveness**, *n.*

corrugate (ko'rəgāt), *v.t.* to contract or bend into wrinkles or folds. *v.i.* to become wrinkled. **corrugated iron**, *n.* sheet iron pressed into folds and galvanized. **corrugation**, *n.* the act of corrugating; a wrinkle, a fold.

corrupt (kərŭpt'), *a.* putrid, decomposed; depraved; perverted by bribery; vitiated by additions or alterations; not genuine. *v.t.* to change from a sound to an unsound state; to make impure or unwholesome; to vitiate or defile; to debauch, to seduce; to bribe; to falsify. *v.i.* to become corrupt. **corrupt practices**, *n.pl.* (*Law*) direct or indirect bribery in connection with an election. **corrupter**, *n.* **corruptibility** (-bil'-), *n.* **corruptible**, *a.* liable to corruption. **corruptibly**, *adv.* **corruption**, *n.* the act of corrupting; the state of being corrupt; decomposition, putrefaction; moral deterioration; misrepresentation; bribery; a corrupt reading or version. **corruptive**, *a.* **corruptly**, *adv.* **corruptness**, *n.*

corsage (kawsahzh'), *n.* the bodice of a woman's dress; a flower worn therein.

corsair (kaw'seə), *n.* a pirate or a privateer, esp. on the Barbary coast.

corse (kaws), *n.* (*poet.*) a corpse.

corselet[1] (kaw'slit), CORSLET.

corselette (kawsəlet'), **corselet**[2] (kaws'lit), *n.* a woman's one-piece foundation undergarment.

corset (kaw'sit), *n.* a close-fitting stiffened or elasticated undergarment worn by women to give a desired shape to the body; a similar undergarment worn by either sex to support a weakened or injured part of the body. *v.t.* to restrain or support with a corset. **corsetry**, *n.*

corslet, corselet (kaws'lit), *n.* body armour; a light cuirass; the thorax of insects.

cortege, cortège (kawtezh'), *n.* a procession, esp. at a funeral.

Cortes (kaw'tez), *n.* the legislative assembly of Spain.

cortex (kaw'teks), *n.* (*pl.* **-tices** (-tisēz)) the layer of plant tissue between the vascular bundles and epidermis; the outer layer of an organ, as the kidney or brain. **cortical** (-ti-), *a.* pertaining to the cortex. **corticate, -cated** (-kət, -kātid), *a.* coated with bark; resembling bark.

corticosteroid (kawtikōstiə'roid), **corticoid** (kaw'tikoid), *n.* a steroid (e.g. cortisone) produced by the adrenal cortex, or a synthetic drug with the same actions.

cortisone (kaw'tizōn, -sōn), *n.* a corticosteroid, natural or synthetic, used to treat rheumatoid arthritis, allergies and skin diseases.

corundum (kərŭn'dəm), *n.* a rhombohedral mineral of great hardness, allied to the ruby and sapphire; a class of minerals including these, consisting of crystallized alumina.

coruscate (ko'rəskāt), *v.i.* to sparkle, to glitter in flashes. **coruscant** (-rŭs'-), *a.* **coruscation**, *n.*

corvée (kawvā', kaw'-), *n.* an obligation to perform a day's unpaid labour for a feudal lord, as the repair of roads etc.; hence, forced labour.

corves CORF.

corvette (kawvet'), *n.* a small, fast escort vessel armed with anti-submarine devices; a flush-decked, full-rigged

ship of war, with one tier of guns.
corvine (kaw'vin), *a.* pertaining to the crows.
Corvus (kaw'vəs), *n.* a genus of conirostral birds, including the raven, jackdaw, rook and crow.
corybant (ko'ribant), *n.* (*pl.* **-tes** (-tēz)) a priest of Cybele, whose rites were accompanied with wild music and dancing. **corybantian** (-ban'shən), *a.* **corybantic** (-ban'-), *a.* **corybantine** (-ban'tin), *a.* **corybantism**, *n.*
corymb (ko'rimb), *n.* a raceme or panicle in which the stalks of the lower flowers are longer than those of the upper. **corymbiate** (-rim'biət), *a.* with clusters of berries or blossoms in the form of corymbs. **corymbiferous** (-bif'-), *a.* **corymbiform** (-rim'bifawm), *a.* **corymbose** (-rim'bōs), *a.*
coryphaeus (korifē'əs), *n.* the leader of a chorus in a classic play.
coryphee (korifā'), *n.* a ballerina, esp. the chief dancer in the corps de ballet.
coryza (kərī'zə), *n.* nasal catarrh; a cold.
COS, (*abbr.*) Chief of Staff.
cos[1] (kos), *n.* a long-leaved variety of lettuce.
cos[2] (koz), (*abbr.*) cosine.
'cos (koz), *conj.* (*coll.*) short for BECAUSE.
Cosa Nostra (kō'zə nos'trə), *n.* the branch of the Mafia operating in the US. [It., our thing]
cosec (kō'sek), (*abbr.*) cosecant.
cosecant (kōsē'kənt), *n.* (*Math.*) the secant of the complement of an arc or angle.
cosech (kō'sech, -shek'), (*abbr.*) hyperbolic cosecant.
coseismal (kōsīz'məl), *a.* relating to the points simultaneously affected by an earthquake. *n.* a coseismal line. **coseismal line, curve**, *n.* a line drawn on a map through all the points simultaneously affected by an earthquake. **coseismic**, *a.*
coset (kō'set), *n.* (*Math.*) a set which forms a given larger set when added to another one.
cosh[1] (kosh), *n.* a short, heavy rod, used as a weapon. *v.t.* to hit with a cosh.
cosh[2] (kosh, kozäch'), (*abbr.*) hyperbolic cosine.
cosher (kō'shə), KOSHER.
cosignatory (kōsig'nətəri), *n.* one who signs jointly with others.
cosine (kō'sīn), *n.* (*Math.*) the sine of the complement of an arc or angle.
cosmetic (kəzmet'ik), *a.* beautifying; used for dressing the hair or skin. *n.* an external application for rendering the skin soft, clear and white, or for improving the complexion. **cosmetic surgery**, *n.* surgery to improve the appearance rather than to treat illness or injury. **cosmetically**, *adv.* **cosmetician** (kozmeti'shən), *n.* one professionally skilled in the use of cosmetics.
cosmic (koz'mik), *a.* pertaining to the universe, esp. as distinguished from the earth; derived from some part of the solar system other than the earth; inconceivably vast; of world-wide importance. **cosmic dust**, *n.* minute particles of matter distributed throughout space. **cosmic radiation, rays**, *n.* very energetic radiation falling on the earth from outer space, consisting chiefly of charged particles. **cosmically**, *adv.*
cosmo-, *comb. form* pertaining to the universe.
cosmogony (kozmog'əni), *n.* a theory, investigation or dissertation respecting the origin of the world. **cosmogonic, -ical** (-gon'-), *a.* **cosmogonist**, *n.*
cosmography (kozmog'rəfi), *n.* a description or delineation of the features of the universe, or of the earth as part of the universe. **cosmographer**, *n.* **cosmographic, -ical** (-graf'-), *a.*
cosmology (kozmol'əji), *n.* the science which investigates the laws of the universe as an ordered whole; the branch of metaphysics dealing with the universe and its relation to the mind. **cosmological** (-loj'-), *a.* **cosmologist**, *n.*
cosmonaut (koz'mənawt), *n.* a (Soviet) astronaut.

cosmopolitan (kozməpol'itən), *a.* common to all the world; at home in any part of the world; free from national prejudices and limitations; composed of people or things from all over the world. *n.* a cosmopolite. **cosmopolitanism**, *n.* **cosmopolitanize, -ise**, *v.t., v.i.* **cosmopolite** (-mop'əlit), *n.* a citizen of the world; one who is at home in any part of the world; an animal or plant found in most parts of the world.
cosmos (koz'mos), *n.* the universe regarded as an ordered system; order, as opp. to chaos; a genus (*Cosmos*) of tropical American plants, grown as garden plants for their showy flowers.
Cossack (kos'ak), *n.* one of a race living on the southern steppes of Russia, and formerly furnishing light cavalry to the Russian army.
cosset (kos'ət), *v.t.* to pet, to pamper.
cost (kost), *v.t.* (*past, p.p.* **cost**) to require as the price of possession or enjoyment; to cause the expenditure of; to result in the loss of or the infliction of. *v.i.* to require expenditure. *v.i., v.t.* (*past, p.p.* **costed**) to fix prices (of commodities). *n.* the price charged or paid for a thing; expense, charge; expenditure of any kind; penalty, loss, detriment; pain, trouble; (*pl.*) expenses of a lawsuit, esp. those awarded to the successful against the losing party. **at all costs**, regardless of the cost. **at cost**, at cost price. **cost of living**, the cost of those goods and services considered necessary to a reasonable standard of living. **prime cost**, the cost of production. **cost accounting**, *n.* recording and analyzing costs incurred by a business. **cost-effective**, *a.* giving a satisfactory return on the initial outlay. **cost-plus**, *a.* used of a contract where work is paid for at actual cost, with an agreed percentage addition as profit. **cost price**, *n.* the price paid by the dealer. **costing**, *n.* calculating the exact cost of production, so as to ascertain the profit or loss entailed. **costless**, *a.* **costly**, *a.* of high price; valuable; entailing considerable loss or sacrifice. **costliness**, *n.*
costa (kos'tə), *n.* (*pl.* **-tae** (-tē)) a rib; any process resembling a rib in appearance or function. **costal**, *a.* **costate** (-tāt), *a.*
co-star (kō'stah), *n.* a star appearing (in a film) with another star. *v.i.* (*past, p.p.* **co-starred**) to be a co-star.
costard (kos'təd), *n.* a large, round apple.
coster, costermonger (kos'təmüng'gə), *n.* a seller of fruit, vegetables etc., esp. from a street barrow. **costermongering**, *n.*
costive (kos'tiv), *a.* constipated; reserved, reticent; niggardly. **costiveness**, *n.*
costly COST.
costmary (kost'meəri), *n.* an aromatic plant of the aster family, used in flavouring.
costo-, *comb. form* (*Anat., Physiol.*) pertaining to the ribs.
costume (kos'tūm, -chəm), *n.* dress; the customary mode of dressing of a particular time or country; fancy dress; the attire of an actor or actress; a set of outer garments; (*dated*) a woman's coat and skirt, usu. tailor-made, a suit; a swimming costume. *v.t.* to furnish or dress with costume. **costume jewellery**, *n.* cheap and showy jewellery worn to set off one's clothes. **costume piece, play**, *n.* a play in which the actors wear historical or foreign costume. **costumer, costumier** (-miə), *n.* a maker of or dealer in costumes.
cosy (kō'zi), *a.* comfortable; snug; complacent. *n.* a padded covering for keeping something warm, esp. a teapot (**tea-cosy**), or a boiled egg (**egg-cosy**). **cosily**, *adv.* **cosiness**, *n.*
cot[1] (kot), *n.* a small house, a hut.
cot[2] (kot), *n.* a light or portable bedstead; a small bedstead with high barred sides for a young child; (*Naut.*) a swinging bed like a hammock. **cot death**, *n.* the sudden and inexplicable death of a baby while sleeping.
cot[3] (kot), (*abbr.*) cotangent.

cotangent (kōtan'jənt, kō'-), *n.* (*Math.*) the tangent of the complement of an arc or angle.
cote (kōt), *n.* a sheepfold; a small house or shelter.
cotenant (kōten'ənt), *n.* a joint tenant.
coterie (kō'təri), *n.* an exclusive circle of people in society; a clique.
coterminous (kōtœ'minəs), CONTERMINOUS.
coth (koth, kotäch'), (*abbr.*) hyperbolic cotangent.
cothurnus (kəthœ'nəs), *n.* the buskin worn by actors in Greek and Roman tragedy.
cotidal (kōti'dəl), *a.* having the tides at the same time as some other place.
cotillion (kətil'yən), **cotillon** (kotēyō'), *n.* (the music for) an 18th-cent. French ballroom dance for four or eight persons.
cotoneaster (kətōnias'tə), *n.* a genus of ornamental shrubs belonging to the order Rosaceae.
Cotswold (kots'wəld), *n.* a breed of sheep, formerly peculiar to the counties of Gloucester, Worcester and Hereford.
cotta (kot'ə), *n.* a short surplice.
cottage (kot'ij), *n.* a small house, esp. for labourers; a small country or suburban residence. **cottage cheese**, *n.* a soft white cheese made from skimmed milk curds. **cottage hospital**, *n.* a small hospital without a resident medical staff. **cottage industry**, *n.* a small-scale industry in which the workers, usu. self-employed, work at home. **cottage loaf**, *n.* a loaf of bread made with two rounded masses of dough stuck one above the other. **cottage piano**, *n.* a small upright piano. **cottage pie**, *n.* shepherd's pie made with beef. **cottager**, *n.* one who lives in a cottage; (*N Am.*) a person living in a country or seaside residence; (*Hist.*) a cottar. **cottagey, -gy**, *a.*
cottar (kot'ə), *n.* a Scottish farm labourer living in a cottage belonging to a farm and paying rent in the form of labour.
cotter (kot'ə), *n.* a key, wedge or bolt for holding part of a machine in place. **cotter pin**, *n.*
cotton (kot'n), *n.* a downy substance, resembling wool, growing in the fruit of the cotton plant, used for making thread, cloth etc.; thread made from this; cloth made of cotton; cotton plants collectively, as a crop. **to cotton on**, to be attracted (to); to begin to understand. **cotton-cake**, *n.* cottonseed pressed into cakes as food for cattle. **cotton candy**, *n.* (*N Am.*) candy floss. **cotton-gin**, *n.* a device for separating the seeds from cotton. **cotton-grass**, *n.* plants with downy heads belonging to the genus *Eriophorum*, growing in marshy ground. **cotton-lord**, *n.* a rich cotton manufacturer. **cotton-picking**, *a.* (*chiefly N Am., sl.*) despicable. **cotton-seed**, *n.* **cottontail**, *n.* any of several common American rabbits. **cotton waste**, *n.* refuse cotton used for cleaning machinery. **cottonwood**, *n.* (*N Am.*) several kinds of poplar, esp. *Populus monilifera* and *P. angulata*; (*Austral.*) the dogwood of Tasmania. **cotton wool**, *n.* cotton in its raw state, used for surgical purposes etc. **in cotton wool**, pampered, protected from hard reality. **cotton-yarn**, *n.* spun cotton ready for weaving. **cottonocracy** (-ok'rəsi), *n.* the great employers in the cotton industry. **cottony**, *a.*
cotyledon (kotilē'dən), *n.* the rudimentary leaf of an embryo in the higher plants, the seed-leaf; a genus of plants, chiefly greenhouse evergreens, including *Cotyledon umbilicus*, the navelwort. **cotyledonal**, *a.* resembling a cotyledon. **cotyledonous**, *a.* possessing cotyledons.
couch (kowch), *v.t.* (*in p.p.*) to cause to lie; to lay (oneself) down; to deposit in a layer or bed; to express in words; to set (a spear) in rest; to operate upon for a cataract; to treat (a cataract) by displacement of the lens of the eye. *v.i.* to lie down, to rest; to lie in concealment; to be laid or spread out. *n.* a bed, or any place of rest; an upholstered seat with a back for more than one person; a similar piece of furniture with a headrest for a doctor's or psychiatrist's patient to lie on; a layer of steeped barley germinating for malting; the frame or floor for this; a preliminary coat of paint, size etc. **couch potato**, *n.* (*sl.*) an inactive person who watches an excessive amount of television instead of taking part in other forms of entertainment or exercise.
couchant (kow'chənt), *a.* (*Her.*) lying down with the head raised.
couchette (kooshet'), *n.* a seat in a continental train which converts into a sleeping berth.
couch-grass (kowch, kooch grahs), *n. Triticum repens*, whose long, creeping root renders it difficult of extirpation.
cougar (koo'gə), *n.* the puma or American lion.
cough (kof), *n.* a convulsive effort, attended with noise, to expel foreign or irritating matter from the lungs; an irritated condition that excites coughing. *v.t.* to drive from the lungs by a cough. *v.i.* to expel air from the lungs with a cough; of an engine, to make a similar noise when malfunctioning. **to cough up**, to eject; (*sl.*) to produce (money or information), esp. under duress. **cough-drop**, *n.* a lozenge taken to cure or relieve a cough. **cough mixture**, *n.* medicine for a cough.
could, *past* CAN [2].
coulee, coulée (koo'li, -lā), *n.* a solidified lava-flow; (*N Am.*) a ravine or gully.
coulomb (koo'lom), *n.* a unit of electrical charge.
coulter (kōl'tə), *n.* the iron blade fixed in front of the share in a plough.
coumarin (koo'mərin), *n.* an aromatic crystalline substance extracted from the Tonka bean, used in flavourings and as an anticoagulant.
council (kown'sl), *n.* a meeting to deliberate or advise on certain matters; a group of people elected or appointed to act in an advisory, administrative or legislative capacity; an ecclesiastical assembly attended by the representatives of various churches; the governing body of a university; (*NT*) the Jewish Sanhedrin; an elected body in charge of local government in a county, parish, borough etc.; (*loosely*) a local bureaucracy. *a.* used by a council; provided or maintained by a council. **British Council**, an official organization for dissemination of British culture abroad. **Council of Europe**, a council set up in 1949 by W European countries to discuss matters of common concern, excluding defence. **Council of Ministers**, an EEC decision-making body consisting of ministers from member states. **council of war**, a council of officers called together in time of difficulty or danger; a meeting to decide on future action. **Privy Council** PRIVY. **councilchamber**, *n.* the room where a council meets. **councilman, -woman**, *n.* (*N Am.*) a councillor. **councillor**, *n.* a member of a council. **councillorship**, *n.*
counsel (kown'sl), *n.* a consultation; advice; opinion given after deliberation; (*Law*) a barrister; (*collect.*) the advocates engaged on either side in a law-suit. *v.t.* (*past, p.p.* **counselled**) to give advice or counsel to; to advise. **counsel of perfection**, a precept aiming at a superhuman standard of righteousness (ref. to Matt. xix.21). **Queen's, King's Counsel**, counsel to the Crown, who take precedence of ordinary barristers. **to keep one's counsel**, to keep a matter secret. **counselling**, *n.* advice and information given in (difficult) personal situations by a qualified adviser. **counsellor**, *n.* one who gives counsel or advice; an adviser; (*N Am.*) a lawyer, esp. one who conducts a case in court. **counsellorship**, *n.*
count[1] (kownt), *v.t.* to add up or check the amount of; to call the numerals in order; to keep a reckoning; to have a certain numerical value; to include in a reckoning; to consider. *v.i.* to name the numbers in order; to have value or significance. *n.* a reckoning or numbering; the sum (of); (*Law*) one of several charges in an indictment; the counting of the seconds after a boxer has been knocked down.

not counting, excluding. **out for the count**, unconscious. **to count down**, to count in reverse order towards zero, zero being the instant at which an event is timed to take place. **to count for** or **against**, to be a factor in favour of, or against. **to count on**, to rely on; to consider as certain. **to count out**, to reckon one by one from a number of units; to declare a boxer defeated upon his failure to stand up within 10 seconds of the referee beginning to count. **to count up**, to calculate the sum of. **to keep, lose count**, to keep or be unable to keep an accurate record of a numerical series. **count-down**, *n.* **counting house**, *n.* the house, room or office appropriated to the business of keeping accounts etc. **countable**[1], *a.* **countless**, *a.* innumerable; beyond calculation.

count[2] (kownt), *n.* a foreign nobleman corresponding to an English earl. **count palatine**, *n.* a high judicial officer under the Merovingian kings; the ruler of either of the Rhenish Palatinates. **countship**, *n.*

countable[2] (kown'təbl), ACCOUNTABLE.

countenance (kown'tənəns), *n.* the face; air, look or expression; composure of look; favour, support. *v.t.* to sanction, to approve, to permit. **to keep one's countenance**, to continue composed in look; to refrain from laughter. **to put out of countenance**, to abash; to cause to feel ashamed. **countenancer**, *n.*

counter[1] (kown'tə), *n.* one who or that which counts; a piece of metal, ivory etc., used for reckoning, as in games; an imitation coin or token; a table or desk over which business is conducted (in a shop, bank, library, cafe etc.). **over the counter**, of medicines, not on prescription. **under the counter**, referring to trade in black market goods; secret(ly); surreptitious(ly).

counter[2] (kown'tə), *n.* the opposite, the contrary; the curved part of a ship's stern; in fencing, a circular parry; in boxing, a blow dealt just as the opponent is striking; the part of a boot or shoe enclosing the wearer's heel. *a.* contrary, adverse, opposed; opposing; duplicate. *adv.* in the opposite direction; wrongly; contrarily. *v.t.* to oppose; to counteract; to return a blow by dealing another one. *v.i.* to take opposing or retaliatory action.

counter- (kowntə-), *comb. form* in return; in answer; in opposition; in an opposite direction.

counteract (kowntərakt'), *v.t.* to act in opposition to, so as to hinder or defeat; to neutralize. **counteraction**, *n.* **counteractive**, *a.*

counterattack (kown'tərətak), *v.t., v.i.* to make an attack after an attack by the enemy. *n.* such an attack.

counterattraction (kown'tərətrak'shən), *n.* attraction in an opposite direction; a rival attraction. **counterattractive**, *a.*

counterbalance (kowntəbal'əns), *v.t.* to weigh against or oppose with an equal weight or effect; to countervail. *n.* (kown'-), an equal weight or force acting in opposition.

counterblast (kown'təblahst), *n.* an argument or statement in opposition.

counterchange (kown'təchānj), *n.* exchange, reciprocation. *v.t.* to exchange, to alternate; to interchange, to chequer.

countercharge (kown'təchahj), *n.* a charge in opposition to another; a counterclaim. *v.t.* to make a charge against in return; to charge in opposition to (a charge of troops).

countercheck (kown'təchek), *n.* an opposing check or restraint; a second check.

counterclaim (kown'təklām), *n.* a claim brought forward by a defendant against a plaintiff.

counterclockwise (kowntəklok'wiz), *adv.* (*N Am.*) anticlockwise.

counterculture (kown'təkŭlchə), *n.* a way of life deliberately contrary to accepted social usages.

counterespionage (kowntəres'piənahzh), *n.* work of an intelligence service directed against the agents and networks of another service.

counterfeit (kown'təfit), *v.t.* to imitate, to mimic; to imitate or copy without right and pass off as genuine; to pretend, to simulate. *v.i.* to make counterfeit copies; to pretend. *a.* made in imitation with intent to be passed off as genuine; sham, simulated. *n.* one who pretends to be what he is not, an impostor; a counterfeit thing. **counterfeiter**, *n.* one who counterfeits.

counterfoil (kown'təfoil), *n.* the counterpart of a cheque, receipt or other document, retained by the giver.

counterinsurgency (kowntərinsœ'jənsi), *n.* measures taken to combat rebellion, guerrilla activity etc.

counterintelligence (kowntərintel'ijəns), *n.* work of an intelligence service designed to prevent or damage intelligence-gathering by an enemy service.

counterirritant (kowntəir'itənt), *n.* an irritant applied to the body to remove some other irritation. *a.* acting as a counterirritant. **counterirritate**, *v.t.* **counterirritation**, *n.*

countermand (kowntəmahnd'), *v.t.* to revoke, to annul; to recall; to cancel. *n.* an order contrary to or revoking a previous order.

countermarch (kown'təmahch), *v.i., v.t.* to march in an opposite direction. *n.* the action of countermarching.

countermeasure (kown'təmezhə), *n.* an action intended to oppose, or counteract the effect of, another action.

countermine (kown'təmīn), *n.* a gallery or mine to intercept or frustrate a mine made by the enemy; a submarine mine employed to explode the mines sunk by the enemy; a stratagem to frustrate any project.

countermove, -movement (kown'təmoov, -moovmənt), *n.* a movement in an opposite or contrary direction.

counteroffensive (kowntərəfen'siv), *n.* a counterattack.

counterpane (kown'təpān), *n.* a coverlet for a bed; a quilt.

counterpart (kown'təpaht), *n.* a correspondent part; a duplicate or copy, esp of a legal document; anything which exactly fits another, as a seal and the impression; one who is exactly like another in person or character.

counterplot (kown'təplot), *v.t.* to oppose or frustrate by another plot. *n.* a plot to defeat another plot.

counterpoint (kown'təpoint), *n.* the art of combining independent melodies in a single harmonic texture; an independent melodic line added above or below another one; a contrasting element. *v.t.* to set off by juxtaposition or contrast.

counterpoise (kown'təpoiz), *n.* a weight in opposition and equal to another; a counterbalancing force, power or influence; equilibrium. *v.t.* to oppose with an equal weight so as to balance; to oppose, check or correct with an equal force, power or influence; to bring into or maintain in equilibrium.

counterproductive (kowntəprədŭk'tiv), *a.* producing an opposite, or undesired, result.

counterproposal (kowntəprəpō'zəl), *n.* one made as an alternative to a previous proposal.

Counter-reformation (kowntərefawmā'shən), *n.* (*Hist.*) the attempt of the Roman Church to counteract the results of the Protestant Reformation.

counter-revolution (kowntərevəloo'shən), *n.* a revolution opposed to a former one, and designed to restore a former state of things. **counter-revolutionary**, *n., a.*

counterscarp (kown'təskahp), *n.* the exterior wall or slope of the ditch in a fortification.

countershaft (kown'təshahft), *n.* an intermediate shaft driven by the main shaft and transmitting motion.

countersign (kown'təsīn), *v.t.* to attest the correctness of by an additional signature; to ratify. *n.* a password, a secret word or sign by which one may pass a sentry, or by which the members of a secret association may recognize each other. **countersignature** (-signəchə), *n.* the signature of an official to a document certifying that of another person.

countersink (kown'təsingk), *v.t.* to chamfer a hole for a screw or bolt head; to sink (the head of a screw etc.) into such a hole. *n.* a chamfered hole; a tool for making such a hole.

countertenor (kown'tətenə), *n.* (a singer with) a voice higher than tenor, an alto; a part written for such a voice.
countervail (kowntəvāl'), *v.t.* to act against with equal effect or power; to counterbalance. *v.i.* to be of equal weight, power or influence on the opposite side.
counterweigh (kowntəwā'), *v.t.* to counterbalance. **counterweight**, *n.*
countess (kown'tis), *n.* the wife of a count or of an earl; a woman holding this rank in her own right.
counting house, countless COUNT¹.
countrified (kŭn'trifīd), *a.* rustic in manners or appearance.
country (kŭn'tri), *n.* a region or state; the inhabitants of any region or state; one's native land; the rural part as distinct from cities and towns; the rest of a land as distinguished from the capital. *a.* rural. **across country**, not using roads etc. **to go to the country**, to hold a general election. **up country**, away from the coast or from the capital city. **country-and-western**, *n.*, *a.* (pertaining to) country music. **country club**, *n.* a sporting or social club in country surroundings. **country cousin**, *n.* a relation of countrified ways or appearance. **country-dance**, *n.* a dance in which the partners are ranged in lines opposite to each other; any rural English dance. **countryman, -woman**, *n.* one who lives in a rural district; a native of the same country as another. **country music**, *n.* a style of popular music based on the folk music of rural areas of the US. **country-seat, -house**, *n.* a gentleman's country mansion. **countryside**, *n.* rural areas or landscapes. **countrywide**, *a.* extending right across a country.
county (kown'ti), *n.* a division of land for administrative, judicial and political purposes; in the British Isles, the chief civil unit, the chief administrative division; in the US, the civil division next below a state; county families collectively. *a.* pertaining to a county; characteristic of county families. **county borough**, *n.* (before 1974) a large borough ranking administratively as a county. **county council**, *n.* the elected council administering the civil affairs of a county. **county councillor**, *n.* **County court**, *n.* a local court dealing with civil cases. **county cricket**, *n.* cricket played between sides representing counties. **county family**, *n.* a family belonging to the nobility or gentry with an ancestral seat in the county. **county palatine**, *n.* a county formerly governed by a count or earl palatine invested with royal privileges, as Cheshire and Lancashire. **county town**, *n.* the chief town of any county.
coup (koo), *n.* a telling or decisive blow; a successful move, piece of strategy or revolution; a coup d'état. **coup de foudre** (də foodr''), a sudden and overwhelming event; love at first sight. **coup de grâce** (də grahs), a finishing stroke. **coup de main** (də mi''), a sudden and energetic attack. **coup d'état** (-dāta'), a sudden and violent change of government, esp. of an illegal and revolutionary nature. **coup de théâtre** (də tāahtr''), a sensational turn of events (in a play). **coup d'oeil** (dœy''), a quick comprehensive glance; a general view. [F.]
coupe (koop), *n.* a dessert made of fruit or ice-cream; the shallow glass dish in which it is served.
coupé (koo'pā), *n.* a four-wheeled closed carriage; a two-door car with an enclosed body.
couple (kŭp'l), *n.* two of the same kind considered together; a pair or brace; (*loosely*) a few; a betrothed or married pair; a pair of dancers; (*Carp.*) a pair of rafters connected by a tie; a pair of equal forces acting in parallel and opposite directions so as to impart a circular movement. *v.t.* to connect or fasten together; to unite persons together, esp. in marriage; to associate. *v.i.* to copulate. **coupler**, *n.* one who or that which couples; a connection between two or more organ manuals. **couplet** (-lit), *n.* two lines of running verse. **coupling** *n.* the action of the verb TO COUPLE; a device for connecting railway carriages etc. together; a device for connecting parts of machinery and transmitting motion. **coupling-box**, *n.* a contrivance for connecting the ends of two shafts and causing them to rotate together.
coupon (koo'pon), *n.* a detachable certificate for the payment of interest on bonds; a detachable ticket or certificate entitling to food ration etc.; a voucher; a detachable slip of paper used as an order form or as an entry form for a competition; a piece of paper which can be exchanged for goods in a shop.
courage (kŭ'rij), *n.* bravery, boldness, intrepidity. **Dutch courage** DUTCH. **the courage of one's convictions**, the courage to act in accordance with one's beliefs. **to pluck up courage**, to summon up boldness or bravery. **courageous** (-rā'-), *a.* **courageously**, *adv.* **courageousness**, *n.*
courante (kurahnt'), *n.* an old dance with a running or a gliding step; the music for this.
courgette (kuəzhet'), *n.* a small kind of vegetable marrow.
courier (ku'riə), *n.* a messenger sent in great haste, an express; a person employed by a travel agency to accompany a party of tourists or to assist them at a resort; an employee of a private postal company offering a fast collection and delivery service usu. within a city or internationally; a person who conveys secret information for purposes of espionage, or contraband for e.g. a drug-smuggling ring.
course (kaws), *n.* the act of passing from one place to another; the track passed over, the route; the bed or the direction of a stream; the ground on which a race is run or a game (as golf) is played; a chase after a hare by one or a brace of greyhounds; continuous progress; one of a series of dishes served at one meal; mode of procedure; a planned programme of study; a medical treatment administered over a particular period; method of life or conduct; a row or tier of bricks or stones in a building. *v.t.* to run after, to pursue; to hunt with dogs which follow by sight. *v.i.* to chase hares with greyhounds; to run or move quickly; to circulate, as the blood. **in due course**, in due, regular or anticipated order. **in (the) course of**, in the process of; during. **matter of course**, a natural event. **of course**, certainly; naturally. **to take its course**, to proceed to its normal conclusion. **courser**, *n.* a swift horse, a war-horse; one who practises coursing; a dog used in coursing; a bird of the genus *Cursorius*, noted for swiftness in running. **coursing**, *a.* that courses. *n.* the sport of hunting hares with greyhounds. **coursing-joint**, *n.* the mortar-joint between two courses of bricks or stones.
court (kawt), *n.* a piece of ground enclosed by buildings; a narrow street; a quadrangle; (a marked-out section of) an area used for games; the residence of a sovereign; the retinue of a sovereign; the body of courtiers; the sovereign and advisers regarded as the ruling power; a State reception by a sovereign; any meeting or body having jurisdiction; the chamber in which justice is administered; the judges or persons assembled to hear any cause; deferential attention paid in order to secure favour or regard. *v.t.* to seek the favour or love of; to seem to invite, as *to court disaster*. *v.i.* to woo; to be in a relationship likely to lead to marriage. **Court of Arches**, ARCHES-COURT under ARCH¹. **Court of St James's**, the court of the British Crown. **Court of Session**, SESSION. **General Court**, GENERAL. **out of court**, not worth considering; without the case being heard in a civil court. **to go to court**, to take legal action. **to hold court**, to preside over a circle of admirers. **court-card**, *n.* any king, queen or knave in a pack of cards. **court circular**, *n.* an official daily report in a newspaper, of the activities and engagements of the royal family. **court-dress**, *n.* the costume proper for a royal levee. **court-house**, *n.* a house or building containing rooms used by any court. **court-martial**, *n.* a court for the trial of service offenders, composed of officers, none of whom must be of inferior rank to the prisoner. *v.t.* to try by court-martial. **drumhead court-martial**, a court held

(orig.) round the drumhead in war-time. **court-plaster**, *n.* silk surfaced with a solution of balsam of benzoin (used in the 18th cent. by fashionable ladies for patches, and since for cuts or slight wounds). **courtroom**, *n.* a room in which a law court sits. **court shoe**, *n.* a woman's low-cut shoe without straps etc. **court tennis**, *n.* real tennis. **courtyard**, *n.* an open area round or within a large building. **courtship**, *n.* the act of soliciting in marriage; the act of seeking after anything.
Courtelle® (kawtel'), *n.* a synthetic acrylic fibre.
courteous (kœ'tiəs), *a.* polite, considerate. **courteously**, *adv.* **courteousness**, *n.*
courtesan, -zan (kaw'tizan), *n.* a prostitute.
courtesy (kœ'təsi), *n.* politeness; graciousness; favour, as opposed to right; an act of civility; a bow, a curtsy. **(by) courtesy of**, through the generosity or kind permission. **courtesy light**, *n.* the interior light in a car. **courtesy title**, *n.* a title to which a person has no legal right.
courtier (kaw'tiə), *n.* one who is in attendance at the court of a prince; one of polished or distinguished manners.
courtly (kawt'li), *a.* polished, elegant, polite; flattering, obsequious. **courtliness**, *n.*
couscous (koos'koos), *n.* a N African dish of pounded wheat steamed over meat or broth.
cousin (kŭz'n), *n.* the son or daughter of an uncle or aunt. **cousin once removed**, the child of one's first cousin. **first cousins**, the children of brothers or sisters. **second cousins**, the children of cousins. **cousin-german**, *n.* a first cousin. **cousinhood**, *n.* **cousinly**, *a.* **cousinship**, *n.*
couture (kətüə', -tuə'), *n.* dressmaking; dress-designing. **couturier** (-riā), *n.* a dress-designer or dress-maker. **couturière** (-rieə), *n.fem.*
couvade (koovahd'), *n.* a custom among primitive races, by which the father on the birth of a child performs certain acts and abstains from certain foods etc.
covalent (kōvā'lənt), *a.* having atoms linked by a shared pair of electrons. **covalence, -ency**, *n.*
cove¹ (kōv), *n.* a small creek, inlet or bay; (also **coving**) a hollow in a cornice-moulding; the cavity of an arch or ceiling. *v.t.* to arch over; to cause to slope inwards. **coved ceiling**, *n.* one with a hollow curve at the junction with the wall. **covelet**, *n.*
cove² (kōv), *n. (dated, sl.)* a man, a fellow, a chap.
coven (kŭv'ən), *n.* an assembly of witches.
covenant (kŭv'ənənt), *n.* a solemn agreement, a compact; a document containing the terms of agreement; *(Law)* a formal agreement under seal; a clause in an agreement; *(Bibl.)* a covenant between the Israelites and Jehovah. *v.t.* to grant or promise by covenant. *v.i.* to enter into a covenant. **Ark of the Covenant** ARK. **New Covenant**, the Christian relation to God. **Old Covenant**, the Jewish dispensation. **Solemn League and Covenant**, the Presbyterian compact of 1643. **covenantal** (-nan'-), **-anted**, *a.* secured by or held under a covenant; bound by a covenant. **covenanter** (-nan'-), **-tor**, *n.* one who enters into a covenant; an adherent of the Scottish National Covenant of 1638 or the Solemn League and Covenant of 1643.
Coventry (kŭv'əntri, kov'-), *n.* **to send to Coventry**, to refuse to have communication or dealings with.
cover (kŭv'ə), *v.t.* to overlay; to overspread with something (so as to protect or conceal); to clothe; to hide or screen; *(Cricket)* to stand behind so as to stop balls that are missed; to lie over so as to shelter or conceal; to incubate; of the lower animals, to copulate with (a female); to include; to be enough to defray; to have as one's field of operations; to extend over; to hold under aim with a firearm; to have within one's field of fire; to protect by insurance; to report on for a newspaper, broadcasting station etc.; *(Mil.)* to protect with troops. *v.i.* to be spread over so as to conceal; to put one's hat on. *n.* anything which covers or hides; a lid; *(often pl.)* the outside covering of a book; one side or board of this; anything which serves to conceal, screen, disguise; pretence, pretext; shelter, protection; a shelter; a thicket, woods which conceal game; *(Comm.)* sufficient funds to meet a liability or ensure against loss; the coverage of an insurance policy; a bed-covering, blanket; an envelope or other wrapping for a packet in the post; a place setting in a restaurant. **to cover for**, to substitute for or replace (an absent fellow worker). **to cover up**, to cover completely; to conceal (esp. something illegal). **under cover**, enclosed in an envelope addressed to another person; concealed; protected. **coverall**, *a.* covering everything. *n.* a one-piece garment covering limbs and body, e.g. a boiler-suit. **cover-charge**, *n.* the amount added to a restaurant bill to cover service. **cover crop**, *n.* one grown between main crops to provide protective cover for the soil. **cover-girl**, *n.* a pretty girl whose photograph is used to illustrate a magazine cover. **cover note**, *n.* a note given to an insured person to certify that he or she has cover. **cover-point**, *n. (Cricket)* a fielder or the position behind point. **cover-up**, *n.* **cover version**, *n.* a version of a song, etc., similar to the original, recorded by a different artist. **coverage**, *n.* the act of covering; the extent to which anything is covered; the area or the people reached by a broadcasting or advertising medium; the amount of protection provided by an insurance policy. **covered**, *a.* **covered wagon**, *n.* a type of large wagon with a tent roof used by American settlers to transport their families and belongings. **covering**, *n.* that which covers; a cover. **covering letter**, *n.* a letter explaining an enclosure.
coverlet (kŭv'əlit), *n.* an outer covering for a bed; a counterpane.
covert (kŭv'ət), *(esp. N Am.* (kō'-)), *a.* covered; disguised, secret, private; *(Law)* under protection. *n.* a place which covers and shelters; a cover for game. **feme-covert** (fem-), **femme-couvert** (famkooveə'), *(Law)* a married woman. **covertly**, *adv.* **covertness**, *n.* **coverture**, *n.* covering, shelter, a hiding-place; secrecy; disguise.
covertical (kōvœ'tikəl), *a. (Geom.)* having common vertices.
covet (kŭv'it), *v.i., v.t.* to desire (someone else's property) inordinately or unlawfully. **covetable**, *a.* **covetous**, *a.* eagerly desirous; eager to obtain and possess; avaricious. **covetously**, *adv.* **covetousness**, *n.*
covey (kŭv'i), *n.* a brood or small flock of birds (prop. of partridges).
coving COVE¹.
cow¹ (kow), *n.* the female of any bovine species, esp. of the domesticated species *Bos taurus;* a female elephant or cetacean; *(sl., derog.)* a woman; *(Austral., sl.)* a difficult or unpleasant situation. **till the cows come home**, *(coll.)* forever. **cow-bane**, *n.* the water-hemlock, *Cicuta virosa.* **cow-bell**, *n.* a bell hung around the neck of a cow. **cowberry**, *n.* the red whortleberry, *Vaccinium vitisdaea.* **cowbird**, *n. (N Am.)* applied to several species of the genus *Molothrus*, from their accompanying cattle. **cowboy**, *n.* a boy who tends cattle; a man in charge of cattle on a ranch; *(sl.)* an unqualified or unscrupulous businessman or workman. **cow-catcher**, *n. (N Am.)* an inclined frame attached to the front of a locomotive etc., to throw obstructions from the track. **cow-fish**, *n.* the sea-cow or manatee; a fish, *Ostracion quadricorne*, with horn-like protuberances over the eyes. **cowgirl**, *n. fem.* **cow-grass**, *n.* a wild trefoil, *Trifolium medium.* **cow-heel**, *n.* the foot of a cow or ox used to make jelly. **cowherd**, *n.* one who tends cattle. **cowhide**, *n.* the hide of a cow; a whip made of cowhide. **cow-house**, *n.* a house or shed in which cows are kept. **cowlick**, *n.* a tuft of hair that grows up over the forehead. **cow-parsley**, *n.* the wild chervil, *Anthriscus sylvestris.* **cowpat**, *n.* a small pile of cow dung. **cowpoke**, *n. (N Am., sl.)* a cowboy. **cow-pony**, *n. (N Am.)* the mustang of a cowboy. **cowpox**, *n.* a vaccine disease affecting the udders of cows, capable of being transferred to

cow-puncher, n. (N Am.) a cowboy. **cowish**, a.
cow² (kow), v.t. to intimidate, to deprive of spirit or courage, to terrify, to daunt. **cowed**, a.
cowage, cowhage (kow'ij), n. (the sharp, stinging hairs of) a tropical climbing plant, *Macuna pruriens*, used as an anthelmintic.
coward (kow'əd), n. someone who lacks courage. a. timid, pusillanimous. **cowardice** (-dis), n. extreme timidity; want of courage. **cowardlike**, a. **cowardly**, adv. in the manner of a coward. a. craven, faint-hearted, spiritless. **cowardliness**, n.
cowboy cow¹.
cower (kow'ə), v.i. to crouch or shrink away through fear.
cowl (kowl), n. a hooded garment, esp. one worn by a monk; a hood-like chimney-top, usu. movable by the wind, to facilitate the exit of smoke. v.t. to cover with a cowl. **cowled**, a. **cowling**, n. a removable metal casing for an aircraft engine.
cowlick, cowpox cow¹.
co-worker (kōwœ'kə), n. a fellow worker.
cowry, cowrie (kow'ri), n. a gasteropod of the genus *Cypraea*, esp. *C. moneta*, a small shell used as money in many parts of southern Asia and Africa.
cowslip (kow'slip), n. a wild plant with fragrant flowers, *Primula veris*, growing in pastures in England.
cox (koks), n. COXSWAIN. v.i. to act as coxswain (to).
coxa (kok'sə), n. (pl. **coxae** (-ē)) the hip; the articulation of the leg to the body in arthropoda. **coxal**, a. **coxalgia** (-al'jə), n. pain in the hip; hip disease. **coxitis** (-ī'tis), n. inflammation of the hip-joint. [L]
coxcomb (koks'kōm), n. the comb resembling that of a cock formerly worn by jesters; a conceited person, a fop, a dandy.
coxitis COXA.
coxswain (kok'sn, kok'swān), n. one who steers a boat, esp. in a race; the petty officer on board ship in charge of a boat and its crew.
coy (koi), a. shrinking from familiarity; modest, shy, reserved; simulating reserve, coquettish; avoiding commitment. **coyly**, adv. **coyness**, n.
coy., (abbr.) company.
coyote (koiō'ti, kiō'-), n. the N American prairie wolf.
coypu (koi'poo), n. a S American aquatic rodent, *Myopotamus*, naturalized in Europe; its fur.
cozen (kūz'n), v.t. to deceive, to cheat. **cozenage**, n. **cozener**, n.
cozy (kō'zi), COSY.
CP, (abbr.) Communist Party; (*Austral.*) Country Party.
cp, (abbr.) candlepower.
cp., (abbr.) compare.
CPAG, (abbr.) Child Poverty Action Group.
cpl, (abbr.) corporal.
CPO, (abbr.) chief petty officer.
CPR, (abbr.) Canadian Pacific Railway.
CPRE, (abbr.) Council for the Preservation of Rural England.
cps, (abbr.) characters per second; cycles per second.
CPSA, (abbr.) Civil and Public Services Association.
CPU, (abbr.) central processing unit.
Cr, (*chem. symbol*) chromium.
cr, (abbr.) created; credit(or); crown.
crab¹ (krab), n. a decapod crustacean of the group Brachyura, esp. the common crab, *Cancer pagurus*, and other edible species; the zodiacal constellation Cancer; a crab-louse. **to catch a crab**, in rowing, to sink an oar too deep or miss a stroke and fall over backwards. **crab-louse**, n. an insect, *Phthirius inguinalis*, found in the pubic area of the human body. **crabwise**, adv. sideways.
crab² (krab), n. a crab-apple; a peevish, morose person. v.i. (*past, p.p.* **crabbed**) (coll.) to complain, grumble.
crab-apple, n. a wild apple, the fruit of *Pyrus malus*; (N Am.) wild apples of other species. **crab-tree**, n.
crabbed (krab'id), a. morose; sour-tempered; perplexing, abstruse; cramped, undecipherable. **crabbedly**, adv. **crabbedness**, n. **crabby**, a. (*coll.*) bad-tempered.
crack (krak), v.t. to break without entire separation of the parts; to cause to give a sharp, sudden noise; to hit hard; to say smartly or sententiously; to open and drink (as a bottle of wine); to break (molecules of a compound) down into simpler molecules by the application of heat; to break into; to solve, decipher. v.i. to break partially asunder; to fail; to give way under psychological pressure; to utter a loud sharp sound; to change (applied to the changing of voices at puberty); to chat. n. a sudden and partial separation of parts; the chink, fissure or opening so made; a sharp sudden sound or report; a smart blow; (*coll.*) a sarcastic joke; (*sl.*) a relatively pure and highly addictive form of cocaine for smoking. a. excellent, superior, brilliant. **a (fair) crack of the whip**, a fair opportunity or chance. **crack of dawn**, the first light of dawn. **crack of doom**, the end of the world. **to crack down (on)**, (*coll.*) to take very strict measures (against). **to crack up**, to extol highly, to puff; to suffer a mental or physical breakdown. **to get cracking**, to start moving quickly. **to have a crack at**, (*coll.*) to have a try, to attempt. **crack-brained**, a. crazy, cracked. **crackdown**, n. **crack-jaw**, a. unpronounceable. **crackpot**, n. (*coll.*) a crazy person. **crack of doom**, the end of the world. **crackup**, n. **crackable**, a. **cracked**, a. (*coll.*) insane, crazy. **cracker**, n. a form of explosive firework; a thin, brittle, hard-baked (savoury) biscuit; a paper tube containing a toy etc., that gives a sharp report in being torn open; an implement for cracking; (*sl.*) one who or that which is exceptional or excellent. **crackerjack**, n. (*sl.*) an excellent person or thing. **crackers**, a. (*sl.*) crazy. **cracking**, a. (*coll.*) vigorous; very good.
crackle (krak'l), v.i. to make short, sharp cracking noises; to be energetic. n. a rapid succession of slight, sharp noises like cracks; a small crack; a series of such cracks. **crackle-china, -glass, -ware**, n. porcelain or glass covered with a delicate network of cracks. **cracklin** (-lin), n. crackle-china. **crackling**, n. making short, sharp, frequent cracks. n. the browned scored skin of roast pork.
cracknel (krak'nəl), n. a hard, brittle biscuit.
-cracy, *suf.* government, rule of; influence or dominance by means of; as in *aristocracy, democracy, plutocracy, theocracy*.
cradle (krā'dl), n. a baby's bed or cot, usu. rocking or swinging; place of birth or early nurture; infancy; a frame to protect a broken or wounded limb in bed; a bed or framework of timbers to support a vessel out of water; (*Mining*) a gold-washing machine; a platform or trolley in which workers are suspended to work on the side of a building or boat. v.t. to lay or place in a cradle; to rock to sleep; to hold protectively in or as in a cradle. **from the cradle to the grave**, throughout one's life. **cradle snatcher**, n. (*coll., derog.*) a person who takes a much younger person as a lover or spouse. **cradlesong**, n. a lullaby. **cradling**, n. the act of laying or rocking in a cradle; (*Build.*) a framework of wood or iron; the framework in arched or coved ceilings to which the laths are nailed.
craft (krahft), n. dexterity, skill; cunning, deceit; an art, esp. one applied to useful purposes, a handicraft, occupation or trade; the members of a particular trade; (*pl.* **craft**) a ship; an aircraft or spacecraft. v.t. to make with skill or by hand. **craft-brother**, n. one of the same craft or guild. **craft-guild**, n. an association of workers in the same occupation or trade. **craftsman**, n. a skilled artisan. **craftsmanship**, n. **-woman**, n.
crafty (krahf'ti), a. artful, sly, cunning, wily. **craftily**, adv. **craftiness**, n.
crag (krag), n. a rugged or precipitous rock. **crag-and-tail**,

crake

n. a rock or hill with a precipitous face on one side and a gradually sloping descent on the other. **cragsman,** *n.* a skilful rock-climber. **cragged,** *a.* **craggedness** (-gid-), *n.* **craggy,** *a.* full of crags, rugged, rough. **cragginess,** *n.*
crake (krāk), *n.* the corncrake; other birds of the same family; their cry.
cram (kram), *v.t.* (*past, p.p.* **crammed**) to stuff, push or press in so as to fill to overflowing; to thrust in by force; to coach for examination by storing the pupil's mind with formulas and answers to probable questions. *v.i.* to eat greedily; to stuff oneself; to get up a subject hastily and superficially, esp. to undergo cramming for examination. *n.* the system of cramming for an examination; information acquired by cramming; a crush, a crowd. **crammer,** *n.* a coach who crams; a school which specializes in cramming.
crambo (kram'bō), *n.* (*pl.* **-boes**) a game in which one selects a word to which another finds a rhyme. **dumb crambo,** a similar game in which the rhymes are expressed in dumb show.
cramp[1] (kramp), *n.* a spasmodic contraction of some limb or muscle, attended with pain and numbness. *v.t.* to affect with cramp.
cramp[2] (kramp), *n.* a cramp-iron; a clamp; restraint, a hindrance. *v.t.* to confine closely; to hinder, to restrain; to fasten with a cramp-iron. **to cramp one's style,** to spoil the effect one is trying to make; to impede a person's actions or self-expression. **cramp-iron,** *n.* an iron with bent ends binding two stones together in a masonry course. **cramped,** *a.* restricted, hindering free movement; difficult to read. **crampedness,** *n.*
crampon (kram'pon), *n.* a hooked bar of iron; a grappling-iron; a plate with iron spikes worn on climbing boots to assist in climbing ice-slopes.
cran (kran), *n.* (*Sc.*) a measure of 37½ gall (170 l) by which herrings are sold.
cranberry (kran'bəri), *n.* the American cranberry, *Vaccinium macrocarpon;* the British marsh whortleberry *V. oxycoccos,* both with a small, red, acid fruit used in sauces etc.
crane[1] (krān), *n.* a bird of the genus *Grus,* esp. *G. cinerea,* a migratory wading bird. *v.t., v.i.* to stretch out (the neck) like a crane, esp. to see over or round an object. **crane-fly,** *n.* the daddy-long-legs, any fly of the genus *Tipula.* **crane's-bill,** *n.* various species of wild geranium. **craner,** *n.*
crane[2] (krān), *n.* a machine for hoisting and lowering heavy weights; a moving platform for a film camera. *v.t.* to raise by a crane.
cranial CRANIUM.
cranio-, *comb. form* (*Anat., Ethn.*) pertaining to the skull. **craniology** (-ol'-), *n.* the scientific study of crania. **craniological** (-loj'-), *a.* **craniologist** (-ol'-), *n.* **craniometer** (-om'itə), *n.* an instrument for measuring the cubic capacity of skulls. **craniometrical** (-met'-), *a.* **craniometry** (-om'-), *n.* **cranioscopy** (-os'kəpi), *n.* the examination of the skull for scientific purposes. **craniotomy** (-ot'əmi), *n.* surgical incision into the skull.
cranium (krā'niəm), *n.* (*pl.* **-niums, -nia** (-ə)) the skull, esp. the part enclosing the brain. **cranial,** *a.* **cranial index,** *n.* the ratio of width to length of the skull, expressed as a percentage. **craniate,** *a.* having a cranium.
crank[1] (krangk), *n.* an arm at right angles to an axis for converting rotary into reciprocating motion, or the converse; an iron elbow-shaped brace for various purposes; (also **crank-handle**) a handle which turns the shaft of a motor until the pistons reach the maximum of compression. *v.t.* to rotate by means of a crank; to crank up. **to crank up,** to start an engine with the crank-handle. **crank-case,** *n.* a metal casing for the crankshaft, connecting-rods etc. in an engine. **crankpin,** *n.* a cylindrical pin parallel to a shaft and fixed at the outer end of a crank.

crawl

crankshaft, *n.* a shaft that bears one or more cranks.
crank[2] (krangk), *n.* a whim, a crotchet; a crotchety person; an eccentric.
crank[3] (krangk), *a.* (*Naut.*) liable to capsize.
cranky (krang'ki), *a.* irritable, fidgety; whimsical; eccentric; full of twists; shaky, sickly; (*Naut.*) liable to capsize.
crannog (kran'əg), *n.* an ancient lake-dwelling in Scotland and Ireland, built up on brushwood and piles, and often fortified.
cranny (kran'i), *n.* a crevice, a chink; a corner, a hole. **crannied,** *a.*
crap[1] (krap), *n.* (*taboo*) excrement; (*sl.*) rubbish, nonsense. *v.i.* (*past, p.p.* **crapped**) (*taboo*) to defecate. **crappy,** *a.* (*sl.*) rubbishy, worthless.
crap[2] (krap), *n.* a losing throw in the game of craps. **crap-shooting, crap-shooter,** *n.* [see CRAPS]
crape (krāp), *n.* a gauzy fabric of silk or other material, with a crisped, frizzly surface, formerly usu. dyed black, used for mourning; a band of this material worn round the hat as mourning. **crapy,** *a.*
crappy CRAP[1].
craps (kraps), *n.sing.* a gambling game played with two dice. **to shoot craps,** to play this game.
crapulent (krap'ūlənt), *a.* drunken; given to intemperance. **crapulence,** *n.* **crapulous,** *a.*
crash[1] (krash), *v.t.* to break to pieces with violence; to dash together violently; to cause to make a loud smashing noise; to cause an aircraft to crash-land or a vehicle to be involved in a collision or serious accident; (*coll.*) to go to (a party) uninvited, to intrude. *v.i.* to make a loud smashing noise; (*Aviat.*) to crash-land; of a vehicle, train etc., to be involved in a collision or serious accident; to fail, to be ruined; to be defeated; of a computer or its program, to cease operating suddenly; (*sl.*) to crash out. *n.* an act or instance of crashing; a loud sudden noise, as of many things broken at once; a violent smash, collision or crash-landing; a sudden failure, collapse, bankruptcy. **to crash out,** (*sl.*) to (go to) sleep. **crash barrier,** *n.* a metal barrier along the edge of a motorway etc. to prevent crashes. **crash course,** *n.* a very rapid and intensive course of study. **crash dive,** *n.* (*Naut.*) a submarine's sudden and rapid dive, usu. to avoid an enemy. **crash helmet,** *n.* a helmet padded with resilient cushions, to protect the head in the event of an accident. **crash-land,** *v.t., v.i.* (to cause) to make a crash-landing. **crash-landing,** *n.* an emergency landing of an aircraft, resulting in damage. **crashing,** *a.* (*coll.*) extreme.
crash[2] (krash), *n.* a coarse linen cloth used for towelling.
crass (kras), *a.* thick, coarse, gross, stupid; obtuse. **crassitude** (-itūd), *n.* crassness; dully. **crassly,** *adv.* **crassness,** *n.*
-crat, *suf.* a partisan, a supporter, a member, as *autocrat, democrat, plutocrat.* **-cratic,** *a.* **-cratically,** *adv.*
crate (krāt), *n.*(the contents of) a box made usu. of wooden slats; (*sl.*) an old and unreliable car, aircraft etc. *v.t.* to pack in a crate. **crateful,** *n.*
crater (krā'tə), *n.* the mouth of a volcano; a funnel-shaped cavity; a large cavity formed by the explosion of a shell or bomb. **crateriform** (-ifawm), *a.*
cravat (krəvat'), *n.* a neckcloth for men.
crave (krāv), *v.t.* to ask for earnestly and submissively; to beg, to beseech, to entreat; to long for; to require. *v.i.* to beg; to long (for). **craver,** *n.* **craving,** *n.* an intense desire or longing. **cravingly,** *adv.*
craven (krā'vən), *n.* a coward, a recreant, a dastard. *a.* cowardly, faint-hearted. **cravenly,** *adv.*
craw (kraw), *n.* the crop or first stomach of fowls or insects.
crawfish CRAYFISH.
crawl[1] (krawl), *v.i.* to move slowly along the ground; to creep; to move slowly; to assume an abject posture or manner; to get on by meanness and servility; to feel as

though insects were creeping over the flesh; to be covered with crawling things; to swim using the crawl. *n.* the act of crawling; a racing stroke in swimming. **crawler,** *n.* **crawlingly,** *adv.* **crawly,** *a.*
crawl² (krawl), *n.* an enclosure in shallow water for keeping fish, turtles etc. alive.
crayfish (krā'fish), **crawfish** (kraw'-), *n.* (*pl.* **crayfish**) the freshwater lobster, *Astacus fluviatilis;* the spiny lobster, *Palinurus vulgaris.*
crayon (krā'ən, -on), *n.* a stick or pencil of coloured material; a drawing made with crayons. *v.t., v.i.* to draw or colour with crayons.
craze (krāz), *v.t.* to make insane; to make cracks or flaws in (china etc.); *v.i.* to become cracked, as the glaze on pottery; to go mad. *n.* a mania, an extravagant idea or enthusiasm; a short-lived fashion. **crazed,** *a.* **crazy,** *a.* unsound, shaky; mad, deranged; ridiculous; (*coll.*) very enthusiastic. **like crazy,** (*sl.*) extremely. **crazy paving,** *n.* a pavement of irregularly-shaped flat stones. **crazily,** *adv.* **craziness,** *n.*
CRE, (*abbr.*) Commission for Racial Equality.
creak (krēk), *v.i.* to make a continued sharp grating noise. *n.* a creaking sound. **creaky,** *a.*
cream (krēm), *n.* the oily part of milk which rises and collects on the surface; a sweetmeat or dish prepared from cream; the best part of anything; a group of the best things or people; a pale yellowish-white colour; a cosmetic preparation with a thick consistency like cream. *v.t.* to skim cream from; to add cream to; (also **to cream off**) to remove the best part from; to make creamy, as by beating. *v.i.* to gather cream; to mantle or froth. *a.* cream-coloured; (of sherry, sweet. **cream of tartar,** purified potassium bitartrate. **cream-bun, -cake,** *n.* a cake with a cream filling. **cream-cheese,** *n.* a soft cheese made of unskimmed milk and cream. **cream cracker,** *n.* an unsweetened crisp biscuit. **cream soda,** *n.* a soft drink flavoured with vanilla. **creamer,** *n.* a flat dish used for skimming the cream off milk; a cream-separator; (*N Am.*) a small jug for cream. **creamery,** *n.* a shop for the sale of dairy produce and light refreshments; an establishment where cream is bought and made into butter. **creamy,** *a.* **creaminess,** *n.*
crease (krēs), *n.* a line or mark made by folding or doubling; (*Cricket*) a line on the ground marking the position of bowler and batsman at each wicket. *v.t.* to make a crease or mark in; (*N Am.*) to graze the skin of with a bullet; (*sl.*) to exhaust; (*sl.*) to crease up. *v.i.* to become creased or wrinkled. **to crease up,** (*coll.*) to double up with laughter. **crease-resistant,** *a.* **creaser,** *n.* **creasy,** *a.*
create (kriāt'), *v.t.* to produce, to bring into existence; to be the occasion of; to originate; to invest with a new character, office or dignity. *v.i.* (*coll.*) to make a fuss; to cause a disturbance. **creative,** *a.* having the ability to create; imaginative; original. **creatively,** *adv.* **creativeness,** *n.* **creativity** (krēə-), *n.*
creatine (krē'ətin, -tēn), *n.* an organic compound found in muscular fibre.
creation (kriā'shən), *n.* the act of creating, esp. creating the world; that which is created or produced; the universe, the world, all created things; the act of appointing or investing; a production of art, craft, or intellect. **creational,** *a.* **creationism,** *n.* the theory that the universe was brought into existence out of nothing by God, and that new forms and species are the results of special creations. **creationist,** *n.* **creator,** *n.* one who or that which creates; a maker; the Maker of the Universe.
creature (krē'chə), *n.* that which is created; a living being; an animal, esp. as distinct from a human being; a person (as an epithet of pity, or endearment); one who owes it or her rise or fortune to another; a tool. *a.* of or pertaining to the body. **creature comforts,** *n.* those pertaining to the body, esp. food and drink. **creaturely,** *a.*

crèche (kresh), *n.* a day nursery in which young children are taken care of.
cred (kred), *n.* (*coll.*) short for CREDIBILITY.
credal CREED.
credence (krē'dəns), *n.* belief, acceptance as true; a credence table. **credence table,** *n.* a small table or shelf near the (south) side of the altar to receive the eucharistic elements before consecration; a small sideboard. **credential** (kriden'-), *n.* anything which gives a title to confidence; (*pl.*) certificates or letters accrediting any person or persons.
credible (kred'ibl), *a.* deserving of or entitled to belief. **credibility** (-bil'-), *n.* **credibility gap,** *n.* the discrepancy between the facts and a version of them presented as true. **credibly,** *adv.*
credit (kred'it), *n.* belief, trust; a reputation (esp. for solvency) inspiring trust or confidence; trust with regard to property handed over on the promise of payment at a future time; the time given for payment of goods sold on trust; a source or cause of honour or reputation; the positive balance in a person's bank account; a sum in excess of deposits available for a person's use; the side of an account in which payment is entered, opposed to debit; an entry on this side of a payment received; (*pl.*) (also **credit titles**) a list of contributors at the beginning or end of a film; an acknowledgment that a student has completed a course of study; the passing of an examination at a mark well above the minimum required. *v.t.* to believe; to set to the credit of (*to* the person); to give credit for (*with* the amount); to believe (a person) to possess something; to ascribe to. **letter of credit,** an order authorizing a person to draw money from an agent. **on credit,** to be paid for later. **credit account,** *n.* a type of account in which goods and services are charged to be paid for later. **credit card,** *n.* a card issued by a bank or credit company which allows the holder to buy goods and services on credit. **credit rating,** *n.* an assessment of credit-worthiness. **credit squeeze,** *n.* government restrictions imposed on banks to limit their loans to clients. **credit-worthy,** *a.* deserving credit because of income-level, past record of debt-repayment etc. **credit-worthiness,** *n.* **creditable,** *a.* bringing credit or honour. **creditability** (-bil'-), *n.* **creditableness,** *n.* **creditably,** *adv.* **creditor,** *n.* one to whom a debt is due.
credo (krā'dō, krē'-), *n.* (*pl.* **-dos**) the Apostles' or the Nicene Creed; a musical setting of the Nicene Creed; a statement of beliefs. [L, I believe]
credulous (kred'ūləs), *a.* disposed to believe, esp. without sufficient evidence; characterized by or due to such disposition. **credulity** (-dū'-), *n.* **credulously,** *adv.* **credulousness,** *n.*
creed (krēd), *n.* a brief summary of the articles of religious belief; any system or solemn profession of beliefs or opinions. **credal, creedal,** *a.*
creek (krēk), *n.* a small inlet, bay or harbour, on the coast; a backwater or arm of a river; (*N Am., Austral.*) a small river, esp. a tributary. **up the creek,** (*sl.*) in trouble or difficulty.
creel (krēl), *n.* an osier basket; a fisherman's basket.
creep (krēp), *v.i.* (*past, p.p.* **crept** (krept)) to crawl along with the body close to the ground; to grow along, as a creeping plant; to move slowly and stealthily, or with timidity; to gain admission unobserved; to behave with servility; to fawn; to have a sensation of shivering or shrinking as from fear or repugnance. *n.* creeping; a slow, almost imperceptible movement; a place for creeping through; a low arch or passage for animals; (*pl.*) a feeling of shrinking horror; (*sl.*) an unpleasant or servile person. **to make someone's flesh creep** FLESH. **creeper,** *n.* one who or that which creeps or crawls; any animal that creeps; a reptile; a plant with a creeping stem; a four-clawed grapnel used in dragging a harbour, pond or well;

(*sl.*) a soft-soled shoe. **creeping**, *n.*, *a.* **creeping jenny** MONEYWORT. **creeping Jesus**, *n.* (*sl.*) a sly or sanctimonious person. **creepingly**, *adv.* **creepy**, *a.* having the sensation of creeping of the flesh; causing this sensation. **creepy-crawly**, *a.* creepy. *n.* a creeping insect.
cremate (krimāt'), *v.t.* to dispose of esp. a corpse by burning. **cremation**, *n.* **cremationist**, *n.* **cremator**, *n.* **crematorium** (kremətaw'riəm), *n.* a place where bodies are cremated. **crematory** (krem'-), *a.* employed in or connected with cremation. *n.* a crematorium.
crème (krem), *n.* cream. **crème de la crème**, (də la), the pick, the most select, the élite. **crème de menthe** (də mēth, mēt), a liqueur made with peppermint. **crème fraîche** (freəsh), a mixture of double cream and soured cream. [F, CREAM]
crenate, -nated (krē'nāt, -nātid), *a.* notched; (*Biol.*) having the edge notched. **crenation**, *n.* **crenature** (kre'nəchə, kren'-), *n.* a scallop; a small rounded tooth on the edge of a leaf.
crenel (kren'əl), **crenelle** (-nel'), *n.* an embrasure in a parapet or battlement. **crenellate** (kren'-), *v.t.* to furnish with battlements or loopholes. **crenellation** (kren-), *n.*
crenulate (kren'ūlət), *a.* finely notched or scalloped. **crenulated** (-lātid), *a.* **crenulation**, *n.*
creole (krē'ōl), *n.* one born of European parentage in the W Indies or Spanish America; in Louisiana, a native descended from French or Spanish ancestors; a native born of mixed European and Negro parentage; the native language of a region, formed from prolonged contact between the original native language and that of European settlers. *a.* relating to the Creoles or a creole language. **creolize, -ise**, *v.t.* creolization, *n.*
creosote (krē'əsōt), *n.* a liquid distilled from coal-tar, used for preserving wood etc. *v.t.* to saturate (as woodwork) with creosote.
crepe, crêpe (krāp, krep), *n.* crape; a crapy fabric other than mourning crape; a thin pancake. **crepe de Chine** (də shēn'), *n.* crape manufactured from raw silk. **crepe paper**, *n.* thin crinkly paper, used in making decorations. **crepe rubber**, *n.* rubber with a rough surface used for shoe soles etc. **crepe suzette** (soozet) *n.* an orange-flavoured crepe flambéed in a liqueur. **crepé** (krāpā'), *a.* frizzled. **creperie** (-əri), *n.* a restaurant or cafe specializing in crepes. **crepey, crepy**, *a.* like crepe, crinkled, as dry skin.
crepitate (krep'itāt), *v.i.* to crackle; to burst with a series of short, sharp reports, as salt in fire; to rattle. **crepitant**, *a.* crackling. **crepitation**, *n.* **crepitus** (krep'itəs), *n.* crepitation; a rattling sound heard in the lungs during pneumonia etc.; the sound of the ends of a broken bone scraping against each other.
crept, *past, p.p.* of CREEP.
crepuscular (krepūs'kūlə), *a.* pertaining to or connected with twilight; glimmering, indistinct; appearing or flying about at twilight.
Cres., (*abbr.*) crescent (of buildings).
crescendo (krishen'dō), *n.* (*pl.* **-dos, -di** (-di)) (a musical passage performed with) a gradual increase in volume; a gradual increase in force or effect. *adv.* with an increasing volume of sound.
crescent (kres'nt), *a.* increasing, growing; shaped like a new moon. *n.* the increasing moon in its first quarter; a figure like the new moon; a row of buildings in crescent form. **Red Crescent** RED.
cresol (krē'sol), *n.* a compound found in coal-tar and creosote, used in antiseptics and as a raw material for plastics.
cress (kres), *n.* a name for various cruciferous plants with a pungent taste (see WATER-CRESS).
cresset (kres'it), *n.* (*Hist.*) a metal cup or basket, usu. on a pole, for holding oil for a light or a beacon fire.
crest (krest), *n.* a plume or comb on the head of a bird; any tuft on the head of an animal; a plume or tuft of feathers, esp. affixed to the top of a helmet; (*Her.*) any figure placed above the shield in a coat-of-arms; the same printed on paper or painted on a building etc.; the summit of a mountain or hill; the top of a ridge; the line of the top of the neck in animals; the ridge of a wave; a ridge on a bone. *v.t.* to ornament or furnish with a crest; to serve as a crest to; to attain the crest of (a hill). *v.i.* to rise into a crest or ridge. **crestfallen**, *a.* dispirited, abashed. **crestfallenly**, *adv.* **crestfallenness**, *n.* **crested**, *a.*
cretaceous (kritā'shəs), *a.* of the nature of or abounding in chalk. **Cretaceous**, *a.*, *n.* (of or formed in) the last period of the Mesozoic era.
cretin (kret'in), *n.* a person mentally and physically deficient because of a (congenital) thyroid malfunction; (*coll.*) a very stupid person. **cretinism**, *n.* **cretinize, -ise**, *v.t.* **cretinous**, *a.*
cretonne (kreton', kret'on), *n.* a cotton fabric with pictorial patterns, used for upholstering, frocks etc.
crevasse (krəvas'), *n.* a deep fissure in a glacier; (*N Am.*) a break in an embankment or levee of a river.
crevice (krev'is), *n.* a crack, a cleft, a fissure. **creviced**, *a.*
crew[1] (kroo), *n.* the company of seamen manning a ship or boat; the personnel on board an aircraft, train or bus; a squad of workers under one foreman or assigned to a specific task; a gang, a mob. *v.t.*, *v.i.* to act as, or serve in, a crew (of). **crew cut**, *n.* a very short style of haircut. **crewman**, *n.* **crew neck**, *n.* a close-fitting round neckline on a jersey.
crew[2], *past of* CROW[2].
crewel (kroo'əl), *n.* fine two-threaded worsted; embroidery worked with such thread. **crewel-work**, *n.*
crib (krib), *n.* a rack or manger; a stall for cattle; a child's cot; a small cottage, a hut, a hovel; a model of the Nativity scene (placed in churches at Christmas); a timber framework lining a mine shaft; (*N Am.*) a bin for grain; cribbage; a hand at cribbage made up of cards thrown out by each player; (*coll.*) anything stolen; a plagiarism; a translation of or key to an author, used by students; something (as a hidden list of dates, formulae etc.) used to cheat in an examination. *v.t.* (*past, p.p.* **cribbed**) to shut up in a crib; (*coll.*) to steal, to appropriate; to plagiarize; to copy from a translation. *v.i.* of horses, to bite the crib; to cheat using a crib. **crib-biting**, *n.* a bad habit in some horses of biting the crib. **cribbing**, *n.*
cribbage (krib'ij), *n.* a game at cards for two, three or four players. **cribbage-board**, *n.* a board on which the progress of the game is marked.
cribriform (krib'rifawm), *a.* (*Anat.*, *Bot.*) resembling a sieve; perforated like a sieve. **cribrate** (-rāt), **cribrose** (-rōs), *a.*
crick (krik), *n.* a spasmodic stiffness, esp. of the neck or back. *v.t.* to cause a crick to.
cricket[1] (krik'it), *n.* any insect of the genus *Acheta*; the house-cricket, well known from its chirp, is *A. domestica*, and the field-cricket *A. campestris*.
cricket[2] (krik'it), *n.* an open-air game played by two sides of 11 each, consisting of an attempt to strike, with a ball, wickets defended by the opponents with bats. **not cricket**, unfair; not straightforward. **cricketer**, *n.*
cricoid (krī'koid), *a.* (*Anat.*) ring-like. **cricoid cartilage**, *n.* the cartilage at the top of the trachea.
cri de coeur (krē də kœ'), *n.* a heartfelt appeal or protest. [F]
crier (krī'ə), *n.* one who cries or proclaims. **town crier**, an officer who makes public proclamation of sales, lost articles etc.
crikey (krī'ki), *int.* (*coll.*) an expression of astonishment.
crim. con., *n.* short for CRIMINAL CONVERSATION under CRIMINAL.
crime (krīm), *n.* an act contrary to law, human or divine; any act of wickedness or sin; wrong-doing, unlawful activity; (*coll.*) a deplorable or foolish act. **capital crime**, a crime punishable with death. **crimeful**, *a.* criminal,

wicked. **crimeless**, *a.*
criminal (krim'inəl), *a.* contrary to duty, law or right; guilty of a crime; relating to crime or its punishment; (*coll.*) deplorable, senseless. *n.* one guilty of a crime; a convict. **Criminal Investigation Department**, the detective branch of a police force. **criminal conversation**, *n.* (*Law*) adultery. **criminality** (-nal'-), *n.* **criminalize, -ise**, *v.t.* **criminally**, *adv.*
criminology (kriminol'əji), *n.* the scientific study of crime and criminals.
crimp (krimp), *v.t.* to curl; to compress into ridges or folds, to frill; to corrugate, to flute, to crease; to join by pinching or folding together. **crimping-iron, -machine**, *n.* an instrument or machine for fluting cap fronts, frills etc.
crimpy, *a.* [cp. CRAMP]
Crimpiene® (krim'plēn), *n.* a kind of crease-resistant synthetic fabric.
crimson (krim'zən), *n.* a deep red colour. *a.* of this colour. *v.t.* to dye with this colour. *v.i.* to turn crimson; to blush.
crinal (krī'nəl), *a.* of or pertaining to the hair.
cringe (krinj), *v.i.* to bend humbly; to pay servile court; to flinch or shrink in fear; to wince in embarrassment; to feel embarrassment or distaste. *n.* an act of cringing. **cringer**, *n.*
cringle (kring'gl), *n.* an iron ring on the bolt-rope of a sail, for the attachment of a bridle.
crinkle (kring'kl), *v.i., v.t.* to wrinkle; to crimp; to rustle. *n.* a wrinkle, a twist; a short bend or turn. **crinkly**, *a.*
crinoid (krin'oid, krī'-), *a.* (*Zool.*) lily-shaped. *n.* any individual of the Crinoidea. **crinoidal** (-noi'-), *a.* pertaining to or containing crinoids. **Crinoidea** (-oi'diə), *n.* a class of echinoderms, containing the sea-lilies and hair-stars.
crinoline (krin'əlin, -lēn), *n.* (a petticoat made of) a stiff fabric of horsehair; any stiff petticoat used to expand the skirts of a dress; a large hooped skirt, *orig.* worn in the mid-19th cent.; the whalebone hoops for such a skirt.
cripes (krīps), *int.* (*coll., dated*) expressing surprise.
cripple (krip'l), *n.* a lame person; one who is disabled or impaired in some way. *v.t.* to make lame; to deprive of the use of the limbs; to deprive of the power of action.
crisis (krī'sis), *n.* (*pl.* **crises** (-sēz)) the turning-point, esp. that of a disease indicating recovery or death; a decisive moment in war, politics, commerce, domestic affairs etc.; a period of instability or distress; an emergency. **crisis-management**, *n.* (ways of) coping with a crisis.
crisp (krisp), *a.* firm but fresh; easily crumbled, brittle; bracing, brisk; curt, sharp, decisive; clear-cut, confident; curled. *v.t.* to curl, to wrinkle; to make crisp. *v.i.* to become curly; to become crisp. **crispbread**, *n.* a thin, dry unsweetened biscuit of rye or wheat flour. **crisped**, *a.* **crisper**, *n.* one who or that which curls or crisps; a compartment in a refrigerator for vegetables to keep them crisp. **crisply**, *adv.* **crispness**, *n.* **crisps**, *n.pl.* very thin slices of potato deep-fried and eaten cold. **crispy**, *a.* curled, curling; wavy; crisp.
crispate (kris'pāt), *a.* (*Biol.*) curled or wrinkled at the edges.
criss-cross (kris'kros), *n., a.* (a network of lines) crossing one another; repeated(ly) crossing to and fro. *v.t., v.i.* to move in, or mark with, a criss-cross pattern.
cristate (kris'tāt), *a.* having a crest; tufted with hairs.
crit., (*abbr.*) critical; criticism; critique; critical mass.
criterion (krītiə'riən), *n.* (*pl.* **-ria** (-ə)) a principle or standard by which anything is or can be judged.
critic (krit'ik), *n.* a judge, an examiner; a censurer, a caviller; one skilled in judging literary or artistic merit; a reviewer. **critical**, *a.* pertaining to criticism; fastidious, exacting; indicating a crisis; decisive; containing or making a negative judgment; hazardous; seriously ill; of a nuclear reactor, capable of sustaining a chain reaction. **critical mass**, *n.* the smallest amount of fissile material that can sustain a chain reaction. **critical path analysis**, *n.* evaluation of a project in terms of the minimum time required for its completion. **critical temperature**, *n.* the temperature below which a gas cannot be liquefied. **critically**, *adv.* **criticaster** (-kahstə), *n.* a petty or contemptible critic. **criticism** (-sizm), *n.* the act of judging, esp. literary or artistic works; a critical essay or opinion; the work of criticizing; an unfavourable judgment. **textual criticism** TEXT. **criticize, -ise**, *v.t.* to examine critically and deliver an opinion upon; to censure. *v.i.* to play the critic. **criticizable**, *a.*
critico-, *comb. form* critically; with criticism.
critique (kritēk'), *n.* a critical essay or judgment; the analysis of the basis of knowledge.
critter (krit'ə), *n.* (*N Am., coll.*) a creature.
CRO, (*abbr.*) cathode ray oscilloscope; criminal records office.
croak (krōk), *v.i.* to make a hoarse low sound in the throat, as a frog or a raven; to grumble, to forbode evil; (*sl.*) to die. *v.t.* to utter in a low hoarse voice; (*sl.*) to kill. *n.* the low harsh sound made by a frog or a raven. **croaker**, *n.* one who croaks; a querulous person; (*sl.*) a dying person. **croaky**, *a.* croaking, hoarse.
Croat (krō'at), *n.* a native of Croatia. **Croatian** (-ā'-), *a., n.*
croc (krok), *n.* (*coll.*) short for CROCODILE.
crochet (krō'shā), *n.* a kind of knitting performed with a hooked needle. *v.t., v.i.* to knit in this manner.
crocidolite (krəsid'əlīt), *n.* a silky fibrous silicate of iron and sodium, also called blue asbestos.
crock[1] (krok), *n.* an earthenware vessel; a pot, a pitcher, a jar; a potsherd; soot or black collected from combustion on pots or kettles etc. *v.t.* to blacken with soot from a pot. **crockery**, *n.* earthenware; earthenware vessels.
crock[2] (krok), *n.* a broken-down horse; a broken-down machine or implement; (*coll.*) a sick person. *v.i.* to break down (often with **up**).
crocket (krok'it), *n.* (*Arch.*) a carved foliated ornament on a pinnacle, the side of a canopy etc.
crocodile (krok'ədīl), *n.* a large amphibian reptile having the back and tail covered with large, square scales; (leather made from) the skin of the crocodile; a string of school children walking two by two. **crocodile bird**, *n.* African plover-like bird which feeds on the insect parasites of the crocodile. **crocodile clip**, *n.* a clip with sprung jaws which have serrated edges, used esp. for making electrical connections. **crocodile tears**, *n.pl.* hypocritical tears like those with which the crocodile is fabled to attract its victims. **crocodilian** (-dil'-), *a.*
crocus (krō'kəs), *n.* (*pl.* **-cuses**) one of a genus (**Crocus**) of small bulbous plants belonging to the Iridaceae, with yellow, white or purple flowers.
Croesus (krē'səs), *n.* a very wealthy man. [king of Lydia, 6th cent. BC]
croft (kroft), *n.* a piece of enclosed ground, esp. adjoining a house; a small farm in the Highlands and islands of Scotland. **crofter**, *n.* one who farms a croft, esp. one of the joint tenants of a farm in Scotland.
croissant (krwa'sō), *n.* a crescent-shaped roll of rich flaky pastry.
Cro-Magnon (krōmag'non), *a.* referring to an early type of modern man, living in late Palaeolithic times, whose remains were found at *Cro-Magnon*, in SW France.
cromlech (krom'lekh), *n.* a prehistoric structure in which a large flat stone rests horizontally on upright ones.
crone (krōn), *n.* an old woman.
cronk (krongk), *a.* (*Austral., coll.*) unwell; unsound. **to cronk up**, to go sick, to be ill.
crony (krō'ni), *n.* a close friend.
crook (kruk), *n.* a bent or curved instrument; a shepherd's or bishop's hooked staff; a curve, a bend, a meander; (*sl.*) a thief, a swindler; a short tube for altering the key on a brass wind instrument. *a.* (*Austral., coll.*) sick; (*Austral.,*

coll.) unpleasant, dishonest. *v.t.* to make crooked or curved. *v.i.* to be bent or crooked. **by hook or by crook** HOOK. **to go crook (on),** (*Austral., coll.*) to become annoyed (with). **crook-back,** *n.* one who has a deformed back. **crook-backed,** *a.* crooked (-kid), *a.* bent, curved; turning, twisting, winding; deformed; not straightforward; perverse; dishonest, illegal. **crookedly,** *adv.* **crookedness,** *n.*

croon (kroon), *v.i.* to sing in a low voice. *v.t.* to mutter. *n.* a hollow, continued moan; a low hum. **crooner,** *n.* one who sings sentimental songs in low tones.

crop (krop), *n.* the craw of a fowl, constituting a kind of first stomach; an analogous receptacle in masticating insects; a short whipstock with a loop instead of a lash; that which is cut or gathered; the harvest yield; a plant for cultivation; a group of things appearing at one time; a short haircut. *v.t.* (*past, p.p.* **cropped**) to cut off the ends of; to harvest; to graze on; to cut off, to cut short; to sow; to plant and raise crops on; to reduce the margin of (a book) unduly, in binding. *v.i.* to yield a harvest, to bear fruit. **neck and crop,** altogether. **to crop out,** to come to light; (*Geol.*) to come out at the surface by the edges, as an underlying stratum of rock. **to crop up,** to come up unexpectedly. **crop-dusting,** *n.* the spreading of crops with powdered insecticide etc., from an aeroplane. **crop-ear,** *n.* a horse with cropped ears. **crop-eared,** *a.* having the hair or ears cut short. **cropper,** *n.* one who or that which crops; a grain or plant which yields a good crop; (*coll.*) a heavy fall; a collapse, as in **to come a cropper.**

croquet (krō'kā), *n.* an open-air game played on a lawn with balls and mallets; the act of croqueting an opponent's ball. *v.t.* to drive an opponent's ball away in this game by placing one's own ball against it and striking. *v.i.* to play croquet.

croquette (krəket'), *n.* a savoury ball made with meat, potato etc. fried in breadcrumbs.

crosier (krō'zhə, -ziə), *n.* the pastoral staff of a bishop or abbot.

cross[1] (kros), *n.* an ancient instrument of torture made of two pieces of timber set transversely at various angles; a monument, emblem, staff or ornament in this form; a sign or mark in the form of a cross; the mixture of two distinct stocks in breeding animals; the animal resulting from such a mixture; a mixture; a compromise; anything that thwarts or obstructs; trouble, affliction; the Christian religion, Christianity; in boxing, a punch delivered from the side; in football, a pass across the field, esp. towards the opposing goal. *a.* transverse, oblique, lateral; intersecting; adverse, contrary, perverse; angry, bad-tempered. **cross of St Anthony,** one shaped like a **T. fiery cross** FIERY. **Greek cross,** an upright cross with arms of equal length. **Latin cross,** one with a long upright below the cross-piece. **Maltese cross,** one with limbs of equal size widening towards the extremities. **on the cross,** diagonally. **St Andrew's cross,** one shaped like an **X. St George's cross,** a Greek cross used on the British flag. **Southern Cross,** a cross-shaped constellation visible in the southern hemisphere. **Victoria Cross** VICTORIA. **cross-bar,** *n.* a transverse bar. **cross-beam,** *n.* a large beam running from wall to wall. **cross-bench,** *n.* (*Parl.*) one of the benches for those independent of the recognized parties. **cross-bencher,** *n.* **cross-bill,** *n.* a bird of the genus *Loxia*, the mandibles of the bill of which cross each other when closed. **cross-bond,** *n.* brick-laying in which points of one course fall in the middle of those above and below. **cross-bones,** *n.pl.* the representation of two thigh-bones crossed as an emblem of mortality. **cross-bow,** *n.* a weapon for shooting, formed by placing a bow across a stock. **cross-bowman,** *n.* **cross-bred,** *a.* of a cross-breed, hybrid. **cross-breed,** *n.* a breed produced from different strains or varieties; a hybrid. *v.t.* to produce a cross-breed. **cross-buttock,** *n.* a wrestling throw over the hip. **cross-check,**

v.t. to check (a fact etc.) by referring to other sources of information. **cross-country,** *a., adv.* across fields etc. instead of along the roads. *n.* a cross-country race. **cross-cultural,** *a.* concerning the differences between two cultures. **cross-current,** *n.* a current in a river or sea flowing across the main current; a conflicting tendency. **crosscut,** *n.* a cut across; (*Mining*) a drift from a shaft. *v.t.* to cut across. **crosscut saw,** *n.* a saw for cutting timber across the grain. **crosscutting,** *n.* moving rapidly from one scene to another and back again in a film. **cross-dresser,** *n.* a transvestite. **cross-dressing,** *n.* **cross-examine,** *v.t.* to examine systematically to elicit facts not brought out in direct examination, or to confirm or contradict the direct evidence. **cross-examination,** *n.* **cross-examiner,** *n.* **cross-eye,** *n.* a squinting eye. **cross-eyed,** *a.* with both eyes squinting inwards. **cross-fade,** *v.t.* in TV or radio, to fade out (one signal) while introducing another. **cross-fertilize, -ise,** *v.t., v.i.* to apply the pollen of one flower to the pistil of a flower of another species. **cross-fertilization,** *n.* **crossfire,** *n.* firing in directions which cross each other; a rapid or lively argument. **cross-grain,** *n.* the grain or fibres of wood running across the regular grain. **cross-grained,** *a.* having the grain or fibres running across or irregular; perverse, peevish; intractable. **cross-hatch,** *v.t.* to shade with parallel lines crossing regularly in drawing or engraving. **cross-head,** *n.* the block at the head of a piston-rod communicating motion to the connecting rod; a heading printed across the page or a column. **cross-infection,** *n.* infection of a hospital patient with an unrelated illness from another patient. **cross-legged** (-gid), *a.* having one leg over the other. **cross-match,** *v.t.* to test (as blood samples from two people) for compatibility. **crossover,** *n.* a point of transition, intersection or interchange; a person or thing (e.g. a record) whose success is not confined to the anticipated audience or market. *a.* being or pertaining to a crossover. **crosspatch,** *n.* (*coll.*) a cross, ill-tempered person. **crosspiece,** *n.* a transverse piece. **cross-ply,** *a.* of motor-vehicle tyres, having the cords crossing each other diagonally to strengthen the tread. **cross-pollination,** *n.* the transfer of pollen from one flower to the stigma of another. **cross-pollinate,** *v.t.* **cross-purpose,** *n.* a contrary purpose. **to be at cross-purposes,** to misunderstand or act unintentionally counter to each other. **cross-question,** *n.* one put in cross-examination. *v.t.* to cross-examine. **cross-reference,** *n.* a reference from one part of a book to another. **cross-refer,** *v.t., v.i.* **cross-roads,** *n.* a point where two or more roads intersect; a crucial point, esp. where a decision has to be made. **cross-section,** *n.* a cutting across the grain, or at right angles to the length; a cutting which shows all the strata; a comprehensive representation, a representative example. **cross-sectional,** *a.* **cross-stitch,** *n.* a kind of stitch crossing others in series; needlework done thus. **cross-stone** HARMOTOME. **cross-talk,** *n.* unwanted signals in a telephone, radio etc. channel, coming in from another channel; repartee. **cross-trainer,** *n.* an all-purpose sports shoe. **cross-training,** *n.* combining two or more sports or forms of exercise in a training programme. **cross-trees,** *n.pl.* timbers on the top of masts to support the rigging of the mast above. **cross-wind,** *n.* an unfavourable wind; a sidewind. **crossword (puzzle),** *n.* a puzzle in which a square divided into blank chequered spaces is filled with words corresponding to clues provided. **crossly,** *adv.* in an ill-humoured manner. **crossness,** *n.* **crossways, -wise,** *adv.* across; in the form of a cross.

cross[2] (kros), *v.t.* to mark or erase with a cross; to make the sign of the cross on or over; to place crossways; to pass across, to traverse; to intersect; to pass over or in front of; to meet and pass; to cause to interbreed; to cross-fertilize; to mark (a cheque) with two lines in order to render payable through another bank; to thwart, to

crosse   191   **cruet**

counteract; to be inconsistent with; in football, to pass (the ball) across the field from the wing. *v.i.* to lie or be across or over something; to pass across something; to move in a zigzag; to be inconsistent; to interbreed. **to cross off, out,** to strike out; to cancel. **to cross one's mind,** to occur to one's memory or attention. **to cross the floor,** used of a member of Parliament leaving one political party to join another. **to cross the path of,** to meet with; to thwart. **crossed,** *a.* of a telephone line, connected in error to more than one telephone. **crossing,** *n.* the intersection of two roads, railways etc.; a place where a road etc. may be crossed; a journey across something, esp. a stretch of water. **crossing-over,** *n.* the interchange of segments of homologous chromosomes during meiotic cell division.

crosse (kros), *n.* the long, netted stick used in lacrosse.

crossing CROSS².

crossways, crosswise CROSS¹.

crotch (kroch), *n.* the parting of two branches; the angle between the thighs where the legs meet the body. **crotched,** *a.*

crotchet (kroch'it), *n.* a perverse fancy, a whim; (*Printing*) a square bracket; (*Mus.*) a note, equal in length to one beat of a bar of 4/4 time. **crotchety,** *a.* irritable, bad-tempered. **crotchetiness,** *n.*

croton (krō'tən), *n.* a genus of euphorbiaceous medicinal plants from the warmer parts of both hemispheres. **croton-oil,** *n.* a drastic purgative oil.

crouch (krowch), *v.i.* to stoop, to bend low; to lie close to the ground; to cringe, to fawn. *n.* the action of crouching.

croup¹ (kroop), *n.* the rump (esp. of a horse).

croup² (kroop), *n.* inflammation of the larynx and trachea, characterized by hoarse coughing and difficulty in breathing, common in infancy. **croupy,** *a.*

croupier (kroo'piā), *n.* one who superintends a gaming-table and collects the money won by the bank.

croûton (kroo'ton), *n.* a small cube of fried or toasted bread, served with soup or salads.

crow¹ (krō), *n.* a large black bird of the genus *Corvus*, esp. *C. cornix*, the hooded crow, and *C. corone*, the carrion crow; a crowbar. **as the crow flies,** in a direct line. **stone the crows,** *int.* (*sl.*) an expression of amazement. **to eat crow,** (*coll.*) to (be made to) humiliate or abase oneself. **crowbar,** *n.* a bar of iron wedge-shaped at one end and used as a lever. **crowberry,** *n.* a heathlike plant, *Empetrum nigrum*, with black berries. **crow-flower** CROWFOOT. **crowfoot,** *n.* name for several species of buttercup, *Ranunculus bulbosus*, *R. acris* and *R. repens*. **crow's-foot,** *n.* a wrinkle at the corner of the eye in old age. **crow's nest,** *n.* a tub or box for the look-out man on a ship's mast. **crow-steps** CORBIE-STEPS.

crow² (krō), *v.i.* (*past* **crew** (kroo), **crowed**) to make a loud cry like a cock; to make a cry of delight like an infant; to exult; to brag, to boast. *v.t.* to proclaim by crowing. *n.* the cry of a cock; the cry of delight of an infant.

crowd (krowd), *v.t.* to press or squeeze closely together; to fill by pressing; to throng or press upon; to press (into or through). *v.i.* to press, to throng, to swarm; to collect in crowds. *n.* a number of persons or things collected closely and confusedly together; the mass, the mob, the populace; (*coll.*) a set, a party, a lot; a large number (of things); any group of persons photographed in a film but not playing definite parts. **to crowd out,** to force (a person or thing) out by leaving no room; to fill to absolute capacity. **crowd-puller,** *n.* an event that attracts a large audience. **crowded,** *a.*

crowdy, -die (krow'di), *n.* (*Sc.*) meal and water (or milk) stirred together cold, to form a thick gruel; a kind of soft unripened cheese made from soured milk.

crown (krown), *n.* a garland of honour worn on the head; the ornamental circlet worn on the head by emperors, kings or princes as a badge of sovereignty; royal power; the sovereign; (*Hist.*) a five-shilling piece; a foreign coin of certain values; a size of paper, 15 in. × 20 in. (381 × 508 mm) (formerly with a crown for a watermark); the top of anything, as a hat, a mountain etc.; the top of the head; the vertex of an arch; the highest part of a road, bridge or causeway; the portion of a tooth above the gum; an artificial crown for a broken or discoloured tooth; (*Naut.*) the part of an anchor where the arms join the shank; the culmination, glory; reward, distinction. *a.* belonging to the Crown or the sovereign. *v.t.* to invest with a crown, or regal or imperial dignity; to surround, or top, as with a crown; to form a crown, ornament or top to; to dignify, to adorn; to consummate; to put a crown or cap on (a tooth); (*Draughts*) to make a king; (*coll.*) to hit on the head. **Crown Colony,** *n.* a colony administered by the home Government. **crown court,** *n.* in England and Wales, a local criminal court. **crown-glass,** *n.* the finest kind of window glass, made in circular sheets without lead or iron; glass used in optical instruments. **crown green,** *n.* a type of bowling green which slopes slightly from the sides up to the centre. **crown imperial,** *n.* a garden flower, *Fritillaria imperialis*, with a whorl of florets round the head. **crown jewels,** *n.pl.* the regalia and other jewels belonging to the sovereign. **Crown lands,** *n.pl.* lands belonging to the Crown as the head of the government. **Crown Office,** *n.* a section of the Court of King's (or Queen's) Bench which takes cognizance of criminal cases. **crown-post,** *n.* a king-post. **crown prince,** *n.* the name given in some countries to the heir apparent to the Crown; **crown princess,** *n.fem.* **crown-wheel,** *n.* a contrate wheel. **crown-witness,** *n.* a witness for the Crown in a criminal prosecution. **crowned,** *a.* **crownless,** *a.*

crozier (krō'zhə, -ziə), CROSIER.

CRT, (*abbr.*) cathode ray tube.

crucial (kroo'shl), *a.* decisive; searching; (*Anat.*) in the form of a cross; (*loosely*) very important; (*sl.*) excellent. **crucially,** *adv.*

crucian (kroo'shn), *n.* the German or Prussian carp, a small fish without barbels.

cruciate (kroo'shiət), *a.* (*Biol.*) cruciform.

crucible (kroo'sibl), *n.* a melting-pot of earthenware, porcelain, or of refractory metal, adapted to withstand high temperatures without softening, and sudden and great alterations of temperature without cracking; a basin at the bottom of a furnace to collect the molten metal; a searching test or trial.

crucifer (kroo'sifə), *n.* a person who carries a cross in a procession; one of the Cruciferae. **Cruciferae** (-sif'ərē), *n.pl.* an order of plants, the flowers of which have four petals disposed crosswise, such as a wallflower. **cruciferous** (-sif'-), *a.*

crucifix (kroo'sifiks), *n.* a cross bearing a figure of Christ. **crucifixion** (-fik'shən), *n.* the act of crucifying; punishment by crucifying; the death of Christ on the cross; a picture of this; torture; mortification. **cruciform** (-fawm), *a.* cross-shaped; arranged in the form of a cross.

crucify (kroo'sifī), *v.t.* to inflict capital punishment by affixing to a cross; to torture; to mortify, to destroy the influence of; to subject to scathing criticism, obloquy or ridicule; to defeat utterly.

crud (krŭd), *n.* (*sl.*) any dirty, sticky or slimy substance; (*sl.*) a contemptible person. **cruddy,** *a.*

crude (krood), *a.* in a natural state, unprocessed; rough, unfinished; of statistics, not classified or analysed. *n.* crude oil. **crude form,** *n.* the original form of an inflected substantive divested of its case ending. **crude oil,** *n.* unrefined petroleum. **crudely,** *adv.* **crudeness,** *n.* **crudity,** *n.*

cruel (kroo'əl), *a.* disposed to give pain to others; inhuman, unfeeling, hard-hearted; causing pain to others, painful. **cruel-hearted,** *a.* **cruelly,** *adv.* **cruelty,** *n.* cruel disposition or temper; a barbarous or inhuman act.

cruet (kroo'it), *n.* a small container for pepper, salt etc. at

cruise — **crystal**

table; a small bottle for holding the wine or water in the Eucharist. **cruet-stand**, *n.* a frame or stand for holding cruets.

cruise (krooz), *v.i.* to sail to and fro for pleasure or in search of plunder or an enemy; of a motor vehicle or aircraft, to travel at a moderate but sustained speed. *n.* the act or an instance of cruising, esp. a pleasure-trip on a boat. **cruise control**, *n.* a device which automatically keeps a vehicle at a pre-set speed. **cruise missile**, *n.* a low-flying subsonic guided missile. **cruiser**, *n.* a person or ship that cruises; a warship designed primarily for speed, faster and lighter than a battleship. **cruiserweight**, *n.* in boxing, a light heavyweight.

crumb (krŭm), *n.* a small piece, esp. of bread; the soft inner part of bread; a tiny portion, a particle; (*sl.*) an unpleasant or contemptible person. *v.t.* to break into crumbs; to cover with crumbs (for cooking). **crumbs**, *int.* (*coll.*) a mild expression of surprise or dismay. **crumby**, *a.* covered with crumbs; (*sl.*) crummy.

crumble (krŭm′bl), *v.t.* to break into small particles. *v.i.* to fall into small pieces; to fall into ruin. *n.* a pudding topped with a crumbly mixture of flour, sugar and butter, as *apple crumble*. **crumbly**, *a.* apt to crumble. *n.* (*coll.*) a very old person.

crummy (krŭm′i), *a.* (*sl.*) unpleasant, worthless; (*sl.*) unwell.

crump (krŭmp), *n.* the sound of the explosion of a heavy shell or bomb. *v.i.* to make such a sound.

crumpet (krŭm′pit), *n.* a thin, light, spongy tea-cake; (*sl.*) the head; (*sl.*) a girl or girls collectively.

crumple (krŭm′pl), *v.t.* to crush or press into wrinkles. *v.i.* to become wrinkled, to shrink, as cloth, paper etc.; to collapse, give way.

crunch (krŭnch), *v.t.* to crush noisily esp. with the teeth. *v.i.* to make a noise, as of crunching; to advance with crunching. *n.* a noise of or as of crunching. **the crunch**, (*coll.*) the decisive or critical moment. **crunchy**, *a.*

crupper (krŭp′ə), *n.* a loop which passes under a horse's tail to keep the saddle from slipping forward; the hindquarters of a horse.

crusade (kroosād′), *n.* one of several expeditions undertaken in the Middle Ages to recover the Holy Land, then in Muslim hands; any enterprise conducted in an enthusiastic or fanatical spirit. *v.i.* to engage in a crusade. **crusader**, *n.*

cruse (krooz), *n.* a small pot, cup or bottle.

crush (krŭsh), *v.t.* to press or squeeze together between two harder bodies so as to break or bruise; to break or grind into small particles; to crumple; to overwhelm by superior power; to oppress, to ruin. *v.i.* to be pressed into a smaller compass by external force or weight. *n.* the act of crushing; a crowd; (*coll.*) a crowded meeting or social gathering; (*coll.*) an infatuation or the object of this; a drink made by or as by crushing fruit. **crush bar**, *n.* a bar in a theatre where patrons may use in the intervals of a play. **crush barrier**, *n.* a temporary barrier to keep back, or to separate, a crowd. **crush-hat**, *n.* an opera-hat. **crushable**, *a.* **crusher**, *n.*

crust (krŭst), *n.* the hard outer part of bread; the crusty end of a loaf; any hard rind, coating, layer, deposit or surface covering; a hard piece of bread; the pastry covering a pie; a scab; a deposit from wine as it ripens; (*Geol.*) the solid outer portion of the earth; a (meagre) living; (*sl.*) impertinence. *v.t.* to cover with a crust; to make into crust. *v.i.* to become encrusted. **crustal**, *a.* pertaining to the earth's crust. **crusted**, *a.* **crusty**, *a.* resembling or of the nature of crust; harsh, surly. **crustily**, *adv.* **crustiness**, *n.*

Crustacea (krŭstā′shə), *n.pl.* a class of Articulata, containing lobsters, crabs, shrimps etc., named from their shelly covering, cast periodically. **crustacean**, *n., a.* **crustaceology** (-shiol′-), *n.* the branch of science dealing with the Crustacea. **crustaceologist**, *n.* **crustaceous**, *a.* of the nature of shell; crustacean.

crutch (krŭch), *n.* a staff, with a crosspiece to fit under the arm-pit, to support a lame person; a support; the crotch of a person; the corresponding part of a garment.

crux (krŭks), *n.* (*pl.* **cruxes, cruces** (kroo′sēz)) the essential point; anything exceedingly puzzling.

cruzeiro (kroozeə′rō), *n.* (*pl.* **-ros**) the monetary unit of Brazil, equal to 100 centavos.

cry (krī), *v.i.* to call loudly, vehemently or importunately; to exclaim; to lament loudly; to weep; of animals, to call; to utter inarticulate sounds; to yelp. *v.t.* to utter loudly; to proclaim, to declare publicly; to announce for sale. *n.* a loud utterance, usu. inarticulate, expressive of intense joy, pain, suffering, astonishment or other emotion; an importunate call or prayer; the characteristic call of a bird or animal; proclamation, public notification; a catchword or phrase; a fit of weeping. **a far cry**, a long way off. **for crying out loud**, (*coll.*) interjection expressing impatience, annoyance. **in full cry**, in hot and vociferous pursuit. **to cry down**, to decry, to depreciate. **to cry mercy**, to beg pardon. **to cry off**, to withdraw from a bargain. **to cry ones eyes** or **heart out**, to weep copiously or inconsolably. **to cry out**, to vociferate, to clamour. **to cry out against**, to exclaim loudly, by way of censure or reproach. **to cry over spilt milk**, to lament or regret something it is too late to do anything about. **to cry quits** QUITS. **to cry stinking fish**, to decry or condemn, esp. one's own wares. **to cry up**, to praise highly. **crybaby**, *n.* (*coll.*) a child or person easily provoked to tears. **crying**, *a.* that cries; calling for notice or vengeance, flagrant.

cryo-, *comb. form* very cold.

cryogen (krī′əjen), *n.* (*Chem.*) a freezing mixture. **cryogenics** (krīəjen′iks), *n.sing.* the branch of physics which studies very low temperatures and the phenomena associated with them. **cryogenic**, *a.*

cryolite (krī′əlīt), *n.* a brittle fluoride of sodium and aluminium from Greenland.

crypt (kript), *n.* a vault, esp. one beneath a church, used for religious services or for burial. **cryptic**, *a.* hidden, secret; obscure, puzzling; of animal coloration, serving as camouflage. **cryptically**, *adv.*

crypt(o)-, *comb. form* secret; inconspicuous; not apparent or prominent.

cryptocrystalline (kriptōkris′təlin), *a.* (*Min.*) having a crystalline structure which cannot be resolved under the microscope.

cryptogam (krip′tōgam), *n.* a plant destitute of pistils and stamens. **Cryptogamia** (-gām′-), *n.pl.* a Linnaean order of plants in which the reproductive organs are concealed or not distinctly visible, containing ferns, lichens, mosses and seaweeds, fungi etc. **cryptogamic** (-gam′-), **-mous** (-tog′ə-), *a.* **cryptogamist** (-tog′ə-), *n.* **cryptogamy** (-tog′-), *n.* concealed or obscure fructification.

cryptogram (krip′tōgram), *n.* a secret symbol. **cryptograph** (-grahf), *n.* a system of writing, or something written, in cipher. **cryptographer, -phist** (-tog′-), *n.* **cryptographic** (-graf′-), *a.* **cryptography** (-tog′-), *n.*

crystal (kris′tl), *n.* a clear transparent mineral; transparent quartz, also called rock crystal; an aggregation of atoms arranged in a definite pattern which often assumes the form of a regular solid terminated by a certain number of smooth plane surfaces; a very pellucid kind of glass; the transparent cover over a watch dial; a crystalline component in various electronic devices, used as an oscillator etc. *a.* clear, transparent, as bright as crystal; made of crystal. **crystal ball**, *n.* one made of glass, used in crystal gazing. **crystal detector**, *n.* a crystal arranged in a circuit so that the modulation on a radio carrier wave becomes audible in earphones etc. **crystal gazing**, *n.* looking into a crystal ball in order to foresee the future. **crystal set**, *n.* an early form of radio receiver using a crystal detector.

crystalline (-lin), *a.* consisting of crystal; resembling crystal; clear, pellucid. **crystalline lens**, *n.* a lenticular, white, transparent solid body enclosed in a capsule behind the iris of the eye, the lens of the eye. **crystallinity** (-lin'-), *n.* **crystallite** (-lit), *n.* one of the particles of definite form observed in thin sections of igneous rock cooled slowly after fusion. **crystallize, -ise**, *v.t.* to cause to form crystals; to coat (fruit) with sugar crystals. *v.i.* to assume a crystalline form; of views, thoughts etc., to assume a definite form. **crystallizable**, *a.* **crystallization**, *n.*

crystallo-, *comb. form* forming, formed of, pertaining to crystal, crystalline structure or the science of crystals.

crystallogeny (kris'təloj'əni), *n.* that branch of science which treats of the formation of crystals. **crystallogenic** (-jen'-), *a.*

crystallography (kristəlog'rəfi), *n.* the science which deals with the forms of crystals. **crystallographer**, *n.* **crystallographic** (-graf'-), *a.* **crystallographically**, *adv.*

crystalloid (kris'təloid), *a.* like a crystal. *n.* a body with a crystalline structure.

CS, (*abbr.*) chartered surveyor; civil service; Court of Session.

Cs, (*chem. symbol*) caesium.

c/s, (*abbr.*) cycles per second.

CSC, (*abbr.*) Civil Service Commission.

CSE, (*abbr.*) certificate of secondary education.

CS gas, *n.* an irritant gas, causing tears, painful breathing etc., used in riot control. [from the initials of its US inventors Corson and Stoughton]

CSIRO, (*abbr.*) Commonwealth Scientific and Industrial Research Organization.

CSM, (*abbr.*) Company Sergeant-Major.

CSO, (*abbr.*) Central Statistical Office; community service order.

CST, (*abbr.*) central standard time.

Ct, (*abbr.*) Connecticut; court.

ct, (*abbr.*) carat; cent.

CTC, (*abbr.*) city technology college; Cyclists' Touring Club.

ctenoid (tē'noid, ten'-), *a.* comb-shaped; pectinated; having ctenoid scales. *n.* a ctenoid fish.

Ctenophora (tinof'ərə, tē'-), *n.pl.* (*Zool.*) a division of the Coelenterata, characterized by fringed or comb-like locomotive organs. **ctenophoral** (-nof'-), *a.* **ctenophore** (ten'-, tē'-), *n.*

CTT, (*abbr.*) capital transfer tax.

Cu, (*chem. symbol*) copper. [L *cuprum*]

cub (kŭb), *n.* the young of certain animals, e.g. lion, bear, fox; (*coll.*) an uncouth, mannerless youth; (also **Cub Scout**) a young Boy Scout. *v.i.* (*past, p.p.* **cubbed**) to bring forth cubs; to hunt young foxes. **cub-hunting**, *n.* **cub reporter**, *n.* an inexperienced newspaper reporter. **cubbing**, *n.* **cubbish**, *a.* **cubhood**, *n.*

cuban heel (kū'bən), *n.* a straight heel of medium height on a boot or shoe.

cubby (kŭb'i), *n.* **cubby-hole**, *n.* a narrow or confined space; a cosy place.

cube (kūb), *n.* a solid figure contained by six equal squares; the third power of a number (as 8 is the cube of 2). *v.t.* to raise to the third power; to cut or shape into cubes. **cube root**, *n.* the number which, multiplied twice by itself, produces the cube, thus 3 is the cube root of 27. **cube sugar**, *n.* lump sugar. **cubic, -ical**, *a.* having the properties or form of a cube; being or equalling a cube, the edge of which is a given unit; (*Math.*) of the third degree. **cubically**, *adv.* **cubiform** (-ifawm), *a.*

cubeb (kū'beb), *n.* the small spicy berry of *Cubeba officinalis*, a Javanese shrub used in medicine and cookery. **cubebic** (-beb'-), *a.* **cubebin**, *n.* a vegetable principle found in the seeds of the cubeb.

cubic CUBE.

cubicle (kū'bikl), *n.* a portion of a bedroom partitioned off as a separate sleeping apartment; a compartment.

cubiform CUBE.

cubist (kū'bist), *n.* one of an early 20th-cent. school of painters who show several aspects of the same object simultaneously using geometrical shapes. **cubism**, *n.*

cubit (kū'bit), *n.* an old measure of length, from the elbow to the tip of the middle finger, varying at different times from 18 to 22 in. (0·46 to 0·5 m).

cuboid (kū'boid), *a.* resembling a cube. *n.* (*Geom.*) a solid like a cube but with the sides not all equal; (*Anat.*) a bone on the outer side of the foot. **cuboidal**, *a.*

cucking-stool (kŭk'ingstool), *n.* a kind of chair, formerly used for ducking scolds, dishonest tradesmen etc.

cuckold (kŭk'əld), *n.* one whose wife is unfaithful. *v.t.* to make (a man) a cuckold. **cuckoldry**, *n.*

cuckoo (kuk'oo), *n.* a migratory bird, *Cuculus canorus*, which visits Britain in the spring and summer and lays its eggs in the nests of other birds; (*sl.*) a fool. *a.* (*sl.*) crazy. **cuckoo in the nest**, an unwanted and alien person, an intruder. **cuckoo clock**, *n.* a clock which announces the hours by emitting a sound like the note of the cuckoo. **cuckoo-flower**, *n.* a local name for many plants, esp. for the lady's smock, *Cardamine pratensis*. **cuckoo-pint**, *n.* the common arum. **cuckoo-spit**, *n.* an exudation on plants from the frog-hopper.

cucullate, -ated (kū'kəlāt, -tid), *a.* (*Bot., Zool.*) hooded; resembling or formed like a hood. **cuculliform** (-kŭl'-ifawm), *a.*

cucumber (kū'kŭmbə), *n.* a trailing plant, *Cucumis sativus*; its elongated fruit, extensively used as a salad and pickle. **cool as a cucumber**, very unemotional, imperturbable. **cucumber tree**, *n.* any of several American magnolias with fruit like small cucumbers. **cucumiform** (-kū'mifawm), *a.*

cucurbit (kūkœ'bit), *n.* a gourd; a gourd-shaped vessel used in distillation. **cucurbitaceous**, *a.*

cud (kŭd), *n.* food deposited by ruminating animals in the first stomach, thence drawn and chewed over again. **to chew the cud**, to ruminate; to reflect.

cuddle (kŭdl), *v.i.* to lie close or snug together; to join in an embrace. *v.t.* to embrace, to hug, to fondle. *n.* a hug, an embrace. **cuddlesome** (-səm), **cuddly**, *a.* attractive to cuddle.

cuddy[1] (kŭd'i), *n.* (*chiefly Sc.*) a donkey; a blockhead.

cuddy[2] (kŭd'i), *n.* a cabin in a ship where officers and passengers take their meals; a small cabin in a boat.

cudgel (kŭj'l), *n.* a short club or thick stick, a bludgeon. *v.t.* (*past, p.p.* **cudgelled**) to beat with a cudgel. **to cudgel one's brains**, to try to recollect or find out something. **to take up (the) cudgels**, to enter a dispute to defend somebody or something.

cudweed (kŭd'wēd), *n.* popular name for the genus *Gnaphalium*, esp. *G. sylvaticum*, formerly administered to cattle that had lost their cud.

cue[1] (kū), *n.* the last words of a speech, a signal to another actor that he or she should begin; any similar signal, e.g. in a piece of music; a hint, reminder. *v.t.* (*pres. p.* **cuing, cueing**) to give a cue to. **on cue**, at the right time. **to cue in**, to give a cue to, to inform.

cue[2] (kū), *n.* a long straight rod used by billiard-players. *v.t., v.i.* (*pres. p.* **cuing, cueing**) to strike (a ball) with a cue. **cue ball**, *n.* the ball which is struck with a cue. **cueist**, *n.*

cuff[1] (kŭf), *v.t.* to strike with the open hand. *n.* a blow of this kind.

cuff[2] (kŭf), *n.* the fold or band at the end of a sleeve; (*N Am.*) a trouser turn-up; (*pl.*) (*coll.*) handcuffs. **off the cuff**, extempore. **cufflink**, *n.* a usu. ornamental device used to fasten a shirt cuff.

cui bono? (kwē bō'nō), for whose advantage? [L]

cuirass (kwiras', kū-), *n.* armour for the body, consisting of a breastplate and a backplate strapped or buckled together. **cuirassier** (-siə'), *n.* a soldier wearing a cuirass.

cuisine   194   cupboard

cuisine (kwizen'), *n*. style of cooking; cookery.

cul-de-sac (kul'disak), *n*. (*pl*. **culs-de-sac** (kŭl-)) a street or lane open only at one end; (*Anat*.) a vessel, tube or gut open only at one end.

-cule, *dim. suf.*, as in *animalcule, corpuscule*.

Culex (kū'lcks), *n*. a genus of dipterous insects, containing the gnat and the mosquito. **culiciform** (-lis'ifawm), *a*.

culinary (kŭl'inəri), *a*. relating to the kitchen or cooking; used in kitchens or in cooking.

cull (kŭl), *v.t*. to pick; to select, to choose (the best); to select (an animal) from a group, esp. a weak or superfluous one, to kill it; to reduce the size of (a group) in this way. *n*. an instance of culling; (*pl*.) (*N Am*.) defective logs or planks picked out from lumber. **culler**, *n*. **culling**, *n*.

cullender COLANDER.

culm[1] (kŭlm), *n*. a stem, esp. of grass or sedge. **culmiferous**, *a*.

culm[2] (kŭlm), *n*. stone-coal; anthracite coal, esp. if in small pieces; coal-dust. **culmiferous** (-mif'-), *a*. abounding in anthracite.

culmen (kŭl'mən), *n*. the ridge on the top of a bird's bill.

culminate (kŭl'mināt), *v.i*. to reach the highest or a climactic or decisive point; (*Astron*.) to come to the meridian. **culminant**, *a*. at the highest point; (*Astron*.) on the meridian. **culmination**, *n*. the highest point; the end of a series of events etc.

culotte(s) (kəlot', kū-), *n*. a divided skirt.

culpable (kŭl'pəbl), *a*. blamable; blameworthy; guilty. **culpability** (-bil'-), *n*. **culpableness**, *n*. **culpably**, *adv*.

culpatory (-təri), *a*. involving or expressing blame.

culprit (kŭl'prit), *n*. an offender; one who is at fault; one who is arraigned before a judge on a charge.

cult (kŭlt), *n*. a system of religious belief; the rites and ceremonies of any system of belief; a sect regarded as unorthodox or harmful to its adherents; an intense devotion to a person, idea etc.; the object of such devotion; an intense fad or fashion. *a*. pertaining to a cult; very fashionable. **cultic**, *a*. **cultish**, *a*. **cultism**, *n*. adherence to a cult. **cultist**, *n*. **cultus** (-təs), *n*. a cult.

cultivar (kŭl'tivah), *n*. a variety of a naturally-occurring species, produced and maintained by cultivation.

cultivate (kŭl'tivāt), *v.t*. to till; to prepare for crops; to raise or develop by tilling; to improve by labour or study, to civilize; to cherish, to foster, to seek the friendship of. **cultivable**, *a*. **cultivation**, *n*. the art or practice of cultivating; the state of being cultivated; a state of refinement or culture. **cultivator**, *n*. one who cultivates; an implement to break up the soil and remove weeds.

culture (kŭl'chə), *n*. the act of tilling; husbandry, farming; breeding and rearing; the experimental growing of bacteria or other microorganisms in a laboratory; the group of microorganisms so grown; intellectual or moral discipline and training; civilized tastes and values; the customary activities, social forms etc. of a particular group; the inherited traditions, beliefs, values etc. of a society or group. *v.t*. to cultivate; to grow (microorganisms) in a laboratory. **culture shock**, *n*. feelings of disorientation caused by the transition from one culture or environment to another. **culture vulture**, *n*. (*coll., often derog*.) a person avidly interested in the arts. **cultural**, *a*. **cultured**, *a*. in a state of intellectual development; grown artificially, as pearls or microorganisms. **culturist**, *n*.

cultus CULT.

culverin (kŭl'vərin), *n*. a long cannon or hand-gun.

culvert (kŭl'vət), *n*. a drain or covered channel for water beneath a road, railway etc.; an underground channel for electric wires or cables.

cum (kŭm, koom), *prep*. combined with; together with. [L]

Cumb., (*abbr*.) Cumberland; Cumbria.

cumber (kŭm'bə), *v.t*. to hamper, to clog; to hinder, to impede; to perplex; to embarrass. **cumbersome**, *a*. unwieldy, unmanageable; burdensome, troublesome. **cumbersomely**, *adv*. **cumbersomeness**, *n*. **cumbrance**, *n*. an encumbrance. **cumbrous**, *a*. **cumbrously**, *adv*.

Cumbrian (kŭm'briən), *n.*, *a*. (a native) of Cumberland or Cumbria.

cumin (kŭm'in), *n*. a plant of the parsley family, the Umbelliferae, with aromatic and carminative seeds; the name of a genus containing this, together with the caraway and other plants. **cumin-oil**, *n*. a volatile extract from the seeds of cumin. **cumic** (kū'-), *a*.

cummerbund (kŭm'əbŭnd), *n*. a waistband or sash, worn esp. by men with evening dress.

cummin (kŭm'in), CUMIN.

cumquat (kŭm'kwot), *n*. a small orange, fruit of *Citrus aurantium*, var. *japonica*.

cumulate (kū'mūlāt), *v.t., v.i*. to accumulate. *a*. (-lət), heaped up, accumulated. **cumulation**, *n*. **cumulative**, *a*. increasing by additions; tending to accumulate; (*Law*) enforcing a point by accumulated proof; used of drugs which, after remaining quiescent, exert their influence suddenly. **cumulatively**, *adv*. **cumulativeness**, *n*.

cumulo-, *comb. form* (*Meteor*.) cumulus.

cumulocirrostratus (kūmūlōsirōstrah'təs), *n*. a combination of cirrus and stratus with or into cumulus, a common form of rain-cloud.

cumulonimbus (kūmūlōnim'bəs), *n*. a very thick, dark cumulus cloud, usu. a sign of thunder or hail.

cumulostratus (kūmūlōstrah'təs), *n*. a mass of cumulus cloud with a horizontal base.

cumulus (kū'mūləs), *n*. (*pl*. **-li** (-lī)) a round billowing mass of cloud, with a flattish base. **cumulous**, *a*.

cuneate (kū'niət), **cuneatic** (-at'-), **cuneiform** (-nifawm), *a*. wedge-shaped. **cuneiform writing**, *n*. wedge-shaped writing used in Babylonian, Hittite, Ninevite and Persian inscriptions.

cunjevoi (kŭn'jəvoi), *n*. an Australian plant grown for its edible rhizome; (*Austral*.) a sea-squirt.

cunnilingus (kūniling'gəs), **cunnilinctus** (-lingk'təs), *n*. stimulation of the female genitals by the lips and tongue.

cunning (kŭn'ing), *a*. ingenious; artful, crafty; (*N Am*.) amusingly interesting, piquant. *n*. skill, knowledge acquired by experience; artfulness, subtlety. **cunningly**, *adv*. **cunningness**, *n*.

cunt (kŭnt), *n*. (*taboo*) the female genitalia; (*taboo*) a contemptible person; (*taboo*) a woman regarded as a sexual object.

CUP, (*abbr*.) Cambridge University Press.

cup (kŭp), *n*. a vessel to drink from, usu. small and with one handle; the liquor contained in it; an ornamental drinking-vessel, usu. of gold or silver, awarded as a prize or trophy; anything shaped like a cup, as an acorn, the socket for a bone; one of two cup-shaped supports for the breasts in a brassière; in golf, the hole or its metal lining; the lot one has to endure; the chalice used in the Holy Communion; an alcoholic mixed drink, usu. with wine or cider as a base; in cooking, a measure of capacity equal to 8 fl oz (0·23 l). *v.t*. (*past, p.p*. **cupped**) (*Med*.) to bleed by means of a cupping-glass; (*Golf*) to strike the ground when hitting the ball; to hold as if in a cup. *v.i*. (*Bot*.) to form a cup or cups. **in one's cups**, intoxicated. **one's cup of tea**, one's preferred occupation, company etc. **cupbearer**, *n*. a person who serves wine, esp. in royal or noble households. **cupcake**, *n*. a small sponge cake baked in a paper or foil case. **cup final**, *n*. the final match of a competition to decide who wins a cup. **cup tie**, *n*. a match in a knockout competition for a cup. **cupful**, *n*. **cupper**, *n*. one who uses a cupping-glass. **cupping**, *n*. **cupping-glass**, *n*. a partially evacuated glass vessel placed over a (usu.) scarified place to excite the flow of blood.

cupboard (kŭb'əd), *n*. a sideboard; an enclosed case with shelves to receive plates, dishes, food etc.; a wardrobe. **cupboard-love**, *n*. greedy or self-interested love.

cupel (kū′pl), *n.* a small shallow vessel used in assaying precious metals. *v.t.* (*past, p.p.* **cupelled**) to assay in a cupel. **cupellation**, *n.*
Cupid (kū′pid), *n.* the Roman god of Love; a picture or statue of Cupid; a beautiful boy.
cupidity (kūpid′əti), *n.* an inordinate desire to possess; covetousness, avarice.
cupola (kū′pələ), *n.* a little dome; a lantern or small apartment on the summit of a dome; a spherical covering to a building, or any part of it; a cupola furnace; a revolving dome or turret on a warship. **cupola furnace**, *n.* a furnace for melting metals.
cuppa, cupper[1] (kŭp′ə), *n.* (*coll.*) a cup of tea.
cupper[2], **cupping** CUP.
cupreous (kū′priəs), *a.* of, like or composed of copper.
cupric, *a.* having bivalent copper in its composition. **cupriferous** (-prif′-), *a.* **cuprite** (-prīt), *n.* red oxide of copper, a mineral with cubic crystal structure. **cupronickel** (-prō-), *n.* an alloy of copper and nickel. **cuprous**, *a.* having monovalent copper in its composition.
Cupressus (kūpres′əs), *n.* a genus of conifers, containing the cypress.
cupule (kū′pūl), *n.* (*Zool.*) a cup-like part or organ. **cupular, -late** (-lət), *a.* **cupuliferous** (-lif′-), *a.*
cur (kœ), *n.* a mongrel, worthless dog; a cowardly or surly fellow. **currish**, *a.* **currishly**, *adv.* **currishness**, *n.*
curable CURE[1].
curaçao (kūrəsah′ō, koo-), *n.* a liqueur flavoured with bitter orange peel, sugar and cinnamon.
curacy CURATE.
curare (kūrah′ri), *n.* the dried extract of the vine *Strychnos toxifera* used by the Indians of S America for poisoning arrows, and employed in physiological investigations as a muscle relaxant. **curarine** (-rin), *n.* an alkaloid from curare. **curarize, -ise** (kūr′-), *v.t.*
curassow (kū′rəsō), *n.* a turkey-like bird found in S and Central America.
curate (kū′rət), *n.* a clergyman of the Church of England who assists the incumbent. **perpetual curate** PERPETUAL. **curate's egg**, *n.* something of which (optimistically) parts are excellent. **curacy**, *n.* the office of a curate; the benefice of a perpetual curate.
curative (kū′rətiv), *a.* tending to cure. *n.* anything that tends to cure.
curator (kūrä′tə), *n.* one who has charge of a library, museum or similar establishment. **curatorial** (-taw′ri-), *a.* **curatorship**, *n.* **curatrix** (-triks), *n. fem.*
curb (kœb), *n.* a chain or strap passing behind the jaw of a horse in a curb-bit; (*N Am.*) a kerb; an injury to the hock-joint of a horse; a check, a restraint. *v.t.* to put a curb on; to restrain or hold back, to keep in check. **curbless**, *a.* **curby**, *a.*
curcuma (kœ′kūmə), *n.* a plant of a genus (*Curcuma*) of tuberous plants of the ginger family; turmeric, which is obtained from its root.
curd (kœd), *n.* the coagulated part of milk, used to make cheese; the coagulated part of any liquid. *v.t.* to curdle. *v.i.* to congeal. **curd-breaker**, *n.* an instrument used to break the cheese-curd into small pieces. **curd-cutter, -mill**, *n.* an instrument with knives to cut the curd. **curdy**, *a.* full of curds; curdled, congealed.
curdle (kœ′dl), *v.t.* to break into curds; to coagulate; to congeal. *v.i.* to become curdled. **to curdle the blood**, to terrify, as with a ghost-story or the like.
cure (kūə), *n.* the act of healing or curing; a remedy; anything which acts as a remedy or restorative; the state of being cured or healed; the care or spiritual charge of souls. *v.t.* to heal, to restore to health, to make sound or whole; to preserve or pickle; to correct a habit or practice. *v.i.* to effect a cure; to be cured or healed. **cure of souls**, a benefice to which parochial duties are annexed. **cure-all**, *n.* a panacea, a universal remedy; a name for the plant *Geum rivale*, water avens. **curable**, *a.* **curability** (-bil′-), *n.* **cureless**, *a.* **curer**, *n.* one who cures or heals; one who prepares preserved food (*often in comb.*, as *fish-curer*). **curing**, *n.* **curing-house**, *n.* a building in which sugar is drained and dried; a house in which articles of food are cured.
curé (kū′rā), *n.* a parish priest, a French rector or vicar. [F]
curette (kūret′), *n.* an instrument used for scraping a body cavity. *v.t.* to scrape or clean with a curette. **curettage** (-ret′ij, -ritahzh′), *n.*
curfew (kœ′fū), *n.* a regulation in the Middle Ages to extinguish fires at a stated hour; the bell announcing or the hour for this; a military or civil regulation to be within doors between stated hours; the hour at which this regulation takes effect or the period for which it is effective.
curia (kū′riə), *n.* (*pl.* **curiae** (-ē)) the Roman See, including Pope, cardinals etc. in their temporal capacity. **curial**, *a.* pertaining to the Papal curia. **curialism**, *n.*
curie (kū′ri), *n.* the standard unit of radioactivity, 3.7×10^{10} disintegrations per second. [Pierre *Curie*, 1859–1906, and Marie *Curie*, 1867–1934, F scientists]
curio (kū′riō), *n.* (*pl.* **curios**) a curiosity, esp. a curious piece of art; a bit of bric-a-brac.
curiosa (kūriō′sə), *n.pl.* unusual (collectable) objects; erotic or pornographic books.
curious (kū′riəs), *a.* inquisitive, desirous to know; given to research; extraordinary, surprising, odd. **curiosity** (-os′-), *n.* a desire to know; inquisitiveness; a rarity, an object of curiosity. **curiously**, *adv.* **curiousness**, *n.* strangeness, oddness.
curium (kū′riəm), *n.* an artificially-produced transuranic metallic element, at.no.96; chem. symbol Cm.
curl (kœl), *n.* a ringlet or twisted lock of hair; anything coiled, twisted or spiral; the state of being curled; a contemptuous curving of the lip; a disease in potatoes of which curled shoots and leaves are a symptom. *v.t.* to twine; to twist into curls; to dress with ringlets; to curve up (the lip) in contempt. *v.i.* to twist, to curve up; to rise in curves or undulations; to play at the game of curling. **out of curl**, limp, out of condition. **to curl up**, to go into a curled position; (*coll.*) to be embarrassed or disgusted. **to make someone's hair curl**, to horrify or scandalize someone. **curl-cloud**, *n.* cirrus. **curl-paper**, *n.* paper round which hair is wound to form a curl. **curler**, *n.* one who or that which curls; a device for curling the hair; one who plays at curling. **curling** (kœ′ling), *n.* a game on the ice in which contending parties slide smooth stones towards a mark. **curling-stone**, *n.* the stone used in the game. **curling-irons, -tongs**, *n.pl.* an instrument for curling the hair. **curlingly**, *adv.* **curly**, *a.* having curls; wavy, undulated; (*Bot.*) having curled or wavy margins. **curliness**, *n.*
curlew (kœ′lū), *n.* a migratory wading bird, esp. the European *Numenius arquatus*.
curlicue (kœ′likū), *n.* a fantastic curl; a flourish in writing.
curmudgeon (kəmŭj′ən), *n.* a miserly or churlish person. **curmudgeonly**, *adv.*
curmurring (kəmœ′ring), *n.* (*Sc.*) a low rumbling, esp. a sound in the bowels from flatulence.
currach, curragh (kū′rə, -əkh), *n.* a skiff made of wickerwork and hides, a coracle. [Ir.]
currant (kū′rənt), *n.* the dried fruit of a dwarf seedless grape from the Levant; the fruit of shrubs of the genus *Ribes*, containing the black, red and white currants.
currency (kū′rənsi), *n.* a continual passing from hand to hand, as of money; the circulating monetary medium of a country, whether in coin or paper; the period during which anything is current; the state of being current.
current (kū′rənt), *a.* passing at the present time; belonging to the present week, month, year; in circulation, as money; generally received or acknowledged; in general circulation among the public. *n.* a flowing stream, a body

of water, air etc., moving in a certain direction; general drift or tendency; electrical activity regarded as the rate of flow of electrical charge along a conductor. **current account**, *n.* a bank account which usu. does not pay interest and on which one may draw cheques. **currentless**, *a.* **currently**, *adv.* with a constant progressive motion; generally; at present.
curricle (kŭ'rikl), *n.* a two-wheeled chaise for a pair of horses.
curriculum (kərik'ūləm), *n.* (*pl.* **-la** (-lə), **-lums**) a fixed course of study at a school etc. **curriculum vitae** (vē'tī), *n.* a brief outline of one's education, previous employment, and other achievements. **curricular**, *a.*
currier (kŭ'riə), *n.* one who curries, dresses and colours leather after it has been tanned. **curriery**, *n.* the trade or work-place of a currier.
currish etc. CUR.
curry[1] (kŭ'ri), *v.i.* to dress a horse with a comb; to dress leather. **to curry favour**, to seek favour by officiousness or flattery. **curry-comb**, *n.* a comb used for grooming horses.
curry[2] (kŭ'ri), *n.* a highly-spiced Indian dish of stewed meat, fish etc., seasoned with turmeric etc.; curry-powder. *v.t.* to season or dress with curry. **curry-paste, -powder**, *n.* a mixture of ginger, turmeric and other strong spices used in curries etc.
curse (kœs), *v.t.* to invoke harm or evil upon; to blast, to injure, vex or torment; to excommunicate. *v.i.* to swear, to utter imprecations. *n.* a solemn invocation of divine vengeance (upon); a sentence of divine vengeance; an oath; an imprecation (upon); the evil imprecated; anything which causes evil, trouble or great vexation; a sentence of excommunication. **the curse**, (*coll.*) menstruation. **cursed** (-sid), **curst**, *a.* execrable, accursed, deserving of a curse; blasted by a curse, execrated; vexatious, troublesome. **cursedly**, *adv.* **cursedness**, *n.* **curser**, *n.*
cursive (kœ'siv), *n., a.* (handwriting or a typeface) with joined-up letters.
cursor (kœ'sə), *n.* the moving part of a measuring instrument, e.g. the slide with the reference line in a slide rule; on a VDU screen, a movable point of light showing the position of the next action, e.g. the beginning of an addition or correction.
cursores (kœsaw'rēz), *n.pl.* an order of birds with rudimentary wings and strong feet well adapted for running, containing the ostrich, the emu, cassowary and apteryx. **cursorial**, *a.*
cursory (kœ'səri), *a.* hasty, superficial, careless. **cursorily**, *adv.* **cursoriness**, *n.*
curst CURSE.
curt (kœt), *a.* short, concise, abrupt; esp. rudely terse and abrupt. **curtly**, *adv.* **curtness**, *n.*
curtail (kœtāl'), *v.t.* to shorten; to cut off the end or tail of; to lessen; to reduce. **curtailer**, *n.* **curtailment**, *n.*
curtain (kœ'tən), *n.* a cloth hanging beside a window or door, or round a bed, which can be drawn across at pleasure; a screen, a cover, a protection; the screen in a theatre separating the stage from the spectators; the end of a scene or play, marked by the closing of the curtains; a curtain wall; a partition or cover of various kinds; a shifting plate in a lock; (*pl.*) (*sl.*) death, the end. *v.t.* to enclose with or as with curtains; to furnish or decorate with curtains. **curtain call**, *n.* applause for an actor which calls for a reappearance before the curtain falls. **curtain-lecture**, *n.* a reproof or lecture from a wife to a husband after they have gone to bed. **curtain-pole**, *n.* a pole for hanging curtains on. **curtain-raiser**, *n.* a short piece given before the main play; any short preliminary event. **curtain-rings**, *n.pl.* rings by which the curtains can be drawn backwards or forwards along the curtain pole or **curtain rods**. **curtain wall**, *n.* a wall that is not load-bearing; a wall between two bastions. **curtainless**, *a.*

curtana (kœtah'nə, -tā'-), *n.* the unpointed Sword of Mercy carried before the English sovereigns at their coronation.
curtilage (kœ'təlij), *n.* a piece of ground included within the same fence as a dwelling-house.
curtsy, -sey (kœt'si), *n.* a bow; an act of respect or salutation, performed by women by slightly bending the body and knees at the same time. *v.i.* to make a curtsy.
curule chair, *n.* the chair of honour, shaped like a campstool with crooked legs, of the old Roman kings, and of the higher magistrates of senatorial rank under the republic.
curvaceous (kœvā'shəs), *a.* (*coll.*) of a woman's body, generously curved.
curvate (kœ'vət), *a.* curved, bent. **curvative** (-və-), *a.* (*Bot.*) having the margins slightly curved. **curvature** (-vəchə), *n.* deflection from a straight line; a curved form; (*Geom.*) the continual bending of a line from a rectilinear direction.
curve (kœv), *n.* a bending without angles; that which is bent; a line on a graph. *v.t.* to cause to bend without angles. *v.i.* to form or be formed into a curve. **curved**, *a.* **curvy**, *a.*
curvet (kœvet'), *n.* a particular leap of a horse raising all four legs off the ground at once. *v.i.* to make a curvet; to frolic, to frisk.
curvi-, *comb. form* curved. **curvilinear** (-lin'iə), *a.* bounded by curved lines; consisting of curved lines. **curvilinearity** (-a'-), *n.* **curvilinearly**, *adv.*
cusec, cu.-sec. (kū'sek), *n.* unit of rate of flow of water, 1 cu. ft. (0·0283 m^3) per second.
cushion (kush'ən), *n.* a kind of stuffed pillow or pad for sitting, kneeling or leaning on; anything padded, as the lining at the side of a billiard-table; a cushion-like organ, part or growth; anything serving as a buffer or protection against shocks. *v.t.* to protect or pad with cushions; to furnish with cushions; to place or leave (a billiard ball) close up to the cushion; to suppress or deaden a blow or shock. **lady's cushion** LADY. **pin-cushion** PIN. **sea cushion** SEA. **cushiony**, *a.*
cushy (kush'i), *a.* (*sl.*) soft, easy, comfortable; well paid and little to do. **cushily**, *adv.*
cusp (kŭsp), *n.* a point; (*Arch.*) a Gothic ornament consisting of a projecting point formed by the meeting of curves; the point in a curve at which its two branches have a common tangent; the pointed end of a leaf or other part; a projection on a molar tooth; either of the two points of a crescent moon; (*Astrol.*) a division between signs of the zodiac. **cusped**, *a.* **cuspid** (-pid), *a.* **cuspidal**, *a.* (*Geom.*) ending in a point. **cuspidate, -dated** (-dāt, -dātid), *a.* (*Bot.*) tapering to a rigid point. **cuspidate teeth**, *n.pl.* canine teeth.
cuspidor (kŭs'pidaw), *n.* a spittoon.
cuss (kŭs), *n.* (*coll.*) a curse; (*coll.*) a worthless fellow. *v.t., v.i.* (*coll.*) to curse. **cussed**, *a.* obstinate; perverse. **cussedly**, *adv.* **cussedness** (-sid-), *n.*
custard (kŭs'təd), *n.* a composition of milk and eggs, sweetened and flavoured; a sweet sauce made of milk, sugar and custard powder; orig., an open pie. **custard-apple**, *n.* a W Indian fruit, *Anona reticulata*, with a soft pulp. **custard pie**, *n.* an open pie filled with custard, thrown in slapstick comedy. **custard powder**, *n.* a composition of cornflour, colouring and flavouring, used in the making of custard.
custodian (kəstō'diən), *n.* one who has the custody or guardianship of anything. **custodial**, *a.* pertaining to custody or guardianship.
custody (kŭs'tədi), *n.* guardianship, care esp. of a minor; imprisonment, detention. **to take into custody**, to arrest.
custom (kŭs'təm), *n.* a habitual use or practice; established usage; familiarity, use; buying of goods, business; frequenting a shop to purchase; (*pl.*) custom-duties; (*pl.*) the

cut   197   **cutty**

government department which collects custom-duties; (*pl.*) the place at a port or airport where goods or baggage are examined for liability to custom-duties; (*Law*) long established practice constituting common law. *a.* (*chiefly N Am.*) made to a customer's specifications. **custom-duties**, *n.pl.* duties imposed on goods imported or exported. **custom(s)-house**, *n.* the office where vessels enter and clear, and where custom-duties are paid. **custom-made, -built**, *a.* (*chiefly N Am.*) made to measure, custom. **customary**, *a.* habitual, usual, wonted; (*Law*) holding or held by custom, liable under custom. **customarily**, *adv.* **customariness**, *n.* **customed**, *a.* usual, accustomed. **customer**, *n.* one who deals regularly at a particular shop; a purchaser; (*coll.*) a person one has to do with, a fellow. **customize, -ise**, *v.t.* to make to a customer's specifications.
cut[1] (kŭt), *v.t.* (*past, p.p.* **cut**) to penetrate or wound with a sharp instrument; to divide or separate with a sharp-edged instrument; to sever, to detach, to hew, to fell, to mow or reap; to carve, to trim or clip; to form by cutting; to have (a tooth) come through the gums; to reduce by cutting; to mutilate or shorten (a play, article or book); to edit (a film); to intersect, to cross; to divide (as a pack of cards); to hit (a cricket ball) with a downward stroke and make it glance to one side; to wound deeply; to absent oneself from; to renounce the acquaintance of; to reduce as low as possible; (*sl.*) to dilute (a drink or drug); to record e.g. a song on (a gramophone record). *v.i.* to make a wound or incision with or as with a sharp-edged instrument; to have a good edge; to divide a pack of cards; (*sl.*) to move away quickly, to run; to intersect; to (be able to) be cut or divided; to change direction; to change abruptly from one scene to another in a film; to stop filming or recording. **cut-and-cover**, *n.* a tunnel made by excavating an open cutting and covering it in. **cut and dry, dried**, already settled, not open to change. **(not) cut out for**, (not) naturally fitted for. **to cut a caper**, to frisk about. **to cut across**, to take a shorter usu. oblique route; to go contrary to (usual procedure etc.). **to cut a dash** DASH. **to cut a figure, a flourish**, etc., to look, appear or perform (usu. qualified by an adjective). **to cut and come again**, to help oneself and take more if one will. **to cut and run**, to depart rapidly. **to cut away**, to detach by cutting; to reduce by cutting. **to cut back**, to prune; to reduce. **to cut both ways**, to have both good and bad consequences. **to cut corners**, to take short cuts. **to cut dead**, to refuse to acknowledge. **to cut down**, to fell; to compress, to reduce. **to cut in**, to drive in front of another person's car so as to affect his driving; to take a lady away from her dancing partner; to interrupt, to intrude; (*coll.*) to allow to have a share in. **to cut it fine**, to reduce to the minimum; to take a risk by allowing little margin. **to cut it out**, to desist from doing something annoying. **to cut no ice** ICE. **to cut off**, to remove by cutting; to intercept; to prevent from access; to obstruct; to sever; to discontinue; to bring to an untimely end, to kill; to disinherit. **to cut one's losses**, to write off as lost, to abandon a speculation. **to cut one's teeth on**, to practise or gain experience by means of. **to cut out**, to shape by cutting; to remove or separate by cutting; to supplant; to cease doing, taking or indulging in something unpleasant or harmful; to cease operating suddenly and unexpectedly or by the automatic intervention of a cut-out device; to relinquish a game as the result of cutting the cards. **to cut short**, to hinder by interruption; to abridge. **to cut to pieces**, to exterminate, to massacre. **to cut up**, to cut in pieces; to criticize severely; to distress deeply. **to cut up rough**, (*sl.*) to become quarrelsome or savage. **cutaway**, *a.* denoting a drawing of an engine etc. in which part of the casing is omitted to show the workings. *n.* a coat with the skirts cut away diagonally to the back of the knee. **cutback**, *n.* a reduction. **cut-glass**, *n.* flint glass in which a pattern is formed by cutting or grinding. **cut-line**, *n.* a caption. **cut-off**, *n.* a passage cut by a river, affording a new channel; a valve to stop discharge. **cut-off road**, *n.* (*N Am.*) a by-pass. **cut-out**, *n.* something, e.g. a shape, which has been cut out; a device for automatic severance of an electric circuit in case the tension becomes too high for the wiring. **cut-price**, *a.* at the lowest price possible; at a reduced price. **cutpurse**, *n.* (*Hist.*) a pickpocket. **cut-rate** CUT-PRICE. **cut-throat**, *n.* a murderer, an assassin. *a.* murderous, barbarous; of competition etc., fierce, merciless. **cutwater**, *n.* (*Naut.*) the fore-part of a ship's prow which cuts the water. **cutworm**, *n.* a caterpillar, esp. (*N Am.*) the larva of the genus of moths *Agrotis*, which cuts off plants near the roots. **cutter**, *n.* one who or that which cuts; one who cuts out men's clothes to measure; a cutting tool; (*Naut.*) a man-of-war's boat smaller than a barge, with from four to eight oars; a one-masted vessel with fore-and-aft sails; a lightly-armed boat used by coastguard or customs patrols. **cutting**, *a.* dividing by a sharp-edged instrument; sharp-edged; wounding the feelings deeply; sarcastic, biting. *n.* the action of the verb TO CUT; a piece cut off or out (of a newspaper etc.); (*Hort.*) a slip; an excavation for a road, railway or canal; the selection of those portions of a film that are finally to be shown. **cutting-bench**, *n.* the table on which a cutter assembles and edits a film in the **cutting-room. cuttingly**, *adv.* in a cutting manner.
cut[2] (kŭt), *n.* the action of cutting; a stroke or blow with a sharp-edged instrument; an opening, gash or wound made by cutting; anything done or said that hurts the feelings; a reduction; the omission of a part of a play; a slit, a channel, a groove, a trench; a part cut off; a gelding; a stroke with a whip; a particular stroke in various games with balls; the act of dividing a pack of cards; the shape in which a thing is cut, style; a degree (from count being formerly kept by notches); a share or portion; the place where one strip of film ends in a picture and another begins. *a.* subjected to the act or process of cutting; severed; shaped by cutting; castrated. **a cut above**, (*fig.*) superior to. **cut and thrust**, cutting and thrusting; a hand-to-hand struggle. **short cut**, a near way; readiest means to an end.
cutaneous (kūtā'niəs), *a.* belonging to or affecting the skin.
cute (kūt), *a.* cunning, sharp, clever; (*chiefly N Am.*) delightful, attractive; pretty. **cutely**, *adv.* **cuteness**, *n.* **cutie** (-ti), *n.* (*sl.*) a bright, attractive person.
cuticle (kū'tikl), *n.* the epidermis or scarf-skin; the outer layer of the integument in the lower animals; the thin external covering of the bark of a plant. **cuticular** (-tik'ū-), *a.* **cuticularize, -ise** (-tik'-), *v.t.*
cutis (kū'tis), *n.* the true skin beneath the epidermis; the peridium of certain fungi.
cutlass (kŭt'ləs), *n.* (*Hist.*) a broad curved sword, esp. that used by sailors.
cutler (kŭt'lə), *n.* one who makes or deals in cutting instruments. **cutlery**, *n.* the business of a cutler; knives and other edged instruments or tools; knives, spoons and forks used for eating.
cutlet (kŭt'lit), *n.* a small slice of meat, usu. from the loin or neck, for cooking; minced meat or meat-substitute shaped to look like a cutlet.
cutter, cutting CUT[1].
cuttle (kŭt'l), *n.* a cuttlefish. **cuttlebone**, *n.* the internal skeleton of the cuttlefish, used as a polishing agent and as a dietary supplement for cagebirds. **cuttlefish**, *n.* a 10-armed cephalopod, *Sepia officinalis*; other members of the genus *Sepia*.
cutty (kŭt'i), *a.* (*Sc., North.*) short, cut short. **cutty sark**, *n.* a short shift. **cutty-stool**, *n.* a bench in old Scottish churches on which women guilty of unchastity were compelled to sit and undergo public rebuke.

cuvée (koovā'), *n.* a batch of blended wine. [F]
cuvette (kūvet'), *n.* a little scoop; a clay crucible.
CV, (*abbr.*) Common Version; curriculum vitae.
Cwith, (*abbr.*) Commonwealth.
cwm (kum), *n.* a valley in Wales; a cirque. [W]
cwo, (*abbr.*) cash with order.
CWS, (*abbr.*) Cooperative Wholesale Society.
cwt, (*abbr.*) hundredweight.
-cy (-si), *suf.* forming nouns of quality from adjectives, and nouns of office (cp. -SHIP) from nouns.
cyan (sī'an), *n.* a bluish-green colour. *a.* of this colour. **cyan-,** *comb. form.* CYANO-.
cyanate (sī'ənāt), *n.* a salt of cyanic acid.
cyanic (sian'ik), *a.* derived from cyanogen; blue. **cyanic acid,** *n.* a compound of cyanogen and hydrogen.
cyanide (sī'ənīd), *n.* a (usu. highly poisonous) compound of cyanogen with a metallic element.
cyanite (sī'ənīt), *n.* a hard, translucent mineral, often blue, occurring in flattened prisms in gneiss and mica-schist. **cyanitic** (-nit'-), *a.*
cyan(o)-, *comb. form.* of a blue colour; pertaining to or containing cyanogen.
cyanogen (sian'əjən), *n.* a colourless, poisonous gas composed of carbon and nitrogen, burning with a peach-blossom flame, and smelling like almond.
cyanosis (sīənō'sis), *n.* (*pl.* **-ses** (-sēz)) a condition in which the skin becomes blue or leaden-coloured owing to the circulation of oxygen-deficient blood.
cybernetics (sībənet'iks), *n. sing.* the comparative study of control and communication mechanisms in machines and living creatures. **cybernetic,** *a.* **cybernate** (sī'-), *v.t.* to control automatically, e.g. by means of a computer. **cybernation,** *n.*
cycad (sī'kad), *n.* a cycadaceous plant. **cycadaceous** (sikədā'shəs), *a.* belonging to the Cycadaceae, an order of gymnosperms, allied to the conifers.
cyclamate (sik'ləmāt, sī'-), *n.* any of several compounds derived from petrochemicals, formerly used as sweetening agents.
cyclamen (sik'ləmən), *n.* the sowbread, a genus of tuberous plants with beautiful flowers.
cycle (sī'kl), *n.* a series of years, events or phenomena recurring in the same order; a series that repeats itself; a complete series or succession; the period in which a series of events is completed; a long period, an age; a body of legend connected with some myth; a series of poems, songs etc. about a central theme or character; a bicycle or tricycle; one complete series of changes in a periodically varying quantity, e.g. an electric current. *v.i.* to revolve in a circle; to ride a bicycle or tricycle. **cycle of the moon, lunar cycle, Metonic cycle,** a period of 19 years, after which the new and full moon recur on the same days of the month. **cycle of the sun, solar cycle,** a period of 28 years, after which the days of the month recur on the same days of the week. **cycle-track, -way,** *n.* a path, often beside a road, reserved for cyclists. **cyclic, -ical,** *a.* pertaining to, or moving or recurring in, a cycle; (*Bot.*) arranged in whorls; of an organic chemical compound, containing a ring of atoms. **cyclically,** *adv.* **cyclist,** *n.* one who rides a bicycle or tricycle.
cyclo-, *comb. form* circular; pertaining to a circle or circles.
cyclo-cross (sī'klōkros), *n.* the sport of cross-country racing on a bicycle.
cycloid (sī'kloid), *n.* the figure described by a point in the plane of a circle which rolls along a straight line till it has completed a revolution. **cycloidal** (-kloi'-), *a.* resembling a circle; (*Zool.*) having concentric striations.
cyclometer (sīklom'itə), *n.* an instrument for recording the revolutions of a wheel, esp. that of a bicycle, and hence the distance travelled.
cyclone (sī'klōn), *n.* a disturbance in the atmosphere caused by a system of winds blowing spirally towards a central region of low barometric pressure; a violent hurricane. **cyclonic** (-klon'-), *a.*
cyclopaedia (sīkləpē'diə), etc. ENCYCLOPAEDIA.
cyclopropane (sīklōprō'pān), *n.* a colourless hydrocarbon gas used as an anaesthetic.
Cyclops (sī'klops), *n.* (*pl.* **Cyclopes** (-ōpēz)) mythical one-eyed giants who dwelt in Sicily. **cyclopean,** *a.*
cyclorama (sīklərah'mə), *n.* a panorama painted on the inside of a large cylinder and viewed from the middle; a large curved screen at the back of a stage onto which backgrounds can be projected.
cyclosis (sīklō'sis), *n.* circulation, as of blood, the latex in plants, or protoplasm in certain cells.
Cyclostomata (sīklostō'mətə), *n.* a subclass of fishes, with a circular suctorial mouth, containing the lampreys and hags. **cyclostomatous** (-stom'-), **cyclostomous** (-klos'-), *a.* **cyclostome** (sī'-), *n.*
cyclostyle (sī'kləstīl), *n.* a machine for printing copies of handwriting or typewriting by means of a sheet perforated like a stencil. *v.t.* to print using this machine.
cyclothymia (sīklōthī'miə), *n.* a psychological condition characterized by swings between elation and depression. [Gr. *thumos*, spirit]
cyclotron (sī'klətron), *n.* a particle accelerator designed to accelerate protons to high energies.
cyder (sī'də), CIDER.
cygnet (sig'nət), *n.* a young swan.
cylinder (sil'ində), *n.* a straight roller-shaped body, solid or hollow, and of uniform circumference; (*Geom.*) a solid figure described by the revolution of a right-angled parallelogram about one of its sides which remains fixed; a cylindrical part in various machines, esp. the chamber in an internal-combustion engine in which the piston is acted upon by the exploding gases; the roller used in machine-printing. **cylinder block,** *n.* the casing which houses the cylinders in an internal-combustion engine. **cylinder head,** *n.* the closed end of an internal-combustion cylinder. **cylindrical** (-lin'dri-), *a.* having the form of a cylinder. **cylindriform** (-lin'drifawm), *a.*
cylindroid (sil'indroid), *n.* (*Geom.*) a solid body differing from a cylinder in having the bases elliptical instead of circular.
cymbal (sim'bl), *n.* one of a pair of disks of brass or bronze more or less basin-shaped, clashed together to produce a sharp, clashing sound. **cymbalist,** *n.*
cymbalo (sim'bəlō), *n.* (*pl.* **-los**) the dulcimer, a stringed instrument played by means of small hammers held in the hands.
cyme (sīm), *n.* an inflorescence in which the central terminal flower comes to perfection first, as in the guelder-rose. **cymoid,** *a.* resembling a cyme. **cymose** (-mōs), *a.*
Cymric (kim'rik), *a.* pertaining to the Welsh. *n.* the Welsh language. [W *Cymru*, Wales]
cynic (sin'ik), *n.* (*Hist.*) one of a rigid sect of Greek philosophers (of which Diogenes was the most distinguished member) founded at Athens by Antisthenes, a pupil of Socrates, who insisted on the complete renunciation of all luxury and the subjugation of sensual desires; one who is habitually morose and sarcastic; one who is pessimistic about human nature. **cynical,** *a.* bitter, sarcastic, misanthropical; contemptuous of accepted standards, unprincipled; (*Hist.*) of or belonging to the cynics. **cynically,** *adv.* **cynicism** (-sizm), *n.*
cynosure (sin'əzūə, -shuə), *n.* the constellation of the Lesser Bear (*Ursa Minor*), containing the north star; a centre of interest or attraction.
cypher CIPHER.
cypress (sī'prəs), *n.* a tree of the coniferous genus *Cupressus,* esp. *C. sempervirens,* valued for the durability of its wood; a branch of this as emblem of mourning.
cyprine (sip'rīn), *n.* of or belonging to the fish genus *Cyprinus,* containing the carp.

Cypriot (sip'riət), *a.* of or belonging to Cyprus. *n.* an inhabitant of Cyprus.

cypripedium (sipripē'diəm), *n.* lady's slipper, an orchid (genus *Cypripedium*) possessing two fertile stamens, the central stamen (fertile in other orchids) being represented by a shield-like plate.

Cyrenaic (sirənā'ik, sī'-), *a.* of or pertaining to Cyrene, an ancient Greek colony in the north of Africa, or to the hedonistic or eudaemonistic philosophy founded at that place by Aristippus. *n.* a philosopher of the Cyrenaic school.

Cyrillic (siril'ik), *a.* a term applied to the alphabet of the Slavonic nations who belong to the Orthodox Church, from the fact that it was introduced by Clement, a disciple of St Cyril.

cyst (sist), *n.* a bladder, vesicle or hollow organ; a sac containing morbid matter; a protective membrane enclosing a cell, larva etc. **cystic**, *a.* pertaining to or enclosed in a cyst, esp. the gall or urinary bladder; having cysts, or of the nature of a cyst. **cystic fibrosis**, *n.* a hereditary disease appearing in early childhood, marked by overproduction of mucus and fibrous tissue, with consequent breathing and digestive difficulties. **cystic worms**, *n.pl.* immature tapeworms, encysted in the tissues of their host. **cystiform** (sis'tifawm), *a.* **cystose** (-tōs), **cystous**, *a.* containing cysts.

cyst(i)-, **cysto-**, *comb. forms* pertaining to the bladder; bladder-shaped.

cystitis (sisti'tis), *n.* inflammation of the urinary bladder.

cysto- CYST(I)-.

cystocele (sis'təsēl), *n.* hernia caused by protrusion of the bladder.

cystoscope (sis'təskōp), *n.* an instrument or apparatus for the exploration of the bladder.

cystose CYST.

cystotomy (sistot'əmi), *n.* the act or practice of opening cysts; the operation of cutting into the bladder to remove calculi.

-cyte, *suf.* (*Biol.*) a mature cell, as in *leucocyte*.

cytherean (sithərē'ən), *a.* (*poet.*) pertaining to Venus, the goddess of love.

cyto-, *comb. form* (*Biol.*) cellular; pertaining to or composed of cells. **cytology** (-tol'-), *n.* the study of cells. **cytological** (-loj'-), *a.* **cytologist** (-tol'-), *n.* **cytolysis** (-tol'isis), *n.* the dissolution of cells. **cytoplasm** (-plaz'm), *n.* the protoplasm of a cell apart from the nucleus. **cytoplasmic** (-plaz'-), *a.* **cytotoxin** (-tok'sin), *n.* a substance which is poisonous to cells.

czar (zar), *n.* the title of the former emperors of Russia. **czarevich**, **-vitch**, *n.* the son of a czar. **czarevna** (-ev'nə), *n.* the daughter of a czar. **czarina** (-rē'nə), **czaritza** (-rit'sə), *n.* an empress of Russia; the wife of a czar.

Czech (chek), *n.* a native or inhabitant of W Czechoslovakia; the slavonic language of the Czechs; loosely, a Czechoslovak. *a.* of or pertaining to the Czechs, their language or (loosely) Czechoslovakia. **Czechoslovak** (chekōslō'vak), **-vakian** (-vak'iən, -vah'kiən), *n.* a native of Czechoslovakia (republic in Central Europe); a Czech or Slovak. *a.* of Czechoslovakia; of the Czechs or Slovaks, or their languages.

D

D, d, the fourth letter in the English alphabet. D is a symbol for the second note of the musical scale of C, corresponding to the Italian *re;* the fourth in numerical series; (*Roman numeral*) 500. **D-day,** *n.* the code name for the date of the invasion of France, 6 June 1944. **D mark** DEUTSCHE MARK. **D region, layer,** *n.* the lowest part of the ionosphere, between 25 and 40 miles (40 and 65 km) above the earth's surface.
D., (*abbr.*) democrat; department; God (L *deus*); Lord (L *dominus*).
d., (*abbr.*) date; daughter; day; dead; depart(s); diameter; penny (before decimalization, L *denarius*).
-d, *suf.* forming past tense and p.p. of weak verbs, as in *died, heard, loved, proved.*
da DAD.
dab[1] (dab), *v.t.* (*past, p.p.* **dabbed**) to strike gently with some moist or soft substance; to pat; to strike or touch tentatively; to apply with short, light strokes. *n.* a gentle blow; a light stroke or wipe with a soft substance; a small amount, patch or piece esp. of a soft substance; (*often pl., sl.*) fingerprints. **a dab hand at,** (*coll.*) an expert at. **dabber,** *n.*
dab[2] (dab), *n.* a small flatfish, *Pleuronectes limanda.*
dabble (dab'l), *v.t.* to keep on dabbing; to wet by little dips; to besprinkle, to splash. *v.i.* to play or splash about in water; to do or practise anything in a superficial manner; to dip into a subject. **dabbler,** *n.* **dabblingly,** *adv.* superficially, shallowly.
dabchick (dab'chik), *n.* the little grebe, *Podiceps minor.*
da capo (da kah'pō), (*Mus. direction*) the player is to begin again.
dace (dās), *n.* a small river fish, *Leuciscus vulgaris.*
dacha, datcha (dach'ə), *n.* a country house or cottage in Russia.
dachshund (daks'hunt, dakhs'-), *n.* a short-legged long-bodied breed of dog.
dacoit (dəkoit'), *n.* one of an Indian or Burmese band of armed robbers. **dacoity,** *n.* robbery by armed gang.
dactyl (dak'til), *n.* a metrical foot consisting of one long followed by two short syllables. **dactylic** (-til'-), *a.* [L *dactylus*, as foll.]
dactyl-, dactylio-, dactylo-, *comb. form* having fingers or digits; pertaining to fingers or digits.
dactyliology (daktiliol'əji), *n.* the study of finger-rings.
dactylology (daktilol'əji), *n.* the art of conversing with the deaf and dumb by means of the fingers.
dad (dad), **da** (dah), **dada** (dad'ah), **daddy** (dad'i), *n.* a child's name for father. **daddy,** *n.* a form of address for an old man. **the daddy of them all,** (*coll.*) the supreme example of something. **daddy-long-legs,** *n.* various species of crane-fly.
Dada[1] (dah'dah), **Dadaism,** *n.* an early 20th-cent. school of art and literature that aimed at suppressing any correlation between thought and expression.
dada[2] DAD.
daddy DAD.
dado (dā'dō), *n.* (*pl.* **-dos, -does**) the cube of a pedestal between the base and the cornice; an arrangement of wainscoting or decoration round the lower part of the walls of a room.
daemon etc. DEMON.

daff (daf), *n.* (*coll.*) short for DAFFODIL.
daffodil (daf'ədil), **daffodilly** (-dil'i), **daffadowndilly** (-downdil'i), *n.* the Lent lily or yellow narcissus, *Narcissus pseudonarcissus;* other species and garden varieties of the genus *Narcissus.*
daffy (daf'i), *a.* (*coll.*) crazy, daft.
daft (dahft), *a.* weak-minded, imbecile; foolish, silly, thoughtless; frolicsome. **daft about,** (*coll.*) very fond of. **daftly,** *adv.* **daftness,** *n.*
dag (dag), *v.t.* to remove the daglock from (sheep). *n.* a daglock.
dagga (dag'ə, dakh'ə), *n.* a type of hemp used as a narcotic.
dagger (dag'ə), *n.* a short two-edged weapon adapted for stabbing; (*Print.*) a reference mark (†). **at daggers drawn,** on hostile terms; ready to fight. **to look daggers,** to look with fierceness or animosity.
daglock (dag'lok), *n.* the dirt-covered clumps of wool around the hindquarters of sheep.
dago (dā'gō), *n.* (*sl., offensive*) a contemptuous term for a Spaniard, Italian or Portuguese.
daguerreotype (dəge'rətip), *n.* the process of photographing on copper plates coated with silver iodide, developed by exposure to mercury vapour, used by *Daguerre* (1789-1851), of Paris; a photograph made by this process. **daguerreotyper, -pist,** *n.* **daguerreotypic, -ical** (-tip'-), *a.* **daguerreotypism,** *n.*
dahl (dahl), DAL.
Dahlia (dāl'yə), *n.* a genus of composite plants from Mexico, cultivated for their beautiful flowers; (dahlia) a plant of this genus or its flower.
Dail Eireann (doil eə'rən), the House of Representatives in the parliament of Eire.
daily (dā'li), *a.* happening, done or recurring every day; published every week-day; necessary for every day; ordinary, usual. *adv.* day by day; often; continually, always. *n.* a newspaper published every week-day; a woman employed daily for house-work. **daily dozen,** *n.* (*coll.*) daily physical exercises.
dainty (dān'ti), *n.* a delicacy; a choice morsel; a choice dish. *a.* pleasing to the taste, choice; pretty, delicate, elegant; fastidious, nice; excessively genteel. **daintily,** *adv.* **daintiness,** *n.*
daiquiri (dak'əri, dī'-), *n.* a cocktail made of rum and lime-juice.
dairy (deə'ri), *n.* the place or building or department of a farm where milk is kept and converted into butter or cheese; a place where milk, cream and butter are sold; a dairy-farm. *a.* belonging to a dairy or its business. **dairy-farm,** *n.* **dairy-maid,** *n. fem.* **dairy-man,** *n.* **dairy products,** *n.pl.* **dairying,** *n.* dairy-farming.
dais (dā'is), *n.* a raised platform usu. at one end of a hall.
daisy (dā'zi), *n.* a small composite flower, *Bellis perennis;* other flowers resembling this. **daisy-chain,** *n.* a string of daisies made by children. **daisy-cutter,** *n.* a ball at cricket bowled so low that it rolls along the ground. **daisy-wheel,** *n.* a wheel-shaped printer with characters on spikes round the circumference. **daisied,** *a.*
dakoit (dəkoit'), DACOIT.
dal (dahl), *n.* a split grain, pulse; a soup or purée made from this, eaten in the Indian subcontinent.

Dalai-lama LAMA [1].

dale (dāl), *n.* a valley, esp. from the English midlands to the Scottish lowlands. **dalesman**, *n.* a native or inhabitant of a dale, esp. in the northern counties of England. **daleswoman**, *n.fem.*

Dalek (dahlek), *n.* fictional creature which lives in a conical, armoured shell and is very aggressive. [from TV series *Dr Who*]

dally (dal'i), *v.i.* to trifle, toy; to exchange caresses; to delay, to waste time. **dalliance**, *n.*

Dalmatian (dalmā'shən), *a.* belonging to Dalmatia, in Yugoslavia. *n.* a Dalmatian dog; a native or inhabitant of Dalmatia. **Dalmatian dog**, *n.* a variety of hound, white with numerous black or brown spots.

dalmatic (dalmat'ik), *n.* an ecclesiastical vestment worn by bishops and deacons in the Roman and Greek Churches at High Mass; a similar robe worn by monarchs at coronation and other ceremonies.

dal segno (dal sen'yō), *adv.* (*Mus.*) repeat from point indicated.

daltonism (dawl'tənizm), *n.* colour-blindness, esp. inability to distinguish between red and green.

dam[1] (dam), *n.* a female parent (chiefly of quadrupeds); used of a human mother in contempt.

dam[2] (dam), *n.* a bank or mound raised to keep back water; the water so kept back; a causeway. *v.t.* (*past, p.p.* **dammed**) to keep back or confine by a dam; to obstruct, to hinder.

damage (dam'ij), *n.* hurt, injury, mischief or detriment to any person or thing; loss or harm incurred; (*sl.*) cost; (*Law, pl.*) reparation in money for injury sustained. *v.t.* to cause damage to. *v.i.* to receive damage. **damage limitation**, *n.* (precautionary measures) restricting the amount of harm which any fault, error or indiscreet revelation can do.

damascene (dam'əsēn, -sēn'), *v.t.* to ornament by inlaying or incrustation, or (as a steel blade) with a wavy pattern in welding. *a.* (**Damascene**) pertaining to Damascus. *n.* (**Damascene**) a native of Damascus. **Damascus blade** (dəmas'kəs), *n.* a sword of fine quality the blade of which is variegated with streaks or veins.

damask (dam'əsk), *n.* a linen fabric, with raised figures woven in the pattern, used for table-cloths, dinner-napkins etc.; the colour of the damask rose; damask steel. *a.* made of damask; red, like the damask rose; of or resembling damask steel. *v.t.* to damascene. **damask rose**, *n.* an old-fashioned rose, *Rosa gallica*, var. *damascena*. **damask steel**, *n.* a laminated metal of pure iron and steel, used for Damascus blades.

dame (dām), *n.* a title of honour (now applied to the wives of knights and baronets) and female members of the Order of the British Empire; a comic old woman in pantomime; (*N Am., sl.*) a woman. **dame-school**, *n.* (*formerly*) an elementary school kept by a woman.

dammar (dam'ə), *n.* a resin of various kinds from eastern conifers.

damn (dam), *v.t.* to condemn; to call down curses on; to condemn to eternal punishment; to reject as a failure. *int.* an expression of annoyance. *a.* (*coll.*) damnable. *adv.* (*coll.*) very, exceedingly. **damn all**, (*sl.*) nothing at all. **not to care, give, a damn**, to be totally unconcerned. **damnable** (-nə-), *a.* deserving damnation or condemnation; odious. **damnably** (-nə-), *adv.* **damnation** (-nā'-), *n.* condemnation to eternal punishment; eternal punishment. *int.* expressing annoyance. **damnatory** (-nə-), *a.* causing or implying condemnation. **damned** (damd), *a.* condemned to everlasting punishment; hateful, execrable; damnable, infernal. *adv.* confoundedly; very. **a damned good try**, (*coll.*) an exceedingly good try. **damnedest, damndest** (dam'dist), *n.* (*coll.*) the best; (*coll.*) extraordinary. **to do one's damnedest**, to do one's very best. **damnify** (-ni-), *v.t.* (*Law*) to cause damage to. **damnification** (-nifikā'-),

n. **damning** (dam'ing), *a.* proving guilty.

damosel (dam'əzel), DAMSEL.

damp (damp), *a.* moist, humid; admitting moisture; clammy. *n.* humidity, moisture in a building or article of use or in the air; discouragement, chill; subterranean gases met with in mines. *v.t.* to moisten; to check, to depress; to discourage, to chill, to deaden. **to damp down**, to make (a fire) burn more slowly. **dampcourse**, *n.* a layer of impervious material put between the courses of a wall to keep moisture from rising. **damp-proof**, *a.* impenetrable to moisture. *v.t.* to render impervious to moisture. **dampen**, *v.t.* to make damp; to dull, to deaden, to deject. *v.i.* to become damp. **damper**, *n.* a valve or sliding plate in a flue for regulating a fire; (*Austral.*) bread or cake baked in hot ashes; (*Mus.*) a padded finger in a piano for deadening the sound; a mute in brass wind instruments. **to put a damper on**, to discourage, to stifle; to reduce the chances of success of. **damping**, *n.* (*Motor*) the deadening of the shock of sudden movement; the rate at which an electrical oscillation dies away. **dampish**, *a.* **damply**, *adv.* **dampness**, *n.*

damsel (dam'zl), *n.* (*Archaic*) a young unmarried woman; a female attendant. **damselfly**, *n.* an insect similar to a dragonfly but which folds its wings above the body when at rest.

damson (dam'zən), *n.* a small black plum, *Prunus domestica*, var. *damascena;* the tree that bears this. *a.* damson-coloured. **damson cheese**, *n.* a conserve of damsons, pressed to the consistency of cheese. **damson plum**, *n.* a large kind of damson.

dan (dan), *n.* in judo and karate, any of the black-belt grades of proficiency; a person who has reached such a level.

dance (dahns), *v.i.* to move or trip, usu. to music with rhythmical steps, figures and gestures; to skip, to frolic; to move in a lively or excited way; to bob up and down; to be dangled; to exult, to triumph. *v.t.* to express or accomplish by dancing; to perform (a particular kind of dance); to toss up and down, to dandle; to cause to dance. *n.* a rhythmical stepping with motions of the body, usu. adjusted to the measure of a tune; the tune by which such movements are regulated; a figure or set of figures in dancing; a dancing-party, a ball. **dance of death** DEATH. **St Vitus's dance**, CHOREA. **to dance attendance on**, to pay assiduous court to; to be kept waiting by. **to lead one a dance**, to cause one trouble or delay in the pursuit of an object. **dancer**, *n.* one who dances, esp. one who earns money by dancing in public. **dancing**, *n., a.* **dancing-girl**, *n.* a professional female dancer. **dancing-master**, *n.* one who teaches dancing.

D and C, dilatation of the cervix and curettage of the uterus, to cure menstrual disorders etc.

dandelion (dan'dilian), *n.* a well-known composite plant, *Taraxacum dens leonis*, with a yellow rayed flower and toothed leaves.

dander (dan'də), *n.* (*coll.*) temper, anger. **to get one's dander up, to have one's dander raised**, to get into a passion.

Dandie Dinmont (dan'di din'mənt), *n.* a type of short-legged, rough-coated terrier.

dandify (dan'difī), *v.t.* to make smart, or like a dandy. **dandification** (-fikā'-), *n.*

dandle (dan'dl), *v.t.* to dance up and down on the knees or toss in the arms (as a child); to pet. **dandler**, *n.*

dandruff (dan'drŭf), *n.* scaly scurf on the head.

dandy (dan'di), *n.* a man extravagantly fond of dress; a fop. *a.* neat, spruce, smart; (*esp. N Am.*) very good, fine. **dandy-brush**, *n.* a hard whalebone brush for cleaning horses. **dandy-roll**, *n.* a roller used to produce watermarks on paper. **dandify**, *v.t.* to dress up like a dandy. **dandyish**, *a.* **dandyism**, *n.*

Dane (dān), *n.* a native of Denmark; (*Hist.*) one of the

Northmen who invaded Britain in the Middle Ages. **Great Dane**, a Danish breed of large, short-haired dogs.

danegeld (dān'geld), *n.* an annual tax formerly levied on every hide of land in England to maintain forces against or furnish tribute to the Danes (finally abolished by Stephen).

Danelaw, Danelagh (dān'law), *n.* the portion of England allotted to the Danes by the treaty of Wedmore (AD 878), extending north-east from Watling Street.

danger (dān'jə), *n.* risk, peril, hazard; exposure to injury or loss; anything that causes peril. **in danger of**, liable to. **on the danger list**, dangerously ill (in hospital). **danger-money**, *n.* money paid in compensation for the risks involved in any unusually dangerous job. **danger-signal**, *n.* a signal on railways directing stoppage or cautious progress. **dangerous**, *a.* **dangerously**, *adv.*

dangle (dang'gl), *v.i.* to hang loosely; to swing or wave about. *v.t.* to cause to dangle; to hold out (as a temptation, bait etc.). **dangler**, *n.*

Danish (dā'nish), *a.* pertaining to Denmark or the Danes. *n.* the Danish language (*coll.*) a Danish pastry. **Danish blue**, *n.* a strong-tasting, blue-veined cheese. **Danish pastry**, *n.* a flaky pastry usu. filled with jam, almonds or apples and often iced.

dank (dangk), *a.* damp, moist; chilly with moisture. **dankish**, *a.* **dankly**, *adv.* **dankness**, *n.*

danse macabre DANCE OF DEATH under DEATH.

danseuse (dāsœz'), *n.* a female professional dancer. [F, fem. of *danseur*]

Dantean (dan'tiən), *a.* relating to Dante; in the style of Dante, esp. of his *Inferno;* sombre, sublime. *n.* a student of Dante. **Dantesque** (-tesk'), *a.* Dantean.

dap (dap), *v.i.* (*past, p.p.* **dapped**) to fish by letting the bait fall gently into the water.

Daphne (daf'ni), *n.* one of Diana's nymphs, fabled to have been changed into a laurel; a genus of shrubs, partly evergreen, allied to the laurel; (**daphne**) a plant of this genus. **daphnin**, *n.* the bitter principle obtained from species of Daphne.

dapper (dap'ə), *a.* spruce, smart, brisk, active. **dapperly**, *adv.* **dapperness**, *n.*

dapple (dap'l), *n.* a spot on an animal; a mottled marking; a horse or other animal with a mottled coat. *a.* spotted; variegated with streaks or spots. *v.t.* to spot, to streak, to variegate. *v.i.* to become dappled. **dapple-grey**, *n.* a horse with a mottled grey coat.

darbies (dah'biz), *n.pl.* (*sl.*) handcuffs.

Darby and Joan (dah'bi ənd jōn), an elderly married couple living in domestic bliss. **Darby and Joan club**, a club for elderly people.

dare (deə), *v.i.* (*past, conditional* **durst** (dœst), **dared**) to venture; to have the courage or impudence; to be able, willing or ready; to be bold or adventurous. *v.t.* to attempt, to venture on; to challenge, to defy. *n.* a challenge esp. to prove one's courage. **I dare say**, I suppose. **dare-devil**, *n.* a fearless, reckless fellow. **dare-devilry**, *n.* **daring**, *a.* courageous, bold; fearless, presumptuous. *n.* boldness, bravery; presumption. **daringly**, *adv.*

Darjeeling (dahjēling), *n.* high-quality tea from the mountains around Darjeeling in N India.

dark (dahk), *a.* destitute of light; approaching black; shaded; swarthy, brown-complexioned; opaque; gloomy, sombre; obscure, ambiguous; hidden, concealed; without spiritual or intellectual enlightenment; wicked, evil; cheerless; sad, sullen, frowning; of a theatre, not presenting a play etc. *n.* darkness; absence of light; night, nightfall; shadow, shade; dark tint, the dark part of a picture. **in the dark**, in ignorance, uninformed. **to keep dark**, to keep secret. **Dark Ages**, *n.pl.* the period from the 5th to the 10th cent. (from an incorrect view of the ignorance then prevailing). **Dark Blues**, *n.pl.* the representatives of Oxford University in sporting events. **Dark Continent**, *n.* Africa, esp. in the period before it was explored. **dark-eyed**, *a.* having dark-coloured eyes. **dark horse**, *n.* one who keeps his/her own counsel; a person of unknown capabilities. **dark lantern**, *n.* a lantern that can be obscured at pleasure. **dark room**, *n.* a room from which actinic light is shut out for photographic work. **dark star**, *n.* one emitting no light, whose existence is known only from its radio waves, infrared spectrum or gravitational effect. **darken**, *v.i.* to become dark or darker; to become gloomy or displeased. *v.t.* to make dark or darker; to render gloomy, ignorant or stupid; to obscure; to sully. **not to darken one's door**, not to appear as a visitor. **darkish**, *a.* **darkling** (-ling), *adv.* (*poet.*) in the dark. *a.* gloomy, dark; in the dark; obscure. **darkly**, *adv.* **darkness**, *n.* the state or quality of being dark; blindness; obscurity; ignorance; wickedness; the powers of hell. **Prince of Darkness**, Satan. **darksome** (-səm), *a.* dark, gloomy.

darky (dah'ki), *n.* (*coll. offensive*) a Negro.

darling (dah'ling), *n.* one who is dearly beloved; a favourite, a pet. *a.* dearly beloved; (*coll.*) charming, delightful.

darn[1] (dahn), *v.t.* to mend by imitating the texture of the material of the garment etc. *n.* a place mended by darning. **darner**, *n.* **darning**, *n., a.* **darning-needle**, *n.* a needle used in darning.

darn[2] (dahn), *v.t.* a mild form of imprecation.

darnel (dah'nəl), *n.* a kind of grass, *Lolium temulentum,* formerly believed to be poisonous, which grows among corn; the genus *Lolium.*

dart (daht), *n.* a small pointed missile used in the game of darts; (*poet.*) an arrow; a javelin; a sudden leap or rapid movement; a needle-like object which one snail fires at another during mating for sexual stimulation; (*Dressmaking*) a V-shaped tuck. *v.t.* to throw; to shoot or send forth suddenly. *v.i.* to run or move swiftly. **dart-board**, *n.* a marked target used in the game of darts. **darter**, *n.* any species of *Plotus,* a genus of long-necked swimming birds; (*pl.*) the order Jaculatores, comprising the kingfishers and bee-eaters; the archer-fish. **darts**, *n.pl.* an indoor game of throwing small darts at a marked target.

Darwinian (dahwin'iən), *a.* pertaining to Charles *Darwin* (1809–82), or to Darwinism. *n.* a believer in Darwinism. **Darwinianism, Darwinism** (dah'-), *n.* the teaching of Charles Darwin, esp. the doctrine of the origin of species by natural selection. **Darwinist**, *n.* **Darwinistic** (-is'-), *a.* **Darwinite** (dah'winit), *n.* **Darwinize, -ise** (dah'-), *v.t., v.i.*

dash (dash), *v.t.* (usu. with *to pieces*) to break by collision; (usu. with *out, down, away* etc.) to strike; to knock; to cause to come into collision; to throw violently or suddenly; to bespatter, to besprinkle; to dilute or adulterate by throwing in some other substance; (with *off*) to compose or sketch hastily; to obliterate with a stroke; to destroy; to frustrate; to discourage, to daunt; (*sl.*) to confound (as a mild imprecation). *v.i.* to rush, fall or throw oneself violently; to strike against something and break; (usu. with *up, off* or *away*) to run, ride or drive smartly; to move or behave showily or spiritedly. *n.* a rapid movement; a rush, an onset; a slight admixture; a splash; activity, daring; brilliancy, display, ostentation; a mark (–) denoting a break in a sentence, a parenthesis or omission; a hasty stroke with a pen etc.; a short stroke placed above notes or chords, directing that they are to be played staccato; the long element in Morse code; (*Athletics*) a sprint; (*coll.*) a dashboard. **to cut a dash**, to make an impression. **dashboard**, *n.* a splashboard; a fascia in front of the driver of a car. **dash-light**, *n.* a light illuminating the dashboard of a car. **dasher**, *n.* **dashing**, *a.* daring, spirited; showy, smart. **dashingly**, *adv.* **dashy**, *a.* ostentatious, showy, smart.

dashiki (dahshē'ki), *n.* colourful, loose-fitting, pull-on shirt worn esp. by black people.

dassie (das'i), *n.* (*S Afr.*) a hyrax.

dastard (das'tǝd), *n.* a cowardly villain. **dastardly**, *a.* cowardly and malicious. **dastardliness**, *n.*
Dasypodidae (dasipod'idē), *n.* the South American family of armadillos.
Dasyure (das'iūǝ), *n.* a genus of small marsupials found in Australia, Tasmania and New Guinea.
data (dā'tǝ), *n.pl.* (*often now sing. in constr.*; *pl. of* DATUM) facts or information from which other things may be deduced; the information operated on by a computer program. **data-bank, -base**, *n.* a large amount of information, usu. stored in a computer for easy access. **data-capture**, *n.* the conversion of information into a form which can be processed by a computer. **data-processing**, *n.* the handling and processing of data in computer files.
datable DATE [1].
date[1] (dāt), *n.* a particular day of the month; the day or year when anything happened or is appointed to take place; the specification of this in a book, inscription, document or letter; (*coll.*) a social or other engagement (usu. with one of the opposite sex); the person thus concerned; period, age, duration; conclusion. *v.t.* to put the date on; to find the date of; to reveal the age of; to make an appointment with (esp. a member of the opposite sex); to go out regularly with. *v.i.* to have originated at; to become old-fashioned; to go out regularly together. **out of date**, dated; obsolete. **to make, have, a date**, (*coll.*) to make or have an appointment. **up to date**, (*coll.*) recent, modern. **date-line**, *n.* the line on either side of which the date differs; line with date and place of sending printed above a newspaper dispatch. **date-mark, -stamp**, *n.* a stamp on perishable goods showing the date before which they are best used or consumed. *v.t.* to mark (goods) in this way. **datable**, *a.* **dated**, *a.* old-fashioned. **dateless**, *a.* **dater**, *n.*
date[2] (dāt), *n.* the fruit of the date-palm, an oblong fruit with a hard seed or stone. **date-palm, -tree**, *n. Phoenix dactylifera*, the palm-tree of Scripture, common in N Africa and Asia Minor.
dative (dā'tiv), *a., n.* (denoting) the grammatical case used to represent the indirect object, or the person or thing interested in the action of the verb. **datival** (-tī'-), *a.* **datively**, *adv.*
datum (dā'tǝm), *n.* (*pl.* **data** (-tǝ)) a quantity, condition, fact or other premise, given or admitted, from which other things or results may be found (cp. DATA). **datum-line, -level, -plane**, *n.* the horizontal line, such as sea-level, from which calculations are made in surveying etc.
Datura (dǝtūǝ'rǝ), *n.* a genus of solanaceous plants, containing the thorn-apple, *D. stramonium*, which yields a powerful narcotic. **daturine** (-rin, -rin), *n.* an alkaloid obtained from the thorn-apple.
daub (dawb), *v.t.* to smear or coat with a soft adhesive substance; to paint coarsely; to stain, to soil. *v.i.* to paint in a crude or inartistic style. *n.* a smear; a coarse painting; a plaster or mud wall-covering. **dauber**, *n.* **daubing**, *n.* **daubster** (-stǝ), *n.*
daughter (daw'tǝ), *n.* a female child with relation to its parents; a female descendant; a female member of a family, race, city etc.; (*Biol.*) a cell formed from another of the same type; a nuclide formed from another by radioactive decay. **daughter-in-law**, *n.* a son's wife. **daughterhood** (-hud), *n.* **daughterly**, *a.* **daughterliness**, *n.*
daunt (dawnt), *v.t.* to intimidate, to dishearten. **dauntless**, *a.* fearless, intrepid. **dauntlessly**, *adv.* **dauntlessness**, *n.*
dauphin (dō'fī, daw'fin), *n.* the title of the heir-apparent to the French throne. **dauphiness**, *n. fem.* **dauphine** (-fēn), *n.* the wife of the dauphin.
davenport (dav'npawt), *n.* a small writing-desk with drawers on both sides; (*esp. N Am.*) a large sofa, a couch.
davit (dav'it), *n.* one of a pair of beams projecting over a ship's side, with tackles to hoist or lower a boat.
Davy Jones (dā'vi jōnz), *n.* an imaginary malign spirit with power over the sea. **Davy Jones's locker**, the sea as the tomb of the drowned.
Davy lamp (dā'vi), *n.* a miner's wire-gauze safety-lamp. [the inventor, Sir Humphry *Davy*, 1778–1829]
dawdle (daw'dl), *v.i.* to move slowly; to lag behind; to waste time. **dawdler**, *n.*
dawn (dawn), *v.i.* to grow light, to break (as day); to begin to open, expand or appear. *n.* the break of day; the first rise or appearance. **to dawn upon**, to be realized gradually by. **dawn-chorus**, *n.* the singing of birds at dawn. **dawning**, *n.* dawn.
day (dā), *n.* the time the sun is above the horizon; the space of twenty-four hours, commencing at midnight and called the *civil day*, as distinguished from a *mean solar day* which begins at noon; the time it takes for the earth to rotate on its axis relative to the stars (about four minutes less than the mean solar day), also called the *sidereal* or *natural day;* daylight, light, dawn, day-time; the part of the day occupied with regular activity; any specified time; an age; (*often pl.*) life, lifetime; period of vigour, prosperity or popularity; a day appointed to commemorate any event; a contest, a battle, the victory. **at the end of the day**, all things considered; when all is said and done. **better days, evil days**, a period of prosperity or of misfortune. **day and night**, throughout both day and night; always; by or in both day and night. **day by day**, daily. **Day of Judgment**, the end of the world, the Last Day. **days of grace**, a customary number of days (in England three) allowed for the payment of a note, or bill of exchange, after it becomes due. **let's call it a day**, that's all we can do today; let's bring this to an end. **one day, one of these days**, shortly; in the near future; at some unspecified time in the future. **present day**, modern times; modern. **some day**, in the future. **the other day** OTHER. **today**, this day, now. **to gain, win, the day**, to come off victor. **to name the day**, to settle the marriage date. **day-bed**, *n.* a couch, a sofa. **day-blindness**, *n.* indistinct vision by day. **daybook**, *n.* one in which the business transactions of the day are recorded. **day-boy, -girl**, *n.* a boy or girl attending a boarding school but not living in. **day-break**, *n.* the first appearance of daylight. **daycare**, *n.* the daytime supervision by trained staff of preschool children or elderly or handicapped people. **daycentre**, *n.* one providing social amenities for the elderly, handicapped etc. **day-dream**, *n.* a reverie, a castle in the air. **day-dreamer**, *n.* **day-dreaming**, *n.* **day-labourer**, *n.* an unskilled worker hired by the day. **day-lily**, *n.* a liliaceous plant of the genus *Hemerocallis*, the flowers of which last one day. **day-long**, *a.* lasting all day. *adv.* the whole day. **day-nursery**, *n.* a crèche. **day-release**, *n.* a system which frees people from work for some hours each week to follow part-time education relevant to their employment. **day-return**, *n.* a special cheap ticket for travel to a place, returning the same day. **day-room**, *n.* a common living-room in a school, hospital, hostel etc.; a ward where prisoners are confined during the day. **day-school**, *n.* a school held in the daytime, distinguished from evening-school, Sunday school or boarding-school. **day-shift**, *n.* work during the day; the group of workers undertaking such work. **day-spring**, *n.* the dawn; day-break. **daytime**, *n.* day as opposed to night. **day-to-day**, *a.* daily, routine. **day-trip**, *n.* an excursion made to and from a place in a single day. **day-tripper**, *n.*
Day-glo® (dā'glō), *n.* a type of fluorescent paint. *a.* being or resembling this type of paint; (*loosely*) of a glowingly bright colour (usu. pink, orange or green).
daylight (dā'līt), *n.* the light of day as opposed to that of the moon or artificial light; dawn; light visible through an opening; hence, an interval, a gap; a visible space; openness, publicity. **to see daylight**, to begin to understand; to draw near to the end of a task. **daylight robbery** ROBBERY under ROB. **daylight saving**, *n.* a system of advancing the

daze (dāz), *v.t.* to stupefy, to confuse, to dazzle. *n.* the state of being dazed. **dazed**, *a.* **dazedly**, *adv.*
dazzle (daz'l), *v.t.* to overpower with a glare of light; to daze or bewilder with rapidity of motion, brilliant display, stupendous number etc. *v.i.* to be dazzled; to be excessively bright. **dazzlement**, *n.* **dazzling**, *a.* that dazzles; brilliant, splendid. **dazzlingly**, *adv.*
dB, (*abbr.*) decibel(s).
DBE, (*abbr.*) Dame Commander of the (Order of the) British Empire.
DBS, (*abbr.*) direct broadcasting by satellite.
DC, (*abbr.*) (*Mus.*) da capo; direct current; (*N Am.*) District of Columbia.
DCM, (*abbr.*) Distinguished Conduct Medal.
DD, (*abbr.*) Doctor of Divinity; direct debit.
DDR, (*abbr.*) German Democratic Republic (East Germany, G *Deutsche Demokratische Republik*).
DDS, (*abbr.*) Doctor of Dental Surgery.
DDT, (*abbr.*) dichlorodiphenyltrichloroethane, an insecticide.
de-, *pref.* from; down; away; out; (*intens.*) completely, thoroughly; (*priv.*) expressing undoing, deprivation, reversal or separation.
deacon (dē'kən), *n.* in the Roman Catholic and Anglican churches, a clergyman ranking below a priest; one who superintends the secular affairs of a Presbyterian church; one who admits persons to membership, and assists at communion in the Congregational Church. **deaconess**, *n.* a female deacon; a member of a Lutheran sisterhood. **deaconship, -ry**, *n.*
deactivate (dēak'tivāt), *v.t.* to render harmless or less radioactive. **deactivation**, *n.* **deactivator**, *n.*
dead (ded), *a.* having ceased to live; having no life, lifeless; benumbed, temporarily deprived of the power of action; resembling death; unconscious or unappreciative; without spiritual feeling; cooled, abated, obsolete, effete, useless; inanimate or inorganic as distinct from organic; extinct; lustreless, motionless, soundless; flat, vapid; of a ball, out of play; certain, unerring. *adv.* absolutely, quite, completely; profoundly. *n.* the time when things are still, stillness. **dead against**, immediately against or opposite; absolutely opposed to. **dead of night**, the middle of the night. **dead on**, (*coll.*) absolutely accurate(ly). **dead to the world**, (*coll.*) fast asleep. **the dead**, dead persons. **dead-alive, dead and alive**, *a.* spiritless. **dead-beat**, *a.* quite exhausted. *n.* (*coll.*) a worthless, lazy fellow. **dead centre**, (in) the very middle. **dead certainty**, *n.* something sure to occur; also (*coll.*) **dead cert**. **dead-drunk**, *a.* helpless from drink. **dead duck**, *n.* (*coll.*) a person or idea doomed to failure. **dead-end**, *n.* a cul-de-sac; a situation offering no hope of progress. *a.* leading nowhere. **dead-eye**, *n.* (*Naut.*) one of the flat, round blocks having eyes for the lanyards, by which the rigging is set up. **dead-fall**, *n.* a trap with a heavy weight which falls to crush the prey. **deadhead**, *n.* (*coll.*) one who has a free pass; a stupid, unimaginative person. *v.t.* to remove withered blooms (from flowers) to encourage future growth. **Dead Heart**, *n.* (*Austral.*) the land of the Central Australian Desert. **dead heat**, *n.* a race resulting in a draw. **dead language**, *n.* a language no longer spoken. **dead letter**, *n.* a letter which cannot be delivered by the post office or returned; a law or anything that has become inoperative. **dead-lights**, *n.pl.* shutters placed over port-holes or cabin windows in rough weather. **deadline**, *n.* the time of newspapers, books etc. going to press; a fixed time or date terminating something. **deadlock**, *n.* a lock worked on one side by a handle, and on the other by a key; a complete standstill, a position whence there is no exit. **dead loss**, *n.* a loss with no compensation whatever; a useless person, thing or situation. **dead man**, *n.* (*sl.*) an empty wine bottle. **dead man's, men's, fingers**, various species of orchis, the wild arum and other flowers; the zoophyte *Alcyonium digitatum*, also called **dead man's hand** or **dead man's toes**. **dead man's handle**, device for automatically cutting off the current of an electrically driven vehicle if the driver releases his pressure on the handle. **dead men's shoes** SHOE. **dead march**, *n.* a piece of solemn music played at funerals (esp. of soldiers). **dead-nettle**, *n.* a non-stinging labiate plant, like a nettle, of several species belonging to the genus *Lamium*. **deadpan**, *a.* (of the face) expressionless. **dead reckoning**, *n.* the calculation of a ship's position from the log and compass, when observations cannot be taken. **dead ringer**, *n.* (*coll.*) a person or thing exactly resembling someone or something else. **dead set**, *n.* a determined attempt (to captivate). **dead set on**, determined (to). **dead-water**, *n.* water that is absolutely still; the eddy under the stern of a ship or boat. **dead weight**, *n.* a mass of inert matter; the difference between a ship's loaded and unloaded weight; any very heavy weight or load. **dead-wood**, *n.* a useless person etc. **deadness**, *n.*
deaden (ded'n), *v.t.* to diminish the vitality, brightness, sensitivity or power of; to soundproof.
deadly (ded'li), *a.* causing or procuring death; fatal; like death; implacable, irreconcilable; unerring; intense; (*coll.*) very boring. *adv.* as if dead; extremely, intensely. **deadly nightshade**, *n.* a poisonous shrub with dark purple berries, *Atropa belladonna*. **deadly sin**, *n.* one of the seven mortal sins. **deadliness**, *n.*
deaf (def), *a.* incapable or dull of hearing; disregarding, refusing to listen, refusing to comply; insensible (to). **deaf-and-dumb alphabet, language**, a system of signs for holding communication with deaf people. **to turn a deaf ear to**, to ignore. **deaf-aid**, *n.* a hearing-aid. **deaf-mute**, *n.* one who is deaf and dumb. **deaf-mutism**, *n.* **deafen**, *v.t.* to make wholly or partially deaf; to stun with noise. **deafening**, *a.* **deafly**, *adv.* **deafness**, *n.*
deal[1] (dēl), *n.* an indefinite quantity; the distribution of cards to the players; a bargain, transaction or agreement (usu. beneficial to both parties); treatment. *v.t.* (*past, p.p.* **dealt** (delt)) to award as his/her proper share to someone; to distribute or give in succession (as cards); to inflict, as a blow. *v.i.* to distribute cards to the players; to sell illegal drugs. **a deal**, (*coll.*) a good amount. **a great, good, deal**, a large quantity; to a large extent; by much, considerably. **a raw deal**, harsh, unfair treatment. **big deal** BIG. **to deal in**, to be engaged in; to trade in. **to deal with**, to have to do with; to punish; to behave towards; to take action in respect of, to handle; to do business with. **dealer**, *n.* a trader, a merchant; a drug-pusher; one who deals the cards. **dealership**, *n.* a dealer's premises. **dealing**, *n.* conduct towards others; intercourse in matters of business; traffic.
deal[2] (dēl), *n.* a plank of fir or pine wood; fir or pine wood.
dean[1] (dēn), *n.* an ecclesiastical dignitary presiding over the chapter of a cathedral or collegiate church; a rural dean; a resident fellow in a university with disciplinary and other functions; the head of a faculty; a doyen. **Dean of Faculty**, (*Sc.*) the president of the Faculty of Advocates. **deanery**, *n.* the office, district or official residence of a dean. **deanship**, *n.* the office or personality of a dean.
dean[2], **dene**[1] (dēn), *n.* a valley; a deep and narrow valley (chiefly in place-names).
dear (diə), *a.* beloved, cherished; precious, valuable; costly, of a high price; characterized by high prices; a conventional form of address used in letter-writing. *n.* a darling, a loved one; a cherished person, a favourite; a term of endearment. *adv.* dearly, at a high price. *int.* expressing distress, sympathy or mild astonishment and protest. **Dear John letter**, (*esp. N Am., coll.*) a letter, esp.

dearth | **decalcify**

from a woman to a man, ending a relationship. **dearly**, *adv.* **dearness**, *n.* **deary, dearie** (-ri), *n.* (*dial., coll.*) a term of endearment.
dearth (dœth), *n.* scarcity.
death (deth), *n.* extinction of life; the act of dying; the state of being dead; decay, destruction; a cause or instrument of death; a skull or skeleton as the emblem of mortality; spiritual destruction, annihilation; capital punishment. **Black Death** BLACK. **civil death**, extinction of one's civil rights and privileges. **like grim death**, tenaciously. **to be at death's door**, to be close to death. **to be in at the death**, to be present at the finish. **to be the death of**, (*coll.*) to make (someone) 'die of laughing'; to be a source of great worry to. **to catch one's death of cold**, to catch a very bad cold. **to do to death**, to kill. **to look like death warmed up**, (*coll.*) to look very ill. **to put to death**, to execute. **death-adder**, *n.* a genus of venomous snakes. **death-angel, -cap, -cup**, *n.* a poisonous fungus. **death-bed**, *n.* the bed on which a person dies; a last illness. *a.* of or pertaining to a death-bed. **death-blow**, *n.* a mortal blow; utter ruin, destruction. **death certificate**, *n.* a document issued by a doctor certifying death and giving the cause, if known. **death duties**, *n.pl.* a tax levied on property when it passes to the next heir. **death knell**, *n.* a passing bell; something that heralds the end or destruction of something. **death-mask**, *n.* a plaster cast of the face after death. **death-rate**, *n.* the proportion of deaths in a given period in a given district. **death-rattle**, *n.* a gurgling sound in the throat of a person just before death. **death's-door**, *n.* a near approach to death. **death's-head**, *n.* a human skull, or a representation of one, as an emblem of mortality. **death's-head moth**, *Acherontia atropos*, the largest European moth, with markings on the back of the thorax faintly resembling a human skull. **death-struggle, -throe**, *n.* the agony of death. **death-trap**, *n.* a place unsuspectedly dangerous to life through insanitary or other conditions. **death-warrant**, *n.* an order for the execution of a criminal; an act or measure putting an end to something. **death-watch-beetle**, *n.* any of a genus of wood-boring beetles that make a clicking sound formerly thought to presage death. **death wish**, *n.* a desire for one's own death. **deathless**, *a.* immortal, imperishable. **deathlessly**, *adv.* **deathlessness**, *n.* **deathlike**, *a.* resembling death. **deathly**, *a.* like death; deadly; pertaining to death. *adv.* so as to resemble death.
deb (deb), short for DEBUTANTE, DEBENTURE.
debacle (dăbah'kl, di-), *n.* a breaking up of ice in a river; breaking up and transport of rocks and gravel by a sudden outburst of water; a rout, a complete overthrow; a fiasco.
debag (dēbag'), *v.t.* (*coll.*) to remove (someone's) trousers by force.
debar (dibah'), *v.i.* (*past, p.p.* **debarred**) to exclude from approach, enjoyment or action; to prohibit.
debark (dibahk'), *v.t., v.i.* to disembark. **debarkation** (dēbahkā'-), *n.*
debase (dibās'), *v.t.* to lower in condition, quality or value; to adulterate; to degrade. **debasement**, *n.* **debasingly**, *adv.*
debate (dibāt'), *v.t.* to contend about by words or arguments; to contend for; to discuss; to consider. *v.i.* to discuss or argue a point; to engage in argument. *n.* a discussion of a question; an argumentative contest; contention. **debatable**, *a.* open to discussion or argument; contentious. **debatement**, *n.* **debater**, *n.* one who takes part in a debate. **debating**, *n., a.* **debating society**, *n.* a society established for holding debates, and to improve the extempore speaking of the members.
debauch (dibawch'), *v.t.* to lead into sensuality or intemperance; to seduce from virtue; to vitiate. *n.* an act of debauchery; a carouse. **debauchee** (-chē'-, -shē'), *n.* a profligate. **debaucher**, *n.* **debauchery**, *n.* indulgence of the sensual appetites.
debenture (diben'chə), *n.* a written acknowledgment of a debt; a deed or instrument issued by a company or a public body as a security for a loan of money on which interest is payable till it is redeemed; a certificate issued by a custom-house to an importer entitling him to a refund of duty. **debenture-stock**, *n.* debentures consolidated or created in the form of stock, the interest on which constitutes the first charge on the dividend. **debentured**, *a.* secured by debenture, entitled to drawback.
debilitate (dibil'ităt), *v.t.* to weaken, to enfeeble; to enervate, to impair. **debilitating, debilitative**, *a.* **debilitation**, *n.* **debility**, *n.* weakness, feebleness.
debit (deb'it), *n.* an amount set down as a debt; the left-hand side of an account, in which debits are entered. *v.t.* to charge to as a debt; to enter on the debit side. **debit card**, *n.* one issued by a bank which enables the holder to debit a purchase to his/her account at the point of purchase.
debonair, debonnaire (debənea'-), *a.* courteous, genial; urbane; carefree. **debonairly**, *adv.* **debonairness**, *n.*
debouch (dibowch'-, -boosh'), *v.i.* to march out from a confined place into open ground; to flow out from a narrow ravine. **debouchment**, *n.*
Debrett (dəbret'), a publication listing the members of the English aristocracy.
debrief (dēbrēf'), *v.t.* to gather information from (someone, e.g. a soldier, diplomat or spy) after a mission.
debris (deb'rē, dā'-), *n.* broken rubbish, fragments; (*Geol.*) fragmentary matter detached by a rush of water.
debt (det), *n.* that which is owing from one person to another, esp. a sum of money that is owing; obligation, liability. **bad debt**, an irrecoverable debt. **debt of honour**, one which is morally but not legally binding; a gambling debt. **debt of nature**, death. **in debt**, under obligation to pay something due. **National Debt**, the debt of a nation in its corporate capacity (**funded debt**, the portion of this converted into bonds and annuities; **floating debt**, the portion repayable at a stated time or on demand). **debtless**, *a.* **debtor**, *n.* one who is indebted to another.
debug (dēbŭg'), *v.t.* to find and remove hidden microphones from; to find and remove the faults in (a system); to remove insects from.
debunk (dēbŭngk'), *v.t.* to dispel false sentiment, to destroy pleasing legends or illusions.
debus (dēbŭs'), *v.t., v.i.* (*Mil.*) to (cause to) alight from a motor vehicle.
debut (dā'bū, deb'ū), *n.* a first appearance before the public; a first attempt. *v.i.* to make a first appearance. *a.* first, constituting a debut. **debutant** (deb'ūtant), *n.* one who makes a debut at court. **debutante**, *n. fem.*
Dec., (*abbr.*) December.
dec., (*abbr.*) declaration; declension; declination; decoration; decorative; deceased; (*Mus.*) decrescendo.
dec(a)-, *comb. form* ten.
decade (dek'ād), *n.* a group of ten; a period of ten years. **decadal** (-kə-), *a.* **decadic** (-kad'-), *a.*
decadence (dek'ədəns), *n.* decay, deterioration, esp. in morality or culture; a falling-off from a high standard of excellence. **decadent**, *a.* in a state of decadence. *n.* a decadent writer or artist, esp. one who is self-indulgent, morally corrupt or unoriginal.
decaffeinate (dikaf'ināt), *v.t.* to remove the caffeine from e.g. coffee. **decaffeinated**, *a.*
decagon (dek'əgon), *n.* a plane figure with ten sides and ten angles.
decagram (dek'əgram), *n.* a weight of 10 grams, 0·353 oz.
decahedron (dekəhē'drən), *n.* a solid figure with ten sides. **decahedral**, *a.*
decal (dēk'l), *n.* a transfer. *v.t.* to transfer (a design).
decalcify (dēkal'sifī), *v.t.* to clear (bone etc.) of calcareous matter. **decalcification** (-fikā'-), *n.*

decalcomania (dikalkəmā'niə), *n.* process of transferring a design; a design so transferred.
decalitre (dek'əlētə), *n.* a liquid measure of capacity containing 10 litres, nearly 2½ gallons.
Decalogue (dek'əlog), *n.* the Ten Commandments.
decametre (dek'əmētə), *n.* a measure of length, containing 10 metres or 393·7 in.
decamp (dikamp'), *v.i.* to break camp; to depart quickly; to take oneself off.
decanal (dikā'nəl), *a.* pertaining to a dean or a deanery, or to the south side of the choir, where the dean has his seat.
decant (dikant'), *v.t.* to pour off by gently inclining, so as not to disturb the sediment; to pour from one vessel into another (as wine); to move (people) from one area to another to provide better housing etc. **decantation** (dēkantā'-), *n.* **decanter**, *n.* a vessel for decanted liquors; an ornamental glass bottle for holding wine or spirits.
decapitate (dikap'itāt), *v.t.* to behead. **decapitable**, *a.* **decapitation**, *n.*
decapod (dek'əpod), *n.* any individual of the Decapoda. *a.* pertaining to the Decapoda; having ten limbs. **Decapoda** (-kap'ədə), *n.pl.* a section of cephalopods, with two tentacles and four pairs of arms, containing the cuttlefishes, squids etc.; an order of crustaceans with five pairs of ambulatory limbs, the first pair chelate, comprising crabs, lobsters etc. **decapodal, -dous**, *a.*
decarbonate (dēkah'bənāt), *v.t.* to remove carbon dioxide from.
decarbonize, -ise (dēkah'bənīz), **decarburize, -ise** (-kah'būriz), *(coll.)* **decarb** (dē'-), *v.t.* to remove the solid carbon deposited in the combustion chamber and on the piston crown of an internal-combustion engine. **decarbonization**, *n.*
decasyllabic (dekəsilab'ik), *a.* having ten syllables. *n.* a line of ten syllables. **decasyllable** (dek'-), *n., a.*
decathlon (dikath'lon), *n.* an athletic contest consisting of ten events. **decathlete**, *n.*
decay (dikā'), *v.i.* to fall away, to deteriorate; to decline in excellence; to waste away; of radioactive matter, to disintegrate. *v.t.* to impair, to cause to fall away. *n.* gradual failure or decline; deterioration; a state of ruin; wasting away, consumption, gradual dissolution; decomposition of dead tissue, rot; decayed matter; disintegration of radioactive matter.
decease (disēs'), *n.* death. *v.i.* to die. **deceased**, *a.* dead. *n.* one lately dead.
deceit (disēt'), *n.* the act of deceiving; propensity to deceive; trickery, deception, duplicity; anything intended to mislead another. **deceitful**, *a.* **deceitfully**, *adv.* **deceitfulness**, *n.*
deceive (disēv'), *v.t.* to mislead; to cheat, to delude; to be unfaithful (of husband or wife). *v.i.* to act deceitfully. **deceivable**, *a.* **deceiver**, *n.*
decelerate (dēsel'ərāt), *v.t.* to slow down. **deceleration**, *n.*
decem-, *comb. form* ten; in or having ten parts.
December (disem'bə), *n.* the twelfth and last month of the year; orig. the tenth and afterwards the twelfth month of the Roman year. **Decemberly**, *a.* **Decembrish**, *a.* **Decembrist**, *n.* one of the conspirators against the Czar Nicholas in Dec. 1825.
decemvir (disem'və), *n. (pl. -viri* (-rī), **-virs**) one of the various bodies of ten magistrates appointed by the Romans to legislate or rule. **decemviral**, *a.* **decemvirate** (-rət), *n.* a governing body of ten persons.
decency DECENT.
decennary (disen'əri), *n.* a period of ten years. *a.* pertaining to a period of ten years. **decenniad** (-ad), **decennium** (-əm), *n. (pl. -nniads, -nnia* (-ə)) a period of ten years.
decennial, *a.* lasting ten years; occurring every ten years. **decennially**, *adv.*
decent (dē'snt), *a.* becoming, seemly; decorous; respectable; passable, tolerable; kind, obliging. **to be decent**, *(coll.)* to be sufficiently clothed to be seen in public. **decentish**, *a.* **decently**, *adv.* **decency** (dē'sənsi), *n.* conformity to accepted (moral) standards in words or behaviour; kindness, humanity.
decentralize, -ise (dēsen'trəlīz), *v.t.* to break up (as a centralized administration); to organize on the principle of local management rather than central government. **decentralization**, *n.*
deception (disep'shən), *n.* the act of deceiving; the state of being deceived; that which deceives; a deceit, a fraud. **deceptive**, *a.* tending or apt to deceive, easy to mistake. **deceptively**, *adv.* **deceptiveness**, *n.*
deci-, *pref.* a tenth part of.
decibel (des'ibel), *n.* a unit to compare levels of intensity of sound.
decide (disīd'), *v.t.* to determine; to adjudge; to settle by adjudging (victory or superiority); to bring to a decision. *v.i.* to come to a decision. **decidable**, *a.* **decided**, *a.* settled; clear, unmistakable; determined, resolute. **decidedly**, *adv.* **decider**, *n.* one who or that which decides; a deciding heat or game.
deciduous (disid'ūəs), *a.* not permanent; shed (as wings) during the lifetime of an animal; falling, not perennial (applied to leaves etc. which fall in autumn, and to trees which lose their leaves annually). **deciduous teeth**, *n.pl.* milk teeth. **decidua** (-ə), *n.* the membrane lining the internal surface of the uterus, coming away after parturition. **deciduate** (-ət), *a.* having a decidua; thrown off after birth. **deciduousness**, *n.*
decigram (des'igram), *n.* a weight equal to one-tenth of a gram, 1·54 grain.
decilitre (des'ilētə), *n.* a fluid measure of capacity of one-tenth of a litre, 0·176 pint.
decillion (disil'yən), *n.* a million raised to the tenth power, represented by 1 followed by 60 ciphers. **decillionth**, *a.*
decimal (des'iməl), *a.* of or pertaining to ten or tenths; counting by tens. *n.* a decimal fraction. **decimal coinage, currency**, *n.* a monetary system in which the basic unit is divided in multiples of ten. **decimal fraction**, *n.* a fraction having some power of 10 for its denominator, esp. when expressed by figures representing tenths, hundredths etc. following a dot (the **decimal point**) to the right of the unit figure. **decimal notation**, *n.* the Arabic system of numerals. **decimal system**, *n.* a system of weights and measures in which the values increase by multiples of ten. **decimalist**, *n.* **decimalize, -ise**, *v.t.* to reduce or adapt to the decimal system. **decimalization**, *n.* **decimally**, *adv.*
decimate (des'imāt), *v.t.* to destroy a tenth or a large proportion of; *(Mil.)* to punish every tenth man with death. **decimation**, *n.*
decimetre (des'imētə), *n.* the tenth part of a metre, 3·937 in.
decipher (disī'fə), *v.t.* to turn from cipher into ordinary language; to discover the meaning of (something written in cipher or hard to make out). **decipherability**, *n.* **decipherable**, *a.* **decipherment**, *n.*
decision (disizh'ən), *n.* the act or result of deciding; the determination of a trial, contest or question; resolution, firmness of character. **decisive** (-sī'-), *a.* having the power of deciding; conclusive, final; characterized by decision. **decisively**, *adv.* **decisiveness**, *n.*
deck (dek), *v.t.* to adorn, to beautify; to cover, to put a deck to. *n.* the plank or iron flooring of a ship; a pack (of cards), a heap, a pile (of cards); the floor of a bus or tramcar a tape deck; the platform supporting the turntable of a record player. **hurricane deck** HURRICANE. **lower deck** LOW. **main deck** MAIN. **the deck**, *(coll.)* the ground. **to clear the decks**, *(Naut.)* to prepare for action; *(fig.)* to make tidy. **to hit the deck**, *(coll.)* to fall down quickly or suddenly. **deck-chair**, *n.* a collapsible chair, camp-stool or long chair for reclining in. **deck-hand**, *n.* a seaman who

deckle

does manual work. **deck-house**, *n.* a room erected on deck. **deck-passenger**, *n.* a passenger who has no right in the cabins. **decked** (dekt), *a.* adorned; furnished with a deck or decks; (*Her.*) edged with another colour, as the feathers of a bird. **-decker**, *comb. form.* having a specified number of decks, as in *double-decker*, a bus, etc. with two decks.

deckle (dek'l), *n.* a frame used in paper-making to keep the pulp within the desired limits. **deckle-edge**, *n.* the rough, untrimmed edge of paper. **deckle-edged**, *a.* of paper or books, uncut.

declaim (diklām'), *v.t.* to utter rhetorically. *v.i.* to speak a set oration in public; to inveigh; to speak rhetorically or passionately. **declaimer**, *n.* **declamation** (dekləmā'-), *n.* the act or art of declaiming according to rhetorical rules; a formal oration; impassioned oratory. **declamatory** (-klam'-), *a.*

declaration (dekləra'shən), *n.* the act of declaring or proclaiming; that which is declared or proclaimed; the document in which anything is declared or proclaimed; a manifesto, an official announcement, esp. of constitutional or diplomatic principles, laws or intentions; (*Cricket*) a voluntary close of innings.

declare (diklēə'), *v.t.* to make known; to announce publicly, to proclaim formally; to pronounce, to assert or affirm positively. *v.i.* to make a declaration, to avow; to state the possession of (dutiable articles); (*Cards*) to name the trump suit; (*Cricket*) to announce an innings as closed. **to declare against, for**, to side against or with. **to declare an interest**, often of a Member of Parliament, to admit to a usu. financial interest in a company about which there is (parliamentary) discussion. **to declare oneself**, to avow one's intentions; to disclose one's character or attitude. **declarant**, *n.* (*Law*) one who makes a declaration. **declarative** (-klā'-), *a.* explanatory, declaratory. **declaratively**, *adv.* **declaratory** (-klā'-), *a.* making declaration; expressive, affirmatory. **declaredly** (-kleə'rid-), *adv.*

déclassé (dāklasā'), *a.* (*fem.* **-ssée**) having lost social position or estimation.

declassify (deklas'ifī), *v.t.* to remove from the security list. **declassification** (-fikā'-), *n.*

declension (diklen'shən), *n.* declining, deterioration; (*Gram.*) the case-inflection of nouns, adjectives and pronouns; the act of declining a noun etc.; a number of nouns declined in the same way.

declinable DECLINE.

declination (deklinā'shən), *n.* the angular distance of a heavenly body north or south of the celestial equator; a formal refusal. **declination of the needle, compass**, the angle between the geographic and the magnetic meridians, also called **magnetic declination**. **declination-compass**, *n.* a declinometer. **declinational**, *a.*

decline (diklīn'), *v.i.* to slope downwards; to deviate; to diminish; to fall off, to deteriorate, to decay; to approach the close; to refuse. *v.t.* to refuse; to reject; (*Gram.*) to inflect (as a noun); to recite the cases of a noun in order. *n.* deterioration, decay, diminution; fall in prices; gradual failure of strength or health; setting; gradual approach to extinction or death. **to go into a decline**, to deteriorate gradually in health. **declinable**, *a.*

declinometer (deklinom'itə), *n.* an apparatus for measuring the declination of the needle of the compass; (*Astron.*) an instrument for registering declinations.

declivity (dikliv'iti), *n.* an inclination, a slope or gradual descent of the surface of the ground; an inclination downward. **declivitous**, *a.* **declivous** (-klī'-), *a.*

declutch (deklŭch'), *v.t.* to release the clutch (of a vehicle); to disconnect the drive.

decoct (dikokt'), *v.t.* to boil down in hot water; to extract the essence of by boiling. **decoction**, *n.* the act of boiling or digesting a substance to extract its virtues; the liquor or substance obtained by boiling.

decrease

decode (dēkōd'), *v.t.* to translate from code symbols into ordinary language. **decoder**, *n.*

decoke (dēkōk'), *v.t.* to remove carbon from, to decarbonize.

décolleté (dākol'tā), *a.* (*fem.* **-tée**) wearing a low-necked dress; low-necked (of a dress). **décolletage** (-tahzh), *n.* the low-cut neckline of a dress.

decolonize, -ise (dēkol'əniz), *v.t.* to grant independence to (a colonial state).

decolour, (*N Am.*) **decolor** (dēkŭl'ə), *v.t.* to deprive of colour. **decolorant**, *a.* bleaching, blanching. *n.* a bleaching substance. **decolorate**, *v.t.* **decoloration**, *n.* **decolorize, -ise**, *v.t.* **decolorization, -isation**, *n.* **decolorizer, -iser**, *n.*

decompose (dēkəmpōz'), *v.t.* to resolve into constituent elements; to separate the elementary parts of, to analyse; to cause to rot. *v.i.* to become decomposed; to putrefy. **decomposable**, *a.* **decomposer**, *n.* **decomposition** (-kompəzish'ən), *n.*

decompress (dēkəmpres'), *v.t.* gradually to relieve pressure, to return to normal atmospheric pressure conditions. **decompression** (-shən), *n.* **decompression chamber**, *n.* one in which a person (e.g. a diver) is gradually returned to normal pressure conditions. **decompression sickness**, *n.* severe pain and breathing problems caused by sudden change in atmospheric pressure. **decompressor**, *n.* a contrivance for relieving pressure on an engine.

decongestant (dēkənjes'tənt), *a.* relieving congestion. *n.* a drug or medicine relieving nasal or chest congestion.

deconsecrate (dēkon'sikrāt), *v.t.* to deprive of consecration; to secularize. **deconsecration**, *n.*

deconstruct, *v.t.* to expose the existing structure in. **deconstruction**, *n.* a procedure in criticism which aims to expose the latent contradictions in a work of art arising from the very structure of language itself. **deconstructionism**, *n.* **deconstructionist**, *n.*, *a.*

decontaminate (dēkəntam'ināt), *v.t.* to clear of a poisonous substance or radioactivity. **decontamination**, *n.*

decontrol (dēkəntrōl'), *v.t.* to terminate government control of (a trade etc.).

decor (dā'kaw), *n.* the setting, arrangement and decoration of a scene on the stage, or of a room. [F]

decorate (dek'ərāt), *v.t.* to adorn, to beautify; to be an embellishment to; to confer a badge of honour on; to paint, paper etc. (a house). **decorated**, *a.* ornamented, possessing a medal or other badge of honour; an epithet applied to the middle pointed architecture in England (*c.* 1300–1400). **decoration**, *n.* the act of decorating; ornamentation, ornament; a badge of honour; (*pl.*) flags, flowers and other adornments put up at a church festival or on an occasion of public rejoicing. **decorative**, *a.* **decorativeness**, *n.* **decorator**, *n.* one whose business it is to paint and paper rooms or houses.

decorous (dek'ərəs), *a.* becoming, seemly; befitting, decent. **decorously**, *adv.* **decorousness**, *n.* **decorum** (-kaw'rəm), *n.* decency and propriety of words and conduct; etiquette, polite usage.

decorticate (dēkaw'tikāt), *v.t.* to strip the bark, skin or husk from. **decortication**, *n.* **decorticator**, *n.* a machine for stripping the hull from grain.

découpage (dākoopahzh'), *n.* the art of decorating furniture with cut-out patterns.

decoy (dē'koi, dikoi'), *v.t.* to lure into a trap or snare; (*fig.*) to entrap, to allure, to entice. *n.* a pond or enclosed water into which wild-fowl are decoyed; a decoy-duck; a person employed to lure or entrap; something used to distract attention. **decoy-duck**, *n.* a tame duck or an imitation of one; a duck used to lure wild-fowl into the decoy.

decrease (dikrēs'), *v.i.* to become less, to wane, to fail. *v.t.* to make less; to reduce in size gradually. *n.* (dē'-, -krēs'), lessening, diminution; the waning of the moon.

decreasingly, *adv.*
decree (dikrē'), *n.* an edict, law or ordinance made by superior authority; (*Law*) a judicial decision. *v.t.* to command by a decree; to ordain or determine; to decide by law or authoritatively. **decree absolute**, *n.* the final decree in divorce proceedings. **decree nisi**, *n.* a provisional decree in divorce proceedings.
decrement (dek'rimənt), *n.* decrease, diminution; the quantity lost by diminution; the wane of the moon; (*Radio*) a measure of the speed of damping out of damped waves. **decremeter** (-mētə), *n.* an instrument for measuring this.
decrepit (dikrep'it), *a.* broken down by age and infirmities; feeble, decayed. **decrepitude** (-tūd), *n.*
decrepitate (dikrep'itāt), *v.t.* to calcine in a strong heat, so as to cause a continual crackling of the substance. *v.i.* to crackle, as salt in a strong heat. **decrepitation**, *n.*
decrescendo (dēkrishen'dō), *n.*, *a.*, *adv.* (*Mus.*) diminuendo.
decrescent (dikres'nt), *a.* waning; (*Bot.*) decreasing gradually from base to summit.
decretal (dikrē'təl), *a.* pertaining to a decree. *n.* a decree, esp. of the Pope; (*pl.*) a collection or body of papal decrees on points of ecclesiastical law or discipline. **decretory**, *a.* judicial, deciding; determining.
decrial, decrier DECRY.
decriminalize, -ise (dēkrim'inəlīz), *v.t.* to make (an action) no longer illegal.
decry (dikrī'), *v.t.* to express strong disapproval of; to depreciate. **decrial**, *n.* **decrier**, *n.*
decumbent (dikŭm'bənt), *a.* lying down, reclining; prostrate; (*Bot.*) lying flat by its own weight. **decumbence, -ncy**, *n.*
decuple (dek'ūpl), *a.* tenfold. *n.* a tenfold number. *v.t.*, *v.i.* to increase tenfold.
decussate (dikŭs'āt), *v.t.*, *v.i.* to intersect (as nerves, lines or rays) in the form of an X. *a.* (-ət), having this form; (*Bot.*) arranged in this manner. **decussated**, *a.* crossed, intersected; of leaves, crossing each other in pairs at right angles. **decussately**, *adv.* **decussation** (dek-), *n.*
dedicate (ded'ikāt), *v.t.* to apply or give up wholly to some purpose, person or thing; to inscribe or address (as a literary work to a friend or patron); to set apart and consecrate solemnly to God or to some sacred purpose. *a.* (-kət), dedicated, consecrated. **dedicated**, *a.* devoting one's time to one pursuit or cause; of computers etc., designed to perform a specific function. **dedicatee** (-tē'), *n.* **dedication**, *n.* the act of dedicating; the words in which a book, building etc. is dedicated. **dedicative**, *a.* **dedicator**, *n.* **dedicatory**, *a.*
deduce (didūs'), *v.t.* to draw as a conclusion by reasoning, to infer. **deducible**, *a.* **deduction**, *n.* the act of deducing; an inference, a consequence. **deductive**, *a.* deduced, or capable of being deduced, from premises. **deductive reasoning**, *n.* (*Log.*) the process of reasoning by which we arrive at the necessary consequences, starting from admitted or established premises. **deductively**, *adv.* a priori.
deduct (didŭkt'), *v.t.* to take away, to subtract. **deduction**, *n.* the act of deducting; that which is deducted; abatement.
dee (dē), *n.* the fourth letter of the alphabet, D, d; anything shaped like a capital D.
deed (dēd), *n.* an action, a thing done with intention; an illustrious exploit, an achievement; fact, reality (see IN-DEED); (*Law*) an instrument comprehending the terms of a contract, and the evidence of its due execution. *v.t.* (*N Am.*) to transfer or convey by deed. **deed-poll**, *n.* (*Law*) a deed made by one person only esp. when that person wishes to change his or her name.
deejay (dē'jā), *n.* a written form of the abbreviation **DJ** (disk jockey).

deem (dēm), *v.t.* to suppose; to think; to judge, to consider. **deemster** (-stə), *n.* one of two officers who officiate as judges in the Isle of Man.
deep (dēp), *a.* extending far down; extending far in from the surface or away from the outside; having a thickness or measurement back or down; dark-coloured, intensely dark; profound, intellectually demanding, abstruse; heartfelt, grave, earnest; intense, extreme, heinous; sonorous, low in pitch; sagacious; (*coll.*) artful, scheming, secretive; near the limits of the playing area. *adv.* deeply, far down; far on; profoundly, intensely. *n.* the sea; (*usu. pl.*) the deep parts of the sea; a deep place, a cavity; the abyss of space; (*Cricket*) the outfield. **deep litter egg**, one produced by hens living in sheds whose floors are thickly covered in straw or peat. **to go off the deep end**, to give way to one's anger. **deep-freeze**, *n.* (a container for) the storage of foods and perishable goods at a very low temperature. *v.t.* to freeze or keep in a deep freeze. **deep-fry**, *v.t.* to fry (food) submerged in fat or oil. **deep-laid**, *a.* profoundly, secretly or elaborately schemed. **deep-rooted**, *a.* firmly established. **deep-sea**, *a.* pertaining to the open sea or the deeper parts of it. **deep-seated**, *a.* profound; situated far in; firmly seated. **deep-set**, *a.* of eyes, deeply set in the face. **deep space**, *n.* that area of space beyond the earth and the moon. **deep structure**, *n.* in generative grammar, the underlying structure of a sentence in which logical and grammatical relationships become clear. **deep-toned**, *a.* emitting a low, full sound. **deepen**, *v.t.* to make deeper. *v.i.* to become deeper. **deeply**, *adv.* **deepmost**, *a.* **deepness**, *n.*
deer (diə), *n.* any of the Cervidae, ruminant quadrupeds, only the males horned, except in the one domesticated species, the reindeer. **deer-forest**, *n.* a tract of wild land on which red deer are bred for stalking. **deer-hound**, *n.* a large greyhound with a rough coat, formerly used for hunting deer. **deer-lick**, *n.* a wet or marshy spot impregnated with salt where deer come to lick. **deerskin**, *n.* the skin of a deer; leather made therefrom. *a.* of this material. **deer-stalker**, *n.* one who hunts deer by stalking; a cap peaked in front and behind.
de-escalate (dēes'kəlāt), *v.t.* to reduce the intensity of. **de-escalation**, *n.*
def (def), *a.* (*sl.*) very good; brilliant.
def., (*abbr.*) defendant; defined; definite; definition.
deface (difās'), *v.t.* to disfigure; to spoil the appearance or beauty of. **defaceable**, *a.* **defacement**, *n.* **defacer**, *n.*
de facto (dā fak'tō), in reality, actually, although not necessarily legally.
defaecation (defəkā'shən), DEFECATION.
defalcate (dē'falkāt), *v.t.* to misappropriate (money etc.) held in trust, to embezzle. *v.i.* to commit embezzlement. **defalcation**, *n.* **defalcator**, *n.*
defame (difām'), *v.t.* to speak evil of maliciously; to slander, to libel. **defamation** (defəmā'-), *n.* **defamatory** (-fam'-), *a.*
default (difawlt'), *n.* want, lack, absence; omission or failure to do any act; neglect; (*Law*) failure to appear in court on the day assigned; failure to meet liabilities; an assumption made by a computer if no alternative instructions are given. *v.i.* to fail in duty; to fail to meet liabilities; (*Law*) to fail to appear in court. *v.t.* (*Law*) to enter as a defaulter and give judgment against, in case of nonappearance. **in default of**, instead of (something lacking). **judgment by default**, (*Law*) decree against a defendant who does not appear. **defaulter**, *n.* one who fails to account for moneys entrusted to him; one who is unable to meet his engagements (esp. on the Stock Exchange or turf); (*Law*) one who fails to appear in court; (*Mil.*) a soldier guilty of a military offence.
defeasance (difē'zəns), *n.* the act of annulling a contract; (*Law*) a condition relating to a deed which being performed renders the deed void. **defeasible**, *a.* that may

be annulled or forfeited. **defeasibility** (bil'-), *n*.
defeat (difēt'), *v.t.* to win a victory over; to resist successfully, to frustrate; to render null; to baffle. *n*. overthrow, discomfiture, esp. of an army; the loss of a contest; a rendering null. **defeatism**, *n*. persistent belief in and resignation to the likelihood of defeat. **defeatist**, *n*.
defecate (def'əkāt), *v.t.* to purify from lees, dregs or other impurities. *v.i.* to become clear by depositing impurities, excrement etc.; to eject faeces from the body. **defecation**, *n*. **defecator**, *n*.
defect[1] (dēfekt'), *n*. absence of something essential to perfection or completeness; blemish, failing; moral imperfection. **defects of one's qualities**, (*coll.*) shortcomings that usually correspond to the particular abilities or good points one possesses. **defective** (difek'-), *a*. imperfect, incomplete, faulty; wanting in something physical or moral; (*Gram.*) lacking some of the forms or inflections. **defectively**, *adv*. **defectiveness**, *n*.
defect[2] (difekt'), *v.i.* to desert one's country or cause for the other side. **defection**, *n*. desertion, apostasy. **defector**, *n*.
defence, (*esp. N Am.*) **defense** (difens'), *n*. the state or act of defending; that which defends; fortifications, fortified posts; military capability and resources; justification, vindication; excuse, apology; (*Law*) the charges; the defending party in defendant's reply to a trial; the defensive players in a team. **line of defence**, a succession of fortified places, forming a continuous line; a thing, institution, argument etc. which acts as a defence. **defence mechanism**, *n*. (*Psych.*) a usually unconscious mental adjustment for excluding from the consciousness matters the subject does not wish to receive. **defenceless**, *a*. **defencelessly**, *adv*. **defencelessness**, *n*. the state of being undefended.
defend (difend'), *v.t.* to protect, to guard; to shield from harm; to keep safe against attack; to support, to maintain by argument, to vindicate; (*Law*) to plead in justification of. *v.i.* to plead on behalf of the defendant; to play in defence. **defendable**, *a*. **defendant**, *n*. (*Law*) one summoned into court to answer some charge; one sued in a law-suit. *a*. holding this relationship. **defender**, *n*. one who defends. **Defender of the Faith**, a title bestowed by Pope Leo X on Henry VIII, in 1521, for his defence of the Roman Church against Luther, and since borne by English sovereigns. **defensible**, *a*. **defensibility** (-bil'-), *n*. **defensibly**, *adv*. **defensive**, *a*. serving to defend; entered into or carried on in self-defence; protective, not aggressive. *n*. an attitude or condition of defence. **to be, act, stand, on the defensive**, to be, act or stand in a position to repel attack. **defensive medicine**, *n*. diagnostic procedures carried out by doctors whether absolutely necessary or not in order to avoid any future legal indictment. **defensively**, *adv*.
defer[1] (difœr'), *v.t.* (*past, p.p.* **deferred**) to put off; to postpone. *v.i.* to delay; to procrastinate. **deferred pay**, *n*. wages or salary, esp. of a soldier, held over to be paid at his discharge or death. **deferment**, *n*.
defer[2] (difœr'), *v.i.* (*past, p.p.* **deferred**) to yield to the opinion of another. **deference** (def'ə-), *n*. submission to the views or opinions of another; compliance; respect, regard; courteous submissiveness. **deferent**[1], *a*. deferential. **deferential** (defəren'-), *a*. showing deference. **deferentially**, *adv*.
deferent[2] (def'ərənt), *n*. that which carries or conveys; (*Physiol.*) a vessel or duct conveying fluids. *a*. (*Physiol.*) conveying fluids.
defeudalize, -ise (dēfū'dəliz), *v.t.* to deprive of feudal character or form.
defiance DEFY.
defibrillator (dēfib'rilātə, -fī'-), *n*. a machine used to apply an electric current to the chest and heart area to stop fibrillation of the heart.
deficiency (difish'ənsi), *n*. a falling short; deficit, lack, want, insufficiency; the amount lacking to make complete or sufficient. **deficiency disease**, *n*. one due to lack or insufficiency of one or more of the essential food constituents. **deficient**, *a*. defective; falling short; not fully supplied. **deficiently**, *adv*. **deficit** (def'isit), *n*. a falling short of revenue as compared with expenditure; the amount of this deficiency; the amount required to make assets balance liabilities. **deficit spending**, *n*. a remedy for economic depression whereby the government increases its expenditure and finances the resulting budget deficit by loans.

defier DEFY.
defile[1] (difīl'), *v.t.* to make foul or dirty; to soil, to stain; to corrupt the chastity of, to violate; to desecrate, to make ceremonially unclean. **defilement**, *n*.
defile[2] (difīl', dē'-), *v.i.* to march in a file or by files. *n*. a long, narrow pass or passage, as between hills, along which men can march only in file; a gorge.
define (difīn'), *v.t.* to determine the limits of; to mark out, to fix with precision (as duties etc.); to state the meaning of, to describe a thing by its qualities and circumstances. **definable**, *a*. **definably**, *adv*. **definite** (def'init), *a*. fixed precisely; exact, distinct, clear; decided, obvious; certain, positive; (*Gram.*) indicating exactly, limiting, defining. **definite article**, *n*. the. **definitely**, *adv*. **definiteness**, *n*. **definition** (definish'ən), *n*. the act of defining; a statement of the meaning (of a word) an exact description of a thing by its qualities and circumstances; (*Log.*) an expression which explains a term so as to distinguish it from everything else; an enumeration of the constituents making up the logical essence; distinctness, clearness of form, esp. of an image transmitted by a lens or a television image. **definitive** (-fin'-), *a*. decisive; conclusive; most reliable and authoritative; of postage stamps, permanently on sale. *n*. a word used to limit the application of a common noun, as an adjective or pronoun. **definitively**, *adv*.
deflagrate (def'ləgrāt), *v.t.* to consume by means of rapid combustion. *v.i.* to be consumed by means of rapid combustion. **deflagration**, *n*.
deflate (diflāt'), *v.t.* to let down (a pneumatic tyre, balloon etc.) by allowing the air to escape; to reduce the inflation of currency; to puncture a person's self-importance or self-confidence abruptly. **deflater, deflator**, *n*. **deflation**, *n*. reduction of size by allowing air to escape; the reduction and control of the issue of paper money, causing prices to fall. **deflationary**, *a*.
deflect (diflekt'), *v.i.* to turn or move to one side, to deviate. *v.t.* to cause to turn or bend. **deflector**, *n*. **deflection**, *n*. **deflexion** (-shən), **deflexure** (-shə), *n*.
deflower (diflow'ə), *v.t.* to deprive of virginity, to ravish; to strip of its bloom; to remove the flowers from. **deflowerer**, *n*.
defoliate (dēfō'liāt), *v.t.* to deprive of leaves. **defoliant**, *n*. a chemical used to remove leaves. **defoliation**, *n*. the fall or shedding of leaves.
deforest (dēfo'rist), *v.t.* to clear of forest. **deforestation**, *n*.
deform (difawm'), *v.t.* to render ugly or unshapely; to disfigure, to distort; to mar, to spoil. **deformation** (defawmā'-), *n*. the act or process of deforming; a disfigurement, perversion or distortion; a change for the worse as opp. to reformation; alteration in the structure and external configuration of the earth's crust through the action of internal forces. **deformed**, *a*. disfigured, ugly, misshapen. **deformer**, *n*. **deformity**, *n*. the state of being deformed; a disfigurement, a malformation; that which mars or spoils the beauty of a thing.
defraud (difrawd'), *v.t.* to deprive of what is right by deception; to cheat. **defrauder**, *n*.
defray (difrā'), *v.t.* to pay; to bear the charge of; to settle. **defrayable**, *a*. **defrayal**, *n*. **defrayment**, *n*.
defrock (dēfrok'), *v.t.* to unfrock (a priest etc.).
defrost (dēfrost'), *v.t.* to remove frost from; to thaw.

deft — 210 — **deliberate**

defroster, *n*. a device for defrosting a windscreen or a refrigerator.
deft (deft), *a*. neat in handling; dexterous, clever. **deftly**, *adv*. **deftness**, *n*.
defunct (difŭngkt'), *a*. dead, deceased; no longer in operation. *n*. a dead person. **defunctive**, *a*. funereal.
defuse (dēfūz'), *v.t.* to render (a bomb) harmless by removing the fuse; to dispel the tension of (a situation).
defy (difī'), *v.t.* to challenge to do or substantiate; to disregard or oppose openly; to resist, to baffle. **defiance**, *n*. challenge to battle, single combat or any contest; contemptuous disregard; opposition; open disobedience. **defiant**, *a*. challenging; openly disobedient; hostile in attitude. **defiantly**, *adv*. **defier**, *n*.
deg., (*abbr.*) degree (of temperature).
dégagé (dăgahzhä'), *fem*. **-gée**, *a*. easy, unembarrassed, unconstrained. [F]
de-gauss (dēgows'), *v.t.* (*Elec.*) to neutralize the magnetization of, e.g. a ship, by the installation of a current-carrying conductor.
degenerate (dijen'ərət), *a*. fallen off from a better to a worse state; sunk below the normal standard; declined in natural or moral growth. *n*. a degenerate person or animal. *v.i.* (-āt), to fall off in quality from a better to a worse physical or moral state; to deteriorate; (*Biol.*) to revert to a lower type; to become wild. **degeneracy**, *n*. **degenerately**, *adv*. **degeneration**, *n*. the act or process of degenerating; the state of being degenerated; the return of a cultivated plant to the wild state; (*Bot.*) transition to an abnormal state; gradual deterioration of any organ or class of organisms. **degenerative**, *a*.
deglutition (dēglootish'ən), *n*. the act or power of swallowing.
degradation (degrədā'shən), *n*. the act of degrading; the state of being degraded; debasement, degeneracy; diminution or loss of strength, efficacy or value; (*Chem.*) decomposition; the wearing away of higher lands, rocks etc.
degrade (digrād'), *v.t.* to reduce in rank; to debase, to lower; to bring into contempt; (*Biol.*) to reduce from a higher to a lower type; to wear away; of a chemical compound, to decompose. *v.i.* to degenerate; to decompose. **degradable**, *a*. capable of decomposing biologically or chemically. **degraded**, *a*. **degrading**, *a*. lowering the level or character; humiliating, debasing. **degradingly**, *adv*.
degree (digrē'), *n*. a step or stage in progression, elevation, quality, dignity or rank; relative position or rank; a certain distance or remove in the line of descent designating proximity of blood; a rank or grade of academic proficiency conferred by universities after examination, or as a compliment to distinguished persons; relative condition, relative quantity, quality or intensity; (*Geom.*) the 90th part of a right angle; the 360th part of the circumference of the earth; the unit of measurement of temperature; one of the three grades of comparison of adjectives and adverbs (POSITIVE, COMPARATIVE, SUPERLATIVE). **by degrees**, gradually, step by step. **degree of freedom**, an independent component of motion of a molecule or atom; (*Chem.*) any of the independent variables which define the state of a system. **honorary degrees**, those conferred by a university without examination. **third degree**, THIRD. **to a degree**, (*coll.*) exceedingly. **degree-day**, *n*. the day on which degrees are conferred at a university. **degreeless**, *a*.
dehisce (dihīs'), *v.i.* to gape, to burst open (of the capsules or anthers of plants). **dehiscence**, *n*. **dehiscent**, *a*.
dehumanize, **-ise** (dēhū'məniz), *v.t.* to divest of human character, esp. of feeling or tenderness; to brutalize.
dehumidify (dēhūmid'ifī), *v.t.* to remove humidity from. **dehumidifier**, *n*.
dehydrate (dēhīdrāt'), *v.t.* to release or remove water or its elements from e.g. the body or tissues. *v.i.* to lose water. **dehydration**, *n*.
de-ice (dēīs'), *v.t.* to disperse ice which has formed on the wings and control surfaces of an aircraft or on the windows of a car. **de-icer**, *n*. an apparatus or liquid used to effect this.
deicide (dē'isīd, dā'-), *n*. the killing of a god; one concerned in this.
deictic (dīk'tik), *a*. (*Gram., Log.*) proving directly; demonstrative, as distinguished from indirect or refutative.
deify (dē'ifī, dā'-), *v.t.* to make a god of; to make godlike; to adore as a god; to idolize. **deific** (-if'-), *a*. making divine. **deification** (-fikā'-), *n*. **deifier**, *n*.
deign (dān), *v.i.* to condescend, to vouchsafe. *v.t.* to condescend to allow or grant.
dei gratia (dā'ē grah'tiə), by the grace of God. [L]
deindustrialize, **-ise** (dēindŭs'triəliz), *v.t.* to make a (country etc.) less industrial. **deindustrialization**, **-isation**, *n*.
deinstitutionalize, **-ise** (dēinstitū'shənəliz), *v.t.* to remove from an institution, esp. from a mental hospital. **deinstitutionalization**, **-isation**, *n*.
deism (dē'izm, dā'-), *n*. belief in the existence of God on purely rational grounds, without accepting divine revelation. **deist**, *n*. **deistic**, **-ical** (-is'-), *a*. **deistically**, *adv*.
deity (dē'əti, dā'-), *n*. divine nature, character or attributes; the Supreme Being; a god or goddess.
déjà vu (dăzhah vü), *n*. an illusion of already having experienced something one is experiencing for the first time. [F, already seen]
deject (dijekt'), *v.t.* to cast down; to depress in spirit; to dishearten. **dejected**, *a*. **dejectedly**, *adv*. **dejection**, *n*. the act of casting down; the state of being dejected; lowness of spirits.
de jure (dā joo'ri), by right. [L]
dekko (dek'ō), *n*. (*pl.* **dekkos**) (*coll.*) a (quick) look at. *v.i.* to look.
Del., (*abbr.*) Delaware (US).
del., (*abbr.*) delegate.
delaine (dilān'), *n*. a kind of untwilled wool muslin; a fabric of wool and cotton.
delay (dilā'), *v.t.* to postpone, to put off; to hinder, to retard. *v.i.* to put off action; to linger. *n*. a stay or stopping; postponement, retardation; detention; hindrance. **delayed-action**, *a*. which operates or takes effect after an interval of time. **delayer**, *n*. **delayingly**, *adv*.
dele (dē'lē), *n*., *v.t.* (*Print.*) (a mark instructing a printer) to delete (a word).
delectable (dilek'təbl), *a*. delightful, highly pleasing. **delectability** (-bil'-), *n*. **delectableness**, *n*. **delectation** (dēlektā'-), *n*. delight, pleasure, enjoyment.
delegate (del'igət), *n*. one authorized to transact business as a representative, esp. at a conference. *v.t.* (-gāt), to depute as delegate, agent or representative, with authority to transact business. **delegation**, *n*. the act of delegating; a body of delegates, a deputation.
delete (dilēt'), *v.t.* to strike out, to erase. **delenda** (-len'də), *n.pl.* things to be deleted. **deletion**, *n*.
deleterious (delitiə'riəs), *a*. harmful; injurious to health or mind.
delf (delf), **delft** (delft), *n*. glazed earthenware, orig. made at Delft, Holland.
deli (del'i), *n*. short for DELICATESSEN.
deliberate (dilib'ərət), *a*. weighing matters or reasons carefully; circumspect, cool, cautious; done or carried out intentionally; leisurely, not hasty. *v.i.* (-āt) to weigh matters in the mind, to ponder; to estimate the weight of reasons or arguments; to consider, to discuss, to take counsel. *v.t.* to weigh in the mind. **deliberately**, *adv*. **deliberateness**, *n*. **deliberation**, *n*. calm and careful consideration; discussion of reasons for and against; freedom from haste or rashness; leisurely, not hasty, movement. **deliberative**, *a*. pertaining to, proceeding from, or acting with, delibera-

delicacy 211 **demean**

tion. **deliberatively**, *adv.*
delicacy (del′ikəsi), *n.* the quality of being delicate; anything that is subtly pleasing to the senses, the taste or the feelings; a luxury, a dainty; fineness of texture, design, tint or workmanship; subtlety and sensitiveness of construction and action; weakness, fragility, susceptibility to injury; need for careful or tactful handling; nicety of perception; fineness, sensitiveness, shrinking from coarseness and immodesty; gentleness, consideration for others.
delicate (del′ikət), *a.* pleasing to senses in a mild or subtle way; dainty or exquisite in form or texture; fastidious, squeamish; sensitive, subtly perceptive or appreciative; subtle in colour, form or style; requiring acuteness of sense to distinguish; precise or sensitive in action; easily injured, fragile, constitutionally weak or feeble; requiring careful treatment; critical, ticklish; refined, chaste, pure; gentle, considerate. *n.* anything choice, esp. food, a dainty. **delicately**, *adv.* **delicateness**, *n.*
delicatessen (delikətes′n), *n.pl.* cooked meats and preserves; (*sing. in constr.*) a shop or part of a shop selling such products.
delicious (dilish′əs), *a.* yielding exquisite pleasure to the senses, to taste or to the sense of humour. **deliciously**, *adv.* **deliciousness**, *n.*
delight (dilīt′), *v.t.* to please greatly, to charm. *v.i.* to be highly pleased; to receive great pleasure (in). *n.* a state of great pleasure and satisfaction; a source of great pleasure or satisfaction. **delightedly**, *adv.* **delightful**, *a.* **delightfully**, *adv.* **delightfulness**, *n.*
Delilah (dili′lə), *n.* a temptress; a loose woman. [the Philistine woman in the Bible who betrayed Samson (Judges xvi)]
delimit (dilim′it), *v.t.* to fix the boundaries or limits of. **delimitate**, *v.t.* to delimit. **delimitation**, *n.*
delineate (dilin′iāt), *v.t.* to draw in outline; to sketch out; to describe, to depict, to portray. **delineation**, *n.* **delineative**, *a.* **delineator**, *n.* **delineatory**, *a.*
delinquent (diling′kwənt), *a.* offending, failing, neglecting. *n.* one who fails in his duty; an offender, a culprit. **juvenile delinquent**, in Britain, an offender under 17 years of age. **delinquency**, *n.* a failure or omission of duty; a fault, an offence; guilt.
deliquesce (delikwes′), *v.i.* to liquefy, to melt away gradually by absorbing moisture from the atmosphere; to melt away (as money). **deliquescence**, *n.* **deliquescent**, *a.*
delirious (dili′riəs), *a.* suffering from delirium, wandering in mind; raving, madly excited; frantic with delight or other excitement. **deliriously**, *adv.*
delirium (dili′riəm), *n.* a mental disturbance characterized by frantic excitement, confusion, hallucinations; frantic excitement or enthusiasm, rapture, ecstasy. **delirium tremens** (trem′ens), *n.* an acute psychotic phase in chronic alcoholism.
deliver (diliv′ə), *v.t.* to save, to rescue; to assist at the birth of a child; to release or send forth, e.g. a blow, a ball; to utter, to pronounce formally or officially; to surrender, to give up; to give over, to hand over or on; to convey, to take to a destination, addressee, consignee etc.; (*N Am.*) to persuade (voters) to support. *v.i.* to convey goods to a destination; (*coll.*) deliver the goods. **to deliver over**, to put into the hands of. **to deliver the goods**, to fulfil a promise, to carry out an undertaking, to live up to expectations. **to deliver up**, to surrender possession of. **deliverable**, *a.* **deliverance**, *n.* release, rescue; a formal expression of an opinion or verdict. **deliverer**, *n.* one who delivers; a saviour, a preserver. **delivery** (-əri), *n.* the act of delivering; setting free; rescue; transfer, surrender; a distribution of letters, goods etc.; the letters, goods, etc. so distributed; the utterance of a speech; style or manner of speaking; childbirth; discharge of a blow or missile; (*Cricket*) the act or style of bowling; (*Law*) the act of putting another in formal possession of property; the handing over of a deed to the grantee. **delivery room**, *n.* a room in a hospital, where babies are delivered.
dell (del), *n.* a hollow or small valley, usually wooded.
delouse (dēlows′), *v.t.* to rid a person or place of vermin, esp. lice.
delph (delf), DELF.
Delphian (del′fiən), **Delphic** (-fik), *a.* of or belonging to *Delphi*, a town in Greece, where there was a celebrated oracle of Apollo; susceptible of two interpretations, ambiguous.
delphinium (del′fin′iəm), *n.* the genus comprising the larkspurs.
delta (del′tə), *n.* the fourth letter of the Greek alphabet (δ, Δ), corresponding to the English *d;* the delta-shaped alluvial deposit at the mouth of the Nile; any similar alluvial deposit at the mouth of a river. **delta rays**, *n.pl.* electrons moving at relatively low speeds. **delta rhythm, wave**, *n.* the normal activity of the brain during deep sleep. **delta wing**, *n.* a triangular-shaped wing on an aeroplane. **deltaic** (-tā′-), *a.* **deltoid**, *a.* shaped like a delta; triangular. *n.* a triangular muscle of the shoulder which moves the arm.
deltiology (deltiol′əji), *n.* the study and collecting of postcards. **deltiologist**, *n.*
delude (dilood′), *v.t.* to deceive, to impose upon. **deluded**, *a.* under a false impression. **deluder**, *n.*
deluge (del′ūj), *n.* a general flood or inundation, esp. the general flood in the days of Noah; a heavy downpour of rain; an overwhelming amount or number. *v.t.* to flood, to inundate; to overwhelm, to swamp.
delusion (diloo′zhən), *n.* the act of deluding; the state of being deluded; an error, a fallacy; an erroneous idea in which the subject's belief is unshaken by facts. **delusional**, *a.* **delusive** (-siv), *a.* deceptive, misleading, unreal. **delusively**, *adv.* **delusiveness**, *n.* **delusory** (-z-), *a.*
de luxe (di lŭks′), *a.* luxurious, of superior quality.
delve (delv), *v.t.* to work with a spade; to burrow; to carry on laborious research; to search, to rummage.
demagnetize, -ise (dēmag′nitīz), *v.t.* to deprive of magnetism. **demagnetization, -isation**, *n.*
demagogue (dem′əgog), *n.* a leader of the people; an agitator who appeals to the passions and prejudices of the people; an unprincipled politician. **demagogic** (-gog′-), *a.* **demagogism**, *n.* **demagoguery**, *n.* **demagogy**, *n.*
demand (dimahnd′), *n.* an authoritative claim or request; the thing demanded; *pl.* the effort, dedication, sacrifices etc. required by e.g. a job; a claim; a peremptory question; desire to purchase or possess; a legal claim. *v.t.* to ask or claim with authority or as a right; to question, to interrogate; to seek to ascertain by questioning; to need, to require; to ask in a peremptory or insistent manner. *v.i.* to ask something as a right; to ask. **demand and supply**, a phrase used to denote the relations between consumption and production: if demand exceeds supply, the price rises; if supply exceeds demand, the price falls. **in demand**, much sought after. **on demand**, whenever requested. **demand feeding**, *n.* feeding a baby when it needs to be fed rather than at set times. **demand note**, *n.* the final notice served for payment of rates, taxes etc. **demandable**, *a.* **demandant**, *n.* a plaintiff; one who demands. **demander**, *n.*
demarcate (dē′mahkāt), *v.t.* to fix the limits of. **demarcation**, *n.* the fixing of a boundary or dividing line; the division between different branches of work done by members of trade unions on a single job.
demarche (dā′mahsh), *n.* a diplomatic approach; method of procedure; announcement of policy. [F]
dematerialize, -ise (dēmətiə′riəlīz), *v.t.* to deprive of material qualities or characteristics. *v.i.* to lose material form; to vanish. **dematerialization, -isation**, *n.*
deme (dēm), *n.* a subdivision or township in Greece; (*Biol.*) an undifferentiated aggregate of cells.
demean (dimēn′), *v.t.* to conduct (oneself), to behave; to

dement

debase, to lower. *n.* behaviour, demeanour. **demeaning**, *a.* humiliating. **demeanour** (-nə), *n.* conduct, manner; behaviour towards others.
dement (diment'), *v.t.* to drive insane. **demented**, *a.* insane. **dementedly**, *adv.* **dementedness**, *n.*
dementia (dimen'shə), *n.* loss or feebleness of the mental faculties. **dementia praecox** (prē'koks), *n.* a mental disorder resulting from a turning inwards into self away from reality, schizophrenia.
demerara (deməroh'rə), *n.* a kind of brown sugar.
demerge (dēmœj'), *v.t.* to split, to separate (of companies formerly acting as one). **demerger**, *n.*
demerit (dēme'rit, di-), *n.* a fault, a defect; a drawback; (*N Am.*) a bad mark given for misconduct etc.
demersal (dimœ'səl), *a.* (*Zool.*) found in deep water or on the ocean bed.
demesne (dimēn'. -mān'), *n.* an estate in land; the manor-house and the lands near it, which a lord keeps in his own hands; (*Law*) possession as one's own; a region, territory.
demi- (demi-), *pref.* half, semi-, partial, partially. **demigod**, *n.* one who is half a god; an inferior deity; the offspring of a god and a human being. **demi-monde**, *n.* persons not recognized in society, kept women; the section of a profession etc. which is not wholly legal or above board. **demi-mondaine** (-mondān'), *n.* a prostitute. **demi-rep** (-rep), *n.* a woman of doubtful chastity. **demi-semiquaver**, *n.* a note of the value of half a semiquaver or one-fourth of a quaver. **demitasse** (-tas), *n.* a small cup of or for black coffee.
demijohn (dem'ijon), *n.* a glass vessel or bottle with a large body and small neck, enclosed in wickerwork.
demilitarize, -ise (dēmil'itəriz), *v.t.* to end military involvement (in) and control (of).
demise (dimīz'), *n.* death, decease, esp. of the sovereign or a nobleman; (*Law*) a transfer or conveyance by lease or will for a term of years or in fee simple. *v.t.* to bequeath; (*Law*) to transfer or convey by lease or will. **demise of the Crown**, transference of sovereignty upon the death or abdication of the monarch. **demisable**, *a.*
demission DEMIT.
demist (dēmist'), *v.t.* to make clear of condensation. **demister**, *n.*
demit (dimit'), *v.t.*, *v.i.* to resign. **demission** (-shən), *n.* the act of resigning or abdicating.
demiurge (dem'iœj), *n.* a name given by the Platonists to the creator of the universe; the Logos of the Platonizing Christians. **demiurgic** (-œ'-), *a.*
demi-veg (dem'ivej), *a.* not completely vegetarian, but including white meat and fish in the diet. *n.* one who follows this diet.
demo (dem'ō), *n.* short for DEMONSTRATION.
demo-, *comb. form* pertaining to the people or the population generally.
demob (dēmob'), *v.t.* (*past, p.p.* **demobbed**) (*coll.*) to demobilize. *n.* demobilization.
demobilize, -ise (dēmō'biliz), *v.t.* to disband, to dismiss (as troops) from a war footing. **demobilization**, *n.*
democracy (dimok'rəsi), *n.* the form of government in which the sovereign power is in the hands of the people, and exercised by them directly or indirectly; a democratic state; any group in which all the members participate in decision-making; social equality. **democrat** (dem'əkrat), *n.* one in favour of democracy; (*US*) a member of the Democratic party. **democratic** (deməkrat'-), *a.* pertaining to a democracy; governed by or maintaining the principles of democracy. **Democratic party**, *n.* one of the two main political parties in the US. **democratically**, *adv.* **democratism**, *n.* **democratize, -ise**, *v.t., v.i.* **democratization**, *n.* the inculcation of democratic views and principles.
démodé (dāmōdā'), *a.* out of fashion. [F]
demodernization, -isation (dēmodənizā'shən), *n.* the mak-

demur

ing less modern of, e.g. modern housing estates, to make them more habitable and less prone to social problems.
demodulate (dēmod'ūlāt), *v.t.* to extract the original audio signal from (the modulated carrier wave by which it is transmitted). **demodulation**, *n.* **demodulator**, *n.*
demography (dimog'rəfi), *n.* the study of population statistics dealing with size, density and distribution. **demographer**, *n.* **demographic** (deməgraf'-), *a.* **demographically**, *adv.*
demolish (dimol'ish), *v.t.* to pull or throw down; to raze; to ruin, to destroy; (*coll.*) to eat up. **demolition** (deməlish'ən), *n.* the act of demolishing; (*Mil.*) destruction using explosives.
demon, daemon (dē'mən), *n.* (*Gr. Myth.*) an attendant spirit supposed to exercise guardianship over a particular individual; an evil spirit having the power of taking possession of human beings; a devil; (*sl., usu. in comb.*) an extremely energetic, enthusiastic or effective person, as *demon-bowler*. **demoness**, *n. fem.* **demoniac** (dimō'niak), *a.* pertaining to or produced by demons; possessed by a demon; devilish; frantic, frenzied. *n.* one possessed by a demon. **demoniacal** (dēməni'əkl), *a.* demoniac. **demoniacally**, *adv.* **demonic, daemonic** (dimon'-), *a.* **demonism**, *n.* belief in demons or false gods. **demonist**, *n.*
demonetize, -ise (dēmŭn'ətiz), *v.t.* to deprive of its character as money; to withdraw (a metal) from currency. **demonetization**, *n.*
demoniac, demonism etc. DEMON.
demonolatry (dēmənol'ətri), *n.* the worship of demons or evil spirits.
demonology (dēmənol'əji), *n.* the study of demons or of evil spirits.
demonstrate (dem'ənstrāt), *v.t.* to show by logical reasoning; to prove beyond doubt; to exhibit, describe and prove by means of specimens and experiments; to display, to indicate. *v.i.* to organize or take part in a military or public demonstration. **demonstrant** (-mon'-), *n.* **demonstrable**, *a.* that may be proved beyond doubt; apparent, evident. **demonstrability** (-bil'-), *n.* **demonstrably**, *adv.* **demonstration**, *n.* the act of demonstrating; clear, indubitable proof; an outward manifestation of feeling etc.; a public exhibition or declaration of principles, feelings etc. by means of a procession, mass-meeting etc.; exhibition and description of objects for the purpose of teaching; (*Mil.*) a movement of troops as if to attack. **demonstrative** (-mon'-), *a.* having the power of exhibiting and proving; proving; conclusive; pertaining to proof; acting as a demonstrative; manifesting the feelings strongly and openly. *n.* (*Gram.*) a class of determiners used to highlight the referent(s), as *this, that*. **demonstratively**, *adv.* **demonstrativeness**, *n.* **demonstrator**, *n.* one who demonstrates; one who teaches by means of exhibition and experiment; one who takes part in a public demonstration of political, religious or other opinions. **demonstratorship**, *n.*
demoralize, -ise (dimo'rəliz), *v.t.* to corrupt the discipline or morale of; to discourage; to throw into confusion. **demoralization, -isation**, *n.*
Demos (dē'mos), *n.* the people; the mob.
demote (dimōt'), *v.t.* to reduce in status or rank. **demotion**, *n.*
demotic (dimot'ik), *a.* of or belonging to the people; popular, common, vulgar. *n.* (**Demotic**) the spoken form of modern Greek. **demotic alphabet**, *n.* that used by the laity and people of Ancient Egypt as distinguished from the hieratic.
demur (dimœ'), *v.i.* (*past, p.p.* **demurred**) to have or express scruples, objections or reluctance; (*Law*) to take exception to any point in the pleading as insufficient. *n.* the act of demurring; scruple, objection. **demurrable**, *a.* liable to esp. legal objection. **demurral**, *n.* **demurrant**, *n.* **demurrer**, *n.* (*Law*) an objection made to a point

demure (dimū·ʹ), *a.* staid; modest; affectedly modest.

demurrage (dimū·ʹrij), *n.* an allowance by the freighter of a vessel to the owners for delay in loading or unloading beyond the time named in the charter-party; the period of such delay; a charge for the detention by one company of trucks etc. belonging to another.

demy (dimī·ʹ, dem·ʹi), *n.* (*pl.* **demies**) a size of paper, 22½ × 17½ in. (444·5 × 571·5 mm) for printing, 20 × 15½ in. (508 × 393·7 mm) for drawing or writing (*N Am.* 21 × 16 in.).

demystify (dēmis·ʹtifī), *v.t.* to remove the mystery from, to clarify. **demystification** (-fikā·ʹ-), *n.*

demythologize, -ise (dēmithol·ʹəjīz), *v.t.* to remove the mythological elements from, e.g. the Bible, to highlight the basic meaning. **demythologization, -isation,** *n.*

den (den), *n.* the lair of a wild beast; a centre of disreputable or illicit activity; a miserable room; (*coll.*) a study, a sanctum; (*Sc.*) a small valley.

denarius (dināhʹriəs), *n.* (*pl.* **-rii** (-riī)) a Roman silver coin, worth 10 asses; a penny.

denary (dēʹnəri), *a.* containing 10; based on the number 10, decimal.

denationalize, -ise (dēnashʹənəlīz), *v.t.* to deprive of the rights, rank or characteristics of a nation; to transfer from public to private ownership. **denationalization,** *n.*

denaturalize, -ise (dēnachʹərəlīz), *v.t.* to render unnatural; to deprive of naturalization. **denaturalization,** *n.*

denature (dēnāʹchə), **denaturize, -ise** (-īz), *v.t.* to change the essential nature or character of (by adulteration etc.); to modify (e.g. a protein) by heat or acid; to render (alcohol) unfit for human consumption; to add non-radioactive material to radioactive material, to prevent the latter being used in nuclear weapons. **denaturant,** *n.* **denaturation,** *n.*

denazify (dēnahtsʹifī), *v.t.* to purge of Nazism and its evil influence on the mind.

dendr(i)-, dendro-, *comb. form.* resembling a tree; branching.

dendrite (denʹdrīt), *n.* a stone or mineral with arborescent markings; one of the branched extensions of a nerve cell which conduct impulses to the body of the cell. **dendritic, -ical** (-drit·ʹ-), *a.* like a tree; arborescent; with tree-like markings.

dendrochronology (dendrōkrənolʹəji), *n.* the study of the annual growth rings in trees, used to date historical events.

dendroid (denʹdroid), *a.* tree-like, arborescent.

dendrology (dendrolʹəji), *n.* the natural history of trees. **dendrologist,** *n.*

dene[1] DEAN[2].

dene[2] (dēn), *n.* a sandy down or low hill, a tract of sand by the sea.

denegation (dēnigāʹshən), *n.* contradiction, denial.

dengue (dengʹgi), *n.* an acute fever common in the E and W Indies, Africa and America, characterized by severe pains, an eruption like erysipelas and swellings.

deniable, denial, denier DENY.

denier (dəniəʹ), *n.* a small French coin, the 12th part of a sou; a coin of insignificant value; (denʹiə) the unit for weighing and grading silk, nylon and rayon yarn, used for women's tights and stockings.

denigrate (denʹigrāt), *v.t.* to belittle, to disparage; to defame. **denigration,** *n.* **denigrator,** *n.*

denim (denʹim), *n.* a coarse, twilled cotton fabric used for overalls, jeans etc.; (*pl.*) jeans made of denim.

denitrate (dēnīʹtrāt), *v.t.* to set free nitric or nitrous acid or nitrate from. **denitrify,** *v.t.* to denitrate. **denitrification** (-fikā·ʹ-), *n.* the liberation of nitrogen from the soil by bacteria.

denizen (denʹizn), *n.* a citizen, an inhabitant, a resident; a foreign word, plant or animal, that has become naturalized. **denizenship,** *n.*

denominate (dinomʹināt), *v.t.* to name; to give a name, epithet or title to; to designate. **denomination** (-nāʹ-), *n.* the act of naming; a designation, title or appellation; a class, a kind, esp. of particular units (as coins, weights etc.); a religious organization or sect. **denominational** (-nāʹ-), *a.* pertaining to a particular denomination, sectarian. **denominationalism,** *n.* **denominationalist,** *n.* **denominationalize, -ise,** *v.t.* **denominationally,** *adv.*

denominative (dinomʹinətiv), *a.* that gives or constitutes a distinctive name. **denominator** (-nā-), *n.* one who or that which denominates; (*Arith.*) the number below the line in a fraction which shows into how many parts the integer is divided.

denote (dinōtʹ), *v.t.* to mark, to indicate; to mean, to signify; to mark out, to distinguish; (*Log.*) to be a name of, to be predicable of (distinguished from CONNOTE). **denotable,** *a.* **denotation** (dēnōtāʹ-), *n.* the act of denoting; separation or distinction by means of a name or names; meaning, signification; a system of marks or symbols. **denotative,** *a.* signifying, pointing out; designating, without implying attributes. **denotatively,** *adv.*

dénouement (dānooʹmē), *n.* the unravelling of a plot or story; the catastrophe or final solution of a plot; an outcome.

denounce (dinownsʹ), *v.t.* to accuse publicly; to charge, to inform against; to give formal notice of termination of (a treaty or convention). **denouncement,** *n.* **denunciation. denouncer,** *n.*

de novo (dī nōʹvō), *adv.* anew.

dense (dens), *a.* thick, compact; having its particles closely united; stupid, obtuse; (*Phot.*) opaque, strong in contrast. **densely,** *adv.* **denseness,** *n.* **densimeter** (-simʹitə), *n.* an apparatus for measuring density or specific gravity. **densimetry** (-simʹ-), *n.* **density,** *n.* denseness; (*Phys.*) the mass per unit volume of a substance measured, for example, in grams per cubic centimetre; a crowded condition; a measure of the reflection or absorption of light by a surface; (*fig.*) stupidity.

dent (dent), *n.* a depression such as is caused by a blow with a blunt instrument; an indentation; a lessening or diminution. *v.t.* to make a dent in; to indent.

dental (denʹtəl), *a.* pertaining to or formed by the teeth; pertaining to dentistry. *n.* a sound formed by placing the end of the tongue against the upper teeth. **dental floss,** *n.* thread used to clean between the teeth. **dental formula,** *n.* a formula used to describe the dentition of a mammal. **dental plaque,** *n.* a deposit of bacteria and food on the teeth. **dental surgeon** DENTIST. **dentalize, -ise,** *v.t.* to pronounce as a dental. **dentate** (denʹtāt), **dentated** (-tāʹ-), *a.* (*Bot., Zool.*) toothed; indented. **dentately,** *adv.* **dentation,** *n.*

dentelle (dentelʹ), *n.* a style of angular decoration like saw-teeth; a lace edging resembling a series of small teeth.

denti-, *comb. form.* of or pertaining to the teeth.

denticle (denʹtikl), *n.* a small tooth; a projecting point, a dentil. **denticular** (-tikʹū-), *a.* **denticulate** (-tikʹūlət), **-lated** (-lātid), *a.* finely toothed; formed into dentils. **denticulately,** *adv.* **denticulation,** *n.*

dentifrice (denʹtifris), *n.* powder, paste or other material for cleansing the teeth.

dentilingual (dentilingʹgwəl), *n., a.* (a consonant) formed by the teeth and the tongue.

dentine (denʹtēn), *n.* the ivory tissue forming the body of a tooth.

dentist (denʹtist), *n.* one skilled and qualified in treating disorders of the teeth. **dentistry,** *n.* **dentition** (-tiʹ-), *n.* teething; the time of teething; the arrangement of the teeth in any animal. **denture** (-chə), *n.* (*often pl.*) set of teeth, esp. artificial.

denuclearize, -ise (dēnūʹkliərīz), *v.t.* to deprive of nuclear

denude | 214 | **deposit**

arms; to prohibit the presence of any nuclear material or any installation using nuclear power. **denuclearization, -isation,** n.
denude (dinūd'), v.t. to make bare or naked; to strip of clothing, attributes, possessions, rank or any covering; (*Geol.*) to lay bare by removing whatever lies above. **denudate** (dē'-), v.t. to denude. a. (dət), made naked, bare, stripped; (*Bot.*) appearing naked. **denudation** (dēnūdā'-), n.
denumerable (dēnū'mərəbl), a. able to be put into a one-to-one correspondence with the positive integers; countable. **denumerably,** adv.
denunciate (dinŭn'siāt), v.t. to denounce. **denunciation,** n. **denunciative,** a. **denunciator,** n. **denunciatory,** a.
deny (dinī'), v.t. to assert to be untrue or non-existent; to disown, to reject, to repudiate; to refuse to grant, to withhold from; to refuse admittance or access to; to say 'no' to. **to deny oneself,** to refrain or abstain from; to practise self-denial. **deniable,** a. **denial,** n. the act of denying, contradicting or refusing; a negation; abjuration, disavowal; self-denial. **denier,** n.
deodar (dē'ədah), n. a large Himalayan tree, *Cedrus deodara,* allied to the cedars of Lebanon.
deodorize, -ise (dēō'dəriz), v.t. to deprive of odour; to disinfect. **deodorant,** n. an agent which counteracts unpleasant smells; a substance used to mask the odour of perspiration. **deodorization,** n. **deodorizer, -iser,** n.
Deo gratias (dā'ō grah'tias), thanks be to God.
deontology (dēəntol'əji), n. the science of duty, ethics. **deontic** (-on'-), a.
Deo volente (dā'ō volen'ti), God willing. [L]
deoxidize, -ise (dēok'sidīz), v.t. to deprive of oxygen; to extract oxygen from. **deoxidization,** n. **deoxidizer,** n.
deoxygenate (-jənāt), v.t. to deoxidize. **deoxygenation,** n. **deoxygenize, -ise,** v.t. to deoxidize.
deoxycorticosterone (dēoksikawtikōstiə'rōn), **deoxycortone** (-kaw'tōn), n. a hormone which maintains the sodium and water balance in the body.
deoxyribonucleic acid (dēoksiribōnūklē'ik), **desoxyribonucleic acid** (des-), n. the full name for DNA.
dep., (abbr.) depart(s); department; deposed; deputy.
depart (dipaht'), v.i. to go away, to leave; to diverge, to deviate, to pass away; to die. v.t. to go away from, to quit. **departed,** a. past, bygone; dead. **the departed,** the dead.
department (dipaht'mənt), n. a separate part or branch of business, administration or duty; a branch of study or science; a sphere of activity or responsibility; one of the administrative divisions of a country, as in France; a ministry, e.g. Department of Trade and Industry. **department store,** n. a shop selling a great variety of goods, organized in departments. **departmental** (dēpahtmen'-), a. **departmentalism** (-men'-), n. a too-rigid adherence to regulations, red tape. **departmentalize, -ise** (-men'-), v.t. **departmentally,** adv.
departure (dipah'chə), n. the act of departing; leaving; starting; quitting; death; divergence, deviation; (*Naut.*) distance of a ship east or west of the meridian she sailed from; the position of an object from which a vessel commences her dead reckoning. **new departure,** a new course of thought or ideas; a new enterprise.
depend (dipend'), v.i. to hang down; to be contingent, as to the issue or result, on something else; to rely, to trust, to reckon (upon); to rely for support or maintenance. **depend upon it,** you may rely upon it, you may be certain. **that depends,** that is conditional; perhaps. **dependable,** a. that may be depended upon. **dependableness,** n. **dependably,** adv. **dependant, -ent,** n. one depending upon another for support or favour; that which depends upon something else; a retainer. **dependence,** n. the state of being dependent; a compulsive physiological or psychological need; connection, concatenation; reliance, trust, confidence; a dependency. **dependence, dependency, culture,** n. a way of life which takes state welfare benefits for granted and relies wholly on them for subsistence. **dependency,** n. dependence; something dependent, esp. a country or state subject to another. **dependent,** a. hanging down; depending on another; subject to, contingent (upon), relying on for support, benefit or favour. **dependent variable,** n. one in a mathematical equation whose value depends on that of the independent variable. **dependently,** adv.
depersonalize, -ise (dēpœ'sənəliz), v.t. to divest of personality; to regard as without individuality. **depersonalization,** n. the divesting of personality; (*Psych.*) the experience of unreality feelings in relation to oneself.
depict (dipikt'), v.t. to paint, to portray; to describe or represent in words. **depicter,** n. **depiction,** n. **depictive,** a.
depilate (dep'ilāt), v.t. to remove hair from. **depilation,** n. **depilator,** n. **depilatory** (dipil'-), n., a. (an application for) removing superfluous hair without injuring the skin.
deplane (dēplān'), v.i. (*N Am.*) to disembark from an aeroplane.
deplete (diplēt'), v.t. to empty, to exhaust; to reduce to very low level. **depletion,** n. **depletive,** a. causing depletion. n. a depleting agent. **depletory,** a.
deplore (diplaw'), v.t. to lament or grieve over; to regret; to express disapproval (of), to censure. **deplorable,** a. **deplorableness, deplorability** (-bil'-), n. **deplorably,** adv. **deploration** (-rā'shən), n.
deploy (diploi'), v.t. (*Mil.*) to open out; to extend from column into line; to organize or distribute ready for use. v.i. to form a more extended front. **deployment,** n. use, esp. of troops or weapons.
deplume (diploom'), v.t. to strip of plumage. **deplumation** (dē-), n.
depolarize, -ise (dēpō'ləriz), v.t. to free from polarization; to deprive of polarity. **depolarization,** n. **depolarizer,** n.
depoliticize, -ise (dēpəlit'isiz), v.t. to make non-political.
depone (dipōn'), v.t., v.i. to testify, esp. on oath. **deponent,** n. a deponent verb; (*Law*) a witness; one who makes an affidavit to any statement of fact. **deponent verb,** n. a Latin verb with a passive form and active meaning.
depopulate (dēpop'ūlāt), v.t. to clear of inhabitants; to reduce the inhabitants of. v.i. to become less populous. **depopulation,** n.
deport (dipawt'), v.t. to expel (an alien) from a country; to conduct, to behave (oneself etc.). **deportation** (dē-), n. expulsion from a country. **deportee,** n. one who is deported.
deportment (dipawt'mənt), n. carriage, posture; conduct, demeanour, manners.
depose (dipōz'), v.t. to remove from a throne or other high office; to bear witness, to testify on oath. v.i. to bear witness. **deposable,** a. **deposal,** n.
deposit (dipoz'it), v.t. to lay down, to place; to entrust; to lodge for safety or as a pledge; to lay (as eggs); to leave behind as precipitation; to bury. n. anything deposited or laid down; that which is entrusted to another; a pledge, an earnest or first instalment; a security; money lodged in a bank; matter accumulated or precipitated and left behind. **on deposit,** when buying on hire-purchase, payable as a first instalment. **deposit account,** n. a bank account earning interest, usu. requiring notice for withdrawals. **depositary,** n. one with whom anything is deposited for safety; a trustee. **deposition** (depəzi'shən, dē-), n. the act of depositing; the act of deposing, esp. from a throne; a statement, a declaration; an affidavit; the act of bearing witness on oath; the evidence of a witness reduced to writing. **the Deposition,** the taking down of Christ from the Cross; a picture of this. **depositor,** n. one who makes a deposit, esp. of money. **depository,** n. a depositary; a place where anything, esp. furniture, is placed for safety.

depot (dep'ō), *n.* a place of deposit, a magazine, a storehouse; a building for the storage and servicing of buses and trains; (*N Am.*) a railway station; (*Mil.*) a magazine for stores; a station for recruits; the headquarters of a regiment.
deprave (diprāv'), *v.t.* to make bad or corrupt; to vitiate. **depravation** (deprəvā'-), *n.* **depraved,** *a.* corrupt. **depravity** (prav'-), *n.* a state of corruption; viciousness, profligacy; perversion, degeneracy.
deprecate (dep'rikāt), *v.t.* to express regret or reluctance about; to express disapproval of or regret for. **deprecatingly,** *adv.* **deprecation,** *n.* **deprecative,** *a.* **deprecatory,** *a.*
depreciate (diprē'shiāt), *v.t.* to lower the value of; to disparage, to decry; to lower the exchange value of (money etc.). *v.i.* to fall in value. **depreciatingly,** *adv.* **depreciation,** *n.* the act of depreciating; the state of becoming depreciated; (allowance for) a fall in value due to wear and tear etc. **depreciatory,** *a.*
depredation (deprədā'shən), *n.* plundering, spoliation. **depredator,** *n.*
depress (dipres'), *v.t.* to press down; to lower; to bring down; to reduce or keep down the energy or activity of; to cast down; to dispirit. **depressant,** *a.* lowering the spirits. *n.* a sedative. **depressed,** *a.* **depressed area,** *n.* an area of very serious unemployment. **depressible,** *a.* **depressing,** *a.* **depressingly,** *adv.* **depression** (-shən), *n.* the act of pressing down or lowering; dejection; lowering of energy or activity; a mental disorder characterized by low spirits, reduction of self-esteem and lowering of energy; slackness of business; an economic crisis; a hollow place on a surface; the angular distance of a heavenly body below the horizon; an area of low barometric pressure. **depressive,** *a.* causing depression; characterized by depression. *n.* one subject to periods of depression. **manic depressive** MANIC under MANIA. **depressor,** *n.* a muscle which depresses the part to which it is attached; an instrument for reducing or pushing back an obtruding part.
depressurize, -ise (dēpresh'əriz), *v.t.* to reduce the atmospheric pressure in (a pressure-controlled area, such as an aircraft cabin). **depressurization,** *n.*
deprive (dipriv'), *v.t.* to take from, to dispossess; to divest of an ecclesiastical office. **deprivable,** *a.* **deprival,** *n.* **deprivation** (deprivā'-), *n.* the act of depriving; the state of being deprived; loss, dispossession; the act of divesting a clergyman of his office. **deprived,** *a.* lacking acceptable social, educational and medical facilities.
de profundis (dā prəfun'dis), *n.* a cry from the depths of penitence or affliction; the title of the 130th Psalm. [L, 'Out of the depths']
deprogram (dēprō'gram), *v.t.* to remove a program from (a computer); (**deprogramme**) to persuade (someone) to reject obsessive beliefs, ideas and fears.
dept., (*abbr.*) department.
depth (depth), *n.* deepness; measurement from the top or surface downwards or from the front backwards; (*usu. pl.*) the deepest, innermost part; the middle or height of a season; (*pl.*) the deep part of the ocean, deep water; abstruseness, profundity; sagacity, mental penetration; intensity of colour, shade, darkness or obscurity; profundity of thought or feeling; (*Mil.*) the number of men in a file. **depth of field,** *n.* the distance in front of and behind an object focused on by a lens (such as a camera or microscope) which will be acceptably sharp. **in depth,** thoroughly; of defences, consisting of several lines. **out of one's depth,** in deep water; puzzled beyond one's knowledge or ability. **depth-bomb, depth-charge,** *n.* a mine or bomb exploded under water, used for attacking submarines. **depth-psychology,** *n.* the study of the unconscious.
depurate (dep'ūrāt), *v.t.* to purify. *v.i.* to become pure. **depuration,** *n.* **depurative** (-pū'-), *a.* **depurator,** *n.*
depute (dipūt'), *v.t.* to appoint or send as a substitute or agent; to give as a charge, to commit. *n.* (*Sc.*) a deputy.

deputation (depūtā'-), *n.* the act of deputing; the person or persons deputed to act as representatives for others, a delegation. **deputational,** *a.* **deputationist,** *n.* **deputize, -ise** (dep'-), *v.t.* to appoint or send as deputy. *v.i.* to act as deputy. **deputy** (dep'ūti), *n.* one who is appointed or sent to act for another or others; a member of the French and other legislative chambers; (*Law*) one who exercises an office in another's right; (*in comb.*) acting for, vice-; acting. **deputy-governor,** *n.* **deputy-speaker,** *n.*
der., deriv., (*abbr.*) derivation; derivative; derived.
deracinate (diras'ināt), *v.t.* to tear up by the roots; to destroy. **deracination,** *n.*
derail (dirāl'), *v.t.* to cause to leave the rails. *v.i.* to run off the rails. **derailer,** *n.* **derailment,** *n.*
derange (dirānj'), *v.t.* to put out of line or order; to disorganize; to disturb, to unsettle (esp. the intellect). **deranged,** *a.* insane; slightly insane. **derangement,** *n.*
Derby (dah'bi), *n.* a race for three-year-old horses, held at Epsom in May or June, founded by the 12th Earl of Derby in 1780; any race; (**derby**) a match between two teams from the same area; (*N Am.*) a bowler hat. **Derby day,** *n.* the day on which the Derby is run.
deregulate (dēreg'ūlāt), *v.t.* to remove legal or other regulations from, often so as to open up to general competition.
derelict (de'rəlikt), *a.* forsaken, abandoned; in ruins; negligent, remiss. *n.* anything abandoned (as a vessel at sea), relinquished or left to decay; a down-and-out. **dereliction** (-lik'-), *n.* abandonment; the state of being abandoned; omission or neglect (as of a duty); the abandonment of land by the sea; land left dry by the sea. **dereliction of duty,** reprehensible neglect or shortcoming.
derestrict (dēristrikt'), *v.t.* to free from restriction, e.g. a road from speed limits. **derestriction,** *n.*
deride (dirīd'), *v.t.* to laugh at, to mock. *v.i.* to indulge in mockery or ridicule. **deridingly,** *adv.* **derision** (-ri'zhən), *n.* ridicule, mockery, contempt. **in derision,** in contempt, made a laughing-stock. **derisive** (-siv), *a.* scoffing, deriding, ridiculing. **derisory,** *a.* ridiculous, ridiculously small; derisive. **derisively,** *adv.* **derisiveness,** *n.*
de rigueur (də rigœ'), prescribed (by etiquette or fashion). [F]
derision, derisive etc. DERIDE.
derive (diriv'), *v.t.* to obtain as by logical sequence; to deduce; to draw, as from a source, root or principle; to trace (an etymology); to deduce or determine from data. *v.i.* to come, to be descended; to originate. **derivable,** *a.* **derivation** (derivā'-), *n.* the act of deriving; deduction, extraction; a derivative; the etymology of a word, the process of tracing a word to its root; (*Math.*) the process of deducing a function from another. **derivative** (diriv'-), *a.* derived; taken from something else; secondary, not original. *n.* anything derived from a source; a word derived from or taking its origin in another; (*Math.*) a differential coefficient. **derivatively,** *adv.* **derived,** *a.* **derived unit,** *n.* a unit of measurement derived from the basic units of a system.
derm (dœm), **dermis** (-mis), *n.* skin; true skin or corium lying beneath the epidermis. **dermal,** *a.* **dermic,** *a.*
derm-, dermo-, dermato-, *comb. form.* pertaining to the skin.
-derm, *comb. form.* skin.
dermatitis (dœmətī'tis), *n.* inflammation of the skin.
dermatology (dœmətol'əji), *n.* the science of the skin and its diseases. **dermatological,** *a.* **dermatologist,** *n.*
dernier (dœ'niā), *a.* last. **dernier cri** (krē), *n.* the last word, the latest fashion. **dernier ressort** (rəsaw'), *n.* the last resort. [F]
derogate (de'rəgāt), *v.i.* to detract (from), to withdraw a part (from); to disparage. **derogation,** *n.* the act of derogating; disparagement. **derogative** (-rog'-), *a.* **derogatively,** *adv.* **derogatory** (-rog'-), *a.* disparaging,

depreciatory. **derogatorily,** *adv.*
derrick (dĕ'rik), *n.* a hoisting machine with a boom stayed from a central post, wall, floor, deck etc.; the framework over an oil-well.
derrière (derieu'), *n.* (*euphem.*) the buttocks, the behind. [F]
derring-do (deringdoo'), *n.* courageous deeds; bravery.
derringer (de'rinjə), *n.* a short-barrelled large-calibre pistol. [Henry Derringer, the 19th-cent. inventor]
derris (de'ris), *n.* an (insecticidal) extract of the root of tropical trees of the genus *Derris*.
derv (dœv), *n.* diesel engine fuel oil. [acronym for *d*iesel *e*ngine *r*oad *v*ehicle]
dervish (dœ'vish), *n.* a member of one of the various Muslim ascetic orders, whose devotional exercises include meditation and often frenzied physical exercises. **whirling dervish,** a member of a Muslim ascetic order whose physical exercises take the form of wild, ecstatic, whirling dances.
DES, (*abbr.*) Department of Education and Science.
desalinate (dēsal'ināt), *v.t.* to remove salt from, usu. sea water. **desalination,** *n.* **desalinator,** *n.*
descant[1] (des'kant), *n.* a counterpoint above the melody; a series of comments. *a.* having usu. the highest range in a family of instruments, as the descant recorder.
descant[2] (deskant'), *v.i.* to comment or discourse at large, to dilate (on); to add a descant to a melody or subject.
descend (disend'), *v.i.* to come or go down, to sink, to fall, to slope downwards; to make a sudden visit or attack (on); to have birth, origin or descent; to be derived; to be transmitted in order of succession; to pass on, as from more to less important matters, from general to particular, or from more remote to nearer times; to stoop; to lower or abase oneself morally or socially. *v.t.* to walk, move or pass along downwards. **descendable, -dible,** *a.* that may be transmitted from ancestor to heir. **descendant,** *n.* one who descends from an ancestor; offspring, issue. **descended,** *a.* derived, sprung (from a race or ancestor). **descender,** *n.* that part of a letter (e.g. j, p, y) which is below the level of the line of type.
descent (disent'), *n.* the act of descending; a declivity, a slope downwards; a way of descending; downward motion; decline in rank or prosperity; a sudden attack, esp. from the sea; pedigree, lineage, origin, evolution; issue of one generation; transmission by succession or inheritance.
describe (diskrīb'), *v.t.* to draw, to trace out; to form or trace out by motion; to set forth the qualities, features or properties of in words. **describable,** *a.* **description** (-skrip'-), *n.* the act of describing; an account of anything in words; a kind, a sort, a species. **descriptive** (-skrip'-), *a.* containing description; capable of describing; given to description; (*Gram.*) applied to a noun modifier which is not limiting or demonstrative, e.g. 'red'. **descriptively,** *adv.*
descry (diskrī'), *v.t.* (*past, p.p.* **descried**) to make out, to espy.
desecrate (des'ikrāt), *v.t.* to divert from any sacred purpose; to profane, to violate the sanctity of. **desecration,** *n.* **desecrator,** *n.*
desegregate (dēseg'rigāt), *v.t.* to end racial segregation in (an institution, e.g. a school). **desegregation,** *n.*
deselect (dēsilekt'), *v.t.* to drop from a group or team; to refuse to readopt as a candidate, esp. as a prospective parliamentary candidate.
desensitize, -ise (dēsen'sitīz), *v.t.* to render insensitive to e.g. a chemical agent. **desensitization, -isation,** *n.* **desensitizer,** *n.*
desert[1] (dez'ət), *a.* uninhabited, waste; untilled, barren. *n.* a waste, uninhabited, uncultivated place, esp. a waterless and treeless region; solitude, dreariness. **cultural desert,** a place completely lacking in cultural activities. **desert boots,** *n.pl.* suede ankle-boots with laces. **Desert Rat,** *n.* (*coll.*) a soldier of the 7th Armoured Division in N Africa (1941-2).
desert[2] (dizœt'), *v.t.* to forsake, to abandon; to quit, to leave; to fail to help. *v.i.* (*Mil. etc.*) to abandon the service without leave. **deserter,** *n.* **desertion,** *n.*
desert[3] (dizœt'), *n.* merit or demerit, meritoriousness; (*pl.*) deserved reward or punishment. **to get one's (just) deserts,** to receive what one's behaviour merits.
deserve (dizœv'), *v.t.* to merit by conduct or qualities, good or bad, esp. by excellence, good conduct or useful deeds. **deservedly** (-vid-), *adv.* **deserver,** *n.* **deserving,** *a.* worthy; meriting financial aid. **deservingly,** *adv.*
desex (dēseks'), *v.t.* to desexualize. **desexualize, -ise** (-sek'sūəlīz), *v.t.* to deprive of sexuality; to castrate or spay.
deshabille (dāzabē'), *n.* undress, state of being partly or carelessly attired; a loose morning dress.[F]
desiccate (des'ikāt), *v.t.* to dry, to exhaust of moisture. **desiccant,** *a.* drying up. *n.* a drying agent. **desiccated,** *a.* **desiccation,** *n.* **desiccative,** *a.* **desiccator,** *n.* an apparatus for drying substances liable to be decomposed by moisture.
desiderate (dizid'ərāt), *v.t.* to feel the loss of; to want, to miss. **desiderative,** *a.* expressing desire.
desideratum (dizidərah'təm), *n.* (*pl.* **-rata** (-tə)) anything desired, esp. anything to fill a gap; a state of things to be desired.
design (dizīn'), *v.t.* to contrive, to formulate, to invent; to plan, to sketch out; to purpose, to intend; to devote or apply to a particular purpose. *v.i.* to make a design or designs. *n.* a plan, a scheme; a purpose, an intention; a preliminary sketch; a working plan; the art of designing; the way a thing is arranged, shaped or made; artistic structure; plot, construction, general idea; a decorative pattern. **by design,** intentionally. **to have designs on,** to intend to acquire, appropriate or seduce. **designed,** *a.* intentional. **designedly** (-nid-), *adv.* **designer,** *n.* one who designs; one who produces detailed plans for a manufacturer; one who makes designs for clothing, stage or film sets etc. *a.* of clothes, produced by a famous designer; applied generally to anything considered extremely fashionable, unusual or expensive. **designer drugs,** *n.pl.* illegal drugs made up from a mixture of existing narcotics. **designer stubble,** *n.* (*coll.*) two or three days' growth of beard. **designing,** *a.* crafty, scheming. **designingly,** *adv.*
designate (dez'ignāt), *v.t.* to point out, to specify by a distinctive mark or name; to cause to be known as, to entitle; to select, to appoint. *a.* (*often placed after the noun*) nominated (to an office), as *president designate*. **designation,** *n.* the act of designating; appointment, nomination; name, title, description. **designative, -tory,** *a.*
desire (dizīə'), *v.t.* to wish (to do); to wish for the attainment or possession of; to request, to beseech; to command. *n.* an eagerness of the mind to obtain or enjoy some object; a request, an entreaty; the object of desire; sensual appetite, lust. **desirable,** *a.* worthy of being desired; agreeable; attractive. **desirability** (-bil'-), *n.* **desirableness,** *n.* **desirably,** *adv.* **desireless,** *a.* **desirous,** *a.* desiring, eager. **desirously,** *adv.*
desist (dizist'), *v.i.* to cease, to leave off. **desistance,** *n.*
desk (desk), *n.* a writing table often with drawers or other compartments; the place from which prayers are read; a pulpit; a counter for information or registration in a public place, e.g. a hotel; a newspaper department, as *news desk;* a music-stand for orchestral players. **desk-top computer,** *n.* one small enough to use on a desk. **desk-top publishing,** *n.* the production of text at a desk equipped with a computer and printer capable of producing high-quality printed copy. **deskwork,** *n.* writing, copying. **deskful,** *n.*
desman (des'mən), *n.* (*pl.* **-mans**) either of two mole-like

desolate — determinate

aquatic mammals, the Russian or the Pyrenean desman.
desolate (des'əlɒt), *a.* forsaken, solitary, lonely; uninhabited, deserted, neglected, ruined; forlorn, wretched. *v.t.* (-lāt), to deprive of inhabitants; to lay waste; to make wretched. **desolately**, *adv.* **desolateness**, *n.* **desolating**, *a.*
desolation, *n.* the act of desolating; the state of being desolated; neglect, ruin; loneliness; bitter grief, affliction. **desolator**, *n.*
despair (dispeə'), *v.i.* to be without hope; to give up all hope. *n.* hopelessness; that which causes hopelessness. **despairer**, *n.* **despairing**, *a.* hopeless, desperate. **despairingly**, *adv.*
despatch DISPATCH.
desperado (despərah'dō), *n.* a desperate or reckless ruffian.
desperate (des'pərət), *a.* reckless, regardless of danger or consequences; fearless; affording little hope of success, recovery or escape; tried as a last resource; extremely dangerous; very bad, awful; feeling despair, hopeless; intensely desirous of or anxious for. **desperately**, *adv.* in a desperate manner; awfully, extremely. **desperateness**, *n.* **desperation**, *n.*
despicable (dispik'əbl, des'-), *a.* meriting contempt; vile, worthless, mean, **despicably**, *adv.*
despise (dispīz'), *v.t.* to look down upon; to regard with contempt; to scorn. **despisingly**, *adv.*
despite (dispīt'), *n.* spite, malice; contemptuous treatment. *prep.* notwithstanding; in spite of.
despoil (dispoil'), *v.t.* to strip or take away from by force; to plunder. **despoiler**, *n.* **despoliation** (-spōli-), *n.* plunder; the state of being despoiled.
despond (dispond'), *v.i.* to be cast down in spririts; to lose hope. *n.* despondency. **despondency**, *n.* **despondent**, *a.* disheartened. **despondently**, *adv.* **despondingly**, *adv.*
despot (des'pot), *n.* an absolute ruler or sovereign; a tyrant, an oppressor. **despotic** (-spot'-), *a.* absolute; arbitrary, tyrannical. **despotically**, *adv.* **despotism**, *n.* absolute authority; arbitrary government, autocracy; tyranny.
despotist, *n.* an advocate of autocracy. **despotize, -ise**, *v.i.*
desquamate (des'kwəmāt), *v.t., v.i.* peel off in scales. **desquamation**, *n.* **desquamative, -tory** (-skwam'-), *a.*
des res (dez rez), (in estate agent's jargon, a desirable residence.
dessert (dizœt'), *n.* (formerly) the last course at dinner, consisting of fruit and sweetmeats; the sweet course. **dessert-spoon**, *n.* a medium-sized spoon holding half as much as a tablespoon and twice as much as a teaspoon.
destabilize, -ise (dēstābilīz), *v.t.* to render unstable; to attempt to undermine (a government, regime).
destination (destinā'shən), *n.* the act of destining; the purpose for which a thing is appointed or intended; the place to which one is bound or to which a thing is sent.
destine (des'tin), *v.t.* to appoint, fix or determine to a use, purpose, duty or position. **destined**, *a.* foreordained; bound for, on the way to.
destiny (des'tini), *n.* the purpose or end to which any person or thing is appointed; fate, fortune, lot, events as the fulfilment of fate; the power which presides over the fortunes of men. **the Destinies**, the three Fates.
destitute (des'titūt), *a.* in want, devoid of the necessities of life; forsaken, forlorn; bereft (of). **destitution**, *n.*
destroy (distroi'), *v.t.* to pull down or demolish; to pull to pieces; to undo, to nullify; to annihilate; to lay waste; to kill; to sweep away; to disprove; to put an end to. **destroyer**, *n.* one who destroys; a fast multi-purpose warship.
destruction (distrŭk'shən), *n.* the act of destroying; the state of being destroyed; demolition, ruin; death, slaughter; that which destroys. **destruct**, *v.t.* to destroy deliberately, e.g. a rocket or missile in flight. *v.i.* to be destroyed. **destructible** (-strŭk'-), *a.* **destructibility** (-bil'-), *n.*

n. **destructive**, *a.* causing or tending to destruction; ruinous, mischievous, wasteful; serving or tending to subvert or confute (arguments or opinions); negative, not constructive. **destructive distillation** DISTILLATION. **destructively**, *adv.* **destructiveness**, *n.* **destructor**, *n.* a furnace for burning up refuse.
desuetude (disū'itūd, des'wi-), *n.* disuse; cessation of practice or habit.
desultory (des'əltəri), *a.* passing quickly from one subject to another; following no regular plan; loose, disjointed, discursive. **desultorily**, *adv.* **desultoriness**, *n.*
Det., (*abbr.*) Detective.
detach (ditach'), *v.t.* to disconnect, to separate; to disengage; (*Mil., Nav.*) to separate from the main body for a special service. **detachable**, *a.* **detached**, *a.* separated, standing on its own; free from prejudice or emotional involvement. **detachedly** (-chid-), *adv.* **detachedness** (-chid-), *n.* **detachment**, *n.* the act of detaching; the state of being detached; a body of troops or a number of ships detached from the main body and sent on a special service or expedition; freedom from prejudice, self-interest or emotional involvement.
detail (dē'tāl), *v.t.* to set forth the particular items of; to relate minutely; (*Mil.*) to appoint for a particular service. *n.* a minute and particular account; a separate, small item or fact; such small items collectively; (*Mil.*) a list of names detailed for particular duties; a body of men selected for a special duty; a minor matter; a minor part of the minor parts of a picture, statue etc., as distinct from the work as a whole. **in detail**, minutely; item by item. **detailed**, *a.* related in detail; minute, complete.
detain (ditān'), *v.t.* to keep back or from; to withhold; to delay, to hinder; to keep in custody. **detainee** (-ē'), *n.* a person held in custody. **detainer**, *n.* one who detains; the holding possession of what belongs to another; (a writ authorizing) the continued detention of a person in custody. **forcible detainer** FORCIBLE under FORCE.
detect (ditekt'), *v.t.* to discover the existence or presence of; to perceive. **detectable**, *a.* **detection**, *n.* the act of detecting; the discovery of crime, guilt etc., or of minute particles. **detective**, *a.* employed in or suitable for detecting. *n.* a police officer employed to investigate special cases of crime etc. (in full, **detective officer**). **private detective**, a private person or an agent of a detective bureau employed privately to investigate cases. **detector**, *n.* one who detects; the part of a radio receiver which demodulates the radio waves.
detent (ditent'), *n.* a pin, catch or lever forming a check to the mechanism in a watch, clock, lock etc.
détente (dātēnt'), *n.* relaxation of tension, esp. between nations, or other warring forces. [F]
detention (diten'shən), *n.* the act of detaining; the state of being detained; arrest; confinement, compulsory restraint; keeping in school after hours as a punishment. **detention centre**, *n.* a place where young offenders are detained.
deter (ditœ'), *v.t.* to discourage or frighten (from); to hinder or prevent. **determent**, **deterrence** (-te'-), *n.* the act of deterring; a deterrent. **deterrent** (-te'-), *a.* tending to deter. *n.* that which deters; a nuclear weapon the possession of which is supposed to deter the use of a similar weapon by another power.
detergent (ditœ'jənt), *n.* a cleansing. *n.* a chemical cleansing agent for washing clothes etc.
deteriorate (ditiə'riərāt), *v.t.* to make worse. *v.i.* to become worse; to degenerate. **deterioration**, *n.* **deteriorative**, *a.*
determinant (ditœ'minənt), *a.* determinative, decisive. *n.* one who or that which determines or causes to fix or decide; the sum of a series of products of several numbers, the products being formed according to certain laws, used in the solution of equations and other processes.
determinate (ditœ'minət), *a.* limited, definite; conclusively determined; (*Math.*) admitting of a finite number of

solutions. **determinately,** *adv.* **determinateness,** *n.* **determination** (minā'shən), *n.* the act of determining or settling; that which is determined on; fixed intention, resolution, strength of mind; direction to a certain end, a fixed tendency; ascertainment of amount etc.; (*Law*) settlement by a judicial decision. **determinative,** *a.* that limits or defines; directive, decisive; defining, serving to limit; tending to determine the genus etc. to which a thing belongs. *n.* that which decides, defines or specifies. **determinator,** *n.*
determine (ditœ'min), *v.t.* (*Law*) to bring to an end; to fix the limits of, to define; to fix, to settle finally, to decide; to direct, to condition, to shape; to ascertain exactly; to cause to decide. *v.i.* to end, to reach a termination; to decide, to resolve. **determinable,** *a.* **determinability** (-bil'-), *n.* **determined,** *a.* resolute; having a fixed purpose; ended; limited, conditioned. **determinedly,** *adv.* **determiner,** *n.* one who or that which determines; (*Gram.*) a word that limits or modifies a noun, as *that, my, his.* **determinism** (-minzm), *n.* the doctrine that the will is not free, but is determined by antecedent causes, whether in the form of internal motives or external necessity, the latter being the postulate of fatalism. **determinist,** *a.* pertaining to determinism. *n.* one who believes in determinism. **deterministic** (-nis'tik), *a.*
deterrent DETER.
detest (ditest'), *v.t.* to hate exceedingly, to abhor. **detestable,** *a.* **detestableness,** *n.* **detestability** (-bil'-), *n.* **detestably,** *adv.* **detestation** (dētestā'-), *n.* extreme hatred; abhorrence, loathing; a person or thing detested.
dethrone (dithrōn'), *v.t.* to remove or depose from a throne; to drive from power or pre-eminence. **dethronement,** *n.* **dethroner,** *n.*
detinue (det'inū), *n.* (*Law*) unlawful detention. **action of detinue,** an action to recover property illegally detained.
detonate (det'ənāt), *v.t.* to cause to explode. *v.i.* to explode. **detonating,** *n.*, *a.* **detonation,** *n.* the act or process of detonating; an explosion with a loud report; a noise resembling this; the spontaneous combustion in a petrol engine of part of the compressed charge after sparking; the knock that accompanies this. **detonator,** *n.* a device which sets off a larger explosive charge; on the railways, a fog signal.
detour (dē'tuə), *n.* a roundabout way; a diversion from the direct route. *v.t.* to send by an indirect route. *v.i.* to make a deviation from a direct route.
detoxicate (dētok'sikāt), **detoxify** (dētok'sifī), *v.t.* to remove poison or toxin from. **detoxicant,** *n.* a detoxifying substance. **detoxication, detoxification** (-fi-), *n.*
detract (ditrakt'), *v.i.* to take some desirable quality away (from). **detraction,** *n.* the act of detracting; depreciation, slander. **detractive,** *a.* **detractor,** *n.* one who detracts; a defamer, a slanderer; a muscle which draws one part from another.
detrain (dētrān'), *v.t.* to cause to alight from a train. *v.i.* to alight from a train. **detrainment,** *n.*
detriment (det'rimənt), *n.* loss; harm, injury, damage. **detrimental** (-men'-), *a.* causing detriment. **detrimentally,** *adv.*
detrited (ditrī'tid), *a.* (*Geol.*) worn away; disintegrated. **detrital,** *a.* **detrition** (-trī'-), *n.* a wearing down or away by rubbing. **detritus** (-təs), *n.* (*Geol.*) accumulated matter produced by the disintegration of rock; debris, gravel, sand etc.
de trop (də trō'), in the way. [F]
detumescence (dētūmes'əns), *n.* the diminution of swelling.
deuce[1] (dūs), *n.* two; a card or die with two spots; (*Tennis*) a score of 40 all, requiring two successive points to be scored by either party to win. **deuce-ace,** *n.* the one and two thrown at dice.
deuce[2] (dūs), *n.* the devil, invoked as a mild oath.

deuced, *a.* confounded, devilish. *adv.* damned, devilishly. **deucedly,** *adv.*
deus (dā'us), *n.* god. **deus ex machina** (eks mak'inə, məshē'nə), in Greek and Roman drama, a god brought on to resolve a seemingly irresolvable plot; a contrived dénouement. [L]
Deut., (*abbr.*) Deuteronomy.
deuterium (dūtə'riəm), *n.* heavy hydrogen, an isotope of hydrogen with double mass.
deutero-, deuto-, *comb. form.* second, secondary.
deuteron (dū'təron), *n.* a heavy hydrogen nucleus.
Deuteronomy (dūtəron'əmi), *n.* the fifth book of the Pentateuch, named from its containing a recapitulation of the Mosaic law. **Deuteronomic, -ical** (-nom'-), *a.*
deuto- DEUTERO-.
deutoplasm (dū'tōplazm), *n.* that portion of the yolk that nourishes the embryo, the food yolk of an ovum or eggcell.
Deutsche Mark, Deutschmark (doitsh'mahk), *n.* the standard monetary unit in West Germany.
deutzia (dūt'siə, doit'-), *n.* a genus of Chinese or Japanese shrubs of the saxifrage family, with clusters of pink or white flowers.
devalue (dēval'ū), **devaluate** (-āt), *v.t.* to reduce the value of currency; to stabilize currency at a lower level. **devaluation,** *n.*
Devanagari (dāvənah'gəri), *n.* the formal alphabet in which Sanskrit and certain vernaculars are usually written, also called simply Nagari.
devastate (dev'əstāt), *v.t.* to lay waste, to ravage; to shock, to overwhelm. **devastation,** *n.* **devastating,** *a.* (*coll.*) shattering, overwhelming. **devastator,** *n.*
develop (divel'əp), *v.t.* to unfold or uncover, to bring to light gradually; to bring from a simple to a complex state; to promote the growth of; to bring to completion or maturity by natural growth; to contract (a disease); to acquire, e.g. a taste or liking for something; (*Mil.*) to carry out the successive stages of an attack; to render visible (as the picture latent in sensitized film); to build on or change the use of (land); to exploit (natural resources); to elaborate upon (a musical theme); to move (a chess piece) into a more effective position. *v.i.* to expand; to progress; to be evolved; to come to light; to come to maturity. **developable,** *a.* **developer,** *n.* one who or that which develops, esp. one who develops land; a chemical agent used to expose the latent image on film or light-sensitive paper. **development,** *n.* the act of developing; the state of being developed; growth and advancement; evolution; maturity, completion; the section of a movement in which a musical theme is elaborated. **development area,** *n.* a region where new industries are being encouraged by Government to combat unemployment. **developmental** (-men'-), *a.* pertaining to development or growth; evolutionary. **developmentally,** *adv.*
deviate (dē'viāt), *v.i.* to turn aside; to diverge or stray from the statistical, psychological or political norm. *n.* (-ət), deviant. **deviant,** *n.* one whose esp. sexual behaviour deviates from what is considered acceptable; *a.* abnormal, deviating. **deviance,** *n.* **deviation,** *n.* the act of deviating; error; the deflection of a compass from the true magnetic meridian; in statistics, the difference between an observed value and the mean; divergence from an accepted standard or the party line. **deviationist,** *n.* one who departs from orthodox Communist doctrine. **deviator,** *n.* **deviatory,** *a.*
device (divīs'), *n.* a plan, a scheme; a trick; a tool, a machine, a contrivance; a design, a pattern; (*Her.*) an emblem or fanciful design, a motto; a fanciful idea, a conceit; a bomb. **to leave (someone) to his/her own devices,** to leave (someone) to do as he/she wishes.
devil (dev'l), *n.* Satan, the chief of the fallen angels, the spirit of evil, the tempter; any evil spirit; the spirit

possessing a demoniac; a wicked person; a malignant or cruel person; a person of extraordinary energy, ingenuity and self-will directed to selfish or mischievous ends; an unfortunate person, a wretch; energy, dash, unconquerable spirit; an expletive expressing surprise or vexation; a printer's errand-boy; a hot grilled dish, highly seasoned; one who does literary work for which another takes the credit; a barrister who prepares a case for another, or who takes the case of another without fee in order to gain reputation; a spiked mill for tearing rags; a Tasmanian marsupial, *Dasyurus ursinis;* various other animals, fish etc. *v.t.* to make devilish; to prepare (food) with highly spiced condiments; to tear up rags with a devil. *v.i.* to act as a literary or legal devil; to do the hard spade-work. **between the devil and the deep blue sea**, torn between two equally undesirable alternatives. **devil's food cake**, (*N Am.*) rich chocolate cake. **speak of the devil!** said when the person who is the subject of conversation arrives. **the devil**, a nuisance; a dilemma, an awkward fix; (*int.*) an expression of surprise or annoyance. **the devil take the hindmost**, one must look after one's own interests. **the devil to pay**, the consequences will be serious. **to give the devil his due**, to allow the worst man credit for his good qualities. **to go to the devil**, to go to ruin; (*imper.*) be off! **you little, young, devil**, a playful, semi-ironical address. **devil-fish**, *n.* the octopus; various other fish, as (*N Am.*) *Lophius piscatorius* and *Cephalopterus vampyrus*. **devil-may-care**, *a.* reckless. **devil-may-careness**, *n.* **devil-may-carish**, *a.* **devil-may-carishness**, *n.* **Devil's advocate** ADVOCATE. **devil's bit**, *n.* a small dark-blue scabious, *Scabiosa succisa*. **devil's coach-horse**, *n.* a large cocktail beetle, *Ocypus olens*. **devilhood** (-hud), *n.* **devilish**, *a.* befitting a devil; diabolical; damnable. *adv.* extraordinarily, damnably, infernally, awfully. **devilishly**, *adv.* **devilishness**, *n.* **devilment**, *n.* mischief, roguery, devilry. **devilry, -iltry** (-tri), *n.* diabolical wickedness, esp. cruelty; diabolism, black magic, wild and reckless mischief, revelry or high spirits.

devious (dē'viəs), *a.* indirect, out of the way, circuitous; erring, rambling; insincere, evasive; deceitful. **deviously**, *adv.* **deviousness**, *n.*

devise (diviz'), *v.t.* to invent, to contrive; to form in the mind, to scheme, to plot; (*Law*) to give or assign (real property) by will. *n.* the act of bequeathing landed property by will; a will or clause of a will bequeathing real estate. **devisable**, *a.* **devisee** (-zē'), *n.* (*Law*) one to whom anything is devised by will. **deviser**, *n.* one who devises. **devisor**, *n.* (*Law*) one who bequeaths by will.

devitalize, -ise (dēvi'təliz), *v.t.* to deprive of vitality or of vital power. **devitalization, -isation**, *n.*

devoice (dēvois'), *v.t.* to pronounce without vibrating the vocal chords. **devoiced**, *a.*

devoid (divoid'), *a.* vacant, destitute, empty (of).

devolution (dē'vəloo'shən), *n.* transference or delegation of authority (as by Parliament to its committees); transference of authority from central to regional government, thus a modified form of Home Rule; passage from one person to another; descent by inheritance; descent in natural succession.

devolve (divolv'), *v.t.* to cause to pass to another, to transfer, e.g. duties, power. *v.i.* to be transferred, delegated or deputed (to); to fall by succession, to descend.

Devonian (divō'niən), *a.* pertaining to *Devonshire;* relating to the fourth period of the Palaeozoic era, between the Silurian and Carboniferous periods. *n.* a native or inhabitant of Devon; (*Geol.*) the Old Red Sandstone formation, well displayed in Devonshire. **Devonshire cream** (dev'ənshə), *n.* clotted cream. **Devonshire split**, *n.* a yeast bun with jam and cream.

devonport (dev'ənpawt), DAVENPORT.

devote (divōt'), *v.t.* to consecrate, to dedicate; to apply; to give wholly up (to). **devoted**, *a.* dedicated, consecrated; wholly given up, zealous, ardently attached. **devotedly**, *adv.* **devotedness**, *n.* **devotee** (devətē'), *n.* a votary, a person devoted (to); an enthusiast. **devotion**, *n.* the act of devoting; the state of being devoted; (*pl.*) prayer, religious worship; deep, self-sacrificing attachment, intense loyalty. **devotional**, *a.* pertaining to or befitting religious devotion. **devotionalism**, *n.* **devotionalist**, *n.* **devotionality** (-nal'-), *n.* **devotionally**, *adv.*

devour (divowə'), *v.t.* to eat up ravenously or swiftly; to consume as a beast consumes its prey; to destroy wantonly, to waste; to swallow up, to engulf; to take in eagerly with the senses; to absorb, to overwhelm. **devourer**, *n.* **devouring**, *a.* that devours; consuming, wasting. **devouringly**, *adv.*

devout (divowt'), *a.* deeply religious; pious, filled with devotion; expressing devotion; heartfelt, earnest, genuine. **devoutly**, *adv.* **devoutness**, *n.*

dew (dū), *n.* moisture condensed from the atmosphere upon the surface of bodies at evening and during the night; anything falling cool and light, so as to refresh; an emblem of freshness; dewy moisture, tears, sweat. *v.t.* to wet with dew. *v.i.* to form as dew; to fall as dew. **mountain-dew**, whisky distilled illicitly. **dewberry**, *n.* a kind of blackberry, *Rubus caesius*. **dew-claw**, *n.* one of the bones behind a deer's foot; the rudimentary upper toe often found in a dog's foot. **dewdrop**, *n.* a drop of dew; a drop at the end of one's nose. **dew-dropping**, *a.* wetting, rainy. **dewfall**, *n.* the falling of dew; the time when dew falls. **dewpoint**, *n.* the temperature at which dew begins to form. **dew pond**, *n.* a shallow, artificial pond formed on high land where water collects at night through condensation. **dewless**, *a.* **dewy**, *a.* **dewy-eyed**, *a.* naive, innocent. **dewily**, *adv.* **dewiness**, *n.*

Dewey Decimal System (dū'i), a library book classification system based on ten main subject classes. [Melvil *Dewey*, 1851–1931, US educator]

dewlap (dū'lap), *n.* the flesh that hangs loosely from the throat of an ox or cow; the flesh of the throat become flaccid through age; the wattle of a turkey etc. **dewlapped**, *a.*

DEW line (dū), *n.* the radar network in the Arctic regions of N America. [*distant early warning*]

dexter (deks'tə), *a.* pertaining to or situated on the right-hand side; situated on the right of the shield (to the spectator's left) etc.

dexterity (dekste'riti), *n.* physical or mental skill, expertness; readiness and ease; cleverness, quickness, tact. **dexterous** (deks'-), *a.* expert in any manual employment; quick mentally; skilful, able; done with dexterity. **dexterously**, *adv.*

dextral (dek'strəl), *a.* inclined to the right; right-handed; having the whorls (of a spiral shell) turning towards the right, dextrorse. **dextrality** (-tral'-), *n.* **dextrally**, *adv.*

dextran (deks'tran), *n.* a carbohydrate produced by the action of bacteria in sugar solutions, used as a substitute for blood plasma in transfusions.

dextrin (deks'trin), *n.* a gummy substance obtained from starch.

dextro-, *comb. form.* on or towards the right.

dextrocardia (dekstrōkah'diə), *n.* a condition in which the heart lies on the right side of the chest instead of the left. **dextrocardiac** (-ak), *n.*, *a.*

dextro-rotary (dekstrōrō'təri), **dextro-rotatory** (-tā'-), *a.* causing to rotate clockwise. **dextro-rotation**, *n.*

dextrorse (dekstraws'), *a.* rising from left to right in a spiral line. **dextrorsely**, *adv.*

dextrose (deks'trōs), *n.* a form of glucose which rotates polarized light clockwise; grape-sugar.

dextrous (deks'trəs), DEXTEROUS.

DF, (*abbr.*) Defender of the Faith; Dean of Faculty; direction-finding.

DFC, (*abbr.*) Distinguished Flying Cross.

DG   220   **diameter**

DG, (*abbr.*) by the grace of God (L *Dei gratia*); Director-General.
dg, (*abbr.*) decigram(me).
dhal (dahl), DAL.
dharma (dŭr'mə), *n.* in Hinduism and Buddhism, the fundamental concept of both natural and moral law, by which everything in the universe acts according to its essential nature or proper station.
dhobi (dō'bi), *n.* an Indian washerman.
dhoti (dō'ti), *n.* a loin-cloth worn by male Hindus.
dhow (dow), *n.* an Arab vessel with one mast, a very long yard, and a lateen sail.
DHSS, (*abbr.*) Department of Health and Social Security.
di-¹, *pref.* twice, two, dis-, double.
di-², *pref.* form of DIA- used before a vowel.
dia-, *pref.* through; thorough, thoroughly; apart, across.
diabetes (dīəbē'tis, -tēz), *n.* a disease marked by excessive discharge of urine. **diabetes insipidus** (insip'idəs), *n.* diabetes caused by a disorder of the pituitary gland. **diabetes mellitus** (meli'təs), *n.* diabetes characterized by a disorder of carbohydrate metabolism, caused by insulin deficiency. **diabetic** (-bet'-), *a.* pertaining to diabetes. *n.* a person suffering from diabetes.
diablerie (diahb'ləri), *n.* dealings with the devil; diabolism, magic or sorcery; rascality, devilry.
diabolic (dīəbol'ik), *a.* pertaining to, proceeding from, or like the devil; outrageously wicked or cruel; fiendish, devilish. **diabolical**, *a.* diabolic; (*coll.*) appalling, dreadful. **diabolically**, *adv.* **diabolism** (-ab'-), *n.* devil-worship; belief in the Devil or in devils; black magic; devilish conduct or character, devilry. **diabolist** (-ab'-), *n.* **diabolize, -ise** (-ab'-), *v.t.* to make diabolical; to represent as a devil.
diabolo (diab'əlō), *n.* a game with a double cone spun in the air by a cord on two sticks.
diachronic (dīəkron'ik), *a.* applied to the study of the historical development of a subject, e.g. language. **diachronically**, *adv.* **diachronism** (-ak'-), *n.* **diachronistic** (-nis'-), *a.* **diachronous** (-ak'-), *a.*
diacid (dīas'id), **diacidic** (-sid'ik), *a.* having two replaceable hydrogen atoms; capable of neutralizing two protons with one molecule.
diaconal (dīak'ənəl), *a.* pertaining to a deacon. **diaconate** (-nət), *n.* the office, dignity or tenure of the office of a deacon; deacons collectively.
diacritic (dīəkrit'ik), **-ical**, *a.* distinguishing, distinctive; serving as a diacritic. **diacritical mark, diacritic,** *n.* a mark (e.g. accent, cedilla, umlaut) attached to letters to show modified phonetic value or stress.
diadelph (dī'ədelf), *n.* a plant of the Linnaean order Diadelphia, in which the stamens are united into two bodies or bundles by their filaments. **diadelphous** (-del'-), *a.*
diadem (diə'dem), *n.* a fillet or band for the head, worn as an emblem of sovereignty; a crown esp. of glory or victory; supreme power, sovereignty.
diaeresis (dīe'rəsis), *n.* (*pl.* **diaereses** (-sēz)) the resolution of one syllable into two; a mark placed over the second of two vowels to show that it must be pronounced separately, as *naïve*. **diaeretic** (-ret'-), *a.*
diagnosis (dīəgnō'sis), *n.* (*pl.* **-noses** (-sēz)), identification of diseases by their symptoms; a statement of the cause of a particular set of symptoms; an analysis of phenomena or problems in order to gain an understanding. **diagnose** (-əgnōz'), *v.t.* to ascertain the nature (of a disease, problem etc.) from symptoms. *v.i.* to make a diagnosis of a disease. **diagnosable**, *a.* **diagnostic** (-nos'-), *a.* serving to distinguish; pertaining to diagnosis. *n.* a sign or symptom by which anything is distinguished from anything else; (*pl.*) diagnosis. **diagnostically**, *adv.* **diagnostician** (-tish'ən), *n.* one who diagnoses, esp. a doctor.
diagonal (dīag'ənəl), *a.* extending from one angle of a quadrilateral or multilateral figure to a non-adjacent angle, or from one edge of a solid to a non-adjacent edge; oblique, crossing obliquely. *n.* a right line or plane extending from one angle or edge to a non-adjacent one; a diagonal row, line, beam, tie etc.
diagram (dī'əgram), *n.* (*Geom.*) a drawing made to demonstrate or illustrate some proposition, statement or definition; a figure, drawn roughly or in outline to illustrate the form or workings of something; a graphical representation of the results of meteorological, statistical or other observations. **diagrammatic** (-mat'-), *a.* **diagrammatically**, *adv.* **diagrammatize, -ise** (-gram'-), *v.t.*
dial., (*abbr.*) dialect.
dial (dī'əl), *n.* a sundial; the graduated and numbered face of a timepiece; a similar plate on which a needle marks revolutions, indicates pressure etc.; the panel or face on a radio showing wavelength and frequency; the rotating, numbered disk on a telephone; (*sl.*, *dated*) the human face. *v.t.* (*past, p.p.* **dialled**) to measure or indicate with or as with a dial; to indicate on the dial of a telephone the number one wishes to call up. **dialling code**, *n.* a group of numbers dialled to obtain an exchange in an automatic dialling system. **dialling tone**, *n.* the sound given by a telephone to show that the line is clear. **dial-plate**, *n.* the face of a timepiece or other instrument with a dial.
dialect (dī'əlekt), *n.* a form of speech or language peculiar to a limited district or people. **dialectal** (-lek'-), *a.* **dialectally**, *adv.* **dialectology** (-tol'-), *n.* the study of dialects. **dialectologist**, *n.*
dialectic (dīəlek'tik), *a.* pertaining to logic; logical, argumentative. *n.* (*often pl.*) logic in general; the rules and methods of reasoning; discussion by dialogue usu. in order to reconcile or resolve contradictions; the investigation of truth by analysis; (*Kant*) critical analysis of knowledge based on science; (*Hegel*) the philosophic process of reconciling the contradictions of experience in a higher synthesis, the world-process which is the objective realization of this synthesis. **dialectical**, *a.* dialectic; dialectal. **dialectical materialism**, *n.* the economic, political and philosophical system developed by Marx and Engels, based on the idea of constant change through a dialectical process of thesis, antithesis and synthesis. **dialectically**, *adv.* in a logical manner; dialectally. **dialectician** (-ti'shən), *n.* one skilled in dialectics; a logician; a reasoner.
dialogue (dī'əlog), *n.* a conversation or discourse between two or more persons; a literary composition in conversational form; the conversational part of a novel, play, film etc.; a political discussion between two groups or nations. *v.i.* to hold a dialogue. **dialogic** (-loj'-), *a.* of the nature of a dialogue. **dialogically**, *adv.* **dialogist** (-al'əjist), *n.* one who takes part in a dialogue; a writer of dialogues.
dialysis (dīal'isis), *n.* the process of separating the crystalloid from the colloid ingredients in soluble substances by passing through moist membranes; the filtering of blood to remove waste products, either by semipermeable membranes in the body, or by a kidney machine. **dialyse**, (*esp. N Am.*) **dialyze** (dī'əliz), *v.t.* **dialyser** (di-), *n.* the apparatus in which the process of dialysis is performed. **dialytic** (-lit'-), *a.*
diam., (*abbr.*) diameter.
diamagnetic (dīəmagnet'ik), *a.* pertaining to or exhibiting diamagnetism. **diamagnetically**, *adv.* **diamagnetism** (-mag'-), *n.* the force which causes certain bodies, when suspended freely and magnetized, to assume a position at right angles to the magnetic meridian, and point due east and west. **diamagnetize, -ise** (-mag'-), *v.t.*
diamanté (dēəman'ti), *n.*, *a.* material covered with glittering particles, such as sequins. **diamantine** (dīəman'tin), *a.* diamond-like. **diamantiferous** (-tif'ərəs), *a.* yielding diamonds.
diameter (dīam'itə), *n.* (the length of) a straight line passing through the centre of any object from one side to the

diamond   221   **dichroic**

other; (the length of) a straight line passing through the centre of a circle or other curvilinear figure, and terminating each way in the circumference; transverse measurement, width, thickness. **diametral**, *a.* **diametrally**, *adv.* **diametrical** (met'-), *a.* pertaining to a diameter; along a diameter; directly opposed; as far removed as possible. **diametrically**, *adv.*

diamond (dī'əmənd), *n.* the hardest, most brilliant and most valuable of the precious stones, a transparent crystal of pure carbon, colourless or tinted; a facet of this when cut; a figure resembling this, a rhomb; a playing-card with figures of this shape; (*pl.*) the suit comprising these cards; a glazier's cutting tool with a diamond at the point; a small rhomboid sheet of glass used in old-fashioned windows; in baseball, the entire playing area or that part of it bounded by the four bases. *a.* made of, or set with, diamonds; resembling a diamond or lozenge. *v.t.* to adorn with or as with diamonds. **black diamonds**, dark-coloured diamonds; coal. **rough diamond**, a diamond in the native state, not yet cut; (*coll.*) a worthy, good-hearted, but uncouth person. **diamond-back**, *n.* the salt-water turtle or terrapin; a deadly N American rattlesnake with diamond-shaped markings. **diamond-drill**, *n.* an annular drill with diamonds set in the cutting edge. **diamond-field**, *n.* a region yielding diamonds. **diamond jubilee**, *n.* the 60th anniversary of a sovereign's accession. **diamond wedding**, *n.* the 60th anniversary of a marriage. **diamondiferous** (-dif'-), *a.* yielding diamonds. **diamond-wise**, *adv.*

Diandria (dīan'driə), *n.pl.* a Linnaean order of plants the flowers of which have only two stamens. **diandrous**, *a.*

dianthus (dīan'thəs), *n.* a genus of caryophyllaceous plants, including the pinks and carnations.

diapason (dīəpā'zən), *n.* a foundation stop of an organ; a harmonious burst of music; a recognized standard of pitch amongst musicians; range, compass.

diapause (dī'əpawz), *n.* a period of suspended growth in insects.

diaper (dī'əpə), *n.* a silk or linen cloth woven with geometric patterns; a towel or napkin made of this; (*esp. N Am.*) a baby's napkin, a nappy; a surface decoration consisting of square or diamond reticulations. *v.t.* to decorate or embroider with this. **diaper-work**, *n.*

diaphanous (diaf'ənəs), *a.* transparent, pellucid; of fabrics, so fine as to be almost transparent. **diaphanously**, *adv.*

diaphoretic (dīəfəret'ik), *a.* having the power of promoting perspiration. *n.* a medicine having this property.

diaphragm (dī'əfram), *n.* the large muscular partition separating the thorax from the abdomen; the straight calcareous plate dividing the cavity of certain shells into two parts; the vibrating disk in the mouthpiece and earpiece of a telephone, or in the loudspeaker of a radio receiver; a dividing membrane or partition; an annular disk excluding marginal rays of light; a thin rubber or plastic cap placed over the mouth of the cervix as a contraceptive. **diaphragmatic** (-fragmat'-), *a.* **diaphragmatitis** (-fragməti'tis), *n.* inflammation of the diaphragm.

diapositive (dīəpoz'itiv), *n.* a positive photographic transparency; a slide.

diarist etc. DIARY.

diarrhoea (dīərē'ə), *n.* the excessive discharge of faecal matter from the intestines. **diarrhoeal, diarrhoeic**, *a.*

diary (dī'əri), *n.* an account of the occurrences of each day; the book in which these are registered; a daily calendar with blank spaces for notes. **diarial** (-eə'riəl), *a.* **diarian** (-eə'riən), *n.* **diarist**, *n.* one who keeps a diary. **diaristic** (-ris'-), *a.* **diarize, -ise**, *v.t., v.i.*

Diaspora (dīas'pora), *n.* (*Hist.*) the dispersion of the Jews after the Babylonian captivity; Jews living outside Palestine, or now, outside Israel; a dispersion or migration of peoples.

diastase (dī'əstās), *n.* a nitrogenous substance produced during the germination of all seeds, and having the power of converting starch into dextrine, and then into sugar. **diastasic** (-sta'-), *a.*

diastole (dīas'təli), *n.* dilatation of the heart and arteries alternating with systole. **systole and diastole**, the pulse; (*fig.*) regular reaction; fluctuation. **diastolic** (-stol'-), *a.*

diastrophism (dīas'trəfizm), *n.* deformation of the earth's crust, giving rise to mountains etc.

diathermancy (dīəthœ'mənsi), *n.* the property of being freely pervious to heat. **diathermaneity** (-məne'əti), *n.* **diathermanous, diathermous**, *a.* **diathermometer** (-mom'itə), *n.*

diathermy (dī'əthœmi), *n.* (*Med.*) the employment of high-frequency currents for the production of localized heat in the tissues.

diathesis (dīath'əsis), *n.* (*pl* **-theses** (-sēz)) a constitution of body predisposing it to certain diseases. **diathetic** (-thet'-), *a.*

diatom (dī'ətəm), *n.* an individual of a group of microscopic algae with siliceous coverings which exist in immense numbers at the bottom of the sea and occur as fossils in such abundance as to form strata of vast area and considerable thickness. **diatomaceous** (-mā'shəs), *a.* **diatomist** (-at'-), *n.* one who studies diatoms. **diatomite** (-at'əmit), *n.* (*Geol.*) any diatomaceous deposit.

diatomic (dīətom'ik), *a.* (*Chem.*) containing only two atoms; containing two replaceable univalent atoms.

diatonic (dīəton'ik), *a.* (*Mus.*) of the regular scale without chromatic alteration; applied to the major and minor scales, or to chords, intervals and melodic progressions. **diatonically**, *adv.*

diatribe (dī'ətrib), *n.* a bitter, abusive or denunciatory criticism or verbal attack.

diazepam (dia'zipam), *n.* a type of tranquillizer and muscle relaxant.

diazo (dia'zō), *a.* of a compound, having two nitrogen atoms and a hydrocarbon radical; of a photocopying technique using a diazo compound exposed to light. *n.* (*pl.* **-zos, -zoes**) a copy made in this way.

dib (dib), *n.* a sheep's knuckle-bone; (*pl.*) a children's game in which these are thrown into the air and caught on the back of the hand; (*pl.*) (*sl.*) money.

dibasic (dibā'sik), *a.* containing two bases or two replaceable atoms.

dibber (dib'ə), *n.* a dibble.

dibble (dib'l), *n.* a pointed instrument used to make a hole in the ground to receive seed. *v.t.* to make holes with a dibble; to plant with a dibble. *v.i.* to use a dibble; to dap as in angling. **dibbler**, *n.*

dice (dis), *n.* (*pl.* **dice**) a small cube marked with a different number of spots on each side, used in games of chance; a gambling game played with dice; a small cube of e.g. meat or vegetable. *v.i.* to play at dice. *v.t.* to gamble (away) at dice; to cut up (food) into small cubes. **no dice**, an expression of refusal or lack of success. **to dice with death**, to take a great risk. **dice-box**, *n.* the case out of which dice are thrown. **dicer**, *n.*

dicephalous (disef'ələs), *a.* two-headed.

dicey (di'si), *a.* (*coll.*) risky, difficult.

dichloride (diklaw'rid), *n.* a compound having two atoms of chlorine with another atom.

dichlorodiphenyltrichloroethane (diklawrōdifeniltriklawroē'thān. -fēnīl-), *n.* the full name for DDT, a white powder used as an insecticide.

dichotomy (dikot'əmi), *n.* a separation into two; (*Log.*) distribution of ideas into two mutually exclusive classes; (*Bot., Zool.*) a continued bifurcation or division into two parts. **dichotomic** (-tom'-), *a.* **dichotomist**, *n.* **dichotomize, -ise**, *v.t., v.i.* **dichotomous**, *a.*

dichroic (dikrō'ik), *a.* assuming two or more colours, according to the direction in which light is transmitted. **dichroism** (dī'-), *n.* **dichroitic**(-it'-), *a.* **dichroscope** (dī'-), *n.*

dichromate (dīkrō'māt), *n.* a salt containing two chromium atoms.
dichromatic (dīkrəmat'ik), *a.* characterized by or producing two colours, esp. of animals able to distinguish two colours only. **dichromatism** (krō'-), *n.*
dichromic (dīkrō'mik), *a.* having two chromium atoms; dichromatic.
dick (dik), *n.* a fellow or person; (*N Am. sl.*) a detective; (*sl., taboo*) the penis. **clever dick**, a know-all. **dickhead**, *n.* (*sl.*) a fool.
dickens (dik'ənz), *n.* (*coll.*) the devil, the deuce.
Dickensian (diken'ziən), *a.* pertaining to or in the style of Charles *Dickens* (1812-70), British novelist; applied to squalid conditions as described in his novels. *n.* an admirer of Dickens.
dicker (dik'ər), *n., v.i.* to barter, to haggle; to hesitate, to dither. *v.t.* to barter or exchange.
dicky[1] (dik'i), *n.* a front separate from the shirt; a seat behind the body of a carriage or a motor-car; a driver's seat; a bird. **dicky-bird**, *n.* (*coll.*) a little bird; (*sl.*) a word. **dicky-bow**, *n.* a bow-tie.
dicky[2] (dik'i), *a.* doubtful, questionable; unsteady, unwell.
diclinous (dī'klinəs), *a.* having the stamens and the pistils on separate flowers, on the same or different plants. **diclinism**, *n.*
dicotyledon (dīkotilē'dən), *n.* a plant with two cotyledons. **Dicotyledones** (-nēz), *n.pl.* the largest and most important class of flowering plants containing all those with two cotyledons. **dicotyledonous**, *a.*
dicrotic (dīkrot'ik), *a.* (*Physiol.*) double-beating (of a pulse in an abnormal state).
dict., (*abbr.*) dictionary.
dicta DICTUM.
Dictaphone® (dik'təfōn), *n.* an apparatus for recording sounds, used for taking down correspondence etc., to be afterwards transcribed.
dictate[1] (diktāt'), *v.t.* to read or recite to another words to be written or repeated; to prescribe, to lay down with authority, to impose, as terms. *v.i.* to give orders; to utter words to be written or repeated by another. **dictation**, *n.* dictating material to be written down or recorded; the material dictated; command. **dictator** (-tə), *n.* one who dictates; a Roman magistrate created in time of emergency, and invested with absolute power; a ruler with supreme and often tyrranical authority, esp. one appointed in a time of civil disorder or securing the supremacy after a revolution; a very overbearing or authoritarian person. **dictatorate, dictature** (-tūə), *n.* **dictatorial** (-tətaw'ri-), *a.* pertaining to a dictator; imperious, overbearing. **dictatorially**, *adv.* **dictatorship**, *n.* **dictatress** (-tris), *n. fem.*
dictate[2] (dik'tat), *n.* an order, an injunction; a direction; a precept.
diction (dik'shən), *n.* choice of words in writing or speech; enunciation of words and sounds.
dictionary (dik'shənəri), *n.* a book containing the words of any language in alphabetical order, with their definitions, pronunciations, parts of speech, etymologies and uses, or with their equivalents in another language; a work of information on any subject under words arranged alphabetically.
dictum (dik'təm), *n.* (*pl.* **-ta** (-tə)) a positive or dogmatic assertion; a judge's personal opinion on a point of law as distinguished from the decision of a court; a maxim, an adage.
did (did), past of DO[1].
didactic (dīdak'tik, di-), *a.* adapted or tending to teach, esp. morally; containing rules or precepts intended to instruct; in the manner of a teacher. *n.pl.* the science or art of teaching. **didactically**, *adv.* **didacticism**, *n.*
didapper (dī'dapə), *n.* a small diving-bird, the dab-chick.

diddicoy DIDICOI.
diddle (did'l), *v.t.* to cheat; to swindle. **diddler**, *n.*
diddums (did'əmz), *int.* expressing commiseration to a baby. [baby-talk, did you/he/she]
didgeridoo (dijəridoo'), *n.* an Austral. instrument, a long, hollow wooden tube that gives a deep booming sound when blown.
didicoi, did(d)icoy (did'ikoi), *n.* an itinerant traveller or tinker, who is not a true Romany.[1]
didst (didst), 2nd sing. past of DO[1].
didymium (dīdim'iəm), *n.* a mixture of the two elements neodymium and praseodymium, orig. thought to be a single element.
didymous (did'iməs), *a.* (*Bot.*) twin, growing in pairs.
die[1] (di), *v.i.* to lose life, to expire; to depart this life; to come to an end; to cease to exist; to wither, to lose vitality; to fail, to become useless; to go out, to disappear; to cease or pass away gradually; to faint, to fade away, to languish with affection; to suffer spiritual death; to perish everlastingly. **to die away**, to become gradually less distinct. **to die down**, of plants, to die off above ground, only the roots staying alive; to become less loud, intense etc., to subside. **to die of laughing**, (*fig.*) to laugh at something immoderately. **to die for**, to sacrifice one's life for; to pine for. **to die off**, to die one by one. **to die out**, to become extinct. **to be dying to do**, to be very eager to do. **die-away**, *a.* fainting or languishing. **die-hard**, *n., a.* (one) resistant to change, or holding an untenable position, esp. in politics.
die[2] (di), *n.* (*pl.* **dice** (dīs)) a dice; (*in foll. senses pl.* **dies**) (*Arch.*) the cube or plinth of a pedestal; a machine for cutting out, shaping or stamping; a stamp for coining money, or for impressing a device upon metal, paper etc. **straight as a die**, completely honest. **the die is cast**, an irrevocable decision has been taken. **die-cast**, *v.t.* to shape an object by forcing molten lead or plastic into a reusable mould. **die-casting**, *n.* **die-stock**, *n.* a handle or stock to hold the dies in screw-cutting.
dieldrin (dēl'drin), *n.* an insecticide containing chlorine.
dielectric (diilek'trik), *n.* any medium, such as glass, through or across which electric force is transmitted by induction; a non-conductor; an insulator. *a.* non-conductive, insulating. **dielectrically**, *adv.*
dieresis DIAERESIS.
dies (dē'āz), *n.* (*pl.* **dies**) day. **dies irae** (ē'rā), *n.* a 13th-cent. Latin hymn describing the Last Judgment, used in the mass for the dead. **dies non** (non), *n.* a Sunday, holiday or other day on which the courts are not open; a day on which business cannot be transacted.
diesel (dē'zl), *n.* any vehicle driven by a diesel engine; diesel oil. **diesel-electric**, *a.* using power from a diesel-operated electric generator. *n.* a locomotive so powered. **diesel engine**, *n.* a type of reciprocating internal-combustion engine which burns heavy oil. **diesel oil, fuel**, *n.* a heavy fuel oil used in diesel engines, also called DERV. **diesel train**, *n.* one drawn by a diesel engine. **dieselize, -ise**, *v.t.* to adapt or convert (an engine) to diesel fuel. *v.t.* to be equipped with a diesel engine. **dieselization**, *n.* [R. *Diesel*, 1858–1913, G engineer]
diesis (dī'əsis), *n.* (*pl.* **-eses** (-sēz)) the double dagger (‡).
diet[1] (dī'ət), *n.* a prescribed course of food followed for health reasons, or to reduce or control weight; the food and drink usually taken; an allowance of food. *v.t.* to put on a diet. *v.i.* to take food, esp. according to a prescribed regimen or to reduce or control weight. **to be, go, on a diet**, to follow a strict plan of eating so as to lose weight. **dietary**, *a.* pertaining to a diet. *n.* a prescribed course of diet; a fixed daily allowance of food (esp. in institutions). **dietary fibre** FIBRE. **dieter**, *n.* one who follows a diet to lose weight. **dietetic, -ical** (-tet'-), *a.* pertaining to diet; prepared according to special dietary needs. **dietetically**, *adv.* **dietetics**, *n.pl.* the science of diet; rules of diet.

dietician, dietitian (-tish'ən), *n.* a professional adviser on dietetics.
diet[2] (dī'ət), *n.* a legislative assembly or federal parliament (esp. as an English name for Continental parliaments); a conference or congress, esp. on international affairs; (*Sc.*) a session of a court or any assembly.
dietetic, dietician DIET[1].
Dieu et mon droit (dyœ ä mô drwa'), God and my right (motto of British sovereigns). [F]
differ (dif'ə), *v.i.* to be dissimilar; to disagree in opinion; to dissent; to quarrel. **to agree to differ**, to give up trying to convince each other.
difference (dif'rəns), *n.* the state of being unlike or distinct; the quality by which one thing differs from another; disproportion between two things; the remainder of a quantity after another quantity has been subtracted from it; a distinction, a differential mark, the specific characteristic or differentia; a point or question in dispute; a disagreement in opinion, a quarrel, a controversy; a figure on a coat-of-arms which distinguishes the younger branch of a family from the elder. **to make a difference**, to have an effect; to behave differently. **to split the difference**, to compromise; to divide the remainder in an equal way. **with a difference**, with something distinctive added; differently; (*Her.*) as a mark of distinction.
different (dif'rənt), *a.* unlike, dissimilar, distinct, not the same. **differently**, *adv.*
differentia (difəren'shə), *n.* (*pl.* **-tiae** (-iē)) that which distinguishes one species from another of the same genus; an essential attribute, which when added to the name of the genus distinctly marks out the species.
differential (difəren'shəl), *a.* differing; consisting of a difference; making or depending on a difference or distinction; relating to specific differences; pertaining to differentials; relating to the difference between sets of motions acting in the same direction, or between pressures etc. *n.* (*Math.*) an infinitesimal difference between two consecutive states of a variable quantity; a differential gear; the amount of difference within a wages structure between rates of pay for different classes of work. **differential calculus**, *n.* (*Math.*) a method of dealing with the rate of change of functions relative to their variables. **differential coefficient**, *n.* (*Math.*) the measure of the rate of change of a function relative to its variable. **differential equation**, *n.* an equation involving differential coefficients. **differential gear**, *n.* a device of bevelled planetary and other wheels which enables the driving wheels of a motor vehicle to rotate at different speeds when rounding a corner. **differential motion**, *n.* a mechanical movement in which a part moves with a velocity equal to the difference between the velocities of two other parts. **differentially**, *adv.*
differentiate (difəren'shiāt), *v.t.* to make different; to constitute difference between, of or in; to discriminate by the differentia, to mark off as different; (*Math.*) to obtain the differential coefficient of; (*Biol.*) to develop variation in; to specialize. *v.i.* to develop so as to become different; to distinguish (*between*). **differentiation**, *n.* **differentiator**, *n.*
difficult (dif'ikəlt), *a.* hard to do or carry out; troublesome; hard to please; not easily managed; hard to understand; cantankerous. **difficulty**, *adv.* **difficulty**, *n.* the quality of being difficult; anything difficult; an obstacle; objection, reluctance, scruple; (*pl.*) pecuniary embarrassment.
diffident (dif'idənt), *a.* distrustful of oneself or of one's powers; bashful, modest, shy. **diffidence**, *n.* **diffidently**, *adv.*
diffract (difrakt'), *v.t.* to break in parts; to bend or deflect a ray of light by passing it close to an opaque object. **diffraction**, *n.* **diffraction grating**, *n.* an array of fine, closely-spaced opaque lines on glass which disperses light into its component colours since the amount of diffraction differs for different-coloured rays of light.

diffuse[1] (difūz'), *v.t.* to spread abroad by pouring out; to disperse widely; to circulate; to cause to intermingle; to dissipate. *v.i.* to be diffused; to intermingle by diffusion. **diffused**, *a.* **diffused lighting**, *n.* light which is softened and spread over an area instead of being concentrated in one spot. **diffusedly** (-zid-), *adv.* **diffuser**, *n.* **diffusible**, *a.* **diffusibility** (-bil'-), *n.*
diffuse[2] (difūs'), *a.* diffused, scattered, spread out; copious, prolix, not concise; (*Bot.*) diverging or spreading widely. **diffusely**, *adv.* **diffuseness**, *n.* **diffusion** (-zhən), *n.* the act of diffusing a liquid, fluid etc.; a spreading abroad of news etc.; the state of being widely dispersed; the mingling of liquids, gases or solids through contact; spread of cultural elements from one community to another; the re-issuing or adaptation of expensive designer items for a broader popular market. **diffusion line, range**, *n.* a cheaper version of designer items, esp. clothes, intended for a mass market. **diffusive**, *a.* diffusing; tending to diffuse; spreading, circulating, widely distributed. **diffusively**, *adv.* **diffusiveness**, *n.*
dig (dig), *v.t.* (*past* **dug** (dŭg), *p.p.* **dug**) to excavate or turn up with a spade or with hands, claws etc.; to thrust or push into something; to obtain by digging; to make by digging; to poke, to pierce; (*dated sl.*) to approve of or like. *v.i.* to work with a spade; to excavate or turn up ground with a spade or other implement; to search, make one's way, thrust, pierce or make a hole by digging. *n.* a piece of digging (esp. archaeological); a thrust, a poke; a cutting remark; (*N Am., coll.*) a plodding student. **to dig in**, to establish defensive positions; to entrench or establish (oneself); to refuse to budge in an argument; (*coll.*) to begin to eat heartily. **to dig one's heels in** HEEL[1]. **to dig out**, to obtain by digging; to obtain by research. **to dig up**, to excavate; to extract or raise by digging; to unearth, to break up (ground) by digging; to unearth, to obtain by research. **to have a dig at**, to make a cutting or sarcastic remark. **digger**, *n.* one who digs, esp. a gold-miner; an implement, machine or part of a machine that digs; (*Austral.*) a fellow, a man. **digger-wasp**, *n.* any of several wasps that dig a hole in the ground for a nest. **digging**, *n.* **diggings**, *n.pl.* a place where mining is or has been carried on; spoil; (*coll.*, also **digs**) lodgings.
digastric (dīgas'trik), *a.* having two fleshy portions joined by a tendon. **digastric muscle**, *n.* a double muscle which depresses the lower jaw.
digest[1] (dijest'), *v.t.* to arrange under proper heads or titles, to classify; to assimilate mentally; to think over; to soften and prepare by heat; to break (food) down in the stomach into forms which can be assimilated by the body; to promote the digestion of; to make a summary of. *v.i.* to be digested; to be prepared by heat. **digester**, *n.* anything which helps to promote digestion; an apparatus for cooking food by exposure to heat above boiling point. **digestible**, *a.* **digestibility** (-bil'-), *n.* **digestibly**, *adv.* **digestion** (-chən), *n.* the act or process of assimilating food in the stomach; the conversion of food into chyme; the power of digesting; concoction for the purpose of extracting the essence from a substance; mental assimilation. **digestive**, *a.* pertaining to or promoting digestion. *n.* any substance which aids or promotes digestion. **digestive biscuit**, *n.* a semi-sweet biscuit made of wholemeal flour. **digestively**, *adv.*
digest[2] (dī'jest), *n.* a compendium or summary arranged under proper heads or titles; a magazine containing summaries of articles etc. in current literature; (*Law*) a collection of Roman laws arranged under proper heads, as the pandects of Justinian.
digger, digging DIG.
digit (dij'it), *n.* a finger or toe; the measure of a finger's breadth, or three-quarters of an inch; any integer under ten. **digital**, *n., a.* **digital clock, watch**, *n.* one without a traditional face, the time being indicated by a display of

digitalin   diminish

numbers. **digital computer**, *n.* an electronic computer which uses binary or decimal notation. **digital disk, digital (audio) tape**, *n.* a record or tape which is recorded using a digital sound signal. **digitalize, -ise**, *v.t.* to digitize. **digitize, -ise**, *v.t.* to put into digital form for use in a computer. **digitization**, *n.*
digitalin (dijitā'lin), **digitalia** (-liə), *n.* an alkaloid obtained from the foxglove.
digitalis (dijitā'lis), *n.* a genus of scrophulariaceous plants, containing the foxglove (*Digitalis purpurea*); the dried leaves of the foxglove, which act as a cardiac sedative.
digitate (dij'itāt), **-tated**, *a.* having finger-like processes; (*Bot.*) branching into distinct leaves or lobes like fingers. **digitately**, *adv.* **digitation**, *n.*
digitigrade (dij'itigrād), *a.* belonging to the Digitigrada, a section of the Carnivora comprising the cats, dogs, hyenas and weasels, which walk on their toes. *n.* a digitigrade animal.
dignify (dig'nifī), *v.t.* to make worthy; to invest with dignity; to make illustrious; to exalt. **dignified**, *a.* invested with dignity; stately; gravely courteous.
dignity (dig'niti), *n.* rank; the importance due to rank or position; gravity, stateliness; a high office, a position of importance or honour. **to be beneath one's dignity**, to be degrading, in one's own opinion. **to stand on one's dignity**, to assume a manner showing one's sense of self-importance. **dignitary**, *n.* one who holds a position of dignity, esp. ecclesiastical.
digraph (dī'grahf), *n.* a combination of two letters to represent one simple sound, as *ea* in *mead*, *th* in *thin*. **digraphic** (-graf'-), *a.*
digress (digres'), *v.i.* to deviate, to wander from the main topic. **digression** (-shən), *n.* a part of a discourse etc. which wanders from the main subject. **digressive**, *a.* **digressively**, *adv.* **digressiveness**, *n.*
digs DIGGINGS under DIG.
dihedral (dīhē'drəl), *a.* (*Cryst.*) having two sides or faces; (*Math.*) of the nature of a dihedron. **dihedral angle**, *n.* that made by the wing of an aeroplane in relation to the horizontal axis. **dihedron** (-drən), *n.* (*Geom.*) a figure with two sides or surfaces.
dik-dik (dik'dik), *n.* a name for several small E African antelopes.
dike, dyke (dīk), *n.* a ditch, a water-course or channel, natural or artificial; a mound or dam to protect low-lying lands from being flooded; (*Sc.*) a wall or fence of turf or stone without cement; a causeway; a barrier, a defence; a wall-like mass of cooled and hardened volcanic or igneous rock, occupying rents and fissures in sedimentary strata; (*Austral.*, *sl.*) a lavatory; (*sl.*, *offensive*) a lesbian. *v.t.* to defend with dikes or embankments. **dike-reeve**, *n.* an officer in charge of dikes, drains and sluices in fen districts.
diktat (dik'tat), *n.* a settlement imposed on the defeated; an order or statement allowing no opposition.
dilapidate (diləp'idāt), *v.t.* to damage, to bring into decay or ruin. *v.i.* to fall into decay or ruin. **dilapidated**, *a.* ruined; shabby. **dilapidation**, *n.* decay for want of repair; a state of partial ruin, decay; disrepair existing at the end of a tenancy or incumbency; charge for making this good. **dilapidator**, *n.*
dilate (dilāt', di-), *v.t.* to expand, to widen, to enlarge in all directions. *v.i.* to expand, to swell; to expatiate, to speak fully and copiously (upon a subject). **dilatable**, *a.* capable of dilatation; elastic. **dilatability** (-bil'-), *n.* **dilatant**, *n.*, *a.* **dilatancy**, *n.* **dilatation** (-lətā'-), *n.* the act of dilating; the state of being dilated; a dilated or expanded form or part; amplification, diffuseness. **dilation** DILATATION. **dilatometer** (-lətom'itə), *n.* **dilator**, *n.* a muscle that dilates the parts on which it acts; (*Surg.*) an instrument for dilating the walls of a cavity.
dilatory (dil'ətəri), *a.* causing or tending to cause delay;

addicted to or marked by procrastination; slow, tardy; wanting in diligence. **dilatorily**, *adv.* **dilatoriness**, *n.*
dildo(e) (dil'dō), *n.* an object serving as an erect penis.
dilemma (dilem'ə, dī-), *n.* (*Log.*) an argument in which a choice of alternatives is presented, each of which is unfavourable; a position in which one is forced to choose between alternatives equally unfavourable. **the horns of a dilemma**, the alternatives presented to an adversary in a logical dilemma. **dilemmatic** (-mat'-), *a.* **dilemmist**, *n.*
dilettante (dilətan'ti), *n.* (*pl.* **-tantes, -tanti** (-ti)) a lover or admirer of the fine arts; a superficial amateur, a would-be connoisseur, a dabbler. *a.* art-loving; amateurish; superficial. **dilettantish**, *a.* **dilettantism**, *n.*
diligence[1] (dil'ijəns, dēlēzhēs'), *n.* a public stage-coach, formerly used in France and adjoining countries. [F]
diligence[2] (dil'ijəns), *n.* steady application or assiduity in business of any kind; care, heedfulness.
diligent (dil'ijənt), *a.* assiduous in any business or task; persevering, industrious, painstaking. **diligently**, *adv.*
dill (dil), *n.* an annual umbellifer, *Anethum graveolens*, cultivated for its carminative seeds, and for its flavour. **dill pickle**, *n.* a pickled cucumber flavoured with dill. **dill-water**, *n.* a popular remedy for flatulence in children, prepared from the seeds of dill.
dilly (dil'i), *n.* (*N Am. coll.*) a remarkable person or thing. *a.* (*Austral. coll.*) silly.
dilly-bag (dil'ibag), *n.* an Australian Aboriginal basket or bag made of rushes or bark.
dilly-dally (dil'idal'i), *v.i.* to loiter about; to waste time; to hesitate.
dilute (diloot', di-), *v.t.* to make thin or weaken (as spirit, acid or colour) by the admixture of water; to reduce the force or effectiveness of. *a.* diluted; weakened, washed out, faded, colourless. **dilutedly**, *adv.* **diluent** (dil'ūənt), *a.* diluting. *n.* that which dilutes; a substance tending to increase the proportion of fluid in the blood. **dilution**, *n.*
diluvial (diloo'viəl), **-vian**, *a.* pertaining to Noah's flood; (*Geol.*) produced by or resulting from a flood. **diluvial clay**, *n.* the boulder clay. **diluvialist**, *n.* one who regards certain physical phenomena as the result of Noah's flood or a series of catastrophic floods.
dim., (*abbr.*) dimension; diminuendo; diminutive.
dim (dim), *a.* lacking in light or brightness; not clear, not bright; faint, indistinct, misty; tarnished, dull; not clearly seen; not clearly understanding or understood; mentally obtuse. *v.t.* (*past, p.p.* **dimmed**) to render dim. *v.i.* to become dim. **dim-dip headlights**, *n.pl.* low-powered dipped headlights which come on automatically with the sidelights when the engine is running. **to take a dim view of**, (*coll.*) to regard pessimistically; to view with suspicion or disfavour. **dimly**, *adv.* **dimmer**, *n.* (*Elec.*) a device whereby an electric lamp can be switched on and off gradually. **dimmish**, *a.* **dimness**, *n.*
dime (dīm), *n.* a US coin worth 10 cents or one-tenth of a dollar. **a dime a dozen**, (*coll.*) cheap, ordinary. **dime novel**, *n.* a sensational story, the equivalent of the penny dreadful.
dimension (dimen'shən, di-), *n.* (*usu. pl.*) measurable extent or magnitude, length, breadth, height, thickness, depth, area, volume etc.; (*usu. pl.*) size, scope, extent; (*Math.*) the number of co-ordinates required to locate a point in space; an aspect. **fourth dimension**, the extra (time) coordinate needed to locate a point in space. **three dimensions**, length, breadth and thickness. **dimensional**, *a.* (*usu. in comb.*) of a specific dimension; having the stated number of dimensions. **dimensioned**, *a.* (*usu. in comb.*) having dimensions of the stated type. **dimensionless**, *a.*
diminish (dimin'ish), *v.t.* to make smaller or less; to reduce in quantity, power, rank etc.; to disparage, to degrade; (*Mus.*) to lessen by a semitone. *v.i.* to become less, to decrease; to taper. **diminishable**, *a.* **diminished**, *a.*

diminuendo 225 **dip**

reduced in size or quality; (*Mus.*) lessened by a semitone.
diminished responsibility, *n.* a plea in law in which criminal responsibility is denied on the grounds of mental derangement. **diminisher**, *n.* **diminishing returns**, *n.pl.* progressively smaller increases in output in spite of increased work or expenditure. **diminishingly**, *adv.* **diminution** (-nū'-), *n.* the act of diminishing; the state of becoming less or smaller; (*Arch.*) the gradual decrease in the diameter of the shaft of a column from the base to the capital.
diminuendo (diminūen'dō), *a., adv.* (*Mus.*) gradually decreasing in loudness. *n.* (*pl.* **-dos, does**) a gradual decrease in loudness; a passage characterized by this.
diminutive (dimin'ūtiv), *a.* small, tiny; (*Gram.*) expressing diminution. *n.* anything of a small size; a word formed from another to express diminution in size or importance, or affection. **diminutival** (-ti'-), *a.* expressing diminution; pertaining to a diminutive word. **diminutively**, *adv.* **diminutiveness**, *n.*
dimity (dim'iti), *n.* a stout cotton fabric with stripes or patterns.
dimmer, dimming DIM.
dimorphic (dimaw'fik), *a.* having or occurring in two distinct forms. **dimorphism**, *n.* the power of assuming or crystallizing into two distinct forms; a difference of form between members of the same species; a state in which two forms of flower are produced by the same species; the existence of a word in more than one form. **sexual dimorphism**, difference in form between the two sexes of a species. **dimorphous**, *a.*
dimple (dim'pl), *n.* a little depression or hollow; a small natural depression on the cheek or chin; a ripple; a shallow dell or hollow in the ground. *v.t.* to mark with dimples. *v.i.* to form dimples; to sink in slight depressions. **dimply**, *a.*
dim sum (dim sŭm), *n.* a Chinese appetizer consisting of small steamed dumplings with various fillings.
dimwit (dim'wit), *n.* (*coll.*) a stupid person. **dimwitted**, *a.* **dimwittedness**, *n.*
DIN (din), *n.* a method of classifying the speed of photographic film by sensitivity to light (the greater the light sensitivity the higher the speed). [acronym for *Deutsche Industrie Norm* (G, German industry standard)]
din (din), *n.* a loud and continued noise; a rattling or clattering sound. *v.t.* (*past, p.p.* **dinned**) to harass with clamour; to stun with a loud continued noise; to repeat or impress with a loud continued noise. *v.i.* to make a din. **to din into**, (*fig.*) to teach by constant repetition.
dinar (dē'nah), *n.* a Persian money of account; the monetary unit of Yugoslavia, Iraq, Jordan, Kuwait, Algeria, South Yemen and Tunisia.
dine (din), *v.i.* to take dinner. *v.t.* to give or provide a dinner for; to afford accommodation for dining. **to dine out on**, to be invited to dinner, to be popular socially, because of (something interesting to recount). **to wine and dine (someone)**, to entertain (someone) to dinner. **diner**, *n.* one who dines; a railway dining-car. **diner-out**, *n.* one who habitually dines away from home; one who is frequently invited out to dinner. **dinette** (-net'), *n.* an alcove or a small part of a room set aside for eating. **dining-car**, *n.* a railway coach in which meals are cooked and served. **dining-chamber, -hall, -room, -table**, *n.* a place or table for taking dinner at.
ding (ding), *v.t.* to beat, to surpass; to knock or drive with violence; to ring. *v.i.* to ring, keep sounding.
dingbat (ding'bat), *n.* (*Austral. sl.*) a stupid person; (*pl.*) an attack of nerves.
ding-dong (ding'dong), *n.* the sound of a bell; (*coll.*) a violent argument. *a.* sounding like a bell; violent, keenly contested. *adv.* like the sound of a bell.
dinges (ding'əs), *n.* (*S Afr. coll.*) a name for any person or thing whose name is forgotten or unknown.

dinghy (ding'gi), *n.* orig. a rowing-boat on the Ganges; a small ship's boat; any small boat.
dingle (ding'gl), *n.* a dell, a wooded valley between hills.
dingo (ding'gō), *n.* (*pl.* **-goes**) the Australian wild dog, *Canis dingo*.
dingy (din'ji), *a.* soiled, grimy; of a dusky, soiled, or dun colour; faded. **dingily**, *adv.* **dinginess**, *n.*
dinkum (ding'kəm), *a.* (*Austral. coll.*) good, genuine, satisfactory.
dinky[1] *ding'ki*), *a.* small, dainty; (*N Am.*) trivial, insignificant.
dinky[2] (ding'ki), *n.* (*coll.*) a socially upwardly mobile couple with two incomes and no children. [acronym for *dual income no kids*]
dinner (din'ə), *n.* the principal meal of the day; a feast, a banquet. **dinner dance**, *n.* a dinner followed by dancing. **dinner-hour**, *n.* the time set apart for dinner. **dinner-jacket**, *n.* a formal jacket, less formal than a dress coat, without tails and worn with black tie. **dinner lady**, *n.* a woman who prepares or supervises school meals. **dinner party**, *n.* invitation of guests to dinner; the guests so invited. **dinner service, set**, *n.* the china plates etc., used for serving dinner. **dinner-table**, *n.* a dining-table. **dinner-time**, *n.* the hour for dinner. **dinnerless**, *a.*
dinosaur (di'nəsaw), *n.* a gigantic Mesozoic reptile. **dinosaurian** (-saw'-), *a.* pertaining to the group Dinosauria. *n.* a dinosaur.
dint (dint), *n.* a blow, a stroke; the mark or dent caused by a blow. *v.t.* to mark with a dint. **by dint of**, by force of; by means of.
dioc., (*abbr.*) diocese; diocesan.
diocese (di'əsis), *n.* the district under the jurisdiction of a bishop. **diocesan** (-os'-), *a.* pertaining to a diocese. *n.* a bishop or archbishop of a diocese; a member of a diocese.
diode (di'ōd), *n.* a simple electron tube in which the current flows in one direction only between two electrodes; a semiconductor with two terminals.
Dioecia (diē'shə), *n.pl.* a Linnaean class of plants, having the stamens on one individual and the pistils on another. **dioecious**, *a.* (*Bot.*) belonging to the Dioecia; (*Zool.*) having the sexes in separate individuals.
dionysian (dioniz'iən), *a.* relating to *Dionysus*, the Greek god of wine; wild.
Diophantine equation (diəfan'tin), *n.* (*Math.*) an indeterminate equation which needs an integral or rational solution.
dioptric (diop'trik), *a.* refractive; pertaining to dioptrics. **dioptric light**, *n.* light produced in lighthouses by refraction through a series of lenses. **dioptre** (-tə), *n.* the unit of refractive power, being the power of a lens with a focal distance of one metre. **dioptrically**, *adv.* **dioptrics**, *n.sing.* that part of optics which treats of the refraction of light in passing through different mediums, esp. through lenses.
diorama (diərah'mə), *n.* a scenic representation viewed through an aperture by means of reflected and transmitted light; a miniature three-dimensional scene with figures; a small-scale set used in films or television; a life-size museum display of animals, figures etc. against a naturalistic background. **dioramic** (-ram'-), *a.*
diorite (di'ərit), *n.* a granite-like rock, consisting principally of hornblende and feldspar. **dioritic** (-rit'-), *a.*
Dioscuri (dioskū'ri), *n.pl.* the twins Castor and Pollux.
diothelism (dioth'əlizm), etc. DYOTHELISM.
dioxide (diok'sid), *n.* one atom of a metal combined with two of oxygen.
dioxin (diok'sin), *n.* a highly toxic substance found in some weedkillers which causes birth defects, cancers and various other diseases.
dip (dip), *v.t.* (*past, p.p.* **dipped**) to plunge into a liquid for a short time; to baptize by immersion; to wash, to dye, to coat by plunging into a liquid; to lower for an instant; to put the hand or a ladle into liquid and scoop out; to lower (the headlights); (*Naut.*) to salute by lowering (the flag)

Dip. 226 **direction**

and hoisting it again. *v.i.* to plunge into liquid for a short time; to sink, as below the horizon; to bend downwards, to bow; to slope or extend downwards, to enter slightly into any business; to read a book cursorily; to choose by chance. *n.* the act of dipping in a liquid; bathing, esp. in a river, sea etc.; a candle made by dipping wicks in melted tallow; the quantity taken up at one dip or scoop; a preparation for washing sheep; a savoury mixture into which biscuits or raw vegetables are dipped before being eaten; depth or degree of submergence; the angle at which strata slope downwards into the earth; a curtsy; (*sl.*) a pickpocket. **dip of the horizon**, the apparent angular depression of the visible horizon below the horizontal plane through the observer's eye. **dip of the needle**, the angle which a magnetic needle makes with the horizontal, also called **magnetic dip**. **to dip into**, to draw upon (e.g. resources); to read in cursorily. **dip-chick**, *n.* the dabchick. **dip-net**, *n.* a small fishing-net with a long handle. **dipstick**, *n.* a rod for measuring the level of liquid in a container. **dip switch**, *n.* a device in a car for dipping headlights. **dipper**, *n.* one who dips; a vessel used for dipping; (*N Am.*) the seven stars of the Great Bear; popular name for several birds, esp. the water-ousel.
Dip., (*abbr.*) Diploma.
Dip.Ed., (*abbr.*) Diploma in Education.
dipetalous (dīpet'ələs), *a.* having two petals.
diphase (dī'fāz), **diphasic** (-fā'zik), *a.* having two phases.
diphtheria (difthiə'riə), *n.* an infectious disease characterized by acute inflammation and the formation of a false membrane, chiefly on the pharynx, nostrils, tonsils and palate. **diphtherial, diphtheric** (-the'-), *a.* **diphtheritis** (-thəri'is), *n.* **diphtheritic** (-thərit'-), *a.* **diphtheroid** (dif'thə-), *a.*
diphthong (dif'thong), *n.* the union of two vowels in one syllable; a digraph or combination of two vowel characters to represent a vowel sound; the vowel ligatures, æ, œ. **diphthongal** (-thong'gəl), **-gic** (-gik), *a.* **diphthongally**, *adv.* **diphthongize, -ise**, *v.t.* **diphthongization**, *n.*
diphyllous (difil'əs), *a.* having two leaves or sepals.
diphysite (dif'izīt), etc. DYOPHYSITE.
dipl., (*abbr.*) diploma; diplomat(ist); diplomatic.
dipl(o)-, *comb. form* double.
Diplodocus (diplod'əkəs), *n.* a genus of sauropod dinosaurs characterized by a large tail and a small head.
diploid (dip'loid), *a.* having the full number of paired homologous chromosomes; double, twofold. **diploidic**, *a.*
diploma (diplō'mə), *n.* a document conveying some authority, privilege or honour; a charter, a state paper; a certificate of a degree, licence etc. **diplomaed** (-məd), *a.* **diplomaless**, *a.* **diplomate** (dip'ləmāt), *n.* **diplomatics** (-mat'-), *n. sing.* the art or science of ascertaining the authenticity, date, genuineness etc. of ancient literary documents.
diplomacy (diplō'məsi), *n.* the art of conducting negotiations between nations; the act of negotiating with foreign nations; skill in conducting negotiations of any kind; adroitness, tact. **diplomat** (dip'ləmat), *n.* a professional diplomatist, one skilled or trained in diplomacy. **diplomatic** (-mat'-), *a.* pertaining to diplomacy or to ambassadors; skilled and tactful in handling people, situations etc. **diplomatic bag**, *n.* one used for sending official mail, free of customs control, to and from embassies and consulates. **diplomatic corps**, *n.* the body of diplomatic representatives accredited to any government. **diplomatic immunity**, *n.* the immunity from taxation and local laws given to diplomats resident in a foreign country. **diplomatic relations**, *n.pl.* official relations between countries marked by the presence of diplomats in each other's country. **Diplomatic Service**, *n.* that part of the Civil Service which provides diplomats to represent Britain abroad. **diplomatically**, *adv.* **diplomatics** see DIPLOMA. **diplomatist**, *n.* one skilled or engaged in diplomacy. **diplomatize, -ise**, *v.i.* to act as a diplomatist; to exert the arts of a diplomatist.

diplopia (diplō'piə), *n.* a disease of the eyes in which the patient sees objects double.
dipole (dī'pōl), *n.* two equal and opposite electric charges or magnetic poles a small distance apart· a molecule in which the centres of positive and negative charge do not coincide; an aerial made of a single metal rod with the connecting wire attached half-way down. **dipolar** (-pō'-), *a.* (*Elec., Opt.*) having two poles. **dipolarize, -ise**, *v.t.* **dipolarization**, *n.*
dipper, dipping DIP.
dippy (dip'i), *a.* (*sl.*) slightly mad.
dipsomania (dipsəmā'niə), *n.* a morbid, irresistible craving for alcohol. **dipsomaniac** (-ak), *n.* **dipsomaniacal** (-məni'əkəl), *a.*
Diptera (dip'tərə), *n.pl.* an order of insects, such as flies and gnats, that have two wings and two small knobbed organs called poisers. **dipteran**, *a.* **dipterous**, *a.* having two wings or wing-like appendages.
diptych (dip'tik), *n.* an ancient writing-tablet with two hinged leaves; an altar-piece or other painting with hinged sides closing like a book.
dire (dīə), *a.* dreadful, disastrous; ominous; desperate, urgent. **direful**, *a.* **direfully**, *adv.* **direfulness**, *n.* **direly**, *adv.*
direct (dīrekt', di-), *a.* straight; in a straight line from one body or place to another; not reflected or refracted; nearest, shortest; tending immediately to an end or result; not circuitous; not collateral in the line of descent; diametrical; immediate; personal, not by proxy; honest, to the point; (*Mus.*) not inverted; (*Gram.*) as spoken, not in reported form; plain, to the point, straightforward; from east to west, applied to the motion of a planet when in the same direction as the movement of the sun. *v.t.* to point or turn in a direct line towards any place or object; to show the right road to; to inscribe with an address or direction; to address, to speak or write to; to guide, to prescribe a course to; to advise; to order, to command; to manage, to control; to plan and supervise the staging of a play, shooting of a film, etc.; to conduct (musicians) esp. from the keyboard. *v.i.* to give orders or instructions; to act as a director. *adv.* (*coll.*) directly; immediately; absolutely. **direct broadcasting by satellite**, a system of broadcasting television programmes direct to the consumer using satellites. **directed-energy weapon**, one that destroys its target with high-energy radiation or sub-atomic particles. **direct grant school**, (before 1979) a school partly funded by a state grant on condition that it accepted some non-fee-paying pupils. **direct access**, *n.* a way of reading data in a computer file without having to read through the whole file. **direct action**, *n.* the use of the strike as a weapon to force political or social measures on a government. **direct current**, *n.* (*Elec.*) a current which flows in one direction only. **direct debit**, *n.* a method by which a creditor is paid directly from the payer's bank account. **direct labour**, *n.* workers belonging to an employer's own workforce, esp. that of a local authority, not supplied by outside contractors. **direct method**, *n.* a way of learning foreign languages involving minimal use of the student's native tongue. **direct object**, *n.* the word or group of words which is acted upon by a transitive verb. **direct primary**, *n.* in the US, a primary election in which voters select the candidates who are to stand for office. **direct speech**, *n.* a report of actual words spoken. **direct tax**, *n.* one levied on the persons who actually pay it, e.g. income tax. **directive**, *n.* an authoritative instruction or direction. *a.* having the power of directing; indicating direction. **directly**, *adv.* in a direct or straight line; in a direct manner; without any intervening space; at once; *conj.* (*coll.*) as soon as, directly that. **directness**, *n.*
direction (dīrek'shən, di-), *n.* the act of directing; the course along which something moves or is aimed; the

point towards which someone or something faces; (*often pl.*) the superscription of a letter or parcel; an order or instruction; management, guidance; a directorate; sphere, subject. **direction finder**, *n.* an apparatus for finding the bearings of a transmitting station. **directional**, *a.* **directional aerial**, *n.* one that transmits or receives radio waves from one direction. **directional drilling**, *n.* non-vertical drilling of oil wells, esp. when several are drilled from the same platform.

directoire (dērektwah'), *a.* characteristic of the Directory period in France, 1795–99. [F]

director (direk'tə, dī-), *n.* one who directs or manages; an instructor, a counsellor; anything which controls or regulates; one appointed to direct the affairs of a company; a confessor; the person who supervises the acting team in a film or play. **director general**, *n.* the head of a large, often non-commercial organization, such as the BBC. **directorate** (-rət), *n.* the position of a director; a body or board of directors. **directorial** (-taw'ri-), *a.* **directorship**, *n.* **directress** (-tris), **directrix**[1] (-triks), *n.fem.* [F *directeur* (as prec.)]

directory (direk'təri, dī-), *n.* a board of directors; a book containing the names, addresses and telephone numbers of the inhabitants etc. of a district; a book of direction for public worship; a list of all the files on a computer disk; (*Hist.*) the executive council of the French Republic in 1795–99.

directrix[1] DIRECTOR.

directrix[2] (direk'triks), *n.* (*pl.* **directrices** (-sēz)) (*Geom.*) a line determining the motion of a point or another line so that the latter describes a certain curve or surface.

dirge (dœj), *n.* a funeral song or hymn; a mournful tune or song; a lament.

dirham (dœ'ham, -ham'), **dirhem** (dœ'hem, -hem'), *n.* the standard monetary unit in Morocco; a coin (of different values) in the monetary systems of several N African and Middle Eastern countries.

dirigible (dirij'ibl), *a.* that may be directed or steered. *n.* a balloon or airship that can be steered. **dirigibility**, *n.*

dirigisme (dērēzhēsm'), *n.* state control of economic and social affairs.

dirk (dœk), *n.* a dagger, esp. that worn by a Highlander. *v.t.* to stab with a dirk.

dirndl (dœn'dl), *n.* a dress like that of Alpine peasant women with tight-fitting bodice and full gathered skirt.

dirt (dœt), *n.* foul or unclean matter; mud, dust; a worthless thing, trash; earth, soil; (*derog.*) land; foul talk, scurrility; (*Mining*) the material put into the cradle to be washed. *v.t.* to make dirty or filthy. **to eat dirt**, to put up with insult and abuse. **dirt-cheap**, *a.* very cheap. **dirt-road**, *n.* (*N Am.*) an unmade-up road. **dirt-track**, *n.* a racing-track with a soft, loose surface, for motor-cycle racing; a speedway. **dirty**, *a.* full of, mixed, or soiled with dirt; foul, nasty, unclean; obscene; sordid, mean; contemptible; of weather, rough, gusty; of nuclear weapons, producing much radioactive fall-out. *v.t.* to make dirty, to soil. *v.i.* to become dirty. **dirty old man**, a lewd old man, or one with the sexual appetites thought proper to a younger man. **to do the dirty on**, (*coll.*) to play an underhand trick on. **dirty dog**, *n.* (*coll.*) a dishonest or untrustworthy person. **dirty linen**, *n.* intimate and usu. unsavoury secrets. **dirty look**, *n.* (*coll.*) a glance of disapproval or dislike. **dirty money**, *n.* extra pay for unpleasant or dirty work. **dirty trick**, *n.* treacherous or underhand manoeuvre; *pl.* illegal, irregular and clandestine activities carried on by organizations in order to eliminate, disable or compromise their opponents. **dirty word**, *n.* (*coll.*) a swear word or taboo word; something currently out of favour or very much disliked. **dirty work**, *n.* work that involves dirtying the hands and clothes; (*coll.*) dishonesty, trickery, foul play. **dirtily**, *adv.* **dirtiness**, *n.* **dirtyish**, *a.*

dis DISS.

dis-, *pref.* asunder, apart; separating; (*intensively*) utterly, exceedingly; (*forming negative compounds*) not, the reverse of; undoing, depriving or expelling from.

disability (disəbil'iti), *n.* want of physical or intellectual power, or pecuniary means; weakness, incapacity; handicap; legal disqualification.

disable (disā'bl), *v.t.* to render unable, incapable or ineffective; to disqualify legally; to injure so as to incapacitate, to cripple. **disabled**, *a.* **disablement**, *n.*

disabuse (disəbūz'), *v.t.* to undeceive.

disadvantage (disədvahn'tij), *n.* injury, detriment; an unfavourable position or condition; a drawback, a handicap. *v.t.* to cause disadvantage to. **disadvantaged**, *a.* deprived of social or economic resources; discriminated against. **disadvantageous** (-advəntā'jəs), *a.* prejudicial, detrimental; unfavourable to one's interest. **disadvantageously**, *adv.* **disadvantageousness**, *n.*

disaffect (disəfekt'), *v.t.* (*chiefly pass.*) to alienate the affection or loyalty of. **disaffected**, *a.* **disaffectedly**, *adv.* **disaffection**, *n.* alienation of affection, esp. from those in authority; disloyalty.

disaffiliate (disəfil'iāt), *v.t.* to end an affiliation to; to detach. *v.i.* to separate oneself (from). **disaffiliation**, *n.*

disaffirm (disəfœm'), *v.t.* to deny to repudiate. **disaffirmation** (-afœmā'-), *n.*

disafforest (disəfo'rist), *v.t.* to reduce from the legal status of forest to that of ordinary land; to strip of forest. **disafforestation**, *n.*

disagree (disəgrē'), *v.i.* to be different or unlike; to differ in opinion; to quarrel, to fall out; to be unsuitable or injurious (to health, digestion etc.). **disagreeable**, *a.* not in agreement or accord; offensive, unpleasant; ill-tempered. **disagreeableness**, *n.* **disagreeably**, *adv.* **disagreement**, *n.* lack of correspondence; difference of opinion; a quarrel, dissension.

disallow (disəlow'), *v.t.* to refuse to sanction or permit; to refuse assent to. **disallowance**, *n.*

disambiguate (disambig'üāt), *v.t.* to remove any ambiguities from.

disappear (disəpiə'), *v.i.* to go out of sight; to become invisible; to be lost; to cease to exist. **disappearance**, *n.*

disappoint (disəpoint'), *v.t.* to defeat of expectation, hope or desire; to frustrate, to hinder, to belie. **disappointed**, *a.* frustrated of one's desires or expectations; sad because of a disappointment. **disappointedly**, *adv.* **disappointing**, *a.* disappointingly, *adv.* **disappointment**, *n.* the failure of one's hopes; that which disappoints.

disapprobation (disaprəbā'shən), *n.* disapproval, condemnation. **disapprobative** (-ap'-), *a.* **disapprobatory** (-ap'-), *a.*

disapprove (disəproov'), *v.i.*, *v.t.* to condemn or censure as wrong; to reject, as not approved of. **disapproval**, *n.* **disapprovingly**, *adv.*

disarm (disahm'), *v.t.* to deprive of weapons; to reduce to a peace footing; (*fig.*) to render harmless; to subdue, to tame (esp. by charm, friendliness etc.). *v.i.* to lay aside arms; to be reduced to a peace footing; to reduce or abandon military and naval establishments. **disarmament**, *n.* reduction of armaments by mutual agreement between nations. **disarmer**, *n.* **disarming**, *a.* tending to allay hostility or criticism; charming. **disarmingly**, *adv.*

disarrange (disərānj'), *v.t.* to put out of order; to derange. **disarrangement**, *n.*

disarray (disərā'), *v.t.* to throw into confusion. *n.* disorder, confusion.

disarticulate (disahtik'ūlāt), *v.t.* to separate the joints of. *v.i.* to become disjointed or separated at the joints. **disarticulation**, *n.*

disassemble (disəsem'bl), *v.t.* to take apart. **disassembler**, *n.* a computer programming tool which facilitates the analysis of how programs are constructed. **disassembly**, *n.*

disassociate (disəsō'siāt), *v.t.* to separate, to disjoin. **disassociation**, *n.*
disaster (dizah'stə), *n.* a catastrophic event; a sudden misfortune, a calamity; misfortune, ill luck; (*coll.*) a fiasco, a flop. **disaster area**, *n.* a locality which has suffered a disaster and needs emergency aid; (*coll.*) one that is very untidy, or a situation that is very unfortunate. **natural disaster**, *n.* a diastrous event produced by the forces of nature, as a flood, earthquake etc. **disastrous**, *a.* occasioning or threatening disaster; ruinous, calamitous. **disastrously**, *adv.*
disavow (disəvow'), *v.t.* to deny the truth of; to disown, to disclaim; to repudiate. **disavowal**, *n.*
disband (disband'), *v.t.* to break up (as a body of men in military service). *v.i.* to be separated or dispersed; to separate, to disperse. **disbandment**, *n.*
disbar (disbah'), *v.t.* (*past, p.p.* **disbarred**) to deprive of status as a barrister; to expel from membership of the bar.
disbelieve (disbilēv'), *v.t.* to refuse to believe in. *v.i.* to be a sceptic. **disbelief**, *n.* **disbeliever**, *n.* **disbelieving**, *a.* **disbelievingly**, *adv.*
disburse (disbœs'), *v.t.* to pay out, to expend. **disbursement**, *n.* **disburser**, *n.*
disc, disk (disk), *n.* a flat, circular shape, object or surface; a gramophone record; (*Astron.*) the face of a celestial body; any round, luminous and apparently flat object; a layer of fibrocartilage between vertebrae; (*Comput.*) a small, circular piece of plastic in a rigid case, coated with a magnetic oxide substance, used for storing information and software. **compact disc** COMPACT[2]. **floppy disk**, a computer disk made of flexible plastic, used in microcomputers. **hard disk**, a computer disk of large capacity, made of inflexible material and held in a sealed container. **parking disc**, one displayed in a parked vehicle to show the time of arrival or the latest time of departure. **slipped disc**, a displacement of one of the discs between the vertebrae. **disc brake**, *n.* one consisting of a metal disc attached to the axle, on the opposite surfaces of which the brake pads press. **disk drive**, *n.* the electromechanical device in a computer which reads information from, and writes it onto, the disk. **disk file, store**, *n.* a random-access device in which information is stored, in tracks, on magnetic disks. **disc harrow**, *n.* one consisting of sharpened saucer-shaped discs for cutting soil. **disc jockey**, *n.* the compere of a programme of popular recorded music, on radio, TV or live. **diskette** (-ket'), *n.* another name for a computer disk.
disc., (*abbr.*) discovered; discoverer.
discard (diskahd'), *v.t.* to throw aside or away as useless; to get rid of, to reject, to cast aside; to play (a particular card) that does not follow suit. *v.i.* to play a non-trump card that does not follow suit. *n.* (dis'kahd) the playing of useless cards; the card or cards so played; anything rejected as useless.
discern (discen'), *v.t.* to perceive distinctly with the senses, to make out; to recognize clearly or perceive mentally; to judge or decide between. *v.i.* to make distinction (between); to discriminate; to see. **discernible**, *a.* **discernibleness**, *n.* **discernibly**, *adv.* **discerning**, *a.* having power to discern; discriminating, acute, penetrating. *n.* discernment. **discerningly**, *adv.* **discernment**, *n.* the act, power or faculty of discerning; clear discrimination, accurate judgment.
discharge[1] (dischahj'), *v.t.* to unload from (a ship, vehicle etc.); to take out or away, as a load; to get rid of, to dismiss, as from employment; to emit, to let fly; to release from confinement; to fire off; to empty; to pay off, to settle; to perform, as duties. *v.i.* to unload a cargo; to empty itself, as a river. **discharger**, *n.* **discharging**, *n., a.*
discharge[2] (dis'chahj), *n.* the act of discharging; (an) unloading, release, emission, firing off; performance, as of duties; payment, satisfaction; dismissal, release, acquittal, liberation; a paper certifying any of these; that which is emitted.
disciple (disī'pl), *n.* a pupil or adherent (of a philosopher, leader etc.); a follower (of a particular cult, area of interest etc.); one of the early followers, esp. one of the twelve personal followers of Christ. **discipleship**, *n.*
discipline (dis'iplin), *n.* instruction, training, exercise, or practice of the mental, moral and physical powers to promote order, regularity and efficient obedience; correction, chastisement; training supplied by adversity. *v.t.* to teach, to train, to drill, esp. in obedience, orderly habits and methodical action; to chastise, to chasten, to bring into a state of order and obedience. **disciplinable**, *a.* **disciplinal** (dis'-, -plī'-), *a.* **disciplinarian** (-neə'ri-), *a.* pertaining to discipline. *n.* one who rigidly enforces discipline. **disciplinary** (dis'iplinəri), *a.* pertaining to or promoting discipline. **discipliner**, *n.*
disclaim (disklām'), *v.t.* to deny, to repudiate; to refuse to acknowledge, to disown, to disavow; (*Law*) to renounce, to relinquish or to disavow. **disclaimer**, *n.* renunciation, disavowal or repudiation.
disclose (disklōz'), *v.t.* to uncover; to make known, to reveal, to divulge. **disclosure** (-zhə), *n.* the act of disclosing; that which is disclosed, a revelation.
disco (dis'kō), *n.* (*pl.* **-cos**) short and more usual form of DISCOTHEQUE. *a.* (*attrib.*) suitable for or adapted for discotheques, as in *disco dancing*.
discography (diskog'rəfi), *n.* the literature and study of gramophone records; a list of gramophone records made by a named person.
discoid (dis'koid), **discoidal** (-koi'-), *a.* having the shape of a disk.
discolour (diskŭl'ə), *v.t.* to alter the colour of; to give an unnatural colour to; to stain, to tarnish, to cause to fade. *v.i.* to become stained or tarnished in colour; to fade, to become pale. **discoloration**, *n.* the act of discolouring; the state of being discoloured; a discoloured appearance, a spot, a stain. **discolourment**, *n.*
discomfit (diskŭm'fit), *v.t.* to defeat, to rout; to disconcert, to confound. **discomfiture** (-chə), *n.* defeat, overthrow; disconcertedness, frustration.
discomfort (diskŭm'fət), *v.t.* to deprive of comfort; to cause pain or uneasiness to. *n.* uneasiness, disquietude, distress.
discommode (diskəmōd'), *v.t.* to incommode.
discompose (diskəmpōz'), *v.t.* to disturb, to destroy the composure of; to agitate, to vex, to disquiet. **discomposedly** (-zid-), *adv.* **discomposingly**, *adv.* **discomposure** (-zhə), *n.* lack of composure; agitation, perturbation, disquiet.
disconcert (diskənsœt'), *v.t.* to throw into confusion, to baffle; to upset the composure of, to disquiet. **disconcertedness**, *n.*
disconformity (diskənfaw'miti), *n.* (a) lack of conformity or agreement; inconsistency.
discongruity (diskəngroo'iti), *n.* (a) lack of congruity.
disconnect (diskənekt'), *v.t.* to separate; to sever (a connection, esp. to electricity, gas etc.). **disconnected**, *a.* severed; incoherent, rambling, as speech. **disconnectedly**, *adv.* **disconnectedness**, *n.* **disconnection**, (*dated*) **disconnexion**, *n.* the act of disconnecting; the state of being separated or incoherent.
disconsolate (diskon'səlat), *a.* dejected, cheerless, forlorn; that cannot be consoled or comforted. **disconsolately**, *adv.* **disconsolateness**, *n.*
discontent (diskəntent'), *n.* want of content; dissatisfaction. *a.* not content, dissatisfied. *v.t.* to make discontented or dissatisfied. **discontented**, *a.* dissatisfied. **discontentedly**, *adv.* **discontentedness**, *n.* **discontentment**, *n.*
discontiguous (diskəntig'ūəs), *a.* not contiguous; having

discontinue (diskəntin'ū), *v.t.* to break off, to interrupt; to leave off, to end; to give up. *v.i.* to cease; to break off.
discontinuance, *n.* (an) interruption in continuance; a break in succession; (a) cessation, interruption, intermission; (*Law*) an interruption or breaking-off of possession.
discontinuous, *a.* not continuous, disconnected; incoherent; intermittent. **discontinuity** (-kontinū'-), *n.* **discontinuously**, *adv.*
discophile (dis'kōfīl), *n.* one who collects gramophone records.
discord[1] (dis'kawd), *n.* want of concord or agreement; disagreement, contention, strife; disagreement or opposition in quality, esp. in sounds, as, a lack of harmony in a combination of notes sounded together; the sounding together of two or more inharmonious or inconclusive notes; the interval or the chord so sounded; a note that is out of harmony with another.
discord[2] (diskawd'), *v.i.* to be out of harmony (with); to disagree; to be inconsistent, to clash (with). **discordance** (-kaw'-), *n.* **discordant** (-kaw'-), *a.* disagreeing, not in accord, unpleasing, esp. to the ear; opposite, contradictory; inconsistent. **discordantly**, *adv.*
discotheque (dis'kətek), *n.* a club or public place where people dance to recorded pop music; mobile apparatus for playing records at such a place.
discount[1] (diskownt'), *v.t.* to deduct a certain sum or rate per cent from (an account or price); to deduct such a sum or percentage from the normal price of (an item); to leave out of account or consideration, to disregard. **discountable** *a.*
discount[2] (dis'kownt), *n.* a deduction from the amount of a price or an account, as for early or immediate payment; the act of discounting; the rate or amount of discount. **at a discount,** below the normal price. **discount house, store,** *n.* a shop which sells most of its merchandise at below the recommended price.
discountenance (diskown'tənəns), *v.t.* to discourage, to express disapprobation of; to put out of countenance, to embarrass.
discourage (diskŭ'rij), *v.t.* to dishearten, to dispirit; to attempt to prevent (something) or deter (someone) by arguing against a proposed action, etc. **discouragement,** *n.* **discourager,** *n.* **discouraging,** *a.* **discouragingly,** *adv.*
discourse[1] (dis'kaws), *n.* talk, conversation, exchange of ideas; a dissertation, a lecture or sermon; a formal treatise.
discourse[2] (diskaws'), *v.t. v.i.* to talk, to speak, to converse (with someone); to talk formally, to hold forth (on or upon).
discourteous (diskœ'tiəs), *a.* not courteous, impolite, rude. **discourteously,** *adv.* **discourteousness,** *n.* **discourtesy,** *n.*
discover (diskŭv'ə), *v.t.* to disclose, to reveal, to make known; to betray; to gain the first sight of, to find by exploration; to ascertain, to realize suddenly, to find out; to detect. **discoverable,** *a.* **discoverer,** *n.* one who discovers; an explorer.
discovery (diskŭv'əri), *n.* the act of discovering; that which is made known for the first time; something that is found out; (a) revelation, disclosure, manifestation; the unravelling of the plot of a play; (*Law*) compulsory disclosure of facts and documents essential to the proper consideration of a case.
discredit (diskred'it), *n.* (the cause of) disrepute or disgrace. *v.t.* to bring into disrepute; to deprive of credibility. **discreditable,** *a.* tending to discredit; disreputable, disgraceful. **discreditably,** *adv.*
discreet (diskrēt'), *a.* prudent, wary, circumspect; judicious, careful in choosing the best means of action to avoid offence, embarrassment, discovery etc. **discreetly,** *adv.* **discreetness,** *n.*

discrepancy (diskrep'ənsi), *n.* a difference; a conflict, esp. between two figures or claims.
discrete (diskrēt'), *a.* distinct, discontinuous, detached, separate; (*Phil.*) not concrete, abstract. **discreteness,** *n.*
discretion (diskresh'ən), *n.* the power or faculty of distinguishing things that differ, or discriminating correctly between opposites; discernment, judgment, circumspection; freedom of judgment and action. **at one's discretion,** at one's judgment or pleasure. **years of discretion,** the age when one is capable of exercising one's own judgment, in English law, the age of 14. **discretionary,** *a.* left to one's discretion or judgment.
discriminate (diskrim'ināt), *v.t.* to distinguish between; to mark or observe the difference or distinction between; to tell apart by marks of difference, to differentiate. *v.i.* to make a distinction or difference; to mark the difference between things. *a.* (-nət), distinctive; having the difference clearly marked. **to discriminate against,** to distinguish or deal with unfairly or unfavourably. **discriminately,** (-nət-), *adv.* **discriminating,** *a.* distinguishing clearly, distinctive; having or exercising good taste or judgment, discerning. **discriminatingly,** *adv.* **discrimination,** *n.* the power or faculty of discriminating; discernment, judgment; unfair treatment of an individual or group on the grounds of race, religion, sex, age or any other characteristic. **positive discrimination,** discrimination in favour of an individual or group of people (previously discriminated against or likely to be discriminated against) in areas such as employment. **discriminative** (-nət-), *a.* serving to distinguish; observing distinctions or differences. **discriminator,** *n.* **discriminatory** (-nət-), *a.* discriminative.
discursive (diskœ'siv), *a.* passing from one subject to another; rambling, desultory; (*Psych.*, *Log.*) rational, argumentative, reasoned as opposed to intuitive. **discursively,** *adv.* **discursiveness,** *n.* **discursory,** *a.*
discus (dis'kŭs), *n.* (*pl.* **-cuses, -ci** (-kī)) a metal disc with a thick, heavy middle, thrown in modern sporting contests.
discuss (diskŭs'), *v.t.* to debate; to consider or examine by argument. **discussible, -able,** *a.* **discussion** (-shən), *n.* (an) act of discussing; consideration or investigation by argument for and against. **discussion group,** *n.* a group (in a school, club etc.) formed to discuss current political or other topics.
disdain (disdān'), *n.* scorn, a feeling of contempt combined with haughtiness and indignation. *v.t.* to regard as unworthy of notice; to despise or repulse as unworthy of oneself, to scorn. **disdained,** *a.* **disdainful,** *a.* **disdainfully,** *adv.*
dis-ease (disēz'), *n.* lack of ease, discomfort.
disease (dizēz'), *n.* any alteration of the normal vital processes of human beings, the lower animals or plants, under the influence of some unnatural or hurtful condition; any disorder or morbid condition, habit or function, mental, moral, social etc. **diseased,** *a.* affected with disease; morbid, unhealthy, deranged.
diseconomy (disikon'əmi), *n.* something which is uneconomic or unprofitable.
disembark (disimbahk'), *v.t., v.i.* to put or come on shore. **disembarkation** (-em-), *n.*
disembarrass (disimba'rəs), *v.t.* to free from embarrassment or perplexity; to relieve of the burden (of). **disembarrassment,** *n.*
disembody (disimbod'i), *v.t.* to divest of body or the flesh. **disembodied,** *a.* parted from the body; not emanating from a body. **disembodiment,** *n.*
disembogue (disimbōg'), *v.t., v.i.* to pour out or discharge at the mouth, as a stream, mouth of a river, bay, gulf etc.
disembowel (disimbow'əl), *v.t.* to take out the bowels of, to eviscerate; to lacerate so as to let the bowels protrude. **disemboweller,** *n.* **disembowelment,** *n.*

disenable (disinäb'l), *v.t.* to disable, to incapacitate (from).
disenchant (disinchahnt'), *v.t.* to free from enchantment or glamour, or from a spell; to disillusion. **disenchanted**, *a.* **disenchanter**, *n.* **disenchantment**, *n.*
disencumber (disinkŭm'bə), *v.t.* to free from burden or encumbrance. **disencumbrance**, *n.*
disenfranchise (disinfran'chīz), *v.t.* to deprive of electoral privileges or rights of citizenship. **disenfranchisement** (-chiz-), *n.*
disengage (disingāj'), *v.t.* to separate; to loosen, to detach; to withdraw (oneself); to release; to disentangle; to set free from any attachment or engagement. **disengaged**, *a.* separated, disjoined; at leisure, having the attention unoccupied; free from any engagement. **disengagement**, *n.* the act of disengaging; extrication; the state of being disengaged; freedom from mental occupation or care, detachment; dissolution of a matrimonial engagement.
disentangle (disintang'gl), *v.t.* to unravel, to free from entanglement; to disengage, to disembarrass. **disentanglement**, *n.*
disenthral (disinthrawl'), *v.t.* to set free from thraldom, to emancipate. **disenthralment**, *n.*
disequilibrium (disekwilib'riəm), *n.* (*pl.* **-ria** (-riə)) a lack of balance or equilibrium, esp. in economic affairs.
disestablish (disistab'lish), *v.t.* to annul the establishment of, esp. to deprive a Church of its connection with the State; to depose from established use or position. **disestablishment**, *n.*
disesteem (disistēm'), *n.* a lack of esteem or regard.
disfame (disfām'), *n.* ill-fame, dishonour.
disfavour (disfā'və), *n.* a feeling of dislike or disapprobation; displeasure, odium; an ungracious, disobliging or harmful act.
disfigure (disfig'ə), *v.t.* to injure the beauty or appearance of; to mar, to spoil, to sully. **disfigurement**, *n.* **disfiguration** (-gū-), *n.* **disfigurer**, *n.*
disgorge (disgawj'), *v.t.* to eject from the mouth or stomach; to vomit; to empty (as a river). *v.i.* to disembogue, to discharge. **disgorgement**, *n.*
disgrace (disgrās'), *n.* the state of being out of favour; disesteem, discredit, ignominy, shame, a fall from honour or favour; a cause or occasion of discredit or shame. *v.t.* to dismiss from favour; to degrade; to dishonour; to bring disgrace on. **disgraceful**, *a.* shameful; dishonourable. *a.* **disgracefully**, *adv.* **disgracefulness**, *n.*
disgruntled (disgrŭn'tld), *a.* (*coll.*) offended, disappointed, discontented. **disgruntle**, *v.t.* to discontent, to disappoint.
disguise (disgīz'), *v.t.* to conceal or alter the appearance of, with a mask or unusual clothing etc. (*fig.*) to hide by a counterfeit appearance; to alter, to misrepresent. *n.* a costume, mask or manner put on to disguise or conceal; a pretence or show. **disguisedly** (-zid-), *adv.* **disguisement**, *n.* **disguiser**, *n.* one who or that which disguises; a masquer, a mummer. **disguising**, *n.* the act of concealing with or wearing a disguise.
disgust (disgŭst'), *v.t.* to excite loathing or aversion in; to offend the taste of. *n.* a strong distaste; aversion, loathing, repulsion. **disgustedly**, *adv.* **disgusting**, *a.* **disgustingly**, *adv.*
dish (dish), *n.* a broad, shallow, open vessel for serving up food at table; the food so served; any particular kind of food; a concave reflector used as a directional aerial for radio or TV transmissions; (*sl.*) an attractive person. *v.t.* to put into or serve in a dish; to make concave. **side-dish**, an extra food at a meal, such as a salad. **to dish out**, (*coll.*) to distribute freely. **to dish up**, (*coll.*) to serve up; to present. **dish-cloth**, *n.* one used for washing up dishes, plates etc. **dish-pan**, *n.* (*N Am.*) a washing-up bowl. **dish-towel**, *n.* a tea-towel. **dish-water**, *n.* water in which dishes have been washed. **dishwasher**, *n.* a machine for washing dishes and cutlery; a person who washes dishes as a job. **dished**, *a.* concave. **dishful**, *n.* the amount in a dish. **dishy**, *a.* (*sl.*) good-looking.
dishabille (disabē'), DESHABILLE.
dishabituate (dis-həbit'ūāt), *v.t.* to make unaccustomed (to).
dishallow (dis-hal'ō), *v.t.* to make unholy; to profane.
disharmony (dis-hah'məni), *n.* lack of harmony, agreement etc.; discord, incongruity. **disharmonious** (-mō'-), *a.* **disharmonize, -ise**, *v.t.*, *v.i.*
dishearten (dis-hah'tn), *v.t.* to discourage, to dispirit. **disheartening**, *a.* **dishearteningly**, *adv.* **disheartenment**, *n.*
dishevel (dishev'l), *v.t.* to disorder (the hair, clothes etc.). **dishevelled**, *a.* **dishevelment**, *n.*
dishonest (dison'ist), *a.* lacking in honesty, probity or good faith; fraudulent, deceitful, insincere, untrustworthy. **dishonestly**, *adv.* **dishonesty**, *n.* lack of honesty or uprightness; fraud, cheating, violation of duty or trust.
dishonour (dison'ə), *n.* lack of honour; disgrace, discredit, ignominy; the cause of this. *v.t.* to bring disgrace or shame on; to damage the reputation of; to violate the chastity of; to refuse to accept or pay (as a bill or cheque). **dishonourable**, *a.* causing dishonour; disgraceful, ignominious; unprincipled, mean, base; lacking in honour. **dishonourableness**, *n.* **dishonourably**, *adv.* (without the customary) honour. **dishonourer**, *n.*
dishorn (dis-hawn'), *v.t.* to deprive of horns.
dishwasher, dishwater, dishy DISH.
disillusion (disiloo'zhən), *v.t.* to free or deliver from a misunderstanding or illusion; to undeceive. *n.* disenchantment; release from an illusion. **disillusioned**, *a.* freed from incorrect or over-optimistic beliefs, or from an illusion. **disillusionment**, *n.*
disincentive (disinsen'tiv), *n.* that which discourages.
disincline (disinklīn'), *v.t.* to make averse or not disposed (to). **disinclination** (-klinā'-), *n.* lack of inclination, desire or propensity; unwillingness.
disincorporate (disinkaw'pərāt), *v.t.* to deprive of the rights, powers or privileges of a corporate body; to dissolve (such a body). **disincorporation**, *n.*
disindividualize, -ise (disindivid'ūəlīz), *v.t.* to take away the individuality of.
disinfect (disinfekt'), *v.t.* to free or cleanse from infection, often by chemical means. **disinfectant**, *n.* a (usu. chemical) substance which removes infection. **disinfection** (-shən), *n.* **disinfector**, *n.*
disinfest (disinfest'), *v.t.* to rid of vermin, rats and insects, esp. lice, from. **disinfestation**, *n.*
disinflation (disinflā'shən), *n.* a return to normal economic conditions after inflation, without a reduction in production. **disinflationary**, *a.*
disinformation (disinfəmā'shən), *n.* the deliberate propagation or leaking of misleading or incorrect information.
disingenuous (disinjen'ūəs), *a.* not ingenuous; wanting in frankness, openness or candour; underhand, insincere. **disingenuously**, *adv.* **disingenuousness**, *n.*
disinherit (disinhe'rit), *v.t.* to cut off from a hereditary right; to deprive of an inheritance; (*fig.*) to dispossess. **disinheritance**, *n.*
disintegrate (disin'tigrāt), *v.t.* to separate into component parts; to reduce to fragments or powder. *v.i.* to fall to pieces, to crumble; to lose cohesion. **disintegrable**, *a.* **disintegration**, *n.*
disinter (disintœ'), *v.t.* to dig up, esp. from a grave; (*fig.*) to unearth, bring to notice. **disinterment**, *n.*
disinterest (disin'trist), *n.* impartiality, lack of bias, prejudice or personal interest. **disinterested**, *a.* without personal interest or prejudice, unbiased, impartial; (*loosely*) uninterested. **disinterestedly**, *adv.* **disinterestedness**, *n.*
disinvestment (disinvest'mənt), *n.* a reduction or cessation of investment, esp. in a country generally disapproved of, as a form of sanction.
disinvolve (disinvolv'), *v.t.* to disentangle; to remove from

involvement.
disjoin (disjoin'), *v.t.* to separate, to put asunder. **disjoinable**, *a.*
disjoint (disjoint'), *v.t.* to put out of joint, to dislocate; to separate at the joints. **disjointed**, *a.* out of joint; broken up, rambling, incoherent. **disjointedly**, *adv.* **disjointedness**, *n.*
disjunction (disjŭngk'shən), *n.* the act of disjoining; separation. **disjunctive**, *a.* separating, disjoining; marking separation. **disjunctively**, *adv.*
diskinesia (diskinē'ziə), *n.* abnormal, involuntary nervous muscular movement.
disleal (dislēl'), *a.* disloyal.
dislike (dislīk'), *v.t.* to regard with repugnance or aversion. *n.* a feeling of repugnance; an aversion; a thing disliked. **dislikable, dislikeable**, *a.* **disliker**, *n.* **disliking**, *n.*, *a.*
dislocate (dis'ləkāt), *v.t.* to put (a bone) out of joint; to disturb, derange; to displace. **dislocation**, *n.*
dislodge (disloj'), *v.t.* to eject from a place of rest, retreat or defence; to drive out, to expel. **dislodgement**, *n.*
disloyal (disloi'əl), *a.* not true to an allegiance, unfaithful, disaffected. **disloyally**, *adv.* **disloyalty**, *n.*
dismal (diz'məl), *a.* dark, cheerless, depressing, dreary. **dismally**, *adv.* **dismalness**, *n.*
dismantle (disman'tl), *v.t.* to take to pieces, demolish. **dismantlement**, *n.*
dismask (dismahsk'), *v.t.* to unmask.
dismast (dismahst'), *v.t.* to deprive (a ship) of a mast or masts.
dismay (dismā'), *v.t.* to deprive of courage, to dispirit; to disappoint; to terrify, to daunt. *n.* utter loss of courage or resolution; complete disappointment; a state of terror.
dismember (dismem'bə), *v.t.* to separate limb from limb; (*fig.*) to tear apart. **dismemberment**, *n.*
dismiss (dismis'), *v.t.* to send away; to dissolve, disband; to give permission to depart; to discharge (from office or employment); to put aside, reject (from one's thoughts); (*Law*) to discharge from further consideration; (*Mil.*, *imper.*) break ranks! disperse! **dismissal**, *n.* **dismissible**, *a.* **dismissive**, *a.* tending to reject (esp. without due consideration).
dismount (dismownt'), *v.i.* to alight from a horse, carriage, bicycle etc. *v.t.* to throw down or remove from (a carriage or support), as cannon; to unhorse. *n.* the act or mode of dismounting.
Disneyesque (dizniesk'), *a.* pertaining to the type of cartoon film or character created by US film producer Walt *Disney* (1901–66); pertaining to fantasy, whimsical.
disobedience (disəbēd'yəns), *n.* refusal to obey; wilful neglect or violation of duty; non-compliance. **disobedient**, *a.* (habitually) refusing or neglecting to obey. **disobediently**, *adv.*
disobey (disəbā'), *v.t.* to neglect or refuse to obey; to violate, transgress. *v.i.* to be disobedient. **disobeyer**, *n.*
disoblige (disəblīj'), *v.t.* to act in a way contrary to the wishes or convenience of. **disobliging**, *a.* not obliging, not disposed to gratify the wishes of another; churlish, ungracious. **disobligingly**, *adv.* **disobligingness**, *n.*
disorder (disaw'də), *n.* confusion, irregularity, lack of order; tumult, commotion; neglect or infraction of laws or discipline; a disease, an illness. *v.t.* to throw into confusion; to derange the natural functions of. **disorderly**, *a.* confused, disarranged; unlawful, irregular; causing disturbance, unruly. **disorderly conduct**, *n.* public misconduct leading to distress or harassment. **disorderly house**, *n.* (*Law*) a term including brothels, illegal gaming-houses, betting-houses and certain unlicensed places of entertainment. **disorderliness**, *n.*
disorganize, -ise (disaw'gənīz), *v.t.* to throw into confusion; to destroy the systematic arrangement of. **disorganization**, *n.* **disorganized**, *a.* lacking order, confused.
disorient (disaw'riənt), **disorientate** (-tāt), *v.t.* to throw out of one's reckoning; to confuse. **disorientation**, *n.*
disown (disōn'), *v.t.* to disclaim, to renounce, to repudiate. **disownment**, *n.*
disparage (dispa'rij), *v.t.* to think little of, to undervalue; to treat or speak of slightingly; to injure by unjust comparison. **disparagement**, *n.* **disparagingly**, *adv.*
disparate (dis'pərət), *a.* quite dissimilar, having nothing in common. *n.* (*usu. pl.*) things so unlike that they admit of no comparison with each other. **disparately**, *adv.* **disparateness**, *n.*
disparity (dispa'riti), *n.* inequality; difference in degree; unlikeness.
dispassionate (dispash'ənət), *a.* free from passion; calm, temperate; impartial. **dispassionately**, *adv.*
dispatch, despatch (dispach'), *v.t.* to send off to some destination, esp. to send with speed or haste; to transact quickly; to settle, to finish; to put to death. *v.i.* to go quickly, to hurry. *n.* the act of dispatching or being dispatched; prompt execution; promptitude, quickness; a message or letter dispatched, esp. an official communication on state affairs; a department in a company which deals with the delivery of letters, or goods; a putting to death. **mentioned in dispatches**, (*Mil.*) cited for bravery or valuable services. **dispatch box**, *n.* one for carrying dispatches and other state papers. **dispatch case**, *n.* (*Mil.*) a leather case for carrying papers. **dispatch-rider**, *n.* a motor-cyclist who carries dispatches. **dispatcher**, *n.*
dispel (dispel'), *v.t.* to dissipate, to disperse; to drive away, to banish. **dispeller**, *n.*
dispensable (dispen'səbl), *a.* that may be dispensed with, inessential. **dispensability** (-bil'-), *n.*
dispensary (dispen'səri), *n.* a place where medicines are dispensed; formerly, an establishment where medicines and medical advice are given gratis to the poor. **dispensatory**, *n.* a book listing medical prescriptions, their composition and use.
dispensation (dispənsā'shən), *n.* the act of dispensing; in the Roman Catholic Church, a licence to omit or commit something enjoined or forbidden by canon law; permission to do something not usually allowed, or not to do something usually required; the act of dispensing with or doing without. **dispensational**, *a.*
dispense (dispens'), *v.t.* to deal out, to distribute; to administer, to prepare and give out (medicine). *v.i.* to dispense medicines. **to dispense with**, to forgo, to do without; to render unnecessary; to suspend, to waive the observance of (as formalities). **dispenser**, *n.* one who or that which dispenses; (*Med.*) one who dispenses medicines. **dispensing optician** OPTICIAN.
dispermatous (dīspœ'mətəs), **dispermous** (-spœ'-), *a.* having only two seeds.
disperse (dispœs'), *v.t.* to scatter; to send, drive or throw in different directions; to dissipate, to cause to vanish; to put (particles) into a colloidal state, to distribute evenly in a fluid. *v.i.* to be widely scattered; to break up, to vanish; to become widely spread. **dispersal**, *n.* dispersion; the spreading of anything to new areas. **dispersedly** (-sid-), *adv.* **dispersive**, *a.* **dispersively**, *adv.* **dispersiveness**, *n.*
dispersion (dispœ'shən), *n.* the act of dispersing; the state of being dispersed; (*Med.*) the removal of inflammation from a part; (*Statistics*) the scattering of variables around the arithmetic mean or median. **dispersion of heat, light, etc.**, the separation produced by the refraction at different angles of rays of different wavelengths.
dispirit (dispi'rit), *v.t.* to deprive of spirit or courage; to discourage, to dishearten, to deject. **dispirited**, *a.* **dispiritedly**, *adv.* **dispiritedness**, *n.* **dispiriting**, *a.* **dispiritingly**, *adv.*
displace (displās'), *v.t.* to remove from the usual or proper place; to dismiss; to take the place of, to put something in the place of, to supersede. **displaced persons**, *n.pl.* refugees who for any reason cannot be repatriated. **dis-**

placement, *n.* the act of displacing; the state of being displaced; supersession by something else; the water displaced by a floating body (as a ship), the weight of which equals that of the floating body at rest; (*Psych.*) the unconscious transferring of strong emotions from the original object of them to another.

display (displā'), *v.t.* to exhibit, to expose, to show; to exhibit ostentatiously, to parade; to make known, to unfold, to reveal. *v.i.* (*Zool.*) to behave in an ostentatious way as part of a courtship or competitive ritual. *n.* displaying, show, exhibition; ostentatious parade. **display type,** *n.* (*Print.*) large type for headings etc.

displease (displēz'), *v.t.* to dissatisfy, to offend; to vex, to annoy, to be disagreeable to. *v.i.* to cause displeasure or offence. **to be displeased at, with,** to be annoyed or vexed (at or with); to disapprove of. **displeasing,** *a.* **displeasingly,** *adv.*

displeasure (displezh'ə), *n.* a feeling of annoyance, vexation or anger.

dispone (dispōn'), *v.t.* (*Sc. Law*) to make over or convey (as property). **disponee,** *n.* (*Sc. Law*) one to whom property is disponed. **disponer,** *n.* (*Sc. Law*) one who dispones property.

disport (dispawt'), *v.t.* to amuse, to divert (oneself); to enjoy (oneself). *v.i.* to play, to amuse or divert oneself; to gambol.

disposal (dispō'zl), *n.* the act of disposing; distributing, bestowing, giving away, selling or dealing with things in some particular way; order or arrangement in which things are set out. **at (someone's) disposal,** in the power of, at the command of (someone).

dispose (dispōz'), *v.t.* to arrange, to set in order, to place; to incline (towards). **to dispose of,** to put into the possession of another by any means; to get rid of; to finish, to kill; to put away, to stow away. **disposable,** *a.* capable of being disposed of; applied to anything designed for disposal after use, as *disposable plates*. **disposable income,** *n.* net income after payment of tax, available for use. **disposed,** *a.* inclined (towards). **well, ill, disposed towards,** viewing with favour, disfavour. **disposedness** (-zid-), *n.* disposition, inclination. **disposer,** *n.*

disposition (dispəzish'ən), *n.* the act of disposing, ordering, arranging or bestowing; (the state or manner of) disposal; arrangement in general; inclination, temperament; propensity, natural tendency; (*Sc. Law*) the disposal of property; any unilateral writing by which a person makes over to another a piece of heritable or movable property; the posting of troops in the most advantageous position. **dispositioned,** *a.*

dispossess (dispəzes'), *v.t.* to oust from possession, esp. of real estate. **dispossessed,** *a.* **dispossession** (-shən), *n.*

dispraise (disprāz'), *v.t.* to censure, to express disapprobation of.

disproof (disproof'), *n.* refutation; (a) proof of error or falsehood.

disproportion (disprəpaw'shən), *n.* want of proportion between things or parts; inadequacy, disparity; lack of symmetry. *v.t.* to make out of proportion; to spoil the symmetry of. **disproportionate,** *a.* not correctly proportioned; unsymmetrical; too great or too small in relation to something; unnecessarily great. **disproportionately,** *adv.*

disprove (disproov'), *v.t.* to prove to be erroneous or unfounded. **disprovable,** *a.* **disproval,** *n.*

dispute (dispūt'), *v.i.* to contend in argument; to argue in opposition to another; to quarrel, to wrangle; to strive against another, to compete. *v.t.* to argue over; to oppose, to question, to challenge or deny the truth of; to reason upon, to discuss, to argue; to contend or strive for, to contest (esp. ownership). *n.* contention or strife in argument; debate, controversy; (a) difference of opinion; a falling out, a quarrel. **disputable** (dis'-. -pū'-), *a.* open to dispute, controvertible; questionable; uncertain. **disputability** (-bil'-), *n.* **disputableness,** *n.* **disputably,** *adv.* **disputant** (dis'-), *n.*, *a.* (a person) engaged in disputation or controversy. **disputation** (-tā'-), *n.* the act of disputing; controversy, discussion; an exercise in arguing both sides of a question for the sake of practice. **disputatious** (-tā'-), *a.* given to dispute or controversy; contentious. **disputatiously,** *adv.* **disputatiousness,** *n.* **disputer,** *n.*

disqualify (diskwol'ifī), *v.t.* to render unfit, to disable, to debar; to render or declare legally incompetent for any act or post; to disbar from a sporting competition on account of an irregularity. **disqualification** (-fikā'-), *n.* (an) act of disqualifying; that which disqualifies.

disquiet (diskwī'ət), *v.t.* to disturb, to make uneasy. *n.* uneasiness, restlessness, anxiety. **disquieting,** *a.* **disquietingly,** *adv.* **disquietness,** *n.* **disquietous,** *a.* **disquietude** (-tūd), *n.* anxiety, uneasiness.

disregard (disrigahd'), *n.* lack or omission of attention or respect. *v.t.* to take no notice of; to ignore as unworthy of respect. **disregarder,** *n.* **disregardful,** *a.* negligent, heedless. **disregardfully,** *adv.*

disrelish (disrel'ish), *n.* a distaste or dislike; aversion, antipathy.

disrepair (disripeə'), *n.* a state of being out of repair; dilapidation.

disreputable (disrep'ūtəbl), *a.* not reputable; of bad reputation, not respectable. **disreputableness,** *n.* **disreputably,** *adv.* **disrepute** (-ripūt'), *n.* a loss or lack of good reputation.

disrespect (disrispekt'), *n.* lack of respect or reverence; rudeness, incivility. **disrespectful,** *a.* lacking in respect; uncivil, rude. **disrespectfully,** *adv.* **disrespectfulness,** *n.*

disrobe (disrōb'), *v.t.* to strip of a robe or dress; to undress (oneself). *v.i.* to undress. **disrober,** *n.*

disrupt (disrŭpt'), *v.t.* to tear apart, to break in pieces; to interrupt. **disruption** (-shən), *n.* (an) act of tearing or bursting asunder; the state of being torn asunder; an interruption; breach, rent, split. **disruptive,** *a.* tending to cause disruption.

diss., (*abbr.*) dissertation.

dissatisfy (disat'isfī), *v.t.* to fall short of the wishes or expectations of; to make discontented, to displease. **dissatisfaction** (-fak'shən), *n.* **dissatisfactory** (-fak'-), *a.* **dissatisfied,** *a.* discontented.

dissect (disekt'. dī-), *v.t.* to cut into pieces; to cut up (an organism) so as to examine its parts and structure; to analyse, to criticize in detail. **dissected,** *a.* (*Bot.*) cut into narrow segments; (*Geol.*) applied to hills and valleys cut by erosion. **dissectible,** *a.* **dissecting microscope,** *n.* one allowing dissection of the object being examined. **dissecting room, table,** *n.* a room or table where anatomical dissection is carried out. **dissection** (-shən), *n.* **dissector,** *n.*

dissemble (disem'bl), *v.t.* to hide under a false appearance. *v.i.* to hide one's feelings, opinions or intentions; to play the hypocrite. **dissembler,** *n.* **dissemblingly,** *adv.*

disseminate (disem'ināt), *v.t.* to scatter widely, as seed, with a view to growth or propagation; to spread, to circulate. **disseminated,** *a.* **dissemination,** *n.* **disseminator,** *n.*

dissension (disen'shən), *n.* disagreement in opinion; discord, contention, strife.

dissent (disent'), *v.i.* to differ or disagree in opinion; to withhold assent or approval; to differ from an established Church, esp. from the Church of England. *n.* difference or disagreement of opinion; refusal of assent; (a declaration of) disagreement or nonconformity; a protest by a minority. **dissenter,** *n.* one who dissents or disagrees, esp. one who dissents from an established Church; (**Dissenter**) a member of a sect that has separated from the Church of England.

dissentient (disen'shənt), *a.* disagreeing or differing in opinion; holding or expressing contrary views. *n.* one who holds or expresses contrary views; a dissenter from the

views of a political or other party. **dissentience**, *n*.
dissertation (disətā'shən), *n*. a formal discourse on any subject; a written treatise or essay.
disserve (dis-sɜːv'), *v.t.* to do a disservice to. **disservice**, *n*. a detriment or disfavour.
dissever (disev'ə), *v.t.* to sever, to separate. **disseverment**, *n*.
dissident (dis'idənt), *a*. not in agreement, esp. with official government views. *n*. one who disagrees, esp. with the government; a dissenter. **dissidence**, *n*.
dissimilar (disim'ilə), *a*. not similar; unlike in nature, properties or appearances. **dissimilarity** (-la'-), *n*. **dissimilarly**, *adv*.
dissimilate (disim'ilāt), *v.t.* (*Philol.*) to make unlike. **dissimilation**, *n*. (*Philol.*) the rendering of two similar sounds unlike, when such sounds come together.
dissimilitude (disimil'itūd), *n*. unlikeness, dissimilarity.
dissimulate (disim'ūlāt), *v.t.* to dissemble, to conceal, to disguise. **dissimulation**, *n*. (an) act of dissimulating; hypocrisy. **dissimulator**, *n*.
dissipate (dis'ipāt), *v.t.* to scatter; to drive in different directions; to disperse, to dispel; to squander, to waste, to fritter away. *v.i.* to be dispersed, to vanish. **dissipated**, *a*. scattered, dispersed; given to dissipation, dissolute; wasted in indulgence. **dissipation**, *n*. the act of dissipating or scattering; the state of being dispersed or scattered; lack of concentration (of one's energies) or perseverance; excessive indulgence in luxury, frivolity, extravagance or vice; (*Phys.*) disintegration, dispersion, diffusion.
dissociate (disō'shiāt, -si-), *v.t.* to separate, to disconnect; (*Chem.*) to decompose, esp. by the action of heat. **dissociation**, *n*. separation, disconnection; (*Psych.*) a loosening of control over consciousness in which the personality is temporarily taken control of by unconscious complexes. **dissociative**, *a*.
dissoluble (disol'ūbl), *a*. that can be dissolved, decomposed, disconnected or (of e.g. an agreement, or partnership) broken off. **dissolubility** (-bil'-), *n*.
dissolute (dis'əloot), *a*. given to dissipation, loose in conduct; licentious, debauched. **dissolutely**, *adv*. **dissoluteness**, *n*.
dissolution (disəloo'shən), *n*. the act or process of dissolving, separating, disintegrating, decomposing; the destruction of any body by the separation of its parts; separation or break-up of a meeting, contract, association etc.; (*Chem.*) separation into the elements or components. **dissolution of parliament**, the end of a parliament, to be followed by a general election. **Dissolution of the Monasteries**, the suppression of the monasteries by Henry VIII.
dissolve (dizolv'), *v.t.* to diffuse the particles of (a substance) in a liquid; to convert from a solid to a liquid state by heat or moisture; to decompose; to separate, to break up; to put an end to (as a meeting etc.); to dismiss, to disperse (as a crowd); to rescind, to annul (as a marriage). *v.i.* to become liquefied; to decompose, to disintegrate; to break up, to separate; to fade away, to melt away; to melt or become liquid by the action of heat or moisture; to vanish; (*coll.*, *fig.*) to collapse (e.g. into laughter); in films, TV and video, to fade out one scene and merge in the next. *n*. a scene in a film or TV programme which is dissolved; a feature on a camera which allows this dissolving effect. **dissolvable**, *a*. **dissolver**, *n*.
dissonant (dis'ənənt), *a*. discordant, inharmonious; disagreeing; harsh, incongruous. **dissonance, dissonancy**, *n*. discordant sounds; want of harmony or agreement; an unresolved chord in music. **dissonantly**, *adv*.
dissuade (diswād'), *v.t.* to persuade not to do some act; to advise against doing something. **dissuader**, *n*. **dissuasion** (-zhən), *n*. **dissuasive**, *a*. **dissuasively**, *adv*.
dissyllable etc. DISYLLABLE.
dissymmetry (disim'ətri), *n*. absence of symmetry between objects or parts. **dissymmetrical** (-met'-), *a*. **dissymmetrically**, *adv*.
dist., (*abbr.*) distant; distinguished; district.
distaff (dis'tahf), *n*. a cleft stick about 3 ft. (0·91 m) long, on which wool or carded cotton was wound for spinning. **distaff side**, *n*. the female side of a family or descent.
distal (dis'təl), *a*. applied to the extremity of a bone or organ farthest from the point of attachment or insertion; situated at the farthest point from the centre; in contradistinction to *proximal*. **distally**, *adv*.
distance (dis'təns), *n*. the space between two objects measured along the shortest line; extent of separation however measured; the quality of being distant, remoteness; the length of a course run, swum etc. in a competition; reserve, coolness, avoidance of familiarity, unfriendliness; remoteness in time (past or future); separation in rank or relationship; ideal space or separation; the remoter parts of a view or the background of a landscape picture; (*Mus.*) a tone interval; a winning margin in horseracing, literally the length between the winning post and a point 204 yd. (219.5 m) from it. *v.t.* to place far off; to leave behind in a race; to outstrip, to outdo; to cause (esp. oneself) to seem distant, unfriendly etc. **angular distance**, the space included between the lines drawn from two objects to the eye. **middle distance**, the central portion of a picture between the foreground and the distance. **to go the distance**, to complete something one has started; to endure to the end of (e.g. a game or bout in sport). **to keep at a distance**, not to become (too) friendly with. **to keep one's distance**, to behave respectfully; to behave with reserve or coldness. **distance learning, teaching**, *n*. learning or teaching carried on by correspondence or with instruction through the media of TV and radio rather than by personal attendance or contact.
distant (dis'tənt), *a*. separated by intervening space; remote in space, time (past or future), succession, consanguinity, resemblance, kind or nature; (*pred.*) at a certain distance (specified numerically); faint, slight; reserved, cool. *n*. (also **distant signal**) a railway signal indicating whether the home signal is at danger or not. **distantly**, *adv*.
distaste (distāst'), *n*. disrelish, aversion of the taste; dislike or disinclination (for). **distasteful**, *a*. unpleasant to the taste; offensive, displeasing. **distastefully**, *adv*. **distastefulness**, *n*.
distemper[1] (distem'pə), *n*. a catarrhal disorder affecting young dogs; mental derangement or perturbation. **distempered**, *a*. disordered in mind or body; intemperate, immoderate.
distemper[2] (distem'pə), *v.t.* to paint or colour with distemper. *n*. a type of water-based paint, for indoor use, which has been mixed with chalk or clay and diluted with size instead of oil.
distend (distend'), *v.t.*, *v.i.* to spread or swell out; to inflate. **distensible** (-si-), *a*. **distensibility** (-bil'-), *n*. **distension** (-shən), *n*. the act of distending; the state or amount of being distended.
disthrone (disthrōn'), *v.t.* to dethrone.
distich (dis'tik), *n*. a couplet; two lines of poetry making complete sense.
distil, distill (distil'), *v.i.* (*past*, *p.p.* **distilled**) to undergo the process of distillation. *v.t.* to extract by means of vaporization and condensation; to extract the essence of; to make or obtain (e.g. whisky) by this process; to purify (e.g. water) by this process; to let fall in drops, to shed. **distillable**, *a*. **distillate** (dis'tilāt), *n*. the end product of distillation. **distillation** (-lā'shən), *n*. the act or process of heating a solid or liquid in a vessel so constructed that the vapours thrown off from the heated substance are collected and condensed; the end product of this process, a distillate. **destructive distillation**, distillation at a temperature sufficiently high to decompose the substance,

and evolve new products possessing different qualities. **dry distillation,** the distillation of a solid substance without the addition of water. **fractional distillation,** the separation of liquids having different boiling-points. **vacuum distillation,** distillation carried out under reduced pressure. **distiller,** *n.* one who distils, esp. a manufacturer of spirits by distillation. **distillery,** *n.* a building where spirits are produced by distillation.

distinct (distingkt'), *a.* clearly distinguished or distinguishable, different, standing clearly apart, not identical; unmistakable, clear, plain, evident, definite. **distinction** (-shən), *n.* a mark or note of difference; a distinguishing quality, a characteristic difference; the act of distinguishing, discrimination; (an) honour, title, rank; eminence, superiority. **without distinction,** indiscriminately. **distinctive,** *a.* serving to indicate or display distinction or difference; separate, distinct. **distinctively,** *adv.* **distinctiveness,** *n.* **distinctly,** *adv.* **distinctness,** *n.*

distingué (dēstĭgā'), *a.* (*fem.* **-guée**) having an air of distinction. [F, p.p. of *distinguer*, to DISTINGUISH]

distinguish (disting'gwish), *v.t.* to discriminate, to differentiate; to indicate the difference of from others by some external mark; to tell apart, to discriminate between; to perceive the existence of by means of the senses; to recognize; to be a mark of distinction or a characteristic of; to separate from others by some token of honour or preference; to make eminent, prominent, or well known. *v.i.* to differentiate (between); to draw distinctions. **distinguishable,** *a.* **distinguishably,** *adv.* **distinguished,** *a.* marked by some distinctive (esp. elegant) sign or property; eminent, celebrated, remarkable; conspicuous, specially marked. **Distinguished Conduct Medal,** a medal awarded to warrant officers and other ranks for gallantry in the field. **Distinguished Flying Cross,** a medal for gallantry awarded to officers and warrant officers of the RAF. **Distinguished Service Cross, Medal,** medals awarded for gallantry to RN officers and warrant officers, and CPOs and other ratings. **Distinguished Service Order,** a medal for meritorious service awarded to officers in the three Services. **distinguishedly,** *adv.* in a distinguished manner; eminently. **distinguishing,** *a.* constituting a difference or distinction. **distinguishingly,** *adv.*

distort (distawt'), *v.t.* to twist or alter the natural shape or direction of; to pervert from the true meaning. **distortedly,** *adv.* **distortion** (-shən), *n.* the act of distorting; the state of being distorted; a distorted part of the body, a deformity; a perversion of meaning, a misrepresentation; (*Radio*) unclear reproduction in a receiver or loudspeaker. **distortional,** *a.* **distortive,** *a.*

distract (distrakt'), *v.t.* to draw or turn aside, to divert the mind or attention of; to confuse, bewilder, perplex; (*usu. p.p.*) to drive mad, to make frantic. **distracted,** *a.* disturbed mentally, crazed, maddened; confounded, harassed, perplexed; preoccupied. **distractedly,** *adv.* **distracting,** *a.* **distractingly,** *adv.* **distraction** (-shən), *n.* diversion of the mind or attention; the thing that diverts; an interruption, a diversion, relaxation, relief, amusement; (a) preoccupation; confusion, perplexity, agitation, violent mental excitement arising from pain, care etc.; mental aberration, madness, frenzy. **distractive,** *a.* **distractively,** *adv.*

distrain (distrān'), *v.t.* to seize for debt; to take (personal chattels) in order to satisfy a demand or enforce the performance of an act. **distrainable,** *a.* **distrainee** (-nē'), *n.* one whose goods are distrained. **distrainer, -or,** *n.* **distrainment,** *n.* **distraint,** *n.* the act of seizing goods for debt.

distrait (distrā'), *a.* absent-minded, abstracted, inattentive. [F]

distraught (distrawt'), *a.* bewildered, agitated, frantic (with worry, grief etc.).

distress (distres'), *n.* anguish or pain of mind or body; misery, poverty, destitution; exhaustion, fatigue; a state of danger; (*Law*) the act of distraining. *v.t.* to afflict with anxiety, unhappiness, grief or anguish, to vex; to exhaust, to tire out. **in distress,** in a condition of grief, anxiety, trouble, exhaustion etc.; (*Naut.*) in a disabled or perilous condition (of a ship). **distress rocket, signal,** *n.* a signal for help from a ship in need of assistance. **distress sale,** *n.* a sale of goods under a distress warrant. **distress warrant,** *n.* a writ authorizing the seizure and compulsory sale of household effects etc. in settlement of a debt. **distressed,** *a.* afflicted with pain or anxiety; distressed; exhausted; (of leather, denim etc. clothing) fashionably worn-looking, perhaps frayed or torn. **distressed areas,** *n.pl.* industrial areas where there is wide unemployment and poverty. **distressful,** *a.* **distressfully,** *adv.* **distressing,** *a.* painful, afflicting; awakening pity or compassion. **distressingly,** *adv.*

distribute (distrib'ūt), *v.t.* to divide or deal out amongst a number; to spread around, to disperse. **distributable,** *a.* **distributary,** *a.* distributive. *n.* a branching-off from a river. **distribution** (-bū'shən), *n.* the act of distributing; apportionment, division; apportionment (of wealth) among the various classes of the community; the dispersal of commodities among consumers; geographical incidence or arrangement of a number of scattered units, e.g. the manner, degree and extent in which the flora and fauna of the world are distributed over the surface of the earth. **distributional,** *a.* **distributive** (-trib'ūtiv), *a.* distributing or allotting the proper share to each; pertaining to distribution; (*Gram.*) expressing distribution, separation or division; *n.* a distributive word as *each, every, either* and *neither.* **distributively,** *adv.* **distributiveness,** *n.* **distributor,** *n.* one who or that which distributes; a wholesaler or middleman who distributes goods to retailers; the device in a petrol engine which distributes current to the spark plugs.

district (dis'trikt), *n.* a portion of territory specially defined for judicial, administrative, fiscal, sales or other purposes; a division having its own representative in a legislature, its own district council, a church or chapel of its own or a separate magistrate; any region or tract of country. **district attorney,** *n.* (*N Am.*) the prosecuting officer of a district. **district court,** *n.* (*N Am.*) a court having jurisdiction over cases arising within a defined district. **district heating,** *n.* a system for heating all the buildings in a given area from a central generating point. **district judge,** *n.* (*N Am.*) the judge of a district court. **district nurse,** *n.* a nurse employed by a local authority to visit and tend patients in their own homes. **district surveyor,** *n.* a local government-appointed officer, usually a civil engineer, who examines buildings, roads etc., superintends repairs etc.

distrust (distrŭst'), *v.t.* to have no trust in; to question the reality, truth or sincerity of. *n.* want of trust, reliance or faith (in); suspicion. **distrustful,** *a.* inclined to distrust. **distrustfully,** *adv.* **distrustfulness,** *n.*

disturb (distœb'), *v.t.* to agitate, to make anxious, to unsettle; to move from any regular course, or habitual or customary position; to hinder, to interrupt; to interfere with, as something sacred. **disturbance,** *n.* interruption of a settled or customary state of things; (an instance of) (public) agitation or excitement, disorder, uproar; (*Radio*) interference; a minor earthquake; a small atmospheric depression; a mental or emotional disorder. **disturbed,** *a.* emotionally or mentally unstable.

disulphate (dīsŭl'fāt), *n.* a salt of sulphuric acid. **disulphide** (-fīd), *n.* a compound in which two atoms of sulphur are united to another element or radical in each molecule.

disunion (disūn'yən), *n.* the state of being disunited; disagreement, discord; in the US, secession from the Union. **disunionist,** *n.*

disunite / divine

disunite (disūnīt'), *v.t.* to separate, to divide, to put at variance, cause to disagree. *v.i.* to become divided. **disunity** (-ū'ni-), *n.* disunion; a state of division or disagreement.
disuse (disūs'), *n.* a cessation of use, practice or exercise; the state of being disused. *v.t.* (-ūz'), to cease to use. **disused**, *a.* no longer in use; unaccustomed.
disyllable, dissyllable (disil'əbl), *n.* a word or metrical foot of two syllables. **disyllabic** (-lab'-), *a.* **disyllabically**, *adv.* **disyllabism**, *n.* **disyllabize, -ise**, *v.t.*
dit[1] (dit), *n.* a word representing the dot in the Morse code when this is spoken.
dit[2] (dē), *a.* named; reputed. [F]
ditch (dich), *n.* a trench made by digging, to form a boundary or for drainage; (*Fort.*) a trench or fosse on the outside of a fortress, serving as an obstacle to assailants. *v.t.* to make a ditch, trench or drain in; to surround with a ditch; (*coll.*) to get rid of, to abandon. **ditch-water**, *n.* stagnant water in a ditch, whence **dull as ditch-water**, very uninteresting or unentertaining. **last-ditch**, *a.* final and desperate, as *a last-ditch attempt*. **ditcher**, *n.* a person who or a machine which makes or repairs ditches.
ditheism (dī'thēizm), *n.* the theory of two co-equal gods or opposing powers of good and evil, the basic principle of Zoroastrianism and Manichaeism. **ditheistic** (-is'-), *a.*
dither (didh'ə), *v.i.* to be distracted or uncertain; to hesitate, to be indecisive. **ditherer**, *n.* **dithery**, *a.*
dithyramb (dith'iram, -ramb), *n.* a choral hymn in honour of Bacchus, full of frantic enthusiasm; hence, any wild, impetuous poem or song. **dithyrambic** (-bik), *a.* of the nature of a dithyramb; wild, enthusiastic.
ditrochee (dītrō'kē), *n.* a metrical foot of two trochees.
ditto (dit'ō), *n.* (*pl.* **-ttos**) that which has been said before; the same thing, a similar thing, as before. **to say ditto**, to repeat, endorse a view; to coincide in opinion. **ditto marks**, *n.pl.* a mark consisting of two dots, placed under a word to show that it is to be repeated below.
ditty (dit'i), *n.* a little poem, song, or air.
diuretic (diūret'ik), *a.* causing or increasing the secretion of urine. *n.* a diuretic medicine.
diurnal (diœ'nəl), *a.* of or pertaining to a day or the daytime; happening each day; of common occurrence; (*Zool.*) active in the daytime, as distinct from *nocturnal*. **diurnal arc**, *n.* (*Astron.*) the arc described by a heavenly body from rising to setting. **diurnally**, *adv.*
div., (*abbr.*) divide(d); dividend; divine; division; divorced.
diva (dē'və), *n.* (*pl.* **-vas, -ve** (-vi)) a famous female (esp. opera) singer, a prima donna.
divalent (dīvā'lənt), *a.* (*Chem.*) having a valency of two; having two valencies.
divan (divan'), *n.* a thickly-cushioned backless seat or sofa; (also **divan bed**,) a bed that can be converted into a sofa by day.
divaricate (dīva'rikāt), *v.i.* to diverge into branches or forks; (*Bot.*) to branch off from the stem at a right or obtuse angle. *a.* (-kət), branching off at a right or obtuse angle. **divarication**, *n.*
dive (dīv), *v.i.* to plunge, esp. head first, under water from above the surface; to descend under water from the surface; to descend quickly through the air; to descend quickly and disappear (not in air or water); to thrust one's hand rapidly (into something); (*fig.*) to enter deeply (into any question, science or pursuit). *v.t.* to explore by skin-diving. *n.* a sudden plunge head foremost into water; a sudden plunge or dart (not in water); a drinking-club or other place of entertainment of an inferior type; a steep descent through the air with the nose (or beak) down. **take a dive**, (*sl.*) intentionally to lose a boxing-match. **dive-bomber**, *n.* a military aeroplane which releases its bombs while in a steep dive. **dive-bombing**, *n.* **diver**, *n.* one who dives; esp. one who dives for pearls, or to work on sunken vessels etc.; a type of bird which dives for food, esp. the loon family. **diving-bell**, *n.* a hollow vessel, orig. bell-shaped, in which persons may remain for a time under water, air being supplied through a flexible tube. **diving board**, *n.* a platform from which one may dive into a swimming-pool. **diving-suit**, *n.* waterproof clothing and breathing-helmet for divers working at the bottom of the sea.
diverge (dīvœj', di-), *v.i.* to tend in different directions from a common point or from each other; to branch off; to vary from a normal form; to deviate, to differ. **divergence, -gency**, *n.* **divergent**, *a.* **divergingly**, *adv.*
divers (dī'vəz), *a.* (*esp. Bibl. or poet.*) several, sundry.
diverse (dīvœs', di-), *a.* different, unlike, distinct; varying, changeable. **diversely**, *adv.* **diverseness**, *n.* **diversiform** (-sifawm), *a.* of many or varied forms. **diversify**, *v.t.* to make different from others; to give variety to. *v.i.* to invest in securities of different types; to be or become engaged in the production of several types of manufactured goods etc.; to spread one's activities or interests (into). **diversification** (-fi-), *n.*
diversion (dīvœ'shən, di-), *n.* the act of diverting or turning aside; that which tends or serves to divert the mind or attention from care, business or study; a relaxation, amusement; a redirection of traffic owing to the temporary closing of a road; something which takes attention away from a criminal act or a military attack which one wishes to remain unseen. **diversity**, *n.* difference, variety or non-identity.
divert (dīvœt', di-), *v.t.* to turn from any course or direction; to turn aside, to deflect, to avert; to distract (one's attention); to entertain, to amuse. *v.i.* to follow a traffic diversion. **divertible**, *a.* **divertimento** (dīvœtimen'tō), *n.* (*pl.* **-ti** (-tē)) a piece of entertaining music; a musical pot-pourri; a ballet-interlude. **diverting**, *a.* entertaining, amusing. **divertingly**, *adv.* **divertissement** (dēvœtēs'mē), *n.* an (esp. musical) interlude or light entertainment.
diverticulum (dīvətik'ūləm), *n.* (*pl.* **-la** (-lə)) a sac or pouch on the wall of a tubular organ, esp. the intestine. **diverticulitis** (-lī'tis), *n.* inflammation of the diverticula.
Dives (dī'vēz), *n.* the popular name for a wealthy man (after the parable of Lazarus and the rich man in Luke xvi.19–31). rich.
divest (dīvest', di-), *v.t.* to strip of clothing; to deprive (of); to rid (oneself of). **divestiture** (-tichə), *n.* divestment.
divestment, *n.* the act of divesting (esp. of property).
divide (divīd'), *v.t.* to cut or sever, to partition; to cause to separate, to break into parts; to distribute, to deal out; to make an opening or passage through; to form the boundary between; to mark divisions on (as on mathematical instruments etc.); to distinguish the different kinds of, to classify (into); to separate (as Parliament, a meeting) by taking opinions on, for and against; to destroy unity amongst, to disunite in feelings; (*Math.*) to separate into factors, to perform the operation of division on. *v.i.* to be parted or separated; to diverge; to be classified by type; to express decision by separating into two parts, as a legislative house; (*Math.*) to be an exact division of; to be disunited in feelings, opinions etc. **divided**, *a.* **dividedly**, *adv.* **divider**, *n.* one who or that which divides; one who causes division or disunion; (*pl.*) compasses used to divide lines into a given number of equal parts.
dividend (div'idend), *n.* the share of the interest or profit which belongs to each shareholder in a company, bearing the same proportion to the whole profit that the shareholder's capital bears to the whole capital; (*Law*) the fractional part of the assets of a bankrupt paid to a creditor, in proportion to the amount of his debt; (*Math.*) a number to be divided by a divisor; a bonus, esp. unexpected.
divine[1] (divīn'), *v.t.* to find out by inspiration, intuition or magic; to foresee, to presage; to conjecture, to guess. *v.i.* to practise divination; to guess. **divination** (-vi-), *n.* the act of predicting or foretelling events, or of discovering

hidden or secret things by real or alleged supernatural means; something divined by these means. **diviner**, *n.* one who divines; a dowser. **divining-rod**, *n.* a forked twig or other staff used by dowsers to discover subterranean waters or minerals.

divine[2] (divīn'), *a.* pertaining to, proceeding from or of the nature of God, a god, or gods; used in the service of the Deity, religious, sacred; above the nature of man, god-like, celestial; pertaining to theology (*coll.*) wonderful. *n.* a clergyman, an ecclesiastic; a theologian. **divine office**, *n.* the service of the Roman church, the recitation of which is made by all persons in Holy Orders. **divine right**, *n.* the claim of kings to hold their office by divine appointment, and hence to govern absolutely without any interference from their subjects. **divine service**, *n.* the worship of God according to established forms. **divinely**, *adv.*

diving DIVE.

divinity (divin'iti), *n.* (the quality of being) a deity, a godhead; the Divine Being, God; theology.

divisible (diviz'ibl), *a.* capable of being divided; (*Math.*) able to be divided into equal parts without a remainder. **divisibility** (-bil'-), *n.* **divisibly**, *adv.*

division (divizh'ən), *n.* the act of dividing; the state of being divided; that which divides or separates, a boundary, a partition; a separate or distinct part; a district, an administrative unit; disunion, disagreement, variance; (*Nat. Hist.*) a separate class, kind, species or variety; the separation of Members of Parliament for the purpose of voting; a formal vote in Parliament; (*Math.*) the process of dividing one number by another; (*Arith.*) a sum in which one number is divided by another; (*Mil.*) a body of men, usu. three brigades, under the command of a general officer, applied loosely to smaller bodies; (*Nav.*) a number of vessels under one command. **division of labour**, distribution of parts of work among different persons in order to secure specialization on particular processes and to save time. **long division**, (*Arith.*) (a sum involving) the process of dividing a number by another number greater than 12. **division lobby**, *n.* either of the two corridors in Parliament in which the members vote. **division sign**, *n.* (*Arith.*) the sign ÷. **divisional**, **-sionary**, *a.* **divisionally**, *adv.* **divisive** (-vī'siv), *a.* causing separation, division or dissension. **divisively**, *adv.* **divisiveness**, *n.* **divisor** (-vī'zə-), *n.* (*Math.*) that number by which a dividend is divided; a number that divides another without a remainder.

divorce (divaws'), *n.* the dissolution of the marriage tie, the separation of husband and wife by judicial sentence of a secular or ecclesiastical court; (*fig.*) a separation of things closely connected. *v.t.* to dissolve by legal process the bonds of marriage between; to obtain a divorce from; (*fig.*) to disunite (things closely connected); (*fig.*) to distance (oneself or one's opinions etc.). *v.i.* legally to dissolve one's marriage. **divorcee** (-sē'), *n.* one who has been divorced.

divot (div'ət), *n.* a piece of turf torn up by the head of a golf club when playing the ball.

divulge (divŭlj'. dī-), *v.t.* to make known, to reveal, to disclose (esp. something secret). **divulgence**, *n.* **divulger**, *n.* **divulgement**, *n.*

divvy (div'i), *n.* (*coll.*) a dividend, esp. on the football pools; a share. *v.t.* to divide (up).

dixie (dik'si), *n.* a pot for cooking over an outdoor fire.

Dixieland (dik'silənd), **Dixie**, *n.* an early type of jazz music played by small combinations of instruments; the southern states of the US from which this music originates.

DIY, (*abbr.*) do-it-yourself.

dizzy (diz'i), *a.* giddy; causing dizziness; high; whirling; (*dated*) foolish, stupid. *v.t.* to make dizzy; to confuse, to muddle. **dizzily**, *adv.* **dizziness**, *n.*

DJ, (*abbr.*) dinner jacket; disc jockey.

djellaba, djellabah (jəlah'bə), *n.* a cloak (*orig.* Arabic) with wide sleeves and a hood. [Arab. *jallabah*]

djinn (jin), JINN.

dl, (*abbr.*) decilitre.

DLitt(t), (*abbr.*) Doctor of Literature, Doctor of Letters (L *Doctor Litteratum*).

DM, (*abbr.*) Deutschmark.

dm, (*abbr.*) decimetre.

DMus, (*abbr.*) Doctor of Music.

DMZ, (*abbr.*) demilitarized zone.

DNA, (*abbr.*) deoxyribonucleic acid, the main constituent of chromosomes, which is self-replicating and transmits hereditary characteristics.

D-notice (dē'nōtis), *n.* an official notice prohibiting publication of information prejudicial to the security of the UK.

do., (*abbr.*) ditto.

do[1] (doo), *v.t.* (*2nd sing.* **do**, *Bibl.* **doest** (doo'ist), *aux.* (*Bibl.*) **dost** (dŭst); *3rd.sing.* **does** (dŭz), *Bibl.* **doth** (dŭth); *past* **did** (did), *Bibl.* **didst** (didst); *p.p.* **done** (dŭn); **don't** (dōnt); **didn't** (did'nt), (*coll.*) for *do not*, *did not*; **doesn't** (dŭz'nt), for *does not*) to execute, perform, effect (a work, thing, service, benefit, injury etc., or the action of any verb understood); to bring about as a result; to produce (to make); to complete, finish, accomplish; to produce, to cause, to render (good, evil, honour, justice, injury etc.) to; to work, act, deal with; hence, to study, to prepare, to cook, to play the part of, etc.; (*coll.*) to cheat; (*coll.*) to injure; (*coll.*) to serve (a given length of prison sentence, as a punishment); (*coll.*) to visit and see the sights of; to travel (a distance) or travel at (a speed); (*sl.*) too arrest; (*sl.*) to convict; (*sl.*) to self-administer (drugs). *v.i.* to act, behave, conduct oneself; to perform deeds; to finish, to cease; to fare, to get on (in an undertaking or in health etc.); to serve, to suffice, to be enough (for). *aux.v.* (in neg. and in interrog. sentences), as *I do not play*, *do you not play?*; with inf. for special emphasis, as *I do believe, they do love him*; in the imper., as *do give it to him, do but ask*; in inverted sentences, as *seldom did it occur*; also poetically, *it did appear*; substitute (for a verb expressing any action, usu. to avoid repetition), as *I walked there in the same time as he did; you play as well as he does; he often comes here, I seldom do. n.* party, a celebration. **anything doing?** anything going on?. **do's and don'ts**, rules for action. **nothing doing**, (*coll.*) no business; no offers; no acceding to a request. **do-do**, bustle, confusion. **to do away with**, to abolish, to get rid of; to make away with, to kill. **to do by**, to treat, to deal with. **to do down**, to get the better of, to cheat. **to do for**, to suit; to put an end to; to ruin, to kill; (*coll.*) to do domestic work for. **to do in**, (*sl.*) to kill. **to do one's best** to make one's best efforts. **to do or die**, to make a last, desperate attempt. **to do (someone) proud**, (*coll.*) to treat (someone) very well. **to do to death**, to (cause to) put to death. **to do up**, to put in repair; to pack (in a parcel), to fasten (a garment). **to do well**, to prosper. **to do with**, to have business or connection with. **to do without**, to dispense with. **to have done with**, to finish with. **to have to do with**, to have business or intercourse with; to concern. **well-to-do**, well off; prosperous. **do-gooder**, *n.* one who tries to help others, often in a meddlesome or ineffectual way. **do-it-yourself**, *n.* decorating, household repairs and building as a hobby. **do-nothing**, *n.* an idler. *a.* lazy, idle. **doable**, *a.* **doer**, *n.*

do[2] (dō), *n.* (*Mus.*) the first of the syllables used in sol-fa; the note C.

DOA, (*abbr.*) of a body, dead on arrival.

dobbin (dob'in), *n.* a draught horse.

Dobermann pinscher (dō'bəmən pin'shə), *n.* a large breed of dog with a smooth usu. black and tan coat, used as a guard dog. [L *Dobermann*, 19th-cent. German dog-breeder, G. *Pinscher*, terrier]

doc (dok), *n.* short for DOCTOR.

docile (dō'sīl), *a.* willing or ready to learn; tractable; easily managed. **docility** (dəsil'-), *n.*
dock[1] (dok), *n.* a common name for various species of perennial herbs, most of them troublesome weeds.
dock[2] (dok), *v.t.* to cut the tail off; to cut short; to deduct a part from (as wages); to deprive of a part of.
dock[3] (dok), *n.* an artificial basin in which ships are built or repaired; (*often pl.*) an artificial basin for the reception of ships to load and unload; (*esp. N Am.*) a wharf. *v.t.* to bring into dock; to place in a dry dock; to join (one spacecraft) to another in space. *v.i.* (of spacecraft) to join together in space. **dry dock**, a dock from which the water can be pumped out, for building and repairing vessels. **floating dock**, a structure into which a vessel can be floated, the internal water then being pumped out to result in a floating dry dock. **to be in, to put into, dock**, to be, to send, away for repairs; (*coll.*) to be unwell. **dockland**, *n.* the area around the docks. **dock-master**, *n.* the officer in charge of docks or of a dockyard. **dockside**, *n.* the area beside a dock. **dockyard**, *n.* a large enclosed area with wharves, docks etc. where vessels are built or repaired, usually in connection with the Navy. **docker**, *n.* a labourer at the docks.
dock[4] (dok), *n.* the enclosure for prisoners in a criminal court. **in the dock**, charged with some offence.
docket (dok'it), *n.* any endorsement of a letter or document summarizing the contents; an internal order-form; a ticket or label showing the address of a package etc.
Doc Martens® (dok mah'tinz), *n.pl.* heavy, strong, lace-up shoes or calf-length boots.
doctor (dok'tə), *n.* a qualified practitioner of medicine or surgery; one who has obtained the highest degree in a faculty at a university either for proficiency or as an honour. *v.t.* to adulterate; to falsify; to castrate or spay (a dog or cat). *v.i.* (*coll.*) to practise as a physician. **(just) what the doctor ordered**, (*coll.*) something much needed or desired. **doctoral**, *a.* **doctorate** (-rət), *n.* the degree, rank or title of a doctor. **doctoring**, *n.* medical treatment; adulteration; falsification. **doctorless**, *a.*
doctrinaire (doktrineə'), *n.*, *a.* (a person who is) determined to follow the exact detail of a theory or doctrine, regardless of whether or not this is practical. **doctrinarian**, *n.*, *a.* **doctrinairism, doctrinarianism**, *n.*
doctrine (dok'trin), *n.* that which is taught; the principles, tenets or dogmas of any church, sect, literary or scientific school, or party. **doctrinal** (-trī'-), *a.* pertaining to doctrine; of the nature of or containing a doctrine. **doctrinally**, *adv.* **doctrinism**, *n.* **doctrinist**, *n.* **doctrinize, -ise**, *v.t.*
docudrama (dok'ūdrahmə), *n.* (a) television film which presents real events in a dramatized style.
document (dok'ūmənt), *n.* a written or printed paper (esp. official) containing information. *v.t.* to furnish with documents necessary to establish any fact; to prove by means of documents. **documentary** (-men'-), *a.* relating to documents; presenting facts or reality. **documentary (film)**, *n.* one which shows real events. **documentation**, *n.* the preparation or use of documents; the documents given, needed etc.
dodder (dod'ə), *v.i.* to shake, to tremble, to totter (along). **dodderer**, *n.* **doddery**, *a.* shaky, trembling because old and feeble.
doddle (dod'l), *n.* (*coll.*) something very easily accomplished.
dodec(a)-, *pref.* twelve.
dodecagon (dōdek'əgon), *n.* (*Geom.*) a plane figure of 12 equal angles and sides.
dodecahedron (dōdekəhē'drən), *n.* a solid figure of 12 equal sides, each of which is a regular pentagon. **dodecahedral**, *a.*
dodecapetalous (dōdekəpet'ələs), *a.* having 12 petals.
dodecaphonic (dōdekəfon'ik), *a.* relating to a 12-tone musical system.

dodecasyllable (dōdekəsil'əbl), *n.* a verse of 12 syllables; an alexandrine. **dodecasyllabic** (-lab'-), *a.*
dodge (doj), *v.i.* to change position by a sudden movement, esp. in order to avoid harm; to move rapidly from place to place so as to elude pursuit, detection etc.; to act trickily. *v.t.* to escape from by moving quickly aside; to evade by craft; to cheat. *n.* a sudden movement to one side; a trick, an artifice, an evasion; a skilful contrivance or expedient. **dodger**, *n.* one who dodges or evades; a trickster, a cheat. **dodgy**, *a.* full of craft, artful, tricky; uncertain, risky, dubious.
Dodgem® (doj'əm), *n.* a bumper car in an amusement ground.
dodo (dō'dō), *n.* (*pl.* **-does, -dos**) a large bird allied to the pigeons, with rudimentary wings, found in Mauritius in great numbers when that island was colonized in 1644 by the Dutch, but soon totally exterminated. **as dead as a dodo**, completely obsolete, defunct or extinct.
DOE, DoE (*abbr.*) Department of the Environment.
doe (dō), *n.* the female of the fallow deer; the female of the rabbit, hare and sometimes of other animals.
doer, does, doest DO[1].
doff (dof), *v.t.* to take off (as clothes, esp. one's hat as a mark of respect); **doffer**, *n.* a part of a carding-machine for stripping the cotton or wool from the cylinder; a person who removes the full bobbins or spindles.
dog (dog), *n.* a wild or domesticated quadruped of numerous breeds, derived from crossing of various species living and extinct; the male of the wolf, fox and other animals; a surly fellow, a contemptible person; a lively young fellow; either of two southern constellations, *Canis major* and *Canis minor*; a name given to various mechanical contrivances acting as searing points; a device with a tooth which penetrates or grips an object and detains it; an andiron or firedog; (*sl.* esp. N Am.) an unattractive woman. *v.t.* to follow like a dog; to track the footsteps of; (*fig.*) to follow or attend closely, as bad luck. **dog in the manger**, one who prevents other people from enjoying what he/she cannot enjoy; a churlish person. **dog-in-the-manger**, *a.* **dressed up like a dog's dinner**, dressed (too) flamboyantly. **give a dog a bad name**, if a person once gets a bad reputation it is difficult to regain people's good opinion. **hair of the dog** HAIR. **hot dog**, a hot sausage sandwich. **lucky dog**, a lucky fellow. **sea-dog**, an old sailor, esp. one of the Elizabethan adventurers. **sick as a dog**, (*coll.*) very sick, vomiting; very disappointed. **sly dog**, an artful fellow. **the dogs**, *n.pl.* (*coll.*) greyhound races. **to go to the dogs**, to go to ruin. **to lead (someone) a dog's life**, to (cause to) live a life of continual wretchedness. **to let sleeping dogs lie**, to leave well alone. **to rain cats and dogs**, to rain very heavily. **to throw, give, to the dogs**, to throw away. **dog-bane**, *n.* a plant with a bitter root, supposed to be poisonous to dogs. **dog-cart**, *n.* a light, two-wheeled, double-seated, one-horse vehicle. **dog-collar**, *n.* a leather or metal collar worn by dogs; a stiff, white collar fastening at the back, as worn by clergymen. **dog-days**, *n.pl.* the period in July and August during which the dog-star rises and sets with the sun, a conjunction formerly supposed to account for the great heat usual at that season. **dog-ear**, *n.* a corner of a leaf of a book turned down like a dog's ear. *v.t.* to turn down the corners of (a book) by careless handling. **dog-eared**, *a.* **dog-eat-dog**, *n.*, *a.* (characterized by) ruthless pursuit of one's own interests. **dog-end**, *n.* (*sl.*) a cigarette-end. **dog-fancier**, *n.* one who keeps and breeds dogs for sale. **dog-fight**, *n.* a fight between dogs; a wrangle; (*Aviat.*) a duel in the air between two aircraft. **dog-fish**, *n.* any species of fish sometimes extended to include small sharks which follow their prey in packs, whence their popular name. **dog-house**, *n.* a dog-kennel. **in the dog-house**, in disfavour, in disgrace. **dog-kennel**, *n.* a house or hut for a dog. **dog-Latin**, *n.* barbarous, ungrammatical Latin.

dog-leg, *n.*, *a.* something bent like a dog's hind leg, applied to e.g. a golf hole with a bent fairway. **dog-, doggy-paddle**, *n.* a simple swimming stroke in which the arms imitate the front legs of a swimming dog. **dog-rose**, *n.* the wild brier. **dogsbody**, *n.* (*coll.*) someone made use of by others for any menial tasks; a useful person treated as a drudge. **dog's chance**, *n.* the slightest chance. **dog sled**, *n.* (*N Am.*) a sled pulled by a team of dogs. **dog's mercury**, *n.* a common poisonous weed. **dog-star**, *n.* Sirius, the principal star in the constellation *Canis major*. **dog's-tooth**, *n.*, *a.* (*Arch.*) (of) a kind of protruding ornament used in Early English mouldings. **dog-tag**, *n.* (*N Am.*) an identification tag for military personnel. **dog-tired**, (*coll.*) *a.* worn out. **dog-trot**, *n.* a gentle easy trot. **dog-violet**, *n.* the scentless wild violet. **dog-watch**, *n.* (*Naut.*) one of two watches of two hours each between 4 and 8 p.m. **doggish**, *a.* **doggishly**, *adv.* **doggishness**, *n.* **doggo** (-gō), *a.* hidden. **to lie doggo**, (*coll.*) to wait silently and motionlessly. **doggy, doggie**, *n.* childish or pet term for a dog. *a.* pertaining to, characteristic of, a dog; fond of dogs. **doggy bag**, *n.* (*coll; esp. N Am.*) a bag for taking home uneaten food after a restaurant meal. **doglike**, *a.* like a dog; unquestioningly obedient.
doge (dōj), *n.* the title of the chief magistate of the republics of Venice and Genoa. **dogate** (dō'gāt), *n.* the position, office or rank of a doge.
dogged (dog'id), *a.* stubborn (like a dog), obstinate, persistent, tenacious. **doggedly**, *adv.* **doggedness**, *n.*
dogger (dog'ə), *n.* a type of Dutch fishing-vessel employed in the North Sea cod and herring fishery.
doggerel (dog'ərəl), *n.*, *a.* orig. applied to loose, irregular verses, such as those in Butler's *Hudibras*; now to verses written with little regard to rhythm or rhyme.
doggone (dog'on), *int.* (*N Am.*) expressing annoyance. **doggoned**, *a.* (*euphem. N Am.*), god-damned.
dogma (dog'mə), *n.* (*pl.* **-mas, -mata** (-tə)) an established principle, tenet or system of doctrines put forward to be received on authority, esp. that of a Church, as opposed to one deduced from experience or reasoning. **dogmatic** (-mat'-), *a.* pertaining to dogma, doctrinal; based on theory not induction; asserted with (sometimes unwarranted) authority; arrogant, dictatorial. **dogmatically**, *adv.* **dogmatism**, *n.* arrogance or undue positiveness in assertion. **dogmatist** *n.* **dogmatize, -ise**, *v.i.* to make dogmatic assertions. *v.t.* to lay down as a dogma.
dogsbody DOG.
doh DO [2].
doily, doyley (doi'li), *n.* a small ornamental mat or napkin on which to place cakes, sandwiches, bottles, glasses etc. [*Doily*, an 18th-cent. London haberdasher]
doings (doo'ingz), *n.pl.* (*coll.*) things done or performed; events, transactions, proceedings, goings-on; behaviour, conduct; applied to any object whose name one has forgotten or does not want to say.
Dolby® (dol'bi), *n.* a system used to cut down interference on broadcast or recorded sound. [R. Dolby, b.1933, US engineer]
dolce (dol'chi), *a.* sweet, soft, usu. of music. *adv.* sweetly, softly. **dolce far niente** (fah nyen'ti), *n.* sweet idleness. **dolce vita** (vētə), *n.* a life of luxury and self-indulgence.
doldrums (dol'drəmz), *n.pl.* low spirits, the dumps; that part of the ocean near the equator between the regions of the trade-winds where calms and variable winds prevail. **in the doldrums**, in low spirits, in the dumps; (*Naut.*) becalmed.
doleful (dōl'fəl), *a.* sorrowful, sad; dismal, gloomy. **dolefully**, *adv.* **dolefulness**, *n.*
dole (dōl), *n.* unemployment benefit. (*coll.*) **on the dole**, in receipt of unemployment benefit. **to dole out**, to distribute in small quantities.
dolerite (dol'ərit), *n.* a variety of rock consisting of feldspar and pyroxene.

dolichocephalic (dolikōsifal'ik), **-cephalous** (-sef'ələs), *a.* long-headed; applied to skulls in which the width from side to side bears a smaller proportion to the width from front to back than 80%. **dolichocephalism** (-sef'-), *n.*
doll (dol), *n.* a child's toy representing a human figure; a pretty (but silly) woman; (*esp. N Am. coll.; sometimes derog.*) a term of endearment to a woman. **to doll up**, (*coll.*) to dress up, to make (oneself) look smart. **dollish**, *a.* **dollishly**, *adv.* **dollishness**, *n.*
dollar (dol'ə), *n.* the unit of currency (orig. a silver coin) in the US and Canada, also Australia, New Zealand etc., equivalent to 100 cents; applied to coins of different values; (*sl.*) five shillings sterling (25p). **dollar area**, *n.* the area in which currency is linked to the US dollar. **dollar diplomacy**, *n.* diplomacy dictated by financial interests abroad; diplomacy which uses financial power as a weapon. **dollar gap**, *n.* the excess of imports over exports in trade with a dollar-area country.
dollop (dol'əp), *n.* (*coll.*) a shapeless lump; a heap, quantity.
dolly (dol'i), *n.* a pet name for a doll; a hoisting platform; a trolley; a simple catch in cricket. *a.* dollish. **dolly camera**, *n.* a cine-camera moving on a type of trolley. **dolly mixture**, *n.* one of a mixture of tiny coloured sweets. **dolly switch**, *n.* one for an electric light etc. consisting of a lever to be pushed up and down.
Dolly Varden (dol'i vah'dn), *n.* a large-patterned print dress; a wide-brimmed woman's hat with one side bent down. [a character in Dickens's *Barnaby Rudge*]
dolman (dol'mən), *n.* a long Turkish robe, open in front, and with narrow sleeves; a woman's loose mantle with hanging sleeves; a hussar's jacket or cape with sleeves hanging loose. **dolman sleeve**, *n.* one which tapers from a wide armhole to a tightly-fitting wrist.
dolmen (dol'mən), *n.* a cromlech; the megalithic framework of a chambered cairn, consisting usually of three or more upright stones supporting a roof-stone.
dolomite (dol'əmit), *n.* a brittle, translucent mineral consisting of the carbonates of lime and magnesia. **dolomitic** (-mit'-), *a.*
dolorous (dol'ərəs), *a.* full of pain or grief; causing or expressing pain or grief, dismal, doleful. **doloroso** (-rō'sō), *a.*, *adv.* (*Mus.*) in a soft, dolorous manner. **dolorously**, *adv.* **dolorousness**, *n.*
dolour (dol'ə), *n.* suffering, distress; grief, sorrow.
dolphin (dol'fin), *n.* a variety of marine mammal, in size between the whale and the porpoise; (*Her. etc.*) a conventional representation of a curved fish. **dolphinarium** (-eə'riəm), *n.* an aquarium for dolphins, often one for public displays.
dolt (dōlt), *n.* a stupid person. **doltish**, *a.* **doltishly**, *adv.* **doltishness**, *n.*
dom., (*abbr.*) domestic; dominion.
dom (dom), *n.* in the Roman Catholic Church, a title given to members of the Benedictine and Carthusian orders; the Portuguese form of DON [1]. [abbr. of L *dominus*, lord]
-dom, *comb. form.* noting power, jurisdiction, office or condition, or a group of people, as in *earldom, kingdom, officialdom, freedom*.
domain (dəmān'), *n.* territory, district or space over which authority, jurisdiction or control is or may be exercised; one's landed property, estate; (*fig.*) sphere, province, field of influence, expertise, thought or action.
dome[1] (dōm), *n.* a roof, usually central, the base of which is a circle, an ellipse or a polygon, and its vertical section a curved line, concave towards the interior; a cupola; a natural vault, arching canopy or lofty covering, as (*poet.*) the sky; any dome-shaped object or structure; (*sl.*) the head, esp. when bald. *v.t.* to cover with, or shape into, a dome. *v.i.* to swell into a domelike shape. **domed**, *a.* having a dome; dome-shaped. **domelike**, *a.*
Domesday Book (doomz'dā buk), *n.* a register of the lands

domestic   239   **doodle**

of England compiled (1084-86) by order of William the Conqueror, from the results of a Great Inquisition or survey, forming a basis for all historical accounts of the economic state of the country at that epoch.
domestic (dəmes'tik), *a.* pertaining to the home or household; made, done or performed at home; employed or kept at home; fond of home; tame, not wild; relating to the internal affairs of a nation; not foreign; made or occurring in one's own country; native-grown (as for wine etc.). *n.* a household servant. **domestic architecture**, *n.* the architecture of dwelling-houses. **domestic economy**, *n.* the economical management of household affairs. **domestic science**, *n.* the study of household skills, inc. cookery, needlework etc. **domesticable**, *a.* **domestically**, *adv.* **domesticate**, *v.t.* to make domestic or familiar; to naturalize (foreigners etc.); to accustom to domestic life and the management of household affairs; to tame; to bring into cultivation from a wild state; to civilize. **domesticated**, *a.* tamed; content with home life; used to, proficient at, household chores. **domestication**,
domesticity (dōməstis'əti, dom-), *n.* the state of being domestic; domestic character, homeliness; (the circumstances of) home life; (*pl.*) domestic affairs, family matters.
domicile (dom'isil), **domicil** (-sil), *n.* a house, a home, a place of abode; (*Law*) a place of permanent residence; length of residence (differing in various countries) necessary to establish jurisdiction in civil actions. *v.t.* to establish in a place of residence. *v.i.* to dwell, to reside. **domiciled**, *a.* **domiciliary** (-sil'i-), *a.* pertaining to a domicile or residence. **domiciliary visit**, *n.* a visit under legal authority to a private house to search for suspected persons or things.
dominant (dom'inənt), *a.* ruling, governing; predominant, overshadowing; supereminent; pertaining to the fifth note of a scale. *n.* the fifth note of the scale of any key, counting upwards; the reciting note of Gregorian chants. **dominant chord**, *n.* a chord formed by grouping three tones rising from the dominant by intervals of a third. **dominance**, *n.* **dominancy**, *n.* **dominantly**, *adv.*
dominate (dom'ināt), *v.t.* to predominate over, to be the most influential or the chief or most conspicuous of; to tower over (as a hill); to influence control, to rule, govern. *v.i.* to be the most influential, etc. **domination**, *n.* the exercise of power or authority; rule, control, dominion.
domineer (dominiə'), *v.i.* to exercise authority arrogantly and tyrannically; arrogantly to take a position of superiority over others in actions, argument etc. *v.t.* to control or tyrannize by taking such a position. **domineering**, *a.* **domineeringly**, *adv.*
Dominican (dəmin'ikən), *n.* one of an order of preaching friars, founded in 1216 by Domingo de Guzman (canonized as St Dominic); a Black Friar; a nun in one of the orders founded by St Dominic; a native of the Dominican Republic or Dominica. *a.* pertaining to the Dominicans; pertaining to the Dominican Republic or Dominica.
dominie (dom'ini), *n.* (*Sc.*) a schoolmaster; a minister or clergyman.
dominion (dəmin'yən), *n.* sovereign authority, lordship; control, rule, government; (*Law*) uncontrolled right of possession or use; the district, region or country governed by a given person; a self-governing country of the British Commonwealth, esp. Canada. **Dominion Day**, *n.* another name for Canada Day, which commemorates Canada receiving dominion status on 1 July 1867, a public holiday in Canada.
domino (dom'inō), *n.* (*pl.* **-noes**) a masquerade dress worn for disguise by both sexes, consisting of a loose black cloak or mantle with a small mask; a kind of half mask; a person wearing a domino; one of 28 oblong dotted pieces, orig. of bone or ivory, used in playing dominoes. **domino effect**, *n.* the fall of a long row of dominoes, all standing on end, caused by pushing the first domino in the row. **domino theory**, *n.* the theory that a single event (esp. the fall of a government) leads to many similar events elsewhere as a chain reaction. **dominoed**, *a.* wearing a domino. **dominoes**, *n.pl.* any of various games played with dominoes, often involving pairing up matching patterns of dots.
don[1] (don), *n.* a title formerly restricted to noblemen and gentlemen, now common to all classes in Spain, equal to sir, Mr; a Spanish gentleman; a Spaniard; a fellow or tutor of a college, esp. at Oxford or Cambridge; one who assumes airs of importance. **Don Juan** (joo'ən, hwahn), *n.* a lady-killer (from the hero of Byron's poem); a male flirt; a would-be rake. **Don Quixote** (kihō'ti. kwik'sōt), *n.* one who is excessively idealistic or chivalrous, esp. one who goes to foolish extremes (from the hero of Cervantes' *Don Quixote de la Mancha*, 1605). **donnish**, *a.* having the air of an academic don. **donnishness**, *n.*
don[2] (don), *v.t.* (*past, p.p.* **donned**) to put on; (*fig.*) to assume.
doña (don'yə), *n.* a Spanish title equivalent to lady, madam, Mrs.
donate (dənāt'), *v.t.* to bestow as a gift, esp. on a considerable scale for philanthropic or religious purposes. **donator**, *n.* a donor.
donation (dənā'shən), *n.* the act of giving; that which is given, a gift, a presentation, a contribution, esp. to a public or religious institution; (*Law*) an act or contract by which any thing, or the use of and the right to it, is transferred as a free gift to any person or corporation.
donative (dō'nətiv), *n.* a gift, a present, a gratuity, esp. an official donation; a benefice directly given by a patron without presentation to or institution by the bishop. *a.* vested or vesting by this form of presentation. **donatory**, *n.* the recipient of such a donation.
done (dūn), *p.p.* performed, executed; (*coll.*) tricked, cheated; cooked (to a given degree); finished. *int.* accepted (used to express agreement to a proposal, as a wager, or a bargain). **to have done**, to have finished. **to have done with**, to have or take no further concern with. **done for**, *a.* ruined, killed, doomed; exhausted.
donee (dōnē'), *n.* a person to whom anything is donated.
doner kebab (dō'nə kibab'), *n.* a dish of lamb sliced from a block grilled on a spit, served with salad in a piece of pitta bread.
donjon (dŭn'jən), *n.* the grand central tower or keep of esp. a mediaeval Norman castle, the lower storey generally used as a prison.
donkey (dong'ki), *n.* a member of the horse family, an ass; (*fig.*) a stupid person. **to talk the hindlegs off a donkey**, to talk constantly and at great length. **donkey-engine**, *n.* an auxiliary engine for light work. **donkey jacket**, *n.* a short, thick workman's jacket. **donkey's years**, *n.pl.* (*coll.*) a long time. **donkey-work**, *n.* drudgery, routine work. **nodding donkey** NOD.
donna (don'ə), *n.* a lady; an Italian title equivalent to Madam.
donor (dō'nə), *n.* a giver; (*Law*) one who grants an estate; one who gives blood for transfusion. *a.* (of organs) given for transplanting. **donor card**, *n.* one carried by a person willing to have parts of their body used for transplant in the event of their death.
don't (dōnt), (*coll.*) short for *do not*. *n.* a prohibition. **don't care**, *a.* careless, reckless. **don't know**, *n.* a person without a firm opinion on a matter, esp. on how to vote; an answer given by one with such a lack of opinion.
donut DOUGHNUT under DOUGH.
doodah (doo'dah), **doodad** (-dad), *n.* (*coll.*) any small decorative article or gadget. **to be all of a doodah**, (*sl.*) to be flustered, in a state of confusion.
doodle (doo'dl), *v.t.* to draw pictures or designs semiconsciously while thinking or listening. *n.* a picture drawn

doodlebug   **dost**

in this way. **doodler**, *n*.
doodlebug (doo'dlbŭg), *n*. (*coll*.) the earliest type of flying bomb used by the Germans in the war of 1939-45, the V-1.
doolally (doolal'i), *a*. (*sl*.) insane, eccentric. [*Deolali*, a town near Bombay, India]
doom (doom), *n*. judgment; judicial decision, sentence or penalty; fate or destiny (usu. in an evil sense); ruin, destruction, perdition. *v.t.* to condemn to punishment; to condemn (to do something); to predestine; to consign to ruin or calamity. **crack of doom**, the dissolution of all things at the time of universal Judgment. **doomsday**, *n*. the Day of Judgment; the end of the world; a day of judgment or dissolution. **till doomsday**, for ever. **Doomsday Book** DOMESDAY BOOK. **doomwatch(ing)**, *n*. pessimism, esp. about the future of the environment; observation of the environment to prevent its destruction by pollution etc. **doomwatcher**, *n*.
door (daw), *n*. a frame of wood or metal, usually on hinges, closing the entrance to a building, room, cupboard etc.; (an opening for) entrance, exit, access; a house, a room, as *two doors down the street/corridor*. **door-to-door**, from one house to the next; of a journey. **front door**, the principal entrance from the street. **next door**, in the next house or room. **next-door**, *a*. (*attrib*.) **next door to**, immediately adjacent to; closely bordering on, almost. **out of door(s)**, outside the house; in the open air. **to lie at one's door**, to be chargeable or attributable to. **to show the door**, to send away unceremoniously. **door-bell**, *n*. a bell inside a building actuated by a button or pull outside. **door-case**, **-frame**, *n*. the structure in which a door swings. **door-keeper**, *n*. a porter, a janitor. **door-knob**, *n*. a round handle on a door. **door-knocker**, *n*. a hinged device attached to a door, for knocking. **doorman**, *n*. a porter, one employed to open doors. **doormat**, *n*. a mat for removing dirt from the boots, placed inside or outside a door; (*coll*.) a submissive person, often imposed on by others. **door-nail**, *n*. a large nail formerly used for studding doors. **door-plate**, *n*. a plate on a door bearing the name of the occupant. **door-post**, *n*. side-piece or jamb of a doorway. **doorstep**, *n*. a step leading up to an outer door; (*coll*.) a thick slice of bread. *v.t.*, *v.i.* to go from door to door (in a neighbourhood) to canvass during a political campaign, or to try to sell goods, often intrusively. **doorstepping**, *n*. **doorstop**, *n*. a device which stops a door from opening too far or from closing. **doorway**, *n*. an opening in a wall fitted with a door; a means of access. **doored**, *a*. (*usu. with a. prefixed*). **doorless**, *a*.
dope (dōp), *n*. any thick liquid or semi-fluid used as a lubricant; (*coll*.) a drug given to a horse or greyhound to make it win a race; any drug, but esp. cannabis; (*coll*.) a fool; a varnish used for waterproofing, protecting and strengthening the fabric parts of an aircraft; (*sl*.) inside information, particulars. *v.t.* to add impurities to a semi-conductor in order to change its properties; to apply aircraft dope to; to drug. **dope-fiend**, *n*. a drug addict. **dopey**, *a*. (*coll*.) stupid; (as if) drugged; sluggish. **doping**, *n*.
doppelganger (dop'lgangə), *n*. the ghostly apparition of a living person; a double. [G *Doppelgänger*, double-goer]
Doppler effect (dop'lə), the apparent change of pitch of sound produced by a body when approaching, passing and receding with considerable velocity. **Doppler shift**, *n*. an observed shift in the frequency of light.
dopplerite (dop'lərīt), *n*. a black substance found in peat beds.
dorado (dərah'dō), *n*. (*pl*. **-dos**) a fish of brilliant golden colour when dying, sometimes called a dolphin; (*Astron*.) a southern constellation, the Swordfish.
Dorcas Society (daw'kəs), *n*. a charitable association for making clothes for the poor.
Dorian (daw'riən), *a*. of or relating to *Doris*, in ancient Greece, or its inhabitants. *n*. an inhabitant of Doris; a member of one of the four great ethnic divisions of the ancient Greeks. **Dorian mode**, *n*. (*Mus*.) a simple, solemn form of music, the first of the authentic Church modes. **Doric** (do'-), *a*. Dorian. *n*. (*Arch*.) the Doric order; any broad rustic dialect. **Doric order**, *n*. the earliest, strongest and most simple of the three Grecian orders of architecture.
Dorking (daw'king), *n*. name of a breed of domestic fowls, orig. from Dorking in Surrey.
dorm (dawm), *n*. short for DORMITORY.
dormant (daw'mənt), *a*. in a state resembling sleep, torpid, inactive (of animals hibernating); undeveloped (as talent); (*esp. Law*), inoperative, not asserted or claimed; (of volcanoes) inactive for the time being. **dormancy**, *n*.
dormer (daw'mə), **dormer-window**, *n*. a window piercing a sloping roof and having a vertical frame and a gable (orig. used in sleeping chambers, whence the name).
dormie, **dormy** (daw'mi), *a*. (*Golf*) applied to a player or his score when he is as many holes ahead of his opponent as there remains holes to play, as *dormie two*.
dormitory (daw'mitri), *n*. a sleeping-room, esp. in a school or public institution, containing a number of beds. **dormitory suburb**, **town**, *n*. one whose inhabitants work elsewhere, often in a nearby city.
Dormobile® (daw'məbēl), *n*. a van equipped for living in while travelling.
dormouse (daw'mows), *n*. (*pl*. **-mice**) a small British hibernating rodent; others of the same genus, animals between the mouse and the squirrel.
dormy DORMIE.
dorp (dawp), *n*. a S African small town.
dorsal (daw'səl), *a*. of or pertaining to the back; situated on the back. **dorsal fin**, *n*. the fin on the back of a fish which aids balance. **dorsally**, *adv*.
dors(i)-, **dorso-**, *comb. form*. belonging to, situated on, the back.
dorsibranchiate (dawsibrang'kiət), *a*. having gills on the back.
dorsiflexion (dawsiflek'shən), *n*. a bending backwards.
dorsigrade (daw'sigrād), *a*. walking on the back of the toes.
dorsispinal (dawsispī'nəl), *a*. belonging to the spine and the back.
dorsum (daw'səm), *n*. (*Zool*., *Anat*.) the back.
dory[1] (daw'ri), *n*. a golden-yellow sea-fish called also the John Dory, eaten as food.
dory[2] (daw'ri), *n*. a small, flat-bottomed boat.
dos-à-dos, **dosi-do** (dōsidō'), *a*. back-to-back. *n*. a seat designed for sitting back-to-back; a square-dance step in which dancers pass each other back-to-back.
dose (dōs), *n*. the quantity of any medicine taken or prescribed to be taken at one time; the quantity of radiation absorbed at one time; (*fig*.) a quantity or amount of anything offered or given, esp. of anything nauseous or unpleasant which one has to take. *v.t.* to administer doses to; to give anything unpleasant to. **like a dose of salts**, very quickly (and thoroughly). **dosage**, *n*. the process or method of dosing; the correct dose to be taken at any one time. **dosimeter** (-sim'itə), *n*. an instrument which measures radiation doses. **dosimetric** (-met'-), *a*.
dosh (dosh), *n*. (*sl*.) money.
dosi-do DOS-À-DOS.
doss (dos), *n*. (*sl*.) a bed or a sleeping-place in a common lodging-house (*sl*.) a sleep. *v.i.* to sleep in this; (*sl*.) to sleep; to go to bed. **dosser**, *n*. one who sleeps in cheap lodging-houses or hostels; (*sl*.) a lazy, idle person. **to doss down**, (*sl*.) to go to sleep in a makeshift bed. **doss-house**, *n*. a cheap lodging-house.
dossier (dos'iā, -iə), *n*. a collection of papers and other documents and information relating to a person, a thing or an event.
dost (dŭst), DO[1].

dot¹ (dot), *n.* a little mark, spot or speck made with a pen or pointed instrument; a similar mark used as a full point, a point over *i* or *j*, a decimal point, or used as a diacritic; (*Mus.*) a point used as a direction to increase the length of a note, play a note staccato, or in other ways; a round mark of one colour repeated on a differently coloured background as a pattern. *v.i.* (*past, p.p.* **dotted**) to make dots or spots. *v.t.* to mark with dots; to mark or scatter over with (small detached objects like dots). **dot and dash**, the system of symbols in Morse telegraphy. **dot matrix printer**, one using a pattern of dots to form characters rather than continuous lines. **on the dot (of)**, (*coll.*) precisely (at). **the year dot**, (*coll.*) as far back as memory reaches. **to dot one's i's and cross one's t's**, to be precisely exact.
dotage (dō'tij), *n.* impairment of the intellect by old age; the period of life characterized by this; infatuation. **dotard**, *n.* a man in his dotage; one who is foolishly and excessively fond. **dotardly**, *a.*
dote (dōt), **to dote on**, to be foolishly fond of. **doter**, *n.*
doth (dŭth), DO¹.
dotterel (dot'ərəl), *n.* a small migratory plover.
dottle (dot'l), *n.* a plug of tobacco left unsmoked in a pipe.
dotty (dot'i), *a.* marked with dots, dot-like; (*sl.*) silly or eccentric.
douane (dooahn'), *n.* a continental custom-house. **douanier** (-iā), *n.* a custom-house officer.
double¹ (dŭb'l), *a.* composed of two, in a pair or in pairs; forming a pair, twofold; folded, bent back or forward; twice as much, as great or as many; of twice the strength or value; of two kinds, aspects or relations; ambiguous; (*fig.*) hypocritical, treacherous, deceitful; an octave lower in pitch; applied to flowers when the stamens become more or less petaloid. *adv.* twice; in two ways; in twice the number, quantity, amount, strength etc.; two together. **double summer time**, the time indicated by clocks advanced one hour more than summer time or two hours in front of Greenwich mean time. **double act**, *n.* an entertainment act by two people; the two entertainers. **double-acting**, *a.* exerting force in two directions. **double agent**, *n.* a spy working for two opposing sides at the same time. **double ale**, *n.* ale of double strength. **double back**, *v.i.* to go back in the direction one has come from. **double bar**, *n.* (*Mus.*) two single bars put together, to denote the end of a section. **double-barrel(led)**, *a.* having two barrels, as a gun; producing a double effect, serving a double purpose; (*coll.*) (of a surname) having two parts, often hyphenated. **double-bass**, *n.* the largest and lowest-toned of the stringed instruments played with a bow. **double bassoon**, *n.* the largest instrument in the oboe class, with the lowest pitch. **double bed**, *n.* a bed for two people. **double-bedded**, *a.* having two beds or a double bed. **double boiler**, *n.* (*N Am.*) a double saucepan. **double bottom**, *n.* an articulated lorry with a second trailer. **double-breasted**, *a.* lapping over and buttoning on both sides, as a coat or waistcoat. **double-check**, *v.t.* to check a second time. **double chin**, *n.* a fold of fat resembling a second or subsequent chin, due to obesity etc. **double concerto**, *n.* one for two solo instruments. **double cream**, *n.* thick cream, with a higher fat content than single cream. **double-cross**, *v.t.* (*coll.*) to betray, to trick, cheat. **double-crosser**, *n.* **double-dagger**, *n.* a reference mark (‡). **double-dealer**, *n.* a double-dealing person. **double-dealing**, *n.*, *a.* (conduct which is) deceitful, tricky. **double-decker**, *n.* a bus or coach with two decks; a sandwich with two layers of filling and three of bread; a novel etc. in two volumes. **double-declutch**, *v.i.* to change to a different gear by moving into neutral and then into the desired gear, disengaging the clutch at each of the two stages. **double-dotted note**, one increased in length by three-quarters, as shown by two dots placed after it. **double-Dutch**, *n.* gibberish, jargon; a language not understood by the hearer. **double-dyed**, *a.* stained or tainted with infamy; doubly infamous. **double-eagle**, *n.* an American gold coin worth 20 dollars; a representation, as in the imperial arms of Russia and Austria, of an eagle with two heads. **double-edged**, *a.* having two cutting edges; (*fig.*) telling for and against; working both for and against the user. **double-entendre** (dooblētēdr''), *n.* a word or phrase with two interpretations, one of which is usually indelicate. **double entry**, *n.* a method of book-keeping in which every transaction is entered twice, once on the credit side of the account that gives, and once on the debit side of the account that receives. **double exposure**, *n.* the recording of two superimposed images on a single frame of film; the picture resulting from this. **double-fault**, *n.* in tennis, two service faults in succession, resulting in the loss of a point. **double feature**, *n.* two full-length feature films shown in a single programme; such a programme. **double figures**, *n.pl.* a number greater than 9 but less than 100. **double first**, *n.* one who attains a first-class pass in two subjects in an examination for a degree. **double flat**, *n.* (*Mus.*) a sign indicating a drop of two semitones. **double-fronted**, *a.* of a house, having the main door in the centre of the front wall, with windows to each side. **double-glazing**, *n.* the fitting of a double layer of glass in a window to act as a form of insulation. **double-glaze**, *v.t.* **double Gloucester**, *n.* a rich hard cheese orig. made in Gloucestershire. **double-handed**, *a.* deceitful, treacherous; in tennis, using both hands. **double-headed**, *a.* having two heads; (*Railway*) having two locomotives at the front. **double-header**, *n.* a train drawn by two locomotives; (*N Am.*) two games played consecutively. **double helix**, *n.* two helices coiled round the same axis, the molecular structure of DNA. **double jeopardy**, *n.* a second trial for the same offence. **double-jointedness**, *n.* abnormal mobility of joints not associated with injury or disease, nor causing symptoms. **double knitting**, *n.* a medium-thickness knitting wool. **double-lock**, *v.t.* to lock by turning the key twice. **double negative**, *n.* (*Gram.*) a construction with two negatives where only one is needed, as in, *I don't need nothing*. **double-park**, *v.t.* to park (a vehicle) outside one already parked at the kerb. **double-parked**, *a.* **double pneumonia**, *n.* pneumonia in both lungs. **double-quick, time**, *n.* (*Mil.*) the quickest pace next to a run. **double reed**, *n.* a set of two reeds in the mouthpiece of a wind instrument, which vibrate against each other. **double-refine**, *v.t.* to refine twice over. **double refraction**, *n.* birefringence. **double salt**, *n.* one which, when dissolved, gives two different salts in solution. **double saucepan**, *n.* two saucepans, one fitting into the other, food being cooked gently in the inner pan by the heat of boiling water in the outer one. **double sharp**, *n.* (*Mus.*) a sign indicating a rise of two semitones. **double-space**, *v.t.* to type with a line-space between the lines. **double-speak**, *n.* double-talk. **double standard**, *n.* a single moral principle applied in different ways to different groups of people and allowing different behaviour, e.g. that of allowing young men but not young women to have sexual experience before marriage. **double-take**, *n.* a second look followed by a delayed reaction. **double-talk**, *n.* talk that sounds sensible though it is actually a compound of sense and gibberish; intentionally misleading talk, meaning the opposite of what it says. **double-think**, *n.* the holding of two contradictory beliefs at the same time. **double time**, *n.* a fast marching pace; overtime paid at twice the rate of normal time. **double vision**, *n.* diplopia. **doubleness**, *n.* **doubly**, *adv.*
double² (dŭb'l), *n.* twice as much or as many, a double quantity; a fold, a plait; a doppelganger; a person exactly resembling someone else; (*Theat.*) an understudy; a turn in running to escape pursuit; (*Lawn Tennis etc., pl.*) a game between two pairs of players; a bet on two races, the stake and winnings on the first being applied to the

doublet • down

second race; (*Darts*) (a throw into) one of the scoring positions between the two outer circles. *v.t.* to increase by an equal quantity, amount, number, value etc.; to multiply by two; to make twice as thick; to fold down or over, to bend, to turn upon itself; to be twice as much as; (*Mus.*) to add the upper or lower octave to; (*Theat.*) to act two (parts) in the same play; (*Theat.*) to understudy. *v.i.* to become twice as much or as great; (*Bridge*) on the strength of one's own hand to double the number of points an opponent may gain or lose; to turn or wind to escape pursuit; to run very quickly. **at, on, the double**, at twice the normal speed; very fast. **double or quits**, a game or bet where the person shall pay twice his debt or nothing, according to whether he loses or wins. **to double up**, to bend one's body into a stooping or folded posture; to (cause to) collapse (in pain, laughter etc.); of paper etc., to become folded or crumpled. **doubler,** *n.*

doublet (dŭb'lĭt), *n.* (one of) a matching pair; one of two words from the same root, but differing in meaning; (*Print.*) a word or passage printed twice by mistake; (*Hist.*) a close-fitting garment covering the body from the neck to a little below the waist, introduced from France in the 14th cent. **doublet and hose**, regular masculine attire in the Tudor period; an undress attire suitable for active exertion (implying the absence of a cloak).

doubleton (dŭb'ltən), *n.* (only) two cards of one suit.

doubloon (dəbloon'), *n.* a Spanish and S American gold coin.

doubt (dowt), *v.i.* to be in uncertainty about the truth, probability or propriety of anything; to hesitate, to waver (esp. in faith). *v.t.* to hold or think questionable; to hesitate to believe or credit. *n.* (an) uncertainty of mind upon any point, action or statement; suspense; distrust, inclination to disbelieve. **beyond a doubt, no doubt, without doubt**, certainly, admittedly, unquestionably. **in doubt**, not settled, not decided. **to give the benefit of the doubt**, to presume innocent, esp. when proof of guilt is lacking or dubious. **doubtable,** *a.* **doubter,** *n.* **doubtful,** *a.* liable to doubt; uncertain, admitting of doubt; uncertain, undecided, hesitating; suspicious, characterized by fear or apprehension; questionable. **doubtfully,** *adv.* **doubtfulness,** *n.* **doubting,** *n.*, *a.* **doubting Thomas,** *n.* one who persists in doubt until he/she has tangible evidence (from Thomas the apostle who would not believe in the Resurrection until he had seen Jesus, John, xx.24–25). **doubtingly,** *adv.* **doubtless,** *adv.* assuredly, certainly; probably. **doubtlessly,** *adv.* **doubtlessness,** *n.*

douce (doos), *a.* (*Sc.*) peaceable. **doucely,** *adv.* **douceness,** *n.*

douche (doosh), *n.* a jet of water or vapour directed upon some part of the body; an instrument for applying this. *v.t.* to apply a douche to, esp. to flush out the vagina or other cavity. *v.i.* to take a douche.

dough (dō), *n.* the paste of bread etc. before baking; anything resembling this in appearance or consistency; (*sl.*) money. **dough-boy,** *n.* (*N Am.*) a private soldier in the US army. **doughnut, donut,** *n.* a cake made of sweetened dough and fried in fat; (*sl.*) a circle of MPs who sit immediately behind and around a speaker during televising of parliament to give the illusion of a crowded house. **doughnutting, donutting,** *n.* (*sl.*) this practice. **doughy,** *a.* soft and pliable like dough. **doughiness,** *n.*

doughty (dow'ti), *a.* brave, valiant, redoubtable. **doughtily,** *adv.* **doughtiness,** *n.*

Douglas fir, pine, spruce (dŭg'ləs), *n.* a tall American conifer, grown for timber.

dour (duə), *a.* (*esp. Sc., North.*) hard, sullen; stern, severe, obstinate. **dourly,** *adv.* **dourness,** *n.*

doura (du'rə), DURRA.

douroucouli (doorəkoo'li), *n.* a nocturnal ape of Central and South America.

douse[1], **dowse** (dows), *v.t.* to plunge into water, to throw water over, to drench; to extinguish.

douse[2] DOWSE[1].

dove (dŭv), *n.* one of several kinds of pigeon, an emblem of gentleness and innocence; the symbol of the Holy Ghost; advocate of peaceable and conciliatory policies towards opponents, opposite to *hawk*; a term of endearment. **dove-coloured,** *a.* grey with a tinge of pink. **dove-cot, -cote,** *n.* a small house or box for domestic pigeons. **dovelike,** *a.*

dovetail (dŭv'tāl), *n.* a mode of fastening boards together by fitting tenons, shaped like a dove's tail spread out, into corresponding mortises; a tenon or a joint of this kind. *v.t., v.i.* to fit together by means of dovetails; (*fig.*) to fit exactly or follow on (from) neatly.

dovetail joint, *n.* one fastened by such tenons.

dowager (dow'əjə), *n.* a widow in possession of a title etc. left to her by her husband; a title given to a widow to distinguish her from the wife of her husband's heir, as *dowager duchess*; (*sl.*) an old lady.

dowdy (dow'di), *a.* awkward, shabby, unfashionable. **dowdily,** *adv.* **dowdiness,** *n.* **dowdyish,** *a.*

dowel (dow'əl), *n.* a pin or peg for connecting two stones or pieces of wood, being sunk into the side of each; a thin wooden rod (esp.) for hanging light curtains. *v.t.* to fasten by dowels. **dowel-joint,** *n.* a joint made by means of a dowel or dowels. **dowel-pin,** *n.* a dowel. **dowelling,** *n.* (lengths of) thin wooden rod.

dower (dow'ə), *n.* the property which a wife brings to her husband in marriage; that part of the husband's property which his widow enjoys during her life. *v.t.* to endow; to give a dower or portion to. **dower house,** *n.* a house on an estate reserved for the widow of the late owner. **dowerless,** *a.*

dowitcher (dow'ichə), *n.* either of two snipe-like birds, found on the shores of arctic and subarctic N America.

Dow-Jones average, index (dow jōnz'), *n.* an index of the prices of stocks and shares on the New York Stock Exchange.

down[1] (down), *n.* (*usu. pl.*) a tract of upland, esp. the chalk uplands of southern England, used for pasturing sheep; a dune. **the Downs,** the downs in the south of England. **downland,** *n.*

down[2] (down), *n.* fine soft plumage, found in young birds or under the feathers of adult birds; fine soft hair, esp. the first hair on the human face; the similar covering on plants; the feather-like substance by which airborne seeds are transported to a distance. **Downie**® (-ni), *n.* a duvet. **downy,** *a.* covered with down; made of down; resembling down. **downiness,** *n.*

down[3] (down), *adv.* towards the ground; from a higher to a lower position; on the ground; from the sky to the earth; below the horizon; (*fig.*) from former to later (years, ages); from north to south; away from the capital or a university; with a stream or current; (*Naut.*) to leeward; into less bulk; to finer consistency; to quiescence; to or in a state of subjection, disgrace or depression; at a low level, prostrate, in a fallen posture or condition; downstairs, out of bed; reduced in price; as a deposit; (a number) behind in scoring; (written etc.) on paper. *prep.* along, through, or into, in a descending direction; from the top or the upper part to the bottom or a lower part of; at a lower part of; along (a river) towards the mouth. *a.* (*superl.* **downmost**) depressed, sad; (*Railway*) pertaining to a journey leaving a more important place than its destination, as *down train/line/signal*. (*Comput.*) not working, inoperative. *v.t.* (*coll.*) to put, strike or throw down, to overcome. *n.* (*esp. pl.*) a reverse; (*coll.*) a grudge, a dislike. **down!** *imper.* (*ellipt.*) get, lie, put or throw down. **down and out,** utterly destitute and without resources. **down-and-out,** *n.* **down at heel,** shabby, poorly dressed. **down in the mouth,** discouraged, depressed. **down on one's luck,** (*sl.*) short of money. **down-to-earth,** realistic,

practical, sensible. **down with!** (*imper.*) swallow; abolish. **to be down on,** to be unfairly severe towards. **to down tools,** to stop work; to strike. **to get down,** to alight; to swallow (something). **to go down,** to sink; to leave the university for the vacation, or at the end of one's term; to prove acceptable to a stated extent, as *go down well/badly*. **to have a down on,** (*coll.*) to have a grudge against. **to put, set, take** or **write down,** to write on paper etc.; hence *down for (Tuesday),* announced to take place on (Tuesday), *down for squash,* announced, booked as playing squash. **to ride, hunt down,** to overtake by pursuit; to bring to bay. **to send down,** (*Univ.*) to expel or suspend (an undergraduate). **to shout down,** to silence with noise. **up and down,** here and there; throughout. **ups and downs,** vicissitudes (of fortune, life etc.). **downbeat,** *n.* a downward movement of a conductor's baton; an accented beat marked in this way. *a.* depressed, not optimistic. **downcast,** *a.* cast downward; dejected, sad. **downdraught,** *n.* a downward current of air. **downfall,** *n.* a fall of rain, snow etc.; a sudden loss of prosperity, rank, reputation, hence ruin, overthrow. **down-grade,** *n.* (down'-) a downward gradient on a railway etc. *v.t.* (-grād') to lower in status, marks, etc. **downhearted,** *a.* dispirited, dejected. **downhill** (down'-), *a.* descending, sloping downwards, declining. *n.* in skiing, a race without obstacles to the foot of a slope. *adv.* (-hil'), on a descending slope; (*fig.*) towards ruin or disgrace. **to go downhill,** to deteriorate physically or morally. **down line,** a railway line for trains going away from a main terminus. **download,** *v.t.* to transfer data directly from one computer to another. **down-market,** *a.* aiming at the poorer-quality, less-well-off end of the consumer spectrum. **down payment,** *n.* a deposit paid on an article bought on hire purchase. **downpipe,** *n.* a drainpipe which carries water from a roof to the ground. **down platform,** *n.* the platform adjoining a down line. **downpour,** *n.* a heavy, persistent fall of rain. **downright,** *a.* plain, absolute. *adv.* thoroughly, absolutely. **downrightness,** *n.* **downside,** *n.* that side of a coin which faces downwards when a coin has been tossed; a negative side, adverse aspect, disadvantage. **downsize,** *v.t.* (*euphem.*) to reduce in size. **downstage,** *a., adv.* at the front of the stage in a theatre. **downstairs,** *adv., a.* down the stairs; on or to a lower floor. *n.* the lower part of a building; the servant's quarters. **downstream,** *a.* towards the mouth of a river. **downswing,** *n.* a downward trend in e.g. trade; that part of a swing in golf when the club is moving downward towards the ground. **downtime,** *n.* the time during a normal working day when a computer, or other machinery, is inoperative. **downtown,** *n.* the business and commercial centre of a city. *a.* situated in, belonging to, this area. *adv.* towards this area. **down train,** *n.* one travelling away from a main terminus. **downtrodden,** *a.* trodden under foot; oppressed, tyrannized. **downturn,** *n.* a downward trend, esp. in business. **down under,** *n., a., adv.* (*coll.*) (in or to) Australia and New Zealand. **downward, -wards,** *adv.* from a higher to a lower position, level, condition or character; from earlier to later; from superior to inferior etc. **downward,** *a.* moving, directed or tending from higher, superior or earlier to lower, inferior or later. **downwind,** *a., adv.* (from a specified point) in the direction in which the wind is blowing.
Down's syndrome (downz), *n.* mongolism.
downy, DOWN².
dowry (dow'ri), *n.* the property which a wife brings on marriage to her husband.
dowse¹ (dows), *v.t.* to use a divining rod for the discovery of subterranean waters or minerals. **dowser,** *n.* **dowsing-rod,** *n.*
dowse² DOUSE¹.
-dox, *comb. form.* pertaining to doctrines or opinions.
doxology (doksol'əji), *n.* a brief formula or hymn of praise to God. **doxological** (-loj'-), *a.*

doxy (dok'si), *n.* (*dated*) a jade, a paramour; a loose woman.
doyen (doien'), *n.* the senior member of a body.
doyley DOILY.
doz., (*abbr.*) dozen.
doze (dōz), *v.i.* to sleep lightly; to be drowsy. *n.* a light sleep; a nap. **dozer,** *n.* **dozily,** *adv.* **doziness,** *n.* **dozy,** *a.* sleepy; (*coll.*) stupid.
dozen (dŭz'n), *n.* an aggregate of 12 things; (*esp. in pl.*) an indefinite number. *a.* 12. **baker's, dozen,** 13. **daily dozen,** daily physical exercises. **to talk nineteen to the dozen,** to talk incessantly.
DP, (*abbr.*) data processing.
DPhil (dēfil'), (*abbr.*) Doctor of Philosophy.
DPP, (*abbr.*) Director of Public Prosecutions.
dpt, (*abbr.*) department.
Dr, (*abbr.*) Doctor; Drive.
dr., (*abbr.*) debit; debitor; drachma; dram (weight).
drab¹ (drab), *n.* (*dated*) a prostitute, a slut. **drabbish,** *a.*
drab² (drab), *a.* of a dull brown or dun colour; (*fig.*) dull, commonplace, monotonous. *n.* drab colour. **drably,** *adv.* **drabness,** *n.*
drachm (dram), *n.* a drachma, a dram (weight); (*Apothecary's weight*) 60 grains (⅛ ounce, 3·542 g); (*Avoirdupois*) 27⅓ grains (1/16 ounce, 1·771 g). **drachma** (drak'mə), *n.* an Attic weight, about 60 gr. avoirdupois (3·542 g); the principal silver coin of the ancient Greeks, worth six obols; the standard unit of currency of modern Greece.
draconian (drəkō'niən), **draconic** (-kon'ik), *a.* inflexible, severe, cruel.
draft (drahft), *n.* the first outline of any writing or document; a rough copy; a rough sketch of work to be executed; a written order for the payment of money; a cheque or bill drawn, esp. by a department or a branch of one bank upon another; (*N Am.*) conscription into the army etc. *v.t.* to draw up an outline of, to compose the first form, or make a rough copy, of; to draw off (a portion of a larger body of men) for some special purpose; (*N Am.*) to conscript. **draftee** (-tē'), *n.* (*N Am.*) a conscript. **drafter, draughter,** *n.* **draftsman,** *n.* one who draws up documents; a draughtsman.
drag (drag), *v.t.* (*past, p.p.* **dragged**) to pull (esp. along the ground) by force; to haul, esp. with difficulty; to search (a river etc.) with a grapnel. *v.i.* to trail along the ground (as a dress); to search a river etc. with a grapnel, nets etc.; (*Mus.*) to move slowly or heavily; to move or go too slowly; to stretch (out) over a too-long period; to draw (on a cigarette). *n.* anything which retards movement; an iron shoe or skid fastened on a wheel of a vehicle to check the speed; a dredge; a four-clawed grapnel for dragging or dredging under water; a drag-net; the total resistance of an aeroplane along its line of flight; (*Hunting*) an artificial scent; the action of dragging; laborious movement, slow progress; an impediment; a drawback; (*sl.*) clothes appropriate to the opposite sex, esp. women's clothes worn by men; (*sl.*) anything or anyone tiresome; (*sl.*) a draw on a cigarette; (*coll.*) a long upward slope. **to drag one's feet,** (*coll.*) to go slow deliberately. **to drag out,** to protract. **to drag (something) out of (someone),** to get (information) from (someone) with difficulty. **to drag the anchor,** to trail it along the bottom when it will not take firm hold (said of a ship). **drag artist,** *n.* an entertainer who works in drag. **drag-hunt,** *n.* a hunt in which a drag is used. **drag-net,** *n.* a net dragged along the bottom of a river etc. for catching fish or over a field to enclose game; (*fig.*) a police search over a wide area, esp. for a fugitive or criminal. **drag race,** *n.* one in which specially modified cars race over a timed course. **drag racing,** *n.* **dragster** (-stə), *n.* a car modified for drag racing. **dragstrip,** *n.* a track for drag racing.
dragée (drazh'ā), *n.* a sweetmeat consisting of a nut, fruit etc. with a hard sugar coating; a small (silver-coloured)

draggle — draw

sugar ball for decorating cakes; a pill with a hard sugar coating.
draggle (drag'l), *v.t.*, *v.i.* to make or become wet and dirty (as if) by dragging on the ground. **draggletail**, *n.* a slut. **draggle-tailed**, *a.* sluttish. **draggling**, *a.*
dragoman (drag'əmən), *n.* (*pl.* **-mans**) one who acts as guide, interpreter and agent for travellers in the East.
dragon (drag'ən), *n.* a fabulous monster found in the mythology of nearly all nations, generally as an enormous winged fire-breathing serpent with formidable claws etc., (*Astron.*) a constellation in the northern hemisphere, Draco; any species of the genus comprising the flying lizard; an unpleasant, hostile woman. **to chase the dragon**, (*sl.*) to smoke heroin. **dragonfly**, *n.* an insect having two pairs of wings of brilliant colour and different sizes, and a long brilliant body. **dragonet** (-nit), *n.* a small, bright-coloured, spiny fish. **dragonish**, *a.*
dragoon (drəgoon'), *n.* a cavalry soldier, orig. a mounted infantryman armed with a short musket or carbine called a dragon; in the British army the name is applied to certain regiments that were formerly mounted infantry. *v.t.* to compel to submit by violent measures.
drain (drān), *v.t.* to draw off (liquid) gradually; to empty by drawing away liquid or moisture from; to drink up; to deprive, to exhaust (of vitality, resources etc.). *v.i.* to flow off gradually; to be emptied of liquid or moisture. *n.* a strain, heavy demand; a channel for conveying water, sewage etc.; (*Surg.*) a tube for drawing off pus etc. **down the drain**, (*coll.*) wasted. **drain-cock**, *n.* a tap for emptying a tank or other vessel. **drainpipe**, *n.* a pipe for draining superfluous or waste water, particularly from a roof or gutter. **drainpipes, drainpipe trousers**, *n.pl.* (*dated coll.*) ones with very narrow legs. **drainable**, *a.* **drainage**, *n.* the act, practice or science of draining; the natural or artificial system by which land or a town is drained. **drainage-area, basin**, *n.* the region drained by a river and its tributaries. **drainage-tube**, *n.* a tube introduced into a suppurating wound or chronic abscess to allow free discharge of putrid accumulations. **drainer**, *n.* one who or that which drains; a vessel in which wet things are put to drain. **draining**, *n.*, *a.* **draining-board**, *n.* a board beside a sink on which washed-up crockery is put to dry.
drake (drāk), *n.* the male of the duck.
Dralon® (drā'lon), *n.* an acrylic fibre usu. used in upholstery.
dram (dram), *n.* a drachm in apothecaries' weight and in avoirdupois weight; (*fig.*) a small quantity; as much spirit as is drunk at once.
drama (drah'mə), *n.* a play or other composition representing life and action, usually intended for performance by living actors on the stage; a series of real-life events invested with the tension, emotions and interest of a play; dramatic art, the composition and presentation of plays; plays collectively; the dramatic literature or theatrical art of a particular country or period. **drama documentary**, *n.* a film, play etc. composed of a mixture of fact and fiction. **dramatic, -ical** (-mat'-), *a.* pertaining to or of the nature of drama; pertaining to the stage, theatrical; intended or suitable for representation on the stage; striking, impressive, full of tension, emotion and excitement; meant for effect. **dramatic irony**, *n.* a situation in a film or play where the irony is clear to the audience but not to the characters. **dramatics**, *n.pl.* (*coll.*) a display of exaggerated behaviour; (*usu. sing. in constr.*) the producing or study of plays. **dramatically**, *adv.*
dramatis personae (dram'ətis pəsō'nē), *n.pl.* the set of characters in a play.
dramatist (dram'ətist), *n.* a writer of plays.
dramatize, -ise (dram'ətīz), *v.t.* to set forth (a story, novel etc.) in the form of a play; to exaggerate. **dramatizable**, *a.* **dramatization**, *n.*
dramaturge (dram'ətœj), *n.* a dramatist, a playwright. **dramaturgic, -ical** (-tœj'-), *a.* **dramaturgist**, *n.* **dramaturgy**, *n.* the technique of writing or producing plays.
drank (drangk), past tense of DRINK.
drape (drāp), *v.t.* to cover, clothe or decorate with cloth etc.; to adjust or arrange the folds of (a dress, curtains etc.). *n.pl.* (*esp. N Am.*) hangings, curtains. **draper**, *n.* one who deals in cloth and other fabrics. **Drapers' Company**, *n.* the third of the 12 great London livery companies, whose charter was granted by Edward III. **drapery**, *n.* the trade of a draper; cloth and other fabrics collectively; items of dress etc. made of these materials; that with which an object is draped, hangings, tapestry etc.; (*esp. N Am.*) (*pl.*) curtains; the arrangement of dress in sculpture, painting etc. **draperied**, *a.*
drastic (dras'tik, drahs'-), *a.* acting vigorously; extreme; effective; (*Med.*) strongly purgative. **drastically**, *adv.*
drat (drat), *v.t.* (*coll.*) confound it, bother, dash (as a mild form of imprecation). **dratted**, *a.*
draught, (*esp. N Am.*) **draft** (drahft), *n.* the act of pulling; the capacity of being pulled; the act of dragging with a net; the quantity of fish taken in one sweep of a net; the quantity of liquor drunk at once; a dose (of medicine); an unwanted current of air; the depth to which a ship sinks in water; a preliminary drawing, design or plan for a work to be executed; the drawing of liquor from a container; (*pl.*; *sing. in constr.*) a game played by two persons on a chess-board with twelve round pieces of different colours on each side. *v.t.* to sketch; to draft. **beast of draught**, an animal for pulling loads. **on draught**, able to be obtained by drawing off (from a cask etc.). **to feel the draught**, (*coll.*) to be aware of, or affected by, adverse (economic) conditions. **draught beer, lager, etc.** *n.* beer drawn from the cask, as distinguished from bottled or canned beer etc. **draught-board**, *n.* a board on which draughts is played. **draught-engine**, *n.* an engine for raising ore, water etc. **draught-horse**, *n.* a horse for pulling heavy loads.
draughtsman, (*esp. N Am.*) **draftsman** (drahfts'mən), *n.* one who draws designs or plans; one skilled in drawing; a piece used in the game of draughts. **draughtsmanship**, *n.* skill at drafting plans etc. **draughtswoman**, *n.fem.*
draughty *a.* full of draughts or currents of air. **draughtiness**, *n.*
Dravidian (drəvid'iən), *a.* of or pertaining to Dravida, an old province of southern India. *n.* one of the dark-skinned peoples of India, comprising the peoples speaking Tamil, Telugu, Canarese and Malayalam.
draw (draw), *v.t.* (*past* **drew** (droo), *p.p.* **drawn**) to drag or pull (along); to haul; to pull out or up (from), as water; to extract by pulling, as teeth; to cause to come forth, to elicit; to take, to receive, as information; to infer, deduce; to take in, to inhale; to draft, to picture, to portray; to lengthen, to stretch, to protract; to disembowel; to take out of a box or wheel (as tickets); to unsheathe; to allure, attract, to cause to follow one; to search for (game); to write (a cheque, draft, order) on a bank etc.; (*Naut.*) to need (a specified depth of water) to float; to leave undecided, as a match; (*Golf*) to impart a slight (intentional) hook to (a shot); (*Bowls*) to play a bowl so that it curls towards the target. *v.i.* to pull, to haul; to attract; to allow a free motion of air, current etc. (as a chimney, pipe etc.); to unsheathe a sword or take a pistol from its holster; to draw lots; to move, to approach as if pulled (towards); to come (out or away, as if pulled); to practise the art of sketching or drafting; to curl towards a target, as a golf ball or a bowl. *n.* a pull, a strain; an attraction, a lure; the act of drawing lots; a lot or chance drawn; the act of drawing out game; a drawn game or contest; (the amount of) a curling towards a target. **draw and quarter**, (*Hist.*) penalty of disembowelling and dismemberment after hanging. **to draw a bead** BEAD. **to draw away**, to get further in front. **to draw back**, to move back; to withdraw; to retreat (from fulfil-

ling a promise). **to draw (a) blank**, to find nothing. **to draw in**, to collect, to contract; to entice; to inveigle; of rights, to extend, making the days shorter. **to draw lots**, to choose (a person, course of action etc.) by the random selection of objects, usu. straws of different lengths. **to draw near, nigh**, to approach. **to draw off**, to cause (liquid) to flow out. **to draw on**, to allure, attract, entice; to approach. **to draw out**, to lengthen, to protract (esp. unnecessarily); to induce to talk; to elicit, as information. **to draw rein**, to slow down and stop. **to draw stumps**, (*Cricket*) to stop playing for the day. **to draw the line at**, to refuse to go any further than. **to draw the teeth of**, (*fig.*) to render harmless. **to draw trumps**, to play trumps (at cards) until the opponents have none left. **to draw up**, to range in order, or in line of battle; to list, to compose, to put into proper form; to put (oneself) into a stiff erect attitude. **drawback**, *n.* a disadvantage, an inconvenience, an obstacle. **draw-bar**, *n.* a bar to connect a locomotive with a tender. **drawbridge**, *n.* a bridge that may be raised on hinges at one or both ends to allow or prevent passage across or under. **draw-gear**, *n.* railway-carriage couplings. **draw-sheet**, *n.* (*Med.*) an extra sheet doubled lengthwise and placed across the bed so that it may be pulled beneath the patient as required. **drawstring**, *n.* a cord or thread, threaded through or otherwise attached to fabric, which can be pulled to gather in the fabric.
drawer (draw'ə), *n.* one who draws; (draw) a sliding receptacle in a table etc.; one who draws a cheque or order for the payment of money. **chest of drawers** CHEST. **out of the top drawer**, belonging to the upper social class. **drawerful**, *n.* **drawers** (drawz), *n.pl.* (*dated or facet.*) an undergarment covering the lower body and part of the legs.
drawing (draw'ing), *n.* the action of the verb to draw; the art of representing objects on a flat surface by means of lines drawn with a pencil, crayon etc. (but not paint); a sketch or draft of this kind; the distribution of prizes in a lottery. **drawing-block**, *n.* a pad of paper for drawing on. **drawing-board**, *n.* a rectangular frame for supporting a sheet of paper while drawing. **drawing-office**, *n.* the department in an engineering works where designs and plans are made. **drawing-pin**, *n.* a flat-headed tack for securing drawing-paper to a board. **drawing-room**, *n.* a room in a house for holding social gatherings; the company assembled in a drawing-room; (*N Am.*) a private compartment in a railway coach. *a.* suitable for a drawing-room.
drawl (drawl), *v.t.* to utter in slow, lengthened tones. *v.i.* to speak with a slow, prolonged utterance. *n.* such a manner of speaking. **drawling**, *a.* **drawlingly**, *adv.*
drawn (drawn), past participle of DRAW. *a.* pulled out (as a sword); depicted, sketched; haggard; eviscerated, disembowelled; neither won nor lost. **at daggers drawn**, of individuals, very hostile, on the point of fighting each other.
dray[1] (drā), *n.* a low cart, generally of strong and heavy construction, used by brewers etc. **dray-horse**, *n.* a strong, heavy horse for drawing a dray, a cart-horse. **drayman**, *n.* a driver in charge of a dray.
dray[2] DREY.
dread (dred), *v.t.* to fear greatly; to anticipate with terror. *n.* great fear or terror; awe, reverence; the person or thing dreaded. *a.* (*poet.*) exciting great fear or terror, frightful; awe-inspiring. **dreadlocks** (dred'loks), *n.pl.* long hair worn in many tight plaits by Rastafarians. **dreadnought**, *n.* a type of battleship, first built 1905–06, with its main armament composed of big guns. **dreadful**, *a.* inspiring dread; terrible; (*coll.*) annoying, disgraceful, frightful, horrid; (*coll.*) dreadfully. **penny dreadful**, (*dated*) a journal or story-book dealing with crude sentiment and horrors. **dreadfully**, *adv.* (*coll.*) very. **dreadfulness**, *n.*
dream (drēm), *n.* a vision; (any one of) the thoughts and images that pass through the mind of a sleeping person; the state of mind in which these occur; a visionary idea, a fancy, (a)waking reverie; something beautiful or enticing; a fervently-held wish, often unlikely to come true. *v.i.* (*past* **dreamt** (dremt), *or* **dreamed**) to have visions; to think, to imagine as in a dream; to waste time in idle thoughts. *v.t.* to see, hear, feel etc. in dream; to imagine or conceive mentally, to picture in hope or imagination. **day-dream**, a romantic scheme or unlikely fancy voluntarily indulged in. **waking dream**, a hallucination. **to dream away**, to spend (time) idly or vainly. **to dream up**, (*coll., often derog.*) to invent (esp. an idea or excuse). **to go like a dream**, (*coll.*) to work very smoothly; to be very successful. **dreamboat**, *n.* (*dated coll.*) a very desirable member of the opposite sex. **dreamland**, *n.* the region of fancy or imagination. **dream ticket**, *n.* (*esp. N Am.*) in politics, the ideal vote-catching combination of candidates. **dream-ticket**, *a.* (*attrib.*) indicating the best possible combination of candidates, participants etc. **dreamtime**, *n.* in Australian aboriginal mythology, a golden age after the creation of the world. **dreamworld**, *n.* a world of illusions. **dreamer**, *n.* **dreamful**, *a.* **dreamingly**, *adv.* **dreamless**, *a.* **dreamlessly**, *adv.* **dreamlike**, *a.* **dreamy**, *a.* full of or causing dreams; addicted to dreaming (*coll.*) attractive to the opposite sex. **dreamily**, *adv.* **dreaminess**, *n.*
dreary (driə'ri), *a.* dismal, gloomy; cheerless, tiresome, dull. **drearily**, *adv.* **dreariness**, *n.* **drearisome** (-səm), *a.*
dredge[1] (drej), *n.* a drag-net for taking oysters or for dragging up objects from the bottom for scientific purposes; a bucket or scoop for scraping mud etc. from the bed of a river etc. to deepen it. *v.t.* to gather or bring up with a dredge; to clean or deepen with a dredger. *v.i.* to use a dredge. **to dredge up**, to lift with a dredge; to bring to light (something) previously obscure or well hidden. **dredger**, *n.* one who fishes with a dredge; a ship for raising silt etc. from the bottom of a river, harbour, channel etc. to deepen it or obtain ballast.
dredge[2], *v.t.* to sprinkle (as flour or sugar) upon or over; to sprinkle with flour etc. **dredger, dredging-box**, *n.* a box with perforated lid for sprinkling.
dregs (dregz), *n.* (*pl.*) the sediment or lees of liquor; the end, the bottom, the last part; worthless refuse; the lowest class (of society, humanity etc.); the most undesirable part. **dreggy**, *a.*
dreich (drēkh), *a.* (*Sc.*) tedious, wearisome, long; bleak. [DREE]
drench (drench), *v.t.* to wet thoroughly; to soak, to saturate. *n.* a liquid medicine for horses or cattle. **drencher**, *n.* one who or that which drenches; an apparatus for drenching cattle. **drenching**, *n.* a soaking.
Dresden china, porcelain, ware (drez'dən), *n.* fine, delicately decorated china made at Meissen, near *Dresden* (now in the German Democratic Republic) from 1710.
dress (dres), *v.t.* (*Mil.*) to form (ranks) into a straight line; to order, arrange, array; to clothe, to attire; to adorn, to deck; (*Naut.*) to decorate with flags etc.; to furnish with costumes; to cleanse, trim, brush, comb etc.; to curry or rub down (as a horse); to cleanse and apply remedies to (as a wound); to prepare for use, to cook; to cover with dressing (as a salad); to manure; to square and give a smooth surface to (as stone); of a shop window, to arrange goods attractively in. *v.i.* to clothe oneself; to put on evening clothes; to attire oneself formally (as for dinner); (*Mil.*) to arrange oneself in proper position in a line; *n.* that which is worn as clothes, esp. outer garments; garments, apparel; a lady's gown, a frock. **evening dress**, formal clothes worn at dinners, evening receptions etc. **full dress**, that worn on state or important occasions. **morning dress**, formal clothes worn during the day. **to dress down**, to chastise, to reprimand severely. **dressing-down** DRESSING. **to dress up**, to clothe elaborately; to deck, to adorn, esp. to look like something else;

dressage 246 **drive**

to adopt a characteristic style of dress (of), as *dress up (as a clown)*. **dress circle**, *n*. the first tier of seats above the stalls in a theatre. **dress-coat**, *n*. a man's coat with narrow pointed tails, worn as evening dress. **dress length**, *n*. enough fabric to make a dress. **dressmaker**, *n*. one who makes women's dresses. **dressmaking**, *n*. **dress parade**, *n*. a formal military parade in full uniform. **dress rehearsal**, *n*. (*Theat*.) the final rehearsal, with costumes and effects. **dress sense**, *n*. a knowledge of style in dress and the ability to pick clothes which suit one. **dress shirt**, *n*. a man's shirt worn with formal evening dress. **dress uniform**, *n*. a full ceremonial military uniform. **dressy**, *a*. fond of showy dress; wearing rich or showy dress; showy; stylish, smart. **dressiness**, *n*.

dressage (dres'ahzh), *n*. (the training of a horse in) deportment, obedience and response to signals given by the rider's body.

dresser[1] (dres'ə), *n*. one who dresses in a stated way, as *an untidy dresser*; one who dresses another, esp. an actor; a surgeon's assistant in operations who dresses wounds. **window dresser**, one who arranges the window displays in a shop.

dresser[2] (dres'ə), *n*. a kitchen sideboard; a set of shelves or an open cupboard for plates etc.; (*N Am*.) a chest of drawers.

dressing (dres'ing), *n*. the action of the verb to DRESS; gum, starch etc. used in sizing or stiffening fabrics; stuffing, sauce, salad-dressing; manure applied to a soil; a combination of ointment, liniment, bandages etc. applied to a wound or sore; (*pl*.) the mouldings and sculptured decorations on a wall or ceiling. **French dressing** FRENCH. **dressing-case**, *n*. a small bag or case for toilet requisites. **dressing-down**, *n*. a severe telling-off. **dressing-gown**, *n*. a loose robe worn over night- or underclothes. **dressing-room**, *n*. a room used for dressing; a room where actors put on costumes and stage make-up. **dressing-station**, *n*. (*Mil. etc*.) a first-aid post. **dressing-table**, *n*. a table fitted with drawers and a mirror, used by women while dressing, making-up etc.

dressmaker etc. DRESS.

drew (droo), past tense of DRAW.

drey, dray[2] (drā), *n*. a squirrel's nest.

dribs (dribz) *v.t.* **in dribs and drabs** (drabz), (*coll*.) little bits at a time.

dribble (drib'l), *v.t., v.i.* to (cause to) drip or trickle; to (cause to) fall or run slowly; to slaver, to drivel; to manoeuvre (a football) in a forward direction by slight kicks from alternate sides. *n*. slavering drizzle; (*Football etc*.) an instance of dribbling. **dribbler**, *n*. **dribbly**, *a*.

drier DRY.

drift (drift), *n*. that which is driven along by a current of air or water; a current, a driving or compelling force; the direction of movement in a current; meaning, tenor; a mass (of snow, leaves, sand etc.) driven together; (*Naut*.) deviation from the course caused by currents; (*Mining*) a horizontal passage following a lode or vein; a cattle-track; a drift-net; (*S Afr*.) a ford; a controlled skid for taking bends at high speed; a gradual change in a supposedly constant piece of equipment, as a tuner. *v.i.* to be driven into heaps; to float or be carried along by, or as by, a current; to be carried along by circumstances or chance; to travel through life aimlessly; (*Mining*) to make a drift. *v.t.* to drive along or into heaps; to carry along (of a current). **drift-ice**, *n*. floating ice drifting on the sea. **drift-net**, *n*. a large fishing net which drifts with the tides. **driftwood**, *n*. wood washed ashore from out at sea. **driftage**, *n*. the extent to which a ship drifts off course. **drifter**, *n*. a trawler or fishing-boat using a drift-net to fish; one who wanders aimlessly from place to place.

drill[1] (dril), *v.t.* to bore or pierce with a pointed tool, to make (holes) by this means; to train by repeated exercise; to train to the use of arms; *v.i.* to go through a course of (esp. military) exercise. *n*. a manual or electric tool for boring holes in hard material; a drill-bit; constant practice or exercise in any art or business; (a) series of repetitive exercises by which soldiers etc. are trained; (*coll*.) correct procedure, the right way to do something. **drill-bit**, *n*. a piece of metal with a spiral cutting edge, inserted into an electric or hand drill to cut a hole of a given diameter. **drillmaster**, *n*. a military drill instructor. **drill-press, drill-stand**, *n*. a device for holding an electric drill vertical. **drilling fluid**, *n*. a mixture of clay and water pumped down during the drilling of an oil-well. **drilling platform**, *n*. one which is either mobile or attached to the sea bed, used as a base for equipment during the drilling of an oil-well. **drilling rig**, *n*. the machinery needed to drill an oil-well; an offshore mobile drilling platform.

drill[2] (dril), *v.t.* to sow (seed) or plant in rows. *v.i.* to sow or plant in this manner. *n*. a small trench or furrow, or a ridge with a trench along the top, for seeds or small plants; a row of plants in such a furrow; a machine for sowing grain in rows. **drill-plough**, *n*. one for sowing grain in drills. **driller**, *n*.

drill[3] (dril), *n*. a heavy cotton twilled cloth used for trousers etc.

drill[4] (dril), *n*. a baboon from the coast of Guinea.

drily DRY.

drink (dringk), *v.t.* (*past* **drank** (drangk), *p.p.* **drunk** (drŭngk)) to swallow (a liquid); to imbibe, absorb, suck in; to pledge, to toast (someone's health, happiness etc.); to waste (money, wages, property) on indulgence in liquor. *v.i.* to swallow a liquid; to take intoxicating liquors (esp. to excess). *n*. something to be drunk; a draught, esp. of intoxicating liquor; excessive indulgence in intoxicating liquors, intemperance. **strong drink**, alcoholic liquor. **the drink**, (*sl*.) the sea. **to drink deep**, to take a long draught. **to drink in**, to absorb readily; to receive greedily, as with the senses; to gaze upon, listen to etc. with delight. **to drink like a fish**, (*coll*.) to be a habitual heavy drinker. **to drink off**, to swallow at a single draught. **to drink (someone) under the table**, to drink and remain comparatively sober while one's drinking companion gets completely drunk. **to drink the health of**, to wish health to by means of a toast. **to drink to**, to salute in drinking; to drink the health of. **to drink up**, to swallow completely. **drink-driver**, *n*. one who drives with (excessive) alcohol in the bloodstream. **drink-driving**, *n*. **drinkable**, *a*. that may be drunk; fit for drinking. **drinkableness**, *n*. **drinkably**, *adv*. **drinker**, *n*. one who drinks; a tippler, a drunkard. **drinking**, *n., a*. **drinking-up time**, the time between the call for last orders and closing time in a public house, in which to finish drinks. **drinking-fountain**, *n*. a fountain erected in a public place to supply drinking water. **drinking-horn**, *n*. a drinking-vessel make of horn. **drinking-song**, *n*. a song in praise of drinking. **drinking water**, *n*. water suitable for drinking.

drip (drip), *v.i.* (*past, p.p.* **dripped**) to fall in drops; to throw off moisture in drops. *v.t.* to let fall in drops. *n*. the act of dripping, a falling in drops; an apparatus for the intravenous administration of liquid, drop by drop; (*coll*.) a stupid or insipid person. **drip-dry**, *a*. of clothing, made of such a material that it dries quickly without wringing and needs no ironing. **drip-feed**, *v.t.* to feed nutrients to (a patient etc.) in liquid form, using a drip. *n*. a drip containing nutrients. **drip-tray**, *n*. the tray in a refrigerator for catching the drops of water and ice during defrosting. **dripless**, *a*. **dripper**, *n*. **dripping**, *n*. the act of falling in drops; the fat which falls from roasting meat; (*pl*.) drops of water, grease etc. falling or trickling from anything. **drippy**, *a*. inclined to drip; insipid, inane.

drive (drīv), *v.t.* (*past* **drove** (drōv), *p.p.* **driven** (driv'n)) to push or urge by force; to urge (in a particular direction), to guide, to direct (as a horse, an engine, a ship); to convey in a carriage etc.; to compel; to prosecute, to carry

drivel   247   **drop**

on; to chase, hunt, esp. to frighten (e.g. animals) into an enclosure or towards guns; to overwork (oneself or others); to propel straight and firmly; (*Golf*) to propel (the ball) with the driver; (*Cricket*) to hit (the ball) to or past the bowler with a swift free stroke; to force (as a nail into wood) with blows; to propel (machinery etc.) to cause (it) to function; to bore (a tunnel etc.). *v.i.* to dash, to rush violently, to hasten (along, on etc.); to travel in a carriage or other vehicle, esp. under one's own direction or control; to control or direct a vehicle, engine etc.; (*Golf*) to hit a ball with a driver. *n.* a ride in a vehicle; the distance travelled by driving; a road for driving on, esp. a private carriageway to a house; (*Golf*) a stroke with a driver; a forward stroke at cricket etc.; a driving of game, cattle, or of an enemy; (*N Am.*) an annual gathering of cattle for branding; energy, motivation; (*Comput.*) a device for writing information on to, and reading it from, magnetic tape; an effort (as *sales drive*); a series of competitive games of whist. **to drive a coach and horses through**, (*coll.*) to demolish (an argument or idea) by pointing out obvious faults. **to drive a hard bargain**, to be tough in negotiations. **to drive at**, to hint at, to imply. **to drive home**, to force (something) completely in; to explain (something) emphatically and make sure it is understood. **to drive up the wall**, (*coll.*) to madden. **drive-belt**, *n.* a flexible belt for connecting parts of a machine and transferring motion from one to another. **drive-in**, *n.* a café, cinema etc. where customers are served or can watch a film without leaving their cars. **drive-shaft**, *n.* a shaft transmitting power from an engine to what it drives. **drive-time**, *n.*, *a.* (*coll.*) (of) the time when people are driving home from work. **driveway**, *n.* a path large enough for a car, from a road to a house. **driver**, *n.* one who or that which drives, esp. one who drives a vehicle or an engine; (*Golf*) a wooden- or metal-headed club used to propel the ball from the tee. **driverless**, *a.* **driving**, *a.* imparting strong motivation; having great force, as rain. **driving licence, driver's licence**, *n.* a permit to drive, granted to one who has passed a **driving test**, or examination in the driving and handling of a vehicle. **driving seat**, *n.* the seat for the driver in a vehicle; (*fig.*) a position of authority or control. **driving-wheel**, *n.* one of the large wheels of a locomotive, one of the pair(s) of wheels in a motor vehicle, to which motive force is applied from the engine.

drivel (driv'l), *v.i.* (*past, p.p.* **drivelled**) to allow spittle to flow from the mouth, as a child, idiot or dotard; (*coll.*) to be weak or silly. *n.* spittle flowing from the mouth; utterly nonsensical talk, twaddle. **driveller**, *n.* a slaverer; an idiot, a dotard, a fool.

driver DRIVE.

drizzle (driz'l), *v.i.* (of rain) to fall in fine drops, to rain slightly. *n.* fine small rain. **drizzly**, *a.*

drogue (drōg), *n.* a wind-sock; a parachute (also **drogue parachute**) which reduces the speed of a falling object or landing aircraft; a cone-shaped device on the end of the refuelling hose of a tanker aircraft into which the probe of the receiving aircraft fits.

droit (droit, drwah), *n.* a right, a due, esp. a legal right.

droll (drōl), *a.* odd, ludicrous; comical, laughable. *n.* a merry fellow, a jester, a buffoon. **drollery**, *n.* (an instance of) idle sportive jocularity, buffoonery. **drollness**, *n.* **drolly**, *adv.*

-drome, *comb. form.* a large area specially prepared for some specific purpose, as in *aerodrome* for aircraft, *hippodrome* for races.

dromedary (drom'idəri), *n.* the Arabian camel, distinguished from the Bactrian camel by its single hump.

drone (drōn), *n.* the male of the bee, larger than the worker by which the honey is made; an idler, a lazy person who lives on the industry of others; a deep humming sound; the humming sound made by a bee; the unchanging bass produced from the three lower pipes of the bagpipe; any of these lower pipes; (a person with) a low, monotonous speaking voice; a radio-controlled aircraft. *v.i.* to make a monotonous humming noise, as a bee or a bagpipe; to talk in a monotonous tone. *v.t.* to read or say in a monotonous tone. **drone-pipe**, *n.* the drone of a bagpipe. **droner**, *n.* **droning**, *a.* **droningly**, *adv.*

drongo (drong'gō), *n.* (*esp. Austral., sl.*) a slow-witted person, a fool.

drool (drool), *v.i.* to drivel, to slaver (over); to show excessive or lascivious pleasure in something.

droop (droop), *v.i.* to hang, lean or bend down languidly; to fail, to flag, to languish, to decline; (*fig.*) to be dejected, tired etc. *n.* the act of drooping; a drooping attitude. **droop snoot**, *n.* (*coll.*) an adjustable nose (of an aircraft). **droopily**, *adv.* **droopiness**, *n.* **droopingly**, *adv.* **droopy**, *a.*

drop (drop), *n.* a globule or small portion of liquid in a spherical form, which is falling, hanging or adhering to a surface; a very small quantity of a fluid; (*Med.*) the smallest quantity separable of a liquid; (*pl.*) a dose measured in such units; (*fig.*) a minute quantity, an infinitesimal particle; (*coll.*) a glass or drink of liquor; anything resembling a drop, or hanging as a drop, as an earring, or other pendent ornament; a type of sweetmeat; the act of dropping, a fall, a descent, a collapse (in position, value, estimation etc.); a thing that drops or is dropped; the unloading of troops by parachute; (*coll.*) a delivery of goods; a falling trap-door; the part of a gallows contrived so as to fall from under the feet of persons to be hanged; the distance they are allowed to fall; an abrupt fall in a surface; the amount of this; (*N Am.*) a slot in a receptacle through which things can be dropped; a drop-kick. *v.t.* (*past, p.p.* **dropped**) to allow or cause to fall in drops, as a liquid; to cause to fall, to fell; to lower, to let down; to set down from a passenger vehicle; to write (a note, postcard etc.) to in an informal manner; to mention casually; to bear (a foal, calf etc.); to omit; to relinquish (as a demand); to stop (doing something), to have done with; (*coll.*) to bring down, to kill; (*coll.*) to lose (a game), to fail to gain (points); (*Golf*) to score (one or more shots) more than par; to score (a goal) with a drop-kick; (*coll.*) to stop seeing or associating with (someone). *v.i.* to fall in drops; to fall (in height, pressure, temperature etc.); to collapse suddenly, to sink as if exhausted, to faint; to disappear (from sight); to die (suddenly); to fall (back or behind). **a drop in the ocean**, a proportionately tiny amount. **a drop too much**, slightly too much to drink. **at the drop of a hat**, immediately; at the slightest provocation. **to drop a brick**, to make a bad (social) mistake, a faux pas. **to drop a curtsy**, to curtsy. **to drop anchor**, to let down the anchor. **to drop away**, to depart; to desert a cause. **to drop in**, to make an informal visit; to call (on someone) unexpectedly. **to drop out**, (*coll.*) to refuse to follow a conventional life style, esp. to leave school or college early. **to let drop**, to disclose, (seemingly) without any intention of so doing. **to drop off**, (*fig.*) to fall gently asleep. **drop-down**, *a.* (of a computer menu) hanging from the top of the screen, superimposed on the work in hand. **drop forge**, *a.* a forge for metal with two dies, one of them fixed, the other acting by force or gravity. **drop goal**, *n.* one scored in rugby with a drop-kick. **drop handlebars**, *n.pl.* curving, lowered handlebars on a racing bicycle. **drop-kick**, *n.* a kick made by dropping the ball from the hands and kicking it on the rise. **drop leaf**, *n.* a hinged flap on a table which can be lowered or raised. **drop-out**, *n.* (*coll.*) one who rejects conventional society. **drop-scene**, *n.* a painted curtain suspended on pulleys which is let down to conceal the stage. **drop scone**, *n.* one made of batter and cooked on a hot griddle. **drop shot**, *n.* a shot in tennis, squash or badminton which falls to the ground immediately after crossing the net or hitting the wall.

droplet (-lit), *n.* a tiny drop. **dropper**, *n.* one who or that which drops; a small glass tube with a rubber bulb at one end, for administering medicinal drops; (*Angling*) an artificial fly set at some distance from the end of a cast. **dropping**, *n.* **droppings**, *n.pl.* that which falls or has fallen in drops; the dung of beasts or birds.

dropsy (drop'si), *n.* (an illness causing) an accumulation of watery fluid in the tissues or body cavities. **dropsied**, *a.* suffering from dropsy.

droshky (drosh'ki), **drosky** (dros'-), *n.* a Russian open four-wheeled carriage in which the passengers ride astride a bench, their feet resting on bars near the ground; a public cab in Berlin and other German towns.

drosophila (drəsof'ilə), *n.* (*pl.* **-las, -lae** (-ē)) any of the small fruit flies used in laboratory genetic experiments.

dross (dros), *n.* the scum or useless matter left from the melting of metals; anything utterly useless, refuse, rubbish. **drossy**, *a.* **drossiness**, *n.*

drought (drowt), *n.* long-continued period of rainless weather; thirst.

drove[1] (drōv), *n.* a collection of animals driven in a body; a road for driving cattle on; (often in *pl.*) a shoal, a crowd, a mass of people, esp. as moving together. *v.i.* to drive cattle in droves. **drove-road**, *n.* an old grassy track formerly used by droves of cattle. **drover**, *n.* one who drives cattle or sheep to market.

drove[2] (drōv), past tense of DRIVE.

drown (drown), *v.i.* to be suffocated and die in water or other liquid. *v.t.* to suffocate by submersion in water or other liquid; to drench, to deluge; (*fig.*) to overpower (as by a volume of sound). **to drown one's sorrows**, to drink alcohol in order to forget one's problems and sorrows. **to drown out**, to overpower or make inaudible by a flood, as of protests, noise etc.

drowse (drowz), *v.i.* to be sleepy or half-asleep; to doze. *n.* the state of being half-asleep; a nap, a doze. **drowsy**, *a.* **drowsy-head**, *n.* a sleepy person. **drowsy-headed**, *a.* sleepy, sluggish in disposition. **drowsiness**, *n.* **drowsily**, *adv.*

drub (drŭb), *v.t.* to beat with a stick; to beat thoroughly in a fight or contest. **drubber**, *n.* **drubbing**, *n.* a heavy beating or defeat.

drudge (drŭj), *v.i.* to perform menial work with little reward; to slave. *n.* one employed in menial work; one who toils at uncongenial work and is ill-paid; a slave. **drudgery**, *n.* hard menial or tedious work.

drug (drŭg), *n.* any substance, mineral, vegetable or animal, used as the basis or as an ingredient in medical preparations, a poison, a potion; a narcotic causing addiction. *v.t.* (*past, p.p.* **drugged**) to mix drugs with, esp. to make narcotic; to administer drugs to, esp. narcotics; to render insensible with drugs. **drug addict, fiend**, *n.* one addicted to the use of narcotics. **drug-pusher**, *n.* one who sells narcotic drugs illegally. **drugstore**, *n.* (*N Am.*) a chemist's shop where pharmaceuticals and many other types of article are sold including refreshments. **drug traffic(king)**, *n.* illegal trading in narcotic drugs. **druggist**, *n.* one who deals in (legal) drugs; (*N Am.*) a pharmacist.

drugget (drŭg'it), *n.* a coarse woollen fabric, felted or woven, used as a covering or as a substitute for carpet.

druid (droo'id), *n.* the name commonly given to the priests and teachers of the early Gauls and Britons or perh. of pre-Celtic peoples, who taught the transmigrating of souls, frequently celebrated their rites in oak-groves, and are stated by Caesar to have offered human sacrifices; a member of the Ancient Order of Druids, a benefit society, established in 1781; an officer of the Welsh Eisteddfod; a member of any of several movements trying to revive druidic practices. **druidess**, *n. fem.* **druidic, -ical** (-id'-), *a.* **druidism**, *n.*

drum[1] (drŭm), *n.* a musical instrument made by stretching parchment over the head of a hollow cylinder or hemisphere; (the membrane across) the tympanum or hollow part of the middle-ear; anything drum-shaped, esp. a reel onto which cable etc. is wound; the quantity contained on such a reel; (*Mach.*) a revolving cylinder over which a belt or band passes; a cylindrical instrument on which computer data can be stored; (*sl.*) a place of abode. *v.i.* (*past, p.p.* **drummed**) to beat or play a tune on a drum; to beat rapidly or thump, as on a table, the floor or a piano; to make a sound like the beating of a drum (as certain insects, birds etc.). *v.t.* to perform on a drum or a substitute for a drum; to summon (up) recruits by the sound of a drum; (with *into*) to din into a person, to drive (a lesson) into by persistence. **to beat the drum for**, to try to raise interest in. **to drum out**, to expel from a regiment with disgrace, to cashier. **to drum up**, to canvass (aid or support). **drum-beat**, *n.* the sound made by a beating drum. **drum brake**, *n.* a type of brake with shoes which rub against a cylindrical brake. **drum-head**, *n.* the membrane stretched at the top of a drum; the membrane across the drum of the ear; the top of a capstan. **drum-major**, *n.* a non-commissioned officer in charge of the drums of a regiment, or who leads the band on the march. **drum majorette**, *n.* a girl who leads a marching band dressed in a uniform and twirling a baton. **drumstick**, *n.* a stick with which a drum is beaten; anything resembling such a stick, as the leg of a fowl. **drummer**, *n.* one who performs on a drum, esp. a soldier whose duty is to beat the various calls etc. on his drum.

drum[2] (drŭm), *n.* a narrow hill or ridge; (*Geol.*) a long narrow ridge of drift or alluvial formation; also called a **drumlin** (-lin).

drunk (drŭngk), *a.* intoxicated, stupefied, inebriated or overcome with alcoholic liquors; (*fig.*) highly excited (with joy etc.) *n.* (*sl.*) a drunken person. **drunkard** (-kəd), *n.* one addicted to the excessive use of alcoholic liquors; one who is habitually or frequently drunk. **drunken**, *p.p.*, *a.* intoxicated; given to drunkenness; caused by drunkenness; characterized by intoxication. **drunkenly**, *adv.* **drunkenness**, *n.*

drupe (droop), *n.* a fleshy fruit containing a stone with a kernel, as the peach, plum etc. **drupel** (-pl), **drupelet** (-lit), *n.* a succulent fruit formed by an aggregation of small drupes, as the raspberry.

Druse, Druz(e) (drooz), *n.* a member of a politico-religious sect of Islamic origin, inhabiting the region of Mt Lebanon in Syria.

dry (drī), *a.* devoid of moisture; arid; without sap or juice, not succulent; lacking rain, having an insufficient rainfall; thirsty; dried up, removed by evaporation, draining or wiping; not giving milk, not yielding juice; not under water (of land, a shore etc.); not sweet (of wines etc.); without butter (as bread); prohibited by law from the sale of alcoholic liquors; (*fig.*) lacking interest, dull; sarcastic, ironical; without sympathy or cordiality, cold, discouraging. *v.t.* to free from or deprive of water or moisture. *v.i.* to lose or be deprived of moisture; to grow dry; (*Theat.*) to dry up. **dry-bulb thermometer**, one of a pair of thermometers the other of which is always kept moist, the two together indicating the degree of humidity of the air. **the Dry**, *n.* (*Austral.*) the dry season. **to dry out**, to make or become dry; to undergo treatment for alcohol abuse. **to dry up**, to deprive totally of moisture; to lose all moisture; to cease to flow, to cease to yield water; (*Theat.*) to forget one's lines; (*sl.*; *usu. imper.*) to cease talking or doing something. **to go dry**, to prohibit the sale of alcoholic liquors. **dry battery**, *n.* one made up of dry cells. **dry cell**, *n.* a battery cell in which the electrolyte is a paste and not a fluid. **dry-clean**, *v.t.* to clean with a petrol-based solvent or other detergent. **dry-cooper**, *n.* a maker of casks for dry goods. **dry-cure**, *v.t.* to cure by drying and salting, as distinguished from pickling. **dry dock**, *n.* a dock which can be emptied of water for ship repairs. *v.t.* to put in dry dock. **dry-eyed**, *a.* without tears. **dry-fly**, *n.* an angler's fly that floats on the surface, as distinguished from one that is allowed to sink. **dry-goods**, *n.pl.* cloths, silks, drapery, haberdashery etc., as distinguished from grocery; sometimes extended to include any non-liquid goods. **dry-goods store**, (*N Am.*) a draper's shop. **dry hole**, *n.* an oil-well which does not produce a viable amount of oil. **dry ice**, *n.* solid carbon dioxide used in refrigeration. **dry-measure**, *n.* a measuring system for dry-goods. **dry-nurse**, *n.* a nurse who rears a child without the breast. *v.t.* to rear without the breast. **dry plate**, *n.* a photographic plate with a hard, dry, sensitized film, adapted for storing and carrying about. **dry riser**, *n.* a vertical, empty pipe with connections on every floor of a building, to which a fireman's hose can be attached. **dry-rot**, *n.* decay in timber caused by fungi which reduce it to a dry brittle mass. **dry run**, *n.* a shooting practice without live ammunition; a practice run, a rehearsal. **dry-salt**, *v.t.* to dry-cure. **drysalter**, *n.* a dealer in dried and salted meat, pickles etc.; a dealer in dye-stuffs, chemical products etc. **drysaltery**, *n.* the goods sold by a drysalter; the shop or business of a drysalter. **dry shaver**, *n.* an electric razor. **dry ski**, *n.* a specially adapted ski for use on a dry surface. **dry skiing**, *n.* **drystone**, *a.* of a wall, built without mortar. **dry(-stone) waller**, *n.* one who builds walls without mortar. **dryish**, *a.* **drily**, **dryly**, *adv.* **dryness**, *n.* **drier, dryer**, *n.* a desiccative; a material added to oil paints and printers' ink to make them dry quickly; a machine for drying clothes after washing etc.; a hair-drier.

dryad (drī'əd), *n.* (*pl.* **dryads, -ades** (-dēz)) a nymph of the woods.

DSc, (*abbr.*) Doctor of Science.
DSC, (*abbr.*) Distinguished Service Cross.
DSM, (*abbr.*) Distinguished Service Medal.
DSO, (*abbr.*) Distinguished Service Order.
DST, (*abbr.*) daylight saving time; double summer time.
DTI, (*abbr.*) Department of Trade and Industry.
DTP, (*abbr.*) desk-top publishing.
DTs, DT's, (*abbr.*) (*coll.*) delirium tremens.

dual (dū'əl), *a.* consisting of two; twofold, binary, double; expressing two; applied to an inflection of a verb, adjective, pronoun or noun, which, in certain languages, expresses two persons or things, as distinct from the plural which expresses more than two. *n.* the dual number. **dual carriageway**, (*N Am.* **divided highway**) *n.* a road which has at least two lanes in each direction, with traffic travelling in opposite directions separated by a central reservation. **dual control**, *a.* that can be operated by either of two people. **dual personality**, *n.* a psychological condition in which a single person has two distinct characters. **dual-purpose**, *a.* having, or intended for, two separate purposes. **dualism**, *n.* duality, the state of being twofold; a system or theory based on a radical duality of nature or animating principle, as mind and matter, good and evil in the universe. **dualist**, *n.* **dualistic** (-lis'-), *a.* **duality** (-al'-), *n.* **dualize, -ise**, *v.t.* **dually**, *adv.*

dub[1] (dŭb), *v.t.* (*past, p.p.* **dubbed**) to confer knighthood upon by a blow of a sword on the shoulder; to confer any dignity, rank, character or nickname upon.

dub[2] (dŭb), *v.t.* to give a new sound-track (esp. in a different language) to (a film).

dubbin (dŭb'in), **dubbing** (-ing), *n.* a preparation of grease for preserving and softening leather.

dubious (dū'biəs), *a.* doubtful; wavering in mind; obscure, vague, not clear; questionable; open to suspicion. **dubiety** (-bī'ə-), *n.* **dubiously**, *adv.* **dubiousness**, *n.*

dubitation (dūbitā'shən), *n.* doubt, hesitation, uncertainty.
dubitative (dū'-), *a.* tending to doubt; expressing doubt.
dubitatively, *adv.*

Dublin Bay prawn (dŭb'lin), *n.* a large prawn usu. cooked as scampi.

ducal (dū'kəl), *a.* of or pertaining to a duke or duchy.

ducally, *adv.*

ducat (dŭk'ət), *n.* a coin, of gold or silver, first minted in the Duchy of Apulia in about 1140, afterwards current in several European countries.

Duce (doo'chi), *n.* the official title of Benito Mussolini (1883–1945) when head of the Fascist state in Italy. [It., leader]

duchess (dŭch'is), *n.* the wife or widow of a duke; a lady who holds a duchy in her own right; a size of roofing slate; a kind of fancy blind with ornamental edging. **duchesse lace**, *n.* Flemish lace with designs in cord outline. **duchesse potatoes**, *n.pl.* mashed potatoes mixed with butter, milk and egg-yolk, piped on to a dish and baked. **duchy**, *n.* the territory, jurisdiction or dominions of a duke; the royal dukedom of Cornwall or Lancaster. **duchy court**, *n.* the court of a duchy, esp. that of Lancaster in England.

duck[1] (dŭk), *n.* a kind of untwilled linen or cotton fabric lighter and finer than canvas, used for jackets, aprons etc.

duck[2] (dŭk), *n.* a web-footed bird esp. the domestic duck, a variety of the wild duck or mallard; the female of this species (as distinguished from a drake); a stone made to skip along the surface of water; (*coll.*) darling; (*Cricket*) a duck's egg, a score of nothing (0); (*Mil.*) an amphibious motor vehicle. **Bombay-duck** BUMMALO. **ducks and drakes**, a game of making a flat stone skip along the surface of water. **lame duck**, a crippled or ineffective person; a defaulter on the Stock Exchange. **like water off a duck's back**, quite without effect. **sitting duck**, an easy target. **to break one's duck**, to score one's first (cricket) run. **duckboard**, *n.* planking used to cover muddy roads or paths. **duck's arse**, *n.* (*sl.*) hair at the back of the neck cut to look like a duck's tail. **duck's-egg**, *n.* (*Cricket*) no score. **duck-shot**, *n.* small shot for shooting wild duck. **duck soup**, *n.* (*sl.*, *N Am.*) anything easy to do. **duckweed**, *n.* a popular name for several floating water-weeds esp. one which is eaten by duck and geese. **duckling** (-ling), *n.* a young duck. **ducky, ducks**, *n.* (*coll.*) a term of familiarity or endearment.

duck[3] (dŭk), *v.i.* to dive, dip or plunge under water, out of sight, out of shot or range etc. *v.t.* to dip under water and suddenly withdraw; to throw into water; quickly to withdraw (one's head etc.) behind an obstacle; to avoid, esp. by keeping out of sight. *n.* a quick plunge or dip under water. **ducking**, *n.* immersion in water; a thorough wetting. **ducking-stool**, *n.* a kind of stool or chair on which petty offenders were tied and ducked.

duct (dŭkt), *n.* a tube, canal or passage by which a fluid or gas is conveyed; (*Anat.*) a tubular passage for conveying chyle, lymph and other fluids; (*Bot.*) a canal or elongated cell holding water, air etc. **ductless**, *a.*

ductile (dŭk'tīl), *a.* that may be drawn out into threads or wire; malleable, not brittle; capable of being moulded, plastic; (*fig.*) pliant, tractable, yielding to persuasion or advice. **ductileness, ductility** (-til'-), *n.*

dud (dŭd), *n.* (*coll.*) a useless thing, esp. a bad coin; a valueless cheque, a forgery; an artillery shell, bomb, bullet etc. that has failed to explode. *a.* useless, worthless. **duds**, *n.pl.* (*coll.*) clothes.

dude (dūd), *n.* (*N Am.*) a city-bred person. **dude ranch**, *n.* (*N Am.*) a ranch run as a pleasure resort by city people.

dudish, *a.* **dudishly**, *adv.*

dudgeon (dŭj'ən), *n.* anger, indignation, as *in high dudgeon*.

due[1] (dū), *a.* owed, owing, that ought to be paid, rendered or done to another; proper, suitable, appropriate; expected, appointed (to arrive), calculated (to happen); ascribable, that may be attributed (to). *adv.* exactly, directly (north, south, east or west). *n.* that which is owed to one; that which one owes; a debt, an obligation, tribute, toll, fee or other legal exaction; (*esp. N Am., pl.*) a club subscription. **dock-, harbour-dues**, charges levied

duel (dū'əl), *n.* a combat between two persons with deadly weapons to decide a private quarrel, usu. an affair of honour; any contest or struggle between two persons, parties, causes, animals etc. *v.i.* (*pres.p.* **duelling**, *past*, *p.p.* **duelled**) to fight in a duel. **duellist**, *n.*

duenna (dūen'ə), *n.* an elderly female employed as companion and governess to young women, a chaperon.

duet (dūet'), **duetto** (-ō), *n.* a composition for two performers, vocal or instrumental; two such performers; any performance by two persons. **duettino** (-tē'nō), *n.* a short duet. **duettist**, *n.*

duff[1] (dŭf), *n.* (*dial., coll.*) a stiff, flour pudding boiled in a bag. **plum duff**, such a pudding made with raisins. **up the duff**, (*Austral., sl.*) pregnant.

duff[2] (dŭf), *v.t.* (*sl.*) to beat (up, over). *a.* useless, not working.

duffel (dŭf'l), *n.* a thick, coarse kind of woollen cloth, having a thick nap or frieze. **duffel bag**, *n.* a sort of kit-bag fastened by a drawstring. **duffel coat**, *n.* a three-quarter-length coat usu. made from duffel, hooded and fastened with toggles.

duffer (dŭf'ə), *n.* a stupid, awkward or useless person.

dug[1] (dŭg), *n.* a teat, a nipple (now used only of the lower animals).

dug[2] (dŭg), past and p.p. of DIG. **dugout**, *n.* a canoe made of a single log hollowed out, or of parts of two logs thus hollowed out and afterwards joined together; a rough cabin cut in the side of a bank or hill; a cellar, cave or shelter used as a protection against enemy shelling; the enclosure at baseball, football, rugby etc. occupied by the trainer and men waiting to play.

dugong (doo'gong), *n.* a large herbivorous aquatic mammal, with two forelimbs only, inhabiting the Indian seas.

duiker, duyker (dī'kə), *n.* a small African antelope; any of various southern African cormorants.

duke (dūk), *n.* a noble holding the highest hereditary rank outside the royal family; the sovereign prince of a duchy; (*pl.*) (*sl.*) fists. **dukedom** (-dəm), *n.* the territory, title, rank or quality of a duke. **dukeship**, *n.*

Dukeries (dū'kəriz), *n.pl.* a district in Nottinghamshire formerly comprising five ducal seats.

dulcet (dŭl'sit), *a.* sweet to the senses, esp. the hearing. **dulcetly**, *adv.*

dulcimer (dŭl'simə), *n.* a musical instrument with strings of wire, which are struck with rods.

Dulcinea (dŭlsinē'ə, -sin'iə), *n.* a sweetheart; an idealized mistress. [a character in Cervantes' *Don Quixote*]

dull (dŭl), *a.* slow of understanding; stupid, not quick in perception; blunt, obtuse, not sharp or acute; wanting keenness in any of the senses; sluggish, inert, slow of movement; not brisk or active (as trade); dim, tarnished; cloudy, overcast, gloomy, uninteresting, tedious, wearisome; hard of hearing, deaf; not loud or clear. *v.t.* to make dull or stupid; to make blunt of edge, to render less acute, sensitive, interesting or effective; to make heavy or sluggish, to deaden; to tarnish, to dim. *v.i.* to become blunt, stupid dim or inert. **dull-brained**, *a.* stupid; of dull intellect. **dull-eyed**, *a.* having a listless or gloomy look. **dull-witted**, *a.* stupid. **dullard** (-ləd), *n.* a blockhead; a dunce. **dullish**, *a.* **dully**, *adv.* **dullness, dulness**, *n.*

dulse (dŭls), *n.* an edible kind of seaweed.

duly (dū'li), *adv.* in suitable manner; properly; punctually; sufficiently.

dumb (dŭm), *a.* unable to utter articulate sounds; unable to speak, mute; silent, refraining from speaking, reticent, taciturn; soundless; (*coll.*) stupid, unintelligent. **to strike dumb**, to astonish; to render speechless by astonishment. **dumb-bells**, *n.pl.* pairs of weights connected by short bars or handles, swung in the hands for exercise. **dumb cluck**, *n.* (*coll.*) a fool. **dumb-piano**, *n.* a keyboard for exercising the fingers. **dumb-show**, *n.* gestures without speech; pantomime. **dumbstruck**, *a.* temporarily shocked into silence. **dumb-waiter**, *n.* a dining-room apparatus with (usu. revolving) shelves for holding dishes etc.; a movable framework for conveying food etc. from one floor to another, a service-lift. **dumbly**, *adv.* **dumbness**, *n.*

dumbfound (dŭm'fownd), *v.t.* to strike dumb; to confound, confuse, perplex.

dumdum bullet (dŭm'dŭm), *n.* a soft-nosed expanding bullet that lacerates the flesh.

dummy (dŭm'i), *n.* one who is dumb; any sham article; a sham package displayed in a shop; a model of a person, for showing off dress etc.; (the player holding) the fourth exposed hand when playing at bridge etc.; a person who appears on the stage without speaking; a doll; a stupid fellow; a rubber teat for a baby to suck; a prototype of a book; the design for a page. **double dummy**, a game at bridge etc., with only two players, the two other hands being exposed. **to sell the dummy**, successfully to feign a pass or move in a football game. **dummy run**, *n.* a trial run, a rehearsal.

dump (dŭmp), *v.t.* to throw into a heap, as rubbish; to unload (as dirt) from wagons etc. by tilting them up; to send surplus produce, esp. manufactured goods that are unsaleable at home, to a foreign market for sale at a low price; to get rid of superfluous or objectionable things or people by sending elsewhere; to record (the data on an internal computer memory) on an external storage device during a computer run. *v.i.* to sit (down) heavily and suddenly. *n.* a pile of refuse; a place for rubbish; an untidy or otherwise undesirable house or room; an army stor- age depot; (*esp. N Am.*) a refuse-tip; the act of dumping computer data. **dump-bin**, *n.* a container in a shop for e.g. sale and bargain items. **dump-truck, dumper-truck**, *n.* a vehicle that tips up in front and so dumps its load. **dumping-ground**, *n.*

dumpling (dŭmp'ling), *n.* a mass of dough or pudding, boiled or baked, often enclosing fruit or meat.

dumps (dŭmps), *n.pl.*, **down in the dumps**, low-spirited, depressed.

dumpy (dŭm'pi), *a.* short and thick or plump. **dumpily**, *adv.* **dumpiness**, *n.*

dun[1] (dŭn), *a.* of a dull brown or brownish-grey colour; (*poet.*) dark, gloomy. *n.* a dun horse.

dun[2] (dŭn), *v.t.* (*past, p.p.* **dunned**) to demand payment from with persistence. *n.* an importunate demand for the payment of a debt.

dun[3] (dŭn), *n.* a hill, a mound, an earthwork (largely used in place-names).

dunce (dŭns), *n.* a dullard, one slow in learning. **dunce's cap**, *n.* a conical paper cap formerly worn by a school pupil to indicate slowness of learning. **Dunciad** (-siad), *n.* the epic of dunces, title of a satire (1728) by Alexander Pope.

Dundee cake (dŭndē'), *n.* a rich fruit cake usu. decorated with almonds.

dunderhead (dŭn'dəhed), *n.* a blockhead, a numskull, a dolt. **dunderheaded**, *a.* **dunderpate**, *n.*

dune (dūn), *n.* a hill, mound or ridge of sand on the seashore. **dune-buggy**, *n.* (*orig. N Am.*) a small open car with wide tyres for driving on beaches.

dung (dŭng), *n.* the excrement of animals; manure. *v.t.* to manure or dress with dung. *v.i.* to void excrement. **dung-beetle**, *n.* a species of beetle the larvae of which develop in dung. **dung-fly**, *n.* a two-winged fly that feeds upon dung. **dung-fork**, *n.* a fork for spreading manure. **dungheap, dunghill**, *n.* a heap of dung; an accumulation of dung and refuse in a farmyard; (*fig.*) a mean, filthy abode. **dungy**, *a.*

dungaree (dŭng·gərē'), *n.* a coarse kind of calico used for overalls; (*pl.*) overalls made of this.
dungeon (dŭn'jən), *n.* a prison or place of confinement, esp. one that is dark and underground.
dunghill etc. DUNG.
dunk (dŭngk), *v.t.*, *v.i.* to dip (cake, biscuits etc.) in what one is drinking, e.g. tea or coffee.
dunlin (dŭn'lin), *n.* the red-backed sand-piper, a common shore-bird.
Dunlop (dŭnlop'), *n.* a kind of rich, white cheese made in Scotland from unskimmed milk.
dunnage (dŭn'ij), *n.* (*Naut.*) loose wood, faggots, boughs etc., laid in the hold to raise the cargo above the bilge-water, or wedged between the cargo to keep it from rolling when stowed.
dunno (dŭnō'), (*coll.*) contr. form of (I) DON`T KNOW.
dunnock (dŭn'ək), *n.* the hedge-sparrow, from its colour.
dunny (dŭn'i), *n.* (*esp. Austral.*) a lavatory; (*esp. Sc.*) an outside lavatory.
duo (dū'ō), *n.* a duet.
duo-, *pref.* two.
duodecagon (dūōdek'əgon), DODECAGON.
duodecahedron (dūōdekəhē'drən), DODECAHEDRON.
duodecennial (dūōdisen'iəl), *a.* occurring once every twelve years.
duodecimal (dūōdes'iməl), *a.* proceeding in computation by twelves; applied to a scale of notation in which the value of digits increases twelvefold as they proceed from right to left. **duodecimally**, *adv.* **duodecimo** (-mō), *a.* consisting of 12 leaves to the sheet. *n.* a book consisting of sheets of 12 leaves or 24 pages; the size of such a book (written 12mo and called 'twelve-mo').
duodenary (dūədē'nəri), *a.* pertaining to the number twelve; proceeding by twelves.
duodenum (dūōdē'nəm), *n.* (*pl.* **-na** (-nə), **-nums**) the first portion of the small intestine, so called from being about the length of twelve fingers' breadths. **duodenal**, *a.* **duodenectomy** (-dinek'təmi), *n.* excision of the duodenum. **duodenitis** (-dini'tis), *n.* inflammation of the duodenum.
duologue (dū'əlog), *n.* a dialogue for two persons; a dramatic composition for two actors.
duomo (dwō'mō), *n.* a Italian cathedral.
duopoly (dūop'əli), *n.* an exclusive trading right enjoyed by two companies.
duotone (dū'ətōn), *n.* a picture in two tones or colours. *a.* in two tones or colours. [L *duo*, two]
dupatta (dupŭt'ə), *n.* a thick fold of cloth worn over one shoulder as part of Indian women's dress.
dupe[1] (dūp), *n.* one who is easily deceived; a credulous person. *v.t.* to trick, to cheat. **dupable**, *a.* **dupability** (-bil'-), *n.*
dupe[2] (dūp), *n.*, *a.*, *v.t.* (*coll.*) short for DUPLICATE.
duple (dū'pl), *a.* double, twofold; (*Mus.*) having two beats to the bar.
duplex (dū'pleks), *a.* double, twofold; compounded of two. *n.* (*N Am.*) a duplex apartment or house. *v.t.* (*Teleg.*) to make (a wire, cable or system) duplex, so that two messages can be sent at once in opposite directions.
duplex apartment, *n.* (*N Am.*) a two-storey apartment.
duplex house, *n.* (*N Am.*) a semi-detached house.
duplicate (dū'plikət), *a.* double, twofold, existing in two parts exactly corresponding; corresponding exactly with another. *n.* one of two things exactly similar in material and form; a reproduction, replica, copy; a copy of an original legal document having equal binding force; a copy made in lieu of a document lost or destroyed. *v.t.* (-kāt), to make a reproduction of; to double; to repeat (often needlessly); to make in duplicate; to make copies of on a machine; to divide and form two parts or organisms. **in duplicate**, in the original plus a copy. **duplication**, *n.* **duplicator**, *n.* a machine for printing copies of original written matter or drawings.

duplicity (dūplis'iti), *n.* double-dealing, dissimulation. **duplicitous**, *a.*
durable (dūə'rəbl), *a.* having the quality of endurance or continuance; lasting, permanent, firm, stable. **durability** (-bil'-), *n.* **durably**, *adv.*
Duralumin® (dūral'ūmin), *n.* an alloy of aluminium, copper and other metals, having great strength and lightness.
dura mater (dūə'rə mā'tə), *n.* the first of three lining membranes of the brain and spinal cord.
duration (dūrā'shən), *n.* (length of time of) continuance; power of continuance. **for the duration**, (*sl.*) so long as the situation lasts. **durative** (dū'-), *a.* denoting the aspect of a verb which implies continuance of action in time, as the imperfect and progressive tenses. *n.* a verb in this aspect.
durbar (dœ'bah), *n.* an Indian ruler's court; a state-reception by an Indian ruler or formerly by a British governor; a hall of audience.
duress (dūres'), *n.* constraint, compulsion, restraint of liberty, imprisonment; (*Law*) restraint of liberty or threat of violence to compel a person to do some act. **under duress**, thus compelled or threatened.
Durex® (dūə'reks), *n.* a make of condom; (**durex**; *coll.*) any condom.
during (dūə'ring), *prep.* in or within the time of; throughout the course or existence of.
durst (dœst), past tense of DARE [1].
durum (wheat) (dūə'rəm), *n.* a variety of spring wheat with a high gluten content, used mainly for the manufacture of pasta.
dusk (dŭsk), *n.* the time of day approaching darkness; partial darkness, twilight. **dusky**, *a.* swarthy, dark-skinned. **duskiness**, *n.*
dust (dŭst), *n.* earth or other matter reduced to such small particles as to be easily raised and carried about by the air; a stirring of such fine particles; pollen; (*poet.*) the decomposed bodies of the dead; a low or despised condition; turmoil, excitement, confusion, commotion. *v.t.* to brush or sweep away the dust from; to sprinkle or cover (as) with dust. **to bite the dust**, (*fig.*, *coll.*) to be beaten; to be humiliated; to die. **to raise, make, kick up a dust**, to make a disturbance. **to throw dust in someone's eyes**, to mislead, to deceive, to delude. **dust bath**, *n.* the rubbing of dust into their feathers by birds, prob. to get rid of parasites. **dustbin**, *n.* a receptacle for household refuse.
dust-bowl, *n.* an area reduced to aridity by drought and over-cropping. **dust-cart**, *n.* a vehicle for removing refuse from houses, streets etc. **dust-colour**, *n.* a light greyish brown. **dust-cover**, **-jacket**, *n.* a paper book-jacket on a hardback book. **dust devil**, *n.* a small whirlwind which whips up dust, leaves and litter. **dust-guard**, *n.* a fitting on a machine to protect a worker, rider etc. from dust.
dustman, *n.* one whose occupation is to remove refuse from dustbins. **dustpan**, *n.* a domestic utensil into which dust is swept. **dust-sheet**, *n.* one thrown over furniture while a room is being dusted or painted, or while it is unused. **dust storm**, *n.* a windstorm which whips up clouds of dust as it travels through arid areas. **dust-up**, *n.* (*coll.*) a row, a heated quarrel. **duster**, *n.* a cloth or brush used to remove dust; a person who dusts; (*Naut. coll.*) a flag. **dusting**, *n.*, *a.* **dusting-down**, *n.* a scolding, a severe reprimand. **dusting powder**, *n.* very fine powder, esp. talcum powder. **dusty**, *a.* covered with or full of dust; like dust; old, dull and uninteresting. **not so dusty**, (*sl.*) pretty good. **dustily**, *adv.* **dustiness**, *n.*
Dutch[1] (dŭch), *a.* pertaining to the Netherlands, its people or language; (*Hist.*) pertaining to the Low Germans, or to the German or Teutonic race; from the Netherlands; made or invented by the Dutch; (*N Am.*) of German extraction. *n.* (*pl.* **Dutch**) the language of the Netherlands; (*pl.*) the Low Germans, esp. the Hollanders. **double Dutch** DOUBLE. **Dutch elm disease**, a fungal disease of elms carried by beetles, causing withering and de-

dutch   252   **dynamometer**

foliation, and often fatal. **Dutch Reformed Church**, the Afrikaans-speaking branch of the Calvinist church in South Africa. **High Dutch**, the southern Germans; their language. **Low Dutch**, the Germans of the coast, esp. of the Netherlands; their language. **to go Dutch**, to pay for oneself when in a group. **Dutch auction** AUCTION. **Dutch barn**, *n.* one for storage, with open sides and a steel frame supporting a curved roof. **Dutch cap**, *n.* a moulded rubber cap fitting over the cervix to act as a contraceptive barrier. **Dutch cheese**, *n.* a small round cheese manufactured in Holland from skim milk. **Dutch clinker**, *n.* a yellow hard brick made in Holland. **Dutch courage**, *n.* false or fictitious courage, esp. inspired by alcohol. **Dutch doll**, *n.* a wooden doll with jointed limbs. **Dutch door**, *n.* (*N Am.*) a stable door. **Dutch hoe**, *n.* a garden hoe with a blade. **Dutchman**, *n.* **if so, I'm a Dutchman**, an emphatic negative. **the Flying Dutchman**, a legendary mariner condemned to sail against the wind till the Day of Judgment; his ghostly ship. **Dutch oven**, *n.* a cooking-chamber suspended in front of a fire so as to cook by radiated heat. **Dutch School**, *n.* a school of painters distinguished for minute realism and for the artistic treatment of commonplace subjects. **Dutch treat**, *n.* (*coll.*) an outing where each person pays his own way. **Dutch uncle**, *n.* one who criticizes in a stern, blunt manner. **Dutchwoman**, *n. fem.*

dutch² (dŭch), *n.* (*Cockney sl.*) a wife.

duty (dū'ti), *n.* that which must be or ought to be paid, done or performed; that which a particular person is bound morally or legally to do; moral or legal obligation; the course of conduct prescribed by ethics or religion; obedience or submission due (to parents, superiors); an act of reverence, respect or deference; toll, tax, impost or custom charged by a government upon the importation, exportation, manufacture or sale of goods; office, function, occupation, work; (any of) the various acts entailed in these; the obligations and responsibilities implied in one's engagement to perform these. **off duty**, not engaged in one's appointed duties. **on duty**, engaged in performing one's appointed duties. **to do duty for**, to serve in lieu of someone or something else; to serve as makeshift. **duty bound**, *a.* obliged by one's sense of duty (to do something). **duty-free**, *a.* not liable to tax or customs fees. **duty-free shop**, one, usu. on a ship or at an airport, where duty-free goods are on sale. **duty officer**, *n.* the officer on duty at any particular time. **duty-paid**, *a.* on which duty has been paid. **duteous**, *a.* obedient, dutiful. **duteously**, *adv.* **duteousness**, *n.* **dutiable**, *a.* liable to the imposition of a duty or custom. **dutiful**, *a.* careful in performing the duties required by law, justice or propriety; obedient, deferential. **dutifully**, *adv.* **dutifulness**, *n.*

duumvir (dūŭm'viə), *n.* (*pl.* **-viri, -virs**) one of two officers or magistrates in ancient Rome appointed to carry out jointly the duties of any public office. **duumvirate** (-rət), *n.* the association of two officers or magistrates in the carrying out of any public duties; a government of two; their term of office.

duvet (doo'vă), *n.* a quilt stuffed with down or man-made fibres, used as a bed-covering instead of blankets.

dux (dŭks), *n.* (*Sc.*) the top pupil of a school.

duyker DUIKER.

DV, (*abbr.*) God willing (L *Deo volente*).

dwaal (dwahl), *n.* (*S Afr.*) a state of bewilderment.

dwale (dwāl), *n.* deadly nightshade.

dwarf (dwawf), *n.* (*pl.* **dwarfs, dwarves** (-vz)) a human being, animal or plant much below the natural or ordinary size; a supernatural being of small stature. *a.* below the ordinary or natural size; stunted, puny, tiny. *v.t.* to cause to look small by comparison; to check the physical or mental development of. **dwarf star**, *n.* any relatively small star with high density and ordinary luminosity, e.g. the sun. **dwarf tree**, *n.* a miniature tree whose branches have been made to shoot near the root. **dwarf wall**, *n.* a low wall serving to surround an enclosure. **dwarfed**, *a.* **dwarfish**, *a.* **dwarfishly**, *adv.* **dwarfishness**, *n.* **dwarfism**, *n.* the condition of being a dwarf. **dwarflike**, *a.*

dwell (dwel), *v.i.* (*past, p.p.* **dwelt** (-t)) to reside, to abide (in a place); to live, spend one's time; to linger, pause, tarry. *v.* (*Mech.*) a pause; a slight regular stoppage of a movement whilst a certain operation is effected. **to dwell on, upon**, to occupy a long time with; to fix the attention upon. **dweller**, *n.* (*usu. in compounds*) a resident, an inhabitant (in). **dwelling**, *n.* the action of the verb to DWELL; a residence, abode, habitation. **dwelling-house**, *n.* a house for residence, in contradistinction to a house of business, office, warehouse etc. **dwelling-place**, *n.* a place of residence.

dwindle (dwin'dl), *v.i.* to shrink, to diminish, to become smaller; to waste or fall away; to degenerate, to decline.

dwt, (*abbr.*) pennyweight.

DX, *n.* long-range radio transmissions. **DXer**, *n.* one whose hobby is listening to such transmissions. **DXing**, *n.*

Dy, (*chem. symbol*) dysprosium.

dybbuk (dib'ək), *n.* (*pl.* **-buks, -bukkim** (-kim)) in Jewish folklore, the soul of a dead sinner that enters the body of a living person and takes control of his/her actions.

dye (dī), *v.t.* (*pres.p.* **dyeing**, *past* **dyed**) to stain, to impregnate with colouring-matter. *v.i.* to follow the business of a dyer; to take a colour (of a material that is being dyed). *n.* (a) fluid used for dyeing, colouring-matter; a colour, tinge, hue, produced by or as by dyeing; (*fig.*) a stain. **to dye in the wool**, to dye the wool before spinning to give a more permanent result. **dye-house, -works**, *n.* a building where dyeing is carried on. **dye-stuffs**, *n.pl.* the colouring materials used in dyeing. **dyed**, *a.* **dyed-in-the-wool**, *a.* fixed in one's opinions, uncompromising. **dyeing**, *n.* **dyer**, *n.* one whose business is dyeing.

dying (dī'ing), *a.* about to die; mortal, perishable; done, given or uttered just before death; associated with death; drawing to an end, fading away. *n.* the act of expiring, death. **dying declaration**, *n.* a legal declaration made by a person on the point of death. **dyingly**, *adv.*

dyke DIKE.

dynamic (dīnam'ik), *a.* of or pertaining to forces not in equilibrium, as distinct from *static*; causing motion, active, energetic, as distinct from *potential*; pertaining to dynamics; involving or dependent upon mechanical activity. *n.* the motive force of any action; (*Mus.*) (a degree of) loudness, or a sign indicating this. **dynamical**, *a.* dynamic; pertaining to dynamism. **dynamically**, *adv.* **dynamics**, *n.sing.* the branch of mechanics which deals with the behaviour of bodies under the action of forces which produce changes of motion in them.

dynamism (dī'nəmizm), *n.sing.* a system or theory explaining phenomena as the ultimate result of some immanent force; the restless energy of a forceful personality.

dynamite (dī'nəmīt), *n.* a powerful explosive compound, extremely local in its action, consisting of nitroglycerine mixed with an absorbent material. *v.t.* to smash or destroy with dynamite. **dynamiter**, *n.* a revolutionary or criminal employing dynamite. **dynamitist**, *n.*

dynamo-, *comb. form.* pertaining to force or power.

dynamo (dī'nəmō), *n.* (*pl.* **-mos**) a dynamoelectric machine. **dynamoelectric**, *a.* pertaining to electrical current; pertaining to the conversion of mechanical into electric energy or the reverse. **dynamoelectric machine**, *n.* a machine for converting mechanical energy into electric by means of electromagnetic induction (usu. applied only to d.c. generators). **dynamoelectrical**, *a.*

dynamograph (dīnam'əgrahf), *n.* a dynamometer used for recording speed, power, adhesion etc. on electric railways.

dynamometer (dīnəmom'itə), *n.* an instrument for the measurement of power, force or electricity.

dynast (din'ast, dī'-), *n.* a ruler, a monarch; a member or founder of a dynasty. **dynastic** (-nas'-), *a.* **dynastically**, *adv.* **dynasty** (din'-), *n.* a line, race or succession of sovereigns of the same family or (*loosely*) members of any rich and powerful family.

dynatron (dī'nətron), *n.* a four-electrode thermionic valve which generates continuous oscillation.

dyne (dīn), *n.* a unit for measuring force, the amount that, acting upon a gram for a second, generates a velocity of one centimetre per second.

dys-, *pref.* bad, badly, depraved; difficult, working badly, painful.

dysentery (dis'əntri), *n.* an infectious tropical febrile disease, seated in the large intestines, accompanied by mucous and bloody evacuations. **dysenteric** (-te'-), *a.*

dysfunction (disfŭngk'shən), *n.* impaired or abnormal functioning (of any organ or part of the body). **dysfunctional**, *a.*

dysgenic (disjen'ik), *a.* unfavourable to the hereditary qualities of any stock or race. **dysgenics**, *n.sing.* the study of racial degeneration.

dysgraphia (disgraf'iə), *n.* inability to write; impaired ability in writing. **dysgraphic**, *a.*

dyslexia (dislek'siə), *n.* word-blindness, impaired ability in reading and spelling caused by a neurological disorder. **dyslexic**, *a.*

dysmenorrhoea (dismenərē'ə), *n.* difficult or painful menstruation. **dysmenorrhoeal, -rrhoeic**, *a.*

dyspepsia (dispep'siə), **dyspepsy** (-si), *n.* indigestion arising from dysfunction of the stomach. **dyspeptic** (-tik), *a.* pertaining to, of the nature of, or suffering from dyspepsia. *n.* one subject to dyspepsia.

dysphagia (disfā'jiə), *n.* (*Path.*) difficulty of swallowing. **dysphagic**, *a.*

dysphasia (disfā'ziə), *n.* difficulty in speaking or understanding speech, caused by injury to or disease of the brain. **dysphasic**, *a.*

dysphonia (disfō'niə), *n.* a difficulty in speaking arising from disease or malformation of the organs.

dyspnoea, (*esp. N Am.*) **dyspnea** (dispnē'ə), *n.* difficulty of breathing. **dyspnoeal, dyspneal**, *a.* **dyspnocic**, *a.*

dysprosium (disprō'ziəm), *n.* a rare metallic element, at no. 66; chem. symbol Dy, of the rare earth group.

dystrophy (dis'trəfi), *n.* any of various disorders characterized by the wasting away of muscle tissue; the condition of lake water when it is too acidic to support life. **dystrophic** (-trof'-), *a.*

dysuria (disū'riə), **dysury** (dis'-), *n.* difficulty and pain in passing urine; morbid condition of the urine. **dysuric**, *a.*

dziggetai (dzig'itī), *n.* a species of wild ass, somewhat resembling the mule, native to Central Asia.

E

E, e, the fifth letter and second vowel of the alphabet; (*Mus.*) the third note of the diatonic scale; a second-class ship in Lloyd's register; (*Math.*) symbol for the base of Napierian logarithms, approximately equalling 2·718. **E-boat,** *n.* a small, fast motor-boat of the German navy armed with guns and torpedoes. **E number,** *n.* a number preceded by the letter E denoting a certain food additive in accordance with EEC regulations.
e-, *pref.* a form of EX-, as in *elocution, emend, evade, evolve.*
each (ēch), *a., pron.* every one (of a number) considered separately. **each way,** of a bet, that a horse will either win or be placed.
eager (ē'gə), *a.* excited by an ardent desire to attain, obtain or succeed; keen, ardent, impatient. **eagerly,** *adv.* **eagerness,** *n.*
eagle (ē'gl), *n.* a large bird of prey, the larger species of the Falconidae; any bird of the genus *Aquila,* esp. the golden eagle; a representation of this, esp. a Roman military ensign; the national emblem of the US and other countries; a gold coin of the US worth 10 dollars; the constellation Aquila in the northern hemisphere; in golf, a score of two under par for a hole. **eagle-eyed,** *a.* sharp-sighted; quick to discern. **eagle-hawk,** *n.* a S American hawk of the genus *Morphuus.* **eagle-owl,** *n.* large European and American owls, esp. the European *Buvo maximus.* **eagle-stone,** *n.* an argillaceous oxide of iron occurring in nodules. **eagle-winged,** *a.* having wings like those of the eagle; soaring high like an eagle. **eaglet** (-lit), *n.* a young eagle.
eagre (ē'gə), *n.* a tidal wave or bore in an estuary.
ealdorman (awl'dəmən), ALDERMAN.
-ean, -aean, -eian, *suf.* belonging to; like.
ear[1] (iə), *n.* the organ of hearing; the external part of this organ; the sense of hearing; a delicate perception of the differences of sounds, and judgment of harmony; attention (esp. favourable consideration); an ear-like projection from a body, usually for support or attachment. **all ears,** (*coll.*) listening carefully. **middle ear,** the ear-drum. **over head and ears, up to the ears,** completely, so as to be overwhelmed. **to bring (down) about one's ears,** to involve oneself in (trouble etc.). **to give, lend, an ear,** to listen. **to go in one ear and out the other,** to make no lasting impression. **to have, keep, one's ear to the ground,** to be well informed about what is happening. **to prick up one's ears,** to begin to listen attentively. **to send away with a flea in one's ear,** to dismiss (someone) angrily or contemptuously. **to set by the ears,** to incite or cause strife between. **to turn a deaf ear** TURN. **earache,** *n.* **ear-drop,** *n.* an ear-ring. **ear-drum,** *n.* the tympanum; the membrane of the tympanum. **earfui,** *n.* (*sl.*) a reprimand. **ear-mark,** *n.* a mark on the ear by which a sheep can be identified; any distinctive mark or feature. *v.t.* to set a distinctive mark upon; to allocate (funds etc.) for a particular purpose. **ear-muffs,** *n.pl.* a pair of pads joined on a band and used to keep the ears warm. **earphone,** *n.* an instrument which is held close to the ear and converts electrical signals into audible sound. **ear-piercing,** *a.* loud and shrill. *n.* the making of a hole in the lobe of the ear, for fastening ear-rings. **ear-plug,** *n.* soft material placed in the ear to block sound or water. **ear-ring,** *n.* a pendant or ornamental ring worn in the lobe of the ear. **earshot,** *n.*

hearing distance. **ear-splitting,** *a.* ear-piercing. **ear-trumpet,** *n.* a tube to aid the sense of hearing. **ear-wax,** *n.* a wax-like substance found in the ear, cerumen. **eared,** *a.* **earless,** *a.* **earlet** (-lit), *n.* a little ear.
ear[2] (iə), *n.* a spike or head of corn. *v.i.* to form ears, as corn.
earache EAR.
earl (œl), *n.* an English nobleman ranking next below a marquess and next above a viscount. **Earl Marshal,** *n.* an English officer of state, head of the College of Arms. **earldom,** *n.*
early (œ'li), *adv.* (*comp.* **earlier,** *superl.* **earliest**) in good time; soon; towards, in or near the beginning. *a.* soon; before or in advance of something else, or the usual time; situated in or near the beginning. **early warning system,** a system of advance notice, esp. of danger such as a nuclear attack. **early-closing,** *n.* a regular weekly half-holiday for shops. **early days,** (*coll.*) *adv.* too soon to take effect, have results etc. **Early English,** *n, a.* (of) the first of the Gothic styles of architecture in England, characterized by lancet windows, clustered pillars and vaulted roofs. **earliness,** *n.*
earn (œn), *v.t.* to gain as the reward of labour; to merit, deserve or become entitled to. **earned income,** *n.* income from paid employment. **earner,** *n.* one who earns; (*sl.*) a (shady) profitable activity. **earnings,** *n.pl.* that which is earned, wages.
earnest[1] (œ'nist), *a.* serious, grave; ardent or zealous in any action; sincere. *n.* seriousness; reality, not a pretence. **in earnest,** seriously; with sincerity. **earnestly,** *adv.* **earnestness,** *n.*
earnest[2] (œ'nist), *n.* a small payment as assurance of something to come. **earnest-money,** *n.* an instalment paid to seal a bargain.
earphone EAR.
earth (œth), *n.* the ground, the visible surface of the globe; the planet on which we live; dry land; this world, as opposed to other possible worlds; soil, mould, as distinguished from rock; dead, inert matter; the human body; the hole of a fox, badger etc.; the part of the ground completing an electrical circuit; a connection to ground; (*Radio*) plates or wires buried in the earth which provide a path to ground for currents flowing in the aerial; an earth-like metallic oxide, such as alumina. *v.t.* (*usu. with up*) to cover with earth; to drive (fox etc.) to his earth; to complete a circuit by connecting with the earth. *v.i.* to retire to an earth (as a fox). **rare earth metals** RARE. **earth-board,** *n.* a mould-board of a plough. **earth-born,** *a.* born from or on the earth; earthly. **earth-bound,** *a.* fixed in or to the earth; (*fig.*) fixed on earthly objects. **earth-closet,** *n.* a lavatory in which earth is used instead of water. **earth-fall,** *n.* a landslide. **earth-light,** *n.* light reflected from the earth upon the dark part of the moon. **earth mother,** *n.* a large, motherly woman. **earth mover,** *n.* a machine for excavating large quantities of earth. **earth-nut,** *n.* the pig-nut or ground-nut; the truffle etc. **earthquake,** *n.* a movement of the earth's crust; (*fig.*) a social or other disturbance. **earth science,** *n.* any science dealing with the earth, e.g. geography or geology. **earth-shine** EARTH-LIGHT. **earth tremor,** *n.* an earthquake. **earth-wolf** AARDWULF under AARDVARK. **earthwork,** *n.* a fortification made of heaped-up earth; an embankment.

earth-worm, *n.* a burrowing worm, esp. belonging to the genus *Lumbricus;* (*fig.*) a grovelling or sordid person.
earthen, *a.* made of baked clay. **earthenware,** *n.* pottery made of baked clay. **earthling** (-ling), *n.* an inhabitant of the earth; an earthly-minded person. **earthly,** *a.* of or pertaining to this world or life; mortal; carnal; corporeal. **not an earthly,** (*coll.*) not a chance. **earthly-minded,** *a.* **earthly-mindedness,** *n.* **earthliness,** *n.* **earthward, -wards,** *adv.* **earthy,** *a.* consisting of or resembling earth; carnal, material; lustreless. **earthiness,** *n.*
earwig (iə'wig), *n.* an insect, *Forficula auricularia,* having curved forceps at its tail; a flatterer or tale-bearer.
ease (ēz), *n.* a state of freedom from work, trouble or pain; freedom from constraint; facility, readiness; absence of effort. *v.t.* to free from pain, anxiety, labour or trouble; to relieve or free from a burden; to make easier or lighter; to mitigate; to make looser, adjust. *v.i.* to relax one's efforts or exertions. **at ease,** in a relaxed or peaceful state of mind or body; (*Mil.*) a command to stand with the legs apart and hands behind the back. **ill at ease,** in a state of mental or bodily disquiet, trouble or pain. **to ease off,** to become less oppressive. **to ease oneself,** to empty the bowels; to urinate. **easeful,** *a.* promoting ease, quiet or repose; comfortable; indolent. **easement,** *n.* the act of easing; a convenience; (*Law*) a right without profit which one proprietor has in or through the estate of another, as a right of way, light, air etc. **easer,** *n.*
easel (ē'zl), *n.* a frame used to support a picture, blackboard etc.
easily EASY.
east (ēst), *a.* situated towards the point where the sun rises when in the equinoctial; coming from this direction. *n.* the point of the compass where the sun rises at the equinox; 90° to the right of north; the eastern part of a country; the countries in the east of Europe or in Asia. *adv.* towards, at or near the east. *v.i.* to move towards the east; to veer from the north or south towards the east. **Far East,** the regions east of India. **Middle East,** Iraq, Iran, Mesopotamia etc. **Near East,** Turkey, the Levant etc. **East End,** *n.* the east and unfashionable end of London. **East-Ender,** *n.* **East Indiaman,** *n.* (*Naut. hist.*) a ship sailing to and from the E Indies. **east wind,** *n.* a wind from the east. **easterly,** *a.* situated in the east; looking towards the east; coming from the east. *adv.* in the direction of the east; in or from the east. *n.* a wind from the east. **easting,** *n.* distance east of a given meridian; movement to the east. **eastward,** *n.*, *a.* **eastwards,** *adv.*
Easter[1] (ēs'tə), *n.* the festival in commemoration of the resurrection of Christ, taking place on the Sunday after the full moon that falls on or next after 21 Mar. **Easter Day,** *n.* Easter Sunday. **Easter-eggs,** *n.pl.* eggs boiled hard and stained or gilded, to symbolize the resurrection; egg-shaped presents given at Easter. **Easter-eve,** *n.* the day before Easter Day. **Eastertide,** *n.* the time round about Easter. **Easter week,** *n.* the week beginning with Easter Day.
eastern (ēs'tən), *a.* situated in the east; pertaining to the east; blowing from the east. **Eastern Church,** *n.* the Greek Church. **Easterner,** *n.* an inhabitant of the east. **easternmost,** *a.*
easting EAST.
easy (ē'zi), *a.* at ease; free from pain, trouble, care or discomfort; well-to-do; not strict; free from embarrassment, constraint or affectation; smooth, fluent; not difficult, not requiring great effort; easily persuaded; indulgent, not exacting; (*Comm.*) easy to obtain; fitting loosely; slight, trivial. *adv.* in an easy manner. **easy!** move or go gently. **honours easy,** (*Cards., coll.*) honours equally divided. **in easy circumstances,** well-to-do, affluent. **take it easy!** take your time! **easy-care,** *n.* of a fabric, requiring little work to look after. **easy chair,** *n.* a comfortable arm-chair for resting in. **easy-going,** *a.* moving easily; taking things in an easy manner; indolent. **easy meat,** *n.* (*coll.*) a gullible fellow. **easy money,** *n.* (*coll.*) money acquired without much effort. **easy-osy** (-ōzi), *a.* indolent, easy-going. **easy street,** *n.* (*coll.*) a position of financial good fortune or security. **easily,** *adv.* **easiness,** *n.*
eat (ēt), *v.t.* (*past* **ate** (āt, et), *p.p.* **eaten**) to chew and swallow as food; to devour; to destroy by eating; (*fig.*) to corrode; to wear away; to use up. *v.i.* to take food; to be eaten; to taste, to relish. **to eat away,** to destroy, to rust, to corrode. **to eat crow, dirt,** to retract or acquiesce humbly. **to eat into,** to corrode. **to eat one's heart out,** to pine away. **to eat one's words,** to retract what one has said. **to eat out,** to eat in a restaurant, cafe or hotel. **to eat out of house and home,** to ruin (someone) by consuming all he/she has. **to eat up,** to eat completely; to wear away. **eatable,** *a.* fit to be eaten. *n.* anything fit or proper for food; (*pl.*) food. **eater,** *n.* one who eats; fruit suitable for eating uncooked. **eatery,** *n.* (*coll.*) a restaurant, cafe etc. **eating-house,** *n.* a restaurant. **eats,** *n.pl.* (*coll.*) food.
eau (ō), *n.* water (used in compounds to designate various spirituous waters and perfumes). **eau-de-Cologne** (də kəlōn'), *n.* a scent consisting of a solution of volatile oils in alcohol, orig. made in Cologne. **eau-de-Nil** (ōdənēl'), *n.* a pale greenish colour, said to be like Nile water. **eau de vie** (ō də vē'), *n.* brandy. [F]
eaves (ēvz), *n.pl.* the lower edge of the roof which projects beyond the wall. **eavesdrop,** *v.i.* to listen secretly (orig. under the eaves) so as to overhear confidences. **eavesdropper,** *n.*
ebb (eb), *n.* the flowing out of the tide; (*fig.*) decline, decay. *v.i.* to flow back; (*fig.*) to decline, decay. **at a low ebb,** weak or in a state of decline. **to ebb and flow,** to rise and fall; (*fig.*) to increase and decrease. **ebb-tide,** *n.* the retiring tide.
ebenezer (ebənē'zə), *n.* a chapel or meeting-house. [Heb.]
ebony (eb'əni), *n.* the wood of various species of *Diospyros,* noted for its solidity and black colour, capable of a high polish, and largely used for mosaic work and inlaying. *a.* made of ebony; intensely black. **ebon,** *n.*, *a.* **ebonist,** *n.* a worker in ebony. **ebonite** (-nīt), *n.* vulcanite. **ebonize, -ise,** *v.t.* to make the colour of ebony.
éboulement (āboōl'mē), *n.* (*Geol.*) a landslide.
ebracteate (ibrak'tiāt), *a.* (*Bot.*) without bracts.
ebriety (ibrī'əti), *n.* drunkenness, intoxication. **ebriate** (ē'bri·ə), *v.t.* to intoxicate. **ebriosity** (ēbrios'-), *n.* habitual drunkenness. **ebriose** (ē'briōs), **ebrious** (-əs), *a.*
ebullient (ibūl'yənt), *a.* boiling over; (*fig.*) overflowing (with high spirits or enthusiasm). **ebullience, -ency,** *n.* **ebullition** (ebə-), *n.* the boiling or bubbling of a liquid; sudden outburst (of feeling).
eburnation (ebənā'shən), **eburnification** (ibœnifikā'shən), *n.* excessive hardness of bone, caused by degenerative disease.
eburnean, -ian (ibœ'niən), **eburnine** (-nīn), *a.* of ivory; ivory-like.
EC, (*abbr.*) East Central; European Community.
écarté (ākah'tā), *n.* a game of cards played by two persons with 32 cards.
ecaudate (ēkaw'dāt), *a.* without a tail or stem.
ecbole (ek'bəli), *n.* (*Rhet.*) a digression. **ecbolic** (-bol'-), *n.* a drug which stimulates uterine contractions, used to induce labour.
ecce homo (ekihō'mō), *n.* a name given to paintings representing Christ crowned with thorns (John xix.5). [L, behold the man]
eccentric (iksen'trik), *a.* deviating from the centre; deviating from the conventional; erratic, irregular; (*Geom.*) of circles, not having the same centre. *n.* a person of odd or peculiar habits; a mechanical contrivance for converting circular into reciprocating rectilinear motion. **eccentric-rod,** *n.* one transmitting the motion of an eccentric-wheel. **eccentric-wheel,** *n.* one whose axis of revolution is

different from its centre. **eccentrically**, *adv.* **eccentricity** (eksəntris'iti), *n.* the state of not being concentric; deviation from the centre; departure from what is usual or regular; whimsical character; oddity, peculiarity.

ecchymosis (ekimō'sis), *n.* a bruise.

Eccles cake (ek'lz), *n.* a cake like a Banbury cake. [from *Eccles* in Lancashire]

ecclesia (iklē'ziə), *n.* a church; a religious assembly, a congregation. **ecclesiastic** (-as'-), *a.* ecclesiastical. *n.* a clergyman. **ecclesiastical**, *a.* pertaining to the Church or the clergy. **ecclesiastical courts**, *n.pl.* courts for administering ecclesiastical law and for maintaining the discipline of the Established Church; courts in the Presbyterian Church for deciding matters of doctrine and discipline. **ecclesiastical modes**, *n.pl.* (*Mus.*) the Ambrosian and Gregorian scales in which plain song and plain chant are composed. **ecclesiastically**, *adv.* **ecclesiasticism** (-sizm), *n.*

ecclesi(o)-, *comb. form.* pertaining to the Church or Churches or to ecclesiastic matters.

ecclesiolatry (ikleziol'ətri), *n.* excessive reverence for ecclesiastical forms and traditions. **ecclesiolater**, *n.*

ecclesiology (ikleziol'əji), *n.* the study of matters connected with churches, esp. architecture, decoration and antiquities. **ecclesiological** (-loj'-), *a.* **ecclesiologist**, *n.*

eccrine (ek'rin, -rīn), *a.* denoting sweat glands, which secrete on to the surface of the skin. **eccrinology** (-nol'-), *n.* the physiological study of secretions.

ecdysis (ek'disis), *n.* the casting of the skin, as by snakes etc.

ECG, (*abbr.*) electrocardiogram; electrocardiograph.

echelon (esh'əlon), *n.* an arrangement of troops with parallel divisions one in advance of another but slightly to the side; (a group of persons in) a level or grade of an organization etc. *v.t.* to form in echelon.

Echidna (ikid'nə), *n.* a genus of monotreme mammals from Australia and New Guinea, popularly known as spiny ant-eaters.

echinoderm (iki'nədœm), *n.*, *a.* (any individual) of the Echinodermata. **Echinodermata** (-dœ'mətə), *a* class of animals containing the sea-urchins, starfish and sea-cucumbers. **echinodermatous**, *a.*

echinus (iki'nəs), *n.* a sea-urchin; the convex projecting moulding below the abacus of an Ionic column. **echinate** (-nāt), **-nated**, *a.* bristly or spiny like a hedgehog or sea-urchin. **echinid** (-nid), **-nidan**, *n.* a sea-urchin. **echinite** (-nīt), *n.* (*Geol.*) a fossil echinoderm or sea-urchin. **echinoid** (-noid), *a.*

echo (ek'ō), *n.* the repetition of a sound caused by its being reflected from some obstacle; close imitation in words or sentiment; (*Mus.*) repetition of a phrase in a softer tone; repetition of the last syllables of a verse in the next line; in whist, a response to a partner's call for trumps. *v.i.* to give an echo; to resound. *v.t.* to return or send back (as a sound); (*fig.*) to repeat with approval; to imitate closely. **echo chamber**, *n.* a room whose walls echo sound for recording or radio effects or for measuring acoustics. **echogram**, *n.* a recording made by an echo sounder. **echo-location**, *n.* determination of the position of an object by measuring reflected sound. **echometer** (ikom'itə), *n.* an instrument for measuring the duration of sounds. **echo sounder**, *n.* (*Naut.*) an apparatus for sounding the depth of water beneath the keel of a ship. **echoer**, *n.* **echoic**, *a.* **echoism**, *n.* onomatopoeia. **echoless**, *a.*

eclair (ikleə'), *n.* an iced, finger-shaped cream cake. [F]

éclaircissement (ikleə'sēsmā), *n.* an explanation or clearing up of a dispute or misunderstanding.

eclampsia (iklamp'siə), *n.* convulsions or fits, particularly occurring with acute toxaemia in pregnancy.

eclat (āklah'), *n.* brilliant success; acclamation; splendour, striking effect. [F]

eclectic (iklek'tik), *a.* selecting, choosing, picking out at will from the (best of) doctrines, teachings etc. of others; broad, not exclusive. *n.* one who derives opinions etc. from various sources. **eclectically**, *adv.* **eclecticism** (-sizm), *n.*

eclipse (iklips'), *n.* the total or partial obscuration of the light from a heavenly body by the passage of another body between it and the eye or between it and the source of its light; a temporary failure or obscuration; loss of brightness, glory, honour or reputation. *v.t.* to cause an eclipse of (a heavenly body) to intercept the light of, to obscure; to outshine, excel. **ecliptic** (-tik-), *a.* pertaining to the ecliptic or an eclipse. *n.* the apparent path of the sun round the earth; a great circle on the terrestrial globe answering to, and falling within, the plane of the celestial ecliptic.

eclogue (ek'log), *n.* an idyll or pastoral poem, esp. one containing dialogue.

eco- (ē'kō-, ek'ō-), *comb. form.* concerned with ecology, habitat or the environment.

ecofriendly (ēkōfrend'li, ek'-), *a.* not wasteful or destructive of the environment.

ecology (ikol'əji), *n.* the branch of biology dealing with the relations between organisms and their environment. **ecological** (ēkəloj'-, ek-), *a.* **ecologist**, *n.*

econometrics (ikonəmet'riks), *n. sing.* statistical and mathematical analysis of economic theories.

economic (ēkənom'ik, ek-), *a.* relating to economics; pertaining to industrial concerns or commerce; maintained for the sake of profit or for the production of wealth; financially viable; economical. **economic zone**, *n.* a coastal area which a country claims as its own territory for purposes of fishing etc. **economical**, *a.* characterized by economic management; thrifty; economic. **economically**, *adv.* **economics**, *n. sing.* the science of the production and distribution of wealth; the condition of a country, community or individual, with regard to material prosperity. **economist**, *n.* one who manages with economy; one skilled in the science of economics. **economize, -ise** (ikon'ōmīz), *v.i.* to manage domestic or financial affairs with economy; (with *on*) to use sparingly. **economization**, *n.* **economizer**, *n.*

economy (ikon'əmi), *n.* the financial management of a community, state etc.; the complex of commercial, industrial and financial activities of a community or state; a frugal and judicious use or expenditure of money, time etc.; carefulness; (*usu. pl.*) a saving or reduction of expense; the disposition, arrangement or plan of any work; a cheap class of air travel. *a.* combining large size and cheapness. **political economy** ECONOMICS.

écorché (ākawshā'), *n.* an anatomical figure with the muscular system exposed for the purpose of study. [F]

ecospecies (ē'kōspēshiz, ek'-), *n.* a taxonomic species regarded as an ecological unit.

ecosphere (ē'kōsfiə, ek'-), *n.* the parts of the universe, esp. the earth, where life can exist.

écossaise (ākosez'), *n.* a Scottish dance or the music to it. [F]

ecostate (ikos'tāt), *a.* having no central rib (as some leaves).

ecosystem (ē'kōsistəm, ek'-), *n.* a system consisting of a community of organisms and its environment.

ecotype (ē'kōtōp, ek'-), *n.* a group of organisms within a species that has adapted itself to a changed environment.

écraseur (ākrazœ'), *n.* an instrument for removing tumours etc. without effusion of blood. [F]

ecru (ek'roo, ākrü'), *n*, *a.* the colour of unbleached linen.

ecstasy (ek'stəsi), *n.* a state of exaltation or rapture; excessive emotion, as delight or distress; a trance; (*sl.*) an illegal stimulant drug of the amphetamine type. **ecstasize, -ise**, *v.t.* to fill with ecstasy. *v.i.* to go into ecstasies.

ecstatic (-stat´-), *a.* pertaining to or producing ecstasy; rapturous; entranced. **ecstatically,** *adv.*
ECT, (*abbr.*) electroconvulsive therapy.
ecthyma (ekthi´mə), *n.* a skin disease characterized by an eruption of pimples.
ecto-, *comb. form.* (*Biol., Zool.*) pertaining to the outside.
ectoblast (ek´təblahst), *n.* ectoderm.
ectoderm (ek´tədœm), *n.* the outermost layer of cells of an embryo; the tissues (skin, nerves etc.) which develop from this.
ectomorph (ek´təmawf), *n.* a person of slight or thin build.
ectopic (ektop´ik), *a.* out of place. **ectopic pregnancy,** *n.* the abnormal development of a foetus outside the womb, usu. in a Fallopian tube.
ectoplasm (ek´təplazm), *n.* the outer layer of protoplasm of a cell; the substance which seems to emanate from the body of a spiritualist medium during a trance.
ectozoon (ektəzō´on), *n.* (*pl.* **-zoa** (-ə)) an external animal parasite.
ectype (ek´tīp), *n.* a copy as distinguished from an original. **ectypal** (-ti-), *a.*
ECU (ā´kū), (*abbr.*) European Currency Unit.
écu (ākū´. ākü´), *n.* an old French silver coin of varying value.
ecumenical (ēkūmen´ikəl, ek-), *a.* belonging to the Christian church, the Christian world or the **ecumenical movement,** encouraging unity between Christian Churches on issues of belief, worship etc. **ecumenically,** *adv.* **ecumenicism** (-sizm), **ecumenism,** *n.*
eczema (ek´simə), *n.* an inflammatory disease of the skin. **eczematous** (-sem´-), *a.*
ed., (*abbr.*) edited; edition; editor.
-ed, *suf.* forming the past tense and p.p. of regular verbs; used also to form adjectives, as in *cultured, talented.*
edacious (idā´shəs), *a.* greedy, voracious. **edacity** (-das´i-), *n.*
Edam (ē´dam), *n.* a kind of pressed, yellow cheese with a red outer skin. [town in Holland]
Edda (ed´ə), *n.* two Icelandic books, the *Elder* or *Poetic Edda* (*c.* 1200) and the *Younger* or *Prose Edda,* (*c.* 1230), containing Norse myths and legends.
eddy (ed´i), *n.* a small whirlpool; a current of air, fog, smoke etc. moving in a circle. *v.i., v.t.* to whirl in an eddy. **eddy current,** *n.* electrical current circulating in the mass of a conductor caused by a change in the magnetic field.
edelweiss (ā´dlvīs), *n.* a small white composite plant, *Gnaphalium leontopodium,* growing in rocky places in the Alps.
edema (ədē´mə), etc. OEDEMA.
Eden (ē´dn), *n.* the region in which Adam and Eve were placed at their creation; a region or a state of perfect bliss. **Edenic** (ēden´-), *a.*
edentate (ēden´tāt), *a.* having no incisor teeth; belonging to the **Edentata,** an order of mammals with no front teeth or no teeth whatsoever, containing the armadillos, sloths and ant-eaters. **edentulous** (-tū´ləs), *a.* toothless.
edge (ej), *n.* the sharp or cutting part of an instrument, as a sword; anything edge-shaped, a boundary-line; the brink, border or extremity of anything; keenness of mind or appetite; bitterness. *v.t.* to sharpen, to make or be an edge or border to; to move or put forward little by little. *v.i.* to move forward or away little by little; to move sideways, to sidle (up). **on edge,** to be irritable. **to edge out,** to get rid of gradually. **to have the edge on,** to have an advantage over. **to set the teeth on edge,** to cause a tingling or grating sensation in the teeth; to cause a feeling of irritation or revulsion. **edge-bone** AITCH-BONE. **edge-tool,** *n.* a cutting-tool; (*fig.*) anything dangerous to deal or play with. **edgeless,** *a.* **edger,** *n.* **edgeways, -wise,** *adv.* with the edge turned up, or forward in the direction of the edge; sideways. **to get a word in edgeways,** to speak with difficulty because of someone else talking. **edging,** *n.* that which forms the border or edge of anything, as lace on a dress, a row of small plants along the edge of a flowerbed. **edgy,** *a.* having or showing an edge; irritable, nervy. **edginess,** *n.*
edible (ed´ibl), *n., a.* (something) eatable. **edibility** (-bil´-), *n.*
edict (ē´dikt), *n.* a proclamation or decree issued by authority. **edictal** (idik´-), *a.*
edifice (ed´ifis), *n.* a building, esp. one of some size and pretension.
edify (ed´ifī), *v.t.* to build up spiritually; to improve the mind of. **edification** (-fi-), *n.* **edificatory,** (-fikā´-), *a.*
edile (ē´dīl), AEDILE.
edit (ed´it), *v.t.* to prepare for publication by compiling, selecting, revising etc.; (*fig.*) to censor, to alter; to act as editor of; to make a final version of (a film) by selection and arrangement of material. **edition** (idish´ən), *n.* the form in which a literary work is published; the whole number of copies published at one time. **edition de luxe** (ādēsyō də lüks´), *n.* a handsomely printed and bound edition of a book. **editor,** *n.* one who edits; one who conducts or manages a newspaper or periodical. **editorial** (-taw´-), *a.* of or pertaining to an editor. *n.* an article written by or proceeding from an editor. **editorialize, -ise,** *v.t., v.i.* to introduce personal opinions into reporting. **editorially,** *adv.* **editorship,** *n.* **editress** (-tris), *n. fem.*
EDP, (*abbr.*) electronic data processing.
educate (ed´ūkāt), *v.t.* to bring up (a child or children); to train and develop the intellectual and moral powers of; to provide with schooling; to train or develop (an organ or a faculty). **educable** (-kəbl), *a.* **educability** (-bil´-), *n.* **education,** *n.* the process of educating or being educated; (a course of) instruction; the result of a systematic course of training and instruction. **educational,** *a.* **educationalist, educationist,** *n.* an advocate of education; one who is versed in educational methods. **educationally,** *adv.* **educative** (-kətiv), *a.* **educator,** *n.*
educe (idūs´), *v.t.* to bring out, evolve, develop; to deduce, infer. **educible,** *a.* **educt** (ē´dūkt), *n.* that which is educed; an inference, deduction. **eduction** (idūk´-), *n.* **eductive,** *a.*
edulcorate (idūl´kərāt), *v.t.* to remove acidity from; to free from acids, salts or impurities, by washing. **edulcoration,** *n.* **edulcorator,** *n.*
Edwardian (edwawd´iən), *a.* referring to the periods of any of the kings of England named Edward, but usu. to that of Edward VII (1901-10).
-ee (-ē), *suf.* denoting the recipient, as in *legatee, payee;* or the direct or indirect object, as in *addressee, employee;* also used arbitrarily, as in *bargee, devotee.*
EEC, (*abbr.*) European Economic Community.
EEG, (*abbr.*) electroencephalogram; electroencephalograph.
eel (ēl), *n.* a snake-like fish, the genus *Anguilla,* esp. the common European species, *A. anguilla;* an eel-like fish; a slippery person; an eel-worm, as the vinegar-eel. **eel-fare,** *n.* the passage of young eels up streams; a brood of young eels. **eel-grass** GRASS-WRACK. **eel-pout,** *n.* a burbot; a blenny. **eel-worm,** *n.* a minute eel-like worm found in vinegar, sour paste etc. **eely,** *a.*
e'en (ēn), EVEN [1].
-eer (-iə), *suf.* denoting an agent or person concerned with or who deals in, as *charioteer, pamphleteer.*
e'er (eə), EVER.
eerie (iə´ri), *a.* causing fear; strange, weird. **eerily,** *adv.* **eeriness,** *n.*
ef- (ef-, if-), *pref.* form of EX- used before *f,* as in *efface, effigy.*
eff (ef), *v.t., v.i.* (*sl.*) euphem. for FUCK. **effing and blinding,** the use of obscene language.
efface (ifās´), *v.t.* to rub out, obliterate; to make not noticeable; to render negligible. **effaceable,** *a.* **effacement,** *n.*

effect (ifekt'), *n.* the result or product of a cause or operation; accomplishment, fulfilment; aim, purpose; an impression created; *(pl.)* goods, personal estate. *v.t.* to produce as a consequence or result; to accomplish. **for effect**, in order to produce a striking impression. **in effect**, in reality, substantially; practically. **special effects**, the creation of lighting, sounds etc. for a film, play, TV or radio. **to give effect to**, to carry out; to make operative. **to no effect**, in vain. **to take effect**, to operate, to produce its effect. **without effect**, invalid, without result. **effecter**, *n.* **effective**, *a.* having an effect; producing a striking impression; fit for duty or service; real, actual. *n.* one who is fit for duty. **effectively**, *adv.* **effectiveness**, *n.* **effectless**, *a.* **effector**, *n.* an organ that effects response to stimulus, e.g. muscle, gland. **effectual** (-chuəl), *a.* productive of an intended effect; efficacious. **effectuality** (-chual'-), *n.* **effectually**, *adv.* **effectualness**, *n.* **effectuate**, *v.t.* to effect, accomplish. **effectuation**, *n.*

effeminate (ifem'inət), *a.* womanish; unmanly, weak. **effeminacy**, *n.* **effeminately**, *adv.*

Effendi (ifen'di), *n. (formerly)* as a Turkish title of respect, bestowed on civil dignitaries and learned men.

efferent (ef'ərənt), *a. (Physiol.)* conveying outwards; discharging. *n.* an efferent vessel or nerve; a stream carrying off water from a lake etc.

effervesce (efəves'), *v.i.* to bubble up, from the escape of gas; to escape in bubbles; to boil over with excitement. **effervescence**, *n.* **effervescent**, *a.*

effete (ifēt'), *a.* worn out or exhausted; weak or decadent. **effeteness**, *n.*

efficacious (efikā'shəs), *a.* able to produce the effect intended. **efficaciously**, *adv.* **efficaciousness**, *n.* **efficacy** (ef'ikəsi), *n.*

efficient (ifish'ənt), *a.* causing or producing effects or results; competent, capable. **efficiency**, *n.* power to produce a desired result; *(Eng.)* the ratio of the output of energy to the input of energy. **efficiently**, *adv.*

effigy (ef'iji), *n.* the representation or likeness of a person, as on coins, medals etc. **to burn** or **hang in effigy**, to burn or hang an image of, to show hatred or contempt.

effloresce (eflares'), *v.i.* to burst into flower; *(Chem.)* to crumble to powder through loss of water or crystallization on exposure to the air; of salts, to form crystals on the surface; *(fig.)* to blossom forth. **efflorescence**, *n.* **efflorescent**, *a.*

effluent (ef'luənt), *a.* flowing or issuing out. *n.* a river or stream which flows out of another or out of a lake; liquid industrial waste or sewage. **effluence**, *n.*

effluvium (ifloo'viəm), *n. (pl.* **-via** (-ə)) a disagreeable smell and vapour as from putrefying substances etc.

efflux (ef'lŭks), *n.* the act of flowing out; that which flows out; a passing away. **effluxion** (iflŭk'shən), *n.*

effort (ef'ət), *n.* an exertion of physical or mental power, a strenuous attempt; an achievement. **effortful**, *a.* **effortless**, *a.* **effortlessness**, *n.*

effrontery (ifrŭn'təri), *n.* impudence, shamelessness, insolence.

effulgent (ifŭl'jənt), *a.* shining brightly; diffusing radiance. **effulgence**, *n.* **effulgently**, *adv.*

effuse (ifūz'), *v.t.* to pour out, to emit; to diffuse. *a.* (ifūs'), *(Bot.)* spreading loosely (of an inflorescence). **effusion** (-zhən), *n.* the act of pouring out; that which is poured out; frank expression of feeling, effusiveness; the escape of any fluid from one part of the body into another. **effusive** (-siv), *a.* gushing, demonstrative. **effusively**, *adv.* **effusiveness**, *n.*

E-fit (ē'fit), *n. (acronym)* electronic facial identification technique, a computerized form of photo-fit.

EFL, *(abbr.)* English as a foreign language.

eft (eft), *n.* the common newt.

EFTA (eft'ə), *(acronym)* European Free Trade Association.

EFTPOS, eftpos, (eft'pos), *(acronym)* electronic funds transfer at *point of sale*, debiting of a bank account directly at the time of purchase.

EFTS, *(abbr.)* electronic funds transfer system.

e.g., *(abbr.)* for example. [L *exempli gratia*]

egad (igad'), *int.* by God (a minced oath).

egalitarian (igalitəa'riən), *a.* believing in the principle of human equality. *n.* **egality** (-gal'-), *n.*

egest (ijest'), *v.t.* to eject; to void as excrement. **egesta** (-tə), *n.pl.* waste matter thrown out; excreta. **egestion** (-chən), *n.*

egg[1] (eg), *n.* the ovum of birds, reptiles, fishes and many of the invertebrates, usually enclosed in a spheroidal shell, and containing the embryo of a new individual; the egg of a bird, esp. of domestic poultry, used as food; anything shaped like this; an ovum or germ-cell; the early stage of anything. *v.t. (N Am.)* to pelt with rotten eggs. *v.i.* to collect eggs. **bad egg**, *(coll.)* a worthless person. **egg and anchor, dart, tongue**, *(Arch.)* various kinds of moulding carved alternately with egg-shapes and anchors etc. **egg and spoon race**, a race in which the runners carry eggs in spoons. **good egg!** *int. (sl.)* excellent! **egg-bird**, *n.* a kind of sea-bird, whose eggs are collected for food etc. **egg-bound**, *a.* applied to the oviduct of birds when obstructed by an egg. **egg-cosy** COSY. **egg-cup**, *n.* a cup-shaped vessel used to hold an egg at table. **egg-flip, -nog**, *n.* a drink compounded of eggs beaten up, sugar, milk and wine or spirits. **egg-head**, *n. (coll.)* an intellectual. **egg-plant**, *n.* aubergine. **egg-shaped**, *a.* **egg-shell**, *n.* the calcareous envelope in which an egg is enclosed. **egg-shell china**, very thin porcelain. **egg-shell paint**, paint with a slightly glossy finish. **egg-slice**, *n.* a kitchen utensil for removing eggs, omelets etc. from the pan. **egg-spoon**, *n.* a small spoon used for eating eggs. **egg-tooth**, *n.* a hard knob on the beak or snout of an embryo bird or reptile, for cracking the containing shell. **egg-whisk**, *n.* a kind of wire utensil used for beating up eggs. **egger**, *n.* one who gathers eggs. **egger-, eggar-moth**, *n.* various British moths with brown wings. **eggy**, *a.*

egg[2] (eg), *v.t. (usu. with* **on**) to incite, to urge.

egis (ē'jis), AEGIS.

eglantine (eg'ləntin, -tin), *n.* the sweet brier.

ego (ē'gō, eg'ō), *n.* individuality, personality; the self-conscious subject, as contrasted with the non-ego, or object; *(Psych.)* the conscious self, which resists on the one hand the threats of the super-ego, and on the other the impulses of the id. **egocentric** (-sen'trik), *a.* self-centred. **egocentricity** (-tris'-), *n.* **egomania** (-mānia), *n.* excessive or pathological egotism. **egomaniac**, *n.* **ego trip**, *n. (coll.)* an action or experience which adds to a person's self-important feelings. **egoism**, *n.* the theory that man's chief good is the complete development and happiness of self; pure self-interest, systematic selfishness; egotism. **egoist**, *n.* **egoistic, -ical** (-is'-), *a.* **egoistically**, *adv.* **egotism** (-tizm), *n.* a too frequent mention of oneself; self-centredness, self-conceit. **egotist**, *n.* **egotistic, -ical** (-tis'-), *a.* **egotistically**, *adv.* **egotize, -tise** (-tiz), *v.i.* [L]

egregious (igrē'jəs), *a.* notorious; conspicuously bad. **egregiously**, *adv.* **egregiousness**, *n.*

egress (ē'gres), *n.* the act or power of going out; departure; a means or place of exit. **egression** (igresh'ən), *n.*

egret (ē'gret), *n.* a heron, esp. the lesser white heron, of those species that have long and loose plumage over the back.

Egyptian (ijip'shən), *a.* of or pertaining to Egypt or the Egyptians. *n.* a native of Egypt. **Egyptian lotus**, *n. Nymphaea lotus.* **Egyptology** (-tol'əji), *n.* the study of the antiquities, language etc. of ancient Egypt. **Egyptological** (-loj'-), *a.* **Egyptologist** (-tol'-), *n.*

eh (ā), *int.* an exclamation expressive of doubt, inquiry, surprise etc.

eider (ī'dər), *n.* a large Arctic sea-duck. **eiderdown**, *n.* the

eidetic (īdet'ik), *a.* able to recall a vivid image of something previously seen or imagined.

eight (āt), *n.* the number or figure 8 or VIII; the age of eight; a set of eight things or people; (*Rowing*) a crew of eight in a boat; (*Skating*) a curved outline resembling the figure 8; articles of attire such as shoes etc. denoted by the number 8; a card with eight pips; a score of eight points; the eighth hour after midday or midnight. *a.* consisting of one more than seven. **one over the eight,** (*coll.*) slightly drunk. **eight-day,** *a.* (of clocks) going for eight days. **eight-fold,** *a.* **eighth,** (ātth), *n.* one of eight equal parts; (*Mus.*) the interval of an octave. *n., a.* (the) last of eight (people, things etc.); the next after the seventh. **eighthly,** *adv.* **eightsome** (-səm), *n.* a form of Scottish reel for eight dancers.

eighteen (ātēn'), *n.* the number or figure 18 or XVIII; the age of 18. *a.* 18 in number; aged 18. **eighteen-mo** (-mō), *n.* (*coll.*) octodecimo. **eighteenth,** *n., a.*

eighty (ā'ti), *n.* the number or figure 80 or LXXX; the age of 80. *a.* 80 in number; aged 80. **eighties,** *n.pl.* the period of time between one's 80th and 90th birthdays; the range of temperature between 80 and 90 degrees; the period of time between the 80th and 90th years of a century. **eightieth,** *n.* one of 80 equal parts. *n., a.* (the) last of 80 equal parts; the next after the 79th.

eikon (ī'kon), ICON.

einsteinium (īnstī'niəm), *n.* a radioactive element, at. no. 99; chem. symbol Es; artificially produced from plutonium and named after Albert *Einstein* (1879–1955).

eisteddfod (īstedh'vod), *n.* (*Welsh*) a competitive congress of bards and musicians held annually to encourage native poetry and music.

either (ī'dhə, ē'-), *a., pron.* one or the other of two; each of two. *adv., conj.* in one or the other case (as a disjunctive correlative); any more than the other (with neg. or interrog., as *If you don't I don't either*).

ejaculate (ijak'ūlāt), *v.t.* to utter suddenly and briefly; to eject. *v.i.* to utter ejaculations; to emit semen. **ejaculation,** *n.* **ejaculative,** *a.* **ejaculatory,** *a.*

eject[1] (ijekt'), *v.t.* to discharge, emit; to expel; (*Law*) to oust or dispossess. *v.i.* to use an ejector seat. **ejection** (-shən), *n.* **ejective,** *a.* **ejectment,** *n.* the act of casting out or expelling; dispossession; (*Law*) an action to recover possession. **ejector,** *n.* **ejector seat,** *n.* a seat that can be shot clear of the vehicle in an emergency.

eject[2] (ē'jekt), *n.* (*Psych.*) something that is not an object of our own consciousness but inferred to have actual existence.

eke[1] (ēk), *v.t.* to make up for or supply deficiencies in (with *out*); to produce, support or maintain with difficulty.

eke[2] (ēk), *adv.* (*formerly*), also, besides, likewise.

el[1] (el), *n.* the 12th letter of the alphabet, L, l; anything shaped like the capital form of this letter.

el[2] (el), *n.* (*US*) an elevated railway.

-el -LE.

elaborate (ilab'ərət), *a.* carefully planned; complicated, detailed. *v.t.* (-rāt), to produce by labour; to develop in detail; to change (something simple) into something complex. **elaborately,** *adv.* **elaborateness,** *n.* **elaboration,** *n.* **elaborative,** *a.* **elaborator,** *n.*

élan (ilan', ālā'), *n.* ardour; dash. [F]

eland (ē'lənd), *n.* a large ox-like antelope from S Africa.

elapse (ilaps'), *v.i.* (of time) to pass away.

elasmobranch (ilaz'məbrangk), *n.* one of a class of fishes, the Elasmobranchii, containing the sharks, rays and chimeras. **elasmobranchiate** (-brang'kiət), *n.*

elastic (ilas'tik), *a.* having the quality of returning to its original form or volume; springy; flexible, adaptable; admitting of extension; readily recovering from depression or exhaustion. *n.* a strip of elastic substance. **elastic band,** *n.* a rubber band. **elastic tissue,** *n.* yellow fibrous tissue occurring in ligaments etc. **elastically,** *adv.* **elasticate,** *v.t.* to elasticize. **elasticin** (-las'tisin), **elastin,** *n.* (*Chem.*) the substance forming the fibres of elastic tissue. **elasticity** (elastis'iti), *n.* **elasticize, -ise** (-īz), *v.t.* to make elastic.

Elastoplast® (ilas'təplahst), *n.* gauze surgical dressing on a backing of adhesive tape, suitable for small wounds, cuts and abrasions.

elate (ilāt'), *v.t.* to raise the spirits of; to make exultant. **elation,** *n.*

Elater (el'ətə), *n.* a genus of coleopterous insects, called click-beetles or skip-jacks; (*Bot.*) an elastic spiral filament attached to spores.

elaterite (ilat'ərit), *n.* a soft elastic mineral.

elaterium (elətiə'riəm), *n.* a powerful purgative obtained from the fruit of the squirting cucumber.

elbow (el'bō), *n.* the joint uniting the forearm with the upper arm; a sharp bend or corner. *v.t.* to push or thrust with the elbows, to jostle. *v.i.* to make one's way by pushing with the elbows. **at one's elbow,** near at hand. **out at elbows,** shabby in dress; in needy circumstances. **to crook** or **lift the elbow,** to drink. **to jog the elbow,** to give a reminder. **up to the elbows,** deeply engaged in business. **elbow-chair,** *n.* an armchair. **elbow-grease,** *n.* hard and continued manual exercise. **elbow-room,** *n.* ample room for action.

eld (eld), *n.* (*old-fashioned*) old age; an old man; former ages.

elder[1] (el'də), *a.* older; pertaining to former times; in card-playing, having the right to play first. *n.* a senior in years; one whose age entitles him to respect; (*pl.*) persons of greater age; a counsellor; an officer in the Jewish synagogue, in the Presbyterian and other churches. **Elder Brethren,** *n.pl.* the masters of Trinity House, London. **elder statesman,** *n.* a retired and respected politician or administrator. **elderly,** *a.* bordering on old age. **eldership,** *n.* **eldest,** *a.* oldest; first born of those surviving.

elder[2] (el'də), *n.* a tree of the genus *Sambucus*; the common elder, a small tree bearing white flowers and dark purple berries. **elderberry,** *n.* another name for the elder tree; the fruit of the elder tree. **elder-wine,** *n.* a wine made from elderberries and elder-flowers.

El Dorado, Eldorado (el dərah'dō), *n.* an imaginary land of gold in South America; any place where money or profit is easily obtained.

eldritch (el'drich), *n.* (*Sc.*) strange, weird, ghastly.

Eleatic (eliat'ik), *a.* pertaining to Elea, a Greek town in Italy, or to the school of philosophy founded there by Xenophanes, Parmenides and Zeno.

elecampane (elikampān'), *n.* a composite plant, *Inula helenium*, formerly used in medicinally.

elect (ilekt'), *a.* chosen; (*placed after the noun*) designated to an office, but not yet installed, as in *president elect*. *v.t.* to choose by vote; to choose to everlasting life; to determine on any particular course of action. **the elect,** those chosen by God etc.; highly select or self-satisfied people. **election** (ilek'shən), *n.* the act of choosing out of a number, esp. by vote; the ceremony or process of electing; power of choosing or selection; (*Theol.*) the selection of certain individuals from mankind to be eternally saved (the characteristic doctrine of Calvinism). **by-election** BY. **general election** GENERAL. **electioneer** (-niə'), *v.i.* to work for the election of a candidate. **elective,** *a.* chosen by election; pertaining to election or choice; having or exercising the power of choice. **electively,** *adv.* **elector** (-tə), *n.* one who is eligible to elect; (*Hist.*) one of the princes of Germany who were entitled to vote in the election of the Emperor. **electoral,** *a.* **electoral college,** *n.* in the US the body of people who elect the President and the Vice-President, having been themselves elected by vote. **electorate** (-rət), *n.* the whole body of electors; (*Hist.*) the

dignity or territory of an elector of the German Empire. **electorship**, *n.* **electress**, *n. fem.*
Electra complex (ilek'trə), *n.* (*Psych.*) attraction of a daughter for her father accompanied by hostility to her mother. [character in Gr. mythology]
electric (ilek'trik), *a.* containing, generating or operated by electricity; resembling electricity, magnetic; startling, thrilling, tense. *n.* a non-conductor, in which electricity can be excited by means of friction. **electric battery** BATTERY. **electric blanket**, *n.* a blanket containing an electrically-heated element. **electric blue**, *n.* a steely blue. **electric cable**, *n.* an insulated wire or flexible conductor for conveying a current. **electric chair**, *n.* a chair in which persons condemned to death are electrocuted. **electric charge**, *n.* the accumulation of electric energy in an electric battery. **electric circuit**, *n.* the passage of electricity from a body in one electric state to a body in another by means of a conductor; the conductor. **electric current**, *n.* continuous transition of electricity from one place to another. **electric eel**, *n.* a large S American eel, *Gymnotus electricus*, able to give an electric shock. **electric eye**, *n.* a photocell; a miniature cathode-ray tube. **electric fence**, *n.* a wire fence charged with electricity, used for purposes of security. **electric field**, *n.* a region in which forces are exerted on any electric charge present in the region. **electric furnace**, *n.* a furnace used for industrial purposes heated by electricity. **electric guitar**, *n.* an electrically amplified guitar. **electric hare**, *n.* an artificial hare made to run by electricity, used in greyhound racing. **electric jar**, *n.* a Leyden jar. **electric ray**, *n.* a flat-fish of the genus *Torpedo*. **electric shock**, *n.* the sudden pain felt from the passing of an electric current through the body. **electric storm**, *n.* a disturbance of electric conditions of the atmosphere. **electric strength**, *n.* the maximum electric field strength that can be applied to an insulator without causing breakdown. **electrical**, *a.* relating to electricity; electric. **electrically**, *adv.* **electrician** (eliktrish'ən), *n.* one who instals or repairs electrical equipment. **electricity** (eliktris'iti), *n.* a form of energy associated with positively and negatively charged subatomic particles and their movements; the science of the laws and phenomena of this energy. **electrics**, *n.pl.* (*coll.*) electric circuits, equipment, etc. **electrify** (ilek'trifi), *v.t.* to charge with electricity; to give an electric shock to; (*fig.*) to thrill or excite; to adapt (a mechanical system) to operation by electricity. **electrification** (-fi-), *n.* **electrize, -ise** (ilek'trīz), *v.t.* to electrify. **electrization**, *n.*
electro (ilek'trō), *n.* an electrotype; electroplate.
electro- (ilektrō-), *comb. form.* having electricity for its motive power; resulting from, or pertaining to, electricity. **electrobiology**, *n.* the science of the electric phenomena of living organisms. **electrobiologist**, *n.* **electrocardiograph**, *n.* an instrument which indicates and records the contractions of the heart muscle. **electrocardiogram**, *n.* a record so produced. **electrochemistry**, *n.* the science of the chemical effects produced by electricity. **electrochemical**, *a.* **electroconvulsive therapy**, *n.* treatment of mental or nervous disorders by the use of electric shocks. **electrodynamics**, *n.* the science of electricity in motion. **electrodynamic**, *a.* **electrodynamometer**, *n.* an instrument for measuring the strength of an electric current. **electroencephalograph**, *n.* an instrument recording electrical activity in the brain. **electroencephalogram**, *n.* such a record. **electroencephalography**, *n.* **electrokinetics**, *n.* electrodynamics. **electromagnet**, *n.* a bar made of soft iron rendered magnetic by the passage of a current of electricity through a coil of wire surrounding it. **electromagnetic**, *a.* **electromagnetic radiation**, *n.* light waves, radiowaves, X-rays etc. **electromagnetism**, *n.* magnetism produced by an electric current; the study of the production of magnetism by electricity, and the relations between magnetism and electricity. **electromechanical**, *a.* pertaining to the use of electricity in mechanical systems. **electrometallurgy**, *n.* the separation of metals from their alloys by means of electrolysis. **electromotive**, *a.* producing an electric current. **electromotive force**, *n.* difference of potential, the force of an electric current, measured in volts. **electromotor**, *n.* a machine for converting electric into mechanical energy. **electronegative**, *a.* having a negative electric charge. **electroplate**, *v.t.* to cover with a coating of metal by electrolysis. *n.* articles so produced. **electropositive**, *a.* having a positive electric charge. **electrostatics**, *n.* the science of static electricity. **electrostatic**, *a.* **electrotherapy**, *n.* treatment of disease with electricity. **electrothermy**, *n.* the science of the relation of electric current and temperature. **electrothermic**, *a.*
electrocute (ilek'trəkūt), *v.t.* to kill by an electric shock; to execute by administering a powerful electric shock. **electrocution**, *n.*
electrode (ilek'trōd), *n.* one of the poles of a galvanic battery, or of an electrical device; an anode, cathode, grid, collector, base etc.
electrograph (ilek'trəgrahf), *n.* the record of an electrometer.
electrolier (ilektrəliə'), *n.* a pendant or bracket for supporting an electric lamp.
electrolyse, -yze, (ilek'trəliz), *v.t.* to decompose by to remove (hair) by electrolysis.
electrolysis (eliktrol'isis), *n.* (the study of) the decomposition of chemical compounds by the passage of an electric current through them; removal of body hair by applying an electrically charged needle to the hair follicles. **electrolytic**, *a.*
electrolyte (ilek'trəlit), *n.* a compound which may be decomposed by an electric current.
electrometer (eliktrom'itə), *n.* an instrument for measuring the amount of electrical force, or for indicating the presence of electricity. **electrometrical** (ilektrəmet'-), *a.*
electron (ilek'tron), *n.* (*Phys.*) a particle bearing a negative electric charge, the most numerous constituent of matter and probably the cause of all electrical phenomena. **electron camera**, *n.* a device which converts an optical image into an electric current by electronic means. **electron microscope**, *n.* a very powerful microscope using electrons instead of light. **electron volt**, *n.* a unit of energy in atomic physics, the increase in energy of an electron when its potential is raised by 1 volt. **electronics** (eliktron'iks), *n. sing.* the science of applied physics that deals with the conduction of electricity in a vacuum, or a semiconductor, and with other devices in which the movement of electrons is controlled. **electronic**, *a.* pertaining to electronics; operated or produced by means of electronics. **electronic data interchange**, a process whereby information is transferred by linked computer terminals, telephones, fax machines etc. **electronic funds transfer at point of sale** EFTPOS. **electronic mail**, *n.* messages sent from one computer or fax machine to another by means of linked terminals.
electrophoresis (ilektrəfərē'sis), *n.* the movement of charged particles in a fluid under the influence of an electric field.
electrophorus (ēlektrof'ərəs), *n.* an instrument for generating static electricity by induction.
electroscope (ilek'trəskōp), *n.* an instrument for detecting the presence of electricity. **electroscopic**, *a.*
electrotype (ilek'trətīp), *n.*, *v.t.* (to make) a copy by the electric deposition of copper upon a mould. **electrotyper**, *n.* **electrotypist**, *n.*
electrum (ilek'trəm), *n.* a natural alloy of gold and silver in use among the ancients.
electuary (ilek'tūəri), *n.* a medicine mixed with some sweet confection.
eleemosynary (eliəmos'inəri), *a.* given or done by way of alms; supported by or dependent on charity.

elegant (el'igənt), *a.* pleasing to good taste; graceful, well-proportioned, refined; simple and effective; excellent. **elegance**, *n.* **elegantly**, *adv.*

elegy (el'əji), *n.* a lyrical poem or a song of lamentation; a poem of a plaintive, meditative kind; a poem written in elegiac verse. **elegiac** (-jī'-), *a.* pertaining to or of the nature of elegies; mournful. *n.pl.* verse consisting of alternate hexameters and pentameters, in which the elegies of the Greeks and Romans were commonly written. **elegiacally**, *adv.* **elegize, -ise**, *v.t.* to compose an elegy upon. *v.i.* to compose an elegy; to write elegiacally. **elegist**, *n.*

element (el'əmənt), *n.* one of the fundamental parts of which anything is composed; any of the 105 a substances which cannot be resolved by chemical analysis; (*pl.*) earth, air, fire and water, formerly considered as simple elements; the natural habitat of a animal or plant; the proper or natural sphere of any person or thing; anything necessary to be taken into account in coming to a conclusion; (*pl.*) the rudiments of any science or art; (*pl.*) the bread and wine used in the Eucharist; the resistance wire of an electric heater; (*pl.*) (bad) weather. **elemental** (-men'-), *a.* pertaining to or arising from first principles; pertaining to the four elements of earth, air, water and fire; pertaining to the primitive forces of nature; ultimate, simple, uncompounded. **elemental spirits**, *n.pl.* disembodied spirits formerly supposed to inhabit the four elements. **elementalism**, *n.* **elementally**, *adv.* **elementary** (-men'-), *a.* consisting of one element; introductory. **elementary particle**, *n.* any of several particles, such as electrons, protons or neutrons, which are less complex than atoms. **elementary school** *n.* (*N Am.*) one attended by children for the first six to eight years of education. **elementarily**, *adv.* **elementariness**, *n.*

elemi (el'əmi), *n.* a gum resin obtained from a tropical tree, used in pharmacy.

elenchus (ileng'kəs), *n.* (*pl.* **-chi** (-i)) an argument by which an opponent is made to contradict himself; a refutation. **elenctic** (-tik), *a.*

elephant (el'ifənt), *n.* a large pachydermatous animal, four-footed, with flexible proboscis and long curved tusks, of which two species now exist, the Indian and African elephants. **white elephant**, a useless and expensive possession. **elephant seal**, *n.* a very large seal, the male having a proboscis-like snout. **elephant's ear**, *n.* the begonia. **elephantiasis** (eləfənti'əsis), *n.* a tropical cutaneous disease in which the skin becomes hardened and the part affected greatly enlarged. **elephantine** (-fan'tīn), *a.* pertaining to or resembling an elephant; immense; clumsy. **elephantoid** (-fan'toid), *n., a.*

Eleusinian (elūsin'iən), *a.* relating to Eleusis, in ancient Attica, or to the mysteries in honour of Ceres annually celebrated there.

elevate (el'əvāt), *v.t.* to raise to a higher place; to exalt in rank or dignity; to make louder or higher; to exhilarate. **elevated**, *a.* raised; at or on a higher level; lofty in style, dignified; (*coll.*) slightly intoxicated. **elevated railway**, *n.* a city railway raised on pillars above the street-level. **elevation**, *n.* the act of elevating; the state of being elevated; an elevated position or ground; height above sea-level; the height of a building; a drawing to scale of the outside of a building; (*Astron.*) the angular altitude of a heavenly body above the horizon; (*Gunnery*) the angle of the line of fire with the plane of the horizon; grandeur, dignity. **elevator**, *n.* one who or that which elevates; a muscle which raises any part of the body; a machine to raise grain from a car or ship to a high level; a lift; a hinged flap on the tailplane to provide vertical control. **elevatory**, *a.*

eleven (ilev'n), *n.* the number or figure 11 or xi; the age of 11; a set of eleven things or people; an article of attire, such as a shoe etc., denoted by the number 11; a score of 11 points; the 11th hour after midday or midnight; (*Cricket, Assoc. Football*) the eleven people selected to play for a particular side. *a.* 11 in number; aged 11. **eleven plus exam**, (*formerly*) a school examination taken by children of about 11 to determine the particular type of secondary education they were suited for. **eleven year period**, (*Astron.*) the cycle of periodic changes in the occurrence of sun-spots. **elevenses** (-ziz), *n.pl.* (*coll.*) a snack taken in the middle of the morning. **eleventh**, *n.* one of 11 equal parts; (*Mus.*) the interval of an octave and a fourth. *n., a.* (the) last of 11 (people, things etc.); (the) next after the 10th. **at the eleventh hour**, at the last moment.

elf (elf), *n.* (*pl.* **elves**, elvz) a tiny supernatural being supposed to be very mischievous; a fairy; a mischievous person; a tiny creature; a pet name for a child. **elf-arrow, -bolt, -dart**, *n.* a Stone-Age flint arrow-head, popularly thought to be shot by fairies. **elf-child**, *n.* a child supposed to be left by fairies in exchange for one taken away by them. **elf-lock**, *n.* a tangled lock of hair. **elfstruck**, *a.* bewitched by elves. **elfin**, *a.* elfish; small, delicately pretty. *n.* a little elf; an urchin. **elfish, elvish**, *a.* like an elf; of the nature of an elf; proceeding from or caused by elves; mischievous. **elfishly, elvishly**, *adv.*

Elgin Marbles (el'gin), *n.pl.* ancient sculptured marbles brought to England in 1812, by the Earl of Elgin (1766–1841), from the Parthenon etc. at Athens.

elicit (ilis'it), *v.t.* to draw out, evoke; to extract. **elicitation**, *n.*

elide (ilīd'), *v.t.* to omit, delete; (*esp. in Gram.*) to cut off (as the last syllable). **elision** (ilizh'ən), *n.* the suppression of a letter or syllable; the suppression of a passage in a book or a discourse.

eligible (el'ijibl), *a.* fit or deserving to be chosen (as for office); desirable, suitable; (*coll.*) desirable for marriage. **eligibility** (-bil'-), *n.* **eligibly**, *adv.*

eliminate (ilim'ināt), *v.t.* to cast out, expel; to remove, get rid of; to ignore (certain considerations); (*Math.*) to cause to disappear from an equation; (*sl.*) to murder. **eliminable**, *a.* **elimination**, *n.* **eliminator**, *n.* (*Radio*) a device for supplying a battery receiving-set with electricity from the mains.

elision ELIDE.

elite (ālēt'), *n.* the pick; a group of the best (or richest or most powerful) people; a type size for typewriters of 12 characters per in. (2.54 cm). **elitism**, *n.* (*often derog.*) the favouring of the creation of an elite. **elitist**, *a.* [F]

elixir (ilik'sə), *n.* the alchemists' liquor for transmuting metal into gold; a potion for prolonging life, usu. called **elixir vitae** or **elixir of life**; a sweetened medicine.

Elizabethan (ilizəbē'thən), *a.* pertaining esp. to Queen Elizabeth I or her time; in the style characterizing the literature, architecture, dress etc., of her time. *n.* a person living at that time.

elk (elk), *n.* the largest animal of the deer family, a native of northern Europe and N America, where it is called the moose; applied also to the wapiti and the eland. **Irish elk**, a large extinct deer. **elkhound**, *n.* a large, thick-haired breed of dog from Norway. **Elks**, *n.pl.* a US fraternal society.

alces, Gr. *alkē*)]

ell (el), *n.* an obsolete measure of length, about 45 in. (114·3 cm). **give him/her an inch and he'll/she'll take an ell**, he/she will take liberties if possible.

ellagic (ilaj'ik), *a.* pertaining to gall-nuts or to gallic acid.

elleborin HELLEBORE.

ellipse (ilips'), *n.* a regular oval, a conic section formed by a plane intersecting a cone obliquely. **ellipsograph**, *n.* an instrument for describing ellipses. **ellipsis** (-sis), (*pl.* **-ses**) omission of one or more words necessary to the complete construction of a sentence; a set of three dots (...) marking such an omission. **ellipsoid** (-soid), *n.* a solid figure of which every plane section through one axis is an ellipse and every other section an ellipse or a circle. **ellipsoid,**

elm 262 **embezzle**

-oidal (elipsoi'-), *a.* **elliptic, -ical** (-tik), *a.* pertaining to an ellipse or to ellipsis; obscure, ambiguous. **elliptically**, *adv.* **ellipticity** (cliptis'-), *n.*
elm (elm), *n.* a tree of the genus *Ulmus;* the common English elm, *U. campestris.* **elmy**, *a.*
Elmo's, St Elmo's fire (el'mōz), *n.* the corposant.
elocution (eləkū'shən), *n.* art, style or manner of speaking; effective oral delivery. **elocutionary**, *a.* **elocutionist**, *n.*
eloge (ālōzh'), *n.* an encomium, a panegyric, esp. a discourse in honour of a deceased person. [F]
Elohim (ilō'him), *n.* ordinary name of God in the Hebrew Scriptures. **Elohist**, *n.* a Biblical writer or one of the writers of parts of the Hexateuch, where the word *Elohim* is habitually used for *Yahveh,* Jehovah. **Elohistic** (eləhis'-), *a.*
elongate (ē'long-gāt), *v.t.* to extend; to make longer. *v.i.* to grow longer. *a.* lengthened, extended; (*Bot., Zool.*) very slender in proportion to length. **elongation**, *n.* the act of lengthening or extending; the state of being elongated; a prolongation, an extension.
elope (ilōp'), *v.i.* to run away with a lover, with a view to marriage; to abscond. **elopement**, *n.* **eloper**, *n.*
eloquence (el'əkwəns), *n.* fluent, powerful and appropriate verbal expression, esp. of emotional ideas; eloquent language; rhetoric. **eloquent**, *a.* having eloquence; full of expression, feeling or interest. **eloquently**, *adv.*
Elsan® (el'san), *n.* a type of chemical lavatory.
else (els), *adv.* besides, other; instead; otherwise, in the other case, if not. **elsewhere**, *adv.* in or to some other place.
ELT, (*abbr.*) English Language Teaching.
elucidate (iloo'sidāt), *v.t.* to throw light on; to render intelligible; to explain. **elucidation**, *n.* **elucidative**, *a.* **elucidator**, *n.* **elucidatory**, *a.*
elude (ilood'), *v.t.* to escape from by artifice or dexterity; to evade, shirk; to remain undiscovered or unexplained by; to baffle. **elusion** (-zhən), *n.* **elusive** (-siv), *a.* **elusively**, *adv.* **elusiveness**, *n.* **elusory**, *a.*
Elul (ē'lūl), *n.* the sixth month of the Jewish ecclesiastical, and the 12th of their civil year, beginning with the new moon of our September.
elute (iloot'), *v.t.* to purify or separate by washing. **elution**, *n.* **eluant** (el'ū-), *n.* a liquid used for elution.
elutriate (iloo'triāt), *v.t.* to purify by straining or washing so as to separate lighter and heavier particles. **elutriation**, *n.*
elvan (el'vən), *n.* intrusive igneous rock penetrating sedimentary strata in Cornwall, Devon and Ireland; a vein or dike of this. **elvanite** (-nīt), *n.* **elvanitic** (-nit'-), *a.*
elver (el'və), *n.* a young eel.
elves, elvish ELF.
Elysium (iliz'iəm), *n.* the abode of the souls of Greek heroes after death; (*fig.*) a place or state of perfect happiness. **Elysian**, *a.* **the Elysian Fields**, the Greek paradise.
elytron (el'itron), *n.* (*pl.* **-tra** (-ə)) one of the horny sheaths constituting the anterior wings of beetles. **elytriform** (ilit'rifawm), *a.*
Elzevir (el'zəvə, -viə), *n.* a book printed by the Elzevirs. *a.* printed by the Elzevirs; pertaining to or resembling the type used by them. [a family of printers, of Amsterdam, 1595–1680]
em (em), *n.* the 13th letter of the alphabet, M, m; (*Print.*) the square of the body of any size of type, used as the unit of measurement for printed matter; a printers' general measure of 12 points or 1/6 in. (0·42 cm).
'em (əm), *pron.* (*coll.*) short for THEM.
em- (em-. im-), *pref.* form of EN- used before *b, p* and *m,* as in *embank, empanoply, emmarble.*
emaciate (imā'siāt), *v.t.* to cause to lose flesh or become (too) lean; to impoverish. **emaciation**, *n.*
emanate (em'ənāt), *v.i.* to issue or flow as from a source, to originate; to proceed (from). *v.t.* to emit, send forth.

emanation, *n.* the act of emanating; that which emanates; the theory that all things are outflowings from the essence of God; a product of radioactive decay. **emanative**, *a.*
emancipate (iman'sipāt), *v.t.* to release from bondage, slavery, oppression, or legal, social or moral restraint; to set free. **emancipation**, *n.* **emancipationist**, *n.* an advocate of emancipation of slaves. **emancipator**, *n.* **emancipatory**, *a.* **emancipist**, *n.* (*Austral. Hist.*) a convict who had served his/her term.
emarginate (imah'jināt), *v.t.* to take away the edge or margin of. **emarginate** (-nət), **-nated** (-nātid), *a.* with the margin notched; (*Bot.*) notched at the apex. **emargination**, *n.*
emasculate (imas'kūlāt), *v.t.* to castrate; to deprive of strength or vigour; to deprive (as language) of force or energy; to enfeeble (a literary work) by undue expurgation or excision. *a.* (-lət), castrated; effeminate, weak. **emasculation**, *n.* **emasculative**, *a.* **emasculatory**, *a.*
embalm (imbahm'), *v.t.* to preserve (as a body) from putrefaction by means of spices and aromatic drugs; (*fig.*) to preserve from oblivion. **embalmer**, *n.* **embalmment**, *n.*
embank (imbangk'), *v.t.* to confine or defend with a bank, dike etc. **embankment**, *n.* the act or process of embanking; a bank or stone structure for confining a river etc. or carrying a road etc.
embargo (imbah'gō), *n.* (*pl.* **-goes**) a government prohibition upon the arrival or departure of vessels from ports; a complete suspension of foreign commerce or of a particular branch of foreign trade; a prohibition or restraint, as on publication. *v.t.* to lay an embargo upon; to requisition, seize; to prohibit, forbid.
embark (imbahk'), *v.t.* to put on board ship. *v.i.* to go on board ship; to engage or enter (upon any undertaking); to invest (in an enterprise). **embarkation** (embah-), *n.* the act of putting or going on board ship.
embarras de (ābara' də), a perplexing amount or number of, as *embarras de richesse* (rēshes), of anything; *embarras de choix* (-shwa), of things to choose from. [F]
embarrass (imbar'əs), *v.t.* to encumber, entangle; to disconcert, make self-conscious; to complicate; to involve in pecuniary difficulties. **embarrassing**, *a.* **embarrassingly**, *adv.* **embarrassment**, *n.*
embassy (em'bəsi), *n.* the function, office or mission of an ambassador; the body of persons sent as ambassadors; an ambassador and his suite; the official residence of an ambassador.
embattle[1] (imbat'l), *v.t.* to array in order of battle; to fortify.
embattle[2] (imbat'l), *v.t.* to furnish with battlements.
embay (imbā'), *v.t.* to enclose (a vessel) in a bay; to confine, enclose; to form into a bay. **embayment**, *n.*
embed (imbed'), *v.t.* (*past, p.p.* **embedded**) to lay as in a bed; to set firmly in surrounding matter; to enclose firmly (said of the surrounding matter).
embellish (imbel'ish), *v.t.* to beautify, adorn; to add incidents or imaginary accompaniments so as to heighten a story. **embellishment**, *n.*
ember (em'bə), *n.* a smouldering piece of coal or wood; (*pl.*) smouldering remnants of a fire or (*fig.*) of passion, love etc.
Ember days (em'bə), *n.pl.* certain days set apart for fasting and prayer; the Wednesday, Friday and Saturday next following the first Sunday in Lent, Whit-Sunday, Holy Cross Day (14 Sept.), and St Lucy's Day (13 Dec.). **Ember-tide**, *n.* the season at which Ember days occur. **Ember weeks**, *n.pl.* the weeks in which the Ember days fall.
ember-goose (em'bəgoos), *n.* the northern diver or loon, also called **ember-diver**.
embezzle (imbez'l), *v.t.* to appropriate fraudulently what is committed to one's care. *v.i.* to commit embezzlement. **embezzlement**, *n.* **embezzler**, *n.*

embitter   263   émeute

embitter (imbit'ə), *v.t.* to make bitter, or more bitter; to aggravate; to render more hostile. **embitterment**, *n.*

emblazon (imblä'zən), *v.t.* to adorn with heraldic figures or armorial designs; to decorate; to make brilliant; to celebrate, praise. **emblazoner**, *n.* **emblazoning, emblazonment**, *n.* **emblazonry** BLAZONRY.

emblem (em'bləm), *n.* a symbolic figure; a representation of an object symbolizing some other class, action or quality, as a crown for royalty; a type, personification; a heraldic device. *v.t.* to represent by an emblem. **emblematic, -ical** (-mat'-), *a.* **emblematically**, *adv.* **emblematist** (-blem'-), *n.* a writer of allegories or inventor of emblems. **emblematize, -ise**, *v.t.* to represent by or as an emblem; to symbolize. **emblematology** (-tol'-), *n.*

emblement (em'bləmənt), *n. (usu. in pl.) (Law)* growing crops annually produced by the cultivator's labour, which belong to the tenant; sometimes extended to the natural products of the soil.

emblossom (imblos'əm), *v.t.* to cover with blossoms.

embody (imbod'i), *v.t.* to incarnate or invest with a material body; to express in a concrete form; to be a concrete expression of; to incorporate, include. **embodier**, *n.* **embodiment**, *n.*

embog (imbog'), *v.t.* to encumber in or as in a bog.

embogue (imbōg'), *v.i.* to disembogue.

embolden (imbōl'dən), *v.t.* to give boldness to; to encourage.

embolism (em'bəlizm), *n.* an interpolation or insertion, e.g. days, months or years into the calendar; anything interpolated or inserted; partial or total blocking-up of a blood-vessel by a clot of blood, air bubble etc. **embolic** (-bol'ik), *a.* **embolus** (-ləs), *n.* the substance which causes embolism.

embonpoint (ēbōpwī'), *n.* plumpness of person or figure. *a.* well-nourished; stout. [F]

emborder (imbaw'də), *v.t.* to adorn or furnish with a border.

embosom (imbuz'm), *v.t.* to place or hold in or as in the bosom of anything; to enclose; to cherish.

emboss (imbos'), *v.t.* to engrave or mould in relief; to decorate with bosses or raised figures; to cause to stand out in relief. **embossment**, *n.*

embouchure (ēbooshooə'), *n.* the mouth of a river etc.; *(Mus.)* the mouthpiece of a wind instrument; the shaping of the lips to the mouthpiece.

embowel (imbow'əl), *v.t.* to disembowel. **embowelment**, *n.*

embower (imbow'ə), *v.t.* to enclose in or as in a bower.

embox (imboks'), *v.t.* to shut in or as in a box.

embrace (imbrās'), *v.t.* to enfold in the arms; to clasp and hold fondly; to enclose, surround; to include; to accept eagerly; to take in with the eye, to comprehend. *v.i.* to join in an embrace. *n.* a clasping in the arms. **embraceable**, *a.* **embracement**, *n.* **embracer**[1], *n.* **embracingly**, *adv.* **embracingness**, *n.* **embracive**, *a.*

embracer[2] (imbrā'sə), *n. (Law)* one who endeavours to corrupt a jury by threats etc. **embracery**, *n.*

embranchment (imbrahnch'mənt), *n.* a branching out; a ramification.

embrangle (imbrang'gl), *v.t.* to entangle, to confuse, perplex. **embranglement**, *n.*

embrasure (imbrā'zhən), *n.* an opening in a parapet or wall to fire guns through; the inward bevelling or splaying of the sides of a window or door.

embrave (imbrāv'), *v.t.* to embolden.

embreathe (imbrēdh'), *v.t.* to breathe into; to breathe in.

embrocate (em'brəkāt), *v.t.* to moisten or foment (as a diseased or injured part of the body). **embrocation**, *n.* the act of bathing or fomenting; the liquid used.

embroglio (imbrō'liō), IMBROGLIO.

embroider (imbroi'də), *v.t.* to ornament with figures or designs in needlework; to embellish with additions, esp. a narrative with exaggerations. **embroiderer**, *n.* **embroidery**,

n. the act, process or art of embroidering; the fabric ornamented; additional embellishment; exaggeration or fiction added to a narrative.

embroil (imbroil'), *v.t.* to throw into confusion; to entangle; to involve (someone) in a quarrel or contention. **embroilment**, *n.*

embrown (imbrown'), *v.t.* to make brown; to darken, obscure.

embrue (imbroo'), IMBRUE.

embryectomy (embriek'təmi), *n.* the operation of removing the foetus through an incision in the abdomen.

embryo (em'briō), *n. (pl.* **-bryos**) the unborn offspring; the developing foetus; the rudimentary plant in the seed after fertilization; the beginning or first stage of anything. *a.* in the germ, undeveloped; rudimentary. **in embryo**, in the first or earliest stage; in a rudimentary or undeveloped state. **embryonic** (-on'ik), *a.*

embry(o)-, *comb. form.* of or pertaining to the embryo or embryos.

embryogenesis (embriōjen'isis, embriōjənē'sis), **embryogeny** (-oj'əni), **embryogony** (embriog'əni), *n.* the formation of an embryo.

embryology (embriol'əji), *n.* the science of the formation and development of the embryo. **embryological** (-loj'-), *a.* **embryologist**, *n.*

embryotomy (embriot'əmi), *n.* a cutting up of an embryo or foetus in the uterus.

embus (imbūs'), *v.t., v.i. (past, p.p.* **embussed**) *(Mil.)* to put (or to mount troops) into omnibuses for transport.

emend (imend'), *v.t.* to correct, remove faults; to improve (as a text by editing). **emendable**, *a.* **emendation** (ēmendā'-), *n.* **emendator** (ē'men-), *n.* **emendatory** (imen'də-), *a.*

emerald (em'ərəld), *n.* a variety of beryl, distinguished by its beautiful green colour; the colour of this. *a.* of a bright green colour. **emerald-copper**, *n.* dioptase. **the Emerald Isle**, Ireland.

emerge (imœj'), *v.i.* to rise up, appear (as from under water, below the horizon or from a place of concealment); to come out (as facts on an enquiry); to become apparent; to issue from a state of depression, suffering or obscurity. **emergence**, *n.* **emergent**, *a.* coming into being; of a country etc., having recently acquired independence. **emerging**, *a.*

emergency (imœ'jənsi), *n.* a sudden occurrence or situation demanding immediate action; a crisis. **emergency exit**, *n.* a door specially provided for exit in case of fire or other contingency. **emergency landing**, *n.* a forced descent by a plane due to engine trouble etc.

emeritus (ime'ritəs), *a. (placed after the noun)* having served one's term of office, as *professor emeritus. n. (pl.* **-ti** (-tī)) one who has served his time and retired from any office.

emerods (em'ərods), *n.pl. (Bibl.)* haemorrhoids.

emersion (imœ'shən), *n.* the act of emerging; the reappearance of a heavenly body at the end of an eclipse or occultation.

emery (em'əri), *n.* a hard, coarse variety of corundum, black or greyish-black in colour, used for polishing. **emery board**, *n.* a strip of card or wood, coated with crushed emery and used to file fingernails. **emery-cloth, -paper**, *n.* cloth or paper dusted with powdered emery. **emery-wheel**, *n.* one faced with emery, used for grinding and polishing metal articles.

emesis EMETIC.

emetic (imet'ik), *a.* inducing vomiting. *n.* a preparation for causing vomiting. **emetically**, *adv.* **emesis** (em'əsis), *n.* the action of vomiting. **emetine** (em'ətin, -tin), *n.* an alkaloid obtained from ipecacuanha, causing vomiting. **emetocathartic** (emitōkəthah'tik), *a.* producing vomiting and purging.

émeute (imūt'), *n.* a riot or popular disturbance. [F]

EMF 264 **empyreuma**

EMF, (*abbr.*) electromotive force.
emiction (imik'shən), *n.* the discharge of urine. **emictory**, *n., a.*
emigrate (em'igrāt), *v.i.* to leave one's native country in order to settle in another. **emigrant**, *a.* **emigration**, *n.* **emigrationist**, *n.* an advocate or promoter of emigration. **emigratory**, *a.*
émigré (em'igrā), *n.* an emigrant, esp. one of the royalists who left France at the time of the French Revolution. [F]
eminent (em'inənt), *a.* rising above others; high, lofty, prominent; distinguished. **eminent domain**, *n.* the right of the State to confiscate private property for public use.
eminence, -nency, *n.* loftiness, height; a part rising above the rest, or projecting above the surface; high rank; distinction (**Eminence**) a title of honour applied to cardinals. **eminence grise** (eminās grēz'), *n.* (F) a person in the background exercising power unofficially. **eminently**, *adv.*
emir (imiə'), *n.* in the Middle East and N Africa, a prince, chieftain; a title given to the descendants of Mohammed through Fatima, his daughter.
emissary (em'isəri), *n.* a messenger or agent, esp. one sent on a secret mission; (*Physiol.*) an excretory vessel. *a.* of, or serving as, an emissary.
emission (imish'ən), *n.* the act or process of emitting or being emitted; that which is emitted; the act of issuing bank-notes etc.; the number and value of the notes etc. sent out; light or other radiation emitted from a source. **emissive, emissory**, *a.* **emissivity** (emisiv'-), *n.*
emit (imit'), *v.t.* to give out, give vent to, discharge; to print and send into circulation (as bank-notes). **emitter**, *n.* an electrode of a transistor.
emmarble (imah'bl), *v.t.* to turn into marble; to decorate with marble.
emmenagogue (imen'əgog), *n.* a medicine that induces or restores the menses. **emmenology** (emənol'-), *n.*
Emmental, -thal (em'əntahl), *n.* a Swiss cheese with holes in it, made in the Emmenthal valley.
emmet (em'it), *n.* an ant.
Emmy (em'i), *n.* (*pl.* **Emmys, Emmies**) the television equivalent of an Oscar, awarded by the American Academy of Television Arts and Sciences.
emollient (imol'iənt), *a.* making soft or supple, esp. the skin; soothing. *n.* an emollient substance; (*fig.*) anything intended to soothe or comfort.
emolument (imol'ūmənt), *n.* the profit arising from any office or employment; remuneration. **emolumentary** (-men'-), *a.*
emotion (imō'shən), *n.* a strong feeling of any kind, whether of pain or pleasure; excitement. **emote**, *v.i.* to show or express exaggerated emotion as in acting. **emotional**, *a.* pertaining to emotion; easily affected with emotion. **emotionalism**, *n.* **emotionalist**, *n.* **emotionality** (-nal'-), *n.* **emotionally**, *adv.* **emotionless**, *a.* **emotive**, *a.* emotional; tending to produce emotion. **emotively**, *adv.*
emove (imoov'), *v.t.* to affect with emotion.
Emp., (*abbr.*) Emperor; Empire; Empress.
empanel (impan'əl), *v.t.* to enter on the list of jurors; to enrol as a jury.
empathy (em'pəthi), *n.* the power of entering into the feelings of another person, and understanding them fully; the losing of one's identity in, e.g. a work of art. **empathetic** (-thet'-), **empathic**, *a.* **empathize, -ise**, *v.t.*
emperor (em'pərə), *n.* the sovereign of an empire; the highest dignity (superior to king). **purple emperor**, *Apatura iris*, a large and handsome British butterfly. **emperor moth**, *n.* a large and beautiful British moth. **emperor penguin**, *n.* the large penguin, *Aptenodytes forsteri*. **emperorship**, *n.* **empress**, *n. fem.*
emphasis (em'fəsis), *n.* (*pl.* **-ses** (-sēz)) a particular stress laid upon a word or words, to indicate special significance; intensity of expression, language, gesture etc.; accent; prominence. **emphasize, -ise**, *v.t.* to pronounce with emphasis; to make more distinct, prominent or impressive. **emphatic, -ical** (imfat'-), *a.* bearing emphasis or special stress; accentuated, forcible, striking; positive. **emphatically**, *adv.*
emphractic (imfrak'tik), *a.* having the quality of closing the pores of the skin. *n.* an emphractic medicine.
emphysema (emfisē'mə), *n.* enlargement of the air-sacs in the lungs, thereby causing breathing difficulties. **emphysematous**, *a.*
empire (em'piə), *n.* a region or group of states ruled by an emperor; a group of industrial or commercial concerns under the control of one person; absolute power. *a.* indicating the style of costume and furniture of the First or Second French Empire. **British Empire**, former name of the British Commonwealth of Nations (see COMMONWEALTH). **Eastern Empire**, the Greek or Byzantine Empire (AD 395-1453). **the Empire**, the British Empire, the first Napoleonic Empire (1804-15); the Holy Roman Empire. **the Second Empire**, the empire of Napoleon III (1852-70). **Western Empire** WEST. **empire-builder**, *n.* a person who seeks added power and authority, esp. by increasing the number of his/her staff. **Empire Day** COMMONWEALTH DAY. **empire gown**, *n.* a high-waisted gown after the style of those worn during the First French Empire. **Empire State**, *n.* the state of New York.
empiric, -ical (impi'rik, -əl), *a.* founded on experience or observation, not theory; acting on this; of medical treatment, based on experience rather than formal training. *n.* one who relies solely on experience or observation; a medical practitioner without scientific training. **empirically**, *adv.* **empiricism** (-sizm), *n.* **empiricist** (-sist), *n.*
emplacement (implās'mənt), *n.* a setting in position; (*Fort.*) a platform for guns.
emplane (implān'), *v.t., v.i.* to go or put on board an aeroplane.
employ (imploi'), *v.t.* to use, exercise; to use the skills or labour of another in return for payment; to spend or pass (time etc.) in any occupation. *n.* employment. **employable**, *a.* **employee** (imploi'ē, emploiē'), *n.* one who is employed regularly for salary or wages. **employer**, *n.* one who employs people for salary or wages. **employment**, *n.* the act of employing; the state of being employed; regular occupation, trade or profession. **employment agency**, *n.* one used by people looking for work and employers seeking employees.
emplume (imploom'), *v.t.* to adorn with or as with plumes.
empoison (impoi'zn), *v.t.* to mix poison with; to envenom; to taint, corrupt; to render hostile.
emporium (impaw'riəm), *n.* (*pl.* **-ria** (-iə)) a mart; (*coll..*) a large shop where many kinds of goods are sold.
empower (impow'ə), *v.t.* to authorize; to enable.
empress EMPEROR.
empressement (ēpres'mē), *n.* cordiality, goodwill, eagerness. [F]
empty (emp'ti), *a.* void, containing nothing; devoid (of); vacant, unoccupied; unloaded; unsubstantial; senseless; without intelligence, ignorant; hungry, unsatisfied. *n.* an empty box, bottle etc. *v.t.* to remove the contents from; to remove from a receptacle (into another); to pour out, discharge. *v.i.* to become empty; to discharge (as a river). **empty-handed**, *a.* bringing nothing; carrying away nothing. **empty-headed**, *a.* silly, witless. **emptier**, *n.* **emptily**, *adv.* **emptiness**, *n.*
empyema (empiē'mə), *n.* a collection of pus, esp. in the chest.
empyrean (empirē'ən), *n.* the highest and purest region of heaven, where the element of fire was supposed by the ancients to exist; the upper sky. *a.* pertaining to the highest heaven. **empyreal** (-pi'ri-), *a.*
empyreuma (empiroo'mə), *n.* (*pl.* **-mata** (-tə)) the disagreeable smell and taste produced when animal or vegetable

EMS

substances are burnt. **empyreumatic, -ical** (-mat'-), *a.* **empyreumatize, -ise,** *v.t.*
EMS, (*abbr.*) European Monetary System.
emu (ē'mū), *n.* a large Australian flightless bird of the genus *Dromaeus*, resembling the cassowary and ostrich.
emu-wren, *n.* a small Australian bird, *Stipiturus malachurus*, having the tail feathers somewhat resembling those of the emu.
emulate (em'ūlāt), *v.t.* to try to equal or excel, esp. by imitation; to rival. **emulation,** *n.* **emulative,** *a.* **emulatively,** *adv.* **emulator,** *n.*
emulgent (imūl'jənt), *a.* milking or draining out.
emulous (em'ūləs), *a.* desirous of emulating; engaged in rivalry or competition; desirous of fame or honour. **emulously,** *adv.* **emulousness,** *n.*
emulsion (imūl'shən), *n.* a colloidal suspension of one liquid in another; a light-sensitive substance held in suspension in collodion or gelatine, used for coating plates or films. *v.t.* to apply emulsion paint. **emulsion paint,** *n.* a water-thinnable paint made from an emulsion of a resin in water. **emulsify** (-sifi), *v.t.* to convert into an emulsion. **emulsification** (-fi-), *n.* **emulsifier,** *n.* an emulsifying agent, esp. one that prevents separation in processed foods. **emulsionize, -ise,** *v.t.* **emulsive,** *a.*
emunctory (imūngk'təri), *a.* serving to wipe the nose; serving to carry noxious or useless particles out of the body. *n.* an excretory duct.
emys (em'is), *n.* (*pl.* **emydes** (-idēz)) (*Zool.*) the freshwater tortoise.
en (en), *n.* the 14th letter of the alphabet, N, n; (*Print.*) the unit of measurement for casting-off copy, an en being the average width of a letter.
en-, *pref.* (1) in, on, into, upon; as in *encamp*; (2) to cause to be, as in *enslave*; (3) with intensive meaning, as in *heighten*.
-en (-ən), *suf.* (1) diminutive, as in *chicken, maiden*; (2) noting the feminine, as in *vixen*; (3) pertaining to, made of, of the nature of, as in *earthen, flaxen*; (4) forming pl., as in *oxen*; (5) forming verbs from adjectives, as *deepen, fatten*; (6) forming p.p. of strong verbs, as *bounden, spoken*.
enable (inā'bl), *v.t.* to make able; to authorize, empower (to); to supply with means (to do any act). **enabling act,** *n.* legislation conferring specified powers on a person or organization.
enact (inakt'), *v.t.* to decree; to pass, as a bill into a law; to represent, play. **enacting clauses,** *n.pl.* clauses in a bill which contain new enactments. **enaction,** *n.* **enactive,** *a.* **enactment,** *n.* **enactor,** *n.* **enactory,** *a.*
enallage (inal'əjē), *n.* a substitution of one word, tense, number etc. for another.
enamel (inam'əl), *n.* a vitreous material with which metal, porcelain etc. are coated by fusion, for decorative or preservative purposes; an article coated with enamel; any smooth, hard, glossy coating; a lacquer, varnish, paint, cosmetic; the ivory-like substance which covers the surface of the teeth. *v.t.* (*past, p.p.* **enamelled**) to coat, paint, encrust or inlay with enamel; to form a smooth glossy surface upon; to decorate with various colours. *v.i.* to practise the art of enamelling. **enameller,** *n.* **enamellist,** *n.*
enamour (inam'ə), *v.t.* to captivate, charm; to inflame with love. **enamoured,** in love; fond (of).
enantiosis (inantiō'sis), *n.* a figure of speech by which one says (usually ironically) the reverse of what one means.
enarch (inahch'), *v.t.* to arch over; (*Hort.*) to inarch.
enarthrosis (enahthrō'sis), *n.* a ball-and-socket joint. **enarthrodial** (-diəl), *a.*
en avant (ën avä'), forward. [F]
en bloc (ē blok'), as a unit, all together. [F]
encaenia (insē'niə), *n.pl.* a festival of commemoration; the annual commemoration of founders and benefactors of Oxford Univ.

enclose

encage (inkāj'), *v.t.* to shut in or as in a cage.
encamp (inkamp'), *v.i., v.t.* to settle or cause to settle temporarily in tents. **encampment,** *n.* the act of encamping; a camp.
encapsulate (inkap'sūlāt), *v.t.* to enclose in a capsule; to capture the essence of; to put in a shortened form. **encapsulation,** *n.*
encarnalize, -ise (inkah'nəliz), *v.t.* to make carnal; to embody.
encase (inkās'), *v.t.* to put into or as into a case; to enclose. **encasement,** *n.*
encash (inkash'), *v.t.* to convert into or exchange for cash. **encashable,** *a.* **encashment,** *n.*
encaustic (inkaw'stik), *n., a.* (of) a mode of painting in which the colours (coloured clay or wax) are fixed by heat (now chiefly of painting on vitreous or ceramic ware). **encaustic brick, tile,** *n.*
-ence (-əns), *suf.* forming abstract nouns, as *existence, corpulence*.
enceinte (ësit'), *a.* pregnant. *n.* the space within the ramparts of a fortification.
encephalitis (insefəli'tis), *n.* inflammation of the brain. **encephalitis lethargica** (lithah'jikə), *n.* acute inflammation of the brain, commonly called sleepy sickness. **encephalitic** (-lit'-), *a.*
encephalocele (insef'ələsēl), *n.* hernia of the brain.
encephalography (insefəlog'rəfi), *n.* radiography of the brain. **encephalograph** (insef'ələgrahf), *n.*, **-gram,** *n.* an X-ray of the brain.
encephalon (insef'əlon), *n.* (*pl.* **-la** (-ə)) the brain. **encephalic** (ensəfal'ik), *a.* **encephaloid** (-loid), *a.* pertaining to or resembling brain matter. **encephalous,** *a.* having a distinct brain or head.
encephalopathy (insefəlop'əthi), *n.* any disease which affects the functioning of the brain. **encephalopathic** (-path'-), *a.*
encephalotomy (insefəlot'əmi), *n.* dissection of the brain.
enchafe (inchāf'), *v.t.* to make hot; to irritate.
enchain (inchān'), *v.t.* to bind with chains; to hold fast, (attention etc.). **enchainment,** *n.*
enchant (inchahnt'), *v.t.* to influence by magic, bewitch; to fascinate, charm, delight. **enchanter,** *n.* a magician; one who delights or fascinates. **enchanter's nightshade,** *n.* a woodland plant of the genus *Circaea*, esp. *C. lutetiana*. **enchantingly,** *adv.* **enchantment,** *n.* **enchantress,** *n. fem.*
encharge (inchahj'), *v.t.* to commission (with).
enchase (inchās'), *v.t.* to set or encase within any other material, as a gem in precious metal; to adorn with embossed work.
enchilada (enchilah'də), *n.* a Mexican dish of a meat-filled tortilla served with chilli sauce.
enchiridion (enkirid'iən), *n.* a small guide or book of reference.
encincture (insingk'chə), *n., v.t.* (to surround with or as with) a ring or girdle.
encircle (insœ'kl), *v.t.* to enclose or surround (with); to take up a position round; to encompass. **encirclement,** *n.*
en clair (ē kleə'), *a.* of telegrams etc., not in code or cipher. [F]
enclasp (inklahsp'), *v.t.* to enfold in a clasp.
enclave (en'klāv), *n.* a territory completely surrounded by that of another power; an enclosure. **enclavement** (-klāv'-), *n.*
enclitic (inklit'ik), *a.* (*Gr. Gram.*) applied to a word which is pronounced as part of the preceding word, on which it throws its accent, e.g. *thee* in *prithee*. *n.* an enclitic word or particle. **enclitically,** *adv.*
enclose (inklōz'), *v.t.* to shut in; to surround or hem in on all sides; to surround by a fence; to put one thing inside another for transmission or carriage; to contain. **enclosed order,** *n.* a Christian contemplative order which does not allow its members to go into the outside world. **encloser,**

n. **enclosure** (-zhə), *n.* the act of enclosing; that which is enclosed; a space of ground enclosed or fenced in; that which encloses, as a fence.
encode (inkōd'), *v.t.* to translate into code.
encomiast (inkō'miast), *n.* one who composes an encomium. **encomiastic** (-as'-), *a.* **encomiastically**, *adv.*
encomium (inkō'miəm), *n.* (*pl.* **-miums, -mia**) a formal eulogy or panegyric; high commendation.
encompass (inkŭm'pəs), *v.t.* to bring about; to surround; to include. **encompassment,** *n.*
encore (ong'kaw), *adv.* again, once more; used as a call for a repetition at a concert, theatre etc. *n.* a demand for a repetition of a song etc.; the repetition itself. *v.t., v.i.* to call for a repetition (of).
encounter (inkown'tə), *v.t.* to meet face to face; to meet in battle; to confront resolutely. *n.* a hostile meeting, fight; an unplanned or unexpected meeting. **encounter group,** *n.* a group of people who meet to develop self-awareness and understanding of others by frank exchange of feelings, opinions and contact. **encounterer,** *n.*
encourage (inkŭ'rij), *v.t.* to give courage or confidence to; to animate; to urge, (to do); to stimulate, promote, foster. **encouragement,** *n.* **encourager,** *n.* **encouragingly,** *adv.*
encrimson (inkrim'zən), *v.t.* to make crimson.
encrinite (en'krinīt), *n.* a fossil crinoid. **encrinal, encrinic** (-krīn'-), **encrinital** (-ni'-), *a.*
encroach (inkrōch'), *v.i.* to intrude gradually or stealthily (upon) what belongs to another; to infringe (upon). **encroacher,** *n.* **encroachingly,** *adv.* **encroachment,** *n.*
encrust (inkrŭst'), *v.t.* to cover with a crust or hard coating; to form a crust upon the surface of; to apply a decorated surface to. **encrustation** (enkrəstā'shən), **encrustment** INCRUSTATION.
encrypt (inkript'), *v.t.* to put into code. **encryption,** *n.*
encumber (inkŭm'bə), *v.t.* to hamper, impede or embarrass by a weight, burden or difficulty; to weigh down with debt. **encumberment,** *n.* **encumbrance,** *n.* a hindrance to freedom of action or motion; a burden, a hindrance; (*Law*) a liability upon an estate, such as a mortgage, a claim etc. **encumbrancer,** *n.* (*Law*) one who holds an encumbrance.
encurtain (inkœ'tin), *v.t.* to enwrap or veil with or as with a curtain.
-ency (-ənsi), *suf.* forming nouns of state or quality.
encyclic (insīk'lik, -sī'-), **-ical,** *a.* sent about to many persons or places. *n.* a circular letter, esp. a letter from the Pope to all bishops.
encyclopaedia, (*esp. N Am.* **-pedia**) (insīklǝpē'diǝ), *n.* a book containing information on all branches of knowledge, or on a particular branch, usually arranged alphabetically; a general system of knowledge or instruction. **encyclopaedian, encyclopaedic, -ical,** *a.* **encyclopaedism,** *n.* the possession of a large range of knowledge and information. **encyclopaedist,** *n.* a compiler of an encyclopaedia; one who has acquired an extensive range of knowledge or information.
encyst (insist'), *v.t.* to enclose in a cyst, bladder or vesicle. **encystation** (en-), *n.* **encystment,** *n.*
end (end), *n.* the extreme point or boundary of a line or of anything that has length; the termination, limit or last portion; the last part of a period; the conclusion of a state or action; a ceasing to exist; abolition; death; the cause of death; a purpose, a designed result; a reason for (a thing's) existence, a final cause; (*usu. in pl.*) a remnant. *a.* final; farthest; last. *v.i.* to cease; to result (in). *v.t.* to bring to an end; to destroy; **at a loose end,** (*coll.*) temporarily disengaged. **at one's wits' end,** bewildered, utterly perplexed, nonplussed. **at the end of one's tether,** unable to do anything more. **end on,** with the end pointing towards one. **in the end,** finally. **no end,** plenty, many. **odds and ends,** odd remnants. **on end,** upright, erect. **the be all and end all,** the sole aim. **the end of the road,** the point beyond which one can no longer go on or survive. **the ends of the earth,** the remotest parts of the earth. **to come to an end,** to be finished, exhausted. **to go off the deep end,** to lose one's temper. **to keep one's end up,** (*coll.*) to stand one's ground. **to make both ends meet,** to keep expenditure within income. **to put an end to,** to terminate; to abolish. **to that end,** for that purpose. **without end,** everlasting; very long; inexhaustible. **wrong end of the stick,** the contrary to what is meant. **end-game,** *n.* the last part of a game of chess etc., when only a few pieces remain in play. **end-paper,** *n.* a blank page placed between the cover and the body of a book. **end product,** *n.* the final product obtained after a series of processes. **end result,** *n.* the final outcome. **end-stopped,** *a.* (*Pros.*) having a pause in sense at the end of a line of poetry. **end-user,** *n.* the person, firm etc. in receipt of a manufactured product being sold. **ending,** *n.* a conclusion, termination; the terminating syllable of a word in grammar. **endless,** *a.* having no end; infinite, unlimited, perpetual; incessant. **endless band, cable** or **chain,** *n.* a band with ends fastened together for conveying mechanical motion. **endless screw,** *n.* a screw conveying motion to a wheel in the teeth of which the threads engage. **endlessly,** *adv.* **endlessness,** *n.* **endlong,** *adv.* lengthwise as distinguished from crosswise; straight along. **endmost,** *a.* the nearest to the end, the furthest. **endways,** *adv.* on end; with the end foremost or uppermost; end to end; lengthwise. **endwise,** *adv.*
end- END(O)-.
endamage (indam'ij), *v.t.* to damage; to prejudice.
endanger (indān'jə), *v.t.* to expose to danger, to put in hazard. **endangered,** *a.* in danger, esp. of extinction.
endear (indiə'), *v.t.* to make dear (to); to cause to be loved. **endearing,** *a.* **endearingly,** *adv.* **endearment,** *n.* an affectionate word or phrase.
endeavour (indev'ə), *v.i.* to strive (after) a certain end; to try (to). *n.* an effort, attempt to attain some object.
endeictic (indīk'tik), *a.* showing, exhibiting. **endeixis** (-ksis), *n.* a symptom.
endemic (indem'ik), *a.* peculiar to a particular locality or people. *n.* an endemic disease. **endemic disease,** *n.* one common from local causes in a particular district or among a particular people. **endemically,** *adv.* **endemicity** (endǝmis'-), *n.* **endemiology** (indēmiol'ǝji), *n.* the study of endemic diseases.
endermic (indœ'mik), *a.* acting upon or through the skin, as an unguent. **endermically,** *adv.*
enderon (en'dǝron), *n.* (*Physiol.*) the inner derm or true skin.
en déshabille (ē dāzabēy''), in a state of undress. [F]
endive (en'div, -dīv), *n.* a kind of chicory, *Cichorium endivia*, much cultivated for use in salads.
end(o)-, *comb. form.* pertaining to the inside of anything.
endocardium (endǝkah'diǝm), *n.* a membrane lining the human heart. **endocardiac** (-ak), *a.* **endocarditis** (-dī'tis), *n.* inflammation of the endocardium.
endocarp (en'dǝkahp), *n.* (*Bot.*) the inner layer of a pericarp.
endocrine (en'dǝkrīn, -krin), *n.* the internal secretion of a gland. **endocrine gland,** *n.* gland secreting directly into the blood-stream. **endocrinology** (-krinol'ǝji), *n.* the study of the secretions of the endocrine glands. **endocrinologist,** *n.*
endoderm (en'dǝdœm), *n.* the inner layer of the blastoderm.
endogamous (indog'ǝmǝs), *a.* marrying within the tribe. **endogamy,** *n.* the custom of taking a wife only within the tribe; pollination between two flowers on the same plant.
endogenous (endoj'inǝs), *a.* growing or developing from within. **endogeny,** *n.*
endometrium (endǝmē'triǝm), *n.* the membrane lining the cavity of the womb. **endometritis** (-mitrī'tis), *n.* in-

flammation of the endometrium.
endomorph (en'dəmawf), *n.* a person of plump, thick-set build. **endomorphic,** *a.*
endoparasite (endōpa'rəsit), *n.* (*Zool.*) a parasite living in the interior of its host. **endoparasitic** (-sit'-), *a.*
endoplasm (en'dəplazm), *n.* the partially fluid inner layer of protoplasm. **endoplasmic,** *a.*
endorhiza (endəri'zə), *n.* (*Bot.*) the sheath-enclosed radical of the embryo in many monocotyledonous plants. **endorhizal, -rhizous,** *a.*
endorphin (indaw'fin), *n.* any of a group of chemicals occurring in the brain which have a similar effect to morphine.
endorse (indaws'), *v.t.* to write one's name, on the back of (a cheque); to transfer ownership by signing the back of a document; to ratify, approve; to record a conviction on (an offender's driving licence). **to endorse over,** to transfer one's rights in (a bill etc.) to another person. **endorsee** (-sē'), *n.* the person to whom a bill etc. is assigned by endorsing. **endorsement,** *n.* **endorser,** *n.*
endosarc (en'dōsahk), *n.* endoplasm.
endoscope (en'dəskōp), *n.* an instrument for inspecting internal parts of the body. **endoscopy** (endos'kəpi), *n.*
endoskeleton (endəskel'itən), *n.* the internal bony and cartilaginous framework of the vertebrates.
endosmose (indoz'mōs), **endosmosis** (-mō'sis), *n.* the passage of a fluid from outside inwards through a porous diaphragm. **endosmotic** (-mot'-), *a.*
endosperm (en'dəspœm), *n.* the albumen of a seed. **endospermic** (-spœ'-), *a.*
endothelium (endəthē'liəm), *n.* a membrane lining blood-vessels, tubes, cavities etc.
endow (indow'), *v.t.* to invest (with qualities etc.); to bestow a permanent income upon; to give a dowry to. **endowment,** *n.* the act of endowing; the fund or property with which an institution etc. is endowed; (*pl.*) natural gifts or ability. **endowment assurance,** *n.* an assurance to provide a fixed sum at a specified age or on death before that age.
endozoic (endəzō'ik), *a.* denoting the method of seed-dispersal by being swallowed by an animal and then passed out in its excreta.
endue (indū'), *v.t.* to provide or invest (with); (*usu.* in *p.p.*) to endow, to furnish.
endure (indūə'), *v.t.* to bear, to stand (a test or strain); to undergo, suffer; to submit to. *v.i.* to last; to bear sufferings with patience and fortitude. **endurability** (-bil'-), *n.* **endurable,** *a.* **endurableness,** *n.* **endurance,** *n.* the act or state of enduring; the ability to endure; continuance, duration. **endurer,** *n.* **enduring,** *a.* bearing; durable, permanent. **enduringly,** *adv.* **enduringness,** *n.*
endways, endwise END.
ENE, (*abbr.*) East North East.
-ene, *suf.* (*Chem.*) denoting a hydrocarbon, such as *benzene, naphthalene.*
en effet (on efā'), in effect. [F]
enema (en'əmə), *n.* an injection of fluid into the rectum; the fluid injected.
enemy (en'əmi), *n.* one who hates or is hated; an adversary; a hostile or harmful person or force. **the Enemy,** the Devil. **how goes the enemy?** (*coll.*) what is the time?
energetic ENERGY.
energumen (enəgū'mən), *n.* one possessed by an evil spirit; an enthusiast, fanatic.
energy (en'əji), *n.* internal or inherent power; force, vigour; capability of action or performing work; active operation; (*Phys.*) a body's power of performing mechanical work. **actual, kinetic** or **motive energy,** the energy of a body in actual motion (measured by the product of half the mass and the square of the velocity). **conservation of energy** CONSERVATION. **latent, potential** or **static energy,** the energy possessed by virtue of the relative condition of parts of a body or of bodies to each other. **energetic** (enəjet'ik), *a.* forcible, powerful; active, vigorously operative. **energetically,** *adv.* **energetics,** *n. sing.* physical, as distinct from vital, dynamics. **energize, -ise** (en'-), *v.t.* to act energetically and vigorously. *v.i.* to give energy to.
enervate (en'əvāt), *v.t.* to deprive of force or strength; to weaken. *a.* (inœ'vət), weakened; wanting in spirit, strength or vigour. **enervation** (enəvā'-), *n.*
enface (infās'), *v.t.* to write, print, stamp the face of.
en famille (ē famēy''), at home with the family. [F]
enfant terrible (ēfã terēbl''), *n.* a child who makes embarrassing remarks; a person who embarrasses people by behaving indiscreetly or unconventionally. [F]
enfeeble (infē'bl), *v.t.* to make feeble or weak. **enfeeblement,** *n.*
enfeoff (infēf', -fef), *v.t.* (*Law*) to invest with a fief; to bestow or convey an estate in fee simple or fee tail. **enfeoffment,** *n.*
en fête (ē fet'), dressed for and/or celebrating a holiday. [F]
enfilade (enfilād'), *n.* a position liable to a raking fire; a fire that may rake a line or body of troops, from end to end. *v.t.* to rake with shot from end to end.
enfold (infōld'), *v.t.* to wrap up, to enclose; to embrace.
enforce (infaws'), *v.t.* to execute strictly; to compel obedience to; to give force to; to press or urge forcibly. **enforceable,** *a.* **enforced,** *a.* forced, not voluntary. **enforcedly** (-sid-), *adv.* **enforcement,** *n.* **enforcer,** *n.*
enframe (infrām'), *v.t.* to set in or as in a frame; to be a frame to.
enfranchise (infran'chīz), *v.t.* to set free; to give (a town, constituency etc.) full municipal or parliamentary rights and privileges; to give (someone) the right to vote. **enfranchisement** (-chiz-), *n.* **enfranchiser,** *n.*
ENG, (*abbr.*) electronic news gathering.
Eng., (*abbr.*) England; English.
eng., (*abbr.*) engineer(ing); engraver; engraving.
engage (ingāj'), *v.t.* to bind by a promise or contract, esp. by promise of marriage; to hire, order, bespeak; to employ, to occupy the time or attention of; to attack, come into conflict with. *v.i.* to pledge oneself (to do something); to undertake; to enter into, embark (on); to begin to fight, to enter into conflict (with); to interlock (with). *n.* (*Fencing*) the order to interlock (swords or foils). **engaged,** *a.* occupied; betrothed. **engaged column,** *n.* (*Arch.*) a column fastened into a wall so that it is partly concealed. **engaged wheels,** *n.pl.* wheels interlocking with each other by means of cogs etc. **engagement,** *n.* the act of engaging or state of being engaged; an obligation, a contract; a mutual promise of marriage; an appointment; a contract to employ; the state of being hired; an enterprise embarked on; an action or battle between armies or fleets. **engagement ring,** *n.* a ring given by a man to a woman to show that she is engaged to be married. **engaging,** *a.* winning, pleasing, attractive (used of manners or address). **engagingly,** *adv.*
en garde (ē gahd'), in fencing, a warning to be ready to receive attack; the stance taken at the start of a fencing bout. [F]
engarland (ingah'lənd), *v.t.* to invest with a garland, to wreathe (with).
engender (injen'də), *v.t.* to beget; to be the cause of, to bring about.
engine (en'jin), *n.* an apparatus consisting of a number of parts for applying mechanical power, esp. one that converts energy into motion; a locomotive; a machine or instrument used in war; an instrument, a tool. *v.t.* to furnish (a ship) with engines. **engine-driver,** *n.* one who drives or manages a locomotive. **engine-turning,** *n.* complex ornamental turning, as on the outside of watch-cases, done by machinery.

engineer (enjinia'), *n.* one who designs or carries out construction work of mechanical, electrical or civic nature; (*N Am.*) an engine-driver; a soldier trained in engineering work. *v.t.* to direct or carry out, as an engineer, the formation or execution of (as railways, canals etc.); (*coll.*) to contrive, to manage by tact or ingenuity. **engineering**, *n.* **civil engineering** CIVIL. **electrical engineering**, construction of electrical engines and equipment. **electronic engineering**, construction of electronic equipment and apparatus. **hydraulic engineering**, the construction of waterworks, the application of water-power, the construction of dams, docks etc. **mechanical engineering** MECHANIC. **military engineering**, the construction of fortifications, and of roads, bridges etc., used for military purposes.
engirdle (ingœ'dl), *v.t.* to surround with or as with a girdle.
English (ing'glish), *a.* pertaining to England or its inhabitants; spoken or written in the English language; characteristic of or becoming an Englishman. *n.* the language of the British Isles, N America, Australasia and other parts of the British Commonwealth; the people of England (sometimes of Britain). *v.t.* to translate into the English language; to render English in style or method. **Basic English** BASIC. **Middle English**, the English language in use from about 1150 to 1500. **Old English**, the English language in use before 1150, also called Anglo-Saxon; (*Print.*) BLACK-LETTER. **plain English**, plain, unambiguous terms. **Queen's, King's English**, correct English as spoken by educated people. **Englishism**, *n.* **Englishman**, *n.* a native or a naturalized inhabitant of England; one of English blood. **Englishness**, *n.* **Englishwoman**, *n. fem.*
englut (inglŭt'), *v.t.* to swallow; to glut, satiate.
engorge (ingawj'), *v.t.* (*in p.p.*) to fill to excess; to congest (with blood). **engorgement**, *n.*
engraft (ingrahft'), *v.t.* to graft upon, insert (a scion of one tree) upon or into another; to incorporate; to implant, instil.
engrail (ingrāl'), *v.t.* (*chiefly Her.*) to indent in curved lines, to make ragged at the edges as if broken with hail. **engrailment**, *n.*
engrain (ingrān'), *v.t.* to dye in fast colours; to dye deeply; (*fig.*) to implant (qualities, esp. vices) ineradicably.
en grande tenue (ē grād tǝnü'), in full evening dress. [F]
en grand seigneur (ē grā senyœr''), like a great lord. [F]
engrave (ingrāv'), *v.t.* to cut (figures etc.) on (a printing surface) with a chisel etc.; to represent on wood, metal etc., by carving; to impress deeply. *v.i.* to practise the art of engraving. **engraver**, *n.* **engraving**, *n.* the act, process or art of cutting figures, letters etc. on wood, stone or metal; an impression from an engraved plate, a print.
engroove (ingroov'), *v.t.* to make a groove in; to set in a groove.
engross (ingrōs'), *v.t.* to write in large, bold letters; to write out in legal form; to occupy the attention entirely. **to be engrossed in**, to be absorbed in (as in reading a book). **engrosser**, *n.* **engrossment**, *n.*
engulf (ingŭlf'), *v.t.* to swallow up, as in a gulf or whirlpool. **engulfment**, *n.*
enhalo (inhā'lō), *v.t.* to encircle with or as with a halo.
enhance (inhahns'), *v.t.* to raise in importance, degree etc.; to augment, intensify; to heighten (in price); *v.i.* to be raised; to grow larger, to increase. **enhanced radiation weapon**, a neutron bomb. **enhancement**, *n.* **enhancer**, *n.* **enhancive**, *a.*
enharmonic (enhahmon'ik), *a.* (*Mus.*) having intervals less than a semitone, as between G sharp and A flat. *n.* enharmonic music. **enharmonic modulation**, *n.* change as to notation, but not as to sound. **enharmonically**, *adv.*
enhearten (inhah'tǝn), *v.t.* to encourage, cheer, strengthen.
enigma (inig'mǝ), *n.* a saying in which the meaning is concealed under obscure language; any inexplicable or mysterious proceeding, person or thing. **enigmatic, -ical** (enigmat'-), *a.* **enigmatically**, *adv.* **enigmatist**, *n.* **enigmatize, -ise**, *v.i.* to speak or write enigmatically.
enjamb(e)ment (injamb'mǝnt), *n.* (*Pros.*) the continuation of a sentence from one verse or couplet into the next.
enjoin (injoin'), *v.t.* to direct, prescribe, impose (an act or conduct); to direct or command (a person to do something); to instruct (that). **enjoiner**, *n.*
enjoy (injoi'), *v.t.* to take pleasure or delight in; to have the use or benefit of; to experience or have. **to enjoy oneself**, (*coll.*) to experience pleasure or happiness. **enjoyable**, *a.* **enjoyableness**, *n.* **enjoyably**, *adv.* **enjoyment**, *n.*
enkephalin (enkef'ǝlin), *n.* a chemical found in the brain, having an effect similar to that of morphine.
enkindle (inkin'dl), *v.t.* to kindle, set on fire; (*fig.*) to inflame, rouse.
enlace (inlās'), *v.t.* to encircle tightly, surround; to embrace, enfold; to entangle. **enlacement**, *n.*
enlarge (inlahj'), *v.t.* to make wider; to increase in size or number; to make more comprehensive. *v.i.* to become bigger; to expatiate (upon). **enlargement**, *n.* the act or process of extending or increasing; increase in size or bulk; an addition; (*Phot.*) a print or negative of a larger size taken from another. **enlarger**, *n.*
enlighten (inlī'tǝn), *v.t.* to give mental or spiritual light to, to instruct; to give (someone) information (on); to release from ignorance, prejudice or superstition. **enlightener**, *n.* **enlightenment**, *n.*
enlink (inlingk'), *v.t.* to join together as with a link.
enlist (inlist'), *v.t.* to enrol, esp. to engage for military service; to gain the interest, assistance or support of. *v.i.* to engage oneself for military service. **enlisted man**, *n.* (*US*) a private soldier, not a conscript. **enlistment**, *n.*
enliven (inlī'vǝn), *v.t.* to give animation to; to stimulate; to brighten, render cheerful. **enlivener**, *n.* **enlivenment**, *n.*
en masse (ē mas'), in a group, all together. [F]
enmesh (inmesh'), *v.t.* to entangle or catch in or as in a net. **enmeshment**, *n.*
enmity (en'miti), *n.* the quality of being an enemy; hatred, hostility.
ennea-, *comb. form.* nine.
ennead (en'iad), *n.* a set of nine, esp. of nine books.
enneahedral (eniǝhē'drǝl), *a.* having nine sides.
Enneandria (enian'driǝ), *n.pl.* (*Bot.*) a Linnaean class of plants distinguished by the nine stamens of the flowers. **enneandrian, -drous**, *a.*
ennoble (inō'bl), *v.t.* to make a noble of; to elevate in character or dignity. **ennoblement**, *n.*
ennui (onwē'), *n.* listlessness; want of interest in things; boredom. **ennuyé** (-yā') (*fem.* **ennuyée**), **ennuied** (-wēd'), *a.* affected with ennui.
enormous (inaw'mǝs), *a.* exceedingly great in size, number or quantity; huge, immense. **enormity**, *n.* the state or quality of being inordinate, outrageous, esp. of being excessively wicked; a monstrous crime, an outrage, atrocity. **enormously**, *adv.* **enormousness**, *n.*
Enosis (en'ōsis), *n.* the proposed political union of Cyprus with Greece. [Gr.]
enough (inŭf'), *a.* (*usu. placed after the noun*) sufficient for or adequate to need or demand. *n.* a sufficiency; a quantity or amount which satisfies requirement or desire; that which is equal to the powers or abilities. *adv.* sufficiently, tolerably, passably. **well enough**, tolerably well.
enounce (inowns'), *v.t.* to enunciate, state definitely; to pronounce. **enouncement**, *n.*
E number E.
en passant (ē pas'ǎ), by the way; applied to the taking of a pawn in chess that has moved two squares as if it has moved only one. [F]
en pension (ē pē'syō), on boarding-house terms. [F]
enprint (en'print), *n.* an enlarged photographic print.
enquire (inkwīǝ'), INQUIRE.

enrage (inrāj'), *v.t.* to put in a rage; to provoke to fury.
en rapport (ē rapaw'), in sympathy with. [F]
enrapture (inrap'chə), *v.t.* to fill with rapture; to delight.
enregiment (inrej'imənt), *v.t.* to form into a regiment; to organize and discipline.
enrich (inrich'), *v.t.* to make rich or richer; to fertilize (soil); to add nutrients to (food).
enring (inring'), *v.t.* to encircle, to surround (with).
enrobe (inrōb'), *v.t.* to put a robe upon, to attire.
enrol (inrōl'), *v.t.* (*past, p.p.* **enrolled**) to write down on or enter in a roll; to record, register; to include as a member, to record the admission of. **enroller**, *n.* **enrolment**, *n.*
enroot (inroot'), *v.t.* to fix by the root; to implant deeply.
en route (ē root'), on the way; on the road. [F]
Ensa (en'sə), *n.* an official organization for entertaining men and women in the armed services during World War II. [acronym for *e*ntertainments *n*ational *s*ervices *a*ssociation]
ensanguine (insang'gwin), *v.t.* (*now only in p.p.*) to smear or cover with blood; to make crimson.
ensate (en'sāt), *a.* (*Bot.*) shaped like a sword.
ensconce (inskons'), *v.t.* to hide; to settle (oneself) comfortably or securely.
ensemble (ēsēbl'', onsom'bl), *n.* all the parts of anything taken together; (*Mus.*) the joint effort of all the performers; a combination of two or more performers or players; a group of supporting players or performers; an outfit of several (matching) garments. **tout ensemble** (toot), the general effect.
ensepulchre (insep'əlkə), *v.t.* to place in a sepulchre.
enshield (inshēld'), *v.t.* to shield, guard, protect.
enshrine (inshrīn'), *v.t.* to place in or as in a shrine; to cherish as if sacred. **enshrinement**, *n.*
enshroud (inshrowd'), *v.t.* to cover with or as with a shroud.
ensiform (en'sifawm), *a.* sword-shaped.
ensign (en'sīn, -sin), *n.* a banner or flag, esp. naval or regimental; a sign or symbol; formerly, the lowest rank of commissioned officers in an infantry regiment; the lowest ranking commissioned officer in the US navy. *v.t.* to distinguish by a badge; to be the distinguishing mark of. **naval ensign**, a flag with a field of white, blue or red, with the union in the upper corner next the staff (white ensign carried by Royal Navy and Royal Yacht Squadrons, blue by naval reserve and red by merchant service). **ensigncy, ensignship**, *n.*
ensilage (en'silij), *n.* a method of preserving green forage crops by storing them en masse in pits or trenches; fodder so preserved, silage. *v.t.* to ensile. **ensile** (insīl'), *v.t.* to put into a silo for this purpose.
enslave (inslāv'), *v.t.* to make a slave of; to bring under the domination of. **enslavement**, *n.* **enslaver**, *n.*
ensnare (insneə'), *v.t.* to entrap.
ensorcell (insaw'səl), *v.t.* to bewitch, fascinate.
ensphere (insfiə'), *v.t.* to place in or as in a sphere.
enstamp (instamp'), *v.t.* to mark as with a stamp.
enstatite (en'stətīt), *n.* a rock-forming mineral, magnesium silicate.
ensue (insū'), *v.i.* to follow in course of time; to result (from). **ensuing**, *a.* coming next after.
en suite (ē swēt'), in succession; forming a unit, as a *bathroom en suite*. [F]
ensure (inshooə'), *v.t.* to make certain (that); to make safe (against or from any risk); to assure or guarantee.
enswathe (inswādh'), *v.t.* to enwrap, to bandage. **enswathement**, *n.*
ENT, (*abbr.*) ear, nose and throat.
-ent (-ənt), *suf.* forming adjectives, e.g. *consistent, frequent;* noting an agent, e.g. *student.*
entablature (intab'ləchə), *n.* (*Arch.*) that part of an order supported upon the columns, the architrave, frieze and cornice.
entablement (intā'blmənt), *n.* the platform supporting a statue, above the dado and base.
entail (intāl'), *v.t.* to bestow or settle a possession inalienably on a certain person and his heirs; to impose (certain duties, expenses etc. upon someone); to involve, necessitate. *n.* an estate in fee limited in descent to a particular heir or heirs; the limitation of inheritance in this way. **entailment**, *n.*
entamoeba (entəmē'bə), *n.* any amoeba of the *Entamoeba* genus, which causes amoebic dysentery in humans.
entangle (intang'gl), *v.t.* to twist together so that unravelling is difficult; to ensnare, as in a net; to involve in difficulties, obstacles, contradictions etc.; to perplex, to embarrass. **entanglement**, *n.*
entasis (en'təsis), *n.* (*Arch.*) the almost imperceptible convex curvature given to a shaft or a column.
entellus (intel'əs), *n.* (*Zool.*) an East Indian monkey.
entente (ētēnt'), *n.* a friendly understanding. **Little Entente**, that between Czechoslovakia, Yugoslavia and Romania. **Triple Entente**, that between Britain, France and Russia, 1907. **Entente Cordiale** (kawdiahl'), *n.* understanding between France and Britain reached in 1904. [F]
enter (en'tə), *v.t.* to go or come into; to penetrate; to associate oneself with, become a member of; to insert; to write (an item, name etc.) in a list (as a candidate etc.); to initiate into a business etc.; to cause to be inscribed upon the records of a court or legislative body; to admit as a pupil or member, to procure admission as such; (*Law*) to take possession of. *v.i.* to go or come in; to become a competitor; (*Theat.*) to appear on the scene. **to enter an appearance**, to show oneself. **to enter a protest**, to make a protest. **to enter into**, to form a part of; to join; to sympathize with; to become a party to (an agreement etc.). **to enter up**, to set down in a regular series; to complete a series of entries. **to enter upon**, to begin; to begin to treat of (a subject etc.); to take legal possession of. **enterable**, *a.*
enterectomy (entərek'təmi), *n.* surgical removal of part of the small intestine.
enteric (inte'rik), *a.* pertaining to the intestines. **enteric fever**, *n.* typhoid fever.
enteritis (entəri'tis), *n.* inflammation of the intestine.
enter(o)-, *comb. form.* pertaining to the intestines.
enteropathy (entərop'əthi), *n.* disease of the small intestine.
enterotomy (entərot'əmi), *n.* surgical incision into the intestine.
enterovirus (entərəvī'rəs), *n.* one which infects the intestinal tract.
enterprise (en'təprīz), *n.* an undertaking, esp. a bold or difficult one; spirit of adventure, boldness, readiness to attempt; a business company. **enterprise culture**, *n.* a social and commercial environment in which entrepreneurs can flourish. **enterprise scheme**, *n.* a government scheme to encourage the setting up of small firms with state financial support. **enterprise zone**, *n.* a depressed area given special government financial etc. backing to encourage commercial etc. improvement. **enterprising**, *a.* energetic, adventurous; full of enterprise. **enterprisingly**, *adv.*
entertain (entətān'), *v.t.* to receive and treat as a guest; to occupy agreeably; to divert, amuse; to hold in mind. *v.i.* to exercise hospitality; to receive company. **entertainer**, *n.* one who entertains, esp. at an entertainment. **entertaining**, *a.* amusing. **entertainingly**, *adv.* **entertainment**, *n.* the act of entertaining; hospitality; the art of entertaining, amusing or diverting; pleasure, amusement; a dramatic or other performance intended to amuse.
enthalpy (en'thəlpi), *n.* (*Phys.*) heat content of a substance per unit mass.
enthral (inthrawl'), *v.t.* (*past, p.p.* **enthralled**) to enslave, enchant, to captivate. **enthralment**, *n.*

enthrone (enthrōn'), *v.t.* to place on a throne; to invest with sovereign power; to induct or instal as an archbishop or bishop. **enthronement**, *n.*
enthronize, -ise (inthrō'nīz), *v.t.* to enthrone, to induct. **enthronization**, *n.*
enthusiasm (enthū'ziazm, -thoo'-), *n.* intense and passionate zeal; ardent admiration; fervour. **enthusiast**, *n.* one filled with or prone to enthusiasm; one whose mind is completely possessed by any subject; a fanatic; **enthusiastic** (-as'-), *a.* **enthusiastically**, *adv.* **enthuse**, *v.i.* (*coll.*) to manifest enthusiasm; to gush.
enthymeme (en'thimēm), *n.* (*Log.*) a syllogism of which one premise is suppressed, and only an antecedent and a consequent expressed in words. **enthymematic** (-mat'-), *a.*
entice (intīs'), *v.t.* to allure, to tempt, seduce (from). **enticement**, *n.* **enticer**, *n.* **enticing**, *a.* alluring, seductive. **enticingly**, *adv.*
entire (intīə'), *a.* whole, complete, perfect; unbroken, undivided; unmixed, pure; unqualified; not castrated (of a horse); (*Bot.*) having the edges (as of a leaf) unbroken or unserrated. **entirely**, *adv.* wholly; fully, completely; exclusively. **entireness**, *n.* **entirety** (-rəti), *n.* entireness, completeness; the entire amount, quantity or extent. **in its entirety**, completely, as a whole.
entitle (intī'tl), *v.t.* to give a certain name or title to, to designate; to dignify (someone) by a title; to give a right, title or claim to anything. **entitlement**, *n.*
entity (en'titi), *n.* essence, existence; anything that has real existence, a being; the essential nature of a thing. **entitative**, *a.*
ent(o)-, *comb. form.* pertaining to the inside of anything.
entoblast (en'təblahst), *n.* (*Biol.*) the nucleus of a cell.
entoil (intoil'), *v.t.* to entrap. **entoilment**, *n.*
entomb (intoom'), *v.t.* to place in a tomb, to bury. **entombment**, *n.*
entom(o)-, *comb. form.* pertaining to insects. **entomic** (intom'ik), *a.* relating to insects. **entomoid** (en'təmoid), *n.*, *a.* (anything) resembling an insect.
entomology (entəmol'əji), *n.* the scientific study of insects. **entomologic, -ical** (-loj'-), *a.* **entomologically**, *adv.* **entomologist**, *n.*
entomophagous (entəmof'əgəs), *a.* feeding on insects.
entomophilous (entəmof'iləs), *a.* attractive to insects; pollinated by insects.
entomostracous (entəmos'trəkəs), *a.* belonging to the Entomostraca, a division of crustaceans, small in size, with the body segments usually distinct, and gills attached to the feet or organs of the mouth.
entoparasite (entōpa'rəsīt), *n.* (*Zool.*) an internal parasite.
entophyte (en'təfīt), *n.* any parasitic plant growing in the interior of animal or vegetable structures.
entourage (ētoorahzh', on-), *n.* surroundings, environment; retinue, attendants.
entozoon (entəzō'on), *n.* (*pl.* **-zoa**, (-ə)) (*Zool.*) an animal living within the body of another animal. **entozoal, -zoic**, *a.* **entozoology** (-ol'əji), *n.* the study of the entozoa. **entozoologist**, *n.*
entr'acte (ētrakt'), *n.* the interval between the acts of a play; music, dancing etc. between acts.
entrails (en'trālz), *n.pl.* the internal parts of animals; the intestines; the internal parts of anything.
entrain[1] (intrān'), *v.t.* to draw after; of a fluid, to carry (particles) along with it.
entrain[2] (intrān'), *v.t.*, *v.i.* to put or get into a railway train. **entrainment**, *n.*
en train (ē trī), in progress, under way. [F]
entrammel (intram'l), *v.t.* to entangle, hamper.
entrance[1] (en'trəns), *n.* the act of entering; the power, right or liberty of entering; the passage or doorway by which a place is entered; the act of coming on to the stage; entrance-fee. **entrance-fee, -money**, *n.* money paid for entrance or admission. **entrant**, *n.* one who enters.

entrance[2] (intrahns'), *v.t.* to throw into a state of ecstasy; to enrapture; to put into a trance. **entrancement**, *n.* **entrancing**, *n.*
entrant ENTRANCE[1].
entrap (intrap'), *v.t.* to catch in or as in a trap; to lure into making a compromising statement or into committing a (criminal) offence. **entrapment**, *n.*
entreat (intrēt'), *v.t.* to beseech, to ask earnestly. *v.i.* to make entreaties. **entreatingly**, *adv.* **entreaty**, *n.* an urgent solicitation; importunity.
entrechat (ē'trəsha), *n.* a leap in dancing with a striking of the heels together several times. [F]
entrecôte (on'trəkōt), *n.* a beefsteak cut from between the ribs.
entrée (ē'trā, on'-), *n.* freedom or right of entrance; a dish served before the main course; (*orig. N Am.*) the main course of a meal. [F]
entremets (ē'trəmā), *n.pl.* side dishes. [F]
entrench (intrench'), *v.t.* to surround with trenches; to defend (oneself) as if with trenches; to trespass, encroach (upon). **entrenchment**, *n.*
entre nous (ē'trə noo'), between ourselves, in confidence. [F]
entrepot (ē'trəpō), *n.* a warehouse for the temporary deposit of goods; a free port where foreign merchandise is kept in bond till re-exported. [F]
entrepreneur (ētrəprənœ'), *n.* one who undertakes a (financial) enterprise, esp. one with an element of risk; a contractor. **entrepreneurial**, *a.* [F]
entresol (ētrəsol'), *n.* a low storey between two higher ones, usually between the first and the ground floor. [F]
entropion (intrō'piən), *n.* introversion of the eyelids.
entropy (en'trəpi), *n.* (*Phys.*) an index of the availability of the thermal energy of a system for mechanical work; (*coll.*) growing disorder in any system.
entrust (intrŭst'), *v.t.* to commit or confide to a person's care; to charge with (a duty, care etc.).
entry (en'tri), *n.* the act of entering; the passage, gate, opening or other way by which anything is entered; the act of entering or inscribing in a book etc.; an item so entered; (*Law*) the act of taking possession by setting foot upon land or tenements; the depositing of a document in the proper office; (*pl.*) a list of competitors etc. **double entry, single entry**, systems of accounts in which each item is entered twice or once in the ledger etc. **entryism**, *n.* the policy of joining a political party etc., in order to influence policy from within. **entryist**, *n.*
Entryphone® (en'trifōn), *n.* a telephonic device at the entrance to a block of flats etc., which allows visitors to communicate with the flat occupier.
entwine (intwīn'), *v.t.* to twine or twist together; to embrace. *v.i.* to become twined or twisted together. **entwinement**, *n.*
entwist (intwist'), *v.t.* to twist around; to form into a twist; to twist (with something else).
enucleate (inū'kliāt), *v.t.* to elucidate, solve; (*Surg.*) to extract (a tumour). **enucleation**, *n.*
enumerate (inū'mərāt), *v.t.* to reckon up one by one, to count; to specify the items of. **enumeration**, *n.* **enumerative**, *a.* **enumerator**, *n.*
enunciate (inŭn'siāt), *v.t.* to pronounce distinctly; state formally. *v.i.* to pronounce words or syllables; to speak. **enunciable**, *a.* **enunciation**, *n.* **enunciative**, *a.* **enunciatively**, *adv.* **enunciator**, *n.*
enure (inūə'), INURE.
enuresis (enūrē'sis), *n.* involuntary urination.
enveigle (invē'gl, -vā'-), INVEIGLE.
envelop (invel'əp), *v.t.* to surround so as to hide; to wrap in or as in an envelope or covering; to surround with troops. **envelopment**, *n.*
envelope (en'vəlōp, on'-), *n.* a wrapper, covering, esp. a paper case to contain a letter; (*Astron.*) the nebulous

envenom   271   epicentre

covering of the head of a comet; (*Biol.*) a shell, membrane etc. the gas-bag of a balloon. **window envelope** WINDOW.

envenom (inven'əm), *v.t.* to impregnate with poison; (*fig.*) to make bitter or spiteful.

enviable etc. ENVY.

environ (inviə'rən), *v.t.* to surround, to encompass; to surround so as to attend or protect; to surround (with persons or things). **environage**, *n.* environment. **environment**, *n.* the act of surrounding; that which encompasses, surrounding objects, scenery, circumstances etc.; the sum of external influences affecting an organism; (*loosely*) living conditions. **environmental**, *a.* **environmental health officer**, an official employed to investigate and prevent potential public health hazards, such as lack of hygiene. **environmentalism**, *n.* the belief that the environment is the main influence on people's behaviour and development; concern for the environment and its preservation from pollution etc. **environmentalist**, *n.* **environs**, *n. pl.* the parts or districts round any place.

envisage (inviz'ij), *v.t.* to contemplate; to visualize; to consider as something possible or likely. **envisagement**, *n.*

envision (invizh'ən), *v.t.* to envisage as a possibility; to foresee.

envoy[1] (en'voi), *n.* a postscript to a collection of poems, or a concluding stanza to a poem.

envoy[2] (en'voi), *n.* a diplomatic agent, next in rank below an ambassador, a messenger, a representative. **envoyship**, *n.*

envy (en'vi), *n.* ill-will at the superiority, success or good fortune of others; a grudging sense of another's superiority to oneself; the object of this feeling. *v.t.* to regard with envy; to feel jealous of; to covet. **enviable**, *a.* capable of exciting envy; greatly to be desired. **enviably**, *adv.* **envier**, *n.* **envious**, *a.* infected with envy; instigated by envy. **enviously**, *adv.*

enwind (inwīnd'), *v.t.* to wind or coil around.

enwrap (inrap'), *v.t.* to wrap or enfold; to envelop; to engross.

enwreathe (inrēth'), *v.t.* to encircle with or as with a wreath.

enzootic (enzōot'ik), *n., a.* (pertaining to) a disease which affects animals in a certain district either constantly or periodically.

enzyme (en'zīm), *n.* a catalyst produced by living cells, esp. in the digestive system. **enzymic** (-zīm'-, -zim'-), *a.* **enzymation**, *n.* **enzymology** (-mol'-), *n.* the scientific study of enzymes.

eoan (ēō'ən), *a.* pertaining to the dawn; eastern.

EOC, (*abbr.*) Equal Opportunities Commission.

Eocene (ē'əsēn), *a.* pertaining to the lowest division of the Tertiary strata.

eod, (*abbr.*) every other day.

eohippus (ēəhip'əs), *n.* the earliest known form of horse-like mammal, now extinct.

Eolian (ēō'liən), etc. AEOLIAN.

eolipyle (ē'əlipil, ēol'-), AEOLIPYLE.

eolith (ē'əlith), *n.* (*Palaeont.*) a rough stone implement anterior in date to the Palaeolithic age, not accepted as artificial by many archaeologists. **eolithic** (-lith'-), *a.*

eon (ē'on), AEON.

eosin (ē'əsin), *n.* a red fluorescent dye, sometimes used in biology.

-eous (-iəs), *suf.* forming adjectives meaning of the nature of, as *ligneous, righteous.*

EP, (*abbr.*) extended play (record); electroplated.

ep-, *pref.* a form of EPI used before a vowel, as in *epact, epoch.*

epact (ē'pakt), *n.* the moon's age at the beginning of the year; the excess of the solar year above the lunar year.

epagoge (epəgō'gi), *n.* (*Log.*) argument by induction.

epan(a)-, *comb. form* (*Rhet.*) denoting repetition, doubling.

epanadiplosis (ipanədiplō'sis), *n.* a figure by which a sentence begins and ends with the same word.

epanalepsis (ipanəlep'sis), *n.* a figure by which a word or clause is repeated after other words intervening.

epanastrophe (epənas'trəfi), *n.* a figure by which the end word of one sentence becomes the first word of the following sentence.

epanodos (ipan'ədos), *n.* a figure in which the second member of a sentence is an inversion of the first.

epanthous (ipan'thəs), *a.* growing upon a flower, as certain fungi.

eparch (ep'ahk), *n.* in the Russian Church, the bishop of an eparchy; a governor of a province in modern Greece. **eparchy**, *n.* a province of modern Greece; a diocese in the Russian Church.

epaulette, epaulet (ep'əlet), *n.* an ornamental badge worn on the shoulder, esp. in military, naval and certain civil full dress uniforms. **epauletted** (-let'-), *a.*

épée (ep'ā), *n.* a duelling sword or fencing foil.

epeirogenesis (ipīrōjen'əsis), **epeirogency** (-roj'ini), *n.* (*Geol.*) the making of a continent.

epenthesis (ipen'thəsis), *n.* (*pl.* **-ses**) (*Gram.*) the addition of a letter or sound in the middle of a word. **epenthetic** (epənthet'-), *a.*

epergne (ipœn'), *n.* an ornamental stand, usu. branched, for the centre of a table etc.

epexegesis (ipeksəjē'sis), *n.* (*pl.* **-ses**) further elucidation of something which has gone before; further statement. **epexegetical** (-jet'-), *a.*

ephah, epha (ē'fə), *n.* a Jewish measure of capacity for dry goods.

Ephedra (ifed'rə), *n.* a genus of jointed, almost leafless, desert plants.

ephedrine (ifed'rin, ef'idrin, -drēn), *n.* a drug used in treating hayfever and asthma.

Ephemera (ifem'ərə), *n.* (*pl.* **-rae**, (-rē), **-ras**) a genus of insects, containing the may-fly; (**ephemera**) the may-fly; anything short-lived. **ephemeral**, *a.* beginning and ending in a day; short-lived, transient. *n.* something ephemeral, esp. a plant or animal. **ephemerality** (-ral'-), *n.* **ephemerid**, *n.* a may-fly. **ephemeris** (-ris), *n.* (*pl.* **-merides** (-mer'idēz)) a collection of tables or data showing the daily position of the planets; an astronomical almanac. **ephemeron** (-rən), *n.* (*pl.* **-ra, -rə**) a mayfly; (*pl.*) anything short-lived, esp. posters, tickets etc. intended to be short-lived. **ephemerist**, *n.* a collector of tickets, handbills and similar ephemera.

Ephesian (ifē'zhən), *n., a.* (an inhabitant) of the ancient city of Ephesus in Asia Minor.

ephod (ef'od, ē'-), *n.* an emblematic short coat formerly worn by Jewish priests.

ephor (ef'aw), *n.* (*pl.* **ephori, -ī**) one of the five magistrates chosen at Sparta and invested with the highest power, controlling even the kings. **ephoralty**, *n.*

epi-, *pref.* above, upon; outer; besides, in addition.

epic (ep'ik), *a.* narrating some heroic event in a lofty style; large-scale; impressive; (*coll.*) very good. *n.* a long poem narrating the history, real or fictitious, of some notable action or series of actions, accomplished by a hero; a work of art associated with some aspect of the epic poem, such as a long adventure novel, a long historical film. **epical**, *a.* **epically**, *adv.*

epicalyx (epikā'liks), *n.* a whorl of leaves forming an additional calyx outside the true calyx.

epicanthus (epikan'thəs), *n.* a fold of skin over the inner corner of the eye, as in the Mongolian peoples.

epicarp (ep'ikahp), *n.* the integument of fruits.

epicene (ep'isēn), *a.* (*Gram.*) having only one form for both sexes; pertaining to both sexes; hermaphrodite; sexless; effeminate. *n.* a noun common to both genders, as *sheep;* a person having the characteristics of both sexes.

epicentre (ep'isentə), *n.* the point over the focus of an earthquake.

epicure (cp'ikūə), *n.* one devoted to sensual pleasures, esp. those of the table. **epicurism**, *n.* **Epicurean** (-rē'ən), *a.* pertaining to Epicurus or his system of philosophy, which taught that pleasure is the supreme good and the basis of morality; (**epicurean**) devoted to pleasure, esp. the more refined varieties of sensuous enjoyment. *n.* a follower of Epicurus; (**epicurean**) a person devoted to pleasure; a gourmet. **epicureanism**, *n.*
epicycle (ep'isīkl), *n.* a small circle the centre of which is carried round upon another circle. **epicyclic** (-sī', -sik'-), *a.* **epicycloid** (-sī'kloid), *n.* a curve generated by the revolution of a point in the circumference of a circle rolling along an exterior of another circle. **epicycloidal** (-sikloi'-), *a.*
epideictic (epidīk'tik), *a.* showing off; displaying (applied to set orations).
epidemic (epidem'ik), *a.* affecting at once a large number in a community. *n.* a disease attacking many persons at the same time, and spreading with great rapidity; a rapidly spreading outbreak. **epidemical**, *a.* **epidemically**, *adv.* **epidemiology** (-dēmiol'-), *n.* the study and treatment of epidemic diseases. **epidemiologist**, *n.*
epidermis (epidœ'mis), *n.* the cuticle or skin constituting the external layer in animals; (*Bot.*) the exterior cellular coating of the leaf or stem of a plant. **epidermal, -mic**, *a.* **epidermoid** (-moid), **-moidal**, *a.*
epidiascope (epidī'əskōp), *n.* a magic lantern which may be used for opaque objects or transparencies.
epididymis (epidid'imis), *n.* a mass of sperm-carrying tubes leading from the back of the testes.
epidote (ep'idōt), *n.* a brittle, lustrous mineral, a silicate of alumina and lime. **epidotic** (-dot'-), *a.*
epidural (epidū'rəl), *a.* situated on, or administered outside, the dura mater. *n.* (also **epidural anaesthetic**) the epidural injection of an anaesthetic into the lower portion of the spinal canal, e.g. in childbirth. **epidurally**, *adv.*
epigean (epijē'ən), **-geal, -geous**, *a.* existing or growing on or above the surface of the ground.
epigene (ep'ijēn), *a.* (applied to rocks) originating on the surface of the earth.
epiglottis (epiglot'is), *n.* a leaf-like cartilage at the base of the tongue which covers the glottis during the act of swallowing. **epiglottic**, *a.*
epigone (ep'igōn), *n.* one belonging to a later and less noteworthy generation. [Gr.]
epigram (ep'igram), *n.* a short poem or composition of a pointed or antithetical character; a pithy or antithetical saying or phrase. **epigrammatic, -ical** (-mat'-), *a.* **epigrammatically**, *adv.* **epigrammatist** (-gram'-), *n.* **epigrammatize, -ise** (-gram'-), *v.t.* to write or express by way of epigrams.
epigraph (ep'igrahf), *n.* a sentence placed at the beginning of a book, chapter etc. as a motto; an inscription on a building, statue, tomb etc. **epigraphic, -ical** (-graf'-), *a.* **epigraphically**, *adv.* **epigraphist** (ipig'-), **-grapher**, *n.* **epigraphy** (ipig'-), *n.* the deciphering and explanation of inscriptions; inscriptions taken collectively.
epigynous (ipij'inəs), *a.* of the stamens or corolla, growing on the top of the ovary, with only the upper portions free.
epilate (ep'ilāt), *v.t.* to remove hair by the roots, by any method. **epilation**, *n.*
epilepsy (ep'ilepsi), *n.* a functional disorder of the brain which involves convulsions of varying intensity, with or without loss of consciousness. **epileptic** (-lep'-), *a.* suffering from epilepsy; pertaining to or indicating the presence of epilepsy. *n.* one who has epilepsy. **epileptoid** (-lep'toid), *a.*
epilogue (ep'ilog), *n.* a short speech or poem addressed to the spectators at the end of a play; the speaker of this; the concluding part of a book, essay or speech. **epilogist** (ipil'əjist), *n.*
epinephrine (epinef'rin), *n.* (*chiefly N Am.*) adrenaline.

Epiphany (ipif'əni), *n.* the manifestation of Christ to the Magi at Bethlehem; the annual festival, held on 6 Jan. to commemorate this; (**epiphany**) the appearance or manifestation of a divinity; a sudden revelation or enlightenment.
epiphenomenon (epifinom'inən), *n.* (*pl.* **-na**) a phenomenon that is secondary and incidental.
epiphysis (ipif'əsis), *n.* (*pl.* **-physes**, -sēz) (*Anat.*) a process formed by a separate centre of ossification; the pineal gland.
epiphyte (ep'ifīt), *n.* a plant growing upon another, usu. not deriving its nourishment from this. **epiphytal** (-fī'-), **epiphytic** (-fit'-), *a.*
episcopacy (ipis'kəpəsi), *n.* government of a Church by bishops, the accepted form in the Latin and Greek communions and the Church of England, prelacy; the bishops taken collectively. **episcopal**, *a.* appertaining to a bishop; constituted on the principles of episcopacy. **episcopal church**, *n.* a Church, like the Anglican, constituted on this basis. **episcopalian** (-pā'-), *n.* (**Episcopalian**) a supporter of episcopal Church government and discipline; a member of an episcopal church. *a.* episcopal. **episcopalianism**, *n.* **episcopalism**, *n.* **episcopally**, *adv.* **episcopate** (-pət), *n.* the office or see of a bishop; the term during which any bishop holds office; bishops collectively.
episiotomy (epēziot'əmi), *n.* cutting of the perineum during childbirth in order to prevent its tearing.
episode (ep'isōd), *n.* orig., the parts in dialogue between the choric parts in Greek tragedy, which were primarily interpolations; an incident or series of events in a story, separable though arising out of it; an incident or closely connected series of events in real life; (*Mus.*) a portion of a fugue deviating from the main theme; one part of a series on radio or television. **episodic, -ical** (-sod'-), *a.* pertaining to or resembling an episode; composed of episodes; sporadic. **episodically**, *adv.*
epistaxis (epistak'sis), *n.* a nose-bleed.
epistemology (ipistəmol'əji), *n.* the science which deals with the origin and method of knowledge.
epistle (ipis'l), *n.* (*formal or facet.*) a written communication, a letter; a literary work (usu. in verse) in the form of a letter; (**Epistle, epistle**) any of the letters written by Apostles to the Churches, now forming part of the New Testament; a lesson in the Church service, so called as being taken from the apostolic Epistles. **epistler** (ipis'-, ipist'-), **-toler** (-tələ), *n.* a writer of letters; the person who reads the Epistle in church service. **epistolary** (-tə-), *a.* pertaining to or suitable for letters; contained in or carried on by means of letters. *n.* a book containing the Epistles.
epistrophe (ipis'trəfi), *n.* (*Rhet.*) a figure in which several sentences or clauses end with the same word.
epistyle (ep'istīl), *n.* (*Arch.*) the architrave.
epitaph (ep'itahf), *n.* an inscription on a tomb; an inscription in prose or verse, as for a tomb or monument. **epitaphic** (-taf'-), *a.* **epitaphist**, *n.*
epitasis (ipit'əsis), *n.* the portion of a play in which the plot is developed, between the protasis or introduction and the catastrophe.
epitaxy (ep'itaksi), *n.* the growth of one layer of crystals on another so that they have the same structure.
epithalamium (epithəlā'miəm), *n.* (*pl.* **-mia**, (-ə)) a nuptial song or poem. **epithalamic** (-lam'-), *a.*
epithelium (epithē'liəm), *n.* (*pl.* **-lia**, (-ə)) the cellular tissue covering the external surfaces of the body and lining cavities and tracts; the thin epidermis lining inner cavities, the stigma etc. of plants.
epithet (ep'ithet), *n.* an adjective or phrase denoting any quality or attribute; a descriptive term; (*coll.*) an abusive expression; a nickname. **epithetic, -ical** (-thet'-), *a.* **epithetically**, *adv.*

epitome (ipit'əmi), *n.* a brief summary of a book, document etc.; a condensation, abridgment, abstract; a representation in miniature; a typical example, embodiment. **epitomist**, *n.* **epitomize, -ise,** *v.t.* to make an abstract, summary or abridgment of; to represent in miniature; to typify, embody.

epizoon (epizō'on), *n.* (*pl.* **-zoa, -ə**) an animal parasitic upon the exterior surface of another. **epizootic** (-ot'-), *a.* pertaining to diseases epidemic among animals. *n.* an epizootic disease.

e pluribus unum (ā pluə'ribəs oo'nəm), (*L*) one out of many (motto of the US).

EPNS, (*abbr.*) electro-plated nickel silver.

epoch (ē'pok), *n.* a fixed point from which succeeding years are numbered, a memorable date; a period characterized by momentous events, an era; a subdivision of geological time; (*Astron.*) the moment when a certain event takes place or a certain position is reached; the longitude of a planet at any given time. **epoch-making,** *a.* of such importance or significance as to mark an epoch. **epochal** (ep'-), *a.*

epode (ep'ōd), *n.* in lyric poetry, the part after the strophe and antistrophe; lyric poetry in which a shorter line follows a longer one. **epodic** (ipod'-), *a.*

eponym (ep'ənim), *n.* a name given to a people, place or institution, after some person; the name of a mythical person made to account for the name of a country or people; a character whose name is the title of a play or book. **eponymic** (-nim'-), **eponymous** (ipon'-), *a.*

epopee (ep'əpē), **epopoeia** (-pē'ə), *n.* an epic or heroic poem; epic poetry.

epos (ep'os), *n.* an epopee; epic poetry; unwritten narrative poetry embodying heroic traditions.

epoxy (ipok'si), *a.* containing oxygen plus two other atoms, frequently carbon, themselves already attached. **epoxy** or **epoxide resin,** *n.* any of a group of synthetic resins containing epoxy groups and used for coatings and adhesives. **epoxide,** *n.* an epoxy compound.

epsilon (ep'silon), *n.* the fifth letter of the Greek alphabet.

Epsom salts (ep'səm), *n.* sulphate of magnesia, a saline purgative, formerly prepared from a mineral spring at Epsom, Surrey.

equable (ek'wəbl), *a.* characterized by evenness or uniformity; even-tempered. **equability** (-bil'-), *n.* **equableness,** *n.* **equably,** *adv.*

equal (ē'kwəl), *a.* the same in magnitude, number, quality, degree etc.; even, uniform, not variable; impartial, unbiased, fair, just; having adequate power, ability or means (to). *n.* one not inferior or superior to another; one of the same or similar age, rank, office, talents or the like. *v.t.* (*past, p.p.* **equalled**) to be equal to, to match. **equal opportunities, equal opportunity,** *n.* a policy of non-discrimination on grounds of race, religion, sex etc. **equal temperament** TEMPERAMENT. **equality** (ikwol'-), *n.* the state of being equal. **equalize, -ise,** *v.t.* to make equal (to, with). *v.i.* to become or make something equal; to reach the same score as an opponent. **equalization, -isation,** *n.* **equalizer,** *n.* one who, or that which, equalizes; esp. the goal etc. that makes the scores equal. **equally,** *adv.* **equalness,** *n.*

equanimity (ekwənim'iti, ē-), *n.* evenness or composure of mind or temper; resignation.

equate (ikwāt'), *v.t.* to equalize; to reduce to an average or common standard. *v.i.* to be equal.

equation (ikwā'shən), *n.* the act of making equal; equality; two algebraic expressions equal to one another, and connected by the sign =; a symbolic expression of a chemical reaction; something involving many factors. **personal equation,** (*Astron.*) the quantity of time by which a person is in the habit of noting a phenomenon wrongly; aberration from strict accuracy, logical reasoning or absolute fairness, due to personal characteristics. **equational,** *a.* **equationally,** *adv.*

equator (ikwā'tə), *n.* a great circle on the earth's surface, equidistant from its poles, and dividing it into the northern and southern hemispheres; (*Astron.*) a great circle of the heavens, dividing it into a northern and a southern hemisphere, constituted by the production of the plane of the earth's equator; a circle dividing a sphere or globular body into two equal parts. **equatorial** (ekwətaw'-), *a.* pertaining to the equator; situated on or near the equator. **equatorial telescope,** *n.* a telescope mounted on an axis parallel to that of the earth, used for noting the course of the stars as they move through the sky. **equatorially,** *adv.*

equerry (ek'wəri, ikwe'ri), *n.* an officer having the care of the horses of nobles or princes; an officer of a royal household.

equestrian (ikwes'triən), *a.* pertaining to horses or horsemanship; mounted on horseback; pertaining to the Roman Equites or Knights. *n.* a rider or performer on horseback. **equestrianism,** *n.* **equestrienne** (-en'), *n. fem.*

equi-, *comb. form.* equal.

equiangular (ēkwiang'gūlə, ek-), *a.* having or consisting of equal angles.

equidistant (ēkwidis'tənt, ek-), *a.* equally distant. **equidistance,** *n.* **equidistantly,** *adv.*

equilateral (ēkwilat'ərəl, ek-), *a.* having all the sides equal. *n.* a figure having all its sides equal. **equilaterally,** *adv.*

equilibrate (ēkwili'brāt, ek-, ikwil'i-), *v.t.* to balance (two things) exactly; to counterpoise. *v.i.* to balance (each other) exactly; to be a counterpoise (to). **equilibrant,** *n.* a force that balances another. **equilibration,** *n.*

equilibrium (ēkwilib'riəm, ek-), *n.* a state of equal balance; esp. a state of rest or balance due to the action of forces which counteract each other; equality of weight or force; the equal balancing of the mind between conflicting motives or reasons; due proportion between parts; the normal state of balance of the animal body. **equilibrist** (ikwil'-), *n.* one who balances in unnatural positions, a rope-dancer, an acrobat.

equine (ek'wīn), *a.* pertaining to or resembling a horse.

equinox (ek'winoks, ē'-), *n.* the moment at which the sun crosses the equator and renders day and night equal throughout the world, now occurring (vernal equinox) on 21 Mar. and (autumnal equinox) on 23 Sept.; (*Astron.*) one of two points at which the sun in its annual course crosses the celestial equator. **equinoctial** (-nok'shəl), *a.* of or pertaining to the equinoxes, or the regions or climates near the terrestrial equator; designating an equal length of day and night; happening at or about the time of the equinoxes. *n.* the equinoctial line; (*pl.*) equinoctial gales. **equinoctial gales,** *n.pl.* gales happening at or near either equinox. **equinoctial line,** *n.* (*Astron.*) the celestial equator, a circle the plane of which is perpendicular to the axis of the earth and passes through the terrestrial equator. **equinoctial points,** *n.pl.* the two points wherein the equator and ecliptic intersect each other. **equinoctial time,** *n.* time reckoned from the moment when the sun passes the vernal equinox. **equinoctially,** *adv.* in the direction of the equinoctial line.

equip (ikwip'), *v.t.* (*past, p.p.* **equipped**) to furnish, accoutre; to fit out, to prepare for any particular duty; to qualify. **equipage** (ek'wipij), *n.* that with which one is equipped; arms and general outfit of a body of troops, including baggage, provisions etc.; a carriage with horses and attendants. **equipaged,** *a.* **equipment,** *n.* the act of equipping; the state of being equipped; that which is used in equipping or fitting out; outfit, furniture, apparatus required for work; intellectual and other qualifications.

equipoise (ek'wipoiz), *n.* a state of equality of weight or force, equilibrium; that which counterbalances. *v.t.* to counterbalance; to hold in equilibrium; (*fig.*) to hold (a person) in mental suspense.

equipollent (ēkwipol'ənt, ek-), *a.* having equal force,

equipotential   274   **errant**

power, significance etc.; equivalent. **equipollence, -lency**, *n.* equality of force etc.; (*Log.*) equivalence between two or more propositions. **equipollently**, *adv.*
equipotential (ēkwipəten'shəl, ek-), *a.* of a line, surface or region, having the same, or at the same, electrical potential at all points.
equisetum (ekwisē'təm), *n.* (*pl.* **-ta, -ə, -turns**) the horsetail. **equisetaceous** (-sitā'shəs), *a.* **equisetic** (-set'-), *a.*
equitation (ekwitā'shən), *n.* the act or art of riding on horseback; horsemanship.
Equites (ek'witēz), *n.pl.* the Knights, the ancient Roman equestrian order of nobility.
Equity (ek'witi), *n.* the actors' trade union.
equity (ek'witi), *n.* justice, fairness; the application of principles of justice to correct the deficiencies of law; (*Law*) the system of law, collateral and supplemental to statute law, administered by courts of equity; an equitable right or claim; the net value of mortgaged property; (*pl.*) stocks and shares not bearing a fixed rate of interest. **equitable**, *a.* fair, just; (*Law*) pertaining to a court or the rules of equity; valid in equity. **equitableness**, *n.* **equitably**, *adv.*
equivalent (ikwiv'ələnt), *a.* of equal value, force or weight; alike in meaning, significance or effect; interchangeable, corresponding; having the same result; (*Geom.*) having equal areas or dimensions; (*Chem.*) having the same combining power. *n.* anything which is equal to something else in amount, weight, value, force etc. **equivalently**, *adv.* **equivalence, -ency**, *n.*
equivocal (ikwiv'əkəl), *a.* doubtful of meaning, ambiguous, capable of a twofold interpretation; of uncertain origin, character etc.; open to doubt or suspicion. †**equivocality** (-kal'-), *n.* **equivocally**, *adv.* **equivocalness**, *n.* **equivocate** (-kāt), *v.i.* to speak ambiguously, esp. so as to deceive; to prevaricate. **equivocation**, *n.* **equivocator**, *n.* **equivocatory**, *a.*
equivoque (ek'wivōk), *n.* an ambiguous term or phrase, an equivocation; a pun or other play upon words. [ME, from late L *aequivocus*, EQUIVOCAL]
ER, (*abbr.*) Elizabeth Regina. [L, Queen Elizabeth]
Er, (*chem. symbol*) erbium.
er (œ), *int.* a sound made when hesitating in speech.
-er[1] (-ə), *suf.* denoting an agent or doer, as *hatter, player, singer;* sometimes doubled, as in *caterer, poulterer;* denoting residence etc., as *Lowlander, Londoner;* denoting a person or thing connected with, as *butler, draper, officer, sampler.*
-er[2] (-ə), *suf.* denoting the comparative, as *richer, taller.*
era (iə'rə), *n.* a historical period or system of chronology running from a fixed point of time marked by an important event such as the birth of Christ, the Hegira etc.; the date from which this is reckoned; a major division of geological time; a period characterized by distinctive events.
eradicate (irad'ikāt), *v.t.* to root up; to extirpate. **eradicable**, *a.* **eradication**, *n.*
erase (irāz', irās'), *v.t.* to rub out; to obliterate, to expunge; to remove (information) from (magnetic tape or other storage medium). **erasable**, *a.* **eraser**, *n.* one who, or that which, erases; esp. a piece of rubber for erasing pencil marks etc. **erasure** (-zhə), *n.* the act of erasing; a place or mark where something has been erased.
erbium (œ'biəm), *n.* a rare metallic element, at. no. 68; chem. symbol Er, forming a rose-coloured oxide. [*Ytterby,* in Sweden]
ere (eə), *prep., conj.* (*poet.*) before. **ere long**, before long; soon.
erect (irekt'), *a.* upright; standing up straight; vertical; pointing straight up or perpendicular to a main axis (as leaves); in a state of physiological erection. *v.t.* to set upright; to raise; to construct, to build; (*fig.*) to elevate, to exalt; to set up. *v.i.* to become erect. **to erect a perpendicular**, (*Geom.*) to draw a line at right angles to another line or plane. **erectile** (-tīl), *a.* susceptible of erection. **erectile tissue**, *n.* tissue formed of blood-vessels intermixed with nervous filaments, and capable of dilatation under excitement. **erection**, *n.* the act of setting upright, building, constructing, establishing etc.; the state of being erected; a building, a structure; the distension of a part consisting of erectile tissue, esp. the penis. **erectly**, *adv.* **erectness**, *n.* **erector**, *n.*
eremite (e'rəmit), *n.* a hermit or anchorite. **eremetic, -ical** (-mit'-), *a.*
erethism (e'rəthizm), *n.* undue excitation of an organ or tissue; abnormal sensitivity to esp. sexual stimuli.
erewhile ERE.
erg[1] (œg), *n.* the unit of work done in moving a body through 1 cm of space against the resistance of 1 dyne. **ergometer** (-gom'itə), *n.* an apparatus that measures the work done by muscles. **ergonomics** (œgənom'iks), *n. sing.* the science concerned with the relationship between workers, their environment and machinery. **ergonomist**, *n.*
erg[2] (œg), *n.* a region of shifting sand dunes in the (Sahara) desert.
ergo (œ'gō), *adv.* therefore; consequently.
ergot (œ'gət), *n.* a disease in various grains and grasses, esp. rye, caused by the presence of a fungus; a preparation of the dried fungus, used medicinally. **ergosterol** (œgos'tərol), *n.* a sterol occurring in ergot. **ergotism**, *n.* poisoning produced by eating grain affected with ergot.
Erica (e'rikə), *n.* a genus of shrubby plants forming the heath family. **ericaceous** (-kā'shəs), *a.*
Erigeron (irij'ərən), *n.* a genus of plants resembling the aster, and including the flea-bane.
eristic (iris'tik), *a.* controversial; given to logical argument or dispute. *n.* a controversialist; the art of disputation.
erk (œk), *n.* (*sl.*) an aircraftsman.
erlking (œl'king), *n.* in German and Scandinavian folklore, a goblin harmful to children.
ermine (œ'min), *n.* an animal of the weasel tribe, *Mustela erminea,* the stoat, whose fur in winter becomes snowy white, with the exception of the tip of the tail which is always black; the fur of this used for the robes of judges, peers etc.; (*fig.*) the office of judge; (*Her.*) a fur represented by triangular black spots on white. **ermined**, *a.* clothed with or wearing ermine.
erne (œn), *n.* an eagle, esp. the golden eagle or the seaeagle.
Ernie (œ'ni), *n.* the device employed for drawing the prize-winning numbers of Premium Bonds. [acronym for *e*lectronic *r*andom *n*umber *i*ndicator *e*quipment]
erode (irōd'), *v.t.* to eat into or away; to corrode; (*Geol.*) to wear away; to eat out (a channel etc.). *v.i.* to become eroded. **erosion** (-zhən), *n.* **erosional**, *a.* **erosive** (-siv), *a.*
erogenous (iroj'inəs), *a.* sensitive to sexual stimulation; producing sexual desire.
EROPS, (*acronnym*) *E*xtended *R*ange *O*perations, with reference to that part of an aircraft's flight which extends beyond a specified maximum flying time from an airport.
erosion ERODE.
erotic (irot'ik), *a.* pertaining to, caused by or arousing sexual love; amatory. **erotica**, *n.sing.* erotic literature or art. **eroticism** (-sizm), **erotism**, *n.* sexual excitement; erotic quality or character; an exaggerated display of sexual feelings. **erotogenic** (-ōjen'ik), *a.* erogenous. **erotomania** (-ōmā'niə), *n.* melancholia or insanity caused by sexual love or desire.
err (œ), *v.i.* to miss the truth, right or accuracy; to be incorrect; to deviate from duty; to sin.
errand (e'rənd), *n.* a short journey to carry a message or perform some other commission; the object or purpose of such a journey. **errand-boy, -girl**, *n.* a boy/girl employed to run errands.
errant (e'rənt), *a.* wandering, roving, rambling, esp. roam-

erratic — **escrow**

ing in quest of adventure as a knight errant; erring. **knight errant** KNIGHT. **errancy, errantry,** *n*.
erratic (irat'ik), *a*. irregular or inconsistent in behaviour, eccentric; wandering, straying, having no fixed course; (*Geol.*) of boulders, transported from their original situation. *n*. an erratic boulder. **erratically,** *adv*.
erratum (irah'tǝm), *n*. (*pl*. **-ta, -ǝ**) an error or mistake in printing or writing; (*pl*.) a list of corrections appended to a book.
erroneous (irō'niǝs), *a*. mistaken, incorrect. **erroneously,** *adv*. **erroneousness,** *n*.
error (e'rǝ), *n*. a mistake; deviation from truth or accuracy; wrong opinion, belief or judgment; a transgression, a sin of a venial kind; a measure of the difference between some quantity and an approximation of it obtained by observation or calculation. **errorless,** *a*.
ersatz (œ'zats, eǝ'-), *n*. a substitute in a pejorative sense. *a*. imitation; artificial.
Erse (œs), *n*. the Gaelic dialect of the Scottish Highlands. *a*. Gaelic; *erron*. Irish.
erst (oest), *adv*. once, formerly, of yore. **erstwhile,** *adv*. some while ago. *a*. former.
erubescent (erǝbes'ǝnt), *a*. reddening, blushing. **erubescence,** *n*.
eructation (ērūktā'shǝn), *n*. the act of belching; that which is ejected by belching; any sudden ejection of gases or solid matter from the earth. **eruct,** *v.t., v.i.*
erudite (e'rǝdit), *a*. learned, well-read. **eruditely,** *adv*. **eruditeness,** *n*. **erudition** (-dish'-), *n*. learning, extensive knowledge gained by study; scholarship.
erupt (irūpt'), *v.i.* to emit lava, steam etc. violently, as a volcano, geyser etc.; to force or break through, as teeth through the gums, a skin rash etc.; to burst out. *v.t.* to force out or emit violently. **eruption,** *n*. the act of bursting forth; a sudden emission; that which breaks out; the breaking out of vesicles, pimples, rash etc. upon the skin; the breaking through of teeth; an outburst of lava etc. from a volcano or other vent. **eruptive,** *a*.
-ery (-ǝri), **-ry** (-ri), *suf*. used with nouns and adjectives, and sometimes with verbs, to form nouns, generally abstract or collective, meaning a business; place of business, cultivation etc.; conduct, things connected with or of the nature of etc.; originally confined to Romance words, but now used with those of Teutonic origin, e.g. *foolery, grocery, pinery, rockery, tannery, witchery*.
eryngo (iring'gō), *n*. a plant of the genus *Eryngium*, esp. sea-holly.
erysipelas (erisip'ilǝs), *n*. an inflammation of the skin in which the affected parts are of a deep red colour, with swelling of the underlying tissue.
erythema (erithē'mǝ), *n*. a superficial skin-disease characterized by redness in patches. **erythematic** (-mat'-), **erythematous,** *a*.
erythr(o)-, *comb. form*. red.
erythrocyte (irith'rǝsit), *n*. a red blood-cell in vertebrates.
erythromycin (irithrōmī'sin), *n*. an antibiotic used to treat bacterial infections.
erythropoiesis (irithrōpoiē'sis), *n*. the formation of red blood-cells.
Es, (*chem. symbol*) einsteinium.
-es (-iz), *suf*. used to form the pl. of most nouns that end in -s; used to form the 3rd pers. sing. pres. of most verbs that end in -s.
ESA, (*abbr*.) European Space Agency; environmentally sensitive area.
escadrille (es'kǝdril), *n*. a squadron of the French air force; a flotilla of ships.
escalade (eskǝlād'), *n*. an attack on a fortified place in which scaling-ladders are used to mount the ramparts etc. *v.t.* to storm by means of scaling-ladders.
escalate (es'kǝlǝt), *vi*, *v.t.* to increase in extent, intensity or magnitude. **escalation,** *n*.

escalator (es'kǝlātǝ), *n*. a moving staircase. **escalater clause,** *n*. a clause in a contract that allows for an upward or downward adjustment in prices, wages etc. in the event of changes in the cost oof living etc.
Escallonia (eskǝlō'niǝ), *n*. a genus of S American flowering trees or shrubs of the saxifrage family. [*Escallon*, a Spanish traveller]
escallop (iskal'ǝp), SCALLOP.
escalope (es'kǝlop), *n*. a thin boneless slice of meat, esp. veal or pork.
escapade (eskǝpād'), *n*. a wild prank or adventure.
escape (iskāp'), *v.t.* to get safely away from; to flee so as to be free from; to evade, to avoid; to slip away from, elude attention or recollection of; to slip from unawares or unintentionally. *v.i.* to get free; to get safely away; to issue, to leak; to evade punishment, capture, danger, annoyance etc. *n*. the act of escaping; the state of having escaped; a means of escaping; a leakage (from a gas or water pipe, electric main etc.); a plant from a garden apparently growing wild. **fire-escape** FIRE. **escape-pipe, -valve,** *n*. an outlet for steam, water etc. in case of necessity. **escape road,** *n*. a track on a hill, bend, etc. for drivers to turn on to if the vehicle is out of control. **escape velocity,** *n*. the minimum velocity that must be attained by a body to escape from the gravitational field of a planet etc. **escapee** (eskǝpē'), *n*. one who has escaped, esp. an escaped prisoner. **escapement,** *n*. a device in a clock or watch for checking and regulating the movement of the wheels; the space between the hammer and the string in a piano, that allows vibration of the string; a vent, an escape. **escapism,** *n*. shirking unpleasant facts and realities by filling the mind with pleasing irrelevancies. **escapist,** *n., a.* **escapologist** (eskǝpol'ǝjist), *n*. a performer whose stage turn is escaping from locked handcuffs, chains, boxes etc. **escapology,** *n*.
escargot (iskah'gō), *n*. an edible snail. [F]
escarp (iskahp'), *v.t.* (*Fort.*) to cut or form into a slope; to scarp. *n*. the slope on the inner side of a ditch, below a rampart; a scarp. **escarpment,** *n*. the precipitous face of a hill or ridge; (*Fort.*) ground cut away precipitously so as to render a position inaccessible.
-esce (-es), *suf*. forming inceptive verbs, as *acquiesce, coalesce, effervesce*. **-escent** (-es'nt), *suf*. forming adjectives from inceptive verbs, as *acquiescent, coalescent, iridescent, opalescent*. **escence** (-es'ns), *suf*. forming abstract nouns from inceptive verbs, as *acquiescence, coalescence, opalescence*.
eschatology (eskǝtol'ǝji), *n*. the doctrine of the final issue of things, death, the last judgment, the future state etc. **eschatological** (-loj'-), *a*.
escheat (ischēt'), *n*. the reverting of property to the lord of the fee, or to the Crown or the state, on the death of the owner intestate without heirs; property so reverting. *v.t.* to confiscate. *v.i.* to revert by escheat. **escheator,** *n*. an officer formerly appointed in every county to register the escheats of the Crown.
eschew (ischoo'), *v.t.* to avoid; to shun; to abstain from. **eschewal,** *n*. **eschewer,** *n*.
Eschscholtzia (esholt'siǝ), *n*. a genus of flowering herbs comprising the California poppy. [J F von *Eschscholtz*, 1793–1831, German naturalist]
escort[1] (es'kawt), *n*. one or more people, ships, etc. accompanying a person or persons, baggage, munitions etc. as a protection against attack or for compulsion or surveillance; a person who accompanies another; esp. of the opposite sex, for company on a social occasion. **escort agency,** *n*. a company which provides people, usu. of the opposite sex, to act as hired escorts on social occasions.
escort[2] (iskawt'), *v.t.* to act as escort to.
escritoire (eskritwah'), *n*. a writing-desk, with drawers etc. for papers and stationery, a bureau.
escrow (eskrō'), *n*. (*Law*) a fully-executed deed or engage-

escudo

escudo (eskoo'dō), *n. (pl.* **-dos**) the unit of currency in Portugal.
Esculapian (ĕskūlā'piən), AESCULAPIAN.
esculent (es'kūlənt), *a.* fit or good for food, edible. *n.* a thing suitable for food.
escutcheon (iskŭch'ən), *n.* a shield or shield-shaped surface charged with armorial bearings; any similar surface or device; a perforated plate to finish an opening, as a keyhole etc.; part of a ship's stern bearing her name. **a blot on the escutcheon**, a stain on the reputation of a person, family etc.
ESE, (*abbr.*) East South East.
-ese (-ēz), *suf.* belonging to a country etc. as inhabitant(s) or language, as *Maltese, Chinese*; pertaining to a particular writer, writing etc. with regard to style, language, theme etc., as *Johnsonese, journalese.*
eskar, esker (es'kə), *n.* a bank or long mound of glacial drift.
Eskimo (es'kimō), *n. (pl.* **-mos**) a member of a race inhabiting Greenland and the adjacent parts of N America; their language. **Eskimo-dog**, *n.* a wolf-like variety of the domestic dog, used by the Eskimos to draw sledges.
Esky® (es'ki), *n.* in Australia, a portable container or chest for cooled drinks.
ESN, (*abbr.*) educationally subnormal.
esophagus (ēsof'əgəs), OESOPHAGUS.
esoteric (esəte'rik, ē-), **-ical**, *a.* meant for or intelligible only to the initiated; recondite, secret, confidential. **esoterically**, *adv.* **esotericism**, *n.*
ESP, (*abbr.*) extrasensory perception.
espadrille (espədril'), *n.* a rope-soled shoe with a cloth upper.
espalier (ispal'iə), *n.* lattice-work on which to train shrubs or fruit-trees; a tree so trained. *v.t.* to train (a tree or shrub) in this way.
esparto (ispah'tō), *n.* a kind of coarse grass or rush, growing in the sandy regions of northern Africa and Spain, largely used for making paper, mats etc.
especial (ispesh'əl), *a.* distinguished in a certain class or kind; pre-eminent, exceptional, particular; pertaining to a particular case, not general or indefinite. **especially**, *adv.*
Esperanto (espəran'tō), *n.* an international artificial language invented by L. L. Zamenhof (1887), based on the chief European languages. **Esperantist**, *n.*
espial ESPY.
espionage (es'piənahzh), *n.* the act or practice of spying; the use of spies.
esplanade (esplənād'), *n.* a level space, esp. a level walk or drive by the seaside etc.; a clear space between the citadel and the houses of a fortified town.
espouse (ispowz'), *v.t.* to marry; to give in marriage (to); to adopt, to support, defend (a cause etc.). **espousal**, *n.* (*usu. pl.*) a betrothal, marriage; adoption or support (of a cause etc.).
espressivo (espresē'vō), *a.* (*Mus.*) with expression. [It.]
espresso (ispres'ō), *n. (pl.* **-ssos**) a coffee-making machine using pressure for high extraction; coffee made in this way.
esprit (isprē'), *n.* wit, sprightliness. **esprit de corps** (də kaw'), the spirit of comradeship, loyalty and devotion to the body or association to which one belongs.
espy (ispī'), *v.t.* to catch sight of; to detect, to discern. **espial** (ispī'əl), *n.* spying, observation.
Esq., (*abbr.*) Esquire.
-esque (-esk), *suf.* like, in the manner or style of, as *arabesque, burlesque, Dantesque, picturesque.*
Esquimau (es'kimō), (*pl*, **-maux**, -mōz) ESKIMO.
esquire (iskwīə'), *n.* the armour-bearer or attendant on a knight, a squire; a title of dignity next in degree below a knight; a title given to professional men, and used as a complimentary adjunct to a person's name in the addresses of letters.
ESRC, (*abbr.*) Economic and Social Research Council.
-ess (-is), *suf.* noting the feminine; as *empress, murderess.*
essay[1] (es'ā), *n.* an attempt; an informal literary composition or disquisition, usu. in prose. **essayist**, *n.* a writer of essays.
essay[2] (esā'), *v.t.* to try, to attempt; to test; to test the quality or nature of.
essence (es'ns), *n.* that which constitutes the nature of a thing; that which makes a thing what it is; (a solution of) an ethereal or immaterial being; essential oil or extract possessing the characteristic properties of a plant etc.; perfume, scent; one who, or that which, epitomizes the nature of a thing.
Essene (esēn', es'-), *n.* a member of an ancient Jewish sect of religious mystics who cultivated poverty, community of goods and asceticism of life. **Essenism** (es'ə-), *n.*
essential (isen'shəl), *a.* of or pertaining to the essence of a thing; necessary to the existence of a thing, indispensable (to); important in the highest degree; containing the essence or principle of a plant etc.; of a disease, idiopathic; of an amino or fatty acid, necessary for the normal growth of the body, but not synthesized by the body. *n.* that which is fundamental, indispensable or of the highest importance. **essential oil**, *n.* a volatile oil containing the characteristic constituent or principle, usually obtained by distillation with water. **essentiality** (-shial'-), *n.* **essentially**, *adv.*
EST, (*abbr.*) Eastern Standard Time; electric shock treatment.
est., (*abbr.*) established; estimated.
-est (-ist), *suf.* forming the superlative degree of adjectives and adverbs, as *richest, tallest, liveliest.*
establish (istab'lish), *v.t.* to set upon a firm foundation, to found, institute; to settle or secure firmly (in office, a job, opinion etc.); to make firm or lasting (as a belief, custom, one's health etc.); to substantiate, verify, put beyond dispute; to ordain officially and settle on a permanent basis (as a Church). **established Church**, the church established by law, the State Church. **establishment** (-mənt), *n.* the act of establishing; the state of being established; a permanent organization such as the army, navy or civil service, a staff of servants etc.; a public institution, business organization or large private household with the body of persons engaged in it. **the Establishment**, a phrase of journalistic use to suggest the unconscious association of the respectable and conventional leaders in education and public affairs. **establishmentarian** (-teə'ri-), *a.* advocating or supporting an established Church or the Establishment. *n.* a person with establishmentarian views.
estaminet (estam'inā), *n.* a cafe in which wine etc. is sold. [F]
estate (istāt'), *n.* property, esp. a landed property; (*Law*) a person's interest in lands and tenements (**real estate**) or movable property (**personal estate**); a person's assets and liabilities taken collectively; land built on either privately or by a local authority for housing (**housing estate**) or for factories and businesses (**industrial** or **trading estate**); state, condition, circumstances, standing; a class or order invested with political rights (in Great Britain the Three Estates are the Lords Spiritual, the Lords Temporal and the Commons). **fourth estate**, the newspaper press. **third estate**, the bourgeoisie of France before the Revolution, as distinguished from the nobles and the clergy. **estate agent**, *n.* the manager of a landed property; an agent concerned with the renting or sale of real estate. **estate car**, *n.* one with a large open space behind the passenger seats, and a rear door. **estate duty**, *n.* death duty.
esteem (istēm'), *v.t.* to hold in high estimation, to regard with respect; to prize; to consider, to reckon. *n.* opinion or judgment as to merit or demerit, esp. a favourable opi-

nion; respect, regard. **estimable** (es'ti-), *a.* worthy of esteem or regard. **estimably,** *adv.*
ester (es'tə), *n.* an organic compound derived by the replacement of hydrogen in an acid by an organic radical.
esthete (es'thēt), etc. AESTHETE.
estimable ESTEEM.
estimate (es'timāt), *v.t.* to compute the value of, to appraise; to form an opinion about. *n.* (-mət) an approximate calculation of the value, number, extent etc. of anything; the result of this; a contractor's statement of the sum for undertaking a piece of work; a judgment respecting character, circumstances etc. **estimation,** *n.* the act of estimating; opinion or judgment; esteem. **estimative,** *a.* **estimator,** *n.*
estival etc. AESTIVAL.
estop (istop'), *v.t.* (*past, p.p.* **estopped**) (*Law*) to bar, preclude, prevent. **estoppage** (-ij), *n.* **estoppel** (-əl), *n.* (*Law*) an act or statement that cannot legally be denied; a plea alleging such an act or statement.
estovers (istō'vəz), *n.pl.* (*Law*) necessaries or supplies allowed by law, esp. wood which a tenant can take from a landlord's estate for repairs etc.
estradiol OESTRADIOL.
estrange (istränj'), *v.t.* to alienate, to make indifferent or distant in feeling; to cut off from friendship; to make (oneself) a stranger to. **estranged,** *a.* having been estranged; of a man and wife, no longer living together. **estrangement,** *n.*
estrogen, estrus etc. OESTROGEN, OESTRUS etc.
estuary (es'chuəri), *n.* the mouth of a river etc. in which the tide meets the current. **estuarine** (-rin), *a.*
esurient (isū'riənt), *a.* hungry; needy. **esurience,** *n.*
-et (-it), *suf.* diminutive, as *coronet, russet, violet.*
ETA, (*abbr.*) estimated time of arrival.
eta (ē'tə, ā'-), *n.* the seventh letter of the Greek alphabet.
et al., (*abbr.*) and others. [L *et alii, aliae* or *alia*]
etc., (*abbr.*) etcetera.
etcetera (etset'ərə), and the rest; and others of like kind; and so forth, and so on, usually written *etc.* or *&c.* **etceteras,** *n.pl.* sundries, extras; things unspecified.
etch (ech), *v.t.* to produce or reproduce (figures or designs) on (metallic plates), for printing copies, by biting with an acid through the previously drawn lines; to imprint, fix deeply. *v.i.* to practise this art. **etcher,** *n.* **etching,** *n.* the act of etching; an impression taken from an etched plate.
ETD, (*abbr.*) estimated time of departure.
eternal (itœ'nəl), *a.* without beginning or end; everlasting, perpetual; unchanging, valid throughout time; (*coll.*) incessant, unintermittent. **the Eternal,** God. **Eternal City,** *n.* Rome. **eternal triangle,** *n.* a sexual or emotional relationship involving three people, usu. two of one sex and one of the other, often resulting in tension or conflict. **eternalize, -ise, eternize, -ise,** *v.t.* to make eternal; to immortalize. **eternally,** *adv.*
eternity (itœ'niti), *n.* eternal duration; endless time; the future life after death; (*coll.*) a seemingly endless time. **eternity ring,** *n.* a ring set all round with stones, signifying continuity.
etesian winds (itē'zhiən), *n.pl.* periodical winds, esp. north-westerly blowing for about six weeks in summer in the Mediterranean.
-eth (-əth), *suf.* used to form ordinal numbers, as *fortieth.*
ethane (ē'thān, eth'-), *n.* a colourless and odourless gaseous compound of the alkane series. **ethanol** (-ənol), *n.* ethyl alcohol.
ethene (eth'ēn, ēth'ēn), ETHYLENE under ETHYL.
ether (ē'thə), *n.* (also **aether**) a fluid of extreme subtlety and elasticity formerly assumed to exist throughout space and between the particles of all substances, forming the medium of transmission of light and heat; (also **aether**) the upper air, the higher regions of the sky, the clear sky; a light, volatile and inflammable fluid, produced by the distillation of alcohol with an acid, esp. sulphuric acid, and formerly used as an anaesthetic; the class of compounds to which this belongs. **ethereal** (ithiə'riəl), *a.* of the nature of or resembling celestial ether; light, airy, tenuous, exquisite, impalpable, spiritual; pertaining to a chemical ether. **ethereality** (-al'-), *n.* **etherealize, -ise,** *v.t.* to convert into ether; to render spiritual. **etherealization,** *n.* **ethereally,** *adv.* **etherize, -ise,** *v.t.* (*Chem.*) to convert into ether; to anaesthetize with ether. **etherization, -isation,** *n.*
ethic (eth'ik), *a.* ethical. *n.* a moral principle or system of values. **ethical,** *a.* treating of or relating to morals or ethics; dealing with moral questions or theory; conforming to a recognized standard of behaviour or conduct; of a drug, available only on prescription; pertaining to the practice of investing money only in those companies which are not involved in racial discrimination or in products causing potential harm to health, life or the environment, as cigarettes, nuclear weapons. **ethically,** *adv.* **ethicize, -ise** (-siz), *v.t.* to make ethical; to treat ethically. **ethicism** (-sizm), *n.* **ethics,** *n.pl.* a system of principles and rules of conduct; the moral correctness of an action etc.; (*sing. in constr.*); the field of moral science, including political and social science, law, jurisprudence etc.
Ethiopian (ēthiō'piən), *n., a.* (an inhabitant) of Ethiopia or Abyssinia. **Ethiopic** (-op'-), *a.* Ethiopian. *n.* the language of Ethiopia.
ethmoid (eth'moid), **-moidal** (-moi'-), *a.* resembling a sieve. *n.* the ethmoid bone. **ethmoid bone,** *n.* a cellular bone situated between the orbital processes at the root of the nose, through which the olfactory nerves pass.
ethnarch (eth'nahk), *n.* the governor of a people or district.
ethnic (eth'nik), *a.* pertaining to or characteristic of a race or people; pertaining to the culture or traditions of a particular race or people; (*coll.*) out of the ordinary; racial, ethnological. *n.* (*chiefly N Am.*) a member of a (minority) ethnic group. **ethnical,** *a.* **ethnically,** *adv.* **ethnicity,** *n.*
ethn(o)- *n. comb. form.* pertaining to race.
ethnobotany (ethnōbot'əni), *n.* the traditional plant love of a people. **ethnobotanical** (-tan'-), *a.* **ethnobotanist,** *n.*
ethnocentrism (ethnōsent'rizm), *n.* the mental habit of viewing the world solely from the perspective of one's own culture. **ethnocentric,** *a.* **ethnocentrically,** *adv.*
ethnography (ethnog'rəfi), *n.* the science which describes different human societies. **ethnographer,** *n.* **ethnographic, -ical** (-graf'-), *a.* **ethnographically,** *adv.*
ethnology (ethnol'əji), *n.* the science which treats of the varieties of the human race, and attempts to trace them to their origin. **ethnologic, -ical** (-loj'-), *a.* **ethnologically,** *adv.* **ethnologist,** *n.*
ethnomusicology (ethnōmūzikol'əji), *n.* the study of the music of different societies. **ethnomusicologist,** *n.*
ethology (ethol'əji), *n.* the science of animal behaviour. **ethologic, -ical** (-loj'-), *a.* **ethologist,** *n.*
ethos (ē'thos), *n.* the characteristic spirit, character, disposition or genius of a people, community, institution, system etc.
ethyl (eth'il, ē'thil), *n.* a monovalent fatty hydrocarbon radical of the alkane series, forming the base of common alcohol and ether, acetic acid etc. **ethyl alcohol,** *n.* the ordinary alcohol of commerce. **ethylene** (eth'-), *n.* a colourless gas occurring in coal gas, used in making polythene.
etiolate (ē'tiəlāt), *v.t.* to blanch (a plant kept in the dark); to render (persons) pale and unhealthy. *v.i.* to become blanched by deprivation of light. **etiolation,** *n.*
etiology (ētiol'əji), AETIOLOGY.
etiquette (et'iket), *n.* the conventional rules of behaviour

in polite society; the established rules of precedence and ceremonial in a court, or a professional or other body.
Etonian (itō′niən), *n.* a person educated at Eton College. **Old Etonian**, an Etonian. **Eton collar**, *n.* a wide, starched collar worn outside the jacket. **Eton crop**, *n.* a fashion of cutting a woman's hair short like a man's. **Eton jacket**, *n.* a boy's untailed dress-coat.
étrier (ā′triā), *n.* a small rope ladder used in mountaineering. [F, a stirrup]
Etrurian (itroo′riən), *a.* pertaining to Etruria, an ancient country in central Italy. *n.* a native of Etruria. **Etruscan** (itrŭs′kən), *a.* Etrurian.
et seq. (et sek), (*abbr.*) and the following (passage). [L *et sequentes*, *et sequentia*]
-ette (-et), *suf.* diminutive, as *palette*, *cigarette*; female, as *brunette*, often offensive, as *jockette*; imitation, as *flannelette*, *leatherette*.
étude (ātüd′), *n.* (*Mus.*) a short composition written mainly to test a player's technical skill. [F, a study]
etui (ātwē′, e-), *n.* a pocket-case for pins, needles etc.
etymology (etimol′əji), *n.* the science that treats of the origin and history of words; the history of the origin and modification of a particular word. **etymologer, etymologist**, *n.* **etymologic, -ical** (-loj′-), **etymologically**, *adv.* **etymologize, -ise**, *v.t.* to give or trace the etymology of. *v.i.* to study etymology; to propose etymologies for words.
etymon (et′imon), *n.* the primitive or root form of a word.
eu-, *comb. form.* good, well, pleasant, as in *eulogy, euphony*.
eucalyptus (ūkəlip′təs), *n.* (*pl.* **-tuses, -ti**, -tī) any of an Australasian genus of evergreen myrtaceous trees comprising the gum-trees.
Eucharist (ū′kərist), *n.* the sacrament of the Lord's Supper; the elements, bread and wine, given in this sacrament. **eucharistic, -ical** (-ris′-), *a.*
euchre (ū′kə), *n.* an American card game for several persons, usu. four, with a pack from which the cards from the twos to the nines have been excluded. *v.t.* to beat by taking three of the five tricks at euchre; (*chiefly N Am. coll.*) to trick, cheat, outwit.
Euclidean (ūklid′iən), *a.* of or pertaining to Euclid, Alexandrian mathematician (fl. 300 BC); according to the axioms and postulates of Euclid's geometry.
eudaemonism, eudemonism (ūdē′mənizm), *n.* the system of ethics which makes the pursuit of happiness the basis and criterion of moral conduct. **eudemonic** (-mon′-), *a.* **eudemonics**, *n.sing.* **eudemonist**, *n.* **eudemonistic** (-nis′-), *a.*
eudiometer (ūdiom′itə), *n.* an instrument for ascertaining the quantity of gases in a mixture or taking part in a chemical reaction. **eudiometric, -ical** (-met′-), *a.* **eudiometrically**, *adv.* **eudiometry**, *n.*
eugenic (ūjen′ik), *a.* pertaining to the development and improvement of offspring, esp. human offspring, through selective breeding. **eugenics**, *n. sing.* the science relating to this. **eugenicist, eugenist**, *n.* **eugenism** (ū′-), *n.*
eukaryon (ūka′riən), *n.* **eukaryote** (ūka′riŏt, -ot), *n.* an organism with a highly organized cell nucleus, surrounded by membrane.
eulogy (ū′ləji), *n.* praise; a writing or speech in praise of a person. **eulogist**, *n.* **eulogistic, -ical** (-jis′-), *a.* **eulogistically**, *adv.* **eulogium** (-ūlō′jiəm), *n.* (*pl.* **-giums, -gia**) (a) eulogy. **eulogize, -ise**, *v.t.* to speak or write of in praise, to commend, to extol.
Eumenides (ūmen′idēz), *n.pl.* a euphemism for the Furies.
eunuch (ū′nək), *n.* a castrated man, esp. an attendant in a harem, or a state functionary in Oriental palaces and under the Roman emperors; (*loosely*) an ineffectual or powerless person. **eunuchal, a. eunuchize, -ise**, *v.t.*
Euonymus (ūon′iməs), *n.* a genus of shrubs containing the spindle-tree.
eupepsia (ūpep′siə), *a.* good digestion. **eupepsy** (-si), *n.* **eupeptic**, *a.* **eupepticity** (-tis′-), *n.*

euphemism (ū′fəmizm), *n.* the use of a soft or pleasing term or phrase for one that is harsh or offensive; such a term or phrase. **euphemistic** (-mis′-), *a.* **euphemistically**, *adv.* **euphemize, -ise**, *v.t.* to speak of euphemistically; to express in euphemism. *v.i.* to speak in euphemism.
euphonium (ūfō′niəm), *n.* (*Mus.*) a brass instrument related to the tuba.
euphony (ū′fəni), *n.* an agreeable sound; smoothness or agreeableness of sound in words and phrases; a pleasing pronunciation; (*Philol.*) the tendency towards greater ease of pronunciation shown in phonetic changes. **euphonic, -ical** (ūfon′), *a.* **euphonically**, *adv.* **euphonious** (ūfō′-), *a.* pleasing in sound. **euphoniously**, *adv.* **euphonize, -ise**, *v.t.*
Euphorbia (ūfaw′biə), *n.* a genus of plants known as the spurges, comprising about 700 species, many of which are poisonous while others have medicinal qualities.
euphoria (ūfaw′riə), *n.* a feeling of well-being, supreme content. **euphoric**, *a.*
euphuism (ū′fūizm), *n.* a pedantic affectation of elegant and high-flown language (from *Euphues* (1578 –80), a work by John Lyly, which brought the style into vogue). **euphuist**, *n.* **euphuistic** (-is′-), *a.* **euphuistically**, *adv.*
Eurasian (ūrā′zhən), *a.* of mixed European and Asian descent; pertaining to both Europe and Asia. *n.* one of European and Asian descent.
Euratom (ūrat′əm), *n.* the European Atomic Energy Community of 1958 in which France, Belgium, West Germany, Italy, the Netherlands and Luxemburg united for the peaceful development of nuclear energy.
eureka (ūrē′kə), *int.* expressing triumph at discovery. [Gr. *heurēka*, I have found, Archimedes' exclamation on discovering a test for the purity of the gold in Hiero's crown, involving the displacement of water]
eurhythmics (ūridh′miks), *n. sing.* the science or art of rhythmical movement, esp. as applied to dancing and gymnastic exercises. **eurhythmic, -ical**, *a.* **eurhythmy** (-mi), *n.*
Euro- (-ū′rō), *comb. form* pertaining to Europe to Europeans; pertaining to the European Economic Community.
euro (ū′rō), *n.* (*Austral.*) a kangaroo, the wallaby of S and Central Australia.
Eurobond (ū′rŏbond), *n.* a bond sold outside the country in whose currency it is issued.
Eurocheque (ū′rōchek), *n.* a type of cheque able to draw on certain international banks on receipt of the appropriate card.
Eurocommunism (ūrōkom′ūnizm), *n.* the form of communism followed by European communist parties, traditionally more pragmatic than, and independent of, the Soviet version. **Eurocommunist**, *n.*
Eurocrat (ū′rəkrat), *n.* an official involved in the administration of any part of the European Economic Community.
Eurocurrency (ū′rōkŭrənsi), *n.* any currency held outside its country of issue and used in the European money market.
Eurodollar (ū′rōdolə), *n.* a US dollar held in European banks to ease the financing of trade.
Euroformat (ū′rōfawmat), *a.* of a passport, issued to a citizen of member countries of the European Economic Community to replace national passports.
Euro-MP (ū′rō-), *n.* a member of the European Parliament.
European (ūrəpē′ən), *a.* of, pertaining to, happening in, or extending over, Europe; native to Europe. *n.* a native or inhabitant of Europe; one of European race. **European (Economic) Community**, an association of European countries as a single economic unit for the purposes of trade. **European Parliament**, *n.* the legislative assembly of the European Economic Community. **European plan**, *n.* (*N Am.*) the system of charging for a hotel room without in-

cluding meals. **Europeanism**, n. **Europeanize, -ise**, v.t. **Europeanization**, n.

europium (ūrō'piəm), n. an extremely rare metallic element, at. no. 63; chem. symbol Eu, discovered in 1901.

Eurovision (ū'rəvizhən), n. the network of European television.

Eustachian tube (ūstā'kiən), n. a duct leading to the cavity of the tympanum of the ear from the upper part of the pharynx. [*Eustachius*, 16th-cent. Italian physician]

eustasy (ūs'təsi), n. changes in the world shore-line level or sea-level. **eustatic** (ūstat'-), a.

eutectic (ūtek'tik), a. (*Chem.*) applying to the mixture of two or more substances whose proportions result in the minimum melting-point.

euthanasia (ūthənā'ziə), n. easy, painless death; a method of producing this; putting to death in this manner, esp. in cases of extreme or terminal suffering.

euthenics (ūthen'iks), n. *sing.* the study of the improvement of human living standards. **euthenist**, n.

Eutheria (ūthiə'riə), n.pl. the subclass of mammals which have a placenta. **eutherian**, n., a. [Gr. *thēr*, a beast]

eutrophic (ūtrof'ik), a. of a body of water, rich in dissolved nutrients and supporting an abundance of plant life.

eV, (*abbr.*) electron-volt.

EVA, (*abbr.*) extravehicular activity.

evacuate (ivak'ūāt), v.t. to make empty, esp. to empty (the excretory passages); to discharge from the body; to form a vacuum in; to withdraw from, as troops; to remove inhabitants from (a danger zone). v.i. to withdraw or move from a place, to discharge waste from the body. **evacuant**, n., a. purgative. **evacuation**, n. **evacuee** (-ē'), n. a person transferred from a danger zone.

evade (ivād'), v.t. to avoid or elude by artifice, stratagem or sophistry; to avoid (doing something), to shirk; to defeat, baffle, foil. **evadable**, a.

evaginate (ivaj'ināt), v.t. to turn inside out, to unsheathe (as a tubular organ). **evagination**, n.

evaluate (ival'ūāt), v.t. to determine the value, amount or worth of, to appraise. **evaluation**, n.

evanesce (evənes'), v.i. to disappear, vanish; to be dissipated in vapour. **evanescence**, n. **evanescent**, a. disappearing gradually, fading; fleeting; **evanescently**, adv.

†**evangel** (ivăn'jəl), n. the Gospel; one of the four Gospels; a gospel, a doctrine of political or social reform. **evangelical** (ēvănjel'-), a. pertaining to the Gospel; according to the doctrine of the Gospel; proclaiming or maintaining the truth taught in the Gospel; accepting for gospel only what Protestants consider the fundamental teaching of Scripture, the doctrines of the Fall, Christ's atonement, and salvation by faith not works; firmly believing in and actively promoting a cause. n. a member of the evangelical party in the Church, esp. in the Church of England, where it corresponds to the Low Church Party. **evangelicalism**, n. **evangelically**, adv. **evangelist**, n. one of the four writers of the Gospels (Matthew, Mark, Luke and John); a preacher of the Gospel; a lay preacher; one who evangelizes or believes in evangelism; an enthusiastic and active supporter of a cause. **evangelism**, n. preaching of the Gospel; evangelicalism; fervent support of a cause. **evangelistic** (-lis'-), a. **evangelize, -ise**, v.t. to preach the Gospel to; to convert to Christianity. v.i. to preach the Gospel; to promote a cause, esp. to gain support on members. **evangelization**, n.

evaporate (ivap'ərāt), v.t. to convert into vapour; to vaporize; to drive off the moisture from by heating or drying. v.i. to become vapour; to pass away in vapour; to give off moisture; (*coll.*) to disappear, to vanish. **evaporated milk**, n. unsweetened tinned milk from which some of the water has been evaporated. **evaporable**, a. **evaporation**, n. **evaporative**, a. **evaporator**, n.

evasion (ivā'zhən), n. the act of evading or escaping (as from a question, argument or charge); a subterfuge, an equivocation. **evasive** (-siv), a. **evasively**, adv. **evasiveness**, n.

Eve (ēv), n. (*Bibl.*) the wife of Adam and mother of mankind; the personification of womankind. **daughter of Eve**, (*often derog.*) a woman, usu. with an implication of curiosity, vanity etc.

eve (ēv), n. the evening before a holiday or other event or date; the period immediately preceding some important event; (*poet.*) evening.

even[1] (ē'vən), n. (*poet.*) evening. **evenfall**, n. (*poet.*) early evening. **evensong**, n. a form of worship for the evening; the time for evening prayer. **eventide**, n. (*poet.*) evening. **eventide home**, n. a home for elderly people.

even[2] (ē'vən), a. level, smooth, uniform; on the same level, in the same plane (with); parallel; regular, unfluctuating; capable of being divided by the number 2 without any remainder; opposed to odd; equal; equally probable; exact in number or amount; showing neither profit nor loss; equally balanced, fair, impartial; unvarying, equable, unruffled. **on an even keel** KEEL. **to be** or **get even with**, to revenge oneself on. **to even out**, to become even or equal. **to even up**, to balance, to make equal. **even chance**, n. an equal likelihood of success or failure. **even date**, n. (*Comm.*) today. **even-handed**, a. impartial, equitable, fair. **even-handedly**, adv. **even-handedness**, n. **even money**, n. an equal amount placed on each side of a bet. **evenly**, adv. **evenness**, n. **evens**, n. odds quoted on a racehorse etc. such that if it wins the person betting gains an amount equal to the stake.

even[3] (ē'vən), v.t. to make smooth or level; to place on a level. **evener**, n.

even[4] (ē'vən), adv. to a like degree, equally; as much as, so much as (expressing unexpectedness, surprise, concession or emphasis, a comparison being implied); evenly; exactly, just, simply, neither more nor less than. **even so**, exactly; yes.

evening (ēv'ning), n. the close or latter part of the day; the period from sunset to dark, or from sunset to bed-time; (*fig.*) the close or decline, as of life; the latter part. **evening dress**, n. the dress prescribed by convention for wearing for a formal occasion in the evening. **evening primrose**, n. a plant belonging to the genus *Oenothera*, the yellow flowers of which usually open in the evening. **evening star**, n. (also called Hesperus or Vesper) Jupiter, Mercury or Venus when visible in the west in the evening.

event (ivent'), n. anything that happens, as distinguished from a thing that exists; an occurrence, esp. one of great importance; the contingency or possibility of an occurrence; the consequence of any action; outcome, conclusion; any item in a programme of games, contests etc. **at all events**, in any case, at any rate. **in the event of**, if so, if it so happens. **three-day event**, an equestrian competition taking place over three days and including dressage, show-jumping and cross-country riding. **eventer**, n. a horse that takes part in three-day events. **eventful**, a. full of events; attended by important changes. **eventing**, n. taking part in three-day events. **eventless**, a.

eventide (ēv'əntīd), EVEN[1].

eventual (iven'chuəl), a. happening as a consequence of something else; finally resulting, ultimate, final. **eventuality** (-al'-), n. **eventually**, adv. **eventuate**, v.i. to happen; to come to pass; to result; to turn out (well or ill).

ever (ev'ə), adv. at all times, always; continually; at any time; in any degree. **ever after**, or **since**, continually after a certain time. **ever and anon**, now and then; at one time and another. **ever so**, to any degree or extent conceivable. **ever such**, (*coll. or dial.*) very. **for ever**, for all future time, eternally; incessantly. **or ever** OR[2]. **everglade** (ev'əglād), n. (*US*) a low, marshy tract of country, interspersed with patches covered with high grass; (*pl.*) the region of this character in Florida. **evergreen**, *a.* retaining foliage throughout the year; (*fig.*) always young or fresh. *n.* a plant which retains its foliage through the year. **everlasting**, *a.* lasting for ever, eternal; continual, unintermittent; (*fig.*) interminable, tiresome; of flowers, not changing colour when dried. *n.* eternity; a plant whose flowers retain their colour when dried. **everlastingly**, *adv.* **everlastingness**, *n.* **evermore**, *adv.* always, eternally.

evert (ivœ't), *v.i.* to turn outwards or inside out. **eversion** (-shən), *n.*

every (ev'ri), *a.* each of a number, all separately; each. **every bit**, quite; the whole. **every now and then, every now and again, every so often**, from time to time; at brief intervals. **every one**, each one. **every other** OTHER. **everybody**, *n.* every person. **everybody else**, (*collect.*) all other persons. **everyday**, *a.* met with or happening daily; worn or used on ordinary occasions; common, usual; commonplace. *adv.* on each or every day; continually. **Everyman**, *n.* a figure in a mediaeval morality play who represents everyone or mankind; (**everyman**) the person in the street. **everyone**, *n.* everybody. **everything**, *n.* (*collect.*) all things; all of the things making up a whole; something of the highest importance. **everyway**, *adv.* in every way; in every respect. **everywhere**, *adv.* in every place.

evict (ivikt'), *v.t.* to dispossess by legal process; to eject from lands or property by law. **eviction**, *n.* **evictor**, *n.*

evidence (ev'idəns), *n.* anything that makes clear or obvious; ground for knowledge, indication, testimony; that which makes truth evident, or renders evident to the mind that it is truth; (*Law*) information by which a fact is proved or sought to be proved, or an allegation proved or disproved; such statements, proofs etc. as are legally admissible as testimony in a court of law. *v.t.* to make evident, to attest. **in evidence**, received or offered as evidence in a court of law; (*coll.*) plainly visible, conspicuous. **to turn King's, Queen's, evidence**, to bear witness against one's accomplice in return for a free pardon. **evident**, *a.* open or plain to the sight; manifest, obvious. **evidential** (-den'-), *a.* affording evidence; proving conclusively. **evidentially**, *adv.* **evidentiary** (-den'shəri), *a.* obviously, manifestly; apparently. **evidently**, *adv.*

evil (ē'vl), *a.* bad, injurious; morally bad, wicked; calamitous, unlucky, producing disastrous results; disagreeable, unpleasant. *adv.* in an evil manner. *n.* an evil thing; that which injures or displeases, calamity, harm; sin, depravity, malignity. **King's evil**, scrofula. **the Evil One**, the Devil. **evil-doer**, *n.* one who does evil, a wrong-doer, a malefactor. **evil eye**, *n.* a supposed power of injuring by the look. **evil-eyed**, *a.* malicious; looking malicious; having the power of the evil eye. **evil-minded**, *a.* unkindly and injuriously disposed. **evilly**, *adv.*

evince (ivins'), *v.t.* to show clearly; to indicate, to make evident; to demonstrate. **evincible**, *a.* **evincive**, *a.*

eviscerate (ivis'ərāt), *v.t.* to disembowel; to empty of all that is vital. **evisceration**, *n.*

evoke (ivōk'), *v.t.* to call up, to summon forth (a memory etc.), esp. from the past; to elicit or provoke; to cause (spirits) to appear. **evocation**, *n.* **evocative** (ivok'ətiv), *a.*

evolute (ev'əloot), *n.* (*Geom.*) a curve from which another is described by the end of a thread gradually wound upon or unwound from the former, thus forming the locus of the centres of curvature of the other, which is called the INVOLUTE.

evolution (-evəloo'shən, ev-), *n.* development, as of a plot, design or political, social or planetary system etc.; the derivation of forms of life from early forms of a simpler character or from a single rudimentary form; the theory based on this principle, opp. to that of special creation (see CREATIONISM); (*Math.*) the extraction of roots from any given power, the reverse of involution; the evolving of gas, heat etc.; (*Mil., Nav.*) doubling of ranks or files, countermarching or other changes of position, by which the disposition of troops or ships is changed; (*pl.*) movements, changes of position etc. in dancing etc. **evolutional, -tionary**, *a.* produced by or pertaining to evolution. **evolutionism**, *n.* the theory or doctrine of evolution. **evolutionist**, *n.* **evolutionistic** (-nis'-), *a.* **evolutive**, *a.*

evolve (ivolv'), *v.t.* to develop; to bring to maturity; to give off (gas, heat etc.); to produce by the process of evolution. *v.i.* to develop; to undergo evolution. **evolvable**, *a.* **evolvement**, *n.* **evolver**, *n.*

evulsion (ivŭl'shən), *n.* the act of forcibly plucking or extracting.

ewe (ū), *n.* a female sheep.

ewer (ū'ə), *n.* a wide-mouthed pitcher or large jug for water.

Ex., (*abbr.*) Exodus.

ex [1] (eks), *prep.* (*Comm.*) from, out of, sold from; without. **ex dividend**, not including the next dividend.

ex [2] (eks), *n.* (*coll.*) a former spouse, boyfriend or girlfriend.

ex [3], (*abbr.*) examined; example; except, exception; exchange; executive.

ex-, *pref.* out, forth, out of; thoroughly; without, -less; formerly, previously occupying the position of; as *exceed, exclude; exacerbate, excruciate; exonerate, expatriate; ex-chancellor, ex-president.*

exacerbate (igzas'əbāt, -sas'-), *v.t.* to irritate, to exasperate, to embitter; to aggravate; to increase the violence of (as a disease). **exacerbation**, *n.*

exact [1] (igzakt'), *a.* precisely agreeing in amount, number or degree; accurate, strictly correct; precise, strict, punctilious; consummate, perfect. **not exactly**, (*iron.*) not at all. **exact sciences**, *n.pl.* those in which mathematical accuracy is attainable. **exactitude** (-titūd), *n.* exactness, precision. **exactly**, *adv.* in an exact manner; quite so; precisely, just so (in answer to a question or affirmation); in express terms. **exactness**, *n.* the quality of being exact.

exact [2] (igzakt'), *v.t.* to compel to be paid or surrendered; to demand of right, to insist on, to require authoritatively. *v.i.* to practise extortion. **exactable**, *a.* **exacting**, *a.* severe or excessive in making demands. **exactingly**, *adv.* **exaction**, *n.* the act of exacting; a forcible, illegal or exorbitant demand; extortion; that which is exacted. **exactor**, *n.*

exaggerate (igzaj'ərāt), *v.t.* to heighten, to overstate, to represent as greater than truth warrants; to increase, intensify, aggravate; to represent (features, colours etc.) in a heightened manner. *v.i.* to use or be given to exaggeration. **exaggeratedly**, *adv.* **exaggeration**, *n.* **exaggerative**, *a.* **exaggeratively**, *adv.* **exaggerator**, *n.*

exalt (igzawlt'), *v.t.* to raise in dignity, rank, character etc. to elate; to praise, extol, glorify; to intensify. **exaltation** (egzawltā'-), *n.* the act of exalting; elevation in rank, dignity etc.; elation, rapture. **exaltedly**, *adv.* **exaltedness**, *n.*

examine (igzam'in), *v.t.* to inquire into, to investigate, scrutinize; to consider critically; to inspect; to question (as a witness); to test the capabilities, qualifications, knowledge of etc., by questions and problems; to inspect ((a part of) a patient's body) with a view to diagnosing possible illness. **examinable**, *a.* **examination**, *n.* the act of examining; careful inspection, scrutiny or inquiry; the process of testing the capabilities or qualifications of a candidate for any post, or the progress, attainments or knowledge of a student; the act of inspecting a patient's body to diagnose possible illness; (*Law*) a careful inquiry into facts by taking evidence. **examination paper**, *n.* a paper containing questions for candidates, pupils etc.; a series of answers to such questions by an examinee. **examinee** (-nē'), *n.* **examiner**, *n.*

example (igzahm'pl), *n.* a sample, a specimen; a model or pattern; any person, fact or thing illustrating a general rule; a warning; a precedent, an instance; a problem or

exanimate 281 **exclude**

exercise (in mathematics etc.) for the instruction of students. *v.t.* to exemplify.
exanimate (igzan'imət), *a.* lifeless, dead; without animation, depressed, spiritless.
exanthema (eksanthē'mə), *n.* (*pl.* **-mata** (-tə)) any disease which is accompanied by a skin rash, e.g. measles.
exarch (ek'sahk), *n.* a governor of a province under the Byzantine Empire; in the E Orthodox Church, a grade in the ecclesiastical hierarchy between a patriarch and metropolitan. **exarchate** (ek'sahkāt, -sah'-), *n.*
exasperate (igzas'pərāt), *v.t.* to aggravate; to anger; to irritate to a high degree. **exasperation**, *n.*
exc., (*abbr.*) except.
ex cathedra (eks kəthē'drə), from the chair, with authority. [L]
excavate (eks'kəvāt), *v.t.* to hollow out; to form by digging or hollowing out; to remove by digging; to uncover by digging, to dig out, esp. for archaeological research. **excavation**, *n.* **excavator**, *n.*
exceed (iksēd'), *v.t.* to go or pass beyond; to be more or greater than; to do more than is warranted or required by; to surpass, to outdo, to excel. *v.i.* to excel. **exceeding**, *a.* very great in amount, duration, extent or degree. **exceedingly**, *adv.* very much.
excel (iksel'), *v.t.* (*past, p.p.* **excelled**) to surpass in qualities; to exceed, to outdo. *v.i.* to be superior, distinguished or pre-eminent (in or at). **excellence** (ek'-), *n.* the state of excelling; superiority, pre-eminence; surpassing virtue, goodness or merit; that in which any person or thing excels; an excellent quality, feature or trait. **excellency**, *n.* excellence; a title of honour given to a governor, an ambassador, a commander-in-chief and certain others of high rank or position. **excellent**, *a.* surpassing others in some good quality; of great virtue, worth etc. **excellently**, *adv.*
except (iksept'), *v.t.* to leave out, to omit, to exclude. *v.i.* to make objection (to or against). *prep.* not including, exclusive of, omitting, but. *conj.* unless; but. **excepting**, *prep.* (*usu. after* not) omitting, with the exception of.
exception, *n.* the act of excepting; that which is excepted; an instance of that which is excluded from or is at variance with a rule, class or other generalization; an objection, disapproval. **to take exception**, to object, to find fault; to find offensive. **exceptionable**, *a.* liable to objection; objectionable. **exceptional**, *a.* forming an exception; unusual, extraordinary, unprecedented. **exceptionality** (-nal'-), *n.* **exceptionally**, *adv.*
excerpt[1] (iksœpt'), *v.t.* to make an extract of or from. **excerptible**, *a.* **excerption**, *n.* **excerptor**, *n.*
excerpt[2] (ek'sœpt), *n.* an extract or selection from a book, play, film etc.
excess (ikses', ek'-), *n.* that which exceeds what is usual or necessary; the quality, state or fact of exceeding the ordinary measure, proportion or limit; the amount by which one number or quantity exceeds another; (*usu. pl.*) transgression of due limits; intemperance, over-indulgence, extravagance. **excess fare**, *n.* the amount paid for travelling beyond the point for which a ticket has been taken or in a higher class. **excess luggage**, *n.* a quantity above the weight allowed free carriage. **excess postage**, *n.* payment due when not enough stamps have been put on a letter or package. **excessive** (-ses'-), *a.* **excessively**, *adv.*
exchange (ikschānj'), *v.t.* to give or receive in return for something else; to hand over for an equivalent in kind; to give and receive in turn, to interchange; to give, resign or abandon (as one state or condition for another). *v.i.* to be given or received in exchange; to pass from one post or office to another by taking the place of another. *n.* the act of exchanging; a parting with one article or commodity for an equivalent in kind; the act of giving and receiving reciprocally, interchange; the act of resigning one state for another; that which is given or received in exchange; exchanging of coin for its value in coins of the same or another country; the system by which goods or property are exchanged and debts settled, esp. in different countries, without the transfer of money; the place where merchants, brokers etc. meet to transact business; the central office where telephone connexions are made; (*coll.*) an exchange student or teacher. **bill of exchange** BILL[3]. **rate of exchange** EXCHANGE RATE. **to exchange words, blows**, to quarrel verbally or physically. **exchange rate**, *n.* the ratio at which the currency of one country can be exchanged for that of another. **exchange student, teacher**, *n.* one who exchanges posts with a corresponding person from another country. **exchangeable**, *a.* **exchangeability** (-bil'-), *n.* **exchanger**, *n.*
exchequer (ikschek'ə), *n.* the State treasury; the Government department dealing with the public revenue; finances or pecuniary resources; the Court of Exchequer. **Chancellor of the Exchequer** CHANCELLOR. **Court of Exchequer**, a court originally intended for the recovery of debts due to the king and to vindicate his proprietary rights etc., but afterwards developed into an ordinary law-court with a jurisdiction in equity, now merged into the Queen's (King's) Bench Division.
excise[1] (ek'sīz, iksīz'), *n.* a tax or duty on certain articles produced and consumed in a country (in the United Kingdom on spirits, beer and tobacco); the branch of the Civil Service which collects and manages the excise duties; a tax, often in the form of a licence to carry on certain trades. *v.t.* (iksīz'), to impose an excise duty on. **excisable**, *a.* subject or liable to excise duty.
excise[2] (iksīz'), *v.t.* to cut out (part of a book or of the body). **excision** (-sizh'ən), *n.*
excite (iksīt'), *v.t.* to rouse, to stir into action, energy or agitation; to stimulate, to bring into activity; to inflame the spirits or emotions of; to arouse sexually; to provoke, to bring about by stimulating; to set up electric activity in; to magnetize the poles of (an electric machine); to raise (an electron, atom etc.) to a higher energy level. **excitable**, *a.* susceptible of excitement; easily excited; responding to stimuli. **excitability** (-bil'-), *n.* **excitant** (ek'si-), *a.* stimulating; tending to excite. *n.* that which excites; a stimulant. **excitation** (eksi-), *n.* **excitative, excitatory**, *a.* **excited**, *a.* **excitedly**, *adv.* **excitement**, *n.* **exciter**, *n.* **exciting**, *a.* stimulating; producing excitement. **excitingly**, *adv.* **excitor**, *n.* an afferent nerve belonging to the spinal group.
exclaim (iksklām'), *v.t., v.i.* to cry out or to utter in an abrupt or passionate manner. **exclamation** (ekskləmā'shən), *n.* the act of exclaiming; an expression of surprise, pain etc. **exclamation mark**, *n.* a sign (!) indicating emotion etc. **exclamatory** (-klam'ə-), *a.*
exclave (eks'klāv), *n.* part of a country disjoined from the main part and surrounded by foreign territory, where it is considered an enclave.
exclosure (iksklō'zhə), *n.* an area shut off from entry or intrusion.
exclude (iksklood'), *v.t.* to shut out, to prevent from coming in; to prevent from participating; to debar; to expel and keep out; to reject, to leave out. **exclusion** (-zhən), *n.* **exclusion order**, *n.* one preventing the entry into Britain of anyone known to be involved in terrorism. **exclusionary**, *a.* **exclusive** (-siv), *a.* shutting out or tending to shut out; desiring to shut out; limited or belonging to a single individual or group; fastidious in the choice of associates or members; snobbish; select, stylish, fashionable; sole; not inclusive (of); excluding all else; excluding all that is not specified. *n.* a story published in only one newspaper or journal. **exclusive zone**, *n.* an area of a country's territorial waters in which exploitation by other countries is officially banned. **exclusively**, *adv.* **exclusiveness**, *n.* **exclusivity** (-siv'-)

excommunicate (ekskəmū'nikāt), *v.t.* to exclude from the communion and privileges of the Church; to expel. **excommunication**, *n.* **excommunicative, -catory**, *a.*

excoriate (ekskaw'riāt), *v.t.* to strip the skin from; to criticize severely.

excrement (eks'krəmənt), *n.* refuse matter discharged from the body after digestion, faeces. **excremental** (-men'-), *a.* **excrementitious** (-tish'əs), *a.*

excrescence (ikskres'əns), **-ency** (-si), *n.* an outgrowth, esp. when unnatural, useless or disfiguring. **excrescent**, *a.* growing abnormally or redundantly.

excrete (ikskrēt'), *v.t.* to separate and discharge or eliminate (superfluous matter) from the organism. **excreta** (-tə) *n.pl.* matter discharged from the body, esp. faeces and urine. **excretal**, *a.* **excretion**, *n.* the ejection of waste matter from the body; that which is excreted. **excretive**, *a.* **excretory**, *a.*

excruciate (ikskroo'shiāt), *v.t.* to torture; to inflict severe pain or mental agony upon. **excruciating**, *a.* extremely painful, agonizing; intense; (*coll.*) very bad, inferior. **excruciatingly**, *adv.* **excruciation**, *n.*

exculpate (eks'kəlpāt, iks'kŭl-), *v.t.* to clear from a charge; to free from blame, exonerate; to vindicate. **exculpation** (ekskŭlpā'-), *n.* **exculpatory** (ikskŭl'pə-), *a.*

excursion (ikskœ'shən), *n.* a journey or trip for pleasure; a wandering from the subject, a digression; a deviation from the fixed course. **excursion fare, ticket**, *n.* a special cheap fare or ticket allowed on some journeys on public transport. **excursional, -ary**, *a.* **excursionist**, *n.* one who goes on an excursion. **excursive**, *a.* rambling, deviating, exploring. **excursively**, *adv.* **excursiveness**, *n.*

excursus (ikskœ'səs), *n.* a dissertation appended to a work, containing an exposition of some point raised or referred to in the text.

excuse[1] (ikskūz'), *v.t.* to free from blame or guilt; to pardon, to acquit; to ask pardon or indulgence for; to serve as a vindication or apology for, to justify; to relieve of or exempt from an obligation or duty; to allow to leave; to dispense with. **to be excused**, (*euphem.*) to go to the lavatory. **excuse me**, *int.* expressing apology or disagreement. **excuse-me**, *n.* a dance during which partners may be changed on request. **excusable**, *a.* **excusableness**, *n.* **excusably**, *adv.* **excusatory**, *a.* **excuser**, *n.*

excuse[2] (ikskūs'), *n.* a plea offered in extenuation of a fault or for release from an obligation, duty etc.; an apology, a justification; the ground or reason for excusing; a pretended reason; the act of excusing.

ex-directory (eksdirek'təri), *a.* of a telephone number, not listed in a telephone directory and not revealed to inquirers; of a person, having such a telephone number.

ex dividend EX[1].

exeat (ek'siāt), *n.* leave of absence, as to a student at university; permission granted by a bishop to a priest to go out of his diocese; permission by a Roman Catholic bishop to one of his subjects to take orders in another diocese. **exeant** (-ant), *n.* leave of absence to several persons.

exec., (*abbr.*) executive; executor.

execrate (ek'sikrāt), *v.t.* to curse, to imprecate evil upon; to detest; to denounce as evil. **execrable**, *a.* detestable, abominable; very bad. **execrably**, *adv.* **execration**, *n.* **execrative**, *a.* **execratory**, *a.*

execute (ek'sikūt), *v.t.* to carry into effect, to put in force; to perform, to accomplish, complete; to perform what is required to give validity to (a legal instrument); as by signing and sealing; to discharge (a duty, function, office etc.); to produce (as a drawing); to play or perform (as a piece of music); to inflict capital punishment on. **executable**, *a.* **executant** (igzek'ū-), *n.* one who performs; (*Mus.*) a performer on any instrument. **execution**, *n.* the act of executing; performance, accomplishment; the infliction of capital punishment; the mode of performing a work of art, skill, technique; the carrying into effect of the judgment of a court; the warrant empowering an officer to carry a judgment into effect, esp. one authorizing the seizure of a debtor's goods in default of payment. **executioner**, *n.* one who inflicts capital punishment; one who kills. **executive** (igzek'ūtiv), *a.* having the function or power of executing; pertaining to performance or carrying into effect; carrying laws, decrees etc. into effect pertaining to or for use by business executives; luxurious expensive. *n.* the person or body of persons carrying laws, ordinances, sentences etc. into effect; the administrative branch of a government, business organization etc.; a person who excersises managerial or administrative control. **executive officer**, *n.* the second in command of a military or naval unit. **executor** (igzek'ūtə), *n.* one who executes, esp. a person appointed by a testator to carry out the provisions of a will. **literary executor**, *n.* a person appointed to deal with the copyrights and unpublished works of a deceased author. **executorial** (-taw'ri-), *a.* **executorship**, *n.* **executrix** (-triks), *n. fem.* (*pl.* **-trices**, -tri'sēz)

exegesis (eksijē'sis), *n.* (*pl.* **-geses**, -sēz) exposition, interpretation, esp. of the Scriptures. **exegete** (ek'sijēt), **exegetist**, *n.* one who is skilled in the exegesis of the Scriptures. **exegetic** (-jet'-), *a.* **exegetics**, *n.sing.* scientific interpretation, esp. of Scripture. **exegetical**, *a.* **exegetically**, *adv.*

exemplar (igzem'plə), *n.* a pattern or model to be copied; a noted example; a typical example; a copy, as of a book. **exemplary**, *a.* serving as a pattern or model; worthy of imitation; serving to exemplify, illustrative; serving as a warning. **exemplary damages**, *n.pl.* damages given in excess of the loss suffered by the plaintiff, in order to act also as punishment to the defendant. **exemplarily**, *adv.* **exemplariness**, *n.*

exemplify (igzem'plifī), *v.t.* to illustrate by example; to be an example of; to make an authenticated copy of. **exemplifiable**, *a.* **exemplification** (-fi-), *n.*

exemplum (igzem'pləm), *n.* an example; a short story or anecdote which illustrates a moral.

exempt (igzempt'), *a.* free (from); not liable or subject to. *n.* one who is exempted or freed, e.g. from a duty. *v.t.* to free or allow to be free; to grant immunity (from). **exemption**, *n.* the state of being exempt; immunity; freedom from the obligation, duty etc.

exequies (ek'sikwiz), *n.pl.* funeral rites; the ceremony of burial. **exequial** (-sē'-), *a.*

exercise (ek'səsīz), *n.* the act of using, employing or exerting; practice (of a function, virtue, occupation, art etc.); systematic exertion of the body for the sake of health; exertion for the training of the body or mind; a task set for this purpose; a composition for the improvement of a player or singer; (*pl.*) military manoeuvres or drill; *v.t.* to employ, to exert, to put in practice or operation; to perform the duties of, to fulfil; to train; to keep employed or busy; to make anxious or solicitous, to perplex, worry; to exert (muscles, brain, memory etc.) so as to develop their power. *v.i.* to take or do exercise for health or training. **the object of the exercise**, the purpose of a particular action or activity. **exercise bike**, *n.* a static machine pedalled like a bicycle for exercise. **exercise book**, *n.* a book for written school work. **exercisable**, *a.* **exerciser**, *n.*

exergue (eks'œg), *n.* the small space beneath the base line of a subject engraved on a coin or medal.

exert (igzœt'), *v.t.* to employ or apply (as strength, power, or ability); to put in action or operation. **to exert oneself**, to strive, to use effort. **exertion**, *n.*

exeunt (ek'siunt), *v.i.* (*stage direction*) they go off the stage, they retire.

exfoliate (eksfō'liāt), *v.i.* of skin, bark, rocks etc., to shed or come off in flakes or scales; to separate into flakes. *v.t.* to remove or shed in flakes; to cause to come off in, or

ex gratia 283 **expand**

separate into, flakes. **exfoliation,** *n.* **exfoliator, exfoliant,** *n.* a substance or device used to remove dead skin cells.
ex gratia (eksgrä'shə), *a., adv.* as an act of favour, and with no acceptance of liabilty.
exhale (ikshāl', igzāl'), *v.t.* to emit, or cause to be emitted, in vapour; to breathe out. *v.i.* to be given off as vapour; to make an expiration, as distinct from inhaling. **exhalant,** *a.* **exhalation** (eksəlā'shən), *n.* the act or process of exhaling; that which is exhaled; a breathing out; vapour, mist; an emanation.
exhaust (igzawst'), *v.t.* to draw off; to empty by drawing out the contents; to create a vacuum in, thus; to use up the whole of, to consume; to wear out by exertion; to drain of resources, strength or essential properties; to study, discuss, treat the whole of (a subject) in a thorough manner. *n.* the discharge or escape of steam, gas, vapour etc. from an engine after it has performed its work; the gases and vapour emitted; exhaust pipe. **exhaust-pipe,** *n.* a pipe conducting spent steam etc. from the cylinder of an engine. **exhauster,** *n.* **exhaustible,** *a.* **exhaustibility** (-bil'-), *n.* **exhausting,** *a.* tending to exhaust or tire out completely. **exhaustion** (-chən), *n.* the act of exhausting; the state of being exhausted; a complete loss of strength. **exhaustive,** *a.* tending to exhaust (esp. a subject), comprehensive. **exhaustively,** *adv.* **exhaustiveness,** *n.*
exhibit (igzib'it), *v.t.* to offer to public view; to present officially or for inspection; to show, to display, to manifest. *v.i.* to display something for public view. *n.* anything exhibited; an article or collection of articles sent to an exhibition; a document or other article produced in court and used as evidence. **exhibition** (eksibi'-), *n.* the act of exhibiting; a display; the act of allowing to be seen, as temper; a public display of works of art or manufacture, natural products etc.; an allowance to a student in college, school etc. **to make an exhibition of oneself,** to behave so as to appear foolish or contemptible. **exhibitioner,** *n.* one who has obtained an exhibition at a college or school. **exhibitionism,** *n.* a tendency to show off, to attract attention to oneself; a tendency to indecent exposure in public. **exhibitionist,** *n.* **exhibitor,** *n.* **exhibitory,** *a.*
exhilarate (igzil'ərāt), *v.t.* to gladden, to enliven, to animate. **exhilarant,** *a.* **exhilarating,** *a.* **exhilaratingly,** *adv.* **exhilaration,** *n.* **exhilarative,** *a.*
exhort (igzawt'), *v.t.* to incite or urge by words; to advise or encourage strongly by argument. *v.i.* to deliver an exhortation. **exhortation** (-eg-), *n.* the act or practice of exhorting; language or an address intended to incite or encourage. **exhortative,** *a.* **exhortatory,** *a.*
exhume (igzūm', ekshūm'), *v.t.* to disinter; to unearth, to discover. **exhumation** (ekshūmā'-, egzū-), *n.*
exigence (ek'sijəns, -sij'-), **-gency** (iksij'-, igzij'-), *n.* urgent need, demand, necessity; a state of affairs demanding immediate action or remedy, an emergency. **exigent,** *a.* urgent, pressing; demanding more than is reasonable, exacting.
exiguous (igzig'ūəs, iksig'-), *a.* small, slender, scanty. **exiguity** (eksi-), **exiguousness,** *n.*
exile (eg'zīl, ek'sīl), *n.* banishment, expatriation; long absence from one's native country, whether voluntary or enforced; one who is banished, or has been long absent from his or her native country. *v.t.* to banish from one's native country. **exilian** (-il'-), **exilic** (-il'-), *a.* pertaining to exile or banishment, esp. to that of the Jews in Babylon.
exist (igzist'), *v.i.* to be, to have actual being; to live; to continue to be; to live or have being under specified conditions. **existence,** *n.* the state of being or existing; continuance of being; life; mode of existing; a thing that exists; all that exists. **existent,** *a.* having being or existence, existing, actual. **existential** (egzisten'shəl, eksi-), *a.* pertaining to or consisting in existence; pertaining to existentialism. **existentialism,** *n.* a philosophy largely deriving from Kierkegaard, implying a special conception of the idea of existence. 'It considers self as a unity of finiteness and freedom, of involvement in natural process and transcendence over process' (R. Niebuhr). **existentialist,** *n.*, *a.* **existentially,** *adv.*
exit (ek'sit, -eg'zit), *n.* the departure of an actor from the stage; departure, esp. from this life; a going out; freedom to go out; a way out. *v.i.* to go out, leave, depart; *(stage direction)* to go off the stage. **exit poll,** *n.* an unofficial poll taken by asking people leaving a polling station how they have voted.
ex-libris (ekslib'ris), *n.* *(often as pl.)* a book-plate, a label bearing an owner's name, crest, device etc. [L, out of books, from the library (of)]
exo- *comb. form.* pertaining to the outside of anything.
exobiology (eksobiol'əji), *n.* astrobiology. **exobiologist,** *n.*
Exocet® (ek'səset), *n.* a French-built surface-skimming missile that can be launched from surface or air.
exocrine (ek'səkrin), *n.*, *a.* (a gland) producing secretions that are released through a duct.
Exod., *(abbr.)* Exodus.
exodus (eks'ədəs), *n.* a departure, esp. of a large body of persons; the departure of the Israelites from Egypt under Moses; (**Exodus**) the second book of the Old Testament, narrating this event; *(coll.)* the departure of many people at once.
ex officio (eksəfish'iō), *adv.* by virtue of one's office. *a.* official. [L]
exogamy (eksog'əmi), *n.* the custom prevalent among some tribes forbidding a man to marry a woman of his own tribe. **exogamic** (-gam'-), **exogamous,** *a.*
exogenous (eksoj'ənəs), *a.* developing externally; having external origins.
exon[1] (ek'son), *n.* one of the four officers of the Yeomen of the Guard.
exon[2] (ek'son), *n.* a small section of a nucleic acid that contains information coding for protein synthesis, opp. to *intron.*
exonerate (igzon'ərāt), *v.t.* to free from a charge or blame; to exculpate; to relieve from a duty, obligation or liability. **exoneration,** *n.* **exonerative,** *a.*
exophthalmia (eksofthal'miə), **exophthalmos** (-məs), *n.* protrusion of the eyeball. **exophthalmic,** *a.*
exorbitant (igzaw'bitənt), *a.* out of all bounds, grossly excessive, inordinate, extravagant. **exorbitance,** *n.* **exorbitantly,** *adv.*
exorcize, ise (ek'sawsīz), *v.t.* to expel (as an evil spirit) by adjurations, prayers and ceremonies; to free or purify from unclean spirits. **exorcizer, -iser,** *n.* **exorcism,** *n.* **exorcist,** *n.*
exordium (igzaw'diəm), *n.* *(pl.* **-diums, -dia)** the beginning of anything, esp. the introductory part of a literary work or discourse. **exordial,** *a.*
exoskeleton (eksōskel'itən), *n.* an external skeleton formed by a hardening of the integument.
exosphere (ek'səsfiə), *n.* the outermost layer of the earth's atmosphere.
exoteric (eksəte'rik), **-ical,** *a.* external, public, fit to be imparted to outsiders; comprehensible to the public, as opposed to *esoteric;* not admitted to esoteric doctrines; ordinary, popular. **exoterically,** *adv.*
exothermal (eksōthoe'məl), **exothermic** (-mik), *a. (Chem.)* involving the evolution of heat.
exotic (igzot'ik), *a.* foreign; introduced from a foreign country; *(coll.)* rare, unusual. *n.* anything foreign; anything introduced from a foreign country, as a plant or animal. **exotica** (-kə), *n.pl.* rare or unusual objects, esp. when forming a collection. **exotically,** *adv.* **exotic dancer,** *n.* a striptease or belly dancer. **exoticism** (-sizm), *n.* **exoticness,** *n.*
exp., *(abbr.)* export; exponential.
expand (ikspand'), *v.t.* to open or spread out; to distend, to cause to increase in bulk; to widen, to extend, to en-

expanse | 284 | **explicate**

large; to write out in full (what is condensed or abbreviated); (*Math.*) to develop into a series, to state in a fuller form. *v.i.* to become opened or spread out, distended, or enlarged in bulk; to speak or write in more detail; to become more talkative or sociable. **expanded,** *a.* of a plastic, produced in the form of a light cellular foam, used for packaging and heat insulation. **expanded metal,** *n.* sheet metal cut and formed into a lattice, used for reinforcing concrete etc. **expander,** *n.* **expanding universe,** *n.* the theory that the universe is ever expanding, based on the Doppler effect in the light from stars and galaxies. **expansible** (-sibl), *a.* **expansibility** (-bil'-), *n.* **expansile,** (-sil), *a.* capable of expanding; expansible.

expanse (ikspans'), *n.* that which is expanded; a wide, open extent or area; expansion.

expansion (ikspan'shən), *n.* the act of expanding; the state of being expanded; enlargement, extension, distension; extension of business, increase of liabilities, extension of the currency; increase of volume, as of steam in a cylinder. **expansionary,** *a.* **expansionism,** *n.* **expansionist,** *n.* one who advocates territorial expansion of a nation.

expansive (ikspan'siv), *a.* having the power of expanding; able or tending to expand; extending widely, comprehensive; frank, effusive done on a grand scale, magnificant, extravagant. **expansively,** *adv.* **expansiveness,** *n.*

ex parte (ekspah'ti), *a.*, *adv.* (*Law*) proceeding from one side only; in the interests of one side. [L, from one side]

expat (ekspat'), *n.* short for EXPATRIATE.

expatiate (ikspā'shiāt), *v.i.* to speak or write copiously on a subject; to wander at large. **expatiation,** *n.* **expatiatory,** *a.*

expatriate (ekspā'triāt), *v.t.* to exile (oneself or another); to emigrate; to deprive (oneself) of citizenship in one's country. *n.* (-ət), one living away from his/her own country. **expatriation,** *n.*

expect (ikspekt'), *v.t.* to look forward to; to regard as certain or likely to happen, to anticipate; to require as due; (*coll.*) to think, to suppose. *v.i.* to wait with anticipation. **expectancy, -ance,** *n.* the act or state of expecting, expectation; the state of being expected; prospect of possessing, enjoying etc.; that which is expected. **expectant,** *a.* expecting, waiting in expectation (of); anticipating, presumptive; pregnant. *n.* one who waits in expectation of something. **expectantly,** *adv.* **expectation** (ekspik-), *n.* the act or state of expecting, anticipation, a confident awaiting (of); (*pl.*) prospects (of inherited wealth, fortune); the ground for confident anticipation (of); the probability of a future event. **expectation of life, life expectancy,** the number of years which a person of a given age may, on the average of chances, expect to live. **expecting,** *a.* pregnant. **expectingly,** *adv.*

expectorate (ikspek'tərāt), *v.t.*, *v.i.* to discharge (matter) from the lungs or air-passages by coughing, hawking or spitting; to spit. **expectorant,** *a., n.* (a medicine) having the quality of promoting expectoration. **expectoration,** *n.* **expectorative,** *a.*

expedient (ikspē'diənt), *a.* promoting the object in view; advantageous, convenient; conducive to personal advantage; politic as opposed to just. *n.* that which promotes an object; an advantageous way or means; a shift, a contrivance. **expedience, -ency,** *n.* **expediential** (-en'-), *a.* **expediently,** *adv.*

expedite (ek'spədīt), *v.t.* to facilitate, to assist or accelerate the progress of; to dispatch. **expediter, -tor,** *n.* **expeditious** (-di'shəs), *a.* speedy, ready, active; done with dispatch. **expeditiously,** *adv.*

expedition (ekspədi'shən), *n.* speed, promptness, dispatch; a journey or voyage by an organized body for some definite object; the persons with their equipment engaged in this. **expeditionary,** *a.* relating to or constituting an expedition.

expel (ikspel'), *v.t.* (*past, p.p.* **expelled**) to drive or force out; to eject, to banish; to turn out formally (as from a school, college, or society). **expellable,** *a.* **expellent, -ant,** *n., a.*

expend (ikspend'), *v.t.* to spend, to lay out; to consume, to use up. **expendable,** *a.* likely to be or intended to be wasted. **expenditure** (-dichə), *n.* the act of expending; disbursement, consumption; the amount expended.

expense (ikspens'), *n.* a laying out or expending; cost, charge, outlay, price paid; (*pl.*) outlay in performance of a duty or commission; (*coll.*) money reimbursed for this; something requiring high or constant expenditure. **at the expense of,** at the cost of; to the discredit or detriment of. **expense account,** *n.* an account of expenses refunded to an employee by an employer. **expensive,** *a.* costly, requiring a large expenditure; extravagant, lavish. **expensively,** *adv.* **expensiveness,** *n.*

experience (ikspiə'riəns), *n.* practical acquaintance with any matter; knowledge gained by observation or trial; a particular instance of such knowledge; something undergone of an affecting or impressive nature. *v.t.* to make trial or proof of; to gain a practical knowledge of by trial or observation; to undergo, to feel, to meet with. **experienced,** *a.* taught by experience; practised, skilled; known from personal trial or observation.

experiential (ikspəriən'shəl), *a.* pertaining to or derived from experience. **experientialism,** *n.* the doctrine that all our ideas are derived from experience. **experientialist,** *a.* **experientially,** *adv.*

experiment (ikspe'rimənt), *n.* a trial, proof or test of anything; an act, operation or process designed to discover some unknown truth, principle or effect, or to test a hypothesis. *v.i.* to make an experiment or trial, test or proof of. **experimental** (-men'-), *a.* pertaining to, derived from, or founded upon experiment; practising experiments; tentative; empirical. **experimentalism,** *n.* reliance on, use of or advocacy of experiment. **experimentalist,** *n.* **experimentally,** *adv.* **experimentation,** *n.* the act or practice of making experiments. **experimenter,** *n.*

expert (ek'spœt), *a.* experienced, dexterous from use and experience; practised, skilful (at or in). *n.* one who has special skill or knowledge; a scientific or professional witness. **expert system,** *n.* a computer system designed to mimic human thought processes so that apparently intelligent dialogue with the machine is possible. **expertise** (-tēz'), *n.* expert skill, opinion or knowledge. **expertly,** *adv.* **expertness,** *n.*

expiate (ek'spiāt), *v.t.* to atone for; to make reparation or amends for. **expiable,** *a.* **expiation,** *n.* **expiator,** *n.* **expiatory,** *a.*

expire (ikspiə'), *v.t.* to breathe out from the lungs; to send forth, to emit. *v.i.* to breathe out; to die; to cease, to come to an end. **expiration** (ekspi-), *n.* the act of breathing out; cessation, termination. **expiratory,** *a.* **expiry,** *n.* expiration, termination.

explain (ikspān'), *v.t.* to make clear, plain, or intelligible; to expound and illustrate the meaning of; to account for. *v.i.* to give explanations. **to explain away,** to get rid of or modify the significance of by explanation. **to explain oneself,** to make one's meaning clear; to give an account of one's motives, intentions, conduct etc. **explainable,** *a.* **explainer,** *n.* **explanation** (eksplənā'shən), *n.* the act of explaining; the sense or definition given by an interpreter or expounder; the process of arriving at a mutual understanding or reconciliation; that which accounts for anything. **explanatory** (-splan'-), *a.* containing an explanation; serving to explain. **explanatorily,** *adv.*

expletive (iksplē'tiv), *a.* serving or introduced to fill out or complete. *n.* a word not necessary to the sense introduced to fill up; an interjection or word added for emphasis, esp. a profane exclamation. **expletory,** *a.* expletive.

explicate (eks'plikāt), *v.t.* to unfold the meaning of, make clear; to develop (the contents of an idea, proposition, etc.). **explicable,** *a.* capable of being explained. **explica-**

tion, *n.* **explicative, -catory** *a.*
explicit (iksplis'it), *a.* plainly expressed, distinctly stated, opposed to implied; definite; unreserved, outspoken; graphic, candid. **explicitly,** *adv.* **explicitness,** *n.*
explode (iksplōd'), *v.t.* to cause to burst or blow up with a loud report; to refute, expose, discredit (a theory, fallacy etc.). *v.i.* to burst or blow up with a loud report; to break forth with violence; to increase rapidly. **exploded,** *a.* burst, blown up; discredited; of a diagram or photograph, showing the constituent parts separately but in their correct relative positions. **exploder,** *n.* **exploding,** *n., a.* **exploding star,** *n.* (*Astron.*) a nova or supernova.
exploit[1] (ek'sploit), *n.* a feat, a great or noble achievement; an adventure.
exploit[2] (eksploit'), *v.t.* to turn to account; to utilize, esp. to make use of for one's own profit. **exploitable,** *a.* **exploitation** (eksploi-), *n.* **exploitative, exploitive,** *a.*
explore (iksplaw), *v.t.* to investigate, to examine; to travel over in order to examine. *v.i.* to travel into unknown country. **exploration** (eks-), *n.* **explorative,** *a.* **explorator,** *n.* **exploratory** (-splo'rə-), *a.* **exploratory operation,** *n.* one carried out for purposes of diagnosis. **explorer,** *n.* one who explores; a traveller into unknown or little-known parts.
explosion (iksplō'zhən), *n.* a bursting or exploding with a loud report; a sudden and violent noise; a sudden and violent outbreak, as of physical forces, anger etc.; a rapid increase. **explosive** (-siv), *a.* bursting or driving forth with great force and noise; liable to explode or cause explosion; of consonants produced by a sudden expulsion of breath, as *p, b, t, d, k, g,* discontinuous, forming a complete vocal stop. *n.* an explosive agent or substance, as gunpowder, dynamite etc.; a mute or non-continuous consonant. **explosively,** *adv.* **explosiveness,** *n.*
expo (eks'pō), *n.* a public exhibition. [*expo*sition]
exponent (ikspō'nənt), *a.* setting forth, explaining or exemplifying. *n.* one who, or that which, sets forth or explains; one who advocates or favours; an interpretive performer; a type, a representative; a number or quantity written to the right of and above another number or quantity, to show how many times the latter is to be multiplied by itself. **exponential** (ekspənen'-), *a.* pertaining to an exponent or exponents; involving exponents.
exponential curve, *n.* a relationship between two quantities such that as one quantity increases by equal steps the other increases by equal percentages of its previous value.
exponential function, quantity, *n.* a quantity with a variable exponent. **exponentially,** *adv.*
export[1] (ikspawt'), *v.t.* to carry or send (goods) to foreign countries; to spread (a custom etc.) abroad. *v.i.* to send out commodities to foreign countries. **exportable** (-spaw'-), *a.* **exportation** (eks'-), *n.* **exporter,** *n.*
export[2] (eks'pawt), *n.* the act of exporting; a commodity or service sold to a foreign country. **export duty,** *n.* a duty paid on goods exported.
expose (ikspōz'), *v.t.* to lay bare or open; to leave unprotected; to subject (to any influence or action); to exhibit (photographic film or plate) to radiant energy; to turn out and abandon (as a child); to exhibit, to display, as for sale; to disclose, reveal; to unmask. **to expose oneself,** to lay bare one's genitals in public. **exposé** (-zā), *n.* a formal declaration or recital of facts; a disclosure, a public exposure of scandal etc. **exposition** (ekspəzi'-), *n.* the act of exposing; an explanation or interpretation of a subject or work, a commentary; the part of a musical composition in which the main themes are introduced; a public exhibition. **expositive** (-spoz'-), *a.* **expositor** (-poz'-), *n.* one who expounds or explains; a commentator. **expository,** *a.* explanatory. **exposure** (-zhə), *n.* the act of exposing; the state of being exposed to view, inconvenience, danger etc.; the state of being unsheltered from cold, heat, sun etc.; esp. the condition of or resulting from being unprotected from cold; public display, esp. of goods for sale; a disclosure, revelation, unmasking; situation with respect to the points of the compass, or free access of light and air; outlook, aspect; (*Phot.*) the act of allowing light from an object to fall upon a sensitized plate or film; the duration of this exposure; an area of film exposed, equal to an individual photograph. **exposure meter,** *n.* a device that indicates suitable camera settings for the intensity of light present. **indecent exposure** INDECENT.
expostulate (ikspos'tūlāt), *v.i.* to reason earnestly (with a person), to remonstrate. **expostulation,** *n.* **expostulator,** *a.* **expostulator,** *n.* **expostulatory,** *a.*
expound (ikspownd'), *v.t., v.i.* to set forth the meaning of; to explain, to interpret. **expounder,** *n.*
express[1] (ikspres'), *a.* set forth or expressed distinctly; direct, explicit, definitely shown or stated, intended, prepared, done, made, sent for a special purpose intended for travel at high speed; pertaining to or designed for rapid delivery by special messenger. *adv.* with speed; by express messenger or mail. *n.* an express train; an express messenger; express mail. **express train,** *n.* a fast train with a few intermediate stops. **expressway,** *n.* (*N Am.*) a motorway. **expressly,** *adv.*
express[2] (ikspres'), *v.t.* to squeeze or press out; to set forth, to make manifest to the understanding; to put into words; to reveal, to exhibit; to represent (by symbols, in terms etc.); to send by express mail. **to express oneself,** to communicate one's opinions, feelings etc. **expressible,** *a.* **expression** (-shən), *n.* the act of expressing; that which is expressed, an utterance, saying, statement of a thought; a word, a phrase; a manifestation of an opinion, feeling etc. without words; a combination of symbols representing a quantity or meaning; the aspect of the face as indicative of feeling and character, purpose etc.; intonation of voice; the exhibition of character and feeling (in a picture, statue etc.); the mode of utterance or performance that expresses the spirit and feeling of a poem, musical passage etc., expressiveness. **expression mark,** *n.* (*Mus.*) a word or sign indicating the way in which a passage is to be performed. **expressional,** *a.* **expressionism,** *n.* a 20th-cent. movement in art and literature devoted to the expression of feeling, character etc. **expressionist,** *n., a.* **expressionistic,** *a.* **expressionless,** *a.* **expressive,** *a.* serving to express; significant; vividly indicating any expression or emotion. **expressively,** *adv.* **expressiveness,** *n.*
expropriate (iksprō'priāt), *v.t.* to take from an owner, esp. for public use; to dispossess. **expropriation,** *n.* **expropriator,** *n.*
expulsion (ikspŭl'shən), *n.* the act of expelling; the state of being expelled; ejection. **expulsive,** *a.*
expunge (ikspŭnj'), *v.t.* to blot or rub out; to efface, to erase. **expunction** (-pŭngk'-), *n.*
expurgate (ek'spəgāt), *v.t.* to free (esp. a book) from anything offensive, obscene or noxious. **expurgation,** *n.* **expurgator,** *n.* **expurgatorial** (-pœgətaw'-), **expurgatory** (-pœ'-), *a.*
exquisite (ek'skwizit, ikskwiz'-), *a.* fine, delicate; very beautiful; of outstandingly quality; delicate or refined in perception, keenly sensitive; nice, fastidious; intensely pleasurable or painful, acute. *n.* a fop; one who dresses or behaves finically. **exquisitely,** *adv.* **exquisiteness,** *n.*
exsert (iksœt'), *v.t.* to thrust out, protrude.
ext., (*abbr.*) extension: external, externally; extinct; extract.
extant (ikstant'), *a.* still existing; surviving.
extemporaneous (ikstempərā'niəs), **extemporary** (-tem'-), *a.* uttered, made, composed or done without preparation. **extemporaneously,** *adv.* **extemporaneousness,** *n.* **extemporarily,** *adv.*
extempore (ikstem'pəri), *adv.* without premeditation or preparation. *a.* unstudied, delivered without preparation. **extemporize, -ise,** *v.t.* to compose or produce without preparation. *v.i.* to speak or perform without notes or

previous study. **extemporization,** n.
extend (ikstend'), v.t. to stretch out; to make greater in space, time or scope; to prolong (as a line, a period etc.); to amplify, to expand; to cause to reach (to, over or across); to enlarge; to put forth; to hold out, offer, grant; of limbs, muscles etc., to stretch out, to unbend. v.i. to stretch; to reach (in space, time or scope). **to extend a welcome (to),** to welcome cordially. **extended,** a. **extended family,** n. a social unit comprising more than a couple and their children, e.g. grandparents, aunts, uncles etc. **extendedly,** adv. **extendible, extensible,** a. **extensibility** (-bil'-), n. **extensile** (-sil), a. capable of being stretched out or protruded.
extension (iksten'shən), n. the act or process of extending; the state of being extended; extent, range, space; prolongation, enlargement; an increase of dimension, an addition, an additional part; extra time given for payment of a debt, completion of work etc.; the property by virtue of which every body occupies a limited portion of space in three dimensions; the pulling of the broken part of a limb in a direction away from the trunk, to bring the ends of the bone into their proper position; (Log.) the extent of the application of a general term, as opposed to *intension;* an additional wing or annexe of a house; an additional telephone using the same line as the main one. **university extension,** a system by which university instruction is extended to non-members of universities by means of lectures, classes and examinations. **extensional,** a.
extensive (iksten'siv), a. widely spread or extended; large; comprehensive; pertaining to farming that depends on amplitude of area, as opposed to *intensive.* **extensively,** adv. **extensiveness,** n.
extensor (iksten'sə), n. a muscle which serves to extend or straighten any part of the body.
extent (ikstent'), n. the space, dimension or degree to which anything is extended; size, width, compass, scope; degree, amount; a large space.
extenuate (iksten'üāt), v.t. to (attempt to) lessen, diminish the gravity of, palliate; to offer excuses for. **extenuating,** a. **extenuating circumstances,** n.pl. those which make an act seem less wrong or less criminal. **extenuation,** n. **extenuator,** n. **extenuatory,** a.
exterior (ikstiə'riə), a. external, outer; situated on the outside; coming from without, extrinsic; outward, visible. n. the outer surface; the external features; the outward or visible aspect, dress, conduct, deportment etc.; an outdoor scene in a film etc. **exterior angle,** n. an angle between any side of a rectilinear figure and the adjacent side produced. **exteriority** (-o'-), n. **exteriorize, -ise,** v.t. in surgery, to move (an internal part) temporarily outside the body; to externalize. **exteriorization,** n. **exteriorly,** adv.
exterminate (ikstœ'minät), v.t. to extirpate, to eradicate, to destroy utterly. **extermination,** n. **exterminator,** n. **exterminatory,** a.
external (ikstœ'nəl), a. situated on the outside; pertaining to the outside, derived from outside; belonging to the world of phenomena as distinguished from the conscious mind, objective; (*Theol.*) consisting in outward acts; of a medicine etc., applied to the outside of the body; pertaining to foreign countries; extraneous, extrinsic. n. an exterior or outer part; (pl.) outward features, aspects, circumstances; (pl.) non-essentials. **external degree,** n. a degree taken without actually attending the university that awards it, studying being done elsewhere. **external examiner,** n. one from another educational institution who ensures examinations are fairly conducted. **externalism,** n. **externality** (ekstœnal'-), n. **externalize, -ise,** v.t. to give external shape or objective existence to; to treat as consisting of externals; to ascribe to external causes. **externalization,** n. **externally,** adv.
exteroceptor (ek'sktərōseptə), n. a sensory organ which receives impressions from outside the body, e.g. the eye.

exterritorial (eksteritaw'riəl), EXTRATERRITORIAL.
extinct (ikstingkt'), a. extinguished, put out; that has ceased eruption; come to an end, that has died out; of a family, species etc.
extinction (ikstingk'shən), n. the act of extinguishing; the state of being extinguished; the act of making extinct; the state of being extinct; extermination, destruction, annihilation.
extinguish (iksting'gwish), v.t. to put out, to quench (as a light, a fire, hope, passion, life etc.); to eclipse, to obscure, to throw into the shade; to destroy, to annihilate; to suppress; to pay off (a debt, mortgage etc.). **extinguishable,** a. **extinguisher,** n. one who or that which extinguishes; esp. a device for putting out a fire. **extinguishment,** n.
extirpate (ek'stəpät), v.t. to root out, to destroy utterly, to exterminate; to cut out or off. **extirpation,** n. **extirpator,** n.
extol (ikstöl'. -tol'), v.t. (past, p.p. **extolled**) to praise in the highest terms, to glorify.
extort (ikstawt'), v.t. to wrest or wring (from) by force, threats, importunity etc.; (*Law*) to exact illegally under colour of a public office. **extorter,** n. **extortion** (ikstawshən), n. the act of extorting; oppressive or illegal exaction; that which is extorted; a gross overcharge. **extortionary,** a. **extortionate** (-ət), a. characterized by extortion; oppressive; of prices, exorbitant. **extortionately,** adv. **extortioner,** n. **extortive,** a.
extra (ek'strə), a. beyond or more than what is absolutely necessary or usual; supplementary, additional; of superior quality. adv. over and above what is usual. n. something beyond what is absolutely necessary or usual, esp. something not covered by the ordinary fee; an addition; an additional edition of a newspaper; (*cricket*) a run scored otherwise than off the bat; an actor temporarily engaged as one of a crowd etc. **extra time,** n. additional time allowed at the end of a sports match to compensate for time lost through injury etc.
extra- (ekstrə-), *comb. form.* on the outside, without.
extract[1] (ikstrakt'), v.t. to draw or pull out; to draw out or separate by mechanical or chemical means; to select a part from, to copy out or quote (as a passage from a book etc.); to derive (from); to deduce (from); to extort. **to extract the root of,** to find the root of (a number or quantity). **extractable,** a. **extraction,** n. the act of extracting; descent, family, lineage, derivation; something extracted. **extractive,** a. tending or serving to extract; capable of extraction. n. an extract. **extractive industries,** n.pl. those (e.g. mining, agriculture, fishing) concerned with obtaining natural productions. **extracter, extractor,** n. **extractor fan,** n. an electric fan which extracts air, gas etc. from a room.
extract[2] (eks'trakt), n. that which is extracted by distillation, solution etc.; a passage quoted from a book or writing; (a preparation containing) the essence or active component of a substance.
extra-curricular (ekstrəkərik'ülə), a. of an activity, outside or in addition to the normal course of school or college study; outside the usual duties or activities.
extradition (ekstrədi'shən), n. the surrender of fugitives from justice by a government to the authorities of the country where the crime was committed. **extraditable** (-di'-), a. subject to extradition, rendering one liable to extradition. **extradite** (ek'-), v.t. to surrender under a treaty of extradition; to secure the extradition of.
extrados (ikstrā'dos), n. the exterior curve of an arch, esp. measured on the top of the voussoirs (cp. INTRADOS).
extragalactic (ekstrəgəlak'tik), a. outside the Milky Way.
extrajudicial (ekstrəjodish'əl), a. taking place outside the court, not legally authorized; outside the ordinary course of law or justice. **extrajudicially,** adv.
extramarital (ekstrəma'ritl), a. esp. of sexual relations, outside marriage.

extramundane (ekstrəmŭndān), *a.* existing in or pertaining to a region outside our world.
extra-mural (ekstrəmū'rəl), *a.* beyond or outside the walls or boundaries; connected with a university or college department though outside its normal programme.
extraneous (ikstrā'niəs), *a.* foreign, not belonging to a class, subject etc.; not intrinsic, external; not essential. **extraneously**, *adv.* **extraneousness, extraneity** (-nē'iti), *n.*
extraordinary (ikstraw'dinəri, ekstraaw'-), *a.* beyond or out of the ordinary course, unusual; of an uncommon degree or kind, remarkable, rare, exceptional, surprising; additional, extra; sent or appointed for a special purpose or occasion. **extraordinarily**, *adv.* **extraordinariness**, *n.*
extrapolate (ikstrap'əlāt), *v.t.* to estimate (the value of a function etc.) beyond the known values by the extension of a curve; to infer, conjecture from what is known. **extrapolation**, *n.* **extrapolative, extrapolatory**, *a.*
extrasensory (ekstrəsen'səri), *a.* beyond the ordinary senses. **extrasensory perception**, *n.* (the ability of) perceiving by means, e.g. telepathy or clairvoyance, other than the ordinary senses.
extraterrestrial (ekstrətəres'triəl), *a.* of or from outside the earth.
extraterritorial (ekstrətəriterm'riəl), *a.* beyond the jurisdiction of the laws of the country in which one resides. **extraterritoriality** (-al'-), *n.* immunity from the laws of a country, such as that enjoyed by diplomats.
extravagant (ikstrav'əgənt), *a.* exceeding due bounds, unrestrained by reason, immoderate; visionary, fantastic; prodigal in expenditure, wasteful; of prices etc., exorbitant. **extravagance**, *n.* the state or quality of being extravagant; an extravagant act, statement or conduct; excessive expenditure. **extravagantly**, *adv.*
extravaganza (ekstrăvəgan'zə), *n.* a spectacular musical, dramatic etc. show; a fantastic piece of conduct or imagination.
extravasate (ikstrav'əsāt), *v.t.* to force or let out of the proper vessels (as blood). *v.i.* to flow out of the proper vessels. **extravasation**, *n.*
extravehicular (ekstrəvehik'ūlə), *a.* taking place outside a spacecraft.
extravert EXTROVERT.
extreme (ikstrēm'), *a.* outermost, farthest; at the utmost limit, at either end; last, final; of the highest degree, most intense; very strict or rigorous; going to great lengths, immoderate. *n.* the utmost or farthest point or limit, the extremity; the utmost or highest degree; the first or the last term of a ratio or series; (*pl.*) things or qualities as different or as far removed from each other as possible. **in the extreme**, in the highest degree; extremely. **to extremes**, (resorting) to the most severe or drastic measures. **extreme unction**, *n.* in the Roman Catholic Church, a sacrament in which those believed to be dying are anointed with holy oil. **extremely**, *adv.* very, greatly, to a great degree. **extremeness**, *n.* **extremism**, *n.* **extremist**, *n.* one ready to go to extremes; one holding extreme opinions and ready to undertake extreme actions. **extremity** (-stren'ē-), *n.* the utmost point, side or limit; the greatest degree; the remotest part, the end; a condition of the greatest difficulty, danger or distress; (*pl.*) the limbs; (*pl.*) extreme measures.
extricate (ek'strikāt), *v.t.* to disentangle, to set free from any perplexity, difficulty or embarrassment. **extricable**, *a.* **extrication**, *n.*
extrinsic (ikstrin'sik), *a.* being outside or external; proceeding or operating from without; not inherent or contained in a body. **extrinsicality** (-kăl'-), *n.* **extrinsically**, *adv.*
extrovert, extravert (ek'strəvœt), *n.* a term to denote a type of temperament which is predominantly engaged with the external world; a person more interested in other people and his/her surroundings etc. than in his/her own thoughts etc. *a.* pertaining to (the personality of) such a person. **extroverted**, *a.* **extroversion**, *n.*
extrude (ikstrood'), *v.t.* to thrust, push or squeeze out; to produce (shaped sections of metal, plastic etc.) by forcing through a die or nozzle. *v.i.* to protrude. **extrusion** (-zhən), *n.* the act or process of extruding; something formed by this process; the pouring out of magma onto the earth's surface; a rock formed from magma on the earth's surface. **extrusive**, *a.*
exuberant (igzū'bərənt), *a.* luxuriant; characterized by abundance or richness; overflowing, copious, superabundant; overflowing with vitality, spirits or imagination. **exuberance**, *n.* **exuberantly**, *adv.* **exuberate**, *v.i.* to be exuberant.
exude (igzūd'), *v.t.* to emit or discharge through pores, as sweat, moisture, or other liquid matter; to give out slowly; to make manifest an air of. *v.i.* to ooze or flow out slowly through pores etc. **exudation** (eksū-), *n.* **exudative**, *a.*
exult (igzŭlt'), *v.i.* to rejoice exceedingly; to triumph (over). **exultant**, *a.* rejoicing, triumphing; feeling or displaying exultation. **exultantly**, *adv.* **exultancy, exultation** (eg-), *n.* **exultingly**, *adv.*
exurbia (eksœ'biə), *n.* residential areas outside the suburbs of a town or city. **exurb** (eks'-), *n.* one such area. **exurban**, *a.*
exuviae (igzū'viē), *n.pl.* the cast or shed skin, shells, teeth etc. of animals; fossil remains of animals in a fragmentary state; things cast off or relinquished. **exuvial**, *n.*, *a.* **exuviate**, *v.t.* to cast off, to shed (an old shell, skin etc.). **exuviation**, *n.*
ex-voto (eksvō'tō), *adv.* in pursuance of a vow. *n.* anything offered (to a divinity) in gratitude for an exemplary favour. **ex-votive**, *a.*
eyas (ī'əs), *n.* an unfledged hawk; (*Falconry*) one taken from the nest for training or whose training is not complete.
eye[1] (ī), *n.* the organ of vision; the eyeball, iris or pupil; the socket or part of the face containing this organ; sight, ocular perception, perception, discernment, acuteness of vision; careful observation, oversight, care, attention; look, mien, expression; mental perception, way of regarding; (*pl.*) estimation, judgment (of conduct etc.); anything more or less eye-shaped; the bud of a plant; a spot on some feathers, as those of the peacock and argus pheasant; the centre of a target, a bull's-eye; a small opening or perforation; the thread-hole of a needle; the loop or catch in which the hook of a dress is fastened; a circular or oval window; the calm centre of a cyclone. **eye for an eye**, strict retaliation; **eyes front, right, left**, turn your head and eyes in front, to right or to left. **in the eye** or **eyes of**, in the regard, estimation or judgment of; from the point of view of. **mind's eye**, mental view or perception. **my eye**, (*sl.*) expressing astonishment. **to be all eyes**, to watch intently. **to catch someone's eye**, to succeed in getting someone's attention. **to find favour in the eyes of**, to be graciously received and treated by. **to give the glad eye**, (*sl.*) to ogle. **to have an eye to**, to regard, to have designs on. **to keep an eye on**, to watch carefully or narrowly. **to make eyes at**, to regard amorously. **to open one's eyes**, to be greatly astonished. **to pull the wool over someone's eyes** WOOL. **to see eye to eye**, to be in complete agreement (with). **to set, lay or clap eyes on**, to have sight of. **to turn a blind eye to** TURN. **up to the eyes**, deeply (immersed, engaged, in debt etc.). **with an eye to**, with the aim of; considering. **with one's eyes open, shut**, aware, unaware of all the facts, problems etc. of a situation. **eyeball**, *n.* the pupil or globe of the eye. *v.t.* (*esp. N Am., sl.*) to stare at. **eyeball to eyeball**, of discussions etc., at close quarters, face to face. **eye-bath**, *n.* a small utensil for bathing the eyes. **eye-bolt**, *n.* (*Naut.*) a bolt having an eye or loop at one end for the reception of a ring, hook etc. **eyebright**, *n.* a plant,

Euphrasia officinalis, formerly much used as a remedy for diseases of the eye. **eyebrow**, *n.* the fringe of hair above the orbit of the eyes. **eyebrow pencil**, *n.* a pencil applied to the eyebrows to alter their shape or colour. **eye-catching**, *a.* striking. **eye contact**, *n.* a direct look between people. **eyeglass**, *n.* a lens to aid the sight; (*pl.*) a pair of these fastened over the nose or held in the hand; the lens nearest the eye in an optical instrument; an eyebath. **eyehole**, *n.* a hole to look through; the cavity containing the eye. **eyelash**, *n.* the row of hairs edging the eyelids; a single hair from the edge of the eyelid. **eyelid**, *n.* a fold of skin covering the eye that can be moved to open or close the eye. **eye-liner**, *n.* a cosmetic used to draw a line along the edge of the eyelid. **eye-opener**, *n.* something that furnishes enlightenment or astonishment. **eyepiece**, *n.* the lens or combination of lenses at the end nearest the eye in an optical instrument. **eye shadow**, *n.* a coloured cosmetic for the eyelids. **eyeshot**, *n.* sight, range of vision, view. **eyesight**, *n.* vision; view, observation. **eyesore**, *n.* anything offensive to the sight. **eye-teeth**, *n.pl.* the upper canine teeth of humans. **eye-wash**, *n.* (*coll.*) deception, humbug; a medicated or soothing lotion for the eyes. **eyewitness**, *n.* one who sees a transaction with his or her own eyes and is able to bear witness. **eyeful**, *n.* as much as the eye can take in at a look; (*sl.*) a beautiful sight, esp. an attractive woman. **eyeless**, *a.* destitute of eyes; blind.

eye² (ī), *v.t.* to watch, to observe (fixedly, suspiciously, jealously etc.). **to eye up**, to look at in a shrewd or assessing manner; esp. to assess sexual attractiveness.

eyelet (ī′lit), *n.* a small hole or opening; esp. one for a cord, lace etc.; a metal or plastic reinforcement for this. **eyelet-hole**, *n.* a hole made for looking or shooting through or for fastening a hook etc.

Eyetie (ītī), *n.*, *a.* (*sl.*, *offensive*) (an) Italian.

eyot (āt, ä′ət), AIT.

eyre (eə), *n.* a journey or circuit; a court of itinerant justices.

eyrie AERIE.

Ez, (*abbr.*) Ezra.

Ezek, (*abbr.*) Ezekiel.

F

F¹, f¹, the sixth letter, is a labiodental spirant, formed by the emission of breath between the lower lip and the upper teeth; (*Mus.*) the fourth note of the diatonic scale of C major. **F clef,** *n.* the bass clef. **f number,** *n.* a number expressing the size of the aperture of a camera lens.
F², (*abbr.*) Fahrenheit; fail, failure; farad; filial generation; force; France. **F₁, F₂,** (*abbr.*) first and second filial generations.
F³, (*chem. symbol*) fluorine.
f², (*abbr.*) fathom; feminine; folio; following; forte; franc(s).
FA, (*abbr.*) Fanny Adams (*euphem.* for *fuck all*); Football Association; (*euphem.*) fuck all.
fa (fah), *n.* the fourth note in the sol-fa notation.
fab (fab), *a.* (*coll.*) short for FABULOUS, used as a term of approbation. **fabby,** *a.* (*coll.*) fab.
Fabian (fā'biən), *a.* of or pertaining to Fabius Maximus Cunctator, who harassed Hannibal in the second Punic war by his cautious and dilatory strategy; hence, cautious, avoiding open conflict. *n.* a member of the Fabian Society, an organization of Socialists relying entirely on moral force. **Fabianism,** *n.* **Fabianist,** *n.*
fable (fā'bl), *n.* a story, esp. one in which lower animals are represented as endowed with speech in order to convey some moral lesson; a legend, a myth; a fabrication, a falsehood. *v.i.* to write fables or fictitious tales; to tell falsehoods. *v.t.* to feign, to invent; to describe or narrate fictitiously or falsely. **fabled,** *a.* fictitious; celebrated in fable. **fabler,** *n.*
fabliau (fab'liō), *n.* (*pl.* **-liaux** (-ōz)) a metrical tale, dealing usually with ordinary life, composed in the 12th and 13th cents.
fabric (fab'rik), *n.* something put together, a system of correlated parts; a building, an edifice; the basic structure of a building, stonework, timbers etc.; woven, felted or knitted material; mode of construction or manufacture; texture.
fabricate (fab'rikāt), *v.t.* to build, to construct; to form by art or manufacture; to forge, to invent, to trump up. **fabrication,** *n.* **fabricator,** *n.*
fabulist (fab'ūlist), *n.* a writer or inventor of fables; a liar.
fabulous, *a.* related or described in fables; mythical, legendary; exaggerated, absurd; amazing, incredible; (*coll.*) wonderful, very good, very enjoyable. **fabulously,** *adv.* **fabulosity** (-los'-), **fabulousness,** *n.*
facade, façade (fəsahd'), *n.* the front of a building, the principal face; outward appearance, esp. one put on for show or to deceive.
face (fās), *n.* the front part of the head, the visage, the countenance; that part of anything which presents itself to the view, the front, the upper or main surface; the plane surface of a solid; an exposed surface of rock on a cliff or mountain, or in a mine or quarry; the dial of a watch, clock etc.; the working side of a tool or instrument; the printed surface of a playing card; the printing surface of type; a design or style of type; the striking surface of a bat, racket or golf-club; the visible state of things, the appearance, aspect; a facial expression, a look; a grimace; dignity, reputation; (*coll.*) impudence, cheek; (*coll.*) make-up. *v.t.* to turn the face towards; to meet in front; to confront boldly, to stand up to; to acknowledge without evasion; to bring face to face with; to stand opposite to; to put a coating or covering on; to put facings on (a garment); to cause to turn in any direction. *v.i.* to look in a certain direction; to be situated with a certain aspect; to turn the face in a certain direction. **about face, left face, right face,** (*Mil. order*) turn right-about, left or right, without moving from the same spot. **face to face (with),** in someone's or each other's actual presence; in confrontation; opposite. **face-to-face,** *a.* **in the face of,** in spite of. **loss of face,** humiliation, loss of personal prestige. **on the face of it,** to judge by appearances. **to face down,** to confront sternly or defiantly; to force to give way. **to face the music,** to meet consequences boldly. **to face up to,** to meet courageously. **to fly in the face of,** to act in direct opposition to. **to look in the face,** to confront steadily and unflinchingly. **to lose face,** to suffer loss of personal prestige. **to make, pull a face,** to grimace. **to put a bold, brave,** or **good face on,** to maintain that all is well with something; to make the best of. **to save (one's) face,** to save oneself from manifest disgrace or discomfiture. **to set one's face against,** to oppose firmly. **to show one's face,** to appear. **to (some)one's face,** openly. **until one is blue in the face,** for ever without success. **face-card,** *n.* a court-card. **face-cloth, -flannel,** *n.* a cloth used to wash the face. **face-harden,** *v.t.* to harden the surface of (as steel). **face-lift,** *n.* an operation to remove wrinkles and make the face look younger and smoother; (*coll.*) renovations, repairs carried out to improve or modernize the appearance of something. **face-off,** *n.* the dropping of the puck or ball between two opposing players to start or restart a game of ice-hockey or lacrosse; (*coll.*) a confrontation. **face pack,** *n.* a cosmetic paste applied to the face to clean and improve the skin. **face powder,** *n.* cosmetic powder for the face. **face-saving,** *a.* intended to prevent humiliation or loss of prestige. **face-value,** *n.* the nominal value shown on coin, bank-notes etc.; the apparent value of anything. **face-worker,** *n.* a miner who works at the face. **-faced,** *comb. form* having a face of a certain kind; having a certain number of faces. **faceless,** *a.* destitute of a face; anonymous; of bureaucrats etc., remote from and unmoved by the concerns of ordinary citizens.
facer, *n.* a sudden check; a dilemma.
facet (fas'it), *n.* a small face or surface; one of the small planes which form the sides of a crystal or cut gem; an aspect. *v.t.* to cut a facet or facets on.
facetiae (fəsē'shiē), *n.pl.* humorous or witty sayings; (*Bibliog.*) curious, comic, esp. indecent books.
facetious (fəsē'shəs), *a.* given to or characterized by levity, flippant; waggish, jocular; intended to be amusing. **facetiously,** *adv.* **facetiousness,** *n.*
facia (fā'shə), FASCIA.
facial (fā'shəl), *a.* of or pertaining to the face. *n.* a beauty treatment for the face.
-facient *comb. form.* producing the action expressed in the verb, as *calefacient, liquefacient.*
facies (fā'shiēz), *n.* the general aspect of an assembly of organisms or rocks characteristic of a particular locality or period of the earth's history.
facile (fas'īl), *a.* easily done; easily led, pliant, yielding; dexterous, skilful, handy; ready, fluent; glib, superficial. **facilely,** *adv.* **facileness,** *n.*

facilitate (fəsil'itāt), *v.t.* to make easy or less difficult; to further, to help forward. **facilitation**, *n.* **facilitative**, *a.* **facility**, *n.* easiness in performing or in being performed; ease, readiness, fluency (of speech etc.); quickness, dexterity, aptitude; readiness to be persuaded or led, pliability; (*usu. pl.*) means or equipment provided to facilitate any activity; a service; (*chiefly N Am.*) a building or plant serving a particular purpose; a sum made available for borrowing.

facing (fā'sing), *n.* the action of the verb TO FACE; a covering or coating for ornament, protection etc.; (*pl.*) the trimmings on the collar, cuffs etc. of a uniform.

facsimile (faksim'ili), *n.* an exact copy of handwriting, printing, a picture etc.; the transmission by wire or radio and reproduction of written or pictorial material. *v.t.* to make a facsimile of. **in facsimile**, exactly like. **facsimilist**, *n.*

fact (fakt), *n.* an act or deed; something that has really occurred or been done; something known to be true or existing, as distinct from an inference or conjecture; reality, actuality; the occurrence of an event, the actual doing of a deed; a piece of (relevant) information. **as a matter of fact**, actually, in fact. **before, after the fact**, before or after the actual event. **facts of life**, the details of, esp. human, reproduction; the (often unpleasant) realities of a situation. **in (point of) fact**, in reality, actually. **fact-finding**, *a.* investigative, appointed to establish the facts of a situation. **factual** (-chuəl), *a.* **factually**, *adv.*

faction[1] (fak'shən), *n.* a body of persons combined or acting in union, esp. a party within a party combined to promote their own views or purposes at the expense of order and the public good; discord, dissension. **factional**, *a.* **factious**, *a.* given to faction; opposed to the established government; seditious, turbulent. **factiously**, *adv.* **factiousness**, *n.*

faction[2] (fak'shən), *n.* literary etc. work which blends factual events and characters with fiction. [from *fact* and *fiction*]

-faction *comb. form.* denoting making, turning or converting, as in *rarefaction, satisfaction*.

factitious (faktish'əs), *a.* made by art, artificial; unnatural, conventional, affected; unreal, bogus. **factitiously**, *adv.* **factitiousness**, *n.*

factitive (fak'titiv), *a.* causing, effecting; (*Gram.*) applied to that relation existing between two words, as applicable to an active verb and its object, when the action expressed by the verb causes a new state or condition in the object, as in *The people made him a king*.

factoid (fak'toid), *n.* a piece of information accepted as true purely on the basis of its appearance in print or its repetition, but not proved.

factor (fak'tə), *n.* an agent, a deputy; (*Sc.*) a steward or agent of an estate; an employee employed to sell goods on commission; one of the quantities that multiplied together make up a given number or expression; any circumstance, fact or influence which contributes to a result. *v.i.* to act as a factor. **Factor 8**, *n.* a blood-clotting agent used in the treatment of haemophiliacs. **factorage**, *n.* the commission given to a factor by his or her employer. **factorial** (-taw'-), *a.* pertaining to a series of mathematical factors; pertaining to a factor or land agent. *n.* the product of an integer multiplied into all its lower integers, e.g. the factorial of 4 = 4 × 3 × 2 = 24. **factoring**, *n.* the work of a factor; the buying up of trade debts or lending money on the security of them. **factorize, -ise**, *v.t.* to express a number in terms of its factors. **factorization**, *n.* **factorship**, *n.*

factory (fak'təri, -tri), *n.* a trading station established in a foreign place by a company of merchants; a building in which any manufacture is carried out. **factory farm**, *n.* a farm practising factory farming. **factory farming**, *n.* the intensive rearing of animals for milk, egg or meat production in a largely man-made environment. **factory ship**, *n.* a vessel in a fishing fleet which processes the catches.

factotum (faktō'təm), *n.* a person employed to do all sorts of work.

factual FACT.

factum (fak'təm), *n.* (*pl.* **-ta** (-tə)) a thing done; an act or deed; (*Law*) a deed, a sealed instrument; a memorial reciting facts or points in a controversy. [L, FACT]

facula (fak'ūlə), *n.* (*pl.* **-lae** (-lē)) (*Astron.*) a luminous spot or streak upon the sun's disc.

faculty (fak'əlti), *n.* power or ability of any special kind; a natural power of the mind, as the will, reason, sense etc.; capacity for any natural action, as seeing, feeling, speaking; the members collectively of any of the learned professions; one of the departments of instruction in a university; the professors and lecturers in such a department; an authorization or licence to perform certain functions, esp. ecclesiastical. **facultative**, *a.* imparting a faculty or power; empowering, permissive, as opposed to compulsory, optional; pertaining to a faculty; able to live under more than one set of environmental conditions. **facultize, -ise**, *v.t.*

fad (fad), *n.* a whim, a passing fancy, taste or fashion, a craze; a favourite theory or idea; an idiosyncratic taste or distaste for something. **faddish**, *a.* **faddishness**, *n.* **faddism**, *n.* **faddist**, *n.* **faddy**, *a.* **faddiness**, *n.*

fade (fād), *v.i.* to wither, as a plant, to lose freshness, brightness, vigour or beauty; to languish; to grow lighter in colour, pale, dim or indistinct; to disappear gradually; of a person, to grow weaker, to decline; of electronic signals, to decrease in strength or volume; of brakes, to lose their effectiveness gradually; of an athlete, team etc., to perform less well, to cease to mount a serious challenge; to perform a fade-in or fade-out. *v.t.* to cause to fade; (*Golf*) to slice (a shot) slightly. *n.* an instance of fading in or out or both simultaneously; a dimming of stage lighting; (*Golf*) a slight (often deliberate) slice. **to fade away**, to fade; (*coll.*) to grow very thin. **to fade in, up**, to cause sound or a picture to appear gradually. **to fade out, down**, to cause sound or a picture to disappear gradually. **fade-in, -up**, *n.* **fade-out, -down**, *n.* **fadeless**, *a.* unfading. **fadelessly**, *adv.* **fadingly**, *adv.*

faeces, (*esp. N Am.*) **feces** (fē'sēz), *n.pl.* excrement from the bowels. **faecal** (fē'kl), *a.*

faerie, faery (feə'ri), *n.* (*poet.*) fairyland; a fairy. *a.* fairy.

faff (faf), *v.i.* (*coll.*) to dither, to fuss (often with *about*).

fag[1] (fag), *v.i.* (*past, p.p.* **fagged**) to toil wearily; to work till one is weary; to act as a fag in a public school. *v.t.* to tire, to exhaust, to weary (often with *out*); to use as a fag or drudge in a public school. *n.* laborious drudgery, toil; (*coll.*) a tiresome or boring task; a junior at a public school who has to perform certain duties for some senior boy; (*coll.*) a cigarette. **fag-end**, *n.* the loose end of a web of cloth; the latter or meaner part of anything; the untwisted end of a rope; (*coll.*) a cigarette butt.

fag[2] (fag), *n.* (*chiefly N Am., offensive*) a (male) homosexual.

faggot (fag'ət), *n.* a bundle of sticks or small branches of trees, used for fuel, filling ditches, road-making etc.; a bundle of steel or wrought-iron rods; a cake or ball of chopped liver, herbs etc.; (*chiefly N Am., offensive*) a (male) homosexual. *v.t.* to bind or tie up in a faggot or bundle; to embroider with faggoting. **faggoting**, *n.* a type of embroidery in which some horizontal threads are tied together in hourglass shapes.

fah (fah), FA.

Fahrenheit (fa'rənhīt), *a.* pertaining to the temperature scale on which the freezing-point of water is marked at 32° and the boiling-point at 212°. [inventor, Gabriel Daniel *Fahrenheit*, 1686–1736]

faience (fayēs'), *n.* tin-glazed earthenware of a particular kind. [F; It. *Faenza*, town in Romagna]

fail (fāl), *v.i.* to come short of the due amount or measure; not to succeed (in); not to succeed in the attainment (of); to lose strength or spirit, to decline; to die away; to be or become ineffective or inoperative; to become bankrupt or insolvent; not to pass an examination. *v.t.* to be insufficient for; to disappoint, to desert; to neglect or omit (to do something); not to pass (an examination); to cause not to pass. *n.* failure, default; a failure grade in an examination; one who fails an examination. **without fail,** assuredly, certainly. **failsafe,** *a.* of a mechanism, incorporated in a system to render it safe in the event of failure or malfunction. **failing,** *n.* an imperfection, a weakness. *prep., pres.p.* in default of. **failure** (-yə), *n.* a failing or coming short; an omission, non-performance, non-occurrence; decay, breaking down; insolvency, bankruptcy; want of success; an unsuccessful person or thing.

fain (fān), *a. (poet.)* glad, well-pleased; in default of something better; desirous. *adv. (poet.)* gladly, readily.

faint (fānt), *a.* weak, feeble; giddy, inclined to faint; timid, fearful; of sound or brightness, dim, indistinct, slight. *v.i.* to lose consciousness suddenly and temporarily, a fainting fit. **faint-hearted,** *a.* cowardly, timid. **faint-heartedly,** *adv.* **faint-heartedness,** *n.* **faintish,** *a.* **faintly,** *adv.* **faintness,** *n.*

fair[1] (feə), *a.* beautiful, comely, pleasing to the eye; just, equitable, legitimate; not effected by unlawful or underhand means, above-board; passably good, of moderate quality; clear, pure, clean; free from spot, blemish or cloud; favourable, auspicious, promising; open, unobstructed; civil, obliging, polite; specious; legible, plain; light in colour or complexion; blond; *adv.* courteously, civilly; openly, honestly, justly; according to the rules, straight, clean. *v.i.* of the weather, to become fair. **by fair means or foul** MEAN[2]. **fair and square,** honourable, above-board. **fair dos** (dooz), *(coll.)* a phrase used when asking for, or consenting to, fair play, equal shares, fair treatment etc. **fair enough,** *(coll.)* (indicating at least partial assent to a proposition, terms etc.) all right, OK. **fair to middling,** *(coll.)* not bad, about average. **fair copy,** *n.* a copy of (a document etc.) not defaced by corrections. **fair game,** *n.* a legitimate target for attack, criticism or ridicule. **fair-haired,** *a.* having hair of a light colour, blond. **fair-minded,** *a.* honest-minded, impartial, just. **fair play,** *n.* equitable conduct; just or equal conditions for all. **fair sex,** *n.* women. **fair-spoken,** *a.* using courteous language. **fairway,** *n.* the navigable part of a river, channel or harbour; (*Golf*) the smooth passage of turf between holes. **fair-weather,** *a.* appearing only in times of prosperity. **fairing,** *n.* a structure to provide streamlining of an aircraft, car etc. **fairish,** *a.* pretty fair; tolerably large. **fairly,** *adv.* in a fair manner; completely, absolutely, utterly; moderately, passably. **fairness,** *n.* **in all fairness,** being strictly honest or just.

fair[2] (feə), *n.* a market or gathering for trade in a particular town or place, held periodically, with shows and entertainments; a funfair; a charity bazaar; a trade show. **fairground,** *n.* open space where fairs, exhibitions etc. are held.

Fair Isle (feə), *a.* applied to woollen articles knitted in coloured patterns typical of *Fair Isle* (one of the Shetland Islands).

fairy (feə'ri), *n.* a small supernatural being having magical powers, supposed to assume human form and to meddle for good or for evil in human affairs; (*offensive*) an effeminate man or homosexual. *a.* pertaining to or connected with fairies; fairy-like; fanciful, imaginary. **fairy cycle,** *n.* a child's bicycle. **fairy godmother,** *n.* an (often unexpected) benefactor. **fairyland,** *n.* the imaginary abode of the fairies; a region of enchantment. **fairy lights,** *n.pl.* small lights of many colours used for decoration. **fairy-ring,** *n.* a circular band of turf greener than the rest caused by the growth of fungi, but formerly supposed to be caused by the dancing of fairies. **fairy story, fairy-tale,** *n.* a tale about fairies; a fanciful or highly improbable story. *a.* as in a fairy-tale; extremely beautiful; extremely fortunate. **fairily,** *adv.* **fairydom** (-dəm), *n.* **fairyhood** (-hud), *n.* **fairyism,** *n.*

fait accompli (fāt, fet əkom'plē), *n.* an accomplished fact. [F]

faith (fāth), *n.* the assent of the mind to what is stated or put forward by another; firm and earnest belief, conviction, complete reliance, trust; spiritual apprehension or voluntary acceptance of divine revelation apart from absolute proof; operative belief in the doctrines and moral principles forming a system of religion; a system of religious belief; a philosophical, scientific or political creed or system of doctrines; fidelity, constancy, loyalty. **bad faith,** intent to deceive. **in faith,** in deed, in truth. **in good faith,** with honest intentions. **the faith,** the Christian religion; the true religion. **to keep faith with,** to be loyal to. **faith-cure, -healing,** *n.* curing of disease by means of prayer and faith, without the use of drugs etc. **faith-curer, -doctor, -healer,** *n.* **faithful,** *a.* loyal to one's promises, duty or engagements; sexually loyal to one's partner; conscientious, trustworthy; upright, honest; truthful, worthy of belief; exact, accurate. **the faithful,** true believers in a particular creed or religious system. **faithfully, to promise faithfully,** with the most emphatic assurances. **yours faithfully,** a conventional mode of subscribing a letter. **faithfulness,** *n.* **faithless,** *a.* destitute of faith, unbelieving; disloyal, unfaithful, not true to promises or duty, unreliable; perfidious, treacherous. **faithlessly,** *adv.* **faithlessness,** *n.*

fake (fāk), *v.t.* to falsify, to doctor, to counterfeit, to contrive, to fabricate; to pretend, to simulate. *v.i.* to simulate something. *n.* a thing prepared for deception, a sham, a person that is not genuine, an imposter. *a.* bogus, sham, counterfeit. **faker,** *n.* **fakery,** *n.*

fakir (fä'kiə, fəkiə'), *n.* a Muslim religious mendicant; often used for a mendicant, ascetic or wonder-worker of other faiths, esp. in India; a very holy man.

Falangist (fəlan'jist), *n.* name adopted by General Franco and his supporters in the revolution against the republican government of Spain (1936–39).

falcate (fal'kāt), **-cated,** *a.* hooked; bent or curved like a sickle or scythe.

falchion (fawl'chən, -shən), *n.* a short, broad sword with a slightly curved blade.

falciform (fal'sifawm), *a.* falcate.

falcon (fawl'kən), *n.* a small diurnal bird of prey, esp. the peregrine falcon and others trained to hawk game; a female falcon, esp. the peregrine (cp. TIERCEL). **falconer,** *n.* one who keeps and trains hawks for hawking; one who hunts with hawks. **falconet** (-nit), *n.* a species of shrike. **falconry,** *n.* the art of training falcons to pursue and attack game; the sport of hawking.

falderal (fal'dəral), *n.* a trifle, a gewgaw; nonsense.

faldstool (fawld'stool), *n.* a portable folding seat, stool or chair, used by a bishop officiating out of his own cathedral; a desk at which the Litany is said; a desk or stool to kneel at during one's devotions.

fall (fawl), *v.i.* (*past,* **fell** (fel), *p.p.* **fallen**) to descend from a higher to a lower place or position by the force of gravity; to descend suddenly, to drop; to sink, to flow down, to be poured down, to become lower in level of surface; to come down, to become prostrate; to be hit or wounded; to be killed (esp. in battle); to be overthrown, to lose power; to be taken by the enemy; (*Cricket*) to be taken by the bowling side; to decrease in number, amount, value, weight, loudness etc.; to become lower in pitch; to subside, to abate, to ebb, to die away; to be degraded or disgraced; to sink into sin, vice, error, to give way to temptation; to lose one's virginity; of the face, to assume a despondent expression; to become, to pass into a specified state, as in *fall asleep, fall ill*; to be transferred by chance, lot, inheritance, or otherwise; to occur at a specified place or time; to be apportioned or assigned; to turn out, to result, to happen; to be uttered or dropped,

fallacy 292 **family**

as a chance remark; to hang down; to droop. *n.* the act of falling; a bout at wrestling or a throw in this; a cataract, a cascade, a waterfall; the degree of inclination, the gradient or slope; a declivity; the amount of descent, the distance through which anything falls; a decrease in value, amount etc.; (*chiefly N Am.*) autumn; the amount of rain, snow etc. in a district; the number of lambs born; downfall, declension from greatness or prosperity, ruin, disgrace; overthrow; the surrender or capture of a town; a lapse from virtue; a yielding to temptation; a veil; that part of the rope in hoisting-tackle to which the power is applied; (*Mus.*) a cadence. **the Fall**, the lapse of Adam and, through him, of his posterity from a state of primeval innocence. **to fall about**, to laugh hysterically. **to fall apart**, to collapse, to become unstitched, unstuck etc. **to fall away**, to desert; to revolt; to apostatize; to decay, to languish; to pine, to become thin. **to fall back**, to recede, to retreat. **to fall back (up)on**, to have recourse to. **to fall behind**, to be passed by, to lag behind; to become in arrears with. **to fall between two stools**, to fail through being unable to choose between two alternatives; to be neither one thing nor the other. **to fall down**, to drop or collapse; to fail, to be inadequate. **to fall flat**, to be a failure; to fail to arouse interest. **to fall for**, (*coll.*) to be impressed by, to fall in love with; to be fooled by. **to fall foul of** FOUL. **to fall from grace**, to fall into sin. **to fall in**, to give way inwards; (*Mil.*) to take one's place in line. **to fall in with**, to meet with accidentally; to agree to, to concur in. **to fall off**, to drop from a place or a position of attachment; to become depreciated, to decrease in quality, quantity or amount. **to fall on**, to make an attack; to set to, to begin eagerly. **to fall out**, to happen, to result; to quarrel; (*Mil.*) to leave the ranks. **to fall over backwards** BACKWARD. **to fall over oneself**, to be eager, or over-eager (to do something). **to fall short**, to be deficient. **to fall short of**, to fail to attain. **to fall through**, to fail, to come to nothing. **to fall to**, to begin hastily or eagerly, to set to; to begin eating. **to fall under**, to be subject to; to come within the range of; to be classed with or reckoned with or under. **to fall (up)on**, to come across; to attack. **fallback**, *a.* which one can retreat to; alternative (and usu. less ambitious). **fall guy**, *n.* a scapegoat; one who is easily duped. **fall-off**, *n.* a decline. **fall-out**, *n.* the deposit of radioactive dust after a nuclear explosion; secondary consequences, by-products. **fallen**, *a.* killed, esp. in battle; seduced; morally degraded; overthrown. **faller**, *n.* a race-horse which falls during a race. **falling**, *n.* **falling-off**, *n.* a fall-off. **falling star**, *n.* a meteor appearing to fall rapidly to the earth.
fallacy (fal'əsi), *n.* an unsound argument or mode of arguing; anything that misleads or deceives the mind; (*Log.*) a delusive mode of reasoning, an example of such; an error, a sophism; unsoundness of reasoning or of belief. **fallacious** (-lā'-), *a.* **fallaciously**, *adv.* **fallaciousness**, *n.*
fallible (fal'ibl), *a.* liable to err or to be mistaken. **fallibility** (-bil'-), *n.*
Fallopian tubes (fəlō'piən), *n.pl.* two ducts or canals by which ova are conveyed to the uterus. [from *Fallopius*, 1523–62, It. anatomist, incorrectly credited with their discovery]
fallow[1] (fal'ō), *a.* of a pale brownish or reddish-yellow colour. **fallow deer**, *n.* a small species of deer, preserved in a semi-domesticated state in many English parks.
fallow[2] (fal'ō), *n.* land ploughed and harrowed but left unsown; land left uncultivated for a period. *a.* ploughed and tilled but not sown; uncultivated, unused, neglected. *v.t.* to plough and harrow and leave unsown.
false (fawls), *a.* not true, contrary to truth, not conformable to fact; deceptive, misleading; erroneous, wrong, incorrect; uttering untruth, lying, deceiving; deceitful; treacherous; feigned, sham, spurious, counterfeit; forced, unconvincing; artificial, man-made; fitting over or replacing a main part; esp. of plants, resembling, though not belonging to, the specified species; (*Mus.*) out of tune. *adv.* falsely; wrongly. **to play one false**, to deceive. **false alarm**, *n.* a needless warning, a cause of unnecessary anxiety or excitement. **false bottom**, *n.* a partition inserted above the true bottom often concealing a secret compartment. **false dawn**, *n.* light appearing just before sunrise. **false imprisonment**, *n.* illegal imprisonment. **false move**, *n.* unwise movement or action. **false pregnancy**, *n.* a psychosomatic condition producing symptoms of pregnancy, pseudocyesis. **false pretences**, *n.pl.* (*Law*) misrepresentations made with intent to deceive or defraud. **false rib**, *n.* a rib not directly attached to the breastbone. **false start**, *n.* a disallowed start to a race, usu. caused by a competitor getting away too early; an abortive beginning to any activity. **false step**, *n.* a stumble; an imprudent action. **falsehood** (-hud), *n.* untruthfulness, falseness; a lie, an untruth; lying, deceitfulness. **falsely**, *adv.* **falseness**, *n.* **falsies** (-siz), *n.pl.* (*coll.*) pads used to improve the shape of the breasts. **falsity**, *n.*
falsetto (fawlset'ō), *n.* (*pl.* **-ttos**) a pitch or range of (usu. the male) voice higher than the natural register; a singer using this range. *a.* pertaining to or produced by such a voice.
falsify (fawl'sifī), *v.t.* to make false; to give a false or spurious appearance to (a document, statement etc.); to misrepresent; to counterfeit, to forge; to disprove, show to be false. **falsification** (-fi-), *n.*
Falstaffian (fawlstahf'iən), *a.* fat, coarsely humorous, convivial. [*Falstaff*, a character in Shakespeare's *Henry IV* and *V* and *Merry Wives of Windsor*]
falter (fawl'tə), *v.i.* to totter, to waver, to be unsteady; to stammer, to stutter; to hesitate in action, to act with irresolution; to fail, flag. *v.t.* to utter with hesitation or stammering. **falteringly**, *adv.*
fam., (*abbr.*) familiar, familiarly; family.
fame (fām), *n.* reputation, esp. good reputation, renown, celebrity. *v.t.* to make famous or renowned. **house of ill fame**, a brothel. **ill fame**, evil reputation. **famed**, *a.* renowned, celebrated.
familial FAMILY.
familiar (fəmil'yə), *a.* of one's own acquaintance, well-known; closely acquainted; (with); unduly or unlawfully intimate; usual, common, ordinary, not novel; easily understood, not abstruse; unconstrained, free, unceremonious. *n.* an intimate or close friend or companion; a demon or spirit supposed to attend at call; a confidential servant in the household of the Pope or a bishop. **familiarity** (-lia'-), *n.* use, habitude; close friendship, intimacy; freedom from constraint, unceremonious behaviour, esp. towards superiors or inferiors; a liberty. **familiarize, -ise**, *v.t.* to make familiar; to habituate, to accustom; to make well acquainted (with). **familiarization**, *n.* **familiarly**, *adv.*
family (fam'ili), *n.* those that live in the same house, including parents, children and servants; father and mother and children; such a group including other relations; children, as distinguished from their parents; those who can trace their descent from a common ancestor; a race, a group of peoples from a common stock; a brotherhood of persons or peoples connected by bonds of civilization, religion etc.; (noble) lineage; a group of related things; (*Biol.*) a group of genera, a subdivision of an order; (*Chem.*) a group of compounds having a common basic radical; a group of languages having the same source language. *a.* esp. of entertainment, deemed suitable for the family with young children, not containing bad language or scenes of sex or violence. **family income supplement**, (*formerly*) in Britain, a social security benefit paid to families with earnings below a set level. **Holy Family** HOLY. **in the family way**, pregnant. **to keep something in the family**, to ensure that something, e.g. a possession or piece of information, does not pass outside

the family or a select group. **family allowance**, *n*. the former name for child benefit. **family Bible**, *n*. a large Bible in which the names and dates of birth of members of a family are entered. **family credit**, *n*. in Britain, a social security benefit paid to low-income people in work who have at least one child. **Family Division**, *n*. a division of the High Court dealing with divorce, the custody of children etc. **family man**, *n*. one who has a (large) family; one who is fond of home life. **family name**, *n*. (*esp. N Am.*) surname. **family planning**, *n*. regulating the number of, and intervals between, children, usu. by means of contraception; (*coll.*) an agency or clinic giving advice on contraception etc. **family tree**, *n*. a genealogical chart. **familial** (fəmil'iəl), *a*. characteristic of a family.

famine (fam'in), *n*. distressing scarcity of food; extreme scarcity of anything; hunger, starvation.

famish (fam'ish), *v.t., v.i.* to reduce to or suffer extreme hunger.

famous (fā'məs), *a*. renowned, celebrated; noted; (*coll.*) first-rate, very good. **famously**, *adv*. **famousness**, *n*.

fan[1] (fan), *n*. an instrument, usu. flat, with radiating sections opening out in a wedge-shape for agitating the air and cooling the face; an implement shaped like an open fan; a winnowing implement or machine; a small sail or vane for keeping the sails of a windmill to the wind; (*Naut.*) the blade of a screw-propeller; a bird's tail, a wing, a leaf shaped like a fan; a rotatory apparatus for causing a current of air for ventilation; a fan-shaped talus. *v.t.* (*past, p.p.* **fanned**) to agitate (the air) with a fan; to stir up; to spread like a fan; to cool with a fan; to move or stimulate with or as with a fan; to stir up; to spread like a fan; to winnow; to winnow or sweep away (as chaff). *v.i.* to move or blow gently; to spread out like a fan. **to fan out**, to radiate outwards. **fanbelt**, *n*. a belt which drives the radiator fan and generator in a car engine. **fan dance**, *n*. a titillating dance by a nude solo performer manipulating a large fan or fans. **fan heater**, *n*. an electric heater in which the heat from the element is dispersed by a fan. **fan-jet**, *n*. (an aircraft with) a jet engine with rotating fans in which some of the air sucked in bypasses the combustion chamber. **fanlight**, *n*. a window with divisions in the shape of an open fan; the light placed over a doorway. **fantail**, *n*. a variety of the domestic pigeon; an Australian flycatcher of the genus *Rhidipura*.. **fan-tailed**, *a*. **fan-vaulting**, *n*. (*Arch.*) vaulting in which the tracery spreads out like a fan from springers or corbels. **fanner**, *n*.

fan[2] (fan), *n*. an enthusiastic admirer; a devotee. **fan club**, *n*. an organized group of admirers. **fan mail**, *n*. adulatory letters to a celebrity from admirers; a group of companies or financiers who buy shares in a company in order to vote, esp. in support of a takeover. **fandom**, *n*. [abbr. of FANATIC]

fanatical (fənat'ikl), *a*. wild or extravagant in opinions or beliefs, esp. as regards religious matters; enthusiastic in the extreme. **fanatic**, *a*. fanatical. *n*. a fanatical person. **fanatically**, *adv*. **fanaticism** (-sizm), *n*. **fanaticize**, **-ise** (-siz), *v.t., v.i.* to render or become fanatical.

fancy (fan'si), *n*. the faculty or the act of forming images, esp. those of a playful, frivolous or capricious kind; imagination as an inventive and comparative power, distinguished from creative imagination; a mental image; a visionary idea or supposition; a delusion, a baseless impression; a caprice, a whim; a personal inclination, liking or attachment; a fad, a hobby. *v.t.* to form as a conception in the mind; to be inclined to think, to suppose; to imagine or believe erroneously; to think a good deal of (oneself etc.); to like, to take a fancy to, to be attracted to; to have a desire or wish for; to breed as a hobby or sport. *a*. adapted to please the fancy rather than for use; ornamental, decorative; not plain; not ordinary; requiring skill; esp. of prices, extravagant, high. **fancy!**, **just fancy!** an expression of surprise. **the fancy**, sporting characters generally, esp. pugilists, pugilism, dog-fanciers etc. **to take a fancy to**, to conceive a liking or an affection for, to desire. **to tickle one's fancy**, to attract. **fancy dress**, *n*. masquerade costume. **fancy-dress ball**, *n*. **fancy-free**, *a*. not in love, not involved in a relationship, hence able to do as one likes. **fancy-goods**, *n.pl*. articles of a showy rather than a useful kind. **fancy man**, *n*. (*derog.*) a woman's lover; a prostitute's pimp. **fancy-woman**, *n*. (*derog.*) a kept mistress. **fancy work**, *n*. ornamental knitting, embroidery, crocheting etc. **fancied**, *a*. imagined; favoured to do well. **fancier**, *n*. one who breeds or sells birds, dogs, rabbits etc. for their special points; one with an interest in or liking for something. **fanciful**, *a*. dictated by or arising in the fancy; baseless, unreal, imaginary; indulging in fancies; whimsical, fantastical. **fancifully**, *adv*. **fancifulness**, *n*.

fandangle (fandang'gl), *n*. a gaudy trinket, a gewgaw; a nonsensical idea or behaviour.

fandango (fandang'gō), *n*. a lively Spanish dance in triple time, for two persons who beat time with castanets; the accompaniment of such a dance.

fanfare (fan'feə), *n*. a flourish of trumpets or bugles; ostentation, parade; any short, prominent passage of the brass.

fang (fang), *n*. a tusk or long pointed tooth: the canine tooth of a dog, wolf or boar; the long, hollow or grooved tooth through which a poisonous snake injects its venom; a curved spike, the point of any device for seizing or holding; the part of a tooth embedded in the gum. **fanged**, *a*. furnished with fangs. **fangless**, *a*.

fanlight FAN[1].

fanny (fan'i), *n*. (*taboo*) the female genitals; (*sl.*) the buttocks.

Fanny Adams (ad'əmz), *n*. (*Naut., sl.*) tinned mutton; (*usu.* **sweet Fanny Adams**) (*sl.*) nothing at all, euphem. for *fuck-all*. [from the name of a young murder victim whose body was cut up into small pieces]

fan-tan (fan'tan), *n*. a Chinese gambling game.

fantasia (fantā'ziə), *n*. (*Mus.*) a composition in which form is subservient to fancy.

fantasm (fan'tazm), PHANTASM.

fantastic (fantas'tik), **-ical**, *a*. illusory, imaginary; fanciful, whimsical, eccentric; odd, grotesque; uncertain, fickle, capricious; extravagant; incredible, amazing; wonderful, very good, very enjoyable. **fantasticality** (-kal'-), *n*. **fantastically**, *adv*. **fantasticalness**, *n*. **fantasticism** (-sizm), *n*.

fantasy, (*esp. formerly*) **phantasy** (fan'təsi), *n*. an extravagant, whimsical or bizarre fancy, image or idea; the faculty of inventing or forming fanciful images; a mental image or daydream which gratifies a psychological need; a fanciful or whimsical invention or design; a novel, drama, film etc. characterized by strange, unrealistic, alien or grotesque characters and settings; such works collectively; (*Mus.*) a fantasia; a visionary idea or speculation; a caprice, a whim. **fantasize**, **-ise**, *v.t., v.i.* to conjure up and indulge in gratifying mental images; to dream up or imagine fantastic (and usu. impracticable) schemes, ideas etc. **fantasist**, *n*.

Fanti, Fante (fan'ti), *n*. (*pl*. **Fantis, Fantes**, *collectively* **Fanti, Fante**) (the language of) a member of a Ghanaian tribe.

fantoccini (fantəchē'ni), *n.pl*. puppets or marionettes made to perform by concealed wires or strings; dramatic representations at which such puppets are made to perform.

fantom (fan'təm), PHANTOM.

fanzine (fan'zēn), *n*. a magazine for fans of a celebrity.

FAO, (*abbr.*) Food and Agriculture Organization (of the United Nations).

far (fah), *a*. (*comp.* **farther, further**, *superl.* **farthest, furthest** (-dh-)) distant, a long way off in space or time; separated by a wide space; extending or reaching a long way; more

farad 294 **fascine**

distant of two, other, opposite; remote from or contrary to one's purpose, intention or wishes; alienated. *adv.* at or to a great distance in space, time or proportion; to a great degree, very greatly, by a great deal; by a great interval, widely. **a far cry** CRY. **as far as**, up to (a certain point); to the extent that. **by far**, in a very great measure; very greatly; exceedingly. **far and away** AWAY. **far and wide**, everywhere. **far be it from me**, I would not even consider; I repudiate the intention (of doing something). **far from**, anything but, not at all; *(followed by pres.p.)* indicates that the speaker's actions or intentions are the opposite of those stated. **far from it**, on the contrary. **so far**, up to a specified point; up to now, hitherto. **so far as** AS FAR AS. **to go far**, to be successful (esp. in one's career); *(esp. in neg.)* to be sufficient for. **to go too far**, to exceed reasonable limits. **far-away**, *a.* remote in time, place or relationship; distant; dreamy, absent-minded. **Far East** EAST. **far-fetched**, *a.* of reasons or arguments, unnatural, improbable, fanciful. **far-flung**, *a.* remote; extending to far-off places. **far-gone**, *a.* in an advanced state (of exhaustion, illness, wear etc.). **far left**, *n.* the extreme left wing of a political party etc. *a.* holding very left-wing views. **Far North**, *n.* the Arctic regions. **far-off**, *a.* distant, remote. **far-out**, *a.* (*sl.*) unconventional, eccentric, weird; (*also int.*) wonderful, great. **far-reaching**, *a.* having broad scope, influence or implications. **far right**, *n.* the extreme right wing of a political party etc. *a.* holding very right-wing views. **far-seeing, -sighted**, *a.* seeing to a great distance; provident for remote issues. **far-sightedly**, *adv.* **far-sightedness**, *n.*

farad (fa'rəd), *n.* the practical unit of capacitance, the capacity of a condenser in which the electrical potential is raised 1 volt by the addition of 1 coulomb. **faradic** (-rad'-), *a.* of an electric current, inductive. [Michael *Faraday*, 1791–1867, British chemist and physicist]

farandole (farəndōl'), *n.* a lively Provençal dance.

farce[1] (fahs), *n.* a short dramatic work in which the action is trivial and the sole purpose to excite mirth; drama of this kind; an absurd proceeding; a pretence, mockery, hollow formality. **farcical**, *a.* of or pertaining to farce; ludicrous, droll, comical; ridiculous, absurd, contemptible. **farcicality** (-kal'-), *n.* **farcically**, *adv.*

farce[2] *v.t.* to stuff (poultry etc.) with forcemeat. *n.* stuffing, forcemeat.

farceur (fahsœ'), *n.* a joker, a jester, a wag. **farceuse**, *n.fem.*

farcy (fah'si), *n.* a disease in horses, closely allied to glanders.

fard (fahd), *n.* paint or rouge for the face, esp. white paint. *v.t.* to paint (the face) with this; to hide the blemishes of.

fare (feə), *v.i.* to go, to travel; to get on, to be in any state, to happen, to turn out (well or ill); to feed or be fed (well etc.). *n.* the sum paid for conveyance on a journey; the person or persons conveyed in a vehicle for hire; food provided. **fare stage**, *n.* one of the sections into which a bus journey is divided for the purpose of setting fares.

farewell (feəwel'), *int.* adieu, good-bye; orig. and properly addressed to one about to start on a journey, now a common formula of leave-taking; used also as expression of simple separation, and in the sense of 'no more of', 'good-bye to'. *n.* a good-bye, an adieu; a departing. *a.* valedictory.

farina (fərē'nə), *n.* flour or meal; the powder obtained by grinding the seeds of gramineous and leguminous plants, nuts, roots etc.; starch. **farinaceous** (farinā'-), *a.* consisting of or high in starch; mealy. **farinaceously**, *adv.* **farinose** (fa'ri-), *a.* producing farina; floury, meally.

farl (fahl), *n.* (*Sc.*) a triangular biscuit or cake of oatmeal or flour.

farm (fahm), *n.* a tract of land used under one management for cultivating crops or rearing livestock; a farm-house; an area of land or water where a particular kind of animal, fish or plant is bred. *v.t.* to till, to cultivate, (land); to rear or cultivate on a farm; to lease or let out (as taxes, offices etc.) at a fixed sum or rate per cent; to take the proceeds of (taxes, offices etc.) for such a fixed sum or rate; to contract for the feeding, lodging etc. of (as children) at so much per head. *v.i.* to be a farmer. **home-farm** HOME. **to farm out**, to delegate, to contract out; to board out, to put into someone's care. **farm hand, labourer**, *n.* an agricultural labourer employed on a farm. **farmhouse**, *n.* a dwelling-house attached to a farm. **farmstead** (-sted), *n.* a farm with the dwelling and other buildings on it. **farmyard**, *n.* a yard or open area surrounded by or adjacent to farm buildings. **farmer**, *n.* one who farms or cultivates land; one who contracts to collect taxes, imposts etc. at a certain rate per cent. **farming**, *n.* the business of cultivating land or rearing livestock.

faro (feə'rō), *n.* a game at cards in which persons play against the dealer.

farouche (fəroosh'), *a.* wild, untamed; (*coll.*) unsociable, unmannerly, brutal. [F]

farrago (fərah'gō), *n.* (*pl.* **-goes**) a confused mixture, a medley. **farraginous** (-raj'i-), *a.*

far-reaching FAR.

farrier (fa'riə), *n.* one who shoes horses; a shoeing smith who is also a horse-doctor. **farriery**, *n.*

farrow (fa'rō), *n.* a litter of pigs. *v.t., v.i.* to bring forth (pigs).

far-sighted FAR.

fart (faht), *v.i.* (*sl., sometimes considered taboo*) to break wind through the anus. *n.* a discharge of wind from the anus; an unpleasant, stupid or boring person. **to fart about**, (*sl.*) to behave foolishly, to waste time.

farther (fah'dhə), *a.* more distant or remote; more extended; additional. *adv.* at or to a greater distance, extent or degree; (now usu. **further**) in addition, moreover, besides, also. **farthermost**, *a.* farthest. **farthest**, *a.* most distant. *adv.* at or to the greatest distance, extent or degree.

farthing (fah'dhing), *n.* the fourth part of an old penny, the smallest British copper coin (withdrawn in 1961); the smallest possible amount.

farthingale (fah'dhing-gāl), *n.* a hooped skirt used to extend the wide gown and petticoat of the 16th cent.

fasces (fas'ēz), *n.pl.* the ancient insignia of the Roman lictors, consisting of a bundle of elm or birch rods, in the middle of which was an axe; an emblem of authority.

fascia (fā'shə), *n.* (*pl.* **fasciae** (-iē)) a thin, tendon-like sheath surrounding the muscles and binding them in their places; a band, belt, fillet; the nameboard above a shop; (*Arch.*) a flat surface in an entablature or elsewhere; the instrument board of a car. **fasciated** (-ātid), *a.* (*Bot.*) flattened by the growing together of several parts; striped. **fasciation**, *n.* union of stems or branches in a ribbon-like form.

fascicle (fas'ikl), **fascicule** (-kūl), *n.* a small bundle, cluster or group; a cluster of leaves, flowers etc., a tuft; (*Anat.*) a bundle of fibres; a serial division of a book sold separately. **fascicled**, *a.* clustered together in a fascicle. **fascicular** (-sik'ū-), *a.* **fasciculate** (-sik'ūlət), **-lated** (-lātid), *a.* (*Nat. Hist.*) collected in clusters, small bundles or bunches. **fasciculation**, *n.* **fasciculus** (fəsik'ūləs), *n.* (*pl.* **-li**, **-lī**)) a fascicle.

fascinate (fas'ināt), *v.t.* to exercise an irresistible influence over; of snakes etc., to deprive of volitional power by magic or by means of look or presence; to captivate, to attract irresistibly, to enchant, to charm. **fascinating**, *a.* irresistibly attractive, charming, bewitching. **fascinatingly**, *adv.* **fascination**, *n.* **fascinator**, *n.*

fascine (fasēn'), *n.* a cylindrical faggot of brushwood used in building earthworks, filling trenches, protecting river-banks etc.

Fascism (fash'izm), *n.* (a movement or regime based on) a theory of government introduced into Italy by Benito Mussolini in 1922. Its object was to oppose socialism and communism by controlling every form of national activity. It was anti-democratic in principle, permitting no other party to exist and tolerating no opposition; (usu. **fascism**) any ideology or system regarded as brutal, repressive, excessively nationalistic or militaristic. **fascist, -istic,** *a.* **Fascist,** *n.* a member of a fascist party; (usu. **fascist**) one who advocates brutal or repressive policies, or (*loosely*) is regarded as holding very right-wing, illiberal views.

fashion (fash'ən), *n.* the form, make, style or external appearance of any thing; mode, manner, way, pattern; the prevailing style or mode of dress; prevailing practice, custom or usage; fashionable society. *v.t.* to give shape and form to; to frame, to mould; to fit, to adapt. **after, in a fashion,** in a way; middling, rather badly; somehow or other. **after the fashion of,** in the same way as; like. **in, out of fashion,** conforming or not conforming to the prevailing mode. **to set the fashion,** to set the example in a new style of dress or behaviour. **fashion-plate,** *n.* a picture illustrating a style in dress; an ultra-fashionably dressed woman. **fashion victim,** *n.* a person for whom wearing the current fashion in dress is of the utmost importance. **fashionable,** *a.* conforming to or observant of the fashion or established mode; characteristic of, approved by, or patronized by people of fashion. **fashionableness,** *n.* **fashionably,** *adv.* **-fashioned,** *comb. form* made or shaped (in a certain way).
-fashion, *comb. form* in the manner of, like.

fast[1] (fahst), *a.* firmly fixed, firm, tight; firmly adhering, faithful, steady, close; lasting, durable, permanent, unfading, not washing out; swift, rapid. (capable of) moving quickly; taking a short time; promoting or permitting quick motion; imparting quick motion, as a bowler, pitcher etc.; of a clock etc., showing a time ahead of the true time; of photographic film, requiring a short exposure time; of a camera shutter, permitting short exposure times; dissipated, rakish, pleasure-seeking, promiscuous; acquired with little effort or by shady means. *adv.* firmly, tightly, securely; quickly, swiftly; in rapid succession; in a dissipated manner, so as to expend one's energies quickly. **fast and furious,** vigorous and eventful, noisy or heated. **fast asleep,** sound or firmly asleep. **fast-breeder reactor,** a nuclear reactor which produces at least as much fissionable material as it consumes. **to play fast and loose** PLAY. **to pull a fast one,** (*coll.*) to trick, to use underhand methods. **fastback,** *n.* (a car with) a back which forms a continuous slope from roof to bumper. **fast food,** *n.* food, e.g. burgers and chicken pieces, which can be prepared and served very quickly. **fast-food,** *a.* serving fast food. **fast-forward,** *n., a.* (a switch) enabling video or recording tape to be wound on very rapidly. **fast lane,** *n.* a part of the carriageway used by fast-moving traffic, esp. the outer lane of a motorway; (*coll.*) a pace of life that is particularly fast, exciting or risky. **fast neutron,** *n.* a neutron with high kinetic energy. **fast-talk,** *n., v.t.* (*chiefly N Am., coll.*) (to persuade by) fluent, forceful (and often facile) speech. **fast worker,** *n.* a person who gets things done quickly; (*coll.*) one who makes rapid progress in relations with the opposite sex. **fastish,** *a.* rather fast or dissipated. **fastness,** *n.* the quality or state of being fast or secure; a fortress, a stronghold, esp. in a remote and inaccessible place.

fast[2] (fahst), *v.i.* to abstain from food; to abstain entirely or partially from food voluntarily as a religious observance. *n.* a (period of) total or partial abstinence from or deprivation of food, esp. from religious motives; a time set apart for fasting. **fast day,** *n.* **fasting,** *n.*

fasten (fah'sn), *v.t.* to fix firmly; to attach; to secure, as by a bolt, a lock, a tie, knot etc.; to fix or set firmly or earnestly. *v.i.* to become fast; to seize, to lay hold (upon). **to fasten on,** to lay hold on; to become aware of and concentrate one's attention on; to attach (blame, responsibility, a nickname etc.) to. **fastener,** *n.* **fastening,** *n.* the act of making fast or secure; anything which makes fast or secure, as a bolt, strap, catch etc.

fastext (fahs'tekst), *n.* a page-searching system used with teletext that facilitates rapid access to the desired information.

fastidious (fəstid'iəs), *a.* difficult to please; extremely careful, delicate, refined, esp. in matters of taste; squeamish, easily disgusted. **fastidiously,** *adv.* **fastidiousness,** *n.*

fastigiate (fəstij'iət), *a.* (*Biol.*) tapering to a point like a pyramid.

fat (fat), *a.* (*comp.* **fatter,** *superl.* **fattest**) plump, fleshy, corpulent; well-filled, thick; oily, greasy, unctuous; prosperous, thriving, affluent; producing a large income; fertile, fruitful; substantial, rewarding. *n.* a substance of a more or less oily character, deposited in vesicles in adipose tissue and in plant tissues; animal tissue containing fat; the fat part of anything; obesity, corpulence; the best or choicest part of anything; that part of anything which is deemed redundant or excessive; an organic compound of glycerine with one of a group of acids. *v.t., v.i.* to make or become fat. **a fat lot,** (*coll., iron.*) very little. **the fat is in the fire,** (*coll.*) there's going to be trouble. **to live off the fat of the land,** to have the best of everything, esp. in terms of food. **fat cat,** *n.* (*chiefly N Am., coll.*) a wealthy person, esp. one who contributes to political campaigns. **fat chance,** *n., int.* (*iron.*) very little or no chance. **fathead,** *n.* a dull, stupid person. **fat-hen,** *n.* a kind of goosefoot or *Chenopodium.* **fat-soluble,** *a.* soluble in fats and various other organic compounds, as ether and chloroform. **fatstock,** *n.* livestock fed up for market. **fatness,** *n.* **fatten,** *v.t.* to make fat; to feed for the table; to make (ground) fruitful, to fertilize. *v.i.* to grow or become fat. **fattish,** *a.* **fatty,** *a.* consisting of or having the qualities of fat; greasy, unctuous; containing (large amounts of) fat. *n.* (*coll.*) a fat person. **fatty acid,** *n.* any of a class of aliphatic carboxylic acids, e.g. palmitic acid, acetic acid. **fatty degeneration,** *n.* the abnormal production of granular fatty matter.

fatal, fatalism etc. FATE.

fate (fāt), *n.* the power by which the course of events is unalterably predetermined; destiny, lot, fortune; one's ultimate condition as brought about by circumstances and events; what is destined to happen; death, destruction. **the Fates,** the three Greek goddesses supposed to preside over the birth, life and fortunes of men, Clotho, Atropos and Lachesis. **fatal,** *a.* decreed by fate, inevitable; fateful, decisive; causing death or ruin; having unwelcome consequences. **fatalism,** *n.* the doctrine that events are predetermined and beyond human control; submission to fate. **fatalist,** *n.* **fatalistic** (-lis'-), *a.* **fatalistically,** *adv.* **fatality** (fətal'-), *n.* a fixed and unalterable course of things; predetermination by fate esp. to death or disaster; deadliness; a (person who suffers) death by accident or violence. **fated,** *a.* decreed by fate, predetermined; doomed to destruction. **fateful,** *a.* having momentous or catastrophic consequences; bringing death or destruction. **fatefully,** *adv.* **fatefulness,** *n.*

father (fah'dhə), *n.* a male parent; he who begets a child; a male ancestor, a patriarch; an originator, author, contriver, an early leader; a respectful mode of address to an old man or any man deserving great reverence; one who exercises paternal care; the senior member of any profession or body; (**Father**) the First Person of the Trinity; (**Father**) a priest, a religious teacher etc.; (*pl.*) elders, senators, the leading men (of a city etc.). *v.t.* to beget; to be or act as father of; to originate; to adopt or assume as one's own child, work etc.; to accept responsibility for. **Conscript Fathers** CONSCRIPT. **father-in-law,** *n.* the father of one's husband or wife. **father of the chapel** CHAPEL. **Fathers of the Church**, the ecclesiastical writers of the early church. **how's your father**, (*coll.*, *facet.*) illicit goings-on, esp. of a sexual nature. **Right, Most Reverend Father in God**, the formal title of a bishop or archbishop. **the Holy Father**, the Pope. **to father (up)on**, to suggest that someone is responsible for. **Father Christmas** SANTA CLAUS. **father confessor**, *n.* a priest who hears confessions; a person to whom one confides intimate matters. **father figure**, *n.* an older man whom one looks to for advice and support. **fatherland**, *n.* one's native country. **Father's day**, *n.* the third Sunday in June. **fatherhood** (-hud), *n.* the condition of being a father; the character or authority of a father. **fatherless**, *a.* **fatherlessness**, *n.* **fatherly**, *a.* like, proper to or becoming a father; kind, tender, loving. **fatherliness**, *n.*

fathom (fa'dhəm), *n.* a measure of length, 6 ft. (1·8 m) used principally in nautical and mining measurements. *v.t.* to ascertain the depth of; (often with *out*) to get to the bottom of, to comprehend. **fathom-line**, *n.* (*Naut.*) a sounding-line. **fathomable**, *a.* **fathomless**, *a.* not to be fathomed. **fathomlessly**, *adv.* **fathometer** (-om'itə), *n.* an instrument for measuring the depth of the sea by sound waves.

fatigue (fətēg'), *n.* weariness, exhaustion from bodily or mental exertion; toil or exertion causing weariness or exhaustion; labour not of a military nature performed by soldiers; a weakening in materials, e.g. metals, due to prolonged strain; temporary reduced response to stimuli caused by over-stimulation; (*pl.*) military overalls, fatigue-dress. *v.t.* to tire, to weary. **fatigue-dress**, *n.* the dress worn by soldiers on fatigue-duty.

fatness, **fatten** etc. FAT.

fatuous (fat'ūəs), *a.* foolish, inane, silly. **fatuity** (-tū'-), *n.* **fatuitous** (-tū'-), *a.* **fatuously**, *adv.* **fatuousness**, *n.*

fauces (faw'sēz), *n.pl.* the hinder part of the mouth, terminated by the pharynx and larynx. **faucal** (-kəl), *a.*

faucet (faw'sit), *n.* (*chiefly N Am.*) a tap; a beer-tap.

faugh (faw), *int.* an exclamation of disgust or abhorrence.

fault (fawlt), *n.* a defect, blemish, imperfection; an error, failing, mistake or blunder; a slight offence or deviation from right or propriety; responsibility for a mistake or mishap, blame; an improper service at tennis; a penalty point in showjumping; (*Teleg.*) a leak through broken insulation etc.; the sudden interruption of the continuity of strata till then upon the same plane, accompanied by a crack or fissure. *v.i.* to commit a fault; (*Geol.*) to undergo a break in continuity. *v.t.* (*Geol.*) to break the continuity of; to find a fault in, criticize. **at fault**, in error, to blame. **to find fault with**, to censure, esp. in a carping manner. **fault-finder**, *n.* **fault-finding**, *n.* **faultless**, *a.* **faultlessly**, *adv.* **faultlessness**, *n.* **faulty**, *a.* **faultily**, *adv.* **faultiness**, *n.*

faun (fawn), *n.* one of a kind of demigods, or rural deities, bearing a strong resemblance in appearance and character to the satyrs, with whom they are generally identified.

fauna (faw'nə), *n.* (*pl.* **-nas**, **-nae**, (-nē)) the animals found in or peculiar to a certain region or epoch; a treatise upon these. **faunal**, *a.* **faunist**, *n.* **faunistic** (-nis'tik), *a.*

Fauvism (fō'vizm), *n.* a 20th-cent. art movement, characterized by vivid use of colour and a free treatment of form. **Fauvist**, *n.*, *a.*

faux pas (fō pah), *n.* a blunder, a slip. [F]

favour, (*esp. N Am.*) **favor** (fā'və), *n.* friendly regard, kindness, goodwill; countenance, approval; partiality, preference, excessive kindness or indulgence; a kind or indulgent act; a token of love or affection, esp. something given by a lady to her lover; a knot of ribbons worn on any festive occasion; a small gift given to a guest at a party; (*pl.*) a woman's consent to sexual activity. *v.t.* to regard or behave toward with kindness; to befriend, to support; to facilitate; to promote; to oblige (with); to show partiality to; to resemble in features; to avoid using, to treat with special care (as an injured limb). **in favour**, approved; approving. **in favour of**, approving, on the side of; to the account of; to the advantage of. **out of favour**, disapproved. **to curry favour** CURRY[1]. **favourable**, *a.* well-disposed, encouraging; propitious; approving, consenting; tending to promote or to encourage; advantageous. **favourableness**, *n.* **favourably**, *adv.* **favoured**, *a.* **-favoured**, *comb. form* having a certain look or appearance, as *ill-favoured*, *well-favoured*. **favouredness**, *n.*

favourite, (*esp. N Am.*) **favorite** (fā'vərit), *n.* a person or thing regarded with special affection, predilection or partiality; one chosen as a companion and intimate by a superior and unduly favoured; (*Sport*) the competitor considered to have the best chance, and against whom or which the shortest odds are offered. *a.* regarded with special favour; preferred before all others. **favouritism**, *n.* showing a special preference for a person or group, partiality.

fawn[1] (fawn), *n.* a young deer, esp. in its first year; the colour of a young deer. *a.* like a fawn in colour, yellowish-brown. *v.t.*, *v.i.* to bring forth (a fawn).

fawn[2] (fawn), *v.i.* of animals, esp. dogs, to show affection by cringing, licking the hand etc.; (usu. with *upon*) to court in a servile manner, to grovel, to cringe. **fawner**, *n.* **fawning**, *a.* **fawningly**, *adv.*

fax (faks), *n.* a system for electronically scanning, transmitting and reproducing documents etc. via a telephone line; a document etc. sent in this way. *v.t.* to send (a document etc.) by fax. **fax machine**, *n.* [abbr. of *facsimile*]

fay (fā), *n.* a fairy.

faze (fāz), *v.t.* (*chiefly N Am.*) to disconcert, to put off one's stroke.

FBA, (*abbr.*) Fellow of the British Academy.

FBI, (*abbr.*) Federal Bureau of Investigation (in US).

FC, (*abbr.*) Football Club; Forestry Commission.

FCA (*abbr.*) Fellow of the Institute of Chartered Accountants.

FD, (*abbr.*) Defender of the Faith. [L *Fidei Defensor*]

Fe, (*chem. symbol*) iron. [L *ferrum*]

fealty (fē'əlti), *n.* fidelity of a vassal or feudal tenant to his lord; fidelity, loyalty, allegiance.

fear (fiə), *n.* a painful apprehension of danger or of some impending evil; dread, a state of alarm; anxiety, solicitude; awe, reverence; an object of fear. *v.t.* to be afraid of, to dread; to shrink from, to hesitate (to do); to reverence, to venerate; to suspect, to doubt. *v.i.* to be afraid; to feel anxiety or solicitude. **for fear**, in dread (that or lest); lest. **no fear**, (*coll.*) not likely; certainly not. **feared**, *a.* regarded with fear. **fearful**, *a.* timid, timorous; apprehensive, afraid (lest); produced by or indicating fear; (*coll.*) terrible, awful, very bad. **fearfully**, *adv.* **fearfulness**, *n.* **fearless**, *a.* **fearlessly**, *adv.* **fearlessness**, *n.* **fearsome** (-səm), *a.* fearful, terrible, alarming. **fearsomely**, *adv.* **fearsomeness**, *n.*

feasible (fē'zibl), *a.* that may or can be done, practicable, possible; (*coll.*) manageable; likely, plausible. **feasibility** (-bil'-), *n.* **feasibly**, *adv.*

feast (fēst), *n.* a sumptuous meal or entertainment, esp. a public banquet; an anniversary or periodical celebration, esp. a religious anniversary; anything giving great enjoyment to body or mind. *v.t.* to entertain sumptuously; to gratify or please greatly, as with something delicious or luscious. *v.i.* to feed sumptuously; to be highly gratified or pleased. **immovable, movable feasts**, festivals or anniversaries occurring on a fixed date, as Christmas, or on varying dates, as Easter. **feast-day**, *n.* a day of feasting; a festival. **feaster**, *n.*

feat (fēt), *n.* a notable act or performance, esp. one displaying great strength, skill or daring; an exploit, an achievement.

feather (fe'dhə), *n.* a plume or quill, one of the dermal appendages forming collectively the covering of a bird; a

strip of a feather attached to an arrow-shaft; (*usu. pl.*) a hairy fringe on a dog's tail or legs, a tuft of long hair on a horse's leg; a tongue on the edge of a board fitting into a groove on the edge of another board; (*Rowing*) the act of feathering. *v.t.* to dress, cover or furnish with feathers; to adorn with or as with feathers; to turn (an oar) so that the blade passes horizontally through the air; to change the angle or allow free rotation of (a propeller blade) to minimize wind resistance. *v.i.* to grow feathers; to have a feathery appearance; (*Rowing*) to feather an oar; to change the angle of a propeller blade. **a feather in one's cap**, an honour, a distinction. **birds of a feather**, people of the same sort, taste, disposition etc. **to be in high feather**, to be in high spirits, to be elated. **to feather one's nest**, to accumulate wealth; to make provision for oneself. **to show the white feather**, to show signs of cowardice or timidity (a white feather in the tail of a game-cock was a sign of cowardice). **feather-bed**, *n.* a mattress stuffed with feathers. *v.t.* to pamper, to spoil; to give financial assistance to (an industry). **feather-bedding**, *n.* the practice of protecting jobs by allowing overmanning. **feather-brain** FEATHER-HEAD. **feather duster**, *n.* a long-handled brush of feathers. **feather-edge**, *n.* the thinner edge of a wedge-shaped board or plank. **feather-edged**, *a.* **feather-head, -brain**, *a.* a silly, frivolous person. **feather-headed, -brained**, *a.* **feather-stitch**, *n.* an embroidery stitch producing a zigzag line somewhat like feathers. **featherweight**, *n.* something very light; a jockey of the lightest weight allowed to be carried by a horse in a handicap; a boxer not above 9 st. (57 kg); one who or that which is of little importance or moment. **feathered**, *a.* covered with feathers (*also in comb.*, as *well-feathered*); fitted, fringed or adorned with a feather or feathers. **feathering**, *n.* the action of the verb TO FEATHER; plumage; feathers on an arrow; a feathery fringe or tuft on a dog, horse etc. **featherless**, *a.* **featherlet** (-lit), *n.* **feathery**, *a.* covered, fringed or adorned with or as with feathers; feather-like, resembling feathers; (*fig.*) light, flimsy, fickle; (*Bot.*) plumose. **featheriness**, *n.*
feature (fē'chə), *n.* (*usu. pl.*) a part of the face, esp. such as gives individual expression and character; a prominent or distinctive part of anything, a striking incident, a mark of individuality; a full-length film, esp. the main film in a programme; a prominent article in a newspaper or magazine on a particular topic; a radio or television documentary. *v.t.* (*coll.*) to resemble in features; to have as a characteristic; to give prominence to, to make a feature of; (*Cinema*) to present in a role less important than that of a star. *v.i.* to be a characteristic, to figure prominently. **featured**, *a.* **-featured**, *comb. form* having a certain kind of features or cast of face. **featureless**, *a.* without any distinct or distinctive features.
Feb, (*abbr.*) February.
febrifuge (feb'rifūj), *n.* a medicine which has the property of dispelling or mitigating fever. **febrifugal** (-brif'ūgəl), *a.*
febrile (fē'brīl), *a.* pertaining to, proceeding from, or indicating fever.
February (feb'rūəri), *n.* the second month of the year, containing in ordinary years 28 days, and in the bissextile or leap-year 29.
feces (fē'sēz), **fecal** (fē'kəl), etc. FAECES.
feckless (fek'lis), *a.* puny, weak, feeble in mind; improvident, irresponsible. **fecklessly**, *adv.* **fecklessness**, *n.*
feculent (fek'ūlənt), *a.* full of dregs, lees or sediment; muddy, turbid; filthy, fetid. **feculence**, *n.*
fecund (fē'kənd, fek'-), *a.* fruitful, prolific, fertile. **fecundate**, *v.t.* to make fruitful or prolific; to impregnate. **fecundation**, *n.* **fecundity** (-kūn'-), *n.* the quality of being fruitful or prolific; the power or property of producing young or germinating; power of creation or invention. **fecundize, -ise**, *v.t.*
Fed., fed[1] (fed), (*abbr.*) Federal; Federation.

fed[2] (fed), *past, p.p.* FEED[1]. **fed up**, *a.* (*coll.*) sick or tired (of). **to be fed up (to the back teeth) with**, to have had more than enough of, to be sick of.
fed[3], *n.* (*coll.*) a federal agent (in the US).
fedayee (fedah'yē), *n.* (*pl.* **-yeen**, (-yēn')) a member of an Arab commando group, esp. against Israel.
federal (fed'ərəl), *a.* relating to, arising from or supporting a polity formed by the union of several states; relating to such a government as distinguished from the separate states; supporting the cause of the Union in the American Civil War. *n.* a supporter of the principle of federation, esp. a supporter of the American Union in the Civil War. **Federal Bureau of Investigation**, a branch of the US Department of Justice concerned with internal security, espionage and sabotage. **Federal Reserve Bank**, one of 12 US banks holding reserves and performing functions similar to those of the Bank of England. **federalism**, *n.* **federalist**, *n., a.* **federalize, -ise**, *v.t., v.i.* to bring or come together in a political confederacy. **federally**, *adv.* **federate**, *v.t., v.i.* to organize as or form a federal group; to federalize; to bring or join together for a common object. *a.* (-ət), united under a federal government; leagued together. **federation**, *n.* the act of uniting or federating; a federated body; a federal government. **federationist**, *n.* **federative**, *a.* **federatively**, *adv.*
fedora (fidaw'rə), *n.* (*N Am., coll.*) a soft felt hat with a curled brim.
fee (fē), *n.* (*Feudal Law*) land and estate held of a superior; a freehold estate of inheritance; payment or remuneration to a public officer or a professional person for the execution of official functions or for the performance of a professional service; a charge paid for a privilege, such as admission to an examination, society, public building etc.; (*Sc.*) wages. *v.t.* (*past, p.p.* **feed**) to pay a fee or reward to; (*Sc.*) to hire. **retaining fee**, a payment made to a professional engaging his or her services for a case etc. **to hold in fee**, (*Law*) to own absolutely. **fee-simple**, *n.* (*Law*) an estate held by a person in his or her own right, without limitation to any particular class of heirs. **fee-tail**, *n.* (*Law*) an estate entailed to the possessor's heirs. **feeless**, *a.*
feeble (fē'bl), *a.* weak, destitute of physical strength; lacking in force, vigour or energy; lacking in moral or intellectual power; ineffective, pointless, insipid; dim, faint; unconvincing, lame. **feeble-minded**, *a.* intellectually deficient, imbecile; wanting in resolution. **feebleness**, *n.* **feeblish**, *a.* **feebly**, *adv.*
feed[1], *v.t.* (*past, p.p.* **fed**, (fed)) to give food to; to put food into the mouth of; to supply with that which is necessary to existence, continuance or development; to serve as food or nourishment for; to cause to pass (as a rope or tape) through or into something; to supply (as a machine) with material; to cue in a response from (another performer); to pass the ball or puck to (another player); to gratify. *v.i.* to take food; to eat; to subsist (on or upon). *n.* food, fodder, pasturage; the act of feeding or giving food; amount of food or provender given to horses, cattle etc. at a time; (*coll.*) a meal, a feast; the operation of supplying a machine with material, or of bringing a tool into operation; the machinery for this; the amount supplied; a performer who supplies cues, esp. a straight man. **off one's feed**, without appetite. **to feed up**, to give plenty to eat, to fatten. **feedback**, *n.* the return of part of the output of a system, circuit or mechanical process to the input; the return of part of the sound output of a loud-speaker to the microphone, producing a high-pitched whistle; reactions and comments from customers, consumers, audience etc. **feed-pipe**, *n.* the pipe carrying water to the boilers of steam-engines. **feed-pump**, *n.* a force-pump for supplying water to boilers. **feeder**, *n.* one who supplies food or nourishment; one who eats, esp. in a certain manner, as a *quick feeder*; a feeding-bottle; a

child's bib; a tributary stream; an artificial channel supplying a canal etc.; a branch railway; a subsidiary road that joins a major route; (*Elec.*) a wire or cable, usu. in pairs, carrying electricity to various points in a system; the apparatus feeding a machine; a theatrical feed; one who nourishes, encourages or supports. **feeding**, *n.* **feeding-bottle**, *n.* a bottle for supplying liquid nutriment to infants.

feed² (fēd), *past, p.p.* FEE.

feel (fēl), *v.t.* to perceive by the touch; to have the sense of touch; to have a sensation of, otherwise than by the senses of sight, hearing, taste or smell; to be conscious of; to have the emotions stirred by; to experience, to undergo; to know in one's inner consciousness, to be convinced (that); to examine or explore by the touch. *v.i.* to have perception by the sense or act of touching; to be conscious of a certain sensation (as cold, wet, hungry or tired); (*reflex.*) to be conscious of (oneself) as in a certain state (as afraid, anxious, busy etc.); to be stirred in one's emotions; to seem to the sense of touch, to produce a certain sensation, as *the air feels damp or cold. n.* the sense of touch; characteristic sensation of something, esp. one related to that of touch; perception, esp. of an emotional kind. **to feel for,** to have sympathy or compassion for. **to feel like,** to wish to, to be in the mood for. **to feel up,** (*sl.*) to touch in such a way as to arouse oneself or another person sexually. **to feel up to,** (*coll.*) to feel able or strong enough to. **feeler,** *n.* one who feels; (*fig.*) any device to ascertain the designs, wishes or opinions of others; a scout; a generic term for various organs of touch in invertebrate animals. **to put out feelers,** to make tentative enquiries. **feeler gauge,** *n.* a thin metal strip of a known thickness used to measure a gap. **feeling,** *n.* the sense of touch; the sensation produced when a material body is touched; a physical sensation of any kind; emotion; an emotional state or reaction; tenderness, sympathy; (*pl.*) susceptibilities, sympathies; an impression, a sense, an intuition; a sentiment, belief or conviction (usu. nonrational); the emotional content or mood of a work of art. *a.* perceiving by the touch; easily affected or moved, sensitive; expressive of or manifesting great sensibility; affecting. **feelingly,** *adv.*

feet FOOT.

feign (fān), *v.t.* to invent, to pretend, to simulate, to counterfeit. *v.i.* to make pretences. **feignedly** (-nid-), *adv.* **feignedness** (-nid-), *n.*

feint¹ (fānt), *n.* a feigned or sham attack; a pretence of aiming at one point while another is the real object; a pretence. *v.i.* to make a feint or pretended attack (upon, against or at).

feint² (fānt), *a.* of ruled lines on paper, faint.

feisty (fīs'ti), *a.* plucky, full of fight.

felafel, falafel (fəlah'fəl), *n.* a thick paste of ground chick peas with spices, onion etc. formed into balls and deep-fried.

feldspar (feld'spah), *n.* a name including several minerals found abundantly in igneous rocks, chiefly silicates of alumina combined with some other mineral. **feldspathic** (-spath'-), *a.* pertaining to feldspar; having feldspar in the composition.

felicity (fəlis'iti), *n.* happiness, blissfulness; a source of happiness, a blessing; appropriateness; a happy turn or expression; a happy way or faculty of expressing, behaving etc. **felicitate,** *v.t.* to confer happiness upon; to congratulate. **felicitation,** *n.* congratulation. **felicitous,** *a.* happy, delightful, prosperous; well-suited, apt, well-expressed; charming in manner, operation etc. **felicitously,** *adv.* **felicitousness,** *n.*

felid (fē'lid), *n.* one of the Felidae, a family of fissiped carnivores, containing lions, tigers, leopards, pumas and cats.

feline (fē'līn), *a.* belonging to the Felidae; of or pertaining to cats, cat-like; sly, stealthy; graceful, sinuous. *n.* one of the Felidae, a cat. **felinity** (-lin'-), *n.*

fell¹ (fel), *v.t.* to knock down; to hew or cut down; (*Sewing*) to finish with a fell. *n.* a quantity of timber felled; a seam or hem in which one edge is folded over another and sewed down. **feller**¹, *n.* one who fells or cuts down trees.

fell² (fel), *n.* the hide or skin of an animal, esp. if covered with hair. **fell-monger,** *n.* a dealer in hides and skins.

fell³ (fel), *n.* a rocky hill; a lofty tract of barren moorland.

fell⁴ (fel), *a.* cruel, savage, fierce; terrible, deadly.

fell⁵ (fel), *past* FALL.

fella (fel'ə), *n.* (*coll.*) fellow, man; a male sweetheart. [alteration of FELLOW]

fellah (fel'ə), *n.* (*pl.* **fellaheen,** (-hēn)) an Egyptian agricultural labourer or peasant.

fellatio (fəlā'shiō), *n.* oral stimulation of the penis.

feller¹ FELL¹.

feller² (fel'ə), FELLA.

felloe (fel'ō), *n.* one of the curved segments of a wheel, joined together by dowels to form the rim; the whole rim of a wheel.

fellow (fel'ō), *n.* an associate; a companion; one of the same kind or species; an equal in rank, a peer; one of a pair; a person or thing like or equal to another, a counterpart; a member of an incorporated society; an incorporated member of a college; the holder of a fellowship or stipendiary position endowed for purposes of research; a man, a boy; (*coll.*) a male sweetheart; a person of little estimation. **fellow-craft,** *n.* a Freemason of the second degree. **fellow-creature,** *n.* one of the same race. **fellow-feeling,** *n.* sympathy; joint interest. **fellow traveller,** *n.* (*usu. derog.*) one who without declaring him- or herself a member sympathizes with the aims of the Communist Party or other similar organization. **fellowship,** *n.* the condition or state of being a fellow; companionship, association, friendliness, cordiality of feeling, community of interest; a body of associates; a brotherhood, a fraternity; the dignity of fellow in a college or learned society; an endowment for maintaining a graduate engaged in research; membership of a community partaking of Holy Communion together; the rule by which profit or loss is divided among partners in proportion to the capital invested.

felly (fel'i), FELLOE.

felo-de-se (fel'ō də sē), *n.* (*pl.* **felos-** (-z)) orig., one who commits felony by self-murder; self-murder, suicide.

felon (fel'ən), *n.* one who has committed a felony; a whitlow or abscess close to the nail. **felonious** (-lō'-), *a.* pertaining to or of the nature of a felony. **feloniously,** *adv.* **feloniousness,** *n.* **felony,** *n.* an offence of a heinous character, conviction for which formerly involved loss of lands and goods; in US law and in Eng. law until 1967, an offence of graver character than a misdemeanour.

felspar (fel'spah), etc. FELDSPAR.

felt¹ (felt), *n.* a kind of cloth made of wool or wool and cotton compacted together by rolling, beating and pressure, with lees or size; an article made of it, as a felt hat. *v.t.* to make into felt; to cover with felt; to press into a compact mass. *v.i.* to become matted together. **felt-tip (pen),** *n.* a pen with a writing point made of pressed felt or similar fibres. **felting,** *n.* **felty,** *a.*

felt² (felt), *past, p.p.* FEEL.

felucca (fəlŭk'ə), *n.* a small vessel used in the Mediterranean, propelled by oars or lateen sails or both.

fem., (*abbr.*) feminine.

female (fē'māl), *a.* denoting the sex which brings forth young or lays eggs from which new individuals are developed; (*Bot.*) having a pistil, but no stamens, capable of being fertilized and producing fruit; of, pertaining to or characteristic of woman or womanhood; womanly, feminine; (*Mech.*) fitted to receive the corresponding male part as a *female screw. n.* (*sometimes derog.*) a woman or girl; a female animal or plant. **female impersonator**, *n.* a male performer who dresses as, and imitates a woman for the purposes of his act. **female screw**, *n.* the spiral-threaded cavity into which another (male) screw works.

feminine (fem'inin), *a.* of, pertaining to or characteristic of women or the female sex; womanly; effeminate, womanish; soft, tender, delicate; (*Gram.*) belonging to the gender denoting females. **femininely**, *adv.* **feminineness**, *n.* **femininity** (-nin'-), *n.* the qualities or manners becoming a woman. **feminism**, *n.* advocacy of the claims of women to political, economic and social equality with men. **feminist**, *n.*, *a.* **feministic** (-nis'-), *a.* **feminize, -ise**, *v.t.*, *v.i.* to make or become feminine.

femme fatale (fam fatahl'), *n.* (*pl.* **femmes fatales**) a seductive woman, esp. one who lures men into ruin. [F, fatal woman]

femto- (femtō-), *pref.* a thousand million millionth (10^{-15}).

femur (fē'mə), *n.* (*pl.* **femurs, femora** (fem'ərə)) the thigh-bone; the third joint of the leg in insects. **femoral** (fem'ərəl), *a.* of or belonging to the thigh. *n.* the femoral artery.

fen (fen), *n.* low, flat and marshy land, esp. (*pl.*) the low-lying districts in the east of England, partially drained and abounding in broads or lakes. **fenland**, *n.* a fen; the fens. **fenlander, -man**, *n.* an inhabitant of the fens. **fenny**, *a.*

fence (fens), *n.* a structure serving to enclose and protect a piece of ground, or to keep cattle from straying, esp. a line of posts joined by rails, panels, wire netting etc.; a guardplate, guide or gauge of various kinds in machinery etc.; (*sl.*) a purchaser or receiver of stolen goods, or a place where such are purchased or deposited. *v.t.* to defend, shield or protect; to ward (off); to enclose, encircle or protect with or as with a fence; to parry. *v.i.* to practise the art of swordplay; to use a sword; to defend oneself or repel attack skilfully; to parry enquiries adroitly, to equivocate; (*sl.*) to deal in stolen goods. **sunk fence**, a fence set along the bottom of a ditch; a ditch forming a fence. **to mend one's fences**, to restore good relations, to make up differences. **to sit on the fence**, to remain neutral in respect to opposing policies. **fenceless**, *a.* **fencer**, *n.* one skilled in fencing; a builder of fences. **fencing**, *n.* the act of making fences; (*collect.*) fences; materials for fences; the act or art of using a sword or foil in attack or defence.

fencible (fen'sibl), *n.* formerly, a soldier enlisted for home defence.

fend (fend), **to fend for**, to provide or to get a living for. **to fend off**, to keep off, ward off. **fender** (fen'də), *n.* a piece of furniture, usu. of iron or brass, placed on the hearth to confine the ashes; a piece of timber or plastic or mass of rope to protect the side of a vessel from injury by collision; (*chiefly N Am.*) a metal frame on the front of a locomotive to reduce injury in a collision. (*N Am.*) the wing or mudguard of a motor vehicle. **fenderless**, *a.*

fenestella (fenistel'ə), *n.* (*Arch.*) a niche on the south side of the altar containing the piscina, and often the credence.

fenestra (fənes'trə), *n.* (*pl.* **-trae** (-trē)) a window-like aperture in a bone or between bones; a transparent spot or aperture in a wing, leaf etc. **fenestral**, *a.* **fenestrated** (-trātid, fen'is-), **fenestrate**, *a.* furnished with windows; (*Anat.*) having pores or fenestrae. **fenestration**, *n.* (*Arch.*) the construction, arrangement or mode of design of windows; (the surgical) formation of a fenestra; the condition of having fenestrae.

Fenian (fē'nyən), *n.* a member of an Irish secret society formed in America about 1858, having for its object the overthrow of the British Government in Ireland, the establishment of an independent republic; (*offensive*) an (esp. Irish) Roman Catholic. *a.* pertaining to this society or to Fenianism; (*offensive*) (Irish) Roman Catholic. **Fenianism**, *n.*

fennec (fen'ik), *n.* a small fox-like animal common in Africa.

fennel (fen'əl), *n.* a fragrant umbelliferous plant with yellow flowers, *Foeniculum vulgare*.

fenugreek (fen'ūgrēk), *n.* a leguminous plant, *Trigonella faenum-Graecum*, the seeds of which are used as a seasoning.

feoff (fēf, fef), *v.t.* (*Law*) to grant possession, to enfeoff. *n.* a fief. **feoffee** (-ē'), *n.* one who is invested with an estate by feoffment. **feoffment**, *n.* the conveyance of any corporeal hereditament to another, accompanied by actual delivery of possession. **feoffor**, *n.* one who grants a fief.

feral (fiə'rəl), *a.* wild, savage; uncultivated; (*fig.*) brutal, savage.

fer-de-lance (feədəlās'), *n.* the yellow viper of Martinique, *Bothrops lanceolatus*.

feretory (fe'ritəri), *n.* the bier or shrine in which relics of saints were borne in procession, a reliquary, a chapel or place in a church in which shrines were kept.

ferial (fiə'riəl), *a.* (*Eccles.*) pertaining to ordinary week-days, such as are not festival or fast days; pertaining to holidays.

fermata (fœmah'tə), *n.* (*Mus.*) a continuation of a note or rest beyond its usual length. [It.]

ferment[1] (fœ'mənt), *n.* any substance which causes fermentation; leaven; fermentation; commotion, tumult, agitation.

ferment[2] (fəment'), *v.t.* to excite fermentation in; to rouse, to agitate, to excite. *v.i.* to be in a state of fermentation, to effervesce; to be agitated, as by violent emotions. **fermentable** (-men'-), *a.* **fermentation**, *n.* chemical decomposition excited in certain organic compounds by living organisms or chemical agents, often with evolution of heat and effervescence; esp., the breakdown of sugar to ethyl alcohol by yeast; commotion, agitation, excitement. **fermentative** (-men'-), *a.*

fermi (fœ'mi), *n.* a unit of length equal to 10^{-15} metre.

fermion (-ən), *n.* any of a group of subatomic particles which behave according to the relations laid down by Fermi and Dirac. **fermium** (-əm), *n.* an element, at. no. 100; chem. symbol Fm, artificially produced from plutonium. [after Enrico *Fermi*, It. physicist, 1901–54]

fern (fœn), *n.* a cryptogamic plant springing from a rhizome, and having the reproductive organs on the lower surface of fronds or leaves, which are often divided in a graceful, feathery form. **fernery**, *n.* a place where ferns are cultivated. **fernless**, *a.* **ferny**, *a.*

ferocious (fərō'shəs), *a.* fierce, savage, cruel, barbarous; extreme, intense. **ferociously**, *adv.* **ferociousness**, *n.* **ferocity** (-ros'-), *n.* the state or quality of being ferocious; wildness, fury.

-ferous (-fərəs), *suf.* bearing, producing, having, as *auriferous, fossiliferous*.

ferrate (fe'rāt), *n.* a salt of ferric acid.

ferret (fe'rit), *n.* a partially tamed variety of polecat used for killing rats and driving rabbits out of their holes; a sharp-eyed searcher or detective. *v.t.* to drive out of a hole with ferrets; to hunt or take with ferrets; to search (out) by persevering means; to worry. *v.i.* to hunt rabbits etc. with a ferret; to search or rummage about (for). **ferreter**, *n.* **ferrety**, *a.*

ferri-, *comb. form* (*Chem.*) denoting a compound of iron in the ferric state (cp. FERRO-).

ferriage (fer'iij), *n.* (the fare paid for) conveyance by a ferry.

ferric (fe'rik), *a.* of, pertaining to or containing (trivalent) iron.

Ferris wheel (fe'ris), *n.* a big, upright fairground wheel with seats suspended from its rim. [after G.W.G. *Ferris*, US engineer, 1859–96]

ferrite (fe'rīt), *n.* a sintered ceramic consisting of a mixture of ferric oxide and other metallic oxides, which possesses magnetic properties.
ferro- (ferō-), *comb. form* (*Min.*) denoting a substance containing iron; (*Chem.*) denoting a compound of iron in the ferrous state (cp. FERRI-).
ferroconcrete (ferōkon'krēt), *n.* concrete strengthened by incorporation of iron bars, strips etc.; reinforced concrete.
ferromagnetic (ferōmagne'tik), *a.* acting magnetically like iron. *n.* a substance acting thus. **ferromagnetism,** *n.*
ferrotype (fe'rōtīp), *n.* a positive photograph on a sensitized film laid on a thin iron plate; the iron plate used in this process.
ferrous (fe'rəs), *a.* of, pertaining to or containing (divalent) iron.
ferruginous (fəroo'jinəs), *a.* containing iron or iron-rust; of the colour of iron-rust.
ferrule (fe'rool, -rəl), *n.* a metallic ring or cap on the handle of a tool, the end of a stick, the joint of a fishing rod, a post etc. to strengthen it; a short piece of pipe screwed into a main to form a connection with a service-pipe. **ferruled,** *a.*
ferry (fe'ri), *v.t.* to transport over a river, strait or other body of water, in a boat, barge etc.; to convey in a vehicle. *v.i.* to pass across narrow water in a boat etc. *n.* the passage where a ferry-boat plies to carry passengers and goods across a river etc.; the right of ferrying and charging toll for so doing; a ferry-boat. **ferry-boat,** *n.* a boat used at a ferry. **ferryman,** *n.*
fertile (fœ'til), *a.* able to sustain abundant growth; able to bear offspring; capable of growing or developing; productive, fruitful; inventive, resourceful; able to be transformed into fissionable material. **fertility** (-til'-), *n.* **fertility drug,** *n.* a drug given to apparently infertile women to stimulate ovulation. **fertility rite,** *n.* a pagan religious ceremony intended to ensure the fertility of the soil, beasts or human population. **fertility symbol,** *n.* **fertilize, -ise** (-ti-), *v.t.* to make fertile or productive; to make rich (as soil); to impregnate or pollinate. **fertilizable,** *a.* **fertilization,** *n.* **fertilizer,** *n.* a fertilizing agent; a chemical applied to the soil to improve its growth-promoting qualities and modify its acidity or alkalinity.
Ferula (fe'rələ), *n.* (*pl.* **-lae** (-lē)) a genus of umbelliferous plants, from the shores of the Mediterranean and Persia, yielding gum-resin, typified by the giant fennel; (**ferula**) a ferule.
ferule (fe'rool, -rəl), *n.* a rod or cane used to punish children in school. *v.t.* to punish with a ferule.
fervent (fœ'vənt), *a.* hot, boiling, glowing; ardent, earnest, zealous, vehement. **fervently,** *adv.* **fervency,** *n.* **fervid** (-vid), *a.* burning, very hot, fervent; impassioned. **fervidly,** *adv.* **fervidness,** *n.* **fervour,** (*esp. N Am.*) **fervor** (-və), *n.* heat, warmth; ardour, intensity of feeling, vehemence; zeal.
fescue (fes'kū), *n.* a small rod or pin with which a teacher pointed out the letters to a child learning to read; a genus of grasses, *Festuca.*
fesse, fess (fes), *n.* (*Her.*) a broad band of metal or colour crossing the shield horizontally, and occupying one-third of it; one of the nine honourable ordinaries, representing a knight's girdle. **fesse-point,** *n.* the centre of an escutcheon.
-fest, *comb. form,* an event or gathering for a particular activity, as *songfest.*
festal (fes'təl), *a.* pertaining to a feast or holiday; festive, joyous, gay, merry. **festally,** *adv.*
fester (fes'tə), *v.i.* to ulcerate or suppurate; to form purulent matter; to rankle; to become corrupted or rotten. *v.t.* to cause to fester or rankle. *n.* a purulent tumour or sore.
festival (fest'tivəl), *a.* pertaining to or characterizing a feast. *n.* a festal day or time, a joyous celebration or anniversary; a cultural entertainment or programme of events on a large scale, usually periodical. **festive,** *a.* of or befitting or used for a feast or festival; joyous, gay, mirthful. **festively,** *adv.* **festivity** (-tiv'-), *n.* a feast, a festival; gaiety, mirth, joyfulness; (*pl.*) merry-making.
festoon (festoon'), *n.* a chain or garland of flowers, foliage, drapery etc. suspended by the ends so as to form a depending curve; a carved ornament in the form of a garland or wreath. *v.t.* to form into or adorn with or as with festoons.
festschrift (fest'shrift), *n.* a collection of learned writings by various authors, published in honour of some person, usu. a scholar. [G]
feta, fetta (fet'ə), *n.* a firm white Greek cheese made from sheep's or goat's milk.
fetal (fē'təl), **fetus** (-təs), FOETUS.
fetch[1] (fech), *v.t.* to go for and bring; to cause to come; to draw forth, to heave (as a sigh); to derive, to elicit; to bring in, to sell for (a price); to reach, to arrive at; (*coll.*) to strike; (*coll.*) to deal (as a blow). *v.i.* (*Naut.*) to reach a place, to bring up. *n.* a stratagem, a trick; the distance over which a wave or the wind travels. **to fetch and carry,** to go to and fro with things; to perform menial tasks. **to fetch up,** to vomit; to come to a stand; (*coll.*) to end up. **fetcher,** *n.* **fetching,** *a.* (*coll.*) fascinating, charming, taking.
fetch[2] (fech), *n.* a wraith or double.
fete, fête (fāt, fet), *n.* a festival, an entertainment; in Roman Catholic countries, the festival of the saint after whom a person is named; an outdoor event with stalls and sideshows, usu. locally organized to raise money for charity. *v.t.* to entertain, to feast; to honour with festivities.
fetid, foetid (fet'id, fē'-), *a.* having an offensive smell, stinking. **fetidly,** *adv.* **fetidness,** *n.* **fetor** (fē'tə), *n.* a strong or offensive smell, a stench.
fetish (fet'ish), *n.* any material object supposed to be the vessel, vehicle or instrument of a supernatural being, the possession of which gives to the possessor power over that being; an object of devotion, an idol; (*Psych.*) an object providing sexual gratification; sexual obsession with such an object; a fixation. **fetishism,** *n.* belief in fetishes; worship of them; (*Psych.*) a form of perversion in which sexual gratification is obtained from an object or body part other than the genitals. **fetishist,** *n.* **fetishistic** (-shis'-), *a.*
fetlock (fet'lok), *n.* (a projection bearing) a tuft of hair behind the pastern joint of a horse, the pastern joint.
fetor (fētə), FETID.
fetta FETA.
fetter (fet'ə), *n.* (*often pl.*) a chain for the feet; anything which restrains or confines. *v.t.* to put fetters upon; to confine, restrain.
fettle (fet'l), *v.t.* to line the walls of (a furnace); to clean or put right; to arrange, order. *n.* condition, order, trim. **in fine, good fettle,** in good form or trim.
fettuccine (fetəchē'ni), *n.* tagliatelle.
fetus (fē'təs), FOETUS.
feu (fū), *n.* (*Sc. Law*) orig., tenure on condition of the performance of certain services or certain returns in money or kind; now, a perpetual lease at a fixed rent; the land, houses or other real estate so held. *v.t.* (*Sc. Law*) to give or take in feu. **feu-duty,** *n.* the annual rent for such a holding. **feu-holding,** *n.*
feud[1] (fūd), *n.* hostility between two tribes or families in revenge for an injury, often carried on for several generations; a state of enmity, a quarrel. *v.i.* to carry on a feud.
feud[2] (fūd), *n.* land held in trust, or on condition of performing certain services, a fee, a fief. **feudal,** *a.* pertaining to, consisting of or founded upon a feud or fief; according to or resembling the feudal system. **feudal system,** *n.* a system of social polity prevailing in Europe during the Middle Ages, by which the ownership of land

feuilleton / fiddle

inhered in the lord, possession or tenancy being granted to the vassal in return for military service. **feudalism,** *n.* the feudal system; a system resembling this. **feudalist,** *n.* **feudalistic** (-lis'-), *a.* **feudality** (dal'-), *n.* the quality or state of being feudal; feudal principles; a fief, a feudal holding. **feudalize, -ise,** *v.t.* to reduce to feudal tenure; to make feudal in character. **feudalization,** *n.* **feudally,** *adv.* **feudatory,** *a.* holding or held by feudal tenure; subject; under foreign overlordship. *n.* one who holds lands of another by feudal tenure; a vassal.

feuilleton (fœy''tō), *n.* that part of a French newspaper which is devoted to light literature, criticism or fiction; a light article or a serial story in a newspaper. [F]

fever (fē'və), *n.* a disease or group of diseases usu. characterized by high temperature, quickened pulse, nervous and muscular prostration and destruction of tissues; a body temperature above normal; a state of nervous excitement, agitation. *v.t.* to put or throw into a fever. *v.i.* to become feverish. **fever pitch,** *n.* a state of intense excitement or agitation. **fevered,** *a.* (*chiefly Sc.*) on the verge of death; (*chiefly Sc.*) in unnaturally high spirits; (*chiefly Sc.*) clairvoyant, psychic; eccentric, odd in a whimsical, other-worldly way.

feverfew (fē'vəfū), *n.* a common European plant supposed to act as a febrifuge.

few (fū), *a.* not many; small, limited or restricted in number. *n.* a small number (of). **a good few,** (*coll.*) a considerable number. **every few days, hours,** once in every series of a few days or hours. **few and far between,** rare, occurring very infrequently. **not a few,** a good many. **the few,** the minority; the elect. **fewness,** *n.*

fey (fā), *a.* (*chiefly Sc.*) fated to die, doomed, on the verge of death; (*chiefly Sc.*) in unnaturally high spirits; (*chiefly Sc.*) clairvoyant, psychic; eccentric, odd in a whimsical, other-worldly way.

fez (fez), *n.* (*pl.* **fezes**) a red cap without a brim, fitting close to the head, with a tassel of silk, wool etc., worn in the Middle East.

ff, (*abbr.*) folios; (and those e.g. pages) following; fortissimo.

fiacre (fiahkr''), *n.* a French hackney-coach invented about 1640. [said to be named after an innkeeper at the Hotel de St *Fiacre,* Paris]

fiancé (fiá'sā -on'-), *n.* one who is betrothed. **fiancée,** *n. fem.*

fiasco (fias'kō), *n.* (*pl.* **-cos, -coes**) a ridiculous or ignominious failure.

fiat (fī'ət, -at), *n.* an order, command, decree, usu. a peremptory one; (*Law*) the order or warrant of a judge or other constituted authority sanctioning or allowing certain processes.

fib (fib), *n.* a harmless or trivial lie. *v.i.* (*past, p.p* **fibbed**) to tell fibs. **fibber, fibster,** *n.* one who tells fibs.

Fibonacci number (fēbonah'chi), *n.* a number in the **Fibonacci sequence** or **series** in which each term is the sum of the preceding two. [after Leonardo *Fibonacci,* c.1170 – c.1250 It. mathematician]

fibre, (*esp. N Am.*) **fiber** (fī'bə), *n.* a slender filament; a thread, string or filament, of which the tissues of animals and plants are constituted; a natural or manmade thread or filament forming the raw material in textile manufactures; a structure composed of filaments; foodstuffs with a high fibre content, roughage; essence, nature, character; nerve, strength. **fibreboard,** *n.* a building-board composed of fibrous material. **fibreglass,** *n.* very fine filaments of molten glass worked into a synthetic material. **fibre-optics,** *n. sing.* a technology based on the transmission of light along bundles of very thin glass or plastic fibres, used esp. in telecommunications and exploratory medicine. **fibre-optic,** *a.* **fibrescope,** *n.* a flexible instrument using fibre-optics which enables the operator to see into otherwise inaccessible areas. **fibred,** *a.* composed of or having fibres (*esp. in comb.,* as *fine-fibred*). **fibreless,** *a.*

fibriform (-brifawm), *a.* fibrous, *a.* made of, containing or resembling fibres. **fibrously,** *adv.* **fibrousness,** *n.*

fibril (fī'bril), **fibrilla** (-bril'ə), *n.* (*pl.* **fibrils, fibrillae** (-ē)) a little fibre; (*Bot.*) one of the minute subdivisions in which a branching root terminates; a minute subdivision of a fibre in a nerve, muscle etc. **fibrillar, -llary, fibrillate** (fī'brilət), **-ated** (-ātid), *a.* **fibrillate,** *v.i.* to form fibrils; of a muscle, to twitch involuntarily; of the heart muscle, to contract rapidly and irregularly. **fibrillation,** *n.* **fibrilliform** (-bril'ifawm), *a.* **fibrillose,** *a.*

fibrin (fī'brin), *n.* an insoluble protein formed from fibrinogen, causing clotting of the blood. **fibrinogen** (-brin'əjən), *n.* a protein in the blood that is converted into fibrin during the process of coagulation. **fibrinolysin** (-ol'isin), *n.* an enzyme that promotes the breakdown of blood clots. **fibrinous,** *a.* composed of or of the nature of fibrin.

fibr(o)- *comb. form* denoting a substance consisting of or characterized by fibres.

fibroid (fī'broid), *a.* of the nature or form of fibre. *n.* a benign tumour.

fibroin (fī'broin), *n.* the chief constituent of silk, cobweb, the horny skeleton of sponges etc.

fibroma (fībrō'mə), *n.* (*Path.*) (*pl.* **-mas -mata** (-tə)) a benign fibrous tumour.

fibrosis (fībrō'sis), *n.* the abnormal formation of fibrous tissue.

fibrositis (fībrosī'tis), *n.* inflammation of fibrous tissue, esp. of muscles.

fibrous etc. FIBRE.

fibula (fib'ūlə), *n.* (*pl.* **-lae** (-lē), **-las**) the outer and smaller bone of the leg; a clasp, buckle or brooch. **fibular,** *a.*

-fic (-fik), *suf.* forming adjectives from nouns, verbs etc., as *honorific, horrific, malefic.*

-fication (-fikā'shən), *suf.* forming nouns from verbs with -FY, as *purification.*

fiche MICROFICHE.

fichu (fē'shoo), *n.* a light covering worn by women over the neck, throat and shoulders.

fickle (fik'l), *a.* changeable, inconstant. **fickleness,** *n.*

fictile (fik'tīl), *a.* capable of being moulded; moulded by art; made of earth or clay; manufactured by or suitable for the potter.

fiction (fik'shən), *n.* the act or art of feigning or inventing; an invented statement or narrative; a story; literature, esp. in prose, consisting of invented narrative; a falsehood; any point or thing assumed for the purposes of justice or convenience. **legal fiction,** an accepted falsehood which averts the raising of an awkward issue. **fictional,** *a.* **fictionalize, -ise,** *v.t.* to introduce fictional elements into (a narrative of real events). **fictionist,** *n.* a writer of fiction. **fictitious** (-tish'əs), *a.* feigned, counterfeit, false; of or pertaining to novels; having no real existence. **fictitiously,** *adv.* **fictitiousness,** *n.* **fictive,** *a.* imaginative, creative; imaginary, fictitious.

fid (fid), *n.* (*Naut.*) a bar of wood or iron to support a top-mast; a pointed wooden pin used to open the strands of a rope in splicing.

-fid *comb. form* divided into parts.

fiddle (fid'l), *n.* a violin; (*Naut.*) a frame of bars and strings, to keep things from rolling off the cabin table in bad weather; (*coll.*) a swindle, a dishonest practice; (*coll.*) an awkward or tricky operation. *v.i.* to play on a fiddle; to make restless movements with the hands or fingers; to waste time in aimless activity. *v.t.* to play (as a tune) on a fiddle; to falsify (accounts etc.); to contrive to do or obtain by underhand means. **fit as a fiddle,** in good condition, ready for anything. **on the fiddle,** (*coll.*) cheating, being dishonest, falsifying accounts etc. for one's own advantage. **to fiddle (about, around) with,** to tinker, to fuss with; to interfere or tamper with. **to play first, second fiddle,** to take a leading or a subordinate part or position. **fiddledeedee** (-didē'), *n., int.* nonsense. **fiddle-**

faddle (-fadl), *n.* trifling talk; nonsense. *a.* trifling; making a fuss about trifles. *v.i.* to trifle; to make a fuss about trifles. **fiddlesticks**, *int.* fiddledeedee. **fiddler**, *n.* one who plays the fiddle; (also **fiddler crab**) a small crab having one large claw and one very small one; one who fiddles. **fiddling**, *a.* trifling, fussy; petty. **fiddly**, *a.* tricky, awkward; small, difficult to manipulate.

fidei defensor (fid'ī difen'saw), *n.* Defender of the Faith. [L]

fidelity (fidel'əti), *n.* careful and loyal observance of duty; faithful adherence to a bond, covenant, engagement or connection; loyalty, faithfulness, esp. to husband or wife; honesty, veracity, reliability; accurate correspondence (of a copy, description, picture etc.) to the original; accurate reproduction of a sound by a radio, record player etc.

fidget (fij'it), *n.* a state of nervous restlessness; one who fidgets; (*pl.*) restless movements. *v.i.* to move about restlessly; to worry, to be uneasy. *v.t.* to worry or make (others) uncomfortable. **fidgety**, *a.* **fidgetiness**, *n.*

fiducial (fidū'shəl), *a.* confident, sure, firm; of the nature of a trust; (*Phys.*, *Surv.* etc.) denoting a fixed point or line used as a basis for measurement or comparison. **fiducially**, *adv.* **fiduciary**, *a.* pertaining to or of the nature of a trust or a trusteeship; held in trust; dependent on public confidence. *n.* a trustee.

fie (fī), *int.* an exclamation indicating contempt, irony, disgust, shame or impatience.

fief (fēf), *n.* an estate held of a superior under feudal tenure.

field (fēld), *n.* an area of open country; a piece of land enclosed for tillage or pasture; a region yielding some natural product abundantly (as an oil- or coal-field); the place where a battle is fought; the battle itself; the scene of military operations; the ground on which cricket, football or other games are played or athletic competitions held; the fielders or the players taken collectively; all the competitors in a race, or all except the favourite; the participants in a hunt; a sphere of activity or knowledge; an interest or speciality; the sphere of practical operations away from the office, laboratory etc.; a wide expanse; the surface on which the figures in a picture are drawn; (*Her.*) the surface of a shield or one of its divisions; a field of force; a field of view; (*Math.*) a set of mathematical elements subject to two binary operations, addition and multiplication, such that the set is a commutative group under addition and also under multiplication if zero is excluded; (*Comput.*) a set of characters comprising a unit of information; (*TV*) one of two interlaced sets of scanning lines. *v.t.* (*Cricket etc.*) to catch or stop (the ball) and return it; to retrieve (something or someone liable to go astray); to deal with (as questions), esp. off the cuff; to assemble ready for action (as a team, an army). *v.i.* to act as fielder in cricket and other games; (*Sporting*) to back the field against the favourite. **field of force**, (*Phys.*) the space within which a certain force is present, as a magnetic field. **field of view, vision**, the space visible in an optical instrument at one view. **to hold the field**, to maintain one's ground against all comers; to surpass all competitors. **to play the field**, to diversify one's interests or activities, esp. not to commit oneself to a steady boy or girl friend. **to take the field**, to commence active military operations; to begin a campaign; to go on to the field of play. **field-artillery**, *n.* light ordnance suitable for use in the field. **field-day**, *n.* a day on which troops are exercised in field evolutions; (*fig.*) a day or time of unusual importance, excitement or activity; a day of races and other sporting competitions held by a school, church etc. **to have a field day**, to take gleeful advantage of. **field events**, *n.pl.* athletic events other than racing, e.g. jumping. **field-glass(es)**, *n.* (*pl.*) a binocular telescope in compact form. **field goal**, *n.* in American football, a score made by kicking the ball over the crossbar from ordinary play; in basketball, a score made while the ball is in play. **field-gun**, *n.* a light artillery piece for service in the field. **field hockey**, *n.* (*N Am.*) hockey played on grass. **field-hospital**, *n.* an ambulance or temporary hospital near a battlefield. **field-ice**, *n.* ice formed in the polar regions in fields or floes, as distinct from icebergs. **field-marshal**, *n.* an officer of highest rank in the British Army. **fieldmouse**, *n.* one of several species of mice living in fields etc. **field-notes**, *n.pl.* notes made on the spot during fieldwork. **field-officer**, *n.* (*Mil.*) an officer above the rank of captain, but below that of general (as a major, a colonel etc.). **field-sports**, *n.pl.* outdoor sports, such as hunting, shooting, coursing etc. **field trial**, *n.* (*often pl.*) a test on a new invention, design etc. carried out under actual operating conditions. **field trip**, *n.* a visit undertaken by schoolchildren or students to study phenomena or collect information in situ. **field-winding**, *n.* a coil of wire wound on iron in order to make a strong electromagnetic field when the current is passing. **fieldwork**, *n.* observations or operations carried out in situ by students, researchers etc.; (*pl.*) temporary fortifications thrown up by besiegers or besieged. **fieldworker**, *n.* **fielder, fieldsman**, *n.* one who fields at cricket etc. **fieldwards**, *adv.*

fieldfare (fēld'feə), *n.* a species of thrush, *Turdus pilaris*, a winter visitant in England.

fiend (fēnd), *n.* a demon, a devil, an infernal being; a person of diabolical wickedness or cruelty; (*coll.*) a devotee, a fan; (*coll.*) an addict. **the fiend**, Satan. **fiendish**, *a.* **fiendishly**, *adv.* **fiendishness**, *n.* **fiendlike**, *a.*

fierce (fiəs), *a.* savage, furiously hostile or combative; raging, violent; vehement, ardent, eager, impetuous; intense, strong. **fiercely**, *adv.* **fierceness**, *n.*

fiery (fī'ri), *a.* consisting of fire, on fire, flaming with fire; hot, like fire; glowing or red, like fire; inflamed; highly inflammable, liable to explosions; of curry etc., hot-tasting; vehement, ardent, eager; passionate, hot-tempered, irascible. **fiery cross**, *n.* a charred and blood-stained wooden cross, formerly sent round in the Highlands to summon a clan to war; a flaming cross used as a means of intimidation by the Ku Klux Klan. **fierily**, *adv.* **fieriness**, *n.*

fiesta (fies'tə), *n.* a saint's day; a holiday or festivity.

fife (fīf), *n.* a small flute-like pipe, chiefly used in martial music. *v.i., v.t.* to play (tunes) on the fife. **fife-rail**, *n.* (*Naut.*) a rail on the quarter-deck and poop or around the mast of a vessel. **fifer**, *n.*

fifteen (fif'tēn, -tēn'), *n.* the number or figure 15 or XV; the age of 15; a set of 15 players, pips on a card, or other things; a Rugby football team; (a shirt with) a neck measuring 15 inches. *a.* 15 in number; aged 15. **fifteenth**, *n.* one of 15 equal parts; (*Mus.*) the interval of a double octave; an organ-stop sounding two octaves above the open diapason. *n.*, *a.* (the) last of 15 (people, things etc.); (the) next after the 14th.

fifth (fifth), *n.* one of five equal parts; (*Mus.*) a diatonic interval of five notes, equal to three tones and a semitone; two notes separated by this interval sounded together; the resulting concord. *n.*, *a.* (the) last of five (people, things etc.); (the) next after the fourth. **the Fifth Amendment**, an amendment to the US constitution allowing a defendant the right to refuse to testify against him- or herself and prohibiting a second trial for an offence of which a person has been acquitted. **fifth column**, *n.* persons in a country who, are ready to give help to an enemy. Origin of the phrase is attributed to General Mola who, in the Spanish Civil War, said that he had four columns encircling Madrid and a fifth column in the city, being sympathizers ready to assist the attacking party. **fifth columnist**, *n.* **Fifth Monarchy**, *n.* the last of the five great empires referred to in Dan. ii. 44, identified with the millennial reign of Christ prophesied in the Apocalypse. **fifth wheel**, *n.* (*chiefly N Am.*) a spare wheel; a superfluous person or thing.

fifthly, *adv.* in the fifth place.
fifty (fif'ti), *n.* the number or figure 50 or L; the age of 50. *a.* 50 in number; aged 50. **fifty-fifty**, *adv.* in equal shares, half each. *a.* even, as likely to be unfavourable as favourable. **fifties**, *n.pl.* the period of time between one's 50th and 60th birthdays; the range of temperatures between 50 and 60 degrees; the period of time between the 50th and 60th years of a century. **fiftieth**, *n.* one of 50 equal parts. *n.*, *a.* (the) last of 50 (people, things etc.); (the) next after the 49th. **fiftyfold** (-fōld), *a.*, *adv.*
fig[1] (fig), *n.* the pear-shaped fleshy fruit of the genus *Ficus*, esp. *F. carica;* the tree bearing this, noted for its broad and handsome leaves; anything valueless, a trifle. **fig-leaf**, *n.* the leaf of a fig-tree; a flimsy covering, from the use made of the fig-leaf in statuary to conceal nakedness. **figwort**, *n.* plants of the genus *Scrophularia*.
fig[2] (fig), *n.* dress, array, outfit, equipment. **in full fig**, in full dress.
fig[3], (*abbr.*) figure; figurative(ly).
fight (fit), *v.i.* (*past, p.p.* **fought**) to contend in arms or in battle, or in single combat (with, against); to strive for victory or superiority; to oppose, to offer resistance; to quarrel, to disagree. *v.t.* to contend with, to struggle against; to engage in combat; to maintain by conflict; to contend over; to engage in, to carry on or wage (a contest, battle, lawsuit, campaign etc.); to gain or win by conflict; to take part in (a boxing match); to set on or cause (as cocks) to fight. *n.* a struggle between individuals or armies, to injure each other or obtain the mastery; a battle, a combat; a contest of any kind; a boxing match; a quarrel, a row; power of or inclination for fighting. **stand-up fight**, an open encounter. **to fight back**, to resist; to counterattack. **fightback**, *n.* **to fight (it) out**, to decide (a contest or wager) by fighting. **to fight off**, to repel. **to fight shy of**, to avoid from a feeling of mistrust, dislike or fear. **fighter**, *n.* one who fights; a boxer; a combative person, one who does not give in easily; an aircraft equipped to attack other aircraft. **fighter-bomber**, *n.* **fighting**, *n.* **fighting chance**, *n.* a chance of success if every effort is made. **fighting-cock**, *n.* a game-cock. **fighting-fish**, *n.* a variety of *Betta pugnax*, a small Thai freshwater fish, kept for fighting. **fighting fit**, *a.* in peak condition. **fighting-man**, *n.*
figment (fig'mənt), *n.* an invented statement, something that exists only in the imagination, a fabrication.
figure (fig'ə, -yə), *n.* the external form or shape of a person or thing; bodily shape, esp. from the point of view of its attractiveness; the representation of any form, as by carving, modelling, painting, drawing etc.; process; a statue, an image; a combination of lines or surfaces enclosing a space, as a triangle, sphere etc.; a diagram, an illustrative drawing, a pattern; an emblem, a type, a simile; a fancy, a creation of the imagination, an idea; a personage, a character; the sensible or mental impression that a person makes, appearance, distinction; a symbol representing a number, esp. one of the 10 Arabic numerals; a step or movement or a combination of these made by a dancer or skater; (*Rhet.*) any mode of speaking or writing in which words are deflected from their literal or ordinary sense, such as metaphor, ellipsis, hyperbole; (*Gram.*) a recognized deviation from the ordinary form or construction; (*Mus.*) a phrase, a short series of notes producing a single impression; (*Log.*) the form of a syllogism with respect to the position of the middle term; a sum, an amount; value, a price. *v.t.* to form an image, likeness or representation of; to represent, to picture; to imagine; to cover, adorn or ornament with figures; to work out in figures, to cipher, to reckon; to mark with numbers or prices; to express by a metaphor or image; (*Mus.*) to mark with figures indicating the harmony; (*chiefly N Am.*) to believe, to consider, to conclude. *v.i.* to cipher; to appear, to be conspicuous; to seem rational, to accord with expectation. **a high, low figure**, high or low price. **double, three, four** etc. figures, number, price or income between 9 and 100, 99 and 1000, 999 and 10,000 etc. **figure of eight**, (*esp. N Am.*) **figure eight**, a shape or movement resembling the Arabic numeral eight (8). **figure of speech**, a figurative use of language. **to cut a figure** CUT[1]. **to figure on**, (*chiefly N Am.*) to plan to; to base one's plans or calculations on. **to figure out**, to ascertain by computation, to work out; to understand, to fathom out. **to keep, lose one's figure**, to remain or cease to be shapely and attractive. **figurehead**, *n.* the ornamental bust or full-length carving on the prow of a ship; a nominal leader or chief personage without real authority. **figure skating**, *n.* skating in prescribed patterns. **figure skater**, *n.* **figural**, *a.* represented by a figure or delineation. **figurant**, *n.* a ballet-dancer who appears as one of a group. **figurante** (-āt), *n. fem.* **figuration**, *n.* the act of giving a certain determinate form to; form, shape, conformation, outline; (a) figurative representation using allergy, symbolism etc.; ornamentation; (*Mus.*) florid or figured counterpoint. **figurative**, *a.* representing something by a figure or type; emblematic, symbolic, metaphorical, not literal; full of figures of speech. **figuratively**, *adv.* **figurativeness**, *n.* **figured**, *a.* adorned with figures or devices; represented by figures, pictured; of wood, with variegated or ornamental grain; shaped in a (certain) fashion. **-figured**, *comb. form* having a certain or specified kind of figure. **figured bass**, *n.* (*Mus.*) a bass having the accompanying chords indicated by numbers above or below the notes. **figured muslin**, *n.* muslin in which a pattern is worked. **figureless**, *a.* shapeless. **figurine** (-ēn), *n.* a carved or moulded statuette.
figwort FIG[1].
Fijian (fējē'ən), *n.* a member of the Melanesian population of the Fiji islands; the language of this people; a citizen of Fiji. *a.* of Fiji or its people.
filament (fil'əmənt), *n.* a slender, thread-like process, a fibre or fibril; the thread of carbon or metal in an incandescent electric lamp; the heater wire of a thermionic valve; that part of the stamen which supports the anther. **filamentary** (-men'-), *a.* of the nature of or formed by a filament or filaments. **filamentose, -tous** (-men'-), *a.* like a filament; composed of filaments; bearing filaments.
filar (fi'lə), *a.* of or pertaining to a thread; furnished with threads.
Filaria (fileə'riə), *n.* a genus of parasitic nematode worms producing live embryos which find their way into the bloodstream of the human host. **filarial**, *a.* **filariasis** (filəri'əsis), *n.* elephantiasis and other manifestations of filarial infection.
filature (fil'əchə), *n.* the reeling of silk from cocoons; the apparatus used; floss-silk; an establishment for reeling silk.
filbert (fil'bət), *n.* the nut of the cultivated hazel, *Corylus avellana*. [from St *Philibert*, whose feast is on 22 Aug. (o.s.), when they are ripe]
filch (filch), *v.t.* to steal, to pilfer. **filcher**, *n.* a petty thief, a pilferer.
file[1] (fīl), *n.* a box or folder, a string or wire, or similar devices in or on which documents are kept in order, for preservation and convenience of reference; the papers so preserved; a collection of papers arranged in order of date or subject for ready reference; (*Comput.*) a block of data with a unique name by means of which it can be accessed; a row of soldiers ranged one behind the other from front to rear; a row of persons or things arranged in this way; (*Chess*) a line of squares extending from player to player. *v.t.* to place in or on a file; to arrange in order; (*Law*) to place on the records of a court; to initiate (charges, a lawsuit); to send in (a story) to a newspaper. *v.i.* to place in file; to initiate a lawsuit; to march in file or line, as soldiers. **Indian, single file**, a single line of

people drawn up or marching thus. **in file**, drawn up or marching in a line or lines of people one behind another. **on file**, preserved and catalogued for reference. **rank and file** RANK. **to file away**, to preserve or catalogue in a file. **to file off**, to wheel off by files and march at right angles to the former direction. **filing cabinet**, *n*. a cabinet with drawers for storing files.

file² (fil), *n*. a steel instrument with ridged surface, used for cutting and smoothing metals, ivory, wood etc. *v.t.* to smooth, polish or cut away (the surface) with a file. **file-fish**, *n*. any fish of the family Balistidae from the toothed character of the dorsal spine. **filings** (fī'lingz), *n.pl.* the fine particles, esp. of metal, cut or rubbed off with a file.

filet (fil'it), FILLET. **filet mignon** (fē'lä mēn'yō), *n*. a small, very tender steak cut from the tail end of a fillet of beef.

filial (fil'iəl), *a*. pertaining to a son or daughter; befitting a child in relation to parents; bearing the relation of a son or daughter. **filial generation**, *n*. (*Genetics*) a generation following a parental generation. **filiality** (-al'-), *n*. **filially**, *adv*. **filiate**, *v.t.* to affiliate. **filiation**, *n*. the relation of a child to its father, the correlative of paternity; descent, transmission (from); genealogical relation; (*Law*) affiliation.

filibeg, fillibeg (fil'ibeg), *n*. a kilt of the modern kind, dist. from the great kilt of olden times, which covered the body.

filibuster (fil'ibŭstə), *n*. a lawless adventurer, esp. one in quest of plunder, a freebooter; one who takes part in an unauthorized military expedition into a foreign state; a parliamentary obstructionist, one who seeks to hinder legislation by prolonged speeches; obstruction of legislation. *v.i.* to act as a filibuster. **filibusterism**, *n*. **filibusterous**, *a*.

filiform (fil'ifawm, fī'-), *a*. having the form of a thread.

filigree (fil'igrē), *n*. ornamental openwork or tracery, executed in fine gold, silver etc. wire; any ornamental tracery or openwork; anything delicate and fantastic, showy and fragile. *a*. pertaining to filigree; composed of or resembling filigree. **filigreed**, *a*. ornamented with filigree.

filings FILE².

Filioque (filiō'kwi), *n*. the clause in the Nicene Creed asserting the procession of the Holy Ghost from the Son as well as from the Father, which is rejected by the Eastern Church.

Filipino (filipē'nō), *n*. (*pl.* **-nos**) an inhabitant of the Philippine Islands. *a*. pertaining to the Philippines or their inhabitants.

fill (fil), *v.t.* to put or pour into till no more can be admitted; to make full (with); to occupy the whole capacity or space of, to pervade, to spread over or throughout; to block up (cracks with putty, hollow tooth with stopping etc.); to satisfy, to glut; to fulfil, to meet; to cause to be filled or crowded; to appoint an incumbent or person to discharge the duties of; to hold; to discharge the duties of; to occupy (time); to distend (as sails); to trim (a sail) to catch the wind; (*N Am.*) to make up (a prescription). *v.i.* to become or grow full; to be distended; to be satisfied. *n*. as much as will satisfy; a full supply; as much as will fill. **to fill in**, to complete (anything that is unfinished, as an outline or a form); (*coll.*) to provide with necessary or up-to-date information; to occupy (time); to act as a temporary substitute (for); (*sl.*) to murder. **to fill out**, to become bigger, more substantial, fatter; (*chiefly N Am.*) to complete (a form etc.). **to fill the bill**, (*coll.*) to do or be all that is required. **to fill up**, to fill or occupy completely; to stop up by filling; to become full. **to have one's fill of**, to have rather too much of. **filler**, *n*. one who or that which fills; a substance added to something to increase bulk, weight etc.; material used to fill cracks and holes in plaster, woodwork etc.; an item used to fill a space between more important items (as in a newspaper, a TV programme, a schedule etc.); the filling orifice of a petrol tank, gearbox, crankcase etc. **filling**, *a*. of food, satisfying. *n*. anything serving to fill up; gold or other material used to fill up a cavity in a tooth; substances used to fill up holes, cavities or defects; rubble and other rough material filling up the interior of a stone- or brick-faced wall; a food mixture filling sandwiches, cakes etc. **filling-station**, *n*. a roadside establishment supplying petrol, oil etc.

fillet (fil'it), *n*. a band of metal, a string or ribbon for binding the hair or worn round the head; a ribbon, a narrow band or strip; a bandage; a fleshy portion or slice of meat; the fleshy part of the thigh of an animal used for meat; a portion of meat or fish removed from the bone and served either flat or rolled together and tied round; a plain liner band on the back of a book; a narrow, flat band between mouldings; the projection between the flutes of a column; a small horizontal division of a shield. *v.t.* to bind with a fillet or bandage; to adorn with a fillet or fillets; to make into fillets (as meat or fish).

fillibeg FILIBEG.

fillip (fil'ip), *v.t.* to strike with the nail of the finger by a sudden jerk from under the thumb; to propel with such a blow; to stimulate, incite, encourage. *n*. a sharp, sudden blow with the finger jerked from under the thumb; a stimulus, an incentive.

filly (fil'i), *n*. a female foal; (*dated*) a young, lively girl.

film (film), *n*. a thin pellicle, skin, coating or layer; a fine thread or filament; a thin, slight covering or veil; a thin sheet of plastic or similar material used for packaging; a series of connected cinematographic images projected on a screen; (*often pl.*) the cinematographic industry generally; (*Phot.*) a thin coating of sensitized material for receiving a negative or positive image; a thin piece or strip of celluloid or other material supporting such a coating. *v.t.* to cover with a film; to record on a cinematographic film; to make a cinematographic film of. *v.i.* to become covered with or as with a film; to make a cinematographic film. **filmgoer** CINEMAGOER. **filmset**, *v.t.* to expose (type characters) on to photographic film from which printing plates are made. **filmsetter**, *n*. **filmsetting**, *n*. **film star**, *n*. a leading cinema actor or actress. **filmstrip**, *n*. a sequence of images on a strip of photographic film, projected as stills. **filmic**, *a*. pertaining to motion pictures. **filmography** (-mog'-), *n*. a list of films by a particular artist or director or on a particular subject. **filmy**, *a*. gauzy, transparent; misted, blurred. **filmily**, *adv*. **filminess**, *n*.

Filofax® (fī'lōfaks), *n*. a small ring-binder with a leather or similar cover into which the owner can insert sheets to make up a diary, an address-list etc., intended as a personal, portable compendium of information.

filoselle (filəsel', fil'-), *n*. floss-silk.

filter (fil'tə), *n*. an apparatus for straining liquids and freeing them from impurities, usu. by means of layers of sand, charcoal or other material through which they are passed; (the layer of) porous material through which the liquids are passed; an apparatus for purifying air by a similar process; a filter-tip; a device for altering the relative intensity of the wavelengths in a beam of light; a circuit for altering or controlling the relative intensity of different frequencies of an alternating current; an auxiliary traffic light at a road junction in the form of a green arrow, which permits a stream of traffic to turn left or right while the main stream is held up. *v.t.* to pass (liquid) through a filter; to remove or separate by means of a filter. *v.i.* to pass through a filter; to pass slowly; to become known gradually, to permeate; of traffic, to move in the direction shown by the filter. **filter-bed**, *n*. a reservoir with a layer of sand or other filtering material at the bottom through which water is allowed to flow. **filter-paper**, *n*. paper used for filtering liquids. **filter-tip**, *n*. (a cigarette with) an attached tip made of a porous substance to trap impurities. **filterable, filtrable**, *a*.

filth (filth), *n*. anything dirty or foul; foulness, corruption,

pollution; anything that defiles morally; foul language, obscenity. **the filth**, (*sl.*, *offensive*) the police. **filthy**, *a.* dirty, foul, unclean; morally impure. **filthy lucre**, *n.* gain obtained by base methods; (*facet.*) money. **filthily**, *adv.* **filthiness**, *n.*
filtrate (fil'trāt), *n.* any liquid that has passed through a filter. *v.t., v.i.* to filter. **filtration**, *n.*
fimbria (fim'briə), *n.* (*pl.* **fimbriae** (-ē)) the radiated fringe of the Fallopian tube. **fimbriate** (-ət), *a.* fringed. **fimbriated** (-ātid), *a.* fringed; (*Her.*) ornamented, as an ordinary, with a narrow border or hem of another tincture. **fimbricate** (-kət), **-cated** (-kātid), *a.* fimbriate.
fin (fin), *n.* the organ by which fishes propel and steer themselves, consisting of a membrane supported by rays, named according to position on the body, as *anal, caudal, dorsal, pectoral*, or *ventral fin;* anything resembling a fin; the flipper of a seal, whale etc.; a projection from the surface of a radiator or engine cylinder by which heat is dissipated; (*Aviat.*) a fixed aerofoil, usu. inserted in or parallel to the plane of symmetry, generally constituting part of the tail structure. *v.i.* (*past, p.p.* **finned**) to beat the water with the fins, as a whale. **fin-back** (**whale**) FINNER. **finless**, *a.* **finlike**, *a.* **finned**, *a.* having fins; having broad edges on either side. **-finned**, *comb. form* having a certain kind of fins, as *prickly-finned, red-finned.* **finner, finner-whale**, *n.* a whale with an adipose fin on its back, as those of the genus *Balaenoptera*, esp. the rorqual. **finny**, *a.* having fins; like a fin; (*poet.*) abounding in fish.
Fin., (*abbr.*) Finland; Finnish.
fin., (*abbr.*) finance; financial.
finable FINE [1].
finagle (fināgl'), *v.t.* to obtain by trickery or dishonest means; to cheat, trick. *v.i.* to use trickery or dishonest means. **finagler**, *n.*
final (fī'nəl), *a.* pertaining to the end or conclusion; ultimate, last; finishing, conclusive, decisive; concerned with the end or purpose. *n.* the deciding heat of a contest; (*usu. pl.*) the last of a series of public examinations. **final cause**, *n.* (*Phil.*) the end or aim contemplated in the creation of the universe. **finalist**, *n.* a competitor in a final. **finality** (-nal'-), *n.* the state or quality of being final; the state of being finally and completely settled; the end of everything, completeness; the final and decisive act or event; (*Phil.*) the doctrine that everything exists or was created for a determinate cause. **finalize, -ise**, *v.t.* to put in final form; to settle; to give final approval to. **finally**, *adv.*
finale (finah'li), *n.* the last part, piece, scene or action in any performance, programme or exhibition; (*Mus.*) the last movement of a musical composition; (*fig.*) the close, end, the final catastrophe.
finality, finalize FINAL.
finance (finans', fī'-), *n.* the science or system of management of revenue and expenditure, esp. public revenue and expenditure; (*pl.*) monetary affairs, the income of a state, sovereign, firm, or individual; obtaining money, esp. to fund purchases etc.; money. *v.t.* to raise money for; to provide with capital. **finance company**, *n.* a company that specializes in making loans, esp. for hire purchase. **financial** (-nan'shəl), *a.* pertaining to finance or revenue; monetary, fiscal. **Financial Times Index**, an indicator of prices on the London Stock Exchange based on the average daily prices of a selected list of ordinary shares. **financial year**, *n.* the period for which public or official accounts are made up. **financialist** (-nan'-), *n.* a financier. **financially**, *adv.* **financier** (-nan'siə), *n.* one who is skilled in finance, esp. the management of public revenues; one engaged in large-scale monetary dealings.
finch (finch), *n.* a popular name for various small birds, many of them of the family Fringillidae; the genus *Fringilla*, see also BULLFINCH, CHAFFINCH, GOLDFINCH.
find (fīnd), *v.t.* (*past, p.p.* **found** (fownd)) to meet with, to come across; to discover, learn or acquire by search, study or other effort; to rediscover (something lost); to ascertain by experience or experiment; to perceive, to recognize; to consider, to be of the opinion that; to reach, to arrive at; to succeed in obtaining; to gain or regain the use of; to supply, to furnish, to provide; (*Law*) to decide, to determine; to declare by verdict. *v.i.* to discover anything by searching or seeking; (*Law*) to arrive at a decision. *n.* the discovery of anything valuable; the thing so found. **to find fault with** FAULT. **to find oneself**, to be or perceive oneself to be (in a certain situation); to be or feel as regards health; to realize one's own capabilities or vocation. **to find one's feet** FEET. **to find out**, to discover; to detect in an offence, lie etc. **findable**, *a.* **finder**, *n.* one who finds; a discoverer, an inventor; a small telescope fixed to the tube and parallel to the axis of a larger one, for finding objects to be examined by the larger telescope; a contrivance for the same purpose attached to a microscope or to a camera. **finders keepers**, (*coll.*) whoever finds something has the right to keep it. **finding**, *n.* the action of the verb TO FIND; a discovery; the act of returning a verdict; a verdict; (*pl.*) the results of an investigation; (*pl., coll.*) things found.
fin de siècle (fī də sye'kl'), *a.* pertaining to or characteristic of the close of the 19th cent.; decadent. [F, end of the age]
fine [1] (fīn), *n.* a sum of money imposed as a penalty for an offence; a fee paid by an incoming tenant to the landlord. *v.t.* to impose a fine upon; to punish by a fine. **in fine**, in conclusion, in short, finally; to sum up. **finable**, *a.* deserving or liable to a fine.
fine [2] (fīn), *a.* excellent in quality, form or appearance; refined, pure, free from dross or extraneous matter; containing a certain proportion of pure metal, as in *22 carats fine*; of feelings, taste etc., also of differences, distinctions etc., delicate, subtle, nice, fastidious; in small grains or particles; thin, small, slender, tenuous; keen, sharp; very small or delicate; finished, consummate, accomplished; handsome, beautiful; showy, smart, pretentious; good, satisfactory, enjoyable, pleasant; well, in good health; free from clouds or rain, sunshiny; (*iron.*) anything but pleasant or satisfactory; (*Cricket*) at or through a position close to the line of the stumps. *adv.* (*coll.*) finely. *v.t.* to refine, purify, clear from impurities; to make finer, to sharpen, to taper; to make less coarse. *v.i.* to become finer, purer, clarified; to taper, to dwindle (away). *int.* Good! all right! Well done! **fine-tooth(ed) comb**, a comb with thin teeth set very close together. **to go over, through with a fine-tooth(ed) comb**, to examine minutely, to investigate very thoroughly. **one of these fine days**, at some unspecified date in the future. **to cut it fine** CUT. **to fine down**, to reduce and improve by the removal of superfluous matter. **fine arts**, *n.pl.* the arts, such as poetry, music, painting, sculpture and architecture, that appeal to our sense of the beautiful. **fine-draw**, *v.t.* to draw together the edges of and mend a rent so that no trace remains visible. **fine-drawn**, *a.* drawn out finely, as wire; excessively subtle; (*Athletics*) reduced by training. **fines herbes**, (fēnz ɛərb'), *n.pl.* a mixture of finely chopped herbs used as flavouring. **fine-spoken**, *a.* using fine phrases. **fine-spun**, *a.* drawn or spun out to minuteness; hence, over-refined or elaborate; delicate, flimsy. **fine-tune**, *v.t.* to make delicate adjustments to. **fine-tuning**, *n.* **finely**, *adv.* **fineness**, *n.* **finery** (-nə-), *n.* fine clothes, showy decorations; a furnace in which cast-iron is made malleable. **fines**, *n.pl.* ore or coal that is too fine or powdery for using in the ordinary way. **fining**, *n.* the process of refining metals; the clarifying of wines, malt liquors etc.; (*usu. pl.*) the preparation, generally a solution of gelatine or isinglass, used to fine or clarify liquors.
fine [3] (fēn), *n.* ordinary quality French brandy. [F]
finesse (fines'), *n.* artifice, stratagem or artful manipula-

tion; a subtle contrivance to gain an end; skill, dexterity, adroitness, esp. in handling difficult situations; elegance, refinement; (*Whist etc.*) an attempt to take a trick with a lower card, so as to retain a higher one for later tricks. *v.i.* to use artifice to gain an end; to try to win a trick with a lower card than one possibly in your opponent's hand, while you have a higher card in your own. *v.t.* to play (a card) in this manner; to manipulate, to manage by means of trickery or stratagem.

finger (fing'gə), *n.* one of the five terminal members of the hand; one of the four longer digits as distinguished from the thumb; anything resembling or serving the purpose of a finger; the part of a glove that covers a finger; the width of a finger, a measure of length or of the quantity of liquid in a glass; (*pl.*) the hand, the instrument of work or art. *v.t.* to touch with or turn about in the fingers; to meddle or interfere with; to touch thievishly, to pilfer; to perform with the fingers; to play with the fingers (as a musical instrument); to mark (a piece of music) so as to indicate which fingers should be used; (*sl.*) to identify (to the police). *v.i.* to use or touch with the fingers. **not to lift a finger**, to do nothing, to stand idly by. **to get, pull one's finger out**, (*coll.*) to start making an effort, to get cracking. **to have a finger in every pie**, to be involved in everything. **to lay, put a finger (up)on**, to touch, to interfere with in the slightest. **to point the finger (at)**, to accuse; to censure. **to put the finger on**, (*sl.*) to identify or inform against. **to twist, wrap around one's little finger**, to have someone in thrall, to be able to do as one likes with someone. **finger-board**, *n.* the board at the neck of a stringed instrument, where the fingers act on the strings. **finger-bowl, -glass**, *n.* a bowl or glass in which to rinse the fingers after dessert. **fingermark**, *n.* a dirty mark left by fingers. **fingernail**, *n.* **finger paint**, *n.* thickish paint for applying with the fingers, hand etc., used esp. by children. **finger painting**, *n.* **finger-plate**, *n.* a plate on the side of a door, near the handle, to preserve the paint from finger-marks. **finger-post**, *n.* a sign-post in the form of a hand or finger pointing out direction. **fingerprint**, *n.* an impression of the whorls of lines on fingers, used for purposes of identification; GENETIC FINGERPRINT under GENETIC. *v.t.* to take the fingerprints of. **fingerprinting**, *n.* **finger-stall**, *n.* a cover for protecting a finger. **fingertip**, *n.* **to have at one's fingertips**, to know familiarly, to be well versed in. **to the fingertips**, completely. **fingered**, *a.* having fingers; (*Bot.*) digitate. **-fingered**, *comb. form* having a certain kind or number of fingers. **fingering**, *n.* the act of touching with the fingers; the management of the fingers in playing upon a keyed, stringed or holed instrument; marks upon a piece of music to guide the fingers in playing. **fingerless**, *a.* **fingerling** (-ling), *n.* the young of the salmon or trout when no longer than a finger.

finial (fin'iəl), *n.* a terminal ornament surmounting the apex of a gable, pediment, roof, canopy etc.

finical (fin'ikl), *a.* finicky. **finicality** (-kal'-), *n.* **finically**, *adv.* **finicalness**, *n.* **finicking**, *a.* (*coll.*) **finicky**, *a.* affecting great nicety, precision or delicacy; over-nice, fastidious; intricate, fiddly.

fining FINE [2].

finis (fin'is), *n.* (*printed at end of book*) the end, finish, conclusion.

finish (fin'ish), *v.t.* to bring to an end; to complete; to arrive at the end of; to perfect; to give the final touches to, to trim, to polish; to consume, to get through; to kill, to defeat, to render powerless; to complete the education of. *v.i.* to come to the end, to cease, to expire; to leave off; to come to the end of a relationship. *n.* the act of finishing; the termination, the final stage; the end of a race, when the competitors are close to the winning-post; the last touches, that which gives the effect of perfect completeness; the final stage of any work, as the last raw coat of plaster on a wall; the appearance of texture of a finished surface; grace, elegance, polish, refinement. **to finish up**, to consume or use up entirely; to arrive, come to rest or end up. **finisher**, *n.* one who or that which finishes; a worker or a machine that performs the final operation in a process of manufacture; a blow that settles a contest. **finishing-coat**, *n.* the last coat in painting or plastering. **finishing school**, *n.* a private school where girls are taught social graces.

finite (fi'nit), *a.* having limits or bounds, opposed to infinite; applied to those moods of a verb which are limited by number and person, as the indicative, subjunctive, imperative. **finitely**, *adv.* **finiteness, finitude** (-nitūd), *n.*

fink (fingk), *n.* (*chiefly N Am., coll.*) an informer; a strikebreaker; a contemptible person.

Finn (fin), *n.* the Teutonic name for the people who inhabit parts of NW Russia and NE Scandinavia; a native or inhabitant of Finland. **Finlander**, *n.* **Finlandization, -isation**, *n.* being under the necessity of accommodating the wishes of a powerful neighbour, esp. the USSR. **Finnic**, *a.* belonging to the Finnish group of peoples. **Finnish**, *a.* pertaining to Finland, the Finns or their language. *n.* the language of the Finns. **Finno-Ugrian** (-oo'griən), **Finno-Ugric**, *n., a.* (pertaining to) a family of languages spoken in Hungary, Lapland, Finland, Estonia, and NW USSR.

finnan (fin'ən), **finnan-haddock**, *n.* a kind of smoke-dried haddock.

finned, finner etc. FIN.

fino (fē'nō), *n.* a dry sherry.

fiord, fjord (fyawd, fē'awd), *n.* a long, narrow inlet of the sea, bounded by high cliffs, as on the coast of Norway.

fioritura (fyoritoo'rə), *n.* (*pl.* **fioriture** (-rā)) (*Mus.*) a decorative phrase or turn, a flourish. [It.]

fipple (fip'l), *n.* an arrangement of a block and a sharp edge, the sound-producing mechanism in e.g. a recorder. **fipple-flute**, *n.*

fir (fœ), *n.* the popular name for many coniferous timber trees of the genus *Abies* or allied genera; the wood of these. **Scotch fir** SCOTS PINE under SCOTS. **silver fir** SILVER. **spruce-fir** SPRUCE [2]. **fir-apple, -ball, -cone**, *n.* the cone-shaped fruit of the fir. **fir-needle**, *n.* the spine-like leaf of the fir. **firry**, *a.*

fire (fīə), *n.* the production of heat and light by combustion; combustion, flame, incandescence; fuel in a state of combustion, as in a furnace, grate etc.; a radiant gas or electric heater; anything burning; a conflagration; a light, glow or luminosity resembling fire; a spark or sparks emitted when certain substances are struck violently; intense heat, fever; the discharge of firearms; ardent emotion, fervour; liveliness of imagination, poetic inspiration; a severe affliction, torture, persecution. *v.t.* to set on fire, to kindle, to ignite; to discharge, to cause to explode; to throw, direct, launch with rapidity and force; to bake (as pottery); to supply with fuel (as a furnace); to inflame, to irritate; to excite, to animate, to inspire; to dismiss, to discharge from employment. *v.i.* to take fire, to be kindled; of an internal-combustion engine, to be in operation; to discharge firearms; to shoot (at) with firearms. *int.* a word of command for soldiers to discharge their firearms. **ball of fire** BALL. **cross-fire** CROSS. **fire away! begin! Greek fire**, an artificial combustible used by the Greeks in their wars with the Saracens for setting hostile ships on fire. **on fire**, burning, in flames; excited, ardent, eager. **running fire**, a discharge of firearms in rapid succession by a line of troops. **St Anthony's fire**, erysipelas. **St Elmo's fire**, the corposant. **to catch, take fire**, to ignite. **to fire away**, to begin, to proceed. **to fire up**, to kindle a fire; to be inflamed with passion, to be irritated. **to play with fire**, to expose oneself to risk. **to set fire to, on fire, a-fire**, to kindle; to excite, to inflame. **to set the Thames on fire**, to do something clever or remarkable. **under fire**, exposed to the enemy's firearms. **fire-alarm**, *n.*

an automatic apparatus for communicating warning of a fire. **firearm,** *n.* a weapon that projects a missile by the explosive force of gunpowder, esp. a rifle or pistol. **fire-back,** *n.* the rear wall of a furnace or fireplace. **fireball,** *n.* globular lightning; a large meteor or shooting star; the luminous cloud of hot gases at the centre of a nuclear explosion. **firebird,** *n.* the Baltimore oriole. **fire-blanket,** *n.* a blanket of a non-inflammable material for throwing over and extinguishing small fires. **fire-blast, -blight,** *n.* a disease in plants, esp. fruit trees. **fire bomb,** *n.* an incendiary bomb. **firebox,** *n.* the chamber in which the fuel is burned in a locomotive etc. **firebrand,** *n.* a piece of wood kindled or on fire; an arsonist; one who inflames passions or kindles strife. **firebreak,** *n.* a strip of land kept clear of trees or vegetation to stop the spread of fire. **firebrick,** *n.* a brick capable of withstanding fire used for fireplaces, furnaces etc. **fire-brigade,** *n.* a body of people organized esp. by a public authority for the extinction of fires. **fire-bucket,** *n.* a bucket (usu. filled with sand or water) kept in readiness in case of fire. **fire-bug,** *n.* (*coll.*) an arsonist. **fire-clay,** *n.* a kind of clay consisting of nearly pure silicate of alumina, capable of standing intense heat, used in the manufacture of fire-bricks. **fire-control,** *n.* (*Nav., Mil.*) the system of controlling gun-fire from one spot. **fire-cracker,** *n.* a small firework that explodes usu. with a series of loud bangs. **firedamp,** *n.* the explosive carburetted hydrogen which accumulates in coal-mines. **fire department,** *n.* (*N Am.*) the fire-brigade. **fire-dog,** *n.* an and-iron. **fire-drill,** *n.* practice in the routine to be observed in case of fire. **fire-eater,** *n.* a juggler who pretends to swallow fire; a belligerent person. **fire-engine,** *n.* a vehicle equipped with fire-fighting equipment. **fire-escape,** *n.* an apparatus for enabling persons to escape from the upper parts of buildings that are on fire. **fire-extinguisher,** *n.* a portable apparatus for extinguishing fires by spraying them with water or chemicals. **firefight,** *n.* an exchange of fire between military units. **fire-fighter,** *n.* a fireman. **fire-fighting,** *n., a.* **firefly,** *n.* a small luminous winged beetle. **fireguard,** *n.* a wire frame placed before an open fire as a safeguard against accidental fire or injury to children etc. **fire-insurance,** *n.* insurance against loss by fire. **fire-irons,** *n.pl.* the implements for tending a fire – poker, tongs and shovel. **fire-light,** *n.* the light from a fire. **fire-lighter,** *n.* an inflammable substance for kindling fuel. **fire-lock,** *n.* an old-fashioned musket or other gun having a lock with a flint and steel, by means of which the priming was ignited. **fireman,** *n.* one who is employed to extinguish fires; a member of a fire-brigade; a stoker; the assistant to the driver of a diesel or electric locomotive. **fire-master,** *n.* (*chiefly Sc.*) the chief of a fire-brigade. **fire-opal** GIRASOL. **fireplace,** *n.* a grate; a hearth. **fire-plug,** *n.* (*chiefly N Am.*) a hydrant for connecting a fire-hose with a water-main. **fire-power,** *n.* the effective capability of weaponry, missiles etc. **fire-proof,** *a.* proof against fire; incombustible. *v.i.* to render proof against fire. **fire-proofing,** *n.* **fire-raising,** *n.* the act of setting on fire; incendiarism, arson. **fire-screen,** *n.* a fireguard; a screen placed between a person and the fire to intercept the direct rays. **fire-ship,** *n.* a vessel freighted with combustibles and explosives, and sent among an enemy's ships in order to set them on fire. **fireside,** *n.* the space around a fire-place, the hearth; hence home, home life. *a.* home, domestic. **fire station,** *n.* a building from which fire-engines and firemen operate. **firestone,** *n.* a stone capable of bearing a high degree of heat, used in furnaces etc. **fire storm,** *n.* a huge fire, one started by bombing, which causes and is kept ablaze by violent inrushing winds. **firethorn,** *n.* a shrub of the genus *Pyracantha* with red or orange berries. **fire-trap,** *n.* (*coll.*) a building without adequate means of exit in case of fire. **fire watcher,** *n.* a person who watches for the outbreak of fires, esp. during an air raid.

fire-water, *n.* the name given by the native Indians of N America to ardent spirit. **fireweed,** *n.* any weed springing up after a fire, esp. rose-bay willow-herb. **firewood,** *n.* wood for burning as fuel. **firework,** *n.* a preparation of various kinds of combustibles and explosives for producing a brilliant display at times of public rejoicing etc.; similar preparations used for illumination, signalling, incendiary purposes or in war; (*pl.*) a display of bad temper; (*pl.*) a spectacular display of virtuosity. **firer,** *n.* one who or that which fires; (*in comb.*) a gun with one or more barrels, as a *single-firer*. **firing,** *n.* the adding of fuel to a furnace or fire; the ignition of an explosive mixture in an internal-combustion cylinder; the act of discharging firearms; fuel; the baking of ceramic products in a kiln. **firing-line,** *n.* a line of troops engaging the enemy with firearms. **to be in the firing-line,** to be at the forefront of any activity and hence exposed to greatest risk. **firing-party,** *n.* a detachment told off to fire over a grave at a military funeral, or to shoot a condemned man. **firing-pin,** *n.* a sliding pin in firearms that strikes upon the detonator and explodes the charge. **firing-squad,** *n.* a detachment which carries out executions by shooting.
firkin (fœ′kin), *n.* a measure of capacity, the fourth part of a barrel or 9 gallons (41 l); a small wooden cask used for butter, tallow etc., of no fixed capacity.
firm[1] (fœm), *a.* fixed, stable, steady; difficult to move or disturb; solid, compact, unyielding; securely established, immutable; steadfast; staunch, enduring, resolute; constant, unwavering; of prices etc., not changing in level. *adv.* firmly. *v.t.* to fix firmly; to make firm, to consolidate. *v.i.* to become firm; to solidify. **firm offer,** *n.* a definite offer. **firmware,** *n.* a computer program or data stored in a read-only memory. **firmly,** *adv.* **firmness,** *n.*
firm[2] (fœm), *n.* a partnership or association of two or more persons for carrying on a business; the business itself.
firmament (fœ′məmənt), *n.* the sky regarded as a solid expanse, the vault of heaven. **firmamental** (-men′-), *a.*
firn (fœn), *n.* névé, snow on the higher slopes of lofty mountains, not yet consolidated into ice.
first (fœst), *a.* foremost in order, time, place, rank, importance or excellence; earliest in occurrence; nearest, coming next (to something specified); chief, highest, noblest. *adv.* before all others in order, time, place, rank, importance or excellence; before some time, act or event (specified or implied); sooner, rather, in preference, for the first time. *n.* that which or the person who comes first; the first mentioned; the beginning; a place in the first class of an examination list, a candidate winning this; the first place in a race, the winner of this; (*pl.*) the best quality of a commodity (such as flour); the upper part in a duet, trio etc.; first gear. **at first,** at the beginning; originally. **at first blush** BLUSH. **first and last,** essentially. **first-day cover,** an envelope postmarked on the first day of issue of new stamps. **first-degree burn,** a mild burn in which the skin is reddened and painful but unblistered. **first off,** (*coll.*) firstly, first of all. **first-past-the-post,** *a.* of an electoral system in which each voter casts a single vote and only the candidate who polls highest is returned. **first thing,** early, as the first action of the day. **from first to last,** throughout; altogether. **not to know the first thing about,** to be entirely ignorant of. **first aid,** *n.* assistance rendered to an injured person before a doctor comes. **first-aider,** *n.* **first-born,** *n., a.* (the) first in order of birth, (the) eldest. **first-class,** *a.* first-rate; of the highest quality or degree; in the first class; of postage, charged at higher rate for quicker delivery. *n.* the highest division in an examination list; a place in this; the first or best class of railway carriage or other accommodation. **first floor,** *n.* the floor or storey of a building next above the ground floor; (*N Am.*) the ground floor (the first floor in Eng. is N Am. second floor). **first-foot,** *n.* (*Sc.*) (also **first-footer**) the first caller at a house on New Year's Day. *v.t.* to en-

ter as first-foot. **first-fruits**, *n.pl.* the fruit or produce first gathered in any season; the first effects or results; the first profits of any office, paid to a superior. **first gear**, *n.* the lowest forward gear on a motor vehicle. **first-hand**, *a.*, *adv.* (obtained) directly from the first or original source. **at first hand** HAND. **first lady**, *n.* the wife of or official hostess for the US president or a state governor; any woman pre-eminent in her field. **first mate**, *n.* the chief officer of a merchant-vessel, next in rank to the captain. **first name**, *n.* christian name or first forename. **first night**, *n.* the first public performance of a theatrical production. **first-nighter**, *n.* one who makes a point of attending first performances of plays. **first offender**, *n.* one not previously convicted. **first-rate**, *a.* of the first or highest class or quality; of the highest excellence. *adv.* excellently, very well. **first refusal**, *n.* the option of accepting or refusing something before it is offered to others. **first school**, *n.* a primary school for children aged 5 to 8. **first strike**, *n.* an initial, unprovoked or preemptive attack with nuclear missiles. **first-strike**, *a.* **first string**, *a.* of regular team members as opposed to substitutes. **first water**, *n.* the purest quality (of diamonds etc.). **firstling** (-ling), *n.* the first-born, the first-born in a season; (*pl.*) the first-fruits. **firstly**, *adv.* in the first place, to begin with.

firth (fœth), *n.* an estuary, an arm of the sea, esp. in Scotland.

fisc (fisk), *n.* the treasury of the State, the public purse or exchequer. **fiscal**, *a.* pertaining to the public revenue or exchequer; financial. *n.* a procurator-fiscal. **fiscally**, *adv.*

fish (fish), *n.* (*pl.* in general **fish**; in particular **fishes**) an aquatic, oviparous, cold-blooded vertebrate animal, provided with permanent gills, usu. covered with scales, and progressing by means of fins; (*loosely*) any of various aquatic animals; the flesh of fish used as food; (*coll.*) a certain kind of person, as an *odd fish*. *v.i.* to try to catch fish, by angling, netting etc.; to search for something under water; to grope or feel around for; to seek to learn or obtain anything by indirect means or finesse. *v.t.* to attempt to catch fish in; to lay hold of and drag up from under water or from inside something; to search (water etc.) by sweeping, dragging etc. **a fish out of water**, anyone out of his or her element, in a strange or bewildering situation. **fish and chips**, fried fish and fried potato chips. **neither fish, flesh nor fowl**, nondescript; of a vague indefinite character. **other fish to fry**, more important matters to attend to. **to drink like a fish**, to drink to excess. **to fish out**, to find and draw out; to ascertain by cunning inquiry. **fish-ball, -cake**, *n.* a fried cake of chopped fish and mashed potatoes. **fish-eye**, *a.* of a wide-angle photographic lens with a convex front which covers almost 180°. **fish farm**, *n.* an installation for the rearing of fish, usu. in ponds or tanks. **fish farmer**, *n.* **fish farming**, *n.* **fish finger**, *n.* a small bar-shaped portion of fish coated in breadcrumbs or batter. **fish-hawk**, *n.* the osprey. **fish-hook**, *n.* a barbed hook for catching fish. **fish-joint**, *n.* a joint made with fish-plates on a railway-line. **fish-kettle**, *n.* a long oval pan for boiling fish. **fish ladder**, *n.* a series of pools arranged in step to enable fish swimming upstream to bypass dams etc. **fishmonger**, *n.* a retail dealer in fish. **fishnet**, *n.* open mesh fabric resembling netting. **fish-plate**, *n.* a plate used to fasten rails end to end. **fish-pond**, *n.* a pond in which fish are kept. **fish-slice**, *n.* a broad-bladed knife for serving fish at table; a similar instrument used by cooks for turning or taking fish out of the pan etc. **fish stick**, *n.* (*N Am.*) a fish finger. **fish-tail**, *a.* shaped like the tail of a fish. **fish-tail burner**, a gas-burner producing a jet like a fish's tail. **fish-way** FISH LADDER. **fish-wife**, *n.* a woman that sells fish; a coarse, foul-mouthed woman. **fisher**, *n.* a fisherman. **fisherman**, *n.* one whose employment is to catch fish; an angler; a boat or vessel employed in catching fish. **fishery**,

n. the business of catching fish; any place where fishing is carried on; (*Law*) permission to fish in reserved water. **fishing**, *n.* the action of the verb TO FISH; the sport of angling; a place where angling is carried on. **fishing-boat**, *n.* **fishing-line**, *n.* a line with hook attached for catching fish. **fishing-net**, *n.* **fishing-rod**, *n.* a long, slender, tapering rod, usu. in sections jointed together, for angling. **fishy**, *a.* like, consisting of, pertaining to, or suggestive of fish; inhabited by or abounding in fish; of a doubtful character, questionable, dubious. **fishily**, *adv.* **fishiness**, *n.*

fissile (fis'īl), *a.* that may be cleft or split, esp. in the direction of the grain, as wood, or along natural planes of cleavage, as rock; capable of undergoing nuclear fission.

fission (fish'ən), *n.* the act or process of cleaving, splitting or breaking up into parts; nuclear fission; a form of asexual reproduction in certain simple organisms, the individual cell dividing into new cells. **fission bomb**, *n.* an atom bomb. **fissionable**, *a.*

fissiparous (fisip'ərəs), *a.* propagating by fission. **fissiparously**, *adv.* **fissiparity** (-pa'-), *n.*

fissure (fish'ə), *n.* a cleft or opening made by the splitting or parting of any substance; a slit or narrow opening, as the deep narrow depression between the anterior and middle lobes of the cerebrum on each side. *v.t., v.i.* to cleave, to split.

fist (fist), *n.* the clenched hand, esp. in readiness to strike a blow; (*coll.*) the hand; (*facet.*) handwriting; (*Typography*) a hand pointing, ☞. *v.t.* to strike or grip with the fist. **-fisted**, *comb. form* having a certain kind of fist. **fisticuffs** (-tikŭfs), *n.pl.* a fight in which the fists are used; a boxing-match.

fistula (fis'tūlə), *n.* (*pl.* **-las, -lae** (-lē)) a kind of ulcer or suppurating swelling, in form like a pipe; a narrow pipe-like passage, duct or spout. **fistular, -ulose, -ulous**, *a.* hollow like a pipe or reed; of the form or nature of a fistula.

fit¹ (fit), *n.* a violent seizure or paroxysm; a sudden transitory attack of illness; esp. a sudden attack of epilepsy or other disease characterized by convulsions; a spasm, a seizure; a transient state of impulsive action, a mood, a caprice. **by fits and starts**, intermittently. **to have a fit**, (*coll.*) to be very angry or upset. **fitful**, *a.* spasmodic, capricious, wavering; acting by fits and starts. **fitfully**, *adv.* **fitfulness**, *n.*

fit² (fit), *a.* (*comp.* **fitter**, *superl.* **fittest**) adapted, suitable, appropriate; becoming, proper, meet; qualified, competent; acceptable, worthy; ready, prepared, in a suitable condition (to do or for); in good physical condition; (*coll.*) as if, in such a mood or condition as (to cry, to do something violent etc.). *v.t.* (*past, p.p.* **fitted**) to adapt to any shape, size or measure; to make suitable, to accommodate; to try on (a garment); to supply, to furnish, to equip; to qualify, to prepare; to be adapted, suitable or proper for; to be of the right size, measure and shape for; to correspond to exactly. *v.i.* to be adjusted or adapted to the right shape, measure, form etc.; to be proper, suitable, convenient or becoming; to accord with what is known, a set of circumstances etc. *n.* exact adjustment, as of a dress to the body; the manner in which anything fits, the style in which a garment fits. **to fit in**, to find room or time for; to prove accommodating or suitable. **to fit out**, to equip. **to fit up**, to furnish with the things suitable or necessary; (*sl.*) to frame. **to think fit to**, to decide to (do something). **fitly**, *adv.* **fitment**, *n.* a piece of furniture; (*usu. pl.*) fittings; an accessory part of a machine. **fitness**, *n.* suitability; good physical condition or health. **fitted**, *a.* adapted, suitable (for); cut or sewn or constructed to fit exactly; of furniture, fitting exactly into a certain space, and usu. permanently attached; furnished with fitted, matching cupboards etc. **fitter**, *n.* one who or that which fits; one who puts together the several parts of machin-

fitch 309 **flag**

ery; one who fits or repairs certain kinds of apparatus, as in *gas-fitter*. **fitting**, *a.* suitable, appropriate, right, proper. *n.* the act of making fit; a small, removable part or attachment, as *light fitting*; (*pl.*) apparatus, furniture employed in fitting up a house, shop etc.; preliminary trying on of a garment. **fittingly**, *adv.*
fitch (fich), **fitchew** (fich'oo), *n.* (the fur of) the polecat; a brush made of this.
five (fīv), *n.* the number or figure 5 or V; the age of five; the fifth hour after midnight or midday; a card, counter etc. with five pips; (*pl.*) articles of attire, such as boots, gloves etc. of the fifth size. *a.* five in number, aged five. **a bunch of fives**, the fist. **five-figure tables**, tables of five-figure logarithms. **five-o'clock shadow**, beard growth which becomes visible on a man's shaven face late in the day. **five-eighth**, *n.* (*Austral.*) a player in rugby football posted between the half-backs and three-quarter backs. **five-penny** (fiv'pəni, fip'ni), *a.* priced at five pence. **five-star**, *a.* of the highest class, esp. of hotels. **fivefold**, *a., adv.* five times as much or as great. **fiver**, *n.* (*coll.*) a five-pound note; anything that counts as five, as a stroke for five at cricket etc. **fives**, *n.* a game in which a ball is struck against a wall by the open hand or a small wooden bat. **fives-court**, *n.* a court with two, three or four walls where the game of fives is played.
fix (fiks), *v.t.* to make fast, firm or stable; to fasten, attach, secure firmly; to establish; to make permanent or stable (as colours, a photographic picture etc.); to solidify; to arrest and hold (as eyes, attention etc.); to direct steadily; to settle, to determine, to decide (on); to appoint a definite position for; (*coll.*) to adjust, to arrange properly, to set to rights, to repair; (*chiefly N Am.*) to prepare; (*usu. pass. coll.*) to be provided with; (*euphem.*) to spay or castrate (an animal); (*sl.*) to punish, to get even with; (*sl.*) to influence illicitly. *v.i.* to become fixed; (*chiefly N Am., coll.*) to be about to, to be set to. *n.* an awkward predicament, a dilemma; the position of a ship, aircraft etc. as determined by radar etc.; the determination of such a position; (*sl.*) an injection of heroin or a similar drug. **to fix on, upon**, to determine on; to choose, to select. **to fix up**, (*coll.*) to arrange, to organize; to settle; to assemble or construct; to provide. **fixable**, *a.* **fixed**, *a.* fast, firm; established, settled, unalterable; not volatile. **fixed assets**, *n.pl.* business assets of a relatively permanent nature, as buildings, plant etc. **fixed idea**, *n.* a rooted idea, one tending to become a monomania. **fixed link**, *n.* a permanent means of crossing a stretch of water, e.g. a bridge or tunnel, as opposed to e.g. a ferry. **fixed-penalty**, *a.* involving the payment of a predetermined and invariable fine. **fixed stars**, *n.pl.* stars which apparently maintain the same relative positions to each other in the sky, as distinct from planets. **fixed-wing**, *a.* having permanently attached wings, as opposed to e.g. a helicopter. **fixedly** (fik'sid-), *adv.* steadfastly, firmly; intently. **fixedness** (fik'sid-), *n.* the quality or state of being fixed; immobility, steadfastness; absence of volatility. **fixer**, *n.* one who or that which fixes; a person adept at finding crafty or illicit solutions to problems. **fixings**, *n.pl.* (*N Am.*) trimmings.
fixate (fik'sāt), *v.t.* to render fixed; to fix the gaze upon; (*Psych.*) to arrest the psychological development of an immature stage; (*usu. pass.*) to be obsessed. **fixation**, *n.* the act of fixing; the process of making nonvolatile, as causing a gas to combine with a solid; (*Psych.*) an emotional arrest of development of the personality; an obsession. **fixative**, *a.* serving to fix. *n.* a fixing agent; a substance used to make colours permanent or prevent crayon or pastel drawings from becoming blurred; a substance added to a perfume to prevent evaporation.
fixer, fixings FIX.
fixity (fiks'iti), *n.* fixedness, stability, permanence.
fixture (fiks'chə), *n.* anything fixed in a permanent position; (*Law*) an article of a personal nature fitted in a building or attached to land and regarded as an integral part; a person or thing regarded as permanently established and immovable; a sporting event arranged for a particular date.
fizgig (fiz'gig), *n.* a gadding, flirting girl; a firework that fizzes.
fizz (fiz), *v.i.* to make a hissing or sputtering sound. *n.* a hissing, sputtering sound; effervescence; (*coll.*) champagne, from its effervescence; (*coll.*) ginger-beer, lemonade; spirit, life. **fizzy**, *a.*
fizzle (fiz'l), *v.i.* to fizz; (*coll.*) to fail. *n.* the sound or action of fizzing or fizzling; (*coll.*) a lame ending, a fiasco. **to fizzle out**, to come to a lame conclusion.
fjord (fyawd, fe'awd), FIORD.
FL, (*abbr.*) Florida.
fl., (*abbr.*) florin; flourished.
Fla, (*abbr.*) Florida.
flab FLABBY.
flabbergast (flab'əgahst), *v.t.* to astound, to stagger with surprise.
flabby (flab'i), *a.* hanging loosely, limp, flaccid; having flab; lacking in fibre or nerve, languid, feeble. **flab**, *n.* (*coll.*) flaccid body tissue, a sign of being overweight or out of condition. **flabbily**, *adv.* **flabbiness**, *n.*
flaccid (flak'sid, flas'id), *a.* lacking firmness or vigour; limp, flabby, drooping; relaxed, feeble. **flaccidity** (-sid'-), *n.* **flaccidly**, *adv.* **flaccidness**, *n.*
flacon (flakō'), *n.* a small bottle, esp. a scent-bottle.
flag[1] (flag), *v.i.* (*past, p.p.* **flagged**) to hang loosely, to droop; to become limp; to lose strength or vigour; to become spiritless or dejected; to lose interest.
flag[2] (flag), *n.* a piece of bunting or other cloth, usu. square or oblong, and plain or bearing a device, typically attached by one edge to a staff or halyard by which it can be hoisted on a pole or mast, and displayed as a banner, ensign or signal; something that marks, signals or is used as a token; (*Naut.*) a flag carried by a flagship to show that the admiral is in command; the flagship itself; the bushy part of a dog's tail, as of a setter; the uncut tuft of hair on a brush. *v.t.* (*past, p.p.* **flagged**) to decorate with flags; to mark out with flags; to signal by means of a flag or flags. **black flag** BLACK. **flag of convenience**, a foreign flag under which a vessel is registered to escape taxation etc. in its real country of origin. **flag of truce**, a white flag indicating that the enemy has some pacific communication to make; an offer of peace. **red flag** RED. **to flag down**, to signal to (a vehicle) to stop. **to flag out**, to register a vessel under a flag of convenience. **to keep the flag flying**, to continue to represent or stand up for e.g. a country or principles. **to show the flag**, to send an official representative or military unit to a place as a courtesy or a means of asserting a claim etc.; (*coll.*) to put in an appearance. **to strike, lower the flag**, to pull the flag down in token of surrender or submission; of an admiral, to relinquish the command. **white flag** FLAG OF TRUCE. **yellow-flag** YELLOW. **flag-captain**, *n.* the commanding officer of a flagship. **flag day**, *n.* a day on which street collections are made for a specific charity, a small flag being worn as a token of having given money. **flag-lieutenant**, *n.* an officer in immediate attendance upon a flag-officer. **flag-officer**, *n.* a commodore, admiral, vice-admiral or rear-admiral; **flag-pole** FLAGSTAFF. **to run (something) up the flagpole**, to sound out an idea etc., to test reactions to something. **flagship**, *n.* the ship which carries the admiral, and on which his flag is displayed; the largest and most important of a set, esp. something regarded as embodying e.g. a company's prestige. **flagstaff**, *n.* (*pl.* **-staffs, -staves**) the pole or staff on which a flag is displayed. **flag-waving**, *n.* (*coll.*) talk or activity intended to stir patriotic feeling.
flag[3] (flag), *n.* one of various herbaceous plants with long

blade-like leaves growing in moist places, chiefly belonging to the genus *Iris*.

flag[4] (flag), **flagstone**, *n.* a broad flat stone used for paving; (*pl.*) a pavement made of such stones; a fine-grained rock which can be split into slabs for paving. *v.t.* (*past, p.p.* **flagged**) to pave with flags. **flagging**, *n.* (a pavement of) flagstones.

flagellate (flaj'əlāt), *v.t.* to whip, to beat, to scourge. *a.* (-lət), (*Zool., Bot. etc.*) having whip-like processes or flagella; resembling a flagellum. *n.* an organism bearing one or more flagella. **flagellant**, *n.* one of a sect of fanatics which arose in Italy about 1260 who sought to avert the divine wrath by scourging themselves till the blood came; (*Psych.*) one who thrashes (himself or others) for sexual gratification. *a.* given to scourging. **flagellation**, *n.* a scourging or flogging. **flagellator**, *n.* **flagellatory**, *a.* **flagelliform** (-jel'ifawm), *a.* **flagellum** (-jel'əm), *n.* (*pl.* **-lla** (-lə)) (*Zool., Biol.*) a minute whip-like appendage; a trailing shoot; a runner.

flageolet[1] (flajəlet', flaj'-), *n.* a small wind instrument blown from a mouthpiece at the end, and producing a shrill sound similar to but softer than that of the piccolo.

flageolet[2] (flajəlet', -lā'), *n.* the French or haricot bean.

flagon (flag'ən), *n.* a vessel with a narrow mouth or spout, used for holding liquors; a flat bottle holding the contents of nearly two bottles, used in the wine-trade.

flagrant (flā'grənt), *a.* glaring, notorious, outrageous, scandalous. **flagrancy**, *n.* **flagrantly**, *adv.*

flagrante delicto (fləgran'ti dilik'tō), in the very act, red-handed. [L]

flagstone FLAG[4].

flail (flāl), *n.* a wooden instrument consisting of a staff or swingle hinged to a longer staff or handle, used for threshing grain by hand. *v.t.* to swing or beat wildly; to strike (as) with a flail. *v.i.* to thresh around.

flair (fleə), *n.* keen perception, discernment; a natural aptitude or gift; stylishness.

flak (flak), *n.* fire from anti-aircraft guns; adverse criticism, dissent. **flak-catcher**, *n.* (*chiefly N Am., sl.*) a subordinate who deals with adverse criticism on a superior's behalf. **flak jacket**, *n.* a reinforced jacket worn by soldiers, police etc. as protection against gunshot etc. [initials of G *Flug abwehr kanone*, anti-aircraft gun]

flake[1] (flāk), *n.* a thin scale-like fragment; a thin piece peeled off; a chip (as of flint); a small fleecy particle (as of snow); (*N Am., sl.*) a flaky person. *v.t.* to form into flakes or loose particles; to chip flakes off or in flakes; to sprinkle with flakes. *v.i.* to peel or scale off in flakes. **flake-white**, *n.* English white lead in the form of scales, used as a pigment. **flaky**, *a.* consisting of flakes; liable to flake; (*N Am., sl.*) unstable, unreliable; (*N Am., sl.*) unattractively unconventional. **flakiness**, *n.*

flake[2] (flāk), *n.* a light platform or rack, e.g. for drying fish.

flam (flam), *n.* a false pretext, a deception, a lie. *v.t.* to deceive.

flambé (flā'bā, flawm'-), *v.t.* (*past, p.p.* **flambéed**) to sprinkle with brandy and ignite. *a.* served as above. [F]

flambeau (flam'bō), *n.* (*pl.* **-beaus, -beaux** (-bō, -boz)) a torch, esp. one made of thick wicks covered with wax or pitch.

flamboyant (flamboi'ənt), *a.* a term applied to the decorated French Gothic (contemporary with the Perpendicular style in England), from the flame-like tracery; florid, highly decorated; gorgeously coloured; exuberant, extravagant, showy.

flame (flām), *n.* a mass or stream of vapour or gas in a state of combustion; a blaze; fire; a glow, a bright light; a blaze of colour; ardour, excitement, passion; the object of one's affection, a sweetheart. *v.t.* to burn; to flambé. *v.i.* to burn with a flame; to send out flame, to blaze, to burst into flames; to break (out) or blaze (up) in violent passion; to shine, to glow. **flame-colour**, *n.* a bright reddish-yellow colour. **flame-coloured**, *a.* **flame-thrower**, *n.* a weapon that projects a stream of burning liquid. **flame-tree**, *n.* the Australian fire-tree. **flameless**, *a.* **flamelet** (-lit), *n.* **flaming**, *a.* burning, blazing; intensely bright; vehement, violent; (*sl.*) bloody. **flaming onion**, *n.* an anti-aircraft projectile having the appearance of a string of yellow fire-balls. **flamingly**, *adv.* **flamy**, *a.*

flamen (flā'men), *n.* an ancient Roman priest devoted to some special deity. **flaminical** (-min'-), *a.*

flamenco (fləmeng'kō), *n.* (*pl.* **-cos**) a Gipsy song or dance from Andalusia.

flamingo (fləming'gō), *n.* (*pl.* **-gos, -goes**) a long-necked bird, with small body and very long legs, its feathers rose or scarlet in colour, belonging to the genus *Phaenicopterus*.

flan (flan), *n.* an open pastry or sponge tart with fruit or savoury filling. **flan-case**, *n.*

flange (flanj), *n.* a projecting rib or rim affixed to a wheel, tool, pipe, rail etc., for strength, as a guide, or for attachment to something else. *v.t.* to supply with a flange. **flanged**, *a.*

flank (flangk), *n.* the fleshy or muscular part of the side between the hips and the ribs; either side of a building, mountain etc.; the side of an army or body of troops. *v.t.* to stand or be at the flank or side of, to border; to attack, turn or threaten the flank of; to secure or guard the flank of. *v.i.* to border, to touch; to be posted on the flank or side. **flanker**, *n.* one who or that which flanks, or is posted, stationed or placed on the flanks; (*Fort.*) a work projecting so as to command the flank of an assailing body; (*Rugby*) a wing forward.

flannel (flan'əl), *n.* a soft woollen stuff of open texture, with a light nap; (*pl.*) garments made of this material, esp. trousers for cricketers etc.; a piece of cloth used for washing the face etc.; (*coll.*) flattery, soft-soap; (*coll.*) evasive waffling, nonsense. *v.t.* (*past, p.p.* **flannelled**) to wrap in or rub with flannel or a flannel; (*coll.*) to flatter. *v.i.* (*coll.*) to waffle on evasively. **flannelette** (flanəlet'), *n.* a cotton fabric made to imitate flannel. **flannelly**, *a.*

flap (flap), *v.t.* (*past, p.p.* **flapped**) to beat, strike or drive away with anything broad and flexible; to move rapidly up and down or to and fro (as wings). *v.i.* to move the wings rapidly up and down or to and fro; to be moved to and fro or up and down, to flutter, swing about or oscillate; to hang down, as the brim of a hat; to strike a loose blow or blows, to beat (as with the wings); to be agitated. *n.* anything broad and flexible, hanging loosely, or attached by one side only, usu. used to cover an opening; the hinged leaf of a table or shutter; a movable control surface on the wing of an aircraft to increase lift on take-off and drag on landing; the motion or act of flapping; a light stroke or blow with something broad and loose; a slap; (*coll.*) a state of anxiety or confusion. **flapdoodle**, *n.* (*coll.*) rubbish, nonsense, bunkum. **flapjack**, *n.* a kind of pancake; a biscuit made of oat flakes and syrup; a flattish circular case for holding a powder-puff and a mirror. **flapper**, *n.* one who or that which flaps; (*coll.*) in the 1920s, a flighty young woman.

flare (fleə), *v.i.* to blaze, to flame up, or to glow, esp. with an unsteady light; to open or spread outwards. *v.t.* to cause to flare up; to burn off (excess gas or oil or flame); to provide with a flare or flares (as a skirt or trousers). *n.* a large unsteady light, a glare; a sudden outburst; (*Dressmaking*) material cut on the cross to give additional fullness; a widening or spreading out; (a thing with) a flared shape; (a device producing) a blaze of light used for illumination, signalling, or to attract attention; a flame for burning off excess gas or oil. **to flare up**, to blaze out; to fly into a passion. **flare-path**, an illuminated path allowing an aircraft to land or take off when visibility is low. **flare-up**, *n.* a sudden outbreak into flame; an outburst of

flash — flaw

anger, violence, hostilities etc. **flared**, a. having a flare or flares, flare-shaped. **flaring**, a. **flaringly**, adv. **flary**, a.

flash (flăsh), v.i. to appear with a sudden and transient gleam; to burst suddenly into flame or light; to send out a rapid gleam; to reflect light, to glitter; to burst forth, appear or occur suddenly; to rush swiftly; to dash, break or splash, as water or waves; to signal using e.g. a torch or the headlights of a car; (sl.) to expose oneself indecently. v.t. to emit or send forth in flashes or like flashes; to cause to gleam; to convey or transmit instantaneously (as news by telegraph); to signal (a message) to (someone) using light; to display or expose suddenly and briefly; to display ostentatiously; to cover (plain glass) with a thin coating, as of coloured glass; to send swiftly along; to send a rush of water down (a river, weir etc.). n. a sudden and transitory blaze or gleam of bright light; the space of time taken by this, an instant; a sudden occurrence or display; a body of water driven along with violence; a label with regimental name etc. sewn on the uniform shoulder; a sticker on goods etc. advertising e.g. a reduction in price; (Phot.) flashlight, an apparatus for producing flashlight; a newsflash; a sudden outburst, as of anger, wit, merriment etc.; show, ostentation. a. occurring or carried out very quickly; gaudy, vulgarly showy; counterfeit, forged; pertaining to thieves or vagabonds. **a flash in the pan**, a flash produced by the hammer of a gun upon a flint which fails to explode the powder; hence, an abortive attempt. **flashback**, n. an interruption in the narrative of e.g. a film or novel to show past events. **flashbulb**, n. (Phot.) a (usu. disposable) bulb used to produce flashlight. **flash burn**, n. a burn suffered as the result of momentary exposure to intense heat. **flash card**, n. a card with e.g. words or numbers printed on it for children to look at briefly as an aid to learning. **flash cube**, n. (Phot.) a plastic cube containing four flashbulbs. **flash flood**, n. a sudden flood, caused by heavy local rainfall. **flash gun**, n. (Phot.) a device which holds and fires a flashbulb. **flashlight**, n. (Phot.) a brilliant light for taking (usu. indoor) photographs; an electric battery torch; a regularly flashing light, as from a lighthouse. **flashpoint**, n. the temperature at which the vapour from oil or spirit ignites; the point at which tension erupts into violence; a place or region where such eruptions are likely to occur. **flasher**, n. one who or that which flashes; a device that causes a light to flash; a vehicle indicator light; (sl.) one who exposes him or herself indecently. **flashing**, n. a lap-joint used in roofing with sheet metal, a strip of lead carrying the drip of a wall into a gutter. **flashy**, a. showy but empty, brilliant but shallow; gaudy, tawdry, cheap and showy. **flashily**, adv. **flashiness**, n.

flask (flahsk), n. a small bottle or similar vessel; a powder-flask; a flat bottle, usu. mounted in metal, for carrying spirits in the pocket; a thin, long-necked bottle, encased in wicker, for wine or oil; a narrow-necked glass vessel used in a laboratory; a vacuum flask; a large reinforced metal container for transporting nuclear waste.

flat¹ (flat), a. having a level and even surface; horizontal, level; even, smooth, having few or no elevations or depressions; having little depth or thickness; level with the ground, lying prone, prostrate; having a surface or side in continuous contact with another surface; of feet, having little or no arch; of shoes, not having a raised heel; of a battery, having little or no charge; of a tyre, deflated; depressed, dejected; monotonous, dull, uninteresting, vapid, insipid, pointless, spiritless; having lost sparkle or freshness; plain, positive, absolute, downright; neither more nor less, as in *ten seconds flat;* of trade, inactive, dull; of prices, low; (*Painting*) wanting relief or prominence of the figures; uniform, without variety of tint or shading; without lustre, matt; lacking contrast; (*Mus.*) below the true pitch; minor (applied to intervals). adv. flatly, positively; prostrate, level with the ground; below the true pitch. n. a flat, plain surface; a level plain or low tract of land; a shoal, a shallow, a low tract flooded at high tide; a flat part of anything; anything that is flat; the palm of the hand; (*Theat.*) scenery on a wooden frame pushed on to the stage from the sides; a note semitone lower than the one from which it is named; the sign indicating this lowering of pitch; a punctured tyre. **flat broke**, (coll.) having no money, skint. **flat out**, at full speed, with maximum effort; completely exhausted. **that's flat!** that is final, irrevocable. **the flat**, the flat-racing season. **to fall flat** FALL. **flat-boat**, n. a large boat with a flat bottom, used for transport on rivers in the US. **flat fish**, n. any fish (such as the sole, plaice, turbot etc.) of the Pleuronectidae, distinguished by their laterally compressed body, absence of coloration on the under side, and the position of both eyes on the upper side. **flatfoot**, n. (derog.) a policeman. **flat-footed**, a. with the feet not arched; awkward; ponderous, unimaginative; (coll.) off guard; (chiefly N Am., sl.) downright, resolute, determined. **flat-head**, n. (Austral.) an edible fish with flattened head and body. **flat-iron**, n. an instrument for smoothing clothes etc. **flat-race**, n. a race on level ground without obstacles. **flat-racing**, n. **flat rate**, n. a rate of payment not varying in proportion with the amount supplied. **flat spin**, n. a spin in which the aircraft is almost horizontal; a confused and frantic state. **flat-top**, n. a haircut in which the hair on top of the head is cut shorter than the sides and back. **flatworm**, n. a worm of the phylum Platyhelminthes having a soft flattened body. **flatly**, adv. **flatness**, n. **flatten**, v.t. to make flat, to level; to make dull or insipid; to deject, to dispirit; (*Mus.*) to depress or lower in pitch; to knock down or out; to defeat resoundingly. v.i. to become flat. **to flatten out**, of a plane, to change from the gliding approach to the position to alight, when approaching to land. **flattie**, n. (coll.) a shoe with a very small or no heel. **flattish**, a.

flat² (flat), n. a floor or storey of a house; a suite of rooms on one floor forming a separate residence. **flatmate**, n. a person with whom one shares a flat. **flatlet** (-lit), n. a small flat.

flatter (flat'ə), v.t. to court, cajole or gratify by compliment, adulation or blandishment; to praise falsely or unduly; to raise false hopes in; to persuade (usu. oneself of some favourable contingency); to represent too favourably; to display to advantage. v.i. to use flattery. **flatterer**, n. **flatteringly**, adv. **flattery**, n. the act or practice of flattering; false or venal praise; adulation, cajolery.

flatulent (fla'chələnt), a. affected with or troubled by wind or gases generated in the alimentary canal; generating or likely to generate wind in the stomach; inflated, empty, vain; pretentious, turgid. **flatulence, -lency**, n. **flatulently**, adv. **flatus** (flā'təs), n. wind in the stomach or bowels; flatulence.

flaunt (flawnt), v.i. to make an ostentatious or gaudy show. v.t. to display ostentatiously or impudently; to parade, to show off; to wave or flutter in the wind. n. the act of flaunting. **flauntingly**, adv. **flaunty**, a.

flautist (flaw'tist), n. a player on the flute.

flavescent (fləves'ənt), a. yellowish; turning yellow.

flavour, (esp. N Am.) **flavor** (flā'və), n. that quality in any substance which affects the taste, or the taste and smell; a characteristic or distinctive quality. v.t. to impart a flavour to; to render pleasing to the palate; to season. **flavour of the month**, (often iron.) a person or thing much in favour at a particular time. **flavorous, flavoursome** (-səm), a. pleasing to taste or smell. **flavoured**, a. having a distinct flavour; (in comb.) having a particular flavour, as *full-flavoured*. **flavouring**, n. an (artificial) substance that gives flavour. **flavourless**, a.

flaw¹ (flaw), n. a crack, a slight fissure; a defect, an imperfection; (*Law*) a defect in an instrument, evidence etc., rendering it invalid. v.t. to break, to crack; to mar;

flaw   312   **flick**

to render invalid. *v.i.* to crack. **flawless**, *a.* **flawlessly**, *adv.* **flawlessness**, *n.*

flaw² (flaw), *n.* a sudden puff or gust; a squall, a violent but transient storm.

flax (flaks), *n.* a plant of the genus *Linum*, esp. *L. usitatissimum*, the common flax, the fibre of which is made into yarn, and woven into linen cloth; the fibrous part of the plant prepared for manufacture; one of various kinds of similar plants, as white flax, false flax or toad-flax. **flax-dresser**, *n.* one who prepares flax for the spinner. **flax-seed**, *n.* linseed. **flaxen**, *a.* made of flax; like flax; light yellow or straw-coloured. **flaxen-haired, -headed**, *a.* **flaxy**, *a.*

flay (flā), *v.t.* to strip the skin from; to whip, flog; to criticize savagely. **flayer**, *n.*

flea (flē), *n.* a blood-sucking insect belonging to the genus *Pulex*, parasitic on mammals and birds, and remarkable for its leaping powers. **sand-flea**, (*N Am.*) **beach-flea, water-flea**, small crustaceans with similar leaping powers. **with a flea in one's ear** EAR. **fleabag**, *n.* (*sl.*) a sleeping bag; (*sl.*) a dirty or neglected person; (*chiefly N Am., sl.*) an inferior lodging house. **flea-bane, -wort**, *n.* compositous plants of the genus *Pulicaria, Erigeron* or *Conyza*, from their supposed efficacy in driving away fleas. **flea-bite**, *n.* the bite of a flea; the red spot caused by the bite; a tiny amount; the smallest trifle; a trifling inconvenience. **fleabitten**, *a.* bitten by a flea; full of fleas; coloured, as some horses, with small red spots on a lighter ground. **flea market**, *n.* an open-air market selling usu. second-hand goods. **fleapit**, *n.* (*coll., facet.*) a shabby cinema or theatre.

flèche (flesh), *n.* a spire, esp. a slender one, usu. of wood covered with lead, over the intersection of nave and transepts.

fleck (flek), *n.* a tiny particle, a spot, a speck; a dot, stain or patch of colour or light. *v.t.* to spot, to streak, to variegate with spots or flecks. **flecker**, *v.t.* to fleck.

flection (flek'shən), FLEXION under FLEXIBLE.

fled (fled), *past, p.p.* FLEE.

fledge (flej), *v.t.* rear until capable of flight; to feather (an arrow); to deck or cover with (anything resembling) feathers. *v.i.* to acquire feathers or plumage for flight. **fledged**, *a.* feathered; able to fly. **fledgeling, fledgling** (-ling), *n.* a young bird just fledged; a raw and inexperienced person. *a.* newly fledged.

flee (flē), *v.i.* (*past, p.p.* **fled** (fled)) to run away, as from danger; to vanish, to disappear, to pass away swiftly. *v.t.* to run away from.

fleece (flēs), *n.* the woolly covering of a sheep or similar animal; the quantity of wool shorn from a sheep at one time; anything resembling a fleece, as a woolly head of hair, a fleecy cloud or fall of snow. *v.t.* to shear the wool from; to cover with anything fleecy; to rob, to plunder, to overcharge. **fleeceable**, *a.* **fleeceless**, *a.* **fleecer**, *n.* **fleecy**, *a.* woolly, wool-bearing; resembling a fleece in appearance or qualities.

fleet¹ (flēt), *n.* a number of ships or smaller vessels in company with a common object or destination, esp. a body of warships under one command; the entire body of warships belonging to one government, a navy; a collection of aircraft or road vehicles used for a common purpose and usu. under one ownership.

fleet² (flēt), *v.i.* to move or pass swiftly. *a.* swift of pace, nimble, rapid, speedy. **fleet-footed**, *a.* able to run with great speed. **fleeting**, *a.* passing quickly, transient. **fleetingly**, *adv.* **fleetly**, *adv.* **fleetness**, *n.*

Fleet-Street (flēt strēt), *n.* (*esp. formerly*) the centre of newspaper offices in London; journalism.

Fleming (flem'ing), *n.* a native of Flanders; a Flemish-speaking Belgian. **Flemish**, *a.* pertaining to Flanders or the Flemings. *n.* the Flemish language, one of the two languages of Belgium. **Flemish bond** BOND¹.

flense (flens), **flench** (flench), **flinch**² (flinch), *v.t.* to strip the blubber or the skin from (a whale or seal).

flesh (flesh), *n.* the soft part of an animal body, esp. the muscular tissue, investing the bones and covered by the skin; animal tissue used as food, as distinct from vegetable, fish, and sometimes from poultry; excess weight, fat, flab; the body, as distinguished from the soul; animal nature; the human race; carnal appetites; the present state of existence; kindred; the soft pulpy part of a fruit or plant. *v.t.* to encourage by giving flesh to, to make eager (from the sportsman's practice of giving hawks, dogs etc. the flesh of the first game they take); to initiate. **flesh and blood**, human nature; one's children or near relations. **in the flesh**, in bodily form. **proud flesh** PROUD. **to be one flesh**, to be closely united as in marriage. **to flesh out**, to elaborate, to give more substance or detail to. **to make someone's flesh creep**, to arouse (a physical sense of) horror in someone. **to press the flesh**, (*chiefly N Am.*) to shake hands. **flesh-colour**, *n.* yellowish-pink. **flesh-coloured**, *a.* **flesh-fly**, *n.* a carnivorous insect of the genus *Sarcophaga*, esp. *S. carnaria*, the larvae of which feed on decaying flesh. **flesh-pot**, *n.* (*usu. pl.*) sumptuous living; (*often pl.*) a night-club etc. offering lavish or sexually titillating entertainment. **flesh-tints**, *n.pl.* the colours which best represent the human skin. **flesh-wound**, *n.* a wound not reaching the bone or any vital organ. **fleshless**, *a.* destitute of flesh, lean, scraggy. **fleshings**, *n.pl.* light flesh-coloured tights to represent the skin, worn by actors, dancers etc. **fleshly**, *a.* sensual, carnal, lascivious; human, as distinct from spiritual; worldly. **fleshliness**, *n.* **fleshy**, *a.* like flesh; fat, plump, corpulent; of fruit etc., pulpy. **fleshiness**, *n.*

fletch (flech), *v.t.* to feather (as an arrow). **fletcher**, *n.* one who feathered arrows, a maker of bows and arrows.

fleur de lis (flœ də lē', lēs'), *n.* (*pl.* **fleurs de lis**) various species of iris; the heraldic lily, a charge borne in the French royal arms.

fleuron (floo'ron, flœrö'), *n.* a flower-shaped ornament, used in architecture, on coins etc.

flew, *past* FLY².

flews (flooz), *n.pl.* the large chaps of a deep-mouthed hound.

flex¹ (fleks), *v.t., v.i.* to bend or cause to bend; to contract (a muscle). **to flex one's muscles**, to contract the muscles, esp. of the arm in order to display them or as a preliminary to a trial of strength; (*fig.*) to put on a show of power or strength.

flex² (fleks), *n.* flexible insulated wire, or a piece of this. [short for FLEXIBLE]

flexible (flek'sibl), *a.* pliant, easily bent; tractable, easily persuaded; adaptable, versatile. **flexibility**, *n.* **flexibly**, *adv.* **flexile** (-īl), *a.* flexible. **flexility**, *n.* **flexion** (-shən), *n.* the act or process of bending; a bend, a curve; (*Gram.*) inflection; bending movement of a joint or limb. **flexional**, *a.* **flexionless**, *a.* **flexor**, *n.* a muscle that causes a limb or part to bend. **flexuous**, *a.* full of bends or turns, winding. **flexuosity** (-os'-), *n.* **flexuously**, *adv.* **flexure** (-shə), *n.* the act, process or manner of bending; the state of being bent; a bend, a curve, a turn, curvature; curving of a line, surface or solid.

flexitime (flek'sitim), *n.* a system of working which allows the worker some freedom to choose when to arrive for and leave work, usu. so long as he or she is present during a stipulated period (core time). [*flexible time*]

flibbertigibbet (flib'ətijibit), *n.* a flighty, thoughtless person.

flick¹ (flik), *n.* a smart, light blow or flip, as with a whip. *v.t.* to touch or strike with such a stroke; to move quickly or jerkily; to remove or cause to move with a flick. *v.i.* to move quickly. **to flick through**, to read through quickly or inattentively. **flick-knife**, *n.* a knife with a blade that springs out when a button in the handle is pressed.

flick² (flik), *n.* (*coll.*) a film, a movie. **the flicks**, the cinema.

flicker (flik'ə), *v.i.* to flutter; to quiver; to burn unsteadily, to waver. *n.* the act of flickering; an unsteady or dying light; a brief awakening (of interest, hope etc.). **flickeringly**, *adv.*

flier (flī'ə), FLYER under FLY[2].

flight[1] (flīt), *n.* the act or power of flying through the air; an air or space journey, esp. a scheduled trip made by a commercial air service; swift movement or passage; a trajectory; a soaring, a sally, an excursion, a sustained effort; a number of birds or insects moving together; a volley (of arrows, spears etc.); a series of steps mounting in one direction; the basic tactical unit of an airforce; (*Racing*) a line of hurdles on a course; a feather or vane attached to the tail of an arrow or dart. *v.t.* to shoot at (wild-fowl flying overhead); to give a high, slow trajectory to (a ball etc.); to put a feather or vane on (an arrow or dart). **flight-deck**, *n.* an aircraft-carrier's deck from which planes take off and land; the compartment at the front of a large aircraft housing the controls, navigation equipment etc. **flight-engineer**, *n.* a member of the crew of an aeroplane in charge of the motors. **flight-feather**, *n.* one of the large wing-quills used in flying. **Flight Lieutenant**, *n.* a commissioned rank in the RAF equivalent to captain in the army. **flight-path**, *n.* the path taken by an aeroplane, spacecraft or projectile through the air. **flight plan**, *n.* the proposed route and schedule of an aircraft flight. **flight recorder**, *n.* an instrument which records details of an aircraft's (performance in) flight. **Flight Sergeant**, *n.* a non-commissioned rank in the RAF. **flightless**, *a.* unable to fly. **flighty**, *a.* capricious, volatile; wild, fickle. **flightily**, *adv.* **flightiness**, *n.*

flight[2] (flīt), *n.* the act of fleeing or running away; a hasty departure, retreat or evasion. **to put to flight**, to cause to run away or disappear. **to take (to) flight**, to run away, to flee.

flighty FLIGHT[1].

flimflam (flim'flam), *n.* nonsense, bosh; humbug, deception; a piece of deception.

flimsy (flim'zi), *a.* thin, slight; frail; without strength or solidity; ineffective, unconvincing. *n.* thin paper used for carbon copies; a copy on this. **flimsily**, *adv.* **flimsiness**, *n.*

flinch[1] (flinch), *v.i.* to shrink from pain, suffering an undertaking etc, to wince; to give way, to fail. **flincher**, *n.* **flinchingly**, *adv.*

flinch[2] FLENSE.

flinder (flin'də), *n.* (*usu. pl.*) a fragment, a piece, a splinter.

fling (fling), *v.i.* (*past, p.p.* **flung** (flŭng)) to rush violently, to flounce; of horses, to kick, struggle, plunge (out). *v.t.* to cast or throw with sudden force; to send or put suddenly and unceremoniously; to apply (oneself) vigorously. *n.* a cast or throw from the hand; a period of unrestrained enjoyment; a lively Highland dance; a try.

flint (flint), *n.* a variety of quartz, usu. grey, smoke-brown or brownish-black and encrusted with white, easily chipped into a sharp cutting edge; a nodule of flint, a flint pebble; a piece of flint shaped for use in a gun, a tinder-box, lighter, or as an implement used by prehistoric man; a piece of iron alloy used to make a spark in a modern lighter; anything extremely hard. **flint-glass**, *n.* a very pure and lustrous kind of glass, orig. made with calcined flints. **flint-lock**, *n.* a lock for firearms, in which the cock holds a piece of flint, and comes down upon the steel cap of the pan containing the priming, which is ignited by the spark thus caused; a firearm having such a lock. **flinty**, *a.* composed of flint; of the nature of or resembling flint; cruel, pitiless, hard-hearted. **flintiness**, *n.*

flip (flip), *v.t.* (*past, p.p.* **flipped**) to fillip, flick or jerk; to toss or propel esp. so as to turn in the air before landing, with a light blow. *v.i.* to flap or flick (at); (*sl.*) to lose control of oneself, to become very angry; (*sl.*) to become wildly enthusiastic. *n.* a quick, light blow; a mixed alcoholic drink containing beaten egg; a somersault. *a.* (*coll.*) flippant; (*coll.*) impertinent. **to flip one's lid**, (*coll.*) to lose self-control. **to flip over**, to (cause to) turn over. **to flip through**, to read through quickly or carelessly. **flip-flap, flip-flop**, *adv.* with (a noise as of) repeated flapping. **flip-flop**, *n.* a backward handspring; an electronic device or circuit capable of assuming either of two stable states; (*N Am.*) a complete reversal (of opinion etc.); a kind of sandal consisting simply of a sole and a strap held between the toes. *v.i.* to move about with a flapping noise. **flipside**, *n.* the B side of a popular single record on which material additional to the title number is recorded. **flipper**, *n.* the broad fin of a fish; the limb or paddle of a turtle, penguin etc.; a paddle-shaped shoe worn for esp. underwater swimming. **flipping**, *a., adv.* (*coll., euphem.*) bloody.

flippant (flip'ənt), *a.* trifling, lacking in seriousness; impertinent, disrespectful. **flippancy**, *n.* **flippantly**, *adv.* [perh. from FLIP]

flipper FLIP.

flirt (flœt), *v.i.* to make sexual advances for amusement or self-gratification; to play at love-making, to coquet. *n.* an act of flirting; a person, esp. a woman, who plays at courtship. **to flirt with**, to treat lightly, to risk carelessly; to entertain thoughts of, to toy with. **flirtation**, *n.* coquetry; a playing at courtship; a casual involvement or interest. **flirtatious** (-tā'-), *a.* **flirtingly**, *adv.* **flirtish**, *a.* **flirty**, *a.*

flit (flit), *v.i.* (*past, p.p.* **flitted**) to move, to pass from place to place; to fly about lightly and rapidly; to depart; (*chiefly Sc.*) to leave one's house, usu. secretly. *n.* a stealthy departure.

flitch (flich), *n.* the side of a pig salted and cured; a board or plank from a tree-trunk, usu. from the outside.

flitter (flit'ə), *v.i.* to flit about; to flutter. **flittermouse**, *n.* a bat.

float (flōt), *v.i.* to be supported on the surface of or in a fluid; to move or glide without effort; to move lightly through or with a fluid, to drift; to move aimlessly; of a currency, to be free to find its own level on foreign exchange markets. *v.t.* to support on the surface of or in a fluid, to convey, to carry on or as on water; to set afloat, to launch; to flood with a liquid; to put into circulation; to allow (a currency) to find its own level; to form (a limited company) with a view to making a public issue of shares; to offer for sale on the Stock Exchange. *n.* anything buoyed up on the surface of a liquid; a buoyant device designed to keep a person afloat; the cork or quill on a fishing-line; a cork on a fishing-net; the bladder supporting fish, animals etc. in the water; the ball of a ballcock regulating a supply-tap; a timber-raft, a floating wharf; the gear of an aircraft for alighting on water; a small delivery vehicle with a flat platform for goods; (a vehicle carrying) a tableau or exhibit in a parade; (*usu. pl.*) the footlights of a theatre; a kind of trowel for smoothing the plastering on walls; a float-board; a drink with a lump of ice-cream floating in it; a small sum of money used to provide change at the start of business; an act of floating. **float-board**, *n.* one of the boards of an undershot waterwheel or a paddle-wheel. **floatable**, *a.* **floatage, flotage**, *n.* anything found floating, flotsam, floating power, buoyancy. **floatation** FLOTATION. **floatel** FLOTEL. **floater**, *n.* one who or that which floats; a spot, composed of dead cells and cell fragments, appearing in one's vision; a vagrant. **floating**, *a.* resting on the surface of a fluid; unattached, free, disconnected; circulating, not fixed; not invested; fluctuating, variable. **floating capital** CAPITAL[2]. **floating debt** DEBT. **floating dock** DOCK[3]. **floating kidney**, *n.* a condition in which the kidney is displaced. **floating ribs**, *n.pl.* the lowest two pairs of ribs, which are not attached to the sternum. **floating voter**, *n.* a person of no fixed party-political allegiance. **floatingly**, *adv.*

floccus (flok'əs), *n.* (*pl.* **-cci** (flok'sī)) a long tuft of hair

flock terminating the tail in some mammals; the down of unfledged birds. **floccose** (-ōs), *a.* covered with little woolly tufts. **flocculate**, *v.t.*, *v.i.* to form into floccules or flocculent masses. **floccule** (-ūl), *n.* a loose tuft; a small woolly or tuft-like portion. **flocculent**, *a.* in small flakes, woolly, tufted. **flocculose, flocculous**, *a.* **flocculus** (-ləs), *n.* (*pl.* **-li** (-lī)) a lobe on the under surface of the human cerebellum; a cloudy marking on the surface of the sun.

flock[1] (flok), *n.* a company or collection of animals, esp. sheep, goats or birds; a crowd, a large body; a congregation, considered in relation to their minister. *v.i.* to come together in a flock; to congregate, to assemble, to go or move in crowds. *v.t.* to crowd; to press by crowding. **flock-master**, *n.* a sheep-farmer. [OE *flocc* (cp. Icel. *flokkr*)]

flock[2] (flok), *n.* a lock or tuft of wool, cotton, hair etc.; (*usu.pl.*) wool-dust used in coating certain portions of the patterns in some wallpapers (**flock-paper**); fibrous material, made by tearing up woollen rags by machinery, used to stuff upholstery, mattresses etc.; (*Chem.*) matter in woolly or loose floating masses precipitated in a solution. **flocking**, *n.* flock used for flock-paper. **flocky**, *a.*

floe (flō), *n.* a large sheet of floating ice.

flog (flog), *v.t.* (*past*, *p.p.* **flogged**) to thrash, esp. with a whip or rod; to urge or drive (as if) by beating; to repeat or labour to the point of tedium; (*sl.*) to sell. **to flog a dead horse**, to try to revive interest in something stale; to pursue a hopeless task. **flogger, flogster**, *n.* **flogging**, *n.* punishment by whipping.

flong (flong), *n.* (*Print.*) prepared paper used for the matrices in stereotyping.

flood (flŭd), *n.* an abundant flow of water; a body of water rising and overflowing land not usually covered with water, an inundation; the inflow of the tide; a downpour, a torrent; an overflowing abundance; a floodlight; excessive menstrual discharge. *v.t.* to overflow, to inundate, to deluge; to supply copiously (with). *v.i.* to rise and overflow; to become inundated or submerged by a flood; to surge (through). to have uterine haemorrhage, to have excessive menstrual discharge. **the Flood**, the Deluge recorded in Genesis. **flood-gate**, *n.* a gate in a waterway arranged to open when the water attains a certain height, and so allow it to escape freely to prevent floods, a sluice; the lower gate of a lock; a restraint against an emotional outburst. **floodlight**, *n.* a powerful beam of artificial light used esp. in the theatre, in sports stadiums or to illuminate buildings; a lamp producing such light. *v.t.* to illuminate with floodlight. **floodlighting**, *n.* **flood plain**, *n.* an area of flat land near a river, formed by sediment deposited during floods. **flood-tide**, *n.* the rising tide. **flooding**, *n.*

floor (flaw), *n.* the bottom surface of a room, on which the inmates walk and which supports the furniture; the boards or other material of which this is made; a storey in a building; the part of the house assigned to members of a legislative assembly; the (area occupied by) people attending a meeting or debate as audience; the right to address an assembly or meeting; any level area corresponding to the floor of a room; a ground surface, as of the sea, a body part etc.; the lowest limit of prices, wages etc. *v.t.* to furnish with a floor; to be or serve as a floor (to); to knock down; to put to silence (as in argument); to pose (a difficult question); to get the better of, to defeat. **to cross the floor**, of an MP etc., to change one's party-political allegiance. **to take the floor**, to rise to speak, to take part in a debate; to get up to dance; **to wipe, mop the floor with**, to defeat completely. **floorboard**, *n.* one of the planks making up a floor. **floorcloth**, *n.* a piece of soft fabric used for washing floors; a substitute for a carpet. **floor manager**, *n.* the stage manager of a television programme; the manager of a floor in a large store. **floor-show**, *n.* a performance on the floor of a restaurant etc.

floorwalker, *n.* a shopwalker. **flooring**, *n.* material for floors; a floor, a platform. **floorless**, *a.*

floozy, floozie, floosie (floo'zi), *n.* (*derog.*) a (young) woman thought to be free with her company and favours.

flop (flop), *v.i.* (*past*, *p.p.* **flopped**) to tumble about or fall loosely and heavily; to sway about heavily, to make a dull sound as of a soft body flapping; to fail dismally; (*chiefly N Am.*) to go to bed. *v.t.* to let fall negligently or noisily; to cause to strike with a heavy dull sound. *n.* the act or motion of flopping; the noise of a soft outspread body falling suddenly to the ground; a complete failure. *adv.* with a flop; suddenly. **flop-house**, *n.* (*N Am.*) a dosshouse, a cheap lodging-house. **floppily**, *adv.* **floppiness**, *n.* **floppy**, *a.* soft and flexible, limp. **floppy disc**, *n.* (*Comput.*) a flexible magnetic disc for data storage.

flor., (*abbr.*) floruit.

flora (flaw'rə), *n.* (*pl.* **-ras, -rae** (-rē) the whole vegetation of a region, country or (geological) period; a book dealing with the vegetation of a country or district. **floral**, *a.* of or pertaining to floras; of or pertaining to flowers; consisting of, or decorated with, flowers. **floral envelope**, *n.* the perianth or parts surrounding the stamens and pistils, generally consisting of calyx and corolla. **florally**, *adv.*

floreat (flo'riat), may it flourish. [L]

Florentine (flo'rəntin), *a.* of or pertaining to Florence. *n.* a native or inhabitant of Florence.

florescence (flərĕs'əns), *n.* the flowering of a plant; the season when a plant flowers. **florescent**, *a.*

floret (flo'rit), *n.* a small flower; a small flower forming part of a composite one.

floriate (flo'riət), **-ated** (-ātid), *a.* adorned with floral ornaments or designs. **floriation**, *n.*

floribunda (floribŭndə), *n.* any of several hybrid roses with flowers in open clusters.

floriculture (flaw'rikŭlchə), *n.* the cultivation of flowers or flowering plants. **floricultural** (-kŭl'-), *a.* **floriculturist** (-kŭl'-), *n.*

florid (flo'rid), *a.* bright in colour; flushed with red, ruddy; flowery, highly embellished, elaborately ornate; showy. **floridity, floridness**, *n.* **floridly**, *adv.*

floriferous (flŏrĭf'ərəs), *a.* bearing flowers.

florin (flo'rin), *n.* a former British coin, orig. silver, worth the equivalent of 10p, a two-shilling piece; a foreign gold or silver coin, of various values according to country and period.

florist (flo'rist), *n.* a cultivator of flowers; one who sells flowers; one skilled in flowers. **floristic**, *a.* of flowers or floras.

floruit (flŏ'ruit), *n.* the period of a person's eminence; the date at which he or she was known to be alive (in the absence of exact dates of birth and death). [L, he flourished]

floss (flos), *n.* the exterior soft envelope of a silkworm's cocoon; the downy substance on the husks of certain plants, as the bean; dental floss. *v.t.*, *v.i.* to use dental floss. **floss-silk**, *n.* untwisted filaments of fine silk, used in embroidery etc. **floss-thread**, *n.* soft cotton yarn or thread for embroidery. **flossy**, *a.*

flotage (flō'tij), FLOATAGE.

flotation, floatation (flōtā'shən), *n.* the act or state of floating; the science of floating bodies; (*Finance*) the floating of a company. **flotative** (flō'-), *a.*

flotel, floatel (flōtel'), *n.* a boat or platform providing accommodation for off-shore oil-rig workers. [*floating-hotel*]

flotilla (flŏtil'ə), *n.* a small fleet; a fleet of small vessels.

flotsam (flot'səm), *n.* goods lost in shipwreck and found floating. **flotsam and jetsam**, wreckage or any property found floating or washed ashore.

flounce[1] (flowns), *v.i.* to move abruptly or violently; to exaggerate one's movements as a means of calling attention to oneself or one's impatience etc. *n.* an abrupt or

flounce / **fluorescence**

impatient movement.
flounce² (flowns), *n.* a gathered or pleated strip of cloth sewed to a petticoat, dress etc., with the lower border hanging loose. *v.t.* to deck or trim with flounces.
flounder¹ (flown'də), *n.* a flatfish resembling the plaice, but with paler spots.
flounder² (flown'də), *v.i.* to struggle or stumble about violently, as when stuck in mire; to struggle along with difficulty; to blunder along, to do things badly. *n.* a stumbling or blundering effort; the motion or act of floundering.
flour (flowə), *n.* the finer part of meal, esp. of wheatmeal; fine soft powder of any substance. *v.t.* to sprinkle or coat with flour; to grind into flour. **flour-dredge, -dredger**, *n.* a perforated tin for sprinkling flour. **flour-mill**, *n.* a mill for grinding and sifting grain to make flour. **floury**, *a.* covered with flour; like flour.
flourish (flŭ'rish), *v.i.* to grow luxuriantly; to thrive, to prosper; to be in good health; to reach a peak of development, condition, activity etc.; to be alive or at work (at or about a certain date); to make bold and fanciful strokes in writing; to move about fantastically; (*Mus.*) to play in a bold, dashing style, with ornamental notes; to sound a fanfare. *v.t.* to brandish, fling or wave about; to flaunt, to show ostentatiously; to embellish with ornamental or fantastic figures. *n.* a figure formed by strokes or lines fancifully drawn; rhetorical display, florid diction, a florid expression; a brandishing or waving of a weapon or other thing; a showy or ostentatious action or gesture; (*Mus.*) a passage played for display, a fanfare of trumpets etc., an improvised prelude or other addition. **flourishing**, *a.* thriving, prosperous; making a show. **flourishingly**, *adv.* **flourishy**, *a.*
flout (flowt), *v.t.* to treat with contempt, to disregard, to defy. *v.i.* to behave with contempt or mockery. **flouter**, *n.* **floutingly**, *adv.*
flow (flō), *v.i.* to move, run or spread, as a fluid; to circulate, as the blood; to rise, as the tide; to issue, to spring, to gush out; to sway, glide or float, to move or proceed easily or freely; to have smooth unbroken lines; to hang loosely; to be poured out abundantly, to abound, to come or go in abundance or great numbers; to discharge blood in excess from the uterus. *v.t.* to overflow, to flood. *n.* the act, state or motion of flowing; the quantity that flows; a flowing liquid, a stream; a copious stream, abundance, a plentiful supply; a smooth uninterrupted progression; the rise of the tide. **flowchart, flowsheet**, *n.* a diagram showing the sequence of operations in a process or computer program. **flowing**, *a.* **flowingly**, *adv.* **flowingness**, *n.*
flower (flow'ə), *n.* the organ or growth comprising the organs of reproduction in a flowering plant; a flowering plant; the blossom, the bloom; the state of flowering; the finest, choicest or best individual, part, period etc.; an embellishment; the prime, the period of youthful vigour; (*pl.*) (*Chem.*) substances of a powdery consistency or form, esp. if produced by sublimation, as in *flowers of sulphur*. *v.i.* to produce flowers, to bloom, to blossom; to reach maturity or complete development; to be in the prime. *v.t.* to embellish with flowers; to cause to blossom. **flower-bed**, *n.* a plot of ground in which flowering-plants are grown. **flower-girl**, *n.* a girl or woman selling flowers. **flower-head** CAPITULUM. **flowerpot**, *n.* an earthenware pot to hold plants. **flower-stalk**, *n.* the peduncle supporting the flowers of a plant. **flowered**, *a.* having or embellished with flowers or figures of flowers; (*in comb.*) bearing flowers, as *blue-flowered, six-flowered*. **flowerer**, *n.* a plant that flowers (at a particular time or in a particular way), as *spring-flowerer*. **flowering**, *a.* that flowers; flowery. **flowering plant**, *n.* an angiosperm; a plant species with conspicuous flowers. **flowerless**, *a.* **flowerlessness**, *n.* **flowery**, *a.* abounding in flowers or blossoms; highly figurative, florid.
flowing FLOW.
flown *p.p.* FLY².
fl.oz., (*abbr.*) fluid ounce.
flu (floo), *n.* short for INFLUENZA.
fluctuate (flŭk'chūāt), *v.i.* to rise and fall like waves; to vary, to change irregularly, to be unsettled; to waver. *v.t.* to cause to fluctuate. **fluctuating**, *a.* unsteady, wavering. **fluctuation**, *n.*
flue (floo), *n.* a passage or tube by which smoke can escape or hot air be conveyed. **flue-pipe**, *n.* an organ pipe in which the sound is produced by air passing through a fissure and striking an edge above.
fluent (floo'ənt), *a.* moving or curving smoothly, graceful; ready in the use of words; eloquent, copious, voluble; effortless, smooth, polished. **fluency**, *n.* the quality of being fluent; readiness and easy flow (of words or ideas). **fluently**, *adv.*
fluff (flŭf), *n.* light down or fur; flocculent matter; the nap of anything; a mistake made esp. in delivering lines, reading a text or playing a piece of music. *v.t.* to make fluffy; to shake or spread (feathers out, as a bird); (*coll.*) to bungle. *v.i.* (*coll.*) to make a mistake in performing. **a bit of fluff**, (*sl.*) a girl. **fluffy**, *a.* **fluffiness**, *n.*
flugelhorn (floo'glhawn), *n.* a valued brass instrument resembling, but slightly larger than, a cornet. [G *Flügel*, wing]
fluid (floo'id), *a.* composed of particles that move freely in relation to each other; capable of flowing, as water; liquid, gaseous; not rigid, stable, or fixed; smooth and graceful. *n.* a liquid or gas, not a solid; a substance whose particles readily move and change their relative positions. **fluid drive**, *n.* (*Eng.*) a system of transmitting power through a change in the momentum of oil. **fluid ounce**, *n.* a British unit of liquid capacity equal to 1/20th of an imperial pint (28·4 ml); a unit equal to 1/16th of a US pint (29·5 ml). **fluidify** (-id'-), **fluidize, -ise**, *v.t.* **fluidity** (-id'-), *n.*
fluke¹ (flook), *n.* a flounder; applied, with distinctive epithet, to other flatfish; a parasitic worm belonging to the Trematoda, found chiefly in the livers of sheep.
fluke² (flook), *n.* the broad holding portion of an anchor; one of the flat lobes of a whale's tail; a barb of a lance, harpoon etc.
fluke³ (flook), *n.* an accidentally successful stroke or act; any lucky chance. *v.t.* to hit or obtain by a fluke. **fluky, flukey**, *a.* obtained by chance, not skill; unsteady, variable. **flukily**, *adv.* **flukiness**, *n.*
flume (floom), *n.* a river; an artificial channel for conveying water to a mill or for some other industrial use; a water chute; (*N Am.*) a deep ravine traversed by a torrent. *v.t.* to carry down a flume.
flummery (flŭm'əri), *n.* a pudding made with oatmeal; anything insipid or out of place; nonsense, humbug; empty compliment.
flummox (flŭm'əks), *v.t.* to perplex, confound.
flung, *past*, *p.p.* FLING.
flunk (flŭngk), *v.t.* (*chiefly N Am.*, *coll.*) to (cause to) fail (a subject, course etc.) *v.i.* to fail, esp. in an examination or course. **to flunk out**, (*chiefly N Am.*, *coll.*) to be expelled for failure.
flunkey (flŭng'ki), *n.* a servant in livery, a footman; one who performs menial duties; a lackey, a toady. **flunkeydom** (-dəm), *n.* **flunkeyish**, *a.* **flunkeyism**, *n.*
fluor (floo'aw), **fluorspar**, (-spah), **fluorite**, (-rīt), *n.* a transparent or subtranslucent, brittle mineral, having many shades of colour, composed of calcium fluoride.
fluor- FLUOR(O)-.
fluorate FLUORINE.
fluorescence (flooəres'əns, flə-), *n.* the emission by certain substances of light or other radiation when bombarded by particles or radiation from another source; the property of

a substance of exhibiting fluorescence; the radiation thus emitted. **fluoresce**, *v.i.* to exhibit fluorescence. **fluorescent**, *a.* having the quality of fluorescence. **fluorescent lamp**, *n.* a lamp consisting of a glass tube with a fluorescent coating inside, which emits light on the passage through the tube of an electric discharge.
fluorine (flooə'rēn), *n.* a non-metallic gaseous element, at. no. 9; chem. symbol F, forming with chlorine, bromine, iodine and astatine the halogen group. **fluoric** (-o'-), *a.* containing fluorine. **fluoric acid**, *n.* **fluoridate**, *v.t.* to add a fluoride to (as drinking water). **fluoridation**, *n.* **fluoride** (-rīd), *n.* a compound of fluorine with an element or radical. **fluorinate**, *v.t.* to treat or cause to combine with fluorine. **fluorination**, *n.*
fluorite (flooə'rīt), FLUOR.
fluor(o)-, *comb. form.* fluorine; fluorescence.
fluorocarbon (flooərōkah'bən), *n.* any of a series of compounds of fluorine and carbon, which are chemically inert and highly resistant to heat.
fluoroscope (flooə'rəskōp), *n.* an apparatus with a fluorescent screen, for directly observing X-ray images. **fluoroscopy** (-ros'kəpi), *n.*
fluorspar FLUOR.
flurry (flŭ'ri), *n.* a squall; a sudden light shower of rain, snow etc.; commotion, agitation, nervous excitement. *v.t.* to agitate, to fluster.
flush[1] (flŭsh), *v.i.* to take wing or start up suddenly. *v.t.* to cause to take wing; force from a place of hiding.
flush[2] (flŭsh), *v.i.* to flow swiftly; to become filled (as pipes) with a sudden rush of water; to become suffused. *v.t.* to cleanse by a rush of water; to flood. *n.* a sudden flow of water; the cleansing of something with a rush of water; a device for flushing e.g. a toilet. **flusher**, *n.* one who flushes drains etc.
flush[3] (flŭsh), *v.i.* to colour as with a rush of blood, to blush; to glow. *v.t.* to cause to colour or become red; to inflame; to encourage, to excite, as with passion. *n.* a sudden flow or rush of blood to the skin, esp. the face causing a redness; any warm colouring or glow; a sudden access of emotion, elation, excitement; a sudden increase or growth; a hot fit, esp. HOT FLUSH under HOT; vigour, bloom, blossoming. **flusher**, *n.*
flush[4] (flŭsh), *a.* full to overflowing; copious, abounding, plentifully supplied, esp. with money; abundant; adjacent, continuous, level, even, on the same plane (with). *adv.* so as to be level. *v.t.* to make even or level. **flushness**, *n.* fullness; abundance.
flush[5] (flŭsh), *n.* a hand of cards all of one suit. **royal flush**, cards in a sequence headed by the ace. **straight flush**, cards in a sequence.
fluster (flŭs'tə), *v.t.* to flurry or confuse; to agitate, to make nervous. *v.i.* to be in an agitated or confused state. *n.* confusion of mind, agitation.
flute (floot), *n.* a tubular wind-instrument with a blow-hole near the end and holes stopped by the fingers or with keys for producing variations of tone, esp. a transverse flute; an organ-stop with a similar tone; a long vertical groove, esp. in the shaft of a column; a long thin French roll of bread; a tall, narrow wine glass. *v.i.* to play a flute; to whistle or sing with a flute-like sound. *v.t.* to play, sing or utter with flute-like tones; to form flutes or grooves in. **fluted**, *a.* **fluting**, *n.* fluted work in pillars etc. **flutist**, *n.* a flautist. **fluty**, *a.* resembling a flute in tone.
flutter (flŭt'ə), *v.i.* to flap the wings rapidly; to hover, flit or move about in a fitful, restless way; to move with quick, irregular motions; to quiver, to vibrate; to beat spasmodically, as the pulse; to be agitated or uncertain. *v.t.* to cause to quiver or flap rapidly; to agitate or alarm. *n.* the act of fluttering; quick, short and irregular vibration; a variation or distortion in pitch occurring at higher frequencies in sound reproduction; potentially dangerous oscillation set up in something, e.g. part of an aircraft, by natural forces; a state of excitement, anxiety, or agitation; disorder, stir; (*coll.*) a gamble, a bet; a venture or speculation. **flutteringly**, *adv.*
fluvial (floo'viəl), **fluviatile** (-til, -tīl), *a.* of or belonging to a river; (*Geol.*) caused by a river; living in rivers.
fluvio-, *comb.form.* relating to a river or rivers.
flux (flŭks), *n.* the act or state of flowing; the motion of a fluid; a state of continuous movement or change; an issue or flowing out, a discharge; the flow of the tide, as opposed to the ebb; an abnormal discharge of fluid matter from the body; any substance which assists the fusion of minerals or metals; the rate of flow of energy, particles or a fluid; the quantity of light falling on an area; the strength of a magnetic field. *v.t.* to melt, to fuse; to facilitate fusion of with a flux. *v.i.* to melt, become fluid.
fluxion (-shən), *n.* continuous variation; (*Math.*) the rate of variation of a varying quantity; (*pl.*) differential calculus. **fluxional, -nary**, *a.*
fly[1] (flī), *n.* (*pl.* **flies**) a two-winged insect, esp. of the genus *Musca*, of which the house-fly, *M. domestica*, is the type; (*loosely*) any winged insect; an artificial fly for fishing. **a fly in the ointment**, a slight flaw, or minor disadvantage, that spoils the quality of something. **a fly on the wall**, an intimate, but unnoticed, observer of events. **like flies**, in vast numbers and offering no resistance. **there are no flies on him/her**, etc., he/she is no fool. **fly-agaric**, *n.* a scarlet-capped mushroom, growing in woods. **fly-blow**, *v.t.* to deposit eggs in, as the blow-fly in meat; to corrupt, to taint. *n.* the egg of a blow-fly. **fly-blown**, *a.* **fly-book**, *n.* a book or case for anglers' flies. **fly-catcher**, *n.* a bird of the genus *Muscicapa*. **fly-fish**, *v.i.* to angle with natural or artificial flies for bait. **fly-fisher**, *n.* **fly-fishing**, *n.* **fly-paper**, *n.* paper prepared to catch or poison flies. **fly-speck**, *n.* the small speck of a fly's excrement; any small speck. **fly-specked**, *a.* **flyspray**, *n.* (an aerosol containing) insecticide. **fly-trap**, *n.* a trap for catching flies; an insectivorous plant, as Venus's fly-trap, *Dionaea muscipula*. **flyweight**, *n.* a professional boxer weighing not more than 112 lb. (50·4 kg); an amateur boxer weighing between 106 and 112 lb. (47·7–50·4 kg); a wrestler weighing not more than 115 lb. (51·7 kg). **fly-whisk**, *n.* a whisk for driving away flies.
fly[2] (flī), *v.i.* (*past* **flew** (floo), *p.p.* **flown** (flōn)) to move through the air with wings; to pilot or ride in an aircraft, spacecraft or balloon; to flutter or wave in the air; to pass or be driven through the air with great speed or violence; to pass, as time, very swiftly; to depart in haste; (*with p.p.* **fled** (fled)) to flee, to run away, to try to escape; to burst or break violently (in pieces); to start, to pass suddenly or violently, to spring (as to arms or into a rage). *v.t.* to cause to fly or float in the air; to pilot (as an aircraft); to travel over by air; to transport by air; to use for air travel (as an airline); to flee from, to avoid, to quit by flight; to make (a hawk, pigeon etc.) fly; to set or keep (a flag) flying. *n.* (*pl.* **flies**) the act or state of flying; a one-horse carriage, a hackney coach; a fly-wheel or a regulating device acting on the same principle; the portion of a vane that shows the direction of the wind; the length of a flag from the staff to the outer edge; the part of a flag farthest from the staff; a flap covering button-holes or a zip; a loose flap for covering the entrance to a tent; (*pl.*) a gallery over the proscenium in a theatre where the curtains or scenes are controlled. **fly-by-night**, *n.* an untrustworthy or irresponsible person; a runaway debtor. *a.* unreliable, untrustworthy. **on the fly**, (*Baseball*) in the air, without bouncing. **to fly a kite** KITE. **to fly at**, to attack suddenly, to rush at with violence or fierceness. **to fly high**, **to be ambitious**. **to fly in the face of** FACE. **to fly off the handle**, HANDLE. **to let fly**, to shoot or throw out; to direct a violent blow (at); to use violent language. **fly-away**, *a.* of hair, tending not to stay in place; flighty, volatile. **flyby** (-bī), *n.* an observation flight, esp. by a

fly spacecraft, past a target or object of investigation at close range. **fly-dumping** FLY-TIPPING. **fly-front,** *n.* a concealed closing on the front of a garment. **fly-half,** *n.* (*Rugby*) the player who acts as a link between the scrum-half and the three-quarter line. **fly-leaf,** *n.* a blank leaf at the beginning or end of a book. **flyover,** *n.* an intersection, esp. of two roads at which the one is carried over the other on a bridge; (*N Am.*) a flypast. **flypast,** *n.* a ceremonial flight by aircraft over a certain point. **fly-posting,** *n.* unauthorized affixing of posters. **fly-sheet,** *n.* a handbill; an extra sheet of canvas that can be fitted over the roof of a tent. **fly-tipping, -dumping,** (*coll.*) *n.* unauthorized dumping of rubbish. **flyway,** *n.* an established bird-migration route. **fly-wheel,** *n.* a heavy-rimmed wheel attached to a machine for regulating the speed by its inertia. **flyer, flier,** *n.* one who flies or flees; a flying jump; (*coll.*) a flying start; (*coll.*) a horse, vehicle, train etc. that goes with exceptional speed; a fly-wheel; (*pl.*) a straight flight of stairs; a speculative attempt or venture.

fly[3] (flī), *a.* (*sl.*) sharp, wide-awake, knowing.

flyer FLY[2].

flying (flī'ing), *a.* moving with or as with wings; moving or adapted to move swiftly; brief, hurried. **flying boat,** *n.* a large seaplane with a buoyant fuselage. **flying bomb,** *n.* a jet-propelled, pilotless aeroplane with a charge of explosive in the head which is detonated when the plane falls with the failure of the propelling jet. **flying bridge,** *n.* a temporary bridge for military purposes; the highest bridge on a ship. **flying buttress,** *n.* an arched or slanting structure springing from solid masonry and serving to support another part of a structure. **flying colours** COLOUR[1]. **flying doctor,** *n.* a doctor in remote areas who uses an aircraft to answer calls. **Flying Dutchman** DUTCHMAN. **flying fish,** *n.* a fish which has the power of sustaining itself in the air for a time by means of its fins. **flying fox,** *n.* an E Indian frugivorous bat belonging to the genus *Pteropus;* (*Austral.*) a conveyor on a suspended wire. **flying-gurnard,** *n.* a fish with large pectoral fins that allow the fish to glide above the surface of the water. **flying jump,** *n.* a jump taken with a running start. **flying-lemur,** *n.* a mammal of SE Asia whose fore and hind limbs are connected by a fold of skin enabling the animal to take flying leaps from tree to tree. **Flying Officer,** *n.* a junior commissioned rank in the RAF equivalent to lieutenant in the army. **flying-phalanger,** *n.* a popular name for the marsupial genus *Petaurus.* **flying picket,** *n.* (a member of) a mobile band of pickets who reinforce local pickets during a strike. **flying saucer,** *n.* a UFO, esp. in the shape of a large disc. **flying squad,** *n.* a mobile detachment of police etc. ready to act swiftly in an emergency. **flying-squirrel,** *n.* a squirrel with a patagium or fold of skin like that of the flying-lemurs, by which it makes flying leaps. **flying start,** *n.* a start in a race in which the competitors are moving at speed as they cross the starting line; a privileged start or beginning.

FM, (*abbr.*) Field Marshal; frequency modulation.

Fm, (*chem. symbol*) fermium.

fm, (*abr.*) fathom.

FO, (*abbr.*) Field Officer; Flying Officer; Foreign Office.

fo., (*abbr.*) folio.

foal (fōl), *n.* the young of an equine animal, as of the horse, ass etc. *v.i., v.t.* to bring forth (a foal).

foam (fōm), *n.* the aggregation of bubbles produced in liquids by violent agitation or fermentation; the similar formation produced by saliva in an animal's mouth or by sweating; froth, spume; chemical froth used in fire-fighting; a light, cellular solid, produced by aerating and then solidifying a liquid; (*poet.*) the sea. *v.i.* to gather, produce or emit foam; to be covered or filled with or as with foam; to pass (away) in foam. *v.t.* to cause to foam; to convert into a foam. **to foam at the mouth,** to be very angry. **foam-rubber,** *n.* rubber of foamlike consistency largely used in upholstery etc. **foamingly,** *adv.* **foamless,** *a.* **foamy,** *a.*

fob[1] (fob), *n.* a watch-pocket, formerly in the waistband of breeches; a chain or strap attaching a watch to a watch-pocket; a seal or ornament on such a chain.

fob[2] (fob), *v.t.* (*past, p.p.* **fobbed**). **to fob off,** to put off with lies or excuses; to pass off as genuine. **to fob off with,** to delude into accepting by a trick.

f.o.b., (*abbr.*) free on board.

focal FOCUS.

fo'c'sle FORECASTLE.

focus (fō'kəs), *n.* (*pl.* **-ci** (-sī), **-cuses**) a point at which rays of light, heat, electrons etc. meet after reflection, deflection or refraction, or from which they appear to diverge; the relation between the eye or lens and the object necessary to produce a clear image; the point from which any activity (as a disease or an earthquake wave) originates; the point on which attention or activity is concentrated; (*Geom.*) one of two points having a definite relation to an ellipse or other curve. *v.t.* (*past, p.p.* **focused, focussed**) to bring (rays) to a focus or point; to adjust (eye or instrument) so as to be at the right focus; to concentrate. *v.i.* to come to a focus; to adjust the eye or an instrument to a particular range. **in focus,** adjusted so as to obtain a clear image; clearly perceived or defined. **focal,** *a.* of, pertaining to or situated at a focus. **focal distance, length,** *n.* the distance between the centre of a lens and the point where initially parallel rays converge. **focal plane,** *n.* a plane containing the foci of the systems of parallel rays passing through a lens. **focalize, -ise,** *v.t.* to focus. **focalization,** *n.*

fodder (fod'ər), *n.* food served to cattle, as hay etc., distinguished from pasture; (*facet.*) food. *v.t.* to feed or supply with fodder.

FoE, (*abbr.*) Friends of the Earth.

foe (fō), *n.* a personal enemy; an opponent, an adversary; an enemy in war; an ill-wisher. **foe-like,** *a.*, *adv.* **foeman,** *n.* an enemy in war.

foehn FÖHN.

foetid (fet'id, fē'-), FETID.

foetus, (*esp. N Am.*) **fetus** (fē'təs), *n.* the young of viviparous animals in the womb, and of oviparous vertebrates in the egg, after the parts are distinctly formed. **foetal,** *a.* pertaining to a foetus. **foetation,** *n.* the formation of a foetus. **foeticide** (-tisid), *n.* the destruction of a foetus.

fog[1] (fog), *n.* coarse, rank grass which has not been eaten off in summer, aftermath; coarse grass remaining through the winter.

fog[2] (fog), *n.* a dense watery vapour rising from land or water and suspended near the surface of land or sea; (a cloud of fine particles causing) murkiness of the atmosphere; (*Phot.*) a cloudiness on a negative; a state of confusion or perplexity. *v.t.* (*past, p.p.* **fogged**) to surround with or as with a fog; to perplex, to bewilder; to make (a negative) cloudy. *v.i.* to become foggy; (*Phot.*) to become cloudy. **fog-bank,** *n.* a dense mass of fog at sea resembling land at a distance. **fog-bound,** *a.* immobilized by fog; covered in fog. **fog-bow,** *n.* a faint bow, resembling a rainbow, produced by light on a fog. **fog-horn,** *n.* an instrument to give warning to ships in a fog. **foglamp,** *n.* a strong light fitted to a vehicle to facilitate driving in fog. **fog-signal,** *n.* a detonator placed on a railway for the guidance of engine-drivers. **foggy,** *a.* thick, murky; full of or subject to fog; obscure, perplexed, indistinct. **not the foggiest,** (*coll.*) not the slightest notion. **foggily,** *adv.* **fogginess,** *n.*

fogy, fogey (fō'gi), *n.* an old-fashioned eccentric person. **fogydom** (-dəm), *n.* **fogyish,** *a.* **fogyism,** *n.*

föhn, foehn (fön), *n.* the warm south wind in the Alps. [G]

foible (foi'bl), *n.* a weak point in one's character; an idiosyncracy, an eccentricity; the part of a sword-blade between the middle and point.

foie gras (fwah grah'), *n.* the fatted liver of an animal, used esp. in the making of pâté. [F]

foil[1] (foil), *n.* an amalgam of quicksilver and tin at the back of a mirror; very thin sheet metal; a thin leaf of metal put under gems to increase their lustre or brighten or alter their colour; that which serves to set off something else to advantage; a rounded leaf-like space or arc in window tracery. *v.t.* to back (glass, crystal etc.) with foil; to set off by contrast; (*Arch.*) to decorate or design with foils. **foiling,** *n.*

foil[2] (foil), *v.t.* to baffle, to frustrate; to throw off the scent; to defeat.

foil[3] (foil), *n.* a straight thin sword, blunted by means of a button on the point, used in fencing.

foist (foist), *v.t.* to introduce surreptitiously or wrongfully; to insert fraudulently; to palm off (on or upon) as genuine.

fol., (*abbr.*) following.

fold[1] (fōld), *n.* a pen or enclosure for sheep; a flock of sheep; (*fig.*) the Church, the flock of Christ. *v.t.* to put or enclose in or as in a fold.

fold[2] (fōld), *v.t.* to double or lay one part of (a flexible thing) over another; to bring together and entwine (as arms, legs); to close (as wings, petals); to clasp (arms etc.) round; to embrace; to enfold, to envelop. *v.i.* to become folded or doubled; to shut in folds; (*Geol.*) to be doubled up; to fail, to cease operations. *n.* a part doubled or laid on another; a bend or doubling, a pleat; a hollow between two parts (as of a fabric); (*Geol.*) a flexure in strata. **to fold in,** (*Cookery*) to mix in gradually and carefully. **fold-away,** *a.* designed to be made compact by folding when not in use. **folding-chair, stool,** *n.* a collapsible chair or stool. **folding-doors,** *n.pl.* two doors hung on opposite side-posts, and meeting in the middle.

-fold, *suf.* forming adjectives and adverbs denoting multiplication, as *fourfold, manifold.*

folder (fōl'də), *n.* one who or that which folds; a holder for loose papers.

foliaceous (fōliā'shəs), *a.* resembling the leaf of a plant; furnished with leaves or leaflike structures; (*Cryst.*) consisting of or splitting into thin laminae.

foliage (fō'liij), *n.* leaves in the aggregate; leaves and branches used for decoration; (*Art, esp. Arch.*) the representation of leaves or clusters of leaves, as ornament. **foliar,** *a.* consisting of or pertaining to leaves.

foliate (fō'liāt), *v.i.* to split or disintegrate into thin laminae. *v.t.* to beat into a leaf or thin plate; to cover over with a thin coat or sheet of tin, quicksilver etc. (as a mirror); to number the leaves of (a book, manuscript). *a.* (-ət), leaf-shaped; furnished with leaves. **foliated,** *a.* foliate; having or splitting into thin layers. **foliation,** *n.* foliating; (*Arch.*) the process of forming leaves; the state of being in leaf; ornamentation by tracery based on the form of a leaf; the foliated texture of a rock.

folic acid (fō'lik. fō'-), *n.* a vitamin of the vitamin B complex found esp. in green vegetables and liver and used in the treatment of anaemia.

folie (fōlē'), *n.* madness; folly. **folie à deux** (a dœ'), the presence of similar delusions in the minds of two closely associated people. [F]

folio (fō'liō), *n.* (*pl.* **-lios** (-ōz)) a sheet of paper folded once; a book of the largest size, whose sheets are folded once, hence, any large volume or work; a page of manuscript; a leaf of paper or other material for writing etc., numbered on the front; a page in an account book, or two opposite pages numbered as one; the number of a page; 72 words of manuscript in legal documents, 90 words in Parliamentary proceedings.

folk (fōk), *n.* (*pl. in constr.*) people, people collectively; (*pl. in constr.*) a particular class of people, as *old folk*; (*pl. in constr.*) members of one's own family; a people, nation or race; folk music; (*in comb.*) people of a specified kind, as *menfolk, kinsfolk. a.* originating among the common people; based on or employing traditional motifs. **folk dance,** *n.* a traditional dance of countryfolk. **folk-etymology,** *n.* a popular but often erroneous derivation of a word. **folk-lore,** *n.* popular superstitions, tales, traditions or legends; the systematic study of such superstitions etc. **folkloric,** *a.* **folklorism,** *n.* **folklorist,** *n.* **folk memory,** *n.* a memory of a distant event passed down through several generations of a community. **folk music,** *n.* the traditional popular music of the common people; modern popular music in the style of this. **folk-singer,** *n.* **folk-song,** *n.* a song or ballad, supposed to have originated among the people and to have been handed down by tradition. **folk-tale,** *n.* a popular myth. **folkways,** *n.pl.* traditional social customs. **folk-weave,** *n.* a fabric with a loose weave. **folks,** *n.pl.* people, folk; family members. **folksy,** *a.* (*chiefly N Am., coll.*) informal, casual, sociable, friendly; (affectedly) traditional in style.

foll., (*abbr.*) following.

follicle (fol'ikl), *n.* a small cavity or sac, as that surrounding the base of a hair; a fruit formed by a single carpel dehiscing by one suture. **follicular** (-lik'ū-), *a.* **folliculate, -lated,** *a.*

follow (fol'ō), *v.t.* to go or come after; to move behind; to pursue, as an enemy; to accompany; to adhere to, to side with, to espouse the cause of; to imitate, to pattern oneself upon; to go after as an admirer or disciple; to go along (a path, road etc.); to engage in, to practise (as a profession); to conform to, act upon (a rule, policy etc.); to come or happen after in point of time, order, rank or importance; to watch the course of; to keep the mind or attention fixed on; to understand, to grasp the meaning of; to result, to be the consequence of; to seek after, to try to attain. *v.i.* to come or go after another person or thing; to pursue; to be the next thing to be done or said; to be a natural consequence, to ensue; to be the logical consequence, to be deducible; to understand. **as follows,** a prefatory formula to a statement, enumeration etc. **follow-my-leader,** (*chiefly N Am.*) **-the-leader,** *n.* a game in which those behind must follow the steps and imitate the actions of the leader. **to follow on,** to continue without break; to continue from where somebody else left off; (*Cricket*) to bat again immediately after completing one's first innings because one is more than a predetermined number of runs behind. **to follow suit** SUIT. **to follow through,** (*Golf, Cricket etc.*) to continue the swing after hitting the ball; to follow to a conclusion. **to follow up,** to pursue closely and steadily; to make further efforts to the same end; to take appropriate action about; to re-examine a patient, or check progress, at intervals after treatment. **follow-on,** *n.* an act of following on. **follow-up,** *n.* a check or checks on a patient's progress; something that reinforces an initial action. **follower,** *n.* one who follows; a disciple, an imitator or adherent; a subordinate, a servant; a fan; an admirer; a Victorian maidservant's sweetheart. **following,** *a.* coming next after, succeeding, now to be mentioned; of wind, blowing in the direction one is travelling. *n.* a body of followers or adherents. *prep.* after.

folly (fol'i), *n.* foolishness, want of understanding or judgment, senselessness; a foolish act, idea or conduct; a structure built for picturesque effect or to gratify the builder's whim; (*derog.*) any building which seems more grand, elaborate or expensive than its purpose warrants; (*pl.*) a theatrical revue featuring girls in glamorous costumes.

foment (fament'), *v.t.* to apply moist heat to; to nourish, to foster, to encourage (trouble etc.). **fomentation** (fōmentā'-), *n.* **fomenter,** *n.*

fond (fond), *a.* tender or loving; doting; cherished. **to be fond of,** to like very much, to love. **fondly,** *adv.* affectionately; foolishly, credulously. **fondness,** *n.*

fondant (fon'dənt), *n.* a soft kind of sweetmeat.
fondle (fon'dl), *v.t.* to caress. *v.i.* to indulge in caresses (with). **fondler**, *n.*
fondue (fon'doo, -dū), *n.* a dish consisting of a hot sauce (usu. of cheese and white wine) into which pieces of bread etc. are dipped, or of cubes of meat which are cooked by dipping into hot oil at table and eaten with a variety of spicy sauces.
font[1] (font), *n.* the vessel or basin to contain water for baptism; the oil-reservoir for a lamp. **fontal**, *a.*
font[2] (font), FOUNT [2].
fontanelle, (*esp N Am.*) **fontanel** (fontənel'), *n.* an interval between the bones of the infant cranium.
food (food), *n.* any substance which, taken into the body, is capable of sustaining or nourishing, or which assists in sustaining or nourishing the living being; victuals, provisions, esp. edibles as distinguished from drink; nutriment for plants; that which nourishes, sustains or is material for. **food additive**, *n.* a substance added to commercially processed food to preserve it, increase its shelf life etc. **food chain**, *n.* a community of organisms thought of as a hierarchy in which each eats the one below and is eaten by the one above. **food poisoning**, *n.* a severe gastrointestinal condition caused by eating food which is naturally poisonous or has been contaminated. **food processor**, *n.* an electrical appliance which chops, shreds, blends etc. food. **foodstuff**, *n.* any thing or material used for food. **food value**, *n.* the degree of nourishment obtained from a particular food. **foodie**, *n.* (*coll.*) a person with an intense interest in (esp. more exotic kinds of) food. **foodism**, *n.* **foodless**, *a.* **foodster**, *n.* a foodie.
fool[1] (fool), *n.* a person without common sense or judgment; a silly person; a dupe; a jester, a buffoon; an idiot, an imbecile. *a.* (*coll.*) foolish, silly. *v.i.* to play the fool; to trifle, to idle. *v.t.* to make a fool of; to dupe, to cheat, to play tricks upon; to waste (time away). **to fool around, about,** to behave foolishly or irresponsibly; to waste time; to trifle (with). **to fool with,** to meddle with in a careless and risky manner. **to make a fool of,** to cause to appear ridiculous; to deceive. **to play, act the fool,** to act like a fool; to act the buffoon. **foolhardy** (fool'hahdi), *a.* daring without sense or judgment, foolishly bold, rash, reckless. **foolhardily**, *adv.* **foodhardihood, foolhardiness**, *n.* **foolproof**, *a.*, *adv.* secure against any ignorant mishandling. **fool's-errand**, *n.* an absurd or fruitless errand or quest; the pursuit of what cannot be found. **fool's gold**, *n.* iron pyrites. **fool's mate**, *n.* the simplest mate in chess. **fool's paradise**, *n.* a state of unreal or deceptive joy or good fortune. **foolery**, *n.* (habitual) folly; the act of playing the fool; absurdity. **fooling**, *n.* **foolish**, *a.* weakminded; lacking judgment; unwise, silly; absurd. **foolishly**, *adv.* **foolishness**, *n.*
fool[2] (fool), *n.* a dish made of fruit, esp. gooseberries, stewed and crushed, then folded together with cream etc.
foolscap (foolz'kap), *n.* a pointed cap with bells, formerly worn by professional jesters; a size of writing-paper 17 × 13½ in. (43·2 × 34·3 cm) or of printing paper, folio, 13½ × 8½ in. (34·3 × 21·6 cm), quarto, 8½ × 6¾ in. (21·6 × 17·1 cm), octavo, 6¾ × 4¼ in. (17·1 × 10·8 cm), named from its original watermark of a fool's cap and bells.
foot (fut), *n.* (*pl.* **feet** (fēt)) the part of the leg which treads on the ground in standing or walking, and on which the body is supported; the part below the ankle; the organ of locomotion or attachment of invertebrate animals; that which serves to support a body; that part of an article of dress which receives the foot; unit of length equalling 12 in. (30·5 cm); the lowest part, the base, the lower end; the bottom; the part of a sewing machine that clamps the fabric in position; foot-soldiers, infantry; (*Pros.*) a set of syllables forming the rhythmical unit in verse; (*pl.* **foots**) sediment, dregs, oil refuse etc. *v.i.* to walk; to dance. *v.t.* to travel over by walking; to perform (a dance); to pay (a bill); to add a new foot to (as to stockings). **a foot in both camps,** connections with two mutually antagonistic groups. **a foot in the door,** a first step towards a desired end; a favourable position from which to advance. **at the feet of,** humbly adoring or supplicating; submissive to; as a disciple or student of. **feet of clay,** initially unsuspected weaknesses. **foot-and-mouth disease,** a contagious eczematous disease chiefly affecting cattle. **my foot!** an exclamation of disbelief. **on foot,** walking; in motion, action or process of execution. **on one's feet,** standing up; in good health; thriving, getting on well. **to catch on the wrong foot,** to take unprepared or at a disadvantage. **to fall on one's feet,** to emerge safely or successfully. **to find one's feet,** to become accustomed to, and able to function effectively in, new circumstances. **to foot it,** to go on foot; to dance. **to get off on the wrong foot,** to make a bad start, esp. in personal relations with someone. **to have one foot in the grave,** to be near death, very old or moribund. **to have one's feet on the ground,** to be realistic, sensible or practical. **to put a foot wrong,** to make a mistake. **to put one's best foot forward,** to step out briskly; to try to show oneself at one's best. **to put one's foot down,** to be firm, determined. **to put one's foot in it,** to blunder. **to sweep off one's feet,** to enrapture, to make a complete and sudden conquest of. **to think on one's feet,** to react to situations as they arise. **under foot,** on the ground. **foot-board**, *n.* a platform for a footman behind a carriage; a step for getting into or out of a vehicle; a foot-plate; a treadle; a board at the foot of a bed. **foot-bridge**, *n.* a narrow bridge for pedestrians. **footfall**, *n.* the sound of a footstep. **foot-fault**, *n.* (*Lawn tennis*) the act of overstepping the baseline when serving. **foothill**, *n.* a hill lying at the base of a range of mountains. **foothold**, *n.* a stable place for the feet; a position of stability or security. **footlights**, *n.pl.* a row of lights, screened from the audience, in front of the stage of a theatre. **footloose**, *a.* free, unbound by ties. **footman**, *n.* a male domestic servant in livery; a foot-soldier. **footmark**, *n.* a footprint. **footnote**, *n.* a note at the bottom of the page of text. **footpad**, *n.* a highwayman who robs on foot. **footpath, -road, -way**, *n.* a narrow path or way for pedestrians only. **footplate**, *n.* a platform for the driver and fireman on a locomotive. **foot-pound**, *n.* a unit of energy, the amount that will raise one pound avoirdupois one foot. **footprint**, *n.* the mark or print of a foot; an area in which something (as a spacecraft) lands; over which something (as a communication satellite) is operational or has an effect; the space occupied by a (desk-top) computer. **foot-race**, *n.* a running-match on foot. **foot-rest**, *n.* a support for the feet. **foot-rot**, *n.* a disease in the feet of sheep and cattle, characterized by an abnormal growth. **foot-soldier**, *n.* an infantry soldier. **footsore**, *a.* having the feet sore or tender. **footstep**, *n.* the act of stepping or treading with the feet; tread; a footprint; the sound of the step of a foot; (*pl.*) traces of a course pursued or actions done. **footstool**, *n.* a stool for supporting the feet. **foot-warmer**, *n.* a device for warming the feet. **footwear**, *n.* shoes, boots etc.; **footwork**, *n.* skilful use of the feet in boxing, dancing etc.; clever manoeuvring, esp. of an evasive kind. **footage**, *n.* length in feet; film on which a scene has been shot; the length of a film (in feet). **footed**, *a.* having feet, usu. in comb. as *swift-footed, four-footed*. **footer** (fut'ə), *n.* (*coll.*) the game of football. **-footer**, *comb. form.* one of a specified length or height in feet. **footsie** (-si), *n.* (*coll.*) erotic or flirtatious touching with the feet.
football (fut'bawl), *n.* an inflated bladder encased in leather used in the game of football: a game between two teams in which a football is kicked, or handled and kicked, to score goals or points, there being many different varieties of the game; a contentious issue, esp. one

footing 320 **forebrain**

which is bandied about between opposing groups. **Football Association**, *n.* the body founded in 1863 to make rules, supervise and preside over Association football in Britain. **Football League**, *n.* an organized collection of Association Football clubs founded in 1888 to arrange matches and supervise the business arrangements of its constituents. **football pools**, (*n.pl.*) a form of gambling based on forecasting the results of football matches. **footballer**, *n.*
footing (fut'ing), *n.* a place for standing or putting the feet on; a firm or secure position; relative position, status or condition; (*Arch.*) a course at the base or foundation of a wall.
footle (foo'tl), *v.i.* (*coll.*) to trifle; to potter about aimlessly. **footling**, *n.*, *a.*
foots FOOT.
foozle (foo'zl), *v.t.* to make a mess of, to bungle. *n.* (*Golf*) a bungled stroke. **foozler**, *n.* **foozling**, *a.*
fop (fop), *n.* a man over-fond of dress, a dandy. **foppery**, *n.* **foppish**, *a.* **foppishly**, *adv.* **foppishness**, *n.*
for (faw, *unstressed* fə), *prep.* in the place of, instead of; in exchange against, as the equivalent of; as the price or requital or payment of; in consideration of, by reason of; because of, on account of, in favour of, on the side of; in order to, with a view to; appropriate or suitable to; toward, tending toward, conducive to; to fetch, to get, to save; to attain, to reach, to arrive at; (*sl.*) against; on behalf of, for the sake of; with regard to, in relation to; as regards; so far as; as, as being, in the character of; to the amount or extent of; at the cost of; in spite of, notwithstanding; in comparison of, contrast with; during; to prevent; because of. *conj.* since, because; seeing that; in view of the reason that. **as for** AS. **to be for it**, (*coll.*) to be marked for reprimand or punishment. **for all that**, nevertheless; in spite of all that. **for all the world**, exactly, completely. **for as much as** FORASMUCH. **for good**, for ever, permanently. **for short**, as an abbreviation or contraction. **once (and) for all**, finally.
for-, *pref.* away, off, as in *forget, forgive;* negative, prohibitive or privative, as in *forbear, forbid, forsake;* amiss, badly, as in *fordo, forshapen;* intensive, as in *forlorn, forspent.*
forage (fo'rij), *n.* food for horses and cattle; the act of foraging. *v.i.* to seek for or to collect forage; to hunt for supplies; to rummage (about). *v.t.* to overrun in order to collect forage; to ravage, to plunder; to obtain for forage; to supply with forage or food. **forage-cap**, *n.* a military undress cap. **forager**, *n.*
foramen (fərā'men), *n.* (*pl.* **foramina** (-ram'inə)) a small natural opening, passage or perforation in parts of plants and animals. **foraminate** (-ram'ināt), **-nated** (-nātid), *a.* **foraminifer** (forəmin'ifə), *n.* one of the Foraminifera. **Foraminifera** (-rāminif'ərə), *n.pl* (*Zool.*) a large group of Protozoa, esp. an order of Rhizopoda, the body of which is contained within a calcareous shell, perforated by numerous foramina. **foraminiferal, -iferous** (-nif'-), *a.* **foraminous** (-ram'i-), *a.*
forasmuch as (forəzmŭch'), *conj.* seeing that; since.
foray (fo'rā), *v.t., v.i.* to ravage, to make a raid (on). *n.* a predatory expedition, a raid.
forbear[1] (fəbeə'), *v.i.* (*past* **-bore** (-baw'), *p.p.* **-borne** (-bawn')) to refrain or abstain from; to bear with, to treat with patience. *v.i.* to refrain or abstain (from); to be patient, to refrain from feelings of resentment. **forbearance**, *n.* **forbearingly**, *adv.*
forbear[2] (faw'beə), FOREBEAR.
forbid (fəbid'), *v.t.* (*past* **-bad, -bade** (-băd), *p.p.* **-bidden**) to order not to do; to interdict, to prohibit; to exclude, to oppose. **forbidden**, *a.* prohibited, interdicted. **forbidden fruit**, *n.* the fruit of the tree of the knowledge of good and evil, which Adam was commanded not to eat (Gen. ii.17); anything desired but pronounced unlawful. **forbidder**, *n.*

forbidding, *a.* uninviting, disagreeable; giving rise to aversion or dislike; threatening, formidable. **forbiddingly**, *adv.* **forbiddingness**, *n.*
forbore, *past*, **-borne**, *p.p.* FORBEAR[1].
force[1] (faws), *n.* strength, energy, active power; military or naval strength; an organized body (of troops, police, workers etc.); (*pl.*) troops; power exerted on a person or object; violence, coercion, compulsion; unlawful violence; efficacy, validity; significance, weight, import, full meaning; persuasive or convincing power; energy, vigour, animation, vividness; that which produces or tends to produce a change of velocity in a body at rest or in motion; that which influences, brings a change, or exerts an effect. *v.t.* to constrain by force (to do or to forbear from); to compel, to constrain; to use violence to achieve or acquire; to rape; to strain, to distort; to impose or impress (upon); to bring about, to accomplish, or to make a way by force; to break open by force; to stimulate artificially, to cause to grow or ripen prematurely; (*Cards*) to compel (a player) to play in a certain way, to compel (a certain card) to be played. **by force**, by compulsion. **in force**, in operation, valid; (*Mil.*) in large numbers. **the Force**, the police. **to force someone's hand** HAND. **to force the pace** PACE[1]. **force-feed**, *v.t.* (*past, p.p.* **force-fed**) to feed forcibly. **force-land**, *v.i.* to make a forced landing. **force-pump, forcing-pump**, *n.* a pump which delivers water under pressure, so as to raise it to an elevation above that attainable by atmospheric pressure. **forced**, *a.* constrained, affected; unnatural. **forced landing**, *n.* a landing of an aircraft elsewhere than at one's destination, owing to mechanical failure or other mishap. **forcedly** (-sid-), *adv.* **forceful**, *a.* full of or possessing force, forcible. **forcefully**, *adv.* **forcefulness**, *n.* **forceless**, *a.* **forcible**, *a.* done or brought about by force; having force, powerful, efficacious, impressive. **forcibleness**, *n.* **forcibly**, *adv.*
force[2] (faws), *n.* (*North.*) a waterfall.
force majeure (faws mazhœ'), *n.* superior power; circumstances not under one's control. [F]
forcemeat (faws'mēt), *n.* meat chopped fine and highly seasoned, used as stuffing or served up alone.
forceps (faw'səps), *n.* (*pl.* **forceps**) a pair of tongs, pincers or pliers for holding or extracting anything; (*Anat., Zool.*) an organ shaped like a pair of forceps.
forcible FORCE.
ford (fawd), *n.* a shallow part of a river where it may be crossed by wading. *v.t.* to cross (as water) by wading. **fordable**, *a.* **fordless**, *a.*
fore (faw), *prep.* before; (*chiefly now in asseverations*) for, in the presence of, as *fore God. adv.* in the front part; (*Naut.*) in or towards the bows. *a.* being in front (of some other thing); being the front part; anterior, prior, former. *n.* the front part; something at the front. *int.* (*Golf*) before, beware in front (warning to persons standing in the direction of a drive). **fore-and-aft**, at, along or over the whole length of a ship from stem to stern. **fore-and-aft rigged**, having sails set lengthwise to the ship, as opposed to square sails set on yards. **to the fore**, to the front, prominent, conspicuous; ready, available, forthcoming.
fore-, *pref.* (*chiefly with verbs*) before, earlier, beforehand, as *foreconceive, foreordain;* in front, as *forecourt;* front part of, as *forearm.*
forearm[1] (fawrahm'), *v.t.* to prepare beforehand for attack or defence.
forearm[2] (faw'rahm), *n.* the anterior part of the arm, between the wrist and elbow.
forebear, forbear (faw'beə), *n.* a forefather, an ancestor.
forebode (fəbōd'), *v.t.* to foretell, predict; to prognosticate, to portend; to feel a presentiment of. *v.i.* to prognosticate, esp. evil. **foreboder**, *n.* **foreboding**, *n.* prophecy, presage or anticipation, esp. of evil. **forebodingly**, *adv.*
forebrain (faw'brān), *n.* the front part of the brain.

forecast (faw′kahst), *v.t.* (*past*, *p.p.* **forecast**, *erron.* **forecasted**) to calculate beforehand; to foresee, to predict; to be an early sign of. *v.i.* to predict or calculate future events. *n.* a prediction or calculation of probable events, esp. regarding future weather. **forecaster**, *n.*

forecastle, fo'c'sle (faw′kahsl, fōk′sl), *n.* a short upper deck forward, formerly raised to command the enemy's decks; in merchant-ships, a forward space below deck where the crew live.

foreclose (fawklōz′), *v.t.* to shut out, exclude or bar; to preclude; to deprive (the mortgagor) of his or her equity of redemption on failure to pay money due on (a mortgage). *v.i.* to foreclose a mortgage. **foreclosure** (-zhə), *n.* the act of foreclosing.

forecourt (faw′kawt), *n.* an open or paved area in front of a building, esp. a filling station.

foredeck (faw′dek), *n.* the forepart of a deck; the deck in the forepart of a ship.

foredoom (fawdoom′), *v.t.* to doom beforehand; to predestinate.

forefather (faw′fahdhə), *n.* an ancestor.

forefend (fawfend′), FORFEND.

forefinger (faw′fing-gə), *n.* the finger next to the thumb, also called the first or index finger.

forefoot (faw′fut), *n.* (*pl.* **-feet**) a front foot of a quadruped; the forward end of a vessel's keel.

forefront (faw′frŭnt), *n.* the extreme front, the foremost part or position.

foregather (fawgä′dhə), FORGATHER.

forego[1] (fawgō′), *v.t., v.i.* (*past* **-went** (went′), *p.p.* **-gone** (-gon)) to go before, to precede in time, order or place. **foregoing**, *a.* preceding, previously mentioned. **foregone**, *a.* past; preceding; determined before. **foregone conclusion**, *n.* a conclusion determined on beforehand or arrived at in advance of evidence or reasoning; a result that might be foreseen.

forego[2] (fawgō′), FORGO.

foreground (faw′grownd), *n.* the nearest part of a view; the part of a picture which seems to lie nearest the spectator; a prominent position.

forehand (faw′hand), *n.* a forehand stroke; the side on which such strokes are made; that part of a horse before the rider; the upper hand, superiority, advantage. *a.* (*Tennis etc.*) with the palm of the hand facing in the direction of the stroke.

forehead (fo′rid, faw′hed), *n.* that part of the face from the eyebrows upwards to the hair.

foreign (fo′rin), *a.* belonging to, connected with, or derived from another country or nation; concerned with or involving other nations; alien, strange, not belonging (to); having no connection with, irrelevant, inappropriate. **foreign body**, *n.* a substance occurring in an organism or tissue where it is not normally found. **foreign correspondent**, *n.* a representative of a newspaper sent to a foreign country to report on its politics etc. **foreign exchange**, *n.* (trading in) foreign currencies. **foreign legion**, *n.* a unit of foreign volunteers serving within a national regular army. **foreign minister, secretary**, *n.* a government minister in charge of relations with foreign countries. **Foreign Office**, *n.* the government department for foreign affairs. **foreigner**, *n.* a person born or belonging to a foreign country or speaking a foreign language; a foreign ship, an import or production from a foreign country; a stranger, an outsider. **foreignism**, *n.* **foreignness**, *n.* **foreignize, -ise**, *v.t., v.i.*

forejudge (fawjŭj′), *v.t.* to judge before trial or decide before hearing the evidence. **forejudgment**, *n.*

foreknow (fawnō′), *v.t.* (*past* **-knew** (-nū′), *p.p.* **-known** (-nōn′)) to know beforehand. **foreknowledge** (-nol′ij), *n.* prescience; knowledge of a thing before it happens.

foreland (faw′lənd), *n.* a point of land extending into the sea, a promontory; a strip of land outside of or in front of an embankment etc.

foreleg (faw′leg), *n.* a front leg.

forelock (faw′lok), *n.* a lock of hair growing over the forehead.

foreman (faw′mən), *n.* the person who acts as chairman and spokesman for a jury; a worker supervising others. **forewoman** (-wumən), *n. fem.*

foremast (faw′mahst), *n.* the mast nearest the bow of a vessel. **foremastman, -hand, -seaman**, *n.* a common sailor.

forementioned (fawmen′shənd), *a.* already mentioned.

foremost (faw′mōst), *a.* first in time, place, order, rank or importance; chief, most notable. *adv.* in the first place; first, before anything else.

forename (faw′nām), *n.* a name preceding the surname; a Christian name. **forenamed**, *a.* named or mentioned before.

forenoon (fawnoon′), *n.* (*chiefly Sc.*) the early part of the day, from morning to noon.

forensic (fəren′sik), *a.* pertaining to courts of judicature, or to public debate; used in debates or legal proceedings. *n.* (*N Am.*) an argumentative thesis at a college. **forensic medicine**, *n.* the science of medicine in its relation to law, medical jurisprudence. **forensically**, *adv.*

foreordain (fawrawdān′), *v.t.* to ordain beforehand, to predestinate. **foreordination** (-dinā′-), *n.*

forepart (faw′paht), *n.* the first or most advanced part; the earlier part.

foreplay (faw′plā), *n.* sexual stimulation preceding intercourse.

forequarter (faw′kwawtə), *n.* the front half of the side of a carcass, as of beef; (*pl.*) the forelegs, shoulders and chest of a horse.

forerun (fawrŭn′), *v.t.* (*past* **-ran** (-ran′), *p.p.* **-run**) to precede; to betoken, to usher in. **forerunner**, *n.* a messenger sent before; a precursor; a predecessor, an ancestor; an omen.

foresail (faw′sāl, -sl), *n.* the principal sail on the foremast.

foresee (fawsē′), *v.t.* (*past* **-saw** (-saw′), *p.p.* **-seen**) to see beforehand; to know beforehand, to have prescience of. **foreseeable**, *a.* **foreseer**, *n.*

foreshadow (fawshad′ō), *v.t.* to typify or indicate beforehand. **foreshadower**, *n.*

foresheet (faw′shēt), *n.* the rope holding the lee corner of a foresail; (*pl.*) the space in a boat forward of the foremost thwart, usu. covered with a grating.

foreshore (faw′shaw), *n.* the part of the shore lying between high- and low-water marks; the ground between the sea and land that is cultivated or built upon.

foreshorten (fawshaw′tn), *v.t.* in drawing or painting, to represent (figures or parts of figures that project towards the spectator) so as to give a correct impression of form and proportions.

foresight (faw′sīt), *n.* prescience, forethought; provident care for the future, prudence, precaution; the muzzle-sight of a gun. **foresighted**, *a.* **foresightedly**, *adv.* **foresightedness**, *n.*

foreskin (faw′skin), *n.* the prepuce, the loose skin covering the end of the penis.

forest (fo′rist), *n.* an extensive wood or tract of wooded country; a wild uncultivated tract of ground partly covered with trees and underwood; a large tract of country set apart for game and hunting, in many cases orig. a royal hunting-ground; something resembling a forest. *v.t.* to plant with trees; to convert into a forest. **forestation**, *n.* **forester**, *n.* one who has charge of a forest; an inhabitant of a forest; one who looks after the trees on an estate; a member of the Forester's Benefit Society. **forestry**, *n.* the act or science of cultivating trees and forests; the management of growing timber; (*poet.*) woodland.

forestall (fawstawl′), *v.t.* to hinder or prevent by anticipation; to anticipate; to be beforehand with; to buy up (commodities) beforehand so as to control the sale. **to**

forestall the market, to engross or buy up commodities, so as to obtain the control of the market. **forestaller**, *n.*
forestay (faw'stā), *n.* (*Naut.*) a strong rope, reaching from the foremast head to the bowsprit end, to support the mast.
forester, forestry FOREST.
foretaste[1] (faw'tāst), *n.* experience or enjoyment (of) beforehand; anticipation.
foretaste[2] (fawtāst'), *v.t.* to taste beforehand; to anticipate enjoyment (of).
foretell (fawtel'), *v.t.* (*past, p.p.* **-told** (-tōld')) to predict, to prophesy; to foreshadow. **foreteller**, *n.*
forethought (faw'thawt), *n.* consideration beforehand; premeditation; foresight, provident care.
foretoken[1] (fawtō'kən), *v.t.* to foreshadow.
foretoken[2] (faw'tōkn), *n.* a token beforehand, an omen.
foretooth (faw'tooth), *n.* (*pl.* **-teeth** (-tēth)) a front tooth.
foretop (faw'top), *n.* (*Naut.*) the top or platform at the head of the foremast. **fore-topmast**, *n.* the mast at the head of the foremast, and surmounted by the **fore-topgallant-mast. fore-topsail, fore-topgallant-sail**, *n.*
forever (fərе'və), *adv.* for ever; always, incessantly. *n.* (*poet.*) eternity; a very long time. **forevermore**, *adv.*
forewarn (faw-wawn'), *v.t.* to warn or caution beforehand; to give notice to beforehand.
forewoman FOREMAN.
foreword (faw'wœd), *n.* a preface, a short introduction.
forfeit (faw'fit), *n.* that which is lost through fault, crime, omission or neglect; a penalty, a fine, esp. a stipulated sum to be paid in case of breach of contract; (*pl.*) a game in which for every breach of the rules the players have to deposit some article, which is subsequently redeemed by the performance of a playful task or ceremony; the article so deposited. *a.* lost or alienated through fault or crime. *v.t.* to lose the right to or possession of by fault, crime, omission or neglect; to lose; to cause to lose, to confiscate. **forfeitable**, *a.* **forfeiter**, *n.* **forfeiture** (-chə), *n.* the act of forfeiting; that which is forfeited; a penalty or amercement.
forfend, forefend (fawfend'), *v.t.* to avert, to ward off.
forgather, foregather (fəga'dhə), *v.i.* to meet or associate (with); to meet together, to assemble.
forgave, *past* FORGIVE.
forge[1] (fawj), *n.* the workshop of a smith; a blacksmith's open fireplace or hearth where iron is heated by forced draught; a furnace or hearth for making wrought iron; a workshop. *v.t.* to shape, form or fabricate by heating and hammering; to make, form or construct; to make, invent, or imitate fraudulently, to counterfeit. *v.i.* to commit forgery. **forgeable**, *a.* **forger**, *n.* **forgery** (-jə-), *n.* the act of forging, counterfeiting or falsifying; a fraudulent imitation; a deception. **forging**, *n.* that which is forged; a piece of forged metal work.
forge[2] (fawj), *v.i.* to move steadily (forward or ahead); to move at an increased speed (forward or ahead).
forget (fəget'), *v.t., v.i.* (*past* **-got** (-got'), *p.p.* **-gotten**, *poet.* **-got**) to lose remembrance of; to put out of mind purposely; to fail to remember through inadvertence; to neglect (to do something). **to forget oneself**, to lose one's self-control, to behave unbecomingly; to act unselfishly. **forgetful**, *a.* **forgetfully**, *adv.* **forgetfulness**, *n.* **forgettable**, *a.* **forgetter**, *n.*
forget-me-not (fəget'minot), *n.* a small plant of the genus *Myosotis*, esp. *M. palustris*, with bright blue flowers.
forgive (fəgiv'), *v.t.* (*past* **-gave** (-gāv'), *p.p.* **-given**) to pardon or remit (as an offence or debt); not to exact the penalty for; to pardon, not to punish (a person or offence, or a person his or her offence); to cease to feel resentment towards. *v.i.* to show forgiveness. **forgivable**, *a.* **forgiveness**, *n.* the act of forgiving; a disposition to forgive; remission, pardon. **forgiver**, *n.* **forgiving**, *a.* disposed to forgive; merciful, gracious. **forgivingly**, *adv.* **forgivingness**, *n.*

forgo, forego (fəgō'), *v.t.* (*past* **forwent** (-went'), *p.p.* **forgone** (-gon')) to go without; to give up, renounce, relinquish.
forgotten, forgot, *p.p.* FORGET.
forint (fo'rint), *n.* the monetary unit of Hungary since 1946, equivalent to 100 fillér.
fork (fawk), *n.* an agricultural implement terminating in two or more prongs, used for digging, impaling, lifting, carrying or throwing; a pronged implement used in cooking or at table; anything of a similar form; a forking or bifurcation; a diverging branch; a confluent, a tributary; a point where a road divides into two; a forked support into which a bicycle wheel fits; (*Chess, Draughts*) a simultaneous attack on two pieces. *v.t.* to raise, pitch, dig or break up with a fork; to make sharp or pointed; (*Chess, Draughts*) to attack two pieces so that only one can escape. *v.i.* to divide into two; to send out branches. **to fork out, over**, (*sl.*) to hand or deliver over; to produce the cash for. **forklift (truck)**, *n.* a vehicle which raises and transports objects on mobile steel prongs. **forked**, *a.* dividing into branches, branching, cleft, bifurcated; terminating in points or prongs.
forlorn (fəlawn'), *a.* deserted, abandoned; helpless, wretched, hopeless; deprived, bereft (of). **forlornly**, *adv.* **forlornness**, *n.*
forlorn hope (fəlawn' hōp), *n.* a detachment of people selected for some service of uncommon danger; a bold, desperate enterprise. [after Dut. *verloren hoop*, lit. lost troop]
form (fawm), *n.* the shape or external appearance of anything apart from colour; configuration, figure, esp. of the human body; particular arrangement, disposition, organization or constitution; established practice or method; a rule of procedure, ceremony or ritual; the mode in which anything is perceptible to the senses or intellect; kind, specific state, species, variety, variation; a specific shape of a word as regards inflection, spelling or pronunciation; a shape, mould or model upon which a thing is fashioned; a customary method or formula, a fixed order of words; a document with blanks to be filled in; (*Art*) style or mode of expression, as opposed to content or subject-matter; orderly arrangement of parts, order, symmetry; behaviour according to accepted rules or conventions; good physical condition or fitness, a good state of health or training; past performance of a horse, athlete etc.; ability to perform; a long seat without a back; a class in a public or secondary school considered as an administrative unit, all the pupils in a particular year or a subdivision of a year group; the seat or bed of a hare; a body of type composed and locked in a chase ready for printing; literary nature of a book etc., as distinct from the subject; that which differentiates matter and generates species; (*Phil.*) the essential nature of a thing; (*coll.*) a criminal record. *v.t.* to give form or shape to; to arrange in any particular manner; to make, construct or create; to model or mould to a pattern; to train, to instruct, to mould or shape by discipline; to conceive, devise, construct (ideas etc.); to articulate; to become; to be the material for; to be or constitute (a part or one of); (*Mil.*) to combine into (a certain order); (*Gram.*) to make by derivation or by affixes or prefixes. *v.i.* to come into existence; to assume a form. **bad, good form**, bad, good manners; ill, good breeding. **in, on form**, showing one's talent to advantage, playing or performing well. **off form**, playing or performing below one's usual standard. **form letter**, *n.* (a copy of) a standard letter sent to many different people, often with relevant individual details added. **form-master, -mistress, -teacher**, *n.* the teacher with general administrative and tutelary responsibility for a form. **formwork**, *n.* framing to hold concrete in place whilst setting.
-form, *suf.* like, having the shape of, as *cruciform, dendriform;* having a certain number of forms, as *multiform*.

formal (faw'məl), *a.* made, performed or done according to established forms; orderly, regular; explicit, definite; observant of established form, ceremonious, punctilious, precise; conventional, perfunctory; of or pertaining to the outward form as opposed to reality, outward; (*Log.*) pertaining to form as opposed to matter; (*Phil.*) pertaining to the formative essence that makes a thing what it is, essential, not material. **formalism**, *n.* the quality of being formal; formality, as in religion. **formalist**, *n.* **formalistic** (-lis'-), *a.* **formality** (-mal'-), *n.* the condition or quality of being formal; conformity to custom, rule or established method; conventionality, mere form; an established order or method, an observance required by custom or etiquette. **formalize, -ise,** *v.t.* to render formal; to formulate. **formalization**, *n.* **formally**, *adv.*
formaldehyde (fəmal'dihīd), *n.* a colourless gas generated by the partial oxidation of methyl alcohol, and used as an antiseptic and disinfectant.
formalin (fawm'əlin), *n.* a solution of formaldehyde used as an antiseptic and as a preservative for biological specimens.
formant (faw'mənt), *n.* a component of a sound which gives it its particular tone colour or quality.
format (faw'mat), *n.* the external form and size of a book or other publication; the general plan, arrangement and style of e.g. a television programme; (*Comput.*) the arrangement of data on a disc etc. *v.t.* (*past, p.p.* **formatted**) to arrange in a specific format; (*Comput.*) to prepare (a disc etc.) for the reception of data.
formation (fəmā'shən), *n.* the act or process of forming or creating; the state of being formed or created; the manner in which anything is formed; conformation, arrangement, disposition of parts, structure; a thing formed, regarded in relation to form or structure; a group of rocks or strata of common origin, structure or physical character; an arrangement of troops, aircraft, ships etc. **formative** (faw'mə-), *a.* having the power of giving form, shaping; pertaining to formation, growth or development; of combining forms, prefixes etc., serving to form words, inflectional, not radical. *n.* a formative affix or element.
forme (fawm), (*Print.*) FORM.
former (faw'mə), *a.* preceding in time; mentioned before something else, first-mentioned (of two); past, earlier, ancient, bygone. *n.* the first mentioned (of two). **formerly**, *adv.* in former times.
formic (faw'mik), *a.* (*Chem.*) pertaining to or produced by ants; derived from formic acid. **formic acid**, *n.* an acid found in the fluid emitted by ants, in stinging-nettles etc., and now obtained from oxalic acid distilled with glycerin. **formicary**, *n.* an ant-hill. **formication**, *n.* irritation of the skin like the crawling of ants. **formyl** (-mil), *n.* (*Chem.*) the radical theoretically constituting the base of formic acid.
Formica® (fəmī'kə), *n.* a laminated plastic used for surfacing materials and other purposes.
formidable (faw'midəbl, -mid'-), *a.* tending to excite fear; to be feared; dangerous to encounter; difficult to resist, overcome or accomplish; awe-inspiring; to be reckoned with. **formidableness, formidability** (-bil'-), *n.* **formidably**, *adv.*
formless (fawm'lis), *a.* without form, shapeless; having no regular form. **formlessly**, *adv.* **formlessness**, *n.*
formula (faw'mūlə), *n.* (*pl.* **-lae** (-lē), **-las**) a prescribed form of words; a formal enunciation of faith, doctrine, principle etc.; a compromise solution to a dispute, an agreed form of words; a fixed rule, a set form, a conventional usage; a prescription, a recipe; a milk mixture or substitute used as baby food; (*Chem.*) an expression by means of symbols of the elements of a compound; the expression of a rule or principle in algebraic symbols; a technical specification which determines the class in which a racing car competes. **formulaic** (-lā'ik), *a.* **formu-**

larize, -ise, *v.t.* to formulate. **formularization**, *n.* **formulary**, *a.* of the nature of a formula. *n.* a collection of formulas; a book containing stated and prescribed forms, esp. relating to religious belief or ritual; a formula. **formulate**, *v.t.* to express in a formula; to set forth in a precise and systematic form; to devise. **formulation**, *n.* **formulize, -ise,** *v.t.* to formulate. **formulization**, *n.*
formwork FORM.
formyl FORMIC.
fornicate (faw'nikāt), *v.i.* to commit fornication. **fornication**, *n.* sexual intercourse outside marriage; (*Bibl.*) applied to idolatry, incest or adultery. **fornicator**, *n.*
forsake (fəsāk'), *v.t.* (*past* **-sook** (-suk'), *p.p.* **-saken**) to leave, to abandon; to renounce, to reject. **forsaker**, *n.*
forsooth (fəsooth'), *adv.* (*chiefly iron.*) in truth, certainly, doubtless.
forswear (fəsweə'), *v.t.* (*past* **-swore** (-swaw'), *p.p.* **-sworn**) to abjure; to renounce upon oath or with protestations. *v.i.* to swear falsely. **to forswear oneself,** to perjure oneself.
Forsythia (fawsī'thiə), *n.* a genus of oleaceous shrubs bearing numerous yellow flowers in early spring before the leaves. [W. *Forsyth*, 1737–1804]
fort (fawt), *n.* a fortified place, esp. a detached outwork or an independent fortified work of moderate extent.
forte¹ (fawt), *n.* the strong part of a sword blade, i.e. from the hilt to the middle; (-ti), a person's strong point; (-ti), that in which one excels.
forte² (faw'ti), *adv.* (*Mus.*) with loudness or force. **forte forte**, *adv.* very loud. **forte piano**, *adv.* loudly, then softly.
fortepiano, *n.* an early form of pianoforte. [It.]
forth (fawth), *adv.* out; out into view; out from home; out of doors; forward in place, time or order; indefinitely forward, in time. **and so forth,** and the rest, and so on, and the like. **back and forth,** to and fro. **forthcoming,** *a.* ready to appear, or to be brought forward; approaching, soon to take place; available; of people, communicative, responsive. **forthright,** *a.* direct, outspoken, to the point. *adv.* straightforward; at once, straightway. *n.* a direct course. **forthwith** (-with', -dh'), *adv.* immediately; without delay.
forties (faw'tiz), *n.pl.* the period of time between one's 40th and 50th birthdays; the range of temperature between 40 and 50 degrees; the period of time between the 40th and 50th years of a century. **the roaring forties,** the stormy part of the Atlantic between 39° and 50° S lat.
fortieth (faw'tiəth), *n.* one of 40 equal parts. *n., a.* (the) last of 40 (people, things etc.); the next after the 39th.
fortify (faw'tifī), *v.t.* to make strong; to give power or strength to; to invigorate; to encourage; to add alcoholic strength to; to enrich (a food) by adding vitamins etc.; to confirm, to corroborate; (*Fort.*) to strengthen or secure by forts, ramparts etc.; to make defensible against the attack of an enemy. *v.i.* to raise fortifications. **fortifiable**, *a.* **fortification** (-fi-), *n.* the act, art or science of fortifying a place or position against the attacks of an enemy; a defensive work, a fort; (*pl.*) works erected to defend a place against attack; increasing the strength of wine with alcohol; something that fortifies. **fortifier**, *n.*
fortissimo (fawtis'imō), *adv.* (*Mus.*) very loud. [It.]
fortitude (faw'titūd), *n.* strength, esp. that strength of mind which enables one to meet danger or endure pain with calmness. **fortitudinous** (-tū'din-), *a.*
fortnight (fawt'nīt), *n.* a period of two weeks or 14 days. **fortnightly,** *a.* happening once a fortnight. *adv.* once a fortnight; every fortnight. *n.* a fortnightly publication.
Fortran, FORTRAN (faw'tran), *n.* a high-level computer language used esp. for mathematical and scientific purposes. [*formula translation*]
fortress (faw'tris), *n.* a fortified place, esp. a large fort or strongly fortified town accommodating a large garrison and forming a permanent stronghold.

fortuitous (fətū'itəs), *a.* happening by chance; fortunate. **fortuitously**, *adv.* **fortuitousness**, *n.* **fortuitist**, *n.* **fortuity**, *n.* a chance occurrence; fortuitousness.

fortunate (faw'chənət), *a.* happening by good luck; bringing or presaging good fortune; auspicious; lucky, prosperous. **fortunately**, *adv.*

fortune (faw'chən), *n.* chance, luck, that which happens as if by chance; that which brings good or ill, a personification of this, a supernatural power supposed to control one's lot and to bestow good or evil; one's future lot; (*pl.*) the progress or history of a person or thing; good luck, prosperity; wealth; a large property or sum of money. *v.i.* to happen, to chance. **a small fortune**, a large sum of money. **fortune cookie**, *n.* (*N Am.*) a biscuit with a slip of paper inside it, which has a prediction, proverb, joke etc. written on it. **fortune-hunter**, *n.* one who seeks to marry a wealthy woman. **fortune-hunting**, *n.*, *a.* **fortune-teller**, *n.* one who claims to reveal future events. **fortune-telling**, *n.*

forty (faw'ti), *n.* the number or figure 40 or XL; the age of 40. *a.* 40 in number; aged 40. **forty-five**, *n.* a record played at 45 r.p.m. **the Forty-five**, the Jacobite rebellion of 1745–46. **forty-niner**, *n.* one of the adventurers who went to California at the time of the gold-rush in 1849. **forty winks**, *n.* a nap.

forum (faw'rəm), *n.* the public place in ancient Rome in which were the courts of law, public offices etc. and where orations were delivered; a place of assembly for public discussion or judicial purposes; a meeting to discuss matters of public interest; a medium for open discussion; a tribunal, a court of law.

forward (faw'wəd), *a.* at or near the forepart of anything; in front; towards the front; onward; in advance, advancing or advanced; well advanced, progressing, early, premature, precocious; eager, prompt; pert, presumptuous; of or preparing for the future. *n.* a mainly attacking player at football etc. stationed at the front of a formation. *v.t.* to help onward, to promote; to hasten the growth of; to send on or ahead, to send to a further destination; to send. *adv.* (*Naut.*) towards, at or in the fore part of a vessel. **forward**, **-wards**, *adv.* towards the front; onward in place or time; towards the future; to an earlier time; ahead, in advance; to the front, to a prominent position. **forward-looking**, *a.* progressive; looking to, or planning for, the future. **forwarder**, *n.* **forwardly**, *adv.* **forwardness**, *n.* the quality or state of being forward; assurance; pertness.

forwent *past* FORGO.

fossa (fos'ə), *n.* (*pl.* **-ssae**, -ē) (*Anat.*) a shallow depression, pit or cavity.

fosse (fos), *n.* a ditch, a trench, esp. around a fortification, commonly filled with water; a canal.

fossick (fos'ik), *v.i.* (*chiefly Austral.*) to search for gold or precious stones, esp. in abandoned workings; to rummage about. **fossicker**, *n.*

fossil (fos'l), *a.* found underground; dug from the earth; preserved in the strata of the earth's crust, esp. if mineralized; antiquated. *n.* an organic body preserved in the strata of the earth's crust; (*coll.*) an antiquated, out-of-date or inflexible person or thing; a word or word element once current but now found only in a few special contexts. **fossil fuel**, *n.* a naturally-occurring fuel formed by the decomposition of prehistoric organisms. **fossiliferous** (-lif'-), *a.* **fossilize**, **-ise**, *v.i.*, *v.t.* to convert or be converted into a fossil; to render or become antiquated or inflexible. **fossilization**, *n.*

fossorial (fosaw'riəl), *a.* (*zool.*) adapted for digging.

foster (fos'tə), *v.t.* to bring up or nurse (esp. a child not one's own); to place in the charge of foster parents; to nourish, to support, to encourage, to promote the growth of; to harbour (as an ill feeling). **foster-brother**, **-sister**, *n.* a brother or sister by fostering, but not by birth. **foster-child**, **-daughter**, **-son**, *n.* a child brought up or nursed by someone other than its natural parent(s). **foster-father**, **-mother**, **-parent**, *n.* one who takes the place of a parent in rearing a child. **fosterage**, *n.* the act of fostering; the state of being a foster-child; the care of a foster-child; fostering or encouraging. **fosterer**, **fosterling** (-ling), *n.* a foster-child.

fought, *past*, *p.p.* FIGHT.

foul (fowl), *a.* dirty, filthy, unclean; loathsome, offensive to the senses; covered or filled with noxious matter; clogged, choked; morally offensive, obscene, disgusting; polluted; unfair, unlawful, dishonest, against the rules; stormy, cloudy, rainy; of a proof, full of printer's errors, dirty, inaccurate; (*coll.*) bad, unpleasant, disagreeable. *adv.* irregularly, against the rules. *n.* (*Sport*) a wilful collision, an interference; any breach of the rules of a game or contest. *v.t.* to make foul; to defile, to soil, to pollute; to dishonour; to come into collision with, to impede, block or entangle; to commit a foul against. *v.i.* to become foul or dirty; to come into collision; to become clogged or entangled; to commit a foul. **to fall**, **run foul of**, to come or run against with force; to come into collision with; to quarrel with. **to foul up**, to make dirty, to pollute; to block, to entangle; to become blocked or entangled; (*coll.*) to blunder; (*coll.*) to spoil or cause to break down by making mistakes etc. **foul-mouthed**, **-spoken**, **-tongued**, *a.* addicted to profane, scurrilous or obscene language. **foul play**, *n.* unfair behaviour in a game or contest, a breach of the rules; dishonest or treacherous conduct; violence, murder. **foully**, *adv.* **foulness**, *n.*

foulard (foolahd'. -lah'), *n.* a soft, thin material of silk or silk mixed with cotton; a silk handkerchief.

found[1] (fownd), *v.t.* to cast by melting (metal) or fusing (material for glass) and pouring it into a mould; to make of molten metal or glass. **founder**[1], *n.* **foundry**, *n.* a building where metals are cast; the act or art of casting metals.

found[2] (fownd), *v.t.* to lay the foundation or basis of; to set up, to establish; to endow; to originate; to conduct or base (on). *v.i.* to rest (on) as a foundation. **founder**[2], *n.* one who founds or originates anything, esp. one who endows a permanent fund for the support of an institution. **founder-member**, *n.* one of the original members who combined to establish a society etc. **founding**, *n.*, *a.* **Founding Father**, *n.* a member of the American Constitutional Convention of 1787; one who establishes or institutes something. **foundress** (-dris), *n. fem.*

found[3] (fownd), *past*, *p.p.* FIND. **all found**, *adv.* with complete board and lodging.

foundation (fowndā'shən), *n.* the act of founding or establishing; that on which anything is established or by which it is sustained; the fund or endowment which supports an institution; the natural or artificial basis of a structure; (*pl.*) the part of a structure below the surface of the ground; a cosmetic used as a base for other facial make-up; the grounds, principles or basis on which anything stands; an endowed institution. **foundation course**, *n.* a basic, general course, taught e.g. in the first year at some universities and colleges. **foundation garment**, *n.* a woman's undergarment that supports the figure, e.g. a corset. **foundation stone**, *n.* a stone laid with ceremony to commemorate the founding of a building. **foundationless**, *a.*

founder[1, 2] FOUND[1, 2].

founder[3] (fown'də), *v.i.* to fill with water and sink, as a ship; of a horse, to fall lame; to fall in, to give way; to fail, to break down; to be ruined. *v.t.* to lame by causing soreness or inflammation in the feet of (a horse); to sink (a ship) by making her fill with water.

foundling (fownd'ling), *n.* a deserted child of unknown parents.

foundry FOUND[1].

fount[1] (fownt), *n.* a spring, a fountain, a well; a source.

fount[2] (fownt), *n.* a set of type of one face and size.

fountain (fown'tin), *n.* a spring of water; the source of a river or stream; a structure producing an ornamental jet of water driven high into the air by pressure; a public structure with a drinking-supply; a reservoir to contain a liquid, as in a lamp, printing-press, fountain-pen etc.; a source, a first principle. **fountain-head**, *n.* an original source or spring. **fountain-pen**, *n.* a pen with an ink reservoir.

four (faw), *n.* the number or figure 4 or IV; the age of four; the fourth hour after midnight or midday; a set of four persons or things, a team of four horses, a four-oared boat or its crew; a card or domino with four spots; (*Cricket*) (a score of four runs from) a shot which crosses the boundary after hitting the ground. *a.* four in number; aged four. **four-in-hand**, *n.* a vehicle drawn by four horses and driven by one driver. **four-letter word**, any of a number of short English words referring to the body, sex or excrement and considered vulgar or obscene. **to be, go, run on all fours**, to crawl on the hands and feet or knees. **four-ale**, *n.* small ale, once sold at fourpence a quart. **four-ball** FOURSOME. **four-colour**, *a.* pertaining to a printing or photographic reproduction process using cyan, magenta, yellow and black to form an image. **four-eyes**, *n.* (*sl.*) a person in spectacles. **four flush**, *n.* a worthless poker hand in which only four of the five cards are of the same suit. **four-footed**, *a.* having four feet; quadruped. **four-handed**, *a.* quadrumanous; of games, for four players; of music, for two performers. **four-leaf, four-leaved**, *a.* applied to a clover leaf with four leaflets instead of three, supposed to bring good luck. **four o'clock**, *n.* the Marvel of Peru, *Mirabilis dichotoma*, so named from its flowers opening at four o'clock in the afternoon. **fourpence**, *n.* the sum of four pennies. **fourpenny**, *n.* a old silver coin worth 4d. *a.* (faw'pəni, fawp'ni), worth fourpence; costing fourpence. **fourpenny one**, *n.* (*dated sl.*) a blow, a cuff. **four-poster**, *n.* a (usu. large) bedstead with four high posts at the corners to support a canopy and curtains. **fourscore**, *n., a.* 4 times 20, 80. **foursquare**, *a.* having four sides and angles equal; square; firmly established; forthright, resolute. **four-stroke**, *a.* term applied to an internal-combustion engine which fires once every four strokes of movement of the piston. **four-wheel, -wheeled**, *a.* having four wheels. **four-wheel drive**, a system whereby power is transmitted to all four wheels of a motor vehicle. **four-wheeler**, *n.* a vehicle having four wheels, esp. a horse-drawn cab. **fourfold**, *a.* four times as many or as much, quadruple. *adv.* in fourfold measure. **foursome** (-səm), *n.* a group of four persons; (*Golf*) a game between two pairs, the partners playing their ball alternately.

Fourierism (fu'riərizm), *n.* a system of social reorganization advocated by the French socialist F.M.C. Fourier (1772–1837), based on the principle of natural affinities. **Fourierist, -ite**, *n.*

fourteen (fawtēn'), *n.* the number or figure 14 or XIV; the age of 14. *a.* 14 in number; aged 14. **fourteenth**, *n.* one of 14 equal parts; an interval of an octave and a seventh; a note separated from another by this interval, two such notes sounded together. *n., a.* (the) last of 14 (people, things etc.); (the) next after the 13th.

fourth (fawth), *n.* one of four equal parts, a quarter; the fourth forward gear of a motor vehicle; (*Mus.*) an interval of four diatonic notes, comprising two whole tones and a semitone; two notes separated by this interval sounded together. *n., a.* (the) last of four (people, things etc.); (the) next after the third. **Fourth of July**, Independence Day in the US, anniversary of the Declaration of Independence, 4 July 1776. **fourth dimension** DIMENSION. **fourth estate**, *n.* the press. **fourthly**, *adv.* in the fourth place.

fovea (fō'viə), *n.* (*pl.* **-veae** (-ē)) (*Anat. etc.*) a small pit or depression. **foveate** (-āt), *a.*

fowl (fowl), *n.* (*pl.* **fowls**; *collect.* **fowl**) a bird hunted or raised for its flesh; a cock or hen of the domestic or poultry kind; their flesh used as food. *v.i.* to hunt, catch or kill wild birds for sport. **barn-door fowl** BARN-DOOR. **fowling-piece**, *n.* a light smooth-bore gun adapted for shooting wild-fowl. **fowl-pest**, *n.* a contagious virus disease of birds. **fowler**, *n.* one who pursues wild-fowl for sport.

fox (foks), *n.* a quadruped, (esp. genus *Vulpes*), of the dog family with a straight bushy tail and erect ears; the fur of the fox; a sly, cunning person. *v.t.* to baffle, to perplex; to trick, to outwit; (*chiefly p.p.*) to discolour (pages of a book etc.). *v.i.* of paper etc., to become discoloured, esp. to turn reddish. **foxglove**, *n.* the genus *Digitalis*, esp. *D. purpurea*, with purple flowers resembling the fingers of a glove, the leaves of which are used as a sedative. **fox-hole**, *n.* (*Mil.*) a small trench. **foxhound**, *n.* a hound trained to hunt foxes. **foxhunt**, *n.* the hunting of a fox with a pack of hounds. *v.i.* to hunt foxes with hounds. **foxhunter**, *n.* **foxhunting**, *a.* **foxtail**, *n.* kinds of grasses, esp. genus *Alopercurus*. **fox-terrier**, *n.* a short-haired dog, orig. employed to unearth foxes, now chiefly as a pet. **fox-trot**, *v.i., n.* (to dance) a kind of ballroom dance with varying steps. **foxed**, *a.* stained with spots, as a book or print; (*sl.*) drunk. **fox-like**, *a.* **foxy**, *a.* fox-like, tricky, crafty; foxed; (*chiefly N Am.*) physically attractive; reddish-brown in colour. **foxiness**, *n.*

foyer (fo'yā, -yə), *n.* a large public room in a theatre; the entrance hall of a hotel, cinema etc.

fp, (*abbr.*) fortepiano; freezing point.

FPA, (*abbr.*) Family Planning Association.

fps, (*abbr.*) feet per second; foot-pound-second; frames per second.

Fr, (*chem. symbol*) francium.

Fr., (*abbr.*) Father; Franc; France; French; Friar.

fr., (*abbr.*) franc; from.

Fra (frah), *n.* brother, a title given to an Italian monk or friar. [It.]

fracas (frak'ah), *n.* (*pl.* **fracas** (-z)) an uproar; a noisy quarrel.

fractal (frak'təl), *n.* an irregular or fragmented figure or surface of a type unsuitable for conventional geometric representation.

fraction (frak'shən), *n.* a part, portion; a fragment, a small piece; (*Math.*) the expression of one or more parts of a unit; the rite of breaking the bread in the Eucharist; a component of a mixture separable by fractionation. **fractional, -nary**, *a.* of or pertaining to fractions; constituting a fraction; forming but a small part, insignificant. **fractional distillation** DISTIL. **fractionally**, *adv.* **fractionate**, *v.t.* to separate (a mixture) into portions having different properties, by distillation or analogous process. **fractionation**, *n.* **fractionize, -ise**, *v.t.* to break up into fractions or divisions.

fractious (frak'shəs), *a.* apt to quarrel; snappish, cross, fretful, peevish. **fractiously**, *adv.* **fractiousness**, *n.*

fracture (frak'chə), *n.* the act of breaking; a break, a breakage; (*Min.*) the irregularity of surface produced by breaking a mineral across, as distinguished from splitting it along the planes of cleavage; the breakage of a bone (when only the bone is broken the fracture is called **simple**, when there is also a wound of the surrounding tissue it is termed **compound**). *v.i., v.t.* to break, to crack.

fraenum, frenum (frē'nəm), *n.* (*pl.* **-na** (-nə)) a band or ligament restraining the action of an organ, as that of the tongue. **fraenulum** (-nūləm), *n.* (*pl.* **-la** (-lə)) a small fraenum.

fragile (fraj'īl), *a.* brittle, easily broken; weak, frail, delicate. **fragility** (-jil'-), *n.*

fragment[1] (frag'mənt), *n.* a piece broken off; a small detached portion; an incomplete or unfinished portion. **fragmental, -tary** (-men'-), *a.* pertaining to or consisting of

fragment 326 **fray**

fragments; disconnected. **fragmentally, fragmentarily**, *adv.* **fragmentariness**, *n.*
fragment² (fragment'), *v.t., v.i.* to (cause to) break into fragments. **fragmentation**, *n.* the breaking into fragments.
fragmentation bomb, *n.* a bomb whose casing is designed to shatter in small, deadly fragments on explosion.
fragmented (-men'-), *a.*
fragrant (frā'grənt), *a.* emitting a pleasant perfume, sweet-smelling. **fragrance**, *n.* (the state of having or emitting) a sweet smell; the particular scent of a perfume, toilet water etc. **fragrantly**, *adv.*
frail (frāl), *a.* fragile, delicate; infirm, in weak health; weak in character or resolution, liable to be led astray. **frailish**, *a.* **frailly**, *adv.* **frailness**, *n.* **frailty**, *n.*
framboesia (frambē'ziə), *n.* (*Path.*) the yaws, a contagious eruption characterized by swellings like raspberries.
frame (frām), *v.t.* to form or construct by fitting parts together; to fit, adapt or adjust; to contrive; to devise, to invent; to compose, to express; to plan, to arrange; to form in the mind, to conceive; to articulate, to form with the lips; to surround with a frame, to serve as a frame to; to (conspire to) incriminate. *n.* a fabric or structure composed of parts fitted together; a structure or fabric of any kind; the skeleton of a structure; the rigid part of a bicycle; the construction, constitution or build of a thing or person; the established order or system (of society or the body politic); disposition of mind; a case or border to enclose or surround a picture, a pane of glass etc.; (*Hort.*) a glazed portable structure for protecting plants from frost; various machines in the form of framework used in manufacturing, mining, building, printing etc.; a structure on which embroidery is worked; a single exposure on a film; a single, complete television picture; a wooden triangle used to set up the balls for a break in snooker etc.; the balls so arranged; a single round of a game of snooker etc.; a frame-up. **frame of reference**, a set of axes used to describe the location of a point; a set or system of standards, derived from an individual's experience, to which he or she refers when making judgments etc. **frame-house**, *n.* a house with a wooden framework covered with boards. **frame-saw**, *n.* a flexible sawblade stretched in a frame to stiffen it. **frame-up**, *n.* (*sl.*) an attempt to incriminate, a false criminal charge. **framework**, *n.* the frame of the structure; the fabric for enclosing or supporting anything, or forming the substructure to a more complete fabric; (*fig.*) structure, arrangement (of society etc.). **framer**, *n.* **frameless**, *a.* **framing**, *n.* a frame or framework.
franc (frangk), *n.* the standard unit of currency in France, Belgium, Switzerland and various other countries.
franchise (fran'chīz), *n.* a right, privilege, immunity or exemption granted to an individual or to a body; a licence to market a company's goods or services in a specified area; the district or territory to which a certain privilege or licence extends; citizenship; the right to vote; the qualification for this. *v.t.* to grant a franchise to. **franchisee** (-zē'), *n.* the holder of a franchise. **franchisement** (-chiz-), *n.* **franchiser**, *n.* one having the elective franchise; one who grants a franchise.
Franciscan (fransis'kən), *a.* of or pertaining to St Francis of Assisi (1182–1226), or the order of mendicant friars founded by him in 1209. *n.* a member of the Franciscan order, a grey friar.
francium (fran'siəm), *n.* a radioactive chemical element of the alkali metal group, at. no. 87; chem. symbol Fr.
Franco- (frang'kō-), *comb. form.* pertaining to the French. **Francophile** (-fīl), *n.* **Francophobe** (-fōb), *n.*
francolin (frang'kəlin), *n.* a bird of the genus *Francolinus*, allied to the partridges, esp. *F. vulgaris*, a richly-coloured species common in India. **Francophone** (frang'kōfōn), *a.* French-speaking, having French as the native or an official language. *n.* a French-speaker.

frangible (fran'jibl), *a.* that may be easily broken. **frangibleness, frangibility** (-bil'-), *n.*
frangipane (fran'jipān), **frangipani** (franjipah'ni), *n.* a kind of pastry made with cream, almonds and sugar; a perfume prepared from the flowers of a W Indian tree. [the inventor of the perfume, the Marquis *Frangipani*]
franglais (frā'glā), *n.* French which contains a high proportion of English words. [F *Français*, French and an*glais*, English]
Frank (frangk), *n.* a member of the ancient German peoples or tribes who conquered France in the 6th cent. **Frankish**, *a.*
frank (frangk), *a.* open, ingenuous, sincere, candid; free, unrestrained, outspoken. *v.t.* to mark (a letter etc.) in such a way as to indicate that postage has been paid; to send or cause to be sent under an official privilege, such as, formerly, the signature of a member of Parliament, so as to pass free; to secure the free passage of (a person or thing). *n.* a signature authorizing a letter to go through the post free of charge; the franked letter or package. **franking machine**, *n.* a machine that franks letters etc. **frankly**, *adv.* **frankness**, *n.*
Frankenstein (frang'kənstīn), *n.* a work that brings disaster to its creator; a destructive monster in human form. [character in the novel by Mary Shelley]
Frankfurter (frangk'fœtə), *n.* a small, smoked sausage of beef and pork.
frankincense (frang'kinsens), *n.* a gum or resin burning with a fragrant smell, used as incense, in the East, olibanum, an exudation from trees of the genus *Boswellia*, is used.
Frankish FRANK.
franklin (frang'klin), *n.* in the 14th and 15th cents., an English freeholder, not liable to feudal service.
frantic (fran'tik), *a.* raving, outrageously excited or demented; marked by extreme haste or agitation. **frantically, -ticly**, *adv.* **franticness**, *n.*
frappé (frap'ā), *a.* iced.
fraternal (frətœ'nəl), *a.* brotherly; pertaining to or becoming brethren; existing between brothers; of twins, from two separate ova. **fraternally**, *adv.* **fraternity**, *n.* the state of being a brother; brotherliness; a body of men associated for a common interest or for religious purposes; a body of men associated or linked together by similarity of rank, profession etc.; (*N Am.*) a college association of students. **fraternize, -ise** (frat'-), *v.i.* to associate or hold fellowship with others of like occupation or tastes; to associate (with) on friendly terms. **fraternization**, *n.* **fraternizer**, *n.*
fratricide (frat'risīd), *n.* the murder of a brother; one who murders a brother. **fratricidal** (-sīd'-), *a.*
Frau (frow), *n.* (*pl.* **Frauen**) a German woman, wife or widow; Mrs. **Fräulein** (fraw'līn, frow'-, froi'-), *n.* (*pl.* **Fräulein**) a young lady, a German spinster; Miss. [G]
fraud (frawd), *n.* an act or course of deception deliberately practised to gain unlawful or unfair advantage; (*Law*) such deception directed to the detriment of another; a deception, a trick, trickery; (*coll.*) a deceitful person, a humbug. **fraudster**, *n.* **fraudulence**, *n.* **fraudulent** (-ūlənt), *a.* practising fraud; characterized by or containing fraud; intended to defraud, deceitful. **fraudulently**, *adv.*
fraught (frawt), *a.* involving, entailing, attended by, charged (with); tense, characterized by or inducing anxiety.
fraxinella (fraksinel'ə), *n.* kinds of rue or dittany, esp. *Dictamnus fraxinella* and *D. albus*, cultivated for their leaves and flowers. **Fraxinus** (-nəs), *n.* a genus of deciduous trees containing the common ash etc.
fray¹ (frā), *n.* an affray; a noisy quarrel, a brawl, a riot; a combat, a contest.
fray² (frā), *v.t.* to wear away by rubbing; to make strained or irritated. *v.i.* of a garment, cloth etc., to become

frazil 327 **free**

rubbed or worn, esp. so as to become unravelled or ragged at the edges.
frazil (frā'zil), *n.* anchor-ice.
frazzle (fraz'l), *v.t.* to reduce to a state of physical or nervous exhaustion. *v.i.* to be worn out, nervous. *n.* an exhausted state. **to a frazzle**, completely, thoroughly.
freak (frēk), *n.* a sudden wanton whim or caprice; a humour, a vagary; an abnormal or deformed person or thing; an unconventional or eccentric person; an unrestrained enthusiast for something. *a.* highly unusual, abnormal, esp. in magnitude or intensity. *v.t.* (*usu. p.p.*) to variegate, to streak. **to freak (out)**, (*coll.*) to (cause to) hallucinate; to (cause to) be in a highly emotional or excited state. **freakful**, *a.* **freakish**, *a.* whimsical; eccentric, unconventional; abnormal. **freakishly**, *adv.* **freakishness**, *n.* **freaky**, *a.* (*coll.*) freakish.
freckle (frek'l), *n.* a yellowish or light-brown spot on the skin, due to sunburn or other causes; any small spot or discoloration. *v.i.*, *v.t.* to mark or become marked with freckles. **freckling**, *n.* **freckly**, *a.*
free[1] (frē), *a.* (*comp.* **freer**, *superl.* **freest**) at liberty; not in bondage or under restraint; living under a government based on the consent of the citizens; of a government, not arbitrary or despotic; of a State, not under foreign domination; released from authority or control; not confined, restricted, checked or impeded; at liberty to choose or act, permitted (to do); independent, unattached, unconnected with the State; released, clear, exempt (from); unconstrained, not bound or limited (by rules, conventions etc.); of a translation, not literal; unconventional, unceremonious, careless, reckless; forward, impudent; indelicate, broad; unreserved, frank; not subject to charges, duties, fees etc.; without restriction, open, gratuitous; liberal, generous; spontaneous, unforced; unoccupied, vacant; clear, unobstructed; not busy, having no obligations or commitments; not fixed or joined; (*Chem.*) not combined with another body; (*Zool.*) unattached; (*Bot.*) not adhering, not adnate. *adv.* freely; without cost or charge; (*Naut.*) not close-hauled. **for free**, (*coll.*) gratis, for nothing. **free alongside ship**, delivered free on the dock or wharf. **free-and-easy**, *a.* unconstrained, unceremonious; careless. **free collective bargaining**, negotiations between trade unions and employers unhampered by government guidelines or legal restraints. **free-for-all**, *n.* a free fight, a disorganized brawl or argument. **free on board**, of goods, delivered on board or into conveyance free of charge. **to make free**, to take liberties (with). **free agent**, *n.* one who is free to act according to his/her own opinions and wishes. **free agency**, *n.* **free association**, *n.* (*Psych.*) the bringing to consciousness of unconscious processes through words and ideas which the subject spontaneously associates with key words provided by a psychoanalyst. **free-base**, *v.t.*, *v.i.* (*sl.*) to purify (cocaine); to smoke (cocaine) so purified. **free-board**, *n.* the space between the water-line on a vessel and the upper side of the deck, or the uppermost full deck. **freebooter**, *n.* a pirate or buccaneer, an adventurer who makes a business of plundering. **freeboot**, *v.i.* **free-born**, *a.* born free; inheriting the right and liberty of a citizen. **Free Church**, *n.* a Church exempt from State control, or one in which there are no enforced payments, esp. the ecclesiastical body founded by those who left the Scottish Presbyterian establishment at the Disruption in 1843; in England, a Nonconformist Church. **freedman** FREE[2]. **free enterprise**, *n.* the conduct of business without state interference or control. **free fall**, *n.* the motion of an unrestrained or unpropelled body in a gravitational field; the part of a parachute jump before the parachute opens; a rapid unrestrained fall in share prices, exchange rates etc. **free-fall**, *v.i.* **free fight**, *n.* a fight in which anyone can join. **free flight**, *n.* the flight of a rocket etc. when its motor has ceased to produce thrust. **free-floating**, *a.* unattached, having no specific object, uncommitted. **free hand**, *n.* (to be given) complete freedom to do. **free-hand**, *a.* (*Drawing*) executed by the hand without the aid of instruments. **free-handed**, *a.* open-handed, liberal. **free-heartedness**, *n.* **freehold**, *n.* an estate held in fee-simple or fee-tail; the tenure by which such an estate is held; an office held for life. *a.* held in fee-simple or fee-tail; of the nature of a free-hold. **freeholder**, *n.* the possessor of a freehold. **free house**, *n.* a public-house free to buy its goods from any supplier. **free kick**, *n.* (*Football*) a kick with which an opponent may not interfere, awarded for a foul or infringement by the other side. **freelance**, *n.* (*Hist.*) a member of one of the free companies of mercenaries in the Middle Ages; (also **freelancer**) a self-employed person hired by others for specific (usu. short-term) assignments. *a.*, *adv.* not bound to a particular employer. *v.i.* to work freelance. **free-liver**, *n.* one who indulges his or her appetites, esp. at table; (*Biol.*) an organism which is neither parasitic nor symbiotic. **free-living**, *n.*, *a.* **freeload**, *v.i.* (*coll.*) to sponge, to live at another's expense. **freeloader**, *n.* **free love**, *n.* sexual intercourse without marriage; the doctrine that the affections should be free to fix on any object to which they are drawn, without restraint of marriage obligation. **freeman**, *n.* one not a slave or serf; one who holds the franchise of a citizen or a particular privilege, esp. the freedom of a city, company etc. **free-woman**, *n.*, *fem.* **free market**, *n.* an economic market in which there is free competition. **Freemason**, *n.* a member of an association of 'Free and Accepted Masons', a secret order or fraternity, stated to have been traced back to the building of Solomon's Temple, but probably originating as a fraternity of skilled masons, with right of free movement, about the 14th cent. **Freemasonry**, *n.* the system, rites and principles of Freemasons; (usu. **freemasonry**) a secret understanding, community of interests, or instinctive sympathy among a number of people. **free pass**, *n.* a ticket that has not been paid for, entitling the holder to travel or to enter an exhibition, theatre etc. **free port**, *n.* a port where ships of all nations may load or unload free of duty. **free radical**, *n.* an atom, or group of atoms, containing at least one unpaired electron. **free-range**, *a.* kept or produced in natural conditions. **free school**, *n.* a school where no fees are charged. **freesheet**, *n.* a newspaper distributed free. **free skating**, *n.* that part of a figure-skating competition in which the competitors have partial or complete freedom to organize their programmes. **free-spoken**, *a.* speaking without reserve; blunt, candid, frank. **free-spokenness**, *n.* **free-standing**, *a.* not attached to, supported by or integrated with other objects. **Free States**, *n.pl.* those States of the American Union in which slavery never existed, or was abolished before the Civil War. **freestone**, *n.* a stone which can be cut freely in any direction. **free-stone**, *n.* a kind of peach easily freed from its stone when ripe. **freestyle**, *n.* a (swimming) race in which each competitor can choose which style to use; all-in wrestling. **free-thinker**, *n.* a rationalist, sceptic or agnostic; one who rejects authority in religious belief. **free-thinking**, *n.*, *a.* **free-thought**, *n.* **free trade**, *n.* unrestricted trade with other countries; free interchange of commodities without protection by customs duties. **free-trader**, *n.* **free verse**, *n.* unrhymed verse with no set metrical pattern. **free vote**, *n.* a vote left to the individual's choice, free from party discipline. **freeway**, *n.* (*N Am.*) a motorway. **free-wheel**, *n.* a driving wheel on a cycle that can be disconnected from the driving gear and allowed to revolve while the pedals are at rest. *v.i.* to coast (on a cycle or motor-car) without employing locomotive power or brakes; to move or live in an unconstrained or irresponsible fashion. **freewheeling**, *n.*, *a.* **free will**, *n.* the power of directing one's own actions without constraint by any external influence. **free-will**, *a.* given freely, voluntary. **Free World**, *n.* the non-

Communist countries collectively. **freely,** *adv.* **freeness,** *n.*
free[2] (frē), *v.t.* to set at liberty, to emancipate; to rid or relieve (of or from); to extricate, to clear, to disentangle; to make available for use (often with *up*). **freedman,** *n.* a manumitted slave.
-free, *comb. form.* free from, not containing.
freebie (frē'bi), *n.* (*coll.*) something for which one does not have to pay.
freedom (frē'dəm), *n.* the state of being free, liberty, independence; personal liberty, non-slavery, civil liberty; liberty of action, free will; exemption, immunity (from); lack of conventionality, frankness, (excessive) familiarity; ease or facility in doing anything; participation in certain privileges, exemptions, and immunities pertaining to citizenship of a city or membership of a company; free use (of). **freedom fighter,** *n.* one who fights (esp. as an irregular soldier) for the liberation of a nation etc. from foreign rule or a tyrannical regime.
freemartin (frē'mahtin), *n.* a sexually imperfect cow, usu. born as twin with a bull-calf.
freesia (frē'ziə, -zhə), *n.* any of a S African genus of bulbous flowering plants allied to the iris.
freeze (frēz), *v.i.* (*past,* **froze** (-frōz) *p.p.* **frozen**) to be turned from a fluid to a solid state by cold; (*impers.*) to be at that degree of cold at which water turns to ice or becomes covered with ice; to become clogged by ice; to become attached (to) or fastened (together) by frost; to feel very cold; to die of cold; to be chilled (by fear); to become motionless or paralysed. *v.t.* to congeal by cold; to form ice upon or convert into ice; to make very cold; to injure, overpower or kill with cold; to preserve (food) by freezing and storing at a temperature below 32° F or 0° C; to chill with fear; to anaesthetize (as if) by cold; to render motionless or paralysed; to stop at a particular stage or state; to stop (a moving film) at a particular frame; (*Finance*) to prohibit the use of or dealings in; to fix or stabilize (prices etc.). *n.* the act or state of freezing; a period of freezing weather; a period of fixing wages, prices etc. at a certain level. **to freeze out,** (*coll.*) to compel the retirement of from business, competition, society etc., by boycotting, contemptuous treatment or similar methods. **freeze-dry,** *v.t.* to dehydrate while in a frozen state in a vacuum, esp. for preservation. **freeze-frame,** *n.* a single frame of a film repeated to give the effect of a still photograph; a single frame of a video recording viewed as a still. **freezable,** *a.* **freezer,** *n.* an apparatus for freezing (meat etc.), a room or cabinet, or a compartment in a refrigerator for the long-term storage of perishable foodstuffs. **freezing,** *a.* very cold; distant, chilling. **freezing-mixture,** *n.* a mixture of salt and snow, or pounded ice, or a combination of chemicals with or without ice, for producing intense cold. **freezing-point,** *n.* the point at which water freezes, marked 32° on the Fahrenheit scale, and 0° on the Centigrade (Celsius) and Réaumur scales; the temperature at which a substance freezes. **freezingly,** *adv.* **frozenly,** *adv.* **frozenness,** *n.*
freight (frāt), *n.* the money due or paid for the transportation of goods, esp. by water; that with which a ship is loaded; a cargo; ordinary transportation, as distinct from express; a goods train. *v.t.* to load (as a ship) with goods for transportation. **freightliner**®, *n.* a train designed for the rapid transportation of containerized cargo. **freight train,** *n.* (*N Am.*) a goods train. **freightage,** *n.* money paid for the hire of a ship or the transportation of goods; the transporting of goods; freight. **freighter,** *n.* one who hires or loads a ship; a cargo-boat; a shipper; one who contracts to receive and forward goods. **freightless,** *a.*
French (french), *a.* pertaining to France or its inhabitants; belonging to or native to France. *n.* the language spoken by the people of France; (*collect.*) the people of France. **excuse, pardon, my French,** excuse my bad language. **to take French leave,** to go away without permission.

French-bean, *n.* the kidney or haricot bean, *Phaseolus vulgaris.* **French bread,** *n.* crusty white bread in thin, long loaves. **French Canadian,** *n.* a French-speaking Canadian. *a.* of the French-speaking part of Canada or its people. **French chalk,** *n.* a variety of talc, steatite, or soapstone used for marking cloth, and in powder as a dry lubricant. **French curve,** *n.* an instrument designed to assist in drawing curved lines. **French dressing,** *n.* a salad dressing made of oil and vinegar or lemon juice with seasoning. **French fries,** *n.pl.* (potato) chips. **French horn,** *n.* a metal wind instrument of circular shape with a gradual taper from the mouthpiece to a large everted bell. **French kiss,** *n.* a kiss in which the tongue is inserted into the partner's mouth. **French knickers,** *n.pl.* wide-legged knickers. **French letter,** *n.* (*coll.*) a condom. **Frenchman,** *n.* a native or naturalized inhabitant of France; a French ship. **French mustard,** *n.* a type of mustard mixed with vinegar etc. **French polish,** *n.* a solution of resin or gum-resin in alcohol or wood naphtha, for polishing cabinet-work etc.; the polish produced. *v.t.* to polish with this. **French polisher,** *n.* **French seam,** *n.* a double seam, stitched first on the wrong, then on the right side, so that the edges are hidden. **French window,** *n.* (*often pl.*) a pair of doors with full-length glazing. **Frenchwoman,** *n. fem.* **Frenchify,** *v.t.* to make French; to influence with French tastes or manners. **Frenchification** (-fi-), *n.*
frenetic (*esp. formerly*) **phrenetic** (frənet'ik), *a.* frantic, frenzied. **frenetically,** *adv.*
frenum FRAENUM.
frenzy (fren'zi), *n.* temporary mental derangement; delirium or unnatural excitement or agitation; wild or excessively intense activity. *v.t.* (*usu. p.p.*) to drive to madness; to infuriate. **frenzied,** *a.* **frenziedly,** *adv.*
Freon® (frē'on), *n.* any of a group of fluorocarbons used as refrigerants.
freq., (*abbr.*) frequent, frequently, frequentative.
frequence (frē'kwəns), **-ency** (-nsi), *n.* the quality of occurring frequently; rate of occurrence; the comparative number of occurrences in a given time; (*Statistics*) the number or proportion of individuals in a single class; (*Elec.*) a term referring to the speed of variations of alternating currents, alternating electromotive forces, and electromagnetic waves; (*Phys.*) rate of repetition or recurrence. **high frequency** HIGH. **frequency distribution,** *n.* (*Statistics*) an arrangement of data which shows the frequency of occurrence of the different values of a variable. **frequency modulation,** *n.* (*Radio*) the varying of the frequency of the carrier wave in accordance with changes in the amplitude of the signal; the broadcasting system using this.
frequent[1] (frē'kwənt), *a.* occurring often, common; repeated at short intervals; occurring near together, abundant. **frequentative** (-kwen'-), *n., a.* (*Gram.*) (a verb) expressing frequent repetition of an action. **frequently,** *adv.* often, commonly, at frequent intervals. **frequentness,** *n.*
frequent[2] (frikwent'), *v.t.* to visit or resort to often or habitually. **frequentage, frequentation** (-tā'-), *n.* **frequenter,** *n.*
fresco (fres'kō), *n.* (*pl.* **-cos, -coes**) a kind of water-colour painting done on fresh plaster or on a wall covered with mortar not quite dry; a picture done in this manner. *v.t.* to paint (a picture) or decorate (a wall etc.) in fresco.
fresh (fresh), *a.* new; not known, met with or used previously, recent; other, different, additional; newly produced, not withered or faded; not stale, decayed or tainted; pure, not salt, drinkable; not preserved with salt, or by pickling, tinning etc.; raw, inexperienced; just arrived (from); looking young or healthy; vividly and distinctly retained in the mind; refreshed, reinvigorated; of a horse, frisky; brisk, active, vigorous, fit; of air, a breeze etc., refreshing, reviving, cool; cheeky, impertinent,

amorously impudent. *adv.* (*esp. in comb.*) freshly, as *fresh-blown*; recently; with fresh vigour. *n.* a freshet; a freshwater river or spring; (*ellipt.*) the fresh part (of the day, season etc.). **fresh out of,** (*chiefly N Am.*) having recently (completely) run out of. **freshman,** *n.* a novice, a beginner, esp. a student in the first year at a university. **freshwater,** *a.* pertaining to, found in or produced by fresh water; used to river or coasting trade, as a sailor. **freshen,** *v.t.* to make fresh; to enliven, to revive. *v.i.* to become fresh; to become brisk, to gain strength. **to freshen up,** to refresh oneself, to have a wash or shower, change one's clothes etc.; to revive, to give a fresher, more attractive appearance to; to replenish (a drink). **fresher,** *n.* (*coll.*) a freshman. **freshet** (-it), *n.* a sudden flood caused by heavy rains or melted snow; a freshwater stream. **freshly,** *adv.* **freshness,** *n.*
fret[1] (fret), *v.t.* (*past, p.p.* **fretted**) to eat away, to corrode; to wear away, to rub or chafe; to make (a way or passage) by rubbing; to irritate, vex or worry. *v.i.* to be worn, rubbed or eaten away; to be irritated, vexed, or troubled. *n.* the act or process of fretting or rubbing away; a spot abraded or corroded; a state of chafing or vexation. **fretful,** *a.* angry, peevish, irritable. **fretfully,** *adv.* **fretfulness,** *n.*
fret[2] (fret), *v.t.* (*past, p.p.* **fretted**) to ornament, to decorate; esp. with interlacing lines; to ornament (esp. a ceiling) with carved work. *n.* fretwork; ornamental work; an ornament formed by small bands or fillets intersecting each other at right angles, used in classical architecture; (*Her.*) a figure composed of bars crossed and interlaced. **fretsaw,** *n.* a small ribbon-saw used in cutting fretwork. **fretwork,** *n.* carved or open woodwork in ornamental patterns and devices; a variegated pattern composed of interlacing lines of various patterns. **fretted,** *a.*
fret[3] (fret), *n.* a small piece of wood, metal etc. placed upon the fingerboard of certain stringed instruments to regulate the pitch of the notes. **fretted,** *a.*
Freudian (froi'diən), *a.* of or pertaining to the psychological theories of the Austrian psychologist Sigmund Freud (1856–1939). *n.* a follower of Freud. **Freudian slip,** *n.* an unintentional action, such as a slip of the tongue, held to betray an unconscious thought.
Fri., (*abbr.*) Friday.
friable (fri'əbl), *a.* capable of being easily reduced to powder; readily crumbled. **friability** (-bil'-), **friableness,** *n.*
friar (fri'ər), *n.* one belonging to a monastic order, esp. one of the four mendicant orders, Augustinians or Austin Friars, Franciscans or Grey Friars, Dominicans or Black Friars, and Carmelites or White Friars. **friar's balsam,** *n.* a tincture of benzoin for application to ulcers and wounds. **friary,** *n.* a monastery of a mendicant order.
fricandeau (frik'ondō), *n.* (*pl.* **-deaus, -deaux** (-z)) a larded veal cutlet, braised or roasted and glazed. *v.t.* to make into a fricandeau.
fricassee (frik'əsē, -sē'), *n.* small pieces of meat, esp. chicken or veal, fried, stewed and served in a usu. white sauce. *v.t.* to cook as a fricassee.
fricative (frik'ətiv), *n.* a consonant, such as *f, sh, th*, produced by the friction of the breath issuing through a narrow opening. *a.* produced by this friction.
friction (frik'shən), *n.* the act of rubbing two bodies together; (*Phys.*) resistance which any body meets with in moving over another body; conflict, disagreement, lack of harmony; chafing or rubbing a part of the body to promote circulation. **friction-clutch, -cone, -coupling, -gear, -gearing,** *n.* contrivances for applying or disconnecting parts of machinery by the use of friction. **frictional,** *a.* **frictionally,** *adv.* **frictionless,** *a.*
Friday (fri'di), *n.* the sixth day of the week, dedicated by Teutonic peoples to Frig, the wife of Odin, as a translation of the late L *dies Veneris*, day of the planet Venus. **Black Friday** BLACK. **Good Friday** GOOD.

fridge (frij), *n.* short for REFRIGERATOR.
fried, *past, p.p.* FRY[1].
friend (frend), *n.* one attached to another by intimacy and affection, as distinguished from sexual love or family relationship; an acquaintance; one of the same nation or party, one who is not an enemy; one on the same side, an adherent, a sympathizer; a patron or promoter (of a cause, institution etc.); (Friend) a member of the Society of Friends; anything that helps one, esp. in an emergency. **a friend at court,** one who has influence to help another. **Society of Friends,** a religious sect (commonly called Quakers), founded by George Fox in the 17th cent., who object to taking oaths, believe the sacraments of baptism and the Lord's Supper to be symbols, and that there is an Inner Light of God in every person. **to make friends,** to become intimate or reconciled (with). **friendless,** *a.* **friendlessness,** *n.* **friendly,** *a.* having the disposition of a friend, good-natured; acting as a friend; characteristic of friends or of kindly feeling; amicable, not hostile; favourable, propitious; played for amusement or entertainment, not as part of a competition. *n.* a game played for entertainment, not a league or competition fixture. **-friendly,** *comb. form.* helpful to; favouring, protecting. **friendly society,** *n.* a society for the purpose of mutual assurance against sickness, distress or old age. **friendlily,** *adv.* **friendliness,** *n.* **friendship,** *n.* mutual attachment between persons, as distinguished from sexual and family affection; the state of being friends; friendliness.
Friesian (frē'zhən), *n.* any of a breed of large black and white dairy cattle from N Holland and Friesland; FRISIAN.
frieze[1] (frēz), *n.* the middle division of an entablature, between the architrave and the cornice, usu. enriched by sculpture; the band of sculpture occupying this; a horizontal band or strip, either plain or decorated, elsewhere in a wall.
frieze[2] (frēz), *n.* a coarse woollen cloth, with a rough nap on one side.
frig (frig), *v.t., v.i.* (*past, p.p.* **frigged**) (*taboo*) to masturbate; (*taboo*) to have sexual intercourse (with). **to frig about,** (*sl.*) to potter or mess about. **frigging,** *a., adv.* (*taboo*) bloody, fucking.
frigate (frig'ət), *n.* a warship of the period *c.* 1650–1840, next in size and strength to a line-of-battle ship; a steam warship of considerably larger size and strength which preceded the ironclad; (*loosely*) a cruiser, a general-purpose escort vessel smaller than a destroyer. **frigate-bird,** *n.* a large tropical raptorial bird, of great swiftness, usu. found at sea near land.
fright (frit), *n.* sudden and violent fear or alarm; an instance of this, a sudden shock; one who presents a ridiculous or shocking appearance in person or dress. *v.t.* (*poet.*) to frighten. **frighten,** *v.t.* to throw into a state of fright; to alarm, terrify, scare; to drive (away, out of, or into) by fright. **frightener,** *n.* **to put the frighteners on,** (*sl.*) to (attempt to) coerce or deter someone with threats (of violence). **frightful,** *a.* dreadful, fearful, shocking; horrible, hideous, very disagreeable; (*coll.*) awful, extraordinary. **frightfully,** *adv.* **frightfulness,** *n.*
frigid (frij'id), *a.* cold; wanting heat or warmth; lacking warmth of feeling or ardour; stiff, formal, forbidding; without animation or spirit, dull, flat; sexually unresponsive. **frigid zones,** *n.pl.* the parts of the earth between the Arctic Circle and the North Pole and the Antarctic Circle and the South Pole. **frigidarium** (-deə'riəm), *n.* (*pl.* **-aria** (-ə)) the cooling-room in a Roman bath; the cold bath itself. **frigidity** (-jid'-), *n.* the state of being frigid; the decrease or absence in a woman of sexual response. **frigidly,** *adv.* **frigidness,** *n.*
frijole (frēkhōl'), *n.* a Mexican bean resembling the kidney-bean.
frill (fril), *n.* a pleated or fluted edging of cloth, used on

fringe clothing, a ruffle, a flounce; a ruff or frill-like fringe of hair, feather etc. on an animal, bird or plant; (*pl.*) (*coll.*) airs, affectations, frippery, decorative non-essentials. *v.t.* (*past*, *p.p.* **frilled**) to furnish with a frill; to serve as a frill to. **with no frills, without frills,** plain, unornamented, no-nonsense. **frilled,** *a.* **frilled lizard,** *n.* a large Australian lizard with an erectile fold of skin around its neck. **frilly,** *a.*

fringe (frinj), *n.* an ornamental border to dress or furniture, consisting of loose threads or tassels; a border, an edging; the front hair cut short with a straight edge along the forehead; (*Bot., Zool.*) a border of hairs or hairlike processes; (*Opt.*) one of the coloured bands seen when a beam of light is transmitted through a slit; something marginal or additional; a group with marginal or extreme views. *v.t.* to border with or as with a fringe; to serve as a fringe to. *a.* existing alongside mainstream or conventional forms, institutions etc.; marginal, secondary. **The Fringe,** that part of an arts festival, the London theatre etc. which presents new, experimental or avant-garde works away from the main venues. **fringe benefit,** *n.* something additional to wages or salary regularly received as part of one's remuneration from one's employer. **fringeless,** *a.* **fringe-like,** *a.* **fringing,** *n.* **fringy,** *a.*

frippery (frip'əri), *n.* worthless, needless or trumpery adornments; tawdry finery; mere display; knick-knacks, gewgaws. *a.* tawdry, trifling.

Frisbee® (friz'bi), *n.* a plastic disc, used in throwing and catching games.

frisé, frisée (frēză'), *n.* endive. [F]

Frisian (friz'iən), **Friesian** (-frē'-), *a.* of, pertaining to or native of Friesland. *n.* the language of Friesland; a native of Friesland.

frisk (frisk), *v.i.* to leap, skip or gambol about; to frolic. *v.t.* (*coll.*) to search (a person) for firearms etc. *n.* a gambol, a frolic; (*coll.*) an act of frisking someone. **frisker,** *n.* **frisky,** *a.* **friskily,** *adv.* **friskiness,** *n.*

frisson (frē'sō), *n.* a shudder, a thrill. [F]

frit (frit), *n.* a calcined mixture of sand and fluxes ready to be melted in a crucible to form glass; applied to other vitreous compositions used in manufactures. *v.t.* (*past, p.p.* **fritted**) to heat so as to decompose and fuse.

frit-fly (frit'flī), *n.* a small fly that arrests the growth of wheat by boring into the bud.

fritillary (fritil'əri), *n.* the liliaceous genus *Fritillaria*, esp. *F. meleagris*, with flowers speckled with dull purple; a butterfly of the genus *Argynnis*, from their wings being marked like this flower.

fritter[1] (frit'ə), *n.* a piece of fruit, meat etc. dipped in a light batter and fried.

fritter[2] (frit'ə), *v.t.* to waste in trifles (often with *away*).

frivolous (friv'ələs), *a.* trifling, of little or no moment; inclined to unbecoming levity or trifling, silly. **frivolity** (-vol'-), *n.* **frivolously,** *adv.* **frivolousness,** *n.*

frizz (friz), *v.t.* to form (as the hair) into a curly, crinkled mass. *v.i.* to curl tightly. *n.* frizzed hair, a mass or row of curls. **frizzy,** *a.*

frizzle[1] (friz'l), *v.t.*, *v.i.* to form (into) crisp, tight curls. *n.* a curled or crisped lock of hair; frizzed hair. **frizzly,** *a.*

frizzle[2] (friz'l), *v.t.* to fry (bacon etc.) with a hissing noise. *v.i.* to make a hissing noise while being fried.

fro (frō), *adv.* away, backwards. **to and fro,** forwards and backwards.

frock (frok), *n.* the long upper garment worn by monks; a loose garment, formerly a loose over-garment worn by men; a woman's dress; a frock-coat; a smock-frock; a woven woollen tunic worn by sailors. **frock-coat,** *n.* a close-fitting body-coat, with broad skirts of the same length before and behind. **frocked,** *a.* **frocking,** *n.* material for smock-frocks.

Froebel (frœ'bl, frō'-), *a.* applied to the Froebel System, a form of kindergarten in which the child's senses are developed by handwork etc. [F.W.A. *Froebel*, 1782–1852]

frog[1] (frog), *n.* a squat, smooth-skinned, tailless amphibian of any species of the genus *Rana*; (usu. **Frog**) (*derog.*) a French person; an iron or steel plate to guide train wheels over an intersection in the track; the hollow in one or both faces of a brick; the block by which the hair is attached to the heel of a violin etc. bow. **a frog in one's throat,** phlegm on the vocal cords impeding speech. **frogfish,** *n.* the angler, *Lophius piscatorius*, and other fish. **froghopper,** *n.* a genus of small insects, remarkable for their leaping powers, living on plants. **frogman,** *n.* an underwater swimmer equipped with rubber suit, flippers, face mask etc. **frogmarch,** *v.t.* to carry face downwards between four people each holding a limb; to move (a person) by force, usu. by seizing from behind and propelling forwards, or by dragging backwards between two people each grasping an arm. **frog-mouth,** *n.* (*Austral.*) a bird of the mopoke family, a variety of goat-sucker. **frogspawn,** *n.* a gelatinous mass of frog's eggs.

frog[2] (frog), *n.* an ornamental fastening of looped braid used with a button or toggle for fastening military cloaks and undress coats, ladies' mantles etc.; the loop of a scabbard. **frogged,** *a.* **frogging,** *n.*

frog[3] (frog), *n.* a tender horny substance in the middle of the sole of a horse's foot.

frolic (frol'ik), *n.* a wild prank; an outburst of gaiety and mirth; a merry-making; a light-hearted entertainment. *v.i.* (*past, p.p.* **frolicked**) to play pranks; to frisk; to indulge in merry-making. **frolicsome** (-səm), *a.* **frolicsomely,** *adv.* **frolicsomeness,** *n.*

from (from), *prep.* away, out of (expressing separation, departure, point of view, distinction or variation); beginning with, after (expressing the starting-point or lower limit in time or space); arriving, coming, deriving (indicating the original location, source or model); by means of, because of, by reason of (expressing instrumentality, cause, source or motive). **from out,** out from, forth from. **from time to time,** at intervals, now and then.

fromage frais (fro'mahzh frā), *n.* a soft or runny curd cheese. [F]

frond (frond), *n.* (*Bot.*) a leaf-like expansion in which the functions of stem and foliage are not entirely differentiated, often bearing the organs of fructification, as in many cryptogams, esp. the ferns.

front (frŭnt), *n.* the forward part or side of anything; the most conspicuous part; the beginning, the first part; the part of a garment covering the chest; a face of a building, esp. the principal face; a frontage; a seaside promenade; a position directly ahead, or in the foremost part of something; the position of leadership; the vanguard; the area where fighting between opposing armies is taking place; the lateral space occupied by a military unit; the direction in which a line of troops faces; a particular sphere of activity; a group of people or organizations who make common cause together; the line of separation between air masses of different density and temperature; outward appearance or bearing; impudence, boldness; something which serves as a cover or disguise for secret or nefarious activities; a front man; a dicky. *a.* relating to or situated in or at the front; articulated at or towards the front of the mouth. *v.i.* to stand or be situated opposite to; to face, to look (to or towards); to confront, to meet face to face, to oppose; to furnish with a front; to be the leader or head of; to be the presenter of (a TV programme etc.). *v.i.* to face, to look, to be situated with the front (towards); to act as a front or cover for. **front of house,** (*Theat.*) those activities which involve direct contact with the public in a theatre, e.g. box office, selling programmes. **front-of-house,** *a.* **in front of,** before; in advance of; in the presence of. **out front,** (*Theat.*) in the audience or auditorium. **to front up,** (*Austral.*) to turn up. **front bench,** *n.* the foremost bench in either house of Parliament, assigned to ministers and leading members. **front**

bencher, *n.* **front door**, *n.* the principal entrance to a building. **front line**, *n.* the positions closest to the enemy in a battle; the most advanced and active and/or most exposed and dangerous positions in any field of activity. **frontline**, *a.* pertaining to or suitable for the front line in battle; neighbouring a hostile state or a scene of (armed) conflict. **front man**, *n.* a nominal leader or figurehead; the presenter of a TV programme. **front-page**, *a.* (worthy of) figuring on the front page of a newspaper. **front room**, *n.* a living room. **front-runner**, *n.* the leader or most favoured contestant in a race, election etc.; a person who runs or performs best when in the lead. **frontage**, *n.* the front part or face of a building; the extent of this; land between this and a road; the direction in which anything faces. **frontal**, *a.* situated on or pertaining to the front; belonging to the forehead. *n.* a small pediment over a door or window; an ornamental hanging or panel in front of an altar. **frontal lobe**, *n.* the front lobe of either side of the brain. **fronted**, *a.* formed with a front, as troops; changed into or towards a front sound. **frontward**, *a.*, *adv.* **frontwards**, *adv.*
frontier (frŭntiə', frŭn'-), *n.* that part of a country which fronts or borders upon another; (*chiefly N Am.*) the margins of settled or developed territory; (*often pl.*) the current limit of knowledge or attainment in a particular sphere. *a.* pertaining to or situated on the frontier. **frontiersman**, *n.*
frontispiece (frŭn'tispēs), *n.* a picture fronting the title-page of a book; a façade, a decorated front or chief entrance.
frontlet (frŭnt'lit), *n.* a small band or fillet worn on the forehead, a phylactery; the forehead in birds.
fronto- *comb. form.* pertaining to the forehead, the frontal bone of the forehead, or the frontal region.
frost (frost), *n.* the act or state of freezing; temperature below freezing-point; the state of the atmosphere that produces freezing; frosty weather; minute crystals of frozen dew or vapour, rime or hoar frost; coldness of manner or attitude; (*sl.*) a disappointment, a fiasco, a failure. *v.t.* to injure or kill by frost; to cover with or as with rime; to give a fine-grained, slightly roughened appearance to (glass, metal etc.); to dredge with fine sugar; (*chiefly Am.*) to ice (a cake). **degrees of frost**, (*with number*) degrees below freezing-point. **Jack Frost**, frost personified. **frost-bite**, *n.* inflammation often resulting in gangrene, usu. of the extremities, caused by exposure to extreme cold. **frost-bitten**, *a.* **frost-work**, *n.* the figures formed by frost on glass etc. **frosted**, *a.* covered with frost or any substance resembling frost; damaged by frost; having a rough, granulated surface; (*chiefly N Am.*) iced; having a shimmering or sparkling sheen. **frosting**, *n.* (*chiefly N Am.*) icing; a rough, granulated surface produced on glass, metal etc. in imitation of frost. **frostless**, *a.* **frosty**, *a.* producing frost; attended with frost; covered with or as with rime; cool, unenthusiastic. **frostily**, *adv.* **frostiness**, *n.*
froth (froth), *n.* foam, spume, the mass of small bubbles caused in liquors by agitation or fermentation; foamy excretion, scum; empty display of wit or rhetoric; light, unsubstantial matter. *v.t.* to cause to foam; to cover with froth. *v.i.* to form or emit froth. **frothless**, *a.* **frothy**, *a.* **frothily**, *adv.* **frothiness**, *n.*
frottage (frotahzh'), *n.* the technique of producing images or textures by rubbing with e.g. a pencil on a sheet of paper placed on top of an object.
frou-frou (froo'froo), *n.* a rustling, as of a silk dress.
frown (frown), *v.i.* to express displeasure or seriousness by contracting the brows; to look gloomy, threatening or with disfavour; to manifest displeasure (at or upon). *v.t.* to express with a frown. *n.* a knitting of the brows in displeasure or mental absorption; any sign of displeasure. **frowningly**, *adv.*
frowst (frowst), *n.* stuffiness, a fug. **frowsty**, *a.*

frowzy (frow'zi), *a.* musty, fusty, close; slovenly, unkempt, dirty. **frowziness**, *n.*
froze, *past* FREEZE.
frozen (frō'zn), *p.p.* FREEZE. *a.* preserved by freezing; very cold; fixed, immobilized; of prices etc., pegged at a certain level; of assets etc., not convertible; frigid, aloof, disdainful. **the frozen mitt**, (*sl.*) hostility, rejection. **frozen shoulder**, *n.* painful stiffness in the shoulder joint.
FRS, (*abbr.*) Fellow of the Royal Society.
fructify (frŭk'tifī), *v.t.* to make fruitful or productive; to fertilize. *v.i.* to bear fruit. **fructiferous** (-tif'-), *a.* bearing fruit. **fructification** (-fikā'-), *n.* the act or process of fructifying; (*Bot.*) the organs of reproduction; the fruit and its parts. **fructiform** (-tifawm), *a.* **fructose**, *n.* fruit-sugar. **fructuous** (-chu-), *a.* fruitful, fertile.
frugal (froo'gəl), *a.* thrifty, sparing; not profuse or lavish; economical in the use or expenditure of food, money etc. **frugality** (-gal'-), *n.* economy, thrift; a sparing use of anything. **frugally**, *adv.*
frugivorous (frəjiv'ərəs), *a.* feeding on fruit.
fruit (froot), *n.* the edible succulent product of a plant or tree in which the seeds are enclosed; (*Bot.*) the matured ovary or seed-vessel with other parts adhering thereto; the spores of cryptogams; (*pl.*) the vegetable products yielded by the earth, serving for food to humans and animals; (*Bibl.*) offspring; product, result or consequence; benefit, profit; (*chiefly N Am.*, *sl.*, *offensive*) a male homosexual. *v.i.*, *v.t.* to (cause to) produce fruit. **fruit bat**, *n.* a large Old World fruit-eating bat found in tropical and subtropical regions. **fruit-cake**, *n.* a cake containing currants etc. **fruit-fly**, *n.* a small fly of the genus *Drosophila*. **fruit-knife**, *n.* a knife with a blade of silver or other material that resists corrosion by acids, for paring and cutting fruit. **fruit machine**, *n.* a coin-in-the-slot gambling machine which spins symbols (as of fruit) past little windows in its front and pays out if certain combinations are visible when it stops. **fruit salad**, *n.* a mixture of fruits cut up. **fruit-sugar**, *n.* laevulose or fructose, obtained from fruit or honey. **fruit-tree**, *n.* a tree cultivated for its fruit. **fruitarian** (-teə'ri-), *n.* one that feeds on fruit. **fruiter**, *n.* a tree that bears fruit. **fruiterer**, *n.* one who deals in fruits. **fruitful**, *a.* producing fruit in abundance; productive, fertile; bearing children, prolific. **fruitfully**, *adv.* **fruitfulness**, *n.* **fruiting**, *a.* bearing fruit. **fruitless**, *a.* not bearing fruit; unsuccessful, unprofitable, useless, idle. **fruitlessly**, *adv.* **fruitlessness**, *n.* **fruitlet** (-lit), *n.* a drupel. **fruity**, *a.* like fruit in taste etc.; of wine, tasting of the grape; rich, full-flavoured; of the voice, round, mellow and rich; salacious, risqué; (*N Am.*, *sl.*) crazy; (*chiefly N Am.*, *sl.*) homosexual. **fruitiness**, *n.*
fruition (frooish'ən), *n.* the condition of bearing fruit; attainment, fulfilment; pleasure or satisfaction derived from attainment of a desire. **fruitive** (froo'-), *a.*
frumenty (froo'mənti), *n.* a dish made of wheat boiled in milk and flavoured with spices.
frump (frŭmp), *n.* an old-fashioned, prim or dowdy-looking woman. **frumpish**, **frumpy**, *a.*
frustrate (frustrāt'), *v.t.* to make of no avail; to defeat, to thwart, to balk; to nullify; to cause feelings of dissatisfaction or discouragement in. **frustrated**, *a.* thwarted; dissatisfied, discouraged. **frustration**, *n.*
frustule (frŭs'tūl), *n.* the covering or shell, usu. in two valves, of a diatom.
frustum (frŭs'təm), *n.* (*pl.* **-tums**, **-ta** (-tə)) the part of a regular solid next to the base, formed by cutting off the top; the part of a solid between two planes.
frutex (froo'teks), *n.* (*pl.* **frutices** (-tisēz)) a shrub. **frutescent** (-tes'-), *a.* shrubby. **frutescence**, *n.* **fruticose** (-kōs), *a.* of the nature of a shrub, shrubby; (*Zool.*) shrub-like in appearance (as certain zoophytes).
fry[1] (frī), *v.i.*, *v.t.* to cook or be cooked with fat in a pan over the fire. *n.* (*pl.* **fries**) a dish of anything fried; the li-

fry   332   **full**

ver, lights, heart etc. of pigs, sheep, calves and oxen; (*pl.*) French fries. **fryer, frier**, *n.* a vessel for frying. **frying-pan**, *n.* a shallow metal pan with a long handle, in which food is fried. **out of the frying-pan into the fire**, out of one trouble into a worse.
fry² (frī), *n.* young fish, esp. those fresh from the spawn; a swarm of young. **small fry**, unimportant, insignificant people or things.
FT, (*abbr.*) the Financial Times. **FT Index**, an indicator of the general trend in share prices, based on the movements of selected shares and published daily in the Financial Times.
ft., (*abbr.*) foot, feet; fort.
fth, fthm, (*abbr.*) fathom.
ft lb, (*abbr.*) foot-pound.
fuchsia (fū'shə), *n.* any of a genus of garden plants with pendulous funnel-shaped flowers. [L *Fuchs*, German botanist, 1501–66]
fuchsine (fook'sēn), *n.* a magenta dye of the rosaniline series.
fuck (fŭk), *v.i.*, *v.t.* (*taboo*) to have sexual intercourse (with). *n.* an act of sexual intercourse; a partner in sexual intercourse. *int.* used to express violent displeasure or (with an object, as *fuck you*) one's disregard or defiance of someone. **fuck all**, nothing at all. **not to give a fuck**, not to care in the least. **to fuck about, around**, to waste time, to mess around; to treat inconsiderately. **to fuck off**, to go away. **to fuck up**, to botch, to damage; to make a mess of. **fuck-up**, *n.* **fucked**, *a.* (*taboo*) broken, damaged, kaput; exhausted. **fucker**, *n.* (*taboo*) a (stupid) person, fellow. **fucking**, *n.* (*taboo*) sexual intercourse. *a.* used to express one's annoyance with something, often a virtually meaningless expletive. *adv.* very, extremely.
Fucus (fū'kəs), *n.* (*pl.* **-ci** (-sī)) a genus of algae, containing some of the commonest seaweeds. **fucoid**, *a.* resembling a fucus. *n.* a fossil plant, like a fucus. **fucoidal** (-koi'-), *a.*
fuddle (fŭd'l), *v.t.* to make stupid with drink, to intoxicate; to confuse. *n.* the state of being muddled.
fuddy-duddy (fŭd'idŭdi), *n.* (*coll.*) an old fogy; a carper. *a.* old-fogyish, old-fashioned; stuffy, pompous; prim, censorious.
fudge¹ (fŭj), *int.* nonsense, stuff, humbug. *n.* nonsense; a made-up or nonsensical story; a soft confection of chocolate, candy etc.
fudge² (fŭj), *v.t.* to patch or make up, to fake; to contrive in a makeshift, careless way; to falsify, to make imprecise, esp. as a means of covering up unpalatable facts; to dodge, to evade. *v.i.* to contrive in a makeshift way; to be evasive and imprecise. *n.* (*Print.*) an attachment on rotary machines for the insertion of a small form giving an item of late news; a makeshift compromise; an evasion.
fuel (fū'əl), *n.* combustible matter, such as wood, coal, peat etc., for fires; fissile material for use in a nuclear reactor; anything which serves to feed or increase passion or excitement. *v.t.* (*past*, *p.p.* **fuelled**) to supply or store with fuel. *v.i.* to take fuel. **fuel cell**, *n.* a cell in which chemical energy is continuously converted into electrical energy. **fuel injection**, *n.* a system whereby fuel is introduced directly into the combustion chamber of an internal-combustion engine, obviating the need for a carburettor. **fuelless**, *a.*
fug (fŭg), *n.* the close atmosphere of an unventilated room. **fuggy**, *a.*
fugacious (fūgā'shəs), *a.* fleeting, lasting but a short time; (*Bot.*) falling off early. **fugacity** (-gas'-), *n.* fleetingness, transience; (*Chem.*) the tendency to expand or escape.
fugal, fugato FUGUE.
-fuge, *suf.* (*Med.*) expelling, driving out, as in *febrifuge*.
fugitive (fū'jitiv), *a.* fleeing, running away, having taken flight, runaway; transient, not stable or durable, volatile; fleeting, evanescent, ephemeral. *n.* one who flees from danger, pursuit, justice, bondage or duty; a runaway, a deserter, a refugee; a person or thing hard to be caught or detained.
fugleman (fū'gəlman), *n.* (*pl.* **-men**) a soldier who takes up a position in front of a company as a guide to the others in their drill; one who sets an example for others to follow, a leader.
fugue (fūg), *n.* a polyphonic composition on one or more short subjects, which are repeated by successively entering voices and developed contrapuntally; (*Psych.*) loss of memory coupled with disappearance from one's usual resorts. **fugal**, *a.* (*Mus.*) in the style of a fugue. **fugally**, *adv.* **fugato** (foogah'tō. fū-), *adv.* in the fugue style but not in strict fugal form.
Führer (fū'rə, fü'-), *n.* the head of the National-Socialist German government, Adolf Hitler (1889–1945). [G, leader]
-ful (-fəl), *suf.* full of, abounding in, having, as in *artful*, *beautiful*; the quantity or number required to fill, as in *cupful, handful*.
Fulah (foo'lah), *n.* a member of one of the dominant races in the Sudan; the language of this race.
fulcrum (ful'krəm, fŭl'-), *n.* (*pl.* **-crums, -cra** (-krə)) the fixed point on which the bar of a lever rests or about which it turns; a means of making any kind of force or influence effective; (*Bot.*, *Zool.*) an additional organ, as a stipule, scale, spine etc. **fulcral, fulcrant**, *a.*
fulfil, (*esp. N Am.*) **fulfill** (fulfil'), *v.t.* (*past, p.p.* **fulfilled**) to accomplish, to carry out, to execute, perform; to satisfy, to correspond to, to comply with; to finish, to complete (a term of office etc.); to realize the potential of. **fulfiller**, *n.* **fulfilment**, *n.*
fulgent (fŭl'jənt), *a.* shining, dazzling, exceedingly bright. **fulgency**, *n.* **fulgently**, *adv.* **fulgorous**, *a.* **fulguration** (-gū-), *n.* (*usu. pl.*) flashing, as of lightning. **fulgurite** (-gūrit), *n.* (*Geol.*) a vitrified tube in sand, supposed to be produced by the action of lightning.
fuliginous (fūlij'inəs), *a.* sooty, smoky, soot-coloured; dusky, gloomy. **fuliginously**, *adv.* **fuliginosity** (-nos'-), *n.*
full¹ (ful), *a.* filled up, replete; having no space empty, containing as much as the limits will allow; well supplied, having abundance (of); filled to repletion, satisfied, esp. with food and drink; charged or overflowing (with feeling etc.); preoccupied or engrossed with; plentiful, copious, ample; having all rights and privileges; complete, perfect, at the height of development; visible in its entire dimensions; of the moon, having the whole disc illuminated; ample in volume or extent, swelling, plump; strong, sonorous; high, as the tide. *adv.* quite, equally; completely, exactly, directly; very. *n.* complete measure or degree; the utmost or fullest extent; the highest state or point. *v.t.* to give fullness to, to make full. *v.i.* to become full. **in full**, completely, without abridgment, abatement or deduction. **to be full of oneself**, to have an exaggerated view of one's own importance. **to the full**, to the utmost extent. **full back**, *n.* (*Football etc.*) a defensive player, usu. the rearmost in any on-field formation. **full-blooded**, *a.* vigorous; sensual; of pure blood. **full-blown**, *a.* fully expanded, as a flower; mature, perfect, fully developed; fully qualified. **full-bodied**, *a.* having a full, rich flavour or quality. **full-bottomed**, *a.* having a large bottom, as distinguished from a bob-wig. **full brother, sister**, *n.* one having both parents in common. **full-cream**, *a.* of milk, not skimmed. **full-cry**, *n.* the state of giving tongue in chorus, as a pack of hounds. **full dress**, *n.* dress worn on ceremonious occasions; evening dress. *a.* at which full dress is to be worn. **full-faced**, *a.* having a broad chubby face; facing directly towards the spectator. **full-flavoured**, *a.* strongly flavoured, highly spiced. **full-fledged** FULLY FLEDGED. **full-frontal**, *a.* of a nude, with the genitals fully revealed; unrestrained, omitting no detail. **full house**, *n.* in poker, three of a kind and a pair; in bingo etc., the set of numbers needed to win; an auditorium filled to capacity.

full-length, *a.* of the entire figure; of the standard length.
full-out, *adv.* at full power. **full-page**, *a.* taking up a whole page. **full pitch**, *n.*, *adv.* (a delivery which reaches the batsman) without touching the ground. **full-pitched**, *a.* **full-scale**, *a.* of the same size as the original; using all available resources; all-out. **full stop**, *n.* a period (.), the longest pause in reading; an abrupt finish. **full term**, *n.* the normal or expected end date for a pregnancy. **full-term**, *a.* **full time**, *n.* the end of play in e.g. a football match. *adv.* for the whole of the (standard) working week. **full-time**, *a.* **full-timer**, *n.* **full-toss** FULL PITCH. **full up**, *adv.* quite full; with no room for more. **fullish**, *a.* **fully**, *adv.* completely, entirely, quite. **fully-fashioned**, *a.* shaped to the lines of the body. **fully-fledged**, *a.* of a bird, having all its feathers; fully qualified, having full status as. **fullness, fulness** (ful'nəs), *n.* the state or quality of being full; completeness, satiety; largeness, richness, volume, force. **in the fullness of time**, at the destined time.
full² (ful), *v.t.* to cleanse and thicken (cloth). **fuller**, *n.* one whose occupation is to full cloth. **fuller's earth**, *n.* an argillaceous earth which absorbs grease, used in fulling cloth.
fullness, fully FULL¹.
fulmar (ful'mə), *n.* a sea-bird, *Fulmaris glacialis*, allied to the petrels, abundant in the Arctic seas.
fulminate (ful'mināt), *v.i.* to explode with a loud noise or report; to thunder out denunciations (at or against). *v.t.* to cause to explode; to utter (threats, denunciations or censures). **fulminant**, *a.* fulminating; of diseases, developing suddenly. **fulminating**, *a.* thundering, explosive. **fulmination**, *n.* **fulminatory**, *a.*
fulness (ful'nəs), FULLNESS under FULL¹.
fulsome (ful'səm), *a.* esp. of compliments, flattery etc., characterized or made disgusting by excess or grossness, coarse, excessive. **fulsomely**, *adv.* **fulsomeness**, *n.*
fulvous (ful'vəs), *a.* (*chiefly Bot. Zool.*) tawny, reddish-yellow.
fumarole (fū'mərōl), *n.* a hole in the ground in a volcanic region forming an exit for subterranean vapours. **fumarolic** (-rol'-), *a.*
fumble (fŭm'bl), *v.i.* to grope about; to act, esp. to use one's hands, in an uncertain aimless, or awkward manner. *v.t.* to handle or manage awkwardly; to fail to catch or hold; to deal with in an uncertain or hesitating manner. **fumbler**, *n.* **fumblingly**, *adv.*
fume (fūm), *n.* (*usu. pl.*) a smoke, vapour or gas, esp. a malodorous or toxic one; (*usu. pl.*) a narcotic vapour, esp. such as is supposed to rise from alcoholic liquors and to affect the brain; mental agitation, esp. an angry mood. *v.i.* to emit smoke or vapour; to pass off in smoke or vapour; to show irritation, to fret, to chafe. *v.t.* to dry, perfume, stain or cure with smoke, esp. to darken (oak, photographic plates etc.) with chemical fumes, as of ammonia; to dissipate in vapour. **fume cupboard, chamber**, *n.* an enclosed area with an extractor, used for chemical experiments in which fumes are given off. **fumy**, *a.*
fumigate (fū'migāt), *v.t.* to subject to the action of smoke or vapour, esp. for the purpose of disinfection. **fumigant**, *n.* a substance used for fumigating. **fumigation**, *n.* **fumigator**, *n.* one who or that which fumigates, esp. an apparatus for applying smoke, gas etc. for the purpose of cleansing or disinfecting, or perfuming.
fumitory (fū'mitəri), *n.* a herb belonging to the genus *Fumaria*, esp. *F. officinalis*, formerly used for skin diseases.
fumy FUME.
fun (fŭn), *n.* (a source of) amusement, merriment, jollity; hectic activity or argument. *a.* enjoyable; amusing, entertaining. **a figure of fun**, a butt of ridicule. **for fun, for the fun of it**, for pleasure simply.. **fun and games**, (*iron.*) hectic activity, trouble. **in fun**, as a joke. **like fun**, (*coll.*) energetically; thoroughly. **to make fun of, to poke fun at**, to hold up to or turn into ridicule. **funfair**, *n.* a usu. outdoor show with rides, sideshows, games of skill and other amusements. **fun fur**, *n.* inexpensive, artificial fur (often dyed) for clothes, seat covers etc. **fun run**, *n.* a long-distance race entered into for enjoyment and usu. to collect money for charity rather than for serious competition.
funambulist (fūnam'būlist), *n.* a performer on the tight or slack rope; a rope-walker or rope-dancer. **funambulism**, *n.*
function (fŭngk'shən), *n.* the specific activity, operation or power belonging to an agent; duty, occupation, office; a public or official ceremony; a religious service of an elaborate kind; (*coll.*) a social entertainment of some importance; (*Physiol.*) the specific office of any animal or plant organ; (*Math.*) a quantity dependent for its value on another or other quantities so that a change in the second correspondingly affects the first. *v.i.* to perform a function or duty; to operate. **functional**, *a.* pertaining to some office or performing a function; practical, utilitarian, eschewing ornament; able to perform (its function), working; (*Physiol.*) pertaining to or affecting the action or functions of an organ, not its substance or structure; (*Math.*) relating to or depending on a function. **functionalism**, *n.* design centred on or determined by the use of a thing; a theory or practice that emphasizes purpose or usefulness. **functionally**, *adv.* **functionary**, *n.* one who holds any office or trust; an official. **functionless**, *a.*
fund (fŭnd), *n.* a sum of money or stock of anything available for use or enjoyment; assets, capital; a sum of money set apart for a specific object, permanent or temporary; (*pl.*) money lent to a government and constituting a national debt; (*pl.*) the stock of a national debt regarded as an investment; (*pl.*) (*coll.*) money, finances, pecuniary resources. *v.t.* to convert into a single fund or debt, esp. to consolidate into stock or securities bearing interest at a fixed rate; to place in a fund; to provide money for. **fund-raiser**, *n.* one who raises money for an (often charitable) organization, project etc. **fund-raising**, *n.*, *a.* **fundable**, *a.* **funded**, *a.* invested in public funds; forming part of the national debt of a country, existing in the form of bonds bearing regular interest. **funded debt**, *n.* **fundless**, *a.*
fundament (fŭn'dəmənt), *n.* (*chiefly euphem. or facet.*) the lower part of the body, the buttocks; the anus.
fundamental (fŭndəmen'təl), *a.* pertaining to or serving as a foundation or base; essential, primary, original, indispensable. *n.* a principle, rule or article forming the basis or groundwork; the lowest note or 'root' of a chord. **fundamental bass**, *n.* (*Mus.*) a bass consisting of a succession of fundamental notes. **fundamental particle** ELEMENTARY PARTICLE under ELEMENT. **fundamentalism**, *n.* (*Christianity*) belief in the literal truth of the Bible; (*Islam*) strict observance of the teachings of the Koran and of Islamic law. **fundamentalist**, *n.*, *a.* **fundamentality** (-tal'-), *n.* **fundamentally**, *adv.*
funeral (fū'nərəl), *a.* pertaining to or connected with the committal of the dead. *n.* the solemn and ceremonious committal of the dead; a funeral service; a procession of persons at a funeral. **it's your** etc. **funeral**, it's your affair. **funeral director**, *n.* an undertaker. **funeral home**, (*esp. N Am.*) **funeral parlour**, *n.* a place where the dead are prepared for burial or cremation and funerals may be held. **funebrial** (-nē'bri-), *a.* funereal. **funerary**, *a.* pertaining to funerals. **funereal** (-niə'ri-), *a.* pertaining to or suitable for a funeral; dismal, sad, mournful; gloomy, dark. **funereally**, *adv.*
funfair FUN.
fungible (fŭn'jibl), *a.* (*Law*) of such a nature that it may be replaced by another thing of the same class. *n.* (*pl.*) movable goods which may be valued by weight or measure.

fungus (fŭng'gəs), *n.* (*pl.* **-gi** (-gī, -jī)) a mushroom, toadstool, mould, mildew, or other cryptogamous plant, destitute of chlorophyll and deriving its nourishment from organic matter; (*Path*.) a morbid growth or excrescence of a spongy nature; something of rapid or parasitic growth. **fungal,** *a.* of, pertaining to or of the nature of a fungus. **fungicide** (-jisid), *n.* anything that destroys fungi or their spores. **fungicidal,** *a.* **fungiform** (-jifawm), **fungiliform** (-jil'-), *a.* having a termination like the head of a mushroom. **fungivorous** (-jiv'-), *a.* feeding on fungi. **fungoid** (-goid), *a.* of the nature of or like a fungus. **fungous,** *a.* like or of the nature of a fungus; springing up suddenly, ephemeral; spongy, unsubstantial.

funicular (fūnik'ūlə), *a.* pertaining to, consisting of, or depending on a rope or cable. *n.* a railway worked by means of a cable, usu. a mountain railway. **funicle** (fū'nikl), *n.* (*Bot.*) a funiculus. **funiculus** (-ləs), *n.* (*pl.* **-culi** (-lī)) the umbilical cord; a number of nerve-fibres enclosed in a tubular sheath; (*Bot.*) a cord connecting the seed with the placenta.

funk[1] (fŭngk), *n.* (*coll.*) a state of fear or panic; a coward. *v.i.* (*coll.*) to be in a state of terror; to flinch, to shrink in fear or cowardice. *v.t.* to be afraid of; to try to evade through fear or cowardice; (*usu. p.p.*) to frighten, to scare. **blue funk,** (*coll.*) abject terror. **funkhole,** *n.* a dug-out; any refuge one can retreat to. **funker,** *n.* **funky,** *a.*

funk[2] (fŭngk), (*sl.*) funky music. **funkster,** *n.* one who plays funk. **funky,** *a.* (*sl.*) of jazz, pop etc., earthy, unsophisticated, soulful, like early blues; (*sl.*) with it.

funnel (fŭn'əl), *n.* a conical vessel usu. terminating below in a tube, for conducting liquids etc. into vessels with a small opening; a tube or shaft for ventilation, lighting etc.; the chimney of a steamship or steam-engine. *v.t.* (*past, p.p.* **funnelled**) to pour or pass (as if) through a funnel. *v.i.* to move (as if) through a funnel. **funnel-web spider,** a large venomous spider found in New South Wales. **funnelled,** *a.* having a funnel or funnels; funnel-shaped.

funny (fŭn'i), *a.* droll, comical, laughable; causing mirth or laughter; strange, curious, puzzling; suspicious; underhand, involving trickery; (*coll.*) slightly unwell. *n.* (*coll.*) a joke; (*pl.*) comic strips or the comics section of a newspaper. **funny-bone,** *n.* the lower part of the elbow over which the ulnar nerve passes, a blow on which causes a curious tingling sensation. **funny business,** *n.* trickery; dubious or suspicious goings on. **funny farm,** *n.* (*coll.*) a mental hospital. **funny man,** *n.* a clown; a buffoon or wag. **funnily,** *adv.* **funniness,** *n.*

fur (fœ), *n.* the soft fine hair growing thick upon certain animals, distinct from ordinary hair; (a piece of) the skin, esp. dressed skin, of such an animal; a garment made of fur; the downy covering on the skin of a peach; a coat of morbid matter collected on the tongue; a crust deposited on the interior of kettles etc. by hard water. *v.t.* (*past, p.p.* **furred**) to cover, line or trim with fur; to cover or coat with morbid matter; to nail pieces of timber to (as joists or rafters) in order to bring them into a level. *v.i.* to become encrusted with fur or scale, as the inside of a boiler. **fur and feather,** fur-bearing animals and game birds. **to make the fur fly,** to create a scene, to start a row. **fur-seal,** *n.* a seal or sea-bear yielding a fur valuable commercially. **furred,** *a.* bearing fur; lined or ornamented with fur; coated with fur or scale. **furrier** (fŭr'iər), *n.* a dealer in furs; one who prepares and sells furs. **furring,** *n.* trimming or lining with furs; a deposit of scale (on the inside of boilers etc.); thin pieces fixed on the edge of timber to make the surface even; a lining on a brick wall to prevent dampness. **furry,** *a.* covered or clad in fur; made of fur; resembling fur; coated with a scale or deposit.

fur., (*abbr.*) furlong.

furbelow (fœ'bilō), *n.* a flounce, a ruffle; (*pl.*) finery. *v.t.* to furnish or trim with furbelows.

furbish (fœ'bish), *v.t.* to rub so as to brighten, to polish up; to renovate, to restore the newness or brightness of. **furbisher,** *n.*

furcate (fœ'kāt, -kət), *a.* forked, dividing into branches like the prongs of a fork. *v.i.* to fork. **furcation** (-kā'-), *n.*

furcula (-kūlə), *n.* (*pl.* **-lae** (-lē)) the two clavicles of birds anchylosed together so as to form one V-shaped bone, the wishbone. **furcular,** *a.*

furioso (fūriō'sō), *adv.* (*Mus.*) with fury or vehemence. [It. furious]

furious (fū'riəs), *a.* full of fury, raging, violent, frantic; marked by vehemence or impetuosity, tempestuous, unrestrained. **furiously,** *adv.* **furiousness,** *n.*

furl (fœl), *v.t.* to roll up (a sail) and wrap about a yard, mast or stay; to roll, wrap, fold or close (up). *v.i.* to become rolled or folded up.

furlong (fœ'long), *n.* a measure of length, the eighth part of a mile, 220 yd. (201 m).

furlough (fœ'lō), *n.* leave of absence, esp. to a soldier. *v.t.* to grant leave of absence to.

furmenty (fœ'mənti), **furmety** (-məti), FRUMENTY.

furnace (fœ'nəs), *n.* a chamber in which fuel is burned for the production of intense heat, esp. for melting ores, metals etc.; a closed fireplace for heating a boiler, hot-water pipes etc.; a time, place or occasion of severe trial or torture.

furnish (fœ'nish), *v.t.* to provide or supply (with); to equip, to fit up, esp. (a house or room) with movable furniture; to supply, to afford, to yield. **furnisher,** *n.* **furnishings,** *n.pl.* furniture, apparatus.

furniture (fœ'nichə), *n.* equipment, equipage, outfit; movable articles, esp. chairs, tables etc. with which a house or room is furnished; (*Print.*) the material, of wood, metal or plastic, which keeps the pages firmly fixed in the chase, and separates them so as to allow a uniform margin when printed; accessories; locks, door and window trimmings etc.; the masts and rigging of a ship.

furore (fūraw'ri), (*esp. N Am.*) **furor** (fū'raw), *n.* great excitement or enthusiasm; a craze, a rage; an uproar, an outburst of public indignation.

furrier, furring FUR.

furrow (fŭ'rō), *n.* a trench in the earth made by a plough; a narrow trench, groove or hollow; a wrinkle on the face. *v.t.* to make grooves, furrows or wrinkles in; to mark (the face) with deep wrinkles. **furrowless,** *a.* **furrowy,** *a.*

furry FUR.

furry dance (fœ'ridahns), *n.* a festival dance through the streets of Helston and certain other Cornish towns on 8 May, called Flora's day.

further (fœ'dhə), *a.* more remote; more advanced; going or extended beyond that already existing or stated, additional (chiefly used when distance in space is not implied, cp. FARTHER). *adv.* to a greater distance, degree or extent; moreover, in addition, also. *v.t.* to help forward, to advance, to promote. **further education,** *n.* formal, post-school education other than at a university or polytechnic. **furtherance,** *n.* promotion, help, assistance. **furtherer,** *n.* **furthermore,** *adv.* moreover, besides. **furthermost,** *a.* furthest, most remote. **furthest,** *a.* most remote in time or place. *adv.* at or to the greatest distance or extent.

furtive (fœ'tiv), *a.* stealthy, sly; secret, surreptitious, designed to escape attention; obtained by or as by theft. **furtively,** *adv.* **furtiveness,** *n.*

furuncle (fū'rŭngkl), *n.* a boil. **furuncular, -culoid, -culous** (-rŭng'-), *a.*

fury (fū'ri), *n.* vehement, uncontrollable anger, rage; a fit of raving passion; impetuosity, violence; intense, ecstatic passion, inspiration, enthusiasm; (**Fury**) one of the three avenging goddesses of classical mythology; hence, a furious woman, a virago. **like fury,** (*coll.*) with furious energy.

furze (fœz), *n.* the gorse or whin, *Ulex europaeus*, a spinous evergreen shrub with bright yellow flowers, common on waste, stony land. **furzy**, *a.*

fuscous (fŭs'kəs), *a.* brown tinged with grey or black; dingy.

fuse[1] (fūz), *v.t.* to melt, to reduce to a liquid or fluid state by heat; to unite by or as by melting together; to cause to fail by blowing a fuse. *v.i.* to melt, to become fluid; to become united by or as by melting together; to fail because of a blown fuse. *n.* (a device containing) a strip of fusible wire or metal which melts if the current in an electric circuit exceeds a certain value. **fuse box**, *n.* a box containing one or more fuses. **fusible**, *a.* capable of being fused or melted. **fusibility** (-bil'-), *n.* **fusion** (-zhən), *n.* the act of melting or rendering liquid by heat; the state of being so melted or liquefied; union by or as by melting together, blending; a product of such melting or blending; nuclear fusion; coalescence or coalition (as of political parties). **fusion bomb**, *n.* a bomb, e.g. the hydrogen bomb, whose energy results from nuclear fusion. **fusion reactor**, *n.* a nuclear reactor operating on the fusion principle. **fusionism**, *n.* **fusionist**, *n.* one who advocates political fusion.

fuse[2] (fūz), *n.* a tube, cord or casing filled or saturated with combustible material, and used for igniting a charge in a mine or projectile; (*esp. N Am.* **fuze**) a detonating device in a bomb or shell. *v.t.* (*esp. N Am.* **fuze**) to furnish with a fuse or fuses.

fusee, fuzee (fūzē'), *n.* the cone round which the chain is wound in a clock or watch; a fuse; a match with a mass of inflammable material at its head, used for lighting pipes etc. in a wind.

fuselage (fū'zəlahzh), *n.* the main body of an aeroplane.

fusel oil (fū'zl), *n.* a poisonous oily product, composed chiefly of amyl alcohol, formed during the manufacture of corn, potato or grape spirits.

fusiform (fū'zēfawm), *a.* shaped like a spindle, tapering at both ends.

fusil (fū'zil), *n.* an obsolete firelock, lighter than a musket.

fusilier (-liə'), *n.* orig. a soldier armed with a fusil, as distinguished from a pikeman or archer, still applied in the British army to certain regiments of the line. **fusillade** (-lād'), *n.* a continuous, rapid discharge of firearms; a rapid succession of blows, critical comments etc. *v.t.* to shoot down or storm by fusillade.

fusion FUSE[1].

fuss (fŭs), *n.* excessive activity, labour or trouble, taken or exhibited; unnecessary bustle or commotion, too much ado; undue importance given to trifles or petty details. *v.i.* to make much ado about nothing; to worry, to be nervous or restless. *v.t.* to worry, to agitate. **to make, kick up a fuss**, to cause a commotion, esp. by complaining. **to make a fuss of**, to lavish attention on as a sign of affection. **fussbudget, fusspot**, *n.* (*coll.*) one who fusses. **fussy**, *a.* nervous, excitable, esp. over small details; finicky, fastidious; overelaborate, overornate. **fussily**, *adv.* **fussiness**, *n.*

fustanella (fŭstənel'ə), *n.* the short white skirt worn by men in Greece and Albania.

fustian (fŭs'chən), *n.* a coarse twilled cotton or cotton and linen cloth, with short velvety pile; applied as an old trade-name to velveteen, corduroy etc.; inflated or pompous writing or speaking, bombast. *a.* bombastic. **fustianed**, *a.*

fustic (fŭs'tik), *n.* a yellow wood used in dyeing, that of *Maclura tinctoria*, a large W Indian tree, sometimes called in distinction **old fustic**, and that of *Rhus cotinus*, a bushy shrub of southern Europe, now usu. called **young fustic**.

fusty (fŭs'ti), *a.* mouldy, musty; old-fashioned, out-dated. **fustiness**, *n.*

fut., (*abbr.*) future.

futhark (foo'thahk), **futhorc, futhork** (-thawk), *n.* the Runic alphabet. [from the first six letters $f\ u\ \þ\ o\ r\ k$]

futile (fū'til), *a.* useless; of no effect; trifling, worthless, frivolous. **futilely**, *adv.* **futility** (-til'-), *n.*

futon (foo'ton), *n.* a Japanese floor-mattress used as a bed.

futtock (fŭt'ək), *n.* (*Naut.*) one of the timbers in the compound rib of a vessel.

future (fū'chə), *a.* that will be; that is to come or happen hereafter; (*Gram.*) expressing action yet to happen; that will be something specified, as *our future king*. *n.* time to come; that which will be or will happen hereafter; prospective condition, state, career, etc.; likelihood of success; (*Gram.*) the future tense; (*pl.*) goods, stocks etc.; bought or sold for future delivery. **future perfect**, *n.*, *a.* (the tense) expressing an action as completed in the future, as *it will have been*. **futureless**, *a.* **futurism, Futurism**, *n.* an early 20th-cent. movement in painting, poetry and sculpture aiming at visualizing the movement and development of objects, instead of the picture they present at a given moment. **futurist**, *n.* one who holds that a great part of Scripture prophecy (esp. of the Apocalypse) is still to be fulfilled; (**futurist, Futurist**) a follower of futurism. **futuristic** (-ris'-), *a.* of the future or futurism; of design, architecture etc., ultramodern, apparently anticipating styles of the future. **futurity** (-tūə'-), *n.* the state of being future; future time, esp. eternity; (*often pl.*) a future event, thing to come. **futurology** (-rol'-), *n.* the prediction of future developments from current, esp. sociological and technological, trends. **futurological** (-loj'-), *a.* **futurologist** (-rol'-), *n.*

fuze FUSE[2].

fuzee FUSEE.

fuzz (fŭz), *v.i.* to fly off in minute particles. *n.* minute light particles of down or similar matter, fluff; fuzziness. **the fuzz**, (*sl.*) the police. **fuzzy**, *a.* covered with fuzz; having many small, tight curls; blurred, indistinct; pertaining to recognizing or operating according to principles that allow for uncertain or in-between states. **fuzzy-wuzzy** (-wŭzi), *n.* (*Austral. offensive*) a native of new Guinea; (*Kipling*) a Sudanese, a Sudanese fighter; (*loosely, derog.*) any black (African) person. **fuzzily**, *adv.* **fuzziness**, *n.*

FWD, (*abbr.*) four-wheel drive.

fwd., (*abbr.*) forward.

-fy (fī), *suf.* to make, to produce; forming verbs, to bring into a certain state, as in *beautify, deify, horrify, petrify*; (*coll.*) as in *argufy, Frenchify, speechify*.

fylfot (fil'fot), *n.* an ancient figure consisting of a great cross with arms continued at right angles, used heraldically, as a mystic symbol, or for decoration; called also gammadion and swastika.

fz, (*abbr.*) sforzando.

G

G, g, gee, *n*. the seventh letter, and fifth consonant, of the Roman and English alphabets; **G** (*pl*. **Gs, G's, Gees**) (*Mus*.) the fifth note of the diatonic scale of C major; the key or scale corresponding to this; the fourth string of a violin, the third of the viola and violoncello, the first of the double-bass; the mark of the treble clef; German **g**, (*Phys*.) a symbol of the acceleration due to gravity, about 32 ft (9.8 m) per second; (*chiefly N Am., sl.*) a symbol for *grand* (1000 dollars or pounds). **G-man**, *n*. a special agent of the US Federal Bureau of Investigation. [initials of Government *man*] **G-string**, *n*. a garment consisting of a small piece of cloth covering the pubic area and attached front and back to a waistband that is worn e.g. by an entertainer when performing striptease.
Ga, (*chem. symbol*) gallium.
gab (gab), *n*. idle talk, chatter; (*Sc*.) the mouth. *v.i.* to talk glibly, to chatter, to prate. **the gift of the gab**, (*coll.*) a talent for speaking, fluency. **gabfest**, *n*. (*chiefly N Am., coll.*) a prolonged session of speeches, discussion or gossip; a gathering for this. **gabby**, *a*. talkative, loquacious. **gabster**, *n*.
gabardine (gab'ədēn, -dēn'), GABERDINE.
gabble (gab'l), *v.i.* to make cackling sounds like a goose; to talk rapidly and incoherently. *v.t.* to say very rapidly or inarticulately. *n*. rapid, incoherent or inarticulate talk; cackle, chatter. **gabbler**, *n*.
gabbro (gab'rō), *n*. rock composed of feldspar and diallage, sometimes with serpentine or mica.
gabelle (gəbel'), *n*. a tax or duty, esp. the tax on salt in France before the Revolution (1789).
gaberdine, gabardine (gab'ədēn, -dēn'), *n*. a long coarse gown or cloak, worn in the Middle Ages by Jews and others; a cloth with a corded effect, used largely for raincoats; a rainproof coat made of this.
gabion (gā'biən), *n*. a cylindrical basket of wicker- or iron-work, filled with earth, used for foundations etc. in engineering work and (*esp. formerly*) for shelter against an enemy's fire while trenches are being dug. **gabionade** (-nād'), *n*. a work composed of gabions. **gabionage**, *n*. gabions collectively. **gabioned**, *a*.
gable (gā'bl), *n*. the triangular portion of the end of a building, bounded by the sides of the roof and a line joining the eaves; a wall with upper part shaped like this; a canopy or other architectural member with this shape. **gable-end**, *n*. the end wall of a building with such an upper part. **gable-roof**, *n*. a ridge roof ending in a gable. **gable-window**, *n*. a window in a gable or with a gable over it. **gabled**, *a*. having gables.
gaby (gā'bi), *n*. a fool, a simpleton.
gad[1] (gad), *v.i.* (*past, p.p.* **gadded**) to rove or wander idly (about, out etc.), esp. in search of pleasure. *n*. gadding or roaming about. **gadabout**, *n*. one who gads about habitually. **gadder**, *n*. **gaddingly**, *adv*.
gad[2] (gad), *int*. (*dated*) an exclamation of surprise etc. **begad** or **by gad, gadzooks** (-zooks'), old-fashioned, euphemistic oaths.
Gadarene (gad'ərēn), *a*. headlong, precipitate, panic-stricken. [in allusion to the Gadarene swine (Matthew 8:28ff)]
gadfly (gad'flī), *n*. an insect of the genus *Tobanidae* or *Oestrus*, which bites cattle and other animals, a breeze-fly; a person, thing or impulse that irritates or torments.
gadget (gaj'it), *n*. a tool, an appliance; a contrivance for making a job easier; a trick of the trade. **gadgetry**, *n*. gadgets collectively.
Gadhelic (gədel'ik), *a*. of or pertaining to the branch of the Celtic race that includes the Gaels of Scotland, the Irish and the people of the Isle of Man. *n*. the language spoken by this branch of the Celtic race.
gadoid (gā'doid), *n., a*. (any fish) belonging to the family *Gadidae*, which comprises the cod-fishes.
gadolinite (gad'əlinīt), *n*. a black, vitreous silicate of yttrium, formed in crystals. **gadolinium** (gadəlin'iəm), *n*. a soft metallic element, at. no. 64; chem. symbol Gd, of the rare-earth group. [J. *Gadolin*, 1760–1852, Finnish mineralogist]
gadroon (gədroon'), *n*. (*usu. pl.*) an ornament consisting of a series of convex curves, used in architecture and metalwork for edgings, mouldings etc. **gadrooned**, *a*.
gadwall (gad'wawl), *n*. a large freshwater duck, *Anas strepera*, of N Europe and America.
gae (gā), (*Sc*.) var. of GO.
Gael (gāl), *n*. a Scottish Celt; (*less commonly*) an Irish Celt. **Gaelic** (gā'-, gal'-), *a*. of or pertaining to the Gaels or their language. *n*. the language spoken by the Gaels; (*less commonly*) the language of the Irish and Manx Celts. **Gaelic coffee**, *n*. IRISH. **Gaelic football**, *n*. a game involving two teams of 15 players, the object of which is to kick, bounce or punch a ball into a net stretched between two posts or over a crossbar above the net. **Gaelic League**, *n*. an association formed to further the revival of the Irish language and ancient culture.
gaff[1] (gaf), *n*. a stick with a metal hook at the end, used by anglers to land heavy fish; the spar which extends the upper edge of fore-and-aft sails not set on stays. *v.t.* to seize or land with a gaff.
gaff[2] (gaf), *n*. **to blow the gaff**, (*sl.*) to let out the secret; to give information.
gaffe (gaf), *n*. a social solecism. **to make a gaffe**, to put one's foot in it. [F]
gaffer (gaf'ə), *n*. an old fellow, esp. an aged rustic (formerly a term of respect, now of familiarity); a foreman, an overseer; a schoolmaster; (*coll.*) the chief lighting electrician on a television or film set.
gag (gag), *v.t.* (*past, p.p.* **gagged**) to stop the mouth (of a person) by thrusting something into it, so as to prevent speech; to silence; to deprive of freedom of speech; to cause to choke or retch. *v.i.* to choke, retch; to tell jokes. *n*. something thrust into the mouth to prevent one from speaking; (*Surg*.) an instrument for holding the mouth open; in Parliament, the closure; a joke; a hoax, a trick. **gag-bit**, *n*. a very powerful bit used in horse-breaking. **gag-rein**, *n*. a rein used for pulling the bit upward or backward. **gagger**, *n*.
gaga (gah'gah), *a*. foolish, senile, fatuous.
gage[1] (gāj), *n*. a pledge, a pawn; something laid down as security, to be forfeited in case of non-performance of some act; a glove or other symbol thrown down as a challenge to combat; hence, a challenge. *v.t.* to deposit as a pledge or security for some act; to stake, to wager.
gage[2] (gāj), *n*. a greengage.
gage[3] GAUGE.

gaggle (gagl), *v.i.* to make a noise like a goose; to cackle, to chatter. *n.* a collection of geese; a straggling or disorderly group.
gaiety (gā′ətī), *n.* the state of being gay; mirth, merriment; gay appearance.
gaily GAY.
gain (gān), *n.* anything obtained as an advantage or in return for labour; profit; increase, growth, accession; amount of this; (*pl.*) profits, emoluments; the acquisition of wealth; the ratio of the output power of an amplifier to the input power usu. measured in decibels, volume. *v.t.* to obtain by or as by effort; to earn, to win, to acquire; to get more of; to reach, to attain to; to win (over); to obtain as a result, to incur. *v.i.* to profit; to gain ground; to increase, esp. in weight; of a clock, to run fast. **to gain ground**, to advance in any undertaking; to make progress. **to gain on** or **upon**, to get nearer to (an object of pursuit); to encroach upon. **to gain the upper hand**, to be victorious. **to gain time**, to obtain delay for any purpose. **gain-control**, *n.* the volume control in an amplifier or receiving set. **gainable**, *a.* **gainer**, *n.* **gainful**, *a.* profitable, advantageous, remunerative. **gainfully**, *adv.* **gainfulness**, *n.* **gainings**, *n.pl.* profits, gains. **gainless**, *a.* unprofitable. **gainlessness**, *n.*
gainsay (gānsā′), *v.t.* (*past, p.p.* **-said** (-sed′)) to contradict, to deny; to controvert, to dispute, †to hinder. **gainsayer**, *n.* **gainsaying**, *n.*
gainst, 'gainst (gānst, genst), AGAINST.
gairfowl (geə′fowl), GAREFOWL.
gait (gāt), *n.* manner of walking or going, carriage; a sequence of foot movements for a horse at a particular speed, e.g. walk, trot etc. **(-)gaited**, *a.* (*usu. in comb.*) having a particular gait.
gaiter (gā′tə), *n.* a covering for the ankle or the leg below the knee, usu. fitting down upon the shoe; (*N Am.*) a half-boot with a cloth top or elastic sides. *v.t.* to dress with gaiters. **gaiterless**, *a.*
gal¹ (gal), GIRL.
gal² (gal), *n.* in physics, a unit of acceleration equal to 1 cm per second per second.
gal., (*abbr.*) GALLON.
gala (gah′lə, gā′-), *n.* a festivity, a celebration; a sporting event.
galact GALACT(O)-.
galactic (gəlak′tik), *a.* pertaining to milk or the secretion of milk; in astronomy, pertaining to a galaxy, esp. the Milky Way. **galactic circle** or **equator**, *n.* the great circle of the celestial sphere which contains the galactic plane. **galactic plane**, *n.* the plane of the galactic circle. **galactic poles**, *n. pl.* the two opposite points on the celestial sphere that are the furthest N and S and which can be joined by an imaginary line perpendicular to the galactic plane.
galact(o)-, *comb. form* milk or milky.
galactogogue (gəlakt′təgog), *n., a.* (a medicine) promoting the flow of milk.
galactose (gəlak′tōs), *n.* a sweet crystalline glucose obtained from milk-sugar by treatment with dilute acid.
galago (gəlā′gō), *n.* (*pl.* **-gos**) an African genus of lemurs.
galah (gəlah′), *n.* the grey, rose-breasted cockatoo; (*Austral., coll.*) a silly person, a simpleton.
Galanthus (gəlan′thəs), *n.* a genus of bulbous plants, containing the snowdrop.
galantine (gal′əntēn), *n.* a dish of white meat, freed from bone, tied up, sliced, boiled, covered with jelly, and served cold. [F]
Galatian (gəlā′shən), *a.* belonging to Galatia. *n.* a native or inhabitant of Galatia in Asia Minor.
galaxy (gal′əksi), *n.* the Milky Way; any similar independent star system of vast extent; a brilliant assemblage of persons or things.
galbanum (gal′bənəm), *n.* a bitter, odorous gum resin obtained from Persian species of *Ferula*, esp. *F. galbaniflua*.
gale¹ (gāl), *n.* a wind stronger than a breeze but less violent than a tempest; a wind with a velocity of 40 mph (64 km) or over, registering force eight on the Beaufort scale; at sea, a storm; a noisy outburst.
gale² (gāl), *n.* the bog-myrtle, *Myrica gale*, a twiggy shrub growing on marshy ground, also called **sweet-gale**.
galea (gā′lə), *n.* (*pl.* **-leae**) a helmet-like organ or part; in botany, the arched upper lip in some labiates. **galeate** (-ət), **galeated** (-ātid), *a.*
galena (gəlē′nə), *n.* native sulphide of lead or lead-ore. **galenic, -ical** (-len′-), *a.* **galenite** (-nīt), *n.* galena. **galenoid**, *n., a.*
Galenic (gālen′ik), **-ical**, *a.* of or according to Galen, a second-century Greek physician, esp. applied to medicines prepared from vegetable substances by infusion or decoction, as opp. to chemical remedies. **Galenism**, *n.* **Galenist**, *n.*
galeopithecus (galiōpithē′kəs, -pith′i-), *n.* a genus of flying lemurs. **galeopithecine** (-sin), **-coid** (-koid), *a.*
galette (gəlet′), *n.* a flat, round cake. [F]
Galician (gəlish′iən), *a.* of or pertaining to Galicia. *n.* a native or inhabitant of Galicia, a province in NW Spain.
Galilean¹ (galilē′ən), *a.* of or according to Galileo, the astronomer, esp. applied to the simple telescope developed and used by him.
galilee (gal′ilē), *n.* a porch or chapel at the entrance of a church. **Galilean**² (-lē′-), *a.* pertaining to Galilee, a Roman province, comprising the north of Palestine west of the Jordan. *n.* a native or inhabitant of Galilee; (*Eccles. Hist.*) (applied contemptuously by pagans) a Christian. **the Galilean**, (*derog.*) Jesus Christ.
galingale (gal′ing-gāl), *n.* the aromatic root-stock of certain E Indian plants of the ginger family and of the genus *Alpinia* and *Kaempferia*, formerly used for culinary purposes; applied to a rare English sedge, *Cyperus longus*.
galiot (gal′iət), GALLIOT.
galipot (gal′ipot), *n.* a yellowish-white, viscid resin exuding from *Pinus maritimus* and hardening into a kind of turpentine, called, after refining, white, yellow or Burgundy pitch. **galipot varnish**, *n.*
galipot² (gal′ipot), GALLIPOT.
gall¹ (gawl), *n.* bile; anything exceedingly bitter; rancour, malignity, bitterness of mind; self-assurance, cheek. **gall and wormwood**, a symbol for all that is hateful, exasperating and unwelcome. **gall-bladder**, *n.* a pear-shaped membraneous sac, lodged on the under surface of the liver, which receives the bile. **gall-stone**, *n.* an abnormal calcareous concretion formed in the gall-bladder. **gall-less**, *a.*
gall² (gawl), *n.* an abnormal excrescence on plants, esp. the oak, caused by the action of some insect. **oak-gall**, **gall-apple** GALL-NUT. **gall-fly, -insect, -louse**, *n.* an insect, chiefly belonging to the genus *Cynips*, that causes the production of galls. **gall-nut, -apple**, *n.* a gall produced on the oak, esp. by the puncture by *C. gallae tinctoria*. **gall-oak**, *n.* the oak, *Quercus infectoria*.
gall³ (gawl), *n.* a sore, swelling, or blister, esp. one produced by friction or chafing on a horse; soreness, irritation; one who or that which causes this. *v.t.* to chafe, hurt or injure by rubbing; to make sore by friction; to annoy, to harass, to vex. **galling**, *a.* vexing, irritating, mortifying. **gallingly**, *adv.*
gall., **gal.**, (*abbr.*) GALLON.
gallant¹ (gal′ənt), *a.* showy, well-dressed; fine, stately; brave, high-spirited, courageous, chivalrous. *n.* a man of fashion; a bold and dashing man. **gallantly**, *adv.* **gallantry**, *n.* bold, dashing, magnanimous courage.
gallant² (gəlant′), *n.* a man attentive and polite to women; a lover, a wooer; a paramour. *a.* specially attentive to women. *v.t.* to attend as a gallant or cavalier, to escort; to pay court to; to flirt with. *v.i.* to play the gallant; to flirt

(with). **gallantly** (gal'-. -lant'-), *adv.* **gallantry** (gal'əntri), *n.* politeness and deference to women, with or without evil intent; amorous intrigue.

galleass (gal'ias), *n.* a heavy, low-built galley, usu. with three masts and about 20 guns.

galleon (gal'iən), *n.* a large sailing ship, with three or four decks, much used in 15th–17th cents., esp. by the Spaniards in their commerce with their American possessions.

gallery (gal'əri), *n.* an elevated floor or platform projecting from the wall toward the interior of a church, hall, theatre, or other large building, commonly used for musicians, singers or part of the congregation or audience; (*Theat.*) the highest and cheapest tier of seats; the persons occupying these, hence the most unrefined of the auditors; a passage open at one side, usu. projecting from the wall of a building and supported on corbels or pillars; a corridor, a passage, a long and narrow room; a room or building used for the exhibition of pictures, hence, a collection of pictures; a portico or colonnade; (*Fort.*) a covered passage in a fortification, either for defence or communication; (*Mining*) an adit, drift, or heading; an underground passage. *v.t.* to furnish or pierce with a gallery or galleries. **to play to the gallery,** to court popular applause. **galleried,** *a.* **galleryful,** *n.*

galley[1] (gal'i), *n.* a low, flat vessel, with one deck, navigated with sails and oars, which were usu. worked by slaves or convicts; an ancient Greek or Roman war-vessel of this type with one or more tiers of oars; a row-boat of large size, esp. one used by the captain of a man-of-war; the cook-house on board ship. **galley-slave,** *n.* a criminal condemned to the galleys; a drudge.

galley[2] (gal'i), *n.* (*Print.*) in hot-metal composition, an oblong tray on which compositors place matter as it is set up; a galley-proof. **galley-press,** *n.* a press at which galley-proofs are pulled. **galley-proof,** *n.* a proof taken from type in a galley, usu. in one column on a long strip of paper as dist. from that arranged in pages.

galley-west (galiwest'), *adv.* (*N Am., coll.*) **to knock galley-west,** to put somebody or something into a state of confusion, unconsciousness or inaction.

galliard (gal'yəd), *n.* a lively dance; the music to this. **galliardise,** *n.* merriment, liveliness.

Gallic (gal'ik), *a.* of or pertaining to ancient Gaul; (*loosely*) French. **Gallice** (-isi), *adv.* in French. **Gallicism,** *n.* a French expression or idiom. **Gallicize, -ise,** *v.t., v.i.* to convert to French practice, idiom etc.

gallic acid, *n.* (*Chem.*) an acid derived from oak-galls and other vegetable sources.

Gallican (gal'ikən), *a.* pertaining to the ancient Church of Gaul or France; ultramontane, claiming autonomy for the Church in France and repudiating papal control. *n.* a member of the French Church who holds these views. **Gallicanism,** *n.* **Gallicanist,** *n.*

galligaskins (galigas'kinz), *n.pl.* loose hose or breeches worn in the 16th and 17th cents.; leather leggings worn in the 19th cent.

gallimaufry (galimaw'fri), *n.* a hash, a hodge-podge; an inconsistent or ridiculous medley.

gallinaceous (galinā'shəs), *a.* of or pertaining to the Gallinae, a group of birds containing pheasants, partridges, grouse, turkeys, domestic fowls, and allied forms. **gallinacean,** *a.* gallinaceous. *n.* one of the Gallinae.

galling GALL[3].

gallinule (gal'inūl), *n.* any bird of the genus *Gallinula,* esp. *G. chloropus,* the moor-hen.

galliot (gal'iət), *n.* a small, swift galley propelled by sails and oars; a one- or two-masted Dutch or Flemish merchant vessel.

gallipot (gal'ipot), *n.* a small glazed earthenware pot used to contain ointments, medicines, preserves etc.

gallium (gal'iəm), *n.* a soft, grey metallic element of extreme fusibility, at. no. 31; chem. symbol Ga, used in semiconductors.

gallivant (galivant'), *v.i.* to gad about, to go pleasure-seeking.

galliwasp (gal'iwosp), *n.* a small harmless W Indian lizard, *Celestus occiduus.*

gallo-[1], *comb. form* pertaining to gallic acid, gallic.

Gallo-[2], *comb. form* French.

Gallomania (galōmā'niə), *n.* a mania for French fashions, habits, or practices, literature etc. **Gallomaniac** (-ak), *n.*

Gallophil (gal'əfil), *n.* a devotee of French customs etc.

Gallophobe (gal'ōfōb), *n.* one who hates French ways or fears the French. **Gallophobia** (-fō'biə), *n.*

galloglass (gal'ōglahs), *n.* an armed soldier or retainer of an ancient Irish chieftain.

gallon (gal'ən), *n.* an English measure of capacity; a dry measure equal to one-eighth of a bushel (4·55 l); a British measure for liquids, containing 277¼ cu. in. (4·55 l) (also called **imperial gallon**); a US measure for liquids containing 231 cu. in. (3·78 l).

galloon (gəloon'), *n.* a narrow braid of silk, worsted, or cotton, with gold or silver thread interwoven, for binding uniforms, dresses etc.; other materials used for binding or edging.

gallop (gal'əp), *v.i.* to run in a series of springs, as a horse at its fastest pace; to ride at a gallop; to go or do anything at a very rapid pace. *v.t.* to make (a horse) gallop. *n.* the motion of a horse at its fastest speed; the act of riding or a ride at this pace; a galop. **gallopade** (-pad'), *n.* a sidelong or curvetting kind of gallop; a brisk dance, of Hungarian origin. *v.i.* to dance this. **galloper,** *n.* a horse that gallops; a person who gallops on a horse, or who makes great haste; an aide-de-camp. **galloping,** *a.* of a disease or something bad, rapidly spreading or increasing.

Galloway (gal'əwā), *n.* a small, hardy variety of horse, or black breed of cattle, orig. bred in Galloway, SW Scotland.

gallows (gal'ōz), *n. sing.* a framework, usu. consisting of timber uprights and a crosspiece, on which criminals are executed by hanging; execution by hanging; a similar framework used for gymnastics, for hanging things on, in printing, cookery etc. **gallows humour,** *n.* macabre, ironic humour. **gallows-tree,** *n.* the gallows.

Gallup Poll[®] (gal'əp), *n.* a method of ascertaining the trend of public opinion by questioning a representative cross-section of the population. [inventor G. *Gallup,* 1901–84, US statistician]

galoche (gəlosh'), GALOSH.

galoot (gəloot'), *n.* an awkward, uncouth person.

galop (gal'əp), *n.* a lively dance in 2/4 time; the music to the dance. *v.i.* to dance this.

galore (gəlaw'), *n.* plenty, abundance. *adv.* in plenty, abundantly.

galosh (gəlosh'), *n.* (*usu. pl.*) an overshoe, usu. of vulcanized rubber, for protecting one's boots or shoes in wet weather.

galt (gawlt), GAULT.

galumph (gəlŭmf'), *v.i.* to prance exultantly. **galumphing,** *a.*

galvan- GALVAN(O)-.

galvanic (galvan'ik), *a.* of, pertaining to or produced by galvanism; forced, spasmodic or violently energetic (of movements, expression etc.) as if caused by the action of an electric current. **galvanic battery** or **pile,** *n.* a number of connected galvanic cells for producing an electric current. **galvanic belt,** *n.* a galvanic apparatus in the form of a belt for applying electricity to the body. **galvanic electricity,** *n.* **galvanically,** *adv.* **galvanism** (gal'vənizm), *n.* electricity produced by chemical action, esp. that of acids on metals; the branch of science dealing with this; its application for medical purposes. **galvanist,** *n.*

galvanize, -ise (gal'vəniz), *v.t.* to apply galvanism to, esp.

to stimulate muscular action etc. by galvanism; to plate with gold or other metal by galvanism; to rouse into life or activity as by a galvanic shock. **galvanization**, *n.* **galvanizer**, *n.* **galvanized iron**, *n.* iron coated with zinc (orig. by galvanic deposition), to protect it from moisture. [L. *Galvani*, 1737–98, Italian physician and its discoverer]
galvan(o)-, *comb. form.* galvanic current.
galvanometer (galvənom'itə), *n.* a delicate apparatus for determining the existence, direction, and intensity of electric currents. **galvanometric, -ical** (met'rik, -əl), *a.* **galvanometry** (-ətri), *n.*
galvanoscope (gal'vənəskōp), *n.* an instrument for detecting the presence and showing the direction of electric currents.
gam (gam), *n.* a herd of whales; a keeping company or exchange of visits among whalers at sea.
gam- GAM(O)-.
gamba[1] (gam'bə), *n.* the metacarpus or metatarsus.
gamba[2] (gam'bə), *n.* a *viola da gamba*, an organ stop with a tone like that of the violin or violoncello.
gambade (gambād'), **gambado** (-dō), *n.* (*pl.* **-bades, -bados, -badoes**) a bound or spring of a horse; a caper, a frolic.
gambier (gam'biə), *n.* an extract from the leaves of *Uncaria gambir*, used in medicine as an astringent, and also for dyeing and tanning.
gambit (gam'bit), *n.* an opening in chess, in which a pawn is sacrificed in order to obtain a favourable position for attack [most of the gambits have distinctive names, as *King's gambit, Queen's gambit, Steinitz gambit*]; the opening move in a concerted plan; a remark intended to initiate a conversation or as the introduction to an anecdote, argument etc.
gamble (gam'bl), *v.i.* to play, esp. a game of chance, for money; to risk large sums or other possessions on some contingency; to speculate financially; to take a chance (on). *v.t.* to bet, to wager; to run the risk of losing, to hazard. *n.* gambling; a gambling venture or speculation; something involving an element of risk. **to gamble away**, to squander or lose in gambling. **gambler**, *n.* **gamblesome** (-səm), *a.*
gamboge (gambōj', -boozh'), *n.* a gum-resin, from Kampuchea, Sri Lanka etc., used as a yellow pigment, and in medicine.
gambol (gam'bl), *v.i.* to frisk or skip about; to frolic. *n.* a frolic; a skipping or playing about.
gambrel (gam'brəl), *n.* a horse's hock; a bent piece of wood used for suspending carcases; (*N Am.*) a gambrel-roof. **gambrel-roof**, *n.* a hipped roof with a steeper, gabled section at the top.
game[1] (gām), *n.* an exercise for diversion, usu. in concert with other players, a pastime; the equipment required for playing e.g. a board game; a contest played according to specified rules and decided by chance, strength, skill or combination of these; (*pl.*) athletic contests, esp. such as are held at periodical dates, as the Olympic Games etc.; a match, e.g. football; a single round in a sporting contest; the number of points required to win a game; a project, plan, or scheme designed to defeat others; success in a game or contest; (*coll.*) trick, dodge, subterfuge; wild animals or birds which are hunted, such as hares, grouse, partridges, pheasants; the flesh of these; an object of pursuit; (*coll.*) a lark, an amusing incident. *a.* pertaining to game; plucky, spirited; ready, willing (to do etc.). *v.i.* to play at games of chance; to play for a stake; to gamble. **game, set and match**, (*coll.*) a final and convincing victory. **the game**, (*coll.*) prostitution. **the game is up**, everything has failed; the game (bird or animal) has started up. **to be off one's game**, to be playing poorly, not giving one's best performance. **to be on the game**, (*coll.*) to be earning a living as a prostitute. **to give the game away**, to reveal a secret or strategy. **to make game of**, to turn into ridicule. **to play the game**, to abide by the rules; to act in an honourable way. **game-bag**, *n.* a bag to hold the game killed or taken by a sportsman. **game ball**, *n.* GAME POINT. **game-bird**, *n.* a bird hunted for sport. **game-book**, *n.* a book for recording game killed. **game-cock**, *n.* a cock bred and trained for fighting. **game fish**, *n.* a large fish that is caught for sport. **game-fowl**, *n.* GAME-BIRD. **gamekeeper**, *n.* one who is employed to look after game, coverts etc., and to prevent poaching on a private estate or game reserve. **game laws**, *n.pl.* laws for the preservation of game; the regulation of the seasons for killing it etc. **game licence**, *n.* one giving the right to kill or deal in game. **game plan**, *n.* the tactics etc. of a football team, prearranged before a match; any carefully planned strategy. **game point**, *n.* a situation in a game of tennis when one point is enough to determine the game. **game reserve**, *n.* an area of land set aside for the protection of wild animals. **gameshow, games show**, *n.* a television programme, esp. a quiz show, in which selected contestants compete for prizes. **gamesmanship**, *n.* the art or practice of winning games by disconcerting the opponent (by talking etc.) but without actually cheating. **game theory, games theory**, *n.* the analysis of all choices and strategies available in a game or military, social etc. conflict in order to choose the best possible course of action. **game warden**, *n.* one who is employed to look after game, esp. on a game reserve. **gamely**, *adv.* **gameness**, *n.* **gamesome** (-səm), *a.* inclined to play; merry, gay. **gamesomely**, *adv.* **gamesomeness**, *n.* **gamester**, *n.* one who is addicted to gaming, a gambler. **gaming**, *n.* gambling. **gaming-house**, *n.* a house where gambling is carried on; a house of ill-repute. **gaming-table**, *n.* a table for gambling games. **gamy**, *a.* having the flavour or odour of game, high; abounding in game; plucky, spirited, game. **gaminess**, *n.* **gamer**, *n.* one who plays a game, esp. a role-playing game or a computer one.
game[2] (gām), *a.* (*coll.*) lame, crippled. **gammy** (gam'i), *a.* lame; injured.
gamelan (gam'əlan), *n.* a SE Asian percussion instrument; an orchestra made up of a number of gamelans.
gamet- GAMET(O)-.
gametangium (gamitan'jiəm), *n.* (*pl.* **-gia**) a cell or organ in which gametes are formed.
gamete (gam'ēt, -mēt'), *n.* a sexual reproductive cell, either of the two germ cells that unite to form a new organism – in the male, a spermatozoon, in the female an ovum. **gamete intra-fallopian transfer**, (GIFT), *n.* a treatment for infertile women in which eggs and sperm (the gametes) are injected directly into the fallopian tubes via a catheter. **gametal** (gam'etəl, -ē'-), **gametic** (-et'-, -ēt'-), *a.*
gamet(o)-, *comb. form.* gamete.
gametocyte (gamē'tōsīt), *n.* a cell that breaks up into gametes.
gametogenesis (gamētōgen'əsis), *n.* the formation of gametes.
gametophyte (gamē'tōfīt), *n.* a plant of the generation that produces gametes, in plant species which show alternation of generations.
gamic (gam'ik), *a.* (*Biol.*) of or pertaining to sex, sexual; capable of development after sexual fertilization (of ova).
gamin (gam'in, -mī), *n.* a street arab, an urchin. [F]
gamine (gam'ēn), *n., a.* (a girl or woman) having a tomboyish or impish charm. [F]
gamma (gam'ə), *n.* the third letter of the Greek alphabet, Γ, γ, G, g, representing 3 in enumerations; (a person or piece of work in) the third grade. **gamma globulin** (gam'ə), *n.* any of a group of proteins that are carried in blood and serum and include most known antibodies. **gamma-rays**, *n.pl.* (*Phys.*) short-wavelength, penetrating electromagnetic rays emitted by radioactive substances; used in treatment of cancer and in radiography of metals.

gammadion, gammation (gəma'diən, -tiən), *n*. (*pl.* **-tia** (-iə)) an ornament composed of the gamma singly or in combination; a cruciform ornament composed of four gammas, placed back to back; a fylfot, swastika.

gammer (gam'ə), *n*. (*esp. dial.*) an old woman.

gammon[1] (gam'ən), *n*. the buttock or thigh of a hog salted and dried; a cured ham. *v.t.* to make into bacon; to salt and dry in smoke.

gammon[2] (gam'ən), (*coll., dated*) *n*. nonsense, humbug; a fraud, a hoax. *int.* nonsense, humbug. *v.t.* to hoax, to impose upon. *v.i.* to pretend.

gammon[3] (gam'ən), *n*. a victory at backgammon in which the winner throws off all his men before his opponent throws off any and scores two games. *v.t.* to win a gammon against (an opponent in backgammon).

gammy GAME[2].

gam(o)-, *comb. form* (*Biol.*) sexual; having certain parts united.

gamogenesis (gamōjen'əsis), *n*. (*Biol.*) sexual reproduction. **gamogenetic** (-net'ik), *a*. **gamogenetically**, *adv*.

gamopetalous (gamōpet'ələs), *a*. (*Bot.*) having the petals united.

gamophyllous (gamōfil'əs), *a*. (*Bot.*) having the leaves united.

gamosepalous (gamōsep'ələs), *a*. (*Bot.*) having the sepals united.

gamp (gamp), *n*. (*coll.*) an umbrella, esp. a large and clumsy one.

gamut (gam'ət), *n*. the major diatonic scale; the whole series of notes recognized by musicians; the whole range, compass or extent.

gamy GAME[1].

-gamy, *comb. form* marriage or kind of marriage, as in *bigamy, endogamy, misogamy*.

gander (gan'də), *n*. the male of the goose; a simpleton, a noodle; (*coll.*) a quick look.

gang[1] (gang), *n*. a number of persons associated for a particular purpose (often in a bad sense); a group of friends or associates; a number of workmen under a foreman, or of slaves or convicts; a set of tools operating in concert; a gangue. *v.i.* to act in concert with. **to gang up**, to join with others (in doing something). **to gang up on**, to join with others to make an attack on somebody. **gang-bang**, *n*. (*sl.*) an occasion on which a number of males have successive sexual intercourse with one female. **gangland**, *n*. the world of organized crime. **gang mill**, *n*. a saw mill with gang saws. **gangsman**, *n*. a ganger. **gang saw**, *n*. a saw with several blades fitted in a frame, producing parallel cuts. **ganger**, *n*. the overseer or foreman of a gang of labourers. **gangster** (-stə), *n*. a member of a criminal gang. **gangsterland**, *n*. gangland.

gang[2] (gang), *v.i.* (*Sc.., past, p.p.* **gaed** (gäd)) to go. **gang-plank, gang-board**, *n*. a plank, usu. with cleats, used for boarding or landing from a vessel. **gangway** (gang'wā), *n*. a passage into or out of a building or between rows of seats; in the House of Commons, a narrow cross passage giving access to the back benches, and dividing the more independent members from the immediate supporters of the Government and the opposition; a temporary bridge affording means of passage from a ship to the shore; an opening in the bulwarks affording entrance to or exit from a vessel; a passage connecting different parts of a vessel; in a mine, a main level. *int.* clear the way!

gangling (gang'gling), *a*. loosely built, lanky, awkward.

ganglion (gang'gliən), *n*. (*pl.* **-lia** (-ə)) an enlargement in the course of a nerve forming a local centre for nervous action; an aggregation of nerve-cells forming a nucleus in the central nervous system; in pathology, a globular growth in the sheath of a tendon. **gangliac** (-ak), **gangliar, ganglionic** (-on'ik), *a*. pertaining to a ganglion or ganglia. **gangliated, ganglionated** (-ātid), *a*. **gangliform**

(-fawm), *a*. **ganglionary**, *a*. composed of ganglia. [Gr.]

gangrene (gang'grēn), *n*. cessation of vitality in a part of the body, the first stage of mortification, usu. followed by decay; corruption, decay. *v.t.* to cause gangrene in. **gangrenescent** (-grənes'-), *a*. **gangrenous** (-grə-), *a*.

gangster GANG[1].

gangue (gang), *n*. the earthy matter or matrix in which ores are embedded.

ganister (gan'istə), *n*. a kind of grit or hard sandstone from the lower coal-measures; a mixture of ground quartz and fire-clay used for lining Bessemer converters.

ganja (gan'jə), *n*. a dried preparation of *Cannabis sativa* or Indian hemp, smoked as an intoxicant and narcotic.

gannet (gan'it), *n*. a large, white sea-bird, *Sula bassana*, with black-tipped wings; a greedy person.

ganoid (gan'oid), *a*. bright, smooth, like enamel of fish-scales; of fish belonging to the Ganoidei. *n*. any fish of the Ganoidei. **ganoidal** (-oi'dəl), **ganoidean** (-oi'diən), *a*. **Ganoidei** (-oi'diī), *n. pl.* a division of fishes comprising the sturgeons and numerous extinct forms, so called from their shining scales. **ganoin** (-ōin), *n*. a calcareous substance that forms a shiny, enamel-like coating on ganoid scales.

gantlet (gant'lit), GAUNTLET[1].

gantry (gan'tri), *n*. a wooden frame for standing a barrel upon; a bridgelike structure for carrying a travelling crane, railway signals etc.

gaol JAIL.

gap (gap), *n*. an opening, a breach, as in a hedge, a fence etc.; a chasm, a break in a mountain ridge; a breach of continuity, a blank, hiatus, interruption; a deficiency, a wide divergence. *v.t.* (*past, p.p.* **gapped**) to make a gap in. **to stop, fill, or supply a gap**, to repair a defect or make up a deficiency. **gap-toothed**, *a*. having spaces between the teeth. **gapped**, *a*. **gappy**, *a*.

gape (gāp), *v.i.* to open the mouth wide; to yawn; to stare with open mouth in wonder, surprise or perplexity; to open in a fissure or chasm, to split open. *n*. the act of gaping; a stare with open mouth, a yawn; the width of the mouth when opened, as of birds etc.; the part of a beak that opens; the opening between the shells of a bivalve that does not shut completely; (*pl.*) a disease in young poultry caused by the gapeworm and characterized by much gaping; a fit of yawning. **to gape at**, to open the mouth and gaze with astonishment. **to gape for** or **after**, to desire eagerly, to crave. **gapeworm**, *n*. a nematode worm, *Syngamus trachea*, that causes gapes in poultry. **gaper**, *n*. one who or that which gapes, esp. various kinds of birds, fish and molluscs. **gapingly**, *adv*.

gar (gah), **garfish**, *n*. a fish with a long pointed snout, esp. *Belone vulgaris*, a European fish called also greenbone, in allusion to the bones of its spine; (*N Am.*) species of the genus *Lepidosteus*, also called **garpike**.

garage (ga'rahzh, -rij), *n*. a building for housing or repairing motor-cars; an establishment where this is done as a business and where motor fuels etc. are sold. *v.t* to put or keep in a garage. *a*. (*coll.*) rough-and-ready, amateurish, improvised. **garage sale**, *n*. a sale of second-hand goods held on the grounds of a private home, esp. in a garage.

garam marsala (gahrəm mahsahlə), *n*. a mixture of spices used in curries.

garb (gahb), *n*. dress, costume; distinctive style of dress; outward appearance. *v.t.* to put garments upon, esp. to put in a distinctive dress.

garbage (gah'bij), *n*. anything worthless or offensive, sordid rubbish; (*N Am.*) kitchen waste. **garbage can**, *n*. (*N Am.*) a dustbin. **garbage disposal unit**, *n*. (*N Am.*) a waste disposal unit.

garble (gah'bl), *v.t.* to mutilate, in such a way as to convey a false impression; to jumble or confuse unintentionally. **garbler**, *n*.

garbo (gah'bō), *n*. (*pl.* **-bos**) (*Austral., coll.*) a dustman.

garboard (gah'bawd), *n.* (*Naut.*) the first plank fastened on either side of a ship's keel. **garboard-strake**, *n.* the row of planks or plates next to the keel on a ship's bottom.

garçon (gah'sō), *n.* a waiter. [F]

garda (gah'də), *n.* (*pl.* **gardai** (-də, -dē)) a member of the Irish police, the **Garda Síochána** (shē'chənə).

gardant (gah'dənt), GUARDANT.

garden (gah'dən), *n.* an enclosed piece of ground for the cultivation of fruit, flowers or vegetables; a place or region particularly fertile, well-cultivated or delightful; (*pl.*) a public pleasure-ground adorned with trees, flower-beds etc. *a.* pertaining to a garden; cultivated, not wild. *v.i.* to cultivate a garden; (*coll.*) in cricket, to smooth out bumps etc. in the pitch with the bat. **common or garden**, (*coll.*) ordinary. **everything in the garden is lovely**, everything appears to be well. **to lead someone up the garden path**, to mislead or deceive someone. **garden centre**, *n.* a place where plants, fertilizers, and garden tools and equipment are sold. **garden city** or **suburb**, *n.* a planned township or suburb in rural surroundings. **garden flat**, *n.* a flat that opens onto a garden. **garden-frame**, *n.* a glazed frame for protecting plants during the winter or for forcing. **garden-glass**, *n.* a bell-glass for protecting plants. **garden-party**, *n.* a social meeting or a company entertained on a lawn or in a garden. **garden-stuff**, *n.* vegetables, herbs, fruit etc. **gardened**, *a.* **gardener**, *n.* one who gardens, esp. one whose occupation is to attend to or to manage gardens. **gardenesque** (-esk'), *a.* **gardening**, *n.* horticulture; work in a garden.

gardenia (gahdē'niə), *n.* a genus of tropical shrubs and trees cultivated in greenhouses for their large fragrant flowers. [Dr Alexander *Garden*, US botanist, *d.* 1791]

garfish GAR [1].

gargantuan (gahgan'tuən), *a.* immense, enormous, incredibly big. **gargantuism**, *n.*

garget (gah'git), *n.* a distemper affecting the throat in cattle; an affection of the udder of cows or ewes.

gargle (gah'gl), *v.t.* to rinse (the mouth or throat) with some medicated liquid. *n.* a liquid used for washing the mouth or throat.

gargoyle (gah'goil), *n.* a grotesque spout, usu. carved to represent a human or animal figure, projecting from a Gothic building to throw rain-water clear of the wall.

garibaldi (garibawl'di), *n.* a loose kind of blouse worn by women or children. **garibaldi biscuit**, *n.* a sandwich-type biscuit with a layer of currants.

garish (geə'rish), *a.* gaudy, showy, flashy; excessively or extravagantly decorated; dazzling, glaring. **garishly**, *adv.* **garishness**, *n.*

garland (gah'lənd), *n.* a wreath, chaplet or festoon of flowers, leaves etc., a similar festoon of metal, stone, ribbons or other material used for decoration etc.; the prize, the chief honour; a collection of choice pieces, esp. of poems. *v.t.* to deck with a garland. **garlandage** (-dij), (*poet.*), **garlandry**, *n.* **garlandless**, *a.*

garlic (gah'lik), *n.* a bulbous-rooted plant, *Allium sativum*, with a strong odour and a pungent taste, used in cookery. **garlic bread**, *n.* bread, sliced, spread with garlic butter and heated. **garlic butter**, *n.* butter flavoured with garlic. **garlic-eater**, *n.*

garment (gah'mənt), *n.* an article of clothing, esp. one of the larger articles, as a coat or gown; an outer covering; (*pl.*) clothes. *v.t.* (*poet.*, *usu. in p.p.*) to attire with or as with a garment. **garmentless**, *a.* **garmenture** (-chə), *n.* dress, apparel, clothing.

garner (gah'nə), *n.* a place for storing grain, a granary; a store, a repository. *v.t.* to store in or as in a garner; to gather.

garnet (gah'nit), *n.* a vitreous mineral of varying composition, colour and quality, the deep red, transparent kinds of which are prized as gems.

garnish (gah'nish), *v.t.* to adorn; to embellish (as a dish) with something laid round or on top of it; to supply, to furnish; (*Law*) to warn, to give notice to. *n.* an ornament; a decoration, especially things put round a dish as embellishment. **garnishee** (-shē'), *n.* one who has received notice not to pay any money which he owes to a third person, who is indebted to the person giving notice. *v.t.* to serve with a garnishment. **garnisher**, *n.* one who garnishes. **garnishing**, *n.* **garnishment**, *n.* an ornament, an embellishment; in law a warning to a party to appear in court, or not to pay money etc. to a defendant. **garnishry**, *n.* embellishment.

garniture (gah'nichə), *n.* ornamental appendages, trimmings, embellishment.

garotte (gərot'), GARROTTE.

garpike GAR [1].

garret (ga'rit), *n.* an upper room or storey immediately under the roof.

garrison (ga'risən), *n.* a body of troops stationed in a town or fortified place; a town or fortified place manned with soldiers. *v.t.* to furnish (a fortress) with soldiers; to occupy as a garrison. **garrison-town**, *n.* a town in which a garrison is stationed.

garrotte (gərot'), *n.* a method of execution in which the victim is fastened by an iron collar to an upright post, and a knob operated by a screw or lever dislocates the spinal column, or a small blade severs the spinal cord at the base of the brain (orig. the method was strangulation by a cord twisted with a stick); hence, robbery by means of strangling. *v.t.* to execute by this means; to render helpless or insensible in order to rob. **garrotter**, *n.*

garrulous (ga'rələs), *a.* talkative, loquacious, wordy; chattering. **garrulity** (-roo'-), *n.* **garrulously**, *adv.* **garrulousness**, *n.*

garter (gah'tə), *n.* a band round the leg for holding the stocking up; (*N Am.*) a sock-suspender. *v.t.* to fasten (a stocking) with a garter; to put a garter upon. **the Garter**, the badge of the highest order of British knighthood, instituted by Edward III, about 1348; the order itself; membership of this. **Garter Principal King-of-Arms**, *n.* the chief herald of this order. **garter-snake**, *n.* a harmless American snake belonging to the genus *Eutaenia*. **garter stitch**, *n.* (knitting made by using) a plain stitch.

garth (gahth), *n.* a close, a yard; the grass-plot surrounded by the cloisters of a religious house.

gas (gas), *n.* (*pl.* **gases**) a substance in an airy form, possessing the condition of perfect fluid elasticity; such a fluid used for lighting and heating, esp. that obtained from coal; a gaseous anaesthetic; (*esp. N Am.*, *coll.*) gasolene, petrol; in coal-mining, an explosive mixture of firedamp and air; (*coll.*) a gas-jet; empty talk, boasting; (*dated coll. or dial.*) something great or wonderful. *v.i.* (*past, p.p.* **gassed**) to indulge in empty talk; to boast. *v.t.* to supply gas to; to subject to the action of burning gas (as lace) in order to free from loose fibres; to attack, to stupefy or kill by means of poison-gas. **to step on the gas**, to accelerate a motor-car; to hurry. **gasbag**, *n.* a bag for holding gas esp. in an airship; (*coll.*) a talkative person. **gas-bottle**, *n.* a steel cylinder for holding compressed gas. **gas-bracket**, *n.* a pipe projecting from a wall and fitted with a burner or burners. **gas-burner**, *n.* the tube or jet at which the gas issues and is ignited. **gas chamber** or **oven**, *n.* an airtight place designed for killing animals or humans by means of a poisonous gas. **gas chromatography**, *n.* a method of analysing a mixture of volatile substances which depends on the relative speeds at which the various components of the mixture pass through a long narrow tube that contains an inert gas and a solvent. **gas-cooled**, *a.* **gas-cooled reactor**, *n.* a nuclear reactor that uses a gas as the coolant. **gas-engine** or **-motor**, *n.* **gas escape**, *n.* a leakage of gas. **gasfield**, *n.* a region in which natural gas occurs. **gas-fire**, *n.* a device for burning gas for heating a room etc. **gas-fired**, *a.*

fuelled by a gas or gases. **gas-fitter**, *n*. a person employed to lay pipes and put up fixtures for gas. **gas gangrene**, *n*. a gangrenous infection in deep wounds caused by bacteria which produce gases in the surrounding tissues. **gas-guzzler**, *n*. (*N Am.*, *coll.*) a (usu. large) car that uses a lot of petrol. **gas-holder**, *n*. a structure for storing gas, a gasometer. **gas-jet**, *n*. a gas burner; a jet of flame from it. **gas-lamp**, *n*. **gaslight**, *n*. the light produced by the combustion of coal-gas; a gas-jet. **gas-main**, *n*. a principal pipe leading from a gas-works and having branches and distributing pipes. **gas-man**, *n*. a gas-fitter; a person employed to read household gas-meters. **gas mantle**, *n*. a chemically-prepared incombustible gauze hood for a gas-lamp that becomes incandescent when heated. **gas-mask**, *n*. a mask with a chemical filter to protect the wearer against poisonous gases and fumes. **gas-meter**, *n*. a machine for measuring and recording the quantity of gas consumed. **gas-oil**, *n*. an oil distilled from crude petroleum used as a fuel for heating etc. **gas-ring**, *n*. a hollow pipe with perforations that serve as gas-jets, used for cooking. **gas-shell**, *n*. an artillery shell that produces or diffuses poison-gas on explosion. **gas station**, *n*. (*N Am.*, *coll.*) a filling-station, petrol station. **gas-tank**, *n*. a gasometer; (*N Am.*, *coll.*) the petrol tank on a motor vehicle. **gas-tight**, *a*. not allowing gas to escape. **gas-trap**, *n*. in plumbing, a double curve or U-shaped section of a pipe in which water remains and forms a seal that blocks the escape of foul gases. **gas turbine**, *n*. an internal-combustion engine in which a turbine is driven by the hot expanding gases from the combustion chamber. **gas-well**, *n*. a well that yields natural gas. **gas-works**, *n.pl.* an industrial plant where gas, esp. coal-gas, is produced. **gaseous** (gā'siəs, gas'-), *a*. in the form of gas; like gas. **gaseity** (-sē'-), *n*. **gasiform**, *a*. **gasify**, *v.t.* to convert into gas. **gasifiable** (-fī'-), *a*. **gasification** (-fikā'-), *n*. **gasless**, *a*. **gassy**, *a*. containing gas; like gas; gaseous; full of empty talk. **gassiness**, *n*.

Gascon (gas'kən), *n*. a native of Gascony, France; a boaster. **gasconade** (-nād'), *n*. boasting, bravado, bragging. *v.i.* to boast, to brag. **gasconader** (-nā'də), **gasconism**, *n*. [F]

gaseous GAS.

gash (gash), *v.t.* to make a long, deep, gaping cut in. *n*. a deep, open cut, especially in flesh; a flesh-wound, a cleft.

gasiform etc. GAS.

gasket (gas'kit), *n*. a plaited cord by which the sails, when furled, are bound close to the yards or gaffs; a strip of leather, rubber, asbestos etc. for packing or caulking joints in pipes, engines etc. to make them air-tight or water-tight. **to blow a gasket**, (*coll.*) to lose one's temper.

gaskin (gas'kin), *n*. the part of a horse's hind leg between the stifle and the hock, lower thigh.

gaskins (gas'kinz), *n.pl.* GALLIGASKINS.

gas(o)- *comb. form* pertaining to or using gas.

gasolene, gasoline (gas'əlēn), *n*. a volatile inflammable product of the distillation of petroleum, used for heating and lighting; (*N Am.*) petrol.

gasometer (gəsom'itə), *n*. a large cylindrical reservoir used at gas-works for the storage of gas, a gas-holder; an instrument for measuring the gases used in chemical experiments etc. **gasometric** (-met'-), *a*. **gasometry**, *n*. the science, art or practice of measuring gases.

gasp (gahsp), *v.i.* to breathe in a convulsive manner, as from exhaustion or astonishment. *v.t.* to emit or utter with gasps. *n*. a short painful catching of the breath. **at the last gasp**, at the last extremity; at the point of death. **to gasp out**, to utter breathlessly. **gasper**, *n*. (*dated sl.*) a cigarette. **gaspingly**, *adv*.

gaspacho (gaspach'ō), GAZPACHO.

Gastarbeiter (gast'ahbītə), *n*. a migrant worker in the Federal Republic of Germany. [G]

gasteral GASTRIC.

gasteropod (gas'tərəpod), *n*. an individual of the Gasteropoda. *a*. gasteropodous. **Gasteropoda** (-rop'-), *n.pl.* a class of molluscs, usu. inhabiting a univalve shell (as the snails), of which the general characteristic is a broad muscular ventral foot. **gasteropodous** (-rop'-), *a*. belonging to or characteristic of the Gasteropoda.

gastral GASTRIC.

gastralgia (gastral'jə), *n*. neuralgia in the stomach. **gastralgic**, *n*., *a*.

gastrectomy (gəstrek'təmi), *n*. the surgical removal of (part of) the stomach.

gastric (gas'trik), *a*. of or pertaining to the stomach. **gastric acid** GASTRIC JUICE. **gastric juice**, *n*. a colourless pellucid acid secreted by the stomach, one of the principal agents in digestion. **gastric ulcer**, *n*. an ulcer of the inner wall of the stomach. **gasteral, gastral**, *a*. **gastrin** (-trin), *n*. a hormone produced in the pyloric mucosa that stimulates the secretion of gastric juice.

gastritis (gastrī'tis), *n*. inflammation of the stomach.

gastr(o)- *comb. form* stomach.

gastrocnemius (gastroknē'miəs), *n*. (*pl.* **-ii** (-ii)) the large muscle in the calf of the leg which helps to extend the foot.

gastroenteric (gastrōente'rik), *a*. pertaining to the stomach and the intestines. **gastroenteritis** (-təri'tis), *n*. inflammation of the stomach and of the intestines.

gastroenterology (gastrōentərol'əji), *n*. the study of diseases of the stomach and the intestines. **gastroenterologist**, *n*.

gastrointestinal (gastrōintestī'nəl), *n*. of or pertaining to the stomach or the intestines.

gastrology (gastrol'əji), *n*. the science of matters pertaining to the stomach; the science of cookery or of eating, gastronomy. **gastrologer, -logist**, *n*. **gastrological** (-loj'-), *a*.

gastronomy (gastron'əmi), *n*. the art or science of good eating, epicurism. **gastronome** (gas'trənōm), **gastronomer, gastronomist**, *n*. one given to good living; an epicure; a gourmet. **gastronomic, -ical** (-nom'-), *a*. **gastronomically**, *adv*.

gastropod (gas'trəpod), GASTEROPOD.

gastroscopy (gastros'kəpi), *n*. an examination of the abdomen in order to discover disease.

gastrotomy (gəstrot'əmi), *n*. the operation of cutting into or opening the abdomen. **gastrotomic** (gastrətom'-), *a*.

gastrula (gas'trələ), *n*. an embryonic stage in the development of a metazoon, consisting of a double-walled sac enclosing a cup-like cavity. **gastrular**, *a*. **gastrulation** (-lā'-), *n*. the formation of a gastrula.

gat (gat), *n*. (*N Am.*, *sl.*) a revolver. [abbr. of GATLING]

gate[1] (gāt), *n*. a movable barrier, consisting of a usu. open-work frame of wood or iron, swinging on hinges or sliding, to close a passage or opening; an opening affording entrance and exit to an enclosure, a gateway; a natural opening, as a strait, a mountain pass etc.; a sluice admitting water to or shutting it off from a lock or dock; a numbered exit in an airport terminal from which passengers board an aircraft; two posts forming an obstacle in a slalom race; either of a pair of barriers that close a road at a level-crossing; in horseracing, a device to start racing usu. consisting of a set of stalls with barriers that are simultaneously removed at the moment of starting, starting gate; the number of people attending a race-meeting, football match etc.; the amount of money taken at the gates; an H-shaped series of slots that controls the position of the gear-lever in a motor vehicle; an electronic circuit (in a computer) that controls the passage of information signals when permitted by another independent source of similar signals. *v.t.* to furnish with a gate; to confine (a student) to the grounds of a school or college. **to gatecrash**, to attend a function or entertainment without an invitation. **gatefold**, *n*. a folded insert in a book or

gate — gazelle

magazine that exceeds the size of the other pages; foldout.
gatehouse, *n.* a lodge, house or defensive structure at or over a gate; a toll-gate cottage. **gate-keeper**, *n.* person in charge of a gate; the lessee or collector of tolls at a toll-gate; a variety of butterfly. **gate-leg, gate-legged**, *a.* descriptive of a folding table with legs that swing in to permit the leaves to be shut down. **gate-money**, *n.* entrance money taken at a sports ground etc. **gate-post**, *n.* a post on which a gate is hung or against which it shuts.
gateway, *n.* an opening or passage that may be closed by a gate; an entrance; a location through which one has access to an area.
gate² (gāt), *n.* (Sc.) one's way, manner of doing; course; (*usu. in comb.* as *Boargate, Friargate*) a street. **any gate, some gate, that gate**, (*dial.*) anywhere, somewhere etc.
-gate, *comb. form* indicating events or actions associated with political scandal. [from *Watergate*]
gâteau (gat'ō), *n.* (*pl.* **-teaux**) a rich cake.
gather (gadh'ə), *v.t.* to bring together, to collect, to cause to assemble; to accumulate, to acquire; to cull, to pluck; to pick (up); to get in, as harvest; to deduce, to conclude; to draw together, to pucker, to draw into folds or pleats; to sum (up); in printing, to arrange (pages) in their proper sequence. *v.i.* to come together, to assemble; to grow by addition, to increase; to generate pus or matter. *n.* a pleat or fold of cloth, made by drawing together. **to gather breath**, to recover one's wind, to have respite. **to gather oneself together**, to concentrate all one's strength or faculties, as for an effort. **to gather way**, of a vessel, to begin to move, to gain impetus, so as to answer to the helm. **gatherable**, *a.* **gatherer**, *n.* **gathering**, *n.* the act of collecting or assembling together; an assembly, a meeting, a party; an abscess, a boil. **gathering-coal, -peat**, *n.* (Sc.) a large piece of coal or peat put on the fire at night to keep it alive. **gathering-cry**, *n.* a rallying-cry, a summons to war. **gathering-ground**, *n.* catchment area. **gathers**, *n.* small pleats.
Gatling (gat'ling), **Gatling-gun**, *n.* an early machine-gun with rotating barrels. [US inventor, Dr R.J. *Gatling*, 1818–1903]
gator (gā'tə), (*abbr.*) alligator.
gauche (gōsh), *a.* awkward, clumsy; tactless, uncouth. **gaucherie** (-əri, -ərē'), *n.* awkwardness; a blunder, esp. a social mistake or awkwardness.
gaucho (gow'chō), *n.* (*pl.* **-chos**) a cowboy of the pampas of Uruguay and Argentina.
gaud (gawd), *n.* a showy ornament or trinket. **gaudy**¹, *a.* vulgarly and tastelessly brilliant and ornate, garish, flashy. **gaudily**, *adv.* **gaudiness**, *n.*
gaudy¹ GAUD.
gaudy² (gawd'i), *n.* a grand festival or entertainment, esp. one held annually at an English college in commemoration of some event.
gauge (gāj), (*Naut.*) **gage**, *v.t.* to ascertain the dimensions, quantity, content, capacity or power of; to test the content or capacity of (casks etc.) for excise purposes; to estimate or appraise (abilities, character etc.); to reduce to a standard size. *n.* a standard of measurement; an instrument for regulating or determining dimensions, amount, capacity etc. according to a fixed standard; a graduated instrument showing the height of a stream, quantity of rainfall, force of the wind, steam-pressure in a boiler etc.; the diameter of the barrel of a gun; the thickness of a sheet of plastic, film, metal etc.; the diameter of wires, screws, needles etc.; the position of a ship with reference to another and the wind, the **weather-gauge** being to windward, and the **lee-gauge** to leeward; in carpentry, an instrument for striking a line parallel to the straight side of a board; the distance between the two rails of a railway track, the **standard gauge** being 4 ft. 8½ in. (1.43 m).
gauge-glass, *n.* a tube to indicate the height of water in a boiler. **gaugeable**, *a.* **gauger**, *n.* one who gauges; esp. one who gauges casks etc., a customs officer. **gauging-rod, -rule, -ruler, -stick**, *n.* a customs officer's measuring instrument.
Gaul (gawl), *n.* an inhabitant of ancient Gaul; (*loosely*) a Frenchman. **Gaulish**, *a.* pertaining to Gaul; hence, French. *n.* the language of ancient Gaul.
Gauleiter (gow'lītə), *n.* the chief official in a district in Nazi Germany; a small-minded, bullying person in a position of minor authority.
Gaullist (gō'list), *n.* one who adheres to the policies and principles associated with General Charles de Gaulle, president of France 1959–69. [F]
gaultheria (gawlthiə'riə), *n.* a genus of evergreen aromatic shrubs of the heath family, containing the wintergreen, *Gaultheria procumbens.* [Dr *Gaultier*, 18th-cent. Canadian botanist]
gaumless, gormless (gawm'lis), *a.* witless, clumsy, stupid.
gaunt (gawnt), *a.* attenuated, thin, emaciated, haggard. **gauntly**, *adv.* **gauntness**, *n.*
gauntlet¹ (gawnt'lit), *n.* a long glove covered with plate-metal, worn with armour; a long stout glove covering the wrists. **to take up the gauntlet**, to accept a challenge. **to throw down the gauntlet**, to challenge, to defy. **gauntleted**, *a.* wearing gauntlets.
gauntlet² (gawnt'lit), *n.* a military punishment, in which the prisoner had to run between two files of men who struck at him with sticks, knotted cords or the like, as he passed. **to run the gauntlet**, to suffer this punishment; to expose oneself to possible attack.
gaup, gawp (gawp), *v.i.* to gape, esp. in astonishment. **gaupy**, *a.*
gaur (gowə), *n.* a large fierce ox, *Bos gaurus*, found in the mountain jungles in India.
gauss (gows), *n.* the cgs unit of magnetic flux density.
gauze (gawz), *n.* a light, transparent silk or cotton stuff; any perforated material resembling this, esp. a surgical dressing of muslin; a thin veil or haze. **wire-gauze**, *n.* a textile fabric made of wire, used for very fine sieves, respirators etc. **gauze-lamp**, *n.* a safety-lamp with gauze surrounding the flame. **gauzy**, *a.* **gauziness**, *n.*
gave (gāv), *past* GIVE.
gavel (gav'l), *n.* a mason's setting-maul; a small mallet, esp. one used by a chairman for demanding attention or by an auctioneer.
gavial (gā'viəl), *n.* an Indian crocodile, *Gavialis gangeticus*, with a long, slender snout.
gavotte (gəvot'), *n.* a dance of a lively yet dignified character resembling the minuet; the music for this; a dance-tune in common time and in two parts, each repeated.
Gawd (gawd), (*facet.*) God.
gawk (gawk), *n.* a simpleton, a booby. *v.i.* to stare (at or about) stupidly. **gawky**, *a.* awkward and usu. lanky. *n.* an awkward or clownish person. **gawkihood** (-hud), **gawkiness**, *n.* **gawkish**, *a.*
gawp GAUP.
gay (gā), *a.* light-hearted, lively, cheerful, merry; given to pleasure; (*euphem.*) wanton, licentious; showy, brilliant in appearance, dressed in bright colours; (*coll.*) homosexual. *n.* (*coll.*) a homosexual. **gay liberation, gay lib**, *n.* (*coll.*) a movement whose aims are to secure rights for homosexuals. **gay-libber**, *n.* (*coll.*) **gaily**, *adv.* **gayness**, *n.*
gaz., (*abbr.*) gazette, gazetteer.
gaze (gāz), *v.i.* to fix the eye intently (at or upon). *n.* a fixed look; a look of curiosity, attention, admiration or anxiety. **gazement**, *n.* **gazer**, *n.* **gazing**, *n.* **gazy**, *a.*
gazebo (gəzē'bō), *n.* an ornamental turret, lantern, or summer-house with a wide prospect, often erected in a garden; a belvedere.
gazel (gaz'əl), GHAZAL.
gazelle (gəzel'), *n.* a swift and very graceful antelope, esp. *Gazella dorcas*, noted for its large, soft black eyes.

gazette (gəzet'), *n.* a newspaper; an official journal containing lists of appointments to any public office or commission, legal notices, lists of bankrupts etc. *v.t.* to publish in a gazette, esp. to announce the appointment or bankruptcy of (*usu. in p.p.*). **gazetteer** (gazətiə'), *n.* a geographical dictionary. *v.t.* to describe in a geographical dictionary.

gazpacho (gaspach'ō), *n.* a spicy iced soup made from uncooked ripe tomatoes, chopped onion, cucumber and green peppers with olive oil, vinegar and water. [Sp.]

gazump (gəzŭmp'), *v.t., v.i.* esp. of a house vendor before entering into a binding contract, to force an intending purchaser to agree a higher price than that originally accepted. *n.* an act or instance of gazumping.

gazunder (gəzŭn'də), *v.t., v.i.* to force an intending seller to agree a lower price than originally accepted.

GB, (*abbr.*) Great Britain.

GBE, (*abbr.*) (Knight or Dame) Grand Cross of the British Empire (a British title).

GBH, gbh, (*abbr.*) grievous bodily harm.

GC, (*abbr.*) George Cross.

GCB, (*abbr.*) (Knight or Dame) Grand Cross of the Bath (a British title).

GCE, (*abbr.*) General Certificate of Education.

G-clef, *n.* (*Mus.*) a sign indicating the position of G above middle C; treble clef.

gcm, (*abbr.*) greatest common measure (in mathematics).

GCMG, (*abbr.*) (Knight or Dame) Grand Cross of the Order of St Michael and St George (a British title).

GCSE, (*abbr.*) General Certificate of Secondary Education.

GCVO, (*abbr.*) (Knight or Dame) Grand Cross of the Royal Victorian Order (a British title).

Gd, (*chem. symbol*) gadolinium.

Gdns, (*abbr.*) Gardens.

GDP, (*abbr.*) gross domestic product.

GDR, (*abbr.*) German Democratic Republic.

gds, (*abbr.*) goods; guards regiments.

Ge, (*chem. symbol*) germanium.

gean (gēn), *n.* (*chiefly Sc.*) the wild cherry, *Prunus avium*.

gear (giə), *n.* apparatus, tools, mechanical appliances, harness, tackle, equipment; dress, esp. (*coll.*) young people's fashionable clothing; a gear-wheel; a mechanism for transmitting motion by means of gear-wheels, links, levers etc.; the arrangement by which the driving-wheel of a cycle, motor-car etc. performs more or fewer revolutions relatively to the pedals, piston etc.; the state of being engaged or connected up; the state of being in working order; personal belongings; (*coll.*) illegal drugs, esp. marijuana. *v.t.* to harness, to put gear on; of a machine or motor vehicle, to put into gear; to furnish with gearing; to adjust or adapt to specific requirements; in company finance, to borrow money in order to increase the amount of total liabilities in relation to the share capital. *v.i.* to come or be in gear (with). **differential gear**, DIFFERENTIAL. **high** or **low gear**, on cars, cycles etc. apparatus for transmitting a high or low number of revolutions to the driving-wheel relatively to the motion of the engine, pedals, etc. **in gear**, of a machine or motor vehicle, connected up and ready for work. **landing gear** LAND. **to change gear**, to select a higher or lower gear; to increase or decrease the tempo of something. **to gear up**, to prepare physically or psychologically. **to throw out of gear**, to disconnect (gearing or couplings); to put out of working order; to disturb, to upset. **gear-box, -case**, *n.* the casing in which gears are enclosed in a motor vehicle or bicycle etc. **gear lever, shift, stick**, *n.* in a motor vehicle, a device for selecting or connecting gears. **gearwheel**, *n.* a wheel with cogs, esp. one transmitting motion to a similar wheel or chain. **gearing**, *n.* gear, working parts; a series of wheels etc. for transmitting motion; in company finance, the ratio of the amount a company has borrowed to its share capital, usu. expressed as a percentage. **gearless**, *a.*

gecko (gek'ō), *n.* (*pl.* **-os, -oes**) a genus of lizards with adhesive toes, by which means they can walk on a wall or ceiling.

gee[1] (jē), **gee-up**, *int.* go on, move faster (command to horse). **gee-gee**, *n.* (*childish and coll.*) a horse.

gee[2] (jē), *int.* (*coll.*) an exclamation expressing surprise, delight etc., also **gee-whizz**.

gee-bung (jē'bŭng), *n.* any shrub or tree of the proteaceous genus *Persoonia*, or its fruit.

geek (gēk), *n.* a sideshow performer who bites the heads off live animals; (*sl.*) a freakish, unattractive or uninteresting person, a flake. **geeky**, *a.*

geep (gēp), *n.* a cross between a goat and a sheep.

geese (gēs), *n.pl.* GOOSE.

gee-string (jē'-), G.

geezer (gē'zə), *n.* (*coll.*) an old man or woman.

gefilte, gefüllte fish (gəfil'tə), *n.* in Jewish cookery, cooked chopped fish mixed with matzo meal, egg and seasonings and then poached, either stuffed back into the skin of the fish or as dumplings.

gegenschein (gā'gənschīn), *n.* a faint glow in the night sky at a position opposite to that of the sun.

Gehenna (gəhen'ə), *n.* a valley near Jerusalem, where (Jer. xix.) men sacrificed their children to Baal or Moloch; whence, hell, a place of torment.

Geiger counter (gī'gə), *n.* a device for the detection and counting of particles from radioactive materials.

geisha (gā'shə), *n.* (*pl.* **-sha, -shas**) a professional female companion for men in Japan.

gel (jel), *n.* the jelly-like material formed when a colloidal solution is left standing; a jelly-like substance used for styling the hair; a sheet of a coloured transparent substance used to give colour esp. to theatrical lighting. *v.i., v.t.* to turn into a gel; to (cause to) assume a definite shape or final form.

gelatine (jel'ətin, -tēn), *n.* a transparent substance forming a jelly in water, obtained from connective animal tissue, such as skin, tendons, bones, horns etc. **gelatigenous** (-tij'i-), *a.* producing gelatine. **gelatinate** (-lat'-), **-ize, -ise**, *v.i., v.t.* to convert into a substance like jelly. **gelatination, -ization**, *n.* **gelatinizable** (-nīz'-), **gelatinoid** (-lat'-), *n., a.* **gelatinous** (-lat'-), *a.* of the nature of or consisting of gelatine, jelly-like. **gelose** (jəlōs'), *n.* a gelatinous substance obtained from Chinese and Japanese moss and seaweeds, used for finishing cotton goods, and in Asian cookery.

gelation (jəlā'shən), *n.* solidification by cooling or freezing.

geld (geld), *v.t.* to castrate (esp. a horse), to emasculate; to deprive of any essential part; to expurgate excessively. **gelder**, *n.* one who gelds (*usu. in comb.*, *a sow-gelder*). **gelding**, *n.* the act of castrating, castration; a castrated animal, esp. a castrated horse.

gelder, gelders rose (gel'dəz), GUELDER ROSE.

gelid (jel'id), *a.* extremely cold; icy. **gelidity**, *n.* **gelidly**, *adv.*

gelignite (jel'ignīt), *n.* an explosive containing nitroglycerine.

gelose (jəlōs'), GELATINE.

GEM, (*abbr.*) Graphics Environment Manager.

gem (jem), *n.* a precious stone, as the diamond, ruby, emerald etc., esp. when cut and polished for ornamental purposes (**gemstone**); an object of great rarity, beauty or value; a treasure, the most prized or the choicest part; in zoology, a gemma. *v.t.* (*past, p.p.* **gemmed**) to adorn with or as with gems. **gemless**, *a.* **gemmeous**, **gemmy**, *a.* full of or set with gems; bright, glittering; (*sl.*) spruce, smart, neat. **gemmily**, *adv.* **gemminess**, *n.* **gem(m)ology**, *n.* the science or study of gems. **gem(m)ological**, *a.* **gem(m)ologist**, *n.*

Gemara (gəmah'rə), *n.* the second portion of the Talmud, consisting of a commentary on the Mishna, or text.

Gemaric, *a.* of or pertaining to the Gemara. **Gemarist**, *n.*
gemeinschaft, *n.* a social group united by kinship, common beliefs etc.
geminate (jem'inət), *a.* united or arranged in pairs. *v.t.* (-nāt), to double, to arrange in pairs. *v.i.* to occur in pairs. **gemination**, *n.* **geminative**, *a.*
Gemini (jem'inī), *n.pl.* a constellation, the Twins, containing the two conspicuous stars, Castor and Pollux; the third sign of the zodiac; a mild oath, **geminy**, *(coll.)* (-ni), **jiminy** (jim'-). **Geminids** (-nidz), *n.pl.* meteoric bodies radiating, usu. in early December, from the constellation Gemini.
gemma (jem'ə), *n.* (*Bot.*) (*pl.* **-mae** (-ē)) a leaf-bud; (*pl.*) minute green cellular bodies in the fructification of Marchantia, and in some mosses and Hepaticae; (*Zool.*) a bud-like outgrowth in polyps, ascidians etc., which separates from the parent organism and develops into an individual. **gemmaceous** (-ā'-), *a.* pertaining to or of the nature of leaf-buds or gemmae.
gemmate (jem'āt), *a.* (*Bot.*) having buds; (*Zool.*) reproducing by gemmation. *v.i.* (jəmāt'), to bud; to reproduce by gemmation. **gemmation** (-ma'-), *n.* the act of budding; (*Zool.*) reproduction by the development of gemmae from the parent body. **gemmative**, *a.*
gemmeous, gemmy GEM.
gemmiferous (jəmif'ərəs), *a.* producing gems; producing or propagating by buds or gemmae.
gemmiparous (jəmip'ərəs), *a.* (*Bot.*) producing buds; (*Zool.*) propagating by gemmation. **gemmiparity** (jemipa'-), *n.* **gemmiparously**, *adv.*
gemmology, etc. GEM.
gemmule (jem'ūl), *n.* (*Biol.*) a small gemma or reproductive bud; (*Bot.*) the plumule or growing point of an embryo; a reproductive cell of a cryptogam; (*Zool.*) the ciliated embryo of many of the Coelenterata; one of the small reproductive bodies thrown off by sponges. **gemmuliferous** (-lif'-), *a.*
gemsbok (gemz'bok), *n.* a large antelope of Southern Africa, *Oryx gazella*, with long straight horns, also **gemsbuck**.
gemstone GEM.
gen (jen), *n.* *(coll.)* full particulars of, information about. **to gen up**, to read up about.
gen., *(abbr.)* gender; general; generally; generic; genitive; genus.
Gen., *(abbr.)* General; Genesis.
-gen, *suf.* producing; produced; growth; as in *hydrogen, nitrogen, oxygen; acrogen, endogen, exogen.*
gendarme (zhē'dahm), *n.* an armed policeman, in France etc. **gendarmerie** (-mərē'), **gendarmery** (-dah'-), *n.* the armed police of France; a body of gendarmes.
gender (jen'də), *n.* (*Gram.*) one of the classes (MASCULINE, FEMININE and NEUTER) into which words are divided according to the sex, natural or grammatical, of the things they represent; classification of words into genders according to their forms etc.; sex. **gender bender**, *n.* *(coll.)* one whose appearance and behaviour is of a kind usu. associated with the opposite sex. **gender bending**, *n.* sexually ambiguous appearance or behaviour. **gender gap**, *n.* lack of communication, understanding etc. between the sexes.
gene (jēn), *n.* the unit of heredity; the factor in a gamete which determines the appearance of an hereditary characteristic. **gene therapy**, *n.* the treatment of certain diseases by the insertion of new genes into non-reproductive cells in a patient. **genetic** (jenet'-), **genic** (jen'-), *a.*
genealogy (jēnial'əji), *n.* the history or investigation of the descent of families; (a record or exhibition of) a person's or family's descent in the natural order of succession; the course of a plant's or an animal's development from earlier forms. **genealogical** (-loj'-), *a.* **genealogical tree**, *n.* the genealogy of a family drawn out in the figure of a tree, with the root, stem, branches etc. **genealogically**, *adv.* **genealogist**, *n.*
genera GENUS.
general (jen'ərəl), *a.* relating to a whole genus, kind, class or order; not special, particular, partial or local; common, universal; ordinary, usual, widespread, prevalent; not limited in scope or application; indefinite, vague; not specialized or restricted; taken or viewed as a whole; of a rank or office, chief or supreme within a certain sphere. *n.* in the Roman Catholic Church, the chief of a religious order, or of all the houses or congregations having the same rule; an officer ranking next below a field-marshal, usu. extended to lieutenant-generals and major-generals; the commander of an army; a strategist; the chief part, the majority. **General Certificate of Education (GCE)**, formerly, in England and Wales a certificate in secondary education obtainable in Ordinary, Advanced and Scholarship levels. **General Certificate of Secondary Education (GCSE)**, a certificate of secondary education replacing GCE O level and the Certificate of Secondary Education. **in general**, in the main, generally; in most cases or in all ordinary cases, for the most part. **general anaesthetic**, *n.* a drug which anaesthetizes the whole body, with loss of consciousness. **General Assembly**, *n.* the governing body of the Church of Scotland; the deliberative body of the United Nations. **general average** AVERAGE. **general confession**, *n.* one in which the whole congregation joins. **general dealer**, *n.* one who sells many articles of daily use. **general election**, *n.* an election for representatives for all constituencies in a state. **general post office**, (GPO) *n.* a chief or head post office. **general practice**, *n.* the work of a general practitioner. **general practitioner**, *n.* a physician or surgeon treating all kinds of cases. **general staff**, *n.* in the army, officers assigned to advise senior officers on operations and policy. **general strike**, *n.* a strike by all or most workers in a city or in most parts of a province or country. **generalism**, *n.* a general conclusion, statement, or opinion. **generalissimo** (jenərəlis'imō), *n.* (*pl.* **-mos**) the chief commander of a combined military force; a commander-in-chief; *(coll.)* any esp. autocratic leader. **generalist**, *n.* a person knowledgeable in many fields as dist. from a specialist. **generality** (jenəral'əti), *n.* the state of being general, as opposed to specific; a general statement or principle; a vague statement, vagueness; the main body, the majority. **generalize, -ise** (jen'ərəlīz), *v.t.* to apply generally, to make of wider or of universal application; to deduce or infer (as a general principle) from many particulars. *v.i.* to form general ideas; to reason inductively; to draw general inferences; to speak vaguely, to employ generalities; to represent typical not particular features. **generalizable**, *a.* **generalization**, *n.* the act or process of generalizing; a general statement, proposition etc., esp. one that does not adequately account for individual cases; a general inference. **generalizer**, *n.* **generally**, *adv.* in general; for the most part, in most cases; ordinarily, commonly, usually; without minute detail, without specifying. **generalship** (jen'ərəlship), *n.* the office or rank of a general; skill in the management of troops and the conduct of war, strategy; skilful leadership, management or organization.
generant GENERATE.
generate (jen'ərāt), *v.t.* to produce or bring into existence; to cause to be; to evolve, to originate; (*Math.*) to trace out or form by the motion of a point, line etc. **generable**, *a.* **generant**, *a.* generating, producing. *n.* that which generates; (*Math.*) a point, line or surface conceived of as, by its motion, generating a line, surface or solid. **generating plant**, *n.* all the equipment needed for generating electrical energy. **generation** (-rā'shən), *n.* the act of generating; reproduction, propagation; production, creation; a single succession or step in natural descent; an age or period between one succession and another; the people of the same period or age; the average time in which the child

takes the place of the parent (usu. estimated at about $\frac{1}{3}$ of a century); offspring. **generation gap**, *n.* the difference in opinions and understanding between members of different generations. **generative** (jen'ərətiv), *a.* having the power of generating; pertaining to generation or production; productive, fruitful. **generative grammar**, *n.* a description of language in terms of a finite set of rules able to generate an infinite number of sentences. **generator** (jen'ərātə), *n.* one who or that which begets, generates, or produces; any apparatus for the production of gas, steam, electricity etc.; (*Elec.*) a dynamo. **generatrix** (-ā'triks), *n.* a female parent; (*Math.*) a generant.

generic (jəne'rik), **-al** (-əl), *a.* pertaining to a genus, class or kind, opp. to specific; comprehensive; not having a trademark. **generic name**, *n.* the name of a genus, as Saxifraga in *Saxifraga longifolia*; a general name for a product, not a brand name or trademark. **generically**, *adv.*

generous (jen'ərəs), *a.* liberal, munificent, open-handed; abundant, fertile; strong, stimulating (as wine); magnanimous, high-spirited. **generosity** (-ros'-), *n.* **generously**, *adv.*

genesis (jen'əsis), *n.* (*pl.* **-ses** (-sēz)) the act of begetting, producing, or giving origin to; creation, beginning, origination. **Genesis**, *n.* the first book of the Old Testament, in which the story of the Creation is told.

genet (jen'it), *n.* a small mammal, *Genetta vulgaris*, allied to the civet; its fur, or cat-skin dressed in imitation of this fur.

genetic (jənet'ik), *a.* of or relating to the origin, generation, or creation of a thing. *n.pl.* the study of heredity and variation. **genetic code**, *n.* the system, based on the molecular arrangement of the chromosomes, that ensures the transmission of hereditary characteristics. **genetic engineering**, *n.* the artificial alteration of the genes of an organism in order to control the transmission of certain hereditary characteristics. **genetic fingerprint**, *n.* the particular DNA pattern that is unique to an individual and can be used to identify that individual or his or her offspring. **genetic fingerprinting**, *n.* taking a genetic fingerprint from an individual's saliva, blood or sperm, in forensic science etc. **genetical**, *a.* **genetically**, *adv.* **geneticist** (-sist), *n.* one who studies genetics.

Geneva[1] (jənē'və), *a.* of, originating from, or pertaining to Geneva, a town in Switzerland. **Geneva bands**, *n.pl.* clerical bands such as those worn by Swiss Calvinist clergy. **Geneva Bible**, *n.* a translation of the Bible into English, made and published at Geneva in 1560. **Geneva Convention**, *n.* a convention made between the great powers (1864–5) to ensure the neutrality of ambulances, military hospitals, and those in charge of them, in time of war. **Geneva cross**, *n.* a red Greek cross on a white ground, the symbol of the Red Cross Society. **Geneva gown**, *n.* the black preaching gown worn by Presbyterian ministers and Low Church clergymen in England. **Genevan**, *a.* of or pertaining to Geneva. *n.* an inhabitant of Geneva. **Genevese** (jənəvēz), *a.*, *n.* Genevan.

geneva[2] (jənē'və), *n.* a spirit distilled from grain flavoured with juniper-berries, also called Hollands.

genial[1] (jē'niəl), *a.* of a cheerful and kindly disposition, cordial, sympathetic; conducive to life and growth, soft, mild. **geniality** (-al'-), *n.* **genially**, *adv.*

genial[2] (jəni'əl), *a.* of, pertaining to, or near the chin.

genic GENE.

-genic, *comb. form* of or pertaining to generation, as in *antigenic*; suitable for, as in *photogenic*.

genie (jē'ni), *n.* (*pl.* **genii** (-niī)) a jinnee.

genio-, *comb. form* chin.

genista (jənis'tə), *n.* a genus of leguminous shrubs and small trees, with yellow flowers.

genital (jen'itəl), *a.* pertaining to generation or procreation. *n.pl.* (also, **genitalia** (-tā'liə)) the external organs of reproduction.

genitive (jen'itiv), *a.* in grammar, indicating origin, possession, or the like (applied to a case in inflected languages roughly corresponding to the Eng. possessive). *n.* the genitive case. **genitival** (-tī'-), *a.*

genito- *comb. form.* genital.

genito-urinary (jen'itōū'rinəri), *a.* pertaining to the genital and urinary organs.

genius (jē'niəs), *n.* (*pl.* **genii** (-niī)) a tutelary deity or spirit, supposed to preside over the destinies of an individual, place, nation etc.; also one of two spirits attendant on a person through life, one good, the other evil; (*fig.*) one who exercises a powerful influence over another for good or ill; a jinnee; (*pl.* **geniuses**) natural bent or inclination of the mind; the dominant character, spirit, or sentiment (of); an extraordinary endowment of intellectual, imaginative, expressive or inventive faculty; a person so endowed; a representative type or impersonation.

Genoa (jen'ōə), *n.* a city in N Italy. **Genoa cake**, *n.* a rich fruit-cake with almonds on the top. **Genoese** (-ēz'), *a.* of or pertaining to Genoa. *n.* an inhabitant of Genoa; (as *pl.*) the people of Genoa.

genocide (jen'ōsīd), *n.* the systematic destruction of a national, racial, ethnical or religious group, e.g. the Jews by the Nazi Germans during World War II.

genome (jē'nōm), *n.* the complete set of chromosomes that is contained in any single cell. **genomic** (-nom'-), *a.*

genotype (jēn'ətīp), *n.* the basic genetic structure of an organism; a group of organisms with the same genetic structure. **genotypic, -typical** (-tip'-), *a.* **genotypically**, *adv.*

-genous, *comb. form* born; bearing; producing; as in *indigenous, polygenous.*

genre (zhé'r'), *n.* kind, sort, class; style, manner; a painting the subject of which is some scene in everyday life; this style of painting, also called **genre-painting**.

gens (jenz), *n.* (*pl.* **gentes** (-tēz)) a clan, house, or sept among the ancient Romans.

gent (jent), *n.* (*coll.*) a gentleman; a would-be gentleman. **gents, gents'**, a public lavatory for men.

genteel (jentēl'), *a.* (*now coll. or iron.*) gentlemanly or ladylike; elegant in manners, or dress, stylish; well-bred, refined, free from vulgarity. **genteelish**, *a.* **genteelly**, *adv.*

gentian (jen'shən), *n.* the English name of *Gentiana*, a genus of bitter herbs, usu. having blue flowers, common in mountain regions, one among which, the yellow gentian, *G. lutea*, yields gentian-root, used in medicine as a tonic. **gentian violet**, *n.* a greenish crystalline substance that forms a violet solution in water and is used in the treatment of burns and boils, as an antiseptic, and as a biological stain.

gentile (jen'tīl), *a.* not a Jew; heathen, pagan; applied by the Mormons to all who are not of their faith. *n.* one who is not a Jew; a heathen, a pagan; one who is not a Mormon.

gentility (jəntil'əti), *n.* the quality of being genteel, assumed social superiority; manners and habits distinctive of good society; gentle birth; genteel people.

gentle (jen'tl), *a.* mild, tender, kindly; not rough, coarse, violent or stern; moderate, not severe, not energetic; not steep; in heraldry, having the right to bear arms. *n.* (*pl.*) gentlefolk; the larva of the flesh-fly, used as bait in angling. *v.t.* to make gentle, amiable, or kind; to tame (as a colt). **the gentle birth**, of honourable birth, belonging to the gentry, having good breeding. **the gentle craft**, angling, also called **the gentle art**. **the gentle or gentler sex**, women. **gentlefolk**, *n.* (*earlier in pl.* **gentlefolks**) people of good position of gentle birth. **gentlehood** (-hud), *n.* gentle birth, rank, or breeding. **gentleness**, *n.* **gently**, *adv.* **gently born**, of gentle birth.

gentleman (jen'tlmən), *n.* a man of good breeding, kindly feelings and high principles, a man of honour; one who by education, occupation or income holds a good social position; used as a polite equivalent for man, esp. (*pl.*) in addressing the male members of an audience; a man of respectable position who follows no occupation. **gentleman-at-arms**, *n.* one of a company forming a bodyguard to the sovereign on state occasions. **gentleman-farmer**, *n.* a man of property who occupies his own farm. **gentleman's gentleman**, *n.* (*facet.*) a valet. **gentleman-usher**, *n.* a gentleman who officiates as usher to a sovereign or other person of high rank. **gentlemanhood** (-hud), *n.* **gentlemanship**, *n.* **gentlemanlike**, *a.* **gentlemanly**, *a.* like a gentleman in appearance, feeling or behaviour; pertaining to or becoming a gentleman. **gentlemanliness**, *n.* **gentlemen's agreement**, *n.* an agreement binding in honour but not legally.

gentlewoman (jen'tlwumən), *n.* a woman of gentle birth or breeding; a lady; a woman who waits upon a lady of high rank. **gentlewomanhood** (-hud), *n.* **gentlewomanlike, -ly**, *a.* **gentlewomanliness**, *n.*

gentry (jen'tri), *n.* the social class below the nobility; (*coll.*) people, folks. **gentrification** (-fikā'-), *n.* the process by which the character of an esp. inner urban area formerly lived in by working-class people is changed by an influx of middle-class people, with a consequent increase in property values. **gentrify**, *v.t.*

gents GENT.

genuflect (jen'üflekt), *v.i.* to bend the knee, esp. in worship. **genuflector**, *n.* **genuflectory** (-flek'-), *a.* **genuflexion, -ection** (-flek'shən), *n.*

genuine (jen'üin), *a.* belonging to or coming from the true stock; real, true; not counterfeit, false, spurious or adulterated; (*Zool.*) true to type, not aberrant. **genuinely**, *adv.* **genuineness**, *n.*

genus (jē'nəs), *n.* (*pl.* **genera** (jen'ərə)) a class or kind of objects containing several subordinate classes or species; (*Zool.* and *Bot.*) a group or class of plants or animals differentiated from all others by certain common characteristics and comprising one or more species; kind, group, class, order, family.

-geny, *comb. form* production or mode of production, as in *ontogeny, philogeny*.

geo-, *comb. form* pertaining to the earth.

geocentric (jēōsen'trik), **-al**, *a.* as viewed from or having relation to the earth as centre; having reference to the centre of the earth, as distinguished from any spot on its surface. **geocentrically**, *adv.* **geocentricism** (-sizm), *n.*

geochronology (jēōkrənol'əji), *n.* the measuring of geological time.

geod., (*abbr.*) geodesy.

geode (jē'ōd), *n.* a hollow nodule of any mineral substance, often lined with crystals; the cavity in such a nodule. **geodic** (-od'-), *a.* **geodiferous** (-dif'-), *a.*

geodesy (jēod'əsi), *n.* the science or art of measuring the earth's surface or large portions of it, as distinguished from surveying, which deals only with limited tracts. **geodesic** (-dē'-, -des'-), **geodetic** (-det'-), *a.* pertaining to geodesy; carried out or determined by means of geodesy. *n.pl.* geodesy. **geodesic dome**, *n.* a light, strong dome built from a lattice-work of polygons so that the pressure load is evenly distributed throughout the structure. **geodetic, geodesic line**, *n.* the shortest line between two points on the earth's surface or that of a geometrical solid. **geodetic surveying**, *n.* a method of surveying large areas which takes into account the curvature of the earth. **geodetically**, *adv.* **geodesist**, *n.*

geodynamic (jēōdīnam'ik), **-al**, *a.* relating to the latent forces of the earth. **geodynamics**, *n.*

geog., (*abbr.*) geographer; geographical; geography.

geography (jēog'rəfi), *n.* the science of the surface of the earth, its physical features, natural productions, inhabitants, political divisions, commerce etc.; a book dealing with this. **mathematical geography**, those parts of the science involving mathematics, such as astronomical geography, geodesy and cartography. **physical geography**, geography treating of the physical features of the earth's surface, the distribution of land and water, climate, and the distribution of plants and animals. **political geography**, geography dealing with countries, states, political, social and economic conditions. **geographer**, *n.* **geographic, -ical** (-graf'-), *a.* of or pertaining to geography; relating to or containing a description of the earth. **geographical mile**, *n.* one minute of longitude measured at the equator, nautical mile. **geographic latitude**, *n.* the angle between the plane of the equator and a perpendicular to the surface of the earth at a given point. **geographic variation**, *n.* the alteration in form, habits etc. of a species or variety of plant or animal due to a change of habitat. **geographically**, *adv.*

geoid (jē'oid), *n.* the surface the earth would have if all parts of it were the same height as the mean sea level of the oceans; the shape of this. **geoidal**, *a.*

geol., (*abbr.*) geological; geologist; geology.

geology (jēol'əji), *n.* the science of the earth's crust, its composition, its structure, and the history of its development. **dynamical geology**, the study of the forces that have brought about geological changes. **structural geology**, the study of the relations between the rock-masses forming the earth's crust and of the physical causes to which they are due. **geologic** (-loj'-), *a.* forming part of the subject-matter of geology. **geological**, *a.* pertaining to geology. **geological time**, *n.* the time occupied by the development of the planet earth to the present. **geologically**, *adv.* **geologist**, *n.* **geologize, -ise**, *v.i.* to study geology; to make geological investigation, esp. in a particular district.

geom., (*abbr.*) geometer; geometrical; geometry.

geomagnetism (jēōmag'nətizm), *n.* the magnetic field of the earth; the study of the earth's magnetism. **geomagnetic** (-net'-), *a.* **geomagnetically**, *adv.* **geomagnetist**, *n.*

geometer (jēom'iətə), *n.* a geometrician; (also **geometrid**) a moth or its caterpillar belonging to the tribe called Geometrae, on account of their seeming to measure the ground as they move along, looper. **geometrid** (-trid), *a.*

geometry (jiom'ətri), *n.* the science of magnitudes, whether linear, superficial, or solid, with their properties and relations in space. **plane geometry**, the branch of geometry dealing with magnitudes and their relations in one plane. **solid geometry**, geometry dealing with all three dimensions of space. **geometric, -al** (jēəmet'-), *a.* pertaining to geometry; done, determined or prescribed by geometry; disposed in mathematical figures. **geometrical progression**, *n.* a progression in which the terms increase or decrease by a common ratio, as 1, 3, 9, 27; 144, 72, 36, 18. **geometrical proportion**, *n.* one based on equal ratios in its two parts, as 2 : 4, 6 : 12. **geometrically**, *adv.* **geometrician** (jēəmətrish'ən), **geometrist**, *n.* **geometrize, -ise**, *v.i.* to work or construct according to the rules or methods of geometry; to proceed geometrically.

geomorphology (jēōmawfol'əji), *n.* the study of the origin, development and characteristics of land forms. **geomorphologist**, *n.* **geomorphologic** (-loj'-), **geomorphological**, *a.* **geomorphologically**, *adv.*

geophysics (jēōfiz'iks), *n.* the science that deals with the physical characteristics of the earth.

geopolitics (jēōpol'itiks), *n.* the study of how the political views and aims of a nation are affected by its geographical position.

geoponics (jēōpon'iks), **geopony** (-op'əni), *n.* the art and science of agriculture.

Geordie (jaw'di), *n.* (*coll.*) a native or the dialect of Tyneside, NE England.

George (jawj), *n.* a jewel bearing the figure of St George worn by the knights of the Garter; (*Aviat., coll.*) an automatic pilot. **by George**, a mild oath. **St George's Cross** CROSS[1]. **George Cross**, *n.* a decoration instituted in 1940, primarily in recognition of acts of heroism or conspicuous

courage performed by civilians. **George Medal**, *n.* similarly awarded for acts of great bravery.
georgette (jawjet'), *n.* a plain semi-transparent dress material. [Mme *Georgette*, a French dressmaker]
Georgian[1] (jaw'jən), *a.* relating to the period of George I–IV in Great Britain (1714–1830); relating to the reign of George V (1910–36).
Georgian[2] (jaw'jən), *a.* of or pertaining to Georgia, a region south of the Caucasus, or to Georgia, one of the southern States of the US. *n.* a native or inhabitant of one of these.
georgic (jaw'jik), *n.* one book of Virgil's *Georgics*, a poem in four books on husbandry.
geoscience (jēōsi'əns), *n.* any of the sciences that are concerned with the earth, e.g. geology, geophysics or geodesy; these sciences collectively.
geosphere (jē'əsfiə), *n.* the solid part of the earth, as distinct from the *atmosphere* or *hydrosphere;* lithosphere.
geostatic (jēəstat'ik), *a.* applied to an arch so constructed as to be in equilibrium under vertical pressure, as in an embankment. **geostatics**, *n.* the branch of physics concerned with the statics of rigid bodies.
geostationary (jēōstā'shənəri), *a.* of a satellite, orbiting the earth at the same speed as the earth rotates so remaining above the same spot on the earth's surface.
geostrophic (jēəstrof'ik), *a.* of or caused by the force produced by the rotation of the earth. **geostrophic wind**, *n.* a wind the direction and force of which are influenced by the earth's rotation.
geosynchronous (jēōsing'krənəs), *a.* of a satellite, geostationary.
geosyncline (jēōsing'klīn), *n.* a part of the earth's crust that has sunk inwards, resulting in a usu. long and broad depression containing deep thicknesses of rock or sediment.
geotaxis (jēōtak'sis), *n.* the response of an organism or a plant to the stimulus of gravity. **geotactic, -ical**, *a.*
geothermal (jēōthœ'mal), *a.* pertaining to the internal heat of the earth. **geothermal energy**, *n.* energy from the natural heat of the earth, e.g. hot springs. **geothermic**, *a.*
geotropism (jēōt'rəpizm), *n.* the tendency exhibited by the organs of a plant to turn towards the centre of the earth. **geotropic** (-trop'-), *a.* **geotropically**, *adv.*
Ger., (*abbr.*) German; Germany.
ger., (*abbr.*) gerund, gerundive.
geranium (jərā'niəm), *n.* a genus, with about 100 species, of hardy herbaceous plants, rarely shrubs, natives of all temperate regions, typified by *Geranium maculatum*, the crane's-bill, so called from the shape of its seed-pod; a plant of this genus; a cultivated plant of the allied genus *Pelargonium.*
gerbil, gerbille (jœ'bl), *n.* any of numerous small, burrowing, mouselike rodents of the subfamily gerbillinae, of desert regions of Asia and Africa.
gerfalcon (jœ'fawlkən), *n.* a large and powerful falcon of northern regions, typified by the Iceland falcon, *Falco islandus.*
geriatrics (jeriat'riks), *n.sing.* the branch of medicine dealing with old age and its diseases. **geriatric**, *a.* **geriatrician** (-trish'ən), *n.*
germ (jœm), *n.* (*Biol.*) the portion of living matter from which an organism develops; the embryo of an animal or plant; a partially-developed organism; a microorganism, esp. such as is supposed to cause disease, a microbe; that from which anything springs; the origin, source or elementary principle. **in germ**, existing in an undeveloped state. **germ-cell**, *n.* the parent cell from which a new individual develops usu. dist. as the female element in reproduction. **germ-line therapy**, *n.* in medicine, the treatment of certain diseases by the insertion of new genes into the reproductive cells of a patient, such genes then being passed on to all future generations, as distinct from *gene therapy.* **germ-plasm**, *n.* the part of the protoplasm in which the power of reproduction is supposed to reside and which is transmitted from one generation to its offspring. **germ theory**, *n.* the theory that certain diseases are caused by the development of microorganisms introduced into the body through germs or spores. **germ-warfare**, *n.* the use of bacterial weapons against enemy troops. **germless**, *a.* **germicide** (-misid), *a.* destroying germs, esp. disease-germs. *n.* a substance used for this purpose. **germicidal** (-si'dəl), *a.*
German (jœ'mən), *a.* pertaining or relating to Germany. *n.* (*pl.* **Germans**) a native or inhabitant of Germany; the language of Germany, High German. **High German**, originally the form of German spoken in the south, but since Luther's translation of the Bible (1450) adopted as the literary language all over Germany. **Low German**, German of the Netherlands, including Dutch, Frisian, Flemish and Old Saxon. **German measles**, *n.pl.* a mild infectious disorder resembling measles which if contracted by a pregnant woman may cause birth deformities in her unborn child, rubella. **German shepherd (dog)**, ALSATIAN. **German silver**, *n.* a white alloy of nickel, copper and zinc, used for mathematical instruments, table-ware, etc. **Germanesque** (-nesk'), *a.* **Germanic** (-man'-), *a.* of or pertaining to Germany; of or pertaining to the Teutonic race. *n.* the primitive Teutonic language. **East Germanic**, the group of extinct Teutonic languages represented by Gothic. **North Germanic**, the Scandinavian group of languages. **West Germanic**, the group comprising High and Low German, Dutch, Frisian, English etc. **Germanism**, *n.* **Germanist**, *n.* **Germanity** (-man'-), *n.* **Germanize, -ise**, *v.t.* to assimilate or make to conform to German ideas, customs, idioms etc. *v.i.* to conform to these. **Germanization**, *n.* **Germanizer**, *n.*
german (jœ'mən), *a.* of blood relationships, full (*usu. in comb.*, as *cousin-german*); closely connected, relevant, pertinent.
germander (jœman'də), *n.* a plant of the genus *Teucrium*, esp. the wall germander, *T. chamaedrys.*
germane (jəmān'), *a.* relevant to, pertaining to, relating to.
germanium (jəmā'niəm), *n.* a metallic element of a greyish-white colour, at. no. 32; chem. symbol Ge, used in the construction of transistors because of its electrical properties.
Germano-, *comb. form* German.
Germanomania (jœmənōmā'niə), *n.* enthusiasm for Germany or German things.
Germanophil (jœman'əfil), *n.* **Germanophilist** (-nof'-), *n.* a lover of Germany or Germans. **Germanophilia**, *n.*
Germanophobe (jœman'əfōb), *n.* one who hates Germany or Germans. **Germanophobia** (-fō'biə), *n.* **Germanophobic**, *a.*
germen (jœ'mən), *n.* the ovary or rudimentary seed-vessel of a plant. **germigenous** (-mij'ə-), *a.* **germiniparous** (-minip'ə-), *a.* **germinal** (-mi-), *a.* pertaining to or of the nature of a germ; germinative; in the earliest stage of development. **germinally**, *adv.*
germicide GERM.
Germinal (zhœmēnal'), *n.* the name given by the French Convention to the seventh month of the republican year, 21 Mar to 19 Apr. [F]
germinal GERMEN.
germinate (jœ'mināt), *v.t.* to sprout, to bud; to develop. *v.i.* to cause to sprout or bud; to produce. **germinable**, *a.* **germinant**, *a.* sprouting, growing, developing. **germination**, *n.* the first act of growth in an embryo plant, ovum etc.; the act or process of germinating. **germinator**, *n.* **germinative**, *a.*
geront(o)-, *comb. form* pertaining to old age.
gerontic (jəron'tik), *a.* pertaining to old people, senile.
gerontocracy (jerəntok'rəsi), *n.* government by old men; a government of old men. **gerontocratic** (jərontōkrat'ik), *a.*

gerontology (jerəntol'əji), *n.* the scientific study of the ageing process.

-gerous, *comb. form* bearing, having; as in *armigerous, florigerous.*

gerrymander (je'rimandə), *v.t.* to tamper with (an electoral district or constituency) so as to secure unfair advantages for a particular candidate, party or class; to misconstrue or garble (a question, argument etc.) so as to arrive at unfair conclusions. *n.* an unfair rearrangement of a constituency. **gerrymanderer,** *n.*

gerund (je'rənd), *n.* in Latin, a part of the verb used as a noun instead of the infinitive in cases other than the nominative; in English, a verbal noun ending in *-ing*, which retains some of the characteristics of a verb. **gerundial** (-rŭn'-), *a.* **gerundive** (-rŭn'div), *a.* pertaining to or of the nature of a gerund. *n.* in Latin, a verbal adjective formed on the gerundial stem giving the sense of *must* and *should* (be done). **gerundival** (-di'-), *a.* **gerundively** (-run'-), *adv.*

gesso (jes'ō), *n.* a prepared ground of plaster of Paris for painting, sometimes for sculpture. **gesso work,** *n.*

gest, geste (jest), *n.* a deed, an exploit; a tale or history of the exploits of a hero or heroes, esp. a mediaeval ballad or metrical romance.

gestalt (gəstalt'), *n.* in psychology, an organized whole in which each part affects every other part. Its exponents have demonstrated that the mind tends to perceive events and situations as a pattern, or whole, rather than as a collection of separate and independent elements. [G, form, pattern]

Gestapo (gəstah'pō), *n.* the body of secret police formed to secure strict obedience to the government of Nazi Germany. [first letters of G *Geheime Staats Polizei*, secret state police]

gestation (jəstā'shən), *n.* the act of carrying or the process of being carried in the uterus from conception to parturition; the period of this; the (period of) development of an idea in the mind. **gestate** (jes'-), *v.t.*

gesticulate (jəstik'ūlāt), *v.i.* to make expressive gestures or motions, as in speaking or instead of speaking. *v.t.* to express or represent by gestures. **gesticulation,** *n.* the act or art of gesticulating to express emotion or illustrate an argument; a gesture. **gesticulator,** *n.* **gesticulative, gesticulatory,** *a.* pertaining to or represented by gesticulation.

gesture (jes'chə), *n.* a motion of the face, body or limbs, used to express emotion or to illustrate or enforce something that is said; the art of using such movements for rhetorical or dramatic purposes; an action which serves as a token of something or indicates one's willingness to do something. *v.i.* to gesticulate. *v.t.* to accompany or represent with gestures or action. **gestural,** *a.* **gestureless,** *a.* **gesturer,** *n.*

gesundheit (gəzunt'hīt), *int.* your health (said after someone has sneezed). [G]

get[1] (get), *v.t.* (*p. get* (got), *p.p.* got, gotten (got'n)), to obtain, to gain possession of by any means; to fetch; to earn, to win; to capture; to receive as one's portion or penalty, to suffer; to understand; to learn, to commit to memory; (*coll. in p.p.*) to have, to possess; (*coll.*) to be obliged (to); to beget; to succeed in obtaining, bringing, putting etc.; to succeed in communicating with; to induce, to persuade (to); to betake (oneself); (*coll.*) to catch, to outwit; to affect or grip emotionally; (*coll.*) to hit; (*coll.*) to take revenge on, to kill. *v.i.* to arrive at any place, condition or posture; to go, to depart; (*coll.*) to succeed, to find the way or opportunity (to); to be a gainer, to profit. **get!** (*imper.*) be off! **get away,** an exclamation of mild disbelief; also, **get lost, get out,** (*imper.*) be off! **has got to be done,** must be done. **to get about,** to be able to move or walk about (after an illness); to become known, to be reported abroad; to travel from place to place. **to get across,** to communicate or be communicated successfully. **to get ahead,** to prosper; to overtake. **to get along,** to proceed, to advance; to succeed, to fare, to manage (well or badly); (*coll.*) to go away. **to get at,** to be able to reach; to ascertain; (*sl.*) to banter, to tease; (*sl.*) to influence, corrupt, bribe (a jockey etc.); to drug or illegally tamper with (a racehorse). **to get away,** to quit; to escape; to disengage oneself (from). **to get away with,** to make off with; to escape discovery in connection with (something wrong or illegal). **to get back,** to receive back, to recover; to return, to come back. **to get behind,** to lag; to fall into arrears; to penetrate, to unravel. **to get by,** to elude; to be good enough; (*coll.*) to manage, to cope. **to get clear,** to disengage oneself; to be released. **to get cracking** CRACK. **to get down,** to dismount, to descend; to depress. **to get down to,** to concentrate upon; to start work on. **to get even with,** to revenge oneself on; to pay back. **to get going,** to begin; to make haste. **to get in,** to be elected; to enter; to collect and place under cover (as crops); to make room for. **to get into,** (*coll.*) to put on (as clothes etc.); (*coll.*) to become involved in; to possess, dominate or take over (a person's mood, personality etc.). **to get into one's head,** to be convinced of. **to get it in the neck** NECK. **to get it together,** to achieve harmony or success. **to get loose or free,** to liberate or disengage oneself. **to get off,** to dismount, to alight (from); to escape, to be released (from); to be acquitted, to be let off (with or for); to start; to take off, to remove; to procure the acquittal of. **to get off on,** (*coll.*) to be impressed by; to enjoy. **to get off with,** (*coll.*) to have a sexual relationship with; to escape blame or punishment for. **to get on,** to put or pull on; to move on; to succeed or prosper; to grow late; to grow old; to have a friendly relationship; to do, fare or manage (with or without); to mount, to board. **to get one's eye in** EYE. **to get one's goat** GOAT. **to get one's own back,** to revenge oneself. **to get onto,** to make contact with; to become aware of, discover. **to get out,** to pull out; to escape from any place of confinement or restraint; to be divulged. **to get out of,** to avoid (doing something). **to get over,** (*coll.*) to persuade; to surmount, overcome (a difficulty etc.); to recover from (illness, surprise, disappointment etc.); to make intelligible. **to get round,** to evade, to circumvent; to cajole; to get one's way with. **to get round to,** to find time or opportunity to. **to get set,** (a command at the start of a race) be ready. **to get stuck in or into,** (*coll.*) to eat hungrily; to start doing a task vigorously; to attack (someone) physically or verbally. **to get the better or best of,** to gain the advantage; to be victorious. **to get the hang of** HANG[2]. **to get there,** (*coll.*) to succeed; to understand. **to get the worst of it,** to be defeated. **to get through,** to reach a point beyond, to reach one's destination; to pass (as a Bill); to succeed in doing, to complete, to finish (with); to pass (an examination); to use up. **to get through to,** to make a telephone connection with; (*coll.*) to make (someone) understand. **to get to,** to reach, to arrive at; to begin (a task etc.); (*coll.*) to annoy or irritate. **to get together,** to meet, to assemble; to bring together, to amass. **to get under one's skin** SKIN. **to get under way,** to start a ship; to start, to begin to move (of a ship). **to get up,** to learn, to work up; to dress up, to disguise; to invent, to devise; to rise (as from a bed etc.); to mount; to begin to rage or be violent (as the wind, waves etc.). **to get up to,** (*coll.*) to be involved in. **to get wind of** WIND. **to get with child,** to make pregnant. **get-at-able,** *n.* accessible. **getaway,** *n.*, *a.* (*coll.*) (of) an escape. **get-together,** *n.* (*coll.*) an informal gathering. **get-up,** *n.* dress and other accessories; the manner in which anything is presented, as on the stage; the style or format (of a book). **get-up-and-go,** *n.* energy and enthusiasm; ambition. **gettable,** *a.* obtainable. **getter,** *n.* **getting,** *n.pl.* gains, profits.

get[2] (get), GIT.

geta (gā'tə), *n.* (*pl.* **geta, getas**) a Japanese wooden sandal.

geum (jē'əm), *n.* a hardy genus of rosaceous plants comprising the avens or herb-bennet.
gewgaw (gū'gaw), *n.* a showy trifle; a toy, a bauble.
geyser (gē'zə), *n.* a hot spring throwing up a column of water at intervals; an apparatus for heating a stream of water supplying a bath, etc.
GG, (*abbr.*) Girl Guides; Governor-General; Grenadier Guards.
Ghanaian (gahnä'ən), *n.* an inhabitant of Ghana. *a.* pertaining to Ghana.
gharry (ga'ri), *n.* a variety of wheeled carriage in India.
ghastly (gahst'li), *a.* pale, death-like, haggard; horrible, frightful, shocking; (*coll.*) awful, unpleasant. *adv.* in a ghastly manner. **ghastlily**, *adv.* **ghastliness**, *n.*
ghaut, ghat (gawt), *n.* a mountain pass; a range of mountains; a flight of steps descending to a river, a landing-place.
Ghazi (gah'zi), *n.* one who has fought for Islam against infidels. **Ghazism**, *n.*
ghee, ghi (gē), *n.* clarified butter, usu. prepared from buffalo-milk.
gherkin (gœ'kin), *n.* a young and green small variety of cucumber, used for pickling.
ghetto (get'ō), *n.* (*pl.* **-tos, -toes**) the quarter of a town formerly inhabited by Jews; a poor, densely populated area of a city, esp. inhabited by a racial minority. **ghetto-blaster**, *n.* a large portable stereo radio-cassette player. **ghettoize, -ise**, *v.t.* to make into a ghetto. **ghettoization**, *n.*
ghi GHEE.
ghillie GILLIE.
ghost (gōst), *n.* the spirit or soul of a deceased person appearing to the living, an apparition; the soul of a dead person in the other world; the soul or spirit, the vital principle; a mere shadow or semblance; the remotest likelihood; one who does literary or artistic work for which another takes the credit; in optics, a spot, gleam or secondary image caused by a defect in a lens; in television reception, a duplicated image. *v.i., v.t.* to ghost write. **Holy Ghost** HOLY. **to give up the ghost**, to die, to expire. **ghost-story**, *n.* a tale concerned with the supernatural, esp. one of a terrifying character. **ghost town**, *n.* a deserted or semi-deserted town, which was formerly flourishing. **ghost-word**, *n.* a word which originated in an error by a scribe, printer etc. **ghost write**, *v.t., v.i.* to write (a speech etc.) for another. **ghost writer**, *n.* one who writes (speeches etc.) for another who is presumed to be the author. **ghosthood** (-hud), *n.* **ghost-like**, *a.* **ghostly**, *a.* pertaining to the spirit or soul, spiritual; pertaining to religious matters; pertaining to ghosts or apparitions; dismal, gloomy. **ghostliness**, *n.*
ghoul (gool), *n.* an evil spirit supposed, in Eastern tales, to devour human corpses; a person who robs graves; a person interested in morbid things. **ghoulish**, *a.* **ghoulishly**, *adv.* **ghoulishness**, *n.*
GHQ, (*abbr.*) General Headquarters.
ghyll (gil), GILL[2].
GI, *n.* (*pl.* **GIs, GI's**) (*N Am., coll.*) a soldier in the US Army, esp. a private. *a.* (of equipment etc.) conforming to US Army regulations. [abbr. of *government issue*]
giant (jī'ənt), *n.* a mythical being of human form but superhuman size; any person, animal, plant etc. of abnormal size; a person of extraordinary powers, ability etc. *a.* gigantic; like a giant. **giant-killer**, *n.* a person, team etc. that defeats a much stronger opponent. **giant panda** PANDA. **giant star**, *n.* a star of great brightness and a very low mean density. **giantess**, *n. fem.* **gianthood** (-hud), *n.* **giantship**, *n.* **giantism**, *n.* abnormal development in size esp. as caused by dysfunction of the pituitary gland. **giant-like**, *a.* **giantly**, *a.*, *adv.* **giantry**, *n.*
giaour (jowə), *n.* an infidel, a name given by the Turks to those who disbelieve in Mohammed, esp. Christians.
Gib. (jib'), (*abbr.*) Gibraltar.

gibber (jib'ə), *v.i.* to jabber, to talk rapidly and inarticulately. **gibberish**, *n.* inarticulate sounds; unmeaning or unintelligible language, jargon. *a.* unmeaning.
gibbet (jib'it), *n.* an upright post with a crosspiece from which criminals were formerly hanged. *v.t.* to execute by hanging; to hang or expose on or as on a gibbet; to expose to public contempt and derision.
gibbon (gib'ən), *n.* any individual of the genus *Hylobates*, long-armed anthropoid apes from E Asia.
gibbous (gib'əs), **gibbose** (-ōs), *a.* hunch-backed, humped; protuberant, convex; a term used when the illuminated portion of the moon or of a planet exceeds a semicircle but falls short of a circle. **gibbosity** (-bos'-), *n.* **gibbously**, *adv.*
gibe (jīb), *v.i.* to use sneering or taunting expressions; to rail, to flout, to jeer, to scoff (at). *v.t.* to use sneering or taunting expressions towards; to mock, to taunt, to sneer at. *n.* a sneer, a scoff, a taunt. **giber**, *n.* **gibingly**, *adv.*
giblets (jib'lits), *n.pl.* the feet, neck, and internal eatable parts of a fowl, such as the heart, liver, gizzard etc., which are removed before cooking.
giddy (gid'i), *a.* having a whirling, swimming or dizziness in the head; reeling, tending to stagger or fall; causing this sensation (as a precipice, a dance, success etc.); inconstant, changeable, fickle, flighty; elated, excited, rash. **to play the giddy goat**, to act the fool. **giddy-up**, *imper.* a command to a horse to make it start moving or go faster. **giddily**, *adv.* **giddiness**, *n.*
gidgee (gij'ē), *n.* a small Australian tree, *Acacia cambagei*, which gives off a foul smell at the approach of rain. Also, **stinking wattle**.
gie (gē), (*Sc.*) GIVE.
GIFT, (*acronym*) gamete intra-fallopian transfer.
gift (gift), *n.* the act, right, or power of giving; that which is given, a present, a contribution; in law, the voluntary bestowal of property without consideration; a natural quality, talent or endowment. *v.t.* to bestow or confer; to endow with gifts; to present (with) as a gift. **must not look a gift-horse in the mouth**, must not criticize what one is given for nothing. **the gift of the gab** GAB. **gift-book**, *n.* a book given as a present, or suitable for so giving. **gift-wrap**, *v.t.* to wrap (a gift) in attractive paper. **gifted**, *a.* largely endowed with intellect, talented.
gig[1] (gig), *n.* a fish-spear.
gig[2] (gig), *n.* a light two-wheeled vehicle drawn by one horse; a light clinker-built boat, 20–28 ft (6–9 m) long, rowed by 4, 6 or 8 alternate oars, usu. reserved for the commanding officer; a somewhat similar boat used for racing; a machine for raising a nap on cloth by passing it over rotary cylinders furnished with wire teeth.
gig[3] (gig), *n.* (*coll.*) a job, esp. a booking for a musician to perform.
giga- (gī'gə-), *pref.* denoting ten to the ninth power (10^9) as in *gigavolt, gigahertz*.
gigantic (jīgan'tik), *a.* huge, enormous, giant-like; immense, extraordinary. **gigantean** (-tē'-), **gigantesque** (-tesk'), *a.* **gigantically**, *adv.* **giganticidal** (-si'-), *a.* **giganticide**, *n.* **gigantify**, *v.t.* **gigantism** (jī'-), *n.* GIANTISM under GIANT.
giggle (gigl), *v.i.* to laugh in a silly or affected manner, to titter; to laugh in a nervous, catchy way, with attempts to restrain oneself. *n.* a laugh of such a kind. **giggler**, *n.* **gigglesome** (-səm), *a.*
gigolo (zhig'əlō), *n.* (*pl.* **-los**) a professional dance-partner or escort; a man who is kept by a much older woman.
gigot (jig'ət), *n.* a leg of mutton. **gigot-sleeve** *n.* a sleeve shaped like a leg of mutton. [F]
gigue (zhēg), *n.* a piece of dance music, usu. in 6/8 time.
Gila monster (hē'lə), *n.* a large poisonous lizard, *Heloderma suspectum*, found in Arizona and New Mexico.
gilbert (gil'bət), *n.* the cgs unit for measuring magnetomotive force. [William *Gilbert*, 1544–1603, Eng. scientist]

Gilbertian (gilbœ'tiən), *a.* absurdly topsy-turvy; of humour, in the style of Sir W.S. Gilbert, 1836–1911, writer of comic operas.

gild[1] (gild), *v.t.* (*past, p.p.* **gilded** or **gilt** (gilt)) to coat, overlay or wash thinly with gold; to impart a golden colour or appearance to; to make brilliant, to brighten; to give a specious or agreeable appearance to. **gilded youth**, young people of wealth and fashion. **to gild the lily**, to spoil beauty by overembellishing. **gilder**, *n.* one whose occupation is to coat articles with gold. **gilding**, *n.* the act, process, or art of overlaying with gold; gilding-metal for application to any surface; outward decoration, covering, or disguise designed to give a fair appearance to anything. **gilding-metal**, *n.* an alloy of copper, brass and tin. **gilding-size**, *n.* sizing used for cementing gold-leaf on a surface.

gild[2] (gild), GUILD.

gilet (zhē'lā), *n.* a woman's garment or the bodice of a dress shaped like a waistcoat.

gill[1] (gil), *n.* (*usu. in pl.*) the organs of respiration or branchiae of fishes and some amphibia; hair or leaf-like respiratory processes projecting from the body of some aquatic insects; the vertical lamellae under the cap of fungi; the wattles of a fowl; (*facet.*) the flesh about a person's jaws and chin. (*coll.*) to **be, go white** or **green at, around, about the gills**, to be pale in the face because of nausea, fear, exhaustion etc. **gill-cover**, *n.* the external bony covering of a fish's gills. **gill-net**, *n.* a net, usu. set vertically, for entangling fish by the gills. **gill-opening**, *n.* the opening by which the water passes into the gills.

gill[2] (gil), *n.* a deep and narrow ravine, often wooded; a gully or stream-bed on a precipitous hillside.

gill[3] (jil), *n.* a liquid-measure, usu. one-fourth of a pint (about 140 cl).

gillie, ghillie (gil'i), *n.* (*Sc.*) a Highland man-servant, esp. one who attends a sportsman in fishing or hunting.

gillion (gil'yən, jil'-), *n.* in Britain, one thousand million (equivalent to US billion).

gillyflower (jil'iflowə), *n.* the clove-pink *Dianthus caryophyllus*; also applied to the white stock, *Matthiola incana*, and the wallflower, *Cheiranthus cheiri*.

gilt[1] (gilt), *a.* gilded; adorned with gold or something resembling gold. *n.* gold laid over the surface of a thing, gilding; superficial attraction; (*pl.*) gilt-edged securities. **to take the gilt off the gingerbread**, to remove the glamour or appeal of something. **gilt-edged**, *a.* having the edges gilded. **gilt-edged securities**, *n.pl.* investments of the most reliable character.

gilt[2] (gilt), *n.* a young sow.

gimbal (jim'bəl, gim'-), *n.* (*usu. in pl.*) a form of universal joint for securing free motion in suspension, or for suspending anything, as a lamp, a compass, a chronometer etc., so that it may always retain a horizontal or other required position, or be in equilibrium.

gimblet (gim'blit), GIMLET.

gimcrack, jimcrack (jim'krak), *n.* a pretty but useless or flimsy article, a gewgaw. *a.* showy but flimsy and worthless. **gimcrackery**, *n.* **gimcracky**, *a.*

gimlet (gim'lit), *n.* a small boring-tool with a worm or screw for penetrating wood, and a wooden crosspiece for handle; a cocktail consisting of gin or vodka, lime juice and soda water.

gimmal (jim'əl), *n.* (*pl.*) a pair or series of interlocking rings, as in machinery, a gimbal; a gemel-ring.

gimmick (gim'ik), *n.* a trick, device or oddity of behaviour used to attract extra interest, attention or publicity. **gimmickry**, *n.* (the use of) gimmicks. **gimmicky**, *a.* characterized by or reliant on gimmicks; in the nature of a gimmick.

gimp (gimp), *n.* silk, wool, or cotton twist interlaced with wire or coarse cord; a silk fishing-line whipped with thin wire to protect it against injury from the teeth of large fish.

gin[1] (jin), *n.* a spirit distilled from grain and flavoured with juniper berries. **gin and it**, a mixture of gin and vermouth. **gin-fizz**, *n.* a drink composed of gin, aerated water, and lemon. **gin rummy**, *n.* a variant of rummy in which the players may go out if their unmatched cards amount to less than ten points. **gin-sling**, *n.* a cold drink, composed of gin, soda-water, lemon, and sugar.

gin[2] (jin), *n.* a trap, a snare for small mammals and birds; a machine for hoisting or moving heavy weights; a machine for separating cotton-fibre from the seeds. *v.t.* to clean (as cotton) of the seeds by means of a gin; to snare, to entrap. **ginning**, *n.*

gin[3] (gin), *v.i., v.t.* (*poet.*) to begin, to commence.

gin[4] (jin), *n.* (*Austral.*) an Aboriginal woman.

gin[5] (gin), *prep.* (*Sc.*) against.

gin[6] (gin), *conj.* (*Sc., North.*) if.

ginger (jin'jə), *n.* a plant, *Zingiber officinale*, with a pungent, spicy root-stock; the root-stock of this, either whole or powdered, used in cookery, as a sweet, or in medicine; a reddish-brown colour; (*coll.*) a red-haired person; mettle, dash, go. *v.t.* to flavour with ginger; to spirit (up). **preserved ginger**, a conserve or sweetmeat made from the immature root. **gingerade**, *n.* ginger-beer. **ginger-ale**, *n.* an aerated non-alcoholic beverage, prepared by dissolving sugar in water, flavouring with ginger or essence of ginger, and colouring with a solution of caramel. **ginger-beer, -pop**, *n.* an effervescing mildly alcoholic beverage prepared from ginger, white sugar, water and yeast. **ginger group**, *n.* a pressure group within a larger body which it aims to liven up or radicalize. **ginger-nut, ginger-snap**, *n.* a crisp ginger-flavoured biscuit. **ginger-wine**, *n.* a wine made by the fermentation of sugar, water and ginger. **gingerous**, *a.* (*coll.*) sandy, carroty (of hair). **gingery**, *a.* spiced with ginger; (*coll.*) red-haired, carroty.

gingerbread (jin'jəbred), *n.* a dark-coloured cake or biscuit made of flour, treacle or molasses, ground ginger and other spices and often cut into shapes.

gingerly (jin'jəli), *adv.* daintily, fastidiously, cautiously, so as to move without noise or risk of hurting oneself or anything trodden upon. *a.* dainty, fastidious, cautious. **gingerliness**, *n.*

gingham (ging'əm), *n.* a kind of linen or cotton fabric woven of dyed yarn, usu. in stripes or checks.

gingili (jin'jili), *n.* an E Indian herb, *Sesamum indicum*, the seeds of which yield a sweet oil.

gingival (jinji'vəl), *a.* pertaining to the gums. **gingivitis** (-jivi'tis), *n.* inflammation of the gums.

gingko (ging'kō), **ginkgo** (gingk'gō), *n.* (*pl.* **-koes, -goes**) a Japanese tree, *Gingko biloba*, with handsome fan-shaped leaves, also called the maidenhair-tree.

gink (gingk), *n.* (*coll.*) fellow, man, person.

ginormous (jīnaw'məs), *a.* (*coll.*) huge.

ginseng (jin'seng), *n.* one of two herbs belonging to the genus *Arabia* or *Panax*, the root of which has a sharp, aromatic taste, and is highly esteemed as a medicine or tonic by the Chinese and others.

gip (jip), GYP[1,3].

gippo (jip'ō), *n.* (*pl.* **-pos**) (*offensive*) Egyptian; gypsy. **gippy tummy** (jip'i), diarrhoea, esp. as afflicting visitors to hot countries.

gipsy (jip'si), *n.* one of a nomad race (calling themselves Romany), prob. of Hindu extraction, dark in complexion and hair, and speaking a corrupt Sanskrit dialect, who live largely by dealing, fortune-telling etc.; one resembling a gipsy, esp. in dark complexion; an itinerant traveller, wanderer. **gipsy-bonnet, -hat**, *n.* a bonnet or hat with a large brim or side flaps, often tied down to the side of the head. **gipsy-cart, -caravan, wagon**, *n.* a large horse-drawn van such as gipsies formerly lived and travelled in from place to place. **gipsy moth**, *n.* a moth whose hairy caterpillar is destructive of trees. **gipsify**, *v.t.*

(*usu. in p.p.*). **gipsyish**, *a.*
giraffe (jirahf′. -raf′), *n.* an African ruminant, *Giraffa camelopardalis*, with an extremely long neck, and two bony excrescences on the head, lawn in colour with darker spots.
girandole (ji′rəndōl), *n.* a branching chandelier or candlestick; a revolving firework discharging rockets; a rotating jet of water; a pendent jewel, usu. for the ears, with a large set encircled by smaller ones.
girasol (ji′rəsol), *n.* a variety of opal with reddish refractions, also called fire-opal.
gird[1] (gœd), *v.t.* (*past*, *p.p.* **girded**, **girt** (gœt)) to bind round (usu. the waist) with some flexible band, esp. in order to secure or confine the clothes; to secure (one's clothes) with a girdle, belt etc.; to fasten (a sword on or to) with a girdle or belt; to invest or equip (with); to surround or encircle with or as with a girdle. **to gird up one's loins**, to get ready to do something; to prepare oneself for (vigorous) action.
gird[2] (gœd), *v.i.* to sneer, to mock (at). *n.* a sarcasm, a sneer.
girder (gœ′də), *n.* a principal beam, esp. a compound structure of iron plates or lattice-work, wood or metal, spanning the distance from wall to wall, or pier to pier, used to support joists, walls, roof, roadway etc. **girder bridge**, *n.* a bridge consisting of girders.
girdle[1] (gœ′dl), *n.* a belt or cord for securing a loose garment round the waist; anything that encircles like a belt; a woman's corset reaching from waist to thigh; the bones by which the limbs are united to the trunk in vertebrate animals; a small circular band or fillet round the shaft of a column; the line of greatest marginal circumference of a brilliant, at which it is grasped by the setting; (*Bot.*) a zone-like ring on a stem, etc. *v.t.* to gird or surround with or as with a girdle, to surround, to environ; to make a cut round (the trunk of a tree) through the bark, so as to kill it or in some cases to make it fruit better. **girdler**, *n.*
girdle[2] (gœ′dl), *n.* (*Sc.*, *North.*) a round flat plate of iron hung over a fire for baking cakes.
girkin (gœ′kin), GHERKIN.
girl (gœl), *n.* a female child, a young and unmarried woman; a female servant; a sweetheart; (*coll.*) a woman of any age. **old girl** OLD. **girl Friday**, *n.* a female secretary and general assistant in an office. **girlfriend**, *n.* a female friend; a regular female companion, esp. one with whom there is a romantic relationship. **girl guide**, *n.* a member of the Girl Guides, an international organization founded with the aim of developing health, character and practical skills. **girl scout**, *n.* (*N Am.*) GIRL GUIDE. **girlhood**, *n.* **girlie**, **girly**, *a.* of magazines etc., showing pictures of nude or scantily clad women. **girlish**, *a.* **girlishly**, *adv.* **girlishness**, *n.*
giro (ji′rō), *n.* (*pl.* **-ros**) in the UK, a system operated by banks and post offices whereby, when the required instructions have been issued, payments can be made by transfers from one account to another. **giro cheque**, *n.* in the UK, a benefit cheque for people who are ill, unemployed etc. that can be cashed at a post office or bank.
Gironde (zhērôd′), *n.* in French history, the name given to the moderate Republican party in the French Assembly (1791-3), from the fact that its leaders represented the department of the Gironde in SW France. **Girondin** (-dī), *n.* a member of the Gironde. **Girondist**, *n.*, *a.*
girr (gœ), *n.* (*Sc.*) a child's hoop; a barrel-hoop.
girt (gœt), *past*, *p.p. of* GIRD.
girth (gœth), *n.* the band by which a saddle or burden is made fast and kept secure on a horse's back by passing round its belly; measure round anything, circumference, waist-measure. *v.t.* to measure the girth of; to surround, to encompass; to fit or secure with a girth.
gismo (giz′mō), GIZMO.
gist (jist), *n.* the essence or main point of a question.

git (git), *n.* (*sl.*) a contemptible person; a bastard.
gite (zhēt), *n.* a sleeping-place, a lodging in France, a privately-owned, self-contained, self-catering apartment or cottage available for holiday lets.
gittern (git′ən), *n.* an instrument like a guitar, a cithern.
give[1] (giv), *v.t.* (*past* **gave** (gāv), *p.p.* **given**) to hand over or transfer the possession of or right to without price or compensation; to bestow, to confer, to present; to grant, to concede, to allow; to hand over, to deliver; to commit, to put in one's keeping; to transfer as price or in exchange, to pay; to administer; to award; to surrender, to relinquish; to devote; to yield as product; to communicate, to impart; to be the source or author of; to occasion, to cause; to perform, to show or exhibit; to present; to act as the host of. *v.i.* to part with freely and gratuitously; to yield as to pressure, to collapse; to move back, to recede; to make way or room; to lead, to open (upon). **give me**, I prefer. **to give and take**, to be fair; to play fair. **to give a dog a bad name** NAME. **to give a miss**, (*coll.*) to avoid. **to give away**, to hand over for nothing; to concede or surrender, esp. through folly or neglect; to give in marriage; (*sl.*) to let out or divulge inadvertently. **to give birth to**, to bring forth. **to give back**, to restore. **to give chase to**, to pursue. **to give ear** EAR. **to give forth**, to publish, to tell. **to give ground** GROUND. **to give in**, to yield; to submit, to hand in. **to give in marriage**, to permit the marriage of (a daughter). **to give it to someone**, (*coll.*) to scold, punish severely, beat. **to give of**, to contribute. **to give off**, to emit. **to give on**, to afford a prospect on or into, to face. **to give out**, to emit; to publish, to proclaim; to distribute; (*coll.*) to show, to profess; to break down; to run short. **to give over**, to hand over, to transfer; to abandon, to despair of; (*in p.p.*) to devote or addict; to cease (from), to desist; to yield. **to give place to** PLACE. **to give rise to** RISE. **to give someone his head** HEAD. **to give the lie to** LIE. **to give the sack** or **boot**, (*coll.*) to dismiss, esp. in a summary fashion. **to give tongue**, to bark. **to give up**, to surrender; to resign; to commit; to despair of; to renounce; to dedicate. **to give way**, to yield, to fail to resist; to make room; to break down; to abandon (oneself to); to be depreciated in value; to begin to row; to row with increased energy. **to give what for**, (*coll.*) WHAT. **what gives?** (*coll.*) what is happening? **give-away**, *n.* (*coll.*) an unintentional revelation; something given free. **giver**, *n.*
give[2] (giv), *n.* the state of yielding or giving way; elasticity. **give and take**, mutual concession or forbearance; fair measure on either side.
given (giv′n), *a.* inclined or addicted (to); fixed, specified; of official documents, issued or executed; granted, assuming (that). **given name**, *n.* a Christian name.
gizmo (giz′mō), *n.* (*pl.* **-mos**) (*coll.*) a gadget. **gizmology** (-mol′-), *n.* (*sl.*) technological gadgetry.
gizzard (giz′əd), *n.* a strong muscular division of the stomach, esp. the second stomach in birds; a thickened muscular stomach in certain fish, insects and molluscs. **it sticks in one's gizzard**, (*coll.*) it is very disagreeable to one.
Gk., (*abbr.*) Greek.
gl., (*abbr.*) glass; gloss.
glabella (gləbel′ə), *n.* the smooth flat area of bone between the eyebrows.
glabrous (glā′brəs), *a.* smooth; devoid of hair or pubescence. **glabrate** (-brāt, -brət), *a.* **glabrescent**, *a.*
glacé (glas′ā), *a.* iced, or with a surface or covering like ice (as confectionery); polished, glossy (as leather goods).
glacial (glā′shl, glā′siəl), *a.* of or pertaining to ice; due to or like ice, icy; of geological formations, due to or characterized by glaciers, ice-sheets or floating ice; (*Chem.*) crystallizing at ordinary temperatures. **glacial drift**, *n.* gravel, sand, clay and other debris transported or deposited by ice. **glacial period**, **epoch**, **era**, *n.* a period during

which a large part of the northern hemisphere was covered with an ice-sheet, called also the ice age. **glacialist**, *n*. one who considers that certain geological phenomena are due to the action of ice. **glacially**, *adv*. **glaciate**, *v.t.* (*Geol.*) to scratch, polish or wear down by means of ice; to cover with ice in the form of sheets or glaciers. *v.i.* to be converted into ice. **glaciation**, *n*. the subjection of an area to glacial conditions.

glacier (glas'iə, glā'-), *n*. a stream-like mass of ice, formed by consolidated accumulations of snow at high altitudes, slowly descending to lower regions. **glacier-lake**, *n*. a lake held back temporarily or permanently by a glacier or its deposits. **glacier-mud, -silt**, *n*. mud, sand or pulverized debris formed underneath glaciers and deposited by glacier streams.

glacio-, *comb. form* glacial; glacier.

glaciology (glāsiol'əji, glas-), *n*. the study of glacial action and its geological effects. **glaciologic, -cal** (-loj'-), *a*. **glaciologist**, *n*.

glacis (glas'is, -i, glā'-), *n*. (*Fort.*) a sloping bank, esp. in front of a rampart, where assailants would be exposed to fire.

glad (glad), *a*. pleased, gratified; indicating pleasure or satisfaction; affording pleasure, joy or satisfaction; bright, gay. **glad-eye**, *n*. (*coll.*) ogling. **glad hand**, *n*. (*coll.*) a warm, but not always entirely disinterested welcome. *v.t.* to welcome, esp. by shaking hands. **gladhanding**, *n*. **gladhander**, *n*. **glad rags**, *n*. (*coll.*) best or smartest clothes; evening dress. **gladden**, *v.t.* to make glad or joyful; to cheer. **gladly**, *adv*. **gladness**, *n*. **gladsome** (-səm), *a*. **gladsomely**, *adv*. **gladsomeness**, *n*.

glade (glād), *n*. an open space in a wood or forest.

gladiator (glad'iǎə), *n*. in Roman times, a man employed to fight in the amphitheatre; a political combatant; a controversialist. **gladiatorial** (-iətaw'riəl), *a*. **gladiatorship**, *n*.

gladiolus (gladiō'ləs), *n*. (*pl.* -li (-lī)) an iridaceous genus of plants with a fleshy bulb, sword-shaped leaves, and spikes of bright-coloured flowers. Also called **sword lily**.

gladius (glā'diəs), *n*. the cuttlebone or pen of a cuttlefish.

gladstone bag (glad'stən), *n*. a light leather bag with flexible sides, opening along the middle and secured with a clasp and straps.

Glagol (glag'əl), *n*. the earliest Slavonic alphabet, principally used in Istria and Dalmatia, in the offices of the Roman Catholic Church. **Glagolitic** (-lit'-), *a*.

glair (gleə), *n*. white of egg, or a preparation made with this, used as size or varnish; any similar viscous, transparent substance. *v.t.* to smear or overlay with glair. **glaireous, glairy,** *a*.

glamour (glam'ə), (*esp. N Am.*) **glamor**, *n*. the influence of some charm on the vision, causing things to seem different from what they are; charm, allure; alluring or exciting personal attractiveness. **glamour-girl**, *n*. an esp. pretty girl or woman regarded as being particularly attractive to men; a girl or woman who has a job that is regarded as glamorous. **glamorize, -ise**, *v.t.* **glamorization**, *n*. **glamorous**, *a*. alluring; beautiful and smart; of a job, lifestyle etc., conspicuous by its desirability.

glance (glahns), *v.i.* to glide off or from (as a blow); to allude, to hint (at); to dart or flash a gleam of light or brightness; to give a quick or cursory look (at); to move about rapidly. *v.t.* to shoot or dart swiftly or suddenly; to direct (a look or the eye) rapidly or cursorily. *n*. an oblique impact of an object on another causing it to be deflected; in cricket, a hit with the bat turned obliquely to the ball; a flash, a gleam; a quick or transient look, a hurried glimpse (at). **at a glance**, immediately. **glancingly**, *adv*.

gland (gland), *n*. an organ secreting certain constituents of the blood, either for specific use or for elimination as waste products; a cellular organ in plants, usu. secreting oil or aroma; a sleeve employed to press packing tight on or around a piston-rod. **glandule** (glan'dūl), *n*. a small gland. **glandular**, *a*. characterized by the presence of a gland or glands; consisting or of the nature of a gland or glands; affecting the glands. **glandular fever**, *n*. an infectious disease characterized by the swelling of the lymph nodes. **glandularly**, *adv*. **glanduliferous** (-lif'-), *a*. **glandulose** (*Bot.*), **glandulous**, (*Physiol.*) *a*. **glandless**, *a*.

glanders (glan'dəz), *n.pl*. a very dangerous and contagious disease in horses, attended with a running of corrupt matter from the nostrils, and enlargement and induration of the glands of the lower jaw. **glandered**, *a*. **glanderous**, *a*.

glandiferous (glandif'ərəs), *a*. bearing acorns or other nut-like fruits. **glandiform** (glan'difawm), *a*. acorn-shaped; in physiology, resembling a gland.

glandule GLAND.

glans (glanz), *n*. the nut-like fruit of some forest trees; a structure of somewhat similar form, as the extremity of the penis.

glare (gleə), *v.i.* to shine with a dazzling or overpowering light; to look with fierce, piercing eyes, to stare; to be garish or gaudy; to be very conspicuous. *v.t.* to shoot or dart forth in or as in intense lustre. *n*. a fierce overpowering light, disagreeable brightness; tawdry splendour; an intense, fierce look or stare. **glaring**, *a*. shining with dazzling brightness; staring; too conspicuous or overcoloured; flagrant, blatant. **glaringly**, *adv*. **glaringness**, *n*. **glary**, *a*.

glareous GLAIR.

glasnost (glaz'nost), *n*. esp. of the USSR government under Mikhail Gorbachev, a willingness to be more open and accountable. [Rus., frankness]

glass (glahs), *n*. (*pl.* -es) a hard, brittle, transparent substance, formed by fusing together mixtures of the silicates of potash, soda, lime, magnesia, alumina and lead in various proportions; a substance of vitreous structure or composition; an article made of glass; a mirror; a drinking-vessel of glass; the quantity which such a vessel will hold; a lens; an optical instrument composed partly of glass, an eye-glass, a telescope; an hour-glass; a barometer; a window-pane; (*pl.*) a pair of spectacles; (*collect.*) ornaments or utensils made of glass, greenhouses, windows. *v.t.* to case in glass; to fit or cover with or as with glass, to glaze. **glass-blower**, *n*. one whose business is to blow and mould glass. **glass-blowing**, *n*. the art or process of shaping molten or softened glass into vessels. **glass case**, *n*. a case or shallow box having a glass lid or sides to show the contents. **glasscloth**, *n*. a cloth for wiping and cleaning glasses; cloth covered with powdered glass, like sand-paper; a fabric woven of fine-spun glass threads. **glass-cutter**, *n*. a worker or a tool that cuts glass. **glass-cutting**, *n*. the art or process of cutting, grinding and polishing glass-ware. **glass-dust**, *n*. powdered glass used for grinding and polishing. **glass eye**, *n*. an artificial eye of glass. **glass fibre**, FIBREGLASS under FIBRE. **glass-grinding**, *n*. glass-cutting. **glass-house**, *n*. a house or building where glass is made; a greenhouse or conservatory; a glass-roofed photographic studio; (*sl.*) a military prison. **glass jaw**, *n*. esp. in boxing, a jaw that is particularly susceptible to injury. **glass-painting**, *n*. the art of painting designs on glass with colours which are burnt in. **glass-paper**, *n*. paper covered with finely-powdered glass used for rubbing down and smoothing rough surfaces of wood etc. **glass-snake**, *n*. an American lizard without limbs, *Ophisaurus ventralis*. **glass-soap**, *n*. oxide of manganese and other substances used in the manufacture of glass to remove colour due to ferrous salts etc. **glass-stainer**, *n*. **glass-staining**, *n*. the art or process of colouring glass during manufacture. **glass-ware**, *n*. (*collect.*) articles made of glass. **glass wool**, *n*. fine, spun glass, used in insulation etc. **glass-work**, *n*. glass manufacture; glass-ware. **glass-worker**, *n*. **glass-works**, *n*. a place or building where glass is manufactured. **glasswort**, *n*. one of various maritime herbs containing alkali formerly used in glass-making. **glassful**, *n*. as much as a glass will hold. **glassine**, *n*. glazed, translucent paper. **glassless**, *a*. **glass-like**, *a*. **glassy**, *a*. like glass, vitreous; lustrous, smooth, mirror-like (of water); hard, dull, lacking fire, fixed (of the eye). **glassily**, *adv*. **glassiness**, *n*.

Glaswegian (glăzwē′jən, glas-), *n*. a native or inhabitant of Glasgow in Scotland.

glauber's salt (glow′bəz, glaw′-), *n*. sodium sulphate, a strong purgative. **glauberite** (-it), *n*. a yellow, grey or brick-red mineral, composed of sulphate of soda and sulphate of lime. [J.R. *Glauber*, 1604–68, German chemist]

glaucescent (glawses′ənt), *a*. tending to become or becoming glaucous. **glaucescence**, *n*.

glaucoma (glawkō′mə), *n*. a disease of the eye, causing opacity in the crystalline humour, tension of the globe, dimness and ultimately loss of vision. **glaucomatous**, *a*. **glaucosis** (-sis), *n*.

glaucous (glaw′kəs), *a*. sea-green, pale greyish-blue; in botany, covered with a bloom or down of this tinge (as grapes).

glaze (glāz), *v.t.* to furnish, fit or cover with glass; to fit with a sheet or panes of glass; to furnish with windows; to overlay (pottery) with a vitreous substance; to cover (a surface) with a thin glossy coating; to make smooth and glossy; to cover (the eyes) with a film. *v.i.* to become glassy (as the eyes). *n*. a smooth, lustrous coating; such a coating, formed of various substances, used to glaze earthenware, pictures, paper, confectionery etc. **double-glazing** DOUBLE. **glaze-kiln**, *n*. a kiln in which glazed biscuit-ware is placed for firing. **glazed**, *a*. having been glazed; esp. of a person's expression, vacant, bored. **glazer**, *n*. one who glazes earthenware. **glazier** (-ziə, -zhə), *n*. one whose business it is to set glass in windows etc. **glazier's diamond**, *n*. a small diamond fixed on a handle, used by glaziers for cutting glass. **glaziery**, *n*. **glazing**, *n*. the act or process of setting glass in window-sashes, picture-frames etc.; covering with a glaze, or giving a glazed or glossy surface to pottery and other articles; the material used for this; glass-work; glazed-windows; the process of applying semi-transparent colours thinly over other colours to tone down asperities. **glazy**, *a*.

gld, (*abbr.*) guilder.

gleam (glēm), *n*. a flash, a beam, a ray, esp. one of a faint or transient kind. *v.i.* to send out rays of a quick and transient kind; to shine, to glitter. **gleamingly**, *adv*. **gleamy**, *a*.

glean (glēn), *v.t.* to gather (ears of corn which have been passed over on the cornfield); to gather ears of corn from; to collect bit by bit, to pick up here and there. *v.i.* to gather the ears of corn left on the ground. **gleaner**, *n*. **gleaning**, *n*. **gleanings**, *n*.

glebe (glēb), *n*. (*poet.*) (a piece of) cultivated ground; the land furnishing part of the revenue of an ecclesiastical benefice. **glebe-house**, *n*. a parsonage-house. **glebe-land**, *n*. **glebeless**, *a*.

glee (glē), *n*. joy, gladness, delight; mischievous pleasure; a musical composition for several voices in harmony, consisting usu. of two or more contrasted movements and without instrumental accompaniment. **glee club**, *n*. (*esp. N Am.*) a choral society. **gleeful**, *a*. merry, gay, joyous; taking mischievous pleasure in. **gleefully**, *adv*.

glen (glen), *n*. a narrow valley, a dale.

glengarry (glengə′ri), *n*. a woollen cap, high in front with ribbons hanging down behind, worn by some Highland regiments; also called **glengarry bonnet**. [valley in Scotland]

glib (glib), *a*. off-hand; voluble, fluent, not very weighty or sincere. **glibly**, *adv*. **glibness**, *n*.

glide (glīd), *v.i.* to move smoothly and gently; to slip or slide along, as on a smooth surface; to pass rapidly, smoothly, and easily; to pass imperceptibly (away); (*Mus.*) to pass from tone to tone without a perceptible break; (*Aviat.*) to fly an engineless heavier-than-air aeroplane which is catapulted or launched from a height, and makes use of rising air currents; to fly without the use of motive power. *n*. the act of gliding; (*Mus.*) a passage from one tone to another without a break; (*Phon.*) a continuous sound produced in passing from one position of the organs of speech to another; a sliding dance step. **glide path**, *n*. the path followed by an aircraft as it descends to a landing. **glider**, *n*. one who or that which glides; a heavier-than-air flying-machine with no motive power. **gliding**, *n*. the art or sport of piloting such an aircraft. **glidingly**, *adv*.

glimmer (glim′ə), *v.i.* to emit a faint or feeble light; to shine faintly. *n*. a faint, uncertain or unsteady light; a faint gleam, an uncertain sign (as of intelligence etc.); a glimpse. **glimmering**, *n*. a glimmer, a twinkle; a faint gleam (as of knowledge, sense etc.); an inkling, a glimpse. **glimmeringly**, *adv*.

glimpse (glimps), *n*. a momentary look, a rapid and imperfect view (of); a passing gleam, a faint and transient appearance; a faint resemblance, a slight tinge. *v.t.* to catch a glimpse of; to see for an instant.

glint (glint), *v.i.* to gleam, to flash; to glitter, to sparkle. *v.t.* to reflect, to flash back. *n*. a gleam, a flash, a sparkle.

glissade (glisahd′, -sād′), *n*. a method of sliding down a steep snow-slope, usu. with an ice-axe or alpenstock held as rudder and support; a gliding step. *v.i.* to slide down a steep snow-slope in this manner.

glissando (glisan′dō), *n*., *a*. (*pl*. **-dos**) (of) a rapid sliding of the finger(s) up and down the musical scale.

glisten (glis′n), *v.i.* to gleam, to sparkle, usu. by reflection. *n*. a glitter or sparkle, esp. by reflection; a gleam. **glistening**, *a*. **glisteningly**, *adv*.

glister (glis′tə), *v.i.* (*poet.*) to glitter, to sparkle. *n*. glitter, lustre, brightness.

glitch (glich), *n*. (*sl.*) an extraneous electric current or false signal, esp. one that disrupts the smooth operation of a system; malfunction.

glitter (glit′ə), *v.i.* to gleam, to sparkle; to shine with a succession of brilliant gleams or flashes; to be brilliant, showy or specious. *n*. a bright sparkling light; brilliancy, splendour; speciousness, attractiveness; tiny glittering particles used for decoration. **glitterati** (-rah′ti), *n. pl.* (*sl.*) fashionable people, as media personalities, artists, jet-setters etc., as a social group. **glitteringly**, *adv*. **glittery**, *a*.

glitz (glits), *n*. (*coll.*) ostentation, conspicuous showiness. *v.i.* to dress in a showy ostentatious way. **glitzy**, *a*. (*coll.*)

gloaming (glō′ming), *n*. evening twilight.

gloat (glōt), *v.i.* to look or dwell (on or over) with exultant feelings of malignity, lust or avarice. **gloatingly**, *adv*.

glob (glob), *n*. a rounded lump of something soft, dollop.

global, globate GLOBE.

globe (glōb), *n*. a ball, a sphere, a round or spherical body; the earth; a sphere on which are represented the heavenly bodies (called a **celestial globe**), or representing the land and sea, and usu. the political divisions of the world (called a **terrestrial globe**); anything of a globular or nearly globular shape; an orb borne as emblem of sovereignty; an almost spherical vessel, as an aquarium, lampshade etc.; the eyeball; (*Austral.*) an electric light bulb. *v.t.* to form into a globe. *v.i.* to become globular. **globe artichoke**, *n*. a type of artichoke *Cynara scolymus*, cultivated for food. **globe-fish**, *n*. a fish having the power of inflating the skin till it becomes nearly globular. **globe-flower**, *n*. the ranunculaceous genus *Trollius*, esp. the British *T. europaeus*, with yellow, almost spherical flowers. **globe-lightning**, *n*. a fire-ball. **globe-trotter**, *n*. a traveller who hurries from place to place sight-seeing or who visits many foreign countries. **globe-trotting**, *n*., *a*. **global**

(glō'bəl), *a.* relating to the globe as an entirety; worldwide; taking in entire groups of classes; across-the-board. **global village**, *n.* the world viewed as an integrated system, esp. as linked by means of instant (mass) communication. **globalism**, *n.* **globalize, -ise,** *v.t.* to make global in scope or application. **globalization,** *n.* **globally,** *adv.* **globate** (glō'bāt), **-bated,** *a.* spherical. **globigerina** (glōbijəri'nə), *n.* (*pl.* **-nae**) a genus of Foraminifera, with a many-chambered shell. **globigerina mud,** or **ooze,** *n.* a light-coloured calcareous mud or ooze in the ocean depths, consisting of shells of globigerinae. **globoid,** *a.* like a globe in shape. *n.* a globular granule. **globose,** *a.* spherical, globular. **globosity** (-bos'-), *n.*
globin (glō'bin), *n.* a colourless protein of the blood.
globoid GLOBE.
globose GLOBE.
globule (glob'ūl), *n.* a particle of matter in the form of a small globe; a minute drop or pill. **globular,** *a.* having the shape of a small globe or sphere; composed of globules. **globularity** (-la'-), *n.* **globularly,** *adv.* **globularness,** *n.* **globuliferous** (-lif'-), *a.* producing, containing or having globules. **globulin** (-lin), *n.* an albuminous protein or class of proteins obtained from animals and plants, insoluble in water but soluble in salt solutions. **globulite** (-it), *n.* a minute globular body representing the most rudimentary stage in the formation of crystals. **globulous,** *a.*
glockenspiel (glok'ənshpēl), *n.* an instrument consisting of hanging metal bars or tubes, to be struck with a hammer.
glom (glom), *v.t.* (*sl.*, *esp. N Am.*) to snatch, seize; to steal. **to glom onto,** to take possession of; to grab hold of.
glomerate (glom'ərət), *a.* in anatomy, compactly clustered (as glands, vessels etc.); in botany, congregated into a head. **glomeration,** *n.* **glomerule,** *n.* a flower-cluster forming a compact head.
gloom (gloom), *v.i.* to look dismal, sullen or frowning; to lour, to be or become cloudy or dark. *n.* obscurity, partial darkness; depression, dejection, melancholy; circumstances that occasion melancholy or despondency. **gloomful,** *a.* **gloomfully,** *adv.* **gloomy,** *a.* dark, obscure; sad, melancholy, dispiriting; sullen, morose. **gloomily,** *adv.* **gloominess,** *n.*
gloria[1] (glaw'riə), *n.* (*pl.* **-as**) a halo.
gloria[2] (glaw'riə), *n.* (*pl.* **-as**) a song or versicle of praise, forming part of the English Church service or the Mass; a doxology; the music to which one of these, esp. the *Gloria in excelsis,* is sung.
glorify (glaw'rifī), *v.t.* to pay honour and glory to in worship, to praise, to extol; to exalt to celestial glory; to make splendid, to beautify; to make appear more splendid or impressive than in reality. **glorifiable,** *a.* **glorification** (-fi-), *n.* **glorifier,** *n.*
gloriole (glaw'riōl), *n.* a glory, halo or nimbus.
glory (glaw'ri), *n.* high honour, honourable distinction; fame, renown; an occasion of praise, a subject for pride or boasting; illustriousness, magnificence, grandeur; brilliance, effulgence, splendour; a state of exaltation; adoration or praise ascribed in worship; the divine presence or its manifestations; the felicity of heaven; a combination of the nimbus and aureola; a halo. *v.i.* to boast, to feel pride, to exult. **crowning glory,** something that is esp. distinctive or worthy of praise. **glory be!,** an exclamation expressing surprise. **in glory,** enjoying the felicity of heaven. **morning glory** MORNING. **to glory in,** to be proud of. **glory box,** *n.* (*Austral.*) a box, chest etc. in which a young woman stores her trousseau etc., bottom drawer. **glory-hole,** *n.* (*coll.*) a room, cupboard etc. where rubbish and odds and ends have been stowed away anyhow; an opening through which one can look into the interior of a furnace. **gloryingly,** *adv.* **glorious** (glaw'riəs), *a.* full of glory, illustrious; worthy of admiration or praise; entitling one to fame or honour; splendid, magnificent; (*coll.*) hilarious, uproarious; very amusing; completely satisfactory. **gloriously,** *adv.* **gloriousness,** *n.*
glose (glōz), GLOZE.
gloss[1] (glos), *n.* an explanatory word or note in the margin or between the lines of a book, as an explanation of a foreign or strange word; a comment, interpretation or explanation; a superficial or misleading interpretation etc.; a glossary, translation, or commentary. *v.t.* to explain by note or comment; to annotate; to comment upon, esp. in a censorious way. *v.i.* to make comments, to annotate, to write glosses. **glossator** (-sā'-), **glosser,** *n.* a writer of glosses.
gloss[2] (glos), *n.* the brightness from a polished surface; polish, sheen; a specious or deceptive outward appearance. *v.t.* to make glossy or lustrous; to render specious or plausible. **to gloss over,** to palliate, to excuse; to make light of by passing rapidly over. **gloss paint,** *n.* paint containing a varnish that gives it a shiny finish. **glosser,** *n.* one who puts a gloss on. **glossy,** *a.* having a smooth, lustrous surface. *n.* a glossy magazine. **glossy magazine,** *n.* a magazine printed on glossy paper with many colour illustrations. **glossily,** *adv.* **glossiness,** *n.*
gloss, (*abbr.*) glossary.
glossa (glos'ə), *n.* (*pl.* **-sae** (-ē), **-sas**) tongue. **glossal,** *a.* of or pertaining to the tongue, lingual. **glossitis** (-sī'tis), *n.* an inflammation of the tongue.
glossary (glos'əri), *n.* a list, vocabulary or dictionary of explanations of obsolete, rare, technical or dialectal words or forms; a collection of glosses or notes. **glossarial** (-seə'ri-), *a.* **glossarist,** *n.*
glosso-, *comb. form* pertaining to the tongue; linguistic.
glossolalia (glosəlā'liə), *n.* speech in an unknown tongue, occurring in religious ecstasy, trances etc.
glossy GLOSS[2].
glottis (glot'is), *n.* (*pl.* **-tises, -tides** (-dēz)) the mouth of the windpipe forming a narrow aperture which contributes, by its dilatation and contraction, to the modulation of the voice. **glottal,** *a.* **glottal stop,** *n.* a speech sound in some languages, e.g. German, produced by closing and suddenly opening the glottis. **glottic,** *a.* **glottologic** (glotəloj'ik), *a.* **glottologist** (-tol'əjist), *n.* **glottology** (-tol'-), *n.*
glove (glǔv), *n.* a covering for the hand, usu. with a separate division for each finger; a padded glove for the hands in boxing, also called **boxing-glove.** *v.t.* to cover with or as with a glove. **hand in glove** HAND. **to fight with the gloves off,** to box without gloves; to fight or contend in earnest, to show no mercy. **to fit like a glove,** to fit perfectly in size and shape. **glove-compartment, -box,** *n.* a small storage compartment in a car, usu. set into the dashboard. **glove-puppet,** *n.* a puppet that fits onto the hand. **gloved,** *a.* **glover,** *n.* one who makes or sells gloves. **gloveress** (-ris), *n. fem.* **gloveless,** *a.* **gloving,** *n.* the occupation of making gloves.
glow (glō), *v.i.* to radiate light and heat, esp. without flame; to be incandescent; to be bright or red with heat, to show a warm colour; to feel great bodily heat; to be warm or flushed with passion; to feel intense pleasure or satisfaction. *n.* incandescence, red or white heat; brightness, redness, warmth of colour; vehemence, ardour; a feeling of well-being or satisfaction; heat produced by exercise. **glow-worm,** *n.* a beetle, *Lampyris noctiluca* or *L. splendidula,* the female of which is phosphorescent. **glowing,** *a.* emitting a glow; bright and warm in colour; flushed; expressing praise and admiration, commendatory. **glowingly,** *adv.*
glower (glow'ə), *v.i* to scowl, to stare fiercely or angrily. *n.* a savage stare, a scowl. **gloweringly,** *adv.*
gloxinia (gloksin'iə), *n.* a genus of plants with large bell-shaped flowers, from tropical America.
gloze (glōz), *v.t., v.i.* to palliate, to extenuate; to explain by note or comment; to flatter, to wheedle.
glucose (gloo'kōs), *n.* a fermentable sugar, less sweet than

cane-sugar, obtained from dried grapes and other fruits, dextrin etc.; any of the group of sweet compounds including dextrose, laevulose etc. **glucic** (-sik), **glucosic** (-kos'ik), *a*. derived from or pertaining to glucose. **glucoside** (-sid), *n*. a vegetable substance yielding glucose when decomposed.

glucosuria (glookəsū'riə), *n*. one form of diabetes, the principal characteristic of which is the occurrence of sugar in the urine. **glucosuric**, *a*.

glue (gloo), *n*. an impure gelatine made of the chippings of hides, horns and hoofs, boiled to a jelly, cooled in moulds, and used hot as a cement; an adhesive or sticky substance. *v.t.* (*pres.p.* **gluing** or **glueing**, past, p.p. **glued**) to join or fasten with or as with glue; to unite, to attach firmly. **glue-pot**, *n*. a vessel for heating glue, with an outer vessel to hold water and prevent burning. **glue-sniffing**, *n*. the inhalation of the fumes of certain glues for their narcotic effects. **glue-sniffer**, *n*. **gluer**, *n*. **gluey**, *a*. **glueyness**, *n*.

glug (glŭg), *n*. (*coll.*) the sound of liquid being poured, esp. out of or into a narrow opening.

gluhwein (gloo'vīn), *n*. mulled wine. [G]

glum (glŭm), *a*. sullen, moody, dejected, dissatisfied. **glumly**, *adv*. **glumness**, *n*.

glume (gloom), *n*. a chaff-like scale or bract forming part of the inflorescence in grasses; a husk. **glumaceous** (-mā'-), *a*. **glumiferous** (-mif'-), *a*. **glumose, -mous**, *a*.

glut (glŭt), *v.t.* to fill to excess, to gorge, to sate; to fill with an over-supply (as a market); to swallow, swallow down. *n*. a surfeit, even to loathing; a superabundance; an over-supply of a market.

glutaeus, gluteus (gloo'tiəs, -tē'-), *n*. (*pl.* **-taei, -tei** (-tii)) one of the three large muscles forming the buttock. **gluteal, -taeal**, *a*.

glutamate (gloo'təmāt), *n*. a salt or ester of glutamic acid.

glutamic acid (glootam'ik), *n*. an amino acid occurring in proteins, which plays an important part in nitrogen metabolism.

gluten (gloo'tən), *n*. an elastic protein present in wheat flour, insoluble in water; a sticky substance, glue. **gluten-bread**, *n*. bread containing a large quantity of gluten, used by those suffering from diabetes. **glutinize, -ise**, *v.t.* to render viscous or gluey. **glutinously**, *adv*. **glutinosity** (-nos'-), *n*.

glutton (glŭt'n), *n*. one who eats to excess; one who overindulges in any activity; the wolverine. **glutton-like**, *a*. **gluttonize, -ise**, *v.i.* to eat to excess. **gluttonous**, *a*. **gluttonously**, *adv*. **gluttony**, *n*.

glycerine (glis'ərin, -rēn), **glycerin** (-rin), *n*. a viscid, sweet, colourless liquid obtained from animal and vegetable fats and oils, used in the manufacture of soaps, medicines, confectionery etc. **glyceric** (-se'-), *a*. **glycerate** (-rāt), *n*. a salt of glyceric acid. **glyceride** (-rid), *n*. **glycerinate**, *v.t.* to treat (esp. vaccine lymph) with glycerine. **glyceroid**, *a*. **glycerol** (-rol), *n*. glycerine. **glyceryl** (-ril), *n*. the radical of glycerine and the glycerides.

glycero-, *comb. form*. glycerine.

glyc(o)-, *comb. form*. sugar.

glycocoll (glī'kōkol), *n*. a crystalline sweetish compound found in bile.

glycogen (glī'kəjən), *n*. a white starch-like compound occurring in the liver and convertible into glucose. **glycogenic** (-jen'-), *a*. **glycogenesis** (-jen'-), *n*.

glycol (glī'kol), *n*. an aliphatic alcohol used as an antifreeze and de-icer. **glycolic** (-kol'-), *a*.

glyconic (glīkon'ik), *a*. applied to varieties of classic verse consisting of three trochees and a dactyl.

glycoprotein (glīkōprō'tēn), *n*. any of a group of complex proteins containing a carbohydrate mixed with a simple protein. Also **glycopeptide** (-pep'tid).

glycosuria (glīkōsū'riə), GLUCOSURIA.

glyph (glif), *n*. (*Arch.*) a fluting or channel, usu. vertical.

glyphic, *a*. carved, sculptured.

glyphograph (glif'əgrahf, -graf), *n*. (an impression from) a plate prepared by glyphography. *v.t., v.i.* to engrave by glyphography. **glyphographer** (-fog'-), *n*. **glyphographic** (-graf'-), *a*. **glyphography** (-fog'-), *n*. the process of making engravings for printing in which an electrotype with the design in relief is obtained from an intaglio etching.

glyptic (glip'tik), *a*. relating to carving or engraving, esp. on gems.

glyptodon (glip'tədon), *n*. a huge fossil quadruped allied to the armadillo, from S America.

glyptography (gliptog'rəfi), *n*. the art of engraving on gems. **glyptograph** (glip'təgraf), *n*. an engraving on a gem. **glyptographer** (-tog'-), *n*. **glyptographic** (-graf'-), *a*.

GM, (*abbr.*) General Manager; George Medal; Grand Master.

Gmc, (*abbr.*) Germanic.

GMT, (*abbr.*) Greenwich Mean Time.

G-man G.

gnamma hole (nam'ə), NAMMA HOLE.

Gnaphalium (nəfā'liəm), *n*. a genus of woolly plants, typified by the cudweed, having small sessile flower-heads.

gnar (nah), KNAR.

gnarl (nahl), *v.t.* (*usu. in. p.p.*) to twist or contort. *n*. a twisted growth or knot in a tree. **gnarled, gnarly**, *a*. rugged, lined, weather-beaten, twisted.

gnash (nash), *v.t.* to strike or grind (the teeth) together; to grind or champ. *v.i.* to grind the teeth together. **gnasher**, *n*. **gnashingly**, *adv*.

gnat (nat), *n*. a small two-winged biting fly.

gnathic (nath'ik), *a*. of or pertaining to the jaw. **gnathal** (nā'), *a*. **gnathion** (nā'-, nath'-), *n*. the lowest point of the midline of the lower jaw.

gnathitis (nathī'tis), *n*. inflammation of the upper jaw or cheek.

gnath(o)-, *comb. form* pertaining to the jaw or cheek.

gnathoplasty (nath'əplasti), *n*. the formation of a cheek by plastic surgery.

gnathopod (nath'əpod), *n*. (*pl*. **gnathopoda** (-thop'ədə)) the foot-jaw of crustaceans.

-gnathous, *comb. form* having a jaw of a certain kind, as in *prognathous*.

gnaw (naw), *v.t.* (*p.p.* **gnawed, gnawn**) to bite repeatedly or persistently; to wear away by biting; of anxiety, rage etc., to distress constantly. *v.i.* to bite repeatedly or persistently (at or into); to cause corrosion or wearing away. **gnawer**, *n*. **gnawing**, *n*. **gnawingly**, *adv*.

gneiss (nīs), *n*. a laminated metamorphic rock consisting of feldspar, quartz and mica. **gneissic, gneissoid, gneissose, gneissy**, *a*.

gnocchi (nok'i, nyok'i), *n*. an Italian dish, small potato or semolina dumplings.

gnome[1] (nōm), *n*. an imaginary being, small and misshapen, supposed to live underground and to be the guardian of mines, quarries etc. **Gnomes of Zurich**, (*coll.*) international financiers. **gnomish**, *a*.

gnome[2] (nōm), *n*. a maxim, an aphorism. **gnomic**, *a*. **gnomic aorist**, *n*. in Greek grammar, a use of the aorist tense to express, not the past, but a general truth. **gnomically**, *adv*.

gnomo-, *comb. form* pertaining to a maxim or saying.

gnomology (nōmol'əji), *n*. a collection of maxims or sententious reflections or sayings. **gnomologic, -al** (-loj'-), *a*. **gnomologist** (-mol'-), *n*.

gnomon (nō'mon), *n*. a pin on a sundial, indicating the time of day by its shadow; in geometry, the figure remaining when a parallelogram has been removed from the corner of a larger one of the same form. **gnomonic, -al** (-mon'-), *a*. **gnomonic projection**, the projection of the lines of a sphere from the centre. **gnomonically**, *adv*. **gnomonics**, *n.pl*. the art or science of making and using dials.

gnosiology (nōziol'əji), *n*. the philosophy dealing with

gnosis

cognition or the theory of knowledge.
gnosis (nō'sis), *n. (pl.* **-ses**) knowledge, esp. of mysteries; gnostic philosophy.
-gnosis, *comb. form* esp. in medicine, recognition, as in *diagnosis.*
gnostic (nos'tik), *a.* relating to knowledge; having esoteric knowledge; of or belonging to the Gnostics or Gnosticism. *n.* an adherent of Gnosticism. **gnostically,** *adv.* **Gnosticism** (-sizm), *n.* a system of religious philosophy flourishing in the first six centuries of the Church, that combined ideas from Greek and Oriental philosophy with Christianity. **gnosticize, -ise** (-siz), *v.t., v.i.* **gnosticizer,** *n.*
GNP, *(abbr.)* Gross National Product.
gnu (noo), *n.* a large-horned antelope, *Catoblepas gnu* of southern Africa.
go[1] (gō), *v.i. (past* **went** (went), *p.p.* **gone** (gon), *2nd sing.* **goest** (-ist), *3rd sing.* **goes** (gōz)), to move, to move from one place or condition to another; to begin to move; to depart; to be moving; to be operating or working; to travel; to advance; to end; to turn out (well or ill); to take a certain course (as for or against); to be habitually (as hungry etc.); to be current; to average; to extend, to reach, to point in a certain direction; to tend; to have a certain tenor; to be applicable, to fit (with); to be harmonious (with a tune etc.); to be released, to get away; to be given up, abolished or lost; to fail, to give way, break down; *(usu. in. p.p.)* to die; to become (as wild, mad etc.); to be sold; to be spent; *(as aux. verb)* to be about (to do), to intend. **go on,** come now; *(iron., remonstr.)* come, come! **to go,** *(esp. N Am. coll.)* of food, for taking away. **to go about,** to set to work at; to go from place to place; to take a circuitous course; of a vessel, to change course. **to go abroad,** to go to a foreign country; to go out of doors. **to go against,** to be in opposition to; to be unfavourable to. **to go ahead,** to proceed in advance; to make rapid progress. **to go aside,** to withdraw apart from others. **to go astray,** to wander from the right path. **to go at,** to attack; to work at vigorously. **to go back on,** to fail to keep (one's word). **to go behind,** to look beyond (the apparent facts etc.). **to go between,** to mediate between. **to go bush,** *(Austral.)* to take to or hide in the bush. **to go by,** to pass by or near to; to pass unnoticed or disregarded; to take as a criterion. **to go down,** to descend; to set; to founder (as a ship); to be recorded; esp. in the UK, to leave a university; to be swallowed, to be acceptable. **to go dry,** *(coll.)* to adopt prohibition; to give up drinking. **to go for nothing,** to count for nothing. **to go far,** *(coll.)* to be very successful. **to go for,** to go somewhere to obtain something; to attack; to be true for; to be attracted by; to be sold for. **to go forth,** to issue or depart; to be spread abroad. **to go hard with,** *(impers.)* to be difficult or dangerous for. **to go ill, well with,** *(impers.)* to happen or fare evil or well with. **to go in,** to enter; to go behind clouds; in cricket, to have an innings. **to go in for,** to enter as a competitor; to follow as a pursuit or occupation. **to go into,** to enter; to take part in; to investigate or discuss. **to go it,** to behave recklessly or outrageously. **to go it alone,** to proceed without help. **to go native,** to adopt the ways and customs of a place. **to go off,** to depart; to be discharged (as a firearm); to cease to function; to succeed (well or ill); to become rotten; to lose affection for. **to go off one's head,** *(coll.)* to become insane. **to go on,** to proceed, continue; *(coll.)* to behave (badly etc.); to grumble, complain; *(coll., in imper.)* rubbish, nonsense!; to appear on the stage. **to go one better,** to surpass. **to go out,** to depart, leave (a room etc.); to be extinguished; to go into society; to go on strike. **to go over,** to cross over; to change one's party or opinions; to read, examine; to rehearse; to retouch. **to go phut** PHUT. **to go round,** to pay a number of visits; to be enough for (the whole party etc.). **to go steady,** to go about regularly with the same boyfriend or girlfriend. **to go the whole hog** HOG. **to go through,** to undergo; to examine; *(coll.)*; to ransack; to discuss thoroughly; to perform (a duty etc.). **to go through with,** to perform thoroughly, to complete. **to go together,** to match each other. **to go under,** to sink; to be submerged or ruined. **to go up,** to climb, pass upwards; to increase; to be constructed; to be destroyed, as by fire or explosion. **to go upon,** to act upon as a principle. **to go west,** *(sl.)* to die. **to go with,** to accompany; to side or agree with; to suit, match. **to go without,** to manage without, to put up with the want of. **go-ahead,** *a.* characterized by energy and enterprise. **go-as-you-please,** *a.* unceremonious. **go-between,** *n.* one who acts as an intermediary between two parties. **go-by,** *n.* intentional failure to notice; evasion, deception. **go-getter,** *n.* a bustling, pushing person. **go-kart,** a small light racing car with a low-powered engine. **go-slow,** *n.* a deliberate curtailment of the rate of production by organized labour in an industrial dispute.
go[2] (gō), *n. (pl.* **goes)** the act of going; an attempt; life, animation; rush, energy, enterprise; *(coll.)* an awkward turn of affairs; a turn, (of doing something); one's turn in a game; in cribbage, a player's turn at which he is unable to play, counting one to his opponent; *(coll.)* fashion, the mode; a spree. **great go, little go,** *(formerly)* in UK universities, the final and preliminary examinations for degrees. **no go,** of no use; not to be done; a complete failure. **on the go,** on the move; vigorously in motion. **to have a go,** *(coll.)* to make an attempt. **to have a go at,** *(coll.)* to attack, physically or verbally.
goad (gōd), *n.* a pointed instrument to urge oxen to move faster; *(fig.)* anything that stings, spurs or incites. *v.t.* to drive or urge on with or as with a goad; to incite. **goadsman, goadster** (-stə), *n.* one who drives with a goad.
goaf (gōf), *n.* a part of a colliery from which the coal has been removed.
goal (gōl), *n.* the mark indicating the end of a race; the end or terminus of one's ambition; destination; in many ball games, the net etc. into which the ball must be driven to win a point; such a scoring. **goal-keeper,** *n.* a player stationed near to guard the goal. **goal kick,** *n.* in soccer, a free kick from or near the corner of the goal area taken by the defending side after the ball has been put out of play by a member of the attacking side. **goal-line,** *n.* a line drawn through the goal-posts to form the boundary at each end of the field of play in football. **goal-post,** *n.* in football etc. either of the two posts marking the goal. **to move the goalposts,** *(coll.)* to change the conditions, regulations, limits etc. applying to a particular matter or action. **goalie,** *n. (coll.)* goal-keeper. **goalless,** *a.*
goanna (gōan'ə), *n. (Austral.)* a large monitor lizard.
goat (gōt), *n.* a hairy, horned and bearded domesticated ruminant belonging to the genus *Capra;* a fool; a lascivious person. **to get one's goat,** to make one angry. **to play the giddy goat,** to play the fool. **goat-god,** *n.* Pan. **goatherd,** *n.* one who tends goats. **goat-moth,** *n.* a large brown and grey moth with black markings. **goat's-beard,** *n.* the meadow-sweet; the salsify. **goat's-rue,** *n.* a leguminous plant, *Galega officinalis.* **goatskin,** *n.* the skin of a goat. **goatsucker,** *n.* any bird of the genus *Caprimulgus,* chiefly nocturnal and insectivorous, fabled to milk goats. **goatish,** *a.* resembling a goat; of a rank smell; lecherous. **goatishly,** *adv.* **goatishness,** *n.* **goatling** (-ling), *n.* **goaty,** *a.*
goatee (gōtē'), *n.* a small beard like a goat's on the point of the chin.
gob (gob), *n.* the mouth; a mouthful; a clot of something slimy, as saliva. *v.i. (past* **gobbed)** to spit. **gob-stopper,** *n.* esp. in UK, a large boiled sweet. **gobsmacked, gob-struck,** *a. (sl.)* amazed, dumbfounded.
gobang (gōbang'), *n.* a game played on a chequer-board, with 50 coloured counters, the object being to get five

gobbet (gob'it), *n.* a mouthful, a lump, piece, esp. of meat.
gobbin, gobbins GOB.
gobble (gob'l), *v.t.* to swallow hastily and greedily. *v.i.* to swallow food thus; to make a noise in the throat as a turkey-cock. *n.* such a noise; in golf, a rapid putt which sends the ball straight into the hole. **gobbler**, *n.* one who gobbles; a turkey-cock.
gobbledegook, gobbledygook (gob'ldigook), *n.* (*coll.*) pretentious language characterized by jargon and circumlocution. [perh. from GOBBLE]
gobelin (gō'bəlin, goblī), *a.* applied to a superior kind of French tapestry. **gobelin blue**, *n.* a blue such as appears a good deal in this tapestry.
go-between GO[1].
gobioid GOBY.
goblet (gob'lit), *n.* a drinking-vessel, with a stem and without a handle.
goblin (gob'lin), *n.* a mischievous spirit of ugly or grotesque shape; a gnome. **goblinism**, *n.*
gobo (gō'bō), *n.* (*pl.* **-bos** or **-boes**) a shield placed around a camera or microphone to exclude unwanted light or sound.
goby (gō'bi), *n.* a small fish of the genus *Gobius* (family Gobiidae), characterized by the union of the ventral fins into a disk or sucker. **gobioid**, *a.*
GOC, (*abbr.*) General Officer Commanding.
god[1] (god), *n.* a superhuman or supernatural being regarded as controlling natural forces and human destinies and worshipped or propitiated by man; a personification of any of the forces of nature; an image worshipped as an embodiment of supernatural power, an idol; (*fig.*) a person or thing greatly idolized; (*pl.*) (the occupants of) the upper gallery in a theatre; (in monotheist religions, **God**) the Supreme Being, the self-existent and eternal Creator and Ruler of the universe. **for God's sake** SAKE. **God almighty**, *int.* expressing surprise or anger. **God knows**, a mild oath expressing apathy or annoyance. **God's (own) country**, (*sometimes iron.*) any country seen as being ideal. Also (*Austral., New Zealand coll.*) **Godzone**. **God willing**, if circumstances permit. **household gods**, the Roman gods of the hearth; one's household treasures. **the blind god**, Cupid. **to play God** PLAY. **ye gods!** (*facet.*) ye gods and little fishes, grandiloquent exclamations of surprise, protest etc. **god-awful**, *a.* (*coll.*) terrible; very unpleasant. **goddam(n), goddamned**, *a.* damned; hateful; complete. *int.* (*esp. N Am. coll.*) expressing annoyance. **goddaughter** GODCHILD. **God-fearing**, *a.* worshipping God, upright. **God forbid**, *int.* expressing the hope that something will not happen. **God-forsaken**, *n.* wretched, miserable; remote. **godmother, -parent** GODFATHER. **godsend**, *n.* an unlooked-for acquisition or gain, a piece of good fortune. **godson** GODCHILD. **God's acre**, *n.* a burial ground. **God-speed**, *n.* the wish 'God speed you' to a person starting on a journey, etc. **goddess** (-is), *n.* a female deity; (*fig.*) a woman of pre-eminent beauty, goodness or charm. **goddess-like**, *a.*, *adv.* **goddess-ship**, *n.* **Godhead**, *n.* divine nature or essence; a deity. **the Godhead**, God. **Godhood**, *n.* **godkin** (-kin), **godlet** (-lit), *n.* **godless**, *a.* irreligious; wicked. **godlessly**, *adv.* **godlessness**, *n.* **godlike**, *a.* **godly**, *a.* God-fearing, pious. **godliness**, *n.* **godship**, *n.* **godward**, *adv.*, *a.* **godwards**, *adv.*
god[2] (god), *v.t.* to deify.
godchild (god'chīld), *n.* one for whom a person stands sponsor at baptism.
goddess GOD[1].
godet (gō'dā, -det'), *n.* in dressmaking, a piece of cloth inserted in a skirt, so that it may hang in folds suggestive of a flare.
godetia (gədē'shə), *n.* a genus of hardy annual flowering herbs allied to the evening primroses. [M. *Godet*, Swiss botanist]
godfather (god'fahdhə), **godmother** (-mūdhə), *n.* one who is sponsor for a child at baptism; the head of a Mafia family or other criminal organization.
godly GOD[1].
godown (gōdown', gō'-), *n.* an E Indian warehouse.
godroon (gədroon'), GADROON.
godwit (god'wit), *n.* a marsh or shore bird resembling the curlew but having a slightly upturned bill.
goel (gō'el), *n.* the next of kin of a murdered man whose duty it was to hunt down and slay the murderer. **goelism**, *n.* [Heb.]
goer (gō'ə), *n.* one who or that which goes (*usu. in comb.*, as *fast-goer*); one who attends regularly (*usu. in comb.*, as in *church-goer*); (*Austral., coll.*) one who or that which is likely to succeed.
Goethian (gœ'tiən), *a.* of, pertaining to or characteristic of *Goethe* (1749–1832) German poet. *n.* a follower or admirer of Goethe.
goety (gō'əti), *n.* black magic. **goetic** (-et'-), *a.*
gofer[1] (gō'fə), *n.* a thin butter-cake with a honeycomb pattern on both sides.
gofer[2] (gō'fə), *n.* (*coll.*) a person employed to run errands etc.
goffer (gō'fə, gof'ə), *v.t.* to crimp (edges of lace etc.) with a heated iron; to emboss (edges of books). *n.* a fluting or ruffle; a tool for goffering. **goffering**, *n.* this process; a ruffle so produced; an embossed design on a book.
Gog and Magog (gog and mā'gog), the last two survivors of a mythical race of giants in ancient Britain.
go-getter GO[1].
goggle (gog'l), *v.i.* to roll the eyes; to squint; to stare; to project (of the eyes). *v.t.* to roll (the eyes). *a.* prominent, staring; rolling from side to side. *n.* a strained or staring rolling of the eyes; (*pl.*) spectacles for protecting the eyes; (*sl.*) spectacles; (*sl.*) the eyes. **goggle-box**, *n.* (*coll.*) a television set. **goggle-eyed**, *a.* **goggled**, *a.* staring, prominent (of the eyes). **goggly**, *a.*
goglet (gog'lit), *n.* a water-cooler used esp. in India.
go-go, gogo (gō'gō), *a.* (*coll.*) active, alert; lively. **go-go dancer**, *n.* a (scantily clad) dancer who performs in nightclubs etc.
Goidelic (goidel'ik), GADHELIC. **Goidel**, *n.*
going (gō'ing), *n.* the act of moving; departure; course of life; the condition of ground as regards walking, riding etc. (*also in comb.* as *slow-going, rough-going*). *a.* working, in actual operation; available. **to be hard going**, to be difficult (to make progress). **going down**, setting, sunset. **going on**, esp. of the time, one's age etc. almost, nearly. **going concern**, *n.* a business etc., in actual operation. **going order**, *n.* order or condition suitable for working. **going-over**, *n.* an examination, check; (*coll.*) a beating. **goings-on**, *n.pl.* behaviour, conduct (usu. in a bad sense).
goitre (goi'tər), *n.* a morbid enlargement of the thyroid gland, causing a deformity of the neck. **goitred**, *a.* **goitrous**, *a.*
go-kart GO[1].
Golconda (golkon'də), *n.* an inexhaustible mine of wealth. [a ruined city NW of Hyderabad, India]
gold (gōld), *n.* a precious metallic element of a bright yellow colour, at. no. 79, chem. symbol Au; the most ductile, malleable, and one of the heaviest of metals, much used for coins, jewellery etc.; this metal in the form of coin, money; wealth, riches; anything very precious or valuable and genuine or pure; a gold medal; the colour of gold. *a.* made of gold, consisting of gold; coloured like gold. **dead gold**, unburnished gold. **old gold**, a dull brownish-gold colour. **as good as gold** GOOD. **gold-amalgam**, *n.* gold combined with mercury in a soft plastic state. **gold-beater**, *n.* one who beats out gold for gilding. **gold-beating**, *n.* **gold-cloth**, *n.* cloth interwoven with gold thread. **gold-digger**, *n.* someone who embarks upon a ro-

mantic association merely for gain. **gold-digging,** *n.* the act of digging for gold; (*usu. in pl.*) a place or district where gold is found. **gold disk,** *n.* a gold-plated record presented to a singer etc. after a certain number of records have been sold. **gold-dust,** *n.* gold in very fine particles. **gold-fever,** *n.* a mania for gold-seeking. **goldfield,** *n.* a district where gold is found. **gold-filled,** *a.* more thickly plated with gold than ordinary gold-plated articles. **goldfinch,** *n.* a yellow-marked singing bird, *Carduelis elegans.* **goldfish,** *n.* a golden-red carp, *Cyprinus auratus,* kept in ponds, aquaria etc. **goldfish-bowl,** *n.* a fishbowl; (*coll.*) a state or situation lacking privacy. **goldfoil,** *n.* a thicker kind of gold leaf. **gold leaf,** *n.* gold beaten into a thin sheet. **gold medal,** *n.* an award for first place in a race or competition. **gold-mine,** *n.* a place where gold is mined; (*fig.*) a source of wealth or profit. **gold-plate,** *n.* vessels, dishes etc. of gold. **gold reserve,** *n.* the total amount of gold held by a central bank to make national and international payments and to protect the value of currency. **gold rush,** *n.* a rush to a place where gold has been discovered. **gold-size,** *n.* a size used in gilding. **goldsmith,** *n.* a worker or dealer in gold. **goldsmithy, -ery, -ry,** *n.* goldsmith's work. **gold standard,** *n.* a system in which a national currency has a set value in gold. **Gold Stick,** *n.* a court official carrying a gilt rod, attending the sovereign on state occasions. **gold thread,** *n.* a flattened silver-gilt wire, laid over a thread of silk. **gold wire,** *n.* gold drawn to the form of wire. **goldless,** *a.* **goldy,** *a.*
golden (gōl'dən), *a.* made or consisting of gold; of the colour of gold; bright, shining; excellent, precious; most favourable; rich in or yielding gold. **golden age,** *n.* a fabled primeval period of perfect human happiness and innocence, the most illustrious period of a people's literature or prosperity. **golden balls,** *n.pl.* the three balls displayed as the emblem of a pawnbroker. **golden calf,** *n.* (*fig.,* see Ex. xxxii.4) money as an aim in itself. **golden-cup,** *n.* various species of *Ranunculus* and other yellow-flowered plants. **Golden Delicious,** *n.* a variety of sweet, green-skinned apple. **golden eagle,** *n.* a large eagle found in the mountainous parts of Britain, esp. Scotland. **golden-eye,** *n.* a sea-duck of the genus *Clangula.* **Golden Fleece,** *n.* the fleece of gold in quest of which the Argonauts sailed under Jason; an order of knighthood instituted in 1429 in Spain and Austria. **golden handcuff, hello,** *n.* (*coll.*) a payment or benefit given to an employee as an inducement to continue working upon joining a company. **golden handshake,** *n.* (*coll.*) a payment or benefit given to an employee when leaving a job, esp. upon retirement. **golden jubilee,** *n.* a 50th anniversary. **golden maidenhair,** *n.* a British moss, *Polytrichum commune.* **golden mean,** *n.* the principle of neither too much nor too little, moderation. **golden mouse-ear,** *n.* mouse-ear hawkweed. **golden-mouthed,** *a.* eloquent, musical. **golden number,** *n.* the number denoting the year's place in a lunar cycle of 19 years, used in calculating the date of Easter. **golden oldie,** *n.* (*coll.*) an old recording or film that is still popular. **golden rain,** *n.* a kind of firework. **golden-rod,** *n.* a tall yellow-flowered plant of the genus *Solidago.* **golden rule,** *n.* the rule that we should do as we would be done by. **golden-samphire,** *n.* a herb, *Inula crithmoides,* of the aster family. **golden share,** *n.* a controlling share held by the government in a privatized company, that can be used to prevent a take-over. **golden syrup** SYRUP. **golden wedding,** *n.* the 50th anniversary of marriage. **goldenly,** *adv.* splendidly, excellently.
goldfinch, goldfish GOLD.
goldilocks (gōl'diloks), *n.* any of several plants with bright yellow flowers.
goldsmith GOLD.
golem (gō'ləm), *n.* in Jewish legend, a human-shaped figure brought to life by supernatural means.
golf (golf), *n.* a game played by two persons or couples with club-headed sticks and small hard balls, on a large grassy space, consisting in driving the balls into a series of small holes in the ground. *v.i.* to play golf. **golf-ball,** *n.* a small, hard, white ball used in playing golf; (an electric typewriter that has) a small metal ball bearing the characters. **golf club,** *n.* the club used in playing golf; a golfing association. **golf-links, -course,** *n.* (*pl.*) the course of 9 or 18 holes on which golf is played. **golfer,** *n.* **golfing,** *n.*
Golgotha (gol'gəthə), *n.* a burial-place, a charnel-house.
Goliath (gəli'əth), *n.* a Biblical giant; a gigantic person or thing. **goliath beetle,** *n.* a huge tropical beetle. **goliath frog,** *n.* the largest living frog, *Rana goliath,* of Africa.
golliwog, gollywog (gol'iwog), *n.* a black-faced doll.
golly (gol'i), *int.* God; by God.
golosh (gəlosh'), GALOSH.
GOM (*abbr.*) Grand Old Man, (*orig.* applied to W.E. Gladstone, 1809–98, British statesman).
gombo (gom'bō), GUMBO.
Gomorrah (gəmo'rə), *n.* a dissolute town. [one of the Biblical cities of the plain]
gomphosis (gomfō'sis), *n.* a kind of articulation by which the teeth are firmly implanted in their sockets.
gon– GON(O)-.
-gon, *comb. form* angled, as in *hexagon, octagon.*
gonads (gō'nadz, gon'-), *n.pl.* a reproductive organ producing gametes, a testis or ovary. **gonadic** (-nad'-), *a.* **gonadotrophic** (-dətrō'fik), *a.* stimulating the gonads. **gonadotrophin** (-dətrō'fin), *n.* a hormone that does this.
gonagra (gənag'rə), *n.* gout in the knee.
gonalgia (gənal'jə), *n.* any painful affection of the knee.
gonarthritis (gonahthri'tis), *n.* inflammation of the knee-joint.
gondola (gon'dələ), *n.* a long, narrow Venetian boat with peaked ends, propelled by one oar; the car of an airship or balloon; (*N Am.*) a kind of freight-boat; a shelved unit in a self-service shop. **gondolier** (-liə'), *n.* one who rows a gondola.
Gondwanaland (gondwah'nəland), **Gondwana,** *n.* a huge land-mass of Palaeozoic and Mesozoic times made up of India, Australia, Antarctica and parts of Africa and S America.
gone (gon), *a.* p.p. of GO; ruined; lost, beyond hope; past, bygone. **gone on,** (*sl.*) infatuated with. **goneness,** *n.* a sensation of weakness. **goner,** *n.* (*coll.*) one who is ruined or ill beyond recovery.
gonfalon (gon'fələn), *n.* a flag with streamers. **gonfalonier** (-niə'), *n.* a gonfalon-bearer. **gonfanon** (-nən), *n.* a gonfalon.
gong (gong), *n.* a tambourine-shaped metal instrument which when struck with a padded stick emits a loud sonorous note; a flattish bell struck with a hammer; (*sl.*) a medal. *v.t.* (*N Am.*) to stop a person or activity by sounding a gong. **gong metal,** *n.* a sonorous metal, 100 parts copper, 25 parts tin.
goniatite (gō'niətit), *n.* a Palaeozoic genus of ammonites.
gonidium (gənid'iəm), *n.* (*pl.* **-dia** (-ə)) a reproductive cell produced asexually in algae. **gonidial, gonidic,** *a.*
goniometer (gōniom'itə), *n.* an instrument for measuring angles, esp. of crystals. **goniometric, -al** (-met'), *a.* **goniometry,** *n.*
gonk (gongk), *n.* a soft round toy with arms and legs.
gon(o)-, *comb. form* sexual or reproductive, as in *gonochorism.*
gonococcus (gonōkok'əs), *n.* (*pl.* **-cocci** (-kok'sī)) the organism that causes gonorrhoea. **gonococcal,** *a.*
gonocyte (gon'ōsit), *n.* an oocyte; spermatocyte.
gonorrhoea (gonərē'ə), *n.* a venereal disease affecting the urethra and other mucous surfaces, accompanied by inflammation and mucopurulent discharge; clap. **gonorrhoeal,** *a.*
goo (goo), *n.* (*coll.*) sticky matter. **gooey,** *a.*
good (gud), *a.* (*comp.* **better,** *superl.* **best**) having such qua-

lities as are useful, proper, and satisfactory; fit; expedient; profitable, serviceable; competent; advantageous, beneficial; genuine, valid; wholesome; complete, thorough; safe, sure; sound financially; ample; possessed of moral excellence; benevolent, friendly, courteous; pleasant, acceptable, palatable. *n.* that which contributes to happiness, advantage etc.; that which is right, useful etc.; welfare; prosperity; benefit, advantage; goodness, good qualities, virtuous and charitable deeds; (*pl.*) movable property, chattels, effects; wares, merchandise. **as good as**, not less than, practically, virtually. **as good as gold**, esp. of children, very well behaved. **as good as one's word**, trustworthy; not to be deterred. **for good, for good and all**, finally, definitely, completely. **a good one**, (*coll.*) a funny joke; an implausible statement or assertion. **goods and chattels**, personal property. **good on or for you!** an exclamation expressing approval, encouragement etc. **to be good for**, to be relied on to pay or bring in (a stated amount). **to be in someone's good books**, to be in favour with someone. **to come good**, (*esp. Austral., coll.*) esp. after a setback, to succeed or improve; to recover one's health after illness etc. **to make good**, to confirm; to fulfil; to supply a deficiency; to replace; to compensate (for). **to the good**, as a balance or profit. **to think good**, to consider good; to be pleased. **Good Book**, *n.* the Bible. **good breeding**, *n.* courteous manners formed by nurture and education. **good day**, *n., int.* a form of salutation at meeting or parting. **good evening**, *n., int.* a form of salutation. **good fellow**, *n.* a genial, sociable person. **good-fellowship**, *n.* **good folk** or **people**, *n.pl.* (*euphem.*) the fairies. **good-for-nothing**, *a.* of no value, worthless. *n.* an idle person, a vagabond. **Good Friday**, *n.* the Friday before Easter, commemorating the Crucifixion. **good grace** GRACE. **good humour**, *n.* a cheerful temper, amiability. **good-humoured**, *a.* **good-humouredly**, *adv.* **good-looking**, *a.* handsome; pretty. **good-looker**, *n.* **good looks**, *n.* **good luck**, *n., int.* good-fortune, prosperity. **goodman**, *n.* (*dated*) the head of a family; the master of a house; a husband. **good morning**, *n., int.* a wish or salutation. **good nature**, *n.* kindness of disposition; freedom from selfishness. **good-natured**, *a.* **good-naturedly**, *adv.* **good-neighbourliness**, *n.* friendliness and kindness between neighbours. **good night**, *n., int.* a wish at parting. **good offices**, *n. pl.* mediating influence. **Good Samaritan**, *n.* a friend in need. **good sense**, *n.* sound judgment. **good temper**, *n.* freedom from irritability. **good-tempered**, *a.* **good-temperedly**, *adv.* **good thing**, *n.* a favourable bargain or speculation; (*pl.*) delicacies, good fare. **good-time girl**, *n.* (*euphem.*) a prostitute. **good turn**, *n.* a kindly helpful act. **goodwife**, *n.* (*dated*) the mistress of a house. **goodwill**, *n.* kindly feeling or disposition; ready consent; the established popularity or custom of a business sold with the business itself. **goodies**, *n. pl.* objects, gifts etc. which are especially desirable. **goodish**, *a.* **goodly**, *a.* large. **goodliness**, *n.* **goodness**, *n.* the quality or state of being good; the virtue or essence of anything; (*euphem.*) God. **goodness gracious!**, *int.* expressing surprise etc. **goodness knows!**, *int.* expressing lack of knowledge etc. **goods**, *n.* merchandise. **the goods**, (*sl.*) just what is wanted; (*esp. N Am. coll.*) evidence (against someone). **to deliver the goods**, to carry out one's promise, keep one's word. **goods train, truck**, *n.* a train, truck carrying merchandise only. **goody**, *int.* (*coll.*) expressing delight. **goody-goody**, *n., a.* (a) priggishly good (person).

good-bye (gudbī'), *n., int.* farewell.
goof (goof), *n.* (*coll.*) a foolish mistake; a stupid person. *v.i., v.t.* to blunder. **goof-ball**, *n.* (*sl.*) a barbiturate pill; a mentally abnormal person. **goofy**, *n.* (*coll.*) silly; infatuated. **goofiness**, *n.*
googly (goo'gli), *n.* in cricket, a ball bowled so as to break a different way from that expected.
goon (goon), *n.* a stupid person; a hired thug.
goondie (goon'di), *n.* (*Austral.*) a hut.
Goorkha (guə'kə), GURKHA.
goosander (goosan'də), *n.* a merganser, *Mergus merganser*.
goose (goos), *n.* (*pl.* **geese** (gēs)) a web-footed bird intermediate in size between the duck and the swan, belonging to the genus *Anser*; the female of this, dist. from gander; a silly person; (*pl.* **gooses**) a tailor's smoothing iron; (*sl.*) a prod between the buttocks. *v.t.* (*sl.*) to prod between the buttocks. **to cook one's goose** COOK. **goose-bump** GOOSE-FLESH. **goose-fish**, *n.* (*N Am.*) the angler-fish. **goose-flesh, -pimples, -skin**, *n.* a bumpiness of the human skin produced by cold, fear etc. **goose-foot**, *n.* any plant of the genus *Chenopodium*, with leaves shaped like a goose's foot. **goose-grass**, *n.* silverweed; cleavers. **goose-grease**, *n.* the melted fat of the goose, formerly used as a remedy. **gooseherd**, *n.* one who tends geese. **goose-neck**, *n.* a piece of iron shaped like the neck of a goose. **goose-pimples** GOOSE-FLESH. **goose-quill**, *n.* a quill-feather of a goose, esp. (formerly) used as a quill pen. **goose-skin** GOOSE-FLESH. **goose-step**, *n.* (*Mil.*) a marching step in which the legs are raised very high without bending the knees. **goose-wing**, *n.* a lower corner of a square mainsail or foresail when the middle part is furled. **goose-winged**, *a.* **goosey**, *n.* **goosy**, *a.*
gooseberry (guz'bəri), *n.* the fruit of a thorny shrub, *Ribes grossularia*; an unwanted third to a pair of lovers. **gooseberry fool**, *n.* stewed gooseberries strained through a sieve and mixed with cream.
gopak (gō'pak), *n.* a folkdance from the Ukraine characterized by high leaps, performed by men.
gopher[1] (gō'fə), *n.* a name given to various American burrowing animals.
gopher[2] (gō'fə), *n.* the wood of which Noah's ark was made, so far unidentified.
gopher[3] (gō'fə), GOFER[2].
goral (gaw'rəl), *n.* a Himalayan goat-like antelope.
goramy (gaw'rəmi), *n.* a nest-building Oriental fish much valued for food.
Gorbachevism (gaw'bəchofizm), *n.* adherence to or support for Mikhail Gorbachev, President of the USSR (1985–) or his policies.
gorcrow (gaw'krō), *n.* the carrion crow.
Gordian (gaw'diən), *a.* intricate, complicated. **Gordian knot**, *n.* any apparently inextricable difficulty or deadlock. **to cut the Gordian knot**, to remove a difficulty by drastic measures.
gordius (gaw'diəs), *n.* a genus of threadlike worms.
gore[1] (gaw), *n.* blood from a wound, esp. thick, clotted blood. **gore-blood**, *n.* **gory**, *a.* covered with gore; bloody; involving bloodshed and killing. **gory-dew**, *n.* a minute freshwater alga coating damp shady walls with rosy gelatinous patches. **gorily**, *adv.* **goriness**, *n.*
gore[2] (gaw), *n.* a triangular piece sewed into a dress, a sail etc. to widen it out; a triangular piece of land. *v.t.* to make into or shape as a gore; to fit with a gore.
gore[3] (gaw), *v.t.* to pierce with or as with a horn or horn-like point.
Gore-tex® (gaw'teks), *n.* a type of waterproof fabric.
gorge (gawj), *n.* the throat; the gullet; that which is swallowed; the act of gorging; a heavy meal; a narrow pass between cliffs or hills; in a fortification, the narrow entrance into an outwork. *v.t.* to swallow, devour greedily; to glut, satiate. *v.i.* to feed greedily. **gorged**, *a.* having a gorge or throat; in heraldry, bearing a crown or the like round the neck.
gorgeous (gaw'jəs), *a.* splendid, magnificent; ornate; (*loosely*) very fine, beautiful etc. **gorgeously**, *adv.* **gorgeousness**, *n.*
gorget[1] (gaw'jit), *n.* a piece of armour for the throat or neck; a part of a wimple covering the chest; a necklace.
gorget[2] (gaw'jit), *n.* a surgical instrument for removing bladder stones.

Gorgio (gaw'jō), *n.* the gipsy name for one not a gipsy.
Gorgon (gaw'gən), *n.* (*Gr. Myth.*) one of three snake-haired female monsters so hideous that the sight of them was supposed to turn beholders to stone; a repulsive-looking woman. **gorgonesque** (-nesk'), *a.* **gorgonize, -ise,** *v.t.* to gaze at so as to paralyse or turn to stone.
Gorgonia (gawgō'niə), *n.* (*pl.* **iae** (-ē), **-ias**) the sea-fan; a genus of flexible polyps growing in the form of shrubs, feathers etc. **gorgonian,** *n., a.*
Gorgonzola (gawgənzō'lə), *n.* a blue-veined, strong-flavoured cheese. [village near Milan]
gorilla (gəril'ə), *n.* a large vegetarian African anthropoid ape, *Gorilla gorilla*; a large, strong man.
gormandize, -ise (gaw'məndīz), *n.* one who indulges in (good) eating. *v.t.* to eat greedily, **gormandizer,** *n.* **gormandizing,** *n.*
gormless GAUMLESS.
gorse (gaws), *n.* furze, whin. **gorsy,** *a.*
Gorsedd (gaw'sedh), *n.* a meeting of bards and Druids. [W]
gory GORE[1].
gosh (gosh), *int.* a mild oath expressing suprise.
goshawk (gos'hawk), *n.* a large, short-winged hawk used in falconry.
Goshen (gō'shən), *n.* a land of plenty.
gosling (goz'ling), *n.* a young goose; a silly or inexperienced person.
go-slow GO[1].
gospel (gos'pl), *n.* the teachings of Jesus Christ as revealed in the four canonical books of Matthew, Mark, Luke and John; one of these books; a selection from these books read in the Church service; anything accepted as infallibly true; a guiding principle; a style of jazz music, orig. US, strongly religious in character. **gospel-book,** *n.* a book containing the Gospels or one of them. **gospel oath,** *n.* an oath sworn on the Gospels. **gospel side,** *n.* the north side of the chancel where the Gospel is read. **gospel truth,** *n.* something as true as the Gospel. **gospeller,** *n.* one who reads the Gospel in the Communion service; a missionary.
gossamer (gos'əmə), *n.* cobweb threads floating in the air in calm weather; thin, filmy gauze; anything exceedingly flimsy or unsubstantial. **gossamered,** *a.* **gossamery,** *a.*
gossan (gos'ən), *n.* in mining, decomposed, ferruginous rock forming the upper part of a metallic vein.
gossip (gos'ip), *n.* idle or casual talk, usu. about other people; one who indulges in such talk; mere rumour; informal chat or writing. *v.i.* to chat idly or maliciously; to talk or write in an informal easy-going way. **gossip-monger,** *n.* a spreader of gossip. **gossiper,** *n.* **gossipry,** *n.* **gossipy,** *a.*
gossypium (gəsip'iəm), *n.* a tropical genus of the mallow family, including three species whence the cotton of commerce is obtained.
got (got), past, p.p. of GET[1], **got-up,** *a.* dressed up or prepared for effect or to take in.
Goth (goth), *n.* one of an ancient Germanic tribe which swept down upon southern Europe in the 3rd–5th cents.[?] establishing kingdoms in Italy, southern France and Spain; a barbarian, a rude, ignorant person.
Goth., (*abbr.*) Gothic.
Gothamist (gō'təmist, got'-), *n.* a foolish or gullible person. **Gothamite** (-it), *n.* a Gothamist; (*US, facet.*) a New Yorker.
Gothic (goth'ik), *a.* pertaining to the Goths or their language; in the style of architecture characterized by pointed arches, clustered columns etc.; rude, barbarous; (*Print.*) black-letter. *n.* the language of the Goths; the Gothic style of architecture; (*Print.*) black-letter. **gothically,** *adv.* **Gothicism** (-sizm), *n.* a Gothic idiom; conformity to the Gothic style of architecture; rudeness of manners. **Gothicist** (-sist), *n.* **gothicize, -ise** (-sīz), *v.t.* to make Gothic.

gotten (got'n), *p.p.* GET[1].
Götterdämmerung (gœtədem'ərung), *n.* in German mythology, the final destruction of the world. [G]
gouache (guahsh'), *n.* a method of painting with opaque colours mixed with water, honey and gum.
Gouda (gow'də), *n.* a round mild cheese made at Gouda, in Holland.
gouge (gowj), *n.* a chisel with a concave blade, used to cut holes or grooves; (*sl.*) a swindle, a fraud. *v.t.* to cut, force or scoop (out) with or as with a gouge; (*N Am.*) to cheat.
goulash (goo'lash), *n.* a stew of meat and vegetables highly seasoned with paprika.
gourami (goo'rəmi), GORAMY.
gourd (guəd), *n.* a large fleshy fruit, the hard outer coat of which serves for vessels to hold water; such a vessel; (*pl.*) hollow dice employed for cheating. **gourdful,** *n.*
gourmand (guə'mənd), *a.* gluttonous, fond of eating. *n.* one who loves good food; a glutton.
gourmet (guə'mā), *n.* a connoisseur in wines and meats; a dainty feeder, an epicure.
gout (gowt), *n.* a metabolic disease affecting the joints, esp. the great toe, with inflammation, pain and irritability; a drop; a disease of wheat caused by the **gout-fly**. **gouty,** *a.* **goutily,** *adv.* **goutiness,** *n.*
goût (goo), *n.* taste, relish; good taste, artistic discernment. [F]
gov., (*abbr.*) Governor; Government.
govern (gŭv'ən), *v.t.* to direct and control; to rule with authority, esp. to administer the affairs of a state; to regulate, to determine; to restrain, curb; in grammar, to require a particular case in a following noun or pronoun (said of a verb or preposition). *v.i.* to exercise authority; to administer the law; to have the control (over). **governable,** *a.* **governability,** *n.* **governably,** *adv.* **governance,** *n.*
governess (gŭv'ənis), *n.* a woman who has the care and instruction of young children, esp. in a private household. **governess-car, -cart,** *n.* a light two-wheeled vehicle with two seats, facing each other. **governessy,** *a.*
government (gŭv'ənmənt), *n.* control, direction, regulation, exercise of authority, esp. authoritative administration of public affairs; the form or system of such administration; the body of persons in charge of the government of a state; the power of controlling; the form of policy in a state; the right of governing; the executive power; (*N Am.*) state administration; (*Gram.*) the influence of a word in determining the case or mood of another. **government issue,** *n.* (*esp. US*) supplied by the government. **governmental** (-men'-), *a.* **governmentally,** *adv.* **governmentalism** (-men'-), *n.* **governmentalist,** *n.*
governor (gŭv'ənə), *n.* one who governs, esp. one invested with authority to administer a state, province etc.; a head of the executive; the Crown representative in a colony or dependency; (*US*) the elective chief magistrate of a state; the commander in a prison or garrison; (*sl.*) (gŭv'nə), one's father or employer; an unceremonious mode of address; a contrivance for regulating the speed of an engine motor etc., or the flow or pressure of a fluid or gas; **governor-general,** *n.* a chief of the executive in a large dependency, having deputy-governors under him. **governor-generalship,** *n.* **governorship,** *n.*
Gov.-Gen., (*abbr.*) Governor-General.
gowk (gowk), *n.* (*Sc., North.*) a fool, a simple or awkward person.
gown (gown), *n.* a woman's loose, long, outer garment; a dress, esp. a handsome or stylish one; a long, loose robe worn by clergymen, judges, lawyers, university graduates etc.; protective garment worn by surgeons during an operation. **town and gown,** the townspeople as opposed to the professors and students in a university town. **gowned,** *a.* **gownsman,** *n.* one whose professional dress is a gown.
goy (goi), *n.* (*pl* **goyim, goys**) Yiddish name for a non-Jewish person. [Heb.]

GP, (*abbr.*) general practitioner.
gp., (*abbr.*) group.
GPO, (*abbr.*) General Post Office.
Gr., (*abbr.*) Grand; Greece; Greek.
gr., (*abbr.*) grain(s).
Graafian (grah'fiən), *a.* named after de Graaf. **Graafian follicle** a small sac in which the ova mature in a mammalian ovary. [Regnier de *Graaf*, 1641-73, Dutch anatomist]
graal (grāl), GRAIL².
grab (grab), *v.t.* (*past, p.p.* **grabbed**) to seize, snatch suddenly; to take possession of violently or lawlessly; (*coll.*) to capture, arrest; (*coll.*) to interest. *v.i.* to grasp, snatch or clutch (at). *n.* a sudden snatch, grasping or seizing (at); an implement for clutching. **grab-bag**, *n.* (*N Am.*) a lucky dip. **up for grabs**, (*coll.*) for sale; ready for taking. **grabber**, *n.*
grabble (grab'l), *v.i.* to grope, to feel about (for).
grace (grās), *n.* that quality which makes form, movement, expression or manner elegant, harmonious, refined and charming; a natural gift or endowment; an acquired accomplishment, charm or attraction; a courteous or affable demeanour; free, unmerited favour or goodwill; clemency, mercy; a benefaction; (*Mus.*) an ornamental note or passage; (*Theol.*) the free, unmerited favour of God; a divine, regenerating and inspiring influence; a spiritual favour or excellence; a short prayer of thanks before or after a meal; a privilege or indulgence, esp. an extension of time allowed after a payment falls due; a licence to take a university degree, a dispensation from statutes etc. **airs and graces**, assumed refinement. **days of grace** DAY. **grace-and-favour**, *a.* of a house, flat etc., granted free of rent by the sovereign as a mark of gratitude. **her, his, your Grace**, forms of address for an archbishop, duke or duchess. **the Graces**, (*Gr. Myth.*) three goddesses embodying and conferring beauty and charm. **to be in the good graces of**, to enjoy the favour of. **to fall from grace** FALL. **with a good, bad grace**, willingly, reluctantly. **year of grace** YEAR. **grace-note**, *n.* (*Mus.*) an extra note introduced for embellishment. **graceful**, *a.* full of grace, elegance or beauty, esp. of form or movement. **gracefully**, *adv.* **gracefulness**, *n.* **graceless**, *a.* void of grace; lacking in propriety or decency, mannerless. **gracelessly**, *adv.* **gracelessness**, *n.*
gracile (gras'īl), *a.* slender, lean. **gracility** (-sil'-), *n.*
gracious (grā'shəs), *a.* exhibiting grace or kindness; benevolent; courteous, condescending, affable; graceful, pleasing; elegant, affluent; proceeding from divine grace; benignant, merciful. **gracious me!** *int.* exclamation of surprise or protest. **graciously**, *adv.* **graciousness**, *n.*
grackle (grak'l), *n.* any bird of the genus *Gracula*, allied to the starlings.
gradate GRADATION.
gradatim (grədā'tim), *adv.* gradually, by degrees. [L]
gradation (grədā'shən), *n.* an orderly succession or progression step by step; (*usu. in pl.*) a step, stage or degree; (*Fine Art*) the gradual blending of one tint, tone etc. with another; (*Mus.*) an ascending or descending succession of chords; (*Philol.*) ablaut. **gradate**, *v.t.* to arrange (colours etc.) by imperceptible gradation. *v.i.* to change by such gradations. **gradational**, *a.* **gradationally**, *adv.* **gradationed**, *a.* **gradatory** (grā'də-), *a.*
grade (grād), *n.* a degree or step in rank, quality, value, order etc.; a class of people of similar rank, ability, proficiency etc.; (*N Am.*) class (at school); a mark showing relative quality; gradient of a road; (*Philol.*) the position of a vowel or root in an ablaut series. *v.t.* to arrange in grades; to gradate; to adjust the rate of slope in, as a road. **at grade**, (*N Am.*) at the same level. **on the down, up grade**, descending or ascending. **to make the grade**, to succeed. **grade cricket**, *n.* (*Austral.*) competitive cricket played between teams arranged in grades. **grade-crossing**, *n.* (*N Am. Rail.*) a level-crossing. **grade school**, *n.* (*N Am.*) elementary school. **gradable**, *a.* **grader**, *n.* one who or that which grades; a machine with a blade for levelling earth, rubble etc.
-grade, *comb. form* of a kind or manner of movement or progression, as in *retrograde*.
gradely (grād'li), *a.* (*dial.*) excellent; respectable; proper; good-looking. *adv.* fine.
gradient (grā'diənt), *n.* degree of slope, inclination; a sloping road; rate of variation or increase or decrease in height of thermometer or barometer over a large area.
gradin (grā'din), **gradine** (grədēn'), *n.* one in a series of rising steps or a tier of seats; a shelf or step at the back of an altar.
gradual (grad'ūəl, graj'l), *a.* proceeding by steps or degrees; regular and slow. *n.* an antiphon sung between the Epistle and the Gospel; a book containing such antiphons. **gradualism**, *n.* the principle of making change slowly and gradually. **gradualist**, *n.* **gradually**, *adv.* **gradualness**, *n.*
graduate (grad'ūāt, graj'-), *v.t.* to mark with degrees; to divide into or arrange by gradations; to apportion (a tax etc.) according to a scale of grades; to temper or modify by degrees; (*N Am.*) to confer an academic degree upon. *v.i.* to alter, change or pass by degrees; to take a degree in a university. **graduand** (-ənd), *n.* a person about to graduate from a university. **graduateship**, *n.* **graduation**, *n.* regular progression by successive degrees; a division into degrees or parts; the conferring or receiving of academic degrees. **graduation exercises**, *n.pl.* (*N Am.*) prize-day at school etc. **graduator**, *n.*
graduction (grədŭk'shən), *n.* in astronomy, the division of circular arcs into degrees, minutes etc.
gradus (grā'dəs), *n.* a dictionary of Greek or Latin prosody.
Graecism (grē'sizm), *n.* a Greek idiom, style or mode of expression; cultivation of the Greek style. **graecize, -ise**, *v.t.* to give a Greek form or character to. *v.i.* to cultivate or follow the Greek spirit, ideas, ways of expression etc.
graeco-, *comb. form* Greek.
Graeco-Roman (grēkōrō'mən), *a.* pertaining to both Greeks and Latins.
graffito (grəfē'tō), *n.* (*pl.* **-ti** (-tē)) (*Archaeol.*) a drawing or inscription scratched on a wall or other surface; (*usu. pl.*) drawings or words, sometimes obscene, sometimes political, painted or written on walls etc. in public view.
graft¹ (grahft), *n.* a small shoot of a plant inserted into another which supplies the sap to nourish it; living tissue transplanted to another part of the body; incorporation with a foreign stock. *v.t.* to insert (a shoot or scion) in or upon another plant; to insert grafts upon; to plant (a tree or stock) thus with another variety; to transplant (as living animal tissue); to incorporate with another stock; to insert or implant (upon) so as to form a vital union; *v.i.* to insert grafts or scions in or on other stocks. **grafter**, *n.* **grafting clay, wax**, *n.* a plastic composition used for covering grafted parts and excluding air. **grafting scissors**, *n.* scissors used by surgeons in skin-grafting.
graft² (grahft), *n.* a spit of earth, the amount thrown up at one dig with the spade.
graft³ (grahft), *n.* a swindle; acquisition of money etc. by taking advantage of an official position; bribery; illicit gains so obtained; hard work. *v.i.* (*coll.*) to work (hard). **grafter**, *n.*
grail (grāl), *n.* a dish or cup said to have been used by Christ at the Last Supper, and employed by Joseph of Arimathea to collect His blood while on the Cross; also called the **Holy Grail, Saint Grail**, and **Sangreal** (san'grāl, -grāl').
grain (grān), *n.* a single seed of a plant; (*collect.*) corn in general or the fruit of cereal plants, as wheat, barley, rye etc.; (*N Am.*) wheat; (*pl.*) the husks or refuse of malt after brewing or of any grain after distillation; any small, hard particle; the smallest amount; the unit of weight in

the English system, 1/7000 lb. avoirdupois (65 mg); in photography, one of the particles in a photographic emulsion; granular texture, degree of roughness or smoothness of surface; the arrangement of the fibres of wood or leather or the particles of stone; the pattern formed by this; a red dye made from cochineal or kermes insects; any fast dye, esp. red, crimson or purple; temper, disposition, natural tendency. *v.t.* to form into grains, to granulate; to treat so as to bring out the natural grain; to paint or stain in imitation of this; to give a granular surface to; to scrape the hair off (hides) with a grainer. *v.i.* to form grains, to become granulated. **against the grain**, against one's natural inclination. **grains of paradise** or **Guinea grains**, the seeds of *Amomum melegueta*, a tropical W African spice, used in stimulants, diuretics and spirituous liquors. **in grain**, downright, thorough, absolute, inveterate. **to dye in grain**, to dye in a fast colour, esp. in kermes; to dye deeply or into the fibre. **grain alcohol**, *n.* alcohol made by the fermentation of grain. **grain elevator**, *n.* ELEVATOR under ELEVATE. **grain leather**, *n.* leather dressed with the grain-side outwards. **grain-side**, *n.* the side (of leather) from which the hair has been removed. **grainage**, *n.* **grained**, *a.* (*esp. in comb.*, as *fine-grained*). **grainer**, *n.* one who paints or stains in imitation of the grain of wood; also the brush he uses; a tanner's knife. **graining**, *n.* the act of producing a grain; a process in tanning; painting in imitation of the grain of wood. **grainless**, *a.* **grainy**, *a.*
grainz (grāns), *n.* (*pl., usu. construed as sing.*) a forked fish-spear, a kind of harpoon.
grakle (grak'l), GRACKLE.
grallatorial (gralətaw'riəl), *a.* of or pertaining to long-legged wading birds.
gralloch (gral'əkh), *v.t.* to disembowel (a deer). *n.* the viscera of a deer. [Gael.]
gram¹ (gram), *n.* the chick-pea or other kinds of pulse.
gram², **gramme** (gram), *n.* a unit of weight in the metric system, equalling a thousandth part of a standard kilogram (about 0.04 oz). **grammetre, gram-centimetre**, *n.* a unit of work, equalling the amount done in raising one gram vertically one centimetre.
gram., (*abbr.*) grammar, grammarian, grammatical.
-gram, *comb. form* forming nouns, meaning something written or recorded, as in *epigram, monogram*.
grama (grah'mə), **gramma grass** (gram'ə), *n.* various species of low pasture grass in W and SW US.
gramarye (gram'əri), *n.* an old name for magic, necromancy.
gramercy (grəmœ'si), *int.* (*old fashioned*) thanks; an exclamation expressive of surprise. *n.* an expression of thanks.
Gramineae (grəmin'iē), *n.pl.* the grass family of plants. **graminaceous** (graminā'shəs), **gramineous** (-min'-), *a.* pertaining to grass or grasses. **graminifolious** (graminifō'liəs), *a.* having leaves like grass. **graminivorous** (graminiv'ərəs), *a.* subsisting on grass.
gramma grass GRAMA.
grammalogue (gram'əlog), *n.* in shorthand, a word represented by a single sign; a logogram.
grammar (gram'ə), *n.* the science of the correct use of language, dealing with phonology, etymology, inflections and syntax; a system of principles and rules for speaking and writing a language; a book containing these principles and rules; speech or writing considered with regard to its correctness; the elements of an art or science, a treatise on these. **grammar school**, *n.* esp. formerly, a secondary school with an academic course. **grammarian** (-mea'ri-), *n.* one who studies, writes on or teaches grammar. **grammarless**, *a.* **grammatical** (-mat'-), *a.* pertaining to grammar; according to the rules of grammar. **grammatical gender**, *n.* gender based on grammar, not sex. **grammatical sense**, *n.* the literal sense. **grammatical subject**, *n.* the literal as dist. from the logical subject. **grammatically**, *adv.* **grammaticize, -ise** (-mat'isiz), *v.t.* to render grammatical.
gramophone (gram'əfōn), *n.* a record-player.
grampus (gram'pəs), *n.* a large delphinoid cetacean belonging to the genus *Orca*, esp. the voracious *O. gladiator;* also the inoffensive cetacean *Grampus griseus* or cow-fish.
granadilla (granədil'ə), *n.* various species of passion-flower, *Passiflora;* their edible fruit.
granary (gran'əri), *n.* a storehouse for grain; a country or district producing much corn.
grand (grand), *a.* great or imposing in size, character or appearance; fine, splendid; dignified, morally impressive; (*Mus.*) for full orchestra, or with all accessory parts and movements; (*coll.*) distinguished, fashionable or aristocratic (society); (*coll.*) excellent; pre-eminent in rank, chief; main, comprehensive, complete, final; in the second degree (of relationships). *n.* (*sl.*) a thousand dollars or pounds. **grandam**, *n.* (*old-fashioned*) a grandmother; an old woman. **grandaunt**, *n.* the sister of a grandfather or grandmother. **grand captain**, *n.* a chief captain, commander or general. **grandchild** (gran'chīld), *n.* the child of a son or daughter. **grand committee**, *n.* one of two standing committees of the House of Commons appointed every session to consider Bills relating to law or trade. **granddaughter** (gran'daw-), *n.* the daughter of a son or daughter. **grandad** (gran'dad), **granddaddy**, *n.* (*coll.*) grandfather. **Grand Duke**, *n.* a sovereign of lower rank than a king, the ruler in certain European states; hence, **Grand Duchy, Grand Duchess. Grand Ducal**, *a.* **grandfather**, *n.* the father of a parent. **grandfather clock**, *n.* a clock worked by weights, in a tall wooden case. **grandfatherly**, *a.* **grand juror, grand jury**, *n.* one which ascertains whether there is sufficient ground for a prisoner to be tried. **grand larceny** LARCENY. **grandma** (gran'mah), **grandmamma**, *n.* (*coll.*) grandmother. **Grand Master**, *n.* the head of a military order of knighthood, the head of the Freemasons etc.; in chess or bridge, an outstanding player, winner of many international tournaments, competitions. **grandmother**, *n.* the mother of a parent. **grandmother clock**, *n.* one similar to but slightly smaller than a grandfather clock. **grandmotherly**, *a.* **Grand National**, *n.* an annual steeplechase run at Aintree, Liverpool. **grandnephew, grandniece**, *n.* the grandson or granddaughter of a brother or sister. **grand opera**, *n.* OPERA¹. **grandpa** (gram'pah, gran'-), **grandpapa**, *n.* (*coll.*) grandfather. **grandparent**, *n.* a grandfather or grandmother. **grand passion**, *n.* an overwhelming love affair. **grand piano**, *n.* a large piano with horizontal framing. **grandsire**, *n.* a grandfather. **grand slam**, *n.* in bridge, the winning of all 13 tricks by a side; in tennis, golf etc., the winning of all the major competitions in a season. **grandson**, *n.* the son of a son or daughter. **grandstand**, *n.* the principal stand for spectators on a race-course etc. *v.i.* (*coll.*) to behave ostentatiously. **grandstand finish**, *n.* a close and exciting finish in a sporting contest. **grand total**, *n.* the total of all subordinate sums. **grand tour**, *n.* one through the countries of continental Europe esp. as formerly part of the education of young people of good family; any extended sightseeing tour. **Grand Turk**, *n.* formerly, the Sultan of Turkey. **Grand Vizier**, *n.* formerly, the prime minister of the Ottoman Empire. **grandly**, *a.* **grandness**, *n.*
grand cru (grā krü), *n.* of a wine, from a famous vineyard. [F]
granddaughter GRAND.
grandee (grandē'), *n.* a Spanish or Portuguese nobleman of the highest rank; a person of high rank or power.
grandeeship, *n.*
grandeur (gran'dyə), *n.* the quality of being grand; greatness, nobility, sublimity; splendid or magnificent appearance or effect.
grandfather GRAND.

grand guignol (grä geenyol'), *n.* melodramatic or horrifying events or stories. [F]
grandiloquent (grandil'əkwənt), *a.* using lofty or pompous language. **grandiloquence**, *n.* **grandiloquently**, *adv.*
grandiose (gran'diōs), *a.* imposing; intended to produce the effect of grandeur, affecting impressiveness, pompous. **grandioseness**, *n.* **grandiosity** (-os'-), *n.*
grand mal (grä mal), *n.* a major epileptic attack, as opp. to *petit mal.* [F]
grand marnier (grämah'niä), *n.* a liqueur somewhat like curaçao. [F]
grand monde (grä mŏd), *n.* highest society. [F]
grandmother, grandparent, grandson GRAND.
Grand Prix (grä prē), *n.* a famous horse race held at Paris; an international race or competition in other sports. [F]
grand seigneur (grä senyœ'), *n.* a person of high rank.
grange (grānj), *n.* a farmhouse with the out-buildings etc.
granger, *n.* (*N Am.*) a farmer.
grangerize, -ise (grān'jəriz), *v.t., v.i.* to illustrate (a book) with illustrations taken from other books (from the practice of so illustrating Granger's *Biographical History of England*, 1769). **grangerism**, *n.* **grangerization**, *n.* **grangerite** (-it), **grangerizer**, *n.* [James Granger, 1716–76]
graniferous (grənif'ərəs), *a.* bearing grain or seed of grain-like form. **graniform** (grä'nifawm), *a.* **granivorous** (-niv'-), *a.* feeding on grain.
granite (gran'it), *n.* a granular, igneous rock consisting of feldspar, quartz and mica. **granite ware**, *n.* an enamelled ironware or hard pottery with speckled surface resembling granite. **granitic, -ical** (-nit'-), *a.* **granitification** (-fikā'), *n.* formation into granite. **granitiform** (-nit'ifawm), *a.* **granitoid**, *a.* resembling granite.
granny (gran'i), **grannie, gran**, *n.* a grandmother; an old woman. **granny flat**, *n.* a self-contained flat added to or part of a house, for an elderly relative. **granny-knot** or **granny's bend**, *n.* a badly-tied reef-knot having the tie crossed the wrong way. **Granny Smith**, *n.* a green-skinned apple.
granolith (gran'əlith), *n.* artificial stone consisting of crushed granite and cement. **granolithic**, *n., a.*
grant (grahnt), *v.t.* to concede or give, esp. in answer to request; to allow as a favour or indulgence; (*Law*) to transfer the title to, to confer or bestow (a privilege, charter etc.); to admit as true. *n.* the act of granting; the thing granted; a gift, assignment, formal bestowal; a sum of money bestowed or allowed; a concession or admission of something as true; (*Law*) a conveyance in writing; the thing conveyed. **to take for granted**, to assume to be true; to neglect, fail to appreciate. **grant-in-aid**, *n.* a sum granted towards the maintenance of a school or other institution. **grantable** *a.* **grantee** (-tē'), *n.* (*Law*) the person to whom a grant or conveyance is made. **granter**, *n.* **grantor**, *n.* (*Law*)
granule (gran'ūl), *n.* a little grain; a small particle. **granular**, *a.* **granularity** (-la'-), *n.* **granularly**, *adv.* **granulate**, *v.t.* to form into granules; to make rough on the surface. *a.* (-lət), granulated. **granulated sugar**, *n.* coarse-grained white sugar. **granulation**, *n.* **granulative**, *n.* **granulator**, *n.* **granuliferous** (-lif'-), *a.* bearing or full of granules. **granuliform** (-ifawm), *a.* **granulitic** (-lit'-), *a.* **granulize, -ise**, *v.t.*
granulo-, *comb. form.* pertaining to granules. **granulous**, *a.*
granulocyte (gran'ūləsit), *n.* a white blood cell with granular cytoplasm. **granulocytic** (-sit'-), *a.*
grape (grāp), *n.* a berry constituting the fruit of the vine; grape-shot. **the grape**, (*coll.*) wine. **sour grapes**, some object of desire disparaged because it is out of reach. **grape-brandy**, *n.* brandy distilled from grapes or wine. **grape-fruit**, *n.* a large round citrus fruit with a pale yellow rind. **grape-house**, *n.* a glass-house for growing vines. **grape-hyacinth**, *n.* a bulbous plant belonging to the genus *Muscari.* **grape-scissors**, *n.pl.* scissors for thinning out bunches of grapes on the vines, or for dividing bunches at the table. **grapeseed**, *n.* the seed of the vine. **grapeseed oil**, *n.* cooking oil expressed from this. **grapeshot**, *n.* shot that scatters when fired. **grape-stone**, *n.* seed of the grape. **grape sugar**, *n.* glucose or dextrose. **grape-vine**, *n.* any species of *Vitis*, esp. *V. vinifera*; (*coll.*) the spread of rumour by unofficial means. **grape-wort**, *n.* the baneberry. **grapeless**, *a.* **grapery**, *n.* **grapy**, *a.*
graph (grahf, graf), *n.* a diagram representing mathematical or chemical relationship and based on two axes. *v.t.* to plot on a graph. **graph paper**, *n.* squared paper used for drawing graphs, diagrams etc.
-graph, *comb. form.* -written, -writing, -writer, as in *autograph, seismograph, telegraph.* **-grapher**, *comb. form.* a person skilled in a specified science or means of communication.
graphic (graf'ik), **-ical**, *a.* pertaining to writing, delineating, engraving, painting etc.; vividly or forcibly descriptive; using diagrams or graphs. *n.* (*pl.*) the art of drawing, esp. in mathematics, engineering etc.; (*pl.*) the production of designs and images by computer; the designs so produced. **graphic formula**, *n.* a chemical formula representing the relations of the atoms of a molecule to each other. **graphic granite**, *n.* a compound of quartz and feldspar, with markings roughly like Hebrew characters. **-graphic, -ical**, *comb. form.* **graphically**, *adv.* **graphicalness, graphicness**, *n.*
graphite (graf'it), *n.* blacklead, plumbago. **graphitic** (-fit'-), *a.* **graphitoid** (-i-), *a.*
graphium (graf'iəm), *n.* (*pl.* **-phia** (-ə)) a stylus, a pencil.
graphiure (graf'iūə), *n.* a southern African rodent resembling the dormouse, with a tufted tail.
grapho-, *comb. form.* of, pertaining to or for writing.
grapholite (graf'əlit), *n.* a kind of slate suitable for writing on.
graphology (grəfol'əji), *n.* the study of handwriting; the art of inferring character from handwriting. **graphologic, -ical** (-loj'-), *a.* **graphologist**, *n.*
graphomania (grafōmā'niə), *n.* a psychological urge to write or scribble.
-graphy (-grəfi), *comb. form.* description; style of writing, as in *geography, stenography.*
grapnel (grap'nəl), *n.* a grappling-iron; an anchor with flukes for mooring boats, balloons etc.
grappa (grap'ə), *n.* a coarse brandy distilled from the residue of a wine press. [It.]
grapple (grap'l), *n.* a grappling-iron or similar clutching device; a close hold or grip; a close struggle. *v.t.* to seize, clutch. *v.i.* to contend or struggle (with or together) in close fight; to strive to accomplish or master. **grappling-iron, -hook**, *n.* an iron instrument with claws or hooks for seizing and holding fast.
graptolite (grap'təlit), *n.* a kind of fossil zoophyte. **graptolitic** (-lit'-), *a.*
grasp (grahsp), *v.t.* to seize and hold fast; to hold on to, esp. eagerly or greedily; to comprehend with the mind. *v.i.* to clutch (at); to attempt to lay hold; to accept eagerly. *n.* a fast grip or hold; ability to seize and hold; forcible possession; intellectual comprehension. **graspable**, *a.* **grasper**, *n.* **grasping**, *a.* greedy. **grasping reflex**, *n.* the response by an infant's fingers or toes to grasp an object that touches them. **graspingly**, *adv.* **graspingness**, *n.*
grass (grahs), *n.* the green-bladed plants on which cattle, sheep etc. feed; any plant of the Gramineae, distinguished by simple, sheathing leaves, a stem usu. jointed and tubular, and flowers enclosed in glumes, including cereals, reeds and bamboos etc.; ground covered with grass; pasture; (*coll.*) marijuana; (*sl.*) an informer; (*Mining*) the surface of the ground. *v.t.* to cover with grass or turf; to land (as a fish); (*sl.*) to fall, to knock down. *v.i.* (*sl.*) to inform against. **as green as grass** GREEN. **grass of Parnassus**, a white-flowered plant belonging to the saxi-

grasshopper　　　　　　　　　　　　　　　　　graze

frage order, growing in moist places. **to bring to grass,** (*Mining*) to bring up to the pit-head. **to go, put, send** or **turn out to grass,** to go or send out to pasture; to go or send out from work, on a holiday, into retirement etc. **to let the grass grow under one's feet,** to waste time and lose opportunities. **grass-box,** *n.* a container attached to a lawn-mower to catch grass cuttings. **grass-cloth,** *n.* a fabric made from the fibres of the ramie. **grass court,** *n.* a tennis court with a grass surface. **grass-green,** *n.*, *a.* verdant, dark green (colour). **grass-grown,** *a.* overgrown with grass. **grassland,** *n.* land kept under grass. **grass-oil,** *n.* a fragrant volatile oil distilled from various Indian grasses. **grass roots,** *n. pl.* (*coll.*) the ordinary people; the basic essentials, foundation, origin. **grass-snake,** *n.* a harmless snake, *Natrix natrix*. **grass-tree,** *n.* an Australasian tree having spear-like stalks etc. **grass widow,** *n.* a wife temporarily separated from her husband. **grass-widower,** *n.* **grass-widowhood,** *n.* **grass-wrack,** *n.* a seaweed belonging to the genus *Zostera*. **grassed,** *a.* of a golf-club, with the face slightly filed back. **grassless,** *a.* **grasslike,** *a.* **grassy,** *a.* covered with grass; like grass; green. **grassiness,** *n.*

grasshopper (grahs'hopə), *n.* an orthopterous insect of various species with hind legs formed for leaping. *a.* of a mind constantly moving from subject to subject. **knee-high to a grasshopper,** young, small. **grasshopper-warbler,** *n.* a small warbler, so called from its note.

grate[1] (grāt), *n.* a grating; a frame of iron bars for holding fuel for a fire. *v.t.* to furnish with a grate or grating. **grated,** *a.* **grateless,** *a.*

grate[2] (grāt), *v.t.* to rub against a rough surface so as to reduce to small particles; to rub, as one thing against another, so as to cause a harsh sound; to grind down; to produce (as a hard, discordant sound) by friction. *v.i.* to rub (upon) so as to emit a harsh, discordant noise; to have an irritating effect (upon). **grater,** *n.* a utensil with a rough surface for reducing a substance to small particles. **grating**[1], *a.* harsh, discordant, irritating. **gratingly,** *adv.*

grateful (grāt'fl), *a.* pleasing, refreshing; thankful, marked by or indicative of gratitude. **gratefully,** *adv.* **gratefulness,** *n.*

graticulation (grətikülā'shən), *n.* division of a design or drawing into squares. **graticule** (grat'i), *n.* the grid pattern of latitude and longitude lines on a map.

gratify (grat'ifī), *v.t.* to please, delight; to satisfy the desire of; to indulge. **gratifying,** *a.* **gratifyingly,** *adv.* **gratification** (-fikā'-), *n.* the act of gratifying; that which gratifies; an enjoyment, satisfaction. **gratifier,** *n.*

gratin (grat'i), *n.* a dish prepared with bread-crumbs or grated cheese and browned on top.[F]

grating[1] GRATE[2].

grating[2] (grā'ting), *n.* an open framework or lattice of metal or wooden bars, parallel or crossed; a series of parallel wires or lines ruled on glass or the like for producing spectra by diffraction.

gratis (grat'is), *adv.*, *a.* for nothing; without charge. [L]

gratitude (grat'itūd), *n.* grateful feeling towards a benefactor; thankfulness.

grattoir (grat'wah), *n.* (*Archaeol.*) a flint implement used as a scraper. [F]

gratuitous (grətü'itəs), *a.* granted without claim or charge; free; without cause, uncalled for, unnecessary. **gratuitously,** *adv.* **gratuitousness,** *n.*

gratuity (grətū'əti), *n.* a present voluntarily given in return for a service, a tip; a payment to soldiers on retirement, discharge etc.

gratulate (grat'ūlāt), etc. CONGRATULATE.

gratulatory (grat'ūlātəri), *a.* congratulatory, expressing joy.

graupel (grow'pl), *n.* soft hail. [G]

gravamen (grəvā'mən), *n.* (*pl.* **-mina** (-minə)) (*Law*) the substantial cause of an action; the most serious part of a charge.

grave[1] (grāv), *v.t.* to clean by scraping or burning, and cover with pitch and tallow (as a ship's bottom). **graving-dock,** *n.* a dry dock.

grave[2] (grāv), *v.t.* (*p.p..* **graved, graven**) to engrave, carve; to produce (a figure, inscription etc.) by engraving or carving; to impress in the wind. *n.* a hole in the earth for burying a dead body in; a place of burial; a tomb; death, destruction; a place of destruction, extinction. **to have one foot in the grave,** to be near death. **to turn in one's grave,** (of a dead person) to be (thought to be) shocked or distressed by some modern event. **grave-clothes,** *n.pl.* wrappings in which the dead are buried. **grave-digger,** *n.* one who digs graves; an insect that buries dead insects etc., to feed its larvae. **grave-mound,** *n.* a barrow, tumulus. **gravestone,** *n.* a memorial stone at a grave. **graveyard,** *n.* a burial ground. **graveless,** *a.* **graven image,** a carved idol. **graver,** *n.* an engraver; an engraving tool.

grave[3] (grāv), *a.* important, serious; sedate, solemn; sombre, plain; (*Mus.*) low in pitch; slow in movement; (*Gram.*) (grahv) low-pitched, not acute (of accents). *n.* a grave accent. **gravely,** *adv.*

gravel (grav'l), *n.* small water-worn stones or pebbles intermixed with sand etc.; a stratum of this; minute concretions in the kidney or bladder. *v.t.* to cover, lay or strew with gravel; to perplex. **gravel-pit,** *n.* a pit out of which gravel is dug. **gravel-walk,** *n.* a path laid with gravel. **gravelling,** *n.* like, or covered with, gravel; of a voice, deep and harsh. **gravelly,** *a.*

graven, etc. GRAVE[2].

Graves (grahv), *n.* a light red or white French wine. [F]

gravestone, graveyard GRAVE[2].

gravid (grav'id), *a.* pregnant.

gravigrade (grav'igrād), *a.* (*Zool.*) walking heavily. *n.* one of the heavy-limbed animals, like the elephant or the megatherium.

gravimeter (grəvim'itə), *n.* an instrument for determining specific gravity or for measuring variations of gravity on the earth's surface. **gravimetric** (-met'-), *a.* **gravimetrically,** *adv.* **gravimetry** *n.*

graving-dock GRAVE[1].

gravitate (grav'itāt), *v.i.* to be acted on by gravity; to be attracted, to tend (towards); to tend downwards, to sink. **gravitater,** *n.* **gravitation** (-tā'-), *n.* the act or process of gravitating; the force of gravity. **gravitational** (-tā'-), *a.* **gravitative,** *a.*

gravity (grav'iti), *n.* heaviness; importance, seriousness; solemnity, sedateness; (*Phys.*) the force causing bodies to tend towards the centre of the earth or another body; the degree of intensity of this force. **specific gravity** SPECIFY. **gravity feed,** *n.* a feed or supply in which the material runs downhill.

gravure (grəvüə'), *n.* an engraving; (short for) photogravure.

gravy (grā'vi), *n.* the fat and juice from meat during and after cooking; a sauce made with this; (*sl.*) money acquired with little effort. **gravy boat,** *n.* a boat-shaped bowl or dish for holding gravy. **gravy dish,** *n.* a meat-dish with a hollow for gravy; a dish in which gravy is served. **gravy train,** *n.* (*sl.*) a source of easy money, benefits etc.

gray (grā), GREY.

grayling (grā'ling), *n.* a freshwater fish of the salmon family, with a large dorsal fin.

graze[1] (grāz), *v.i.* to eat growing grass; to supply grass for grazing (of land, fields etc.); (*coll.*) to eat in small snacks rather than set meals; *v.t.* to feed (cattle etc.) on growing grass; to supply with pasturage; to tend (cattle etc.) at pasture. **grazer,** *n.* **grazing,** *n.*

graze[2] (grāz), *v.t.* to touch or brush slightly in passing; to scrape or abrade in rubbing past. *v.i.* to touch some person or thing lightly in passing; *n.* a slight touch in passing; (*coll.*) to watch only short excerpts of television

programmes whilst constantly changing channels.
grazier (grā'ziə), *n*. one who pastures cattle, and rears and fattens them for market. **graziery**, *n*.
grazioso (grätsiō'sō), *a*. (*Mus*.) graceful, elegant. **graziosamente** (-səmen'tā), *adv*. elegantly. [It.]
grease (grēs), *n*. animal fat in a melted or soft state; oily or fatty matter of any kind. *v.t*. (grēs, grēz), to smear, lubricate or soil with grease. **like greased lightning**, (*coll*.) very quickly; **to grease someone's palm** or **hand**, (*coll*.) to bribe. **greasebox**, *n*. a holder on a wheel or axle for grease as a lubricant. **greasegun**, *n*. a syringe for injecting grease or oil into machinery. **grease monkey**, *n*. (*coll*.) a mechanic. **greasepaint**, *n*. a paste used in theatrical make-up. **greasetrap**, *n*. a contrivance fixed in drains for catching grease from sinks etc. **greaser**, *n*. one who or that which greases, a mechanic; (*N Am*., *offensive*) a Mexican or Spanish-American.
greasy (grē'zi, -si), *a*. smeared, saturated or soiled with or as with grease; made of or like grease; oily; unpleasantly unctuous. **greasy spoon**, *n*. (*coll*.) a cheap restaurant, esp. one specializing in mainly fried foods. **greasily**, *adv*. **greasiness**, *n*.
great (grāt), *a*. large in bulk, number, amount, extent or degree; large or important beyond the ordinary; preeminent, the chief; of exceptional ability; (*coll*.) very skilful, experienced or knowing (at); having lofty moral qualities, noble; grand, sublime; teeming; (with); excessive; notorious; (*coll*.) excellent; denoting a step of ascending or descending consanguinity (as **great-grandfather**, the father of a grandfather; **great-grandson**, the son of a grandson etc.). *n*. (*collect*.) great people; (*pl*.) GREATS (*pl*.); **great at**, to be skilful at. **great ape**, *n*. one of the larger apes, such as the gorilla, chimpanzee etc. **great auk** AUK. **great circle**, *n*. a circle on a sphere (such as the earth) formed by a plane passing through the centre of the sphere. **great-coat**, *n*. an overcoat. **greatcoated**, *a*. **Great Dane**, *n*. a breed of large smooth-coated dog. **Great Britain** BRITAIN. **great gross**, 144 dozen. **great-hearted**, *a*. high-spirited, magnanimous; brave. **greatheartedness**, *n*. **great organ** ORGAN. **Great Powers**, *n.pl*. the leading states of the world collectively. **great primer**, *n*. 18-point type. **Great Scot(t)**, *int*. expressing surprise etc. **Great Seal** SEAL². **Great Spirit**, *n*. the name given by the N American Indians to their deity. **great toe**, *n*. the big toe. **Great War**, *n*. World War I. **greatly**, *adv*. **greatness**, *n*. **greats**, *n.pl*. the course of study in classics and philosophy at Oxford Univ.; the final examination for this.
greave (grēv), *n*. (*usu. pl*.) armour for the legs.
greaves (grēvz), *n.pl*. fibrous scraps or refuse of melted tallow.
grebe (grēb), *n*. a diving-bird of the genus *Podiceps*, with lobed feet and no tail.
Grecian (grē'shən), *a*. of or pertaining to Greece. *n*. a Greek; one who adopted Greek manners or habits; a Greek scholar; a senior boy at Christ's Hospital. **Grecian knot**, *n*. a knot of hair at the back of the head. **Grecian nose**, *n*. a nose continuing the line of the forehead. **Grecianize, -ise**, *v.t., v.i*.
Grecism (grē'sizm), **Grecize** (-sīz), etc. GRAECISM.
grecque (grek), *n*. an ornamental Greek fret; a coffee-strainer or a coffee-pot fitted with a strainer. [F]
greed (grēd), *n*. insatiable desire for food or possessions.
greedy, *a*. having an inordinate desire for food or drink; eager to obtain, desirous (of). **greedily**, *adv*. **greediness**, *n*.
Greek (grēk), *n*. a native of Greece; one of the Greek race; the language of Greece; something one does not understand. *a*. pertaining to Greece or its people or to the Hellenic race. **when Greek meets Greek**, describing an equal encounter of champions. **Greek Church**, *n*. the Orthodox or Eastern Church, including most of the Christians in Greece, Russia and Eastern Europe, which

separated from Rome in the 9th cent. **Greek cross** CROSS. **Greek fire** FIRE. **Greek gift**, *n*. a gift bestowed with some treacherous motive (in alln. to Virgil's *Aeneid* ii.49).
Greekness, *n*.
green (grēn), *a*. having a colour like growing grass; of the colour in the spectrum between yellow and blue; unripe; inexperienced, gullible; fresh, not withered, dried, seasoned, cured, dressed or tanned; pale, sickly; envious, jealous; concerned with environmental protection, esp. by political means. *n*. the colour of growing grass; a colour composed of blue and yellow; a green pigment or dye; a grassy plot or piece of land; (*coll*.) a person who is concerned about environmental issues; (*pl*.) fresh leaves or branches of trees; (*pl*.) green vegetables used for food. *v.i*. to become or grow green. *v.t*. to make green; to plant trees etc. in (urban areas); to make environmentally conscious. **as green as grass**, naive, inexperienced, immature. **greenback**, *n*. (*coll*.) a note issued by any national bank in US. **green belt**, *n*. an area around a city in which building is restricted. **green card**, *n*. an international insurance certificate for motorists. **green cheese**, *n*. unripened cheese, whey cheese; cheese coloured with sage. **green-coloured**, *a*. pale, sickly. **green-crop**, *n*. a crop of food-stuff in the green state. **green drake**, *n*. the mayfly. **green-earth**, *n*. glauconite. **green eye**, *n*. jealousy. **green-eyed**, *a*. **the green-eyed monster**, jealousy. **green fat**, *n*. the green gelatinous part of the turtle, much esteemed by epicures. **greenfield**, *a*. applied to a development site which has not previously been built on. **greenfinch**, *n*. a common British singing-bird with green and gold plumage. **green fingers**, *n.pl*. skill at growing plants. **greenfly**, *n*. the green aphid, destructive to plants. **greengrocer**, *n*. a retailer of fresh vegetables and fruit. **greengrocery**, *n*. **green heart**, *n*. a hard-timbered W Indian tree. **greenhorn**, *n*. an inexperienced person. **greenhouse**, *n*. a glass-house for cultivating and preserving tender plants. **greenhouse effect**, *n*. the increased temperature of the earth caused by man-made gases in the atmosphere trapping the sun's heat. **greenkeeper**, *n*. the person in charge of a golf-course. **green laver**, *n*. an edible seaweed. **green light**, *n*. a signal to proceed. **greenmail**, *n*. a business tactic whereby a company buys a large number of shares in another company with the threat of a takeover, thereby forcing the threatened company to repurchase the shares at a higher price. *v.t*. to practise this tactic. **greenmailer**, *n*. **green manuring**, *n*. the cultivation and ploughing-in of a crop of vetch, rape etc. **Green Paper**, *n*. a set of policy proposals issued by the government. **green pepper**, *n*. the green unripe fruit of the sweet pepper eaten raw or cooked. **green pound**, *n*. an adjustable financial unit of account used in agricultural dealings with EEC. **green-room**, *n*. a room in which actors or musicians wait during the intervals of their parts. **Green-sand**, *n*. sandstone, largely consisting of green-earth. **greenshank**, *n*. a large European sandpiper. **green-sickness**, *n*. chlorosis. **green stick**, *n*. a form of fracture occurring in children in which one side of the bone is broken and the other bent. **green-stone**, *n*. a greenish igneous rock; a kind of jade. **green-stuff**, *n*. green vegetables. **greensward**, *n*. an area of turf. **green tea**, *n*. tea prepared by drying with steam. **greenthumb** GREEN FINGERS. **greenwood**, *n*. a wood in summer; wood rendered greenish by the fungus *Chlorosplenium aeruginosum*. **greener**, *n*. (*sl*.) a novice; a black-leg. **greenery**, *n*. **greenie** (-ni), *n*. (*coll*.) a conservationist. **greening**, *n*. the act of becoming green; greenness; a kind of apple which is green when ripe. **greenish**, *a*. **greenishness**, *n*. **greenly**, *adv*. **greenness**, *n*. **greeny**, *a*.
greengage (grēn'gāj), *n*. a green, fine-flavoured variety of plum.
Greenwich (grin'ij, gren'-, -ich), *a*. pertaining to Greenwich, in SE London, or its meridian. **Greenwich mean**

greet 367 **grilse**

time, mean time for the meridian of Greenwich, adopted as the standard time in Great Britain and several other countries.

greet[1] (grēt), *v.t.* to address with a salutation at meeting; to hail; to meet. **greeting**, *n.* the act of saluting or welcoming; a salutation, welcome.

greet[2] (grēt), *v.i.* (*now chiefly Sc.*) to weep, to cry, to lament.

greffier (gref'iə), *n.* a registrar, clerk or notary.

gregarious (grigeə'riəs), *a.* living in flocks or herds; tending to associate, not solitary; growing in clusters or in association with others; sociable. **gregariously,** *adv.* **gregariousness,** *n.*

Gregorian (grigaw'riən), *a.* pertaining to or established or produced by Gregory. *n.* a Gregorian chant. **Gregorian calender,** *n.* the reformed calendar introduced by Pope Gregory XIII in 1582; hence **Gregorian epoch, style, year. Gregorian chant,** *n.* plainsong, introduced by Pope Gregory I, 590–604. **Gregorian telescope,** *n.* the first form of reflecting telescope, invented by James Gregory, *c.* 1663.

Gregory powder (greg'əri), *n.* the compound powder of rhubarb, magnesium carbonate and ginger, formerly used as an aperient. [James *Gregory*, 1758–1822, Scottish physician]

gremial (grē'miəl), *a.* of or pertaining to the lap or bosom; resident. *n.* in the Roman Catholic Church, an episcopal vestment covering the lap.

gremlin (grem'lin), *n.* (*sl.*) a supposed goblin that accompanies an aviator in the air, performing ill-natured tricks; any source of mischief.

grenade (grənād'), *n.* a small explosive shell thrown by hand or fired from a rifle; a glass shell which is thrown and shatters to release chemicals.

grenadier (grenədiə'), *n.* a member of a battalion chosen for long service and approved courage; now, the Grenadier Guards; a southern African weaver-bird with vivid red and black plumage.

grenadine[1] (gren'ədēn), *n.* a thin, gauzy, silk or woollen fabric for women's dresses etc.

grenadine[2] (gren'ədēn), *n.* a dish of veal or poultry fillets, larded and glazed; a pomegranate syrup.

gressorial (grəsaw'riəl), *a.* adapted for walking, applied to the feet of some birds.

Gretna Green see MARRIAGE.

grew (groo), *past* GROW.

grey (grā), *a.* of a colour between black and white; dull, dark, dismal; hoary with age; ancient; mature, experienced. *n.* a grey colour, grey pigment; twilight, cold, sunless light; grey clothes; a grey animal, esp. a horse; (*pl.*) GREYS. *v.t., v.i.* to make or become grey. **grey area,** *n.* the area midway between two extremes; an issue or situation that is not clear-cut. **greybeard,** *n.* an old man; a large earthen jar for spirit. *a.* having a grey beard. **greybearded,** *a.* grey drake, *n.* a species of *Ephemera.* **grey eminence** EMINENCE. **grey falcon,** *n.* the hen-harrier; also the peregrine falcon. **Grey Friar,** *n.* a Franciscan friar. **grey goose,** *n.* the grey lag. **grey-haired, -headed,** *a.* having grey hair; old, time-worn. **greyhead,** *n.* a person with grey hair; an old male sperm whale. **grey-hen,** *n.* the female of the black grouse. **grey lag,** the European wild goose, the original of the domestic goose. **grey market,** *n.* the unofficial, but not necessarily illegal, selling of products, alongside the official market. **grey matter,** *n.* the greyish tissue of the brain and spinal cord containing the nerve cells; (*coll.*) intellect, intelligence. **grey nurse,** *n.* an E Australian shark. **grey squirrel,** *n.* a N American squirrel, *Sciurus carolinensis*, now established in Britain. **greystone,** *n.* a compact volcanic grey or greenish rock. **grey wether,** *n.* (*usu. pl.*) detached blocks of sarsen or sandstone occurring chiefly in SW England. **grey wolf,** *n.* the N American timber wolf. **greyish,** *a.* **greyly,** *adv.* **greyness,** *n.* **Greys,** *n.pl.* a British cavalry regiment, the 2nd Dragoons (orig. Scottish).

greyhound (grā'hownd), *n.* a swift-running dog, slender and keen-sighted. **ocean greyhound,** a swift ship. **greyhound racing,** *n.* racing greyhounds in pursuit of an electric hare.

greywacke (grā'waki), *n.* (*Geol.*) a gritstone or conglomerate, usu. consisting of small fragments of quartz, flinty slate etc.

grid (grid), *n.* a grating; a gridiron for cooking; a perforated or ridged plate used in a storage battery; a system of power transmission lines; a gridiron for docking ships; an electrode placed between an anode and a cathode to control the flow of current between them; a network of vertical and horizontal lines covering a map or plan. **grid bias,** *n.* voltage applied to the grid of a valve. **grid current,** *n.* the current passing between grid and cathode. **grid potentiometer,** *n.* a mechanism to facilitate critical adjustment of grid potential or grid bias.

griddle (grid'l), *n.* a circular iron plate for baking scones; (*Mining*) a wire-bottomed sieve or screen. *v.t.* (*Mining*) to screen with a griddle. **griddle-cake,** *n.* a scone baked on a griddle.

gride (grīd), *v.i.* to grind, scrape or jar (along, through etc.); to grate. *n.* a grating sound.

gridelin (grid'əlin), *n.* a grey-violet or purple colour.

gridiron (grid'īən), *n.* a grated iron utensil for broiling fish, flesh etc.; any similar framework; (*Theat.*) a framework above the stage supporting the apparatus for drop-scenes etc.; a series of parallel lines for shunting goods trains; wire network between cathode and anode; (*N Am.*) a football field. **gridiron manoeuvre,** *n.* a naval movement in which ships in two parallel columns cross each to the opposite column.

grief (grēf), *n.* deep sorrow due to loss or disaster or disappointment; regret, sadness; that which causes sorrow or sadness. **to come to grief,** to fail; to come to ruin. **grief-stricken,** *a.* suffering great sorrow. **griefless,** *a.* **grieflessness,** *n.*

grievance (grē'vəns), *n.* that which causes grief; a wrong, an injustice; a ground for complaint. **to air a grievance,** to state a cause of complaint.

grieve[1] (grēv), *v.t.* to annoy; to cause pain or sorrow to. *v.i.* to feel grief, to mourn, to sorrow. **griever,** *n.* **grievingly,** *adv.* **grievous,** *a.* causing grief; hard to be borne, distressing, oppressive; hurtful, injurious; atrocious, heinous. **grievous bodily harm (gbh),** in law, a serious injury to a person caused by another person. **grievously,** *adv.* **grievousness,** *n.*

grieve[2] (grēv), *n.* an overseer, steward or bailiff.

grievous GRIEVE[1].

griffin (grif'in), **-on**[1] (-ən), *n.* a fabulous creature, with the body and legs of a lion, the head and wings of an eagle and listening ears, emblematic of strength, agility and watchfulness. **griffin-like,** *a.*

griffon[2] (grif'ən), *n.* a large vulture, *Gyps fulvus*, usu. called **griffon-vulture.**

griffon[3] (grif'ən), *n.* a variety of dog like a terrier, with short, coarse hair.

grig (grig), *n.* a sand-eel or a young eel; a cricket or grasshopper; a lively or merry person.

grill (gril), *v.t.* to cook on a gridiron by radiant heat; to bake or torture as if by fire; (*coll.*) to interrogate severely. *n.* food broiled; a gridiron. **grill-room,** *n.* a room in a restaurant where meat etc. is grilled and served. **griller,** *n.*

grillage (gril'ij), *n.* a structure of sleepers and cross-beams forming a foundation in marshy soil.

grille (gril), *n.* an open grating, railing or screen of latticework, to enclose or shut a sacred or private place, or to fill an opening in a door etc.; (*Real Tennis*) a square opening in the end wall on the hazard side of the court.

grilse (grils), *n.* a young salmon when it first returns from the sea, usu. in its second year.

grim (grim), *a.* stern, relentless, harsh, savage, cruel; hideous, ghastly. **like grim death**, with determination, unyieldingly. **grimly**, *adv.* **grimness**, *n.*

grimace (grim′əs), *n.* a distortion of the features, a wry face, expressing disgust, contempt, affectation etc. *v.i.* to make grimaces. **grimacer**, *n.*

grimalkin (grimal′kin), *n.* an old cat, esp. a she-cat.

grime (grīm), *n.* dirt, smut; dirt deeply engrained. *v.t.* to dirty; to begrime. **grimy**, *a.* **grimily**, *adv.* **griminess**, *n.*

Grimm's law (grimz), *n.* a law formulated by Grimm respecting the modification of consonants in the most important of the Indo-European languages. [Jakob Grimm, 1785–1863, German philologist]

grin (grin), *v.i.* to show the teeth as in laughter, derision or pain; to smile broadly or maliciously; to stand wide open (as a joint). *v.t.* to express by grinning. *n.* the act of grinning; a smile with the teeth showing. **grinningly**, *adv.*

grind (grīnd), *v.t.* (*past*, *p.p.* **ground** (grownd)) to reduce to powder or fine particles by crushing and friction; to produce (flour etc.) by this process; to sharpen, smooth or polish by friction; to grate; to oppress; to work (a machine) by turning a handle; to study laboriously. *v.i.* to perform the act of grinding; to be ground; to drudge; to study laboriously. *n.* the act or process of grinding; hard and monotonous work or study; a turn at the handle of a machine or instrument. **grinder**, *n.* one who or that which grinds; a molar tooth; (*sl.*) a crammer; one who studies hard. **grindery**, *n.* a place where tools etc. are ground. **grindingly**, *adv.* **grindstone**, *n.* a flat circular stone, used for grinding tools. **to keep one's nose to the grindstone**, to stick to one's work.

gringo (gring′gō), *n.* (*pl.* **-gos**) (*esp. N Am., offensive*) a contemptuous name for an English-speaking foreigner.

grip[1] (grip), *n.* the act of seizing or holding firmly; a firm grasp, clutch; the power of grasping; a particular mode of clasping hands; the part of a weapon, instrument etc. that is held in the hand; a grasping or clutching part of a machine; a grappling-tool; (*pl.*) on a film set, a person employed to carry camera equipment; power of holding the attention; a hold-all. *v.t.* (*past*, *p.p.* **gripped**) to seize hold of; to grasp or hold tightly; to hold the attention of. *v.i.* to take firm hold. **to come, get to grips with**, to deal with, tackle (a problem etc.). **grip-brake**, *n.* a brake that is worked by gripping with the hand. **grip-sack**, *n.* (*N Am.*) a travelling bag, suitcase. **gripper**, *n.* **gripping**, *a.* having the power of holding the attention.

grip[2] (grip), *n.* a small ditch or furrow.

gripe (grīp), *v.t.* to affect the bowels of with colic pains. *v.i.* to get money by extortion; (*coll.*) to complain. *n.* a pinch, squeeze; the part by which anything is grasped; a handle or hilt; a brake applied to the wheel of a crane or derrick; (*pl.*) pains in the abdomen; (*coll.*) a complaint. **gripe water**, *n.* a solution given to a baby to ease the pain of colic. **griper**, *n.* **griping**, *a.* grasping, greedy; pinching the bowels. **gripingly**, *adv.*

grippe (grēp), *n.* a former name for influenza.

gripper GRIP[1].

Griqua (grē′kwə), *n.* (*pl.* **-qua, -quas**) one of a mixed ancestry people, descended from Dutch settlers and the Hottentot of southern Africa.

grisaille (grizäl′, -zī), *n.* a style of painting or staining in grey monochrome, esp. on stained glass.

grise[1] (grēs), GREE[2].

grise[2] (grīs), GRICE.

Griselda (grizel′də), *n.* a woman of great meekness and patience.

griseous (griz′iəs, gris′-), *a.* bluish-grey.

grisette (grizet′), *n.* a lively and attractive young French working woman.

griskin (gris′kin), *n.* the lean part of the loin of a bacon pig.

grisled (griz′ld), GRIZZLED under GRIZZLE.

grisly (griz′li), *a.* horrible, terrible, fearful, grim. **grisliness**, *n.*

grist[1] (grist), *n.* corn to be ground; corn which has been ground; malt for a brewing. **grist to the mill**, profitable business or gain. **grist-mill**, *n.* a mill for grinding corn.

grist[2] (grist), *n.* a size of rope as denoted by the number and thickness of the strands.

gristle (gris′l), *n.* cartilage, esp. when found in meat. **gristly**, *a.*

grit (grit), *n.* coarse rough particles such as sand; gritstone; the character of a stone as regards texture or grain; (*coll.*) courage, firmness, determination. *v.i.* (*past*, *p.p.* **gritted**) to be ground together; to give out a grating sound; to grate. *v.t.* to grind or grate (as the teeth); to cover with grit. **gritstone**, *n.* a coarse-grained sandstone. **gritter**, *n.* **gritty**, *a.* **grittiness**, *n.*

grits (grits), *n.pl.* husked and granulated but unground meal, esp. coarse oatmeal, or (*N Am.*) maize.

grizzle[1] (griz′l), *n.* a grey-haired man; grey hair; a grey colour. **grizzled**, *a.* grey, grey-haired; interspersed with grey. **grizzly**, *a.* grey, greyish. *n.* a grizzly-bear. **grizzly-bear**, *n.* a N American bear, *Ursus ferox*, of great size and strength.

grizzle[2] (griz′l), *v.i.* to worry, to fret; to whimper. *n.* one who grizzles. **grizzler**, *n.*

gro., (*abbr.*) gross.

groan (grōn), *v.i.* to utter a deep moaning sound, as in pain or grief; to suffer hardship; to complain. *v.t.* to utter with groans. *n.* a low moaning sound, expressing pain or sorrow; any low rumbling sound. **groaner**, *n.* **groaningly**, *adv.*

groat (grōt), *n.* (*formerly*) silver fourpenny-piece; any trifling sum. **not worth a groat**, worthless. **groatsworth**, *n.*

groats (grōts), *n.pl.* husked and crushed oats.

grocer (grō′sə), *n.* a dealer in foodstuffs and miscellaneous household supplies. **grocery**, *n.* (*usu. in pl.*) grocers' wares; a grocer's shop; (*N Am.*) a grog-shop.

grog (grog), *n.* a mixture of spirit and cold water; spirituous liquor; (*Austral., coll.*) any esp. cheap alcoholic drink. *v.i.* to drink grog. **to grog on**, (*Austral., coll.*) to take part in a session of heavy drinking. **grog-blossom**, *n.* a redness or eruption on the nose or face, due to excessive drinking. **grog on**, *n.* (*Austral., coll.*) a drinking party. **grog-shop**, *n.* a place where spirits are sold. **groggery**, *n.* (*N Am.*) a grog-shop. **groggy**, *a.* staggering; acting like one stupefied with drink; moving uneasily, as with tender feet or forelegs (said of a horse); unwell. **grogginess**, *n.*

grogram (grog′rəm), *n.* a coarse stuff of silk and mohair or silk and wool.

groin[1] (groin), *n.* the hollow in the human body where the thigh and the trunk unite; (*Arch.*) the edge formed by an intersection of vaults; the fillet or moulding covering this. *v.t.* to form (a roof) into groins; to furnish with groins. **groin-centring**, *n.* the centring of timber during construction. **groined**, *a.* **groining**, *n.*

groin[2] (groin), GROYNE.

Grolier (grō′liə, -iā), *n.* a book or binding from Grolier's collection. **Grolier design**, *n.* geometrical or arabesque ornament such as characterized Grolier's bindings. **Grolieresque** (-əresk′), *a.* [Jean Grolier, 1479–1565, French bibliophile]

grommet (grom′it), GRUMMET.

gromwell (grom′wəl), *n.* a genus of trailing herbs of the borage family, esp. *Lithospermum officinale*.

groom (groom), *n.* a person in charge of horses or a stable; one of several officers in the royal household; a bridegroom. *v.t.* to tend or care for, as a groom does a horse; to curry and brush; to prepare (someone) for a job or position. **well-groomed**, *a.* neatly or smartly got up. **groomsman**, *n.* an unmarried friend who attends the bridegroom.

groove (groov), *n.* a channel, furrow or long hollow, such

grope

as may be cut with a tool; the long spiral furrow cut into a gramophone record; a rut, routine; (*dated sl.*) an exalted state; (*dated sl.*) a satisfying experience. *v.t.* to cut or form a groove or grooves in. *v.i.* (*dated sl.*) to be delighted, pleased, satisfied etc. **grooved**, *a*. **groover**, *n*. **groovy**, *a*. of a groove; up-to-date; excellent; very good. **grooviness**, *n*.
grope (grōp), *v.i.* to search (after) something as in the dark, by feeling about with the hands; to feel one's way; to seek blindly. *v.t.* to seek out by feeling; (*sl.*) to fondle for sexual gratification. **groper**[1], *n*. **gropingly**, *adv*.
groper[2] (grō'pə), GROUPER[2].
grosbeak (grōs'bēk), *n*. a name given to several birds having thick bills, esp. the hawfinch.
groschen (grō'shən), *n*. an Austrian coin worth onehundredth of a schilling; a German 10-pfennig piece. [G]
grosgrain (grō'grān), *n*. a heavy ribbed silk or rayon fabric or ribbon. [F]
gros point (grō point), *n*. a stitch in embroidery covering two horizontal and two vertical threads; work done in this stitch. [F]
gross (grōs), *a*. big; fat, bloated, overfed; coarse, uncleanly; lacking fineness, dense, thick; unrefined; indelicate, obscene; flagrant, total, not net; general, not specific; *n*. 12 dozen; the main body, the mass; the sum total. *v.t.* to bring in as total revenue. **gross domestic product**, *n*. the total annual value of all goods and services produced domestically in a country. **gross national product**, *n*. the total annual value of all goods and services produced in a country, including net investment incomes from foreign nations. **in (the) gross**, in the bulk, wholesale; in a general way, on the whole. **gross up**, *v.t.* to convert a net figure to a gross figure. **gross weight**, *n*. the total weight of goods with the container. **grossly**, *adv*. **grossness**, *n*.
grossular (gros'ūlər), *a*. of or belonging to a gooseberry. *n*. a Siberian variety of garnet, sometimes called the gooseberry garnet.
grot[1] (grot), *n*. a grotto.
grot[2] (grot), *n*. (*coll.*) dirt, filth.
grotesque (grətesk'), *a*. irregular, extravagant or fantastic in form; ludicrous, absurd, bizarre. *n*. ornamentation consisting of fanciful or distorted figures of plants and animals; (*pl.*) whimsical figures or scenery; (*Print.*) a square-cut type without serifs. **grotesquely**, *adv*. **grotesqueness**, *n*. **grotesquerie** (-kəri), *n*.
grotto (grot'ō), *n*. (*pl.* **-ttoes**, **-ttos**) a small cave, esp. one that is picturesque; an artificial cave or cave-like room. **grottoed**, *a*.
grotty (gro'ti), *a*. (*coll.*) inferior, substandard; unattractive. **grottily**, *adv*. **grottiness**, *n*.
grouch (growch), *v.i.* to grumble. *n*. a discontented mood; a grumbler. **grouchy**, *a*. **grouchiness**, *n*.
ground[1] (grownd), *n*. the surface of the earth; a floor, pavement etc.; a region or tract of land; (*pl.*) private land attached to a house; firm, solid earth; the base or foundation; the background of a picture; (*Painting*) the first layer of paint; motive, origin, cause; (*pl.*) basis, valid reason; the first or fundamental principles; the extent of an inquiry or survey; (*pl.*) sediment, dregs, esp. of coffee; the position occupied by an army; in sport, the area allotted to a single player or to a side. *v.t.* to set upon or in the ground; to base or establish (on); to instruct thoroughly (in) the elementary principles of; to run (a ship) aground; to prevent an aeroplane from taking off. *v.i.* of a vessel, to strike the ground. **above ground** ABOVE. **below ground** BELOW. **common ground** COMMON. **down to the ground**, (*coll.*) thoroughly; in every respect. **forbidden ground**, an area or subject that must be avoided. **home ground** HOME. **to break (new) ground**, to take the first step; to make a start. **to cut the ground from under someone('s feet)**, (*coll.*) to anticipate someone's argu-

group

ments or actions etc., and render them meaningless or ineffective. **to get off the ground**, (*coll.*) to make a successful start. **to gain ground**, to advance, to meet with success. **to give ground**, to give way, to yield. **to have one's feet on the ground** FOOT. **to lose ground**, to be driven back, to give way; to lose advantage or credit; to decline. **to shift one's ground**, to change the basis or premises of one's reasoning. **to stand one's ground**, not to yield or give way. **ground-angling**, *n*. angling without a float, with the weight placed close to the hook. **groundash**, *n*. an ash sapling. **ground-bait**, *n*. bait thrown into the water to attract fish. **ground-bass**, *n*. a bass passage constantly repeated. **ground-box**, *n*. small box shrubs for edging garden plots and paths. **ground-colour**, *n*. the first coat of paint; the general colour or tone on which a design is painted. **ground control**, *n*. control of an aircraft or spacecraft by information transmitted from the ground. **groundcover**, *n*. low-growing plants and shrubs, esp. as used to cover a whole area; air support for ground troops. **ground floor**, *n*. the storey or rooms level with the exterior ground. **ground frost**, *n*. a ground temperature of 0° C or under. **ground game**, *n*. running game, as hares, rabbits etc. **ground-hog**, *n*. the aardvark; the American marmot. **ground-ice**, *n*. ice formed at the bottom of the water. **ground-ivy**, *n*. a labiate creeping plant with purple-blue flowers. **ground-note**, *n*. (*Mus.*) the note or fundamental bass on which a common chord is built. **ground-nut**, *n*. the pea-nut. **ground-oak**, *n*. an oak sapling. **ground-pine**, *n*. a herb with a resinous odour. **ground-plan**, *n*. a horizontal plan of a building at the ground level; an outline or general plan. **ground plane**, *n*. the horizontal plane of projection in perspective drawing. **ground-plot**, *n*. the ground upon which a building is placed. **ground-rent**, *n*. rent paid to a landlord for a building-site. **ground rule**, *n*. (*often pl.*) a basic rule of a game, procedure etc. **ground-sea**, *n*. a heavy sea or swell without apparent cause. **ground-sheet**, **-cover**, *n*. a waterproof sheet spread on the ground to protect against dampness. **groundsman**, *n*. a person employed to look after a sports field. **ground speed**, *n*. the speed of an aircraft relative to the ground. **ground-squirrel**, *n*. a genus of American burrowing squirrels, esp. the chipmunk. **ground staff**, *n*. the non-flying staff of an airport. **ground stroke**, *n*. a stroke (as in tennis) made by hitting a ball that has rebounded from the ground. **ground-swell**, *n*. a long, deep rolling of the sea. **ground-tackle**, *n*. of a vessel, the ropes and tackle concerned with mooring. **ground-tier**, *n*. the lower range of boxes in a theatre. **ground-torpedo**, *n*. one laid at the bottom of the sea. **ground water**, *n*. underground water consisting mainly of surface water that has seeped down. **groundwork**, *n*. that which forms the foundation or basis; a fundamental principle; **groundage**, *n*. dues paid for space occupied by a ship in port. **groundedly**, *adv*. **grounding**, *n*. instruction in the elements of a subject. **groundless**, *a*. without foundation, reason or warrant, baseless. **groundlessly**, *adv*. **groundlessness**, *n*. **groundling** (-ling), *n*. a spectator who stood on the floor of a theatre; a fish that keeps at the bottom of a river etc.; a creeping plant. **groundy**, *a*.
ground[2] (grownd), *a*. having been ground. **ground glass**, *n*. glass with the surface ground to make it nontransparent.
groundsel[1] (grownd'sl), *n*. a composite plant with pinnatifid leaves and small yellow flowers, esp. the common weed, *Senecio vulgaris*.
groundsel[2] (grownd'sl), **-sill** (-sil) *n*. the lowest timber of a building.
groundwork, groundy GROUND[1].
group (groop), *n*. the combination of several figures or objects to form a single mass; a cluster, assemblage; a number of persons or things classed together on account of certain resemblances; a grade in classification not corre-

sponding precisely to any regular division or sub-division; a series of minerals agreeing essentially in chemical composition; in the RAF, the highest subdivision of a Command; a pop group; a number of companies under one owner. *v.t.* to form into or place in a group; to put (an object) in close relation or contact (with); to bring together so as to produce a harmonious whole or effect. *v.i.* to form or fall into a group. **Group Captain**, *n.* an RAF officer equivalent to Colonel in the Army. **group practice**, *n.* a medical practice run by a partnership of general practitioners. **group therapy**, *n.* in psychiatry, the treatment of a group of patients in regular sessions where problems are shared in group discussion. **groupage**, *n.* **Grouper**, *n.* a member of the Oxford Group. **groupy, -ie**, *n.* (*sl.*) a female fan who travels with and is sexually available to the members of a pop group. **grouping**, *n.*
grouper[2] (groo'pə), *n.* name of certain Californian, Atlantic and Australian fish.
grouse (grows), *n.* (*pl.* **grouse**) a gallinaceous game-bird with feet more or less feathered, esp. *Lagopus scoticus*, the red grouse; also the black grouse, capercailzie and ptarmigan etc.; the flesh of these. *v.i.* to hunt or shoot grouse. **grousy**, *a.*
grouse[2] (grows), *v.i.* to grumble. *n.* a grievance. **grouser**, *n.*
grouse[3] (grows), *a.* (*Austral., coll.*) very good.
grout (growt), *n.* (*pl.*) dregs, grounds; a thin mortar to run into the joints of tiles, masonry etc.; a finishing coat of fine plaster. *v.t.* to fill up with grout. **grouter**, *n.* **grouting**, *n.* **grouty**, *a.* muddy, dirty.
grouter (grow'tə), *n.* (*Austral., coll.*) a bet in the game of two-up.
grove (grōv), *n.* a small wood; a cluster of trees. **groved**, *a.* **groveless**, *a.* **grovy**, *a.*
grovel (grov'l), *v.i.* to crawl; to prostrate oneself, to be humble or abject. **groveller**, *n.* **grovellingly**, *adv.*
grow (grō), *v.i.* (*past* **grew** (groo), *p.p.* **grown**) to increase in bulk by the assimilation of new matter into the living organism; to develop; to increase in number, degree, power etc.; to exist as a living thing; to spring up, to be produced; to pass into a certain state; to become rooted; *v.t.* to raise by cultivation; to produce. **to grow on one**, to impress one more and more. **to grow out of**, to issue from; to develop or result from; to outgrow. **to grow together**, to become closely united, to become incorporated in each other. **to grow up**, to arrive at maturity; to arise. **grow-**, **growing-bag**, *n.* a large bag, containing a growing medium (as compost) in which seeds can be germinated and plants grown to full size. **growable**, *a.* **grower**, *n.* one who or that which grows (*usu. in comb.*, as *free-grower*); a producer of corn, vegetables etc.; a cultivator. **growing**, *n., a.* **growing pains**, *n.pl.* pains in the limbs felt by young children. **growingly**, *adv.* **grown**, *a.* **grown-up**, *n., a.* (an) adult. **growth** (grōth), *n.* the act or process of growing; increase in number, extent, bulk, stature etc.; cultivation of vegetable produce; that which grows or is grown; (*Path.*) an abnormal formation, as a tumour.
growl (growl), *v.i.* to make a deep guttural sound as of anger; to speak angrily or gruffly; to rumble. *v.t.* to utter or express by a growl. *n.* a deep guttural sound like that made by an angry dog; a grumbling, complaint. **growler**, *n.* one who growls; a grumbler; an American fish, *Grystis salmonides*, from the sound it emits when landed; a small iceberg. **growlery**, *n.* a place to grumble in, a private room. **growlingly**, *adv.*
grown (grōn), etc. GROW.
groyne (groin), *n.* a breakwater on a foreshore, preventing erosion.
grub (grŭb), *v.i.* (*past*, *p.p.* **grubbed**) to dig by scratching or tearing up the ground superficially; to rummage; to drudge, toil. *v.t.* to dig (up or out); to clear (ground) of roots etc.; to find by searching. *n.* the larva of an insect, esp. of bees and wasps; a drudge; in cricket, a ball bowled along the ground; (*sl.*) food; **grub-axe, -hoe, -hook** etc. GRUBBING-AXE etc. **grub-screw**, *n.* a small headless screw. **grub-stake**, *n.* provisions etc., given to a prospector in return for a share of the finds. **grubber**, *n.* one who or that which grubs; an instrument for stirring up and clearing the soil; (*Austral., coll.*) in football, a kick that sends the ball along the ground, also **grub-, grubber-kick**. **grubbing**, *n., a.* **grubbing-axe, -hoe, -hook, -machine, -tool** etc., *n.* implements for grubbing up roots, stumps etc. **grubby**, *a.* full of grubs; dirty, grimy. **grubbily**, *adv.* **grubbiness**, *n.*
Grub Street (grŭb), *n.* (*collect.*) poor, mean or needy authors, or the region they live in (former name of Milton Street, Moorfields, London). *a.* of or pertaining to this kind of writer.
grudge (grŭj), *v.i.* to be unwilling or reluctant; to be envious, to cherish ill-will. *v.t.* to feel discontent or envy at; to give or take unwillingly or reluctantly. *n.* ill-will, a feeling of resentment; unwillingness. **grudger**, *n.* **grudgingly**, *adv.*
gruel (groo'əl), *n.* semi-liquid food made by boiling oatmeal or other meal in water or milk; any food of like consistency. **gruelling**, *n.* severe or harsh treatment. *a.* exacting, requiring fortitude.
gruesome (groo'səm), *a.* horrible, grisly, repulsive. **gruesomely**, *adv.* **gruesomeness**, *n.*
gruff (grŭf), *a.* of a rough, surly or harsh aspect or voice. **gruffish**, *a.* **gruffly**, *adv.* **gruffness**, *n.*
grumble (grŭm'bl), *v.i.* to murmur with discontent; to complain in a surly or muttering tone; to growl, to mutter, rumble. *v.t.* to express or utter in a complaining manner. *n.* the act of grumbling; a complaint. **grumbler**, *n.* **grumbling**, *n., a.* causing intermittent discomfort or discontent. **grumblingly**, *adv.* **grumbly**, *a.*
grume (groom), *n.* a thick, viscid fluid; a clot. **grumous**, *a.* thick; concreted; clotted, coagulated (of blood); (*Bot.*) divided into little clustered grains. **grumousness**, *n.*
grummet (grŭm'it), *n.* a ring or eyelet of metal, rubber or plastic designed to strengthen or protect the opening of a hole; a tube inserted through the eardrum to drain the middle ear.
grumous GRUME.
grump (grŭmp), *n.* a bad-tempered person; (*pl.*) a fit of bad-temper or sulkiness. **grumpy**, *a.* surly, cross, peevish, ill-tempered. **grumpily**, *adv.* **grumpiness**, *n.* **grumpish**, *a.*
Grundyism (grŭn'diizm), *n.* prudishness; a slavish respect for conventions in matters of sex. **Grundyish**, *a.* **Grundyist, Grundyite** (-it), *n.* [Mrs *Grundy*, a character in Morton's *Speed the Plough*, 1798, adopted as the type of conventional respectability]
grungy (grŭn'ji), *a.* (*sl.*) squalid, seedy.
grunsel (grŭn'sl), GROUNDSEL[1] and [2].
grunt (grŭnt), *v.i.* to make a deep guttural noise like a pig; to grumble, growl, complain. *v.t.* to express or utter in a grunting manner. *n.* a deep guttural sound, as of a pig. **grunter**, *n.* **gruntingly**, *adv.* **gruntling** (-ling), *n.* a young pig or hog.
gruntled (grŭn'təld), *a.* (*coll.*) pleased, contented.
Gruyère (groo'yeə, grē'-), *n.* a Swiss cheese made from cows' milk, pale-coloured, firm and full of cavities. [town in Switzerland]
gr. wt., (*abbr.*) gross weight.
grype (grīp), GRIPE.
gryphon (grif'ən), GRIFFIN[1].
grysbok (grĭs'bok), *n.* a speckled, reddish-brown southern African antelope.
GS, (*abbr.*) General Secretary; General Staff.
G-string, G-suit G.
GT., (*abbr.*) gran turismo, a touring car, usu. a fast sports car.
GU, (*abbr.*) genitourinary.

guacamole (gwakəmō′li), *n.* a Mexican dish of mashed avocado, citrus juice and seasonings.

guacharo (gwah′chərō), *n.* (*pl.* **-ros**) the oil-bird, a S American goatsucker, feeding on fruit.

guacho (gwah′chō), GAUCHO.

guaco (gwah′kō), *n.* (*pl.* **-cos**) a tropical American plant said to cure snake-bites.

guaiacum (gwī′əkəm), *n.* a genus of tropical American trees and shrubs, one of which furnishes *lignum vitae*; the wood of this genus; a drug made from the resin used medicinally.

guan (gwahn), *n.* a S American genus *Penelope*, gallinaceous birds allied to the curassou.

guana (gwah′nə), IGUANA.

guanaco (gwənah′kō), *n.* (*pl.* **-cos**) a wild llama inhabiting the Andes.

guano (gwah′nō), *n.* (*pl.* **-nos**) a manure, composed chiefly of the excrement of sea-fowl, brought from S America and the Pacific; a similar artificial manure. **guaniferous** (-nif′-), *a.* producing guano. **guanine** (-nen), *n.* a white amorphous substance found in guano, a constituent of nucleic acids.

guarana (gwərah′nə), *n.* the powdered seeds of *Paullinia sorbilis*, a Brazilian shrub. **guarana-bread, -paste,** *n.* bread or paste made from this.

guarantee (garəntē′), *n.* an engagement to see an agreement, duty or liability fulfilled; the act of guaranteeing; any security, warranty or surety given; the person to whom the guarantee is given; one who acts as a guarantor. *v.t.* to become guarantor or surety for; to undertake responsibility for the fulfilment of a promise, contract etc.; to pledge oneself or engage (that); to assure the continuance or permanence of; to undertake to secure (to another). **guaranteed,** *a.* warranted. **guarantor,** *n.* one who guarantees.

guaranty (gar′ənti), *n.* the act of guaranteeing, esp. an undertaking to be responsible for a debt or obligation of another person; that which guarantees, that on which a guarantee or security is based.

guard (gahd), *v.t.* to secure the safety of; to watch over, protect, defend (from or against); to prevent the escape of; to secure (against criticism etc.). *v.i.* to be cautious or take precautions (against). *n.* defence, protection, a state of vigilance, watch against attack, surprise etc.; a state, posture or act of defence, esp. in boxing, fencing, cricket etc.; a protector; a man or body of men on guard; an escort; (*N Am.*) prison warder; a contrivance to prevent injury, accident or loss; a man in charge of a railway train or a coach; the part of a sword-hilt which protects the hand; a watch-chain; a screen to prevent accident placed in front of a fire-place etc. **on** or **off one's guard,** prepared or unprepared for attack, surprise etc. **to mount guard,** to go on duty as a guard or sentinel. **to stand guard,** of a sentry, to keep watch. **guard cell,** *n.* in botany, either of the two cells that border the pore of a stoma and cause it to open and close. **guard-chain,** *n.* a chain for securing a watch, brooch etc. **guard-house, -room,** *n.* a house or room for those on guard or for prisoners. **guard-rail,** *n.* a rail to protect against falling off a deck etc.; a rail fixed inside the inner rail at curves, points etc., to prevent derailment. **guard-ring,** *n.* a keeper for a wedding-ring etc. **guard's van,** *n.* a carriage usu. at the rear of a train for the use of the guard. **guardant** (-dənt), *a.* (*Her.*) presenting the full face to the spectator. **guarded,** *a.* wary; reserved. **guardedly,** *adv.* **guardedness,** *n.* **guarder,** *n.* **guardfully,** *adv.* **guardless,** *a.*

guardian (gah′diən), *n.* one who has the charge, care or custody of any person or thing; a protector; in law, one who has the charge, custody and supervision of a person not legally capable of managing his own affairs. *a.* acting as a guardian or protector. **guardian angel,** *n.* an angel or spirit supposed to be assigned to a person as guardian and protector; (*pl.*) (**Guardian Angels**) an orig. American group which organizes vigilante partrols on underground trains. **guardianship,** *n.*

Guards (gahdz), *n.pl.* British household troops consisting of the Coldstream, Grenadier, Irish, Welsh and Scots Guards. **Guardsman,** *n.* a soldier in the Guards.

guava (gwah′və), *n.* the luscious fruit of various species of the tropical American myrtaceous genus *Psidium*; the trees on which they grow.

gubbins (gŭb′inz), *n.sing.* (*coll.*) an anonymous or trifling object; a gadget; rubbish.

gubernatorial (gūbənətaw′riəl), *a.* pertaining to governor, esp. of a US state.

guddle (gŭd′l), *v.t.*, *v.i.* (*Sc.*) to catch (fish) by groping with the hands. *n.* (*coll.*) a muddle, confusion.

gude (gid), (*Sc.*) GOOD.

gudgeon[1] (gŭj′ən), *n.* a small freshwater fish, *Gobio fluviatilis*, easily caught and largely used as bait; one easily taken in.

gudgeon[2] (gŭj′ən), *n.* the metallic journal-piece let into the end of a wooden shaft; the bearing of a shaft; an eye or socket in which a rudder turns. **gudgeon pin,** *n.* a metal pin that links the piston of an internal combustion engine to the little-end bearing of the connecting rod.

guelder rose (gel′də), *n.* a shrubby plant, *Viburnum opulus,* bearing ball-shaped bunches of white flowers.

Guelph, Guelf (gwelf), *n.* a member of the popular party in mediaeval Italy which aimed at national independence, and supported the Pope against the Ghibellines. **Guelphic,** *a.*

guenon (gənon′), *n.* any of various long-tailed African monkeys of the genus *Cercopithecus*.

guerdon (gœ′dən), *n.* (*poet.*) a reward, recompense. **guerdonless,** *a.*

guereza (ge′rizə), *n.* a black Abyssinian monkey, *Colobus guereza,* with a fringe of white hair and a bushy tail.

guerilla GUERRILLA.

guerite (gārēt′), *n.* (*Mil.*) a small loopholed tower, usu. on the point of a bastion, to hold a sentinel.

Guernsey, (gœn′zi), *n.* a close-fitting knitted or woven sweater usu. blue, worn by seamen; (*Austral.*) a similar garment, sometimes sleeveless, worn by football players. **Guernsey cow,** *n.* one of a breed of dairy cattle originating from Guernsey. **Guernsey lily,** *n.* a pink amaryllis, *Nerine sarniensis,* cultivated in Guernsey for the market. [one of the Channel Islands]

guerrilla, guerilla (gəril′ə), *n.* a member of a small irregular army; an irregular, petty war. *a.* belonging to or consisting of guerrillas; carried on in an irregular manner (of a war). **guerrilla strike,** *n.* a sudden industrial strike.

guess (ges), *v.t.* to judge or estimate on imperfect grounds, conjecture; to suppose on probable grounds; to conjecture rightly; (*N Am.*, *coll.*) to suppose; to believe. *v.i.* to form a conjecture, to judge at random. *n.* a conjecture; an opinion, estimate or supposition based on imperfect grounds. **guesstimate** (-timət), *n.* (*coll.*) an estimate made by guessing. *v.t.* (-māt), to estimate in this way. **guess-rope** GUEST-ROPE. **guess-work,** *n.* action or calculation based on guess; procedure by guessing. **guessable,** *a.* **guesser,** *n.* **guessingly,** *adv.*

guest (gest), *n.* a person received and entertained in the house or at the table of another; one who resides temporarily at a hotel or boarding-house; a parasitic animal or vegetable. *v.i.* to be a guest; to appear as a guest on a television or radio show etc. **paying guest,** a boarder. **guest house,** *n.* a boarding-house, a small hotel. **guest-night,** *n.* a night when visitors are entertained by a club etc. **guest-room,** *n.* a room for the accommodation of a guest. **guestship,** *n.*

guest- (-gest′-), **guess-rope** (ges′rōp), *n.* a rope hanging over the side of a ship for making fast a boat; also called **guest-, guess-warp.**

guff (gŭf), *n.* (*coll.*) nonsense, humbug.
guffaw (gəfaw'), *n.* a burst of loud or coarse laughter. *v.i.* to laugh loudly or coarsely.
guggle (gŭg'l), GURGLE.
guide (gīd), *v.t.* to direct, lead or conduct; to regulate, govern; to steer; to be the motive or criterion of (action, opinion etc.). *n.* one who leads another or points the way; a leader, conductor, esp. a person employed to conduct a party of tourists etc.; an adviser; a girl guide; anything adopted as a model or criterion; a guide-book; (*pl.*) a company formed for reconnoitring etc.; a ship by which a squadron or fleet regulate their movements; a device acting as indicator or regulating motion in a medicine. **guide-book**, *n.* a book for tourists, describing places of interest etc. **guide-dog**, *n.* a dog trained to lead a blind person. **guide-line**, *n.* a line drawn as a guide for further drawing or writing; a statement setting out future policy, courses of action etc. **guide-post**, *n.* a finger-post to show the way. **guide-rope** GUY[1]. **guideway**, *n.* on a machine, a groove, track or frame directing the motion of a part. **guidable**, *a.* **guidage**, *n.* guidance. **guidance**, *n.* the act of guiding; direction; government. **guided**, *a.* **guided missile**, *n.* a rocket- or jet-propelled projectile with a warhead, electronically guided to its target by remote control. **guideless**, *a.* **guider**, *n.* **guiding**, *a.* **guiding light, star**, *n.* person or thing used as a guide or model.
guidon (gī'dən), *n.* the forked or pointed flag of a troop of light cavalry; a standard-bearer.
guignol GRAND GUIGNOL.
guild (gild), *n.* a society or corporation belonging to the same trade or pursuit, combined for mutual aid and protection of interests; a club or fellowship. **guild-brother**, *n.* a fellow-member of a guild. **guild-hall**, *n.* a hall where a guild or corporation meets; a town-hall. **guildsman, guildswoman**, *n.* **Guild Socialism**, *n.* a form of socialism under which every industry would be organized as an autonomous guild. **guildry**, *n.* (*Sc.*) a guild, the corporation of a burgh royal.
guilder (gil'də), *n.* a coin formerly current in the Netherlands; GULDEN.
guile (gīl), *n.* deceit, craft, cunning. **guileful**, *a.* **guilefully**, *adv.* **guilefulness**, *n.* **guileless**, *a.* **guilelessly**, *adv.* **guilelessness**, *n.*
guillemot (gil'imot), *n.* any swimming bird of the genus *Alca* or *Uria*, with a short tail and pointed wings.
guilloche (gilōsh'), *n.* an ornament of intertwisted or interlaced bands.
guillotine (gil'ətēn), *n.* an apparatus for beheading persons consisting of an upright frame, down which a weighted blade slides in grooves; a machine for cutting thicknesses of paper etc.; (*Surg.*) an instrument for cutting tonsils; in Parliament, the curtailment of debate by fixing beforehand the hours when parts of a Bill must be voted on. *v.t.* to execute by guillotine; to cut with a guillotine. **guillotiner**, *n.* [F]
guilt (gilt), *n.* the state of having committed a crime or offence; criminality, culpability. **guilt complex**, *n.* (real or imagined) obsessive feeling of guilt or responsibility. **guiltless**, *a.* free from guilt; innocent; having no knowledge (of), inexperienced. **guiltlessly**, *adv.* **guiltlessness**, *n.* **guilty**, *a.* having committed a crime; criminal, culpable (of); characterized by guilt. **guiltily**, *adv.* **guiltiness**, *n.*
guimp GIMP[1].
guinea (gin'i), *n.* a former British gold coin with the nominal value of 21s.; a sum of money equivalent to a guinea £1.05p. **Guinea corn**, *n.* Indian millet, called also durra. **guinea-fowl, -hen**, *n.* a gallinaceous bird of the genus *Numida*, dark-grey with white spots, orig. from Africa. **Guinea grains** GRAIN[1]. **guinea-pig**, *n.* a small domesticated cavy, *Cavia cobaya*, native to Brazil; a person used as a subject for an experiment. **Guinea worm**, *n.* a nematode worm, *Filaria medinensis*, parasitic in the skin of the human feet etc.

guipure (gipūə'), *n.* a lace without a ground or mesh, the pattern being held in place by threads; a kind of gimp.
guise (gīz), *n.* external appearance; semblance, pretence. *v.t.* to dress up. *v.i.* to play the mummer. **guiser**, *n.* (*chiefly Sc.*) a person in fancy dress esp. at Halloween. **guising**, *n.*
guitar (gitah'), *n.* a (usu. six-) stringed instrument, somewhat like the violin in shape, the strings being plucked with the fingers or a plectrum. **guitar-fish**, *n.* a tropical sea-fish, one of the rays. **guitarist**, *n.*
gulag (goo'lag), *n.* the system of forced labour camps in the USSR, esp. as used to correct dissidents. [Rus. *G(lavnoye) U(pravleniye Ispravitelno-Trudovykh) Lag(erei)*, Main Administration for Corrective Labour Camps]
gulch (gŭlch), *n.* a deep rocky ravine.
gulden (gul'dən), *n.* one of various obsolete coins of Germany, Austria and Hungary; the monetary unit of Holland.
gules (gūlz), *n.* (*Her.*) a red colour, represented by vertical lines. *a.* red. **guly**, *a.*
gulf (gŭlf), *n.* a deep narrow bay; a deep hollow; a whirlpool, anything that swallows or engulfs; a profound depth, as of the ocean; an impassable chasm or difference. **Gulf Stream**, *n.* a warm ocean current flowing from the Gulf of Mexico across the Atlantic to the British Isles and Scandinavia. **gulf-weed**, *n.* a seaweed with berry-like air-vessels, found in the Gulf Stream, the Sargasso Sea etc. **gulfy**, *a.*
gull (gŭl), *n.* a long-winged, web-footed bird of the genus *Larus*, mostly marine in habitat; a simpleton, a dupe. *v.t.* to fool, trick. **gull wing**, *n.* of an aircraft wing, having a short inner section that slopes up from the fuselage and a long horizontal outer section; of a car door, opening upwards. **gullery**, *n.* a breeding-place for gulls. **gullish**, *a.*
gullet (gŭl'it), *n.* the throat; the oesophagus.
gullible (gŭl'ibl), *a.* credulous, easily deceived. **gullibility**, *n.* **gullibly**, *adv.*
gully (gŭl'i), *n.* a channel or ravine worn by water; a ditch, drain or gutter; a gully-hole; in cricket, (a fielder in) the position between slips and point. *v.t.* to wear a gully or gullies in. **gully-drain**, *n.* one connecting a gully-hole with a sewer. **gully-hole**, *n.* an opening into a drain at the side of a street. **gully-trap**, *n.* a grated trap to receive the discharge from rainwater pipes etc.
gulosity (gūlos'əti), *n.* gluttony, greediness.
gulp (gŭlp), *v.t.* to swallow (down) eagerly or in large draughts. *v.i.* to make a noise in swallowing, to gasp or choke. *n.* the act of gulping; a large mouthful; an effort to swallow, a catching or choking in the throat. **to gulp back**, esp. of tears, to keep back or suppress. **gulper**, *n.* **gulpingly**, *adv.*
gum[1] (gŭm), *n.* the fleshy tissue surrounding the necks of the teeth. **gumboil**, *n.* a boil or small abscess on the gums. **gum-rash**, *n.* a teething rash frequent in children. **gumshield**, *n.* a pad worn by sportsmen to protect the gum and teeth. **gummy**, *a.* toothless.
gum[2] (gŭm), *n.* a viscid substance which exudes from certain trees, and hardens, but is more or less soluble in water, used as an adhesive etc.; a plant or tree exuding this; (*coll.*) chewing-gum. *v.t.* (*past, p.p.* **gummed**) to cover or stiffen with gum; to fasten or stick with or as with gum. *v.i.* to exude gum; to become sticky or clogged. **to gum up the works**, (*coll.*) to interfere with, spoil or delay (something). **gum ammoniac** AMMONIAC under AMMONIA. **gum arabic**, *n.* a gum that exudes from certain acacias. **gum-boots**, *n.pl.* knee-high rubber boots. **gum-digger**, *n.* (*New Zealand*) one who digs for fossilized gum. **gum-dragon** TRAGACANTH. **gum-drop**, *n.* a gelatinous sweet containing gum arabic. **gum-elastic**, *n.* rubber. **gum-juniper**, *n.* sandarac. **gumnut**, *n.* (*Austral.*) the woody seed capsule of the eucalyptus. **gum-resin**, *n.* a vegetable

gum   373   **gut**

secretion consisting of a gum and a resin, e.g. gamboge. **gum shoe,** *n.* (*esp. N Am.*) a rubber overshoe; (*N Am., coll.*) a policeman or detective. *v.i.* (*N Am., coll.*) to move or act silently or stealthily. **gum-tree,** *n.* (*Austral.*) one of several species of eucalyptus. **to be up a gum-tree,** cornered, in a fix, brought to bay. **gummiferous** (-mif'-), *a.* producing gum. **gumming,** *n.* **gummous,** *a.* of the nature of gum. **gummosity** (-mos'-), *n.* **gummy,** *a.* sticky, viscous, adhesive; productive of or covered with gum. **gumminess,** *n.*

gum[3] (gŭm), **by gum,** a mild oath or expletive.

gumbo (gŭm'bō), *n.* (*pl.* **-bos**) the okra, *Hibiscus esculentus;* a soup or a dish made of okra pods; a Negro patois in Louisiana and the W Indies; a silty soil of the W and S US prairies that becomes very sticky when wet.

gumma (gŭm'ə), *n.* (*pl.* **-as, -ata** (-ah'tə)) a syphilitic tumour with gummy contents. **gummatous,** *a.*

gummiferous etc. GUM[2].

gumption (gŭmp'shən), *n.* common sense, practical shrewdness; capacity for getting on. **gumptious,** *a.*

gun (gŭn), *n.* a tubular weapon from which projectiles are shot by means of gunpowder or other explosive force, a cannon, rifle etc.; any device which projects something by force; a person with a gun, a member of a shooting party. *v.t.* to shoot; (*coll.*) to accelerate (a car) sharply. *a.* (*Austral., coll.*) of a person, skilled, expert. **great gun,** an important person. **son of a gun,** a rascal. **to beat or jump the gun,** (*coll.*) to begin prematurely. **to go great guns,** (*coll.*) to make vigorous and successful progress. **to gun for,** (*coll.*) to seek to kill, harm or destroy; to strive to obtain. **to spike someone's guns,** to frustrate someone's aims, spoil someone's chances. **to stick to one's guns,** to maintain an opinion in face of opposition. **gun-barrel,** *n.* the barrel or tube of a gun. **gunboat,** *n.* a warship of small size carrying heavy guns. **gunboat diplomacy,** *n.* the use of naval or military threats as part of international negotiations. **gun-carriage,** *n.* the apparatus upon which a cannon is mounted for service. **gun-cotton,** *n.* a highly explosive substance made by soaking cotton in nitric and sulphuric acids. **gun dog,** *n.* a dog trained to locate and retrieve game. **gunfight,** *n.* a fight using firearms. **gunfighter,** *n.* **gun-fire,** *n.* discharge of guns. **gun-harpoon,** *n.* a harpoon shot from a gun. **gun-house,** *n.* a shelter for a gun and the gunners against the enemy's fire. **gun-layer,** *n.* the gunner who sights and elevates a gun. **gun-lock,** *n.* the mechanism by which the charge in a gun is exploded. **gunman,** *n.* an armed gangster. **gun-metal,** *n.* an alloy of copper and tin or zinc from which cannon were formerly cast; a dark grey colour. **gun play,** *n.* the use of guns. **gunpoint,** *n.* the muzzle of a gun. **at gunpoint,** under the threat of being shot. **gunpowder,** *n.* a powdered mixture of saltpetre, carbon and sulphur, used as an explosive; gunpowder-tea. **Gunpowder Plot,** *n.* a plot to blow up the Houses of Parliament by gunpowder on 5 Nov. 1605. **gunpowder-tea,** *n.* a fine kind of green tea, each leaf of which is rolled up. **gun-reach,** *n.* (*N Am.*) gunshot. **gun-room,** *n.* a junior officers' mess on board a war vessel; a room where guns are stored. **gun-runner,** *n.* one who smuggles firearms into a country. **gun-running,** *n.* **gunshot,** *n.* a shot fired from a gun; the range of a gun. **gun-shy,** *a.* frightened at the report of firearms. **gunslinger** GUNFIGHTER. **gunsmith,** *n.* one who makes or repairs small firearms. **gun-stock,** *n.* the shaped block of wood to which the barrel of a gun is fixed. **gunless,** *a.* **gunner,** *n.* in the navy, a warrant officer in charge of ordnance or ordnance stores; in the army, an artilleryman, esp. a private. **gunnery,** *n.* the art of managing heavy guns; the science of artillery; practice with heavy guns. **gunnery-ship,** *n.* a vessel for training officers and men in gunnery. **gunning,** *n.* shooting game with a gun.

gunge (gŭnj), *n.* (*coll.*) an unpleasant sticky or dirty incrustation. **gungy,** *a.*

gung ho (gŭng hō'), excessively enthusiastic, over-zealous.

gunk (gŭngk), *n.* (*sl.*) an unpleasant sticky or slimy substance.

gunnel[1] (gŭn'l), *n.* the butter-fish, a blenny common on N Atlantic shorees.

gunnel[2] GUNWALE.

gunny (gŭn'i), *n.* a heavy coarse sackcloth, usu. of jute or hemp.

Gunter (gŭn'tə), *n.* a Gunter's scale; **Gunter's chain,** *n.* an ordinary surveyor's chain, 22 yd. (approx. 20 m) in length. **Gunter's line,** *n.* (*N Am.*) a logarithmic line on Gunter's scale, used for performing the multiplication or division of numbers. **Gunter's scale,** *n.* a flat, 2-ft. (60-cm) rule having scales of chords, tangents etc. and logarithmic lines, engraved on it, for solving navigational and surveying questions.

gunwale, gunnel[2] (gŭn'l), *n.* the upper edge of the side of a ship or boat.

gunyah (gŭn'yah), *n.* (*Austral.*) a hut usu. built of twigs and bark.

Guppie (gŭp'i), *n.* (*coll.*) an environmentally-conscious young professional person. [Green and *Yuppie*]

guppy (gŭp'i), *n.* a small brightly-coloured W Indian freshwater fish, now a common aquarium fish.

gurdwara (gœd'wahrə), *n.* a Sikh place of worship.

gurgle (gœ'gl), *v.i.* of running water, to make a purling or bubbling sound; to utter a similar noise, as a baby. *v.t.* to utter with such a sound. *n.* a gurgling sound.

gurgoyle (gœ'goil), GARGOYLE.

Gurkha (gœ'kə), *n.* a member of the dominant ethnic group in Nepal, of Hindu descent; (*pl.*) Indian soldiers of this ethnic group.

gurnard (gœ'nəd), **-net** (-nət), *n.* any fish of the genus *Trigla,* characterized by a large angular head, covered with bony plates, and three free pectoral rays.

gurrah (gŭ'rə), *n.* a plain, coarse Indian muslin.

gurry (gŭ'ri), *n.* in whale-fishing, fish-offal.

guru (gu'roo), *n.* a Hindu spiritual teacher or guide; a mentor.

gush (gŭsh), *v.i.* to flow or rush out copiously or with violence; to be uttered rapidly and copiously; to be effusive or affectedly sentimental. *v.t.* to pour (out) rapidly or copiously. *n.* a violent and copious issue of a fluid; the fluid thus emitted; an outburst; extravagant affectation of sentiment. **gusher,** *n.* one who or that which gushes; an oil-well that discharges without requiring pumps. **gushing,** *n., a.* **gushingly,** *adv.* **gushy,** *a.*

gusset (gŭs'it), *n.* a small angular piece of cloth inserted in a garment to enlarge or strengthen some part. **gusseted,** *a.*

gust (gŭst), *n.* a short but violent rush of wind; a squall; an outburst of passion. *v.i.* of wind, to blow in gusts. **gusty,** *a.* **gustily,** *adv.*

gustation (gŭstā'shən), *n.* the act of tasting; the sense of taste. **gustative, gustatory,** *a.* of or pertaining to gustation. **gustatory nerve,** *n.* the lingual nerve upon which taste depends.

gusto (gŭs'tō), *n.* zest, enjoyment, pleasure; flavour, relish.

gusty etc. GUST.

gut (gŭt), *n.* the intestinal canal; (*pl.*) the intestines; an intestine or a part of the alimentary canal; (*pl.*) the belly or the stomach as symbol of gluttony; (*pl.*) the core or essential part of something; catgut; fibre drawn from a silkworm before it spins its cocoon, used for fishing-lines; a narrow sound or strait; (*pl.*) (*coll.*) stamina, courage, persistence. *v.t.* (*past, p.p.* **gutted**) to eviscerate; to draw the entrails out of; to remove or destroy the contents of. *a.* of feelings etc., very strong, basic; instinctive. **to hate someone's guts,** (*coll.*) to dislike someone intensely. **to work** etc., **one's guts out,** to work etc. extremely hard.

gut-scraper, *n.* (*coll.*) a fiddler. **gutful,** *n.* (*Austral., coll.*)

gutta | 374 | **gyrostat**

more than enough of an unacceptable situation etc. **gutless**, *a*. cowardly. **guts**, (*coll*.) *n*. a glutton. *v.i.* to gormandize. **gutsy**, *a*. greedy; plucky. **gutsiness**, *n*. **gutser**, *n*. (*coll*.) a glutton. **gutted**, *a*. (*sl.*) fed up, disappointed. **gutty**, *a*. corpulent.
gutta (gŭt'ə), *n*. (*pl*. **-ttae** (-ē)) a drop; an ornament resembling a drop. **gutta serena** (sərē'nə), *n*. amaurosis. **guttate** (-tāt), *a*. speckled; containing drops. **gutté** (-ā), **guttee** (-tē'), *a*. (*Her*.) sprinkled with drops. **guttiferous** (-tif'-), *a*. (*Bot*.) yielding gum or resinous sap. **guttiform** (-fawm), *a*. drop-shaped. [L]
gutta-percha (gŭtəpœ'chə), *n*. the inspissated juice of the Malayan gutta-percha tree, forming a rubbery substance used for insulators etc. **gutty**, *n*. a gutta-percha golf-ball.
gutté GUTTA.
gutter (gŭt'ə), *n*. a channel at the side of a street or a trough below eaves for carrying away water; a channel worn by water; a trench, conduit etc. for the passage of water or other fluid; in printing, the space between the printed matter in two adjacent pages; a poor, sordid environment or lifestyle. *v.t.* to form channels or gutters in. *v.i.* to become channelled or worn with hollows, as a burning candle; to stream (down). **gutter press**, *n*. cheap and sensational newspapers. **guttersnipe**, *n*. a street urchin. **guttering**, *n*. the act of forming gutters; a gutter or arrangement of gutters; material for gutters; the act of falling in drops.
guttiform etc. GUTTA.
guttle (gŭt'l), *v.t., v.i.* to eat voraciously, to gobble. **guttler**, *n*.
guttural (gŭt'ərəl), *a*. pertaining to the throat; produced or formed in the throat. *n*. a sound or combination of sounds produced in the throat or the back part of the mouth. **gutturize, -ise**, *v.t.* to form in the throat. **gutturalism**, *n*. **gutturally**, *adv*.
gutturo- *comb. form* of the throat. **gutturo-nasal** (gŭt'ərō), *a*. pertaining to or produced by the throat and the nose.
gutty SEE GUT, GUTTA-PERCHA.
guv (gŭv), *n*. (*coll., esp. dial*.) used as a term of address to a man (in authority).
guy[1] (gī), *n*. a rope, chain etc., to steady a load in hoisting or to act as a stay. *v.t.* to guide or steady by means of a guy or guys. **guy-rope**, *n*.
guy[2] (gī), *n*. an effigy of *Guy* Fawkes burnt on 5 Nov. in memory of GUNPOWDER PLOT; a fright, a dowdy, a fantastic figure; (*coll*.) a man, a person. *v.t.* to ridicule. **a regular guy**, (*N Am., coll.*) a good fellow.
guzzle (gŭz'l), *v.t., v.i.* to eat or drink greedily. *n*. a debauch. **guzzler**, *n*.
gwyniad (gwin'iad), *n*. a salmonoid fish, *Coregonus pennantii*, found in Bala Lake and the English Lakes.
gybe (jīb), *v.i.* to swing from one side of the mast to the other (of a fore-and-aft sail); to take the wind on the other quarter (of a vessel). *v.t.* to shift (a sail) in this way; to make (a vessel) take the wind on the opposite quarter. *n*. the act or process of gybing.
gym (jim), *n*. short for GYMNASIUM; short for GYMNASTICS under GYMNAST. **gym shoe**, *n*. a plimsoll. **gymslip**, *n*. a tunic worn by schoolgirls as part of a school uniform.
gymkhana (jimkah'nə), *n*. a meeting for equestrian sports and games.
gymnasium (jimnā'ziəm), *n*. (*pl*. **-ia** (-ə), **-iums**) a building or room where athletic exercises are performed; **Gymnasium** (gimnah'ziəm), in Germany, a school of the highest grade preparatory to the universities. **gymnasial**, *a*. **gymnasiast** (-ast), *n*.
gymnast (jim'nast), *n*. an expert in gymnastic exercises. **gymnastic** (-nas'-), *a*. of or pertaining to gymnastics; involving athletic effort; involving great mental agility. *n*. (*usu. in pl.*) a course of instruction, discipline or exercise for the development of body or mind; exercises for the development of bodily strength and agility. **gymnastically**, *adv*.
gymno-, *comb. form* naked; destitute of protective covering.
gymnocarpus (jimnōkah'pəs), *a*. of plants, having the fruit or spore-bearing parts bare.
gymnorhinal (jimnōri'nəl), *a*. of birds, having the nostrils naked or unfeathered.
gymnosperm (jim'nōspœm), *n*. one of a class of plants having naked seeds, as the pine. **gymnospermous**, *a*.
gymnospore (jim'nōspaw), *n*. of plants, a naked spore. **gymnosporous** (-spaw'-), *a*.
gymnotus (jimnō'təs), *n*. (*pl*. **-ti** (-ī)) an electric eel.
gymp (jimp), JIMP.
gymslip GYM.
gynaeceum (jīnisē'əm, gī-), *n*. (*Bot*.) the female organs in a plant.
gynaeco- *comb. form* pertaining to women.
gynaecocracy (jīnikok'rəsi, gī-), *n*. government by women. **gynaecocrat** (-krat), *n*. **gynaecocratic**, *a*.
gynaecology (gīnikol'əji), *n*. the branch of medicine dealing with women's diseases. **gynaecological** (-loj'-), *a*. **gynaecologist**, *n*.
Gynandria (jīnan'driə), *n.pl.* a Linnaean class of plants, in which the stamens and pistils are united. **gynandrian**, **-drous**, *a*. **gynandromorph** (-drəmawf), *n*. an animal with both male and female characteristics. **gynandromorphically**, *adv*. **gynandromorphism**, *n*. **gynandromorphous**, *a*. **gynandry**, *n*.
gyneco- GYNAECO-.
gyn(o)-, *comb. form*. pertaining to women or females; pertaining to female reproductive organs.
gynobase (jī'nōbās, gī'-), *n*. enlargement of the receptacle of a flower, bearing the gynaeceum.
gynoecium (jīnisē'əm, gī-), GYNAECEUM.
gynophobia (jīnōfō'biə, gī-), *n*. a morbid fear of women.
gynophore (jī'nōfaw, gī-), *n*. (*Bot*.) the stalk of the ovary, as in the passion-flower.
-gynous, *comb. form* pertaining to women or to female reproductive organs.
gyp[1] (jip), *n*. a college servant at Cambridge and Durham Univs.
gyp[2] (jip), *v.t.* (*coll*.) to cheat, swindle. *n*. a swindle.
gyp[3] (jip), *n*. (*coll*.) pain. **to give someone gyp**, to cause (someone) pain.
gypsum (jip'səm), *n*. a mineral consisting of hydrous sulphate of lime, used to make plaster of Paris. **gypseous, gypsous**, *a*. **gypsiferous**, *a*.
gypsy (jip'si), GIPSY.
gyrate (jī'rət), *a*. (*Bot*.) coiled into a circle. *v.i.* (-rāt'), to revolve, whirl, in a circle or spiral. **gyration**, *n*. **gyrational**, *a*. **gyrator**, *n*. **gyratory**, *a*. **gyre**, *n*. a gyration, a revolution. *v.t., v.i.* to turn or move in a circle. **gyral**, *a*. **gyrally**, *adv*.
gyrfalcon (jœ'fawlkən), GERFALCON.
gyro (jī'rō), *n*. gyro-compass; gyroscope.
gyro-, *comb. form* round, curved; relating to revolutions.
gyro-compass (jī'rō-), *n*. a navigating compass consisting of an electrically driven gyroscope the axle of which orientates the sensitive element.
gyroidal (jīroi'dəl), *a*. arranged or moving spirally.
gyron (jī'ron), *n*. (*Her*.) a triangular charge formed by two lines meeting at the fesse-point. **gyronny**, *a*.
gyroplane (jī'rōplān), *n*. an aeroplane with freely rotating rotors in a horizontal plane, a rota-plane, a helicopter.
gyroscope (jī'rəskōp), *n*. a heavy fly-wheel rotated at very high speed and supported on an axis at right angles to the plane of the wheel, used as a compass etc. Any alteration of direction of the axis of rotation is resisted by the turning movement. **gyroscopic** (-skop'-), *a*.
gyrose (jī'rōz), *a*. (*Bot*.) marked with wavy lines.
gyrostabilizer, -iser (jīrōstāb'ilīzə), *n*. a gyroscopic device for steadying the roll of a vessel.
gyrostat (jī'rōstat), *n*. a modification of the gyroscope, for

illustrating the dynamics of rotating bodies. **gyrostatic,** *a.*
gyrus (jī'rəs), *n.* (*pl.* **gyri** (-rī)) (*Anat.*) a convolution of the brain.

gyve (jīv), *n.* (*usu. in pl.*) a fetter, a shackle.

H

H¹, h, the eighth letter of the English alphabet (*pl.* **aitches, Hs, H's**). **to drop one's hs**, to fail to give the breathing in words beginning with the letter *h;* to speak incorrectly.
H², (*chem. symbol*) hydrogen. **H-bomb**, *n.* a hydrogen bomb.
ha¹, hah (ha, hah), *int.* an exclamation denoting surprise, joy, suspicion or other sudden emotion; an inarticulate sound expressive of hesitation; when repeated, **ha ha!, hah, hah!** it denotes laughter. *n.* the exclamation so defined, or the sound of it. *v.i.* to express surprise, wonder etc.; to hesitate.
ha², (*abbr.*) hectare.
haar (hah), *n.* (*dial. esp. Sc.*) a wet mist, esp. a sea-fog.
habanera (habənea'rə), *n.* a Cuban dance in slow duple time.
habeas corpus (hä'biəs kaw'pəs), *n.* a writ to produce a prisoner before a court, with particulars of the day and cause of his arrest and detention, in order that the justice of this may be determined. **Habeas Corpus Act**, Act 31 Charles II, c. 2 (1679), authorizing this. [L, thou mayest have the body]
Habenaria (habənea'riə), *n.* a large genus of low terrestrial orchids bearing spikes of brilliant flowers.
haberdasher (hab'ədashə), *n.* a seller of small articles of apparel, as ribbons, laces, silks etc. **haberdashery**, *n.* a haberdasher's shop or business; the type of goods sold by a haberdasher.
habiliment (həbil'imənt), *n.* (*usu. pl.*) an item of clothing.
habilitate (həbil'itāt), *v.t.* to furnish with means, to finance. *v.i.* to become qualified (for). **habilitation**, *n.* **habilitator**, *n.* one who supplies means.
habit (hab'it), *n.* a permanent tendency to perform certain actions; a settled inclination, disposition or trend of mind; manner, practice, use or custom, acquired by frequent repetition; (*Bot., Zool.*) a characteristic manner of growth; an addiction; garb, dress costume, esp. one of a distinctive kind, as of a religious order. *v.t.* to inhabit; to habituate; to dress, to clothe. **habit-forming**, *a.* tending to become a habit or an addiction.
habitable (hab'itəbl), *a.* that may be dwelt in or inhabited. **habitability** (-bil'-), **habitableness**, *n.* **habitably**, *adv.* **habitant**, *n.* an inhabitant; (abētā'), an inhabitant of Lower Canada of French origin. **habitation**, *n.* the act of inhabiting; the state of being inhabited; a place of abode; natural region or locality.
habitat (hab'itat), *n.* the natural abode or locality of an animal or plant.
habitual (həbit'üəl), *a.* formed or acquired by habit; according to habit, usual; customary, constant; rendered permanent by use. **habitually**, *adv.* **habitualness**, *n.* **habituate**, *v.t.* to accustom; to make familiar by frequent repetition. **habituation**, *n.* **habitude** (hab'itūd), *n.* customary manner or mode, habit, aptitude, tendency, propensity; customary relation, familiarity.
habitué (həbit'üā), *n.* one who habitually frequents a place, esp. a place of amusement. **habituée**, *n.fem.*
haboob (həboob'), *n.* a high wind charged with sand that blows from the desert in the Sudan.
haček (hah'chek), *n.* a diacritical mark (ˇ) placed above a letter to modify its pronunciation, esp. in Slavonic languages.

hachure (hashooə'), *n.* (*usu. pl.*) short lines employed to represent half-tints and shadows, and on maps to denote hill-slopes. *v.t.* to cover or mark with hachures.
hacienda (hasien'də), *n.* (*Sp. Am.*) an estate; a farm or plantation, an establishment in the country for stock-raising etc., esp. with a residence for the proprietor.
hack¹ (hak), *v.t.* to cut irregularly or into small pieces; to chop, to notch; to cut unskilfully; to kick (a player's shins) at football; to mangle in uttering; (*sl.*) tolerate; cope with. *v.i.* to cut or chop away at anything; to emit a short dry cough. (*Comput.*) to use computers as a hobby, esp. in order to manipulate another computer system illegally. *n.* an irregular cut, a gash, a notch, a dent; the result of a kick (on the shins etc.); a mattock or large pick. **hacksaw**, *n.* a hand-saw used for cutting metal. **hacked (up), (off)**, *a.* **to be hacked up, off**, to be fed up, disgruntled, sickened. **hacker**, *n.* one who writes computer programs as a hobby; one who uses a computer to gain access to another computer system, often for illegal purposes; **hacking**, *a.* slashing, chopping, mangling; short, dry and intermittent (of a cough). **hacking jacket, coat**, *n.* a short jacket with a vent or vents at the back, worn for riding.
hack² (hak), *n.* a hackney, a horse for hire; a horse for general purposes, esp. as dist. from a hunter or racer; (*N Am.*) a hackney-carriage; one who earns money from routine literary or journalistic work. *v.t.* to let out for hire; to make a hack of; to make common, to hackney. *v.i.* to be let out for hire; to ride a hack or (*N Am.*) in a hack; to ride at the pace of an ordinary hack; to be common or vulgar; to live as a prostitute. **hackwork**, *n.* work done by a literary or journalistic hack. **hackery**, *n.* hackwork.
hack³ (hak), *n.* a rack or grated frame, a hatch; a drying-frame for fish; a frame for drying bricks; a feeding-rack or manger; (*Hawking*) a feeding-board for hawks, also the state of partial liberty in which young hawks are kept. *v.t.* to keep young hawks at hack.
hackamore (hak'əmaw), *n.* a rope with a loop used instead of a bit on a horse unused to a bridle. [Sp. *jáquima*]
hackberry (hak'beri), *n.* a N American tree of the genus *Celtis*, related to the elms; the hagberry, called also the nettle-tree, sugarberry, hog berry.
hackle (hak'l), *n.* an instrument with sharp steel spikes for dressing or combing (flax etc.); fibrous substance unspun, as raw silk; a long shining feather on a cock's neck; a fly for angling, dressed with this; the hairs on a cat's or dog's neck. *v.t.* to dress or comb (flax or hemp) with a hackle; to tie a hackle on (an artificial fly). **to make one's hackles rise**, to anger.
hackney (hak'ni), *n.* a horse kept for riding or driving; a horse kept for hire; a hackney-carriage. *v.t.* (*usu. in p.p.*) to make stale, trite or commonplace by overuse. **hackney-carriage**, *n.* a passenger road-vehicle licensed for hire. **hackneyed**, *a.* stale, overused, trite.
had (had), *past, p.p.* HAVE. **you've had it**, (*coll.*) there's no chance of your getting it now; you've had your chance and lost it; something unpleasant is going to happen to you.
haddock (had'ək), *n.* a sea-fish, *Gadus aeglefinus*, allied to the cod and fished for food.
hade (hād), *n.* the inclination of a fault or vein from the

vertical, complementary to the dip. **hading**, *n.*
Hades (hā'dēz), *n.* the lower world, the abode of the spirits of the dead. [Gr. *Hadēs*, *Aidēs*, the god of the lower world]
hadith (had'ith), *n.* the body of tradition relating to the sayings and doings of Mohammed.
hadj, ha, *n.* (*pl.* **-s**) a pilgrimage to Mecca. **hadji** (haj'i), *n.* (*pl.* **-es**) a Muslim who has performed the pilgrimage to Mecca; a title conferred on such a man.
hadron (had'ron), *n.* an elementary particle taking part in strong nuclear interactions.
haecceity (heksē'iti), *n.* (*Phil.*) the quality of being a particular thing, individuality.
haem (hēm), *n.* red organic compound containing iron, found in haemoglobin.
haema-, haemat-, haemato-, (*esp. N Am.*) **hema-** etc. *comb. form.* consisting of or containing blood; pertaining to or resembling blood.
haemal (hē'məl), *a.* of or pertaining to the blood; on or pertaining to the side of the body containing the heart and great blood-vessels.
haematic (himat'ik), *a.* of or pertaining to the blood; acting on the blood; containing blood; blood-coloured. *n.* a medicine acting on the blood. **haematics**, *n.* the branch of physiology which treats of the blood.
haematite (hē'mətīt, hem'-), (*esp. N Am.*) **hematite**, *n.* native sesquioxide of iron, occurring in two forms, red and brown, a valuable iron-ore.
haematology, (*esp. N Am.*) **hematology** (hēmətol'əji, hem-), *n.* the branch of physiology dealing with diseases of the blood.
haemo-, (*esp. N Am.*) **hemo-,** short form of HAEMATO-.
haemocyte (hē'məsīt, hem'-), (*esp. N Am.*) **hemocyte**, *n.* a blood cell, esp. of an invertebrate animal.
haemoglobin (hēməglō'bin), *n.* (*Chem.*) the colouring matter of the red corpuscles of the blood.
haemophilia, (*esp. N Am.*) **hemophilia** (hēməfil'iə), *n.* a constitutional tendency to haemorrhage. **haemophiliac** (-ak), *n., a.* (a person) suffering from this.
haemorrhage, (*esp. N Am.*) **hemorrhage** (hem'ərij), *n.* abnormal discharge of blood from the heart, arteries, veins or capillaries; (*fig.*) a serious, continuous depletion or loss.
haemorrhoids, (*esp. N Am.*) **hemorrhoids** (hem'əroidz), *n.pl.* (*Path.*) piles. **haemorrhoidal** (-roi'-), *a.*
haemostatic (hēməstat'ik), (*esp. N Am.*) **hemostatic**, *a.* serving to stop haemorrhage. *n.* a medicine for doing this; (*pl.*) the branch of physiology relating to the hydrostatics of blood. **haemostasia** (-stā'ziə), *n.* congestion of blood; stoppage of the flow of blood by means of constriction or compression of an artery.
hafiz (hah'fiz), *n.* one knowing the Koran by heart (a Muslim title).
hafnium (haf'niəm), *n.* metallic element occurring in zirconium ores.
haft (hahft), *n.* a handle, esp. of a dagger, knife or tool. *v.t.* to set in or fit with a handle.
hag (hag), *n.* a witch; a fury; an ugly old woman; an eel-like fish, *Myxine glutinosa,* of low organization, parasitic within the bodies of other fishes. **hag-ridden**, *a.* suffering from nightmares. **hag-weed**, *n.* the broom, *Cytisus scoparius.* **haggish**, *a.* **haggishly**, *adv.*
hagberry (hag'beri), *n.* the bird-cherry, *Prunus padus;* (*N Am.*) the hackberry.
Haggadah (həgah'də), *n.* the legendary part of the Talmud. **Haggadic, -ical** (-gad'-), *a.* **Haggadist**, *n.* **Haggadistic** (-dis'-), *a.*
haggard (hag'əd), *a.* wild-looking; anxious, careworn or gaunt from fatigue, trouble etc. *n.* a wild or untrained hawk. **haggardly**, *adv.* **haggardness**, *n.*
haggis (hag'is), *n.* a Scottish dish, made of liver, lights, heart etc., minced with onions, suet, oatmeal etc., boiled in a sheep's stomach.
haggle (hag'l), *v.t.* to hack, to mangle. *v.i.* to wrangle, esp. over a bargain. *n.* a wrangle about terms. **haggler**, *n.*
hagiarchy (hag'iahki), *n.* government by priests; the order of priests or holy men.
hagio-, *comb. form.* pertaining to saints or to holy things.
hagiocracy (hagiok'rəsi), *n.* government by priests or holy persons.
Hagiographa (hagiog'rəfə), *n.pl.* the third and last of the Jewish divisions of the Old Testament, comprising the books not included in 'the Law' and 'the Prophets,' i.e. consisting of the Psalms, Proverbs, Job, Song of Songs, Ruth, Lamentations, Ecclesiastes, Esther, Daniel, Ezra, Nehemiah and Chronicles. **hagiographical** (graf'-), *a.*
hagiography (hagiog'rəfi), *n.* biography of saints; a series of lives of saints; any biography that treats its subject as excessively good, noble etc.; the Hagiographa. **hagiographer, hagiographist,** *n.* **hagiographic, -ical** (-graf'-), *a.*
hagiolatry (hagiol'ətri), *n.* the worship of saints. **hagiolater**, *n.*
hagiology (hagiol'əji), *n.* literature relating to the lives and legends of saints; a work on the lives of saints. **hagiologic, -ical** (-loj'-), *a.* **hagiologist**, *n.*
hah, ha ha HA [1].
ha-ha (hah'hah), *n.* a hedge, fence or wall sunk between slopes.
haik (hīk, hāk), *n.* a strip of woollen or cotton cloth worn as an upper garment by Arabs over the head and body.
haiku (hī'koo), *n.* (*n.pl.* **haiku**) a Japanese verse of 17 syllables.
hail[1] (hāl), *n.* frozen rain or particles of frozen vapour falling in showers; (*fig.*) a great number of violent or abusive words etc. *v.i.* (*impers.*) to pour down hail; to come down with swiftness or violence. *v.t.* to pour down or out, as hail. **hailstone**, *n.* a single pellet of hail. **hailstorm**, *n.*
hail[2] (hāl), *v.t.* to call to (a person at a distance); to greet, designate (as); to welcome, to salute. *v.i.* to come (as a ship). *int.* an address of welcome or salutation. *n.* a salutation; a shout to attract attention. **hail fellow well met,** on easy, familiar terms. **hail Mary** AVE MARIA. **to hail a ship,** to call to those on board. **to hail from,** to come from (a place designated). **within hailing distance,** within the reach of the voice.
hair (heə), *n.* a filament composed of a tube of horny, fibrous substance, with a central medulla enclosing pigment cells, growing from the skin of an animal; (*collect.*) the mass of such filaments forming a covering for the head or the whole body; (*Bot.*) hair-like cellular processes on the surface of plants; something very small or fine, a hair's breadth; **by a hair,** by a very small margin. **keep your hair on,** don't lose your temper. **to get in one's hair,** to become a nuisance, cause irritation. **hair of the dog,** small amount of what has proven harmful, esp. of alcohol during a hangover. **to make one's hair curl, stand on end,** to shock extremely, to terrify. **to let one's hair down,** to talk without restraint; to forget ceremony. **not to turn a hair,** not to show any sign of fatigue or alarm. **to split hairs,** to quibble about trifles. **hair-breadth, hair's breadth,** *n.* a very minute distance. **hairbrush,** *n.* a brush for the hair. **haircut,** *n.* the act or style of cutting a man's hair. **hairdo,** *n.* (*pl.* **-dos**) a style of hairdressing. **hairdresser,** *n.* one who styles and cuts hair. **hairdressing,** *n.* **hair-dryer, -drier,** *n.* an electric device for drying the hair with warm air. **hair-grass,** *n.* tall, tufted grass of the genus *Aira.* **hairline,** *n.* the up-stroke of a letter; a fishing-line of horse-hair; the edge of the hair on a person's head, esp. the forehead. **hairpencil,** *n.* a fine brush made of hair for painting. **hairpiece,** *n.* a piece of false hair worn to change the style of or add to the natural hair. **hairpin,** *n.* a pin for fastening the hair. **hairpin bend,** a U-shaped turn in a road. **hair-raising,** *a.* inspiring fear. **hair-raisingly,** *adv.* **hair-shirt,** *n.* a shirt made of

horse-hair, worn as a penance. **hairsplitting**, *n*. the practice of making minute distinctions. *a*. quibbling. **hairspring**, *n*. the fine steel spring regulating the balance-wheel in a watch. **hair-stroke**, *n*. a hair-line in penmanship or on type, a serif. **hairstyle**, *n*. a particular way of arranging the hair. **hair-trigger**, *n*. a secondary trigger for releasing a main trigger by very slight pressure. **-haired**, *a*. (*comb. form*), as in *grey-haired, fair-haired*. **hairless**, *a*. **hairy**, *a*. covered with hair; consisting of or resembling hair; (*coll*.) difficult, exciting or dangerous. **hairiness**, *n*.
hajj, hajji (haj, haj′i), HADJ.
haka (hah′kə), *n*. a ceremonial Maori dance.
hake (hāk), *n*. a fish, *Merlucius vulgaris*, allied to the cod.
hakeem, hakim (həkēm′), *n*. (in Muslim countries) a physician; a governor,, a judge.
Halachah, Halakah (hәlah′kә), *n*. a body of traditional laws, supposed to be of Mosaic origin, included in the Mishna. **Halachic**, *a*. **Halachist**, *n*.
halal (həlahl′), *n*. meat which is prepared in accordance with Muslim law. *v.t*. to prepare (meat) in this way. *a*. of meat, prepared in this way.
halation (hәlā′shәn), *n*. (*Phot*.) a blurring in a negative caused by the reflection of a strong light from the back of the plate during exposure.
halberd (hal′bәd), *n*. a weapon consisting of a combination of spear and battle-axe, mounted on a pole 5 to 7 ft. in length.
halcyon (hal′siәn), *n*. the kingfisher; calm, peace; (*Zool*.) the genus of birds containing the Australasian kingfishers. *a*. peaceful, happy, pleasant. **halcyon days**, *n.pl*. a time of prosperity, peace and happiness.
hale[1] (hāl), *a*. sound and vigorous, robust, esp. in **hale and hearty**.
hale[2] (hāl), *v.t*. to drag, to draw violently.
half (hahf), *n*. (*pl*. **halves**) one of two equal parts into which a thing is or may be divided; a moiety; a halfback; (*coll*.) a half-pint. *a*. consisting of or forming a half. *adv*. to the extent or degree of a half; to a certain extent or degree; partially, imperfectly (*often in comb*.). *v.t*. to halve. **by halves**, badly, imperfectly. **half (past) one, two etc.**, half an hour past as in *half past one, 1.30*. **not half**, (*sl*.) not at all; (*iron*.) rather. **one's other half**, one's spouse or partner. **to cry halves**, to claim an equal share. **to go halves**, to share equally (with or in). **too clever, cocky etc. by half**, far too clever, cocky etc. **half-and-half**, *n*. a mixture of two substances in equal parts. **halfback**, *n*. (*Football, hockey etc.*) a position behind the forwards; one who plays in this position. **half-baked**, *a*. undercooked; not thought out, silly. **half-blood**, *n*. relationship between two persons having but one parent in common; one so related; a half-breed. **half board**, *n*. in hotels etc. the provision of bed, breakfast and one main meal per day. **half-breed, -caste**, *n*. an offspring of parents of different races. **half-brother, -sister**, *n*. sharing one parent only. **half-cock**, *n*. the position of the cock of a fire-arm when retained by the first notch, so that it cannot be moved by the trigger. **to go off half-cocked, at half-cock**, to fail as a result of being too impetuous. **half-cocked**, *a*. inadequately prepared. **half-crown**, *n*. (*formerly*) a British silver coin, value 12½p. **half-dozen**, *n*., *a*. six. **half-guinea**, *n*. (*formerly*) an English gold coin, value *c*. 50p. **half-hardy**, *a*. (of a plant) able to survive outside except in the severest frosts. **half-hearted**, *a*. luke-warm, indifferent. **half-heartedly**, *adv*. **half-heartedness**, *n*. **half-hour**, *n*. thirty minutes. **half-hourly**, *adv*., *a*. **half-landing**, *n*. a landing half-way up a flight of stairs. **half-length**, *n*. a portrait showing only the upper half of the body. *a*. consisting of only half the full length. **half-life**, *n*. the time taken for the radiation from a radioactive substance to decay to half its initial value. **half-light**, *n*. dim light as at dawn or dusk. **half mast**, *n*. the middle of or half-way up the mast, the position of a flag denoting respect for a dead person. **half-moon**, *n*. the moon at the quarters when but half is illuminated; a crescent-shaped thing. *a*. crescent-shaped. **half-nelson**, *n*. (*Wrestling*) a grip in which one arm is driven through the corresponding arm of an opponent and the hand pressed on the back of his neck. **half-note**, *n*. (*N Am*., *Mus*.) a minim. **half-pay**, *n*. a reduced allowance to an officer retired or not in active service. *a*. entitled to half-pay, on half-pay. **half-pint**, *n*., *a*. (someone) of small stature, size or consequence. **half-price**, *n*. reduced price, esp. for children when travelling or gaining admittance. **half-round**, *a*. semicircular. *n*. (*Arch*.) a semicircular moulding. **half-starved**, *a*. poorly fed, not having sufficient food. **half-step**, *n*. (*N Am*., *Mus*.) a semitone. **half-sword**, *n*. half-sword, *n*. half the length of a sword. **half term**, *n*. a short holiday half-way through a school term. **half-tide**, *n*. half the time of a tide, about six hours; the tide midway between flow and ebb. **half-timbered**, *a*. (*Build*.) having the foundations and principal supports of timber, and the interstices of the walls filled with plaster or brickwork. **half-time**, *n*. the interval between two halves of a game. **half-title**, *n*. a short title of a book, printed on the recto preceding the title page; a title printed on the recto preceding a separate section of a book. **halftone**, *a*. of or pertaining to a process by which printing blocks are made with the shaded portions in small dots, by photographing on to a prepared plate through a finely-ruled screen or grating. **half-track**, *n*. a vehicle running on or system consisting of one pair of wheels and one pair of caterpillar tracks. **half-truth**, *n*. a statement suppressing part of the truth. **half-volley**, *n*. a stroke in tennis in which a ball is hit immediately after it bounces. **halfway**, *adv*., *a*. in the middle; at half the distance. **halfway house**, *n*. an inn half-way between two towns etc.; a compromise; short-term accommodation provided for people leaving institutions such as prisons or mental hospitals as rehabilitation for going back into the community. *a*. equidistant from two extremes. **halfwit**, *n*. an idiot, a stupid person. **half-witted**, *a*. weak in the intellect, imbecile. **half-yearly**, *a*. happening every six months. *adv*. twice in every year.
halfpenny (hāp′ni), *n*. (*pl*. **-pennies**) (*formerly*) an English copper coin, half the value of an old penny. *a*. of the value or price of a halfpenny; trumpery, almost worthless. **halfpennyworth, ha'p'worth** (hā′pәth), *n*. as much as can be bought for a halfpenny; a very small amount.
halibut (hal′ibәt), *n*. a large flat-fish, *Hippoglossus vulgaris*, sometimes weighing from 300 to 400 lb. (135–180 kg), much esteemed for food. **halibut oil**, *n*. (*Med*.) extract from the liver of this fish, rich in vitamins A and D.
halicore (hәlik′әri), *n*. (*Zool*.) a genus of sirenians, comprising the dugong.
halide (hā′lid, hal′-), *n*. a binary salt of halogen.
halidom (hal′idәm), *n*. a holy relic or sacred thing; a holy place, a sanctuary; lands belonging to a religious foundation.
halitosis (-tō′sis), *n*. offensive breath.
hall (hawl), *n*. a large room, esp. one in which public meetings are held, the large public room in a palace, castle etc.; a large building in which public business is transacted; the building occupied by a guild etc.; (*Univ. etc.*) a large room in which scholars dine in common, hence the dinner itself; a manor-house or mansion; a room or passage at the entrance of a house; (*N Am*.) a connecting passage between rooms, a landing; a room forming the entry area of a house; the room in a mansion in which the servants dine etc.; (*Univ*.) a residential building for undergraduates or other students; a college or department of a university. **hallmark**, *n*. an official stamp stamped by the Goldsmiths' Company and Government assay offices on gold, silver and platinum articles to guarantee the standard; any mark of genuineness, worth or distinction. *v.t*. to stamp with this. **hallway**, *n*. (*esp. N Am*.) an entrance hall.

hallelujah, halleluiah (haliloo'yə), *n., int.* an ascription of praise to God, sung at the commencement of many psalms and in hymns of praise.
halliard (hal'iahd), HALYARD.
hallo HELLO.
halloo (həlō'), *n.* a call for attention; a call to cheer on dogs. *v.i.* to cheer dogs on with cries; to call out loudly. *v.t.* to shout loudly to; to cheer, or urge on; to chase with shouts.
hallow (hal'ō), *v.t.* to make sacred or worthy of reverence; to revere; to consecrate, to sanctify. **hallowed,** *a.* **Hallowe'en** (-ēn'), *n.* the eve of All-Hallows or All Saints' Day (31 Oct.).
Hallstatt (hawlstat), *a.* denoting the first period of the Iron Age, typified by weapons found in the necropolis of Hallstatt, Austria which illustrate the transition from the use of bronze to that of iron.
hallucinate (həloo'sināt), *v.i.* to experience hallucinations. *v.t.* to affect with hallucination. **hallucination,** *n.* an apparent sense perception or appearance of an external object arising from disorder of the brain, an illusion. **hallucinatory,** *a.* **hallucinogen** (-əjen), *n.* a drug etc. that induces hallucinations. **hallucinogenic** (-jen'-), *a.* inducing hallucinations.
hallux (hal'əks), *n.* (*pl.* **-uces**) the great toe; the digit corresponding to this (as in some birds).
halm (hahm), HAULM.
halo (hā'lō), *n.* (*pl.* **-loes, -los**) a luminous circle round the sun or moon caused by the refraction of light through mist; a nimbus or bright disk surrounding the heads of saints etc.; (*fig.*) an ideal glory investing an object. *v.t.* to surround with or as with a halo. *v.i.* to be formed into a halo.
hal(o)- *comb. form.* pertaining to salt or the sea; pertaining to a halogen.
halogen (hal'əjən), *n.* an element or other radical which by combination with a metal forms a salt; fluorine, chlorine, bromine and iodine.
halser (haw'zə), HAWSER.
halt[1] (hawlt), *v.i.* to doubt, to hesitate; to be defective, to fall or come short; to be faulty in measure or rhyme. **halting,** *a.* **haltingly,** *adv.*
halt[2] (hawlt), *n.* a stop or interruption in activity or motion; a minor stopping-place for trains, without a siding. *v.i.* to come to a stand, esp. of soldiers; (*Mil.* command) cease marching, come to a stand. *v.t.* to cause to stop. **to call a halt (to something),** to cause something to end. **halting-place,** *n.*
halter (hawl'tə), *n.* a headstall and strap or rope by which an animal is fastened; a rope to hang malefactors; hence, death by hanging; a halter neck. *v.t.* to put a halter upon; to tie up with a halter. **halter neck,** *n.*, *a.* (an item) of sleeveless female apparel with a strap around the neck resembling a halter.
haltere (hal'tiə), **halter** (-tə), *n.* (*pl.* **halteres** (-tiə'rēz)) either of two modified hind wings on dipterous insects, used for maintaining balance in flight.
halva(h) (hahl'vah), *n.* an Eastern sweetmeat often made with nuts, sesame seeds, honey, semolina etc.
halve (hahv), *v.t.* to divide into two equal parts; to share equally; to lessen by half, to reduce to half; (*Golf*) to win the same number of holes, or to reach a hole in the same number of strokes, as the other side.
halyard, halliard (hal'yəd), *n.* a rope or tackle for hoisting or lowering yards, sails or flags.
ham[1] (ham), *n.* the hind part of the thigh; (*usu. in pl.*) the thigh and buttock; the thigh of an animal, esp. of a pig, salted and dried in smoke, or otherwise cured. **hamstring,** *n.* a tendon behind the knee. *v.t.* to cripple by severing this.
ham[2], *n.* an amateur radio operator; a ham actor; the acting of a ham actor. *v.t., v.i.* to act in a clumsy or exaggerated way. **ham actor,** *n.* a bad, inexperienced or amateur actor with a tendency to overact. **ham-fisted, -handed,** (*adj.*) (*coll.*) clumsy; inept. **hammy,** *a.*
hamadryad (haməḍri'əd, -ad), *n.* (*pl.* **-ads, -des,** (-dēz)) (*Gr. Myth.*) a dryad or wood-nymph, who lived and died with the tree in which she lived; an Indian venomous snake; an Arabian and Abyssinian baboon.
hamate (hā'māt), *a.* hooked; furnished with a hook; hook-shaped.
hamburger, *n.* a flat cake of minced beef, fried and often served in a bun.
ham-fisted, -handed HAM[2].
Hamite (ham'it), *n.* a descendant of Ham, second son of Noah; belonging to the Hamitic stock, comprising the Egyptians and other African races. **Hamitic** (-mit'-), *a.* of or belonging to Ham, his supposed descendants, or the languages spoken by them.
hamlet (ham'lit), *n.* a small village; a little cluster of houses in the country.
hammer (ham'ə), *n.* a tool for driving nails, beating metals etc., consisting of a head, usu. of steel, fixed at right angles on a handle; a machine, part of a machine or other appliance, performing similar functions; the part of a gun-lock for exploding the charge; the striker of a bell etc.; an auctioneer's mallet; a metal ball attached to a handle by a long wire and thrown in an athletics contest; the contest in which it is thrown. *v.t.* to strike, beat or drive with or as with a hammer; to forge or form with a hammer; to work (out) laboriously in the mind; (*coll.*) to defeat easily. *v.i.* to work or beat with or as with a hammer. **hammer and sickle,** the emblem symbolic of worker and peasant adopted on the flag etc. of USSR. **hammer and tongs,** with great noise and vigour; violently. **to come under the hammer,** to be sold by auction. **to hammer home,** to stress greatly. **hammer-head,** *n.* the head of a hammer; a S African bird; a shark with a head like a hammer. **hammer-toe,** *n.* a malformation of the foot consisting of permanent angular fixing of one or more toes. **hammerwort,** *n.* the common pellitory (*Parietaria*). **hammering,** *n.* (*coll.*) a clear defeat.
hammock (ham'ək), *n.* a swinging or suspended bed made of canvas or network, and hung by hooks or other contrivance from a roof, ceiling, tree etc.
hamose (hā'mōs), **-mous,** *a.* (*Bot.*) curved like a hook; having hooks.
hamper[1] (ham'pə), *n.* a large, coarsely made wickerwork basket, with a cover; a package of groceries etc. put together for a special occasion.
hamper[2] (ham'pə), *v.t.* to impede the movement or free action of; to obstruct or impede (movement etc.); to hinder, to shackle, to fetter.
hamster (ham'stə), *n.* a rat-like rodent with large cheek-pouches in which it carries grain for food during hibernation, a common domestic pet.
h & c, (*abbr.*) hot and cold (water).
hand[1] (hand), *n.* the part used for grasping and holding, consisting of the palm and fingers, at the extremity of the human arm; a similar member terminating the limbs of monkeys; the end of a limb, esp. a fore-limb, in other animals, when serving as a prehensile organ; power of execution, skill, performance, handiwork; a pledge of marriage; possession, control, authority, power (*often in pl.*); source, person; (*pl.*) operatives, labourers, crew of a ship, players, persons engaged in a game etc.; a part, a share, a turn, an innings; an act of helping; a game at cards; the cards held by a player; a part in a game of cards; one of the players in a game of cards; style of workmanship, handwriting etc.; signature; a lineal measure of 4 ins. (10 cm), a palm (measuring horses); the pointer or index finger of a watch, clock or counter; side direction (right or left); a round of applause. **all hands,** (*Naut.*) the entire crew. **at first, second hand,** as the origi-

nal purchaser, owner, hearer etc., or as one deriving or learning through another party. **at hand**, near, close by; available. **by hand**, with the hands (as dist. from instruments or machines); by messenger or agent; by artificial rearing (of children or the young of the lower animals). **clean hands**, innocence, freedom from guilt. **from hand to hand**, from one person to another, bandied about. **from hand to mouth**, without provision for the future. **hand in glove**, on most intimate terms (with). **hand to hand**, at close quarters; in close fight. **hands off!** stand off! don't touch! **hands up!** show hands, those who assent etc.; show hands to preclude resistance. **in hand**, in a state of preparation or execution; in possession; under control. **on hand**, in present possession; in stock. **on one's hands**, (left) to one's responsibility; (left) unsold. **on the one hand, on the other**, from this point of view, from that. **out of hand**, done, ended, completed; at once, directly, extempore; out of control. **(the) upper hand**, dominance; mastery. **to bite the hand that feeds one**, to be ungrateful to a benefactor. **to change hands**, to become someone else's property. **to come cap in hand**, to come humbly, to come seeking a favour. **to force one's hand**, to make someone take action against his or her will. **to hand**, near; available. **to have a hand in**, to have a share in; to be mixed up with. **to have one's hands full**, to be fully occupied. **to know like the back of one's hand**, to be very familiar with. **to lay hands on**, to touch; to assault; to seize; to lay the hands on the head of (in ordination, confirmation etc.). **to lend a hand**, to help, to give assistance. **to show hand**. **one's hand**, to reveal one's plans, resources etc. **to take in hand**, to undertake, to attempt. **to take one in hand**, to deal with, to manage; to discipline. **to tie one's hands**, to prevent one from taking action. **to wash one's hands of**, to declare oneself no longer responsible for; to renounce for ever. **to win hands down**, without an effort, easily (of a jockey). **with a high hand**, arbitrarily, arrogantly. **handbag**, *n*. a small bag for carrying things with the hand. **handball**, *n*. a ball played with the hand; a game played with this between goals. **handbell**, *n*. a small bell rung with the hand, esp. one of a series played musically. **handbill**, *n*. a small printed sheet for circulating information. **handbook**, *n*. a small book or treatise on any subject, a compendium, a manual. **hand-brake**, *n*. a brake in a motor vehicle worked by a hand lever. **hand-car**, *n*. (*Rail.*) a small hand-propelled truck running on the rails, used by workers on the line. **handcuff**, *n*. (*usu. pl.*) a manacle for the wrists, consisting of a chain and locking-rings. *v.t.* to secure with handcuffs. **hand-grenade**, *n*. a grenade for throwing, by hand. **hand-gun**, *n*. a gun that can be held and fired in one hand. **handhold**, *n*. something for the hand to hold on by (in climbing etc.). **hand-line**, *n*. a line worked by the hand, esp. a fishing-line without a rod. **hand-made**, *a*. produced by hand, not by machinery. **hand-me-downs**, *n.pl.* (*coll.*) second-hand clothes. **handout**, *n*. information handed out to the press; financial help given, esp. to the poor. **hand-pick**, *v.t.* to choose carefully. **hand-picked**, *a*. **handrail**, a rail protecting stairs, landings etc. **handset**, *n*. the receiver of a telephone. **handshake**, *n*. a shake of another's hand as a greeting. **hands-off**, *a*. allowing things or people to follow their own course without intervening. **hands-on**, *a*. having, through practical experience. **handstand**, *n*. the act of balancing upright on one's hands. **handwriting**, *n*. writing done by hand; the style of writing peculiar to a person. **handed**, *a*. having a hand of a certain kind (*in comb.*, as *free-handed*). **-hander**, *comb. form.* blow, stroke etc. using the stated hand. **handful**, *n*. as much as can be held in the hand; a small number or quantity; (*coll.*) a troublesome person or task. **handless**, *a*.

hand[2] (hand), *v.t.* to give, deliver or transmit with the hand; to assist or conduct with the hand (into, out of etc.). **to hand down**, to transmit, to give in succession; to pass on. **to hand in**, to deliver to an office etc. **to hand it to**, to give someone credit, to acknowledge someone's superiority, victory etc.

handicap (han'dikap), *n*. a race or contest in which an allowance of time, distance or weight is made to the inferior competitors; the heavier conditions imposed on a superior competitor; any physical or mental disability; a disadvantage. *v.t.* to impose heavier weight or other disadvantageous conditions on a competitor; to put at a disadvantage. **handicapped**, *a*. having a physical or mental disability. [from the drawing of lots out of a hat or cap]

handicraft (han'dikrahft), *n*. skill in working with the hands; manual occupation or trade.

handily HANDY.

handiwork (han'diwœk), *n*. work done by the hands; the product of one's hands, labour or effort.

handkerchief (hang'kəchif), *n*. (*pl.* **-chiefs**, **-chieves** (-chēvz)) a piece of cloth for wiping the nose, face etc.

handle (han'dl), *v.t.* to touch, to feel with, to wield or use with the hands; to treat (well, ill etc.); to deal with, to manage, to treat of; to deal in. *v.i.* to work with the hands; to be handled (of a vehicle) to respond in a specified way to control by a driver. *n*. that part of a vessel, tool or instrument, by which it is grasped and held in the hand; an instrument or means by which anything is done; (*sl.*) name, title. **to fly off the handle**, to become angry, to go into a rage. **to give a handle**, to furnish an occasion or advantage that may be utilized. **handlebar**, *n*. a horizontal bar with grips at each end for steering a bicycle, motorcycle etc. **handlebar moustache**, *n*. a thick, wide moustache that curls upwards at each end. **handler**, *n*. one who handles; a coach or trainer. **handling**, *n*. the carriage shipment etc. of an item; the responsiveness of a vehicle or animal to a driver, rider, trainer's control; the manner in which something is treated.

handshake etc. HAND[1].

handsome (han'səm), *a*. well formed, finely featured, good-looking; noble; liberal, generous; ample, large. **handsomely**, *adv*. **handsomeness**, *n*.

handwriting HAND[1].

handy (han'di), *a*. ready or convenient to the hand; close at hand; dexterous, skilful with the hands; near, convenient. **handyman**, *n*. a man who does odd jobs; a man who is good at DIY. **handily**, *adv*. **handiness**, *n*.

handywork HANDIWORK.

hang[1] (hang), *v.t.* (*past, p.p.* **hung** (hŭng); for put to death and as imprecation **hanged**) to suspend; to attach loosely to a point of support above the centre of gravity; to fasten so as to leave movable (as a bell, gate, the body of a coach etc.); to suspend by the neck on a gallows as capital punishment; to cause to droop; to cover or decorate with anything suspended, as wallpaper; to attach, to fasten. *v.i.* to be suspended; to depend, to dangle, to swing; to cling; to be executed by hanging; to droop, to bend forwards; to project (over), to impend; to be fixed or suspended with attention; to depend (as on a basis etc.); to be in suspense. **hang! hang it! I'll be hanged!** forms of imprecation or exclamation. **to hang about, around**, to loiter, to loaf; to stay close. **to hang back**, to act reluctantly, to hesitate. **to hang down**, to decline, to droop. **to hang fire**, said of a fire-arm when the charge does not ignite immediately; to hesitate; to be delayed. **to hang heavy**, to go slowly (as time). **to hang in (there)**, (*esp. N Am. coll.*) to persist. **to hang on**, to grasp or hold; to persist; to depend on; (*coll.*) to wait. **to hang on to**, keep holding; retain, to spend time. **to hang on, upon**, to adhere closely to; to be a weight or drag on; to dwell upon, to listen closely to. **to hang out**, to suspend from a washing line etc.; to protrude loosely (of a tongue); (*sl.*) to frequent (a place). **to hang together**, to be closely united; to be consistent. **to hang up**, to suspend; to re-

place a telephone receiver and so end the call. **hangman**, *n.* a public executioner. **hangdog**, *a.* sullen, furtive, guilty-looking. **hang-glider**, *n.* a type of large kite controlled by the person suspended beneath it in a harness; the person who flies it. **hang-gliding**, *n.* **hangnail**, a sore at the foot of a finger- or toe-nail; see AG-NAIL. **hangout**, *n.* (*coll.*) haunt. **hangover**, *n.* someone or thing remaining, left over (from); (*coll.*) the after-effects of the excessive drinking of alcohol. **hang-up**, *n.* (*coll.*) a source of neurosis or anxiety. **hung over**, *a.* suffering from a hangover. **hung up**, *a.* nervous, tense, obsessed. **hanger**, *n.* one who hangs or causes to be hanged; that on which a thing is hung or suspended. **hanger-on**, *n.* one who hangs on or sticks to a person, place etc.; a dependant, a parasite. **hanging**, *n.* the act of suspending; an execution by the gallows; an exhibition; (*pl.*) fabrics hung up to cover or drape a room or window. *a.* suspended, dangling; steep, inclined. **hanging garden**, a garden rising in terraces one above the other.
hang² (hang), *n.* a slope, a declivity; mode of hanging; general tendency, drift or bent. **to get the hang of**, to understand the drift or connexion of; to get the knack of.
hangar (hang'ə), *n.* a large shed, esp. for aircraft.
hank (hangk), *n.* a coil or skein; two or more skeins of yarn, silk, wool or cotton, tied together; (*Naut.*) one of the hoops or rings to which a fore-and-aft sail is bent.
hanker (hang'kə), *v.i.* to have strong desire or longing (after). **hankering**, *n.* **hankeringly**, *adv.*
hanky, hankie (han'kē), *n.* (*pl.* **-kies**) (*coll.*) handkerchief.
hanky-panky (hang'kipang'ki), *n.* jugglery, trickery, fraud; (*coll.*) improper activity, esp. of a sexual kind.
Hanoverian (hanəvə'riən, -veə'-), *a.* of or pertaining to Hanover. *n.* a native or inhabitant of Hanover; an adherent of the House of Hanover, the dynasty that came to the throne of Great Britain and Ireland in 1714.
Hansard (han'sahd), *n.* the official report of the proceedings of the British Parliament, from the name of the compilers and printers (1774–1889).
Hanseatic League, *n.* a celebrated confederacy formed in the 13th cent. between certain German towns for the protection of commerce.
hansom (han'səm), *n.* a two-wheeled cab in which the driver's seat is behind the body, the reins passing over the hooded top.
Hants. (hants), (*abbr.*) Hampshire.
hanuman (han'uman), *n.* entellus monkey. **Hanuman**, *n.* Hindu monkey-god.
hapless, *a.* unhappy, unfortunate, luckless. **haplessly**, *adv.*
haphazard (haphaz'əd), *a.* happening by chance; random. **haphazardly**, *adv.* **haphazardness**, *n.*
ha'p'orth HALFPENNY.
happen (hap'n), *v.i.* to fall out; to happen; to chance (to); to light (upon). **happening**, *n.* (*usu. in pl.*) something that happens, a chance occurrence; an event.
happy (hap'i), *a.* lucky, fortunate; prosperous, successful; enjoying pleasure from the fruition or expectation of good; contented, satisfied; apt, felicitous; favourable; (*coll.*) slightly drunk. **happy-go-lucky**, *a.* carefree, thoughtless, improvident. **happy hour**, *n.* a period when a bar etc. sells drinks at reduced prices to attract customers. **happy hunting-ground**, (*coll.*) an area of activity offering easy rewards. **happily**, *adv.* **happiness**, *n.*
hara-kiri (hah'rəkē'ri, -ki'-), *n.* a Japanese method of suicide by disembowelling.
haram HAREM.
harangue (hərang'), *n.* a declamatory address to a large assembly; a noisy and vehement speech, a tirade. *v.i.* to make an harangue. *v.t.* to address in an angry, vehement way.
harass (ha'rəs), *v.t.* to torment by or as by importunity; to worry, to molest; to tire out with care, worry or attacks.

harassed, *a.* **harassment**, *n.* the act of harassing; the state of being harassed.
harbinger (hah'binjə), *n.* a precursor; one who or that which goes before and foretells what is coming.
harbour, (*N Am.*) **harbor** (hah'bə), *n.* a refuge, esp. a refuge or shelter for ships; a port or haven; an asylum, shelter, security. *v.t.* to shelter, to entertain, to cherish, to foster. *v.i.* to take shelter, to lodge. **harbour-master**, *n.* an official having charge of the berthing and mooring of ships in a harbour. **harbourage** (rij), *n.* shelter, harbour, refuge. **harbourer**, *n.* one who harbours another.
hard (hahd), *a.* firm, solid, compact; not yielding to pressure; difficult of accomplishment, comprehension or explanation; laborious, fatiguing, toilsome; intricate, perplexing; harsh, severe, galling, inflexible, cruel, unfeeling; sordid, miserly, stingy; difficult to bear, oppressive, unjust; (of a drug) highly addictive and harmful; (of a drink) alcoholic; coarse, unpalatable; rough and harsh to the palate, the touch etc.; containing mineral salts unfitting it for washing (of water); (*Phon.*) sounded gutturally (as *c* and *g* when not pronounced like *s* and *j*), aspirated (as *k, t, p,* compared with *g, d, b*). *adv.* forcibly, violently; strenuously, severely; with effort or difficulty; close, near. **hard and fast**, strict; that must be strictly adhered to. **hard by**, close by; close at hand. **hard put to it**, in straits, in difficulties. **hard upon**, close behind. **hard of hearing**, rather deaf. **to go hard with**, to fare ill with. **hardback**, *n., a.* (relating to or denoting) a book with a stiff binding. **hard-bitten**, *a.* tough, resolute. **hardboard**, *n.* a form of compressed fibreboard. **hard-boiled**, *a.* (*sl., coll.*) hard, sophisticated, unemotional, callous; shrewd, hard-headed. **hard case**, *n.* a tough or violent person. **hard cash**, *n.* money in the form of coins and notes. **hard cheese**, (*coll.*) hard luck. **hard copy**, (*Comput.*) copy that can be ready without the use of a word processor etc. **hard-core**, *n.* refuse stone, brickbats etc., crushed to form the substratum of a road; members of a group devoted to their beliefs and resistant to change. *a.* loyal to beliefs and resistant to change; of pornography, sexually explicit. **hard currency**, *n.* coin, metallic money; currency unlikely to depreciate suddenly or fluctuate in value. **hard drinker**, *n.* a drunkard. **hard-earned**, *a.* earned with difficulty. **hard-fought**, *a.* closely contested. **hard-got, -gotten**, *a.* hard-earned. **hard hat**, *n.* a protective helmet, such as worn on construction sites. **hard-head**, *n.* a hard-headed person; the menhaden and other fishes. **hard-headed**, *a.* matter-of-fact, practical, not sentimental. **hard-hearted**, *a.* cruel, unfeeling, pitiless. **hardheartedly**, *adv.* **hardheartedness**, *n.* **hard hit**, seriously damaged, especially by monetary losses; smitten with love. **hard-hitting**, *a.* forceful; effective. **hard labour**, *n.* enforced labour, esp. when added to imprisonment. **hard-liner**, *a.* (of a policy) uncompromising; extreme. **hard-liner**, *n.* a person following a hard-line policy. **hard lines**, *n.pl.* hard luck. **hard luck**, *n.* misfortune, lack of success. **hard-nosed**, *a.* (*coll.*) unsentimental; tough. **hard-on**, *n.* (*sl.*) an erect penis. **hard porn**, *n.* (*coll.*) sexually explicit pornography. **hard pressed**, *a.* closely pressed; in straits. **hard sell**, *n.* (*coll.*) aggressive selling, advertising etc. **hard shoulder**, *n.* an extra lane beside the nearside lane of a motorway etc. used for stopping in emergencies. **hard-up**, *a.* in great want, esp. of money; very poor. **hardware**, *n.* articles of metal, iron mongery etc.; items of machinery, weaponry etc. (*Comput.*) the physical apparatus of a computer system, contrasted with the programs for it (cp. SOFTWARE). **hard-wearing**, *a.* durable. **hard wheat**, *n.* a type of wheat with hard kernels that are high in gluten, used for making bread and pasta. **hard-won**, *a.* won with difficulty. **hardwood**, *n.* close-grained wood from deciduous trees, as dist. from pines etc. **hard-working**, *a.* working hard and diligently. **hard water**, *n.* water which from holding mineral salts in solution is unfit for washing

purposes. **hardish**, *a.* **hardly**, *adv.* with difficulty; harshly, rigorously; unfavourably; scarcely, not quite. **hardness**, *n.* **hardly earned**, *a.* earned with difficulty.
harden (hah'dn), *v.t.* to make hard or harder; to temper (tools); to confirm (in effrontery, wickedness, obstinacy etc.); to make firm; to make insensible, unfeeling or callous. *v.i.* to become hard or harder; to become unfeeling or inured; to become confirmed (in vice); (of prices) become stable. **hardener**, *n.*
hardship (hahd'ship), *n.* that which is hard to bear, as privation, suffering, toil, fatigue, oppression, injury, injustice.
hardy (hah'di), *a.* bold, over-confident, audacious; inured to fatigue, robust; (of plants) capable of bearing exposure to winter weather. **hardy annual**, an annual plant that may be sown in the open; a question that crops up annually or periodically. **hardihood** (-hud), *n.* boldness, daring; audacity, effrontery. **hardily**, *adv.* **hardiness**, *n.*
hare (heə), *n.* a long-eared short-tailed rodent of the genus *Lepus*, with cleft upper lip, similar to but larger than the rabbit. **hare and hounds**, a paper-chase. **to run with the hare and hunt with the hounds**, to keep in with both sides. **hare-bell**, *n.* the blue-bell of Scotland, the round-leaved bell-flower. **hare-brained**, *a.* (*coll.*) rash, giddy, flighty. **hare-lip**, *n.* a congenital fissure of the upper lip. **hare-lipped**, *a.* **hare's-foot**, *n.* a species of clover; a tropical American cork-tree.
harem (heə'rəm, hah'rēm, -rēm'), *n.* the apartments reserved for the women in a Muslim household; the occupants of these; a Muslim sanctuary (usu. **haram** (hərahm')).
haricot (har'ikō), *n.* a stew or ragout of meat, usu. mutton, with beans and other vegetables; the kidney or French bean, *Phaseolus vulgaris*. **haricot-bean**, *n.*
hari-kari (hahrikah'ri), HARA-KIRI.
hark (hahk), *v.i.* to listen. **to hark back**, to return to some point or matter from which a temporary digression has been made.
harken (hah'kən), HEARKEN.
harl (hahl), *n.* filaments of flax; fibrous substance; a barb of a feather, esp. one from a peacock's tail used in making artificial flies.
harlequin (hah'likwin), *n.* the leading character in a pantomime or harlequinade, adopted from Italian comedy; supposed to be invisible to the clown, he is dressed in a mask, parti-coloured and spangled clothes, and bears a magic wand. **harlequina** (-kwi'nə), **harlequiness** (-nes'), *n. fem.* **harlequinade** (-nād'), *n.* that part of a pantomime in which the harlequin and clown play the principal parts; an extravaganza; a piece of fantastic conduct. **harlequinesque** (-nesk'), *a.*
harlot (hah'lət), *n.* a prostitute. **harlotry**, *n.* the practices or trade of a harlot.
harm (hahm), *n.* hurt, injury, damage, evil. *v.t.* to injure, hurt or damage. **out of harm's way**, safe. **harmful**, *a.* hurtful, injurious, detrimental. **harmfully**, *adv.* **harmfulness**, *n.* **harmless**, *a.* not hurtful or injurious; uninjured, unharmed. **harmlessly**, *adv.* **harmlessness**, *n.*
harmala (hah'mələ), **harmel** (-məl), *n.* wild rue. **harmaline** (-lin), *n.* a white crystalline alkaloid obtained from the seeds of this.
harmattan (hah'mətən), *n.* a dry hot wind blowing from the interior of Africa to the upper Guinea coast.
harmonic (hahmon'ik), *a.* pertaining to harmony or music; concordant, harmonious. *n.* a harmonic tone; an overtone; (*pl.*) the science of musical sounds. **harmonic progression**, *n.* a series of numbers whose reciprocals are in arithmetical progression, as 1/5, 1/7, 1/9 etc. **harmonic quantities**, *n.pl.* numbers or quantities having this relation. **harmonic proportion**, *n.* the relation of three consecutive terms of a harmonic progression. **harmonic tones**, *n.pl.* tones produced by the vibration of aliquot parts of a string, column of air etc. **harmonical**, *a.* **harmonically**, *adv.* **harmonica**, *n.* a mouth-organ. **harmonious** HARMONY. **harmonist** HARMONY.
harmonium (hahmō'niəm), *n.* a keyed musical wind-instrument whose tones are produced by the forcing of air through free reeds.
harmony (har'məni), *n.* the just adaptation of parts to each other, so as to form a complete, symmetrical or pleasing whole; the agreeable combination of simultaneous sounds, music; an arrangement of musical parts for combination with an air or melody; the science dealing with musical combination of sounds; concord or agreement in views, sentiments etc.; a literary work showing the agreement between parallel or corresponding passages of different authors, esp. of the Gospels. **harmonious** (-mō'-), *a.* concordant, having harmony; having parts adapted and proportioned to each other, symmetrical; without discord or dissension; musical, tuneful. **harmoniously**, *adv.* **harmoniousness**, *n.* **harmonist** (har'mənist), *n.* one skilled in harmony; one who treats of and shows the agreement between corresponding passages of different authors. **harmonize, -ise**, *v.t.* to make harmonious; to arrange in musical concord; to add the proper accompaniment to; to adjust in proper proportions; to cause to agree (with). *v.i.* to agree in sound or effect; to live in peace and concord; to correspond, to be congruous (with). **harmonization**, *n.* **harmonizer**, *n.*
harness (hah'nis), *n.* the working gear of a horse or other draught-animal; an arrangement of straps etc. to hold a person or thing safely, e.g. in a pram, car seat etc.; *v.t.* to put harness on (a horse etc.); to utilize natural forces, e.g. water, for motive power. **in harness**, at work. **to die in harness**, to continue to the last in one's business or profession. **harness race, racing**, *n.* a type of trotting with horses harnessed to a two-wheeled trap. **harnesser**, *n.*
harp (hahp), *n.* a musical instrument of triangular shape, with strings which are plucked by the fingers. *v.i.* to play upon a harp. **to harp on**, to dwell incessantly upon anything. **harp-seal**, *n.* an Arctic seal with dark bands on its back resembling the former saddle shape of a harp. **harp-shell**, *n.* a tropical genus of molluscs. **harper, harpist**, *n.* a player on the harp.
harpoon (hahpoon'), *n.* a barbed, spearlike missile weapon with a line attached, used for striking and killing whales etc. *v.t.* to strike, catch or kill with a harpoon. **harpoon-gun**, *n.* a gun for firing a harpoon. **harpooner**, *n.*
harpsichord (hahp'sikawd), *n.* a stringed instrument with a keyboard actuating quills that pluck instead of hammers that strike, similar in form to the pianoforte, by which it was superseded.
harpy (hah'pi), *n.* a fabulous monster represented with the face of a woman, the body of a vulture and fingers armed with sharp claws; an extortioner, a rapacious person or animal.
harquebus (hah'kwibəs), *n.* an arquebus.
harridan (ha'ridən), *n.* a worn-out haggard old woman; an ill-tempered woman.
harrier[1] (ha'riə), *n.* a variety of dog, smaller than the fox-hound, used for hare-hunting by mounted huntsmen; a cross-country runner.
harrier[2] (ha'riə), *n.* one who harries or plunders; a falconoid bird of the genus *Circus*.
Harris Tweed® (ha'ris), *n.* a type of tweed woven in the Outer Hebrides.
Harrovian (hərō'viən), *a.* of or pertaining to Harrow School. *n.* a person educated there.
harrow[1] (ha'rō), *n.* a large rake or frame with teeth, drawn over ground to level it, stir the soil, destroy weeds or cover seed. *v.t.* to draw a harrow over; to torment, to cause anguish or suffering to. **harrowing**, *a.* causing anguish or torment.
harrow[2] (ha'rō), *v.t.* to plunder, to spoil, to harry, to

pillage.
harrumph (hərŭmf'), *v.i.* to make a sound as if clearing one's throat, often to indicate disapproval.
harry (ha'ri), *v.t.* to plunder, to pillage, to lay waste; to harass.
harsh (hahsh), *a.* rough to the touch or other senses; discordant, irritating; austere, morose, severe; rigorous, inclement; unfeeling. **harshly**, *adv.* **harshness**, *n.*
hart (haht), *n.* a stag, esp. a male red deer, from its fifth year onwards. **hart's-tongue**, *n.* a fern, with tongue-shaped leaves.
hartal (hah'tǎl), *n.* a boycott or protest in India, carried out by closing shops and suspending work, a strike.
hart(e)beest (hah'tibĕst), *n.* the S African *Alcephalus caama*, the commonest of the larger antelopes.
hartshorn (hahts'hawn), *n.* a preparation from shavings or chippings of the horns of the hart; smelling salts.
harum-scarum (heərəmskeə'rəm), *a.* giddy, hare-brained. *n.* a giddy, hare-brained person.
haruspex (hərŭs'pĕks), *n. (pl.* **-pices** (-pisēz)) an ancient Etruscan or Roman soothsayer who divined the will of the gods by inspecting the entrails of victims. **haruspicy** (-si), *n.*
harvest (hah'vist), *n.* the season of reaping and gathering crops, esp. of corn; ripe corn or other agricultural products gathered and stored; the yield of any natural product for the season; the product or result of any labour or conduct. *v.t., v.i.* to reap and gather in, as corn, grain etc. **harvest-bug, -louse, -mite, -tick**, *n.* a minute tick, mite or acaridan which burrows in or attaches itself to the skin during late summer and autumn, setting up an irritating itch. **harvest festival**, *n.* a religious service of thanksgiving for the harvest. **harvest home**, *n.* the close of harvesting; a merry-making in celebration of this. **harvest moon**, the moon at its full about the time of the autumnal equinox. **harvest mouse**, a very small fieldmouse, *Mus messorius*, which makes a nest usually among wheat-stalks. **harvest queen**, *n.* a person or image representing Ceres, the goddess of fruits, flowers etc. on the last day of harvest. **harvester**, *n.* a reaper; a reaping and binding machine; a harvest-bug.
has (haz), HAVE. **has-been**, *(coll.)* one whose days of success, fame etc. are past; a not-so-young person.
hash¹ (hash), *n.* meat, specially such as has already been cooked, cut into small pieces, mixed with vegetables and stewed etc.; a second preparation of old matter; *(coll.)* a mess, a muddle. *v.t.* to cut or chop up in small pieces; to mince. **to make a hash of**, *(coll.)* to make a mess of, to spoil utterly.
hash² (hash), *n.* a race resembling hare and hounds run by expatriate Britons throughout the world. **hashing**, *n.*
hash³ (hash), *n. (coll.)* hashish.
hashish (hash'ēsh), *n.* a resinous substance extracted from the tender tops and sprouts of Indian hemp used as an intoxicant for smoking, chewing etc.
haslet (haz'lit), **harslet** (hahs'-), *n.* a part of the entrails, liver, heart etc. of an animal, usu. a hog, eaten as a cold meat loaf.
hasp (hahsp), *n.* a fastening, esp. a clamp or bar hinged at one end, the other end passing over a staple, where it is secured by a pin, key or padlock.
hassle (has'l), *n. (coll.)* an argument; something causing difficulty or problems. *v.i.* to argue; to behave in a difficult or destructive way. *v.t.* to cause difficulty or problems for; to harass.
hassock (has'ək), *n.* a small stuffed footstool or cushion for kneeling on in church; a matted tuft of rank grass, a tussock.
hast (hast), HAVE.
hastate (has'tāt), *a.* triangular, like the head of a spear.
haste (hāst), *n.* hurry, speed of movement of action, urgency, precipitation. *v.i.* to make haste. **to make haste**, to be quick; to be in a hurry. **hasten** (hā'sn), *v.t.* to

cause to hurry; to urge or press on; to expedite. *v.i.* to move with haste or speed. **hasty**, *a.* hurried, quick; eager, precipitate; rash, inconsiderate; irritable; ripening early. **hastily**, *adv.* **hastiness**, *n.*
hat (hat), *n.* a covering for the head, usu. having a crown or top and a continuous brim; the dignity of a cardinal, from the broad-brimmed scarlet hat worn by cardinals; a specified function. *v.t.* to provide, fit or cover with a hat. **at the drop of a hat** DROP. **old hat**, outdated, old-fashioned; familiar and dull. **to pass, send round the hat**, to ask for subscriptions, charity etc. **to raise the hat to**, to salute. **to talk through one's hat**, to talk about something one does not understand. **hat-rack, -stand**, *n.* a contrivance or piece of furniture for hanging hats on. **hat trick**, *n.* the feat of taking three wickets with consecutive balls; the feat of one player scoring three goals etc. in one match; three successes in any area of activity. **hatless**, *a.* **hatter**, *n.* a maker of hats; (*Austral.*) a miner who works by himself; a bush recluse.
hatch¹ (hach), *n.* a half-door, a wicket; an opening in a roof for access to the outside; an opening in a wall between two rooms, a serving hatch; a flood-gate or a grated opening in a weir; (*Naut.*) a hatch-way or a trap-door or shutter to cover this; the door in a spacecraft or aircraft. **down the hatch!** cheers! drink up! **to be under hatches**, (*Naut.*) to be confined below; to be in a state of bondage or repression. **hatchback**, *n.* a car with a door at the back that opens upwards. **hatchway**, *n.* a large opening in the deck of a ship for lowering cargo etc.
hatch² (hach), *v.t.* to produce from eggs by incubation or artificial heat; to produce young from (eggs); to evolve, to contrive, to devise. *v.i.* to produce young (of eggs); to come out of the egg; to be developed from ova, cells of a brood-comb etc. *n.* act of hatching; a brood hatched. **to count one's chickens before they are hatched** CHICKEN. **hatchery**, *n.* a place where fish ova are hatched artificially.
hatch³ (hach), *v.t.* to mark with fine lines, parallel or crossing each other; to engrave, to chase with these. *n.* a fine line in drawing or engraving. **hatched moulding**, *n.* (*Arch.*) ornamentation with a series of cuts or grooves crossing each other, common in Norman work. **hatching**, *n.* shading produced by lines crossing each other at more or less acute angles.
hatchet (hach'it), *n.* a small axe with a short handle for use with one hand. **to bury, take up the hatchet**, to make peace or war. **hatchet-faced**, *a.* having a narrow face with sharp, prominent features. **hatchet job**, *n. (coll.)* a devastating attack on someone's reputation, argument, proposals etc. **hatchetman**, *n. (coll.)* a person hired to carry out violent or illegal tasks; a person appointed to sack people in an organization.
hatchment (hach'mənt), *n.* a funeral escutcheon or panel bearing the coat of arms of a deceased person placed on the front of his house, in a church etc.
hate (hāt), *n.* extreme dislike or aversion; detestation. *(coll.)* a hated thing or person. *v.t.* to dislike exceedingly; to abhor, to detest. **hatable**, *a.* **hateful**, *a.* causing hate; odious, detestable; feeling hatred. **hatefully**, *adv.* **hatefulness**, *n.* **hater**, *n.* **hatred** (-rid), *n.* exceeding dislike or aversion; active malevolence, animosity, enmity.
hath (hath), HAVE.
hatred HATE.
hatha yoga (hath'ə), YOGA.
hauberk (haw'bœk), *n.* a coat of mail, sometimes without sleeves, formed of interwoven steel rings.
haughty (haw'ti), *a.* proud, arrogant, disdainful, supercilious; proceeding from or expressing disdainful pride. **haughtily**, *adv.* **haughtiness**, *n.*
haul (hawl), *v.t.* to pull or drag with force; to transport or move by dragging. *v.i.* to pull or drag (at or upon) with force; to alter the course of a ship. *n.* a hauling, a pull; the drawing of a net; the amount that is taken or stolen at

haulm 384 **hay**

once; take, acquisition. **a long haul**, (*coll.*) a long and wearisome task, journey etc. **to haul over the coals** COAL. **to haul up**, to bring for trial in a court of law. **to haul the wind**, to turn the head of the ship nearer to that point from which the wind blows. **haulage** (ij), *n.* **hauler, haulier** (-liə), *n.* one who hauls, esp. a workman who hauls trucks to the bottom of the shaft in a coal-mine; a person or business that transports goods by lorry.

haulm, halm (hawm), *n.* a stem, a stalk; (*collect.*) the stems or stalks of peas, beans, potatoes etc.

haunch (hawnch), *n.* that part of the body between the ribs and the thigh; the buttock, the basal joint; the leg and loin of an animal as meat; (*Arch.*) the shoulder of an arch. **haunch bone**, *n.* **haunched**, *a.* having haunches.

haunt (hawnt), *v.t.* to frequent, to resort to often; to frequent the company of; to visit frequently, to recur to the mind of frequently in an irritating way; to frequent as a ghost or spirit. *v.i.* to stay or be frequently (about, in etc.). *n.* a place to which one often or customarily resorts; habit of frequenting a place. **haunted**, *a.* frequented by a ghost or spirit; troubled, anxious, obsessed. **haunter**, *n.* **haunting**, *a.* **hauntingly**, *adv.*

Hausa (how'zə), *n.* (*pl.* **Hausa**) a member of the Negroid race occupying a large area of W Africa, esp. N Nigeria; their language.

hausfrau (hows'frow), *n.* a housewife.

haustellum (hawstel'əm), *n.* (*pl.* **-lla** (-ə)) (*Zool.*) the sucking organ of certain insects and crustaceans. **haustellate** (haw'stəlat), *a.*

hautboy (hō'boi), *n.* (*formerly*) an oboe; a tall species of strawberry.

haute couture (ōt kutüə', -tüə'), *n.* the designing and making of exclusive trend-setting fashions; the designers and houses creating such fashions.

haute cuisine (ōt kwēzēn'), *n.* cooking of a very high standard.

hauteur (ōtœ'), *n.* haughtiness, lofty manners or demeanour.

hautton (ōtō'), *n.* high fashion; people of the most approved fashion, high society.

haut monde (ō mŏd'), *n.* high society.

Havana (həvan'ə), *n.* a cigar made at Havana or elsewhere in Cuba.

have (hav), *v.t.* (*2nd sing.* **hast** (hast), *3rd sing.* **has** (haz), **hath** (hath); *past* **had** (had), *2nd sing.* †**hadst** (hadst), *p.p.* **had**) to possess, to hold as owner; to enjoy, to suffer, to experience; to receive, to get, to obtain; to require, to claim; to hold mentally, to retain; to entertain; to maintain; to hold as part, appurtenance, quality etc., to contain, to comprise; to know, to understand, to be engaged in; to vanquish, to hold at one's mercy; to circumvent, to cheat; to bring forth, to bear; (*sl.*) to engage in sexual intercourse with. *v.i.* (*usu. in imper.*) to go, to betake oneself, to get (at, after, with etc.). *aux.* used with past participles to denote the completed action of verbs. **had I known**, if I had known. **have done**, stop, cease. **let him have it**, (*coll.*) punish, censure or abuse him; give it him. **you've had it** HAD. **the haves and the have-nots**, the propertied classes and the unpropertied. **to be had**, to be taken in. **to have a care**, to be cautious. **to have had it**, to be at the end of one's tether, exhausted, defeated; to have missed a chance, lost out. **to have it in one**, to be capable, have the ability. **to have it in for**, to want to harm somebody. **to have it off, away**, (*sl.*) to have sexual intercourse. **to have it out**, to settle a quarrel or dispute by fighting, debate etc. **to have it that**, to maintain or argue that. **to have nothing for it**, to have no alternative. **to have on**, to wear; to have planned; to deceive, trick. **to have someone up**, (*coll.*) to cause someone to be prosecuted in court; to censure someone.

havelock (hav'lok), *n.* a light covering for the cap hanging over the neck, worn as a protection against sunstroke.

haven (hā'vən), *n.* a port, a harbour; a station or refuge for ships; a refuge, an asylum.

haver (hāv'ə), *n.* (*Sc. and North.*) (*usu. pl.*) nonsense, foolish talk. *v.i.* to talk nonsense. **haverel** (rəl), **haverer**, *n.*

haversack (hav'əsak), *n.* a strong canvas bag to hold rations etc. on march or journey.

havildar (hav'ildah), *n.* a sergeant of a regiment of infantry in India.

having HAVE.

havoc (hav'ək), *n.* widespread destruction; devastation, waste; chaos. **to play havoc with**, to damage; to upset.

haw[1] (haw), *n.* the berry or fruit of the hawthorn; a hedge, an enclosed field or yard.

haw[2] (haw), *int.*, *n.* a sound expressive of hesitation in speaking. *v.i.* to utter this sound, to speak with hesitation.

haw[3] (haw), *n.* the nictitating membrane or third eyelid (of a horse etc.); (*often in pl.*) a disease of this characterized by inflammation, enlargement etc.

Hawaiian (həwi'ən), *a.* of or pertaining to Hawaii, an island in the N Pacific, its inhabitants or its language. *n.* a native or inhabitant of Hawaii, the language spoken by them.

hawfinch (haw'finch), *n.* the common grosbeak, *Coccothraustes coccothraustes*.

haw-haw (hawhaw'), HA-HA.

hawk[1] (hawk), *n.* a name for many species of raptorial birds allied to the falcons; a bird of prey with short, rounded wings used in falconry; a rapacious person, a sharper. *v.i.* to hunt birds etc. by means of trained hawks or falcons; to attack on the wing, to soar (at). **hawk-bell**, *n.* a small bell on the foot of a hawk. **hawk-eyed**, *a.* having sharp sight; quick to notice. **hawk-moth**, *n.* a moth of the family Sphingidae, the flight of which resembles a hawk in quest of prey. **hawk's-beard**, *n.* the composite genus *Crepis*, related to the hawkweeds. **hawkweed** (hawk'wēd), *n.* any plant of the composite genus *Hieracium*. **hawker**[1], *n.* are who practises the sport of hawking. **hawking**, *n.* falconry.

hawk[2] (hawk), *v.i.* to clear or try to clear the throat in a noisy manner. *v.t.* to force (up) phlegm from the throat. *n.* an effort to force up phlegm from the throat.

hawk[3] (hawk), *v.t.* to carry about for sale, to cry for sale; to carry or spread about. **hawker**[2] (haw'kə), *n.* one who travels around offering goods for sale in the street or from house to house.

hawk[4] (hawk), *n.* a plasterer's board with handle underneath, for carrying plaster, mortar etc.

hawse (hawz), *n.* (*Naut.*) that part of the bow in which the hawseholes are situated; the distance between a ship's head and the anchors by which she rides; the situation of the cables when a ship is moored from the bows with two anchors. **hawsehole**, *n.* a hole in each bow through which a cable or hawser can be passed.

hawser (haw'zə), *n.* a cable, used in warping and mooring.

hawthorn (haw'thawn), *n.* a thorny, rosaceous shrub or tree belonging to the genus *Crataegus*, bearing white or pink flowers which develop into haws. Other names are whitethorn and may.

hay[1] (hā), *n.* grass cut and dried for fodder. *v.t.* to make (grass etc.) into hay; to supply or feed with hay. *v.i.* to make hay. **to hit the hay**, (*sl.*) to go to bed. **to make hay**, to turn, toss and expose mown grass to the sun for drying. **to make hay while the sun shines**, to take advantage of every favourable opportunity. **haybox**, *n.* an air-tight box, with a thick layer of hay, used for keeping food hot, and for continuing the process of slow cooking after the food has been removed from the fire. **haycock**, *n.* a conical heap of hay. **hayfever**, *n.* a severe catarrh with asthmatic symptoms caused by an allergic reaction to the inhalation of pollen. **hayfork**, *n.* a fork for turning over or pitching hay. **hayloft**, *n.* a loft for storing hay. **haymaker**,

n. one employed in making hay; a machine for tossing hay; (*coll.*) a swinging punch. **haymaking**, *n.* **hayrick**, **haystack**, *n.* a pile of hay in the open air, built with a conical or ridged top, and thatched to keep it dry. **haywire**, *a.* (*coll.*) crazy, mad; chaotic, disordered.
hay² (hā), *n.* a country dance with a winding movement.
hazard (haz'əd), *n.* a game at dice; danger, risk; chance, casualty; the stake in gaming; one of the winning openings in a tennis-court; difficulties, obstacles, bunkers etc. on a golf-course; in billiards, a stroke putting a ball into a pocket. *v.t.* to risk; to expose to chance or danger; to run the risk of; to venture (an act, statement etc.). **chicken-hazard** CHICKEN. **to (run the) hazard**, to (run the) risk. **hazardous**, *a.* full of hazard, danger, or risk. **hazardously**, *adv.* **hazardousness**, *n.*
haze (hāz), *n.* want of transparency in the air, a very thin mist or vapour, usu. due to heat; obscurity or indistinctness of perception. *v.t.* to make hazy. **hazy**, *a.* misty; thick with haze; dim, vague, indistinct, obscure; (*sl.*) rather drunk; muddled. **hazily**, *adv.* **haziness**, *n.*
hazel (hā'zl), *n.* a shrub or small tree of the genus *Corylus*, esp. the European *C. avellana*, bearing the hazel nut; a reddish-brown colour. *a.* reddish-brown. **hazel-eyed**, *a.* having light-brown eyes. **hazelnut**, *n.* the fruit of the hazel, the cob-nut.
hazy HAZE.
HB, (*abbr.*) (of pencils) hard and black.
H-bomb H.
HC, (*abbr.*) House of Commons.
hdqrs, (*abbr.*) headquarters.
He, (*chem. symbol*) helium.
he (hē), *pron.* (*obj.* **him** (him), *poss.* **his** (hiz), *pl.* **they** (dhā), *obj.* **them** (dhem), *poss.* **their** (dheə)), the male person or animal referred to. *n.* a male person; a children's game of chasing to touch another player; (*in comb.*) male, as in *he-goat*. **he-man**, *n.* (*coll.*) a virile man.
head¹ (hed), *n.* the foremost part of the body of an animal, the uppermost in man, consisting of the skull, with the brain and the special sense-organs; any part, organ or thing of an analogous kind; a measure of length equal to a head, esp. in a horse race; the upper part of anything, the top; the upper end of a valley, lake, gulf etc.; the front part of a ship, plough, procession, column of troops etc.; a ship's toilet; a promontory; the capital of a pillar etc.; the part of a bed where the head rests; the more honourable end of a table etc.; the obverse of a coin or medal; the knobbed end of a nail etc.; the striking part of a tool; the part of a machine tool etc. that holds a drill or cutter; the device on a tape recorder that can record sound, or play back or erase recorded sound; the globular cluster of flowers or leaves at the top of a stem; the first or most honourable place, the forefront, the place of command; a chief, a ruler, a principal or leader; a head teacher of a school; a person, an individual; a single one (as of cattle); a main division, a topic, a category; a culmination, a crisis, a pitch; the ripened part of an ulcer or boil; froth on liquor; pressure of water available for driving mills; available steam-pressure; liberty, licence, freedom from restraint; an aptitude for something specified; the mind, the understanding, the intellect, esp. as distinguished from the feelings; one's life; (*sl.*) addict, devotee, fan. **from head to foot**, over the whole person. **head and shoulders**, by the height of the head and shoulders; by a great margin. **head over heels**, turning upside down; completely (in love). **off one's head**, out of one's mind; wildly excited, demented. **over someone's head**, beyond someone's understanding; appealing to a higher authority than someone. **to come to a head**, to suppurate (of an ulcer or boil); hence, to ripen; to reach a crisis or culminating point. **to give someone, let someone have his head**, to give liberty or licence to; to let (a horse) go as he pleases. **to go to one's head**, (of a success etc.) make one vain, arrogant etc. **to have one's head screwed on the right way**, to be sensible, well-balanced. **to hold one's head high**, to retain one's dignity. **to keep one's head**, to remain calm. **to keep one's head above water** WATER. **to lose one's head**, to be carried away by excitement; to lose one's presence of mind; to be decapitated. **to make neither head nor tail of**, to fail to understand. **to raise, rear its (ugly) head**, to become apparent, esp. in an ominous way. **to turn someone's head**, to cause someone to be vain or infatuated. **headache**, *n.* a neuralgic or other persistent pain in the head; (*coll.*) a source of worry. **headachy**, *a.* suffering from or tending to cause headache. **headband**, *n.* a fillet or band for the hair; a band at the top and bottom inside the back of a book; the band connecting a pair of receivers or ear-phones. **headbanger**, *n.* (*sl.*) a person who makes violent head movements in time to pop music; a stupid, crazy or violent person. **headboard**, *n.* a panel at the head of a bed. **head case**, *n.* (*coll.*) a mad or foolish person. **head count**, *n.* a count of all the people etc. present. **head-dress**, *n.* covering and ornaments for the head, esp. of a woman. **head-first**, *adv.* with the head in front (of a plunge); precipitately. **headgear**, *n.* the covering, dress or ornaments of the head; a bridle; machinery at the top of a mine shaft or boring. **headhunt**, *v.t., v.i.* to seek and recruit business executives. **head-hunters**, *n.pl.* several races or tribes, notably the Dyaks of Borneo and Celebes, so called from their practice of making hostile raids in order to secure human persons and heads as trophies; an agency that specializes in seeking and recruiting business executives. **head-hunting**, *n.* **headland**, *n.* a point of land projecting into the sea, a cape, a promontory; a ridge or strip of unploughed land at either end of a field, where the plough is turned. **headlight, -lamp**, *n.* the lamp carried at the front of a locomotive, motor vehicle etc. **headline**, *n.* the line at the head of a page or paragraph giving the title etc.; news set out in large, heavy type. **to hit the headlines**, to gain notoriety, to get notice in the press. **headlong**, *adv.* head first; violently, hastily, rashly. *a.* steep, precipitous; violent, precipitate; rash, thoughtless. **headman**, *n.* a chief, a leader, a head worker. **headmaster, headmistress**, *n.* the principal master or mistress at a school. **headmastership**, *n.* **headmost**, *a.* most forward, most advanced. **head-on**, *a.* head to head; (of a collision) with the front of one vehicle hitting that of another. **headphone**, *n.* a telephone receiver to fit to the head; (*pl.*) a set of earphones joined by a band over the head, a headset. **headrace**, *n.* a race that leads water to a waterwheel. **headrest**, *n.* a padded support for the head, esp. at the top of a seat in a vehicle. **headroom**, *n.* room or space for the head in a low tunnel etc. **headset**, *n.* a set of earphones joined by a band over the head; headphones. **headspring**, *n.* the source of a stream; source, origin. **headstall**, *n.* the bridle without the bit and reins. **head start**, *n.* an advantage given or taken at the beginning of a race etc.; an advantageous beginning to any enterprise. **headstone**, *n.* a stone at the head of a grave; the principal stone in a building; a cornerstone. **headstrong**, *a.* ungovernable, obstinate, intractable, self-willed. **head teacher**, *n.* a headmaster or headmistress. **head-water**, *n.* (*usu. pl.*) the upper part of a stream near its source. **headway**, *n.* motion ahead, rate of progress; head-room. **headwind**, *n.* a contrary wind. **headword**, *n.* one constituting a heading, esp. in a dictionary. **headwork**, *n.* brain-work. **headed**, *a.* having a head; having intellect or mental faculties (*esp. in comb.*, as *hard-headed*). **headless**, *a.* without a head; having no leader. **headship**, *n.* the office of a head teacher. **heady**, *a.* headstrong, precipitate; violent, impetuous, intoxicating, inflaming, exhilarating. **headily**, *adv.* **headiness**, *n.*
head² (hed), *v.t.* to lead, to be the leader to, to direct; to move, travel in a specified direction; to be or form a head

to; to provide with a head; to put or to be a heading to a chapter etc.; to get ahead of; to lop (as trees); to oppose, to check; to strike (a ball) with the head. *v.i.* to go or tend in a direction; to form a head. **to head back, off,** to intercept; to get ahead of and turn back or aside. **heading,** *n.* the action of the verb TO HEAD; an inscription at the head of an article, chapter etc.; a running title; (*Mining*) the end or the beginning of a drift or gallery; a gallery, drift or adit; (*Football*) the act of hitting the ball with the head; the compass bearing of an aircraft etc.

-head (-hed), **-hood** (-hud), *suf.* denoting state or quality, as in *godhead, maidenhead, childhood, manhood.*

header (hed'ə), *n.* one who puts or fixes a head on anything; a plunge or dive head-foremost; a brick or stone laid with its end in the face of the wall; (*Mech.*) a reaper that clips off the corn heads only; a machine for heading nails, rivets etc.; a tube or water-chamber in a steam boiler into which either end of a stack of water tubes is secured in such a manner that the steam and water can go from one tube or coil to another; (*coll.*) an act of heading a ball.

headquarters (hedkwaw'təz), *n.pl.* the residence of the commander-in-chief of an army; the place whence orders are issued; the centre of authority.

heal (hēl), *v.t.* to make whole, to restore to health; to cure (of disease etc.); to cause to cicatrize; to reconcile; to free from guilt, to purify. *v.i.* to grow or become sound or whole. **healer,** *n.* **healing,** *a.* tending to heal; soothing, mollifying.

health (helth), *n.* a state of bodily or organic soundness, freedom from bodily or mental disease or decay; physical condition (good, bad etc.); a toast wishing that one may be well, prosperous etc. **health centre,** *n.* a centre containing doctors' surgeries and administrative offices. **health farm,** *n.* an establishment, often in the country, where clients can diet, exercise, relax etc. **health food,** *n.* types of food, e.g. organically grown or with no synthetic ingredients, regarded as promoting health. **health resort,** *n.* a place where sick, delicate or convalescent people stay for the benefit of their health. **health visitor,** *n.* a nurse specializing in preventive medicine, who visits people in their own homes. **healthful,** *a.* promoting health, either physical or spiritual; salubrious; healthy. **healthfully,** *adv.* **healthfulness,** *n.* **healthy,** *a.* enjoying good health; hale, sound; promoting health, salubrious, salutary. **healthily,** *adv.* **healthiness,** *n.*

heap (hēp), *n.* a pile or accumulation of many things placed or thrown one on another; (*coll.*) a large number, a lot, a crowd, a good many times, a good deal. *v.t.* to throw (together) or pile (up) in a heap; to load or overload (with); to pile (upon).

hear (hiə), *v.t.* (*past, p.p.* **heard** (hœd)) to perceive by the ear, to perceive the sound of; to listen to, to attend to; to listen to as a judge etc.; to understand by listening; to be a hearer of; to pay regard to, to heed, to obey; to be informed of by report; to receive a communication (from). *v.i.* to have the sense of hearing; to be told, to be informed (of, about etc.). **hard of hearing,** *a.* having defective hearing. **hear! hear!** a form of applause or ironical approval. **hearer,** *n.* one who hears; one of an audience. **hearing,** *n.* the act of perceiving sound; the sense by which sound is perceived; audience, attention; a judicial trial or investigation; earshot. **hearing aid,** *n.* a device for assisting the deaf to hear. **hearing-impaired,** *a.* (*euphem.*) having defective hearing.

hearken (hah'kən), *v.i.* (*poet.*) to listen attentively (to).

hearsay (hiə'sā), *n.* common talk, report or gossip. *a.* told or given at second-hand.

hearse (hœs), *n.* a vehicle in which the dead are taken to the place of burial.

heart (haht), *n.* the central organ of circulation in the body; the mind, the soul; the emotions or affections, esp. the passion of love; sensibility, tenderness, courage, spirit; zeal, ardour; the breast as seat of the affections; the central part; strength, efficacy, fertility; anything heart-shaped; (*pl.*) a suit of cards marked with figures like hearts. **after one's own heart,** exactly as one desires. **a heart of gold,** a quality of kindness, helpfulness etc. **at heart,** in reality, truly, at bottom; in the inmost feelings. **by heart,** by rote, by or from memory. **from (the bottom of) one's heart,** with absolute sincerity; fervently. **in (good) heart,** in good spirits. **in one's heart,** inwardly, secretly. **to break the heart of,** to cause the greatest grief to. **to cross one's heart,** to promise or aver something solemnly. **to eat one's heart out,** to brood over or pine away through trouble; be envious. **to find in one's heart,** to be willing. **to give, lose one's heart to,** to fall deeply in love with. **(not) to have one's heart in,** (not) to be fully committed or devoted to. **to have one's heart in one's mouth,** to be violently frightened or startled. **(not) to have the heart to,** (not) to be able or have the courage to (do something unkind or unpleasant). **to lose heart,** to become discouraged. **to make one's heart bleed,** (*iron.*) to distress. **to one's heart's content,** as much as one likes. **to set the heart on,** to want very much. **to take heart,** to pluck up courage. **to take to heart,** to be greatly affected by. **to wear one's heart upon one's sleeve,** to be excessively frank and unreserved; to reveal one's inmost feelings and thoughts. **with all one's heart,** very willingly; completely, utterly. **heartache,** *n.* anguish of mind. **heart attack,** *n.* an acute loss of normal function in the heart. **heartbeat,** *n.* a pulsation of the heart. **heart block,** *n.* a condition in which the atria and the ventricles of the heart do not beat in coordination. **heartbreak,** *n.* overpowering sorrow. **heartbreaker,** *n.* one who or that which breaks the heart; a kind of curl; a love-lock. **heart-breaking,** *a.* **heart-broken,** *a.* **heartburn,** *n.* a burning pain in the stomach arising from indigestion. **heart disease,** *n.* a generic term for various affections of the heart. **heart failure,** *n.* a condition in which the heart fails to function normally, often leading to death. **heartfelt,** *a.* deeply felt, sincere. **heartland,** *n.* (*often pl.*) the central or most important part of a country. **heart–lung machine,** *n.* a machine that adopts the function of a patient's heart and lungs during heart surgery. **heart-rending,** *a.* heart-breaking, intensely afflictive. **heart-searching,** *n.* an anguished examination of one's feelings etc. **heart-sore,** *n.* a cause of deep sorrow. *a.* grieved at heart. **heart-strings,** *n.pl.* the sensibilities; pity, compassion; one's deepest affections. **heart-throb,** *n.* a person, e.g. a filmstar, adulated by many. **heart-to-heart,** *n., a.* (a conversation) of a searching and intimate nature. **heartwarming,** *a.* inspiring emotional approval. **heartwood,** *n.* duramen. **-hearted,** *comb.form.* having emotions of the specified kind.

heartburn HEART.

hearten (hah'tn), *v.t.* to encourage, to inspirit, to stir up. *v.i.* to cheer (up). **heartening,** *a.*

hearth (hahth), *n.* the floor of a fireplace; that part of a reverberatory furnace in which the ore is laid, or in a blast furnace the lowest part through which the metal flows; the fireside, the domestic circle, the home. **hearth-rug,** *n.* a rug placed in front of a fireplace. **hearth-stone,** *n.* the stone forming the hearth; (*formerly*) a soft kind of stone for whitening hearths etc.

heartless (haht'lis), *a.* destitute of feeling or affection; insensible, pitiless, cruel. **heartlessly,** *adv.* **heartlessness,** *n.*

heartsease (hahts'ēz), *n.* peace of mind; the wild pansy.

hearty (hah'ti), *a.* cordial, good-natured, kindly; healthy; of keen appetite; full, abundant, satisfying; boisterous; irritatingly cheerful. **my hearties,** (*Naut.*) a friendly mode of address. **heartily,** *adv.* **heartiness,** *n.*

heat (hēt), *n.* a form of energy, probably consisting in the vibration of the ultimate molecules of bodies or of the ether, capable of melting and decomposing matter, and transmissible by means of radiation, conduction or convection; hotness, the sensation produced by a hot body; hot weather; an inflamed condition of the skin, flesh etc.; redness, flush, high colour; hotness or pungency of flavour; violence, vehemence, fury; anger; intense excitement; warmth of temperament; animation, fire; sexual excitement in animals, esp. in females; a single course in a race or other contest; (*sl.*) coercive pressure; (*sl.*) searches etc. by police after a crime. *v.t.* to make hot; to inflame, to cause to ferment; to excite. *v.i.* to become hot; to become inflamed or excited. **in, on heat,** of a female when sexually excited. **to take the heat out of,** to make less emotional or vehement. **heat exchanger,** *n.* a device that transfers heat from one fluid to another. **heat-shield,** *n.* a shield that protects from high temperatures, e.g. those produced by a spacecraft re-entering the earth's atmosphere. **heatspot,** *n.* a freckle; an urticarious pimple attributed to heat. **heatstroke,** *n.* prostration from excessive heat. **heat-treat,** *v.t.* to heat and cool (metals) in order to change their properties. **heat treatment,** *n.* **heat wave,** *n.* a wave of radiant heat; an unbroken spell of hot weather. **heated,** *a.* passionate, angry. **heatedly,** *adv.* **heater,** *n.* a heating-apparatus; (*N Am., sl., dated*) a pistol. **heating,** *a.* promoting warmth or heat; exciting; stimulating.

heath (hēth), *n.* an open space of country, esp. if covered with shrubs and coarse herbage; any plant belonging to the genus *Erica*, or the allied genus *Calluna*, consisting of narrow-leaved evergreen shrubs with wiry stems and red or reddish flowers. **one's native heath,** one's home country or area.

heathen (hē'dhən), *n.* (*pl.* (**the**) **heathen, -ns**) a Gentile; one who is not Christian, Jewish or Muslim; a pagan, an idolater; an unenlightened or barbarous person. *a.* gentile; pagan; unenlightened; barbarous. **heathenish,** *a.* of or belonging to the heathens; barbarous, rapacious, cruel. **heathenism, heathenry,** *n.* **heathenize, -ise,** *v.t.* to render heathen.

heather (hē'dhə), *n.* heath, esp. *Calluna vulgaris,* called in the north ling. **to set the heather on fire,** to create a disturbance. **heather-bell,** *n.* the cross-leaved heather. **heather-mixture, -tweed, -wool,** *n.* a fabric or garment of a speckled colour supposed to resemble heather. **heathery,** *a.* abounding in heather.

Heath Robinson (hēth rob'insən), *a.* (of an apparatus) ingenious and extremely complex. [*Heath Robinson,* 1872–1944, English cartoonist who drew such devices]

heave (hēv), *v.t.* (*past, p.p.* **heaved**) to lift, to raise, with effort; to utter or force from the breast; (*coll., orig. Naut.*) to throw, to cast (something heavy); (*Naut.*) to hoist (as the anchor), to haul. *v.i.* to rise; to rise and fall with alternate or successive motions; to pant; to retch, to vomit. *n.* an upward motion or swelling; the act of heaving; a sigh; an effort to vomit. **heave ho!** sailor's cry in hauling up the anchor. **the (old) heave-ho,** (*sl.*) dismissal, expulsion, abandonment (in a relationship). **to heave out,** to throw out. **to heave to,** (*Naut.*) to bring the head (of a ship) to the wind and so stop her motion; to bring a ship to a standstill. **heaver,** *n.* one who or that which heaves.

heaven (hev'n), *n.* the sky, the firmament (*often in pl.*); the abode of God and the blessed; the place of supreme felicity; any place or state of extreme joy or pleasure. **Good heavens!** an exclamation. **seventh heaven,** a state of supreme felicity. **to move heaven and earth,** to overcome very great difficulties. **heaven-sent,** *a.* (of an opportunity etc.) coming at an opportune moment. **heavenly,** *a.* pertaining to the heavens, celestial; inhabiting heaven; situated in the heavens (as the planets, stars etc.); divine; (*coll.*) highly pleasing, delicious. **heavenly body,** *n.* a sun, star, planet or other mass of matter, distinct from the earth. **heavenly host,** *n.* the angels. **heavenliness,** *n.* **heavenward,** *a., adv.* **heavenwards,** *adv.*

heavy (hev'i), *a.* having great weight, weighty, ponderous; of a large and ponderous kind (as metal, artillery etc.); of great density or specific gravity, dense; great, powerful, forcible, violent; unwieldy, clumsy; large in amount; weighed down, loaded (with); not easily borne; oppressive, grievous, severe; difficult; drowsy, dull, sluggish, stupid; tedious; doleful, depressing, depressed; excessively serious, sombre; threatening, louring. *n.* (*coll.*) a thug, villain; (*Sc.*) a type of strong beer. **time hangs heavy,** time passes tediously. **to make heavy weather,** to make a labour of a task. **heavy breather,** *n.* a person who makes obscene telephone calls. **heavy-duty,** *a.* designed to sustain more than usual wear. **heavy-handed,** *a.* clumsy, awkward; oppressive. **heavy-hearted,** *a.* dejected. **heavy metal,** *n.* a type of loud rock music with a strong beat. **heavy water,** *n.* deuterium oxide. **heavyweight,** *n.* a person or animal of more than average weight, esp. a boxer weighing over 12 st. 10 lb (80·74 kg); (*coll.*) a person of great power, influence or intellect. **heavily,** *adv.* **heaviness,** *n.*

Heb., Hebr., (*abbr.*) Hebrew; (*bible*) Hebrews.

hebdomad (heb'dəmad), *n.* a week. **hebdomadal** (-dom'-), *a.* weekly.

Hebe (hē'bi), *n.* the goddess of youth, cupbearer to the gods of Olympus; (*Astron.*) the sixth asteroid.

Hebraic, -ical (hibrā'ik), *a.* pertaining to the Hebrews, their culture, or language. **Hebraism** (hē'-), *n.* the thought or religion of the Hebrews; a Hebrew characteristic; a Hebrew idiom or expression. **Hebraist,** *n.* one learned in the Hebrew language and literature. **hebraize, -ise,** *v.t.* to convert into a Hebrew idiom. *v.i.* to become Hebrew; to act according to Hebrew manners or fashions. **Hebrew** (hē'broo), *n.* a Jew, an Israelite; the language of the ancient Jews and of the State of Israel. *n.pl.* a book in the New Testament. *a.* pertaining to the Jews or their language.

Hebridean (hebridē'ən), *a.* of or pertaining to the Hebrides, islands off the West coast of Scotland.

Hecate (hek'ət), *n.* (*Gr. Myth.*) a mysterious goddess holding sway in earth, heaven and the underworld; a hag, a witch.

heck (hek), *int.* (*coll. euphem.*) an exclamation of irritation, used instead of *hell*.

heckle (hek'l), *v.t., v.i.* to worry (a public speaker) by inconvenient questions. **heckler,** *n.* **heckling,** *n.*

hectare (hek'teə, -tah'), *n.* a measure of area equal to 10,000 sq. metres or 2·471 acres.

hectic (hek'tik), *a.* relating to or symptomatic of fever; full of excitement or activity, exciting, wild.

hecto-, *comb. form.* a hundred.

hectogram (hek'təgram), *n.* a weight of 100 grams or 3·52 oz av.

hectolitre (hek'təlētə), *n.* a liquid measure containing 100 litres or 3·531 cu. ft.

hectometre (hek'təmētə), *n.* a measure of length equal to 100 metres or 109·3633 yds.

hector (hek'tə), *v.t.* to bully, to treat with insolence. *v.i.* to play the bully, to bluster. **hectorer,** *n.* **hectoring,** *n., a.*

heddle (hed'l), *n.* (*Weaving*) one of the sets of parallel cords or wires forming loops for the warp-threads of a loom.

hedera (hed'ərə), *n.* a genus of climbing plants containing two species, the common and the Australian ivy.

hedge (hej), *n.* a fence of bushes or small trees; a barrier of any kind; a means of securing oneself against loss; a shifty or non-committal statement. *v.t.* to fence (in) with or separate (off) by a hedge; to surround or enclose with or as with a hedge; to secure oneself against loss (on a speculation etc.) by transactions that would compensate one. *v.i.* to plant or repair hedges; to act in a shifty way, to avoid making a decisive statement. **hedgehop,** *v.i.* to fly very low over fields etc. **hedgehopping,** *n., a.* **hedge-**

row, *n.* a row of shrubs planted as a hedge. **hedge sparrow**, *n.* a common European bird, *Accentor modularis*, one of the warblers. **hedging**, *n.*

hedgehog (hej'hog), *n.* a small insectivorous mammal, *Erinaceus europaeus*, covered above with spines, and able to roll itself up into a ball.

hedonic (hēdon'ik), *a.* of or pertaining to pleasure. *n.pl.* the science of pleasure; the branch of ethics dealing with the relations of duty and pleasure. **hedonism** (hed'-, hē'-), *n.* (*Phil.*) the doctrine that pleasure is the chief good; the pursuit of sensual pleasures. **hedonist** (hed'-, hē'-), *n.* **hedonistic** (-nis'-), *a.*

-hedral, *comb. form.* having the specified number of sides.

-hedron, *comb. form.* a solid figure having the stated number of sides.

heebie-jeebies (hēbijē'biz), *n. pl.* (*coll.*) a feeling of anxiety or apprehension.

heed (hēd), *v.t.* to regard, to take notice of. *n.* care, attention; careful consideration. **to take** or **pay heed to**, to take notice of, pay regard to. **heedful**, *a.* circumspect, wary; attentive, regardful (of). **heedfully**, *adv.* **heedfulness**, *n.* **heedless**, *a.* careless; thoughtless; negligent (of). **heedlessly**, *adv.* **heedlessness**, *n.*

heehaw (hē'haw), *v.i.* to bray like an ass. *n.* an ass's bray; a loud and foolish laugh.

heel[1] (hēl), *n.* the rounded hinder part of the human foot; the corresponding part of the hind limb in quadrupeds, often above the foot; the hinder part of a shoe, stocking etc. covering the heel; a block built up of pieces of leather to raise the hinder part of a boot or shoe from the ground; a heel-like protuberance, knob or part; (*coll.*) the crusty end of a loaf of bread, the latter part, the tail-end of anything; (*sl.*) a contemptible person. *v.t.* to add a heel to; (*Football*) to pass the ball out from a scrimmage with the heels; (*Golf*) to hit the ball with the heel of a club. **at** or **on someone's heels**, close behind someone. **come to heel, to heel**, come close behind, so as to be under control (direction to a dog). **head over heels** HEAD. **on the heels of**, following closely after. **to be down at heel**, to be slipshod or slovenly; to be in unfortunate circumstances. **to cool one's heels**, to be made to wait. **to dig one's heels in**, to be obstinate. **to show a clean pair of heels, to take to one's heels**, to run away. **heel ball**, *n.* a composition of hard wax and lamp-black, used to give a smooth surface to heels, and for taking rubbings of inscriptions etc. **heel tap**, *n.* a thickness of leather in a shoe-heel; a small quantity of liquor left in the bottom of a glass. **heeler**, *n.* **-heeled**, *comb. form.* having heels of the specified type, e.g. *high-heeled*.

heel[2] (hēl), *v.i.* (*Naut.*) to incline or cant over to one side. *v.t.* to make (a vessel) do this. *n.* an inclination to one side (of a ship etc.).

heft (heft), *v.t.* to try the weight of by lifting.

hefty (hef'ti), *a.* (*coll.*) strong, muscular, powerful; big.

Hegelian (higā'liən, hāgē'-), *a.* pertaining to the German philosopher George Frederick William Hegel (1770-1831) or his philosophy. *n.* one who accepts the teaching of Hegel. **Hegelianism**, *n.* the philosophical system of Hegel.

hegemony (higem'əni), *n.* leadership, predominance, esp. applied to the relation of one state to another or to a confederation. **hegemonic** (hegəmon'-), *a.*

Hegira, Hejira (hej'irə), *n.* the flight of Mohammed from Mecca to Medina, 19 July 622, from which the Muslim era is computed; a hurried escape from a dangerous situation.

heifer (hef'ə), *n.* a young cow that has not yet calved.

height (hīt), *n.* the quality or state of being high; the distance of the top of an object above its foot, basis or foundation; altitude above the ground, sea-level or other recognized level; an elevated position; an eminence, a summit; stature; elevation in rank, office, society etc.; the fullest extent or degree. **heighten**, *v.t.* to make high or higher, to raise, to elevate; to increase, to enhance, to intensify, to accentuate, to emphasize; to exaggerate.

heinous (hā'nəs), *a.* abominable, flagrant, atrocious; wicked in the highest degree. **heinously**, *adv.*

heir (eə), *n.* one who by law succeeds or is entitled to succeed another in the possession of property or rank; one who succeeds to any gift, quality etc. **heir apparent**, *n.* the heir who will succeed on the death of the present possessor. **heir presumptive**, *n.* one whose actual succession may be prevented by the birth of someone else nearer akin to the present possessor of the title, estate etc. **heirdom** (-dəm), **heirship**, *n.* **heiress** (-rəs), *n. fem.* **heirless**, *a.*

heirloom (eə'loom), *n.* a chattel which descends with an estate to an heir; any possession that has remained in a family for several generations.

heist (hīst), *n.* (*chiefly N Am.*) a robbery. *v.t.* to commit a robbery, steal.

Hejira (hej'irə), HEGIRA.

held (held), *past, p.p.* HOLD[1].

heliacal (hilī'əkəl), *a.* closely connected with the sun; rising just before the sun. **heliacal rising, setting**, the apparent rising or setting of a star when it first becomes perceptible or invisible in rays of the sun.

helianthus (hēlian'thəs), *n.* a genus of plants containing the sunflower.

helical (hel'ikəl), *a.* like a helix; spiral. **helical gears**, *n.pl.* gear-wheels in which the teeth are set at angle to the axis. **helically**, *adv.* **helicograph** (-kəgraf), *n.* an instrument for describing spirals. **helicoid** (-koid), *a.* **helicoidal**, *a.*

helicopter (hel'ikoptə), *n.* an aircraft with one or more powerdriven airscrews mounted on vertical axes with the aid of which it can take-off or land vertically. **helicopter pad, helipad**, *n.* an area, e.g. on the roof of a building, where helicopters can take off and land. **heliport**, *n.* airport for the landing and departure of helicopters.

heli(o)-, *comb. form.* pertaining to the sun; produced by the rays of the sun.

heliocentric (hēliōsen'trik), *a.* having reference to the sun as centre; regarded from the point of view of the sun. **heliocentrically**, *adv.*

heliograph (hē'liəgrahf), *n.* an apparatus for signalling by reflecting flashes of sunlight. *v.i.* to signal with this, to photograph by a heliographic process. **heliography** (-og'-), *n.* the operation of signalling with the heliograph.

heliolatry (hēliol'ətri), *n.* sun-worship. **heliolater**, *n.* **heliolatrous**, *a.*

heliology (hēliol'əji), *n.* the science of the sun.

heliosis (hēliō'sis), *n.* spots caused on leaves etc. by the concentration of the sun's rays shining through glass, water-drops etc.; sunstroke.

heliostat (hē'liəstat), *n.* an instrument, comprising a turning mirror, by which the rays of the sun are continuously reflected in a fixed direction.

heliotherapy (hēliəthe'rəpi), *n.* (*Med.*) curative treatment by exposing the body to the rays of the sun.

heliotrope (hē'liətrōp), *n.* a genus of tropical or subtropical plants belonging to the borage family, whose flowers turn with the sun; a purple tint characteristic of heliotrope flowers; a red-spotted variety of quartz, also called bloodstone. **heliotropic** (-trop'-), *a.* pertaining to or manifesting heliotropism. **heliotropism** (-ot'rə-), *n.* movement of leaves or flowers towards the sun.

helipad (hel'ipad), **heliport** (hel'ipawt), HELICOPTER.

helium (hē'liəm), *n.* a gaseous inert element, at. no. 2; chem. symbol He, discovered in the atmosphere of the sun and afterwards found in the atmosphere and occluded to certain minerals.

helix (hē'liks), *n.* (*pl.* **-lices, -lixes**) a spiral line, as of wire or rope in coil; the rim or fold of the external ear; a genus of molluscs, containing the common snails.

hell (hel), *n.* the place of punishment for the wicked after

hellebore 389 **henchman**

death; the place or state of the dead; a place of extreme misery, pain or suffering; torment, torture. *int.* an exclamation expressing anger, annoyance etc. **hell of a, helluva,** (*coll.*) very good, bad, remarkable etc. **as hell,** (*coll.*) extremely. **come hell or high water,** (*coll.*) whatever may happen. **for the hell of it,** for amusement. **hell for leather,** (*coll.*) very fast. **hell to pay,** unpleasant consequences. **like hell,** (*coll.*) used to deny a statement made by another. **to give someone hell,** (*coll.*) to scold severely. **to play hell with,** (*coll.*) to harm or damage. **what the hell,** what does it matter? **what, where, why etc. the hell?** (*coll.*) used as an intensifier. **hell bent,** *adj.* recklessly intent (on). **hellcat,** *n.* a spiteful or fractious woman. **hell fire,** *n.* the torments of hell. *a.* describing sermons, preachers etc. emphasizing these. **hellraiser,** *n.* (*coll.*) a troublemaker; one given to violent carousing. **Hell's Angel,** *n.* a member of an often violent gang wearing leather and riding motorcycles. **hell's bells, teeth,** *int.* an exclamation expressing anger, annoyance etc. **hellion** (-'yŭn), *n.* (*chiefly N Am.*) a wild or unruly person, esp. a child; a hellraiser. **hellish,** *a., adv.* **hellishness,** *n.*
hellebore (hel'ibaw), *n.* any plant of the ranunculaceous genus *Helleborus*, containing the Christmas rose.
Hellene (hel'ēn), *n.* (*pl.* **-lenes**) a Greek. **Hellenic** (-len'-, -lē'-), *a.* **Hellenism** (hel'ə-), *n.* cultivation of Greek ideas, language, style etc.; Greek civilization or culture; Greek nationalism. **Hellenist,** *n.* one who adopted the Greek language, dress, customs etc., esp. a Greek Jew in the early days of Christianity; one who is learned in the Greek language and literature. **hellenize, -ise** (hel'ə-), *v.i.* to adopt or follow Greek habits; to use or study the Greek language. *v.t.* to permeate with Greek ideas, culture etc.; to make Greek. **hellenization, -isation,** *n.*
hello, hallo, hullo (həlō'), *int.* an informal greeting; an exclamation of surprise; a call for attention.
helm (helm), *n.* the instrument or apparatus by which a vessel is steered; the rudder and its operative parts, such as the tiller or wheel; the tiller; a position of management or direction. **helmless,** *a.* **helmsman,** *n.* the man who steers.
helmet (hel'mit), *n.* a piece of defensive armour for the head; a hat of similar form, worn as a protection. **helmeted,** *a.* wearing a helmet.
helminth (hel'minth), *n.* a worm, esp. a parasitic intestinal worm. **helminthic** (-min'-), **helminthoid,** *a.* shaped like a worm.
helot (hel'ət), *n.* a serf or bond slave in ancient Sparta; a slave or serf. **helotism,** *n.* the system of serfdom in Sparta or elsewhere. **helotry,** *n.* helots collectively; bond slaves or serfs.
help (help), *v.t.* to assist, to aid; to further; to supply succour or relief to in time of distress; to remedy, to prevent; to serve (e.g. food, a customer). *v.i.* to lend aid or assistance; to be of use; to avail. *n.* aid or assistance; succour, relief; escape, remedy; a helper; a domestic servant. **a helping hand,** assistance. **it cannot be helped,** there is no remedy; it cannot be prevented or avoided. **so help me God,** a strong oath or asseveration. **to help oneself,** to improve one's situation unassisted; to steal; to take without permission, authority etc. **to help out,** to help to complete or to get out of a difficulty; to assist. **to help to,** to supply with, to furnish with. **to help up,** to raise, to support. **helpline,** *n.* a telephone line offering an advice service to callers. **helper,** *n.* **helpful,** *a.* giving help, useful, serviceable, beneficial. **helpfully,** *adv.* **helpfulness,** *n.* **helping,** *n.* a portion of food given at table. **helpless,** *a.* wanting power to help oneself. **helplessly,** *adv.* **helplessness,** *n.*
helpmate (help'māt), **-meet** (-'mēt), *n.* a helper; a partner or helpful companion, esp. a spouse.
helter-skelter (hel'təskel'tə), *adv., n., a.* (in) great hurry and confusion. *n.* a fun-fair amusement consisting of a tower with a spiral slide.
helve (helv), *n.* the handle of a weapon or tool.
Helvetian (helvē'shən), *a.* Swiss. *n.* a Swiss native or citizen. **Helvetic** (-vet'-), *a.*
hem[1] (hem), *n.* the edge or border of a garment or piece of cloth, esp. when doubled and sewn in to strengthen it. *v.t.* (*past, p.p.* **hemmed**) to double over and sew in the border of; to enclose or shut (in, about or round). **hemline,** *n.* the hemmed bottom edge of a skirt or dress. **hemstitch,** *n.* an ornamental stitch made by drawing out parallel threads and fastening the cross threads. *v.t.* to hem with this. **hemmer,** *n.* one who or that which hems; an attachment to a sewing-machine for hemming.
hem[2] (hem), *int., n.* a voluntary short cough, uttered by way of warning, encouragement etc. *v.i.* to make a 'hem' sound; to hesitate.
hema-, hemat-, hemato- etc. HAEMA-.
he-man HE.
hematite (hem'ətīt, hē'-), HAEMATITE.
hemeralopia (hemərəlō'piə), *n.* a condition in which the eyes see badly by daylight and better by night or artificial light, day blindness.
hemi-, *pref.* half, halved; pertaining to or affecting one half.
hemidemisemiquaver (hemidemisem'ikwāvə), *n.* a musical note equal in time to half a demisemiquaver.
hemihedral (hemihē'drəl), *a.* (*Cryst.*) having only half the normal number of planes or facets. **hemihedron,** *n.*
hemimetabola (hemimitab'ələ), *n.* a section of insects that undergo incomplete metamorphosis.
hemione (hem'iōn), **hemionus** (himī'-), *n.* the dziggetai, a species of wild ass.
hemiplegia (hemiplē'jə), *n.* paralysis of one side of the body. **hemiplegic** (-plej'-, -plē'-), *a.*
hemipteran, *n.* one of an order of insects with suctorial mouth-organ, and usually having four wings, the upper pair partly horny and partly membranous, comprising bugs, lice etc. **hemipteral, -terous, -teran,** *a.*
hemisphere (hem'isfiə), *n.* the half of a sphere or globe, divided by a plane passing through its centre; half of the terrestrial or the celestial sphere; a map or projection of either of these; cerebral hemisphere. **hemispheric, -ical** (-sfe'-), *a.*
hemistich (hem'istik), *n.* half a verse, usu. as divided by the cæsura; an imperfect verse.
hemlock (hem'lok), *n.* the poisonous umbelliferous genus *Conium,* esp. *C. maculatum,* the common hemlock; a poison obtained from it; (*N Am.*) the hemlock fir or spruce. **hemlock fir, spruce,** or **tree,** *n.* a N American conifer.
hemorrhage (hem'ərij), etc. HAEMORRHAGE.
hemorrhoids (hem'əroidz), etc. HAEMORRHOIDS.
hemp (hemp), *n.* an Indian herbaceous plant, *Cannabis sativa*; the fibre of this, used for making ropes, coarse fabrics etc. (applied also to other vegetable fibres used for cloth or cordage); the drug of this, also known as hashish, cannabis, marijuana. **hempseed,** *n.* the seed of hemp, much used as food for cage-birds. **hempen,** *a.* made of or resembling hemp.
hen (hen), *n.* the female of any bird, esp. the domestic fowl; a female bird (*in comb.,* as *guinea-hen, pea-hen*); a female crab, lobster, fish etc.; in parts of Scotland a term of endearment or friendliness to a woman. **henbane,** *n.* a plant of the genus *Hyoscyamus,* esp. *H. niger*; a poisonous drug obtained from *H. niger*. **henbit,** *n.* a species of deadnettle; the ivy-leaved speedwell. **hen-coop,** *n.* a coop or cage for fowls. **hen-party,** *n.* a party for women only. **henpeck,** *v.t.* to nag (esp. of a wife who domineers her husband). **henpecked,** *a.* **hen-toed,** *a.* having the toes turned in.
hence (hens), *adv.* from this place, time, source or origin; in consequence of this, consequently, therefore. **henceforth, henceforward(s),** *adv.* from this time on.
henchman (hench'mən), *n.* (*pl.* **-men**) a faithful follower; a

political supporter.

hendeca-, *comb. form.* eleven.

hendecagon (hendek'əgon), *n.* a plane rectilinear figure of 11 sides or angles.

hendecasyllable (hendekəsil'əbl), *n.* a verse or line of 11 syllables. **hendecasyllabic** (lab'-), *a.* containing 11 syllables; *n.* a hendecasyllabic verse.

hendiadys (hendī'ədis), *n.* a rhetorical figure representing one idea by two words connected by a conjunction, e.g. 'by hook or crook'.

henequen (hen'ikin), *n.* sisal hemp.

henge (henj), *n.* a circle of stones or staves of prehistoric date. [backformation from *Stonehenge*]

henna (hen'ə), *n.* the Egyptian privet; a dye obtained from this plant used largely for dyeing hair, also in the East for dyeing parts of the body. *v.t.* to dye with henna.

henpecked HEN.

henry (hen'ri), *n.* (*pl.* **-ries**) unit of inductance; inductance of a circuit in which a change of current of 1 ampere per second induces e.m.f. of 1 volt.

hep (hep), HIP [4].

heparin (hep'ərin), *n.* (*Med.*) a substance which prevents blood-clotting.

hepatic (hipat'ik), *a.* of or belonging to the liver; resembling the liver in colour or form.

hepatica (hipat'ikə), *n.* (*pl.* **-cae** (-sē)) a sub-genus of the genus *Anemone* containing the liverleaf; the common liverwort; (*pl.*) a sub-class of cryptogams comprising the liverworts.

hepatitis (hepətī'tis), *n.* (*Med.*) inflammation or congestion of the liver.

hepat(o)-, *comb. form.* liver.

hepatology (hepətol'əji), *n.* the branch of medical science relating to the liver. **hepatologist**, *n.*

hepta-, *comb. form.* consisting of seven.

heptachord (hep'təkawd), *n.* (*Mus.*) a series of seven notes; the interval of a seventh; an instrument with seven strings.

heptad (hep'tad), *n.* a sum, group or series of seven.

heptagon (hep'təgon), *n.* a plane rectilinear figure having seven sides and seven angles. **heptagonal** (-tag'-), *a.*

heptahedron (heptəhē'drən), *n.* (*Geom.*) a solid figure having seven sides. **heptahedral**, *a.*

heptameter (heptam'itə), *n.* a verse of seven metrical feet.

heptane (hep'tān), *n.* a hydrocarbon of the methane series.

heptarchy (hep'tahki), *n.* a government by seven. **heptarchic, -al** (-tah'-), *a.*

Heptateuch (hep'tətūk), *n.* the first seven books of the Old Testament.

heptathlon (heptath'lon), *n.* women's athletic contest consisting of seven separate events.

her (hœ), *pron.* the possessive, dative or accusative case of the personal pronoun SHE; used in the possessive as an adj., and absolutely in the form **hers** (hœz), when the noun is not expressed.

herald (he'rəld), *n.* an officer whose duty was to proclaim peace or war, to challenge to battle and to carry messages between sovereigns and princes; an officer whose duty it is to superintend state ceremonies, to grant, record and blazon arms, trace genealogies etc.; a messenger; a harbinger, a precursor. *v.t.* to act as herald to; to proclaim; to announce; to introduce, to usher in. **heraldmoth**, *n.* a noctuid moth which appears in the autumn. **Heralds College**, *n.* a royal corporation, founded in 1483 whose duty now is to record pedigrees and grant armorial bearings.

heraldry (he'rəldri), *n.* the art and study of armorial bearings etc.; pomp, ceremony etc.; the office of a herald; heraldic bearings, emblazonment. **heraldic** (-ral'-), *a.* pertaining to heraldry or heralds. **heraldically**, *adv.*

herb (hœb), *n.* a plant producing shoots of only annual duration; herbage, grass and other green food for cattle; a plant having medicinal, culinary or aromatic properties, a simple. **herb bennet**, *n.* the wood avens. **herb Paris**, *n.* a herb growing in woods, with four leaves in the form of a cross and a terminal green flower. **herb Robert**, *n.* a species of crane's-bill. **herbaceous** (bā'shəs), *a.* pertaining to herbs; of the nature of herbs; (of plants) flowering annually. **herbaceous border**, *n.* a bed of perennial plants. **herbage** (-ij), *n.* herbs collectively; grass, pasture. **herbal**, *a.* pertaining to, consisting of or made from herbs. *n.* a book containing the names of plants, with a description of their properties, medicinal and other virtues etc. **herbalist**, *n.* one skilled in the knowledge of herbs and their qualities; a dealer in medicinal herbs. **herbarium** (-beə'riəm), *n.* (*pl.* **-ia** (-iə)) a systematic collection of dried plants; a case or room for the preservation of dried plants. **herbicide** (-sīd), *n.* a chemical that destroys vegetation, used to control weeds. **herby**, *a.* of the nature of or like herbs; abounding in herbs.

herbivore (hœ'bivaw), *n.* an animal which feeds on grass or plants. **herbivorous**, *a.*

Hercules (hœ'kūlēz), *n.* a mythical Greek hero celebrated for his bodily strength, which enabled him to perform twelve labours of superhuman magnitude; a man of enormous strength. **Hercules beetle**, *n.* a Brazilian arboreal beetle 5–6 in. (12–15 cm) long. **Herculean** (-lē'ən), *a.* pertaining to Hercules; exceedingly strong or powerful; exceedingly great, difficult or dangerous (as the labours of Hercules).

herd[1] (hœd), *n.* a number of beasts or cattle feeding or driven together; a crowd of people, a rabble. *v.i.* to go in herds or companies; to associate; to act as a herd or shepherd. *v.t.* to tend or watch (cattle etc.); to form or bring into a herd; to drive in a herd. **the herd**, *n.* the masses. **herd-book**, *n.* a book containing the pedigrees of high-bred cattle. **herd instinct**, *n.* the instinct that urges men and animals to react to contagious impulses and follow their leader. **herdsman**, *n.* (*pl.* **-men**) one who breeds or tends herds, esp. cattle.

herd[2] (hœd), *n.* a keeper of a herd (*in comb.*, as *shepherd, goatherd*).

Herdwick (hœd'wik), *n.* a hardy breed of sheep raised in the mountainous parts of Cumberland and Westmorland.

here (hiə), *adv.* in this place; to this place, hither, in this direction; in the present life or state; at this point; on this occasion; from this, hence. *n.* this place, point or time. **here and now**, right now, the present. **here and there**, in this place and that; hither and thither. **here goes**, *int.* said by a speaker who is about to do something. **here we go again**, *int.* meaning the same unpleasant, predictable etc. thing is about to happen again. **neither here nor there**, without reference to the point; irrelevant. **hereabout(s)**, *adv.* somewhere about or near this place. **hereafter**, *adv.* for the future; in a future state. *n.* a future state; the future life. **hereby**, *adv.* by this, by means or by virtue of this. **herein**, *adv.* in this; here. **hereinafter**, later or below in this (writing, book, document etc.). **hereinbefore**, *adv.* **hereto**, *adv.* up to this place, point or time; (attached) to this. **heretofore**, *adv.* below in this (document etc.). **hereupon**, *adv.* upon this, after this, at this, in consequence of this. **herewith**, *adv.* with this.

hereditable (hired'itəbl), *a.* heritable. **hereditability** (-bil'-), *n.*

hereditament (herədit'əmənt), *n.* any property that may be inherited.

hereditary (hired'itəri), *a.* descending or passing by inheritance; transmitted by descent from generation to generation; holding or deriving by inheritance. **hereditarily**, *adv.* **hereditariness**, *n.*

heredity (hired'iti), *n.* the tendency to transmit individual characteristics to one's offspring; the tendency in an organism to resemble the parent.

Hereford (he'rifəd), *n.*, *a.* (of) a breed of red cattle with

white markings and faces.
heresiarch (hirē'ziahk), *n*. a leading heretic.
heresy (he'rəsi), *n*. departure from what is held to be true doctrine, esp. when such opinions lead to division in the Christian Church.
heretic (he'rətik), *n*. one who holds unorthodox opinions, esp. in religious matters. **heretical** (-ret'-), *a*. **heretically**, *adv*.
hereto, heretofore, hereupon, herewith HERE.
heritable (he'ritəbl), *a*. capable of being inherited; (*Law*) passing by inheritance, esp. of lands and appurtenances as dist. from movable property; capable of inheriting by descent. **heritably**, *adv*. by inheritance.
heritage (he'ritij), *n*. land or other property that passes by descent or course of law to an heir; (*Sc. Law*) heritable estate, realty; share, portion, lot; anything passed from one generation to another. **heritor**, *n*. one who inherits; (*Sc. Law*) a landholder in a parish; heiress.
herl (hœl) HARL.
hermandad (œmandad'), *n*. (*Sp. Hist*.) a popular league or association formed to resist oppression, esp. by the cities of Castile against the nobles.
hermaphrodite (hœmaf'rədit), *n*. a human being or an animal combining in itself both male and female organs; a plant having the stamens and pistils in the same floral envelope; a person or thing in which opposite qualities are embodied. *a*. possessing to a greater or less extent the characteristics of both sexes, or other opposite attributes, in a single individual. **hermaphroditic, -ical** (-dit'-), *a*. **hermaphroditism** (-dit-), *n*.
hermeneutic (hœmənü'tik), **-ical**, *a*. interpreting, explaining, explanatory. *n.pl*. the art or science of interpretation, esp. of Scripture. **hermeneutically**, *adv*.
Hermes (hœ'mēz), *n*. (*pl*. **-mae, -mai** (-mē, -mī)) messenger of the gods of Olympus, god of science, commerce etc., identified by the Romans with Mercury.
hermetic (hœmet'ik), **-ical**, *a*. fitting by or as by fusion so as to be air-tight. **hermetically**, *adv*.
hermit (hœ'mit), *n*. a person who retires from society to live in solitary contemplation or devotion, esp. an early Christian anchorite. **hermit crab**, *n*. the genus *Pagurus*, esp. *P. bernhardus*, named thus because they live in abandoned univalve shells. **hermitage** (-ij), *n*. the cell or habitation of a hermit. **hermitical** (-mit'-), *a*.
hernia (hœ'niə), *n*. (*pl*. **-nias, -niae**) rupture; the protrusion of any organ, or part of an organ, from its natural place. **hernial, herniated**, *a*. **herniotomy** (-ot'-), *n*. operation for strangulated hernia.
hero (hiə'rō), *n*. (*pl*. **heroes**) a person of extraordinary valour, fortitude or enterprise; the principal male character in a novel, play, poem etc.; orig., in Greek mythology, a man of superhuman powers, often deified or regarded as a demigod. **hero worship**, *n*. the deification of a hero; excessive devotion shown to a person who is regarded as a hero. *v.t*. to regard or treat as a hero. **hero-worshipper**, *n*.
heroine (her'ōin), *n. fem*. **heroism** (he'-), *n*. the quality, character or conduct of a hero; extreme bravery.
heroic (hirō'ik), *a*. pertaining to or becoming a hero; having the qualities or attributes of a hero; producing heroes; relating to or describing the deeds of heroes; bold, vigorous, attempting extreme deeds or methods. *n.pl*. heroic verses; high-flown or bombastic language or sentiments. **heroic age**, *n*. the age in which heroes or demigods were supposed to have lived. **heroic size**, *n*. of sculpture, between life-size and colossal. **heroic verse**, the metre of heroic or epic poetry, e.g. the five-foot iambic, the Alexandrine, and the hexameter. **heroical**, *a*. **heroically**, *adv*.
heroin (he'rōin), *n*. a derivative of morphine, a white crystalline powder, used as an anodyne, a sedative and as an addictive drug.
heron (her'ən), *n*. a long-legged, long-necked wading bird. **heronry**, *n*. a place where herons breed.

herpes (hœr'pēz), *n*. a skin complaint consisting of vesicles grouped on an inflamed surface such as the lip. **herpes simplex**, *n*. an acute viral disease, often transmitted sexually, resembling herpes. **herpes zoster**, *n*. SHINGLES. **herpetic** (-pet'-), *a*.
herpestes (hœpes'tēz), *n*. a genus of small carnivorous mammals containing the mongooses.
herpet(o)-, *comb. form*. pertaining to reptiles; pertaining to herpes.
herpetology (hœpitol'əji), *n*. the natural history of reptiles. **herpetologic, -ical** (-loj'-), *a*. **herpetologist**, *n*.
Herr (heə), *n*. (*pl*. **Herren**) German title corresponding to the English Mr.
Herrenvolk (he'rənfolk), *n*. the master race, esp. the Aryan race as conceived by Nazi ideology.
herring (he'ring), *n*. a clupeoid marine fish of the N Atlantic, moving in large shoals and spawning near the coast. **herringbone**, *a*. like the spine and bones of a herring; denoting a kind of masonry in which the stones etc. are set obliquely in alternate rows. **herringbone stitch**, *n*. a kind of cross-stitch. **herring gull**, *n*. a large common seagull feeding on herrings.
hers HER.
herself (həself'), *pron*. the reflexive form of SHE, used to give emphasis in either the nominative or the objective case; her usual self. **by herself**, alone, unaided.
Herts (hahts), (*abbr*.) Hertfordshire.
hertz (hœts), *n*. (*pl*. **hertz**) a standard unit of frequency equal to one cycle per second.
Heshvan (hesh'vahn), *n*. the second month of the Jewish civil year and the eighth month of the Jewish ecclesiastical year.
hesitate (hez'itāt), *v.i*. to stop or pause in action; to be doubtful or undecided; to be reluctant (to); to stammer. **hesitant**, *a*. hesitating, dubious, vacillating, undecided. **hesitance, -tancy**, *n*. **hesitantly**, *adv*. **hesitatingly**, *adv*. **hesitation**, *n*. **hesitative**, *a*. **hesitator**, *n*.
Hesperides (hespe'ridēz), *n.pl*. (*Gr. Myth*.) the daughters of Hesperus, possessors of the garden of golden fruit watched over by a dragon at the western extremity of the earth; the garden so watched over.
hesperidium (hespərid'iəm), *n*. (*pl*. **-dia** (-ə)) a citrus fruit, e.g. the orange, with a leathery rind and a pulp divided into sections.
Hesperus (hes'prəs), *n*. the evening star. **Hesperian** (-piə'ri-), *a*. (*poet*.) situated at or in the west, western.
hessian (hes'iən), *n*. a coarse cloth made of hemp and jute. **Hessian fly**, *n*. a small fly or midge, the larva of which attacks wheat in the US.
hest (hest), *n*. a command, an injunction.
hetaera (hitiə'rə), *n*. (*pl*. **-rae** (-rē)) one of a class of highly educated courtesans in ancient Athens. **hetaerism**, *n*. recognized concubinage; community of women within the limits of the tribe.
heter(o)-, *comb. form*. different, dissimilar; irregular, abnormal; erroneous.
heterauxesis (hetərawksē'sis), *n*. (*Bot*.) irregular or unsymmetrical growth.
heteroblastic (hetəröblas'tik), *a*. (*Biol*.) derived from unlike cells, dist. from homoblastic. **heteroblasty** (het'-), *n*.
heterocarpous (hetərōkah'pəs), *a*. producing fruit of more than one kind.
heterochromous (hetərōkrōm'əs), *a*. of different colours.
heteroclite (het'ərəklīt), *a*. deviating from the ordinary rules or forms; anomalous, irregular. *n*. a word that deviates from the ordinary forms of inflexion; a person or thing deviating from the ordinary forms. **heteroclitic** (-klit'-), *a*.
heterocyclic (hetərōsī'klik), *a*. of organic chemical compounds with a ring structure of atoms of different kinds in the molecules.
heterodox (het'ərədoks), *a*. contrary to received or estab-

heterodyne 392 **hiccup**

lished doctrines, principles or standards; heretical; not orthodox. **heterodoxy**, *n.*
heterodyne (het'ərədin), *n.* a beat frequency caused in a radio receiver by the interplay of two alternating currents of similar frequencies.
heterogamous (hetərog'əməs), *a.* having flowers or florets sexually different. **heterogamy**, *n.*
heterogeneous (hetərəjē'niəs), *a.* diverse in character, structure or composition; (*Math.*) of different kinds, dimensions or degrees; incommensurable. **heterogeneousness, heterogeneity** (-nē'-), *n.*
heterogenesis (hetərəjen'əsis), *n.* the production of offspring differing from the parent; abiogenesis, spontaneous generation; alternation of generations. **heterogenetic** (-net'-), *a.*
heterogonous (hetərog'ənəs), *a.* of certain flowers, stamens and pistils dimorphous or trimorphous so as to ensure cross-fertilization. **heterogony**, *n.*
heterograft (het'ərəgrahft), *n.* a tissue graft from a member of one species onto a member of another.
heterologous (hetərol'əgəs), *a.* consisting of different elements, or of the same elements combined in different proportions; (*Path.*) differing in structure from normal tissue. **heterology** (-ji), *n.*
heteromerous (hetərom'ərəs), *a.* differing in number, form or character of parts.
heteromorphic (hetərōmaw'fik), **-ous**, *a.* differing from the normal form; having dissimilar forms; of insects, having different forms at different stages of development. **heteromorphism, heteromorphy** (het'-), *n.* the quality of being heteromorphic; existence in different forms.
heteronomous (hetəron'əməs), *a.* subject to the law or rule of another, not autonomous; (*Biol.*) having different laws of growth, diverging from the type. **heteronomy**, *n.*
heteronym (het'ərənim), *n.* a word spelt the same way as another but differing in sound and meaning, as *gill* (gil), a breathing-organ, and *gill* (jil), a measure. **heteronymous** (-on'-), *a.*
heterophyllous (hetərəfil'əs), *a.* having leaves of different form on the same plant. **heterophylly** (-of'-), *n.*
heteropod (het'ərəpod), *a.* belonging to the Heteropod, a group of Gasteropoda having the foot modified into a swimming-organ. *n.* one of the Heteropoda.
Heteroptera (hetərop'tərə), *n.pl.* a sub-order of Hemiptera in which the wings are of dissimilar parts, comprising the bugs. **heteropterous**, *a.*
heterosexual (hetərəsek'shəl), *a.* having or concerning sexual attraction to the opposite sex. *n.* a heterosexual person. **heterosexuality** (-al'-), *n.* **heterosexism**, *n.* prejudice against those who are not heterosexual (cp. HOMOSEXUAL).
heterosis (hetərō'sis), *n.* abnormal vigour or strength typical of a hybrid plant or animal.
heterosporous (hetəros'pərəs, -spaw'-), *a.* having two kinds of spores.
heterostyled (het'ərəstild), *a.* heterogonous, the styles or pistils on different plants of the species differing in length so as to promote cross-fertilization. **heterostylism** (-sti'-), *n.*
heterotaxy (het'ərətaksi), *n.* deviation of organs or parts from ordinary arrangement.
hetman (het'mən), *n.* a commander or leader of the Cossacks.
het up (het), *a.* (*coll.*) excited, agitated, annoyed.
heuchera (hū'kərə), *n.* a genus of herbaceous plants of the saxifrage family, with roundish leaves and scapes of red, white or green flowers rising directly from the rootstock.
heulandite (hū'ləndīt), *n.* a monoclinic, transparent brittle mineral, consisting chiefly of silica, alumina and lime.
heuristic (hūris'tik), *a.* serving or tending to find out; not correct or provable, but aiding the discovery of truth; based on trial and error. *n.* the branch of logic dealing with discovery and invention, also called **heuretic** (-ret'-).

heurism, *n.* **heuristically**, *adv.*
hew (hū), *v.t.* (*p.p.* **hewed, hewn**) to cut (down, away, off etc.) with an axe or similar tool; to hack, to chop; to make or fashion with toil and exertion. **hewer**, *n.* **hewn**, *a.*
hex (heks), *v.i.* (*N Am.*) to practise witchcraft. *v.t.* to cast a spell on; jinx. *n.* a person who practices witchcraft; a spell.
hex(a)-, *comb. form.* six.
hexachord (hek'səkawd), *n.* (*Mus.*) an interval of four tones and a semitone; a scale or diatonic series of six notes with a semitone between the third and the fourth.
hexad (hek'sad), *n.* a group of six; an atom with a valency of six.
hexadecimal (heksa'desməl), *n.*, *a.* (of) a number system (esp. used in comput.) with a base of 16. **hexadecimal notation**, *n.*
hexagon (hek'səgən), *n.* a plane figure having six sides and six angles. **hexagonal** (-sag'-), *a.*
hexagram (hek'səgram), *n.* a star-shaped figure formed by two equilateral triangles whose points coincide with those of a regular hexagon.
hexahedron (heksəhē'drən), *n.* a solid body of six sides, esp. a regular cube. **hexahedral**, *a.*
hexameter (heksam'itə), *n.* a line of heroic verse consisting of six metrical feet. **hexametric, -ical** (-met'-), *a.*
hexane (hek'sān), *n.* a hydrocarbon of the methane series.
hexangular (heksang'gūlə), *a.* having six angles.
hexapetalous (heksəpet'ələs), *a.* having six petals.
hexaphyllous (heksəfil'əs), *a.* having six leaves or sepals.
hexapod (hek'səpod), *n.* one of the Hexapoda or insects.
hexastich (hek'səstik), *n.* a poem or poetical passage of six lines or verses.
hexastyle (hek'səstīl), *n.*, *a.* (a portico or temple) having six columns.
Hexateuch (hek'sətūk), *n.* the first six books of the Old Testament.
hey (hā), *int.* an exclamation of joy, surprise, interrogation, encouragement etc.; (*esp. N Am.*) an exclamation used to attract someone's attention, often used meaninglessly.
hey presto, *int.* PRESTO².
heyday (hā'dā), *n.* the prime, the time of unexhausted spirits, vigour, prosperity etc.
Hf, (*chem. symbol*) hafnium.
Hg, (*chem. symbol*) mercury.
HH, (*abbr.*) Her/His Highness; His Holiness (the Pope); of pencils, extra hard.
hi (hī), *int.* (*coll.*) hello.
hiatus (hiā'təs), *n.* a gap, a break, a lacuna in a manuscript, connected series etc.; the coming together of two vowels in successive syllables or words. **hiatus hernia**, *n.* a hernia caused when part of the stomach protrudes through the oesophagal opening in the diaphragm.
hibachi (hi'batshi), *n.* a portable grill or barbecue for cooking out of doors.
hibernate (hī'bənāt), *v.i.* to pass the season of winter in sleep or torpor, as some animals; (*fig.*) to live in seclusion or remain inactive at a time of stress. **hibernal** (-bœ'-), *a.* pertaining to winter. **hibernant** (-bœ'-), *a.* hibernating. **hibernation**, *n.*
Hibernian (hībœ'niən), *a.* pertaining to Ireland. *n.* a native or inhabitant of Ireland. **Hibernianism, Hibernicism**, *n.* a phrase, mode of speech, or other peculiarity of the Irish.
hibiscus (hibis'kəs), *n.* any plant belonging to a genus of mostly tropical mallows, with large showy flowers.
hic (hik), *int.* a sound like a hiccup, denoting interruption, as in the speech of a drunken person.
hiccup (hik'ŭp), *n.* a short, audible catching of the breath due to spasmodic contraction of the diaphragm and the glottis; a series of sudden, rapid and brief inspirations, followed by expiration accompanied by noise; (*coll.*) a disruption, a problem. *v.i.* to have or utter a hiccup. *v.t.* to utter with a hiccup. **hiccupy**, *a.*

hick (hik), *n. (esp. N Am.)* a farmer, countryman, yokel. *a.* rustic, rural; parochial, provincial.
hickey (hik'i), *n. (N Am., coll.)* a lovebite.
hickory (hik'əri), *n.* a name for several N American trees allied to the walnuts, the timber of which is tough and elastic.
hid (hid), *past, p.p.* HIDE¹.
hidalgo (hidal'gō), *n.* a Spanish nobleman of the lowest class, a gentleman by birth.
hidden (hid'n), *p.p.* HIDE¹.
hide¹ (hīd), *v.t. (past* **hid** (hid), *p.p.* **hidden** (hid'n), **hid**) to conceal; to put out of or withhold from sight; to secrete, to cover up; to keep secret, to withhold from the knowledge (of); to suppress. *v.i.* to lie concealed, to conceal oneself. *n.* a place of concealment for observing wild life. **neither hide nor hair of someone,** nothing at all of someone. **hide-and-seek,** *n.* a children's game in which one hides and the others try to find; evasion. **hideaway,** *n.* a concealed or secluded place. **hideout,** *n.* a place where someone can hide or take refuge. **hidden,** *a.* **hider,** *n.* one who hides. **hiding,** *n.* concealing, lying in concealment. **hid(e)y-hole,** *n.* a secret chamber, priest's hidingplace; a hiding place.
hide² (hīd), *n.* the skin of any animal, raw or dressed; *(coll.)* the human skin. *v.t. (coll.)* to thrash. **hidebound,** *a.* narrow-minded, bigoted, obstinate. **hiding,** *n. (coll.)* a thrashing, a beating.
hideous (hid'iəs), *a.* horrible, frightful or shocking to eye or ear; ghastly, grim. **hideously,** *adv.* **hideousness,** *n.*
hidrosis (hidrō'sis), *n. (esp.* excessive) sweating. **hidrotic** (-drot'-), *a.* causing perspiration. *n.* a sudorific.
hieracium (hiərā'shiəm), *n.* any plant of the hawk-weed genus of *Compositae.*
hierarch (hī'ərahk), *n.* the chief of a sacred order, one who has authority in sacred things, a chief priest, prelate or archbishop; one who heads a hierarchy. **hierarchal, hierarchic, -ical** (-ah'-), *a.* of or pertaining to a hierarch or hierarchy. **hierarchism,** *n.* hierarchical principles, power or character. **hierarchy** (hīə'rahki), *n.* a rank or order of sacred persons (orig. of angels); priestly or ecclesiastical government; organization in classes, grades or orders (e.g. of plants); the body so organized.
hieratic (hīərat'ik), *a.* pertaining to the priesthood, priestly; applied to the written characters employed in Egyptian records and to early styles in Egyptian and Greek art. **hieratically,** *adv.*
hier(o)-, *comb. form.* sacred; pertaining to sacred things.
hierocracy (hīərok'rəsi), *n.* government by priests, hierarchy. **hierocratic,** *a.*
hieroglyph (hīə'rəglif), *n.* the figure of an animate or inanimate object used to represent a word, sound etc., a kind of writing practised by the ancient Egyptians; a character or symbol employed to convey a secret meaning; *(usu. in pl., facet.)* illegible writing. **hieroglyphic** (-glif'-), *a.* written in or covered with hieroglyphs; written in characters difficult to decipher; mysterious, emblematic, esoteric. *n. (usu. in pl.)* hieroglyphs; hieroglyphic writing. **hieroglyphical,** *a.* **hieroglyphically,** *adv.* **hieroglyphist** (-og'-), *n.* one skilled in deciphering hieroglyphs.
hierolatry (hīərol'ətri), *n.* the worship of sacred persons or things, esp. the worship of saints.
hierology (hīərol'əji), *n.* the science of hieroglyphics, esp. of the ancient writings of the Egyptians; the science or study of religious or of sacred literature. **hierologist,** *n.*
Hieronymite (-on'imit), *n.* one of a monastic order named after St Jerome.
hierophant (hīə'rəfant), *n.* one who teaches or explains the mysteries of religion; a priest who acted as initiator to the Eleusinian mysteries. **hierophantic** (-fan'-), *a.*
hi-fi (hī'fī), *n.* any equipment for high-quality sound reproduction. *a.* HIGH FIDELITY.

higgledy-piggledy (higlidipig'ldi), *adv. (coll.)* in confusion, topsy-turvy. *a.* confused, jumbled about anyhow.
high (hī), *a.* lofty, elevated; situated at a great elevation; rising or extending upwards for or to a specified extent; upper, inland; exalted in rank, position, or office; chief; of noble character or purpose; proud, lofty in tone or temper, arrogant; great, extreme, intense; full, complete, consummate; far advanced (of time); expensive, costly (in price); lively, animated; boisterous, violent; *(Mus.)* sharp, acute in pitch; tainted, approaching putrefaction, strongsmelling; chief, principal; *(coll.)* under the influence of alcohol or drugs; in a nervous or excited state. *adv.* to a great altitude, aloft; in or to a high degree; eminently, greatly, powerfully; at a high price; at or to a high pitch. **high and dry,** *(Naut.)* out of the water; aground; left behind, stranded, of no account in affairs. **high and low,** of people, all sorts and conditions; everywhere. **high and mighty,** arrogant. **on high,** aloft; to or in heaven. **on one's high horse,** arrogant, affecting superiority, giving oneself airs. **to be for the high jump,** to be in for severe admonishment or punishment. **to be riding high,** to be in a state of good fortune or prosperity. **to hightail it,** *(sl.)* to run away. **to play high,** to play or gamble for heavy amounts; to play a high card. **with a high hand,** in an arrogant or arbitrary manner. **high altar,** *n.* the principal altar. **high-ball,** *n.* iced whisky and soda in a tall glass. **highborn,** *a.* of noble birth. **high-boy,** *n. (N Am.)* a tall-boy. **highbrow,** *(coll.)* an intellectually superior person; a person who takes an intellectual or academic line in conversation. *a.* intellectual, superior. **high camp,** *n.*, *a.* (of or displaying) sophisticated camp style, behaviour etc. **high-chair,** *n.* a baby's chair with a tray, raised on long legs to table height. **High Church,** *n.* one of the three great schools in the Anglican Church, distinguished by its maintenance of sacerdotal claims and assertion of the efficacy of the sacraments. *a.* belonging to the High Church party, hence, **High Churchism, High Churchman. high-class,** *a.* of high quality, refinement, sophistication etc. **high command,** *n.* the supreme headquarters of the armed forces. **high commissioner,** *n.* the chief representative of one Commonwealth country in another. **high-end,** *a.* (of goods) at the expensive end of the market, high-quality, expensive. **high-energy,** *a.* concerning elementary particles accelerated in a particle accelerator. **high-energy, hi-NRG music,** up-tempo dance music, usu. recorded. **higher education,** *n.* education after secondary schools, e.g. at a college or university. **higher-up,** *n.* a person in a position of greater authority or higher rank. **high explosive,** *n.* an explosive of extreme rapidity and great destructive energy. **highfalutin,** *a.* bombastic, affected. **high fidelity,** *n.* reproducing sound with very little distortion. **high-flier, -flyer,** *n.* one with high qualifications, or who is likely to achieve high position. **high-flown,** *a.* proud, turgid, bombastic. **high frequency,** *n.* any frequency of alternating current above the audible range, from about 12,000 cycles per second upward. **High German,** *n.* the form of German spoken in central and southern Germany, regarded as standard speech. **high-handed,** *a.* overbearing, domineering, arbitrary. **highjack** HIJACK. **high jinks,** *n.pl.* high festivities or revelry; great sport. **high-level,** *a.* placed, done etc. at a high level; having a high rank. **high-level language,** *n. (Comput.)* a language in which each word is equal to several machine instructions, making it closer to human language. **high life,** *n.* the style of living or the manners of the fashionable world. **highlight,** *n.* the most brilliantly lit spot in a photograph or picture; *(pl.)* streaks of artificial light colour in dark hair; a moment or event of particular importance or interest. *v.t.* to put emphasis on; to put highlights in (hair). **high living,** *n.* living in extravagance and luxury. **High Mass,** *n.* a Mass in which the celebrant is attended by deacon and sub-deacon, usually, but not necessarily, sung at the high

highlands   394   **hip**

altar. **high-minded**, *a.* having or revealing lofty ideals. **high-mindedness**, *n.* **high noon**, *n.* the time when the sun is in the meridian. **high-octane**, *a.* (of petrol) of high efficiency. **high-pitched**, *a.* (of a musical note, sound, cry, etc.) acute, tuned high. **high point**, *n.* the most pleasurable, significant etc. moment or time. **high-power(ed)**, *a.* (*coll.*) having or showing great energy or vigour. **high priest**, *n.* a chief priest, esp. the head of the Jewish hierarchy. **high-reaching**, *a.* reaching to a great height; aspiring, ambitious. **high relief** ALTO-RILIEVO. **high-rise**, *a.* (in a building) having many storeys. **high-risk**, *a.* of a person, group etc., esp. vulnerable to a particular danger. **high road**, *n.* a main road, a highway. **high school**, *n.* a secondary school. **high seas**, *n.* the open sea or ocean beyond a country's territorial waters. **high season**, *n.* peak holiday time. **high-sounding**, *a.* pompous, ostentatious. **high-speed**, *a.* moving or operating at a high speed; (of photographic film) requiring brief exposure. **high-spirited**, *a.* having a lofty or courageous spirit; bold, daring. **high street**, *n.* the principal street (often used as the proper name of a street). **high-strung**, *a.* highly strung. **high tea**, *n.* tea at which meat is served. **high tech** (tek), *n.* advanced technology. **high-tech**, *a.* **high tension**, *n.*, *a.* (providing, carrying or operating at a) steady and high voltage. **high tide**, *n.* high water; the tide at its full. **high time**, *n.* fully time. **high-toned**, *a.* high in pitch; strong in sound; morally or culturally elevated. **high-top**, *a.* (of training shoes) lacing up over and above the ankle. **high-tops**, *n.pl.* training shoes reaching above the ankle. **high treason** TREASON. **high-up**, *n.* a person of high rank or authority. **high-velocity**, *a.* applied to projectiles with a low trajectory and long range; applied to guns firing such projectiles. **high water**, *n.* the utmost flow of the tide; the time when the tide is at its full. **high-water mark**, *n.* the level reached by the tide at its utmost height. **high wire**, *n.* a tightrope high above the ground. **highly**, *adv.* in a high degree, extremely, intensely; honourably, favourably. **highly-strung**, *a.* of a nervous and tense disposition. **highness**, *n.* the quality or state of being high; a title of honour given to princes and others of high rank (used with a possessive pronoun).

highlands (hī'ləndz), *n.pl.* a mountainous region. **the Highlands**, *n.pl.* the northern mountainous parts of Scotland. **Highland**, *a.* pertaining to the Highlands of Scotland. **Highland cattle**, *n.* a long-haired, long-horned breed of cattle of a red-brown colour. **Highland fling**, *n.* a hornpipe, peculiar to the Sc. Highlanders. **Highlander**, *n.* an inhabitant of the Highlands of Scotland.

highway (hī'wā), *n.* a public road open to all passengers; a main route either by land or by water. **highway code**, *n.* the official guide and instructions for proper behaviour on the road to avoid accidents etc. **highwayman**, *n.* one who robs on the highway.

HIH, (*abbr.*) His/Her Imperial Highness.

hijack (hī'jak), *v.t.* to steal goods in transit; to take over a vehicle, aircraft etc. by force, esp. to divert it from its route. *n.* an act of hijacking. **hijacker**, *n.*

hike (hīk), *n.* a ramble, a walking-tour; (*N Am.*, *coll.*) an increase, e.g. in prices. *v.i.* to go for a hike. *v.t.* to hoist, lift, throw up; (*N Am.*, *coll.*) to increase. **hiker**, *n.*

hilarious (hilεə'riəs), *a.* cheerful, mirthful, merry; enjoying or provoking laughter. **hilariously**, *adv.* **hilariousness**, **hilarity** (-la'-), *n.*

Hilary Term (hil'əri), *n.* one of the four terms of the High Court of Justice etc. in England (11 Jan. – 31 Mar.); the spring term at Oxford and Dublin universities.

hill (hil), *n.* a natural elevation on the surface of the earth, a small mountain; a heap, a mound; (*N Am.*) a cluster of plants, roots etc., with earth heaped round them. *v.t.* to form into hills, heaps or mounds; to heap (up). **(as) old as the hills**, (*coll.*) very old. **over the hill**, of an age when one has lost one's vigour, energy etc. **hillbilly**, *n.* (*N Am.*)

a rustic from the mountain country. **hillside**, *n.* the slope or declivity of a hill. **hill station**, *n.* a settlement in the hills, esp. of N India, used as a retreat during hot weather. **hilltop**, *n.* **hillock** (-ək), *n.* a little hill or mound. **hillocky**, *a.* **hilly**, *a.* **hilliness**, *n.*

hilt (hilt), *n.* the handle of a sword or dagger. **to the hilt**, to the fullest extent.

hilum (hī'ləm), *n.* the spot upon a seed where it was attached to the placenta; a small aperture or a small depression in a body organ.

him (him), *pron.* the objective or accusative case of HE. **himself**, *pron.* an emphatic or reflexive form of the personal pronoun of the 3rd pers. sing. masc.; his usual self. **by himself**, alone, unaccompanied; unaided.

Himalayan (himəlā'ən), *a.* pertaining to the Himalayas, a lofty range of mountains in the north of India; vast, gigantic. **Himalayan pine**, *n.* the Nepal nutpine. **Himalayan primrose**, **cowslip**, *n.* a large yellow primula.

himation (himat'ion), *n.* (*pl.* **-tia**) the ordinary outer garment in ancient Greece, an oblong piece of cloth thrown over the left shoulder.

himself HIM.

hinau (hin'ow), *n.* a New Zealand tree the bark of which yields a black dye.

hind[1] (hīnd), *n.* the female of the deer, esp. the red deer.

hind[2] (hīnd), **hinder**,[1] *a.* pertaining to or situated at the back or rear. **hindquarters**, *n.pl.* the posterior of an animal. **hindsight**, *n.* wisdom after the event, the reverse of foresight. **hindermost**, **hindmost**, *a.* the last; that is or comes last of all.

hinder[2] (hin'də), *v.t.* to obstruct, to impede; to prevent from proceeding or moving. *v.i.* to cause a hindrance; to interpose obstacles or impediments. **hinderer**, *n.* **hindrance**, *n.* the act of hindering; that which hinders; an impediment, an obstacle.

Hindi (hin'di), *n.* the group of Indo-European languages spoken in northern India.

hindrance HINDER[2].

hindsight HIND[2].

Hindu, **Hindoo** (hindoo', hin'-), *n.* a native of India adhering to Hinduism. **Hinduism** (hin'-), *n.* the Hindu polytheistic system of Brahminism modified by Buddhism and other accretions. **Hindustani** (-stah'ni), *a.* of or belonging to Hindustan (properly India north of the Nerbudda), Indian. *n.* a native of Hindustan proper; the form of the Hindu language adopted by the Muslim conquerors of Hindustan, Urdu.

hinge (hinj), *n.* the joint or mechanical device on which a door or lid turns; a natural articulation fulfilling similar functions; a piece of gummed paper for sticking a stamp in an album etc.; the point on which anything depends or turns. *v.t.* to furnish with or as with a hinge. *v.i.* to turn on or as on a hinge; to depend (upon). **hinged**, *a.* **hingeless**, *a.*

hinny (hin'i), *n.* the offspring of a stallion and a she-ass.

hint (hint), *n.* a slight or distant allusion; an indirect (usu. pointed) mention or suggestion. *v.t.* to mention indirectly, to suggest, to allude to. *v.i.* to make remote allusion. **to hint at**, to make slight but pointed allusion to. **hinter**, *n.* **hintingly**, *adv.*

hinterland (hin'təland), *n.* the region situated behind that on the coast or that along a navigable river.

hip[1] (hip), *n.* the projecting fleshy part covering the hipjoint; the haunch; the external angle formed by the meeting sides of a roof; a rafter along the edge of this; a truncated gable. **hip bath**, *n.* a bath in which the body can be immersed to the hips. **hip flask**, *n.* a flask, usu. containing spirits, carried in a pocket at the hip. **hip-hop**, *n.* a form of music and dancing originating among black and Hispanic youngsters in New York. **hip joint**, *n.* the articulation of the femur and the thigh-bone. **hip-roof**, *n.* a roof rising directly from the walls on every side and conse-

quently having no gable. **hipped**[1], *a.* (*Arch.*) furnished with a hip; (*in comb.*) having hips of the specified kind (as *wide-hipped*). **hipped roof** HIP-ROOF. **hipsters** (-stəz), *n.pl.* trousers that start at the hips, not the waist.
hip[2] (hip), *n.* the fruit of the dog-rose.
hip[3] (hip), *int.* an exclamation, usu. twice or three times repeated, introducing a hurrah.
hip[4] (hip), *adv.* (*esp. N Am.*) (*dated*), *sl.*) aware, in the know. **hippie, hippy** (-i), *n.* a member of the youth culture of the 1960s, which stressed universal love and rejected middle-class values. **hipster**, *n.* one who knows what's what, one in the know.
hippie HIP[4].
hippo (hip'ō), *n.* (*coll.*) short for HIPPOPOTAMUS.
hipp(o)-, *comb. form.* pertaining to or resembling a horse.
hippocampus (hipōkam'pəs), *n.* (pl. **-pi** (-pi)) any of various small teleostean fishes with a head resembling that of a horse, the sea horse; one of two eminences on the floor of the lateral ventricle of the brain.
Hippocratic (hipəkrat'ik), **-ical** (-krat'i-), *a.* of or pertaining to Hippocrates, the Gr. physician born about 460 BC. **Hippocratic oath**, *n.* an oath taken by a physician binding him to observe the code of medical ethics, secrecy etc., first drawn up in the 4th or 5th cent. BC, possibly by Hippocrates.
hippocrene (hip'əkrēn), *n.* a spring on Mount Helicon in Greece, a supposed source of poetic inspiration.
hippodrome (hip'ədrōm), *n.* (*Gr. and Rom. Ant.*) a circus for equestrian games and chariot races; a circus.
hippogriff, hippogryph (hip'əgrif), *n.* a fabulous creature, half horse and half griffin; a winged horse.
hippology (hipol'əji), *n.* the study of the horse. **hippologist**, *n.*
hippophile (hip'əfil), *n.* a lover of horses.
hippophobia (hipəfō'biə), *n.* dislike, fear of horses.
hippopotamus (hipəpot'əməs), *n.* (*pl.* **-es, -mi** (-mī)) a gigantic African pachydermatous quadruped of amphibious habits, with a massive, heavy body, short, blunt muzzle and short limbs and tail.
Hippuris (hipū'ris), *n.* a genus of plants containing the mare's-tail, common in pools and marshes.
hippy HIP[4].
hipster HIP[1], HIP[4].
hircine (hœ'sīn), *a.* goatish; strong smelling.
hire (hīə), *n.* the price paid for labour or services or the use of things; the engagement of a person or thing for such a price. *v.t.* to procure at a certain price or consideration for temporary use; to employ (a person) for a stipulated payment; to grant the use or service of for a stipulated price. **on** or **for hire**, available for hiring. **hire car**, *n.* a car hired usu. for a short period. **hire purchase**, *n.* a method by which payments for hire are accepted as instalments of the price and the article eventually becomes the property of the hirer. **hir(e)able**, *a.* **hireling** (-ling), *n.* (*usu. derog.*) one who serves for money. **hirer**, *n.* one who hires or lets on hire.
hirrient (hi'riənt), *a.* (*Phon.*) trilled. *n.* a trilled sound.
hirsute (hœ'sūt), *a.* rough, hairy, unshorn; (*Bot.*) covered with bristles. **hirsuteness**, *n.*
hirudin (hiroo'din), *n.* a substance secreted by the salivary gland of the leech, preventing blood-clotting.
his (hiz), *pron., a.* of or belonging to him; used absolutely as in *this is his*, this belongs to him.
Hispanic (hispan'ik), *a.* pertaining to Spain or the Spanish people. *n.* (*N Am.*) a citizen or inhabitant of the US of Latin American descent. **Hispanicism** (-sizm), *n.* a Spanish idiom.
hispid (his'pid), *a.* rough, bristly.
hiss (his), *v.i.* to make a sound like that of the letter *s*, as do geese, or by rapid motion through the air, as an arrow etc.; to express disapprobation by making such a sound. *v.t.* to utter with a hissing sound; to condemn by hissing.
n. a hissing sound; an expression of derision or disapprobation.
hist (hist), *int.* silence! hush! listen! *v.t.* to attraction attention with this sound.
hist(o)-, *comb. form.* pertaining to organic tissues.
histamine (his'tamēn), *n.* a substance released from body tissue causing an allergic reaction.
histochemistry (histəkem'istri), *n.* the application of chemistry to organic tissue.
histocompatibility (histōkəmpatibil'iti), *n.* the compatibility of tissues that allows one to be grafted successfully onto another.
histogenesis (histəjen'əsis), *n.* the science of the origin of tissues. **histogenetic** (-net'-), **histogenic**, *a.* **histogeny** (-toj'-), *n.* histogenesis; the formation and development of the organic tissues.
histogram (his'təgram), *n.* a pictorial method of showing the distribution of various quantities, e.g. rainfall month by month.
histology (histol'əji), *n.* the science of organic tissues. **histologic, histological** (-loj'-), *a.* **histologically**, *adv.* **histologist**, *n.*
histolysis (histol'isis), *n.* the decay and dissolution of organic tissue. **histolytic** (-lit'-), *a.*
historian (histaw'riən), *n.* a writer of history; one versed in history. **historiated** (-ātid), *a.* ornamented with figures (as illuminated capitals etc.).
historic (histo'rik), *a.* celebrated in history, associated with historical events, important, momentous. **historic tenses**, *n.pl.* (*Gram.*) the tenses normally employed to express past events. **historical**, *a.* pertaining to or of the nature of history, distinguished from legendary, fictitious etc. **historical novel**, *n.* a novel set in the past, using actual historical events and characters as background. **historically**, *adv.* **historicism**, *n.* a theory that all political and social events are historically determined. **historicist**, *n.* **historicity** (-ris'-), *n.* historical existence.
historiographer (histawriog'rəfə), *n.* a writer of history, esp. an official historian; an expert on historical method and practice. **historiographic, -ical** (-graf'-), *a.* **historiography**, *n.*
history (his'təri), *n.* a systematic record of past events, esp. those of importance in the development of men or peoples; a study of or a book dealing with the past of any country, people, science, art etc.; past events, esp. regarded as material for such a study; an eventful past, an interesting career; an historical play; a story; a record, e.g. of someone's past medical treatment. **to make history**, to do something momentous.
histrionic (-on'-), *a.* (*rare*) pertaining to actors or acting; theatrical; stagey, affected, unreal. *n.pl.* (*rare*) the art of theatrical representation; theatricals; an ostentatious display of usu. false emotion. **histrionically**, *adv.*
hit (hit), *v.t.* (*past, p.p.* **hit**) to strike; to strike or touch with a blow or missile after taking aim; to reach, attain; to experience; to score (in cricket); to guess; to affect, to wound; (*esp. N Am., sl.*) to kill; to encounter, meet; to arrive in or at. *v.i.* to strike (at, against etc.); to come into collision (against). *n.* a blow, a stroke; a touch with the sword or stick in fencing; a lucky chance; a felicitous expression or turn of thought; a successful effort; a best-selling book, record etc. **hit and, or miss**, succeeding and failing in a haphazard way. **hit-and-run**, *a.* of a driver, causing an accident and not stopping to help the injured; of an accident, involving a hit-and-run driver. **to hit below the belt** BELT. **to hit it off with, together**, to come to a quick understanding, sympathy, compatibility etc. **to hit off**, to represent or describe rapidly or cleverly. **to hit on, upon**, to light or chance on; to discover by luck. **to hit on someone**, to demand, attempt to extract (esp. money) from someone. **to hit out**, to strike out straight from the shoulder; to attack verbally. **to hit the bottle** BOTTLE. **to**

hit the nail on the head NAIL. **to hit the road** ROAD. **to hit the roof**, to explode with anger. **to make a hit**, to be a sudden success, to become popular. **hit list**, *n.* (*coll.*) a list of people to be killed, punished, sacked etc.; a list of coal mines, companies etc. targetted for closure etc. **hit man**, *n.* (*coll.*) a hired professional killer; a person who undertakes unpleasant tasks. **hit parade**, *n.* a list of the currently most popular recordings of pop music.

hitch (hich), *v.t.* to fasten loosely; to make fast by a hook, loop etc.; to pull up with a jerk; (*coll.*) to obtain (a lift) by hitch-hiking. *v.i.* to move with jerks; (*coll.*) to hitch-hike. *n.* a catch, a stoppage; an impediment, a temporary difficulty; the act of catching, as on a hook; a pull or jerk up; (*Naut.*) various species of knot by which a rope is bent to a spar or to another rope. **to get hitched**, (*sl.*) to get married. **hitchhike**, *v.i.* to travel by obtaining lifts from passing motorists. **hitchhiker**, *n.* **hitchhiking**, *n.* **hitcher**, *n.* a hitchhiker.

hi-tech (hī tek), HIGH TECH.

hither (hi'dhə), *adv.* to this place, end or point; in this direction. **hither and thither**, to this place and that; here and there. **hitherto**, *adv.* up to this limit or time.

Hitlerism *n.* the ideology of National Socialism as propounded by Adolf Hitler.

hi-top (hī top), HIGH-TOP. **hi-tops** HIGH-TOPS.

Hittite (hit'īt), *a.* of or pertaining to the Hittites, a people of doubtful origin inhabiting parts of Asia Minor and Syria before 1000 BC.

HIV, (*abbr.*) human immunodeficiency virus, the virus which causes AIDS.

hive (hīv), *n.* an artificial structure for housing bees; a swarm of bees inhabiting a hive; a place swarming with busy occupants. *v.t.* to put into or secure in a hive; to house as in a hive. *v.i.* to enter or live in a hive; to take shelter or swarm together, as bees. **to hive off**, to assign part of a firm's work to a subsidiary company; to divert (assets) from one concern to another.

hives (hīvz), *n.* an eruptive disease characterized by scattered vesicles filled with a fluid.

hiya (hī'ya), *int.* (*coll.*) a greeting, short for *how are you*.

hl, (*abbr.*) hectolitre.

HM, (*abbr.*) His (or Her) Majesty.

HMC, (*abbr.*) His (or Her) Majesty's Customs.

HMI, (*abbr.*) His (or Her) Majesty's Inspector, Inspectorate (of schools).

HMS, (*abbr.*) His (or Her) Majesty's Ship or Service.

HMSO, (*abbr.*) His (or Her) Majesty's Stationery Office.

HNC, (*abbr.*) Higher National Certificate.

HND, (*abbr.*) Higher National Diploma.

Ho, (*chem. symbol*) holmium.

ho (hō), *int.* an exclamation to call attention, or to denote exultation, surprise etc.; a cry used by teamsters to stop their teams. **ho! ho!** *int.* expressing amusement, derision etc.

hoar (haw), *a.* white, grey or greyish-white, esp. with age, foam or frost. *n.* hoarfrost. **hoarfrost**, *n.* frozen dew, white frost. **hoary**, *a.* white or whitish-grey as with age; white- or grey-headed; of great antiquity; venerable.

hoard (hawd), *n.* a stock, a store, a quantity of things, esp. money, laid by; an accumulated stock of anything. *v.t.* to collect and lay by; to store up. *v.i.* to amass and store up anything of value. **hoarder**, *n.*

hoarding (haw'ding), *n.* a temporary screen of boards round or in front of a building where erections or repairs are in progress; a large screen for posting bills on.

hoarse (haws), *a.* of the voice, harsh, rough; grating, discordant; having such a voice, as from a cold. **hoarse-sounding**, *a.* **hoarsely**, *adv.* **hoarseness**, *n.*

hoary HOAR.

hoax (hōks), *n.* a deception meant as a practical joke. *v.t.* to play a practical joke upon, to take in for sport. **hoaxer**, *n.*

hob (hob), *n.* the projecting side of a grate, or the top of this, on which things are placed to be kept warm; the top part of a cooker consisting of two burners or rings.

Hobbesian (hob'ziən), *a.* concerning the philosopher Hobbes Thomas (1588–1679) or his political philosophy. **Hobbism**, *n.* the system of philosophy contained in or deduced from the writings of Hobbes, esp. his teachings with regard to absolute monarchy.

hobbit (hob'it), *n.* a member of a fictional race of small people created by J.R.R. Tolkien (1892–1973).

hobble (hob'l), *v.i.* to walk lamely or awkwardly; to walk with unequal and jerky steps; to move in a halting or irregular way. *v.t.* to cause to hobble; to shackle the legs of (horses etc.) to prevent straying. *n.* an awkward, uneven or limping gait; a rope, shackle etc. for hobbling an animal.

hobby[1] (hob'i), *n.* any recreation or pursuit, plan or object. **hobby-horse**, *n.* orig. a figure rudely imitating a horse used in morris-dances, pantomime etc.; a toy horse's head on a stick; a horse on a merry-go-round; a topic to which one constantly reverts.

hobby[2] (hob'i), *n.* a small species of falcon.

hobgoblin (hobgob'lin), *n.* a kind of goblin, elf or fairy, esp. one of a frightful appearance.

hobnail (hob'nāl), *n.* a short thick nail with a large head, used for heavy boots. **hobnailed**, *a.* set with hobnails.

hobnob (hob'nob), *v.i.* to associate familiarly (with); to chat intimately (with).

hobo (hō'bō), *n.* (*pl.* **-boes**) (*esp. N Am.*) a vagrant, a tramp.

Hobson's choice CHOICE.

hock[1] (hok), *n.* the joint between the knee and the fetlock in the hind leg of quadrupeds. *v.t.* to hamstring.

hock[2] (hok), *n.* a kind of light white wine, still or sparkling, of the Rhine region.

hock[3] (hok), *v.t.* (*coll.*) to pawn. **in hock**, the state of being pawned or pledged; prison.

hockey (hok'i), *n.* a team game played with a club having a curved end (cp. *ice hockey* under ICE).

hocus pocus (hō'kəs pō'kəs), *n.* an expression used by jugglers in playing tricks; a trick, a fraud, a hoax. *v.t.* to cheat, to trick.

hod (hod), *n.* a wooden holder shaped like a trough and fixed on a long handle, for carrying mortar or bricks on the shoulder; a coal-scuttle.

hodge-podge (hoj'poj), *n.* a hotchpotch; a mixture or medley.

Hodgkin's Disease (hoj'kinz), *n.* a disease causing progressive anaemia and enlargement of the liver, lymph glands etc.

hodiernal (hōdiœ'nəl), *a.* pertaining to the present day.

hodograph (hod'əgrahf, -graf), *n.* the curve traced by the end of lines, drawn from a fixed point, representing in magnitude and direction the velocity of a moving point.

hodometer (hədom'itə), ODOMETER.

hodoscope (hod'əskōp), *n.* any device for tracing the path of a charged particle.

hoe (hō), *n.* a tool used to scrape or stir up earth around plants, cut weeds up from the ground etc. *v.t.* (*pres.p.* **hoeing**) to scrape or loosen (ground), cut (weeds), or dig (up) with a hoe. *v.i.* to use a hoe. **hoedown**, *n.* (*esp. N Am.*) a social gathering for square-dancing.

hog (hog), *n.* a swine, esp. a castrated boar meant for killing; (*N Am.*) any kind of pig; a young sheep or bullock, usu. of a year old, a hogg; a dirty, gluttonous or low person. *v.t.* (*past, p.p.* **hogged**) to cut short like the bristles of a hog; to keep greedily to oneself. **to go the whole hog**, to do anything completely; to make no compromise or reservations. **hogback, hog's back**, *n.* a long ridged hill; (*Geol.*) a monocline; an eskar. **hogfish**, *n.* a fish with dorsal spine or bristles on the head. **hogmane**, *n.* a horse's mane cut so as to stand erect. **hog-plum**, *n.* a name for several species of W Indian trees and their fruit,

hogg

which is used for feeding hogs. **hog's back** HOGBACK.
hogskin, *n*. tanned pig's skin. **hogtie**, *v.t.* to tie the feet of (an animal or person); to make helpless. **hog-wash**, *n*. the refuse of a kitchen or brewery, used for feeding hogs; (*coll.*) anything worthless. **hogweed**, *n*. a name applied to many coarse plants, esp. the cow-parsnip. **hoggish**, *a*. having the qualities or manners of a hog; brutish, gluttonous, filthy, selfish. **hoggishly**, *adv*. **hoggishness**, *n*.
hogg (hog), **hoggerel** ('ərəl), **hogget** (-it), HOG.
hogmanay (hogmənā'), *n*. in Scotland, the last day of the year; celebrations held on that day.
hogshead (hogz'hed), *n*. a measure of capacity containing 52½ imperial gal. (238.7 l); a large cask; a butt.
ho hum (hō hŭm), *int*. used to express a feeling of tedium, lack of interest, resignation etc.
hoi polloi (hoi pəloi'), *n*. (*often derog.*) the common herd; the masses.
hoick (hoik), *v.t.* (*coll.*) to pull up or out of with sudden force.
hoist (hoist), *v.t.* to raise up; to lift by means of tackle; to run up (a sail or flag); an apparatus for hoisting or raising. **hoist with his own petard** PETARD.
hoity-toity (hoititoi'ti), *a*. (*usu. derog.*) haughty, superior.
hokey (hō'ki), *a*. (*N Am.*) sentimental, corny; false, phoney.
hokum (hō'kəm), *n*. (*esp. N Am.*, *coll.*) bunkum; a foolish stage or book plot.
hol- HOL(O)-.
Holarctic (hōlahk'tik), *a*. of or pertaining to the entire northern region of the globe.
hold[1] (hōld), *v.t.* (*past*, *p.p.* **held** (held)) to grasp and retain; to keep in, to confine; to enclose, to contain; to be able to contain, to keep from running or flowing out; to keep back, to restrain; to keep in a certain manner or position; to retain possession or control of; to reserve; to occupy, to possess; to regard, to believe; to maintain (that); to judge, to assert (that); to carry on; to celebrate; to have use of, title to. *v.i.* to maintain a grasp or attachment; to continue firm, not to break; to adhere (to); to maintain a course; to be valid or true, to stand; to be fit or consistent; to wait to be connected (on the telephone). **hold it!** stop! **to hold back**, to restrain; to retain in one's possession; to keep oneself in check. **to hold by**, to hold to, to adhere to. **to hold forth**, to stretch or put forward; to propose, to offer; to speak in public; to harangue, to dilate. **to hold good, true**, to remain valid; to apply; to be relevant. **to hold in**, to restrain, to restrain oneself; to keep quiet, to keep silent. **to hold in (high) esteem** etc., to regard with esteem etc. **to hold off**, to keep at a distance; to remain at a distance; to delay. **to hold on**, to continue or proceed without interruption; (*coll.*) to wait. **to hold one's head (high)**, to conduct oneself proudly or arrogantly. **to hold one's own**, to maintain one's position. **to hold one's tongue**, to be silent. **to hold out**, to hold forward; to offer; to bear, to endure; to persist, not to yield. **to hold out on**, (*coll.*) not to tell someone about something. **to hold over**, to keep back or reserve, to defer; (*Law*) to keep possession of after the expiration of one's term. **to hold to**, to bind by (bail, one's statement etc.); to adhere to. **to hold together**, to keep in union, cause to cohere; to continue united; to cohere. **to hold up**, to raise or lift up; to support, to encourage; to sustain; to show forth, to exhibit (to ridicule etc.); to rob; to keep from falling; of the weather, to keep fine. **to hold water** WATER. **to hold with**, to approve of, to side with. **holdall**, *n*. a bag or soft case for carrying clothes etc. **holdfast**, *n*. a means by which something is clamped to another; a support. **hold-up**, *n*. a delay; a robbery, esp. when armed. **holder** (hōl'də), *n*.
hold[2] (hōld), *n*. the act of seizing or grasping in the hands; a grasp, a clutch; mental grasp; a support, any-

Holocene

thing to hold by or support oneself by; influence; custody, possession; a pause. **no holds barred**, observing no rules. **on hold**, of a telephone call(er), waiting to be connected; deferred until later. **to get hold of**, to grasp; to get in contact with.
hold[3] (hōld), *n*. the interior cavity of a ship or aircraft, in which the cargo is stowed.
holding (hōl'ding), *n*. tenure or occupation; that which is held, esp. land, property, stocks or shares. **holding company**, *n*. a company formed to acquire the majority of shares in one or more subsidiary companies. **holding pattern**, *n*. the course an aircraft takes while waiting to land.
hole (hōl), *n*. a hollow place or cavity; (*Austral.*) a pool; an aperture, an orifice, a perforation; a wild animal's burrow; a mean habitation; a small pit or hollow into which the ball has to be driven in various games; in golf, one of the points made by the player who drives his ball from one hole to another with the fewest strokes, the distance between two consecutive holes; a dingy, disreputable place; a difficulty, a fix. *v.t.* to form a hole or holes in; to put or drive into a hole. *v.i.* to go into a hole; in golf, to drive one's ball into a hole. **to hole up**, to go into hiding. **to make a hole in**, to take or consume a large part of. **to pick holes in**, to find fault with. **hole-and-corner**, *a*. secret, clandestine. **hole-in-the-wall**, *a*. hard to find, out of the way. **holey**, *a*.
holiday (hol'idā), *n*. a day of exemption from work; a day of amusement or pleasure; any period devoted to this; a vacation. *a*. pertaining to or befitting a holiday. **holiday camp**, *n*. an enclosed area with accommodation, entertainment facilities etc. **holidaymaker**, *n*. a person taking a holiday away from home.
holier-than-thou HOLY.
holiness HOLY.
holism (hō'lizm), *n*. (*Phil.*) the tendency in nature to evolve wholes that are more than the sum of the parts; (*Med.*) a form of treatment concerned with the whole person. **holistic**, *a*.
holla (hol'ə), HELLO.
holland (hol'ənd), *n*. coarse unbleached linen with a glazed surface. **hollands**, *n*. a kind of gin made in Holland.
hollandaise (holəndāz'), *a*. of sauce, made with butter, egg-yolk and lemon-juice or vinegar.
holler (hol'ə), (*N Am.*) *v.i.*, *v.t.* to shout. *n*. a loud call, a shout.
hollo, holloa HELLO.
hollow (hol'ō), *a*. containing a cavity or empty space; not solid; excavated, sunken, concave; empty, vacant; meaningless; deep, low (of sounds); insincere, not genuine. *n*. a depression or unoccupied space; a cavity, a hole, a basin; a valley. *v.t.* to make hollow, to excavate. **to beat someone hollow**, (*coll.*) to beat someone completely. **hollow-eyed**, *a*. having sunken eyes. **hollowly**, *adv*. **hollowness**, *n*.
holly (hol'i), *n*. a shrub or tree with glossy, prickly leaves and scarlet or, more rarely, yellow berries. **holly oak** HOLM.
hollyhock (hol'ihok), *n*. a tall garden plant with red, pink and yellow flowers.
Hollywood (hol'iwud), *n*., *a*. (of or pertaining to) the films, styles and practices of the big US cinema studios situated in Hollywood, a suburb of Los Angeles.
holm (hōm), *n*. the ilex or evergreen oak often called **holm-oak**.
holmium (hōl'miəm), *n*. a metallic element of the rare-earth group, at. no. 67; chem. symbol Ho.
hol(o)-, *comb. form*. entire, complete; completely.
holocaust (hol'əkawst), *n*. a wholesale sacrifice of life, or general destruction, esp. by fire; the wholesale slaughter of Jews in Europe by the Nazis in the 1940s.
Holocene (hol'əsēn), *n*. the most recent period of geologi-

hologram 398 **homeopathy**

cal time. *a.* of or concerning this period.
hologram (hol'əgram), *n.* (a photographic reproduction of) a pattern produced by the interference between a beam of coherent light (e.g. from a laser) and a direct beam of such light reflected off an object; a three-dimensional image produced by illuminating this reproduction.
holograph (hol'əgrahf), *n.* a document, letter etc. wholly in the handwriting of the author or signatory. **holographic** (-graf'-), *a.* **holography** (-log'-), *n.* the technique of making or using a hologram.
holohedral (holəhē'drəl), *a.* of crystals, having the full possible number of planes symmetrically arranged.
holometabolic (-metabol'-), **holometabolous**, *a.* of insects which undergo complete metamorphosis. **holometabolism**, *n.*
holophrastic (holofras'tik), *a.* expressing a whole sentence in a single word.
holophyte (hol'əfit), *n.* a plant that obtains food like a green plant, esp. by photosynthesis. **holophytic** (-fit'-), *a.*
holothurian (holəthū'riən), *a.* belonging to the Holothuroidea, a class of echinoderms comprising the sea-slugs. *n.* an animal of this class.
holotype (hol'ətip), *n.* the original specimen from which a new species is derived.
hols (holz), *n.pl.* (*coll.*) school holidays.
holster (hōl'stə), *n.* a leather case to hold a pistol. **holstered**, *a.*
holt[1] (hōlt), *n.* a wood, a grove, a copse.
holt[2] (hōlt), *n.* a burrow, a hole; a covert, a shelter.
holy (hō'li), *a.* sacred; set apart for the service of God or other sacred use; morally pure; free from sin or sinful affections; of high spiritual excellence. **Holy Communion** COMMUNION. **Holy of holies**, the innermost and most sacred apartment of the Jewish Tabernacle and the Temple, where the ark was kept; the inmost shrine. **Holy Roman Empire** ROMAN. **Holy City**, *n.* Jerusalem. **holy cross**, *n.* the cross on which Christ was put to death. **holy day**, *n.* a day commemorating some religious event. **Holy Family**, *n.* the infant Jesus with Joseph and Mary. **Holy Ghost, Holy Spirit**, *n.* the third Person of the Trinity. **Holy Grail** GRAIL. **Holy Land**, *n.* Palestine. **holy orders** ORDERS. **holy rood**, *n.* a cross or crucifix, esp. one on the rood-beam in churches. **Holy Saturday**, *n.* the Saturday before Easter. **Holy See**, *n.* the bishopric of Rome, the Pope's see. **Holy Thursday**, *n.* in the English Church, Ascension Day; in the Roman Catholic Church, Maundy Thursday, the Thursday in Holy Week. **holy war**, *n.* a war waged on behalf of a religion. **holy water**, *n.* water blessed by a priest, used in the Roman and Greek ritual. **Holy Week**, *n.* the week from Palm Sunday to Holy Saturday inclusive. **Holy Willie**, *n.* a hypocritically pious person. **holy writ**, *n.* sacred scriptures, esp. the Bible. **holier**, *comp. a.* more holy. **holier-than-thou**, *a.* convinced of one's moral superiority, sanctimonious. **holily**, *adv.* **holiness**, *n.* the state of being holy, sanctity; moral purity or integrity; the state of being consecrated to God or His worship; that which is so consecrated. **his Holiness**, a title of the Pope.
hom- HOM(O)-.
homage (hom'ij), *n.* the service paid and fealty professed to a sovereign or superior lord; respect paid by external action; deference, obeisance, reverence, worship.
hombre (om'bri), *n.* (*N Am., coll.*) man.
Homburg (hom'bœg), *n.* a trilby hat.
home[1] (hōm), *n.* one's own house or abode; the abode of the family to which one belongs; one's own country; the place of constant residence, of commonest occurrence, or where anything is indigenous; a place or state of rest or comfort; an institution of rest or refuge for orphans, the destitute, the elderly etc.; in various games, the goal or den. *a.* connected with, carried on or produced at home or in one's native country; domestic, opposed to foreign;

personal, touching the heart or conscience; describing a football match won by a home team. *adv.* to one's home or country; to the point, pointedly, closely, intimately. **at home**, in one's own house; accessible to visitors; in one's own area, country etc.; at one's ease, comfortable; conversant with. **at-home**, *n.* a gathering or party held in one's own home. **home and dry**, safe after having successfully come through an experience. **nothing, not much etc. to write home about**, (*coll.*) not very impressive, great etc. **to bring home to one**, to convince. **to come home to one**, to reach one's heart or conscience. **home base, plate**, *n.* the rubber plate on which the batter stands in baseball. **home banking**, *n.* a system of banking using home computer terminals. **home-brew**, *n.* a beverage (esp. beer) brewed at home. **home-brewed**, *a.* **homecoming**, *n.*, *a.* (relating to) a return to, or arrival at home. **Home Counties**, *n.pl.* the counties nearest London: Middlesex, Surrey, Kent, Essex, Herts, Bucks, Berks. **home economics**, *n.* the study of how to run a home, including cookery, child-care etc. **home farm**, *n.* a farm attached to and run by the owner of a large country estate. **home ground**, *n.* a familiar topic or subject. **home-grown**, *a.* grown in one's own garden, area, country etc. **Home Guard**, *n.* the citizen army formed in Britain in World War II. **home help**, *n.* a local authority employee who cleans the home of a disabled or elderly person. **homeland**, *n.* one's native land; in S Africa, a semi-autonomous state reserved for Black Africans. **home-made**, *a.* made at home; not manufactured abroad; roughly made. **Home Office**, *n.* the department of the Secretary of State for Home Affairs, dealing with police administration, prisons, factories, licensing etc.; the building occupied by this. **Home Rule**, *n.* the government of a country, esp. Ireland, by a separate parliament. **home run**, *n.* a hit in baseball that allows the batter to make a complete circuit and score a run. **home shopping**, *n.* shopping by mail order or telephone. **homesick**, *a.* **homesickness**, *n.* a vehement desire to return home, causing depression of spirits and affecting physical health. **homespun**, *a.* home-made; plain, unaffected. *n.* cloth spun at home. **homestead** (-sted), *n.* a house, esp. a farmhouse, with the buildings attached; (*N Am.*) a lot granted for the residence and maintenance of a family, under the Homestead Act of 1862; (*Austral.*) the owner's house on a sheep station. *v.t.* to occupy as a homestead. **homesteader**, *n.* **homesteading**, *n.* **home straight, stretch**, *n.* the last section of a race-course before the winning-post is reached; the last phase of any enterprise. **home truth**, *n.* an unwelcome truth expressed in a pointed way. **home unit**, *n.* (*Austral.*) one of a number of separate apartments in the same building. **homework**, *n.* study exercises to be done at home. **to do one's homework**, to prepare well. **homeless**, *a.* **homelessness**, *n.* **homely**, *a.* without affectation, unpretending; unadorned, unvarnished; (*chiefly N Am.*) plain in looks. **homeliness**, *n.* **homeward(s)**, *adv.* towards home. *a.* being or going in the direction of home. **homeward-bound**, *a.* returning home from abroad. **hom(e)y**, *a.*
home[2] (hōm), *v.i.* of pigeons, to fly home; to go home; to dwell. *v.t.* to send (pigeons) home; to provide with a home; to direct onto a target, e.g. with a navigational device. **homer**, *n.* a homing pigeon. **homing device**, *n.* the mechanism for the automatic guiding of missiles.
homelyn (hō'məlin), *n.* the spotted ray, a European seafish.
homeo-, *comb. form.* similar.
homeopathy (hōmiop'əthi, hom-), *n.* the system which aims at curing diseases by administering in small doses medicines which would produce in healthy persons symptoms similar to those they are designed to remove. **homeopath** (hō'-. hom'-), *n.* a homeopathist. **homeopathic** (-path'-), *a.* **homeopathically**, *adv.* **homeopathist**,

n. one who practises or believes in homeopathy.
homeostasis (hōmiəstā'sis, hom'-, -os'-), *n.* the keeping of an even level. **homeostatic** (-stat'-), *a.*
homeozoic (homiəzō'ik), *a.* containing similar forms of life (of regions of the earth).
homer HOME [2].
Homeric (hōme'rik), *a.* pertaining to Homer or his poems; resembling Homer's poems in style.
homestead HOME [1].
homicide (hom'isid), *n.* the killing of a human being; one who kills another. **homicidal** (-sī'-), *a.*
homiletic (homilet'ik), *a.* pertaining to homilies. *n.pl.* the art of preaching; the art or method of presenting spiritual truths to an audience in the most effective form. **homilist** (hom'-), *n.*
homily (hom'ili), *n.* a religious discourse; a sermon, esp. on some practical subject; a tedious moral exhortation. **Books of Homilies**, two books published in England by authority in 1547 and 1562, to be read in churches when no sermon was prepared.
hominid (hom'inid), *n.* a creature of the genus *Homo;* a man-like fossil.
hominy (hom'ini), *n.* (*chiefly N Am.*) maize hulled and coarsely ground, boiled with water or milk for food.
Homo (hō'mō), *n.* (*pl.* **homines** (hom'ināz)) man, the genus of which man is the only living species. **Homo sapiens** (sap'ienz), *n.* man as a species.
homo (hō'mō), *n.* (*coll. derog.*; *often offensive*) short for HOMOSEXUAL.
homo-, *comb. form.* noting likeness or sameness.
homocyclic (hōməsī'klik, hom-), *a.* of an organic compound, having a closed chain of atoms of the same kind.
homoeo- HOMEO-.
homoerotic (hōmōirot'ik), *a.* of or concerning sexual attraction to the same sex. **homoeroticism** (-sizm), *n.*
homogeneous (homəjē'niəs), *a.* composed of the same or similar parts or elements; of the same kind or nature throughout; (*Math.*) having all its terms of the same degree; commensurable. **homogeneousness, homogeneity** (həmojənē'-, -nā'-), *n.*
homogenesis (homəjen'əsis), *n.* reproduction characterized by the likeness of the offspring to the parent and correspondence in the course of its development.
homogenetic (homəjənet'ik), *a.* pertaining to or characterized by homogenesis; corresponding in structure so as to show community of descent; (*Geol.*) similar in structural relations prob. owing to community of origin. **homogenetical, homogenous** (-moj'-), *a.* **homogenize, -ise** (-moj'-), *v.t.* to break up to the same size and distribute evenly (esp. of fat particles in milk). **homogeny** (-moj'-), *n.*
homograph (hom'əgraf, -grahf), *n.* a word which has the same form as another, but a different origin and meaning.
homoio- HOMEO-.
homologate (həmol'əgāt), *v.t.* to admit, to concede; to approve, to confirm. **homologation,** *n.*
homologous (həmol'əgəs), *a.* having the same relative position, proportion, value, structure etc. **homological** (-loj'-), *a.* characterized by homology; homologous. **homologically,** *adv.* **homologize, -ise,** *v.t.* to be homologous. *v.t.* to make homologous. **homologue,** (*N Am.*) **homolog** (hom'əlog), *n.* something that is homologous; the same organ in different animals under every variety of form and function. **homology,** *n.* correspondence; identity of relation between parts developed from the same embryonic structures, as the arm of a man, the foreleg of a quadruped, and the wing of a bird.
homomorphic (homəmaw'fik), **homomorphous,** *a.* analogous, identical or closely similar in form. **homomorphism,** *n.*
homonomous (həmon'əməs), *a.* subject to the same law of growth. **homonomy,** *n.*

homonym (hom'ənim), *n.* a word having the same sound and perhaps the same spelling as another, but differing in meaning. **homonymic** (-nim'-), **homonymous** (-mon'-), *a.* **homonymy** (-mon'əmi), *n.* the state of being homonymous; a sameness of name with difference of meaning; ambiguity.
homophone (hom'əfōn), *n.* a letter or word agreeing in sound with another, but having a different meaning, as *heir* and *air.* **homophonic** (-fon'-), *a.* (*Mus.*) having the same pitch; in unison, opp. to polyphonic. **homophony** (-mof-'), *n.*
Homoptera (həmop'tərə), *n.pl.* a suborder of Hemiptera having the wings uniform throughout. **homopterous,** *a.*
homosexual (hōməsek'shəl, hom'-), *n., a.* (a person) sexually attracted by those of the same sex. **homosexuality** (-shual'-), *n.*
homunculus (həmūng'kūləs), **homunc(u)le** (-kl), *n.* a little man; a dwarf; a manikin. **homuncular,** *a.*
Hon., (*abbr.*) honourable; honorary.
hon (hūn), *n.* (*coll.*) a term of endearment short for HONEY.
honcho (hawn'chō), *n.* (*pl.* **-chos**) (*coll.*) a boss, a chief; a controller.
hone (hōn), *n.* a stone for giving an edge to a cutting tool. *v.t.* to sharpen on a hone.
honest (on'ist), *a.* upright, fair, truthful, trustworthy in dealings, business or conduct; just, equitable; open, frank, candid, sincere, honourable; unimpeached, unstained. *int.* used to affirm the honesty of a statement. **honest to goodness,** absolutely genuine. **to make an honest woman of,** to marry (a woman). **honestly,** *adv.* **honesty,** *n.* the quality or state of being honest; integrity, sincerity, uprightness; chastity; a cruciferous garden plant bearing flat, round, semi-transparent seed-pods.
honey (hūn'i), *n.* a sweet viscid product collected from plants by bees, and largely used as an article of food; sweetness; a term of endearment. **honey-bear,** *n.* a S American quadruped, also called the kinkajou, which destroys the nests of bees. **honey buzzard,** *n.* a British raptorial bird which feeds on the larvae of bees and wasps. **honeycomb,** *n.* a waxy substance formed in hexagonal cells by the hive-bee, for the reception of honey and for the eggs and larvae; anything similarly perforated. *v.t.* to fill with holes or cavities. **honeydew,** *n.* a saccharine substance found on the leaves of some plants; something extremely sweet, nectar; a kind of tobacco moistened with molasses. **honeydew melon,** *n.* a type of melon with sweet flesh and a greenish rind. **honey guide,** *n.* the S African cuckoo whose cry is supposed to indicate the nests of bees. **honey-stalk,** *n.* the flower of clover. **honeysuckle,** *n.* the woodbine, a wild climbing plant with sweet-scented flowers; (*Austral.*) any one of the Banksia shrubs. **honeytongued,** *a.* smooth in speech. **honey-wort,** *n.* two cultivated plants of the borage family both attractive to bees. **honeyed,** *a.* sweetened with honey; of words, ingratiating.
honeymoon (hūn'imoon), *n.* the period immediately following marriage spent by the married couple by themselves away from home. *v.i.* to spend the honeymoon (in, at etc.). **honeymooner,** *n.* **honeymoon period,** *n.* a period of goodwill and harmony at the start of a new business appointment, relationship etc.
honi soit qui mal y pense (on'i swah kē mal ē pēs'), shame be to him who thinks evil of it (motto of the Order of the Garter).
honk (hongk), *n.* the cry of the wild goose; any similar cry or noise, esp. that of a vehicle's horn. *v.t., v.i.* to (cause to) make this noise.
honky, honkie (hong'ki), *n.* (*N Am., derog., sl.*) a white person.
honky-tonk (hongkitongk'), *n.* (*esp. N Am.*) a disreputable nightclub, bar etc.; a type of ragtime piano-playing, esp. on a cheap upright piano. *a.* of a piano, of such a kind; of music, of this type of ragtime.

honor, (*N Am.*) HONOUR.

honorarium (onərea'riəm), *n.* (*pl.* **honorariums, -ria** (-ə) a fee or payment for the services of a professional person.

honorary (on'ərəri), *a.* done, made, or conferred as a mark of honour; holding a title or an office without payment or without undertaking the duties; depending on honour, not enforceable by law (of duties or obligations).

honour (on'ə), *n.* respect, esteem, reverence; reputation, glory, distinction, a mark or token of distinction; high rank; nobleness of mind, probity, uprightness; conformity to the accepted code of social conduct; chastity; (*pl.*) courteous attentions paid to guests etc.; (*Univ.*, *pl.*) a distinction awarded for higher proficiency than that required for a pass; marks of respect; a title of address given to certain officers, as a county court judge etc.; in golf, the right of driving off first; (*pl.*) the four highest trump cards. *v.t.* to treat with reverence or respect; to bestow honour upon; to dignify, to glorify, to exalt; to acknowledge; to accept and pay when due (as a bill). **honours of war**, a distinction or privilege granted to an enemy who has surrendered on terms. **in honour of**, to celebrate. **on, upon one's honour**, a declaration pledging one's honour or reputation to the accuracy or good faith of a statement. **to do the honours**, to perform the courtesies required of a host at a dinner, reception etc. **funeral** or **last honours**, marks of respect paid to the deceased at a funeral. **honours list**, *n.* the list of people who have received honours, e.g. knighthoods etc., from the Queen. **honorific** (-rif'-), *a.* conferring or doing honour. *n.* an honorific title etc.

honourable (on'ərəbl), *a.* worthy of honour; conferring honour; actuated by principles of honour, upright; accompanied or performed with or as with marks of honour; proceeding from a laudable cause; not base. **(the) Honourable**, a title of respect or distinction borne by the children of peers below the rank of marquess, maids of honour, Justices of the High Court etc. **honourably**, *adv.*

hooch, hootch (hooch), *n.* (*N Am.*) crude alcoholic liquor.

hood¹ (hud), *n.* a loose covering for the head and back of the neck, separate, or an appendage to a cloak or coat; an appendage to an academic gown marking a degree; anything more or less resembling a hood, as the folding roof of a convertible car, a pram top etc.; (*N Am.*) the bonnet of a motor vehicle. *v.t.* to dress in a hood; to put a hood on; to cover. **hoodwink**, *v.t.* to deceive, to take in. **hooded**, *a.* covered with a hood; blinded; (*Bot.*) hood-shaped, cucullate; (of a bird, snake or other animal) having a hood-like part.

hood² (hood), *n.* (*N Am., coll.*) short for HOODLUM.

-hood (-hud), -HEAD.

hoodlum (hood'ləm), *n.* (*N Am., coll.*) a street rowdy, a hooligan, esp. one of a gang.

hoodoo (hoo'doo), *n.* bad luck; the cause of bad luck, a Jonah. *v.t.* to bring bad luck.

hoodwink HOOD.

hooey (hoo'i), *n.* bosh, nonsense.

hoof (hoof), *n.* (*pl.* **hoofs, hooves** (hoovz)) the horny sheath covering the feet of horses, oxen etc. **on the hoof**, of livestock, alive. **to hoof it**, (*coll.*) to walk, to tramp it. **hoofed**, *a.*

hoo-ha (hoo'hah), *n.* (*coll.*) fuss, noisy excitement.

hook (huk), *n.* a curved piece of metal or other material by which an object is caught or suspended; a bent and pointed wire, usu. barbed, for catching fish; a trap, a snare; a sickle; a sharp bend; a cape, a headland; a type of blow in boxing; (*sl.*) a repetitive catchy musical phrase. *v.t.* to catch, grasp or hold with or as with a hook; to fasten with a hook or hooks; (*esp. passive; coll.*) to attract or cause to become addicted; in golf, to drive (the ball) widely to the left; in football, to pull (the ball) in with the foot in a certain manner. *v.i.* to fit or fasten (on) with or as with hooks. **by hook or by crook**, by fair means or foul; somehow. **hook and eye**, a metal hook and corresponding loop for fastening a dress etc. **off the hook**, freed from punishment, guilt, responsibility etc.; of a telephone receiver, not on its rest. **to sling one's hook**, (*sl.*) to decamp; to run away. **to swallow something hook, line and sinker**, to believe something completely. **hook-up**, *n.* a radio network, a series of connected stations; a connection, contact or link-up. **hookworm**, *n.* a parasite infesting both humans and animals. **hooked**, *a.* bent; furnished with hooks; (*sl.*) addicted to (a drug), obsessed by; caught, trapped.

hooka(h) (huk'ə), *n.* a tobacco or marijuana pipe in which the smoke passes through water.

hooker (huk'ə), *n.* (*N Am., sl.*) a prostitute; in rugby, the player who takes possession of the ball in a scrum.

hookey, hooky (huk'i), *n.* (*N Am., coll.*) truant, esp. in the phrase **to play hookey**.

hooligan (hoo'ligən), *n.* a street rough given to violent attacks on persons. **hooliganism**, *n.*

hoop¹ (hoop), *n.* a strip of wood or metal bent into a band or ring to bind the staves of casks, to expand the skirts of woman's dresses, as a play thing, a circus prop etc.; a small iron arch used in croquet. *v.t.* to bind or fasten with hoops; to encircle. **to go, be put etc. through the hoop(s)**, to go, be put etc. through an ordeal. **hoop-la** (-lah), *n.* game of winning small objects by throwing rings over them. **hooped**, *a.*

hoop² (hoop), **hooping-cough** WHOOP.

hoopoe (hoo'poo), *n.* a bird with large crest and fine plumage, a rare British visitant.

hoorah, hooray (hurah', -rā'), *int.* HURRAH. **Hooray Henry** (hoo'rah hen'ri), *n.* an extrovert, loud youth with an upper crust accent.

hoot (hoot), *v.i.* to shout or make loud cries in derision or contempt; to cry as an owl; to make a sound like this. *v.t.* to shout (down, out, away etc.) in contempt or derision; express by hooting. *n.* a cry like that of an owl; an inarticulate shout in contempt or derision. *n.* (*coll.*) something or someone very amusing. **not to give two hoots**, (*coll.*) not to care at all. **hooter**, *n.* one who or that which hoots; a steam-whistle or siren; (*coll.*) a noise.

Hoover® (hoo'və), *n.* a vacuum-cleaner. *v.t., v.i.* to clean with a vacuum-cleaner.

hop¹ (hop), *v.i.* (*past, p.p.* **hopped**) to spring, leap or skip on one foot; to skip with both feet (as birds) or with all four feet (as quadrupeds). *v.t.* to jump lightly or skip over; (*esp. N Am.*) to ride on (a bus etc.). *n.* a jump, spring, or light leap on one foot; (*coll.*) a dance; a short trip by aircraft, a short run, a quick passage; a distance easily covered in a few paces. **hop it!** go away. **hop, skip and a jump**, a short distance. **to catch on the hop**, to catch by surprise or when unprepared. **hopping mad**, (*coll.*) very angry.

hop² (hop), *n.* a perennial climbing plant the mature cones of which are used in brewing beer. *v.i.* to pick hops. **hop-picker**, *n.* one who gathers hops; a machine for this purpose. **hop-pillow**, *n.* a pillow stuffed with hops for inducing sleep. **hoppy**, *a.* tasting of hops.

hope (hōp), *n.* an expectant desire; confidence in a future event; a ground for expectation, trust or confidence; that in which one confides; a person or thing that is the object of someone's hopes. *v.i.* to have confidence; to trust with confidence; to look (for) with desire or expectation, to trust (in). *v.t.* to expect with desire; to look forward to with trust; (*coll.*) to think, to suppose. **hope chest**, *n.* (*N Am.*) bottom drawer. **to hope against hope**, to cling to a slight chance. **hopeful**, *a.* full of hope; giving rise to hope. **hopefully**, *adv.* in a hopeful way; one hopes. **hopefulness**, *n.* **hopeless**, *a.* destitute of hope, despairing; affording no hope, desperate, incurable; (*coll.*) incompetent or showing incompetence. **hopelessly**, *adv.* **hopelessness**, *n.*

Hopi (hōpi), *n. (pl.* **-pis, -pi**) a member of a N Am. Indian people inhabiting NE Arizona; the language of this people.
hoplite (hop'līt), *n.* a heavy-armed soldier in ancient Greece.
hopper (hop'ə), *n.* a funnel-shaped vessel for feeding material to a machine; a funnel or trough for passing grain etc. through a mill into vehicles; a barge for receiving and dumping mud, sand etc. from a dredging-machine; a tilting bottom in a barge, car etc. for discharging refuse.
hopscotch (hop'skoch), *n.* a children's game in which a stone is driven by the foot of a player hopping from one compartment to another of a figure traced on the ground.
Horatian (hərā'shən), *a.* pertaining to or resembling the Latin poet Horace or his poetry.
horde (hawd), *n.* a nomadic tribe or clan; (*usu. derog.*) a multitude.
Hordeum (haw'diəm), *n.* a genus of grasses typified by wild barley. **hordein** (-diin), *n.* a protein found in barley grains.
horehound, hoarhound (haw'hownd), *n.* a labiate herb with woolly stem and leaves and aromatic juice, formerly used as a tonic and a remedy for colds etc.
horizon (həri'zən), *n.* the circular line where the sky and the earth seem to meet; the great circle parallel to it, the centre of which is the centre of the earth; the boundary of one's mental vision, experience etc.
horizontal (horizon'təl), *a.* pertaining or relating to the horizon; situated at or near the horizon; parallel to the horizon, level, flat, plane; measured or contained in a plane of the horizon. *n.* a horizontal line, plane, bar etc. **horizontally**, *adv.*
hormone (haw'mōn), *n.* a secretion from an internal gland having the property of stimulating vital and functional activity. **hormonal** (-mō'-), *a.* **hormone replacement therapy**, treatment involving replacement of deficient hormones in menopausal women.
horn (hawn), *n.* a projecting bony growth, usu. pointed and in pairs on the heads of certain animals; the substance of which such growths are composed; anything made of or like a horn in shape; a sounding device as in a motor vehicle; an organ or growth resembling horns, as the feeler of a snail etc.; an extremity of a curved object, piece of land, stretch of water etc.; a metal wind instrument, orig. of horn; one of the alternatives of a dilemma. *a.* made of horn. **horn of plenty** CORNUCOPIA. **to draw, pull in one's horns**, to repress one's ardour; to curtail one's expenses; to draw back, to check oneself. **hornbeak**, *n.* the garfish. **hornbeam**, *n.* a small tree yielding tough timber. **hornbill**, *n.* a bird with bone-crested bills from India and the Indian Archipelago. **horn-fish**, *n.* the garfish; the sand-pike, and other fishes. **hornpipe**, *n.* an old wind instrument; a lively dance, usu. for one person, popular among sailors; the music for such a dance. **horn-rimmed**, *a.* esp. of spectacles, having rims made of (a material resembling) horn. **horn-rims**, *n.pl.* horn-rimmed spectacles. **hornstone**, *n.* chert. **horned**, *a.* furnished with horns; having projections or extremities like horns. **horned screamer**, *n.* a S American grallatorial bird with a horn on its forehead, and a piercing voice. **horned viper**, *n.* an Indian or African viper with horns over the eyes. **horned toad**, *n.* a small American toadlike lizard covered with spines. **hornless**, *a.* **horny**, *a.* made of or like horn; callous; having or abounding in horns; (*esp. N Am., sl.*) sexually excited; causing sexual excitement; lustful. **hornily**, *adv.* **horniness**, *n.*
hornblende (hawn'blend), *n.* a dark-coloured mineral consisting of silica, magnesia, lime and iron.
hornet (haw'nit), *n.* a large social wasp with a formidable sting. **to stir up a hornet's nest**, to excite (often unintentionally) the animosity of a large number of people.

horo-, *comb. form.* pertaining to times or seasons, or to the measurement of time.
horologe (hor'əloj), *n.* (*rare*) an instrument for showing the hour, a time-piece. **horologist** (-rol'-), *n.* one skilled in horology; a maker of horologes. **horology** (-rol'-), *n.* the art of measuring time, or of constructing instruments to indicate time.
horoscope (ho'rəskōp), *n.* an observation of the sky and the configuration of the planets at a particular time, esp. at the moment of one's birth, in order to foretell one's future; fortune-telling by this method.
horrendous (hərən'dəs), *a.* (*coll.*) awful; horrifying.
horrible (ho'ribl), *a.* causing or tending to cause horror; dreadful, shocking, harrowing; (*coll.*) extremely unpleasant, awful. **horribleness**, *n.* **horribly**, *adv.*
horrid (ho'rid), *a.* causing horror; shocking; (*coll.*) nasty, unpleasant, frightful. **horridly**, *adv.* **horridness**, *n.*
horrify (ho'rifī), *v.t.* to strike with horror; (*coll.*) to scandalize. **horrific** (-rif'-), *a.* **horrifically**, *adv.*
horripilation (həripilā'shən), *n.* a sensation of a creeping or motion of the hair of the body, caused by disease, terror etc. **horripilant** (-rip'-), *a.* **horripilate** (-rip'-), *v.t., v.i.*
horror (ho'rə), *n.* dread or terror, mingled with detestation or abhorrence; that which excites terror or repulsion; (*coll.*) an unpleasant person or thing. *a.* esp. of a cinema film, depicting gruesome, frightening, often paranormal events. **the horrors**, *n.pl.* the blues; delirium tremens. **horror-stricken, -struck**, *a.* overwhelmed with horror.
hors d'oeuvre (dœvr''), *n.* (*pl.* **hors d'oeuvres** (dœvr'')) a dish not forming part of the regular course, served as relish before or during a meal.
horse (haws), *n.* a solid-hoofed quadruped with mane and tail of long coarse hair, domesticated and employed as beast of draught and burden or for riding; the adult male of the species; (*collect.*) cavalry; a frame or other device used as a support; a vaulting-block; (*sl.*) heroin. *v.t.* to provide with a horse or horses. **a dark horse**, (*coll.*) a person who is secretive or reserved. **a Trojan horse**, something apparently innocuous that introduces potential danger, harm etc. **from the horse's mouth**, from the original source. **to be, get on one's high horse**, to be arrogant, to put on consequential airs. **to change horses in midstream**, to alter plans, views, loyalties etc. in the middle of a project. **to eat like a horse**, (*coll.*) to eat very much. **to flog a dead horse** FLOG. **to hold one's horses**, (*coll.*) to stop; hesitate; refrain from acting. **to horse about or around**, to engage in horseplay. **to look a gift horse in the mouth**, to criticize something freely offered. **horseback**, *n.* **on horseback**, (mounted on) the back of a horse. **horse bean**, *n.* a coarse variety of bean, the broad bean. **horse block**, *n.* a block or stage to assist a person in mounting on horseback. **horse box**, *n.* a closed trailer for taking horses by road. **horse brass**, *n.* a brass decoration originally hung on a horse's harness. **horse-breaker**, *n.* one whose occupation it is to break in or to train horses. **horse chestnut**, *n.* a large variety of chestnut with coarse, bitter fruit; its fruit. **horse-cloth**, *n.* a rug to cover a horse. **horse-flesh**, *n.* the flesh of the horse, used as food; (*collect.*) horses. **horsefly**, *n.* any large fly that irritates horses. **Horse Guards**, *n.pl.* the brigade of cavalry of the English household troops; their barracks or headquarters. **horse-hair**, *n.* the long hair of the mane and tail of horses. *a.* made of this. **horse laugh**, *n.* a loud, coarse laugh. **horse mackerel**, *n.* the cavally and other fishes. **horseman, -woman**, *n.* one skilled in riding or the management of horses. **horsemanship**, *n.* **horseplay**, *n.* rough, boisterous play. **horsepower**, *n.* the power a horse can exert, used as a unit of measurement of the rate of doing mechanical work, equivalent to 33,000 foot-pounds (44.7 kJ) per minute. **horseradish**, *n.* a plant with a pungent, acrid root, used as a condiment. **horse sense**, *n.* (*coll.*) rough, practical common sense. **horseshit**, *n.* (*esp.*

N Am., taboo, sl.) nonsense; rubbish. **horseshoe,** *n.* a shoe for horses; anything resembling this in shape. *a.* shaped like this. **horseshoe crab,** *n.* any of several types of crab of N America and Asia with a heavily armoured crescent-shaped body. **horse trading,** *n.* hard bargaining. **horse-whip,** *v.t.* to thrash, to flog. **hors(e)y,** *a.* pertaining to or fond of horses or horse-racing; resembling a horse; coarse in behaviour. **horsiness,** *n.*
horst (hawst), *n.* a raised block of land separated by faults from the surrounding land.
hortative (haw'tətiv), **hortatory,** *a.* giving or containing advice or encouragement. **hortation,** *n.*
horticulture (haw'tikŭlchə), *n.* the art or science of cultivating or managing gardens. **horticultural** (-kŭl'-), *a.* **horticulturist,** *n.*
hosanna (hōzan'ə), *n.* a shout of praise and adoration.
hose (hōz), *n.* (*collect.*) stockings or socks; (*sing. with pl.* **hoses**) flexible tubing for water or other fluid, a **hose pipe.** *v.t.* to water or drench with a hose. **hosier** (-ziə), *n.* one who deals in hosiery. **hosiery,** *n.* (*collect.*) stockings, socks and tights.
hospice (hos'pis), *n.* a convent or other place for the reception and entertainment of travellers; a nursing home or hospital for the terminally ill.
hospitable (hos'pitəbl, -pit'-), *a.* entertaining or disposed to entertain strangers or guests with kindness. **hospitableness,** *n.* **hospitably,** *adv.* **hospitality** (-tal'-), *n.* liberal entertainment of strangers or guests.
hospital (hos'pitəl), *n.* an institution for the reception and treatment of the sick or injured. **hospitalize, -ise,** *v.t.* to send to hospital; to admit for hospital treatment. **hospitalization,** *n.*
hospitality HOSPITABLE.
hospitaller (hos'pitələ), *n.* (*Hist.*) one of a religious brotherhood whose office was to relieve the poor, strangers and the sick. **Knights Hospitallers,** *n.pl.* a military and charitable religious brotherhood established in the Middle Ages, esp. the Knights Hospitallers of St John of Jerusalem founded *c.* 1048.
host[1] (hōst), *n.* one who entertains another; the landlord of an inn; the compere of a TV or radio show; an animal or plant on which another is parasitic; an organism into which an organ or tissue is grafted or transplanted. *v.t.* to be the compere of; to entertain at. **hostess** (-is), *n.* a female host; the landlady of an inn or hotel; an airhostess; a woman paid to entertain customers in a bar, nightclub etc.
host[2] (hōst), *n.* a great number, a multitude. **the heavenly host,** *n.* the angels and archangels.
host[3] (hōst), *n.* the consecrated bread or wafer used in the Eucharist.
hostage (hos'tij), *n.* a person given or seized in pledge for the performance of certain conditions or for the safety of others. **to give a hostage to fortune,** put oneself at a disadvantage by risking the loss of someone or something valued highly. **hostage-taker,** *n.* **hostage-taking,** *n.*
hostel (hos'tl), *n.* a house or extra-collegiate hall for the residence of students etc.; a place of residence not run commercially, esp. for the homeless; youth hostel. **hostelling,** *n.* the practice of staying at youth hostels when travelling. **hostelry,** *n.* an inn.
hostess HOST[1].
hostile (hos'til), *a.* pertaining to an enemy; showing enmity; unfriendly; inimical. **hostile witness,** *n.* a witness whose evidence is unfavourable to the party which has called him. **hostilely,** *adv.* **hostility** (-til'-), *n.* enmity; antagonism; state of war; (*pl.*) acts of war.
hostler (hos'lə), OSTLER.
hot (hot), *a.* having a high temperature; having much sensible heat; producing a sensation of heat; burning, acrid, pungent; ardent, impetuous, passionate, fierce; (*coll.*) exciting, excited; (*Hunting*) of scent, strong; (*coll.*) super efficient, clever, quick; of news, fresh, recent; (*sl.*) stolen; wanted by the police; (*coll.*) very good; (*coll.*) radioactive. **hot under the collar,** indignant, angry. **in hot water** WATER. **in the hot seat** SEAT. **the hots,** (*esp. N Am., sl.*) strong (sexual) desire. **to hot up,** to become more intense, exciting etc. **to make a place too hot to hold one,** to make it too uncomfortable for one to stay. **hot air,** *n.* (*coll.*) boastful, empty talk. **hotbed,** *n.* a bed of earth heated by means of fermenting manure, used for raising early and tender plants; (*fig.*) of disease, vice etc., any place which favours rapid growth. **hot-blooded,** *a.* excitable, irritable, passionate. **hot-cross bun,** *n.* a spicy yeast bun with a cross marked on the top, eaten esp. on Good Friday. **hot dog,** *n.* a cooked frankfurter sausage sandwiched in a roll. **hot favourite,** *n.* the horse, runner etc. thought most likely to win in a race etc. **hot flush,** a sudden feeling of warmth accompanied by blushing, usu. associated in women with the menopause. **hot-foot,** *adv.* very hastily, swiftly. **to hotfoot it,** (*sl.*) to run; go quickly. **hot-head,** *n.* **hot-headed,** *a.* fiery, impetuous, passionate. **hothouse,** *n.* a plant-house where a relatively high artificial temperature is maintained to facilitate growth. *a.* (*coll.*) too sensitive, delicate. **hot housing,** *n.* intensive training or teaching to achieve high results. **hot line,** *n.* a telephone line for swift communication in emergencies, esp. the one between Washington and Moscow. **hot plate,** *n.* a round plate, electrically heated, on top of a cooker; a portable heatable plate for keeping food warm. **hotpot,** *n.* meat cooked with potatoes in a closed pot. **hot potato** POTATO. **hot-press,** *n.* a machine for giving a gloss to paper or linen by pressure between heated metal plates and glazed boards. *v.t.* to subject to this process. **hot rod,** *n.* (*coll.*) a car with an engine considerably modified to increase its performance greatly. **hot-rodder,** *n.* **hotshot** *n.* (*esp. N Am., coll.*) an important, often ostentatious, person. **hotspot,** *n.* a point in an engine etc. with an (excessively) high temperature; a lively nightclub or similar; a place of potential trouble; (*coll.*) a warm, sunny place, esp. a holiday resort. **hot stuff,** *n.* (*sl.*) an impressive, excellent or alluring thing or person; a sexually explicit film, book, play etc. **hot-tempered,** *a.* quick to anger; irascible. **hot tub,** *n.* a Jacuzzi. **hot-water bottle,** *n.* a usu. rubber vessel containing hot water, used for warming a bed. **hotly,** *adv.* **hotness,** *n.*
hotchpotch (hoch'poch), *n.* a confused mixture, a jumble; a dish composed of various ingredients, esp. thick broth made with mutton or other meat and vegetables.
hotel (hətel'), *n.* a commercial establishment providing accommodation, meals etc. for travellers. **hotelier** (-iə), *n.* a hotel manager or owner.
Hottentot (hot'əntot), *n.* a member of a great aboriginal people formerly inhabiting the region near the Cape of Good Hope; the language spoken by this people.
hound (hownd), *n.* a dog used in hunting (*usu. in comb.,* as *bloodhound, deerhound, foxhound* etc.); one of those who chase the hares in hare and hounds; a mean, contemptible fellow; (*coll.*) an enthusiastic seeker, as a *newshound. v.t.* to hunt or chase mercilessly with or as with hounds. **hound's tongue,** *n.* a coarse, hairy plant, of the borage family, with dull-red flowers. **hound-fish,** *n.* a dog-fish.
hour (owə), *n.* the 24th part of a natural day, the space of 60 minutes; the point of time indicated by a clock etc.; a particular time; 15° of longitude; (*pl.*) times appointed for work, attendance at office etc.; in the Roman Catholic Church, certain prayers to be said at fixed times of the day; the distance travelled in an hour. **at all hours,** at all times. **at the eleventh hour,** at the last moment. **on the hour,** at exactly one, two etc. o'clock. **the hour,** the present time. **the small hours,** the early hours of the morning. **hourglass,** *n.* a glass having two bulbs and a connecting opening through which the sand in one bulb runs into the other, formerly used for measuring time. *a.* of a wo-

houri   403   hub

man's figure, having a narrow waist and large bust and hips. **hour hand,** *n.* that hand which shows the hour on a clock or watch. **hourly,** *a.* happening or done every hour; continual. *adv.* hour by hour; frequently.

houri (hoo′ri), *n.* a nymph of the Muslim paradise; a beautiful woman.

house[1] (hows), *n.* a building for shelter or residence; a dwelling, a place of abode; a building used for a specified purpose (as *coffee-house, farmhouse, public house, warehouse*); the abode of a religious fraternity, a monastery; the fraternity itself; a household; a family or stock, esp. a noble family; an assembly, esp. one of the legislative assemblies of a country; a quorum of a legislative body; a theatre; the audience at a place of entertainment; manner of living, table; a commercial establishment; a square on a chess-board; the game of lotto; the station of a planet in the heavens; a twelfth part of the heavens; (*usu.* **House**) a type of disco or pop music characterized by electronically synthesized effects. **a halfway house,** a compromise; a mid point; a place or residence where people formerly institutionalized in prisons, mental hospitals etc. can adjust to living outside these. **house and home,** an emphatic expression for home. **house of cards,** any scheme or enterprise of an insecure or precarious kind. **house of correction,** a prison; a penitentiary. **house of God,** a church, a place of worship. **house of ill fame,** a brothel. **house of the ascendant** ASCENDANT. **house-to-house,** performed at every house (of an enquiry etc.). **like a house on fire,** very quickly and successfully. **on the house,** esp. of alcoholic drinks, given for no payment. **(as) safe as houses,** completely safe. **to bring down the house** BRING. **to keep house,** to maintain or manage a household. **to keep open house,** to provide hospitality for all comers. **to put, set etc. one's house in order,** to settle one's affairs. **house agent,** *n.* one who sells and lets houses, collects rents etc. **house arrest,** *n.* detention in one's own home under guard. **houseboat,** *n.* a boat or barge with a cabin or house for living in. **housebound,** *a.* unable to leave one's house, e.g. because of a disability. **housebreaker,** *n.* one who breaks into and robs houses. **housebreaking,** *n.* **house-broken,** *a.* HOUSE-TRAINED. **housecoat,** *n.* a woman's long over-garment, worn in the house, a dressing gown. **house-dog,** *n.* a dog kept to guard the house. **housefather, -mother,** *n.* a man or woman in charge of children in an institution. **housefly,** *n.* the common fly. **house guest,** *n.* a guest in a private house. **house-husband,** *n.* a married man who stays at home to run a household instead of having a paid job. **houseleek,** *n.* a plant with thick, fleshy leaves growing on the tops of walls and houses in Britain. **houseman,** *n.* a junior doctor in a hospital. **house martin,** *n.* a black and white bird with a forked tail, resembling a swallow. **housemaster, -mistress,** *n.* a teacher in charge of a house of residence at a boarding-school. **houseparent,** *n.* a housefather or housemother. **house party,** *n.* a party of guests at a country house. **house plant,** *n.* a plant for growing indoors. **house-proud,** *a.* taking a pride in the care and embellishment of a home. **houseroom,** *n.* accommodation or storage space in a house. **house-sitter,** *n.* a person who stays in a house to look after it while the occupier is away. **house-sit,** *v.i.* **house sparrow,** *n.* the common sparrow. **house-trained,** *a.* of an animal, trained not to foul places indoors; of a person, well-mannered. **house-warming,** *n.* a celebration held on moving into a new house. **housework,** *n.* work connected with housekeeping. **houseful,** *n.* as many or as much as a house will hold.

house[2] (howz), *v.t.* to place or store in a house; to lodge, contain; to shelter; to provide housing for.

household (hows′hōld), *n.* those who live together under the same roof; a domestic establishment. *a.* pertaining to the house and family, domestic. **household troops,** *n.pl.* troops specially employed to guard the person of the sovereign. **household name, word,** *n.* a familiar name or word. **householder,** *n.* the head of a household, the occupier of a house.

housekeeper (hows′kēpə), *n.* one (usu. a woman) who is employed to manage the affairs of a household. **housekeeping,** *n.* the care of a household; domestic economy.

housemaid (hows′mād), *n.* a female servant employed to keep a house clean etc. **housemaid's knee,** inflammation of the knee-cap, due to much kneeling.

housewife (hows′wīf), *n.* a married woman who stays at home to run a household instead of having a paid job. **housewifely,** *a.* **housewifery** (-wifəri), *n.*

housey-housey (howsihow′si), BINGO[2].

housing (how′zing), *n.* lodging, shelter, accommodation; part of a mechanism or structure intended to contain, protect, cover etc. **housing association,** *n.* a non profit-making body which builds or renovates dwellings and lets them at a reasonable rent. **housing estate,** *n.* a planned residential area; such an estate built by a local authority.

houyhnhnm (win′im), *n.* one of the race of horses with the finer human characteristics, in Jonathan Swift's *Gulliver's Travels.*

hove (hōv), *past* HEAVE.

hovel (hov′l, huv′l), *n.* a miserable dwelling-house.

hover (hov′ə), *v.i.* to hang or remain (over or about) fluttering in the air or on the wing; to loiter (about); to be irresolute, to waver. **Hovercraft**®, *n.* an aircraft supported above land or water on a cushion of air which it generates itself. **hoverfly,** *n.* any brightly-coloured fly of the family Syrphidae, which hover and dart. **hoverport,** *n.* a place where passengers enter and leave hovercraft.

how (how), *adv.* in what way or manner; by what means; to what extent, degree etc.; in what proportion; in what condition. **and how!** *int.* (*sl.*) and how much more! **how about?** used to suggest a possible choice. **how come?** (*coll.*) how does it, did that etc. happen? **how-do-you-do?** how are you? a conventional form of greeting. **how-d'ye-do,** *n.* (*coll.*) an awkward situation. **how's that?** used in cricket to ask for the batsman to be given out. **however,** *adv.* in whatever manner or degree; nevertheless, notwithstanding. **howsoever,** *adv.* in whatever manner; however.

howdah, houdah (how′də), *n.* a seat, usu. canopied, carried on an elephant's back.

howdy (how′di), *n.* (*esp. N Am.*) a greeting short for HOW-DO YOU-DO, see HOW.

however HOW.

howitzer (how′itsə), *n.* a short, light or heavy piece of ordnance with a high trajectory and low muzzle velocity.

howl (howl), *v.i.* to utter a protracted hollow cry; to cry as a dog or wolf; to weep; to laugh; to make a wailing sound like the wind. *v.t.* to utter in wailing or mournful tones. *n.* the cry of a wolf or dog; a protracted, hollow cry, esp. one of anguish, distress or derision. **howler,** *n.* one who howls; a S American monkey; (*coll.*) a ludicrous blunder. **howling,** *a.* that howls; (*sl.*) extreme, glaring.

hoy (hoi), *int.* an exclamation to draw attention etc.

Hoya (hoi′ə), *n.* a genus of tropical climbing shrubs with pink, white or yellow flowers, commonly called the wax flowers.

hoyden (hoi′dən), *n.* a boisterous girl. **hoydenish,** *a.*

HP, hp, (*abbr.*) high pressure; horsepower; hire purchase; Houses of Parliament.

HQ, (*abbr.*) headquarters.

HR, (*abbr.*) Home Rule; House of Representatives.

hr., hr, (*abbr.*) hour.

HRH, (*abbr.*) His/Her Royal Highness.

HRT, (*abbr.*) hormone replacement therapy.

hrw, (*abbr.*) heated rear window (of a car).

ht, (*abbr.*) high tension.

hub (hŭb), *n.* the central part of a wheel from which the spokes radiate, the nave; a place of central importance.

hubcap, *n*. a (decorative) plate or disk covering the hub of a wheel.
hubble bubble (hŭblbŭb′l), *n*. a kind of hookah; a bubbling noise; a hubbub; a jabbering or chattering.
hubbub (hŭb′ŭb), *n*. a confused noise; a noisy disturbance; a tumult, an uproar.
hubby (hŭb′i), (*coll.*) husband.
hubris (hū′bris), *n*. insolent pride or security, arrogance. **hubristic** (-bris′-), *a*.
huckaback (hŭk′əbak), *n*. a coarse linen or cotton cloth, with a rough surface, used for table-cloths and towels.
huckleberry (hŭk′lberi), *n*. the edible fruit of low shrubs bearing dark-blue berries; the fruit of the blueberry and other allied species.
huckster (hŭk′stə), *n*. a retailer of small goods, a pedlar, a hawker; a mean, trickish, mercenary fellow; (*N Am.*) a person who produces advertising material for radio or TV. *v.i.* to deal in petty goods; to bargain, to haggle.
huddle (hŭd′l), *v.t.* to throw or crowd (together, up etc.) promiscuously; to do or make hastily and carelessly; to hunch (oneself up); to put (on) hurriedly or anyhow. *v.i.* to gather or crowd (up or together) promiscuously. *n.* a confused crowd; disorder, confusion; (*coll.*) a secretive discussion between a group of people.
hue (hū), *n*. colour, tint; a compound colour, esp. one in which a primary predominates. **hued**, *a*. having a particular hue (*esp. in comb.*, as *light-hued*).
hue and cry, *n*. a cry or general summons to pursue a felon or offender; a clamour or outcry (against); a great stir or alarm.
huff (hŭf), *v.i.* to make a puffing noise; *v.i., v.t.* to give or take offence. *n.* a sudden fit of anger or petulance. **huffy**, *a*. **huffily**, *adv*. **huffiness**, *n*.
hug (hŭg), *v.t.* (*past, p.p.* **hugged**) to embrace closely; to clasp or squeeze tightly; to hold fast or cling to, to cherish; of a ship, to keep close to (the shore). *n.* a close embrace. **huggable**, *a*.
huge (hūj), *a.* very large; enormous, immense. **hugely**, *adv.* exceedingly, extremely. **hugeness**, *n*.
huggermugger (hŭg′əmŭg′ə), *n.* (*rarely*) secrecy, privacy; disorder, confusion. *a., adv.* (*rarely*) clandestine(ly); confused(ly).
Huguenot (hū′gənō), *n.* a name formerly applied to the Protestants of France.
huh (hŭ), *int.* used to express surprise, contempt, disbelief etc.
hula (hoo′lə), **hula-hula**, *n.* a Hawaiian dance performed by women. **Hula Hoop**®, *n.* a light hoop kept in motion by swinging round the waist. **hula skirt**, *n.* a grass skirt worn by hula dancers.
hulk (hŭlk), *n.* the hull or body of a ship, especially an unseaworthy or unwieldy one; (*derog.*) any unwieldy object or person. **the hulks**, *n.pl.* old dismasted ships formerly used as convict prisons. **hulking**, *a.* bulky, unwieldy, awkward.
hull[1] (hŭl), *n.* the outer covering of anything, especially of a nut or seed; the pod, shell or husk. *v.t.* to strip the hull or husk off.
hull[2] (hŭl), *n.* the body of a ship. *v.t.* to pierce the hull of.
hullabaloo (hŭləbəloo′), *n.* an uproar.
hullo HELLO.
hum (hŭm), *v.i.* (*past, p.p.* **hummed**) to make a prolonged murmuring sound like a bee; to sing with the lips closed; to make an inarticulate sound in speaking, from embarrassment or hesitation; (*sl.*) to smell unpleasant; (*sl.*) to be very active. *v.t.* to utter in a low murmuring voice. *n.* a low droning or murmuring sound; the act of humming; an inarticulate expression of hesitation, disapproval etc. **to hum and ha**, to hesitate in speaking; to refrain from giving a decided answer. **to make things hum**, to stir (people etc.) into activity.
human (hū′mən), *a.* pertaining to man or mankind; having the nature, qualities or characteristics of man; of or pertaining to mankind as dist. from divine, animal or material. *n.* a human being. **human being**, *n.* a member of the human race; a person. **human interest**, *a.* (of a newspaper story etc.) having appeal to human emotions. **human nature**, *n.* all those characteristics considered typical of human beings, esp. the weaknesses. **human rights**, *n.pl.* the rights of an individual to freedom of speech, freedom of movement, justice etc. **humankind**, *n.* the human race. **humanly**, *adv.* after the manner or according to the knowledge or capacity of human beings. **humanness**, *n.* **humanoid** (-oid), *n., a.* (a being) resembling a human in form or attributes.
humane (hūmān′), *a.* tender, compassionate, kind, gentle; relieving distress, aiding those in danger etc. **humane killing**, *n.* a method of slaughtering animals painlessly; the killing of an animal by such a method. **Humane Society**, *n.* a society that campaigns for humane behaviour, esp. in the treatment of animals. **humanely**, *adv.* **humaneness**, *n.*
humanism (hū′mənizm), *n.* a moral or intellectual system that regards the interests of mankind as of supreme importance, in contradistinction to theism; devotion to humanity or human interests; culture derived from literature, esp. the Greek and Latin classics. **humanist**, *n.* one versed in human history or the knowledge of human nature; one versed in the humanities, esp. one of the classical scholars of the Renaissance. **humanistic** (-is′-), *a.* **humanistically**, *adv.*
humanitarian (hūmaniteə′riən), *a.* humane. *n.* a philanthropist. **humanitarianism**, *n.*
humanity (hūman′iti), *n.* human nature; (*collect.*) the human race; kindness, benevolence, humaneness; humanism. **the humanities**, *n.pl.* the study of literature, music, history, esp. of Ancient Rome and Greece etc. distinguished from social or natural sciences. **humanize, ise** (hū′-), *v.t.* to render human; to give human character or expression to; to render humane. *v.i.* to become human or humane. **humanization**, *n.*
humanoid HUMAN.
humble (hŭm′bl), *a.* having or showing a sense of lowliness or inferiority, modest; of lowly condition, kind, dimensions etc.; submissive, deferential. *v.t.* to lower; to bring to a state of subjection or inferiority; to abase. **to eat humble pie**, to submit oneself to humiliation or insult; to apologize humbly. **humbleness**, *n.* **humbly**, *adv.*
humbug (hŭm′bŭg), *n.* a hoax, a sham; nonsense; an impostor; a sweet highly flavoured with peppermint. *int.* nonsense. *v.t.* (*past, p.p.* **humbugged**) to hoax, to take in.
humdinger (hŭmding′ə), *n.* (*coll.*) an excellent person or thing.
humdrum (hŭm′drŭm), *a.* dull, commonplace, tedious. **humdrumness**, *n.*
Humean (hū′miən), *a.* of or pertaining to the philosophical doctrines of David Hume (1711–76). **Humism**, *n.*
humectant (hūmek′tənt), *a.* moistening. *n.* a substance that increases the fluidity of another substance.
humerus (hū′mərəs), *n.* the long bone of the upper arm; the corresponding bone in the foreleg of quadrupeds. **humeral** (hū′mərəl), *a.*
humic (hū′mik), *a.* pertaining to mould or earth. **humify**, *v.i.* to turn into humus. **humification** (-fi-), *n.*
humid (hū′mid), *a.* moist, damp. **humidify** (-mid′-), *v.t.* **humidification** (-fi-), *n.* **humidifier**, *n.* a device for increasing the amount of moisture in the air. **humidity** (-mid′-), *n.* the state of being humid; a measure of the amount of moisture in the atmosphere. **humidor** (-daw), *n.* a box or room for keeping cigars moist.
humiliate (hūmil′iāt), *v.t.* to lower in self-esteem, to mortify; to humble, to lower in condition, to abase. **humiliating**, *a.* **humiliation**, *n.*
humility (hūmil′iti), *n.* the state of being humble; modesty, a sense of unworthiness; self-abasement.

humming (hŭm'ing), *a.* that hums. *n.* a low, prolonged murmuring sound. **hummingbird**, *n.* one of a family of diminutive birds, mostly tropical, of brilliant plumage and very rapid flight.
hummock (hŭm'ək), *n.* a mound or hillock; a protuberance formed by pressure in an icefield; (*N Am.*) an elevation in a swamp or bog, esp. if wooded. **hummocky**, *a.*
hummum (hŭm'əm), HAMMAM.
hummus (hŭm'əs, hu'məs), *n.* a kind of Middle Eastern hors d'oeuvre consisting of pureed chick-peas, tahini, garlic and lemon.
humor (hū'mə), HUMOUR.
humoresque (hūməresk'), *n.* a musical composition of a humorous or capricious character.
humorist (hū'mərist), *n.* one who displays humour in his conversation, writings etc.
humorous (hū'mərəs), *a.* full of humour; tending to excite laughter; jocular. **humorously**, *adv.*
humour, (*esp. N. Am.*) **humor** (hū'mə), *n.* mental disposition, frame of mind, mood; drollery, comicality; the capacity of perceiving the ludicrous elements in life or art. *v.t.* to fall in with the humour of; to indulge, to give way to, to make concessions to. **out of humour**, in an ill-temper, displeased. **humoured**, *a.* having a certain humour (*usu. in comb.*, as **good-humoured**). **humouredly**, *adv.* (*usu. in comb.*) **humourless**, *a.*
hump (hŭmp), *n.* a swelling or protuberance, esp. on the back; a rounded hillock; (*coll.*) a fit of annoyance, ill-temper or the blues. *v.t.* to make hump-shaped; (*coll.*) to carry; (*sl.*) to have sexual intercourse with. *v.i.* (*sl.*) to have sexual intercourse. **over the hump**, (*coll.*) past the difficult or critical stage of something. **humpback**, *n.* a hunchback; an American whale also called the humpbacked whale. **humpback bridge**, *n.* a small, narrow bridge with steep inclines on either side leading to its centre. **humpbacked**, *a.* **humped**, *a.* having a hump. **humpy**, *a.*
humph (hŭmf), *int.* expressing doubt, disapproval etc.
humpy (hŭm'pi), *n.* (*Austral.*) an Aborigine hut; a shack, a lean-to.
humus (hū'məs), *n.* soil or mould, esp. that largely composed of decayed vegetation.
Hun (hŭn), *n.* one of an ancient people from Asia, that overran Europe in the 4th and 5th cents., and gave their name to Hungary; (*coll., derog.*) a German; a barbarian, a destroyer. **Hunnish**, *a.*
hunch (hŭnch), *n.* a hump; a lump, a thick piece; an intuition. *v.t.* to draw (oneself) in, to bring together parts of the body (e.g. the shoulders). **hunchback**, *n.* a person with a humped back. **hunchbacked**, *a.* **hunched**, *a.*
hundred (hŭn'drid), *n.* the cardinal number representing 10 times 10; the product of 10 multiplied by 10. **hundreds and thousands**, tiny strips or balls of sugar coated with different bright colours, used esp. for cake decoration. **hundredweight**, *n.* a weight of 112 lb av. (50·8 kg).
hundredth, *a.* the ordinal of a hundred. *n.* one of a hundred equal parts; the one after the ninety-ninth in a series.
hung (hŭng), *past* HANG. *a.* of an election, not resulting in a clear majority for any party; of a Parliament, produced by such an election; of a jury, unable to reach a verdict. **to be hung over**, (*coll.*) to be suffering the after effects of too much alcohol. **to be hung up on**, to be obsessively interested (in a person) or concerned with (a matter).
Hungarian (hŭng-geə'riən), *a.* pertaining to Hungary. *n.* a native or inhabitant of Hungary; the Hungarian language.
hunger (hŭng'gə), *n.* a craving for food; a painful sensation caused by the want of food; any strong desire. *v.i.* to desire or long eagerly. **hunger march**, *n.* a march of the unemployed to protest against their lot. **hunger strike**, *n.*, *v.i.* a refusal to take food, usu. as a political protest. **hunger-striker**, *n.*

hungry (hŭng'gri), *a.* feeling a sensation of hunger; having a keen appetite; showing hunger, emaciated, thin; causing hunger; longing or craving eagerly; barren, poor (of soil). **hungrily**, *adv.* **hungriness**, *n.*
hunk (hŭngk), *n.* (*coll.*) a large piece; a big, strong, sexually attractive man.
hunker (hŭng'kə), *n.pl.* the haunches. **on one's hunkers**, squatting down.
hunky-dory (hŭngkidaw'ri), *a.* (*esp. N Am., coll.*) satisfactory, fine.
hunt (hŭnt), *v.t.* to chase (as wild animals) for the purpose of catching and killing; to employ (horses, dogs etc.) in hunting; to pursue or chase in or over (a district etc.); to search for, to seek after. *v.i.* to follow the chase; to pursue game or wild animals; to search (after or for); of a machine etc., to vary in speed of operation. *n.* hunting, the chase; a search; a pack of hounds; a group of people who regularly go hunting together; a district hunted by a pack of hounds. **to hunt down**, to bring to bay; to search out and destroy. **to hunt out**, to track down, to find by searching. **to hunt up**, to search for. **hunt ball**, *n.* a ball given by the members of a hunt. **hunter**, *n.* one who follows the chase; a huntsman; a horse trained for hunting; one who searches or seeks for anything (*usu. in comb.*, as *fortune-hunter*); a watch with a hinged metal cover over the face (or a **half-hunter** with a glass disc). **hunter-killer**, *n.* a naval craft designed to pursue and destroy enemy craft. **hunter's moon**, *n.* the full moon after harvest moon. **hunting**, *a.* chasing game or wild animals; pertaining or given to hunting. **hunting horn**, *n.* a bugle or horn used in the chase; the second pommel of a side-saddle. **hunting knife**, *n.* a knife used for killing game at bay, or skinning it. **huntress** (-tris), *n. fem.* **huntsman**, *n.* one who hunts; a person employed to manage the hounds, esp. the foxhounds. **huntsmanship**, *n.*
Huon pine (hū'on), *n.* a large Tasmanian yew, valued for its finely-marked wood.
hurdle (hœ'dl), *n.* a movable framework of twigs or split timber serving for gates, enclosures etc.; a barrier for jumping over in racing; (*pl.*) a race over hurdles; a barrier or obstacle. *v.i.* to participate in a hurdle race. **hurdler**, *n.* one who runs in a race with hurdles.
hurdy-gurdy (hœdigœ'di), *n.* a barrel-organ, or other similar instrument which is played with a handle.
hurl (hœl), *v.t.* to throw with violence; to utter or emit with vehemence.
hurley (hœ'li), **hurling** (hœ'ling), *n.* an Irish game resembling hockey.
hurly-burly (hœlibœ'li), *n.* a tumult, commotion, uproar.
hurrah, hooray (hərah'), **hurray, hoorah** (-rā'), *int.* an exclamation of joy, applause etc.
hurricane (hŭ'rikən), *n.* a storm with violent wind; an extremely violent gale, orig. a W Indian cyclone; anything that sweeps along violently. **hurricane lamp**, *n.* a glass-covered lamp designed to keep alight in a wind.
hurry (hŭ'ri), *v.t.* to impel to greater speed, to accelerate; to push forward; to drive or cause to act or do carelessly or precipitately. *v.i.* to hasten; to move or act with excessive haste. *n.* the act of hurrying; urgency, bustle, precipitation; eagerness (to do etc.); (*coll.*) need for haste. **hurried**, *a.* impelled to speed; done in a hurry, hasty. **hurriedly**, *adv.* **hurriedness**, *n.*
hurt (hœt), *v.t.* (*past, p.p.* **hurt**) to cause pain, injury, loss or detriment to; to damage; to grieve or distress (as the feelings). *v.i.* (*usu. impers.*) to be painful, to cause pain. *n.* a wound; an injury, damage, harm. **hurtful**, *a.* causing hurt; mischievous, noxious. **hurtfully**, *adv.* **hurtfulness**, *n.*
hurtle (hœ'tl), *v.t.* to strike or dash against with violence; to move or whirl with great force. *v.i.* to rush with great force and noise; to make a crashing noise.
husband (hŭz'bənd), *n.* a man joined to a woman in marriage. *v.t.* to manage with frugality, to economize.

husbandry (hŭz'bəndri), *n.* the business of a farmer, agriculture; economy, esp. domestic; frugality, careful management.
hush (hŭsh), *v.t.* to make silent; to repress the noise of. *v.i.* to be still or silent. *n.* silence, stillness. *int.* silence! be still! **to hush up**, to keep concealed, to suppress. **hush-hush**, *a. (coll.)* very secret. **hush money**, *n.* a bribe paid to secure silence (about a scandal etc.). **hushed**, *a.*
husk (hŭsk), *n.* the dry external integument of certain fruits or seeds; a mere frame, shell or worthless part. *v.t.* to strip the husk from. **husked**, *a.* having or covered with a husk; stripped of a husk. **husker**, *n.* **husking**, *n.* the act of stripping off husks.
husky[1], *a.* dry, hoarse, rough and harsh in sound (of the voice); *(esp. N Am., coll.)* strong, stalwart. **huskily**, *adv.* **huskiness**, *n.*
husky[2] (hŭs'ki), *n.* a Canadian sledge-dog; *(Canada coll.)* an Eskimo; the Eskimo language.
hussar (həzah'), *n.* originally a light horseman in the national cavalry of Hungary; now, a soldier of a light cavalry regiment in European armies.
hussy (hŭs'i), *n.* a pert, forward girl; a worthless woman.
hustings (hŭs'tĭng), *n.pl. (formerly)* a platform from which candidates addressed the electors during parliamentary elections; proceedings at an election.
hustle (hŭs'l), *v.t.* to shake together in confusion; to jostle, to push violently; to hurry or cause to move quickly; *(sl.)* to acquire (something) by aggressive or dishonest means. *v.i.* to press roughly; to hurry; to push one's way in an unceremonious or unscrupulous way; *(sl.)* to make a living by aggressive or dishonest means; *(esp. N Am., sl.)* to engage in prostitution. *n.* hustling. **hustler**, *n.*
hut (hŭt), *n.* a small, rude house, a mean dwelling; a cabin, a hovel.
hutch (hŭch), *n.* a coop or box-like pen for small animals esp. rabbits; in mining, a truck, for carrying ore.
huzza (həzah'), *int.* a cry of joy, applause etc. *v.i.* to shout 'huzza'. *v.t.* to applaud or greet with this cry. *n.* a shout of 'huzza'.
hw, *(abbr.)* hit wicket.
HWM, *(abbr.)* n. high water mark.
hwyl (hū'əl), *n.* passion or fervour, esp. in rhetoric.
hyacinth (hī'əsinth), *n.* a bulbous-rooted flowering plant of the order Lilaceae; a brownish, orange or reddish gem stone; a colour ranging from purplish-blue to violet.
Hyades (hī'ədēz), **Hyads** (-adz), *n.pl.* a cluster of stars, including Aldebaran, in the head of Taurus, supposed by the ancients to bring rain when they rose with the sun.
hyaena (hĭe'nə), HYENA.
hyaline (hī'əlin, -lin), *a.* glassy, transparent, crystalline; vitreous. **hyaline cartilage**, *n.* a translucent cartilage found in joints and respiratory passages, containing little fibrous tissue. **hyalite** (-līt), *n.* a glassy variety of opal.
hyaloid (hī'əloid), *a.* glassy, vitriform. *n.* the hyaloid membrane. **hyaloid membrane**, *n.* the transparent membrane enclosing the vitreous humour of the eye.
hybrid (hī'brid), *a.* produced by the union of two distinct species, varieties etc.; produced by cross-fertilization or interbreeding; derived from incongruous sources. *n.* an animal or plant produced by the union of two distinct species, varieties etc.; a word compounded from different languages; anything composed of heterogeneous parts or elements. **hybridity** (-brĭd'-), *n.* the state or quality of being hybrid. **hybridism**, *n.* hybridity; the act or process of interbreeding, hybridization. **hybridize**, **-ise**, *v.t.* to produce by the union of different species or varieties; to produce by cross-fertilization or interbreeding. *v.i.* to produce hybrids, to be capable of cross-fertilization or interbreeding. **hybridization**, *n.*
hydatid (hī'dətid), *n.* a watery cyst occurring in animal tissue, esp. one resulting from the development of the embryo of a tapeworm. **hydatic** (-dat'-), *a.*

hydatoid (hī'dətoid), *a.* resembling water.
hydr- HYDR(O)-.
Hydra (hī'drə), *n.* in Greek mythology, a water-serpent with many heads, each of which, when cut off, was succeeded by two, destroyed by Hercules. **hydra**, *n.* an evil or calamity difficult to extinguish; *(Astron.)* one of the 15 ancient southern constellations; a genus of freshwater polyps which multiply when divided. **hydra-headed**, *a.* having many heads; hence, difficult to get rid of; spreading.
hydracid (hīdrăs'id), *n., a. (Chem.)* (of or pertaining to) an acid containing hydrogen but no oxygen.
Hydrangea (hīdrān'jə), *n.* a genus of flowering shrubs of the saxifrage family, from Asia and America.
hydrant (hī'drənt), *n.* a spout or discharge pipe, usu. with a nozzle for attaching hose, connected to a water-main for drawing water.
hydrargyrum (hīdrah'jirəm), *n.* mercury, quicksilver. **hydrargyral** (-ral), *a.*
hydrate (hī'drāt), *n.* a compound of water with an element or another compound. *v.t.* to combine with water to form a hydrate. **hydration**, *n.* **hydrator**, *n.*
hydraulic (hīdrol'ik), *a.* pertaining to fluids in motion, or to the power exerted by water conveyed through pipes or channels; operating or operated by such power. *n.pl.* the science of water or other liquids both at rest and in motion, esp. the conveyance of water through pipes etc., and the practical application of water-power. **hydraulic lift**, *n.* a lift worked by means of water-power. **hydraulic press**, *n.* a heavy pressing machine worked by water-power. **hydraulic ram**, *n.* a machine by which the fall of a column of water supplies power to elevate a portion of the water to a greater height than that at the source. **hydraulically**, *adv.*
hydrazine (hī'drəzēn, -zin), *n.* a colourless corrosive liquid that is a strong reducing agent, used esp. in rocket fuel.
hydric (hī'drik), *a.* of, pertaining to or containing hydrogen in chemical combination; containing or related to moisture. **hydride** (-drīd), *n.* a compound of hydrogen with another element or radical. **hydriodic** (-driod'-), *a.* of, pertaining to, or containing hydrogen and iodine in chemical combination.
hydro[1] (hī'drō), *n. (coll.)* a hydropathic establishment.
hydro[2] (hī'drō), *a.* hydroelectric. *n.* hydroelectricity.
hydr(o)- (hīdrō-), *comb. form.* pertaining to or connected with water; containing hydrogen in chemical combination; of a mineral, containing water as a constituent; belonging to the genus *Hydra* or the class Hydrozoa.
hydrobromic (hīdrōbrō'mik), *a.* composed of hydrogen and bromine. **hydrobromic acid**, *n.*
hydrocarbon (hīdrōkah'bən), *n.* a compound of carbon and hydrogen.
hydrocele (hī'drəsēl), *n.* an accumulation of fluid in a saclike cavity, esp. in the scrotum.
hydrocephalus (hīdrəsef'ələs), **-cephaly** (-li), *n.* water on the brain. **hydrocephalic** (-fal'-), **hydrocephalous**, *a.* pertaining to or akin to hydrocephalus.
hydrochloric (hīdrəklaw'rik), *n.* a compound of chlorine and hydrogen. **hydrochloric acid**, *n.* a solution of hydrogen chloride in water, a strong corrosive acid.
hydrochloride (hīdrəklaw'rīd), *n.* a compound of hydrochloric acid, esp. with an organic base.
hydrocortisone (hīdrōkaw'tizōn), *n.* the steroid hormone naturally secreted by the adrenal cortex, synthesized to treat, e.g. rheumatoid arthritis, skin diseases.
hydrocyanic (hīdrōsīan'ik), *a.* formed by the combination of hydrogen and cyanogen. **hydrocyanic acid**, *n.*
hydrodynamics (hīdrōdinam'iks), *n.* the science which deals with water and other liquids in motion, hydromechanics. **hydrodynamic**, **-ical**, *a.* pertaining to hydrodynamics; derived from the force of water.
hydroelectric (hīdrōilek'trik), *a.* pertaining to electricity

hydro-extractor   407   **hymn**

generated from water-power. **hydroelectricity** (-eliktris'-), *n.*
hydro-extractor (hīdrōikstrak'tə), *n.* an apparatus for removing moisture.
hydrofluoric (hīdrōfloo·o'rik), *a.* consisting of fluorine and hydrogen. **hydrofluoric acid**, *n.*
hydrofoil (hī'drəfoil), *n.* a fast vessel with one or more pairs of vanes attached to its hull which lift it out of the water at speed; such a vane.
hydrogen (hī'drəjən), *n.* (*Chem.*) an invisible, inflammable, gaseous element, the lightest of all known bodies, which in combination with oxygen produces water. **hydrogen bomb**, *n.* an exceedingly powerful bomb in which an immense release of energy is obtained by the conversion by fusion of hydrogen nuclei into helium nuclei, the H-bomb. **hydrogen chloride**, *n.* a colourless pungent corrosive gas obtained from the interaction of sulphuric acid and sodium chloride. **hydrogen cyanide**, *n.* a colourless poisonous liquid faintly redolent of bitter almonds. **hydrogen peroxide** PEROXIDE. **hydrogen sulphide**, *n.* a colourless poisonous gas smelling of rotten eggs. **hydrogenate, -nize, -nise** (hīdroj'-, hī'-), *v.t.* to cause to combine with hydrogen; to charge with hydrogen. **hydrogenation, -genization**, *n.* **hydrogenous** (-droj'-), *a.*
hydrography (hīdrog'rəfi), *n.* the science and art of studying, surveying and mapping seas, lakes, rivers and other waters, and their physical features, tides, currents etc. **hydrographer**, *n.* **hydrographic** (-graf'-), *a.*
hydrokinetics *n.* the kinetics of liquids.
hydrology (hīdrol'əji), *n.* the science of water, its properties, phenomena, laws and distribution. **hydrological** (-loj'-), *a.* **hydrologist**, *n.*
hydrolysis (hīdrol'isis), *n.* the formation of an acid and a base from a salt by the action of water. **hydrolyse, -lyze**, *v.t.* to subject to hydrolysis. **hydrolytic** (-lit'-), *a.*
hydromechanics (hīdrōmikan'iks), *n.* the mechanics of liquids.
hydrometer (hīdrom'itə), *n.* an instrument for determining the specific gravity of liquids or solids by means of flotation. **hydrometric, -ical** (-met'-), *a.* **hydrometry**, *n.* the art or process of measuring the specific gravity of fluids etc.
hydropathy (hīdrop'əthi), *n.* the treatment of disease by the internal and external application of water, hydrotherapy. **hydropathic** (-path'-), *a.* **hydropathically**, *adv.*
hydrophane (hī'drəfān), *n.* an opal which becomes translucent when immersed in water.
hydrophilic (hīdrəfil'ik), *a.* having a great affinity for water.
hydrophobia (hīdrəfō'biə), *n.* an unnatural dread of water, a symptom of rabies; rabies. **hydrophobic**, *a.*
hydrophone (hī'drōfōn), *n.* an instrument for detecting sound by water.
hydrophyte (hī'drəfit), *n.* an aquatic plant. **hydrophytic** (-fit'-), *a.*
hydroplane (hī'drəplān), *n.* a light motor-boat capable of rising partially above the surface of water; a flat fin for governing the vertical direction of a submarine; a plane for lifting a boat partially from the water, so as to diminish the resistance and increase the speed.
hydroponics (hīdrəpon'iks), *n.pl.* the cultivation of plants without soil in water containing chemicals. **hydroponic**, *a.* **hydroponically**, *adv.*
hydropower, *n.* hydroelectric power.
hydroquinone (hīdrōkwin'ōn), *n.* a compound derived from quinone, employed in the development of photographs.
hydrosphere (hī'drəsfiə), *n.* the watery envelope of the earth.
hydrostatic (hīdrəstat'ik), *a.* pertaining or relating to hydrostatics; pertaining to the pressure and equilibrium of liquids at rest. **hydrostatics**, *n.pl.* the science concerned with the pressure and equilibrium of liquids at rest.

hydrotherapeutic (hīdrōthcrəpū'tik), *a.* pertaining to the therapeutic application of water; hydropathic. **hydrotherapist** (-the'-), *n.* **hydrotherapy** (-the'-), *n.*
hydrothermal (hīdrəthœ'məl), *a.* relating to the action of heated water, esp. on the materials of the earth's crust.
hydrotropism (hīdrot'rəpizm), *n.* the tendency in the growing parts of plants to turn towards or away from moisture. **hydrotropic** (-trop'-), *a.* **hydrotropically**, *adv.*
hydrous (hī'drəs), *a.* containing water.
hydroxide (hīdrok'sīd), *n.* a compound formed by the union of a basic oxide with the molecules of water.
hydroxyl (hīdrok'sil), *n.* the monad radical formed by the combination of one atom of hydrogen and one of water occurring in many chemical compounds.
Hydrozoa (hīdrəzō'ə), *n.pl.* a class of coelenterates, principally marine, comprising the hydra, medusa, jellyfish etc. **hydrozoan**, *n.*, *a.*
hyena, hyaena (hīē'nə), *n.* a genus of carnivorous quadrupeds allied to the dog, including the laughing hyena.
hyetal (hī'ətəl), *a.* of or belonging to rain; relating to the rainfall of different countries.
Hygeia (hījē'ə), *n.* the goddess of health.
hygiene (hī'jēn), *n.* the science of the prevention of disease; practices that promote health; sanitary science. **hygienic** (-jēn'-), *a.* **hygienics**, *n.* hygiene. **hygienically**, *adv.* **hygienist** (-jēn'-), *n.*
hygr(o)-, *comb. form.* moist, pertaining to or denoting the presence of moisture.
hygrometer (hīgrom'itə), *n.* an instrument for measuring the moisture of the air etc. **hygrometric** (-met'-), *a.* **hygrometry**, *n.* the measurement of moisture, esp. of the air.
hygrophilous (hīgrof'iləs), *a.* living or growing in moist places.
hygroscope (hī'grəskōp), *n.* an instrument for indicating the degree of moisture in the atmosphere. **hygroscopic**, *a.* pertaining to or indicated by the hygroscope; imbibing moisture from the atmosphere (of bodies). **hygroscopically**, *adv.* **hygroscopicity** (-skəpis'-), *n.*
hyleg (hī'leg), *n.* the planet ruling, or in the sign of the zodiac above the eastern horizon, at the hour of a person's nativity.
hylic (hī'lik), *a.* of or relating to matter; material.
hylo-, *comb. form.* of matter; pertaining to wood.
hylogenesis (hīləjen'əsis), *n.* the origin of matter.
hylomorphism (hīləmaw'fizm), *n.* the philosophy that finds the first cause of the universe in matter.
hylotheism (hīləthē'izm), *n.* the system which regards God and matter as identical; pantheism. **hylotheist**, *n.*
hylotomous (hīlot'əməs), *a.* of certain insects, wood-cutting.
hylozoism (hīləzō'izm), *n.* the doctrine that matter is necessarily endowed with life.
Hymen[1] (hī'mən), *n.* the god of marriage. **hymeneal** (-nē'əl), *a.* pertaining to marriage. *n.* a marriage song.
hymen[2] (hī'mən), *n.* a membrane stretched across the vaginal entrance; the fine pellicle enclosing a flower in the bud.
hymenium (hīmē'niəm), *n.* (*pl.* **-nia** (-ə), **-niums**) (*Bot.*) the spore-bearing stratum or surface in fungi.
hymenomycete (hīmənōmī'sēt), *n.* one of the Hymenomycetae, an order of fungi characterized by an exposed hymenium.
Hymenoptera (hīmənop'tərə), *n.pl.* an order of insects having four membranous wings, as the bee, wasp, ant etc. **hymenopterous**, *a.* **hymenopteran, hymenopteron**, *n.*
hymn (him), *n.* a song or ode in praise or adoration of God or some deity; a sacred or solemn song or ode, esp. a religious song not taken from the Bible. *v.t.* to praise or worship in hymns. *v.i.* to sing hymns. **hymn book**, *n.* a book of hymns. **hymnal** (him'nəl), *n.* a collection of hymns, esp. for public worship. **hymnary** (-nə-), *n.* a hymnal.

hymnic (nik), *a*. **hymnist** (-nist), *n*. a composer of hymns. **hymnology** (-nol'-), *n*. the composition or the study of hymns; hymns collectively. **hymnologist,** *n*.
hyoid (hī'oid), *a*. pertaining to the hyoid bone. **hyoid bone**, *n*. the bone supporting the tongue.
hyoscyamine (hīəsi'əmen, -min), *n*. a white crystalline alkaloid obtained from the seeds of henbane, highly poisonous, used as a sedative. **hyoscine** (hī'əsēn, -sin), *n*. a strong narcotic drug, scopolamine.
hyp (hip), hip^3.
hypaethral, hynethral (hipē'thrəl, hī-), *a*. open to the sky, roofless.
hype[1] (hīp), *n*. (*coll*.) exaggerated or false publicity used to sell or promote; a deception, a swindle. *v.t*. (*sometimes with up, coll*.) to sell or promote something or somebody by using exaggerated or false publicity.
hype[2] (hīp), *n*. (*coll*.) short for HYPODERMIC NEEDLE, see HYPODERM. **hyped up,** *a*. (*sl*.) full of nervous excitement.
hyper (hīpər), *a*. (*sl*.) full of nervous excitement, overwrought.
hyper-, *comb. form*. above, beyond; excessive, beyond measure.
hyperacidity (hīpərəsid'iti), *n*. excessive acidity in the digestive tract, esp. in the stomach.
hyperactive (hīpərak'tiv), *a*. abnormally active. **hyperactivity** (-tiv'-), *n*.
hyperaemia (hīpərē'miə), *n*. morbid or excessive accumulation of blood.
hyperaesthesia (hīpərəsthē'ziə), *n*. morbid or excessive sensibility, esp. of the nerves. **hyperaesthetic** (-thet'-), *a*.
hyperalgesia (hīpəraljē'ziə), *n*. a condition of exaggerated sensibility to pain.
hyperbaton (hīpœ'bətɒn), *n*. a figure of speech by which words are transposed or inverted from their natural and grammatical order.
hyberbola (hīpœ'bələ), *n*. (*pl*. **-las, lae**) a plane curve formed by cutting a cone when the intersecting plane makes a greater angle with the base than the side of the cone makes. **hyperbolic** (-bol'-), *a*.
hyperbole (hīpœ'bəli), *n*. a figure of speech expressing much more than the truth; rhetorical exaggeration. **hyperbolic, -ical** (-bol'-), *a*. of the nature of hyperbole. **hyperbolically,** *adv*. **hyperbolism,** *n*. the use of hyperbole; a hyperbolic expression. **hyperbolist,** *n*. **hyperbolize, -ise,** *v.i*. to use hyperbolical language. *v.t*. to express in hyperbolical language.
hyperborean (hīpəbaw'riən), *a*. belonging to or inhabiting the extreme north. *n*. one living in the extreme north.
hypercharge (hī'pəchahj), *n*. an interaction between elementary particles that is a weak force tending to oppose gravitational attraction between objects.
hypercritical, *a*. unreasonably critical or censorious. **hypercritically,** *adv*. **hypercriticism** (-sizm), *n*.
hyperfocal distance (hīpəfō'kəl), *n*. the distance beyond which objects appear sharply defined through a lens focused at infinity.
hyperglycaemia (hīpəglīsē'miə), *n*. an excessive level of sugar in the blood.
Hypericum (hīpε'rikəm), *n*. a genus of herbaceous plants or shrubs typified by the St John's wort.
hyperinflation (hīpərinflā'shən), *n*. a very high level of inflation in an economy.
hypermarket (hī'pəmahkit), *n*. a very large self-service store, usually on the outskirts of a town or city.
hypermetropia (hīpəmitrō'piə), **hypermetropy** (-trō'pi), *n*. an abnormal state of the eye characterized by longsightedness, opposed to myopia. **hypermetropic** (-trop'-), *a*. **hyperopia** (-ō'-), *n*. **hyperopic** (-op'-), *a*.
hyperon (hī'pəron), *n*. an elementary particle of the baryon group with a greater mass than a proton or a neutron.
hyperphasia (hīpəfā'ziə), *n*. (*Path*.) lack of control over the organs of speech. **hyperphasic** (-faz'-), *a*.

hyperphysical (hīpəfiz'ikəl), *a*. supernatural.
hyperplasia (hīpəplā'ziə), *n*. (*Path*.) excessive growth caused by abnormal multiplication of cells. **hyperplastic** (-plas'-), *a*.
hypersensitive (hīpəsen'sitiv), *a*. excessively or morbidly sensitive. **hypersensitivity** (-tiv'-), *n*.
hypersonic (hīpəson'ik), *a*. of speeds, higher than Mach 5. **hypersonically,** *adv*. **hypersonics,** *n.pl*.
hyperspace (hī'pəspās), *n*. space that has more than three dimensions.
hypersthene (hī'pəsthēn), *n*. (*Min*.) an orthorhombic, foliated, brittle mineral allied to hornblende, with a beautiful pearly lustre.
hypertension (hīpəten'shən), *n*. abnormally high blood pressure. **hypertensive,** *a*. suffering from hypertension.
hyperthermia (hīpəthœ'miə), *n*. abnormally high body temperature.
hyperthyroidism (hīpəthi'roidizm), *n*. excessive activity of the thyroid gland, causing an accelerated metabolic rate, nervousness etc.
hypertrophy (hīpœ'trəfi), *n*. excessive development or enlargement. *v.t*. to affect with hypertrophy. *v.i*. to be affected by hypertrophy. **hypertrophic,** *a*.
hyperventilation (hīpəventilā'shən), *n*. excessive breathing, causing excessive loss of carbon dioxide in the blood.
hyphen (hī'fən), *n*. a short stroke (-) joining two words or parts of words. *v.t*. to hyphenate. **hyphenate, hyphenated,** *a*. **hyphenation,** *n*.
hypn(o)-, *comb. form*. sleep.
hypnagogic, hypnogogic (hipnəgoj'ik), *a*. of or concerning the state of drowsiness before sleep.
hypnology (hipnol'əji), *n*. the study of the phenomena of sleep. **hypnologist,** *n*.
hypnopompic (hipnəpom'pik), *a*. of or concerning the state of drowsiness between sleep and waking.
hypnosis (hipnō'sis), *n*. inducement of a state resembling sleep in which the sub-conscious mind responds to external suggestions.
hypnotherapy (hipnəthe'rəpi), *n*. treatment by hypnotism.
hypnotic (hipnot'ik), *a*. causing sleep; soporific; of, pertaining to or inducing hypnotism. *n*. a medicine that produces sleep; a person who is susceptible to hypnotism.
hypnotism (hip'nətizm), *n*. an artificial method of inducing sleep or hypnosis; the study or practice of this. **hypnotist,** *n*. **hypnotize, -ise,** *v.t*. to affect (as) with hypnotism. **hypnotization, -isation,** *n*. **hypnotizer,** *n*.
hypnum (hip'nəm), *n*. (*pl*. **-nums, -na** (-nə)) a genus of pleurocarpous mosses known as feather-moss.
hypo[1] (hī'pō), *n*. common term for sodium thiosulphate, the normal fixing solution in photography.
hypo[2] (hī'pō), *n*. (*coll*.) short for HYPODERMIC NEEDLE.
hyp(o)-, *comb. form*. under, below; less than; (*Chem*.) denoting compounds having a lower degree of oxidation in a series.
hypoallergenic (hīpoaləjenik), *a*. (of soaps, cosmetics etc.) containing substances unlikely to cause an allergic reaction.
hypocaust (hī'pəkawst), *n*. in ancient Roman buildings, an underfloor heating system.
hypochlorite (hīpəklaw'rīt), *n*. a salt or ester of hypochlorous acid. **hypochlorous acid,** *n*. an unstable acid formed when chlorine dissolves in water, used as a bleach, disinfectant etc.
hypochondria (hīpəkon'driə), *n*. a morbid condition characterized by exeessive anxiety with regard to one's health. **hypochondriac** (-ak), *n*., *a*. **hypochondriacal,** *a*. **hypochondriasis** (-drī'əsis), *n*. hypochondria.
hypocorism (hīpok'ərizm), *n*. a pet name. **hypocoristic** (-ris'-), *a*.
hypocrisy (hipok'rəsi), *n*. dissimulation; a feigning to be what one is not. **hypocrite** (hip'əkrit), *n*. one who practises hypocrisy; a dissembler. **hypocritical** (hīpəkrit'-),

a. **hypocritically,** *adv.*
hypodermic (-dœ'-), *a.* pertaining to parts underlying the skin; pertaining to an injection under the skin. *n.* (a drug introduced into the system by) an injection under the skin; (*coll.*) a hypodermic syringe. **hypodermic injection,** *n.* an injection (of narcotics, antitoxins etc.) beneath the skin. **hypodermic needle,** *n.* (the hollow needle of) a hypodermic syringe. **hypodermic syringe,** *n.* a small syringe with a hollow needle for giving hypodermic injections. **hypodermically,** *adv.*
hypogastrium (hīpəgas'triəm), *n.* (*Anat.*) the middle part of the lowest zone into which the abdomen is divided. **hypogastric,** *a.*
hypogeum (hīpəjē'əm), *n.* (*pl.* **-gea** (-ə)) (part of) a building below the level of the ground.
hypoglossal (hīpəglos'əl), *a.* under the tongue. **hypoglossal nerve,** *n.* the motor nerve of the tongue. **hypoglossus** (-səs), *n.* (*Anat.*) the hypoglossal nerve.
hypoglycaemia (hīpōglīsē'miə), *n.* an abnormally low level of sugar in the blood.
hypomania (hīpəmā'niə), *n.* the mental state of overexcitability. **hypomanic** (-man'-), *a.*
hypophosphate (hīpəfos'fāt), *n.* a salt of hypophosphoric acid. **hypophosphite** (-fīt), *n.* a salt of hypophosphorous acid. **hypophosphoric** (-fo'rik), **hypophosphorous** (-fos'-), *a.* **hypophosphoric acid,** *n.* an acid formed by action of water and oxygen on phosphorus. **hypophosphorous acid,** *n.* a weak acid composed of hydrogen, phosphorus and oxygen.
hypophysis (hīpof'isis), *n.* (*pl.* **-physes** (-sēz)) the pituitary gland. **hypophyseal, -physial** (-fīz'iəl, -sē'əl), *a.*
hypoplasia (hīpəplā'ziə), *n.* underdevelopment of an organ or part.
hypostasis (hīpos'təsis), *n.* (*pl.* **-stases** (-sēz)) that which forms the basis of anything; in metaphysics, that by which a thing subsists, substance as distinguished from attributes; the essence or essential principle; the personal subsistence, as opposed to substance, of the Godhead; one of the persons of the Trinity; congestion of the blood (in an organ). **hypostatic, -ical** (-stat'-), *a.* **hypostatic union,** *n.* union of the divine and human natures in Christ.
hypostyle (hī'pəsūl), *a.* having the roof supported by pillars. *n.* a building with a roof or ceiling supported by pillars.
hyposulphite (hīpəsūl'fīt), *n.* a thiosulphate, a salt of hyposulphurous acid. **hyposulphuric acid** (-fū'-), *n.* acid containing two more atoms of oxygen per molecule than sulphuric acid. **hyposulphurous acid,** *n.* an unstable acid containing one more sulphur atom per molecule than sulphuric acid.
hypotension (hīpōten'shən), *n.* abnormally low blood pressure.
hypotenuse (hīpot'ənūz), *n.* the side of a right-angled triangle opposite to the right angle.
hypothalamus (hīpəthal'əməs), *n.* a region at the base of the brain controlling autonomic functions, e.g. hunger, thirst.
hypothec (hīpoth'ik), *n.* (*esp. Sc., Law*) a security in favour of a creditor over the property of his debtor, while the property continues in the debtor's possession. **hypothecate,** *v.t.* to pledge or mortgage in security for some debt or liability. **hypothecation,** *n.* **hypothecator,** *n.*
hypothermia (hīpəthœ'miə), *n.* subnormal body temperature.
hypothesis (hīpoth'əsis), *n.* (*pl.* **-theses** (-thəsēz)) a proposition assumed for the purpose of argument; a theory assumed to account for something not understood; a mere supposition or assumption. **hypothesize, -ise,** *v.i.* to form hypotheses. *v.t.* to assume. **hypothetical** (-thet'-), *a.* founded on or of the nature of a hypothesis; conjectural, conditional. **hypothetically,** *adv.*
hypothyroidism (hīpəthī'roidizm), *n.* underactivity of the thyroid gland.
hypoxia (hīpok'siə), *n.* a deficiency of oxygen reaching the body tissues.
hypozoic (hīpəzō'ik), *a.* (*Geol.*) situated beneath the strata that contain organic remains. **Hypozoa** (-ə), *n.pl.* Protozoa. **hypozoan,** *n.*, *a.*
hyps(o)-, *comb. form.* height.
hypsography (hipsog'rəfi), *n.* the branch of geography concerned with the altitudes above sea-level. **hypsographical** (-graf'-), *a.*
hypsometer (hipsom'itə), *n.* an instrument for measuring heights above sea-level. **hypsometric, -ical** (-met'-), *a.* **hypsometry,** *n.* the art of measuring heights by observing differences in barometric pressures at different altitudes.
hypural (hīpū'rəl), *a.* (*Ichthyol.*) situated below the tail, as the bones supporting the fin-rays.
hyrax (hīr'aks), *n.* a genus of small hare-like quadrupeds, comprising the Syrian rock-rabbit or cony of Scripture and the S African rock-badger. **hyracid** (-ras'id), *a.* **hyracoid** (hī'rəkoid), *n.*, *a.*
hyson (hī'sən), *n.* a kind of green tea.
hyssop (his'əp), *n.* a labiate plant with blue flowers; in Biblical times, an unidentified plant used in Jewish rites of purification.
hyster- HYSTERO.
hysterectomy (histərek'təmi), *n.* the removal of the womb by surgery.
hysteresis (histərē'sis), *n.* the tendency of a magnetic substance to remain in a certain magnetic condition, 'the lag of magnetic effects behind their causes'.
hysteria (histiə'riə), *n.* a nervous disorder, occurring in paroxysms, and often simulating other diseases; (extreme over-excitement characterized by) a fit of laughing or crying. **hysteric** (-te'-), *n.* one subject to hysteria. *a.* hysterical. *n.pl.* a fit or fits of hysteria, hysteria; (*coll.*) emotional paroxysm of crying, laughing etc. **hysterical,** *a.* **hysterically,** *adv.*
hysteritis (histəri'tis), *n.* inflammation of the uterus.
hystero-, hyster-, *comb. form.* womb; hysteria.
hysteron proteron (his'təron prot'əron), *n.* a figure of speech in which what should follow comes first; an inversion of the natural or logical order.
Hz, (*abbrev.*) hertz.

I

I¹, i, the ninth letter and the third vowel in the English alphabet (*pl.* **Is, I's**); the Roman numeral symbol for one; (*Math.*) the symbol for the square root of minus one.
I² (ī), *nom. sing. 1st pers. pron.* in speaking or writing denotes oneself. *n.* (*Metaph.*) the self-conscious subject, the ego.
I³, (*abbr.*) Institute; Island; Italy.
I⁴, (*chem. symbol*) iodine.
-i, *suf.* indicating plural of L nouns in *-us* or *-er*, as *fungi, hippopotami;* also of It. nouns and adjectives in *-o* or *-e*, as *banditti, literati.*
Ia., (*abbr.*) Iowa.
-ia, *suf.* forming abstract nouns, as *mania, militia;* names of countries etc., as *Australia, Bulgaria;* names of diseases, as *hysteria, malaria;* names of botanical genera etc., as *Begonia, Saponaria;* names of alkaloids, as *morphia, strychnia;* (*pl.* of L *-ium*, Gr. *-ion*) *bacteria, mammalia, regalia, reptilia.*
IAEA, (*abbr.*) International Atomic Energy Agency.
-ial (-iəl), *suf.* forming adjectives, as *celestial, terrestrial.*
iambus (īam′bəs), **iamb** (i′-), *n.* (*pl.* **-buses**) a poetic foot of one short and one long, or one unaccented and one accented syllable. **iambic**, *a.* of or pertaining to the iambus; composed of iambics. *n.* an iambic foot; an iambic verse.
-ian (-iən), *suf.* forming nouns or adjectives, as *Athenian, Baconian, Bristolian.*
-iasis, *comb. form.* indicating a disease, as *elephantiasis, phthiriasis.*
IATA (īah′tə), (*abbr.*) International Air Transport Association.
iatric, -ical (īat′rik), *a.* pertaining to physicians or medicine. **iatrogenic** (-jen′ik), *a.* resulting unintentionally from medical treatment. **iatrogenically**, *adv.* **iatrogenicity** (-nis′-), *n.* **-iatrics**, *comb. form.* indicating medical care, as *paediatrics.* **-iatry**, *comb. form.* indicating healing treatment, as *psychiatry.*
IBA, (*abbr.*) Independent Broadcasting Authority.
Iberian (ībiə′riən), *a.* of or pertaining to ancient Iberia in Europe, comprising modern Spain and Portugal, or ancient Iberia in Asia, now Transcaucasian Georgia. *n.* one of the inhabitants of ancient Iberia in Europe, or in Asia; one of an ancient race, chiefly dolichocephalic, who inhabited western Europe and probably entered the British Isles early in the Neolithic period; the language of ancient Iberia. **Iberian Peninsula**, *n.* Spain and Portugal.
Iberis (ībiə′ris), *n.* (*Bot.*) a genus of crucifers comprising the candytufts.
ibex (ī′beks), *n.* the name given to several species of wild goats with large backward-curving horns inhabiting the mountain regions of Europe and Asia.
ibid. (ibid′), (*abbr.*) ibidem.
ibidem (ib′idem), *adv.* in the same place (as in a book, page etc.).
ibis (ī′bis), *n.* any of a genus (*Ibis*) of heron-like wading birds, esp. *I. religiosa*, the sacred ibis, venerated by the ancient Egyptians.
-ible (-ibl), *suf.* as in *edible, risible.* **-ibility** (-ibil′iti), *suf.* **-ibly** (-ibli), *suf.*
Iblees, *pl.* EBLIS.
Ibo (ē′bō), *n.* a black African people living in SE Nigeria; a member of this people; their language. *a.* of the Ibo.
-ic (-ik), *suf.* of, pertaining to, like, as in *alcoholic, algebraic, domestic, Miltonic, plutonic;* (*Chem.*) in acids etc., denoting a higher state of oxidation than the suffix *-ous;* forming names of sciences, arts etc., as *arithmetic, epic, logic, music.*
ICA, (*abbr.*) Institute of Chartered Accountants; Institute of Contemporary Arts.
-ical, *suf.* forming adjectives, as *algebraical, comical, historical, political.* **-ically**, *suf.* forming adverbs, as *historically, politically.*
ICBM, (*abbr.*) intercontinental ballistic missile.
ice (īs), *n.* water congealed by cold; a frozen confection of cream, syrup etc., ice-cream; icing; (*sl.*) diamonds. *v.t.* to cover or cool with ice; to convert into ice; to coat with concreted sugar; to frost; to freeze; (*sl.*) to kill. *v.i.* to freeze; to become covered with ice. **dry ice**, frozen carbon dioxide. **on ice**, in abeyance. **on thin ice**, in a vulnerable or dangerous situation. **to break the ice** BREAK. **to cut no ice**, (*coll.*) to fail to make an impression, to be unimportant. **ice age**, *n.* a glacial period. **ice-axe**, *n.* an axe shaped like a pickaxe, used by mountain-climbers for cutting steps on glaciers etc. **ice-blink**, *n.* a luminous reflection over the horizon from snow- or ice-fields. **ice-boat**, *n.* a boat for travelling on ice; a heavily-built boat for breaking a passage through ice. **ice-bound**, *a.* completely surrounded with ice; fringed or edged with ice; unable to get out because of ice. **icebox**, *n.* the freezing compartment of a refrigerator; a portable insulated box containing ice; (*chiefly N Am.*) a refrigerator. **ice-breaker**, *n.* a ship with a reinforced hull for forcing a channel through ice; (*coll.*) something that encourages a relaxed atmosphere among a group of people. **ice bucket, pail**, *n.* a bucket containing ice, for keeping wine etc. cool. **ice-cap**, *n.* a mass of ice and snow permanently covering an area. **ice-cream**, *n.* cream or custard flavoured and artificially frozen. **ice dancing**, *n.* a form of ice-skating with movements based on ballroom dancing. **ice-fall**, *n.* a shattered part of a glacier where it descends a steep slope. **ice-field**, *n.* a large expanse of ice, esp. such as exist in the Polar regions. **ice-floe, -pack**, *n.* a sheet of floating ice. **ice-foot**, *n.* a hill or wall of ice along the shore in Polar regions. **ice hockey**, *n.* a type of hockey played on ice by teams of skaters. **ice-house**, *n.* a repository for the storage of ice. **ice lolly**, *n.* (*coll.*) a flavoured piece of ice or ice-cream on a stick. **ice-pack**, *n.* ICE-FLOE; a bag etc. containing ice applied to a part of the body to reduce swelling or ease pain; a gel-filled sachet which can be frozen and used to cool the contents of e.g. a cool bag. **ice pail** ICE BUCKET. **ice pick**, *n.* a pointed tool for splitting ice. **ice-plant**, *n.* a creeping plant, *Mesembryanthemum crystallinum*, whose leaves have a glistening lustre somewhat like ice. **ice rink**, *n.* a rink for ice-skating. **ice-show**, *n.* (*Theat.*) a performance on ice by actors wearing skates. **ice skate**, *n.* a boot with a blade attached for skating on ice. **ice-skate**, *v.i.* **ice-skater**, *n.* **Ice station**, *n.* a scientific research station in polar regions. **ice-water, iced water**, *n.* water from melted ice; water cooled by ice. **icing**, *n.* a coating of concreted sugar for cakes. **icing sugar**, *n.* powdered sugar used for icing cakes etc. **icy**, *a.* pertaining to or consisting of ice; like ice, frozen; (*fig.*) frigid, chil-

ling. **icily,** *adv.* **iciness,** *n.*

-ice, *suf.* forming nouns, as *justice, malice, novice, service.*

iceberg (īs'bɛrg), *n.* a large mass of ice, usu. floating on the sea at high latitudes, usu. formed by detachment from a glacier; a cold and unresponsive person. **tip of the iceberg,** the part of an iceberg visible above the water; the most obvious part of a huge problem etc.

Iceland (īs'lənd), *n.* an island in the N Atlantic between Scandinavia and Greenland. **Iceland lichen, moss,** *n.* an edible moss or lichen, *Cetraria islandica,* growing in the northern and mountainous parts of Europe, used as a medicine. **Iceland poppy,** *n.* the yellow Arctic poppy. **Iceland spar,** *n.* a transparent variety of calcite. **Icelander,** *n.* a native or inhabitant of Iceland. **Icelandic** (-lan'-), *a.* pertaining to Iceland. *n.* the language of Iceland.

I Ching (ī ching), *n.* an ancient Chinese method of divination employing a set of symbols, together with the text known as the *I Ching* which serves to interpret them.

ichneumon (iknū'mon), *n.* a small carnivorous animal, *Herpestes ichneumon,* related to the mongoose, found in Egypt; (also **ichneumon-fly**) a hymenopterous insect which lays its eggs in or upon the larvae of other insects, upon which its larvae will feed.

ichnite (ik'nīt), ICHNOLITE.

ichnography (iknog'rəfi), *n.* the art of drawing groundplans etc. **ichnograph** (ik'nəgraf, -grahf), *n.* a groundplan. **ichnographic, -ical** (-graf'-), *a.* **ichnographically,** *adv.*

ichnolite (ik'nəlīt), **ichnite,** *n.* a stone with the impression of a footprint.

ichor (ī'kaw), *n.* (*Gr. Myth.*) the ethereal fluid which took the place of blood in the veins of the gods; a thin watery humour like serum; a watery acrid discharge from a wound etc. **ichorous,** *a.*

ichthy-, ichthyo-, *comb. form.* pertaining to fish; fish-like.

ichthyic (ik'thiik), *a.* pertaining to fishes; having the characteristics of a fish.

ichthyography (ikthiog'rəfi), *n.* a description of or a treatise on fishes. **ichthyographer,** *n.*

ichthyoid (ik'thioid), *a.* resembling fish. *n.* a vertebrate of fishlike form.

ichthyology (ikthiol'əji), *n.* the branch of zoology concerned with fishes; the natural history of fishes. **ichthyologic, -ical** (-loj'-), *a.* **ichthyologist,** *n.*

ichthyophagy (ikthiof'əji), *n.* the practice of eating fish; fish diet. **ichthyophagist,** *n.* **ichthyophagous** (-gəs), *a.*

ichthyosaurus (ikthiəsaw'rəs), **ichthyosaur** (ik'-), *n.* a gigantic fossil marine reptile, resembling a porpoise and with paddle-like limbs.

ichthyosis (ikthiō'sis), *n.* a hereditary skin disease, marked by thick, hard, imbricated grey scales. **ichthyotic** (-ot'-), *a.*

ichthys (ik'this), *n.* a symbol in the form of a fish, connected with Christ.

ICI, (*abbr.*) Imperial Chemical Industries.

-ician (-ishən), *suf.* indicating a specialist in a subject, as in *beautician.*

icicle (ī'sikl), *n.* a hanging conical point of ice, formed by dripping water freezing.

icily, icing etc. ICE.

icky (ik'i), *a.* (*coll.*) cloying; over-sentimental.

-icle, *suf.* diminutive, as in *particle, versicle.*

icon, ikon (ī'kon), *n.* in the Eastern Church, a sacred image, picture, mosaic, or monumental figure of a holy personage, usu. regarded as endowed with miraculous attributes; a symbol; a hero-figure; a pictorial representation of a facility available to the user of a computer system. **iconic** (-kon'-), *a.* pertaining to or consisting of figures or pictures; (*Art*) following a conventional pattern or type, as busts, memorial effigies etc. **iconic memory,** *n.* the continuation of a sense impression after the stimulus has disappeared.

icono-, *comb. form.* of or pertaining to images or idols.

iconoclast (īkon'əklast), *n.* a breaker of images, esp. one of the religious zealots in the Eastern Empire who attacked the worship of images during the 8th and 9th cents.; an assailant or despiser of established practices etc. **iconoclasm,** *n.* **iconoclastic** (-klas'-), *a.* **iconoclastically,** *adv.*

iconography (īkənog'rəfi), *n.* pictorial matter relating to a subject; the symbols used in a work of art, esp. those traditionally or conventionally associated with the subject; the meaning of such symbols. **iconographer,** *n.* **iconographic, -ical** (-graf'-), *a.*

iconology (īkənol'əji), *n.* the science or study of images, pictures etc. **iconological** (-loj'-), *a.* **iconologist,** *n.*

iconomatic (īkonəmat'ik), *a.* denoting a kind of writing in which pictures represent phonetic elements.

iconometer (īkənom'itə), *n.* an instrument for measuring the size or distance of an object; a direct-vision viewfinder. **iconometry,** *n.*

iconoscope (īkon'əskōp), *n.* a type of electron camera.

iconostasis (īkənos'təsis), *n.* (*pl.* **-ses** (-sēz)) in the Eastern Church, a screen on which icons are placed separating the sanctuary from the rest of the church.

icosahedron (īkəsəhē'drən), *n.* (*pl.* **-dra** (-drə), **-drons**) (*Geom.*) a solid figure having 20 plane sides; a regular solid contained by 20 equilateral triangles. **icosahedral,** *a.*

-ics, *suf.* (*usu. sing. in constr.*) indicating a science or art, as *linguistics;* indicating specified activities, as *acrobatics;* indicating matters etc. relating to, as *mechanics.*

icterus (ik'tərəs), *n.* jaundice; a disease of plants characterized by yellowness of the leaves. **icteric** (-te'-), *a.* affected with jaundice; good against jaundice. *n.* a remedy for jaundice.

ictus (ik'təs), *n.* (*pl.* **ictuses, ictus**) the stress, beat, or rhythmical accent in metre; the beat of the pulse.

icy ICE.

ID, (*abbr.*) identification.

Id., (*abbr.*) Idaho.

I'd (īd), *contr. form. I had* or *I would.*

id[1] (id), *n.* (*Psych.*) the instinctive impulses of the individual.

id.[2] (id), (*abbr.*) idem, the same.

-id, *suf.* forming adjectives denoting the quality orig. expressed by a Latin verb, as *acid, frigid, morbid, tepid;* (*Bot.*) denoting a member of an order, as orchid (Orchidaceae); (*Zool.*) member of a family, as arachnid (Arachnida).

-idae, *suf.* indicating membership of a specified zoological family.

-ide, *suf.* indicating chemical compounds of an element with another element or a radical, as *chloride, fluoride, oxide.*

idea (īdē'ə), *n.* a mental image, form, or representation of anything; a notion, a conception, a supposition; a more or less vague opinion, belief, or fancy; a plan, an intention or design; the purpose, the aim; a view, a way of thinking or conceiving (something); (*Platonic*) the archetype or perfect and eternal pattern of which actual things are imperfect copies. **not my** etc. **idea of,** not what I etc. expect something or someone to be like. **the very idea!,** that is ridiculous. **to get ideas,** (*coll.*) to become overambitious; to develop the wrong expectations or impressions. **to have no idea,** to be unaware of what is going on; (*coll.*) to be innocent, ignorant or stupid. **what's the big idea?** (*coll.*) what is the meaning of this?, what is going on?

ideal (īdē'əl), *a.* consisting of, existing in, or pertaining to ideas; visionary, fanciful; reaching one's standard of perfection; (*Phil.*) of or pertaining to idealism or the Platonic ideas. *n.* a standard of perfection; an actual thing realizing this. **idealism,** *n.* belief in or the pursuit of ideals; (*Phil.*) the doctrine that in external perceptions the objects immediately known are ideas. **idealist,** *n.* **idealistic** (-lis'-), *a.* **idealistically,** *adv.* **ideality** (-al'-), *n.* the

idée fixe (ē'dā fēks), n. a fixed idea, monomania. [F]
idem (id'em), n. the same (word, author, book etc.). [L]
identical (īden'tikl), a. absolutely the same, not different; similar in essentials; (*Math.*) expressing identity. **identical twins** TWIN. **identically**, *adv.* **identicalness**, n.
identify (īden'tifī), v.t. to consider or represent as precisely the same (with); to determine or prove the identity of; to unite or associate (oneself) closely (with a party, interests etc.). v.i. to associate or consider oneself to be at one (with). **identifiable**, a. **identifiably**, adv. **identification** (-fi-), n. a proof of identity; the assumption of the characteristics of another, esp. of an admired person. **identification parade**, n. a number of persons assembled by the police and among whom a witness is invited to identify a suspect. **identifier**, n.
identikit (īden'tikit), n. a set of facial features on transparent slips, used to compose a likeness, esp. of a criminal suspect; a portrait built up in this way. *adj.* pertaining to such portraits; (*coll.*) conforming to an unimaginative pattern.
identity (īden'titi), n. the state of being identical; one's individuality; who or what a particular person or thing is; (*Alg.*) absolute equality between two expressions; an equation expressing such equality. **identity card, disc,** n. an object bearing the owner or wearer's name and used as proof of his or her identity. **identity crisis,** n. a state of psychological confusion resulting from a failure to reconcile discordant elements in one's personality. **identity element,** n. a mathematical element belonging to a set and which leaves any other member of that set unchanged when combining with it.
ideo-, *comb. form.* pertaining to or expressing ideas.
ideograph (id'iəgraf, -grahf), **ideogram** (-gram), n. a symbol, figure etc., suggesting or conveying the idea of an object, without expressing its name. **ideographic, -ical** (-graf'-), a. **ideographically,** adv. **ideography** (-og'-), n.
ideology (īdiol'əji), n. the science of ideas; the political or social philosophy of a nation, movement, group etc. **ideological** (-loj'-), a. **ideologically,** adv. **ideologist, ideologue** (ī'diəlog), n. a supporter of an ideology; one who treats of ideas; a theorist, a visionary. **ideologize, -ise,** v.t.
Ides (īdz), n.pl. in the ancient Roman calendar, the 15th of March, May, July, October, and 13th of the other months.
id est (id est), that is, that is to say (usu. written i.e.). [L]
idio-, *comb. form.* individual, peculiar.
idiocy IDIOT.
idiolect (id'iəlekt), n. a form of speech or language peculiar to an individual. **idiolectal, -tic** (-lek'-), a.
idiom (id'iəm), n. a mode of expression, esp. an irregular use of words, peculiar to a language; an expression whose meaning cannot be determined from its constituent parts; a dialect, a particular variety of a language; a mode of artistic expression characteristic of a particular person or school. **idiomatic, -ical** (-mat'-), a. **idiomatically,** adv.
idiopathy (idiop'əthi), n. a primary disease, one not occasioned by another. **idiopathic, -ical** (-path'-), a. **idiopathically,** adv.
idiosyncrasy (idiəsing'krəsi), n. individual quality, habit, or attitude of mind; a characteristic peculiar to an individual; (*Med.*) an abnormal sensitivity to a particular food, drug etc. **idiosyncratic, -ical** (-krat'-), a. **idiosyncratically,** adv.
idiot (id'iət), n. a person of weak or defective understanding; one belonging to the lowest grade of mental defectives; a stupid, silly person. a. idiotic. **idiot board,** n. (*coll.*) an autocue. **idiot box,** n. (*sl.*) a television set. **idiot savant** (ē'dyō savä', id'iət sav'ənt), n. a mentally retarded person who possesses a remarkable talent for a single specialized activity. **idiot tape,** n. a computer tape printing out information in an unbroken stream, lacking any line breaks. **idiocy** (-si), n. **idiotic, -ical** (-ot'-), a. resembling or characteristic of an idiot; foolish, silly, absurd. **idiotically,** adv.
idle (ī'dl), a. doing nothing; inactive, not occupied; not in use; averse to work, lazy; useless, vain, ineffectual; unfruitful, barren; trifling, without foundation. v.i. to spend time in idleness; to move about aimlessly or lazily; of machinery, to run slowly without the transmission being engaged. v.t. to spend (time) in idleness; to cause to idle. **to idle away,** to spend in idleness. **idle-pulley,** n. a pulley able to rotate freely as a means of guiding or controlling the tension of a belt. **idle-tongs** LAZY-TONGS. **idle-wheel,** n. a cogged wheel between two others for transmitting motion. **idleness,** n. **idler,** n. one who spends his or her time in idleness. **idly,** adv.
Ido (ē'dō), n. an artificial international language based on Esperanto.
idol (ī'dl), n. an image, esp. one worshipped as a god; a false god; a person or thing loved or honoured excessively. **idolater** (īdol'ətə), n. one who worships idols; a pagan; an adorer, an extravagant admirer. **idolatress** (-tris), n. *fem.* **idolatrous,** a. **idolatrously,** adv. **idolatry,** n. **idolism,** n. idolatry; idolization; a vain opinion or fancy. **idolize, -ise,** v.t. to worship as an idol; to make an idol of; to love or venerate to excess. **idolization,** n. **idolizer,** n.
idolon (īdō'lon), **idolum** (-əm), n. (*pl.* **-lia** (-iə)) an image, an appearance; a phantom, an apparition; (*Phil.*) a fallacious appearance or misconception.
idyll (id'l), n. a brief narrative or description of harmonious rustic life, either in verse or prose; a work of art, esp. a musical piece, of a similar character; a scene, episode, or situation suitable for the tone of such a composition. **idyllic** (idil'-), a. pertaining to or suitable for an idyll; perfect in harmony, peace, beauty etc. **idyllically,** adv. **idyllist,** n. **idyllize, -ise,** v.t.
i.e., (*abbr.*) id est.
-ie, *suf.* -Y.
-ier, *suf.* denoting occupation, profession etc., as in *bombardier, brigadier, chevalier, financier.*
if (if), *conj.* on the supposition that, providing that, in case that; even on the supposition, allowing that; whenever, at the time when; whether; also used in an exclamatory sense, as *if only you were here!.* n. (*coll.*) an uncertain or doubtful factor; a condition. **as if,** as it would be if. **ifs and ans,** things that might have been. **ifs and buts,** objections.
-iferous (-if'ərəs), *suf.* -FEROUS.
-iform (-ifawm), *suf.* -FORM.
iffy (if'i), a. (*coll.*) doubtful, uncertain; risky.
igloo (ig'loo), n. a dome-shaped Eskimo hut, often built of blocks of snow.
igneous (ig'niəs), a. containing or of the nature of fire; emitting fire; (*Geol.*) produced by volcanic action.
ignis fatuus (ig'nis fat'ūəs), n. (*pl.* **ignes fatui** (ig'nēz fat'ūī)), an apparent flame probably due to the spontaneous combustion of inflammable gas, floating above the ground in marshes etc.; a delusive object or aim.
ignite (ignīt'), v.t. to set on fire; to render luminous or red with heat. v.i. to take fire; to become red with heat. **ignitable, ignitible,** a. **ignitability, ignitibility** (-bil'-), n. **ignition** (-nish'-), n. the act of igniting; the state of being ignited; the mechanism for igniting the explosive mixture in the internal-combustion engine; (*coll.*) the slot into which the ignition key is inserted. **ignition key,** n. the key that operates the ignition system in a motor vehicle. **igniter,** n.
ignoble (ignō'bl), a. of humble or mean birth; mean, base, despicable, unworthy, dishonourable. **ignobility** (-bil'-), n. **ignobleness,** n. **ignobly,** adv.
ignominy (ig'nəmini), n. public disgrace or shame; dis-

ignoramus 413 **illustrate**

honour, infamy; an act deserving disgrace. **ignominious** (-min'-), *a.* **ignominiously,** *adv.* **ignominiousness,** *n.*
ignoramus (ignərā'məs), *n.* (*pl.* **-muses**) an ignorant person; a stupid person, a fool.
ignorance (ig'nərəns), *n.* the state of being ignorant; want of knowledge (of). **ignorant,** *a.* destitute of knowledge, unconscious (of); illiterate, uninstructed, uneducated. **ignorantly,** *adv.*
ignore (ignaw'), *v.t.* to pass over without notice, to disregard; deliberately to pay no attention. **ignorable,** *adj.* **ignoration,** *n.*
iguana (igwah'nə), *n.* any of a genus (*Iguana*) of large tropical American lizards, esp. *I. tuberculata.*
iguanodon (igwah'nədon), *n.* a genus of extinct gigantic lizards.
ikebana (ikibah'nə), *n.* the Japanese art of arranging flowers.
ikon ICON.
-il, -ile, *suf.* that may be, capable of being, pertaining to etc., as in *civil, fossil, docile, fragile, Gentile, puerile, senile.*
ileac (il'iak), etc. ILIAC.
ileo-, *comb. form.* (*Anat., Path.*) ileum.
ileum (il'iəm), *n.* (*pl.* **ilea** (-ə)) the portion of the small intestine communicating with the larger intestine.
ilex (ī'leks), *n.* (*pl.* **-lexes**) the holm-oak; a genus of trees or shrubs with coriaceous leaves, typified by the holly.
iliac (il'iak), *a.* of or belonging to the ileum or smaller intestines; pertaining to the ilium or hip-bone. **iliac region,** *n.* the part of the abdomen between the ribs and the hips.
ilio-, *comb. form.* pertaining to or situated near the ilium.
ilium (il'iəm), *n.* (*pl.* **-ia** (-ə)) the upper part of the hip-bone.
ilk (ilk), *a.* the same. **of that ilk,** of the same name (used when the surname of a person is the same as the name of his estate). **that ilk,** (*coll., erron.*) that family or kind.
ilka (il'kə), *a.* (*Sc.*) each, every.
I'll (il), *contr. form.* of *I will* or *I shall.*
Ill., (*abbr.*) Illinois.
ill (il), (*comp.* **worse** (wœs), *superl.* **worst** (wœst)) *a.* unwell, sick; bad morally, evil; malevolent, hostile, adverse; noxious, mischievous, harmful; unfortunate, unfavourable, unlucky; not right, faulty, inferior, incorrect; cross (in temper). *adv.* (*comp.* **worse,** *superl.* **worst**) not well, badly; not rightly; not easily; imperfectly, scarcely; unfavourably, in bad part or humour. *n.* evil; injury, harm; wickedness; (*pl.*) misfortunes. **ill at ease,** uncomfortable, anxious. **to be taken ill,** to fall sick. **to speak ill,** to speak (of or about) unfavourably. **to take ill, to take in ill part,** to take offence at. **ill-advised,** *a.* imprudent; injudicious. **ill-advisedly,** *adv.* **ill-affected,** *a.* not friendly disposed. **ill-assorted,** *a.* poorly matched; not compatible. **ill blood, ill feeling,** *n.* resentment, enmity. **ill-bred,** *a.* brought up badly; rude, unmannered, offensive. **ill breeding,** *n.* **ill-conditioned,** *a.* having a bad temper or disposition; in a bad physical condition. **ill-considered,** *a.* done without careful thought; misconceived. **ill-defined,** *a.* poorly defined; lacking a clear outline. **ill-disposed,** *a.* wickedly or maliciously inclined; unfavourably inclined (towards). **ill fame,** *n.* disrepute. **ill-fated,** *a.* unfortunate, unlucky. **ill-favoured,** *a.* ugly, deformed; unattractive; objectionable. **ill-favouredness,** *n.* **ill-feeling** ILL BLOOD. **ill-founded,** *a.* lacking any foundation in fact, not substantiated. **ill-got, ill-gotten,** *a.* obtained in an improper way. **ill humour,** *n.* bad temper. **ill-humoured,** *a.* **ill-humouredly,** *adv.* **ill-judged,** *a.* not well-judged; injudicious, unwise. **ill luck,** *n.* bad luck, misfortune. **ill-mannered,** *a.* rude, boorish. **ill-matched,** *a.* not well-matched or suited. **ill nature,** *n.* evil disposition; lack of kindness or good feeling. **ill-natured,** *a.* of a churlish disposition, bad-tempered; expressive of or indicating ill nature. **ill-naturedly,** *adv.* **ill-naturedness,** *n.* **ill-omened,** *a.* unlucky, inauspicious. **ill-starred,** *a.* born under the influence of an unlucky planet, hence unlucky. **ill temper,** *n.* **ill-tempered,** *a.* having a bad temper, sour, peevish. **ill-timed,** *a.* done, said, or attempted, at an unsuitable time. **ill-treat** ILL USAGE. **ill treatment** ILL USAGE. **ill turn,** *n.* an ill-natured act or treatment. **ill usage,** *n.* unkind treatment. **ill use,** *v.t.* to treat badly. **ill-versed,** *a.* uninstructed, lacking skill (in). **ill will,** *n.* malevolence, enmity.
illation (ilā'shən), *n.* deduction; a deduction, an inference.
illative (il'-), *a.* denoting, expressing, or of the nature of an inference; of some Finno-Ugrian languages, denoting a noun case expressing motion or direction. *n.* an illative particle; the illative case. **illatively,** *adv.*
illegal (ilē'gəl), *a.* not according to law; contrary to law, unlawful. **illegality** (-gal'-), *n.* **illegalize, -ise,** *v.t.* to render illegal. **illegally,** *adv.*
illegible (ilej'ibl), *a.* that cannot be read or deciphered. **illegibility** (-bil'-), **-ibleness,** *n.* **illegibly,** *adv.*
illegitimate (iləjit'imət), *a.* born out of wedlock; contrary to law or recognized usage; irregular, improper; illogical, unsound. *n.* an illegitimate child. **illegitimacy** (-si), *n.* **illegitimately,** *adv.* **illegitimation,** *n.* **illegitimatize, -ise,** *v.t.*
illiberal (ilib'ərəl), *a.* not generous, petty, sordid; narrow-minded, niggardly, stingy; uncultured. **illiberality** (-ral'-), *n.* **illiberalize, -ise,** *v.t.* **illiberally,** *adv.*
illicit (ilis'it), *a.* not allowed or permitted; unlawful. **illicitly,** *adv.* **illicitness,** *n.*
illimitable (ilim'itəbl), *a.* boundless, limitless. **illimitability** (-bil'-), **illimitableness,** *n.* **illimitably,** *adv.*
illing (il'ing), *a.* (*sl.*) bad, wicked.
illiquid (ilik'wid), *a.* of assets, not easily convertible into cash; of a company etc., lacking liquid assets. **illiquidity** (-kwid'-), *n.*
illiterate (ilit'ərət), *a.* unable to read or write; ignorant in a specific subject, esp. literature; rude, uncultivated. *n.* an ignorant or uneducated person, esp. one unable to read. **illiterately,** *adv.* **illiteracy, illiterateness,** *n.*
illness (il'nəs), *n.* the state of being ill, sickness, physical indisposition.
illogical (iloj'ikəl), *a.* ignorant or careless of the rules of logic; contrary to reason. **illogically,** *adv.* **illogicalness, illogicality** (-kal'-), *n.*
illuminate (iloo'mināt), *v.t.* to throw light upon; to light up; to adorn (buildings, streets etc.) with festal lamps; to adorn (a manuscript etc.) with coloured pictures, letters etc.; to enlighten mentally or spiritually; to make illustrious. **illuminable,** *a.* **illuminant,** *a.* illuminating. *n.* that which illuminates. **illuminating,** *a.* lighting up; enlightening. **illumination,** *n.* **illuminative,** *a.* **illuminator,** *n.*
illuminati (iloominah'tē), *n.pl.* a name given to several religious sects and secret societies professing to have superior enlightenment; hence any persons who affect to possess extraordinary knowledge or gifts.
illumine (iloo'min), *v.t.* to illuminate.
illus., (*abbr.*) illustrated; illustration.
illusion (iloo'zhən), *n.* the act of deceiving; that which deceives; a delusion; a conjuring trick; an unreal image presented to the vision; esp. a deceptive sensuous impression; (*Psych.*) a wrong interpretation of what is perceived through the senses. **illusionism,** *n.* a theory that regards the external world as a mere illusion of the senses; the artistic practice of aiming to give an illusion of reality. **illusionist,** *n.* **illusive** (-siv), **illusory** (-zəri), *a.* delusive, deceptive. **illusively, illusorily,** *adv.* **illusiveness, illusoriness,** *n.*
illustrate (il'əstrāt), *v.t.* to make clear, to explain or elucidate by means of examples, figures etc.; to be an example of; to embellish or elucidate by pictures etc. **illustration,** *n.* the act of illustrating; the state of being illustrated; that which illustrates, an example, a typical instance; an engraving or drawing illustrating a book or article in a periodical; an embellishment. **illustrational, illustrative,** *a.* **illustratively,** *adv.* **illustrator,** *n.*

illustrious (ilŭs'triəs), *a.* distinguished, famous; conferring lustre, renown, or glory; brilliant. **illustriously,** *adv.* **illustriousness,** *n.*

ILP, (*abbr.*) the International Labour Party.

I'm (im), contr. form of *I am.*

image (im'ij), *n.* the visible representation or similitude of a person or thing; a likeness, a statue, esp. one intended for worship, an idol; a copy, a counterpart; the living embodiment of a particular quality; an idea, a conception; a mental picture; the impression given to others of a person's character etc.; an expanded metaphor or simile; the figure of an object formed (through the medium of a mirror, lens etc.) by rays of light; a mental representation of a sense impression. *v.t.* to make an image of; to mirror; to portray; to represent mentally; to conceive in the mind; to typify, to symbolize; to represent a part of the body pictorially for medical purposes. **image intensifier,** *n.* a device which enables the user to see objects at night. **image-maker,** *n.* a public relations expert employed to improve the impression that someone, e.g. a politician, makes on the general public. **image orthicon,** *n.* a type of television camera tube. **imageable,** *a.* **imageless,** *a.* **imagery** (-əri), *n.* (*collect.*) images, statues; figures evoked by the fancy; rhetorical figures, figurative description. **imagism,** *n.* **imagist,** *n.* a follower of a poetical school that seeks to express itself through clear and precise images of nature etc. **imagistic** (-jis'-), *a.*

imaginable etc. **IMAGINE.**

imaginal IMAGO.

imagine (imaj'in), *v.t.* to form an image of in the mind, to conceive, to form an idea of; (*coll.*) to suppose, to think; to believe without any justification; to conjecture, to guess; to plot, to devise. *v.i.* to form images or ideas in the mind. **imaginable,** *a.* **imaginably,** *adv.* **imaginary,** *a.* existing only in imagination or fancy; not real, esp. a mathematical quantity or value assumed as real for the purposes of an equation etc. **imaginary number,** *n.* a number involving the square root of a negative number. **imaginarily,** *adv.* **imaginariness,** *n.* **imagination,** *n.* the act or process of imagining; the power of imagining; the mental faculty that forms ideal images or combinations of images from the impressions left by sensuous experience; fancy, fantasy; the constructive or creative faculty of the mind; mental scope or resourcefulness. **imaginative,** *a.* endowed with imagination; creative, constructive; produced or characterized by imagination. **imaginatively,** *adv.* **imaginativeness,** *n.* **imaginer,** *n.* **imagining,** *n.* imagination; a conception, an idea. **imaginal,** *a.*

imago (imā'gō), *n.* (*pl.* **-goes, -gines** (-jinēz)) the adult, fully-developed insect as manifested after its metamorphoses; an idealized type of a parent or other person exercising a persistent influence in the subconscious. **imaginal** (-maj'-), *a.*

imam (imahm'), *n.* a person who leads congregational prayer in a mosque; the title of various Muslim rulers and founders. **imamate** (-āt), **imamship,** *n.*

imbalance (imbal'əns), *n.* a lack of balance.

imbecile (im'bəsēl), *a.* mentally weak, half-witted; stupid, fatuous. *n.* one mentally weak; one who, though mentally deficient, shows signs of rudimentary intelligence; a stupid or foolish person. **imbecilely,** *adv.* **imbecility** (-sil'-), *n.*

imbibe (imbīb'), *v.t.* to drink in; to absorb; to receive into the mind. *v.i.* (*facet.*) to drink; to indulge in a drinking session. **imbiber,** *n.*

imbricate (im'brikāt), *v.t.* to lap (leaves, scales on fish etc.) the one over the other like tiles. *v.i.* to be arranged in this position. *a.* (-kət) overlapping. **imbrication,** *n.* **imbricative,** *a.*

imbroglio (imbrō'liō), *n.* (*pl.* **-lios**) a complicated plot, as of a play or novel; a perplexing or confused state of affairs; a disorderly heap; a misunderstanding.

imbrue (imbroo'), *v.t.* to steep, to soak or moisten (in or with blood, carnage etc.); to stain, to dye (in or with).

imbue (imbū'), *v.t.* to dye (with); to tinge strongly (with); to inspire, to impregnate (with).

IMF, (*abbr.*) International Monetary Fund.

imide (im'id), *n.* a compound derived from ammonia by the replacement of two atoms of hydrogen by a metal or organic radical.

imit., (*abbr.*) imitation; imitative.

imitate (im'itāt), *v.t.* to produce a likeness of in form, colour, or appearance; to follow the example of; to mimic, to ape. **imitable,** *a.* **imitability** (-bil'-), *n.* **imitation,** *n.* the act of imitating; a copy or likeness; (*Mus.*) the repetition of a phrase or subject by another part or key. **imitative,** *a.* given to or aiming at imitation; done in imitation (of); counterfeit. **imitatively,** *adv.* **imitativeness,** *n.* **imitator,** *n.*

immaculate (imak'ūlət), *a.* spotlessly clean or tidy; pure; free from blemish; absolutely faultless; (*Biol.*) not spotted. **Immaculate Conception,** *n.* in the Roman Catholic Church, the doctrine that the Virgin Mary was conceived and born free from original sin. **immaculacy** (-si), **-lateness,** *n.* **immaculately,** *adv.*

immanent (im'ənənt), *a.* inherent, in-dwelling; (*Theol.*) present throughout the universe as an essential sustaining spirit. **immanence, -ency,** *n.* **immanently,** *adv.*

immaterial (imətiə'riəl), *a.* not consisting of matter; incorporeal; irrelevant, unimportant. **immaterialism,** *n.* the doctrine that there is no material substance, and that all being may be reduced to mind and ideas in mind. **immaterialist,** *n.* **immateriality** (-al'-), *n.* **immaterialize, -ise,** *v.t.* **immaterially,** *adv.*

immature (imətūə'), *a.* not mature, not ripe, imperfect; not fully developed; lacking the appropriate maturity of character etc. **immaturely,** *adv.* **immaturity, immatureness,** *n.*

immeasurable (imezh'ərəbl), *a.* that cannot be measured; immense. **immeasurability** (-bil'-), **-ableness,** *n.* **immeasurably,** *adv.*

immediate (imē'diət), *a.* situated in the closest relation; not separated by any space etc.; acting or acted upon by direct agency, direct; proximate, next, present; done or occurring at once, instant. **immediacy** (-si), *n.* **immediately,** *adv.* without delay, at once, closely or directly; just close by. *conj.* as soon as. **immediateness,** *n.*

immemorial (imimaw'riəl), *a.* extending beyond the reach of memory, record or tradition. **immemorially,** *adv.*

immense (imens'), *a.* huge, vast, immeasurable; (*coll.*) very great, very large; (*sl.*) very good, excellent. **immensely,** *adv.* **immenseness, immensity,** *n.*

immerse (imœs'), *v.t.* to plunge, to dip (into or under water or other fluid); to baptize in this manner; to involve or absorb deeply (in difficulty, debt, study, etc.). **immersible,** *a.* **immersion** (-shən), *n.* the act of immersing; the state of being immersed; baptism by plunging completely under water; the state of being deeply involved (in thought etc.); the disappearance of a celestial body behind or into the shadow of another; a language-teaching method involving the exclusive use of the language concerned in the learning situation. **immersion heater,** *n.* an electrical appliance that is immersed in a tank etc. to heat the water contained therein. **immersionist,** *n.* one who believes in baptism by immersion.

immigrate (im'igrāt), *v.i.* to come into a foreign country for settlement there. **immigrant,** *n.* one who immigrates. **immigration,** *n.*

imminent (im'inənt), *a.* impending; close at hand. **imminence, -ency,** *n.* **imminently,** *adv.*

immiscible (imis'ibl), *a.* not capable of being mixed. **immiscibility** (-bil'-), *n.* **immiscibly,** *adv.*

immobile (imō'bīl), *a.* not mobile, immovable; impassible; (*coll.*) not moving. **immobility** (-bil'-), *n.* **immobilize, -ise** (-bi-), *v.t.* to render immovable; to withdraw (specie) from circulation; to render (troops) incapable of being moved. **immobilization,** *n.*

immoderate (imod'ərət), *a.* excessive; unreasonable. **immoderacy** (-si), **immoderateness, immoderation,** *n.* **immoderately,** *adv.*

immodest (imod'ist), *a.* not modest, forward; unchaste, indelicate, indecent. **immodestly,** *adv.* **immodesty,** *n.*

immolate (im'əlāt), *v.t.* to kill in sacrifice, to offer up; to sacrifice (to). **immolation,** *n.* **immolator,** *n.*

immoral (imo'rəl), *a.* not moral; inconsistent with or contrary to (esp. sexual) morality; licentious, vicious. **immoralism,** *n.* the rejection of morality. **immoralist,** *n.* **immorality** (-ral'-), *n.* **immorally,** *adv.*

immortal (imaw'təl), *a.* not mortal, not subject to death; imperishable; relating to immortality; eternally famous. *n.* one who is immortal, esp. one of the ancient gods; a person whose works assure him or her of lasting fame. **immortality** (-tal'-), *n.* **immortalize, -ise,** *v.t.* to make immortal; to perpetuate the memory of. **immortalization,** *n.* **immortally,** *adv.*

immortelle (imawtel'), *n.* a plant with flowers that keep their shape and colour for a long period after being gathered.

immovable (imoo'vəbl), *a.* that cannot be moved; firmly fixed; steadfast; unchanging, unalterable; unfeeling; (*Law*) not liable to be removed. **immovability** (-bil'-), **-ableness,** *n.* **immovably,** *adv.*

immune (imūn'), *a.* free or exempt (from); highly resistant to (a disease etc.); pertaining to immunity. *n.* one who is not liable to infection. **immunist,** *n.* **immunity,** *n.* freedom or exemption from any obligation, duty, or office; exemption from a penalty, taxation etc.; freedom from liability to infection. **immunize, -ise** (im'-), *v.t.* **immunization,** *n.* the conferring of immunity to a disease by artificial means.

immuno-, *comb. form.* immunity; immune.

immunodeficiency (imūnōdifish'ənsi), *n.* a deficiency in, or breakdown of, a person's immune system.

immunoglobulin (imūnōglob'ūlin), *n.* one of five classes of proteins showing antibody activity.

immunology (imūnol'əji), *n.* the scientific study of immunity. **immunological** (-loj'-), *a.* **immunologically,** *adv.* **immunologist,** *n.*

immunosuppressive (imūnōsəpres'iv), *a.* pertaining to a drug that minimizes the body's natural reactions to a foreign substance, esp. the rejection of a transplanted organ. **immunosuppression,** *n.*

immunotherapy (imūnōthe'rəpi), *n.* the treatment of disease through the stimulation of the patient's own natural immunity.

immure (imūə'), *v.t.* to shut in or up; to surround, as with a wall; to confine. **immurement,** *n.*

immutable (imū'təbl), *a.* unchangeable, not susceptible to change or variation. **immutability** (-bil'-), **-ableness,** *n.* **immutably,** *adv.*

imp (imp), *n.* a young or little devil; a little malignant spirit; a mischievous child. *v.t.* to supply (esp. the wing of a falcon) with new feathers. **impish,** *a.* having the characteristics of an imp; mischievous. **impishly,** *adv.* **impishness,** *n.*

imp., (*abbr.*) imperative; imperfect; imperial; impersonal.

impact (im'pakt), *n.* a forcible striking (upon or against), a collision; effect, influence. *v.t.* (pakt'), to press or drive firmly together, to pack firmly in. **impacted** (-pak'-), *a.* of a tooth, wedged in such a way as to be unable to come through the gum; of a fracture, having jagged ends that are wedged into each other. **impaction** (-pak'-), *n.*

impair (impeə'), *v.t.* to diminish in excellence, value, strength etc.; to damage, to injure. **impairment,** *n.*

impala (impah'lə), *n.* a large antelope of southern and eastern Africa.

impale (impāl'), *v.t.* to transfix, esp. to put to death by transfixing with a sharp stake; (*Her.*) to arrange two coats of arms on one shield, divided by a vertical line; to render helpless, as though by impaling. **impalement,** *n.*

impalpable (impal'pəbl), *a.* not perceptible to the touch; not coarse; not to be readily apprehended by the mind, intangible. **impalpability** (-bil'-), *n.* **impalpably,** *adv.*

impanel (impan'l), EMPANEL.

impart (impaht'), *v.t.* to grant or bestow a share of; to communicate the knowledge of; to give, to bestow. **impartation, impartment,** *n.* **imparter,** *n.*

impartial (impah'shəl), *a.* not partial; not favouring one party or one side more than another; equitable, disinterested. **impartiality** (-al'-), **impartialness,** *n.* **impartially,** *adv.*

impartible (impah'tibl), *a.* not subject to or capable of partition. **impartibility** (-bil'-), *n.*

impassable (impahs'əbl), *a.* that cannot be passed. **impassability** (-bil'-), **-ableness,** *n.* **impassably,** *adv.*

impasse (am'pas, im'-), *n.* a blind alley; an insurmountable obstacle; deadlock.

impassible (impas'ibl), *a.* insensible to pain or suffering; incapable of being injured; not subject to feeling or passion. **impassibility** (-bil'-), **-ibleness,** *n.* **impassibly,** *adv.*

impassion (impash'ən), *v.t.* to rouse the deepest feelings of, to stir to ardour or passion. **impassionable,** *a.* **impassioned,** *a.* charged with passion.

impassive (impas'iv), *a.* not affected by pain, feeling, or passion; apathetic; unmoved, serene. **impassively,** *adv.* **impassiveness, impassivity** (-siv'-), *n.*

impasto (impas'tō), *n.* the application of a thick layer or body of pigment, to give relief etc.; paint so applied.

impatient (impā'shənt), *a.* not able to wait or to endure; fretful; not patient or tolerant (of); eager (for or to). **impatience,** *n.* **impatiently,** *adv.*

impawn (impawn'), *v.t.* to deposit as security; to pledge.

impeach (impēch'), *v.t.* to charge with a crime or misdemeanour; to bring a charge of maladministration or treason against; to accuse, to charge, to find fault; to call in question; to bring discredit upon. **impeachable,** *a.* **impeacher,** *n.* **impeachment,** *n.* the act of impeaching; the arraignment before a proper tribunal for maladministration or treason; an accusation; a calling in question.

impeccable (impek'əbl), *a.* not liable to fall into sin; blameless; faultless. **impeccability** (-bil'-), *n.* **impeccably,** *adv.* **impeccant,** *a.* sinless, impeccable.

impecunious (impikū'niəs), *a.* destitute of money; short of money. **impecuniosity** (-os'-), **impecuniousness,** *n.* **impecuniously,** *adv.*

impede (impēd'), *v.t.* to hinder, to obstruct. **impedance,** *n.* resistance to alternating current, esp. due to inductance or capacitance together with ohmic resistance. **impediment** (-ped'-), *n.* that which impedes; hindrance, obstruction; a speech defect; an obstacle to lawful marriage. **impedimenta** (-men'tə), *n.pl.* baggage, esp. for an army; things that impede progress. **impedimental,** *a.*

impel (impel'), *v.t.* (*past, p.p.* **impelled**) to drive or push forward; to drive or urge (to an action or to do). **impellent,** *a., n.* (one who or that) which impels. **impeller,** *n.* a rotor.

impend (impend'), *v.i.* to threaten, to be imminent; to hang (over), to be suspended (over). **impendence, -ency,** *n.* **impendent, impending,** *a.*

impenetrable (impen'itrəbl), *a.* that cannot be penetrated or pierced; inscrutable, incomprehensible; dull, obtuse, stupid; (*Phys.*) preventing any other substance from occupying the same place at the same time. **impenetrability** (-bil'-), **-ableness,** *n.* **impenetrably,** *adv.*

impenitent (impen'itənt), *a.* not penitent, not contrite. **impenitence, -ency,** *n.* **impenitently,** *adv.*

imper., (*abbr.*) imperative.

imperative (impe'rətiv), *a.* (*Gram.*) expressive of command; authoritative, peremptory; obligatory, essential; urgent, vital. *n.* that mood of a verb which expresses command, entreaty, or exhortation; something absolutely essential or very urgent. **categorical imperative**

CATEGORICAL. **imperatival** (-tī'-), *a.* (*Gram.*) **imperatively**, *adv.* **imperativeness**, *n.*

imperator (impərah'taw), *n.* (*Rom. Hist.*) a title originally bestowed upon a victorious leader by his soldiers; afterwards the equivalent of 'emperor'. **imperatorial** (-perətaw'-), *a.* **imperatorially**, *adv.* **imperatrix** (-triks), *n.* an empress.

imperceptible (impəsep'tibl), *a.* not easily apprehended, indistinguishable; insignificant, extremely slight, small, or gradual. **imperceptibility** (-bil'-), **-ibleness**, *n.* **imperceptibly**, *adv.* **imperceptive**, *a.*

impercipient (impəsip'iənt), *a.* not perceiving; not having power to perceive.

imperf., (*abbr.*) imperfect.

imperfect (impœ'fikt), *a.* defective; incomplete, not fully made, done etc.; lacking some part or member; (*Gram.*) expressing action as continuous and not completed; (*Mus.*) diminished; (of a cadence, passing to a dominant chord from another, esp. a tonic, chord. *n.* the imperfect tense; a verb in this tense. **imperfectible** (-fek'-), *a.* incapable of being perfected. **imperfectibility** (-bil'-), *n.* **imperfect tense**, *n.* a tense expressing or denoting an uncompleted action or state, usu. relating to past time. **imperfection** (-fek'-), *n.* a moral or physical fault; a defect; a deficiency. **imperfective** (-fek'-), *a.*, *n.* (denoting) that aspect of a verb which shows that the action is in progress. **imperfectively**, *adv.* **imperfectly**, *adv.* **imperfectness**, *n.*

imperforate (impœ'fərət), *a.* not perforated; not separated by rows of perforations, as stamps; (*Anat.*) having no opening or normal orifice etc. **imperforable**, *a.* **imperforation**, *n.*

imperial (impiə'riəl), *a.* of or pertaining to an empire or an emperor; sovereign, supreme; lordly, majestic; of weights and measures, conforming to official British nonmetric standards. *n.* a baggage-case on a travelling carriage; an outside seat on a diligence or coach; a size of paper about 22 × 30 in. (57 × 76 cm); a tuft of hair on a man's chin (named from Napoleon III). **Imperial City**, *n.* Rome; an independent city in the Holy Roman Empire. **imperialism**, *n.* government by an emperor; imperial spirit, state, or authority; the policy of extending the authority of a nation by means of territorial acquisition; the extension of power or authority in any sphere. **imperialist**, *n.* **imperialistic** (-is'-), *a.* **imperialistically**, *adv.* **imperialize**, **-ise**, *v.t.* **imperialization**, *n.* **imperially**, *adv.*

imperil (impe'ril), *v.t.* (*past*, *p.p.* **imperilled**) to endanger.

imperious (impiə'riəs), *a.* arbitrary, overbearing; haughty, arrogant; urgent, pressing. **imperiously**, *adv.* **imperiousness**, *n.*

imperishable (impe'rishəbl), *a.* enduring permanently; not subject to decay. **imperishability** (-bil'-), **-ableness**, *n.* **imperishably**, *adv.*

imperium (impə'riəm), *n.* absolute command, authority, or rule. **imperium in imperio** (-ō), an independent authority within the dominion of another authority.

impermanent (impœ'mənənt), *a.* not permanent. **impermanence**, *n.* **impermanently**, *adv.*

impermeable (impœ'miəbl), *a.* not allowing passage, esp. of a fluid, impervious. **impermeability** (-bil'-), **-ableness**, *n.* **impermeably**, *adv.*

impermissible (impəmis'ibl), *a.* not permissible. **impermissibility** (-bil'-), *n.*

impers., (*abbr.*) impersonal.

impersonal (impœ'sənəl), *a.* without personality; not relating to any particular person or thing; lacking in human warmth; (*Gram.*) applied to verbs used only in the third person singular. **impersonality** (-nal'-), *n.* **impersonalize**, **-ise**, *v.t.* **impersonally**, *adv.*

impersonate (impœ'sōnāt), *v.t.* to pretend to be; to imitate the mannerisms of, esp. for entertainment. **impersonation**, *n.* **impersonator**, *n.*

impertinent (impœ'tinənt), *a.* not pertaining to the matter in hand; trifling, frivolous; offensive, impudent, insolent. **impertinence**, *n.* **impertinently**, *adv.*

imperturbable (impətœ'bəbl), *a.* that cannot be easily disturbed or excited; unmoved, calm, cool. **imperturbability** (-bil'-), **-ableness**, *n.* **imperturbably**, *adv.*

impervious (impœ'viəs), *a.* not penetrable; not receptive or open (to). **imperviously**, *adv.* **imperviousness**, *n.*

impetigo (impəti'gō), *n.* (*pl.* **-tigines** (-tij'inēz)), a clustered yellow-scaled pustular eruption on the skin. **impetiginous** (-tij'-), *a.*

impetuous (impet'ūəs), *a.* moving with violence or great speed; acting violently or suddenly, hasty, impulsive, precipitate. **impetuously**, *adv.* **impetuosity** (-os'-), **impetuousness**, *n.*

impetus (im'pitəs), *n.* the force with which a body moves or is impelled; impulse, driving force; stimulus.

impi (im'pi), *n.* a body of southern African native fighters.

impiety (impī'əti), *n.* the quality of being impious; an impious act; want of filial affection or of reverence towards God.

impinge (impinj'), *v.i.* to come into collision, to strike (on, against etc.); to encroach (on); to have an effect (on). **impingement**, *n.*

impious (im'piəs), *a.* wanting in piety or reverence, esp. towards God; irreverent, profane. **impiously**, *adv.* **impiousness**, *n.* impiety.

impish IMP.

impiteous (impit'iəs), *a.* (*poet.*) pitiless, ruthless.

implacable (implak'əbl), *a.* not to be appeased; inexorable, unrelenting. **implacability** (-bil'-), **-ableness**, *n.* **implacably**, *adv.*

implant (implahnt'), *v.t.* to plant for the purpose of growth; to set or fix (in); to graft or insert into the body surgically; to inculcate, to instil. *n.* something engrafted, esp. surgically. **implantation**, *n.*

implausible (implaw'zibl), *a.* not having an appearance of truth and credibility. **implausibility** (-bil'-), **-ibleness**, *n.* **implausibly**, *adv.*

implement (im'plimənt), *n.* a tool, a utensil; an instrument, an agent; (*pl.*) things that serve for equipment, furniture, use etc. *v.t.* to fulfil; to carry into effect; to complete, to supplement. **implemental** (-men'-), *a.* **implementation**, *n.* **implementiferous** (-tif'-), *a.* of strata, containing stone implements.

implicate (im'plikāt), *v.t.* to entangle, to entwine; to involve, to bring into connection with; to show to be involved. **implication**, *n.* the act of implicating; the state of being implicated; something that is implied; a logical relationship between two propositions such that if the first is true, the second must necessarily be true as well. **implicative**, *a.*

implicit (implis'it), *a.* implied; understood or inferable; tacitly contained but not expressed; depending upon complete belief or trust in another; hence, unquestioning, unreserved. **implicitly**, *adv.* **implicitness**, *n.*

implied etc. IMPLY.

implode (implōd'), *v.t.*, *v.i.* to burst inwards; to sound by implosion. **implosion** (-zhən), *n.* imploding; the inward release of obstructed breath involved in the articulation of certain stop consonants. **implosive**, *n.*, *a.*

implore (implaw'), *v.t.* to call upon in earnest supplication; to ask for earnestly; to entreat. **imploration**, *n.* **implorer**, *n.* **imploringly**, *adv.* **imploringness**, *n.*

implosion IMPLODE.

imply (impli'), *v.t.* to involve or contain by implication; to signify; to import; to mean indirectly, to hint. **implied**, *a.* contained in substance or essence, though not actually expressed. **impliedly** (pli'əd-), *adv.*

impolarizable, **-isable** (impō'lərizəbl), *a.* (*Elec.*) incapable of polarization (as some voltaic batteries).

impolder, **empolder** (impōl'də), *v.t.* to form into a polder;

impolite

to reclaim (land) from the sea.
impolite (impəlīt'), *a.* not polite, ill-mannered. **impolitely**, *adv.* **impoliteness**, *n.*
impolitic (impŏl'itik), *a.* not politic; injudicious, inexpedient. **impoliticly**, *adv.*
imponderable (impon'dərəbl), *a.* not having sensible weight; impossible to assess or evaluate. *n.* a body or agent without sensible weight (as light, heat, electricity); an element or factor whose importance cannot be assessed or evaluated. **imponderabilia** (bil'iə), *n.pl.* imponderables. **imponderability** (-bil'-), **-ableness**, *n.*
import[1] (impawt'), *v.t.* to bring (goods) from a foreign country (into); to introduce; to imply, to signify, to mean. **importable**, *a.* **importability** (-bil'-), *n.* **importation**, *n.* the act or practice of importing; that which is imported. **importer**, *n.* one who imports goods.
import[2] (im'pawt), *n.* that which is imported from abroad (*usu. pl.*); importation; that which is signified or implied; importance, moment, consequence. **importance** (-paw'-), *n.* the quality of being important; weight, authority, consequence; personal consideration, self-esteem.
important (-paw'-), *a.* of great moment or consequence; of great personal consequence, pretentious; notable, eminent. **importantly**, *adv.*
importunate (impaw'tūnət), *a.* unreasonably and pertinaciously solicitous or urgent; troublesome for this reason. **importunately**, *adv.* **importunacy, importunateness, importunity** (-tū'-), *n.*
importune (impaw'tūn), *v.t.* to solicit pertinaciously or urgently; to harass with persistent requests; to solicit for immoral purposes. *v.i.* to be importunate. **importuner**, *n.*
importunity IMPORTUNATE.
impose (impōz'), *v.t.* to set, to attach; to lay (as a burden, tax, toll etc.) upon; to force (views etc.) upon; to palm off (upon); to arrange (pages of type) in a forme for printing. **to impose on, upon**, to cheat, to deceive; to take advantage of. **imposer**, *n.* **imposing**, *a.* commanding; impressive, majestic. **imposingly**, *adv.* **imposingness**, *n.*
imposition (impəzish'ən), *n.* the act of imposing or placing upon; that which is laid or placed upon; an unfair and excessive burden; an exercise enjoined as a punishment in schools etc.; a duty, a tax, an impost; a deceit, an imposture, a fraud; the process of assembling pages in type and then locking them into a chase. **imposition of hands**, (*Eccles.*) the laying on of hands in the ordination ceremony etc.
impossible (impos'ibl), *a.* not possible; (*loosely*) impracticable, not feasible; that cannot be done, thought, endured etc.; outrageous, monstrous; (*Math.*) imaginary. **impossibility** (-bil'-), *n.* **impossibly**, *adv.*
impost[1] (im'pōst), *n.* that which is imposed or levied as a tax, a tribute, a duty (esp. on imported goods); (*Racing*) a weight carried by a horse in a handicap.
impost[2] (im'pōst), *n.* the upper member of a pillar or entablature on which an arch rests.
impostor (impos'tə), *n.* one who falsely assumes a character; a deceiver by false pretences. **impostorship**, *n.*
impostrous, *a.* **imposture** (-chə), *n.* deception by the assumption of a false character, imposition; a fraud, a swindle.
impotent (im'pətənt), *a.* wanting in physical, intellectual, or moral power; of the male, lacking the power of sexual intercourse. **impotence, -ency**, *n.* **impotently**, *adv.*
impound (impownd'), *v.t.* to shut up (cattle) in a pound; to confine; to collect and confine or retain (water) in a reservoir, mill-pond etc.; to take possession of or confiscate (a document etc.). **impoundable**, *a.* **impoundage** (-ij), *n.* the act of impounding. **impounder, impoundment**, *n.*
impoverish (impov'ərish), *v.t.* to make poor; to exhaust the strength, fertility, or resources of. **impoverisher**, *n.* **impoverishment**, *n.*
impracticable (imprak'tikəbl), *a.* not possible to be

improbable

effected by the means at command; not feasible; unsuitable for a particular purpose; impassable. **impracticability** (-bil'-), **-ableness**, *n.* **impracticably**, *adv.*
impractical (imprak'tikl), *a.* unpractical. **impracticality** (-kal'-), **impracticalness**, *n.*
imprecate (im'prikāt), *v.t.* to invoke (as an evil on); to invoke a curse on. *v.i.* to curse. **imprecation**, *n.* the act of imprecating; a prayer for evil to fall on anyone; a curse. **imprecatory**, *a.* involving a curse.
imprecise (imprəsīs'), *a.* not precise, vague; inexact. **imprecision** (-sizh'ən), *n.*
impregnable[1] (impreg'nəbl), *a.* that cannot be stormed or taken by assault; able to resist all attacks, invincible. **impregnability** (-bil'-), *n.* **impregnably**, *adv.*
impregnate (im'pregnāt), *v.t.* to make pregnant; to render fruitful or fertile; to infuse the particles or qualities of any other substance into; to saturate (with); to imbue, to inspire (with). **impregnable**[2] (-preg'-), *a.* able to be impregnated. **impregnation**, *n.*
impresario (imprizah'riō), *n.* (*pl.* **-rios**) one who organizes or manages a concert, an opera company etc.
imprescriptible (impriskrip'tibl), *a.* that cannot be lost or impaired by usage or claims founded on prescription. **imprescriptibility** (-bil'-), *n.*
impress[1] (impres'), *v.t.* to press or stamp (a mark etc., in or upon); to produce (a mark or figure) by pressure; to fix deeply (in or on the mind); to affect strongly and favourably. **to impress on**, to emphasize to (someone); to urge, insist. **impressible**, *a.* capable of being impressed; yielding to pressure; susceptible. **impression** (-shən), *n.* the act of impressing; the mark made by impressing; a copy taken from type, an engraved plate etc.; the visible or tangible effect of an action etc.; (*collect.*) copies constituting a single issue of a book, engraving etc., esp. a reprint from standing type, as dist. from an edition; effect produced upon the senses, feelings etc.; an indistinct notion, a slight recollection, belief etc.; a mental effect of a previous experience; an imitation or impersonation. **impressionable**, *a.* easily impressed, impressible. **impressionability** (-bil'-), *n.* **impressionism**, *n.* an artistic movement that began in France acting on the principle that the hand should paint what the eye sees, thus ruling out all conventions of lighting and composition. **impressionist**, *n.* a painter of the impressionist school; an entertainer who does impersonations. *a.* pertaining to impressionism. **impressionistic** (-nis'-), *a.* **impressionistically**, *adv.* **impressive**, *a.* adapted to make an impression on the mind; commanding; inspiring; leaving a deep impression. **impressively**, *adv.* **impressiveness**, *n.*
impress[2] (im'pres), *n.* the act of marking by pressure; a mark or stamp made by pressure; a stamp, an impression; a characteristic mark.
impress[3] (impres'), *v.t.* to compel (seamen) to enter the public service; to seize or set apart (goods, property etc.) for the public service. **impressment**, *n.*
imprest (im'prest), *n.* a loan, an advance, esp. for carrying on any of the public services.
imprimatur (imprimah'tə), *n.* a licence to print a book, granted by the authorities, esp. of the Roman Catholic Church; (a mark of) sanction or approval.
imprint (imprint'), *v.t.* to impress, to stamp; to print; to impress (on or in the mind). *n.* (im'-), a mark, stamp, or impression; the name of the printer or publisher of a book, periodical etc., with the place and usu. the date of publication (on the title-page or at the end of a book). **imprinting**, *n.* the process by which young animals develop the tendency to recognize and be attracted to members of their own species.
imprison (impriz'n), *v.t.* to put into prison; to confine, to hold in custody or captivity. **imprisonment**, *n.*
improbable (improb'əbl), *a.* not likely to be true; not likely to happen. **improbability** (-bil'-), *n.* **improbably**, *adv.*

improbity (impro'biti), *n.* want of probity; dishonesty.
impromptu (impromp'tū), *adv.* off-hand, without previous study. *a.* done or said off-hand, extempore. *n.* (*pl.* **-tus**) an extemporaneous composition, performance, act etc.
improper (improp'ə), *a.* not proper; unsuitable, unfit; unbecoming, indecent; not accurate, erroneous. **improper fraction,** *n.* a fraction the numerator of which is equal to or greater than the denominator. **improperly,** *adv.*
impropriety (impropri'əti), *n.* the quality of being improper; an unbecoming act, expression etc.; indecency.
improve (improov'), *v.t.* to make better; to increase the value, goodness, or power of; to turn to profitable account; to take advantage of, to utilize. *v.i.* to grow or become better; to recover from illness, to regain health or strength; to increase in value, to rise, to be enhanced. **to improve on, upon,** to make something better than. **improvable,** *a.* **improvability** (-bil'-), **-ableness,** *n.* **improvement,** *n.* the act of improving; advancement in value, goodness, knowledge etc.; profitable use or employment; progress, growth, increase; that which is added or done to anything in order to improve it; a beneficial or valuable addition or substitute; the practical application of a discourse. **improver,** *n.* **improving,** *a.* tending to improve; morally edifying. **improvingly,** *adv.*
improvident (improv'idənt), *a.* not provident; neglecting to make provision for future exigencies; thriftless; careless, heedless. **improvidence,** *n.* **improvidently,** *adv.*
improvise (im'prəvīz), *v.t.* to compose and perform as one goes along, to extemporize; to do, produce, or prepare on the spur of the moment. **improvisation** (imprəvīz-), *n.* the act of improvising; something improvised. **improvisatorial** (-vīzətaw'-), **improvisatory** (-vīz'ə-), *a.* pertaining to improvisation. **improviser,** *n.*
imprudent (improo'dənt), *a.* wanting in foresight or discretion; rash, incautious, indiscreet. **imprudence,** *n.* **imprudently,** *adv.*
impudent (im'pūdənt), *a.* wanting in shame or modesty; impertinent, insolent. **impudence,** *n.* **impudently,** *adv.*
impudicity (-dis'-), *n.* immodesty, shamelessness.
impugn (impūn'), *v.t.* to call in question, to contradict, to gainsay. **impugnable,** *a.* **impugner,** *n.* **impugnment,** *n.*
impulse (im'pŭls), *n.* the application or effect of an impelling force; influence acting suddenly on the mind tending to produce action; a sudden tendency to action; stimulus, inspiration; a large force acting for an extremely short time, the momentum due to such a force; a disturbance passing along a nerve or muscle. **on impulse,** spontaneously, on a whim. **impulse buying,** *n.* making purchases on the spur of the moment. **impulsion** (-pŭl'shən), *n.* the act of impelling; the state of being impelled; impetus; an impelling force; a compulsion; instigation, incitement. **impulsive** (-pŭl'-), *a.* communicating impulse, urging forward; resulting from or liable to be actuated by impulse rather than reflection; acting momentarily, not continuous. **impulsively,** *adv.* **impulsiveness,** *n.*
impunity (impū'niti), *n.* exemption from punishment, penalty, injury, damage, or loss.
impure (impūə'), *a.* not pure; mixed with foreign matter, adulterated; defiled, unclean, unchaste; mixed with other colours. **impurely,** *adv.* **impureness,** *n.* **impurity,** *n.*
impute (impūt'), *v.t.* to ascribe, to attribute esp. blame (to); to set to the account or charge of. **imputable,** *a.* **imputability** (-bil'-), *n.* **imputation,** *n.* the act of imputing; that which is imputed as a charge or fault; reproach, censure, insinuation, *a.* coming by imputation. **imputatively,** *adv.* **imputer,** *n.*
in (in), *prep.* within, inside of, contained or existing within; denoting presence or situation within the limits of time, place, circumstance, reason, tendency, ratio, relation etc.; pregnant with. *adv.* within or inside some place; indoors, at home; in office; in favour; in fashion; in season; into the bargain, over and above; of a fire, alight; (*Cricket*) at the wicket. *a.* directed inwards; internal, living inside (as a hospital); fashionable; understood by a select group. **in absentia** (absen'tiə), not being present. **in as much as, inasmuch as,** seeing that, since; in so far as. **in itself,** by itself, apart from other things or considerations, absolutely. **in on,** (*coll.*) sharing in. **ins and outs,** (*coll.*) windings; complications, details. **in so** (or **as**) **far as, insofar as,** in such measure as. **in that,** seeing that; since. **to be in for,** to be committed to or involved in; to be entered for (a race etc.); to be heading for. **to be in for it,** to be certainly heading for trouble. **to be in with,** to be on intimate terms with. **to have it in for,** to intend harm or trouble to. **in-and-in,** *a.*, *adv.* from closely related parents. **in between,** *prep.* between. **in-between,** *a.* intermediate. **in-built,** *a.* built in, inherent. **in-depth,** *a.* detailed, thorough, comprehensive. **in-fighting,** *n.* behind-the-scenes squabbling or jockeying for power within a group etc. **in-flight,** *a.* available during an aeroplane flight. **in-house,** *a.* pertaining to, or employed within, a particular organization, company etc. **in-off,** *n.* a billiards or snooker shot that falls into a pocket after striking another ball. **in-patient,** *n.* a person residing inside a hospital and receiving regular treatment. **in-phase,** *a.* of two electric currents, alternating simultaneously. **in-service,** *a.* performed whilst remaining in one's ordinary employment. **in-tray,** *n.* a tray holding letters and documents still to be dealt with.
in., (*abbr.*) inch, inches.
in-[1], *pref.* in; into; within; on; against, towards; as in *indicate, induce.*
in-[2], *pref.* un-, not, without, as in *incomprehensible.*
inability (inəbil'iti), *n.* the state of being unable (to do, understand etc.); lack of power or means.
inaccessible (inəkses'ibl), *a.* that cannot be reached, attained, or approached; not affable, unapproachable. **inaccessibility** (-bil'-), **-ibleness,** *n.* **inaccessibly,** *adv.*
inaccurate (inak'ūrət), *a.* not accurate. **inaccuracy,** *n.* want of accuracy; an inaccurate statement, an error **inaccurately,** *adv.*
inaction (inak'shən), *n.* idleness, sloth; sluggishness, supineness. **inactive,** *a.* sluggish, inert; idle, indolent; chemic- ally or biologically lacking in reactivity; not in active service. **inactively,** *adv.* **inactivity** (tiv'-), *n.*
inadaptable (inədap'təbl), *a.* not adaptable. **inadaptability** (-bil'-), *n.* **inadaptation** (adəp-), *n.*
inadequate (inad'ikwət), *a.* not adequate; insufficient, unequal; unable to cope. **inadequately,** *adv.* **inadequacy, inadequateness,** *n.*
inadmissible (inədmis'ibl), *a.* that cannot be admitted, allowed, or received. **inadmissibility** (-bil'-), *n.*
inadvertent (inədvœ'tənt), *a.* not paying attention; of actions, unintentional, accidental. **inadvertence, -ency,** *n.* **inadvertently,** *adv.*
inadvisable (inədvī'zəbl), UNADVISABLE.
inalienable (inā'liənəbl), *a.* that cannot be alienated or transferred. **inalienability** (-bil'-), *n.* **inalienably,** *adv.*
inalterable (inawl'tərəbl), *a.* incapable of alteration. **inalterability** (-bil'-), *n.* **inalterably,** *adv.*
inamorato (inamərah'tō), *n.* a lover. **inamorata** (-tə), *n. fem.*
inane (inān'), *a.* empty; senseless; silly, fatuous. **inanely,** *adv.* **inanition** (inənish'ən), *n.* emptiness, voidness; exhaustion from want of food or nourishment. **inanity** (inan'-), *n.*
inanimate (inan'imət), *a.* not animate, not living; not endowed with animal life; void of animation, dull, lifeless. **inanimately,** *adv.* **inanimateness, inanimation,** *n.*
inanition, inanity INANE.
inapplicable (inap'likəbl, -əplik'-), *a.* not applicable; irrelevant. **inapplicability** (-bil'-), **-ableness,** *n.* **inapplicably,** *adv.*
inapposite (inap'əzit), *a.* not apposite; not pertinent. **inappositely,** *adv.* **inappositeness,** *n.*
inappreciable (inəprē'shəbl), *a.* not appreciable, not

inappropriate 419 **incident**

perceptible; too insignificant to be considered. **inappreciably,** *adv.* **inappreciation,** *n.* lack of appreciation; inability to appreciate properly. **inappreciative,** *a.*
inappropriate (inəprō'priət), *a.* not appropriate, unsuitable. **inappropriately,** *adv.* **inappropriateness,** *n.*
inapt (inapt'), *a.* not apt; unsuitable; unfit, unqualified. **inaptitude, inaptness,** *n.* **inaptly,** *adv.*
inarch (inahch'), *v.t.* to graft by inserting a scion, without separating it from the parent tree, into a stock growing near.
inarticulate (inahtik'ūlət), *a.* not articulated, not jointed; not uttered with distinct articulation, indistinct, dumb, speechless; unable to express oneself clearly. **inarticulacy** (si), **inarticulateness,** *n.* **inarticulately,** *adv.*
inartistic (inahtis'tik), *a.* not designed, done etc., according to the principles of art; not having artistic taste or ability. **inartistically,** *adv.*
inasmuch (inəzmŭch'), IN AS MUCH as under IN.
inattention (inətenshən), *n.* lack of attention; heedlessness, negligence; disregard of courtesy. **inattentive,** *a.* **inattentively,** *adv.* **inattentiveness,** *n.*
inaudible (inaw'dibl), *a.* not audible, so low as not to be heard. **inaudibility** (-bil'-), *n.* **inaudibly,** *adv.*
inaugurate (inaw'gūrāt), *v.t.* to install or induct into an office solemnly or with appropriate ceremonies; to commence, introduce, or celebrate the opening of with some degree of formality or solemnity. **inaugural,** *a.* pertaining to or performed at an inauguration; marking a commencement. *n.* an inaugural address. **inauguration,** *n.* **inaugurator,** *n.* **inauguratory,** *a.*
inauspicious (inəspish'əs), *a.* unlucky; ill-omened, unfavourable. **inauspiciously,** *adv.* **inauspiciousness,** *n.*
inboard (in'bawd), *adv.* within the sides or towards the middle of a ship, aircraft, or vehicle. *a.* situated thus.
inborn (in'bawn), *a.* innate, naturally inherent.
inbred (inbred'), *a.* innate, inborn; produced by inbreeding.
inbreed (inbrēd'), *v.t.* to breed or produce within; to breed from animals nearly related. **inbreeding,** *n.*
inc. (ingk), (*abbr.*) including; inclusive; incorporated.
Inca (ing'kə), *n.* the title given to the sovereigns of Peru up to the conquest under Pizarro, AD 1531; one of the royal race formerly dominant in Peru.
incalculable (inkal'kūləbl), *a.* not calculable, not to be reckoned or estimated in advance; too vast or numerous to be calculated; unpredictable, uncertain. **incalculability** (-bil'-), **-ableness,** *n.* **incalculably,** *adv.*
in camera CAMERA.
incandesce (inkandes'), *v.i.* to glow with heat. **incandescence,** *n.* **incandescent,** *a.* glowing with heat; intensely luminous with heat; strikingly radiant or bright. **incandescent lamp,** *n.* an electric or other lamp in which a filament or mantle is made intensely luminous by heat.
incantation (inkantā'shən), *n.* a formula, said or sung, supposed to add force to magical ceremonies, a charm. **incantatory** (-kan'tə-), *a.*
incapable (inkā'pəbl), *a.* not physically, intellectually, or morally capable (of); not susceptible (of); legally incapacitated; unable to take care of oneself; incapacitated by drink. *n.* one who is incapable. **incapability** (-bil'-), *n.* **incapably,** *adv.*
incapacitate (inkəpas'itāt), *v.t.* to render incapable, to disable; to render unfit, to disqualify (for, from etc.). **incapacitated,** *a.* **incapacitation,** *n.*
incapacity (inkəpas'iti), *n.* want of capacity; inability, incompetency; legal disqualification.
incarcerate (inkah'sərāt), *v.t.* to imprison; to shut up or confine. **incarceration,** *n.* **incarcerator,** *n.*
incarnadine (inkah'nədin), *a.* (*poet.*) of a flesh or carnation colour. *v.t.* (*poet.*) to dye this colour; to tinge with red.
incarnate (inkah'nət), *a.* invested or clothed with flesh, embodied in flesh, esp. in human form; typified, personified; (*esp. Bot.*) flesh-coloured, pink. *v.t.* (in'kahnāt), to clothe with flesh; to embody in flesh; to embody (an idea) in a living form; to be the embodiment of. **incarnation,** *n.* the act of assuming flesh; embodiment, esp. in human form; Christ's assumption of human nature; a vivid exemplification or personification; carnation, flesh-colour; the process of healing wounds, and filling or covering the damaged part with new flesh.
incautious (inkaw'shəs), *a.* wanting in caution; rash, unwary. **incautiously,** *adv.* **incautiousness,** *n.*
incavo (inkah'vō), *n.* the incised portion of an intaglio. [It.]
incendiary (insen'diəri), *a.* pertaining to the malicious burning of property; exciting or tending to excite factions, seditions or quarrels; inflammatory; igniting readily. *n.* one who maliciously sets fire to property etc.; an incendiary bomb; one who excites factions, seditions etc. **incendiary bomb,** *n.* a bomb containing violently incendiary materials that are scattered in flames on detonation. **incendiarism,** *n.*
incense[1] (in'sens), *n.* a mixture of fragrant gums, spices etc. used for producing perfumes when burnt, esp. in religious rites; the smoke of this; flattery; an agreeable perfume. *v.t.* to perfume with or as with incense; to offer incense to. **incense-boat,** *n.* a small boat-shaped vessel for holding incense. **incensation,** *n.* the offering of incense as an act of divine worship, or as a ceremonial adjunct. **incenser, -sory,** *n.* a censer.
incense[2] (insens'), *v.t.* to exasperate, to provoke, to enrage.
incentive (insen'tiv), *a.* inciting, urging. *n.* that which acts as a motive, incitement or spur.
incept (insept'), *v.i.* at Cambridge University, to be finally admitted to the degree of Master or Doctor. *v.t.* (*Biol.*) to receive, to take in. **inception** (insep'shən), *n.* a commencement; at Cambridge University, the act or ceremony of incepting. **inceptive,** *a.* beginning, commencing; (*Gram.*) denoting the beginning of an action. *n.* a verb that denotes the beginning of an action.
incertitude (insœ'titūd), *n.* uncertainty.
incessant (inses'ənt), *a.* unceasing, perpetual. **incessantly,** *adv.* **incessancy,** †**incessantness,** *n.*
incest (in'sest), *n.* sexual intercourse between persons related within the prohibited degrees of matrimony. **incestuous** (-tūəs), *a.* guilty of or involving incest; of a group etc., inward-looking, closed to external influences etc. **incestuously,** *adv.* **incestuousness,** *n.*
inch[1] (inch), *n.* the 12th part of a linear foot; the least quantity or degree; the unit of measurement of the rainfall, the quantity that would cover the surface of the ground to the depth of one inch (2·54 cm); the pressure, atmospheric or other, equivalent to the weight of a column of mercury one inch (2.54 cm) high in a barometer; (*pl.*) stature. *v.t.* to drive by inches or small degrees. *v.i.* to move thus. **by inches, inch by inch,** bit by bit; gradually, by very small degrees. **every inch,** entirely, from head to foot. **(by) inchmeal,** by inches, bit by bit. **inch-measure, -rule, -tape,** *n.* a measure divided into inches. **inchworm,** *n.* a looper caterpillar. **incher,** *n.* (*usu. in comb.*), as *six-incher*.
inch[2] (inch), *n.* an island.
inchoate (inkō'āt, -ət), *a.* only begun, commenced; existing only in elements, incomplete, undeveloped. **inchoately,** *adv.* **inchoateness,** *n.* **inchoation,** *n.* **inchoative,** *a.* incipient; indicating the beginning of an action. *n.* an inchoative verb.
incident (in'sidənt), *a.* falling or striking (on or upon); likely to happen; naturally appertaining or belonging (to); consequent (on). *n.* a fortuitous event; a concomitant or subsidiary event; an occurrence, esp. one of a picturesque or striking nature; a minor event causing a public disturbance. **incidence** (in'sidəns), *n.* the act or state of fall-

ing on or upon; (*Phys.*) the direction in which a body, or a ray of light, heat etc. falls upon any surface; scope, bearing, range; frequency of occurrence. **angle of incidence** ANGLE. **line of incidence**, the line in which a ray of light, heat etc. moves to strike a plane. **incidental** (insiden'təl), *a.* casual, accidental, contingent; undesigned, fortuitous, not essential; concomitant, naturally connected with or related (to); occasional. *n.* something that is incidental; (*pl.*) casual expenses. **incidental music,** *n.* music accompanying the action of a play or film. **incidentally,** *adv.*
incinerate (insin'ərāt), *v.t.* to reduce to ashes. **incineration,** *n.* **incinerator,** *n.* a receptacle in which refuse etc. is burned.
incipient (insip'iənt), *a.* beginning, in the first stages. **incipiently,** *adv.* **incipience, -ency,** *n.*
incise (insīz'), *v.t.* to cut into, to carve (with an inscription, pattern etc.); to engrave. **incision** (sizh'ən), *n.* the act of incising; a cut, a gash made by surgery in the body. **incisive** (-siv), *a.* having the quality of cutting into; having a sharp cutting edge; sharp, penetrating; trenchant, acute. **incisively,** *adv.* **incisiveness,** *n.*
incisor (insī'zə), *n.* a tooth adapted for cutting or dividing the food, one of those between the canines.
incite (insīt'), *v.t.* to stir up, to urge; to prompt, to encourage (to action, to do etc.). **incitation** (insi-), *n.* the act of inciting; an incitement. **incitement,** *n.* a stimulus, an incentive, a motive. **inciter,** *n.* **incitingly,** *adv.*
incivility (insivil'iti), *n.* rudeness, impoliteness; an act of rudeness.
incl., (*abbr.*) including.
inclement (inklem'ənt), *a.* merciless; rough, severe; boisterous, stormy. **inclemently,** *adv.* **inclemency** (-si), *n.*
incline[1] (inklīn'), *v.i.* to deviate from any direction that is regarded as the normal one; to lean, to bend down or forwards; to be disposed (to); to have a propensity, proneness or inclination. *v.t.* to cause to deviate from a line or direction; to give an inclination or leaning to; to direct; to cause to bend (the head or body) down, to bow or stoop; to dispose, to turn. **inclinable** (klīn'-), *a.* having a tendency; inclined, disposed, willing (to). **inclinableness,** *n.* **inclination** (inkli-), *n.* the act of inclining or bending; a deviation from any direction regarded as the normal one; leaning or bent of the mind or will; disposition, proclivity, propensity (to, for etc.); liking, affection (for); (*Geom.*) the mutual approach or tendency of two bodies, lines or planes towards each other, esp. as measured by the angle between them. **inclinational,** *a.* **inclined,** *a.* **inclined plane,** *n.* a plane set at an acute angle to the horizon. **incliner,** *n.* **inclinometer** (inklinom'itə), *n.* an instrument for detecting the vertical intensity of the magnetic force, a dipping-compass; an instrument that indicates the angle an aircraft is making with the horizon.
incline[2] (in'klīn), *n.* a slope, a gradient.
inclose (inklōz'), ENCLOSE.
include (inklood'), *v.t.* to contain, to comprise, to comprehend as a component part, member etc.; to put in or classify as part of a set etc.; to enclose, to confine within. **includable, -dible,** *n.* **included,** *a.* enclosed; contained, comprehended; of the style and stamens of a plant, not projecting beyond the mouth of the corolla. **inclusion** (inkloo'zhən), *n.* **inclusive** (inkloo'siv), *a.* including, containing, comprehending (usu. with *of*); comprehending in the total sum or number; including everything; including the limits specified. **inclusively,** *adv.* **inclusiveness,** *n.*
incog., (*abbr.*) incognito.
incognito (inkogne'tō), *a., adv.* living or going under an assumed name or character. *n.* (*pl.* **-tos**) a person who is unknown or under an assumed name or character; the state of being unknown or in disguise; an assumed identity. **incognita** (-tə), *n., a., adv., fem.*
incognizable, -isable (inkog'nizəbl), *a.* not capable of being perceived or apprehended. **incognizance, -isance,** *n.* **incognizant, -isant,** *a.* not aware, unknowing.
incoherent (inkəhiə'rənt), *a.* lacking cohesion; loose, disconnected, inconsistent; inarticulate, rambling. **incoherence, -ency, incohesion** (-zhən), *n.* **incoherently,** *adv.* **incohesive,** *a.*
incombustible (inkəmbŭs'tibl), *a.* incapable of being burnt or consumed by fire. **incombustibility** (-bil'-), *n.* **incombustibly,** *adv.*
income (in'kəm), *n.* the amount of money (usu. annual) accruing as payment, profit, interest etc. from labour, business, profession, or property. **income support,** *n.* in Britain a social security payment made to the unemployed or people on low incomes. **income-tax,** *n.* a tax levied on incomes above a certain amount.
incomer (in'kŭmə), *n.* one who comes in, an immigrant. **incoming,** *a.* coming in or entering into possession; accruing; succeeding.
incommensurable (inkəmen'shərəbl), *a.* having no common measure (with another integral or fractional number or quantity); not fit or worthy to be measured (with). **incommensurability** (-bil'-), *n.* **incommensurably,** *adv.* **incommensurate,** *a.* not commensurate; incommensurable; inadequate (to or with). **incommensurately** (-rət-), *adv.* **incommensurateness,** *n.*
incommode (inkəmōd'), *v.t.* to cause trouble or inconvenience to; to embarrass, to disturb, to hinder. **incommodious,** *a.* inconvenient; cramped, too small. **incommodiously,** *adv.* **incommodiousness,** *n.*
incommunicable (inkəmū'nikəbl), *a.* that cannot be communicated to, or shared with another. **incommunicability** (-bil'-), **-ableness,** *n.* **incommunicably,** *adv.* **incommunicative,** *a.* taciturn, reserved. **incommunicativeness,** *n.*
incommunicado (inkəmūnikah'dō), *a.* with no means of communication with the outside world; in solitary confinement.
incomparable (inkom'pərəbl), *a.* not to be compared (to or with); unequalled, peerless. **incomparableness,** *n.* **incomparably,** *adv.*
incompatible (inkəmpat'ibl), *a.* inconsistent with something else; incapable of subsisting with something else; unable to cooperate or work together; mutually intolerant; not suited for use together because of harmful effects. *n.* an incompatible person or thing. **incompatibility** (-bil'-), **-ibleness,** *n.* **incompatibly,** *adv.*
incompetent (inkom'pitənt), *a.* lacking adequate power, means, capacity, or qualifications (to do); grossly lacking in ability or fitness for a task. *n.* an incompetent person. **incompetence, -ency,** *n.* **incompetently,** *adv.*
incomplete (inkəmplēt'), *a.* not complete, not perfect; (*Bot.*) destitute of calyx, corolla, or of both. **incompletely,** *adv.* **incompleteness, incompletion,** *n.*
incomprehensible (inkomprihen'sibl), *a.* that cannot be comprehended, conceived or understood. **incomprehensibility** (-bil'-), **-ibleness,** *n.* **incomprehensibly,** *adv.* **incomprehension** (-shən), *n.* lack of comprehension; failure to understand.
incompressible (inkəmpres'ibl), *a.* not compressible; strongly resisting compression. **incompressibility** (-bil'-), *n.*
inconceivable (inkənsē'vəbl), *a.* not conceivable, incomprehensible; hence, incredible, most extraordinary. **inconceivability** (-bil'-), **-ableness,** *n.* **inconceivably,** *adv.*
inconclusive (inkənkloo'siv), *a.* not conclusive; of evidence etc., not cogent or decisive. **inconclusively,** *adv.* **inconclusiveness,** *n.*
incongruous (inkong'grūəs), *a.* not congruous, not agreeing or harmonizing; unsuitable, not fitting, out of place. **incongruent,** *a.* **incongruity** (-groo'-), **incongruousness,** *n.* **incongruously,** *adv.*
inconsequent (inkon'sikwənt), *a.* not following regularly from the premises, irrelevant; illogical; disconnected. **inconsequence,** *n.* **inconsequential** (-kwen'-), *a.* inconse-

quent; of no consequence, trivial. **inconsequentiality** (-kwenshial'-), *n.* **inconsequentially**, *adv.* **inconsequently**, *adv.*

inconsiderable (inkənsid'ərəbl), *a.* not deserving consideration or notice; insignificant, unimportant, trivial; small. **inconsiderableness**, *n.* **inconsiderably**, *adv.*

inconsiderate (inkənsid'ərət), *a.* hasty, incautious; having no consideration for the feelings of others. **inconsiderately**, *adv.* **inconsiderateness**, **-ation**, *n.*

inconsistent (inkənsis'tənt), *a.* discordant, incompatible (with); self-contradictory, not agreeing with itself or oneself; not uniform, changeable, unsteady. **inconsistency**, *n.* **inconsistently**, *adv.*

inconsolable (inkənsō'ləbl), *a.* of a person, grief etc., not to be consoled. **inconsolability** (-bil'-), **-ableness**, *n.* **inconsolably**, *adv.*

inconsonant (inkon'sənənt), *a.* not consonant, discordant (with). **inconsonance**, *n.*

inconspicuous (inkənspik'ūəs), *a.* not conspicuous; not easy to see; (*Bot.*) small in size, obscure in colour etc. **inconspicuously**, *adv.* **inconspicuousness**, *n.*

inconstant (inkon'stənt), *a.* not constant, changeable, fickle; variable, unsteady; unfaithful. **inconstancy**, *n.* **inconstantly**, *adv.*

incontestable (inkəntes'təbl), *a.* indisputable, undeniable. **incontestability** (-bil'-), *n.* **incontestably**, *adv.*

incontinent (inkon'tinənt), *a.* not restraining (esp. sexual) the passions or appetites; (*Med.*) not able to restrain natural evacuations. **incontinence**, *n.* **incontinently**, *adv.* unchastely; at once, straightway, immediately.

incontrollable (inkəntrō'ləbl), *a.* not controllable. **incontrollably**, *adv.*

incontrovertible (inkontrəvœ'tibl), *a.* that cannot be controverted; incontestable, indisputable. **incontrovertibility** (-bil'-), **-ibleness**, *n.* **incontrovertibly**, *adv.*

inconvenience (inkənvēn'yəns), *n.* the quality or state of being inconvenient; a cause of difficulty. *v.t.* to put to inconvenience; to incommode. **inconvenient** (-nyənt), *a.* not convenient, incommodious; causing or tending to cause trouble, uneasiness or difficulty; inopportune, awkward. **inconveniently**, *adv.*

inconvertible (inkənvœ'tibl), *a.* incapable of being converted into or exchanged for something else, esp. money. **inconvertibility** (-bil'-), *n.* **inconvertibly**, *adv.*

incorporate¹ (inkaw'pərət), *a.* combined into one body or corporation, closely united; of a society, company etc., made into a corporation.

incorporate² (inkaw'pərāt), *v.t.* to unite, combine or mingle into one mass or body (with); to form into a legal corporation; to receive into a corporation; to embody. *v.i.* to become united or incorporated (with another substance, society etc.) so as to form one body; (*N Am.*) to form a limited company. **incorporated**, *a.* (*N Am.*) of a joint stock company, limited. **incorporation**, *n.* the act of incorporating; the state of being incorporated; embodiment; formation of or reception into a corporate body; a corporate body, a corporation. **incorporative**, *a.* **incorporator**, *n.*

incorporeal (inkawpaw'riəl), *a.* not corporeal; immaterial; (*Law*) lacking material existence of itself but based on something material.

incorrect (inkərekt'), *a.* not in accordance with truth, propriety etc.; wrong, inaccurate; improper, unbecoming. **incorrectly**, *adv.* **incorrectness**, *n.*

incorrigible (inko'rijibl), *a.* incapable of being amended or improved; bad beyond hope of amendment. *n.* one who is incorrigible. **incorrigibility** (-bil'-), **-ibleness**, *n.* **incorrigibly**, *adv.*

incorrupt (inkərŭpt'), *a.* not corrupt; not decayed, marred or impaired; pure, untainted; not depraved; above the influence of bribery. **incorruptible**, *a.* incapable of corruption, decay or dissolution; eternal; not to be bribed; high-principled. **incorruptibility** (-bil'-), *n.* **incorruptibly**, *adv.* **incorruption**, *n.* **incorruptly**, *adv.* **incorruptness**, *n.*

increase¹ (inkrēs'), *v.i.* to grow; to become greater in bulk, quantity, number, value, degree etc.; to multiply by the production of young. *v.t.* to make greater in number, bulk, quantity etc.; to add to, to extend, to enlarge, to intensify. **increasable**, *a.* **increasingly**, *adv.*

increase² (in'krēs), *n.* the act, state or process of increasing; growth, multiplication; that which is added; increment; produce, crops; progeny; profit.

incredible (inkred'ibl), *a.* not credible; passing belief; (*coll.*) extraordinarily great, astounding. **incredibility** (-bil'-), **-ibleness**, *n.* **incredibly**, *adv.*

incredulous (inkred'ūləs), *a.* indisposed to believe, sceptical (of); unbelieving. **incredulity** (-dū'-), **incredulousness**, *n.* **incredulously**, *adv.*

increment (in'krimənt), *n.* the act or process of increasing; an addition, an increase; the amount of increase; (*Math.*) the finite increase of a variable. **incremental** (-men'-), *a.*

incriminate (inkrim'ināt), *v.t.* to charge with a crime; to suggest or indicate (a person's) guilt. **incrimination**, *n.* **incriminatory**, *a.*

incrustation (inkrŭstā'shən), *n.* the act or process of encrusting; a crust or hard coating on a surface etc.; a facing or lining of foreign material, as marble, stone etc., on masonry etc. **incrust** (inkrŭst'), ENCRUST.

incubate (ing'kūbāt), *v.t.* to sit on (eggs) in order to hatch; to hatch by sitting on or by artificial means; to cause (bacteria etc.) to develop; to evolve (a plan etc.) by meditation. *v.i.* to sit on eggs for hatching, to brood; to undergo incubation. **incubation**, *n.* the act or process of incubating or hatching; brooding, as of a hen upon eggs; meditation on a scheme etc.; (*Path.*) the period between infection and the development of symptoms of a disease. **incubative**, **-atory**, *a.* **incubator**, *n.* an apparatus for hatching eggs by artificial heat, for developing bacteria etc., or rearing a child prematurely born; one that incubates, esp. a brooding hen.

incubus (ing'kūbəs), *n.* (*pl.* **-bi**, (-ī), **-buses**) a demon supposed (esp. in the Middle Ages) to have sexual intercourse with women; any person, thing or influence that oppresses, harasses or restrains.

inculcate (in'kəlkāt), *v.t.* to impress (upon the mind) by emphasis or frequent repetition; to enforce, to instil. **inculcation**, *n.* **inculcator**, *n.*

inculpate (in'kəlpāt), *v.t.* to charge with participation in a crime, to incriminate. **inculpation**, *n.* **inculpatory** (-kŭl'-), *a.*

incumbent (inkŭm'bənt), *a.* lying or resting (on); imposed (upon) as a duty or obligation; currently holding a post or office. *n.* a person in possession of an office etc., esp. a clergyman holding a benefice. **incumbency** (-si), *n.* the act, state, sphere or period of holding a benefice as incumbent; an ecclesiastical benefice.

incunabula (inkūnab'ūlə), *n.pl.* (*sing.* **-lum** (-ləm)) the beginning (of a race, art, development etc.); examples of books etc., printed before AD 1500. **incunabular**, *a.*

incur (inkœ'), *v.t.* (*past, p.p.* **incurred**) to render oneself liable to (risk, injury, punishment etc.); to bring upon oneself, to run into. **incurrable**, *a.*

incurable (inkū'rəbl), *a.* that cannot be cured or healed; irremediable, irreparable. *n.* one suffering from an incurable disease. **incurability** (-bil'-), **-ableness**, *n.* **incurably**, *adv.*

incurious (inkū'riəs), *a.* not curious or inquisitive; indifferent, heedless. **incuriosity** (-os'-), **incuriousness**, *n.* **incuriously**, *adv.*

incursion (inkœ'shən), *n.* a sudden inroad, a raid; an irruption; a brief and temporary exploration (into a subject etc.). **incursive**, *a.*

incus (ing'kəs), *n.* (*pl.* **-cudes** (-kū'dēz)) one of the small bones of the middle ear supposedly shaped like an anvil.

incuse (inkūz'), *v.t.* to impress (a device etc.) by stamping; to stamp with a device etc. *a.* stamped or impressed (on a coin etc.). *n.* an impression made by stamping.

Ind.[1], (*abbr.*) Independent; India; Indiana.

ind.[2], (*abbr.*) independent.

indaba (indah'ba), *n.* (*S Afr.*) a council; a conference.

indebted (indet'id), *a.* being under a debt or obligation (to or for); owing money (to). **indebtedness**, *n.*

indecent (indē'sənt), *a.* unbecoming, unseemly; offensive to modesty or propriety; immodest, grossly indelicate, obscene. **indecent exposure**, *n.* the offence of exposing a part of the body, esp. the genitals, publicly. **indecency** (-si), *n.* **indecently**, *adv.*

indecipherable (indisī'fərəbl), *a.* not decipherable, illegible.

indecision (indisizh'ən), *n.* lack of decision; wavering of the mind, irresolution. **indecisive** (-sī'-), *a.* not decisive, final, or conclusive; irresolute, vacillating, hesitating. **indecisively**, *adv.* **indecisiveness**, *n.*

indeclinable (indiklī'nəbl), *a.* (*Gram.*) not varied by inflections. *n.* an indeclinable word. **indeclinably**, *adv.*

indecorous (indek'ərəs), *a.* violating propriety, decorum or good manners. **indecorously**, *adv.* **indecorousness**, *n.*

indecorum (-dikaw'rəm), *n.* violation of decorum or propriety.

indeed (indēd'), *adv.* in reality, in truth, in point of fact, actually (expressing emphasis, interrogation, concession etc.). *int.* expressing surprise, irony, interrogation etc.

indef., (*abbr.*) indefinite.

indefatigable (indifat'igəbl), *a.* unflagging, unwearied; unremitting. **indefatigability** (-bil'-), **-ableness**, *n.* **indefatigably**, *adv.*

indefeasible (indifē'zibl), *a.* incapable of being annulled or forfeited. **indefeasibility** (-bil'-), *n.* **indefeasibly**, *adv.*

indefensible (indifen'sibl), *a.* incapable of being defended, excused or justified. **indefensibly**, *adv.* **indefensibility** (-bil'-), *n.*

indefinable (indifī'nəbl), *a.* that cannot be defined. **indefinably**, *adv.*

indefinite (indef'init), *a.* not limited or defined, not determinate; vague, uncertain; not certain, not settled; of certain adjectives, adverbs and pronouns, not defining or determining the persons, things etc. to which they apply. **indefinitely**, *adv.* **indefiniteness**, **-finitude** (-fin'itūd), *n.*

indehiscent (indihis'ənt), *a.* of seed-capsules etc., not splitting open to set free the seeds.

indelible (indel'ibl), *a.* that cannot be blotted out or effaced. **indelible pencil**, *n.* a pencil that makes ineffaceable marks. **indelibility** (-bil'-), **-ibleness**, *n.* **indelibly**, *adv.*

indelicate (indel'ikət), *a.* coarse, unrefined; offensive, embarrassing. **indelicacy**, *n.* **indelicately**, *adv.*

indemnify (indem'nifī), *v.t.* to secure from or compensate for damage, loss, penalty or responsibility. **indemnification** (-fi-), *n.*

indemnity (indem'niti), *n.* security against damage, loss or penalty; compensation for damage, loss or penalties incurred; a sum paid as such compensation, esp. by a defeated state to the conqueror as a condition of peace; legal exemption from liabilities or penalties incurred.

indemonstrable (indimon'strəbl), *a.* that cannot be demonstrated; assumed as self-evident, axiomatic. **indemonstrability** (-bil'-), *n.*

indent[1] (indent'), *v.t.* to notch or cut into as with teeth; (*Print.*) to set in farther from the margin than the rest of the paragraph; to indenture; to order by an indent; to execute or draw up (a contract etc.) in exact duplicate. *v.i.* to make an indent or order (upon). **indentation**, *n.* the act of indenting; a notch, dent or incision, esp. in a margin; a deep recess, esp. in a coast-line; a zigzag moulding. **indented**, *a.* **indenter**, *n.* **indention**, *n.* (*Print.*) the setting in of a line of print farther from the margin; indentation.

indent[2] (in'dent), *n.* a notch in the margin of anything; an indentation or recess; an official order for stores; an order for goods, esp. one from abroad.

indent[3] (indent'), *v.t.* to dent; to make a dent in; to mark with a dent. *n.* a dent.

indenture (inden'chə), *n.* (*Law*) an agreement or contract under seal, esp. one binding an apprentice to a master; an official voucher, certificate, register etc.; an indentation. *v.t.* to bind (esp. an apprentice) by an indenture.

independence (indipen'dəns), *n.* the quality or state of being independent; income sufficient to make one independent of others, a competency. **Independence Day**, *n.* a day set apart for publicly celebrating the attainment of national independence; esp. 4 July, the day on which the American colonies declared their independence in 1776.

independent, *a.* not dependent upon or subject to the control, power or authority of another, not subordinate; free to manage one's own affairs without the interference of others; not affiliated with or part of a larger organization; not depending on anything for its value, cogency etc.; having or affording the means of independence; self-asserting, self-reliant; free from bias or prejudice. *n.* one who exercises his or her judgment and choice of action without dependence on any person, party etc. **independently**, *adv.*

in-depth IN.

indescribable (indiskrī'bəbl), *a.* not describable, too fine or too bad for description, passing description. **indescribability** (-bil'-), *n.* **indescribably**, *adv.*

indestructible (indistrŭk'tibl), *a.* incapable of being destroyed. **indestructibility** (-bil'-), *n.* **indestructibly**, *adv.*

indeterminable (inditœ'minəbl), *a.* that cannot be determined or defined; that cannot be terminated as a dispute. **indeterminably**, *adv.*

indeterminate (inditœ'minət), *a.* not determinate; indefinite, undefined, not precise; (*Math.*) having no fixed value. **indeterminately**, *adv.* **indeterminacy** (-si), **indeterminateness**, *n.*

index (in'deks), *n.* (*pl.* **indexes**, (*Math.*) **indices** (-disēz)) that which serves to point out or indicate; the forefinger; a hand (as of a watch etc.), an arm or a pointer, that directs to anything; a table of the contents of a book in alphabetical order with page-references; anything that indicates or denotes (an inner meaning, character etc.); (*Alg.*) the exponent of a power; the decimal number expressing the ratio between the length and breadth of a skull; a numerical scale indicating the relative changes in the cost of living etc., by reference to a given base level. *v.t.* to provide with an index; to enter in an index; to relate to an index, index-link. **index librorum prohibitorum** (libraw'rəm prōhibitaw'rəm), a list of books forbidden to be read by Roman Catholics on pain of excommunication. **index of a logarithm**, the integral part of the logarithm. **index of refraction**, (*Opt.*) the ratio of the sines of the angles of incidence and refraction. **the Index** INDEX LIBRORUM PROHIBITORUM. **index-finger**, *n.* the forefinger, from its being used in pointing. **index-linked**, *a.* increasing or decreasing in direct relation to changes in an index, esp. the cost of living index. **index number**, *n.* an indicator of the relative change in the price or value of something by reference to an earlier period, usu. taken to be 100. **index point**, *n.* a sub-division of a track on a compact disc. **indexation**, *n.* the act of linking wages, rates of interest etc. to the cost of living index. **indexer**, *n.* one who makes an index. **indexical** (-dek'-), *a.* pertaining to or of the form of an index. **indexless**, *a.*

India (in'diə), *n.* a great peninsula in the south of Asia; (also **Republic of India**) the non-Muslim portion of India (including the princely states) declared independent in the political partition of 1947. **India ink** INDIAN INK. **Indiaman**, *n.* (*pl.* **-men**) a large ship employed in the Indian trade. **India paper**, *n.* a fine paper, imported from China,

used by engravers for taking proofs. **india-rubber,** *n.* a soft, elastic substance obtained from the coagulated juice of certain tropical plants, usu. called rubber. **india-rubbery,** *a.* **Indian,** *a.* belonging to the East or West Indies, to the natives of India, or to the aboriginal inhabitants of America. *n.* a native of India; one of the aboriginal inhabitants of America or the West Indies. **Red Indian** RED. **Indian club,** *n.* a bottle-shaped club used in gymnastic exercises. **Indian corn,** *n.* maize. **Indian file,** *n.* single file. **Indian ink,** *n.* (ink made from) a black pigment, composed of lamp-black and animal glue, used for writing and in water-colour painting. **Indian rope-trick,** *n.* the supposed Indian feat of climbing an unsupported rope. **Indian-rubber** INDIA-RUBBER. **Indian summer,** *n.* summerlike weather, occurring late in autumn.
Indic, *a.* originating or existing in India; pertaining to the Indian branch of the Indo-European languages. *n.* this group of languages.
indic., (*abbr.*) indicative.
indicate (in'dikāt), *v.t.* to show, to point out; to be a sign or token of; (*Med.*) to point out or suggest (as a remedy); to state briefly, to suggest. **indication,** *n.* the act of indicating; that which indicates; intimation; a symptom suggesting certain treatment. **indicative** (-dik'ə-), *a.* applied to that mood of a verb which expresses matters of fact; (*sometimes pron.* in'dikātiv) indicating; denoting something not visible or obvious. *n.* the indicative mood. **indicatively,** *adv.* **indicator,** *n.* one who or that which indicates; a reagent used to indicate, by change of colour, the presence of an acid, alkali etc.; an instrument attached to apparatus, machinery, a vehicle etc., to indicate or record pressure, speed, number etc.; a device for indicating the times of departure etc. of trains etc.; a device, esp. a flashing light, on a vehicle to show an intention to change direction; a statistic such as the level of industrial production that indicates the condition of a national economy. **indicatory,** *a.*
indices INDEX.
indicium (indish'iəm), *n.* (*pl.* **-cia** (-ə)) an indicating sign or mark; a symptom.
indict (indīt'), *v.t.* to charge with a crime or misdemeanour, esp. by means of an indictment. **indictable,** *a.* of a person, liable to be indicted; of an offence, forming a ground of indictment. **indictably,** *adv.* **indicter,** *n.* **indictment,** *n.* the act of indicting; a formal accusation of a crime or misdemeanour, presented upon oath by the grand jury to a court; the document embodying this; (*Sc. Law*) a process by which a criminal is brought to trial at the instance of the Lord Advocate; (grounds for) condemnation.
Indies (in'diz), *n.pl.* India and the neighbouring regions, also called the East Indies; the West Indies.
indifferent (indif'rənt), *a.* impartial; having no inclination or disinclination (to); unconcerned, apathetic; neither good nor bad; of no importance, of little moment (to); of a barely passable quality, not good; (*Chem., Elec. etc.*) neutral, not active. **indifference,** *n.* the quality or state of being indifferent; impartiality; absence of inclination or disinclination; lack of interest or attention (to or towards); unconcern, inattention; mediocrity; unimportance, insignificance. **indifferential** (-en'-), *a.* **indifferentiated** (-en'shiātid), *a.* **indifferentism,** *n.* systematic indifference, esp. with regard to religious belief. **indifferentist,** *n.* **indifferently,** *adv.*
indigenous (indij'ənəs), *a.* native, not exotic; natural, innate (to). **indigene** (in'dijēn), *n.* **indigenously,** *adv.*
indigent (in'dijənt), *a.* in want, poor, needy, necessitous; in need (of); destitute (of). **indigence,** *n.* **indigently,** *adv.*
indigestible (indijest'əbl), *a.* not easily digested; hard to understand or to follow; not acceptable. *n.* an indigestible substance or thing. **indigestibility** (-bil'-), *n.* **indigestibly,** *adv.* **indigestion** (-chən), *n.* difficulty of digestion, dyspepsia. **indigestive,** *a.*
indignant (indig'nənt), *a.* feeling or showing indignation, esp. at meanness, injustice etc., or with a person acting meanly etc. **indignantly,** *adv.* **indignation,** *n.* a mingled feeling of anger and disdain; the feeling excited by that which is unworthy, mean, base or unjust.
indignity (indig'niti), *n.* undeserved contemptuous treatment; a slight, an insult.
indigo (in'digō), *n.* (*pl.* **-gos, -goes**) a beautiful and very durable blue dye obtained from the indigo-plant, largely used in calico printing etc.; a deep-blue colour. *a.* of a deep-blue colour. **indigo-bird,** *n.* a N American finch, *Cyanospiza cyanea.* **indigo-blue,** *n.* the colour or the colouring-matter of indigo. **indigo-plant,** *n.* a plant of the genus *Indigofera*, esp. *I. tinctoria.* **indigotic** (-got'-), *a.*
indirect (indirekt'. -dī-), *a.* not direct, deviating from a direct line; not straight or rectilinear; not resulting directly or immediately from a cause; of taxes, not paid directly to the Government, but in the form of increased prices etc.; (*Gram.*) in oblique oration or reported speech; not fair, not honest, not open or straightforward. **indirect evidence, r testimony,** *n.* evidence deduced from collateral circumstances. **indirect object,** *n.* (*Gram.*) the person or thing indirectly affected by an action though not the direct object of the verb. **indirect speech,** *n.* the reporting of spoken or written discourse by indicating what was meant rather than by repetition of the exact words. **indirectly,** *adv.* **indirectness,** *n.*
indiscernible (disœ'nibl), *a.* not distinguishable; not visible. **indiscernibleness,** *n.* **indiscernibly,** *adv.*
indiscipline (indis'iplin), *n.* lack of discipline. **indisciplinable,** *a.*
indiscreet (indiskrēt'), *a.* lacking in discretion; injudicious, incautious; foolish, rash. **indiscreetly,** *adv.* **indiscreetness,** *n.* **indiscretion** (-kresh'ən), *n.* lack of discretion; imprudence, rashness; an indiscreet act, indiscreet conduct.
indiscrete (indiskrēt'), *a.* not discrete or separated.
indiscriminate (indiskrim'inət), *a.* lacking in discrimination; making no distinction; confused; random, promiscuous. **indiscriminately,** *adv.* **indiscriminateness, -ation,** *n.* **indiscriminating, -ative,** *a.*
indispensable (indispen'səbl), *a.* that cannot be dispensed with; absolutely necessary or requisite. **indispensability** (-bil'-), **-ableness,** *n.* **indispensably,** *adv.*
indispose (indispōz'), *v.t.* to make disinclined or unfavourable; to render unfit or unable (for or to); to make slightly ill. **indisposed,** *a.* **indisposition** (-pəzish'ən), *n.*
indisputable (indispū'təbl), *a.* incontestable, beyond question or dispute. **indisputability** (-bil'-), **-ableness,** *n.* **indisputably,** *adv.*
indissoluble (indisol'ūbl), *a.* not to be dissolved or disintegrated; stable, binding, subsisting and binding for ever. **indissolubility** (-bil'-), **indissolubleness,** *n.* **indissolubly,** *adv.*
indistinct (indistingkt'), *a.* obscure; not readily distinguishable; confused, faint. **indistinctive,** *a.* not distinctive. **indistinctively,** *adv.* **indistinctly,** *adv.* **indistinctness,** *n.*
indistinguishable (indisting'gwishəbl), *a.* not distinguishable. **indistinguishably,** *adv.*
indite (indīt'), *v.t.* to put in words, to compose; to write.
indium (in'diəm), *n.* a rare soft, silver-white metallic element, at. no. 49; chem. symbol In.
individual (individ'ūəl), *a.* subsisting as a single indivisible entity; single, particular as opp. to general; separate or distinct; characteristic of a particular person or thing, distinctive. *n.* a single person, animal or thing, esp. a single human being; a single member of a species, class etc.; a person. **individualism,** *n.* conduct or feeling centred in self, egoism, self-interest, selfishness; idiosyncrasy, personal peculiarity; an attitude, tendency or system in which each individual works for his or her own ends; inde-

indivisible 424 **ineffective**

pendent action as opposed to cooperation, or as opp. to collectivism or Socialism. **individualist,** *n.* **individualistic** (is'-), *a.* **individualistically,** *adv.* **individuality** (-al'-), *n.* separate or distinct existence; distinctive character, strongly-marked personality. **individualize, -ise,** *v.t.* to distinguish from other individuals; to connect with one particular individual; to package separately; to make so as to suit the needs of a particular person. **individualization,** *n.* **individually,** *adv.* **individuate** (individ'ūāt), *v.t.* to give the character of individuality to; to make an individual or a distinct entity. **individuation,** *n.*
indivisible (indiviz'ibl), *a.* not divisible; that cannot be exactly divided. **indivisibility** (-bil'-), *n.* **indivisibly,** *adv.*
Indo- (in'dō), *comb.form* Indian; derived from, belonging to, or connected with India.
Indo-Chinese (indōchīnēz'), *a.* pertaining to Indo-China, the south-eastern peninsula of Asia, its people or their languages.
indoctrinate (indok'trināt), *v.t.* to imbue with the distinctive principles of any system; to brainwash. **indoctrination,** *n.*
Indo-European (indōūrəpē'ən), **-Germanic,** *a.* of or pertaining to the family of languages spoken over most of Europe and over Asia as far as northern India.
indole (in'dōl), *n.* a white or yellowish crystalline heterocyclic compound derived from coal tar.
indolent (in'dələnt), *a.* habitually idle or lazy; (*Path.*) causing no pain. **indolence,** *n.* **indolently,** *adv.*
indomitable (indom'itəbl), *a.* untamable, unconquerable; indefatigable. **indomitably,** *adv.*
Indonesian (indənē'zhən), *a.* pertaining to the East Indian islands forming the Republic of Indonesia. *n.* an inhabitant of Indonesia; the language.
indoor (in'daw), *a.* being or done within doors. **indoors** (-dawz'), *adv.* within a house or building.
indraught (in'drahft), *n.* an inward flow, draught or current.
indri (in'dri), **indris** (-dris), *n.* the babacoote, a Madagascan lemur.
indubitable (indū'bitəbl), *a.* not doubtful, unquestionable; too evident to admit of doubt. **indubitability** (-bil'-), **-ableness,** *n.* **indubitably,** *adv.*
induce (indūs'), *v.t.* to lead by persuasion or reasoning, to prevail on; to bring about, to cause; to bring on or speed up (labour) by artificial means, as by the use of drugs; (*Elec.*) to produce by induction; (*Log.*) to derive inductively, opp. to *deduce*. **inducement,** *n.* the act of inducing; that which induces; a motive, a reason, an incentive; (*Law*) a preamble or statement of facts introducing other material facts. **inducer,** *n.* **inducible,** *a.*
induct (indŭkt'), *v.t.* to introduce (as into a benefice or office); to put in actual possession of an ecclesiastical benefice or of any office, with the customary forms and ceremonies. *v.i.* (*N Am.*) to enlist for military training. **inductance,** *n.* the property of an electric circuit which produces an electromotive force when the current is varied. **induction** (indŭk'shən), *n.* (*Log.*) the process of inferring a law or general principle from particular instances, as dist. from *deduction;* a general statement or conclusion attained by this kind of reasoning; the production of an electric or magnetic state by the proximity or movement of an electric or magnetized body; instalment in an office or benefice; an introduction, a prologue; (*N Am.*) enlistment for military training. **induction coil,** *n.* an apparatus for producing currents by electromagnetic induction. **induction motor,** *n.* an electric motor in which an electromagnetic flux set up by currents in a primary winding induces currents in a secondary winding, such that interaction of currents with flux produces rotation. **inductional,** *a.* **inductive,** *a.* (*Log.*) proceeding or characterized by induction; (*Elec.*) pertaining to, producing or susceptible of induction; leading or drawing on. **inductive**

method, *n.* (*Log.*) the process of reasoning from particular instances to general principles. **inductively,** *adv.* **inductivity** (-tiv'-), *n.* **inductor** (indŭk'tə), *n.* one who inducts a clergyman into office; any part of an electrical apparatus acting inductively.
indulge (indŭlj'), *v.t.* to yield, esp. unduly, to the desires, humours or wishes of, to humour (in or with); to favour; to gratify (one's desires, weakness etc.). *v.i.* to yield to one's desires (in); to take alcoholic drink, esp. in excess. **indulgence** (indŭl'jəns), *n.* the act or practice of indulging, yielding or complying to desires etc.; an indulgent act, a favour or privilege granted; a pleasurable thing or habit indulged in; liberality, tolerance, leniency; in the Roman Catholic Church, a remission of the punishment still due to sin after sacramental absolution. **indulgent,** *a.* indulging or disposed to indulge the wishes, humours or caprices of others; not exercising restraint or control. **indulgently,** *adv.* **indulger,** *n.*
indurate (in'dūrāt), *v.t.* to make hard, to harden; to render obdurate or unfeeling. *v.i.* to become hard; to become fixed or inveterate, as a custom. **induration,** *n.* insensibility. **indurative,** *a.*
indusium (indū'ziəm), *n.* (*pl.* **-sia** (-iə)) (*Bot.*) a hairy cup enclosing a stigma; a shield or scale covering the fruit-cluster in some ferns; the larval case of an insect.
industry (in'dəstri), *n.* diligence, assiduity, steady application to any business or pursuit; useful work, esp. mechanical and manufacturing pursuits as dist. from agriculture and commerce; any branch of these; (*Polit. Econ.*) the employment of labour in production; any field of activity as organized for economic gain. **industrial** (indŭs'triəl), *a.* pertaining to industry, to productive occupations or to produce; characterized by advanced and sophisticated industries. *n.* a person engaged in an industrial occupation; (*pl.*) shares or securities relating to industrial enterprises. **industrial action,** *n.* action taken by employees to try to coerce their employer into complying with demands or as a protest, esp. a strike or go-slow. **industrial archaeology,** *n.* the study of the remains of past industrial activity. **industrial estate,** *n.* an industrial area specially planned to provide employment in factories of different kinds. **industrial exhibition,** *n.* an exhibition of industrial products, machinery, appliances etc. **industrial relations,** *n.* a general term covering the relationships between employer and employees. **Industrial Revolution,** *n.* the changes brought about in the way of life by the extensive introduction of machinery after about 1760. **industrialism,** *n.* a state of society characterized by the preeminence of large-scale manufacturing industries. **industrialist,** *n.* a person engaged in management or ownership in industry. **industrialize, -ise,** *v.t.* **industrialization,** *n.* **industrially,** *adv.* **industrious** (indŭs'triəs), *a.* characterized by industry; diligent and assiduous in business or study. **industriously,** *adv.* **industriousness,** *n.*
indwell (indwel'), *v.t., v.i.* (*past, p.p.* **indwelt** (-dwelt')) to abide in; (*usu. fig.*) to inhabit. **indweller,** *n.*
-ine, *suf.* pertaining to, of the nature of; forming adjectives, as *crystalline, divine, marine;* forming feminine nouns, as *heroine,* abstract nouns, as *discipline, medicine,* (*Chem.*) names of alkaloids and basic substances, as *cocaine, morphine.*
inebriate (inē'briāt), *v.t.* to make drunk; to intoxicate or exhilarate. *a.* (-ət), (*lit.* or *fig.*) intoxicated, drunk. *n.* a habitual drunkard. **inebriant,** *a.* intoxicating. *n.* anything which intoxicates. **inebriated,** *a.* **inebriation,** *n.*
inedible (ined'ibl), *a.* not edible. **inedibility** (-bil'-), *n.*
ineffable (inef'əbl), *a.* unutterable, beyond expression. **ineffableness,** *n.* **ineffably,** *adv.*
ineffaceable (inifā'səbl), *a.* that cannot be rubbed out. **ineffaceably,** *adv.*
ineffective (inifek'tiv), *a.* not producing any or the desired effect; inefficient; useless. **ineffectively,** *adv.*

ineffectual 425 **inferior**

ineffectiveness, n.
ineffectual (inifek'chuəl), a. not producing any effect; powerless, vain. **ineffectualness,** n. **ineffectually,** adv.
inefficacious (inefikā'shəs), a. not efficacious; producing no result or effect. **inefficacy** (-ef'-), n.
inefficient (inifish'ənt), a. not efficient; lacking ability or capacity. **inefficiently,** adv. **inefficiency,** n.
inelastic (inilas'tik), a. lacking elasticity. **inelasticity** (-elastis'-), n.
inelegant (inel'igənt), a. not elegant; lacking grace, polish, refinement etc. **inelegance,** n. **inelegantly,** adv.
ineligible (inel'ijibl), a. not eligible; not qualified or permitted to take part, be elected etc. **ineligibility** (-bil'-), n. **ineligibly,** adv.
ineluctable (inilŭk'təbl), a. inescapable.
inept (inept'), a. not apt, fit or suitable; clumsy, incompetent; fatuous. **ineptitude** (-titūd), **ineptness,** n. **ineptly,** adv.
inequable (inek'wəbl), n. changeable; not uniform.
inequality (inikwol'iti), n. lack of equality; difference, diversity, unevenness (of dimensions, position, intensity etc.); disparity; unfairness, partiality.
inequitable (inek'witəbl), a. not equitable, not fair or just. **inequitably,** adv. **inequity,** n.
ineradicable (inirad'ikəbl), a. that cannot be eradicated. **ineradicably,** adv.
inert (inœt'), a. lacking inherent power of motion or active resistance to motive power applied; motionless, slow, sluggish; indisposed to move or act; (*Chem.*) chemically unreactive, neutral. **inert gas,** n. any of a group of gaseous elements that react very little with other elements and include helium, neon, argon, krypton, xenon and radon. **inertia** (-shə), n. inertness; that property of a body by which it persists in an existing state of rest or of uniform motion in a straight line, unless an external force changes that state. **vis inertiae** (-ī), the resistance of matter to a force operating to move it. **inertia-reel seat belt,** n. a seat belt in which the belt unwinds freely except when the violent deceleration of the vehicle causes it to lock. **inertia selling,** n. the practice of sending unsolicited goods to householders and requesting payment if the goods are not returned. **inertial,** a. **inertial navigation,** n. a system of gyroscopic guidance for aircraft, missiles etc., that dispenses with magnetic compass or ground-based radio direction. **inertly,** adv. **inertness,** n.
inescapable (iniskā'pəbl), a. inevitable, not to be escaped.
inessential (inisen'shəl), a. unessential; not vitally necessary.
inestimable (ines'timəbl), a. that cannot be estimated; of surpassing worth or excellence. **inestimably,** adv.
inevitable (inev'itəbl), a. that cannot be avoided or prevented; certain to happen; utterly predictable. **inevitability** (-bil'-), **-ableness,** n. **inevitably,** adv.
inexact (inigzakt'), a. not exact, not precisely accurate. **inexactitude** (-titūd), **inexactness,** n. **inexactly,** adv.
inexcusable (inikskū'zəbl), a. not to be excused or justified. **inexcusability** (-bil'-), **-ableness,** n. **inexcusably,** adv.
inexhaustible (inigzaws'tibl), a. that cannot be exhausted; unfailing, unceasing. **inexhaustibility** (-bil'-), **-ibleness,** n. **inexhaustibly,** adv.
inexorable (inek'sərəbl), a. incapable of being persuaded or moved by entreaty or prayer; inflexible, relentless. **inexorability** (-bil'-), n. **inexorably,** adv.
inexpensive (inikspen'siv), a. not expensive; cheap. **inexpensively,** adv. **inexpensiveness,** n.
inexperience (inikspiə'riəns), n. lack of knowledge gained by experience. **inexperienced,** a.
inexpert (inek'spœt), a. not expert, unskilful. **inexpertly,** adv.
inexplicable (iniksplik'əbl, -ek'-), a. not capable of being made plain or intelligible; not to be explained. **inexplicability** (-bil'-), **-ableness,** n. **inexplicably,** adv.
inexplicit (iniksplis'it), a. not definitely or clearly stated.

inexplicitly, adv. **inexplicitness,** n.
inexpressible (inikspres'ibl), a. incapable of being expressed or described; unutterable, unspeakable. **inexpressibly,** adv. **inexpressive,** a. **inexpressively,** adv. **inexpressiveness,** n.
inextinguishable (iniksting'gwishəbl), a. incapable of being extinguished. **inextinguishably,** adv.
in extremis (in ikstrē'mis), at the point of death. [L]
inextricable (inek'strikəbl, -strik'-), a. that cannot be disentangled or solved; inescapable. **inextricably,** adv.
inf., (*abbr.*) infinitive.
infallible (infal'ibl), a. exempt from liability to error or to failure; certain not to fail. **infallibility** (-bil'-), n. **papal infallibility,** the dogma that the Pope, speaking *ex cathedra*, is infallible when he defines a doctrine regarding faith or morals. **infallibly,** adv.
infamous (in'fəməs), a. having or deserving a bad reputation; detestable, scandalous. **infamously,** adv. **infamy,** n. total loss of reputation or character; public reproach; extreme baseness; an infamous act.
infant (in'fənt), n. a child during the earliest years of its life (*usu.* a babe, *also*, a child less than seven years old); (*Law*) a minor. *a.* young, tender; pertaining to or designed for infants; in its earliest stages. **infancy** (-si), n. early childhood; an early stage of development. **infantile** (in'fəntīl), a. pertaining to infants or infancy; puerile, childish. **infantile paralysis,** n. poliomyelitis. **infantilism** (-fan'til-), n.
infanta (infan'tə), n. (in Spain and Portugal) any royal princess (usu. the eldest) except an heiress-apparent. **infante** (-tā), n. any son of the king except the heir-apparent.
infanticide (infan'tisid), n. murder of a new-born infant; the practice of killing new-born children; the murderer of an infant. **infanticidal** (-sī'-), a.
infantry (in'fəntri), n. (*collect.*) foot-soldiers, usu. armed with small arms or rifle and bayonet. **infantryman,** n. a soldier in an infantry regiment.
infarct (infahkt'), n. an area of tissue that is dying from lack of a blood supply. **infarction,** n.
infatuate (infat'ūāt), v.t. to cause to act foolishly; to inspire with an extravagant or foolish passion. **infatuatedly,** adv. **infatuation,** n.
infect (infekt'), v.t. to affect with a communicable disease; to contaminate with micro-organisms; to corrupt, to taint, e.g. morally; to affect adversely. **infectedly,** adv. **infection,** n. the act or process of infecting, esp. the communication of disease by means of water, the atmosphere etc., as distinct from *contagion;* that which infects, infectious matter; an infectious disease; moral contamination. **infectious,** a. infecting or capable of infecting; likely to communicate disease; liable to be communicated by the atmosphere, water etc.; of feelings etc., apt to spread, catching. **infectiously,** adv. **infectiousness,** n. **infective,** a. infectious. **infectiveness, infectivity** (-tiv'-), n.
infelicitous (infilis'itəs), a. not felicitous; unfortunate; inappropriate, inept. **infelicitously,** adv. **infelicity,** n.
infer (infœ'), v.t. (*past, p.p.* **inferred**) to deduce as a fact, consequence or result; to conclude; to prove, to imply. **inferable, inferrable,** a. **inference** (in'-), n. the act of inferring; that which is inferred from premises, a conclusion or deduction. **inferential** (-ren'-), a. **inferentially,** adv.
inferior (infiə'riə), a. lower in place, rank, value, quality, degree etc.; subordinate; of mediocre or poor quality; (*Astron.*) within the earth's orbit; below the horizon; (*Bot.*) growing below another organ, as the calyx or the ovary; (*Print.*) set below ordinary letters or below the line, as the figures in H_2SO_4. *n.* a person who is inferior to another in station etc.; a subordinate. **inferiority** (-o'-), n. **inferiority complex,** n. a suppressed sense of inferiority which produces as compensation some abnormal reaction such as megalomania, assertiveness, or the like.

inferiorly, *adv.*
infernal (infœ'nəl), *a.* pertaining to hell or the lower regions; worthy of hell, hellish; detestable, diabolical; (*coll.*) abominable, confounded. **infernally,** *adv.* **inferno** (nō), *n.* (*pl.* **-nos**) hell, esp. as conceived by Dante; any place supposed to resemble hell; a blaze or conflagration.
inferrable INFERABLE under INFER.
infertile (infœ'til), *a.* not fertile; unfruitful. **infertility** (-til'-), *n.*
infest (infest'), *v.t.* to overrun, to swarm over or about, in vast numbers. **infestation,** *n.* **infester,** *n.*
infidel (in'fidel), *a.* disbelieving in a given form of faith (that of the person using the epithet), esp. rejecting the Christian religion or Islam; rejecting revelation, agnostic, sceptical. *n.* one who disbelieves in a given form of faith; (*Hist.*) a Turk, a pagan, a Jew; an agnostic, a sceptic. **infidelity** (-del'-), *n.* disbelief in a religion (as Christianity); (an act of) disloyalty or deceit, esp. unfaithfulness to the marriage vow.
infield (in'fēld), *n.* (*Cricket*) the part of the field close to the wicket; (*Baseball*) the ground within the base lines. **infielder,** *n.* (*Baseball*) one of the players in the infield.
infighting (in'fīting), *n.* boxing at close quarters; IN.
infill (in'fil), *v.t.* to fill in; to fill up. *n.* (also **infilling**) closing up gaps, esp. between houses; material for filling up holes etc.
infiltrate (in'filtrāt), *v.t.* to (cause to) enter by penetrating the pores or interstices of; to (cause to) pass secretly through (enemy lines etc.); to (cause to) gain access secretly to. *v.i.* to pass or percolate (into) thus. **infiltration,** *n.*
infin., (*abbr.*) infinitive.
infinite (in'finit), *a.* limitless, endless; indefinitely great or numerous; (*Math.*) greater than any assignable quantity. *n.* infinite space, infinity; a vast or infinite amount; (*Math.*) an infinite quantity. **infinitely,** *adv.* **infiniteness,** *n.* **infinitesimal** (-tes'iməl), *a.* infinitely small; (*coll.*) insignificant; negligible; (*Math.*) less than any assignable quantity. *n.* an infinitesimal quantity. **infinitesimally,** *adv.*
infinity (infin'iti), **infinitude** (-tūd), *n.* boundlessness; an infinite quantity or distance; a boundless expanse, vastness, immensity.
infinitive (infin'itiv), *a.* (*Gram.*) applied to that mood of a verb which expresses the action without regard to any person etc. *n.* the infinitive mood; a verb in this mood. **infinitival** (-ti'-), *a.* **infinitively, infinitivally,** *adv.*
infinitude, infinity INFINITE.
infirm (infœm'), *a.* lacking bodily strength or health, esp. through age or disease; weak-minded, irresolute; uncertain, unstable. **infirmary** (infœ'məri), *n.* a hospital or establishment for treating the sick or injured. **infirmity,** *n.* **infirmly,** *adv.*
infix (infiks'), *v.t.* to fasten or fix in; to implant firmly; to insert (an infix) in a word. *n.* (in'-), (*Gram.*) a modifying element inserted in the body of a word.
in flagrante (delicto) (in flagran'ti dilik'tō), whilst actually committing the misdeed. [L]
inflame (inflām'), *v.t.* to cause to blaze, to kindle; to cause inflammation in; to excite, to stir up to passion etc.; to intensify, to aggravate.
inflammable (inflam'əbl), *a.* that may be easily set on fire; easily excited or aroused. **inflammability** (-bil'-), **-ableness,** *n.*
inflammation (infləmā'shən), *n.* an abnormal condition characterized by heat, redness, swelling, pain and loss of function in the part affected; the act of inflaming or the state of being inflamed. **inflammatory** (-flam'-), *a.* tending to inflame; exciting or arousing passions.
inflate (inflāt'), *v.t.* to cause to swell with air or gas; to swell, to puff up; to elate; to raise (prices, reputation etc.) artificially or excessively. *v.i.* to expand by being filled with gas or air. **inflatable,** *a.* that can be inflated. *n.* an inflatable toy, esp. an imitation castle etc. for children to jump or

climb on; anything inflatable. **inflated,** *a.* distended with air; bombastic, turgid; exaggerated; expanded or raised artificially; (*Bot.*) hollow and distended. **inflation,** *n.* the act of inflating, the state of being inflated; progressive increase in prices; an increase in the amount of money in circulation accompanied by a decrease in its buying power. **inflationary,** *a.* **inflationism,** *n.* **inflationist,** *n.* one who favours an increased issue of paper money. **inflator,** *n.*
inflect (inflekt'), *v.t.* to bend; to modulate (as the voice); (*Gram.*) to change the terminations of (words) for purposes of declension or conjugation. **inflectedness,** *n.* **inflection, inflexion** (inflek'shən), *n.* a bend; modulation of the voice; the variation of the termination of nouns etc. in declension, and of verbs in conjugation; change from concave to convex in a curve. **inflectional, inflexional,** *a.* pertaining to or having grammatical inflections. **inflectionless, inflexionless,** *a.* **inflective,** *a.* capable of bending; (*Gram.*) inflectional. **inflector,** *n.*
inflexible (inflek'sibl), *a.* incapable of being bent or curved; unyielding, obdurate; firm of will or purpose. **inflexibility** (-bil'-), *n.* **inflexibly,** *adv.*
inflexion INFLECT.
inflict (inflikt'), *v.t.* to impose upon as a penalty or punishment; to cause to feel or experience (something of an unpleasant nature). **inflictable,** *a.* **inflicter,** *n.* **infliction,** *n.* the act of inflicting; a punishment inflicted; (*coll.*) a trouble, an annoyance.
in-flight IN.
inflorescence (inflɔres'əns), *n.* the act or process of flowering; the arrangement of flowers upon a branch or stem; the collective flower or flowers of a plant.
inflow (in'flō), *n.* flowing in; influx; something which flows in.
influence (in'fluəns), *n.* power serving or tending to affect, modify or control; the effect of such power; an ethereal fluid supposed to flow from the stars and to affect character and control human destinies; power resulting from wealth, position, contacts etc.; a person, thing, feeling etc., exercising moral power (over). *v.t.* to exercise influence upon; to modify (motives etc.) to any end or purpose; to bias, to sway. **influencer,** *n.* **influential** (-en'-), *a.* possessing or exercising influence. **influentially,** *adv.*
influent (in'fluənt), *a.* flowing in; influential. *n.* a tributary.
influenza (influen'zə), *n.* a catarrhal inflammation of the mucous membranes of the air-passages, attended by fever and nervous prostration.
influx (in'flŭks), *n.* a flowing of or as of water (into); the point of inflow (of a stream); the arrival of many people or things. **influxion** (-shən), *n.* **influxive** (-flŭk'siv), *a.*
info (in'fō), *n.* (*coll.*) short for INFORMATION.
infold (infōld'), ENFOLD.
infopreneur (infōprənœ'), *n.* (*coll.*) a person engaged in business activity in the field of information technology. **infopreneurial,** *adj.*
inforce (infaws'), ENFORCE.
inform (infawm'), *v.t.* to animate, to imbue (with feeling, vitality etc.); to communicate knowledge to, to tell; to give form or shape to. *v.i.* to disclose facts, to bring a charge (against). **to inform against,** or **on,** to give information to the police about (a criminal). **informant,** *n.* **informatics,** *n.* information science or technology. **information,** *n.* the act of informing or communicating knowledge etc.; intelligence communicated; notice, knowledge acquired; facts, data; a complaint or accusation presented to a court or magistrate as a preliminary to criminal proceedings. **information retrieval,** *n.* the recovery of data stored in computerized form. **information science,** *n.* the computerized processing and communication of data; the study of this. **information technology,** *n.* the gathering, processing and communication of information through computing and telecommunications combined. **information**

informal (infaw'məl), *a.* not in accordance with official, proper or customary forms; without formality. **informality** (-mal'-), *n.* **informally**, *adv.*

theory, *n.* mathematical theory on the subject of the transmission, storage, retrieval and decoding of information. **informational**, *a.* **informative**, *a.* conveying information or instruction. **informed**, *a.* having information; apprized of the facts; educated, enlightened. **informer**, *n.* one who informs, esp. one who gives information about a person to the police.

infotainment (infōtān'mənt), *n.* television or radio shows which deal with serious subjects.
infra (in'frə), *adv.* of a passage in a book etc., below, further on.
infra-, *pref.* below, beneath.
infraction (infrak'shən), *n.* the act of breaking or violating; violation, infringement. **infract**, *v.t.* (*N Am.*) to infringe. **infractous**, *a.* (*Bot.*).
infra dig (infrə dig'), *phr.* beneath one's dignity, undignified.
infrangible (infran'jibl), *a.* unbreakable, that cannot be infringed or violated. **infrangibility** (-bil'-), *n.* **infrangibly**, *adv.*
infra-red rays (infrəred'), *n.pl.* invisible radiations beyond the visible spectrum at the red end.
infrasonic (infrəson'ik), *a.* having a frequency below the usual audible limit.
infrastructure (in'frəstrŭkchə), *n.* underlying structure or basic framework; the network of communications etc. systems essential for industry, military operations etc.
infrequent (infrē'kwənt), *a.* rare, uncommon, unusual. **infrequency**, *n.*
infringe (infrinj'), *v.t.* to break (a law, compact, contract etc.); to violate, to neglect to obey. *v.i.* to encroach, to intrude (upon). **infringement**, *n.* **infringer**, *n.*
infundibulum (infŭndib'ūləm), *n.* (*pl.* **-bula** (-ə)) (*Anat.*) any funnel-shaped part. **infundibular, -ulate** (-lət), **-uliform** (-fawm), *a.* funnel-shaped.
infuriate (infū'riāt), *v.t.* to provoke to madness or fury. **infuriatingly**, *adv.*
infuse (infūz'), *v.t.* to pour (into); to inculcate, to implant; to steep in liquid so as to obtain an extract or infusion. **infuser**, *n.* **infusion** (infū'zhən), *n.* instillation, inculcation; the act or process of steeping; the liquid extract obtained by steeping any substance; that which is instilled or implanted, an admixture, a tincture. **infusive** (-siv), *a.* having the power of infusing.
ingathering (in'gaḥdhering), *n.* the act of gathering or collecting, esp. of getting in the harvest.
ingenious (injēn'yəs), *a.* skilful, clever, esp. in inventing or contriving; cleverly designed. **ingeniously**, *adv.* **ingeniousness, ingenuity** (-jənū'-), *n.*
ingénue (i'zhänū), *n.* an ingenuous or naive girl, esp. such a character on the stage.
ingenuous (injen'ūəs), *a.* candid, frank, sincere; artless, naive. **ingenuously**, *adv.* **ingenuousness**, *n.*
ingest (injest'), *v.t.* to take (food) into the stomach. **ingestible**, *a.* **ingestion** (-chən), *n.* **ingestive**, *a.*
ingle (ing'gl), *n.* a fire on the hearth; a fireplace. **inglenook**, *n.* a chimney-corner. **ingle-side**, *n.*
inglorious (inglaw'riəs), *a.* not glorious; shameful, ignominious. **ingloriously**, *adv.* **ingloriousness**, *n.*
ingoing (in'gōing), *a.* going in, entering. *n.* entrance.
ingot (ing'gət), *n.* a mass of cast metal, esp. steel, gold or silver; a bar of gold or silver for assaying.
ingrain[1] (in'grān. *predicatively* -grān'), *a.* dyed in the grain or yarn before manufacture; thoroughly imbued, inherent, inveterate. *n.* a yarn or fabric dyed with fast colours before manufacture. **ingrain carpet**, *n.* a carpet manufactured from wool dyed in the grain, the pattern showing through the fabric. **ingrained** (in'-. *predicatively* -grānd'), *a.* deeply imprinted; complete, total; of dirt etc., worked into the fibres, pores etc.
ingrain[2] (ingrān'), ENGRAIN.
ingrate (in'grāt), *a.* ungrateful; unpleasant. *n.* (*chiefly N Am.*) an ungrateful person.
ingratiate (ingrā'shiāt), *v.t.* to insinuate (oneself) into goodwill or favour (with) another. **ingratiating**, *a.* **ingratiatingly**, *adv.*
ingratitude (ingrat'itūd), *n.* lack of gratitude.
ingredient (ingrē'diənt), *n.* that which enters into a compound as an element, a component part; any of the separate items required for a recipe in cooking.
ingress (in'gres), *n.* the act of entering, entrance; power or liberty of entrance. **ingression** (-shən), *n.* **ingressive**, *a.*
ingrowing (in'grōing), *a.* growing inwards; of a toe-nail etc., growing abnormally into the flesh. **ingrown**, *a.* ingrowing; innate, native. **ingrowth**, *n.*
inguinal (ing'gwinəl), *a.* of, pertaining to or situated near the groin. **inguino-**, *comb. form.*
inhabit (inhab'it), *v.t.* to live or dwell in; to occupy as a place of settled residence; to reside in. **inhabitable**, *a.* fit for habitation. **inhabitancy**, *n.* domiciliation or residence for a considerable period, esp. such as confers the rights of an inhabitant. **inhabitant**, *n.* a person or animal that lives in a particular place. **inhabitation**, *n.* the act of inhabiting; the state of being inhabited. **inhabiter**, *n.*
inhale (inhāl'), *v.t.* to breathe in, to draw into the lungs; to inspire, as distinct from *exhale*. **inhalant**, *n.*, *a.* **inhalation** (-hə-), *n.* **inhaler**, *n.* one who inhales; a respirator; an instrument for enabling the inhalation of medicated vapours etc.
inharmonious (inhahmō'niəs), *a.* not harmonious; unmusical. **inharmonic, -ical** (-mon'-), *a.* **inharmoniously**, *adv.*
inhere (inhiə'), *v.i.* to be an essential or necessary part (in); to be vested (in). **inherence, -ency** (-hiə'-, -he'-), *n.* **inherent** (inhe'rənt, -hiə'-), *a.* permanently belonging or intrinsic (in or to); innate, inborn. **inherently**, *adv.*
inherit (inhe'rit), *v.t.* to receive by legal succession from a former possessor; to derive from one's ancestors by genetic transmission; to take over (a position etc.) from a predecessor. *v.i.* to take or come into possession as an heir. **inheritable**, *a.* capable of inheriting or being inherited. **inheritability** (-bil'-), *n.* **inheritably**, *adv.* **inheritance**, *n.* the act of inheriting; that which is inherited; a hereditary succession to an estate etc.; the right of an heir to succeed; the hereditary derivation of characteristics of one generation from another. **inheritor**, *n.* **inheritress** (-tris), **-trix** (-triks), *n. fem.*
inhibit (inhib'it), *v.t.* to restrain, to hinder, to put a stop to (an action, desire, chemical reaction etc.); to prohibit, to forbid; to hamper free and spontaneous activity, expression etc. in. **inhibiter, -tor**, *n.* **inhibition** (inibish'ən), *n.* the act of inhibiting; the state of being inhibited; the stopping or retardation of a chemical reaction; (*Psych.*) habitual shrinking from some action which is instinctively thought of as a thing forbidden; the partial or complete stoppage of a physical process by a nervous influence. **inhibitory, inhibitive** (-hib'-), *a.*
inhospitable (inhəspit'əbl), *a.* not inclined to show hospitality to strangers; affording no shelter, desolate. **inhospitableness**, *n.* **inhospitably**, *adv.*
in-house IN.
inhuman (inhū'mən), *a.* brutal, cruel, unfeeling; not human. **inhumanity** (-man'-), *n.* **inhumanly**, *adv.*
inhumane (inhūmān'), *a.* lacking in humanity.
inhume (inhūm'), *v.t.* to bury, to inter. **inhumation**, *n.*
inimical (inim'ikəl), *a.* hostile, unfriendly; adverse, unfavourable (to). **inimicality** (-kal'-), **inimicalness**, *n.* **inimically**, *adv.*
inimitable (inim'itəbl), *a.* that cannot be imitated; superb. **inimitability** (-bil'-), **-ableness**, *n.* **inimitably**, *adv.*
iniquity (inik'witi), *n.* gross injustice; (act of) unright-

initial   428   **innumerable**

eousness, wickedness. **iniquitous**, *a*. **iniquitously**, *adv*. **iniquitousness**, *n*.
initial (inish'əl), *a*. beginning; first; placed at or pertaining to the beginning. *n*. the first letter of a word; (*pl*.) the first letters of a Christian name and surname. *v.t.* (*past, p.p.* **initialled**) to mark with one's initials, as a guarantee of correctness, a sign of ownership etc. **initially**, *adv*. at the beginning; at first.
initiate (inish'iāt), *v.t.* to begin or originate; to start; to instruct in the rudiments or principles; to admit (into a society or association or mysteries or secret science), usu. with ceremonial rites. *n*. (-ət), one who has been initiated; a novice. **initiation**, *n*. the act of initiating; admission into a new society or association; the ceremony by which one is so admitted. **initiative**, *a*. serving to begin or initiate; introductory. *n*. the first step or action in any business; power or right to take the lead or originate (esp. legislation); the energy and resourcefulness typical of those able to initiate new projects etc. **on one's own initiative**, without being prompted by others. **initiator**, *n*. **initiatrix** (-triks), *n.fem*. **initiatory**, *n*., *a*.
inject (injekt'), *v.t.* to introduce (as a liquid) by mechanical means; to charge (with a liquid) by injection; to interject; to add, insert. **injection**, *n*. the act of injecting; that which is injected; the introduction of a therapeutic agent into the body; the spraying of oil fuel into the cylinder of a compression ignition engine. **injection moulding**, *n*. the manufacture of rubber or plastic items by the injection of heated material into a mould. **injector**, *n*.
injudicious (injudish'əs), *a*. ill-judged, indiscreet, unwise. **injudicial**, *a*. not judicial. **injudiciously**, *adv*. **injudiciousness**, *n*.
Injun (in'jən), *n*. (*facet.; sometimes offensive*) a N American Indian.
injunction (injŭngk'shən), *n*. the act of enjoining; (*Law*) a writ or process whereby a party is required to do or (more usually) to refrain from doing certain acts; an admonition, direction or order. **injunctive**, *a*. **injunctively**, *adv*.
injure (in'jə), *v.t.* to cause physical harm to; to hurt, to damage; to impair or diminish. **injurer**, *n*. **injurious** (-joo'ri-), *a*. that injures or tends to injure; wrongful, hurtful, pernicious, detrimental; insulting, abusive. **injuriously**, *adv*. **injuriousness**, *n*.
injury (in'jəri), *n*. a wrong; that which occasions loss or detriment; (an instance of) physical damage or harm. **injury time**, *n*. time added on to normal playing time in soccer, rugby etc. to compensate for interruptions to play on account of injuries.
injustice (injŭs'tis), *n*. lack of right or equity, unjustness, unfairness; violation of justice, a wrong.
ink (ingk), *n*. a coloured liquid or viscous material used in writing or printing; the dark fluid exuded by a cuttle-fish to cover its escape. *v.t.* to blacken, daub or cover with ink (as type etc.); to mark (in or over) with ink. **printer's ink** PRINTER. **ink-blot**, *n*. **ink-blot test** RORSCHACH TEST. **ink-cap**, *n*. a mushroom of the genus *Coprinus*. **ink-eraser**, *n*. india-rubber treated with fine sand, used for rubbing out ink-marks. **inkstand**, *n*. a stand for one or more inkpots, usu. with a place for pens. **inkwell**, *n*. a container for ink often let into a school desk. **inker**, *n*. a roller for inking type. **inkless**, *a*. **inky**, *a*. of the nature of or resembling ink; discoloured with ink; black as ink. **inkiness**, *n*.
inkling (ing'kling), *n*. a hint, an intimation; a mere suspicion (of).
inlaid INLAY.
inland (in'lənd), *a*. remote from the sea; situated in the interior of a country; carried on within a country, domestic, not foreign. *adv*. in or towards the interior of a country. **inland revenue**, *n*. taxes and duties levied on home trade etc., not foreign. **Inland Revenue**, *n*. in Britain, the government department responsible for collecting these. **inlander**, *n*. **inlandish**, *a*.
in-law (in'law), *n*. (*pl*. **in-laws**) (*coll*.) a relation by marriage.
inlay (inlā'), *v.t.* (*past, p.p.* **inlaid** (-lād) to lay or insert in; to decorate by inserting different materials into a groundwork, leaving the surfaces even; to fasten a print, picture etc. evenly (into a page or sheet). *n*. (in'-), material inlaid or prepared for inlaying. **inlayer** (in'-), *n*. **inlaying** (in'-), *n*. the business of an inlayer; inlaid work.
inlet (in'lət), *n*. a means of entrance; a passage allowing fuel etc. into a machine; a small arm of the sea; a creek.
inlier (in'liə), *n*. (*Geol*.) an isolated portion of an underlying bed, which has become surrounded by a later formation.
in loco parentis (in lō'kō pəren'tis), in place of or having the responsibilities of a parent. [L]
inly (in'li), *adv*. inwardly, internally; closely, deeply.
inmate (in'māt), *n*. a resident or occupant, esp. of a prison, hospital etc.
in memoriam (in mimaw'riam), in memory of; as a memorial. [L]
inmost (in'mōst), **innermost** (in'ə-), *a*. remotest from the surface; most inward; deepest, most heartfelt, most secret.
inn (in), *n*. a public house esp. one providing board and lodging mainly for travellers. **Inns of Court**, four corporate societies in London (*Inner Temple, Middle Temple, Lincoln's Inn, Gray's Inn*), which have the exclusive right of admitting persons to practise at the bar; the buildings belonging to such societies. **innkeeper**, *n*.
innards (in'ədz), *n.pl*. (*coll*.) entrails; the components of a machine etc.
innate (ināt'), *a*. inborn, natural; native, not acquired; instinctive; (*Phil*.) present in the mind previous to any experience. **innately**, *adv*. **innateness**, *n*.
inner (in'ə), *a*. interior; farther inward or nearer the centre; internal; spiritual; dark, hidden, esoteric. *n*. that part of a target immediately outside the bull's eye; a shot striking that part. **inner city**, *n*. the central, usu. densely populated part of a city; esp. in relation to the social problems found there. **inner man**, *n*. the inner or spiritual part of a person; (*coll*.) the stomach, the appetite for food. **inner tube**, *n*. an inflatable tube inside a tyre. **innerliness**, *n*. **innermost** INMOST.
innervate (in'œvāt), *v.t.* to give a nerve impulse to; to supply with nerves or nerve filaments. **innervation**, *n*.
inning (in'ing), *n*. (*Baseball*) a turn at batting. **innings**, *n*. (*pl*. **innings**) (*Cricket*) the time or turn for batting of a player or a side; the time during which a party or person is in possession, in power etc.
innkeeper INN.
innocent (in'əsənt), *a*. free from moral guilt; guiltless (of); blameless, sinless; pure, unspotted; guileless; naive or credulous; devoid (of). *n*. an innocent person, esp. a child. **Innocents' Day**, *n*. the festival (28 Dec.) commemorating the massacre of the children of Bethlehem by Herod (Matt. ii.16). **innocence**, *n*. **innocently**, *adv*.
innocuous (inok'ūəs), *a*. having no injurious qualities, harmless. **innocuously**, *adv*. **innocuousness**, *n*.
innominate (inom'inət), *a*. not named; nameless. **innominate bone**, *n*. the hip-bone.
innovate (in'əvāt), *v.i.* to introduce alterations (in anything); to introduce or invent something new. *v.t.* to alter or change, by the introduction of something new. **innovation**, *n*. the act of innovating; a new device, method, procedure etc. **innovative, -atory**, *a*. **innovator**, *n*.
innuendo (inūen'dō), *n*. (*pl*. **-dos, -does**) an indirect or oblique hint or intimation; an insinuation.
Innuit, Inuit (in'ūit), *n*. (*pl*. **Innuit, Innuits**) (one of) a N American or Greenland Eskimo people.
innumerable (inū'mərəbl), *a*. countless, numberless; inde-

finitely numerous. **innumerably,** *adv.*
innumerate (inū'mərət), *a.* ignorant of or unskilled in mathematics or science.
inoculate (inok'ūlāt), *v.t.* to communicate a disease to (humans or the lower animals) by the introduction of infectious matter, in order to render the subject immune against further attack; to imbue (with); (*Hort.*) to graft on by the insertion of buds. *v.i.* to graft trees by budding; to practise inoculation. **inoculable,** *a.* **inoculation,** *n.* **inoculative,** *a.* **inoculator,** *n.* **inoculum,** *n.* the material used in giving an inoculation.
in-off (inof'), IN.
inoffensive (inəfen'siv), *a.* giving no offence; unobjectionable, harmless. **inoffensively,** *adv.* **inoffensiveness,** *n.*
inoperable (inop'ərəbl), *a.* that cannot be operated on. **inoperability** (bil'-), **-ableness,** *n.* **inoperably,** *adv.*
inoperative (inop'ərətiv), *a.* not in operation; producing no effect.
inopportune (inop'ətūn, -tūn'), *a.* inconvenient; unseasonable. **inopportunely,** *adv.* **inopportuneness, -tunity** (-tū'-), *n.*
inordinate (inaw'dinət), *a.* excessive, immoderate. **inordinately,** *adv.* **inordinateness,** *n.*
inorganic (inawgan'ik), *a.* not organic, not having the organs or instruments of life; not having organic structure, e.g. rocks, metals etc.; not resulting from natural growth. **inorganic chemistry** CHEMISTRY. **inorganically,** *adv.* without organization.
inosculate (inos'kūlāt), *v.i., v.t.* of e.g. two vessels in the body, to unite by the mouth of one fitting into the mouth of the other, or by a duct; to anastomose. **inosculation,** *n.*
inositol (inos'itol), *n.* a member of the vitamin B complex, found in most plant and animal tissues.
inotropic (inətrop'ik), *a.* of or directing contraction of the heart muscle.
in-patients, in-phase IN.
input (in'put), *n.* the amount put into (a machine, the body etc.); a place where energy, information etc. goes into a system; data fed into a computer; the process of entering such data; a contribution. *a.* pertaining to computer input. *v.t.* to put into (esp. a computer).
inquest (in'kwest), *n.* a judicial inquiry or investigation esp. a coroner's inquest; the jury itself; an inquiry, an investigation. **coroner's inquest,** a judicial inquiry before a coroner and a jury into death occurring suddenly, from violence, an unknown cause or in a prison; also into cases of treasure trove.
inquiline (in'kwilin), *n.* an animal living in the abode of another, as certain beetles in ants' nests. **inquilinous** (-li'-), *a.*
inquire (inkwīə'), *v.i.* to ask questions (of); to seek information by asking questions (about or after); to investigate (into). *v.t.* to ask information about; to ask (what, whether, how etc.). **inquirer,** *n.* **inquiring,** *a.* given to inquiry; inquisitive. **inquiringly,** *adv.* **inquiry** (inkwī'ri), *n.* the act of inquiring; a question, an interrogation; a searching for truth, information or knowledge; examination of facts or principles; a judicial investigation. **court of inquiry,** a court appointed to make a legal investigation into charges against soldiers, usu. before proceedings are instituted before a court-martial.
inquisition (inkwizish'ən), *n.* inquiry, search, investigation; a searching examination; a judicial inquiry; (*often* **Inquisition**) a tribunal in the Roman Catholic Church for inquiring into offences against the canon law, aimed especially at the suppression of heresy, also called the Holy Office. **inquisitional,** *a.* **inquisitive** (inkwiz'itiv), *a.* unduly given to asking questions; prying, curious. **inquisitively,** *adv.* **inquisitiveness,** *n.* **inquisitor** (inkwiz'itə), *n.* one who inquires, esp. searchingly or ruthlessly; a functionary of the Inquisition. **Grand Inquisitor,** the president of a court of the Inquisition. **inquisitorial** (-taw'ri-), *a.* of a system of

criminal procedure where the judge is also the prosecutor. **inquisitorially,** *adv.* **inquisitress** (-tris), *n. fem.*
inquorate (inkwaw'rət), *a.* insufficient for, not constituting, a quorum.
INRI, (*abbr.*) Jesus of Nazareth King of the Jews. [L *Iesus Nazarenus Rex Iudaeorum*]
inroad (in'rōd), *n.* a raid; an encroachment.
inrush (in'rūsh), *n.* an irruption; a sudden influx.
insalubrious (insəloo'briəs), *a.* unhealthy. **insalubriously,** *adv.* **insalubrity,** *n.*
insane (insān'), *a.* mad; exceedingly rash or foolish. **insanely,** *adv.* **insanity** (-san'-), **insaneness,** *n.*
insanitary (insan'itəri), *a.* not sanitary. **insanitation,** *n.*
insatiable (insā'shəbl), *a.* that cannot be satisfied or appeased; immoderately greedy (of). **insatiability** (-bil'-), **-ableness,** *n.* **insatiably,** *adv.* **insatiate** (-ət), *a.* (*poet.*) insatiable.
inscribe (inskrīb'), *v.t.* to write, carve or engrave (in or upon a stone, paper or other surface); to mark (a stone etc. with writing or letters); to dedicate (as a book to a friend); to enter in or on a book, list etc.; (*Geom.*) to delineate (a figure) within another so that it touches the boundary surfaces of the latter. **inscribable,** *a.* **inscriber,** *n.*
inscription (inskrip'shən), *n.* the art or act of inscribing; that which is inscribed, as a dedicatory address or the words on the reverse of some coins and medals. **inscriptional, inscriptive,** *a.*
inscrutable (inskroo'təbl), *a.* impenetrable, unfathomable, mysterious. **inscrutability** (-bil'-), **-ableness,** *n.* **inscrutably,** *adv.*
insect (in'sekt), *n.* one of a class of articulate, usu. winged animals, with three pairs of legs, and divided into three distinct segments, the head, thorax and abdomen; used incorrectly of other articulated animals resembling these, as a spider or centipede; a small or contemptible person or creature. **insect-powder,** *n.* a powder for destroying insects. **insectarium** (-teə'riəm), *n.* (*pl.* **-riums, -ria**) an insectary. **insectary** (in'sektəri), *n.* a place for keeping or breeding insects. **insecticide** (-sek'tisid), *n.* a preparation for killing insects. **insecticidal** (-si'-), *a.* **insectifuge** (-sek'tifūj), *n.* a substance for keeping insects away. **insectile** (-sek'til), *a.* of the nature of insects. **insectology** (-tol'-), *n.* entomology. **insectologist,** *n.*
insectivore (insek'tivaw), *n.* an animal, e.g. mole or hedgehog, or plant that feeds on insects. **insectivorous** (-tiv'-), *a.*
insecure (insikūə'), *a.* not safe; apprehensive of danger; not effectually guarded; not strongly fixed or supported; lacking in self-confidence. **insecurely,** *adv.* **insecurity,** *n.*
inseminate (insem'ināt), *v.t.* to impregnate, esp. by artificial means; to implant (in the mind etc.). **insemination,** *n.*
insensate (insen'sət), *a.* lacking sensation, unconscious; unfeeling; foolish, mad. **insensately,** *adv.*
insensible (insen'sibl), *a.* imperceptible; unconscious; unaware; indifferent, heedless (of, how etc.); unfeeling, callous, apathetic. **insensibility** (-bil'-), *n.* **insensibilize, -ise,** *v.t.* **insensibilization,** *n.* **insensibly,** *adv.* imperceptibly, gradually.
insensitive (insen'sitiv), *a.* not sensitive (to). **insensitiveness,** *n.*
insentient (insen'shiənt), *a.* not sentient, inanimate.
inseparable (insep'ərəbl), *a.* incapable of being separated; (*Gram.*) incapable of being employed separately (as the prefixes DIS-, RE-). *n.* (*usu. pl.*) things which cannot be separated; persons who are constantly together. **inseparability** (-bil'-), **-ableness,** *n.* **inseparably,** *adv.*
insert (insœt'), *v.t.* to set or place (in, amongst etc.); to introduce (in or into). *n.* something inserted; a printed sheet etc. placed inside the leaves of a newspaper, periodical etc. **inserted,** *a.* **inserter,** *n.* **insertion** (insœ'shən), *n.* the act of inserting; that which is inserted, an intercalation, a passage etc. introduced (in or into); a band of lace

or embroidery inserted in a dress, handkerchief, fancy work etc.; (*Anat., Bot. etc.*) the manner in which one part is inserted into or adheres to another.
in-service IN.
insessorial (insesaw'riəl), *a.* of birds, with feet adapted for perching and walking.
inset¹ (inset'), *v.t.* (*past, p.p.* **inset**) to set or fix (in), to insert (in).
inset² (in'set), *n.* that which is set or fixed in; an insertion, as a piece let into a dress etc., a small map or diagram set within a larger one, a page or number of pages inserted in a book, newspaper etc.
inshore (inshaw'), *a., adv.* on, near or towards the shore.
inside¹ (in'sid), *a.* situated within; interior, internal, inner; indoor. *n.* the inner or interior part; the inner side, surface, part etc. (of); (*Print.*) the side of a sheet containing the second page; the middle part (of); (*pl.*) the contents; (*pl.*) the bowels. **inside information**, *n.* confidential knowledge not generally accessible. **inside job**, *n.* (*coll.*) a crime, organized with the help of someone trusted or employed by the victim. **inside out**, *adv.* having the inner side turned out and vice versa. **to know inside out**, (*coll.*) to have thorough knowledge of. **inside track**, *n.* the inner lane of a race track; an advantageous position. **insider** (-sī'-), *n.* one who belongs to a society, clique etc.; one who has inside information. **insider dealing, trading**, *n.* the criminal practice of conducting share deals on the basis of inside information.
inside² (insid'), *adv.* in or into the interior; within; indoors; (*coll.*) in or into prison. *prep.* within, on the inner side of, into. **inside of a mile, an hour etc.**, (*coll.*) within or in less than a mile, an hour etc.
insidious (insid'iəs), *a.* treacherous, sly; working secretly or deceptively; harmful but attractive; working gradually but dangerously. **insidiously**, *adv.* **insidiousness**, *n.*
insight (in'sit), *n.* (something which demonstrates) power of observation or discernment of the real character of things; penetration; awareness, esp. self-awareness. **insightful**, *a.*
insignia (insig'niə), *n.pl.* (*in N Am. often sing. in constr.*) badges of office or honour; distinguishing marks or signs (of).
insignificant (insignif'ikənt), *a.* unimportant, trivial; contemptible; tiny; without meaning. **insignificantly**, *adv.* **insignificance, -ancy**, *n.*
insincere (insinsiə'), *a.* not sincere; false, dissembling; hypocritical, deceitful. **insincerely**, *adv.* **insincerity** (-se'-), *n.*
insinuate (insin'ūāt), *v.t.* to introduce (into favour, office etc.) by gradual and artful means; to hint or suggest obliquely or by remote allusion. **insinuatingly**, *adv.* **insinuation**, *n.* the art or power of insinuating; a hint, an indirect suggestion. **insinuative**, *a.* **insinuator**, *n.*
insipid (insip'id), *a.* tasteless, savourless; dull, vapid. **insipidity** (-pid'-), **insipidness**, *n.* **insipidly**, *adv.*
insist (insist'), *v.i.* to be emphatic, positive, urgent or persistent (on or upon). *v.t.* to maintain emphatically; to urge strongly or without accepting any refusal. **to insist on**, to demand emphatically; to assert positively. **insistence, -ency**, *n.* **insistent**, *a.* **insistently**, *adv.*
in situ (in sit'ū), in the proper, appropriate or destined position. [L]
insobriety (insəbrī'əti), *n.* intemperance (usu. in drinking).
insofar IN.
insolation (in'səlāshən), *n.* exposure to the sun; sunstroke; solar radiation falling on a given surface.
insole (in'sōl), *n.* the inner sole of a boot or shoe; a strip of waterproof or other material placed inside a shoe.
insolent (in'sələnt), *a.* showing overbearing contempt; impudent, offensive, insulting. **insolently**, *adv.* **insolence**, *n.*
insoluble (insol'ūbl), *a.* that cannot be dissolved; that cannot be solved; inexplicable. **insolubility** (-bil'-), **insolubleness**, *n.* **insolubly**, *adv.*
insolvable (insol'vəbl), *a.* that cannot be solved or explained, insoluble. **insolvability** (-bil'-), *n.* **insolvably**, *adv.*
insolvent (insol'vənt), *a.* not able to discharge all debts or liabilities; pertaining to insolvents. *n.* a bankrupt. **insolvency**, *n.*
insomnia (insom'niə), *n.* sleeplessness; chronic inability to sleep or sleep well. **insomniac** (-ak), *n.*
insomuch (insōmŭch'), *adv.* so, to such a degree (that).
insouciant (insoo'siənt), *a.* careless, unconcerned. **insouciance**, *n.*
Insp., (*abbr.*) Inspector.
inspect (inspekt'), *v.t.* to look closely into; to scrutinize carefully; to view and examine officially. **inspectable**, *a.* **inspection** (inspek'shən), *n.* **inspector** (inspek'tə), *n.* one who inspects; an overseer, a superintendent; a police officer usu. ranking next below a superintendent. **inspectoral, -orial** (-taw'ri-), *a.* **inspectorate** (-ət), *n.* the office of inspector; a body of inspectors; a district overseen by an inspector. **inspectorship**, *n.* **inspectress** (-tris), *n. fem.*
inspire (inspiə'), *v.t.* to inhale; to instil or infuse (ideas, feelings etc.) into, esp. by or as by supernatural agency; to imbue or animate (with); to infuse or instil (as emotion in or into); to convey privately suggestions or material for. **inspirable**, *a.* **inspiration** (inspirā'shən), *n.* inhalation; an act of inspiring, breathing in or infusing feelings, ideas etc.; a person that inspires others; supernatural influence, esp. that exerted by the Holy Spirit on certain teachers and writers; the feeling, ideas or other influences imparted by or as by divine agency; an inspiring idea. **inspirational**, *a.* **inspirationally**, *adv.* **inspiratory** (-spi'-), *a.* pertaining to inspiration; aiding in the process of inspiration. **inspired**, *a.* so brilliant or accurate as to seem to be produced by supernatural agency. **inspirer**, *n.* **inspiringly**, *adv.*
inspirit (inspi'rit), *v.t.* to infuse spirit, life or animation into; to inspire. **inspiriting**, *a.*
inspissate (inspis'āt), *v.t.* to thicken, to render more dense. **inspissation**, *n.*
inst. (inst), (*abbr.*) instant (this month).
instability (instəbil'iti), *n.* lack of stability or firmness; lack of mental or emotional consistency.
install (instawl'), *v.t.* (*past, p.p.* **installed**) to induct or invest (in an office etc.) with customary ceremonies; to establish in a specified place, condition etc.; to put (apparatus etc.) in position for use. **installation** (-stə-), *n.* the act of installing; a piece or complex of machinery; a military base etc. **installer**, *n.*
instalment (instawl'mənt), *n.* a part of a debt paid at successive periods; a part (of anything) supplied at different times; part of a serial story etc. **instalment plan**, *n.* (*N Am.*) the hire-purchase system.
instance (in'stəns), *n.* an example, illustrative case or precedent; (*Law*) a process or suit. *v.t.* to bring forward as an instance or example. **at the instance of**, at the suggestion or desire of. **for instance**, for example. **in the first instance**, at the first stage, in the first place.
instant (in'stənt), *a.* pressing, urgent; immediate; esp. of food, processed so as to be quickly and easily prepared; of the current month. *n.* a particular point of time; a moment, a very brief space of time. **instantaneous** (instəntā'niəs), *a.* happening or done in an instant or immediately. **instantaneously**, *adv.* **instantaneity** (-nē'i-), **instantaneousness**, *n.* **instantly**, *adv.* immediately; without delay.
instanter (instan'tə), *adv.* at once, immediately. [L]
instar (in'stah), *n.* (*Zool.*) a stage in the development of an insect or other arthropod between successive moults.
instate (instāt'), *v.t.* to place in a particular office, condition etc., to install.

instead   431   **integral**

instead (insted'), *adv.* as an alternative or substitute. **instead of**, in the place of; rather than.

instep (in'step), *n.* the arched upper side of the human foot, near the ankle; the part of a shoe, stocking etc., corresponding to this; the front part of the hind leg of a horse reaching from the ham to the pastern-joint.

instigate (in'stigāt), *v.t.* to urge on (to an action or to do); to provoke or bring about (an action, esp. of an evil kind). **instigation**, *n.* **instigator**, *n.*

instil (instil'), *v.t.* (*past, p.p.* **instilled**) to pour by drops (into); to infuse slowly and gradually (into the mind of a person). **instillation, instilment**, *n.* **instillator** (in'-), *n.*

instinct (in'stingkt), *n.* a natural impulse, esp. in the lower animals, leading them without reasoning or conscious volition to respond to certain stimuli in a particular way; an innate or intuitive impulse, tendency or aptitude; intuition. **instinctive** (-stingk'-), *a.* prompted by instinct; spontaneous, impulsive. **instinctively**, *adv.* **instinctual** (-stingk'-), *a.* pertaining to instinct.

institute (in'stitūt), *v.t.* to set up, to establish; to start, to begin; to nominate, to appoint (to or into). *n.* a society established for the promotion or furtherance of some particular object (usu. literary or scientific); the building in which such a society meets; an established law, precept or principle; (*pl.*) a book of elements or principles, esp. of jurisprudence or medicine. **institution** (institū'shən), *n.* the act of instituting; that which is instituted; an established order, law, regulation or custom; an organization founded for the promotion of some particular object, esp. a charitable or educational one; the building occupied by such an organization; a mental hospital; (*coll.*) a familiar custom, person etc. **institutional**, *a.* pertaining to an institution; routine or unimaginative. **institutional religion**, *n.* the form of religion that expresses itself through ritual and church services. **institutionalism**, *n.* **institutionalist**, *n.* **institutionalize, -ise**, *v.t.* to make an institution of; to confine to an institution; to subject to the effects of confinement in an institution. **institutionalization**, *n.* **institutionally**, *adv.*

instruct (instrŭkt'), *v.t.* to teach, to educate (in a subject); to inform; to give orders or directions to; to supply (a solicitor, counsel etc.) with information relating to a case. **instruction** (-shən), *n.* teaching, education; a code directing a computer to perform a certain operation; a command; (*pl.*) directions regarding use, assembly etc.; directions to a solicitor, counsel etc. **instructional**, *a.* **instructive**, *a.* conveying instruction. **instructively**, *adv.* **instructiveness**, *n.* **instructor**, *n.* one who instructs; (*N Am.*) a college teacher ranking below a professor. **instructress**, *n. fem.*

instrument (in'strəmənt), *n.* a means by which work is done or any object or purpose effected; a tool, a mechanical implement, esp. one for scientific and other delicate operations; an indicator or controlling device, esp. in an aircraft; a contrivance for producing musical sound; (*Law*) a document giving formal expression to an act; an agent, a person used as a means by another. *v.t.* (*Mus.*) to orchestrate. **instrumental** (instrəmen'təl), *a.* serving as instrument or means (to some end or in some act); pertaining to an instrument; pertaining to or produced by musical instruments. *n.* a piece of music for instruments as opposed to voices. **instrumentalist**, *n.* one who plays an instrument. **instrumentality** (-tal'-), *n.* **instrumentally**, *adv.* **instrumentation** (instrəmentā'shən), *n.* orchestration; the art or manner of using an instrument or instruments.

insubordinate (insəbaw'dinət), *a.* not submissive to authority; disobedient, disorderly. **insubordinately**, *adv.* **insubordination**, *n.*

insubstantial (insəbstan'shəl), *a.* unsubstantial, unreal; flimsy or slight. **insubstantiality** (-al'-), *n.*

insufferable (insŭf'ərəbl), *a.* unendurable; detestable, intolerable. **insufferably**, *adv.*

insufficient (insəfish'ənt), *a.* not sufficient; deficient, inadequate. **insufficiency** (-si), **-ence**, *n.* **insufficiently**, *adv.*

insufflate (in'səflāt), *v.t.* (*Med.*) to blow or breathe (air, vapour, powder etc.) into an opening, cavity etc.; to treat (a person, organ etc.) by insufflation. **insufflator**, *n.* an instrument used for this purpose. **insufflation**, *n.*

insular (in'sūlə), *a.* pertaining to or of the nature of an island; remote, detached; narrow-minded, inward-looking. **insularism, -larity** (-la'-), *n.* **insularly**, *adv.*

insulate (in'sūlāt), *v.t.* to place in a detached position; to isolate; (*Phys.*) to separate from other bodies by a non-conductor, so as to prevent the passage of electricity, heat, sound etc. **insulation**, *n.* **insulator**, *n.*

insulin (in'səlin), *n.* a hormone produced in the pancreas which regulates the metabolism of sugar and fat and is employed in the treatment of diabetes.

insult[1] (insŭlt'), *v.t.* to treat with gross indignity, insolence or contempt; to affront. **insulting**, *a.* **insultingly**, *adv.*

insult[2] (in'sŭlt), *n.* an affront, an indignity; an insulting act or speech.

insuperable (insū'pərəbl), *a.* insurmountable. **insuperability** (-bil'-), *n.* **insuperably**, *adv.*

insupportable (insəpaw'təbl), *a.* insufferable, intolerable; incapable of being sustained or justified. **insupportably**, *adv.*

insure[1] (inshua'), *v.t.* to secure compensation, whole or partial, for loss or injury of (property, life etc.) by paying a periodic premium; to furnish (a person) with an insurance policy. *v.i.* to take out an insurance policy. **the insured**, the person covered by insurance. **insurability** (-bil'-), *n.* **insurable**, *a.* **insurance** (inshua'rəns, -shaw'-), *n.* the act of insuring against damage or loss; a contract by which a company, for a sum of money, becomes bound to indemnify the insured against loss by fire, shipwreck etc.; the sum so insured; the premium so paid. **insurance policy** POLICY[2]. **insurer**, *n.*

insure[2] (inshua'), (*chiefly N Am.*) ENSURE.

insurgent (insœ'jənt), *n.*, *a.* (one) who rises up against established government or authority; (a) rebel. **insurgence, -ency**, *n.*

insurmountable (insəmown'təbl), *a.* that cannot be surmounted, passed over or overcome. **insurmountability** (-bil'-), *n.* **insurmountably**, *adv.*

insurrection (insərek'shən), *n.* the act of rising in open opposition to established authority; uprising, rebellion in the initial stage. **insurrectional, -tionary**, *a.* **insurrectionist**, *n.*

int., (*abbr.*) interjection.

intact (intakt'), *a.* untouched; unimpaired, uninjured; entire. **intactness**, *n.*

intaglio (intahl'yō), *n.* (*pl.* **-ios**) a figure cut or engraved in a hard substance; the act or process of producing this; a gem with a figure cut or engraved into it; (*N Am.*) a rotogravure. **intaglioed** (-ātid), *a.*

intake (in'tāk), *n.* that which is taken in; the point where a tube or woven article narrows; a place where water is taken in, an inlet; an air-shaft in a mine; the point at which fuel enters an engine; a quantity of new members (of a school etc.).

intangible (intan'jibl), *a.* imperceptible to the touch, impalpable; not to be grasped mentally; of assets, saleable but without intrinsic productive value. *n.* an intangible thing. **intangibility** (-bil'-), *n.* **intangibly**, *adv.*

integer (in'tijə), *n.* the whole of anything; a whole number as distinguished from a fraction.

integrable (in'tigrəbl), *a.* capable of being integrated.

integral (in'tigrəl, -teg'-), *a.* whole, entire, complete; necessary to completeness, an essential part of a whole; (*Math.*) pertaining to or constituting an integer; pertaining to or produced by integration. *n.* (*Math.*) the limit of the sum of a series of values of a differential $f(x) dx$ when x varies by indefinitely small increments from one given

integrate 432 **intercommunicate**

value to another (cp. DIFFERENTIAL); a whole, a total. **integral calculus**, *n*. a method of summing up differential quantities. **integrality** (-gral'-), *n*.
integrate (in'tigrāt), *v.t.* to make into a whole, to complete by addition of the parts; to combine into a whole; to end the racial segregation of; (*Math.*) to find the integral of. **integrated circuit**, *n*. a minute electronic circuit in or on a slice of semiconductor material. **integration**, *n*. the making into a whole; the unification of all elements in a society, esp. of white and coloured; (*Math.*) the act or process of integrating. **integrative**, *a.* **integrator**, *n*. a device or instrument for determining the value of an integral, as an area, rate of speed etc.
integrity (integ'riti), *n*. entireness, completeness; soundness; genuine, unadulterated state; probity, rectitude, high principle.
integument (integ'ūmənt), *n*. a covering, esp. a natural one; the skin; the outer covering of a seed, the husk, rind etc. **integumentary** (-men'-), *a*.
intellect (in'tilekt), *n*. the faculty of the human mind by which it understands and reasons; a mind, esp. a brilliant one; a person who possesses a brilliant mind. **intellection**, *n*. thought; reasoning. **intellectual** (intilek'tūəl), *a*. possessing intellect in a high degree; pertaining to or performed by the intellect; appealing to or perceived by the intellect. *n*. an intellectual person. **intellectualism**, *n*. the cultivation of the intellect; the doctrine that knowledge is exclusively or principally derived from pure reason. **intellectualist**, *n*. **intellectuality** (-al'-), *n*. **intellectualize, -ise**, *v.t.* to make intellectual; to treat intellectually; to give an intellectual character or significance to. **intellectualization**, *n*. **intellectually**, *adv*.
intelligence (intel'ijəns), *n*. intellectual power; capacity for the higher functions of the intellect; quickness or sharpness of intellect; news, information, notice, notification; a department concerned with gathering secret or little-known information of importance for military activity; such information; an intelligent being, esp. an incorporeal or spiritual being regarded as pure intellect. **intelligence quotient (IQ)**, *n*. a number denoting a person's intelligence by dividing the mental age by the age in years. **intelligence test**, *n*. a psychological test to determine a person's relative mental capacity.
intelligent (intel'ijənt), *a*. endowed with understanding, sensible, clever, quick; of computerized processes, able to modify action in the light of ongoing events. **intelligently**, *adv*. **intelligentsia, -gentzia** (-jent'siə), *n*. people who claim or possess enlightenment or culture.
intelligible (intel'ijibl), *a*. capable of being understood, comprehensible; plain, clear; apprehensible only by the intellect, as distinct from *sensible*. **intelligibility** (-bil'-), **-ibleness**, *n*. **intelligibly**, *adv*.
intemperate (intem'pərət), *a*. not exercising due moderation or self-restraint; given to overindulgence in alcohol; immoderate, excessive; violent, inclement. **intemperance**, *n*. **intemperately**, *adv*. **intemperateness**, *n*.
intend (intend'), *v.t.* to propose, to plan; to signify, to mean; to design (for); to destine (for); to mean, to have a certain intention. **intendancy**, *n*. a body of intendants; the position or office of intendant. **intendant**, *n*. a superintendent or manager. **intended**, *a*. (*coll.*) a person whom one is expecting to marry. **intendedly**, *adv*. **intendment**, *n*. (*Law*) true intent or meaning as determined by the law.
intense (intens'), *a*. violent, vehement; extreme in degree; severe, immoderate, excessive; ardent, eager, fervent; strongly or deeply emotional. **intensely**, *adv*. **intenseness**, *n*. **intensify** (-fī), *v.t.* to render more intense; (*Phot.*) to increase the density of (a negative) so as to produce stronger contrasts. *v.i.* to become more intense. **intensification** (-fi-), *n*. **intensifier**, *n*. **intensity**, *n*. the condition or quality of being intense; an extreme degree of force or strength; magnitude of force per unit.

intension (inten'shən), *n*. intensity, high degree (of a quality), as distinct from *extension*; (*Log.*) the sum of the qualities which distinguish the referents of a given word. **intensive** (inten'siv), *a*. concentrated, thorough, as opp. to extensive; unremitting; characterized by intensity; (*chiefly in comb.*) utilizing one specified element in production proportionately more than others; (*Econ.*) conducive to high productiveness within a narrow area; pertaining to methods (of inoculation etc.) in which injections, doses etc. are successively increased; (*Gram.*) serving to intensify, or to add force or emphasis. **intensive cultivation**, *n*. the system whereby land is kept under cultivation by a rotation of crops and manuring. **intensively**, *adv*.
intent[1] (intent'), *a*. ·concentrated; determined, directing one's mind or energy (on). **intently**, *adv*. **intentness**, *n*.
intent[2] (intent'), *n*. purpose, intention; meaning, drift. **to all intents and purposes**, practically, really, in reality.
intention (inten'shən), *n*. determination to act in some particular manner; purpose, design, intent; (*pl.*) (*coll.*) designs with regard to marriage; (*Log.*) a general concept; ultimate aim or object. **intentional**, *a*. done with design or purpose. **intentionality** (-nal'-), *n*. **intentionally**, *adv*. **intentioned**, *a*. (*chiefly in comb.*, as *well-intentioned*).
inter (intœ'), *v.t.* (*past, p.p.* **interred**) to bury; to place in a grave or tomb. **interment**, *n*.
inter-, *pref.* between, among; with, into or upon each other; as *intercede, intercostal, international, interstellar, intertexture, interwoven*.
interact (intərakt'), *v.i.* to act reciprocally; to act on each other. **interaction**, *n*. **interactive**, *a*. capable of mutual action; permitting continuous mutual communication between computer and user.
inter alia (in'tə ah'liə, ā'liə), among other things. [L]
interbreed (intəbrēd'), *v.t., v.i.* (*past, p.p.* **interbred**) to breed from within a closed population; to crossbreed.
intercalary (intœ'kələri), **intercalar**, *a*. of a day, month etc., inserted in the calendar to make this correspond with the solar year; of a year, containing such an addition; inserted, interpolated. **intercalate**, *v.t.* to insert between or amongst others (esp. a day etc. into a calendar); to interpolate, to insert anything in an unusual or irregular way. **intercalation**, *n*. **intercalative**, *a*.
intercede (intəsēd'), *v.i.* to plead (with someone) in favour of another; to mediate. **interceder**, *n*.
intercept (intəsept'), *v.t.* to stop, take or seize by the way or in passage; to obstruct, to stop, to shut off; (*Math.*) to mark off or include between two points etc. *n*. (*Math.*) the part of a line that is intercepted. **interception**, *n*. **interceptive**, *a*. **interceptor** (in'-), *n*. one who or that which intercepts; a fighter used to intercept enemy aircraft.
intercession (intəsesh'ən), *n*. the act of interceding; a prayer offered for others. **intercessional**, *a*.
intercessor (intəses'ə), *n*. one who intercedes; a mediator. **intercessory**, *a*.
interchange (intəchānj'), *v.t.* to exchange with each other, to give and take; to put each (of two things) in the place of the other, to cause to alternate. *v.i.* to alternate. *n*. (in'-), reciprocal exchange; alternate succession, alternation; a junction of two or more roads designed to prevent traffic streams crossing one another. **interchangeable**, *a*. **interchangeability** (-bil'-), **-ableness**, *n*. **interchangeably**, *adv*. **interchanger** (in'-), *n*.
intercity (intəsit'i), *a*. existing or carried on between different cities.
intercollegiate (intəkəlē'jət), *a*. existing or carried on between colleges.
intercom (in'təkom), *n*. a system of intercommunication in aircraft etc. [*internal c*ommunication]
intercommunicate (intəkəmū'nikāt), *v.i.* to hold or enjoy mutual communication; to have free passage to and from each other. *v.t.* to give or communicate mutually. **intercommunicable**, *a*. **intercommunication**, *n*. **intercommu-**

nion, *n.* the partaking of communion in common between members of different Churches or sects.
interconnect (intəkənekt'), *v.i.* to connect (with) by links or parts acting reciprocally. **interconnectedness**, *n.* **interconnection**, *n.*
intercontinental (intəkontinen'təl), *a.* existing between or connecting different continents or persons belonging thereto.
intercostal (intəkos'təl), *a.* situated between the ribs. *n.pl.* the intercostal muscles.
intercourse (in'təkaws), *n.* reciprocal dealings, association, communication etc., between persons, nations etc.; sexual intercourse.
intercrop (in'təkrop), *n.* a crop raised between the rows of another crop; a quickly-maturing crop between crops grown in a regular series. *v.t.* (*past, p.p.* **intercropped**) to raise (a crop) in this way. *v.i.* to plant intercrops.
intercurrent (intəkŭ'rənt), *a.* occurring between or among; intervening; occurring during the progress of another disease. **intercurrence**, *n.*
intercut (intəkŭt'), *v.t.* (*pres.p.* **intercutting**, *past, p.p.* **intercut**) to alternate (contrasting camera shots) by cutting.
interdenominational (intədinominā'shənəl), *a.* existing or carried on between different denominations.
interdepartmental (intədēpahtmen'təl), *a.* involving or carried on between different departments.
interdepend (intədipend'), *v.i.* to depend upon each other. **interdependent**, *a.* **interdependently**, *adv.* **interdependence**, *n.*
interdict (in'tədikt), *n.* a prohibitory decree; (*Sc. Law*) an order of the Court of Session equivalent to an injunction; in the Roman Catholic Church, a sentence by which places or persons are debarred from ecclesiastical functions and privileges. *v.t.* (-dikt'), to prohibit; to restrain (from); to lay under an interdict. **interdiction**, *n.* **interdictory** (-dik'-), *a.*
interdisciplinary (intədisiplin'əri), *a.* involving two or more disciplines or fields of study.
interest[1] (in'trist), *n.* lively, sympathetic or curious attention; the power of eliciting such attention; personal concern, sympathy; something in which one has a personal concern; participation in advantages, benefits or profits; (*often pl.*) benefit, advantage; a share, a portion or stake (in); (*collect.*) those having a concern in a particular business etc.; influence with or over others; payment for the use of borrowed money or on a debt. **compound interest** COMPOUND. **simple interest** SIMPLE. **to take an interest in**, to pay sympathetic or curious attention to. **interest group**, *n.* a group of people concerned to defend a common interest.
interest[2] (in'trist), *v.t.* to arouse or hold the attention or curiosity of; to concern; to cause to participate (in). **interested**, *a.* having the interest excited; concerned (in); having an interest, concern or share in; liable to be biased through personal interest, not disinterested. **interestedly**, *adv.* **interesting**, *a.* arousing interest, attention or curiosity. **to be in an interesting condition**, to be pregnant. **interestingly**, *adv.*
interface (in'təfās), *n.* (*Geom., Cryst.*) a surface lying between two spaces; the point at which independent systems meet and act on each other; an electrical circuit linking computers or other devices. **interfacial** (-fā'shəl), *a.* included between two faces of a crystal etc.; pertaining to an interface. **interfacing**, *n.* stiffening material inserted between layers of fabric.
interfere (intəfiə'), *v.i.* to come into collision, to clash (with); to meddle (with); (*coll.*) to assault sexually; to interpose, to intervene (in); (*Phys.*) to act reciprocally, to modify each other; of a horse, to strike the hoof against the opposite fetlock. **interference**, *n.* the act of interfering; meddling; hindrance, esp. of an opponent in certain games; (*Radio.*) the spoiling of reception by atmospherics or by other signals. **interferer**, *n.* **interfering**, *a.* inclined to interfere; officious. *n.* interference. **interferingly**, *adv.*
interferometer (-om'itə), *n.* an optical instrument for accurate measuring, esp. of the wavelength of light.
interferon (intəfiə'ron), *n.* an antiviral substance produced in living cells in humans and other creatures in response to infection from various viruses.
interfuse (intəfūz'), *v.t.* to commix or intersperse; to blend together. *v.i.* to blend into each other. **interfusion** (-zhən), *n.*
intergalactic (intəgəlak'tik), *a.* between galaxies.
interglacial (intəglā'shəl), *a.* occurring or formed between two of the glacial periods.
interim (in'tərim), *n.* the meantime; the intervening time or period. *a.* temporary, provisional.
interior (intiə'riə), *a.* inner; inland; remote from the coast, frontier or exterior; domestic, as dist. from foreign; pertaining to the inner consciousness, the soul or spiritual matters. *n.* the internal part of anything, the inside; the central or inland part of a country; the inside of a building or room, esp. as portrayed in a picture, photograph etc.; the domestic affairs of a country; the government department dealing with these; the inward nature, the soul. **interior angle**, *n.* the angle between two sides of a polygon. **interior-sprung**, *a.* of a mattress etc., having springs. **interiority** (-o'-), *n.* **interiorly**, *adv.*
interject (intəjekt'), *v.t.* to throw in (an abrupt remark etc.); to insert, to interpose. **interjection**, *n.* the act of interjecting; an exclamation, a word thrown in to express feeling, and which is differentiated as a separate part of speech. **interjectional, -jectory, -jectural**, *a.* **interjectionally**, *adv.*
interlace (intəlās'), *v.t.* to lace or weave together; to entangle together; to intermix. *v.i.* to be interwoven (with each other); to intersect in a complicated fashion. **interlacement**, *n.*
interlard (intəlahd'), *v.t.* to diversify (a conversation, passage in a book etc., with unusual phrases etc.); to intersperse.
interleaf (in'təlēf), *n.* a leaf, usu. blank, inserted among others for purposes of illustration etc. **interleave** (-lēv'), *v.t.* to insert (a blank leaf or leaves) between the leaves of.
interline (intəlin'), *v.t.* to write or print between the lines of; to write or print in alternate lines; to insert a lining between the outer cloth and the lining of (a garment). **interlineal** (-lin'iəl), **-linear** (-lin'iə), *a.* **interlineation**, *n.* **interlining** (in'-), *n.*
interlink (intəlingk'), *v.t.* to connect (together or with) by links. *n.* an intermediate link.
interlock (intəlok'), *v.t.* to connect firmly together by reciprocal engagement of parts; to link or lock together. *v.i.* to engage with each other by reciprocal connections. *n.* (in'-), the state of being interlocked; a device in a logic circuit preventing the initiation of an activity in the absence of certain preceding events.
interlocutor (intəlok'ūtə), *n.* one who takes part in a conversation; the compere of a minstrel show; (*Sc. Law*) an interlocutory or interim decree in a case. **interlocutory** (-ləkū'-), *a.* consisting of dialogue; (*Law*) intermediate, not final. **interlocutress** (-tris), **-trice**, (-tris), **-trix** (-triks), *n. fem.*
interloper (in'təlōpə), *n.* one who thrusts him- or herself into a place, office, affairs etc., without a right; an intruder; one who trades without a licence or infringes upon another's business. **interlope**, *v.i.*
interlude (in'təlood), *n.* an interval; a piece of instrumental music played between the acts of a drama, between the verses of a hymn, portions of a church service etc.; a period of time of a contrasting character to those preceding and following it; a dramatic representation, usu. farcical, intervening between the acts of the mystery-plays and moralities.

intermarriage (intəmă'rəj), *n.* marriage between persons of different families, tribes, castes or nations; marriage between persons closely akin. **intermarry**, *v.i.*
intermediary (intəmē'diəri), *a.* being, coming or acting between; intermediate; mediatory. *n.* an intermediate agent, a go-between; intermediation. **intermediate** (ət), *a.* coming or being between; intervening, interposing. *n.* an intermediate thing. *v.i.* (-āt), to act as intermediary; to mediate (between). **intermediate technology**, *n.* technology as adapted for the conditions and requirements of developing nations. **intermediately**, *adv.* **-mediateness**, **-mediation**, *n.*
interment INTER.
intermezzo (intəmet'sō), *n.* (*pl.* **-mezzi** (-ē), **-mezzos**) a short movement connecting the main divisions of a large musical composition; a short self-contained musical composition.
interminable (intœ'minəbl), *a.* endless; tediously protracted. **interminableness**, *n.* **interminably**, *adv.*
intermingle (intəming'gl), *v.t.* to mingle together, to intermix. *v.i.* to be mingled (with).
intermit (intəmit'), *v.t.* (*past, p.p.* **intermitted**) to cause to cease for a time; to suspend. *v.i.* to cease or relax at intervals (as a fever, pain etc.). **intermittence, intermission** (-shən), *n.* the act or state of intermitting; temporary cessation of a paroxysm; a pause; an interlude; an interval between acts of a play etc.; (*N Am.*) school break. **intermittent**, *a.* ceasing or relaxing at intervals, periodic. *n.* an intermittent fever. **intermittently**, *adv.*
intermix (intəmiks'), *v.t.* to mix together, to intermingle. *v.i.* to be intermingled. **intermixture** (-chə), *n.*
intermolecular (intəmələk'ūlə), *a.* between molecules.
intern (intœn'), *v.t.* to send to or confine in the interior of a country; to keep under restraint; to confine aliens (in time of war), political opponents, prisoners of war etc. *n.* (in'-), (also **interne**, (*N Am.*)) an assistant surgeon or physician resident in a hospital. **internee** (-nē'), *n.* one who is interned. **internment**, *n.* **internment camp**, *n.* a camp for the internment of aliens in time of war, or of prisoners of war. **internship** (in'-), *n.* (*N Am.*).
internal (intœ'nəl), *a.* situated in the inside; of or pertaining to the inside, inherent, intrinsic; domestic as opp. to foreign; pertaining to the inner being, inward. *n.* a medical examination of the vagina or uterus; (*pl.*) the inner parts. **internal-combustion engine**, *n.* an engine in which mechanical energy is produced by the combustion or explosion of a mixture of air and gas, oil-vapour etc. in its cylinder. **internal evidence**, *n.* evidence derived from what the thing itself contains. **internality** (-nal'-), *n.* **internalize, -ise**, *v.t.* to assimilate (an idea etc.) into one's outlook, to contain (an emotion) within oneself instead of expressing it. **internally**, *adv.*
internat., (*abbr.*) international.
international (intənash'ənəl), *a.* pertaining to, subsisting or carried on between, or mutually affecting different nations; known or famous in more than one country. *n.* a match between two national teams; one who has taken part in such a match; (**International**) any of three international socialist organizations intended to promote the joint political action of the working classes throughout the world; a member of these. **International Court of Justice**, the principal judicial organ of the United Nations. **International Date Line**, a line roughly along the 180th meridian, east and west of which the date is one day different. **International Labour Organization**, an independent body established with the object of raising the standard of labour conditions throughout the world. **International Phonetic Alphabet**, a series of symbols intended to give an accurate representation of human speech sounds. **international law**, *n.* an accepted system of laws or jurisprudence regulating intercourse between nations. **Internationale** (-nahl), *n.* the French socialist hymn adopted by the International. **internationalism**, *n.* the promotion of community of interests between nations; international interests or outlook. **internationalist**, *n.* **internationality** (-nal'-), *n.* **internationalize, -ise**, *v.t.* to make international; to bring under the joint protection or control of different nations. **internationalization**, *n.* **internationally**, *adv.*
interne (in'tœn), *n.* an intern.
internecine (intənē'sīn), *a.* mutually destructive; involving conflict within a group.
internee INTERN.
internist (intœ'nist), *n.* a specialist in internal medicine.
internment INTERN.
interoceptive (intərōsep'tiv), *a.* of or being stimuli developing inside the viscera.
interpellate (intœ'pəlāt), *v.t.* to interrogate, esp. to interrupt discussion etc. in order to demand a statement or explanation from (a minister). **interpellant**, *n.*, *a.* **interpellation**, *n.* **interpellator**, *n.*
interpenetrate (intəpen'itrāt), *v.t.* to penetrate thoroughly, to permeate; to penetrate (each other). *v.i.* to penetrate each other. **interpenetration**, *n.* **interpenetrative**, *a.*
interplanetary (intəplan'itəri), *a.* pertaining to the regions or to communication between planets.
interplay (in'təplā), *n.* reciprocal action between parts or things.
interpleader (in'təplēdə), *n.* a suit to determine the claims of two parties to money or property, so that a third party, on whom the claim is made, may know which party to pay. **interplead**, *v.i.*
Interpol (in'təpol), *n.* the *Inter*national *Pol*ice Commission, that ensures cooperation between police forces in the suppression and detection of crime.
interpolate (intœ'pəlāt), *v.t.* to insert (esp. a spurious word or passage) in (a book or document); to insert or intercalate; (*Math.*) to estimate (values of a function) between two values already known. *v.i.* to make interpolations. **interpolation**, *n.* **interpolator**, *n.*
interpose (intəpōz'), *v.t.* to place between or among; to put forward (as an objection, veto, obstruction etc.) by way of intervention or interference. *v.i.* to intervene, to intercede, to mediate between; to remark by way of interruption, to interrupt. **interposal**, *n.* **interposer**, *n.* **interposition** (intəpəzish'ən), *n.*
interpret (intœ'prit), *v.t.* to explain the meaning of; to translate esp. orally from one language into another; to expound, to make intelligible; to find out the meaning of, to construe or understand (in a particular way); to represent the meaning of, or one's idea of, artistically. *v.i.* to act as an interpreter. **interpretable**, *a.* **interpretation**, *n.* **interpretative** (-prətātiv), *a.* **interpretatively**, *adv.* **interpreter**, *n.* one who interprets, esp. one employed to translate orally to persons speaking a foreign language. **interpretership**, *n.* **interpretress** (-tris), *n. fem.*
interracial (intərā'shəl), *a.* between different races.
interregnum (intəreg'nəm), *n.* (*pl.* **-nums, -na**) the period between two reigns, ministries or governments; a suspension or interruption of normal authority, succession etc.
interrelation (intərəlā'shən), *n.* mutual relation. **interrelationship**, *n.*
interrog., (*abbr.*) interrogative.
interrogate (intœ'rəgāt), *v.t.* to put questions to; to examine in a formal manner. *v.i.* to ask questions. **interrogable**, *a.* **interrogation**, *n.* the act of interrogating; a question put; (*Gram.*) the sign (?) marking a question. **interrogational**, *a.* **interrogative** (-rog'-), *a.* denoting a question; expressed in the form or having the character of a question. *n.* (*Gram.*) a word used in asking questions. **interrogatively**, *adv.* **interrogator**, *n.* **interrogatory** (-rog'-), *a.* interrogative. *n.* a question; an inquiry; (*Law*) a question or set of questions put formally to a defendant etc.
interrupt (intərŭpt'), *v.t.* to stop or obstruct by breaking in

intersect | 435 | **intrapreneur**

upon; to break the continuity of; to cause a break or gap in; to obstruct (a view etc.); (*coll.*) to disturb. *v.i.* to make interruption. **interruptedly,** *adv.* **interrupter,** *n.* **interruptible,** *a.* **interruption,** *n.* **interruptive, -tory,** *a.* **interruptively,** *adv.*
intersect (intəsekt'), *v.t.* to pass or cut across; to divide by cutting or passing across. *v.i.* to cut or cross each other. **intersection,** *n.* the act or state of intersecting; (in'-), a crossroads; (*Geom.*) the point or line in which two lines or planes cut each other. **intersectional,** *a.*
intersex (in'təseks), *n.* an individual developing certain characters of the opposite sex. **intersexual** (sek'-), *a.* intermediate in sexual characters between male and female. **intersexuality** (-al'-), *n.* **intersexually,** *adv.*
interspace¹ (in'təspās), *n.* intervening space; an interval between two things or occurrences. **interspatial** (-shəl), *a.* **interspatially,** *adv.*
interspace² (intəspās'), *v.t.* to put a space or spaces between; to fill the intervals between.
interspecific (intəspisif'ik), *a.* subsisting between different species.
intersperse (intəspœs'), *v.t.* to scatter here and there (among etc.); to diversify or variegate (with scattered objects, colours etc.). **interspersion** (-shən), *n.*
interstate (in'təstāt), *a.* (*N Am.*) subsisting, maintained or carried on between states. *n.* (*N Am.*) an interstate highway.
interstellar (intəstel'ə), **-ary,** *a.* situated between or passing through the regions between the stars.
interstice (intœ'stis), *n.* a space, opening, crevice etc., between things near together or between the component parts of a body. **interstitial** (-sti'shəl), *a.*
interstratify (intəstrat'ifi), *v.t.* (*usu. in p.p.*) (*Geol.*) to stratify between or among other strata. **interstratification** (-fi-), *n.*
intertidal (intəti'dəl), *a.* situated between the low-water and high-water marks.
intertie (in'təti), *n.* a horizontal timber framed between two posts to tie them together; a binding joist.
intertribal (intətri'bəl), *a.* occurring or carried on between different tribes.
intertrigo (intətri'gō), *n.* (*pl.* **-gos**) inflammation of the skin through the rubbing of two parts together.
intertwine (intətwin'), *v.t.* to entwine or twist together. *v.i.* to be twisted together. **intertwinement,** *n.* an intertwining. **intertwiningly,** *adv.*
interurban (intərœ'bən), *a.* between cities.
interval (in'təvəl), *n.* intermediate space, distance or time; a break, a gap; a pause or interlude; the extent of difference between two things, persons etc.; the difference of pitch between two sounds; the break between scenes or acts of a play etc. *v.t.* to separate or interrupt at intervals. **at intervals,** from time to time; with spaces in between. **intervallic** (-val'-), *a.*
intervene (intəvēn'), *v.i.* to come in as an extraneous feature or thing; to come or be situated (between); to occur between points of time or events; to happen or break in so as to interrupt or disturb, to interfere, to interpose; to practise intervention. **intervener,** *n.* **intervenient,** *a.* **intervention** (-ven'-), *n.* the act of intervening; violating a sovereign state's independence by interfering in its domestic or external affairs; the practice of the EEC of buying and storing surplus products when the market price is low; the action of a central bank in buying large quantities of a currency to prevent its international value from falling. **interventionism,** *n.* **interventionist,** *n.*
interview (in'təvū), *n.* a formal meeting between some person and a press representative employed to obtain information or opinions for publication; the article describing this or recording the result; a similar meeting broadcast on television or radio; a meeting in which an employer questions a candidate for a job in order to test the candidate's suitability. *v.t.* to have an interview with, esp. for purposes of publication or broadcasting or to test a candidate's suitability for a post. **interviewee** (-ē'), *n.* **interviewer,** *n.*
interwar (intəwaw'), *a.* occurring in the period between World Wars I and II.
interweave (intəwēv'), *v.t.* (*p.p.* **-woven**) to weave together; to blend or mingle closely together.
interwork (intəwœk'), *v.t.* (*past, p.p.* **-wrought** (-rawt'), **-worked**) to work things together or into each other. *v.i.* to work reciprocally, to interact.
intestacy (intes'təsi), *n.* lack of a will or testament.
intestate (intes'tāt), *a.* dying without having made a will; not disposed of by will.
intestine (intes'tin), *a.* internal, domestic. *n.* (*usu. pl.*) the long membranous tube from the stomach to the anus; the bowels, the guts. **intestinal** (-tī'-, -tes'-), *a.* pertaining to the intestines. **intestinally,** *adv.*
intifadeh, intifada (intifah'də), *n.* an uprising esp. of Palestinians in Israel.
intimate¹ (in'timāt), *v.t.* to make known, to announce; to indicate, to hint. **intimation,** *n.*
intimate² (in'timət), *a.* close in friendship or fellowship; familiar, confidential; private, personal; having an atmosphere conducive to close personal relationships; pertaining to one's inner being; having a profound knowledge of; having sexual relations. *n.* a familiar friend or associate. **intimacy** (-si), *n.* **intimately,** *adv.*
intimidate (intim'idāt), *v.t.* to frighten, to cow; to deter (from an action or doing). **intimidation,** *n.* **intimidator,** *n.* **intimidatory,** *a.*
intituled (intit'ūld), *a.* chiefly of Acts of Parliament, entitled.
into (in'tu), *prep.* expressing motion or direction towards the interior, or change from one state to another; entrance; penetration; insertion; inclusion or comprehension; (*sl.*) very keen on; indicating the dividend in division.
intolerable (intol'ərəbl), *a.* not tolerable, unendurable. **intolerableness,** *n.* **intolerably,** *adv.*
intolerant (intol'ərənt), *a.* not tolerant (of); not enduring or allowing difference of opinion, teaching or worship; bigoted. **intolerance, intoleration,** *n.* **intolerantly,** *adv.*
intone (intōn'), *v.i., v.t.* to recite or chant in a monotone; to give a musical tone to one's delivery. **intonate** (in'-), *v.i., v.t.* to intone. **intonation** (intənā'shən), *n.* modulation of the voice, accent; intoning; the opening phrase of a plain-song melody; the mode of producing sound from a voice or an instrument, esp. as regards correctness of pitch.
in toto (in tō'tō), *adv.* completely.
intoxicate (intok'sikāt), *v.t.* to make drunk; to excite to enthusiasm; to make delirious, as with joy; (*Med.*) to poison. **intoxicant,** *n., a.* **intoxicatedly,** *adv.* **intoxicating,** *a.* tending to intoxicate. **intoxicatingly,** *adv.* **intoxication,** *n.*
intra- (intrə-), *pref.* within, on the inside.
intracommunity (intrəkəmū'niti), *a.* situated or occurring inside the European Economic Community (EEC).
intractable (intrak'təbl), *a.* unmanageable, indocile, refractory. **intractability** (-bil'-), **-ableness,** *n.* **intractably,** *adv.*
intrados (intrā'dos), *n.* (*pl.* **intrados, intradoses**) the under surface or curve of an arch.
intramural (intrəmū'rəl), *a.* taking place within the confines of an educational establishment.
intransigent (intran'sijənt), *a.* irreconcilable; uncompromising, inflexible. *n.* an irreconcilable; an uncompromising adherent of any creed (political, artistic etc.). **intransigency,** *n.* **intransigently,** *adv.*
intransitive (intran'sitiv), *a.* (*Gram.*) not having a direct object. *n.* an intransitive verb. **intransitively,** *adv.* **intransitiveness, intransitivity,** *n.*
intrapreneur (intrəprənœ'), *n.* one who initiates or manages

a new business or division within an existing firm. **intrapreneurial**, *a*. **intrapreneurship**, *n*.
intraspecific (intrəspisif'ik), *a*. relating to the internal development of a species.
intra-urban (intræ'bən), *a*. existing or carried on within a city.
intrauterine (intrəū'tərin), *a*. situated inside the uterus. **intrauterine device**, *n*. a metal or plastic coil, loop or ring, placed in the uterus to prevent conception.
intravascular (intrəvǎs'kūlə), *a*. situated or occurring within a vessel, esp. a blood-vessel.
intravenous (intrəvē'nəs), *a*. into a vein or veins. **intravenously**, *adv*.
intrepid (intrep'id), *a*. fearless, brave, bold. **intrepidity** (-pid'-), *n*. **intrepidly**, *adv*.
intricate (in'trikət), *a*. entangled, involved, complicated; obscure, complex. **intricacy** (-si), *n*. **intricately**, *adv*.
intrigue (intrēg'), *v.t.* to plot or scheme to effect some object by underhand means; to carry on a secret love affair. *v.t.* to perplex; to fascinate. *n*. (-trēg', in'-), the act of intriguing; a plot to effect some object by underhand means; secret love; a liaison. **intriguer**, *n*. **intriguingly**, *adv*.
intrinsic (intrin'sik), *a*. inward, inherent; belonging to the nature of a thing; essential; genuine; situated within the body. **intrinsically**, *adv*.
intro (in'trō), *n*. (*pl*. **intros**) (*coll*.) introduction.
intro- (intrō-), *pref*. in, into; inward.
introduce (intrədūs'), *v.t.* to bring or lead in; to usher in; to insert; to bring into use or notice; to cause (a person) to discover; to make known, esp. (a person) in a formal way (to another); to bring before the public; to bring out into society; to bring before Parliament; to preface; to present (a programme etc.). **introducer**, *n*. **introduction** (intrədŭk'shən), *n*. the act of introducing; formal presentation of a person to another; a preface or preliminary discourse; an elementary treatise. **letter of introduction**, a letter introducing a friend to a third person. **introductive, -tory**, *a*. **introductively, -torily**, *adv*.
introit (in'troit), *n*. a psalm or antiphon sung or recited as the priest approaches the altar to begin the Mass.
introject (intrəjekt'), *v.t.* to assimilate unconsciously into one's personality. **introjection**, *n*.
intromit (intrəmit'), *v.t.* (*past, p.p.* **intromitted**) to send in; to insert. **intromission** (-shən), *n*. **intromittent**, *a*.
intron (in'tron), *n*. a section of a nucleic acid not coding information for protein synthesis.
introrse (intraws'), *a*. (*Bot*.) turned towards the axis.
introspect (intrəspekt'), *v.t.* to look into or within; to examine one's own mind and its working. **introspection**, *n*. **introspectionist**, *n*. one who introspects; one who employs introspection as a psychological instrument. **introspective**, *a*. **introspectively**, *adv*. **introspectiveness**, *n*.
introvert¹ (intrəvœt'), *v.t.* to turn inwards; to turn (the mind or thoughts) inwards; to turn (an organ or a part) in upon itself; to turn inside out. **introversible**, *a*. **introversion** (-shən), *n*.
introvert² (in'trəvœt), *n*. a part or organ that is introverted or introversible; a person who is interested chiefly in his or her own mental processes and standing with other people, this making the person shy and unsociable. **introversive, -vertive**, *a*.
intrude (introod'), *v.t.* to thrust or force (into); to force (volcanic rock etc.) into sedimentary strata. *v.i.* to thrust oneself or force one's way (into); to force oneself (upon others); to enter without invitation. **intruder**, *n*. someone who intrudes, esp. a person discovered on someone's property with suspected criminal intent; an aircraft encroaching on restricted airspace.
intrusion (intrū'zhən), *n*. the act of intruding; an encroachment; (*Geol*.) the penetration of volcanic rocks into sedimentary strata; (*Law*) unlawful entry by a stranger upon lands or tenements, invasion, usurpation.

intrusive (intrū'siv), *a*. tending to intrude; entering without invitation or welcome; of rocks, which have forced their way into sedimentary strata. **intrusively**, *adv*. **intrusiveness**, *n*.
intubate (in'tūbāt), *v.t.* to insert a tube into (e.g. the larynx). **intubation**, *n*. **intubator**, *n*. an instrument for inserting a tube thus.
intuition (intūish'ən), *n*. immediate perception by the mind without reasoning; the power of the mind for such perception; instinctive knowledge; a truth so perceived. **intuit** (-tū'it), *v.t.* to know by intuition. *v.i.* to acquire knowledge by means of intuition. **intuitional**, *a*. **intuitionalism**, *n*. the doctrine that the perception of truth, or of certain truths, is by intuition. **intuitionalist**, *n*. **intuitionism**, *n*. intuitionalism; an extreme form of this which holds that the objects of sense-perception are known intuitively as real. **intuitionist**, *n*. **intuitive** (-tū'-), *a*. perceived by intuition; perceiving by intuition; seeing immediately and clearly. **intuitively**, *adv*. **intuitiveness**, *n*. **intuitivism**, *n*. the doctrine that ideas of right and wrong are intuitive.
intumesce (intūmes'), *v.i.* to swell up, to become tumid. **intumescence**, *n*. **intumescent**, *a*.
intussuscept (intəsəsept'), *v.t.* (*Path*.) to receive within itself or another part; to invaginate. **intussusception**, *n*. the taking in of anything; the reception of foreign matter (as food) by an organism and its conversion into living tissue; the accidental insertion or protrusion of an upper segment of the bowels into a lower. **intussusceptive**, *a*.
Inuit INNUIT.
inundate (in'əndāt), *v.t.* to overflow; to flood; to submerge; to deluge; to overwhelm. **inundation**, *n*. a flood, a deluge.
inure (inūə'), *v.t.* to accustom, to habituate, to harden (to). *v.i.* (*Law*) to come into operation; to take or have effect. **inurement**, *n*.
inurn (inœn'), *v.t.* to place in a cinerary urn; to bury.
in utero (in ū'tərō), in the uterus. [L]
in vacuo (in vak'ūō), in a vacuum. [L]
invade (invǎd'), *v.t.* to enter (a country) as an enemy; to enter by force; to assail; to encroach on, to violate. *v.i.* to make an invasion. **invader**, *n*.
invaginate (invaj'ināt), *v.t.* to put into or as into a sheath; to introvert or turn (a tubular sheath) upon itself. **invaginable**, *a*. **invagination**, *n*.
invalid¹ (inval'id), *a*. of no force, weight or cogency; null. **invalidate**, *v.t.* to weaken or destroy the validity of, to render not valid, to annul. **invalidation**, *n*. **invalidator**, *n*. **invalidity** (-lid'-), **invalidness**, *n*. **invalidly**, *adv*.
invalid² (in'vəlid), *a*. infirm or disabled through ill-health or injury. *n*. an infirm or disabled person. *v.t.* to disable by illness or injury; to register or discharge as unfit for military or naval duty on account of illness etc. **invalidism**, *n*. chronic ill health, esp. neurotic. **invalidity** (-lid'-), *n*. invalidity pension, benefit, *n*. money paid by the government to someone who is chronically ill or disabled.
invaluable (inval'ūəbl), *a*. precious above estimation; priceless. **invaluably**, *adv*.
Invar® (in'vah), *n*. a nickel-steel alloy with small coefficient of expansion.
invariable (inveə'riəbl), *a*. not variable, uniform; not liable to change; (*Math*.) fixed, constant. *n*. (*Math*.) a constant quantity. **invariability** (-bil'-), **-ableness**, *n*. **invariably**, *adv*. **invariant**, *n., a*. (*Math*.) (that) which remains fixed and unchanged though its constituents may vary.
invasion (invā'zhən), *n*. the act of invading; a hostile attack upon or entrance into the territory of others; infringement, violation; the approach or assault of anything dangerous or pernicious. **invasive** (-siv), *a*.
invective (invek'tiv), *n*. a violent expression of censure or abuse; vituperation. *a*. abusive. **invectively**, *adv*.
inveigh (invā'), *v.t.* to utter or make use of invectives; to

declaim censoriously and abusively (against). **inveigher**, n.
inveigle (invē'gl, -vā'-), v.t. to wheedle, to entrap (into an action, deed etc.) by cajolery. **inveiglement**, n. **inveigler**, n.
invent (invent'), v.t. to devise or contrive (a new means, instrument etc.); to concoct, to fabricate. **invention**, n. the act of inventing; the production of something new; the faculty or power of inventing, inventiveness; that which is invented, a contrivance; a fabrication, a fiction; a short piece of music, usu. in double counterpoint. **Invention of the Cross**, the finding of the true Cross by Helena, the mother of Constantine the Great, AD 326; the festival (3 May) commemorating this. **inventive**, a. able to invent; resourceful, ingenious; imaginative; characterized by creative skill. **inventively**, adv. **inventiveness**, n. **inventor**, n. **inventress** (-tris), n. fem.
inventory (in'vəntəri), n. a detailed list or catalogue of goods and chattels; the articles enumerated in such a list; (chiefly N Am.) (the quantity or value of) a firm's current assets in terms of raw materials and stock; the material in a nuclear reactor. v.t. to enter in an inventory; to make a list, catalogue or schedule of. **inventorial** (-taw'ri-), a. **inventorially**, adv.
Inverness, Inverness cape (invənes'), n. a kind of sleeveless cloak with a cape hanging loosely over the shoulders.
inverse (invœs'. in'-), a. opposite in order or relation; contrary, inverted. n. that which is inverted; the direct opposite of; (Math.) the result of inversion. **inversely**, adv.
inversion (invœ'shən), n. the act of inverting; reversal of order, place or relation; (Gram.) reversal of the natural order of words in a sentence; (Mus.) the process or result of altering or reversing the relative position of the elements of a chord etc.; the rearrangement of molecular structure taking place when starch, dextrin or sugar is boiled with a dilute acid; (Math.) the operation of changing the order of the terms, so that the antecedent takes the place of the consequent and the reverse in both ratios; the assumption of the characteristics of the other sex. **inversive**, a.
invert[1] (invœt'), v.t. to turn upside down; to place in a contrary position or order; to reverse. **inverted commas** QUOTATION MARKS. **invertedly** (-vœ'-), adv. **inverter, -tor** (-vœ'-), n. a device that converts direct current into alternating current. **invertible** (-vœ'-), a.
invert[2] (in'vœt), n. an inverted arch, esp. forming the bottom of a sewer etc.; one with inverted sexual instincts, a homosexual. **invert sugar**, n. a mixture of laevulose and dextrose. **invertase** (invœ'tāz), n. an enzyme able to convert sucrose into invert sugar.
invertebrate (invœ'tibrət), a. destitute of a backbone or vertebral column; lacking strength or firmness. n. an invertebrate animal.
invest (invest'), v.t. to clothe (esp. in the robes of office); to install ceremoniously (in an office, rank etc.); to surround, besiege; to employ (money in remunerative property, business, stocks etc.); to devote (effort etc.) to a project etc. for future rewards. v.i. to make an investment; (coll.) to spend money (as on a small purchase). **investable**, a. **investive**, a. **investment**, n. the act of laying out money; money invested; that in which money is invested; the act of surrounding or besieging. **investment trust**, n. a financial enterprise which invests its subscribers' capital in securities and distributes the net return among them. **investor**, n.
investigate (inves'tigāt), v.t. to examine or inquire into closely. v.i. to research or make investigation. **investigation**, n. **investigative, -gatory**, a. **investigator**, n.
investiture (invest'ichə), n. the act of investing, esp. the ceremonial of investing (with office, rank etc.).
investment INVEST.
inveterate (invet'ərət), a. long-established; deeply-rooted, obstinate, confirmed by long continuance; habitual; determinedly settled in a habit. **inveteracy** (-si), **inveterateness**, n. **inveterately**, adv.

invidious (invid'iəs), a. tending to incur or provoke envy or ill-will; offending through real or apparent unfairness or injustice. **invidiously**, adv. **invidiousness**, n.
invigilate (invij'ilāt), v.i. to keep a watch over students during an examination. v.t. to supervise. **invigilation**, n. **invigilator**, n.
invigorate (invig'ərāt), v.t. to give vigour or strength to; to animate. **invigoratingly**, adv. **invigoration**, n. **invigorative**, a. **invigorator**, n.
invincible (invin'sibl), a. unconquerable. **invincibility** (-bil'-), **-ibleness**, n. **invincibly**, adv.
inviolable (invī'ələbl), a. not to be violated, profaned or dishonoured; not to be broken or disturbed. **inviolability** (-bil'-), n. **inviolably**, adv. **inviolate** (-lət), a. not violated or profaned; unbroken. **inviolacy** (-si), **inviolateness**, n. **inviolately**, adv.
invisible (inviz'ibl), a. not visible; imperceptible to the eye; too small, distant, misty etc. to be seen; not recorded in published accounts; not showing in statistics; pertaining to services as opposed to goods, as *invisible earnings*. **invisible ink**, n. ink that does not show until heated or otherwise treated. **invisibility** (-bil'-), **-ibleness**, n. **invisibly**, adv.
invite[1] (invīt'), v.t. to solicit the company of (to or in); to request courteously (to do something); to solicit; to tempt; to bring on, provoke, esp. unintentionally. **invitation** (-vi-), n. the act of inviting; words, written or oral, with which one is invited. **invitee** (-tē'), n. one invited. **inviter**, n. **inviting**, a. tempting; physically attractive. **invitingly**, adv. **invitingness**, n.
invite[2] (in'vīt), n. (coll.) an invitation.
in vitro (in vit'rō), in an artificial environment outside the body. [L]
in vivo (in vē'vō), in the body. [L]
invocation (invəkā'shən), n. the act of invoking; a supplication or call, esp. to God; a petition addressed to a muse, saint etc., for help or inspiration; the calling up of a spirit by incantation. **invocable** (in'-), a. **invocatory** (-vok'-), a.
invoice (in'vois), n. a list of goods dispatched, with particulars of quantity and price, sent to a consignee. v.t. to enter (goods) in an invoice; to send an invoice to.
invoke (invōk'), v.t. to address in prayer; to solicit earnestly for assistance and protection; to call upon solemnly; to call on as a witness, to appeal to as an authority; to summon by magical means.
involucre (in'vəlookə), n. a whorl of bracts surrounding the flowers of certain Compositae and other plants. **involucral** (-loo'krəl), **involucrate** (-loo'krət), a.
involuntary (invol'əntəri), a. done unintentionally, not from choice; independent of will or volition. **involuntarily**, adv. **involuntariness**, n.
involute (in'vəloot), a. rolled up, folded; rolled inward at the margin, as certain leaves, petals etc.; complicated, involved. n. (Math.) a curve traced by the end of a string unwinding itself from another curve, which is called the *evolute*. **involuted, -lutive**, a.
involution (invəloo'shən), n. the act of involving; the state of being involved; complication, entanglement, intricacy; a rolling up or curling of parts; anything folding up or enveloping; a complicated grammatical construction; the shrinking of a bodily organ, e.g. of the uterus after pregnancy; (Math.) the act or process of raising a quantity to any power.
involve (involv'), v.t. to enwrap, to enfold or envelop (in); to entangle (in); to implicate (in); to include (in); to commit (as oneself) emotionally; to comprise as a logical or necessary consequence; to imply, to entail; to complicate, to make intricate; (Math.) to raise to any power. **involvedness**, n. **involvement**, n.
invulnerable (invŭl'nərəbl), a. incapable of being wounded

inward

or injured; proof against attack. **invulnerably,** *adv.* **invulnerability** (-bil'-), **-ableness,** *n.*
inward (in'wəd), *a.* internal; situated or being within; towards the interior, connected with the mind or soul. *adv.* inwards. **inwardly,** *adv.* internally, within; towards the centre; in one's thoughts and feelings, mentally, secretly.
inwardness, *n.* the inner quality or essence (of); the quality of being inward; the mental and spiritual nature. **inwards,** *adv.* towards the interior, internal parts or centre; in the mind or soul.
inweave (inwēv'), *v.t.* to weave in or together; to interlace (with).
inwrap (inrap'), ENWRAP.
inwrought (in'rawt, -rawt'), *a.* of a pattern etc., wrought or worked in among other things; of a fabric, adorned with work or figures.
inyala (inyah'lə), *n.* a spiral-horned S African antelope.
iodic (īod'ik), *a.* belonging to, or containing, iodine.
iodide (ī'ədid), *n.* a compound of iodine with an element or radical.
iodine (ī'ədēn, -din), *n.* a non-metallic bluish-black element, at. no. 53; chem. symbol I, yielding violet fumes when heated, and resembling bromine and chlorine in chemical properties, used in photography and for its antiseptic and disinfectant qualities.
iodize, -ise (ī'ədiz), *v.t.* to treat with iodine; to prepare with iodine. **iodization,** *n.*
iodoform (īod'əfawm), *n.* an iodine compound resembling chloroform in its antiseptic effects. **iodoformin,** *n.*
iodopsin (iədop'sin), *n.* a light-sensitive pigment in the retinal cones.
ion (ī'ən), *n.* an electrically charged atom or group of atoms formed by the loss or gain of electrons. **ion exchange,** *n.* a process by which ions are exchanged between a solution and a solid or another liquid, as used in the softening of water etc. **ionic** (īon'-), *a.* **ionic bond,** *n.* a bond within a chemical compound produced by the transfer of electrons, such that the resulting ions are held together by electrostatic attraction. **ionize, -ise,** *v.t.* to convert into an ion or ions. **ionization,** *n.*
Ionian (īō'niən), *a.* pertaining to Ionia, a district of Asia Minor, or to the Ionians. *n.* a member of the division of the Hellenic race which founded colonies on the shores of the Mediterranean and Euxine and esp. in Asia Minor. **Ionic** (īon'-), *a.* Ionian. **Ionic dialect,** the Greek dialect spoken in Ionia. **Ionic order,** *n.* one of the five orders of architecture, distinguished by the volute on both sides of the capital.
ionic, ionize, ionization ION.
ionosphere (īon'əsfiə), *n.* the region surrounding the earth at a height of from 6 miles (about 9·5 km) to about 250 miles (400 km) in which ionized layers of gas occur. **ionospheric** (-sfe'-), *a.*
iota (īō'tə), *n.* the ninth letter, ι, of the Greek alphabet; a jot, a very small quantity.
IOU, *n.* a formal acknowledgment of debt, bearing these letters. [*I owe you*]
IOW, (*abbr.*) Isle of Wight.
IPA, (*abbr.*) International Phonetic Alphabet.
ipecacuanha (ipikakūan'ə), **ipecac,** *n.* the dried root of *Cephaelis ipecacuanha*, a cinchonaceous plant from Brazil, used as an emetic and purgative. **ipecacuanhic,** *a.*
ipomoea (ipəmē'ə), *n.* a genus of Convolvulaceae, with many species.
ipse dixit (ip'si dik'sit), *n.* a mere assertion; a dogmatic statement. [L, he himself has said it]
ipso facto (ip'sō fak'tō), by that very fact. [L]
IQ, (*abbr.*) Intelligence Quotient.
IR, (*abbr.*) Inland Revenue.
Ir, (*chem. symbol*) iridium.
Ir., (*abbr.*) Ireland; Irish.
IRA, (*abbr.*) Irish Republican Army.

iron

Iranian (irā'niən), *a.* of or belonging to Iran in SW Asia, formerly Persia; pertaining to the inhabitants or language of Iran. *n.* a member of the Iranian race; a native of Iran; a branch of the Indo-European family of languages including Persian; the modern Persian language.
Iraqi (irah'ki), *a.*, *n.* (pertaining to) a native or inhabitant of Iraq; (pertaining to) the form of Arabic spoken in Iraq.
irascible (iras'ibl), *a.* easily excited to anger; passionate, irritable. **irascibility** (-bil'-), **-ibleness,** *n.* **irascibly,** *adv.*
irate (īrāt'), *a.* angry, enraged.
IRBM, (*abbr.*) Intermediate Range Ballistic Missile.
ire (īə), *n.* anger, passion. **ireful,** *a.* **irefully,** *adv.*
irenic (īrē'nik, -ren'-), **-ical,** *a.* pacific; promoting peace. **irenicon** (-kon), EIRENICON.
Iricism (ī'risizm), IRISHISM under IRISH.
iridescent (irides'nt), *a.* exhibiting changing colours like those of the rainbow. **iridescence,** *n.* **iridescently,** *adv.*
iridium (irid'iəm), *n.* a shining white metallic element belonging to the platinum group, at. no. 77; chem. symbol Ir. **iridize, -ise** (ī'-), *v.t.* to tip (a pen) with iridium.
iridology (iridol'əji), *n.* a diagnostic technique in alternative medicine involving studying the iris of the eye. **iridologist,** *n.*
iris (ī'ris), *n.* (*pl.* **irises, irides** (-dēz)) (*Gr. Myth.*) the rainbow; an appearance resembling the rainbow; the circular coloured membrane or curtain surrounding the pupil of the eye; a genus of plants of the family Iridaceae, with tuberous roots, sword-shaped leaves, and large variously-coloured flowers; a flower of this genus, a fleur-de-lis. **iris diaphragm,** *n.* an adjustable diaphragm regulating the entry of light into an optical instrument. **irised,** *a.* containing colours like the rainbow.
Irish (īə'rish), *a.* of or pertaining to Ireland or its inhabitants; like an Irishman; ludicrous, illogical. *n.* a native of Ireland; the Irish language; (*N Am. sl.*) temper, contentiousness; (*collect.*) the people of Ireland. **Irish coffee,** *n.* a drink made of sweetened coffee mixed with Irish whiskey and topped with cream. **Irishman, -woman,** *n.* a native of Ireland; one of Irish race. **Irish moss** CARRAGEEN. **Irish stew,** *n.* a stew of vegetables and meat boiled together. **Irishism,** *n.* a mode of expression or idiom peculiar to the Irish, esp. a 'bull'. **Irishize, -ise,** *v.t.*
iritis (īri'tis), *n.* inflammation of the iris of the eye.
irk (œk), *v.t.* to tire, to bore; to annoy. **irksome** (-səm), *a.* wearisome, tedious, annoying. **irksomely,** *adv.* **irksomeness,** *n.*
iron (ī'ən), *n.* a malleable tenacious metallic element, at. no. 26; chem. symbol Fe, used for tools etc.; an article, tool, utensil etc., made of iron; an implement for smoothing clothes; a metal-headed golf club used for lofting; (*pl.*) fetters. *a.* made or composed of iron; like iron, robust, strong, inflexible, or unyielding, merciless. *v.t.* to smooth with a smoothing-iron. **flat-iron, smoothing-iron,** an iron implement that is heated for smoothing cloth. **in irons,** in fetters. **to have (too) many irons in the fire,** to be attempting or dealing with (too) many projects at the same time. **to iron out,** *v.t.* to correct (defects etc.); to find a solution to (problems etc.). **to pump iron** PUMP. **to strike while the iron is hot,** to seize a favourable opportunity without delay. **iron age,** *n.* the late prehistoric age when weapons and many implements began to be made of iron; (*Gr. Myth.*) the last of the four ages of the world, in which oppression and vice prevailed. **iron-bark,** *n.* an Australian eucalyptus with a hard, firm bark. **iron-bound,** *a.* bound with iron; of a coast, surrounded with rocks; unyielding, hard and fast. **ironclad,** *n.* a warship plated with iron. *a.* covered or protected with iron. **Iron Cross,** *n.* a German war-medal. **Iron Curtain,** *n.* the imperceptible barrier to communication formerly existing between the USSR with its satellites and the rest of Europe; (**iron curtain**) any similar barrier to communication. **iron-founder,** *n.* one who makes iron castings. **iron-**

foundry, *n.* **iron-grey**, *a.*, *n.* (of) a grey colour like that of iron freshly broken. **iron hand**, *n.* strict control, often tyranny. **the iron fist** or **hand in the velvet glove**, strict control which is at first concealed. **iron horse**, *n.* (*dated*, *coll.*) a railway locomotive. **iron lung**, *n.* a mechanical device employed for maintaining or assisting respiration. **iron maiden**, *n.* an instrument of torture in the form of a lidded box lined with iron spikes. **iron-master**, *n.* a manufacturer of iron. **ironmonger**, *n.* one who deals in ironware or hardware. **ironmongery**, *n.* **iron rations**, *n.pl.* complete emergency rations packed in a sealed case. **ironstone**, *n.* an iron-ore containing oxygen and silica. **ironware**, *n.* goods made of iron, hardware. **iron-wood**, *n.* the popular name given to several very hard and heavy woods. **ironwork**, *n.* anything made of iron; (*pl.*, *often sing. in constr.*) an establishment where iron is manufactured, wrought or cast. **ironer**, *n.* **ironing**, *n.* smoothing with an iron; clothes to be ironed. **ironing-board**, *n.* **irony**[1], *a.* consisting of, containing, or resembling iron.
iron., (*abbr.*) ironically.
ironic, ironical IRONY.
Ironside (ī'ənsīd), **Ironsides**, *n.* one of Cromwell's troopers; a hardy veteran.
irony[1] IRON.
irony[2] (ī'rəni), *n.* an expression intended to convey the opposite to the literal meaning; the use of such expressions; language having a meaning or implication, for those who understand it, different from the ostensible one or that of which the speaker is conscious. **Socratic irony** SOCRATIC. **ironic** (īron'-), **-ical**, *a.* **ironically**, *adv.* **ironist**, *n.* **ironize, -ise**, *v.i.*
irradiate (irā'diāt), *v.t.* to shed light upon; to light up (a subject etc.); to brighten up (a face, expression etc.); to subject to sunlight or ultraviolet rays; to expose food to low levels of gamma radiation in order to sterilize and preserve it. **irradiant**, *a.* **irradiance**, **irradiation**, *n.* **irradiative**, *a.*
irrational (irash'ənəl), *a.* incapable of reasoning; illogical, contrary to reason, absurd; not expressible by a whole number or common fraction. *n.* an irrational number. **irrationalism**, *n.* **irrationalist**, *n.* **irrationality** (-nal'-), *n.* **irrationalize, -ise**, *v.t.* **irrationally**, *adv.*
irreclaimable (irəklā'məbl), *a.* incapable of being reclaimed; obstinate, inveterate. **irreclaimability** (-bil'-), *n.* **irreclaimably**, *adv.*
irreconcilable (irekənsī'ləbl), *a.* incapable of being reconciled; implacably hostile; incompatible, inconsistent, incongruous. *n.* one who cannot be reconciled, appeased or satisfied. **irreconcilability** (-bil'-), **-ableness**, *n.* **irreconcilably**, *adv.*
irrecoverable (irikŭv'ərəbl), *a.* that cannot be recovered; irreparable. **irrecoverableness**, *n.* **irrecoverably**, *adv.*
irredeemable (iridē'məbl), *a.* not terminable by payment of the principal (as an annuity); not convertible into cash (as a banknote); irreclaimable; beyond redemption or improvement. **irredeemability** (-bil'-), **-ableness**, *n.* **irredeemably**, *adv.*
irredentist (iriden'tist), *n.* one of a party formed about 1878 to bring about the inclusion of all Italian-speaking districts in the kingdom of Italy. **irredentism**, *n.*
irreducible (iridū'sibl), *a.* not reducible; not to be lessened; not to be brought to a required condition etc.; (*Math.*) not giving way to treatment; (*Math.*) not to be simplified. **irreducibility** (-bil'-), **-ibleness**, *n.* **irreducibly**, *adv.*
irrefragable (iref'rəgəbl), *a.* incapable of being refuted; undeniable. **irrefragability** (-bil'-), **-ableness**, *n.* **irrefragably**, *adv.*
irrefrangible (irifran'jibl), *a.* inviolable; not susceptible of refraction.
irrefutable (irifū'təbl), *a.* incapable of being refuted. **irrefutability** (-bil'-), *n.* **irrefutably**, *adv.*
irreg., (*abbr.*) irregular, irregularly.

irregular (ireg'ūlə), *a.* not according to rule or established principles or custom; lawless, disorderly; uneven; abnormal, asymmetrical; not occurring at expected intervals; (*Gram.*) deviating from the common form in inflection; not belonging to the regular army. *n.* an irregular soldier. **irregularity** (-la'-), *n.* **irregularly**, *adv.*
irrelevant (irel'əvənt), *a.* not applicable or pertinent, not to the point; having no application (to the matter in hand). **irrelevance, -ancy**, *n.* **irrelevantly**, *adv.*
irreligion (irəlij'ən), *n.* indifference or hostility to religion. **irreligionist**, *n.* **irreligious**, *a.* **irreligiously**, *adv.* **irreligiousness**, *n.*
irremediable (irəmē'diəbl), *a.* incurable, irreparable; incapable of being remedied or corrected. **irremediableness**, *n.* **irremediably**, *adv.*
irremissible (irəmis'ibl), *a.* that cannot be remitted or pardoned. **irremissibility** (-bil'-), *n.* **irremissibly**, *adv.*
irremovable (irəmoo'vəbl), *a.* that cannot be removed or displaced, permanent, immovable. **irremovability** (-bil'-), *n.* **irremovably**, *adv.*
irreparable (irep'ərəbl), *a.* incapable of being repaired, remedied or restored. **irreparableness, -ability** (-bil'-), *n.* **irreparably**, *adv.*
irreplaceable (irəplā'səbl), *a.* not to be made good in case of loss. **irreplaceably**, *adv.*
irrepressible (iripres'ibl), *a.* not to be repressed; always lively and good-humoured. **irrepressibility** (-bil'-), *n.* **irrepressibly**, *adv.*
irreproachable (iriprō'chəbl), *a.* blameless, faultless. **irreproachability** (-bil'-), **-ableness**, *n.* **irreproachably**, *adv.*
irresistible (irəzis'tibl), *a.* that cannot be resisted; not to be withstood; extremely attractive or alluring. **irresistibility** (-bil'-), **-ibleness**, *n.* **irresistibly**, *adv.*
irresolute (irez'əloot), *a.* not resolute; undecided, hesitating. **irresolutely**, *adv.* **irresoluteness, -lution**, *n.*
irrespective (irispek'tiv), *a.* regardless of, without reference to; irrespectively. **irrespectively**, *adv.* without regard to circumstances or conditions.
irresponsible (irispon'sibl), *a.* not responsible; not trustworthy; performed or acting without a proper sense of responsibility; lacking the capacity to bear responsibility. **irresponsibility** (-bil'-), *n.* **irresponsibly**, *adv.*
irresponsive (irispon'siv), *a.* not responsive (to). **irresponsiveness**, *n.*
irretrievable (iritrē'vəbl), *a.* not to be retrieved; irreparable. **irretrievability** (-bil'-), *n.* **irretrievably**, *adv.*
irreverent (irev'ərənt), *a.* lacking in reverence; disrespectful; proceeding from irreverence. **irreverence**, *n.* **irreverential** (-ren'shəl), *a.* **irreverently**, *adv.*
irreversible (irivœ'sibl), *a.* not reversible; irrevocable. **irreversibility** (-bil'-), **-ibleness**, *n.* **irreversibly**, *adv.*
irrevocable (irev'əkəbl), *a.* incapable of being revoked or altered, unalterable. **irrevocability** (-bil'-), **-ableness**, *n.* **irrevocably**, *adv.*
irrigate (i'rigāt), *v.t.* to water (land) by causing a stream to flow over it; of streams, to supply (land) with water; to moisten (a wound etc.) with a continuous jet or stream of antiseptic fluid; to refresh or fertilize the mind as with a stream. **irrigable**, *a.* **irrigative**, *a.* **irrigation**, *n.* **irrigator**, *n.*
irritate (i'ritāt), *v.t.* to excite to impatience or ill-temper; to fret, to annoy, to exasperate; to stir up, to excite; to cause an uneasy sensation in (the skin, an organ etc.); to stimulate (an organ) artificially. **irritable**, *a.* easily provoked, fretful; easily inflamed or made painful, highly sensitive; of nerves, muscles etc., responsive to artificial stimulation. **irritability** (-bil'-), **-ableness**, *n.* **irritably**, *adv.* **irritancy**, *n.* **irritant**, *n.*, *a.* **irritation**, *n.* **irritative**, *a.*
irruption (irŭp'shən), *n.* a bursting in; a sudden invasion or incursion. **irrupt**, **irruptive**, *a.*
Is., (*abbr.*) Isaiah; Island(s).
is (iz), *3rd. pers. sing. pres. ind.* [see AM, BE]
is- IS(O)-.

isagogic (isəgoj'ik), *a.* introductory. *n.pl.* preliminary investigation regarding the Scriptures, the department of biblical study concerned with literary history, authorship etc.

isatin (ī'sətin), *n.* a compound obtained by oxidizing indigo, crystallizing in yellowish-red prisms.

ISBN, (*abbr.*) International Standard Book Number.

ischaemia, (*esp. N Am.*) **ischemia** (iskē'miə), *n.* a shortage of blood in part of the body. **ischaemic,** *a.*

ischium (is'kiəm), *n.* (*pl.* **ischia** (-ə)) one of the posterior bones of the pelvic girdle. **ischial,** *a.*

-ise -IZE.

isenergic (īsənœ'jik), *a.* (*Phys.*) of or indicating equal energy.

isentropic (īsəntrop'ik), *a.* having equal entropy.

-ish (-ish), *suf.* of the nature of, pertaining to, as in *childish, English, outlandish*; rather, somewhat, as in *reddish, yellowish.*

Ishmael (ish'māl), *n.* an outcast (Gen. xvi.12); one whose hand is against every person. **Ishmaelite** (-īt), *n.* a descendant of Ishmael; one at war against society. **Ishmaelitish,** *a.*

isinglass (ī'zing-glahs), *n.* a gelatinous substance prepared from the swimming-bladders of the sturgeon, cod, and other fish, used for making jellies, glue etc.

Islam (iz'lahm), *n.* the Muslim religion; the Muslim world or culture. **Islamic,** *a.* **Islamicize, -ise,** *v.t.* Islamize. **Islamism,** *n.* **Islamist, Islamite** (-īt), *n.* **Islamitic** (-mit'-), *a.* **Islamize, -ise,** *v.t.* to convert to Islam.

island (ī'lənd), *n.* a piece of land surrounded by water; anything isolated or resembling an island; an area in the middle of a highway which divides the traffic and affords a refuge for the pedestrian; (*N Am.*) wood surrounded by prairie; a cluster of cells, mass of tissue etc., different in formation from those surrounding it. **islander,** *n.*

isle (īl), *n.* an island, esp. a small island. **islesman,** *n.* an islander, esp. belonging to the Hebrides, Orkneys or Shetlands. **islet** (-lit), *n.* a little island. **islets or islands of Langerhans,** groups of endocrine cells in the pancreas that secrete insulin, discovered by Paul *Langerhans* (1847–88), German anatomist.

ism (izm), *n.* (*usu. derog.*) a doctrine or system of a distinctive kind. **ismatic, -ical** (-mat'-), *a.* **ismaticalness,** *n.*

-ism (-izm), *suf.* forming abstract nouns denoting doctrine, theory, principle, system etc., as *altruism, Conservatism, Socialism, spiritualism, Gallicism, scoundrelism.*

Ismaili (izmahē'li), *n.* one of a sect of Shiite Muslims whose spiritual leader is the Aga Khan. **Ismailism** (iz'-), *n.* **Ismailitic** (-lit'-), *a.*

isn't (iz'nt), is not.

is(o)-, *comb. form* equal; having the same number of parts; indicating an isomeric substance.

isobar (ī'sōbah), *n.* a line on a map connecting places having the same mean barometric pressure, or the same pressure at a given time; any of two or more atoms having the same atomic mass but being of different chemical elements. **isobaric** (-ba'-), **isobarometric** (īsōbarəmet'rik), *a.* of equal barometric pressure; pertaining to isobars.

isocheim (ī'sōkīm), *n.* a line connecting places having the same mean winter temperature. **ischeimal** (-kī'-), **ischeimenal** (īsōkī'mənəl), *a.* marking equal winters. *n.* an isocheimenal line.

isochor (ī'sōkaw), *n.* a line (on a diagram representing relations between pressure and temperature) connecting the points denoting equal volumes. **isochoric** (-ko'-), *a.*

isochromatic (īsōkrəmat'ik), *a.* of the same colour.

isochronal (īsok'rənəl), **-chronous, -chronic** (īsōkron'-), *a.* denoting or occupying equal spaces of time; having regular periodicity (as the swinging of a pendulum). **isochronism** (īsok'-), *n.* **isochronously,** *adv.*

isoclinal (īsōklī'nəl), **-clinic** (-klin'-), *a.* having the same inclination or dip; having the same magnetic inclination; (*Geol.*) having the same angle or dip.

isocrymal (īsōkrī'məl), *a.* connecting points having the same temperature at the coldest season. *n.* an isocrymal line. **isocryme** (ī'-), *n.*

isodiametric (īsōdīəmet'rik), *a.* (*Bot., Cryst.*) equal in diameter.

isodont (ī'sōdont), *a.* having the teeth all alike. **isodontous** (-don'-), *a.*

isodynamic (īsōdīnam'ik), *a.* having equal force, esp. of terrestrial magnetism.

isoelectric (īsōilek'trik), *a.* having identical electric potential.

isogeny (īsoj'əni), *n.* general similarity of origin; general correspondence or homology. **isogeneic, -genous,** *a.*

isogeotherm (īsōjē'əthœm), *n.* a line connecting places having the same mean temperature below the surface. **isogeothermal, -thermic** (-thœ'-), *a.*

isogloss (ī'sōglos), *n.* a line on a map separating regions differing in a specific dialectal feature. **isoglossal** (-glos'-), **-glottic** (-glot'-), *a.*

isogon (ī'sōgon), *n.* a geometrical figure having the angles all equal. **isogonal** (īsog'-), *a.* equiangular; isogonic. **isogonic** (-gon'-), *a.* connecting points (on the earth's surface) having the same magnetic declination or variation from true north. *n.* an isogonic line.

isohel (ī'sōhel), *n.* a line connecting places having equal amounts of sunshine.

isohyet (īsōhī'it), *n.* a line connecting places having equal amounts of rainfall. **isohyetal,** *a.*

isolate (ī'səlāt), *v.t.* to place in a detached situation; (*Elec.*) to insulate; (*Chem.*) to obtain in an uncombined form; to subject to quarantine. **isolability** (-ləbil'-), *n.* **isolable,** *a.* **isolation,** *n.* **isolationism,** *n.* **isolationist,** *n.* one who believes in the policy of holding aloof from all political entanglements with other countries. **isolator,** *n.*

isoleucine (īsōloo'sēn), *n.* an essential amino acid.

isomeric (īsəme'rik), **-ical,** *a.* (*Chem.*) having identical elements, molecular weight and proportions, with difference in physical characteristics or chemical properties owing to different grouping; of atomic nuclei, having the same numbers of protons and neutrons but different energy states. **isomer** (ī'sōmə), *n.* a compound, chemical group, atom etc. isomeric with one or more other compounds etc. **isomerism** (īsom'-), *n.* **isomerous** (īsom'-), *a.* (*Chem.*) isomeric; (*Bot., Zool. etc.*) having the parts or segments equal in number.

isometric (īsōmet'rik), **-ical,** *a.* of equal measure. **isometric line,** *n.* a line on a graph representing variations of pressure and temperature at a constant volume. **isometric projection,** *n.* (*Eng.*) a drawing in approximate perspective from which lengths can be scaled. **isometrics,** *n. sing.* a system of exercises in which the muscles are strengthened as one muscle is opposed to another or to a resistant object.

isomorphism (īsōmaw'fizm), *n.* (*Cryst.*) the property of crystallizing in identical or nearly identical forms; (*Math.*) identity of form and construction between two or more groups. **isomorphic, -phous,** *a.*

isopleth (ī'sōpleth), *n.* a line on a map connecting points at which a variable such as humidity has a constant value. **isoplethic** (-pleth'-), *a.*

isopod (ī'sōpod), *n.* one of the Isopoda or sessile-eyed crustaceans characterized by seven pairs of thoracic legs almost of the same length. *a.* isopodous. **isopodan** (īsop'-), *n. a.* **isopodous** (īsop'-), *a.*

isoprene (ī'sōprēn), *n.* a hydrocarbon of the terpene group used esp. in synthetic rubber.

isoprinosine (īsōprinō'zēen), (*esp. N Am.*) **-nocine,** *n.* an antiviral drug used in treatment of the early symptoms of AIDS.

isosceles (īsos'əlēz), *a.* of a triangle, having two sides equal.

isoseismal (īsōsīz'məl), *a.* connecting points at which an earthquake has been of the same intensity. *n.* an isoseismal line. **isoseismic**, *a.*
isostatic (īsōstat'ik), *a.* (*Geol.*) in equilibrium owing to equality of pressure on every side, as that normally prevailing in the crust of the earth. **isostasy** (īsos'təsi), *n.*
isotheral (īsoth'ərəl), *a.* connecting points having the same mean summer temperature. *n.* an isotheral line.
isotherm (ī'sothœm), *n.* a line on a globe or map passing over places having the same mean temperature. **isothermal** (-thœ'-), *n.*, *a.*
isotonic (īsōton'ik), *a.* having equal tones; of muscles, having equal tension or tonicity; having the same concentration as a surrounding or other liquid. **isotonicity** (-nis'i-), *n.*
isotope (ī'sətōp), *n.* one of a set of species of atoms of a chemical element having the same atomic number but differing in atomic weight etc. **isotopic**, **-ical** (-top'-), *a.* **isotopy** (īsot'-), *n.*
isotropic (īsōtrop'ik), *a.* manifesting the same physical properties in every direction. **isotropism, -tropy** (-sot'-), *n.* **isotropous** (īsot'-), *a.*
I-spy (īspī'), *n.*, HY-SPY.
Israel (iz'rəl), *n.* (*collect.*) the Israelites, the Jewish people; an autonomous country founded in Palestine in 1948. **Israeli** (-rā'-), *n.*, *a.* (an inhabitant) of the State of Israel. **Israelite** (-īt), *n.* a descendant of Israel, a Jew. **Israelitic** (-lit'-), **Israelitish** (-lī'-), *a.*
Issei (ē'sā), *n.* Japanese immigrant in the US.
issue (ish'oo, is'ū), *n.* the act of passing or flowing out; egress, outgoing, outflow; that which passes or flows out; a discharge, as of blood; way or means of exit or escape; outlet; the mouth of a river; progeny, offspring; the produce of the earth; profits from land or other property; result, consequence; the point in debate; (*Law*) the point between contending parties; the act of sending, giving out or putting into circulation; publication; that which is published at a particular time; the whole quantity or number sent out at one time. *v.i.* to pass or flow out; to be published; to emerge (from); to be descended; to proceed, to be derived (from); to end or result (in). *v.t.* to send out; to publish; to put into circulation. **at issue**, in dispute; at variance. **side issue**, a less important issue arising from the main business or topic. **to join, take issue**, to take opposite sides upon a point in dispute. **issuable**, *a.* **issuance**, *n.* the act of issuing. **issueless**, *a.* **issuer**, *n.*
-ist (-ist), *suf.* denoting an agent, adherent, follower etc., as *Baptist, botanist, fatalist, Socialist.*
isthmus (is'məs), *n.* (*pl.* **-muses**) a neck of land connecting two larger portions of land; (*Anat. etc.*) a narrow passage or part between two larger cavities or parts. **Isthmian**, *a.* pertaining to an isthmus, esp. to the Isthmus of Corinth in Greece.
istle (ist'li), *n.* a species of Mexican agave, or the tough wiry fibre of its leaves, used for cordage etc.
It., (*abbr.*) Italy; Italian.
it[1] (it), *3rd pers. neut. pron.* (*poss.* **its**) a non-human creature, inanimate thing or a small baby; the thing spoken about (ref. to noun mentioned or understood); used as subject of a verb the actual subject of which follows, usu. in apposition or introduced by 'that'; the grammatical subject of an impersonal verb; the indefinite object of an intransitive or transitive verb (as *to rough it, to fight it out*); the player in a children's game chosen to oppose the others; that which corresponds precisely to what one has been seeking; personal magnetism, charisma, sex appeal; (*coll.*) sexual intercourse.
it[2] (it), *n.* (*coll.*) Italian vermouth.
ital., (*abbr.*) italics.
Italian (ital'yən), *a.* pertaining to Italy. *n.* a native of Italy; the Italian language. **Italianate**, *v.t.* to render Italian. *a.* (-āt), Italianized. **Italianism**, *n.* **Italianize, -ise**, *v.i.*, *v.t.*

italic (ital'ik), *a.* applied to a sloping type (*thus*), introduced by the Venetian printer Aldus Manutius, *c.* 1500; pertaining to ancient Italy or the Italian races or their languages, esp. as distinguished from Roman. *n.pl.* italic letters or type. **italicize, -ise** (-sīz), *v.t.* to print in italics; to emphasize. **italicism** (-sizm), **italicization, -isation**, *n.*
Italo-, *comb. form* Italian.
ITC, (*abbr.*) Independent Television Commission.
itch (ich), *v.i.* to have a sensation of uneasiness in the skin exciting a desire to scratch the part; to feel a constant teasing desire (for etc.). *n.* a sensation of uneasiness in the skin causing a desire to scratch; an uneasy desire or craving (for etc.); a contagious skin-disease produced by the itch-mite *Sarcoptes scabiei*. **itchiness**, *n.* **itchy**, *a.* **to have itchy feet**, to be restless, to have a desire to travel.
-ite (-īt), *suf.* belonging to, a follower of, as *Pre-Raphaelite, Spinozite;* denoting fossils, minerals, chemical substances, explosives etc., as *belemnite, ichnite, dolomite, quartzite.*
item (ī'təm), *n.* a separate article or particular in an enumeration; an individual entry in an account, schedule etc.; a paragraph or detail of news in a newspaper; (*coll.*) an acknowledged couple; a well-established grouping. *adv.* likewise, also. *v.t.* to make a note or memorandum of. **itemize, -ise**, *v.t.* to set forth in detail.
iterate (it'ərāt), *v.t.* to repeat, to say, make or do over and over again. **iterant**, *a.* repeating, iterating. **iteration**, *n.* **iterative**, *a.*
ithyphallic (ithifal'ik), *a.* of or pertaining to the erect phallus, esp. as carried in Bacchic processions; grossly indecent.
itinerant (itin'ərənt), *a.* passing or moving from place to place; travelling on a circuit. *n.* one, esp. a worker, who journeys from place to place. **itineracy, itinerancy**, *n.*
itinerary (itin'ərəri), *n.* a guide-book; a route taken or to be taken; an account of travels. *a.* pertaining to roads or to travel. **itinerate**, *v.i.* to journey from place to place; to preach on circuit. **itineration**, *n.*
-itis, *suf.* denoting inflammation, as *gastritis, peritonitis.*
ITN, (*abbr.*) Independent Television News.
its (its), *poss.* [IT]
it's (its), contr. form of *it is.*
itself (itself'), *pron.* (*usu. in apposition*) used emphatically; used reflexively. **by itself**, alone, separately. **in itself**, independently of other things; in its essential qualities.
itsy-bitsy (itsibit'si), *a.* (*coll.*) tiny.
IUD, (*abbr.*) intrauterine device.
-ium, *suf.* used chiefly to form names of metals, as *aluminium, lithium, sodium.*
I've (īv), contr. form of *I have.*
-ive (-iv), *suf.* disposed, serving or tending to; of the nature or quality of; as *active, massive, pensive, restive, talkative;* forming nouns, as *captive, detective.*
ivied IVY.
ivory (ī'vəri), *n.* the hard white substance composing the tusks of the elephant, the narwhal etc.; the colour of ivory; (*pl.*) (*sl.*) teeth, billiard-balls, dice, keys of a piano etc. *a.* consisting, made of or resembling ivory. **vegetable ivory**, the hard albumen of ivory-nuts. **ivory-nut**, *n.* the seed of a tropical American palm, *Phytelephas macrocarpa.* **ivory tower**, *n.* a shelter from realities.
IVP, (*abbr.*) intravenous pyelogram (an X-ray of the kidneys).
ivy (ī'vi), *n.* (*pl.* **ivies**) an evergreen climbing plant, *Hedera helix*, usu. having five-angled leaves, and adhering by aerial rootlets. **Ivy League**, *n.* a group of eight long-established and prestigious US universities. **ivy-mantled**, *a.* overgrown with ivy. **ivied**, *a.*
ixia (ik'siə), *n.* a genus of S African bulbous flowering plants of the iris family.
Ixion's wheel (ik'siən), *n.* (*Gr. Myth.*) the wheel on which Ixion was condemned to revolve forever in Hades.
Iyar (iyah'), *n.* the eighth month of the Jewish civil, and

the second of the ecclesiastical year.

izard (iz'əd), *n.* a kind of antelope related to the chamois, inhabiting the Pyrenees.

-ize, -ise (-iz), *suf.* forming verbs denoting to speak or act as; to follow or practise; to come to resemble; to come into such a state; (*transitively*) to cause to follow, resemble or come into such a state; as *Anglicize, Christianize, evangelize, Hellenize*.

Izod Test (ī'zod), *n.* a test to determine particular characteristics of structural materials.

Izvestia (izves'tiə), *n.* the official organ of the legislature of the USSR.

J

J, j, the 10th letter in the English alphabet (*pl.* **jay's, Js, J's**).
J, symbol for current density; joule.
J., (*abbr.*) Journal; Judge; Justice.
JA, (*abbr.*) Judge Advocate.
jab (jab), *v.t.*, *v.i.* to poke violently; to stab; to thrust (something) roughly (into). *n.* a sharp poke, stab, thrust; a short punch; (*coll.*) an injection.
jabber (jab'ə), *v.i.* to talk volubly and incoherently; to talk unintelligibly. *v.t.* to utter rapidly and indistinctly. *n.* rapid, indistinct, or nonsensical talk; gabble. **jabberer**, *n.*
jabbernowl (jab'ənōl), JOBBERNOWL.
jabberwock (jab'əwok), *n.* a fabulous monster created by Lewis Carroll in the poem *Jabberwocky*. **jabberwocky**, *n.* nonsense, gibberish.
jabiru (jab'iroo), *n.* a bird of the genus *Mycteria*, S American stork-like wading-birds.
jaborandi (jabəran'di), *n.* a sudorific and diuretic drug got from certain tropical American shrubs.
jabot (zhab'ō), *n.* a lace frill worn at the neck of a woman's bodice; a ruffle on a shirt front.
jacamar (jak'əmah), *n.* any bird of the tropical American genus *Galbula*, resembling the kingfisher.
jacana (jak'ənə), *n.* any bird of the grallatorial genus *Parra*, from the warmer parts of N and S America.
jacaranda (jakəran'də), *n.* a genus of tropical American trees of the order Bignoniaceae yielding fragrant and ornamental wood.
Jacchus (jak'ŭs), *n.* a small squirrel-like S American monkey.
jacinth (jas'inth), *n.* a variety of zircon.
jack[1] (jak), *n.* familiar or diminutive for John; a labourer, odd-job man; a sailor; the male of certain animals; the knave of cards; a contrivance for turning a spit, or lifting heavy weights, e.g. a car; a lever or other part in various machines; a wooden frame on which wood or timber is sawn; (*Mining*) a gad, a wooden wedge; a small flag; a small white ball at which bowlers aim; (*N Am. sl.*) money; a jack-plug. *v.t.* to lift, hoist, or move with a jack; (*sl.*) to resign, to give (up). **before one can say Jack Robinson**, quite suddenly and unexpectedly. **every man jack**, every individual. **jack-by-the-hedge**, *n.* hedge garlic. **Jack-in-office**, *n.* one who assumes authority on account of holding a petty office. **jack-in-the-box**, *n.* a grotesque figure that springs out of a box when the lid is raised; a kind of firework. **Jack of all trades**, *n.* one who can turn his hand to any business. **jack o' lantern**, *n.* an ignis fatuus. **jack the lad**, *n.* an adventurous stylish young man. **yellow jack**, yellow fever. **jack-a-dandy**, *n.* a little foppish fellow. **jack-bean**, *n.* a climbing plant of the *Camavali* genus. **jack-block**, *n.* a block for raising and lowering the top gallant mast. **jack-boot**, *n.* a military boot reaching to the thigh; harsh military government. **jack-chain**, *n.* (*Forestry*) an endless spiked chain which carries logs from one point to another. **jack-flag**, *n.* a flag hoisted at the spritsail top-mast head. **Jack Frost**, *n.* frost personified. **jack-hammer**, *n.* a hand-held compressed-air hammer used for drilling rock. **jack-high**, *a.*, *adv.* (*Bowls*) from the green to the jack. **Jack Ketch**, *n.* the public hangman. **jack-knife**, *n.* a large clasp-knife, esp. orig. one with a horn handle, carried by seamen. *v.i.* to double up like a jack-knife; of an articulated vehicle, to turn or rise and form an angle of 90° or less when out of control. **jack-plane**, *n.* the coarsest of the joiner's bench-planes. **jack-plug**, *n.* a one-pronged electric plug. **jackpot**, *n.* the money pool in card games and competitions; a fund of prize-money. **jack-rabbit**, *n.* a N American hare with long ears. **Jack Russell**, *n.* a breed of small terrier. **jack-screw**, *n.* a lifting implement worked by a screw. **jack-snipe**, *n.* a small European species of snipe. **jack-staff**, *n.* a flagstaff at the bow for flying the jack. **jack-stay**, *n.* (*Naut.*) a rib or plate with holes, or a rod running through eye-bolts, passing along the upper side of a yard, to which the sail is bent. **jack-straw**, *n.* a person of no weight or substance, also a straw or twig used in jack-straws or spillikins. **jack-tar** TAR[2]. **jack-towel**, *n.* a long round towel on a roller. **Jacky**, *n.* (*Austral. coll.*) an Aboriginal man.
jack[2], (jak), *n.* an E Indian fruit, like a coarser breadfruit. **jack-tree**, *n.*
jackal (jak'əl, -awl), *n.* a gregarious animal, *Canis aureus*, closely allied to the dog; one who does dirty work or drudgery for another.
jackanapes (jak'ənāps), *n.* an impudent person; a coxcomb.
jackaroo (jakəroo'), *n.* (*Austral. sl.*) a new-comer, a novice. **jillaroo** (jil-), *n.fem.*
jackass (jak'as), *n.* a male ass; a stupid fellow. **laughing jackass**, the Australian giant kingfisher, so called from its discordant cry. **jackass fish**, *n.* the edible 'morwong' of Australia and New Zealand. **jackass rabbit**, *n.* a male rabbit.
jackdaw (jak'daw), *n.* the smallest of the British crows, *Corvus monedula*.
jacket (jak'it), *n.* a short coat or sleeved outer garment; the coat of an animal; a wrapper, cover; a paper covering for a hardback book; the skin of a potato; an exterior covering or casing esp. an insulating covering round a boiler, steam-pipe, cylinder of an internal-combustion engine etc. *v.t.* to envelop in a jacket. **to dust one's jacket** DUST. **jacketed**, *a.* **jacketing**, *n.*
jack-flag etc. JACK[1].
jacko (jak'ō), JOCKO.
jack o' lantern etc. JACK[1].
Jacobean (jakəbē'ən), *a.* belonging to the reign of King James I.
Jacobin (jak'əbin), *n.* a Dominican friar; a member of a revolutionary republican club, that met in the hall of the Jacobin friars in Paris, 1789–94; an extreme revolutionist, a violent republican; a variety of hooded pigeon. **Jacobinic** (-bin'-), *a.* **Jacobinism** *n.* **Jacobinize, -ise**, *v.t.*
Jacobite (jak'əbit), *n.* a partisan of James II after his abdication, or of the Stuart pretenders to the throne. *a.* pertaining to or holding the opinions of the Jacobites. **Jacobitic, -ical** (-bit'-), *a.* **Jacobitism**, *n.*
Jacob's ladder (jā'kəbz), *n.* a garden plant with closely pinnate leaves; a rope ladder with wooden rounds.
Jacob's staff (jā'kəbz), *n.* a mediaeval instrument for measuring distances and heights.
jaconet (jak'ənit), *n.* a fine, close, white cotton cloth.
Jacquard loom (jak'ahd), *n.* a loom for weaving figured fabrics. **jacquard** *n.* fabric woven in such a manner. [French inventor J.M. *Jacquard*, 1752–1834]

jacquerie (zhakərē'), *n.* a revolt of the peasants against the nobles in France, in 1357–8; any peasant revolt.

jactation (jaktā'shən), *n.* the act of throwing; agitation of the body in exercise, as in riding; jactitation. **jactitation** (-ti-), *n.* restlessness, a tossing or twitching of the body during illness.

Jacuzzi® (jəkoo'zi), *n.* a type of bath with a mechanism which makes the water swirl round; this mechanism itself; a bathe in such a bath.

jade¹ (jād), *n.* a broken-down, worthless horse; (*playfully or in contempt*) an immoral woman. *v.t., v.i.* to make or become tired. **jadedly**, *adv.* **jadedness**, *n.* **jadish**, *a.*

jade² (jād), *n.* a green, massive, sometimes cryptocrystalline, silicate of lime and magnesia, used for ornamental purposes; the green colour of jade.

j'adoube (zhadoob'), *int.* in chess, expressing an intention to adjust, but not move, a piece.

Jaeger® (yā'gə), *n.* a woollen dress material, orig. one containing no vegetable fibre. [Dr Gustav *Jaeger*, the manufacturer]

jaeger (yā'gə), *n.* a huntsman; a sharpshooter; an attendant waiting on a person of quality. [G]

Jaffa, Jaffa orange (jaf'ə), *n.* a type of orange from *Jaffa* in Israel.

jag (jag), *n.* a notch; a ragged piece, tooth or point; a stab, prick; (*sl.*) a bout of drinking or drug-taking. *v.t.* to cut or tear raggedly; to cut into notches. **jagged** (jag'id), *a.* having notches; sharply uneven. **jaggedly**, *adv.* **jaggedness**, *n.* **jagger**, *n.* one who or that which jags; a toothed chisel. **jaggy**, *a.*

jaguar (jag'ūə), *n.* a S American feline animal resembling the leopard.

jaguarundi (jagwərūn'di), *n.* a S American wild cat.

Jah (yah), *n.* Jehovah. **Jahveh** (-wä), *n.* (*form adopted by Bibl. critics*).

Jahad (jəhad'), JIHAD.

jai-alai (hī'əlī), *n.* a game played by two or four players on a court, who wear woven baskets tied to their wrists and using these hurl a ball at the walls. [Sp.]

jail, gaol (jāl), *n.* a prison, a place of confinement for persons charged with or convicted of crime. **jail-bird**, *n.* one who has been to prison; an inveterate criminal. **jail-break**, *n.* an escape from jail. **jail-fever**, *n.* an old name for typhus. **jailer, gaoler**, *n.* the keeper of a prison. **jaileress, gaoleress** (-ris), *n. fem.*

Jain (jīn), **Jaina** (-nə), *n.* an adherent of Jainism. *a.* of or belonging to the Jains or Jainism. **Jainism**, *n.* an Indian non-Brahminical religion akin to Buddhism. **Jainist**, *n.*

jake (jāk), *a.* (*coll.*) honest; correct; very good.

jalap (jal'əp), *n.* the dried root of a Mexican plant used as a purgative. **jalapin**, *n.* an amorphous glucoside existing in jalap root.

jalopy (jəlop'i), *n.* (*coll.*) a much-worn automobile.

jalousie (zhal'uzi), *n.* a louvre blind, a Venetian shutter. **jalousied**, *a.*

jam¹ (jam), *v.t.* (*past, p.p.* **jammed**) to wedge or squeeze (in, into, between, together); to block up by crowding into; to make (a machine etc.) immovable or unworkable by forcible handling; to prevent clear radio reception of a signal by transmitting an interfering signal on the same wavelength. *v.i.* to become immovable or unworkable by rough handling; of a jazz musician to improvise freely; to take part in a jam session. *n.* a crush, squeeze; a stoppage in a machine due to jamming; a crowd, press; congestion as in traffic jam; a predicament. **jam-packed**, *a.* very crowded; filled to capacity. **jam session**, *n.* (*coll.*) an improvised performance by jazz musicians.

jam² (jam), *n.* a conserve of fruit boiled with sugar. **jam-jar**, *n.* **jam-pot**, *n.* **jammy**, (*coll.*) sticky (with jam); lucky; desirable.

Jamaica pepper (jəmā'kə), *n.* allspice; pimento.

jamb (jam), *n.* one of the upright sides of a doorway, window, or fireplace.

jambalaya (jambəlī'ə), *n.* a spicy Southern US dish made of meat or fish with rice, onions, etc.

jambok (jam'bok), SJAMBOK.

jamboree (jambərē'), *n.* a Scout rally; a frolic.

jampan (jam'pan), *n.* a sedan-chair used in India. **jampanee** (-nē'), *n.* one of the bearers of a jampan.

Jan., (*abbr.*) January.

jane (jān), *n.* (*Am., Austral. sl.*) a woman.

jangle (jang'gl), *v.i.* to sound harshly or discordantly. *v.t.* to cause to sound discordantly; to utter harshly. *n.* discordant sound, as of bells out of tune. **jangler**, *n.*

janissary JANIZARY.

janitor (jan'itə), *n.* a doorkeeper; caretaker, porter; (*Sc.*) a school caretaker. **janitorial** (-taw'ri-), *a.* **janitorship**, *n.* **janitress** (-tris), **-trix** (-triks), *n. fem.*

janizary, janissary (janizəri), *n.* a soldier of the old Turkish infantry forming the Sultan's bodyguard.

Jansenist (jan'sənist), *n.* a follower of Cornelius Jansenius, bishop of Ypres, Flanders (*d.* 1638), who denied the freedom of the human will and that Christ died for all mankind. *a.* pertaining to or characteristic of Jansenism. **Jansenism**, *n.* **Jansenistic** (-nis'-), *a.*

janty (jan'ti), jantily etc. JAUNTY.

January (jan'ūəri), *n.* the name of the first month of the year.

Janus (jā'nəs), *n.* an ancient Italian deity presiding over doors and gates, having two faces looking in opposite directions.

Jap (jap), *n., a.* (*derog. or offensive*) short form of JAPANESE. **Jap silk**, *n.* a pure silk fabric plainly woven from net silk yarns.

japan (jəpan'), *n.* a hard, black varnish orig. from Japan; work varnished and figured in the Japanese style. *v.t.* (*past, p.p.* **japanned**) to cover with or as with japan. **Japan earth**, catechu. **Japanese** (jəpənēz'), *a.* pertaining to Japan or its inhabitants. *n.* a native or inhabitant of Japan; the language of Japan. **Japanese cedar**, *n.* a tall Japanese conifer. **Japanize, -ise**, *v.t.* **Japanization**, *n.* **japanner**, *n.* one whose business is to japan goods.

jape (jāp), *v.i.* to jest, play tricks. *n.* a jest, trick, joke. **japer**, *n.*

Japhetic (jəfet'ik), *a.* descended from Japheth, the third son of Noah.

Japonic (jəpon'ik), *a.* Japanese. **japonica** (-kə), *n.* the Japanese quince, a common garden shrub. **japonically**, *adv.* **japonicize, -ise**, *v.i.*

jar¹ (jah), *v.i.* (*past, p.p.* **jarred**) to emit a harsh or discordant sound; to vibrate harshly; to be discordant, disagreeable or offensive; to disagree, to clash, to be inconsistent (with). *v.t.* to cause to shake or tremble; to give a shock to. *n.* a harsh vibration as from a shock; a harsh discordant sound; a disagreement, conflict. **jarringly**, *adv.*

jar² (jah), *n.* a wide-mouthed vessel of glass or earthenware; (*coll.*) (a glass of) alcoholic drink. **jarful**, *n.*

jar³ (jah), **on the jar**, partly closed, ajar.

jardiniere (zhahdinycə'), *n.* an ornamental pot or stand for growing flowers in a room etc. [F]

jargon¹ (jah'gən), *n.* unintelligible talk; gabble; any professional, technical or specialized language. *v.i.* to talk unintelligibly; (of birds) to twitter. **jargoner**, *n.* **jargonesque** (-nesk'), **jargonic** (-gon'-), *a.* **jargonist**, *n.* **jargonize, -ise**, *v.i.* **jargonization**, *n.*

jargon² (jah'gən), *n.* a transparent, colourless or smoky variety of zircon.

jargonelle (jahgənel'), *n.* a kind of early pear.

jarl (yahl), *n.* (*Hist.*) a Norse nobleman or chieftain. [Icel.]

jarrah (ja'rə), *n.* the W Australian mahogany gum-tree.

jarringly JAR¹.

Jas, (*abbr.*) James.

jasmine (jaz'min), *n.* any plant of the genus *Jasminum*,

jaspé | 445 | **Jerusalem artichoke**

climbers with sweetscented white or yellow flowers, esp. the common white *J. officinale*.
jaspé (jas'pä), *a.* (*Ceram.*) having an appearance like jasper; mottled.
jasper (jas'pə), *n.* an opaque, impure variety of quartz, of many colours and shades; a greenish marble, with small red spots. **jasperite** (rit), *n.* a red variety of jasper. **jasperize, -ise,** *v.t.* **jasperous,** *a.* **jaspoid** (-poid), *a.* resembling jasper.
jataka (jah'takə), *n.* the birth story of Buddha.
jato (jä'tō), *n.* (acronym) *jet assisted take-off*.
jaundice (jawn'dis), *n.* yellowness of the skin caused by obstruction of the bile or absorption of the colouring matter into the blood; caused by jealousy, prejudice. *v.t.* to affect with or as with jaundice; to poison the mind with jealousy, prejudice etc. **jaundiced,** *a.*
jaunt (jawnt), *v.i.* to ramble or rove about; to take a short excursion. *n.* an excursion, a short outing. **jaunting-car,** *n.* an Irish horse-drawn vehicle having two seats, back to back.
jaunty (jawn'ti), *a.* sprightly, airy, perky. **jauntily,** *adv.* **jauntiness,** *n.*
Javanese (jahvənēz'), *a.* of or pertaining to Java. *n.* a native of Java; the language of Java.
javelin (jav'əlin), *n.* a light spear thrown by hand; (**the javelin**) the competitive sport of javelin-throwing. *v.t.* to wound or pierce with or as with a javelin.
Javelle water (zhəvel'), *n.* a solution of sodium hypochlorite used in disinfecting and bleaching.
jaw (jaw), *n.* one of two bones or bony structures in which the teeth are fixed, forming the framework of the mouth; (*pl.*) the mouth; one of two opposing members of a vice or similar implement or machine; (*pl.*) a narrow opening or entrance; (*sl.*) long-winded talk; a lecture. *v.i.* (*coll.*) to talk lengthily. *v.t.* to abuse; to lecture. **hold your jaw,** (*sl.*) shut up. **jaw-breaker,** *n.* (*coll.*) an unpronounceable word. **jaw-lever,** *n.* an instrument for opening the mouths of cattle for the administration of medicine. **jaw-tooth,** *n.* a molar. **jawed,** *a.*
jay (jä), *n.* a chattering bird of brilliant plumage related to the crows; (*fig.*) an impudent chatterer. **jay-walker,** *n.* (*coll.*) a pedestrian who crosses the street heedless of traffic. **jay-walk,** *v.i.*
jazz (jaz), *n.* syncopated music of Negro origin; the form of dancing that goes to this music; vividness; garishness; liveliness; (*sl.*) rigmarole; (*sl.*) insincere talk. **to jazz up,** to quicken the tempo of; to make more attractive, livelier, colourful etc. **jazz-man,** *n.* a jazz musician. **jazz rock,** *n.* music which is a mixture of jazz and rock. **jazzy,** *a.* **jazzily,** *adv.* **jazziness,** *n.*
JC, (*abbr.*) Jesus Christ; Julius Caesar; Justice Clerk.
JCB[b], *n.* a type of excavating machine. [Joseph Cyril Bamford, British manufacturer]
jealous (jel'əs), *a.* suspicious or apprehensive of being supplanted in the love or favour (of a wife, husband, lover or friend); suspicious or apprehensive (of a rival); solicitous or anxiously watchful (of one's honour, rights etc.); envious (of another or another's advantages etc.); (*Bibl.*) requiring exclusive devotion (of God). **jealously,** *adv.* **jealousy,** *n.*
Jeames (jēmz), *n.* a footman, a flunkey.
jean (jēn), *n.* a twilled undressed cloth with cotton warp; (*pl.*) a garment or garments made of this; close-fitting casual trousers usu. made of this.
jeep (jēp), *n.* (*US Mil.*) a small, tough vehicle with four-wheel drive.
jeer (jiə), *v.i.* to scoff, mock (at). *v.t.* to scoff at, to make a mock of, to deride. *n.* a scoff, a gibe, a taunt, mockery. **jeerer,** *n.* **jeeringly,** *adv.*
jeffersonite (jef'əsənīt), *n.* a greenish-black variety of pyroxene.
Jehovah (jihō'və), *n.* the name given in the Old Testament to God. **Jehovah's Witnesses,** *n.* an evangelistic sect who believe that the end of the world is near.
Jehu (jē'hū), *n.* a coachman, a driver, esp. one who drives fast or furiously.
jejune (jijoon'), *a.* meagre, scanty; wanting in substance; devoid of interest or life. **jejunely,** *adv.* **jejuneness,** *n.*
jejunum (jijoo'nəm), *n.* the second portion of the small intestine between the duodenum and the ileum. **jejuno-,** *comb. form* pertaining to the jejunum.
Jekyll and Hyde (jek'l), a person with a split personality, one side evil the other good.
jelly (jel'i), *n.* any gelatinous substance; a conserve made of fruit juice boiled with sugar; a dessert made with gelatine and fruit flavourings; (*sl.*) gelignite. *v.i.* to turn into jelly. *v.t.* to convert into jelly. **jelly-baby,** *n.* a gelatinous sweet shaped like a baby. **jelly-bag, -cloth,** *n.* a bag or cloth for straining jelly. **jelly bean,** *n.* a sugar-coated, bean-shaped sweet filled with jelly. **jelly-fish,** *n.* a name of the medusas and other marine coelenterates; a weak-willed person. **jell, gell** (jel), *v.i.* (*coll.*) to congeal; to take distinct form. **jellify,** *v.t.,* *v.i.*
jemadar (jem'ədah), *n.* an officer in the Indian army.
jemimas (jimi'məs), *n.pl.* (*coll.*) elastic-sided boots; long galoshes for boots.
jemmy (jem'i), *n.* a short, stout crowbar, used by burglars. *v.t.* to open with a jemmy.
jennet (jen'it), *n.* a small Spanish horse.
jenneting (jen'iting), *n.* an early kind of apple.
jenny (jen'i), *n.* a name for certain female animals; a spinning-jenny; (*Billiards*) a stroke pocketing the ball from an awkward position. **jenny-ass,** *n.* **jenny-wren,** *n.*
jeopardy (jep'ədi), *n.* exposure to danger, loss, or injury; risk, hazard, danger, peril. **jeopardize, -ise,** *v.t.* to put in jeopardy.
jequirity (jikwi'riti), *n.* a tropical twining shrub with parti-coloured seeds used ornamentally. **jequirity-beans,** *n.pl.*
jerboa (jœbō'ə), *n.* a small mouse-like rodent with long hind legs adapted for leaping.
jeremiad (jerəmī'əd), *n.* a long lamentation or complaint in the style of the prophet Jeremiah.
Jeremiah (jerəmī'ə), *n.* a prophet of doom; a pessimistic person.
jerfalcon (jœ'fawlkən), GERFALCON.
jerk[1] (jœk), *v.t.* to pull, push or thrust sharply or spasmodically; to throw sharply. *v.i.* to move with jerks. *n.* a sharp, sudden push or tug; a twitch, a spasmodic movement due to involuntary contraction of a muscle; (*sl.*) a stupid, ignorant or contemptible person. **jerk off,** (*sl.*) to masturbate. **jerker,** *n.* **jerky,** *a.* **jerkily,** *adv.* **jerkiness,** *n.*
jerk[2] (jœk), *v.t.* to cut (beef) into long pieces and dry in the sun. **jerked beef,** *n.*
jerkin[1] (jœ'kin), *n.* a short jacket or waistcoat.
jerkin[2] (jœ'kin), GERFALCON.
jerkin-head (jœ'kinhed), *n.* (*Arch.*) a combination of truncated gable and hipped roof.
jeroboam (jerəbō'əm), *n.* a wine-bottle holding 10–12 quarts (about 12 litres).
Jerry (je'ri), *n.* (*esp. war sl.*, *often derog.*) a German soldier.
jerry (je'ri), *n.* (*sl.*) a chamber-pot. **jerry-can,** *n.* a flat-sided rectangular can for carrying petrol etc.
jerry-builder (je'ri-), *n.* a speculative builder of cheap and inferior houses. **jerry-building,** *n.* **jerry-built,** *a.* **jerry-shop,** *n.* a beerhouse.
jerrymander (je'rimandə), GERRYMANDER.
jersey (jœ'zi), *n.* a knitted garment worn on the upper part of the body; fine wool yarn and combed wool; a fine elastic knitted fabric. **Jersey cow,** *n.* one of a breed of dairy cattle originating from the island of Jersey.
jerupigia GEROPIGIA.
Jerusalem artichoke (jəroo'sələm), *n.* a species of sunflower with edible tuberous roots; the tuber eaten as a vegetable.

jess (jes), *n.* (*Falconry*) a short leather or silk strap tied round each leg of a hawk, to which the leash was usually attached. **jessed,** *a.*
jessamine (jes'əmin), JASMINE.
jessant (jes'ənt), *a.* (*Her.*) issuing or springing (from).
Jesse (jes'i), *n.* a tree representing the genealogy of Christ, esp. in the form of a large many-branched candlestick. **Jesse-window,** *n.* one whose tracery and glazing represent a tree of Jesse.
jest (jest), *n.* a joke, something ludicrous said or done to provoke mirth; a jeer, taunt; a laughing-stock; a prank, a frolic. *v.i.* to joke; to utter jests; to jeer (at). **in jest,** as a jest. **jester,** *n.* one who jests or jokes, (*Hist.*) a professional buffoon. **jesting,** *n.*, *a.* **jestingly,** *adv.*
Jesuit (jez'ūit), *n.* a member of the Society of Jesus, a Roman Catholic order founded in 1534 by Ignatius Loyola; (*fig.*) a crafty, insidious person, a subtle casuist or prevaricator. **Jesuits' bark,** *n.* cinchona bark. **Jesuitic, -ical** (-it'-), *a.* crafty, cunning, designing. **Jesuitically,** *adv.* **Jesuitism, Jesuitry** (-ri), *n.* **Jesuitize, -ise,** *v.t.*, *v.i.*
Jesus (je'zəs), *n.* the founder of Christianity. **Society of Jesus,** the Jesuits.
jet[1] (jet), *n.* a black variety of lignite susceptible of a brilliant polish, used for ornaments. *a.* the colour of jet. **jet-black,** *a.* **jetty**[2], *a.*
jet[2] (jet), *v.i.* (*past, p.p.* **jetted**) to spurt or shoot out; to travel by jet-plane. *v.t.* to send out in a jet or jets. *n.* a sudden spurt or shooting out; a spout or nozzle for the discharge of water etc.; (*coll.*) a jet-propelled plane. **jetfoil,** *n.* a hydrofoil powered by a jet of water. **jet-lag,** *n.* the exhaustion caused by the body's inability to adjust to the time-zone changes involved in long-distance air-travel. **jet-plane,** *n.* a jet-propelled plane. **jet-propelled,** *a.* propelled by heating and expanding air which is directed in a jet from the rear of a vehicle. **jet propulsion,** *n.* **jet-set,** *n.* fashionable people who can afford constant travel by jet-plane. **jet-setter,** *n.* **jet-ski,** *n.* a small powered water vehicle with a flat heel shaped like a water-ski. **jet-skiing,** *n.* **jet-stream,** *n.* very strong winds blowing at high altitude; the exhaust of a jet engine.
jeté (zhət'ā-, -tā'), *n.* a leap from one foot to another in ballet. [F]
jetsam (jet'səm), *n.* goods, cargo etc., thrown overboard in order to lighten a ship. **flotsam and jetsam** FLOTSAM.
jettison (jet'isən), *v.t.* to throw (goods) overboard; to discard (anything unwanted).
jetty[1] (jet'i), *n.* a structure of stone or timber projecting into water and serving as a mole, pier or wharf; (*Arch.*) a part of a building which juts beyond the ground-plan.
jetty[2] JET[1].
jeu (zhœ), *n.* (*pl.* **jeux**) a game, play, jest. **jeu de mots** (dəmō), *n.* a pun. **jeu d'esprit** (desprē'), *n.* a witticism, a play of wit.
jeunesse dorée (zhœnes'dorā'), *n.* gilded youth. [F]
Jew (joo), *n.* a member of a Semitic people descended from the ancient Israelites; one whose religion is Judaism; (*offensive*) a usurer, a person who drives a hard bargain. **(jew)** *v.t.* (*offensive*) to drive a hard bargain, to cheat. **jew's-ear,** *n.* a tough edible fungus growing on elder and elm-trees. **jew's harp,** *n.* a musical instrument held between the teeth, with a metal tongue set in motion by the forefinger. **jew's-mallow,** *n.* a plant, *Corchorus capsularis,* used in the East as a potherb. **Jewess,** *n. fem.* **Jewish,** *a.* **Jewishness,** *n.*
jewel (joo'əl), *n.* a precious stone, gem; a personal ornament containing a precious stone or stones; (*fig.*) a person or thing of very great value or excellence. *v.t.* to adorn with or as with jewels; to fit (a watch) with jewels in the pivot-holes. **jewel in the crown,** (*fig.*) the most highly-prized, beautiful etc. one of a collection or group. **jewel-case,** *n.* **Jewel-house, -office,** *n.* the place (in the Tower of London) where the Crown Jewels are deposited. **jewel-like,** *a.* **jeweller,** *n.* a maker of or dealer in jewels. **jewellery, jewelry** (-əlri), *n.* (*collect.*) jewels in general; the art or trade of a jeweller.
Jewry (joo'ri), *n.* (*collect.*) the Jews; Jewish religion or culture; (*Hist.*) Judaea; the Jews' quarter in a town.
jewstone (joo'stōn), *n.* the fossil spine of a sea-urchin or echinus.
Jezebel (jez'əbel), *n.* a wicked, bold, or vicious woman.
jib[1] (jib), *n.* a triangular sail set between the topmast and bowsprit of a vessel; the extended arm of a crane or derrick. **the cut of one's jib,** (*orig. Naut. sl.*) one's physical appearance. **jib-boom,** *n.* a movable spar running out beyond the bowsprit. **jib-door,** *n.* a door flush with the wall on both sides, and usu. papered or painted over so as to be concealed.
jib[2] (jib), *v.t.* (*past, p.p.* **jibbed**) to shift (a boom, yard or sail) from one side of a vessel to the other. *v.i.* to swing round (of a sail etc.).
jib[3] (jib), *v.i.* of a horse, to refuse to move forwards; (*with at*) of a person, to refuse to do (something). **jibber,** *n.*
jibbah (jib'ə), *n.* long, loose coat worn by Muslims; a loose overall or pinafore. [Arab.]
jibber[2] (jib'ə), GIBBER.
jibe (jib), GIBE.
jiblet (jib'lit), GIBLET.
jiff (jif'), **jiffy** (-i), *n.* (*coll.*) a moment, instant.
Jiffy-bag* (jif'ibag), *n.* a kind of strong padded envelope.
jig (jig), *n.* a lively dance for one or more performers; the music for such a dance; a fish-hook with a weighted shank, used for snatching at fish; a device for holding an object and guiding a cutting-tool in a machine. *v.i.* (*past, p.p.* **jigged**) to dance a jig; to skip about. *v.t.* to sing or play in jig time; to jerk up and down rapidly; to separate finer and coarser qualities of (ore etc.) with a jigger. **jig-saw,** *n.* a vertically-reciprocating saw moved by a vibrating lever or crank-rod, used for cutting scrolls, fretwork etc. **jigsaw puzzle,** *n.* a puzzle to put together a picture cut into irregularly shaped pieces. **jigging,** *n.* **jigging-machine,** *n.* a jigger.
jigger[1] (jig'ə), *n.* one who or that which jigs; a sieve shaken vertically in water to separate the contained ore; the man using such a sieve; (*Naut.*) a small lifting tackle; a small sail, usu. set on a jigger-mast; a rest for a billiard-cue; (*coll.*) any kind of mechanical contrivance, implement etc.; (*Golf*) an iron club coming between a mid-iron and a mashie; a small measure of spirits. **jigger-mast,** *n.* a small mast at the stem of a yawl, a small mizzen-mast.
jigger[2] (jig'ə), CHIGOE.
jiggered (jig'əd), *a.* very surprised, confounded.
jiggery-pokery (jigəripō'kəri), *n.* underhand goings-on.
jiggle (jig'l), *v.t.* to jerk or rock lightly to and fro.
jig-jog (jig'jog), *n.* a jogging, jolting motion.
jigot (jig'ət), GIGOT.
jihad, jehad (jēhad'), *n.* a holy war proclaimed by Muslims against unbelievers or the enemies of Islam; (*fig.*) a war or crusade on behalf of a principle etc. [Arab.]
jillaroo JACKAROO.
jilliflower (jil'iflowə), GILLYFLOWER.
jilt (jilt), *n.* a woman who capriciously or wantonly throws over a lover. *v.t.* to throw over or discard (one's lover). *v.i.* to play the jilt.
jimcrack (jim'krak), GIMCRACK.
Jim Crow (jimkrō'), *n.* (*N Am., offensive*) a Negro (from the refrain of a negro-minstrel song); the policy of segregating Negroes; (*Mach.*) an implement for bending or straightening rails.
jim-jams (jim'jamz), *n.pl.* (*coll.*) fluster, jumpiness; delirium tremens.
jimmy (jim'i), JEMMY.
jingal GINGAL.
jingko (jing'kō), GINGKO.
jingle (jing'gl), *v.i.* to make a clinking or tinkling sound;

to correspond in sound, rhyme etc. (in a depreciative sense). *v.t.* to cause to make such a clinking or tinkling sound. *n.* a tinkling metallic sound; a correspondence or repetition of sounds in words, esp. of a catchy inartistic kind; doggerel; a simply rhythmical verse, esp. one used in advertising. **jingle-jangle**, *n.*

Jingo (jing'gō), *n.* (*pl.* **-goes**) a word used as a mild oath; a person given to (excessive) belligerent patriotism. **jingoish**, *a.* **jingoism**, *n.* (excessive) belligerent patriotism; a foreign policy based on this. **jingoist**, *n.* **jingoistic** (-is'-), *a.*

jink (jingk), *v.i.* to move nimbly; to dance; to dodge. *v.t.* to dodge. *n.* a slip, evasion, a dodging turn. **high jinks**, pranks, frolics.

jinker (jing'kə), *n.* (*Austral.*) a sort of two-wheeled bogey for transporting heavy logs and timber.

jinnee (jinē'), *n.* (*pl.* **jinn**, *often taken for sing.*) one of a race of spirits or demons in Muslim mythology supposed to have the power of assuming human or animal forms.

jinrickshaw, jinrickisha (jinrik'shaw), RICKSHAW.

jinx (jingks), *n.* (*sl.*) a person or thing that brings ill luck.

jit (jit), *n.* a type of beat music that originated in Zimbabwe.

jitney (jit'ni), *n.* (*N Am. sl.*) a motor-car.

jitters (jit'əz), *n.pl.* (*sl.*) nervous apprehension. **jitter-bug**, *n.* a type of fast dance to jazz music.

jiu-jitsu (joojit'soo), JU-JITSU.

jive (jīv), *n.* a style of lively, jazz-style music; dancing to such music; (*sl.*) misleading talk. *v.i.* to dance to jive music; (*sl.*) to mislead. **jiver**, *n.* **jiving**, *n.*

joanna (jōan'ə), *n.* (*sl.*) a piano.

Job (jōb), *n.* (*fig.*) an uncomplaining sufferer or victim. **Job's comforter**, *n.* one who lacerates one's feelings whilst pretending to sympathize. **jobe**, *v.t.* to reprove, to reprimand. **jobation**, *n.* a long-winded reproof, a lecture.

job[1] (job), *n.* a piece of work, esp. one done for a stated price; an occupation; a responsibility or duty; (*coll.*) a difficult task; (*coll.*) a situation; (*sl.*) a crime, esp. a robbery. *v.t.* (*past, p.p.* **jobbed**) to let out (as work) by the job; to buy up in miscellaneous lots and retail; to deal in (stocks); to deal with in an underhand way for one's private benefit. *v.i.* to buy and sell as a broker; to do job-work; to let or hire by the job; to make profit corruptly out of a position of trust, esp. at public expense. **a bad, good job**, (*coll.*) an unfortunate, satisfactory turn of affairs. **to job out**, to sublet a piece of work. **job centre**, *n.* a government-run employment agency. **job lot**, *n.* a miscellaneous lot of goods bought. **job-sharing**, *n.* the division of one job by two or more people who work hours complementary to each other. **job-work**, *n.* work done or paid for by the job. **jobber**, *n.* one who does small jobs; one who deals in stocks and shares on the Stock Exchange. **jobbery**, *n.* **jobbing**, *a.* doing job-work. **jobless**, *a.* **joblessness**, *n.*

job[2] (job), *v.t.* to stab, poke or prod with a sharp instrument; to drive (a sharp instrument) in. *v.i.* to stab or thrust (at). *n.* a sudden stab, poke or prod.

jobber JOB[1].

jobbernowl (job'ənōl), *n.* a blockhead.

Jock (jok), *n.* (*coll.*) a soldier of a Scottish regiment.

jockey (jok'i), *n.* a professional rider in horse-races; one given to sharp practice. *v.t.* to outwit, out-manoeuvre etc.; to cheat; (*Horse-racing*) to jostle by riding against. *v.i.* to be tricky; to play a tricky game. **disc-jockey** DISC. **to jockey for position**, to try by skill to get an advantageous position. **jockeydom** (-dəm), *n.* **jockeyism, jockeyship**, *n.*

jocko (jok'ō), *n.* a chimpanzee.

jockstrap (jok'strap), *n.* support for the genitals worn by men engaged in athletic or sporting activity.

jocose (jəkōs'), *a.* humorous, facetious; given to jokes or jestings; containing jokes. **jocosely**, *adv.* **jocoseness**, *n.* **jocosity** (-kos'-), *n.*

jocular (jok'ūlə), *a.* merry, facetious, amusing; embodying a joke. **jocularity** (-la'-), *n.* **jocularly**, *adv.*

jocund (jok'ənd), *a.* merry; inspiring mirth. **jocundity** (-kūn'-), *n.* **jocundly**, *adv.*

jodel (jō'dəl), YODEL.

jodhpurs (jod'pəz), *n.pl.* long riding-breeches fitting closely from knee to the ankle.

Joe (jō), a male christian name. **Joe Bloggs** (blogz), *n.* a typical or ordinary person. **Joe Public**, *n.* the general public. **Joe Soap**, *n.* one who does menial tasks; one who is taken advantage of.

jog (jog), *v.t.* (*past, p.p.* **jogged**) to push or jerk lightly, usually with the hand or elbow; to nudge, esp. to excite attention; to stimulate (one's memory or attention). *v.i.* to move with an up-and-down leisurely pace; to walk or plod idly (on, along etc.); to run at a steady, slow pace for exercise. *n.* a light push or nudge to arouse attention; a leisurely trotting or jogging motion. **jog-trot**, *n.* a slow, easy, monotonous trot; humdrum progress. **jogger**, *n.* one who jogs (for exercise). **jogger's knee, nipple** etc., *n.* an injured or damaged knee, nipple etc. caused by jogging. **jogging**, *n.* the act of jogging, esp. as a form of exercise. **jogging-suit**, *n.* a garment like a track-suit worn when jogging.

joggle (jog'l), *v.t.* to shake, push, nudge or jerk slightly; (*Build.*, perh. from JAG) to unite by means of joggles. *v.i.* to shake slightly, to totter. *n.* an act of joggling; a joint in stone or other material consisting of a projection which fits into a notch in another piece.

Johannisberger (jəhan'isbœgə), *n.* a fine white Rhenish wine.

John (jon), *n.* a male Christian name; (**john**) (*sl.*) a lavatory; (*sl.*) a prostitute's client. **John Barleycorn** BARLEY[1]. **John Bull** BULL[1]. **John Chinaman**, *n.* (*offensive*) a Chinaman. **John Collins** (kol'inz), *n.* an alcoholic drink based on gin. **John Company**, *n.* a familiar name for the East India Company. **John Doe**, *n.* the fictitious plaintiff in an (obsolete) action for ejectment, the defendant being called Richard Roe. **John Dory** DORY[1].

Johnian (jō'niən), *n., a.* (a member or student) of St John's College, Cambridge.

Johnny (jon'i), *n.* (*sl.*) a fellow, chap; (*sl.*) a condom. **Johnny cake**, *n.* (*N Am.*) a maize cake baked on the hearth; (*Austral.*) a similar wheat-meal cake. **johnny-come-lately**, *n.* a newcomer. **Johnny Crapaud** (krapō'), *n.* (*offensive*) a Frenchman. **Johnny Raw**, *n.* a raw beginner, a novice.

Johnsonian (jonsō'niən), *a.* pertaining to Dr Samuel Johnson or his style; pompous, inflated, abounding in words of classical origin. **Johnsonism**, *n.* **Johnsonese** (-nēz'), *n.*

join (join), *v.t.* to connect, fasten together; to couple, to associate; to unite (two persons, or a person or persons with or to) in marriage etc.; to engage in (battle etc.); to become a member (of a club etc.). *v.i.* to be contiguous or in contact; to become associated or combined (with etc.) in views, partnership, action etc. *n.* a joint; a point, line, or mark of junction. **to join issue** ISSUE. **to join up**, (*coll.*) to enlist. **joinant**, *a.* (*Her.*) conjoined. **joinder** (-də), *n.* the coupling of two things in one suit or action, or two or more parties as defendants in a suit. **joiner**, *n.* one who joins; a carpenter who makes articles of furniture, finishes woodwork etc.; (*N Am.*) a carpenter; a person who likes joining clubs etc. **joinery**, *n.*

joint (joint), *n.* a junction or mode of joining parts together; the union of two bones in an animal body; an analogous point or mechanical device connecting parts of any structure, whether fixed or movable; one of the pieces into which a butcher cuts up a carcass; this piece as served at table; a node; an internode; a crack traversing rocks in a straight and well-determined line; (*sl.*) a bar,

joist 448 **jubilate**

club etc.; (*sl.*; *often derog.*) a place, building etc. (*sl.*) a marijuana cigarette. *a.* of, belonging to, performed or produced by different persons in conjunction; sharing or participating (with others). *v.t.* to form with joints or articulations; to connect by joints; to plane and prepare (boards etc.) for joining; to point (masonry); to divide or cut (meat) into joints. **out of joint**, dislocated, out of order. **to put someone's nose out of joint**, to upset, disconcert, or supplant a person. **universal joint**, one in which one part is able to swivel in all directions, as in a ball and socket joint. **joint account**, *n.* a bank account held jointly by two or more persons, any of whom may make transactions. **joint-action**, *n.* the joining of several actions in one. **joint-heir**, *n.* **joint-stock**, *n.* stock or capital divided into shares and held jointly by several persons, hence **joint-stock company, firm** etc. **joint-stool**, *n.* a stool made with parts jointed (orig. *joined*) together. **jointtenancy**, *n.* (*Law*) tenure of an estate by unity of interest, title, time, and possession. **joint-tenant**, *n.* **jointweed**, *n.* the mare's-tail or *Equisetum*; (*N Am.*) a herb of the buckwheat family. **jointed**, *a.* **jointedly**, *adv.* **jointer**, *n.* one who or that which joints; (*Carp.*) a long plane used to true the edges of boards to be joined; a pointing tool used by masons and bricklayers. **jointing**, *n.*, *a.* **jointless** (-lis), *a.* **jointly**, *adv.* **jointress** (-tris), *n. fem.* **jointure** (-tūə), *n.* property settled upon a woman in consideration of marriage, which she is to enjoy after her husband's decease. *v.t.* to settle a jointure upon.
joist (joist), *n.* one of a series of parallel horizontal timbers to which floor-boards or the laths of a ceiling are nailed. *v.t.* to furnish with joists.
jojoba (həhō'bə), *n.* a desert shrub of south-western US whose edible seeds provide waxy oil similar to spermaceti.
joke (jōk), *n.* something said or done to excite laughter or merriment; a jest a ridiculous incident, circumstance etc. *v.i.* to make jokes, to jest. *v.t.* to crack jokes upon; to rally. **practical joke**, *n.* a trick played on a person in order to raise a laugh at his expense. **jokee** (-kē'), *n.* one on whom a joke is played. **jokeless** (-lis), *a.* **jokelet** (-lit), *n.* **joker**, *n.* one who jokes, a jester; (*sl.*) a fellow; (*Cards*) an extra card (often printed with a comic device) used with various values in some games. **jokesman, jokesmith, jokester** (-stə), **jokist**, *n.* **jokesome** (-səm), *a.* **jokingly**, *adv.* **joky**, *a.*
jolly (jol'i), *a.* merry, jovial, festive; inspiring or expressing mirth; (*coll.*) pleasant, agreeable, charming; remarkable, extraordinary; (*iron.*) nice, precious; (*sl.*) slightly drunk. *adv.* (*coll.*) very, exceedingly. *v.i.* to be jolly, to make merry. *v.t.* (*sl.*) to banter, to rally; to treat agreeably so as to keep in good humour. **jolly Roger**, *n.* a pirate's flag with skull and cross-bones. **jollify** (-fi), *v.i.* to make merry; to tipple. **jollification** (-fi-), *n.* **jollily**, *adv.* **jolliness, jollity**, *n.*
jolly-boat (jol'ibōt), *n.* a small boat for the general work of a ship.
jolt (jōlt), *v.t.* to shake with sharp, sudden jerks; to move in a carriage along a rough road; to disturb, to shock. *v.i.* to move thus. *n.* a sudden shock or jerk. **jolter**, *n.* **joltingly**, *adv.*
Jonah (jō'nə), *n.* a bringer of bad-luck. [from the prophet *Jonah*]
Jonathan (jon'əthən), *n.* the American people; a typical American; a kind of late-ripening red apple. [prob. from *Jonathan* Trumbull, 1710–85, Governor of Connecticut]
jongleur (zhŏglœ'), *n.* an itinerant minstrel of the Middle Ages, esp. in France. [F]
jonquil (jong'kwil), *n.* the rush-leaved narcissus, with two to six flowers on a stem.
jorum (jaw'rəm), *n.* a large bowl or drinking-vessel; its contents.
Joseph (jō'zif), *n.* a man of invincible chastity (alln. to Gen. xxxix.12).

josh (josh), *v.t.* (*N Am. sl.*), to make fun of, to ridicule. *n.* a friendly joke. **josher**, *n.*
joskin (jos'kin), *n.* (*sl.*) a bumpkin, yokel.
joss (jos), *n.* a Chinese idol. **joss-house**, *n.* a Chinese temple. **joss-stick**, *n.* a stick of perfumed material burnt as incense, orig. in China.
jostle (jos'l), *v.t.* to push against, hustle; to elbow. *v.i.* to push (against, along etc.); to hustle, to crowd. *n.* a hustling; a collision, conflict.
jot (jot), *n.* a tittle, an iota. *v.t.* to write (down a brief note or memorandum of). **jotter**, *n.* a pad or exercise-book for taking notes etc. **jotting**, *n.* a note or memorandum.
joul (jowl), JOWL.
joule (jool), *n.* the SI unit of work and energy, equal to the work done when a force of 1 newton advances its point of application 1 metre. **joulemeter**, *n.* [Dr J.P. *Joule*, 1818–89, British physicist]
jounce (jowns), *v.t.*, *v.i.* to jolt or shake. *n.* a jolt, shake.
journal (jœ'nəl), *n.* an account of daily transactions; (*Book-keeping*) the book from which daily entries are posted up in the ledger; a daily record of events, a diary; newspaper or other periodical published at regular intervals; the transactions of a learned society etc.; a ship's log-book; the part of a shaft that rests on the bearings. **journal-box**, *n.* the case in which the journal moves. **journalese** (-lēz'), *n.* (*derog.*) a superficial style of writing full of cliches etc., regarded as typical of writing in newspapers etc. **journalist**, *n.* an editor of or contributor to a newspaper or other journal; one who keeps a diary. **journalism**, *n.* **journalistic**, *a.* **journalize, -ise**, *v.t.* (*Book-keeping*) to enter in a journal; to enter in a diary. *v.i.* to follow the profession of a journalist; to keep a journal or diary.
journey (jœ'ni), *n.* passage or travel from one place to another, esp. by land; the distance travelled in a given time. *v.i.* to travel. **journeyman**, *n.* a craftsman who has served his apprenticeship and works for an employer; a hack or hireling. **journey-work**, *n.* work performed for hire.
joust (jowst), **just**[2] (jŭst), *v.i.* to tilt, to encounter on horseback with lances. *n.* a combat between knights or men-at-arms on horseback.
Jove (jōv), *n.* Jupiter. **Jovian**, *a.* **jovial**, *a.* mirthful, merry, convivial. **joviality** (-al'-), **jovialness**, *n.* **jovially**, *adv.*
jowl (jowl), *n.* the (lower) jaw; (*often pl.*) the cheek; the throat or neck, esp. of a double-chinned person; the dewlap; the crop or wattle of a fowl. **cheek by jowl**, with the cheeks close together; close together. **jowler**, *n.* a dog with heavy jowls.
joy (joi), *n.* the emotion produced by gratified desire, success, happy fortune, exultation etc.; gladness, happiness, delight; a cause of joy or happiness. *v.i.* to rejoice. *v.t.* to gladden; to congratulate. **no joy**, (*coll.*) lack of success; no news. **joy-bells**, *n.pl.* peals rung on festive occasions. **joy-ride**, *n.* (*coll.*) a ride in a car for pleasure, especially when unauthorized. **joy-rider**, *n.* **joy-stick**, *n.* the control-lever of an aeroplane; a lever for controlling the movement of a cursor on a computer screen. **joyful**, *a.* **joyfully**, *adv.* **joyfulness**, *n.* **joyless** (-lis), *a.* **joylessly**, *adv.* **joylessness**, *n.* **joyous**, *a.* joyful; causing joy. **joyously**, *adv.* **joyousness**, *n.*
JP, (*abbr.*) Justice of the Peace.
Jr., **jr.**, (*abbr.*) junior.
juba[1] (joo'bə), *n.* a mane, as of a horse. **jubate** (-bāt), *a.* maned; fringed.
juba[2] (joo'bə), *n.* a characteristic Negro dance.
jube (joo'bi), *n.* a rood-loft in a church.
jubilate[1] (joo'bilāt), *v.i.* to exult; to express intense joy. **jubilance, jubilation**, *n.* **jubilant**, *a.* **jubilantly**, *adv.*
jubilate[2] (joobilah'ti, yoo-), *n.* the 100th Psalm from its Latin commencing words *Jubilate Deo*; (*fig.*) a shout of joy or exultation.

jubilee (joo'bilē), *n.* a Jewish festival proclaimed by the sound of a trumpet, and celebrated every 50th year to commemorate their deliverance from Egyptian slavery; the 50th anniversary of an event; a season of public festivity; in the Roman Catholic Church, a year of special indulgence or remission of the guilt of sin. **silver, golden, diamond jubilee,** a 25th, 50th, 60th anniversary.
Judaeo-, *comb.form* of or relating to the Jews or Judaism.
Judaeophobe (judē'əfōb), *n.* one who fears or dislikes Jews. **Judaeophobia** (-fō'biə), *n.*
Judaic (joodā'ik), **-ical**, *a.* pertaining to Jews, Jewish. **Judaically**, *adv.* **Judaism** (joo'-), *n.* the religion of the Jews, according to the law of Moses; the Jews colllectively; Jewish culture. **Judaist**, *n.* **Judaize, -ise,** *v.t., v.i.* **Judaization,** *n.* **Judaizer,** *n.*
Judas (joo'dəs), *n.* the disciple who betrayed Jesus Christ; a traitor. **Judas-coloured,** *a.* red, reddish (from a tradition that Judas had red hair). **Judas-hole, -window,** *a.* a spy-hole cut in a door. **Judas-tree,** *n.* a leguminous tree which flowers before the leaves appear (traditionally the tree on which Judas hanged himself).
judder (jŭd'ə), *v.i.* to wobble; to vibrate; in singing to make rapid changes in intensity during the emission of a note. *n.* a wobble; the vibration of an aircraft.
judge (jŭj), *n.* a civil officer invested with power to hear and determine causes in a court of justice; one authorized to decide a dispute or contest; one skilled in deciding on relative merits, a connoisseur; (*Jewish Hist.*) a chief civil and military magistrate among the Jews, from the death of Joshua to the Kings. *v.t.* to decide (a question); to hear or try (a cause); to pass sentence upon; to examine and form an opinion upon (an exhibition etc.); to criticize; (*coll.*) to consider, to estimate, decide. *v.i.* to hear and determine a case; to give sentence; to form or give an opinion; to come to a conclusion; to be censorious. **judge advocate,** *n.* an officer in charge of proceedings at a court martial. **judges' rules,** *n. pl.* in English law, a set of rules governing the behaviour of the police towards suspects. **judger,** *n.* **judgeship,** *n.* **judgingly,** *adv.* **judgment, judgement,** *n.* the act of judging; a judicial verdict; discernment, discrimination; the capacity for arriving at reasonable conclusions; criticism; opinion, estimate; a misfortune regarded as sent by God. **judgment of Solomon,** any judgment designed to reveal the false claimant, after Solomon in I Kings iii.16–28. **Last Judgment,** the judgment of mankind by God at the end of the world. **Judgment Day,** *n.* the day of this. **judgment debt,** *n.* a debt secured by a judge's order, under which an execution can be levied at any time. **judgmental,** *a.* (severely) critical.
judicature (joo'dikəchə), *n.* the administration of justice by trial and judgment; the authority of a judge; a court of justice; the jurisdiction of a court; a body of judges. **judicatory,** *a.* pertaining to the administration of justice. *n.* a court of justice; the administration of justice.
judicial (joodish'əl), *a.* pertaining to courts of law or the administration of justice; having or exercising the qualities of a judge; impartial. **judicial murder,** *n.* a legal but unjust sentence of capital punishment. **judicial separation,** *n.* separation of married persons by order of the Divorce Court. **judicially,** *adv.* **judiciary,** *a.* judicial; passing judgment. *n.* judges collectively. **judicious,** *a.* sagacious, discerning; wise, prudent; done with reason or judgment. **judiciously,** *adv.* **judiciousness,** *n.*
judo (joo'dō), *n.* a modern sport derived from a form of ju-jitsu.
Judy (joo'di), *n.* the name of Punch's wife in the Punch and Judy show; (*derog.*) a woman.
jug[1] (jŭg), *n.* a vessel, usually with a handle and a spout, for holding liquors; (*sl.*) a prison. *v.t.* (*usu. in p.p.*) to stew (a hare) in a jug or jar; (*sl.*) to imprison. **jugful,** *a.* **jug-jug,** *n.*
jug[2] (jŭg), *v.i.* of the nightingale etc. to make a sound like 'jug'. **jug-jug,** *n.*

jugal (joo'gəl), *a.* pertaining to a yoke or a cheek-bone.
jugate (joo'gət), *a.* (*Bot.*) having leaflets in pairs.
juggernaut (jŭg'ənawt), *n.* Vishnu in his eighth avatar; his idol carried on a large cart under which fanatics are said to have thrown themselves; (*fig.*) a belief, institution etc. to which one is ruthlessly sacrificed or by which one is ruthlessly destroyed; a very large articulated lorry.
juggins (jŭg'inz), *n.* (*coll.*) a blockhead, dolt.
juggle (jŭg'l), *v.i.* to play tricks by sleight of hand, to conjure; to throw in the air and catch several objects, such as balls, continuously so that some are in the air all the time. *v.t.* to deceive by trickery; to manipulate (facts, figures etc.) in order to deceive; to try to keep several activities going at the same time. *n.* an act of juggling. **juggler,** *n.* **jugglery,** *n.*
Juglans (joo'glənz), *n.* a genus of trees containing the walnuts. **juglandaceous** (-dā'shəs), *a.*
Jugoslav (yoo'gəslahv), YUGOSLAV.
jugular (jŭg'ūlə), *a.* belonging to the neck or throat. *n.* a jugular vein. **to go for the jugular,** to attack someone where he/she is most likely to be harmed. **jugular veins,** *n.pl.* the veins of the neck which return the blood from the head. **jugulate,** *v.t.* to kill.
juice (joos), *n.* the watery part of vegetable or animal tissues; (*sl.*) electric current, petrol; the essence or characteristic element of anything. **to juice up,** (*sl.*) to make more lively. **to step on the juice,** (*sl.*) to accelerate a motor-car. **juiceless** (-lis), *a.* **juicer,** *n.* a machine for extracting the juice from fruit. **juicy,** *a.* abounding in juice, succulent; interesting, titillating; (*coll.*) profitable. **juiciness,** *n.*
ju-jitsu (joojit'soo), *n.* the Japanese art of wrestling, based on the principle of making one's opponent exert his strength to his own disadvantage.
ju-ju (joo'joo), *n.* a fetish, an idol credited with supernatural power; the power residing in this.
jujube (joo'joob), *n.* the berry-like fruit of several spiny shrubs of the buckthorn family, dried as a sweetmeat; a sweetmeat flavoured with or imitating this.
juke-box (jook'boks), *n.* a kind of large automatic record player, in which coins are inserted and buttons pressed to select the relevant tunes.
Jul., (*abbr.*) July.
julep (joo'ləp), *n.* a sweet drink, esp. medicated; a stimulant composed of spirit, usu. flavoured with mint.
Julian (joo'liən), *a.* pertaining to or originated by Julius Caesar. **Julian calendar,** *n.* the calendar instituted by him in 46 BC. **Julian year,** *n.* the year of this, containing 365¼ days.
julienne (joolien'), *n.* a clear soup with shredded vegetables; a variety of pear.
July (jəli'), *n.* the seventh month of the year.
jumble (jŭm'bl), *v.t.* to mix confusedly; to throw or put together without order. *n.* a confused mixture; disorder, confusion; articles suitable for a jumble-sale. **jumble-sale,** *n.* a sale of miscellaneous articles at a bazaar etc. **jumbleshop,** *n.* **jumbly,** *a.*
jumbo (jŭm'bō), *n.* (*pl.* **-bos**) a huge, unwieldy person, animal or thing, orig. the proper name of an elephant; an over-sized object. **jumbo jet,** *n.* a very large jet-propelled aircraft. **jumbo-size(d),** *a.* of much larger than usual size. **jumboesque** (-esk'), *a.* **jumboism,** *n.*
jumbuk (jŭm'bŭk), *n.* (*austr.*) a sheep.
jumelle (joomel', zhoo-), *a.* twin, paired. *n.* a gimmal; a pair of opera-glasses. [F]
jump[1] (jŭmp), *v.i.* to throw oneself from the ground by a sudden movement of the legs and feet; to spring, bound; to move suddenly (along, off, up, out); to start or rise (up) abruptly; (*fig.*) to agree, tally (with or together). *v.t.* to pass over or cross by leaping; to cause to leap over; to skip (a chapter, pages etc.). *n.* the act of jumping; a leap, spring, bound; an involuntary nervous movement, esp.

jumper — justice

(*pl.*) convulsive twitching as in delirium tremens; a sudden rise (in price, value etc.); a break, a gap. **to jump at**, to accept eagerly; to reach hastily (as a conclusion). **to jump down one's throat**, to answer or interrupt violently. **to jump on**, to reprimand, abuse, or assail violently. **to jump one's bail**, to abscond. **to jump ship**, of a sailor etc., to leave a ship without permission, to desert. **to jump the gun**, to get off one's mark in a race too soon; to take action prematurely. **to jump the queue**, to get ahead of one's turn. **to jump to it**, (*coll.*) to act swiftly. **jump-jet**, *n.* a jet aeroplane which can take off and land vertically. **jump-jockey**, *n.* one who rides in steeplechases. **jump-leads**, *n.pl.* two cables which can connect a flat car battery to an outside battery in order to start the car. **jump-seat**, *n.* a folding seat in a vehicle. **jump-start**, *v.t.* to start (a car) by pushing it and then engaging gear. **jump-suit**, *n.* a one-piece garment consisting of combined trousers and top. **jumpable**, *a.* **jumped-up**, *a.* up-start. **jumper**[1], *n.* one who or that which jumps or leaps; a jumping insect; a tool or implement worked with a jumping motion; a quarryman's boring-tool. **jumping**, *n.*, *a.* **jumping bean** or **seed**, *n.* the seed of a Mexican plant which jumps about through the movements of larvae inside it. **jumping-deer**, *n.* the black-tailed deer found west of the Mississippi. **jumping-jack**, *n.* a toy figure whose limbs move when a string is pulled. **jumping-rope**, *n.* (*N Am.*) a skipping-rope. **jumpy**, *a.* moving or proceeding with jumps and jerks; (*coll.*) nervous, easily startled. **jumpily**, *adv.* **jumpiness**, *n.*
jumper[2] (jŭm'pə), *n.* a woman's knitted upper garment; (*N Am.*) a pinafore dress.
juncaceous (jŭngkā'shəs), *a.* of or resembling rushes. **juncal** (jŭng'-), *a.*
junco (jŭng'kō), *n.* the snow-bird, a genus of N American finches.
junction (jŭnk'shən), *n.* the act of joining or the state of being joined; a point or place of union, esp. the point where lines of railway meet. **junction box**, *n.* an earthed box in which wires and cables can be safely connected.
juncture (jŭnk'chə), *n.* a junction, union; a point of time marked by important events.
June (joon), *n.* the sixth month of the year. **June-bug**, *n.* an insect or beetle that appears about June, chiefly in the US.
Jungian (yoong'iən), *a.* pertaining to the psychoanalytical style of Carl *Jung* (1875–1961).
jungle (jŭng'gl), *n.* land covered with forest trees or dense, matted vegetation; a place of ruthless competition; anything difficult to negotiate, understand etc.; a confusing mass. **jungle-bear**, *n.* the Indian sloth-bear. **jungle-cat**, *n.* the marsh lynx. **jungle-cock, -hen**, *n.* **jungle-fever**, *n.* a remittent tropical fever. **jungle-fowl**, *n.* an E Indian gallinaceous bird. **jungle juice**, *n.* (*sl.*) alcoholic liquor. **jungled**, *a.* **jungli** (-gli), *a.* uncouth, unrefined. **jungly**, *a.*
junior (joon'yə), *a.* the younger (esp. as distinguishing two of the same surname); lower in rank. *n.* one younger or of lower rank than another; (*N Am.*) a son. **junior common room**, (in some colleges and universities) a common room for the use of students. **junior school**, *n.* (in England and Wales), a school for pupils aged about 7 to 11. **junior service**, *n.* the Army. **juniorship**, *n.*
juniper (joo'nipə), *n.* a genus of prickly evergreen shrubs, berries of which are used to flavour gin.
junk[1] (jŭngk), *n.* a flat-bottomed vessel with lugsails, used in the Chinese seas.
junk[2], *n.* rubbish, valueless odds and ends; a lump or chunk of anything; (*sl.*) a narcotic drug. *v.t.* (*sl.*) to discard, abandon. **junk bond**, *n.* a bond giving a high yield but low security. **junk-dealer**, *n.* a marine-store dealer. **junk food**, *n.* food of little nutritional value, quick to prepare. **junk mail**, *n.* unsolicited mail, usu. advertising material. **junk-ring**, *n.* a steam-tight packing round a piston. **junk-shop**, *n.* a shop where second-hand goods of all kinds are sold. **junkie, junky**, *n.* (*sl.*) a drug addict.
junker (yung'kə), *n.* a young German noble; a member of the German reactionary aristocratic party. **junkerdom** (-dəm), **junkerism**, *n.*
junket (jŭng'kit), *n.* a dish of curds sweetened and flavoured with cream and set with rennet; a feast, an entertainment; a supposed business trip (at public expense), really for pleasure. *v.i.* to feast, to picnic. *v.t.* to regale at a feast. **junketer**, *n.* **junketing**, *n.*
Juno (joo'nō), *n.* the wife of Jupiter, a beautiful queenly woman; the third asteroid.
junta (jŭn'tə), *n.* a legislative or administrative council, esp. in Spain, Italy and S America; (also **junto** (-tō) a group, esp of military officers who take control of a country e.g. after a coup. [Sp.]
jupati-palm (joopətē'pahm), *n.* the S American palm yielding raffia fibre.
jupe (joop), *n.* a woman's skirt.
Jupiter (joo'pitə), *n.* the supreme Roman deity; the largest plane of the solar system.
jupon (joo'pən, zhoo'-), *n.* a skirt or petticoat.
jural (joo'rəl), *a.* of or relating to law or jurisprudence, esp. with regard to rights and obligations.
Jurrassic (jooras'ik), *a.* belonging to the oolitic limestone formation well developed in the Jura Mts; belonging to the second period of the Mesozoic era. *n.* the Jurassic system or period.
jurat (joo'rat), *n.* a person under oath; a municipal officer of the Cinque Ports; a magistrate in the Channel Islands. **jurant**, *a.* taking an oath. *n.* one who takes an oath. **juratory**, *a.* containing an oath.
juridical (joorid'ikəl), *a.* pertaining to the administration of justice, to courts of justice, or to jurisprudence. **juridically**, *adv.*
jurisconsult (jooriskənsŭlt'), *n.* one learned in law, esp. civil or international law; a jurist.
jurisdiction (joorisdik'shən), *n.* the legal power or right of administering justice; the district or extent within which such power may be exercised. **jurisdictional, jurisdictive**, *a.*
jurisprudence (joorisproo'dəns), *n.* the science or philosophy of law; the science of the laws, constitutions and rights of men; the legal system of a particular country. **jurisprudent**, *n.*, *a.* **jurisprudential** (-den'-), *a.*
jurist (joo'rist), *n.* one learned in the law; a writer on legal subjects; a student of law. **juristic, -ical** (-ris'), *a.* **juristically**, *adv.*
juror (joo'rə), *n.* one who serves on a jury; one who takes an oath.
jury (joo'ri), *n.* a body of persons (in England usu. 12) selected according to law and sworn to try, and give a true verdict upon, questions put before them; a committee selected to award prizes at public shows, exhibitions etc. **grand jury** GRAND. **jury-box**, *n.* the enclosure in a court where the jury sits. **juryman**, *n.* **jurywoman**, *n.fem.*
jury-mast (joo'rimahst), *n.* a temporary mast erected in place of one carried away. **jury-rigged**, *a.* **jury-rudder**, *n.*
jussive (jŭs'iv), *n.*, *a.* (*Gram.*) (a form or construction) expressing command.
just[1] (jŭst), *a.* acting according to what is right and fair; equitable, impartial, upright, honest; exact, precise; fit, proper, suitable; deserved. *adv.* exactly, precisely; barely, with nothing to spare; precisely at the moment; a very little time ago; (*coll.*) perfectly, quite. **just about**, nearly; more or less. **just now**, a very little time since; at this instant. **just so**, exactly; that is right; with great precision. **justly**, *adv.* **justness**, *n.*
just[2] JOUST.
justice (jŭs'tis), *n.* the quality of being just; fairness in dealing with others; uprightness, honesty; just requital of deserts; the authoritative administration or maintenance

justify (sjŭs'tifī), *v.t.* to prove or show to be just or right; to vindicate, to make good, to show grounds for; to exonerate; (*Theol.*) to declare free from the penalty of sin; to adjust and make (lines of type) even in length. *v.i.* to coincide or range uniformly (of lines of type). **justifiable**, *a.* **justifiability** (-bil'-), **justifiableness**, *n.* **justifiably**, *adv.* **justification** (-fi-), *n.* **justificative, justificatory**, (jŭs'-), *a.* **justifier**, *n.*

Context before: ...of law and right; a magistrate; a judge. **Justice of the Peace**, a local magistrate commissioned to keep the peace and try cases of felony and other misdemeanours. **Lord Chief Justice**, the chief judge of the King's Bench Division. **to do justice to**, to treat fairly; to treat appreciatively. **to do oneself justice**, to acquit oneself worthily of one's ability. **Justice-Clerk**, *n.* (*Sc. Law*) the President of the Outer House or Second Division of the Court of Session, and Vice-President of the High Court of Justiciary. **Justice-General**, *n.* the highest judge in Scotland, Lord President of the Court of Session. **justiceship**, *n.* **justiciable** (-tish'i-), *a.* liable to be tried in a court of justice. *n.* one subject to (another's) jurisdiction. **justiciary** (-əri), *n.* an administrator of justice. *a.* pertaining to the administration of justice. **High Court of Justiciary**, the supreme court of Scotland in criminal causes.

justle (jŭs'l), JOSTLE.

justly, justness JUST [1].

jut (jŭt), *v.i.* to project, protrude; to stick (out). *n.* a protruding point or part. **jut-window**, *n.*

jute (jōōt), *n.* the fibre from the inner bark of two tropical plants, from which fabrics, paper and rope are prepared.

juvenescent (jōōvənes'ənt), *a.* growing or being young. **juvenescence**, *n.*

juvenile (jōō'vənil), *a.* young, youthful; immature; befitting or characteristic of youth. *n.* a young person; a book for children; an actor who usually performs the part of a young person. **juvenile court**, *n.* a court for **juvenile offenders** (under 17 years of age). **juvenile delinquent**, *n.* **juvenileness**, *n.* **juvenilely**, *adv.* **juvenilia** (-nil'iə), *n.pl.* writings etc., produced in youth. **juvenility** (-nil'-), *n.*

juxtapose (jŭkstəpōz'), *v.t.* to place (a thing) next to or (things) side by side. **juxtaposition** (-zish'ən), *n.*

K

K¹, k¹, the 11th letter of the English alphabet, is a voiceless guttural mute (*pl.* **Ks, K's, Kays**)
K², (*abbr.*) the solar constant; kaon; Kelvin scale; (*Chess*) king; Knight; Köchel (catalogue of Mozart's work); one thousand; 1024 words, bytes or bits.
K³, (*chem. symbol*) potassium.
k², (*abbr.*) kilo-.
Kaaba (kah'bə, -əbə), CAABA.
kabbala, kabala (kəbah'lə), CABBALA.
kabuki (kəboo'ki), *n.* a highly-stylized, traditional and popular form of Japanese drama, based on legend and acted only by men, in elaborate costumes.
Kabyle (kəbil'), *n.* one of the agricultural branch of the Berber people inhabiting the highlands of Algeria; the Berber dialect spoken by the Kabyles.
kadi (kah'di), CADI.
Kafir, Kaffir (kaf'ə), *n.* (*now offensive*) one of a S African Bantu people; their language. *a.* of or pertaining to the Kafirs. **kaffir corn**, *n.* a variety of sorghum cultivated in S Africa.
kafir (kaf'ə), *n.* a native of Kafiristan in E Afghanistan; (*offensive*) an infidel.
Kafkaesque (kafkəesk'), *a.* of or like the ideas and work of the Czech novelist Franz Kafka (1883–1924), esp. his ideas on the alienation of man.
kaftan (kaf'tan, -tən), CAFTAN.
kagool, kagoule (kəgool'), CAGOULE.
kahawai (kah'həwī), *n.* the New Zealand salmon.
kai (kī), *n.* a general word for 'food' in New Zealand and the South Sea Islands. [Maori]
kaiak (kī'ak), KAYAK, CAÏQUE.
kail (kāl), KALE.
kainite (kī'nīt), *n.* hydrous chlorosulphate of magnesium and potassium, used as a fertilizer.
Kaiser (kī'zə), *n.* an emperor; the Emperor of Germany or Austria; the head of the Holy Roman Empire. **the Kaiser's war**, the 1914–18 war. **Kaiserdom** (-dəm), *n.* **Kaiserin** (-rin), *n.* the wife of the Kaiser. **Kaiserism**, *n.* **Kaisership**, *n.*
kaka (kah'kə), *n.* a New Zealand parrot belonging to the genus *Nestor*. **kakapo** (-pō), *n.* (*pl.* **-pos**) the ground- or owl-parrot of New Zealand.
kakemono (kakimō'nō), *n.* (*pl.* **-nos**) a Japanese wall-picture mounted on rollers for putting away.
kala-azar (kahləəzah'), *n.* a chronic tropical disease with a high mortality, caused by a protozoan.
kalanchoe (kalənkō'i), *n.* a succulent plant grown indoors or in a greenhouse, with pink, red or yellow flowers.
kalashnikov (kəlash'nikof), *n.* a submachine gun made in the USSR.
kale, kail (kāl), *n.* (*Sc., North.*) cabbage; a cabbage with crinkled leaves; (*Sc.*) cabbage soup. **Scotch kale**, kale with purplish leaves. **kale-yard**, *n.* a kitchen-garden. **kale-yard school**, *n.* a group of novelists and writers depicting the homely life of Scottish lowlanders, with liberal use of broad dialect.
kaleidoscope (kəlī'dəskōp), *n.* an instrument showing by means of bits of coloured glass and a series of reflecting surfaces, an endless variety of symmetrical forms; any complex, changing pattern. **kaleidoscopic** (-skop'-), **-ical**, *a.* **kaleidoscopically**, *adv.*
kalends CALENDS.
kalevala (kah'ləvahlə), *n.* in Finnish legend, the land of the hero Kaleva; the epic which recounts his exploits.
kali (kal'i, kä'-), *n.* the salt-wort, *Salsola kali*, from which soda-ash was obtained. **kaligenous** (kəlij'-), *a.* **kalinite** (kal'init), *n.* (*Min.*) native potash alum.
Kalmia (kal'miə), *n.* a genus of smooth, evergreen N American flowering shrubs. [Peter *Kalm*, 1715–79, Swed. naturalist]
Kalmuck (kal'mŭk), *n.* one of a Mongol people living in a region extending from W China to the Volga; their language.
kalong (kah'long), *n.* the Malay fox-bat, *Pteropus edulis*.
kalpa (kal'pə), *n.* a day of Brahma, or a period of 4,320,000 years, constituting the age or cycle of a world.
Kama (kah'mə), *n.* the god of love in the puranas; impure or sensual desire. **Kamasutra** (-soo'trə), *n.* an ancient Hindu book on erotic love.
kame (kām), *n.* a long mound of glacial detritus, an eskar.
kameez (kəmēz'), *n.* a type of loose tunic with tight sleeves worn by women in S Asia.
kami (kah'mi), *n.* a Japanese title, equivalent to lord, given to nobles, ministers, governors etc.; in Shinto, a divinity, a god.
kamikaze (kamikah'zi), *n.* a Japanese airman or plane performing a suicidal mission in World War II. *a.* pertaining to a kamikaze; (*coll.*) suicidal, self-destructive.
kampong (kam'pong), *n.* a Malay village.
Kampuchean (kəmpuchē'ən), *n.*, *a.* Cambodian.
kamseen, kamsin KHAMSIN.
Kanaka (kənak'ə, kan'-), *n.* a native Hawaiian; a South Sea islander; one of these employed as an indentured labourer on the Queensland sugar-plantations.
Kanarese, Canarese (kanərēz'-), *n.* (*pl.* **-rese**) a member of a Kannada-speaking people living largely in Kanara in southern India; the Kannada language. *a.* of or from the Kanara area.
kanga, khanga (kang'gə), *n.* a piece of brightly coloured cotton worn as a woman's dress in E Africa.
kangaroo (kang.gəroo'), *n.* a name for several marsupial quadrupeds peculiar to Australia, Tasmania, New Guinea and adjacent islands, distinguished by their large hind limbs, used for leaping, and short limbs, almost useless for walking. **kangaroo closure**, *n.* the parliamentary procedure whereby the chairman or speaker decides what shall be discussed (e.g. which clauses of a Bill) and what passed over. **kangaroo court**, *n.* an irregular court, set up by e.g. the mob, prisoners, or strikers; a court where a fair trial is impossible. **kangaroo paw**, *n.* any of several Australian plants with green and red flowers. **kangaroo-rat**, *n.* a small Australian marsupial, an American pouched burrowing-mouse.
kanji (kan'ji), *n.* (*pl.* **-ji, -jis**) a script for representing Japanese syllables derived from Chinese orthography.
Kannada (kan'ədə), *n.* an important Dravidian language spoken in the Mysore area of southern India.
Kantian (kan'tiən), *a.* pertaining to the philosophy of Immanuel Kant (1724–1804). *n.* a Kantist. **Kantianism, Kantist**, *n.*
kaolin (kā'əlin), *n.* a porcelain clay (also used medicinally as a poultice or internally) derived principally from the

kaon 453 **keep**

decomposition of feldspar, China clay. **kaolinic** (-lin'-), *a.*
kaon (kā'on), *n.* an unstable type of meson, also called K-meson.
kapellmeister (kəpel'mī'stə), *n.* the musical director of a choir, band or orchestra. [G]
kapok (kā'pok), *n.* a fine woolly or silky fibre enveloping the seeds of a tropical silk-cotton tree, used for stuffing cushions etc.
Kaposi's sarcoma (kapō'siz sahkō'mə), *n.* a form of skin cancer associated esp. with AIDS victims. [M J *Kaposi*, 1837–1902, Austrian dermatologist]
kappa (kap'ə), *n.* the 10th letter of the Greek alphabet.
kaput (kəput'), *adv.* finished, done for, smashed up. [G]
karabiner (karəbē'nə), *n.* a metal clip with a spring inside it, for attaching to a piton, belay etc., used in mountaineering.
Karaite (keə'rəit), *n.* a member of a Jewish sect who hold by the literal inspiration of the Scriptures, rejecting rabbinical tradition. **Karaism,** *n.*
karakul, caracul (ka'rəkul), *n.* a breed of sheep from the Bukhara district of Central Asia; the fleece prepared as fur from the lambs of these sheep. [*Karakul*, a village in Bukhara, USSR]
karaoke (karəō'ki), *n.* the technique or entertainment of singing in conjunction with a machine which provides a prerecorded backing track and mixes the voice to blend with the accompaniment. [Jap.]
karat (ka'rət), CARAT.
karate (kərah'ti), *n.* a traditional Japanese martial art, based on blows and kicks. **karate chop,** *n.* a downward blow with the side of the hand.
karma (kah'mə), *n.* in Buddhism, the results of action, ethical causation as determining future existence, esp. the cumulative consequence of a person's acts in one stage of existence as controlling his or her destiny in the next. **karmic,** *a.*
karoo, karroo (kəroo'), *n.* a waterless S African tableland.
kaross (karos'), *n.* a S African native mantle or jacket made of skins with the hair left on.
karri (ka'ri), *n.* a W Australian timber tree.
kars(e)y (kah'zi), CARSEY.
karst (kahst), *n.* the characteristic scenery of a limestone region with underground streams, caverns and potholes forming a drainage system. [*Karst*, limestone plateau east of the Adriatic]
kart (kaht), *n.* a go-kart. **karting,** *n.* go-kart racing.
karyo-, *comb. form* cell nucleus.
karyokinesis (kariōkinē'sis), *n.* the series of changes that take place in mitotic cell-division. **karyokinetic** (-net'-), *a.*
kasbah, casbah (kaz'bah), *n.* the castle or fortress in a N African city, or the area around it.
kashmiri (kashmiə'ri), *n.* a native or inhabitant of Kashmir, India; the language of Kashmir.
kashruth, kashrut (kashroot'), *n.* the state of being kosher; the Jewish dietary rules.
katabasis (kətab'əsis), *n.* a moving down. **katabatic** (katəbat'-), *a.*
katabolism, catabolism (kətab'əlizm), *n.* the process of change by which complex organic compounds break down into simpler compounds, destructive metabolism. **katabolic,** *a.*
katalysis (kətal'isis), **katalytic** (-lit'-) etc. CATALYSIS.
katydid (kā'tidid), *n.* a large green orthopterous insect common in N America, so-called from its stridulating cry.
kauri (kow'ri), *n.* a New Zealand coniferous tree. **kauri-gum,** *n.* a resinous gum from the kauri. **kauri-pine,** *n.*
kava (kah'və), *n.* a beverage prepared from the chewed or pounded roots of a Polynesian shrub.
kayak (kī'ak), *n.* the Eskimo and Alaskan canoe, made of sealskins stretched upon a light wooden framework.
kayo (kāō'), *n.* the spoken form of K.O., knockout. *v.t.* to knock (someone) out.

kazoo (kəzoo'), *n.* (*pl.* **-zoos**) a tube of metal or plastic with a membrane covering a hole in the side, through which one sings or hums to produce sound.
KB, (*abbr.*) King's Bench; Knight Bachelor; Knight of the Bath.
KBE, (*abbr.*) Knight Commander of the Order of the British Empire.
kbyte, (*abbr.*) kilobyte.
KC, (*abbr.*) King's Counsel.
kc, (*abbr.*) kilocycle.
kcal, (*abbr.*) kilocalorie.
KCB, (*abbr.*) Knight Commander of the Order of the Bath.
kea (kā'ə), *n.* a green and blue mountain parrot, *Nestor notabilis*, of New Zealand, feeding on carrion and attacking living sheep for their kidney-fat.
kebab (kibab'), *n.* small pieces of meat, with vegetables, cooked on skewers, also called shish kebab.
keck (kek), *v.i.* to retch, to heave; to make a retching sound.
ked (ked), *n.* a sheep-tick.
kedge (kej), *n.* a small portable anchor, used in warping. *v.t.* to move (a ship) by a light cable attached to kedge. *v.i.* of a ship, to move in this way. **kedger,** *n.* a kedge.
kedgeree (kej'erē), *n.* a stew of rice, pulse, onions etc., a common dish in India; a dish of fish, rice etc.
keek (kēk), *v.i.* (Sc., *North.*) to peep, to pry. *n.* a peep. **keeker,** *n.*
keel[1] (kēl), *n.* the principal timber of a ship, extending from bow to stern and supporting the whole structure; a structure corresponding to this, as in an airship; the two lower petals of a papilionaceous corolla; a projecting ridge or longitudinal process. *v.i.* of a ship, to roll on her keel; to turn (over), to careen. *v.t.* to turn up the keel of, to turn over or keel upwards. **on an even keel,** calm, steady, well-balanced. **to keel over,** to capsize; to faint (*coll.*) to fall over. **keelhaul,** *v.t.* to punish by dragging under water one side of the ship and up again on the other; to rebuke severely. **keeled,** *a.* having a keel; (*Bot. etc.*) carinate. **keelless** (-lis), *a.*
keel[2] (kēl), *n.* a lighter or flat-bottomed barge, esp. one of those used for loading colliers in the Tyne. **keeler, keelman,** *n.*
keelhaul KEEL[1].
keelson KELSON.
keen[1] (kēn), *a.* having a sharp edge or point; of an edge, sharp; sensitive, acute, penetrating; of cold etc., biting, piercing; intense; enthusiastic, eager, ardent; competitive. **keen on,** interested in; **keen prices,** *n.pl.* low, competitive prices. **keen-witted,** *a.* **keenly,** *adv.* **keenness,** *n.*
keen[2] (kēn), *n.* lamentation over the body of a deceased person. *v.i.* to raise the keen. **keener,** *n.* a professional mourner.
keep[1] (kēp), *v.t.* (*past, p.p.* **kept** (kept)), to hold, to retain; to have in charge; to guard, preserve, protect; to maintain; to support financially, provide for; to observe, to pay proper regard to; to fulfil, to celebrate; to supply with the necessaries of life; to protect; to tend, to look after; to remain in; to cause to continue, remain or adhere to; to have in pay; to make regular entries in (a ledger, diary, log etc.); to have regularly on sale; to restrain (from); to detain (in custody etc.); to reserve (for); to refrain from divulging; to preserve; to associate with; to store. *v.i.* to continue or retain one's place in, on etc.); to remain; to continue to be (in a specified condition etc.); to remain unspoiled, untainted etc.; to adhere (to); to restrict oneself (to). **keep your hair on** HAIR. **to keep at,** to persist in. **to keep away,** to prevent from approaching. **to keep back,** to restrain, to hold back; to reserve; to keep secret. **to keep body and soul together,** to survive, to maintain life. **to keep company with** COMPANY. **to keep down,** to repress, to subdue; to keep (expenses etc.) low; not to vomit. **to keep from,** to abstain or refrain from; not

keep — **to tell** (someone about something). **to keep house,** to manage a household. **to keep in,** to repress, to restrain; to confine, esp. after school-hours. **to keep in touch with,** to maintain connection with. **to keep in with,** to remain on friendly terms with. **to keep off,** to hinder from approach; to avoid; to avert; to remain at a distance. **to keep on,** to continue to employ etc.; to continue (doing etc.), to persist. **to keep on about,** to continue talking about. **to keep on at,** to nag. **to keep oneself to oneself,** to avoid other people. **to keep one's hand in,** to keep oneself in practice. **to keep out,** to hinder from entering or taking possession (of). **to keep out of,** to stay away from, to avoid. **to keep school, shop etc.,** to conduct a school, shop etc., on one's own account. **to keep tabs on** TAB[1]. **to keep time,** to go accurately; to go rhythmically. **to keep to,** to adhere strictly to. **to keep up,** to maintain; to keep in repair or good condition; to prevent from falling or diminishing; to cause to stay up at night; to bear up; to go on at the same pace (with). **to keep up with the Joneses,** (*coll.*) to keep on the same social level as one's friends and neighbours. **keep fit,** *n.* physical exercises to keep one fit and healthy. **keepnet,** *n.* a net kept in the water by anglers, where they put the fish they have caught to keep them alive. **keepsake** (kēp'sāk), *n.* anything kept or given to be kept for the sake of the giver. **keeper** (kē'pə), *n.* one who or that which keeps; one who retains others in custody or charge; one who has the charge, care or superintendence of anything, as a museum or park; a gamekeeper; a person in charge of animals in a zoo; a ring worn to protect another; the bar of soft iron used to prevent permanent magnets from losing magnetism; a position in some games. **Keeper of the Great Seal,** the officer of State who holds the Great Seal; the Lord Chancellor. **keepership,** *n.* **keeping** (kē'ping), *n.* the action of holding, guarding, preserving etc.; charge, custody, guardianship; harmony, accord. *a.* that can be kept; as fruit. **in, out of keeping,** in or not in harmony (with). **kept,** *a.* **kept woman,** *n.* a woman supported financially by the man whose mistress she is.

keep[2] (kēp), *n.* subsistence, maintenance; food required for subsistence; a donjon; the main tower or stronghold of a mediaeval castle. **keeps,** *n.pl.* **for keeps,** permanently.

keeshond (kās'hond, kēs'-), *n.* a small breed of dog, with a heavy coat, pointed muzzle, and erect ears.

kef, keif KIEF.

keffiyeh (kefē'yə), *n.* a Bedouin Arab's kerchief headdress.

keg (keg), *n.* a small cask or barrel. **keg beer,** *n.* any beer kept in pressurized kegs.

keir (kiə), *n.* a vat for bleaching-liquor, in cloth-, paper-making etc.

kelim KILIM.

keloid (kē'loid), *n.* a hard, pinkish growth of scar tissue, usu. occurring in dark-skinned people.

kelp (kelp), *n.* the calcined ashes of seaweed, from which carbonate of soda was obtained for glass- and soap-making, now chiefly used for obtaining iodine; the large, coarse seaweed from which kelp is produced.

kelpie (kel'pi), *n.* a water-spirit usu. in the form of a horse, supposed to haunt fords, and to rejoice in the drowning of wayfarers; (*Austral.*) a smooth-haired variety of sheep-dog.

kelson (kel'sən), **keelson** (kēl'-), *n.* a longitudinal piece placed along the floor-timbers of a ship binding them to the keel.

Kelt[1] **Keltic** etc. CELT.

kelt[2] (kelt), *n.* (*Sc.*) a spent salmon or sea-trout.

ketter KILTER.

Kelvin (kel'vin), *a.* referring to a thermometer scale in which zero is absolute zero. **Kelvin,** *n.* the basic SI unit of temperature. [Lord *Kelvin,* 1824–1907, British physicist]

ken (ken), *v.t.* (*past, p.p.* **kenned, kent**) (*chiefly Sc.*) to be acquainted with; to understand; to know. *n.* view, sight;

range of sight or knowledge, apprehension. **beyond one's ken,** beyond the limits of one's knowledge or experience. **in one's ken,** within the limits of one's knowledge. **kenning,** *n.* a metaphorical name or phrase for something, in Old English and Old Norse poetry.

Kendo (ken'dō), *n.* the Japanese martial art of fencing, usu. with pliable bamboo staves, occasionally with swords.

kennel (ken'l), *n.* a house or shelter for a dog or hounds; a place where dogs are bred or boarded; a hovel, a wretched haunt or den; a pack of hounds. *v.i.* (*past, p.p.* **kennelled**) to lie or lodge in or as in a kennel. *v.t.* to confine in or as in a kennel. **kennel-maid, -man,** *n.* one who works in a kennel looking after the dogs.

kennel-coal (ken'lkōl), CANNEL.

kenosis (kənō'sis), *n.* Christ's relinquishment of the divine nature at the incarnation. **kenotic** (-not'-), *a.* **kenoticist, kenotist,** *n.*

kenspeckle (ken'spekl), *a.* (*Sc.*) conspicuous; easily recognized.

kent (kent), *past, p.p.* KEN.

Kentish (ken'tish), *a.* pertaining to the county of Kent. **Man of Kent,** a native of Kent born east of the Medway. **Kentish man,** *n.* a native of Kent born west of the Medway.

kentledge (kent'lij), *n.* pigs of iron used for permanent ballast, laid over the kelson-plates.

képi (kā'pē), *n.* a flat-topped military hat with a horizontal peak. [F]

Kepler's laws (kep'ləz), *n.pl.* laws formulated by the astronomer Johann Kepler (1571–1630) concerning the revolution of planets round the sun. **keplerian** (-liə'ri-), *a.* pertaining to Kepler or his laws.

kept (kept), *past, p.p.* KEEP[1].

keramic CERAMIC.

keratin (ke'rətin), *n.* a nitrogenous substance, the chief constituent of hair, feathers, claws and horns. **keratinous,** *a.* **keratinization, -isation,** *n.* the formation of keratin; the state of becoming horny. **keratinize,** *v.i.* **keratitis** (kerəti'tis), *n.* inflammation of the cornea of the eye. **keratose** (ke'rətōs, -tōz), *n.* the substance of the skeleton of horny sponges. *a.* horny. **keratosis** (-tō'sis), *n.* a horny growth on the skin; the skin condition causing this.

kerb (kœb), *n.* a row of stones set as edging to a pavement etc. **kerb-crawling,** *n.* the act of driving along slowly with the intention of enticing someone into the car for sexual purposes. **kerb-crawler,** *n.* **kerb-drill,** *n.* a pedestrian's procedure, as looking to the left and right, for crossing a road in safety, esp. as taught to and used by children. **kerbside,** *n., a.* **kerb-stone,** *n.*

kerchief (kœ'chif), *n.* (*pl.* **-chiefs**) a cloth to cover the head; a handkerchief, a napkin. **kerchiefed,** *a.*

kerel (ker'əl), *n.* (*S Afr.*) a young man.

kerf (kœf), *n.* the slit, notch or channel made by a saw or axe in cutting; the spot where something has been cut or lopped off.

kerfuffle (kəfŭf'l), *n.* commotion, fuss.

kermes (kœ'mēz), *n.* the dried bodies of female scale insects, yielding a red or scarlet dye. **kermes oak,** *n.* a shrubby, dwarf mediterranean oak.

kermis, kermess, kirmess (kœmis), *n.* in the Netherlands, a fair or outdoor festival or merrymaking, orig. a church festival.

kern[1], **kerne** (kœn), *n.* a light-armed Irish foot-soldier; a country lout. **kernish,** *a.*

kern[2] (kœn), *n.* the projecting part of a piece of printing type. **kerned,** *a.*

kernel (kœ'nəl), *n.* the substance, usu. edible, contained in the shell of a nut or the stone of a fruit; the seed, with its husk, of a cereal; that which is enclosed in a shell, husk, integument etc.; the nucleus, core, gist or essence.

kernelled, *a.* having a kernel. **kernel-less** (-lis), *a.*
kerosene, kerosine (ke'rəsēn), *n.* an oil distilled from petroleum, coal or bituminous shale, chiefly used as a fuel.
kerry (ke'ri), *n.* any of a breed of small black dairy cattle, from Ireland. [County Kerry]
Kerry blue (ke'ri), *n.* a large, grey-blue, longhaired breed of terrier.
kersey (kœ'zi), *n.* a coarse woollen cloth, usu. ribbed. [place in Suffolk]
kerseymere (kœ'zimiə), CASSIMERE.
kerygma (kərig'mə), *n.* in the early Christian church, the teaching of the Gospel. **kerygmatic**, *a.*
Kesp[R] (kesp), *n.* a textured vegetable protein used as a meat substitute.
kestrel (kes'trəl), *n.* a small species of hawk, *Falco tinnunculus.*
ketch (kech), *n.* a fore-and-aft rigged two-masted vessel.
ketchup (kech'ŭp), *n.* a sauce, usu. prepared from mushrooms, tomatoes etc.; tomato sauce.
ketone (kē'tōn), *n.* one of a class of organic compounds, usu. formed by oxidation of a secondary alcohol. **ketone body**, *n.* a compound produced in the liver from fatty acids, found in the blood and urine in abnormal amounts in people unable to use glucose, such as diabetics. **ketonic** (-ton'-), *a.* **ketosis** (-tō'sis), *n.* the excessive formation of ketone bodies, as in diabetes.
kettle (ket'l), *n.* a metallic vessel for heating water or other liquid, esp. one with a lid, handle and spout; a hollow or hole in rock. **a pretty kettle of fish**, a muddle, a troublesome state of affairs. **kettledrum**, *n.* a drum made of a thin hemispherical shell of copper or brass, with a parchment head. **kettle-drummer**, *n.* **kettleholder**, *n.* a thick piece of cloth for protecting the hand in holding a hot kettle.
Kewpie doll[R] (kū'pi), *n.* a plump baby-doll with hair in a top-knot.
key[1] (kē), *n.* a portable instrument, usu. of metal, for working the bolt of a lock to and fro; a tool or instrument by which something is screwed up or turned; that which gives access to or opportunity for something; a place whose military occupation gives control over a region of land or sea; that which explains anything difficult; a solution, an explanation; a translation; a series of solutions of problems etc.; a piece of wood or metal let transversely into the back of a board to prevent warping; a keystone; the first coat of plaster on a wall or ceiling which goes between the laths and binds the whole together; a small lever actuated by the fingers in operating certain instruments, machines etc.; one of several systems of musical notes having definite tonic relations among themselves and to one fundamental note called the key-note; the general tone or style (of a picture, literary composition, speech etc.); a dry, winged fruit, as of the ash or elm. *v.t.* to fasten (on, in etc.) with a key, bolt, wedge etc.; to provide with an identifying or explanatory key; to attune (to); to keyboard. *v.i.* to keyboard. *a.* of great importance, essential, fundamental. **House of Keys**, the representative branch of the legislature in the Isle of Man. **to key in**, to enter data into a computer using a keyboard. **to key up**, to brace up, to incite, to encourage. **keyboard**, *n.* the range of keys on a piano, organ, typewriter etc. *v.t.*, *v.i.* to set (text) in type using a keyboard. **keyboarder**, *n.* **key fruit**, *n.* a winged fruit. **key-grip**, *n.* the person in a television studio or on a film set responsible for setting up scenery and camera tracks. **keyhole**, *n.* the hole in a lock, door, cover etc., by which a key is inserted. **key-money**, *n.* a premium demanded, in addition to rent, for the granting or renewal of a tenancy. **key-note**, *n.* (*Mus.*) the fundamental note of a key; the general tone or spirit (of a picture, poem etc.); a central point or principle. *a.* addressing or pertaining to issues of primary importance. **keypad**, *n.* a small device with a push-button keyboard for operating, e.g. a television or teletext system. **key-person**, *n.* an indispensable worker. **key punch**, *n.* a keyboard operated manually and used to put data onto punched cards. *v.t.* to transfer (data) in this way. **key-ring**, *n.* a ring for carrying keys upon. **key signature**, *n.* the sharps and flats on the musical stave, showing the key of a piece of music. **keystone**, *n.* the central stone of an arch locking the others together; the fundamental element, principle etc. **key-stroke**, *n.* the operation of a key on a keyboard-operated machine. **keyed**, *a.* **keyless** (-lis), *a.* not having a key; wound without a key (as a clock or watch).
key[2] (kē), *n.* a low island, esp. of coral, on the coast of Florida.
keystone KEY[1].
KG, (*abbr.*) Knight of the Order of the Garter.
kg, (*abbr.*), keg; kilogram.
KGB, (*abbr.*) Soviet secret police. [Rus. *komitet gosudarstvennoi bezpasnosti* State Security Committee]
khaddar (kah'də), **khadi** (-di), *n.* Indian hand-woven cloth.
khaki (kah'ki), *a.* dust-coloured, dull-yellow. *n.* cloth or cotton material of this colour, used for army uniforms.
khalif (kā'lif, kəl-), CALIPH.
Khalka (kal'kə), *n.* the official language of the Mongolian People's Republic.
khamsin, kamseen, kamsin (kam'sin), *n.* a hot southerly wind blowing in Egypt in March to May.
khan[1] (kan), *n.* orig. a prince, a lord, a chief; now a title (in India, Central Asia etc.) equivalent to 'esquire'; in mediaeval times, a king or emperor, esp. the chief rulers of Tartar, Turkish, and Mongol tribes. **khanate** (-āt), *n.*
khan[2] (kan), *n.* a caravanserai.
khanga KANGA.
Khedive (kidēv'), *n.* the official title of the Governor of Egypt, conferred upon Ismail Pasha in 1867 by the Porte. **Khediva** (-və), **Khediviah** (-viə), *n.* the wife of the Khedive. **khedival, khedivial**, *a.* **khedivate** (-vāt), *n.*
Khmer (kmeə, kmɛ), *n.* a member of a people inhabiting Cambodia; the official language of Cambodia. *a.* pertaining to this people or their language. **Khmerian**, *a.*
Khoisan (koi'sahn, -sahn'), *n.* a family of African languages which includes Hottentot and Bushman languages.
kHz, (*abbr.*) kilohertz.
kiang (kiang'), *n.* an Asian wild ass.
kia ora (kēə aw'rə), *int.* (*New Zealand*) your health!
kibble[1] (kib'l), *n.* a strong iron (formerly wooden) bucket for raising ore from a mine.
kibble[2] (kib'l), *v.t.* to grind (grain, beans etc.) coarsely.
kibbutz (kibuts'), *n.* (*pl.* **kibbutzim** (-im)) a communal agricultural settlement in Israel. **kibbutznik** (-nik), *n.* someone who lives and works on a kibbutz.
kibe (kib), *n.* a chap occasioned by cold; an ulcerated chilblain.
kibitzer (kib'itsə), *n.* (*N Am. coll.*) an interfering looker-on, a meddling spectator. **kibitz**, *v.i.*
kiblah (kib'lə), *n.* the direction of the Caaba at Mecca, to which Muslims turn during prayer.
kibosh, kybosh (kībosh), *n.* (*sl.*) bosh, humbug. **to put the kibosh on**, to do for, to put an end to.
kick (kik), *v.t.* to strike with the foot; to push, move, or drive, by kicking; to strike in recoil; to achieve or score by a kick; (*coll.*) to free oneself of (a bad habit, addiction etc.). *v.i.* to strike out with the foot or feet; to recoil, as a gun; to show opposition, dislike etc. (against, at etc.); to be alive and well; to make a sudden violent movement. *n.* the act of kicking; a blow with the foot; a recoil (of a gun); power, force; a sudden movement or acceleration; a stimulating reaction to alcohol, a drug etc.; a sudden thrill of excitement; an enthusiastic, short-lived interest. **a kick in the teeth**, a rebuff. **to get a kick out of**, to get enjoyment from. **to kick about, around**, (*coll.*) to wander

from place to place; to lie unnoticed or unused; to consider or discuss (an idea etc.). **to kick ass**, (*sl.*) to make one's presence felt in a forceful manner, to show who's boss. **to kick off**, to throw off by kicking; (*Football*) to give the ball the first kick. **to kick one's heels**, to stand idly waiting. **to kick out**, to eject or dismiss contumeliously or with violence. **to kick over the traces**, to throw off any means of restraint or control. **to kick the bucket** BUCKET. **to kick up a dust, fuss, rumpus** etc. DUST. **to kick up one's heels**, to enjoy oneself with no inhibitions. **to kick upstairs**, (*coll.*) to promote, often to a less active or less powerful post. **kickback**, *n.* a strong reaction to something; a sum paid to another person, confidentially, for favours past or future. **kickdown**, *n.* a way of changing gear in an automatic car, by pressing the accelerator pedal right down. **kick-off**, *n.* (*Football*) the first kick in the game. **kick-pleat**, *n.* a pleat at the back of a tight skirt. **kickstand**, *n.* a metal bar attached to the frame of a bicycle or motorcycle for supporting the vehicle when stationary. **kickstart**, *n.* the starting of an engine by kicking down a pedal. *v.t.* to start (an engine) thus. **kick-starter**, *n.* a pedal for kickstarting, e.g. a motorcycle. **kickable**, *a.* **kicker**, *n.* one who or that which kicks; a horse given to kicking.

kickshaw (kik′shaw), *n.* a trinket, a trifle; a light, unsubstantial dish.

kid[1] (kid), *n.* the young of the goat or of a related animal; leather from the skin of this; (*coll.*) a child. *v.i.* (*past, p.p.* **kidded**) to bring forth a kid or kids. **kid glove**, *n.* a glove made of kid. *a.* too fastidious for common tasks etc.; tactful. **with kid gloves**, very carefully or tactfully. **kidskin**, *n.* a smooth, soft leather from a young goat. **kids' stuff**, *n.* (*coll.*) something suitable for children; something childish or very easy. **kiddy**, *n.* a little child. **kiddywink, kiddiewink**, *n.* a child, kiddy. **kidling** (-ling), *n.*

kid[2] (kid), *v.t.* (*past, p.p.* **kidded**) (*coll.*) to humbug, to hoax; to pretend; to deceive for fun. *v.i.* to play around, joke. *n.* a deception, a fraud. **kidder**, *n.* **kiddingly**, *adv.* **kidology**, *n.* (*coll.*) the art or practice of kidding, bluffing.

Kidderminster (kid′əmənstə), *n.* two-ply ingrain carpet orig. made at Kidderminster in Hereford and Worcester.

kidnap (kid′nap), *v.t.* to carry off by force or illegally, to abduct, esp. for ransom. **kidnapper**, *n.*

kidney (kid′ni), *n.* an oblong flattened glandular organ embedded in fatty tissue in the lumbar region on each side of the spine in humans, and serving to secrete urine and remove nitrogenous matter from the blood; an organ with a similar function in other animals; temperament, kind, fashion. **kidney bean**, *n.* the name of two species of *Phaseolus*, the dwarf French bean and the scarlet runner; the seed of these. **kidney machine**, *n.* a machine used to carry out blood dialysis in cases of kidney failure. **kidney stone**, *n.* a hard mass in the kidney. **kidney-vetch**, *n.* a leguminous plant, *Anthyllis vulneraria*.

kief, kif, kef (kēf. kēif), *n.* the drowsy, dreamy, trance-like condition produced by the use of bhang etc.; Indian hemp, smoked in Morocco and Algeria to produce this condition.

kiekie (kē′kē), *n.* a New Zealand climber, the berries of which are eaten and the leaves used for baskets etc.

kieselguhr (kē′zlguə), *n.* diatomite.

kike (kīk), *n., a.* (*offensive*) Jew. [possibly from *-ki* ending of many Jewish immigrants names in the US at the end of the 19th cent.]

Kikuyu (kikoo′ū), *n.* a Bantu-speaking people of Kenya, E Africa; a member of this people; its language.

kilderkin (kil′dəkin), *n.* a small barrel, usu. of 18 gals. (81·8 l); an obsolete liquid measure of this capacity.

kilim, kelim (kēlim′), *n.* a pileless rug woven in tapestry-style patterns, from the Middle East.

kill (kil), *v.t.* to deprive of life; to put to death; to slay; to put an end to, to destroy, to quell; to deaden, to still (pain etc.); to neutralize (effects of colour etc.); to pass or consume (time) idly; to discard, to cancel; to switch off; to cause pain or discomfort; (*coll.*) to overwhelm with admiration, astonishment, personal charms etc.; in lawn tennis, to strike (the ball) so forcibly that it cannot be returned. *v.i.* to put to death; to slaughter, esp. in sport. *n.* the act of killing; an animal or number of animals killed, esp. in sport; (*Lawn Tennis, Rackets*) the hitting of a ball in such a manner that it cannot be returned. **to be in at the kill**, to be present at the end or conclusion of something. **to kill off**, to get rid of by killing. **to kill oneself**, (*coll.*) to over-exert oneself. **to kill two birds with one stone**, to achieve two things with a single action. **kill-joy**, *n.* a person who sheds a general depression on company, a wet blanket. **killer**, *n.* **killer bee**, *n.* an African honeybee which is very aggressive when disturbed. **killer whale**, *n.* a black-and-white toothed whale, *Orcinus orca*, found in most seas. **killing**, *n.* the act of depriving of life, slaughter, *a.* that kills; (*coll.*) exhausting; (*coll.*) excruciatingly funny. **to make a killing**, to make a large profit. **killingly**, *adv.*

killdee (kil′dē), **killdeer** (-diə), *n.* a N American ringplover.

killick (kil′ik), *n.* a stone or small anchor used for mooring a fishing-boat.

killifish (kil′ifish), *n.* a minnow-like fish of the genus *Fundulus* used as bait and to control mosquitoes.

kiln (kiln), *n.* a furnace, oven or stove for calcining, drying, hardening etc. *v.t.* to dry or bake in a kiln. **brick-kiln**, a kiln for baking bricks. **lime-kiln**, a kiln for calcining lime. **kiln-dry**, *v.t.* to dry in a kiln. **kiln-dried**, *a.*

kilo- (kilō), *comb. form.* one thousand. **kilobit**, *n.* (*Comput.*) 1024 bits. **kilobyte**, *n.* (*Comput.*) 1024 bytes. **kilocalorie** CALORIE. **kilocycle**, *n.* (*Elec.*) 1000 cycles per second, a unit for measuring the frequency of alternating current. **kilogram**, *n.* 1000 grams or 2·2046 lb. av., the SI base unit of mass. **kilogram-metre**, *n.* a unit of measurement of work, the energy expended in raising one kilogram to the height of one metre. **kilohertz**, *n.* 1000 hertz, a unit used to measure the frequency of radio waves. **kilojoule**, *n.* 1000 joules. **kilolitre**, *n.* 1000 litres. **kilometre** (kil′əmētə, kilom′itə), *n.* 1000 metres or 0·621 mile. **kilometrical** (-met′-), *a.* **kiloton**, *n.* a measure of explosive power, equivalent to 1000 tons of TNT. **kilovolt**, *n.* 1000 volts. **kilowatt**, *n.* 1000 watts, a unit of measurement of electrical energy. **kilowatt hour**, *n.* a unit of energy or work equivalent to that performed by 1 kilowatt acting for 1 hour.

kilt (kilt), *v.t.* to tuck up (the skirts of a dress); to gather together (the material of a dress) into vertical pleats. *n.* a kind of short skirt usu. of tartan cloth gathered in vertical pleats, worn as part of male dress by the Highlanders of Scotland. **kiltie** (-ti), *n.* (*coll.*) a soldier of a kilted regiment. **kilted**, *a.*

kilter (kil′tə), **kelter** (kel′-), *n.* (*coll.*) good condition, fitness, form. **out of kilter**, not working properly.

kimberlite (kim′bəlīt), *n.* a diamond-bearing clay-like substance, called by miners 'blue earth' or 'blue ground', found in S Africa. [*Kimberley*, S Afr.]

kimono (kimō′nō), *n.* (*pl.* **-nos**) a loose robe fastened with a sash, the principal outer garment of Japanese costume.

kin (kin), *n.* stock, family; relations or connections collectively, kindred; a relation, a connection. *a.* of the same family, nature, or kind; akin. **kith and kin** KITH. **next of kin** the nearest blood relation. **kinsfolk**, *n.* (*collect.*) family relations, kindred. **kinsman, -woman**, *n.* **kinless** (-lis), *a.*

-kin (-kin), *dim. suf.* as in *bumpkin, buskin, cannikin, catkin.*

kinaesthesis, (esp. N Am.) **kinesthesis** (kinəsthē′sis, kī-), **kinaesthesia** (-thē′ziə), *n.* the muscular sense, the perception of muscular movement. **kinaesthetic** (-thet′-), *a.*

kinase (kī′nāz. kin-), *n.* a chemical in the body which converts a zymogen into an enzyme; an enzyme that facili-

tates the transfer of phosphates from ATP.
kincob (king'kob), *n.* a rich E Indian fabric interwoven with gold or silver thread.
kind (kīnd), *n.* race, genus, species, natural group; sort, class, variety, category; manner, fashion, way; fundamental nature. *a.* disposed to do good to others; sympathetic, benevolent, tender; proceeding from or characterized by goodness of heart; cordial, friendly; mild, pleasant, not harmful. **after its kind**, according to its nature. **a kind of**, a sort of; roughly or approximately of the description or class expressed. **in kind**, of payment, wages etc.; in produce or commodities; in the same way or manner. **kind-hearted**, *a.* sympathetic. **kind-heartedly**, *adv.* **kind-heartedness**, *n.* **kindly**, *a.* kind, good-natured, benevolent, genial, beneficial; favourable, auspicious. *adv.* in a considerate or tolerant way; used to express polite formality or impatience when making requests or commands. **to take kindly**, to react favourably. **kindliness**, *n.* **kindness**, *n.*
kindergarten (kin'dəgahtən), *n.* a school or class for infants and young children.
kindle (kin'dl), *v.t.* to set fire to; to light; to inflame, to inspire (the passions etc.); to excite, to stir up (to action or feeling); to light up or illumine. *v.i.* to take fire, to begin to burn or flame; to become inflamed or excited; to become illumined. **kindler**, *n.* **kindling**, *n.* the act of setting on fire; wood, shavings etc., for lighting fires.
kindly KIND.
kindred (kin'drid), *n.* relationship by blood or marriage; affinity or likeness of character; *(collect.)* relatives, kin. *a.* related by blood; congenial, sympathetic; of like nature or qualities.
kine (kīn), *n.pl.* cows, cattle.
kinematics, *n.sing.* the science of pure motion, admitting conceptions of time and velocity but excluding that of force. **kinematic** (kinəmat'ik), *a.* pertaining to movement or to kinematics. **kinematical**, *a.* **kinematically**, *adv.*
kinesics (kinēsiks), *n.sing.* the study of body movements as non-verbal communication.
kinesiology (kinēsiol'əji), *n.* the study of human movement and anatomy. **kinesiologist**, *n.*
kinesis (kinē'sis), *n.* movement under stimulus.
kinesthesis KINAESTHESIS.
kinetic (kinet'ik), *a.* of or producing motion; due to or depending upon motion. **kinetic art**, *n.* art, e.g. sculpture, which has moving parts. **kinetic energy**, *n.* the energy possessed by a body by virtue of its motion. **kinetic theory**, *n.* a theory which accounts for the behaviour of gases, vapours, liquids etc. in terms of the motions of molecules or atoms comprising them. **kinetically**, *adv.* **kinetics**, *n.sing.* that branch of dynamics which treats of forces imparting motion to or influencing motion already imparted to bodies.
kineto-, *comb. form* pertaining to motion; pertaining to kinetics.
kinetograph (kinet'əgraf), *n.* a camera for obtaining photographs of objects in motion. **kinetographer**, *n.* **kinetographic** (-graf'-), *a.* **kinetography** (-tog'-), *n.*
kinetoscope (kinet'əskōp), *n.* a device for exhibiting pictures taken by the kinetograph, an early form of cinematograph; an instrument for combining arcs of different radii into continuous curves.
king (king), *n.* the male sovereign of a nation, esp. a hereditary sovereign of an independent State; a chief, a ruler; one who or that which is pre-eminent in any sphere; a card bearing a representation of a king, usu. ranking next to the ace and before the queen; *(Chess)* a piece which has to be protected from checkmate; *(Draughts)* a piece which has been crowned and is entitled to move in any direction. *v.i.* to act as king; to play the king. *v.t.* to make a king of. **King Charles's spaniel** SPANIEL. **King James Version**, the authorized version of the Bible. **King-of-Arms**, *n.* a senior herald. **king of beasts**, the lion. **king of birds**, the eagle. **King of Kings**, God, the title of various Oriental monarchs. **King of the Castle**, a children's game; the most important person in a group. **king-bird**, *n.* an American tyrant flycatcher. **king-bolt**, *n.* a main or central pin, bolt or pivot. **king-cobra**, *n.* a large, venomous Asian cobra. **king-crab**, *n.* a large crustacean with a carapace shaped like a horseshoe. **king-craft**, *n.* the art of governing; kingly statesmanship. **king-cup**, *n.* the marsh marigold, *Caltha palustris*, and some allied species. **kingfish**, *n.* the opah. **kingfisher**, *n.* any bird of the genus *Alcedo*, esp. a small British bird with brilliant blue and green plumage, subsisting on fish. **kingklip** (-klip), *n.* an eel-like fish. **kingmaker**, *n.* one who sets up kings, esp. Richard Neville, Earl of Warwick, who supported the Houses of York and Lancaster alternately in the Wars of the Roses. **king penguin**, the largest of the penguins, *Aptenodytes longirostris* or *A. patagonica*, also called the emperor penguin. **king-pin**, *n.* the centre pin in ninepins; *(coll.)* a most important person. **kingpost**, *n.* the middle post of a roof, reaching from the ridge to the tie-beam. **king prawn**, *n.* a large prawn. **King's Bench** BENCH. **King's Counsel** COUNSEL. **King's evidence** EVIDENCE. **King's evil**, *n.* scrofula, formerly believed to be cured by the royal touch. **King's highway**, *n.* a public road, a right-of-way. **king-size, king-sized**, *a.* of beds etc., larger than the standard double size. **kingdom** (-dəm), *n.* the territory under rule of a king or queen; the position or attributes of a king; sovereign power or authority; a domain, a territory; the highest and most comprehensive of the divisions into which natural objects are arranged. **kingdom come**, the world to come. **United Kingdom**, Great Britain and Northern Ireland. **kinghood** (-hud), *n.* **kingless** (-lis), *a.* **kinglet** (-lit), *n.* a petty king; the golden-crested wren, *Regulus cristatus*. **king-like**, *a.* **kingly**, *a.*, *adv.* **kingliness**, **kinglihood** (-hud), *n.* **Kings**, *n.sing.* the title of two books of the Old Testament. **kingship**, *n.*
kinin (ki'nin), *n.* a hormone which causes dilation of the blood vessels; a hormone which promotes cell division and slows down the aging process in plants.
kink (kingk), *n.* a twist or abrupt bend in a rope, thread, wire etc.; a prejudice, a crotchet, a whim; a crick, as in the neck; a flaw. *v.t., v.i.* to (cause to) twist or run into kinks. **kinky**, *a.* twisted; curly; given to abnormal sexual practices; provocative; unusual, idiosyncratic. **kinkily**, *adv.* **kinkiness**, *n.*
kinkajou (king'kəjoo), *n.* an arboreal carnivorous quadruped of S and Central America, allied to the racoon, with long body and prehensile tail.
kino (kē'nō), *n.* an astringent gum used for tanning or dyeing and in medicine, obtained from certain Indian, African and Australian trees.
kinsfolk KIN.
kiosk (kē'osk), *n.* an open pavilion or summerhouse; a light ornamental structure for the sale of newspapers etc.; a public telephone booth.
kip¹ (kip), *n.* the hide of a calf or of small cattle, used for leather; leather made from such skins. **kip-leather, kipskin**, *n.*
kip² (kip), *n.* (*sl.*) a lodging-house; a bed; (a period of) sleep. *v.i.* (*past, p.p.* **kipped**) to lie down to sleep; to sleep. **to kip down**, to go to bed.
kip³ (kip), *n.* (*Austral.*) a wooden bat for tossing coins in the game of two-up.
kipper (kip'ə), *n.* a male salmon during the spawning season; a salmon or herring split open, salted, and smoke-dried *v.t.* to cure and preserve (salmon, herrings etc.) by rubbing with salt, pepper etc., and drying or smoking. **kipperer**, *n.*
kir (kiə), *n.* a drink, made from white wine and cassis. [Felix *Kir*, 1876–1968, mayor of Dijon, France, who invented it]

Kirbigrip*, **kirby grip** (kœ´bigrip), *n.* a type of hair-grip.
kirk (kœk), *n.* (*Sc.*) a church, the Established Church of Scotland, esp. in contradistinction to the Church of England or the Scottish Episcopal Church. **kirkman**, *n.* **kirk-session**, *n.* the lowest court in the Kirk of Scotland and other Presbyterian Churches consisting of the minister and elders.
kirmiss KERMIS.
kirsch (kiəsh), **kirschwasser** (-vasə), *n.* an alcoholic liqueur distilled from the fermented juice of the black cherry. [G]
kirtle (kœ´tl), *n.* a woman's gown or petticoat; a man's short jacket, tunic or coat.
kismet (kiz´mət, kis´-), *n.* fate, destiny.
kiss (kis), *n.* a caress or salute with the lips; in billiards, a mere touch of the moving balls; a confection of sugar, white of eggs etc. *v.t.* to salute or caress by pressing or touching with the lips; to touch or graze in passing. *v.i.* to join lips in affection or respect; of moving billiard balls, to come in contact. **kiss and sell**, the practice of revealing one's sexual adventures for money, as to a newspaper. **kiss-and-tell**, *a.* pertaining to a memoir etc. that relates one's sexual adventures or other private or confidential matters. *n.* the practice of revealing such matters; a book or article that contains such details. **kiss-me-quick**, *n.* the wild pansy or heartsease, *Viola tricolor*; a small old-fashioned bonnet. **kiss of death**, something which will inevitably lead to failure. **kiss of life**, mouth-to-mouth resuscitation. **kiss of peace**, a ceremonial embrace in the Christian church. **to kiss away**, to wipe away by kissing. **kissagram** (-əgram), *n.* a greetings service where the person employed to deliver the greeting kisses the person who is celebrating. **kiss-curl**, *n.* a curl hanging over the forehead, in front of the ear, or at the nape of the neck. **kisser**, *n.* one who kisses; (*sl.*) the mouth. **kissing**, *n.*, *a.* **kissing-cousin**, *n.* a relation familiar enough to be kissed on meeting. **kissable**, *a.*
kit[1] (kit), *n.* a wooden tub, as for pickled fish, butter etc.; the necessaries, tools etc. for a particular purpose assembled in a container; such a container; an outfit, esp. the equipment of a soldier; pieces of equipment, sold as a set, and ready for assembly. **to kit out, up**, to fit out with the necessary clothes or equipment. **kit-bag**, *n.* a strong bag for holding a person's gear, esp. a serviceman's.
kit[2] (kit), *n.* a small violin used by dancing-masters.
kit[3] (kit), *n.* a kitten.
kitchen (kich´ən), *n.* the room in a house etc. where food is cooked. **Kitchen Dutch**, *n.* a mixture of Dutch or Kaffir with English. **kitchen-garden**, *n.* a garden in which fruit and vegetables are cultivated for the table. **kitchen-maid**, *n.* a female servant whose business it is to assist the cook. **kitchen-midden**, *n.* a prehistoric refuse-heap, or shellmound, first noticed on the coast of Denmark, and since found in the British Isles etc. **kitchen range**, *n.* a kitchen grate with oven, boiler etc., for cooking. **kitchen sink**, *a.* of a type of British drama which depicts the reality and often sordid quality of family life. **kitchen tea**, *n.* (*Austral.*, *New Zealand*) a party held before a wedding to which the guests bring gifts of kitchenware. **kitchen unit**, *n.* a modern kitchen fitment, as a cupboard. **kitchenware**, *n.* the pots, pans and utensils used in the kitchen. **kitchener**, *n.* a cooking-range; one employed in a kitchen, esp. that of a monastery. **kitchenette** (-net´), *n.* a small kitchen.
kite (kīt), *n.* a medium-sized bird of the hawk family, esp. *Milvus ictinus*, the common or European kite; a greedy or rapacious person, a device consisting of a light frame of wood and paper constructed to fly in the air by means of a string; (*sl.*) an aircraft; (*Comm. sl.*) an accommodation note or bill; (*pl.*) light sails, set only in very light winds, above the other sails. *v.i.* to fly like a kite. **to fly a kite**, to try how the wind blows; (*Comm. sl.*) to raise money on an accommodation bill; to enquire about a situation, public opinion etc. **kite-balloon**, *n.* an observation-balloon moored to the ground. **kite-flyer**, *n.* **kite-flying**, *n.* flying and controlling a kite; the circulation of rumours to test public opinion. **kite-mark**, *n.* a kite-shaped mark indicating that goods conform in all particulars with the specifications of the British Standards Institution.
kith (kith), *n.* kindred. **kith and kin**, close friends and relations; relatives only.
kitsch (kich), *n.* art or literature that is inferior or in bad taste, and designed to appeal to popular taste. **kitschy**, *a.*
kitten (kit´n), *n.* the young of the cat. *v.i.* to bring forth young, as a cat. **to have kittens**, (*coll.*) to be over-excited, very annoyed etc. **kittenish**, *a.* **kitty**[1], *n.* a pet-name for a kitten.
kittiwake (kit´iwāk), *n.* a seagull of the genus *Rissa*, esp. *R. tridactyla*, common on the British coasts.
kitty[1] KITTEN.
kitty[2] (kit´i), *n.* the pool into which each player puts a stake in poker and other games; a common fund of money.
kiwi (kē´wē), *n.* the New Zealand apteryx or wingless bird; (*coll.*) a New Zealander. **kiwi fruit** (kē´wē), *n.* the edible green fruit of the Chinese gooseberry, an Asiatic climbing plant.
kJ, (*abbr.*) kilojoule.
KKK, (*abbr.*) Ku Klux Klan.
kl., (*abbr.*) kilolitre.
Klan KU KLUX KLAN.
Klaxon* (klak´sən), *n.* a loud horn formerly used on cars.
Kleenex* (klē´neks), *n.* (*pl.* **Kleenex**, **Kleenexes**) soft paper tissue used as a handkerchief etc.
Klein bottle (klīn), *n.* in mathematics, a one-sided surface surrounding a three-dimensional space, formed by putting the narrow end of tapered tube through the surface of the tube, then stretching it to fit into the other end. [Felix Klein, 1849–1925, G mathematician]
kleptomania (kleptəmā´niə), *n.* a form of insanity or mental aberration displaying itself in an irresistible propensity to steal. **kleptomaniac** (-ak), *n.*
Klieg light (klēg), *n.* a powerful arc lamp used as floodlighting in a film studio. [John *Kliegl*, 1869–1959, and Anton *Kliegl*, 1872–1927, inventors]
klipspringer (klip´springə), *n.* a small South African antelope.
kloof (kloof), *n.* (*S Afr.*) a ravine, gully or mountain gorge.
klystron (klis´tron, klī´-), *n.* an electron tube used to amplify or generate microwaves.
km, (*abbr.*) kilometre.
K-meson KAON.
knack (nak), *n.* a trick or adroit way of doing a thing; dexterity, adroitness, ability, aptitude.
knacker (nak´ə), *n.* a dealer in worn-out horses; a horse-slaughterer; a dealer in second-hand goods, houses, ships etc. *v.t.* (*sl.*) to tire out.
knackwurst (nak´wœst), *n.* a spicy sausage.
knap (nap), *v.t.* (*past, p.p.* **knapped**) to break into pieces, esp. with a sharp snapping noise; to break, flake, or chip (flint). **knapper**, *n.*
knapsack (nap´sak), *n.* a case or bag for clothes etc., carried on the back during a march by soldiers, tourists etc.
knapweed (nap´wēd), *n.* a composite plant with purple globular flowers of the genus *Centaurea*, esp. *C. nigra*, the black knapweed and *C. scabiosa*, the great knapweed.
knar, **gnar** (nah), *n.* a knot in wood; a protuberance on the trunk or branch of a tree.
knave (nāv), *n.* a deceitful, cunning fellow, a rogue; a court-card with a representation of a soldier or servant, the jack. **knavery** (-vəri), *n.* dishonesty. **knavish**, *a.* fraudulent. **knavishly**, *adv.* **knavishness**, *n.*
knead (nēd), *v.t.* to work up (flour, clay etc.) with the

hands into a plastic mass; to work or incorporate into dough: to shape, fashion, mingle or blend by this method; to work thus on (the muscles etc.) in massage. **kneadable**, *a.* **kneader**, *n.* **kneading-trough**, *n.* a trough in which dough is worked up.
knee (nē), *n.* (the area surrounding) the joint of the thigh or femur with the lower leg; a joint roughly corresponding to this in animals other than humans; the part of a garment covering the knee; the lap; a piece of timber or metal cut or cast with an angle like that of the knee to connect beams etc.; anything resembling a knee in shape or function. *v.t.* to touch or strike with the knee; (*coll.*) to cause (trousers) to bag at the knees. **to bring to one's knees**, to reduce to submission. **knee-cap**, *n.* a padded cover for the knee; the heart-shaped sesamoid bone in front of the knee-joint; *v.t.* to shoot or injure someone in the knees. **knee-capping**, *n.* **knee-deep**, *a.* sunk in as far as the knees. **knee-high**, *a.* coming up to the knee. **knee-hole**, *n.* the hole between the pedestals of a writing-table or desk. **knee-jerk**, *n.* a reflex kick of the lower part of the leg; (*coll.*) a reflex, an automatic reaction. **knee-length**, *a.* reaching down to, or up to, the knee. **knee-pan**, *n.* the knee-cap or socket of the knee. **knees-up**, *n.* (*coll.*) a party. **kneed**, *a.* (*usu. in comb.*, as *loose-kneed*).
kneel (nēl), *v.i.* (*past, p.p.* **kneeled, knelt** (nelt), to fall or support the body on the knees. **kneeler**, *n.* one who kneels; a stool or cushion for kneeling on.
knell (nel), *v.i.* to ring, to toll, as a funeral bell; to sound in a mournful or ominous manner. *v.t.* to proclaim or summon by or as by a knell. *n.* the sound of a bell when struck, esp. at a death or funeral; an evil omen, a death-blow.
knelt (nelt), *past, p.p.* KNEEL.
Knesset (knes'it), *n.* the single-chamber parliament of the state of Israel.
knew (nū), *past* KNOW.
Knickerbocker (nik'əbokə), *n.* a New Yorker of original Dutch descent. **knickerbockers**, *n.pl.* loose breeches gathered in below the knee. **knickerbocker glory**, *n.* a large ice-cream sundae, with fruit and jelly. [imag. author of Washington Irving's *History of New York*]
knickers (nik'əz), *n.pl.* women's underpants. **to get one's knickers in a twist**, (*coll.*) to be over-anxious, upset etc.
knick-knack (nik'nak), *n.* any little ornamental article; a showy trifle.
knife (nīf), *n.* (*pl.* **knives** (nīvz)) a blade with one edge sharpened, usu. set in a handle; a cutting-blade forming part of a machine. *v.t.* to stab or cut with a knife; to defeat or betray by underhand means. **to have one's knife in someone**, to be vindictive towards someone. **under the knife**, (*coll.*) undergoing a surgical operation. **war to the knife**, mortal combat. **knife-edge**, *n.* the edge of a knife; a hard steel edge used as fulcrum for a balance, pendulum etc.; a sharp ridge; a difficult situation where things could go either right or wrong. **knife-pleat**, *n.* a single, narrow pleat.
knight (nīt), *n.* a man of gentle birth, usu. one who had served as page and esquire, admitted to an honourable degree of military rank, with ceremonies or religious rites; one who holds a corresponding non-hereditary dignity conferred by the sovereign or his or her representative, and entitling the possessor to the title of 'Sir' prefixed to his name; (*Chess*) a piece shaped like a horse's head entitled to move two squares straight and one at right-angles; a chivalrous or quixotic person; one acting as chevalier to a lady. *v.t.* to create or dub (a person) a knight. **knight of the road**, a highwayman; a tramp; a commercial traveller. **knight of the shire**, (*Hist.*) a representative of an English county in Parliament. **knight-bachelor** BACHELOR. **knight-errant**, *n.* a mediaeval knight who wandered about in quest of adventures to show his prowess and generosity. **knight-errantry**, *n.* **knight's progress**, *n.* in chess, a combination of moves which allow a knight to visit every square on the board. **Knight Templar**, *n.* (*pl.* **Knights Templars, Knights Templar**) TEMPLAR. **knightage** (-ij), *n.* knights collectively. **knighthood** (-hud), *n.* **knightlike**, *a.* **knightly**, *a., adv.* **knightliness**, *n.*
kniphofia (nifō'fiə), *n.* the red-hot poker. [J.H. *Kniphof*, 1704–63, G professor of medicine]
knit (nit), *v.t.* (*past, p.p.* **knitted, knit**) to form into a fabric or form (a fabric, garment etc.) by looping or knotting a continuous yarn or thread; to join closely together, to unite; to make close or compact; to contract into folds or wrinkles. *v.i.* to make a textile fabric by interweaving yarn or thread; to grow together; to become closely united. *n.* style of knitting; a knitted fabric or garment. **knitwear**, *n.* knitted clothes. **knitter**, *n.* **knitting**, *n.* the action of one who knits; knitted work. **knitting-machine**, *n.* an apparatus for mechanically knitting jerseys etc. **knitting-needle, -pin**, *n.* a long eyeless needle of metal, wood etc., used in knitting.
knives (nīvz), *pl.* KNIFE.
knob (nob), *n.* a rounded protuberance, usu. at the end of something; a rounded handle of a door, lock, drawer etc.; (*N Am.*) a rounded hill, a knoll; an ornamental terminal boss; a small lump (of coal, sugar etc.); (*sl.*) the penis. *v.t.* (*past, p.p.* **knobbed**) to furnish with a knob or knobs. *v.i.* to bulge or bunch (out). **with knobs on**, even more so. **knobstick**, *n.* a knobbed stick used as a weapon; (*sl.*) a worker who refuses to join a strike. **knobbed, knobby**, *a.* **knobbiness**, *n.* **knobble** (nob'l), *n.* a small knob. **knobbly**, *a.* **knoblike**, *a.*
knobkerrie (nob'keri), *n.* the round-headed club used as a weapon by S African tribesmen.
knock (nok), *v.t.* to strike, to hit, to give a hard blow to; to drive or force by striking; (*coll.*) to be disparaging about, censure. *v.i.* to strike hard or smartly (at, against, together etc.); to collide; to make a sharp rapping sound; of an internal combustion engine, to pink. *n.* a blow; a rap, esp. on a door for admission. **knock-for-knock**, *a.* of an agreement between vehicle insurance companies by which each company pays for the damage sustained to a vehicle insured by them irrespective of legal liability. **to knock about, around**, to strike with repeated blows; to handle violently; (*coll.*) to wander about; to lead an irregular life; to associate (with). **to knock back**, to drink quickly; to cost; to reject; to shock. **to knock cold**, to shock. **to knock down**, to fell with a blow; to prostrate (with astonishment etc.); to demolish; to sell (with a blow of the hammer) to a bidder at an auction; to lower in price, quality etc. **to knock off**, to strike off; to dispatch, to do or finish quickly; to cease work; to deduct; (*sl.*) to murder; (*sl.*) to steal. **to knock on**, in rugby, to play (the ball) with the hand or arm. **to knock on the head**, to stun or kill with a blow on the head; to frustrate, to spoil, to defeat. **to knock out**, to force or dash out with a blow; to disable by a particular blow; to make unconscious by a blow; to eliminate from a competition; (*coll.*) to overwhelm with astonishment or admiration. **to knock sideways**, to knock off course. **to knock someone into the middle of next week**, (*coll.*) to butt someone very hard. **to knock the bottom out of**, to refute (an argument). **to knock together**, to put hastily or roughly into shape. **to knock up**, to arouse by knocking; to fatigue, to wear out, to exhaust; to put together or make up hastily; to make (a score of runs) at cricket; to engage in a preliminary practice before a tennis, squash etc. match; (*sl.*) to make (someone) pregnant. **knock-about**, *a.* noisy, rough, violent; suitable for rough usage, as clothes; noisy, boisterous. **knockback**, *n.* a rejection. **knock-down**, *a.* of a blow, overwhelming; of a price at auction, reserve or minimum. *n.* a knock-down blow; a free fight. **knock-knees**, *n.pl.* knees bent inwards in walking. **knock-kneed**, *a.* **knock-on**, *n.* in rugby, playing the ball with the hand or arm.

knock-on effect, *n.* an indirect result of an action. **knock-out,** *a.* of a blow, disabling. *n.* the act of knocking out; a knock-out blow; (*coll.*) a marvel, wonder; a competition in which losers are eliminated after each round. **knock-out drops,** *n.pl.* (*coll.*) a drug put into someone's drink secretly. **knocker,** *n.* one who knocks; a hammer-like attachment to an outer door to give notice that someone desires admittance; one who finds fault; (*pl.*) (*sl.*) a woman's breasts. **knocking,** *n.* **knocking copy,** *n.* publicity aimed at undermining a competing product. **knocking-shop,** *n.* (*sl.*) a brothel.

knoll (nōl), *n.* a small rounded hill; a mound, a hillock. **knolly,** *a.*

knot[1] (not), *n.* the interlacement or intertwining of a rope or ropes, cords etc., so as to fasten one part to another part of the rope etc. or to another object; an ornamental bow or interlacement of a ribbon etc., as on a dress; a bond; (usu. **porters' knot, shoulder-knot**) a kind of double shoulder-pad, with a loop passing round the forehead, used by London market-porters for carrying burdens; a sense of constriction caused by muscular tension; a difficulty, a perplexity, a problem; something not easily solved; the gist or kernel of a matter; anything resembling a knot; an irregular or twisted portion in a tree caused by branches, buds etc.; a tangle; a node or joint in a stem; a protuberance or excrescence; a hard cross-grained part in a piece of wood, caused by interlacing fibres; a hard lump in the body of an animal; a group, a cluster; a division of the log-line marked off by knots, used as a unit for measuring speed; (*loosely*) a nautical mile per hour. *v.t.* (*past, p.p.* **knotted**) to tie in a knot or knots; to fasten with a knot; to intertwine; to make (fringe) by means of knots; to knit (the brows); to join together closely or intricately; to entangle, to perplex. *v.i.* to form knots; to make knots for fringe. **at a rate of knots,** very fast. **to tie someone up in knots,** to confuse someone completely. **knot-garden,** *n.* a formal garden laid out according to an intricate pattern. **knot-grass,** *n.* a prostrate plant, *Polygonum aviculare*, with internodes and white, pink, crimson or green inconspicuous flowers. **knot-hole,** *n.* a hole in wood where a knot used to be. **knot-work,** *n.* ornamental fringe made by knotting cords together; representation of this in painting or carving; a kind of ornamental needlework. **knotless** (-lis), *a.* **knotted,** *a.* **get knotted!,** an expression of anger, exasperation etc. **knotter,** *n.* **knotting,** *n.* fancy knotted work; the removal of knots from textile fabrics. **knotty,** *a.* full of knots; intricate, perplexing, difficult of solution. **knottiness,** *n.*

knot[2] (not), *n.* a small wading-bird of the snipe family, visiting Britain in the late summer and autumn.

knout (nowt), *n.* a whip or scourge formerly used as an instrument of punishment in Russia. *v.t.* to punish with the knout.

know (nō), *v.t.* (*past* **knew** (nū), *p.p.* **known**) to have a clear and certain perception of; to recognize from memory or description, to identify; to be convinced of the truth or reality of; to be acquainted or familiar with; to have personal experience of; to be on intimate terms with; to be aware of; to understand from learning or study; (*formerly*) to have sexual intercourse with. *v.i.* to have knowledge; to be assured (of). **in the know,** in the secret; acquainted with what is going on. **to know how many beans make five,** to have one's wits about one. **to know the ropes,** (*coll.*) to be acquainted with the particular conditions of any affair or proceeding. **to know what's what,** to be wideawake; to know the ways of the world; to appreciate a good thing. **to know which side one's bread is buttered,** to appreciate what is in one's best interests. **to not know someone from Adam,** to have no idea at all who somebody is. **what do you know?,** an expression of incredulity. **you never know,** things are never certain. **know-all,** *n.* (*derog.*) someone who thinks they know everything. **know-how,** *n.* (*coll.*) specialized skill, expertise. **know-it-all,** *n.* (*derog.*) a know-all. **know-nothing,** *n.* an ignorant person. **knowable,** *a.* **knowability** (-bil'-), **knowableness,** *n.* **knower,** *n.* **knowing,** *a.* intelligent; skilful; sharp, cunning; deliberate, conscious. **there is no knowing,** one can never tell. **knowingly,** *adv.* **knowingness,** *n.*

knowledge (nol'ij), *n.* the result of knowing; that which is known; certain or clear apprehension of truth or fact; cognition, the process of knowing; familiarity gained by actual experience; erudition, science, the sum of what is known; information, notice; range or scope of information; (*formerly*) sexual intercourse. **to the best of my etc. knowledge,** as far as I etc. know. **knowledge engineering,** *n.* the area of artificial intelligence concerned with producing expert systems. **knowledge engineer,** *n.* **knowledgeable,** *a.* (*coll.*) well-informed, intelligent. **knowledgeably,** *adv.*

known (nōn), *p.p.* KNOW.

knub (nŭb), *n.* a lump, a knob; (*usu. pl.*) the waste silk produced in winding off from the cocoon.

knuckle (nŭk'l), *n.* one of the joints of a finger, esp. at the base; the middle or tarsal joint of a quadruped; a joint of meat comprising this and adjoining parts; a knuckle-shaped joint or part in a structure, machinery etc. *v.i.* to hit or press with the knuckles. *v.i.* to keep the knuckles on the ground in a game of marbles. **near the knuckle,** verging on the indecent. **to knuckle down,** to get down to some hard work. **to knuckle under,** to bow to the pressure of authority. **knuckle-bone,** *n.* a bone forming the knuckle of a sheep or other animal; (*pl.*) a game played with such bones. **knuckle-duster,** *n.* an iron instrument to protect the knuckles, and to add force to a blow. **knuckle-head,** *n.* (*coll.*) an idiot. **knuckleheaded,** *a.* **knuckle sandwich,** *n.* (*sl.*) a punch. **knuckly,** *adv.*

knur (nœ), **knurr, knar** (nah), *n.* a hard swelling on the trunk of a tree; a knot; a hard concretion.

knurl, nurl (nœl), *n.* a knot, a lump, an excrescence; a bead or ridge produced on a metal surface as a kind of ornamentation. *v.t.* to make knurls, beadings or ridges. **knurly,** *a.*

k.o., (*abbr.*) knock-out.

koa (kō'ə), *n.* a Hawaiian acacia used for cabinet-work and building.

koala (kōah'lə), **koala bear,** *n.* an Australian marsupial, with dense fur, which feeds on eucalyptus leaves.

koan (kō'an), *n.* a problem with no logical answer, used for meditation by Zen Buddhists.

kobold (kob'ōld), *n.* a German house-spirit, corresponding to the English Robin Goodfellow, and the Scottish brownie; a gnome or goblin haunting mines and hidden lodes.

Köchel number (kœ'khəl), *n.* a number given to the works of Mozart in the Köchel catalogue of his compositions. [Ludwig von *Köchel*, d. 1877, Austrian cataloguer of Mozart's work]

kodiak (kō'diak), **kodiak bear,** *n.* a brown bear found in Alaska and the neighbouring Aleutian Islands, esp. Kodiak Island.

koeksister (kuk'sistə), *n.* (*S Afr.*) a cake made with sweetened dough.

kohl (kōl), *n.* fine powder of antimony used, orig. by Asian women, to darken the eyelids.

kohlrabi (kōlrah'bi), *n.* (*pl.* **-bies**) the turnip-stemmed cabbage, *Brassica oleracea caulorapa*.

Koine (koi'nē), *n.* a Greek dialect used as a common language in the E Mediterranean during the Hellenistic and Roman periods; a lingua franca.

kola COLA.

kolinsky (kəlin'ski), *n.* a type of Asian mink; the fur from this mink.

kolkhoz (kolkhoz'), *n.* (*pl.* **-hozy** (-zi), **-hozies** (-ziz')) a

cooperative or collective farm in the USSR.
Kol Nidre (kol nid′ri), *n.* the service marking the beginning of Yom Kippur; the opening prayer of this service.
komodo dragon (kəmō′dō), *n.* the largest known lizard *Veranus komodoensis*, from Indonesia.
Komsomol (kom′səmol), *n.* the Young Communist League of USSR. [Rus. abbr. of *Kommunisticheskii Soyuz Molodezhi*]
koodoo, kudu (koo′doo), *n.* a S African antelope with white stripes.
kook (kook), *n.* (*coll.*) an eccentric, mad, or foolish person. **kookie** (-i), **kooky,** *a.*
kookaburra (kuk′əbur′ə), *n.* the laughing jackass, an Australian kingfisher.
kopeck, kopek, copeck (kō′pek), *n.* a Russian coin, the hundredth part of a rouble.
kopje, koppie (kop′i), *n.* a small hill in South Africa.
Koran (kərahn′), *n.* the Muslim sacred scriptures consisting of the revelations delivered orally by Mohammed and collected after his death. **Koranic,** *a.*
Korean (kərē′ən), *a.* pertaining to Korea, its people, or its language. *n.* a person living in Korea; the language spoken in N and S Korea.
korfball (kawf′bawl), *n.* a game not unlike basket-ball, with teams of six men and six women.
korma (kaw′mə), *n.* a mild Indian dish of braised meat or vegetables cooked in spices and a yoghurt or cream sauce.
kosher (kō′shə), *a.* permitted, right; (*coll.*) genuine, above-board; of food or a shop where food is sold, fulfilling the requirements of the Jewish law. *n.* a kosher shop or food. *v.t.* to make kosher.
koto (kō′tō), *n.* (*pl.* **-tos**) a Japanese stringed instrument with a wooden body and 13 silk strings.
kotow, kowtow (kōtow′), (kow-), *n.* the ancient Chinese method of obeisance by kneeling or prostrating oneself, and touching the ground with the forehead. *v.i.* to perform the kotow; to act obsequiously.
kotuku (kō′tukoo), *n.* a white heron found in New Zealand.
koumiss KUMISS.
kowhai (kō′hī), *n.* (*pl.* **-hais**) a small shrub with clusters of golden flowers found in Australasia and Chile.
kowtow KOTOW.
Kr, (*chem. symbol*) krypton.
kr, (*abbr.*) krona; krone; kreutzer.
kraal (krahl), *n.* a S African village or group of huts enclosed by a palisade; a hut; an enclosure for cattle or sheep.
kraft (krahft), *n.* strong, brown, wrapping paper.
kragdadige (krahkh′dahdikhə), *n.* (*S Afr.*) someone who advocates hard-line policies. **kragdadigheid** (-hit), *n.* a hard-line attitude, esp. of a government towards demands for liberalization.
krait (krīt), *n.* a poisonous rock snake.
kraken (krah′kən), *a.* a fabulous sea-monster, said to have been seen at different times off the coast of Norway.
krantz (krahnts), **krans** (krahns), **kranz** (krahnts), *n.* (*S Afr.*) a precipitous acclivity, esp. of crags walling in a valley.
krasis (krāsis), CRASIS.
kraut (krowt), *n.* (*offensive*) a German. [from *sauerkraut*]
Kremlin (krem′lin), *n.* the citadel of a Russian town, esp. that of Moscow enclosing the old imperial palace, now government buildings etc.; the Soviet Government. **Kremlinologist,** *n.* **Kremlinology,** *n.* the study of the Soviet government and Soviet politics.
krill (kril), *n. collect.* tiny shrimplike crustaceans, the main food of whales.
krimmer (krim′ə), *n.* the tightly curled black or grey fleece from a type of lamb found in the Crimean.
kris (krēs), CREESE.

Krishnaism (krish′nəizm), *n.* the worship of the Hindu divinity Krishna. **Krishnaist, Krishnaite** (-it), *n.*
kromesky (krəmes′ki), *n.* chicken minced and rolled in bacon, then fried.
krona (krō′nə), *n.* (*pl.* **kronor**) a silver coin of Sweden.
krone (krō′nə), *n.* a silver coin of Denmark and Norway.
Krugerrand (kroo′gərand), *n.* a coin minted in S Africa containing 1 oz. of gold. [President *Kruger* 1825–1904]
krummhorn (krum′hawn), *n.* a mediaeval wind instrument with a curved tube, and a tone like that of a clarinet; an organ stop consisting of reed pipes, with a similar tone.
krypton (krip′tən), *n.* an inert gaseous element, at no. 38; chem. symbol Kr, discovered in 1898 as a constituent of the atmosphere.
Kshatriya (kshah′triyə), *n.* the warrior caste in the Hindu caste system.
kt, (*abbr.*) karat; knot; knight.
kudos (kū′dos), *n.* glory, fame, credit.
kudu (koo′doo), KOODOO.
Ku-Klux-Klan (kooklūksklan′), *n.* a secret society formed in the Southern States after the American Civil War of 1861–65 to keep down the black population. Suppressed by the US government in 1871 but revived since then with the aim of preserving white supremacy. **klanism,** *n.* **klansman,** *n.* **Ku Klux Klanner,** *n.*
kukri (kuk′ri), *n.* a curved knife broadening at the end, used by the Gurkhas.
kulak (koo′lak), *n.* a prosperous Russian peasant.
kumara, kumera (koo′mərə), *n.* (*New Zealand*) the sweet potato.
kumiss, koumiss (koo′mis), *n.* a spirituous liquor made, orig. by Tartars, from fermented mare's milk.
kümmel (kum′l), *n.* a liqueur flavoured with caraway-seeds made in Germany and Russia. [G]
kumquat (kŭm′kwot), CUMQUAT.
kung fu (kŭng foo′), *n.* a Chinese martial art.
kuo-yü (kwöyü′. gwaw-), *n.* a form of Mandarin taught all over China. [Chin., national language]
kurchatovium (kocchatō′viəm), *n.* the chemical element at no. 104, whose discovery was claimed by the Soviets in 1966, also called rutherfordium. [I. V. Kurchatov, 1903–1960, Soviet physicist]
Kurd (kœd), *n.* a native or inhabitant of Kurdistan. **Kurdish,** *a.* pertaining to the Kurds or Kurdistan; *n.* their language.
kurrajong, currajong (kŭ′rəjong), *n.* (*Austral.*) any of several trees and shrubs with fibrous bark.
kurta (kuə′tə), *n.* a loose tunic worn in India.
kurtosis (kətō′sis), *n.* the distribution and density of points around the mean.
kuru (koo′roo), *n.* a disease, usu. fatal, of the nervous system occurring in the inhabitants of eastern New Guinea.
kv, (*abbr.*) kilovolt.
kvass (kvahs), *n.* beer made in the USSR from rye.
kvetch (kvech), *v.i.* (*coll.*) to whine, to complain, **kvetcher,** *n.* a complainer. [Yiddish]
kW, (*abbr.*) kilowatt.
kwacha (kwah′chə), *n.* the unit of currency in Zambia and Malawi.
kwashiorkor (kwashiaw′kə), *n.* a nutritional disease caused by lack of protein.
kWh, (*abbr.*) kilowatt hour.
KWIC (kwik), (*acronym*) keyword in context.
KWOC (kwok), (*acronym*) keyword out of context.
kyanite (kī′ənit), CYANITE.
kyat (kyaht), *n.* the unit of currency in Burma.
kybosh KIBOSH.
kyle (kīl), *n.* (*Sc.*) a narrow channel.
kylie (kī′li), *n.* a boomerang.
Kymric (kim′rik), CYMRIC.
kymograph (kī′məgraf), *n.* an instrument for recording

wave-like oscillations, as of the pulsation of the blood in a living body. **kymographic** (-graf'-), *a.*
kyphosis (kīfō'sis), *n.* a curvature of the spine resulting in a hunched back. **kyphotic** (fot'-), *a.*

Kyrie (ki'riä, kiə'ria), **Kyrie eleison** (ilā'ison, -zon), *n.* this phrase used as a short petition in the liturgies of the Eastern and Western Churches, at the beginning of the Mass; a musical setting of this.

L

L¹, l¹, the 12th letter of the English alphabet (*pl.* **Els, Ls, L's**), an L-shaped thing, part or building; a rectangular joint; the Roman numeral for 50.
L², (*abbr.*) lady; lake; Latin; learner (driver); Liberal; libra (pound); licentiate; lira; London; longitude.
l², (*abbr.*) latitude; league; left; length; line; litre(s).
LA, (*abbr.*) Los Angeles.
La, (*chem. symbol*) lanthanum.
la¹ (lah), *int.* expressing surprise.
la², **lah** (lah), *n.* the name for the sixth note of the scale in solmization.
laager (lah'gə), *n.* (*S Afr.*) a defensive encampment. *v.t.* to form into a laager. *v.i.* to encamp.
Lab (lab), (*abbr.*) Labour; Labrador.
lab (lab), short for LABORATORY.
labarum (lab'ərəm), *n.* the imperial standard of Constantine the Great; a banner resembling this used in religious processions.
labdacism (lab'dəsizm), LAMBDACISM under LAMDA.
labdanum (lab'dənəm), LADANUM.
labefaction (labifak'shən), *n.* a weakening; decay; downfall, ruin.
label (lā'bl), *n.* a narrow strip of paper etc. attached to an object to indicate contents, destination, ownership etc.; a descriptive phrase associated with a person, group etc.; an addition to a document, as a codicil; an adhesive stamp; a drip-moulding; a firm's tradename (esp. of a record company); character(s) indicating the start of an instruction in a computer program. *v.t.* (*past, p.p.* **labelled**) to affix a label to; to describe, to categorize. **labeller**, *n.*
labellum (ləbel'əm), *n.* (*pl.* **-lla** (-lə)) the lower part of the corolla in an orchidaceous flower.
labial (lā'biəl), *a.* of or pertaining to the lips or labium; serving as or resembling a lip; having lips or lip-like edges; formed or modified in sound by the lips. *n.* a sound or letter representing a sound formed with the lips. **labialism**, *n.* **labialization**, *n.* **labialize, -ise**, *v.t.* **labially**, *adv.* **labiate** (-ət), *a.* (*Bot.*) having a corolla with an upper and lower part like a pair of lips; belonging to the mint family.
labile (lā'bil), *a.* unstable, liable to chemical or other change. **lability** (-bil'-), *n.*
labio-, *comb. form* labial.
labiodental (lābiōden'təl), *n., a.* (a sound) produced by the agency of lips and teeth.
labium (lā'biəm), *n.* (*pl.* **-bia** (-biə)) (*Anat.*) a lip or lip-like part, as of the female genitals.
laboratory (ləbo'rətri, *esp. N Am.* lab'rətəri), *n.* a place in which scientific experiments are conducted; a factory for chemical articles, explosives etc.
laborious LABOUR.
labour (lā'bə), (*esp. N. Am.*) **labor**, *n.* physical or mental exertion, esp. in obtaining the means of subsistence; the performance of work; a task, esp. one requiring great effort; the pains of childbirth; the element contributed by toil to production, esp. in opp. to capital; (**Labour**) the Labour Party, its members, causes or ideals; (*collect.*) workers. *a.* pertaining to labour or to the Labour Party. *v.i.* to work hard; to move or proceed with difficulty; to be burdened or oppressed with difficulties; to be in the pains of childbirth. *v.t.* to work out laboriously; to deal with in detail or at great length. **labour of love**, work done without expectation of payment. **labour camp**, *n.* a penal establishment where prisoners are forced to labour. **Labour Day**, *n.* a public holiday honouring working people, esp. 1 May; (*N Am.*) the first Monday in September. **Labour Exchange**, *n.* the former name for a state employment agency. **labour market**, *n.* the supply of unemployed labour in relation to the demand. **Labour Party**, *n.* a British political party representing 'workers by hand or brain', composed of the chief socialist organizations and supported by the trade unions. **labour-saving**, *a.* **laborious** (ləbaw'riəs), *a.* working hard or perseveringly; laboured; difficult, fatiguing. **laboriously**, *adv.* **laboriousness**, *n.* **laboured**, *a.* showing signs of effort, not spontaneous. **labourer**, *n.* one who labours; esp. one who performs work requiring manual labour but little skill. **labourite** (-it), *n.* a follower or member of the Labour Party.
Labrador (lab'rədaw), *n.* a mainland province of Newfoundland, Canada; a Labrador retriever. **Labrador retriever**, *n.* a type of retriever dog of either a golden or black colour. **labradorite** (-rit), *n.* a brightly-coloured feldspar from Labrador.
labrum (lā'brəm), *n.* (*pl.* **-bra**) a lip or lip-like part. **labret** (-brit), *n.* a plug of stone, shell etc. inserted into the lip as an ornament.
laburnum (ləbœ'nəm), *n.* a poisonous tree or shrub with racemes of yellow flowers.
labyrinth (lab'irinth), *n.* a structure composed of intricate winding passages, a maze; an intricate combination, arrangement etc.; the internal portion of the ear. *v.t.* to enclose in or as in a labyrinth. **labyrinthal** (-rin'-), **labyrinthine** (-rin'thin), *a.* **labyrinthitis** (-thitis), *n.* an inflammation of the inner ear.
lac¹ (lak), *n.* a resinous substance secreted, chiefly on the banyan-tree, by the parasitic insect *Coccus lacca*, used to make shellac. **lac-dye, lac-lake**, *n.* colouring matters obtained from lac and used in dyeing scarlet or purple. **laccic** (lak'sik), **laccin** (lak'sin), *n.* the colouring principle in lac.
lac² LAKH.
laccolite (lak'əlit), **-lith** (-lith), *n.* an intrusive mass of lava penetrating between strata and raising the surface into domes.
lace (lās), *n.* a cord or string used to bind or fasten, esp. by interweaving, as a shoe-lace etc.; a kind of ornamental network of threads forming a fabric of open texture; an ornamental braid or edging for uniforms etc. *v.t.* to fasten or compress by means of a lace or string; to intertwist or interweave (with thread etc.); to trim or adorn with lace; to embellish with or as with stripes; to flavour or fortify by adding spirits to; to beat, thrash. *v.i.* to compress the waist by tightening laces; of boots etc., to fasten with laces; to lash (into). **lace-bark**, *n.* the inner bark of a W Indian shrub which resembles coarse lace. **lace-frame**, *n.* a machine used in lacemaking. **lace-glass**, *n.* Venetian glass decorated with lace-like patterns. **lacemaker**, *n.* **lacemaking**, *n.* **laceman**, *n.* one dealing in lace. **lace-pillow**, *n.* a cushion on which various kinds of lace are made. **lace-ups**, *n.pl.* shoes or boots fastened by laces. **lace-wing**, *n.* any of various flying insects with veiny wings. **lace-**

lacerate   464   **laevo-**

winged, *a*. **laced**, *a*. **lacing**, *n*. (a fastening by) a cord passing through holes etc.; a dose of spirit added to a liquor to strengthen it; a thrashing. **lacy**, *a*.

lacerate (las'ərāt), *v.t.* to tear, to rend, harrow. **lacerable**, *a*. **lacerant**, *a*. agonizing, traumatic. **lacerate, -ated** (rāt), *a*. torn, mangled; (*Bot.*) having a jagged edge, as if torn. **laceration**, *n*. **lacerative**, *a*.

Lacerta (ləsœ'tə), *n*. the typical genus of the Lacertilia, the lizards, iguanas etc.; the Lizard, a northern constellation. **lacertian** (-sœ'shən), *n*., *a*. **lacertilian** (lasətil'-), **-tine** (-tīn), **-toid** (-toid), *a*.

laches (lach'iz), *n*. in law, culpable negligence or remissness.

Lachesis (lak'isis), *n*. one of the three Fates in Greek mythology; a genus of venomous rattlesnakes, with the rudiments of a rattle.

Lachryma Christi (lak'rimə kris'tē), *n*. a sweet white wine from S Italy. [L, Christ's tears]

lachrymal, lacrimal, lacrymal (lak'riməl), *a*. pertaining to tears or tear-glands. *n*. a bone near the tear-producing glands; (*pl.*) lachrymal organs. **lachrymation**, *n*. **lachrymatory**, *a*. of, pertaining to or causing tears. **lachrymose** (-mōs), *a*. ready to shed tears; mournful. **lachrymosely**, *adv.*

lacing LACE.

lacinia (ləsin'iə), *n*. (*pl.* **-niae** (-niē)) an incision or slash in a leaf or petal; a slender lobe like the result of slashing or cutting. **laciniate** (-ət), **-ated** (-ātid), *a*. **laciniation**, *n*.

lack (lak), *n*. deficiency, need (of); that which is needed. *v.t.* to be in need of, to be deficient in; to be without. *v.i.* to be deficient (in); to be wanting. **lacklustre**, *a*. wanting brightness or lustre. **lacking**, *a*.

lackadaisical (lakədā'zikəl), *a*. affectedly pensive; listless, absent-minded. **lackadaisically**, *adv.* **lackadaisicalness**, *n*.

lackaday (lak'ədā), *int.*

lackey (lak'i), *n*. a footman; a servile follower. *v.t.*, *v.i.* to follow or attend as a servant; to attend servilely.

lacmus (lak'məs), LITMUS.

laconic (ləkon'ik), **-ical**, *a*. brief, sententious, pithy, concise. **laconically**, *adv.* **laconicism** (-sizm), **laconism** (lak'-), *n*. a concise, pithy or sententious style; a laconic saying.

lacquer (lak'ə), *n*. a varnish composed of shellac dissolved in alcohol; (also **hair lacquer**) a similar substance used to keep a hairstyle in place; a hard glossy varnish made from black resin; woodwork coated with such a varnish. *v.t.* to cover with lacquer. **lacquerer**, *n*.

lacrimal LACHRYMAL.

lacrosse (ləkros'), *n*. a ball-game resembling hockey, but played with a pouched stick. **lacrosse stick**, *n*.

lacrymal LACHRYMAL.

lact- LACT(O)-.

lactate (lak'tāt), *v.i.* (lak'-. -tāt'), to secrete or produce milk. *n*. a salt of lactic acid. **lactation**, *n*.

lacteal (lak'tiəl), *a*. pertaining to milk; conveying chyle. *n.pl.* the vessels which convey chyle from the alimentary canal. **lacteous**, *a*. lacteal.

lactescent (laktes'ənt), *a*. milky; turning to milk; yielding milky juice. **lactescence** (-tes'ns), *n*.

lactic (lak'tik), *a*. pertaining to milk; contained in or derived from milk. **lactic acid**, *n*. a colourless liquid acid produced in tissue and sour milk.

lactific (laktif'ik), **lactiferous**, *a*. carrying or producing milk or milky juice.

lact(o)-, *comb. form* pertaining to milk.

lactoflavin (laktōflā'vin), *n*. earlier name for riboflavin.

lactogenic (laktōjen'ik), *a*. inducing lactation.

lactometer (laktom'itə), *n*. a hydrometer for showing the specific gravity of milk.

lactoscope (lak'təskōp), *n*. an instrument for determining the relative opacity of milk.

lactose (lak'tōs), *n*. milk-sugar, the form in which sugar occurs in milk.

Lactuca (laktū'kə), *n*. a genus of plants containing the lettuce. **lactucic** (-sik), *a*.

lacuna (ləkū'nə), *n*. (*pl.* **-nae** (ē), **-nas**) a gap, hiatus; a small pit or depression. **lacunal, -nar, -nary, -nate** (-nāt), **-nose** (-nōs), *a*.

lacustrine (ləkūs'trin), *a*. of or pertaining to or living on or in a lake. **lacustral**, *a*. **lacustrian**, *n*., *a*.

lacy LACE.

lad (lad), *n*. a boy, youth; (*coll.*) a fellow. **laddie** (-i), *n*. familiar or affectionate term.

ladanum (lad'ənəm), **labdanum** (lab'-), *n*. an odorous resin exuded from the leaves and twigs of various kinds of cistus.

ladder (lad'ə), *n*. a device consisting of two long uprights, connected by rungs or cross-pieces, which form steps by which one may ascend; a vertical rent in a stocking or tights; anything serving as a means of ascent. *v.t.* to equip with a ladder. *v.t.* of stockings etc., to form a ladder. **ladder-back**, *n*. a chair with a back of two uprights joined by several horizontal slats. **ladder-proof**, *a*. of fabrics that are unlikely to ladder. **ladder-stitch**, *n*. a cross-bar stitch used in embroidery and fancy-work. **laddered, laddery**, *a*.

laddie LAD.

lade (lād), *v.t.* (*p.p.* **laden**) to put a load or burden on; to put a cargo or freight on board; to ship (goods) as cargo; to lift (as water) with a ladle etc. **laden**, *a*. weighed down, loaded; encumbered. **lading**, *n*. cargo. **bill of lading** BILL.[3]

la-di-da, lah-di-dah (lahdidah'), *a*. (*sl.*) affectedly genteel, pretentious. *n*. such a person.

ladify LADY.

lading LADE.

ladle (lā'dl), *n*. a large spoon with which liquids are lifted out or served from a vessel; a pan or bowl with a long handle to hold molten metal. *v.t.* to serve out or transfer with a ladle. **to ladle out**, to give or hand out freely. **ladleful**, *n*.

lady (lā'di), *n*. (*pl.* **ladies**) a gentlewoman; a woman of refinement or social standing; a wife; a mistress, girlfriend, sweetheart; the mistress of a house or family; (**Lady**) a title used by women of various ranks in the peerage; also in such titles as lady mayoress; (*pl.*) a public lavatory for women. **lady-in-waiting**, one attending a queen or princess. **my lady, your ladyship**, forms of address for those holding the title. **Our Lady**, the Virgin Mary. **painted lady** PAINT. **Ladies' Gallery**, *n*. a gallery in the House of Commons, formerly screened off by a grille. **ladies' man**, *n*. one attentive to women; one who enjoys the company of women. **lady-altar**, *n*. the altar of a chapel dedicated to the Virgin Mary. **ladybird, -bug**, *n*. a small red coleopterous insect with black spots. **Lady chapel**, *n*. a chapel within a church dedicated to the Virgin Mary. **lady-cow** LADYBIRD. **Lady Day**, *n*. the Feast of the Annunciation of the Virgin Mary, 25 Mar. **lady-fern**, *n*. a tall slender fern, *Asplenium filix-foemina*. **ladyfly** LADYBIRD. **lady-killer**, *n*. (*facet.*) one who (believes he) is irresistibly fascinating to women. **lady-love**, *n*. a female sweetheart. **lady's bedstraw**, *n*. the herb bedstraw. **lady's cushion**, *n*. the thrift or sea-pink. **lady's finger**, *n*. okra. **lady's maid**, *n*. a female attendant on a lady. **lady's mantle**, *n*. the rosaceous herb, *Alchemilla vulgaris*. **lady's slipper**, *n*. an orchid of the genus *Cypripedium*. **lady's-smock, lady-smock**, *n*. the cuckoo flower, *Cardamine pratensis*. **ladified**, *a*. affecting the manners and air of a fine lady. **ladify** (-fī), *v.t.* to make a lady of; to treat as a lady. **ladyhood** (-hud), *n*. **ladyish**, *a*. **ladyism**, *n*. **ladylike**, *a*. **ladyship**, *n*. the title of a lady.

laevo-, lev(o)-, *comb. form.* left, as opposed to right. **laevoglucose** (lēvōgloo'kōs), *n*. laevulose. **laevo-gyrate** (-jī'rət), **-gyrous**, *a*. **laevorotatory** (lēvōrō'tətəri), *a*. turning the plane of polarization to the left. **laevorotation**, *n*.

laevulose (lē'vūlōs), *n.* a sugar or glucose distinguished from dextrose by its turning the plane of polarization to the left.

lag[1] (lag), *v.i.* (*past, p.p.* **lagged**) to loiter, move slowly; to fall behind. *n.* retardation of current or movement; delay in response; an interval; the grey lag. **lag of the tide**, the interval by which the tide lags behind the mean time during the first and third quarters of the moon. **laggard** (-əd), *a.* sluggish, backward; wanting in energy. *n.* a slow, sluggish fellow; a loiterer. **lagger**, *n.* **lagging**, *n.*, *a.* **laggingly**, *adv.*

lag[2] (lag), *v.t.* (*sl.*) to arrest; to send to penal servitude. *n.* a convict; a long-term prisoner, a gaol-bird.

lag[3] (lag), *n.* a stave, lath or strip of wood, felt etc.; one of the pieces of the non-conducting jacket of a boiler or cylinder. *v.t.* to cover or encase with lags or lagging, esp. to preserve against freezing. **lagger**, *n.* **lagging**, *n.* insulating material.

lagan (lag'ən), *n.* wreckage or goods lying at the bottom of the sea, usu. marked by a float or buoy.

lagena (ləjē'nə), *n.* (*pl.* **-nae** (-nē)) a Roman amphora. **lageniform** (-nifawm), *a.*

lager, lager beer (lah'gə), *n.* a light beer, the ordinary beer of Germany. **lager lout**, *n.* (*sl.*) a youth who behaves like a hooligan, esp. when having drunk too much alcohol, esp. lager or beer. [G]

laggard etc. LAG[1].

lagging LAG[1,3].

lagomorph (lag'ōmawf), *n.* any gnawing mammal with two pairs of upper incisors (e.g. hares, rabbits). **lagomorphic, -phous** (-maw'-), *a.*

lagoon (ləgoon'), *n.* a shallow lake near a river or the sea; the water enclosed by an atoll or coral island.

lagrimoso (lahgrēmō'sō), *a., adv.* (*Mus.*) solemnly, plaintively. [It.]

lah LA[2].

lah-di-dah LA-DI-DA.

laic, laical etc. LAY[2].

laid (lād), *past, p.p.* LAY[1]. *a.* lying down; placed or pressed down. **laid back**, *a.* (*coll.*) relaxed, casual. **laid paper**, *n.* paper made with a ribbed surface, opp. to *wove paper*.

lain (lān), *past, p.p.* LIE[2].

lair[1] (leə), *n.* the den or retreat of a wild beast; a hiding-place. *v.i.* to go to or lie in a lair; to make one's lair (in). *v.t.* to place in a lair. **lairage** (-rij), *n.*

lair[2] (leə), *n.* (*Austral. coll.*) an over-dressed man. **laired up**, dressed in a flashy manner. **lairy**, *a.*

laird (leəd), *n.* (*Sc.*) the owner of a landed estate. **lairdship**, *n.*

laissez-aller (lesäal'ä), *n.* unrestraint; absence of conventionality. [F]

laissez-faire (-feə'), *n.* the principle of non-interference, esp. by the Government. [F]

laity, LAY[2].

lake[1] (lāk), *n.* a large sheet of water entirely surrounded by land; surplus of a liquid commodity, as wine. **Lake District, Lakeland**, *n.* the mountainous district occupied by the English lakes, in Cumbria. **lake-dwellers**, *n.pl.* the prehistoric inhabitants of dwellings built on piles on the shallow edges of lakes. **lake-dwellings**, *n.pl.* **Lake Poets**, *n.pl.* **Lake School**, *n.* Coleridge, Southey and Wordsworth, who lived in the Lake District. **lake-settlement**, *n.* **lake-trout**, *n.* a Canadian fish of the salmon family living in lakes. **lakeless** (-lis), *a.* **lake-like, lakelet** (-lit), *n.* **laker**, *n.* a lake fish; a boat for lakes; one who boats on lakes. **laky**, *a.*

lake[2] (lāk), *n.* a crimson pigment, orig. derived from lac or cochineal.

lakh, lac (lahk), *n.* the number 100,000 (usu. of rupees). [Hind.]

laky LAKE[1].

Lallan (lal'ən), **Lallans**, *n.* the Lowlands of Scotland; the broad Scots dialect, esp. its modern literary use. [Sc.]

lallation (ləlā'shən), *n.* pronunciation of *r* as *l.* **lalling**, *n.* continuous repetition of a single sound.

lalopathy (lalop'əthi), *n.* any speech disorder.

lam[1] (lam), *v.t.* (*past, p.p.* **lammed**) (*coll.*) to thrash, wallop.

lam[2] (lam), *n.* (*N Am. sl.*) a quick escape, esp. from the law. *v.i.* (*past, p.p.* **lammed**) to depart quickly, escape.

lama[1] (lah'mə), *n.* a Tibetan or Mongolian Buddhist priest or monk. **Dalai Lama**, the chief lama of Tibet. **Teshu Lama**, that of Mongolia. **lamaism**, *n.* **lamaist**, *n., a.* **lamaistic**, *a.* **lamaserai, -sery** (-səri), *n.* a lamaist monastery.

lama[2] (lah'mə), LLAMA.

lamantin (ləman'tin), *n.* the manatee.

Lamarckian (ləmah'kiən), *a.* of or pertaining to *Lamarck* (1744–1829, French naturalist). *n.* an adherent of the theory of Lamarck that the development of species has been due to inheritable modifications caused by efforts at adaptation to environment etc. **Lamarckianism, Lamarckism**, *n.*

lamasery LAMA[1].

lamb (lam), *n.* the young of a sheep; its flesh used for food; an innocent and gentle person; a term of endearment. *v.i.* to bring forth lambs; to tend ewes at lambing. **like a lamb to the slaughter**, defenceless, unresisting. **the Lamb, Lamb of God**, Christ. **lambskin**, *n.* the skin of a lamb dressed as a leather with the fleece on. **lamb's tails**, *n.pl.* catkins of hazel and filbert. **lamb's-wool**, *n.* wool from lambs used for hosiery. **lambhood** (lam'hud), **lambkin** (-kin), *n.* **lambing**, *n.* **lamb-like**, *a.*

lambast (lambast'), **lambaste** (-bāst'), *v.t.* to beat; to scold severely.

lambda (lam'də), *n.* the 11th letter of the Greek alphabet (λ) transliterated as Roman *l*; a symbol denoting wavelength. **lambda particle**, *n.* an elementary particle, a hyperon, that has no charge. **lambdoid** (-doid), **-oidal**, *a.* resembling the Greek letter lambda (λ) in form, as the suture between the parietal and the occipital bones of the skull.

lambent (lam'bənt), *a.* playing or moving about, touching slightly without burning, as flame or light; softly radiant; light, sparkling, as wit. **lambency**, *n.* **lambently**, *a.*

lambert (lam'bət), *n.* a former measure of the luminous intensity or brightness of a surface, one lumen per square centimetre. [G scientist J.H. *Lambert*, 1728–77]

Lambeth (lam'bəth), *n.* a London borough south of the Thames, where the Archbishop of Canterbury has his palace. **Lambeth degree**, *n.* an honorary degree conferred by the Archbishop of Canterbury.

lambrequin (lam'brikin, -bəkin), *n.* an ornamental strip of drapery over a door, window, mantelshelf etc.

lame (lām), *a.* disabled in esp. the foot or leg; limping; not running smoothly or evenly; unsatisfactory; imperfect. *v.t.* to make lame. **lame duck**, *n.* a defaulter on the Stock Exchange; a weak, ineffective or disabled person. **lamely**, *adv.* **lameness**, *n.* **lamish**, *a.*

lamé (lah'mā), *n.* a fabric containing metallic threads. [F]

lamella (ləmel'ə), *n.* (*pl.* **-llae** (-ē)) a thin plate, layer or scale. **lamellar, lamellate** (-ət), **-ated** (lam'əlātid), **lamellose** (-ōs), *a.* **lamellarly**, *adv.*

lamelli-, *comb. form* pertaining to thin layers, scales etc.

lamellibranch (ləmel'ibrangk), *n.* one of a class of molluscs breathing by two pairs of plate-like gills. **lamellibranchiate** (-brang'kiət), *n., a.*

lamellicorn (ləmel'ikawn), *n., a.* (a beetle) of the Lamellicornia, having short antennae terminated by a short lamellated club. **lamellicornate** (-kaw'nət), **-cornous**, *a.* **lamelliferous** (laməlif'-), **lamelliform** (-fawm), *a.*

lamellirostral (ləmeliros'trəl), *a.* having a lamellose bill, as ducks, geese etc.

lament (ləment'), *v.i.* to mourn, wail; to feel or express sorrow. *v.t.* to mourn over; to deplore. *n.* sorrow expressed in cries or complaints; an elegy, dirge. **lamentable** (lam'-), *a.* mournful; deplorable. **lamentably,** *adv.* **lamentation** (lam-), *n.* the act of lamenting; an audible expression of grief; a wail. **Lamentations,** *n. sing.* the book of the Old Testament containing the lamentations of Jeremiah. **lamented,** *a.* mourned for; deceased. **lamenter,** *n.* **lamenting,** *n.,* *a.* **lamentingly,** *adv.*

lamia (lā'miə), *n.* in classical mythology, a female demon; a sorceress, witch.

lamina (lam'inə), *n.* (*pl.* **-nae** (-nē), **-nas**) a thin plate, layer, coat, leaf, flake, stratum etc. **laminable,** *a.* **laminal, laminar, -nary,** *a.* **laminar flow,** *n.* a smooth liquid flow following the shape of a streamlined surface. **laminarian** (-neə'ri-), *a.* pertaining to the genus *Laminaria*, algae with a flat ribless expansion in place of leaves. **laminarize, -ise,** *v.t.* to form (a surface) for a laminar flow. **laminate,** *v.t.* to beat, press or roll into thin plates; to cut or split into thin layers or sheets; to produce by joining successive layers or sheets; to cover with a thin sheet (of plastic etc.). *v.i.* to split into thin plates. *a.* (-nət), consisting of, or having laminae, laminated. *n.* an article or material produced by laminating. **laminated,** *a.* **lamination,** *n.* **laminator,** *n.* **laminitis** (-ni'tis), *n.* an inflammation of the sensitive tissue lining a horse's hoof. **laminiferous** (-nif'-), **laminose** (-nōs), *a.*

Lammas, Lammas Day (lam'əs), *n.* 1 Aug., esp. formerly a harvest festival. **Lammas-tide,** *n.* the season around Lammas, harvest time.

lammergeyer, -geier (lam'əgiə), *n.* the great bearded vulture, inhabiting lofty mountains of S Europe, Asia and N Africa.

lamp (lamp), *n.* a device for producing light by consuming oil, gas or electricity; an electric device which emits, esp. infrared or ultraviolet light waves, as a *sun lamp*; any source of light; (*pl.*) (*sl.*) the eyes. *v.t.* to supply with lamps; to illuminate. **to smell of the lamp,** to show signs of laborious preparation (as a sermon, speech etc.). **lampblack,** *n.* amorphous carbon, obtained by the imperfect combustion of oil or resin used as a pigment or filler. **lamp-chimney, -glass,** *n.* the upright chimney surrounding the wick and flame of an oil lamp. **lamp-light,** *n.* **lamp-lighter,** *n.* one employed to light public lamps. **lamp-post, -standard,** *n.* a pillar supporting a street lamp. **lampshade,** *n.* a cover for a lamp which softens or directs the light emitted.

lampad (lam'pəd), *n.* a lamp, a torch.

lampadomancy (lampad'əmansi), *n.* divination by the flame of a torch or lamp.

lampas (lam'pəs), *n.* a flowered silk or woollen cloth used in upholstery.

lampern (lam'pən), *n.* the river lamprey.

lampion (lam'piən), *n.* a small coloured globe or cup with wick etc., used in illuminations.

lampoon (lampoon'), *n.* a scurrilous personal satire. *v.t.* to write lampoons upon; to abuse with personal satire. **lampooner, -nist,** *n.*

lamprey (lam'pri), *n.* an eel-like fish with a suctorial mouth.

lana (lā'nə), *n.* the close-grained, tough wood of a S American tree, *Genipa americana.*

lanate (lā'nāt), **lanose** (-nōs), *a.* woolly, covered with curly hairs.

Lancastrian (langkəs'triən), *a.* pertaining to the family descended from John of Gaunt, Duke of Lancaster. *n.* an adherent of this, one of the Red Rose party in the Wars of the Roses; a native of Lancashire.

lance (lahns), *n.* a thrusting weapon consisting of a long shaft with a sharp point; a lancet; a lancer. *v.t.* to pierce with or as with a lance; (*Surg.*) to open with a lancet. **lance-corporal,** *n.* a private who performs the duties and holds the rank of a corporal. **lance-sergeant,** *n.* an acting sergeant. **lance-snake,** *n.* a venomous American snake allied to the rattlesnake. **lancewood,** *n.* the tough, elastic wood of various S American and W Indian trees. **lancer,** *n.* a cavalry soldier armed with a lance; (*pl.*) a set of quadrilles; the music for this. **lanciform** (-ifawm), *a.*

lancelet (lahn'silit), *n.* a small transparent iridescent fish, *Amphioxus lanceolatus.*

lanceolate (lahn'siələt), **-ated,** *a.* tapering to a point at each end.

lancet (lahn'sit), *n.* a sharp-pointed surgical knife; a lancet window or arch. *a.* lancet-shaped. **lancet arch,** *n.* an arch with a sharply pointed top. **lancet window,** *n.* a high narrow window with a sharply pointed arch. **lanceted,** *a.*

lancinate (lahn'sināt), *v.t.* to tear, lacerate. **lancinating,** *a.* of a pain, piercing, keen. **lancination,** *n.*

Lancs. (langks), (*abbr.*) Lancashire.

land (land), *n.* the solid portion of the earth, esp. of the earth's surface; the ground, the soil; a tract of country, esp. a rural area; a country, district, region; a nation; landed property; (*pl.*) estates. *v.t.* to bring to or place on shore; to set down from a vehicle; to bring to or place in a certain position; to deal (a blow); to bring (fish) to land; to win, capture or secure (e.g. a prize, a business deal). *v.i.* to come or go ashore; to disembark; to find oneself in a certain position (with *up*); to alight. *a.* belonging to the land; terrestrial. **land of milk and honey,** any extremely fertile land; a place, country offering wealth and ease. **land of nod,** the state of being asleep. **land of the living,** the present life. **land of the midnight sun,** Norway. **to land with,** to burden with. **to make (the) land,** of a ship, to come in sight of land. **to see how the land lies,** to assess how matters stand (before acting). **land-agent,** *n.* one employed to manage land for the proprietor; an agent for the sale of land. **Land Army,** *n.* a national organization of war-time volunteer farm-workers. **land bank,** *n.* a bank lending money on the security of land. **land-breeze,** *n.* a wind blowing seawards off the land. **land-crab,** *n.* a crab living mainly on land visiting the sea chiefly for breeding. **landfall,** *n.* approach to land after a voyage; the first land seen after a voyage. **landfill,** *n.* the burying of rubbish under layers of earth; a rubbish dump where refuse is buried; the rubbish so buried. **landfilling,** *n.* **land-flood,** *n.* an overflow of water on land. **land-force,** *n.* a military force employed on land. **land-girl,** *n.* a girl or woman employed in farm-work during the two World Wars. **land-grabber** GRAB. **landholder,** *n.* one who owns or (usu.) rents land. **landholding,** *n.,* *a.* **land-hunger,** *n.* desire to acquire land. **land-hungry,** *a.* **land-jobber,** *n.* one who speculates in land. **landlady,** *n.* a woman who keeps an inn or public house; a woman who lets houses, lodgings etc. **land-line,** *n.* an overland telecommunications cable or wire. **landlocked,** *a.* enclosed by land; isolated from the sea. **landlord,** *n.* a man who lets houses, lodgings etc.; the keeper of an inn or a lodging-house. **landlordism,** *n.* the system under which land is owned by individuals to whom tenants pay a fixed rent. **land-lubber,** *n.* (*derog.*) a landsman, one unused to the sea or ships. **landmark,** *n.* anything marking the boundaries of land; a conspicuous object in a place or district; an important event in history etc. **landmass,** *n.* a large area of land uninterrupted by the sea. **land-measuring,** *n.* **land-mine,** *n.* a mine set in the ground to explode under advancing troops etc. *v.t.* to place land-mines. **landowner,** *n.* **landownership,** *n.* **land-owning,** *a.* **land-rail,** *n.* the corn-crake. **land-rat,** *n.* a rat living on land. **land reform,** *n.* a redistribution of land. **land-roll, -roller,** *n.* a roller for crushing clods. **land-scrip,** *n.* (*N Am.*) a certificate entitling the holder to acquire a specified amount of public land. **land-shark,** *n.* a land-grabber. **landslide,** *n.* a landslip; an election debacle. **landslip,** *n.* the sliding down of a portion of ground from a higher to a lower level; the ground thus slipping. **lands-**

man, *n.* one who lives on land; one unused to the sea.
land-spring, *n.* a spring of water appearing intermittently.
land-steward, *n.* one who manages a landed estate. **land-surveying,** *n.* measuring and mapping of land. **land-surveyor,** *n.* **land-tax,** *n.* one assessed upon land and property. **land-waiter,** *n.* a customs officer who watches the landing of dutiable goods. **landwind,** *n.* a wind blowing off the land. **landed,** *a.* having an estate in land; consisting of real estate. **lander,** *n.* one who lands or disembarks. **landing,** *n.* the act of going or setting on land, esp. from a vessel or aircraft; a place for disembarking or alighting; (*N Am.*) the platform of a railway-station; a level space at the top of a flight of stairs or between flights. **landing beam,** *n.* a radio beam guiding an aircraft to ground. **landing-craft,** *n.* a small naval vessel for landing troops etc. **landing-field,** *n.* an area for the landing or take-off of aircraft. **landing-gear,** *n.* an aircraft undercarriage. **landing-net,** *n.* a small bag-net used to take hooked fish from the water. **landing-place,** *n.* **landing-stage,** *n.* a platform on which passengers and goods are disembarked. **landing-strip,** *n.* an airstrip. **landless,** *a.* **landward** (-wəd), *a.*, *adv.* **landwards,** *adv.*
landamman(n) (lan'dəmən), *n.* the chief magistrate in some Swiss cantons.
landau (lan'daw, -dow), *n.* a four-wheeled horse-drawn carriage with a folding top. **landaulet, -lette** (-let'), *n.* a small landau; a motorcar with a covering or hood, fixed in front, movable behind.
lande (läd), *n.* a heathy and sandy plain; a moor. [F]
landfall, landfill etc. LAND.
landgrave (land'gräv), *n.* a German title, orig. distinguishing a governor of a province from inferior counts. **landgraviate** (-viət), **landgraveship,** *n.* **iandgravine** (-grəvēn), *n. fem.*
landing, landlady, landmark etc. LAND.
landscape (land'skāp), *n.* (a picture representing) a view of country scenery. *v.t.* to develop the natural beauty of (an area) by landscape-gardening. **landscape architecture, landscape-gardening,** *n.* the art of laying out grounds so as to develop their natural beauties. **landscape-gardener,** *n.* **landscape-marble,** *n.* a marble with dendriform markings. **landscape-painter, landscapist,** *n.*
lane (lān), *n.* a narrow road, way or passage, esp. between hedges or buildings; a passage between persons or objects; a prescribed route, as for boats or aircraft; a division of a road for a single stream of traffic.
lang., (*abbr.*) language.
langlauf (lahng'lowf), *n.* cross country skiing. **langlaufer,** *n.* [G]
langouste (lägoost'), *n.* the spiny lobster. **langoustine** (-tēn'), *n.* the smaller Norway lobster.
langsyne (langsīn'), *adv.* (*Sc.*) long since, long ago. *n.* time long ago.
language (lang'gwij), *n.* human speech; the vocabulary peculiar to a nation or people; the vocabulary appropriate to a particular profession etc.; the manner of expression peculiar to an individual; literary style; the phraseology or wording (of a book etc.); any method of communicating ideas by symbols, gestures etc. **bad language,** swearing. **to speak the same language,** to have similar background or habits of mind, tastes etc. **language laboratory,** *n.* a place where languages are taught with the aid of tape recorders, headphones etc. **language-teacher,** *n.* **languaged,** *a.* (*usu. in comb.* as *well-languaged*). **languageless** (-lis), *a.*
langue (läg), *n.* in linguistics, language regarded as an abstract system tacitly shared by a speech community. **langue d'oc** (dok'), *n.* (*collect.*) mediaeval Southern French dialects, esp. the Provençal language. **langue d'ocian** (-dō'shən), *a.* **langue d'oïl** (doi), **d'oui** (dwē), *n.* Northern French in the Middle Ages. [F, see prec.]
languet, -ette (lang'gwet), *n.* a tongue-shaped part; the tongue of an organ flue-pipe or the reed of a harmonium;
the tongue of a balance.
languid (lang'gwid), *a.* relaxed, lacking energy; indisposed to exertion; spiritless, lacking animation, listless; sluggish. **languidly,** *adv.* **languidness,** *n.*
languish (lang'gwish), *v.i.* to lose vitality, energy or animation; to grow slack; to droop, to pine (for); to put on a languid expression. **languisher,** *n.* **languishingly,** *adv.* **languishment,** *n.*
languor (lang'gə), *n.* languidness, lassitude; debility; softness of mood or expression; oppressive stillness (of the air etc.). **languorous,** *a.* **languorousness,** *n.*
langur (lŭng-guə'), *n.* an entellus monkey.
laniard LANYARD.
laniary (lan'iəri), *a.* adapted for tearing. *n.* a canine tooth in the Carnivora.
laniferous (lənif'ərəs), **lanigerous** (-nij'-), *a.* bearing wool.
lank (langk), *a.* lean, long and thin; of hair, long and straight. **lankly,** *adv.* **lankness, lankiness,** *n.* **lanky,** *a.*
lanner (lan'ə), *n.* the female of a kind of falcon. **lanneret** (-ret), *n.* the male lanner.
lanolin, -line (lan'əlin), *n.* a fatty substance forming the basis of ointments etc., extracted from wool.
lansquenet (lans'kənet), *n.* a cardgame consisting largely of betting.
Lantana (lantah'nə), *n.* a genus of shrubs of the Verbena family, able to bloom continuously.
lantern (lan'tən), *n.* a case with transparent sides or panes for holding a light; the upper chamber of a lighthouse containing the light; a glazed structure on the top of a dome or roof, for the admission of light and air; a magic lantern. **Chinese lantern** CHINA. **dark lantern** DARK. **magic lantern** MAGIC. **lantern-fly,** *n.* a tropical insect formerly believed to produce light. **lantern-jawed,** *a.* having a long, thin face. **lantern-jaws,** *n.pl.* **lantern slide,** *n.* the glass slide holding the image projected by a magic lantern. **lantern-wheel,** *n.* a form of cog-wheel acting as a pinion to a spur-wheel.
lanthanum (lan'thənəm), *n.* a metallic divalent element, at. no. 57; chem. symbol La, usu. occurring in cerite. **lanthanides** (-nidz), **-noids** (-noidz), **-nons** (-nons), *n.pl.* a group of rare metallic elements, at. nos. 58 to 71.
lanthorn (lant'hawn, -ən), old spelling of LANTERN.
lanugo (lənū'gō), *n.* (*pl.* **-gos**) pre-natal hair; a fine down. **lanuginose** (-jinōs), *a.* covered with soft downy hair.
lanx (langks), *n.* (*Rom. Ant.*) a large dish or platter.
lanyard, laniard (lan'yəd), *n.* (*Naut.*) a short cord for seizing or lashing; cord to which a whistle or knife is attached.
Lao (low), *a.* the people or language of Laos in SE Asia. **Laotian** (lāō'shən, low'-), *n.* a native of Laos. *a.* pertaining to Laos or its people.
Laodicean (lāodisē'ən), *n., a.* (a person) lukewarm in religion, politics etc.
lap[1] (lap), *n.* the part of the person (or a garment) from the waist to the knees in sitting; a place where anything rests or lies securely; that part of anything that extends over something else, the (amount of) overlap; the length of rope, thread etc., making one turn round a wheel, roller etc.; one round of a race-course, running track etc.; a rotating wheel, disc etc. for polishing gems, metal articles etc. *v.t.* (*past, p.p.* **lapped**) to wrap, twist, roll (around, about etc.); to lay (one thing) partly over another; to fold, bend over; to enwrap, surround; to involve; to bind; to get ahead of by a lap or laps; to polish with a lap. *v.i.* to be turned over; to overlap. **in the lap of the gods,** outside human control. **lap of honour,** a victory circuit of a race-track made by a winning contestant. **lap of luxury,** a state of wealth and ease. **the last lap,** the closing stages. **to drop in someone's lap,** to give someone responsibility for something. **lap-dog,** *n.* a small pet dog. **lap-joint,** *n.* a joint in which one part laps over the other. **lap-jointed,** *a.* **lap-robe,** *n.* (*N Am.*) a travelling rug. **lap-streak,** *n., a.*

clinker-built (boat). **lap-top**, *a.* of a portable computer etc., small enough to be held and operated on a person's lap. **lap-work**, *n.* work constructed with lap-joints; work polished by lapping. **lapful**, *n.* **lapper**, *n.*, *a.*

lap[2] (lap), *v.i.* (*past*, *p.p.* **lapped**) to take up liquid with the tongue; to drink by lifting with the tongue; to beat gently (as waves on the shore) with a sound as of lapping. *v.t.* to drink or consume by lapping. *n.* the act of lapping; the amount taken up by this; food or drink that can be lapped up; a weak kind of drink. **to lap up**, to consume or absorb greedily. **lapper, lapping,** *n.*

laparo-, *comb. form* pertaining to the intestines or abdomen. **laparoscope** (lap'ərəskŏp), *n.* an optical instrument for the internal examination of the body's organs. **laparoscopy** (-ros'kəpi), *n.*

laparotomy (lapərot'əmi), *n.* surgical incision into the cavity of the abdomen. **laparotomist**, *n.* **laparotomize, -ise,** *v.t.*

lapel (ləpel'), *n.* the fold on the front of a coat or jacket below the collar. **lapelled,** *a.*

lapidary, (lap'idəri), *n.* one who cuts, polishes or deals in gems. *a.* pertaining to the art of cutting, engraving or polishing gems; inscribed on or suitable for inscription on stones; hence, formal or monumental in style. **lapidary-bee**, *n.* one which nests in or among stones. **lapidarian** (-deə'ri-), *a.* **lapideous** (-pid'i-), *a.* stony. **lapidicolous** (-dikələs), *a.* dwelling under or among stones. **lapidify** (-pid'ifī), *v.t.*, *v.i.* to turn into stone. **lapidific, -ical** (-dif-), *a.* **lapidification** (-fī-), *n.* **lapidose** (-dōs), *a.* stony; growing in stony soil.

lapilli (ləpil'ī), *n.pl.* volcanic ashes, consisting of small, angular, stony or slaggy fragments. **lapilliform** (-ifawm), *a.*

lapis lazuli (lap'is laz'ūlī), *n.* a rich blue silicate of alumina, lime and soda; its colour.

Laplander (lap'landə), **Lapp,** *n.* a native or inhabitant of Lapland. **Laplandish, Lapp, Lappish,** *a.*

lappel (ləpel'), LAPEL.

lapper LAP[1,2].

lappet (lap'it), *n.* a little lap, fold or loose part of a garment or headdress; a flap; a loose, fleshy process, a wattle. **lappeted,** *a.*

lapse (laps), *v.i.* to slide, to pass insensibly or gradually; to fall back or away; to fall into disuse or decay; to make a slip or fault, to fail in duty; to pass from one proprietor to another by omission, negligence or failure; to become void. *n.* the act of lapsing; a gradual decline; easy, smooth and almost imperceptible movement; the imperceptible passage of time; a mistake, a fault, deviation from what is right; a falling into disuse, neglect, decay or ruin; termination of a right or privilege through desuetude. **lapse rate**, *n.* the rate of change of atmospheric factors (e.g. temperature, humidity) with changing altitude. **lapsable, -sible,** *a.* **lapser**, *n.* **lapsus** (-əs), *n.* (*pl.* **lapsus**) a lapse, a slip. **lapsus calami** (kal'əmi, -mē), a slip of the pen. **lapsus linguae** (ling'gwī, -gwē), a slip of the tongue.

Laputan (ləpū'tən), **-tian** (-shən), *a.* pertaining to Laputa the flying island in Swift's *Gulliver's Travels*; visionary, chimerical. *n.* an inhabitant of Laputa; a visionary.

lapwing (lap'wing), *n.* a bird of the plover family, esp. the peewit.

lar (lah), *n.* (*pl.* **lares** (leə'rēz)) a tutelary Roman divinity; (*pl.* **lars**) the white-handed gibbon. **lares and penates**, the home or the valued household possessions contained in it; household gods or their representations.

larboard (lah'bəd), *n.*, *a.* former name for port or left (side of a vessel to a person facing the bow).

larceny (lah'səni), *n.* a legal term for theft. **larcener, -nist,** *n.* **larcenous,** *a.* **larcenously,** *adv.*

larch (lahch), *n.* a coniferous tree having deciduous bright-green foliage and tough, durable timber.

lard (lahd), *n.* the rendered fat of pigs. *v.t.* to fatten; to cover or smear with lard; to insert strips of bacon in (a fowl etc.) before roasting; to intermix or garnish. **lardaceous** (-dā'shəs), *a.* of the nature or consisting of lard. **lardon** (-dən), **-doon** (-doon'), *n.* a strip of bacon for larding fowls etc. **lardy,** *a.* **lardy cake,** *n.* a rich cake made from yeast, lard, flour, dried fruits etc.

larder (lah'də), *n.* a room where meat and other provisions are kept.

lares (leə'rēz), *n.pl.* LAR.

Largactil® (ləgak'til), *n.* chlorpromazine, a tranquillizer.

large (lahj), *a.* great in size, number, quantity, extent or capacity; bulky; extensive; abundant, copious; generous, lavish; wide in range, comprehensive, far-seeing. **as large as life**, unmistakably present or real. **at large**, at liberty; freely, without restraint; diffusely, with ample detail. **by and large** BY. **larger than life**, remarkably vivid or eye-catching. **large calorie**, *n.* 1000 calories, a kilocalorie. **large-handed,** *a.* profuse. **large-hearted,** *a.* having a liberal heart or disposition. **large-heartedness**, *n.* **large intestine**, *n.* that part of the intestine comprising the caecum, colon and rectum. **large-minded,** *a.* generous. **large-mindedness**, *n.* **large-paper,** *a.* of books, prints etc., having wider margins than the ordinary. **large-scale,** *a.* extensive; detailed. **largely,** *adv.* to a large extent. **largeness,** *n.* **largish,** *a.*

largess, -gesse (lahjes'), *n.* a present, a generous bounty (usu. from a superior to inferiors); liberality, esp. in giving.

largo (lah'gō), *adv.* (*Mus.*) slowly, broadly, in an ample, dignified style. *n.* (*pl.* **-gos**) a piece of music played in this manner. **larghetto** (-get'ō), *adv.* somewhat slow. **larghissimo** (-gis'imō), *adv.* very slowly. [It.]

lariat (la'riət), *n.* a rope for tethering horses; a lasso.

lark[1] (lahk), *n.* any bird of the genus *Alauda*, esp. the skylark. **to get up with the lark**, to rise very early in the morning. **lark('s)-heel,** *n.* the larkspur; the nasturtium. **lark-heeled,** *a.* having long back claws. **larkspur,** *n.* a plant with spur-shaped calyx.

lark[2] (lahk), *n.* a prank, frolic. *v.i.* to frolic. **larker,** *n.* **larkish, larky,** *a.* (*coll.*)

larn (lahn), *v.i.* (*dial.*, *facet.*) to learn. *v.t.* to teach.

larrigan (la'rigən), *n.* a high leather boot worn by woodsmen etc.

larrikin (la'rikin), *n.* (*chiefly Austral.*) a rowdy youngster, a hooligan. **larrikinism,** *n.*

larrup (la'rəp), *v.t.* (*coll.*) to thrash, to flog, to lash. **larruper,** *n.* **larruping,** *n.*

Larus (leə'rəs), *n.* a genus of swimming-birds, containing the seagulls. **larine** (-rīn), **laroid** (-roid), *a.*

larva (lah'və), *n.* (*pl.* **-vae** (-vē)) an animal in its immature form, e.g. a grub, caterpillar or maggot. **larval,** *a.* **larvicidal** (-sī'-), *a.* killing larvae. **larvicide** (lah'-), *n.* a preparation for this purpose. **larviform** (-ifawm), *a.* **larvigerous** (-vij'-), **larviparous** (-vip'-), *a.* producing larvae.

laryngectomy (larinjek'təmi), *n.* surgical removal of the larynx.

laryngitis (larinjī'tis), *n.* inflammation of the larynx. **laryngitic** (-rinjit'-), *a.*

laryng(o)-, *comb. form.* pertaining to the larynx.

laryngology (laring-gol'əji), *n.* the medical study of the windpipe and its diseases. **laryngological** (-loj'-), *a.* **laryngologist,** *n.*

laryngoscope (ləring'gəskōp), *n.* an instrument with a reflecting mirror for examining the larynx. **laryngoscopic** (-skop'-), *a.* **laryngoscopist** (-gos'-), *n.* **laryngoscopy,** (-gos'-), *n.*

laryngotomy (laring-got'əmi), *n.* the operation of making an incision into the larynx in order to aid breathing.

larynx (la'ringks), *n.* (*pl.* **larynges** (-rin'jēz), **larynxes**) the upper part of the windpipe, containing the vocal cords. **laryngeal** (-rin'jiəl), **-gal,** *a.*

lasagna, lasagne (ləsan'yə), *n.* a baked dish, wide flat

strips of pasta layered with bolognese and béchamel sauces; such strips of pasta. [It.]
lascar (lasʹkə), *n.* an E Indian sailor.
lascivious (ləsivʹiəs), *a.* wanton, lustful; exciting or provoking lust. **lasciviously,** *adv.* **lasciviousness,** *n.*
laser (lāʹzə), *n.* an instrument which amplifies light waves by stimulation to produce a powerful, coherent beam of monochromatic light; a similar instrument for producing other forms of electromagnetic radiation (e.g. infrared).
lase, *v.i.* to be capable of functioning as a laser. **laser printer,** *n.* a computer printer using a laser beam on special paper.
lash (lash), *n.* the thong or flexible part of a whip; a whip; a stroke with a whip; flogging; an eyelash; a scolding or vituperation. *v.t.* to strike or drive with or as with a whip; to beat or dash against; to fasten or bind with a rope or cord; to assail fiercely with satire. *v.i.* to use a whip; to strike, fling or kick violently (at, out etc.). **to lash out,** to kick, strike out physically or verbally; to spend lavishly. **lasher,** *n.* **lashing,** *n.* a rope etc. by which anything is secured; a whipping; (*pl.*) (*coll.*) a plentiful supply. **lashless** (-lis), *a.*
lasque (lahsk), *n.* a thin, flat diamond; an ill-formed or veiny diamond.
lass (las), *n.* a girl; a sweetheart. **lassie** (-i), *n.* (*Sc.*).
Lassa fever (lasʹə), *n.* an often fatal tropical viral disease symptomized by fever and muscle pain and transmitted by rats etc.
lassitude (lasʹitūd), *n.* weariness, lack of energy or animation.
lasso (lasooʹ), *n.* (*pl.* **lasso(e)s**) a rope with a running noose, used for catching cattle, horses etc. *v.t.* to catch with a lasso.
last[1] (lahst), *a.* coming after all others or at the end; final; pertaining to the end, esp. of life or of the world; definitive; utmost, extreme; lowest; only remaining; least likely, suitable etc.; most recent. *n.* the end, the conclusion; the last moment, hour, day etc.; death; (*ellipt.*) the last thing done, mentioned etc., or the last doing, mention etc. *adv.* on the last time or occasion; for the last time; after all others. **at last,** ultimately. **at long last,** after long delay. **on one's last legs,** very tired; on the verge of ruin. **to breathe one's last,** to die. **to the last,** to the end. **Last Day,** *n.* the Day of Judgment. **last ditch,** *a.* done or made at the final moment or as a last resort. **last minute,** *a.* made or done at the latest possible time. **last post,** *n.* (*Mil.*) the bugle-call signalling the time of turning-in; a bugle-salute at military funerals. **last rites,** *n.pl.* religious rites for the dying. **last straw,** *n.* the limit of endurance or patience. **Last Supper,** *n.* the supper shared by Christ and his disciples the evening before his crucifixion. **last word,** *n.* a concluding statement; a final decision; the most up-to-date model. **lastly,** *adv.* at last; finally.
last[2] (lahst), *n.* a shaped wooden block on which boots and shoes are fashioned or repaired.
last[3] (lahst), *v.i.* to continue in existence, to go on; to hold out, to continue unexhausted or unimpaired, to endure. **to last out,** to endure to the end, to persevere, to survive. **laster,** *n.* **lasting,** *a.* continuing, enduring, permanent. **lastingly,** *adv.* **lastingness,** *n.*
Lat., (*abbr.*) Latin.
lat., (*abbr.*) latitude.
latch (lach), *n.* a fastening for a door, gate etc., consisting of a bolt and catch; a spring-lock fastening with the shutting of a door and opened with a key. *v.t.* to fasten with a latch. **on the latch,** fastened by the latch only, not locked. **to latch on to,** to understand the meaning of; to attach oneself to. **latchkey,** *n.* key of latch on front door.
latchkey child, *n.* one who lets him- or herself into the house after school, one with working parents.
latchet (lachʹit), *n.* a string for a shoe or sandal.
late (lāt), *a.* coming after the proper or usual time; slow, tardy, long delayed; far on towards the close or end; far advanced; existing at a previous time; deceased, departed; lately or recently alive, in office etc.; recent in date. *adv.* after the proper or usual time; at or till a late hour, season, stage etc.; lately, recently. **at the latest,** no later than. **of late,** recently. **the latest,** *n.* (*coll.*) the most recent news. **lately,** *adv.* **lateness,** *n.* **latish,** *a.*, *adv.*
lateen (lətēnʹ), *a.* applied to a triangular sail used principally in the Mediterranean. *n.* a vessel so rigged.
La Tène (la tenʹ), *a.* of the later European Iron Age from the 5th cent. to 1st cent. BC. [*La Tène,* near Neuchâtel, Switzerland]
latent (lāʹtənt), *a.* hidden or concealed; not apparent; dormant, potential. **latent heat** HEAT. **latent period,** *n.* the length of time between stimulation and reaction. **latency,** *n.* **latently,** *adv.*
-later, *comb. form.* worshipper, as *idolater* etc.
lateral (latʹərəl), *a.* of, pertaining to, at, from or towards the side. *n.* a part, member, shoot etc., situated or developing at the side. **lateral axis,** *n.* the cross-wise axis of an aircraft. **lateral line,** *n.* a sensory organ on the side of fish for detecting changes in water pressure or movement. **lateral thinking,** *n.* a way of solving problems by finding new perspectives rather than following conventional lines of thought. **laterality** (-ralʹ-), *n.* physical one-sidedness. **laterally,** *adv.* **lateri-, latero-,** *comb. form.*
Lateran (latʹərən), *n.* a cathedral church at Rome, dedicated to St John the Baptist. *a.* pertaining to this. **Lateran Council,** *n.* name given to five general ecumenical councils held in this church.
laterite (latʹərīt), *n.* a red porous rock, composed of silicate of alumina and oxide of iron. **lateritic** (-ritʹ-), *a.*
latescent LATENT.
latex (lāʹteks), *n.* (*pl.* **-texes, -tices** (-tisēz)) the milky juice of plants, esp. rubber trees; a similar emulsion of a polymer in a watery liquid. **laticiferous** (latisifʹ-), *a.* conveying or producing latex.
lath (lahth, lath), *n.* (*pl.* **laths** (lahdhz, lahths, laths)) a thin strip of wood, esp. one supporting tiles or plastering; anything of similar dimensions or used for the same purposes. *v.t.* to cover or line with laths. **lath-work,** *n.* **lathen,** *a.* **lathing,** *n.* **lathy,** *a.*
lathe (lādh), *n.* a machine for turning and polishing wood, ivory, metal etc.; a potter's wheel. *v.t.* to work on a lathe.
lather (lahʹdhə, ladhʹə), *n.* froth or foam made by soap moistened with water or caused by profuse sweating; (*coll.*) a flustered or excited state. *v.i.* to form a lather; of a horse, to become covered with lather. *v.t.* to cover with lather; (*coll.*) to thrash. **lathering,** *n.* a beating. **lathery,** *a.*
lathi, lathee (lahʹti), *n.* in India, a long, heavy stick.
laticiferous LATEX.
Latin (latʹin), *a.* of or pertaining to ancient Latium or ancient Rome, the inhabitants or their language; pertaining to one or any of the (Romance) languages derived from the Latin language or the peoples who speak them; of the Roman Catholic Church. *n.* the Latin language, the language or inhabitants of ancient Latium and Rome; one belonging to a people whose language derives from Latin; (*N Am.*) a Latin American; a Roman Catholic. **Classical Latin,** that of the golden age of Latin literature (*c.* 75 BC to AD 175). **dog Latin,** barbarous or illiterate Latin. **Late Latin,** that of the period *c.* AD 175–600. **Low Latin,** Mediaeval Latin. **Mediaeval, Middle Latin,** that of the Middle Ages (*c.* AD 600–1500). **Modern, New Latin,** that of periods after AD 1500. **thieves' Latin,** cant or jargon employed by thieves. **Vulgar Latin,** colloquial Latin. **Latin America,** *n.* the parts of America where the official language is derived from Latin (e.g. Spanish, Portuguese). **Latin-American,** *n.*, *a.* **Latin Church,** *n.* the Roman Catholic Church. **Latin cross** CROSS. **Latin peoples,** *n.pl.* those whose language is of Latin origin, the French, Spanish, Portuguese and Italians. **Latin Quarter,** *n.* a left-

bank district of Paris surrounding the Sorbonne, famous for bohemianism. **Latian** (lā'shən), *a.* belonging to Latium, Italy. **Latinate** (-nāt), *a.* imitating or derived from Latin. **Latinism,** *n.* **Latinist,** *n.* **Latinity** (latin'iti), *n.* quality of Latin style or idiom, or of Latin scholarship. **Latinize, -ise,** *v.t.* to translate into Latin; to bring into conformity with the ideas, customs, forms etc., of the Romans or the Roman Catholic Church. *v.i.* to use Latin words, idioms or phrases. **Latinization,** *n.* **Latinizer,** *n.*
latipennate (latipen'āt), *a.* broad-winged. **latirostral** (-ros'trəl), **-trate** (-trāt), **-trous,** *a.* broad-beaked.
latish LATE.
latitude (lat'itūd), *n.* width; scope, comprehensiveness; looseness of application or meaning; absence of strictness; extent of deviation from a standard or rule; the angular distance of a celestial body from the ecliptic; angular distance of a place north or south of the equator; (*pl.*) regions, climates, esp. with reference to distance from the equator or the tropics. **latitudinal** (-tū'-), *a.* **latitudinally,** *adv.* **latitudinarian** (-neə'ri-), *n.* one who does not attach great importance to (religious) dogmas. *a.* wide in range or scope; free from prejudice, attaching little importance to speculative opinions; lax. **latitudinarianism,** *n.* **latitudinous** (-tū'-), *a.*
latria (lat'riə, lətri'ə), *n.* in the Roman Catholic Church, that supreme worship which can lawfully be offered to God alone.
latrine (lətrēn'), *n.* a lavatory, esp. in an army or prison camp.
-latry, *comb. form.* worship, as in *bibliolatry, idolatry.*
latten (lat'ən), *n.* a fine kind of brass; metal in thin sheets.
latter (lat'ə), *a.* coming or happening after something else; modern, present; lately done or past; second, second-mentioned; pertaining to the end of a period, life etc. **latter-day,** *a.* modern, recent. **Latter-day Saints,** the Mormons. **latter-end,** *n.* death; the end. **latterly,** *adv.*
lattice (lat'is), *n.* a structure of strips of metal or wood crossing and forming open work; in a crystal, the geometric pattern of molecules, atoms or ions, or of the points around which they vibrate; in a nuclear reactor, the geometrical arrangement of fissile and non-fissile material. *v.t.* to furnish with a lattice or lattices. **lattice bridge,** *n.* one built of lattice girders. **lattice girder,** *n.* a beam or girder consisting of bars connected together by iron lattice-work. **lattice-window,** *n.* one consisting of small panes set in strips of lead. **latticed,** *a.* **latticing,** *n.*
Latvian (lat'viən), *n.* a native or inhabitant of Latvia (a Soviet Baltic Republic); the language of Latvians. *a.* of or pertaining to Latvia, its people or its language.
laud (lawd), *v.t.* to praise, to extol. *n.* praise; (*pl.*) the psalms immediately following matins. **laudable,** *a.* praiseworthy, commendable. **laudableness, laudability** (-bil'-), *n.* **laudably,** *adv.* **laudation,** *n.* the act of praising; praise. **laudative, laudatory,** *a.* **lauder,** *n.*
laudanum (law'dənəm), *n.* opium prepared in alcohol.
laugh (lahf), *v.i.* to express amusement or exultation by inarticulate sounds and the convulsive movements of the face which are the involuntary effects of such emotions; to scoff (at). *v.t.* to express by laughing; to utter with laughter; to move or influence by ridicule or laughter. *n.* the action or an act of laughing; manner of laughing; entertainment. **to have the last laugh,** to be triumphant after a setback. **to laugh at,** to mock, deride, ridicule. **to laugh away,** to dismiss with a laugh; to pass (time) away in jesting. **to laugh down,** to suppress or silence with derisive laughter. **to laugh in, up, one's sleeve,** to be inwardly amused. **to laugh in someone's face,** to show someone open contempt or ridicule. **to laugh off,** to treat as of trifling importance. **to laugh on the other side of one's face, on the wrong side of one's mouth,** to be made to feel vexation or disappointment after mirth or satisfaction; to cry. **laughable,** *a.* exciting laughter; comical, ridiculous.

laughableness, *n.* **laughably,** *adv.* **laugher,** *n.* **laughing,** *n., a.* **no laughing matter,** not a proper subject for levity. **laughing-gas,** *n.* nitrous oxide, used as an anaesthetic (so-called because when inhaled it produces laughter). **laughing hyena** HYENA. **laughing jackass** JACKASS. **laughing-stock,** *n.* an object of ridicule. **laughingly,** *adv.* **laughter,** *n.* **laughy,** *a.* (*coll.*) prone to laughing.
launce (lans), *n.* a sand-eel.
launch[1] (lawnch), *v.t.* to throw, to propel; to cause to glide into the water (e.g. a vessel), or take off from land (e.g. a space rocket); to start or set (a person etc.) going; to introduce a new product or publication onto the market, usu. with a publicity campaign. *v.i.* of a ship, rocket etc., to be launched; to put to sea; to enter on a new sphere of activity; to expatiate. *n.* the act or occasion of launching. **to launch into,** to propel oneself into a new activity, career etc. with vigour and enthusiasm; to embark on a speech etc. **launch(ing)-pad, -site,** *n.* a platform or place from which a rocket is launched; a starting point for a new activity etc. **launcher,** *n.*
launch[2] (lawnch), *n.* the largest boat belonging to a man-of-war; a large power-driven pleasure-boat.
launder (lawn'də), *v.t.* to wash and iron (clothing, linen etc.); (*coll.*) to legitimize illegally-acquired money by transferring it through banks, foreign institutions etc. *v.i.* to wash and iron clothing, linen etc.; to be suitable for washing and ironing. **launderer, laundress** (-dris), *n.* one who washes and irons (clothes, linen etc.). **laund(e)rette** (-dret'), *n.* (*orig. trademark*) an establishment containing coin-operated washing-machines etc., for public use. **Laundromat**® (-drəmat), *n.* a launderette. **laundry** (-dri), *n.* a place where clothes are washed and ironed; a batch of washing. **laundry-man, -woman,** *n.* one who is employed in a laundry or who delivers washing.
laureate (law'riət, lo'-), *a.* crowned or decked with laurel. *n.* one who has been awarded a prize or other honour; a Poet Laureate. **Poet Laureate** POET. **laureateship,** *n.*
laurel (lo'rəl), *n.* a glossy-leaved evergreen shrub, the bay-tree; (*pl.*) the foliage of this, esp. in the form of a wreath, conferred as a distinction on victors, heroes, poets etc.; (*pl.*) the honours conferred by this; any other species of the genus *Laurus;* the common laurel or cherry laurel; any of various trees and shrubs resembling the laurel. **to look to one's laurels,** to guard against rivalry, to take care not to lose one's pre-eminence. **to rest on one's laurels,** to cease from one's efforts. **laurelled,** *a.* crowned with laurel.
lauric acid, *n.* an insoluble crystalline substance used in cosmetics and detergents.
Laurentian (lawren'shiən), *a.* designating a vast series of old rocks north of the St Lawrence River; relating to the St Lawrence River; relating to Lorenzo de'Medici or the library he established.
laurestine (law'rəstin), **laurustinus** (-tī'nəs), *n.* an ornamental evergreen shrub with pinkish-white winter flowers and dark-blue berries.
Laurus (law'rəs), *n.* a genus of plants containing the laurels, bay-tree etc. **lauraceous** (-rā'shəs), *a.*
lav (lav), *n.* (*coll.*) short for LAVATORY.
lava (lah'və), *n.* (*pl.* **-vas**) molten matter flowing in streams from volcanic vents or solidified by cooling. **lava-cone,** *n.* a volcanic cone formed by successive outflows of lava. **lava-flow, -stream,** *n.* **lavaform** (-fawm), *a.* having the form of lava. **lava-like,** *a.*
lavabo, lavatory LAVE.
lave (lāv), *v.t., v.i.* to wash; to bathe. **lavabo** (ləvah'bō, -vā'-), *n.* (*pl.* **-bos**) the washing of the celebrant's hands, in the Roman Catholic and other churches, after the offertory and before the Eucharist; a wash-basin. **lavage,** *n.* (*Med.*) washing out a hollow part of the body with water etc. **lavation,** *n.* the act of washing. **lavatory** (lav'ətri), *n.* a room or place for washing; a room with a toilet and usu. a washhand basin; a toilet; a ritual vessel

lavender 471 **lay**

for washing. **lavatory paper,** n. **lavatorial** (-taw'ri-), a. **lavement,** n. **laver**[1], n. a vessel containing water for the Jewish priests to wash when they offered sacrifices.
lavender (lav'əndə), n. a sweet-scented flowering shrub, cultivated for its oil which is used in perfumery; the flower and stalks or the oil used for perfuming linen etc.; the colour of the flowers, a pale lilac. a. of this colour. **lavender-water,** n. a perfume made with essential oil of lavender.
laver[1] LAVE.
laver[2] (lä'və), n. a name given to various seaweeds esp. *Porphyra laciniata, P. vulgaris* and other edible species.
laverock (lav'ərək), var. of LARK[1].
lavish (lav'ish), a. spending or giving with profusion; prodigal, unrestrained; existing or produced in profusion; excessive, super-abundant. v.t. to expend or bestow profusely; to squander. **lavisher,** n. **lavishly,** adv. **lavishness,** n.
law (law), n. a rule of conduct imposed by authority or accepted by the community as binding; a system of such rules regulating the intercourse of individuals within a State, or of States with one another; the controlling influence of this; the condition of order and stability it secures; the practical application of these rules, esp. by trial in courts of justice, litigation, judicial process; jurisprudence; legal knowledge; the legal profession; rules governing the conduct of a profession, sport etc.; a generalized statement of the orderly recurrence of natural phenomena and their consequences; the will of God as set forth in the Pentateuch, esp. in the Commandments; (*Ethics*) a principle of conduct emanating from the conscience. v.t., v.i. (*coll.*) to go to law, to take legal proceedings. **canon, civil, common, international, martial law** CANON, CIVIL, COMMON, INTERNATIONAL, MARTIAL. **law of averages,** the principle that extremes cancel one another out, thereby reaching a balance. **law of supply and demand,** the principle that the price of a commodity or service is governed by the relationship between the amount of demand for it and the quantity which can be supplied. **law of the jungle,** rules necessary for survival in adverse conditions or circumstances. **laws of motion** MOTION. **the law,** the police; a policeman. **to go to law,** to take legal proceedings. **to have the law of, on,** to take legal proceedings against. **to lay down the law,** to talk or direct in a dictatorial manner. **to take the law into one's own hands,** to try to secure satisfaction by one's own actions. **law-abiding,** a. obedient to the law. **law-abidingness,** n. **lawbook,** n. a treatise on law. **law-breaker,** n. one who violates the law. **law centre,** n. an office where free legal advice is available to the public. **law-French,** n. Anglo-Norman terms and phrases used in law. **lawgiver,** n. a legislator. **law-Latin,** n. the Latin used in legal documents. **Law-Lord,** n. a member of the House of Lords qualified to deal with the judicial business of the House. **lawmaker,** n. a legislator. **law-making,** n. **lawman,** n. (*N Am.*) a law enforcement officer. **lawmonger,** n. a pettifogging lawyer. **law officer,** n. a public legal functionary, esp. the Attorney-General and Solicitor-General. **law stationer,** n. one who deals in stationery used in legal work. **lawsuit,** n. an action in a court of law. **law-term,** n. a word or phrase used in law; one of the periods appointed for the sitting of the Law Courts. **lawful,** a. conformable to law; allowed by law; legitimate; valid, rightful. **lawfully,** adv. **lawfulness,** n. **lawless** (-lis), a. regardless of or unrestrained by the law, unbridled, licentious; not subject to or governed by law; illegal; anomalous, irregular. **lawlessly,** adv. **lawlessness,** n. **lawyer** (-yə), n. one who practises law, esp. an attorney or solicitor. **lawyer-like, lawyerly,** a.
lawks (lawks), *int.* an old exclamation of surprise or wonder.
lawn[1] (lawn), n. a grassy space kept smooth and closely mown in a garden or pleasure-ground. **lawn-mower,** n. a machine for mowing a lawn. **lawn-sprinkler,** n. a device with a perforated revolving collar for watering lawns. **lawn-tennis,** n. a game somewhat resembling real tennis, orig. played on a lawn but now frequently on a hard court. **lawny,** a.
lawn[2] (lawn), n. a fine cotton or linen fabric. **lawny,** a.
lawrencium (lawren'siəm), n. a radioactive element, at. no. 103; chem. symbol Lr, with a short half-life. [after Ernest O. *Lawrence,* 1901–58]
lawyer LAW.
lax (laks), a. slack, loose; porous; not exact, not strict; careless; ambiguous, vague. **laxative,** a. opening or loosening the bowels. n. a laxative medicine. **laxity, laxness,** n. **laxly,** adv.
lay[1] (lā), v.t. (*past, p.p.* **laid** (lād)) to cause to lie; to place in a prostrate or recumbent position; to bury; to drop (as eggs); to put down, to place, deposit; to wager; to apply; to put in proper position; to spread on a surface; to beat down, to prostrate; to overthrow; to cause to settle (as dust); to cause to be still, to allay; to exorcize; to put or bring into a certain state or position; to put forward, present; to impose, inflict; to bring down (a weapon, blows etc., on); to think out, to plan, to prepare; (*sl.*) to have sexual intercourse with. v.i. to drop or deposit eggs; (*nonstandard or Naut.*) to lie; to make a bet. n. the way, direction or position in which a region or object is situated; the direction in which the strands of a rope are twisted; (*sl.*) particular business, occupation, job etc.; (*sl.*) an act of sexual intercourse; (*sl.*) a sexual partner. **in lay,** of hens, laying eggs. **to lay about one,** to hit out on all sides. **to lay a cable,** to bury or sink an electric cable; to twist the strands of a cable. **to lay aside,** to abandon, to put away. **to lay bare,** to reveal; to strip. **to lay before,** to exhibit to; to bring to the notice of. **to lay by,** to save; to reserve for a future occasion. **to lay by the heels** HEEL[1]. **to lay down,** to put down; to surrender; to delineate; to assert; to formulate; to pay; to wager; to sacrifice; to put down the main structural parts of; to store (wine etc.); to stipulate. **to lay down the law** LAW[1]. **to lay fast,** to seize and keep fast, to prevent from escaping. **to lay hold of, on,** to grasp; to make a pretext of. **to lay in,** to store. **to lay into,** to assault physically or verbally. **to lay it on,** to speak or flatter extravagantly; to charge exorbitantly. **to lay low,** to fell or destroy; to cause to become weak or ill. **to lay off,** to suspend from employment; to desist; to avoid. **to lay on,** to impose; to deal (blows etc.); to supply (as water or gas); to prepare or arrange for printing. **to lay oneself open to,** to expose oneself to (criticism, attack etc.). **to lay oneself out,** to busy or exert oneself to do something. **to lay open,** to cut so as to expose the interior of; to expose. **to lay out,** to arrange according to plan; to spread out; to explain; to expend; to dress in graveclothes and dispose for burial; to knock to the ground or render unconscious. **to lay over,** (*N Am.*) to stop over during a journey; to postpone. **to lay to,** to check the motion of a ship. **to lay to sleep, rest,** to bury. **to lay under,** to subject to. **to lay up,** to store, to save; of illness, to confine to one's bed or room; to dismantle and place in dock. **to lay waste,** to ravage. **layabout,** n. an idle person, a lounger. **lay-by,** n. a widening of a road to enable vehicles to stop without holding up traffic. **lay-out,** n. the make-up of a printed page; a planned arrangement of buildings etc.; that which is set out or displayed. **layer,** n. one who or that which lays; a stratum; a shoot laid with part of its length on or beneath the surface of the ground in order that it may take root; an artificial oyster-bed; the areas between contours on a map marked by distinctive colouring. v.t. to propagate by layers; to place, cut or form in layers. **layering,** n. a method of propagating plants by layers; any method employing layers. **laying,** n.
lay[2] (lā), a. pertaining to the people as distinct from the clergy; non-professional, lacking specialized knowledge;

(*Cards*) other than trumps. **lay brother, sister**, *n.* a brother or sister in a monastery, under vows and wearing the habit of the order, engaged chiefly in manual labour and exempt from other duties. **lay communion**, *n.* membership of the church as a layman. **lay elder**, *n.* a ruling elder in the Presbyterian Church. **layman, laywoman**, *n.* a non-professional, one not an expert. **lay reader**, *n.* a member of the Church of England laity authorized to conduct certain religious services; a layman in the Roman Catholic Church who reads the epistle at Mass. **laic** (lā'ik), **-ical**, *a.* lay, not clerical, secular. *n.* a layman. **laically**, *adv.* **laicization** (-siz-), *n.* **laicize, -ise**, *v.t.* to render lay or secular. **laity** (lā'iti), *n.* (*collect.*) the people, as distinct from the clergy; those not belonging to a particular profession.

lay[3] (lā), *n.* a lyric song or ballad; a short narrative poem for singing or recitation.

lay[4] (lā), *past* LIE[2].

lay[5] (lā), LEY.

layer LAY[1].

layette (lăet') *n.* the outfit for a new-born infant.

lay figure (lā'), *n.* a jointed figure of the human body used by artists for hanging drapery on etc.; a nonentity; an unreal character in a story etc.

laying LAY[1].

layman LAY[2].

laystall LAY[1].

lazar (laz'ə), *n.* a person infected with a loathsome disease, esp. a leper. **lazar-house**, *n.* a lazaretto. **lazarus-house** (-rəs-), *n.*

lazaretto (lazəret'ō), **lazaret** (-ret'), *n.* (*pl.* **-ttos**) a hospital for contagious disease; a ship or other place of quarantine; a store-room for provisions in large merchant-vessels.

lazarone (lahtsərō'nä), LAZZARONE.

laze LAZY.

lazuli LAPIS LAZULI.

lazulite (laz'ūlīt), *n.* an azure-blue to pale greenish-blue mineral.

lazy (lā'zi), *a.* idle, indolent, slothful, disinclined for labour or exertion; disposing to idleness or sloth. **laze**, *v.i.* to be lazy; to live in idleness. *v.t.* to waste or spend in idleness. *n.* a time or spell of idleness. **lazy-bones**, *n.* a lazy fellow, an idler. **lazy daisy**, *n.* a type of embroidery stitch. **lazy Susan**, *n.* a revolving tray for a dining table with compartments for various condiments. **lazy-tongs**, *n.pl.* tongs consisting of levers, in pairs, crossing one another and turning on a pin like scissors, for picking up distant objects. **lazily**, *adv.* **laziness**, *n.*

lazzarone (lahtsərō'nä), *n.* (*pl.* **-ni** (-nē)) a Neapolitan beggar.

lb, (*abbr.*) in cricket, leg-bye; pound(s).

lbw, (*abbr.*) in cricket, leg before wicket.

lc, (*abbr.*) left centre; letter of credit; *loco citato*, in the place cited; lower case (*type*).

lcd, LCD, (*abbr.*) liquid crystal display; lowest common denominator.

LCJ, (*abbr.*) Lord Chief Justice.

lcm, LCM, (*abbr.*) least, lowest common multiple.

L Cpl, (*abbr.*) Lance-Corporal.

LDS, (*abbr.*) Latter-day Saints; *laus Deo semper*, praise be to God for ever.

lea[1] (lē), *n.* a meadow; grassland; open country.

lea[2] (lē), *n.* land left untilled, fallow land, grassland. *a.* fallow, unploughed.

LEA, (*abbr.*) Local Education Authority.

leach[1] (lēch), *v.t.* to wash out or separate (a soluble constituent) by percolation; to strain or drain (liquid) from some material (*usu.* out or away). *v.i.* of liquid in any material, to drain out. *n.* a tub, vat or other vessel used for leaching; a leachate. **leach-tub**, *n.* a tub for leaching ashes in. **leachate** (-āt), *n.* the substance obtained by leaching;

the percolating liquid used in leaching. **leachy**, *a.*

leach[2] (lēch), LEECH[1].

lead[1] (lĕd), *n.* a soft malleable and ductile, bluish-grey, heavy metal, at. no. 82; chem. symbol Pb; in printing, a thin plate of type-metal used to separate lines; graphite used in lead-pencils; (*pl.*) strips of lead used for covering a roof; a roof, esp. a flat roof covered with lead; a lead weight on a line used for sounding; a metal strip holding the glass in diamond-paned windows; lead bullets or (fishing) weights. *a.* pertaining to or made of lead. *v.t.* to cover, fasten, weight, frame or fit with lead; to space out (as lines of type) by inserting leads. **blacklead** BLACK. **red, white lead**, RED, WHITE. **to swing the lead**, to malinger. **lead-glance**, *n.* galena. **leadglass**, *n.* glass containing lead oxide. **lead-line**, *n.* a sounding-line. **lead-oxide**, *n.* litharge, a yellow crystalline material substance used in the manufacture of glass and paint. **lead paint**, *n.* paint with a lead base. **lead-pencil**, *n.* a pencil containing a slip of graphite. **lead-poisoning**, *n.* poisoning caused by the prolonged absorption of lead into the system. **leaded**, *a.* set in or fitted with lead; separated by leads, as lines of printing. **leaden**, *a.* made of lead; dark-coloured; heavy as lead; slow, burdensome; inert. **leadenly**, *adv.* **leadenness**, *n.* **leading**, *n.* (*Print.*) leads; the space introduced between lines of type by inserting leads; the lead strips framing panes of glass or covering a roof. **leadless** (-lis), *a.* **leady**, *a.*

lead[2] (lēd), *v.t.* (*past, p.p.* **led** (lĕd)) to conduct, to show the way; to direct the movements of; to be in command of; to keep in front of; to be at the head of, to direct by example; to indicate, esp. by going in advance; to pass or spend (time etc.); to cause to spend or pass; to draw or drag after one; to begin a round at cards with. *v.i.* to act as conductor or guide; to go in advance; to be the commander or foremost person in any undertaking; to be the first player in a game of cards; to go towards, to extend; to tend (to) as a result. *n.* guidance, direction, esp. by going in front; the first place, precedence; the leading role; an example; a clue; a cord for leading a dog; a principal conductor for distribution of electric current; in cards, the first play or the right to this; the main story in a newspaper. *a.* principal, chief, main, leading. **to lead astray**, to lead into error. **to lead off**, to make a start. **to lead on**, to entice further; to fool or trick. **to lead the way**, to go first so as to point the way. **to lead up to**, to conduct conversation towards (some particular subject); to pave the way for. **lead-in**, *n.* an introduction to a topic; the electric conductor connecting a radio transmitter or receiver with an outside aerial or transmission cable. **lead-off**, *n.* a leading motion. **lead time**, *n.* the interval between the design and manufacture of a product. **leadable**, *a.* **leader**, *n.* one who or that which leads; a guide, conductor; a chief, commander; the leading counsel in a case; a chief editorial article in a newspaper; the principal first violin of an orchestra; (*N Am.*) a conductor of an orchestra; a blank strip of film or tape preceding or following the recorded material; the foremost horse in a team; in printing, a row of dots to lead the eye across a page or column; a trace on a fishing-line; the terminal bud or shoot at the apex of a stem or branch. **leaderless** (-lis), *a.* **leadership**, *n.* **leading**, *a.* guiding, conducting; principal. *n.* the action of the verb TO LEAD; guidance, influence. **leading aircraftman, aircraftwoman**, *n.* a rank in the British Air Force below senior aircraftman. **leading article**, *n.* a leader in a newspaper. **leading case**, *n.* in law, a case that forms a precedent for the decision of others. **leading edge**, *n.* (*Aviat.*) the foremost edge of an aerofoil (e.g. of a wing, propeller blade). **leading lady, man**, *n.* persons taking the chief role in a play. **leading light**, *n.* an influential or prominent member of a movement, group etc. **leading note**, *n.* the subtonic. **leading question**, *n.* a question (esp. in cross-examination) that

suggests a certain answer. **leading-rein**, *n*. a rein for leading a horse by. **leading-reins, -strings**, *n.pl*. a harness by which small children are controlled when learning to walk; a state of dependence on others. **led**, *a*. under another's influence or leading.
leaden LEAD [1].
leading LEAD [2].
leaf (lēf), *n*. (*pl*. **leaves** (lēvz)) one of the (usu.) green, flat organs of plants whose function is photosynthesis and transpiration; anything resembling this; a petal, scale or sepal; (*collect*.) foliage; a sheet of paper in a book or manuscript; a thin sheet of metal or other material; a valved, hinged, sliding or detachable member of a bridge, table, door, shutter, screen etc. *v.i*. to shoot out or produce leaves or foliage. **to leaf through**, to turn the pages of a book, magazine etc., in a casual way. **to turn over a new leaf**, to change one's mode of life or conduct for the better. **leaf-bridge**, *n*. one with a rising leaf or leaves swinging vertically on hinges. **leaf-bud**, *n*. one developing into a leaf. **leaf curl**, *n*. a disease of plants which causes curling of the leaves. **leaf-cutter**, *n*. an insect (as a bee or ant) which cuts out sections of leaves. **leaf-hopper**, *n*. a jumping insect that sucks plant juices. **leaf-insect**, *n*. one having camouflaged wing covers resembling leaves. **leaf-metal**, *n*. **leaf miner**, *n*. any of various insects that as larvae bore into and eat leaf tissue. **leaf-mould**, *n*. decayed leaves reduced to mould and used as compost. **leaf-roll**, *n*. a form of potato virus. **leaf-spring**, *n*. a spring consisting of several flat strips of metal. **leaf-stalk**, *n*. a petiole supporting a leaf. **leaf-work**, *n*. **leafage** (-ij), *n*. **leafed, leaved**, *a*. **leafless** (-lis), *a*. **leaflessness**, *n*. **leaflet** (-lit), *n*. a small leaf; a one-page handbill, circular etc.; a pamphlet; one of the primary divisions of a compound leaf. *v.i*. to distribute leaflets. **leaflike**, *a*. **leafy**, *a*. **leafiness**, *n*.
league [1] (lēg), *n*. a union for mutual help or protection or the pursuit of common interests; a category, class or group; an association of clubs that play matches against one another, as *football league*. *v.t., v.i*. to combine together (with). **in league with**, having formed an alliance with, usu. for a dubious purpose. **League of Nations**, an international organization (1920–46), pledged to cooperate in securing peace and the rigorous observance of treaties by its member states. **not in the same league**, not on the same level of excellence etc. **league match**, *n*. a match between clubs in the same league. **league table**, *n*. a list of competitors in a league in order of performance; a list showing the order of achievement, merit, performance etc. **leaguer**, *n*. a league member.
league [2] (lēg), *n*. an old measure of distance (in England usu. about three and four nautical miles, about 4·8 km).
leak (lēk), *v.i*. to let liquid, gas etc. pass in or out through a hole, fissure etc.; to ooze through a hole or fissure; (*sl*.) to urinate. *v.t*. to allow to enter or pass out; to divulge (confidential information). *n*. a crevice or hole which admits water or other fluid; the oozing of a fluid through such crevice; (*sl*.) urination; the divulgence of confidential information; a loss of electric current from a conductor. **to leak out**, to become gradually known or public. **to spring a leak**, to begin to leak. **leakage** (-ij), *n*. a leak; the quantity that leaks; an allowance at a certain rate for loss by leaking etc. **leaker**, *n*. **leakiness**, *n*. **leaky**, *a*.
leal (lēl), *a*. (*Sc*.) loyal, true. **leally**, *adv*. **lealty** (lē'əl-), *n*.
lean [1] (lēn), *v.i*. (*past, p.p*. **leaned, leant** (lent)) to incline one's body from an erect attitude; to rest (against or upon); to deviate from a straight or perpendicular line or direction; to depend (upon); to have a tendency or propensity (to or towards). *v.t*. to cause to incline. *n*. a leaning, inclination, slope or deviation. **to lean on**, (*coll*.) to coerce, threaten (someone). **lean-to**, *n*. a building with a roof supported by another building or wall. **leaning**, *n*. inclination, propensity (towards or to).

lean [2] (lēn), *a*. thin; of meat, not fat, consisting of muscular tissue; wanting in plumpness; meagre, unproductive. *n*. meat without fat. **lean-burn**, *a*. of an internal-combustion engine, burning a low proportion of fuel to air, to reduce consumption and exhaust emissions. **leanly**, *adv*. **leanness**, *n*.
leap (lēp). *v.i*. (*past, p.p*. **leapt** (lept), **leaped**) to jump, to spring upwards or forwards; to rush, dart; to pass over an interval, esp. in music; to make a sudden transition. *v.t*. to jump over or across; to cause to jump. *n*. the act of leaping; a jump, a spring; the space passed over by leaping; a space or interval; a sudden transition; an increase; a place of leaping. **a leap in72 the dark**, an action whose consequences cannot be foreseen. **by leaps and bounds** BOUND [1]. **leap-day**, *n*. 29 February. **leap-frog**, *n*. a game in which one stoops down and another vaults over. *v.t., v.i*. to vault in this way. **leap year**, *n*. an intercalary year of 366 days, which adds one day to February every four years. **leaping**, *n*. **leapingly**, *adv*.
learn (lœn), *v.t*. (*past, p.p*. **learnt, learned**) to acquire knowledge of or skill in by study, experience or instruction; to fix in the memory; to find out, to be informed of. *v.i*. to acquire knowledge or skill; to receive instruction. **learnable**, *a*. **learned** (-nid), *a*. having acquired learning by study; skilled, skilful (in); erudite; of words etc., introduced or chiefly used by learned people. **learnedly**, *adv*. **learnedness**, *n*. **learner**, *n*. **learning**, *n*. the act of learning; knowledge acquired by study; scholarship.
lease (lēs), *n*. a letting or renting of property for a specified period; the written contract for, the term of or the rights of tenure under such letting. *v.t*. to grant or to take or hold under lease. **a new lease of life**, an anticipated spell of renewed life or enjoyment. **leaseback**, *n*. an arrangement whereby the seller of a property leases it back from the buyer. **leasehold**, *n*. tenure by lease; property held by lease. *a*. held thus. **leaseholder**, *n*. **leasable**, *a*. **leaser**, *n*.
leash (lēsh), *n*. a lead for controlling a dog or other animal; that which controls or restrains as if by a leash. *v.t*. to bind, hold or fasten (as) by a leash. **straining at the leash**, anxious or impatient to begin.
least (lēst), *a*. smallest; less than all others in size, quantity, importance etc. *adv*. in the smallest or slightest degree. *n*. the smallest amount, degree etc. **at (the) least**, at or in the lowest degree; at any rate. **in the least**, in the slightest degree, at all.
leather (ledh'ə), *n*. the tanned or dressed skin of an animal; an article or part made of leather (*often in comb*., as *stirrup-leather*); (*pl*.) a pair of leather breeches or leggings; (*sl*.) a cricketball or football; (*facet*.) one's skin. *a*. made of leather. *v.t*. to cover or furnish with leather; to thrash. **fair, white leather**, leather with its natural colour. **patent leather** PATENT. **leather-back**, *n*. a leathery, soft-shelled turtle. **leather-head**, *n*. a blockhead; an Australian bird without head feathers. **leather-jacket**, *n*. an Australian tree, *Eucalyptus resinifera;* the larva of a crane-fly; one of various fishes. **leather-neck**, *n*. (*Austral*.) a handyman; (*sl*.) a US marine. **Leatherette**[R] (-ret'), *n*. a kind of imitation leather. **leatheriness**, *n*. **leathering**, *n*. a beating. **leathern** (-ən), *a*. **leathery**, *a*.
leave [1] (lēv), *n*. liberty or permission; permission to be absent from duty; the period of this; the act of departing, a formal parting; a holiday. **by, with your leave**, with your permission. **French leave** FRENCH. **on leave**, absent from duty by permission; on holiday. **to take leave of one's senses**, to think or act contrary to reason. **to take (one's) leave**, to say good-bye; to depart. **leave-taking**, *n*. parting; a farewell.
leave [2] (lēv), *v.t*. (*past, p.p*. **left** [2] (left)) to allow to remain, to go without taking; to bequeath; to refrain from removing, consuming or interfering with; to depart from, to

leaven 474 **leg**

quit; to withdraw from; to abandon; to desist from, discontinue; to commit; to refer for consideration, approval etc. *v.i.* to depart, to go away; to cease, discontinue. **to leave alone**, not to interfere with; to have no dealings with. **to leave be**, to avoid disturbing or interfering. **to leave behind**, to go away without; to outstrip; to leave as a record, mark, consequence etc. **to leave off**, to stop, discontinue; to cease to wear. **to leave out**, to omit. **to leave over**, to leave for future consideration etc. **to leave well alone**, to leave be. **leaver**, *n.* **leaving**, *n.* the act of departing; (*pl.*) remnant, refuse.

leaven (lev'ən), *n.* a substance (e.g. yeast) mixed with dough in order to cause fermentation and make it lighter; any influence tending to cause a general change. *v.t.* to raise and make light (as) with leaven; to pervade with an influence causing change. **leavening**, *n.* leaven.

leaves (lēvz), LEAF.

Lebanese (lebənēz'), *a.* pertaining or belonging to the Mediterranean country of Lebanon. *n.* the people of the Lebanon.

Lebensraum (lā'bənzrowm), *n.* territory necessary for a country's expanding population. [G]

lecher (lech'ə), *n.* a lascivious or promiscuous man. **lech, letch**, *v.i.* (*sl.*) to lust (after); to act lecherously. *n.* a lecher; a lascivious act. **lecherous**, *a.* **lecherously**, *adv.* **lechery**, *n.*

lecithin (les'ithin), *n.* a nitrogenous fatty substance containing phosphorus found in animal and vegetable tissue.

lectern (lek'tən), *n.* a reading-desk from which the lessons are read in church; any similar reading desk.

lection (lek'shən), *n.* a portion of Scripture to be read in church. **lectionary**, *n.* a collection of passages of Scripture for daily services. **lector** (-tə), *n.* a cleric in minor orders; a reader, esp. in a German university. **lectorate** (-rət), **lectorship**, *n.*

lecture (lek'chə), *n.* a formal expository or instructive discourse on any subject, before an audience or a class; a reprimand. *v.i.* to deliver, or instruct by, a lecture or lectures. *v.t.* to instruct by lectures; to reprimand. **lecturer**, *n.* **lectureship**, *n.* the academic office of a lecturer.

LED, (*abbr.*) light-emitting diode.

led (led), LEAD ².

lederhosen (lā'dəhōzən), *n.* leather shorts with braces, the trad. male dress of Austria and Bavaria. [G]

ledge (lej), *n.* a shelf or shelf-like projection; a shelf-like ridge or outcrop of rock; a metal-bearing stratum of rock. **ledged**, *a.* **ledgeless** (-lis), *a.* **ledgy**, *a.*

ledger (lej'ə), *n.* the principal book in a set of account-books, containing a record of all trade transactions; (*Angling*) a ledger-line or tackle. *v.i.* to fish with a ledger-tackle. **ledger-bait**, *n.* fishing bait that remains on the bottom; hence **ledger-hook, -line, -tackle. ledger-, legerline**, *n.* (*Mus.*) an additional short line above or below the stave to express ascending or descending notes.

Ledum (lē'dəm), *n.* a genus of low shrubs of the heath family.

lee (lē). *n.* the side or quarter towards which the wind blows, opp. to windward or weather side; the sheltered side; shelter, protection. *a.* pertaining to the side or quarter away from the wind; sheltered. **under the lee of**, protected from the wind by. **lee-board**, *n.* a board let down on the lee-side of a flat-bottomed vessel to prevent a leeward drift. **lee-gage, -gauge**, *n.* position to leeward of another ship. **lee shore**, *n.* the shore on the lee side of a vessel. **lee side**, *n.* the lee of a vessel. **lee tide**, *n.* a tide running in the same direction as the wind blows. **leeward** (-wəd, loo'əd), *a.* relating to, in or facing the lee side. *adv.* towards the lee side. *n.* the lee side or direction. **leeway**, *n.* the leeward drift of a vessel; scope or toleration inside defined limits. **to make up leeway**, to recover lost ground or time.

leech¹ (lēch), *n.* (*old-fashioned*) a physician, a healer; an aquatic bloodsucking worm, formerly used medicinally; one who abstracts or absorbs the gains of others; one who clings tenaciously to another. *v.t.* to apply leeches to, as in phlebotomy; to cling to or prey on. **leechcraft**, *n.* the art of healing.

leech² (lēch), *n.* the perpendicular ledge of a square sail; the after edge of a fore-and-aft sail.

leek (lēk), *n.* a culinary vegetable, *Allium porrum*, allied to the onion, with a cylindrical bulb, the national emblem of Wales.

leer (liə), *n.* an oblique, sly or arch look; a look expressive of malice, lasciviousness or triumph. *v.i.* to look with a leer. **leeriness**, *n.* **leering**, *n.*, *a.* **leeringly**, *adv.* **leery**, *a.* knowing, sly. **leery of**, wary of.

lees (lēz), *n.pl.* the sediment of liquor which settles to the bottom.

leet (lēt), *n.* (*Sc.*) a list of candidates for any office. **short leet**, the final select list of such candidates.

leeward, leeway LEE.

left¹ (left), *a.* of, pertaining to or situated on the side that is to the east when one faces south, opp. to right; correspondingly situated in relation to the front or the direction of anything; radical, politically innovative; of or pertaining to socialism or communism. *adv.* on or towards the left. *n.* the side opposite to the right; the left hand; a left-handed blow; the progressive, democratic or socialist party, wing or faction. **left hand**, *a.* situated on or pertaining to the left side; executed by the left hand. *n.* the left side, direction or region. **left-handed**, *a.* using the left hand more readily than the right; moving from right to left; done with the left hand; awkward, clumsy; ambiguous, equivocal. **left-handedly**, *adv.* **left-handedness**, *n.* **left-hander**, *n.* a left-handed person or blow. **leftward** (-wəd), *adv.*, *a.* **leftwards**, *adv.* **left-wing**, *n.* the left side of an army or sports pitch. *a.* pertaining to, active in or sympathetic to the political left (of a party); playing on the left-wing. **left-winger**, *n.* **leftism**, *n.* the policies and principles of the political left. **leftist**, *n.*, *a.* **lefty**, *n.*, *a.* (*derog.*) (a) leftist.

left² (left), *past, p.p.* LEAVE ². **left-luggage**, *n.* luggage deposited temporarily at a railway station etc. **left-luggage office**, *n.* **left-over**, *n.* (*usu. pl.*) a remainder, esp. of uneaten food.

leg (leg), *n.* one of the limbs by which humans and other animals walk; an animal's hind leg (esp. the upper portion) which is eaten as meat; the part of a garment that covers the leg; one of a set of posts or rods supporting a table, chair etc.; a limb of a pair of compasses etc.; the course and distance run by a vessel on one tack; a stage in a long-distance journey; in cricket, (a fielder in) that part of the field to the rear and left of a batsman; in a contest, any of a series of events, games etc. **a leg up**, a boost. **leg before wicket**, in cricket, stoppage by the batsman's leg of a ball when it would have hit the wicket. **leg-of-mutton**, *a.* of a sleeve etc., tapering sharply. **not to have a leg to stand on**, to have no support or basis for one's position (e.g. in a controversy). **to find one's legs**, to attain ease or mastery. **to leg it**, (*coll.*) to walk. **to pull someone's leg**, to hoax, to tease. **to shake a leg**, (*often int.*) to hurry up. **to show a leg**, to get out of bed. **to stretch a leg, one's legs**, to take exercise, esp. after inactivity. **leg-break**, *n.* in cricket, a ball which breaks from the leg side. **leg-bye**, *n.* in cricket, a run scored for a ball that touches the batsman. **leg-guard**, *n.* in cricket, baseball etc., a pad to protect the leg from knee to ankle. **leg-iron**, *n.* a fetter for the leg. **leg-man**, *n.* (*coll.*) one who runs errands etc. for another. **leg-rest**, *n.* a support for an injured leg. **leg-room**, *n.* space for the legs (e.g. in a car). **leg show**, *n.* entertainment involving the exhibition of women's legs. **legwarmers**, *n.pl.* long footless stockings usu. worn over outer garments. **legwork**, *n.* work involving

leg. 475 **lemon**

much travel on foot. **legged,** *a. (usu. in comb.,* as *four-legged).* **legging,** *n. (usu. pl.)* gaiters; a covering of leather, stretch cotton etc. for the legs. **leggy,** *a. (coll.)* having long legs. **legginess,** *n.* **legless** (-lis), *a.* without legs; *(coll.)* very drunk.
leg., *(abbr.)* legal; legate; legato; legislation.
legacy (leg'əsi), *n.* a bequest; property bequeathed by will; anything left or handed on by a predecessor. **legacy-hunter,** *n.* one who pays court to another in the hope of receiving a legacy. **legatee** (-tē'), *n.* one to whom a legacy is bequeathed. **legator** (-gā'-), *n.*
legal (lē'gəl), *a.* of, pertaining to, or according to law; lawful; recognized, sanctioned or appointed by the law; characteristic of law or lawyers. **legal aid,** *n.* financial assistance for legal proceedings granted to those with low incomes. **legal tender,** *n.* money which a creditor is bound to accept in discharge of a debt. **legalese** (-ēz), *n.* the language of legal documents. **legalism,** *n.* strict adherence to law and formulas; respect for the letter rather than the spirit of law. **legalist,** *n.* **legalistic** (-lis'-), *a.* **legalistically,** *adv.* **legality** (ligal'-), *n.* **legalization,** *n.* **legalize, -ise,** *v.t.* **legally,** *adv.*
legate (leg'ət), *n.* a papal emissary; an ambassador, envoy. **legateship,** *n.* **legatine** (-tīn), *a.* **legation** (ligā'shən), *n.* the act of sending a legate or deputy; a diplomatic mission; a diplomatic representative and his or her delegates; the official residence of a diplomatic representative.
legatee etc. LEGACY.
legato (ligah'tō), *adv., a. (Mus.)* in an even, gliding manner without a break. *n. (pl.* **-tos)** this style of playing. **legatissimo** (legətis'imō), *adv., a.* as smoothly as possible.
legend (lej'ənd), *n.* a traditional story, esp. one popularly accepted as true; a myth, fable; traditional or non-historical story-telling or literature; an inscription or caption, as on a coat of arms, a coin or an illustration; one who is renowned for outstanding deeds or qualities, whether real or fictitious. **legendary,** *a.* **legendist,** *n.* **legendry,** *n.*
leger (lej'ə), LEDGER.
legerdemain (lejədimān'), *n.* sleight of hand, a trick in which the eye is deceived by the quickness of the hand; jugglery, sophistry.
legging, leggy LEG.
leghorn (leg'hawn, ligawn'), *n.* a plait of the straw of bearded Italian wheat; a hat made of this; a breed of domestic fowl.
legible (lej'ibl), *a.* that may be read; easily decipherable; clear. **legibleness, legibility** (-bil'-), *n.* **legibly,** *adv.*
legion (lē'jən), *n.* a division of the ancient Roman army, varying from 3000 to 6000 men; a military force, esp. in France and other foreign countries; a large number, multitude. **American, British Legion,** associations of ex-Service men and women. **Foreign Legion,** corps of foreign volunteers in the French army. **Legion of Honour,** a French order of merit. **legionary,** *a.* pertaining to, consisting of one or more legions. *n.* a member of a legion. **legioned,** *a.* formed or drawn up in legions. **legionnaire** (-neə'), *n.* a legionary. **Legionnaire's disease,** *n.* a serious, sometimes fatal disease resembling pneumonia, caused by *Legionella* bacteria (named because of its occurrence at an American Legion convention in 1976).
legislate (lej'islāt), *v.i.* to make or enact a law or laws; to make allowance (for). **legislation,** *n.* (the act or process of making) laws. **legislative,** *a.* enacting laws; having power to legislate; enacted by or pertaining to legislation. **legislatively,** *adv.* **legislator,** *n.* a member of a legislative assembly. **legislatorial** (-taw'ri-), *a.* **legislatress** (-tris), *n.fem.* **legislature** (-sləchə), *n.* a law-making assembly.
legist (lē'jist), *n.* one learned in the law.
legit (lijit'), *a. (coll.)* short for LEGITIMATE[1]. *n.* the legitimate theatre.
legitimate[1] (lijit'imət), *a.* lawful; legal; born in wedlock; legally descended; of a title to sovereignty, derived from strict hereditary right; proper, regular; conformable to accepted usage; following by logical sequence; pertaining to formal or serious theatre rather than television, cinema, variety etc. **legitimacy,** *n.* **legitimately,** *adv.* **legitimateness,** *n.* **legitimation,** *n.* **legitimism,** *n.* the doctrine of hereditary monarchical government and divine right. **legitimist,** *n., a.* **legitimization,** *n.* **legitimize, -ise,** *v.t.* to render legitimate.
legitimate[2] (lijit'imāt), *v.t.* to make lawful; to render legitimate; to serve as justification for.
Lego® (leg'ō), *n.* a building toy mainly consisting of connecting plastic bricks.
legume (leg'ūm), *n.* the edible fruit or pod of a leguminous plant (as pea or bean); any of various vegetables used as food, esp. pulses. **leguminous** (-gū'-), *a.* producing legumes; pertaining to the Leguminosae, an order of plants bearing legumes.
lei (lā'i), *n.* a Hawaiian garland or necklace of flowers.
Leibnitzian, -nizian (lībnit'siən), *n., a.* (a follower) of the German philosopher Gottfried Leibnitz (1646–1716) or his philosophy, esp. his doctrine of pre-established harmony and the optimism based on this. **Leibnitzianism,** *n.*
Leicester (les'tə), *n.* a type of cheese (orig. made in Leicestershire) resembling cheddar; a breed of sheep with a long fleece.
leiotrichous (līot'rikəs), *a.* smooth-haired.
leipoa (līpō'ə), *n.* a genus of mound-birds, containing the native pheasant of Australia.
leister (lēs'tə), *n., v.t.* (to spear with) a pronged fishing-spear.
leisure (lezh'ə), *n.* freedom from business, occupation or hurry; time at one's own disposal; opportunity, convenience. *a.* unoccupied, free, idle. **at leisure,** at one's ease or convenience; without hurry. **leisure centre,** *n.* a building containing facilities for sports, entertainments, meetings etc. **leisure wear,** *n.* casual clothing. **leisurable,** *a.* **leisurably,** *adv.* **leisured,** *a.* **leisureless** (-lis), *a.* **leisureliness,** *n.* **leisurely,** *a., adv.*
leitmotiv, -motif (līt'mōtēf), *n.* a recurring theme in a composition, orig. a musical theme invariably associated with a certain person, situation or idea throughout an opera etc.
lek[1] (lek), *n.* the Albanian unit of currency.
lek[2] (lek), *n.* an area where certain species of birds (esp. black grouse) assemble for sexual display and courtship.
LEM, *(abbr.)* lunar excursion module.
lemma (lem'ə), *n. (pl.* **-mmas, -mmata** (-mətə)) an auxiliary proposition taken to be valid in order to demonstrate some other proposition; a theme, esp. when prefixed as a heading.
lemming (lem'ing), *n.* a small mouselike rodent of northern Europe, remarkable for migrating at certain periods in immense multitudes; someone who dashes headlong into situations without forethought.
lemniscate (lemnis'kət), *n.* a curve of the general form of a figure 8 (∞).
lemniscus (lemnis'kəs), *n. (pl.* **-ci** (-nis'ī)) *(Anat., Zool.)* a bundle of fibres or ribbon-like appendages.
lemon (lem'ən), *n.* an oval acid citrus fruit; the tree bearing this; its pale yellow colour; *(sl.)* one who or that which is disappointing, unpleasant, useless. *a.* of the colour of a lemon; lemon flavoured. **salt of lemon** SALT. **lemon cheese, curd,** *n.* a spread made from lemon, butter, eggs and sugar. **lemon-dab** LEMON-SOLE. **lemon-drop,** *n.* a lemon-flavoured hard sweet. **lemon grass,** *n.* a lemon-scented hardy grass from the tropics which yields an essential oil. **lemon-peel,** *n.* **lemon-plant, -verbena,** *n.* a S American shrub cultivated for its lemon-scented foliage. **lemon-squash,** *n.* a sweet concentrated lemon drink. **lemon-squeezer,** *n.* **lemon-wood,** *n.* a small New Zealand tree. **lemonade** (-nād'), *n.* lemon-juice or lemon flavour-

lemon-sole / **lese-majesty**

ing mixed with still or aerated water and sweetened. **lemony**, *a.*

lemon-sole, lemon-dab *n.* a flat-fish with brown markings valued as a food.

lemur (lē'mə), *n.* any member of a genus of arboreal nocturnal animals allied to the monkeys, common in Madagascar. **lemurid** (lem'ūrid), *n.* **lemuroid** (lem'ūroid), *n., a.*

lend (lend), *v.t.* (*past, p.p.* **lent** (lent)) to grant the use of on condition of repayment or compensation; to let out (money) at interest; to contribute, esp. for temporary service; to accommodate (oneself). *v.i.* to make loans. **to lend a hand** HAND [1]. **to lend an ear**, to listen. **to lend oneself, itself, to**, to have the right qualities for. **lendable**, *a.* **lender**, *n.* **lending**, *n.* **lending library**, *n.* one from which books can be borrowed freely or for a subscription.

length (length), *n.* measure or extent from end to end, as distinguished from breadth or thickness; a definite portion of the linear extent of anything; the state of being long; extent of time; the distance anything extends; extent or degree of action etc.; the quantity of a vowel or syllable; the distance traversed by a cricket ball before striking the ground; in racing, the linear measure of the body of a horse, boat etc. **arm's length** ARM [1]. **at length**, to the full extent, in full detail; at last. **to go to any length**, to be restrained by no scruples. **lengthen**, *v.t.* to make long or longer; to extend; to protract. *v.i.* to grow longer. **lengthener**, *n.* **lengthways, lengthwise**, *adv., a.* **lengthy**, *a.* long and tedious; prolix. **lengthily**, *adv.* **lengthiness**, *n.*

lenient (lē'niənt), *a.* mild, gentle; merciful. **lenience, -ency**, *n.* **leniently**, *adv.* **lenitive** (len'-), *a.* having the power or quality of softening or mitigating. *n.* a lenitive medicine. **lenity** (len'-), *n.*

Leninism (len'inizm), *n.* the economic and political theory and practice of the Russian statesman and Marxist revolutionary Lenin (pseud. of Vladimir I!yıch Ulyanov, 1870–1924). **Leninist**, *n., a.*

leno (lē'nō), *n.* (*pl.* **-nos**) an open cotton fabric resembling fine muslin.

lens (lenz), *n.* (*pl.* **lenses**) a piece of transparent substance, usu. glass, or a combination of such (**compound lens**), with the surface or both surfaces curved so as to change the direction of rays of light; the crystalline body (**crystalline lens**) in the eye through which rays of light are focused on the retina; a device (**electrostatic** or **electromagnetic lens**) for converging beams of electrons and other charged particles; a device for directing sound waves. **lensed**, *a.* **lensless** (-lis), *a.*

Lent (lent), *n.* a fast of 40 days (excluding Sundays) from Ash Wednesday to Easter Eve, in commemoration of Christ's fasting in the wilderness. **Lent-lily**, *n.* the daffodil. **Lent term**, *n.* the spring school and university term. **lenten**, *a.* of, pertaining to or used in Lent; sparing, meagre.

lent (lent), *past, p.p.* LEND.

-lent (-lənt), *suf.* full, as in *corpulent, opulent.*

lentamente LENTO.

lenten LENT.

lent(i)-, *comb. form* pertaining to a lens.

lenticel (len'tisel), *n.* a pore in the bark of a plant, through which respiration takes place. **lenticellate** (-sel'ət), *a.*

lenticular (lentik'ūlə), *a.* resembling in shape a lentil or lens doubly convex; of or pertaining to the lens of the eye. **lenticularly**, *adv.*

lentiform (len'tifawm), *a.* shaped like a lens.

lentigo (lenti'gō), *n.* (*pl.* **lentigines**, (-tij'inēz)) a freckle, freckly eruption. **lentiginous, -nose** (-nōs), *a.*

lentil (len'tl), *n.* a small branching leguminous plant, *Ervum lens;* (*pl.*) the seeds of this plant, largely used for food.

lentisk (len'tisk), *n.* the mastic tree.

lentivirus (len'tivirəs), *n.* any of a family of viruses including the AIDS virus.

lento (len'tō), *adv., a.* (*Mus.*) slow(ly). *n.* (*pl.* **-tos, -ti** (-tē)) a piece of music played thus. **lentamente** (-əmen'tā), *adv.* slowly; in slow time. **lentando** (-tan'də), *adv.* with increasing slowness. **lentissimo** (-tis'imō), *adv., a.* very slow(ly).

lentoid (len'toid), *a.* pertaining to a lens.

l'envoy (lenvoi'), ENVOY [1].

Leo (lē'ō), *n.* one of the 12 zodiacal constellations, the Lion; the fifth sign of the zodiac. **leonid** (-nid), *n.* one of the meteors that appear in numbers radiating from the constellation Leo. **leonine** (-nīn), *a.* pertaining to or like a lion; (**Leonine**) of or pertaining to one of the Popes Leo, esp. Leo I; describing pentameter or hexameter Latin verse with internal rhyme. **Leonine City**, *n.* the portion of Rome comprising the Vatican which was walled by Leo IV.

leopard (lep'əd), *n.* a large mammal of the cat family from Africa and S Asia, having a pale coat with dark spots, the panther; a leopard-like animal, as the **American leopard** or jaguar, the **hunting leopard** or cheetah, and the **snow leopard** or ounce; in heraldry, a lion passant guardant. **leopard's bane**, *n.* a plant of the composite genus *Doronicum*. **leopardess** (-dis), *n.fem.*

leotard (lē'ətahd), *n.* a close-fitting garment resembling a swimsuit, worn during exercise, dance practice etc.

leper (lep'ə), *n.* one affected with leprosy; one who is deliberately avoided by others. **leprosarium** (-prəsee'riəm), **leproserie** (-prəsəri), **-sery**, *n.* a leper hospital. **leprose** (-rōs), *a.* scaly. **leprosity** (-pros'-), *n.* **leprosy** (-rəsi), *n.* a chronic disease, usu. characterized by tubercles of various sizes, thickening of the skin, loss of feeling, and ulceration and necrosis of parts. **leprous**, *a.*

lepid(o)-, *comb. form* having scales; scaly.

lepidolite (lipid'əlīt), *n.* a pinky-violet mica containing lithium.

Lepidoptera (lepidop'tərə), *n.pl.* an order of insects, having four wings clothed with minute scales, the butterflies and moths. **lepidopteral, -an**, *a.* **lepidopterist**, *n.* **lepidopterology** (-rol'-), *n.* **lepidopterous**, *a.*

lepidosaurian (lepidōsaw'riən), *n., a.* (a member) of the Lepidosauria, a subclass of reptiles having a scaly integument.

lepidosiren (lepidōsī'rən), *n.* the S American mud-fish, from the river Amazon.

lepidote (lep'idōt), *a.* scaly.

leporine (lep'ərīn), *a.* of or pertaining to hares; resembling a hare.

leprechaun (lep'rəkawn), *n.* in Irish folklore, a small sprite who performs domestic tasks, mends shoes etc.

leprosy LEPER.

-lepsy, *suf.* a seizure, as in *epilepsy, catalepsy*. **-leptic**, *a.*

lepto-, *comb. form* fine, small, thin, delicate; narrow, slender.

leptocephalic (leptōsifal'ik), *a.* having a long and narrow skull. **leptocephalous** (-sef'-), *a.*

leptocercal (leptōsœ'kəl), *a.* slender-tailed.

leptodactyl (leptōdak'til), *n., a.* (a bird) having long, slender toes. **leptodactylous**, *a.*

lepton (lep'ton), *n.* (*pl.* **lepta** (-tə)) a small Greek coin, the mite of the New Testament parable, now worth one-hundredth of a drachma; (*pl.* **leptons**) any of various elementary particles (e.g. electron, muon) insensitive to the strong interaction.

leptorrhine (lep'tərīn), *a.* having a long, narrow nose.

leptosome (lep'təsōm), *n.* someone of slender build, narrow chested etc.

leptospirosis (leptōspīrō'sis), *n.* any of various infectious diseases transmitted by animals and caused by bacteria.

lesbian (lez'biən), *n., a.* (of) a female homosexual. **lesbianism**, *n.*

lese-majesty (lēzmaj'əsti), *n.* an offence against the sover-

eign power or its representative, high treason. [F]
lesion (lē'zhən), *n.* a hurt, injury; physical change in a tissue or organ due to injury or disease.
less (les), *a.* smaller; of smaller size, extent, amount, degree, importance, rank etc. *prep.* minus, with deduction of. *adv.* in a smaller or lower degree; not so much. *n.* a smaller part, quantity or number; the smaller, inferior, junior etc., of things compared; (*coll.*) enough. *conj.* unless. **less and less**, gradually diminishing. **nothing less**, anything else (than), anything rather; (*coll.*) nothing of a smaller or milder kind. **lessen**, *v.t.* to make less or diminish; to reduce, to depreciate, degrade. *v.i.* to become less; to decrease, shrink. **lesser**, *a.* less, smaller; inferior.
-less (-lis), *suf.* devoid of, free from, as in *fearless, godless, tireless.*
lessee (lesē'), *n.* one to whom a lease is granted. **lesseeship**, *n.*
lessen, lesser LESS.
lesson (les'ən), *n.* the amount or duration of instruction given to a pupil at one time; a piece of homework; (*pl.*) a course of instruction (in any subject); a portion of Scripture read in divine service; a reprimand or lecture; an occurrence or example taken as a warning or caution.
lessor (les'ə), *n.* one who grants a lease.
lest (lest), *conj.* for fear that; in case; so that not.
let[1] (let), *v.t.* (*pres.p.* **letting**, *past, p.p.* **let**) to permit, allow (to be or do); to give leave to; to cause to; to grant the use, occupation or possession of for a stipulated sum; to give out on contract. *aux.v.* used in the imperative mood, with the force of prayer, exhortation, assumption, permission or command. *v.i.* to be let or leased, for rent. *n.* a letting. **let alone**, not to mention; much less. **to let alone**, to leave without interference; not to do or deal with. **to let be**, not to interfere with. **to let down**, to allow to sink or fall; to humiliate; to fail (someone). **to let drop, fall**, to drop; to mention by or as if by accident. **to let fly** FLY[2]. **to let go**, to release; to relinquish hold of; to cease to retain. **to let in**, to allow to enter; to insert; to cheat, defraud. **to let in for**, to involve (someone) in something unpleasant, difficult etc. **to let in on**, to allow to be involved in. **to let into**, to admit to; to admit to knowledge of. **to let off**, to suffer to go free; to punish lightly; to pardon; to fire off (an arrow, gun etc.). **to let on**, to divulge; to pretend. **to let oneself go**, to give way to any impulse; to lose interest in maintaining one's appearance. **to let out**, to suffer to escape; to divulge; to enlarge (as a dress); to lease or let on hire. **to let rip**, (*coll.*) to act or speak without restraint. **to let slip**, to allow to escape; to lose; to reveal inadvertently. **to let up (on)**, to become less (severe). **let-down**, *n.* a disappointment. **let up**, *n.* a cessation, an alleviation. **lettable**, *a.* **letter**, *n.* **letting**, *n.*
let[2] (let), *n.* in tennis etc., a stoppage, hindrance etc., requiring the ball to be served again; a rally or service affected by this.
-let (-lit), *suf.* diminutive, as in *bracelet, tartlet.*
letch (lech), LECH, under LECHER.
lethal (lē'thəl), *a.* deadly, fatal, mortal. **lethality** (-thal'-), *n.* **lethally**, *adv.*
lethargy (leth'əji), *n.* unnatural sleepiness; a state of torpor, apathy or inactivity. **lethargic, -ical** (-thah'-), *a.* **lethargically**, *adv.* **lethargied**, *a.* **lethargize, -ise**, *v.t.*
Lethe (lē'thē), *n.* in Greek mythology, a river of Hades, the drinking of whose waters produced forgetfulness; forgetfulness, oblivion. **Lethean** (-thē'-), *a.*
lethiferous LETHAL.
Lett (let), *n.* a member of a people largely inhabiting Latvia (Lettland), a Latvian. **Lettic**, *n., a.* (of) the group of languages containing Latvian (Lettish) Lithuanian and Old Prussian. **Lettish**, *n., a.* Latvian.
letter (let'ə), *n.* a mark or character employed to represent a sound in speech; one of the characters in the alphabet; a written message or communication; the literal or precise meaning of a term or terms; a character used in printing; (*pl.*) literature, literary culture; erudition; a degree, membership, title etc. abbreviated after a surname. *v.t.* to impress, mark or stamp with letters. **letter of attorney** ATTORNEY[2]. **letter of credit** CREDIT. **letter of marque (and reprisal)**, a privateer's commission to seize and plunder the merchant ships of a hostile state. **letters of administration**, a document issued by a court authorizing a person to administer an intestate estate. **letters of credence**, a document held by a diplomat presenting his or her credentials to a foreign government. **man of letters**, an author; a scholar. **letter bomb**, *n.* an explosive device contained in an envelope, which detonates when opened. **letter-box**, *n.* a box for the reception of letters. **letterhead**, *n.* (notepaper with) a printed heading. **letter-perfect**, *a.* wordperfect. **letterpress**, *n.* printed matter other than illustrations. **letters patent** PATENT. **lettered**, *a.* marked or impressed with letters; learned; literary. **lettering**, *n.* the act or technique of impressing or marking with letters; an inscription. **letterer**, *n.* **letterless** (-lis), *a.* illiterate, ignorant.
lettre de cachet CACHET.
lettuce (let'is), *n.* a crisp-leaved garden plant of the genus *Lactuca*, much used for salad.
leu (lioo'), *n.* (*pl.* **lei** (lā)) the monetary unit of Romania.
leuc(o)-, leuk(o)-, *comb. form* white, pale.
leucaemia, -chaemia, leukaemia (lookē'miə), (*esp. N Am.*) **-emia**, *n.* a cancerous disease in which leucocytes multiply causing loss of red corpuscles, hypertrophy of the spleen etc.
leucin, -cine (loo'sin), *n.* a white crystalline substance obtained from the decomposition of animal fibre. **leucic, leucinic** (-sin'-), *a.*
leucite (loo'sīt), *n.* a dull, glassy silicate of aluminium and potassium. **leucitic** (-sit-), *a.*
leucocyte (loo'kōsīt), *n.* a white corpuscle or blood cell. **leucocytic** (-sit'-), *a.*
leucocythaemia (lookōsīthē'miə), *n.* leucaemia.
leucocytosis (lookōsītō'sis), *n.* a condition characterized by an increase in the number of white corpuscles in the blood.
leucoma (lookō'mə), *n.* a white opaque spot in the cornea.
leucorrhoea (lookərē'ə), *n.* a mucous discharge from the vagina. **leucorrhoeal, -rrhoeic**, *a.*
leucotomy (lookot'əmi), LOBOTOMY.
leukaemia LEUCAEMIA.
lev (lef), *n.* (*pl.* **leva** (lev'ə)) the monetary unit of Bulgaria.
Lev., (*abbr.*) Leviticus.
Levant (livant'), *n.* the eastern part of the Mediterranean with the adjoining countries; a levanter wind; LEVANT MOROCCO under MOROCCO. **levanter**, *n.* a native or inhabitant of the Levant; an easterly wind in the Mediterranean. **levantine** (lev'əntin), *n., a.*
levant (livant'), *v.t.* to run away, esp. with debts undischarged. **levanter**, *n.*
levator (livā'tə), *n.* a muscle that raises some part of the body.
levee[1] (lev'i), *n.* a morning or early afternoon reception held by a sovereign or distinguished person.
levee[2] (lev'i), *n.* a raised bank of a river, natural or artificial; a quay.
level (lev'əl), *n.* a horizontal line or surface; an instrument for determining whether a surface or a series of objects are horizontal; the altitude of any point or surface; level country; a stage of progress or rank; a position on a scale of values; a horizontal gallery or passage in a mine. *a.* horizontal, flat; of an evenness with something else (e.g. the top of a cup, spoon); equal in rank or degree; equable, uniform, well-balanced. *v.t.* (*past, p.p.* **levelled**) to make horizontal; to reduce to a horizontal plane; to bring (up or down) to the same level (as); to make smooth or even; to point (a gun) in taking aim; to aim (an attack,

satire etc.); to knock down. **on the level**, honest, genuine. **one's level best**, the best one can. **to find one's level**, to settle in a position, office, rank etc. suitable for one's abilities etc. **to level off**, to make flat; to reach and stay in a state of equilibrium. **to level with**, (*sl.*) to be honest with. **level crossing**, *n.* a place where a road crosses a railway line at the same level. **level-headed**, *a.* sensible, shrewd, untemperamental. **level-headedly**, *adv.* **level-headedness**, *n.* **level pegging**, *a.* of contestants etc., equal. **leveller**, *n.* one who or that which levels; one who wishes to destroy all social distinctions (esp. during the English Civil War); that which does so. **levelling-rod, -staff**, *n.* a pole used in surveying. **levelly**, *adv.* **levelness**, *n.*
lever (lē'və), *n.* a rigid bar having a fixed point of support used to overcome a certain resistance (or weight); a part of a machine, instrument etc., acting on the same principle; anything that brings power or influence to bear. *v.t.* to move or lift with or as with a lever. **leverage** (-rij), *n.* the action of a lever; the mechanical power or advantage gained by using a lever; means of accomplishing, influencing etc.
leveret (lev'ərit), *n.* a hare in its first year.
leviable LEVY.
leviathan (livī'əthən), *n.* a huge aquatic monster in Job xli; anything huge or monstrous, esp. a huge ship, a whale, the state.
levigate (lev'igāt), *v.t.* to make smooth; to grind or rub down to a powder, esp. in liquid or a moist state. *a.* (-gət), smooth. **levigable**, *a.* **levigation**, *n.*
levirate (lē'virət), *n.* an ancient Hebrew law binding a man to marry the childless widow of his dead brother. *a.* **leviratical. leviratic, -ical** (-rat'-), *a.* **leviration**, *n.*
Levis[x] (lē'vīz), *n.pl.* a type of (blue) denim jeans.
levitate (lev'itāt), *v.t., v.i.* to (cause to) rise or float in the air through supernatural causes. **levitation**, *n.* **levitational**, *a.* **levitator**, *n.*
Levite (lē'vīt), *n.* one of the tribe of Levi, who acted as assistant priests; a clergyman. **Levitic, -ical** (-vit'-), *a.* pertaining to the Levites or the book of Leviticus. **Levitical degrees**, *n.pl.* degrees of relationship which according to the Levitical law precluded marriage. **Levitically**, *adv.* **Leviticus** (-vit'ikəs), *n.* the third book of the Pentateuch, containing the Levitical law. **levitism**, *n.*
levity (lev'iti), *n.* lightness of conduct or manner; want of seriousness or earnestness, frivolity.
levo- LAEVO-.
levulose (lē'vūlōs), LAEVULOSE.
levy (lev'i), *v.t.* to collect together, to enlist (as an army); to begin to wage (war); to impose and collect (as a tax or forced contribution); to seize (property) by a judicial writ etc. *n.* the levying of a tax or an army; the tax or troops levied. **leviable**, *a.*
lewd (lood, lūd), *a.* lascivious, indecent; depraved. **lewdly**, *adv.* **lewdness**, *n.*
lewis (loo'is), *n.* a hoisting device for heavy stone blocks employing curved metal pieces which fit into and grasp the stone.
Lewis gun (loo'is), *n.* a light machine-gun invented by Col. Isaac Newton Lewis (1858–1931).
lewisite (loo'isīt), *n.* a poisonous liquid used in chemical warfare.
lexicon (lek'sikən), *n.* a dictionary; the vocabulary of a language, subject etc. **lexical** (-kəl), *a.* **lexically**, *adv.* **lexicography** (leksikog'rəfi), *n.* the art or process of compiling dictionaries. **lexicographer**, *n.* **lexicographic, -ical** (-graf'-), *a.* **lexicographist** (-kog'-), *n.* **lexicology** (-kol'-), *n.* the study of the derivation, meaning and application of words. **lexicologist**, *n.* **lexigram** (-gram), *n.* a sign representing a word. **lexigraphy** (-sig'rəfi), *n.* a system of writing using lexigrams. **lexigraphic, -ical** (-graf'-), *a.* **lexis**, *n.* the complete vocabulary of a language, individual or subject.

ley (lā), *n.* LEA[2]; a ley-line. **ley-line**, *n.* a straight line across the landscape joining two landmarks, supposed to be of prehistoric origin.
Leyden jar (lī'dən), *n.* a glass bottle or jar coated inside and out with tinfoil used as an electrical condenser. [invented in *Leyden*, Holland, in 1745]
leze-majesty (lēzmaj'əsti, lezmazhəstä'), LESE-MAJESTY.
LF, (*abbr.*) low frequency.
lf, (*abbr.*) light face (type).
LG, (*abbr.*) Low German.
LH, lh, (*abbr.*) left hand.
lherzolite (lœ'zəlīt), *n.* peridotite.
Li, (*chem. symbol*) lithium.
li (lē), *n.* a Chinese measure of distance, rather more than one-third of a mile (0·5 km).
liable (lī'əbl), *a.* legally obliged; responsible (for); subject (to); exposed or open (to); tending, apt or likely (to). **liability** (-bil'-), *n.* the state of being liable; a debt, a hindrance. **limited liability**, responsibility for debts of a company only to a specified amount, in proportion to the amount of stock held; hence, **limited (liability) company**.
liaison (liā'zon), *n.* an illicit intimacy between a man and woman; a bond, a connection; (*Cookery*) a thickening, usu. made of yolk of egg; the carrying on of the sound of a final consonant to a succeeding word beginning with a vowel or *h* mute; communication between (military) units. **liaison officer**, *n.* a person in charge of communication between units, groups etc. **liaize, liaise**, *v.i.* to maintain communication and contact.
liana (liah'nə), **liane** (-ahn'), *n.* any of the climbing and twining plants common in tropical forests. **lianoid**, *a.*
liar (lī'ə), *n.* one who knowingly utters falsehoods, esp. one addicted to lying.
Lias (lī'əs), *n.* the lowest series of rock strata of the Jurassic system. **Liassic** (-as'-), *a.*
Lib[1] (lib), (*abbr.*) Liberal.
Lib[2], **lib** (lib), (*coll.*) short for LIBERATION.
libation (lībā'shən), *n.* an offering to a deity involving the pouring of oil or wine; the liquid poured; (*usu. facet.*) (the drinking of) an (alcoholic) beverage. **libatory** (lī'bə-), *a.*
libber (lib'ə), *n.* (*coll.*) short for LIBERATIONIST.
libel (lī'bl), *n.* a defamatory writing or publication of any kind, tending to bring any person into ridicule, contempt or disrepute; the act or crime of publishing a libel; an unfair representation or defamatory statement. *v.t.* (*past, p.p.* libelled) to publish a libel; to defame. **libellee** (-lē'), *n.* **libeller, libel(l)ist**, *n.* one who libels; **libellous**, *a.* **libellously**, *adv.*
liberal (lib'ərəl), *a.* generous; ample; open, candid; favourable to liberty and progress; not strict or literal; broadminded, unprejudiced; favourable to democratic government, opposed to aristocratic privileges; esp. of education, not technical, tending to free mental development; (**Liberal**) of or pertaining to a Liberal Party. *n.* one who advocates political and social progress and reform; (**Liberal**) a member or supporter of a Liberal Party. **liberal arts**, *n.pl.* non-technical or non-professional studies including the fine arts, history, languages, literature, philosophy etc. **Liberal Party**, *n.* a former political party, the successor of the Whig Party; a party having liberal policies; (*Austral.*) a major political party supporting conservative policies. **liberalism**, *n.* **liberalist**, *n.* **liberalistic** (-lis'-), *a.* **liberality** (-ral'-), *n.* the quality of being liberal; generosity; largeness or breadth of views; freedom from prejudice. **liberalization**, *n.* **liberalize, -ise**, *v.t.* **liberally**, *adv.* **liberalness**, *n.*
liberate (lib'ərāt), *v.t.* to set at liberty; to release from domination, injustice or confinement; (*euphem. or facet.*) to steal; to set free from chemical combination. **liberated**, *a.* pertaining to peoples freed from foreign domination, of women freed from trad. sexual roles etc. **liberation**, *n.* **animal, gay, women's liberation** ANIMAL. GAY. WOMAN.

liberation theology, *n.* a Christian theory that political involvement to effect social equality and justice is a necessary part of Christianity. **liberationism**, *n.* **liberationist**, *n.* one who seeks or supports the causes of equality, freedom or liberty (e.g. *women's liberationist*). **liberator**, *n.* **liberatory**, *a.*
libertarian LIBERTY.
libertine (lib'ətēn), *n.* a debauchee, profligate. *a.* licentious, dissolute. **libertinage** (nij), **libertinism**, *n.*
liberty (lib'əti), *n.* the quality or state of being free from captivity, bondage or subjection; freedom of choice, opinion or action; permission granted to do any act; free time; a breach of decorum. **at liberty**, free; having the right (to do etc.); unoccupied. **cap of liberty** CAP. **civil liberty**, the freedom of the individual as embodied in the law. **liberty of the press**, freedom of the press to publish without government interference. **to take liberties (with)**, to be unduly familiar or presumptuous (with); to falsify. **liberty bodice**, *n.* a sleeveless bodice worn as an undergarment, esp. by children. **liberty hall**, *n.* a place where one may do as one pleases. **liberty horse**, *n.* a riderless circus horse. **liberty ship**, *n.* a prefabricated, mass-produced cargo ship produced during World War II. **libertarian** (teə'ri-), *a.* inculcating the doctrine of free will. *n.* a believer in freedom or free will. **libertarianism**, *n.* **liberticide** (-bœ'tisid), *n.* destruction of liberty; one who destroys liberty.
libidinous (libid'inəs), *a.* characterized by lust; lascivious. **libidinously**, *adv.* **libidinousness**, *n.*
libido (libē'dō), *n.* (*pl.* **-dos**) in psychoanalysis, the life force deriving from biological impulses; the sexual drive. **libidinal** (-bid'inəl), *a.*
Libra (lē'brə), *n.* (*pl.* **-rae** (-rē)) the Balance, the seventh sign of the zodiac; one of the 12 ancient zodiacal constellations.
libra (li'brə), *n.* (*pl.* **-rae** (-rē)) an ancient Roman pound; hence, a pound weight (*lb*), a pound sterling (£).
library (li'brəri), *n.* a collection of books for use either by the public or by private persons; a building or room containing such a collection, or an institution for its formation or maintenance; a set of books issued (usu. in similar format) by a publisher; a collection of computer software, films, records, tapes etc. **circulating, lending library**, CIRCULATE, LEND. **free, public library**, a library open to members of the public. **reference library** REFERENCE. **library science**, *n.* (*N Am.*) librarianship. **librarian** (-breə'ri-), *n.* one who has charge of a library. **librarianship**, *n.*
librate (li'brāt), *v.i.* to be poised; to oscillate, to swing or sway. **libration**, *n.* **librational, libratory** *a.*
libretto (libret'ō), *n.* (*pl.* **-tti** (-tē), **-ttos**) (a book containing) the words of an opera, oratorio etc. **librettist**, *n.* one who writes a libretto.
Librium (lib'riəm), *n.* a tranquillizing drug containing chlordiazepoxide.
Libyan (lib'iən), *a.* of or pertaining to the N African country of Libya, its language or its people. *n.* a native or inhabitant of Libya.
lice (lis), (*pl.*) LOUSE.
licence (li'səns), *n.* authority, leave, permission; (document containing) permission granted by a constituted authority (to marry, drive a motor vehicle, possess a firearm, own a dog, carry on a business etc.); liberty of action, disregard of law or propriety; abuse of freedom, licentiousness; in literature or art, deviation from the ordinary rules or mode of treatment; permitted freedom of thought or action. **special licence**, *n.* a licence authorizing a marriage without banns.
license (li'səns), *v.t.* to authorize by a legal permit; to allow, permit. **licensed**, *a.* **licensable**, *a.* **licensed victualler**, *n.* one who holds a licence to sell spirits, wines, beer etc. **licensee** (-sē'), *n.* one holding a licence (esp. a publican). **licenser, -sor**, *n.* one who grants a licence or has the authority to do so. **licentiate** (-sen'shiət), *n.* one holding a certificate of competence in some profession; in the Presbyterian Church, one who has a licence to preach. **licentious** (-sen'shəs), *a.* lascivious, dissolute, profligate. **licentiously**, *adv.* **licentiousness**, *n.*
lich (lich), *n.* a corpse. **lich-gate, lych-gate**, *n.* a churchyard gate with a roof, under which a coffin could be rested. **lich-owl**, *n.* the screech-owl.
lichee LITCHI.
lichen (li'kən), *n.* a symbiotic organism consisting of an alga and a fungus, living on stone, wood etc. **lichened**, *a.* **lichenoid** (-oid), **-nose** (-ōs), **-nous**, *a.* **lichenology**, *n.* the botanical study of lichens. **lichenologist** (-ol'-), *n.*
licit (lis'it), *a.* lawful, allowed. **licitly**, *adv.*
lick (lik), *v.t.* to draw or pass the tongue over; to take in or lap (up) with the tongue; of flame etc., to stroke or pass lightly over; (*sl.*) to beat, to overcome. *v.i.* of flames etc., to make a licking motion; (*sl.*) to beat, to win; to lap. *n.* the act of licking; a slight smear or coat (as of paint); a salt-lick; (*coll.*) a blow or slap; (*coll.*) great speed. **a lick and a promise**, (*coll.*) a quick or superficial wash. **salt-lick** SALT. **to lick into shape**, to give form or method to. **to lick one's lips**, to anticipate or remember something with pleasure. **to lick one's wounds**, to withdraw after a defeat to recuperate physically or mentally. **to lick the dust**, to be beaten, to be killed. **lickspittle**, *n.* an abject parasite or toady. **licker**, *n.* **licking**, *n.* a beating, a defeat.
lickerish, liquorish (lik'ərish), *a.* greedy; pleasing to the taste; lecherous. **lickerishly**, *adv.* **lickerishness**, *n.*
lickety-split (likətisplit'), *adv.* (*chiefly N Am. coll.*) speedily; very quickly.
lictor (lik'tə), *n.* a civil officer who attended the chief Roman magistrates, and bore the fasces.
lid (lid), *n.* a hinged or detachable cover or cap, usu. for shutting a container; an eyelid. (*coll.*) a hat. **to blow, lift, take the lid off**, (*coll.*) to reveal, uncover. **to flip one's lid**, (*sl.*) to go berserk. **lidded**, *a.* **lidless** (-lis), *a.*
lido (lē'dō), *n.* (*pl.* **-dos**) a bathing-beach, an out-door swimming pool.
lie[1] (li), *v.i.* (*pres.p.* **lying** (li'ing), *past, p.p.* **lied**) to say or write anything with the deliberate intention of deceiving; to convey a false impression, to deceive. *n.* a false statement deliberately made for the purpose of deception; a deception, an imposture. **to give the lie to**, to disprove; to accuse of lying. **white lie**, a pardonable fiction or misstatement. **lie-detector**, *n.* a device for monitoring physiological changes taken as evidence of mental stress accompanying the telling of lies.
lie[2] (li), *v.i.* (*pres.p.* **lying** (li'ing), *past* **lay** (lā), *p.p.* **lain** (lān)) to rest or place oneself in a reclining or horizontal posture; to be situated or fixed in a specified place, condition or direction; to rest, to remain; of an action, objection etc., to be sustainable. *n.* position, arrangement, direction, manner of lying; the lair (of an animal). **lie of the land**, the present state of affairs. **to lie by**, to be or stay near; to be put aside; to rest; to be quiet; to remain unused. **to lie down**, to go to rest; (*in pres.p.*) to submit tamely. **to lie hard, heavy on**, to oppress, to be a weight upon. **to lie in**, to be in childbed; to remain in bed later than normal. **to lie in one**, to be in one's power or capacity. **to lie in the way**, to be an obstacle or impediment. **to lie in wait**, to wait in ambush. **to lie low**, to remain in hiding. **to lie off**, of a vessel, to stay at a distance from the shore or another ship. **to lie on, upon**, to be incumbent upon. **to lie on one's hands**, of time, to hang heavy. **to lie over**, to be deferred. **to lie to**, of a ship, to be checked or stopped in her course. **to lie under**, to be subject to or oppressed by. **to lie up**, to rest in order to recuperate; of a ship, to go into dock. **to lie with**, to lodge or sleep with; to have sexual intercourse with. **lie-a-bed**, *n.* a late riser.

lie-in, *n*. a longer than normal stay in bed. **lying-in**, *n*. (*pl*. **lyings-in**) confinement in childbirth.
Lied (lēd), *n*. (*pl*. **Lieder** (-ə)) a German song or ballad. [G]
lief (lēf), *adv*. willingly, gladly, freely.
liege (lēj), *n*. bound by some feudal tenure, either as a vassal or as a lord; pertaining to such tenure. *n*. a vassal; a lord, a superior, a sovereign; a law-abiding citizen. **liegedom** (-dəm), *n*. **liegeless** (-lis), *a*. **liegeman**, *n*. a liege vassal.
lien¹ (lē'ən), *n*. (*Law*) a right to detain the goods of another until some claim has been satisfied; (*coll*.) an option.
lientery (li'əntəri), *n*. diarrhoea in which the food passes rapidly through the bowels undigested. **lienteric** (-te'-), *a*.
lierne (licen'), *n*. a cross-rib connecting the main ribs in Gothic vaulting.
lieu (lū, loo), *n*. place, stead. **in lieu of**, instead of.
lieutenant (ləften'ənt, *N Am*. loo-), *n*. an officer acting as deputy or substitute to a superior; an army officer ranking next below a captain; a naval officer ranking next below a lieutenant-commander. **Lord-Lieutenant** LORD. **second-lieutenant**, the lowest commissioned rank in the British army. **lieutenant-colonel**, *n*. an officer next in rank below a colonel, in actual command of a battalion. **lieutenant-commander**, *n*. a naval officer ranking between a lieutenant and a commander. **lieutenant-general**, *n*. an army officer next in rank below a general and above a major-general. **lieutenant-governor**, *n*. a deputy governor; the acting governor in subordination to a governor-general; (*N Am*.) the deputy to a state governor. **lieutenancy**, *n*.
life (līf), *n*. (*pl*. **lives** (līvz)) the state or condition of being alive; the state of an organism in which it is capable of performing its animal or vegetable functions; the period of such existence; any specified portion of a person's existence; the average period which a person of a given age may expect to live; the period of time for which an object functions or operates; the living form; (*collect*.) living things; mode, manner of living; the essential or inspiring idea (of a movement etc.); animation, vivacity, spirit; one who or that which imparts spirit or animation; the active side of existence; human affairs; a biography; (*coll*.) a life sentence. *a*. for the duration of one's life; in drawing, sculpture etc., taken from life. **a matter of life and death**, one of utmost urgency. **for dear life**, with extreme vigour, in order to escape death. **for the life of me**, as if my life depended upon it. **high life**, the habits of fashionable society. **not on your life**, under no circumstances. **the life and soul**, one who is the chief source of amusement or interest. **the life of Riley**, (*coll*.) an easy, carefree existence. **the time of one's life**, an experience of unequalled pleasure. **to bring to life**, to revive (an unconscious person). **to the life**, of a portrait etc., as if the original stood before one. **life-assurance, insurance**, *n*. insurance providing for the payment of a specified sum to a beneficiary on the policy holder's death, or to the policy holder on reaching a certain age. **life-belt**, *n*. a buoyant belt for supporting a person in the water. **life-blood**, *n*. the blood necessary to life; that which is essential to existence. **lifeboat**, *n*. a boat for rescuing people in storms and heavy seas. **lifebuoy** BUOY. **life cycle**, *n*. the series of changes in the form and function of an organism during its lifetime. **life-giving**, *a*. inspiriting, invigorating, animating. **lifeguard**, *n*. a bodyguard; an attendant at a bathing beach or pool who renders aid to swimmers in difficulties. **life-insurance** LIFE-ASSURANCE. **life-jacket**, *n*. a sleeveless jacket used as a life-belt. **life-line**, *n*. a rope used for saving life; a vital line of communication. **lifelong, livelong** (-liv'-), *a*. lasting throughout life. **life-peer**, *n*. **life-peerage**, *n*. a peerage lapsing with the death of the holder. **life-preserver**, *n*. (*N Am*.) a life-belt, life-jacket etc.; a loaded stick or club. **life raft**, *n*. a raft kept on board ships etc. for use in emergencies. **life-saver**, *n*. one who saves a person's life; (*coll*.) one who or that which provides help in distress. **life-saving**, *a*. **life science**, *n*. one that deals with the structure and function of living organisms. **life-size(d)**, *a*. representing the actual size of an object. **life span**, *n*. the length of time during which an organism, machine etc. lives or functions. **life style**, *n*. the attitudes, behaviour, surroundings etc. characteristic of an individual or group. **life support**, *a*. pertaining to a device or system which maintains a person's life. **lifetime**, *n*. the duration of life, function or existence. **life-work**, *n*. the work to which one devotes the best part of one's life. **lifeless** (-lis), *a*. without life; dead, inanimate; inert; deprived of physical energy; dull, spiritless. **lifelessly**, *adv*. **lifelessness**, *n*. **lifelike**, *a*. like a living being. **lifelikeness**, *n*. **lifer**, *n*. one sentenced to imprisonment for life.
lift (lift), *v.t*. to raise to a higher position, to hold or support on high; to raise or take up from the ground; (*coll*.) to steal, to plagiarize; to rescind or remove; to exalt, to elate. *v.i*. to perform or attempt to perform the act of raising something; to rise; to rise and disperse, as a mist. *n*. the act of lifting; the degree of elevation; a rise in the height of the ground; a hoisting-machine, an elevator for persons, goods or material; assistance in lifting; a helping hand; a layer inserted in the heel of a shoe, to increase the height of the wearer; a rise in spirits, morale; a rise in condition; a ride in a vehicle for part or all of a journey; the component of the aerodynamic force on an aircraft or aerofoil acting upwards at right angles to the airflow and opposing the pull of gravity. **lift-off**, *n*. the take-off of an aircraft, rocket or missile; the instant at which this occurs. **lift off**, *v.i*. **lift-pump**, *n*. a pump that lifts to its own level, distinguished from a *force-pump*. **lifter**, *n*.
ligament (lig'əmənt), *n*. anything which binds; a short band of fibrous tissue by which bones are bound together; any tough bands or tissues holding parts together. **ligamental, -tary, -tous** (-men'-), *a*.
ligand (li'gənd, lig'-), *n*. a single atom, molecule, radical or ion attached to a central atom to form a coordination complex.
ligate (li'gāt), *v.t*. to tie with a ligature. **ligation**, *n*. **ligature** (lig'əchə), *n*. that which binds, esp. a thread or cord to tie arteries or veins; anything that unites, a bond; (*Print*.) two or more letters cast on one shank, as ff, ffi; (*Mus*.) a tie connecting notes, a slur. *v.t*. to bind with a ligature.
liger (li'gə), *n*. a cross between a lion and a tigress.
light¹ (līt), *n*. electromagnetic radiation which, by acting on the retina, stimulates the sense of sight; the sensation produced by the stimulation of the visual organs; the state or condition in which things are visible, opp. to darkness; the amount of illumination in a place; a source of light, a lamp, a candle, the sun etc.; daylight; that by which light is admitted, a window, a division of a window; publicity; point of view, aspect; mental illumination, enlightenment; one who enlightens, an example; brightness of the face or eyes; something that kindles or ignites. *a*. having light, bright, clear, not dark; pale-coloured, fair. *v.t*. (*past, p.p*. **lit** (lit), **lighted**) to kindle to give light to; to conduct with a light; to brighten. *v.i*. to take fire, to begin to burn; to be illuminated; to brighten (up); (*coll*.) to decamp, to hurry away. **according to one's lights**, according to one's information or knowledge. **in the light of**, considering, allowing for. **light-emitting diode**, a semiconductor junction which emits light when an electric current passes through it, used in calculators, watches etc. **lighting-up time**, the time of day when vehicles are required by law to show their lights. **lights out**, (a signal indicating) the time when residents in an institution (e.g. a boarding school) are expected to retire for the night. **to bring to light**, to discover, disclose. **to come to light**, to become known. **to light up**, (*coll*.) to light a cigarette, pipe etc.; to illuminate; to switch on (car) lights; to become cheerful or animated suddenly. **to see (the) light**, to be born; to be published; to be enlightened. **to shed, throw light (up)on**, to

elucidate, explain. **light-bulb**, *n.* a glass bulb filled with a low density gas and containing a metal filament which glows when an electric current is passed through it. **lighthouse**, *n.* a tower supporting a powerful light for the warning and guidance of ships at sea. **light pen**, *n.* a pen-shaped photoelectric device used for creating or entering information on a computer; a device for reading bar-codes. **lightship**, *n.* a moored vessel carrying a light to give warning or guidance to ships. **light show**, *n.* a display of multi-coloured lights for visual effects, esp. at a pop concert. **light-year**, *n.* the distance (about 6,000,000,000,000 miles or 9460×10^9 km) travelled by light in one year. **lit-up**, *a.* (*coll.*) slightly drunk. **lighting**, *n.* **lightish**, *a.* **lightless** (-lis), *a.*
light[2] (līt), *a.* of small weight, not heavy; easy to be lifted, carried, moved, handled etc.; not burdensome; easy to be performed; of troops, lightly armed and equipped; nimble, quick; of low specific gravity; below the standard weight; not heavily laden; adapted for rapid movement; employed in or adapted for easy work; not heavy in construction or appearance; graceful, elegant; of fabrics, thin, delicate; loose or sandy, as soil; of bread, not dense; of wine, beer etc., not strong; easily digested; not forcible or violent, gentle, slight; not intense or emphatic; unimportant, trivial; thoughtless, frivolous; volatile, fickle; unchaste; cheerful, airy. **to make light of**, to treat as pardonable or excusable. **light engine**, *n.* one with no train attached. **light-fingered**, *a.* given to thieving. **light-footed**, *a.* nimble, active. **light-handed**, *a.* light of touch, light in handling. **light-handedly**, *adv.* **light-handedness**, *n.* **light-headed**, *a.* delirious. **light-headedness**, *n.* **light-hearted**, *a.* free from care or anxiety; merry, cheerful. **light-heartedly**, *adv.* **light-heartedness**, *n.* **light-heeled**, *a.* nimble, quick-moving. **light literature**, *n.* books intended for entertainment. **light-minded**, *a.* fickle, unsteady, volatile. **light-mindedly**, *adv.* **light-mindedness**, *n.* **light railway**, *n.* adapted for light traffic. **light-spirited**, *a.* cheerful, merry. **light-weight**, *n.* an animal or person below average weight; a professional boxer weighing not more than 135 lb (61·2 kg) or 132 lb (60 kg) if amateur; (*coll.*) a person of small importance or ability. *a.* light in weight; trivial. **lightish**, *a.* **lightly**, *adv.* **lightness**, *n.*
light[3] (līt), *v.i.* of a bird, to descend as from flight, to settle; to alight; to chance (upon). **to light into**, (*sl.*) to attack physically or verbally. **to light out**, (*sl.*) to leave in a hurry.
lighten[1] (līt′n), *v.i.* to become light; to emit lightning; to shine out. *v.t.* to illuminate.
lighten[2] (līt′n), *v.t.* to make less heavy or burdensome; to cheer. *v.i.* to grow lighter; to become less burdensome.
lighter[1] (līt′ər), *n.* a pocket appliance for lighting cigarettes, pipe etc.; one who or that which ignites.
lighter[2] (līt′ər), *n.* an open, flat-bottom boat, used in loading and unloading ships etc. **lighterage** (-rij), *n.* **lighter-man**, *n.*
lightning (līt′ning), *n.* the dazzling flash caused by the discharge of electricity between clouds or between a cloud and the earth. *a.* very fast or sudden. **lightning-bug**, *n.* a fire-fly. **lightning-conductor, -rod**, *n.* a wire or rod for carrying the electric discharge to earth and protecting a building, mast etc., against damage. **lightning strike**, *n.* workers' strike without notice being given.
lights (līts), *n.pl.* the lungs of animals, esp. as food.
lightsome (līt′səm), *a.* light-hearted; airy, graceful. **lightsomely**, *adv.* **lightsomeness**, *n.*
lign- LIGN(O)-.
lign-aloes (līnal′ōz), *n.* the bitter drug aloe; a fragrant Mexican wood.
ligneous (lig′niəs), *a.* made or consisting of wood; resembling wood. **lignescent** (-nes′ənt), *a.* **ligniferous** (-nif′-), *a.* **ligniform** (-nifawm), *a.* **lignify** (-fī), *v.t., v.i.* **lignification** (-fi-), *n.*

lignin (lig′nin), *n.* an organic material which forms the woody cell walls of certain plants.
lignite (lig′nīt), *n.* a partially carbonized coal showing fibrous woody structure. **lignitic** (-nit′-), *a.*
lign(o)-, *comb. form* pertaining to wood.
lignocaine (lig′nōkān), *n.* a local anaesthetic.
lignum (lig′nəm), *n.* wood. **lignum vitae** (vī′tē, vē′tī), *n.* the very hard and heavy wood of various tropical American trees.
ligule (lig′ūl), *n.* one of the rays of a composite plant. **ligula** (-lə), *n.* (*pl.* **-lae** (-lē)) a tongue-like organ or part. **ligular**, *a.* **ligulate** (-lət), **-ated** (-lā′tid), *a.*
likable LIKE[2].
like[1] (līk), *a.* resembling, similar; equal or nearly equal in quantity, quality or degree; characteristic of. *adv.* (*used ellipt. as prep. or conj., coll.*) as, in the manner of, to the same extent or degree as. *n.* a counterpart; a similar or equal thing, person or event. **something like**, (*with emphasis on* like) in some way or nearly resembling; first-rate, highly satisfactory. **the likes of**, (*coll., usu. derog.*) people like (you or me). **to feel like**, to feel as if one resembled; to feel disposed or inclined to. **to look like**, to resemble in appearance; to seem likely. **like-minded**, *a.* having similar disposition, opinions, purpose etc. **likely**, *a.* probable, credible, plausible; liable, to be expected (to); promising, suitable, well-adapted. *adv.* probably. **likelihood** (-hud), **likeliness**, *n.* **liken**, *v.t.* to compare, to represent as similar (to). **likeness**, *n.* similarity, resemblance; a picture or other representation of a person or thing; form, appearance, guise. **likewise**, *adv., conj.* in like manner; also, moreover.
like[2] (līk), *v.t.* to be pleased with; to be inclined towards or attracted by; to enjoy; to be fond of. *v.i.* to be pleased; to choose. *n.* liking; a desire; (*usu. pl.*) predilection. **lik(e)able**, *a.* **lik(e)ableness**, *n.* **liking**, *n.* the state of being pleased; inclination, fondness. **to one's liking**, to one's taste.
-like (-līk), *suf.* forming adjectives, as in *childlike, warlike*.
lilac (lī′lək), *n.* a shrub of the genus *Syringa*, with fragrant pale violet, purple or white flowers. *a.* of a pale violet colour.
liliaceous, lilied LILY.
Lilliputian (lilipū′shən), *a.* of or pertaining to Lilliput, an imaginary country in Swift's *Gulliver's Travels*; diminutive. *n.* a native of Lilliput; a pygmy; a very small person.
Lilo® (lī′lō), *n.* a type of inflatable mattress.
lilt (lilt), *v.i.* to sing in cheerful, lively style; to spring. *v.t.* to sing in a lively style. *n.* a lively melody, rhythm or cadence of a song.
lily (lil′i), *n.* a flower or plant of bulbous genus producing white or coloured flowers of great beauty, esp. the Madonna lily; applied to various plants having resemblances, as the Lent-lily or daffodil, the water-lily etc.; (*Her.*) the fleur-de-lis; a person or thing of unsullied whiteness or purity. *a.* pure white; pure, unsullied. **lily of the valley**, a fragrant spring-flowering plant of the genus *Convallaria*, with white hanging cuplike flowers. **lily-livered**, *a.* cowardly. **lily-pad**, *n.* the broad floating leaf of the water-lily. **lily-white**, *a.* **liliaceous** (-ā′shəs), *a.* pertaining to lilies. **lilied** (-id), *a.* lilylike in complexion.
lima bean (lī′mə), *n.* an edible tropical American bean.
limaceous (līmā′shəs), *a.* pertaining to slugs. **limaciform** (-mas′ifawm), *a.* **limacoid** (-koid), *n., a.* **limaçon** (lim′əson, lēmasō′), *n.* (*Math.*) a particular curve based on the union of two ovals.
limb[1] (lim), *n.* one of the articulated extremities of an animal, an arm, leg or wing; a main branch of a tree; branch or arm of a larger group or institution; (*coll.*) an impish child. **out on a limb**, in a predicament, isolated. **limbed**, *a.* (*usu. in comb.*) having limbs, as *large-limbed*. **limbless** (-lis), *a.*
limb[2] (lim), *n.* the edge or border of the sun, moon etc.;

the graduated arc of a sextant etc.; the expanded portion of a leaf, petal etc. **limbate** (lim'bət), *a.* bordered. **limbation**, *n.* **limbic, limbiferous** (bif'-), **limbous**, *a.*

limber[1] (lim'bə), *n.* the detachable part of a gun-carriage consisting of two wheels and ammunition-box. *v.t.* to attach the limber to the gun (usu. with *up*).

limber[2] (lim'bə), *n.* (*Naut.*) a gutter on each side of the kelson for draining.

limber[3] (lim'bə), *a.* flexible, lithe. **to limber up**, to stretch and flex the muscles in preparation for physical exercise.

limbo[1] (lim'bō), *n.* the abode of infants who died unbaptized and of the just who died before Christ; confinement; a place of neglect or oblivion; an uncertain or transitional state.

limbo[2] (lim'bō), *n.* a West Indian dance in which the participants bend backwards and pass under a bar.

Limburger (lim'bœgə), *n.* a white cheese with a strong taste and smell. [*Limburg*, Belgium]

lime[1] (līm), *n.* a caustic earth, mainly calcium oxide (**quicklime**), obtained by burning calcium carbonate (usu. in limestone form), used in building and agriculture; bird-lime; calcium hydroxide (**slaked lime**), a white powder obtained by the action of water on quicklime. *v.t.* to smear with bird-lime; to ensnare. **lime-cast**, *n.*, *a.* (of) a building covered with lime in the form of mortar. **lime-kiln**, *n.* one in which limestone is calcined and reduced to lime. **limelight**, *n.* a brilliant white light produced by burning lime; the glare of publicity. **lime-pit**, *n.* a pit for liming hides. **limestone**, *n.* any rock whose basis is carbonate of lime. **lime-twig**, *n.* a twig smeared with bird-lime to catch birds. **limewash**, *n.*, *v.t.* (to) whitewash. **limewater**, *n.* a solution of lime in water. **lime-wort**, *n.* the brook-lime. **limy**, *a.* viscous; of the nature of, resembling or containing lime. **liminess**, *n.*

lime[2] (līm), *n.* the linden-tree. **lime-tree**, *n.*

lime[3] (līm), *n.* a small tropical citrus tree; the greenish-yellow fruit of this tree with acid, juicy flesh. **lime-juice**, *n.* **limey**, *n.* (*N Am. sl.*) a British sailor (from the former use of lime juice on British ships to prevent scurvy); any British person.

limelight LIME[1].

limen (lī'mən), *n.* (*Psych.*) the threshold of consciousness, at which a given stimulus begins to produce sensation. **liminal, liminary** (lim'i-), *a.*

Limerick (lim'ərik), *n.* a nonsense verse, usu. of five lines, the first, second and fifth, and the third and fourth of which rhyme together respectively.

limit (lim'it), *n.* a boundary, a line, point or edge marking termination or utmost extent; a restraint, a check; that which has bounds, a district, a period etc. *v.t.* to set a limit or bound to; to confine within certain bounds; to serve as boundary or restriction to. **the limit**, (*coll.*) a very irritating person, happening etc. **limitary**, *a.* stationed at the limits (of a guard); circumscribed; confining. **limitation**, *n.* the act of limiting; the state of being limited; that which limits; a restriction; (*Law*) the period within which an action must be brought and beyond which it may not lie. **statute of limitation**, a statute fixing such periods. **limitative**, *a.* **limited**, *a.* narrow; restricted; confined; (*coll.*) not very clever or well-read. **limited edition**, *n.* an edition of a book, print etc. of which only a small number is issued. **limited liability** LIABLE. **limited monarchy**, *n.* one in which the power of the sovereign is limited by a constitution. **limitedly**, *adv.* **limitedness**, *n.* **limiter**, *n.* **limitless** (-lis), *a.*

limitrophe (lim'itrōf), *a.* on the border, adjacent (to).

limn (lim), *v.t.* to paint or draw, to depict, to portray. **limner** (lim'nə), *n.*

limnology (limnol'əji), *n.* the study of the physical, biological, geographical etc. features of lakes and other freshwater bodies. **limnological** (-loj'-), *a.* **limnologist**, *n.*

limonite (lī'mənīt), *n.* a hydrated sesquioxide of iron, orig. bog iron-ore.

limousine (lim'əzēn), *n.* a large opulent car esp. one with a glass partition dividing the driver from the passengers.

limp[1] (limp), *v.i.* to walk lamely; of verse, logic etc., to be irregular; to proceed with difficulty. *n.* the act of limping; a limping step or walk. **limper**, *n.* **limpingly**, *adv.*

limp[2] (limp), *a.* wanting in stiffness or firmness, pliable; of book covers, not stiffened by boards. **limply**, *adv.* **limpness**, *n.*

limpet (lim'pit), *n.* a marine gastropod having an open conical shell, found adhering firmly to rocks; a tenacious person or thing. **limpet mine**, *n.* an explosive device which clings to a ship's hull, tank etc. by magnetic or adhesive means.

limpid (lim'pid), *a.* clear, pellucid, transparent; lucid. **limpidly**, *adv.* **limpidness, limpidity** (-pid'-), *n.*

limp-wort (limp'wœt), LIME-WORT, under LIME[1].

limy LIME[1].

lin (lin), LINN.

linage (lī'nij), *n.* amount of printed matter reckoned by lines; payment by the line.

linchpin (linch'pin), *n.* one serving to hold a wheel on the axle; someone or something essential to an organization etc.

Lincoln green (ling'kən), *n.* bright green cloth formerly made at Lincoln; its colour.

lincrusta (linkrŭs'tə), *n.* a canvas-backed embossed wallpaper.

Lincs (lingks), (*abbr.*) Lincolnshire.

linctus (lingk'təs), *n.* (*pl.* **-tuses**) a syrupy cough medicine.

linden (lin'dən), *n.* a tree of the genus *Tilia* with soft timber, heart-shaped leaves, and small clusters of delicately-scented flowers, the lime-tree.

line[1] (līn), *n.* a thread or string; a rope, esp. used for sounding etc.; a cord, string, wire etc. used for specific purposes, as with hooks for fishing, with a plumb for testing verticality; a clothes-line; a cord for measuring etc.; a wire or cable for telegraph or telephone; the route traversed by this; a thread-like mark; such a mark drawn by a pencil or other instrument; a narrow band, furrow, wrinkle etc. resembling this; (*Math.*) that which has length without breadth or thickness; the track of a moving point; the equator; shape of contour, outline; a plan, design; a limit, boundary; a row or continuous series of letters, words, people etc.; a short letter, a note; a single verse of poetry; (*pl.*) a piece of poetry, a specified quantity of verse or prose; (*pl.*) a certificate of marriage; (*pl.*) a series of trenches, ramparts etc.; a row of men ranged as in order of battle; the aggregate of troops in an army apart from support units etc.; a row of ships drawn up in order; a series of persons related in direct descent or succession; a series of public conveyances plying between certain places or under one management; a railway track; a certain branch of business, class of goods, a stock of these; field of activity; (*coll.*) pertinent facts; (*sl.*) smooth talk; one of the horizontal bands on a television screen which creates the picture; (*N Am.*) a queue. *v.t.* to draw lines upon, to cover with lines; to mark (in, off etc.) with lines. *v.i.* to come or extend into line. **one's line of country**, one's special field of interest. **to line up**, to arrange, array; to align; to queue. **to read between the lines**, to detect the hidden or unexpressed meaning. **line block, drawing**, *n.* a printing block or drawing using lines only, with no shading. **line-engraving**, *n.* **line frequency**, *n.* (*TV*) the frequency with which the lines in a scanned image are repeated. **lineman**, *n.* one who maintains and repairs a line of railway, telegraph etc. **line-out**, *n.* a method of restarting a match in Rugby Union when the ball has gone out of play, by throwing it in between the forwards of each team lined up facing the touchline. **line printer**, *n.* a high-speed output device, used esp. in conjunction with a computer, which prints copy a whole line at a time. **lines-**

line-man, *n.* a lineman; in various sports, an official who notes when and where a ball crosses a line. **line-up**, *n.* a row or group of persons assembled for a particular purpose; an identification parade.

line² (līn), *n.* the fine long fibre of flax separated from the tow. *v.t.* to put a covering of different material on the inside of (a garment, box etc.); to serve as such a covering for; to fill the inside of. **liner**, *n.* one who makes or fits linings; anything which serves as a lining. **lining**, *n.* the covering of the inside of anything; that which is within.

lineage (lin'ij), *n.* descendants in a direct line from a common progenitor; ancestry.

lineal (lin'iəl), *a.* ascending or descending in the direct line of ancestry; linear. **lineality** (-al'-), *n.* **lineally**, *adv.*

lineament (lin'iəmənt), *n.* (*usu. pl.*) characteristic lines or features, esp. of the face.

linear (lin'iə), *a.* composed of or having the form of lines; having a straight or lengthwise direction; of one dimension; of mathematical functions, expressions etc., able to be represented on a graph as a straight line; narrow with parallel sides. **Linear A,B**, *n.* syllabic scripts used in ancient Crete and mainland Greece (2nd millennium BC). **linear accelerator**, *n.* an apparatus for accelerating charged particles along a straight line by applying high-frequency potential between electrodes placed at intervals along their path. **linear motor**, *n.* an electric motor producing direct thrust without the use of gears. **linear perspective**, *n.* perspective dealing with the apparent positions, magnitudes and forms of objects. **linear programming**, *n.* a method of solving practical problems in economics etc., using mathematical models involving complex interactions of linear equations. **linearity** (-ar'-), *n.* **linearly**, *adv.* **lineate** (-ət), *a.* of leaves, marked with lines. **lineation**, *n.* **lineolate** (-əlāt), *a.* marked with minute lines.

lineman LINE¹.

linen (lin'in), *n.* a cloth made of flax; (*collect.*) articles orig. chiefly made of linen, esp. underclothing, sheets, table cloths etc. *a.* made of flax or linen. **linen-draper**, *n.*

liner¹ LINE².

liner² (lī'nə), *n.* one of a regular line of passenger ships or aircraft; colouring material for outlining the eyes.

linesman LINE¹.

ling¹ (ling), *n.* a long slender food-fish found in northern seas.

ling² (ling), *n.* heather or heath.

-ling¹ (-ling), *suf.* forming nouns (with a diminutive force), as *darling, gosling.*

-ling² (-ling), *suf.* forming adverbs, as *darkling.*

lingam (ling'gəm), **linga**, *n.* the phallus representative of the god Siva, in Hindu mythology.

linger (ling'gə), *v.i.* to delay going, to tarry; to hesitate; to be protracted. **lingerer**, *n.* **lingeringly**, *adv.*

lingerie (li'zhəri), *n.* women's underwear and nightclothes.

lingo (ling'gō), *n.* (*pl.* **-goes**) a foreign language, unfamiliar dialect or phraseology.

lingua (ling'gwə), *n.* the tongue.

lingua franca (frang'kə), *n.* a mixture of Italian with another language used in Mediterranean ports; a language or hybrid language serving as a medium of communication between different peoples.

lingual (ling'gwəl), *a.* pertaining to the tongue; (*phon.*) formed by the tongue. *n.* a letter or sound produced by the tongue, as *t, d, n, l, r.* **lingually**, *adv.* **linguiform** (-gwifawm), *a.* having the form of a tongue.

linguist (ling'gwist), *n.* one skilled in languages. **linguistic** (-gwis'-), *a.* of or pertaining to language or linguistics. **linguistically**, *adv.* **linguistics**, *n.sing.* the science of languages.

lingula (ling'gūlə), *n.* (*pl.* **-lae**) a tongue-shaped part. **lingular**, *a.* **lingulate** (-lət), *a.*

linhay (lin'i), *n.* a shed, usu. a lean-to, open at the sides.

liniment (lin'imənt), *n.* a liquid preparation for rubbing on bruised or inflamed parts, embrocation.

lining LINE².

link¹ (lingk), *n.* a ring or loop of a chain; a connecting part in machinery etc. or in a series, sequence, argument etc.; one-hundredth of a surveyor's chain equal to 7·92 in. (about 20 cm); a unit in a communications system. *v.t.* to connect or attach (to, together, up etc.) by or as by a link or links. *v.i.* to be connected. **missing link**, a conjectured creature linking man and the anthropoid ape; anything required to complete a chain of connection or argument. **linkman**, *n.* a television or radio presenter who provides continuity between separate items in a broadcast. **link-up**, *n.* a connection, joint. **linkage** (-ij), *n.* the act or manner of linking or being linked; a system of links; the product of magnetic flux and the total number of turns in a coil; the occurrence of two genes close together on the same chromosome so that they tend to be inherited together. **linker**, *n.* (*Comput.*) a program which joins separately assembled or compiled modules into a single executable programme.

link² (lingk), *n.* a torch made of tow and pitch. **link-boy, -man**, *n.* one carrying a link.

links (lingks), (*Sc.*) *n.pl.* flattish or undulating sandy ground near the seashore; a golf-course.

linn (lin), *n.* a waterfall; a pool, esp. one below a fall; a precipice or ravine.

Linnaean, Linnean (linē'ən, -nā'), *a.* of or pertaining to Linnaeus (or Linné (1707–78), Swedish naturalist) or his system of classification and naming of plants and animals. *n.* a follower of Linnaeus.

linnet (lin'it), *n.* a common finch with brownish plumage.

lino (lī'nō), *n.* short for LINOLEUM. **linocut**, *n.* an engraving on linoleum in the manner of a woodcut.

linoleum (linō'liəm), *n.* a preparation of oxidized linseed-oil mixed with ground cork and laid upon fabric, used as a floor covering.

Linotype® (lī'nətīp), *n.* a typesetting machine for producing castings or slugs of whole lines of words; type produced by such a method.

linseed (lin'sēd), *n.* the seed of the flax-plant. **linseed-cake**, *n.* the solid mass left after the oil has been pressed out of flax-seed. **linseed-oil**, *n.* the oil expressed from linseed.

linsey-woolsey (lin'ziwul'zi), *n.* a coarse fabric of linen or cotton warp with wool filling.

lint (lint), *n.* the down of linen cloth scraped on one side, or cotton substitute, used in bandages for dressing wounds etc. **linty**, *a.*

lintel (lin'tl), *n.* the horizontal beam or stone over a door or window. **lintelled**, *a.*

lion (lī'ən), *n.* a large and powerful carnivorous feline mammal, usu. brown or tawny, with tufted tail and (in the adult male) a long mane, inhabiting southern Asia and Africa; the sign of the zodiac and constellation Leo; the British national emblem; a courageous person; a celebrity, an object of general attention. **the lion's mouth**, a dangerous place. **the lion's share**, the largest part. **lion-heart**, *n.* **lion-hearted**, *a.* having great courage. **lioncel** (-sl), *n.* (*Her.*) a small lion. **lionesque** (-nesk'), **lion-like**, *a.* **lioness** (-nis), *n.fem.* **lionet** (-nit), *n.* a young lion. **lionhood** (-hud), **-ship**, *n.* **lionize, -ise**, *v.t.* to treat as an object of interest or curiosity.

lip (lip), *n.* one of the two fleshy parts enclosing the opening of the mouth; the edge or margin of an orifice, chasm etc.; (*pl.*) the projecting lobes of a bilabiate corolla; (*pl.*) the mouth, as organ of speech; (*sl.*) impudence, cheek. *v.t.* (*past, p.p.* **lipped**) to touch with the lips; to kiss; of water, to lap against. *v.i.* of water, to lap. **to bite one's lips**, to express vexation, to repress anger, laughter or other emotion. **to hang on one's lips**, to listen eagerly for every word spoken. **to keep a stiff upper lip**, to be self-reliant, inflexible, unflinching. **to smack one's lips**, to

anticipate or recall with relish. **lip-gloss**, *n.* a cosmetic which makes the lips glossy. **lip-read**, *v.i.* **lip-reading**, *n.* the practice of following what is said by observing the movements of the speaker's lips. **lipsalve**, *n.* ointment for the lips. **lip service**, *n.* flattery, servile agreement etc. expressed but not put into practice. **lipstick**, *n.* a stick of cosmetic for colouring the lips. **lip-sync, -synch**, *v.t.* to synchronize the movement of the lips with a prerecorded soundtrack (of words, music etc.) on film or television. **lipless** (-lis), *a.* **lipped** *a.* (*usu. in comb.*, as *thick-lipped*).
lip- LIP(O)-.
lipase (lip'äs), *n.* an enzyme which decomposes fats.
lipid (lip'id), *n.* any of various organic compounds, esters of fatty acids, important structural components of living cells. **lipidic**, *a.*
lip(o)-, *comb. form* fat, fatty.
lipogenesis (lip'əjen'əsis), *n.* the formation of fat. **lipogenic**, *a.*
lipogram (lip'əgram), *n.* a writing in which a particular letter is omitted. **lipogrammatic** (-mat'-), *a.* **lipogrammatism** (-gram'-), *n.* **lipogrammatist** (-gram'-), *n.*
lipography (lipog'rəfi), *n.* the accidental omission of a letter or letters in writing.
lipoid (lip'oid), *a.* fat-like. *n.* a fat-like substance; a lipid.
lipoma (lipō'mə), *n.* (*pl.* **-mata** (-tə)) a fatty tumour. **lipomatosis** (-tō'sis), *n.* excessive growth of fatty tissue. **lipomatous**, *a.*
lipoprotein (lipōprō'tēn), *n.* a protein which includes a lipid.
Lippizaner (lipitsah'nə), *n.* a breed of horses (usu. white or grey in colour) used esp. by the Spanish Riding School in Vienna for dressage displays.
liq., (*abbr.*) liquid; liquor.
liquate (lī'kwāt, likwāt'), *v.t.* to melt; to liquefy (metals) in order to purify. **liquation**, *n.*
liquefy, liquify (lik'wifī), *v.t.*, *v.i.* to make or become liquid. **liquefacient** (-fā'shənt), *n.* that which liquefies. *a.* serving to liquefy. **liquefaction** (-fak'-), *n.* **liquefactive**, *a.* **liquefiable**, *a.* **liquefier**, *n.* **liquescence**, *n.* **liquescent** (-kwes'ənt), *a.*
liqueur (likūə'), *n.* an alcoholic cordial sweetened or flavoured with aromatic substances. **liqueur brandy**, *n.* brandy of special quality drunk as a liqueur. **liqueurglass**, *n.*
liquid (lik'wid), *a.* fluid; flowing or capable of flowing, watery; transparent; of vowels, not guttural, smooth, easily pronounced; of assets, readily convertible into cash; of principles etc., changeable. *n.* a substance whose molecules are incompressible and inelastic and move freely among themselves, but cannot escape as in a gaseous state; a smooth consonant sound, as *l, r*, and sometimes *m, n.* **liquid crystal**, *n.* a liquid with optical, properties analogous to crystals. **liquid crystal display**, a display, esp. in electronic calculators, using liquid crystal cells which change their reflectivity in an electric field. **liquid paraffin**, *n.* an oily liquid obtained from petroleum distillation and used as a laxative. **liquidate**, *v.t.* to pay off (a debt etc.); to wind up (a bankrupt estate etc.); to assassinate. *v.i.* of a company, to have its debts, liabilities and assets liquidated. **liquidation**, *n.* **liquidator**, *n.* **liquidity** (-kwid'-), **liquidness**, *n.* **liquidize, -ise**, *v.t.* to reduce to liquid; to pulverize (food) into a liquid. **liquidizer**, *n.* a kitchen appliance for chopping or puréeing vegetables, blending soup etc. **liquidly**, *adv.*
liquidambar (likwidam'bə), *n.* any of a genus of tropical trees, several species of which yield a fragrant resin or balsam called storax; the resin so produced.
liquor (lik'ə), *n.* a liquid or fluid substance, esp. the liquid part of anything as of a solution, a secretion, food etc.; a solution or dilution; an alcoholic drink, esp. spirits; an aqueous solution of a drug. **liquor-up**, *v.i.* to take a lot of drink.

liquorice (lik'əris), *n.* the root of a bean-like Mediterranean plant; its dried root; an extract from the root used in medicine and confectionery; liquorice-flavoured sweets.
liquorish LICKERISH.
lira (liə'rə), *n.* (*pl.* **lire** (rä), **liras**) the standard unit of currency in Italy and Turkey.
liriodendron (liriəden'drən), *n.* a genus of N American trees containing the tulip tree.
lis FLEUR-DE-LIS.
lisle (līl), *n.* a fine, hard cotton thread.
lisp (lisp), *v.i.* to pronounce *s* and *z* with the sound of *th* or *dh;* to speak affectedly or imperfectly as a child. *v.t.* to pronounce with a lisp. *n.* the act or habit of lisping; the speech-defect which causes one to lisp. **lisper**, *n.* **lispingly**, *adv.*
lissom, lissome (lis'əm), *a.* lithe, supple, nimble. **lissomness**, *n.*
list[1] (list), *n.* the border, edge or selvedge of cloth; a strip of this used as material; (*pl.*) the palisades enclosing a piece of ground for a tournament, the ground so enclosed; a scene of contest. *v.t.* to cover or line with list; (*N Am.*) to plough (land) with a lister. **to enter the lists**, to enter into a contest. **lister**, *n.* (*N Am.*) a plough designed for throwing up ridges.
list[2] (list), *v.t.* (3rd *sing.* **list**, *past* **list, listed**) to be pleasing to. *v.i.* to please, to be disposed. **listless** (-lis), *a.* indifferent; inattentive, languid. **listlessly**, *adv.* **listlessness**, *n.*
list[3] (list), *n.* a leaning over (of a ship, building etc.). *v.i.* to lean over. *v.t.* to heel (a ship) over.
list[4] (list), *v.t.*, *v.i.* to listen (to).
list[5] (list), *n.* a number of names (of people, places, objects) written down in order. *v.t.* to enter in a list; to arrange as a list. **list price**, *n.* a price as listed in a catalogue. **listed**, *a.* entered in a list; of buildings, architecturally important and protected from demolition and alteration. **listing**, *n.* the act or fact of entering in a list; (*pl.*) a published list of current plays, films, radio and TV programmes etc.
listen (lis'n), *v.i.* to give ear or attention (to); to heed, follow. *n.* an act of listening. **to listen in**, to listen to (a discussion) without contributing; to tap a telephone message. **listener**, *n.* **listening**, *a.* **listening post**, *n.* a position where people are posted to overhear what the enemy is saying or planning.
Listeria (listiə'riə), *n.* a genus of bacteria found in the environment and in contaminated food; (**listeria**) a member of this genus. **listeriosis** (-ō'sis), *n.* a disease caused by the presence of Listeria in contaminated foods.
lit (lit), *past, p.p.* LIGHT[1], LIGHT[3].
lit., (*abbr.*) literal(ly); literature; litre.
litany (lit'əni), *n.* a form of prayer, used in public worship, esp. a series of invocations with fixed responses; a long, usu. boring, list or catalogue.
litchi (lēchē'), **lichee, lychee** (lī-), *n.* a Chinese tree bearing an edible fruit; the fruit of this tree with a hard, scaly skin and a soft white pulp.
-lite (-līt), *suf.* forming names of minerals, as *aerolite, coprolite.*
liter LITRE.
literacy LITERATE.
literal (lit'ərəl), *a.* strictly according to the verbal meaning; not figurative or metaphorical; following the exact words (as a translation); consisting of or expressed by letters; unimaginative, prosaic. *n.* a misprint or misspelling. **literalism**, *n.* the interpretation of words and statements in a literal sense; realistic or unimaginative portrayal in art or literature. **literalist**, *n.* **literality** (-ral'-), **literalness**, *n.* **literalize, -ise**, *v.t.* **literally**, *adv.*
literary (lit'ərəri), *a.* of or pertaining to literature or writing; derived from, versed or engaged in literature; well-read; consisting of written or printed compositions; of language, formal in style. **literary agent**, *n.* one who

manages the business affairs of an author.
literate (lit'ərət), *a.* instructed in letters or literature, esp. able to read and write. *n.* one who is able to read and write; a person of liberal education. **literacy**, *n.* **literati** (-rah'ti), *n.pl.* men and women of letters. **literation**, *n.* representation (of a language etc.) by means of letters. **literator**, *n.* a literary man. **literature** (lit'rəchə), *n.* (*collect.*) the written or printed productions of a country, period or particular subject; printed matter; the literary profession. **literose** (-rōs), *a.* affecting literary tastes. **literosity** (-ros'-), *n.*
literatim (litərah'tim), *adv.* letter for letter, literally. [L]
lith (lith), (*abbr.*) lithograph; lithography.
-lith (-lith), *suf.* stone or rock, as in *monolith*.
lithaemia, *esp. N Am.* **-emia** (lithē'miə), *n.* excess of lithic or uric acid in the blood. **lithaemic**, *a.*
lithagogue (lith'əgog), *n.* (*Surg.*) an agent used in expelling kidney or gallstones.
litharge (lith'ahj), *n.* lead oxide.
lithe (līdh), *a.* flexible, supple. **lithely**, *adv.* **litheness**, *n.*
lithesome (-səm), *a.* **lithesomeness**, *n.*
lithia (lith'iə), *n.* oxide of lithium.
lithiasis (lithī'əsis), *n.* the formation of calculi in internal organs.
lithic[1] (lith'ik), *a.* pertaining to or composed of stone or calculi.
lithium (lith'iəm), *n.* the lightest metallic element, at. no. 3; chem. symbol Li, a member of the alkali series, used, esp. in alloys and batteries. **lithic**[2], *a.*
litho, *n., a., adv.* (*pl.* **lithos**) short for LITHOGRAPH, -GRAPHIC, -GRAPHY.
lith(o)-, *comb. form* pertaining to stone; calculus.
lithodome (lith'ədōm), *n.* a small mollusc which excavates and lives in rocks, shells etc.
lithogenous (lithoj'inəs), *a.* stone-producing; forming coral.
lithoglyph (lith'əglif), *n.* a carving on stone, esp. a gem. **lithoglyphic** (-glif'-), *a.*
lithograph (lith'əgraf), *v.t.* to engrave or draw on stone or metal and transfer to paper etc. by printing; to print by lithography. *n.* an impression from a drawing on stone or metal. **lithographer** (-thog'-), *n.* **lithographic**, **-ical** (-graf'-), *a.* **lithographically**, *adv.* **lithography** (-thog'-), *n.*
lithoid (lith'oid), **lithoidal** (-thoi'-), *a.* resembling a stone in nature or structure.
lithology (lithol'əji), *n.* the science of the composition, structure and classification of rocks; the branch of medical science dealing with calculus. **lithologic**, **-ical** (-loj'-), *a.* **lithologist**, *n.*
lithomancy (lith'əmansi), *n.* divination by means of stones.
lithophagous (lithof'əgəs), *a.* eating or perforating stones (as some molluscs).
lithophane (lith'əfan), *n.* ornamental porcelain suitable for lamps, windows and other transparencies.
lithophotography (lithōphətog'rəfi), PHOTOLITHOGRAPHY.
lithophyte (lith'əfit), *n.* a calcareous polyp, as some corals; a plant that grows on stone.
lithosis (lithō'sis), *n.* a disease of the lungs caused by tiny particles of stone.
lithosphere (lith'əsfiə), *n.* the rocky crust of the earth.
lithotome (lith'ətōm), *n.* an instrument used in lithotomy.
lithotomy (-thot'əmi), *n.* the surgical removal of stone in the bladder. **lithotomic** (-tom'-), *a.* **lithotomist** (-thot'-), *n.*
lithotripsy LITHOTRITY.
lithotrity (lithot'riti), **lithotripsy** (lith'ətripsi), *n.* the operation of crushing stones in the bladder, kidney or gallbladder to small fragments. **lithotripter** (-trip'tə), **-triptor**, *n.* a device which uses ultrasound to crush kidney etc. stones without the need for surgery. **lithotritic** (-trit'-), **lithotriptic** (-trip'-), **lithontriptic** (-əntrip'-), *n., a.*
Lithuanian (lithūā'niən), *a.* pertaining to Lithuania, a Baltic republic of the USSR. *n.* a native or inhabitant of Lithuania; the language of Lithuania.

litigate (lit'igāt), *v.t.* to contest in a court of law. *v.i.* to go to law; to carry on a lawsuit. **litigable**, *a.* **litigant**, *n., a.* **litigation**, *n.* **litigator**, *n.* **litigious** (-tij'əs), *a.* fond of litigation; quarrelsome; open to legal dispute; pertaining to litigation. **litigiously**, *adv.* **litigiosity** (-ios'-), **litigiousness**, *n.*
litmus (lit'məs), *n.* a substance obtained from certain lichens, turned red by acids or blue by alkalis. **litmus-paper**, *n.* unsized paper stained with litmus, used to test the acidity or the alkaline nature of a solution.
litotes (lī'tōtēz, lī'tə-), *n.* (*Rhet.*) affirmation expressed by negation of the contrary, or a weaker expression used to suggest a stronger one, as, 'Something has happened to him', meaning 'He is dead'.
litre (lē'tə), (*esp. N Am.*) **liter**, *n.* the unit of capacity in the metric system, equal to a cubic decimetre, or about 1¾ pints.
LittD, LitD, (*abbr.*) Doctor of Letters, Doctor of Literature.
litter (lit'ə), *n.* a couch or stretcher in which a person may be carried by animals or on men's shoulders; straw, hay or other soft material used as a bed for horses, cattle etc.; refuse, odds and ends scattered about; hence, disorder or untidiness; the young brought forth by a sow, bitch, cat etc. at one birth. *v.t.* to supply (beasts) with litter; to scatter (things) about carelessly; to make (a place) untidy with articles scattered about; to bring forth (said esp. of the sow, dog, cat etc.). *v.i.* to bring forth a litter of young. **litter-bug**, **-lout**, *n.* (*coll.*) one who drops rubbish in public places. **littery**, *a.*
littérateur (litərətœ'), *n.* an author, professional writer.
little (lit'l), *a.* (*comp.* **less** (les), *lesser*, (*coll.*) **littler**, *superl.* **least** (lēst), (*coll.*) **littlest** (-list) small, not great in size, extent, amount or quantity; short in duration; short in distance; of small dignity or importance; petty; narrow, contemptible, paltry; smaller than normal, short in stature; young like a child, weak; *adv.* in a small degree; not much, slightly; not at all. *n.* a small amount, quantity, space, distance, time etc.; only a trifle. **little by little**, by small degrees. **in little**, in miniature. **not a little**, a great deal; extremely. **to make little of**, to treat as insignificant; to disparage. **Little Bear** (*N Am.* **Little Dipper**), *n.* Ursa Minor. **little Englander**, *n.* an opponent of British expansion overseas in the 19th cent. **Little-go** GO[2]. **little people**, *n.pl.* the fairies. **little woman**, *n.* (*facet.*) one's wife. **littleness**, *n.*
littoral (lit'ərəl), *a.* pertaining to the shore, esp. the zone between high- and low-water marks. *n.* a coastal region.
liturgy (lit'əji), *n.* a ritual or form of service for public worship; the Mass, the formulary of the Eucharist; **liturgic** (-tœ-), *a.* liturgical. *n.pl.* the study or doctrine of liturgies. **liturgical**, *a.* **liturgically**, *adv.* **liturgiology** (-tœjiol'-), *n.* the study of liturgy. **liturgiologist**, *n.* **liturgist**, *n.*
live[1] (līv), *a.* alive; burning; ready for use; charged with electricity (as a wire); unexploded (as a shell); full of energy, of present interest etc.; of a radio, television broadcast, transmitted at the actual time of an event, not recorded; relating to a living (not recorded) performance of a play, concert etc. *adv.* as a live performance. **live-bait**, *n.* living animals used as fishing bait. **live birth**, *n.* the birth of a living animal. **live-box**, *n.* a case in which living microscopic objects are confined for observation. **live-cartridge**, *n.* one containing a bullet. **live-oak**, *n.* a N American evergreen tree, valuable for shipbuilding. **live rail**, *n.* one charged with an electric current. **livestock**, *n.* animals kept for farming or domestic purposes. **live wire**, *n.* one through which an electric current is flowing; (*coll.*) an energetic person. **liven**, *v.t., v.i.* to make or become lively.
live[2] (liv), *v.i.* to have life; to be alive; to remain in operation or as an active principle; to reside (at, in etc.); to subsist, depend for subsistence (upon); to receive or gain

a livelihood (by); to pass or conduct one's life in a particular condition, manner etc.; to enjoy life intensely; to continue alive, survive. *v.t.* to pass, spend (a specified kind of life); to survive; to manifest, express or effect, by living. **to live and let live**, to give and receive toleration of deficiencies. **to live down**, to efface the recollection of (scandal etc.) by one's conduct. **to live in, out**, to reside or not at one's place of work. **to live it up**, (*coll.*) to live extravagantly. **to live together**, to cohabit. **to live up to**, to conform to a prescribed standard. **to live with**, to cohabit; to accept or tolerate. **liveable**, *a.* worth living (of life); fit to live in; fit to live with. **liveableness**, *n.* **lived**, *a. in comb.* as *long-lived*. **lived-in**, *a.* shabby; untidy; comfortable. **liver**[1], *n.* one who lives (in a specified way, as a *good liver*).
livelihood (liv'lihud), *n.* means of subsistence.
livelong (liv'long), *a.* long-lasting; the whole, entire.
lively (liv'li), *a.* vivid; full of life, brisk; vivacious, bright; striking, exciting. **livelily**, *adv.* **liveliness**, *n.*
liven LIVE[1].
liver[1] LIVE[2].
liver[2] (liv'ə), *n.* a glandular organ in the abdominal cavity of vertebrates which secretes the bile and purifies the blood; this organ from certain animals used as food; (*coll.*) a disordered liver. *a.* liver-coloured. **liver-colour**, *n.* the colour of the liver; dark reddish-brown. **liver-coloured**, *a.* **liver-fluke**, *n.* a parasitic worm causing disease in the human liver. **liver-leaf**, *n.* (*N Am.*) one of the Hepaticae or anemones. **liver salts**, *n.pl.* a preparation of mineral salts used to relieve indigestion. **liver sausage, liverwurst**, *n.* sausage made from liver. **liver spot**, *n.* a liver-coloured spot which appears on the skin in old age. **liver-wing**, *n.* the right wing of a cooked fowl. **liverwort**, *n.* any plant of the cryptogamic Hepaticae family. **livered**, *a. in comb.*, as *white-livered*, cowardly. **liverish**, *a.* having a disordered liver; irritable.
liver[3] (li'və), *n.* a fabulous bird, supposed to have given its name to Liverpool.
Liverpudlian (livəpŭd'liən), *n*, a native or inhabitant of Liverpool.
livery (liv'əri), *n.* a distinctive dress worn by the servants of a particular person or the members of a city company; any distinctive dress, guise or outward appearance; the privileges of a city company or guild. **at livery**, kept at a stable for the owner at a fixed charge. **livery company**, *n.* one of the guilds or companies of the City of London. **liveryman**, *n.* one who belongs to a livery company. **livery- servant**, *n.* a servant wearing a livery. **livery-stable**, *n.* one where horses are kept at livery or let out on hire. **liveried**, *a.*
livid (liv'id), *a.* of a leaden colour; discoloured (as by a bruise); (*coll.*) very angry. **lividity** (-vid'-), *n.* **lividly**, *adv.*
living (liv'ing), *a.* alive, having life; flowing, running; vivifying, quickening; operative, efficient; alive now, contemporary; true to life (of a portrait). *n.* the state of being alive; livelihood; the benefice of a clergyman; manner of life. **living death**, *n.* a life of unmitigated suffering. **living rock**, *n.* rock in its native state or location. **living-room**, *n.* a family sitting-room. **living wage**, *n.* the lowest wage on which it is possible to maintain oneself and family.
lixiviate (liksiv'iāt), *v.t.* to leach, to dissolve out by washing or filtering; to impregnate with salts by lixiviation. **lixiviation**, *n.*
lizard (liz'əd), *n.* any member of the reptilian order Lacertilia, having a long, scaly body and tail, and four limbs, each with five toes of unequal length.
LJ, (*abbr.*) Lord Justice.
LL, (*abbr.*) Late Latin; Low Latin; Lord-Lieutenant.
ll., (*abbr.*) of print, lines.
llama (lah'mə), *n.* a S American wool-bearing animal like a small camel, used as a beast of burden; its wool, material made from this.

llano (lah'nō), *n.* a level, treeless plain in northern S America. **llanero** (-neə'rō), *n.* one who lives on the llanos.
LLB, (*abbr.*) Bachelor of Laws.
LLD, (*abbr.*) Doctor of Laws.
LLM, (*abbr.*) Master of Laws.
Lloyd's (loidz), *n.* a corporation dealing with insurance, the classification and registration of vessels etc. **Lloyd's List**, *n.* a newspaper devoted to shipping news. **Lloyd's Register**, *n.* an annual alphabetical list of world shipping classified according to seaworthiness. [Edward *Lloyd*, who kept a coffee-house frequented by shippers in the 17th cent.]
lm, (*symbol*) lumen.
lo (lō), *int.* see! look!
loach (lōch), *n.* a small British river-fish of the carp family.
load (lōd), *n.* that which is put on or in anything for conveyance; as much as can be carried at a time; on that which is borne with difficulty; that which presses upon, obstructs or resists; the downward pressure of a superstructure; the resistance to an engine or motor apart from friction; the power output of a machine, circuit etc.; a device which receives power; any mental burden; (*coll.*, *pl.*) a large amount. *v.t.* to put a load on or in; to put (a load or cargo) on or in a ship, vehicle etc.; to add weight to, to weight; to weigh down, encumber, oppress; to charge (a gun etc.); to put a film, cartridge etc. in (a camera); to fill to overflowing; to overwhelm (with abuse, honours etc.); to add charges to an insurance premium; (*Comput.*) to transfer a program into the memory. *v.i.* to take in a load or cargo (usu. with *up*); to charge a firearm. **to get a load of**, (*sl.*) to listen to. **load-line**, *n.* the line to which a ship sinks when loaded. **loaded**, *a.* biased; (*coll.*) wealthy; (*sl.*) drunk or drugged. **loaded question**, *n.* one with hidden implications designed to trap the answerer. **loader**, *n.* one who or that which loads; one employed to load a sportsman's gun; a loading-machine; *in comb.*, as *muzzle-loader*. **loading**, *n.* a load, burden; also *in comb.*, as *breech-loading*. **loading-coil**, *n.* an extra coil inserted in an electrical circuit to increase the inductance. **loading-gauge**, *n.* one indicating the height to which railway-trucks can be loaded.
loadsa (lōd'zə), *n.* (*sl.*) short for *a load of*; also *comb. form*, e.g. *loadsamoney*.
loadstar, loadstone LODE.
loaf[1] (lōf), *n.* (*pl.* **loaves** (lōvz)) a shaped mass of bread; a moulded mass of any material; (*sl.*) the head or brains.
loaf[2] (lōf), *v.i.* to lounge or idle about. *v.t.* to spend or pass (time away) idly. **loafer**, *n.* one who loafs; a low shoe similar to a moccasin.
loam (lōm), *n.* soil consisting of sand and clay with some organic matter or humus; in brickmaking etc., a mixture of sand and clay with chopped straw, used for making moulds. *v.t.* to cover with loam. **loamy**, *a.* **loaminess**, *n.*
loan (lōn), *n.* the act of lending; the state of being lent; that which is lent, esp. a sum of money lent at interest; permission to make use of; a word, myth, custom etc., adopted from another people. *v.t.* to grant the loan of. **loan-collection**, *n.* a private art collection lent for public exhibition. **loan shark**, *n.* (*coll.*) one who lends money at excessive or illegal interest rates. **loan translation**, *n.* a compound word or phrase which is a literal translation of the corresponding elements of a foreign expression (e.g. *Superman* from G *Übermensch*). **loan-word**, *n.* a word borrowed from another language. **loanable**, *a.* **loanee** (-nē'), *n.* **loaner**, *n.*
loath, loth (lōth), *a.* unwilling, averse, reluctant. **nothing loath**, quite willing; willingly. **loathness**, *n.*
loathe (lōdh), *v.t.* to feel disgust at; to detest. **loather**, *n.* **loathing**, *n.* disgust, aversion, abhorrence. **loathingly**, *adv.* **loathly**, *a.* (*old-fashioned*) loathsome. **loathsome** (-səm), *a.* causing loathing or disgust; odious, detestable. **loath-

somely, *adv.* **loathsomeness,** *n.*
loaves (lōvz), LOAF[1].
lob (lob), *n.* (*Cricket*) a slow underhand ball; (*Lawn-tennis*) a ball pitched high into the air. *v.t.* (*past, p.p.* **lobbed**) (*Cricket or Lawn-tennis*) to bowl a lob. *v.i.* to make a lob.
lobar, lobate LOBE.
lobby (lob'i), *n.* (*pl.* **-bbies**) a small hall or ante-room; that part of a hall of a legislative assembly to which the public are admitted; one of two corridors to which members go to vote (also **division lobby**); a group of people who try to influence legislators on behalf of special interests. *v.i.* to solicit the votes of members. *v.t.* to influence or solicit (members). **lobby correspondent,** *n.* a reporter working in the lobby system. **lobby system,** *n.* the system which allows correspondents access to political information on condition that the source remains anonymous. **lobbyist,** *n.*
lobe (lōb), *n.* any rounded and projecting or hanging part; a division of a bodily organ; the soft lower part of the ear; a rounded division of a leaf. **lobar, lobate** (-āt), *a.* **lobed,** *a.* **lobelet** (-lit), *n.*
lobectomy (ləbek'təmi), *n.* (*pl.* **-mies**) the surgical removal of a lobe from an organ or gland.
lobelia (ləbē'lyə), *n.* a genus of herbaceous and brilliant flowering plants.
loblolly (lob'loli), *n.* any of various US pine trees.
lob(o)-, *comb. form.* pertaining to a lobe.
lobotomy (ləbot'əmi), *n.* a surgical incision into the lobe of an organ or gland; an operation in which the fibres connecting the frontal lobes to the rest of the brain are cut, formerly used to treat severe depression. **lobotomize, -ise,** *v.t.* to perform a lobotomy on; (*coll.*) to render dull or harmless.
lobscouse (lob'skows), *n.* (*Naut.*) a hash of meat with vegetables and ship's biscuit.
lobster (lob'stə), *n.* a large marine long-tailed and stalk-eyed decapod crustacean esteemed for food. **lobster-pot,** *n.* a wickerwork trap for lobsters.
lobule (lob'ūl), *n.* (a subdivision of) a small lobe. **lobular, lobulated** (-lātid), *a.*
lobworm (lob'wœm), *n.* a large earthworm, used as bait by anglers; a lugworm.
local (lō'kəl), *a.* pertaining to, existing in or peculiar to a particular place or places; pertaining to a part (of the body), not the whole. *n.* an inhabitant of a particular place; a train serving a suburban district; an item of local news; a public-house. **local anaesthesia,** *n.* anaesthesia affecting only a particular area of the body. **local anaesthetic,** *n.* **local authorities,** *n.pl.* the elected bodies which administer local government. **local colour,** *n.* features characteristic of a place or district. **local government,** *n.* administration of towns, districts etc. by elective councils. **local time,** *n.* time calculated on the noon of the meridian of a place, as against standard time. **locale, local** (ləkahl'), *n.* the scene or locality of an event etc. **localism,** *n.* the state of being local; provincialism; a local idiom, custom etc. **locality** (-kal'-), *n.* particular place or region, site, geographical position. **localize, -ise** (lō'kəliz), *v.t.* to make local; to ascertain or indicate the exact place or locality of; to restrict to a particular place. **localizable,** *a.* **localization,** *n.* **locally,** *adv.*
locate (ləkāt'), *v.t.* to set or place in a particular locality; (*in p.p.*) to situate; to discover or determine the site of. **location,** *n.* situation or position; (*Cinema*) a site outside the studio grounds where a scene is shot; (*Comput.*) a specific area in memory capable of holding a unit of information, e.g. a word. **on location,** outside the studio (of filming etc.). **locative** (lok'ətiv), *n., a.* (*Gram.*) (a case) denoting place.
loc. cit. (lok sit'), (*abbr.*) loco citato. [L, in the place cited]
loch (lokh), *n.* a lake, an arm of the sea in Scotland.
lochan (-ən), *n.* a small lake.
lochia (lok'iə), *n.pl.* a uterine discharge following child-birth. **lochial,** *a.*
loci (lō'sī, -kī), *pl.* of LOCUS.
lock[1] (lok), *n.* a device for fastening doors etc. securely; a mechanical device for checking or preventing movement; the firing-apparatus of a gun; an enclosure in a canal, between gates, for raising and lowering vessels by the introduction or liberation of water; the oblique position of a fore-axle to a rear-axle in turning; (*fig.*) a fastening together or interlocking; a block, jam; a hug or grapple in wrestling; in Rugby, a player in the second row of a scrum (**lock forward**). *v.t.* to fasten with a lock; to shut (up a house, box, contents of these etc.) thus; to prevent passage (in, out etc.) by fastening doors etc. with locks; to shut (in); to fasten (together) securely; (*in p.p.*) to embrace, to tangle together; to furnish with locks (as a canal). *v.i.* to become fastened by or as by a lock; to intertwine. **lock, stock and barrel,** the whole lot. **to lock on (to),** to track automatically by means of a radar beam or sensor. **to lock up, in, away,** to close, fasten or secure with lock and key; to invest (money) so that it cannot be readily realized; (*fig.*) to make unavailable. **lock-gate,** *n.* the gate of a canal-lock. **lockjaw,** *n.* a violent spasm of the jaw muscles; tetanus. **lock-keeper,** *n.* one who attends to a canal-lock. **lock-out,** *n.* the temporary discharge of workers by employers to bring them to terms. *v.t.* to try to coerce workers thus. **lock-sill,** *n.* a piece of timber at the bottom of a canal-lock, against which the gates shut. **locksmith,** *n.* a maker and repairer of locks. **lock-spring,** *n.* a spring for closing a watch-case. **lock-stitch,** *n.* a sewing-machine stitch which locks two threads together. **lock-up,** *n.* a place where prisoners are temporarily confined; time for locking up; a small garage; *a.* that may be locked. **lock-up shop,** one having access only from the street, with no living quarters. **lockable,** *a.* **lockage** (-ij), *n.* a toll for passing through locks.
lock[2] (lok), *n.* a tuft of hair, wool or similar substance; a tress, ringlet; (*pl.*) hair.
lockage LOCK[1].
locker (lok'ə), *n.* one who or that which locks; a cupboard, chest etc. with lock and key. **locker room,** *n.* a room with lockers for storing clothes and other belongings.
locket (lok'it), *n.* a small ornamental case, worn on a chain and containing hair, a miniature etc.
Lockian (lok'iən), *a.* characteristic of the teaching of Locke or his followers. **Lockist,** *n.* [John *Locke*, 1632–1704, English philosopher]
lockjaw LOCK[1].
loco[1] (lō'kō), *n.* short for LOCOMOTIVE.
loco[2] (lō'kō), *a.* (*esp. N Am. sl.*) insane; affected with loco disease. *n.* (also **loco-plant, -weed**) any of several leguminous plants of NW America which cause loco disease in livestock when ingested. **loco disease,** *n.* a disease of livestock characterized by paralysis of the limbs and impaired vision and caused by eating loco-weed. [Sp.]
locomobile (lōkəmō'bil), *a.* able to change place. *n.* a locomotive vehicle. **locomobility** (-bil'-), *n.*
locomotion (lōkəmō'shən), *n.* the act or power of moving from place to place; travel, travelling. **locomotive** (lō'-), *v.i.* to move from one place to another. **locomotive,** *a.* pertaining to, capable of or causing locomotion; moving from place to place, not stationary. *n.* a self-propelling machine, esp. a railway engine. **locomotively,** *adv.* **locomotivity** (-tiv'-), **locomotiveness,** *n.* **locomotor,** *a.* of or pertaining to locomotion. *n.* one who or that which is capable of locomotion. **locomotor ataxy** *n.* a nervous disorder characterized by inability to coordinate the movements of the limbs. **locomotory,** *a.*
loculus (lok'ūləs), *n.* (*pl.* **-li** (-lī)) a small cavity, a cell; (*Biol. etc.*) one of numerous cavities in various organisms. **locular, loculate** (-lət), *a.*
locum (lō'kəm), **locum tenens** (tē'nenz), *n.* (*pl.* **-tenentes** (tinen'tēz)), *n.* a deputy or substitute, esp. one acting in the

place of a doctor or clergyman. **locum-tenency** (ten'ənsi), *n.*

locus (lō'kəs), *n. (pl.* **-ci** (-sī)) the exact place, the locality (of); the location of a particular gene on a chromosome. **locus classicus** (klas'ikəs), the best or most authoritative passage quoted. **locus standi** (stan'dī), recognized place or position authorizing intervention, etc. [L]

locust (lō'kəst), *n.* a winged insect of various species allied to the grasshopper, which migrates in vast swarms and is very destructive to vegetation; a locust-tree. **locust-bean**, *n.* the carob bean. **locust-bird, -eater**, *n.* one of various species of birds that feed on locusts. **locust-tree**, *n.* the carob; the N American acacia; applied to various W Indian trees.

locution (ləkū'shən), *n.* style of speech, mode of delivery; a phrase or expression considered with regard to style or idiom.

lode (lōd), *n.* a vein bearing metal. **lodestar**, *n.* a guiding star or one that is steered by, usu. the pole-star; one's guiding principle. **lodestone**, *n.* magnetic oxide of iron, a natural magnet; something that attracts.

loden (lō'dən), *n.* a thick soft waterproof woollen cloth used for making coats; a greyish-green colour typical of this cloth.

lodge (loj), *n.* a temporary residence; a cottage, hut, cabin; a gate-keeper's or gardener's cottage; a room or apartment for a porter in a college etc.; a local branch or place of meeting of certain societies; a beaver's or otter's lair; a N Am. Indian tent or hut. *v.t.* to supply with temporary quarters, esp. for sleeping; to receive as an inmate, usu. for a fixed charge; to deposit, to leave for security (in, with etc.); to deposit in court etc. (as a complaint); to implant, to fix. *v.i.* to reside temporarily, esp. to have sleeping quarters; to reside as an inmate at a fixed charge; to stay or become fixed (in). **lodger**, *n.* one who rents and occupies furnished rooms. **lodging**, *n.* a temporary residence; (*usu. in pl.*) a room or rooms hired in another's house. **lodging-house**, *n.* **lodg(e)ment**, *n.* the act of lodging; the state of being lodged; an accumulation of matter, a deposit; an entrenchment hastily constructed to defend captured enemy territory.

loess (lō'is, lœs), *n.* a deposit of clay, loam, sand etc. formed by wind action.

loft (loft), *n.* the room or air space under a roof; a gallery in a church or hall; a room over a barn or stable; a pigeon-house; (*Golf*) a backward inclination of the face of a club; a lofting stroke. *v.t.* to strike (the ball) so that it rises high in the air; to provide (pigeons) with a loft. **lofter**, *n.* (*Golf*) a club for lofting. **lofty**, *a.* very high, imposing; elevated in character, sentiment, style etc.; grandiose; arrogant. **loftily**, *adv.* **loftiness**, *n.*

log[1] (log), *n.* a bulky piece of unhewn timber; a block; a device consisting of a float attached to a line, used for ascertaining the speed of a ship; a detailed record of the voyage of a ship or flight of an aircraft; a log-book; any record of performance; a dolt. *v.t.* (*past, p.p.* **logged**) to cut into logs; to enter in the log-book; to travel (a specified distance). **to log on**, to identify oneself to a computer system in order to gain access to one's files; (*coll.*) to begin work etc. **to log off**, to close one's files at the end of a session of work; (*coll.*) to finish work etc. **to sleep like a log**, to be in a deep sleep. **log-book**, *n.* an official diary of events occurring in a ship's voyage or aircraft's flight; the registration documents of a motor vehicle. **log-cabin, -house, -hut**, *n.* a dug-out. **log-canoe**, *n.* (*N Am.*) a blockage in a river caused by floating logs; a deadlock, standstill. **log-line**, *n.* a knotted line, fastened to the log for finding a ship's speed. **log-roll**, *v.i.* **log-roller**, *n.* **log-rolling**, *n.* (*N Am.*) mutual political assistance in carrying legislative measures; a sport in which two opponents attempt to spin each other off a floating log on which both are standing. **logwood**, *n.* the wood of a tropical US tree, used then as a dark-red dye-stuff. **logger**, *n.* a lumberman. **logging**, *n.*

log[2] (log), *n.* short for LOGARITHM.

loganberry (lō'gənbəri), *n.* a permanent hybrid obtained by crossing the raspberry and a species of blackberry; the fruit of this.

logan-stone (log'ənstōn), *n.* a rocking-stone.

logarithm (log'əridhm), *n.* the exponent of the power to which a fixed number, called the base, must be raised to produce a given number (used as a means of simplifying multiplication and division). **logarithmic** (-ridh'-), *a.* **logarithmic scale**, *n.* one in which an increase of one unit represents a tenfold increase in the quantity measured. **logarithmically**, *adv.*

loge (lōzh), *n.* a box in the theatre. [F]

loggerhead (log'əhed), *n.* a large marine turtle; a tool consisting of a long handle with a bulbous iron head for heating liquids, melting tar etc. **at loggerheads**, locked in dispute.

loggia (loj'iə), *n. (pl.* **-ggias, -ggie** (-je)) an open arcade along the front of a building.

logic (loj'ik), *n.* the science of reasoning, correct thinking, proving and deducing; a particular system of reasoning; reasoning, argument etc. considered with regard to correctness or incorrectness; force of argument; force of circumstances etc.; (*Comput.*) the elementary principles for performing arithmetical and logical operations. **logic bomb**, *n.* an instruction programmed into a computer that will later trigger a breakdown. **logic circuit**, *n.* (*Comput.*) an electronic circuit which performs logical operations on its two or more inputs. **logical**, *a.* pertaining to, used in or according to the rules of logic, consistent or accurate in reasoning; reasonable; versed or skilled in accurate reasoning; (*Comput.*) of, performed by or used in logic circuits. **logical positivism**, *n.* a philosophical school based on linguistic analysis which demands that meaningful statements must be empirically verifiable, so rejecting metaphysics etc. **logicality** (-kal'-), **logicalness**, *n.* **logically**, *adv.* **logician** (-jish'ən), *n.* one skilled in logic.

-logist (-ləjist), *suf.* as in *anthropologist.*

logistics (ləjis'tiks), *n.pl.* or *sing.* the branch of strategy concerned with the moving and supply of troops; the planning and organization of any complex enterprise. **logistic**, *a.*

loglog (log'log), *n.* the logarithm of a logarithm.

logo (log'ō, lō'-), *n. (pl.* **-gos**) short for LOGOTYPE.

logo-, *comb. form* pertaining to words; wordy.

logogram (log'əgram), *n.* a sign representing a word, esp. in shorthand. **logograph** (-graf), *n.* a logogram; a logotype. **logographic, -ical** (-graf'-), *a.* **logography** (-gog'-), *n.* a method of printing in which a type represents a word instead of a letter.

logorrhoea (logərē'ə), *n.* excessive or uncontrollable talkativeness.

logos (log'os), *n.* in Greek philosophy, the divine reason implicit in and governing the cosmos; (*Theol.*) the Divine Word, the Son of God, the Second Person of the Trinity.

logotype (log'ətip), *n. (Print.)* a type having two or more letters cast in one piece, but not as a ligature, as *are, was* etc.; a symbol or simple design identifying a company, organization etc.

-logue (-log), *suf.* speech, discourse, as *epilogue, prologue.*

logwood LOG[1].

-logy (-ləji), *suf.* forming names of sciences and departments of knowledge, and nouns denoting modes of speaking, as *astrology, eulogy, tautology.*

loin (loin), *n.* the part of the body lying between the lower ribs and the hip-joint; (*pl.*) strength or generative power; a joint of meat from this part. **to gird up the loins**, to prepare oneself for a great effort. **loincloth**, *n.* a cloth worn round the loins.

loiter (loi'tə), *v.i.* to linger, dawdle; to spend time idly; to be dilatory. *v.t.* to idle (time) away. **loiterer,** *n.* **loiteringly,** *adv.*

Lok Sabha (lŏk sab'ə), *n.* the lower chamber of the Indian parliament. [Hind.]

loligo (ləlī'gō), *n.* a genus of cephalopods containing the squids.

loll (lol), *v.i.* of the tongue, to hang from the mouth; to stand, sit or lie in a lazy attitude, to lounge. *v.t.* to allow to or recline lazily. **loller,** *n.* **lollingly,** *adv.*

Lollard (lol'əd), *n.* one of a sect of English religious reformers in the 14th and 15th cents., followers of John Wyclif (?1330–84). **Lollardism, Lollardy,** *n.*

lollipop (lol'ipop), *n.* a flat or round boiled sweet stuck on the end of a stick; an ice lollipop; a piece of popular classical music. **lollipop man, woman, lady,** *n.* (*coll.*) one who conducts children safely across roads by controlling traffic using a pole with a disk on the top.

lollop (lol'əp), *v.i.* to loll about; to roll or flop about heavily; to go or do in a lounging or idle way.

lolly (lol'i), *n.* a lollipop, a sweet on a stick; an ice lolly; (*sl.*) money.

Lombard (lom'bəd, -bahd), *n.* one of the Teutonic Longobardi who conquered Italy in the 6th cent.; a native of Lombardy; *a.* of or pertaining to the Lombards or to Lombardy. **Lombard Street,** the banking centre of the City of London fomerly occupied by Lombard merchants and money-lenders. **Lombardic** (-bah'-), *a.* **Lombardy poplar** (-bədi), *n.* a variety of poplar tree with erect branches.

loment (lō'mənt), *n.* a kind of pod separating by a transverse articulation between each seed. **lomentaceous** (-tā'shəs), *a.* **lomentum** (-men'təm), *n.* (*pl.* **-ta** (-tə)).

London (lŭn'dən), *n.* the capital of England. **London particular,** *n.* (*old coll.*) dense yellow fog. **London pride,** *n.* an Irish saxifrage cultivated in gardens. **Londoner,** *n.* a native, inhabitant or citizen of London.

lone (lōn), *a.* (*chiefly poet. or rhet.*) solitary, uninhabited; without company or a comrade; unmarried, widowed. **lone hand,** *n.* (*Cards*) one played without help from one's partner's cards. **loneness,** *n.* **loner,** *n.* one who prefers solitude or independence. **lonesome** (-səm), *a.* lonely, unfrequented; adapted for solitude. **lonesomeness,** *n.*

lonely (lōn'li), *a.* solitary, unfrequented; without companion. **lonely hearts,** *a.* of or for people seeking friendship or marriage. **loneliness,** *n.*

long[1] (long), *a.* of considerable or relatively great linear extent; of great extent in time; of a specified linear extent or duration in time; protracted in sound; stressed (of vowels or syllables); delayed in coming; far-reaching; lengthy, tedious. *adv.* to a great extent in distance or time; for a long time; throughout a specified period; having a large holding of securities etc. in anticipation of a price rise. *n.* anything that is long, esp. a period, interval etc.; (*Pros.*) a long syllable; (*Mus.*) a note equal in common time to two breves. **before long,** soon. **in the long run,** eventually. **long-drawn-out,** prolonged, overextended. **no longer,** formerly but not now. **so long!,** (*coll.*) goodbye. **long ago,** *n., a.* (of) the distant past. **longboat,** *n.* the largest boat on a sailing vessel. **longbow,** *n.* a long powerful bow drawn by hand. **longcase clock,** *n.* a grandfather clock. **longcloth,** *n.* a fine, soft cotton cloth made in strips. **long-dated,** *a.* of securities, not due for redemption in less than 15 years. **long-distance,** *a.* from or at long range. **long dozen,** *n.* thirteen. **long drink,** *n.* a well-diluted alcoholic drink in a tall glass. **long face,** *n.* a gloomy expression. **long firm** FIRM[2]. **longhand,** *n.* ordinary writing opp. to shorthand. **long haul,** *n.* transport over a great distance; a difficult or extended period of time. **long-headed,** *a.* shrewd, sensible, farsighted. **longhorn,** *n.* any animal with long horns or antennae. **long hundred,** *n.* one hundred and twenty. **long johns,** *n.pl.* underpants with long legs. **long jump,** *n.* an athletic event involving a horizontal jump for distance. **long-lived,** *a.* enjoying long life. **long metre,** *n.* a hymn stanza of four 8-syllable lines. **long odds,** *n.* unequal or unfavourable odds (in betting). **long off, on,** *n.* in cricket, the fielder to the left or right rear of the bowler. **long-playing,** *a.* of or relating to a fine-grooved gramophone record. **long-range,** *a.* involving or fit for an extended distance or time. **longship,** *n.* a long open boat used esp. by the Vikings. **long shot,** *n.* a camera shot from a long distance; a random guess. **long-sighted,** *a.* able to see to a great distance; shrewd. **long-standing,** *a.* of long duration. **long stop,** *n.* in cricket, a fielder positioned to stop balls which pass the wicket-keeper. **long-suffering,** *a.* patient, enduring. **long suit,** *n.* the most numerous suit in a hand of cards; one's special interest or skill. **long-term,** *a.* of a policy, looking to the future rather than the immediate present. **long vacation,** *n.* the long summer holidays of universities etc. **long wave,** *n.* a radio wave with a wavelength of 1000 m or more. **long weekend,** *n.* a holiday of several days including a weekend. **long-winded,** *a.* wordy, tiresome. **longish,** *a.* **longs,** *n.pl.* long trousers; long-dated securities. **longways, -wise,** *adv.*

long[2] (long), *v.i.* to have an earnest desire (to or for). **longing,** *n., a.* **longingly,** *adv.*

long., (*abbr.*) longitude.

longanimity (long-gənim'iti), *n.* long-suffering, forbearance. **longanimous** (-gan'-), *a.*

longe LUNGE[2].

longeron (lon'jəron), *n.* a longitudinal spar of an aeroplane's fuselage.

longevity (lonjev'iti), *n.* great length of life. **longeval** (lonjē'-), *a.* long-lived.

longi-, *comb. form.* long.

longicorn (lon'jikawn), *n., a.* (a member) of a division of beetles with large filiform antennae.

longitude (long'gitūd, lon'ji-), *n.* angular distance of a place E or W of a given meridian, usu. that of Greenwich. **longitudinal** (-tū'-), *a.* pertaining to longitude or length; running lengthwise. **longitudinal wave,** *n.* a wave in which the particles of the medium vibrate in the same direction as the advance of the wave. **longitudinally,** *adv.*

Longobard (long'gəbahd), LOMBARD[1].

longshore (long'shaw), *a.* of or belonging to, existing or working on the shore. **longshoreman,** *n.* a landsman working on the shore; (*N Am.*) a docker.

loo[1] (loo), *n.* (*coll.*) a lavatory.

loo[2] (loo), *n.* a card game in which penalties are paid into a pool. **loo-table,** *n.* a round table.

looby (loo'bi), *n.* an awkward, clumsy fellow.

loof (loof), LUFF.

loofah (loo'fə), *n.* the fibre of the sponge-gourd, used as a flesh-brush.

look (luk), *v.i.* to direct the eye (towards, at etc.) in order to see an object; to exercise the sight; to gaze, stare; to direct one's attention; to face, to be turned (towards, to, into etc.); to have a particular tendency; to appear; to watch; to take care. *v.t.* to express or show by the looks; to view, to inspect. *n.* the act of looking or seeing, a glance; (*usu. in pl.*) appearance, esp. of the face, aspect; expression of the eye and countenance; general appearance. **look before you leap,** be cautious before acting. **look here!** pay attention! **look sharp!** be quick! **on the look out,** on the watch. **to look after,** to attend to; to take care of. **to look down on,** to assume superiority over. **to look for,** to seek; to hope for; to be on the watch for. **to look forward to,** to hope for with pleasure. **to look in,** to pay a brief visit. **to look in the face** FACE. **to look into,** to investigate; to examine the inside of. **to look on,** to be a mere spectator; to regard, consider (as, with etc.). **to look out,** to be on the watch, to be prepared (for); to seek and find. **to look over,** to examine; to overlook or excuse. **to look through,** to penetrate with one's sight or insight; to

loom 490 **lose**

examine the contents of. **to look to**, to expect to (do); to rely upon (for). **to look up**, to search for; to pay a visit to; to improve; (*with* **to**) to admire or respect. **look-alike**, *n.* somebody or something who or that closely resembles another; a double. **look alive** ALIVE. **look in**, *n.* a call, a short visit; a chance, as of winning in a game. **look lively**, *v.i.* (*coll.*) to make haste. **look-out**, *n.* a watch; a person keeping watch; a place from which watch is kept; a view; (*fig.*) future prospect; one's personal affair or concern. *v.i.* to be careful. **look-see**, *n.* (*coll.*) an inspection. **looker**, *n.* an observer; (*coll.*) an attractive person, esp. a woman. **looker-on**, *n.* a mere spectator. **looking-glass**, *n.* a mirror.

loom[1] (loom), *n.* a machine in which yarn is woven into fabric.

loom[2] (loom), *v.i.* to appear indistinctly or faintly in the distance; to appear larger than the real size; to seem very close orrr threatening. *n.* the first indistinct appearance, as of land at sea.

loon[1] (loon), *n.* (*coll.*) a daft or eccentric person.

loon[2] (loon), *n.* the great northern diver; the grebe; the guillemot.

loony (loo'ni), *n., a.* (*sl.*) a lunatic; a foolish person. **loony-bin**, *n.* (*sl.*) a mental hospital.

loop (loop), *n.* a doubling of a string, rope etc. across itself to form a circle or oval; a noose; anything resembling this; a ring etc. by which anything is hung up, fastened etc.; a stitch in crochet or knitting; a loop-line; a length of film or tape joined end to end to form a continuous strip; a flight manoeuvre comprising a complete revolution in a vertical plane, the upper surface of the aircraft being on the inside of the circle; a loop-shaped intrauterine contraceptive device; (*Comput.*) a set of instructions repeated in a program until a specific condition is met. *v.t.* to form into a loop or loops; to fasten or secure with loops. *v.t.* to make a loop. **to loop the loop**, to travel round in a vertical loop in an aeroplane etc. **loop-line**, *n.* a railway, telegraph-line etc. diverging from the main line and joining it again. **looped**, *a.* **looper**, *n.* **loopy**, *a.* (*coll.*) slightly mad.

loop-hole (loop'hōl), *n.* a small hole in a wall for shooting through etc.; a means of evasion or escape. *v.t.* to make loop-holes in.

loose (loos), *a.* not tied, fastened or confined; freed; detachable, hanging partly free; not fixed or tight; not compact or dense; relaxed, slack; careless; rambling; not strict; indefinite; incorrect; ungrammatical; dissolute, wanton; lax in the bowels, opp. to costive. *v.t.* to undo, untie, unfasten; to release, to unbind; to dissolve; to relax; to free from obligation or burden; to discharge. *n.* release, discharge. **at a loose end**, with nothing to do. **on the loose**, on the spree. **to break loose**, to escape from captivity. **to let loose** LET[1]. **to loosen up**, to relax. **to set loose**, to set at liberty. **loose box** BOX[2]. **loose change**, *n.pl.* coins kept for small items of expenditure. **loose cover**, *n.* an easily removable cloth cover for a chair, sofa etc. **loose-leaf**, *a.* bound so that pages may be inserted or removed. **loose-limbed**, *a.* having flexible or supple limbs. **loosely**, *adv.* **loosen**, *v.t., v.i.* **loosener**, *n.* **looseness**, *n.* **loosish**, *a.*

loosestrife (loos'strīf), *n.* any of a genus of plants of the primrose family, with yellow flowers; a water-side plant with red or purple flowers.

loot (loot), *n.* booty, plunder, esp. from a conquered city; stolen money, jewellery etc.; (*coll.*) money. *v.t.* to plunder, esp. a city; to carry off as plunder. *v.i.* to plunder. **looter**, *n.*

lop[1] (lop), *v.t.* (*past, p.p.* **lopped**) to cut off the top or extremities of; to trim (trees, shrubs etc.) by cutting; to omit a part of. *n.* (*usu. in pl.*) that which is lopped. **lopper**, *n.*

lop[2] (lop), *v.i.* to hang down limply; to flop, droop; to hang or idle (about). *v.t.* to allow to hang down. *n.* a lop-eared rabbit. **lop-ear**, *n.* **lop-eared**, *a.* having hanging ears. **lop-sided**, *a.* heavier on one side than the other; not symmetrical. **lop-sidedly**, *adv.* **lop-sidedness**, *n.*

lope (lōp), *v.i.* to gallop or run (along) with long strides or leaps. *n.* motion of this kind. **loper**, *n.*

loquacious (lǝkwā'shǝs), *a.* talkative, garrulous, chattering; apt to disclose secrets. **loquaciously**, *adv.* **loquaciousness, loquacity** (-kwas'-), *n.*

loquat (lō'kwot), *n.* a Chinese and Japanese evergreen tree; its yellow edible fruit.

loral, lorate LORE[2].

lord (lawd), *n.* a ruler, master; one possessing supreme power, a sovereign; (**Lord**) God; (**Lord**) Jesus Christ; a feudal superior, the holder of a manor; (*facet.*) one's husband; a nobleman, a peer of the realm; (*pl.*) the members of the House of Lords. *v.i.* to play the lord (over). *int.* an exclamation of surprise or dismay ((*coll.*) **lor, lordy**). **drunk as a lord**, very drunk. **House of Lords**, the upper legislative chamber in the UK comprising the lords spiritual and temporal. **lord of misrule**, (*Hist.*) one who superintended the games and revels at Christmas. **lords and ladies**, the wild arum lily. **my lord**, a formula for addressing a nobleman (not a duke), bishop, lord mayor or judge of the Supreme Court. **to live like a lord**, to live affluently. **Lord-Lieutenant**, *n.* an official representing the sovereign, in a county. **Lord Mayor**, *n.* the chief magistrate of London, York and certain other large towns. **Lord Rector**, *n.* the elected head officer of certain Scottish universities. **Lord's Cricket Ground**, the headquarters of Marylebone Cricket Club and of cricket generally. **Lord's day**, *n.* Sunday. **Lord's Prayer**, *n.* the prayer taught by Jesus Christ to his disciples (Math. vi.9–13, Luke xi.2–4). **lords spiritual**, *n.pl.* the archbishops and bishops having seats in the House of Lords. **Lord's Supper**, *n.* the Eucharist. **Lord's table**, *n.* the altar in a Christian church. **lords temporal**, *n.pl.* lay peers having seats in the House of Lords. **lordless**, (-lis), *a.* **lordlet** (-lit), **lordling** (-ling), *n.* **lord-like**, *a.*, *adv.* **lordly**, *a.* becoming or befitting a lord; noble, magnificent; superb, haughty, insolent. *adv.* proudly; imperiously. **lordliness**, *n.* **lordship**, *n.* **your, his lordship**, a formula used in speaking deferentially to or of a lord.

lordosis (lawdō'sis), *n.* forward curvature of the spine. **lordotic** (-dot'-), *a.*

lore[1] (law), *n.* learning; the collective traditions and knowledge on a given subject.

lore[2] (law), *n.* the surface between the eye and the beak in birds. **loral, lorate** (-rǝt), *a.*

lorgnette (lawnyet'), *n.* a pair of eye-glasses with a long handle; an opera-glass.

lorica (lǝri'kǝ), *n.* a cuirass; the carapace of a crustacean. **loricate** (lo'rikǝt), *a.* **lorication**, *n.*

lorikeet (lo'rikēt), *n.* a genus of brightly-coloured parrots belonging to the Malay Archipelago.

loriot (law'riǝt), *n.* the golden oriole.

loris (law'ris), *n.* a lemur of Sri Lanka, usu. called the slender loris; also the slow lemur or E Indian loris.

lorn (lawn), *a.* (*old-fashioned*) lost, abandoned, forlorn.

lorry (lo'ri), *n.* (*pl.* **-rries**) a large motor vehicle for carrying heavy loads.

lory (law'ri), *n.* a brilliantly coloured parrot-like bird of SE Asia and Australia.

lose (looz), *v.t.* (*past, p.p.* **lost** (lost)) to be deprived of; to part with accidentally or as a forfeit, penalty etc.; to be freed from; to miss, to be unable to find; to fail to gain, win, hear, obtain or enjoy; to fail to keep possession of; to spend uselessly, to waste; (*in p.p.*) to cause to disappear, die or perish; to cause one the loss of; to make (oneself or itself) disappear. *v.i.* to fail to be successful, to be beaten; to suffer loss; to be worse off (by); of a clock etc., to run slow. **to lose ground** GROUND[1]. **to lose oneself**, to lose one's way; to be bewildered. **to lose out,**

loss   491   **love**

(*coll.*) to make a loss; to fail to take advantage of. **losable**, *a.* **loser**, *n.* one who loses; a person, horse, boat etc. failing to win a race; (*coll.*) a failure; (*pl.*) the beaten party in a game, battle etc. **losing**, *pres.p.* **losing game**, *n.* a hopeless game or contest. **losingly**, *adv.* **lost**, *a.* unable to find the way; no longer possessed or known; missing; confused, helpless; ruined, destroyed; insensible; engrossed. **to be lost in**, to be engrossed in; to merge or be obscured in. **lost cause**, *n.* a futile endeavour. **lost soul**, *n.* one who is beyond redemption.

loss (los), *n.* the act or state of losing or being lost; failure to win or gain; that which is lost or the amount of this; detriment, disadvantage; wasted expenditure, effort etc. **at a loss**, embarrassed or puzzled. **to bear a loss**, to sustain a loss without giving way; to make good a loss. **loss adjuster**, *n.* one who assesses losses through fire, theft etc. for an insurance company. **loss leader**, *n.* an article sold at a loss to attract customers.

löss (lœs), LOESS.

lost (lost), *past, p.p.* LOSE¹.

lot (lot), *n.* anything, such as a die, paper or other object, used in determining chances; choice or decision by random drawing of these; one's fortune, destiny or condition in life; a thing or set of things offered for sale at auction; a parcel of land; a number or quantity of things or persons; (*often in pl.*) a considerable quantity or amount, a great deal; a film studio. *v.t.* to divide into lots. **a bad lot**, a person of bad or doubtful character. **the lot**, the whole quantity. **to cast, draw lots**, to determine by the throw of a die or other contrivance. **lotsa**, *n.* (*sl.*) short for *lots of*; also *comb. form*, as *lotsalolly*. **lotta**, *n.* (*sl.*) short for *a lot of*.

loth LOATH.

Lothario (lothah'riō), *n.* a libertine, seducer.

lotion (lō'shən), *n.* a medicinal or cosmetic liquid application for external use.

lottery (lot'əri), *n.* a method of allotting valuable prizes by chance among purchasers of tickets; a mere hazard. **lottery-wheel**, *n.* a drum-like wheel used for shuffling lottery-tickets. **lotto** (-ō), *n.* a game of chance, played with disks placed on cards divided into numbered squares.

lotus (lō'təs), *n.* (*pl.* **lotuses**) in Greek legend, a name for several plants the eating of whose fruit was said to induce a dreamy languor; the Egyptian or Indian water-lily; an architectural representation of this; a genus of leguminous plants containing the bird's-foot trefoil. **lotus-eater**, *n.* one who gives himself up to dreamy ease. **lotus-eating**, *n.*, *a.* **lotus-land**, *n.* **lotus position**, *n.* a yoga position in which one sits cross-legged with each foot against the opposite thigh.

louche (loosh), *a.* seedy; sinister. [F]

loud (lowd), *a.* powerful in sound; noisy, clamorous; conspicuous, ostentatious, flashy (of attire, manners etc.). **loud-hailer**, *n.* a megaphone with a built-in amplifier and microphone. **loudmouth**, *n.* (*coll.*) someone who brags or talks offensively in a loud voice. **loudspeaker**, *n.* a device which converts electrical signals into audible sound. **louden**, *v.t., v.i.* to make or become louder. **loudish**, *a.* **loudly**, *adv.* **loudness**, *n.*

lough (lokh), *n.* a lake, an arm of the sea in Ireland. [Ir.]

louis (loo'i), **louis d'or** (daw'), *n.* (*unchanged in pl.*) an old French gold coin issued from Louis XIII to Louis XVI. **Louis Treize** (trez), **Quatorze** (katawz'), **Quinze** (kiz), or **Seize** (sez), Louis XIII, XIV, XV or XVI (denoting styles of furniture fashionable in those reigns).

lounge (lownj), *v.i.* to idle about, to saunter; to loll or recline. *n.* the act of lounging; a saunter; a place for lounging; the sitting-room in a house; (also **lounge-bar**) a more comfortable and expensive bar of a public house; a sofa with a back and one raised end. **lounge-lizard**, *n.* a gigolo. **lounge suit**, *n.* a man's suit for daily wear.

lounger, *n.* one who lounges; a comfortable sofa or extending chair for relaxing on; a loose-fitting garment. **loungingly**, *adv.*

loupe (loop), *n.* a small magnifying glass used by jewellers, watchmakers etc.

lour, lower³ (lowə), *v.i.* to appear dark or gloomy; to frown, scowl; to look threatening (of clouds, weather etc.). *n.* a scowl; sullenness; gloominess (of weather etc.). **louring, loury**, *a.* **louringly**, *adv.*

louse (lows), *n.* (*pl.* **lice** (līs)) a bloodsucking insect parasitic on man; applied to various parasites infesting animals, birds, fish and plants; (*sl.*) a mean, contemptible person. *v.t.* (louz), to clean from lice; (*with up* (*sl.*)) to spoil, make a mess of. **lousy** (-'zi), *a.* infested with lice; (*sl.*) low, mean, or obscene; (*sl.*) bad, inferior; (*sl.*) swarming or excessively supplied (with). **lousily**, *adv.* **lousiness**, *n.*

lout (lowt), *n.* an awkward, crude person; an oaf. **loutish**, *a.* **loutishly**, *adv.* **loutishness**, *n.*

louvre (loo'və), *n.* an opening in a chimney pot etc. to let out smoke; (*pl.*) louvre-boards. **louvre-boards**, *n.pl.* sloping overlapping boards across a door or window to exclude rain but allow the passage of air. **louvre-door, -window**, *n.* **louvred**, *a.*

lovage (lŭv'ij), *n.* an umbelliferous herb, used in salads and for flavouring food.

love (lŭv), *n.* a feeling of deep regard, fondness and devotion (for, towards etc.); deep affection, usu. accompanied by yearning or desire for; affection between persons of the opposite sex, more or less founded on or combined with desire or passion; a personification of this or of Cupid; a beloved one (as a term of endearment); (*coll.*) a delightful person, a charming thing; in games, no points scored, nil. *v.t.* to have strong affection for, to be in love with; to delight in, to have a strong partiality or predilection for. *v.i.* to be in love. **for love or money**, by some means or other. **for the love of**, for the sake of. **love all**, in games, nothing scored on either side. **love-in-a-mist**, the fennel-flower, *Nigella damascena*. **love-in-idleness**, *n.* the pansy or heartsease, *Viola tricolor*. **love-lies-bleeding**, a species of amaranth, esp. *Amaranthus caudatus*. **to give, send one's love**, to give, send an affectionate message. **to fall in love**, to become enamoured. **to make love to**, to woo, to pay court or attentions to; to have sexual intercourse with. **love-affair**, *n.* a romantic or sexual attachment between two people, often temporary. **love-apple**, (*old-fashioned*) the tomato. **love-bird**, *n.* a short-tailed African parrot so called from the attachment it shows to its mate. **love-bite**, *n.* a temporary red or purple mark on the skin caused by a partner biting or sucking it during lovemaking. **love-child** *n.* an illegitimate child. **love-feast**, *n.* a religious meeting such as the agape. **love game**, *n.* a game in which the loser has not scored. **love-god**, *n.* Cupid. **love-knot**, *n.* an intricate bow or knot (a token of love). **love-letter**, *n.* a letter between lovers or professing love. **love-lock**, *n.* a curl or tress hanging at the ear or on the forehead. **love-lorn**, *a.* pining away for love. **love-making**, *n.* courtship; sexual play or intercourse between partners. **love-match**, *n.* a marriage for love, not other considerations. **love-nest**, *n.* a secret place where lovers meet. **love seat**, *n.* a small sofa for two people. **lovesick**, *a.* languishing with love. **love-sickness**, *n.* **love-song**, *n.* a song expressing love. **love-story**, *n.* a story dealing mainly with romantic love. **love-token**, *n.* a present in token of love. **lovable, loveable**, *a.* worthy of love; amiable. **lovableness**, *n.* **lovably**, *adv.* **loveless** (-lis), *a.* destitute of love; not loving; not loved. **lovelessly**, *adv.* **lovelessness**, *n.* **lovely**, *a.* beautiful and attractive, inspiring admiration and affection, tempting, delightful. *adv.* so as to excite love or admiration. *n.* a beautiful woman. **lovelily**, *adv.* **loveliness**, *n.* **lover**, *n.* one who loves, one fond of anything; a person having a sexual relationship, often extramarital; (*pl.*) a pair of sweethearts. **loverless** (-lis), *a.*

loverlike, *a.*, *adv.* **loverly**, *a.*, *adv.* **lovey** (i), *n.* (*coll.*) a term of endearment. **lovey-dovey** (-dŭv'i), *a.* (*coll.*) loving, (over-)affectionate. **loving**, *a.* **loving-cup**, *n.* a large two- or three-handled drinking-vessel passed round with wine at a banquet. **lovingly**, *adv.* **lovingness**, *n.*

low[1] (lō), *a.* (*comp.* **lower**[1], *superl.* **lowest**) not reaching or situated far up; not high or tall, below the usual or normal height; below or little above a given surface or level; not elevated; of the sun, moon etc., near the horizon; below the common standard in rank, condition, quality, character etc.; humble, degraded; dishonourable; not sublime; coarse, vulgar; not advanced in civilization; not high in organization; lacking in vigour, feeble; of sounds, not raised in pitch, deep, not loud or intense, soft; not large in amount, nearly exhausted; moderate, cheap; of or pertaining to the Low Church. *adv.* not on high; in or to a low position; deeply; at a low price; in a humble rank or position; with a subdued voice; on a poor diet. *n.* a low position or level; an area of low atmospheric pressure. **to bring low**, to reduce in wealth, position, health etc. **to lay low**, to overthrow. **to lie low**, (*coll.*) to keep quiet, to do nothing for the moment. **low birth**, *n.* humble parentage. **low born**, *a.* **low-brow**, *n.* a person making no claims to intellectuality. *a.* unintellectual. **low-browed**, *a.* having a low brow or forehead. **Low Church**, *n.* the evangelical party in the Church of England. **Low Churchman**, *n.* **low comedy**, *n.* comedy bordering on farce, hence **low comedian**. **Low Countries**, *n.pl.* a collective name for Belgium, Luxemburg and the Netherlands. **low-cut**, *a.* of a dress etc., cut low at the neck. **low-down**, *a.* degraded, mean, despicable. *n.* (*coll.*) the real facts. **Low Dutch** DUTCH. **low frequency**, *n.* a radio frequency lying between 300 and 30 kHz. **Low German** GERMAN. **low-key, -keyed**, *a.* of low intensity; undramatic; restrained. **lowland**, *n.* low-lying or level country. *a.* pertaining to a lowland or the Lowlands. **Lowlands**, *n.pl.* the eastern and southern or less mountainous parts of Scotland. **Lowlander**, *n.* **Low Latin** LATIN. **low latitudes**, *n.pl.* latitudes near the Equator. **low-level language**, *n.* computer programming language that corresponds more to machine code than to human language. **low life**, *n.* (*pl.* **lifes**) persons of a low position in life. **low-loader**, *n.* a road or rail vehicle with a low platform for heavy loads. **Low Mass**, *n.* Mass said without music and without elaborate ritual. **low-minded**, *a.* having a crude mind and character. **low-neck, -necked**, *a.* low-cut. **low-pitched**, *a.* having a low tone or key; of a roof with low angular elevation. **low profile**, *n.* a reserved or inconspicuous attitude or manner to avoid attention or publicity. **low-profile**, *a.* **low relief**, *n.* bas-relief. **low-rise**, *a.* of buildings, having only one or two storeys. **low-spirited**, *a.* dejected. **low-start**, *a.* of a mortgage with a lower than normal rate of repayment for the first few years. **Low Sunday, Week**, *n.* the Sunday or week next after Easter. **low-tech**, *a.* using simple technology; uncomplicated, unsophisticated. **low-tension**, *a.* having, generating or operating at a low voltage. **low tide**, *n.* the lowest point of the ebb tide; the level of the sea at ebb tide. **low-velocity**, *a.* applied to projectiles propelled at a comparatively low velocity and having a high trajectory. **low-voiced**, *a.* having a soft, gentle voice. **low water**, *n.* low tide; hence, **low-water mark**, *n.* **lower**[1], *comp. a.* **lower case**, *n.* (*Print.*) the small letters. **Lower Chamber, House**, *n.* the second of two legislative chambers, as the House of Commons. **lower deck**, *n.* the deck just above the hold of a ship; petty officers and men of a Navy. **lower**[2], *v.t.* to bring down in height, force, intensity, amount, price, estimation etc.; to haul or let down; to reduce the condition of. *v.i.* to become lower or less; to sink, fall. **lowermost**, *a.* **lowish**, *a.* **lowly**, *a.* humble, modest, unpretentious; low in size, rank or condition; inferior. *adv.* humbly, modestly. **lowliness**, *n.* **lowlily**, *adv.* **lowness**, *n.*

low[2] (lō), *v.i.* of cattle, to utter a mooing sound. *v.t.* to utter with such a sound. *n.* the moo of a cow. **lowing**, *n.*

lower[1,2] LOW[1].

lower[3] LOUR.

lox[1] (loks), *n.* a kind of smoked salmon.

lox[2] (loks), *n.* liquid oxygen, used in rocket fuels.

loxodrome (lok'sədrōm), *n.* RHUMBLINE; also **loxodromic curve, line, spiral** (-drom'-).

loyal (loi'əl), *a.* faithful, true, constant; faithful to one's sovereign, government or country. **loyalism**, *n.* **loyalist**, *n.* a patriotic supporter of sovereign or government; (**Loyalist**) in Northern Ireland, a Protestant who supports Ulster's union with Britain; in the American War of Independence, a colonial supporter of Britain; a republican supporter in the Spanish Civil War. **loyalize, -ise**, *v.t.* **loyally**, *adv.* **loyalty**, *n.*

lozenge (loz'inj), *n.* a rhombus or oblique-angled parallelogram; (*Her.*) a diamond-shaped bearing, appropriated to the arms of spinsters and widows; a confection or medicated sweet etc. **lozenge-shaped**, *a.* **lozenged**, *a.* shaped like a rhomb or diamond; arranged in series of lozenges in alternate colours; having diamond panes. **lozengewise**, *adv.* **lozengy**, *a.* (*Her.*) divided lozengewise.

LP, *n.* a *l*ong-*p*laying record, usu. 12 in. (30 cm) in diameter and designed to rotate at 33·3 revolutions per minute.

LPG, (*abbr.*) liquid petroleum gas.

Lr, (*chem. symbol*) lawrencium.

LSD, *n.* lysergic acid diethylamide, a hallucinogenic drug.

L.S.D., £.s.d., l.s.d., (*abbr.*) librae, solidi, denarii. [L, pounds, shillings, pence]

LSE, (*abbr.*) London School of Economics.

LSO, (*abbr.*) London Symphony Orchestra.

Lt, (*abbr.*) Lieutenant.

Ltd, (*abbr.*) limited liability.

Lu, (*chem. symbol*) lutetium.

lubber (lŭb'ə), *n.* a lazy, clumsy fellow; an awkward lout; a bad seaman. **lubber's line**, *n.* the mark inside a compass-case which shows the direction of the ship's head. **lubberlike**, *a.* **lubberly**, *a.*, *adv.* **lubberliness**, *n.*

lubra (loo'brə), *n.* (*Austral.*) an Aboriginal woman.

lubricate (loo'brikāt), *v.t.* to make smooth or slippery by means of grease, oil etc. in order to reduce friction; to bribe. **lubricant**, *n.*, *a.* **lubrication**, *n.* **lubricator**, *n.* one who or that which lubricates. **lubricity** (loobris'iti), *n.* smoothness; slipperiness; lasciviousness. **lubricious** (-brish'əs), *a.*

Lucan (loo'kən), *a.* pertaining to the evangelist St Luke.

lucarne (lookahn'), *n.* a dormer or garret window.

luce (loos), *n.* a pike (fish).

lucent (loo'sənt), *a.* shining, bright, luminous. **lucency**, *n.*

lucerne (loosœn'), *n.* a clover-like fodder-plant.

lucid (loo'sid), *a.* bright, radiant; clear, transparent, easily understood; sane. **lucidity** (-sid'-), **lucidly**, *adv.*

Lucifer (loo'sifə), *n.* the morning star; Satan, the chief of the rebel angels; a match tipped with combustible substance and ignited by friction. **lucifer-match**, *n.* **Luciferian** (-fiə'ri-), *a.*

lucifugous (loosif'ūgəs), *a.* of certain animals, shunning the light.

luck (lŭk), *n.* chance, as bringer of fortune, good or bad; what happens to one, fortune; good fortune, success; something supposed to bring good luck. **down on one's luck**, not having much luck. **to lockout**, (*sl.*) to be successful or fortunate, esp. by chance. **to luck into**, (*sl.*) to acquire or achieve by good fortune or chance. **to push one's luck** PUSH. **to try one's luck**, to attempt something. **tough luck**, an expression of sympathy. **luck-money, -penny**, *n.* a small sum returned to the buyer 'for luck' by the seller. **luckily**, *adv.* fortunately for. **luckiness**, *n.* **luckless** (-lis), *a.* unfortunate. **lucklessly**, *adv.* **lucky**, *a.* characterized or usually attended by good-luck; favoured by fortune; successful, esp. by chance; bringing luck.

lucky-bag, *n.* a lucky dip; a bag (of sweets etc.) bought without knowledge of the contents. **lucky dip,** *n.* a receptacle containing an assortment of articles, for one of which one dips blindly.
lucrative (loo'krətiv), *a.* profitable, bringing in money. **lucratively,** *adv.*
lucre (loo'kə), *n.* money; financial gain.
lucubrate (loo'kūbrāt), *v.i.* to study by lamplight; to produce lucubrations. **lucubration,** *n.* that which is composed at night; composition of a learned or too elaborate and pedantic character.
Luddite (lŭd'īt), *n.* a member of a band of workmen who organized riots (1811–16) for the destruction of machinery as a protest against unemployment; any opponent of technological change. **Luddism,** *n.* [supposedly after Ned Ludd, fl. 1779, a Leicestershire workman who destroyed machinery]
ludicrous (loo'dikrəs), *a.* exciting laughter or derision; ridiculous. **ludicrously,** *adv.* **ludicrousness,** *n.*
ludo (loo'dō), *n.* a game played with counters on a specially chequered board.
luff (lŭf), *n.* that part of a ship's bows where the timbers begin to curve in towards the stem; the weather-edge of a fore-and-aft sail; the part of a ship facing towards the wind. *v.i.* to steer a ship nearer the wind. *v.t.* to bring (a ship's head) or the head of (a ship) nearer the wind; to turn (the helm) so as to do this. **luff-tackle,** *n.* a large tackle composed of a double and single block.
lufta (lŭf'ə), LOOFAH.
Luftwaffe (looft'vahfə), *n.* the German Air Force before and during World War II.
lug[1] (lŭg), *n.* a large marine worm, burrowing in the sand, used for bait. **lugworm,** *n.*
lug[2] (lŭg), LUG-SAIL.
lug[3] (lŭg), *v.t.* (*past, p.p.* **lugged**) to drag, pull, esp. roughly or with exertion; (*fig.*) to drag in, to insert unnecessarily. *v.i.* to drag. *n.* a drag or tug.
lug[4] (lŭg), *n.* a projecting part, esp. of a machine, to hold or grip another part; (*coll.* or *Sc.*) the ear; an unlooped handle of a pot.
luge (loozh), *n.* a small one-man toboggan. *v.t.* to toboggan in one of these. [F]
Luger® (loo'gə), *n.* a type of German automatic pistol.
luggage (lŭg'ij), *n.* a traveller's suitcases etc. **luggage-van,** *n.* a railway carriage for luggage, bicycles, etc.
lugger (lŭg'ə), *n.* a small vessel rigged with lug-sails.
lug-sail (lŭg'sl), *n.* a four-cornered sail bent to a yard lashed obliquely to the mast.
lugubrious (ləgoo'briəs), *a.* mournful, dismal, funereal. **lugubriously,** *adv.* **lugubriousness,** *n.*
lukewarm (look'wawm), *a.* moderately warm; tepid; indifferent. *n.* one who is indifferent or unenthusiastic. **lukewarmly,** *adv.* **lukewarmness,** *n.*
lull (lŭl), *v.t.* to sooth to sleep, to calm, to quiet. *v.i.* to subside, become quiet. *n.* a temporary calm; an intermission or abatement. **lullaby** (-əbī), *n.* a song for lulling a child to sleep. *v.t.* to sing to sleep. **lullingly,** *adv.*
lulu (loo'loo), *n.* (*coll.*) an extremely good or bad person or thing.
lumbago (lŭmbā'gō), *n.* rheumatism in the lumbar region.
lumbar (lŭm'bə), *a.* pertaining to the portion of the body between the lower ribs and the upper part of the hipbone. **lumbar puncture,** *n.* the insertion of a needle between two lumbar vertebrae to withdraw cerebrospinal fluid.
lumber[1] (lŭm'bə), *v.i.* to move heavily, cumbrously or clumsily; to rumble. **lumbering,** *a.* **lumberingly,** *adv.* **lumbersome** (-səm), *a.*
lumber[2] (lŭm'bə), *n.* discarded articles of furniture etc. taking up room; useless and cumbersome things; (*N Am.*) timber sawn into marketable shape. *v.t.* to fill with lumber; to encumber, obstruct; to heap up in a disorderly way; to cut and prepare timber for the market. **lumber camp,** *n.* a lumberman's camp. **lumber-dealer,** *n.* **lumberjack, lumberman,** *n.* one who is employed in cutting trees etc. **lumber-jacket,** *n.* a loose-fitting jacket in a heavy, usu. chequered material that fastens up to the neck. **lumber-mill,** *n.* a saw-mill. **lumber-room,** *n.* a room for the storage of lumber. **lumber-yard,** *n.* a timber yard.
lumbrical (lŭm'brikəl), *a.* of or like an earthworm. **lumbriciform** (-bris'ifawm), *a.* resembling a worm.
lumen (loo'mən), *n.* (*pl.* **-mens, -mina** (-minə)) the SI unit of luminous flux, being the quantity of light emitted per second in a solid angle of one steradian by a uniform point-source having an intensity of one candela; (*Anat.*) the cavity of a tubular organ; a cavity within a plant cell wall. **luminal,** *a.* **luminance,** *n.* luminousness; a measure of the luminous intensity of any surface. **luminant,** *n., a.*
luminary (loo'minəri), *n.* any body yielding light, esp. a heavenly body; a famous or influential person. **luminesce** (-nes'), *v.i.* **luminescence,** *n.* the emission of light at low temperatures by processes other than incandescence, e g. by chemical action. **luminescent,** *a.* **luminosity** (-nos'-), *n.* **luminous,** *a.* emitting light; shining brightly, brilliant; lucid, enlightening. **luminous flux,** *n.* a measure of the rate of flow of luminous energy. **luminous intensity.** *n.* a measure of the amount of light radiated in a given direction from a point source. **luminous paint,** *n.* a paint containing phosphorescent compounds which cause it to glow in the dark after exposure to light. **luminously,** *adv.* **luminousness,** *n.*
lumme, lummy (lŭm'i), *int.* expressing surprise.
lummox (lŭm'əks), *n.* (*coll.*) a clumsy person.
lump[1] (lŭmp), *n.* a small mass of matter of no definite shape; a mass, quantity, heap; a swelling, protuberance; a heavy, stupid person. *v.t.* to put together in a lump, to form into a mass; to take collectively, to treat as all alike. *v.i.* to form or collect into lumps; to move (about) heavily or clumsily. **in the lump,** collectively; altogether. **the lump,** the collective group of self-employed workers in the building trade. **lump-sugar,** *n.* loaf sugar broken into small lumps. **lump sum,** *n.* the whole amount of money taken together. **lumpectomy** (-ek'təmi), *n.* the removal by surgery of a cancerous lump in the breast. **lumper,** *n.* one who lumps things together; a labourer who loads or unloads ships; a small contractor who takes work in the lump and puts it out. **lumping,** *a.* large, heavy; big, bulky. **lumpish,** *a.* like a lump; gross; inert; stupid. **lumpishly,** *adv.* **lumpishness,** *n.* **lumpy,** *a.* full of lumps; (*Naut.*) rough (of the sea). **lumpily,** *adv.* **lumpiness,** *n.*
lump[2] (lŭmp), *v.t.* (*coll.*) to put up with. **like it or lump it,** put up with it as there is no alternative.
lump[3], **lumpfish** (lŭmp), *n.* a suctorial fish of northern seas.
lumpen (lŭm'pən), *a.* (*coll.*) stupid, oafish; denoting a degraded section of any social group. **lumpenproletariat,** *n.* the very poorest section of the urban population, composed of criminals, vagabonds etc.
lunacy (loo'nəsi), *n.* unsoundness of mind, insanity; gross folly, senseless conduct.
luna moth (loo'nə), *n.* a large N American moth with crescent-shaped markings on its forewings.
lunar (loo'nə), *a.* of, pertaining to, caused or influenced by the moon; resembling the moon. *n.* a lunar distance or observation. **lunar cycle** CYCLE. **lunar distance,** *n.* the angular distance of the moon from the sun, a planet or a star, used at sea in finding longitude. **lunar month,** *n.* the period of a complete revolution of the moon, 29½ days, (*pop.*) four weeks. **lunar observation,** *n.* observation of the moon's distance from the sun or a star to find the longitude. **lunar year,** *n.* a period of twelve lunar months. **lunarian** (-neə'ri-), *n.* an inhabitant of the moon. **lunarist,** *n.* an investigator of the moon. **lunary,** *a.* lunar. *n.* the moonwort. **lunate** (-nāt), **luniform** (-nifawm), *a.* crescent-shaped. **lunation,** *n.* the period between two returns of

the moon, a lunar month.
lunatic (loo'nətik), *a.* insane; frantic, crazy, extremely foolish. *n.* an insane person. **lunatic asylum**, *n.* (*offensive*) formerly the name for a hospital for the care and treatment of the mentally ill. **lunatic fringe**, *n.* members of society or of a group holding extreme or fanatical views.
lunation LUNAR.
lunch (lŭnch), *n.* a midday meal; (*N Am.*) a snack. *v.i.* to take lunch. *v.t.* to provide lunch for. **luncheon** (-chən), *n.* lunch (in more formal usage). **luncheon meat**, *n.* a type of pre-cooked meat, usu. pork minced with cereal, served cold. **luncheon voucher**, *n.* a voucher given to employees which can be used to pay for food.
lune (loon), *n.* anything in the shape of a half-moon.
lunette (loonet'), *n.* a semicircular aperture in a concave ceiling; a crescent-shaped or semicircular space or panel for a picture or decorative painting; (*Fort.*) an advanced work of two faces and two flanks; a flattened watch-glass.
lung (lŭng), *n.* one of the two organs of respiration in vertebrates, situated on each side of the chest; an analogous organ in invertebrates. **lung-fish**, *n.* a dipnoan, having lungs as well as gills. **lung-power**, *n.* strength of voice.
lungwort, *n.* a lichen growing on the trunks of trees; a genus of the borage family, formerly held to be good for pulmonary diseases. **lunged**, *a.* **lungless** (-lis), *a.*
lunge[1] (lŭnj), *n.* a sudden thrust with a sword etc.; a sudden forward movement, a plunge. *v.i.* to make a lunge; to plunge or rush forward suddenly.
lunge[2], **longe** (lŭnj), *n.* a long rope or rein used in training horses. *v.t.* to drive a horse round in a circle at the end of a lunge.
lungi (lung'gi), *n.* a long cloth used in India as a loincloth or sash, sometimes as a turban.
luniform LUNAR.
lunisolar (loonisō'lə), *a.* pertaining to, or compounded of the revolutions of, the sun and the moon. **lunisolar period, year**, *n.* a period of 532 years found by multiplying the cycle of the sun by that of the moon.
lunula (loon'ūlə), *n.* a crescent-shaped mark, spot or part, esp. at the base of a fingernail. **lunular, lunulate** (-lət), **-lated** (-lātid), *a.* **lunule**, *n.*
Lupercal (loo'pəkəl), *n.* (*pl.* **-calia** (-kā'liə)) a Roman fertility festival in honour of the god Lupercus celebrated on 15 Feb. **Lupercalian**, *a.*
lupin[1], **lupino** (loo'pin), *n.* a leguminous plant with spikes of white or coloured flowers, grown in flower-gardens and for fodder. **lupinin** (-nin), *n.* (*Chem.*) a bitter glucoside obtained from lupins.
lupine[2] (loo'pin), *a.* pertaining to wolves; like a wolf.
lupoid, lupous LUPUS.
lupulin (loo'pūlin), *n.* the bitter essence of hops; a yellow granular aromatic powder containing that essence. **lupulite** (-līt), *n.* lupulin.
lupus (loo'pəs), *n.* a spreading tuberculous or ulcerous inflammation of the skin, usually of the face. **lupoid** (-poid), **lupous**, *a.*
lurch[1] (lœch), *n.* a losing position in the game of cribbage and some other games. **to leave in the lurch**, to leave in difficulties.
lurch[2] (lœch), *v.i.* of a ship, to roll suddenly to one side; to stagger. *n.* a sudden roll sideways, as of a ship; a stagger.
lurcher (lœ'chə), *n.* a dog supposed to be a cross between a collie and a greyhound.
lure (luə), *n.* an object resembling a fowl, used to recall a hawk; hence, an enticement, allurement. *v.t.* to attract or bring back by a lure; to entice.
Lurex[®] (loo'reks), *n.* (a fabric made from) a thin plastic-coated metallic thread.
lurid (loo'rid), *a.* of a pale yellow colour, wan, gloomy; ghastly, unearthly; shocking in detail; of a story etc., sensational; (*Bot.*) of a dirty brown colour. **luridly**, *adv.*

luridness, *n.*
lurk (lœk), *v.i.* to lie hid; to lie in wait; to be latent, to exist unperceived. **lurker**, *n.* **lurking**, *a.*
luscious (lŭsh'əs), *a.* delicious; sweet to excess; cloying, fulsome, over-rich in imagery, sensuousness etc.; voluptuous. **lusciously**, *adv.* **lusciousness**, *n.*
lush[1] (lŭsh), *a.* luxuriant in growth; succulent, juicy; luxurious. **lushness**, *n.*
lush[2] (lŭsh), *n.* (*sl.*) a heavy drinker, an alcoholic. *v.i.* to drink. **lushy**, *a.* drunk.
lust (lŭst), *n.* a powerful desire for sexual pleasure, lasciviousness; sensual appetite; passionate desire for. *v.i.* to have powerful or inordinate desire (for or after). **lustful**, *a.* **lustfully**, *adv.* **lustfulness**, *n.* **lustiness**, *n.* **lustily**, *adv.* **lusty**, *a.* full of health and vigour.
lustre[1] (lŭs'tə), *n.* brightness, splendour, luminousness, gloss, sheen; the reflection of a light; a chandelier ornamented with pendants of cut glass; a cotton, woollen or other fabric with a glossy surface; a glossy enamel on pottery etc.; illustriousness, radiant beauty. **lustreless** (-lis), *a.* **lustrous**, *a.* **lustrously**, *adv.*
lustre[2] (lŭs'tə), **lustrum** (-trəm), *n.* a purification; a period of five years. **lustral**, *a.* **lustrate**, *v.t.* to purify. **lustration**, *n.*
lustrine (lŭs'trin), **lustring** (-tring), *n.* a glossy silk fabric.
lustrum LUSTRE[2].
lusty LUST.
lute[1] (loot), *n.* a stringed instrument with a pear-shaped body and a long fretted fingerboard. **lute-string**, *n.* a string of a lute; a noctuid moth with string-like markings on its wings. **lutanist, lutenist** (-tənist), **lutist**, *n.* a luteplayer.
lute[2] (loot), *n.* a composition of clay or cement used to secure the joints of vessels and tubes, or as a covering to protect retorts etc. from fire. *v.t.* to seal up or coat with lute.
luteal (loo'tiəl), *a.* of or pertaining to the corpus luteum. **luteinize, -ise** (-tiə-), *v.t.*, *v.i.* to produce or form corpora lutea. **luteinization**, *n.* **luteinizing hormone**, *n.* a hormone secreted from the front lobe of the pituitary gland which stimulates, in females, ovulation and the development of corpora lutea and, in males, maturation of the interstitial cells of the testes and androgen production.
lute-string LUTE[1].
lutestring (loot'string), *n.* lustrine.
Lutetian (lootē'shiən), *a.* Parisian.
lutetium (lootē'shiəm), **lutecium** (-si-), *n.* an extremely rare metallic element, at. no. 71; chem. symbol Lu, of the lanthanides.
Lutheran (loo'thərən), *a.* of or belonging to Luther or his doctrines. *n.* a follower of Luther; a member of the Church based on Luther's religious doctrines. **Lutheranism, -therism**, *n.* [Martin *Luther*, 1483–1546, German Protestant reformer]
Lutine bell (loo'tēn), *n.* a bell recovered from the ship *Lutine* and rung at Lloyd's in London before important announcements, such as the loss of a vessel.
lutist LUTE[1].
lux (lŭks), *n.* (*pl.* **lux, luxes**) the SI unit of illumination equal to one lumen per square metre.
luxate (lŭk'sāt), *v.t.* to put out of joint, to dislocate. **luxation**, *n.*
luxe (luks), *n.* luxury, sumptuousness. [F]
luxury (lŭk'shəri), *n.* habitual indulgence in expensive pleasures; luxurious living; an expensive item or service, pleasant but unnecessary; luxuriousness. **luxuriant** (lŭgzū'riənt), *a.* abundant in growth; plentiful, profuse, exuberant; prolific, rank; ornate, extravagant. **luxuriance, -iancy**, *n.* **luxuriantly**, *adv.* **luxuriate**, *v.i.* to feed or live luxuriously; to revel, to indulge oneself to excess. **luxurious**, *a.* **luxuriously**, *adv.* **luxuriousness**, *n.*
luzerne (loozœn'), LUCERNE.

LV, (*abbr.*) luncheon voucher.
LW, (*abbr.*) long wave; low water.
lx, (*phys. symbol*) lux.
-ly (-li), *suf.* forming adjectives, as *ghastly, godly, manly*, or adverbs, as *badly, heavily, mightily*.
lycanthropy (likan'thrəpi), *n.* insanity in which the patient believes himself a wolf or other animal; belief in a magical transformation of people into wolves. **lycanthrope** (-kənthrōp), *n.* a werewolf, one suffering from lycanthropy. **lycanthropic** (-throp'-), *a.* **lycanthropist**, *n.*
lycée (lē'sä), *n.* a French State secondary school. [F]
lyceum (līsē'əm), *n.* a place devoted to instruction; an institution for literary instruction or mutual improvement by means of lectures, libraries etc.
lych, lych-gate etc. LICH.
lychee LITCHI.
lychnis (lik'nis), *n.* a genus of plants comprising the campions.
lycopod (lī'kəpod), *n.* a club-moss. **lycopodiaceous** (-pōdiā'shəs), *a.* **lycopodium** (-pō'diəm), *n.* a genus of perennial plants comprising the club-mosses; an inflammable yellow powder in the spore-cases of some species, used for making fireworks and as an absorbent in surgery.
Lycra[*] (lī'krə), *n.* a synthetic elastic fibre and material used in swimwear and other tight-fitting garments.
lyddite (lid'it), *n.* a powerful explosive composed mainly of picric acid.
Lydian (lid'iən), *a.* (*Mus.*) applied to one of the modes in Greek music, and the third ecclesiastical mode.
lye (lī), *n.* an alkaline solution leached from wood ashes or other alkaline substance; a detergent.
lying[1] (lī'ing), *n.* the act or habit of telling lies. *a.* telling lies; false, deceitful. **lyingly**, *adv.*
lying[2] (lī'ing), *n.* the act or state of being recumbent. **lying-in**, *n.* child-birth. **low-lying**, *a.* situated at a low level.
lyke-wake (līk'wāk), *n.* a night watch over a dead body.
lyme-grass (līm'grahs), *n.* a coarse grass grown in sand in order to bind it.
Lymeswold[*] (līmz'wōld), *n.* a mild, blue, soft cheese.
lymph (limf), *n.* water or any clear transparent fluid; the comparatively transparent, colourless, alkaline fluid in the tissues and organs of the body, bearing a strong resemblance to blood without the red corpuscles. **lymph gland, node**, *n.* any of the small localized masses of tissue distributed along the lymphatic vessels that produce lymphocytes. **lymphatic** (-fat'-), *a.* pertaining to, containing, secreting or conveying lymph; phlegmatic, sluggish. *n.* a vessel that conveys lymph. **lymphatic system**, *n.* the network of capillary vessels that conveys lymph to the venous system. **lymphocyte** (-fəsit), *n.* a type of white blood cell formed in the lymph nodes, which forms part of the body's immunological defence against infection. **lymphoid** (-foid), *a.* containing or resembling lymph. **lymphoma** (-fō'mə), *n.* (*pl.* **-phomas, -phomata** (-tə)) a tumour of lymphoid tissue. **lymphomatous**, *a.* **lymphomatoid** (-toid), *a.*

lynch (linch), *v.t.* to judge and punish, esp. to execute, by lynch law. **lynch law**, *n.* summary punishment without trial or upon trial by a self-appointed court. **lynching**, *n.*
lynx (lingks), *n.* one of several species of feline mammals characterized by tufted ear-tips, short tail and extremely sharp sight. **lynx-eyed**, *n.*, *a.* having sharp sight. **lyncean** (linsē'ən), *a.*
Lyon (lī'ən), *n.* the chief of the Scottish heralds, also called **Lyon King of Arms**.
lyophil (lī'əfil), **lyophilic** (-fil'-), *a.* of a colloid, easily dispersed in a solvent. **lyophilize, -ise** (-of'-), *v.t.* to freeze-dry. **lyophilization**, *n.* **lyophobe** (-fōb), **lyophobic** (-fō'-), *a.* of a colloid, not easily dispersed in a solvent.
lyre (līə), *n.* a stringed musical instrument of the harp kind, anciently used as an accompaniment to the voice. **lyre-bird**, *n.* an insectivorous Australian bird having the 16 tail-feathers of the male disposed in the form of a lyre.
lyrate (-rət), **-rated** (-rātid), *a.* shaped like a lyre. **lyric** (li'rik), *a.* suitable for singing; of poetry, expressing the individual emotions of the poet. *n.* a lyric poem; a song; (*pl.*) verses used in lyric poetry; the words of a popular song. **lyrical**, *a.* lyric; effusive. **lyrically**, *adv.* **lyricism** (-sizm), *n.* **lyrico-**, *comb. form.* **lyrist**, *n.*
lysis (lī'sis), *n.* the gradual decline in the symptoms of a disease; the destruction of cells by the action of a lysin. **lyse** (līz), *v.t.* to cause to undergo lysis. **-lysis** (-lisis), *comb. form.* denoting a breaking down, loosening or disintegration. **-lyse, -lyze** (-līz), *comb. form.* to cause or undergo loosening or decomposition through lysis. **lysergic acid diethylamide** LSD. **lysin** (-sin), *n.* a substance, esp. an antibody, which causes the disintegration of cells.
lysol (-sol), *n.* a mildly astringent solution of cresol and soap, used as a disinfectant.
-lyst (-list), **-lyte** (-līt), *comb. form* denoting a substance capable of being broken down.
-lytic, -lytical (-lit'i-), *comb. form* of or producing decomposition.

M

M¹, m¹, n. (pl. **M's, Ms, ms**) the 13th letter of the alphabet.
M², (abbr.) Mach; mega-; Majesty; (G) mark(s); million; (F) Monsieur; UK motorway; Roman numeral 1000.
m², (abbr.) male; married; masculine; maiden over; metre(s); mile(s); milli-; minute(s); month(s); moon; meridian.
ma (mah), n. childish shortening of MAMMA¹.
MA, (abbr.) Master of Arts; Massachusetts.
ma'am (mam, mahm, məm), n. contraction of madam (used by servants etc., and at Courts in addressing the queen or a royal princess).
MAC (mak), n. a European standard for satellite TV broadcasting. [acronym for *m*ultiplex *a*nalogue *c*omponents]
mac, mack (mak), n. (coll.) short for MACKINTOSH.
macabre (məkah'br), a. gruesome.
macadam (məkad'əm), n. broken stone for macadamizing; a road made by macadamizing. v.t. to macadamize. **macadamize, -ise**, v.t. to make, cover or pave (a road) with layers of broken stone so as to form a smooth hard surface.
macaroni (makərō'ni), n. an Italian pasta made of fine wheaten flour formed into long slender tubes.
macaroon (makəroon'), n. a small sweet cake or biscuit made of flour, almonds, sugar etc.
macassar oil (məkas'ə), n. an oil for the hair, orig. brought from *Macassar*, in the island of Celebes.
macaw (məkaw'), n. a S American parrot, of various species distinguished by their large size and beautiful plumage.
Mace®¹ (mās), n. a liquid causing the eyes to run and a feeling of nausea, used in self-defence, riot control etc.
mace² (mās), n. a mediaeval weapon shaped like a club with a heavy metal head, usu. spiked; an ornamented staff of office of analogous shape.
mace³ (mās), n. a spice made from the dried covering of the nutmeg.
macedoine (masədwan'), n. a dish of mixed vegetables.
macerate (mas'ərāt), v.t. to soften by steeping; to separate the parts of a digestive process; to make lean, to cause to waste away. v.i. to undergo maceration. **maceration**, n.
Mach MACH NUMBER.
machete (məshet'i, -shāt'i), n. a broad knife or cutlass used in tropical America as a weapon, to cut down sugar canes etc.
Machiavellian (mak'iəvel'yən), a. politically amoral; unscrupulous; devious, crafty, subtle. n. one who imitates the political principles of Niccolo *Machiavelli* (1469–1527), Florentine writer and statesman.
machinate (mak'ināt, mash'-), v.i. to contrive, to plot, to intrigue. **machination**, n. (often pl.). **machinator**, n.
machine (mashēn'), n. a mechanical apparatus by which motive power is applied; any mechanism, simple (as a lever or tool) or compound, for applying or directing force; a person who acts mechanically and without intelligence; any organization of a complex character designed to apply power of any kind. v.t. to effect by means of machinery; to print by machinery; to sew with a sewing-machine. v.i. to be employed in or upon machinery. **machine code, language**, n. a set of instructions for coding information in a form usable by a computer. **machine-gun**, n. a light piece of ordnance loaded and fired automatically. **machine-gunner**, n. **machine head**, n. (*Mus.*) a simple worm and tooth-wheel mechanism fitted to the head of a bass viol or other instrument for stretching the strings to the required pitch. **machine-made**, a. made by machinery, as distinct from *hand-made*. **machine-readable**, a. of data, in a form usable by a computer. **machine-shop**, n. a large workshop where machines are made or repaired. **machine-tool**, n. a machine for doing work with a tool, such as a chisel, plane, drill etc. **machine-work**, n. **machineable**, a. **machinery** (-nəri), n. (collect.) machines; the parts or mechanism of a machine; mechanical combination; any combination to keep anything in action or to effect a purpose; the means and combinations, esp. supernatural, employed to develop a plot in a poem etc. **machinist**, n. one who constructs machines; one versed in the principles of machinery; one who works or tends a machine, esp. a sewing-machine.
machismo MACHO.
Mach number (mak, mahkh, mahk), n. a number representing the ratio of the velocity of a body in a certain medium to the velocity of sound in the same medium.
macho (mach'ō, mah'hō), a. showing machismo. **machismo** (məkiz'mo, -kēz'-, chiz'-, chēz'-), n. aggressive arrogant assertiveness, often associated with masculinity.
mackerel (mak'ərəl), n. a well-known edible sea-fish, blue-green with dark stripes. a strong fresh breeze good for mackerel-fishing. **mackerel-sky**, n. a sky with small roundish masses of cirrocumulus, frequent in summer.
mackintosh, macintosh (mak'intosh), n. a water-proof material made of rubber and cloth; a coat or cloak made of this, a raincoat.
macramé (məkrah'mä, mak'rə-), n. a fringe or trimming of knotted thread or cord; knotted work.
macro (mak'rō), n. (pl. **macros**) a computer instruction that represents a sequence of instructions.
macr(o)- comb. form great, large (as distinct from small).
macrobiotic (-o'tik), a. of a diet, consisting chiefly of whole grains or of vegetables grown without chemical additives; concerning such a diet. **macrobiotics**, n.sing.
macrocephalic, -lous (makrōsifal'ik, -sefal'), a. large-headed. **macrocephalism** (-sef'-), n.
macrocosm (mak'rəkozm), n. the great world, the universe, as distinct from *microcosm;* the great whole of any body etc. **macrocosmic** (-koz'-), a.
macroeconomics (makrōekənom'iks, -ēkə-), n. sing. the study of economics on a large scale, e.g. of national economies.
macro instruction (mak'rō), MACRO.
macromolecule (mak'rōmol'ikūl), n. a large complex molecule formed from a number of simple molecules.
macron (mak'ron), n. a short horizontal line put over a vowel (as ē) to show that it is pronounced with a long sound.
macroscopic (makrəskop'ik), a. visible with the naked eye, as distinct from *microscopic*. **macroscopical**, a. **macroscopically**, adv.
macula (mak'ūlə), n. (pl. **-lae**, (-lē)) a spot, as on the skin, the surface of the sun etc. **macula lutea** (loo'tiə), n. a small yellowish spot near the centre of the retina of the

eye, where vision is especially acute. **macular**, *a*. **maculate** (lāt), *v.t.* to spot, to stain. *a.* (-lət), spotted, stained, impure. **maculation**, *n*. **macule** (mak'ūl), *n*. a spot, a stain; a mackle.

MAD, (*abbr.*) mutual assured destruction, a theory of nuclear deterrence based on the ability of each side to inflict an unacceptable level of damage on the other.

mad (mad), *a.* (*comp.* **madder**[1], *superl.* **maddest**) disordered in mind, lunatic, insane, crazy; furious, frantic, wildly excited; of animals, rabid; extravagant, infatuated, inflamed, wild, frolicsome; exceedingly foolish, very unwise; (*coll.*) enraged, annoyed, vexed. *v.i.* to be or go mad; to act madly. *v.t.* to make mad. **like mad**, (*coll.*) violently, wildly, excitedly. **madcap**, *a*. mad, eccentric. *n*. a person of wild and eccentric habits. **madhouse**, *n*. a lunatic asylum; a scene of confusion or uproar. **madman**, *n*. **madwoman**, *n. fem.* **madden**, *v.t., v.i.* **maddening**, *a*. **maddeningly**, *adv.* **madly**, *adv.* in an insane manner; (*coll.*) extremely. **madness**, *n*.

madam (mad'əm), *n*. a polite form of address to a woman; the formal opening of a letter to a woman; (*coll.*) a brothel keeper; (*coll.*) an impertinent girl.

madame (mədahm'), *n*. the French title for married women and mode of address to a woman.

madder[1] MAD.

madder[2] (mad'ə), *n*. a shrubby climbing-plant, the root of which is used in dyeing; the dye obtained from this plant.

made (mād), *a.* p., p.p. of MAKE. **made to measure**, clothes, footwear etc., made according to the customer's measurements. **to be, have it, made**, (*coll.*) to be certain of success; to be secure or comfortable. **made man, woman**, *n*. a person whose success is assured. **made-up**, *a*. of complexion etc., artificial; of a story etc., invented, coined.

Madeira (mədiə'rə), *n*. a fortified white wine made in Madeira. **madeira cake**, *n*. a light, spongy cake without fruit.

madeleine (mad'əlin), *n*. a small sponge cake, often coated with jam and coconut.

mademoiselle (madəmwəzel', -məzel', mamzel'), *n*. (*pl.* **mesdemoiselles** (mādəmwəzel'), a title given to an unmarried Frenchwoman; a French teacher or governess.

Madonna (mədon'ə), *n*. the Virgin Mary; a picture or statue of the Virgin Mary. **Madonna lily**, *n*. the white lily.

madras (mədras'), *n*. a large bright-coloured handkerchief worn on the head by Afro-Caribbeans; a fine cotton or silk fabric.

madrepore (mad'ripaw), *n*. a perforated coral or the animal producing such.

madrigal (mad'rigl), *n*. a short amorous poem; an unaccompanied vocal composition in five or six parts; (*loosely*) a part-song, a glee.

madroño (mədrō'nyō), *n*. a large evergreen tree, *Arbutus menziesii*, of N California, with hard wood, and edible berries.

Maecenas (mēsē'nəs, mī-), *n*. a munificent patron of literature or art. [a Roman citizen, *c.* 70–8 BC, patron of Horace and Virgil]

maelstrom (māl'strəm, -om), *n*. a dangerous whirlpool; a turmoil, an overwhelming situation.

maenad (mē'nad), *n*. (*pl.* **-nads**) in classical literature, a woman who took part in the orgies of Bacchus, a bacchante; a frenzied woman.

maestoso (mīstō'sō), *adv.* (*Mus.*) with dignity, grandeur and strength.

maestro (mīs'trō), *n*. (*pl.* **-tros, -tri**, (-strē)) a master in any art, esp. in music; a great composer or conductor.

mae west (mā west'), *n*. an airman's life-jacket [because, when inflated, it resembles the bust of the US actress Mae West, 1892–1980]

maffled (maf'ld), *a.* confused, muddled. **maffling**, *n*. a simpleton.

Mafia (maf'iə), *n*. a secret criminal society engaged in international organized crime, esp. in the US. **Mafioso** (ō'sō, -zō), *n*. (*pl.* **Mafiosi**, (-sē, -zē)) a member of the Mafia.

mag (mag), *n*. (*coll.*) short for MAGAZINE.

magazine (magəzēn', mag'-), *n*. a place for storage, a depot, a warehouse; a building or apartment for military stores, esp. ammunition; a storeroom for explosives etc. aboard ship; the chamber holding cartridges in an automatic firearm; a light-tight receptacle or enclosure for holding exposed or unexposed films or plates; a periodical publication or broadcast containing miscellaneous articles by different people.

magenta (məjen'tə), *n*. a brilliant purplish-crimson colour.

maggot (mag'ət), *n*. a grub, a worm, esp. the larva of the housefly, a whim, a crotchet; **maggoty**, *a*. **maggotiness**, *n*.

magi, magian etc. MAGUS.

magic (maj'ik), *n*. the pretended art of employing supernatural power to influence or control events; sorcery, witchcraft; any agency, power or action that has extraordinary results. *a*. pertaining to or used in magic; using magic; exercising supernatural powers; produced by magic; (*coll.*) used as a form of approval. *v.t.* to affect or move by magic. **black magic** BLACK. **white magic** WHITE. **magic lantern**, *n*. an apparatus which projects magnified images onto a screen. **magic mushroom**, *n*. a type of fungus containing a hallucinogenic substance. **magic square**, *n*. a series of numbers so disposed in a square that the totals, taken perpendicularly, horizontally or diagonally, are equal. **magical**, *a*. **magically**, *adv*. **magician** (jish'ən), *n*.

magisterial (majistē'riəl), *a*. pertaining to or befitting a teacher or magistrate; authoritative, commanding; dictatorial, domineering; oracular. **magisterially**, *adv*.

magistrate (maj'istrāt, -strət), *n*. a public officer, commissioned to administer the law, a Justice of the Peace. **magistrates' court**, *n*. a court of summary jurisdiction for minor offences and preliminary hearings. **magistracy**, **magistrateship**, **magistrature**, *n*.

Maglemosian (maglimō'ziən), *a*. of a transitional culture between the Palaeolithic and Neolithic, represented by finds at Maglemose in Denmark.

magma (mag'mə), *n*. (*pl.* **-mas, -mata** (-tə)) a thin paste or suspension of mineral or organic matter dispersed in a liquid; the molten semi-fluid rock below the earth's crust.

Magna Carta (mag'nə kah'tə), *n*. the Great Charter of English liberties sealed by King John on 15 June 1215; any fundamental constitution guaranteeing rights and privileges.

magna cum laude (mag'nə kum law'di, low'dā), with great distinction.

magnanimous (magnan'iməs), *a*. great-minded, elevated in soul or sentiment; brave, generous. **magnanimity** (nim'-), *n*. **magnanimously**, *adv*.

magnate (mag'nāt), *n*. a person of rank, distinction or great wealth.

magnesia (magnē'shə, -zhə), *n*. oxide of magnesium; a white alkaline antacid earth; hydrated carbonate of magnesia, used as an antacid and laxative. **magnesian**, *a*.

magnesium (magnē'ziəm, -zhəm, -shəm), *n*. a divalent metallic element, at. no.12; chem. symbol Mg, which burns with a dazzling white flame.

magnet (mag'nit), *n*. the lodestone; a body, usu. of iron or steel, to which the properties of the lodestone, of attracting iron and pointing to the poles, have been imparted; a thing or person exercising a powerful attractive influence. **magnetic** (-net'-), *a*. pertaining to a magnet or magnetism; having the properties of a magnet; attractive; mesmeric. *n*. any metal capable of receiving the properties of the lodestone; (*pl.*, sing. in constr.) the science or principles of magnetism. **magnetic battery**, *n*. a combination of magnets with their poles similarly arranged. **magnetic dip**, *n*. the angle between the earth's magnetic field and the horizontal. **magnetic disk** DISK. **magnetic equator**, *n*. a line round the globe where the magnetic

needle has no dip. **magnetic field**, *n.* a field of force surrounding a permanent magnet or electric current. **magnetic fluid**, *n.* a fluid formerly supposed to account for magnetism. **magnetic flux**, *n.* a measure of the strength of a magnetic field over a given area. **magnetic friction**, *n.* the reaction of a strong magnetic field on an electric discharge. **magnetic iron**, *n.* magnetite. **magnetic mine**, *n.* a mine detonated by the approach of a metal ship. **magnetic needle**, *n.* a slender poised bar of magnetized steel, as in the mariner's compass, pointing north and south. **magnetic north, south, or pole**, *n.* two nearly opposite points of the earth's surface where the magnetic needle dips vertically. **magnetic resonance**, *n.* the vibration of electrons, atoms, molecules or nuclei in a magnetic field in response to various radiation frequencies. **magnetic storm**, *n.* a disturbance of the earth's magnetic field. **magnetic tape**, *n.* plastic tape impregnated with magnetic particles used for the recording and reproduction of sound and television pictures or computer data. **magnetical** (-net'-), *a.* **magnetically**, *adv.* **magnetism** (mag'nitizm), *n.* the property whereby certain bodies, esp. iron and its compounds, attract or repel each other according to certain laws; the science treating of this property, its conditions or laws; the attractive power itself; personal attractiveness, charm. **magnetite** (magnitīt), *n.* magnetic oxide of iron. **magnetize, -ise**, *v.t.* to communicate magnetic properties to; to attract as with a magnet; to mesmerize. *v.i.* to become magnetic. **magnetization**, *n.* **magnetizer, -iser**, *n.*
magneto (magnē'tō), *n.* (*pl.* **-tos**) a magneto-electric machine (esp. the igniting apparatus of an internal-combustion engine). **magneto-**, *comb. form.* pertaining to magnetism or magnetic properties. **magnetoelectricity**, electricity generated by the inductive action of magnets; the science treating of such electricity. **magnetometer** (magnitom'itə), *n.* a device for measuring the intensity or direction of a magnetic field, esp. of the earth. **magnetometry**, *n.* **magneton** (mag'niton, magnē'ton), *n.* the unit of magnetic moment. **magnetron** (mag'nitron), *n.* a thermionic tube for generating very high frequency oscillations.
Magnificat (magnif'ikat), *n.* the song of the Virgin Mary (Luke i.46–55), so called from the first word in the Latin version; a setting of the same to music.
magnification MAGNIFY.
magnificent (magnif'isənt), *a.* grand in appearance, majestic, splendid; characterized by sumptuousness, luxury, splendour or generous profusion; (*coll.*) first-rate, excellent. **magnificence**, *n.* **magnificently**, *adv.*
magnifico (magnif'ikō), *n.* a grandee, orig. of Venice.
magnify (mag'nifī), *v.t.* to increase the apparent size of (an object) as with an optical instrument; to make greater, to increase; to extol, to glorify; to exaggerate. *v.i.* to increase the apparent size of objects. **magnifying glass**, *n.* an optical lens for magnifying objects. **magnification**, *n.* **magnifier**, *n.*
magniloquent (magnil'əkwənt), *a.* using high-flown, pompous or bombastic language. **magniloquence**, *n.* **magniloquently**, *adv.*
magnitude (mag'nitūd), *n.* size, bulk, extent, quantity, amount; anything that can be measured; importance; the order of brilliance of a star. **of the first magnitude**, among the best, worst, most important etc. of its kind.
Magnolia (magnō'liə), *n.* a genus of beautiful flowering trees or shrubs, chiefly N American.
magnox (mag'noks), *n.* an alloy of magnesium and aluminium used in fuel containers in certain nuclear reactors. **magnox reactor**, *n.*
magnum (mag'nəm), *n.* (*pl.* **-nums**) a wine bottle containing the equivalent of two normal bottles (about 1½ litres).
magnum opus (magnəm ō'pəs, op'-), *n.* the greatest work of a writer, painter etc.

magpie (mag'pī), *n.* a well-known chattering bird with black and white plumage; a chatterer; a person who collects and hoards trifles; in rifle-shooting. a shot that hits the outermost division but one of the target.
maguey (mag'wā), *n.* a type of tropical agave plant whose leaves yield fibre used to make an alcoholic drink.
magus (mā'gəs), *n.* (*pl.* **-gi**, (-jī)) a member of the priestly caste among the Medes and Persians; a magician. **the Magi**, the three holy men of the East who brought presents to the infant Christ. **magian** (-jiən), *n.*, *a.* **magianism**, *n.*
Magyar (mag'yah), *n.* one of the dominant ethnic groups in Hungary; the Hungarian language; a Magyar blouse or bodice. *a.* pertaining to the Magyars or their language. **Magyarism**, *n.*
maharajah, maharaja (mah·hərah'jə), *n.* a title assumed by some Indian princes. **maharani, maharanee** (-ni), *n.* a princess; the wife of a maharajah.
maharishi (mah·hərish'i, -rē'shi), *n.* a Hindu religious teacher.
mahatma (məhat'mə), *n.* in Buddhism, an adept of the highest order.
Mahdi (mah'di), *n.* the Muslim messiah (*hist.*) a title assumed by leaders of insurrection in the Sudan. **Mahdism**, *n.* **Mahdist**, *n.*
mahjong, mahjongg (mahjong'), *n.* a Chinese table game played with 144 pieces called tiles.
mahlstick (mawl'stik), MAULSTICK.
mahoe (məhō'i), *n.* the New Zealand white-wood tree.
mahogany (məhog'əni), *n.* the hard, fine-grained wood of *Swietenia mahogani*, a tree of tropical America, largely used in making furniture; the tree itself; applied also to other trees yielding similar wood; a dining-table; the colour of mahogany, reddish-brown.
Mahomedan, -etan, (məhom'idən, -tən), etc. MOHAMMEDAN.
mahout (məhowt'), *n.* an elephant-driver or keeper.
Mahratta (məraht'ə), MARATHA.
mahseer (mah'siə), *n.* a large and powerful East Indian river-fish somewhat like the barbel.
maid (mād), *n.* a girl, a young unmarried woman, a virgin; a female servant. **maid of all work**, a general servant. **maid of honour**, an unmarried lady attending upon a royal personage; (*esp. N Am.*) an unmarried attendant on a bride; a variety of cheese-cake. **old maid**, an elderly spinster. **maid-servant**, *n.*
maiden (mā'd(ə)n), *n.* a girl; a spinster; an apparatus for washing linen. *a.* of or pertaining to a maid; unmarried; of female animals, unmated; first, new, unused, untried; of a city or fortress, never captured; of a horse, never having won a prize; of a race, open to such horses; **maiden assize**, *n.* an assize at which there are no cases. **maiden name**, *n.* the surname of a woman before marriage. **maiden over**, *n.* (*Cricket*) an over in which no runs are scored. **maiden speech**, *n.* the first speech made by a member of Parliament in the House. **maiden voyage**, *n.* a first voyage (of a ship). **maidenhead, maidenhood**, the state of being a maid or virgin, virginity; the virginal membrane. **maidenish**, *a.* **maidenlike**, *a.* **maidenly**, *a.*, *adv.* **maidenliness**, *n.*
maidenhair (mā'dnheə), *n.* a fern with delicate fronds. **maidenhair tree**, *n.* a gingko.
maigre (mā'gə), *a.* of food, esp. soup, suitable for fast days, not made from meat nor containing gravy; applied to fast days. *n.* a large Mediterranean fish.
mail [1] (māl), *n.* defensive armour for the body, formed of rings, chains or scales; any defensive covering. *v.t.* to invest in or as in mail. **mailed**, *a.* clad in mail. **mailed fist**, *n.* the application of physical force.
mail [2] (māl), *n.* a bag for the conveyance of letters etc.; the letters etc. conveyed by the post; the system of conveying letters etc., the post, esp. for abroad; a mail-train or ship.

maillot   499   **make**

v.t. to send by mail, to post. **mail-coach**, *n.* **mailman**, *n.* (*N Am.*) a postman. **mail-merge**, *n.* the automatic merging of names and addresses from a computer file with the text of a letter etc. **mail order**, *n.* the ordering of goods to be sent by post. **mail-train**, *n.* **mailable**, *a.* **mailing list**, *n.* a list of names and addresses of people to whom letters, advertising material are to be posted.
maillot (mī'ō), *n.* tights for a ballet-dancer; a tight-fitting swimsuit.
maim (mām), *v.t.* to deprive of the use of a limb; to cripple, to mutilate.
main[1] (mān), *a.* of force, concentrated or fully exerted; principal, chief, most important. **main-boom**, *n.* the lower spar of a small vessel on which the mainsail is extended. **mainbrace**, *n.* a brace attached to the mainyard of a sailing ship. **to splice the mainbrace**, to serve an extra rum ration (on a ship). **main chance** CHANCE. **mainframe**, *n.* a large, powerful computer; the central processing and storage unit of a computer. **mainland**, *n.* the principal body of land as opposed to islands etc. **mainlander**, *n.* **main line**, *n.* a primary railway route. **mainline**, *v.t.*, *v.i.* (*sl.*) to inject (a narcotic drug etc.) into a vein. **mainliner**, *n.* **mainmast**, *n.* the principal mast of a ship. **mainsail** (-sāl, -sl), *n.* a sail bent to the main-yard of a square-rigged ship; the sail set on the after part of the mainmast of a fore-and-aft rigged vessel. **mainstay**, *n.* the stay from the main-top to the foot of the foremast; the chief support. **main store**, *n.* the central storage facility of a computer. **mainly**, *adv.* principally, chiefly; in the main; greatly, strongly.
main[2] (mān), *n.* strength, force, violent effort; the main or high sea, the ocean; a chief sewer, conduit, conductor, electric cable etc. **in the main**, for the most part. **Spanish Main** SPANISH.
main[3] (mān), *n.* a throw at dice, or a number (5–9) called by the caster before throwing; a match at cock-fighting; a match in various sports.
mainstream (mān'strēm), *n.* the most important aspects of a culture, society etc. *a.* concerning the mainstream; of jazz music, of the type prevalent between early and modern jazz.
maintain (māntān'), *v.t.* to hold, preserve or carry on in any state; to sustain, to keep up; to support, to provide with the means of living; to keep in order, proper condition or repair; to assert, to affirm, to support by reasoning, argument etc. **maintainable**, *a.* **maintainer**, *n.* **maintenance**, (mān'tənəns), *n.* the act of maintaining; means of support; (*Law*) an officious intermeddling in a suit in which the person has no interest. **cap of maintenance** CAP[1]. **maintenance man**, *n.* a workman employed to keep machines etc., in working order.
maiolica MAJOLICA.
maisonette, maisonnette (māzənet', -sə-), *n.* part of a house let separately; a small house.
maître d'hôtel (metr dōtel'), *n.* (*pl.* **maitres d'hôtel**, metrə) a head waiter; a major-domo.
maize (māz), *n.* an American ceral with yellow edible grains, also called Indian corn.
Maj., (*abbr.*) Major.
majesty (maj'əsti), *n.* the quality of inspiring awe or reverence; impressive dignity, grandeur, stateliness; sovereign power and dignity, esp. (*with poss. pron.*); a title of kings, queens and emperors; in religious art, a representation of God. **majestic -al** (-jes'-), *a.* **majestically**, *adv.*
Maj.-Gen., (*abbr.*) Major-General.
majolica (məjol'ikə), *n.* a fine enamelled Italian pottery, said to have come orig. from Majorca, or an imitation of this.
major (mā'jə), *a.* greater in number, quantity, extent or importance; of considerable importance; serious; main, principal, (*Mus.*) standard, normal, applied to a third consisting of four semitones; of full legal age (18 years).
n. the first premise of a regular syllogism containing the major term; a person of full legal age; an officer next above captain and below lieutenant-colonel; (*N Am.*) a subject of specialization at a college or university; a person specializing in such a subject. **major axis**, *n.* the axis passing through the foci (in a conic section). **major-domo**, (-dō'mō), *n.* (*chiefly It. and Sp.*) the chief officer of a royal or princely household; one who takes charge of a household, a steward. **major-general**, *n.* an officer commanding a division, ranking next below lieutenant-general. **major interval**, *n.* (*Mus.*) an interval greater by a semitone than the minor interval of the same denomination. **major league**, *n.* a league of the highest classification in US sport, esp. baseball. **major mode**, *n.* (*Mus.*) the mode in which the third and sixth tones of the scale form major intervals with the key-note. **major premise**, *n.* in logic, the premise containing the major term. **major suit**, *n.* in contract bridge, spades or hearts, which have a higher value than clubs and diamonds. **major term**, *n.* in logic, that term which forms the predicate of the conclusion.
majorette (-ret'), *n.* one of a group of girls who march in parades twirling batons, playing instruments etc. **majority** (-jor'-), *n.* the greater number; the greater part, more than half; the amount of the difference between the greater and the less number, esp. of votes in an election; full age; rank of major. **the majority**, the dead. **to join the majority**, to die. **majority verdict**, *n.* one reached by a majority of a jury.
majuscule (maj'əskūl), *n.* (*Palaeont.*) a capital or large letter, as in Latin MSS. before the introduction of minuscules.
make (māk), *v.t.* (*past, p.p.,* **made**) to frame, construct, produce; to bring into existence, to create; to give rise to, to effect, to bring about; to execute, to perform, to accomplish (with nouns expressing action); to result in, to cause to be or become; to compose (as a book, verses etc.); to prepare for use; to establish, to enact; to raise to a rank or dignity; to constitute, to form, to become, to turn out to be; to gain, to acquire; to move or proceed (towards etc.); (*Cards*) to win (a trick) or cause (a card) to win, to shuffle; to score; to cause, to compel (to do); to cause to appear, to represent to be; to reckon, to calculate or decide to be; to conclude, to think; to reach the end of; to amount to, to serve for; to travel over (a distance etc.); to fetch, as a price, (*Naut.*) to come near; to arrive at; to infuse (tea); (*sl.*) to succeed in seducing. *v.i.* to go, move, tend or lie (in a specified direction); to contribute, to have effect (for or to); to rise, to flow (of the tide); (*usu. with a.*) to do, to act in a specified way, as *make bold*. *n.* form, shape; arrangement of parts; making; style; disposition, mental or moral constitution; making of electrical contact, completion of a circuit. **on the make**, (*coll.*) intent on personal profit, after the main chance. **to make account of**, to esteem; to consider. **to make against**, to be unfavourable to, to tend to injure. **to make as if**, to pretend, to feint. **to make at**, to attack. **to make away**, to hurry away. **to make away with**, to get rid of, to kill; to waste, to squander. **to make believe** BELIEVE. **to make bold** BOLD. **to make do (with)**, to be satisfied with (something) not completely adequate. **to make for**, to conduce to; to corroborate; to move toward; to attack. **to make free**, to venture (to). **to make free with**, to treat without ceremony. **to make good** GOOD. **to make hay of** HAY[1]. **to make headway**, to advance. **to make it**, (*coll.*) to reach an objective; to succeed. **to make light of** LIGHT[2]. **to make like**, (*esp. N Am.*) to pretend; to imitate. **to make love** LOVE. **to make merry**, to feast, to be jovial; to make much of, to treat with fondness or favour; to treat as of great importance. **to make no doubt**, to be sure. **to make of**, to understand, interpret; to attach a specified degree of importance to. **to make off**, to run away; to abscond. **to make out**, to understand, to decipher; to prove, to estab-

lish; to claim or allege; to draw up; (coll.) to be successful; (N Am., coll.) to engage in necking or petting; (sl.) to have sexual intercourse. **to make over**, to transfer. **to make place, room**, to move so as to leave space (for). **to make sail**, to set more sails; to set sail. **to make sure of**, to consider as certain. **to make the grade** GRADE. **to make the most of**, to use to the best advantage. **to make up**, to compose; to compound; to collect together; to complete; to supply (what is wanting); to compensate; to settle, to adjust; to repair; of an actor, to dress up, to prepare the face to represent a character; to apply cosmetics to the face; to fabricate, to concoct; to arrange (as type) in columns or pages. **to make up one's mind**, to decide, to resolve. **to make up to**, to make advances to. **to make water**, to urinate; (Naut.) to leak. **to make way**, to make room, to open a passage; to progress. **to make with**, (N Am., coll.) to show, produce. **to make words**, to multiply words; to raise a quarrel. **make-believe**, n. a pretending, a pretence, a sham. a. unreal; counterfeit. v.t., v.i. to pretend. **makeshift**, n. a temporary expedient. a. used as a makeshift. **make-up**, n. the arrangement of type into columns or pages; the manner in which an actor's face is made to represent a character; the material used for this; a made-up story, a fiction; cosmetics for use on the face; a person's character or temperament. **make-weight**, n. that which is thrown into a scale to make weight; a stop-gap; anything that counterbalances, a counterpoise. **maker**, n. one who makes; (**Maker**) the Creator, God. **to meet one's Maker**, to die. **making**, n. the act of constructing, producing, causing etc.; possibility or opportunity of success or full development; (pl.) composition, essential qualities; (pl.) profits, earnings. **in the making**, gradually developing or being made. **making-up**, n. balancing of accounts.

mako (mah'kō), n. a small New Zealand tree; a kind of shark.

mal(e)- comb. form. bad(ly); evil; faulty; abnormal.

malacca, malacca cane (məlak'ə), n. a palm-stem used as a walking-stick.

malachite (mal'əkīt), n. a bright green monoclinic carbonate of copper.

maladjusted (malǝjŭs'tid), a. unable to adjust oneself to the physical or social environment. **maladjustment**, n.

maladministration (malǝdministrā'shǝn), n. defective of vicious management, esp. of public affairs. **maladminister**, v.t.

maladroit (malǝdroit'), a. awkward, clumsy. **maladroitly**, adv.

malady (mal'ǝdi), n. (pl. **-dies**) a disease, an ailment, esp. a lingering or deep-seated disorder; a moral defect or disorder.

Malaga (mal'ǝgǝ), n. sweet white wine imported from Malaga in Spain.

Malagasy (malǝgas'i), a. of or pertaining to Madagascar or its inhabitants or language, n. (pl. **Malagasy**) a native or the language of Madagascar.

malaise (malāz'), n. a feeling of uneasiness, mild depression or sickness.

malamute MALEMUTE.

malaprop (mal'ǝprop), **malapropism** (-izm), n. grotesque misapplication of words; a word so misapplied. **malapropian** (-prop'-, -prō'-), a. [Mrs *Malaprop* in Sheridan's *The Rivals*.

malapropos (mal'aprǝpō'), adv. unseasonably, unsuitably, out of place. a. unseasonable etc. n. an unseasonable or inopportune thing, remark, event etc.

malar (mā'lǝ), a. pertaining to the cheek or cheek-bone. n. the bone which forms the prominence of the cheek.

malaria (mǝleǝ'riǝ), n. the unpleasant, harmful air arising from marshy districts, formerly believed to produce fevers etc.; applied to various kinds of fever of an intermittent and remittent nature, now known to be due to a parasite introduced by the bite of mosquitoes. **malarial, -rian,**

-rious, a.

malark(e)y (mǝlah'ki), n. (esp. N Am., coll.) foolish or insincere talk; nonsense.

malassimilation (malǝsimilā'shǝn), n. imperfect assimilation, esp. of nutriment.

malate MALIC.

Malathion[R] (malǝthī'ǝn), n. an insecticide used for houseflies and garden pests.

Malay (mǝlā'), a. of or pertaining to the predominant race in Malaysia and Indonesia. n. a member of this race; their language. **Malayan,** n., a.

Malayala(a)m (malǝyah'lǝm, mǝli'ǝlǝm), n. the language of Malabar, a Dravidian dialect akin to Tamil. **Malayalim,** n.pl. the Dravidians of Malabar speaking this.

malcontent (mal'kǝntent), a. discontented, esp. with the government or its administration. n. one who is discontented, esp. with the government. **malcontented** (-tent'-), a. **malcontentedly**, adv. **malcontentedness**, n.

mal de mer (mal dǝ mœ, meǝ), seasickness. [F]

male (māl), a. pertaining to the sex that begets young or has organs for impregnating ova; of organs, adapted for fertilization; of flowers, having stamens but no pistil; consisting of or pertaining to individuals of this sex; (*Mech.*) designed for entering a correlative female part; masculine, virile. n. one of the male sex; a plant, or part of a plant, that bears the fecundating organs. **male chauvinist (pig)**, n. a man with an arrogant belief in the superiority of the male sex. **male fern**, n. a fern with the fronds clustered in a crown. **male menopause**, n. a (supposed) period in a man's middle life when he experiences an emotional crisis focused on diminishing sexual prowess. **male screw**, n. one whose threads enter the grooves of a corresponding screw.

malediction (malǝdik'shǝn), n. a curse, an imprecation. **maledictory,** a.

malefactor (mal'ǝfaktǝ), n. an evil-doer, a criminal. **malefaction** (fak'-), n. **maleficent** (mǝlef'isǝnt), a. hurtful, mischievous, causing evil (to). **maleficence** (-lef'-), n.

maleic (mǝlē'ik), a. applied to an acid obtained by the dry distillation of malic acid.

malemute, n. an Eskimo dog.

malevolent (mǝlev'ǝlǝnt), a. wishing evil or injury to others; ill-disposed, envious, malicious, spiteful. **malevolence,** n. **malevolently,** adv.

malfeasance (malfē'zǝns), n. evil-doing, esp. illegal conduct by a public official.

malformation (malfǝmā'shǝn), n. faulty formation; a faulty structure or irregularity of form. **malformed** (-fawmd'), a.

malfunction (malfŭngk'shǝn), n. defective function or operation. v.i. to operate defectively.

malic (ma'lik), a. of malic acid, derived from fruit. **malate** (mal'āt), n. a salt or ester of malic acid.

malice (mal'is), n. a disposition to injure others, active malevolence; (*Law*) a premeditated design to do evil or injure another.

malice aforethought, n. (*Law*) a premeditated wish to commit an illegal act, esp. murder. **malicious,** a. **maliciously,** adv. **maliciousness,** n.

malign (mǝlīn'), a. unfavourable, pernicious, malignant, hurtful; malevolent. v.t. to speak evil of, to slander. **maligner,** n. **malignly,** adv.

malignant (mǝlig'nǝnt), a. actuated by extreme enmity or malice; exercising a pernicious influence, virulent; of a disease, tumour etc., resisting treatment and threatening life. **malignancy,** n. **malignantly,** adv. **malignity,** n.

malinger (mǝling'gǝ), v.i. to pretend illness in order to shirk work. **malingerer,** n.

mall (mawl, mal), n. a public walk, orig. a place where pall-mall was played; a street or area of shops reserved for pedestrians.

mallard (mal'ǝd, -lahd), n. a wild drake; a wild duck; the flesh of this.

malleable (mal'iəbl), *a.* capable of being rolled out or shaped by hammering without being broken; easily influenced by outside forces, pliant. **malleability** (-bil'-), **malleableness**, *n.*
mallee (mal'i), *n.* one of various dwarf species of eucalyptus growing in the deserts of Victoria and S Australia. **mallee-bird, -fowl, -hen**, *n.* a mound-bird. **mallee-scrub**, *n.*
mallemuck (mal'imŭk), *n.* the fulmar.
malleolus (məlē'ələs), *n.* one of two bony processes extending either side of the ankle. **malleolar** (-lē'-, mal'-), *a.*
mallet (mal'it), *n.* a light hammer, usu. of wood; a long-handled wooden one for striking the ball in croquet or polo.
malleus (mal'iəs), *n.* one of the small bones of the middle ear. **malleiform** (mal'iə-), *a.*
mallow, -lows, (mal'ō. -z), *n.* a plant of various species belonging to the genus *Malva*, usu. with pink or mauve flowers and hairy stems and foliage.
malm (mahm), *n.* a soft, friable chalky rock or loam, used with clay and sand for brick-making. *v.t.* to mix (clay, chalk etc.) to make malm for bricks; to cover brick-earth with this.
malmsey (mahm'zi), *n.* a strong sweet white wine now chiefly made in the Canaries and Spain.
malnutrition (malnūtrish'ən), *n.* insufficient or defective nutrition.
malodorous (malō'dərəs), *a.* having an unpleasant smell. **malodour**, *n.* an offensive odour.
malpractice (malprak'tis), *n.* illegal or immoral conduct, esp. improper treatment of a case by a physician, lawyer etc.
malpresentation (malprezəntā'shən), *n.* an abnormal position of the foetus at birth.
malt (mawlt), *n.* grain, usually barley, steeped in water and fermented, dried in a kiln and used for brewing and distilling; malt-liquor. *a.* pertaining to, containing or made of malt. *v.t.* to convert into malt; to treat with malt. *v.i.* to be converted into malt; (*facet.*) to drink malt-liquor. **malt extract**, *n.* a thick, sticky liquid made from malt, taken as a health food. **malt-floor**, *n.* the floor in a malthouse on which the grain is spread to germinate. **malt-horse**, *n.* a horse employed in grinding malt; a dull, stupid fellow. **malt-house, maltings**, *n.* building where malt is prepared and stored. **malt-liquor**, *n.* liquor made from malt by fermentation, beer, stout etc. **malt whisky**, *n.* whisky distilled from malted barley. **maltings** MALT-HOUSE. **maltster**, *n.* a man whose occupation is to make malt. **malty**, *a.*
Maltese (mawtēz'), *a.* pertaining to Malta or its inhabitants. *n.* a native of Malta; the Maltese language or the people; a Maltese dog. **Maltese cross** CROSS[1]. **Maltese dog**, *n.* a small variety of spaniel with long silky hair.
Malthusian (malthū'ziən), *a.* pertaining to or supporting the teachings of Malthus. *n.* a follower of Malthus; one who holds that some check is necessary to prevent overpopulation. **Malthusianism**, *n.* [T.R. *Malthus*, 1766–1834, economist]
maltose (mawl'tōs), *n.* a sugar obtained by the action of malt or diastase on starch paste.
maltreat (maltrēt'), *v.t.* to ill-treat; to abuse. **maltreater**, *n.* **maltreatment**, *n.*
maltster, malty MALT.
malvaceous (malvā'shəs), *a.* belonging to or resembling the genus *Malva* or the family Malvaceae, including the mallows, cotton etc.
malversation (malvəsā'shən), *n.* fraudulent conduct or corruption in a position of trust, esp. corrupt administration of public funds.
mam (mam), *n.* (*dial.*) mother.
mama (məmah'), MAMMA[1].
mamba (mam'bə), *n.* any of various African poisonous snakes of the genus *Dendroaspis*.
mambo (mam'bō), *n.* a W Indian syncopated dance or dance tune, like the rumba. *v.i.* to dance the mambo.
Mameluke (mam'əlook), *n.* one of the mounted soldiers of Egypt (orig. Circassian slaves) who formed the ruling class in that country, destroyed by Mehmet Ali in 1811.
mamilla (məmil'ə), (*esp N. Am.*) **mammilla**, *n.* a nipple or teat; a nipple-shaped organ or part. **mamillary** (mam'-)
mamma[1] (məmah'), *n.* mother (used chiefly by young children). **mammy** (mam'i), *n.* (*dial.*) mother; (*N Am.*) a black woman working as a children's nurse in a white family.
mamma[2] (mam'ə), *n.* (*pl.* **-mae, -ē**) the milk-secreting organ in mammals. **mammary**, *a.* of or concerning the mammae. **mammary gland**, *n.* a mamma.
mammal (mam'əl), *n.* (*Zool.*) any individual of the Mammalia. **Mammalia** (-mā'liə), *n.pl.* the class of animals having milk-secreting organs for suckling their young, the highest division of vertebrates. **mammalian** (-mā'-), *a.* **mammography** (-mog'-), *n.* examination of the breasts by X-ray. **mammogram**, *n.*
mammee (məmē'), *n.* a tropical American tree bearing edible pulpy fruit.
mammon (mam'ən), *n.* riches personified as an idol or an evil influence. **mammonish**, *a.* **mammonism**, *n.* **mammonist, -nite**, *n.* **mammonize, -ise**, *v.t.*
mammoth (mam'əth), *n.* a large extinct species of elephant. *a.* gigantic, huge.
mammy (*dial.*) MAMMA[1].
man (man), *n.* (*pl.* **men**, men) a human being, a person; (*collect.*) mankind, the human race; an adult male of the human race; an individual, one; one with manly qualities; manhood; (*dial.*, *coll.*) a husband; a man-servant, a valet, a workman; a person under one's control; a vassal, a tenant; (*pl.*) soldiers, esp. privates; (*pl.*) pieces used in play- ing chess or draughts; (*in comb.*) a ship, as *man-of-war, merchantman* etc. *v.t.* (*past*, *p.p.* **manned**) to furnish with a man or men, esp. for defence or other military service; to fortify the courage of (esp. oneself). **as one man**, all together, in unison. **inner man** INNER. **man about town**, a fashionable idler. **man and boy**, from boyhood upwards. **man in the street**, an ordinary person. **man of letters**, a writer, literary critic etc. **man of straw**, a man of no substance; a false argument or adversary put forward for the sake of being refuted. **man of the world**, an experienced person, sophisticated and urbane. **man to man**, as between individual men, one with or against the other; with complete frankness. **man-to-man**, *adj.* **to a man**, without exception. **to be one's own man**, to be of independent mind. **man-at-arms**, *n.* a heavily-armed mounted soldier, esp. in the Middle Ages. **man-child**, *n.* a male child. **man-day**, *n.* the amount of work done by one person in one day. **man-eater**, *n.* a cannibal; a tiger, shark etc., that devours human beings; a horse that bites. **man-eating**, *a.* **Man Friday**, *n.* a personal servant, factotum. **man-handle**, *v.t.* to move by man-power alone; (*coll.*) to handle roughly, to maltreat. **man-hater**, *n.* one (usu. a woman) who hates men. **manhole**, *n.* a hole in a floor, drain or parts of machinery etc., to allow entrance for cleansing and repairs. **man-hour**, *n.* the amount of work done by one person in one hour. **manhunt**, *n.* a large-scale search for a person, e.g. an escaped prisoner. **man-jack**, *n.* a person. **man-made**, *a.* made by man, not natural, artificial. **man-of-war, man-o'-war**, *n.* a warship belonging to a navy. **manpower**, *n.* amount of men available for any purpose. **man-rope**, *n.* a rope at the side of a gangway etc. **manservant**, *n.* **man-sized**, *a.* of a suitable size for a man; (*coll.*) large. **manslaughter**, *n.* the killing of a human being or beings; (*Law*) the unlawful killing of a person but without malice. **man-slayer**, *n.* one who kills a human being or commits manslaughter. **man-trap**, *n.* a trap set for poachers etc. **manful**, *a.* brave, courageous; resolute,

mana — mango

manly. manfully, *adv.* **manfulness,** *n.* **manhood,** *n.* the state of being a man; the state of being a male person of full age; manliness, courage, resolution. **mankind** (mankīnd'), *n.* the human species; (man'kīnd), men collectively as distinct from humanity. **manlike,** *a.* **manly,** *a.* having the finer qualities characteristic of a man, courageous, resolute, magnanimous; befitting a man; mannish. **manliness,** *n.* **manned,** *a.* furnished with a crew, workers etc., of a spacecraft, having a human pilot or crew. **mannish,** *a.* esp. of a woman, masculine, characteristic of a man. **mannishly,** *adv.* **mannishness,** *n.*
mana (mah'nə), *n.* spiritual power exerted through man or inanimate objects; power, authority. [Polynesian]
manacle (man'əkl), *n.* (*usu. pl.*) a handcuff, a fetter. *v.t.* to put manacles on; to fetter.
manage (man'ij), *v.t.* to conduct, to direct, to carry on, to control; to conduct the affairs of; to handle, to wield; to bring or keep under control; to lead or guide by flattery etc.; to break in, to train (as a horse); to deal with, to make use of; to husband, to use cautiously. *v.i.* to direct affairs; to contrive (to do etc.); to get on (with or without); to succeed (with). *n.* management; manège. **manageable,** *a.* **manageability** (-bil'-), **manageableness,** *n.* **manageably,** *adv.* **management,** *n.* the act of managing; conduct, administration; those who manage, a board of directors etc.; skilful employment of means; skill, ingenuity. **management consultant,** *n.* a person who advises on the efficient management of a business company or institution. **manager,** *n.* one who manages, esp. a business, institution etc.; (*Law*) one appointed to administer a business in chancery etc.; (*usu. with* good, bad etc.) one skilled in economical management; (*pl.*) a committee appointed by either House of Parliament to perform a duty concerning both Houses. **manageress** (-es'), *n.* a female manager, esp. of a retail shop, canteen, restaurant etc. **managerial** (-nəjē'-), *a.* **managership,** *n.* **managing,** *a.* having the management or control of a business, department etc.; careful, economical.
mañana (mənyah'nə), *n.* tomorrow, presently, later on; procrastination.
manatee (man'ətē), *n.* the sea-cow, a large herbivorous sirenian.
manche (mahnsh), *n.* a sleeve, with long hanging ends; in heraldry, a bearing representing such a sleeve; the neck of a violin etc.
Manchester (man'chəstə), *a.* belonging to or made in Manchester. **Manchester goods,** *n.pl.* cotton textiles.
manchineel (manchinēl'), *n.* a W Indian tree with a poisonous sap and apple-like fruit; its timber used for cabinet work.
Manchu (man'choo), *n.* (*pl.* **-chus, -chu**) one of the Mongoloid people of Manchuina who governed China from the 17th to the 20th cent.; their language. *a.* of or relating to the Manchu dynasty.
manciple (man'sipl), *n.* a steward, a purveyor of stores, esp. for a college, inn of court etc.
Mancunian (mankū'niən, mang-), *n.* a native or citizen of Manchester. *a.* of Manchester.
-mancy (-mənsi), *suf.* divination by, as in *necro-*.
mandala (man'dələ, -dah'-), *n.* any of various symbols used to represent the universe in Buddhism or Hinduism, used as an aid to meditation.
mandamus (mandā'məs), *n.* a writ issued from a higher court directed to a person, corporation or inferior court, requiring them to do some particular thing therein specified which appertains to their office or duty.
mandarin (man'dərin), *n.* a Chinese official under the Chinese Empire; a grotesque ornament or statuette in Chinese costume; (**Mandarin**) the chief dialect of the Chinese language; a mandarin orange; a dye the colour of this; a mandarin duck; a liqueur flavoured with juice of the mandarin orange; a high-ranking public servant; an influential, often reactionary (literary) figure. **mandarin collar,** *n.* a stiff, narrow stand-up collar. **mandarin duck,** *n.* a brightly-coloured Asiatic duck. **mandarin orange,** *n.* a small flattish sweet orange, of a dark-yellow colour. **mandarinate,** *n.* the office of a mandarin; mandarins collectively.
mandate (man'dāt), *n.* an authoritative charge, order or command; (*Law*) a judicial command to an officer or a subordinate court; a contract of bailment by which the mandatary undertakes to perform gratuitously a duty regarding property committed to him; a rescript of the Pope; a direction from electors to a representative or a representative body to undertake certain legislation etc.; the authority given (esp. formerly by the League of Nations) to a larger power to govern another country in trust for its native inhabitants; a country ruled in this way. **mandatary,** *n.* **mandator** (-dā'-), *n.* **mandatorily,** *adv.* **mandatory,** *a.* containing, or of the nature of a mandate; bestowing a mandate; obligatory, compulsory.
mandible (man'dibl), *n.* the jaw, the under jaw in vertebrates, the upper or lower in birds, and the pair in insects. **mandibular, -late, -lated** (-dib'-), *a.*
mandolin (man'dəlin), *n.* a musical instrument with a deep almond-shaped body and two or three pairs of metal strings.
mandrake (man'drāk), *n.* the plant *Mandragora officinarum*, the root of which was anciently believed to be like the human form and to shriek when pulled up.
mandrel, mandril (man'drəl), *n.* an arbor or axis on which work is fixed for turning; the revolving spindle of a circular saw; a cylindrical rod or core round which metal or other material is forged or shaped; a miner's pick.
mandrill (man'dril), *n.* a large W African baboon with a red and blue nose.
mane (mān), *n.* the long hair on the neck of some animals, as the horse; long, thick hair on a person's head. **manesheet,** *n.* a covering for the upper part of a horse's head. **maned,** *a.* (*usu. in comb.*) having a mane, as *thick-maned*. **maneless,** *a.*
manège (manäzh', -nezh'), *n.* a school for training horses or teaching horsemanship; the training of horses; horsemanship. *v.t.* to manage; to break in and train (a horse).
manes (mā'nēz, mah'nāz), *n.pl.* the spirits of the dead, esp. of ancestors worshipped as tutelary divinities; the shade of a deceased person regarded as an object of reverence.
maneuver (mənoo'və, -nū'-), (*N Am.*) MANOEUVRE.
manful, etc. MAN.
mangabey (mang'gəbā), *n.* an African monkey of the genus *Cercocebus*.
manganese (mang'gənēz, -nēz'), *n.* a metallic element, at. no.25; chem. symbol Mn, of a greyish-white colour; the oxide of this occurring as a black mineral, used in glassmaking. **manganate** (man'-), *n.* a salt of manganic acid.
mange (mānj), *n.* a skin disease occurring in cattle, dogs etc. **mangy,** *a.* infected with the mange; mean, squalid. **manginess,** *n.*
mangel-wurzel, mangold-wurzel (mang'g(ə)lwœ'zəl), *n.* a large-rooted variety of the common beet cultivated as fodder for cattle.
manger (mān'jə), *n.* a trough for horses or cattle to eat out of. **dog in a manger** DOG.
mangetout (mäzh'too), *n.* a type of pea which is eaten complete with the pod.
mangle[1] (mang'gl), *v.t.* to lacerate; to mutilate; to disfigure by hacking; to mar, to ruin, to destroy the symmetry or completeness of, by blundering etc.
mangle[2] (mang'gəl), *n.* a rolling-machine for pressing and smoothing damp clothes. *v.t.* to press and smooth with a mangle; to calender. **mangler,** *n.*
mango (mang'gō), *n.* (*pl.* **-goes**) an E Indian tree or its fruit; (*N Am.*) a green musk-melon pickled.

mangonel (mang'gənel), *n.* a mediaeval engine for throwing missiles.
mangosteen (mang'gəstēn), *n.* an E Indian tree or its orange-like fruit, with a sweet, juicy pulp.
mangrove (mang'grōv), *n.* a tropical tree of the genus *Rhizophora*, growing in muddy places by the coast.
mangy etc. MANGE.
manhandle, manhole, manhood, manhunt MAN.
Manhattan (manhat'n), *n.* a cocktail containing whisky, vermouth and sometimes a dash of bitters.
mania (mā'niə), *n.* a form of mental disorder characterized by hallucination, emotional excitement and violence; (*coll.*) an infatuation, a craze. **maniac** (-ak), *a.* affected with mania, insane, raving. *n.* a madman, a raving lunatic. **maniacal** (məni'əkəl), *a.* **maniacally,** *adv.* **manic** (man'ik), *a.* of or affected by mania; (*coll.*) over-excited, wildly energetic. **manic-depressive,** *n., a.* (a person) suffering from alternating bouts of mania and depression.
-mania, *comb. form* denoting special kinds of derangement, hallucination, infatuation or excessive enthusiasm, as in *erotomania, kleptomania, megalomania, monomania*. **-maniac,** *suf.* forming nouns and adjectives.
manicure (man'ikūə), *n.* one who undertakes the treatment of the hands and finger-nails as a business; the care of the hands, nails etc. *v.t.* to treat the hands and finger-nails. **manicurist,** *n.*
manifest (man'ifest), *a.* not concealed; plainly apparent, clear, obvious. *v.t.* to make manifest, to show clearly; to display, to exhibit, to evince; to be evidence of; to reveal or exhibit (itself); to record in a ship's manifest. *v.i.* to make a public demonstration of opinion; of a spirit, to reveal its presence. *n.* a list of a ship's cargo for the use of the custom-house officers; a list of passengers on an aircraft. **manifestable,** *a.* **manifestation,** *n.* manifesting or being manifested; a public demonstration. **manifestative** (-fes'-), *a.* **manifester,** *n.* **manifestly,** *adv.* **manifestness,** *n.*
manifesto (manifes'tō), *n.* (*pl.* **-toes**) a public declaration, esp. by a political party, government, sovereign or other authoritative body, of opinions, motives or intentions.
manifold (man'ifōld), *a.* of various forms or kinds; many and various, abundant; shown, applied or acting in various ways. *n.* that which is manifold; a carbon-copy; a tube or system of tubes for conveying steam, gas etc., in an engine, motor etc. *v.t.* to duplicate, multiply. **manifolder,** *n.* **manifoldly,** *adv.* **manifoldness,** *n.* the state of being manifold.
manikin, mannikin (man'ikin), *n.* a little man, a dwarf; an anatomical model exhibiting the parts, organs and structure of the human body; a lay figure; a small tropical American passerine bird.
Manila, Manilla[1] (mənil'ə), *n.* a kind of cheroot made at *Manila* (capital of Philippine Islands); Manila hemp; a rope of this. **Manila hemp,** *n.* hemp made from the fibre of *Musa textilis*, used for making rope. **Manila paper,** *n.* a strong brown paper, orig. made from Manila hemp.
manilla[2] (manil'ə), *n.* a metal ring worn by Africans on the legs or arms; a piece of metal shaped like a ring horseshoe formerly used as a medium of exchange.
manioc (man'iok), *n.* the cassava, *Manihot utilissima*; meal made from the root of this.
maniple (man'ipl), *n.* a strip worn as a eucharistic vestment on a priest's left arm; a subdivision of the Roman legion consisting of 60 to 120 men with their officers. **manipular** (-nip'-), *a.*
maniplies (men'iplīz), MANYPLIES.
manipulate (mənip'ūlāt), *v.t.* to operate on with or as with the hands, to handle, to treat, esp. skilfully or dexterously; to manage, influence or tamper with by artful or sly means. *v.i.* to use the hands skilfully, as in scientific experiments etc. **manipular,** *a.* **manipulation,** *n.* **manipulative, -tory** *a.* **manipulator,** *n.*

manis (mā'nis), *n.* a genus of edentate mammals, containing the scaly ant-eaters.
manitou (man'itoo), *n.* among certain American Indians a spirit or being endowed with supernatural power; an amulet, a fetish.
mankind, manlike, etc. MAN.
manna (man'ə), *n.* (*Bibl.*) the food miraculously supplied to the Israelites in the wilderness; divine food, spiritual nourishment, as the Eucharist; a sweetish exudation, of a slightly laxative nature, from certain species of ash, chiefly *Fraxinus ornus*. **manna from heaven,** anything very advantageous and unexpected. **manna gum,** *n.* dried sap of the eucalyptus, lerp. **manniferous** (-nif'-), *a.* bearing or yielding manna.
mannequin (man'ikin), *n.* a woman employed to wear and display clothes; a dummy used to model clothes.
manner (man'ə), *n.* the mode in which anything is done or happens; method, style, mannerism; practice, habit, use, custom; demeanour, bearing, address; sort, kind; (*pl.*) conduct in social intercourse, behaviour, deportment; politeness, habits showing good breeding; general modes of life, social conditions. **all manner of,** all kinds of. **by no manner of means,** under no circumstances. **in a manner,** in a certain way, somewhat, so to speak. **to the manner born,** (*Shak.*) born to follow a certain practice or custom; (as if) accustomed to something from birth. **what manner of,** what kind of. **mannered,** *a.* (*usu. in comb.*) having manners, as *ill-mannered;* having or betraying mannerism; affected. **mannerism,** *n.* excessive adherence to the same manner or peculiarity; peculiarity of style. **mannerist,** *n.* **manneristic, -ical** (-is'-), *a.* **mannerless,** *a.* devoid of manners or breeding. **mannerly,** *a.*
mannikin (man'ikin), MANIKIN.
mannish etc. MAN.
manoeuvre (mənoo'və, -nū'-), (*esp. N Am.*) **maneuver,** *n.* a tactical movement or change of position by troops or warships; (*pl.*) tactical exercises in imitation of war; skilful or artful management; a trick, a stratagem. *v.i.* to perform manoeuvres; to manage with skill; to employ stratagem. *v.t.* to cause (troops) to perform manoeuvres; to move, drive or effect by means of strategy or skilful management; to manipulate. **manoeuvrable,** *a.* **manoeuvrability** (-bil'-), *n.* **manoeuvrer,** *n.*
manometer (mənom'itə), *n.* an instrument for measuring the pressure of a gas. **manometric** (manəmet'-), **-al,** *a.* **manometry** (-nom'-), *n.*
manor (man'ə), *n.* a landed estate belonging to a lord; (*N Am.*) a tract of land occupied in perpetuity or for long terms by tenants who pay a fee-farm rent to the proprietor; (*sl.*) a police district. **lord of the manor,** a person or corporation holding the rights of a manor. **manor-house,** *n.* **manorial** (-naw'riəl), *a.* pertaining to a manor.
manpower MAN.
manqué (mũ'kā), *a.* having the potential to be, but not actually being, something specified, as in *actor manqué*. [F, having failed]
mansard roof (man'sahd), *n.* a roof with two sets of rafters on each side, the lower nearly vertical, the upper much inclined, giving space for attics.
manse (mans), *n.* the residence of a clergyman, esp. a Presbyterian minister.
mansion (man'shən), *n.* a residence of considerable size and pretensions; a manor-house; (*pl.*) a large building or set of buildings divided into residential flats. **mansion-house,** *n.* a manor-house; an official residence, esp. of the Lord Mayors in London and Dublin.
man-sized, manslaughter MAN.
manta (ray) (man'tə), *n.* any of various very large rays of the family *Mobulidae*.
mantel (man'tl), *n.* the ornamental facing round a fireplace with the shelf above it. **mantel-board,** *n.* a mantelshelf or a shelf resting on it, formerly draped. **mantelpiece,** *n.* a

mantel; a mantel-tree. **mantelshelf**, *n*. the shelf above a fireplace.
mantelet (man'tlət), *n*. a short mantle; a bullet-proof shield, enclosure or shelter.
mantic (man'tik), *a*. pertaining to prophecy or divination.
manticore (man'tikaw), *n*. a fabulous monster with a human head, a lion's body and the tail of a scorpion.
mantilla (mantil'ə), *n*. a woman's light cloak or cape; a veil for the head and shoulders, worn in Spain and Italy.
mantis (man'tis), *n*. (*pl*. **-tises, tes**) a genus of carnivorous orthopterous insects, which hold their forelegs as if in prayer lying in wait for other insects as prey.
mantissa (mantis'ə), *n*. the decimal or fractional part of a logarithm.
mantle (man'tl), *n*. a sleeveless cloak or loose outer garment; a covering; a conical or tubular network coated with refractory earth placed round a gas-jet to give an incandescent light; a covering or concealing skin, part or organ, as the fold enclosing the viscera in the Mollusca; a symbol of leadership, power or authority; the layer of the earth between the crust and the core. *v.t.* to clothe in or as in a mantle; to cover, to envelop, to conceal; to suffuse. *v.i.* to be overspread or suffused (as with a blush); to suffuse the cheeks; of a blush, of liquids, to become covered or coated; to stretch the wings (as a hawk on its perch). **mantle rock**, *n*. unconsolidated rock at the earth's surface.
mantra (man'trə), *n*. a Hindu formula or charm; a Vedic hymn of praise; a word or phrase chanted inwardly in meditation.
mantrap MAN.
mantua (man'tūə), *n*. a woman's loose gown worn in the 17th and 18th cents. **mantua-maker**, *n*.
manual (man'ūəl), *a*. pertaining to or performed with the hands; involving physical exertion; not mechanical or automatic. *n*. a small book or handy compendium, a handbook; a fire-engine worked by hands; a service book, esp. that used by priests in the mediaeval church; an organ keyboard played by the hands; the drill by which soldiers are taught to handle their rifles etc. properly. **manually**, *adv*.
manubrium (mənü'briəm), *n*. (*pl*. **-bria, -ə, -briums**) (*Anat., Zool.*) a handle-like part or process, as the presternum in mammals, the peduncle hanging from the umbrella in medusae; the handle of an organ-stop. **manubrial**, *a*.
manufacture (manūfak'chə), *n*. the making of articles by means of labour or machinery, esp. on a large scale; industrial production; any particular branch of this; (*pl*.) the products of industry or any particular industry. *v.t.* to make or work up into suitable forms for use; to produce or fashion by labour or machinery, esp. on a large scale; to produce (pictures, literature etc.) in a mechanical way; to fabricate, to invent (a story, evidence etc.). *v.i.* to be occupied in manufacture. **manufacturer**, *n*. **manufacturing**, *n*.
manuka (mahn'ukə), *n*. the New Zealand tea-tree.
manumit (manūmit'), *v.t.* (*past, p.p.* **manumitted**) to release from slavery. **manumission**, *n*.
manure (mənūə'), *v.t.* to enrich (a soil) with fertilizing substances. *n*. any substance, as dung, compost or chemical preparations, used to fertilize land. **manurer**, *n*.
manus (mä'nəs), *n*. (*pl*. **manus**) the hand or a corresponding part in an animal.
manuscript (man'ūskript), *a*. written by hand. *n*. a book or document written by hand, not printed; copy for a printer.
Manx (mangks), *a*. pertaining to the Isle of Man, or its inhabitants or its language. *n*. the Celtic language spoken by natives of Man; the people of the Isle of Man. **Manx cat**, *n*. a tailless variety of domestic cat. **Manxman**, *n*. **Manxwoman**, *n. fem*.

many (men'i), *a*. numerous; comprising a great number. *n*. a multitude; a great number. **the many**, the majority; the multitude, the common crowd. **too many**, superfluous, not wanted, in the way. **many-sided**, *a*. having many sides, aspects etc.; widely sympathetic, versatile, liberal. **many-sidedness**, *n*.
manyplies (men'ipliz), *n*. the third stomach of a ruminant, the omasum.
manzanilla (manzənil'ə, -thənē'yə), *n*. a very dry sherry.
Maoism (mow'izm), *n*. the political thought expounded by the Chinese communist leader *Mao* Tse-tung. **Maoist**, *n., a*. (an adherent) of Maoism.
Maori (mowr'i), *n*. (*pl*. **-ris, -ri**) one of the Polynesian original inhabitants of New Zealand; their language. *a*. pertaining to them. **Maori chief**, *n*. a New Zealand flat fish. **Maori hen**, *n*. the flightless wood hen of New Zealand. **Maoriland**, *n*. New Zealand. **Maorilander**, *n*. a white person native of New Zealand. **Maoritanga**, (-tahng'ə), *n*. Maori culture.
map (map), *n*. a representation of a portion of the earth's surface or the stars, upon a plane; any delineation; a mathematical function. *v.t.* (*past, p.p.* **mapped**) to represent or set down in a map; to plan (out) in exact detail; (*Math.*) to assign (each of the elements of a set) to each of the elements in a different set. **off the map**, *adv*. (*coll.*) of no account, not worth consideration, remote; out-of-the-way. **to put on the map**, to cause to become important or well-known. **maplike**, *a*. **mapper, mappist**, *n*.
maple (mā'pl), *n*. a tree or shrub of the genus *Acer*; the wood of this. **maple leaf**, *n*. the emblem of Canada. **maple-sugar**, *n*. a coarse sugar obtained from maples.
maquette (maket'), *n*. a sculptor's preliminary model in clay, wax etc.; a preliminary sketch.
maqui (makē'), *n*. a Chilean evergreen shrub, the berries of which produce a medicinal wine.
maquillage (makēyahzh'), *n*. (the technique of applying) make-up, cosmetics.
maquis (makē'), *n*. scrub or bush in Corsica; the name taken by those surreptitiously resisting the German invaders of France etc., in 1940–45.
mar (mah), *v.t.* (*past, p.p.*. **marred**) to spoil, to ruin; to disfigure. *n*. a blemish, a drawback. **marrer**, *n*.
Mar., (*abbr.*) March.
marabou (ma'rəboo), *n*. a W African stork, the downy feathers from under the wings and tail of which are used for trimming hats etc.; the adjutant-bird.
marabout (ma'rəboot, -boo), *n*. a Muslim hermit or saint, esp. one of a priestly caste in N Africa; the tomb or dwelling of such a saint.
maraca (mərak'ə), *n*. a hollow gourd or shell containing beads, shot etc., shaken as a percussive accompaniment to music, esp. in Latin America.
maranatha (marənath'ə), ANATHEMA.
maraschino (mərəskē'nō, -shē'-), *n*. a cordial or liqueur distilled from bitter cherries grown in Dalmatia. **maraschino cherry**, *n*. a cherry preserved in maraschino or imitation liqueur, used in cocktails, etc.
marasmus (mərəz'məs), *n*. wasting away of the body. **marasmic**, *a*.
Maratha (mərah'tə), *n*. a member of a people of SW India, esp. the state of Maharashtra. **Marathi**, *n*. their Sanskritic language.
marathon (ma'rəthən), *n*. a foot-race of 26 miles 385 yards (42.1 km); any task or contest; requiring great endurance.
maraud (mərawd'), *v.i.* to rove in quest of plunder; to make a raid (on). *v.t.* to plunder. **marauder**, *n*.
marble (mah'bl), *n*. a fine-grained or crystalline limestone capable of taking a fine polish; (*usu. pl.*) a piece of sculpture in this material; a type of smoothness, hardness or inflexibility; a small ball of marble, glass or other hard substance used as a toy; (*pl., coll.*) one's sanity, one's wits. *v.t.* to stain or vein (end-papers of books etc.) to

look like marble. *a.* composed of marble; veined like marble; hard, unfeeling. **marbled**, *a.* **marbling**, *n.* the veined or speckled appearance of marble.

marc (mahk), *n.* the compressed residue of grapes left after pressing, in the making of wine or oil; liqueur-brandy made from this.

marcasite (mah'kəsit), *n.* pyrites, esp. a white orthorhombic form of iron pyrites, used for making ornaments.

marcato (mahkah'tō), *a.* of musical notes, heavily accented. *adv.* (played) with a heavy accent.

marcescent (mahses'nt), *a.* of blooms, leaves etc., withering without falling. **marcescence**, *n.*

March (mahch), *n.* the third month of the year. **March hare**, *n.* a hare excited by the breeding season in March, hence **mad as a March hare**.

march[1] (mahch), *v.i.* to move with regular steps as soldiers; to walk in a grave, deliberate or determined manner. *v.t.* to cause to move (on, off etc.) in military order. *n.* the act of marching; a stately, deliberate or measured movement, esp. of soldiers; the distance marched in a day; progress, advance; (*Mus.*) a composition for accompanying a march. **on the march**, advancing steadily; making progress. **to steal a march on**, to gain an advantage over. **march past**, *n.* a marching of troops in a review past a superior officer etc.

march[2] (mahch), *n.* (*pl.* **-ches**) the frontier or boundary of a territory; (*often pl.*) a borderland or debatable land between two countries, as the border country of England and Wales. *v.i.* to border (upon) or have a common frontier (with). **Lord Marcher**, (*pl.* **Lords Marchers**) Lords holding jurisdiction and privileges on the Welsh border. **marchman**, *n.* **marcher**, *n.* an officer or warden having jurisdiction over marches; an inhabitant of a march.

marchioness (mah'shənis), *n.* the wife or widow of a marquis, or a woman holding this rank in her own right.

marchpane (mahch'pān), MARZIPAN.

Mardi Gras (mahdi grah'), *n.* Shrove Tuesday; the carnival celebrated at this time.

mare[1] (meə), *n.* the female of the horse or other equine animal. **mare's-nest**, *n.* a discovery that turns out a hoax or a delusion. **mare's-tail**, *n.* an aquatic plant long fibrous cirrus-clouds, supposed to prognosticate rain.

mare[2] *n.* (*pl.* **maria**) vast plains on the moon visible as dark patches; similar areas on Mars.

maremma (mərem'ə), *n.* (*pl.* **-me**, **-mā**) a marshy and usu. malarious region by the seashore.

margaric (mahga'rik), *a.* pertaining to pearl, pearly.

margarine (mahjərēn', -gə-, mah'gə-, mah'jə-), *n.* an emulsion of edible oils and fat with water or skimmed milk or other substances with or without the addition of colouring matter, as a substitute for butter.

margarita (mahjərē'tə), *n.* a cocktail made from tequila and lemon (or other fruit) juice.

margay (mah'gā), *n.* a S American tiger-cat.

marge[1] (mahj), MARGIN.

marge[2] (mahj), *n.* (*coll.*) short for MARGARINE.

margin (mah'jin), *n.* an edge, a border, a brink; the blank space round the printed matter on a page; the space of time or the range of conditions within which a thing is just possible; an allowance of time, money, space etc. for contingencies, growth etc.; the difference between cost and selling price; a sum deposited with a broker to protect him against loss; the lowest amount of profit allowing an industry etc. to continue. *v.t.* to furnish with a margin; to enter on the margin. *v.i.* to deposit margin on stock. **marginal**, *a.* of, pertaining to or at the margin; written or printed on the margin; near the limit; (*coll.*) small, slight; of land, difficult to cultivate. *n.* a marginal constituency. **marginal constituency**, *n.* a parliamentary constituency where there is only a small difference between the totals of votes cast for the two leading candidates. **marginalia** (-ā'liə), *n.pl.* marginal notes. **marginalize, -ise**, *v.t.* to reduce in influence, power, importance etc.; to cause to seem irrelevant. **marginally**, *adv.* **marginate, -ated** (-nət, -nātid), *a.* having a margin; edged. *v.t.* (-nāt), to furnish with a margin. **margination**, *n.*

margrave (mah'grāv), *n.* orig. a lord or governor of a march or border province, now a German title of nobility. **margravate** (-grəvət), **margraviate** (-ət), *n.* **margravine** (-grəvēn), *n. fem.*

marguerite (mahg'gərēt), *n.* the ox-eye daisy and other wild or cultivated varieties of chrysanthemum.

maria MARE [2].

Marian (meə'riən), *a.* pertaining to the Virgin Mary, to Mary I of England or Mary Queen of Scots. *n.* an adherent or defender of either of the two last.

marigold (ma'rigōld), *n.* a plant bearing a bright yellow flower; applied to other composite yellow-flowered plants. **marsh marigold** MARSH.

marihuana, marijuana (mariwah'nə), *n.* dried leaves of Indian hemp, used to make cigarettes smoked as a narcotic.

marimba (mərim'bə), *n.* a musical instrument of the nature of a xylophone.

marina (mərē'nə), *n.* a docking area for yachts and pleasure boats.

marinade (marinād'), *n.* a pickle of vinegar, oil etc. flavoured with wine and spices; fish or meat pickled in this. **marinate** (ma'rināt), *v.t.* to pickle in marinade.

marine (mərēn'), *a.* pertaining to, found in or produced by the sea; used at sea or in navigation, nautical, naval; serving on shipboard. *n.* the shipping, fleet or navy of a country; (*pl.*) troops for service on board warships; a member of the Royal Marines; a specialist in commando and amphibious operations; a seascape. **tell it to the marines**, an expression of incredulity and derision (from the sailor's contempt for landsmen). **mariner** (ma'rinə), *n.* a seaman, a sailor. **master mariner**, the captain of a merchant ship.

Mariolatry (meəriol'ətri), *n.* (excessive) worship of the Virgin Mary. **Mariolater**, *n.*

marionette (mariənet'), *n.* a puppet moved by strings on a mimic stage.

marital (ma'ritl), *a.* pertaining to marriage or to a husband. **maritally**, *adv.*

maritime (ma'ritim), *a.* pertaining to, connected with or bordering on the sea; of countries, cities etc., having a navy or commerce by sea.

marjoram (mah'jərəm), *n.* a herb of the genus *Origanum* of the mint family, esp. *O. vulgare*, the wild marjoram, and *O. majorana*, sweet marjoram, a fragrant plant used as a herb in cooking.

mark[1] (mahk), *n.* a visible sign or impression, as a stroke, cut, dot etc.; an indication, symbol, character, brand, device or token; a target, an object to aim at; (*coll.*) a victim, esp. of fraud; the point to be reached; a limit, a standard; a starting-line in a race; a distinguishing sign, a seal etc.; a character made by one who cannot write; a number or sign indicating merit in an examination; a distinguishing feature, a characteristic, a symptom; (*Boxing*) the pit of the stomach; (*Rugby*) an indentation made in the ground by the heel of a player who has secured a fair catch; a boundary, frontier or limit. *v.t.* to make a mark on; to distinguish or designate or indicate, by a mark or marks; to select, to single out; to pay heed to; to indicate or serve as a mark to; to characterize, to be a feature of; to express or produce by marks; to record (points in games); to award (merit in examination); in football, hockey etc., to keep close to an opponent so as to be ready to tackle him. *v.i.* to observe something critically, to take note. **below, not up to, the mark**, not equal to a desired standard. **beside, wide, of the mark**, not hitting the object; not to the point, irrelevant. **on your marks**, an order from the starter in a race to the runners to take their position on the starting-line. **to make one's**

mark, to do something that brings fame, recognition etc. **to mark down, up,** *v.t.* to lower or raise the price. **to mark out,** to set out boundaries and levels for a proposed building; to set out lines and marks on material as a guide for cutting, drilling or other operations. **to mark time,** to move the feet alternately as in marching, without changing position; to pause until further progress can be made. **to toe the mark,** to touch a chalk line with the toes so as to be in rank abreast with others; to do one's duty, to perform one's obligations. **mark-down,** *n.* the amount by which a price is reduced. **mark-up,** *n.* the amount by which a price is increased. **marked,** *a.* noticeable, definite; of a person, destined to suffer misfortune, attack, suspicion etc. **markedly** (kid-), *adv.* **markedness** (-kid-), *n.* **marker,** *n.* one who marks; a counter used in card-playing; one who notes the score at billiards; a bookmark.
mark² (mahk), *n.* the name of several coins of various values, esp. that of the Federal Republic of Germany and of the German Democratic Republic.
market (mah'kit), *n.* a meeting for buying and selling; the place for this; an open space or large building in which cattle, provisions or other commodities are offered for sale; a county or locality regarded as a place for buying and selling commodities in general or a particular form of merchandise; demand for a commodity, value as determined by this. *v.i.* to buy or sell in a market. *v.t.* to sell in a market. **to come into** or **put on the market,** to be offered or to offer for sale. **market-garden,** *n.* a garden in which vegetables and fruit are raised for market. **marketplace,** *n.* a market square etc.; the sphere of commercial trading. **market research,** *n.* research into public demand, need etc. for particular commercial goods. **market town,** *n.* a town having the privilege of holding a public market. **marketable,** *a.* **marketability, marketableness,** *n.* **marketer,** *n.* **marketing,** *n.* the processes involved in selling goods, e.g. promotion, distribution etc.
markhor (mah'kaw), *n.* a wild mountain goat inhabiting the border-land of India, Iran and Tibet.
marking (mah'king), *n.* producing a mark. *n. (often in pl.)* marks or colouring, esp. on natural objects.
markka (mah'kə), *n.* the *(pl.* **-kaa)** Finnish unit of currency.
marksman (mahks'mən), *n.* one skilled in aiming at a mark; one who shoots well. **marksmanship,** *n.* **markswoman,** *n. fem.*
marl (mahl), *n.* clay containing various minerals, much used as a fertilizer; *(poet.)* earth. *v.t.* to manure with marl. **marly,** *a.*
marlin (mah'lin), *n.* any of various large oceanic fishes with a long upper jaw.
marline (mah'lin), *n. (Naut.)* a small two-stranded line, used for lashing etc. **marline-spike,** *n.* a pointed iron pin for opening the strands of rope in splicing.
marmalade (mah'məlād), *n.* a jam or preserve prepared from fruit, esp. oranges or lemons, boiled with the sliced rind. *a.* of cats, having streaks of orange and brown.
Marmiteᴿ (mah'mīt), *n.* a savoury yeast extract used as a spread or for flavouring.
marmoreal (mahmaw'riəl), *a.* like marble, esp. cold, smooth or polished, pure white; made of marble.
marmose (mah'mōs), *n.* one of various S American pouchless opossums.
marmoset (mah'məzet), *n.* a small tropical American monkey of various species, called squirrel-monkeys from their bushy tails.
marmot (mah'mət), *n.* a burrowing rodent about the size of a rabbit.
marocain (ma'rəkăn), *n.* a cloth similar to crêpe de Chine, but coarser.
Maronite (ma'rənīt), *n.* a member of a Christian sect in the Lebanon region.

maroon¹ (məroon'), *a.* of a brownish-crimson colour. *n.* this colour; a detonating firework.
maroon² (məroon'), *n.* one of a group of descendants of fugitive slaves in the W Indies and Guyana; one who has been marooned. *v.t.* to put ashore and abandon on a desolate island.
maroquin (ma'rəkin, -kēn'), *n.* Morocco leather.
marque (mahk), *n.* a brand, model or type. **letter of marque** LETTER.
marquee (mahkē'), *n.* a large field-tent.
marquetry (mah'kətri), *n.* work inlaid with different pieces of fine wood, ivory, plates of metal, steel etc.
marquis, marquess (mah'kwis), *n.* a title or rank of nobility in England, ranking next below a duke and above an earl. **marquessate, marquisate** (sət), *n.* **marquise** (mahkēz'), *n.* (F) a marchioness. **marquise-ring,** *n.* a finger-ring set with gems in a pointed oval cluster.
marquisette (mahkizet'), *n.* a finely-woven mesh fabric used for clothing, curtains and mosquito-nets.
marram (ma'rəm), *n.* various grasses common along sandy shores.
marriage (ma'rij), *n.* the legal union of a man and woman, wedlock; the act or ceremony of marrying, a wedding, a nuptial celebration; sexual union; close conjunction or union; in bezique etc., the declaration of a king and queen of the same suit. **civil marriage** CIVIL. **marriage of convenience,** a marriage contracted for advantage rather than for love; any union that is made to secure an advantage. **marriage guidance,** *n.* counselling and advice given to couples with marital problems. **marriage licence,** *n.* a licence for the solemnization of a marriage without the proclamation of banns. **marriage settlement,** *n.* an arrangement made before marriage securing a provision for the wife and sometimes for future children. **marriageable,** *a.* fit of or age for marriage.
marrons glacés (marō glas'ā), *n.pl.* chestnuts coated with sugar.
marrow (ma'rō), *n.* a fatty substance contained in the cavities of bones; the essence, the pith; the pulpy interior of a fruit etc.; a vegetable marrow. **marrow-bone,** *n.* a bone containing marrow; *(pl.)* the knees. **marrowfat,** *n.* a large variety of pea. **marrowy,** *a.*
marry (ma'ri), *v.t.* to unite as man and wife; to give in marriage; to take for one's husband or wife; to join closely together, to unite intimately. *v.i.* to enter into the state of wedlock. **to marry into,** to gain (esp. money) by marrying; to join (a family) by marrying. **married,** *a.* united in marriage; pertaining to married persons, conjugal.
Mars (mahz), *n.* the Roman god of war; war; the fourth planet in order of distance from the sun.
Marsala (mahsah'lə), *n.* a white fortified wine somewhat like sherry, made at *Marsala* in Sicily.
Marseillaise (mahsāez', -səlāz'), *n.* the national anthem of the French Republic, composed by Rouget de l'Isle and introduced into Paris by the Marseillaise contingent in 1792.
marseilles, marseilles (mahsālz'), *n.* a stiff and heavy cotton fabric quilted in the loom.
marsh (mahsh), *n.* a tract of low land covered wholly or partially with water. **marsh-gas,** *n.* carburetted hydrogen evolved from stagnant water. **marsh-harrier,** *n.* a hawk. **marshland,** *n.* **marsh-mallow,** *n.* a shrubby herb growing near salt marshes; a confection formerly made from its root. **marsh-marigold,** *n.* a ranunculaceous plant with bright yellow flowers. **marshy,** *a.*
marshal (mah'shəl), *n.* an officer regulating ceremonies and directing processions; an officer of state with functions varying by country and period; an earl-marshal; a provost-marshal; a military officer of the highest rank; a field-marshal; *(N Am.)* a civil officer corresponding to an English sheriff. *v.t. (past, p.p.* **marshalled**) to arrange or

rank in order; to conduct in a ceremonious manner; (*Her.*) to dispose in order, as the coats in a shield. *v.i.* to assemble, to take up a position (of armies, processions etc.). **Marshal of the Air**, the highest rank in the RAF, corresponding in rank to Field-Marshal in the Army. **marshaller**, *n.* **marshalling yard**, *n.* a place where goods trucks are sorted according to their destination, and goods trains made up. **marshalship**, *n.*

marsupial (mahsū'piəl, -soo'-), *a.* of or resembling a pouch; belonging to the order Marsupialia, carrying the young in a pouch, as the kangaroos and opossums. *n.* any individual of the Marsupialia. **marsupium**, *n.* (*pl.* **marsupia**) a pouch for carrying the imperfectly developed young of marsupial animals; a pouch-like part or organ in other animals.

mart (maht), *n.* a market, a marketplace; an auction-room; traffic, purchase and sale.

martello (mahtel'ō), *n.* a martello tower. **martello tower**, *n.* a circular, isolated tower of masonry, erected on the coast to oppose the landing of invaders.

marten (mah'tin), *n.* a small carnivorous mammal allied to the weasel, with a valuable fur.

martial (mah'shəl), *a.* pertaining to or suited for war; military; warlike, courageous, bellicose; under the influence of the planet Mars. **martial art**, *n.* any of the various forms of single combat pursued as a sport, e.g. judo, karate. **martial law**, *n.* military law abrogating ordinary law for the time being, proclaimed in time of war, insurrection or like emergency. **martialism**, *n.* **martialist**, *n.* **martially**, *adv.*

Martian (mah'shən), *n.* an inhabitant of the planet Mars. *a.* of the planet or god Mars.

martin (mah'tin), *n.* a bird of the swallow family.

martinet (mahtinet'), *n.* a strict disciplinarian.

martingale (mah'ting-gāl), *n.* a strap fastened to a horse's girth to keep the head down; (*Naut.*) a lower stay for the jib-boom or flying jib-boom; the system of doubling stakes after every loss in gambling.

Martini® (mahtē'ni), *n.* Italian vermouth; a cocktail based on this.

Martinmas (mah'tinməs), *n.* the feast of St Martin, 11 Nov.

martyr (mah'tə), *n.* one who suffers death or persecution in defence of his faith or principles. *v.t.* to put to death for adherence to one's religion or principles; to persecute, to torture. **a martyr to**, a continual sufferer from. **martyrdom**, *n.* **martyrize**, **-ise**, *v.t.* **martyrolatry**, *n.* worship of martyrs. **martyrology**, *n.* a list or history of martyrs. **martyrological**, *a.* **martyrologist**, *n.*

marvel (mah'vl), *n.* a wonderful or astonishing thing; a prodigy. *v.i.* (*past, p.p.* **marvelled**) to be astonished (at or that); to be curious to know (why etc.). **marvellous**, *a.* **marvellously**, *adv.* **marvellousness**, *n.*

Marxian (mahk'siən), *a.* of or pertaining to Karl *Marx* (1818–83), G. socialist, or his political and economic theories. **Marxism**, *n.* the theory that human and political motives are at root economic, and that the class struggle explains the events of history; state socialism as taught by Marx. **Marxism–Leninism**, *n.* the political ideology developed by Lenin from the theories of Marx. **Marxist**, *n.*, *a.*

marzipan (mah'zipan. -pan'), *n.* a confection of almonds, sugar and white of egg.

Masai (masī'. mas'ī), *n.pl.* a dark Hamito-Negroid people inhabiting Kenya and Tanzania.

masc., (*abbr.*) masculine.

mascara (maskah'rə), *n.* a dark cosmetic for eyelashes etc.

mascon (mas'kon), *n.* one of the concentrations of dense material just beneath the moon's surface. [*mass concentration*]

mascot (mas'kət), *n.* an object or person that acts as a talisman and brings luck.

masculine (mas'kūlin), *a.* belonging to or having the characteristic qualities of the male sex; strong, robust, vigorous; manly, spirited; mannish, forward, coarse; (*Gram.*) denoting the male gender. *n.* the masculine gender; a masculine word. **masculine rhyme**, *n.* a rhyme on a word ending with a stressed syllable. **masculinely**, *adv.* **masculineness**, **masculinity** (-lin'-), *n.*

maser (mā'zə), *n.* a device similar to a laser used for amplifying microwave radiation. [acronym for *m*icrowave *a*mplification by *s*timulated *e*mission of *r*adiation]

mash (mash), *n.* a mass of ingredients crushed and mixed into a pulp; a mixture of bran and hot water for horses; crushed or ground grain or malt steeped in hot water to form wort. *v.t.* to crush into a pulpy mass; to make an infusion of (malt) in hot water.

mashie, mashy (mash'i), *n.* in golf, an iron club with a deep short blade, lofted.

masjid (məs'jid), *n.* a mosque.

mask (mahsk), *n.* a covering for the face, for protection or to conceal one's identity; a face-guard; an impression of a face in plastic material; a reproduction of a face used as a gargoyle or part of a moulding; a disguise, a pretence, a subterfuge; a masque; in photography, an opaque screen for framing the image in lantern slides, a silhouette used in printing to cover part of the plate; the head of a fox. *v.t.* to cover with a mask; (*in p.p.*) to disguise with a mask; to hide, screen or disguise; to watch (a hostile force) so as to hinder its effective action. *v.i.* to go in disguise. **masked ball**, *n.* a ball attended by guests wearing masks. **masker**, *n.*

maskinonge (mas'kinonj. -nonj'), *n.* a large pike inhabiting the Great Lakes of N America.

masochism (mas'əkizm), *n.* a variety of sexual perversion in which a person takes delight in being dominated or cruelly maltreated by another. **masochist**, *n.* **masochistic**, *a.* **masochistically**, *adv.*

mason (mā'sn), *n.* a craftsman who works in stone; a Freemason. *v.t.* to build with masonry. **masonic** (-son'-), *a.* pertaining to Freemasonry. **masonry**, *n.* the art or occupation of a mason; mason's work, stonework; Freemasonry.

Mason–Dixon line (māsndik'sn), *n.* the boundary drawn between Pennsylvania and Maryland in 1763–67 by Charles Mason and Jeremiah Dixon, regarded as the dividing line between the Northern states and the Southern slave states prior to the American civil war.

masque (mahsk), *n.* a play or dramatic entertainment, usu. presented by amateurs at court or in noblemen's houses, the performers wearing masks, orig. in dumb show, later with dialogue, poetical and musical accompaniments. **masquer**, *n.*

masquerade (maskərād', mahs-), *n.* a ball or assembly at which people wear masks; disguise, pretence. *v.i.* to wear a mask or disguise, to pass oneself off in a false guise. **masquerader**, *n.*

Mass (mas), *n.* the celebration of the Eucharist in the Roman Catholic Church (also applied by some to the Anglican communion service); the office for this; a setting of certain portions of this to music. **black mass** BLACK. **High Mass** HIGH [1]. **Low Mass** LOW [1]. **mass-book**, *n.* a missal.

mass (mas), *n.* a body of matter collected, concreted or formed into a coherent whole of indefinite shape; a compact aggregation of things; a great quantity or amount; the greater proportion, the principal part or the majority (of); volume, bulk, magnitude; (*Phys.*) the quantity of matter which a body contains. *v.t.* to form or gather into a mass; to concentrate (as troops). *v.i.* to gather into a mass. **in the mass**, in the aggregate. **the masses**, the ordinary people; the populace. **mass defect**, *n.* the difference in mass between a nucleus and its constituent particles. **mass media**, *n.pl.* the means of communication with large numbers of people, i.e. radio, TV,

newspapers etc. **mass observation,** *n.* method of obtaining public opinion by observing and interviewing people of various modes of life. **mass-produce,** *v.t.* **mass-producer,** *n.* **mass production,** *n.* the production of standardized articles in large quantities in which the processes are reduced to simple, usually mechanical, operations performed often along a conveyor belt. **mass spectrograph,** *n.* an instrument for separating charged particles into a ray spectrum according to their mass and for detecting them photographically. **mass spectrometer,** *n.* an instrument like a mass spectrograph which detects particles photographically or electrically. **massive,** *a.* heavy, weighty, ponderous; bulky; substantial, solid; (*coll.*) very large; (*Psych.*) applied to sensations of large magnitude; (*Min.*) without definite crystalline form. **massively,** *adv.* **massiveness,** *n.* **massless,** *a.* **massy,** *a.* **massiness,** *n.*
Mass., (*abbr.*) Massachusetts.
massacre (mas'əkə), *n.* indiscriminate slaughter; carnage, wholesale murder. *v.t.* to kill or slaughter indiscriminately.
massage (mas'ahzh. -ahj. -sahzh'), *n.* treatment by rubbing or kneading the muscles and body, usu. with the hands. *v.t.* to subject to this treatment; to manipulate or misrepresent (esp. statistics). **massage parlour,** *n.* a place where massages are administered; (*euphem. coll.*) a kind of brothel. **masseur** (masœ'), *n.* one skilled in massage. **masseuse,** (masœz'), *n.fem.*
massé (mas'ā), *n.* in billiards, a stroke with the cue held vertically.
massif (mas'ēf. masēf'), *n.* the main or central mass of a mountain or range.
mast[1] (mahst), *n.* a long pole of timber, or iron or steel tube, placed upright in a ship to support the yards, sails etc.; a tall, slender structure carrying a TV or radio aerial. **masthead,** *n.* the top of a mast, usu. of the lowermost as a place for a look-out etc., or of the topmast; the name of a newspaper or periodical as printed at the top of the front page. *v.t.* to send to the masthead as a punishment. **masted,** *a.* furnished with a mast or masts.
mast[2] (mahst), *n.* the fruit of the oak and beech or other forest trees.
mastaba (mas'təbə), *n.* an ancient Egyptian tomb or chapel covering a sepulchral pit, used for the deposit of offerings.
mastectomy (məstek'təmi), *n.* (*pl.* **-mies**) surgical removal of the breast. **radical mastectomy,** surgical removal of the breast including some of the pectoral muscles and the lymph nodes of the armpit.
master (mahs'tə), *n.* one who has control or authority over others; an employer; the head of a household; the owner of a slave, dog, horse etc.; one who has secured the control or upper hand; one thoroughly acquainted with or skilled in an art, craft etc., a great artist; a schoolmaster, a teacher, a tutor, an expert, a proficient; the highest degree in arts and surgery; a title given to the head of certain colleges, corporations etc.; a title of certain judicial officers; a title prefixed to the names of young gentlemen; the captain of a merchant vessel; an officer who navigates a ship of war under the direction of the captain. *a.* having control or authority; employing workmen; in charge of work or of workmen. *v.t.* to become the master of; to overpower; to defeat; to subdue, to bring under control; to become thoroughly conversant with or skilled in using; to be the master of, to rule as a master. **Master of Arts** ART[2]. **master of ceremonies** CEREMONY. **Old Masters,** the great painters of the 13th–17th cents.; their pictures. **to be one's own master,** to be free to do as one likes. **master-at-arms,** *n.* a first-class petty officer acting as head of the ship's police. **master-builder,** *n.* a builder who employs workmen; the chief builder, the architect. **master-class,** *n.* the class that exerts control in a society;

a lesson, esp. in music, given by a leading expert to gifted students. **master-key,** *n.* a key which opens all the locks of a set, opened each by a separate key. **master-mason,** *n.* a Freemason who has attained the third degree. **mastermind,** *n.* the ruling mind or intellect. *v.t.* to direct, plan. **masterpiece,** *n.* a performance superior to anything of the same kind; an achievement showing surpassing skill. **master race,** *n.* the Aryan race, regarded by Nazi ideology as superior to all others; any race regarded as superior. **master sergeant,** *n.* a senior non-commissioned officer in the US army. **masterstroke,** *n.* an instance of great skill, mastery etc. **masterwork,** *n.* a masterpiece. **masterful,** *a.* expressing mastery; domineering, self-willed. **masterfully,** *adv.* **masterfulness,** *n.* **masterly,** *a.* **masterliness,** *n.* **mastery,** *n.*
mastic (mas'tik), *n.* a resin exuding from a Mediterranean evergreen tree chiefly used for varnish; a putty-like preparation used for bedding windowframes etc. in buildings; a liquor flavoured with gum mastic used in Greece and the Levant.
masticate (mas'tikāt), *v.t.* to grind and crush with the jaw, to chew. **masticable,** *a.* **mastication,** *n.* **masticator,** *n.* **masticatory,** *a.*
mastiff (mas'tif), *n.* a large dog of great strength and courage, used as a watch-dog.
mastitis (masti'tis), *n.* inflammation of the breast or udder.
mastodon (mas'tədon), *n.* an extinct mammal closely allied to the elephant.
mastoid (process) (mas'toid), *n.* a process of bone behind the ear.
masturbate (mas'təbāt), *v.i.* to excite one's genitals, usu. with the hand, to obtain sexual pleasure. *v.t.* to do this for (oneself or another). **masturbation,** *n.* **masturbator,** *n.* **masturbatory,** *a.*
mat[1] (mat), *n.* a piece of fabric rubber etc., used as a carpet, to wipe shoes on, for packing etc.; (*Naut.*) a mass of old rope etc. to prevent chafing; a flat piece of cork, wood etc. placed under a dish or similar object; a tangled mass of anything. *v.t.* (*past, p.p.* **matted**) to cover or lay with mats; to twist or twine together. *v.i.* of hair etc., to become twisted into a mat. **matting,** *n.* matwork; mats; material for mats; the making of mats; a coarse fabric esp. for packing and covering.
mat[2] (mat), *a.* dull, lustreless, not glossy. *n.* a dull, lustreless surface, groundwork, border etc., esp. in metal roughened or frosted. *v.t.* to dull; to give a wet surface or appearance to.
matador (mat'ədaw), *n.* in Spanish bullfights the man who has to kill the bull; one of the three principal cards in ombre and quadrille; a game played with dominoes.
match[1] (mach), *n.* a person or thing, equal, like, or corresponding to another; a counterpart, a facsimile; one able to cope with another; a contest of skill, strength etc.; a pairing or alliance by marriage; one eligible for marrying. *v.t.* to be a match for; to compare as equal; to oppose as equal; to oppose (against or with) as a rival, opponent etc.; to be the equal of, to correspond, to join. *v.i.* to agree, to be equal, to tally (of different things or persons); to be married. **to meet one's match,** to encounter someone who is equal to or better than one in combat, skill, argument etc. **matchboard,** *n.* a board having a tongue along one edge and a corresponding groove on the other for fitting into similar boards. **matchmaker,** *n.* one fond of planning and bringing about marriages. **matchmaking,** *n.,* *a.* **match play,** *n.* (*Golf*) scoring according to the number of holes won and lost instead of strokes taken. **match point,** *n.* the point that needs to be won in order for a match to be won in tennis, squash etc. **matchable,** *a.* **matcher,** *n.* **matchless,** *a.* without equal, incomparable. **matchlessly,** *adv.*
match[2] (mach), *n.* a small strip of wood or taper tipped with combustible material for producing or communica-

ting fire; a fuse burning at a uniform rate for firing charges. **matchbox**, *n.* a box for holding matches. **matchlock**, *n.* the lock of an obsolete musket fired by means of a lighted match; a musket so fired. **matchstick**, *n.* the wooden part of a match. *a.* very thin; (of drawn figures, etc) straight and thin. **matchwood**, *n.* wood suitable for making matches; wood reduced to small splinters.

mate[1] (māt), *n.* a companion, a comrade, a fellow-worker, an equal, a match; a spouse; a suitable partner, esp. in marriage; one of a pair of the lower animals, esp. birds, associated for breeding; an officer in a merchant ship ranking below the captain; an assistant to the surgeon, cook etc.; an assistant to a plumber etc. *v.t.* to match, to couple; to join together in marriage; to pair (birds); to vie with. *v.i.* to pair. **mat(e)y**, *a.* (*coll.*) friendly.

mate[2] (māt), *v.t.* to checkmate; to confound, to paralyse. *a.* confounded, paralysed. *n.* a checkmate. **fool's mate** FOOL. **smothered mate** SMOTHER.

matelot (mat'əlō), *n.* (*coll. facet.*) a sailor.

mater (mā'tə), *n.* (*usu. facet.*) a mother. **materfamilias** (mahtəfəmil'ias), *n.* the mother of a family.

material (mətiə'riəl), *a.* pertaining to or consisting of matter; corporeal, substantial; pertaining to or concerning the matter or essence of a thing, not to the form; important, momentous, essential. *n.* the substance or matter from which anything is made; stuff, fabric; elements or component parts (of); notes, ideas etc. for a written or oral composition; a person or persons suitable to fulfil a specified function after training etc. **raw material** RAW. **materialism** (-iəlizm), *n.* the theory that there is nothing in the universe but matter, that mind is a phenomenon of matter, and that there is no ground for assuming a spiritual First Cause; regard for secular to the neglect of spiritual interests; (excessive) devotion to the pursuit of material wealth and physical well-being. **materialist**, *n.* materialistic (-lis'-), *a.* **materialistically**, *adv.* **materialize, -ise**, *v.t.* to make material, to invest with matter or corporeity; to cause (a spirit) to become material or to appear; to make materialistic. *v.i.* of a spirit, to appear; to become actual fact. **materialization**, *n.* **materially**, *adv.* in a material way; to a significant extent.

materia medica (mətiə'riə med'ikə), *n.* a general term for the different substances employed in medicine; the scientific study of such substances.

materiel (mətiəriel'), *n.* the material, supplies, machinery or instruments, as distinguished from the personnel or persons, employed in an art, business, military or naval activity, etc.

maternal (mətœ'nəl), *a.* motherly; pertaining to a mother or to maternity; connected or related on the mother's side. **maternally**, *adv.* **maternity**, *n.* motherhood; motherliness. **maternity leave**, *n.* paid leave granted to a woman having a baby.

mat(e)y (mā'ti), MATE[1].

math (math), *n.* (*N Am., coll.*) short for MATHEMATICS.

mathematical (mathəmat'ikl), *a.* pertaining to mathematics; rigidly precise or accurate. **mathematically**, *adv.* **mathematician** (-tish'-), *n.*

mathematics (mathəmat'iks), *n.* the science of quantity, magnitude as expressed by numbers; the mathematical calculations involved in a particular problem, area of study etc. **applied mathematics**, the application of pure mathematics to branches of physical research, as mechanics, astronomy, etc. **pure mathematics**, the abstract science of magnitudes, etc.

maths (maths), *n.* (*coll.*) short for MATHEMATICS.

Matilda (mətil'də), *n.* (*Austral., coll.*) a swag, a bag of belongings. **waltzing Matilda**, carrying the swag (q.v.).

matinal (mat'inəl), *a.* of, pertaining to or occurring in the afternoon.

matinée (mat'inā), *n.* an afternoon performance. **matinée jacket**, *n.* an infant's top garment of wool or material.

matins (mat'inz), *n.pl.* one of the canonical hours of the Roman Catholic breviary, properly recited at midnight but also at daybreak; the daily office of morning prayer in the Anglican Church; a morning song as of birds.

matriarch (mā'triahk), *n.* a woman regarded as at once ruler and mother; a venerable or patriarchal lady. **matriarchal** (-ah'-), *a.* **matriarchalism** (ah'-), *n.* **matriarchy**, *n.* a social system in which the mother is head of the family, or in which descent is reckoned through the female line.

matric (mətrik'), *n.* (*coll.*) short for MATRICULATION.

matricide (mā'trisid, mat'-), *n.* one who murders his mother; the murder of a mother. **matricidal** (-sid'-), *a.*

matriculate (mətrik'ūlāt), *v.t.* to enter in a register, to admit to membership of a body or society, esp. a college or university. *v.i.* to be admitted as a member or student; to pass the examination formerly required to ensure such admission. *a.* matriculated. *n.* one who has matriculated. **matricular**, *n.*, *a.* **matriculation**, *n.* the examination that must be passed to matriculate; the act of matriculating.

matrilineal (matrilin'iəl), *a.* by succession through the mother. **matrilineally**, *adv.*

matrimony (mat'riməni), *n.* the act of marrying; the state of being married, marriage, wedlock; a card-game; the combination of king and queen of one suit in this and other games. **matrimonial** (-mō'-), *a.*

matrix (mā'triks), *n.* (*pl.* **matrices**, **matrixes**) the womb; a place where anything is generated or developed; (*Biol.*) the formative part from which a structure is produced, intercellular substance; a mould in which anything, esp. type or a die, is cast or shaped; the concave bed into which a stamp or die fits; a mass of rock in which a mineral or fossil is embedded, also the impression left by a fossil, crystal etc. after its removal from the rock; an array of numbers or symbols with special mathematical properties.

matron (mā'trən), *n.* a married woman, esp. an elderly one; (*formerly*) the head of the nursing staff in a hospital; the female superintendent of an institution. **matron of honour**, a bride's principal married attendant at a wedding. **matronage**, *n.* **matronal**, *a.* **matronize, -ise**, *v.t.* to render matronlike; to chaperon; (*facet.*) to patronize. **matronly**, *a.*, *adv.*

matt (mat), MAT[2].

matter (mat'ə), *n.* that which constitutes the substance of physical things; that which has weight or mass, occupies space and is perceptible to the senses; physical substance as distinguished from thought, mind, spirit etc., meaning, sense or substance (of a book, discourse etc.); (*Log.*) content as opposed to form; a subject for thought or feeling; an object of or for attention; an affair, a business; the cause or occasion of or for difficulty, regret etc.; importance, moment; an indefinite amount, quantity or portion; (*Print.*) type set up; (*Law*) a statement or fact forming the ground of an action etc.; purulent substance in an abscess, pus. *v.i.* to be of moment, to signify. **a matter of course**, what may be expected in the natural course of events. **a matter of fact**, a reality, a fact. **for that matter**, so far as that is concerned. **in the matter of**, as regards. **no matter**, it does not matter; regardless of. **matter-of-fact**, *a.* treating of or adhering to facts or realities; not fanciful or imaginary; commonplace, prosaic, plain, ordinary.

matting MAT[1].

mattock (mat'ək), *n.* a kind of pick with one broad adze-edged end, for loosening ground, severing roots etc.

mattress (mat'ris), *n.* a case of strong material stuffed with straw, foam rubber, etc, often with springs, used for the bottom of a bed.

maturate (mat'ūrāt), *v.t.* to mature; to promote suppuration in. *v.i.* to ripen, to suppurate perfectly. **maturation**,

mature *n.* **maturative** (tū'-), *n.*, *a.*
mature (mətūə'), *a.* ripe; ripened; completely developed; fully grown; fully elaborated, considered etc.; become payable (as a bill); in a state of perfect suppuration. *v.t.* to bring to a state of ripeness or complete development; to bring to a state of suppuration. *v.i.* to become ripened or fully developed; of a bill, to become payable. **maturation,** *n.* the attainment of maturity, the completion of growth. **maturely,** *adv.* **matureness, maturity,** *n.*
matutinal (mətūtī'nəl), **matutine** (mat'ūtin), *a.* pertaining to the morning; early.
maty (mā'ti), MATE².
matzo (mat'sō), *n.* (*pl.* **matzoth, -sət, -os**) (a thin wafer of) unleavened bread, eaten esp. at the Passover.
maudlin (mawd'lin), *a.* muddled with drink; characterized by sickly sentimentality, mawkish. *n.* mawkish sentimentality.
maul (mawl), *n.* a heavy wooden hammer; a loose scrum in Rugby; a tussle, struggle. *v.t.* to beat, to bruise (as with a maul); to handle roughly; to damage.
maulstick (mawl'stik), *n.* a light stick with a round pad at the end used as a rest for the right hand by painters.
maunder (mawn'də), *v.i.* to grumble, to mutter; to talk incoherently, to ramble; to act or move about aimlessly. *v.t.* to utter in a grumbling or incoherent manner. **maunderer,** *n.*
maundy (mawn'di), *n.* the ceremony of washing the feet of poor people in commemoration of Christ's performing this office for His disciples; a distribution of alms following this. **maundy money, penny,** *n.* silver money specially struck and distributed on Maundy Thursday. **Maundy Thursday,** *n.* the day before Good Friday, when the royal alms or maundy money is distributed by the royal almoner.
mausoleum (mawsəlē'əm), *n.* (*pl.* **-lea, -leums**) the stately tomb of Mausolus, king of Caria, erected by his widow Artemisia, and reckoned one of the seven wonders of the world; a sepulchral monument of considerable size or architectural pretensions.
mauve (mōv), *n.* a purple- or lilac-coloured aniline dye; the colour of this. *a.* of this colour.
maverick (mav'ərik), *n.* (*N Am.*) an unbranded beast; anything got hold of dishonestly; an irresponsible or independent person. *v.t.* to brand (a stray beast); hence, to seize or appropriate illegally.
mavis (mā'vis), *n.* the song-thrush.
maw (maw), *n.* the stomach of lower animals, esp. the fourth stomach of ruminants; the crop of birds; (*facet.*) the human stomach; the mouth. **mawworm,** *n.* an intestinal worm.
mawkish (maw'kish), *a.* apt to cause satiety or loathing; sickly, insipid; falsely or feebly sentimental. **mawkishly,** *adv.* **mawkishness,** *n.*
maxi (mak'si), *n.*, *a.* short for MAXIMUM, esp. (*n.*) a coat, skirt etc. reaching the ankles.
maxi-, *comb. form* very large or long.
maxilla (maksil'ə), *n.* (*pl.* **-lae, -lē**) one of the jaw-bones, esp. the upper in mammals. **maxillary,** *a.* the part of the skull that forms the upper jaw. *a.* pertaining to a jaw or maxilla.
maxim (mak'sim), *n.* a general principle of a practical kind; a rule derived from experience; (*Law*) an established or accepted principle.
maximal (mak'siml), *a.* of the greatest, largest etc. size, rate etc.; of an upper limit. **maximally,** *adv.*
maximin (mak'simin), *n.* the maximum of a set of minima, esp. of minimum gains in game theory. [*maximum, minimum*]
maximum (mak'siməm), *n.* (*pl.* **-ma**) the greatest quantity or degree attainable in any given case. *a.* greatest; at the greatest or highest degree. **maxima and minima,** the greatest and least values of a variable quantity. **maximize, -ise,** *v.t.* to raise to a maximum; to increase to the utmost extent; to hold rigorous opinions in matters of faith. *v.i.* to interpret doctrines in the most rigorous way.
maxwell (maks'wəl), *n.* a cgs unit of magnetic flux.
May (mā), *n.* the fifth month of the year; the springtime of life, youth; hawthorn blossom, from its appearing in May; Mayday festivities. **May-apple,** *n.* a N American herb with a single white flower and an edible egg-shaped fruit. **may-blossom,** *n.* hawthorn bloom. **May-bug,** *n.* the cockchafer. **Mayday,** *n.* the first of May as a spring festival or, in some countries, as a public holiday in honour of workers. **mayduke,** *n.* a variety of cherry said to have been introduced from Médoc. **mayflower,** *n.* a flower blooming in May, as the cowslip, lady's smock or hawthorn; (*N Am.*) the trailing arbutus. **mayfly,** *n.* an ephemeral insect, esp. *Ephemera vulgata* or *E. dania*; an angler's fly made in imitation of this; the caddis-fly. **maypole,** *n.* a pole decorated with garlands etc., round which people dance on Mayday. **May-queen,** *n.* a young girl chosen to act as queen of the games on Mayday.
may (mā), *aux.v.* (*past* **might** (mīt)) expressing possibility, ability, permission, desire, obligation, contingency or uncertainty. **maybe,** *adv.* perhaps, possibly.
Maya (mī'ə), *n.* an Indian of the native tribes of Yucatan, Honduras etc.; the language of these tribes.
maya (mī'ə), *n.* in Hinduism, the world as perceived by the senses, regarded as illusory.
Mayday (mā'dā), *n.* the international radiotelephone distress signal. [F *m'aider*, help me]
mayhem (mā'hem), *n.* (*formerly*) the offence of maiming a person; wilful damage; a state of disorder or confusion.
mayonnaise (māənāz'), *n.* a thick sauce or salad-dressing made of egg-yolk, vinegar etc.; a dish with this as a dressing, as *egg mayonnaise*.
mayor (meə), *n.* the chief officer of a city or borough. **Lord Mayor** LORD. **mayoral,** *a.* **mayoralty,** *n.* **mayoress,** *n.* a female mayor; the wife of a mayor, or a woman who assists the mayor in official duties.
mayweed (mā'wēd), *n.* the stinking camomile, *Anthemis cotula*; other composite plants, esp. the feverfew.
mazarine (maz'ərēn), *n.*, *a.* a deep rich blue.
maze (māz), *n.* a labyrinth, a confusing network of winding and turning passages; a state of bewilderment, uncertainty, perplexity. **mazy,** *a.* involved, winding, perplexing, intricate; giddy, dizzy.
mazurka (məzœ'kə), *n.* a lively Polish dance like the polka; the music for this.
MB, (*abbr.*) Bachelor of Medicine; (*Comput.*) megabyte.
MBA, (*abbr.*) Master of Business Administration.
MBE, (*abbr.*) Member of (the Order of) the British Empire.
mbx, *n.* (*Comput.*) a message transferred from one computer terminal to another. *v.t.* to send messages, electronic mail etc. between computer terminals. [acronym for *mailbox*]
MC, (*abbr.*) Master of Ceremonies; Member of Congress; Military Cross.
MCC, (*abbr.*) Marylebone Cricket Club.
MCh, (*abbr.*) Master of Surgery.
MCP, (*abbr.*) male chauvinist pig.
MD, (*abbr.*) Doctor of Medicine; Managing Director.
Md., (*abbr.*) Maryland (US).
ME, (*abbr.*) Mechanical Engineer; Military Engineer; Middle English; myalgic encephalomyelitis.
Me., (*abbr.*) Maine (US).
me[1] (mē, mi), *pers. pron.* the dative and objective of the first personal pronoun.
me[2] (mē), MI.
mea culpa (māə kŭl'pə, kul'-), by my fault. [L]
mead[1] (mēd), *n.* a fermented liquor made from honey, water and spices.
mead[2] (mēd), *n.* (*poet.*) a meadow.
meadow (med'ō), *n.* a tract of land under grass, esp. if

grown for hay; low, rich, moist ground, esp. near a river. **meadow-lark**, *n.* an American songbird. **meadow pipit**, *n.* a brown and white European songbird. **meadow-saffron**, *n.* a plant of the genus *Colchicum*, also called autumn crocus. **meadow-sweet**, *n.* a rosaceous plant with white, plumy, fragrant flowers. **meadowy**, *a.*

meagre (mē'gə), *a.* lean, thin, wanting flesh; destitute of richness, fertility or productiveness; poor, scanty. **meagrely**, *adv.* **meagreness**, *n.*

meal[1] (mēl), *n.* food taken at one of the customary times of eating, a repast; the occasion or usual time of this. **to make a meal of**, to exaggerate the importance, difficulty etc. of. **meals-on-wheels**, *n.* a scheme by which precooked meals are delivered by vehicles to the housebound, needy etc. **meal ticket**, *n.* a ticket given in exchange for a meal, often at a subsidized price; (*coll.*, *often derog.*) a person upon whom one can depend for financial support. **meal-time**, *n.*

meal[2] (mēl), *n.* the edible portion of grain or pulse ground into flour. **meal-worm**, *n.* the larva of a beetle that infests meal. **mealy**, *a.* of, containing or resembling meal; powdery, friable, floury; farinaceous; besprinkled with or as with meal, spotty; pale (of the complexion); mealy-mouthed. **mealy bug**, *n.* an insect infesting vines and hot-house plants. **mealy-mouthed**, *a.* soft-spoken, hypocritical.

mealie (mē'li), *n.* (*usu. pl.*) maize.

mean[1] (mēn), *v.t.* (*past*, *p.p.* **meant**, ment) to have in the mind; to purpose, to intend; to design, to destine (for); to denote, to signify; to intend to convey or to indicate. *v.i.* to have a specified intention or disposition. **to mean business** BUSINESS. **to mean well**, to have good intentions. **meaning**, *n.* that which is meant, significance, import. *a.* significant, expressive. **meaningful**, *a.* **meaningfully**, *adv.* **meaningfulness**, *n.* **meaningless**, *a.* **meaninglessly**, *adv.* **meaninglessness**, *n.* **meaningly**, *adv.*

mean[2] (mēn), *a.* occupying a middle position; equidistant from two extremes; not extreme, moderate, not excessive; intervening; (*Math.*) intermediate in value between two extremes, average. *n.* the middle point, state, course, quality or degree between two extremes; (*Math.*) a quantity intermediate between two extremes, an average. **means**, *n.pl.* that by which anything is done or a result attained; available resources, income, wealth. **(a man etc.) of means**, a wealthy man etc. **by all means**, certainly, undoubtedly. **by any means**, in any way possible, somehow; at all. **by fair means or foul**, by any means whatsoever. **by means of**, by the agency or instrumentality of. **by no means**, certainly not, on no account whatever. **means-test**, *n.* the official investigation into the means of a person applying for pension, dole etc. **meantime**, **-while**, *adv.* in the intervening time. *n.* the interval between two given times.

mean[3] (mēn), *a.* low in quality, capacity, value, rank etc.; inferior, poor, inefficient, shabby; low-minded, petty, stingy; shabby, contemptible, miserly; ignoble, of no account, disreputable; despicable; (*coll.*) having or showing great skill, excellent. **mean-spirited**, *a.* **mean-spiritedly**, *adv.* **meanie** (-ni), *n.* (*coll.*) a petty-minded or miserly person. **meanly**, *adv.* **meanness**, *n.*

meander (mian'də), *n.* (*usu. pl.*) a tortuous or intricate course or bend; (*usu. pl.*) a winding, a circuitous path or movement, a deviation; a decorative pattern, fretwork etc. composed of intricately interlacing lines. *v.i.* to wander, wind or flow in a tortuous course. **meandering**, *a.*

meant, *past*, *p.p.* MEAN[1].

measles (mē'zlz), *n.pl.* a contagious viral disease, indicated by a red papular rash, usu. attacking children; applied to the effects of a cystic worm in swine and oxen. **German measles** GERMAN. **measled**, *a.* **measly**, *a.* infected with measles; (*coll.*) worthless, paltry, meagre.

measure (mezh'ə), *n.* the extent or dimensions of a thing as determined by measuring; the measurements necessary to make an article of dress; a standard of measurement; a definite unit of capacity or extent; an instrument for measuring, as a rod, tape etc., or a vessel of standard capacity; a system of measuring; the act of measuring, measurement; a quantity measured out taken as a rule or standard; prescribed or allotted extent, length or quantity; limit, moderation, just degree or amount; metre, poetical rhythm; an action to achieve a purpose; a law, a statute, an Act of Parliament; (*Geol.*) (*pl.*) a series of beds, strata; (*Mus.*) time, pace, the contents of a bar. *v.t.* to determine the extent or quantity of by comparison with a definite unit or standard; to take the dimensions of; to weigh, to judge, to value or estimate by comparison with a rule or standard; to serve as the measure of; to allot or apportion by measure; to travel over, to cover; to survey, look up and down; to bring into competition (with). *v.i.* to take measurements; to be in extent, to show by measurement. **beyond measure**, exceedingly, excessively. **for good measure**, as an additional amount. **in a measure**, to some extent, in a certain degree. **short measure**, less than the due amount. **to measure up to**, to be adequate for. **to take measures**, to adopt means, to take steps (to). **to take someone's measure**, to measure someone for clothes; to find out what kind of a person someone is. **within measure**, in moderation. **without measure**, immoderately. **measurable**, *a.* **measurably**, *adv.* **measured**, *a.* of definite measure; deliberate and uniform; rhythmical; well-considered, carefully weighed. **measureless**, *a.* **measurement**, *n.* **measurer**, *n.* **measuring**, *n.*, *a.* **measuring jug**, *n.* a graduated jug used for measuring ingredients in cooking.

meat (mēt), *n.* the flesh of animals, usu. excluding fish and fowl, used as food; solid food of any kind; the partaking of food, a meal; the edible part of a nut, egg, shell-fish etc.; the substance of something, the pith. **meat-ball**, *n.* a ball of minced meat, eaten e.g. with a sauce and spaghetti. (*esp. N Am.*) a stupid person. **meat loaf**, *n.* a loaf-shaped mass of minced or chopped meat, cooked and often eaten cold. **meaty**, *a.* containing much meat; of or like meat; substantial, pithy. **meatiness**, *n.*

Mecca (mek'ə), *n.* (*fig.*) a holy place; the object of one's aspirations; a place frequently visited.

Meccano[®] (mikah'nō), *n.* a set of toy engineering parts that can be built up into various mechanical models.

mechanic (mikan'ik), *n.* an artisan; a skilled workman; one who is employed or skilled in repairing or maintaining machines; (*pl.*) the branch of physics treating of the motion and equilibrium of material bodies; also the science of machinery; (*pl.*) the practical details of an operation, project etc. *a.* mechanical; industrial; pertaining to or of the nature of machinery, machine-like. **mechanical**, *a.* pertaining to mechanics; in accordance with physical laws; acting or affected by physical power without chemical change; pertaining to or acting as machinery or mechanism; produced by machinery; of or pertaining to handicraft; working with tools or machinery; machine-like, automatic, done from force of habit; slavish, unoriginal. **mechanical engineering**, *n.* the branch of engineering concerned with the design and production of machinery. **mechanically**, *adv.* **mechanicalness**, *n.* **mechanician** (mek-), *n.*

mechanism (mek'ənizm), *n.* the structure or correlation of parts of a machine; machinery; a system of correlated parts working reciprocally together, as a machine; in art, mechanical execution as distinguished from style etc., technique; the philosophical doctrine that phenomena can be explained purely in terms of physical interactions. **mechanist** (mek'-), *n.* **mechanistic** (-nis'-), *a.* **mechanistically**, *adv.*

mechanize, **-ise** (mek'əniz), *v.t.* to make mechanical; to equip (troops) with armoured vehicles. **mechanization**, *n.*

mechanotherapy (mikanəthe'rəpi), *n.* the treatment of disease through the use of mechanical appliances.
Mechlin (mek'lin), *n.* a light lace made at Mechlin (Malines), near Brussels.
meconic (mikon'ik), *a.* contained in or derived from the poppy. **meconin** (mek'ənin, mē'-), *n.* a neutral substance existing in opium. **meconium** (mikō'niəm), *n.* inspissated poppy juice; the first faeces of infants consisting of excretions from the liver etc. **Meconopsis** (měkənop'sis), *n.* a genus of flowering plants related to and resembling the poppy.
MEd, (*abbr.*) Master of Education.
med, (*abbr.*) medical, medicine; mediaeval; medium.
medal (med'l), *n.* a piece of metal, often in the form of a coin, stamped with a figure and inscription to commemorate some illustrious person or event. **medallion** (-dal'yən), *n.* a large medal; (*Arch.*) a tablet or panel, usually round or oval, containing painted or sculptured figures, decorations etc. **medallist,** *n.* one who designs or engraves medals; a collector of or dealer in medals; one who has gained a medal.
meddle (med'l), *v.i.* to interfere (in) officiously; to concern or busy oneself (with) unnecessarily. **meddler,** *n.* **meddlesome,** *a.* **meddlesomeness,** *n.*
media[1], *pl.* MEDIUM.
media[2] MASS[2].
mediacy MEDIATE.
mediaeval, medieval (mediē'vl), *a.* of, or pertaining to, or characteristic of the Middle Ages. *n.* one who lived in the Middle Ages. **medi(a)evalism,** *n.* **medi(a)evalist,** *n.* **medi(a)evally,** *adv.*
medial (mē'diəl), *a.* pertaining to or situated in the middle, intermediate; mean or average.
median (mē'diən), *a.* situated in the middle, esp. in the median plane, dividing anything longitudinally into two equal halves. *n.* a straight line joining the vertex of a triangle to the mid-point of the opposite side; in statistics, the middle value in a number sequence. **medianly,** *adv.*
mediant (mē'diənt), *n.* (*Mus.*) the third tone of any scale.
mediastinum (mēdiosti'nəm), *n.* (*pl.* **-na**) a membranous septum or cavity between the two main parts of an organ etc., esp. the folds of the pleura between the right and left lung. **mediastinal,** *a.*
mediate (mē'diət), *a.* situated in the middle or between two extremes; intervening, indirect, secondary; serving or acting as an intervening or indirect means or agency; effected or connected by such means. *v.t.* (-āt), to interpose between (parties) in order to reconcile them; to effect by means of intervention. *v.i.* to interpose (between) in order to reconcile parties etc.; to serve as connecting link or medium (between). **mediately,** *adv.* **mediation,** *n.* **mediator,** *n.* **mediatory,** *a.,* *n.*
medic (med'ik), *n.* (*coll.*) a medical student; a physician; a doctor.
medical (med'ikl), *a.* pertaining to, connected with or employed in medicine; curative, healing, medicinal; pertaining to medicine as opposed to surgery etc. *n.* (*coll.*) a medical student; an examination to ascertain a person's state of physical fitness. **medical certificate,** *n.* a document issued by a doctor stating that a person is unfit for work etc. **medical examiner,** *n.* (*N Am.*) a public official, usu. a physician, appointed to inquire into cases of sudden or suspicious death. **medical jurisprudence** FORENSIC MEDICINE. **medicable,** *a.* able to be treated or cured. **medically,** *adv.* **medicament** (medik'ə-, med'-), *n.* a healing substance or application. **medicate,** *v.t.* to impregnate with anything medicinal; to treat medically. **medication,** *n.* a medicine or drug; treatment with medicine or drugs. **medicative** (-kə-), *a.*
medicine (med'sin. -isin), *n.* a substance, usu. taken internally, used for the alleviation or removal of disease; the art or science of preserving health and curing or alleviating disease, esp. as distinguished from surgery and obstetrics; a term applied by the N American Indians to anything supposed to possess supernatural powers or influence, a charm, a fetish. *v.t.* to treat or cure with or as with medicine. **a taste of one's own medicine,** unpleasant treatment given to one in retaliation. **medicine ball,** *n.* a heavy ball thrown from one person to another as physical exercise. **medicine chest,** *n.* a box, cupboard etc. containing medicine, bandages etc. **medicine-man,** *n.* a witch-doctor; a magician. **medicinal** (-dis'-), *a.* **medicinally,** *adv.* **medico** (med'ikō), *n.* (*facet.*) a physician; a doctor; a medical student.
medieval MEDIAEVAL.
mediocre (mēdiō'kə), *a.* of middling quality; indifferently good or bad, average, commonplace. **mediocrity** (-ok'-), *n.* the state of being mediocre; a mediocre person.
meditate (med'itāt), *v.i.* to ponder, to engage in thought (upon); to muse, to cogitate; to engage in contemplation, esp. on religious or spiritual matters. *v.t.* to dwell upon mentally; to plan, to design, to intend. **meditation,** *n.* **meditative** (-tə-), *a.* **meditatively,** *adv.* **meditator,** *n.*
Mediterranean (mediterā'niən), *a.* pertaining to the Mediterranean Sea or the countries surrounding it. *n.* the sea between Europe and Africa; a native of a Mediterranean country.
medium (mē'diəm, měd'yəm), *n.* (*pl.* **-dia, -diums**) anything serving as an intermediary, agent or instrument; instrumentality, agency; an intervening substance or element, such as the air or ether, through which forces act, impressions are conveyed etc.; a substance in which germs are developed; a means of communication; an instrument of exchange, as money; a middle or intermediate object, quality, degree etc.; (*Painting*) a liquid vehicle for dry pigments; the middle term of a syllogism; (*pl.* **-diums**) a person claiming to receive communications from the spirit world. *a.* intermediate in quantity, quality or degree; average, moderate; middling, mediocre. **medium waves,** *n. pl.* radio waves of wavelength between 100 and 1000 metres.
medlar (med'lə), *n.* a rosaceous tree the fruit of which is eaten when beginning to decay.
medley (med'li), *n.* a mixed or confused mass, esp. of incongruous objects, persons, materials etc.; a musical or literary miscellany. *a.* mixed, multifarious, motley.
Médoc (mādok'), *n.* a red wine from *Médoc,* a district in Gironde, SW France.
medulla (midŭl'ə), *n.* (*pl.* **-llas, -llae, -ē**) the marrow of bones, esp. that of the spine; the spinal cord; the inner part of certain organs, as the kidneys; the pith of hair; the internal tissue or pith of plants. **medulla oblongata** (oblong·gah'tə), *n.* the elongated medulla or continuation of the spinal cord forming the hindmost segment of the brain. **medullary, medullar,** *a.*
medusa (midū'zə, -sə), *n.* (*pl.* **-sas, -sae, -zē, sē**) a jellyfish. **medusan, -oid,** *n., a.*
meek (mēk), *a.* mild, submissive, humble, tame, gentle, forbearing. **meekly,** *adv.* **meekness,** *n.*
meerkat (miə'kat), *n.* a small, carnivorous lemur-like mammal of southern Africa.
meerschaum (miə'shəm), *n.* a white compact hydrous silicate of magnesia, used for tobacco-pipes; a pipe made of this.
meet[1] (mēt), *v.t.* (*past, p.p.* **met,** met) to come face to face with; to go to a place so as to join or receive; to reach and touch or unite with (of a road, railway, etc.); to encounter, to confront, to oppose; to experience; to refute; to answer, to satisfy; to pay, to discharge. *v.i.* to come together; to assemble; to come into contact; to be united. *n.* a meeting of persons and hounds for hunting, or of cyclists, athletes etc.; the persons assembled or the place appointed for a meet; (*Austral.*) an appointment. **to meet halfway,** to compromise with. **to meet one's maker** MAKE.

to meet the eye or **ear**, to be seen or heard. **to meet with**, to come across; to experience; to encounter, to engage.
meeting, *n*. a coming together, an assembly; the persons assembled; a duel; a race-meeting; a conflux, intersection.
meeting-house, *n*. a dissenting place of worship, esp. of Quakers.
meet[2] (mēt), *a*. fit, proper, suitable. **meetly**, *adv*. **meetness**, *n*.
meg(a)-, megal(o)-, *comb*. *form* great, large; one million; (*coll*.) great in number, significance, impressiveness etc.
mega (meg'ə), *a*. (*coll*.) very large in number; very important.
megabit (meg'əbit), *n*. (*Comput*.) one million bits; 2^{20} bits.
megabyte (meg'əbīt), *n*. (*Comput*.) one million bytes; 2^{20} bytes.
megacephalic (megəsifal'ik), **megacephalous** (-sef'ələs), *a*. large-headed.
megacycle (meg'əsīkl), *n*. a frequency of a million cycles per second, a megahertz.
megadeath (meg'ədeth), *n*. one million deaths, esp. in nuclear war.
megahertz (meg'əhœts), *n*. a unit of frequency equal to one million hertz.
megalith (meg'əlith), *n*. a great stone; a megalithic monument, as a cromlech, stone circle etc. **megalithic** (-lith'-), *a*.
megalomania (megəlǝmā'niə), *n*. a form of mental disorder characterized by self-exaltation; a craze for over-statement etc. **megalomaniac** (-ak), *n*., *a*.
megalopolis (megǝlop'əlis), *n*. a large, densely-populated urban area.
Megalosaurus (megǝlǝsaw'rǝs), *n*. an extinct genus of gigantic carnivorous lizards from the Oolite.
megaphone (meg'əfōn), *n*. an apparatus for enabling persons to converse at a long distance; a large speaking-trumpet.
megapod (meg'əpod), *n*. an Australian or Malaysian mound-bird.
megastar (meg'əstah), *n*. a very popular, internationally-known star of the cinema, theatre etc. **megastardom**, *n*.
megaton (meg'ətūn), *n*. one million tons; a unit of explosive power in nuclear weapons, equal to a million tons of TNT.
megavolt (meg'əvōlt), *n*. one million volts.
megawatt (meg'əwot), *n*. one million watts.
Megger® (meg'ə), *n*. an instrument for measuring high resistances.
megilp (məgilp'), *n*. a vehicle for colours, consisting of a compound of linseed-oil and mastic varnish.
megohm (meg'ōm), *n*. one million ohms.
meiosis (miō'sis, mi-), *n*. litotes, depreciative hyperbole; the stage of a malady when the symptoms tend to abate; the diminution of the number of chromosomes in the cell nucleus.
Meissen (mī'sn), *n*. a type of fine porcelain first produced at *Meissen* near Dresden in the 18th century.
meistersinger (mī'stəsingə), *n*. a German burgher poet and musician of the 14th–16th cent., one of the successors of the minnesingers.
melamine (mel'əmēn), *n*. a white crystalline compound used for making synthetic resins; a resin made from this, used in moulded products, adhesives, coatings etc.
melan- MELAN(O)-.
melancholia (melənkō'liə), *n*. a mental disorder, often preceding mania, characterized by lowness of spirits, frequently with suicidal tendencies (formerly supposed to be due to excess of black bile).
melancholy (mel'ənkəli), *n*. a gloomy, dejected state of mind; sadness, gloom, depression, despondency; (*poet*.) pensive contemplation; melancholia. *a*. sad, gloomy, depressed in spirits; mournful, saddening; pensive; afflicted with melancholia. **melancholic** (-kol'-), *a*. **melancholically**, *adv*.
Melanesian (melənē'zhən), *a*. of or pertaining to Melanesia, the group of islands in the Pacific ocean lying to the east of New Guinea. *n*. a native or inhabitant of Melanesia.
mélange (mālǎzh'), *n*. a mixture, medley or miscellany; a mixed worsted yarn.
melanic (milan'ik), *a*. black, dark-complexioned; applied to the black pigment characteristic of melanosis. **melanin** (mel'-), *n*. a black or dark brown pigment occurring in the hair and skin of dark-skinned races. **melanism** (mel'-), *n*. excess of colouring-matter in the skin, hair and tissues; a disease producing blackness in plants. **melanistic** (melənis'-), *a*. **melanoid** (mel'-), *a*. **melanoma** (-nō'mə), *n*. a malignant tumour with dark pigmentation, esp. on the skin. **melanosis** (-nō'-), *n*. an organic affection, characterized by a deposit of black pigment in the tissues. **melanotic** (-not'-), *a*.
melanite (mel'ənīt), *n*. a black variety of garnet.
Melba toast (mel'bə), *n*. very thin crisp toast.
meld[1] (meld), *v.t.*, *v.i*. (*Cards*) to declare for a score.
meld[2] (meld), *v.t.*, *v.i*. to mix, blend, combine.
mêlée (mel'ā), *n*. a confused hand-to-hand fight, an affray.
meliorate (mē'liərāt), *v.t*. to make better. *v.i*. to grow better. **melioration**, *n*.
meliphagous (məlif'əgəs), *a*. belonging to the family of birds Meliphagidae or honey-eaters.
melisma (məliz'mə), *n*. (*pl*. **-mata**, -tə) a melodic embellishment; a group of notes sung to a single syllable. **melismatic** (-mat'-), *a*.
melliferous (məlif'ərəs), *a*. producing or yielding honey.
mellifluous (məlif'luəs), *a*. flowing smoothly and sweetly. **mellifluent**, *a*. **mellifluence**, *n*.
mellite (mel'īt), *n*. native mellitate of aluminium, honeystone.
mellow (mel'ō), *a*. fully ripe, pulpy, sweet; of earth, rich, friable; of tones and colours, soft and rich; ripened or softened by age and experience; genial, kindly; (*coll*.) jolly, half tipsy. *v.t*. to ripen, mature, soften. *v.i*. to become ripe, mature or softened, by age etc. **mellowly**, *adv*. **mellowness**, *n*. **mellowy**, *a*.
melodeon, -dion (məlō'diən), *n*. a wind-instrument with a row of reeds and a keyboard; a type of small accordion.
melodic, melodious etc. MELODY.
melodrama (mel'ədrahmə), *n*. a sensational play, film, novel etc. with a plot characterized by startling situations; orig. a dramatic composition with songs intermixed; sensational and extravagant events, behaviour or speech. **melodramatic** (-mat'-), *a*. **melodramatically**, *adv*. **melodramatist** (-dram'-), *n*. **melodramatize, -ise** (-dram'-), *v.t*.
melody (mel'ədi), *n*. an agreeable succession of sounds, esp. of simple tones in the same key, an air or tune; a simple setting of words to music; the chief part in harmonic music, the air; music. **melodic** (-lod'-), **melodious** (-lō'-), *a*. of, characterized by or producing melody; musical, sounding sweetly. **melodiously**, *adv*. **melodiousness**, *n*. **melodist** (mel'ədist), *n*. **melodize, -ise**, *v.t*. *v.i*.
melon (mel'ən), *n*. a kind of gourd, esp. *Cucumis melo*, the musk-melon, and *Citrullus vulgaris*, the water-melon. **melon-cactus, -thistle**, *n*. a tropical American cactaceous plant.
Melpomene (melpom'ini), *n*. (*Gr. Myth*.) the Muse of tragedy.
melt (melt), *v.i*. (*p.p*. **melted, molten** mōl'tən), to pass from a solid to a liquid state by heat; to dissolve; to be dissipated, to disappear, to vanish (away); to be softened to kindly influences, to give way; to dissolve in tears; to dissolve or blend (into); *v.t*. to make liquid by heat; to dissolve; to soften to tenderness; to dissipate. *n*. a period of melting, a thaw. **meltdown**, *n*. the melting of fuel rods in a nuclear reactor, often causing the escape of radiation

into the environment; an economic collapse. **melting-point**, *n.* the temperature at which a solid begins to melt. **melting-pot**, *n.* a crucible; a situation or place where there is a mixture of races, cultures, ideas etc. **melter**, *n.* **meltingly**, *adv.*

melton (mel'tən), *n.* a jacket worn in hunting; a stout make of cloth without nap, used largely for overcoats.

member (mem'bə), *n.* a limb, a part or organ of the body; the penis; a component part or element of an organism or complex whole; one belonging to a society or body; a branch or division of a society or organization; a set of figures or symbols forming part of a mathematical expression. **Member of Parliament**, one representing a constituency in the House of Commons. **memberless**, *a.* **membership**, *n.* the state of being a member; (a number of) members.

membrane (mem'brān), *n.* a thin sheet of tissue lining or covering parts of an organism; a morbid tissue produced in certain diseases; a skin of parchment or vellum. **membraneous** (-brā'-), **membranous** (-brə-), *a.* of or like a membrane; very thin, translucent.

memento (mimen'tō), *n.* (*pl.* **-os, -oes**), a memorial, a souvenir, a reminder. **memento mori** (maw'ri), *n.* an emblem of mortality, esp. a skull (L., remember you must die).

memo (mem'ō), *n.* (*pl.* **memos**) short for MEMORANDUM.

memoir (mem'wah), *n.* (*usu. pl.*) an account of events or transactions in which the narrator took part; an autobiography or a biography; a communication to some learned society on a special subject. **memoirist**, *n.*

memorabilia (memərəbil'iə), *n.pl.* things worthy to be remembered.

memorable (mem'ərəbl), *a.* worthy to be remembered; notable, remarkable. **memorability** (-bil'-), *n.* **memorably**, *adv.*

memorandum (memərandəm), *n.* (*pl.* **-dums, -da**, -də) a note to help the memory; a brief record or note; a short informal letter, usu. unsigned, with the sender's name etc. printed at the head; (*Law*) a summary, outline or draft of an agreement etc.

memorial (məmaw'riəl), *a.* preservative of memory; commemorative; preserved in memory. *n.* that which preserves the memory of something; a monument, festival etc. commemorating a person, event etc.; a written statement of facts, esp. of the nature of a petition, remonstrance etc.; an informal diplomatic paper; (*usu. pl.*) a chronicle or record. **memorialize, -ise**, *v.t.* **memorially**, *adv.*

memorize, -ise (mem'əriz), *v.t.* to commit to memory; to learn by heart.

memory (mem'əri), *n.* the mental faculty that retains and recalls previous ideas and impressions; the exercise of this faculty, remembrance, recollection; something that is remembered; the state of being remembered; posthumous reputation; the period during which anything is remembered; (*Comput.*) a device for storing data in a computer; the capacity of a material to return to its former condition after distortion.

mem sahib (mem'sahb), *n.* a term formerly applied to European married women living in India.

men, *pl.* MAN.

menace (men'əs), *n.* a threat; (*coll.*) a nuisance. *v.t.* to threaten. **menacer**, *n.* **menacing**, *a.* **menacingly**, *adv.*

ménage (mānahzh'), *n.* a household; housekeeping, household management. **ménage à trois** (a trwa), an arrangement whereby a couple live together with the lover of one or both of them (F, household of three).

menagerie (mənaj'əri), *n.* a collection of wild animals; a place or enclosure where wild animals are kept.

menarche (mənah'ki), *n.* the first onset of menstruation in a woman's life. **menarcheal**, *a.*

mend (mend), *v.t.* to repair, to restore, to make good; to improve, to make better; to correct, to amend. *v.i.* to grow better, to improve; to amend, to recover health. *n.* the act or process of mending; improvement; a repaired part (in a garment etc.). **on the mend**, improving, recuperating. **mendable**, *a.* **mender**, *n.*

mendacious (mendā'shəs), *a.* given to lying, untruthful. **mendaciously**, *adv.* **mendacity** (-das'-), *n.*

mendelevium (mendəlē'viəm), *n.* an artificially-produced transuranic element, at. no. 101; chem. symbol Md.

Mendelism (men'dəlizm), *n.* a theory of heredity based on researches and generalizations by G. J. *Mendel* (1822–84), Austrian botanist, showing that the characters of the parents of cross-bred offspring reappear by certain proportions in successive generations according to definite laws. **Mendelian** (-dē'-), *a.*

mendicant (men'dikənt), *a.* begging; reduced to beggary. *n.* a beggar; a member of a mendicant order. **mendicant orders**, *n.pl.* monastic orders subsisting on alms. **mendicancy**, *n.* **mendicity** (-dis'-), *n.*

menfolk (men'fōk), *n.pl.* the men, esp. of a particular family or community.

menhaden (menhā'dn), *n.* a N American sea-fish allied to the herring.

menhir (men'hiə), *n.* a prehistoric monument consisting of a tall upright stone.

menial (mē'niəl), *a.* pertaining to or suitable for servants; servile, low, mean. *n.* a domestic servant; one doing servile work. **menially**, *adv.*

meningitis MENINX.

meninx (mē'ningks), *n.* (*pl.* **meninges**, mənin'jēz) one of the three membranes enclosing the brain and spinal cord, comprising the dura mater, arachnoid and pia mater. **meningeal** (mənin'jiəl), *a.* **meningitis** (meninji'tis), *n.* inflammation of the meninges.

meniscus (mənis'kəs), *n.* (*pl.* **-sci**, -sī) a lens convex on one side and concave on the other; the top of a liquid column made convex or concave by capillarity (as mercury in a barometer). **meniscal**, *a.*

Mennonite (men'ənīt), *n.* a member of a Protestant sect originating in Friesland in the 16th cent., with principles similar to those of the Anabaptists.

menopause (men'əpawz), *n.* final cessation of menstruation. **menopausal**, *a.*

menorah (minaw'rə), *n.* a candelabrum with several branches, used in Jewish worship.

menorrhagia (menərā'jiə), *n.* excessive bleeding during menstruation.

menorrhoea (menərē'ə), *n.* ordinary bleeding during menstruation.

menses (men'sēz), *n.pl.* the periodic flow of blood from the uterus of women, usu. occurring once every lunar month; menstruation. **menstrual, menstruous**, *a.* monthly; pertaining to the menses. **menstruant**, *a.* **menstruate**, *v.i.* to undergo menstruation. **menstruation**, *n.* the menses.

menstruum (men'struəm), *n.* (*pl.* **-trua**) any fluid that dissolves a solid, a solvent.

mensurable (men'sūrəbl, -shə-), *a.* measurable; (*Mus.*) having rhythm and measure. **mensurability** (-bil'-), *n.* **mensural**, *a.*

mensuration (mensūrā'shən), *n.* the act or practice of measuring; the branch of mathematics concerned with the determination of lengths, areas and volumes.

-ment (-mənt), *suf.* forming nouns denoting result, state, action etc., as in *agreement, bereavement, enticement, impediment, ornament.*

mental (men'tl), *a.* pertaining to the mind, intellectual; due to or done by the mind; of or concerning psychiatric illness; slightly deranged in mind. **mental age**, *n.* the intellectual maturity of an individual expressed in terms of the average intellectual maturity of a child of a specified age. **mental arithmetic**, *n.* arithmetic done in the head, without writing it down or using a calculator. **mentality**

menthol   515   **merry**

(-tal'-), *n.* mental attitude or disposition. **mentalize, -ise,** *v.t.* **mentally,** *adv.*
menthol (men'thol), *n.* a waxy crystalline substance obtained from oil of peppermint, used as a local anaesthetic. **mentholated,** *a.* esp. of cigarettes, treated with menthol.
mention (men'shən), *n.* a concise notice, allusion to (or of); a naming. *v.t.* to refer to, to allude to; to indicate by naming without describing. **honourable mention,** a distinction sometimes awarded to a competitor who has just failed to win a prize. **mention in dispatches,** (*Mil.*) reference by name (in official dispatches) to an officer who has done well in battle. **mentionable,** *a.*
mentor (men'taw), *n.* a faithful guide, a wise counsellor.
menu (men'ū), *n.* a bill of fare; (*Comput.*) a list of options, topics etc. which the operator can choose from.
meow (miow'), MIAOW.
MEP, (*abbr.*) Member of the European Parliament.
mephitis (məfī'tis), *n.* a foul, offensive or pestilential exhalation. **mephitic, -ical** (-fit'-), *a.*
mercantile (mœ'kəntīl), *a.* commercial, pertaining to buying and selling; mercenary. **mercantile marine** MERCHANT SERVICE. **mercantilism** (-til-), *n.* **mercantilist** (-til-), *n.*
Mercator's projection PROJECTION.
mercenary (mœ'sənəri, -sənri), *a.* hired or serving for money; done from or actuated by motives of gain; venal. *n.* (*pl.* **-aries**) one who is hired, esp. a soldier hired in foreign service.
mercer (mœ'sə), *n.* one who deals in silk, cotton, woollen and linen goods. **mercery,** *n.*
mercerize, -ise (mœ'səriz), *v.t.* to treat cotton fabrics with an alkaline solution in preparation for dyeing. **mercerization,** *n.*
merchandise (mœ'chəndīz), *n.* articles of commerce; commodities for purchase. *v.t. v.i.* to trade, to barter. **merchandiser,** *n.* **merchandising,** *n.* promotion and advertising of goods for sale.
merchant (mœ'chənt), *n.* one who carries on trade on a large scale, esp. with foreign countries; (*N Am., Sc.*) a shopkeeper, a tradesman; (*coll.*) one given to acting in a certain (often disreputable) manner, such as *speed merchant*. *a.* mercantile, commercial. **merchant bank,** *n.* a private bank whose business chiefly involves dealing in bills of exchange and underwriting new security issues. **merchantman,** *n.* a merchant ship. **merchant navy,** *n.* collective name for sea-going vessels other than those of the Royal Navy. **merchant prince,** *n.* a wealthy merchant. **merchant service,** *n.* personnel etc. of shipping employed in commerce. **merchant ship,** *n.* ship for conveying merchandise.
merciful etc. MERCY.
mercury (mœ'kūri), *n.* a liquid, silvery, toxic, metallic element, at. no. 80; chem. symbol Hg; (**Mercury**) the Roman god of commerce, identified with the Greek Hermes, the messenger of the gods; the planet nearest the sun; a messenger; a common title for a newspaper. **mercurial** (məkū'riəl), *a.* pertaining to the god Mercury; flighty, volatile, fickle; pertaining to, consisting of or caused by mercury. *n.* a preparation containing mercury, used as a drug. **mercurially,** *adv.* **mercuric,** *a.* containing mercury in the divalent state. **mercurous** (-kūr'əs), *a.* containing mercury in the monovalent state.
mercy (mœ'si), *n.* a disposition to temper justice with mildness; forbearance, clemency, compassion; an act of clemency, pity or compassion; pardon, forgiveness; control, discretion, liberty to punish or spare; (*coll.*) something to be thankful for. **at the mercy of,** wholly in the power of. **for mercy's sake,** an exclamation or appeal for mercy, or of expostulation. **sister of mercy** SISTER. **mercy dash, mercy flight,** etc., *n.* a trip, flight etc. to bring help to a sick or injured person. **mercy killing,** *n.* euthanasia. **merciful,** *a.* **mercifully,** *adv.* **mercifulness,** *n.* **merciless,** *a.* **mercilessly,** *adv.* **mercilessness,** *n.*

mere [1] (miə), *n.* a lake, a pool.
mere [2] (miə), *a.* such and no more; absolute, unqualified. **merely,** *adv.* purely, only, solely.
mere [3] MERI.
-mere, *comb. form* part, segment.
meretricious (merətrish'əs), *a.* pertaining to or befitting a prostitute; alluring by false or empty show; unreal, tawdry. **meretriciously,** *adv.* **meretriciousness,** *n.*
merganser (mœgan'sə), *n.* the goosander and other diving or fish-eating ducks belonging to the genus *Mergus*.
merge (mœj), *v.t.* to cause to be swallowed up or absorbed; to fuse or cause to blend. *v.i.* to be absorbed or swallowed up; to lose individuality or identity (in); to combine. **merger,** *n.* the merging of an estate, limited company etc. into another; extinction, absorption.
meri, mere (me'ri), *n.* a war-club; a greenstone trinket shaped like this.
meridian (mərid'iən), *a.* pertaining to midday or to a geographical or astronomical meridian, or to the point or period of highest splendour or vigour. *n.* a great circle drawn through the poles and the zenith of any given place on the earth's surface; the line in which the plane of this circle intersects the earth's surface; the time when the sun or other heavenly body crosses this; midday, noon; culmination, zenith, point of highest splendour or vigour. **first, prime, meridian,** a meridian from which longitude is reckoned, usu. that of Greenwich. **meridional,** *a.* pertaining to a meridian; highest; culminating; pertaining to the south, esp. of Europe; running north and south, as a mountain range. *n.* an inhabitant of the south, usu. of the south of France.
meringue (mərang'), *n.* a confection of white of eggs, sugar etc., used as icing; a cake made of this.
merino (mərē'nō), *n.* a breed of sheep introduced from Spain, valuable for their fine wool; a fine woollen dress-fabric, orig. of this wool; a fine woollen yarn used for hosiery. *a.* pertaining to this breed of sheep; made of merino. **pure merino,** (*Austral., coll.*) descendant of an early settler with no convict connection.
meristem (me'ristem), *n.* vegetable tissue or cells in process of growth. **meristematic** (-mat'-), *a.*
merit (me'rit), *n.* the quality of deserving, desert; excellence deserving honour or reward; worth, worthiness; a reward or recompense, a mark or award of merit; (*pl.*) the essential rights and wrongs of a case. *v.t.* to deserve, to earn; to be entitled to receive as a reward; to have a just title to. *v.i.* to acquire merit. **Order of Merit** ORDER. **merited,** *a.* **meritocracy** (-tok'-), *n.* (a society ruled by) those who have gained their positions through talent, intellect or industriousness, not through their family background, inherited wealth etc.; the rule of such people. **meritocrat,** *n.* **meritocratic,** *a.* **meritorious** (-taw'riəs), *a.* deserving reward; praiseworthy. **meritoriously,** *adv.* **meritoriousness,** *n.*
merle (mœl), *n.* (*poet.*) the blackbird.
merlin (mœ'lin), *n.* the smallest of the European falcons.
mermaid (mœ'mād), *n.* an imaginary marine creature, having the upper half like a woman and the lower like a fish. **mermaid's purse** SEA-PURSE. **merman,** *n. masc.*
Merovingian (merəvin'jiən), *a.* a term applied to the Frankish dynasty reigning in Gaul and Germany, founded by Clovis in AD 486. *n.* a sovereign of this dynasty.
merry (me'ri), *a.* joyous, gay, jovial, mirthful; causing merriment; (*coll.*) slightly tipsy. **the more the merrier,** the pleasure will be greater, the more people are involved. **to make merry** MAKE [2]. **merry-andrew,** *n.* a buffoon, a jester, esp. one assisting a mountebank or quack. **merry England,** *n.* an idealized image of England as it used to be, esp. in Elizabethan times. **merry-go-round,** *n.* a revolving frame with seats or wooden horses on which persons ride at fairs etc.; (*coll.*) a traffic roundabout. **merry-make,** *v.i.* to make merry. *n.* a merry-making. **merry-maker,** *n.*

merry-making, *a.* making merry, jovial. *n.* merriment; a festivity. **merrythought**, *n.* the furcula or forked bone in the breast of a bird. **merrily**, *adv.* merriment, **merriness**, *n.*
mes- MISO-.
mesa (mä´sə), *n.* a tableland; a plateau with steep sides.
mésalliance (mäzalias´, -zal´-), *n.* marriage with one of inferior social position.
mesaraic (mesərā´ik), *a.* mesenteric.
mescal (meskal´), *n.* a small globular cactus of the southern US and Mexico, the tubercles of which are chewed for their hallucinogenic effects; an alcoholic liquor distilled from Agave. **mescaline** (mes´kalin), *n.* a hallucinogenic substance derived from mescal.
Mesdames (mädahm´), MADAME.
Mesdemoiselles, *pl.* MADEMOISELLE.
Mesembrianthemum (məzembrian´thiməm), *n.* a genus of very succulent plants, with thick, fleshy leaves and brilliant flowers, containing the ice-plant or fig-marigold.
mesencephalon (mesənsef´əlon), *n.* the mid-brain. **mesencephalic** (-fal´-), *a.*
mesentery (mes´əntəri, mez´-), *n.* a fold of the peritoneum investing the small intestines and connecting them with the wall of the abdomen. **mesenteric** (-ter´-), *a.* **mesenteritis** (-i´tis), *n.* inflammation of the mesentery.
mesh (mesh), *n.* the space or interstice between the threads of a net; (*pl.*) network; a trap, a snare; the engagement of gear-teeth etc.; interlacing structure. *v.t.* to catch in a net, to ensnare; to engage (of gear-teeth etc.). *v.i.* to coordinate (with); of gear-teeth etc., to engage.
mesial (mē´ziəl), *a.* pertaining to, situated or directed towards the middle, esp. the middle line of the body; median.
mesmerism (mez´mərizm), *n.* the art or power of inducing an abnormal state of the nervous system, in which the will of the patient is controlled by that of the agent; the hypnotic state so induced. **mesmeric** (-mer´-), *a.* **mesmerist**, *n.* **mesmerize, -ise**, *v.t.* to hypnotize; to occupy (someone's attention) totally. **mesmerization**, *n.* **mesmerizer**, *n.*
mesne (mēn), *a.* middle, intermediate. **mesne lord**, *n.* in feudal law, one holding of a superior lord. **mesne profits**, *n.pl.* the profits of an estate received by a person wrongfully in possession.
mes(o)-, *comb. form* intermediate, in the middle; pertaining to the middle.
mesoblast (mē´səblahst, -z-), *n.* the intermediate layer of the blastoderm of the embryo. **mesoblastic** (-blahs´-), *a.*
mesocarp (mes´əkahp, -z-), *n.* the middle layer of a pericarp.
mesocephalic (mēsōsifal´ik, -z-), *a.* intermediate between dolichocephalic and brachycephalic (of skulls). **mesocephalism** (-sef´-), **mesocephaly** (-sef´-), *n.* **mesocephalous** (-sef´-), *a.*
mesoderm (mes´ədœm), *n.* the mesoblast; (*Bot.*) the middle layer of the bark, of the wall of a spore-case etc. **mesodermal, -dermic** (-dœ´-), *a.*
Mesolithic (mēsəlith´ik, -z-), *a.* intervening between the Neolithic and Palaeolithic divisions of the stone age.
mesomorphic (mēsəmaw´fik, -z-), *a.* having a muscular physique. **mesomorph** (mē´-), *n.* **mesomorphy** (mē´-), *n.*
meson (mē´zon), *n.* a particle intermediate in mass between a proton and an electron.
mesophloeum (mēsəflē´əm, -z-), *n.* the middle or green layer of bark in exogens.
mesophyll (mes´əfil, -z-), *n.* the inner parenchymatous tissue of a leaf.
mesophyte (mes´əfīt, -z-), *n.* a plant that grows in conditions where there is a moderate supply of water.
mesoplast (mē´səplast, -z-), *n.* the nucleus of a cell.
mesosphere (mē´səsfiə, -z-), *n.* the region of the earth's atmosphere above the stratosphere.
mesothelioma (mēsəthēliō´mə, -z-), *n.* a tumour of the lining of the lungs, heart or stomach, often caused by blue asbestos dust.
mesothorax (mēsəthaw´raks, -z-), *n.* in insects, the middle segment of the thorax bearing the anterior legs and the middle wings.
Mesozoic (mēsəzō´ik, -z-), *a.* belonging to the second great geological epoch.
mesquit (meskēt´), *n.* either of two leguminous shrubs or trees growing in the SW United States and as far south as Peru, the larger yielding the sweetish screw-pod used for fodder. **mesquit-bean**, *n.* **mesquit-grass**, *n.*
mess (mes), *n.* a dish or a portion of food sent to table at one time; liquid or semi-liquid food, esp. for animals; a quantity of food taken; a number of persons who sit down to table together (used esp. of soldiers and sailors); a meal taken thus; officers' living quarters; a state of dirt and disorder; a muddle, a difficulty. *v.i.* to take a meal or meals in company, esp. of soldiers etc.; to muddle or potter (about). *v.t.* to mix together, to muddle, to jumble; to dirty, to soil. **to mess about**, (*coll.*) to tumble about; to treat roughly; to treat improperly or inconsiderately; to potter about. **to mess up**, (*coll.*) to ruin, spoil. **to mess with**, (*coll.*) to interfere with. **messmate**, *n.* a member of the same mess; an associate; a parasite which does not actually feed on the body of its host, a commensal. **messy**, *a.* dirty, muddled; complicated and difficult to handle. **messiness**, *n.* a state of dirt or disorder.
message (mes´ij), *n.* a communication, oral or written, from one person to another; the truths, ideas or opinions of a writer or inspired person; the chief theme of a play, novel etc.
messenger (mes´injə), *n.* one who carries a message or goes on an errand. **queen's, king's, messenger**, official bearer of Foreign Office dispatches to foreign countries. **messenger RNA**, *n.* a type of RNA that carries genetic information from DNA to the ribosomes for the synthesis of protein.
Messiah (misī´ə), **Messias** (-əs), *n.* Christ, as the promised deliverer of the Jews; an expected saviour or deliverer. **messianic** (mesian´ik), *a.* of, or inspired by the hope of, a Messiah; marked by great zeal in support of a cause.
Messieurs (mesyœ´), *n.pl.* sirs; gentlemen (pl. of Mr, usu. abbr. to **Messrs** (mes´əz)).
messmate MESS.
Messrs MESSIEURS.
mestizo (mestē´zō), *n.* (*pl.* **-zos, -zoes**) one of mixed Spanish or Portuguese and American Indian blood; applied also to one of mixed Chinese and Philippine blood. **mestiza** (-zə), *n. fem.*
Met (met), *n.* (*coll.*) the (London) Metropolitan Police; (*coll.*) Meteorological Office in London; the Metropolitan Opera, New York.
met, *past, p.p.* MEET [2].
met(a)-, meth-, *comb. form* on; with, among or between; after (implying change or transposition).
metabolism (mitab´əlizm), *n.* the continuous chemical change going on in living matter, either constructive, by which nutritive material is built up into complex and unstable living matter, or destructive, by which protoplasm is broken down into simpler and more stable substances. **metabolic** (metəbol´-), *a.* **metabolite**, *n.* a substance involved in or produced by metabolism. **metabolize, -ise**, *v.t.*
metacarpus (metəkah´pəs), *n.* (*pl.* **-pi**) the part of the hand between the wrist and the fingers. **metacarpal**, *a.*
metacentre (met´əsentə), *n.* the point in a floating body slightly out of equilibrium where the vertical drawn through the centre of gravity when it is in equilibrium intersects the vertical passing through the centre of buoyancy.
metachrosis (metəkrō´sis), *n.* change of colour, as in

certain lizards.
metagalaxy (metəgal'əksi), *n.* the universe beyond our galaxy. **metagalactic** (-lak'-), *a.*
metage (mē'tij), *n.* official measurement, esp. of coal; toll charged for measuring.
metal (met'l), *n.* one of a class of elementary substances which usu. present in various degrees certain physical characters, as lustre, malleability and ductility, possessed by the six metals known to the ancients, viz. gold, silver, copper, iron, lead and tin; a compound of the elementary metals, an alloy; broken stone for road-making etc.; molten glass ready for blowing or casting; the effective power of the guns of a warship; (*pl.*) rails of a railway etc.; mettle, essential quality. *v.t.* (*past, p.p.* **metalled**) to furnish or fit with metal; to cover or repair (a road) with metal. **metallic** (-tal'-), *a.* **metallic currency**, *n.* money composed of gold, silver etc., as opp. to paper. **metalliferous** (-lif'-), *a.* bearing or yielding metal. **metalling**, *n.* broken stones etc. used in making or mending roads. **metallize, -ise**, *v.t.* to form into a metal; to give metallic properties to; to vulcanize. **metallization**, *n.*
metalanguage (met'əlang-gwij), *n.* a language or system of symbols used to speak about another language.
metallography (-log'-), *n.* the science of metals, esp. the microscopic study of their internal structure.
metallurgy (mital'əji, met'əlœji), *n.* the science of metals; the art of separating metals from ores; the art of working in metal. **metallurgic, -ical** (metəlœ'-), *a.* **metallurgist**, *n.*
metamere (met'əmiə), *n.* one of a series of similar parts of a body. **metameric** (-me'-), *a.* of, pertaining to, or of the nature of a metamere; having the same composition and molecular weight, isomeric but different in chemical properties. **metamerism** (-tam'ə-), *n.*
metamorphose (metəmaw'fōz), *v.t.* to change into a different form; to transmute. *v.i.* to undergo change into a different form. **metamorphic** (-maw'fik), *a.* causing or showing the results of metamorphosis; transforming or transformed. **metamorphism**, *n.* change in the structure of rocks, caused usu. by heat. **metamorphology** (-fol'-), *n.* the science of the metamorphoses of organisms. **metamorphosis** (-fəsis), *n.* a change of form; the result of such a change; transformation, as of a chrysalis into a winged insect; a complete change of character, purpose etc.
metaphor (met'əfə, -faw), *n.* a figure of speech by which a word is transferred from one object to another, so as to imply comparison. **metaphoric, -ical** (-fo'-), *a.* **metaphorically**, *adv.*
metaphrase (met'əfrāz), *v.t.* to translate literally. *n.* (also **metaphrasis** (-taf'rə-)) a literal translation. **metaphrastic** (-fras'-), *a.* **metaphrist**, *n.*
metaphysics (metəfiz'iks), *n. sing.* the philosophy of being and knowing; the theoretical principles forming the basis of any particular science; the philosophy of mind; anything vague, abstract and abstruse. **metaphysical**, *a.* of or pertaining to metaphysics; transcendental, dealing with abstractions; abstruse, over-subtle; imaginary, fantastic; of the group of 17th cent. poets noted for their intellectual tone and ingenious imagery. **metaphysically**, *adv.* **metaphysician** (-zish'-), *n.* **metaphysicize, -ise**, *v.t., v.i.*
metapsychology (metəsikol'əji), *n.* the body of philosophical theory on psychological matters, beyond experiment or reasoning. **metapsychological** (-loj'-), *a.*
metastable (metəstā'bl), *a.* seemingly stable because passing slowly from one state to another. **metastability** (-stəbil'-), *n.*
metastasis (mitas'təsis), *n.* metabolism; a change in the seat of a disease, esp. cancer, from one organ to another. **metastasize, -ise**, *v.i.* **metastatic** (metəstat'ik), *a.*
metatarsus (metətah'səs), *n.* (*pl.* **-si**) that part of the foot between the tarsus and the toes, in man consisting of five long bones. **metatarsal**, *a.*
metathesis (mitath'əsis), *n.* the transposition of sounds or letters in a word; the surgical removal of a morbific agent etc. from one place to another; interchange of radicals or groups of atoms in a compound with others. **metathetic** (metəthet'ik), *a.*
Metazoa (metəzō'ə), *n.pl.* a primary division of the animal kingdom including all animals which have many-celled bodies and differentiated tissues, as distinct from *Protozoa.* **metazoan**, *a.* pertaining to the Metazoa. *n.* any individual of the Metazoa. **metazoic**, *a.*
mete (mēt), *v.t.* to measure; to allot, to apportion (out); to appraise; to be the measure of.
metempsychosis (mitempsikō'sis), *n.* the passage of the soul after death from one animal body to another.
meteor (mē'tiə), *n.* a luminous body appearing for a few moments in the sky and then disappearing, a shooting-star; a small meteoroid; any atmospheric phenomenon, as rain, hail etc.; anything which transiently dazzles or strikes with wonder. **meteoric** (-o'-), *a.* pertaining to or consisting of meteors; resembling a meteor; brilliant but fading quickly, dazzling; of or pertaining to the atmosphere or its phenomena. **meteorically**, *adv.* **meteorite**, *n.* a fallen meteor; stone, metal or a compound of earth and metal, that has fallen upon the earth from space. **meteoroid**, *n.* small celestial bodies that orbit the sun and fall to earth as meteors. **meteoroidal** (-oi'-), *a.*
meteorology (mētiərol'əji), *n.* the science of the atmosphere and its phenomena, esp. for the purpose of forecasting the weather; the general character of the weather in a particular place. **meteorologic, -ical** (-loj'-), *a.* **Meteorological Office**, *n.* a government department responsible for issuing weather forecasts, storm warnings etc. **meteorologically**, *adv.* **meteorologist**, *n.*
meter[1] (mē'tə), *n.* one who or that which measures, esp. an instrument for registering the quantity of gas, water, electric energy etc. supplied. *v.t.* to measure by means of a meter. **meterage**, *n.*
meter[2] METRE[1].
meter[3] METRE[2].
-meter, *suf.* a measuring instrument, as *barometer, thermometer.*
meth- MET(A)- before aspirates.
methadone (meth'ədōn), *n.* a synthetic drug similar to morphine, but less addictive, often used in the treatment of addiction.
methane (mē'thān), *n.* (*Chem.*) a light, colourless gas, methyl hydride or carburetted hydrogen, produced by the decomposition or dry distillation of vegetable matter, one of the chief constituents of coal-gas, and also of fire-damp and marsh-gas.
methanol (meth'ənol), *n.* a colourless, volatile liquid used as a solvent or as fuel.
methinks (mithingks'), *v.impers.* (*past* **-thought, -thawt'**) it seems to me; I think.
method (meth'əd), *n.* mode of procedure, way or order of doing; an orderly, systematic or logical arrangement; orderliness; system; a system or the basis of a system of classification. **method acting**, *n.* an actor's identification of himself with the part rather than giving just a technical performance. **methodical** (-thod'-), *a.* done according to a method; habitually proceeding in a systematic way. **methodically**, *adv.* **Methodism**, *n.* the doctrines, practices or Church system of the Methodists. **Methodist**, *n.* a strict observer of method in philosophical inquiry or medical practice; a member of any of the religious bodies that have grown out of the evangelical movement begun in the middle of the 18th cent. by John Wesley and followers. **methodize, -ise**, *v.t.* to reduce to order; to arrange systematically. **methodizer**, *n.* **methodology** (-dol'-), *n.* the branch of logic dealing with the methods of accurate thinking; the methods used in a particular project, discipline etc.
methought METHINKS.

meths (meths), *n.pl.* (*coll.*) short for METHYLATED SPIRIT(S).
methyl (meth'l, mē'thil), *n.* the hypothetical radical of wood spirit, formic acid and many other organic compounds. **methylic** (-thil'ik), *a.* [F *méthyle*, from *methylène* (Gr. *methu*, wine, *hulē*, wood)]
methylate (meth'əlāt), *v.t.* to mix or saturate with methyl alcohol. **methylated spirit**, *n.* spirit of wine, mixed with 10% of methyl alcohol so as to be rendered unfit to drink and accordingly duty-free.
methylene (meth'əlēn), *n.* a hypothetical organic radical in which two atoms of hydrogen are in chemical combination with one atom of carbon, occurring in numerous compounds.
meticulous (mətik'ūləs), *a.* cautious or over-scrupulous about trivial details, very careful. **meticulously**, *adv.* **meticulousness**, *n.*
métier (met'iā. mä'-), *n.* trade, profession; one's particular strength or specialization.
métis (mä'tēs. mātēs'), *n.* one of mixed blood, esp. (in Canada) the offspring of a European and an American Indian.
metonic (miton'ik), *a.* pertaining to *Meton*, Athenian astronomer, applied to the cycle of 19 Julian years at the end of which the new and full moons recur on the same dates.
metonymy (miton'əmi), *n.* a figure in which one word is used for another, as the effect for the cause, the material for the thing made etc., e.g. 'bench' for 'magistrates'. **metonymic, -ical** (-nim'-), *a.* **metonymically**, *adv.*
metope (met'əpi, -ōp), *n.* the space between the triglyphs in a Doric frieze.
metre[1], (*esp. N Am.*) **meter** (mē'tə), *n.* the rhythmical arrangement of syllables in verse; verse; any particular form of poetic rhythm. **metric** (met'-), *a.* metrical. *n.* (*usu. pl.*) the science or art of metre, prosody. **metrical** (met'-), *a.* of, pertaining to or composed in metre; of or pertaining to measurement. **metrically**, *adv.*
metre[2], (*esp. N Am.*) **meter** (mē'tə), *n.* the standard measure of length in the metric system, orig. the ten-millionth part of the quadrant of a meridian, 39·37 in., now defined as the distance travelled by light in a vacuum in 1/299, 792, 458 of a second. **metric** (met'rik), *a.* **metric system**, *n.* a system of weights and measures in which ascending units carry Greek prefixes and descending units Latin prefixes. Units are multiples of ten times the basic unit. **metric ton** TONNE. **metricate** (met'-), *v.t.* to convert to the metric system. **metrication** (met-), *n.*
metro (met'rō), *n.* (*pl.* **metros**) an underground railway network in a city.
metronome (met'rənōm), *n.* an instrument for indicating and marking time in music by means of a pendulum. **metronomic** (-nom'-), *a.*
metronymic (metrənim'ik), *a.* of names, derived from the name of a mother or maternal ancestor. *n.* a name so derived. **metronymy** (-tron'-), *n.*
metropolis (mitrop'əlis), *n.* (*pl.* **-lises**) the chief town or capital of a country; the seat or see of a metropolitan bishop; a centre or focus of activity etc.; a large town.
metropolitan (metrəpol'itən), *a.* pertaining to a capital city or to an archbishopric; forming part of a sovereign state as distinct from its colonies. *n.* a bishop having authority over other bishops in a province, in the Western Church an archbishop, in the ancient and the modern Greek Church ranking above an archbishop and next to a patriarch. **metropolitanate** (-nət), *n.*
-metry, *suf.* science of measuring, as *geometry, trigonometry.*
mettle (met'l), *n.* quality of temperament or disposition; constitutional ardour; spirit, courage. **to put on one's mettle**, to test one's courage, determination etc. **mettled**, *a.* **mettlesome**, *a.* high-spirited, fiery, ardent.
mew[1] (mū), *n.* a kind of sea-gull.
mew[2] (mū), *v.i.* to cry 'mew' as a cat. *n.* this cry of the cat.

mew[3] (mū), *n.* a cage for hawks, esp. whilst moulting; a place of confinement; a den; (*pl.*) royal stables in London (built on the spot where the royal hawks were formerly mewed); (*pl.*) stables for carriage-horses etc.; a row of dwellings, garages etc. converted from these; (*pl., N Am.*) a back alley.
mewl (mūl), *v.i.* to cry, whine or whimper, as a child; to mew, as a cat. **mewler**, *n.*
Mexican (mek'sikən), *a.* of or pertaining to Mexico. *n.* a native or inhabitant of Mexico.
MEZ, (*abbr.*) Central European Time (G Mitteleuropäische Zeit).
mezuza(h) (məzoo'zə), *n.* a small case containing extracts from Scripture fixed to the doorpost by Jews as a sign of their piety.
mezzanine (mez'ənēn), *n.* a storey intermediate in level between two main storeys, usu. between the ground and first floors; a window in such a storey; a floor beneath the stage of a theatre from which the traps etc. are worked. **mezzanine-floor, -window**, *n.*
mezza voce (met'sə vō'chi), *a., adv.* (singing or sung) softly; quiet(ly).
mezzo (met'sō), *a.* half or medium.
mezzo forte (metsōfaw'ti), *a., adv.* in music, moderately loud(ly).
mezzo-soprano (metsōsəprah'nō), *n.* a voice lower than a soprano and higher than a contralto; a singer with such a voice.
mezzotint (met'sōtint), *n.* a process of engraving in which a copper plate is uniformly roughened so as to print a deep black, lights and half-lights being then produced by scraping away the burr; a print from this. *v.t.* to engrave in mezzotint.
mf, (*abbr.*) mezzo forte.
Mg, (*chem. symbol*) magnesium.
mg, (*abbr.*) milligram.
Mgr, (*abbr.*) Monsignor.
MHz, (*abbr.*) megahertz.
mi (mē), *n.* the third note of the diatonic scale.
MI5, (*abbr.*) the British government agency for counterespionage.
MI6, (*abbr.*) the British government agency for espionage.
miaow (miow'), *n.* the cry of a cat. *v.i.* of a cat, to cry 'miaow'.
miasma (miaz'mə), *n.* (*pl.* -mata, -tə) poisonous or infectious atmosphere. **miasmal, miasmatic** (miazmat'-), *a.*
mica (mī'kə), *n.* a name for a group of silicates having a perfect basal cleavage into thin, tough and shining plates, formerly used instead of glass.
mice, *pl.* MOUSE.
Mich., (*abbr.*) Michigan (US).
Michaelmas (mik'lməs), *n.* the feast of St Michael the Archangel, 29 Sept.; autumn. **Michaelmas daisy**, *n.* the wild aster, *Aster tripolium*, also various perennial cultivated asters.
Mick (mik), *n.* (*sl. often offensive*) an Irishman.
Mickey Finn (mik'i fin'), *n.* (*esp. N Am.*) a doped drink.
mickle (mik'l), *a.* (*chiefly Sc.*) much, great. *n.* a large amount.
micky (mik'i), *n.* (*Austral. sl.*) a young wild bull; (*N Am. sl.*) an Irish lad. **to take the micky out of**, (*coll.*) to debunk; to tease. **micky-taking**, *n.*
micro (mī'krō), *n.* (*coll.*) short for MICROCOMPUTER.
micr(o)-, *comb. form* noting smallness; pertaining to small things (as opposed to large ones).
microbe (mī'krōb), *n.* any minute organism, esp. a bacterium or microzyme causing disease or fermentation. **microbial, -ian, -bic** (mīkrō'biəl, -biən, -bik), *a.* **microbiology** (-biol'-), *n.* **microbiologist** (-ol'-), *n.*
microcephalic (mīkrəsifal'ik), *a.* having an unusually small skull. **microcephalous** (-sef'-), *a.* **microcephaly**, *n.*
microchip (mī'krəchip), *n.* a chip of silicon etc. bearing

many integrated circuits. **microcircuit** (mī'krəsækit), *n*. a very small integrated circuit on a semiconductor.
microclimate (mī'krəklimət), *n*. the climate of a very small area.
microcomputer (mī'krəkəmpūtə), *n*. a small computer with one or more microprocessors.
microcosm (mī'krəkozm), *n*. the universe on a small scale; man as an epitome of the macrocosm or universe; a little community; a representation (of) in little. **microcosmic** (-koz'-), *a*.
microdot (mī'krədot), *n*. a photographic image reduced to the size of a dot, e.g. for espionage purposes.
microeconomics (mīkrŏekənom'iks), *n*. the branch of economics dealing with particular companies, products, individuals, etc.
microelectronics (mīkrŏeliktron'iks), *n. sing*. electronics as applied to microcircuits.
microfiche (mī'krəfēsh), *n*. a sheet of film bearing miniature photographs of documents etc.
microfilm (mī'krəfilm), *n*. a strip of film on which successive pages of a document or book are photographed for purposes of record.
microinstruction (mī'krŏinstrŭkshən), *n*. a computer instruction that activates a particular circuit to execute part of an operation specified by a machine instruction.
microlight (mī'krəlit), *n*. a very small light aircraft for one or two people.
microlite (mī'krəlit), *n*. a native salt of calcium found in small crystals; microlith.
microlith (mī'krəlith), *n*. one of the microscopic bodies found in vitreous feldspar, hornblende etc. **microlithic** (-lith'-), *a*. applied to a particular style of funeral monuments, in which extremely small stones are used.
micrometer (mikrom'itə), *n*. an instrument to measure small distances or objects. **micrometric, -ical** (-met'-), *a*.
micron (mī'kron), *n*. one millionth of a metre, the unit of length in microscopic research.
microorganism (mīkrŏaw'gənizm), *n*. an organism of microscopic size.
microphone (mī'krəfōn), *n*. an instrument for converting sound into electrical waves; the mouthpiece for broadcasting. **microphonic** (-fon'-), *a*.
microphotography (mīkrŏfətog'rəfi), *n*. the photography of objects on a minute scale; the photography of microscopic objects. **microphotograph** (-fō'təgrahf), *n*.
microprocessor (mīkrŏprö'sesə), *n*. an integrated circuit operating as the central processing unit of a microcomputer.
microscope (mī'krəskōp), *n*. an optical instrument by which objects are so magnified that details invisible to the naked eye are clearly seen. **microscopic** (-skop'-), *a*. pertaining to the microscope; too small to be visible except by the aid of a microscope. **microscopically**, *adv*.
microscopy (-kros'-), *n*.
microsecond (mī'krəsekənd), *n*. one millionth of a second.
microsurgery (mī'krəsœjəri), *n*. surgery performed using a microscope and special small instruments.
microtome (mī'krətōm), *n*. an instrument for cutting thin sections for the purpose of microscopic examination.
microtomy (-krot'-), *n*.
microwave (mī'krəwāv), *n*. an electromagnetic wave with a wavelength between 30 cm and 1 mm; (*coll*.) a microwave oven. **microwave oven**, *n*. one that cooks food with microwaves.
micturition (miktūrish'ən), *n*. a frequent desire to urinate; (*loosely*) the act of urinating. **micturate** (mik'-), *v.i.* to urinate.
mid (mid), *a*. (*superl*. **midmost**) (*usu. in comb*.) middle. *prep*. (*poet*.) amid. **midday**, *n*. noon. *a*. pertaining to noon. **mid-Atlantic**, *a*. indicating a blend of British and American characteristics. **mid-iron**, *n*. an iron golf club with a moderate amount of loft. **mid-off**, *n*. (*Cricket*) the fieldsman to the left of the bowler. **mid-on**, *n*. the fieldsman to the right of the bowler. (These definitions apply only when the batsman is right-handed.)
midden (mid'ən), *n*. a dunghill.
middle (mid'l), *a*. placed equally distant from the extremes; intervening; intermediate; of verbs, between active and passive, reflexive. *n*. the point equally distant from the extremes; the waist; the midst, the centre. *v.t.* to place in the middle; (*Naut*.) to fold or double in the middle; in football, to pass or return the ball to midfield from one of the wings. **in the middle of**, during, while. **middle age**, *n*. the period of life between youth and old age, or about the middle of the ordinary human life (35–55). **middle-aged**, *a*. **Middle Ages**, *n.pl.* the period from the 5th to the 15th cent. inclusive. **Middle America**, *n*. the midwestern region of the US; the middle class of the US. **middlebrow**, *n*. (*derog*.) a person of modest and conventional tastes in art, music etc. **middle class**, *n*. the class between the wealthy and working people, the bourgeoisie. **middle-class**, *a*. **middle-distance** DISTANCE. **Middle English** ENGLISH. **middle finger**, *n*. the second finger (third from the little finger inclusive). **Middle Kingdom**, *n*. China. **middleman**, *n*. an agent, an intermediary; one through whose hands a commodity passes between the producer and the consumer. **middle name**, *n*. any name between a person's first given name and his or her family name; (*coll*.) one's most typical quality. **middle-of-the-road**, *a*. moderate; (of music) having a wide appeal. **middle term**, *n*. the term of a syllogism that appears in both major and minor premises. **middling**, *a*. of middle size, capacity or condition; mediocre; moderately good, second-rate. *adv*. moderately, tolerably. **fair to middling** FAIR [1]. **middlingly**, *adv*.
middy [1] (mid'i), MIDSHIPMAN.
middy [2] (mid'i), *n*. (*Austral*., *coll*.) a glass of beer; a 10 oz. (approx. 300 ml.) pot or container.
midfield (mid'fēld), *n*. the central area of a sports pitch, esp. a football pitch; the players with positions between the attackers and defenders, esp. in football.
midge (mij), *n*. a gnat or other minute fly; a tiny person.
midget (mij'it), *n*. a very small person. *a*. very small.
MIDI (mid'i), *n*. (*Comput*.) an interface which allows computer-control of electronic keyboards and other musical instruments. [acronym for Musical Instrument Digital Interface]
Midi (mē'dē), *n*. the South of France. [F]
midi-, *comb. form* of middle size; of a skirt etc., reaching to the mid-calf. **midi** (mid'i), *n*. a midi-skirt, midi-dress etc. **midi-bus**, *n*. a bus larger than a minibus, but smaller than a standard bus. **midi system**, *n*. a compact hi-fi stacking unit.
midland (mid'lənd), *a*. situated in the middle or interior of a country; surrounded by land. *n*. the interior of a country; (*pl*.) the midland counties of England.
midnight (mid'nit), *n*. the middle of the night, twelve o'clock; intense darkness. *a*. pertaining to or occurring in the middle of the night; very dark. **midnight sun**, *n*. the sun visible around midnight in summer in the polar regions.
mid-off, -on MID.
midrib (mid'rib), *n*. the continuation of the petiole to the apex of a leaf.
midriff (mid'rif), *n*. the middle section of the human body; the diaphragm.
midship (mid'ship), *n*. the middle part of a ship or boat. *a*. situated in or belonging to this. **midshipman**, *n*. formerly an officer ranking between a cadet and a sub-lieutenant, a young officer under instruction on shipboard. **midshipmite**, *n*. (*facet*.) a very young or small midshipman. **midships** AMIDSHIPS.
midst (midst), *n*. the middle. *prep*. in the middle of,

amidst. in the midst of, among, surrounded by or involved in.
midstream (midstrēm'), *n.* the middle of a stream. *adv.* in the middle of a stream. **to change horses in midstream** HORSE.
midsummer (midsŭm'ə, mid'-), *n.* the middle of summer, esp. the period of the summer solstice, about 21 June. **midsummer day,** *n.* 24 June.
midway (mid'wā), *a.* situated in the middle or the middle of the way. *adv.* (-wā'), in the middle; half-way.
midweek (midwēk'), *n.* the middle of the week, i.e. Tuesday, Wednesday, Thursday.
midwife (mid'wīf), *n.* (*pl.* **-wives,** -wivz) a woman who assists at childbirth; any person who helps to bring something forth. *v.t.* to perform the office of a midwife. *v.t.* to assist in childbirth. **midwifery** (-wif'əri), *n.*
midwinter (midwin'tə), *n.* the middle of winter, esp. the winter solstice, 21 Dec.
mien (mēn), *n.* air or manner; appearance, deportment, demeanour, bearing, carriage.
miff (mif), *n.* a petty quarrel; a huff. *v.i.* to be vexed (with or at). *v.t.* to vex, to annoy slightly, to offend.
might[1] (mīt), *n.* strength, force; power, esp. to enforce will or arbitrary authority. **mighty,** *a.* strong, forceful, powerful; very great, huge, immense; (*coll.*) great, considerable. *adv.* (*coll.*) exceedingly, very. **mightily,** *adv.* **mightiness,** *n.*
might[2], *past* MAY[1].
mignon (mēn'yon), *a.* delicate and small, dainty.
mignonette (minyənet'), *n.* an annual plant with fragrant greenish flowers.
migraine (mē'grān, mī'-), *n.* a recurrent severe headache, esp. on one side of the head only, often accompanied by nausea and visual disturbances.
migrate (mīgrāt'), *v.i.* to remove from one country, place or habitation to another; of birds, fishes etc., to pass from one region to another according to the season; to pass from one part of the body to another. **migrant** (mī'-), *n.*, *a.* **migration,** *n.* **migratory** (mī'grətəri), *a.*
mikado (mikah'dō), *n.* (*Hist.*) the Emperor of Japan.
mike (mīk), *n.* (*coll.*) short for MICROPHONE.
mil, (*abbr.*) military.
mil (mil), *n.* a unit of length, a thousandth part of an inch (0·0254 mm), in measuring wire; in pharmacy, a millilitre.
miladi, milady (milā'di), *n.* (formerly used in France) my lady (used as address or appellation).
milage (mī'lij), MILEAGE under MILE.
milch (milch), *a.* giving milk. **milch-cow,** *n.* a cow kept for milk; a person from whom money is easily obtained.
mild (mīld), *a.* gentle in manners or disposition; tender, pacific, clement, placid, bland, pleasant; of fruit, liquor etc., soft, not harsh, sharp or strong; of beer, not bitter, not strongly flavoured with hops; moderate, not extreme, tame; moderate in degree; of medicines, operating gently. **mildly,** *adv.* **to put it mildly,** without exaggerating. **mildness,** *n.*
mildew (mil'dū), *n.* a deleterious fungoid growth on plants, cloth, paper, food etc. after exposure to damp. *v.t.* to taint with mildew. *v.i.* to be tainted with mildew. **mildewy,** *a.*
mile (mīl), *n.* a measure of length or distance, 1760 yds. (1·609 km); orig. a Roman measure of 1000 paces, about 1620 yds. (1·481 km). **geographical, nautical, mile,** one sixtieth of a degree, acc. to the British Admiralty 6080 ft., or 2026¾ yds. (1·853 km). **mil(e)ometer** (-lom'-), *n.* a device for recording the number of miles travelled by a vehicle. **mileage,** *n.* the number of miles concerned; an allowance paid for the number of miles travelled; the distance travelled by a vehicle on one gallon or litre of petrol; the benefit to be derived from something. **miler,** *n.* a person, animal or thing qualified to run or travel a mile,

or (*in comb.*) a specified number of miles (as *ten-miler*). **miles,** *adv.* (*coll.*) considerably, very much.
milfoil (mil'foil), *n.* the yarrow.
milieu (mēlyœ'. mēl'-), *n.* environment, surroundings, setting.
militant (mil'itənt), *a.* fighting; combative, warlike, military. *n.* a militant person. **militancy,** *n.* **militantly,** *adv.*
military (mil'itəri), *a.* pertaining to soldiers, arms or warfare; soldierly, warlike, martial; engaged in war. *n.* (*collect.*) soldiers generally; the army; troops. **militaria** (-teə'riə), *n.pl.* military uniforms, medals etc. of the past that are of interest to collectors. **militarism,** *n.* military spirit; military or warlike policy; domination by the military or the spirit of aggression. **militarist,** *n.* **militarize, -ise,** *v.t.* **militarization,** *n.*
militate (mil'itāt), *v.i.* to be or stand opposed; to have weight or influence, to tell (against).
militia (milish'ə), *n.* a military force consisting of the body of citizens not enrolled in the regular army. **militiaman,** *n.*
milk (milk), *n.* the whitish fluid secreted by female mammals for the nourishment of their young, esp. that of the cow; the white juice of certain plants or fruits; an emulsion made from herbs, drugs etc., esp. for cosmetic purposes. *v.t.* to draw milk from; to plunder (creditors); to exploit or get money out of (a person) in an underhand or disreputable way; (*sl.*) to tap (a telegraph wire or message). *v.i.* to yield milk. **milk-and-water,** *n.* nambypamby or mawkish talk, sentiment etc. *a.* namby-pamby, weak, twaddling. **milk chocolate,** *n.* chocolate made with milk. **milk float,** *n.* a usu. electrically-propelled vehicle for delivering milk to houses. **milkmaid,** *n.* a woman employed in dairywork. **milkman,** *n.* a man who sells milk; a dairy worker. **milk-run,** *n.* a regular journey to deliver milk; a routine trip, flight etc. **milk-shake,** *n.* a drink of milk, flavouring and ice-cream, shaken up in a machine. **milksop,** *n.* an effeminate person. **milk-sugar,** *n.* (*Chem.*) lactose. **milk-tooth,** *n.* one of the temporary teeth in young mammals; the foretooth of a foal. **milkweed,** *n.* a plant, of various species, with milky juice. **milkwort,** *n.* a plant formerly believed to promote the secretion of milk, with blue, white or pink flowers. **milker,** *n.* **milky,** *a.* consisting of, mixed with or resembling milk; mild, effeminate; white, opaque, clouded (of liquids); of cattle, yielding milk; timid. **Milky Way,** *n.* a luminous zone, composed of innumerable stars, stretching across the heavens, being the stars composing the galaxy of which our solar system is a part.
mill (mil), *n.* a machine for grinding corn to a fine powder; a building with machinery for this purpose; a machine for reducing substances of any kind to a finer consistency; a building fitted up with machinery for any industrial purpose, a factory; (*sl.*) a fight with fists. *v.t.* to grind (as corn); to produce (flour) by grinding; to serrate the edge of (a coin); to full (cloth); (*sl.*) to thrash, to pummel. *v.i.* to move slowly (around). **to go, be put through the mill,** to undergo a harrowing, exhausting etc. experience. **millboard,** *n.* thick pasteboard used by bookbinders for book-covers. **mill-dam,** *n.* a wall or dam built across a stream to divert it to a mill; a millpond. **millpond,** *n.* **mill-race,** *n.* the canal or the current of water for driving a mill-wheel. **millstone,** *n.* one of a pair of circular stones for grinding corn; a very burdensome person or thing. **mill-wheel,** *n.* a large wheel moved by water, flowing over or under it, for driving the machinery in a mill. **millwright,** *n.* one who constructs or repairs the machinery of mills. **milled,** *a.* passed through a mill; of coins, having the edges serrated; of cloth, fulled. **miller,** *n.* one who keeps or works in a flour mill; one who works any mill; applied to various moths and other insects with white or powdery wings etc.
millennium (milen'iəm), *n.* (*pl.* **-nniums, -nnia,** -iə) a period of 1000 years, esp. that when Christ shall reign on

millepede 521 **mingle**

earth (in allusion to Rev. xx.1–5). **millenarian** (milənea'riən), *a.* consisting of 1000 years; pertaining to the millennium. *n.* one who believes in this. **millenarianism**, *n.* **millenary**, *n.*, *a.* **millennial**, *a.* pertaining to the millennium. *n.* a thousandth anniversary. **millennialist**, *n.*
millepede (mil'ipēd), MILLIPEDE.
miller MILL.
millesimal (miles'iməl), *a.* consisting of one-thousandth parts. *n.* a thousandth.
millet (mil'it), *n.* an E Indian grass, or its nutritive seeds; applied to some other species of grasses bearing edible seeds. **millet-grass**, *n.* a tall N American grass.
milli-, *comb. form* one-thousandth.
milliard (mil'iahd), *n.* one thousand millions.
millibar (mil'ibah), *n.* one-thousandth of a bar, equivalent to the pressure exerted by a column of mercury about 0·03 in. (0·762 mm) high.
milligram (mil'igram), *n.* one-thousandth of a gram, 0·0154 of an English grain.
millilitre (mil'ilētə), *n.* one-thousandth of a litre, 0·06103 cu. in.
millimetre (mil'imētə), *n.* one-thousandth of a metre, or 0·03937 in.
milliner (mil'inə), *n.* one who makes and sells hats, bonnets etc. for women. **millinery**, *n.*
milling (mil'ing), *n.* the act or process of working a mill or mills; the serrated edging of a coin.
million (mil'yən), *n.* one thousand thousand, esp. of pounds, francs or dollars; an indefinitely great number. **millionaire** (-neə'), *n.* a person who has a million pounds, francs, dollars etc.; one immensely rich. **millionairess** (-neə'res), *n. fem.* **millionth**, *n.*, *a.*
millipede (mil'ipēd), *n.* a segmented myriapod, esp. of the genus *Iulus*; any articulate animal with numerous feet.
millisecond (mil'isekənd), *n.* one-thousandth of a second.
millwright MILL[1].
milometer MILE.
milord (milawd'), *n.* my lord (applied to rich Englishmen).
milt (milt), *n.* the spleen; the spermatic organ of a male fish; the soft roe of fishes. *v.t.* to impregnate with milt (as fish ova). **milter**, *n.*
mime (mīm), *n.* in classical theatre, a simple kind of farce characterized by mimicry and gesture; any communication through facial expression, gesture etc. and without words; an actor in mime; a mimic, a clown or buffoon. *v.i.* to act in mime; to play the mime. *v.t.* to mimic.
mimeograph (mim'iəgraf, -grahf), *n.* a duplicating machine which uses a stencil for reproducing written or typewritten matter. *v.t.* to reproduce by means of this.
mimesis (mimē'sis), *n.* mimicry; imitation of or close natural resemblance to the appearance of another animal or of a natural object; the imitation of nature in art. **mimetic** (-met'-), *a.* **mimetically**, *adv.*
mimic (mim'ik), *a.* given to imitation; imitative; imitating, counterfeit. *n.* one who mimics. *v.t.* (*past*, *p.p.* **mimicked**) to imitate, esp. in order to ridicule; to ape, to copy; to resemble closely (of animals, plants etc.). **mimicry**, *n.*
Mimosa (mimō'sə), *n.* a genus of leguminous shrubs, including the sensitive plant, *Mimosa pudica*.
Mimulus (mim'ūləs), *n.* a genus of plants with a mask-like corolla, comprising the monkey-flower.
mina (mī'nə), *n.* one of various Eastern and Australian passerine birds. **minabird**, *n.* Also **myna**, **mynah**.
minaret (minəret', mī'-), *n.* a lofty slender turret on a mosque, from which the muezzin summons the people to prayers.
minatory (min'ətəri), *a.* threatening, menacing.
minauderie (minawdərē'), *n.* affectation, coquettish airs. [F, from *minauder*, to put on airs, from *mine*, MIEN]
mince (mins), *v.t.* to cut or chop into very small pieces; to utter or pronounce with affected delicacy; to minimize, to palliate, to gloss over; to restrain (one's words) for politeness' sake. *v.i.* to talk with affected elegance; to walk in a prim and affected manner. *n.* minced meat; mincemeat. **not to mince matters**, to speak plainly. **mincemeat**, *n.* meat chopped into very small pieces; a filling for pies etc. composed of suet, raisins, currants, candied-peel etc. chopped fine; very fine or small pieces or fragments. **to make mincemeat of**, to crush or destroy completely. **mince-pie**, *n.* a small pie filled with mincemeat. **mincing**, *a.* affectedly elegant. **mincingly**, *adv.*
mind (mīnd), *n.* a person's intellectual powers; the understanding, the intellect; the soul; intellectual capacity; recollection, memory; one's candid opinion; sanity; disposition, liking, way of feeling or thinking; intention, purpose; desire, inclination. *v.t.* to heed, to regard; to pay attention to, to apply oneself to; (*coll.*) to object to; (*coll.*) to look after; to remember, to bear in mind. *v.i.* to take care, to be on the watch. **absence, presence, of mind** ABSENT[2], PRESENCE. **in one's mind's eye**, in one's imagination. **in two minds**, unable to choose between two alternatives. **to bring, call, to mind** CALL[1]. **to cast one's mind back** CAST[1]. **to cross one's mind** CROSS[2]. **to have (half) a mind**, to be inclined (to). **to make up one's mind** MAKE[2]. **to one's mind**, in one's opinion. **to put in mind**, to remind (of). **to put out of one's mind**, to stop thinking about. **to speak one's mind**, to express one's candid opinion (of or about). **mind-blowing**, *a.* of or inducing a state like that produced by psychedelic drugs. Also **mind-bending**. **mind-boggling**, *a.* amazing, astonishing. **mind-reader**, *n.* a person who claims to know what others are thinking. **mind-reading**, *n.* **minded**, *a.* (*usu.* in *comb.*, as *evilminded*). **minder**, *n.* one who watches over someone or something; (*sl.*) a bodyguard; (*sl.*) an aide to a politician or public figure. **mindful**, *a.* attentive, heedful. **mindfully**, *adv.* **mindfulness**, *n.* **mindless**, *a.* done for no reason; done without need for thought; heedless, regardless. **mindlessly**, *adv.* **mindlessness**, *n.*
mine[1] (mīn), *poss. pron.* belonging to me.
mine[2] (mīn), *v.t.* to dig into or burrow in; to obtain by excavating in the earth; to make by digging; to undermine, to sap; to set with mines. *v.i.* to dig a mine, to engage in digging for ore etc.; to burrow; to practise secret methods of inquiry. *n.* an excavation in the earth for the purpose of obtaining minerals; a rich deposit of minerals suitable for mining; crude ironstone; an excavation under an enemy's works for blowing them up, formerly to form a means of entering or to cause a collapse of the wall etc.; a receptacle filled with explosive, floating in the sea or buried in the ground, which is exploded by contact; a rich source of wealth, or of information etc. **mine-layer**, *n.* a ship or aircraft employed to lay mines. **mine-sweeper**, *n.* a naval vessel employed to clear mines laid by the enemy. **miner**, *n.* one who digs for minerals; one who works in mines; a soldier employed to lay mines; (*Austral.*) a variety of honey-eater bird. **mining**, *n.*
mineral (min'ərəl), *n.* an inorganic body, homogeneous in structure, with a definite chemical composition, found in the earth; any inorganic substance found in the ground; (*pl.*) mineral waters; *a.* pertaining to or consisting of minerals; impregnated with mineral matter. **mineral oil**, *n.* any oil derived from minerals; (*N Am.*) liquid paraffin. **mineral water**, *n.* water naturally impregnated with mineral matter; an artificial imitation of this. **mineralize, -ise**, *v.t.* to convert into a mineral; to give mineral qualities to; to impregnate with mineral matter. *v.i.* to become mineralized; to study mineralogy. **mineralization, -isation**, *n.* **mineralizer**, *n.* **mineralogy** (-al'əji), *n.* the science of minerals, their nature and properties. **mineralogical** (-loj'-), *a.* **mineralogically**, *adv.* **mineralogist**, *n.*
minestrone (mini'strōni), *n.* a thick soup of vegetables, pasta etc.
minge (minj), *n.* (*taboo sl.*) the female genitals.
mingle (ming'gl), *v.t.* to mix up together; to blend (with).

mingy — **miracle**

v.i. to be mixed, blended or united (with). **mingler,** *n.* **minglingly,** *adv.*
mingy (min'ji), *a.* (*coll.*) mean, stingy. **minginess,** *n.*
mini (min'i), *n.* (*coll.*) a mini-skirt; a small car.
mini-, *comb. form* smaller than the usual size.
miniature (min'əchə), *n.* a small-sized painting, esp. a portrait on ivory, vellum etc., orig. a small picture in an illuminated manuscript; the art of painting on a small scale; an image on a greatly reduced scale; a reproduction on a small scale. *a.* represented on a very small scale. *v.t.* to portray in miniature. **miniaturist,** *n.* **miniaturize, -ise,** *v.t.* to make or construct on a smaller scale; to reduce the size of. **miniaturization,** *n.*
minibus (min'ibūs), *n.* a small bus for 8 to 12 passengers.
minicab (min'ikab), *n.* a taxi that can be ordered by telephone, but may not cruise in search of passengers.
minicomputer (min'ikəmpūtə), *n.* a small digital computer.
minim (min'im), *n.* a musical note of the value of two crotchets or half a semi-breve; fluid-measure, one drop, or one sixtieth of a drachm (0·059 g); a down-stroke in writing; an insignificant person, a dwarf, a pigmy.
minimal (min'iməl), *a.* pertaining to or being a minimum; least possible; smallest, very small. **minimal art,** *n.* abstract art showing simple geometric forms and ordinary objects in an inexpressive style. **minimalism,** *n.* **minimalist,** *n.* a person ready to accept the minimum; one who practises minimal art.
minimize, -ise (min'imīz), *v.t.* to reduce to the smallest possible amount or degree; to belittle. **minimization,** *n.*
minimum (min'iməm), *n.* (*pl.* -mums, -ma, -mə) the smallest amount or degree possible or usual. *a.* least possible. **minimum lending rate,** the minimum rate at which the Bank of England will discount bills. **minimum wage,** *n.* the rate of wages established by law or collective bargaining below which workers cannot be employed.
minion (min'yən), *n.* a darling, a favourite; a servile dependant.
minipill (min'ipil), *n.* a low-dose oral contraceptive pill without oestrogen.
miniscule (min'iskūl), MINUSCULE.
miniseries (min'isiə'riz), *n.* a television programme in several parts, usu. shown on consecutive nights.
mini-skirt (min'iskœt), *n.* a skirt with the hem far above the knees.
minister (min'istə), *n.* one charged with the performance of a duty, or the execution of a will etc.; a person entrusted with the direction of a state department; a person representing his government with another state, an ambassador; the pastor of a church, esp. a Nonconformist; one who acts under the authority of another, a subordinate, an instrument; a servant. *v.i.* to render aid, service or attendance; to contribute, to be conducive (to); to serve as minister. †*v.t.* to furnish, to supply. **ministerial** (-tē'-), *a.* pertaining to a minister of state or of religion; pertaining to the ministry, esp. in contradistinction to the opposition; subsidiary, instrumental; pertaining to the execution of a legal mandate etc. **ministerially,** *adv.* **ministrant,** *n.*, *a.* **ministration** (-trā'-), *n.* **ministry,** *n.* the act of ministering; administration; the ministers of state or of religion collectively.
miniver (min'ivə), *n.* a kind of fur used for ceremonial robes; applied to the Siberian squirrel and its fur.
mink (mingk), *n.* a name for several species of amphibious stoat-like animals esteemed for their fur. *n.*, *a.* (of) this fur.
Minn., (*abbr.*) Minnesota (US).
minnesinger (min'isingə), *n.* one of a body of German lyric poets and singers (1138–1347) whose chief theme was love.
minnow (min'ō), *n.* a small fish common all over Europe; (*loosely*) any tiny fish; an insignificant person or thing.
Minoan (minō'ən), *a.* pertaining to ancient Crete or its people. *n.* an inhabitant of ancient Crete; their language. **Minoan period,** *n.* the Bronze Age of Crete, loosely 2500–1200 BC.
minor (mī'nə), *a.* less, smaller (not used with *than*); petty, comparatively unimportant; (*Mus.*) less by a semitone; *n.* a person under age; a minor term or premise in logic; a minor key or a composition or strain in this. **minor key,** *n.* (*Mus.*) a key in which the scale has a minor third. **minor premise,** *n.* the premise containing the minor term of a syllogism. **minor suit,** *n.* in bridge, clubs or diamonds. **minor term,** *n.* the subject of the conclusion of a categorical syllogism. **minority** (-nor'-), *n.* the smaller number, esp. the smaller of a group or party voting together in an election, on a Bill etc.; the state of being under age; the period of this.
Minotaur (min'ətaw, mī'-), *n.* in Greek mythology, a monster having the head of a bull and the rest of the body human, devouring human flesh.
minster (min'stə), *n.* the church of a monastery; a cathedral or other large and important church.
minstrel (min'strəl), *n.* one of a class of men in the Middle Ages who lived by singing and reciting; a travelling gleeman, musician, performer or entertainer; a poet; a musician; one of a troupe of entertainers with blackened faces.
mint[1] (mint), *n.* a place where money is coined, usu. under state authority; a source of invention or fabrication; a great quantity, supply or amount. *v.t.* to coin, to stamp (money); to invent; to coin (a phrase etc.). *a.* of a book, coin etc., in its unused state; as new. **in mint condition,** as perfect as when first produced. **minter,** *n.*
mint[2] (mint), *n.* any plant of the aromatic genus *Mentha,* esp. *M. viridis,* the garden mint, from which an essential oil is distilled. **mint-julep,** *n.* spirits, sugar and pounded ice flavoured with mint. **mint-sauce,** *n.* mint chopped up with vinegar and sugar, used as a sauce with roast lamb.
minuend (min'ūend), *n.* the quantity from which another is to be subtracted.
minuet (minūet'), *n.* a slow stately dance in triple measure; music for this or in the same measure.
minus (mī'nəs), *prep.*, *a.* less by, with the deduction of; (*coll.*) short of, lacking; negative. *n.* the sign of subtraction (−).
minuscule (min'iskūl), *a.* small; miniature (esp. applied to mediaeval script). *n.* a minute kind of letter in cursive script of the 7th–9th cent.; a small or lower-case letter; anything very small.
minute[1] (minūt'), *a.* very small; petty, trifling; particular, exact, precise. **minutely,** *adv.* **minuteness,** *n.*
minute[2] (min'it), *n.* the 60th part of an hour; a very small portion of time, an instant; an exact point of time; the 60th part of a degree; a memorandum; an official memorandum of a court or other authority; (*pl.*) official records of proceedings of a committee etc. *v.t.* to write minutes of; to take a note of; to time to the exact minute. **up to the minute,** very modern. **minute-gun,** *n.* a gun fired at intervals of one minute as a signal of distress or mourning. **minute-hand,** *n.* the hand pointing to minutes in a clock or watch. **minute steak,** *n.* a thin steak that can be cooked quickly.
minutia (minū'shiə, -mī-), *n.* (*usu. pl.* -tiae, iē) small and precise or trivial particulars.
minx (mingks), *n.* a pert girl, a hussy.
Miocene (mī'əsēn), *a.* a term applied to the middle division of the Tertiary strata or period.
miosis (miō'sis, mi-), MEIOSIS.
MIPS (mips), (*acronym*) (*Comput.*) million instructions per second.
miracle (mir'əkl), *n.* a wonder, a marvel, a prodigy; a marvellous event or act due to supernatural agency; an extraordinary occurrence; of cleverness etc., an extraordinary example; a miracle play. **miracle play,** *n.* a mediaeval dramatic representation, usu. dealing with historical or

mirador

traditional events in the life of Christ or of the Saints.
miraculous (-rak'ū-), *a.* **miraculously**, *adv.* **miraculousness**, *n.*
mirador (mirədaw'), *n.* a belvedere turret or gallery, commanding an extensive view.
mirage (mirahzh'), *n.* an optical illusion by which images of distant objects are seen as if inverted, esp. in a desert where the inverted sky appears as a sheet of water.
MIRAS (mī'rəs), (*acronym*) *m*ortgage *i*nterest *r*elief *at s*ource.
mire (mīə), *n.* wet, clayey soil, swampy ground, bog; mud, dirt. *v.t.* to plunge in mire; to soil with mire; to involve in difficulties. *v.t.* to sink in mire.
mirepoix (miəpwah', miə'-), *n.* (*pl.* **mirepoix**) a sauce of sautéed root vegetables.
mirk (mœk) etc. MURK.
mirror (mi'rə), *n.* an appliance with a polished surface for reflecting images; a looking-glass; anything that reflects objects; an exemplar, a pattern, a model. *v.t.* to reflect in or as in a mirror. **mirror writing,** *n.* handwriting from right to left, as if reflected in a mirror.
mirth (mœth), *n.* merriment, jollity, gaiety, hilarity. **mirthful,** *a.* **mirthfully,** *adv.* **mirthfulness,** *n.* **mirthless,** *a.* **mirthlessness,** *n.*
MIRV (mœv), *n.* a missile with two or more warheads designed to strike separate targets. [acronym for *m*ultiple *i*ndependently-targetable *r*e-entry *v*ehicle]
mis- *pref.* wrongly, badly, amiss, unfavourably.
misadventure (misədven'chə), *n.* bad luck; ill fortune; an unlucky chance or accident.
misalign (misəlīn'), *v.t.* to align wrongly.
misalliance (misəlī'əns), *n.* an improper alliance, esp. by marriage. **misallied,** *a.* **misally,** *vb.*
misanthrope (mis'ənthrōp, miz'-), *n.* a hater of mankind; one who has a morbid dislike of other people. **misanthropic, -ical** (-throp'-), *a.* **misanthropist** (-an'-), *n.* **misanthropy** (-an'-), *n.*
misapply (misəplī'), *v.t.* to apply wrongly. **misapplication,** *n.*
misappreciate (misəprē'shiāt), *v.t.* to fail to appreciate rightly or fully. **misappreciation,** *n.*
misapprehend (misəprihend'), *v.t.* to misunderstand. **misapprehension,** *n.* **misapprehensive,** *a.* **misapprehensively,** *adv.*
misappropriate (misəprō'priāt), *v.t.* to apply to a wrong use or purpose (esp. funds to one's own use). **misappropriation,** *n.*
misbecome (misbikŭm'), *v.t.* to be improper or unseemly to, to ill become. **misbecomingly,** *adv.*
misbegotten (misbigot'n), *a.* begotten unlawfully, illegitimate, bastard; hideous, despicable.
misbehave (misbihāv'), *v.i.* to behave (oneself) ill or improperly. **misbehaved,** *a.* guilty of misbehaviour, ill-mannered. **misbehaviour** (-yə), *n.*
misbelief (misbilēf'), *n.* false or erroneous belief. **misbelieve,** *v.t.* **misbeliever,** *n.*
misc, (*abbr.*) miscellaneous; miscellany.
miscalculate (miskal'kūlāt), *v.t.* to calculate wrongly. **miscalculation,** *n.*
miscall (miskawl'), *v.t.* to misname; to abuse, to call (someone) names.
miscarry (miska'ri), *v.i.* to be carried to the wrong place; to fail, to be unsuccessful; to be delivered of a child prematurely. **miscarriage,** *n.* an act or instance of miscarrying; failure; the (accidental) premature expulsion of a foetus before it can survive outside the womb. **miscarriage of justice,** a mistake or wrong committed by a court of justice.
miscast (miskahst'), *v.t.* to cast or add up wrongly; to cast (a play or an actor) inappropriately.
miscegenation (misəjənā'shən), *n.* intermarriage or interbreeding between people of different races.

miscellaneous (misəlā'niəs), *a.* consisting of several kinds; mixed, multifarious, diversified; various, many-sided. **miscellanea** (-niə), *n.pl.* a collection of miscellaneous literary compositions. **miscellaneously,** *adv.* **miscellaneousness** (-niə-), *n.* **miscellany** (-sel'əni), *n.* a mixture of various kinds, a medley, a number of compositions on various subjects in one volume.
mischance (mischahns'), *n.* misfortune, ill-luck.
mischief (mis'chif), *n.* harm, injury, damage; vexatious action or conduct, esp. a vexatious prank. **mischiefmaker,** *n.* one who stirs up ill-will. **mischief-making,** *n.,* *a.* **mischievous,** *a.* making mischief; naughty; of a child, full of pranks, continually in mischief; arch, roguish; vexatious. **mischievously,** *adv.* **mischievousness,** *n.*
miscible (mis'ibl), *a.* that may be mixed (with). **miscibility** (-bil'-), *n.*
miscomprehend (miskomprihend'), *v.t.* to comprehend wrongly, to misunderstand. **miscomprehension,** *n.*
miscompute (miskəmpūt'), *v.t.* to compute wrongly. **miscomputation** (-kom'-), *n.*
misconceive (miskənsēv'), *v.t.* to have a wrong idea of, to misapprehend. **misconception** (-sep'-), *n.*
misconduct[1] (miskon'dŭkt), *n.* improper conduct, esp. adultery; mismanagement.
misconduct[2] (miskəndŭkt'), *v.t.* to mismanage. *v.i.* to misbehave.
misconstrue (miskənstroo'), *v.t.* to mistake the meaning of; to put a wrong interpretation or construction upon. **misconstruction** (-strŭk'-), *n.*
miscount[1] (miskownt'), *v.t.* to count wrongly; to estimate or regard wrongly. *v.i.* to make a false account.
miscount[2] (mis'kownt), *n.* a mistake in counting, esp. of votes.
miscreant (mis'kriənt), *n.* a heretic; a vile wretch, a scoundrel. *a.* evil; vile.
miscue (miskū'), *n.* in billiards, snooker etc., failure to strike a ball properly with the cue. *v.i.* to make a miscue.
misdate (misdāt'), *v.t.* to date wrongly. *n.* a wrong date.
misdeal (misdēl'), *v.t.* to deal wrongly (as cards). *v.i.* to make a misdeal. *n.* a wrong or false deal.
misdeed (misdēd'), *n.* an evil deed, a crime.
misdemeanour, *n.* misbehaviour, misconduct; (*Law*) an indictable offence of less gravity than a felony.
misdirect (misdirekt'. -dī-), *v.t.* to direct wrongly; to address (a letter etc.) wrongly. **misdirection,** *n.*
mise en scène (mēz ē sen'), the scenery and general setting of a play; the visible surroundings of an event. [F]
miser (mī'zə), *n.* one who denies himself the comforts of life for the sake of hoarding; an avaricious person; a wretched person. **miserly,** *a.* **miserliness,** *n.*
miserable (miz'ərəbl), *a.* very wretched or unhappy, distressed; causing misery, distressing; sorry, despicable, worthless; very poor or mean. *n.* a miserable person, a wretch. **miserably,** *adv.*
misère (mizeə'), *n.* a declaration in solo whist etc. by which a player undertakes not to take a single trick.
miserere (mizəreə'ri), *n.* the 51st Psalm, beginning with this word in the Vulgate; a musical setting of this psalm; a prayer or cry for mercy; a misericord.
misericord (mize'rikawd), *n.* an apartment in a monastery for monks to whom special indulgences were granted; a bracketed projection on the underside of the seat of a choir-stall, to afford rest to a person standing; a small, straight dagger for giving the coup de grâce.
miserly MISER.
misery (miz'əri), *n.* great unhappiness or wretchedness of mind or body; affliction, poverty; (*coll.*) an ill-tempered, gloomy person.
misfeasance (misfē'zəns), *n.* (*Law*) a trespass, a wrong, esp. negligent or improper performance of a lawful act.
misfire[1] (mis'fīə), *n.* failure to go off or explode (of a gun, charge etc.); in a motor-vehicle engine, failure to fire in

the correct ignition sequence.
misfire² (misfīə'), *v.i.* to fail to go off; to fail to achieve the intended effect; of a motor-vehicle engine, to fail to fire correctly.
misfit¹ (mis'fit), *n.* a bad fit; a garment that does not fit properly; an awkward person.
misfit² (misfit'), *v.t., v.i.* to fail to fit.
misfortune (misfaw'chən), *n.* ill luck, calamity; a mishap, a disaster. **misfortuned,** *a.*
misgive (misgiv'), *v.t. (impers.)* to fill (one's mind) with doubt or suspicion. **misgiving,** *n.*
misgovern (misgŭv'ən), *v.t.* to govern ill; to administer unfaithfully. **misgovernment,** *n.* **misgoverned,** *a.* badly governed, rude.
misguide (misgīd'), *v.t.* to guide wrongly; to lead astray. **misguided,** *a.* foolish. **misguidedly,** *adv.*
mishandle (mis·hand'l), *v.t.* to handle roughly; to ill-treat.
mishap (mis'hap), *n.* a mischance; ill luck.
mishit (mis·hit'), *v.t.* to hit wrongly. *n.* (mis'hit), an instance of hitting wrongly.
mishmash (mish'mash), *n.* a hotchpotch, a jumble.
Mishna (mish'nə), *n.* the second or oral law, the collection of traditions etc. forming the text of the Talmud. **Mishnic,** *a.*
misinform (misinfawm'), *v.t.* to give erroneous information to. **misinformation,** *n.*
misinterpret (misintœ'prit), *v.t.* to interpret wrongly; to draw a wrong conclusion from. **misinterpretation,** *n.* **misinterpreter,** *n.*
misjudge (misjŭj'), *v.t.* to judge erroneously; to form an erroneous opinion of. **misjudgment,** *n.*
mislay (mislā'), *v.t. (past, p.p.* **-laid)** to lay in a wrong place or in a place that cannot be remembered; to lose.
mislead (mislēd'), *v.t. (past, p.p.* **misled,** -led') to lead astray; to cause to go wrong, esp. in conduct; to deceive, to delude. **misleader,** *n.* **misleading,** *a.* **misleadingly,** *adv.*
mismanage (misman'ij), *v.t., v.i.* to manage badly. **mismanagement,** *n.*
mismatch¹ (mismach'), *v.t.* to match unsuitably.
mismatch² (mis'mach), *n.* an unsuitable match.
misnomer (misnō'mə), *n.* a mistaken or misapplied name or designation; an incorrect term.
miso (mē'sō), *n.* a food paste made from soya beans fermented in brine, used for flavouring.
mis(o)-, *comb. form* dislike; hatred.
misogamy (misog'əmi, mī-), *n.* hatred of marriage. **misogamist,** *n.*
misogyny (misoj'əni, mī-), *n.* hatred of women. **misogynic** (-jin'-), *a.* **misogynist,** *n.*
misplace (misplās'), *v.t.* to mislay; to set on or devote to an undeserving object. **misplacement,** *n.*
misplay (misplā'), *n.* wrong or foul play.
misprint¹ (misprint'), *v.t.* to print incorrectly.
misprint² (mis'print), *n.* a mistake in printing.
misprision (misprizh'ən), *n.* (*Law*) an offence within the degree of capital but borderline thereon, esp. one of neglect or concealment; †mistake, misconception. **misprision of treason** or **felony,** concealment of treason or felony without actual participation.
mispronounce (mispronowns'), *v.t.* to pronounce wrongly. **mispronunciation** (-nŭn-), *n.*
misquote (miskwōt'), *v.t.* to quote erroneously. **misquotation,** *n.*
misread (misrēd'), *v.t. (past, p.p.* **-read,** -red) to read incorrectly; to misinterpret. **misreading,** *n.*
misrepresent (misreprizent'), *v.t.* to represent falsely or incorrectly. **misrepresentation,** *n.* **misrepresentative,** *n., a.*
misrule (misrool'), *n.* bad government; disorder, confusion, tumult, riot. *v.t.* to rule incompetently, misgovern.
Miss (mis), *n. (pl.* **misses)** a title of address for an unmarried woman or girl; *(coll.)* a girl; used before the name of a place, activity etc. to refer to a young woman who represents that place, activity etc., often in beauty contests. **missy,** *n. (coll.)* a form of address to a young woman or little girl.
miss (mis), *v.t.* to fail to reach, hit, meet, perceive, find, or obtain; to fall short of, to let slip, to overlook; to fail to understand; to omit; to escape, to dispense with; to feel or perceive the want of. *v.i.* to fail to hit the mark; to be unsuccessful. *n.* a failure to hit, reach, obtain etc. **a miss is as good as a mile,** escape or failure, no matter how narrow the margin, is the point of importance. **to give a miss,** in billiards, to avoid hitting the object ball in order to leave one's own in a safe position; not to take an opportunity to see, visit, enjoy etc. **to go missing,** to disappear or be lost. **to miss fire,** to fail to go off (of a gun, explosive etc.). **to miss out,** to omit; to fail to receive or enjoy. **to miss the boat, bus,** to miss an opportunity. **missing,** *a.* that misses; lost, wanting; absent, not in its place. **missing link,** *n.* something required to complete a series; a hypothetical form connecting types that are probably related, as man and the anthropoid apes.
Miss., *(abbr.)* Mississippi (US).
missal (mis'l), *n.* the book containing the service of the Mass for the whole year; a mediaeval illuminated manuscript.
missel-thrush MISTLE.
misshape (mis·shāp'), *v.t. (p.p.* **misshapen)** to shape badly; to deform. *n.* deformity. **misshapen,** *a.*
missile (mis'īl), *n.* a weapon or other object projected or propelled through the air, esp. a rocket-propelled weapon, often with a nuclear warhead. **missil(e)ry,** *n.* (the design, use or study of) missiles.
mission (mish'ən), *n.* a sending or being sent; the commission, charge or office of a messenger, agent etc.; a person's appointed or chosen end, a vocation; a body of persons sent on a diplomatic errand, an embassy or legation; a body of missionaries established in a district at home or sent to a foreign country to spread religious teaching; their field of work; a missionary station; a religious organization in the Roman Catholic Church ranking below that of a regular parish; a series of special services for rousing spiritual interest. **missionary** (-əri), *a.* pertaining to missions, esp. those of a religious nature; pertaining to the propagation of religion or other moral, social or political influence. *n.* one sent to carry on such work. **missionary position,** *n. (coll.)* the conventional position for sexual intercourse, lying down with the woman on her back and the man on top of her. **missioner,** *n.* a missionary; one in charge of a parochial mission.
missis, missus (mis'iz), *n. (coll.)* the mistress of a household; a wife.
missive (mis'iv), *n.* a message, a letter.
misspell (mis·spel'), *v.t.* to spell incorrectly. **misspelling,** *n.*
misspend (mis·spend'), *v.t. (past, p.p.* **-spent)** to spend badly; to waste.
misstate (mis·stāt'), *v.t.* to state wrongly. **misstatement,** *n.*
misstep (mis·step'), *n.* a false step.
mist (mist), *n.* visible watery vapour in the atmosphere at or near the surface of the earth; a watery condensation dimming a surface; a suspension of a liquid in a gas; a watery film before the eyes; anything which dims, obscures or darkens. *v.t.* to cover as with mist. *v.i.* to be misty. **misty,** *a.* characterized by or overspread with mist; vague, dim, indistinct, obscure. **mistily,** *adv.* **mistiness,** *n.*
mistake (mistāk'), *v.t. (past* **-took,** -tuk, *p.p.* **-taken)** to take or understand wrongly; to take in a wrong sense; to take one person or thing for another. *v.i.* to be in error; to err in judgment or opinion. *n.* an error of judgment or opinion; a misunderstanding, a blunder. **mistakable,** *a.* **mistakably,** *adv.* **mistaken,** *a.* wrong in judgment, opinion etc. **mistakenly,** *adv.* **mistakenness,** *n.*
Mister (mis'tə), *n.* the common form of address prefixed to men's names or certain official titles (abbr. in writing to

MR). *v.t.* to speak of or address (someone) as 'Mister'. [var. of MASTER]
mistime (mistīm'), *v.t.* to say or do inappropriately or not suitably to the time or occasion.
mistitle (mistī'tl), *v.t.* to call by a wrong title.
mistle, mistle-thrush (mis'lthrŭsh), *n.* the largest of the European thrushes.
mistletoe (mis'ltō), *n.* a plant bearing white glutinous berries.
mistral (mistrahl'), *n.* a cold dry NW wind of S France.
mistranslate (mistranslāt'. -tranz-), *v.t.* to translate wrongly. **mistranslation,** *n.*
mistreat (mistrēt'), *v.t.* to ill-treat. **mistreatment,** *n.*
mistress (mis'tris), *n.* a woman who has authority or control; the female head of a family, school etc.; a woman having the control or disposal (of); a woman who has mastery (of a subject etc.); a female teacher; a woman beloved and courted, a sweetheart; a woman with whom a man has a long-term extramarital relationship; a title of address to a married woman (abbr. in writing to MRS, mis'iz).
mistrial (mistrī'əl), *n.* an abortive or inconclusive trial.
mistrust (mistrŭst'), *v.t.* to regard with doubt or suspicion. *n.* distrust, suspicion. **mistrusted,** *a.* **mistrustful,** *a.* **mistrustfully,** *adv.* **mistrustfulness,** *n.*
misunderstand (misŭndəstand'), *v.t.* (*past, p.p.* **-stood,** stud) to misconceive, to misapprehend, to mistake the meaning or sense of. **misunderstanding,** *n.* **misunderstood,** *a.*
misuse (misūz'), *v.t.* to use or treat improperly; to apply to a wrong purpose; to ill-treat. *n.* (-ūs'), improper use; abuse. **misusage,** *n.*
MIT, (*abbr.*) Massachusetts Institute of Technology.
mite[1] (mīt), *n.* a very small coin, orig. Flemish; a small contribution; a minute amount, a tiny thing, esp. a child.
mite[2] (mīt), *n.* a name common to the minute arachnids of the Acarida, esp. those infesting cheese. **mity,** *a.*
miter (mī'tə), MITRE.
Mithra, Mithras (mith'rə, -ras), *n.* the Persian god of light or sun-god. **Mithraic** (-rā'-), *a.* **Mithraism** (-rā-), *n.*
mithridate (mith'ridāt), *n.* an antidote against poison. **mithridatic** (-dat'-), *a.* **mithridatism,** *n.* **mithridatize, -ise** (-thrid'-), *v.t.* to render immune against poison by taking larger and larger doses of it.
mitigate (mit'igāt), *v.t.* to make less rigorous or harsh; to relax (severity); to alleviate (pain, violence etc.); to soften, to diminish, to moderate. *v.i.* to become assuaged, relaxed or moderated. **mitigation,** *n.* **mitigative, mitigatory,** *a.* **mitigator,** *n.*
mitochondrion (mītōkon'driən), *n.* a spherical or rodlike organism, found in cytoplasm, whose function is energy production.
mitosis (mītō'sis), *n.* (*pl.* **-oses**) indirect cell-division involving the reproduction of nuclei. **mitotic** (-tot'-), *a.*
mitral (mī'trəl), *a.* of or resembling a mitre. **mitral valve,** *n.* the valve between the left auricle and ventricle of the heart.
mitre, (*esp. N Am.*) **miter** (mī'tə), *n.* a tall ornamental cap shaped like a cleft cone rising into two peaks, worn as symbol of office by bishops; the dignity of a bishop; a joint at an angle (usu. of 90°), as the corner of a picture-frame, each jointing surface being cut at an angle to the piece on which it is formed; hence, an angle of 45°. *v.t.* to confer a mitre upon; to join with a mitre; to shape off at an angle of 45°. **mitre-block, -box,** *n.* a block or box used to guide the saw in cutting mitres. **mitred,** *a.*
mitt (mit), *n.* a kind of glove or covering, usu. of lace or knitting, for the wrist and palm; a thick mitten worn by the catcher in baseball; (*sl.*) a hand.
mitten (mit'n), *n.* a glove with a thumb but no fingers; (*sl.*) a boxing-glove.
mittimus (mit'iməs), *n.* (*Law*) a warrant of commitment to prison.
mity MITE[2].
mix (miks), *v.t.* to put together or blend into one mass or compound; to mingle or incorporate (several substances, quantities or groups) so that the particles of each are indiscriminately associated; to compound by mingling various ingredients; to cross (breeds); to join; (*Mus.*) to combine individual instruments, voices, etc. electronically. *v.i.* to become united; to be mingled (with or together); to be associated or have intercourse (with); to copulate. *n.* an act or process of mixing; a mixture, a combination; mixed ingredients, e.g. for a cake, sold pre-packed; (*Mus.*) the sound produced by electronic mixing. **to mix up,** to mix thoroughly; to confuse, to bewilder; to involve (in an (esp. dubious) undertaking or with an (esp. undesirable) person). **mixed-up,** *a.* confused, chaotic, muddled; in emotional turmoil. **mixable,** *a.* **mixed** (mikst), *a.* consisting of various kinds or constituents; promiscuous, of company, not select; not wholly good or bad, not of consistent quality; (*coll.*) confused, bewildered, muddled. **mixed bag,** *n.* a miscellaneous collection. **mixed blessing,** *n.* something that has advantages and disadvantages. **mixed doubles,** *n.pl.* (*Tennis*) matches with a man and woman player as partners on each side. **mixed farm,** *n.* a farm combining arable and livestock production. **mixed marriage,** *n.* one in which the contracting parties are of different creeds or races. **mixed mathematics** MATHEMATICS. **mixed metaphor,** *n.* one that brings together incongruous concepts. **mixer,** *n.* a person or thing that causes mixing, that mixes; a person with social tact; one who gets on well with all sorts of people; (*coll.*) a person who stirs up trouble; a non-alcoholic drink suitable for mixing with alcoholic drinks; a kitchen appliance for mixing food ingredients; an electronic device which combines input signals into a single output.
mixture (miks'chə), *n.* that which is mixed; a mixing, compound; gas or vaporized oil mixed with air to form the explosive charge in an internal-combustion engine.
mizzen, mizen (miz'n), *n.* a fore-and-aft sail set on the mizzen-mast, also called the **mizzen-sail. mizzen-mast,** *n.* the aftermost mast of a three-masted ship.
Mk, (*abbr.*) mark; markka.
mks units, *n.pl.* the metric system of units based on the metre, Kilogram and second.
MLA, (*abbr.*) Member of the Legislative Assembly; Modern Languages Association.
MLF, (*abbr.*) multilateral (nuclear) force.
MLitt, (*abbr.*) Master of Letters (*L Magister Litterarum*).
Mlle(s), (*abbr.*) Mademoiselle; Mesdemoiselles.
MLR, (*abbr.*) minimum lending rate.
MM, (*abbr.*) Messieurs.
mm, (*abbr.*) millimetre.
Mme(s), (*abbr.*) Madame; Mesdames.
Mn, (*chem. symbol*) manganese.
mnemonic (nimon'ik), *a.* pertaining to or aiding the memory. *n.* an aid to memory; (*pl.*) the art of or a system for aiding or strengthening memory. **mnemonically,** *adv.*
MO, (*abbr.*) Medical Officer; modus operandi; money order.
Mo, (*chem. symbol*) molybdenum.
Mo., (*abbr.*) Missouri (US).
mo (mō), *n.* (*coll.*) moment, as in *just a mo,* (*coll.*) *half a mo.*
mo., (*abbr.*) month(s).
moa (mō'ə), *n.* an extinct, flightless bird of New Zealand.
moan (mōn), *n.* a low prolonged sound expressing pain or sorrow; a complaint. *v.i.* to utter a moan or moans; to complain, grumble. *v.t.* to lament, to deplore; to mourn; to utter moaningly. **moaner,** *n.* **moanful,** *a.* **moanfully, moaningly,** *adv.*
moat (mōt), *n.* a ditch round a castle, fort etc., usu. filled with water. *v.t.* to surround with or as with a moat.

mob (mob), *n.* a disorderly or riotous crowd, a rabble; the masses, the lower orders; a gang of criminals engaged in organized crime; (*coll., derog.*) a group or class (of people of a specified kind). *v.t.* (*past, p.p.* **mobbed**) to attack in a mob; to crowd roughly round and annoy. *v.i.* to gather together in a mob. **the heavy mob,** (*coll., facet.*) (a group of) people with the power to frighten or coerce. **the mob, Mob,** (*chiefly N Am.*) the Mafia; organized crime. **mob law,** *n.* the rule of the mob; lynch law. **mobster,** *n.* a member of a criminal mob.
mobcap, *n.* a plain indoor cap or headdress for women, usu. tied under the chin.
mobile (mō'bīl), *a.* movable, free to move; easily moved; easily changing (as expression); that may be moved from place to place (as troops). *n.* that which moves or causes motion; an artistic concoction of dangling wires etc. **mobility** (-bil'-), *n.* **mobilize, -ise** (-bil-), *v.t.* to make mobile; to put into circulation; to put (troops, a fleet etc.) in a state of readiness for active service; to put into action. **mobilizable** (-bil-), *a.* **mobilization,** *n.*
Möbius strip (mœ'biəs), *n.* a long, rectangular strip of paper twisted through 180° and joined at the ends, to form a one-sided surface bounded by one continuous curve. [A.F. *Möbius*, German mathematician, 1790–1868]
moccasin (mok'əsin), *n.* a foot-covering, usu. of deer-skin or soft leather in one piece, worn by N American Indians; a bedroom slipper of soft leather made of one piece.
mocha (mo'kə), *n.* a dark brown coffee, orig. from Mocha; in Arabia; this flavour or colour.
mock (mok), *v.t.* to deride, to laugh at; to mimic, esp. in derision; to defy contemptuously; to delude, to take in. *v.i.* to express ridicule, derision or contempt. *a.* sham, false, counterfeit; imitating reality. *n.* a derision, a sneer; that which is derided; an imitation; an examination taken as practice prior to an official one. **mock-turtle soup,** a soup prepared from calf's head, veal etc. to imitate turtle soup. **mock-heroic,** *a.* of a literary work, burlesquing the heroic style. *n.* a burlesque of the heroic style. **mock-orange,** *n.* a shrub, the flowers of which smell like orange-blossoms. **mock-up,** *n.* a full-size dummy model; an unprinted model of a book. **mocker,** *n.* **to put the mockers on,** (*sl.*) to cause to fail; to make impossible. **mockery,** *n.* the act of mocking; ridicule, derision; a subject of ridicule; a delusive imitation; a futile effort. **mocking,** *n., a.* **mocking-bird,** *n.* an American song-bird, with great powers of mimicry; the lyre-bird of Australia; a small New Zealand bird that imitates voices etc. **mockingly,** *adv.*
mod[1] (mod), *n.* a Highland gathering analogous to a Welsh eisteddfod.
mod[2] (mod), *n.* a member of a youth group of the 1960s, who wore smart casual clothes, rode motor-scooters and were often involved in fights with gangs of rockers, a rival youth group. [*modern*]
modal (mō'dl), *a.* pertaining to mode, form or manner, as opp. to substance; of a verb, pertaining to mood or denoting manner. *n.* a modal proposition or verb. **modal proposition,** *n.* a proposition that affirms or denies with some qualification. **modality** (-dal'-), *n.* **modally,** *adv.*
mod. con. (mod kon'), *n.* (*coll.*) a modern device or appliance that gives comfort, convenience etc. [*modern convenience*]
mode (mōd), *n.* manner, method, way of doing, existing etc.; style; common fashion, prevailing custom; an operational state; (*Mus.*) one of the systems of dividing the octave, the form of the scale; the character of the connection in or the modality of a proposition.
model (mod'l), *n.* a representation or pattern in miniature, in three dimensions, of something to be made on a larger scale; a figure in clay, plaster etc. for execution in durable material; a thing or person to be represented by a sculptor or painter; one employed to pose as subject to an artist; one employed to wear clothes to display their effect; a standard, an example regarded as a canon of artistic execution; a particular style or type, e.g. of a car or a garment; a description or representation of something that cannot be observed directly; a set of postulates, mathematical equations etc. used e.g. to predict developments in the economy; (*euphem.*) a prostitute. *a.* serving as a model or example; worthy of imitation, perfect. *v.t.* (*past, p.p.* **modelled**) to shape, mould or fashion in clay etc.; to form after or upon a model; to give a plan or shape to (a document, book etc.); to display (clothes) by wearing them. *v.i.* to make a model or models; to act as a mannequin. **modeller,** *n.*
modem (mō'dem), *n.* a device used to transmit and receive data, esp. between two computers over a telephone line. [*modulator demodulator*]
moderate (mod'ərət), *a.* keeping within bounds; temperate, reasonable, mild; not extreme or excessive; of medium quantity or quality. *n.* one of moderate views in politics, religion etc. *v.t.* (-rāt), to reduce to a calmer, less violent, energetic or intense condition; to restrain from excess; to temper, to mitigate. *v.i.* (-rāt), to become less violent; to quiet or settle down; to preside as a moderator. **moderately,** *adv.* with moderation; not excessively, fairly. **moderation,** *n.* the act of moderating; the quality or state of being moderate; temperance; self-restraint; (*pl.*) the first public examination for a degree at Oxford. **moderato** (-rah'tō), *adv.* (*Mus.*) in moderate time. **moderator** (-rā-), *n.* one who or that which moderates; one who presides at a meeting, esp. the presiding officer at a court of the Presbyterian Church; one whose task is to ensure fairness and consistency in the way examination papers are set and marked; one who superintends certain examinations for degrees and honours at Oxford and Cambridge Univs.. **moderatorship** (-rā-), *n.*
modern (mod'ən), *a.* pertaining to the present or recent time; late, recent; not ancient, old-fashioned or obsolete; being or concerning the present or most recent form of a language. *n.* a person of modern times; an exponent of modernism (as artist, writer etc.). **modern languages,** *n.pl.* those that are still in current use, esp. as a subject of study. **modern pentathlon,** *n.* a sports contest involving swimming, cross-country running, fencing, equestrian steeplechasing and shooting. **modernism,** *n.* a modern mode of expression or thought; a modern term or idiom; in art and literature, the conscious rejection of traditional forms and use of new forms of expression; a tendency towards freedom of thought and the acceptance of the results of modern criticism and research in religious matters. **modernist,** *n.* **modernity** (-dœ'-), *n.* **modernize, -ise,** *v.t., v.i.* **modernization,** *n.* **modernizer,** *n.*
modest (mod'ist), *a.* humble, unassuming or diffident in regard to one's merits or importance; not presumptuous, forward or arrogant; bashful, retiring; restrained by a sense of propriety; decorous, chaste; moderate, not extreme or excessive. **modestly,** *adv.* **modesty,** *n.* the quality of being modest; a sense of propriety; delicacy; chastity.
modicum (mod'ikəm), *n.* a little; a small amount, a scanty allowance.
modify (mod'ifī), *v.t.* to alter, to make different; to change to a moderate extent the form, character or other qualities of; to reduce in degree or extent; to moderate, to tone down; (*Gram.*) to qualify the sense of, to alter (a vowel) by umlaut. **modifiable,** *a.* **modifiability** (-bil'-), *n.* **modification** (-fikā'-), *n.* **modificatory** (-fi-), *a.* **modifier,** *n.* a word or phrase that modifies another.
modish (mō'dish), *a.* fashionable; stylish. **modishly,** *adv.* **modishness,** *n.*
modiste (mōdēst'), *n.* a milliner or dressmaker.
Mods (modz), *n.pl.* short for MODERATIONS.
modulate (mod'ūlāt), *v.t.* to adjust, to regulate; to vary or

module 527 **momentum**

inflect the sound or tone of; (*Mus.*) to change the key of. *v.i.* (*Mus.*) to pass from one key to another. **modulation,** *n.* the act of modulating or being modulated; alterations in the amplitude or frequency of an electrical wave at a different frequency, usually at a lower. **modulator,** *n.* one who or that which modulates; (*Mus.*) a chart of the modulations in the tonic sol-fa system.
module (mod'ūl), *n.* a measure or unit of proportion, esp. (*Arch.*) the semidiameter or other unit taken as a standard for regulating the proportions of a column; any element or unit that forms part of a larger system, e.g. a spacecraft, an educational course. **modular,** *a.*
modulus (mod'ūləs), *n.* (*pl.* **-li,-lī**) (*Math. etc.*) a constant number or coefficient expressing a force, effect, function, etc.; a constant multiplier in a function of a variable; the numerical value of quantity. **modular,** *a.*
modus operandi (mō'dəs opəran'di, -dē), the way one does something; the way a thing works.
modus vivendi (mō'dəs viven'dē, -dī), way of living; a compromise or temporary arrangement pending a final settlement of matters in dispute.
mog (mog), **moggy** (-i), *n.* (*coll.*) a cat.
Mogadon (mog'ədon), *n.* a proprietary drug used to treat insomnia.
Mogul (mō'gl), *n.* a Mongolian; a follower of Baber, descendant of Tamerlane, or of Genghis Khan; a powerful and influential entrepreneur.
MOH, (*abbr.*) Medical Officer of Health.
mohair (mō'heə), *n.* the hair of the angora goat; a fabric made from it; an imitation of this fabric in cotton and wool.
Mohammedan (məham'idən), *a.* pertaining to Mohammed or Mohammedanism. *n.* a follower of Mohammed, a Muslim; an adherent of Mohammedanism, Islam. **Mohammedanism,** *n.* the Muslim religion founded by Mohammed (*c.* 570–632).
Mohawk (mō'hawk), *n.* the name of a tribe of N American Indians; their language; (*Skating*) a stroke from either edge to the same edge on the other foot, but in the opposite direction.
Mohican (mōhē'kən), *n.* a N American Indian of a tribe living in the Hudson river valley; the language of this tribe. *a.* of this tribe or its language.
mohican (mōhē'kən), *n.* a punk hairstyle in which the head is bald apart from a narrow central strip of erect hair from front to rear, often brightly coloured.
moidore (moi'daw), *n.* (*Hist.*) a Portuguese gold coin.
moiety (moi'əti), *n.* a half; a part or share.
moire (mwah'), *n.* watered silk; a watered appearance on textile fabrics or metals. **moiré** (-rā), *v.t.* to give a watered appearance to. *a.* watered (of silk, surfaces of metal etc.). *n.* a surface or finish like watered silk.
moist (moist), *a.* moderately wet, damp, humid; rainy. **moisten** (-sn), *v.t., v.i.* **moistness, moisture,** *n.* **moistureless,** *a.* **moisturize, -ise,** *v.t.* to add moisture to. **moisturizer,** *n.* anything which moisturizes, esp. a cosmetic cream or lotion.
moke (mōk), *n.* (*sl.*) a donkey.
moki (mō'ki), *n.* a variety of New Zealand fish.
moko (mō'kō), *n.* tattooing; the Maori method of doing this.
mol, (*chem. symbol*) MOLE [4].
mol., (*abbr.*) molecular; molecule.
molar[1] (mō'lə), *a.* having power to grind; grinding. *n.* one of the back or grinding teeth.
molasses (məlas'iz), *n.pl.* (*usu. sing. in constr.*) the viscid, dark-brown uncrystallizable syrup drained from sugar during the refining process; treacle.
mold (mōld), MOULD.
mole[1] (mōl), *n.* a spot on the human skin, usu. dark-coloured and sometimes covered with hair.
mole[2] (mōl), *n.* a pile of masonry, such as a breakwater, pier or jetty before a port; a port, a harbour.
mole[3] (mōl), *n.* a small soft-furred burrowing mammal; a spy or subversive person working within an organization on behalf of a rival organization, enemy etc. *v.t.* to burrow or ferret (something out). **molehill,** *n.* a hillock thrown up by a mole burrowing underground; an unimportant or very small matter, problem etc. **moleskin,** *n.* the skin of the mole used as fur; a kind of fustian, dyed after the surface has been shaved; (*pl.*) clothes, esp. trousers, of this material.
mole[4] (mōl), *n.* the basic SI unit of substance, being the amount of substance of a system which contains as many specified elementary entities as there are atoms in 0·012 kg of carbon-12.
molecule (mol'ikūl), *n.* one of the structural units of which matter is built up; the smallest quantity of substance capable of separate existence without losing its chemical identity with that substance; a particle. **molecular** (-lek'-), *a.* **molecular attraction,** *n.* the force by which molecules of bodies act upon each other; cohesion. **molecular biology,** *n.* the study of the structure and chemical organization of living matter, esp. of nucleic acids and protein synthesis. **molecular weight,** *n.* the weight of a molecule of any substance in terms of one-twelfth of the mass of an atom of the isotope carbon-12.
moleskin MOLE [3].
molest (məlest'), *v.t.* to trouble, to disturb, to harm; to assault or attack, esp. for sexual purposes. **molestation** (mol-, mōl-), *n.* **molester,** *n.*
moll (mol), *n.* (*sl.*) a wench, a prostitute; a gangster's girlfriend.
mollify (mol'ifī), *v.t.* (*past, p.p.* **-fied**) to soften, to assuage; to pacify, to appease. **mollifiable,** *a.* **mollification** (-fikā'-), *n.* **mollifier,** *n.*
mollusc (mol'əsk), *n.* any of various invertebrates with soft bodies and sometimes shells, as snails, mussels, cuttlefishes, squid, etc. **molluscan** (-lŭs'-).
molly (mol'i), *n.* an effeminate boy, one who likes to be coddled, a milksop; (*dated sl.*) a wench, a prostitute. **molly-coddle,** *n.* a milksop. *v.t.* to coddle.
Moloch (mō'lok), *n.* (*Bibl.*) an idol of the Phoenicians to which human sacrifices were offered; a devouring influence such as overbearing wealth, tyranny etc.
Molotov cocktail (mol'ətov), *n.* a home-made incendiary device consisting of a bottle containing an inflammable liquid, with a rag for a wick. [V. M. *Molotov,* 1890–1986, Russian statesman]
molten (mōl'tən), *a.* made of melted metal; melted by heat. [p.p. of MELT]
molto (mol'tō), *adv.* (*Mus.*) much, very.
moly (mō'li), *n.* a fabulous herb with white flower and black root, given to Ulysses to counteract the spells of Circe; wild garlic.
molybdenum (məlib'dinəm), *n.* (*Chem.*) a rare metallic element, at. no.42; chem. symbol Mo, found in combination as molybdenite. **molybdenite,** *n.* a sulphide or native disulphide of molybdenum.
mom (mom), *n.* (*N Am., coll.*) mother.
moment (mō'mənt), *n.* a minute portion of time, an instant; importance, consequence; the measure of a force by its power to cause rotation. **at the moment,** at the present, just now. **moment of truth,** a moment when something important must be decided, a difficult task undertaken etc. **the moment,** the right time for anything, the opportunity. **this moment,** at once. **momentary,** *a.* lasting only for a moment; done or past in a moment; transient, ephemeral. **momentarily,** *adv.* for a moment; (*N Am.*) immediately. **momentous** (-men'-), *a.* weighty, important. **momentously,** *adv.* **momentousness,** *n.*
momentum (məmen'təm), *n.* (*pl.* **-ta, -tə**) impetus, power of overcoming resistance to motion; the quantity of motion in a body, the product of the mass and the velocity.

momma (mom'ə), *n.* (*N Am.*, *coll.*) mother.
Mon., (*abbr.*) Monday.
mon- MON(O)-.
monad (mon'ad, mō'-), *n.* a simple, indivisible unit; one of the primary elements of being, esp. according to the philosophy of Leibnitz; a univalent atom, radical or element; an elementary, single-celled organism. **monadic** (-nad'-), *a.*
monadelphous (monədel'fəs), *a.* having the stamens united by their filaments; of stamens, having the filaments united.
monandry (monan'dri), *n.* that form of marriage in which one woman has only one husband at a time; the quality of being monandrous. **monandrous** (monan'drəs), *a.* (*Bot.*) having only one stamen.
monarch (mon'ək), *n.* a sole ruler; a hereditary sovereign, as emperor, empress, king or queen; the chief of its class; a large red and black butterfly. **monarchic, -ical**, (-ah'-), **monarchal** (-ah'-), *a.* **monarchically**, *adv.* **monarchism**, *n.* **monarchist**, *n.* **monarchy**, *n.* government in which the supreme power is vested in a monarch; a state under this system, a kingdom; supreme control.
monastery (mon'əstəri), *n.* a residence for a community, esp. of monks, living under religious vows of seclusion. **monasterial** (-stiə'-), *a.* **monastic** (-nas'-), *a.* **monastical**, *a.* **monastically**, *adv.* **monasticism** (-nas'tisizm), *n.* the theory and system of the monastic life.
monatomic (monətom'ik), *a.* having one atom in the molecule; univalent.
monaural (monaw'rəl), *a.* having or using one ear; of reproduced sound, not stereophonic. **monaurally**, *adv.*
Monday (mŭn'dā, -di), *n.* the second day of the week, following Sunday.
monetary (mŭn'itəri), *a.* of or pertaining to money or the coinage. **monetarism**, *n.* the economic theory that advocates strict control of the money supply as the best method of regulating the economy. **monetarist**, *n.*, *a.* **monetize, -ise**, *v.t.* to give a standard value to (a metal) as currency; to form into coin. **monetization**, *n.*
money (mŭn'i), *n.* (*pl.* **moneys**, erroneously **monies**) coin or other material used as medium of exchange; banknotes, bills, notes of hand and other documents representing coin; wealth, property, regarded as convertible into coin; (*with pl.*) coins of a particular country or denomination; (*pl.*) sums of money, receipts or payments. **for my etc. money**, in my etc. opinion. **money-bag**, *n.* a bag for money; (*pl.*) wealth; (*pl.*, *sing. in constr.*) a rich or miserly person. **money-changer**, *n.* one who changes foreign money at a fixed rate. **money-grubber**, *n.* a person who saves or amasses money in sordid ways. **money-lender**, *n.* a person whose business is to lend money at interest. **money-making**, *n.*, *a.* highly profitable (business). **money-market**, *n.* the field of operation of dealers in stocks etc., the financial world. **money-order**, *n.* an order for money, granted at one post-office and payable at another. **moneyed**, *a.* rich; consisting of money. **moneyless**, *a.*
monger (mŭng'gə), *n.* a trader, a dealer (now only in comb., as *ironmonger*, *scandalmonger*).
Mongol (mong'gl), *n.* one of an Asiatic race now inhabiting Mongolia; (**mongol**) a person suffering from mongolism. *a.* of or pertaining to the Mongols; of or being a sufferer from Down's syndrome. **Mongolian** (-gō'-), *a.* pertaining to the straight-haired yellow-skinned peoples of Asia. *n.* a Mongol; one belonging to the Mongolian races; the language of the Mongols or of the Mongolian stock. **mongolism**, *n.* DOWN'S SYNDROME. **Mongoloid, mongoloid**, *n.*, *a.*
mongoose (mong'goos), *n.* (*pl.* **-gooses**) a small mammal found in Africa, S Europe and SE Asia, which feeds on venomous snakes.
mongrel (mŭng'grəl), *a.* of mixed breed, arising from the crossing of two varieties; of mixed nature or character. *n.* anything, esp. a dog, of mixed breed. **mongrelism**, *n.* **mongrelize, -ise**, *v.t.* **mongrelly**, *adv.*
moni(c)ker (mon'ikə), *n.* (*sl.*) name.
moniplies (mon'ipliz), MANYPLIES.
monism (mon'izm), *n.* the doctrine that all existing things and activities are forms of manifestations of one ultimate principle or substance; any philosophic theory such as idealism, pantheism or materialism, opposed to dualism. **monist**, *n.* **monistic** (-nis'-), *a.*
monition (mənish'ən), *n.* a warning; an intimation or notice; (*Civil Law*) a summons or citation; a formal letter from a bishop or court warning a clergyman to abstain from certain practices.
monitor (mon'itə), *n.* one who warns or admonishes; a senior pupil appointed to keep order in a school or to look after junior classes; one whose duty it is to listen to foreign or other broadcasts; a detector for radioactivity; an ironclad of low draught having revolving turrets; (**Monitor**) a genus of large tropical lizards found in Asia, Africa and Australia; a television screen used e.g. in a studio or with a computer for displaying and checking pictures or information. *v.t.*, *v.i.* to listen to (radio broadcasts) in order to glean information. **monitorial** (-taw'-), *a.* **monitorially**, *adv.* **monitorship**, *n.* **monitory**, *a.* giving warning or admonition. *n.* a warning or admonition from a bishop, pope etc. **monitress**, *n. fem.*
monk (mŭngk), *n.* a member of a religious community of men, living apart under vows of poverty, chastity and obedience. **monkfish**, *n.* any of various anglerfishes. **monk's-hood**, *n.* a plant of the genus *Aconitum*, esp. *A. napellus* (from its hooded sepals). **monkish**, *a.*
monkey (mŭng'ki), *n.* a quadrumanous mammal of various species and families ranging from the anthropoid apes to the lemurs; (*coll.*) a rogue, an imp; an ape, a mimic; a pile-driving machine; (*sl.*) a sum of £500 or $500. *v.t.* to mimic, to ape; to meddle with, to interfere with. *v.i.* to play foolish or mischievous tricks. **to make a monkey of**, to cause to seem foolish. **monkey-business**, *n.* devious or underhand behaviour; mischievous behaviour. **monkey-engine**, *n.* a pile-driving machine. **monkey-flower**, *n.* a plant of the genus *Mimulus*. **monkey-jacket**, *n.* a short jacket. **monkey-puzzle**, *n.* the Chilean pine with spiny leaves and branches. **monkey-wrench**, *n.* a spanner with a movable jaw.
mono (mon'ō), *n.*, *a.* monophonic (sound).
mon(o)-, *comb. form* alone, single; as in *monograph, monosyllable*.
monobasic (monəbā'sik), *a.* (*Chem.*) with one base or replaceable atom.
monocarp (mon'əkahp), *n.* a monocarpic plant. **monocarpic, -pous** (-kah'-), *a.* bearing fruit but once, and dying after fructification.
monochromatic (monəkrəmat'ik), *a.* of light, presenting rays of one colour only; painted etc. in monochrome. **monochrome** (mon'əkrōm), *n.* a painting in tints of one colour only; any representation in one colour. *a.* monochromic. **monochromic** (-krō'-), *a.* executed in one colour. **monochromist, monochromy** (mon'-), *n.*
monocle (mon'əkl), *n.* an eye-glass for one eye.
monoclinal (monəklī'nəl), *a.* (*Geol.*) of strata, dipping continuously in one direction. **monocline**, *n.* a monoclinal fold, a hogback. **monoclinic** (-klin'-), **-clinate** (-nət), *a.* (*Cryst.*) having two oblique axes and a third at right angles to these.
monoclinous (monəklī'nəs), *a.* (*Bot.*) hermaphrodite; (*Geol.*) monoclinal.
monoclonal antibody (mon'əklōnəl), *n.* an antibody composed of cells derived from a single cell.
monocotyledon (monəkotilē'dən), *n.* a plant having a single cotyledon. **monocotyledonous**, *a.*
monocracy (monok'rəsi), *n.* government by a single

person. **monocrat** (mon'əkrat), *n.* **monocratic**, *a.*
monocular (monok'ūlə), *a.* one-eyed; for use with one eye only. **monocularity** (-lar'-), *n.*
monoculture (mon'əkŭlchə), *n.* the cultivation of a single type of crop; the area where it is grown.
monocycle (mon'əsikl), *n.* a unicycle.
monocyte (mon'əsit), *n.* the largest white blood-cell in vertebrate blood.
monody (mon'ədi), *n.* an ode, usu. of a mournful character, for a single actor; a song for one voice, or a musical composition in which one voice predominates; a mournful or plaintive song or poetical composition, a threnody. **monodic** (-nod'-), *a.* **monodist**, *n.*
Monoecia (mənē'shiə), *n.pl.* a Linnaean class, comprising plants in which the stamens and pistils are in distinct flowers. **monoecious**, *a.* belonging to the Monoecia; having separate male and female flowers on the same plant; (*Zool.*) hermaphrodite.
monofil (mon'əfil), *n.* a single strand of synthetic fibre.
monogamy (mənog'əmi), *n.* marriage to one wife or husband only; (*rare*) the practice of marrying only once; (*Zool.*) the habit of pairing with a single mate. **monogamic** (-gam'-), *a.* **monogamist**, *n.* **monogamous**, *a.*
monogenesis (monəjen'əsis), *n.* generation from one parent, asexual reproduction; development of an organism from a parent resembling itself. **monogenetic** (-net'-), **monogenic**, **monogenous** (-noj'-), *a.* **monogenism**, **-geny** (-noj'-), *n.* the doctrine that all human beings are descended from a single pair. **monogenist** (-noj'-), *n.*
monoglot (mon'əglot), *a.* speaking only one language. *n.* a monoglot person.
monogram (mon'əgram), *n.* a character composed of two or more letters interwoven; a single character representing a word etc. **monogrammatic** (-mat'-), *a.*
monograph (mon'əgrahf), *n.* a treatise on a single thing or class of things. *v.t.* to treat of in a monograph. **monographer**, **-phist** (-nog'-), *n.* **monographic**, **-ical** (-graf'-), *a.* **monographically**, *adv.*
Monogynia (monəjin'iə), *n.pl.* a Linnaean order containing plants having flowers with one pistil. **monogyn** (mon'-), *n.* a plant of this kind. **monogynian**, **-gynous** (-noj'-), *a.*
monogyny (-noj'-), *n.* the practice of mating with only one female.
monohull (mon'əhŭl), *n.* a vessel with a single hull, contrasted with a catamaran, trimaran etc.
monokini (mon'əkēni), *n.* a one-piece bathing garment for a woman, usu. similar to the bottom half of a bikini.
monolatry (mənol'ətri), *n.* worship of one god, esp. among many. **monolater**, **-trist**, *n.* **monolatrous**, *a.*
monolayer (mon'əleə), *n.* a single layer of atoms or molecules adsorbed on a surface.
monolingual (monəling'gwəl), *a.* using or expressed in only one language. **monolingualism**, *n.*
monolith (mon'əlith), *n.* a monument or other structure formed of a single stone. **monolithic**, *a.* of or like a monolith; consisting of a large and undifferentiated whole, often entailing inflexibility.
monologue (mon'əlog), *n.* a dramatic scene in which a person speaks by himself; a dramatic piece for one actor; a soliloquy; a long speech in conversation. **monologic**, **-logical** (-loj'-), *a.* **monologist** (-nol'əjist), *n.* **monologize**, **-gise** (-nol'əjiz, -giz), *v.i.*
monomania (monəmā'niə), *n.* mental obsession one subject only. **monomaniac** (-ak), *n.*, *a.* **monomaniacal** (-məni'-), *a.*
monomark (mon'əmahk), *n.* a system of registered combinations of numbers, serving to identify property or manufactured goods.
monomer (mon'əmə), *n.* a chemical compound that can undergo polymerization. **monomeric** (-nom'-), *a.*
monometallism (monəmet'əlizm), *n.* a one-metal standard of value for coinage. **monometallic** (-tal'-), *a.* **monometallist**, *n.*

monomial (mənō'miəl), *n.* a mathematical expression consisting of a single term. *a.* consisting of a single term.
monomorphic, **-phous** (monəmaw'fik, -fəs), *a.* having the same structure or morphological character, esp. throughout successive stages of development. **monomorphism**, *n.*
monophobia (monəfō'biə), *n.* morbid dread of being alone.
monophone (mon'əfōn), *n.* a monophonous sound; a homophone. **monophonic** (-fon'-), *a.* of sound, reproduced through only one electronic channel; homophonic. **monophonous** (-nof'-), *a.* homophonous; (*Mus.*) producing only one tone at a time.
monophthong (mo'nəfthong), *n.* a simple or single vowel sound; two written vowels pronounced as one.
monoplane (mon'əplān), *n.* an aircraft with one pair of wings.
monopoly (mənop'əli), *n.* an exclusive trading right in a certain commodity or class of commerce or business, usu. conferred by government; a company or combination enjoying this; the subject of such a right; exclusive possession, control or enjoyment (of); (**Monopoly**ᴷ) a board game for two or more people, who throw dice to move their pieces round a board marked with the names of streets etc., the object being to accumulate capital through buying the property on which the pieces land. **monopolist**, *n.* **monopolistic** (-lis'-), *a.* **monopolize**, **-ise**, *v.t.* to obtain or possess a monopoly of; to engross the whole of (attention, conversation etc.). **monopolization**, *n.*
monorail (mon'ərāl), *n.* a railway with a track consisting of a single rail. *a.* consisting of one rail.
monosaccharide (monəsak'ərid), *n.* a sugar that cannot be hydrolysed to form simpler sugars.
monosodium glutamate (monəsōdiəm gloo'təmāt), *n.* a salt of glutamic acid used as a flavour-enhancing food additive.
monosyllable (mon'əsiləbl), *n.* a word of one syllable. **monosyllabic** (-lab'-), *a.* of one syllable; speaking in words of a single syllable. **monosyllabically**, *adv.*
monotheism (mon'ōthēizm), *n.* the doctrine that there is only one God. **monotheist**, *n.* **monotheistic** (-is'-), *a.* **monotheistically**, *adv.*
monotint (mon'ətint), *n.* a picture or other representation in one colour.
monotone (mon'ətōn), *n.* continuance of or repetition in the same tone; a succession of sounds of the same pitch; intoning of words on a single note; monotony; monotint. *a.* monotonous. *v.i.*, *v.i.* to chant, recite or speak in the same tone or note. **monotonous** (mənot'-), *a.* wearisome through sameness, tedious; unvarying in pitch. **monotonously**, *adv.* **monotonousness**, **monotony**, *n.*
Monotremata (monətrē'mətə), *n.pl.* a sub-class of mammals having only one aperture or vent for the genital organs and the excretions. **monotrematous**, *a.* **monotreme** (mon'ətrēm), *n.*, *a.*
Monotypeᴷ (mon'ətip), *n.* a type-setting machine that casts and sets single printing-types.
monovalent (monəvā'lənt), UNIVALENT.
monoxide (mənok'sid), *n.* an oxide containing one atom of oxygen in combination with a radical.
Monseigneur (monsenyœ'. mō-), *n.* (*pl.* **Messeigneurs**, mā-) a French title of honour given to high dignitaries, esp. in the Church.
Monsieur (məsyœ', mis-), *n.* (*pl.* **Messieurs**, mes-) the French title of address, Mr or Sir; a Frenchman.
Monsignor (monsēn'yə), *n.* (*pl.* **Monsignori**, -yaw'ri) a title given to Roman Catholic prelates, officers of the Pope's court and others.
monsoon (monsoon'), *n.* a wind in SW Asia and the Indian Ocean, blowing from the south-west from April to October accompanied by heavy rainfall, and from the north-east the rest of the year; applied to other periodical winds.
monster (mon'stə), *n.* something misshapen, abnormal,

out of the ordinary course of nature; an abortion, a deformed creature; an imaginary animal, usually compounded of incongruous parts, such as a centaur, griffin, mermaid, gorgon etc.; an abominably cruel or depraved person; a person, animal or thing of extraordinary size. *a.* of extraordinary size, huge.
monstrance (mon'strəns), *n.* an open or transparent vessel in which the Host is carried in procession or exposed for adoration, esp. in a Roman Catholic church.
monstrous (mon'strəs), *a.* unnatural in form; out of the ordinary course of nature; enormous, huge; shocking, atrocious, outrageous; absurd, incredible. **monstrosity** (-stros'-), *n.* the quality of being monstrous; a monster, an abortion; a deformity, a distortion. **monstrously,** *adv.* **monstrousness,** *n.*
mons veneris (monz ven'əris), *n. (pl.* **montes veneris,** montēz) the pad of fatty tissue over the pubic bone of the human female.
Mont., *(abbr.)* Montana (US).
montage (montahzh'), *n.* cutting and assembling of shots taken when making a cinema picture; an artistic, literary or musical work consisting of heterogeneous elements in juxtaposition.
montane (mon'tān), *a.* of or pertaining to mountainous regions.
montbretia (montbrē'shiə), *n.* a bulbous-rooted plant with orange flowers.
monte (mon'ti), *n.* a Spanish card game.
Montessori method (montisaw'ri), *n.* a system of teaching the very young, in which physical activity, individual tuition and early attention to writing are main features. [Dr Maria *Montessori*, 1870-1952]
month (mŭnth), *n.* one of the twelve parts into which the year is divided, orig. the period of one revolution of the moon round the earth; four weeks. **month of Sundays,** an indefinitely long period. **monthly,** *a.* done in or continuing for a month; happening or payable once a month. *adv.* once a month. *n.* a periodical published every month; *(pl.)* the menses.
montre (mōtr'), *n.* in an organ, a flue-stop the pipes of which are visible in the external case. [F, from *montrer*, L *monstrāre*, to show]
monument (mon'ūmənt), *n.* anything by which the memory of persons or things is preserved, esp. a building or permanent structure; anything that serves as a memorial of a person, event or of past times; a document, a record; a distinctive mark; (*N Am.*) a natural or artificial landmark. **the Monument,** a column in London commemorating the Great Fire of 1666. **monumental** (-men'-), *a.* serving as a monument; stupendous (as of ignorance). **monumentalize, -ise** (-men'-), *v.t.* to commemorate with a monument. **monumentally,** *adv.*
monyplies (mon'iplīz), MANYPLIES.
moo (moo), *v.i.* to make a noise like a cow. *n.* the sound 'moo'.
mooch (mooch), *v.i. (coll.)* to wander aimlessly, amble; to cadge. *v.t. (coll.)* to cadge; to steal. **moocher,** *n.*
mood[1] (mood), *n.* temper of mind, disposition, humour; a morbid state of mind; a favourable state of mind; the expression of mood in art, literature etc. *a.* expressing a mood. **in the mood,** inclined (to or for). **moody,** *a.* indulging in moods or humours; peevish, sullen, out of temper. **moodily,** *adv.* **moodiness,** *n.*
mood[2] (mood), *n.* a verb-form expressing the manner in which the act, event or fact is conceived, whether as actual, contingent, possible, desirable etc.; the nature of the connection between antecedent and consequent in a proposition, modality; the form of a syllogism with regard to the quantity and quality of the propositions; (*Mus.*) mode.
Moog® (moog, mōg), *n.* a type of music synthesizer. [Robert *Moog*, born 1934, US engineer]

moolah (moo'lə), *n. (sl.)* money.
moon (moon), *n.* the earth's satellite revolving round it monthly; the satellite of any planet; a lunar month; anything shaped like a moon or crescent. *v.i.* to wander (about) or stare in a listless manner; *(sl.)* to expose one's buttocks to others. *v.t.* to pass (time) in this way. **blue moon** BLUE. **cycle of the moon** CYCLE. **over the moon,** *(coll.)* very pleased or happy. **moonbeam,** *n.* **moon-calf,** *n.* a blockhead; a born fool. **moonlight,** *n.*, *a.* **moonlight flit,** *n.* a removal of household furniture after dark to escape paying rent. *v.t.* see below. **moonlighter,** *n.* a person in full-time work who has a second, part-time job (in the evening). **moonlighting,** *n.* **moonlight,** *v.i.* **moonlit,** *a.* **moonshine,** *n.* moonlight; unreality, visionary ideas, nonsense; smuggled or illicitly-distilled spirits. **moonshiner,** *n.* an illicit distiller; a smuggler, esp. of spirits. **moonshot,** *n.* the launching of a space-craft to the moon. **moonstone,** *n.* a variety of feldspar with whitish or opalescent reflections. **moonstruck, -stricken,** *a.* affected by the moon; deranged, lunatic; fanciful, sentimental. **mooned,** *a.* shaped like the moon, crescent-shaped; moonlit. **moonless,** *a.* **moony,** *a.* like the moon; crescent-shaped; like moonlight; moonstruck, listless, dreamy, silly; *(sl.)* tipsy.
Moonie (moo'ni), *n. (coll.)* a member of the Unification Church, whose followers give all their possessions to it and live in communes. [Sun Myung *Moon*, Korean industrialist, who founded the church in 1954]
Moor (maw, muə), *n.* a member of a mixed Berber and Arab race inhabiting Morocco and the adjoining parts of NW Africa. **Moorish,** *a.*
moor[1] (maw, muə), *v.t.* to secure (a ship, boat etc.) with chains, ropes or cable and anchor. *v.i.* to secure a ship in this way, to anchor; to lie at anchor or secured by cables etc. **moorage,** *n.* **mooring,** *n. (usu. pl.)* the place where a ship is moored; anchors, chains etc. by which a ship is moored.
moor[2] (maw, muə), *n.* a tract of wild open land, esp. if overgrown with heather. **moor-cock, -fowl,** *n.* the male of the red grouse. **moor game,** *n.* red grouse. **moor-hen,** *n.* the red female of this; the water-hen. **moorland,** *n.*, *a.*
moose (moos), *n. (pl.* **moose)** a large animal allied to the elk, inhabiting the colder parts of N America.
moot (moot), *v.t.* to raise for discussion; to suggest. *v.i.* to argue or plead on a supposed case. *n.* formerly, an assembly of freemen in a township, tithing etc.; a law students' debate on a supposed case. *a.* open to discussion or argument. **moot case, point,** *n.* a debatable case or point; an open question. **moot court,** *n.* a meeting in an inn of court for discussing points of law.
mop (mop), *n.* a bundle of rags, coarse yarn etc. fastened to a long handle, and used for cleaning floors etc.; applied to various similar implements; a thick mass, as of hair; a mop-fair. *v.t. (past, p.p.* **mopped)** to wipe, clean or dry with or as with a mop. **to mop up,** to wipe up with or as with a mop; to clear (a place) of enemy troops etc.; *(sl.)* to seize, to appropriate, to get hold of; to worst, to dispatch.
mope (mōp), *v.i.* to be dull or dispirited. *v.t.* to make dull or dispirited (*usu. refl., p.p.*). *n.* one who mopes; *(pl.)* ennui, the blues. **mope-eyed,** *a.* purblind, short-sighted. **moper,** *n.* **mopy,** *a.*
moped (mō'ped), *n.* a motorized motorcycle, less than 50cc.
mopoke (mō'pōk), **morepork** (maw'pawk), *n. (Austral.)* a night-jar; applied to other birds; (*New Zealand*) a small owl.
moppet (mop'it), *n.* a pet, a darling (applied to children, young girls etc.); a variety of lap-dog.
moquette (moket'), *n.* a woven fabric of wool and hemp or linen with a velvety pile, used for carpets.
MOR, *(abbr.)* middle-of-the-road (music), used esp. in

broadcasting parlance.

mora (maw'rə), *n.* an Italian game in which one has to guess the number of fingers held up by another player, popular also in China and other countries.

moraine (mərān'), *n.* the debris of rocks brought down by glaciers. **morainal, morainic,** *a.*

moral (mo'rəl), *a.* pertaining to character and conduct as regards the distinction between right and wrong; conforming to or regulated by right, good, virtuous, esp. in sexual relations; subject to the rules of morality, distinguishing between right and wrong; based on morality; concerned with or treating of conduct or morality; conveying a moral; probable, virtual; (esp. of support) psychological rather than practical. *n.* the moral lesson taught by a story, incident etc.; (*pl.*) moral habits, conduct, behaviour, esp. in sexual relations; (*pl.*) ethics, moral science; (*sl.*) counterpart, likeness, double (prob. corr. of MODEL). **moral certainty,** *n.* probability that leaves little doubt. **moral courage,** *n.* fortitude in matters of life and conduct, esp. in resisting unjust or iniquitous opposition, odium and abuse, as opp. to physical courage. **moral defeat** MORAL VICTORY. **moral majority,** *n.* (*esp. N Am.*) the majority of the country's population, regarded as acting on and favouring adherence to strict moral principles. **moral philosophy, science,** *n.* ethics. **Moral Rearmament** BUCHMANISM. **moral victory,** *n.* an indecisive result or a partial success the moral effects of which are equivalent to victory. **moralism,** *n.* morality distinguished from religion or divested of religious teaching; †a moral maxim. **moralist,** *n.* a person who teaches morality or behaves in accordance with moral rules. **moralistic** (-lis'-), *a.* **moralistically,** *adv.* **morality** (-ral'iti), *n.* the doctrine, principles or practice of moral duties; moral science, ethics; morals, moral conduct, esp. in sexual relations; moralizing; a kind of drama (popular in the 16th cent.) in which the characters represent virtues, vices etc. **moralize, -ise** (mo'-), *v.t.* to interpret or apply in a moral sense; to provide with moral lessons; to render moral. *v.i.* to make moral reflections (on). **moralization,** *n.* **moralizer,** *n.* **morally,** *adv.* according to morality; practically, virtually.

morale (mərahl'), *n.* mental or moral condition; courage and endurance in supporting fatigue and danger, esp. of troops in war.

morass (məras'), *n.* a swamp, a bog; anything that is confused or complicated, esp. when it impedes progress.

moratorium (morətaw'riəm), *n.* a legal act authorizing a debtor or bank to defer or suspend payment for a time; any deferment, delay or temporary suspension.

Moravian (mərā'viən), *a.* pertaining to Moravia, the Moravians or their dialect of Czech. *n.* a native of Moravia; (*pl.*) a Protestant sect founded in the 18th cent.

moray (mo'rā), *n.* a brightly-patterned coastal eel of the family Muraenidae.

morbid (maw'bid), *a.* sickly, unhealthy, diseased; pathological; unhealthily preoccupied with unpleasant matters, esp. with death. **morbidity** (-bid'-), *n.* unhealthiness, prevalence of morbid conditions; morbidness. **morbidly,** *adv.* **morbidness,** *n.*

morceau (mawsō'), *n.* (*pl.* **-eaux,** -ō') a small piece, a short literary or musical composition.

mordant (maw'dənt), *a.* biting, caustic, pungent; causing pain or smarting; serving to fix colours etc. *n.* a substance for fixing colouring-matter in dyeing; an adhesive substance used in applying gold-leaf; acid or other corrosive used by etchers. **mordancy,** *n.* **mordantly,** *adv.*

mordent (maw'dənt), *n.* (*Mus.*) a rapid alternation of a note with the one immediately below it, a kind of trill by the character indicating this.

more (maw), *a.* (*superl.* **most,** mōst) greater in quantity, extent degree, number, importance etc.; additional, extra. *adv.* in or to a greater degree, extent, or quantity (used to form compar. of most adjectives and adverbs of more than one syllable); further, besides, again. *n.* a greater quantity, amount, number or degree; an additional quantity. **more and more,** with continual increase. **more or less,** to a greater or less extent; about; thereabouts. **more's the pity** PITY. **more than,** very. **no more,** nothing in addition; no longer existing, dead. **mor(e)ish,** *a.* (*coll.*) of food, causing one to want more; delicious.

morel (mərel'), *n.* an edible fungus, *Morchella esculenta*, and other species of *Morchella*.

morello (mərel'ō), *n.* a bitter dark-red cherry.

moreover (mawrō'və), *adv.* besides, in addition, further.

morepork MOPOKE.

mores (maw'rāz), *n.pl.* the customs and conduct which embody the fundamental values of a social group.

moresque (mawresk'), *a.* Moorish in style and decoration. *n.* Moorish decoration, as the profusely ornamented work in the Alhambra.

morganatic (mawgənat'ik), *a.* applied to a marriage between a man of high rank and a woman of inferior station, by virtue of which she does not acquire the husband's rank and neither she nor the children of the marriage are entitled to inherit his title or possessions. **morganatically,** *adv.*

morgen (maw'gən), *n.* a unit of land measurement based on area that can be ploughed by one team in one morning. In SE Africa, Holland and parts of the US it is slightly over two acres (8094 sq.m).

morgue (mawg), *n.* a mortuary; a building or room where the bodies of unknown persons found dead are exposed for identification; a stock of files, clippings etc. kept by a newspaper for reference.

MORI® (maw'ri), (*abbr.*) Market and Opinion Research Institute.

moribund (mo'ribŭnd), *a.* in a dying state; lacking vitality and energy.

Morisco (məris'kō), *a.* Moorish. *n.* a Moor, esp. one of the Moors remaining in Spain after the conquest of Granada; a morris dance; Moresque ornament or architecture.

morish MORE [1].

Mormon (maw'mən), *n.* a member of an American religious body, founded by Joseph Smith in 1830, now calling themselves the Latter-day Saints. **Mormonism,** *n.* [from a mythic personage, author of the *Book of Mormon*, containing the alleged divine revelations on which their creed was based]

morn (mawn), *n.* (*poet.*) morning. **the morn,** (*Sc.*) tomorrow.

mornay (maw'nā), *a.* served with a cheese sauce.

morning (maw'ning), *n.* the first part of the day, beginning at twelve o'clock at night and extending to twelve noon, or from dawn to midday; the early part of a period or epoch; (*poet.*) dawn. *a.* pertaining to or meant to be taken or worn in the morning. **morning coat,** *n.* a tail-coat with cutaway front. **morning dress,** *n.* men's clothes worn on formal occasions during the day, esp. for weddings etc. **morning-glory,** *n.* various climbing or twining plants. **morning prayer,** *n.* in the Anglican Church, matins. **morning sickness,** *n.* nausea and vomiting frequently accompanying early pregnancy. **morning star,** *n.* the planet Venus when visible in the east at dawn.

Moro (maw'rō), *n.* a member of a Muslim people of the S Philippines; the language of this people. *a.* of this people or its language.

morocco (mərok'ō), *n.* a fine leather from goat- or sheepskin, tanned with sumach and dyed (formerly made in Morocco).

moron (maw'ron), *n.* a feeble-minded person; an adult with the mentality of the average child aged between eight and twelve; (*coll.*) a very stupid or foolish person. **moronic** (-ron'-), *a.* **moronically,** *adv.* **moronism,** *n.*

morose (mərōs'), *a.* peevish, sullen; gloomy, churlish; given to morbid brooding. **morosely,** *adv.* **moroseness,** *n.*

-morph, *comb. form* denoting shape or structure. **-morphic, -morphous**, *a*. **-morphism, -morphy**, *n*.
morpheme (maw'fēm), *n*. a linguistic element that can carry meaning and cannot be divided into smaller such elements. **morphemic** (-fēm'-), *a*. **morphemically**, *adv*.
morphine, morphia (maw'fēn, -fiə), *n*. the alkaloid constituting the narcotic principle of opium, used in medicine as a sedative.
morphogenesis (mawfəjen'əsis), *n*. the development of the form of an organism during its growth to maturity. **morphogenetic** (-net'-), *a*.
morphology (mawfol'əji), *n*. the branch of biology dealing with the form of organisms; the science of the forms of words; the science of the forms of rocks etc. **morphologic, -ical**(-loj'-), *a*. **morphologically**, *adv*. **morphologist**, *n*.
morris (mo'ris), *n*. a grotesque dance; a rustic dance in which the performers formerly represented characters from folk legends; any similar dancing performance. *v.i. (sl.)* to decamp. **morris dance**, *n*.
Morris chair (mo'ris), *n*. an armchair with an adjustable back. [William *Morris* (1834-96), British painter, writer and craftsman]
morro (mo'rō), *n. (pl. -rros)* a small promontory or hill.
morrow (mo'rō), *n*. the day next after the present, the following day; the succeeding period.
Morse code (maws), *n*. a system of expressing messages for transmission invented by Morse in combinations of dots and dashes. [G.F.B. *Morse* (1791-1872), US inventor]
morsel (maw'sl), *n*. a mouthful, a bite; a small piece of food; a small quantity, a piece.
mortal (maw'tl), *a*. subject to death, causing death, deadly, fatal; inveterate, implacable; involving physical or spiritual death (as a sin or crime); pertaining to death; liable to death, hence human; *(coll.)* extreme, excessive; long and tedious; *(coll.)* very drunk. *n*. a being subject to death; a human being; *(facet.)* a person. *adv. (coll.)* exceedingly, extremely. **mortality** (-tal'-), *n*. the quality of being mortal; human nature; *(collect.)* human beings; loss of life, esp. on a large scale; the number of deaths in a given period, the death-rate. **mortally**, *adv*. in a fatal way; *(coll.)* exceedingly, greatly.
mortar (maw'tə), *n*. a vessel in which substances are pounded with a pestle; a short piece of ordnance used for throwing shells at a high angle; a device for firing pyrotechnic shells; a cement, made of lime, sand and water, for joining bricks etc. in building. *v.t.* to join, plaster or close up with mortar. **mortar-board**, *n*. a square board for holding mortar; a square-topped academic cap.
mortgage (maw'gij), *n*. the grant of an estate or other immovable property in fee as security for the payment of money, to be voided on the discharge of the debt or loan; the loan thus made. *v.t.* to grant or make over property on mortgage; to pledge, to plight (oneself etc. to or for). **mortgage rate**, *n*. the rate of interest charged by building societies, banks etc. for mortgage loans. **mortgageable**, *a*. **mortgagee** (-jē'-), *n*. the one who accepts a mortgage. **mortgager, mortgagor** (-jə'), *n*. the one who mortgages his property.
mortice (maw'tis), MORTISE.
mortician (mawtish'ən), *n. (N Am.)* an undertaker.
mortier (mawtyā', maw'-), *n*. a cap of state formerly worn by legal and other functionaries in France. [F, as MORTAR]
mortify (maw'tifi), *v.t.* to subdue (the passions etc.) by abstinence or self-discipline; to humiliate, to chagrin, to wound. *v.i.* to lose vitality, to decay, to gangrene. **mortification**, *n*. **mortifier**, *n*. **mortifyingly**, *adv*.
mortise (maw'tis), *n*. a hole cut in timber or other material to receive a tenon. *v.t.* to cut a mortise in; to join by means of mortise and tenon. **mortise-chisel**, *n*. one with a stout blade for cutting mortises. **mortise lock**, *n*. one set into a mortise in the edge of a door, so that the lock mechanism is enclosed by the door.

mortmain (mawt'mān), *n*. possession or tenure of lands or tenements by an ecclesiastical or other corporation who cannot alienate.
mortuary (maw'chuəri, -chəri), *n*. a building for the temporary reception of the dead. *a*. pertaining to death or the burial of the dead.
morwong (maw'wong), *n*. an edible fish found off the coasts of Australia and New Zealand.
Mosaic (məzā'ik), *a*. pertaining to Moses or to the law given through him.
mosaic (məzā'ik), *a*. a term applied to any work in which a pattern or representation is produced by the junction of small pieces of differently-coloured marble, glass or stone; tesselated, inlaid. *n*. a pattern, picture etc. produced in this style; a viral disease of plants in which the leaves display a mottled yellowing; an organism, or part of one, consisting of tissues with different genetic constitutions. *v.t.* to decorate with mosaic; to combine into or as into a mosaic. **mosaically**, *adv*. **mosaicist** (-sist), **mosaist**, *n*.
moschatel (moskətel'), *n*. a small perennial herb with yellowish-green flowers and a musky scent.
moselle (məzel'), *n*. a white wine made in the Moselle district.
mosey (mō'zi), *v.i. (esp. N Am., coll.)* to walk, amble.
Moslem (moz'ləm), MUSLIM.
mosque (mosk), *n*. a Muslim place of worship.
mosquito (məskē'tō), *n. (pl. -toes)* an insect of the genus *Culex* or allied genera, with a proboscis for piercing the skin of animals and sucking their blood. **mosquito-net**, *n*. a fine-mesh netting round a bed, over windows etc. to ward off mosquitoes.
moss (mos), *n*. a bog, a peat-bog, wet, spongy land; a low, tufted, herbaceous plant usually growing on damp soil or the surface of stones, trees etc. *v.t.* to cover with moss. **moss-rose**, *n*. a variety of *Rosa centifolia*, with moss-like calyx. **moss stitch**, *n*. a stitch in knitting consisting of alternating plain and purl stitches. **mossy**, *a*. **mossiness**, *n*.
most (mōst), *a*. greatest in amount, number, extent, quality, degree etc. *adv*. in the greatest or highest degree (forming the superl. of most adjectives and adverbs of more than one syllable); *(coll.)* very. *n*. the greatest number, quantity, amount etc.; the best, the worst etc.; the majority. **at most**, at the utmost extent; not more than. **for the most part**, in the main; usually. **to make the most of**, to use to the best advantage. **mostly**, *adv*. chiefly, mainly; on most occasions, usually.
-most, *suf*. forming superlatives of adjectives and adverbs denoting position, order etc., as in *hindmost, utmost*.
MOT, *(abbr.)* Ministry of Transport; *(coll.)* short for MOT test. **MOT test**, *n*. a test of the roadworthiness of a motor vehicle more than three years old, administered under the authority of the Department of the Environment.
mote (mōt), *n*. a particle of dust, a speck, a spot; anything proverbially small.
motel (mōtel'), *n*. a roadside hotel or furnished cabins where motorists may put up for the night.
motet (mōtet'), *n*. a vocal composition in harmony, of a sacred character.
moth (moth), *n*. one of a group of nocturnal or crepuscular Lepidoptera, distinguished from butterflies by not having knotted antennae, comprising a small insect breeding in cloth, furs etc., on which the larvae feed; that which gradually eats, consumes or wears away anything. **moth-ball**, *n*. a ball of naphthalene or similar substance that keeps away clothes-moths. *v.t.* to lay up in moth-balls; to lay up for later use; to spray with a plastic for laying-up and preserving. **to put into moth-balls**, to defer execution of (a plan, project etc.). **moth-eaten**, *a*. eaten into holes by moths; ragged. **mothy**, *a*. full of moths; moth-eaten.
mother[1] (mŭ'dhə), *n*. a female parent; the source or origin of anything; a motherly woman; a woman performing the

mother — function of a mother; the head of a religious community; a contrivance for rearing chickens artificially; short for MOTHERFUCKER. *v.t.* to act as mother towards; to adopt as a son or daughter; (*lit. or fig.*) to profess oneself to be mother of; to give birth to. *a.* holding the place of a mother; giving birth or origin; native, natural, inborn, vernacular. **motherboard** (mŭ'dhəbawd), *n.* (*Comput.*) the main circuit board holding the CPU, memory chips, etc. **mother country,** *n.* one's native country; a country in relation to its colonies. **mother earth,** *n.* the earth regarded as parent of all that lives on her surface; (*facet.*) the ground. **motherfucker,** *n.* (*N Am., taboo sl.*) an offensive or unpleasant person or thing. **Mothering Sunday,** *n.* the mid-Lent Sunday, when mothers traditionally receive presents from their children. **mother-in-law,** *n.* (*pl.* **mothers-in-law**) the mother of one's wife or husband. **motherland,** *n.* one's native country. **mother language** MOTHER TONGUE. **mother lode,** *n.* in mining, the main lode of a system. **mother-of-pearl,** *n.* the iridescent nacreous or pearly substance forming the internal layer of many shells. *a.* (*usu. in comb.*) made of this. **Mother's Day,** *n.* (*US and Canada*) the second Sunday in May, set apart for the remembrance of one's mother; Mothering Sunday. **mother ship,** *n.* one which supplies a number of other ships with stores, ammunition etc. **mother superior,** *n.* a woman having charge of a community of women in religious orders. **mother-to-be,** *n.* a pregnant woman. **mother tongue,** *n.* one's native language; a language from which others have sprung. **mother wit,** *n.* natural sagacity, common sense. **motherhood,** *n.* **motherless,** *a.* **motherlike,** *a.*, *adv.* **motherly,** *a.*, *adv.* **motherliness,** *n.*
mother[2] (mŭ'dhə), *n.* a thick slimy substance forming in various liquids during fermentation. *v.i.* to become mothery (as vinegar). **mothery,** *a.*
mothy MOTH.
motif (mōtēf'), *n.* the dominant feature or idea in a literary, musical or other artistic composition; a theme in music; an ornamental piece of lace etc. sewn on a dress.
motile (mō'tīl), *a.* capable of motion. **motility** (-til'-), *n.*
motion (mō'shən), *n.* the act, process or state of moving; passage of a body from place to place; change of posture; a gesture; an evacuation of the bowels; a combination of moving parts in a machine etc.; a proposal, esp. in a deliberative assembly; (*Law*) an application to a court for a rule or order; impulse, instigation. *v.t.* to direct by a gesture. *v.i.* to make significant gestures. **angular motion,** motion of a body as measured by the increase of the angle made with some standard direction by a line drawn from the body to a fixed point. **in motion,** moving; not at rest. **laws of motion,** three axioms laid down by Sir Isaac Newton: (1) every body remains in a state of rest or of uniform motion in the same direction, unless it is compelled to change that state; (2) change of motion is proportional to the force applied, and takes place in the direction of the straight line in which the force acts; (3) to every action there is always an equal and contrary reaction. **to go through the motions,** to do something without enthusiasm or conviction. **to put in motion,** to set going or in operation. **motion picture,** *n.* a cinema film. **motion sickness,** *n.* nausea induced by travelling in a car, ship, aircraft etc. **motion study,** *n.* the study of repetitive movement in industrial work with a view to the elimination of unnecessary movement. **motionless,** *a.*
motive (mō'tiv), *a.* causing or initiating motion; tending to cause motion; pertaining to movement. *n.* that which incites to action, or determines the will; cause, ground, incentive; in art, the predominant idea, feeling etc., motif. *v.t.* (*usu. pp.*) to furnish with an adequate motive (as a story, play etc.). **motive power,** *n.* the power by which mechanical motion is imparted; any impelling force; the act of motivating. **motivate,** *v.t.* to motive; to instigate; to provide an incentive. **motivation,** *n.* **motiveless,** *a.*

mot juste (mō zhoost'), *n.* the appropriate or felicitous word or phrase.
motley (mot'li), *a.* variegated in colour; dressed in particoloured clothes; heterogeneous. *n.* the particoloured dress of fools or jesters; a fool, a jester; a heterogeneous mixture.
motocross (mō'təkros), *n.* the sport of racing on motorcycles over rough ground. [*motor, cross-country*]
motor (mō'tə), *n.* that which imparts motive power, esp. a machine imparting motion to a vehicle or vessel (usu. excluding steam-engines); a device that converts electrical energy into mechanical energy; (*coll.*) a motor-car; a muscle for moving some part of the body; a nerve exciting muscular action. *a.* causing or imparting motion. *v.i.* to drive or ride in a motor-car; (*coll.*) to move fast. *v.t.* to convey in a motor-car. **motor torpedo boat,** a light, fast naval vessel equipped chiefly with torpedoes. **motor-boat,** *n.* a boat propelled by a motor. **motor-bus, -cab, -car, -coach, -cycle, -truck,** etc., *n.* various kinds of vehicle propelled by their own motor. **motorcade** (-kād), *n.* a procession of cars. **motor caravan,** *n.* a motor-vehicle fitted with sleeping accommodation, cooking facilities etc. **motor launch,** *n.* a motor-driven small boat, for plying between vessels and the shore. **motorman,** *n.* a man in charge of a motor, esp. of an electric tram or train. **motormouth,** *n.* (*sl.*) a garrulous person. **motor nerve,** *n.* an efferent nerve that excites muscular activity. **motorscooter,** *n.* a small motor-cycle, usu. with fairing reaching from below the handlebars in a curve to under the rider's feet. **motorway,** *n.* a road for fast motor traffic, usu. with a relatively high speed limit. **motorist,** *n.* a driver of a motor-car. **motorize, -ise,** *v.t.* to equip (troops) with petrol-driven vehicles.
Motown (mō'town), *n.* a form of music containing elements of rhythm and blues and pop, often combined with the rhythm of gospel music. [*motor town,* the nickname of Detroit, Michigan, where this originated]
mottle (mot'l), *v.t.* to blotch, to variegate with spots of different colours or shades of colour. *n.* a blotch or patch of colour; a spotted, blotched or variegated appearance on a surface.
motto (mot'ō), *n.* (*pl.* **-toes**) a short pithy sentence or phrase expressing a sentiment or maxim; a principle or maxim adopted as a rule of conduct; a joke, verse or maxim contained in a paper cracker; (*Her.*) a word or sentence used with a crest or coat of arms.
moue (moo), *n.* a small pouting grimace.
mouflon, moufflon (moo'flon), *n.* a wild sheep of Sardinia and Corsica.
moujik (moo'zhik), MUZHIK.
mould[1] (mōld), *n.* fine soft earth, easily pulverized, suitable for tillage; the earth, the ground; the grave. **mouldboard,** *n.* the curved plate in a plough which turns the furrow-slice over.
mould[2] (mōld), *n.* a hollow shape into which molten metal or other substance is poured in a fluid state to cool into a permanent shape; a templet used by plasterers for shaping cornices etc.; various analogous appliances used in trades and manufactures; a vessel for shaping puddings etc.; (*Arch.*) a moulding or group of mouldings; physical form, shape, build; character, nature. *v.t.* to form into a particular shape; to fashion, to make, to produce; to give a particular character to; to shape (bread) into loaves. **mouldable,** *a.* that may be moulded. **moulder**[1], *n.* **moulding,** *n.* the act or process of shaping anything in or as in a mould; anything formed in or as in a mould; an ornamental part of a cornice, capital, arch, woodwork etc., usu. in the form of continuous grooves and projections, showing in profile a complex series of curves.
mould[3] (mōld), *n.* a minute fungoid growth forming a woolly or furry coating on matter left in the damp. **mouldy,** *a.* covered with mould; (*coll.*) bad, poor, nasty;

(coll.) mean, shabby. **mouldiness**, n.
moulder[1] MOULD[2].
moulder[2] (mōl'də), v.i. to turn to dust by natural decay; to crumble; to waste away gradually.
moulding MOULD[2].
mouldy MOULD[3].
moult (mōlt), v.i. to cast the feathers, hair, skin, horns etc. (of certain birds and animals). v.t. to shed or cast. n. the act of moulting.
mound[1] (mownd), n. an artificial elevation of earth, stones etc.; a hillock, a knoll; a barrow, a tumulus. v.t. to heap up in a mound or mounds; to furnish, enclose or protect with a mound.
mound[2] (mownd), n. a ball or globe representing the earth, usu. of gold and surmounted by a cross, used as part of regalia, an orb.
mount (mownt), n. a high hill; a mountain (in poetry, or as first part of a proper name); in palmistry, one of the fleshy protuberances on the palm of the hand; a figure of a green hill occupying the base of a shield; that upon which anything is mounted; a gummed hinge for affixing stamps, photographs etc. to a page; the margin round a picture; a cardboard etc. upon which a drawing is placed; a slide upon which something is placed for microscopic examination; the parts by which various objects are prepared for use, strengthened or ornamented; a horse with the appurtenances necessary for riding; a horse-block or other means of mounting on horseback. v.i. to rise, to ascend; to soar; to get on horseback; to rise in amount. v.t. to ascend, to climb; to ascend upon, to get on; to form a path up; to copulate with; to raise; to prepare for use; to put into working order; to put (a picture) on a mount; to affix (a stamp, photograph etc.) with mounts; to stage (a play); to put (someone) on a horse; to furnish with a horse or horses. **to mount guard**, to go on duty as sentry. **mountable**, a. **mounted**, a. on horseback; placed on a mount. **mounter**, n. **Mountie, Mounty** (-ti), n. (coll.) a member of the Royal Canadian Mounted Police. **mounting**, n.
mountain (mown'tin), n. a natural elevation of the earth's surface rising high above the surrounding land; a large heap or pile; something of very great bulk; an excessive supply, esp. of an agricultural product. **mountain ash**, n. the rowan, *Pyrus aucuparia*; (Austral.) various kinds of Eucalyptus. **mountain bicycle, bike**, n. an ATB (q.v.) for use on steep or rugged terrain. **mountain biking**, n. **mountain sickness**, n. a feeling of indisposition, varying in different people, brought on by ascending into rarefied mountain air. **mountain-side**, n. **mountaineer** (-niə'), n. one who dwells among mountains; one who climbs mountains for amusement or scientific purposes. **mountaineering**, n. **mountainous**, a. full of mountains; exceedingly large. **mountainously**, adv.
mountebank (mown'tibangk), n. a quack doctor, orig. one who proclaimed his nostrums from a platform; a boastful pretender, a charlatan. v.t. to cheat by false boasts or pretences. **mountebankery, -kism**, n.
Mountie MOUNT.
mourn (mawn), v.i. to express or feel sorrow or grief; to wear mourning. v.t. to grieve or sorrow for; to deplore; to utter mournfully. **mourner**, n. **mournful**, a. **mournfully**, adv. **mournfulness**, n. **mourning**, a. grieving, sorrowing; expressive of grief or sorrow. n. grief, sorrow, lamentation; the customary dress, usu. black, worn by mourners. **in mourning**, wearing mourning garments. **mourning-band**, n. a band of black material worn esp. round the sleeve to show that one is in mourning. **mourning-dove**, n. the Carolina turtle dove so called from its plaintive note. **mourningly**, adv.
mousaka MOUSSAKA.
mouse (mows), n. (pl. **mice**, mis) a small rodent quadruped of various species belonging to the genus *Mus*, esp.
M. musculus, the common house mouse; applied to similar animals, as the shrews, voles etc.; (sl.) a black eye; (coll.) a shy or inconspicuous person; (*Comput*.) a device that allows manual control of the cursor and selection of computer functions without use of the keyboard. v.i. (mowz) to hunt for or catch mice; to hunt, to watch craftily, to prowl (about). v.t. to hunt for persistently; to rend or pull about as a cat does a mouse. **mouse-trap**, n. **mouser**, n. a cat good at catching mice. **mousy**, a. of a drab grey or brown colour; (coll.) shy, timid or inconspicuous. **mousiness**, n.
moussaka, mousaka (moosahkə'), n. a Greek dish of layered minced meat, aubergines, tomatoes topped with cheese or white sauce.
mousse (moos), n. a dish of flavoured cream whipped and frozen; any of various light, stiff liquid preparations, e.g. used for hair-styling or cosmetic purposes.
mousseline (mooslēn'), n. fine French muslin.
moustache (məstahsh', (*esp. N Am*.) mŭs'tash), n. the hair on the upper lip of men; applied to growths of hair on various animals, esp. round the mouth. **moustache-cup**, n. a drinking-cup with a guard to keep liquid from wetting the moustache. **moustached**, a.
mouth (mowth), n. the opening at which food is taken into the body with the cavity behind containing the organs of mastication, insalivation and speech; anything analogous to a mouth; a person regarded as needing to be fed; the opening of a vessel, pit, cave or the like; the outfall of a river. v.t. (mowdh), to utter pompously or in an elaborate or constrained manner, to declaim; to take up or seize with the mouth; to chew or roll with the mouth; to train (a horse) to the use of the bit. v.i. to talk pompously or affectedly; to make grimaces. **down in the mouth** DOWN[3]. **to keep one's mouth shut**, not to speak, esp. not to reveal secrets. **to laugh on the wrong side of the mouth** LAUGH. **to shoot one's mouth off**, to speak boastfully or ill-advisedly. **to stop the mouth of**, to put to silence. **mouth-organ**, n. a small musical instrument, played by blowing on metallic reeds. **mouthpiece**, n. a tube by which a cigar or cigarette is held in the mouth; that part of a musical instrument put between the lips; a spokesman for others. **mouth-to-mouth**, a. of resuscitation, carried out by breathing air into someone's mouth directly, with mouths in contact. **mouthwash**, n. an antiseptic liquid used to cleanse the mouth. **mouth-watering**, a. (appearing to be) delicious. **mouth-wateringly**, adv. **mouthed**, a. (usu. in comb.) having a mouth, as *big-mouthed*. **mouthful**, n. an amount that fills the mouth; (coll.) a word or phrase that is pompous or difficult to say; (coll.) an abusive tirade.
mouthy (-dhi), a. talkative; ranting, bombastic.
move (moov), v.t. to cause to change position or posture; to carry, lift, draw or push from one place to another; to put in motion, to stir; to cause (the bowels) to act; to incite, to incline, to prompt, to rouse (to action); to excite, to provoke (laughter etc.); to prevail upon; to affect with feelings, usu. of tenderness, to touch; to propose, to submit for discussion. v.i. to change place or posture; to go from one place to another; to advance, to progress; to change one's place of residence; to change the position of a piece at chess etc.; to make an application, appeal etc.; to begin to act; to take action, to proceed; to be moved, to have an evacuation (of the bowels); to live, to exercise one's activities (in or among); to bow. n. the act of moving; the right to move (in chess etc.); proceeding, action, line of conduct; a step, a device to obtain an object; a change of abode. **on the move**, stirring; moving from place to place, travelling about. **to get a move on**, (coll.) to hurry. **to make a move**, to go, to leave the table etc.; to start; to begin to go; to move a piece at chess etc. **to move heaven and earth**, to make every effort, to leave no stone unturned (to secure an object). **to move the goalposts** GOAL. **movable** (moov'əbl), a. capable of being

moved; occurring at varying times (as a festival). *n.* anything that can be moved or removed, esp. a movable or portable piece of furniture etc. that is not a fixture; *(pl.)* goods, furniture, chattels etc., as distinct from houses and lands, personal as distinct from real property; *(Sc. Law)* not heritable, as distinguished from heritable property. **movable feast,** *n.* a festival the date of which varies; *(facet.)* a meal taken at irregular times. **movability** (-bil'-), **movableness,** *n.* **movement,** *n.* the act or process of changing position, place or posture; a military evolution; change in temper, disposition, feeling etc.; manner or style of moving; action, incident or process of development in a story etc.; the working mechanism of a watch, clock, machine etc., or a connected group of parts of this; a connected series of impulses, efforts and actions, directed to a special end; a tendency in art, politics, literature etc., either actively promoted or occurring spontaneously; the people involved in this; activity in a market, esp. change of value; the mode or rate of a piece of music, also a section of a large work having the same general measure or time. **mover,** *n.* one who or that which moves; a cause or source of motive power; a proposer (of a resolution etc.); one who originates or instigates. **moving,** *a.* causing motion; in motion; impelling, persuading; pathetic, affecting. **moving staircase,** *n.* an escalator. **movingly,** *adv.*
movie (moo'vi), *n.* (*esp. N Am.*) a cinema film.
mow (mō), *v.t.* to cut down (grass, corn etc.) with a scythe, mowing-machine etc.; to cut the grass off (a lawn etc.); to destroy indiscriminately; to cut (down) in great numbers. *v.i.* to cut grass by mowing. **mower,** *n.* **mowing,** *n.* the act of cutting with a scythe or mowing-machine; land from which grass is cut. *a.* of land, crops etc., intended to be mown.
mozzarella (motsərel'ə), *n.* a soft white unsalted curd cheese.
MP, *(abbr.)* Member of Parliament; Military Police.
mp, *(abbr.)* melting-point; *(Mus.)* mezzo piano.
mpg, *(abbr.)* miles per gallon.
mph, *(abbr.)* miles per hour.
MPhil, *(abbr.)* Master of Philosophy.
Mr MISTER.
MRC, *(abbr.)* Medical Research Council.
Mrs MISTRESS.
MS[1], *(abbr.)* (*pl.* **MSS**) manuscript.
MS[2], *(abbr.)* multiple sclerosis.
Ms (miz. məz), *(abbr.)* a title used before the name of a woman in order not to distinguish her marital status.
MSc, *(abbr.)* Master of Science.
Mt, *(abbr.)* Mount.
mu (mū), *n.* the Greek letter M, μ.
much (mŭch), *a.* great in quantity or amount; long in duration. *adv.* in or to a great degree or extent; almost, nearly, about. *n.* a great quantity, a great deal; something uncommon. **a bit much,** *(coll.)* rather excessive, unreasonable etc. **as much,** an equal quantity. **not much,** *(sl.)* certainly not, not likely. **not up to much,** *(coll.)* not very good, of poor quality. **to make much of** MAKE[2]. **too much,** more than enough. **muchness,** *n.* **much of a muchness,** practically the same, very nearly alike.
mucilage (mū'silij), *n.* a gummy or viscous substance from the seeds, bark or roots of various plants; gum prepared for use; a viscous lubricating secretion in animal bodies. **mucilaginous** (-laj'-), *a.*
mucivorous (mūsiv'ərəs), *a.* of some insects, feeding on the juices of plants.
muck (mŭk), *n.* dung or manure; refuse, filth; anything filthy, disgusting or nasty; *(coll.)* untidiness; *(coll.)* money. *v.t.* to make dirty; *(sl.)* to bungle, to make a mess of. **to muck about** MESS ABOUT. **to muck in,** *(coll.)* to help others to do something. **to muck out,** to clean (esp. a stable). **muckrake,** *v.i.* to stir up scandal. **muckraker,** *n.*

muckraking, *n.* **mucker,** *n.* *(sl.)* a bad fall, esp. in the mud; *(coll.)* a friend. **mucky,** *a.* **muckiness,** *n.*
mucous (mū'kəs), *a.* pertaining to, like or covered with mucus; secreting mucus; slimy, viscid. **mucous membrane,** *n.* the membranous lining of the cavities and canals of the body.
mucus (mū'kəs), *n.* the viscid secretion of the mucous membrane; applied to other slimy secretions in animals and fishes; gummy matter found in all plants, soluble in water but not in alcohol. **mucosity** (-kos'-), *n.*
mud (mŭd), *n.* moist, soft earth, or earthy matter; mire; anything that is worthless or defiling. **mud in your eye!** used facetiously as a toast. **to throw mud,** to make disgraceful imputations. **mud-bath,** *n.* a bath of mineral water and mud in which patients are immersed for medicinal purposes. **mudfish,** *n.* a New Zealand fish that burrows in the mud at a distance from water. **mudflap,** *n.* a flap hanging behind the road-wheel of a vehicle to prevent mud etc. being thrown behind. **mud-flat,** *n.* a flat expanse of mud revealed by the ebb-tide. **mudguard,** *n.* a board or strip of metal fastened over a wheel of a carriage or cycle to protect persons riding from mud. **mudlark,** *n.* one who cleans out sewers, or fishes up pieces of coal, metal etc. from the mud of tidal rivers; a street urchin; *(Austral.)* the pee-wee. **mudpack,** *n.* a cosmetic containing fuller's earth, applied in paste form to the face. **mudslinger,** *n.* one who throws mud, a slanderer. **mudslinging,** *n.* **muddy,** *a.* covered or foul with mud; of the colour of mud; resembling mud; turbid, cloudy; confused, muddled, obscure. *v.t.* to make muddy or foul; to confuse. **muddily,** *adv.* **muddiness,** *n.*
muddle (mŭd'l), *v.t.* to confuse, to bewilder, to stupefy; to make half drunk; to mix (up), to jumble (together) confusedly; to make a mess of, to bungle, to waste, to squander. *v.i.* to act or proceed in a confused or bungling way. *n.* a mess; a state of confusion or bewilderment. **to muddle on, along,** to get along somehow. **to muddle through,** to attain a desired result without any efficiency or organization. **muddle-headed,** *a.* **muddle-headedly,** *adv.* **muddle-headedness,** *n.*
muesli (mooz'li, müz'li), *n.* a breakfast cereal of rolled oats, nuts, dried fruit, etc.
muezzin (mooez'in), *n.* a Muslim crier of the hour of prayer.
muff[1] (mŭf), *n.* a covering, usu. cylindrical, of fur or other material, carried by women, in which the hands are placed to keep them warm.
muff[2] (mŭf), *n.* an awkward or stupid person; a bungling action, esp. failure to catch the ball at cricket. *v.t.* to miss (a catch) or to fail to catch (the ball) at cricket; to bungle or fail in. *v.i.* to fail, to bungle badly.
muffin (mŭf'in), *n.* a plain, light, spongy, round cake, usu. toasted and eaten hot with butter.
muffle[1] (mŭf'l), *v.t.* to wrap or cover (up) closely and warmly; to wrap up the head of so as to silence; to wrap up (oars, bells etc.) so as to deaden the sound; to dull, to deaden. *n.* a muffler, a boxing-glove; a large mitten; anything employed to deaden sound; an oven or receptacle placed in a furnace used in operations in which the pottery etc. is not in direct contact with the products of combustion. **muffler,** *n.* a wrapper or scarf for the throat; a boxing-glove; a mitten, a thick stuffed glove; a pad or other contrivance for deadening sound, as in a piano; (*N Am.*) the silencer on a motor-vehicle; a bandage for blindfolding.
muffle[2] (mŭf'l), *n.* the thick, naked upper lip and nose of ruminants and rodents.
mufti (mŭf'ti), *n.* (*with Caps.*) an official interpreter or expounder of the Koran and Muslim law; civilian dress worn by servicemen off duty, ordinary dress as distinguished from that worn on state or ceremonial occasions.

mug[1] (mŭg), *n.* a drinking-cup, usu. cylindrical without a lip; the contents of this; a cooling drink; (*sl.*) the face or mouth; (*coll.*) a dupe, a gullible person. *v.i.* to make faces, to grimace. **mug shot,** *n.* (*coll.*) a photograph of a person's face, esp. one taken for police records.
mug[2] (mŭg), *v.i.* (*coll.*) to study hard, to grind. *v.t.* to work or get up (a subject). *n.* one who works hard for examinations, esp. one who neglects outdoor sports.
mug[3] (mŭg), *v.t.* to rob (someone) violently or by threatening violence, esp. in the street. **mugger,** *n.*
muggins (mŭg'inz), *n.* a children's card-game; a game of dominoes; (*sl.*) a fool, a simpleton.
muggy (mŭg'i), *a.* damp and close, sultry; moist, damp, mouldy (of hay etc.). **mugginess,** *n.*
mujaheddin, mujahedeen (moojəhədēn'), *n.pl.* Islamic fundamentalist guerillas.
mulatto (mŭlat'ō, mə-), *n.* the offspring of a white and a black. *a.* of this colour, tawny, esp. when intermediate in colour between the parents.
mulberry (mŭl'bəri), *n.* any tree of the genus *Morus*, bearing a collective fruit like a large blackberry; its fruit; the colour of this.
mulch (mŭlch), *n.* a surface layer of dead vegetable matter, manure etc. to keep the ground or the roots of plants moist. *v.t.* to cover with mulch.
mulct (mŭlkt), *n.* a fine, esp. for an offence or misdemeanour. *v.t.* to punish with a fine or forfeiture; to deprive (a person of); to swindle.
mule[1] (mūl), *n.* the offspring of a male ass and a mare; also a hinny; a stupidly stubborn or obstinate person; a hybrid between different animals or plants; an instrument for cotton-spinning. **mule-bird, -canary,** *n.* a cross between a canary and a goldfinch. **mulish,** *a.* like a mule; obstinate, sullen. **mulishly,** *adv.* **mulishness,** *n.*
mule[2] (mūl), *n.* a backless shoe or slipper.
mulga (mŭl'gə), *n.* an Australian acacia, used as fodder.
mull[1] (mŭl), *v.t.* to warm (wine, beer etc.), sweeten and flavour with spices.
mull[2] (mŭl), *n.* a thin soft muslin.
mull[3] (mŭl), *v.t.* (*usu. followed by over*) to ponder.
mull[4] (mŭl), (*Sc.*) a promontory.
mullah (mŭl'ə), *n.* an honorary title in Muslim countries for persons learned in theology and sacred law, and for ecclesiastical and civil dignitaries.
mullein (mŭl'in), *n.* a herbaceous plant with woolly leaves and tall spikes of yellow flowers, sometimes called Aaron's rod.
muller (mŭl'ə), *n.* a stone with a flat surface, used to grind and mix pigment etc. on a slab.
mullet (mŭl'it), *n.* a food fish living near coasts and ascending rivers, belonging either to the genus *Mullus* or *Mugil* the former distinguished as red and the latter as grey mullet.
mulligatawny (mŭligətaw'ni), *n.* an E Indian highly-flavoured curry-soup.
mullion (mŭl'yən), *n.* a vertical bar separating the compartments of a window. *v.t.* to divide or separate by mullions.
mullock (mŭl'ək), *n.* (*Austral.*) rock containing no gold; mining refuse from which the gold has been extracted; (*dial.*) rubbish; a muddle.
multangular (mŭltang'gŭlə), *a.* having many angles.
mult(i)-, *comb. form* many, several.
multicolour, -ed (mŭl'tikŭlə, -d), *a.* of or in many colours; many-coloured.
multicultural (mŭltikŭlchər'əl), *a.* of a society, containing diverse cultural groups.
multifaceted (mŭltifas'itid), *a.* of a gem, having many facets; having many aspects or factors.
multifarious (mŭltifə'riəs), *a.* having great multiplicity, variety or diversity. **multifariously,** *adv.* **multifariousness,** *n.*
multiform (mŭl'tifawm), *a.* having many forms. **multiformity** (-fawm'-), *n.*
multigym (mŭl'tijim), *n.* a versatile exercise apparatus used for toning various muscle groups.
multilateral (mŭltilat'ərəl), *a.* many-sided; of an agreement or treaty in which more than two states participate. **multilaterally,** *adv.*
multilingual (mŭltiling'gwəl), *a.* in many languages; able to speak, or speaking, several languages.
multimedia (mŭl'timēdiə), *a.* combining different media, such as television, video, computer graphics, etc.
multinational (mŭltinash'ənəl), *n., a.* (a company) operating in several countries.
multiparous (mŭltip'ərəs), *a.* bringing forth many at a birth; bearing or having borne more than one child.
multiple (mŭl'tipl), *a.* manifold; numerous and multifarious; having many parts, components or relations. *n.* a quantity that contains another a number of times without a remainder. **multiple-choice,** *a.* of a test etc., giving a number of different answers, from which the candidate must choose the correct one. **multiple personality,** *n.* a condition occasioned by the splitting of the normal organization of mental life into a number of distinct parts, each of which is comparable with an individual personality. **multiple sclerosis,** *n.* a progressive disease causing paralysis, speech and visual defects etc. **multiple store,** *n.* a number of retail stores under the same ownership. **multiplicity** (-plis'-), *n.* the quality of being many or manifold; many of the same kind.
multiplex[1] (mŭl'tipleks), *a.* manifold; multiple; of a channel, cable etc., allowing more than one signal to be transmitted simultaneously.
multiplex[2] (mŭl'tipleks), *n.* a cinema complex with a large number of separate screens showing different films.
multiply (mŭl'tipli), *v.t.* to add (a quantity called the multiplicand) to itself a certain number of times (called the multiplier) so as to produce a quantity called the product; to make more numerous, to increase in number or quantity. *v.i.* to increase in number or extent; to increase by propagation. **multiplicable** (-plik'-), **multipliable** (-pli-), *a.* **multiplicand** (-plikand'), *n.* the quantity to be multiplied. **multiplication** (-pli-), *n.* **multiplication table,** *n.* a table exhibiting the products of quantities taken in pairs, usually to 12 times 12. **multiplier,** (-pli-), *n.* one who or that which multiplies or increases; the number by which the multiplicand is multiplied; (*Phys.*) an instrument for intensifying an effect. **multiplying,** *n., a.*
multipurpose (mŭltipœ'pəs), *a.* serving several purposes.
multiracial (mŭltirā'shl), *a.* incorporating several racial groups.
multistage (mŭl'tistāj), *a.* having many stages; of a rocket, having several sections which fall off in series at set points during flight.
multistorey (mŭltistaw'ri), *a.* having several storeys, esp. of a car-park. *n.* a multistorey car-park.
multitasking (mŭltitahs'king), *n.* of a computer, the carrying out of several tasks simultaneously.
multitrack (mŭl'titrak), *a.* of a sound recording, using several different tracks blended to produce the final sound.
multitude (mŭl'titūd), *n.* the state of being numerous; a great number; a very large crowd or throng of people; the common people. **multitudinous** (-tū'-), *a.* very numerous. **multitudinously,** *adv.* **multitudinousness,** *n.*
multiuser (mŭltiū'zə), *a.* of a computer system, designed for use by several people simultaneously.
multivalent (mŭltivā'lənt), *a.* having several degrees of valency; (*Chem.*) having a valency greater than unity. **multivalence, -valency,** *n.*
mum[1] (mŭm), *a.* silent. *int.* silence, hush! *v.i.* to act in mime-show; to play as a mummer. **mum's the word,** a phrase used to ask for silence or discretion.
mum[2] (mŭm), *n.* an informal term for MOTHER. **mummy,**

mumble 537 musk

n. a child's word for MOTHER.
mumble (mŭm'bl), *v.i.* to speak indistinctly; to mutter; to speak with the lips closed. *v.t.* to mutter indistinctly or inarticulately; to chew or mouth gently. *n.* indistinct utterance; a mutter. **mumbler,** *n.* **mumbling,** *n.*, *a.* **mumblingly,** *adv.*
mumbo-jumbo (mŭmbōjŭm'bō), *n.* a W African idol, deity or malignant spirit; an absurd object of popular veneration; (*coll.*) incomprehensible or nonsensical language.
mummer (mŭm'ə), *n.* an actor in mime-show, esp. one of a number of people who formerly went from house to house at Christmas in fantastic disguises performing a kind of play; (*derog.* or *facet.*) an actor. **mummery,** *n.* the act or performance of mumming; tomfoolery, hypocritical parade of ritual etc.
mummy[1] (mŭm'i), *n.* a body of a person or animal preserved from decay by embalming, esp. after the manner of the ancient Egyptians; a bituminous pigment giving a rich brown tint. *v.t.* to mummify. **mummify,** *v.t.* **mummification,** *n.*
mummy[2] MUM[3].
mumps (mŭmps), *n.pl.* (*sing. in constr.*) a contagious disease characterized by a swelling and inflammation in the parotid and salivary glands. **mumpish,** *a.*
munch (mŭnsh, mŭnch), *v.t.*, *v.i.* to chew audibly; to eat with much movement of the jaws.
mundane (mŭndān'), *a.* belonging to this world, earthly, worldly; matter-of-fact; prosaic, everyday, banal. **mundanely,** *adv.* **mundaneness, mundanity** (-dan'-), *n.*
mung bean (mŭng), *n.* (the seed of) an E Asian bean plant used as a forage plant and as the main source of beansprouts.
municipal (mūnis'ipəl), *a.* pertaining to the government of a town or city or to local self-government in general; of a state, kingdom or nation. **municipality** (-pal'-), *n.* a town, city or district having a charter of incorporation or enjoying local self-government. **municipalize, -ise,** *v.t.* **municipalization,** *n.* **municipally,** *adv.*
munificent (mūnif'isənt), *a.* liberal, generous, bountiful; characterized by splendid liberality. **munificence,** *n.* **munificently,** *adv.*
muniment (mū'nimənt), *n.* a title-deed, charter or record kept as evidence or defence of a title; help, protection, defence.
munition (mūnish'ən), *n.* (*usu. pl.*) military stores of all kinds; anything required for an undertaking. *v.t.* to furnish with munitions.
muntjak (mŭnt'jak), *n.* a small Asiatic deer.
muon (mū'on), *n.* a subatomic particle, an unstable lepton with a mass approx. 207 times that of the electron.
mural (mū'rəl), *a.* pertaining to, on or like a wall. *n.* a large painting, mosaic etc. on a wall.
murder (mœ'də), *n.* premeditated homicide. *v.t.* to kill (a human being) intentionally and unlawfully; to slay barbarously; to spoil, to mar, by blundering or clumsiness; to mangle, to ruin (*coll.*) to defeat decisively. **murder will out,** a hidden matter will certainly come to light. **to get away with murder,** (*coll.*) to do something criminal, outrageous etc. without being punished. **murderer,** *n.* **murderess,** *n. fem.* **murderous,** *a.* **murderously,** *adv.*
murk, mirk (mœk), *n.* darkness. **murky,** *a.* dark, gloomy; unclear, hazy. **murkily,** *adv.* **murkiness,** *n.*
murmur (mœ'mə), *n.* a low, confused, continuous or repeated sound, as of running water; a half-suppressed protest or complaint, a grumble; a subdued speech; an abnormal sound heard on auscultation of the heart, lungs or arteries. *v.i.* to make a low continued noise, like that of running water; to mutter in discontent; to find fault. *v.t.* to utter in a low voice. **murmurer,** *n.* a grumbler, a complainer. **murmuringly,** *adv.* **murmurous,** *a.*
murrain (mū'rən), *n.* an infectious disease among cattle. *a.* affected with murrain.

muscatel, muscadel (mŭskətel', -del'), *n.* a kind of rich wine made from muscadine grapes; the grapes from which such wine is made; a sweet fragrant pear. **muscadine** (mŭs'kədin, -din), *n.* one of several varieties of grape with a musky flavour or odour.
muscle (mŭs'l), *n.* an organ consisting of a band or bundle of contractile fibrous tissue serving to effect movement of some part of the animal body; the tissue of which this is composed; muscular strength; power or influence. **to muscle in,** to force one's way in; to interfere. **muscle-bound,** *a.* stiff and inflexible as a result of over-developed muscles. **muscle-man,** *n.* a man with very developed muscles, often used to intimidate. **muscly,** *a.*
Muscovite (mŭs'kəvit), *n.* a native or inhabitant of Moscow.
muscular (mŭs'kūlə), *a.* pertaining to, consisting of or performed by the muscles; having well-developed muscles; strong, brawny. **muscular dystrophy,** *n.* a genetic disease causing progressive deterioration of the muscles. **muscularity** (-la'-), *n.* **muscularly,** *adv.* **musculature** (-ləchə), *n.* the arrangement or disposition of the muscles in the body or an organ.
Muse (mūz), *n.* in Greek mythology, one of nine goddesses who presided over the liberal arts; (**muse**) the inspiring power of poetry, poetical genius; a person, esp. a woman, who inspires or influences a poet or poem.
muse (mūz), *v.i.* to ponder, to meditate (upon); to study or reflect (upon) in silence; to dream, to engage in reverie. *v.t.* to meditate on; to think or say meditatively. *n.* abstraction of mind; reverie.
musette (mūzet'), *n.* a small bagpipe formerly used in France; a soft pastoral melody imitating the sound of the bagpipe; a reed-stop on the organ.
museum (mūzē'əm), *n.* a room or building for the preservation or exhibition of objects illustrating antiquities, art, natural science etc. **museum-piece,** *n.* an object so splendid or old-fashioned that it should be on display in a museum.
mush (mŭsh), *n.* a mash; a soft pulp, pulpy mass; (*N Am.*) porridge made of maize-meal boiled; (*sl.*) sentimental nonsense. **mushy,** *a.*
mushroom (mŭsh'room, -rum), *n.* a quick-growing edible fungus. *v.i.* to gather mushrooms; to expand and flatten out in a mushroom shape; to grow or increase quickly. **mushroom cloud,** *n.* a cloud shaped like a mushroom, esp. that produced by a nuclear explosion.
music (mū'zik), *n.* the art of combining vocal and instrumental tones in a rhythmic form; such an artistic combination of tones; any pleasant combination of sounds; melody, harmony; musical taste; a musical score. **music to one's ears,** something that one is pleased to hear. **to face the music** FACE. **music centre,** *n.* a unit incorporating several devices for sound reproduction, e.g. a turntable, tape-deck. **music-hall,** *n.* a theatre devoted to variety entertainments. **music-stand,** *n.* a light frame for supporting a sheet of music. **music-stool,** *n.* a stool with a revolving adjustable seat. **musical,** *a.* of or pertaining to music; fond of or skilled in music; harmonious, melodious. *n.* a stage show, film etc. with much singing and dancing. **musical-box,** *n.* a box with a barrel-organ mechanism for playing different tunes. **musical chairs,** *n. sing.* a parlour game. **musicality** (-kal'-), *n.* **musically,** *adv.* **musician** (-zish'ən), *n.* one skilled in music, esp. in playing an instrument. **musicianship,** *n.* **musicology** (-kol'-), *n.* the science of musical lore and history. **musicologist,** *n.* a writer on this.
musk (mŭsk), *n.* an odoriferous, resinous substance obtained from a sac in the male musk-deer; the odour of this; similar perfumes; the muskplant, *Mimulus moschatus*; applied to other plants. **musk-bag,** *n.* the bag or sac containing musk in various animals, esp. the musk-deer. **musk-deer,** *n.* a small hornless deer of Central Asia, from

musket 538 myalgia

which musk is obtained. **musk-duck,** *n.* a tropical American duck erroneously called the Muscovy or Barbary duck; an Australian duck, *Biziura lobata.* **musk-ox,** *n.* an Arctic-American bovine ruminant. **musk-rat,** *n.* any of several rodents emitting a musky odour, esp. the musquash. **musk-rose,** *n.* a rambling rose with large white flowers and a musky odour. **musky,** *a.* **muskiness,** *n.*
musket (mŭs'kit), *n.* (Hist.) a fire-arm by the infantry. **musket-shot,** *n.* the distance a musket will carry; a ball or shot from a musket. **musketeer** (-tiə'), *n.* a soldier armed with a musket. **musketry,** *n.* muskets collectively; the art of using the musket; fire from small-arms.
Muslim (muz'lim, mŭz'-), *n.* a person of the Islamic faith. *a.* of or pertaining to the Islamic faith, culture etc. **Muslimism,** *n.*
muslin (mŭz'lin), *n.* a fine, thin, cotton fabric.
musquash (mŭs'kwosh), *n.* a N American aquatic rodent yielding a valuable fur and secreting a musky substance in a large gland, also called the musk-rat.
muss (mŭs), *n.* a state of confusion or disorder, a mess. *v.t.* to disarrange, to throw into disorder. **mussy,** *a.* untidy; disordered.
mussel (mŭs'l), *n.* any mollusc of the bivalve genus *Mytilus,* esp. the edible mussel.
Mussulman (mŭs'lmən), *n. (pl.* **-mans**) formerly used for a Muslim.
must[1] (mŭst), *aux. v.* to be obliged to, to be under a necessity to; to be requisite, to be virtually or logically necessary to; to be certain to; (used also with p.p. as a kind of historic present). *n.* a thing that must not be missed; an essential thing.
must[2] (mŭst), *n.* new wine, the expressed juice of the grape before fermentation; mustiness, mould. *v.t.* to make mouldy. *v.i.* to grow mouldy.
must[3] (mŭst), *n.* mustiness, mould.
mustachio (məstah'shiō), *n.* a moustache, esp. a large one. **mustachioed,** *a.*
mustang (mŭs'tang), *n.* the wild horse of the American prairies.
mustard (mŭs'təd), *n.* the seeds of various plants ground and used as a condiment; a brownish-yellow colour; (*N Am., coll.*) zest. *a.* brownish-yellow. **keen as mustard,** (*coll.*) very keen. **mustard and cress,** white mustard and cress used in the seed-leaf as salad herbs. **mustard gas,** *n.* an irritant poison gas. **mustard plaster,** *n.* a mixture of powdered mustard seeds applied to the skin as a stimulant, counter-irritant etc.
Mustela (mŭstē'lə), *n.* a genus of small Carnivora containing the weasels or martens. **musteline** (mŭs'təlin), *n., a.*
muster (mŭs'tə), *n.* the assembling of troops for parade or review; a register of forces mustered; a collection, a gathering; a collection of peacocks. *v.t.* to collect or assemble for review, checking of rolls etc.; (*N Am.*) to enrol in the army; to bring together; to summon (up strength, courage etc.). *v.i.* to meet in one place. **to muster out,** (*N Am.*) to discharge (a soldier) from the army. **to pass muster,** to pass inspection without censure; to be accepted as satisfactory.
mustn't (mŭs'nt), contr. form of *must not.*
musty (mŭs'ti), *a.* mouldy; sour, stale; vapid, antiquated, spiritless. **mustily,** *adv.* **mustiness,** *n.*
mutable (mū'təbl), *a.* liable to change; inconstant, fickle, unstable. **mutability** (-bil'-), **mutably,** *adv.*
mutagen (mū'təjən), *n.* a substance that causes or assists genetic mutation. **mutagenesis** (-jen'ə-), *n.* **mutagenic,** *a.*
mutant (mū'tənt), *n.* an organism that has undergone mutation.
mutate (mŭtāt'), *v.i.* to change; to undergo motation. *v.t.* to change or modify (a sound), esp. by umlaut. **mutation,** *n.* the act or process of changing; umlaut; the change of an initial consonant in Celtic languages; a permanent variation in organisms giving rise to a new species; a species

so produced. **mutational,** *a.* **mutationally,** *adv.*
mutatis mutandis (mŭtah'tis mŭtan'dis), the necessary changes being made. [L]
mute (mūt), *a.* silent, uttering no sound, speechless; not having the power of speech, dumb; of hounds, not giving tongue; not spoken; (*Phon.*) not sounded, unpronounced; produced by complete closure of the organs of the mouth or interruption of the passage of breath (as *h, p, ph, d, t, th, k* and *g*). *n.* one who is silent or speechless; a dumb person; a hired attendant at a funeral; an actor in dumb show or whose part is speechless; a contrivance for deadening sound (as in a piano); (*Philol.*) a letter which is not pronounced; a consonant that stops the sound entirely. *v.t.* to deaden or muffle the sound of. **mute swan,** *n.* a Eurasian swan with white plumage and an orange bill. **muted,** *a.* unassertive, subdued. **mutely,** *adv.* **muteness,** *n.* **mutism,** *n.* muteness; silence; inability to hear, dumbness.
mutilate (mū'tilāt), *v.t.* to cut off a limb or an essential part of; to maim, to mangle; to disfigure; to injure (literary and other work) by excision. **mutilation,** *n.* **mutilative,** *a.* **mutilator,** *n.*
mutineer (mūtiniə'), *n.* one who mutinies. *v.t.* to mutiny.
mutinous (mū'-), *a.* given to mutiny; rebellious. **mutinously,** *adv.* **mutiny** (mū'tini), *n.* open resistance to or revolt against constituted authority, esp. by sailors or soldiers against their officers. *v.i.* to rise or rebel against authority (esp. in the army or navy).
mutt (mŭt), *n.* (*sl.*) a fool, a silly ass; a dog, esp. a mongrel.
mutter (mŭ'tə), *v.i.* to speak, in a low voice or with compressed lips; to grumble, to murmur (at or against); to make a low, rumbling noise. *v.t.* to utter in a low or indistinct voice; to say in secret. *n.* a low or indistinct utterance; a low rumbling sound; a murmur, a grumble. **muttering,** *n., a.*
mutton (mŭ'tən), *n.* the flesh of sheep used as food; (*facet.*) a sheep. **mutton dressed (up) as lamb,** (*coll.*) an old woman dressed or made up to look younger. **mutton-chop,** *n.* a rib or other small piece of mutton for cooking; (*pl.*) side whiskers of this shape. **mutton-head,** *n.* (*coll.*) a stupid person. **mutton-headed,** *a.* **muttony,** *a.*
mutual (mū'tūəl), *a.* reciprocal, reciprocally given and received; possessed, done, felt etc. by each of two persons, parties etc., to or towards the other; shared by or common to two or more persons, as *mutual friend.* **mutual insurance,** *n.* a system of insurance in which parties agree to indemnify each other for specified losses; insurance under a company granting a certain share of the profits to policy-holders. **mutuality** (-al'-), *n.* **mutually,** *adv.*
muu-muu (moo'moo), *n.* a loose dress worn by women in Hawaii.
Muzak® (mū'zak), *n.* a type of recorded background music played in shops, restaurants etc.
muzhik (moo'zhik), *n.* a Russian peasant; a serf.
muzzle (mŭ'zl), *n.* the projecting mouth and nose of an animal, as of a horse, dog etc.; the snout; the mouth of a gun or cannon; a guard put over an animal's muzzle to prevent biting. *v.t.* to put a muzzle on; to silence. **muzzle-loader,** *n.* a gun loaded at the muzzle. **muzzle velocity,** *n.* the velocity of a projectile as it leaves the muzzle.
muzzy (mŭ'zi), *a.* muddled, dazed; dull; fuddled, tipsy. **muzzily,** *adv.* **muzziness,** *n.*
MW, (*abbr.*) megawatt; medium wave.
Mx, (*chem. symbol*) maxwell.
my (mī, *unstressed* mi), *poss. a.* (*absol.* **mine,** min) belonging to me; used as a vocative in some forms of address (as *my boy, my dear*). *int.* a mild ejaculation of surprise. **my word!** used to express surprise, admiration etc.
my- MY(O)-.
myalgia (mīal'jiə), *n.* a condition of the muscles char-

myalism   539   myxovirus

acterized by pain and cramp. **myalgic**, *a.* **myalgic encephalomyelitis**, *n.* a viral disease affecting the nervous system and characterized by excessive fatigue, general malaise, lack of coordination, depression etc.
myalism (mī'əlizm), *n.* a species of witchcraft practised in the W Indies.
myall (mī'əl), *n.* any of various Australian acacias yielding scented wood used in making tobacco-pipes.
mycelium (mīsē'liəm), *n.* (*pl.* **-lia**, -ə) the vegetative parts of fungi, mushroom spawn. **mycelial**, *a.*
Mycenaean (mīsinē'ən), *a.* pertaining to Mycenae, an ancient city of Argolis, Greece.
mycete-, *comb. form.* fungus. **myceto-, myco-**, *comb. form.* fungus.
mycology (mīkol'əji), *n.* the science of fungi. **mycologic, mycological** (-loj'-), *a.* **mycologically**, *adv.* **mycologist**, *n.*
mycosis (mīkō'sis), *n.* the presence of parasitic fungi in the body. **mycotic** (-kot'-), *a.*
mycotoxin (mīkətok'sin), *n.* a poisonous substance produced by a fungus. **mycotoxology**, *n.*
mydriasis (midrī'əsis, mi-), *n.* an abnormal dilatation of the pupil of the eye. **mydriatic** (-at'-), *a.* of mydriasis. *n.* a drug that causes the pupils to dilate.
myelin (mī'əlin), *n.* a soft, white, fatty tissue forming a sheath round certain nerve fibres. **myelinated**, *a.* having a myelin sheath.
myelitis (mīəlī'tis), *n.* inflammation of the spinal cord. **myelitic** (-lit'-), *a.*
myel(o)-, *comb.form* spinal cord.
myelomeningitis (mīəlōmeninjī'tis), *n.* spinal meningitis.
myna, mynah MINA.
Mynheer (mīnhiə'), *n.* a Dutch form of address equivalent to *Sir* or *Mr*.
my(o)-, *comb.form* pertaining to muscles.
myocarditis (mīōkahdī'tis), *n.* inflammation of the myocardium. **myocardium** (-diəm), *n.* the muscular substance of the heart.
myology (mīol'əji), *n.* the science dealing with the muscles. **myologic, -ical** (-loj'-), *a.* **myologist**, *n.*
myope (mī'ōp), *n.* a short-sighted person. **myopia** (-ō'piə), *n.* short-sightedness. **myopic** (-op'-), *a.*
myosin (mī'əsin), *n.* the main protein in muscular tissue.
myosis (mīō'sis), *n.* contraction of the eye-pupil.
myositis (mīəsī'tis), *n.* inflammation of a muscle.
myosotis (mīəsō'tis), *n.* a genus of hardy plants comprising the forget-me-not. **myosote** (mī'ə-), *n.* the forget-me-not.
myotomy (mīot'əmi), *n.* dissection of muscles.
myriad (mi'riəd), *a.* innumerable, countless. *n.* a very great number.
myriapod (mi'riəpod), *a.* having numerous legs. *n.* one of the Myriapoda. **Myriapoda** (-ap'-), *n.pl.* a class of Arthropoda, comprising the centipedes and millipedes, characterized by a very large indeterminate number of jointed feet.
Myrmidon (mœ'midon), *n.* one of a warlike people of Thessaly, led by Achilles to the siege of Troy; a faithful follower, esp. an unscrupulous underling.
myrobalan (mīrob'əlan), *n.* the dried plum-like fruit of species of *Terminalia*, used in calico-printing, and for dyeing and tanning; a variety of plum-tree largely used as a stock for budding.
myrrh (mœ), *n.* a gum resin from various trees growing in Africa and Asia, used in the manufacture of incense, perfumes etc. **myrrhic** (mœ'-, mi'-), **myrrhy**, *a.*
myrtle (mœ'tl), *n.* a tree or shrub with glossy evergreen leaves and sweet-scented white or rose-coloured flowers, anciently sacred to Venus.

myself (mīself', mi-), *pron.* used in the nominative after 'I', to express emphasis; in the objective reflexively.
mystery (mis'təri), *n.* (*pl.* **-teries**) something beyond human comprehension; a secret or obscure matter; secrecy, obscurity; a form of mediaeval drama the characters and events of which were drawn from sacred history, a miracle-play; a divine truth partially revealed; (*pl.*) secret rites and ceremonies known to and practised only by the initiated; the esoteric rites practised by the ancient Greeks, Romans etc.; the Eucharist. **mystery tour**, *n.* an excursion to a destination that is kept secret until it is reached. **mysterious** (-tiə'ri-), *a.* not plain to the understanding; obscure, mystic, occult; fond of mystery. **mysteriously**, *adv.* **mysteriousness**, *n.*
mystic (mis'tik), *a.* pertaining to or involving mystery or mysticism; occult, esoteric; allegorical, emblematical. *n.* one addicted to mysticism; a supporter of the doctrine of mysticism. **mystical**, *a.* **mystically**, *adv.* **mysticism**, *n.* the belief that by self-surrender and spiritual apprehension one may attain to direct communion with and absorption in God, or that truth may be apprehended directly by the soul without the intervention of the senses and intellect.
mystify (mis'tifī), *v.t.* to involve in mystery; to bewilder, to puzzle, to hoax. **mystification** (-fikā'-), *n.* **mystifyingly**, *adv.*
mystique (mistēk'), *n.* the mystery surrounding some creeds, professions etc.; any mysterious aura surrounding a person or thing.
myth (mith), *n.* a fictitious legend or tradition, accepted as historical, usu. embodying the beliefs of a people on the creation, the gods, the universe etc.; a parable, an allegorical story; a fictitious event, person, thing etc. **mythic, -ical**, *a.* of myths; legendary; imaginary, untrue. **mythically**, *adv.* **mythicize, -ise**, *v.t.* **mythicizer**, *n.*
mythogenesis (mithəjen'əsis), *n.* the creation or production of myths.
mythogony (mithog'əni), *n.* the study of the origin of myths.
mythography (mithog'rəfi), *n.* the writing or narration of myths, fables etc. **mythographer, -phist**, *n.*
mythology (mithol'əji), *n.* a system of myths in which are embodied the beliefs of a people concerning their origin, deities, heroes etc.; the science of myths, a treatise on myths. **mythologer, -gist**, *n.* **mythologic, -ical** (-loj'-), *a.* **mythologically**, *adv.* **mythologize, -ise**, *v.t.* to make the basis of a myth. *v.i.* to invent myths; to study or interpret myths.
mythomania (mithəmā'niə), *n.* an abnormal tendency to lie or exaggerate.
mythopoeic (mithəpē'ik), **mythopoetic** (-pōet'-), *a.* myth-making; pertaining to a stage of culture when myths were developed.
myx(o)-, *comb.form* pertaining to or living in slime; pertaining to or consisting of mucus.
myxoedema, (*esp. N Am.*) **myxedema** (miksədē'mə), *n.* a cretinous disease characterized by atrophy of the thyroid gland and conversion of the connective tissue throughout the body into gelatinous matter. **myxoedematous**, *a.*
myxoma (miksō'mə), *n.* (*pl.* **-mata**, -tə), a tumour composed of mucous tissue. **myxomatous**, *a.*
myxomatosis (miksəmətō'sis), *n.* a contagious and fatal virus disease in rabbits.
Myxomycetes (miksəmī'sētēz), *n.pl.* the slime moulds or fungi, a group of organisms by some regarded as belonging to the Mycetozoa, by others as related to the fungi.
myxovirus (mik'səvirəs), *n.* any of a group of viruses causing such illnesses as influenza, mumps etc.

N

N¹, n¹, the 14th letter of the alphabet (*pl.* **Ns, N's, Ens**) *n*, **en** (en) *n*. (*Print.*) a unit of measurement; an indefinite number.
N², (*chem. symbol*) nitrogen.
N³, (*abbr.*) north(ern).
n², (*abbr.*) neuter; noun.
NA, (*abbr.*) National Academy; Nautical Almanac; North America.
Na, (*chem. symbol*) sodium.
na (nah), *adv.* (*coll.*) NO ¹.
Naafi (naf'i), *n.* an organization for supplying the services with canteens. [acronym for the Navy, Army and Air Force Institutes]
nab (nab), *v.t.* (*sl.*) (*past, p.p.* **nabbed**) to catch, to seize; to apprehend, to arrest.
nabob (nā'bob), *n.* a deputy-governor or prince under the Mogul empire in India; a very rich man, esp. one who amassed wealth in India.
nacelle (nəsel'), *n.* the basket suspended from a balloon; a small, streamlined body on an aircraft, distinct from the fuselage etc.
nacho (nah'chə), *n.* a crisp corn chip used as an appetizer in Mexican cuisine, often served with melted cheese and/ or a chilli dip.
Nachtmusik (nahkht'moozēk), *n.* a piece of music of a serenade-like character. [G]
nacre (nā'kə), *n.* the pinna, sea-pen or other fish yielding mother-of-pearl; mother-of-pearl. **nacred**, *a.* **nacreous, nacrous**, *a.* **nacrite** (-rīt), *n.* a pearly variety of mica.
nadir (nā'diə), *n.* the point of the heavens directly opposite to the zenith or directly under our feet; the lowest point or stage (of decline, degradation etc.).
naevus (nē'vəs), *n.* (*pl.* **-vi** (-vī)) a congenital discoloration of the skin, a birth-mark.
nag¹ (nag), *n.* a small horse or pony for riding; (*coll.*) a horse.
nag² (nag), *v.i.*, (*past, p.p.* **nagged**) to be continually finding fault; to scold (at). *v.t.* to find fault with or scold continually; to be continually pestering with complaints or fault-finding. *n.* one who nags; nagging.
nagor (nā'gaw), *n.* a small brown antelope from Senegal.
naiad (nī'ad), *n.* (*pl.* **-ades, -ads**) a water-nymph. (-dēz), *n.pl.* water-nymphs; an order of aquatic plants; a family of freshwater shellfish.
naif (nahēf'), NAÏVE.
nail (nāl), *n.* the horny substance at the tip of the human fingers and toes; a claw, a talon; a horny plate on the soft bill of certain birds; a measure of 2¼ in. (5·7 cm); a pointed spike, usu. of metal, with a head, for hammering into wood or other material to fasten things together. *v.t.* to fasten or secure (as) with nails; to stud with nails; to hold, to fix; to seize, to catch, to arrest; to engage (attention); to clinch (a bargain); (*coll.*) to hit, stop (as with a bullet); (*coll.*) to expose, discover incriminating information etc. **hard as nails**, callous, unsympathetic. **on the nail**, on the spot; at once (of payments). **to hit the nail on the head**, to hit upon the true facts of a case; to do exactly the right thing. **nail-biting**, *n.* chewing off the ends of one's finger-nails. *a.* of an event or experience, which creates an amount of tension. **nailbrush**, *n.* a small brush for cleaning the finger-nails. **nail polish or varnish**,

n. a type of varnish, often coloured, for decorating finger or toe-nails. **nailed**, *a.* (*usu. in comb.*, as *long-nailed, nailed-on*); caught, arrested.
nainsook (nān'suk), *n.* a thick muslin or jaconet, formerly made in India.
naïve (nīēv'), *a.* artless, ingenuous, simple, unaffected; (*usu. derog.*) over simple (e.g. of an argument). **naïvely**, *adv.* **naïveté** (-tā'), **naïvety**, *n.*
naked (nā'kid), *a.* destitute of clothing, uncovered, nude; without natural covering, as leaves, hair, shell etc.; not sheathed; exposed, unsheltered; defenceless, unarmed; stripped, destitute, devoid (of); unfurnished; not ornamented; bare, plain, undisguised; unsupported, uncorroborated, unconfirmed; unassisted, as without a telescope (of the eye). **the naked eye**, *n.* the eye unassisted by any optical instrument. **naked lady**, *n.* the meadow saffron or autumn crocus. **nakedly**, *adv.* **nakedness**, *n.*
NALGO (nal'gō), *acronym.* National and Local Government Officers' Association.
namby-pamby (nambipam'bi), *a.* weakly and insipidly sentimental; affectedly pretty or simple; one who has these characteristics.
name (nām), *n.* a word denoting any object of thought, esp. that by which a person, animal, place or thing is known, spoken of or addressed; a mere term as distinct from substance, sound or appearance, as opp. to reality; reputation, honourable character, fame, glory; authority, countenance. *v.t.* to give a name to, to call, to style; to call by name; to nominate, to appoint; to mention, to specify, to cite. **in the name of**, by the authority of; in reliance upon (esp. as an invocation); under the designation of. **no names, no pack-drill**, (*coll.*) mention no names, then no one gets into trouble. **the name of the game**, (*coll.*) the central or important thing; what it is all about. **to call names** CALL ¹. **to name names**, to mention people by name usu. in order to accuse or blame them. **to name the day**, (*coll.*) to fix the date for a wedding. **to take a name in vain**, to use it profanely. **name day**, *n.* the day sacred to a saint after whom one is named. **name-dropper**, *n.* one who tries to impress by mentioning the names of important or famous people as if they were close friends. **name-drop**, *v.i.* **name-dropper**, *n.* **name-dropping**, *n.* **namesake**, *n.* a person or thing having the same name as or named after another. **nam(e)able**, *a.* **nameless** (-lis), *a.* having no name; anonymous; illegitimate, unknown, obscure, inglorious; inexpressible, indefinable; unfit to be named, abominable, detestable. **namely**, *adv.* that is to say.
nan (nan), *n.* a type of slightly leavened bread originating in India or Pakistan cuisine.
nancy (nan'si), *n.* (*derog.*) an effeminate young man; a homosexual. **nancy boy**, *n.*
nanism (nā'nizm), *n.* dwarfishness; being stunted.
nankeen (nankēn'), **nankin** (-kin), *n.* a cotton fabric exported from Nankin, usu. of a buff or yellow colour; a fabric made in imitation of this colour.
nanna, nana (nan'ə), *n.* a child's word for grandmother.
nanny (nan'i), *n.* a children's nurse. **nanny goat**, *n. fem.* a she-goat.
nannygai (nan'igī), *n.* an edible red fish found in Australian rivers.

nano-, *comb. form.* one thousand millionth, as in **nanogram, nanosecond**.
nap[1] (nap), *v.i.* to sleep lightly or briefly, to doze; to be careless or unprepared. *n.* a short sleep, a doze, esp. in the day-time. **to catch napping**, to take unawares; to catch unprepared or at a disadvantage.
nap[2] (nap), *n.* the smooth surface produced on cloth or other fabric by cutting and smoothing the pile; a smooth, woolly, downy or hairy growth on a surface. *v.t.* to raise a nap on by brushing.
nap[3] (nap), *n.* a card-game in which five cards are dealt to each player, short for *napoleon*.
napalm (nä'pahm), *n.* a highly inflammable petroleum jelly which is produced from naphthalene and coconut palm oil, largely used for bombs.
nape (näp), *n.* the back of the neck.
napery (nä'pəri), *n.* (*rarely*) linen, esp. table-linen.
naphtha (naf'thə), *n.* an inflammable oil produced from organic substances, as bituminous shale or coal. **naphthalic** (-thal'-), *a.* **naphthaline** (naf'thəlēn), *n.* a white crystalline product of the dry distillation of coal-tar, used as a disinfectant and in the manufacture of dyes and explosives. **naphthene** (-thēn), *n.* a liquid hydrocarbon obtained from Caucasian naphtha. **naphthol** (-thol), *n.* either of two phenols derived from naphthaline.
napkin (nap'kin), *n.* a small cloth usu. of linen, esp. one used at table to wipe the hands on, protect the clothes, etc., a serviette; a baby's nappy; a small towel. **tablenapkin**, *n.* a serviette. **napkin-ring**, *n.* a ring used to enclose a table-napkin.
Naples yellow (nä'plz), *n.* a yellow pigment made from antimony; the colour of this.
Napoleonic (nəpölion'ik), *a.* resembling Napoleon I; dominating, masterful.
nappy (nap'i) *n.* a square of towelling etc., placed between a baby's legs and kept in place by a fastening at the waist, to absorb urine and faeces. **nappy-rash**, *n.* a rash on a baby's body created by its nappy, especially by ammonia from its urine.
narcissism (nah'sisizm), *n.* an abnormally heightened state of self-love. **narcissist**, *n.*
narcissus (nahsis'əs), *n.* (*pl.* **-es, -i** (-ī)) a genus of ornamental bulbous plants, containing the daffodils and jonquils; a plant of this genus.
narco-, *comb. form.* pertaining to torpor or narcotics.
narcolepsy (nah'kəlepsi), *n.* a nervous disease characterized by fits of irresistible drowsiness.
narcosis (nahkō'sis), *n.* narcotic poisoning, the effect of continuous use of narcotics; a state of stupor.
narcotic (nahkot'ik), *a.* producing torpor or coma; soporific; causing sleep or dullness. *n.* a substance that allays pain by inducing sleep or torpor; any of a group of addictive drugs such as opium and morphine that induce numbness and stupor. **narcotism** (nah'-), *n.* **narcotize, -ise** (nah'-), *v.t.* **narcotization, -isation**, *n.*
nares (neə'rēz), *n.pl.* the nostrils.
nark (nahk), *n.* (*sl.*) a police spy, a decoy, an informer; (*coll.*) a complainer, a nag. *v.t.* (*coll.*) to complain; to annoy.
narrate (nərāt'), *v.t.* to tell, to relate, to give an account of the successive particulars of in speech or writing. **narration**, *n.* **narrative** (na'rə-), *a.* in the form of narration; relating to an event or story. *n.* a recital of a series of events; a tale, a story. **narrator**, *n.*
narrow (na'rō), *a.* of little breadth or extent from side to side; constricted, limited, restricted; illiberal in views or sentiments; prejudiced, bigoted; close, near, within a small distance, with little margin; precise, accurate. *v.t.* to make narrow or narrower; to contract in range, views or sentiments; to confine, to limit, to restrict. *v.i.* to become narrow or narrower; to take too little ground (said of a horse). *n.* (*usu. pl.*) a strait; the contracted part of an ocean current. **a narrow escape**, *n.* an escape only just managed. **narrow boat**, *n.* a canal boat. **narrow circumstances**, *n.* poverty. **narrow-gauge**, *n.* a railway gauge of less than 4 ft. 8½ in. (1·43 m). *a.* of a railway, having a narrow gauge. **narrow-minded**, *a.* illiberal, bigoted. **narrow-mindedness**, *n.* **narrowly**, *adv.* **narrowness**, *n.*
narthex (nah'theks), *n.* a vestibule or porch across the west end in early Christian churches.
narwhal (nah'wəl), *n.* an Arctic delphinoid cetacean with a long tusk (or tusks).
NASA (nah'sə), the US space exploration authority [acronym for National Aeronautics and Space Administration].
nasal (nä'zəl), *a.* of or pertaining to the nose; sounded or produced with the nasal passage open. **nasalize, -ise**, *v.t.* **nasalization, -isation**, *n.* **nasally**, *adv.*
nascent (nas'ənt), *a.* coming into being; beginning to develop; immature. **nascency**, *n.*
nashi (nah'shi), *n.* a variety of Japanese pear.
nas(o)-, *comb. form.* pertaining to the nose as in *nasobronchial, naso-labial*.
nasturtium (nəstœ'shəm), *n.* (*pl.* **-ums**) a plant belonging to a genus of Cruciferae containing the watercress; a trailing plant with vivid orange flowers, also called Indian cress.
nasty (nahs'ti), *a.* dirty, foul, filthy to a repulsive degree; indecent, obscene; repellent, nauseous; objectionable, annoying, spiteful, odious, vicious, trying. **nastily**, *adv.* **nastiness**, *n.*
nat (nat), *n.* (*coll.*) short for *nationalist*.
natal (nät'əl), *a.* of, from or pertaining to one's birth.
natant (nä'tənt), *a.* swimming; (*Bot.*) floating. **natation**, *n.* **natatorial** (-taw'ri-), **natatory**, *a.*
natch (nach), *int.* (*sl.*) of course, short for *naturally*.
nation (nä'shən), *n.* a people under the same government and inhabiting the same country; a people belonging to the same ethnological family and speaking the same language. **nation-wide**, *a.* covering the whole nation. **national** (nash'ə-), *a.* of or pertaining to the nation, esp. to the whole nation; public, general, as opp. to local; peculiar to a nation. *n.* a member or subject of a particular nation; a national newspaper. **National Health Service**, in Britain, the system of state-provided medical service. **national anthem**, *n.* a hymn or song embodying the patriotic sentiments of a nation. **National Curriculum**, *n.* the courses of study offered by state-run schools nationally. **national grid**, *n.* a country-wide network of high-voltage electric power-lines linking major power-stations; the co-ordinate system in Ordnance Survey maps. **National Guard**, *n.* a force which took part in the French Revolution; organized militia of individual states in the US. **national insurance**, *n.* a system of compulsory insurance paid for weekly by the employer and employee and yielding benefits to the sick, retired or unemployed. **national park**, *n.* an area owned by the nation and set aside to preserve beauty, wildlife etc. **national service**, *n.* compulsory service in the armed forces. **National Socialism**, *n.* the political doctrine of the National Socialist German Workers' Party, which came into power in Germany under Adolf Hitler in 1933. **National Trust**, *n.* an organization in the UK concerned with the preservation of historic buildings and areas of countryside. **nationalism**, *n.* devotion to the nation, esp. the whole nation as opp. to sectionalism; the policy of national independence, esp. in Ireland; patriotic effort, sentiment etc. **nationalist**, *n.* **nationalistic**, *a.* **nationality** (-nal'-), *n.* **nationalize, -ise**, *v.t.* to make national; to naturalize; to bring (an industry etc.) under state control. **nationalization, -isation**, *n.* **nationally**, *adv.* **nationhood** (-hud), *n.*
native (nä'tiv), *a.* pertaining to a place or country by birth, indigenous not exotic; belonging to a person, animal or thing, by nature; inborn, innate, natural not acquired; natural (to); plain, simple, unaffected; occurring in a pure

or uncombined state (of metals); of or pertaining to the natives of a place or region. *n.* one born in a place; a plant or animal indigenous to a district or country; a member of an indigenous people of a country; (*offensive*) a coloured person. **natively,** *adv.* **nativeness,** *n.*
nativity (nətiv'iti), *n.* birth, esp. that of Jesus Christ, the Virgin or St John the Baptist; a festival in commemoration of this; a picture of the birth of Christ; a horoscope.
NATO (nā'tō), *n.* an international treaty for mutual defence, dating from 1949. [acronym for *N*orth *A*tlantic *T*reaty *O*rganization]
natron (nā'trən), *n.* native sesquicarbonate of soda.
natter (nat'ə), *v.i.* to chatter idly; to chat, exchange gossip. *n.* idle chatter; a chat, gossip.
natterjack (nat'əjak), *n.* a European toad with a yellow stripe down the back.
natty (nat'i), *a.* (*coll.*) neat, tidy, spruce. **nattily,** *adv.* **nattiness,** *n.*
natural (nach'ərəl), *a.* of, pertaining to, produced or constituted by nature; innate, inherent, uncultivated, not artificial; inborn, instinctive; in conformity with the ordinary course of nature, normal, not irregular, exceptional or supernatural; pertaining to physical things, animal, not spiritual; true to life; unaffected, not forced or exaggerated; undisguised; ordinary, to be expected, not surprising; coming by nature, easy (to); related by nature only, illegitimate; (*Mus.*) applied to the diatonic scale of C; (*Theol.*) unregenerate. (*Mus.*) a sign cancelling the effect of a preceding sharp or flat; a certainty, something by its very nature certain. **natural childbirth,** *n.* a method of childbirth involving breathing and relaxation exercise with no anaesthetic. **natural death,** *n.* death owing to disease or old age, not violence or accident. **natural gas,** *n.* gas from the earth's crust, specif. a combination of methane and other hydrocarbons used mainly as a fuel and as raw material in industry. **natural history,** *n.* the science or study of animal life, zoology; the study or description of the earth and its productions, loosely applied to botany, zoology, geology and mineralogy. **natural law,** *n.* the sense of right and wrong implanted by nature; a law governing the operations of physical life etc. **natural numbers,** *n.* the whole numbers starting at one upwards. **natural order,** *n.* an order of plants in a system of classification based on the nature of their sexual organs or their natural affinities. **natural philosophy,** *n.* the study of natural phenomena; physics. **natural religion,** *n.* religion not depending upon revelation. **natural resources,** *n.pl.* features or properties of the land such as minerals, water, timber etc. that occur naturally. **natural scale,** *n.* a scale without sharps or flats. **natural science,** *n.* the science of physical things as distinguished from mental and moral science; natural history. **natural selection,** *n.* the process by which plants and animals best fitted for the conditions in which they are placed survive and reproduce, while the less fitted leave fewer or no descendants. **natural theology,** *n.* theology based on principles established by reason, not derived from revelation. **natural wastage,** *n.* the loss of employees due to retirement, resignation etc. rather than redundancy, sacking etc. **naturalness,** *n.* **naturally,** *adv.* according to nature; spontaneously; as might be expected, of course.
naturalism (nach'ərəlizm), *n.* condition or action based on natural instincts; a philosophical or theological system that explains the universe as produced and governed entirely by physical laws; strict adherence to nature in literature and art, realism.
naturalize, -ise, *v.t.* to adopt; to acclimatize; to confer the rights and privileges of a natural-born subject on. *v.i.* to become naturalized. **naturalization,** *n.*
naturalist (nach'ərəlist), *n.* one versed in natural history; a believer in naturalism. **naturalistic** (-lis'-), *a.* in accordance with nature; realistic, not conventional or ideal; of or pertaining to natural history. **naturalistically,** *adv.*
nature (nā'chə), *n.* the essential qualities of anything; the physical or psychical constitution of a person or animal; natural character or disposition; kind, sort, class; vital or animal force; the whole sum of things, forces, activities and laws constituting the physical universe; the physical power that produces the phenomena of the material world; this personified; the sum of physical things and forces regarded as distinct from man; the natural condition of man preceding social organization; the undomesticated condition of animals or plants; unregenerate condition as opp. to a state of grace. **by nature,** innately. **from nature,** directly from the living model or natural landscape. **in nature,** in actual existence, in the sphere of possibility. **nature-worship,** *n.* worship of natural objects or phenomena or of the powers of nature. **natured,** *a.* (*usu. in comb.* as *ill-natured.*) **naturism,** *n.* nudism. **naturist,** *n., a.*
naught (nawt), *n.* nothing. **to set at naught,** to disregard.
naughty (naw'ti), *a.* perverse, mischievous; disobedient, ill-behaved; disagreeable; mildly indecent. **naughtily,** *adv.* **naughtiness,** *n.*
nauplius (naw'pliəs), *n.* (*pl.* **-plii** (-ī)) a larval stage of development in certain of the lower crustaceans.
nausea (naw'ziə), *n.* a feeling of sickness, with a propensity to vomit; loathing. **nauseate,** *v.t.* to cause to feel nausea; to reject with loathing. **nauseating,** *a.* causing nausea. **nauseation,** *n.* **nauseous,** *a.* causing nausea; disgusting, distasteful. **nauseously,** *adv.* **nauseousness,** *n.*
nautical (naw'tikəl), *a.* pertaining to ships, navigation or sailors; naval. **nautical mile** MILE. **nautically,** *adv.*
nautilus (naw'tiləs), *n.* (*pl.* **-li** (-lī)) a genus of cephalopods comprising the pearly nautilus, with a many-chambered shell; the paper nautilus or argonaut.
naval (nā'vəl), *a.* consisting of or pertaining to ships or a navy. **navally,** *a.*
nave[1] (nāv), *n.* the central block of a wheel in which the axle and spokes are inserted, the hub.
nave[2] (nāv), *n.* the body of a church.
navel (nā'vəl), *n.* the cicatrix of the umbilical cord, forming a depression on the surface of the abdomen. **navel orange,** *n.* a variety of orange with a navel-like depression and a smaller orange enclosed. **navelwort,** *n.* applied to the marsh pennywort and other plants.
navicular (nəvik'ūlə), *a.* shaped like a boat. *n.* the navicular bone. **navicular bone,** *n.* the scaphoid bone of the foot or (rarely) the hand.
navigate (nav'igāt), *v.i.* to sail, to pass from place to place by water or air; to direct and plot the route or position of (a ship, aircraft, car etc.); to manage a ship. *v.t.* to pass over or up or down, in a ship etc.; to manage, to conduct (a ship, etc.). **navigable,** *a.* **navigability** (-bil'-), *n.* **navigation,** *n.* the act, art or science of navigating. **navigator,** *n.* one who navigates; an explorer by sea; a navigational instrument.
navvy (nav'i), *n.* (*pl.* **-vvies**) a labourer employed in construction.
navy (nā'vi), *n.* the shipping of a country; the warships of a nation; their officers, men, dockyards etc. **navy-blue,** *a.* the dark-blue colour used for naval uniforms. **Navy List,** *n.* an official list of naval officers.
nawab (nəwahb'), *n.* an Indian governor or nobleman; a nabob.
nay (nā), *int.* (*dated*) no. *n.* a denial, a refusal.
Nazi (naht'si), *n.* a member of the German National-Socialist Party. *a.* pertaining to that party.
NB, (*abbr.*) note well.
Nb, (*chem. symbol*) niobium.
NCO, (*abbr.*) Non-Commissioned Officer.
Nd, (*chem. symbol*) neodymium.
NE, (*abbr.*) North-East(ern).

Ne, (*chem. symbol*) neon.
Neanderthal (nian'dətahl), *a.* of a Palaeolithic species of man; (*coll.*) extremely old-fashioned, reactionary; (*coll.*) boorish. **Neanderthal man,** *n.*
neap (nēp), *a.* low or lowest (applied to tides). *n.* a neap tide.
Neapolitan (nēəpol'itən), *a.* pertaining to or distinctive of Naples or its inhabitants. *n.* an inhabitant of Naples. **Neapolitan ice-cream,** ice-cream made of different ices in distinct layers.
near (niə), *adv.* at or to a short distance, at hand; not far off, not remote in place, time or degree. *prep.* close to in place, time, condition etc. *a.* close at hand, not distant in place, time or degree; closely related; close, narrow; on the left (of horses, parts or sides of vehicles etc.); parsimonious, niggardly. *v.t.* to approach. **a near thing,** a narrow escape. **as near as dammit,** (*coll.*) very nearly. **near-,** *comb. form.* close; almost, as in *near-white.* **nearby,** *adv.* close at hand. **near-miss,** *n.* a miss that is almost a hit. **nearside,** *n., a.* (of) the left side of a horse etc.; (of) the side of a vehicle nearer the kerb. **near-sighted,** *a.* short-sighted. **near-sightedness,** *n.* **nearly,** *adv.* almost; intimately. **nearness,** *n.*
neat (nēt), *a.* tidy, trim; simply but becomingly ordered; nicely proportioned, well made; elegantly and concisely phrased; adroit, dexterous, clever; of e.g. alcoholic drink, undiluted, pure. (*N Am.*) excellent, admirable. **neaten,** *v.t.* to tidy. **neatly,** *adv.* **neatness,** *n.*
'neath (nēth) *prep.* (*poet.*) beneath.
NEB, (*abbr.*) New English Bible; National Enterprise Board.
neb (neb), *n.* a beak or bill; a nose or snout; the tip or point of anything; a spout; a nib; (*Sc.*) the face, the mouth.
nebbuk (neb'ək), *n.* a thorny shrub supposed to have furnished the thorns for Christ's crown.
nebula (neb'ūlə), *n.* (*pl.* **-lae** (-lē) a cloudy patch of light in the heavens produced by groups of stars or by a mass of gaseous or stellar matter; a speck on the cornea causing defective vision. **nebular,** *a.* of or pertaining to nebulae.
nebulous, *a.* indistinct, obscure; muddled, bewildered; (*Astron.*) belonging to or resembling a nebula.
necessary (nes'isəri), *a.* needful, requisite, indispensable, requiring to be done; inevitable; happening or existing by necessity; resulting from external causes or determinism; determined by natural laws; (*rarely*) compulsory. *n.* that which is indispensably requisite; (*pl.*) things that are essentially requisite, esp. to life; (*sl.*) money. **necessarily** (nes'-, -se'rə-), *adv.* of necessity; inevitably. **necessitarian** (nisesiteə'ri-), *n.* one believing in the doctrine that man's will is not free, but that actions and volitions are determined by antecedent causes. **necessitarianism,** *n.*
necessitate (nises'itāt), *v.t.* to make necessary or unavoidable; to constrain, to compel. **necessity** (nises'-), *n.* the quality of being necessary; inevitableness; absolute need, indispensability; (*rarely*) constraint, compulsion; the compelling force of circumstances; that which is necessary, an essential requisite (*often in pl.*); want, poverty. **necessitous,** *a.* needy, destitute, in poverty.
neck (nek), *n.* the narrow portion of the body connecting the trunk with the head; this part of an animal used for food; anything resembling this; the slender part of a bottle near the mouth; the lower part of a capital; the part of a garment that is close to the neck. *v.t.* (*sl.*) to kiss, to fondle. **neck and neck,** equal, very close (in a race). **to get it in the neck,** (*coll.*) to be hard hit; to be reprimanded severely. **neckband,** *n.* a part of a garment fitting round the neck. **neckerchief** (-əchēf), *n.* a kerchief worn round the neck. **necklace** (-ləs), *n.* a string of beads or gems worn round the neck; in S Africa, a tyre soaked in petrol, put round a person's neck and set alight in order to kill by burning. *v.t.* in S Africa, to kill a person by this means. **necktie,** *n.* (*N Am.*) a strip of silk or other material encircling or worn as if encircling the neck and collar and tied in front, a tie. **necking,** *n.* the hollow part of a column between the shaft and the capital; (*sl.*) kissing, cuddling. **necklet** (-lit), *n.* a small ornament or garment for the neck.
necro-, *comb. form.* pertaining to dead bodies or the dead.
necrolatry (nikrol'ətri), *n.* worship of the dead, esp. ancestors.
necromancy (nek'rəmansi), *n.* the art of revealing future events by communication with the dead; enchantment, magic. **necromancer,** *n.*
necrophagous (nikrof'əgəs), *a.* eating or feeding on carrion.
necrophilia (nekrəfil'iə), *n.* an obsession with, and usu. an erotic interest in, corpses. **necrophiliac** (-ak), *a.*
necropolis (nikrop'əlis), *n.* a cemetery, esp. one on a large scale.
nectar (nek'tə), *n.* the drink of the gods; any delicious drink; the honey or sweet fluid secretion of plants. **nectarine** (-rin), *n.* (-rēn) a smooth-skinned and firm variety of the peach. **nectary,** *n.* the organ or part of a plant or flower secreting honey.
ned (ned), *n.* (*sl.*) a lout, a hooligan.
NEDC, (*abbr.*), National Economic Development Council (also known as **Neddy** (ned'i)).
neddy (ned'i), *n.* a donkey.
née (nā), *a.* born (used with the maiden name of a married woman as in *neé Smith*).
need (nēd), *n.* a state of urgently requiring something; lack of something; a state requiring relief, urgent want; emergency; that which is wanted, requirement. *v.i.* (3rd sing. **need** or **needs**) to be wanting or necessary, to require, to be bound, to be under necessity or obligation to; to be in want. *v.t.* to be in want of, to require. **in need,** poor or in distress. **needful,** *a.* **needfully,** *adv.* **needfulness,** *n.* **needless** (-lis), *a.* unnecessary, not required; useless, superfluous. **needlessly,** *adv.* **needlessness,** *n.* **needs,** *adv.* of necessity, necessarily, indispensably (*usu. with* must). **needy,** *a.* in need.
needle (nē'dl), *n.* a small, thin, rod-shaped, pointed steel instrument with an eye for carrying a thread, used in sewing; analogous instruments of metal, bone, wood etc., used in knitting, etc.; a piece of magnetized steel used as indicator in a compass, guage etc.; a pointed peak or pinnacle of rock; an obelisk; a needle-like leaf of a pine-tree; (*coll.*) a hypodermic syringe; a stylus in a record player. *v.t.* (*coll.*) to irritate; to force into action. **to look for a needle in a haystack,** to engage in a hopeless search. **needle-beam,** *n.* a cross-beam in the flooring of a bridge etc. **needlecord,** *n.* a cotton material with closer ribs and flatter pile than corduroy. **needle-exchange,** *n.* a place where drug addicts may exchange their used hypodermic needles for new. **needle-point,** *n.* point-lace; a type of embroidery done on canvas. **needlewoman,** *n.* a seamstress. **needlework,** *n.* **needled,** *a.* (*coll.*) annoyed, irritated.
needless etc. NEED.
neep (nēp), *n.* (*Sc.*) a turnip.
ne'er (neə), *adv.* (*poet.*) never. **n'er-do-well,** *a.* good for nothing. *n.* a good-for-nothing.
nefarious (nifeə'riəs), *a.* wicked, abominable. **nefariously,** *adv.* **nefariousness,** *n.*
neg., (*abbr.*) negative.
negate (nigāt'), *v.t.* to render negative, to nullify; to deny. **negation,** *n.* denial; a declaration of falsity; refusal, contradiction; nullity, voidness. **negative** (neg'ə-), *a.* containing, declaring or implying negation; denying, contradicting, prohibiting, refusing; lacking positive qualities such as optimism or enthusiasm; denoting the opposite to positive, denoting that which is to be subtracted (expressed by the minus sign −); (of electricity) having an excess of electrons; showing the lights dark and the sha-

neglect — nephr(o)-

dows light. *n.* a proposition, reply, word etc., expressing negation; a veto; a negative quality, lack or absence of something; an image or plate bearing an image in which the lights and shades of the object are reversed. *v.t.* to veto, to reject, to refuse to accept, sanction or enact; to reprove; to contradict. **in the negative**, indicating dissent or refusal. **negative feedback**, *n.* interconnexion of input and output terminals of an amplifier in such a manner that the output opposes the input. **negative pole**, *n.* the pole of a freely swinging magnet that swings to the south. **negatively**, *adv.* **negativeness, negativity** (-tiv'-), *n.* **negativism**, *n.* the quality of being negative; a religious or philisophical system based on denial or scepticism. **negativist**, *n.*, *a.* **negatory**, *a.*
neglect (niglekt'), *v.t.* to treat carelessly; to slight, to disregard; to pass over; to leave undone; to omit (to do or doing). *n.* disregard (of); carelessness, negligence; the state of being neglected. **neglectful**, *a.* **neglectfully**, *adv.* **neglectfulness**, *n.*
negligée (neg'lizhā), *n.* a woman's loose dressing gown of flimsy material.
negligence (neg'lijəns), **negligency**, *n.* disregard, neglectfulness; (*Law*) failure to exercise proper care and precaution. **negligent**, *a.* careless, neglectful. **negligently**, *adv.* **negligible**, *a.* that can be ignored, not worth notice.
negotiate (nigō'shiāt), *v.i.* to treat (with another) in order to make a bargain, agreement, compromise etc.; to traffic. *v.t.* to arrange, bring about or procure by negotiating; to carry on negotiations concerning; to transfer (a bill, note etc.) for value received; to obtain or give value for; to accomplish, to get over successfully. **negotiable**, *a.* **negotiability** (-bil'-), *n.* **negotiation**, *n.* **negotiator**, *n.*
Negress NEGRO.
Negro (nē'grō), *n.* (*pl.* **-oes**, *fem.* **Negress** (-gris)) a person belonging to, or descended from, one of the black-skinned African peoples. *a.* of or pertaining to these peoples; black or dark-skinned. **Negroid** (-groid), *a.* of Negro type; having the physical characteristics associated with the Negro peoples.
negus (nē'gəs), *n.* a beverage of wine, hot water, sugar and spices.
neigh (nā), *v.i.* to utter the cry of a horse; to whinny. *n.* the cry of a horse.
neighbour (nā'bə), *n.* one who lives near one. *v.t.* to adjoin; to lie near to. *v.i.* to border (upon). **neighbourhood** (-hud), *n.* the locality round or near; the vicinity; nearness; (*collect.*) those who live near, neighbours. **in the neighbourhood of**, approximately. **neighbouring**, *a.* situated or living near. **neighbourly**, *a.*, *adv.* **neighbourliness**, *n.*
neither (nī'dhə, nē'-), *a.* not either. *pron.* not the one nor the other. *conj.* not either, not on the one hand (usu. preceding one of two alternatives and correlative with *nor* preceding the other). *adv.* (*coll.*, *at end of sentence*) either.
nelly (nel'i), **not on your nelly**, *int.* (*sl.*) not likely, certainly not.
nelson (nel'sən), *n.* (**full nelson**) a wrestling hold in which the arms are passed under both the opponent's arms from behind, and the hands joined so that pressure can be exerted with the palms on the back of the neck. **half nelson**, *n.* such a hold applied from one side only.
nemato-, *comb. form.* thread-like; filamentous.
nematode (nem'ətōd), *n.* any member of a class of worms comprising the parasitic round-worm, thread-worm etc., a nematode worm.
Nembutal® (nem'būtal), *n.* proprietary name for sodium ethyl methylbutyl barbiturate, used as a sedative, hypnotic and anti-spasmodic.
nemertean (nimœ'tiən), **-tine** (nem'ətin), *a.* belonging to the Nemertea, a division of flat- or ribbon-worms, chiefly marine. *n.* a worm of this class.
Nemesis (nem'əsis), *n.* retribution, justice.

nemophila (nimof'ilə), *n.* an annual trailing plant with blue and white flowers.
nenuphar (nen'ūfah), *n.* the white water-lily.
neo-, *comb. form.* new, recent, modern, later, fresh.
neoclassic (nēōklas'ik), *a.* belonging to the 18th-cent. revival of classicism. **neoclassic, -ical**, *a.* **neoclassicism** (-sizm), *n.*
neocolonialism (nēōkələ'niəlizm), *n.* the policy of a strong nation gaining control over a weaker through economic pressure etc.
neocracy (nēok'rəsi), *n.* government by new or upstart persons.
neo-Darwinism (nēōdah'winizm), *n.* Darwinism as modified by later investigators, esp. those who accept the theory of natural selection but not that of the inheritance of acquired characters. **neo-Darwinian** (-win'-), *a.*, *n.*
neodox (nē'ōdoks), *a.* holding new views. **neodoxy** (-i), *n.*
neodymium (nēōdim'iəm), *n.* a metallic element, at. no. 60; chem. symbol Nd; of the cerium group of rare earth elements.
neofascism (nēōfash'izm), *n.* a movement attempting to reinstate the policies of fascism. **neofascist**, *n.*
neo-Gothic (nēōgoth'ik), *n.*, *a.* the Gothic revival of the mid-19th cent., or pertaining to this.
neolite (nē'əlit), *n.* a dark-green hydrous silicate of aluminium and magnesium.
Neolithic (nēəlith'ik), *a.* pertaining to the later Stone Age characterized by ground and polished implements and the introduction of agriculture.
neology (niol'əji), *n.* the introduction or use of new words; a neologism; the adoption of or the tendency towards rationalistic views in theology. **neological** (nēəloj'-), *a.* **neologically**, *adv.* **neologism**, *n.* a new word or phrase, or a new sense for an old one; the use of new words; neology. **neologistic** (-jis'-), *a.* **neologize, -ise**, *v.i.*
neon (nē'on), *n.* a gaseous element, at. no. 10; chem. symbol Ne; existing in minute quantities in the air, isolated from argon in 1898.
neonatal (nēōnā'təl), *a.* pertaining to the first few weeks of life in human babies. **neonate** (nē'-), *n.* a baby at this stage in its development.
neomycin (nēōmi'sin), *n.* an antibiotic effective against some infections that resist ordinary antibiotics.
neophyte (nē'əfīt), *n.* one newly converted or newly baptized; one newly admitted to a monastery or to the priesthood; a beginner, a novice.
neoplasm (nē'əplazm), *n.* an abnormal growth of new tissue in some part of the body, a cancer.
Neoplatonism (nēōplā'tənizm), *n.* a system of philosophy combining the Platonic ideas with the theosophy of the East, originating in Alexandria in the 3rd cent. AD. **Neoplatonic** (-ton'), *a.* **Neoplatonist**, *n.*
neoteric (nēəte'rik), *a.* new; of recent origin. *n.* one of modern times.
Nepalese (nepal'ēz), **Nepali** (nepawli), *a.*, *n.* (of or pertaining to) the people or language of Nepal. *n.* from Nepal.
nephalism (nef'əlizm), *n.* total abstinence from intoxicants, teetotalism.
nepheline (nef'əlin), **-lite** (-līt), *n.* a vitreous silicate of aluminium and sodium found in volcanic rocks.
nephew (nef'ū), *n.* the son of a brother or sister; extended to the son of a brother- or sister-in-law, also to a grand-nephew.
nephoscope (nef'əskōp), *n.* an instrument for observing the elevation, direction and velocity of clouds.
nephr(i)- *comb. form.* pertaining to the kidney.
nephrite (nef'rīt), *n.* a variety of jade.
nephritic (-frit'-), *a.* pertaining to the kidneys; suffering from kidney disease. **nephritis** (nifri'tis), *n.* a disease or disorder of the kidneys.
nephr(o)-, *comb. form.* pertaining to the kidneys.

nephrology (nifrol'əji), *n.* the study of the kidneys.
ne plus ultra (nä plus ul'trə), the most perfect or uttermost point.
nepotism (nep'ətizm), *n.* favouritism (as in bestowing patronage) towards one's relations. **nepotist,** *n.*
Neptune (nep'tūn), *n.* the Roman god of the sea; one of the sun's planets. **neptunium** (-tū'niəm), *n.* a radio-active element, at. no. 93, chem. symbol Np, obtained by the bombardment of uranium with neutrons.
NERC, (*abbr.*) Natural Environment Research Council.
nerd (nœd), *n.* (*sl.*) an ineffectual or stupid person, a fool. **nerdish,** *a.*
nereid (niə'riid), *n.* (*pl.* **-ids**) a sea-nymph; a sea-worm or marine centipede.
nerine (niri'nē), *n.* a S African amaryllid genus with scarlet or rose coloured flowers, including the Guernsey Lily.
neroli (niə'rəli), **neroli oil,** *n.* an essential oil distilled from the flowers of the bitter or Seville orange, used in perfume.
nerve (nœv), *n.* one of the fibres or bundles of fibres conveying sensations and impulses to and from the brain or other organ; a tendon or sinew; strength, coolness, resolution, pluck; one of the ribs or fibrovascular bundles in a leaf; (*pl.*) (*coll.*) the source or state of anxiety, extreme nervousness. (*coll.*) impudence, cheek, audacity. *v.t.* to give strength or firmness to. **a bundle of nerves,** (*coll.*) a very timid, anxious person. **to get on one's nerves,** (*coll.*) to become oppressively irritating. **to lose one's nerve,** to lose confidence; to become afraid. **nerve cell,** *n.* any cell forming part of the nervous system. **nerve centre,** *n.* an aggregation of nerve cells from which nerves branch out; the central or most important part of a business, organization etc. **nerve gas,** *n.* any one of a number of gases affecting the nervous system, used in warfare etc. **nerve-racking,** *a.* causing anxiety, fear, nervous tension. **nervate** (-vāt), *a.* having nerves or ribs. **nerved,** *a.* (*usu. in comb.*, as *strong-nerved*). **nerveless** (-lis), *a.* destitute of strength, energy or vigour; feeble, flabby. **nervine** (-vin), *a.* capable of acting upon the nerves. *n.* a medicine that acts on the nerves. **nervous,** *a.* abounding in nervous energy; having weak or sensitive nerves, excitable, highly strung, timid. **nervous breakdown,** *n.* a loose term for a mental illness or disorder often characterized by depression, agitation or excessive anxiety. **nervous system,** *n.* a network of nerve cells, including the spinal cord and brain, which collectively controls the body. **nervously,** *adv.* **nervousness,** *n.*
nervy, *a.* nervous, jerky, jumpy; full of nerve, cool, confident.
nescience, *n.* **nescient** (nes'iənt), *n.*
ness (nes), *n.* a promontory, a cape.
-ness (-nəs, -nis), *suf.* forming abstract nouns denoting state or quality, as *goodness, holiness, wilderness*.
nest (nest), *n.* the bed or shelter constructed or prepared by a bird for laying its eggs and rearing its young; any place used by animals or insects for similar purposes; a snug place of abode, shelter or retreat; a series or set, esp. a number of boxes each inside the next larger. *v.t.* to put, lodge or establish in or as in a nest; to pack one inside another. *v.i.* to build and occupy a nest; to hunt for or take birds' nests. **nest egg,** *n.* something laid by, as a sum of money, as a nucleus for saving or a reserve. **nestling** (-ling), *n.* a bird too young to leave the nest.
nestle (nes'l), *v.i.* to nest; to be close or snug; to settle oneself (down, in or among); to press closely (up to).
Nestor (nes'tə), *n.* a wise counsellor; a sage; a venerable senior.
net[1] (net), *n.* a fabric of twine, cord etc., knotted into meshes, for catching fish, birds, or other animals, or for covering, protecting, carrying etc. *v.t.* (*past, p.p.* **netted**) to cover, hold or confine with a net; to catch in a net; to fish with nets; to ensnare. **netball,** *n.* a game in which a ball has to be thrown into a suspended net. **network,** *n.*

netting; a system of intersecting lines, a reticulation; a system of stations for simultaneous broadcasting; any system of lines, roads etc., resembling this; (*Comput.*) a system of communication between different computers, terminals, circuits etc.; a group of people who are useful to each other because of the similarity of their aims, background etc. as in *old boy network*. *v.t., v.i.* to connect; to broadcast (a television or radio programme) nationwide; to form or be part of a network, e.g. through business contact. (*Comput.*) to create or use a system of communication between computers, circuits etc. **netted,** *a.* reticulated. **netting,** *n.*
net[2] (net), *a.* free from all deductions; obtained or left after all deductions; final, conclusive. *v.t.* to yield or realize as clear profit. **net profit,** *n.* total profit reached after the deduction of overheads from the gross profit.
nether (nedh'ə), *a.* lower; belonging to the region below the heavens or the earth. **nether regions** or **world,** Hell; (*rarely*) the earth. **nethermost,** *a.*
Netherlander (nedh'ələndə), *n.* a native or inhabitant of the Netherlands.
netsuke (net'suki), *n.* a small piece of carved wood or ivory worn or attached to various articles, as a toggle or button, by the Japanese.
nett (net), NET[2].
nettle (net'l), *n.* a plant of the genus *Urtica*, with inconspicuous flowers and minute stinging hairs. *v.t.* to irritate, to provoke. **nettle rash,** *n.* an eruption on the skin resembling the sting of a nettle, urticaria.
network NET[1].
Neufchâtel (nœshatel'), *n.* a soft white cheese similar to cream cheese but with less fat.
neur- NEUR(O)-.
neural (nū'rəl), *a.* of or pertaining to the nerves or the nervous system. **neural network,** *n.* (*Comput.*) a computer designed to imitate the functions of the human brain.
neuralgia (nūral'jə), *n.* an acute pain in a nerve or series of nerves, esp. in the head or face. **neuralgic,** *a.*
neurasthenia (nūrəsthē'niə), *n.* weakness of the nervous system, nervous debility.
neuric (nū'rik), *a.* of or pertaining to the nerves. **neuricity** (-ris'-), *n.*
neuritis (nūri'tis), *n.* inflammation of a nerve.
neur(o)-, *comb. form.* pertaining to a nerve cell; pertaining to nerves; pertaining to the nervous system.
neurochemistry (nūrōkem'istri), *n.* the biochemistry of the transmission of impulses down nerves.
neurocomputing (nūrōkəmpūt'ing), *a., n.* (*Comput.*) computing using a neural network. *a.* related to, involved in neural network computing.
neurology (nūrol'əji), *n.* the scientific study of the anatomy, physiology and pathology of nerves. **neurological** (-loj'-), *a.* **neurologist,** *n.*
neuron(e) (nū'ron), *n.* a nerve cell with its processes and ramifications, one of the structural units of the nervous system.
neuropath (nū'rəpath), *n.* a person suffering from a nervous disorder or having abnormal nervous sensibility; a physician who regards nervous conditions as the main factor in pathology. **neuropathic** (-path'-), *a.* relating to or suffering from a nervous disease. **neuropathology** (-thol'-), *n.* the pathology of the nervous system. **neuropathy** (-rop'əthi), *n.* any nervous disease.
neurophysiology (nūrōfiziol'əji), *n.* the physiology of the nervous system.
neuroptera (nūrop'tərə), *n.pl.* (*Ent.*) an order of insects with four reticulated membranous wings. **neuropterous, neuropteran,** *a., n.*
neurosis (nūrō'sis), *n.* (*loosely*) a mild mental disorder, usu. with symptoms of anxiety. **neurotic** (-rot'-), *a.* suffering from neurosis. *n.* a person suffering from neurosis; a person of abnormal nervous excitability.

neurosurgery (nūrōsœ'jəri, nū'-), *n.* the branch of surgery dealing with the nervous system. **neurosurgical,** *a.*
neurotomy (nūrot'əmi), *n.* an incision in a nerve, usu. to produce sensory paralysis.
neurotransmitter (nūrōtranzmit'ə), *n.* a chemical substance by means of which nerve cells communicate with each other.
neuter (nū'tə), *a.* neither masculine, nor feminine; (of verbs) intransitive; neither male nor female, without pistil or stamen; undeveloped sexually, sterile. *n.* a neuter noun, adjective or verb; the neuter gender; a flower having neither stamens nor pistils; a sterile female insect, as a working bee; a castrated animal.
neutral (nū'trəl), *a.* taking no part with either side; belonging to a state that takes no part in hostilities; indifferent, impartial; having no distinct or determinate character, colour etc.; neither good nor bad, indefinite, indeterminate; the position of parts in gear mechanism when no power is transmitted; neither acid nor alkaline; neither positive nor negative. *n.* a state or person that stands aloof from a contest; a subject of a neutral state. **neutrality** (-tral'-), *n.* **neutralize, -ise,** *v.t.* to render neutral; to render inoperative or ineffective, to counteract; to declare (a state or territory) neutral either permanently or during hostilities. **neutralization,** *n.* **neutralizer,** *n.* **neutrally,** *adv.*
neutrino (nūtrē'nō), *n.* a sub-atomic particle with almost zero mass, zero charge but specified spin.
neutron (nū'tron), *n.* a particle that is neutral electrically with approximately the same mass as a proton.
névé (nev'ā), *n.* consolidated snow above the glaciers, in process of being converted into ice.
never (nev'ə), *adv.* not ever, at no time; on no occasion; not at all; none; (*ellipt. in exclamations*) surely not. **never a one,** not a single person etc., none. **never more,** at no future time; never again. **never-never land,** an imaginary place with conditions too ideal to exist in real life. **the never-never,** (*coll.*) the hire-purchase system. **never so,** (*loosely* **ever so**) to an unlimited extent; exceedingly. **The Never-never,** term applied to areas in North and West Queensland. **never-ending, -failing,** *a.* **nevertheless,** *conj.* but for all that; notwithstanding; all the same.
new (nū), *a.* not formerly in existence; lately made, invented or introduced; not before known; recently entered upon or begun; never before used, not worn or exhausted; fresh, unfamiliar, unaccustomed; (of bread) newly baked; fresh (from), not yet accustomed (to). *adv.* newly, recently (*in comb.*, as *new-blown, new-born*); anew, fresh. **the New World,** the Western Hemisphere. **New Age,** *a.*, *n.* (of) a movement or philosophy characterized by concern for the ecology, spiritual well-being etc. **new-born,** *a.* just born; regenerate. **new chum,** *n.* (*Austral. coll.*) a recent immigrant. **newcomer,** *n.* a person who has recently arrived in a place or who has just begun to take part in something. **New Deal,** *n.* economic and social measures introduced by President Roosevelt in the US in 1933, after the economic crisis of 1929. **new man,** *n.* one in sympathy with the Women's Liberation Movement and who undertakes tasks traditionally associated with women, such as housework. **newspeak,** *n.* a form of official language which is ambiguous, misleading and verbose, coined by George Orwell in his novel *Nineteen Eighty-Four*. **new town,** *n.* a town planned by the government to aid housing development in nearby large cities, stimulate development etc. **new woman,** *n.* (*formerly*) a woman of advanced ideas, esp. one who claimed equality with men in the social, economic and political spheres. **new wave,** *a.*, *n.* (of) a movement in the arts concerned with overturning traditional ideas, perspectives etc. **newly,** *adv.* recently (*usu. in comb.*). **newness,** *n.*
newel (nū'əl), *n.* the central column from which the steps of a winding stair radiate; an upright post at the top or bottom of a stair supporting the hand-rail.

newfangled, *a.* different from the accepted fashion; fond of novelties, inconstant. **newfangledness,** *n.*
Newfoundland (nū'fəndlənd, -fownd'-), *n.* a large breed of dog orig. from Newfoundland.
Newgate (nūgət, -gāt), *n.* a London prison, demolished in 1902.
Newmarket (nū'mahkit), *n.* a card game.
news (nūz), *n.pl.* (*usu. as sing.*) recent or fresh information, tidings; (*prec. by* **the**) a regular radio or television broadcast of up-to-date information on current affairs. **news agency,** *n.* an organization for supplying information to newspapers etc. **newsagent,** *n.* a dealer in newspapers and other periodicals. **newscaster,** *n.* a news broadcaster. **newsflash,** *n.* a short important news item, esp. one which interrupts a television or radio programme. **newshound,** (*coll.*) *n.* a reporter. **newsletter,** *n.* any news sent out regularly to a particular group. **newspaper,** *n.* a printed publication, usu. issued daily or weekly, containing news usu. with articles on special topics, advertisements and reviews of literature, plays etc. **newspaper man, -woman,** *n.* a journalist. **newsprint,** *n.* the cheap-quality paper upon which newspapers are printed. **news-reader,** *n.* a newscaster, one who presents the news on television or radio. **newsreel,** *n.* a film giving the day's news. **news-room,** *n.* a room where news is edited. **news-stand,** *n.* a newspaper kiosk. **newsworthy,** *a.* of sufficient interest or note to be included in a newspaper, newscast, etc. **newsless** (-lis), *a.*
newt (nūt), *n.* a small tailed amphibian like the salamander, an eft.
newton (nū'tən), *n.* a unit of force equal to 100,000 dynes.
next (nekst), *a.* nearest in place, time or degree; nearest in order or succession, immediately following. *adv.* nearest or immediately after; in the next place or degree. *n.* the next person or thing. **next best,** second best. **next but one,** the one next to that immediately preceding or following. **next door,** the house adjoining. **next to nothing,** scarcely anything. **what next?** *int.* can anything exceed or surpass this?
nexus (nek'səs), *n.* (*pl.* **nexus**) a link, a connection.
NFU, (*abbr.*) National Farmers' Union.
ngaio (nī'ō), *n.* a New Zealand tree noted for its fine white wood.
NHS, (*abbr.*) National Health Service.
NI, (*abbr.*) National Insurance; Northern Ireland.
Ni, (*chem. symbol*) nickel.
nib (nib), *n.* the point of a pen; a pen-point for insertion in a pen-holder; the point of a tool etc.; the beak of a bird. **his nibs,** (*coll.*) an important or self-important person.
nibble (nib'l), *v.t.* to bite little by little; to bite little bits off; to bite at cautiously (as a fish at a bait). *v.i.* to take small bites or to bite cautiously (at). *n.* the act of nibbling; a little bite. **nibbler,** *n.*
niblick (nib'lik), *n.* (*Golf*) a club with a small cup-shaped iron head.
niccolite (nik'əlīt), *n.* native arsenide of nickel.
nice (nīs), *a.* (rarely) fastidious, over-particular, hard to please, dainty, punctilious; discerning, discriminating, delicate, subtle, minute; pleasing or agreeable; satisfactory; delightful, attractive, friendly, kind. **nicely,** *adv.* **niceness,** *n.* **nicety** (-siti), *n.* exactness, precision; a minute point, a delicate distinction; a small detail. **to a nicety,** exactly, with precision.
niche (nich, nēsh), *n.* a recess in a wall for a statue, vase etc.; one's proper place or natural position.
nick (nik), *n.* a small notch, cut or dent; the critical moment; the exact point or moment; (*sl.*) prison. *v.t.* to cut or make a nick or nicks in; (*sl.*) to steal. **in good nick,** (*coll.*) in good condition. **in the nick of time,** only just in time.
nickel (nik'l), *n.* a lustrous silvery-white ductile metallic element, at. no. 28; chem. symbol Ni; usu. found in asso-

ciation with cobalt, used in manufacture of German silver and other alloys; a US 5-cent piece (formerly a 1-cent piece). **nickel plate,** *v.t.* to cover with nickel. **nickel plating,** *n.* **nickel silver,** *n.* an alloy like German silver but containing more nickel.
nickelodeon (nikəlō'diən), *n.* an early form of juke-box, especially one operated by a 5-cent piece.
nicker (nik'ə), *n.* (*sl.*) a pound (money), £1.
nicknack (nik'nak), KNICK-KNACK.
nickname (nik'nām), *n.* a name given in derision or familiarity. *v.t.* to give a nickname to; to call by a nickname.
nicotine (nik'əten), *n.* an acrid, poisonous alkaloid contained in tobacco.
nictate (nik'tāt), **nictitate** (-ti-), *v.i.* to blink. **nictation, nictitation,** *n.* **nictitating membrane,** a third or inner eyelid possessed by birds, fishes and many animals.
nide (nīd), *n.* a nest, esp. of young pheasants; a collection of pheasants.
nidificate (nid'ifikāt), **nidify** (-fī), *v.i.* to build a nest or nests. **nidification,** *n.*
nidus (nī'dəs), *n.* (*pl.* **-di** (-dī)) a nest, a place for the deposit of eggs laid by birds, insects etc.; a place in which spores or bacteria develop; a source or origin, a place of development.
niece (nēs), *n.* the daughter of one's brother or sister, or one's brother-in-law or sister-in-law.
niello (niel'ō), *n.* (*pl.* **-li** (-lē)) a black alloy used to fill the lines of incised designs on metal plates; an example of this work.
Neirsteiner (niə'shtīnə, -stī-), *n.* a white Rhenish hock.
Nietzschean (nē'chiən), *a.* of Friedrich *Nietzsche*, 1844–1900, or his philosophy. *n.* a follower of Nietzsche.
niff (nif), *n.* (*sl.*) a stink, a bad smell. **niffy,** *a.*
nifty (nif'ti), *a.* (*coll.*) smart, stylish; quick, slick.
nigella (nījel'ə), *n.* a genus of ranunculaceous plants comprising love-in-a-mist.
niggard (nig'ərd), *n.* a stingy person, a miser; one who is grudging (of). **niggardly,** *a.* **niggardliness,** *n.*
nigger (nig'ə), *n.* (*offensive*) a negro; (*offensive*) one of any dark-skinned people. **the nigger in the woodpile,** a person or thing that spoils something good.
niggle (nig'l), *v.i.* to busy oneself with petty details; to fiddle, to trifle. **niggler,** *n.* **niggling,** *a.*, *n.* **niggly,** *a.*
nigh (nī), *adv.*, *a.*, *prep.* near.
night (nīt), *n.* the time of darkness from sunset to sunrise; the darkness of this period; the end of daylight, nightfall; a period of state of darkness; (*fig.*) ignorance; intellectual and moral darkness; death; a period of grief or sorrow; an evening set aside for a particular event or activity. *a.* occurring, belonging to, working during, etc., the night. **to make a night of it,** to spend an evening in festivity. **night-bird,** *n.* the owl or nightingale; (*fig.*) a person who routinely stays up late. **night blindness,** *n.* nyctalopia. **nightcap,** *n.* an alcoholic drink taken at bed-time. **nightclothes,** *n.* clothes worn in bed. **nightclub,** *n.* a club open late at night and in the early hours of the morning. **nightdress, -gown,** *n.* a woman's night attire. **nightfall,** *n.* the beginning of night, the coming of darkness; dusk. **nightflower,** *n.* a flower that opens at night and shuts in the day. **nightgown** NIGHTDRESS. **nighthawk,** *n.* the nightjar; an American bird, *Chordeiles virginianus.* **nightjar,** *n.* the goatsucker. **night-light,** *n.* a dim light for keeping alight at night. **nightlife,** *n.* late evening entertainment or social life. **night-line,** *n.* a line with baited hooks left in the water at night to catch fish. **nightlong,** *a.*, *adv.* (lasting) through a night. **nightmare** (-meə), *n.* a terrifying dream; (*fig.*) a haunting sense of dread or anything inspiring such a feeling. **nightmarish,** *a.* **night owl,** *n.* an exclusively nocturnal owl; a person who habitually stays up late. **night school,** *n.* an evening school. **nightshift,** *n.* a period of work performed at night; (*collect.*) the workers on such a shift. **nightshirt,** *n.* a long shirt worn in bed by men.

nightspot, *n.* (*coll.*) a nightclub. **night-stick,** (*N Am.*) a truncheon. **night-time,** *n.* **night-terrors,** *n.pl.* a nightmare of childhood. **night watch,** *n.* a watch or guard on duty at night; one of the periods into which the Jews and Romans divided the night; a nightwatchman. **nightwatchman,** *n.* a person who keeps watch on a building at night, a night security guard. **nightwear,** *n.* nightclothes. **nightly,** *a.* **nighty** (-i), *a.* (*coll.*) a nightgown.
nightingale (nī'tinggāl), *n.* a small migratory bird singing at night as well as by day.
nightmare NIGHT.
nightshade (nīt'shād), *n.* one of several plants of the genus *Solanum,* esp. the black nightshade with poisonous black berries, the woody nightshade with purple flowers and brilliant red berries and the deadly nightshade.
nigrescent (nīgres'ənt), (*adj.*) growing black; blackish. **nigrescence,** *n.*
nigritude (nīg'ritūd), blackness *n.*
nigrosine (nīg'rəsēn, sin), *n.* a blue-black dye-stuff obtained from aniline hydrochlorates.
nihilism (nī'ilizm), *n.* any theological, philosophical or political doctrine of a negative kind; denial of all existence, or of the knowledge of all existence; (*Hist.*) a Russian form of anarchism aiming at the subversion of all existing institutions. **nihilist,** *n.* **nihilistic** (-lis'-), *a.* **nihility** (-hil'-), *n.* the state of being nothing, of nothingness; a mere nothing.
-nik *comb. form.* a person who practices something, e.g. *beatnik, kibbutznik, peacenik.*
nil (nil), *n.* nothing; zero.
nilgai (nīlgī), **nilghau** (nīl'gaw), *n.* a large Indian antelope.
Nilotic (nīlot'-), *a.* pertaining to the Nile etc.
nimble (nim'bl), *a.* light and quick in motion; agile, swift, dexterous; alert, clever, brisk, lively. **nimble-fingered,** *a.* **nimble-footed,** *a.* **nimbleness,** *n.* **nimbly,** *adv.*
nimbus (nim'bəs), *n.* (*pl.* **-buses**) a halo or glory surrounding the heads of divine or sacred personages in paintings etc.; a rain-cloud, a dark mass of cloud from which rain is falling or likely to fall.
nimby (nim'bi), *a.* an expression indicating the views of those who support the dumping of nuclear waste, the construction of ugly buildings etc. as long as they or their property are not affected. **nimbyism,** *n.* [acronym for *not in my back yard*]
Nimrod (nim'rod), *n.* a great hunter; military jet aircraft equipped with radar.
nincompoop (ning'kəmpoop), *n.* a blockhead, a fool.
nine (nīn), *n.* the number or figure 9; the age of 9; nine o'clock. **dressed (up) to the nines,** dressed elaborately. **nine-days' wonder,** an event, person or thing that is a novelty for the moment but is soon forgotten. **nine times out of ten,** usually, generally. **ninepins,** *n.* a game with nine skittles set up to be bowled at. **ninefold,** *a.* nine times repeated. **nineteen** (-tēn'), *n.* the number or figure 19; the age of 19. *a.* 19 in number; aged 19. **nineteen to the dozen,** volubly. **nineteenth** (-th), *n.* one of nineteen equal parts. *n.*, *a.* (the) last of (people, things etc.); the next after the eighteenth. **nineteenth hole,** *n.* (*colloq. Golf*) the clubhouse bar. **ninety** (-ti), *n.* the number or figure 90; the age of 90; (*pl.*) the period of time between one's ninetieth and one hundredth birthdays; the range of temperature between ninety and one hundred degrees; the period of time between the ninetieth and final years of a century. **ninetieth** (-tiəth), *n.* one of ninety equal parts. *n.*, *a.* (the) last of ninety (people, things etc.); the next after the eighty-ninth.
Ninja (nin'ja), *n.* a member of an ancient Japanese society of assassins.
ninjutsu (ninjut'soo), *n.* a martial art based on the killing techniques of the Ninjas.
ninny (nin'i), *n.* a fool, a simpleton.
ninon (nē'non), *n.* a semi-diaphanous light silk material.

ninth (nīnth) *n.* one of the nine equal parts; an interval of an octave and a second. *n., a.* (the) last of nine (people, things etc.); the next after the eighth. **ninthly,** *adv.*
niobium (nīō'biəm), *n.* a metallic element, at. no. 41; chem. symbol Nbg; occurring in tantalite etc. **niobite** (nī'əbīt), *n.* a niobic salt; a variety of tantalite.
Nip NIPPON.
nip[1] (nip), *v.t.* (*past, p.p.* **nipped**) to pinch, to squeeze or compress sharply; to cut or pinch off the end or point of; to bite; to sting, to pain; to check the growth of; to blast, to wither. *v.i.* to cause pain; (*coll.*) to move, go, or step quickly (in, out, etc.). *n.* a pinch, a sharp squeeze or compression; a bite; a check to vegetation, esp. by frost. **nipper,** *n.* one who or that which nips; a chela or great claw of a crab or other crustacean; a fish of various kinds; (*sl.*) a boy, a lad; (*pl.*) a pair of pincers, forceps or pliers. **nippily,** *adv.* **nippy,** *a.* cold; active; agile; sharp in temper; quick, alert.
nip[2] *n.* a small drink, esp. of spirits *v.i.* to take a nip or nips. *v.t.* to take a nip of.
nipa (-nē'pə, nī'-), *n.* a palm tree of tropical SE Asia and the islands of the Indian Ocean; an intoxicating beverage made from the sap of this.
nipper, etc. NIP[1].
nipple (nip'l), *n.* the small prominence in the breast of female mammals, esp. women, by which milk is sucked or drawn, a teat; a similar contrivance attached to a baby's feeding bottle; a nipple-shaped projection prominence. **nipple-shield,** *n.* a protection worn over the nipple by nursing mothers. **nipplewort,** *n.* a slender weed with small yellow flowers.
Nippon (nip'on), *n.* the Japanese name for Japan. **Nip,** *n.* (*offensive*) a Japanese.
nippy NIP[1].
Nirvana (nœvah'nə), *n.* absorption of individuality into the divine spirit with extinction of personal desires and passions, the Buddhist state of beatitude; (*coll.*) bliss, heaven.
nisi (nī'sī), *conj.* (*Law*) unless, if not. **decree nisi** DECREE.
nissen (nis'ən), *n.* a long hut of corrugated iron with semicircular roof.
nit[1] (nit), *n.* the egg or larva of a louse. **nit-picking,** *n.* (*coll.*) petty criticism of minor details. **nit-pick,** *v.i.* **nit-picker,** *n.* **nitty,** *a.*
nit[2] (nit), *n.* (*coll.*) a fool.
nit[3] *n.* the unit of luminance, one candela per square metre.
niton (nīton), *n.* gaseous radioactive element.
nitrate (nī'trāt), *a.* a salt of nitric acid; sodium or potassium nitrate. *v.t.* to treat or combine with nitric acid. **nitration,** *n.*
nitre (nī'tə), *n.* saltpetre, potassium nitrate, occurring as an orthorhombic mineral. **nitrify** (-fī), *v.t.* to turn into nitre; to make nitrous. *v.i.* to become nitrous. **nitrification** (-fi-), *n.*
nitric (nī'trik), *a.* pertaining to nitre. **nitric acid,** *n.* a colourless, corrosive acid liquid based on the ingredients of nitre, aqua fortis.
nitrogen (nī'trəjən), *n.* a colourless, tasteless, gaseous element, at. no. 7; chem. symbol N; forming 80% of the atmosphere, the basis of nitre and nitric acid. **nitrogenous** (-troj'-), *a.*
nitroglycerine (nītrōglis'ərēn, -rin), *n.* a highly explosive colourless oil, obtained by adding glycerine to a mixture of nitric and sulphuric acids.
nitrous (nī'trəs), *a.* obtained from, impregnated with, or resembling nitre. **nitrous oxide,** *n.* nitrogen monoxide used as an anaesthetic, laughing-gas.
nitty NIT[1].
nitty gritty (nit'i grit'), *n.* the basic facts, the realities of a situation.
nitwit (nit'wit), *n.* (*coll.*) a foolish or stupid person.

nix[1] (niks), *n.* (*sl.*) nothing, nobody.
nix[2] (niks), **nixie** (-i), *n.* a water-sprite.
nizam (nizahm'), *n.* (*pl.* **nizam**) a man in the Turkish regular army; the title of the ruler of Hyderabad.
NNE, (*abbr.*) north north-east.
NNW, (*abbr.*) north north-west.
No[1], (*chem. symbol*) nobelium.
No, no, no[2], (*abbr.*) number.
No[3], **Noh** (nō), *n.* the Japanese drama developed out of religious dance.
n.o., (*abbr.*) not out.
no[1] (nō), *adv.* a word of denial or refusal, the categorical negative. *n.* (*pl.* **noes**) the word 'no'; a negative reply, a denial, a refusal; (*pl.*) voters against a motion. **no less,** *adv.* as much, as much as. **no more,** *adv.* not any more; nothing further; no longer; dead, gone; never again; just as little as.
no[2] (nō) *a.* not any; not one, not a; quite other, quite opposite; not the least; hardly any; absent, lacking. **no ball,** *n.* (*Cricket*) a ball not delivered according to the rules, counting for one to the other side. *int.* a declaration of this by the umpire. **no-claims bonus,** *n.* a reduction in the price of an insurance policy because no claims have been made on it. **no go** GO[2]. **nohow,** *adv.* in no way, not by any means. **no-man's-land,** *n.* waste or unclaimed land; the contested land between two opposing forces. **no-nonsense,** *a.* practical, down-to-earth, sensible. **no one,** *pron.* nobody, no person. **no trump,** *int.* in bridge, the call for the playing of a hand without any trump suit. **noway, nowise,** *adv.* in no way, not at all.
Noah's ark ARK.
nob[1] *n.* a person of rank or distinction; a swell.
nob[2] (nob), *n.* (*Cribbage*) a point scored for holding the knave of the same suit as the turn-up.
nobble (nob'əl), *v.t.* (*coll.*) to dose, lame or otherwise tamper with (a horse) to prevent its winning a race; to circumvent, to over-reach; to get hold of dishonestly; to catch, to nab; to buttonhole (a person); to persuade, influence or win over by dishonest means.
nobbut (nob'ət), (*dial*) no more than; only.
nobelium (nōbē'liəm), *n.* an artificially produced radioactive element, at. no. 102; chem. symbol No.
nobility (nəbil'iti), *n.* the quality of being noble; magnanimity, greatness, dignity; nobleness of birth or family; (*collect.*) the nobles, the peerage.
noble (nō'bəl), *a.* lofty or illustrious in character, worth or dignity; magnanimous, high-minded; of high rank, of ancient or illustrious lineage; belonging to the nobility; magnificent, stately, imposing; excellent, admirable. *n.* a nobleman, a peer, an obsolete gold coin. **nobleman, -woman,** *n.* a peer. **nobleness,** *n.* **nobly,** *adv.* **noblesse oblige** (nō'bles əblēzh'), rank imposes obligations.
nobody (nō'bədi), *n.* no one, no person; a person of no importance. **like nobody's business,** very energetically or intensively.
nock (nok), *n.* a notched tip at the butt-end of an arrow; the notch in this; the notched tip at each end of a bow. *v.t.* to fit (an arrow) to the bowstring.
noctambulation, *n.* somnambulism, *n.* **noctambulism,** *n.*
noctilucent (noktiloo'sənt), *a.* shining by night.
Noctua (nok'tūə). *n.* a genus of moths typical of the Noctuidae, the largest family of Lepidoptera. **noctuid** (-id), *a.* and *n.*
noctule (nok'tūl), *n.* the great bat *Vesperugo noctula.*
Nocturnae (noktœr'nē), *n.pl.* the owls.
nocturnal (noktœr'nəl), *a.* relating to or occurring, performed, or active, by night. **nocturnally,** *adv.*
nocturne (nok'tœn), *n.* a painting or drawing of a night scene; a dreamy piece of music suited to the night or evening.
nod (nod), *v.i.* (*past, p.p.* **nodded**) to incline the head with a slight, quick motion in token of assent, command, indi-

cation, or salutation; to incline, to totter; to let the head fall forward; to be drowsy, to sleep; to make a careless mistake, *v.t.* to bend or incline (the head etc.); to signify by a nod. *n.* a quick bend of the head; a bending downwards; (*fig.*) command. **Land of Nod**, sleep. **to nod off**, (*coll.*) to fall asleep. **to nod through**, in Parliament, to allow to vote by proxy; to pass (a motion etc.) without formal discussion, voting etc. **nodding,** *n.*, *a.* **nodding acquaintance,** *n.* a slight acquaintance.
nodal NODE.
noddle (nod'l), *n.* (*sl.*) the head.
node (nōd), *n.* a knot, a knob; the point of a stem from which leaves arise; a thickening or swelling e.g. of a joint of the body; the point at which the orbit of a planet intersects the ecliptic, or in which two great circles of the celestial sphere intersect; the point at which a curve crosses itself and at which more than one tangent can be drawn; a similar point on a surface; a point of rest in a vibrating body. **nodal,** *a.* **nodule** (nod'ūl), *n.* a small knot, node or lump; a rounded lump or mass of irregularly rounded shape. **nodular, noduled, nodulose** (-lōs), **nodulous,** *a.*
nodus (nō'dəs), *n.* a knotty point, a complication, a difficulty; a node.
Noel (nōel'), *n.* Christmas.
noetic (-et'ik), **-al,** *a.* relating to the intellect; performed by or originating in the intellect.
nog, *n.* a pin or peg; a wooden block; a snag or stump.
nogging (nog'ing), *n.* a timber part in a framed construction.
noggin (nog'in), *n.* a small cup or mug; a measure, usu. a gill (125 ml); the contents of such a measure; (*coll.*) the head.
noise (noiz), *n..* a sound of any kind, esp. a loud, discordant, harsh or disagreeable one; clamour, din, loud or continuous talk. *v.t.* to make public, to spread (about or abroad). **big noise,** (*coll.*) a person of importance. **noiseless** (-lis), *a.* **noiselessly,** *adv.* **noisy,** *a.* causing noise; making much noise. **noisily,** *adv.* **noisiness,** *n.*
noisette[1] (nwazet'), *n.* a variety of rose, a cross between the China rose and the musk-rose.
noisette[2] (nwazet'), *n.* (*pl.*) small pieces of mutton, veal, etc.; a nutlike or hazelnut-flavoured sweet.
noisome (noi'səm), *a.* hurtful, noxious; unwholesome, offensive, (especially of smells). **noisomeness,** *n.*
nolle prosequi (nol'i pros'ikwī), *n.* (*Law*) the record of a decision to proceed no further with part of a prosecution. [L]
nolo contendere (nō'lō konten'dəri), *n.* (*Law*) I will not contest it, a plea which accepts conviction without admitting guilt. [L]
nomad (nō'mad), *n.* one of a people that wanders about seeking pasture for their flocks, a wanderer. *a.* wandering. **nomadic** (-mad'-), *a.* **nomadism,** *n.*
no-man's-land NO[2].
nom de guerre (nōdəgeə', nom-), *n.* an assumed name, a pseudonym. **nom de plume** (ploom) *n.* a pen-name. [F]
nome (nōm), *n.* a province of a country, esp. in modern Greece and Egypt.
nomenclature (nō'men'klǎchə), *n.* a system of names for the objects of study in any branch of science; a system of terminology.
nomic (nom'ik), *a.* ordinary, customary (of spelling).
nominal (nom'inəl), *a.* existing in name only; opp. to real; trivial, inconsiderable; of, pertaining to or consisting of a name or names; of or pertaining to a noun, substantival. **nominalism,** *n.* the doctrine that general or abstract concepts have no existence but as names or words. cp. REALISM. **nominalist,** *n.* **nominally,** *adv.* in name only.
nominate (nom'ināt), *v.t.* to name; to designate; to mention by name; to appoint to an office or duty; to propose as a candidate; to call, to denominate. **nomination,** *n.* **nominative,** *a.* (*Gram.*) applied to the case of the subject. *n.* the case of the subject; the subject of the verb. **nominatival** (-tī'-), *a.* **nominator,** *n.* **nominee** (-nē'), *n.* person named or appointed by name.
nomology (nəmol'əji), *n.* the science of law. **nomologist,** *n.*
non- (non-), *pref.* freely prefixed to indicate a negative; only a selection of such words is given below. **non-ability,** *n.* a want of ability. **nonacceptance,** *n.* **nonaggression pact,** *n.* an agreement between two states to settle differences by negotiation rather than by force. **non-alcoholic,** *a.* not containing alcohol. **non-aligned,** *a.* not taking any side in international politics, esp. not belonging to the Warsaw Pact or NATO. **nonappearance,** *n.* default of appearance, esp. in court. **nonattendance,** *n.* **nonbelligerent,** *n.* a neutral; a country that remains neutral in name only, supporting a belligerent country with everything save armed force. **noncombatant,** *a.* not in the fighting line. *n.* a civilian, esp. a surgeon, chaplain etc. attached to troops. **non-com.** (non'kom), *n.* (*sl.*) a noncommissioned officer. **noncommissioned,** *a.* not holding a commission, applied to military officers below the rank of 2nd lieutenant. **noncommittal,** *a.* not committing one, impartial. **non-compliance,** *n.* (*Law*) failure to comply. **nonconductive,** *a.* not conducting heat or electricity. **nonconductor,** *n.* a substance or medium that offers resistance to heat or electricity. **nonconformist,** *n.* one who does not conform, esp. a member of a Protestant Church or sect dissenting from the Church of England. **nonconformity,** *n.* **non-contagious,** *a.* not contagious. **noncooperation,** *n.* refusal to cooperate; inactive opposition. **noneffective,** *a.* not qualified for active service. **nonessential,** *a.*, *n.* **nonevent,** *n.* an event or occasion which fails to come up to expectation. **nonexistence,** *n.* **nonexistent,** *a.* **nonferrous,** *a.* containing no iron. **non-fiction,** *n.*, *a.* (of) a literary work, containing no deliberate fictitious element. **nonflammable,** *a.* not capable of supporting flame, though combustible, difficult or impossible to set alight. **non-interference, non-intervention,** *n.* the principle or policy of keeping aloof from the disputes of other nations. **noniron,** *a.* not requiring ironing. **nonmember,** *n.* one who is not a member. **non-membership,** *n.* **nonmetallic,** *a.* **nonnuclear,** *a.* not having, or using, nuclear power or weapons. **non-observance,** *n.* **nonpartisan,** *a.* adhering to no one political party. **non-payment,** *n.* **non-performance,** *n.* **nonprofessional,** *a.* not professional, amateur; unskilled. **nonproliferation,** *a.*, *n.* (pertaining to) the limitation of the manufacture or distribution of something, esp. nuclear weapons. **nonrepresentational,** *a.* of art, abstract. **non-residence,** *n.* the state of not residing in a place. **nonresident,** *a.*, *n.* **nonresistance,** *n.* passive obedience or submission even to power unjustly exercised. **nonreturnable,** *a.*, *n.* (of) a bottle jar or other container on which a returnable deposit has not been paid. **nonsexual,** *a.* asexual, sexless. **nonskid,** *n.*, *a.* (a tyre) designed to prevent skidding. **nonsmoker,** *n.* someone who does not smoke; a part of a train etc. in which it is not permitted to smoke. **non-starter,** *n.* in a race, a horse which is entered but does not start; (*coll.*) an idea or person with no chance whatsoever of success. **nonstick,** *a.* of a cooking pan, treated so that food will not stick to it. **nonstop,** *a.* without a pause; not stopping at certain stations etc. **nonunion,** *a.* not connected with a trade union. **nonviolence,** *n.* the practice of refraining from violence on principle. **nonviolent,** *a.*
nonage (nō'nij), *n.* the state of being under age; minority; a period of immaturity.
nonagenarian (nonəjinə'riən), *a.* ninety years old. *n.* a person 90 years old, or between 90 and 100.
nonagon (non'əgon), *n.* a figure having nine sides and nine angles.
nonce (nons), *n.* the present time, occasion, purpose, etc. **nonce-word,** *n.* a word coined for the occasion.
nonchalant (non'shələnt), *a.* careless, cool, unmoved, in-

different. **nonchalance**, *n*. **nonchalantly**, *adv*.
nondescript (non'diskript), *a*. not easily described or classified, without distinguishing features; bland, uninteresting. *n*. such a person or thing.
none (nŭn), *pron*. no one, no person; (*coll*.) no persons; not any, not any portion (of).
nonentity (nonen'titi), *n*. a thing not existing, a mere figment, an imaginary thing; (*fig*.) an unimportant person or thing.
nonet (nonet'), *n*. musical composition for nine players or singers.
nonillion (nənil'yən), *n*. a million raised to the ninth power, denoted by a unit with 54 ciphers annexed; (*esp. F and N Am*.) the tenth power of a thousand, denoted by a unit with 30 ciphers.
nonpareil (nonpərel', -rāl'), *n*. a paragon, a thing of unequalled excellence.
nonplus (nonplŭs'), *v.t*. (*past, p.p*. **nonplussed**) to puzzle, to confound, to bewilder.
nonsense (non'səns), *n*. unmeaning words, ideas etc.; foolish or extravagant talk, conduct etc.; foolery, absurdity; rubbish, worthless stuff, trifles. **nonsense verse**, *n*. verse having no meaning, used for mnemonic purposes; verse intentionally absurd written to amuse. **nonsensical** (-sen'-), *a*. **nonsensicality** (-kal'-), **nonsensicalness**, *n*. **nonsensically**, *adv*.
non sequitur (nonsek'witə), *n*. an inference not warrantable from the premises. [L]
noodle[1] (noo'dl), *n*. a simpleton, a fool.
noodle[2] (noo'dl), *n*. a strip of dough made of wheat-flour and eggs, served with soup etc.
nook (nuk), *n*. a corner; a cosy place, as in an angle; a secluded retreat.
nookie, nooky (nuki), *n*. (*sl*.) lovemaking.
noon (noon), *n*. the middle of the day, 12 o'clock; (*fig*.) the culmination of height. *a*. pertaining to noon. **noonday**, *n., a*. **noontide**, *n*.
noose (noos), *n*. a loop with a running knot binding the closer the more it is pulled, as in a snare, a hangman's halter etc.; (*fig*.) a tie, a bond, a snare. *v.t.* to catch in a noose; to entrap; to tie a noose on; to tie in a noose. **to put one's head in a noose**, to put oneself into a dangerous or exposed situation.
nopal (nō'pəl), *n*. an American genus of cacti grown for the support of the cochineal insect.
nope (nōp), (*sl; orig. N Am*.) an emphatic form of no.
nor (naw), *conj*. and not (a word marking the second or subsequent part of a negative proposition); occasionally used without the correlative.
noraghe (nōrah'gä), *n*. (*pl*. **noraghi**) a prehistoric stone structure common in Sardinia.
noradrenaline (norədren'əlin), *n*. an amine related to adrenalin, used as a heart resuscitant.
Nordic (naw'dik), *n*. a tall, blond dolichocephalic racial type inhabiting Scandinavia, parts of Scotland and other parts of N Europe.
Norfolk jacket, *n*. a man's loose-jacket with vertical pleats in the back and front and a waist-band. [English county]
noria (naw'riə), *n*. an endless chain of buckets on a wheel for raising water from a stream or similar.
norm (nawm), *n*. a standard, model, pattern or type; a typical structure etc.
normal (naw'məl), *a*. according to rule, standard, or established law; regular, typical, usual; perpendicular. *n*. the usual state, quality, quantity etc.; a perpendicular line; the average or mean value of observed quantities; the mean temperature, volume etc. **normality** (-mal'-), *n*. **normalize, -ise**, *v.t.* to make normal; to cause to conform to normal standards etc. **normalization**, *n*. **normally**, *adv*.
Norman (naw'mən), *n*. a native or inhabitant of Normandy; Norman-French; a member of a mixed people of Scandinavians and Franks established there. *a*. of or pertaining to Normandy or the Normans. **Norman architecture**, *n*. a massive Romanesque style of architecture prevalent in Normandy (10th–11th cents.) and England (11th–12th cents.). **Norman conquest**, *n*. CONQUEST.
Norn (nawn), **Norna** (-nə), *n*. one of the Norse Fates.
Norse (naws), *a*. pertaining to Norway or its inhabitants, Norwegian. *n*. the Norwegian language; the language of Norway and its colonies (till the 14th cent.); (*loosely*) the Scandinavian languages, including early Swedish and Danish. **Norseman**, *n*.
North (nawth), **nor'** (naw), *n*. one of the four cardinal points, that to the right of a person facing the setting sun at the equinox; a region or part north of any given point; the northern part (of any country); *a*. situated in or towards the north; belonging or pertaining to the north, northern. *adv*. towards or in the north. **the North**, northern England; Scotland; the U.S. states north of the Mason-Dixon line; the countries of the world lying in the northern hemisphere. **north by east, west**, one compass point east, west of north. **north country**, *n*. the part of a country to the north, esp. northern England or the northern part of Great Britain. *a*. of, pertaining to or characteristic of this. **north-countryman**, *n*. **north-east, -west**, *n*. the point midway between the north and east, or north and west, a region lying in this quarter. *a*. pertaining to or proceeding from the north-east etc. **north-easter, nor'-easter**, *n*. a north-east wind. **north-easterly**, *a*. towards or from the north-east. **north-eastern**, *a*. in or towards this. **north-eastward**, *a*. **north-eastwardly**, *a., adv*. **north pole** POLE. **north-star**, *n*. the pole-star. **north-west, -wester, -westerly, -western, -westward, -westwardly** NORTH-EAST. **northerly** (-dhə-), *a., adv,.* **northerliness**, *n*. **northern** (-dhən), *a*. pertaining to, situated or living in, or proceeding from the north; towards the north. *n*. a northerner. **northern hemisphere**, *n*. the half of the globe north of the equator. **northern lights**, *n.pl*. the aurora borealis. **northerner**, *n*. a native or inhabitant of the north. **northernmost**, *a*. **northward** (-wəd), *a., adv., n*. **northwardly**, *a., adv*. **northwards**, *adv., n*.
Northumbrian (nawthŭm'briən), *n*. a native or inhabitant of ancient Northumbria (England north of the Humber) or of Northumberland; the old English dialect of Northumbria. *a*. of or pertaining to one of these districts.
Norwegian (nawwē'jən), *a*. pertaining to Norway or its inhabitants. *n*. a native or inhabitant of Norway; the language of the Norwegians.
nose (nōz), *n*. the projecting part or the face between the forehead and mouth, containing the nostrils and the organ of smell; a part of things resembling a nose, as a point, a prow etc. *v.t.* to perceive, trace or detect by smelling; to rub or push with the nose; to push (one's way). *v.i.* to smell, to sniff (about, after, at etc.). **to follow one's nose**, to go straight ahead. **to lead by the nose**, to cause to follow blindly. **to pay through the nose**, to be charged an exorbitant price. **to put one's nose out of joint** JOINT. **to stick one's nose into**, to meddle officiously. **to turn up the nose**, to show contempt (at). **under one's nose**, in one's actual presence or sight. **nosebag**, *n*. a bag containing for hanging over a horse's head. **noseband**, *n*. the part of a bridle passing over the nose and attached to the cheek-straps. **nosebleed**, *n*. bleeding from the nose. **nosedive**, *n*. of an aircraft, a sudden plunge towards the earth; (*coll*.) to plummet, decline suddenly e.g. of prices. *v.i.* to make this plunge. **nose job**, *n*. (*sl*.) plastic surgery to remodel the nose. **nose-piece**, *n*. a noseband; the end-piece of a microscope to which the lens is fastened. **nose rag**, *n*. (*sl*.) a handkerchief. **nose ring**, *n*. a ring worn in the nose as ornament; a leading-ring for a bull etc. **nosed**, *a*. (*usu. in comb.*, as *red-nosed*). **noseless** (-lis), *a*. **nosy**, *a*. (*coll*.) very inquisitive. **nosy parker**, *n*. (*coll*.) a prying person.
nosegay (nōz'gā), *n*. a bunch of flowers, especially fragrant flowers.

nosh (nosh), *n.* (*sl.*) food; a meal. *v.i.* to eat. **nosh-up**, *n.* (*sl.*) a large meal, a feast.
nosology (nosol'əji), *n.* a systematic classification of diseases; the branch of medical science treating of such a classification. **nosological** (-loj'-), *a.* **nosologist**, *n.*
nostalgia (nəstal'jiə), *n.* home-sickness; a yearning for the past. **nostalgic**, *a.*
nostoc (nos'tok), *n.* a genus of gelatinous freshwater algae, also called star-jelly or witches' butter.
Nostradamus (nostrədah'məs), *n.* one who predicts or professes to predict. [Michel de *Nostredame*, 1503-66, French physician, astrologer and professional prophet]
nostril (nos'tril), *n.* one of the apertures of the nose.
nostrum (nos'trəm), *n.* (*pl.* **-ums**) a medicine based on a secret formula; a quack remedy.
nosy NOSE.
not[1] (not), *adv.* (*encliticalIy* **n't**) a particle expressing negation, denial, prohibition or refusal. **not a few** FEW. **not in the running**, standing no chance, not worth considering. **not once or twice**, many times, often. **not on**, *n.* (*sl.*) not possible; not morally, socially etc. acceptable. **not out**, *a.* having reached the end of a cricket innings or of play for the day without being dismissed. **not that**, *conj.* it is not meant however that.
nota bene (nōtə ben'i), mark well; take notice; often abbreviated to **NB.** [L]
notable (nō'təbl), *a.* worthy of note; remarkable, memorable, distinguished; eminent, conspicuous. *n.* a person or thing of note or distinction. **notabilia** (-bil'iə), *n.pl.* notable things. **notability** (-bil'), *n.* **notably**, *adv.*
notary (nō'təri), *n.* a public official appointed to attest deeds, contracts etc., administer oaths etc., frequently called a **notary public. notarial** (-teə'ri-), *a.*
notation (nətā'shən), *n.* the act or process of representing by signs, figures etc.; a system of signs, figures etc., employed in any science or art. **notational** (nōtət), *a.*
notch (noch), *n.* a nick, a cut, a V-shaped indentation; a tally-point. *v.t.* to cut a notch or noches in; to score by notches. **a notch above**, a level or step higher; superior to. **notched**, *a.*
note (nōt), *n.* a distinctive feature, a characteristic; a mood, atmosphere; a brief record, a memorandum, a short or informal letter; a diplomatic communication; a bank-note or piece of paper money; an annotation, a comment, explanation or gloss, appended to a passage in a book etc.; attention, observation; distinction, importance; a sign representing the pitch and duration of a sound; a musical sound. *v.t.* to observe, to take notice of; to make a memorandum of; to annotate. **of note**, distinguished. **to compare notes** COMPARE. **to strike the right (a false, wrong) note**, to act, behave etc. in an appropriate (inappropriate), or suitable (unsuitable) manner. **notebook**, *n.* a book for entering memoranda in. **notepad**, *n.* a pad of paper for writing notes. **notepaper**, *n.* a small size of paper for letters. **noted**, *a.* eminent, remarkable; well-known. **notelet**, *n.* a sheet of notepaper folded into a decorative greeting card for short letters. **noteworthy**, *a.* outstanding, famous; worth attention.
nothing (nŭth'ing), *n.* no thing that exists; not anything, naught; no amount, a cipher, a naught; nothingness, non-existence; an insignificant or unimportant thing, a trifle. *adv.* in no way, not at all. **next to nothing**, almost nothing. **nothing doing**, (*sl.*) nothing happening; a refusal to do something. **nothing for it but**, no alternative but. **nothing to you**, not your business. **sweet nothings**, words of endearment. **to come to nothing**, to turn out a failure; to result in no amount or naught. **to make nothing of**, to fail to understand or deal with. **to stop at nothing**, to be totally ruthless. **nothingness**, *n.*
notice (nō'tis), *n.* intelligence, information, warning; a written or printed paper giving information or directions; an intimation or instruction; intimation of the termination of an agreement, contract of employment etc., at a specified date; a review of a book, play etc. in a newspaper; observation, regard, attention; the act of noting. *v.t.* to take notice of, to perceive; to remark upon. **at short notice**, with little advance warning. **to give notice**, to intimate the termination of an agreement, particularly a contract of employment. **to take no notice of**, to pay no attention to; to ignore. **notice board**, *n.* a board exposed to public view on which notices are posted. **noticeable**, *a.* **noticeably**, *adv.*
notify (nō'tifi), *v.t.* to make known, to accounce, to declare, to publish; to give notice to, to inform (of or that). **notifiable**, *a.* to be notified (esp. of cases of diseases that must be reported to the health authorities). **notification** (-fi-), *n.*
notion (nō'shən), *n.* an idea, a conception; an opinion, a view; a theory, a scheme, a device; (*coll.*) an inclination, desire, intention or whim; (*pl.*, *N Am.*) fancy goods, haberdashery, novelties etc. **notional**, *a.* pertaining to notions or concepts; abstract, imaginary, hypothetical; speculative, ideal. **notionally**, *adv.*
notitia (nətish'iə), *n.* a list, register, or catalogue. [L]
notochord (nō'tōkawd), *n.* the elastic cartilaginous band constituting a rudimentary form of the spinal column in the embryo and some primitive fishes.
notorious (nətaw'iəs), *a.* widely or publicly or commonly known (now used only in a bad sense). **notoriety** (nōtəri'ə-), **notoriously**, *adv.*
notornis (nətaw'nis), *n.* a gigantic New Zealand coot, now very rare.
notturno (notuə'nō), *n.* a title given by some 18th cent. composers to music written for evening performance; another word for *nocturne*.
Notus (nō'təs), *n.* the south wind.
notwithstanding (notwidhstan'ding), *prep.* in spite of, despite. *adv.* nevertheless; in spite of this. *conj.* although; in spite of the fact (that).
nougat (noo'gah, nug'ət), *n.* a chewy confection of almonds or other nuts and sugar.
nought (nawt) NAUGHT.
noun (nown), *n.* a word used as the name of anything, a substantive. **nounal**, *a.*
nourish (nŭ'rish), *v.t.* to feed, to sustain, to support; to foster, to cherish, to nurse. *v.i.* to promote growth. **nourisher**, *n.* **nourishing**, *a.* **nourishment**, *n.* the act of nourishing; the state of being nourished; that which nourishes, food, sustenance.
nous (nows), *n.* mind, intellect; (*coll.*) sense, wit, intelligence.
nouveau (noovō), *fem.* **nouvelle** (-vel), new. **nouveau riche** (rēsh), *a.* of a person who has recently acquired wealth but who has not acquired good taste or manners. **nouvelle cuisine** (kwēzēn'), *a.* a style of simple French cooking which does not involve rich food, creamy sauces etc. and relies largely on artistic presentation.
nova (nō'və), *n.* a star that flares up to great brightness and subsides after a time.
novel (nov'əl), *a.* new, recent, fresh; unusual, strange. *n.* a fictitious narrative in prose, usu. of sufficient length to fill a volume, portraying characters and actions from real life; this type of literature. **novelette** (-let'), *n.* a short novel; a short novel of a sentimental nature; (*Mus.*) a kind of romance dealing freely with several themes. **novelettish**, *a.* characteristic of a novelette; cheaply sentimental. **novelist**, *n.* a writer of novels. **novelistic** (-lis'-), *a.* **novella** (nəvel'ə), *n.* a tale, a short story; a short novel. **novelty**, *n.* newness, freshness; something new, unusual.
November (nəvem'bə), *n.* the 11th month of the year; code word for the letter N.
novena (nəvē'nə), *n.* in the Roman Catholic Church, a devotion consisting of a prayer or service repeated on nine successive days.

novennial (nəven'iəl), *a.* happening every ninth year.
novice (nov'is), *n.* one entering a religious house on probation before taking the vows; a new convert; one who is new to any business, an inexperienced person, a beginner.
novitiate (nəvish'iət), *n.* the term of probation passed by a novice; the part of a religious house allotted to novices; (*fig.*) a period of probation or apprenticeship.
Novocaine® (nō'vəkān), *n.* a synthetic produce derived from coal tar, used as a local anaesthetic, known as procaine.
now (now), *adv.* at the present time; at once, immediately; very recently; at this point or time, then (in narrative); in these circumstances; used as an expletive in explaining, remonstrating, conciliating etc. *conj.* since, seeing that, this being the case (that). *n.* the present time. *a.* (*coll.*) present, existing; fashionable. **just now**, a little time or a moment ago. **now and then** or **again**, from time to time; occasionally. **now or never**, at this moment or the chance is gone for ever. **nowadays**, *adv.* at the present time; in these days.
noway, NO ².
nowel (nōel'), *int.* a shout of joy in Christmas carols.
nowhere (nō'weə), *adv.* not in, at, or to any place or state. **nowhere near**, not at all near. **to be** or **come in nowhere**, (*coll.*) to be badly defeated (esp. in a race or other contest).
nowt (nowt), *n.* (*dial.*) nothing.
noxious (nok'shəs), *a.* hurtful, harmful, unwholesome; pernicious, destructive. **noxiously**, *adv.* **noxiousness**, *n.*
noyau (nwahyō'), *n.* brandy cordial flavoured with bitter almonds etc.
nozzle (noz'l), *n.* a spout, projecting mouthpiece, or end of pipe or hose.
NP, (*abbr.*) Notary Public.
n.p. (*abbr.*) new paragraph.
NPL, (*abbr.*) National Physical Laboratory.
NSPCC, (*abbr.*) National Society for the Prevention of Cruelty to Children.
NSU, (*abbr.*) Non-specific urethritis.
NSW, (*abbr.*) New South Wales.
NT, (*abbr.*) National Trust; New Testament.
NTS, (*abbr.*) National Trust for Scotland.
nu (nu), *n.* the 13th letter of the Greek alphabet.
nuance (nū'ās), *n.* a delicate gradation in colour or tone; a fine distinction between things, feelings, opinions etc.
nub (nŭb), *n.* a small lump; the pith or gist (of). **nubbly** (*prov.*), **nubby**, *adv.*
nubile (nū'bīl), *a.* marriageable (usu. of women); sexually mature; sexually attractive. **nubility** (-bil'-), *n.*
nuchal (nū'kəl), *a.* pertaining to the nape of the neck.
nuclear (nūkliə), *a.* relating to atomic nuclei; pertaining to the nucleus of a biological cell; of or using nuclear power or weapons. **nuclear charge**, *n.* the positive electric charge in the nucleus of an atom. **nuclear disarmament**, *n.* the reduction or giving up of a country's nuclear weapons. **nuclear energy**, *n.* energy released during a nuclear reaction, whether by nuclear fission or nuclear fusion. **nuclear family**, *n.* the basic family unit consisting of a mother and father and their children. **nuclear fission**, *n.* the breaking up of a heavy atom, as of uranium, into atoms of smaller mass, thus causing a great release of energy. **nuclear fuel**, *n.* uranium, plutonium and other metals consumed to produce nuclear energy. **nuclear fusion**, *n.* the creation of a new nucleus by merging two lighter ones, with release of energy. **nuclear physics**, *n.pl.* the study of atomic nuclei. **nuclear power**, *n.* power obtained from a controlled nuclear reaction. **nuclear reaction**, *n.* reaction in which the nuclei of atoms are transformed into isotopes of the element itself, or atoms of a different element. **nuclear reactor**, *n.* a structure of fissile material such as uranium, with a moderator such as carbon or heavy water, so arranged that nuclear energy is continuously released under control. Also called an atomic pile. **nuclear warfare**, *n.* the use of atomic or hydrogen bombs in warfare. **nuclear waste**, *n.* radioactive waste. **nuclear winter**, *n.* a period of coldness and darkness predicted as likely to follow a nuclear war.
nucleate (nū'kliāt), *v.i.* to form a nucleus. (-ət), *a.* having a nucleus, nucleated.
nucleic acids (nūklē'ik), *n.pl.* complex organic acids forming part of nucleo proteins.
nucle(o)- *comb. form.* pertaining to a nucleus.
nucleonics (nūklion'iks), *n.* the science of the nucleus of the atom.
nucleus (nū'kliəs), *n.* (*pl.* -**clei** (-ī)) a central part about which aggregation, accretion or growth goes on; a kernel; the charged centre of an atom consisting of protons and neutrons; the central body in an ovule, seed, cell etc., constituting the organ of vitality, growth, or other functions; (*fig.*) a centre of growth, development, activity etc.; the brightest part of the head of a comet.
nude (nūd), *a.* bare, naked, uncovered, unclothed, undraped; (*Law*) made without any consideration and consequently void. *n.* an undraped figure in painting or sculpture. **in the nude**, the undressed human figure or the state of being undressed. **nudely**, *adv.* **nudeness, nudity**, *n.* **nudism**, *n.* belief in the physical and spiritual benefit of being nude; the practice of nudity. **nudist**, *n.* **nudist camp, colony**, *n.* an open-air camp inhabited by nudists.
nudge (nŭj), *v.t.* to push gently, as with the elbow; to draw attention or give a hint with, or as with such a push. *n.* such a push. **nudge-nudge**, *int.* (*coll.*) suggesting some secret or underhand behaviour, esp. sexual.
nudibranch (nū'dibrangk), *n.* a mollusc characterized by naked gills or the absence of a shell.
nugatory (nū'gə-), *a.* trifling, insignificant; futile, ineffective, inoperative.
nugget (nŭg'it), *n.* a lump of native metal, esp. of gold; something small but valuable.
nuisance (nū'səns), *n.* anything that annoys, irritates or troubles; an offensive or disagreeable person, action, experience etc.; (*Law*) anything causing annoyance, inconvenience or injury to another. **nuisance value**, *n.* the capacity to cause irritation, obstruction etc.
NUJ, (*abbr.*) National Union of Journalists.
nuke (nūk), *n.* (*sl.*) a nuclear weapon *v.t.* to attack with nuclear weapons; (*fig.*) to destroy completely, ruin.
null (nŭl), *a.* void, having no legal force or validity; (*fig.*) without character, expression, or individuality; (*Math.*, *Log.*) amounting to nothing, equal to zero, nil.
nullah (nŭl'ə), *n.* a ravine, gully or water-course in India.
nulla-nulla (nŭl'ənŭl'ə), *n.* a club-shaped weapon of hard wood used by the Australian Aborigines.
nullify (nŭl'ifī), *v.t.* to make void; to cancel; to annul, to invalidate; to efface, to destroy. **nullification** (-fi-), *n.* **nullifier**, *n.*
nullity (nŭl'iti), *n.* invalidity; an invalid act, instrument, etc.; nothingness, non-existence; a nonentity, a mere cipher.
NUM, (*abbr.*) National Union of Mineworkers.
numb (nŭm), *a.* deprived of sensation and motion; torpid, stupefied, dulled. *v.t.* to benumb, to paralyse. **numbfish**, *n.* the electric ray or torpedo. **numbly**, *adv.* **numbness**, *n.*
number (nŭm'bə), *n.* a measure of discrete quantity; a name or symbol representing any such quantity, a numeral; a sum or aggregate of persons, things or abstract units; one of a numbered series; a single issue of a periodical; a division of an opera etc.; plurality, multitude, numerical preponderance (*usu. in pl.*); the distinctive form of a word according as it denotes unity or plurality; a song or piece of music forming part of a performance; (*sl.*) a position or job, esp. an advantageous or lucrative one; (*coll.*) an article or object of special interest (e.g. a designer dress). *v.t.* to count, to reckon; to ascertain the number of; to

numbskull 553 **nymph**

assign a number to, to distinguish with a number; to include, to comprise (among etc.). **Numbers**, *n.* the fourth book of the Old Testament. **by numbers**, performed in simple stages esp. orig. with each one numbered. **his/her number is up**, he/she is going to die. **to have someone's number**, (*coll.*) to understand someone's intentions, motives or character. **without number**, *n.* innumerable. **number-crunching**, *n.* the large-scale processing of numbered information esp. by computer. **number one**, *n.* (*coll.*) the first in a series; the most senior person in an organization; (*coll.*) oneself; the product which is at the top of a sales chart esp. a pop record. **number plate**, *n.* the plate on a motor vehicle showing its registration number. **Number Ten**, *n.* 10 Downing Street, the British Prime Minister's residence. **number two**, *n.* a deputy.
numbskull NUMSKULL.
numdah (nŭn'dah), *n.* an embroidered felt rug found in India.
numeral (nū'mərəl), *n.* a word, symbol or group of symbols denoting number; Roman figures, e.g. I, III, V, X, C. *a.* pertaining to, consisting of, or denoting number. **numerical** (-me'ri-), *a.* **numerically**, *adv.*
numerate (nū'mərāt), *v.t.* to count, to number. *a.* able to count; competent in mathematics. **numeracy**, *n.* **numeration**, *n.* **numerator**, *n.* the part of a vulgar fraction written above the line.
numerous (nū'mərəs), *a.* many in number; consisting of a great number of individuals. **numerously**, *adv.* **numerousness**, *n.* **numerology** (-rol'-), *n.* the study of the alleged significance of numbers.
numinous (nū'minəs), *a.* pertaining to divinity; feeling or arousing awe of the divine.
numismatic (nūmizmat'ik), *a.* pertaining to coins or coinage. **numismatics**, *n.* the science or study of coins and medals. **numismatist** (-miz'-), *n.*
nummulite (nŭm'ūlīt), *n.* a fossil foraminifer resembling a coin.
numskull (nŭm'skŭl), *n.* a blockhead, a dunce.
nun (nŭn), *n.* a woman devoted to a religious life and living in a convent under vows, usu. of poverty, chastity and obedience. **nunhood** (-hud), *n.* **nunlike**, *a.*, *adv.*
nunnery, *n.* a religious home for women.
nuncio (nŭn'shiō, siō), *n.* (*pl.* **-ios**) a papal envoy to a foreign power. **nunciature** (-shəchə), *n.*
nunnery, NUN.
NUPE (nū'pi), (*abbr.*) National Union of Public Employees.
nuphar (nū'fah) *n.* the yellow water-lily.
nuptial (nŭp'shəl, -chəl), *a.* pertaining to, done at, or constituting a wedding. *n.pl.* a wedding. **nuptial flight**, *n.* the flight of a virgin queen-bee, during which she is impregnated.
NUR, (*abbr.*) National Union of Railwaymen.
nurd (nūrd), *n.* (*sl.*) a foolish or ineffective person. **nurdish**, *a.*
nurse (neos), *n.* a wet nurse; a nursemaid; one who tends the sick, wounded or infirm; a sexually imperfect bee, ant etc., which tends the young brood. *v.t.* to suckle, to feed (an infant); to hold or clasp, esp. on one's knees or lap; to nurture; to foster, to tend, to promote growth in; to tend in sickness; to manage with care; to cherish. (*Billiards*) to keep (the balls) in a good position for cannons. *v.i.* to act as nurse; to suckle a baby. **wet nurse** WET. **nursemaid**, *n.* a woman in charge of young children. **nursing**, *n.* the act of nursing, the profession of being a nurse. *a.* **nursing home**, *n.* a hospital or home where care is provided for the elderly or chronically sick. **nursing officer**, *n.* any of several grades of nurse having administrative duties.
nursery (nœ'səri), *n.* a room set apart for young children; a daycare establishment for young children; a place or garden for rearing plants; a place where animal life is developed; a place or atmosphere conducive to development; a handicap or race for two-year-old horses. **nurserymaid**, *n.* a nursemaid. **nurseryman, -woman**, *n.* one who raises plants in a nursery. **nursery rhyme**, *n.* a traditional rhyme known to children. **nursery school**, *n.* a school for young children aged two to five. **nursery slopes**, ski slopes set apart for novices.
nurture (nœ'chə), *n.* the act of bringing up, training, fostering; nourishment; education, breeding. *v.t.* to nourish, to rear, to train, to educate.
NUS, (*abbr.*) National Union of Seamen; National Union of Students.
NUT, (*abbr.*) National Union of Teachers.
nut (nŭt), *n.* the indehiscent fruit of certain trees, containing a kernel in a hard shell; a metal block with a hole for screwing on and securing a bolt, screw etc.; (*sl.*) the head; (*sl.*) a crazy person, a fanatic; (*pl.*) small lumps of coal; (*pl., sl.*) testicles. *v.i.* (*past, p.p.* **nutted**) to gather nuts. **a hard nut to crack**, a difficult problem to solve. **can't do it for nuts**, (*sl.*) can't succeed in doing it even in the most favourable circumstances. **nuts and bolts**, the basic essential facts. **off his nut**, (*sl.*) mad. **to be nuts about**, (*sl.*) to delight in; to be very fond of. **nutbrown**, *a.* brown as a hazelnut. **nutcase**, (*sl.*) a crazy person. **nutcracker**, *n.* (*usu. pl.*) an instrument for cracking nuts; a European bird of the genus *Nucifraga*. **nut cutlet**, *n.* a preparation of crushed nuts etc., eaten by vegetarians as a substitute for meat. **nut-hatch**, *n.* a small bird allied to the woodpecker. **nutshell**, *n.* the hard shell enclosing the kernel of a nut. **in a nutshell**, contained or expressed in a very concise statement. **nutlet** (-lit), *n.* **nuts**, *a.* (*sl.*) crazy. **nutter**, *n.* (*sl.*) a mad person. **nutty**, *a.* tasting like nuts; spicy; (*sl.*) crazy.
nutation (nū'tāshən), *n.* a nodding or oscillation; a movement of the tips of growing plants, usu. towards the sun; a periodical oscillation of the earth's axis due to the attractive influence of the sun and moon on the greater mass round the equator; (*Path.*) morbid oscillation of the head.
nutmeg (nŭt'meg), *n.* the hard aromatic seed of the fruit of species of *Myristica* used for flavouring.
nutria (nū'triə), *n.* a S American beaver; its skin, formerly much used in hat-making.
nutrient (nū'triənt), *a.* nourishing; serving as or conveying nourishment. *n.* a nutritious substance. **nutriment**, *n.* that which nourishes or promotes growth, esp. food. **nutrimental** (-men'-), *a.* **nutrition** (-trish'-), *n.* the function or process of promoting the growth of organic bodies; the study of this. **nutritional**, *a.* **nutritionist**, *n.* **nutritious**, *a.* affording nourishment, efficient as food. **nutritiously**, *adv.* **nutritiousness**, *n.* **nutritive**, *a.*, *n.*
nux vomica (nŭksvom'ikə), *n.* the seed of an East Indian tree which yields strychnine.
nuzzle (nŭz'əl), *v.t.* to rub or press the nose against; to fondle; to root up with the nose. *v.i.* to root about with the nose; to nestle, to hide the head, as a child in its mother's bosom.
NW, (*abbr.*) North-West.
NY, (*abbr.*) New York.
NYC, (*abbr.*) New York City.
nyctalopia (niktəlō'piə), *n.* a disease of the eyes in which vision is worse in shade or twilight than in daylight, night blindness.
nyctitropic (niktitrop'ik), *a.* changing position or direction at night (of leaves). **nyctitropism** (-tit'-), *n.*
nycturia (niktū'riə), *n.* nocturnal incontinence of urine, bed-wetting.
nylon (nī'lən), *n.* name applied to a group of thermoplastics, used largely for hosiery, shirts, dress fabrics, imitation furs, ropes, brushes etc.
nymph (nimf), *n.* one of a class of youthful female divinities inhabiting groves, springs, mountains, the sea etc.; a

beautiful or attractive young woman. **nymphet** (-fit), *n.* a young girl who is very sexually attractive and precocious. **nymphlike**, *a.* **Nymphaea** (nimfē'ə), *n.* a genus of aquatic plants, containing the white water-lily. **nympho** (nimf'ō), *n.* (*coll.*) a nymphomaniac.

nympholepsy (nim'fəlepsi), *n.* a wild desire for the unattainable. **nympholept**, *n.* **nympholeptic** (-lep'-), *a.* **nymphomania** (nimfəmā'niə), *n.* excessive sexual desire in woman. **nymphomaniac** (-ak), *n.* **nymphomaniacal** (-məni'ə-), *a.*
NZ, (*abbr.*) New Zealand.

O

O¹, o, the 15th letter of the alphabet, and the fourth vowel (*pl.* **Os, O's, Oes**); a circle, oval, or any round or nearly round shape; a naught, zero. **O-grade**, *n.* (a pass in) an examination in the Scottish Certificate of Education corresponding to O-level. **O-level**, *n.* (a pass in) an examination at the (now defunct) ordinary level of the General Certificate of Education in England and Wales.
O², (*chem. symbol*) oxygen.
O³, o, (*abbr.*) ocean; octavo; old; Ohio; ohm; only; order; ordinary.
O⁴, oh (ō), *int.* an exclamation of entreaty, invocation, pain, surprise, wonder etc., or used in addressing someone formally.
O' (ō), *pref.* descendant of, in Irish surnames.
o' (ō. ə), *prep.* (*coll.* or *dial.*) short for OF, ON.
-o, (ō), *suf.* to *n.*, *a.* (*coll.*) serving as a diminutive, as in *boyo, laddo*; (*coll.*) forming an interjection; as in *cheerio, righto*.
oaf (ōf), *n.* (*pl.* **oafs**) a silly, stupid person, a lout. **oafish**, *a.* **oafishly**, *adv.* **oafishness**, *n.*
oak (ōk), *n.* any tree or shrub of the genus *Quercus*, esp. *Q. robur*, a forest tree much valued for its timber; the wood of this; applied to various trees and plants bearing a real or fancied resemblance to the oak; one of various species of moth. **the Oaks**, a classic horse-race for three-year-old fillies. **oak-apple, -gall, -nut**, *n.* a gall or excrescence of various kinds produced on oaks by various gall-flies. **oaken**, *a.* made of oak. **oaky**, *a.*
oakum (ō'kəm), *n.* old rope, untwisted and pulled into loose fibres, used for stopping leaks etc. in ships.
O&M, (*abbr.*) organization and method(s).
OAP, (*abbr.*) old age pension(er).
oar (aw), *n.* a long pole with a flattened blade, for rowing or steering a boat; an oarsman. *v.i.* to row. *v.t.* to propel (as) by rowing. **to lie, rest on one's oars**, to cease rowing without shipping the oars; to stop for rest, to cease working. **to put, stick one's oar in**, to interfere, esp. with unasked-for advice. **oarlock**, *n.* (*N Am.*) a rowlock. **oarsman, -woman**, *n.* a (skilled) rower. **oarsmanship**, *n.* **oared**, *a.* having oars. **oarless** (-lis), *a.* **oar-like**, *a.*
OAS, (*abbr.*) on active service; Organization of American States.
oasis (ōā'sis), *n.* (*pl.* **oases** (-sēz)) a fertile spot in a waste or desert; a thing or place offering peace, pleasure, refuge etc.
oast-house (ōst'hows), *n.* a building for drying of hops.
oat (ōt), *n.* (*usu. in pl.*) a variety of cereal plant; (*pl.*) the grain of this, used for food. **to get one's oats**, (*sl.*) to have (regular) sexual intercourse. **to sow one's wild oats**, to indulge in youthful (esp. sexual) excess. **wild oats**, a tall grass of the same genus as the oat-plant; youthful dissipation. **oatcake**, *n.* a flat cake or biscuit made from oatmeal. **oatmeal**, *n.* oats ground into meal, used chiefly for making porridge or oatcakes; a fawny-grey colour (as porridge). **oaten**, *a.*
oath (ōth), *n.* (*pl.* **oaths** (ōdhz)) (the form of) a solemn swearing by God or some holy or revered person or thing in witness of the truth of a statement or of the binding nature of a promise, esp. in a court of law; a profanity, an expletive, a blasphemy, a curse. **on, under, upon oath**, sworn to tell the truth; having taken an oath. **to take an oath**, to swear formally to the truth of one's statements. **oath-breaking**, *n.*
oatmeal OAT.
OAU, (*abbr.*) Organization of African Unity.
OB, (*abbr.*) old boy; outside broadcast.
ob., (*abbr.*) *obiit*, he/she died.
ob-, *pref.* toward, to, in, facing; against, opposing, hostile; reversely, obversely, contrary to the normal; as in *object n.*, *objurgate, oblique, obovate*.
Obad., (*abbr.*) Obadiah.
obang (ō'bang), *n.* an oblong gold coin formerly current in Japan.
obbligato (obligah'tō), *a.* (*Mus.*) not to be omitted; indispensable to the whole. *n.* an instrumental part or accompaniment that forms an integral part of the composition or is independently important (usu. by a single instrument).
obdurate (ob'dūrət), *a.* hardened in attitude, esp. against moral influence; stubborn. *v.t.* (-rāt) to make obdurate. **obduracy**, *n.* **obdurately**, *adv.* **obduration**, *n.*
OBE, (*abbr.*) (Officer of the) Order of the British Empire.
obeah (ō'biə), **obi²** (ō'bi), *n.* a form of sorcery practised by some Blacks, esp. in the W Indies; a magical charm.
obedience (əbēd'yəns), *n.* the act or practice of obeying; dutiful submission to authority; compliance with law, command or direction. **obedient**, *a.* **obediently**, *adv.*
obeisance (əbā'səns), *n.* a bow, a curtsy, or any gesture signifying deference, submission, respect or salutation; homage. **obeisant**, *a.*
obelisk (ob'əlisk), *n.* a four-sided, erect stone shaft, usually monolithic and tapering, with a pyramidal apex; an obelus. **obeliscal** (-lis'-), **obeliscoid** (-lis'koid), *a.*
obelus (ob'ələs), *n.* (*pl.* **-li** (-lī)) a mark (−, ÷, or †), used to mark spurious or doubtful passages in ancient MSS; in printing, a dagger symbol (†) indicating a cross-reference, footnote or death date. Also **double obelisk** (‡).
obese (əbēs'), *a.* excessively fat. **obeseness, obesity**, *n.*
obey (əbā'), *v.t.* to perform or carry out (a command or instruction); to be obedient to (a person); to act according to (a law etc.). *v.i.* to do what is directed or commanded. **obeyer**, *n.*
obfuscate (ob'fəskāt), *v.t.* to obscure, to confuse (an issue, etc.). **obfuscation**, *n.* **obfuscatory**, *a.*
obi¹ (ō'bi), *n.* (*pl.* **obi, obis**) a coloured sash worn around a Japanese kimono.
obi² OBEAH.
obit (ob'it), *n.* a memorial service or commemoration of a death; (*coll.*) short for obituary. **obituarist**, *n.* **obituary** (əbit'ū-), *a.* relating to or recording a death or deaths. *n.* a notice of a death, usu. in the form of a short biography of the deceased.
obiter dictum (ob'itə dik'təm), *n.* (*pl.* **obiter dicta** (-tə)), a passing remark.
object¹ (əbjekt'), *v.i.* to express opposition (to); to allege (a fact; usu. with *that*) in criticism, disapproval, or condemnation.
object² (ob'jikt), *n.* anything presented to the senses or the mind, esp. anything visible or tangible; a person or thing to which an action or feeling is directed; an aim, end, ultimate purpose; a noun, or word, phrase or sentence equivalent to a noun, governed by a transitive

verb or preposition; (*coll.*) a person or thing of pitiable or ridiculous appearance. **no object**, no obstacle or problem. **object-ball**, *n.* in billiards etc., the ball at which a player aims with the cueball. **object-finder**, *n.* an eye-piece on a microscope for enabling one to locate the object on a slide; an analogous device on a large telescope. **object-glass**, *n.* the lens or combination of lenses at the end of an optical instrument nearest the object. **object-lesson**, *n.* a lesson in which the actual object described or a representation of it is used for illustration; an excellent practical illustration. **objectification** (-fi-), *n.* **objectify** (-jek'tifi), *v.t.* to render objective. **objection** (-jek'-), *n.* the act of objecting; an adverse argument, reason or statement; disapproval, dislike, or the expression of this. **objectionability** (-bil'-), *n.* **objectionable**, *a.* liable to objection; offensive, unpleasant. **objectionableness**, *n.* **objectionably**, *adv.* **objective** (-jek'-), *a.* actual, real, substantive; as distinct from the subjective, uninfluenced by emotion, impulse, prejudice etc.; denoting the case of the object of a transitive verb or preposition. *n.* an aimed-for point, e.g. of military operations; the objective case; an object-glass. **objective case**, *n.* the case in English of the object of a transitive verb or a preposition. **objectival** (objiktī'-), *a.* **objectively**, *adv.* **objectiveness**, *n.* **objectivism**, *n.* the tendency to give priority to what is objective; the theory that stresses objective reality. **objectivist**, *n.* **objectivistic** (-vis'-), *a.* **objectivity** (objiktiv'-), *n.* **objectivize**, **-ise**, *v.t.*, *v.i.* **objectless** (ob'-), *a.* **objector** (-jek'-), *n.* one who objects. **conscientious objector** CONSCIENTIOUS.
objet (ob'zhā), *n.* (*pl.* **objets** (-zhā)) an object. **objet d'art** (dah), *n.* (*pl.* **objets d'art**) an object of artistic merit or value.
oblast (ob'lahst), *n.* (*pl.* **-lasts**, **-lasti** (-sti)) an administrative district or province in the USSR.
oblate[1] (ob'lāt), *a.* of a spheroid, flattened at the poles. **oblately**, *adv.* **oblateness**, *n.*
oblate[2] (ob'lāt), *n.* one not under vows but dedicated to monastic or religious life or work, esp. one of a congregation of secular priests or sisters who live in community. **oblation**, *n.* the act of offering or anything offered in worship such as the bread and wine of the Eucharist; a religious sacrifice. **oblational**, **oblatory**, *a.*
obligation (obligā'shən), *n.* the binding power of a promise, contract, vow, duty, law etc.; a duty, responsibility, commitment; (*Law*) a binding agreement, esp. one with an attendant penalty. **under an obligation**, indebted for some benefit, favour etc. **obligate** (ob'-), *v.t.* to place under an obligation, legal or moral. **obligatory** (əblig'ə-), *a.* mandatory.
obligato (obligah'tō) OBBLIGATO.
oblige (əblīj'), *v.t.* to bind or constrain by legal or physical force or by moral debt; to do a favour to. *v.i.* to perform a favour or task. **obliged**, *a.* under a legal or moral obligation (to someone, for something). **obliger**, *n.* **obliging**, *a.* kind, helpful, accommodating. **obligingly**, *adv.* **obligingness**, *n.*
oblique (əblēk'), *a.* slanting, deviating from the vertical or horizontal; deviating from the straight or direct line, roundabout; evasive, not to the point; (inclined at an angle) other than a right angle. *n.* acute or obtuse; *n.* that which is at an angle, slanting, esp. a geometric figure, military advance; a solidus. **oblique angle**, an angle greater or less than a right angle. **obliquely**, *adv.* **obliqueness**, *n.* **obliquity** (əblik'wi-), *n.* obliqueness; (an instance of) mental or moral deviation.
obliterate (əblit'ərāt), *v.t.* to erase; to destroy; to reduce to an illegible or imperceptible state. **obliteration**, *n.* **obliterative**, *a.* **obliterator**, *n.*
oblivion (əbliv'iən), *n.* forgetfulness, unconsciousness, unawareness; the state of being forgotten; (*coll.*) nothingness. **oblivious**, *a.* forgetful, unaware (of); lost in thought. **obliviously**, *adv.* **obliviousness**, *n.*
oblong (ob'long), *a.* of greater length or breadth than height esp. of rectangles with adjoining sides unequal; (of leaves) elliptical. *n.* an oblong figure or object.
obloquy (ob'ləkwi), *n.* abusive language; discredit, disgrace.
obnoxious (obnok'shəs), *a.* offensive, objectionable. **obnoxiously**, *adv.* **obnoxiousness**, *n.*
oboe (ō'bō), *n.* a woodwind instrument with a double reed, usu. of soprano pitch; a person who plays this instrument in an orchestra; a reed organ-stop of similar tone. **oboist**, *n.*
obs., (*abbr.*) obsolete.
obscene (əbsēn'), *a.* repulsive, disgusting, indecent, lewd; in law, that which is liable to corrupt or deprave. **obscenely**, *adv.* **obscenity** (-sen'-), *n.*
obscurant (əbskū'rənt), *n.* an opponent of intellectual progress. **obscurantism**, *n.* **obscurantist**, *n.*, *a.*
obscure (əbskūə'), *a.* dark, dim; not clear, indistinct, unexplained; abstruse, esoteric; hidden, secluded, remote from public observation; unknown, lowly, humble. *v.t.* to make dark, to cloud; to make less intelligible, visible or legible; to dim, to overshadow, to outshine; to conceal. **obscurely**, *adv.* **obscurity**, *n.* the quality or state of being obscure; an obscure person or thing.
obsecration (obsikrā'shən), *n.* the act of imploring, entreaty.
obsequies (ob'sikwiz), *n.pl.* funeral rites. **obsequial** (-sē'-), *a.*
obsequious (əbsē'kwiəs), *a.* servile, fawning, over-ready to comply with the desires of others. **obsequiously**, *adv.* **obsequiousness**, *n.*
observe (əbzœv'), *v.t.* to regard attentively, to watch; to scrutinize; to note, to perceive; to examine and note scientifically; to heed; to perform (duties etc.); to comply with; to celebrate (religious festivals etc.); to remark, to express as an opinion. *v.i.* to make a remark or remarks (upon); to watch etc. **observable**, *a.*, *n.* **observableness**, *n.* **observably**, *adv.* **observance**, *n.* the act of complying with, celebrating, performing etc.; a customary rite, form or ceremony; a rule or practice, esp. in a religious order. **observant**, *a.* watchful, attentive; quick or strict in observing, esp. rules etc. **observantly**, *adv.* **observation** (ob-), *n.* the act, habit or faculty of observing; scientific watching and noting of phenomena as they occur, as dist. from experiment; the result of such a scrutiny, a fact scientifically noted or taken from an instrument; experience and knowledge gained by systematic observing; a remark, an incidental comment or expression of opinion or reflection. **observational**, *a.* **observationally**, *adv.* **observation car**, *n.* a railway carriage designed so that one can view passing scenery. **observation-post**, *n.* a post from which an observer can watch the effect of artillery fire. **under observation**, being watched carefully, undergoing scrutiny. **observatory**, *n.* an institution, building, room etc. for observation of astronomical or meteorological phenomena. **observer**, *n.* one who observes; an official looker-on at proceedings; formerly, the member of an aircraft's crew employed on reconnaissance. **observedly** (-vid-), *adv.*
obsess (əbses'), *v.t.* to haunt, to preoccupy the mind of (as a fixed idea). **obsession** (-shən), *n.* (the condition of having) an unhealthily deep-rooted or persistent fixation. **obsessional**, *a.* **obsessionally**, *adv.* **obsessive**, *n.* **obsessively**, *adv.* **obsessiveness**, *n.*
obsidian (obsid'iən), *n.* a black or dark-coloured vitreous lava.
obsolescent (obsəles'ənt), *a.* becoming obsolete; gradually disappearing, going out of use. **obsolescence**, *n.* **built-in**, **planned obsolescence**, the prearranged demise of a commodity through deterioration or supersedence by a newer model. **obsolete** (ob'səlēt), *a.* no longer used, practised, current or accepted. **obsoletely**, *adv.* **obsolete-**

ness, obsoletism, *n.*
obstacle (ob'stəkl), *n.* an impediment, an obstruction.
obstacle-race, *n.* a race in which the competitors have to surmount or avoid a series of natural or artificial obstacles.
obstetric (əbstet'rik), **-ical,** *a.* of or pertaining to childbirth or obstetrics; (*pl.*) the branch of medical science dealing with childbirth and ante- and postnatal care of women. **obstetrically,** *adv.* **obstetrician** (obstətrish'-), *n.*
obstinate (ob'stinət), *a.* tenaciously adhering to one's opinion or purpose, stubborn, headstrong; not easily remedied, persistent. **obstinacy, obstinateness,** *n.* **obstinately,** *adv.*
obstreperous (əbstrep'ərəs), *a.* noisy, boisterous, unruly. **obstreperously,** *adv.* **obstreperousness,** *n.*
obstruct (əbstrŭkt'), *v.t.* to block up, to close by means of obstacles; to hinder, to impede, to stop. **obstructer,** *n.* **obstruction,** *n.* **obstructional,** *a.* **obstructive,** *a.*, *n.* (someone or something) causing or tending to cause obstruction. **obstructively,** *adv.* **obstructiveness,** *n.* **obstructor,** *n.*
obtain (əbtān'), *v.t.* to gain, to acquire, to get; to attain, to reach. *v.i.* to be prevalent or accepted, to be in common use. **obtainable,** *a.* **obtainability** (-bil'-), *n.* **obtainer,** *n.* **obtainment,** *n.*
obtrude (əbtrood'), *v.t., v.i.* to thrust out, forward or upon; to introduce or thrust in without warrant or invitation. **obtruder,** *n.* **obtrusion** (-zhən), *n.* **obtrusive,** *a.* **obtrusively,** *adv.* **obtrusiveness,** *n.* a thrusting forward, a desire to be noticed; undue prominence.
obtuse (əbtūs'), *a.* blunt or rounded, not pointed or acute; denoting an angle greater than a right angle; dull, stupid, slow in comprehending. **obtuse-angled,** *a.* **obtusely,** *adv.* **obtuseness,** *n.*
obverse (ob'vœs), *a.* turned towards one; of leaves, having a base narrower than the apex. *n.* the face or front; the side of a coin or medal bearing the main design; the counterpart or complementary side or aspect (of a statement, fact etc.). **obversely** (-vœs'-), *adv.* **obvert** (obvœt'), *v.t.* to turn the front towards one.
obviate (ob'viāt), *v.t.* to overcome, to remove, counteract or neutralize (as dangers, difficulties etc.). **obviation,** *n.*
obvious (ob'viəs), *a.* plain to the eye, immediately evident; unsubtle. **the obvious,** what is obvious, needing no explanation. **obviously,** *adv.* **obviousness,** *n.*
Oc., (*abbr.*) Ocean.
OC, (*abbr.*) Officer Commanding.
oc-, *pref.* OB- (before *c*).
ocarina (okərē'nə), *n.* a musical instrument of terracotta with finger-notes and mouthpiece, giving a mellow whistling sound.
Occam's, Ockham's razor (ok'əmz), *n.* the philosophic principle (taught by William of *Occam* or *Ockham,* c.1270–c.1349) that interpretations should include as few assumptions as possible.
occas., (*abbr.*) occasional(ly).
occasion (əkā'zhən), *n.* an event, circumstance or time of affairs, giving an opportunity, reason, or motive for doing something; motive, reason, need; a time or event, esp. of special interest or importance. *v.t.* to cause, directly or indirectly; to induce; to influence. **as the occasion arises,** when necessary, circumstances demand. **on occasion,** now and then. **to rise to the occasion,** to be equal to a demanding event or situation. **to take (the) occasion to,** to take the opportunity to. **occasional,** *a.* happening, made, employed or done as opportunity arises; irregular, infrequent, incidental, casual; of or made for a special occasion. **occasional table,** *n.* a small moveable table (used as occasion demands). **occasionally,** *adv.*
Occident, occident (ok'sidənt), *n.* the west; western Europe, Europe and America; the western hemisphere. **Occidental, occidental** (-den'-), *a.* western; characteristic of western culture, thoughts etc. *n.* a westerner. **occidentally,** *adv.*
occiput (ok'sipət), *n.* the back part of the head. **occipital** (-sip'i-), *n.* short for **occipital bone,** the bone forming the back and part of the base of the skull. **occipitally,** *adv.*
occlude (əklood'), *v.t.* to shut or stop up; to close, to bring together (the eyelids, the teeth); to (cause to) form an occlusion; of metals etc., to absorb (a gas). **occluded front,** *n.* (*Meteor.*) an occlusion. **occlusal** (-zəl), *a.*, relating to dental occlusion. **occlusion** (-zhən), *n.* a shutting or stopping up; in meteorology, the closing of a cold front upon a warm one (which is narrowed and raised up), an occluded front; (*Chem.*) absorption; the manner in which teeth come together, the bite; (*Phon.*) closure of the breath passage prior to articulating a sound. **occlusive** (-siv), *n.*, *a.* (*Phon.*) (of) a sound made when the breath passage is completely closed.
occult (əkŭlt'), *a.* concealed, kept secret, esoteric; supernatural, magical, mystical. **the occult,** that which is hidden, mysterious, magical; the supernatural. **occultism,** *n.* **occultist,** *n.* **occultly,** *adv.* **occultness,** *n.*
occupant (ok'ūpənt), *n.* one who occupies; one who resides or is in a place; a tenant in possession as dist. from an owner; one who establishes a claim by taking possession. **occupancy,** *n.* **occupation** (oküpā'shən), *n.* the act of occupying or taking possession (e.g. of a country by a foreign army); the state of being occupied; occupancy, tenure; employment, business, trade, calling, job. **occupational,** *a.* caused by or related to employment. **occupational disease, hazard,** *n.* a disease, injury, risk etc. resulting directly from, or common to, an occupation. **occupational therapy,** *n.* treatment of various illnesses by providing a creative occupation or hobby. **occupational therapist,** *n.* **occupationally,** *adv.* **occupier,** *n.* an occupant of a house etc.
occupy (ok'ūpī), *v.t.* to take possession of; to hold in possession, to be the tenant of; to reside in, to be in; to take up, to fill; to employ, to engage (often in p.p.).
occur (əkœ'), *v.i.* (*past, p.p.* **occurred**) to happen, to take place; to be found, to exist; to present itself (often suddenly) to the mind. **occurrence** (əkŭ'rəns), *n.* an event, an incident; the taking place of anything.
ocean (ō'shən), *n.* the vast body of water covering about two-thirds of the surface of the globe; any one of its principal divisions, the Antarctic, Atlantic, Arctic, Indian and Pacific Oceans; the sea; an immense expanse or quantity. *a.* pertaining to the ocean. **ocean-going,** *a.* (suitable for) travelling on the ocean. **Oceania** (ōshiah'niə), *n.* the island-region of the Pacific and adjoining seas. **Oceanian,** *a.* **oceanic** (ōshian'-), *a.* of, pertaining to, occurring in, or like the ocean; pertaining to Oceania. **oceano-,** *comb. form.* **oceanographer,** *n.* **oceanographic, -ical** (-graf'-), *a.* **oceanography** (-nog'-), **oceanology** (-nol'-), *n.* the branch of science concerned with oceans and their biological, geographic, chemical features etc.
ocellus (əsel'əs), *n.* (*pl.* **-lli** (-lī)) a simple eye, as opposed to the compound eye of insects; a spot or ring of colour, as on feathers, wings etc. **ocellar, ocellate** (-lāt), **-ted,** *a.*
ocelot (ō'səlot), *n.* a small American feline, the tiger-cat or leopard-cat; its fur.
och (okh), *int.* (*dial.*) expressing impatience, contempt, regret, surprise etc.
oche (ok'i), *n.* in darts, the line or mark behind which a player must stand when throwing at the dartboard.
ochre (ō'kə), *n.* (the colour of) a native earth consisting of hydrated peroxide of iron with clay in various proportions, used as a red or yellow pigment.
-ock (-ək), *suf.* indicating smallness or youngness, as in *bullock, hillock.*
ocker (ok'ə), *n.*, *a.* (*Austral. sl.*) (characteristic of) a boorish, chauvinistic Australian.

Ockham's razor OCCAM'S RAZOR.
o'clock CLOCK [1].
Oct., (*abbr.*) October.
oct., (*abbr.*) octavo.
oct(a)-, octo-, *comb. form.* having eight; consisting of eight.
octad (ok'tad), *n.* a group or series of eight. **octadic** (-tad'-), *a.*
octagon (ok'təgən), *n.* a plane figure of eight sides and angles; any object or building of this shape. **octagonal** (-tag'-), *a.* **octagonally,** *adv.*
octahedron (oktəhē'drən), *n.* (*pl.* **-dra** (-drə), **-drons**) a solid figure contained by eight plane faces. **regular octahedron,** one contained by eight equal equilateral triangles. **octahedral,** *a.*
octal (ok'təl), *a.* referring to or based on the number eight.
octameter (oktam'itə), *n.* (*poet.*) a line of eight metrical feet.
octane (ok'tān), *n.* a colourless liquid hydrocarbon that occurs in petroleum. **octane number, rating,** *n.* a percentage measure of the anti-knock quality of a liquid motor fuel.
octangular (oktang'gūlə), *a.* having eight angles.
octaroon OCTOROON.
octastyle, octostyle (ok'təstīl), *n., a.* (a building) having eight columns in front.
octateuch (ok'tətūk), *n.* the first eight books of the Old Testament.
octave (ok'tiv), *n.* the interval between any musical note and that produced by twice or half as many vibrations per second (lying eight notes away inclusively); a note at this interval above or below another; two notes separated by this interval; the scale of notes filling this interval; any group of eight, as the first eight lines of a sonnet, or a stanza of eight lines.
octavo (oktā'vō), *n.* (*pl.* **-vos**) a book in which a sheet is folded into 8 leaves or 16 pages; the size of such a book or paper (written *8vo*). *a.* of this size, having 8 leaves per sheet.
octennial (okten'iəl), *a.* recurring every eighth year; lasting eight years. **octennially,** *adv.*
octet (oktet'), *n.* a musical composition of eight parts or for eight instruments or singers; a group of eight, esp. musicians or singers; an octave (of verse).
octile (ok'tīl), *n.* an eighth part.
octillion (oktil'yən), *n.* the number produced by raising a million to the eighth power, represented by 1 followed by 48 ciphers; (*N Am.* and *Fr.*) the eighth power of a thousand, 1 followed by 27 ciphers. **octillionth,** *a., n.*
oct(o)- OCT(A)-.
October (oktō'bə), *n.* the 10th month of the year (the eighth of the Roman year). **Octobrist,** *n.* a member of a moderate reforming party in Czarist Russia (after the Czar's liberal manifesto pubd. Oct. 1905).
octocentenary (oktōsentē'nəri, -ten'-), *n.* an 800th anniversary. **octocentennial** (-ten'iəl), *n.*
octodecimo (oktōdes'imō), *n.* (*pl.* **-mos**) a book having 18 leaves or 36 pages to the sheet; the size of such a book (written *18mo*). *a.* of this size, having 16 leaves per sheet.
octogenarian (oktōjənea'riən), *n.* one who is 80, or between 80 and 90, years old.
octoped (ok'təped), *n.* an eight-footed animal. **octopetalous** (oktōpet'ələs), *a.* having eight petals.
octopod (ok'təpod), *n., a.* (an animal) having eight feet. **octopodous** (-top'-), *a.*
octopus (ok'təpəs), *n.* (*pl.* **octopuses**) one of a genus of molluscs having eight arms furnished with suckers, a cuttlefish; (*fig.*) an organization or influence having far-extending powers (for harm).
octoroon, octaroon (oktəroon'), *n.* the offspring of a quadroon and a white person (having one-eighth Black ancestry).

octosepalous (oktōsep'ələs), *a.* having eight sepals.
octosyllabic (oktōsilab'ik), *a.* having eight syllables. *n.* a line of eight syllables. **octosyllable** (ok'-), *n.* a word of eight syllables.
octuple (ok'tūpl), *a.* eightfold. *n.* the product of multiplication by eight. *v.t., v.i.* to multiply by eight. **octuplet** (-plit), *n.* one of eight offspring produced at one birth.
ocular (ok'ūlə), *a.* of, pertaining to, by or with the eye or eyes, visual; known from actual sight. *n.* an eyepiece. **ocularist,** *n.* a maker of artificial eyes. **ocularly,** *adv.* **oculate** (-lət), **-ated** (-lātid), *a.* having eye-like spots. **oculist,** *n.* a former name for an ophthalmologist or optician. **oculo-,** *comb. form.*
OD [1], (*abbr.*) Officer of the Day; Old Dutch; ordinance data, datum; outside diameter.
OD [2], *n.* (*coll.*) an overdose of a drug. *v.i.* (*pres.p.* **OD'ing,** *past, p.p.* **OD'd**) to take an overdose (of a drug). [*over dose*]
O/D, o/d, (*abbr.*) on demand; overdraft; overdrawn.
odalisque, odalisk (ō'dəlisk), *n.* an Oriental female slave or concubine, esp. in a harem.
odd (od), *a.* remaining after a number or quantity has been divided into pairs; not even; not divisible without remainder by two; bearing such a number; wanting a match or pair; singular, strange, eccentric; occasional, casual; (*ellipt.*) and more, with others thrown in (added to a round number in enumeration, as *two hundred odd*). **odd man out,** one who is left when a number pair off; one who is at variance with, excluded from, or stands out as dissimilar to a group etc. **oddball,** *n., a.* (*coll.*) (one who is) eccentric or peculiar. **odd job,** *n.* a casual, irregular or occasional piece of work, esp. a domestic repair. **Oddfellows,** *n.* a member of an 18th-cent. Friendly Society known as the Order of Oddfellows. **odd-looking,** *a.* **oddish,** *a.* **oddity,** *n.* oddness; a peculiar feature or trait; an odd person or thing. **oddly,** *adv.* **oddment,** *n.* a remnant. *n.pl.* odds and ends. **oddness,** *n.* **odds,** *n.pl.* (*usu. as sing.*) balance of superiority, advantage; the chances (in favour of or against a given event). **at odds,** at variance; in dispute. **odds and ends, odds and sods,** miscellaneous remnants, trifles, scraps etc. **over the odds,** higher, more than is acceptable, necessary, usual etc. **to make no odds,** to make no difference, to not matter. **to shout the odds,** (*coll.*) to talk loudly, stridently, vehemently etc. **what's the odds?** (*coll.*) what difference does it make? **odds-on,** *a.* having a better than even chance (to happen, win, succeed etc.).
ode (ōd), *n.* formerly, a lyric poem meant to be sung; a lyric poem in an elevated style, rhymed or unrhymed, of varied and often irregular metre, usu. in the form of an address or invocation.
-ode [1] (-ōd), *suf.* denoting a thing resembling or of the nature of, as *geode, sarcode.*
-ode [2] (-ōd), *suf.* denoting a path or way, as *anode, cathode, electrode.*
odeon (ō'diən), **odeum** (-əm), *n.* (*pl.* **odea** (-ə)) in ancient Greece or Rome, a theatre in which poets and musicians contended for prizes; hence, a concert-hall, a theatre used for musical performances; (a name for) a cinema.
odious (ō'diəs), *a.* hateful, repulsive; offensive. **odiously,** *adv.* **odiousness,** *n.* **odium** (-əm), *n.* general dislike, disfavour; blame; repulsion.
odometer (ōdom'itə), *n.* an instrument attached to a vehicle for measuring and recording the distance travelled. **odometry,** *n.*
-odont (-ədont), *comb. form.* -toothed.
odont(o)-, *comb. form.* having teeth, or processes resembling teeth.
odontoid (odon'toid), *a.* toothlike; of or relating to an odontoid process. **odontoid process,** *n.* a toothlike projection from the axis or second cervical vertebra in certain mammals and birds.

odour, (*N Am.*) **odor** (ō'də), *n.* a smell, whether pleasant or unpleasant; scent, fragrance; (*fig.*) repute, esteem. **in bad, good odour,** in, out of, favour. **odorant, odoriferous** (-rif'-), *a.* diffusing fragrance; (*coll.*) (unpleasantly) smelly. **odoriferously,** *adv.* **odoriferousness,** *n.* **odorizer, -iser,** *n.* **odorous,** *a.* having a sweet scent, fragrant. **odorously,** *adv.* **odorousness,** *n.* **odoured, odourless** (-lis), *a.*

Odyssey (od'isi), *n.* an epic poem attributed to Homer, describing the wanderings of Ulysses after the fall of Troy; (*also* **odyssey**; a story of) a long, eventful journey containing a series of adventures and difficulties. **Odyssean** (-sē'ən), *a.*

OE, (*abbr.*) Old English.

OECD, (*abbr.*) Organization for Economic Cooperation and Development.

oecumenical ECUMENICAL.

OED, (*abbr.*) Oxford English Dictionary.

oedema, (*esp. N Am.*) **edema** (idē'mə), *n.* (*Path.*) swelling due to accumulation of serous fluid in the cellular tissue; dropsy; in plants, swelling due to water accumulation in the tissues. **oedematose** (-tōs), **-tous, edematose, -tous,** *a.* **oedematously,** *adv.*

Oedipus complex (ēd'ipəs), *n.* a psychical impulse in offspring comprising excessive love or sexual desire for the parent of the opposite sex and hatred for the parent of the same sex. **oedipal, oedipean** (-pē'ən), ELECTRA COMPLEX.

oen(o)-, oin(o)-, (*esp. N Am.*) **en(o)-,** *comb. form.* wine.

oenology, (*esp. N Am.*) **enology** (ēnol'əji), *n.* the science or study of wines. **oenological, enological** (-loj'-), *a.* **oenologist, enologist,** *n.*

oenomania (ēnōmā'niə), *n.* dipsomania; mania due to intoxication. **oenomaniac** (-ak), *n.*

oenometer (ēnom'itə), *n.* an instrument for testing the alcoholic strength of wines.

oenophil(e) (ēnəfil, -fīl), **oenophilist** (-nof'i-), *n.* a wine connoisseur. **oenophily** (-nof'ili), *n.* love or knowledge of wines.

o'er (aw), *prep.*, *adv.* (*poet.*) short for OVER.

oesophagectomy (ēsofəjek'təmi), *n.* excision of part of the oesophagus.

oesophagitis (ēsofəji'tis), *n.* inflammation of the oesophagus.

oesophag(o)-, (*esp. N Am.*) **esophag(o)-,** *comb. form.* oesophagus.

oesophagotomy (ēsofəgot'əmi), *n.* the operation of opening the oesophagus.

oesophagus, (*esp. N Am.*) **esophagus** (ēsof'əgəs), *n.* (*pl.* **-gi** (-gī)) the gullet, the canal by which food passes to the stomach. **oesophageal, esophageal** (-jē'əl), *a.*

oestradiol (ēstrədī'ol, es-), *n.* the major oestrogenic hormone in human females (used to treat breast cancer and oestrogen deficiency).

oestriol (ēs'triol, es'-), *n.* an oestrogenic hormone often used to treat menopausal symptoms.

oestrogen, (*esp. N Am.*) **estrogen** (ēs'trəjən, es'-), *n.* any of the female sex hormones which induce oestrus and encourage the growth of female secondary sexual characteristics. **oestrogenic** (ēstrəjen'ik, es-), *a.* **oestrogenically,** *adv.*

oestrus, (*esp. N Am.*) **estrus** (ēs'trəs, es'-), **-trum** (-trəm), *n.* a violent impulse, a raging desire or passion, esp. the sexual impulse in animals; the period when this regularly occurs; heat. **oestral,** *a.* **oestrous,** *a.* **oestr(o)us cycle,** *n.* the hormonal changes occurring in a female mammal from one oestrus to the next.

OF, (*abbr.*) Old French.

of (ov, əv), *prep.* denoting connection with or relation to, in family (*son of David*), point of departure (*relieve of one's duties/wallet*), separation, origin (*Henry VIII of England*), motive or cause (*died of old age*) agency (*works of Shakespeare*), substance or material (*made of cotton*), possession (*owner of a car*): inclusion, partition (*part of my collection*; *1lb. of apples*); equivalence or identity (*city of Leeds*): with reference, regard to (*full of ideas*; *dislike of liars*); quality, condition (*man of scholarship/fifty*).

of-, *pref.* OB- (before *f,* as in offence).

off (of), *adv.* away, at a distance or to a distance in space or time, expressing removal (*move off*), separation (*close off*), suspension (*hold off*), discontinuance (*shut off*), decay (*go off*) or termination; to the end, utterly, completely (*finish off*). *prep.* from, denoting deviation (*turn off a road*), distance (*a mile off the coast*), disjunction (*jump off a cliff*), removal (*pull off a sock*), etc. *a.* more distant, farther, opp. to *near*; (of a vehicle, etc.) right, opp. to *left*; (*coll.*) unacceptable, unfair; on bad form (as an *off day*); applied to that part of a cricket field to the left side of the bowler when the batsman is right-handed, or vice versa. *n.* the off side of a cricket field; the beginning, start. *v.i.* (*coll.*) to go away. *v.t.* (*sl.*) to kill (someone). *int.* away! begone! **badly off,** in bad circumstances; poor. **in the offing,** likely to occur soon. **off and on,** intermittently, now and again. **to be off,** to leave. **to be off one's head,** to lose one's reason. **to come off** COME. **to go off** GO¹. **to take off,** to divest oneself of; to mimic; of a plane, to leave the ground. **to tell off** TELL. **well off** WELL¹. **offbeat,** *a.* (*coll.*) unconventional, unusual. **off-Broadway,** *a.* (of a theatre) situated outside the Broadway theatrical area of New York; relating to New York fringe theatre, experimental, low-cost, non-commercial etc. **off-off-Broadway,** *n.*, *a.* **off-centre,** *a.*, *adv.* not (quite) central(ly). **off-chance,** *n.* a remote possibility. **on the off chance,** in the slim hope that, just in case. **off-colour** COLOUR¹. **offcut,** *n.* a section cut off the main piece of a material (fabric, meat, paper, wood etc.). **off-drive,** *n.* in cricket, a drive to the off side. **offhand,** *adv.* without consideration, preparation, or warning; casually; summarily, curtly, or brusquely. *a.* impromptu; casual; summary, curt or brusque. **offhanded,** *a.* **offhandedly,** *adv.* **offhandedness,** *n.* **off-key,** *a.*, *adv.* (*Mus.*) not in tune; (*fig.*) not entirely suitable (for the occasion). **off-licence,** *n.* a shop holding a licence to sell intoxicating liquors to be consumed off the premises. **off-limits,** *a.* out of bounds; forbidden. **off-line,** *a.* of a computer peripheral, switched off, not under the control of a central processor. **off-load,** *v.i.*, *v.t.* to unload; to get rid of something (on to someone else). **off-peak,** *a.* of a service, (supplied) during a period of low demand. **off-print,** *n.* a reprint of an article or separate part of a periodical etc. **off-putting,** *a.* disconcerting; unattractive, displeasing. **off-screen,** *adv.*, *a.* (appearing or happening) out of sight of the viewer of a film, television programme etc. **off-season,** *a.*, *adv.* out of high season. *n.* a period of low (business) activity. **off-set,** *n.* a lateral shoot or branch that takes root and is used for propagation; anything allowed as a counterbalance, equivalent, or compensation; an amount set off (against); a bend or fitting bringing a pipe past an obstacle. *v.t.* to balance by an equivalent; to counterbalance, to compensate. **offset printing,** *n.* a (lithographic) printing process in which the image is first transferred from a plate on to a cylinder before it is printed on to paper. **off-shoot,** *n.* a branch or shoot from a main stem; a side-issue. **off-shore,** *a.*, *adv.* (blowing) off the land; (situated) a short way from the land. **offside,** *n.* a foul caused by a player being offside. *a.*, *adv.* in(to) of a player, a position to kick a goal when no defender is between (the attacker) and the goalkeeper. **offspring,** *n.* issue, progeny; (*collect.*) children, descendants; a child; (*fig.*) a production or result of any kind. **off-stage,** *a.*, *adv.* out of sight of the theatre audience. **off-the-cuff,** *a.*, *adv.* spontaneous(ly); impromptu. **off-the-peg,** *a.* ready-made. **off-the-record,** *a.*, *adv.* unofficial(ly), confidential(ly). **off-the-wall,** *a.* eccentric, bizarre. **off-white,** *a.* not pure white.

offal (of'əl), *n.* parts of the carcass of an animal sometimes regarded as waste, including the head, tail, kidneys,

offend (əfend'), *v.t.* to wound the feelings of, to hurt; to make angry, to cause displeasure or disgust in. *v.i.* to transgress or violate any human or divine law; to cause anger, distaste etc., to scandalize. **offence** (əfens'), (*N Am.*) **offense**, *n.* the act of offending; an offensive act; an affront, an insult; a breach of custom, a transgression, a sin; an illegal act (*N Am.*; *usu.* (of'-)) the attacking part of a team (as dist. from *defensive*). **to give offence,** to cause umbrage, to affront or insult. **to take offence,** to be offended, to feel a grievance. **offenceless** (-lis), *a.* **offendedly,** *adv.* **offender,** *n.* **offensive,** *a.* pertaining to or used for attack, aggressive; causing or meant to cause offence; disgusting, disagreeable, repulsive. *n.* (*Mil.*) a strategic attack. **to be, go on, take the offensive,** to go onto the attack. **offensively,** *adv.* **offensiveness,** *n.*
offer (of'ə), *v.t.* to present as an act of worship; to sacrifice, to immolate; to present, to put forward, to tender for acceptance or refusal; to bid (as a price); to evince readiness (to do something); to proffer, to show (e.g. for sale); to present or show (itself). *n.* an act of offering; an expression of willingness or readiness (to); a tender, or proposal, to be accepted or refused; a price or sum bid. **on offer,** presented for sale, consumption etc., esp. at a bargain price. **special offer,** an article or service proffered at a bargain price. **under offer,** provisionally sold, subject to the signing of a contract. **offerer,** *n.* **offering,** *n.*
offertory (-təri), *n.* that part of the Mass or liturgical service during which offerings or donations are made; in the Church of England, an anthem sung or the text spoken while these are being made; the gifts offered; any collection made at a religious service.
office (of'is), *n.* duty, charge, function, the tasks or services attaching to a particular post or job; a post of service, trust or authority, esp. under a public body; an act of worship of prescribed form; a room, building or other place where business is carried on; (*collect.*) persons charged with such business, the official staff or organization as a whole; a government department or agency. **divine office** DIVINE². **holy office** HOLY. **office-bearer,** *n.* person who holds (an) office. **office-block,** *n.* a large building containing offices. **office-boy, -girl,** *n.* one employed to perform minor tasks in an office. **officer,** *n.* person holding a post or position of authority, esp. a government functionary, one elected to perform certain duties by a society, committee etc., or appointed to a command in the armed services or merchant navy; a policeman. **officer of the day,** *n.* an army officer responsible for security on a particular day. **official** (əfish'əl), *a.* of or pertaining to an office or public duty; holding an office, employed in public duties; derived from or executed under proper authority; duly authorized; characteristic of persons in office. *n.* person who holds a public or committee post; a judge, or a presiding officer in an ecclesiastical court. **Official Receiver,** *n.* an officer appointed to administer a bankrupt's estate. **officialdom** (-dəm), *n.* **officialese** (-ēz'), *n.* official jargon. **officially,** *adv.* **officiate,** *v.i.* to perform official duties, to act in an official capacity; to perform a prescribed function or duty; (*coll.*) to act in any given capacity. **officiation,** *n.* **officiator,** *n.*
officious (əfish'əs), *a.* forward in doing or offering unwanted kindness, meddling, intrusive. **officiously,** *adv.* **officiousness,** *n.*
off-print, offshore, offside etc. OFF.
OFT, (*abbr.*) Office of Fair Trading.
often (of'n. -tən), (*poet.* or in compounds) **oft,** *adv.* frequently, many times; in many instances. **as often as not, more often than not,** (quite) frequently. **oft-recurring,** *a.* frequently recurring.
Oftel (of'tel), *n.* the Government body which monitors British Telecom. [*O*ffice of *Tel*ecommunications]
ogam OG(H)AM.

ogee (ō'jē), *n.* a wave-like moulding having an inner and outer curve like the letter S; short for **ogee arch,** a pointed arch each side of which is formed of a concave and a convex curve. *a.* having such a double curve (of arches, windows etc.).
Ogen melon (ō'gən), *n.* a variety of small sweet melon resembling a cantaloupe.
og(h)am (og'əm), *n.* an ancient Celtic system of writing consisting of an alphabet of twenty characters derived from the runes; any character in this; an inscription in this. **og(h)amic** (-am'-, og'-), *a.*
ogle (ō'gl), *v.t.* to look or stare at with admiration, wonder etc., esp. amorously. *v.i.* to cast amorous or lewd glances. **ogler,** *n.*
ogre (ō'gə), *n.* a fairytale giant living on human flesh; a monster, a barbarously cruel person. **ogr(e)ish,** *a.* **ogress** (ō'gris), *n. fem.*
oh, o⁴.
OHG, (*abbr.*) Old High German.
ohm (ōm), *n.* the unit of electrical resistance, being the resistance between two points on a conductor when a potential difference of one volt produces a current of one amp. **Ohm's law,** *n.* **ohmage** (-ij), *n.* **ohmmeter,** *n.* [Georg S. Ohm, 1787-1854, German electrician]
OHMS, (*abbr.*) On Her/His Majesty's Service.
oho (əhō'), *int.* expressing surprise, irony, or exultation.
oi (oi), *int.* used to give warning, attract attention etc.
-oid (-oid), **-oidal,** *comb. form.* denoting resemblance, as in *colloid, cycloid, rhomboid*.
-oidea, *comb. form.* denoting zoological classes or families.
oik (oik), *n.* (*coll.*) a stupid or inferior person; a fool.
oil (oil), *n.* a greasy liquid, insoluble in water, soluble in ether and usually in alcohol, obtained from various animal, vegetable and mineral substances; (*pl.*) oil, paints. *v.t.* to smear, anoint, rub, soak, treat or impregnate with oil; to lubricate (as) with oil. **essential, fixed, mineral oils,** ESSENTIAL. FIX. MINERAL. **oil of vitriol** VITRIOL. **to burn the midnight oil,** to study or work far into the night. **to oil the wheels,** to facilitate matters, to help things go smoothly. **to oil someone's palm,** to bribe someone. **to strike oil** STRIKE. **oilcake,** *n.* the refuse after oil is pressed or extracted from linseed etc., used as fodder. **oil-can,** *n.* a can for holding oil, esp. one used for oiling machinery. **oilcloth,** *n.* a fabric coated with white lead ground in oil; an oilskin. **oil-fired,** *a.* burning oil as fuel. **oil-field,** *n.* a region where mineral oil is obtained. **oil-gauge,** *n.* an instrument showing the level of oil in a tank etc. **oilman,** *n.* one who works in the oil industry; one who owns an oil well. **oil-nut,** *n.* any of various oil-yielding nuts including the butternut and buffalo-nut. **oil paint,** *n.* a paint made by grinding a pigment in oil. **oil painting,** *n.* the art of painting with oil paints; a painting in oil paints. **oil palm,** *n.* a palm tree bearing fruits yielding palm-oil. **oil platform,** *n.* a floating or fixed offshore structure which supports an oil-rig. **oil-press,** *n.* a machine for pressing the oil from seeds, nuts etc. **oil-rig,** *n.* an installation for drilling and extracting oil and natural gas. **oil seed,** *n.* any of various oil-yielding seeds including sesame and sunflower seeds. **oil shale,** *n.* a shale from which mineral oils can be distilled. **oilskin,** *n.* cloth rendered waterproof by treatment with oil; a garment of this cloth; (*pl.*) a suit of such garments. **oil slick,** *n.* a patch of floating oil, usu. pollutive. **oilstone,** *n.* a fine-grained whetstone lubricated with oil before use. **oil-tanker,** *n.* a large sea-going ship for transporting oil. **oil-well,** *n.* **oiled,** *a.* greased with, lubricated by, saturated or preserved in oil. **well oiled,** (*coll.*) drunk. **oily,** *a.* consisting of, containing, covered with, or like oil; of a person's manner, unctuous, smooth, insinuating. **oiliness,** *n.*
oink (oingk), *n.* the grunt of a pig. *v.i.* to make such a sound.
ointment (oint'mənt), *n.* a soft oily preparation applied to

Ojibwa diseased or injured parts or used as a cosmetic.
Ojibwa, -way (ōjib'wä), *n.* (*pl.* **-wa(s), -way(s)**) (a member of) a N American people living in the westerly region of the Great Lakes; their language.
OK, okay (ōkä'), (*coll.*) *a., int., adv.* quite correct, all right. *v.t.* (*past, p.p.* **OK'd, OKed, okayed**) to authorize, to endorse, to approve. *n.* approval, sanction, agreement.
okeydoke(y) (ōkidōk', -dō'ki), (*coll.*) *int.* a casual or jokey form of assent or agreement.
okapi (ōkah'pi), *n.* a deer-like animal related to and partially marked like a giraffe.
okra (ō'krə), *n.* an African plant cultivated for its green pods used in curries, soups, stews etc.; gumbo.
-ol (-ol), *suf.* denoting a chemical compound containing an alcohol, or (loosely) an oil, as *benzol, menthol, phenol*.
old (ōld), *a.* (*comp.* **older**, *superl.* **oldest**, cp. ELDER [1], -EST) advanced in years or long in existence; not young, fresh, new or recent; like an old person, experienced, wise; decayed by process of time, worn, dilapidated; stale, trite; customary, wonted; obsolete, out-of-date, antiquated; of any specified duration; (*also* **olden**) belonging to a former period, made or established long ago, ancient, bygone; early, previous, former; of a language, denoting the earliest known form; (*coll.*) expressing familiarity or endearment; (*coll.*) used to emphasize (as in e.g. *a right old carry-on*). **of old**, in or from ancient times; long ago, formerly. **the good old days**, better or happier former times; the past viewed nostalgically. **the Old Bill**, (*sl.*) the police. **the Old Country**, the country of origin of an immigrant or his/her ancestors. **old age**, *n.* the latter part of life. **old age pension**, *n.* in the UK, a weekly state pension paid to a man at 65 and a woman at 60 years of age. **old age pensioner**, *n.* **Old Bailey**, *n.* the Central Criminal Court of the City of London. **old boy, old girl**, *n.* a former pupil of a school; (*coll.*) a friendly form of address; an elderly man or woman. **old boy, girl network** NETWORK. **old dear**, *n.* (*coll.*) an elderly person, esp. a woman. **Old English** ENGLISH. **Old English Sheepdog**, a large English breed of sheepdog with a bob tail and a shaggy coat. **old-fashioned**, *a.* out-of-date, outmoded; quaint. **old fogy**, *n.* (*derog.*) an older person with conservative, eccentric or old-fashioned ideas or ways. **Old Glory**, *n.* (*N Am.*) the US flag. **old gold** GOLD. **old guard**, *n.* the old or conservative members of a party, organization etc. **old hand**, *n.* one who is skilled from long practice at a trade, craft or practice of any kind. **old hat** HAT. **old lady, man**, *n.* (*coll.*) a sexual partner, a husband or wife; a father or mother; (*esp.* **old man**) a friendly form of address; an elderly man or woman. **old maid**, *n.* (*derog.*) an unmarried woman, of advanced years or unlikely ever to marry; a variety of card game; a precise, prudish, fidgety person of either sex. **old-maidish**, *a.* **old master**, *n.* a great painter or painting of the 16th–18th cents. **Old Nick**, *n.* (*coll.*) the devil.
old penny, *n.* a former unit of British money of which there were 240 to the £. **old salt**, *n.* an old, experienced or retired sailor. **old school**, *a.*, *n.* (of) those who adhere to past traditions or principles. **old school tie**, a tie sporting a public school's colours, worn by its old boys; (a symbol of) the mutual allegiance of a group of (esp. privileged or upper-class) people. **Old Style** STYLE [1]. **Old Testament** TESTAMENT. **old-time**, *a.* old, ancient; old-fashioned. **old-timer**, *n.* (*coll.*) an old man; a veteran; a person who has remained in a situation for a long time. **old woman**, *n.* (*coll.*) a wife or mother; a timid, fidgety or fussy man. **old-womanish**, *adj.* **Old World**, *n.* the eastern hemisphere. **old(e) world(e)**, *a.* (emphatically or for effect) old-fashioned and quaint. **old year**, *n.* the year just ended or on the point of ending. **olden** OLD. **oldie** (-di), *n.* (*coll.*) an old person or thing (e.g. an old song, film). **oldish**, *a.* **oldness**, *n.* **oldster** (-stə), *n.* an old or oldish person.
olé (ōlā'), *int.* expressing approval or victory at a bull fight.

ole-, OLE(O)-.
oleaginous (ōliaj'inəs), *a.* oily, greasy. **oleaginously**, *adv.* **oleaginousness**, *n.*
oleander (ōlian'də), *n.* a poisonous evergreen shrub with lanceolate leaves and pink or white flowers.
olefin (ō'lifin), **-fine** (-fēn), *n.* any one of a group of hydrocarbons containing two atoms of hydrogen to one of carbon. **olefinic** (-fin'-), *a.*
ole(o)-, *comb. form.* oil.
olfactory (olfak'təri), *a.* pertaining to or used in smelling. *n.* (*usu.* pl.) an organ of smell. **olfaction**, *n.* the sense or process of smelling. **olfactive**, *a.*
OLG, (*abbr.*) Old Low German.
oligarch (ol'igahk), *n.* a member of an oligarchy. **oligarchal, oligarchic, -ical** (-gah'-), *a.* **oligarchically**, *adv.* **oligarchist** (ol'-), *n.* **oligarchy** (-ki), *n.* a form of government in which power is vested in the hands of a small exclusive class; the members of such a class; a state so governed.
Oligocene (ol'igəsēn), *n., a.* tertiary, (of) the age or strata between the Eocene and Miocene.
oligopoly (oligop'əli), *n.* a situation in the market in which a few producers control the supply of a product. **oligopolist**, *n.* **oligopolistic** (-lis'-), *a.*
olio (ō'liō), *n.* (*pl.* **olios**) a mixed dish; a mixture, a variety, a potpourri.
olive (ol'iv), *n.* an evergreen tree with narrow leathery leaves and clusters of oval drupes yielding oil when ripe and eaten unripe as a relish; the fruit of this tree; its wood; the colour of the unripe olive, a dull, yellowish green or brown. *a.* of an olive colour; (made) of olive wood. **olive-branch**, *n.* a branch of the olive-tree as an emblem of peace; something which indicates a desire for peace (e.g. a goodwill gesture, an offer of reconciliation). **olive drab**, *n.* (*N Am.*) a dull olive-green colour; fabric or garments of this colour, esp. a US army uniform. **olive-green**, *a.* of a dull yellowish-green colour. **olive-oil**, *n.* **olivaceous** (-vā'shəs), *a.* olive-coloured.
ology (ol'əji), *n.* a science; one of the sciences whose names end thus.
-ology -LOGY.
oloroso (olərō'sō), *n.* a medium-sweet golden sherry.
olympiad (əlim'piad), *n.* a period of four years, being the interval between the celebrations of the Olympic Games; (a staging of) an international contest, esp. the modern Olympic games. **Olympian**, *a.* pertaining to Mt Olympus, the home of the gods, celestial; magnificent, lofty, superb; Olympic. *n.* a dweller on Olympus; one of the Greek gods; (*chiefly N Am.*) a contestant in the Olympic games. **Olympic**, *a.* pertaining to Olympia or the Olympic Games. **Olympic flame**, the flame which burns throughout the Olympic Games, lit by the Olympic torch. **Olympic Games**, the greatest of the Greek national games, held every four years at Olympia, in honour of Zeus; a revival (since 1896) of this festival, an international four-yearly sports meeting. **Olympic torch**, *n.* the lighted torch brought by a runner from Olympia to light the Olympic flame. **the Olympics**, *n.pl.* the Olympic Games.
-oma (-ō'mə), *comb. form.* (*pl.* **-omas, -omata** (-tə)) denoting a tumour or growth.
omasum (ōmä'səm), *n.* (*pl.* **-sa** (-sə)) the third stomach of a ruminant.
ombre (om'bə), *n.* a game of cards, for two, three or five players, popular in the 17th and 18th cents.
ombro-, *comb. form.* denoting rain.
ombrology (ombrol'əji), *n.* the branch of meteorology concerned with rainfall. **ombrometer** (-brom'itə), *n.* a rain-gauge.
Ombudsman (om'budzmən), *n.* (*pl.* **-men**) an official investigator of complaints against government bodies or employees; a Parliamentary Commissioner. **Ombudswoman**, *n. fem.*

-ome (-ōm), *comb. form.* denoting a mass or part.

omega (ō'migə, om'-), *n.* the last letter of the Greek alphabet, Ω, ω, ō; the last of a series; the conclusion, the end, the last stage or phase.

omelet, -lette (om'lit), *n.* a flat dish made with beaten eggs cooked in fat or oil, eaten plain or seasoned and filled with various combinations of herbs, cheese, meat etc.

omen (ō'mən), *n.* an incident, object or appearance taken as indicating a good or evil event, fortune etc., a portent; a prognostication or prophetic signification. *v.t.* to portend. **ill-omened**, *a.* attended by bad indications; portending bad fortune etc.

omertà (omœtah'), *n.* a conspiracy of silence, part of the Mafia code of honour.

omicron (ōmi'kron, om'i-), *n.* the 15th letter of the Greek alphabet, the short o.

ominous (om'inəs), *a.* portending evil; of evil omen, inauspicious. **ominously**, *adv.* **ominousness**, *n.*

omit (əmit'), *v.t.* to leave out; not to include, insert, or mention; to neglect (to do), to leave undone. **omissible**, *a.* **omission** (-shən), *n.* **omitter**, *n.*

omni- (omni-), *comb. form.* universally, in all ways, of all things.

omnibus (om'nibəs), *n.* a large passenger vehicle for transporting numbers of people by road, a bus; a volume containing reprints of a number of works, usually by the same author. *a.* inclusive, embracing several or various items, objects etc. **omnibus Bill, clause, resolution**, *n.* one dealing with several subjects.

omnidirectional (omnidirek'shənl), *a.* (capable of) moving, sending, receiving, etc. in every direction (of e.g. radio waves, a radio transmitter or receiver).

omnifarious (omnifeə'riəs), *a.* of all kinds. **omnifariously**, *adv.* **omnifariousness**, *n.*

omnipotent (omnip'ətənt), *a.* almighty; having unlimited power. **omnipotence, omnipotency**, *n.* **The Omnipotent**, God. **omnipotently**, *adv.*

omnipresent (omniprez'ənt), *a.* present in every place at the same time. **omnipresence**, *n.*

omniscience (omnis'iəns), *n.* infinite knowledge. **omniscient**, *a.* **omnisciently**, *adv.*

omnivore (om'nivaw), *n.* a creature that eats any type of available food (i.e. vegetable matter and meat). **omnivorous** (-niv'ərəs), *a.* feeding on anything and everything available; *(fig.)* reading anything and everything. **omnivorously**, *adv.* **omnivorousness**, *n.*

omophagic (ōməfaj'ik), **omophagous** (ōmof'əgəs), *a.* eating raw flesh.

omphalos (om'fəlos), *n.* the boss of an ancient Greek shield; a stone in the temple of Apollo at Delphi, believed to be the middle point or navel of the earth; a central point, a hub. **omphalic** (omfal'ik), **omphaloid** (om'fəloid), *a.* pertaining to or resembling the navel.

ON, *(abbr.)* Old Norse.

on (on), *prep.* (as) in(to) contact with, esp. as supported by, covering, environing or suspended from, the upper surface or level of; very close to; about, concerning; attached to; after, added to; as a member of; carried with; taking (e.g. a drug); at the expense of; at the date, time or occasion of; in a condition or state of; sustained by; by means of; after receiving (advice, orders, etc.); because of; in the manner of. *adv.* so as to be in contact with and supported by, covering, environing, suspended from or adhering to something; in advance, forward, in operation, action, movement, progress, persistence *(coll.)*, or continuance of action or movement; taking place; arranged; planned. *a.* applied to that part of a cricket field to the right side of the bowler when the batsman is right-handed, or vice versa; operating; *(coll.)* wagered; performing, broadcasting, playing (e.g. of a batsman); definitely happening as arranged; acceptable, possible, tolerable. *n.* the on side of a cricket field. **... and so on**, etcetera. **off and on, on and off**, intermittently. **on and on**, ceaselessly, continuously. **on-board computer**, a computer installed in a car or other vehicle. **on time**, punctual(ly). **to be on to**, to be aware of, to have tumbled to. **to get on** GET. **to go on at**, to nag at. **to go on to**, to advance, progress, move or travel to a further level, position or place. **oncoming**, *n.*, *a.* (the) coming on, advancing or approaching. **oncost**, *n.* a supplementary or additional expense, an overhead. **on-flow**, *n.* onward flow. **ongoing**, *n.* procedure, progress; *(pl.)* goings-on, misbehaviour. *a.* unceasing, continuous; in progress. **on-line**, *a.* of a computer peripheral, under the control of the central processor. **onlooker**, *n.* a spectator. **onrush**, *n.* a rushing forwards (in time or place), an attack, an onset. **onset**, *n.* an attack, an assault, an onslaught; (the) outset, beginning. **onshore**, *a.*, *adv.* on or towards (the) land. **onside**, *a.*, *adv.* not offside. **onslaught**, *n.* a furious attack. **on-stream**, *a.*, *adv.* of a manufacturing plant, industrial installation etc., in operation. **onward**, *a.* moving, tending or directed forward; advancing, progressive. **onwards**, *adv.* **on to, onto**, *prep.* to and upon, to a position on or upon.

onager (on'əjə), *n.* the wild ass of the Asiatic deserts; an ancient and mediaeval war machine resembling a large catapult for hurling rocks at the enemy.

onanism (ō'nənizm), *n.* masturbation; the withdrawal of the penis from the vagina prior to ejaculation. **onanist**, *n.*

ONC, *(abbr.)* Ordinary National Certificate.

once (wŭns), *adv.* one time (only); formerly, at some past time; (not) at any time, ever, at all; as soon as. *conj.* as soon as. *n.* one time. *a.* former. **all at once**, all together, simultaneously; suddenly. **at once**, immediately, without delay; simultaneously. **for once**, for one time or occasion only or at least. **once (and) for all**, finally; definitively. **once in a while** or **way**, very seldom. **once or twice**, a few times. **once upon a time**, at some past date or period (usu. beginning a fairytale). **once-over**, *n.* a look of appraisal. **oncer**, *n. (coll.)* a £1 note.

onco-, *comb. form.* denoting a tumour.

oncology (ongkol'əji), *n.* the study of tumours and cancers. **oncologist**, *n.* **oncological, oncologic** (-loj'-), *a.*

OND, *(abbr.)* Ordinary National Diploma.

one (wŭn), *a.* single, undivided; being a unit and integral; a or an; single in kind, the only, the same; this, some, any, a certain. *pron.* a person or thing of the kind implied; someone or something, anyone or anything; a person unspecified; any person, esp. the speaker. *n.* the number 1; a thing or person so numbered; a single thing or person; *(coll.)* a joke or story. **all in one**, combined. **all one** ALL. **at one**, in accord or agreement. **many a one**, many people. **never a one**, none. **one and all**, everyone; jointly and severally. **one and only**, unique. **one-armed bandit**, a fruit machine operated by a single lever. **one by one**, singly, individually, successively. **one-man-band**, a sole musician playing a variety of instruments simultaneously; a company, enterprise etc., consisting of, or run by, a single person. **one or two**, a few. **one-, single-parent family**, a family having only one parent present to tend the child or children. **one way and another**, altogether, on balance. **to be a one**, *(coll.)* to be a joker. **to be one for**, to be an enthusiast for. **one another**, *pron.* each other. **one-armed**, *a.* having, or done by, one arm. **one-dimensional**, *a.* having only one dimension; of a person, superficial, shallow. **one-handed**, *a.* having, or done by, one hand. **one-horse**, *a.* drawn by a single horse; *(sl.)* of meagre capacity, resources or efficiency. **one-legged** (-legid), *a.* **one-liner**, *n.* a short punchy joke or witticism. **one-man**, *a.* employing, worked by, or consisting of one man. **one-night stand**, a single performance at one venue; *(coll.)* (a person having) a sexual encounter or relationship lasting one evening or night. **one-off**, *n.* a unique object, product, event etc. *a.* unique. **one-piece**, *a.*, *n.* (of) a gar-

ment, esp. a swimsuit, consisting of one piece of material. **one-sided**, *a.* having or happening on one side only; partial, unfair, favouring one side of an argument, topic etc.; more developed on one side than another. **one-sidedly**, *adv.* **one-sidedness**, *n.* **one-step**, *n.*, *a.* form of a quick-stepping type of ballroom dancing. **one-time**, *a.* former; sometime in the past. **one-to-one**, corresponding, esp. in mathematics, pairing each element of a set with only one of another; of one person to or with another, as in a *one-to-one relationship*. **one-track**, *a.* single-track; (*coll.*) capable of, or obsessed by, only one idea at a time. **one-two**, *n.* in boxing, two successive blows rapidly delivered; a type of passing movement in football. **one-up**, *a.* (*coll.*) having, or in, a position of advantage. **one-upmanship**, *n.* the art of gaining or keeping an advantage over someone. **one-way**, *a.* denoting a traffic system which allows vehicles to go in one direction only through certain streets; unidirectional. *n.*, *a.* (of) a single ticket or fare (e.g. for a bus, train, aeroplane etc.). **oneness**, *n.* singleness; singularity, uniqueness; unity, agreement, harmony; sameness. **oneself**, *pron.* the reflexive form of *one*. **to be by oneself**, to be alone. **to be oneself**, to behave naturally, without constraint, pretension, artifice etc.
onerous (ō'nərəs), *a.* burdensome, heavy, weighty, troublesome. **onerously**, *adv.* **onerousness**, *n.*
oneself ONE.
onflow, ongoing etc. ON.
onion (ŭn'yən), *n.* a plant with an underground bulb of several layers and a pungent smell and flavour, much used in cookery. **to know one's onions**, (*coll.*) to be knowledgeable in one's subject, competent in one's job, etc. **oniony**, *a.*
on-line, onlooker ON.
only (ōn'li), *a.* solitary, single or alone in its or their kind; the single, the sole. *adv.* solely, merely, exclusively, alone; with no other, singly; wholly; not otherwise than; not earlier than. *conj.* except that; but. **if only**, expressing a desire or wish. **only too willing**, extremely willing. **only too true**, regrettably, completely true. **only-begotten**, *a.* born as the sole issue. **only child**, *n.* one without brothers or sisters.
o.n.o., (*abbr.*) or near(est) offer.
onomastic (onəmas'tik), *a.* pertaining to a name. **onomastics**, *n. sing.* the study of proper names.
onomatopoeia (ŏnəmatəpē'ə), *n.* the formation of words in imitation of the sounds associated with or suggested by the things signified; the rhetorical use of a word so formed. **onomatopoeic, -poetic** (-pōet'-), *a.* **onomatopoeically, -poetically**, *adv.*
onrush, onset, etc. ON.
onus (ō'nəs), *n.* a burden; a duty, obligation or responsibility.
-onym, *comb. form.* denoting a name or word, as in *pseudonym, antonym*.
onyx (on'iks), *n.* a variety of quartz resembling agate, with variously-coloured layers. **onyx marble**, *n.* a calcium carbonite mineral resembling marble.
oo-, *comb. form.* pertaining to ova, or to an egg.
oodles (oo'dlz), (*dated coll.*) *n. pl.* a great quantity, superabundance.
oogamous (ōog'əməs), *a.* reproducing by the union of male and female cells. **oogamete** (-mēt), *n.* one of such cells. **oogamy** (-mi), *n.*
oogenesis (ōəjen'əsis), **oogeny** (ōoj'əni), *n.* the origin and development of an ovum. **oogenetic** (-net'-), *a.*
ooh (oo), *int.* expressing delight, surprise, pain, admiration etc. *v.i.* to say 'ooh'.
Oolite (ō'əlit), *n.* a limestone composed of grains or particles of sand like the roe of a fish; the upper portion of the Jurassic strata in England, composed in great part of oolitic limestone. **oolitic** (-lit'-), *a.*
oology (ōol'əji), *n.* the study of birds' eggs. **oological**

(-loj'-), *a.* **oologically**, *adv.* **oologist**, *n.*
oolong (oo'long), *n.* a kind of China tea.
oompah (oom'pah), *n.* an imitation or representation of the sound of a large brass musical instrument. *v.i.* to make such a sound.
oomph (umf), *n.* (*coll.*) vigour, energy.
oops (oops), *int.* expressing surprise, dismay or apology.
ooze (ooz), *n.* wet mud, slime; a slimy deposit consisting of shells and other remains found on ocean-beds; a gentle, sluggish flow, an exudation. *v.i.* to flow, to pass gently; to percolate (through the pores of a body etc.). *v.t.* to emit or exude. **oozily**, *adv.* **ooziness**, *n.* **oozy**, *a.*
o.p., (*abbr.*) out of print.
op. (*abbr.*) opera; operation; operator; optical; opposite; opus.
op-, *pref.* OB- (before *p*).
opacity OPAQUE.
opal (ō'pəl), *n.*, *a.* (of or like) an amorphous, transparent, vitreous form of hydrous silica, several kinds of which have iridescent colours and used as gems; (of) the colour of opal. **opaled**, *a.* **opalescence** (-les'əns), *n.* iridescence. **opalescent**, *a.* **opaline** (-lin), *a.* pertaining to or like opal. *n.* a translucent variety of glass.
opaque (əpāk'), *a.* impervious to rays of light; not transparent or translucent; impenetrable to sight or (*fig.*) understanding. **opaquely**, *adv.* **opaqueness**, **opacity** (əpas'-), *n.*
op art (op), *n.* a type of abstract art employing shapes arranged to produce an optical illusion, esp. that of movement.
op. cit. (op sit), (*abbr.*) *opere citato*, in the work cited.
ope (ōp), *v.t.*, *v.i.* (*poet.*) to open.
OPEC (ō'pek), (*abbr.*) Organization of Petroleum-Exporting Countries.
open (ō'pən), *a.* not closed, obstructed or enclosed; affording entrance, passage, access or clear view; unclosed, unshut, having any barrier, gate, cover etc. removed, withdrawn or unfastened; uncovered, bare, unsheltered, exposed; unconcealed, undisguised, manifest; unrestricted, not exclusive or limited; ready to admit, receive or be affected; liable, subject (to); unoccupied, vacant; widely spaced; spread out; loosely woven, or latticed, or having frequent gaps or spaces; free, generous, liberal; frank, candid; not closed or decided, debatable, moot; (*Mus.*) not stopped, or produced from an unstopped pipe, string etc.; enunciated with the vocal organs comparatively unclosed; (of a vowel or syllable) not ended by a consonant. *n.* (*Sport*) a tournament open to any class of player. *v.t.* to make open; to unclose; to unfasten, unlock; to remove the covering from; to unfold, spread out, expand; to free from obstruction or restriction, to give free access to; to reveal, to make manifest or public; to declare open ceremonially; to widen, to enlarge, to develop; to make a start in, to begin. to set going; in law, to state (a case) before calling evidence. *v.i.* to become unclosed or unfastened; to crack, to fissure; to unfold, to expand; to develop; to make a start, to begin. **the open**, unenclosed space or countryside; public or plain view. **to bring into the open**, to disclose what was hitherto hidden or secret. **to open fire**, to begin firing ammunition. **to open out**, to unfold; to develop. **to open someone's eyes**, to disillusion someone. **to open up**, to make accessible; to reveal; to become (more) communicative, candid, etc., to discover, to explore, to colonize; to make ready for trade. **open air**, *n.* outdoor(s). **open-air**, *a.* **open-and-shut**, *a.* needing little deliberation, easily solved, simple. **open-armed**, *a.* very welcoming. **open book**, *n.* a person or thing easily understood. **open-cast**, *n.*, *a.* in mining, (of) a surface excavation. **open circuit**, *n.* an electrical circuit that has been broken so that no current can flow. **Open College**, *n.* a British college established (1986) for similar motives to the Open University, teaching non-vocational

courses. **open court,** *n.* a court to which the public are admitted. **open day,** *n.* a day when an institution (e.g. a school or university) is open to the public. **open door,** *n.* free admission or unrestricted access. **open-ended,** *a.* having no set limit or restriction on duration, amount etc. **open-eyed,** *a.* watchful, aware; astonished, surprised. **open field,** *n.* undivided arable land. **open fire,** *n.* a domestic hearth fire burning coal or wood, not gas or electricity. **open-handed,** *a.* generous, liberal. **open-handedly,** *adv.* **open-handedness,** *n.* **open-heart surgery,** *n.* surgery on a heart while its functions are temporarily performed by a heart-lung machine. **open-hearted,** *a.* frank, sincere, unsuspicious. **open-heartedly,** *adv.* **open-heartedness,** *n.* **open-hearth,** *a.* of, made by or used in an open-hearth furnace or process. **open-hearth furnace,** *n.* a furnace for producing steel from pig iron. **open-hearth process,** *n.* the process of making steel in such a furnace. **open house,** *n.* hospitality proffered to all comers. **open letter,** *n.* a letter addressed to an individual but (also) published in a newspaper. **open market,** *n.* a market situation of unrestricted commerce and free competition. **open-minded,** *a.* accessible to ideas, unprejudiced. **open-mindedly,** *adv.* **open-mindedness,** *n.* **open-mouthed** (-mowdhd), *a.* gaping with stupidity, surprise etc.; greedy, ravenous; clamorous. **open-plan,** *a.* having no, or few, dividing partitions or walls. **open prison,** *n.* one allowing great freedom of movement. **open question,** *n.* one that is unresolved. **open sandwich,** *n.* a sandwich without an upper slice of bread, exposing its filling. **open season,** *n.* a period during which it is legal to hunt or fish for various species; (*fig.*) an unrestricted period in which to attack etc. **open secret,** *n.* an apparently undivulged secret which is however generally known. **open sesame** SESAME. **open-skies policy,** *n.* one where an outline has free choice of which airports it uses. **Open University,** *n.* a British university established (1971) to teach those not usu. able to attend a university, through broadcasts and by correspondence. **open verdict,** *n.* a verdict which names no criminal or records no cause of death. **openable,** *a.* **opener,** *n.* a person or thing that opens; the first episode of a new series of broadcast programmes. **for openers,** (*coll.*) to begin with, as a first consideration. **opening,** *a.* that opens; beginning, first in order, initial. *n.* the act of making or becoming open; a gap, a breach, an aperture; a beginning, the first part or stage, a prelude; in law, a counsel's statement of a case before evidence is called; in chess etc., a series of moves beginning a game; a vacancy, an opportunity; the two facing pages of an open book, a spread. **opening time,** *n.* the time at which bars and public houses can legally begin selling alcohol. **openly,** *adv.* **openness,** *n.*
opera[1] (op'ərə), *n.* a dramatic entertainment in which music forms an essential part; a composition comprising words and music for this; the branch of the musical and dramatic arts concerned with this; the theatre in which it is performed; the company which performs it; this form of dramatic art. **comic opera** COMIC. **light opera,** operetta. **grand opera,** a type of opera having large or tragic dramatic themes, and without spoken dialogue. **opera bouffe** (boof), **buffa** (-ə), *n.* a farcical variety of opera. **opéra comique** (komēk'), *n.* a type of opera having some spoken dialogue; comic opera. **opera-glasses,** *n. pl.* small binoculars for use in theatres. **opera-hat,** *n.* a collapsible tall hat for men. **opera seria** (seə'riə), *n.* a serious type of opera having a heroic or mythical plot. **operatic** (-rat'-), *a.* **operatically,** *adv.* **operetta** (-ret'ə), *n.* a short opera of a light character. **operettist,** *n.*
opera[2] (op'ərə), *pl.* OPUS.
operate (op'ərāt), *v.i.* to work, to act; to produce effect; (*Med.*) to produce a certain effect on the human system; to perform a surgical operation; (*Mil.*) to carry out strategic movements; to trade, to carry on a business. *v.t.* to work or control the working of; to control, manage, run (a business etc.). **operable,** *a.* suitable or capable of being operated (on); practicable. **operability** (-bil'-), *n.* **operand** (-rənd), *n.* that which is operated on, esp. a quantity in mathematics. **operating table,** *n.* a table on which a patient lies during a surgical operation. **operating theatre,** *n.* a specially-fitted room where surgery is performed. **operation,** *n.* the act or process of operating; action, (mode of) working; activity, performance of function; effect; a planned campaign or series of military or naval movements; a surgical act performed with or without instruments upon the body; in mathematics, the act of altering the value or form of a number or quantity by a process such as multiplication or division; a commercial or financial business; a procedure, a process. **operational,** *a.* ready for or capable of action or use; of or pertaining to military operations. **operational, operations research,** *n.* the application of mathematical techniques to problems of military or naval strategy, industrial planning, economic organization etc. **operations room,** *n.* a room from where (esp. military) operations are controlled or directed. **operative,** *a.* acting, exerting force; relevant, significant; of or pertaining to a surgical operation. *n.* a (skilled) workman, an operator. **operatively,** *adv.* **operativeness,** *n.* **operator,** *n.* one who runs or operates a machine, esp. a telephone switchboard; in mathematics or logic, a symbol etc. representing a function to be performed; (*coll.*) a skilled manipulator.
opere citato (op'ere kitah'tō), in the work cited.
operetta OPERA[1].
ophthalmia (ofthal'miə), *n.* inflammation of the eye. **ophthalmic** (ofthal'mik), *a.* pertaining to the eye. **ophthalmic optician,** *n.* an optician qualified to test eyesight and prescribe and dispense spectacles or lenses. **ophthalm(o)-,** *comb. form.* pertaining to the eye. **ophthalmology** (ofthalmol'əji), *n.* the science of the eye, its structure, functions and diseases. **ophthalmologic, -ical** (-loj'-), *a.* **ophthalmologically,** *adv.* **ophthalmologist,** *n.* **ophthalmoscope** (ofthal'məskōp), *n.* an instrument for examining the inner structure of the eye. **ophthalmoscopic, -ical** (-skop'-), *a.* **ophthalmoscopy** (-mos'kəpi), *n.* **-opia,** *comb. form.* denoting a condition or defect of the eye, as in *myopia.*
opiate (ō'piət), *n.* a medicine compounded with opium; a narcotic; anything serving to dull sensation, relieve uneasiness, or induce sleep.
opine (əpīn'), *v.i., v.t.* to think, to suppose; to express an opinion.
opinion (əpin'yən), *n.* a judgment, conviction, or belief falling short of positive knowledge; views, sentiments, esp. those generally prevailing; one's judgment, belief or conviction with regard to a particular subject; the formal statement of a judge, counsel, physician or other expert. **a matter of opinion,** a matter open to debate or question. **opinion poll** POLL[1]. **opinionated** (-ātid), *a.* stiff or obstinate in one's opinions; dogmatic, stubborn. **opinionatedly,** *adv.* **opinionatedness,** *n.* **opinionless** (-lis), *a.*
opium (ō'piəm), *n.* a narcotic drug prepared from the dried juice of the white poppy. **opium-den,** *n.* a place where opium is smoked. **opium-eater, -smoker,** *n.* one who habitually eats or smokes opium as a stimulant or narcotic.
opossum (əpos'əm), *n.* (*pl.* **-ssums, -ssum**) an American marsupial quadruped with a prehensile tail and a thumb on the hind-foot; applied to similar marsupials of Australia and Tasmania.
opponent (əpō'nənt), *n.* one who opposes, esp. in a debate, argument or contest; an adversary, an antagonist. **opponency,** *n.*
opportune (op'ətūn, -tūn'), *a.* situated, occurring, done etc. at a favourable moment; timely; fit, suitable. **opportunely,** *adv.* **opportuneness,** *n.* **opportunism,** *n.* uti-

lizing circumstances or opportunities to gain one's ends, esp. the act or practice of shaping policy according to the needs or circumstances of the moment; adaptation to circumstances through sacrifice of principle to expediency. **opportunist**, *n.* **opportunity** (-tū'-), *n.* an opportune or convenient time or occasion; a chance, an opening.

oppose (əpōz'), *v.t.* to set against, to place or bring forward as an obstacle, adverse force, counterpoise, contrast or refutation (to); to set oneself against or act against, to resist; to object to, to dispute; (*in pres.p. or p.p.*) opposite, contrasted. *v.i.* to offer resistance or objection. **opposable**, *a.* **opposability** (-bil'-), *n.* **opposer**, *n.*

opposite (op'əzit), *a.* situated contrary in position (to); fronting, facing; contrary, diametrically different (to or from); (*Bot.*) placed in pairs on contrary sides on the same horizontal plane (of leaves on a stem). *n.* one who or that which is opposite; a contrary thing or term; (*Log.*) a contradictory. *adv.* in an opposite place or direction. *prep.* facing; as a co-star with (in a play, film etc.). **opposite number**, *n.* a person in the corresponding position on another side; a counterpart. **oppositely**, *adv.* **oppositeness**, *n.*

opposition (opəzish'ən), *n.* the act or state of opposing; those people oppposing; antagonism, resistance, hostility; the state of being opposite; antithesis, contrast; an obstacle, a hindrance; the chief parliamentary party not in office; (*Astron.*) the situation of two heavenly bodies when their longitudes differ by 180°. **oppositional**, *a.* **oppositionist**, *a.*, *n.*

oppress (əpres'), *v.t.* to overburden; to weigh down; to inflict hardships upon, to govern cruelly or unjustly; to tyrannize. **oppression** (-shən), *n.* **oppressive**, *a.* overbearing, tyrannous; (of the weather) close, muggy, sultry. **oppressively**, *adv.* **oppressiveness**, *n.* **oppressor**, *n.*

opprobrium (əprō'briəm), *n.* disgrace, infamy, obloquy. **opprobrious**, *a.* abusive, vituperative, scornful. **opprobriously**, *adv.* **opprobriousness**, *n.*

oppugn (əpūn'), *v.t.* to oppose; to dispute, to call in question.

opt (opt), *v.i.* to make a choice (for). **to opt out (of)**, to choose not to be involved (in something); (of a school, hospital etc., no longer to be under the control or management of a local authority.

opt., (*abbr.*) optative; optical; optician; optimum; optional.

optative (op'tətiv, optā'-), *a.* (*Gram.*) expressing a wish or desire. *n.* the optative mood; a verbal form expressing this. **optatively**, *adv.*

optic (op'tik), *a.* pertaining to vision or the eye. *n.* a device fixed to the neck of a bottle to measure out spirits. **optical character reader**, a device which scans and reads printed characters optically, translating them into binary code which can then be processed by a computer. **optical disk**, *n.* a small disk which can be read by a laser beam and can hold large quantities of information in digital form, used as a mass storage medium for computers. **optical fibre**, *n.* a thin, hollow glass fibre which can transmit light, used in communications and in fibre optics. **optical glass**, *n.* a high-quality glass used for making lenses. **optical microscope, telescope**, *n.* one used to view objects by the light they reflect or emit, as opposed to an electron microscope or radio telescope. **optic nerve**, *n.* a nerve of sight connecting the retina with the brain. **optically**, *adv.* **optician** (-tish'ən), *n.* person who prescribes or dispenses spectacles and contact lenses to correct eye defects. **optics**, *n.sing.* the science of the nature, propagation, behaviour and function of light.

optimism (op'timizm), *n.* disposition to take a hopeful view of things; the view that the existing state of things is the best possible; the view that good must ultimately prevail. **optimal**, *a.* optimum. **optimally**, *adv.* **optimize, -ise**, *v.t.* to make the most of; to organize or perform with maximum efficiency. **optimist**, *n.* **optimistic** (-mis'-), *a.* **optimistically**, *adv.* **optimum**, *n.* (*pl.* **-ma** (-mə), **-mums**) *a.* the most favourable (condition).

option (op'shən), *n.* the right, power or liberty of choosing; a choice, preference; the thing chosen or preferred; the right to purchase something at a specified rate within a specified time. **soft option**, the easy choice between alternatives. **to keep, leave one's options open**, to refrain from committing oneself. **optional**, *a.* open to choice; not compulsory. **optionally**, *adv.*

opt(o)-, *comb. form.* pertaining to sight or optics.

optometer (optom'itə), *n.* an instrument for ascertaining the range of vision and other powers of the eye. **optometry** (-tri), *n.* **optometrist**, *n.* a fully-qualified optician.

opulent (op'ūlənt), *a.* rich, wealthy, affluent; abundant, esp. (of language) overly so. **opulence**, *n.* **opulently**, *adv.*

opus (ō'pəs, op'-), *n.* (*pl.* **opera** (op'ərə), **opuses**) a work, esp. a musical composition (*usu. written* **op.**, *pl.* **opp.**), esp. one numbered in order of publication.

or[1] (aw), *conj.* a disjunctive particle introducing an alternative; used also to connect synonyms, words explaining, correcting etc.

or[2] (aw), *n.* (*Her.*) gold.

-or (-ə), *suf.* denoting agency or condition as in *actor, author, creator, equator, terror;* (*N Am.*) *favor, vigor.*

OR, (*abbr.*) operational research; other ranks.

oracle[1] (o'rəkl), *n.* the answer of a god or inspired priest to a request for advice or prophecy; the agency or medium giving such responses; the seat of the worship of a deity, where these prophecies were sought; the sanctuary or holy of holies in the Jewish Temple; a person of profound wisdom, knowledge, or infallible judgement; an utterance regarded as profoundly wise, authoritative or infallible. *v.i.* to speak as an oracle. *v.t.* to utter as an oracle. **to work the oracle**, to secure a desired answer from the mouthpiece of an oracle by craft; to obtain some object by secret influence; to gain one's point by stratagem. **oracular** (orak'ū-), **-ulous**, *a.* **oracularly**, *adv.* **oracularity** (-la'-), *n.*

Oracle®[2] (o'rəkl), *n.* the teletext service of British Independent Television. [acronym for *O*ptional *R*eception of *A*nnouncements by *C*oded *L*ine *E*lectronics]

oral (aw'rəl), *a.* spoken, not written, by word of mouth; of, at or near the mouth; pertaining to the early stage of infantile sexual development when gratification is obtained from eating and sucking. *n.* an oral (scholastic) examination. **oracy** (-si), *n.* skill in spoken communication and self-expression. **oral contraceptive**, *n.* a contraceptive pill. **oral history**, *n.* interviews with living people about past events, recorded and written down as historical evidence. **orally**, *adv.*

orange (o'rənj), *n.* a large roundish red-yellow cellular pulpy fruit the evergreen tree on which it grows; the colour of the fruit, reddish-yellow. *a.* of the colour of an orange. **orangeade** (-jād'), *n.* a fizzy drink made from orange-juice. **orange-blossom**, *n.* the blossom of the orange-tree. **orange marmalade**, *n.* marmalade made from oranges. **orange-peel**, *n.* the rind of an orange; a pitted effect on porcelain. **orange squash**, *n.* a concentrated orange drink. **orange-stick**, *n.* a thin piece of orange-tree wood used for manicure purposes. **orange-tip**, *n.* a variety of butterfly. **orangery** (-jəri), *n.* a building designed for the cultivation of orange trees in a cool climate.

Orange (o'rənj), *a.* pertaining to the Irish extreme-Protestant party or to the Society of Orangemen formed 1795 to uphold the Protestant ascendancy in Ireland.

orang-utan (orangutan'), **-outang** (-tang'), *n.* a large, red-haired, arboreal anthropoid ape of Borneo and Sumatra.

oration (ərā'shən), *n.* a formal speech, treating of some important subject in elevated language. *v.i.* to make an oration. **orator** (o'rə-), *n.* one who delivers an oration; an eloquent speaker; an officer at a University

oratorio   566   **ordure**

who acts as public speaker on ceremonial occasions. **oratorical** (-to'ri-), *a.* **oratorically**, *adv.* **oratress** (-tris), **-trix** (-triks), *n. fem.* **oratory**[1] (o'rǝ-), *n.* the art of public speaking, rhetoric; eloquence. **oratorio** (orǝtaw'riō), *n.* (*pl.* **-ios**) a musical composition for voices and instruments, usually semi-dramatic in character, and having a scriptural theme. **oratory**[1] ORATION. **oratory**[2] (o'rǝtǝri), *n.* a small chapel, esp. one for private devotions.
orb (awb), *n.* a sphere, a globe; a heavenly body; (*poet.*) an eye or eyeball; the globe forming part of the regalia. **orbicular** (-bik'ū-), *a.* **orbicularity** (-la'ri-), *n.*
orbit (aw'bit), *n.* (*Anat., Zool., etc.*) the bony cavity of the eye; the ring or border round the eye in insects, birds etc.; the path of one celestial body around another; a sphere of action; the path of an electron around the nucleus of an atom. *v.t.* to move in a curved path around; to circle (a planet etc.) in space. *v.i.* to revolve in an orbit. **orbital**, *a.* **orbiter**, *n.* a spacecraft designed to orbit a planet etc.
orc (awk), *n.* a whale of the genus *Orca*; a sea-monster. **Orcadian** (awkā'diǝn), *a.* pertaining to the Orkney Islands. *n.* a native or inhabitant of these.
orchard (aw'chǝd), *n.* an enclosure containing fruit trees, or a plantation of these. **orchardman**, **orchardist**, *n.* **orcharding**, *n.*
orchestra (aw'kistrǝ), *n.* in an ancient Greek theatre, the semicircular space between the stage and the seats for the spectators, where the chorus danced and sang; the place for the musicians, or musicians and chorus, in modern concert-rooms, theatres etc.; an organized body of musicians playing together. **orchestra stalls**, *n.pl.* seats just behind the orchestra in a theatre. **orchestral** (-kes'-), *a.* **orchestrate**, *v.t.* to compose or arrange (music) for an orchestra; (*fig.*) to organize (e.g. events) in the desired way. **orchestration**, *n.*
orchid ((aw'kid), *n.* one of a large order of plants characterized by tuberous roots and flowers usually of a fantastic shape and brilliant colours. **orchidaceous** (-dā'shǝs), *a.* **orchidist**, *n.* **orchidology** (-dol'-), *n.*
ord., (*abbr.*) ordained; ordinal; ordinance; ordinary.
ordain (awdān'), *v.t.* to appoint and consecrate, to confer Holy Orders on; to decree, to establish, to destine. **ordainable**, *a.* **ordainer**, *n.* **ordainment**, *n.*
ordeal (aw'dēl, -dēl'), *n.* the ancient practice of referring disputed questions of criminality to supernatural decision, by subjecting a suspected person to physical tests by fire, boiling water, battle etc.; an experience testing endurance, patience, courage etc.
order (aw'dǝ), *n.* regular or methodical disposition or arrangement; sequence, succession, esp. as regulated by a system or principle; normal, proper or right condition; a state of efficiency, a condition suitable for working; tidiness, absence of confusion or disturbance; established state of things, general constitution of the world, as opp. to *chaos*; customary mode of procedure, esp. the rules and regulations governing an assembly or meeting; a rule, regulation; a mandate, an authoritative direction; a direction to supply specified commodities or to carry out specified work; a signed document instructing a person or persons to pay money or deliver property; a social class, rank or degree; kind, sort, quality; a class or body of persons united by some common purpose, e.g. a fraternity of monks or friars, or formerly of knights, bound by the same rule of life; a grade of the Christian ministry; (*pl.*) clerical office or status; a body usu. instituted by a sovereign, organized in grades like the mediaeval orders of knights, to which distinguished persons are admitted as an honour; the insignia worn by members of this; any of the nine grades of angels and archangels; a system of parts, ornaments and proportions of columns etc.,

distinguishing styles of architecture, esp. Classical, as the Doric, Ionic, Corinthian, Tuscan and Composite; (*Math.*) degree of complexity; (*Biol.*) a division below that of class and above that of family and genus. *v.t.* to put in proper sequence; to regulate; to direct, to command; to instruct (a person, firm etc.) to supply goods or perform work; to direct the supplying, doing or making of (goods etc.). *v.i.* to give orders. **by order (of)**, according to direction by (proper authority). **Holy Orders**, the different ranks of the Christian ministry; clerical office. **a tall order**, a difficult or demanding task. **in order**, properly or systematically arranged; in due sequence. **in, of the order of**, approximately (the size or quantity specified). **in order to**, so as to. **on order**, having been ordered but not yet arrived. **order of the day**, business arranged beforehand, esp. the programme of business in a legislative assembly. **out of order**, disarranged; untidy; not consecutive; not sequentially or systematically arranged; not (fit for) working. **to order**, according to, or in compliance with, an order. **to order about**, to send from one place to another; to domineer over. **to order arms**, to bring rifles vertically against the right side of the body with the butts resting on the ground. **to take orders**, to be ordained to religious office. **order-book**, *n.* a book in which orders for goods, work etc. are written; a book in which motions to be submitted to the House of Commons must be entered. **order paper**, *n.* a paper on which the order of business, esp. in Parliament, is written or printed. **orderer**, *n.* **ordering**, *n.* **orderly**[1], *a.* in order; methodical, regular; keeping or disposed to keep order, free from disorder or confusion. **orderliness**, *n.*
orderly[2], *n.* a soldier who attends on an officer to carry orders, messages etc.; a male hospital attendant. **orderly officer**, *n.* the officer of the day. **orderly-room**, *n.* a room in barracks used as the office for company or regimental business.
ordinaire (awdineǝ'), *n.* wine of ordinary grade, *vin ordinaire*.
ordinal (aw'dinǝl), *a.* (of a number) denoting order or position in a series. *n.* a number denoting this, e.g. *first*, *second*; a book containing rules, rubrics etc., for ordination in the Church of England.
ordinance (aw'dinǝns), *n.* an order, decree or regulation laid down by a constituted authority; an established rule, rite or ceremony etc. **ordinand** (-nǝnd), *n.* one preparing for Holy Orders.
ordinary (aw'dinǝri), *a.* usual, habitual, customary, regular, normal, not exceptional or unusual; commonplace; mediocre; (of a judge, having immediate or ex officio jurisdiction. *n.* a rule or order, as of the Mass; (*Her.*) one of the simplest and commonest figures. **out of the ordinary**, exceptional. **ordinary level** O-LEVEL at O[1]. **ordinary seaman**, *n.* a sailor not fully qualified as able seaman. **ordinary shares**, *n.pl.* shares in a company which pay dividends according to profit only after the claims of preference shares have been met, cp. PREFERENCE SHARES. **ordinarily**, *adv.* **ordinariness**, *n.*
ordinate (aw'dinǝt), *n.* the vertical coordinate (y-coordinate) of a point. **ordination**, *n.* the act of ordaining; the state of being ordained or appointed; appointment, ordainment.
ordnance (awd'nǝns), *n.* heavy guns, cannon, artillery; a department dealing with military stores and equipment, esp. munitions. **Ordnance datum**, *n.* the level taken as the basis for the Ordnance Survey, since 1921 the mean sea level at Newlyn, Cornwall. **Ordnance Survey**, *n.* the (Government map-making body responsible for the) survey of Great Britain and Northern Ireland.
Ordovician (awdōvish'iǝn), *n.* the middle period of the lower Palaeozoic era, which followed the Cambrian period.
ordure (aw'dyǝ), *n.* excrement, dung, filth.

ore (aw), *n.* a natural mineral substance from which metal may be extracted.

öre (œ′rə), *n.* (*pl.* **öre**) a monetary unit in Sweden and (**øre**) Norway and Denmark.

oregano (origah′nō), *n.* one of a genus of aromatic labiate herbs and shrubs comprising the wild marjoram.

organ (aw′gən), *n.* a musical wind-instrument composed of an assemblage of pipes sounded by means of a bellows and played by keys; an electronic musical instrument producing a sound imitative of this; a means (of doing something); a medium or agent of communication etc., as a newspaper or other periodical; a part of an animal or vegetable body performing some definite vital function. **barrel-organ** BARREL. **mouth-organ** MOUTH. **organ-grinder**, *n.* a player on a barrel-organ. **organ-screen**, *n.* a screen or partition, usu. between the nave and the choir, on which the organ is placed in a large church. **organ-stop**, *n.* the handle by which a set of pipes in an organ is put in or out of action; the set of pipes or reeds of a certain quality controlled by this. **organelle** (-nel), *n.* a unit in a cell having a particular structure and function. **organist**, *n.* one who plays a church or other organ.

organdie (aw′gəndi), *n.* a stiff, light, transparent muslin.

organza (-gan′zə), *n.* a thin transparent fabric of silk, rayon or nylon.

organic (awgan′ik), *a.* of or pertaining to a bodily organ or organs; of, pertaining to, or of the nature of organisms or plants or animals; (*Anat.*) pertaining to or affecting an organ or organs. (*Chem.*) containing or derived from hydrocarbons and their derivatives; organized, systematic, coordinated; structural, inherent, not accidental; vital, not mechanical; of vegetables etc., grown without artificial fertilizers, pesticides etc. **organic chemistry**, *n.* the study of the compounds of carbon. **organically**, *adv.*

organism (aw′gənizm), *n.* an organized body consisting of mutually dependent parts fulfilling functions necessary to the life of the whole; an animal, a plant; a whole having mutually related parts analogous to those of a living body.

organist ORGAN.

organize, -ise (aw′gəniz), *v.t.* to make organic, into an organism, into a living part, structure or being; to correlate the parts of and make into an organic whole; to put into proper working order; to arrange or dispose things or a body of people in order to carry out some purpose effectively. *v.i.* to become organic; to unite into an organic or effective, purposeful whole. **organizable**, *a.* **organization** (-zā′shən), *n.* the act of organizing; the state of being organized; an organized system, body or society. **organizational**, *a.* **organizer**, *n.*

organo-, *comb. form.* organ (*Anat.*); organic.

organometallic (awgənōmital′ik), *a.* of, being or relating to a compound containing linked carbon and metal atoms.

organotherapy (awgənōthe′rəpi), *n.* the treatment of disease by the administration of one or more hormones in which the body is deficient.

organza ORGANDIE.

orgasm (aw′gazm), *n.* a paroxysm of excitement or passion; the culminating excitement in the sexual act. **orgasmic** (-gaz′-), *a.*

orgy (aw′ji), *n.* (*pl.* **-gies**) secret and licentious rites, orig. the worship of Dionysus or Bacchus etc.; (*fig.*) a bout of indulgence, revelry, or (esp. sexual) debauchery. **orgiastic** (-as′-), *a.*

oribi (o′ribi), *n.* (*pl.* **-bis**) a small fawn-coloured antelope of S and E Africa.

oriel (aw′riəl), *n.* a projecting polygonal recess with a window or windows, usu. built out from an upper storey and supported on corbels or a pier. **oriel window**, *n.* the window of such a structure.

orient (aw′riənt), *n.* the East, the countries east of S Europe and the Mediterranean. *v.t.* to orientate. **Oriental** (-en′-), *a.* situated in or pertaining to the East or the (esp. Asiatic) countries east of S Europe and the Mediterranean; derived from or characteristic of the civilization etc. of the East; (*poet.*) easterly. *n.* a native or inhabitant of the East. **Orientalism**, *n.* an idiom or custom peculiar to the East; knowledge of Oriental languages and literature. **Orientalist**, *n.* **Orientally**, *adv.*

orientate (aw′riəntāt), *v.t.* to determine the position of, lit. with reference to the east and accordingly to all points of the compass; to determine one's physical or mental situation. *v.i.* to turn or face towards the east. **orientation**, *n.* the act of orientating oneself; the determination of one's position, mental or physical, with regard to the situation or surroundings. **orienteer** (-tiə′), *v.i.* to take part in orienteering. *n.* one who orienteers. **orienteering**, *n.* a sport in which the contestants race on foot cross-country, following checkpoints located by a map and compass.

orifice (o′rifis), *n.* an opening or aperture, esp. (as) of the mouth, nose, ears etc.

orig., (*abbr.*) origin; original(ly).

origami (origah′mi), *n.* the (traditionally Japanese) art of paper folding.

origin (o′rijin), *n.* beginning, commencement or rise (of anything); derivation, source; extraction, ancestry; (*Math.*) the point where coordinate axes intersect; (*Anat.*) the point of attachment of a muscle. **original** (ərij′-), *a.* of or pertaining to the origin, beginning, or first stage; first, primary, primitive; innate; not copied or imitated, not produced by translation; fresh, novel; able to devise, produce, think or act for oneself; inventive, creative. *n.* the pattern, the archetype, the first version; that from which a work is copied or translated; the language in which a work is written; an eccentric person. **original sin**, the sin of Adam in eating the forbidden fruit; the innate depravity of man. **originality** (-nal′-), *n.* **originally**, *adv.* **originate** (ərij′-), *v.t.* to be the origin of; to cause to begin, to bring into existence. *v.i.* to rise, to begin; to have origin (in, from or with). **origination**, *n.* **originative**, *a.* **originator**, *n.*

orinasal (awrinā′zəl), *a.* of or pertaining to or sounded by the mouth and nose.

oriole (aw′riōl), *n.* a European bird with bright-yellow and black plumage (**golden oriole**): an American orange and black bird of similar genus (**Baltimore oriole**).

Orion (əri′ən), *n.* one of the best-known constellations, a group of stars representing a hunter with belt and sword. **Orion's belt**, *n.* a row of three bright stars across the middle of this constellation.

orison (o′rizən), *n.* a prayer, a supplication.

Oriya (orē′yə), *n.* a member of a people living in Orissa in India; the language of Orissa.

Orlon® (aw′lon), *n.* (a fabric made from) acrylic fibre.

orlop (deck) (aw′lop), *n.* the lowest deck of a vessel having three or more decks.

ormer (aw′mə), *n.* a sea-ear or ear-shell, an edible gasteropod mollusc.

ormolu (aw′məloo), *n.* orig., gold leaf ground and used as a pigment for decorating furniture etc.; a gold-coloured alloy of copper, zinc and tin, used for cheap jewellery; metallic ware, furniture etc. decorated with this.

ornament (aw′nəmənt), *n.* a thing or part that adorns, an embellishment, a decoration; ornamentation; a person, possession or quality that reflects honour or credit; (*Mus.*) decorations such as trills, mordents etc. to be improvised. *v.t.* (-ment), to adorn, to decorate, to embellish. **ornamental** (-men′-), *a.* **ornamentalist**, *n.* **ornamentally**, *adv.* **ornamentation**, *n.* **ornamenter**, *n.*

ornate (awnāt′), *a.* adorned, ornamented, richly embellished; florid, elaborate (of literary style etc.). **ornately**, *adv.* **ornateness**, *n.*

ornery (aw′nəri), *a.* (*N Am., coll.*) stubborn; mean, low.

ornith(o)-, *comb. form.* pertaining to birds.

ornithology (awnithol′əji), *n.* the branch of zoology dealing with birds. **ornithological** (-loj′-), *a.* **ornithologist**, *n.*

ornithopter (awnithop'tə), *n.* an aeroplane driven by power supplied by the aviator and not by an engine.
orotund (o'rətūnd), *a.* characterized by fullness and resonance; rich and musical (said of the voice and utterance); pompous, inflated.
orphan (aw'fən), *n.* a child bereft of one parent or both. *a.* bereft of one parent or both. **orphanage** (-ij), *n.* an institution for bringing up orphans. **orphaned**, *a.* **orphanhood** (-hud), **orphanism,** *n.*
Orphean (aw'fiən), *a.* pertaining to Orpheus, a celebrated mythical musician of Thrace, or his music.
orphrey (aw'fri), *n.* a band of gold and silver embroidery decorating an ecclesiastical vestment.
Orpington (aw'pingtən), *n.* a variety of domestic fowl.
orrery (o'rəri), *n.* a mechanical model for illustrating the motions, magnitudes and positions of the planetary system.
orris (o'ris), *n.* a kind of iris. **orris-root,** *n.* the root of one of three species of iris, used as a perfume and in medicine.
ortho-, *comb. form.* straight; upright; perpendicular; correct.
orthocephalic (awthōsifal'ik), *a.* having a breadth of skull from 70 to 75 per cent of the length, between brachycephalic and dolichocephalic.
orthochromatic (awthōkrəmat'ik), *a.* giving the correct values of colours in relations of light and shade.
orthoclase (aw'thəklās, -klāz), *n.* common or potash feldspar having a right-angled cleavage.
orthodontics (awthədon'tiks), *n.sing.* dentistry dealing with the correction of irregularities of the teeth. **orthodontic,** *a.* **orthodontist,** *n.*
orthodox (aw'thədoks), *a.* holding correct or accepted views, esp. in matters of faith and religious doctrine; in accordance with sound or accepted doctrine; approved, accepted, conventional, not heretical, heterodox or original. **Orthodox Church,** *n.* the Eastern or Greek Church. **orthodoxly,** *adv.* **orthodoxy,** *n.*
orthoepy (awthō'əpi), *n.* the branch of grammar dealing with pronunciation, phonology; correct speech or pronunciation. **orthoepic, -ical** (-ep'-), *a.* **orthoepically,** *adv.* **orthoepist,** *n.*
orthogenesis (awthəjen'əsis), *n.* a theory of evolution that postulates that variation is determined by the action of environment. **orthogenetic** (-net'-), *a.*
orthognathous (awthog'nəthəs), *a.* (*Craniology*) straight-jawed, having little forward projection of the jaws. **orthognathic** (-nath'-), *a.* **orthognathism,** *n.*
orthogonal (awthog'ənəl), *a.* right-angled. **orthogonally,** *adv.*
orthography (awthog'rəfi), *n.* correct spelling; that part of grammar which deals with letters and spelling; a writing system; the art of drawing plans, elevations etc., in accurate projection (**orthographic projection**), as if the object were seen from an infinite distance. **orthographer, -phist,** *n.* **orthographic, -ical** (-graf'-), *a.* **orthographically,** *adv.*
orthopaedic, -ical, *comb.* (*esp. N Am.*) **orthopedic, -ical,** *a.* **orthopaedics, orthopedics** (awthəpē'diks), *n.* the act or art of curing muscular or skeletal deformities by surgery, esp. in children. **orthopaedist, -pedist,** *n.*
Orthoptera (awthop'tərə), *n.pl.* an order of insects with two pairs of wings, the hind wings membranous and those in front stiff and usually straight. **orthopteral, -ous,** *a.* **orthopteran,** *n.*, *a.*
orthorhombic (awthōrom'bik), *a.* having three planes of dissimilar symmetry at right angles to each other.
orthoscope (aw'thəskōp), *n.* an instrument for examining the interior of the eye.
ortolan (aw'tələn), *n.* the garden bunting or ortolan bunting, esteemed as a delicacy; applied to several W Indian and American birds.
-ory[1] (-əri), *comb. form.* denoting place where or instrument, as in *dormitory, lavatory, refectory.*
-ory[2] (-əri), *comb. form.* forming adjectives, as *amatory, admonitory, illusory.*
oryx (o'riks), *n.* (*pl.* **-yxes, -yx**) a genus of straight-horned African antelopes.
oryza (orī'zə), *n.* a tropical genus of grasses comprising rice.
OS, (*abbr.*) Old Style; Ordinary Seaman; Ordnance Survey; outsize.
Os, (*chem. symbol*) osmium.
os (os), *n.* (*pl.* **ossa** (-ə)) a bone.
Oscar (os'kə), *n.* a gold-plated statuette awarded by the American Academy of Motion Picture Arts and Sciences to the actor, director, film-writer etc. whose work is adjudged the best of the year.
oscillate (os'ilāt), *v.i.* to swing, to move like a pendulum; to vibrate; to fluctuate, to vacillate, to vary. **oscillation,** *n.* the movement of oscillation; the regular variation in an alternating current; a single cycle (of something oscillating). **oscillative, -tory,** *a.* **oscillator,** *n.* someone or something that oscillates; a device for producing alternating current. **oscilloscope** (osil'əskōp), *n.* an instrument which registers the oscillations of an alternating current or the fluorescent screen of a cathode-ray tube; an instrument to facilitate the detection of vibrations and other faults in machinery.
osculate (os'kūlāt), *v.t.* (*facet.*) to kiss; (*Geom.*) to touch at three or more poiints. *v.i.* (*facet.*) to kiss; (*Geom.*) to touch at three or more points; (*Biol.*) to be related to, having characters in common or forming an intermediate species etc. **osculant,** *a.* **osculation,** *n.* **osculatory** (os'-), *a.* (*facet.*) kissing; (*Geom.*) osculating.
-ose[1] (-ōs), *comb. form.* denoting fullness, abundance, as in *grandiose, jocose, verbose.*
-ose[2] (-ōs), *comb. form.* denoting the carbohydrates and isomeric compounds, as in *lactose.*
osier (ō'ziə), *n.* a species of willow, the pliable shoots of which are used for basket-making.
-osis (-ōsis), *comb. form.* denoting condition, esp. morbid states, as *necrosis, psyclosis.*
-osity (-ositi), *comb. form.* forming nouns from adjectives in -OSE or -OUS, as *grandiosity, luminosity.*
Osmanli (ozman'li), *a.* of or pertaining to the Ottoman Empire, the W branch of the Turkish peoples or their language. *n.* a member of the Ottoman dynasty; a Turk.
osmiridium (ozmirid'iəm), *n.* a very hard natural alloy of osmium and iridium used esp. in pen nibs.
osmium (oz'miəm), *n.* the heaviest known metallic element, at. numb. 76; chem. symbol Os, usu. found in association with iridium. **osmic,** *a.*
osmosis (osmō'sis. oz-), *n.* the diffusion of a solvent through a semipermeable membrane into a more concentrated solution until both solutions are of like concentration; (*fig.*) unconscious, gradual assimilation (of knowledge etc.). **osmose,** *v.t., v.i.* to (cause to) diffuse by osmosis. **osmograph** (oz'məgraf), **osmometer** (-mom'itə), *n.* an instrument for measuring osmotic pressures. **osmotic** (-mot'-), *a.* **osmotic pressure,** *n.* the pressure required to prevent osmosis. **osmotically,** *adv.*
osnaburg (oz'nəbœg), *n.* a coarse kind of linen originally made in Osnabrück.
osprey (os'prā), *n.* a large bird, preying on fish, also known as the sea-eagle or sea-hawk; an egret plume used for trimming hats and bonnets (a term used erroneously by milliners).
ossein (os'iin), *n.* the gelatinous tissue left when mineral matter is eliminated from a bone.
osseous (os'iəs), *a.* of the nature of or like bone, bony; consisting of bone, ossified.
ossicle (os'ikl), *n.* a small bone; a bony part or outgrowth in various animals.
ossify (os'ifī), *v.t.* to turn into bone. *v.i.* to become bone;

to become inflexible in attitudes, habits etc. **ossification** (-fi-), *n.*

ossuary (os'ūəri), *n.* a charnel-house; a bone-urn; a deposit of bones (as in a cave).

oste- OSTE(O)-.

osteology (ostiol'əji), *n.* the study of bones.

ostensible (əsten'sibl), *a.* put forward for show or to hide the reality; professed, pretended, seeming. **ostensibly,** *adv.* **ostensive,** *a.* exhibiting, showing; ostensible. **ostensively,** *adv.* **ostentation** (os-), *n.* pretentious or ambitious display; parade, pomp. **ostentatious,** *a.* **ostentatiously,** *adv.* **ostentatiousness,** *n.*

oste(o)-, *comb. form.* bone.

osteoarthritis (ostiōahthrī'tis), *n.* degenerative arthritis, esp. of the weight-bearing joints of the spine, hips and knees. **osteoarthritic** (-thrit'-), *a.*

osteoblast (os'tiəblast), *n.* a cell concerned in the development of bone.

osteoclasis (ostiok'ləsis), *n.* the operation of breaking a bone to remedy a deformity etc. **osteoclast** (os'tiōklast), *n.* a surgical tool for osteoclasis.

osteoid (os'tioid), *a.* like bone.

osteology (ostiol'əji), *n.* the branch of anatomy dealing with bones, osseous tissue etc. **osteologic, -ical** (-loj'-), *a.* **osteologist,** *n.*

osteomyelitis (ostiōmiəli'tis), *n.* inflammation of bone marrow.

osteopathy (ostiop'əthi), *n.* a method of treating diseases by manipulation, mainly of the spinal column. **osteopathist, osteopath** (os'tiəpath), *n.* a practitioner of oestopathy.

osteoplasty (os'tiəplasti), *n.* transplantation of bone. **osteoplastic** (-plas'-), *a.*

osteoporosis (ostiōpawrō'sis), *n.* development of porous or brittle bones due to lack of calcium in the bone matrix.

osteotome (os'tiətōm), *n.* an instrument used in the dissection of bones. **osteotomy** (-ot'əmi), *n.* (an instance of) the dissection of bones.

ostinato (ostinah'tō), *n.* (*pl.* **-tos**) a musical figure continuously reiterated throughout a composition.

ostium (os'tiəm), *n.* (*pl.* **-tia**) (*Anat.*) the mouth or opening of a passage.

ostler, hostler (os'lə), *n.* a man who looks after horses at an inn, a stableman.

ostmark (ost'mahk), *n.* the standard unit of currency in East Germany.

ostracize (os'trəsiz), *v.t.* (*Gr. Ant.*) in ancient Greece, to banish by popular vote; to exclude from society, to ban, to send to Coventry. **ostracism,** *n.*

ostrich (os'trich), *n.* a large African bird with rudimentary wings, but capable of running at great speed, and greatly valued for its feathers, which are used as plumes; a person who refuses to recognize unpleasant facts.

Ostrogoth (os'trəgoth), *n.* an eastern Goth, one of the division of the Gothic peoples who conquered Italy in the 5th cent. **ostrogothic** (-goth'-), *a.*

OT, (*abbr.*) occupational therapy; Old Testament; overtime.

ot-, oto-, *comb. form.* pertaining to the ear.

otalgia (ōtal'jiə), *n.* earache.

other (ŭdh'ə), *a.* not the same as one specified or implied; different, distinct in kind; alternative or additional; second, only remaining (of two alternatives); opposite, contrary. *n., pron.* an or the other person, thing, example, instance etc. *adv.* otherwise. **every other,** every alternate (day, week etc.). **other ranks,** non-commissioned servicemen. **someone, something or other,** an unspecified or vaguely defined person or thing. **the other day** etc., on a day etc. recently. **otherness,** *n.* **otherwise,** *adv.* in a different way or manner; in other respects; by or from other causes; in quite a different state. **other world,** *n.* the future life; a world existing outside of or in a different mode from this; fairy-land. **other-worldly,** *a.* preoccupied with imagination rather than reality. **other-worldliness,** *n.*

-otic (-otik), *comb. form.* forming adjectives corresponding to nouns in -osis, as *neurotic, osmotic.*

otiose (ō'tiōs), *a.* not wanted, useless, superfluous. **otiosely,** *adv.* **otioseness,** *n.*

otitis (ōti'tis), *n.* inflammation of the ear.

otology (ōtol'əji), *n.* the science of the ear or of diseases of the ear; anatomy of the ear. **otologist,** *n.*

otorhinolaryngology (ōtōrinōlaringol'əji), *n.* ear, nose and throat medicine. Also **otolaryngology,** *n.*

otoscope (ō'təskōp), *n.* an instrument for inspecting the ear and ear-drum.

ottava rima (otah'və rē'mə), *n.* a form of versification consisting of stanzas of eight lines, of which the first six rhyme alternately, and the last two form a couplet.

otter (ot'ə), *n.* a furred, web-footed, aquatic European mammal feeding exclusively on fish; its fur; the sea-otter.

otto (ot'ō), ATTAR.

Ottoman (ot'əmən), *a.* of or pertaining to the dynasty of Othman or Osman I; pertaining to the Turks. *n.* a Turk.

ottoman (ot'əmən), *n.* a cushioned seat or sofa without back or arms, introduced from Turkey.

OU, (*abbr.*) Open University; Oxford University.

oubliette (ooblict'), *n.* an underground dungeon opening only from above, in which persons condemned to perpetual imprisonment or secret death were confined.

ouch (owch), *int.* used to express sudden pain.

oud (ood), *n.* a Middle Eastern stringed musical instrument resembling a lute.

ought[1] (awt), *v.aux.* to be found in duty or rightness; to be necessary, fit or proper; to behove.

ought[2] (awt), AUGHT.

Ouija®**, Ouija-board**[®] (wē'jə), *n.* a board inscribed with the letters of the alphabet, used for receiving messages etc. in seances.

ounce[1] (owns), *n.* a unit of weight, the 12th part of a pound troy, and 16th part of a pound avoirdupois (about 28 g); (*loosely*) a small quantity.

ounce[2] (owns), *n.* a lynx or (*loosely*) other leopard-like animal; the mountain-panther of S and Central Asia, also called the snow leopard.

our (owə), *a.* of, pertaining to or belonging to us; used instead of 'my' by royalty, editors, reviewers etc. **ours** (owəz), *a.* belonging to us. *pron.* that or those belonging to us. **ourself** (-self'), *pron.* (*pl.* **-selves** (-selvz'), or **-self** when a sovereign) myself (used in regal or formal style); (*pl.*) we, not others, we alone (usu. emphatically in apposition with *we*); (*reflex.*) the persons previously spoken of as we.

-our (-ə), *comb. form.* forming nouns, as *amour, ardour, clamour.*

ourang-outang (orangutang'), ORANG-UTAN.

-ous (-əs), *comb. form.* full of, abounding in, as *glorious, lustrous;* (*Chem.*) denoting a compound having much of the element indicated in the stem than those whose names end in -ic; as *ferrous, nitrous, sulphurous.*

oust (owst), *v.t.* to eject, to expel, to turn out (from); to dispossess, to deprive (of). **ouster,** *n.*

out (owt), *adv., adj.* (*usu. pred.*) from the inside, from within; not in, not within; from among; forth or away (from); not at home, not in office; not engaged or employed; on strike; (*Cricket*) not batting, dismissed from the wicket; (*Boxing*) denoting defeat through inability to rise within the ten seconds allowed after being knocked down; (*coll.*) not in fashion; not in practice; in error, wrong; at odds, not in agreement; not to be thought of; not permissible or possible; (written, read etc.) so as to be visible, audible, revealed, published etc.; introduced to society; exhausted or extinguished; (to speak etc.) clearly; at full extent; asleep, or otherwise no longer conscious; to an end or conclusion, completely, thoroughly; of a card game etc., solved; on loan; having lost or spent (a quant-

ity of money etc.); flowering. *prep. (coll.)* from inside of. *int. (ellipt.)* begone! away! an expression of impatience, anger or abhorrence. **all out**, striving to the uttermost. **from out**, out of. **murder will out**, the guilt will be disclosed; the secret is bound to be revealed. **out and about**, (able to get up and go) outside. **out and away** AWAY. **out and out**, completely, unreservedly. **out-and-out**, *a. (attrib.)* **out at elbows** ELBOW. **out of date** DATE¹. **out of doors**, outdoors. **out of hand** HAND. **out of it**, not included, neglected; at a loss. **out of one's head**, delirious. **out of pocket** POCKET. **out of print**, sold out and no longer on sale by the publisher (of books). **out of sorts**, indisposed, unwell. **out of temper**, irritated, vexed. **out of the way**, unusual; remote. **out-of-the-way**, *a. (attrib.)* **out of trim**, not in good order. **out for**, *a.* striving for; (*Cricket*) dismissed having scored (a number of runs). **out of**, *adv.* from the inside of; from among; beyond the reach of; from (material, source, condition etc.); born of (esp. horses); lacking stock of; denoting deprivation or want of (work, fitness, patience, ideas etc.). **outer**, *a.* being, worn etc. on the exterior side, external; farther from the centre or the inside; objective, material, not subjective or psychical. *n.* the part of a target outside the rings round the bull's-eye the external packaging (of something). **outer space**, *n.* the vast, immeasurable region beyond the Earth. **outermost**, *a.* **outing**, *n.* an excursion, a pleasure-trip, an airing.
out- (owt-), *pref.* out, towards the outside, external; from within, forth; separate, detached, at a distance; denoting issue or result; expressing excess, exaggeration, superiority, surpassing, defeating, enduring, getting through or beyond. **outback** (owt'-), *n., a., adv.* (*Austral.*) (in, of) the hinterland, the bush, the interior. **outbalance**, *v.t.* to outweigh, to exceed. **out-bargain**, *v.t.* to get the better of in a bargain. **outbid**, *v.t.* (*past* **-bad, -bade**, *p.p.* **-bidden**) to bid more than; to outdo by offering more. **outboard** (owt'-), *a.* situated on or directed towards the outside of a ship; of a motor, having an engine and propeller outside the boat. *adv.* out from a ship's side. **outbound** (owt'-), *a.* outward bound, departing. **out-brag**, *v.t.* to outdo in bragging. **outbreak** (owt'-), **-breaking**, *n.* a sudden bursting forth, an eruption; an epidemic; a riot or insurrection. **outbreed**, *v.t.* **outbreeding** (owt'-), *n.* interbreeding of unrelated plants or animals. **outbuilding** (owt'-), *n.* a detached building, an outhouse. **outburst** (owt'-), *n.* an outbreak, an explosion; an outcry. **outcast** (owt'-), *a.* rejected, cast out; exiled. *n.* an exile, a person rejected. **outclass**, *v.t.* to be of a superior class, kind or standard to; to surpass as a competitor. **outcome** (owt'-), *n.* result, consequence, effect. **outcrop** (owt'-), *n.* (*Geol.*) the exposure of a stratum at the surface. *v.i.* to show at the surface. **outcry** (owt'-), *n.* a vehement or loud cry; noise, clamour, esp. in indignation. **outdare**, *v.t.* to exceed in daring. **outdated**, *a.* obsolete, out of date. **outdistance**, *v.t.* to outstrip. **outdo**, *v.t.* to excel, to surpass. **outdoor** (owt'-), *a.* living, existing, being, happening etc. in the open air. **outdoors**, *adv.* in the open air, out of the house. **outdweller** (owt'-), *n.* one who lives outside of or beyond certain limits. **outface**, *v.t.* to brave; to confront boldly; to stare down. **outfall** (owt'-), *n.* the point of discharge of a river, drain etc. **outfield** (owt'-), *n.* (*Cricket, Baseball*) the part of the field at a distance from the batsman. **outfielder**, *n.* **outfit** (owt'-), *n.* the act of equipping for a journey, expedition etc.; the tools and equipment required for (a trade, hobby, profession etc.); a set of (esp. selected) clothes; (*coll.*) a set or group of people who work as a team or a business concern. *v.t.* to provide with an outfit. **outfitter**, *n.* one who deals in outfits for journeys, athletic sports, ceremonies, schools etc. **outflank**, *v.t.* (*Mil.*) to go or pass beyond the flank of; to use strategy to get the better of. **outflow** (owt'-), *n.* the process of flowing out; that which flows out; a place of flowing out, an outfall. **outfly**, *v.t.* to fly faster than; to outstrip. **outfox**, *v.t.* to outwit; to surpass in cunning. **out-general**, *v.t.* to surpass in generalship; to manoeuvre so as to get the better of. **outgoing** (owt'-), *a.* departing; vacating office. *n.* departure, termination; (*usu. in pl.*) outlay, expenditure. **outgrow**, *v.t.* to surpass in growth; to grow too much or too great for; to grow too old for or too large for. **outgrowth** (owt'-), *n.* something, or the process of, growing out from a main body; a result or by-product (from). **outgun**, *v.t.* to defeat with superior weaponry; (*fig.*) to surpass. **out-Herod**, *v.t.* to outdo, to exaggerate; to surpass any kind of excess [*Herod*, Tetrarch of Galilee, represented in the old miracle-plays as a swaggering tyrant] **outhouse** (owt'-), *n.* a smaller building away from the main building. **outlander**, *n.* a foreigner, a stranger; an alien settler. **outlandish** (-lan'-), *a.* foreign-looking, strange, extraordinary; bizarre, unconventional. **outlast**, *v.t.* to last longer than; to surpass in duration, endurance etc. **outlaw** (owt'-), *n.* a lawless person; a fugitive. *v.t.* to declare illegal. **outlay** (owt'-), *n.* expenditure. *v.t.* to expend, esp. in anticipation. **outleap**, *v.t.* to surpass in leaping, to leap farther than. **outlet** (owt'lit), *n.* a passage outwards, a vent; a means of expression; (a point of sale via which one reaches) a market (for something). **outlier** (owt'-), *n.* a person or thing detached from a main body or group; (*Geol.*) a portion of a bed detached from the main mass by denudation of the intervening parts. **outline** (owt'-), *n.* the line or lines enclosing and defining a figure; a drawing of such lines without shading; the first general sketch, rough draft or summary; (*pl.*) general features, facts, basic principles etc. *v.t.* to draw the outline of; to sketch; to detail the main features of. **outlive**, *v.t.* to survive; to outlast. **outlook**, *n.* (owt'-), prospect, general appearance of things, esp. as regards the future; a view, a prospect. **outlying** (owt'-), *a.* situated at a distance, or on the extreme of a frontier. **outmanoeuvre**, *v.t.* to get the better of by manoeuvring or strategy. **out-march**, *v.t.* to march faster than, outstrip by marching. **outmoded**, *a.* out of fashion. **outnumber**, *v.t.* to exceed in number. **outpace**, *v.t.* to walk, run etc. faster than. **out-patient** (owt'-), *n.* a patient receiving treatment at a hospital without being a resident. **out-perform**, *v.t.* to do (much) better than. **outplace** (owt'-), *v.t.* **outplacement**, *n.* professional relocation of redundant employees arranged by their former employer. **outplacer**, *n.* person or agency providing this service. **outplay**, *v.t.* to play better than or defeat an opponent in a game. **outpoint**, *v.t.* to score more points than. **outpost** (owt'-), *n.* a post or station at a distance from the main body; a remote habitation. **outpour**, *v.t.*, *v.i.* to pour out; to discharge. **outpouring** (owt'-), *n.* **output** (owt'-), *n.* the produce of a factory etc.; the aggregate amount produced; the data produced by a computer; the signal, voltage etc. delivered by an electronic system or device. *v.t.* to produce output. **outrank**, *v.t.* to have higher rank than. **outreach**, *v.t.* to exceed in reach, to surpass. *n.*, *a.* (owt'-), (of) welfare work which actively seeks out its potential beneficiaries. **outride**, *v.t.* to ride faster than. **outrider** (owt'-), *n.* an escort who rides ahead of or beside a carriage; one sent in advance as a scout, or to discover a safe route etc. **outrigger** (owt'-), *n.* a projecting spar or framework extended from the sides of a ship for various purposes, esp. balance; a bracket carrying a rowlock projecting from the sides of a boat to give increased leverage in rowing; a boat with these. **outrigged**, *a.* **outright**, *adv.* completely, entirely; openly. *a.* (owt'-), downright, extreme, unrestrained. **outrightness** (owt'-), *n.* **outrival**, *v.t.* to surpass as a rival. **outroar**, *v.t.* to roar louder than. **outrun**, *v.t.* to run faster or farther than, to outstrip; to escape by running. **outsell**, *v.t.* to exceed in price or value; to sell more or faster than. **outset** (owt'-), *n.* commencement, beginning, start. **outshine**, *v.t.* to

excel in lustre; to surpass in splendour. **outsize**, *n.*, *a.* (a person or thing) abnormally large; (a ready-made garment) much larger than the standard size. **outskirt** (owt'-), *n.* (*usu. in pl.*) the outer border. **outsmart**, *v.t.* (*coll.*) to outwit; to get the better of. **outspoken**, *a.* (sometimes too) open, candid, frank in speech. **outspokenly**, *adv.* **outspokenness**, *n.* **outspread**, *a.* (owt'-), spread wide. **outstanding**, *a.* remaining unpaid; projecting outward; salient, conspicuous, prominent; superior, excellent. **outstation** (owt'-), *n.* (*Austral.*) a distant (sheep) station. **outstay**, *v.t.* to stay longer than (a specified time, esp. that for which one is welcome, or another person). **outstretched**, *a.* extended, esp. in welcome. **outstrip**, *v.t.* to outrun, to leave behind; to escape by running; to surpass in progress. **out-take** (owt'-), *n.* an unreleased piece of recorded film, television or music. **out-talk**, *v.t.* to outdo in talking. **out-thrust** (owt'-), *n.* outward thrust or pressure. *a.* thrust or projected forward or outward. *v.t.* (-thrŭst') to thrust forth or forward. **out-tray** (owt'-), *n.* a tray in an office for outgoing documents, correspondence etc. **outvote**, *v.t.* to out-number in voting; to cast more votes than. **out-walk**, *v.t.* to outdo or outstrip in walking. **outwear**, *v.t.* to wear out; to exhaust, to weary out; to last longer than. **outweigh**, *v.t.* to weigh more than; to be of more value, importance etc. than. **outwell**, *v.i.* to pour or flow forth. **outwit**, *v.t.* to defeat by superior ingenuity or cunning. **outwork** (owt'-), *n.* a fortification of a place outside the parapet; work done by an outworker. *v.t.* (-wœk') to work faster than. **outworker** (owt'-), *n.* one who works outside (a factory, shop etc.) i.e., at home, often under sub-contract. **outworn**, *a.* worn out; of usefulness, past; obsolete.

outer, outing OUT.

outrage (owt'răj), *n.* wanton injury to or violation of the rights of others; a gross offence against order or decency; a flagrant insult. *v.t.* commit an outrage on; to injure or insult in a flagrant manner. **outraged**, *a.* furious, grossly offended. **outrageous** (-rā'-), *a.* flagrant, heinous, atrocious; extravagant; grossly offensive or abusive. **outrageously**, *adv.* **outrageousness**, *n.*

outré (oo'trā), *a.* eccentric; outraging convention or decorum.

outside (owtsīd', owt'-), *n.* the external part or surface, the exterior; external superficial appearance; that which is on the exterior; external space, region etc.; the utmost limit, the extreme. *a.* pertaining to, situated on or near(er) to the outside, outer; external; highest or greatest possible (estimate etc.); remotest, most unlikely (e.g. chance). *adv.* to or on the outside; without, not within; (*sl.*) not in prison. *prep.* at, on, or to the exterior of; out from, forth from; beyond the limits of. **outside in**, having the outer side turned in, and vice versa. **outside of**, *prep.* (*coll.*) outside; apart from. **outside broadcast**, *n.* a radio or television broadcast from outside the studio. **outside-left, -right**, *n.* (*Football*, *Hockey*) a member of a team who plays on the extreme left or right. **outside seat**, *n.* one at the open end of a row. **outsider** (-sī'-), *n.* one who is not a member of a profession, party, circle, coterie, district etc.; one not acquainted with or interested in something that is going on; one not admissible to decent society; (*Racing etc.*) a horse or competitor not among the favourites.

outward (owt'wəd), *a.* exterior, outer; tending or directed toward the outside; external, visible, superficial. *adv.* outwards. **outward bound**, *a.* going away from home; **outward-bound, Outward-Bound**) involving outdoor physical activities and training. **outward form**, *n.* appearance; physical form. **outwardly**, *adv.* **outwardness**, *n.* **outwards**, *adv.*

ouzel (oo'zəl), *n.* one of various thrush-like birds, including the dipper or water-ouzel.

ouzo (oo'zō), *n.* (*pl.* **ouzos**) an aniseed-flavoured alcoholic drink from Greece.

ova OVUM.

oval (ō'vəl), *a.* egg-shaped, roughly elliptical. *n.* a closed convex curve with one axis longer than the other; an egg-shaped figure or thing, e.g. a sports field. **ovally**, *adv.* **ovalness**, *n.*

ovary (ō'vəri), *n.* one of the organs (two in number in the higher vertebrates) in a female in which the ova are produced; (*Bot.*) the portion of the pistil in which the ovules are contained. **ovarian** (-veə'ri-), *a.* **ovariectomy** (-ek'təmi), *n.* the removal of the ovary by surgery, or of a tumour from the ovary. **ovate** (ō'vāt), *a.* egg-shaped.

ovation (ōvā'shən), *n.* a display of popular favour, an enthusiastic reception or bout of applause.

oven (ŭv'ən), *n.* a closed chamber in which substances (esp. foods) are baked etc.; a furnace or kiln for assaying, annealing, firing etc. **Dutch oven** DUTCH. **oven glove**, *n.* a thick glove for handling hot dishes. **oven-ready**, *a.* of food, already prepared for immediate cooking in an oven. **ovenware**, *n.* heat-resistant dishes used for cooking food in an oven.

over (ō'və), (*poet.*) **o'er** (aw), *prep.* above, in a physically higher position than; above or superior to in excellence, dignity or value; more than, in excess of; in charge of; across from side to side of; through the extent or duration of; having recovered from the effect of. *adv.* so as to pass from side to side or across some space, barrier etc.; in width, in distance across; on the opposite side; so as to be turned down or upside down from an erect position; so as to be across or down from a brink, edge, brim etc.; so as to traverse a space etc.; from end to end, throughout; in excess, in addition; excessively, with repetition, again. *a.* at an end. *int.* in radio signalling etc., indicating that a reply is expected. *n.* (*Cricket*) the interval between the times when the umpire calls 'over'; the number of balls (usually 6 or 8) delivered by one bowler during this. **all over**, completely, everywhere; (him or her) typically; finished. **over again**, *adv.* afresh, anew. **over against**, *prep.* opposite; in front of. **over and above**, in addition to; besides (something). **over and over**, so as to turn completely round or upside down several times; repeatedly. **over one's head**, beyond one's comprehension or ability. **to give over** GIVE. **to turn over** TURN.

over- (ōvə'-), *pref.* above; across; outer, upper; as a covering; past, beyond; extra; excessively, too much, too great(ly). **overabound**, *v.i.* to be superabundant; to abound too much (in or with). **overact**, *v.t.* to act (a part) in an exaggerated way. *v.i.* to act more than is necessary. **overall**, *a.*, *adv.* from end to end, (in) total. **overall**, *n.sing.* **overalls** (ō'-), *n.pl. or sing.* (a pair of) trousers, a one-piece suit or other garment(s) worn over normal clothing as a protection against dirt etc. **overarch**, *v.t.* to form an arch over. *v.i.* to form an arch overhead. **overarm** (ō'-), *a.* in sports, esp. cricket, bowled or thrown with the arm raised above the shoulder. *adv.* with the arm raised above the shoulder. **overawe**, *v.t.* to keep in awe; to control or restrain by awe. **overbalance**, *v.t.* to outweigh; to destroy the equilibrium of. *v.i.* to lose one's equilibrium; to topple over. **overbearing**, *a.* arrogant, haughty, imperious. **overbearingly**, *adv.* **overbid**, *v.t.*, *v.i.* to outbid; to bid more than the value of (one's hand of cards). *n.* (ō-), a higher bid. **overblown**, *a.* inflated, pretentious. **overboard** (ō'-), *adv.* over the side of a ship; out of a ship. **to go overboard about, for**, (*coll.*) to go to extremes of enthusiasm about, for. **over-bold**, *a.* bold to excess. **over-boldly**, *adv.* **over-boldness**, *n.* **overbook**, *v.t.*, *v.i.* to make bookings for more (people) than there are places available (e.g. in a hotel, plane, ship etc.). **overburden**, *v.t.* to overload, physically or mentally. **overcall**, *v.t.* to bid higher than (a previous bid or player) at bridge. *n.* (ō'-), a higher bid than the preceding one. **over-capitalize, -ise**, *v.t.* to fix the actual, or rate the nominal value of the capital of (a company etc.) too high. **over-careful**, *a.* careful to excess. **overcast**, *v.t.* to sew (an edge etc.) with long stitches to prevent unravelling etc., or as embroidering. *a.* clouded all over (of the sky); sewn or embroidered by overcasting. **overcasting**, *n.* **over-caution**, *n.* excess of caution. **over-cautious**, *a.* **over-cautiously**, *adv.* **overcharge**, *v.t.* to charge (someone) more than is properly due; to overburden, to overload. *v.i.* to charge too much. *n.* (ō'-) an excessive charge, load or burden. **overcoat** (ō'-), *n.* a great-coat, a top-coat. **overcoated**, *a.* **overcoating**, *n.* material for overcoats. **overcome**, *v.t.* to overpower; to vanquish, to conquer. **over-compensate**, *v.t.* to provide too much in compensation. *v.i.* to react to feelings of one kind with excessive demonstration of the opposite. **overcompensation**, *n.* **over-confidence**, *n.* excessive confidence. **over-confident**, *a.* **over-confidently**, *adv.* **over-credulous**, *a.* too credulous. **over-credulously**, *adv.* **over-credulity**, *n.* **overcrop**, *v.t.* to crop (land) to excess; to exhaust by continual cropping. **overcrowd**, *v.t.*, *v.i.* to throng with too many people. **overcrowding**, *n.* **over-cunning**, *a.* too cunning. **over-curious**, *a.* too curious. **overdevelop**, *v.t.* to develop a photographic negative too much so that the image is too dense; to nurture too much. **overdo**, *v.t.* (*past* **-did**, *p.p.* **-done**) to do to excess; to exaggerate; to cook to excess. **to overdo things**, to wear oneself out, to fatigue oneself. **overdose** (ō'-), *n.* an excessive dose. *v.t.*, *v.i.* (dōs´) to give or take too large a dose. **overdraft** (ō'-), *n.* a withdrawal of money from a bank in excess of the amount to one's credit; an arrangement by which this is permitted. **overdraw**, *v.t.* (*past* **-drew**, *p.p.* **-drawn**) **overdress**, *v.t.*, *v.i.* to dress too formally or ostentatiously. **overdrive**, *n.* (ō'-) an extra high gear in a motor car which drives the propeller shaft at a higher speed than the engine crankshaft. **overdue**, *a.* remaining unpaid after the date on which it is due; not arrived at the time it was due. *n.* a debt or account that is overdue. **over-earnest**, *a.* too earnest. **over-eat**, *v.i.* to eat to excess. **overestimate**, *v.t.* to give too high a value to; a too-high value. **over-excitable**, *a.* too readily excited. **over-expose**, *v.t.* (*Phot.*) to expose (a film or paper) to light too long. **over-exposure**, *n.* **overfeed**, *v.t.* to surfeit with food. *v.i.* to eat to excess. **overflow**, *v.t.* to flow over, to flood, to inundate; to cover as with a liquid. *v.i.* to abound; to overflow the banks (of a stream). *n.* (ō'-) a flood, an inundation; any outlet for surplus liquid. **overflowing**, *a.* **overflowingly**, *adv.* **over-fond**, *a.* too fond; doting. **over-fondly**, *adv.* **over-fondness**, *n.* **over-full**, *a.* too full; surfeited. **over-greedy**, *a.* excessively greedy. **overground** (ō'-), *a.*, *adv.* situated or running above ground, opp. to underground. **overgrow**, *v.t.* (*past* **-grew**, *p.p.* **-grown**) to cover with vegetation. **overgrowth** (ō'-), *n.* **overhang**, *v.i.*, *v.t.* (*past*, *p.p.* **-hung**) to hang (over), to jut out (over); to threaten. *n.* (ō'-). the act of overhanging; the part or thing that overhangs. **over-happy**, *a.* too happy. **overhaul**, *v.t.* to examine thoroughly; to overtake, to gain upon and pass. *n.* (ō'-), (an) inspection, thorough examination. **overhead**, *adv.* above the head, aloft; in the sky, ceiling, roof etc. *a.* situated overhead; (*Mach.*) working from above downwards. *n.* (ō'-) a stroke in racket games made above head height; (*usu. pl.*) expense(s) of administration etc. **overhead projector**, *n.* a device that projects an enlarged image of a transparency onto a screen behind the operator. **overhear**, *v.t.* to hear (words not meant for one) by accident or design. **overheat**, *v.t.* to heat too much or to too high a temperature; to agitate. *v.i.* to become too hot or over-agitated. **over-indulge**, *v.t.* (*often reflex.*) to indulge to excess. **over-indulgence**, *n.* **over-indulgent**, *a.* **over-indulgently**, *adv.* **overjoyed**, *a.* extremely happy. **overkill** (ō'-), *n.* destructive capability, esp. in nuclear weapons, in excess of military requirements; unnecessary thoroughness. **over-labour**, *v.t.* to

work upon excessively, to elaborate too much. **overlaid** OVERLAY. **overland** (ō'-), *a.* lying, going, made or performed by land. *adv.* across the land. **overlap**, *v.t.*, *v.i.* to fold over; to extend so as to lie or rest upon (one another). *n.* (ō'-) an act, case, or the extent of overlapping; the part that overlaps something else. **over-lavish**, *a.* too lavish. **overlay**, *v.t.* (*past*, *p.p.* **-laid**) to cover or spread over the surface of; to cover with a layer; (*Print.*) to put overlays on. *n.* (ō'-) something laid over (as a covering, layer etc.). **overleaf**, *adv.* on the next page, on the other side of the leaf (of a book etc.). **over-leap**, *v.t.* to leap over, beyond; (*fig.*) to omit. **overlie**, *v.t.* (*past* **-lay**) to lie above or upon. **overload**, *v.t.* to load too heavily. *n.* (ō'-) an excessive load. **overlook**, *v.t.* to view from a high place; to be situated so as to command a view of from above; to superintend, to oversee; to inspect or peruse, esp. in a cursory way; to disregard, to slight. **overlooker** (ō'-), *n.* **overlord** (ō'-), *n.* one who is supreme over another or others. **overly** (ō'-), *adv.* excessively, too. **overman** (-man´), *v.t.* to staff with too many men. **over-many**, *a.* too many. **overmuch**, *a.* too much, more than is sufficient or necessary. *adv.* in or to too great a degree. *n.* more than enough. **over-nice**, *a.* too nice. **overnight**, *a.* done or happening the night before. *adv.* in the course of the night; during or through the night. *n.* an overnight stay. **over-officious**, *a.* too officious. **over-officiously**, *adv.* **over-officiousness**, *n.* **overpass**, *n.* (ō'-) a flyover. **over-pay**, *v.t.*, *v.i.* to pay more than is sufficient. **over-payment**, *n.* **over-people**, *v.t.* to overstock with people. **over-peopled**, *a.* **overplay**, *v.t.* to exaggerate the importance of; to over-emphasize. **to overplay one's hand**, to try to obtain, achieve etc. more than one's capabilities. **overpower**, *v.t.* to be too strong or powerful (a sensation etc.) for; to overcome, conquer, vanquish; to overcome the feelings or judgment of. *n.* to overwhelm. **overpoweringly**, *adv.* **over-praise**, *v.t.* to praise too highly. **overprint** (ō'-), *n.* printed matter added to a previously printed surface, esp. a postage stamp. **overprint**, *v.t.* **over-prize**, *v.t.* to over-value. **over-production**, *n.* production in excess of demand. **over-produce**, *v.t.*, *v.i.* **over-proud**, *a.* excessively proud. **overrate**, *v.t.* to rate too highly. **overreach**, *v.t.* to reach or extend beyond; (*reflex.*) to try unsuccessfully to go beyond one's capabilities. **over-refine**, *v.t.* to refine too much, to be over-subtle. **over-refinement**, *n.* **override**, *v.t.* (*past* **-rode**, *p.p.* **-ridden**) to ride over; to disregard, to set aside; to take manual control of (an automatic system). *n.* (ō'-) a device used to override automatic control. **overrider** (ō'-), *n.* an attachment to the bumper of a motor vehicle to prevent it becoming interlocked with the bumper of another vehicle. **overriding** (-rī'-), *a.* dominant, taking precedence. **over-ripe**, *a.* **over-ripen**, *v.t.*, *v.i.* **over-roast**, *v.t.* to roast too much. **overrule**, *v.t.* to control by superior power or authority; to set aside, to reject, to disallow. **overrun**, *v.t.* to grow or spread rapidly and haphazardly over; to invade or harass by hostile incursions; to extend beyond. (*Print.*) to carry over and change the arrangement of (type set up). *v.i.* to overflow; to extend beyond the proper limits. **overseas**, *adv.*, *a.* to (or from) foreign parts. **oversee**, *v.t.* to superintend. **overseer** (ō'-), *n.* a superintendent, an inspector; a parish officer charged with the care of the poor. **oversell**, *v.t.* to sell more than; to sell more of (stocks etc.) than one can deliver; to exaggerate the merits of (a commodity). **oversew**, *v.t.* to sew (two pieces or edges) together by taking the needle through from one side only so that the thread between the stitches lies over the edges. **over-sexed**, *a.* obsessed with sexual activity; having an abnormally active sex life. **overshade**, *v.t.* to cover with shade. **overshadow**, *v.t.* to throw a shadow over, to shade, to obscure (as) with cloud; (*fig.*) to cast doubt etc. over; to tower high above; to exceed in

importance, influence etc. **overshoe** (ō'-), *n.* a waterproof cover worn over a shoe. **overshoot**, *v.t.* to go beyond, to overstep, to exceed; to shoot more game than is good for (a moor etc.). *v.i.* to go beyond the mark, target, accepted boundary. **overshot**, *a.* driven by water sent over the top; projecting, overlapping. **oversight** (ō'-), *n.* superintendence, supervision, care; a mistake, an unintentional error or omission. **oversimplify**, *v.t.* to depict as simpler than in reality. **oversimplification**, *n.* **oversize**, *a.* overly or abnormally large. **oversleep**, *v.i..* to sleep too long. **overspend**, *v.t.*, *v.i.* (*past*, *p.p.* **-spent**) to spend too much of (income etc.). **overspill** (ō'-), *n.* people who have moved from crowded cities into surrounding areas; a quantity spilled over. **overspread**, *v.t.* to spread over; to cover (with); to be spread over. **overstate**, *v.t.* to state too strongly, to exaggerate. **overstatement**, *n.* **overstay**, *v.t.* to stay longer than or beyond the limits of (esp. one's welcome). **overstep**, *v.t.* to exceed (accepted boundaries). **overstock** (ō'-), *n.* superabundance, excess. *v.t.* (-stok') to stock to excess; to fill too full. **over-strain**, *v.i.*, *v.t.* to strain or exert too much. **overstrike**, *v.t.* to strike out. **over-stuff**, *v.t.* to cover (furniture) with too-thick upholstery. **oversubscribe**, *v.t.* to sell, or apply for more (seats, shares etc.) than are available. **oversubscription**, *n.* **oversubtle**, *a.* too subtle. **oversubtlety**, *n.* **over-sure**, *a.* too confident. **overswell**, *v.t.*, *v.i.* to overflow. **over-swift**, *a.* too swift. **overtake**, *v.t.* (*past* **-took**, *p.p.* **-taken**) to catch up with and pass; to take by surprise, to come up on suddenly. **over-tax**, *v.t.* to tax too heavily; to overburden. **over-tedious**, *a.* too tedious. **overthrow**, *v.t.* (*past* **overthrew**, *p.p.* **overthrown**) to overturn, throw down, demolish; to overcome, conquer. *n.* (ō'-) defeat, ruin, destruction; (*Cricket*) a ball returned to but missed by the wicket-keeper, allowing further runs to be made. **overtime** (ō'-), *n.* time during which one works beyond the regular hours; work done during this period; the rate of pay for such work. **overtone** (ō'-), *n.* (*Acoustics*) a harmonic; a secondary meaning, a nuance. **overtop**, *v.t.* to tower over; to surpass. **overturn**, *v.t.* to reverse, to upset. *v.i.* to be upset or reversed. *n.* (ō'-) the act of overturning; the state of being overturned. **overturner** (-tœ'-), *n.* **over-value**, *v.t.* to value too highly. **over-valuation**, *n.* **overview** (ō'-), *n.* a general survey. **overweening**, *a.* arrogant, conceited, presumptuous. **overweeningly**, *adv.* **over-weight**, *v.t.* to weigh down too heavily; to give too much emphasis to. **overweight**, *a.* exceeding the normal or accepted weight. **overwhelm**, *v.t.* to cover completely, to submerge; to destroy utterly; to overcome, to overpower. **overwhelming**, *a.* overpowering; (*coll.*) enormous, huge (majority etc.). **overwhelmingly**, *adv.* **overwind**, *v.t.* (*past*, *p.p.* **-wound**) to wind too much or too tight. **overwise**, *a.* too wise. **overwork**, *v.t.* to impose too much work upon; to exhaust with work; to over-stimulate (the imagination). *v.i.* to work to excess. *n.* excessive work. **overwrite**, *v.t.* to write data into computer memory, or onto magnetic tape or disk, thereby erasing (the existing contents). **overwrought**, *a.* overworked; excited, agitated, nervous; too-elaborately decorated.
overt (ōvɛt'), *a.* open, plain, public, visible; (*Her.*) spread open (of wings). **overtly**, *adv.*
overture (ō'vətūə), *n.* (*usu. in pl.*) a preliminary proposal, an offer to negotiate, or of suggested terms; (*Mus.*) an introductory piece for instruments, a prelude to an opera or oratorio; a single-movement orchestral piece.
ov(i)-, ovo-, *comb. form.* pertaining to an egg or ovum.
Ovidian (ovid'iən), *a.* of or in the manner of the Roman poet Ovid.
oviduct (ō'vidŭkt), *n.* a passage through which ova pass from the ovary. **oviducal** (-dū'-), **oviductal** (-dŭk'-), *a.*
oviferous (ōvif'ərəs), *a.* egg-bearing; applied to the receptacle for ova in certain crustaceans.
oviform (ō'vifəm), *a.* egg-shaped.

ovigerous (ōvij'ərəs), *a.* egg-bearing, carrying eggs.
oviparous (ōvip'ərəs), *a.* producing young by means of eggs that are expelled and hatched outside the body. **oviparity** (-pa'-), *n.* **oviparously**, *adv.* **oviparousness**, *n.*
oviposit (ōvipoz'it), *v.i.* to deposit eggs, esp. with an ovipositor. **oviposition** (-zi'-), *n.* **ovipositor**, *n.* a tubular organ in many insects serving to deposit the eggs.
ovisac (ō'visak), *n.* a closed receptacle in the ovary in which ova are developed.
ovoid (ō'void), *a.* egg-shaped, oval with one end larger than the other; ovate. *n.* an ovoid body or figure. **ovoidal** (-voi'-), *a.*
ovoviviparous (ōvōvīvip'ərəs), *a.* producing young by eggs hatched within the body of the parent.
ovule (ov'ūl), *n.* the rudimentary seed; the body in the ovary which develops into the seed after fertilization. **ovular**, *a.* **ovulate**, *v.i.* to form ova; to discharge ova from the ovary. **ovulation**, *n.* the periodical discharge of the ovum or egg-cell from the ovary.
ovum (ō'vəm), *n.* (*pl.* **ova** (ō'və)) the female egg cell, or gamete, produced within the ovary and capable, usu. after fertilization by the male, of developing into a new individual; applied to the eggs of oviparous animals when small; (*Bot.*) an ovule.
owe (ō), *v.t.* to be indebted to for a specified amount; to be under obligation to pay or repay (a specified amount); to be obliged or indebted for (a favour). *v.i.* to be indebted or in debt. **owing**, *a.* due as a debt; attributable (to), ascribable (to), resulting from, on account of.
Owenism (ō'ənizm), *n.* the principles of humanitarian and communistic cooperation taught by Robert *Owen*, 1771–1858, British socialist. **Owenist, -ite** (-it), *n.*
owl (owl), *n.* a nocturnal raptorial bird with large head, short neck and short hooked beak, of various species; a solemn-looking person. **owl-like**, *a.*, *adv.* **owlery**, *n.* an owl nesting-site. **owlet** (-lit), *n.* a young owl. **owlish**, *a.* like an owl, solemn, with large eyes. **owlishly**, *adv.*
own[1] (ōn), *a.* belonging or proper to, particular, individual, not anyone else's (usu. appended as an intensive to the poss. pronoun, adjective etc.). **on one's own**, without aid from other people, independently; alone. **to come into one's own**, to gain what one is due; to have one's talents or potential used to the full, acknowledged. **to get one's own back**, to get even with. **to hold one's own** HOLD.
own[2] (ōn), *v.t.* to possess; to have as property by right; to acknowledge as one's own; to recognize (the authorship, paternity etc. of); to admit, to concede as true or existent. *v.i.* to confess (to). **to own up**, to confess (to). **own-brand**, *a.* denoting goods on sale which display the name or label of the retailer rather than the producer. **own goal**, *n.* in soccer, a goal scored by a player against his own side by accident; (*coll.*) any action which results in disadvantage to the person taking it. **owner**, *n.* a lawful proprietor. **owner-occupied**, *a.* **owner-occupier**, *n.* someone who owns the house he or she lives in. **ownerless** (-lis), *a.* **ownership**, *n.*
ox (oks), *n.* (*pl.* **oxen**) the castrated male of the domesticated bull; any bovine animal, esp. of domesticated species of cattle. **ox-bow** (-bō), *n.* the bow-shaped piece of wood in an ox-yoke; a similarly-shaped bend in a river. **ox-eye (daisy)**, *n.* the moon-daisy, and other composite plants. **ox-head**, *n.* a dolt, a blockhead. **ox-hide**, *n.* the skin of an ox. **oxtail**, *n.* the tail of an ox, esp. when used for making soup; the soup so made.
Oxalis (ok'səlis, -sal'-), *n.* a genus of plants containing the wood-sorrel. **oxalic** (-sal'-), *a.* belonging to or derived from oxalis. **oxalic acid**, *n.* a sour, highly-poisonous acid found in numerous plants.
Oxbridge (oks'brij), *n.*, *a.* (of) the Universities of Oxford and Cambridge, esp. seen as elitist educational establishments conferring unfair social, economic and political advantages.

Oxfam (oks'fam), n. a British charity. [acronym for *Oxf*ord *Com*mittee for *Fam*ine Relief]
Oxford (oks'fəd), a. of, pertaining to, or derived from Oxford. **Oxford bags**, n.pl. trousers very wide at the ankles. **Oxford blue**, n. a dark shade of blue; a representative of Oxford University in sporting competitions. **Oxford shoe**, n. a low shoe laced over the instep.
oxide (ok'sīd), n. a binary compound of oxygen with another element or an organic radical. **oxidant** (-si-), n. a substance used as an oxidizing agent. **oxidation** (-si-), n. the process of oxidizing. **oxidize, -ise** (-si-), v.t. to combine with oxygen; to cover with a coating of oxide, to make rusty. v.i. to enter into chemical combination with oxygen; to rust. **oxidizable**, a. **oxidization**, n. **oxidizer**, n.
Oxon (ok'sən), (abbr.) Oxfordshire; of Oxford University (used for degrees etc.).
Oxonian (oksō'niən), n. a student or graduate of Oxford Univ. a. belonging to Oxford.
oxy-, comb. form. sharp, keen; denoting the presence of oxygen or its acids, or of an atom of hydroxyl substituted for one of hydrogen.
oxyacetylene (oksiəset'ilēn), a. yielding a very hot flame from the combustion of oxygen and acetylene, used for welding metals etc.
oxycarpous (oksikah'pəs), a. having pointed fruit.
oxygen (ok'sijən), n. a colourless, tasteless, odourless divalent element, at. no. 8; chem. symbol O, existing in a free state in the atmosphere, combined with hydrogen in water, and with other elements in most mineral and organic substances. **oxygen mask**, n. an apparatus for supplying oxygen in rarefied atmospheres to aviators etc. **oxygen tent**, n. an oxygen-filled tent placed over a patient to assist breathing. **oxygenate** (-nāt), v.t. to treat or impregnate with oxygen; to oxidize. **oxygenation**, n. **oxygenator**, n. **oxygenize, -ise**, v.t.
oxymoron (oksimaw'ron), n. a rhetorical figure in which an epithet of a quite contrary signification is added to a word for the sake of point or emphasis, e.g. a clever fool, a cheerful pessimist.
oyez, oyes (ōyez', ō'yez), int. thrice repeated as introduction to any proclamation made by an officer of a court of law or public crier.
oyster (oi'stə), n. an edible bivalve mollusc found in salt or brackish water, eaten as food. **oyster-bank, -bed**, n. a part of a shallow sea-bottom forming a breeding-place for oysters. **oyster-catcher**, n. a wading-bird with black and white plumage and red bill and feet. **oyster-farm, -field, -park**, n. a part of the sea-bottom used for breeding oysters commercially. **oyster-knife**, n. a knife specially shaped for opening oysters. **oyster-patty**, n. a small pie made from oysters.
oz, (abbr.) ounce.
Oz (oz), n. (Austral. sl.) Australia.
ozone (ō'zōn), n. a form of oxygen, having three atoms to the molecule, with a slightly pungent odour, found in the atmosphere, probably as the result of electrical action. **ozoniferous** (-nif'-), a. **ozonosphere** (ōzō'nəsfiə, -zon'-), **ozone layer**, n. a layer of ozone in the stratosphere which protects the earth from the sun's ultraviolet rays. **ozone-friendly**, a. of sprays etc., not damaging the ozone layer, not containing chlorofluorocarbons (cfcs).

P

P¹, p¹, the 16th letter, and the 12th consonant (*pl.* **Pees, P's, Ps**), is a voiceless labial mute, having the sound heard in *pull*, *cap*, except when in combination with *h* it forms the digraph *ph*, sounded as *f*. **to mind one's Ps and Qs**, to be careful over details, esp. in behaviour.
P², (*chem. symbol*) phosphorus.
P³, (*abbr.*) parking; in chess, pawn; Portugal.
p², (*abbr.*) page; penny, pence; (*Mus.*) piano, used as an instruction to play softly; pint; power; pressure.
PA, (*abbr.*) Panama; personal assistant; Press Association; public address (system).
Pa¹, (*abbr.*) Pennsylvania.
Pa², (*chem. symbol*) protactinium.
pa (pah), *n.* a child's name for father.
p.a., (*abbr.*) per annum.
pabulum (pab'ūləm), *n.* food; nutriment of a physical, mental or spiritual kind. **pabular**, *a.*
PABX, (*abbr.*) private automatic branch (telephone) exchange.
paca (pah'kə, pak'ə), *n.* a large Central and South American semi-nocturnal rodent.
pace¹ (pās), *n.* a step, the space between the feet in stepping (about 30 in., 76 cm); gait, manner of going, either in walking or running; the carriage and action of a horse etc.; an amble, rate of speed or progress. *v.i.* to walk with slow or regular steps; to walk with even strides or in a slow, deliberate manner; to amble. *v.t.* to measure by carefully regulated steps; to traverse in slow and measured steps; to set the pace for. **to be put through one's paces**, to be examined closely, to be tested. **to force the pace**, to try to increase the speed or tempo of any activity. **to go the pace**, to go very fast; to lead a life of dissipation or recklessness. **to keep pace with**, to go or progress at equal rate with. **to set, make the pace**, to fix the rate of going in a race or any other activity. **pacemaker**, *n.* a rider or runner who sets the pace in a race; a person who sets the pace in any form of activity; a small area of muscle tissue in the heart wall that controls the rhythm of the heartbeat; a small device, usu. implanted in the chest, that corrects irregularities in the heartbeat. **pace-setter**, *n.* a pacemaker. **paced**, *a.* having a particular pace or gait (*in comb.*, as *thorough-paced*). **pacer**, *n.* one who paces; a horse trained in pacing. **pacey, pacy**, *a.* (*coll.*) moving at a fast, exciting pace.
pace² (pā'si, pah'ke, -che), *prep.* with the permission of; with due respect to (someone who disagrees). [L]
pacha (pah'shə), PASHA.
pachisi (pəchē'si), *n.* an Indian game played on a board with cowries for dice.
pachyderm (pak'idœm), *n.* any individual of the Pachydermata, an order of mammals containing hoofed non-ruminant animals with thick integuments; a thick-skinned person. **pachydermatoid** (-dœ'mətoid), **-tous**, **pachydermoid** (-dœ'-), *a.*
pacific (pəsif'ik), *a.* inclined or tending to peace, conciliatory; tranquil, quiet, peaceful; (**Pacific**) of the Pacific Ocean. **the Pacific**, the ocean between America and Asia, so named by Magellan. **pacifically**, *adv.* **pacification** (-kā'-), *n.* the act of pacifying. **pacificator** (-sif'-), *n.* **pacificatory**, *a.* **pacifier**, *n.* one who or that which pacifies; (*N Am.*) a baby's dummy or similar object for sucking. **pacifism** (pas'i-), *n.* the doctrine of non-resistance to hostilities and of total non-cooperation with any form of warfare. **pacifist**, *n.*, *a.* **pacify** (pas'ifī), *v.t.* to appease, to calm, to quiet; to restore peace to.

pack (pak), *n.* a bundle of things tied or wrapped together for carrying; a parcel, a burden, a load; a quantity going in such a bundle or parcel, often taken as a measure, varying with different commodities; a small packet, package or container, e.g. of cigarettes; a set, a crew, a gang; a set of playing-cards; a number of dogs kept together; a number of wolves or other beasts or birds, esp. grouse, going together; pack ice; a quantity of something packed, esp. for sale; in rugby, the forwards of a team; a group of Cub Scouts or Brownies; a face-pack; *v.t.* to put together into a pack or packs; to stow (articles) in (a bundle, box, barrel, bag, tin etc.), for keeping, carrying etc.; to crowd or press closely together, to compress; to fill completely; to cram (with); to wrap tightly, to cover or surround with some material to prevent leakage, loss of heat etc.; to load with a pack; to arrange in a pack; to select or bring together (a jury etc.) so as to obtain some unfair advantage; to send off or dismiss without ceremony; (*N Am.*, *coll.*) to carry (a gun). *v.i.* to put things in a pack, bag, trunk etc., for sending away, carrying or keeping; of animals, to crowd together; to form a firm compacted mass; to leave with one's belongings; to depart hurriedly. **to pack a punch**, (*coll.*) to be able to punch hard; to be strong or forceful. **to pack in**, (*sl.*) to stop doing (something); to stop going out with (someone). **to pack up**, (*sl.*) to stop functioning; to break down; to pack in. **to send packing**, to dismiss summarily. **packdrill**, *n.* a form of military punishment consisting of high-speed drill in full kit. **pack-horse**, *n.* a horse employed in carrying goods. **pack-ice**, *n.* large pieces of ice floating in the polar seas. **pack rat**, *n.* a rat of western N America, with a long tail that is furry in some species. **packsaddle**, *n.* one for supporting packs. **packthread**, *n.* strong thread for sewing or tying up parcels. **package** (-ij), *n.* a parcel, a bundle; the packing of goods, the manner in which they are packed; the container, wrapper etc. in which a thing is packed; a number of items offered together; a set of computer programs for carrying out a generalized operation, e.g. word processing. *v.t.* to place in a packet; to bring (a number of items) together as a single unit; to present in an appealing way. **package deal**, *n.* a deal in which a number of items are offered, and all must be accepted. **package holiday**, *n.* a holiday where travel, accommodation, meals etc. are all included in the price. **packager**, *n.* one who packages; a person or company who undertakes the complete design and production of a book, television programme, video game etc. on behalf of or for sale to, another company under whose name it is distributed. **packaging**, *n.* the container etc. in which something is packaged; the presentation of a person or thing to the public in a particular, esp. favourable, way; the occupation of a packager. **packer**, *n.* one who or that which packs, esp. one employed to pack meat, fish, fruit etc. for sale. **packing**, *n.* that which is used for packing; material closing a joint or helping to lubricate a journal. **packing-case**, *n.* a large box made of unplaned wood.

packet (pak'it), *n.* a small package; a packet-boat; (*sl.*) a

large sum of money; (*Comput.*) a block of data of fixed length sent as a unit. *v.t.* to make up in a packet.
packet-boat, *n.* a vessel conveying mails, goods and passengers at regular intervals. **packet switching**, *n.* the sending of packets of data, that together form a message, via a digital communications network.
pact (pakt), *n.* an agreement, a compact.
pacy PACE¹.
pad¹ (pad), *v.i.* (*past, p.p.* **padded**) to travel on foot; to walk softly. *v.t.* to tramp or travel over.
pad² (pad), *n.* a soft cushion; a bundle or mass of soft stuff; a soft saddle; a cushion-like package, cap, guard etc., for stuffing, filling out, protecting parts of the body etc.; a block of absorbent material saturated with ink for inking a stamp; a number of sheets of paper fastened together at the edge for writing or drawing upon and then detaching; a rocket-launching platform; an area for take-off and landing, esp. for helicopters; the cushion-like sole of the foot, or the soft cushion-like paw of certain animals; the floating leaf of a water lily; (*coll.*) one's home or room. *v.t.* (*past, p.p.* **padded**) to stuff or line with padding; to furnish with a pad or padding; to fill out (a sentence, article etc.) with unnecessary words. **padsaw**, *n.* a small narrow saw for cutting curves. **padded**, *a.* **padded cell, room**, *n.* a room with padded walls for confining violent patients. **padding**, *n.* material used for stuffing; unnecessary matter inserted to fill out an article, magazine or book.
paddle¹ (pad'l), *n.* a broad short oar used without a rowlock; the blade of this or of an oar; a paddle-board; a paddle-wheel; a spell of paddling; a paddle-like implement used for mixing, in washing clothes etc.; a table-tennis bat; a creature's broad, flat limb for swimming. *v.t.* to propel by means of paddles; (*N Am.*) to spank. *v.i.* to ply a paddle; to move along by means of a paddle; to row gently; to swim with short, downward strokes. **paddle-board**, *n.* one of the floats or blades of a paddle-wheel. **paddle-wheel**, *n.* a wheel with floats or boards projecting from the periphery for pressing against the water and propelling a vessel.
paddle² (pad'l), *v.i.* to dabble in the water with the hands or, more usually, the feet; to toddle.
paddock (pad'ək), *n.* a small field or enclosure, usu. under pasture and near a stable; (*Austral.*) any pasture land enclosed by a fence; a turfed enclosure adjoining a racecourse where horses are kept before racing; an area beside a motor-racing circuit where cars are parked, repaired etc.
Paddy (pad'i), *n.* (*sometimes derog.*) an Irishman.
paddy¹ (pad'i), *n.* rice in the straw or in the husk; a paddy-field. **paddy-field**, *n.* a field planted with rice.
paddy² (pad'i), *n.* (*coll.*) a rage, temper.
paddymelon (pad'imelən), *n.* (*Austral.*) a small bush kangaroo or wallaby.
padishah (pah'dishah), *n.* the title of the Shah of Iran, also formerly in India of the British sovereign and of the Great Mogul.
padlock (pad'lok), *n.* a detachable lock with a bow or loop for fastening to a staple etc. *v.t.* to fasten with this.
padre (pah'drā), *n.* used in addressing a priest in Italy, Spain and Spanish America; a chaplain in the armed forces.
padsaw PAD².
paean (pē'ən), *n.* a choral song addressed to Apollo or some other deity; a song of triumph or rejoicing.
paed- PAED(O)-.
paederast (pēd'ərast), PEDERAST.
paediatrics (*esp. N Am.*) **pediatrics** (pēdiat'riks), *n. sing.* the branch of medicine dealing with children's diseases. **paediatric**, *a.* **paediatrician** (-iətrish'ən), *n.* a specialist in paediatrics.
paed(o)-, (*esp. N Am.*) **ped(o)-**, *comb. form.* relating to children.
paedophilia (*esp. N Am.*) **pedophilia** (pēdəfil'iə), *n.* sexual desire directed towards children. **paedophile** (-fīl), *n.*
paella (pīel'ə), *n.* a Spanish dish of rice, seafood, meat and vegetables, flavoured with saffron.
paeony (pē'əni), PEONY.
pagan (pā'gən), *n.* a heathen; a person who has no religion; one who disregards Christian, Jewish or Muslim beliefs. *a.* heathen, irreligious. **pagandom** (-dəm), **paganism**, *n.* **paganish**, *a.* **paganize, -ise**, *v.t.*, *v.i.*
page¹ (pāj), *n.* a young male attendant on persons of rank; hence, a title of various functionaries attached to the royal household; a boy acting as an attendant at a wedding; a boy in livery employed to go on errands, attend to the door etc.; *orig.*, a youth in training for knighthood attached to a knight's retinue. *v.t.* to attend on as a page; to summon (a person in a hotel etc.) by calling the name aloud; to summon by transmitting an audible signal on an electronic device. **page boy**, *n.* a page; a (woman's) medium-length hairstyle, with the ends curled under. **pagehood, pageship**, *n.* **page-three**, *a.* pertaining to material of a mildly sexually titillating nature, as photographs of topless female models that appear in some British tabloid newspapers, traditionally on page three. **pager**, *n.* one who or that which pages; esp. a bleeper.
page² (pāj), *n.* (one side of) a leaf of a book, newspaper etc.; a record, a book; an episode; a subdivision of a computer memory. *v.t.* to put numbers on the pages of (a book). **paginal** (paj'-), *a.* **paginate** (paj'-), *v.t.* to page. **pagination**, *n.*
pageant (paj'ənt), *n.* a brilliant display or spectacle, esp. a parade or procession of an elaborate kind; a theatrical exhibition, usu. representing well-known historical events; a tableau or allegorical design, usu. mounted on a car in a procession; empty and specious show. **pageantry**, *n.*
paginate, pagination PAGE².
pagoda (pəgō'də), *n.* a sacred temple, usu. in the form of a pyramidal tower in many receding storeys, in India, China and other Eastern countries; a building imitating this. **pagoda-tree**, *n.* the name of several kinds of Indian and Chinese trees shaped more or less like pagodas.
pah (pah), *int.* an exclamation of disgust etc.
Pahlavi (pah'lavi), *n.* the characters used for the sacred writings of the Iranians; the literary language of Iran under the Sassanian kings, old Persian.
paid (pād), *past, p.p.* PAY¹.
pail (pāl), *n.* an open, usu. round vessel, esp. of metal or wood, for carrying liquids; a pailful. **pailful**, *n.* (*pl.* **-fuls**).
paillasse PALLIASSE.
pain (pān), *n.* bodily or mental suffering; a disagreeable sensation in animal bodies; (*pl.*) labour, trouble; (*coll.*) a nuisance. *v.t.* to inflict pain upon, to afflict or distress bodily or mentally. **a pain in the neck**, (*coll.*) a nuisance. **on, under pain of**, subject to the penalty of. **to take pains**, to take trouble, to be exceedingly careful. **painkiller**, *n.* a drug that alleviates pain. **painkilling**, *a.* **painstaking**, *a.* taking pains, laboriously careful. **painstakingly**, *adv.* **pained**, *a.* having or showing distress, embarrassment etc. **painful**, *a.* attended with or causing mental or physical pain; laborious, toilsome, difficult; irksome, troublesome. **painfully**, *adv.* **painfulness**, *n.* **painless**, *a.* **painlessly**, *adv.* **painlessness**, *n.*
paint (pānt), *v.t.* to cover or coat with paint; to portray or represent in colours; to adorn with painting; to produce (a picture) with paint; to apply (esp. a liquid) to with a brush; to apply make-up to; to depict vividly in words. *v.i.* to practise painting. *n.* a solid colouring-substance or pigment, usu. dissolved in a liquid vehicle, used to give a coloured coating to surfaces; a coat of this; (facial) make-up. **to paint the town red**, (*sl.*) to go out on a noisy spree. **paint-box**, *n.* a box in which oil- or water-colours are kept

painter in compartments. **paint-brush,** *n.* **painted lady,** *n.* an orange-red butterfly spotted with black and white. **painter¹,** *n.* one whose occupation is to colour walls, woodwork etc. with paint; an artist who paints pictures. **painting,** *n.* the act, art or occupation of laying on colours or producing representations in colours; a picture. **painty,** *a.* (*coll.*) like paint in smell etc.; covered in paint.
painter² (pān'tə), *n.* a bow-rope for fastening a boat to a ring, stake etc.
pair (peə), *n.* two things of a kind, similar in form, or applied to the same purpose or use; a set of two, a couple, usu. corresponding to each other; an implement or article having two corresponding and mutually dependent parts, as scissors, spectacles; two playing-cards of the same value; an engaged or married couple; (*Parl.*) two members of opposite views abstaining from voting by mutual agreement. *v.t.* to make or arrange in pairs or couples; to cause to mate. *v.i.* to be arranged in pairs; to mate; to make a pair (with). **to pair off,** to separate into couples; to make a pair (with). **pair-bond,** *n.* a lasting, exclusive relationship between a male and a female.
paisa (pī'sah), *n.* in India and Pakistan a coin worth 1/100th of a rupee.
paisley (pāz'li), *n.* (a fabric bearing) a colourful pattern of small intricate curves, a shawl made of this fabric. *a.* of this fabric or pattern. [*Paisley*, town in Scotland]
pajamas (pəjam'əz), PYJAMAS.
pakeha (pah'kihah), *n.* a white man in New Zealand.
Paki (pak'i), *n., a.* (*sl., offensive*) Pakistani.
Pakistani (pakistah'ni), *a.* pertaining to Pakistan. *n.* a native or inhabitant of Pakistan; a descendant of Pakistanis.
pakora (pəkaw'rə), *n.* an Indian dish consisting of balls of chopped chicken, vegetable etc. coated in batter and deep-fried.
PAL (pal), (*acronym*) *p*hase *a*lternation *l*ine, a system of colour television broadcasting used in Europe.
pal (pal), *n.* (*coll.*) a friend, chum, mate. **to pal up with,** (*coll.*) to become friendly with. **pally,** *a.* (*coll.*) friendly.
palace (pal'is), *n.* the official residence of an emperor, king, bishop or other distinguished personage; a splendid mansion; a large building for entertainments, a music-hall, cinema, theatre etc.
paladin (pal'ədin), *n.* one of Charlemagne's 12 peers; a knight-errant.
palae- PALAEO-.
Palaearctic (paliahk'tik), *a.* pertaining to the northern parts of the Old World, including Europe, N Africa, and Asia north of the Himalayas, esp. as a zoogeographical region.
palae(o)-, (*esp. N Am.*), **pale(o)-,** *comb. form* pertaining to or existing in the earliest times.
palaeobotany (paliōbot'əni), *n.* the study of extinct or fossil plants. **palaeobotanical** (-tan'-), *a.* **palaeobotanist,** *n.*
Palaeocene (pal'iəsēn), *n., a.* (of) the oldest epoch of the Tertiary period.
palaeoclimatology (paliōklīmətol'əji), *n.* the science of the climates of the geological past.
palaeography (paliog'rəfi), *n.* the art or science of deciphering ancient inscriptions or manuscripts; ancient inscriptions or manuscripts collectively. **palaeographer,** *n.* **palaeographic, -ical** (-graf'-), *a.*
Palaeolithic (paliəlith'ik), *n., a.* (pertaining to) the earlier Stone Age.
palaeomagnetism (paliōmag'nətizm), *n.* the study of the magnetic properties of rocks.
palaeontology (paliəntol'əji), *n.* the science or the branch of biology or geology dealing with fossil animals and plants. **palaeontological** (-loj'-), *a.* **palaeontologist,** *n.*
Palaeozoic (paliəzō'ik), *n., a.* (denoting) the lowest fossiliferous strata or the era in which these were formed.
palaestra (pəles'trə, -lē'-), *n.* in ancient Greece, a place where athletic exercises were taught and practised; a gymnasium or wrestling-school. **palaestral, -tric,** *a.*
palankeen, palanquin (palənkēn'), *n.* a couch or litter borne by four or six men on their shoulders.
palatable (pal'ətəbl), *a.* pleasing to the taste; agreeable, acceptable. **palatability** (-bil'-), **palatableness,** *n.* **palatably,** *adv.*
palate (pal'ət), *n.* the roof of the mouth; the sense of taste; liking, fancy. **hard, bony palate,** the anterior or bony part of the palate. **soft palate,** the posterior part consisting of muscular tissue and mucous membrane terminating in the uvula. **palatal,** *a.* pertaining to or uttered by the aid of the palate. *n.* a sound produced with the palate, esp. the hard palate, as *k, g, ch, y, s, n.* **palatalize, -ise, palatize, -ise,** *v.t.* to pronounce as a palatal. **palatine¹** (-tin), *a.* of or pertaining to the palate. *n.pl.* (also **palatine bones**) the two bones forming the hard palate.
palatial (pəlā'shəl), *a.* pertaining to or befitting a palace; magnificent, splendid. **palatially,** *adv.*
palatine¹ PALATE.
palatine² (pal'ətin), *a.* pertaining to or connected with a palace, orig. the palace of the Caesars, later of the German Emperors; possessing or exercising royal privileges (as the counties of Chester, Durham and Lancaster); of or pertaining to a count palatine. *n.* one invested with royal privileges; a feudal lord having sovereign power within a territory; COUNT PALATINE under COUNT². **The Palatine,** the territory of the Count Palatine of the Rhine, an elector of the Holy Roman Empire. **palatinate** (-lat'ināt), *n.* the office or territory of a palatine.
palaver (pəlah'və), *n.* a discussion, a conference, a parley; talk, chatter; cajolery, flattery; tedious activity. *v.t.* to flatter. *v.i.* to confer; to talk idly and profusely. **palaverer,** *n.*
palazzo (pəlat'sō), *n.* (*pl.* **-azzi** (-at'sē)) a large palatial building in Italy.
pale¹ (pāl), *n.* a pointed stake or narrow board used in fencing; a limit or boundary; a district, a territory; (*Her.*) a vertical band down the middle of a shield. *v.t.* to enclose with or as with pales. **beyond the pale,** unacceptable. **the Pale,** formerly, the part of Ireland in which English authority was recognized. **paling,** *n.* a fence made with pales; a single pale; material for making fences.
pale² (pāl), *a.* whitish, ashen, wanting in colour, not ruddy; of colours or light, dim, faint; poor, feeble. *v.t.* to make pale. *v.i.* to turn pale, to be pale, dim or poor in comparison. **pale ale,** *n.* light-coloured ale. **paleface,** *n.* a name supposed to be given by N American Indians to white persons. **pale-faced,** *a.* **palely,** *adv.* **paleness,** *n.* **palish,** *a.*
palea (pā'liə), *n.* a bract or scale resembling chaff, at the base of the florets in composite flowers, enclosing the stamens and pistil in grass-flowers, or on the stems of ferns. **paleaceous** (-ā'shəs), *a.*
pale(o)- PALAEO-.
Palestinian (paləstin'iən), *a.* of or pertaining to Palestine, a former country of SW Asia, the Holy Land. *n.* a native of Palestine, or a descendant of such a native.
palette (pal'it), *n.* a flat board used by artists for mixing colours on; the colours or arrangement of colours used for a particular picture or by a particular artist; the range of colours a computer can reproduce on a visual display unit. **palette-knife,** *n.* a thin, flexible knife for mixing and applying esp. paint or food.
palfrey (pawl'fri), *n.* a small saddle-horse, esp. for a woman.
Pali (pah'li), *n.* the canonical language of Buddhist literature, akin to Sanskrit.
palimony (pal'iməni), *n.* (*chiefly N Am.*) alimony paid to an unmarried partner after the end of a long-term relationship.
palimpsest (pal'impsest), *n.* a manuscript from which the original writing has been erased to make room for another

palindrome 578 **pan**

record, *a.* treated in this manner.
palindrome (pal'indrŏm), *n.* a word, verse, or sentence that reads the same backwards or forwards, e.g. 'Madam I'm Adam'. **palindromic** (-drom'-), *a.* **palindromist** (-lin'-, pal'-), *n.*
paling PALE¹.
palingenesis (palinjen'əsis), *n.* a new birth, a regeneration; the form of ontogeny in which the development of the ancestors is exactly reproduced. **palingenetic** (-net'-), *a.* **palingenetically**, *adv.*
palisade (palisād'), *n.* a fence or fortification of stakes; such a stake. *v.t.* to enclose or to fortify with stakes. **palisade cells**, *n.pl.* a layer of cells rich in chloroplasts situated beneath the epidermis of leaves.
palish PALE².
pall¹ (pawl), *n.* a large cloth, usu. black, purple or white, thrown over a coffin, hearse or tomb; a pallium; anything that covers or shrouds; an oppressive atmosphere. **pallbearer**, *n.* one who attends the coffin at a funeral, or who holds up the funeral pall.
pall² (pawl), *v.i.* to become vapid or insipid; to become boring. *v.t.* to make vapid, insipid or spiritless; to cloy, to dull.
Palladian (pəlā'diən), *a.* of or according to the Italian architect Andrea Palladio (1518–80) or his school of architecture; pertaining to the free and ornate classical style modelled on the teaching of Vitruvius. **Palladianism**, *n.* **Palladianize, -ise**, *v.t.*
palladium¹ (pəlā'diəm), *n.* a statue of Pallas on the preservation of which, according to tradition, the safety of Troy depended; a defence, a safeguard.
palladium² (pəlā'diəm), *n.* a greyish-white metallic element of the platinum group, at. no. 46; chem. symbol Pd, used as an alloy with gold and other metals.
pallet¹ (pal'it), *n.* a palette; a tool, usu. consisting of a flat wooden blade with handle, used for mixing and shaping clay in pottery-making, or for taking up gold-leaf and for gilding or lettering in bookbinding; a pawl or projection on a part of a machine, for converting reciprocating into rotary motion or vice versa; a valve regulating the admission of air from the wind-chest to an organ-pipe; a flat wooden structure on which boxes, crates etc. are stacked or transported. **palletize, -ise**, *v.t.*, *v.i.* to use pallets for storing and transporting (goods). **palletization**, *n.*
pallet² (pal'it), *n.* a small bed; a straw mattress.
palliasse, paillasse (pal'ias), *n.* a mattress or under-bed of straw or other material.
palliate (pal'iāt), *v.t.* to cover with excuses; to extenuate; to mitigate, to alleviate (a disease etc.) without entirely curing. **palliation**, *n.* **palliative**, *n.*, *a.* (a substance) serving to alleviate a disease etc. without curing it. **palliatively**, *adv.*
pallid (pal'id), *a.* pale, wan; feeble, insipid. **pallidly**, *adv.* **pallidness**, *n.*
pallium (pal'iəm), *n.* (*pl.* **-llia** (-ə)) a man's square woollen cloak, worn esp. by the ancient Greeks; a scarf-like vestment of white wool with red crosses, worn by the Pope and certain metropolitans and archbishops. **pallial**, *a.*
pallor (pal'ə), *n.* paleness, want of healthy colour.
pally PAL.
palm¹ (pahm), *n.* a tree or shrub belonging to the Palmae, a family of tropical or subtropical plants, usu. with a tall branched stem and head of large fan-shaped leaves; a palm-branch or leaf as a symbol of victory or triumph; victory, triumph, the prize, the pre-eminence. **palm-oil**, *n.* an oil obtained from the fruit of certain kinds of palm. **Palm Sunday**, *n.* the Sunday immediately preceding Easter, commemorating the triumphal entry of Christ into Jerusalem. **palmaceous** (palmā'shəs), *a.* **palmy**, *a.* abounding in palms; victorious, flourishing.
palm² (pahm), *n.* the inner part of the hand; the part of a glove etc. covering this; a measure of breadth (3–4 ins., 7·5–10 cm) or of length (8–8½ ins., 20–21·5 cm); the broad flat part of an oar, tie, strut, antler etc.; the fluke of an anchor. *v.t.* to conceal in the palm; to impose or pass off fraudulently. **in the palm of one's hand**, under one's control, in one's power. **to palm off**, to foist. **palm top**, *n.* a portable computer smaller than a lap-top computer. **palmar** (pal'mə), *a.* of, pertaining to, in or connected with the palm. **palmate** (pal'māt), **-ated**, *a.* (*Bot.*, *Zool.*) resembling a hand with the fingers spread out; of the foot of a bird, webbed. **palmately**, *adv.* **palmed**, *a.* having a palm or palms (*usu.* in *comb.*, as *full-palmed*). **palmistry**, *n.* fortune-telling by the lines and marks on the palm of the hand. **palmist**, *n.*
palmer (pah'mə), *n.* a pilgrim who carried a palm-branch in token of his having been to the Holy Land; a pilgrim, devotee, itinerant monk etc.
palmetto (palmet'ō), *n.* (*pl.* **-ttos, -ttoes**) any of several small varieties of palm with fan-shaped leaves, esp. *Sabal palmetto*, a fan-palm of the Southern US.
palmistry PALM².
palmitic (palmit'ik), *a.* of or derived from palm-oil. **palmitate** (pal'mitāt), *n.* a salt or ester of palmitic acid. **palmitin** (pal'mitin), *n.* a natural fatty compound contained in palm-oil etc.
palmy PALM¹.
palmyra (palmī'rə), *n.* an E Indian palm with fan-shaped leaves used for mat-making.
palomino (palǝmē'nō), *n.* (*pl.* **-nos**) a cream, yellow or gold horse with a white mane and tail.
palp (palp), **palpus** (-pəs), *n.* (*pl.* **palps, palpi** (-pī)) one of a pair of jointed sense-organs developed from the lower jaws of insects etc., a feeler. **palpal, palped**, *a.* **palpiform** (pal'pifawm), *a.*
palpable (pal'pəbl), *a.* perceptible to the touch; easily perceived, plain, obvious. **palpability** (-bil'-), *n.* **palpably**, *adv.*
palpate (pal'pāt), *v.t.* to feel, to handle, to examine by feeling. **palpation**, *n.*
palpebral (pal'pibrəl), *a.* pertaining to the eyelid.
palpi PALP.
palpitate (pal'pitāt), *v.i.* to throb, to pulsate; to flutter; of the heart, to beat rapidly. **palpitation**, *n.*
palpus PALP.
palsy (pawl'zi), *n.* paralysis. *v.t.* to paralyse, to affect with palsy. **palsied** (-zid), *a.*
paltry (pawl'tri), *a.* mean, petty, despicable; trivial. **paltrily**, *adv.* **paltriness**, *n.*
paludal (pəlū'dəl), **-udic**, *a.* pertaining to marshes or fens, marshy; malarial. **paludism** (pal'-), *n.* malaria.
pampas (pam'pəs), *n.pl.* (*sing.* **pampa**) the open, far-extending, treeless plains in S America, south of the Amazon. **pampas-grass** (-pas-), *n.* a lofty grass with large feathery flower-heads, originally from the pampas.
pampean (-piən), *a.*
pamper (pam'pə), *v.t.* to indulge (a person, oneself etc.), often excessively; to gratify (tastes etc.) to excess. **pamperedness**, *n.* **pamperer**, *n.*
pampero (pampeə'rō), *n.* (*pl.* **-ros**) a violent westerly or south-westerly wind blowing over the pampas.
pamphlet (pam'flit), *n.* a small book of a few sheets, stitched, but not bound, usu. on some subject of temporary interest. **pamphleteer** (-tiə'), *v.i.* to write pamphlets, esp. controversial ones. *n.* a writer of pamphlets. **pamphleteering**, *n.*
Pan (pan), *n.* the chief rural divinity of the Greeks, represented as horned and with the hindquarters of a goat. **Pan-pipes**, *n.* (*pl.*) a musical instrument made of a number of pipes or reeds, a mouth-organ.
pan¹ (pan), *n.* a broad shallow vessel of metal or earthenware, usu. for domestic uses; a panful; a vessel for boiling, evaporating etc.; a natural hollow in land; a hollow

in the ground for evaporating brine in salt-making; a sheet-iron dish used for separating gold from gravel etc., by shaking in water; the part of a flint-lock that holds the priming; hardpan; a lavatory bowl; either of the two shallow receptacles of a balance; the act or process of panning a camera; the brain-pan. *v.t. (past, p.p.* **panned**) to wash (gold-bearing earth or gravel) in a pan; (*coll.*) to criticize severely; to move (a camera) in panning. *v.i.* to move the camera while taking a picture of a moving object. **to pan out**, to yield gold; to yield a specified result (esp. well or badly). **pancake**, *n*. a thin flat cake of batter fried in a frying-pan; a cake or stick of compressed make-up; an aircraft landing made from a low altitude in an approximately horizontal position. *v.t., v.i.* to (cause to) alight from a low altitude at a large angle of incidence, remaining on an even keel. **Pancake Day** SHROVE TUESDAY. **panful**, *n*.

pan², (pan), *n*. a betel leaf; such a leaf wrapped around sliced betel-nut mixed with spices, used in India for chewing.

pan-, *comb. form.* all.

panacea (panəsē'ə), *n*. a universal remedy.

panache (pənash'), *n*. a tuft or plume, esp. on a headdress or helmet; show, swagger, bounce; style.

panada (pənah'də), *n*. a thickening, as for a sauce, made with flour or breadcrumbs.

Pan-African (panaf'rikən), *a*. of the whole of the African continent; of Pan-Africanism. **Pan-Africanism**, *n*. the advocacy of cooperation between or unification of all the African nations.

panama (hat) (pan'əmah), *n*. a hat made from the undeveloped leaves of the S American screw-pine.

Pan-American (panəme'rikən), *a*. of or pertaining to the whole of both N and S America; of Pan-Americanism. **Pan-Americanism**, *n*. the advocacy of cooperation between the nations of N and S America.

Pan-Arab(ic) (pana'rəbik), *a*. relating to the movement for political unity between the Arab nations. **Pan-Arabism**, *n*.

panatella (panətel'ə), *n*. a type of long slender cigar.

pancake PAN¹.

panchromatic (pankrəmat'ik), *a*. uniformly sensitive to all colours. **panchromatism** (-krōm'-), *n*.

pancratium (pankrā'shiəm), *n*. in ancient Greece, an athletic contest including both boxing and wrestling. **pancratiast, -cratist** (pan'krə-), *n*. **pancratic** (-krat'-), *a*.

pancreas (pang'kriəs), *n*. a gland near the stomach secreting a fluid that aids digestive action and insulin. **pancreatic** (-at'-), *a*. **pancreatic juice**, *n*. the fluid secreted by the pancreas into the duodenum to aid the digestive process. **pancreatin** (-tin), *n*. a protein compound found in the pancreas and pancreatic juice. **pancreatitis** (-tī'tis), *n*. inflammation of the pancreas.

panda (pan'də), *n*. a small racoon-like animal, *Aelurus fulgens*, from the SE Himalayas and Tibet; a giant panda. **giant panda**, the *Ailuropus melanoleucus*, linking the panda with the bears. **panda car**, *n*. a police patrol car.

Pandanus (pandā'nəs), *n*. a genus of trees or bushes containing the screw-pines.

pandect (pan'dekt), *n*. (*usu. pl.*) the digest of the Roman civil law made by direction of the emperor Justinian in the 6th cent.; any complete system or body of laws; a comprehensive treatise or digest.

pandemic (pandem'ik), *a*. widely epidemic, affecting a whole country or the whole world. *n*. a pandemic disease.

pandemonium (pandimō'niəm), *n*. a place or state of lawlessness, confusion or uproar; confusion, uproar.

pander (pan'də), *n*. a procurer, a pimp, a go-between in an amorous intrigue; one who ministers to base passions, prejudices etc. *v.i.* to act as pander; to act as an agent (to) for the gratification of base desires, lusts, weaknesses etc. **panderer**, *n*. a pander. **panderess** (-ris), *n. fem.* [*Pandare*, who procured for Troilus the favour of Criseyde, in Chaucer's *Troilus and Criseyde*]

pandit (pŭn'dit), PUNDIT.

P&O, (*abbr.*) Peninsular and Oriental (Steam Navigation Company).

Pandora (pandaw'rə), *n*. the first woman. **Pandora's box**, *n*. a box containing all human ills and blessings, which Pandora brought with her from heaven; hence, a source of great troubles.

pandowdy (pandow'di), *n*. (*N Am.*) a deep-dish dessert of sweetened apple slices topped with a cake crust.

p&p, (*abbr.*) postage and packing.

pane (pān), *n*. a sheet of glass in a window etc.; a piece, part or division; a panel; a side, face or surface.

panegyric (panəji'rik), *n*. a eulogy written or spoken in praise of some person, act or thing; praise, commendation. **panegyrical**, *a*. **panegyrically**, *adv*. **panegyrism**, *n*. **panegyrist**, *n*. **panegyrize, -ise** (pan'-, -nej'-), *v.t., v.i.*

panel (pan'l), *n*. a usu. rectangular piece of wood or other material inserted in or as in a frame, forming a compartment of a door, wainscot etc.; a section, as of a car body, formed from a sheet of metal, plastic etc.; a thin board on which a picture is painted; a picture, photograph etc., the height of which is much greater than the width; a flat section of metal, plastic etc. into which switches and instruments are set; a control panel; a piece of material let in lengthwise in a woman's dress etc.; a list of persons summoned as jurors; a jury; the team in a quiz game, brains trust etc.; an official list of persons; (*formerly*) persons receiving medical treatment under the National Insurance Act; (*Sc. Law*) a prisoner or the prisoners at the bar. *v.t. (past, p.p.* **panelled**) to fit or furnish (a door, wall etc.) with panels. **panel beater**, *n*. a person who repairs the damaged body panels of motor vehicles. **panel-game**, *n*. a quiz game in which a panel of experts etc. answer questions from an audience or a chairman. **panel pin**, *n*. a short, slender nail with a small head. **panel-plane, -saw**, *n*. a plane or saw used in panel-making. **panel truck**, *n*. (*N Am.*) a delivery van. **panel-work, panelling**, *n*. **panellist**, *n*. a member of a team in a quiz-game etc.

pang (pang), *n*. a sudden paroxysm of extreme pain, physical or mental; a throe, agony.

Pan-German (panjœ'mən), *a*. relating to Germans collectively or to Pan-Germanism. **Pan-Germanism**, *n*. the movement to unite all Teutonic peoples into one nation.

pangolin (pang·gō'lin), *n*. a scaly ant-eater, of various species belonging to the genus *Manis*.

panhandle (pan'handl), *n*. (*N Am.*) a strip of territory belonging to one political division extending between two others. *v.i.* (*N Am., coll.*) to beg. **panhandler**, *n*.

Pan-Hellenic (panhəlen'ik), *a*. of, characteristic of, including, or representing all Greeks. **Pan-Hellenism** (-hel'-), *n*.

panic¹ (pan'ik), *n*. sudden, overpowering, unreasoning fear, esp. when many persons are affected; a general alarm causing ill-considered action. *a*. sudden, extreme, unreasoning; caused by fear. *v.t., v.i. (past, p.p.* **panicked**) to affect or be affected with panic. **panic button**, *n*. a button, switch etc. operated to signal an emergency. **panic buying**, *n*. buying goods in panic, e.g. in anticipation of a shortage. **panic-monger**, *n*. **panic stations**, *n.pl.* a state of alarm or panic. **panic-stricken, -struck**, *a*. struck with sudden fear. **panicky**, *a*.

panic² (pan'ik), *n*. a common name for several species of the genus *Panicum*, esp. the Italian millet. **panic-grass**, *n*.

panicle (pan'ikl), *n*. a loose and irregular compound flower-cluster. **panicled, paniculate** (-nik'ŭlət), *a*.

Panicum (pan'ikəm), *n*. a genus of grasses with numerous species, some (as the millet) valuable for food.

Pan-Islam (paniz'lahm), *n*. the whole of Islam; a union of the Muslim peoples. **Pan-Islamic** (-lam'-), *a*. **Pan-Islamism** (-iz'-), *n*.

panjandrum (panjan'drəm), *n*. a high and mighty func-

tionary or pompous pretender. [the Grand *Panjandrum*, a character in a nonsense work by the Eng. dramatist Samuel Foote, 1720-77]
panne (pan), *n.* a soft, long-napped fabric.
pannier (pan'iə), *n.* a large basket, esp. one of a pair slung over the back of a beast of burden; one of a pair of bags fixed on either side of the wheel of a bicycle, motorcycle etc.; a framework, usu. of whalebone, formerly used for distending a woman's skirt at the hips; an arrangement of material formerly worn, or attached to the skirt, at the hip to give extra width; (*Arch.*) a sculptured basket, a corbel. **panniered**, *a.*
pannikin (pan'ikin), *n.* a small drinking-cup of metal; a small saucepan. **pannikin boss**, *n.* (*Austral.*) a suboverseer on a station.
pannose (pan'ōs), *a.* (*Bot.*) like cloth in texture.
panophobia (panəfō'biə), *n.* excessive fear (literally of everything).
panoply (pan'əpli), *n.* a complete suit of armour; complete defence; a full, impressive array. **panoplied**, *a.*
panopticon (panop'tikon), *n.* a prison constructed on a circular plan with a central well for the warders so that the prisoners would be always under observation; an exhibition-room for novelties etc. **panoptic** (panop'tik), *a.* viewing all aspects; all-embracing, comprehensive.
panorama (panərah'mə), *n.* a continuous picture of a complete scene on a sheet unrolled before the spectator or on the inside of a large cylindrical surface viewed from the centre; complete view in all directions; a general survey. **panoramic** (-ram'-), *a.* **panoramically**, *adv.*
panotitis (panəti'tis), *n.* inflammation of both the middle and internal ear. [Gr. *ous ōtos*, ear]
panotrope (pan'ətrōp), *n.* an electrical reproducer of disc gramophone records operating one or more loudspeakers. [Gr. *tropos*, a turn]
Pan-pipes (pan pīps), *n.sing.* a musical wind instrument made of a number of pipes or reeds, a mouth organ.
Pan-Slavism (panslah'vizm), *n.* a movement for the union of all the Slavic races. **Pan-Slavic**, *a.* **Pan-Slavist**, *n.* **Pan-Slavistic** (-vis'-), *a.* **Pan-Slavonic** (-von'-), *a.*
pansophy (pan'səfi), *n.* universal knowledge. **pansophic, -ical** (-sof'-), *a.* **pansophically**, *adv.*
panspermia (panspœ'miə), **-spermatism**, **-spermism**, **-spermy**, *n.* the theory that the atmosphere is pervaded by invisible germs which develop on finding a suitable environment. **panspermatist**, *n.* **panspermic**, *a.*
pansy (pan'zi), *n.* a species of viola, with large flowers of various colours in the cultivated varieties, the heartsease; (*coll.*) an effeminate man or boy; (*coll.*) a homosexual male.
pant (pant), *v.i.* to breathe quickly; to gasp for breath; to throb, to palpitate; to long, to yearn (after, for etc.). *v.t.* to utter gaspingly or convulsively. *n.* a gasp; a throb.
pant- PANT(O)-.
pantagraph (pan'təgrahf), PANTOGRAPH.
pantalets, -lettes (pantəlets'), *n.pl.* loose drawers extending below the skirts, with frills at the bottom, worn by children and women in the early 19th cent.; detachable frilled legs for these; drawers, cycling knickerbockers etc. for women.
pantaloon (pantəloon'), *n.* a character in pantomime, the butt of the clowns' jokes; (*pl.*) tight trousers fastened below the shoe, as worn in the Regency period; trousers, esp. loose-fitting ones.
pantechnicon (pantek'nikən), *n.* a storehouse for furniture; a place where all sorts of manufactured articles are exposed for sale; a pantechnicon van. **pantechnicon van**, *n.* a large van for removing furniture.
pantheism (pan'thiizm), *n.* the doctrine that God and the universe are identical; the heathen worship of all the gods. **pantheist**, *n.* **pantheistic, -ical** (-is'-), *a.* **pantheistically**, *adv.*

Pantheon (pan'thion), *n.* a temple to all the gods, esp. one with a circular dome at Rome, built about 27 BC; (*collect.*) the divinities of a nation; a treatise on all the gods; a building dedicated to the illustrious dead.
panther (pan'thə), *n.* the leopard; (*N Am.*) the puma.
panties (pan'tiz), *n.pl.* (*coll.*) women's or children's short knickers.
pantihose (pan'tihōz), *n.pl.* (*esp. N Am.*) tights.
pantile (pan'tīl), *n.* a tile curved transversely to an ogee shape.
pantisocracy (pantisok'rəsi), *n.* a Utopian scheme of communism in which all are equal in rank, and all are ruled by all. **pantisocrat** (-tī'səkrat), *n.* **pantisocratic** (-krat'-), *a.*
panto (pan'tō), *n.* (*coll.*) short for PANTOMIME.
pant(o)-, *comb. form.* all.
pantograph, pantagraph (pan'təgrahf), *n.* an instrument used to enlarge, copy or reduce plans etc.; a framework similar in appearance attached to the roof of an electrically-driven vehicle, for collecting electrical power from an overhead cable. **pantographic, -ical** (-graf'-), *a.* **pantography** (-tog'-), *n.*
pantomime (pan'təmīm), *n.* a dramatic performance in dumb show; one who performed in dumb show; representation in dumb show; a theatrical entertainment, usu. produced at Christmas-time, consisting largely of farce and burlesque; a muddled or farcical situation. *v.t., v.i.* to mime. **pantomimic, -ical** (-mim'-), *a.* **pantomimically**, *adv.* **pantomimist**, *n.*
pantothenic acid (pantəthen'ik), *n.* an oily acid, a member of the vitamin B complex.
pantry (pan'tri), *n.* a room or closet in which provisions or tableware are kept. **pantryman**, *n.* a butler or his assistant.
pants (pants), *n.pl.* drawers for men and boys; (*N Am.*) men's or women's trousers; women's knickers. **with one's pants down**, in an embarrassing or ill-prepared position. **pants suit**, *n.* (*N Am.*) a trouser suit for women.
panty hose (pan'ti hōz), *n.pl.* tights.
panzer (pan'zə, -tsə), *n.* term applied to armoured bodies, esp. an armoured division, in the German army; a vehicle in such a division, esp. a tank.
pap[1] (pap), *n.* soft or semi-liquid food for infants etc.; pulp; trivial or insubstantial ideas, talk etc. **pappy**, *a.*
pap[2] (pap), *n.* a teat, a nipple; a conical hill or small peak.
papa[1] (pəpah'), *n.* a dated children's word for father.
papa[2] (pah'pə), *n.* the Pope; a parish priest or one of the inferior clergy of the Greek Church.
papacy (pā'pəsi), *n.* the office, dignity or tenure of office of the Pope; the papal system of government; the Popes collectively. **papal**, *a.* pertaining to the Pope or his office, or to the Roman Catholic Church. **papal cross**, *n.* a cross with three horizontal shafts. **papalism**, *n.* **papally**, *adv.*
papain (pəpā'in), *n.* a protein-digesting enzyme found in the milky juice of the papaya.
paparazzo (paparat'sō), *n.* (*pl.* **-zzi** (-tsē)) a freelance professional photographer who specializes in photographing celebrities at private moments, usu. without their consent. [It.]
papaverous (pəpā'vərəs), *a.* resembling or allied to the poppy. **papaveraceous** (-ā'shəs), *a.*
papaw (pəpaw'), **pawpaw** (paw-), *n.* a N American tree the milky juice of which, obtained from the stem, leaves or fruit, makes meat tender; its fruit; the papaya.
papaya (pəpah'yə), *n.* a tropical American tree yielding papain, that bears large oblong edible yellow fruit; its fruit.
paper (pā'pə), *n.* a thin flexible substance made of rags, wood-fibre, grass or similar materials, used for writing and printing on, wrapping etc.; a piece, sheet or leaf of this; a written or printed document; an essay, a dissertation; a lecture; a newspaper; a set of questions for an examination; negotiable instruments, as bills of exchange;

paper money; paper-hangings, wallpaper; (*sl.*) a free pass; (*pl.*) documents establishing identity etc.; (*pl.*) a ship's documents. *a.* made of paper; like paper; stated only on paper, having no real existence. *v.t.* to cover with or decorate with paper; to rub with sandpaper; to furnish with paper; (*sl.*) to admit a large number to (a theatre etc.) by free passes. *v.i.* to apply wallpaper. **on paper**, theoretically, rather than in reality. **to commit to paper**, to write down, to record. **to paper over**, to disguise, cover up (a dispute, mistake etc.). **paperback**, *n.* a book with a soft cover of flexible card. *a.* being or relating to a paperback or paperbacks. **paper-boy, -girl**, *n.* a boy or girl who delivers newspapers. **paper-chase**, *n.* a game in which one or more persons (called the hares) drop pieces of paper as scent for pursuers (called the hounds) to track them by. **paper-clip**, *n.* a small clip of looped wire used to fasten pieces of paper together. **paper-currency** PAPER-MONEY. **paper-cutter, -knife**, *n.* a flat piece of wood, ivory etc., for cutting open the pages of a book, opening letters etc. **paper-girl** PAPER-BOY. **paper-hanger**, *n.* a person whose occupation is hanging wallpaper. **paper-hangings**, *n.pl.* paper ornamented or prepared for covering the walls of rooms etc. **paper-knife** PAPER-CUTTER. **paper-making**, *n.* **paper-money**, *n.* bank-notes or bills used as currency, opp. to coin. **paper nautilus**, *n.* a cephalopod with a thin, papery external shell, the argonaut. **paper-profits**, *n.pl.* hypothetical profits shown on a company's prospectus etc. **paper tape**, *n.* a strip of paper containing data in the form of perforations, used in computers, telex machines etc. **paper tiger**, *n.* a person or thing that is apparently threatening or powerful, but is not so in reality. **paper-weight**, *n.* a weight for keeping loose papers from being displaced. **paperwork**, *n.* clerical work, e.g. writing letters. **paperer**, *n.* **papery**, *a.* **paperiness**, *n.*
Paphian (pā'fiən), *a.* pertaining to Paphos, a city of Cyprus sacred to Venus; pertaining to Venus or her worship. *n.* a native of Paphos; a courtesan.
papier mâché (pap'yā mash'ā), *n.* a material made from pulped paper, moulded into trays, boxes etc. *a.* made of papier mâché.
papilionaceous (pəpiliənā'shəs), *a.* resembling a butterfly; used of plants with butterfly-shaped flowers, as the pea.
papilla (pəpil'ə), *n.* (*pl.* **-llae** (-ī)) a small pap, nipple or similar process; a small protuberance on an organ or part of the body or on plants. **papillary** (pap'-), **-llate** (-lət), **-llose** *a.* **papilliferous** (-lif'-), *a.* **papilliform** (-pil'ifawm), *a.* **papilloma** (papilō'mə), *n.* (*pl.* **mas, -mata** (-tə)) a tumour formed by the growth of a papilla or group of papillae, as a wart, corn etc. **papillomatous**, *a.*
papilion (pap'iyō, -lon), *n.* a breed of toy spaniel with large butterfly-shaped ears.
papillote (pap'ilōt), *n.* a paper frill used to decorate a cutlet etc.; an oiled paper in which food is enclosed for cooking.
papist (pā'pist), *n.* (*chiefly derog.*) a Roman Catholic. **papism, papistry**, *n.* **papistic, -ical** (-pis'-), *a.*
papoose (pəpoos'), *n.* a N American Indian baby or young child.
pappus (pap'əs), *n.* (*pl.* **pappi** (-ī)) the calyx of composite plants, consisting of a tuft of down or fine hairs or similar agent for dispersing the seed. **pappous, pappose**, *a.*
pappy PAP[1].
paprika (pap'rikə), *n.* a powdered condiment made from a sweet variety of red pepper.
Pap smear, test (pap), *n.* a test for the early diagnosis of cancer in which cells are scraped from a bodily organ, esp. the cervix, and examined under a microscope. [George N. *Pap*anicolaou, 1883–1962, US anatomist]
Papuan (pap'ūən), *a.* of or pertaining to Papua or New Guinea. *n.* a member of the black peoples native to the Melanesian archipelago.
papula (pap'ūlə), *n.* **papule** (*pl.* **-lae** (-lī), **-les**) a pimple.

papular, *a.*
papyrus (pəpīə'rəs), *n.* (*pl.* **-ri** (-rē), **-ruses**) a rush-like plant of the genus *Cypereae*, common formerly on the Nile and still found in Ethiopia, Syria etc.; a writing-material made from this by the Egyptians and other ancient peoples; a manuscript written on this material. **papyraceous** (-pirā'shəs), *a.* **papyrology**, *n.* the study of ancient papyrus manuscripts. **papyrologist**, *n.*
par[1] (pah), *n.* state of equality, parity; equal value, esp. equality between the selling value and the nominal value expressed on share certificates and other scrip; average or normal condition, rate etc.; in golf, the number of shots which a good player is expected to play in order to complete a hole. **above par**, at a price above the face value, at a premium. **below par**, at a discount; out of sorts. **on a par with**, of equal value, degree etc. to. **par for the course**, what is to be expected or usual. **up to par**, of the required standard.
par[2] (pah), PARR.
par[3], (*abbr.*) paragraph; parallel; parish.
par- PAR(A)-.
para[1] (pa'rə), (*abbr.*) paragraph.
para[2] (pa'rə), *n.* (*coll.*) short for PARATROOPER.
par(a)-, *comb. form* denoting closeness of position, correspondence of parts, situation on the other side, wrongness, irregularity, alteration etc.; (*Chem.*) denoting substitution or attachment at carbon atoms directly opposite in a benzene ring.
parabasis (pərab'əsis), *n.* (*pl.* **-bases** (-sēz)) a choral part in ancient Greek comedy in which the chorus addressed the audience, in the name of the poet, on personal or public topics.
parabiosis (pərəbīō'sis), *n.* the anatomical union of two organisms with shared physiological processes. **parabiotic** (-ot'-), *a.*
parable (pa'rəbl), *n.* an allegorical narrative of real or fictitious events from which a moral is drawn; an allegory, esp. of a religious kind.
parabola (pərab'ələ), *n.* a plane curve formed by the intersection of the surface of a cone with a plane parallel to one of its sides. **parabolic, -ical** (-bol'-), *a.* pertaining to or of the form of a parabola; pertaining to or of the nature of a parable, allegorical. **parabolically**, *adv.* **parabolist**, *n.* **parabolize, -ise**, *v.t.* **paraboloid** (-loid), *n.* a solid of which all the plane sections parallel to a certain line are parabolas, esp. that generated by the revolution of a parabola about its axis. **paraboloidal** (-loi'-), *a.*
paracentesis (parəsentē'sis), *n.* the operation of perforating a cavity of the body, or tapping, for the removal of fluid etc.
paracetamol (parəsē'təmol), *n.* a mild painkilling drug.
parachronism (pərak'rənizm), *n.* an error in chronology, esp. post-dating of an event.
parachute (pa'rəshoot), *n.* an umbrella-shaped contrivance by which a descent is made from a height, esp. from an aircraft; a part of an animal or an appendage to a fruit or seed serving for descent or dispersion by the wind. *v.t., v.i.* (to cause) to land by means of a parachute. **parachutist**, *n.*
paraclete (pa'rəklēt), *n.* an advocate, esp. as a title of the Holy Ghost, the Comforter.
paracrostic (parəkros'tik), *n.* a poetic composition in which the first verse contains, in order, all the letters which commence the remaining verses.
parade (pərād'), *n.* show, ostentatious display; a muster of troops for inspection etc.; a procession; ground where soldiers are paraded, drilled etc.; a public promenade; a row of shops. *v.t.* to make display of; to assemble and marshal (troops) in military order for or as for review; to cause to march up and down or in procession. *v.i.* to be marshalled in military order for display or review; to show oneself or walk about ostentatiously; to march in

paradigm 582 **parang**

procession. **parade-ground,** *n.* an area where soldiers parade.
paradigm (par'ədim), *n.* an example, a pattern; an example of a word in its grammatical inflections. **paradigmatic** (-digmat'-), *a.* **paradigmatically,** *adv.*
paradise (pa'rədīs), *n.* the garden of Eden; heaven; a place or condition of bliss; a park or pleasure-ground, esp. one in which animals are kept, a preserve. **paradise-fish,** *n.* an E Indian fish, sometimes kept in aquariums for its brilliant colouring. **paradisaic, -ical** (-sā'-), **paradisiac, -acal** (-dis'iak), **paradisial, -ian, paradisic, -ical** (-dis'-), *a.*
parados (pa'rədos), *n.* (*Mil.*) a rampart or earthwork to protect against fire from the rear.
paradox (pa'rədoks), *n.* a statement, view or doctrine contrary to received opinion; an assertion seemingly absurd but really correct; a self-contradictory or essentially false and absurd statement; a person, thing or phenomenon at variance with normal ideas of what is probable, natural or possible. **paradoxical** (-dok'si-), *a.* **paradoxical sleep,** *n.* sleep that is apparently deep but is marked by rapid eye movement, increased brain activity etc. **paradoxicalness, paradoxy,** *n.* **paradoxically,** *adv.*
paradoxure (parədok'sūə), *n.* a civet-like animal with a long, curving tail, the palm-cat of S Asia and Malaysia.
paraesthesia, (*esp. N Am.*) **paresthesia** (parəsthē'ziə), *n.* disordered perception or hallucination.
paraffin (pa'rəfin), *n.* an alkane; a mixture of liquid hydrocarbons used as a lubricant or fuel, kerosene; paraffin wax. **liquid paraffin** LIQUID. **paraffin wax,** *n.* a colourless, tasteless, odourless, fatty substance consisting primarily of a mixture of alkanes, and obtained from distillation of coal, bituminous shale, petroleum, peat etc., used for making candles, waterproofing etc.
paragliding (pa'rəglīding), *n.* the sport of gliding while attached to a device like a parachute, in which one is pulled by an aircraft, then allowed to drift to the ground. **paraglider,** *n.*
paragon (pa'rəgən), *n.* a pattern of perfection; a model, an exemplar; a person or thing of supreme excellence.
paragraph (pa'rəgrahf, -graf), *n.* a distinct portion of a discourse or writing marked by a break in the lines; a reference mark [¶] or other mark used to denote a division in the text; an item of news in a newspaper etc. *v.t.* to form into paragraphs; to mention or write about in a paragraph. **paragrapher, -phist,** *n.* **paragraphic, -ical** (-graf'-), *a.* **paragraphy,** *n.*
paragraphia (parəgraf'iə), *n.* the habitual writing of words or letters other than those intended, often a sign of brain disorder.
Paraguayan (pa'rəgwiən), *n., a.* (a native or inhabitant) of the S American republic of Paraguay.
Paraguay tea *n.* an infusion of the leaves of *Ilex paraguayensis,* maté.
para-influenza virus (parəinfluen'zə), any of various viruses causing influenza-like symptoms.
parakeet, parrakeet (pa'rəkēt), *n.* a popular name for any of the smaller long-tailed parrots.
paraldehyde (pəral'dihīd), *n.* a hypnotic used in asthma, respiratory and cardiac diseases, and epilepsy.
paralipomena (parəlipom'inə), *n.pl.* things omitted in a work; (*Bibl.* **paralipomenon** (-on), *gen. pl.*) the Books of Chronicles as giving particulars omitted in the Books of Kings.
paralipsis (parəlip'sis), *n.* a rhetorical figure by which a speaker pretends to omit mention of what at the same time he really calls attention to.
parallax (pa'rəlaks), *n.* apparent change in the position of an object due to change in the position of the observer; angular measurement of the difference between the position of a heavenly body as viewed from different places on the earth's surface or from the earth at different positions in its orbit round the sun. **parallactic** (-lak'-), *a.*

parallel (pa'rəlel), *a.* of lines etc., having the same direction and everywhere equidistant; having the same tendency, similar, corresponding. *n.* a line which throughout its whole length is everywhere equidistant from another; any one of the parallel circles on a map or globe marking degrees of latitude on the earth's surface; direction parallel to that of another line; parallel connection; a comparison; a person or thing corresponding to or analogous with another, a counterpart; in printing, a reference mark (||) calling attention to a note etc. *v.t.* (*past, p.p.* **paralleled**) to be parallel to, to match, to rival, to equal; to put in comparison with; to find a match for; to compare. **parallel bars,** *n.pl.* a pair of horizontal bars used for various exercises in gymnastics. **parallel connection,** *n.* the arrangement of pieces of electrical apparatus across a common voltage supply. **parallel processing,** *n.* the processing, by a computer, of several items of data simultaneously. **parallel processor,** *n.* a computer capable of parallel processing. **parallel rule,** *n.* a draughtsman's instrument consisting of two rulers movable about hinged joints, but always remaining parallel. **parallelism,** *n.* the state of being parallel; correspondence; repetition of grammatical construction in successive sentences or paragraphs.
parallelepiped (parəlelep'iped, -ip'ī-), **-pipedon** (-pip'ədon), *n.* a regular solid bounded by six parallelograms, the opposite pairs of which are parallel.
parallelogram (parəlel'əgram), *n.* a four-sided rectilinear figure whose opposite sides are parallel and equal. **parallelogram rule,** *n.* a rule for finding the resultant of two vectors, by constructing a parallelogram in which two sides represent the vectors, and the diagonal originating at their point of intersection represents the resultant. **parallelogrammatic, -ical** (-mat'-), **parallelogrammic, -ical** (-gram'-), *a.*
paralogism (pəral'əjizm), *n.* a fallacious argument, esp. one of which the reasoner is unconscious.
paralyse, (*esp. N Am.*) **paralyze** (par'əlīz), *v.t.* to affect with paralysis; to render powerless or ineffective; to render immobile or unable to function. **paralysation,** *n.*
paralysis (pəral'isis), *n.* total or partial loss of the power of muscular contraction or of sensation in the whole or part of the body; palsy; complete helplessness or inability to act; inability to move or function properly. **paralytic** (-lit'-), *a.* of paralysis; characterized by paralysis; afflicted with paralysis; (*sl.*) very drunk. *n.* a paralysed or paralytic person. **paralytically,** *adv.*
paramagnetic (parəmagnet'ik), *a.* having the property of being weakly magnetized and attracted by the poles of a magnet, weakly magnetic, distinguished from *diamagnetic.* **paramagnetism** (-mag'-), *n.*
paramatta PARRAMATTA.
paramecium, paramoecium (parəmē'siəm), *n.* (*pl.* **-cia** (-ə)) one of a genus (*Paramecium*) of slipper-shaped Protozoa.
paramedical, -medic (parəmed'ik), *a.* auxiliary to the work of medical doctors. *n.* a paramedical worker.
parameter (pəram'itə), *n.* (*Math.*) a quantity remaining constant for a particular case, esp. a constant quantity entering into the equation of a curve etc.; (*coll.*) a limiting factor, a constraint. **parametric** (parəmet'rik), *a.*
paramilitary (parəmil'itəri), *a.* having a similar nature or structure to military forces. *n.* a member of a paramilitary force.
paramnesia (paramnē'zhə), *n.* déjà vu.
paramoecium PARAMECIUM.
paramount (pa'rəmownt), *a.* supreme above all others, pre-eminent; superior (to). *n.* the highest in rank or authority; a lord paramount. **paramountcy** (-si), *n.* **paramountly,** *adv.*
paramour (pa'rəmooə), *n.* a lover, usu. an illicit one.
parang (pah'rang), *n.* a heavy, Malaysian sheath-knife.

paranoia (parənoi'ə), *n.* mental derangement, esp. in a chronic form characterized by delusions etc.; (*coll.*) a sense of being persecuted. **paranoiac** (-ak), *n.*, *a.* **paranoic** (-nō'-), *a.* **paranoid**, *a.*
paranormal (parənaw'məl), *a.* not rationally explicable. *n.* paranormal events.
paranthelion (parənthē'lion), *n.* a diffuse image of the sun at the same altitude and at an angular distance of 120° due to reflection from ice-spicules in the air.
parapet (pa'rəpit), *n.* a low or breast-high wall at the edge of a roof, balcony, bridge etc.; a breast-high wall or rampart for covering troops from observation and attack. **parapeted**, *a.*
paraph (pa'raf), *n.* a flourish after a signature, orig. intended as a protection against forgery. *v.t.* to sign; to initial.
paraphernalia (parəfənä'liə), *n.pl.* (*Law*) personal property allowed to a wife over and above her dower, including her personal apparel, ornaments etc.; miscellaneous belongings, ornaments, trappings, equipment.
paraphrase (pa'rəfräz), *n.* a free translation or rendering of a passage; a restatement of a passage in different terms. *v.t.* to express or interpret in other words. *v.i.* to make a paraphrase. **paraphraser, paraphrast** (-frast), *n.* one who paraphrases. **paraphrastic** (-fras'-), *a.* **paraphrastically**, *adv.*
paraplegia (parəplē'jə), *n.* paralysis of the lower limbs and the lower part of the body. **paraplegic**, *n.*, *a.*
parapodium (parəpō'diəm), *n.* one of the jointless lateral locomotory organs of an annelid.
parapophysis (parəpof'isis), *n.* a process on the side of a vertebra, usu. serving as the point of articulation of a rib.
parapsychical (parəsi'kikl), *a.* denoting phenomena such as telepathy which appear to be beyond explanation by the ascertained laws of science. **parapsychology**, *n.* the study of such phenomena. **parapsychologist**, *n.*
Paraquat® (pa'rəkwot), *n.* a very poisonous weedkiller.
parascending (pa'rəsending), *n.* a sport in which a parachutist is towed at ground level until he or she ascends, then descends by parachute.
paraselene (parəsilē'ni), *n.* (*pl.* -nae) a mock moon appearing in a lunar halo. **paraselenic** (-len'-), *a.*
parasite (pa'rəsīt), *n.* one who lives at the expense of others, a hanger-on, a sponger; an animal or plant subsisting at the expense of another organism. **parasitic, -ical** (-sit'-), *a.* **parasitically**, *adv.* **parasiticide** (-sit'isid), *n.* a preparation for destroying parasites. **parasitism**, *n.* **parasitize, -ise**, *v.t.* to live as a parasite on; to infest with parasites. **parasitization**, *n.* **parasitology** (-tol'-), *n.* **parasitologist**, *n.*
parasol (pa'rəsol), *n.* a small umbrella used to give shelter from the sun, a sunshade.
parasympathetic (parəsimpəthet'ik), *a.* of that part of the autonomic nervous system that slows the heartbeat, stimulates smooth muscle, as of the digestive tract, constricts the bronchi of the lungs etc. and thus counteracts the sympathetic nervous system.
parasynthesis (parəsin'thəsis), *n.* the principle or process of forming derivatives from compound words. **parasynthetic** (-thet'-), *a.* **parasyntheton** (-ton), *n.* (*pl.* -ta (-tə)) a word so formed.
parataxis (parətak'sis), *n.* an arrangement of clauses, sentences etc., without connectives indicating subordination etc. **paratactic** (-tak'-), *a.* **paratactically**, *adv.*
paratha (pərah'tə), *n.* a thin cake of unleavened bread, eaten with Indian food.
parathion (parəthī'on), *n.* a highly toxic insecticide.
parathyroid (parəthī'roid), *a.* a small endocrine gland, one of which is situated on each side of the thyroid.
paratrooper (pa'rətroopə), *n.* a soldier belonging to a unit borne in aeroplanes and gliders and dropped by parachute, with full equipment, usu. behind enemy lines. **paratroops**, *n.pl.*
paratyphoid (parəti'foid), *n.* an infectious fever of the enteric group, similar in symptoms to typhoid but of milder character.
paravane (pa'rəvān), *n.* a mine-sweeping appliance for severing the moorings of submerged mines.
par avion (par avyō'), *adv.* by air mail. [F, by aeroplane]
parazoan (parəzō'ən), *n.* a member of the Parazoa, multicellular invertebrates, such as sponges.
parboil (pah'boil), *v.t.* to boil partially until semi-cooked; orig., to boil thoroughly.
parbuckle (pah'bŭkl), *n.* a double sling usu. made by passing the two ends of a rope through a bight, for hoisting or lowering a cask or gun. *v.t.* to hoist or lower by a parbuckle.
parcel (pah'sl), *n.* a number or quantity of things dealt with as a separate lot; a distinct portion, as of land; a group of things or people; a bundle, a package. *v.t.* (*past*, *p.p.* **parcelled**) to divide (*usu.* out) into parts or lots; to make into a parcel; **parcel-office**, *n.* an office for the receipt or dispatch of parcels. **parcel-post**, *n.* a branch of the postal service for the delivery of parcels.
parcenary (pah'sənəri), *n.* coheirship, coparcenary. **parcener**, *n.*
parch (pahch), *v.t.* to scorch or roast partially dry; to dry up. *v.i.* to become hot or dry. **parched**, *a.* dried up; (*coll.*) very thirsty.
Parcheesi® (pahchē'zi), *n.* a modern board game based on pachisi.
parchment (pahch'mənt), *n.* the skin of calves, sheep, goats etc., prepared for writing upon, painting etc.; a manuscript on this, esp. a deed; a stiff strong paper resembling parchment. *a.* made of or resembling parchment. **parchmenty**, *a.*
parclose (pah'klōz), *n.* a screen or railing enclosing an altar, tomb etc. in a church.
pard[1] (pahd), *n.* (*poet.*) a panther, a leopard.
pard[2] (pahd), **pardner** (pahd'nə), *n.* (*N Am. sl.*) a partner.
pardon (pah'dn), *v.t.* to forgive, to absolve from; to remit the penalty of; to excuse, to make allowance for. *n.* the act of pardoning; a complete or partial remission of the legal consequences of crime; an official warrant of a penalty remitted; a papal indulgence; courteous forbearance. **I beg your pardon, pardon me**, excuse me, a polite apology for an action, contradiction or failure to hear or understand what is said. **pardon my French** FRENCH. **pardonable**, *a.* **pardonableness**, *n.* **pardonably**, *adv.* **pardoner**, *n.* one who pardons; a person licensed to sell papal pardons or indulgences.
pare (peə), *v.t.* to cut or shave (away or off); to cut away or remove the rind etc. of (fruit etc.); to trim by cutting the edges or irregularities of; to diminish by degrees. **parer**, *n.* **paring**, *n.* the act of cutting off, pruning or trimming; that which is pared off; a shaving, rind etc.
paregoric (parigo'rik), *a.* assuaging or soothing pain. *n.* a camphorated tincture of opium for assuaging pain.
pareira (pərea'rə), *n.* a drug used in urinary disorders, obtained from the root of a Brazilian climbing plant.
parenchyma (pəreng'kimə), *n.* the soft cellular tissue of glands and other organs, distinguished from connective tissue etc.; thin cellular tissue in the softer part of plants, pith, fruit pulp etc. **parenchymal** (-ki'-), **-matous** (-kim'-), **-mous**, *a.*
parent (peə'rənt), *n.* a father or mother; a forefather; an organism from which others are produced; a source, origin, cause. *v.t.* to be a parent or the parent of. **parent-teacher association**, an association formed by the parents and teachers of a school, esp. for social and fund-raising purposes. **parent company**, *n.* a company having control of one or more subsidiaries. **parent power**, *n.* the right of parents to influence the way in which their children are educated. **parentage** (-ij), *n.* birth, extrac-

tion, lineage, origin; the position or state of being a parent, parenthood. **parental** (pəren'-), *a*. **parentally**, *adv*.
parenthood (-hud), *n*. **parenting**, *n*. the skills or activity of being or acting as a parent. **parentless**, *a*.
parenteral (pəren'tərəl), *a*. situated or occurring outside the digestive tract, esp. being the means of administering a drug other than via the digestive tract.
parenthesis (pəren'thəsis), *n*. (*pl*. **-theses** (-sēz)) a word, phrase or sentence inserted in a sentence that is grammatically complete without it, usu. marked off by brackets, dashes or commas; (*pl*.) round brackets () to include such words; an interval, interlude, incident etc. **parenthesize**, **-ise**, *v.t*. to insert as a parenthesis; to place (a clause etc.) between parentheses. **parenthetic**, **-ical** (-thet'-), *a*. **parenthetically**, *adv*.
parergon (pərœ'gon), *n*. (*pl*. **-ga** (-gə)) a subsidiary work, apart from one's regular employment.
paresis (pa'rəsis), *n*. incomplete paralysis, affecting muscular movement but not sensation. **paretic** (-ret'-), *a*. [Gr., from *parienai* (*hienai*, to let go)]
paresthesia (parəsthē'ziə), PARAESTHESIA.
par excellence (par eks'əlös), *adv*. above all others, preeminently. [F]
parfait (pahfā'. pah-), *n*. a rich, cold dessert made with whipped cream, eggs, fruit etc.
parget (pah'jit), *v.t*. to cover with plaster; esp. pargeting. *n*. plaster for covering surfaces, esp. walls; pargeting. **pargeter**, *n*. **pargeting**, *n*. plasterwork, esp. decorative plaster-work.
parhelion (pah-hē'liən), *n*. (*pl*. **-lia** (-ə)) a mock-sun or bright spot in a solar halo, due to ice-crystals in the atmosphere. **parheliacal** (-lī'-), **parhelic** (-hē'-, -hel'-), *a*. **parhelic circle**, *n*. a circle of light parallel to the horizon at the altitude of the sun, caused by the reflection of sunlight by ice-crystals in the atmosphere.
pariah (pəri'ə), *n*. one of a people of very low caste in S India and Burma; one of low caste or without caste; a social outcast. **pariah-dog**, *n*. a vagabond mongrel dog, esp. in India.
Parian (peə'riən), *a*. pertaining to the island of Paros, celebrated for its white marble. *n*. a white variety of porcelain, used for statuettes etc.
parietal (pəri'ətəl), *a*. pertaining to the walls of the body and its cavities; pertaining to a parietal bone; (*Bot*.) pertaining to or attached to the wall of a structure, esp. of placentae or ovaries; (*N Am*.) pertaining to residence within the walls of a college. **parietal bone**, *n*. either of the two bones forming part of the top and sides of the skull. **parieto-**, *comb. form*. (*Anat*.).
pari mutuel (pari mū'tūəl), *n*. a system of betting in which the winners divide the losers' stakes less a percentage for management. [F, mutual bet]
paring PARE.
pari passu (pari pas'oo), *adv*. (esp. in legal contexts) with equal pace, in a similar degree, equally. [L]
Paris (pa'ris), *a*. used attributively of things derived from Paris, capital of France. **plaster of Paris** PLASTER. **paris blue**, *n*. a bright Prussian blue; a bright-blue colouring-matter obtained from aniline. **paris green**, *n*. a light-green pigment obtained from arsenite of copper.
parish (pa'rish), *n*. an ecclesiastical district with its own church and clergyman; a subdivision of a county; a civil district for purposes of local government etc.; the people living in a parish or attending the church of a parish. *a*. pertaining to or maintained by a parish. **on the parish**, (*formerly*) being financially supported by the parish. **parish clerk**, *n*. a subordinate lay official in the church, formerly leading the congregation in the responses. **parish council**, *n*. a local administrative body elected by the parishioners in rural districts. **parish pump**, *n*. (*coll*.) of local interest; parochial. **parish register**, *n*. a register of christenings, marriages, burials etc., kept at a parish church.

parishioner (-rish'-), *n*. one who belongs to a parish.
parisyllabic (parisilab'ik), *a*. of Greek and Latin nouns, having the same number of syllables, esp. in all the cases.
parity (par'iti), *n*. equality of rank, condition, value etc. parallelism, analogy; the amount of a foreign currency equal to a specific sum of domestic currency; equivalence of a commodity price as expressed in one currency to its price expressed in another; (*Math*.) the property or state of being odd or even. **parity check**, *n*. a check of computer data which uses the state of oddness or evenness of the number of bits in a unit of information as a means of detecting errors.
park (pahk), *n*. a piece of land, usu. for ornament, pleasure or recreation, with trees, pasture etc., surrounding or adjoining a mansion; a piece of ground, ornamentally laid out, enclosed for public recreation; a large tract or region, usu. with interesting physical features, preserved in its natural state for public enjoyment; (*Mil*.) a space occupied by the artillery, stores etc. in an encampment; a place for temporarily leaving vehicles; a large area designed to accommodate businesses etc. of a similar type or having a similar purpose, as in *science park*. (*coll*.) a sports pitch; (*N Am*.) a sports stadium or arena. *v.t*. to enclose in or as in a park; to mass (artillery) in a park; to leave (a vehicle) in a place allotted for the purpose; to leave temporarily. *v.i*. to leave a vehicle in a place allotted for the purpose. **parking lot**, *n*. (*N Am*.) a car-park. **parkingmeter**, *n*. a coin-operated appliance on a kerb that charges for the time cars are parked there. **parking ticket**, *n*. a document issued for a parking offence requiring payment of a fine or appearance in court. **park-keeper**, *n*. an official who supervises a public park. **parkland**, *n*. cultivated grassed land with trees, used, or suitable for use, as a park.
parka (pah'kə), *n*. a hooded jacket often edged or lined with fur.
parkin (pah'kin), *n*. a cake made of gingerbread or oatmeal and treacle.
Parkinson's disease (pah'kinsənz), **parkinsonism**, *n*. a chronic disorder of the central nervous system causing loss of muscle coordination and tremor. [James *Parkinson*, 1755–1824, British surgeon who first described it]
Parkinson's Law (pah'kinsənz), *n*. the principle in office management etc. that work expands to fill the time available for its completion. [adumbrated facetiously by C. Northcote *Parkinson*, 1909–88, British historian]
parky (pah'ki), *a*. (*coll*.) chilly, uncomfortable.
parl., (*abbr*.) parliament(ary).
parlance (pah'ləns), *n*. way of speaking, idiom.
parley (pah'li), *v.i*. to confer with an enemy with pacific intentions; to talk, to dispute. *n*. a conference for discussing terms, esp. between enemies.
parliament (pah'ləmənt), *n*. a deliberative assembly; a legislative body, esp. the British legislature, consisting of the Houses of Lords and Commons, together with the sovereign. **parliamentarian** (-teə'ri-), *n*. one versed in parliamentary rules and usages or in parliamentary debate; a member of parliament; (*Hist*.) one who supported the Parliament against Charles I in the time of the Great Civil War. *a*. parliamentary. **parliamentary** (-men'-), *a*. of, pertaining to or enacted by a parliament according to the rules of (a) parliament; esp. of language, admissible in (a) parliament, civil. **parliamentary private secretary**, an ordinary member of the British parliament appointed to assist a Minister. **parliamentary agent**, *n*. a person employed by a private person or persons to draft bills or manage the business of private legislation. **Parliamentary Commissioner**, *n*. (in Britain) the official designation of an ombudsman.
parlour, (*esp. N Am*.) **parlor** (pah'lə), *n*. orig. a room in a convent for conversation; the family sitting-room in a private house; a room in an inn, hotel etc. away from the

parlous — **partake**

public rooms; any of various commercial establishments, e.g. a beauty-parlour; a building used for milking cows. **parlour-maid**, *n.* a maid-servant waiting at table.
parlous (pah'ləs), *a.* perilous, awkward, trying; shrewd, clever, venturesome. *adv.* extremely.
Parmesan (pahmizan'), *n.* a kind of hard, dry cheese made at Parma and elsewhere in N Italy, used grated as a topping for pasta dishes.
Parnassus (pahnas'əs), *n.* poetry, literature. [mountain in Greece, anciently famous as the favourite resort of the Muses]
parochial (pərō'kiəl), *a.* relating to a parish; petty, narrow. **parochialism, -ality** (-al'-), *n.* **parochialize, -ise,** *v.t.* **parochially,** *adv.*
parody (pa'rədi), *n.* a literary or musical composition imitating an author's or composer's work for the purpose of ridicule; a feeble imitation, a mere travesty. *v.t.* to turn into a parody, to burlesque. **parodic** (-rod'-), *a.* **parodist,** *n.*
parole (pəröl'), *n.* a word of honour, esp. a promise by, e.g. a prisoner of war to fulfil certain conditions in return for liberty; the release of a prisoner under certain conditions; the daily password used by officers etc., as distinguished from the countersign; actual speech, by contrast with language as an abstract system (cp. *performance*). *v.t.* to put or release on parole. **on parole,** of a prisoner, released under certain conditions.
paronomasia (parənəmā'ziə, -siə), *n.* a play on words, a pun. **paronomastic, -ical** (-mas'-), *a.*
paronym (pa'rənim), *n.* a paronymous word. **paronymous** (-ron'-), *a.* having the same root, cognate; alike in sound, but differing in origin, spelling and meaning. **paronymy** (-ron'-), *n.*
parotid (pərot'id), *a.* situated near the ear. *n.* a parotid gland. **parotid gland,** *n.* one of a pair of salivary glands situated on either side of the cheek in front of the ear, with a duct to the mouth. **parotitis** (-titis), *n.* inflammation of the parotid gland, mumps.
-parous, *comb. form.* producing, bringing forth.
parousia (pəroo'ziə), *n.* Christ's second coming, to judge the world.
paroxysm (pa'rəksizm), *n.* a sudden and violent fit; the exacerbation of a disease at periodic times; a fit of laughter or other emotion. **paroxysmal, -mic** (-siz'-), *a.* **paroxysmally,** *adv.*
paroxytone (pərok'sitōn), *a.* in classical Greek, applied to a word having an acute accent on the penultimate syllable. *n.* a word having such an accent. **paroxytonic** (-ton'-), *a.*
parquet (pah'kā), *n.* a flooring of parquetry; (*N Am.*) the part of the floor of a theatre between the orchestra and the row immediately under the front of the gallery. *v.t.* to floor (a room) with parquetry. **parquetry** (-tri), *n.* inlaid woodwork for floors.
parr (pah), *n.* a young salmon.
parrakeet (pa'rəkēt), PARAKEET.
par(r)amatta (parəmat'ə), *n.* a light twilled dress-fabric of merino wool and cotton, orig. from Parramatta in New South Wales.
parricide (pa'risid), *n.* one who murders, or the murder of, a parent or a revered person. **parricidal** (-si'-), *a.*
parrot (pa'rət), *n.* one of a group of tropical birds with brilliant plumage, esp. the genus *Psittacus*, remarkable for their faculty of imitating the human voice; one who repeats words or imitates actions mechanically or unintelligently. *v.t.* to repeat mechanically or by rote. *v.i.* to repeat words or to chatter as a parrot. **parrot-fashion,** *adv.* by rote. **parrot-fish,** *n.* a fish of the genus *Scarus*, or some allied genera, from their brilliant coloration, and the beak-like projection of the jaws. **parroter,** *n.* **parrotism,** *n.* **parrotry,** *n.*
parry (pa'ri), *v.t.* to ward off (a blow or thrust); to evade. *n.* a defensive movement in fencing, the warding off of a blow etc.; an evasive answer or action.
parse (pahz), *v.t.* to describe or classify (a word) grammatically, its inflectional forms, relations in the sentence etc.; to analyse (a sentence) and describe its component words and their relations grammatically. *v.i.* of a word or sentence, to be conformable to grammatical rules. **parser,** *n.* **parsing,** *n.*
parsec (pah'sek), *n.* a unit of length in calculating the distance of the stars, being $1 \cdot 9 \times 10^{13}$ miles (3×10^{13} km) or 3·26 light years. [*parallax second*]
Parsee, Parsi (pah'sē), *n.* a Zoroastrian, a descendant of the Persians who fled to India from the Muslim persecution in the 7th and 8th cents; the language of the Persians under the Sassanian kings, before it was corrupted by Arabic. **Parseeism, Parsiism,** *n.* the Parsee religion, Zoroastrianism.
parsimonious (pahsimō'niəs), *a.* sparing in the expenditure of money; frugal, niggardly, stingy, mean. **parsimoniously,** *adv.* **parsimoniousness, parsimony** (pah'-), *n.*
parsley (pahs'li), *n.* an umbelliferous herb cultivated for its aromatic leaves used for seasoning and garnishing dishes.
parsnip (pah'snip), *n.* an umbelliferous plant, with an edible root used as a culinary vegetable.
parson (pah'sən), *n.* a rector, vicar or other clergyman holding a benefice; (*coll.*) a clergyman. **parson-bird,** *n.* the poe-bird. **parson's nose,** *n.* the rump of a fowl. **parsonage** (-ij), *n.* the dwelling-house of a parson, provided by a parish. **parsonic** (-son'-), **-ical, parsonish,** *a.*
part (paht), *n.* a portion, piece or amount of a thing or number of things; an essential or integral constituent; a portion separate from the rest or considered as separate; a member, an organ; a proportional quantity; one of several equal portions, quantities or numbers into which a thing is divided, or of which it is composed; a section of a book, periodical etc., as issued at one time; a share, a lot; interest, concern; share of work etc., act, duty; side, party; the role, character, words etc. allotted to an actor; a copy of the words so allotted; a person's allotted duty or responsibility; (*pl.*) qualities, accomplishments, talents; quarters; (*pl.*) region, district; one of the constituent melodies of a harmony; a melody allotted to a particular voice or instrument; (*N Am.*) a parting in the hair. *v.t.* to divide into portions, shares, pieces, etc.; to separate; to brush (the hair) with a division along the head. *v.i.* to divide; to separate (from); to leave. *adv.* partly. *a.* partial. **for my part,** so far as I am concerned. **for the most part** MOST. **in good part,** with good temper, without offence. **in ill part** ILL. **in part,** partly. **on the part of,** done by or proceeding from. **part and parcel,** an essential part or element. **part of speech,** a grammatical class of words of a particular character, comprising noun, pronoun, adjective, verb, adverb, preposition, conjunction, interjection. **to part company,** to separate. **to part with,** to relinquish, to give up. **to play a part,** to assist or be involved; to act deceitfully, pretend. **to take part,** to assist; to participate. **to take the part of,** to back up or support the cause of. **part-exchange,** *n.* a form of purchase in which one item is offered as partial payment for another, the balance being paid as money. *v.t.* to offer in part-exchange. **part-owner,** *n.* one who has a share in property with others. **part-song,** *n.* a composition for at least three voices in harmony, usu. without accompaniment. **part-time,** *a.,* *adv.* (working or done) for less than the usual number of hours. **part-timer,** *n.* a part-time worker, student etc. **part-way,** *adv.* (*esp. N Am.*) to some extent; partially. **part-work,** *n.* a magazine etc. issued in instalments intended to be bound to form a complete book or course of study.
part., (*abbr.*) participle.
partake (pahtāk'), *v.t.* (*past* **partook** (-tuk), *p.p.* **partaken**) to take or have a part or share in common with others. *v.i.* to take or have a part or share (of or in, with another

or others); to have something of the nature (of); to eat and drink (of). **partaker**, *n*.
parterre (pahte͡ə'), *n*. an ornamental arrangement of flower-beds, with intervening walks; the ground-floor of a theatre or the part of this behind the orchestra; (*N Am*.) the part under the galleries.
parthenogenesis (pahthənöjen'əsis), *n*. generation without sexual union. **parthenogenetic** (net'-), *a*. **parthenogenetically**, *adv*.
Parthian (pah'thiən), *a*. of or pertaining to Parthia, an ancient kingdom in W Asia. **Parthian arrow, glance, shaft, shot**, *n*. a look, word etc. delivered as a parting blow, like the arrows shot by the Parthians in the act of fleeing.
partial (pah'shəl), *a*. affecting a part only, incomplete; biased in favour of one side or party, unfair; having a preference for. **partiality** (-al'-), *n*. **partially**, *adv*.
participate (pahtis'ipāt), *v.i*. to have or enjoy a share, to partake (in); to have something (of the nature of). **participable**, *a*. **participant**, *n*., *a*. **participation**, *n*. **participative**, *a*. **participator**, *n*.
participle (pah'tisipl), *n*. a word partaking of the nature of a verb and of an adjective, a verbal adjective qualifying a substantive. **participial** (-sip'-), *a*. **participially**, *adv*.
particle (pah'tikl), *n*. a minute part or portion; a minute speck of matter; a subdivision of matter of negligible size, as an atom or electron; a word not inflected, or not used except in combination. **elementary particle** ELEMENT. **particle accelerator**, *n*. a device for accelerating charged particles, used in high-energy physics. **particulate** (-tik'ūlət), *a*.
parti-coloured (pah'tikūləd), *a*. partly of one colour, partly of another; variegated.
particular (pətik'ūlə), *a*. pertaining to a single person or thing as distinguished from others; special, peculiar, characteristic; minute, circumstantial; fastidious, exact, precise; remarkable, noteworthy. *n*. an item, a detail, an instance; (*pl*.) a detailed account. **in particular**, especially, particularly. **particularism**, *n*. devotion to private interests or those of a party, sect etc.; the policy of allowing political independence to the separate states of an empire, confederation etc.; (*Theol*.) the doctrine of the election or redemption of particular individuals of the human race. **particularist**, *n*. **particularistic** (-is'-), *a*. **particularity** (-la'-), *n*. the quality of being particular; attention to detail; a minute point or instance; a peculiarity. **particularize, -ise**, *v.t*. to mention individually; to specify; to give the particulars of. *v.i*. to be attentive to particulars or details. **particularization, -isation**, *n*. **particularly**, *adv*. in a particular manner; in detail; to a high degree. **particularness**, *n*.
particulate PARTICLE.
parting (pah'ting), *a*. serving to part; departing; given or bestowed on departure or separation. *n*. separation, division; a point of separation or departure; a dividing-line, esp. between sections of hair combed or falling in opposite directions; a departure; leave-taking.
partisan (pahtizan'. pah'-), *n*. an adherent of a party, faction, cause etc., esp. one showing unreasoning devotion; one of a body of irregular troops carrying out special enterprises, such as raids. *a*. pertaining, attached or (unreasonably) devoted to a party. **partisanship**, *n*.
partita (pahtē'tə), *n*. a suite of music.
partite (pah'tīt), *a*. (*Bot*., *Ent*. etc.) divided nearly to the base.
partition (pahtish'ən), *n*. division into parts, separation, distribution; a separate part; that which separates into parts, esp. a wall, screen or other barrier; (*Law*) division of property among joint-owners etc. *v.t*. to separate (off); to divide into parts or shares. **partitioned**, *a*. **partitioner**, *n*. **partitionist**, *n*. **partitive** (pah'-), *a*. denoting a part. *n*. a word denoting partition, as *some*, *any* etc. **partitively**, *adv*.
partly (paht'li), *adv*. in part; to some extent; not wholly.

partner (paht'nər), *n*. one who shares with another, esp. one associated with others in business; an associate or companion; one of two persons who dance together; one of two playing on the same side in a game; a husband, wife, spouse, lover. *v.t*. to join in partnership, to be a partner (of). **partnerless**, *a*. **partnership**, *n*. the state of being a partner or partners; a contractual relationship between a number of people involved in a business enterprise; the people involved in such a relationship.
parton (pah'ton), *n*. an elementary particle postulated as a constituent of neutrons and protons.
partook (pahtuk'), past PARTAKE.
partridge (pah'trij), *n*. a gallinaceous bird of the genus *Perdix*, esp. *P. cinerea*, preserved for game.
part-song, part-time PART.
parturient (pahtū'riənt), *a*. about to bring forth young; of parturition; of the mind etc., learned, fertile. **parturition** (-rish'ən), *n*. the act of bringing forth young.
partwork PART.
party (pah'ti), *n*. a number of persons united together for a particular purpose; a body of persons organized for a political aim; a social gathering or entertainment; each of the actual or fictitious personages on either side in a legal action, contract etc.; an accessory, one concerned in any affair; (*coll*.) a person. *v.i*. to attend parties, entertainments etc. **the party's over**, (*coll*.) something enjoyable, pleasant etc. is at an end. **party line**, *n*. a telephone exchange line used by a number of subscribers; the policy laid down by a political party. **party-spirit**, *n*. zeal for a party. **party-spirited**, *a*. **party-wall**, *n*. a wall separating two buildings etc., the joint-property of the respective owners.
parvenu (pah'vənū, -noo), *n*. a person who has risen socially or financially, an upstart. **parvenue**, *n. fem*.
parvis (pah'vis), *n*. the name given in the Middle Ages to the vacant space before a church where the mysteries were performed.
PASCAL (paskal'), *n*. a computer language suitable for many applications. [Blaise *Pascal*, 1623–62, French scientist and thinker]
pascal (paskal'), *n*. a unit of pressure, 1 newton per square metre.
paschal (pas'kəl), *a*. pertaining to the Passover or to Easter. **pasch**, *n*. the Passover; Easter.
pas de deux (pah də dœ'), (*pl.* **pas de deux**) a dance performed by two people, esp. in ballet. [F, dance of two]
pash (pash), *n*. (*coll*.) a violent infatuation, a crush. [contracted form of *passion*]
pasha, pacha (pah'shə, pash'ə), *n*. a former Turkish title of honour, usu. conferred on officers of high rank, governors etc. **pashalic** (pash'-. -shah'-), *n*. the jurisdiction of a pasha.
Pashto (pŭsh'tō), *n*. a language spoken in Afghanistan and NW India. *a*. of or using this language.
paso doble (pas'ō dō'blä), *n*. (the music for) a Latin-American ballroom dance in fast 2/4 time, based on a march step.
pasque-flower (pask'flowə), *n*. a species of anemone, *Anemone pulsatilla*, with bell-shaped purple flowers.
pasquinade (paskwinād'), *n*. a lampoon, a satire. *v.t*. to lampoon, to satirize.
pass (pahs), *v.i*. to move from one place to another, to proceed, to go (along, on, swiftly etc.); to move past; to overtake a vehicle; to circulate, to be current; to be changed from one state to another; to change gradually; to change hands; to be transferred; to disappear, to vanish; to die; to go by, to elapse; to go through, to be accepted without censure or challenge; to be enacted (as a bill before parliament); to receive current recognition; to meet with acceptance; to be approved by examining; to take place, to happen, to occur; (*Cards*) to give up one's

pass.

option of playing, making trumps etc.; to choose not to do something, esp. to answer a question; (*Law*) to be transferred or handed on. *v.i.* to go by, beyond, over, or through; to transfer, to hand round; to transfer (as a ball) to a teammate by throwing, kicking etc.; to circulate, to give currency to; to spend (time etc.); to admit; to approve; to enact; to satisfy the requirements of (an examination etc.); to outstrip, to surpass; to move, to cause to move; to cause to go by; to allow to go through (as a bill, a candidate etc.) after examination; to pronounce, to utter; to void, to discharge; to overlook, to disregard, to reject; (*N Am.*) to omit. *n.* the act of passing; a passage, avenue or opening, esp. a narrow or difficult way; a narrow passage through mountains, a defile; a navigable passage, as at the mouth of a river; a written or printed permission to pass, or do something; a ticket authorizing one to travel (on a railway etc.) or to be admitted (to a theatre etc.) free; a critical state or condition of things; the act of passing an examination, esp. without special merit or honours; a thrust in fencing; a sexual advance; a passing of hands over anything (as in mesmerism); the act of passing a ball etc. in various games. **to bring to pass** BRING. **to come to pass** COME. **to make a pass at**, to attempt to seduce. **to pass as, for**, to be taken for. **to pass away**, to die, to come to an end. **to pass by**, to omit, to disregard. **to pass off**, to proceed (without a hitch etc.); to disappear gradually; to circulate as genuine, to palm off. **to pass out**, (*coll.*) to faint; of an officer cadet, to complete training at a military academy. **to pass over**, to allow to go by without notice, to overlook; to omit; to die. **to pass the time of day**, to exchange greetings. **to pass up**, to renounce; to neglect, disregard. **pass band**, *n.* a band of frequencies that is transmitted with minimum attenuation. **pass-book**, *n.* a book in which withdrawals from, and payments into, an account (e.g. with a building society) are recorded; a bank-book. **pass degree**, *n.* a bachelor's degree without honours. **pass-key**, *n.* a master-key; a key for passing in when a gate etc. is locked. **pass laws**, *n.pl.* in South Africa, laws that restrict Blacks' freedom of movement. **passman**, *n.* a candidate in an examination obtaining only a pass not honours. **password**, *n.* a word by which to distinguish friends from strangers; a word, phrase, set of characters etc. by which a person gains admission, e.g. to a computer file; a watchword. **passable**, *a.* that may be passed; acceptable, allowable, tolerable, fairly good. **passably**, *adv.* **passer**, *n.* one who passes. **passer-by**, *n.* (*pl.* **passers-by**) one who passes by or near, esp. casually.
pass., (*abbr.*) passive.
passage (pas'ij), *n.* the act of passing; movement from one place to another, transit, migration; transition from one state to another; a journey, a voyage, a crossing; a way by which one passes, a way of entrance or exit; a corridor or gallery giving admission to different rooms in a building; right or liberty of passing; a separate portion of a discourse etc., esp. in a book; a section of a musical composition; the passing of a bill etc., into law. **bird of passage** BIRD. **to work one's passage**, to work as a sailor etc., receiving a free passage in lieu of wages; to work one's way without help from influence etc. **passageway**, *n.* a corridor.
passant (pas'ənt), *a.* (*Her.*) walking and looking towards the dexter side with the dexter fore-paw raised.
passé (pas'ā, -ā'), *a.* (*fem.* **-sée**) past the prime, faded; old-fashioned, behind the times. [F]
passenger (pas'injə), *n.* one who travels on a public conveyance; (*coll.*) a person who benefits from something without contributing to it. **passenger-pigeon**, *n.* an extinct N American migratory pigeon.
passepartout (paspahtoo'), *n.* a paper frame for a picture, photograph etc.; a master-key. [F, pass everywhere]
passer, passer-by PASS.

pasta

passerine (pas'ərin), *a.* pertaining to the order Passeriformes or suborder Passeres or perchers, which contains the great mass of the smaller birds. *n.* a passerine bird.
passible (pas'ibl), *a.* capable of feeling or suffering; susceptible to impressions from external agents. **passibility** (-bil'-), **-ibleness**, *n.*
passim (pas'im), *adv.* here and there, throughout (indicating the occurrence of a word, allusions etc. in a cited work). [L]
passing (pah'sing), *a.* going by, occurring; incidental, casual, cursory; transient, fleeting. *n.* passage, transit, lapse. **in passing**, casually, without making direct reference. **passing-bell**, *n.* a bell tolled at the hour of a person's death to invite prayers on his or her behalf. **passing-note**, *n.* (*Mus.*) a note forming a transition between two others, but not an essential part of the harmony. **passing shot**, *n.* a stroke in tennis that wins the point by passing an opponent beyond his or her reach.
passion (pash'ən), *n.* intense emotion, a deep and overpowering affection of the mind, as grief, anger, hatred etc.; violent anger; ardent affection; sexual love; zeal, ardent enthusiasm (for); the object of this; the last agonies of Christ; an artistic representation of this; a musical setting of the Gospel narrative of the Passion. **passion-flower**, *n.* a plant of the genus *Passiflora*, chiefly consisting of climbers, with flowers bearing a fancied resemblance to the instruments of the Passion. **passion fruit**, *n.* the edible fruit of a passion-flower; a granadilla. **passion-play**, *n.* a mystery-play representing the Passion. **Passion Sunday**, *n.* the fifth Sunday in Lent. **Passiontide**, *n.* the last two weeks of Lent. **Passion Week**, *n.* the week following Passion Sunday; (*loosely*) Holy Week. **passionate** (-nət), *a.* easily moved to strong feeling, esp. anger; excited, vehement, warm, intense; manifesting or characterized by intense sexual feeling. **passionately**, *adv.* **passionateness**, *n.* **passioned**, *a.* impassioned. **passionless**, *a.* passionlessly, *adv.* **passionlessness**, *n.*
passive (pas'iv), *a.* suffering, acted upon, not acting; capable of receiving impressions; of a verb form, expressing an action done to the subject of a sentence; esp. of a metal, not chemically reactive; inactive, inert, submissive, not opposing. *n.* the passive voice of a verb. **passive resistance**, *n.* inert resistance, without active opposition. **passive smoking**, *n.* the inhalation of others' cigarette smoke by non-smokers. **passive smoker**, *n.* **passive voice**, *n.* the form of a transitive verb representing the subject as the object of the action. **passively**, *adv.* **passiveness, passivity** (-siv'-), *n.*
Passover (pah'sōvə), *n.* a Jewish feast, on the 14th day of the month Nisan, commemorating the destruction of the first-born of the Egyptians and the 'passing over' of the Israelites by the destroying angel (Exod. xii); Christ, the paschal lamb.
passport (pahs'pawt), *n.* an official document authorizing a person to travel in a foreign country and entitling him or her to legal protection; anything ensuring admission (to society etc.).
password PASS.
past (pahst), *a.* gone by, neither present nor future; just elapsed; finished, ended; in grammar, denoting action or state belonging to the past; former. *n.* past times; one's past career or the history of this, esp. a disreputable one; the past tense of a verb. *adv.* so as to go by. *prep.* beyond in time or place, after; beyond the influence or range of. **past it**, (*coll.*) no longer young and vigorous. **pastmaster**, *n.* one who has been master of a Freemasons' lodge, a guild etc.; a thorough master (of a subject etc.). **past participle**, *n.* a participle derived from the past tense of a verb, with a past or passive meaning. **past perfect** PLUPERFECT.
pasta (pas'tə), *n.* a flour and water dough, often shaped and eaten fresh or in processed form, e.g. as spaghetti.

paste (pāst), *n.* a mixture of flour and water, usu. with butter, lard etc., kneaded and used for making pastry etc.; sweetmeats of similar consistency; a relish of pounded meat, fish etc.; an adhesive compound of flour, water, starch etc. boiled; any doughy or plastic mixture, esp. of solid substances with liquid; a vitreous composition used for making imitations of gems. *v.t.* to fasten or stick with paste; to stick (up) with paste; (*sl.*) to thrash. **scissors and paste** SCISSORS. **pasteboard**, *n.* a board made of sheets of paper pasted together or of compressed paper pulp. *a.* made of pasteboard; thin, flimsy, sham. **paste-up**, *n.* a sheet of paper on to which proofs, drawings etc. are pasted prior to being photographed for a printing process. **pasting**, *n.* (*coll.*) a thrashing.
pastel (pas'tl), *n.* a dry paste composed of a pigment mixed with gum-water; a coloured crayon made from this; a picture drawn with such crayons; the art of drawing with these; a subdued pale colour. *a.* of pastel colour. **pastellist**, *n.*
pastern (pas'tən), *n.* the part of a horse's leg between the fetlock and the hoof. **pastern-joint**, *n.*
Pasteurism (pas'tərizm), *n.* a method of preventing or curing certain diseases, esp. hydrophobia, by progressive inoculation. **pasteurize, -ise** (tū-), *v.t.* to subject (as milk) to treatment by heat in order to destroy the organisms which may be present. **pasteurization, -isation**, *n.* [Louis Pasteur, 1822–95, French chemist and biologist]
pastiche (pastēsh'), **pasticcio** (-tich'ō), *n.* a musical work, painting etc. composed of elements drawn from other works or which imitates the style of a previous work.
pastille (pastēl'), *n.* a roll, cone or pellet of aromatic paste for burning as a fumigator or disinfectant; an aromatic lozenge.
pastime (pahs'tīm), *n.* that which serves to make time pass agreeably, a game, a recreation.
pasting PASTE.
pastis (pastēs'), *n.* an aniseed-flavoured alcoholic drink.
pastor (pahs'tə), *n.* a minister having charge of a church and congregation; one acting as a spiritual guide; the crested starling. **pastorate** (-rāt), *n.* **pastorship**, *n.*
pastoral (pahs'tərəl), *a.* pertaining to shepherds; of land, used for pasture; of poetry etc., treating of country life, esp. in an idealized manner; rural, rustic; relating to the cure of souls or the duties of a pastor; befitting a pastor. *n.* a poem, romance, play, picture etc. depicting shepherds or rustics or country life; a letter or address from a pastor, esp. from a bishop to his diocese; a pastorale. **pastoralism**, *n.* **pastoralist**, *n.* (*Austral.*) a sheep- or cattle-raiser as distinct from an agriculturalist. **pastorality** (-ral'-), *n.* **pastorally**, *adv.*
pastorale (pastərahl'), *n.* a simple rustic melody; a cantata on a pastoral theme; a symphony dealing with a pastoral subject.
past participle, past perfect PAST.
pastrami (pəstrah'mi), *n.* a highly seasoned smoked beef, esp. cut from the shoulder.
pastry (pās'tri), *n.* (articles, or an article of food made with a crust of) baked flour-paste. **pastry-cook**, *n.*
pasture (pahs'chə), *n.* ground fit for the grazing of cattle; grass for grazing. *v.t.* to put (cattle etc.) on land to graze; of sheep, to eat down (grass-land). *v.i.* to graze. **pasturable**, *a.* **pasturage** (-ij), *n.*
pasty[1] (pās'ti), *a.* of or like paste; pale, unhealthylooking. **pasty-faced**, *a.* having a pale, dull complexion. **pastiness**, *n.*
pasty[2] (pas'ti), *n.* a small pie, usu. of meat and vegetables, baked without a dish.
pat (pat), *n.* a light quick blow with the hand; a tap, a stroke; a small mass or lump (of butter etc.) moulded by patting; the sound of a light blow with something flat. *v.t.* (*past*, *p.p.* **patted**) to strike gently and quickly with something flat, esp. the fingers or hand; to tap, to stroke gently. *v.i.* to strike gently; to run with light steps. *a.* exactly suitable or fitting; opportune, apposite, apt; glib. *adv.* aptly, opportunely; fluently, glibly. **off pat**, fluently, exactly. **pat on the back**, a sign of encouragement or approval. **patly**, *adv.* **patness**, *n.*
pat., (*abbr.*) patent(ed).
patagium (pətā'jiəm), *n.* (*pl.* **-gia** (-ə)) the wing membrane of a bat, flying-lemur etc.
patch (pach), *n.* a piece of cloth, metal or other material put on to mend anything; a piece put on to strengthen a fabric etc.; a piece of cloth worn over an injured eye; a dressing covering a wound etc.; a small piece of black silk or court-plaster worn (esp. in the 17th and 18th cents.) to conceal a blemish or to set off the complexion; a differently coloured piece of a surface; a small piece of ground, a plot; a scrap, a shred; a temporary correction to a computer program. *v.t.* to put a patch or patches on; to mend with a patch or patches (usu. with *up*); to mend clumsily; to make (up) of or as of shreds or patches; to put together or arrange hastily; to serve as a patch for; to connect (electrical circuits or components) temporarily by means of a patch-board. **a bad, good** etc. **patch**, (*coll.*) a sequence of bad, good etc. experiences or achievements. **not a patch on**, (*coll.*) not to be compared with. **to patch up**, to mend. **to patch up a quarrel**, to be reconciled temporarily. **patchboard**, *n.* a board with a number of electrical sockets used for making temporary circuits. **patch pocket**, *n.* one consisting of a flat piece of cloth sown to the outside of a garment. **patch test**, *n.* a test for allergy in which small amounts of potentially allergenic substances are applied to the skin. **patchwork**, *n.* work composed of pieces of different colours, sizes etc., sewn together; clumsy work. **patcher**, *n.* **patchery**, *n.* **patchy**, *a.* of inconsistent quality, frequency etc.; covered with patches; appearing in patches. **patchily**, *adv.* **patchiness**, *n.*
patchouli (pəchoo'li), *n.* an Indian plant yielding a fragrant oil; a perfume prepared from this.
pate (pāt), *n.* the head, esp. the top of the head. **pated**, *a.* (*usu. in comb.*).
pâté (pat'ā), *n.* a pie, a patty; a paste made of cooked, diced meat, fish or vegetables blended with herbs etc. **pâté de foie gras** (də fwah grah), pâté made of fatted goose liver. [F]
patella (pətel'ə), *n.* the knee-cap. **patellar, patellate** (-lət), *a.*
paten (pat'ən), *n.* a plate or shallow dish for receiving the eucharistic bread.
patent (pā'tənt, pat'-), *a.* open to the perusal of all; protected or conferred by letters patent; protected by a trade mark, proprietary plain, obvious, manifest. *n.* a grant from the Crown by letters patent of a title of nobility, or of the exclusive right to make or sell a new invention; an invention so protected; anything serving as a sign or certificate (of quality etc.). *v.t.* to secure a patent for. **letters patent**, an open document from the sovereign or an officer of the Crown conferring a title, right, privilege etc., esp. the exclusive right to make or sell a new invention. **patent-leather**, *n.* a leather with a japanned or varnished surface. **patent medicine**, *n.* a medicine sold under a licence with a registered name and trade mark. **patent office**, *n.* a government department responsible for granting patents. **patent rolls**, *n.pl.* the rolls or register of patents granted by the Crown since 1201. **patency** (-si), *n.* **patentable**, *a.* **patentee** (-tē'), *n.* person granted a right or privilege by patent. **patently**, *adv.* obviously, clearly.
pater (pā'tə), *n.* (*coll.*) father.
paterfamilias (patəfəmil'ias), *n.* the head or father of a family or household. [L]
paternal (pətœ'nəl), *a.* of or pertaining to a father; fatherly; connected or related through the father. **paternalism**, *n.* the exercise of benign, overprotective

paternoster   589   **pauper**

authority, esp. in a form of government, often seen as interference with individual rights. **paternalistic** (-lis'-), *a*.
paternally, *adv*. **paternity**, *n*. fatherhood; ancestry or origin on the male side, descent from a father; authorship, source. **paternity leave**, *n*. official leave from work granted to a man when his wife is in childbirth or recovering from it.
paternoster (patənos'tə), *n*. the Lord's Prayer, esp. in Latin; every 11th bead of a rosary, indicating that the Lord's Prayer is to be repeated; hence, a rosary; a fishing-line with a weight at the end and short lines with hooks extending at intervals. [L, our Father]
path (pahth), *n*. a footway, esp. one beaten only by feet; a course or track; course of life, action etc. **pathfinder**, *n*. an explorer or pioneer; a radar device used for navigational purposes or for targeting missiles. **pathless**, *a*. **pathway**, *n*.
path., (*abbr.*) pathology, pathological.
path- PATH(O)-.
-path, *comb. form*. a person suffering from a pathological disorder; a medical practitioner. **-pathy**, *comb. form*. suffering, feeling; disease, treatment of this, as in *sympathy*, *homoeopathy*.
Pathan (pətahn'), *n*. an Afghan belonging to independent tribes on the NW frontier of India.
pathetic (pəthet'ik), *a*. affecting or moving the emotions, esp. those of pity, sympathy or contempt; (*coll.*) poor, inadequate, contemptible. **pathetic fallacy**, *n*. in literature, the attribution of human feelings to objects associated with nature. **pathetically**, *adv*.
path(o)-, *comb. form*. disease.
pathogen (path'əjen), *n*. any disease-producing substance or micro-organism.
pathogenesis (pathəjen'əsis), *n*. the origin and development of disease. **pathogenetic** (-net'-), *a*. **pathogeny** (-thoj'-), *n*.
pathology (pəthol'əji), *n*. the science of diseases. **pathologic**, *a*. pertaining to pathology; caused by or involving disease; (*coll.*) driven or motivated by compulsion. **pathologically**, *adv*. **pathologist**, *n*.
pathos (pā'thos), *n*. a quality or element in events or expression that excite emotions of pity, compassion, sympathy etc.
pathway PATH.
-pathy -PATH.
patience (pā'shəns), *n*. the quality of being patient; calm endurance of pain, provocation or other evils, fortitude; a card game played by one person. **to have no patience with**, to be unable to stand or put up with; to be irritated by.
patient (pā'shənt), *a*. capable of bearing pain, suffering etc. with fortitude; not easily provoked, indulgent; persevering. *n*. a person under medical treatment. **patiently**, *adv*.
patina (pat'inə), *n*. the green incrustation that covers bronzes and copper exposed to the air; a thin covering or layer; a sheen gained from constant use or age. **patinated** (-nātid), *a*.
patio (pat'iō), *n*. (*pl.* **-tios**) the open inner court of a Spanish or Spanish-American residence; a paved area beside a house, used for outdoor meals, sunbathing etc.
patisserie (pətē'səri), *n*. a shop selling sweet pastries; such pastries.
Patna rice (pŭt'nə), *n*. a variety of long-grain rice used for savoury dishes.
patois (pat'wah), *n*. a non-standard dialect of a district.
patr(i)-, **patro-**, *comb. form*. father.
patriarch (pā'triahk), *n*. the head of a family or tribe, ruling by paternal right (esp. in the Bible); the highest grade in the hierarchy of the Roman Catholic Church; in the Eastern and early Churches, a bishop; the founder of a religion, science etc.; a venerable old man; the oldest man (in an assembly, community etc.). **patriarchal** (-ah'-) *a*. **patriarchate** (-kət), *n*. **patriarchy**, *n*. a patriarchal system of government or social organization, esp. as distinguished from matriarchy; a society in which this system operates.
patrician (pətri'shən), *a*. noble, aristocratic (orig. ancient Rome). *n*. a noble, an aristocrat, a member of the highest class of society.
patricide (pat'risīd), *n*. (the act of) one who kills his or her father. **patricidal** (-sī'-), *a*.
patrilineal (patrilin'iəl), *a*. by descent through the father.
patrimony (pat'riməni), *n*. an estate or right inherited from one's father or ancestors. **patrimonial** (-mō'-), *a*.
patriot (pā'triət, pat'-), *n*. one who loves his or her country and is devoted to its interests, esp. its freedom and independence. **patriotic** (-ot'ik), *a*. **patriotically**, *adv*. **patriotism**, *n*.
patristic (pətris'tik), **-ical**, *a*. pertaining to the ancient Fathers of the Church or their writings. *n.pl.* (*sing.* in *constr.*) the study of patristic writings.
patrol (pətrōl'), *v.i.* (*past, p.p.* **patrolled**) to go on a patrol. *v.t.* to go round. *n*. the action of moving around an area for the maintenance of order and for security or to reconnoitre; the detachment of soldiers, police, firemen etc., or the soldier, constable etc., doing this; a Scout or Girl Guide unit. **patrol car**, *n*. a car in which police officers patrol an area. **patrolman**, *n*. (*N Am.*) a police officer. **patrol wagon**, *n*. (*N Am., Austral., New Zealand*) a Black Maria.
patron[1] (pā'trən), *n*. one who supports, fosters or protects a person, cause, art etc.; a tutelary saint; one who holds the gift of a benefice; a regular customer (at a shop etc.). **patron saint**, *n*. a saint regarded as the patron of a particular group, country etc. **patronage** (pat'rənij), *n*. support, fostering, encouragement, or protection; the right of presentation to a benefice or office; the act of patronizing; support by customers (of a shop etc.). **patroness**, *n. fem*. **patronize**, **-ise** (pat'rəniz), *v.t.* to act as a patron towards; to treat in a condescending way; to frequent as a customer. **patronizing**, **-ising**, *a*. condescending. **patronizingly**, *adv*.
patron[2] (patrō'), *n*. the proprietor of a French hotel, restaurant etc.
patronymic (patrənim'ik), *n.*, *a*. (a name) derived from a father or ancestor.
patsy (pat'si), *n*. (*chiefly N Am., coll.*) a person who is easily deceived, cheated etc., a sucker; a scapegoat.
patter[1] (pat'ə), *v.i.* to strike, as rain, with a quick succession of light, sharp sounds; to move with short, quick steps. *n*. a quick succession of sharp, light sounds or taps.
patter[2] (pat'ə), *v.t.* to say (one's prayers) in a mechanical, singsong way. *v.i.* to pray in this manner; to talk glibly. *n*. the patois or slangy lingo of a particular area; glib talk, chattering, gossip.
pattern (pat'ən), *n*. a model or original to be copied or serving as a guide in making something; a sample or specimen; a decorative design for a carpet, wallpaper, etc.; hence type, style; an arrangement or way of doing things which is repeated or followed regularly. *v.t.* to copy, to model (after, from or upon); to decorate with a pattern.
patty (pat'i), *n*. a little pie; a small, flat cake of minced food.
patulous (pat'ūləs), *a*. open, having a wide aperture; of boughs etc., spreading, expanding.
paua (pow'ə), *n*. a New Zealand abalone; its handsome iridescent shell.
paucity (paw'siti), *n*. fewness in number; scarcity.
paunch (pawnch), *n*. the belly, the abdomen; (*coll.*) a fat or protruding belly. **paunchy**, *a*. **paunchiness**, *n*.
pauper (paw'pə), *n*. one without means of support, a poor or destitute person, one entitled to public assistance.

pauperism, *n*. **pauperize, -ise**, *v.t.* **pauperization**, *n*.
pause (pawz), *n*. a cessation or intermission of action, speaking etc.; a break in reading, speaking, music etc.; hesitation; (*Mus*.) a mark ⌒ or ⌣ over a note etc., indicating that it is to be prolonged. *v.i*. to make a pause or short stop; to wait; to linger (upon or over). **to give pause**, to cause to hesitate, esp. for reconsideration.
pavan, pavane (pəvan', pav'-), *n*. a slow and stately dance, usu. in elaborate dress, in vogue in the 16th and 17th cents.; music for this.
pave (pāv), *v.t*. to make a hard, level surface upon, with stone, bricks etc.; to cover with or as with a pavement. **to pave a way for, to**, to prepare for. **pavement**, *n*. that with which anything is paved; a hard level covering of stones, bricks, tiles, wood-blocks etc.; a paved footway at the side of a street or road. **pavement artist**, *n*. a person drawing figures etc. on a pavement in order to obtain money from passers-by. **paving**, *n*. **paving-stone**, *n*.
pavilion (pəvil'yən), *n*. a tent, esp. a large one, of conical shape; a temporary or movable structure for entertainment, shelter etc.; an ornamental building, usu. for light construction, for amusements etc., esp. one for spectators and players on a cricket-ground etc.
paving PAVE.
pavlova (pavlō'və), *n*. a dessert consisting of a meringue base topped with fruit and whipped cream. [after the Russian ballerina Anna *Pavlova*, 1885–1931]
Pavlovian (pavlō'viən), *a*. of or relating to conditioned reflexes as first described by I. P. Pavlov, Russian physiologist (1849–1936).
paw (paw), *n*. the foot of a quadruped having claws, as dist. from a hoof. *v.t*. to scrape or strike with the forefoot; of a horse, to strike the ground with the hoofs; (*coll*.) to handle roughly, familiarly, sexually or clumsily.
pawl (pawl), *n*. a hinged piece of metal or lever engaging with the teeth of a wheel etc., to prevent it from running back etc.
pawn[1] (pawn), *n*. a piece of the lowest value in chess; an insignificant person used in the plans of a cleverer one.
pawn[2] (pawn), *n*. something deposited as security for a debt or loan, a pledge; the state of being held as a pledge. *v.t*. to deposit as a pledge for the repayment of a debt or loan. **pawnbroker**, *n*. one who lends money on the security of goods pawned. **pawnbroking**, *n*. **pawnshop**, *n*. the place where this is carried on.
pawpaw (paw'paw), PAPAW.
pax (paks), *n*. a tablet or plaque bearing a representation of the Crucifixion or other sacred subject which was formerly kissed by the priest and congregation at Mass, an osculatory. *int*. calling for a truce.
pay (pā), *v.t*. (*past, p.p.* **paid** (pād)) to hand over what is due in discharge of a debt or for services or goods; to discharge (a bill, claim, obligation etc.); to spend; to compensate, to recompense; to be remunerative or worthwhile to; to bestow, to tender (a compliment, visit etc.). *v.i*. to make payment; to discharge a debt; to make an adequate return (to); to be remunerative or worthwhile. *n*. payment, compensation, recompense; wages, salary. **pay as you earn**, a method of collecting income tax by deducting it before payment of the earnings. **to pay back**, to repay; to return (a favour etc.); to take revenge on. **to pay for**, to make a payment for; to suffer as a result of. **to pay off**, to pay in full and discharge; to be profitable or rewarding. **to pay one's way**, to keep out of debt. **to pay out**, to punish; to disburse; to cause (a rope) to run out. **to pay the piper**, to bear the cost. **to pay through the nose**, to pay an exorbitant price. **pay bed**, *n*. a bed for a private patient in a National Health Service hospital. **pay-day**, *n*. the day on which wages are paid; (*coll*.) a windfall, a money-spinner, a pay-off; (*coll*.) the period when a large profit, prize, sum of money etc. is realized. **paydirt**, *n*. (*chiefly N Am*.) a deposit containing enough gold to make mining worth while; anything profitable or useful. **paying guest**, *n*. a lodger, usu. in a private house. **payload**, *n*. the part of a transport vehicle's load that brings profit; the passengers, cargo or weaponry carried by an aircraft; the explosive capacity of a bomb, missile warhead etc.; that which a spacecraft carries as the purpose of its mission. **paymaster**, *n*. one who regularly pays wages etc.; (*Mil., Nav*.) an officer whose duty it is to pay the wages. **Paymaster-General**, *n*. the officer at the head of the treasury department concerned with the payment of civil salaries and other expenses. **pay-off**, *n*. (*coll*.) the final result or outcome; the final payment of a bill etc.; (*coll*.) a bribe; an act of revenge. **pay-packet**, *n*. (an envelope containing) a person's wages. **pay-phone**, *n*. a public telephone operated by coins. **payroll**, *n*. a list of employees; the total of their wages. **payslip**, *n*. a slip of paper giving details of one's pay, income tax deductions etc. **payable**, *a*. that can or must be paid. **payee** (-ē'), *n*. person to whom money is paid. **payer**, *n*. **payment**, *n*.
PAYE (*abbr*.) pay as you earn.
payola (pāō'lə), *n*. (*sl*.) clandestine reward paid for illicit promotion of a commercial product, e.g. of a record by a disc jockey.
Pb, (*chem. symbol*) lead.
PC, (*abbr*.) parish council; personal computer; police constable.
pc, (*abbr*.) per cent; personal computer; postcard.
PD, (*abbr*.) (*N Am*.) police department.
pdq, (*abbr*.) (*coll*.) pretty damn quick.
PDSA, (*abbr*.) People's Dispensary for Sick Animals.
PE, (*abbr*.) physical education.
pea (pē), *n*. a leguminous plant, the seeds of which are used as food; the seed of this. **pea-green**, *a., n*. (of) a colour like that of fresh green peas. **peapod**, *n*. the pericarp of the pea. **peashooter**, *n*. a tube through which dried peas are shot by blowing. **peasoup**, *n*. soup made with peas, esp. dried and split peas. **peasouper**, *n*. (*coll*.) a dense yellowish fog.
peace (pēs), *n*. a state of quiet or tranquillity; absence of civil disturbance or agitation; freedom from or cessation of war or hostilities; a treaty reconciling two hostile nations; a state of friendliness; calmness of mind. **at peace**, in a state of harmony or tranquillity. **Justice of the Peace** JUSTICE. **to hold one's peace**, to keep silent. **to keep the peace**, to abstain from strife; to prevent a conflict. **to make peace**, to reconcile or be reconciled (with); to bring about a treaty of peace. **Peace Corps**, *n*. a US government body that sends volunteers to help developing countries. **peacekeeper**, *n*. one who preserves peace between hostile parties. **peacekeeping**, *n*. **peacekeeping force**, *n*. a military force whose objective is to prevent hostilities breaking out between opposing factions. **peacemaker**, *n*. one who reconciles. **peace-offering**, *n*. a gift to procure peace or reconciliation. **peace pipe**, *n*. a pipe smoked by N American Indians as a sign of peace. **peace sign**, *n*. a hand gesture in which the index and middle fingers open to form a V, signifying peace. **peacetime**, *n*. a time when there is no war. **peaceable**, *a*. peaceful, quiet; disposed to peace. **peaceableness**, *n*. **peaceably**, *adv*. **peaceful**, *a*. in a state of peace; free from noise or disturbance; quiet, pacific, mild. **peacefully**, *adv*. **peacefulness**, *n*.
peach[1] (pēch), *n*. a sweet, fleshy, downy fruit; (*coll*.) anything or person superlatively good or attractive; a pinkish-yellow colour. *a*. pinkish-yellow. **peach-bloom**, *n*. the delicate powder on a ripe peach; a soft, pink colour on the cheeks. **peach brandy**, *n*. a spirit distilled from peach-juice. **peach melba**, *n*. an ice-cream dessert with peaches and raspberries. **peachy**, *a*. soft and downy like a peach; having the colour of a peach; (*coll*.) excellent, good. **peachiness**, *n*.
peach[2] (pēch), *v.i*. to turn informer against an accomplice; to inform (against or upon).

peacock (pē'kok), *n.* the male, peafowl, a bird with gorgeous plumage and long tail capable of expanding like a fan; a vain person. *v.i.* to strut about ostentatiously. **peacock blue**, *a.*, *n.* (of) a brilliant deep blue/green colour. **peacock-butterfly**, *n.* one of various butterflies with ocellated wings. **peacock-fish**, *n.* a brilliantly variegated fish. **peacockish**, *a.*
peafowl (pē'fowl), *n.* a pheasant of the genus *Pavo*, of which the peacock is the male. **pea-chick**, *n.* the young of the peafowl. **peahen**, *n.* the female peafowl.
pea jacket, (pē'jakit), **coat** (kōt), *n.* a coarse, thick, loose overcoat worn by seamen.
peak (pēk), *n.* a sharp point or top, esp. of a mountain; the projecting brim in front of a cap; the point of greatest activity, use, demand etc. *v.i.* to reach a peak. *a.* of or relating to the point of greatest activity, use, demand etc. **peak viewing time**, *n.* the period when the largest number of people watch television. **peaked**, *a.* having or resembling a peak.
peaky (pēkē), *a.* wan, sickly looking.
peal (pēl), *n.* a loud, esp. a prolonged or repercussive sound, as of thunder, bells etc.; a set of bells tuned to each other; a series of changes rung on these. *v.i.* to sound a peal; to resound.
peanut (pē'nūt), *n.* a plant of the bean family with pods ripening underground which are edible and are used for their oil; a monkey-nut; (*pl.*) (*sl.*) an insignificant sum of money. **peanut butter**, *n.* a paste made from ground peanuts.
pear (peə), *n.* the fleshy obovoid fruit of the pear-tree. **pear-shaped**, *a.* **pear-tree**, *n.*
pearl[1] (pœl), *n.* a smooth, white or bluish-grey lustrous and iridescent calcareous concretion, found in several bivalves, the best in the pearl oyster, prized as a gem; mother-of-pearl; something resembling a pearl, any thing or person exceedingly valuable or attractive; *a.* pertaining to, containing or made of pearls. *v.i.* to fish for pearls. **pearl ash**, *n.* crude carbonate of potash. **pearl-barley** BARLEY. **pearl diver**, *n.* one who dives for pearl oysters. **pearl fisher**, *n.* a pearl diver. **pearl-fishing**, *n.* **pearl-spar**, *n.* a variety of dolomite. **pearl-stone**, *n.* perlite. **pearled**, *a.* **pearlies** (-liz), *n.pl.* costermonger's festal dress covered with pearl buttons. **pearliness**, *n.* **pearly**, *a.* **pearly gates**, *n.pl.* (*coll.*, *facet.*) the entrance to heaven. **pearly king**, **queen**, *n.* a costermonger wearing pearlies.
pearl[2] (pœl), *n.* a fine loop, a row of which forms an ornamental edging on various fabrics; a purl stitch in knitting. *v.t.* to purl.
peasant (pez'nt), *n.* one living on or off the land esp. a labourer; (*coll.*) a rough, uncouth person. *a.* rustic, rural; base. **peasantry**, *n.*
pease-pudding (pēz pud'ing), *n.* a pudding made of peas.
peat (pēt), *n.* decayed and partly carbonized vegetable-matter found in boggy places and used as fuel. **peaty**, *a.*
pebble (peb'l), *n.* a small stone rounded by the action of water; a transparent rock-crystal, used for lenses etc.; a lens made of this. *v.t.* to pave with pebbles; to impart a rough indented surface or grain to (leather). **pebble-dash**, *n.* a coating for external walls consisting of small stones imbedded in mortar. **pebble-ware**, *n.* a variety of Wedgwood ware having different coloured clays worked into the paste. **pebbled**, **pebbly**, *a.*
pecan (pikan'), *n.* a N American hickory or its fruit or nut. **pecan pie**, *n.* a pie made with pecan nuts.
peccable (pek'əbl), *a.* liable to sin. **peccability** (-bil'-), *n.*
peccadillo (pekədil'ō), *n.* (*pl.* **-lloes**, **-llos**) a slight fault or offence.
peccary (pek'əri), *n.* one of two small American species of pig-like mammals.
pêcher (pe'shā), *n.* an alcoholic drink, esp. sparkling wine, flavoured with peach.
peck[1] (pek), *n.* a measure of capacity for dry goods, 2 gallons (about 9 l); the fourth part of a bushel; a large quantity.
peck[2] (pek), *v.t.* to strike with a beak or a pointed instrument; to pick up with or as with the beak; (*coll.*) to eat in small amounts; to kiss lightly and quickly. *v.i.* to strike or aim with a beak or pointed implement. *n.* a sharp stroke with or as with a beak; a sharp kiss. **pecker**, *n.* (*sl.*) spirits, courage; (*N Am.*, *taboo sl.*) the penis. **keep your pecker up**, keep cheerful. **pecking order**, *n.* the hierarchical order of importance in any social group. **peckish**, *a.* (*coll.*) hungry.
pectin (-tin), *n.* a white, amorphous compound found in fruits and certain fleshy roots, formed from pectose by the process of ripening. **pectic**, *a.*
pectoral (pek'tərəl), *a.* pertaining to or for the breast. *n.* something worn on the breast; a pectoral fin; a medicine to relieve chest complaints; the pectoral muscle. **pectoral muscle**, *n.* either of the two muscles at the top of the chest on each side, controlling certain arm and shoulder movements.
peculate (pek'ūlāt), *v.t.*, *v.i.* to appropriate to one's own use (money or goods entrusted to one's care). **peculation**, *n.* **peculator**, *n.*
peculiar (pikū'lyə), *a.* belonging particularly and exclusively (to); particular, special; singular, strange, odd. **peculiarity** (-ia'ri-), *n.* the quality of being peculiar; a characteristic; an idiosyncrasy. **peculiarly**, *adv.*
pecuniary (pikū'niəri), *a.* relating to or consisting of money. **pecuniarily**, *adv.*
ped- PAED(O)-.
-ped -PED(E).
pedagogue (ped'əgog), *n.* a teacher (usu. in contempt, implying pedantry). **pedagogic**, **-ical** (-goj'-), *a.* **pedagogy** (-goji), *n.* the science or practice of teaching.
pedal (ped'l), *n.* a lever operated by the foot, e.g. in a car, on a bicycle, piano or organ; (*Mus.*) a sustained note, usu. in the bass. *v.t.* (*past*, *p.p.* **pedalled**) to work (a bicycle, etc.) by pedals; to play on (an organ) by pedals. *v.i.* to play an organ or work a bicycle etc. by pedals. *a.* of or pertaining to a foot or foot-like part (esp. of molluscs). **pedal-pushers**, *n.pl.* women's calf-length trousers, usu. close-fitting below the knee. **pedalist**, *n.* **pedalo** (-ō), *n.* (*pl.* **-loes**, **-los**) a small boat propelled by paddles operated with pedals.
pedant (ped'ənt), *n.* one who makes a pretentious show of learning, or lays undue stress on unimportant details or petty rules or strict formulas. **pedantic** (-dan'-), *a.* **pedantically**, *adv.* **pedantry**, *n.*
peddle (ped'l), *v.i.*, *v.t.* to travel about selling small articles or small amounts of goods. **peddler**, *n.* one who sells illegal drugs; (*chiefly N Am.*) a pedlar.
-ped(e), *comb. form.* foot.
pederast (ped'ərast), *n.* a man who has sexual relations with a boy. **pederasty**, *n.*
pedestal (ped'istəl), *n.* a base for a column, statue etc.; either of the supports of a knee-hole desk. **on a pedestal**, in a position of (excessive) respect or devotion.
pedestrian (pədes'triən), *a.* going or performed on foot; pertaining to walking; prosaic, dull, commonplace. *n.* one who journeys on foot. **pedestrian crossing**, *n.* a marked strip across a road where vehicles must stop to allow pedestrians to cross. **pedestrian precinct**, *n.* an area closed to traffic, for the use of pedestrians only (usu. around shops). **pedestrianize**, **-ise**, *v.t.* to convert (a road etc.) so that it may only be used by pedestrians.
pedi-, *comb. form.* foot.
pediatric (pēdiat'rik), PAEDIATRIC.
pedicel (ped'isel), *n.* (*Bot.*, *Zool.*) the stalk supporting a single flower etc.; any small foot-stalk or stalk-like structure. **pedicle** (-ikl), *n.* a pedicel or peduncle.
pedicure (ped'ikūə), *n.* the medical treatment of the feet; cosmetic care of the feet.

pedigree (ped'igrē), n. genealogy, lineage, esp. ancient lineage; a genealogical table or tree. a. of cattle, dogs etc., pure-bred, having a known ancestry. **pedigreed**, a.
pediment (ped'imənt), n. the triangular part surmounting a portico, crowning doorways, windows etc. in classical buildings. **pedimental** (-men'-), a.
pedlar, (esp. N Am.) **peddler** (ped'lə), n. a travelling hawker of small wares.
ped(o)- PAED(O)-.
pedology (pədol'əji), n. the science of soils. **pedological** (pedəloj'-), a. **pedologist**, n.
pedometer (pidom'itə), n. an instrument for measuring the distance covered on foot by registering the number of the steps taken.
peduncle (pidüng'kl), n. a flower-stalk; a stalk-like process for the attachment of an organ or an organism. **peduncular** (-kū-), **-culate** (-kūlət), **-lated** (-lātid), a.
pee (pē), v.i. (coll. euphem.) to urinate. n. an act of urinating; urine.
peek (pēk), v.i. to peep. n. a peep. **peekaboo** (pē'kəboo), n. a game used for amusing babies, in which the face is hidden, then suddenly revealed. a. (of a garment) revealing.
peel[1] (pēl), v.t. to strip the skin, bark or rind off. v.i. to lose the skin or rind, to become bare; of paint etc., to flake off; to shed skin, esp. after sunburn. n. skin or rind. **to peel off**, (coll.) to undress; to leave and move away from (e.g. a column of marchers). **peeler**, n. **peeling**, n. the skin of a fruit etc. that has been peeled off.
peel[2], **pele** (pēl), n. a square fortified tower, esp. those built about the 16th cent. in the border counties of Scotland and England.
peeling PEEL[1].
peep[1] (pēp), v.i. to cry, chirp or squeak, as a young bird, a mouse etc. n. a chirp, squeak etc.; (coll.) any spoken sound.
peep[2] (pēp), v.i. to look through a crevice or narrow opening; to look slyly or furtively; to come (out) gradually into view. n. a furtive look; a glimpse; the first appearance. **peep-hole**, n. **Peeping Tom**, n. one guilty of prurient curiosity; a voyeur. **peep-show**, n. an exhibition of pictures etc., shown through a small aperture. **peeper**, n. one who peeps; (sl.) an eye.
peer[1] (piə), n. one of the same rank; an equal in any respect; a noble; in the UK, a member of one of the degrees of nobility, comprising dukes, marquesses, earls, viscounts and barons. **peer group**, n. a group of people equal in status, age etc. **peerage** (-rij), n. the rank of a peer; the body of peers, the nobility, the aristocracy; a book containing particulars of the nobility. **peeress** (-res), n. fem. **peerless**, a. without an equal.
peer[2] (piə), v.i. look (at) closely or intently; to squint; to peep out; to appear, to come into sight.
peevish (pē'vish), a. fretful, irritable, petulant; expressing discontent. **peeved**, a. (sl.) irritated, annoyed. **peevishly**, adv. **peevishness**, n.
peewit (pē'wit), PEWIT.
peg (peg), n. a pin or bolt, usu. of wood, for holding parts of a structure or fastening articles together, hanging things on, supporting, holding, marking etc.; a step, a degree; an occasion, pretext, excuse, or topic for discourse etc.; (coll.) a drink. v.t. (past, p.p. **pegged**) to fix or fasten (down, in, out etc.) with a peg or pegs; to mark (a score) with pegs on a cribbage-board; to mark (out) boundaries; (coll.) to fix (esp. prices) at an agreed level. **a square peg in a round hole, a round peg in a square hole**, a person in an unsuitable job, function etc. **off the peg**, ready-made. **to peg away**, to work at or struggle persistently. **to peg down**, to fasten down with pegs; to restrict (to rules etc.). **to peg out**, (sl.) to die. **to take someone down a peg (or two)**, to humiliate, to degrade.
pegboard, n. a board with holes into which pegs can be fixed, used for scoring in games, or placed on a wall and used for hanging things. **peg-leg**, n. a crude wooden leg; a person who has a wooden leg.
Pegasus (peg'əsəs), n. in Greek mythology, a winged steed.
pegmatite (peg'mətit), n. a coarse-grained variety of granite, with a little mica.
peignoir (pān'wah), n. a loose robe or dressing-gown worn by women.
pejorative (pijo'rətiv), a. depreciatory. n. a word or form expressing depreciation. **pejoratively**, adv.
pekan (pek'ən), n. a N American carnivorous animal of the weasel family, prized for its fur.
peke (pēk), n. (coll.) a Pekinese dog.
Pekinese, Pekingese (-nēz'), n. a rough-coated variety of Chinese pug. **Pekinese, Pekingese**, a. of or pertaining to Peking (now Beijing). **Peking man** (pēking'), n. a fossil man of the Lower Palaeolithic age, first found SW of Beijing (Peking) in 1929.
pekoe (pē'kō), n. a fine black tea.
pelage (pel'ij), n. the coat or hair of an animal, esp. of fur.
pelagic (pəlāikj'), a. of or inhabiting the deep sea.
pelargonium (peləgō'niəm), n. any plant of a large genus of ornamental plants popularly called geraniums.
pelf (pelf), n. money, wealth, gain.
pelican (pel'ikən), n. a large piscivorous water-fowl with an enormous pouch beneath the mandibles for storing fish when caught. **pelican crossing**, n. a type of pedestrian crossing controlled by pedestrian-operated traffic lights.
pellagra (pəlag'rə, -lā'-), n. a virulent disease attacking the skin and causing nervous disorders and mania, caused by deficiency of vitamins B.
pellet (pel'it), n. a little ball, esp. of bread, paper or something easily moulded; a small pill; a small shot; a rounded boss or prominence. v.t. to form into pellets; to hit with pellets.
pellicle (pel'ikl), n. a thin skin; a membrane or film.
pellitory (pel'itəri), n. a herb of the genus Parietaria, esp. the wall-pellitory also applied to a herb of the aster family or pellitory of Spain.
pell-mell (pelmel'), adv. in a confused or disorderly manner, anyhow; in disorderly haste.
pellucid (pəloo'sid), a. clear, limpid, transparent; clear in thought, expression or style. **pellucidity** (peləsid'-), n. **pellucidly**, adv. **pellucidness**, n.
pelmet (pel'mit), n. a canopy, built-in or detachable, which conceals the fittings from which curtains hang; a valance.
pelota (pilō'tə), n. a game somewhat like squash played with a ball and a curved racket fitting upon the hand, popular in Spain and the Basque country.
pelotherapy (pelәthe'rәpi), n. treatment of disease by the application of mud.
pelt[1] (pelt), n. a hide or skin of an animal with the hair on; a raw skin stripped of hair or wool. **peltry**, n. pelts.
pelt[2] (pelt), v.t. to strike or assail by throwing missiles; to shower (with insults etc.); to strike repeatedly. v.i. of rain etc., to beat heavily; (sl.) to hurry (along). n. a blow from something thrown. **full pelt**, at full speed, with violent impetus. **pelter**, n.
pelvis (pel'vis), n. the lower portion of the great abdominal cavity; the bony walls of this cavity; the interior cavity of the kidney. **pelvic**, a. **pelvic girdle**, n. the arrangement of bones which supports the hind-limbs of vertebrates, the lower limbs in humans.
pemphigus (pem'figəs), n. a disease characterized by the eruption of watery vesicles on the skin.
PEN, (abbr.) International Association of Poets, Playwrights, Essayists and Novelists.
pen[1] (pen), n. a small enclosure, esp. for cattle, sheep, poultry etc. v.t. (past, p.p. **penned**) to enclose, to confine; to shut up or coop (up or in). **playpen** PLAY.

pen² (pen), *n.* an instrument for writing with ink; writing, style of writing. *v.t.* (*past, p.p.* **penned**) to write, to compose and write. **pen and ink**, instruments for writing; writing. **pen-and-ink**, *a.* written or drawn with these. **pen-friend, -pal**, *n.* a person, usu. living abroad, whom one has not usu. met and with whom one corresponds. **penknife**, *n.* a small knife with one or more folding blades, a pocket-knife. **penmanship**, *n.* the art of writing; style of writing. **pen-name**, *n.* a literary pseudonym. **pen-pusher**, *n.* (*coll.*) a person doing dull, routine, clerical work.

pen³ (pen), *n.* (*N Am., sl.*) short for penitentiary, prison.

pen⁴ (pen), *n.* a female swan.

pen., (*abbr.*) peninsula.

penal (pē'nəl), *a.* enacting, inflicting, used for, prescribing, or pertaining to punishment. **penal code**, *n.* a body of laws concerned with crime and its punishment. **penalize, -ise**, *v.t.* to make or declare penal; (*Sport*) to subject to a penalty or handicap; to put under an unfair disadvantage. **penalization**, *n.* **penally**, *adv.*

penalty (pen'əlti), *n.* legal punishment for a crime, offence or misdemeanour; a fine, a forfeit for non-performance or breach of conditions; (*Sport*) a handicap imposed for a breach of rules. **penalty area**, *n.* a rectangular area in front of the goal in soccer, where a foul against the attacking team results in a penalty and where the goalkeeper may handle the ball. **penalty box**, *n.* the penalty area; an area to which penalized players are confined in ice hockey. **penalty kick**, *n.* in football, a kick allowed to the opposite side when a penalty has been incurred.

penance (pen'əns), *n.* sorrow for sin, repentance; in the Roman Catholic and Greek Churches, a sacrament consisting of contrition, confession and satisfaction, with absolution by the priest; an act of self-mortification undertaken as a satisfaction for sin.

pence, *n.pl.* PENNY.

penchant (pen'chənt, pē'shä), *n.* a strong inclination or liking.

pencil (pen'sl), *n.* a cylinder or slip of graphite, crayon etc., usu. enclosed in a casing of wood, used for writing, drawing etc.; something resembling this; (*Opt.*) a system of rays diverging from or converging to a point; *v.t.* (*past, p.p.* **pencilled**) to draw, write or mark with or as with a pencil. **pencilled**, *a.* **pencilling**, *n.*

pendant (pen'dənt), *n.* anything hanging down or suspended by way of ornament etc., as an earring, a locket, a tassel etc.; a pendant light esp. a chandelier, a boss hanging from a ceiling or roof.

pendent (pen'dənt), *a.* hanging; overhanging; (*Gram.*) incomplete in construction, having the sense suspended. **pendency**, *n.* **pendentive** (-den'-), *n.* one of the triangular pieces of vaulting resting on piers or arches and forming segments of a dome. **pending**, *a.* imminent; awaiting settlement, undecided.

pendulous (pen'dūləs), *a.* hanging, suspended; swinging, oscillating. **pendulously**, *adv.* **pendulousness**, *n.*

pendulum (pen'dūləm), *n.* (*pl.* **-lums**) a body suspended from a fixed point and oscillating freely by the force of gravity, as the weighted rod regulating the movement of the works in a clock. **the swing of the pendulum**, (the regular pattern of) change in public opinion, political power etc.

peneplain (pē'niplān), *n.* an area of flat land produced by erosion.

penetrate (pen'itrāt), *v.t.* to enter, to pass into or through; to pierce; to permeate; to saturate or imbue (with); to move or affect the feelings of; to reach or discern by the senses or intellect. *v.i.* to make way, to pass (into, through, to etc.). **penetrable**, *a.* capable of being penetrated; impressible, susceptible. **penetrability** (-bil'-), *n.* **penetrating**, *a.* sharp, piercing; subtle, discerning. **penetratingly**, *adv.* **penetration**, *n.* penetrating or being penetrated; acuity, discernment. **penetrator**, *n.* **penetrative**, *a.*

penguin (peng'gwin), *n.* a swimming bird belonging to the southern hemisphere, with rudimentary wings or paddles and scale-like feathers.

penicillin (penisil'in), *n.* a ether-soluble substance produced from mould and having an intense growth-inhibiting action against various bacteria, esp. in wounds etc.

peninsula (pinin'sūlə), *n.* a piece of land almost surrounded by water. **the Peninsula**, Spain and Portugal. **peninsular**, *a.* of, pertaining to or resembling a peninsula.

penis (pē'nis), *n.* (*pl.* **penises, penes** (-nēz)) the copulatory and urethral organ of a male mammal. **penis envy**, *n.* in Freudian theory, the female's subconscious desire to be male. **penile** (-nīl), *a.*

penitent (pen'itənt), *a.* contrite, repentant, sorry. *n.* one who is penitent; a contrite sinner; one submitting to penance under the direction of a confessor. **penitence**, *n.* **penitential** (-ten'-), *a.* pertaining to or expressing penitence; relating to or of the nature of penance. *n.* a book containing rules relating to penitence. **penitentially, penitently**, *adv.* **penitentiary** (-ten'-), *a.* penitential. *n.* (*N Am.*) a prison; in the Roman Catholic Church, a papal court granting dispensations and dealing with matters relating to confessions.

penknife etc. PEN².

pennant (pen'ənt), *n.* a pennon; (*N Am.*) a flag indicating championship, e.g. in baseball.

penniless (pen'ilis), *a.* without money; destitute.

pennon (pen'ən), *n.* a small pointed or swallow-tailed flag.

penny (pen'i), *n.* (*pl.* **pennies** (-niz), denoting the number of coins; **pence** (pens), denoting the amount) a bronze coin, a 100th part of a pound sterling, formerly a 12th part of a shilling; (*N Am.*) a one-cent piece. **a pretty penny**, considerable sum, cost or expense. **the penny drops**, (*coll.*) the truth is realized; something is made clear. **to spend a penny**, (*coll.*) to urinate. **to turn an honest penny**, to earn money by honest work. **penny dreadful** DREAD. **penny-farthing**, *n.* an early type of bicycle with a large front wheel and small back wheel. **penny-pinch**, *v.i.* to save money by being niggardly. **penny-pinching**, *a.* miserly, niggardly. **penny-wise**, *a.* saving small sums at the risk of larger ones. **pennywort**, *n.* one of several plants with round peltate leaves. **pennyworth** (pen'iwœth, pen'əth), *n.* as much as can be bought for a penny; a small amount, a trifle.

pennyroyal (peniroi'əl), *n.* a kind of mint formerly and still popularly used for medicinal purposes.

penology (pēnol'əji), *n.* the science of punishment and prison management. **penological** (-loj'-), *a.* **penologist**, *n.*

pen-pusher PEN².

pension¹ (pen'shən), *n.* a periodical allowance for past services, disablement, old age, widowhood, etc. paid by the government, employers or through a personal scheme. *v.t.* to grant a pension to. **to pension off**, to cease to employ and to give a pension to; to discard as useless, worn etc. **pensionable**, *a.* **pensioner**, *n.* one in receipt of a pension.

pension² (pēsyō'), *n.* a continental boarding-house.

pensive (pen'siv), *a.* thoughtful; serious, anxious, melancholy; expressing sad thoughtfulness. **pensively**, *adv.* **pensiveness**, *n.*

penstemon (pentstē'mən), PENTSTEMON.

pent (pent), *a.* penned in or confined; shut (up or in). **pent-up**, *a.* not openly expressed; suppressed.

pent(a)-, *comb. form.* five.

pentacle (pen'təkl), *n.* a figure like a star with five points; a pentagram, a pentangle used as a symbol by the mystics and astrologers of the Middle Ages.

pentad (pen'tad), *n.* the number five; a group of five; a chemical element or radical having a valency of five.

pentadactyl (pentədak'til), *a.* having five fingers or toes.

pentagon (pen'təgon), *n.* a plane (usu. rectilineal) figure having five sides and five angles. **the Pentagon**, the Defence Department of the US in Washington, DC. **pentagonal** (-tag'-), *a.*
pentagram (pen'təgram), *n.* a pentacle.
pentahedron (pentəhē'drən), *n.* a figure having five sides, esp. equal sides. **pentahedral**, *a.*
pentamerous (pentam'ərəs), *a.* of a flower-whorl, composed of five parts; (*Zool.*) five-jointed.
pentameter (pentam'itə), *n.* a verse of five feet; (*Gr. and L Pros.*) a dactylic verse consisting of two halves each containing two feet; (*Eng. Pros.*) the iambic verse of ten syllables.
pentane (pen'tān), *n.* a volatile fluid paraffin hydrocarbon contained in petroleum etc.
pentangle (pen'tang-gl), *n.* a pentagram. **pentangular** (-tang'gū-), *a.* having five angles.
pentastich (pen'təstik), *n.* a stanza or group of five lines of verse.
Pentateuch (pen'tətūk), *n.* the first five books of the Old Testament, usu. ascribed to Moses.
pentathlon (pentath'lon), *n.* in ancient Greece, an athletic contest comprising leaping, running, wrestling, throwing the discus and hurling the spear; a modern athletics event based on this. **pentathlete** (-lēt), *n.*
pentatomic (pentətom'ik), *a.* containing five atoms in the molecule, esp. five replaceable atoms of hydrogen.
pentatonic (pentəton'ik), *a.* (*Mus.*) consisting of five tones. **pentatonic scale,** *n.*
pentavalent (pentəvā'lənt), *a.* having a valency of five.
Pentecost (pen'tikost), *n.* a solemn Jewish festival at the close of harvest, held on the 50th day from the second day of the Passover. **Pentecostal** (-kos'-), *a.* of Pentecost; of or relating to any of various fundamentalist Christian sects which stress the powers of the Holy Spirit, e.g. in healing. **Pentecostalism,** *n.* **Pentecostalist,** *n.*
penthouse (pent'hows), *n.* a dwelling or other structure built on, or occupying, the roof or top floor of a large block of flats, offices etc.
pentode (pen'tōd), *n.* a five-electrode thermionic valve.
pentose (pen'tōs), *n.* any of various sugars containing five carbon atoms in the molecule.
pentstemon (pentstē'mən), *n.* a plant belonging to a genus of scrophulariaceous plants with showy tubular flowers.
penult (pinŭlt'), *n.* the last syllable but one of a word. **penultimate** (-mət), *n., a.* (the) last but one.
penumbra (pinŭm'brə), *n.* (*pl.* **-bras**) the partly-shaded zone around the total shadow caused by an opaque body intercepting the light from a luminous body, esp. round that of the earth or moon in an eclipse; the blending or boundary of light and shade in a painting etc. **penumbral,** *a.*
penury (pen'uri), *n.* extreme poverty, destitution; lack of scarcity (of). **penurious** (-nū'-), *a.* niggardly, stingy; poor, scanty.
peon (pē'on), *n.* in India, a foot-soldier, a native constable, an office messenger or attendant; formerly in Spanish America, a bondman serving his creditor in order to work off a debt; in Spanish America, a day-labourer etc. **peonage** (-nij), *n.*
peony (pē'əni), *n.* a plant of the genus *Paeonia*, with large globular terminal flowers, usu. double in cultivation.
people (pē'pl), *n.* (*collect. sing. with pl.* **peoples**) the persons composing a nation, community or race; persons generally or indefinitely; one's family, kindred or tribe; followers, retinue, workpeople etc. *v.t.* to stock with inhabitants, to populate; to occupy, to inhabit. **the people,** the commonalty, the populace, as dist. from the self-styled higher orders. **people carrier, mover,** *n.* any of various methods of or vehicles for moving a number of people, e.g. moving pavements, driverless shuttles, estate cars etc. **people's front** POPULAR FRONT.

PEP, (*abbr.*) personal equity plan; political and economic planning.
pep (pep), *n.* (*coll.*) vigour, spirit, energy. **to pep up,** to give energy, vigour etc. to; to cheer up. **pep pill,** *n.* (*coll.*) a tablet containing a stimulant. **pep-talk,** *n.* (*coll.*) a talk intended to encourage or stimulate. **peppy,** *a.* (*coll.*) full of vitality, energetic; fast.
peplum (pep'ləm), **peplus** (-ləs), *n.* (*pl.* **-lums, -la** (-lə)) a flared extension attached to the waist of a tight-fitting jacket or bodice.
pepo (pē'pō), *n.* (*pl.* **-pos**) any of various fruits of the gourd family, e.g. cucumber, melon, with a hard rind, watery pulp, and many seeds.
pepper (pep'ə), *n.* a pungent aromatic condiment made from the dried berries of the pepper-plant; the pepper-plant, *P. nigrum,* or other species; applied also to plants of the genus *Capsicum,* or to various strong spices, e.g. cayenne pepper, prepared from the fruit of capsicums. *v.t.* to sprinkle or season with pepper; to sprinkle; to season with pungent remarks; to pelt with missiles. **black pepper,** *Piper nigrum,* the common pepper. **cayenne pepper** CAYENNE. **pepper-and-salt,** *a., n.* (of) cloth of grey and black or black and white closely intermingled and having a speckled appearance; (of) hair, black mingled with grey. **white pepper,** pepper made by removing the skin before grinding. **peppercorn,** *n.* the dried fruit of the pepper-tree; anything of little value. **peppercorn rent,** *n.* a nominal rent. **peppermill,** *n.* a small hand-operated device for grinding peppercorns. **pepper-pot,** *n.* a small container with a perforated top for sprinkling pepper on food; a W Indian dish of meat or fish with okra, chillies etc., flavoured with cassareep. **pepper-tree,** *n.* (*Austral.*) a shrub with leaves and bark having a biting taste like pepper.
peppery, *a.* having the qualities of pepper; pungent; hot-tempered; irascible, hasty.
peppermint (pep'əmint), *n.* a pungent aromatic herb; an essential oil distilled from this plant; a lozenge flavoured with this. **peppermint tree,** *n.* (*Austral.*) a eucalyptus with fragrant leaves.
peppy PEP.
pepsin (pep'sin), *n.* a protein-digesting enzyme contained in gastric juice. **peptic** (-tik), *n.* promoting or pertaining to digestion; caused by or relating to pepsin. **peptide** (-tid), *n.* a group of two or more amino acids, in which the carbon of one amino acid is linked to the nitrogen of another.
peptone (pep'tōn), *n.* any of the soluble compounds into which the proteins in food are converted by the action of pepsin.
per (pœr), *prep.* by; through; by means of; according to.
per-, *comb. form.* through, completely; very, exceedingly; to the extreme; denoting the highest degree of combination or of valency in similar chemical compounds.
perambulate (peram'būlāt), *v.t.* to walk over or through, esp. for the purpose of surveying or inspecting. *v.i.* to walk about. **perambulation,** *n.* **perambulator,** *n.* a pram. **perambulatory,** *a.*
per annum (pəran'əm), *adv.* yearly, by the year. [L]
per ardua ad astra (pœrah'dūə ad as'trə), through difficulties to the stars (motto of the Royal Air Force). [L]
percale (pəkāl', -kahl'), *n.* a closely woven cotton cambric. **percaline** (pœ'kəlēn), *n.* a glossy cotton cloth.
per capita (pœ kap'itə), *adv.* by the head, for each person. [L]
perceive (pəsēv'), *v.t.* to apprehend with the mind; to discern, to understand; to have cognizance of by the senses. **perceivable,** *a.*
per cent (pə sent'), *a.* calculated in terms of 100 parts of a whole. *n.* a percentage. **percentage** (-tij), *n.* a proportion expressed as a per cent figure; allowance, commission, duty; (*coll.*) advantage, profit.
percept (pœ'sept), *n.* that which is perceived, the mental

product of perception. **perceptible** (-sep'-), *a.* that may be perceived by the senses or intellect. **perceptibility** (-bil'-), *n.* **perceptibly**, *adv.* **perceptive** (-sep'-), *a.* having the faculty of perceiving; discerning, astute. **perceptively**, *adv.* **perceptiveness, perceptivity** (-tiv'-), *n.*
perception (pəsep'shən), *n.* the act, process or faculty of perceiving; the mental action of knowing external things through the medium of sense presentations; intuitive apprehension, insight or discernment; way of seeing, viewpoint.
perch[1] (pœch), *n.* a spiny-finned freshwater fish.
perch[2] (pœch), *n.* a pole or bar used as a rest or roost for birds; anything serving this purpose; an elevated seat or position; (*formerly*) a land measure of $5\frac{1}{2}$ yd. (5·03 m). *v.i.* to alight or rest as a bird; to alight or settle on or as on a perch. *v.t.* to set or place on or as on a perch.
perchance (pəchahns'), *adv.* perhaps, by chance.
percipient (pəsip'iənt), *a.* perceiving, apprehending, conscious, discerning. *n.* one who or that which perceives. **percipience**, *n.*
percolate (pœ'kəlāt), *v.i.* to pass through small interstices, to filter (through). *v.t.* to ooze through, to permeate; to make (coffee) in a percolator. **percolation**, *n.* **percolator**, *n.* one who or that which strains or filters; a coffee pot in which the boiling water filters through the coffee.
per contra (pœ kon'trə), *adv.* on the contrary. [L]
percuss (pəkŭs'), *v.t.* to strike quickly or tap forcibly, esp. (*Med.*) to test or diagnose by percussion.
percussion (pəküsh'ən), *n.* forcible striking or collision; the shock of such collision; the effect of the sound of a collision on the ear; medical examination by gently striking some part of the body; the production of sound by striking on an instrument; musical instruments struck in this way. **percussion-cap**, *n.* a detonator consisting of a small metal cap containing fulminating powder. **percussionist**, *n.* a person who plays a percussion instrument. **percussive**, *a.*
percutaneous (pœkütā'niəs), *a.* acting or done through the skin.
per diem (pœ dē'em), *a.*, *adv.* by the day, for each day. [L]
perdition (pədish'ən), *n.* the loss of the soul or of happiness in a future state, damnation.
peregrination (perəgrinā'shən), *n.* a travelling about; a sojourning in foreign countries. **peregrinate** (pe'-), *v.i.*
peregrine (per'əgrin), *n.* a peregrine falcon. **peregrine falcon**, *n.* a widely-distributed species of falcon used for hawking.
peremptory (pəremp'təri), *a.* precluding question or hesitation; absolute, positive, decisive, determined; imperious, dogmatic, dictatorial; (*Law*) final, determinate. **peremptorily**, *adv.* **peremptoriness**, *n.*
perennial (pəren'iəl), *a.* lasting throughout the year; unfailing, unceasing, lasting long, never ceasing; of plants, living for more than two years. *n.* a perennial plant. **perennially**, *adv.*
perestroika (perəstroi'kə), *n.* the policy of restructuring and reforming Soviet institutions initiated by Mikhail Gorbachev. [Rus.]
perf., (*abbr.*) perfect; perforated.
perfect[1] (pœ'fikt), *a.* complete in all its parts, qualities etc., without defect or fault; finished, thoroughly versed, trained, skilled etc.; of the best, highest and most complete kind; entire, complete, unqualified; (*Gram.*) expressing action completed. *n.* a perfect tense of a verb. **future perfect** FUTURE. **perfect number**, *n.* an integer, e.g. 6, that is equal to the sum of all its possible factors, excluding itself. **perfection** (-fek'-), *n.* the act of making or the state of being perfect; supreme excellence; complete development; faultlessness; a perfect person or thing; the highest degree, the extreme (of). **to perfection**, completely, perfectly. **perfectionism**, *n.* **perfectionist**, *n.* one believing in the possibility of attaining (moral or religious) perfection; a person who tolerates no fault or imperfection. **perfective** (-fek'-), *a.* tending to make perfect; (*Gram.*) expressing completed action (cp. *imperfective*). **perfectly**, *adv.*
perfect[2] (pəfekt'), *v.t.* to finish or complete, to bring to perfection; to render thoroughly versed or skilled (in). **perfecter** (-fek'-), *n.* **perfectible** (-fek'-), *a.* **perfectibility** (-bil'iti), *n.*
perfecto (pəfek'tō), *n.* (*pl.* **-tos**) a large cigar that tapers at both ends.
perfidy (pœ'fidi), *n.* violation of faith, allegiance or confidence. **perfidious** (-fid'-), *a.* treacherous, faithless, deceitful, false. **perfidiously**, *adv.* **perfidiousness**, *n.*
perfoliate (pəfō'liət), *a.* applied to leaves so surrounding the stem as to appear as if perforated by it.
perforate (pœ'fərāt), *v.t.* to bore through, to pierce; to make a hole or holes; to pass or reach through. *v.i.* to penetrate (into or through). *a.* (-rət) perforated. **perforated**, *a.* having perforations, pierced with small holes. **perforation**, *n.* a perforating or being perforated; a hole made by piercing, e.g. one of many made in paper so that it is easy to tear.
perforce (pəfaws'), *adv.* of necessity; compulsorily.
perform (pəfawm'), *v.t.* to carry through, to execute, to accomplish; to discharge, to fulful; to represent, as on the stage; to play, to render (music etc.). *v.i.* to act a part; to play a musical instrument etc.; to do what is to be done. **performable**, *a.* **performance**, *n.* execution, carrying out, completion; the performing of a play, an entertainment etc.; the capacity (as of a vehicle) to function (well); (*coll.*) an elaborate or laborious action. **performer**, *n.* one who performs, esp. an actor, musician, gymnast etc. **performing**, *a.* **performing art**, *n.* an art form requiring performance before an audience.
perfume (pəfūm', pœ'-), *n.* (pœ'-), a substance emitting a sweet odour; fragrance, scent. *v.t.* to fill or impregnate with a scent or sweet odour; to scent. **perfumer** (-fū'-), *n.* one who makes or sells perfumes. **perfumery** (-fū'-), *n.* (the preparation of) perfumes; a place where perfumes are made or sold.
perfunctory (pəfüngk'təri), *a.* done merely for the sake of having done with, done in a half-hearted or careless manner; careless, negligent, superficial, mechanical. **perfunctorily**, *adv.* **perfunctoriness**, *n.*
pergola (pœ'gələ), *n.* a covered walk or arbour with climbing plants trained over posts, trellis-work etc.
perh., (*abbr.*) perhaps.
perhaps (pəhaps', praps), *adv.* it may be, possibly.
peri-, *pref.* around, near.
perianth (pe'rianth), *n.* a floral envelope.
pericardium (perikah'diəm), *n.* the membrane enveloping the heart. **pericardiac** (-ak), **-dial**, *a.* **pericarditis** (-dī'tis), *n.* inflammation of the pericardium.
pericarp (per'ikahp), *n.* the seed-vessel or wall of the developed ovary of a plant. **pericarpial** (-kah'-), *a.*
perichondrium (perikon'driəm), *n.* the membrane investing the cartilages except at joints.
periclase (pe'riklās, -klāz), *n.* a greenish mineral composed of magnesia and protoxide of iron.
pericranium (perikrā'niəm), *n.* the membrane investing the skull.
peridium (pərid'iəm), *n.* (*pl.* **-dia** (-ə)) the outer envelope of certain fungi enclosing the spores.
peridot (pe'ridot), *n.* a yellowish-green chrysolite; olivine.
perigee (pe'rijē), *n.* the nearest point to the earth in the orbit of the moon, one of the planets or an artificial satellite. **perigean** (-jē'-), *a.*
perihelion (perihē'liən), *n.* the part of the orbit of a planet, comet etc. nearest the sun.
peril (pe'rəl), *n.* danger, risk, hazard, jeopardy; exposure to injury, loss or destruction. **perilous**, *a.* **perilously**, *adv.*

perilousness, *n*.
perilune (pe'riloon), *n*. the point in the orbit of a body round the moon where the body is closest to the centre of the moon.
perimeter (pərim'itə), *n*. the bounding line of a plane figure; the length of this, the circumference; an instrument for measuring the field of vision; the boundary of a camp etc. **perimetric** (-met'-), *a*.
perinatal (perinā'təl), *a*. occurring at or relating to the period shortly before and after birth.
perineum (perinē'əm), *n*. the part of the body between the genital organs and the anus. **perineal**, *a*.
period (piə'riəd), *n*. a portion of time marked off by some recurring event, esp. an astronomical phenomenon; the time taken up by the revolution of a planet round the sun; any specified portion of time; a definite or indefinite portion of time, an age, an era, a cycle; length of duration, existence or performance; a complete sentence; a full stop (.) marking this; an end, a limit; the time of menstruation; *a*. descriptive of a picture, object etc. characteristic of a certain period, belonging to a historical period. **period piece**, *n*. an objet d'art, piece of furniture etc. belonging to a historical period, esp. one of value; a person of outdated views, dress etc. **periodic** (-od'-), *a*. pertaining to a period or periods; performed in a regular revolution; happening or appearing at fixed intervals. **periodic table**, *n*. a table showing the chemical elements in order of their atomic number, and arranged to show the periodic nature of their properties. **periodical**, *a*. periodic. *a.*, *n*. (a publication) appearing at regular intervals. **periodically**, *adv*. **periodicity** (-dis'-), *n*.
periodontal (periədon'təl), *a*. of tissues, around a tooth.
periodontics, *sing*. **periodontology**, *n*. the branch of dentistry concerned with the treatment of periodontal disorders.
periosteum (perios'tiəm), *n*. a dense membrane covering the bones. **periosteal, periostitic** (-tit'-), *a*. **periostitis** (-ti'tis), *n*. inflammation of the periosteum.
peripatetic (peripətet'ik), *a*. walking about, itinerant.
peripeteia (peripəti'ə), *n*. a reversal of circumstances or sudden change of fortune in a play.
periphery (pərif'əri), *n*. the outer surface; the perimeter or circumference of a figure or surface. **peripheral**, *a*. of a periphery; additional, supplementary; incidental, relatively unimportant. *n*. an additional or auxiliary device, esp. in computing, e.g. a printer, a VDU.
periphrasis (pərif'rəsis), *n*. (*pl.* **-phrases**) roundabout speaking or expression, circumlocution; a roundabout phrase. **periphrastic** (-fras'-), *a*. of or using periphrasis; using two words instead of an inflected form of one word. **periphrastically**, *adv*.
periscope (pe'riskōp), *n*. an apparatus enabling persons inside a submarine, trench etc. to look about above the surface of the water etc. **periscopic** (-skop'-), *a*.
perish (pe'rish), *v.i.* to be destroyed; to die; to decay, to wither. **perishable**, *a*. liable to perish; subject to rapid decay. *n.pl.* foodstuffs and other things liable to rapid decay or deterioration. **perishability** (-bil'-), **-bleness**, *n*. **perished**, *a*. (*coll*.) of a person, very cold; of rubber etc., in poor condition, rotting due to age, damp etc. **perisher**, *n*. (*coll*.) an irritating person, esp. a child. **perishing**, *a*. (*sl*.) infernal, damned; (*coll*.) freezing cold. **perishingly**, *adv*.
perissodactyl (pərisədak'til), *a*. of or belonging to the Perissodactyla, a division of Ungulata in which all the feet are odd-toed. *n*. any individual of the Perissodactyla.
peristalith (pəris'təlith), *n*. a group of stones standing round a burial-mound etc.
peristalsis (peristal'sis), *n*. the automatic vermicular contractile motion of the alimentary canal and similar organs by which the contents are propelled along. **peristaltic**, *a*.
peristome (pe'ristōm), *n*. the fringe round the mouth of the capsule in mosses; the margin of the aperture of a mollusc, the oval opening in insects, Crustacea, Infusoria etc.
peristyle (pe'ristīl), *n*. a row of columns about a building, court etc.; a court etc. with a colonnade around it.
peritoneum (peritənē'əm), *n*. a serous membrane lining the abdominal cavity and enveloping all the abdominal viscera. **peritoneal**, *a*. **peritonitis** (-nī'tis), *n*. inflammation of the peritoneum.
periwig (pe'riwig), *n*. a peruke, a wig.
periwinkle [1] (pe'riwingkl), *n*. a small, edible univalve mollusc.
periwinkle [2] (pe'riwingkl), *n*. a plant of the genus *Vinca*, comprising trailing evergreen shrubs with blue or white flowers.
perjure (pœ'jə), *v.t.* (*Law*) **perjured**, *a*. **perjurer**, *n*. **perjurious** (-joo'-), *a*.
perjury (pœ'jəri), *n*. (*Law*) the act of wilfully giving false evidence.
perk [1] (pœk), *a*. pert, brisk, smart, trim. **to perk up**, (*coll*.) (to cause) to be more cheerful, lively etc.; to make smart or trim; to hold or prick up. **perky**, *a*. lively; cheerful, jaunty. **perkily**, *adv*. **perkiness**, *n*.
perk [2] (pœk), *n*. (*coll*.) a benefit enjoyed by an employee over and above his or her salary, short for *perquisite*.
perk [3] (pœk), *v.t.* to percolate (coffee).
perlite (pœ'līt), *n*. a glassy igneous rock characterized by spheroidal cracks formed by contractile tension in cooling, pearlstone. **perlitic** (-lit'-), *a*.
perm [1] (pœm), *n*. a hairstyle in which hair is shaped and then set by chemicals, heat etc., formerly a *permanent (wave)*. *v.t.* to put a perm in (hair).
perm [2] (pœm), *n*. a forecast of a number of football match results selected from a larger number of matches, short for *permutation*.
permafrost (pœ'məfrost), *n*. a layer of permanently frozen earth in very cold regions, *perma*nent *frost*.
permalloy (pœm'əloi), *n*. an alloy with high magnetic permeability.
permanent (pœ'mənənt), *a*. lasting, remaining or intended to remain in the same state, place or condition. **permanent press**, *n*. treatment with chemicals or heat to give resistance to creasing and soften to impart permanent creases or pleats. **permanent wave** PERM [1]. **permanent way**, *n*. the finished road-bed of a railway. **permanence, -ency**, *n*. **permanently**, *adv*.
permanganate (pəmang'gənət), *n*. a salt of permanganic acid.
permeate (pœ'miāt), *v.t.* to pass through the pores or interstices of; to penetrate and pass through; to pervade, to saturate. *v.i.* to be diffused (in, through etc.). **permeability** (-bil'-), *n*. being permeable; the degree to which a magnetizable medium affects the magnetic field surrounding it. **permeable**, *a*. yielding passage to fluids; penetrable. **permeably**, *adv*. **permeance, permeation**, *n*. **permeant**, *a.*, *n*. **permeative**, *a*.
Permian (pœ'miən), *a.*, *n*. (of) the uppermost strata of the Palaeozoic series.
permit [1] (pəmit'), *v.t.* (*past*, *p.p.* **permitted**) to allow by consent, to tolerate; to give permission to or for, to authorize. *v.i.* to allow, to admit (of). **permissible** (-mis'-), *a*. **permissibly**, *adv*. **permission** (-shən), *n*. the act of permitting; leave or licence given. **permissive**, *a*. permitting, allowing; granting liberty, leave or permission; allowing great licence in social and sexual conduct. **permissively**, *adv*. **permissiveness**, *n*. **permitter**, *n*. **permittivity** (-tiv'-), *n*. a measure of a substance's ability to store potential energy in an electric field.
permit [2] (pœ'mit), *n*. an order to permit, a written authorization, a licence.
permutate (pœmūtāt), **permute** (pəmūt'), *v.t.* to change thoroughly the arrangement of; (*Math*.) to subject to permutation. **permutable**, *a*. interchangeable. **permutable-**

ness, *n.* **permutably,** *adv.*
permutation (pœmūtā'shən), *n.* (*Math.*) change of the order of a series of quantities; each of the different arrangements, as regards order, that can be made in this; alteration, transmutation; PERM².
pernicious (pənish'əs), *a.* destructive, ruinous, deadly, noxious, hurtful; malicious, wicked. **pernicious anaemia,** *n.* a very severe, sometimes fatal, form of anaemia. **perniciously,** *adv.* **perniciousness,** *n.*
pernickety (pənik'əti), (*esp. N Am.*) **persnickety** (-snik'-), *a.* (*coll.*) fastidious, fussy, over-particular, awkward to handle, ticklish.
peroneal (pe'rənē'-), *a.* of or pertaining to the fibula.
perorate (pe'rərāt), *v.i.* to deliver an oration; (*coll.*) to speechify. **peroration,** *n.* (the concluding part or winding up of) an oration.
peroxide (pərok'sīd), *n.* the oxide of a given base that contains the greatest quantity of oxygen; hydrogen peroxide. *v.t.* (*coll.*) to bleach (hair) with hydrogen peroxide. **hydrogen peroxide,** a bleaching compound, used mainly for lightening the hair and as an antiseptic.
perp., (*abbr.*) perpendicular (style).
perpendicular (pœpəndik'ūlə), *a.* at right angles to the plane of the horizon; upright or vertical. *n.* a perpendicular line; perpendicular attitude or condition; a plumb-rule or other instrument for determining the vertical. **Perpendicular style,** *n.* the style of architecture in 14th and 15th cent. England characterized by the predominance of vertical lines, esp. in tracery. **perpendicularity** (-la'-), *n.* **perpendicularly,** *adv.*
perpetrate (pœ'pətrāt), *v.t.* to perform, to commit; to be guilty of. **perpetration,** *n.* **perpetrator,** *n.*
perpetual (pəpet'ūəl), *a.* unending, eternal; always continuing, persistent, continual, constant; of a plant, blooming continually throughout the season. **perpetual calendar,** *n.* a calendar adjustable to any year. **perpetually,** *adv.*
perpetuity (pœpətū'-), *n.* the number of years' purchase to be given for an annuity; a perpetual annuity. **in perpetuity,** for ever.
perpetuate (pəpet'ūāt), *v.t.* to cause to go on existing or continue. **perpetuation** (-ā'shən), *n.* **perpetuator,** *n.*
perplex (pəpleks'), *v.t.* to puzzle, to bewilder; to complicate, confuse or involve. **perplexedly** (-id-), *adv.* **perplexing,** *a.* **perplexity,** *n.*
per pro (pœ prō), (*abbr.*) on behalf of. [L *per pro*curationem]
perquisite (pœ'kwizit), *n.* gain, profit or emolument, or something of value, allowed over and above regular wages or salary, a perk.
perron (pe'rən), *n.* a platform with steps in front of a large building.
perry (pe'ri), *n.* a fermented liquor made from the juice of pears.
per se (pœ sā'), *adv.* by itself, in itself. [L]
persecute (pœ'sikūt), *v.t.* to afflict with suffering or loss of life or property, esp. for adherence to a particular opinion or creed; to harass, to worry. **persecution,** *n.* **persecution complex,** *n.* an irrational conviction that others are conspiring against one. **persecutor,** *n.*
persevere (pœsiviə'), *v.i.* to persist in or with any undertaking, design or course. **perseverance,** *n.* persistence in any design or undertaking; sedulous endeavour; (*Theol.*) continuance in a state of grace.
Persian (pœ'shən), *a.* pertaining to Persia, now Iran, its inhabitants or language. *n.* a native of Persia; the Persian language; a Persian cat. **Persian carpet, rug,** *n.* a carpet or rug made of knotted twine etc., finely decorated, from Persia. **Persian cat,** *n.* a variety of cat with long silky hair. **Persian lamb,** *n.* a karakul lamb; its fleece prepared as fur.
persienne (pœsien'), *n.pl.* window blinds or shutters like Venetian blinds.

persiflage (pœ'siflahzh), *n.* banter, raillery; frivolous treatment of any subject.
persimmon (pəsim'ən), *n.* the plum-like fruit of the American date-plum.
persist (pəsist'), *v.i.* to continue steadfast in the pursuit of any design; to remain, to continue, to endure. **persistence, -ency,** *n.* **persistent,** *a.* persisting, persevering; lasting, enduring; unrelenting; obstinate; insistently repeated, e.g. of questions. **persistently,** *adv.* **persister,** *n.*
person (pœ'sən), *n.* a human being, an individual; the body of a human being; a human being or body corporate having legal rights and duties; one of the three relations of the subject or object of a verb, as speaking, spoken to or spoken of. **in person,** by oneself; not by deputy. **personable,** *a.* handsome, pleasing, attractive. **personableness,** *n.* **personage** (-nij), *n.* a person of rank, distinction or importance; a person. **personal,** *a.* relating to or affecting an individual; individual, private; of criticism etc. reflecting on an individual, esp. disparaging, hostile; relating to the physical person, bodily, corporeal; transacted or done in person; (*Gram.*) denoting or indicating one of the three persons. **personal equity plan,** a scheme under which individuals may invest a fixed sum each year in UK shares without paying tax on capital gains or dividend income. **personal column,** *n.* a newspaper column containing personal messages, requests for donations to charity etc. **personal computer,** *n.* a small computer designed for business or home use. **personal effects,** *n.pl.* articles of property intimately related to the owner. **personal property,** *n.* personalty, as distinguished from real property. **personal organizer,** *n.* a portable personal filing system, usu. in the form of a small looseleafed book, containing appointments, telephone numbers, memoranda etc.; **personal pronoun,** *n.* a pronoun, e.g. *I, you, we* etc., used to refer to a person or thing. **personal stereo,** *n.* a very small, portable stereo set with headphones, designed to be attached to a belt. **personality** (-nal'-), *n.* individual existence or identity; the sum of qualities and characteristics that constitute individuality; a distinctive personal character; an important or famous person, a celebrity; (*pl.*) disparaging personal remarks. **multiple personality** MULTIPLE. **personality disorder,** *n.* any of various psychological disorders marked by a tendency to do harm to oneself or others. **personalize, -ise,** *v.t.* to make personal; to cater for the needs of a particular person; to take as referring to a particular person; to mark (something) so that it is identifiable as belonging to a particular person; to inscribe with a person's name, initials etc.; to personify. **personalization,** *n.* **personally,** *adv.* in person; particularly, individually; as regards oneself.
persona (pəsō'nə), *n.* (*pl.* **-nae**) (*often pl.*) a character in a play, novel etc.; (*pl. also* **-nas**) a person's social façade. **persona non grata** (non grah'tə), an unacceptable person, esp. in diplomatic parlance. **persona grata,** *n.* acceptable person.
personalty (pœ'sənəlti), *n.* personal estate, movable property as distinguished from real property.
personate (pœ'sənāt), *v.t.* to act the part of; to impersonate, esp. for fraudulent purpose. **personation,** *n.* **personator,** *n.*
personify (pəson'ifī), *v.t.* to regard or represent (an abstraction) as possessing the attributes of a living being; to symbolize by a human figure; to embody, to exemplify, to typify, in one's own person. **personification** (-fi-), *n.*
personnel (pœsənel'), *n.* the body of persons engaged in some service, esp. a public institution, military or naval enterprise etc.; the staff of a business firm etc.; the department of a business firm etc. that deals with the welfare, appointment and records of personnel.
perspective (pəspek'tiv), *n.* the art of representing solid objects on a plane exactly as regards position, shape and

Perspex — **petition**

dimensions, as the objects themselves appear to the eye at a particular point; a representation of objects in perspective; the relation of facts or other matters as viewed by the mind; a view, a vista, a prospect; a point of view from which something is considered. **in perspective**, according to the laws of perspective; in due proportion. **perspectively**, *adv.*
Perspex® (pœ'speks), *n.* a transparent plastic, very tough and of great clarity.
perspicacious (pœspikā'shəs), *a.* quick-sighted; mentally penetrating or discerning. **perspicacity** (-kas'-), *n.*
perspicuous (pəspik'ūəs), *a.* free from obscurity or ambiguity, clearly expressed, lucid. **perspicuity** (kū'-), *n.*
perspire (pəspīə'), *v.i.* to sweat. *v.t.* to give out (the excretions of the body) through the pores of the skin, to sweat. **perspiration** (-spi-), *n.* **perspiratory**, *a.* **perspiringly**, *adv.*
persuade (pəswād'), *v.t.* to influence or convince by argument, advice, entreaty, or expostulation; to induce; to try to influence, to advise. **persuadability** PERSUASIBILITY. **persuadable** PERSUASIBLE. **persuader**, *n.* one who or that which persuades. **persuasible** (-si-), *a.* capable of being persuaded. **persuasibility** (-bil'-), *n.* **persuasion** (-swā'zhən), *n.* the act of persuading; power to persuade, persuasiveness; a settled conviction; creed, belief, esp. in religious matters; a sect, denomination, faction, party etc. **persuasive** (-siv), *a.* able or tending to persuade; winning. *n.* that which persuades, a motive, an inducement. **persuasively**, *adv.* **persuasiveness**, *n.*
pert (pœt), *a.* sprightly, lively; saucy, forward. **pertly**, *adv.* **pertness**, *n.*
pertain (pətān'), *v.i.* to belong (to) as attribute, appendage, part etc.; to relate, to apply, to have reference (to).
pertinacious (pœtinā'shəs), *a.* obstinate; stubborn, persistent. **pertinaciously**, *adv.* **pertinacity** (-nas'-), *n.*
pertinent (pœ'tinənt), *a.* related to the matter in hand; relevant, apposite; fit, suitable. **pertinence, -ency** (-), *n.* **pertinently**, *adv.*
perturb (pətœb'), *v.t.* to disturb; to disquiet, to agitate; to throw into confusion or disorder; to cause (a planet, electron etc.) to deviate from a regular path. **perturbation**, **perturbment**, *n.*
pertussis (pətus'is), *n.* whooping-cough. **pertussal**, *a.*
peruke (pərook'), *n.* a periwig.
peruse (pərooz'), *v.t.* to read with attention; to read in a leisurely fashion. **perusal**, *n.* **peruser**, *n.*
Peruvian (pəroo'viən), *a.* pertaining to Peru. *n.* a native of Peru.
perv (pəv), *n.* (*sl.*) PERVERT 2.
pervade (pəvād'), *v.t.* to pass through; to permeate, to saturate; to be diffused throughout. **pervasion** (-zhən), *n.* **pervasive** (-siv), *a.* **pervasively**, *adv.* **pervasiveness**, *n.*
perverse (pəvœs'), *a.* wilfully or obstinately turned against what is considered good, normal, reasonable or fitting; unreasonable, perverted, intractable. **perversely**, *adv.* **perverseness, perversity**, *n.*
perversion (pəvœ'shən), *n.* the act of perverting; a misinterpretation, misapplication or corruption; abnormal sexual proclivity.
pervert[1] (pəvœt'), *v.t.* to turn aside from the proper use; to put to improper use; to misapply, to misinterpret; to lead astray, to mislead, to corrupt. **perversive**, *a.* **perverted**, *a.* marked by (esp. sexual) perversion. **perverter**, *n.* **pervertible**, *a.*
pervert[2] (pœ'vœt), *n.* one who has been perverted, esp. a person with abnormal sexual proclivities.
pervious (pœ'viəs), *a.* allowing passage (to); permeable, accessible (to facts, ideas etc.). **perviousness**, *n.*
peseta (pəsā'tə), *n.* the monetary unit in Spain.
pesky (pes'ki), *a.* (*coll.*) annoying; plaguy, troublesome. **peskily**, *adv.*
peso (pā'sō), *n.* (*pl.* **-sos**) the monetary unit of Mexico, the Philippines, Uruguay and some other S American countries.

pessary (pes'əri), *n.* a device introduced into the vagina to prevent or remedy prolapse of the uterus or as a contraceptive; a medicated plug or suppository introduced into the vagina.
pessimism (pes'imizm), *n.* the habit of taking a gloomy and despondent view of things; the doctrine that pain and evil predominate over good. **pessimist**, *n.* **pessimistic** (-mis'-), *a.* **pessimistically**, *adv.*
pest (pest), *n.* one who or that which is extremely destructive, hurtful or annoying; any plant or animal that harms crops, livestock or humans. **pesticide** (pes'tisīd), *n.* a chemical used to kill pests that damage crops etc.
pester (pes'tə), *v.t.* to bother, to worry, to annoy.
pesticide PEST.
pestilence (pes'tiləns), *n.* any contagious disease that is epidemic and deadly, esp. bubonic plague. **pestilent**, *a.* noxious to health or life, deadly; fatal to morality or society; vexatious, troublesome. **pestilential** (-len'shəl), *a.*
pestle (pes'l), *n.* an implement used in pounding substances in a mortar; any appliance used for pounding or crushing things.
pet[1] (pet), *n.* a tame, well-loved animal kept in the house; a darling, a favourite. *a.* of or for a pet animal; kept as a pet; indulged; favourite. *v.t.* (*past, p.p.* **petted**) to make a pet of; to fondle; to pamper. *v.i.* to engage in amorous fondling. **pet aversion, hate,** *n.* a thing especially disliked. **pet name**, *n.* an affectionate or informal name.
pet[2] (pet), *n.* a fit of peevishness or ill temper. **pettish**, *a.* peevish, fretful; inclined to ill-temper. **pettishly**, *adv.*
petal (pet'l), *n.* one of the divisions of a corolla of a flower, consisting of several pieces. **petal-shaped**, *a.* **petaline** (-līn), *a.* **petalled**, *a.*
pétanque (pātăk'), *n.* a Provençal word for the French game of boules.
petard (pətahd'), *n.* a case containing explosives, formerly used for blowing open gates or barriers. **hoist with one's own petard**, caught in one's own trap.
pataurist (pitaw'rist), *n.* a genus of arboreal marsupials or flying phalangers.
petechiae (pətē'kiē), *n.pl.* spots on the skin formed by extravasated blood etc., **petechial**, *a.*
peter (pē'tə), *v.i.* **to peter out**, to thin or give (out); to come to an end, to die (out). **Peter Pan** (pē'tə pan), *n.* a youthful or immature man, one who never grows up. **Peter Pan collar**, a collar on a round-necked garment having two rounded ends meeting at the front.
petersham (pē'təshəm), *n.* a thick corded-silk ribbon used for belts, hatbands etc. [Viscount *Petersham*, died 1851]
pethidine (peth'idēn), *n.* a synthetic analgesic drug with sedative effects similar to but less powerful than morphine.
pétillant (pā'tēyă), *a.* of wine, slightly sparkling.
petiole (pet'iōl), *n.* the leaf-stalk of a plant; a small stalk. **petiolate**, *a.*
petit (pet'i), *a.* (*fem.* **-tite** (-tēt')) small, petty; inconsiderable, inferior. **petit bourgeois** (pet'i buəzh'wah), *n.*, *a.* (a member) of the petite bourgeoisie. **petite bourgeoisie** (pet'i buəzhwahzē'), *n.* the lower middle class. **petit four** (pet'i foo'ə), *n.* (*pl.* **petits fours, petit fours**) a small fancy cake or biscuit. **petit mal** (pet'i mal), *n.* a mild form of epilepsy. **petit point** (pet'i pwī'), *n.* a kind of fine embroidery. **petits pois** (pet'i pwah'), *n.pl.* small, sweet green peas. [F]
petite (pətēt'), *a.* of a woman, slight, small, dainty, graceful.
petition (pitish'ən), *n.* an entreaty, a request, a supplication, a prayer; a single article in a prayer; a formal written supplication to persons in authority, esp. to Parliament etc.; the paper containing such supplication; (*Law*) a formal written application to a court, as for a writ of habeas corpus, in bankruptcy etc. *v.t.* to solicit, to

ask humbly (for etc.); to address a formal supplication to. **petitionary**, *a.* **petitioner**, *n.*
petrel (pet'rəl), *n.* any individual of the genus *Procellaria* or the family Procellariidae, small dusky sea-birds, with long wings and great power of flight.
petri- PETR(O)-.
Petri dish (pē'tri), *n.* a shallow, circular, flat-bottomed dish used for cultures of microorganisms. [Julius *Petri*, 1852-1921, German bacteriologist]
petrify (pet'rifī), *v.t.* to convert into stone or stony substance; to stupefy, as with fear, astonishment etc.; to make hard, callous, benumbed or stiffened. *v.i.* to be converted into stone or a stony substance; to become stiffened, benumbed, callous etc. **petrifaction** (-fak'-), **petrification** (-fi-), *n.*
petr(o)-, petri-, *comb. form.* stone; petrol.
petrochemical (petrəkem'ikl), *n.*, *a.* (a chemical) obtained from petroleum; relating to such chemicals.
petrocurrency (petrəkŭ'rənsi), *n.* the currency of a country which exports significant quantities of petroleum.
petrodollar (petrədol'ər), *n.* a dollar earned from the exporting of petroleum.
petroglyph (pet'rəglif), *n.* a rock carving.
petrography (petrog'rahf-), *n.* descriptive petrology. **petrographer** (-trog'-), *n.*
petrol (pet'rəl), *n.* a refined form of petroleum used in motor vehicles etc. **petrol bomb**, *n.* a bottle or other container full of petrol, used as an incendiary. **petrol station**, *n.* a retail outlet for the sale of petrol to motorists. **petrolatum** (-lā'təm), *n.* petroleum jelly.
petroleum (pitrō'liəm), *n.* an inflammable oily liquid exuding from rocks or pumped from wells, used to make fuel oil, lubricating oil, paraffin, petrol, etc. **petroleum jelly**, *n.* used as a lubricant and ointment base.
petrology (pitrol'əji), *n.* the study of the origin, structure and mineralogical and chemical composition of rocks. **petrological** (-loj'-), *a.* **petrologist**, *n.*
petticoat (pet'ikōt), *n.* a woman's loose under-skirt; (*sl. offensive*) a woman, a girl. *a.* (*coll., facet., offensive*) of or by women. **petticoat-government**, *n.* government by women, esp. in domestic affairs.
pettifogger, *n.* a petty, second-rate lawyer, esp. one given to sharp practices; a petty quibbler. **pettifoggery**, *n.* **pettifogging**, *a., n.* petty, trivial, quibbling (activity or speech).
pettily etc. PETTY.
pettish PET[2].
petty (pet'i), *a.* little, trifling, insignificant; minor, inferior, subordinate, on a small scale; small-minded, mean. **petty cash**, *n.* minor items of receipt and expenditure. **petty larceny**, *n.* in the US and formerly in the UK, the theft of property worth less than a specified figure. **petty officer**, *n.* a naval officer corresponding in rank to a non-commissioned officer. **petty sessions**, *n.pl.* a magistrates' court for trying minor offences. **pettily**, *adv.* **pettiness**, *n.*
petulant (pet'ūlənt), *a.* given to fits of ill temper; peevish, irritable. **petulance, -ancy**, *n.* **petulantly**, *adv.*
petunia (pitū'niə), *n.* one of a genus of S American plants, allied to the tobacco, cultivated for their showy funnel-shaped flowers.
petuntse (pitŭnt'si), *n.* a fusible substance of feldspathic nature used for the manufacture of porcelain.
pew (pū), *n.* a box-like enclosed seat in a church for a family etc.; a long bench with a back, for worshippers in church; (*coll.*) a seat, a chair. **to take a pew**, (*coll.*) to sit down.
pewit, peewit (pē'wit), *n.* the lapwing; its cry.
pewter (pū'tə), *n.* an alloy usu. of tin and lead, sometimes of tin with other metals; vessels or utensils made of this. *a.* made of pewter. **pewterer**, *n.*
peyote (pāyō'tā), *n.* a small globular cactus of the southern US and Mexico, MESCAL.

P.F., (*abbr.*) Procurator Fiscal.
Pfennig (fen'ig), *n.* (*pl.* **-ige** (-gə)) a small copper coin of Germany, worth the hundredth part of a mark.
PFLP, (*abbr.*) Popular Front for the Liberation of Palestine.
PG, (*abbr.*) parental guidance (used to classify the content of cinema films).
pH (pēach'), *n.* a measure of the acidity or alkalinity of a solution on a scale from 0 to 14, with 7 representing neutrality, and figures below it denoting acidity, those above it alkalinity. **pH balanced**, *a.* (of shampoo, soap, etc.) having a formula in which the acidity and alkalinity is balanced for a specific use.
phaeton (fī'tən, fā'-), *n.* a light four-wheeled open carriage, usu. drawn by two horses.
-phage, *comb. form.* eater. **-phagia, -phagy**, *comb. form.* eating.
phag(o)-, *comb. form.* eating.
phagocyte (fag'əsit), *n.* a leucocyte that absorbs microbes etc., protecting the system against infection. **phagocytic** (-sit'-), *a.* **phagocytosis** (-tō'sis), *n.* the destruction of microbes etc. by phagocytes.
-phagous, *comb. form.* eating, devouring, as in *anthropophagous, sarcophagous.*
-phagy -PHAGE.
phalange PHALANX.
phalanger (fəlan'jə), *n.* any individual of a sub-family of small Australian woolly-coated arboreal marsupials, comprising the flying squirrel and opossum.
phalanx (fal'angks), *n.* (*pl.* **-xes**, *Anat., Bot.* **-ges** (-lan'jēz)) the close order in which the heavy-armed troops of a Greek army were drawn up, esp. a compact body of Macedonian infantry; hence, any compact body of troops or close organization of persons; (also **phalange** (-anj)) each of the small bones of the fingers and toes; one of the bundles of stamens in polyadelphous flowers. **phalangeal** (-lan'ji-), *a.*
phalarope (fal'ərōp), *n.* a small wading bird of the family Phalaropodidae, related to the snipes.
phallus (fal'əs), *n.* (*pl.* **-lli**) a figure of the male sexual organ, venerated as a symbol of the fertilizing power in nature; a penis. **phallic**, *a.* **phallicism** (-sizm), **phallism**, *n.* the worship of the phallus.
phanerogam (fan'ərəgam), *n.* a plant having pistils and stamens, a flowering plant. **phanerogamic** (-gam'-), **-gamous** (-rog'-), *a.*
phantasm (fan'tazm), *n.* a phantom; an optical illusion; a deception, a figment; an illusory mental representation of an object, person, etc. **phantasmal, -mic** (-taz'-), *a.*
phantasmagoria (fantəzməgaw'riə), **phantasmagory** (-taz'məgəri), *n.* a series of fantastic appearances or illusions. **phantasmagorial, -goric** (-go'-), *a.*
phantasy (fan'təsi), FANTASY.
phantom (fan'təm), *n.* a ghost, a spectre; a vision, an illusion, an imaginary appearance; an empty show or mere image (of). *a.* seeming, apparent; illusory, imaginary, fictitious.
Pharaoh (feə'rō), *n.* any one of the ancient Egyptian kings. **Pharaonic** (-rəon'-), *a.*
Pharisee (fa'risē), *n.* one of an ancient Jewish sect who rigidly observed the rites and ceremonies prescribed by the written law; a self-righteous person, an unctuous hypocrite. **pharisaic** (-sā'ik), **-ical**, *a.* **pharisaism** (fa'risā-), *n.* the doctrines of the Pharisees as a sect; hypocrisy in religion, self-righteousness.
pharm., (*abbr.*) pharmaceutical; pharmacy.
pharmaceutical (fahməsū'tikl), *a.* of or pertaining to pharmacy or drugs. **pharmaceutically**, *adv.* **pharmaceutics**, *n.sing.* pharmacy. **pharmacist** (fah'məsist), *n.* one trained in pharmacy; one legally enutled to sell drugs and poisons.
pharmaco-, *comb. form.* pertaining to chemistry or to drugs.

pharmacology (fahməkol'əji), *n.* the science of drugs and medicines. **pharmacological** (-loj'-), *a.* **pharmacologically**, *adv.* **pharmacologist**, *n.*
pharmacopoeia (fahməkəpe'ə), *n.* a book, esp. an official publication containing a list of drugs, formulas, doses etc.
pharmacy (fah'məsi), *n.* the art or practice of preparing, compounding and dispensing drugs, esp. for medicinal purposes; a chemist's shop; a dispensary.
pharos (feə'ros), *n.* a lighthouse, a beacon.
pharyngal, -geal, pharyngitis PHARYNX.
pharynx (fa'ringks), *n.* (*pl.* **-ringes** (-rin'jēz)) the canal or cavity opening from the mouth into the oesophagus and communicating with the air passages of the nose. **pharyngal** (-ring'gəl), **-geal** (-rin'jəl), *a.* **pharyngitis** (farinji'tis), *n.* inflammation of the pharynx.
phase (fāz), *n.* (*pl.* **-ses** (-sēz)) a particular aspect or appearance of something going through a series of events or changes, applied e.g. to the successive quarters etc. of the moon; a stage of change or development. *v.t.* to carry out in phases; to organize in phases (*see* SINGLE PHASE, THREE-PHASE). **to phase in**, to introduce gradually. **to phase out**, to discontinue gradually. **phasic** (-zik, -sik), *a.* **-phasia**, *comb. form.* speech disorder.
phatic (fat'ik), *a.* of speech, used to express feelings, sociability etc., rather than to express meaning.
Ph.D., (*abbr.*) Doctor of Philosophy. [L *Philosophiae Doctor*]
pheasant (fez'ənt), *n.* a game bird naturalized in Britain and Europe, noted for its brilliant plumage and its delicate flesh.
phen- PHEN(O)-.
phenacetin (fənas'ətin), *n.* a white crystalline compound used as an antipyretic.
phen(o)-, *comb. form.* applied to substances derived from or relating to benzene; showing, visible, manifest.
phenobarbitone (fēnōbah'bitōn), *n.* a white, crystalline powder used as a sedative or hypnotic drug.
phenol (fē'nol), *n.* carbolic acid; any of various weakly acidic compounds derived from benzene, and containing a hydroxyl group. **phenolic** (-nol'-), *a.*
phenology (fənol'əji), *n.* the study of the times of recurrence of natural phenomena, esp. of the influence of climate on plants and animals. **phenological** (-loj'-), *a.*
phenomenon, (*esp. formerly*) **phaenomenon** (finom'inən), *n.* (*pl.* **-na** (-nə)) that which appears or is perceived by observation or experiment; (*coll.*) a remarkable or unusual occurrence or person; that which is apprehended by the mind, as distinguished from real existence. **phenomenal**, *a.* of or pertaining to phenomena, perceptible, cognizable by the senses; (*coll.*) extraordinary, prodigious. **phenomenalism, phenomenism**, *n.* the doctrine that phenomena are the sole material of knowledge, and that underlying realities and causes are unknowable. **phenomenalist, phenomenist**, *n.*
phenotype (fē'nōtīp), *n.* the observable characteristics of an organism produced by the interaction of the genotype and the environment.
phenyl (fē'nīl, fen'il), *n.* the organic radical found in benzene, phenol, aniline etc.
pheromone (fe'rəmōn), *n.* any chemical substance secreted by an animal that stimulates responses from others of its species.
phew (fū), *int.* expressing surprise, disgust, tiredness, impatience etc.
phi (fī), *n.* the 21st letter of the Greek alphabet.
phial (fī'əl), *n.* a small glass vessel or bottle, esp. for medicine or perfume.
Phi Beta Kappa (fī bē'tə kap'ə), the oldest of the American college fraternities. [initials of Gr. *Philosophia Biou Kubernētēs*, Philosophy is the guide of life]
phil., (*abbr.*) philosophy.
phil- PHIL(O)-.
philander (filan'də), *v.i.* to make love in a trifling way; to flirt. **philanderer**, *n.* a man who does this.
philanthropy (filan'thrəpi), *n.* love of mankind; active benevolence towards one's fellow-humans. **philanthrope** (fil'ənthrōp), *n.* **philanthropic, -ical** (-throp'-), *a.* **philanthropically**, *adv.* **philanthropist**, *n.*
philately (filat'əli), *n.* the study and collecting of postage stamps. **philatelic** (-tel'-), *a.* **philatelist**, *n.*
-phile, *comb. form.* a lover or friend of; loving, as in *bibliophile, gastrophile, Germanophile*. **-philic, -philous**, *a.*
philharmonic (filəmon'ik), *a.* loving music. *n.* a person fond of music; a musical society.
philhellene (filhe'lēn), *n.* a friend or lover of the Greeks; (*Hist.*) a supporter of Greek independence.
-philia, *comb. form.* love of. **-philiac**, *n.*, *a.*
philibeg (fil'ibeg), FILIBEG.
-philic -PHILE.
philippic (filip'ik), *n.* speech or declamation full of acrimonious invective.
Philistine (fil'istīn), *n.* one of an ancient warlike race in S Palestine who were hostile to the Jews; a person of narrow or materialistic views or ideas, a boor; one hostile to liberal culture; *a.* pertaining to the Philistines; uncultured, boorish. **Philistinism** (-tin-), *n.*
phillumeny (filoo'məni), *n.* the collecting of matchboxes or matchbox labels. **phillumenist**, *n.*
phil(o)-, *comb. form.* fond of, loving.
philodendron (filəden'drən), *n.* (*pl.* **-drons, -dra** (-drə)) any of various plants of the arum family, cultivated for their showy foliage.
philology (filol'əji), *n.* the historical or comparative study of language. **philologer, -logist**, *n.* **philological** (-loj'-), *a.* **philologically**, *adv.*
philosopher (filos'əfə), *n.* one who studies or devotes himself or herself to natural or moral philosophy or to the investigation of the principles of being or of knowledge; one who regulates conduct and actions by the principles of philosophy; one of philosophic temperament. **philosopher's stone**, *n.* an imaginary stone, sought for by the alchemists in the belief that it would transmute the baser metals into gold. **philosophical** (-sof'-), *a.* pertaining to or according to philosophy; devoted to or skilled in philosophy; wise, calm, temperate, unimpassioned, stoical. **philosophically**, *adv.* **philosophize, -ise**, *v.t., v.i.* **philosophizer, -iser**, *n.*
philosophy (filos'əfi), *n.* the knowledge or investigation of ultimate reality or of general principles of knowledge or existence; a particular system of philosophic principles; the fundamental principles of a science etc.; practical wisdom; calmness and coolness of temper; serenity, resignation.
-philous -PHILE.
philtre, (*esp. N Am.*) **philter** (fil'tə), *n.* a love-potion, a love-charm.
phimosis (fīmō'sis), *n.* constriction of the opening of the prepuce.
phiz (fiz), **phizog** (-zog'), *n.* (*sl.*) the face, the visage; the expression, the countenance.
phleb- PHLEB(O)-.
phlebitis (flibī'tis), *n.* inflammation of the inner membrane of a vein. **phlebitic** (-bit'-), *a.*
phleb(o)-, *comb. form.* vein.
phlebotomy (flibot'əmi), *n.* the opening of a vein, bloodletting.
phlegm (flem), *n.* viscid mucus secreted in the air passages or stomach and discharged by coughing etc.; self-possession, coolness; sluggishness, apathy. **phlegmatic, -ical** (flegmat'-), *a.* cool, sluggish, apathetic, unemotional. **phlegmatically**, *adv.* **phlegmy** (flem'i), *a.* abounding in or of the nature of phlegm.
phloem (flō'əm), *n.* the softer cellular portion of fibrovascular tissue in plants, the bark and the tissues closely connected with it.

phlox (floks), *n.* a plant of a genus of N American plants with clusters of showy flowers.

-phobe, *comb. form.* fearing, as in *Anglophobe, Gallophobe*; one who fears or hates. **-phobia**, *comb. form* fear, morbid dislike, as in *Anglophobia, hydrophobia*. **-phobic**, *a.*

phobia (fō'biə), *n.* an irrational fear or hatred. **phobic**, *a.*

Phoenician (fənish'ən), *a.* of or pertaining to Phoenicia, an ancient Semitic country on the coast of Syria; its people or their language. *n.* a native or inhabitant of Phoenicia or its colonies; the (extinct) language of Phoenicia.

phoenix (fē'niks), *n.* a mythical Arabian bird, said to live for 500 or 600 years and to immolate itself on a funeral pyre, from which it rose again in renewed youth; a person or thing of extreme rarity or excellence.

phon (ton), *n.* a unit of loudness.

phon., (*abbr.*) phonetics; phonology.

phon- PHON(O)-.

phonate (fō'nāt), *v.i.* to make a vocal sound. **phonation**, *n.*

phone[1] (fōn), *v.t., v.i.* to telephone. *n.* a telephone. **phone book** TELEPHONE DIRECTORY. **phonecard**, *n.* a plastic card inserted into a slot in a public telephone, which cancels out the prepaid units on the card as the call is made. **phone-in**, *n.* a radio or TV programme in which members of the audience at home telephone to make comments, ask questions etc., as part of a live broadcast.

phone[2] (fōn), *n.* an articulate sound, as a simple vowel or consonant sound.

-phone, *comb. form.* sound; voice; a device producing sound; (a person) speaking a specified language, as in *Francophone.*

phoneme (fō'nēm), *n.* the smallest distinctive group of phones (PHONE[2]) in a language. **phonemic** (-nē'-), *a.* **phonemics**, *n.sing.* the study of phonemes.

phonetic (fənet'ik), *a.* pertaining to the voice or vocal sounds; representing sounds, esp. by means of a distinct letter or character for each. **phonetically**, *adv.* **phonetician**, *n.* **phoneticist** (-sist), *n.* **phonetics**, *n.sing.* the science of articulate sounds. **phonetist** (fō'-), *n.* one versed in phonetics, a phoneticist.

phoney, (*esp. N Am.*) **phony** (fō'ni), *a.* (*coll.*) bogus; false; fraudulent, counterfeit; of a person, pretentious. *n.* a phoney person or thing.

phonic (fon'ik), *a.* pertaining to sounds, acoustic; pertaining to vocal sounds. **phonics**, *n.sing.* a method of teaching people to read by associating sounds with their phonetic values.

phonogram (fō'nəgram), *n.* a written character indicating a particular spoken sound. **phonograph** (-grahf), *n.* an early instrument for automatically recording and reproducing sounds; (*N Am.*) a gramophone. **phonographic** (-graf'-), *a.*

phonology (fənol'əji), *n.* the science of the vocal sounds; the sounds and combinations of sounds in a particular language. **phonologic, -ical** (fonəloj'-), *a.* **phonologist**, *n.*

phonon (fō'non), *n.* a quantum of vibrational energy in a crystal lattice.

phony PHONEY.

phooey (foo'i), *int.* used to express contempt, disbelief etc.

-phore, *comb. form.* bearer, as in *gonophore, gynophore, semaphore.* **-phorous**, *comb. form.* bearing, -ferous, as in *electrophorous, galactophorous.*

-phoresis, *comb. form.* transmission.

phosgene (foz'jēn), *n.* gaseous carbon oxychloride, used as a poison gas.

phosph- PHOSPH(O)-.

phosphate (fos'fāt), *n.* a salt of phosphoric acid; (*pl.*) phosphates of calcium, iron and alumina etc., used as fertilizing agents. **phosphatic** (-fat'-), *a.*

phosphene (fos'fēn), *n.* a luminous image produced by pressure on the eyeball, caused by irritation of the retina.

phosphide (fos'fīd), *n.* a combination of phosphorus with another element or radical. **phosphite** (-fīt), *n.* a salt of phosphorous acid.

phosph(o)-, phosphor(o)-, *comb. form.* phosphorus.

phosphor (fos'fə), *n.* phosphorus; a substance that exhibits phosphorescence. **phosphoresce** (-res'), *v.i.* to give out a light unaccompanied by perceptible heat or without combustion. **phosphorescence**, *n.* the emission of or the property of emitting light under such conditions; such emission of light caused by radiation bombardment, and continuing after the radiation has ceased. **phosphorescent**, *a.*

phosphorate, phosphoric etc. PHOSPHORUS.

phosphorus (fos'fərəs), *n.* a non-metallic element, at.no. 15; chem. symbol P, occurring in two allotropic forms. White phosphorus is waxy, poisonous, spontaneously combustible at room temperature and appears luminous. Red phosphorus is non-poisonous and ignites only when heated. **phosphorate** (fos'fərāt), *v.t.* to combine, impregnate with phosphorus. **phosphoric** (-fo'-), *a.* pertaining to phosphorus in its higher valency; phosphorescent. **phosphorous**, *a.* pertaining to, of the nature of or obtained from phosphorus, esp. in its lower valency.

phot (fot, fōt), *n.* a unit of illumination, one lumen per square centimetre.

phot., (*abbr.*) photographic; photography.

phot- PHOT(O)-.

photic (fō'tik), *a.* pertaining to light; accessible to the sun's light; pertaining to the upper layers of the sea that receive the sun's light.

photo (fō'tō), *n.* (*pl.* **-tos**) a photograph. *v.t.* to photograph. *a.* photographic. **photo-finish**, *n.* a close finish of a race etc., in which a photograph enables a judge to decide the winner. **photo-opportunity**, *n.* an event arranged for or by a politician, group etc. primarily in order to get favourable media coverage through good pictures; a photocall.

phot(o)-, *comb. form.* pertaining to light or to photography.

photocall (fō'tōkawl), *n.* an occasion when someone is photographed by arrangement for publicity purposes.

photocell (fō'tōsel), *n.* a photoelectric cell.

photochemical (fōtəkem'ikl), *a.* of, pertaining to or produced by the chemical action of light. **photochemically**, *adv.* **photochemistry**, *n.*

photocomposition (fōtōkompəzish'ən), *n.* the composition of print etc. directly on film or photosensitive paper for reproduction.

photoconductivity (fōtōkondŭktiv'iti), *n.* electrical conductivity that varies with the incidence of radiation, esp. light.

photocopy (fō'təkopi), *n.* a photographic reproduction of matter that is written, printed etc. *v.t.* to make a photocopy of. **photocopier**, *n.* a device for making photocopies.

photoelectric (fōtōilek'trik), *a.* of or pertaining to photoelectricity, or to the combined action of light and electricity. **photoelectric cell**, *n.* a device for measuring light by a change of electrical resistance when light falls upon a cell, or by the generation of a voltage. **photoelectricity**, *n.* electricity produced or affected by light.

photoelectron (fōtōilek'tron), *n.* an electron emitted during photoemission.

photoemission (fōtōimish'ən), *n.* the emission of electrons from a substance on which radiation falls.

photo-engraving (fōtōingrā'ving), *n.* any process for producing printing plates by means of photography; the plate or image produced by this process.

photo-finish PHOTO.

Photo-fit® (fō'tōfit), *n.* (a method of composing a likeness of someone's face consisting of photographs of parts of faces, used for the identification of criminal suspects.

photogenic (fōtəjen'ik), *a.* produced by the action of light; producing light, phosphorescent; descriptive of one who comes out well in photographs or on film.

photogrammetry (fōtəgram'ətri), *n.* the technique of taking

photograph 602 **physique**

measurements from photographs, e.g. making maps from aerial photographs.
photograph (fō'təgrahf, -graf), *n.* a picture etc. taken by means of photography. *v.t.* to take a picture of by photography. *v.i.* to practise photography; to appear in a photograph (well or badly). **photographer** (-tog'-), *n.* **photographic** (-graf'-), *a.* **photographically**, *adv.* **photography** (-tog'-), *n.* the process or art of producing images or pictures of objects by the chemical action of light on certain sensitive substances.
photogravure (fōtəgrəvūə'), *n.* the process of producing an intaglio plate for printing by the transfer of a photographic negative to the plate and subsequent etching; a picture so produced.
photojournalism, *n.* journalism characterized by the predominance of photographs over written text. **photojournalist**, *n.*
photolithography (fōtōlithog'rəfi), *n.* a mode of producing by photography designs upon plates etc., from which impressions may be taken.
photolysis (fətol'isis), *n.* decomposition resulting from the incidence of radiation. **photolytic** (fōtəlit'-), *a.*
photometer (fətom'itə), *n.* an instrument for measuring the relative intensity of light. **photometry**, *n.*
photomicrography (fōtōmikrog'rəfi), *n.* the process of making magnified photographs of microscopic objects. **photomicrographer**, *n.* **photomicrography** (-mī'-), *n.*
photomontage (fōtōmontahzh'), *n.* a means of producing pictures by the montage of many photographic images; the picture thus produced.
photon (fō'ton), *n.* the unit of light intensity; quantum of radiant energy.
photosensitive (fōtōsen'sitiv), *a.* sensitive to the action of light. **photosensitivity** (-tiv'-), *n.* **photosensitize**, **-ise**, *v.t.* to make photosensitive.
photosetting (fō'təseting), *n.* photocomposition, filmsetting. **photoset**, *v.t.*
photosphere (fō'təsfiə), *n.* the luminous envelope of the sun or a star. **photospheric** (-sfe'-), *a.*
Photostat® (fō'təstat), *n.* a machine for making photographic copies of prints, documents etc.; a copy so produced. *v.t.* to make a copy (of) using a Photostat.
photosynthesis (fōtōsin'thəsis), *n.* the process by which carbohydrates are produced from carbon dioxide and water through the agency of light; esp. this process occurring in green plants. **photosynthesize**, **-ise**, *v.t.* **photosynthetic** (-thet'-), *a.* **photosynthetically**, *adv.*
phototropism (fōtōtrō'pizm), *n.* tropism due to the influence of light.
phrase (frāz), *n.* an expression denoting a single idea or forming a distinct part of a sentence; a brief or concise expression; mode, manner or style of expression, diction; idiomatic expression; noun; (*Mus.*) a short, distinct passage forming part of a melody. *v.t.* to express in words or phrases. **phrase-book**, *n.* a handbook of phrases or idioms in a foreign language with their translations. **phrasal**, *a.* **phrasing**, *n.* the way in which a speech etc. is phrased, phraseology; the manner or result of grouping phrases in music.
phraseogram (frā'ziəgram), *n.* a character standing for a whole phrase, as in shorthand.
phraseology (frāziol'əji), *n.* choice or arrangement of words; manner of expression, diction. **phraseological** (-loj'-), *a.*
phrenic (fren'ik), *a.* of or pertaining to the diaphragm. *n.* the phrenic nerve.
phrenology (frənol'əji), *n.* the theory that the mental faculties and affections are located in distinct parts of the brain denoted by prominences on the skull. **phrenological** (frenəloj'-), *a.* **phrenologist**, *n.*
Phrygian (frij'iən), *a.* pertaining to Phrygia, an ancient country in Asia Minor, or its language or inhabitants. *n.* a native of Phrygia.
phthisis (thī'sis, fthī'-, tī'-), *n.* a wasting disease, esp. pulmonary tuberculosis.
phut (fŭt), *n.* (*coll.*) a representation of the sound of something deflating, collapsing, etc. **to go phut**, to collapse, to stop.
phyl- PHYL(O)-.
phylactery (filak'təri), *n.* a small leather box in which are enclosed slips of vellum inscribed with passages from the Pentateuch, worn on the head and left arm by Jews during morning prayer, except on the Sabbath.
phyletic (filet'ik), *a.* pertaining to a phylum, racial.
-phyll, *comb. form.* leaf, as in *chlorophyll, xanthophyll.*
phyll(o)-, *comb. form.* leaf.
phyllode (fil'ōd), **phyllodium** (-lō'diəm), *n.* (*pl.* **-odes**, **-odia** (-diə)) a petiole having the appearance and functions of a leaf.
phyllotaxis (filətak'sis), **phyllotaxy** (fil'-), *n.* the arrangement of the leaves etc. on the stem or axis of a plant.
phylloxera (filoksiə'rə, -lok'sərə), *n.* an aphid very destructive to vines.
phyl(o)-, *comb. form.* tribe, race.
phylogeny (philoj'əni), **-genesis** (philəjen'isis), *n.* the evolution of a group, species or type of plant or animal life; the history of this. **phylogenetic** (-net'-), **-genic** (-jen'-), *a.*
phylum (fī'ləm), *n.* (*pl.* **phyla** (-lə)) a primary group consisting of related organisms descended from a common form.
phys., (*abbr.*) physical; physics.
Physalia (fisā'liə), *n.* a genus of large oceanic Hydrozoa comprising the Portuguese man-of-war.
Physeter (fisē'tə), *n.* a genus of Cetacea, containing the sperm-whales.
physi- PHYSI(O)-.
physical (fiz'ikəl), *a.* of or pertaining to matter; obvious to or cognizable by the senses; pertaining to physics, as opposed to chemical; material, bodily, corporeal, as opposed to spiritual. *n.* an examination to ascertain physical fitness. **physical education**, *n.* athletic and sporting training or practice, esp. in schools. **physical geography**, *n.* the study of the earth's natural features. **physical jerks**, *n.pl.* (*coll.*) physical exercises to promote fitness. **physicality** (-kal'-), *n.* excessive concern with physical matters.
physically, *adv.*
physician (fizish'ən), *n.* a legally-qualified medical practitioner.
physicist (fiz'isist), *n.* one versed in physics.
physics (fiz'iks), *n.sing.* the science dealing with the phenomena of matter, esp. as affected by energy, and the laws governing these, excluding biology and chemistry.
physio (fiz'iō), *n.* (*coll.*) short for PHYSIOTHERAPIST.
physi(o)-, *comb. form.* pertaining to nature; physical.
physiognomy (fizion'əmi), *n.* the art of reading character from features of the face or the form of the body; the face or countenance as an index of character; the lineaments or external features (of a landscape etc.); aspect, appearance, look (of a situation, event etc.). **physiognomic**, **-ical** (-nom'-), *a.* **physiognomically**, *adv.* **physiognomist**, *n.*
physiography (fiziog'rəfi), *n.* the scientific description of the physical features of the earth, and the causes by which they have been modified; physical geography. **physiographer**, *n.* **physiographic**, **-ical** (-graf'-), *a.*
physiol., (*abbr.*) physiologist; physiology.
physiology (fiziol'əji), *n.* the science of the organic functions of animals and plants. **physiologic**, **-ical** (-loj'-), *a.* **physiologist**, *n.*
physiotherapy (fiziōthe'rəpi), *n.* a form of medical treatment in which physical agents such as movement of limbs, massage, etc. are used in place of drugs or surgery. **physiotherapist**, *n.*
physique (fizēk'), *n.* physical structure of a person.

-phyte, *comb. form.* denoting a vegetable organism, as in *lithophyte, zoophyte.* **-phytic**, *a.*
phyto-, *comb. form. plant.*
phytogenesis (fītōjen'isis), **phytogeny** (-toj'əni), *n.* the origin, generation or evolution of plants.
phytography (fītog'rəfi), *n.* the systematic description and naming of plants.
phytopathology (fītōpəthol'əji), *n.* the science of the diseases of plants.
phytotoxic (fītōtok'sik), *a.* poisonous to plants.
pi[1] (pī), *n.* the Greek letter π, *pl.* the symbol representing the ratio of the circumference of a circle to the diameter, i.e. 3·14159265.
pi[2] (pī), PIE[3].
pi[3] (pī), *a. (coll.)* short for PIOUS.
piacular (pīak'ūlə), *a.* expiatory; requiring expiation; atrociously bad.
pia mater (pī'əmā'tə), *n.* a delicate membrane, the innermost of the three meninges investing the brain and spinal cord.
pianissimo (piənis'imō), *adv. (Mus.)* very softly. *a.* very soft.
piano[1] (piah'nō), *adv. (Mus.)* softly. *a.* played softly.
piano[2] (pian'ō), **pianoforte** (-faw'ti), *n. (pl.* **-nos**) a musical instrument the sounds of which are produced by blows on the wire strings from hammers acted upon by levers set in motion by keys. **piano accordion**, *n.* an accordion equipped with a keyboard resembling that of a piano. **pianism**, *n.* piano-playing; the technique of this. **pianist** (pi'ənist), *n.* a performer on the piano. **Pianola**® (piənō'lə), *n.* a type of mechanical piano.
piastre, *(esp. N Am.)* **piaster** (pias'tə), *n.* (a coin representing) a monetary unit (1/100th of a pound) of various Middle Eastern countries (e.g. Egypt, Sudan, Syria).
piazza (piat'sə), *n.* a square open space or public square, esp. in Italian towns; applied to a colonnade, or an arcaded or colonnaded walk, and *(N Am.)* to a verandah about a house.
pibroch (pē'brəkh), *n.* a series of variations, chiefly martial, played on a bagpipe; *(erron.)* the bagpipe.
pic (pik), *n. (pl.* **pics**, **pix** (piks)) *(coll.)* a photograph.
pica[1] (pī'kə), *n.* a vitiated appetite causing the person affected to crave for things unfit for food, as coal, chalk etc.
pica[2] (pī'kə), *n.* a size of type, the standard of measurement in printing.
picador (pik'ədaw), *n.* in Spanish bullfights, a horseman with a lance who rouses the bull.
picaresque (pikəresk'), *n., a.* being or relating to a style of fiction describing the episodic adventures of a usu. errant rogue.
piccalilli (pik'əlil'i), *n.* a pickle of various chopped vegetables with pungent spices.
piccaninny (pikənin'i), *n. (offensive)* a little child. esp. of blacks or Australian Aborigines.
piccolo (pik'əlō), *n. (pl.* **-los**) a small flute, with the notes one octave higher than the ordinary flute.
pick[1] (pik), *v.t.* to break, pierce, or indent with a pointed instrument; to make (a hole) or to open thus; to strike at with something pointed; to remove extraneous matter from (the teeth etc.) thus; to clean by removing that which adheres with the fingers etc.; to pluck, to gather; to take up with a beak etc.; to eat in little bits; to choose, to cull, to select carefully; to make (one's way) carefully on foot; to find an occasion for (a quarrel etc.); to steal the contents of; to open (a lock) with an implement other than the key; to pluck, to pull apart; to twitch the strings of, to play (a banjo). *v.i.* to strike at with a pointed implement; to eat in little bits; to make a careful choice; to pilfer. *n.* choice, selection; the best (of). **to pick and choose**, to make a fastidious selection. **to pick at**, to criticize in a cavilling way; to eat sparingly. **to pick off**, to shoot with careful aim one by one. **to pick on**, to single out for unpleasant treatment, esp. to bully. **to pick out**, to select; to distinguish (with the eye) from surroundings; to relieve or variegate with or as with distinctive colours; to gather (the meaning of a passage etc.); to gather by ear and play (a tune) on the piano etc. **to pick someone's brains**, to consult someone with special expertise or experience. **to pick to pieces**, to analyse or criticize spitefully. **to pick up**, to take up with the beak, fingers etc.; to gather or acquire here and there or little by little; to collect; to accept and pay (a bill); to detain (a suspect etc.); to receive (an electronic signal etc.); to make acquaintance (with), esp. with the intention of having sexual relations; to regain or recover (health etc.). **pick-me-up**, *n.* a drink or medicine taken to restore the tone of the system. **pickpocket**, *n.* one who steals from pockets. **pick-up**, *n.* the act of picking up, esp. at cricket; a person or thing picked up; a casual acquaintance, esp. one made for the purpose of having sexual intercourse; the act of making such an acquaintance; a vehicle with a driver's cab at the front, and an open back with sides and a tailboard; the arm of a record player. **picked**, *a.* gathered, culled; chosen, selected, choice. **picker**, *n.* **pickings**, *n.pl.* gleanings, odds-and-ends; profit or reward, esp. when obtained dishonestly. **picky**, *a. (coll.)* excessively fastidious, choosy.
pick[2] (pik), *n.* a tool with a long iron head, usu. pointed at one end and pointed or having a chisel-edge at the other, fitted in the middle on a wooden shaft, used for breaking ground etc.; one of various implements used for picking.
pick-a-back (pik'əbak), PIGGY-BACK.
pickaxe (pik'äks), *n.* an instrument for breaking ground etc., a pick.
pickerel (pik'ərəl), *n.* a young or small pike.
picket (pik'it), *n.* a pointed stake, post, or peg, forming part of a palisade or paling; a small body of troops posted on the outskirts of a camp etc., as a guard; a person or group posted in a certain place as part of a protest or demonstration, or to dissuade or prevent employees entering an establishment during a labour dispute. *v.t.* to fortify or protect with stakes etc., to fence in; to tether to a picket; to post as a picket; to set a picket or pickets at (the gates of a factory etc.). *v.i.* to act as a picket. **picket-line**, *n.* a group of people picketing a factory etc.
pickle (pik'əl), *n.* a liquid, as brine, vinegar etc., for preserving fish, meat, vegetables etc.; *(often pl.)* vegetables or other food preserved in pickle; *(coll.)* a disagreeable or embarrassing position; *(coll.)* a troublesome child. *v.t.* to preserve in pickle; to treat with pickle; to immerse in diluted acid used for cleaning, etc. **pickled**, *a. (coll.)* drunk.
pickpocket etc. PICK[1].
Pickwickian (pikwik'iən), *a.* relating to or characteristic of Mr Pickwick in Dickens's *Pickwick Papers*; *(facet.)* of the sense of words, merely technical or hypothetical.
picky PICK[1].
picnic (pik'nik), *n.* an outdoor excursion the members of which carry with them provisions; an informal meal, esp. one eaten outside; *(often in neg; coll.)* an easy or pleasant undertaking. *v.i. (past, p.p.* **picnicked**) to go on a picnic. **picnicker**, *n.*
pico-, (pēkō-), *comb. form.* one millionth of a millionth part (10^{-12}).
picot (pēkō), *n.* a small loop of thread forming part of an ornamental edging.
picotee (pikətē'), *n.* a hardy garden variety of the carnation, with a spotted or dark-coloured margin.
picric acid (pik'rik), *n.* an acid obtained by the action of nitric acid on phenol etc., used in dyeing and in certain explosives.
Pict (pikt), *n.* one of a people who anciently inhabited parts of northern Britain. **Pictish**, *a.* of the Picts. *n.* their

pictograph 604 **pigeon**

language.
pictograph (pik'təgraf), **pictogram** (pik'təgram), *n.* a pictorial character or symbol; a diagram showing statistical data in pictorial form. **pictographic** (graf'-), *a.* **pictography** (-tog'-), *n.*
pictorial (piktaw'riəl), *a.* pertaining to, containing, expressed, or illustrated by pictures. *n.* an illustrated journal etc. **pictorially,** *adv.*
picture (pik'chə), *n.* a painting, drawing, photograph, etc. representing a person, natural scenery, or other objects; an image, copy; a vivid description; a perfect example; a beautiful object; a scene, a subject suitable for pictorial representation; a motion picture, a film; the image on a television screen. *v.t.* to represent by painting; to depict vividly; to form a mental likeness of, to imagine vividly. **in the picture,** having all the relevant information. **the pictures,** *n.pl.* (*coll.*) the cinema. **to get the picture,** to understand the situation. **picture-book,** *n.* an illustrated book, esp. one for children. **picture-card,** *n.* a court card. **picture-gallery,** *n.* a gallery or large room in which pictures are exhibited. **picture-hat,** *n.* a lady's hat with wide drooping brim. **picture-house,** *n.* (*dated*) a cinema. **picture postcard,** *n.* a postcard with a picture on the back. *a.* picturesque. **picture window,** *n.* a large window, usu. with a single pane, framing an attractive view. **picture-writing,** *n.* a primitive method of recording events etc., by means of pictorial symbols, as in hieroglyphics, pictography.
picturesque (pikchərəsk', -tyə-), *a.* having those qualities that characterize a good picture, natural or artificial; of language, graphic, vivid. **picturesquely,** *adv.* **picturesqueness,** *n.*
piddle (pid'l), *v.i.* (*coll.*) to urinate. *n.* an act of urinating; urine. **piddling,** *a.* trifling; squeamish.
piddock (pid'ək), *n.* a bivalve mollusc of burrowing genus.
pidgin (pij'in), *n.* a language that is a combination of two or more languages, used esp. for trading between people of different languages. **pidgin English,** pidgin in which one of the languages is English.
pie[1] (pī), *n.* meat, fruit etc., baked in pastry. **pie in the sky,** an unrealistic aspiration. **to have a finger in every pie** FINGER. **pie chart,** *n.* a pictorial representation of relative quantities, in which the quantities are represented by sectors of a circle. **pie-eyed,** *a.* (*coll.*) drunk.
pie[2] (pī), *n.* a confused mass of printers' type; a jumble, disorder, confusion. *v.t.* to mix or confuse (type).
piebald (pī'bawld), *a.* of a horse or other animal, having patches of two different colours, usu. black and white. *n.* a piebald horse or other animal.
piece (pēs), *n.* a distinct part of anything; a detached portion, a fragment (of); a division, a section; a plot or enclosed portion (of land); a definite quantity or portion in which commercial products are made up or sold; an example, an instance; an artistic or literary composition or performance, usu. short; a coin; a gun, a firearm; a man at chess, draughts etc.; (*offensive sl.*) a woman; (*Sc.*) a thick slice of bread with butter, jam, or cheese. *v.t.* to add pieces to, to mend, to patch; to put together so as to form a whole; to join together, to reunite; to fit (on). *v.i.* to come together, to fit (well or ill). **a piece of cake,** something very easy. **a piece of the action,** (*coll.*) active involvement. **in pieces,** broken. **of a piece,** of the same sort, uniform. **piece of eight,** old Spanish dollar of eight *reals*, worth about 22½p. **to go to pieces,** to collapse; to have a breakdown. **to piece on,** to fit on (to). **to say one's piece,** to express one's opinion. **piecemeal,** *adv.* piece by piece, part at a time; in pieces. *a.* fragmentary. **piecework,** *n.* work paid for by the piece or job.
pièce de résistance (pyes də rezis'täs), *n.* an outstanding item; the main dish of a meal. [F]
pie chart PIE[1].
pied (pīd), *a.* parti-coloured, variegated, spotted.

pied-à-terre (pyädəteə'), *n.* a city apartment or other dwelling kept for occasional use.
pie-eyed PIE[2].
pier (piə), *n.* a mass of masonry supporting an arch, the superstructure of a bridge, or other building; a pillar, a column; a solid portion of masonry between windows etc.; a buttress; a breakwater, mole, jetty; a structure projecting into the sea etc., used as a landing-stage, promenade etc. **pier-glass,** *n.* a looking-glass orig. placed between windows; a large ornamental mirror. **pier-table,** *n.* a low table placed between windows.
pierce (piəs), *v.t.* to penetrate or transfix with or as with a pointed instrument; of such an instrument, to penetrate, to transfix, to prick; to make a hole in; to move or affect deeply; to force a way into, to explore; of light, to shine through; of sound, to break (a silence etc.). *v.i.* to penetrate (into, through etc.). **piercing,** *a.* penetrating; affecting deeply. **piercingly,** *adv.*
pierrot (pyə'rō. piə'-), *n.* a buffoon or itinerant minstrel, orig. French and usu. dressed in loose white costume and with the face whitened. **pierrette** (-et'), *n. fem.*
pietà (piətah'), *n.* a pictorial or sculptured representation of the Virgin and the dead Christ.
pietism (pī'ətizm), *n.* a display of strong religious feelings. **pietism,** *n.* **pietistic, -ical** (-tis'-), *a.*
piety (pī'əti), *n.* the quality of being pious; reverence towards God.
piezochemistry (pīēzōkem'istri), *n.* the study of the effect of high pressures on chemical reactions.
piezoelectricity (pīēzōeliktris'iti), **piezoelectric effect** (-lek'-), *n.* a property possessed by some crystals, e.g. those used in gramophone crystal pick-ups, of generating surface electric charges when mechanically strained, they also expand along one axis and contract along another when subjected to an electric field.
piezometer (pīizom'itə), *n.* an instrument for determining the compressibility of liquids or other forms of pressure.
piffle (pif'əl), *v.i.* to talk or act in a feeble, ineffective, or trifling way. *n.* trash, rubbish, twaddle, nonsense. **piffler,** *n.* **piffling,** *a.*
pig (pig), *n.* a swine, a hog, esp. when small or young; the flesh of this, pork; (*coll.*) a greedy, gluttonous, filthy, obstinate, or annoying person; an oblong mass of metal (esp. iron or lead) as run from the furnace. (*sl.*) a police officer; (*sl.*) a very difficult or unpleasant thing; *v.i.* to bring forth pigs; to be huddled together like pigs. *v.t.* (*past, p.p.* **pigged**) to bring forth (pigs); to overindulge oneself in eating. **a pig in a poke,** goods purchased without being seen beforehand. **to make a pig's ear of,** (*sl.*) to make a mess of, to botch. **to pig it,** (*sl.*) to live in squalor; to behave in an unmannerly way. **pig-eyed,** *a.* having small sunken eyes. **pigheaded,** *a.* having a large, ill-shaped head; stupid; stupidly obstinate or perverse. **pigheadedly,** *adv.* **pigheadedness,** *n.* **pig-iron,** *n.* iron in pigs. **pigskin,** *n.* the skin of a pig; leather made from this; (*sl.*) a saddle; *a.* made of this leather. **pigsticking,** *n.* the sport of hunting wild boars with a spear; pig-killing. **pigsticker,** *n.* **pigsty,** *n.* a sty or pen for pigs; a dirty place, a hovel. **pig's wash,** *n.* swill or refuse from kitchens, etc., for feeding pigs. **pig-tail,** *n.* the tail of a pig; the hair of the head tied in a long queue like a pig's tail; tobacco prepared in a long twist. **pigtailed,** *a.* **piggery,** *n.* **piggish,** *a.* like a pig, esp. in greed. **piggishly,** *adv.* **piggishness,** *n.* **piggy,** *n.* a little pig. **piggy-bank,** *n.* a container for saved coins, usu. in the shape of a pig. **piggy-wiggy,** *n.* a little pig; a term of endearment applied to children. **piglet** (-lit), **pigling** (-ling), *n.* **piglike,** *a.*
pigeon (pij'ən), *n.* a bird of the dove family; a greenhorn, a simpleton. *v.t.* to fleece, to swindle, esp. by tricks in gambling. **not my pigeon,** (*coll.*) not my business, not concerning me. **pigeon-breast,** *n.* a deformity in which the breast is constricted and the sternum thrust forward.

pigeon-breasted, *a.* **pigeon-English** PIDGIN-ENGLISH under PIDGIN. **pigeonhole**, *n.* a hole in a dove-cot, by which the pigeons pass in or out; a nesting compartment for pigeons; a compartment in a cabinet etc., for papers, etc.; a category, esp. an over-simplified one. *v.t.* to put away in this; to defer for future consideration, to shelve; to give a definite place to in the mind, to label, to classify. **pigeon-post**, *n.* the conveyance of letters etc., by homing pigeons. **pigeon-toed**, *a.* having the toes turned in.
piggery etc. PIG.
piggy-back (pig'ibak), *adv.* on the back or shoulders, like a pack. *v.t.* to carry (a person) piggy-pack an act of carrying a person piggy-back.
pigment (pig'mənt), *n.* colouring-matter used as paint or dye; a substance which gives colour to animal or vegetable tissues. **pigmental** (-men'-), **pigmentary**, *a.* **pigmentation**, *n.*
pigsty, pigtail PIG.
pika (pi'kə), *n.* a small burrowing mammal related to the rabbit, a native of Asia and N America.
pike¹ (pīk), *n.* a military weapon, consisting of a narrow, elongated lance-head fixed to a pole; a pickaxe, a spike; a peak, a peaked or pointed hill, esp. in the English Lake District; (prob. short for *pike-fish*) a large slender voracious freshwater fish of the genus *Esox*, with a long pointed snout; a diving position in which the legs are straight, the hips bent, and the hands clasp the feet or knees. *v.t.* to run through or kill with a pike. **pikeman**, *n.* a miner working with a pickaxe; a soldier armed with a pike. **pikestaff**, *n.* the wooden shaft of a pike; a pointed stick carried by pilgrims etc. **plain as a pikestaff** [earlier PACKSTAFF], perfectly clear or obvious. **piked**, *a.* pointed, peaked.
pike² (pīk), *n.* a toll-bar; a turnpike road. **pikeman**, *n.* a turnpike-man. **piker**, *n.* a tramp; (*N Am. sl.*) a poor sport; a timid gambler; (*Austral.*) a wild bullock; a trickster, a sharp fellow.
pilaf(f) (pil'af), PILAU.
pilaster (pilas'tə), *n.* a rectangular column engaged in a wall or pier. **pilastered**, *a.*
pilau, pilaw (pilow'), *n.* a mixed dish of Eastern origin consisting of rice boiled with meat, fowl, or fish, together with raisins, spices etc.
pilchard (pil'chəd), *n.* a small sea-fish allied to the herring, and an important food-fish.
pile¹ (pīl), *n.* a heap, a mass of things heaped together; a funeral pyre, a heap of combustibles for burning a dead body; a very large, massive, or lofty building; an accumulation; (*coll.*) a great quantity or sum, a fortune; an atomic pile. *v.t.* to collect or heap up or together, to accumulate; to load; to stack (rifles) with butts on the ground and muzzles together. *v.i.* to move in a crowd. **to pile it on**, (*coll.*) to exaggerate. **to pile up**, to accumulate; to be involved in a pile up. **to pile up the agony** [AGONY]; **pile-up**, (*coll.*) *n.* a crash involving several vehicles. **piler**, *n.*
pile² (pīl), *n.* a sharp stake or post; a column of timber, concrete, steel, etc. driven into the ground, esp. under water, to form a foundation. *v.t.* to drive piles into; to furnish or strengthen with piles. **pile-driver**, *n.* a device for driving piles into the ground.
pile³ (pīl), *n.* soft hair, fur, down, wool; the nap of velvet, plush, or other cloth, or of a carpet.
pile⁴ (pīl), *n.* (*usu. pl.*) small tumours formed by the dilatation of the veins about the anus, haemorrhoids. **pilewort**, *n.* the lesser celandine or figwort supposed to be a remedy for this.
pileate (pi'liət), **-ated** (-iātid), *a.* having a pileus or cap.
pileum (pī'liəm), *n.* the top of the head, from the base of the bill to the nape, in a bird.
pileus (pī'liəs), *n.* (*pl.* **-lei**) in classical antiquity, a brimless felt cap; the cap of a mushroom; the pileum.
pilfer (pil'fə), *v.t.* to steal in small quantities. **pilferage** (-rij), *n.* **pilferer**, *n.* **pilfering**, *n.* **pilferingly**, *adv.*
pilgrim (pil'grim), *n.* one who travels to a distance to visit some holy place, in performance of a vow etc.; a traveller, a wanderer. *v.i.* to go on a pilgrimage; to wander as a pilgrim. **Pilgrim Fathers**, *n.pl.* the English Puritan colonists who sailed in the *Mayflower* to N America, and founded Plymouth, Massachusetts in 1620. **pilgrimage** (-mij), *n.* a pilgrim's journey to some holy place; the journey of human life. *v.i.* to go on a pilgrimage.
piliferous (pilif'ərəs), *a.* bearing hairs. **piliform** (pī'lifawm), *a.*
pill (pil), *n.* a little ball or capsule of some medicinal substance to be swallowed whole; something unpleasant which has to be accepted or put up with; (*sl.*) a black balloting-ball; (*pl.*) billiard balls. *v.i.* (*sl.*) to blackball, to reject. **the pill**, (*coll.*) the contraceptive pill. **to gild, sugar, sweeten the pill**, to make something unpleasant more acceptable. **pill-box**, *n.* a small box for holding pills; a small carriage or building; a concrete blockhouse, used as a machine-gun emplacement or for other defensive purposes; small round brimless hat, formerly part of some military uniforms, now worn esp. by women.
pillage (pil'ij), *n.* the act of plundering; plunder, esp. the property of enemies taken in war. *v.t.* to strip of money or goods by open force; to lay waste. *v.i.* to rob, to ravage, to plunder. **pillager**, *n.*
pillar (pil'ər), *n.* an upright structure of masonry, iron, timber etc., of considerable height in proportion to thickness, used for support, ornament, or as a memorial; a column, a post, a pedestal; an upright mass of anything analogous in form or function; a mass of coal, stone etc., left to support the roof in a mine or quarry; person or body of persons acting as chief support of an institution, movement etc. *v.t.* to support with or as with pillars; to furnish or adorn with pillars. **from pillar to post**, from one place to another; from one difficult situation to another. **pillar-box**, *n.* a short hollow pillar in which letters may be placed for collection by the post office. **pillar-box red**, *a.* vivid red. **pillared**, *a.*
pillau (pilow'), PILAU.
pillion (pil'yən), *n.* a low light saddle for a woman; a cushion for another to sit on behind a person on horseback; a passenger-seat on a motorcycle. *a.*, *adv.* (riding) on a pillion.
pillory (pil'əri), *n.* a wooden frame supported on a pillar and furnished with holes through which the head and hands of a person were put, so as to expose him or her to public derision (abolished in 1837). *v.t.* to set in the pillory; to hold up to ridicule or execration. **pillorize, -ise**, *v.t.*
pillow (pil'ō), *n.* a cushion filled with feathers or other soft material, used as a rest for the head of a person reclining, esp. in bed; a block used as a cushion or support on a machine; anything resembling a pillow in form or function. *v.t.* to lay or rest on a pillow; to prop up with a pillow or pillows. *v.i.* to rest on a pillow. **pillow-case**, **-slip**, a washable cover of linen etc., for drawing over a pillow. **pillow fight**, *n.* a game in which the participants strike each other with pillows. **pillow talk**, *n.* intimate conversation in bed. **pillowy**, *a.*
pilose (pī'lōs), **-lous** (-ləs), *a.* covered with or consisting of hairs. **pilosity** (-los'-), *n.*
pilot (pīlət), *n.* a steersman, esp. one qualified to conduct ships into or out of harbour or along particular coasts, channels etc.; a person directing the course of an aeroplane, spacecraft etc.; a guide, a director, esp. in difficult or dangerous circumstances; a radio or TV programme made to test its suitability to be extended into a series. *a.* serving as a preliminary test or trial. *v.t.* to act as pilot or to direct the course of (esp. a ship, aircraft etc.) **pilot balloon**, *n.* a small, free hydrogen-filled balloon sent up to obtain the direction and velocity of the upper winds.

pilot-fish, *n.* a small sea-fish said to act as a guide to sharks. **pilot-light**, *n.* a small jet of gas kept burning in order to light a cooker, geyser etc. **pilot officer**, *n.* a junior commissioned rank in the RAF corresponding to second lieutenant in the Army. **pilotage** (tij), *n.*
pilous PILOSE.
pilsner (pil'snə), *n.* a pale beer with a strong flavour of hops.
pilule (pil'ūl), *n.* a pill, esp. a small pill. **pilular, pilulous**, *a.*
pilum (pī'ləm), *n.* (*pl.* **-la** (-lə)) the heavy javelin used by the ancient Roman infantry.
pimento (pimen'tō), *n.* the dried unripe aromatic berries of a W Indian tree, allspice; the tree itself. **pimiento** (piměen'tō), *n.* a Spanish sweet red pepper used as a vegetable.
pimp (pimp), *n.* a man who finds customers for a prostitute or lives from her earnings. *v.i.* to act as a pimp.
pimpernel (pim'pənel), *n.* a plant of the genus *Anagallis* belonging to the family *Primulaceae*, esp. the common red pimpernel, a small annual found in sandy fields etc., with scarlet flowers that close in dark or rainy weather.
pimple (pim'pl), *n.* a small pustule, or inflamed swelling on the skin. **pimpled, pimply**, *a.*
pin (pin), *n.* a short, slender, pointed piece of wood, metal etc., used for fastening parts of clothing, papers etc., together; a peg or bolt of metal or wood used for various purposes, as the bolt of a lock, a peg to which the strings of a musical instrument are fastened, a hairpin, a ninepin etc.; an ornamental device with a pin used as a fastening etc., or as a decoration for the person; (*pl.*) (*sl.*) legs; anything of slight value; a badge pinned to clothing. *v.t.* (*past, p.p.* **pinned**) to fasten (to, on, up etc.) with or as with a pin; to pierce, to transfix; to seize, to make fast, to secure; to enclose; to bind (down) to a promise of obligation; to place (the blame) (on). **not to care a pin**, not to care in the slightest. **pins and needles**, a tingling sensation when a limb has been immobile for a long time. **to pin one's faith to, on**, to place full reliance upon. **pinball**, *n.* a game played on a machine with a sloping board down which a ball runs, striking targets and thus accumulating points. **pinball machine**, *n.* a machine for playing this. **pin-case**, *n.* **pincushion**, *n.* a small cushion for sticking pins into. **pin-feather**, *n.* an incipient feather. **pin-head**, *n.* the head of a pin; a very small object; (*coll.*) a very stupid person. **pin-hole**, *n.* a very small aperture; a hole into which a pin or peg fits. **pinhole camera**, *n.* a camera with a pinhole instead of a lens. **pin-money**, *n.* an allowance of money for dress or other private expenses; money earned or saved, esp. by a woman, for personal expenditure. **pinpoint**, *n.* the point of a pin; anything sharp, painful, or critical; *v.t.* to locate accurately and precisely. **pin-prick**, *n.* a prick or minute puncture with or as with a pin; a petty annoyance. *v.t.* to prick with or as with a pin; to molest with petty insults or annoyances. **pin-stripe**, *n.* (a cloth with) a very narrow stripe. **pintail**, *n.* a duck with a pointed tail; applied also to some species of grouse. **pin-up**, *n.* a person, esp. a girl, whose face or figure is considered sufficiently attractive for his or her photograph to be pinned on the wall. **pin-wheel**, *n.* a wheel with pins set in the face instead of cogs in the rim; a small catherine-wheel. **pin-worm**, *n.* a small threadworm.
pinafore (pin'əfaw), *n.* a sleeveless apron worn to protect the front of clothes. **pinafore dress**, *n.* a sleeveless dress usu. worn over a blouse or sweater.
pinaster (pīnas'tə, pin-), *n.* a pine indigenous to the Mediterranean regions of Europe.
pince-nez (pīsnā'), *n.* a pair of eye-glasses held in place by a spring clipping the nose.
pincers (pin'səz), *n.pl.* a tool with two limbs working on a pivot as levers to a pair of jaws, for gripping, crushing, extracting nails etc.; a nipping or grasping organ, as in crustaceans. **pincer movement**, *n.* a military manoeuvre in which one army encloses another on two sides at once.
pinch (pinch), *v.t.* to nip or squeeze, to press so as to cause pain or inconvenience; to take off or remove by nipping or squeezing; to afflict, to distress, esp. with cold, hunger etc.; to straiten, to stint; to extort, to squeeze (from or out of); (*coll.*) to steal, to rob; (*coll.*) to arrest, to take into custody. *v.i.* to nip or squeeze anything; to be niggardly; to be straitened; to cavil. *n.* a sharp nip or squeeze, as with the ends of the fingers; as much as can be taken up between the finger and thumb; a pain, a pang; distress, straits, a dilemma, stress, pressure. **at a pinch**, in an urgent case; if hard pressed. **with a pinch of salt**, SALT. **pinchpenny**, *n.* a niggard; one who stints his or her own and other people's allowances. **pincher**, *n.* **pinchingly**, *adv.* sparingly, stingily.
pinchbeck (pinch'bek), *n.* an alloy of copper, zinc, etc., formerly used for cheap jewellery. *a.* specious and spurious.
pincushion PIN.
pine[1] (pin), *n.* any tree of the coniferous genus *Pinus*, consisting of evergreen trees with needle-shaped leaves; timber from various coniferous trees; a pineapple. *a.* of pines or pine-timber. **pineapple**, *n.* the large multiple fruit of the ananas, so-called from its resemblance to a pine-cone; (*sl.*) a hand-grenade. **pine-marten**, *n.* a European marten, *Mustela martes*. **pine-needle**, *n.* the needle-shaped leaf of the pine. **piny**, *a.*
pine[2] (pin), *v.i.* to languish, waste away; to long or yearn (for etc.).
pineal (pin'iəl, pī'-), *a.* shaped like a pine-cone. **pineal gland**, *n.* a dark-grey conical organ situated behind the third ventricle of the brain that secretes melatonin into the blood stream.
pineapple, etc. PINE [1].
pin-feather, etc. PIN.
pinfold (pin'fōld), *n.* a pound in which stray cattle are shut up; a narrow enclosure. *v.t.* to shut up in a pound.
ping (ping), *n.* a sharp ringing sound as of a bullet flying through the air. *v.i.* to make such a sound; to fly with such a sound.
Ping-pong®, (ping'pong), *n.* table tennis.
pinion[1] (pin'yən), *n.* a wing-feather; a wing; the joint of a bird's wing remotest from the body. *v.t.* to cut off the first joint of the wing to prevent flight; to shackle, to fetter the arms of; to bind (the arms etc.); to bind fast (to).
pinion[2] (pin'yən), *n.* the smaller of two cog-wheels in gear with each other; a cogged spindle or arbor engaging with a wheel.
pink[1] (pingk), *n.* a plant or flower of the genus *Dianthus*, largely cultivated in gardens; applied to several allied or similar plants; a pale rose colour or pale red slightly inclining towards purple, from the garden pink; the supreme excellence, the very height (of); a fox-hunter's scarlet coat; a fox-hunter. *a.* of the colour of the garden pink, pale red or rose; (*coll.*) moderately left-wing. **in the pink**, in fine condition. **tickled pink** TICKLE. **pink elephants**, *n.pl.* hallucinations induced by intoxication with alcohol. **pink-eye**, *n.* a contagious influenza among horses, cattle and sheep, characterized by inflammation of the conjunctiva; a form of conjunctivitis in humans; the herb *Spigelia marilandica* and other N American plants. **pink gin**, *n.* gin mixed with angostura bitters. **pink slip**, *n.* (*N Am.*) a note given to an employee, terminating employment. **pinkness**, *n.* **pinkish**, *a.* **pinkness**, *n.* **pinko** (-ō), *n.* (*pl.* **-kos, -koes**) (*usu. derog.*) a person with (moderately) left-wing views. **pinky**[1], *a.*
pink[2] (pingk), *v.t.* to pierce, to stab; to make small round holes in for ornament; to decorate in this manner. **pinking-iron**, *n.* a tool for pinking. **pinking shears**, *n.pl.* a pair of shears with zig-zag cutting edges, used to cut cloth

to prevent fraying.
pinkie, pinky² (ping'ki), *n.* (*coll.*) the little finger.
pinna (pin'ə), *n.* (*pl.* **-nae, -nas**) a leaflet of a pinnate leaf; a wing a fin, or analogous structure; the projecting upper part of the external ear.
pinnace (pin'is), *n.* a ship's tender; (*loosely*) any small boat.
pinnacle (pin'əkəl), *n.* a turret, usu. pointed or tapering, placed as an ornament on the top of a buttress etc., or as a termination on an angle or gable; a pointed summit; the apex, the culmination (of). *v.t.* to furnish with pinnacles; to set on or as on a pinnacle; to surmount as a pinnacle.
pinnate (pin'ət), **-ated** (-ātid), *a.* having leaflets arranged featherwise along the stem; divided into leaflets; (*Zool.*) having lateral processes along an axis. **pinnately**, *adv.* **pinnation**, *n.*
pinniped (pin'iped), *a.* having feet like fins. *n.* any individual of an order of marine carnivores containing the seals, sea-lions and walruses.
pinnule (pin'ūl), *n.* one of the smaller or ultimate divisions of a pinnate leaf; a small fin, fin-ray, wing, barb of a feather etc. **pinnulate** (-lət), **-lated** (-lātid), *a.*
pinny (pin'i), (*coll.*) PINAFORE.
pinochle, pinocle, penuchle (pē'nŭkl), *n.* a card-game similar to bezique, played with a 48-card pack by two or four players.
pinpoint PIN.
pint (pīnt), *n.* a measure of capacity, the eighth part of a gallon (0.568 l). **pinta** (pīnt'ə), *n.* (*coll.*) a pint of milk. **pint-size(d)**, *n.* (*coll.*) small.
pintail PIN.
pintle (pin'tl), *n.* a pin or bolt, esp. one used as a pivot; one of the pins on which a rudder swings.
pinto (pin'tō), *n.* (*pl.* **-tos, -toes**) a horse or pony with patches of white and another colour.
pin-up PIN.
pin-wheel, etc. PIN.
piny PIN.
piolet (pyōlā'), *n.* a climber's ice-axe.
pion (pī'on), *n.* a meson with positive or negative or no charge, chiefly responsible for nuclear force.
pioneer (pīəniə'), *n.* one of a body of soldiers whose duty it is to clear and repair roads, bridges etc., for troops on the march; one who goes before to prepare or clear the way; an explorer; an early leader. *v.t.* to prepare the way for; to act as pioneer to; to lead, to conduct.
pious (pī'əs), *a.* reverencing a god or gods; religious, devout; feeling or exhibiting filial affection; dutiful; sanctimonious. **pious fraud**, *n.* a deception in the interests of religion or of the person deceived; a sanctimonious hypocrite. **piously**, *adv.*
pip¹ (pip), *n.* the seed of an apple, orange etc. *v.t.* to remove the pips from (fruit). **pipless**, *a.*
pip² (pip), *n.* a disease in poultry etc., consisting of a secretion of thick mucus in the throat; applied facetiously to various human disorders. **to have, get the pip**, (*sl.*) to be out of sorts or dejected.
pip³ (pip), *n.* a spot on a playing-card, domino, die etc.; one of the segments on the rind of a pineapple; a small flower in a clustered inflorescence etc.; a star on an officer's uniform indicating rank.
pip⁴ (pip), *v.t.* (*past, p.p.* **pipped**) (*sl.*) to blackball; to beat; to hit with a shot; to get the better of. **to pip at the post**, to beat, outdo etc. at the last moment, e.g. in a race or contest.
pip⁵ (pip), *v.i.* to chirp, as a bird. *v.t.* to break through (the shell) in hatching. *n.* a short, high-pitched sound.
pipe (pīp), *n.* a long hollow tube or lines of tubes, esp. for conveying liquids, gas etc.; a musical wind-instrument formed of a tube; a tube producing a note of a particular tone in an organ; a tubular organ, vessel, passage etc., in an animal body; the windpipe; the voice, esp. in singing;

a shrill note or cry of a bird etc.; a tube with a bowl for smoking tobacco; a pipeful (of tobacco); a vein containing ore or extraneous matter penetrating rock; (*pl.*) a bagpipe. *v.t.* to play or execute on a pipe; to whistle; to utter in a shrill tone; to lead or bring (along or to) by playing or whistling on a pipe; to furnish with pipes; to propagate (pinks) by slips from the parent stem; to trim or decorate with piping; to convey or transmit along a pipe or wire. *v.i.* to play on a pipe; to whistle, to make a shrill high-pitched sound; (*sl.*) to smoke the drug crack. **pipe of peace**, a pipe smoked in token of peace, a calumet. **to pipe down**, to fall silent. **to pipe up**, to begin to sing; to sing the first notes of; (*coll.*) to begin to speak. **pipe-clay**, *n.* a fine, white, plastic clay used for making tobacco-pipes, and for cleaning military accoutrements etc. *v.t.* to whiten with pipe-clay. **piped music**, *n.* music recorded for playing in shops, restaurants, etc. as background music. **pipe-dream**, *n.* a fantastic notion, a castle in the air. **pipe-fish**, *n.* a fish of the family Syngnathidae, from their elongated form. **pipe-line**, *n.* a long pipe or conduit laid down from an oil-well, or oil region, to convey the petroleum to a port etc. **in the pipeline**, under preparation, soon to be supplied, produced etc. **pipe-major**, *n.* a non-commissioned officer in charge of pipers. **pipe-rack**, *n.* a stand for tobacco-pipes. **pipeful**, *n.* **pipeless**, *a.* **piper**, *n.* one who plays upon a pipe, esp. a strolling player or a performer on the bagpipes. **to pay the piper** [PAY¹].
piperaceous (pipərā'shəs), *a.* pertaining to or derived from pepper. **piperine** (pip'ərin, -rēn, -rin), *n.* an alkaloid obtained from black pepper.
pipette (pipet'), *n.* a fine tube for removing quantities of a fluid, esp. in chemical investigations.
piping (pī'ping), *n.* the action of one who pipes; a shrill whistling or wailing sound; a fluting; a covered cord for trimming dresses; a cord-like decoration of sugar etc., on a cake; a quantity, series, or system of pipes. *a.* playing upon a pipe; shrill, whistling. **piping hot**, *a.* hissing hot; fresh, newly out.
pipistrelle (pip'istrel), *n.* a small, reddish-brown bat, the commonest British kind.
pipit (pip'it), *n.* a lark-like bird belonging to the genus *Anthus*.
pipkin (pip'kin), *n.* a small earthen pot, pan, or jar.
pippin (pip'in), *n.* a name for several varieties of apples.
pipsqueak (pip'skwēk), *n.* (*coll.*) a small, contemptible or insignificant person.
piquant (pē'kənt), *a.* having an agreeably sharp, pungent taste; interesting, stimulating, racy, lively, sparkling. **piquancy**, *n.* **piquantly**, *adv.*
pique (pēk), *v.t.* to irritate; to touch the envy, jealousy, or pride of; to stimulate or excite (curiosity etc.); to plume or value (oneself on). *n.* ill-feeling, irritation, resentment.
piqué (pē'kā), *n.* a heavy cotton fabric with a corded surface, quilting.
piquet (piket'), *n.* a game of cards for two persons, with a pack of cards from which all below the seven have been withdrawn.
piracy (pī'rəsi), *n.* the crime of a pirate; robbery on the seas; unauthorized publication; infringement of copyright.
piragua (pirag'wə), *n.* a long narrow boat or canoe made from one or two trunks hollowed out, a dug-out; a pirogue.
piranha (pirah'nə), *n.* a small, voracious, flesh-eating S American tropical fish.
pirate (pī'rət), *n.* a robber on the seas; a piratical ship; a marauder; one who infringes the copyright of another; an unauthorized radio station. *v.i.* to practise piracy. *v.t.* to plunder; to publish (literary or other matter belonging to others) without permission or compensation. **piratic, -ical** (-rat'-), *a.* **piratically**, *adv.*
pirogue (pirōg'), *n.* a large canoe formed of a hollowed trunk of a tree; a large flat-bottomed boat or barge for

shallow water, usu. with two masts rigged fore-and-aft.
pirouette (piruet'), *n*. a rapid whirling round on the point of one foot, in dancing; a sudden short turn of a horse. *v.i.* to dance or perform a pirouette.
piscatory (pis'kətəri), *a*. pertaining to fishers or fishing.
piscatorial (-taw'ri-), *a*. piscatory; fond of or pertaining to angling.
Pisces (pīs'ēz), *n.pl.* the Fishes, the 12th sign of the zodiac.
pisciculture (pis'ikŭlchə), *n*. the artificial breeding, rearing, and preserving of fish. **piscicultural** (-kŭl'-), *a*. **pisciculturist** (-kŭl'-), *n*.
pisc(i)-, *comb. form*. fish.
piscina (pisē'nə), *n*. (*pl*. **-nas, -nae** (-nē)) a stone basin with outlet beside the altar in some churches to receive the water used in purifying the chalice etc.; in Roman antiquity, a fish-pond; a bathing pond.
piscine (pis'īn), *a*. of or pertaining to fish.
piscivorous (pisiv'ərəs), *a*. living on fish.
pish (pish), *int*. an exclamation expressing contempt, disgust etc. *v.i.* to express contempt by saying 'pish'.
pisiform (pī'sifawm, piz'-), *a*. pea-shaped.
pismire (pis'mīə), *n*. (*dated*) an ant.
piss (pis), *n*. (*taboo sl.*) urine. *v.i.* to discharge urine. *v.t.* to discharge in the urine; to wet with urine. **piss artist**, *n*. (*sl.*) a habitual heavy drinker; a drunk. **to piss down**, (*sl.*) to rain heavily. **to piss off**, (*sl.*) to go away; to annoy; to bore, to make discontented. **to take the piss**, (*sl.*) to make fun of someone, to tease. **piss-up**, *n*. (*sl.*) a bout of drinking. **pissed**, (*sl.*) drunk.
pissoir (pēs'wah), *n*. a public urinal.
pistachio (pistah'shio), *n*. the nut of a W Asiatic tree, *Pistacia vera*, with a pale greenish kernel; the flavour of this. **pistachio-nut**, *n*.
piste (pēst), *n*. a slope prepared for skiing; a rectangular area on which a fencing contest is held.
pistil (pis'til), *n*. the female organ in flowering plants, comprising the ovary and stigma, usu. with a style supporting the latter.
pistol (pis'təl), *n*. a small firearm for use with one hand. *v.t.* (*past, p.p.*) **pistolled**) to shoot with a pistol. **pistol-shot**, *n*. the range of this. **pistol-whip**, *v.t.* to strike with a pistol.
piston (pis'tən), *n*. a device fitted to occupy the sectional area of a tube and be driven to and fro by alternating pressure on its faces, so as to impart or receive motion, as in a steam-engine or a pump; a valve in a musical wind-instrument; in an internal combustion engine, a plunger which passes on the working pressure of the burning gases via the connecting rod to the crankshaft. **piston-ring**, *n*. a split ring encircling the piston in a groove. **piston-rod**, *n*. a rod attaching a piston to machinery.
pit[1] (pit), *n*. a natural or artificial hole in the ground, esp. one of considerable depth in proportion to its width; one made in order to obtain minerals or for industrial or agricultural operations; a coal mine; a hole dug and covered over as a trap for wild animals or enemies; an abyss; hell; a hollow or depression in the surface of the ground, of the body etc.; a hollow scar, esp. one left by smallpox; the ground floor of the auditorium in a theatre, esp. behind the stalls; the part of an audience occupying this; (*pl.*) an area on a motor-racing course where cars are repaired, their tyres are changed etc.; an area for cockfighting, a cockpit; a trap, a snare. *v.t.* (*past, p.p.*) **pitted**) to put into a pit, esp. for storage; to mark with pits or hollow scars, as with smallpox (*usu. in p.p.*); to match (against) in a pit; to match, to set in competition (against). **the pits**, (*sl.*) a very unpleasant person, thing, place or situation. **pitfall**, *n*. a pit slightly covered so that animals may fall in; a trap; (*fig.*) a hidden danger. **pit-head**, *n*. (the area or buildings near) the top of a mineshaft. **pitman**, *n*. one who works in a pit, a collier. **pit-stop**, *n*. a stop by a racing car in the pits for tyre-changes etc. **pitting**, *n*. the uneven wearing of valve-seatings and other surfaces in an internal combustion engine.
pit[2] *pit*), *n*. (*N Am.*) the stone of a fruit. *v.t.* to remove the pit from (fruit). [PITH]
pita (bread) PITTA.
pit-a-pat (pit'əpat), *n*. a tapping, a flutter, a palpitation. *adv*. with this sound, palpitatingly, falteringly.
pitch[1] (pich), *n*. a dark-brown or black resinous substance obtained from tar, turpentine, and some oils, used for caulking, paving roads etc. *v.t.* to cover, coat, line, or smear with pitch. **pitch-black**, *a*. brownish black; as dark as pitch. **pitchblende** (-blend), *n*. native oxide of uranium, the chief source of radium. **pitch-dark**, *a*. as dark as pitch, very dark. **pitch-darkness**, *n*.
pitch[2] (pich), *v.t.* to fix or plant in the ground; to fix; to set in orderly arrangement, to fix in position; to throw, to fling, esp. with an upward heave or underhand swing; to toss (hay) with a fork; to pave with cobbles or setts; to expose for sale; to set to a particular pitch or keynote; (*coll.*) to put or relate in a particular way; (*Baseball etc.*) to deliver or throw (the ball) to the batsman; (*Golf*) to strike (the ball) with a lofted club. *v.i.* to encamp; to light, to settle; to plunge, to fall; of a ball, esp in cricket, to bounce; to fall headlong; to plunge at the bow or stern, as opposed to *rolling*. *n*. the act of pitching; mode of pitching; the delivery of the ball in various games; height, degree, intensity; extreme height, extreme point; point or degree of elevation or depression; degree of inclination or steepness; degree of slope in a roof; the place or station taken up by a person for buying and selling, residence etc.; an attempt at persuasion, usu. to induce someone to buy something; (*Cricket*) the place in which the wickets are placed or the distance between them; (*Cricket*) the point at which a bowled ball bounces; any area marked out for playing sports, e.g. football; the lineal distance between points etc., arranged in series, as between teeth on the pitchline of a cog-wheel, between floats on a paddle-wheel, between successive convolutions of the thread of a screw etc.; (*Mus.*) the degree of acuteness or gravity of a tone. **to pitch in**, to begin or set to vigorously; (*coll.*) to participate or contribute. **to pitch into**, (*coll.*) to assail with blows, abuse, etc.; to attack vigorously. **to queer the pitch**, to spoil a plan, to thwart. **pitch-and-toss**, *n*. a game in which coins are pitched at a mark, the player getting nearest having the right to toss all the others' coins into the air and take those that come down with heads up. **pitched battle**, *n*. a battle for which both sides have made deliberate preparations. **pitchfork**, *n*. a large fork, usu. with two prongs, with a long handle, used for lifting hay sheaves of corn etc. *v.t.* to lift or throw with or as with a pitch-fork; to place unexpectedly or unwillingly in a certain situation. **pitchpipe**, *n*. a small pipe for sounding with the mouth to set the pitch for singing or tuning. **pitchblende, pitch-cap** etc. PITCH[1].
pitcher[1] (pich'ə), *n*. one who or that which pitches; a player delivering the ball in baseball and other games.
pitcher[2] (pich'ə), *n*. a large vessel, usu. of earthenware, with a handle and a spout, for holding liquids; (*N Am.*) a jug; a pitcher-shaped leaf, usu. closed with an operculum. **pitcher-plant**, *n*. one of various insectivorous plants with such leaves.
pitchy (pich'i), *a*. of the nature of or like pitch; dark, dismal. **pitchiness**, *n*.
piteous (pit'iəs), *a*. exciting or deserving pity; lamentable, sad, mournful; †compassionate; †mean, pitiful. **piteously**, *adv*. **piteousness**, *n*.
pitfall PIT[1].
pith (pith), *n*. a cellular spongy substance occupying the middle of a stem or shoot in dicotyledonous plants; the soft, white tissue under the skin of a lemon, grapefruit etc.; the spinal cord; the essence, the essential part, the

main substance; strength, vigour, energy; cogency, point; importance. *v.t.* to remove the pith of; to sever the spinal cord of; to kill in this way. **pith hat, helmet,** *n.* a lightweight sun-hat made of pith. **pithy,** *a.* consisting of, like, or abounding in pith; forcible, energetic; condensed, sententious. **pithily,** *adv.* **pithiness,** *n.*
pithead PIT [1].
pithecanthrope (pithiəkan'thrōp), *n.* former name for *Homo erectus*, an extinct human species of which remains have been found in Java and elsewhere.
pitiable, pitiful, pitiless, etc. PITY.
piton (pē'ton), *n.* a metal spike used for fixing ropes on precipitous mountain-sides etc.; a peak, a cone.
pitpan (pit'pan), *n.* a narrow, long, flat-bottomed dug-out canoe, used in Central America.
pitta (bread), pita (pit'ə), *n.* a flat, round, slightly leavened bread, hollow inside so that it can be filled with food.
pittance (pit'əns), *n.* orig. a gift or bequest to a religious house for food etc.; a dole, an allowance, esp. of a meagre amount.
pitted PIT [2].
pitter-patter (pitəpat'ə), PIT-A-PAT.
pituitary (pitū'itəri), *a.* containing or secreting phlegm, mucus; of the pituitary gland. **pituitary gland,** *n.* a small gland attached by a pedicle to the base of the brain, secreting hormones which regulate growth, the production of other hormones etc.
pity (pit'i), *n.* a feeling of grief or tenderness aroused by the sufferings or distress of others, compassion; a subject for pity, a cause of regret, an unfortunate fact. *v.t.* to feel pity for. *v.i.* to be compassionate. **more's the pity,** it is unfortunate. **to take pity on,** to be compassionate; to act compassionately towards. **what a pity,** how unfortunate! **pitiable** (pit'iəbl), *a.* deserving or calling for pity; piteous. **pitiableness,** *n.* **pitiably,** *adv.* **pitiful,** *a.* full of pity, compassionate; calling for pity; pitiable, contemptible. **pitifully,** *adv.* **pitifulness,** *n.* **pitiless,** *a.* destitute of pity; merciless, unfeeling, hard-hearted. **pitilessly,** *adv.* **pitilessness,** *n.* **pityingly,** *adv.*
pityriasis (pitirī'əsis), *n.* squamous inflammation of the skin, dandruff.
piu, *adv.* (*Mus.*) more.
pivot (piv'ət), *n.* a pin, shaft, or bearing on which anything turns or oscillates; of a body of troops, a soldier at the flank on whom a company wheels; a thing or event on which an important issue depends. *v.i.* to turn on or as on a pivot; to hinge (upon) *v.t.* to place on or provide with a pivot. **pivotal,** *a.* of a pivot; of crucial importance, critical.
pix PIC.
pixel *pik'sl*), *n.* one of the minute units which together form an image, e.g. on a computer monitor.
pixilated, pixillated (piksilātid), *a.* (*N Am.*) mentally unbalanced; eccentric.
pixy, pixie (pik'si), *n.* a supernatural being akin to a fairy or elf.
pizza (pēt'sə), *n.* a flat, round piece of baked dough covered usu. with cheese and tomatoes, and also often with anchovies, mushrooms, slices of sausage etc. **pizzeria** (-rē'ə), *n.* a place where pizzas are made or sold.
pizzazz, pizazz (pizaz'), *n.* (*esp. N Am., coll.*) vigour, élan, panache, glamour.
pizzicato (pitsikah'tō), *a.* (*Mus.*) played by plucking the strings (of a violin etc.) with the fingers. *adv.* in this manner. *n.* a passage or work so played.
pizzle (piz'l), *n.* the penis of some quadrupeds, esp. a bull.
pl., (*abbr.*) place, plate, plural.
placable (plak'əbl), *a.* that may be appeased; ready to forgive, mild, complacent. **placability** (-bil'-), **placably,** *adv.*
placard (plak'ahd), *n.* a written or printed paper or bill posted up in a public place, a poster. *v.t.* to post placards on; to announce or advertise by placards; to display as a placard.
placate (pləkāt'), *v.t.* to appease, to pacify, to conciliate. **placation,** *n.* **placatory,** *a.* tending to placate.
place (plās), *n.* a particular portion of space; a spot, a locality; a city, a town, a village; a residence, an abode; a building, esp. as devoted to some particular purpose; a residence with its surroundings, esp. in the country; a fortified post; an open space in a town; a passage in a book etc.; position in a definite order, as of a figure in a relation to others in a series or group; a stage or step in an argument, statement etc.; a suitable juncture; stead, lieu; space, room for a person; rank, station in life, official position; situation, employment, appointment, esp. under government; a vacancy, e.g. for a student at a university; duty, sphere, province; a position among the competitors that have been placed. *v.t.* to put or set in a particular place; to put, to set, to fix; to arrange in proper places; to identify; to assign to class; to put in office, to appoint to a post; to find an appointment, situation, or living for; to put out at interest, to invest, to lend; to dispose of (goods) to a customer; to arrange (esp. a bet); to set or fix (confidence etc., in or on); the assign a definite date, position etc., to, to locate; in racing, to indicate the position of (a horse etc.), usu. among the first three passing the winning-post; to get a goal by a place-kick. **in place,** suitable, appropriate. **in place of,** instead of. **out of place,** unsuitable, inappropriate. **to go places,** to be successful. **to put in one's place,** to humiliate someone who is (regarded as) arrogant, presumptuous etc. **to take place,** to come to pass, to occur. **to take the place of,** to be substituted for. **place-kick,** *n.* (*Football*) a kick after the ball has been placed for the purpose by another player. **place-mat,** a table-mat. **place setting,** *n.* the plate, knife, fork etc. set for one person at a table.
placebo (pləsē'bō), *n.* (*pl.* **-bos, -boes**) in the Roman Catholic Church, the first antiphon in the vespers for the dead; a medicine having no physiological action, given to humour the patient or to provide psychological comfort, or as a control during experiments to test the efficacy of a genuine medicine.
placenta (pləsen'tə), *n.* (*pl.* **-tas, -tae** (-tē)) the organ by which the foetus is nourished in the higher mammals; in plants, the part of the ovary to which the ovules are attached. **placental** *n.*, *a.*
placer (plas'ə, plā'-), *n.* a place where deposits are washed for minerals; an alluvial or other deposit containing valuable minerals; any mineral deposits not classed as veins.
placet (plā'set), *n.* permission, assent, sanction.
placid (plas'id), *a.* gentle, quiet; calm, peaceful, serene, unruffled. **placidity** (-sid'-), **placidness,** *n.* **placidly,** *adv.*
placket (plak'it), *n.* the opening or slit in a petticoat or skirt; a woman's pocket; a petticoat.
placoid (plak'oid), *a.* of fish scales, plate-shaped, *n.* one of a group of fish with plate-like scales.
plafond (pla'fō), *n.* a ceiling, esp. one of a richly decorated kind.
plagal (plā'gəl), *a.* of the Gregorian modes, having the principal notes between the dominant and its octave; denoting the cadence formed when a subdominant chord immediately precedes the final tonic chord.
plagiarize, -ise (plā'jəriz), *v.t.* to appropriate and give out as one's own (the writings, inventions, or ideas of another). **plagiarism,** *n.* **plagiarist,** *n.*
plague (plāg), *n.* a blow, a calamity, an affliction; a pestilence, an intensely malignant epidemic, esp. the bubonic or pneumonic forms of infection by *Pasteurella pestis;* a nuisance, a trouble. *v.t.* to visit with plague; to afflict with any calamity or evil; to vex, to tease; to annoy.
plaice (plās), *n.* a flat-fish much used for food.
plaid (plad, pläd), *n.* a long rectangular outer garment of woollen cloth, usu. with a checkered or tartan pattern, worn by Scottish Highlanders; plaiding. *a.* like a plaid in

pattern. **plaided**, *a.* wearing a plaid; made of plaid cloth.
plaiding, *n.* cloth for making plaids.
plain¹ (plān), *a.* clear, evident, manifest; simple, free from difficulties; easily seen, easy to understand; not intricate; of knitting, consisting of plain stitches; devoid of ornament; unvariegated, uncoloured; not luxurious, not seasoned highly; of flour, having no raising agent; homely, unaffected, unsophisticated; straightforward, sincere, frank; direct, outspoken; ugly. *adv.* plainly; totally, utterly. *n.* a tract of level country. **plain-chant** PLAINSONG. **plain chocolate**, *n.* dark chocolate with a slightly bitter flavour. **plain clothes**, *n.pl.* private clothes, as opp. to uniform, mufti. **plain-clothes**, *a.* wearing such clothes. **plain-dealer**, *n.* one who speaks his or her mind plainly. **plain-dealing**, *n.*, *a.* **plain-hearted**, *a.* sincere; free from hypocrisy. **plain Jane**, *n.* (*coll.*) an unattractive woman or girl. **plain sailing** [cp. **plane sailing** under PLANE³], *n.* a simple course of action. **plainsong**, *n.* a variety of vocal music according to the ecclesiastical modes of the Middle Ages, governed as to time not by metre but by word-accent, and sung in unison. **plain-spoken**, *a.* speaking or said plainly and without reserve. **plain stitch**, *n.* a simple stitch in knitting, in which a loop is made by passing wool round the right-hand needle and pulling it through a loop on the left-hand needle. **plainly**, *adv.* **plainness**, *n.*
plain² (plān), *v.i.* to mourn, to lament, to complain; to make a mournful sound.
plaint (plānt), *n.* an accusation, a charge; (*poet.*) a lamentation, a mournful song.
plaintiff (plān'tif), *n.* one who brings a suit against another in a court of law.
plaintive (plān'tiv), *a.* expressive of sorrow or grief. **plaintively**, *adv.* **plaintiveness**, *n.*
plait (plat), *n.* a braid of several strands of hair, straw, twine etc., esp. a braided tress of hair; a flat fold, a doubling over, as of cloth, a pleat. *v.t.* to braid, to form into a plait or plaits; to fold.
plan (plan), *n.* a delineation of a building, machine etc., by projection on a plane surface, usu. showing the relative positions of the parts on one floor or level; a map of a town, estate, on a large scale; a scheme; a project, a design; an outline of a discourse, sermon etc.; method of procedure; habitual method, way, custom. *v.t.* to draw a plan of; to design; to contrive, to scheme, to devise. *v.i.* to make plans. **planner**, *n.* **planning**, *n.* the making of plans, esp. the laying down of economic, social etc. goals and the means of achieving them, or the allocation of land for specific purposes. **planning permission**, *n.* official permission from a local authority etc. to erect or convert a building or change its use.
planar PLANE³.
planarian (plənea'riən), *a.* belonging to the genus of minute, flat, aquatic worms found in salt or fresh water and in moist places. *n.* a flatworm.
planchet (plahn'chit), *n.* a disc of metal for making into a coin.
planchette (plahnshet'), *n.* a small, usu. heart-shaped, board resting on wheels, and a pencil which makes marks as the board moves under the hands of the person resting upon it, believed by spiritualists to be a mode of communicating with the spirit world.
Planck's constant (planks), *n.* a constant (*h*) which expresses the ratio of a quantum of energy to its frequency.
plane¹ (plān), *n.* a tree of various species of the genus *Platanus*, consisting of large spreading branches with broad angular leaves palmately lobed. **plane-tree**, *n.*
plane² (plān), *n.* a tool for smoothing boards and other surfaces. *v.t.* to smooth or dress with a plane; to make flat and even; to remove (away) or pare (down) irregularities.
plane³ (plān), *a.* level, flat, without depressions or elevations; lying or extending in a plane. *n.* a surface such that a straight line joining any two points in it lies wholly within it; such a surface imagined to extend to an indefinite distance, forming the locus for certain points or lines; a level surface; an even surface extending uniformly in some direction; one of the natural faces of a crystal; a main road in a mine; an imaginary surface for determining points in a drawing; level (of thought, existence etc.). *v.i.* to glide, to soar; to skim across water. **plane geometry**, *n.* the geometry of plane figures. **plane sailing**, *n.* the art of determining a ship's position on the supposition that she is moving on a plane; plain sailing, a simple course of action. **plane-table**, *n.* a surveying instrument marked off into degrees from the centre for measuring angles in mapping. *v.t.* to survey with this. **planar** (-nə), *a.*
plane⁴ (plān), *n.* an aeroplane; one of the thin horizontal structures used as wings to sustain an aeroplane in flight.
planet (plan'it), *n.* a heavenly body revolving round the sun, either as a primary planet in a nearly circular orbit or as a secondary planet or satellite revolving round a primary; in ancient astronomy, one of the major planets, Mercury, Venus, Mars, Jupiter, Saturn, together with the sun and moon, distinguished from other heavenly bodies as having an apparent motion of its own. **planetarium** (-teə'riəm), *n.* an apparatus for exhibiting the motions of the planets, an orrery; a building in which this is exhibited on a large scale. **planetary**, *a.* pertaining to the planets or the planetary system. **planetesimal** (- tes'iməl), *n.* a small body of matter in solar orbit existing at an earlier stage in the formation of the solar system. **planetoid** (-toid'), *n.*, *a.* **planetoidal** (-toi'-), *a.*
plane-table PLANE³.
plane-tree PLANE¹.
plangent (plan'jənt), *a.* sounding noisily; resounding sorrowfully. **plangency**, *a.*
planimeter (plənim'itə), *n.* an instrument for measuring the area of an irregular plane surface. **planimetry**, *n.*
planish (plan'ish), *v.t.* to flatten, smooth, or toughen (metal) by hammering or similar means; to reduce in thickness by rolling; to polish (metal plates, photographs etc.) by rolling; to polish by hammering.
planisphere (plan'isfiə), *n.* a plane projection of a sphere, esp. of part of the celestial sphere. **planispheric** (-sfe'-), *a.*
plank (plangk), *n.* a long piece of sawn timber; an article or principle of a political programme. *v.t.* to cover or lay with planks; (*coll.*) to lay down (money, etc.) as if on a board or table. **to walk the plank**, to be compelled to walk blindfold along a plank thrust over a ship's side (a pirates' mode of putting to death). **planking**, *n.*
plankton (plangk'tən), *n.* fauna and flora, esp. minute animals and plants or those of low organization, floating at any level of a sea or lake.
plano-, *comb. form.* flat, level.
plano-concave (plā'nōkon'kāv), *a.* plane on one side and concave the other.
plano-convex (plā'nōkon'veks), *a.* plane on one side and convex on the other.
plano-horizontal (plānōhorizon'təl), *a.* having a level horizontal surface or position.
plant (plahnt), *n.* any vegetable organism, usu. one of the smaller plants distinguished from shrubs and trees; a sapling, a shoot, a slip, a cutting; a growth or crop of something planted; the tools, machinery, apparatus, and fixtures used in an industrial concern; mobile mechanical equipment used for earth-moving, road-building etc.; a factory; any place where an industrial process is carried on; a person or thing used to entrap another, esp. an article secretly left so as to be found in a person's possession and provide incriminating evidence. *v.t.* to set in the ground for growth; to put (young fish, spawn etc.) into a river etc.; to furnish or lay out with plants; to fix firmly, to station; to settle, to found, to introduce, to establish;

plantain 611 **plate**

to implant (an idea etc.); to aim and deliver (a blow etc.); to put into position secretly in order to observe, deceive or entrap. *v.i.* to sow seed; to perform the act of planting. **to plant out**, to transplant (seedlings) from pots etc. to open ground. **plantation**, *n.* a large quantity of trees or growing plants that have been planted; a growing wood, a grove; a large estate for the cultivation of sugar, cotton, coffee, etc.; the act of planting; a colony or settlement, settling of colonists, colonization. **planter**, *n.* one who plants; an implement or machine for planting; one who owns or works a plantation; a settler in a colony; an ornamental pot for plants; (*sl.*) a well-directed blow. **plantlike**, *a.*
plantain[1] (plan'tin), *n.* any plant of the genus *Plantago*, esp. *P. major*, a low perennial weed with broad flat leaves and a spike of dull green flowers.
plantain[2] (plan'tin), *n.* a tropical American herbaceous tree closely akin to the banana, and bearing similar fruit; its fruit.
plantar (planta'), *a.* pertaining to the sole of the foot.
plantation, planter PLANT.
plantigrade (plan'tigrād), *a.* walking on the sole of the foot; of or pertaining to the Plantigrada, a section of the Carnivora embracing the bears, badgers etc. *n.* a plantigrade animal.
planxty (plangk'sti), *n.* (*Ir.*) a melody of a sportive and animated character for the harp.
plaque (plahk), *n.* a plate, slab, or tablet, of metal, porcelain, ivory etc., usu. of an artistic or ornamental character; a small plate worn as a badge or personal ornament; a patch or spot on the surface of the body; a filmy deposit on the surface of the teeth consisting of mucus and bacteria.
-plasm (plazm), *comb. form.* (*Biol.*) denoting material forming cells.
plasma (-mə), **plasm**, *n.* the viscous living matter of a cell, protoplasm; the fluid part of the blood, lymph, or milk; sterilized blood plasma used for transfusions; a hot, ionized gas containing approximately equal numbers of positive ions and electrons; a green variety of quartz allied to chalcedony. **plasma torch**, *n.* a device in which gas is converted to a plasma by being heated electrically and which is used for melting metals etc. **plasmatic** (-mat'-), **plasmic**, *a.* **plasmid** (-mid), *n.* a small circle of DNA found esp. in bacteria, which exists and replicates itself independently of the main bacterial chromosome. *a.* **plasmin** (-min), *n.* an enzyme in blood plasma that dissolves fibrin.
plasmodium (plazmō'diəm), *n.* (*pl.* **dia** (-ə)) a mass of mobile, naked protoplasm resulting from the fusion or aggregation of numerous amoeboid cells, as in the vegetative stage of Myxomycetes and Mycetozoa; a genus of Protozoa found in the blood in malaria. **plasmodial** (-mod'-), *a.*
plasmolysis (plazmol'isis), *n.* the contraction of the protoplasm in active cells under the influence of a reagent or of disease.
-plast, *comb. form.* a living cell or subcellular particle.
plaster (plahs'tə), *n.* a mixture of lime, sand etc., for coating walls etc.; white powder which sets hard when mixed with water, for coating or for moulding into ornaments, figures etc.; an adhesive application of some curative substance, usu. spread on linen, muslin, or a similar fabric, placed on parts of the body; a strip of sticking plaster; a surgical plaster cast. *a.* made of plaster. *v.t.* to cover or overlay with plaster or other adhesive substance; to apply a plaster to (a wound etc.); to daub, to smear over, to smooth over; to cause to lie flat or adhere to; to cover with excessive quantities of, to use excessively and/or tastelessly; to stick (on) as with plaster; (*coll.*) to inflict heavy damage, injury, or casualties on. **plaster of Paris**, *n.* a white powder, orig. made from gypsum, esp. used for making casts of statuary etc. **plasterboard**, *n.* (a thin, rigid board consisting of) a layer of plaster compressed between sheets of fibreboard, used in making partition walls, ceilings etc. **plaster cast**, *n.* a plaster copy, made from a mould, of any object, esp. a statue; a covering of plaster of Paris used to immobilize and protect e.g. a broken limb. **plastered**, *a.* (*coll.*) drunk. **plasterer**, *n.* **plastering**, *n.* the act of coating or treating with plaster; a covering or coat of plaster.
plastic (plas'tik), *a.* having the power of giving form or fashion; capable of being modelled or moulded; pertaining to or produced by modelling or moulding; continuously extensible or pliable without rupturing; capable of adapting to varying conditions; formative, causing growth; forming living tissue; made of plastic; (*coll., derog.*) outwardly and conventionally attractive but lacking substance or reality; synthetic, insincere. *n.* any of a group of usu. synthetic, polymeric substances which, though stable in use at ordinary temperatures, are plastic at some stage in their manufacture and can be shaped by the application of heat and pressure; (*coll.*) a credit card, or credit cards collectively. **plastic art**, *n.* art which is concerned with moulding or shaping or representation in three dimensions; any visual art. **plastic bullet**, *n.* a cylinder of plastic approximately 4 in. (10 cm) long, less lethal than an ordinary bullet, used mainly for riot control. **plastic explosive**, *n.* an adhesive, jelly-like explosive substance. **plastic surgery**, *n.* the branch of surgery concerned with the restoration of lost, deformed, or disfigured parts of the body or with the cosmetic improvement of any feature. **plastically**, *adv.* **plasticity** (-tis'-), *n.* the state of being plastic; (apparent) three-dimensionality. **plasticize, -ise** (-siz), *v.t.* **plasticizer**, *n.* a substance which renders rubber, plastic, etc. more flexible. **plasticization**, *n.*
Plasticine® (plas'tisēn), *n.* a modelling substance used by children.
plastid (plas'tid), *n.* a small particle in the cells of plants and some animals containing pigment, starch, protein etc.
plastron (plas'tron), *n.* a padded leather shield worn by fencers to protect the breast; an ornamental front to a woman's dress; a shirtfront; the under part of the buckler of a tortoise or turtle, an analogous part in other animals. **plastral**, *a.*
-plasty, *comb. form.* formation or replacement by plastic surgery.
plat[1] (plat), *n.* a small plot, patch, or piece of ground; a small bed (of flowering-plants etc.); a map, a chart.
plat[2] (plat) PLAIT.
platan (plat'ən), *n.* a plane-tree.
plate (plāt), *n.* a flat, thin piece of metal etc., usu. rigid and uniform in thickness; a very thin coating of one metal upon another; a flat, rigid layer of bone, horn etc. forming part of an animal's body or shell; a huge plate-like section of the earth's crust; a piece of metal with an inscription for attaching to an object; a piece of metal used for engraving; a print taken from this; a sheet of glass or other material coated with a sensitized film for photography; an electrotype or stereotype cast of a page of type, to be used for printing; a whole-page illustration separately inserted into book and often on different paper from the text; a type of armour; a device for straightening teeth; the plastic base of a denture, fitting the gums and holding the artificial teeth; a horizontal timber laid on a wall as base for framing; a small shallow vessel, now usu. of crockery, for eating from; a plateful; the contents of a plate, a portion served on a plate; any shallow receptacle esp. for taking a collection in church; (*collect.*) domestic utensils, as spoons, forks, knives, cups, dishes etc., of gold, silver, or other metal; plated ware; a cup or other article of gold or silver offered as a prize in a race etc.; a race for such a prize. *v.t.* to cover or overlay with plates of metal for defence, ornament, etc.; to coat with a layer

of metal, esp. gold, silver, or tin; to beat into thin plates; to make an electrotype or stereotype from (type). **handed to one on a plate**, (*coll.*) obtained without effort. **to have on one's plate**, (*coll.*) to have waiting to be done, to be burdened with. **plate-glass**, *n.* a superior kind of glass made in thick sheets, used for mirrors, large windows etc. **plate-layer**, *n.* one who fixes or repairs railway track. **plate-rack**, *n.* a frame for holding plates and dishes. **plate tectonics**, *n.sing.* (the study of the earth's crust based on) the theory that the lithosphere is made up of a number of continually moving and interacting plates. **platelet** (-lit), *n.* a minute blood particle involved in clotting. **plater**, *n.* one who plates articles with silver, etc.; one who works upon plates; one who fits plates in shipbuilding; an inferior race-horse that runs chiefly in selling plates. **plating**, *n.* the act, art, or process of covering articles with a coating of metal; a coating of gold, silver, or other metal; (*collect.*) the plates covering a ship, fort etc.; (*Racing*) competing for plates.

plateau (platō), *n.* (*pl.* **-teaus, -teaux** (-ōz)) a table-land, an elevated plain; a period of stability after or during an upward progression, a levelling-off.

platen (plat'ən), *n.* the part of a printing-press that presses the paper against the type to give the impression; the roller in a typewriter serving the same purpose.

plater PLATE.

platform (plat'fawm), *n.* any flat or horizontal surface raised above some adjoining level; a stage or raised flooring in a hall etc., for speaking from etc.; a landing-stage; a raised pavement etc., beside the line at a railway station etc.; a vehicle or emplacement on which weapons are mounted and fired; a raised metal structure moored to the sea-bed and used for off-shore drilling. marine exploration etc.; (a shoe with) a thick sole; platform oratory; a political programme; the principles forming the basis of a party; (*N Am.*) a declaration of principles and policy issued by a party before an election. *v.t.* to place on or as on a platform. *v.i.* to speak from a platform.

plating PLATE.

platinum (plat'inəm), *n.* a heavy, ductile and malleable metallic element of a silver colour, fusing only at extremely high temperatures, immune to attack by most chemical reagents. *a.* of a record album, selling more than 300,000 in one year. **platinum-black**, *n.* finely divided platinum in the form of a black powder. **platinum blonde**, *n.* (*coll.*) a woman while hair so fair as to be almost white. **platinum metals**, *n.pl.* the platinoids. **platinize, -ise**, *v.t.* to coat with platinum. **platinoid** (-noid), *a.* like platinum. *n.* a name for certain metals found associated with platinum.

platitude (plat'itūd), *n.* flatness, commonplaceness, insipidity, triteness; a trite remark, esp. of a didactic kind. **platitudinize, -ise** (-tū'diniz), *v.i.* **platitudinous** (-tū'-), *a.* **platitudinously**, *adv.*

Platonic (pləton'ik), *a.* of or pertaining to Plato, the Greek philosopher, c. 427–347 BC, or to his philosophy or school; (*usu.* **platonic**) not involving sexual desire or activity. **Platonically**, *adv.* **Platonist** (plā'tə), *n.*

platoon (plətoon'), *n.* a subdivision, usu. half, of a company, formerly a tactical unit under a lieutenant; a body or set of people.

platter (plat'ə), *n.* a large shallow dish or plate.

platting (plat'ing), *n.* slips of bark, cane, straw etc., woven or plaited, for making hats etc.

platy-, *comb. form.* broad, flat.

platypus (plat'ipəs), *n.* a small, aquatic, egg-laying mammal of E Australia having a broad bill and tail, thick fur and webbed feet.

platyrrhine (plat'irin), *n.* of monkeys, broad-nosed.

plaudit (plaw'dit), *n.* (*usu. pl.*) an expression of applause; praise or approval. **plauditory**, *a.*

plausible (plaw'zibl), *a.* apparently right, reasonable, or probable; specious; apparently trustworthy; ingratiating.

plausibility (-bil'-), *n.* **plausibly**, *adv.*

play (plā), *n.* free, light, aimless movement or activity; freedom of movement or action; space or scope for this; a state of activity; a series of actions engaged in for pleasure or amusement; sport, exercise, amusement, fun; playing in a game; manner or style of this; the period during which a game is in progress; (*esp. N Am.*) a manoeuvre, esp. in a game; style of execution, playing (as on an instrument); exercise in any contest; gaming, gambling; a dramatic composition or performance, a drama; conduct or dealing towards others (esp. as fair or unfair). *v.i.* to move about in a lively, light, or aimless manner, to dance, frisk, shimmer etc.; to act or move freely; of instruments, machinery, guns etc. to perform a regular operation; of a part of a machine etc., to move loosely or irregularly; to be discharged or directed onto something, as water, light etc.; to sport, to frolic; to do something as an amusement; to toy, to trifle; to take part in a game; to take one's turn at performing an action specific to a game; to perform in a specified position or manner in a game; to perform on a musical instrument; to emit or reproduce sound; to take part in a game of chance; to game, to gamble; to behave, to act, to conduct oneself in regard to others; to personate a character; to act a part (esp. on a stage); of a drama, show etc., to be in performance; of an actor or company, to be performing. *v.t.* to engage in (a game or sport); to execute (a stroke, a shot etc.); to proceed through (a game, a rubber etc.); to oppose, to compete against; to make use of (as a player or an implement) in a game; to bring (as a card) into operation in a game; to cause (a ball etc) to move in a certain direction by striking, kicking etc.; (*coll.*) to gamble on; to perform (a trick etc.) esp. in jest or mockery; to give a performance or performances of (a musical or dramatic work, the works of a specified composer or author); to perform music on; to emit or reproduce sounds, esp. music; to act the role of; to act, or stage a play, in (a specified theatre or town); to pretend to be; to handle; to deal with, to manage; to give (a fish) freedom to exhaust itself; to discharge (as guns, a hose) continuously (on or upon); to cause to move lightly or aimlessly over. **play on words**, punning, a pun. **to bring, call into play**, to make operative. **to make a play for**, (*coll.*) to try to get. **to make great play with**, to make much of, to parade, to flourish ostentatiously. **to play about, around**, to act in a frivolous or irresponsible manner; to have casual sexual relationships. **to play along (with)**, to (seem to) agree or cooperate (with). **to play at**, to engage (in a game); to perform or execute in a frivolous or half-hearted way. **to play back**, to replay something just recorded. **to play ball**, to cooperate, to comply. **to play down**, to treat as unimportant, not to stress. **to play false**, to betray. **to play fast and loose**, to be fickle; to act recklessly. **to play God**, (seek to) control other people's destinies, to affect omnipotence. **to play hard to get**, (*coll.*) to act coyly, esp. as a come-on. **to play into the hands of**, to play or act so as to give the advantage to one's partner or opponent. **to play it by ear**, not to plan one's actions in advance, to improvise a response as situations develop. **to play it cool**, not to get excited. **to play (it) safe**, to take no risks. **to play off**, to pass (a thing) off as something else; to oppose (one person) against another, esp. for one's own advantage; to show off; to tee off; to take part in a play-off. **to play on**, to perform upon; to play the ball on to one's own wicket; (also **upon**) to exploit. **to play the game**, to play according to the rules of a game and accept defeat without complaint; to act honestly and courageously in any undertaking. **to play up**, to cause trouble or suffering to; to misbehave; to malfunction or function erratically; to give prominence to; to play more vigorously. **to play up to**, to humour, to draw out. **to play with**, to amuse oneself or sport with; to treat with levity. **play-act**, *v.i.* to make believe; to behave insincerely or overdramatically. **play-acting**, *n.* **playback**, *n.* a reproduction, esp. immediately after the recording has been made, of recorded sound or vision; a device for producing the above. **play-bill**, *n.* a bill or programme announcing or giving the cast of a play. **playbook**, *n.* a book of dramatic compositions; a book of games for children. **playboy, playgirl**, *n.* a young man or woman who lives for pleasure, a social parasite. **playfellow**, *n.* a companion in play. **playgoer**, *n.* one who frequents theatres. **playgoing**, *a.* **playground**, *n.* a piece of ground used for games, esp. one attached to a school; a favourite district for tourists, mountain-climbers etc. **playgroup**, *n.* a group of pre-school children who meet regularly for supervised and usu. creative play. **playhouse**, *n.* a theatre. **playmate**, *n.* a playfellow. **play-off**, *n.* a game to decide the final winner of a competition, esp. an extra game when two competitors are tied. **playpen**, *n.* a portable wooden framework inside which young children can play in safety. **playschool**, *n.* a nursery school or playgroup. **plaything**, *n.* a toy; a person or article used for one's amusement. **playtime**, *n.* time allotted for play. **playwright**, *n.* a dramatist. **playable**, *a.* able to be played or performed (on); (*Cricket*) able to be struck by the batsman. **played-out**, *a.* tired out; worn out, used up. **player**, *n.* one who plays; one engaged in a game; a person skilled in a particular game; a professional player, esp. a professional cricketer; an actor; a performer on a musical instrument; an automatic device for playing a musical instrument; a gambler; (*Billiards, Croquet, etc.*) the ball coming next into play. **playful**, *a.* froliscome, sportive; sprightly, humourous, jocular, amusing. **playfully**, *adv.* **playfulness**, *n.* playing, *n.* **playing-cards**, *n.pl.* cards used for games. **playing field**, *n.* a field or open space used for sports.

plaza (plah′zə), *n.* a public square or open paved area; (*chiefly N Am.*) a shopping centre.

PLC, plc (*abbr.*) public limited company.

plea (plē), *n.* the accused's answer to an indictment; something alleged by a party to legal proceedings in support of a claim or defence; an excuse; an urgent entreaty. **plea-bargaining**, *n.* (*chiefly N Am.*) the practice of arranging more lenient treatment by the court in return for an admission of guilt by the accused.

plead (plēd), *v.i.* (*past, p.p.* **pleaded**) to speak or argue in support of a claim, or in defence against a claim; to urge arguments for or against a claim; to urge arguments for or against; to supplicate earnestly; to put forward a plea or allegation, to address a court on behalf of; to answer to an indictment. *v.t.* to discuss, maintain, or defend by arguments; to allege in pleading or argument; to offer in excuse, to allege in defence. **to plead guilty, not guilty**, to admit or deny guilt or liability. **to plead with**, to entreat or supplicate (for, against etc.). **pleadable**, *a.* **pleader**, *n.* one who pleads in a court of law; one who offers reasons for or against; one who draws up pleas. **pleading**, *n.* the act of making a plea; entreating, imploring; a written statement of a party in a suit at law; the art or practice of drawing up such statements. *a.* imploring, appealing. **special pleading** SPECIAL. **pleadingly**, *adv.*

pleasant (plez′ənt), *a.* pleasing, agreeable, affording gratification to the mind or senses; affable, friendly; good-humoured; cheerful, gay, jocular, merry. **pleasantly**, *adv.* **pleasantness**, *n.* **pleasantry** (-tri), *n.* jocularity, facetiousness; a jest, a joke, an amusing trick; an agreeable remark, made esp. for the sake of politeness.

please (plēz), *v.t.* to afford pleasure to; to be agreeable to; to satisfy; to win approval from. *v.i.* to afford gratification; to like, to think fit, to prefer; a polite formula used in making requests or expressing acceptance. **if you please**, if it is agreeable to you; with your permission; (*iron*) expressing sarcasm or protest. **please yourself** *int.* do as you wish. **pleased**, *a.* gratified; delighted. **pleasedly**

(-zid-), *adv.* **pleasing**, *a.* **pleasingly**, *adv.*

pleasure (plezh′ə), *n.* the gratification of the mind or senses; enjoyment, gratification, delight; sensual gratification; a source of gratification; choice, wish, desire. *v.t.* to give pleasure to. *v.i.* to take pleasure (in). **pleasurable**, *a.* affording pleasure; pleasant, gratifying. **pleasurably**, *adv.*

pleat (plēt), *v.t.* to fold or double over, to crease. *n.* a flattened fold, a crease.

pleb (pleb), *n.* short for PLEBEIAN; (*derog.*) a common, vulgar person.

plebeian (pləbē′ən), *a.* pertaining to the ancient Roman commoners; pertaining to the common people; common, vulgar, low. *n.* a commoner in ancient Rome; one of the common people. **plebianism, plebeianness**, *n.*

plebiscite (pleb′isīt, -sit), *n.* a law enacted by a vote of the commonalty in an assembly presided over by a tribune of the people; a direct vote of the whole body of citizens in a state on a definite question, a referendum; an expression of opinion by the whole community.

plectrum (plek′trəm), *n.* (*pl.* **-tra** (-trə), **-trums**) a small implement of plastic, etc., with which players pluck the strings of guitar, etc.

pled (pled), (*Sc.*) *p.p.* of PLEAD.

pledge (plej), *n.* anything given or handed over by way of guarantee of security for the repayment of money borrowed, or for the performance of some obligation; a thing put in pawn; an agreement, promise, or binding engagement; the state of being pledged; a health, a toast. *v.t.* to give as a pledge or security; to deposit in pawn; to engage solemnly; to drink a health to. **to take the pledge**, to pledge oneself to abstain from alcoholic drink. **pledgeable**, *a.* **pledgee** (-ē′), *n.* **pledger**, *n.*

pledget (plej′it), *n.* a compress of lint for laying over an ulcer, wound etc.

-plegia, *comb. form.* paralysis.

pleiad (plī′əd), *n.* a cluster of brilliant persons, esp. seven (as the French poets Ronsard, Du Bellay, and their associates, in the 16th cent.). **Pleiades** (-dēz), *n.pl.* a cluster of small stars in the constellation Taurus, seven of which are discernible by the naked eye.

Pleiocene (plī′əsēn), PLIOCENE.

Pleistocene (plīs′təsēn), *a., n.* (*Geol.*) (pertaining to) the strata or epoch overlying or succeeding the Pliocene formation.

plenary (plē′nəri), *a.* full, complete, entire, absolute; attended by all members. **plenary indulgence**, in the Roman Catholic Church, an indulgence remitting all the temporal penalties due to sin. **plenarily**, *adv.*

plenipotentiary (plenipəten′shəri), *a.* invested with full powers; full, absolute. *n.* an ambassador or envoy to a foreign court, with full powers.

plenitude (plen′itūd), *n.* fullness; completeness, abundance.

plenty (plen′ti), *n.* abundance, copiousness; fruitfulness; (*sing. or pl. in constr.*) a large quantity or number, an ample supply, lots. *a.* (*coll.*) plentiful, abundant. *adv.* (*coll.*) quite, abundantly. **plenteous** (-tiəs), *a.* **plenteously**, *adv.* **plenteousness**, *n.* **plentiful**, *a.* existing in abundance; yielding abundance; copious. **plentifully**, *adv.* **plentifulness**, *n.*

plenum (plē′nəm), *n.* (*pl.* **-nums, -na**) space, as considered to be full of matter, opposed to vacuum; an enclosure containing gas at a higher pressure than the surrounding environment; a condition of fullness, plethora; a full meeting.

pleomorphism (plēōmaw′fizm), *n.* the occurrence of more than one different form in the life cycle of a plant or animal; polymorphism. **pleomorphic**, *a.*

pleonasm (plē′ənazm), *n.* redundancy of expression in speaking or writing. **pleonastic** (-nas′-), *a.* **pleonastically**, *adv.*

plesiosaur (plē′siəsaw-), *n.* any of a genus of extinct mar-

ine reptiles with long necks, small heads and four paddles.
plethora (pleth'ərə), *n.* superabundance; excessive fullness of blood. **plethoric** (-tho-), *a.* having an excess of blood in the body; superabundant.
pleura (ploor'ə), *n.* (*pl.* **-rae** (-re), **-ras**) a thin membrane covering the interior of the thorax and investing the lungs; a part of the body-wall in arthropods; a part to which the secondary wings are attached in insects; a part on each side of the rachis of the lingual ribbon in molluscs. **pleural**, *a.* **pleurisy** (-risi), *n.* (*Path.*) inflammation of the pleura, usu. attended by fever, pain in the chest or side, etc. **pleuritic** (-rit'-), *a.*
pleurenchyma (plooreng'kimə), *n.* the woody tissue of plants.
pleur(o)-, *comb. form.* pertaining to the side or ribs; pertaining to the pleura.
pleuronectid (plooronek'tid), *n.* a fish of the family Pleuronectidae, or the flat-fishes or flounders; a flat-fish.
pleuropneumonia (plooronūmō'niə), *n.* inflammation of the lungs and pleura, esp. as contagious disease among cattle.
Plexiglass® (plek'siglahs), *n.* a transparent plastic.
plexor (plek'sə), *n.* an instrument used as a hammer for chest examination, reflex testing, etc.
plexus (plek'səs), *n.* (*pl.* **plexuses, plexus**) a network of veins, fibres, or nerves; a network, a complication.
pliable (plī'əbil), *a.* easily bent; flexible, pliant; supple, limber; yielding readily to influence or arguments. **pliableness, pliability** (-bil'-), *n.* **pliably**, *adv.* **pliant**, *a.* pliable, flexible; yielding, compliant. **pliancy**, *n.* **pliantly**, *adv.*
plicate (plīkāt, -kət), **-cated** (-kā'tid), *a.* plaited; folded like a fan.
plié (plē'ā), *n.* a ballet movement in which the knees are bent outwards while the back remains straight.
pliers (plī'əz), *n.pl.* small pincers with long jaws for bending wire etc.
plight[1] (plīt), *v.t.* to pledge, to promise, to engage (oneself, one's faith etc.). *n.* a engagement, a promise. **to plight one's troth**, to become engaged to be married.
plight[2] (plīt), *n.* condition, state, case, esp. one of distress or disgrace.
plimsoll (plim'səl), *n.* a rubber-soled canvas shoe worn for sports etc.
Plimsoll line, mark (plim'səl), *n.* a line, required to be placed on every British ship, marking the level to which the authorized amount of cargo submerges her.
plinth (plinth), *n.* a square member forming the lower division of a column etc.; a block serving as a pedestal; the plain projecting face at the bottom of a wall.
Pliocene (plī'əsēn), *n.*, *a.* (*Geol.*) the most modern epoch of the Tertiary period. *a.* pertaining to this epoch.
plissé (plē'sa), *n.* (a fabric having) a wrinkled finish.
PLO, (*abbr.*) Palestine Liberation Organization.
plod (plod), *v.i.* (*past, p.p.* **plodded**) to walk painfully, slowly, and laboriously; to trudge; to toil, to drudge; to study with steady diligence. *v.t.* to make (one's way) thus. *n.* a laborious walk, a trudge; a wearisome piece of work. **plodder**, *n.* **plodding**, *a.* **ploddingly**, *adv.*
plonk[1] (plongk), *v.t.*, *v.i.* to (be) put down or drop heavily, forcefully or with a plonk. *n.* a heavy, hollow sound. **plonker**, *n.* (*sl.*) a (big and) stupid person. **plonking**, *a.* (*sl.*) large, unwielding.
plonk[2] (plongk), *n.* (*coll.*) cheap (and inferior) wine.
plop (plop), *n.* the sound of something falling heavily into water. *adv.* suddenly; heavily, with the sound 'plop'. *v.i.* (*past, p.p.* **plopped**) to fall thus into water.
plosive (plō'siv), *n.*, *a.* (*Phon.*) explosive. **plosion**, *n.*
plot (plot), *n.* a small piece of ground; a plan of a field, farm, estate etc.; a complicated plan, scheme, or stratagem; a conspiracy; the plan or skeleton of the story in a play, novel etc.; a graphic representation. *v.t.* (*past, p.p.* **plotted**) to make a plan, map, or diagram of; to mark on a map (as the course of a ship or aircraft); to locate and mark on a graph by means of coordinates; to draw a curve through points so marked; to lay out in plots; to plan, to devise, to contrive secretly. *v.i.* to form schemes or plots against another; to conspire. **plotter**, *n.*
plotty (plot'i), *n.* a hot drink made of wine, water, spices etc.
plough, (*esp. N Am.*) **plow** (plow), *n.* an implement for cutting, furrowing, and turning over land for tillage; tillage, agriculture; arable land; an implement or machine resembling a plough in form or function, a snow-plough etc.; (*cap.*) the seven brightest stars in Ursa Major, also called Charles's Wain; (*sl.*) failure or rejection in an examination. *v.t.* to turn up (ground) with a plough; to make (a furrow) with a plough; to furrow, groove, or scratch, with or as with a plough; to wrinkle; (*sl.*) to reject at an examination; (*sl.*) to fail in an examination subject. *v.i.* to advance laboriously. **to plough back profits**, to reinvest profits. **to plough into**, to collide with violently. **to plough through**, to smash a way through; to work or read through laboriously. **to put one's hand to the plough**, to begin a task or undertaking. **ploughman**, *n.* one who ploughs; a husbandman, a rustic. **ploughman's lunch**, *n.* a cold snack of bread and cheese with pickle, served esp. in a pub. **ploughshare**, *n.* the blade of a plough. **ploughable**, *a.* **plougher**, *n.*
plover (pluv'ə), *n.* the common English name for several shore birds, esp. the golden, yellow or green plover.
plow etc. PLOUGH.
ploy (ploi), *n.* employment, an undertaking; a game, a pastime; a prank; a manoeuvre, a tactic, a stratagem.
PLP, (*abbr.*) Parliamentary Labour Party.
PLR, (*abbr.*) Public Lending Right.
pluck (plŭk), *v.t.* to pull off or out, to pick; to pull, to twitch; to pull, to drag (away etc.); to strip by pulling out feathers; to fleece, to swindle; (*sl.*) to reject (as a candidate for a degree etc.). *v.i.* to pull, drag, or snatch (at). *n.* the act of plucking; a pull, a twitch; the heart, lungs and liver of an animal; courage, spirit. **to pluck up courage** COURAGE. **plucked**, *a.* (*usu. in comb., as well-plucked*). **plucker**, *n.* **plucky**, *a.* having pluck, spirit, or courage. **pluckily**, *adv.* **pluckiness**, *n.*
plug (plŭg), *n.* a piece of wood or other substance used to stop a hole; a stopper, a peg, a wedge; anything wedged in or stopping up a pipe, or used to block the outlet to a waste-pipe; a piece of wood etc. inserted into masonry to take a nail or screw; a mass of volcanic rock stopping a vent; (a cake, stick, or small piece of) compressed tobacco; a sparking-plug; a fire-plug; a device with a non-conducting case, having usu. three pins, which are attached to an electric cable to make an electrical connection with a suitable socket; (*coll.*) an electrical socket; (*coll.*) a piece of favourable publicity, esp. one inserted into other material. *v.t.* (*past & p.p.* **plugged**) to stop with a plug; to insert as a plug; (*sl.*) to shoot; (*N Am. sl.*) to strike with the fist; (*coll.*) to give favourable publicity to; (*coll.*) to allude to repeatedly, esp. as a form of publicity. **to plug away at**, to work at doggedly and persistently. **to plug in**, to establish an electrical connection (with). **to pull the plug on**, (*coll.*) to bring to an (abrupt) end. **plughole**, *n.* the outlet for waste water in a sink, bath etc. which can be closed with a plug. **plug-ugly**, *a.* (*N Am. sl.*) a hooligan, a rowdy. **plugging**, *n.*
plum (plŭm), *n.* the fleshy stone fruit of *Prunus domestica* or other trees of the same genus; a tree bearing this; applied to other fruits, esp. to the raisin used in cakes, puddings etc.; plum-colour; the best part of anything, the choicest thing of a kind; any handsome perquisite, windfall etc. *a.* plum-coloured; choice, cushy. **plum-cake**, *n.* a cake containing raisins, currants etc. **plum-colour**, *n.* dark

purple. **plum-duff**, *n*. a plain boiled flour pudding with raisins etc. **plum-pudding**, *n*. a pudding containing raisins, currants, etc., esp. a rich one with spices etc., eaten at Christmas. **plummy**, *a*. full of or rich in plums; luscious, inviting; desirable; of the voice, rich and round to the point of affectation.
plumage PLUME.
plumb (plŭm), *n*. a weight, usu. of lead, attached to a line, used to test perpendicularity; a position parallel to this, the vertical; a sounding-lead, a plummet. *a*. perpendicular, vertical; downright, sheer, perfect, complete; (*Cricket*) level. *adv*. vertically; exactly, correctly, right; completely. *v.t.* to adjust by a plum-line; to make vertical or perpendicular; to sound with a plummet, to measure the depth of; to fathom, to understand. **out of plumb**, not exactly vertical. **to plumb (in)**, to connect to a water main and/or drainage system. **plumb bob**, *n*. a conical weight used in a plum-rule or on a plumb-line. **plumb line**, *n*. the cord by which a plumb is suspended for testing perpendicularity; a vertical line. **plumb rule**, *n*. a builder's or carpenter's rule with a plumb-line attached.
plumbago (plŭmbā'gō), *n*. a form of carbon used for making pencils etc., blacklead, graphite; a genus of perennial herbs, with blue, rose, or violet flowers.
plumber (plŭm'ėr), *n*. a person who fits and repairs cisterns, pipes, drains, gas fittings etc. **plumbing**, *n*. the work of a plumber; the arrangement of water-pipes, gas installations etc. in a building.
plume (ploom), *n*. a feather, esp. a large or conspicuous feather; a feather-bunch or tuft of feathers, or anything resembling this worn as a ornament; (*Zool.*) a feather-like part or form; a feathery appendage to a seed etc. *v.t.* to trim, dress, or arrange (feathers), to preen; to adorn with or as with feathers, esp. in borrowed plumage; to pride (oneself on); to strip of feathers. **plumage**, (-mij), *n*. a bird's entire covering of feathers. **plumate, plumose** (-mōs'), *a*. resembling a feather or feathers, feathery. **plumy**, *a*. covered with feathers; adorned with plumes.
plummet (plŭm'it), *n*. a weight attached to a line used for sounding; a ball of lead for a plumb line; (*fig.*) a weight, an encumbrance. *v.i.* to fall sharply or rapidly.
plummy PLUM.
plump[1] (plŭmp), *a*. well-rounded, fat, fleshy, filled out, chubby; of a purse etc, well-filled; rich, abundant. *v.t.* to make plump; to fatten, to distend. *v.i.* to grow plump; to swell (out or up). **to plump up**, to make (pillows, cushions) rounded and soft by shaking. **plumply**, *adv*. **plumpiness**, *n*. **plumpy**, *a*.
plump[2] (plŭmp), *v.i.* to plunge or fall suddenly and heavily; to vote for one candidate when more might be voted for; to give all one's votes to a single candidate. *v.t.* to fling or drop suddenly and heavily. *n*. a sudden plunge, a heavy fall; the sound of this. *adv*. suddenly and heavily; directly, straight down; flatly, bluntly. *a*. downright, plain, blunt.
plumule (ploo'mūl), *n*. the rudimentary stem in an embryo; a little feather, one of the down feathers; a downy scale on the wings of butterflies, etc.
plumy PLUME.
plunder (plŭn'dȧ), *v.t.* to pillage, to rob, to strip; to take by force, to steal, to embezzle. *n*. forcible or systematic robbery; spoil, booty; (*sl.*) profit, gain. **plunderer**, *n*.
plunge (plŭnj), *v.t.* to force or thrust into water or other fluid; to immerse; to force, to drive (into a condition, action etc.); to sink (a flower-pot) in the ground. *v.i.* to throw oneself, to dive (into); to rush or enter impetuously (into a place, condition, etc.); to fall or descend very steeply or suddenly; of a horse, to throw the body forward and the hind legs up; of a ship, to pitch; (*coll.*) to gamble or bet recklessly, to spend money or get into debt heavily. *n*. the act of plunging; a dive, a sudden and violent movement; a risky or critical step. **to take the plunge**, (*coll.*) to commit oneself after hesitating. **plunger**, *n*. one who plunges; (*coll.*) a reckless gambler, speculator, or spendthrift; a part of a machine working with a plunging motion, as the long solid cylinder used as a piston in a force-pump; a rubber suction cup on a handle, used to unblock drains, etc. **plunging**, *a*.
plunk (plŭngk), *v.t.* to pluck the strings of (a banjo etc.); to emit a plunk; to plonk. *v.i.* to plonk. *n*. a dull, metallic sound.
pluperfect (ploopœ'fikt), *a*. (*Gram.*) expressing action or time prior to some other past time. *n*. the pluperfect tense; a pluperfect form.
plural (plua'rȧl), *a*. denoting more than one; consisting of more than one. *n*. the form of a word which expresses more than one, or (in languages having a dual number) more than two. **pluralism**, *n*. the state of being plural; the holding of more than one office, esp. an ecclesiastical benefice, at the same time; (*Phil.*) the doctrine that there is more than one ultimate principle in the universe, opp. to monism; a social system in which members of diverse ethnic, cultural etc. groups coexist, preserving their own customs and lifestyle but having equal access to power. **pluralist**, *n*. **pluralistic** (-lis'-), *a*. **plurality** (-ral'-), *n*. a number consisting of two or more; a majority, or the excess of (votes etc.) over the next highest number; pluralism; a benefice or other office held by a pluralist. **pluralize, -ise**, *v.t.* pluralization, *n*. **plurally**, *adv*.
plus (plŭs), *n*. a character (+) used as the sign of addition; an addition; a positive quantity; an advantage, a positive feature; a surplus. *prep*. with the addition of. *a*. above zero, positive; additional, extra, esp. additional and advantageous electrified positively. **plus fours**, *n.pl.* long, baggy knickerbockers.
plush (plŭsh), *n*. a cloth of various materials with a pile or nap longer than that of velvet. *a*. of plush; (also **plushy**) rich, luxurious, lavishly appointed.
Pluto (ploo'tō), *n*. Greek god of the underworld; (*Astron.*) the ninth planet in the solar system in order of distance from the sun, from which it is distant 3666 million miles (15,902 million km). **Plutonian** (-tō'-), *a*. pertaining to Pluto or the lower regions; infernal, subterranean, dark; igneous. **Plutonic rocks**, *n.pl.* igneous rocks, as granite, basalt etc.
plutocracy (plootok'rȧsi), *n*. the rule of wealth or the rich; a ruling class of rich people; (*coll.*) the wealthy class. **plutocrat** (ploo'tȧkrat), *n*. **plutocratic** (-krat'-), *a*. **plutocratically**, *adv*.
plutonium (plootō'niȧm), *n*. a radioactive element, at no. 94; chem. symbol Pu; formed by the radioactive decay of neptunium.
pluvial (ploo'viȧl), *a*. of or pertaining to rain; rainy; humid; (*Geol.*) due to the action of rain. **pluviometer** (-om'itȧ), *n*. a rain-gauge. **pluviometric** (-met'-), *a*.
ply[1] (plī), *n*. a fold, a plait, a twist, a strand (of a rope, twine etc.); a thickness, a layer; a bent, a bias. **plywood**, *n*. board consisting of three or more thin layers of wood glued together in such a manner that the grain of each is at right-angles to that of its neighbour.
ply[2] (plī), *v.t.* (*pres. p.* **plying**, *past, p.p.* **plied**) to use (a tool) vigorously or busily; to work at, to employ oneself in; to pursue, to press, to urge; to supply (with) or subject (to) repeatedly; to travel regularly along. *v.i.* to go to and fro, to travel or sail regularly; to be busy, to be employed; to stand or wait for custom; (*Naut.*) to work to windward.
Plymouth Brethren (plim'ȧth), *n.pl.* a strict evangelical sect that rose at Plymouth about 1830, having no regular ministry, and formulating no creed.
plywood PLY[1].
PM[1], (*abbr.*) Prime Minister; Provost Marshal.
PM[2], **pm**[1], (*abbr.*) post meridiem (after noon); postmortem.

Pm, (*chem. symbol*) promethium.
pm², (*abbr.*) premium.
PMG, (*abbr.*) Paymaster General; Postmaster General.
PMT, (*abbr.*) pre-menstrual tension.
pneumatic (nūmat'ik), *a.* pertaining to or consisting of air; gaseous; containing or filled with air; actuated by means of compressed air or a vacuum; having air-filled cavities. *n.* a pneumatic tyre. **pneumatic drill**, *n.* a drill in which compressed air reciprocates a loose piston which hammers a steel bit. **pneumatically**, *adv.* **pneumatics**, *n.sing.* the science treating of the mechanical properties of air and other gases.
pneumatology (nūmətol'əji), *n.* the science of spiritual existence; the doctrine of the Holy Spirit.
pneumatophore (nū'mətəfaw, -mat'-), *n.* a respiratory organ in the roots of some tropical trees growing in mud.
pneumococcus (nūmōkok'əs), *n.* (*pl.* **-cocci** (-kok'sī)) a bacterium which causes pneumonia.
pneumoconiosis (nūmōkōniō'sis), *n.* any disease of the lungs or bronchi caused by habitually inhaling metallic or mineral dust.
pneumogastric (nūmōgas'trik), *n.* of or pertaining to the lungs and the stomach.
pneumonia (nūmō'niə), *n.* acute inflammation of a lung or the lungs. **pneumonic** (-mon'-), *a.*
pneumothorax (nūmōthaw'raks), *n.* accumulation of air in the pleural cavity, usu. associated with pleurisy.
PO, (*abbr.*) Personnel Officer; Petty Officer; Pilot Officer; postal order; Post Office.
Po, (*chem. symbol*) polonium.
po (pō), *n.* (*pl.* **pos**) (*coll.*) chamber pot.
poach¹ (pōch), *v.t.* to cook (as an egg, fish) in simmering liquid esp. water or milk. **poacher**, *n.* a vessel for poaching eggs in.
poach² (pōch), *v.i.* to encroach or trespass on another's lands), esp. to take game etc.; to take game, fish etc. by illegal or unsportsmanlike methods; to intrude or enroach upon another's rights, area of responsibility etc.; to take an advantage unfairly, as in a race or game; (of ground, to become soft, swampy, or miry; (*Lawn Tennis*) to hit the ball when in the court of one's partner. *v.t.* to take (game, fish etc.) from another's preserves or by illegitimate methods; to take game from (another's preserves); to trample, to tread into mire. **poacher**, *n.* **poachy**, *a.* wet and soft; swampy; easily trodden into holes by cattle.
pochard (pō'chəd), *n.* a European diving sea-duck.
pock (pok), *n.* a pustule in an eruptive disease, as in smallpox. **pockmark**, *n.* the pit or scar left by a pock; any similar mark or indentation. **pockmarked, pocky**, *a.*
pocket (pok'it), *n.* a small bag, sack, or pouch, esp. a small bag inserted in the clothing, to contain articles carried about the person; pecuniary means; a small netted bag in billiards or snooker to receive the balls; (*Geol.*) a cavity containing gold or other ore; an isolated area or patch; an air pocket. *a.* for the pocket; small. *v.t.* to put into a pocket; to keep in or as in the pocket; to hem in (a horse etc.) in a race; to appropriate, esp. illegitimately; to put up with; to repress or conceal (one's feelings); in billiards or snooker, to drive (a ball) into a pocket. **in, out of pocket**, the richer, the poorer. **out-of-pocket**, *a.* of expenses etc., unbudgeted and paid for in cash. **to have in one's pocket**, to have complete control over; to be assured of winning. **to line one's pockets**, (to abuse a position of trust in order) to enrich oneself. **to pocket one's pride, to put one's pride in one's pocket**, not to stand on one's dignity, to adopt an amenable attitude. **to put one's hand in one's pocket**, to spend or give money. **pocket battleship**, *n.* a small battleship, esp. one built to conform with treaty limitations. **pocketbook**, *n.* a note-book of book or case for carrying papers etc., in the pocket; (*N Am.*) a handbag. **pocket-borough** BOROUGH. **pocket-edition**, *n.* a pocket-size edition of a book; a smaller version of anything. **pocketknife**, *n.* a knife with blades shutting into the handle, for carrying in the pocket. **pocket-money**, *n.* money for occasional expenses or amusements; a small, regular allowance given to a child.
pockmark, pocky POCK.
pod (pod), *n.* a long capsule or seed-vessel, esp. of leguminous plants; applied to similar receptacles, as the case enclosing the eggs of a locust etc.; a streamlined container, housing an engine, fuel, armaments etc., attached to the outside of an aircraft; any protective (external) housing; a detachable compartment on a spacecraft. *v.i.* (*past, p.p.* **podded**) to produce pods; to swell into pods. *v.t.* to shell (peas etc.).
-pod, *comb. form.* foot.
podagra (pədag'rə), *n.* gout, esp. in the foot. **podagral, -gric, -grous**, *a.*
poddy (pod'i), *n.* (*Austral.*) a hand-fed calf or foal.
podge (poj), *n.* a short and stout person. **podgy**, *a.*
podium (pō'diəm), *n.* (*pl.* **-diums, -dia** (-ə)) a low projecting wall or basement supporting a building; a platform encircling the arena in an amphitheatre; a continuous structural bench round a hall etc.; a small raised platform (for a conductor, lecturer etc.). **podial**, *a.*
podophyllum (podəfil'əm, -dof'-), *n.* a genus of plants, containing the may-apple. **podophyllic** (-fil'-), *a.* **podophyllin** (-fil'-), *n.* a purgative resin extracted from the root of *Podophyllum peltatum*.
Podura (pədū'rə), *n.* a genus of apterous insects comprising the springtails.
poe-bird (pō'iboed), *n.* a New Zealand bird, larger than a thrush, with dark metallic plumage and a tuft of white feathers on the neck, also called the parson-bird or tui.
poem (pō'əm), *n.* a metrical composition, esp. of an impassioned and imaginative kind; an artistic and imaginative composition in verse or prose; anything supremely beautiful, well-executed or satisfying.
poesy (pō'isi), *n.* (*pl.* **-sies**) (*dated*) the art of poetry; (*collect.*) metrical compositions.
poet (pō'it), *n.* a writer of poems or metrical compositions, esp. one possessing high powers of imagination and rhythmical expression; one possessed of high imaginative or creative power. **Poet Laureate**, *n.* an officer of the British royal household whose nominal duty is to compose an ode every year for the sovereign's birthday, for any great national victory, etc. **poetaster** (-tas'tə), *n.* an inferior or petty poet; a pitiful versifier. **poetess** (-tis), *n. fem.* **poetic, -ical** (-et'-), *a.* pertaining to or suitable for poetry; expressed in poetry, written in poetry; having the finer qualities of poetry; fit to be expressed in poetry. **poetic justice**, *n.* punishment or reward ideally (often ironically) fitted to deserving. **poetic licence**, *n.* the latitude in grammar etc., allowed to poets. **poetically**, *adv.* **poeticize, -ise** (-siz), *v.t.* **poetics**, *n.sing.* the theory or principles of poetry. **poetize, -ise**, *v.i.* to compose verses, to write poetry. *v.t.* to poeticize. **poetry** (-ri), *n.* the art or work of the poet; that one of the fine arts which expressed the imagination and feelings in sensuous and rhythmical language, usu. in metrical forms; imaginative, impassioned, and rhythmical expression whether in verse or prose; imaginative or creative power; a quality in anything that powerfully stirs the imagination or the aesthetic sense; (*collect.*) metrical compositions, verse, poems.
po-faced (pōfāst'), *a.* (*coll.*) deadpan; humourless, stolid; stupidly solemn.
pogo stick, *n.* a toy consisting of a strong pole attached to a spring and having a handle at the top and a crossbar on which one stands to bounce along.
pogrom (pog'rəm), *n.* an organized attack, usu. with pillage and massacre, upon an ethnic group, esp. Jews.
pohutakawa (pōhootəkah'wə), *n.* the Christmas bush.
poignant (poin'yənt), *a.* sharp; stimulating to the palate, pungent; keen, piercing; bitter, painful. **poignancy**, *n.*

poignantly, *adv.*
poikilothermal (poikilǝthœ'mǝl), **-thermic**, *a.* having a body temperature which varies with the surrounding temperature. **poikilothermy**, *n.*
poind (poind), *v.t. (Sc.)* to distrain upon; to seize and sell (a debtor's goods); to impound. *n.* the act of poinding, distraint.
poinsettia (poinset'iǝ), *n.* a genus of S American and Mexican plants with gorgeous red leaf-like bracts and small greenish-yellow flowerheads.
point (point), *n.* a mark made by the end of anything sharp, a dot; a dot used as a mark of punctuation, to indicate vowels etc.; (*Print.*) a full stop, or decimal mark to separate integral from fractional digits in decimal numbers, etc.; a particular item, a detail; a particular place or position; a specific position or stage in a development or on a scale; a state or condition; a particular moment; the precise moment for an event, action, etc., the instant; the verge; a step or stage in an argument, discourse etc.; a unit used in measuring or counting, in assessing superiority etc., in appraising qualities of an exhibit in a show, a racehorse etc., in reckoning odds given to an opponent in a game, in betting, or in scoring in games; a salient quality, a trait, a characteristic; the essential element, the exact object (of a discussion, joke etc.), the main purport, the gist; the aim, the purpose; a conclusion; a suggestion, a tip; the sharp end of a tool, weapon etc., the tip; a nib; a cape, a promontory (esp. in place-names); point-lace; a sharp-pointed tool, as an etcher's needle, glass-cutter's diamond, various implements or parts of machinery etc.; a tapering rail moving on a pivot for switching a train from one line to another; hence, (*pl.*) a railway switch; (*pl.*) the contact-breakers in the distributor of an internal-combustion engine; a power point; a tine of a deer's horn; pungency, effectiveness, force; (*Geom.*) that which has position but not magnitude; the unit of measurement for printing type-bodies, approx. 1/72 in. (0·351 mm); a position on a shield; a fielder or position on the off-side wicket, of and close in to, the batsman in cricket; (*Hunting*) a spot to which a straight run is made; a musical passage or subject to which special importance is drawn; the leading party of an advanced guard; (*pl.*) the extremities of a horse; (*Ballet*) pointe; a unit of increase or decrease in the price of stocks or the value of currencies; the act of pointing by a setter etc.; (*Fencing*) a twist; a point of the compass; a short cord for reefing sails; a buckling-strap on harness. *v.t.* to sharpen; to mark with points, to punctuate; to mark off (as a psalm) in groups of syllables for singing; to give force or point to; to fill (the joints of masonry) with mortar or cement pressed in with a trowel; to indicate, to show; to direct (a finger etc., at); to turn in a particular direction, to aim; to give effect or pungency to (a remark, jest etc.); to indicate the meaning or point of by a gesture; to turn in (manure etc.) with the point of a spade. *v.i.* to direct attention to; of a pointer or setter, to draw attention to game by standing rigidly and looking at it; to aim (at or towards); to face or be directed (towards); to sail close to the wind. **at all points**, in every part or direction; completely, perfectly. **at** or **on the point**, on the verge (of). **beside the point**, irrelevant. **in point**, apposite, relevant. **in point of fact** FACT. **not to put too fine a point on it**, to speak bluntly. **point of honour**, a matter of punctilio, a matter involving personal honour, involving a demand for satisfaction by a duel etc. **point of no return**, the point in a flight where shortage of fuel makes it necessary to go on as return is impossible; a critical point (at which one must commit oneself irrevocably to a course of action). **point of order**, a question of procedure. **point of sale**, the place, esp. a retail shop, where the sale of an article physically takes place. **point of the compass**, one of the 32 angular divisions of the compass; the angle of 11° 15′ between two such points. **point of view**, the position from which a thing is looked at; way of regarding a matter. **to carry one's point**, to prevail in an argument or dispute. **to make a point**, to score a point; to establish a point in argument. **to make a point of**, to attach special importance to; to take special care to. **to point out**, to indicate. **to point up**, to emphasize, to highlight. **to score points off**, to score off. **to stretch a point**, STRETCH. **to the point**, appropriate, apposite, pertinent. **point-blank**, *a.* fired horizontally; aimed directly at the mark making no allowance for the downward curve of the trajectory; hence very close (permitting such aim to be taken); direct, blunt. *adv.* horizontally, with direct aim; directly, bluntly. **a point-blank shot**. correct, precise, finical, neat. *adv.* correctly, precisely, to a nicety. **point-duty**, *n.* the work of a constable stationed at a junction of streets or other point to regulate traffic. **point-lace**, *n.* lace made with the point of a needle. **pointsman**, *n.* a policeman on point-duty; a person in charge of the switches on a railway. **point-to-point**, *a.* denoting a steeplechase or other race in a direct line from one point of a course to another. **pointed**, *a.* having a sharp point; having point, penetrating, cutting; referring to some particular person or thing; emphasized, made obvious. **pointedly**, *adv.* with special meaning. **pointedness**, *n.* **pointer**, *n.* one who or that which points; the index-hand of a clock etc.; a rod used for pointing on a blackboard etc.; an indication, a hint, a tip; a dog trained to point at game; (*pl.*) two stars of the Plough a line drawn through which points nearly to the pole-star; (*Austral.*) a trickster, a swindler; (*Austral.*) one of the bullocks next to the pole in a team. **pointing**, *n.* the act of indicating, directing, sharpening etc.; punctuation; division into groups of words or syllables for singing; the act of finishing or renewing a mortar-joint in a wall. **pointless**, *a.* having no point; purposeless, futile. **pointlessly**, *adv.* **pointlessness**, *n.*
pointe (pwit), *n.* (*Ballet*) the extreme tip of the toe; a position in which the dancer balances on this.
pointillism (pwi'tilizm), *n.* (*Art*) delineation by means of dots of various colours which merge into a whole. **pointillist**, *n.*, *a.*
poise (poiz), *v.t.* to balance to hold or carry in equilibrium; to place in a carefully balanced position; to counterpoise; to ponder. *v.i.* to be balanced or in equilibrium; to hang (in the air) over, to hover. *n.* equipoise, equilibrium; a counterpoise; a state of suspense, indecision etc.; composure, assurance, self-possession; physical balance. **poised**, *a.* balanced; having or showing composure etc.; in a state of readiness (to), all set (to).
poison (poi'zǝn), *n.* a substance that injures or kills an organism into which it is absorbed; anything noxious or destructive to health or morality; a substance which retards catalytic activity or a chemical reaction, or, by absorbing neutrons, the course of a nuclear reaction; (*sl.*) liquor, drink. *v.t.* to put poison in or upon; to infect with poison; to administer poison; to kill or injure by this means; to taint, to corrupt, to vitiate, to pervert. **poison gas**, *n.* (*Mil.*) poisonous or stupefying gas or liquid used in warfare. **poison ivy**, *n.* any of various N American shrubs or climbing plants which cause an intensely itching skin rash. **poison-pen letter**, *n.* one written maliciously and usu. anonymously, to abuse or frighten the recipient. **poisoner**, *n.* **poisonous**, *a.* **poisonously**, *adv.*
poke[1] (pōk), *v.t.* to thrust, to push (in, out, through etc.) with the end of something; to jab, to prod; to stir (a fire) with a poker; to cause to protrude; to put, move etc. by poking; to make (a hole etc.) by poking; (*coll.*) to punch; (*sl.*) to have sexual intercourse with. *v.i.* to thrust, to jab; to protrude; (*coll.*) to pry, to search; to dawdle, to busy oneself without any definite object. *n.* a poking, a push, a thrust, a prod, a nudge; a collar with a drag attached to

prevent animals from breaking through fences, etc.; (*coll.*) a punch; (*sl.*) an act of sexual intercourse. **to poke fun at,** FUN. **to poke one's nose into,** NOSE. **pokeweed,** *n.* a N American medicinal herb. **poky,** *a.* of a room etc., cramped, confined, stuffy; (*coll.*) shabby; petty, dull.
poke² (pōk), *n.* (*esp. Sc.*) a bag, a sack. **a pig in a poke** PIG.
poke³ (pōk), *n.* a projecting front on a woman's hat or bonnet, formerly a detachable rim. **poke-bonnet,** *n.*
poker¹ (pō'kə), *n.* a metal rod used to stir a fire; an instrument employed in poker-work; (*Univ. sl.*) one of the bedells carrying a mace before the Vice-Chancellor at Oxford or Cambridge. *v.t.* to adorn with poker-work. *v.i.* to carry out (a design) in this. **red-hot poker,** *n.* popular name of plant of the genus *Tritoma*, or flame-flower. **poker-work,** *n.* the production of decorative designs on wood by burning or scorching with a heated instrument.
poker² (pō'kə), *n.* a cardgame in which the players bet on the value in their hands. **poker face,** *n.* an expressionless face. **poker-faced,** *a.*
poky POKE².
Pol., (*abbr.*) Poland; Polish.
pol., (*abbr.*) political; politics.
pol, (pol), *n.* (*sl.*) a politician.
Polack (pō'lak), *n.* (*now offensive*) a Pole.
polar (pō'lə), *a.* pertaining to or situated near the poles of the earth; coming from the regions near the poles; pertaining to a magnetic pole, having polarity, magnetic; having two opposite elements or tendencies, esp. positive and negative electricity; pertaining to the poles of a cell; relating to or of the nature of a polar; remote or opposite as the poles; resembling the polestar, attracting, guiding. **polar bear,** *n.* a white arctic bear. **polar circles** CIRCLE. **polar distance,** *n.* the angular distance of a point from the nearest pole.
polarimeter (pōlərim'itə), *n.* an instrument for measuring the polarization of light **polarimetric** (-met'-), *a.* **polarimetry** *n.*
Polaris (pəlah'ris), *n.* the pole star.
polariscope (pōla'riskōp), *n.* an instrument for showing the phenomena of polarized light.
polarity, *n.* the state of having two opposite poles, or of having different or opposing properties in opposite parts or directions; the quality (in electricity) of being attracted to one pole and repelled from the other; the disposition in a body to place its mathematical axis in a particular direction; diametric opposition.
polarization, -isation (pōlərīzā'shən), *n.* the state of possessing or causing polarity; the restriction of light rays to a single plane of vibration. **polarize, -ise** (pō'-), *v.t., v.i.* to (cause to) acquire polarity or polarization; to divide into sharply opposed groups or factions; of opinions etc., to (cause to) become more sharply opposed or more radically different. **polarizer,** *n.*
Polaroid® (pō'ləroid), a light-polarizing material used esp. in sunglasses; a type of camera which produces a finished print from inside itself within seconds of the picture's being taken.
polder (pōl'də), *n.* a tract of land below sea or river level, that has been drained and cultivated.
pole¹ (pōl), *n.* a long slender piece of wood or metal, usu. rounded and tapering, esp. fixed upright in the ground as a flagstaff, support for a tent, etc.; the shaft to which horses are yoked; an instrument for measuring; a measure of length, a rod or perch, 5½ yd. (5 m); a mast. *v.t.* to furnish or support with, to convey or impel by poles. **up the pole,** (*sl.*) crazy, mad; mistaken, wrong. **pole position,** *n.* the most advantageous position at the start of a race, esp. a motor or horse race.
pole² (pōl), *n.* one of the extremities of the axis on which a sphere, esp. the earth, revolves; one of the two points in a body where the attractive or repelling force is great-

est, as in a magnet; a terminal of an electric cell, battery etc.; the extremity of the axis of a cell nucleus, etc.; either of the polar regions; either of two opposite extremes. **magnetic pole** MAGNETIC. **poles apart,** as far apart as possible; having widely divergent views, attitudes etc. **pole-star,** *n.* a bright star, Polaris in Ursa Minor, within a degree and a quarter of the northern celestial pole; a guiding principle, a lodestar. **pole vault,** *n.* a field event in which the competitor attempts to clear a very high bar with the aid of a long flexible pole. **pole-vault,** *v.i.* **pole-vaulter,** *n.*
Pole (pōl), *n.* a native of Poland or one of Polish race.
poleaxe (pōl'aks), *n.* a form of battle-axe consisting of an axe set on a long handle; a long-handled butcher's axe with a hammer at the back, used for slaughtering cattle. *v.t.* to strike kill or destroy (as if) with a poleaxe.
polecat (pōl'kat), *n.* a small carnivorous European weasel-like mammal, with two glands emitting an offensive smell.
polemic (pəlem'ik), *a.* pertaining to controversy; controversial, disputatious. *n.* a controversy or controversial discussion; a controversialist. **polemical,** *a.* **polemically,** *adv.* **polemicist** (-sist), *n.* **polemics,** *n.sing.* the art or practice of controversial discussion.
polenta (pəlen'tə), *n.* a kind of porridge made of maize-meal or chestnut-meal, common in Italy.
police (pəlēs'), *n.* a civil force organized for the maintenance of order, the detection of crime, and the apprehension of offenders; (*pl.*) the members belonging to this force. *v.t.* to control by the use of police; to supervise, to regulate, to discipline. **police dog,** *n.* a dog trained to assist police. **police force,** *n.* a separately organized body of police. **policeman,** *n.* any (male) member of a police force, esp. a constable. **police officer,** *n.* a policeman or policewoman. **police state,** *n.* a totalitarian state maintained by the use of political police. **police station,** *n.* the headquarters of a local section of the police. **police trap,** *n.* an ambush by police to trap offenders against road regulations. **policewoman,** *n.* a female member of the police force. **policing,** *n.* the policies and methods employed by the police in maintaining law and order etc.
policy¹ (pol'isi), *n.* prudence, foresight, or sagacity in managing or conducting, esp. state affairs; prudent conduct; a course of action or administration recommended or adopted by a party, government, firm, organization or individual.
policy² (pol'isi), *n.* a document containing a contract of insurance; a warrant, voucher etc.; (*N Am.*) a method of gambling by betting on numbers drawn in a lottery. **policy-shop,** *n.* (*N Am.*) an office where drawings take place in connection with such lotteries.
poliomyelitis (-mīəlī'tis), **polio** (pō'liō), *n.* a viral disease of the spinal cord; infantile paralysis.
polish¹ (pol'ish), *v.t.* to make smooth or glossy, usu. by friction; to refine, to free from rudeness or coarseness; to bring to a fully finished state. *v.i.* to take a polish. *n.* a smooth glossy surface, gloss. produced by friction; friction applied for this purpose; a substance applied to impart a polish; refinement, elegance of manners. **to polish off,** (*coll.*) to finish speedily; to get rid of. **to polish up,** to give a polish to; to improve, or refresh, (one's knowledge of something) by study. **polished,** *a.* accomplished; impeccably executed. **polisher,** *n.* **polishing,** *n.*
Polish (pō'lish), *a.* pertaining to Poland or its inhabitants. *n.* the language of the Poles; (*collect.*) the Polish people.
Politburo (pol'itbūrō), *n.* the Political Bureau of the Central Committee of the Communist Party of the USSR.
polite (pəlīt'), *a.* refined in manners; courteous; well-bred; cultivated. **politely,** *adv.* **politeness,** *n.*
politesse (politès'), *n.* formal politeness.
politic (pol'itik), *a.* prudent and sagacious; prudently devised, judicious, expedient; crafty, scheming, artful;

specious; consisting of citizens. **body politic** BODY. **politicly**, *adv*. artfully, cunningly.
political (pəlit'ikəl), *a*. relating to civil government and its administration; of or relating to politics, esp. party politics; interested or involved in politics; having an established system of government. **political economy** ECONOMY. **political geography** GEOGRAPHY. **political prisoner**, *n*. a person imprisoned for his or her political beliefs or activities. **political science**, *n*. the study of government and politics. **political scientist**, *n*. **politically**, *adv*.
politician (politish'ən), *n*. a person versed in politics, a statesman; one engaged in or devoted to party politics, usually as a career; (*N Am.*) one who makes use of politics for private ends, a spoilsman.
politicize, -ise (pəlit'isiz), *v.t*. to give a political tone or scope to; to make politically aware. *v.i*. to engage in or discuss politics. **politicization**, *n*.
politicking (pol'itiking), *n*. political activity, esp. vote-getting.
politico (pəlit'ikō), *n*. (*pl*. **-cos, -coes**) (*coll*.) a politician.
politics (pol'itiks), *n. sing*. the art or science of civil government; the profession of politics; (*sing. or pl. in constr*.) political affairs; (*pl*.) the political dimension to any action or activity; (*pl*.) any activities concerned with the acquisition, apportionment or exercise of power within an organization, manoeuvring, intrigue; (*pl*.) a person's political views of sympathies.
polity (pol'iti), *n*. (*pl*. **-ties**) the form, system, or constitution of the civil government of a State; the State; an organized community, a body politic; the form of organization of an institution, etc.
polka (pōl'kə), *n*. (*also* (pol'-)) a lively round dance of Bohemian origin; a piece of music in duple time for this; a woman's tight-fitting jacket, usu. made of knitted wool. *v.i*. to dance a polka. **polka dots**, *n.pl*. small dots arranged in a regular pattern esp. as a textile design. **polka-dot**, *a*.
poll[1] (pōl), *n*. a register or enumeration of heads or persons, esp. of persons entitled to vote at elections; the voting at an election, the number of votes polled, or the counting of these; the time or place of election; an attempt to ascertain public opinion by questioning a few individuals. *v.t*. to remove the top of (trees etc.); to cut off the horns of; to clip, to shear; to take the votes of; to receive a specified number of votes); to give (one's vote); to question in a poll; (*Comput*.) to interrogate (computer terminals) in sequence to ascertain whether they have any data on them for use. *v.i*. to give one's vote. **deed-poll** DEED[1]. **to poll a jury**, (*N Am.*) to examine each juror as to his or her concurrence in a verdict. **poll-tax**, *n*. a capitation tax or one levied on every person. **pollable**, *a*. **poller**, *n*. **polling**, *n*. **polling booth**, *n*. a semi-enclosed structure in which a voter marks his or her ballot paper. **polling station**, *n*. a building designated as a place where voters should go to cast their votes. **pollster** (-stə), *n*. one who conducts an opinion poll.
poll[2] (pōl), *a*. polled; hornless. **poll-beast, -cow, -ox**, *n*. a polled beast, esp. one of a breed of hornless cattle.
pollack (pol'ək), *n*. a sea-fish allied to the cod.
pollan (pol'ən), *n*. a herring-like Irish freshwater fish.
pollard (pol'əd), *n*. a tree with its top cut off so as to have a dense head of young branches; a stag or other animal that has cast its horns; a polled or hornless ox, sheep or other animal; the chub; a mixture of fine bran with a small quantity of flour, orig. bran sifted from flour. *v.t*. to lop the top of (a tree). **pollarded**, *a*. lopped, cropped; wanting horns.
pollen (pol'ən), *n*. the fertilizing powder discharged from the anthers of flowers and causing germination in the ovules. **pollen count**, *n*. a measure of the pollen present in the air, published to assist hay-fever sufferers etc. **pollinate**, *v.t*. to sprinkle with pollen so as to cause fertilization. **pollination**, *n*.
pollicitation (pəlisitā'shən), *n*. a voluntary promise or engagement, or a paper containing such engagement; (*Law*) a promise not yet accepted, an offer.
polliwog (pol'iwog), *n*. a tadpole.
pollock (pol'ək), POLLACK.
pollute (pəloot'), *v.t*. to make foul or unclean; to contaminate (an environment), esp. with man-made waste; to defile; to corrupt the moral purity of; to dishonour, to ravish; to profane. **polluter**, *n*. **pollution**, *n*.
polo (pō'lō), *n*. a game of Asian origin resembling hockey but played on horseback. **polo-jumper**, *n*. a knitted jumper with a fold-over collar. **polo neck**, *n*. (a jumper with) close-fitting, doubled-over collar.
polonaise (polənāz'), *n*. an article of dress for women, consisting of a bodice and short skirt in one piece; a similar garment for men worn early in the 19th cent.; a slow dance of Polish origin; a piece of music in 3/4 time for this.
polonium (pəlō'niəm), *n*. a radioactive element, at. no. 84; chem. symbol Po.
polony (pəlō'ni), *n*. a sausage of partly-cooked pork.
poltergeist (pōl'təgīst, pol'-,), *n*. an alleged spirit which makes its presence known by noises and violence.
poltroon (pəltroon'), *n*. an arrant coward.
poly (pol'i), *n*. (*pl*. **polys**) short for POLYTECHNIC; short for POLYTHENE, as in *poly bag*.
poly-, *comb. form*. several, many; excessive, abnormal; denoting a polymer.
polyamide (poliam'īd), *n*. a synthetic, polymeric material such as nylon.
polyandrous (polian'drəs), *a*. having numerous stamens; pertaining to or practising polyandry. **polyandrist**, *n*. a woman having several husbands. **polyandry** (pol'-), *n*. the practice or condition of a woman having more than one husband at once; plurality of husbands.
polyanthus (polian'thəs), *n*. (*pl*. **-thuses**) a garden variety of primula, prob. a development from the cowslip or oxlip.
polyarchy (pol'iahki), *n*. government by many.
polyatomic (poliətom'ik), *a*. applied to elements having more than one atom in their molecules, esp. replaceable atoms of hydrogen.
polybasic (polibā'sik), *a*. of acids etc., having two or more equivalents of a base.
polychaete (pol'ikēt), *n*. belonging to the Polychaeta, a class of worms with setae, mostly marine. *n*. a polychaete worm. **polychaetan, -tous** (-kē'-), *a*.
polychromatic (polikrəmat'ik), *a*. exhibiting many colours or a play of colours. **polychrome** (pol'ikrōm), *n*. a work of art executed in several colours, esp. a statue. *a*. having or executed in many colours. **polychromic, -chromous** (-krō'-), *a*. **polychromy** (polyikrōmi), *n*. the art of decorating (pottery, statuary or buildings) in many colours.
polyclinic (poliklin'ik), *n*. a clinic dealing with various diseases; a general hospital.
polycotyledon (polikotilē'dən), *n*. a plant with seeds having more than two cotyledons. **polycotyledonous**, *a*.
polydactyl (polidak'til), *a*. having more than the normal number of fingers or toes. *n*. a polydactyl animal. **polydactylism**, *n*. **polydactylous**, *a*.
polyester (polies'tə), *n*. any of a group of synthetic polymers made up of esters, used esp. in making fibres for cloth, plastics, and resins.
polyethylene (polieth'ilēn), POLYTHENE.
polygamy (pəlig'əmi), *n*. the practice or condition of having a plurality of wives or husbands at the same time; the state of having more than one mate; the state of being polygamous. **polygamist**, *n*. **polygamous**, *a*.
polygastric (poligas'trik), *a*. having many stomachs. **polygastrian**, *n*., *a*.
polygene, *n*. any of a group of genes which determine a

particular characteristic, e.g. height. **polygenesis**, *n.* (*Biol.*) the evolution of organisms from multiple ancestral lines; the theory suggesting that different races of mankind are descended from different original ancestors. **polygenetic**, *a.*

polyglot (pol'iglot), *a.* expressed in or speaking many languages. *n.* a person speaking many languages; a book, esp. the Bible, written in many languages. **polyglottal, -ttic** (-glot'-), *a.*

polygon (pol'igon), *n.* a closed plane figure, usu. rectilinear and of more than four angles or sides. **polygonal** (-lig'-), *a.* **polygonally**, *adv.*

Polygonum (pəlig'ənəm), *n.* a genus of plants comprising the snakeweed, knot-grass etc., belonging to the family Polygonaceae.

polygraph (-graf), *n.* an apparatus for multiplying copies of writing, drawings etc.; an instrument which registers several small physiological changes simultaneously, e.g. in pulse rate, body temperature, often used as a lie detector. **polygraphic** (-graf'-), *a.* **polygraphy** (-lig'-), *n.*

Polygynia (polijin'iə), *n.pl.* a Linnaean class of plants containing those having flowers with many styles. **polygynous** (-lij'-), *a.* pertaining to or practising polygyny. **polygyny** (-lij'-), *n.* plurality of wives.

polyhedron (polihē'drən), *n.* (*pl.* **-drons, -dra** (-drə)) a solid bounded by many (usu. more than four) plane sides. **polyhedral**, *a.* having many sides.

polyhybrid (polihī'brid), *n.* a cross between parents of differing heritable characters.

polymath (pol'imath), *n.* a person of great and varied learning. **polymathic** (-math'-), *a.* **polymathist** (-lim'-), *n.* **polymathy** (-lim'-), *a.* wide and multifarious learning.

polymer (pol'imə), *n.* a chemical compound, formed by polymerization, which has large molecules made up of many comparatively simple repeated units. **polymeric** (-me'rik), *a.* of or constituting a polymer. **polymerization, -isation**, *n.* a chemical reaction in which two or more small molecules combine as repeating units in a much larger molecule. **polymerize**, *v.t.* to render polymerous or polymeric. *v.i.* to become polymeric. **polymerous** (-lim'-), *a.* (*Biol.*) consisting of many parts; polymeric.

polymorphic (polimaw'fik), **-morphous**, *a.* having, assuming, or occurring in many forms. **polymorph** (-pol'-), *n.* **polymorphism**, *n.*

polyneme (pol'inēm), *n.* any fish belonging to the genus *Polynemus*, consisting of tropical spiny sea-fishes having the pectoral fin divided into free rays.

Polynesia (polinē'zhə), *n.* a collective name for the islands of the central and southern Pacific. **Polynesian**, *a.* pertaining to Polynesia. *n.* a native or inhabitant of Polynesia.

polyneuritis (polinūrī'tis), *n.* simultaneous inflammation of many nerves.

polynia, polynya, (pəlin'iə), *n.* an open place in water that is for the most part frozen over, esp. in the Arctic.

polynomial (polinō'miəl), *a.* (*Alg.*) multinomial. *n.* a multinomial. **polynomialism**, *n.*

polyp (pol'ip), *n.* one of various aquatic animals of low organization, as the hydra, the sea-anemone etc., an individual in a compound organism of various kinds; a polypus. **polypary**, *n.* the calcareous or chitonous structure supporting a colony of polyps.

polypeptide (polipep'tid), *n.* any of a group of polymers made up of long amino-acid chains.

polypetalous (polipet'ələs), *a.* having many or separate petals.

polyphagous (pəlif'əgəs), *a.* feeding on various kinds of food; voracious.

polyphase (pol'ifāz), *a.* having two or more alternating voltages of equal frequency, the phases of which are cyclically displaced by fractions of a period.

polyphone (pol'ifōn), *n.* (*Philol.*) a character or sign standing for more than one sound. **polyphonic** (-fon'-), *a.* representing different sounds; contrapuntal; having several sounds or voices, many-voiced. **polyphony**, *n.* the state of being polyphonic; composition in parts, each part having an independent melody of its own, counterpoint.

polyploid (pol'iploid), *a.* having more than twice the basic (haploid) number of chromosomes. **polyploidy**, *n.*

polypod (pol'ipod), *a.* having numerous feet. *n.* a millipede, e.g. a wood-louse.

polypody (pol'ipədi), *n.* a fern of the genus *Polypodium*, esp. *P. vulgaris*, the common polypody, growing on rocks, walls, trees etc. **polypodiaceous** (-ā'shəs), *a.*

polypoid (pol'ipoid), *a.* like a polyp or polypus.

polyporous (pəlip'ərəs), *a.* having many pores.

Polyporus (pəlip'ərəs), *n.* a genus of hymenomycetous fungi growing on the decaying parts of trees, the spores of which are borne on the inner surface of pores or tubes. **polyporaceous** (-rā'shəs), *a.* **polyporoid** (-roid), *a.*

polypropylene (poliprop'ilēn), *n.* any of various plastics or fibres that are polymers of propylene.

polypus (pol'ipəs), *n.* (*pl.* **-pi** (-pī)) a tumour growing in a mucous cavity; a polyp.

polyrhizous (polirī'zəs), *a.* having many roots.

polysemous (polisē'məs), *a.* having several meanings. **polysemy** (pol'isēmi), *n.*

polysepalous (polisep'ələs), *a.* having the sepals distinct.

polyspermal (polispœ'məl), **-mous**, *a.* having many seeds.

polyspore (pol'ispaw), *n.* a compound spore; a spore-case containing many spores. **polysporous** (-spaw'-), *a.*

polystigmous (polistig'məs), *a.* having several carpels each bearing a stigma.

polystome (pol'istōm), *a.* having many mouths. *n.* an animal with many mouths or suckers. **polystomatous** (-stom'ətəs), **-stomous** (-lis'tə-), *a.*

polystyrene (polistī'rēn), *n.* a polymer of styrene used esp. as a transparent plastic for moulded products or in expanded form, as a rigid white foam, for packaging and insulation.

polysyllabic (polisilab'ik), *a.* consisting of many syllables; characterized by polysyllables. **polysyllable** (pol'-), *n.*

polysyndeton (polisin'ditən), *n.* a figure of speech in which the conjunction or copulative is repeated several times.

polytechnic (politek'nik), *a.* connected with, pertaining to, or giving instruction in many subjects, esp. technical ones. *n.* a college where degree and other advanced courses are given in technical, vocational, and academic subjects.

polytetrafluoroethylene (politetraflooərōeth'ilēn), **PTFE**, *n.* a tough, translucent plastic used esp. for moulded articles and as a non-stick coating.

polytheism (pol'ithēizm, -thē'-), *n.* the doctrine or worship of a plurality of gods. **polytheist**, *n.* **polytheistic** (-is'-), *a.* **polytheistically**, *adv.*

polythene (pol'ithēn), *n.* any of various thermoplastics that are polymers of ethylene.

polytocous (pəlit'əkəs), *a.* multiparous, producing several at a birth.

polytomous (pəlit'əməs), *a.* pinnate, the divisions not articulated with a common petiole.

polyunsaturated (poliŭnsat'ūrātid), *a.* of certain animal and vegetable fats, having long carbon chains with many double bonds.

polyurethane (poliū'rəthān), *n.* any of a class of polymeric resins used esp. as foam for insulation and packing.

polyvalent (polivā'lənt), MULTIVALENT.

polyvinyl (polivī'nəl), *n., a.* (of or being) a polymerized vinyl compound. **polyvinyl chloride**, *n.* a plastic used esp. as a rubber substitute, e.g. for coating wires and cables and as a dress and furnishing fabric.

Polyzoa (polizō'ə), *n.pl.* (*sing.* **-zoon** (-zō'on), a class of invertebrate animals, mostly marine, produced by gemma-

tion, existing in coral-like or plant-like compound colonies. **polyzoal**, *a*. **polyzoan**, *n.*, *a*.
pom (pom), (*Austral.*, *New Zealand*, *derog. sl.*) POMMY.
pomace (pŭm'is), *n*. the mashed pulp of apples crushed in a cider-mill, esp. the refuse after the juice has been pressed out. **pomaceous** (-mā'shəs), *a*. of the nature of a pome or of trees producing pomes, as the apple, pear, quince etc.
pomade (pəmād', -mahd'), *n*. pomatum. *v.t.* to apply this to (the hair etc.).
pomander (pōman'də, pom-), *n*. a perfumed ball or powder kept in a box, bag etc., used as a scent and formerly carried about the person to prevent infection; the perforated box or hollow ball, in which this is kept or carried.
pomatum (pəmā'təm), *n*. a perfumed ointment (said to have been prepared partly from apple-pulp) for dressing the hair. *v.t.* to apply pomatum to.
pome (pōm), *n*. a compound fleshy fruit, composed of the walls of an adnate inferior calyx enclosing carpels containing the seeds, as the apple, pear, quince etc. **pomiculture** (-mikŭlchə), *n*. fruit-growing. **pomiferous** (-mif'-), *a*. bearing apples or pomes. **pomiform** (pō'mifawm), *a*. shaped like a pome or apple. **pomology** (pəmol'-), *n*. the art or science of the cultivation of fruit. **pomological** (-loj'-), *a*. **pomologist**, *n*.
pomegranate (pom'igranit), *n*. the fruit of a N African and W Asiatic tree resembling an orange, with a thick, tough rind and acid red pulp enveloping numerous seeds; the tree bearing this fruit.
Pomeranian (pomərā'niən), *a*. of or pertaining to Pomerania. *n*. a native or inhabitant of Pomerania; a Pomeranian dog, esp. a dog about the size of a spaniel, with a fox-like pointed muzzle and long, silky hair.
Pomfret-cake (pom'frit), *n*. a flat cake of liquorice made in Pomfret, now Pontefract, in Yorkshire.
pomiculture POME.
pommel (pom'l, pŭm'l), *n*. a round ball or knob, esp. on the hilt of a sword; the upward projection at the front of a saddle; either of the two handles on top of a gymnastics horse. *v.t.* to pummel.
pomology POME.
pommy (pom'i), *n*. (*Austral.*, *New Zealand*, *derog. sl.*) a British person, esp. an immigrant to Australia or New Zealand. *a*. British.
pomp (pomp), *n*. a pageant; state, splendour; ostentatious display or parade.
pompadour (pom'pəduə), *n*. method of wearing the hair brushed up from the forehead or (in women) turned back in a roll from the forehead, named after the Marquise de *Pompadour*, 1721–64, mistress of Louis XV.
pompano (pom'pənō), *n*. a W Indian food-fish of various species belonging to the genus *Trachinotus*.
pom-pom (pom'pom), *n*. an automatic quick-firing gun, usu. mounted for anti-aircraft defence.
pompom (pom'pom), **pompon** (-pon), *n*. an ornament in the form of a tuft or ball of feathers, ribbon etc. worn on women's and children's hats, shoes etc. a small compact chrysanthemum.
pompous (pom'pəs), *a*. displaying pomp; grand, magnificent; ostentatious, pretentious; exaggeratedly solemn or portentous, self-important. **pomposo** (-pō'sō), *adv*. (*Mus.*) in a stately or dignified manner. **pompously**, *adv*. **pomposity** (-pos'-), **pompousness**, *n*.
ponce (pons), *n*. (*sl.*) a prostitute's pimp; an effeminate man. *v.i.* to pimp. **to ponce about, around**, (*sl.*) to act in an ostentatious or effeminate manner; to fool about, to waste time. **poncy**, *a*. (*sl.*)
poncho (pon'chō), *n*. a woollen cloak, orig. worn in S America, with a slit through which the head passes; a cycling-cape of this pattern.
pond (pond), *n*. a body of still water, often artificial, smaller than a lake; (*facet.*) the sea, esp. the Atlantic. *v.t.* to dam back; to make into a pond. *v.i.* to form a pool or pond (of water). **pond-lily**, *n*. a water-lily, esp. the yellow *Nuphar lutea*. **pond-weed**, *n*. an aquatic plant growing on stagnant water, esp. species of *Potamogeton*.
ponder (pon'də), *v.t.* to weigh carefully in the mind; to think over, to consider deeply, to reflect upon. *v.i.* to think, to deliberate, to muse (on, over etc.). **ponderable**, *a*. capable of being weighed, having appreciable weight, opp. to *imponderable*. **ponderer**, *n*. **ponderingly**, *adv*.
ponderous, *a*. very heavy or weighty; bulky, unwieldy; dull, tedious; pompous, self-important. **ponderously**, *adv*. **ponderosity** (-os'-), **ponderousness**, *n*.
pone[1] (pōn), *n*. a kind of bread made by the N American Indians of maize-meal; similar bread made with eggs, milk etc.; a loaf of this.
pone[2] (pōn. -ni), *n*. the player to the dealer's right who cuts the cards.
pong (pong), *n*. (*coll.*) a bad smell, a stink. *v.i.* to stink.
pongee (pŭnjē'), *n*. a soft unbleached kind of Chinese silk.
pongo (pong'gō), *n*. (*pl.* **-gos**, (*Austral.*) **-goes**) a large African anthropoid ape; erroneously applied to the orang-outan etc.
poniard (pon'yəd), *n*. a dagger. *v.t.* to stab with a poniard.
pons (ponz), *n*. a bridge; (*Anat.*) a connecting part, esp. the pons Varolii. **pons Varolii** (vərō'lii), *n*. a band of fibres connecting the two hemispheres of the cerebellum.
pontifex (pon'tifeks), *n*. (*pl.* **-tifices** (-tif'isēz)) a member of the highest of the ancient Roman colleges of priests. **Pontifex Maximus** (mak'siməs), *n*. the president of this; the Pope.
pontiff (pon'tif), *n*. the Pope; a pontifex, a high priest.
pontifical (-tif'-), *a*. of, pertaining to, or befitting a pontiff, high priest or pope; papal, popish; with an assumption of authority. *n*. a book containing the forms for rites and ceremonies to be performed by bishops; (*pl.*) the vestments and insignia of a pontiff or bishop. **pontifically**, *adv*. **pontificate** (-kət), *n*. *v.t.* (-kāt), to celebrate (Mass etc.) as a bishop. *v.i.* to officiate as a pontiff or bishop, esp. at Mass; to speak or behave in a pompous and dogmatic manner.
pontoon[1] (pontoon'), VINGT-ET-UN.
pontoon[2] (pontoon'), *n*. a flat-bottomed boat, cylinder, or other buoyant structure supporting a floating bridge; a caisson; a barge or lighter; a pontoon-bridge; *v.t.* to bridge with pontoons. **pontoon-bridge**, *n*. a platform or roadway laid across pontoons.
pony (pō'ni), *n*. a small horse, esp. one of a small breed; a small glass; (*sl.*) 25 pounds sterling; (*N Am.*, *sl.*) a crib used in getting up lessons. **pony express**, *n*. a postal and delivery system across the western US (1860–61), using relays of horses and riders. **ponytail**, *n*. a hair style in which the hair is gathered at the back and hangs down over the nape of the neck like a tail. **pony trekking**, *n*. cross-country pony-riding in groups as a pastime.
pooch (pooch), *n*. (*sl.*) a dog, esp. a mongrel.
poodle (poo'dl), *n*. a breed of pet dog with long woolly hair, often clipped in a fanciful style; a servile follower. *v.t.* to clip the hair of (a dog) thus.
poof (puf), **pooftah** (-tə), **poofter**, **poove** (poov), **pouf**, *n*. (*sl.*, *offensive*) a male homosexual.
pooh (poo), *int*. an exclamation of contempt or impatience.
pooh-pooh (-poo'), *v.t.* to laugh or sneer at; to make light of.
Pooh-bah (poobah'), *n*. (*facet.*) a pompous person holding many offices.
pool[1] (pool), *n*. a small body of water, still or nearly still; a deep, still part of a stream; a puddle; a pond, natural or ornamental; a swimming pool; an underground accumulation of oil or gas.
pool[2] (pool), *n*. the receptacle for the stakes in certain

games of cards; the collective amount of stakes, forfeits etc.; a game on a billiard-table in which the players aim to drive different balls into the pockets in a certain order; the collective stakes in a betting arrangement; (*pl.*) football pools; a combination of persons, companies etc., for manipulating prices and suppressing competition; any common stock or fund; a group of people or things which can be called upon when required. *v.t.* to put (funds, risks etc.) into a common fund or pool. **poolroom**, *n.* (*N Am.*) a billiards saloon.

poop[1] (poop), *n.* the stern of a ship; a deck over the after part of a spar-deck. *v.t.* of waves, to break heavily over the poop of; to take (a wave) over the stern (of a ship). **pooped**, *a.* having a poop (*usu. in comb.*) struck on the poop.

poop[2] (poop), *v.t.* (*coll.*) to make out of breath, to exhaust. *v.i.* to become exhausted. **to poop out**, to poop; to give up.

poor (puə), *a.* wanting means of subsistence, needy, indigent; badly supplied, lacking (in); barren, unproductive; scanty, meagre, inadequate in quantity or quality, unsatisfactory; lean, thin, wasted; unhealthy; uncomfortable; inferior, sorry, paltry, miserable, contemptible; insignificant, humble, meek; unfortunate, pitiable, used as a term of slight contempt, pity or endearment. **poor box**, *n.* a money-box, esp. in a church for charitable contributions. **poorhouse**, *n.* a workhouse. **poor law**, *n.* the body of laws formerly relating to the maintenance of paupers. **poor relation**, *n.* a person or thing looked down on, considered inferior, or shabbily treated in comparison to others. **poor-spirited**, *a.* timid, cowardly; mean, base. **poor-spiritedness**, *n.* **poor white**, *n.* (*usu. derog.*) a member of a class of poverty-stricken and socially inferior white people in the southern US or S Africa. **poorly**, *adv.* with poor results, with little success; defectively, imperfectly; meanly, despicably; in delicate health; unwell, indisposed. **poorness**, *n.*

poove POOF.

pop[1] (pop), *v.i.* (*past, p.p.* **popped**) to make a short, sharp, explosive noise as of the drawing of a cork; to burst open with such a sound; esp. of the eyes, to protrude as with amazement; to enter or issue forth with a quick, sudden motion; to dart; to move quickly; to shoot (at) with a gun, pistol etc. *v.t.* to push or thrust (in, out, up) suddenly; to put (down etc.) quickly or hastily; to fire off (a gun etc.); to cause (a thing) to pop by breaking etc.; (*sl.*) to take (drugs) orally or by injection; (*sl.*) to consume habitually. *adv.* with a pop; suddenly. *n.* a short, sharp, explosive noise; a dot, spot or other mark, esp. used in marking sheep etc.; (*coll.*) an effervescing drink, esp. ginger-beer or champagne. **to go pop**, to make, or burst with, a popping sound. **to pop off**, (*coll.*) to leave hastily; to die. **to pop the question**, (*coll.*) to propose marriage. **to pop up**, to appear suddenly. **popcorn**, *n.* maize kernels burst and puffed up by heating; the kind of maize suitable for this. **popeyed**, *a.* with bulging eyes. **popgun**, *n.* a small toy gun used by children, shooting a pellet or cork with air compressed by a piston; a poor or defective firearm. **pop socks**, *n.pl.* knee-length nylon stockings. **pop-up**, *a.* of books, having illustrations etc. which stand up off the page when the book is opened to give a quasi-three-dimensional effect; having a device which causes the contents to spring up; (*Comput.*) of or being a computer facility which can be accessed during the running of a program. **popper**, *n.* something that pops; (*coll.*) a press-stud; (*sl.*) an amyl nitrate capsule inhaled by drug users. **popping-crease**, *n.* (*Cricket*) a line four feet in front of the stumps parallel with the bowling crease.

pop[2] (pop), *n.* pop music; (*coll.*) a popular piece of (usu. light) classical music. *a.* popular. **top of the pops**, (a record, singer etc.) currently (among) the most popular in terms of sales; one who or that which is currently enjoying great popularity. **pop art**, *n.* art incorporating everyday objects from popular culture and the mass media. **pop concert**, *n.* a concert at which pop music is played. **pop festival**, *n.* **pop group**, *n.* a (usu. small) group of musicians who play pop music. **pop music**, *n.* modern popular music, post 1950, esp. as characterized by a simple, heavy rhythmic beat and electronic amplification. **pop record**, *n.* **pop singer**, *n.* **popster**, *n.* (*sl.*) one who is engaged in or interested in pop music.

pop[3] (pop), *n.* (*coll.*) father; a familiar form of address to an old man.

pop., (*abbr.*) popular(ly); population.

popadum (pop′ədŭm), POPPADOM.

pope (pōp), *n.* the bishop of Rome as the head of the Roman Catholic Church; (*fig.*) a person claiming or credited with infallibility; a priest in the Orthodox Church, esp. in Russia. **pope's eye**, *n.* the gland surrounded with fat in the middle of the thigh of an ox or sheep. **popedom** (-dəm), *n.* **popery** (-pəri), *n.* (*derog.*) the religion or ecclesiastical system of the Church of Rome. **popish**, *a.* (*often derog.*) of or pertaining to the Pope; (*derog.*) pertaining to popery, papistical. **popishly**, *adv.*

popinjay (pop′injā), *n.* (*dated*) a parrot; a mark like a parrot set up on a pole to be shot at by archers etc.; a conceited chattering fop.

poplar (pop′lə), *n.* a large tree of the genus *Populus*, of rapid growth, and having a soft, light wood.

poplin (pop′lin), *n.* a silk and worsted fabric with a ribbed surface, a cotton or other imitation of the above; (*N Am.*) broadcloth.

popliteal (poplit′iəl), **poplitic**, *a.* pertaining to the ham or hollow behind the knee-joint.

poppa (pop′ə), *n.* (*N Am.*, *coll.*) father.

poppadom, -dum (pop′ədəm), *n.* a crisp, thin Indian bread, fried or roasted and served with curry.

popper POP[1].

poppet (pop′it), *n.* a framework bearing the hoisting-gear at a pithead; one of the timbers on which a vessel rests in launching; a puppet, a marionette; a darling, a term of endearment.

popple (pop′l), *v.i.* of floating bodies or water, to bob up and down, to toss, to heave. *n.* a tossing or rippling; the sound of this.

poppy (pop′i), *n.* a plant or flower of the genus *Papaver*, containing plants with large showy flowers chiefly of scarlet colour, with a milky juice having narcotic properties. **opium-poppy**, *n.* the species, *P. somniferum*, from which opium is collected. **poppycock**, *n.* (*sl.*) nonsense, balderdash. **Poppy Day**, *n.* remembrance Sunday. **poppyhead**, *n.* the seed-capsule of a poppy; a finial of foliage or other ornamental top to ecclesiastical woodwork, esp. a bench-end.

popsy (pop′si), *n.* (*dated sl.*, *often derog.*) an attractive young woman.

populace (pop′ūləs), *n.* the common people; the masses.

popular (pop′ūlə), *a.* pleasing to or esteemed by the general public or a specific group or an individual; pertaining to, or carried on by, or prevailing, among the general public or the common people; suitable to or easy to be understood by ordinary people, not expensive, not abstruse, not esoteric; democratic. **popular front**, *n.* a coalition of socialist and other parties in a common front against dictatorship. **popularity** (-la′-), *n.* **popularize, -ise**, *v.t.* to make popular; to treat (a subject etc.) in a popular style; to spread (knowledge etc.) among the people; to extend (the suffrage etc.) to the common people. **popularization**, *n.* **popularly**, *adv.*

populate (pop′ūlāt), *v.t.* to furnish with inhabitants, to people; to form the population of, to inhabit. **population**, *n.* the inhabitants of a country etc., collectively; the number of such inhabitants; the (number of) people of a certain class and/or in a specified area; (*Statistics*) the aggregate of individuals or items from which a sample is taken; the group of organisms, or of members of a particular species, in a particular area; the act of populating.

populist (pop'ūlist), *n.* a person claiming to represent the interests of the common people; a person who believes in the rights, virtues, or wisdom of the common people. **populism**, *n.* **populist, -istic** (-lis'-), *a.*

populous (pop'ūləs), *a.* full of people; thickly populated. **populously**, *adv.* **populousness**, *n.*

porbeagle (paw'bēgl), *n.* a shark of the genus *Lamna*, a mackerel-shark.

porcelain (paw'səlin, -slin), *n.* a fine kind of earthenware, white, thin, and semi-transparent; ware made of this. *a.* pertaining to or composed of porcelain; fragile, delicate. **porcellaneous** (-lā'ni-), *a.* **porcellanite** (-sel'ənīt), *n.* a rock composed of hard metamorphosed clay.

porch (pawch), *n.* a covered structure before or extending from the entrance to a building; a covered approach to a doorway; (*N Am.*) a verandah. **porched**, *a.*

porcine (paw'sīn), *a.* pertaining to or resembling a pig.

porcupine (paw'kūpīn), *n.* any individual of the genus *Hystrix*, rodent quadrupeds covered with erectile, quill-like spines; one of various appliances or machines armed with pins, knives, teeth etc. **porcupine-fish**, *n.* a tropical fish, *Diodon hystrix*, covered with spines. **porcupine grass**, *n.* a coarse, spiky, tussocky grass, *Triodia*, that covers many areas in Australia, also known as spinifex. **porcupinish, porcupiny**, *a.*

pore[1] (paw), *n.* a minute opening, esp. a hole in the skin for absorption, perspiration etc.; one of the stomata or other apertures in the cuticle of a plant; a minute interstice between the molecules of a body. **porous**, *a.* having (many) pores; permeable to liquids etc. **porosity** (-ros'-), *n.* **porously**, *adv.* **porousness**, *n.*

pore[2] (paw), *v.i.* to look with steady, continued attention and application (at); to meditate or study patiently and persistently (over, upon etc.). *v.t.* to fatigue (the eyes) by persistent reading. **porer**, *n.*

porge (pawj), *v.t.* in the Jewish faith, to extract the sinews of (slaughtered animals) in order that they may be ceremonially clean. **porger**, *n.*

porgy (paw'ji), *n.* the name of a number of N American sea-fishes, including various species of *Calemus* and *Sparus*.

Porifera (pərif'ərə), *n.pl.* the sponges; the Foraminifera. **poriferal**, *a.* **poriferan**, *n., a.*

pork (pawk), *n.* the flesh of pigs, esp. fresh, as food. **pork pie**, *n.* a pie made of minced pork, usu. round with vertical sides. **pork-pie hat**, *n.* a round hat with flat crown and rolled-up brim. **porker, porklet**, (-lit), *n.* a pig raised for killing, esp. a young fattened pig. **porkling** (-ling), *n.* **porky**, *a.* like pork; (*coll.*) fat, fleshy.

porn(o) (pawn'nō), *n.* (*coll.*) short for PORNOGRAPHY, PORNOGRAPHIC.

pornocracy (pawnok'rəsi), *n.* the rule or domination of courtesans, as in the government of Rome during the 10th cent.

pornography (pawnog'rəfi), *n.* the obscene and exploitive depiction of erotic acts; written, graphic etc. material consisting of or containing the above. **pornographer**, *n.* **pornographic** (-graf'-), *a.* **pornographically**, *adv.*

poromeric (pōrəme'rik), *a.* permeable to water vapour, as certain synthetic leathers. *n.* a substance having this characteristic.

porphyria (pawfi'riə), *n.* one of a group of hereditary metabolic disorders characterized by an abnormal pigment in the urine, severe pain, photosensitivity, and periods of mental confusion.

porphyry (paw'firi), *n.* an igneous rock consisting of a felsitic or crypto-crystalline groundmass full of feldspar or quartz crystals; a rock quarried in Egypt having a purple groundmass with enclosed crystals of feldspar. **porphyritic** (-rit'-), *a.*

porpoise (paw'pəs), *n.* any of various dolphine-like cetaceans, about 5 ft. (1.5 m) long, with a blunt snout.

porridge (po'rij), *n.* a soft or semi-liquid food made by boiling meal etc. in water or milk till it thickens; a broth or stew of vegetables or meal; (*sl.*) a term of imprisonment.

porringer (po'rinjə), *n.* a small basin or bowl out of which soup etc. is eaten by children.

port[1] (pawt), *n.* a harbour, a sheltered piece of water into which vessels can enter and remain in safety; a town or other place having a harbour, esp. where goods are imported or exported under the customs authorities. **free port** FREE. **port of call**, a port at which a ship stops during a voyage; any stopping place on an itinerary. **port of entry**, an airport, harbour etc. having customs facilities through which goods or persons may enter or leave a country.

port[2] (pawt), *n.* a porthole; an aperture in a wall or the side of an armoured vehicle for firing through; (*Mach.*) an opening for the passage of steam, gas, water etc. a connector on a computer into which a peripheral can be plugged. **port-bar**, *n.* a bar to secure the ports of a ship in a gale. **porthole**, *n.* an aperture in a ship's side for light, air etc.; a small window in the side of a ship or aircraft; (*Mach.*) a passage for steam, gas etc. in a cylinder. **portlanyard, -rope**, *n.* a rope for drawing up a port-lid. **port-lid**, *n.*

port[3] (pawt), *n.* carriage, deportment. *v.t.* to carry or hold (a rifle etc.) in a slanting position across the body in front. **port-crayon**, *n.* a pencil-case; a handle to hold a crayon.

port[4] (pawt), *n.* a fortified dessert wine (usu. dark-red or tawny) made in Portugal.

port[5] (pawt), *n.* the left-hand side of a ship as one looks forward. *a.* towards or on the left. *v.t.* to turn or put (the helm) to the left side of a ship. *v.i.* to turn to port (of a ship).

Port., (*abbr.*) Portugal; Portuguese.

porta (paw'tə), *n.* the portal or aperture where veins, ducts etc. enter an organ, esp. the transverse fissure of the liver.

portable (paw'təbl), *a.* capable of being easily carried, esp. about the person; not bulky or heavy; of a pension, transferable as the holder changes jobs. *n.* a portable version of anything. **portability** (-bil'-), *n.* **portably**, *adv.*

portage (paw'tij), *n.* the act of carrying, transport; the cost of carriage; a break in a line of water-communication over which boats, goods etc. have to be carried; transportation of boats etc. over this. *v.t.* to carry over a portage. *v.i.* to make a portage.

Portakabin® (paw'təkabin), *n.* a portable building delivered intact to, or speedily erected on, a site as temporary offices etc.

portal[1] (paw'təl), *n.* a door, gate, gateway, an entrance, esp. one of an ornamental or imposing kind.

portal[2] (paw'təl), *a.* of or connected with the porta. **portal vein**, *n.* the large vein conveying blood to the liver.

portamento (pawtəmen'tō), *n.* (*pl.* **-menti** (-tē)), a smooth, continuous glide from one note to another across intervening tones.

portative (paw'tətiv), *a.* pertaining to or capable of carrying or supporting.

port-crayon PORT[3].

portcullis (pawtkŭl'is), *n.* a strong timber or iron grating, sliding in vertical grooves over a gateway, and let down to close the passage in case of assault. **portcullised**, *a.*

Porte (pawt), *n.* the old Imperial Turkish Government. [F *Sublime Porte*, sublime gate]

porte-cochère (pawtkoshēə'), *n.* a carriage-entrance leading

into a courtyard; a roof extending from the entrance of a building over a drive to shelter people entering or alighting from vehicles.
portend (pawtend'), *v.t.* to indicate by previous signs, to presage, to foreshadow; to be an omen of. **portent** (paw'tent), *n.* that which portends; an omen, esp. of evil; prophetic significance; a prodigy, a marvel. **portentous** (-ten'-), *a.* ominous; impressive; solemn; self-consciously solemn or meaningful. **portentously,** *adv.* **portentousness,** *n.*
porter[1] (paw'tə), *n.* a person employed to carry loads, esp. parcels, luggage etc. at a railway station, airport or hotel, or goods in a market; a person who transports patients and does other manual labour in a hospital; a dark-brown beer made from charred or chemically-coloured malt etc. (perh. so called from having been made specially for London porters). **porter-house,** *n.* a tavern at which porter etc. is sold; an eating-house, a chop-house. **porterhouse steak,** *n.* a choice cut of beef-steak next to the sirloin, and including part of the tenderloin. **porter's knot,** *n.* a pad worn on the shoulders by porters when carrying heavy loads. **porterage** (-rij), *n.* **porterly,** *a.*
porter[2] (paw'tə), *n.* a gatekeeper, a door-keeper esp. of a large building, who usu. regulates entry and answers enquiries. **porter's lodge,** *n.* a room, apartment or house beside a door or gateway for the porter's use. **porteress** (-rəs), **portress** (-tris), *n. fem.*
portfire (pawt'fīə), *n.* a slow match, formerly used for firing guns, now chiefly in mining etc.
portfolio (pawtfō'liō), *n.* a portable case for holding papers, drawings etc.; a collection of such papers; the office and duties of a minister of state; the investments made, or securities held, by an investor.
porthole PORT[2].
portico (paw'tikō), *n.* (*pl.* **-coes, -cos**) a colonnade, a roof supported by columns at systematic intervals; a porch with columns. **porticoed,** *a.*
portière (pawtyeə'), *n.* a door-curtain; a portress. [F]
portion (paw'shən), *n.* a part; a share, a part assigned, an allotment; a helping; a wife's fortune, a dowry; the part of an estate descending to an heir; one's destiny in life. *v.t.* to divide, to distribute; to allot, to endow. **portioner,** *n.* one who portions; (*Eccles.*) a portionist. **portionist,** *n.* at Merton College, Oxford, one of the scholars on the foundation; (*Eccles.*) a joint incumbent of a benefice. **portionless** (-lis), *a.*
Portland (pawt'lənd), *a.* of or derived from Portland, a peninsula in Dorset. **Portland cement,** *n.* a cement having the colour of Portland stone. **Portland stone,** *n.* a yellowish-white freestone, quarried in Portland, much used for building.
portly (pawt'li), *a.* dignified or stately in mien or appearance; stout, corpulent. **portliness,** *n.*
portmanteau (pawtman'tō), *n.* (*pl.* **-teaus, -teaux** (-tōz)) a long leather trunk or case for carrying apparel etc. in travelling. *a.* combining several uses or qualities. **portmanteau word,** *n.* an artificial word combining two distinct words as *chortle*, from *chuckle* and *snort*, coined by Lewis Carroll.
portrait (paw'trit), *n.* a likeness or representation of a person or animal, esp. from life; a vivid description. **portrait-gallery,** *n.* **portrait-painter, portraitist,** *n.* **portraiture** (-chə), *n.* a portrait; portraits collectively; the art of painting portraits; vivid description. **portray** (pətrā'), *v.t.* to make a portrait of; to describe; to play the role of; to present (as). **portrayal,** *n.* **portrayer,** *n.*
portress PORTER[2].
Portuguese (pawtūgēz'), *a.* of or pertaining to Portugal. *n.* (*pl.* **Portuguese**) a native or inhabitant of Portugal; the Portuguese language. **Portuguese man-of-war,** *n.* a large jellyfish with a painful sting.
portulaca (pawtūlă'kə), *n.* a genus of low succulent herbs with flowers opening only in direct sunshine, comprising the purslane.
posada (pəsah'də), *n.* a Spanish inn. [Sp.]
posaune (pōzow'ni), *n.* (*Organ*) a rich and powerful reed-stop. [G]
pose[1] (pōz), *v.t.* to place, to cause to take a certain attitude; to affirm, to lay down; to put forward, to ask; to present, to be the cause of. *v.i.* to assume an attitude or character; to attempt to impress by affecting an attitude or style, to behave affectedly; to appear or set up (as). *n.* a bodily or mental attitude or position, esp. one put on for effect; (*Dominoes*) the first play. **posé** (pō'zā), *a.* (*Her.*) applied to a lion, horse etc. standing still, with all its feet on the ground. **poser,** *n.* one who poses; a poseur. **poseur** (-zœ'), *n.* an affected person.
pose[2] (pōz), *v.t.* to puzzle, to cause to be at a loss. **poser,** *n.* one who or that which puzzles; a puzzling question or proposition. **posingly,** *adv.*
posh (posh), *a.* (*coll.*) smart, elegant, fashionable; (*sometimes derog.*) genteel, upper-class.
posit (poz'it), *v.t.* to place, to set in position; to lay down as a fact or principle, to assume, to postulate.
position (pəzish'ən), *n.* a location, the place occupied by a person or thing; the place belonging or assigned to a person or thing; (*Mil.*) an occupied and defended or a defensible point or area; (*Sport*) a player's place in a team formation or usual area of operation on the field of play; a posture; arrangement, disposition; a point of view, a stance; a situation, a state of affairs; situation relative to other persons or things; social rank; an office, an appointment; status, rank; a principle laid down, a proposition; the act of positing. *v.t.* to place in position; to locate. **in a position to,** able to. **positional,** *a.*
positive (poz'itiv), *a.* definitely, explicitly, or formally laid down or affirmed; explicit, express, definite; intrinsic, inherent, absolute, not relative; existing, real, actual; authoritatively laid down, prescribed by artificial enactment as distinguished from natural; incontestable, certain, undoubted; fully convinced; confident, cocksure, dogmatic; (*coll.*) downright, thorough; tending to emphasize the good or laudable aspects; constructive, helpful; (*Gram.*) simple, not comparative or superlative; (*Phil.*) practical, positivist; (*Phys.*) denoting the presence of some quality, not negative; having the same polarity as the charge of a proton; having relatively higher electrical potential; denoting the north-seeking pole of a magnet or the south pole of the earth; (*Math.*) denoting increase or progress, additive, greater than zero; (*Med.*) indicating the presence of a suspected condition or organism; (*Phot.*) exhibiting lights and shades in the same relations as in nature. *n.* that which may be affirmed; (*Gram.*) the positive degree, a positive adjective; (*Math.*) a positive quantity; a photograph in which the lights and shades are shown as in nature. **positive discrimination** DISCRIMINATE. **positive philosophy,** *n.* positivism. **positive sign,** *n.* the sign +, denoting addition. **positive vetting,** *n.* active investigation of a person's background etc. to check whether he or she is suitable for work involving national security. **positively,** *adv.* **positiveness, positivity** (-tiv'-), *n.* **positivism,** *n.* the philosophical system of Auguste Comte (1798–1857), which recognizes only observed phenomena and rejects speculation or metaphysics. **positivist,** *n.* **positivistic** (-vis'-), *a.*
positron (poz'itron), *n.* a positive electron.
posnet (pōs'nit), *n.* a small basin or pot used for boiling.
posology (pəsol'əji), *n.* the art or science treating of doses or the quantities to be administered; the science of quantity, mathematics. **posological** (-loj'-), *a.*
poss., (*abbr.*) possession; possessive; possible.
posse (pos'i), *n.* a body or force (of persons); a posse co-mitatus; (*coll.*) a gang; (*coll.*) a group of people of similar interests or background; (*Law*) possibility. **in posse,** within possibility, possible. **posse comitatus** (komi-

possess 625 **posterity**

tah'tas), *n.* (*Hist.*) a force which the sheriff of a county is empowered to raise in case of riot etc. [L]
possess (pəzes'), *v.t.* to have the ownership of, to own as property, to have full power over, to control (oneself, one's mind etc.); to occupy, to dominate; to imbue, to impress (with); to acquire; to gain, to hold; to inhabit. **to be possessed of**, to own. **to possess oneself of**, to acquire, to obtain as one's own. **possessed**, *a.* owned; owning; dominated (by an idea etc.); controlled (as by a devil), mad. **possessor**, *n.* **possessory**, *a.*
possession (pəzesh'ən), *n.* the act or state of possessing; holding or occupancy as owner; the exercise of such control as attaches to ownership, actual detention, or occupancy; that which is possessed; territory, esp. a subject dependency in a foreign country; (*pl.*) property, goods, wealth; self-possession; the state of being possessed or under physical or supernatural influence; (*Sport*) control of the ball. **in possession**, in actual occupancy, possessed (of); holding, possessing; position of a bailiff in a house. **to give possession**, to put another in possession. **to take possession of**, to enter on; to seize. **writ of possession**, an order directing a sheriff to put a person in possession.
possessive (pəzes'iv), *a.* of or pertaining to possession; showing a strong urge to possess or dominate, unwilling to share; (*Gram.*) denoting possession, genitive. *n.* the possessive case; a word in this case. **possessively**, *adv.* **possessiveness**, *n.*
posset (pos'it), *n.* a drink made of hot milk curdled with ale, wine etc.
possible (pos'ibl), *a.* that may happen, be done, or exist; not contrary to the nature of things; feasible, practicable; having a specified potential use or quality; that may be dealt with or put up with, tolerable, reasonable. *n.* that which is possible; a possibility; (*Shooting*) the highest score that can be made. **possibilist**, *n.* a member of a political party which aims only at reforms that are actually practicable. **possibility** (-bil'-), *n.* the state of being possible; a possible thing; a contingency; one who or that which has an outside to moderate chance of success, selection etc.; (*usu. pl.*) potential. **possibly**, *adv.* by any possible means; perhaps; by remote chance.
possum (pos'əm), *n.* (*coll.*) an opossum. **to play possum**, to feign death, sleep etc.; to dissemble.
post[1] (pōst), *n.* a piece of timber, metal etc. set upright, and intended as a support to something; a stake; a starting or winning post; an upright forming part of various structures, machines etc. *v.t.* to fix (*usu.* up) on a post or in a public place; to fasten bills etc. upon (a wall etc.); to advertise, to make known; to enter (a name) in a list posted up; to publish (the name of a ship) as overdue or missing.
post[2] (pōst), *n.* a fixed place, position, or station; a military station; the troops at such station; a fort; a place established for trading purposes, esp. in relatively unpopulated areas; a situation, job, appointment; an established system of letter-conveyance and delivery; orig. one of a series of men stationed at points along a road whose duty was to ride forward to the next man with letters; a courier, a messenger; a mail-cart; a post-office; a postal letter-box; a dispatch of mails; (*collect.*) the letters, packages etc. taken from a post-office or letter-box at one time; the letters delivered at a house at one time; a relay of horses; a bugle-call announcing the time of retiring for the night etc. *adv.* in relays of horses; express, with speed. *v.t.* to station, to place in a particular position; to transfer to another unit or location; to transmit by post; to put into a postal letter-box for transmission; to send by or as by post-horses; to send with speed; to transfer (accounts) to a ledger, to enter in this from a day-book etc. *v.i.* to travel with post-horses; to travel rapidly, to hurry. **first, last post**, (*Mil.*) the first or second of two bugle-calls announcing the time for retiring for the night.

to keep posted, to keep supplied with up-to-date information. **to post up**, to complete (a ledger) with entries of accounts from a day-book etc.; to supply with full information. **post-bag**, *n.* a mail-bag; mail received (esp. by a public figure, magazine, radio show etc.). **post-bill**, *n.* a post-office way-bill of letters etc. transmitted by mail. **post-boat**, *n.* a boat employed in postal work, a mail-boat, or one conveying passengers, a stage-boat. **post-boy, -rider**, *n.* a boy who carries the post; a boy who rides a post-horse, a postilion. **postcard**, *n.* one for sending by post unenclosed. **post-chaise**, *n.* a vehicle for travelling by post. **postcode**, *n.* a code of letters and numbers denoting a particular subsection of a postal area, used to help in sorting mail. **post-free**, *a.* carried free of charge for postage. **post-haste**, *adv.* with all speed. **post-horn**, *n.* a long straight horn formerly blown to signalize the arrival of a mail-coach. **post-horse**, *n.* a horse kept as a relay at an inn etc. for post or for travellers. **post-house**, *n.* a house where post-horses were kept for relays. **postman, -woman**, *n.* one who delivers letters brought by post. **postman's knock**, *n.* a children's game in which a kiss is the reward for delivering an imaginary letter. **postmark**, *n.* a mark stamped by post-office officials on letters etc., usu. stating place and date of dispatch, and serving to deface the postage-stamp. *v.t.* to stamp (an envelope etc.) with this. **postmaster**, *n.* the superintendent of a post-office. **postmaster general**, *n.* the executive head of a national post office. **postmastership**, *n.* **postmistress**, *n. fem.* **Post Office**, *n.* the public postal authority. **post-office**, *n.* a place for the receipt and delivery of letters etc. **post-office box**, a private numbered box at a post-office in which the holder's mail is deposited awaiting collection. **post-paid**, *a.* having the postage prepaid. **post-rider** POST-BOY. **post-road**, *n.* a road on which relays of horses were available for posting. **postwoman**, *n. fem.* POSTMAN. **posting**, *n.* an appointment to a particular job or place of work.
post- (pōst), *pref.* after; behind.
postage (pōs'tij), *n.* the fee for conveyance of a letter etc. by post. **postage-stamp**, *n.* an embossed or printed stamp or an adhesive label to indicate that postage has been paid.
postal (pōs'təl), *a.* pertaining to the mail service; carried on by post. **postal code** POSTCODE under POST[2]. **postal order**, *n.* an order for a sum of money (specified on the document) issued at one post-office for payment at another. **postal vote**, *n.* a vote submitted by post.
post-boy etc. POST[2].
post-classical (pōstklas'ikəl), *a.* later than the classical writers, artists etc., esp. those of Greece and Rome.
post-communion (pōstkəmūn'yən), *n.* that part of the eucharistic service which follows after the act of communion.
post-costal (pōstkos'təl), *a.* behind a rib.
post-date (pōstdāt'), *v.t.* to assign or mark with a date later than the actual one. *n.* a date later than the actual one.
postdiluvial (pōstdiloo'viəl), *a.* being or happening after the Flood. **postdiluvian**, *n., a.*
post-doctoral (pōstdok'tərəl), *a.* pertaining to studies, research etc. carried out after obtaining a doctorate.
post-entry (pōsten'tri), *n.* an additional or subsequent entry; a late entry (for a race etc.).
poster (pōs'tə), *n.* a large placard or advertising bill; one who posts this. **poster paint, colour**, *n.* an opaque, gum-based watercolour paint.
poste restante (pōst res'tät), *n.* a department in a post-office where letters are kept until called for. [F]
posterior (pəstiə'riə), *a.* coming or happening after; later; hinder. *n.* the buttocks. **posteriority** (-o'ri-), *n.* **posteriorly**, *adv.*
posterity (pəste'riti), *n.* one's descendants; succeeding generations.

postern (pos'tən), *n.* a small doorway or gateway at the side or back; a private entrance, esp. to a castle, town etc.; a way of escape.
poster paint POSTER.
post-exilian (pōstegzil'iən), **-exilic**, *a.* later than the Babylonian exile.
post-fix (pōstfiks'), *v.t.* to append (a letter etc.) at the end of a word. *n.* (pōst'-), a suffix.
post-glacial (pōstglā'shəl), *a.* later than the glacial period.
postgraduate (pōstgrad'ūət), *a.* carried on or awarded after graduation; working for a postgraduate qualification. *n.* a graduate who pursues a further course of study.
poste-haste etc. POST [2].
posthumous (pos'tūməs), *a.* (of a child) born after the death of the father; happening after one's decease; published after the death of the author or composer. **posthumously**, *adv.*
posthypnotic suggestion (pōst·hipnot'ik), *n.* giving instructions to a hypnotic subject which the latter will act on after emerging from the trance.
postiche (postēsh'), *a.* artificial, superadded (applied to superfluous ornament). *n.* an imitation, a sham; a hairpiece.
posticous (postī'kəs), *a.* (*Bot.*) on the hinder side; turned away from the axis.
postilion, postillion (pəstil'yən), *n.* one who rides on the near horse of the leaders or of a pair drawing a carriage.
post-impressionism (pōstimpresh'ənizm), *n.* the doctrines and methods of a school of (esp. French) painters of the late 19th cent. who rejected the naturalism and momentary effects of impressionism in favour of a use of pure colour for more formal or subjective ends. **post-impressionist**, *n.*, *a.*
posting POST [2].
postliminy (pōstlim'ini), *n.* (*Internat. Law*) the right of restoration of things taken in war to their former civil status or ownership on their coming back into the power of the nation to which they belonged.
postlude (pōst'lood), *n.* (*Mus.*) a closing piece or voluntary.
postman, -mark, -master etc. POST [2].
postmeridian (pōstmərid'iən), *a.* of or belonging to the afternoon; late. **post meridiem** (-diem), after midday (applied to the hours from noon to midnight, usu. abbr. *p.m.*).
post-millennial (pōstmilen'iəl), *a.* of or pertaining to a period after the millennium. **post-millennialism**, *n.* the doctrine that the second advent of Christ will follow the millennium. **post-millennialist**, *n.*
postmodernism (pōstmod'ənizm), *n.* a movement which rejects the basic tenets of 20th century modernism in art and architecture. **postmodernist**, *n.*, *a.*
post-mortem (pōstmaw'təm), *adv.* after death. *a.* made or occurring after death. *n.* an examination of a dead body; a subsequent analysis or review, esp. after defeat or failure. [L]
post-natal (pōstnā'təl), *a.* happening after birth.
post-nuptial (pōstnŭp'shəl), *a.* made or happening after marriage.
post-obit (pōstob'it), *a.* taking effect after death; postmortem. *n.* a bond securing payment of a sum of money to a lender on the death of a specified person from whose estate the borrower has expectations.
post-office etc. POST [2].
post-operative (pōstop'ərətiv), *a.* pertaining to the period just after a surgical operation.
post-oral (pōstaw'rəl), *a.* behind the mouth.
post-orbital (pōstaw'bitəl), *a.* behind the orbit of the eye.
post-partum (pōstpah'təm), *a.* after childbirth.
post-Pliocene (pōstplī'əsēn), *a.* (*Geol.*) pertaining to the formation immediately above the Pliocene.
postpone (pəspōn'), *v.t.* to put off, to defer, to delay; to regard as of minor importance to something else. *v.i.* (*Path.*) to be late in recurring. **postponement**, *n.* **postponer**, *n.*
postposition (pōstpəzish'ən), *n.* the act of placing after; the state of being placed after or behind; a word or particle placed after a word, esp. an enclitic. **postpositional**, *a.* **postpositive**, *a.* (*Gram.*) placed after something else. *n.* a postpositive word or particle.
postprandial (pōstpran'diəl), *a.* after dinner.
post-rider, post-road POST [2].
postscript (pōst'skript), *n.* a paragraph added to a letter after the writer's signature; any supplement added on to the end of a book, document, talk etc.; (later) additional information. **postscriptal** (-skrip'-), *a.*
post-Tertiary (pōst·tœ'shəri), *a.* pertaining to formations later than the Tertiary.
postulant (pos'tūlənt), *n.* a candidate for entry into a religious order or for an ecclesiastical office; one who demands.
postulate [1] (pos'tūlət), *n.* a position assumed without proof as being self-evident; a fundamental assumption; a necessary condition, an indispensable preliminary; a statement of the possibility of a simple operation such as a geometrical construction.
postulate [2] (pos'tūlāt), *v.t.* to demand, to claim; to assume without proof, to take as self-evident; to stipulate. **postulation**, *n.* **postulator**, *n.*
posture (pos'chə), *n.* a pose, attitude or arrangement of the parts of the body, the manner of holding the body; a mental attitude; situation, condition, state (of affairs etc.). *v.t.* to arrange the body and limbs of in a particular posture. *v.i.* to assume a posture, to pose; to endeavour to look or sound impressive. **posture-master**, *n.* one who teaches or practises artificial postures of the body. **postural**, *a.* **posturer**, *n. pōnere*, for put]
postviral syndrome (pōstvī'rəl), *n.* a condition occurring after a viral infection, characterized by prolonged fatigue, depression etc.
posy (pō'zi), *n.* a motto or short inscription, esp. in a ring; a bunch of flowers, a nosegay.
pot (pot), *n.* a round vessel of earthenware or metal, usu. deep relatively to breadth, for holding liquids etc.; a vessel of this kind used for cooking; a large drinking-cup of earthenware, pewter etc.; the quantity this holds; (*loosely*) a quart (1·14 l); a vessel used for various domestic or industrial purposes; a chamber-pot, a coffee-pot, a flower-pot, a teapot etc.; a chimney-pot; a wicker trap for catching lobsters etc.; (*Racing etc.*) a cup offered as a prize; (*often pl., coll.*) a large sum; the money or stakes in the pool in gambling games; a potbelly; (*sl.*) marijuana. *v.t.* (*past, p.p.* **potted**) to put into a pot or pots; to plant in pots; to season and preserve in pots etc.; (*Billiards*) to pocket; (*coll.*) to bring down, esp. with a pot-shot; to secure; to sit (a young child) on a potty; to shape clay as a potter. *v.i.* (*coll.*) to shoot (at). **big pot** BIG [1]. **to go to pot**, (*sl.*) to be ruined, to degenerate. **to keep the pot boiling** KEEP [1]. **pot-ale**, *n.* fermented grain as refuse from a distillery. **potbelly**, *n.* a protuberant belly; a pot-bellied person. **potbellied**, *a.* **pot-boiler**, *n.* a work of art or literature produced merely for money; one who produces this. **potbound**, *a.* of a plant, filling the pot with its roots, not having room to grow. **pot-boy, -man**, *n.* one employed in a public-house to clean pots etc. **pot-hanger**, *n.* a pothook. **pot-herb**, *n.* a culinary herb. **pothole**, *n.* a cauldron-shaped cavity in the rocky bed of a stream; a pit-like cavity in mountain limestone etc., usu. produced by a combination of faulting and water-action; a cavity in a roadway caused by wear or weathering. **potholer**, *n.* **potholing**, *n.* the exploration of underground caverns as a sport. **pot-hook**, *n.* an S-shaped hook for suspending a pot or kettle over a fire; a letter like a pot-hook, esp. in clumsy handwriting. **pot-house**, *n.* a low public-house.

pot-hunter, *n.* one who kills game, fish etc. for food or profit rather than sport; one who competes merely to win prizes. **pot-hunting**, *n.* **pot-lead**, *n.* blacklead, esp. as used on the hulls of racing yachts to reduce friction. **pot-lid**, *n.* **pot-luck**, *n.* whatever food may be available without special preparation; whatever luck or chance may offer. **pot-man** POT-BOY. **pot-roast**, *n.* a piece of meat stewed in a closed receptacle. **pot-shot**, *n.* a shot at game etc. that happens to be within easy range; a random shot; a shot for filling the pot, esp. one of an unsportsmanlike kind. **pot-still**, *n.* one in which the heat is applied directly to the pot containing the liquid to be distilled, used to make whisky. **potstone**, *n.* a granular variety of steatite; a large mass of flint found in chalk. **pot-valiant**, *a.* made courageous by drink. **potful**, *n.* **potted**, *a.* put in a pot; preserved in the form of a paste; condensed, abridged; of music, recorded for reproduction by gramophone etc.; (*coll.*) drunk.
potable (pō'təbl), *a.* drinkable. *n.* (*usu. pl.*) anything drinkable. **potability, potableness**, *n.*
potage (potahzh'), *n.* thick soup. [F]
potamic (pətam'ik), *a.* of or pertaining to rivers. **potamology** (potəmol'-), *n.*
potash (pot'ash), *n.* a powerful alkali, consisting of potassium carbonate in a crude form, orig. obtained from the ashes of plants; caustic potash; potassium or a potassium compound.
potassium (pətas'iəm), *n.* a bluish or pinkish white metallic element, at. no. 19; chem. symbol K. **potassium hydroxide** CAUSTIC POTASH under CAUSTIC. **potassic**, *a.*
potation (pōtā'shən), *n.* the act of drinking; a draught; a beverage; (*usu. pl.*) tippling. **potatory** (pō'-), *a.*
potato (pətā'tō), *n.* (*pl.* **-toes**) a plant, *Solanum tuberosum*, with edible farinaceous tubers; a tuber of this. **hot potato**, a controversial issue, something difficult or dangerous to deal with. **small potatoes**, (*sl.*) something very inferior and contemptible. **to drop like a hot potato**, to cease to have anything to do with (a person or subject which suddenly becomes risky or controversial). **potato-beetle, -bug**, *n.* the Colorado beetle. **potato-box, -trap**, *n.* (*sl.*) the mouth. **potato chip**, *n.* a long slice of potato fried in deep fat; (*N Am., S Afr.*) a potato crisp. **potato crisp**, *n.* a very thin slice of potato thus fried. **potato fern**, *n.* (*Austral.*) a fern with edible tubers, also called the native potato.
poteen (pətēn'), *n.* Irish whiskey illicitly distilled.
potence[1] (pō'təns), *n.* (*Her.*) a cross with ends resembling the head of a crutch or a T; (*Eng.*) a T-shaped framework. **potent**[1], *a.* (*Her.*) having the arms (of a cross) terminating in cross-pieces or crutch-heads. **potented, potentée**, *a.* (*Her.*).
potent[2] (pō'tənt), *a.* powerful, mighty; having great force or influence; cogent; strong, intoxicating; of a male, capable of having sexual intercourse. **potency, potence**[2], *n.* **potently**, *adv.* **potentate** (-tāt), *n.* one who possesses great power; a monarch, a ruler. **potential** (-ten'shəl), *a.* of energy, existing but not in action, latent; existing in possibility not in actuality; (*Gram.*) expressing possibility. *n.* anything that may be possible; a possibility; as yet undeveloped value, resources or ability; (*Gram.*) the potential mood; the voltage of a point above zero or earth; the work done in transferring a unit (of mass, electricity etc.) from infinity to a given point. **potential difference**, *n.* the work required to move an electrical charge between two points, measured in volts. **potential energy**, *n.* energy possessed by a body as a result of its position. **potential function**, *n.* (*Math.*) a quantity by the differentiation of which the value of the force at any point in space arising from any system of bodies can be obtained. **potentiality** (-shial'-), *n.* **potentialize, -ise**, *v.t.* to transform into a potential condition. **potentially**, *adv.* **potentiate** (-ten'shi-), *v.t.* esp. of drugs, to make potent; to make more effective. **potentiation**, *n.* **potentiator**, *n.* **potentiometer** (-om'itə), *n.* an instrument for measuring electromotive force or potential difference.
Potentilla (pōtəntil'ə), *n.* a genus of Rosaceae, comprising the cinquefoil, tormentil etc.
pother (podh'ə), *n.* bustle, confusion. *v.i.* to make a bustle or stir; to make a fuss. *v.t.* to harass.
pothole, pot-hook, pot-hunter etc. POT.
potion (pō'shən), *n.* a drink, a draught; a liquid mixture intended as a medicine, poison, or a magic charm.
potlatch (pot'lach), *n.* a ceremonial feast of Indians of the northwestern US involving emulation in the giving of extravagant gifts.
potoroo (potōroo'), *n.* (*Austral.*) (*pl.* **-roos**) the marsupial kangaroo-rat.
potpourri (pōpərē'), *n.* a mixture of dried flower-petals and spices, usu. kept in a bowl for perfuming a room; a literary miscellany, a musical medley etc. [F]
potsherd (pot'shœd), *n.* a broken piece of earthenware.
pottage (pot'ij), *n.* a kind of thick soup; porridge.
potter[1] (pot'ə), *n.* a maker of pottery. **potter's asthma, bronchitis, consumption**, *n.* an acute form of bronchitis caused by dust in pottery-manufacture. **potter's clay**, *n.* a tenacious clay containing kaolin, used for pottery. **potter's field**, *n.* (*N Am.*) a public burying-place for the poor or strangers. **potter's lathe**, *n.* a machine for moulding clay. **potter's wheel**, *n.* a horizontal wheel used in this. **pottern-ore**, *n.* an ore vitrifying with heat, used by potters to glaze their ware. **pottery**, *n.* earthenware; a place where this is manufactured, a potter's workshop; the making of earthenware.
potter[2] (pot'ə), *v.i.* to busy oneself in a desultory but generally agreeable way; to proceed in a leisurely and often random fashion. *v.t.* to waste or pass (time away) in a desultory way.
pottle (pot'l), *n.* a liquid measure of 4 pt. (2·3 l); a large tankard; a vessel or basket for holding fruit. **pottle-pot**, *n.* a 4 pt. (2·3 l) pot or tankard.
potto (pot'ō), *n.* (*pl.* **-ttos**) a W African lemuroid, *Perodicticus potto*.
potty[1] (pot'i), *a.* insignificant; crazy, foolish. **pottiness**, *n.*
potty[2] (pot'i), *n.* (*coll.*) a chamber pot, esp. one for use by small children.
pouch (powch), *n.* a small bag; a purse, a detachable pocket; a leather bag for holding cartridges etc.; the bag-like part in which marsupials carry their young; a pouch-like sac in plants. *v.t.* to put into a pouch; to pocket; to swallow; (*fig.*) to put up with; (*sl.*) to supply the pocket of, to tip. *v.i.* to form or resemble a pouch. **pouched, pouchy**, *a.*
pouf[1], **pouffe** (poof), *n.* a mode of dressing women's hair fashionable in the 18th cent.; a large, solid cushion used as a seat. [F]
pouf[2], **pouffe** POUF[1], POOF.
Poujadist (poozhah'dist), *n., a.* a follower of, or characteristic of this champion of the small man and tax reduction. [Pierre *Poujade*, F politician]
poulard (poo'lahd), *n.* a fattened or spayed hen.
poulpe (poolp), *n.* an octopus or cuttle-fish, esp. *Octopus vulgaris*.
poult (pōlt), *n.* a young pullet, partridge, turkey etc.
poulterer, *n.* one who deals in poultry for the table. **poultry** (-tri), *n.* domestic fowls, including barn-door fowls, geese, ducks, turkeys etc. **poultry-house, poultry-yard**, *n.*
poultice (pōl'tis), *n.* a usu. heated, moist preparation for applying to sores or inflamed parts of the body to reduce pain or inflammation. *v.t.* to apply a poultice to.
pounce[1] (powns), *n.* the claw of a bird of prey; a pouncing, an abrupt swoop, spring etc. *v.i.* to sweep down or spring upon and seize prey with the claws; to seize (upon), to dart or dash (upon) suddenly; to speak abruptly. **pounced**, *a.* (*Her.*) furnished with claws.
pounce[2] (powns), *n.* a fine powder formerly used to dry

up ink on manuscript; a powder used for sprinkling over a perforated pattern in order to transfer the design. *v.t.* to smooth with pounce or pumice; to mark out (a pattern) by means of pounce. **pounce-box**, *n.* a box out of which pounce is sprinkled. **pouncet-box** (-sit-), *n.* a box with a perforated lid for holding perfumes.

pound[1] (pownd), *n.* an avoirdupois unit of weight divided into 16 ounces and equal to approx. 0·454 kg; a troy unit of weight divided into 12 ounces and equal to approx. 0·373 kg; the basic monetary unit of the UK, divided into 100 (new) pence (formerly 20 shillings); the standard monetary unit of various other countries. *v.t.* to test the weight of (coins). **pound of flesh**, the exact amount owing to one, esp. when recovering it involves one's debtor in considerable suffering or trouble. **pound-cake**, *n.* a rich sweet cake, from the ingredients being pound for pound of each. **pound-foolish**, *a.* neglecting the care of large sums, esp. through trying to make small economies. **pound Scots**, *n.* (*Hist.*) 1s. 8d. **poundage** (-dij), *n.* an allowance, fee, commission etc. of so much in the pound; a percentage of the aggregate earnings of an industrial concern paid as or added to wages; a payment or charge per pound weight; the charge on a postal order. **pounder**, *n.* (*usu. in comb.*) a gun firing a shot of a specified number of pounds weight; a person worth or possessing a specified sum in pounds sterling; something weighing a pound, or a specified number of pounds, as a fish.

pound[2] (pownd), *n.* an enclosure for confining stray cattle etc.; an enclosure, a pen; a trap, a prison; a place whence there is no escape, esp. in hunting; a pond, a part between locks on a canal. *v.t.* to confine in or as in a pound; (*usu. in p.p.*) to shut in, to enclose in front and behind. **pound-keeper**, *n.* **pound-net**, *n.* a series of nets, set in shoal water, to form a trap. **poundage** (-dij), *n.* confinement in a pound.

pound[3] (pownd), *v.t.* to crush, to pulverize, to comminute; to strike heavily; to thump, to pommel. *v.i.* to strike heavy blows, to hammer (at, upon etc.); to fire heavy shot (at); to walk or go heavily along. **pounder**, *n.*

poundage POUND[1], POUND[2].
pounder POUND[1], POUND[3].

pour (paw), *v.t.* to cause (liquids etc.) to flow; to discharge, to emit copiously; to send (forth or out) in a stream or great numbers; to shed freely; to utter, to give vent to. *v.i.* to flow in a stream of rain, to fall copiously; to rush in great numbers; to come in a constant stream. *n.* a heavy fall, a downpour; the amount of molten material poured at one time. **to pour oil on troubled waters**, to exercise a calming or conciliatory influence. **pourer**, *n.*

pourboire (puəbwah'), *n.* a gratuity, a tip. [F]
pour-parler (puəpah'lā), *n.* a preliminary discussion with a view to formal negotiation. [F]
poussette (pooset'), *v.i.* to swing partners with hands joined as in a country dance. *n.* this figure.
poussin (poo'si), *n.* a small chicken for eating. [F]
pout[1] (powt), *n.* any of various fishes that have a pouting appearance (see EEL-POUT, WHITING-POUT).
pout[2] (powt), *v.i.* to thrust out the lips in sullenness, displeasure or contempt; of lips, to be protruded or prominent. *v.t.* to thrust out. *n.* a protrusion of the lips; a fit of sullenness. **in the pouts**, sullen. **pouter**, *n.* one who pouts; a variety of pigeon, from its way of inflating its crop. **poutingly**, *adv.*
poverty (pov'əti), *n.* the state of being poor; destitution; scarcity, meagreness, dearth (of); deficiency (in); inferiority. **poverty-stricken**, *a.* poor; inferior, mean. **poverty trap**, *n.* a situation in which any increase in one's earned income is immediately offset by a decrease in one's entitlement to state benefit, thus making it impossible to raise one's standard of living.
POW, (*abbr.*) prisoner of war.
pow (pow), *int.* an exclamation imitating the sound of an impact, blow etc.
powan (pow'ən), *n.* a freshwater fish found in some Scottish lochs.
powder (pow'də), *n.* any dry substance in the form of fine particles; dust; a cosmetic in the form of fine dust; a medicine in the form of powder; gun-powder. *v.t.* to reduce to powder; to put powder on; to sprinkle or cover with powder; to sprinkle with fine spots or figures for decoration. *v.i.* to become powder or like powder; to powder one's hair. **powder-blue**, *n.* a pale blue colour. *a.* having the colour of smalt. **powder-box**, *n.* a box for cosmetic powder etc. **powder-cart**, *n.* a cart for conveying ammunition for artillery. **powder-closet**, *n.* a small room where women's hair used to be powdered. **powder-down**, *n.* a peculiar kind of down-feathers disintegrating into fine powder, occurring in definite patches on herons etc. **powder-flask**, **-horn**, *n.* a case or horn fitted to hold gunpowder. **powder keg**, *n.* a small barrel to hold gunpowder; a potentially explosive place or situation. **powder-magazine**, *n.* a storing-place for gunpowder. **powder-mill**, *n.* works in which gunpowder is made. **powder-monkey**, *n.* (*Naut.*) a boy formerly employed to bring powder from the magazine to the guns. **powder-puff**, *n.* a soft pad or brush for applying powder to the skin. **powder-room**, *n.* the apartment in a ship where gunpowder is kept; (*coll.*) a ladies' cloakroom. **powdering-tub**, *n.* a tub in which meat is salted or pickled. **powderer**, *n.* **powderiness**, *n.* **powdery**, *a.*
power (pow'ə), *n.* ability to do or act so as to effect something; a mental or bodily faculty, or potential capacity; strength, force, energy, esp. as actually exerted; influence, dominion, authority (over); right or ability to control; legal authority or authorization; political ascendancy; a person or body invested with authority; a state having influence on other states; (*coll.*) a great deal; the product obtained by multiplication of a quantity or number into itself; the index showing the number of times a factor is multiplied by itself; mechanical energy as distinguished from hand labour; electricity; the capacity (of a machine etc.) for performing mechanical work; the rate at which energy is emitted or transferred, esp. the rate of doing work, measured in watts (joules per second), foot-pounds per second, or ergs per second; the magnifying capacity of a lens etc. *a.* concerned with power; worked by mechanical power; involving a high degree of physical strength or skill. *v.t.* to supply with, esp. motive power. *v.t., v.i.* (*coll.*) to (cause to) move with great force or speed. **in power**, in office. **in someone's power**, within the limits of a person's capabilities or authority; under someone's control, at someone's mercy. **more power to your elbow**, (*coll.*) a way of showing one's approval of someone's efforts by urging that person to continue and even intensify them. **power behind the throne**, a person with no official position in government who exercises a strong personal influence on a ruler. **power of attorney** ATTORNEY[2]. **the powers that be**, (*often facet.*) established authority. **power amplifier**, *n.* a low-frequency amplifier for powerful loudspeakers. **powerboat**, *n.* a boat propelled by a motor, esp. a speedboat. **power cut**, *n.* an interruption or reduction in the supply of electricity. **power dive**, *n.* (*Aviat.*) a steep dive under engine power. **power-dive**, *v.i.* **power drill, lathe, loom** etc., *n.* a drill, lathe, loom etc. worked by mechanical or electrical power. **power factor**, *n.* that fraction which is less than unity by which the produce of amperes and volts in an alternating-current circuit has to be multiplied in order to estimate the true power. **power-house**, *n.* a power station; (*coll.*) a very forceful and dynamic person or thing. **power pack**, *n.* a unit for converting a power supply to the voltage required by an electronic circuit. **power plant**, *n.* a power station; the machinery etc. used to generate power; the engine and related parts which power a car, aircraft etc. **power**

point, *n.* an electrical socket by which an appliance can be connected to the mains. **power politics**, *n.* diplomacy backed by armed force or the threat of it. **power-rail**, *n.* an insulated rail conveying current to the motors on electric railways of certain kinds. **power-station**, *n.* a building for the generation of power, esp. electrical power. **power steering**, *n.* a system in which the torque applied to a vehicle's steering wheel is augmented by engine power. **power transmission**, *n.* the transmission of electrical power from the generating system to the point of application. **powerful**, *a.* having great strength or energy; mighty, potent; impressing the mind, forcible, efficacious; producing great effects; (*coll.*) great, numerous, extreme. **powerfully**, *adv.* **powerfulness**, *n.* **powerless** (-lis), *a.* **powerlessly**, *adv.* **powerlessness**, *n.*
powwow (pow'wow), *n.* a N American Indian medicineman or wizard; magic rites for the cure of diseases; a meeting, talk or conference. *v.i.* to practise sorcery, esp. for healing the sick; to hold a powwow.
pox (poks), *n.* any disease characterized by the formation of pustules; syphilis. **poxy**, *a.* syphilitic; (*sl.*) bloody.
pozzolana (potsōlah'nə), **pozzuolana** (-swō-), *n.* a volcanic ash used in hydraulic cements. [It.]
PP, (*abbr.*) Parish Priest; past President.
pp, (*abbr.*) pages; for and on behalf of, by proxy; pianissimo.
p.p., (*abbr.*) past participle.
ppd, (*abbr.*) post-paid; prepaid.
PPE, (*abbr.*) Philosophy, Politics and Economics.
PPS, (*abbr.*) Parliamentary Private Secretary; further postscript.
PQ, (*abbr.*) Parti Québecois; Province of Quebec.
PR, (*abbr.*) proportional representation; public relations; Puerto Rico.
Pr[1], (*chem. symbol*) praseodymium.
Pr[2], (*abbr.*) priest; prince.
pr, (*abbr.*) pair; present; price.
praam (prahm), PRAM [1].
prabble (prab'l), *n.* a squabble, a quarrel. *v.i.* to chatter.
practicable (prak'tikəbl), *a.* capable of being done, feasible; of roads etc., usable, passable; of stage properties, functioning, real, not simulated; (*sl.*) easily taken in, gullible. **practicability** (-bil'-), **-ableness**, *n.* **practicably**, *adv.*
practical (prak'tikl), *a.* of, pertaining to, or governed by practice; derived from practice, experienced; capable of being used, available, serving, or suitable for use; pertaining to action, not theory or speculation; realistic, downto-earth; such in effect, virtual. **practical joke**, *n.* a joke or trick entailing some action and intended to make the victim look foolish. **practicality** (-kal'-), *n.* **practically**, *adv.* in a practical manner; virtually, in effect, as regards results.
practice[1] (prak'tis), *n.* habitual or customary action or procedure; habit, custom; the continued or systematic exercise of any profession, art, craft etc.; business, professional connection; actual performance, doing or execution, as opposed to theory or intention; conduct, dealings; regular, repeated exercise in order to gain proficiency in something; a rule for multiplying quantities of various denominations; legal procedure, the rules governing this. **in practice**, in the sphere of action; in training, in condition for working, acting, playing etc., effectively. **out of practice**, out of training. **practician** (-tish'ən), *n.* one who works or practises, a practitioner.
practice[2] PRACTISE.
practise, (*esp. N Am.*) **practice** (prak'tis), *v.t.* to do or perform habitually; to carry out; (to exercise a profession etc.); to exercise oneself in or on; to instruct, to exercise, to drill (in a subject, art etc.); to accustom. *v.i.* to exercise oneself; to exercise a profession or art; to do a thing or perform an act habitually. **practisant**, *n.* an agent, a plotter. **practised**, *a.* **practiser**, *n.*

practitioner (praktish'ənə), *n.* one who regularly practises any profession, esp. of medicine. **general practitioner**, one practising both medicine and surgery; a physician, not a specialist.
prad (prad), *n.* (*sl.*) a horse.
prae-, *pref.* PRE-.
praecipe (prē'sipē), *n.* a writ requiring something to be done, or a reason for its non-performance. [L]
praecocial (prikō'shəl), PRECOCIAL.
praecordia (prikaw'diə), *n.pl.* the chest and the parts it contains, the region about the heart.
praemunire (prēmūni'ri), *n.* (*Law*) a writ or process against a person charged with obeying or maintaining the papal authority in England; an offence against the Statute of Praemunire (1393) on which this is based; the penalty incurred by it.
praenomen (prēnō'men), PRENOMEN.
praesidium (prisid'iəm), PRESIDIUM.
praetor, (*esp. N Am.*) **pretor** (prē'tə), *n.* (*Rom. Hist.*) a Roman magistrate; orig. a consul as leader of the army; later a curule magistrate elected yearly to perform various judicial and consular duties. **praetorial** (pritaw'ri-), *a.* **praetorian**, *a.* of or pertaining to a praetor; pertaining to the body-guard of a Roman general or emperor, esp. the imperial body-guard established by Augustus. *n.* a soldier in this body-guard; a man of praetorian rank. **praetorian gate**, *n.* the gate of a Roman camp in front of the general's quarters towards the enemy.
pragmatic (pragmat'ik), *a.* pertaining to the affairs of a state; concerned with the causes and effects and the practical lessons of history; concerned with practicalities or expediency rather than principles; pragmatical. *n.* one busy in affairs; a busybody; a sovereign decree. **pragmatic sanction**, *n.* an ordinance made by a sovereign and constituting a fundamental law, esp. that of the Emperor Charles VI settling the succession to the throne of Austria. **pragmatical**, *a.* busy, diligent, officious, given to interfering in the affairs of others; dogmatic; relating to pragmatism; pragmatic. **pragmaticality** (-kal'-), **pragmaticalness**, *n.* **pragmatically**, *adv.* **pragmatism** (prag'-), *n.* pragmaticalness, officiousness; treatment of things, esp. in history, with regard to causes and effects; a practical approach to problems and affairs; (*Phil.*) the doctrine that our only test of the truth of human cognitions or philosophical principles is their practical results. **pragmatist**, *n.* **pragmatistic** (-tis'-), *a.* **pragmatize**, **-ise**, *v.t.* to represent (an imaginary thing) as real.
prairie (preə'ri), *n.* an extensive tract of level or rolling grassland, usu. destitute of trees, esp. in the western US. **prairie-chicken, -hen**, *n.* a N American grouse. **prairiedog**, *n.* a small rodent which lives in large communities on the prairies of N America. **prairie oyster**, *n.* a pickme-up of raw egg, Worcester sauce etc. **prairie-schooner**, *n.* (*N Am.*) an emigrants' name for the covered wagons used in crossing the western plains. **prairie-squirrel**, *n.* a N American ground-squirrel. **prairie value**, *n.* the value of land before labour has been expended on it. **prairie-wolf**, *n.* the coyote.
praise (prāz), *v.t.* to express approval and commendation of, to applaud; to extol, to glorify. *n.* praising, approbation, encomium; glorifying, extolling; an object of praise. **praisable**, *a.* **praiser**, *n.* **praiseful**, *a.* laudable, commendable. **praisefulness**, *n.* **praiseless** (-lis), *a.* **praiseworthy**, *a.* deserving of praise; laudable, commendable. **praiseworthily**, *adv.* **praiseworthiness**, *n.*
Prakrit (prah'krit), *n.* any of a group of Indian languages or dialects no longer used, based on Sanskrit.
praline (prah'lēn), *n.* a confection of almond or other nut with a brown coating of sugar.
pram[1] (prahm), *n.* a flat-bottomed barge or lighter used in Holland and the Baltic.
pram[2] (pram), *n.* a usu. four-wheeled carriage for a baby,

with a box-like body in which the child can lie down.
prance (prahns), *v.i.* to spring or caper on the hind legs, as a horse in high mettle; to walk or strut in a pompous or swaggering style. *n.* the act of prancing. **prancer**, *n.* **prancing**, *a.*
prandial (pran'diəl), *a.* (*facet.*) relating to dinner.
prang (prang), *v.t.* (*sl.*) to bomb heavily; to strike; to crash. *n.* a bombing raid; a crash.
prank[1] (prangk), *v.t.* to dress up in a showy fashion; to deck (out); to adorn (with). *v.i.* to make a show.
prank[2] (prangk), *n.* a trick, a playful act, a practical joke; a gambol, a capricious action. **prankful, prankish, pranky,** *a.* **prankishness,** *n.*
prase (präz), *n.* a dull leek-green translucent quartz. **prasinous,** *a.* of a light-green colour.
praseodymium (präziōdim'iəm), *n.* a rare metallic element, at. no. 59; chem. symbol Pr, occurring in certain rare-earth minerals.
prat (prat), *n.* a stupid or contemptible person.
prate (prāt), *v.i.* to chatter; to talk much and without purpose or reason; to babble. *v.t.* to utter foolishly; to boast idly about. *n.* idle or silly talk; unmeaning loquacity. **prater,** *n.* **prating,** *n.*, *a.* **pratingly,** *adv.*
pratfall (prat'fawl), *n.* (*esp. N Am.*) a fall on one's buttocks; a humiliating blunder or mishap.
pratique (prat'ek, -tēk'), *n.* licence to a ship to hold communication with a port after quarantine, or upon certification that the vessel has not come from an infected place. [F]
prattle (prat'l), *v.i.* to talk in a childish or foolish manner. *v.t.* to utter or divulge thus. *n.* childish or idle talk; a babbling sound, as of running water. **prattler,** *n.* **prattling,** *n.*, *a.*
Pravda (prahv'də), *n.* the official newspaper of the Central Committee of the Communist Party in the USSR.
pravity (prav'iti), *n.* depravity; corruption.
prawn (prawn), *n.* a small marine crustacean like a large shrimp.
praxinoscope (prak'sinəskōp), *n.* an optical toy in which successive positions of moving figures are blended in a rotating cylinder etc., so as to give reflections in a series of mirrors as of objects in motion.
praxis (prak'sis), *n.* use, practice, application as distinguished from theory; (*Gram.*) a collection of examples for practice.
pray (prā), *v.t.* to ask for with earnestness or submission; to entreat, to supplicate; to petition for, to beg; to address devoutly and earnestly. *v.i.* to address God with adoration or earnest entreaty; to make supplication, to beseech or petition (for). *int.* (*often iron.*) may I ask, I ask you. **prayer**[1] (prā'ə), *n.* one who prays. **praying mantis** MANTIS. **prayingly,** *adv.*
prayer[2] (preə), *n.* the act of praying; a solemn petition addressed to God or any object of worship; the practice of praying, a formula for praying; a prescribed formula of divine worship; a liturgy; (*often pl.*) a religious service; an entreaty; a memorial or petition. **not to have a prayer,** (*coll.*) to have not the slightest chance or hope. **prayer-book,** *n.* a book containing prayers and forms of devotion, esp. the Anglican Book of Common Prayer. **prayer-meeting,** *n.* a meeting for divine worship in which prayer is offered by several persons. **prayer rug,** *n.* a small carpet on which a Muslim kneels and prostrates himself while praying. **prayer-wheel, praying-machine,** *n.* a revolving wheel or cylinder on which written prayers are inscribed or fastened by Tibetan Buddhists. **prayerful,** *a.* given to prayer; devotional, devout. **prayerfully,** *adv.* **prayerfulness,** *n.* **prayerless** (-lis), *a.* **prayerlessly,** *adv.*
PRB, (*abbr.*) Pre-Raphaelite Brotherhood.
pre (prē), *prep.* (*coll.*) before.
pre-, *pref.* before, earlier than; in advance; in front of, anterior to; surpassingly.

preach (prēch), *v.i.* to deliver a sermon or public discourse on some religious subject; to give earnest religious or moral advice, esp. in an obtrusive or persistent way. *v.t.* to proclaim, to expound in a common or public discourse; to deliver (a sermon); to teach or advocate in this manner. *n.* (*coll.*) a preaching, a sermon. **to preach down,** to denounce or disparage by preaching; to preach against. **preachable,** *a.* **preacher,** *n.* **preachership,** *n.* **preachify** (-ifī), *v.i.* to hold forth in a sermon, esp. tediously; to sermonize. **preachification** (-fi-), *n.* **preachment,** *n.* (*usu.* derog.) a discourse or sermon. **preachy,** *a.* fond of preaching or sermonizing, disposed to preach. **preachiness,** *n.*
preacquaint (prēəkwānt'), *v.t.* to acquaint beforehand. **preacquaintance,** *n.*
pre-Adamite (prēad'əmīt), *n.* an inhabitant of this world before Adam; one who holds that there were persons in existence before Adam. *a.* existing before Adam; pertaining to the pre-Adamites. **pre-Adamic** (-dam'-), **pre-Adamitic** (-mit'-), *a.*
preadolescent (prēadəles'ənt), *n.*, *a.* (a person) in late childhood, not yet adolescent.
preamble (prēam'bl), *n.* an introductory statement, esp. the introductory portion of a statute setting forth succinctly its reasons and intentions; a preliminary event, fact etc. *v.i.* to make a preamble. **preambulary** (-bū-), *a.*
preambulate, *v.i.* to make a preamble. **preambulatory,** *a.*
preamplifier (prēam'plifiə), *n.* an amplifier used to boost a low-level signal and often to equalize it before feeding it to the main amplifier.
preapprehension (prēaprihen'shən), *n.* a preconceived opinion; a foreboding.
prearrange (prēəranj'), *v.t.* to arrange in advance. **prearranged,** *a.* **prearrangement,** *n.*
preaudience (prēaw'diəns), *n.* (*Law*) the right of being heard before another, precedence at the bar.
prebend (preb'ənd), *n.* the stipend or maintenance granted to a canon of a cathedral or collegiate church out of its revenue; the land or tithe yielding this; a prebendary's prebendaryship. **prebendal,** *a.* **prebendal stall, prebendary stall,** *n.* a prebendary's stall in a cathedral or his benefice. **prebendary,** *n.* the holder of a prebend. **prebendaryship,** *n.*
prec., (*abbr.*) preceding.
Precambrian (prēkam'briən), *n.*, *a.* (of or pertaining to) the earliest geological era.
precancel (prēkan'səl), *v.t.* to cancel (a postage stamp) before use.
precancerous (prēkan'sərəs), *a.* of a growth, not malignant, but likely to become so if left untreated.
precarious (prikeə'riəs), *a.* held by a doubtful tenure; not well-established, insecure, unstable; doubtful, dependent on chance, uncertain, hazardous. **precariously,** *adv.* **precariousness,** *n.*
precast (prēkahst'), *a.* of concrete blocks, panels etc., cast before being put in position.
precative (prek'ətiv), *a.* (*Gram.*) expressing a wish or entreaty. **precatory,** *a.* of the nature of or expressing a wish or recommendation.
precaution (prikaw'shən), *n.* previous caution, prudent foresight; a measure taken beforehand to guard against or bring about something. *v.t.* to caution or warn beforehand. **precautionary,** *a.* **precautiously,** *adv.*
precede (prisēd'), *v.t.* to go before in time, order, rank or importance; to walk in front of; to cause to come before, to preface or prelude. *v.i.* to go or come before; to have precedence. **precedence** (pres'i-), *n.* the act or state of preceding; priority; superiority; the right to a higher position or a place in advance of others at public ceremonies, social functions etc. **precedent,** *n.* something done or said which may serve as an example to be followed in a similar case, esp. a legal decision, usage etc.; a necessary antecedent. *a.* (*also* prisē'dənt) going before in time, order,

rank etc.; antecedent. **preceded,** *a.* having or warranted by a precedent. **precedently,** *adv.* **preceding,** *a.* going before in time, order etc.; previous; immediately before.

precentor (prisen'tə), *n.* a cleric who directs choral services in a cathedral; a person who leads the singing of choir or congregation; in Presbyterian churches, the leader of the psalmody. **precent,** *v.i.* to act as precentor. *v.t.* to lead the singing of (psalms etc.). **precentorship,** *n.* **presentress** (-tris), **precentrix** (-triks), *n. fem.*

precept (prē'sept), *n.* a command, a mandate; an injunction respecting conduct; a maxim; a writ, a warrant; a sheriff's order to hold an election; an order for the levying or collection of a rate. **preceptive** (-sep'-), **preceptual** (-sep'chəl), *a.* of the nature of a precept; containing or giving moral instruction; didactic. **preceptor** (-sep'-), *n.* a teacher, an instructor; the head of a preceptory among the Knights Templars. **preceptoral, preceptorial** (-taw'ri-), *a.* **preceptorship,** *n.* **preceptory** (-sep'-), *n.* a subordinate house or community of the Knights Templars; the estate, manor etc. pertaining to this. **preceptress** (-sep'tris), *n. fem.*

precession (prisesh'ən), *n.* precedence in time or order. **precession of the equinoxes,** (*Astron.*) a slow but continual shifting of the equinoctial points from east to west, occasioned by the earth's axis slowly revolving in a small circle about the pole of the ecliptic, causing an earlier occurrence of the equinoxes in successive sidereal years. **precessional,** *a.*

pre-Christian (prēkris'chən), *a.* of or pertaining to the times before Christ or before Christianity.

precinct (prē'singkt), *n.* the space enclosed by the walls or boundaries of a place, esp. a church; a boundary, a limit; a pedestrianized area of a town set aside for a particular activity, usu. shopping; (*N Am.*) a municipal police district; (*N Am.*) a polling district; (*pl.*) the environs or immediate surroundings (of).

precious (presh'əs), *a.* of great price or value; very costly; highly esteemed, dear, beloved; affected, over-refined in manner, style, workmanship etc.; (*iron.*) worthless, rascally. *adv.* very, extremely. **preciosity** (-ios'-), *n.* overfastidiousness or affected delicacy in the use of language, in workmanship etc. **precious metals,** *n.pl.* gold, silver and (sometimes) platinum. **precious stone,** *n.* a gem. **preciously,** *adv.* **preciousness,** *n.*

precipice (pres'ipis), *n.* a vertical or very steep cliff; the edge of a cliff, hence a situation of extreme danger.

precipitate (prisip'itāt), *v.t.* to throw headlong; to urge on with eager haste or violence; to hasten; to bring on, esp. prematurely. (*Chem.*) to cause (a substance) to be deposited at the bottom of a vessel, as from a solution; to cause (moisture) to condense and be deposited, as from vapour; to cause to fall as rain, snow etc. *v.i.* of a substance in solution, to fall to the bottom of a vessel; of vapour, to condense and be deposited in drops; to fall as rain, snow etc. *a.* (-tət), headlong; flowing or rushing with haste and violence; hasty, rash, inconsiderate; adopted without due deliberation. *n.* a solid substance deposited from a state of solution. **precipitable,** *a.* **precipitability** (-bil'-), *n.* **precipitant,** *a.* falling or rushing headlong; headlong, precipitate. *n.* any substance that, being added to a solution, causes precipitation. **precipitately,** *adv.* **precipitance, -ancy, precipitateness,** *n.* **precipitation,** *n.* the act of precipitating, the state of being precipitated; violent speed; rash haste; (the amount of) rain, snow, sleet etc. **precipitator,** *n.*

precipitous (prisip'itəs), *a.* like or of the nature of a precipice, very steep. **precipitously,** *adv.* **precipitousness,** *n.*

précis (prā'sē), *n.* an abstract, a summary. *v.t.* to make a précis of. [F]

precise (prisīs'), *a.* definite, sharply defined or stated; accurate, exact; strictly observant of rule, punctilious, over-nice, over-scrupulous; particular, identical. **precisely,** *adv.* in a precise manner; exactly, quite so. **preciseness,** *n.* **precisian** (-sizh'ən), *n.* a punctilious person; one rigidly observant of rules etc., a formalist, a stickler. *a.* precise, punctilious; formal. **precisianism,** *n.* **precision** (-sizh'ən), *n.* accuracy, exactness. *a.* characterized by great accuracy in execution; intended for very accurate measurement or operation. **precisionist,** *n.* **precisionize, -ise,** *v.t.* **precisive,** *a.*

preclassical (prēklas'ikl), *a.* belonging to or characteristic of the period before the classical (of Greece or Rome).

preclude (priklood'), *v.t.* to shut out, to exclude; to prevent; to render inoperative; to neutralize. **preclusion** (-zhən), *n.* **preclusive** (-siv), *a.* **preclusively,** *adv.*

precocial (prikō'shəl), *a.* (having young which are) hatched with a complete covering of down and capable of leaving the nest within a very short time. **precocious,** *a.* developing or ripe before the normal time; prematurely developed intellectually; characteristic of such development; forward, pert. **precociously,** *adv.* **precociousness,** **precosity** (-kos'-), *n.*

precognition (prēkəgnish'ən), *n.* foreknowledge; clairvoyance. **precognitive,** *a.*

pre-Columbian (prēkəlŭm'biən), *a.* referring to American cultures, their history and artefacts, before the arrival of Christopher Columbus (1492).

preconceive (prēkənsēv'), *v.t.* to form an opinion or idea of beforehand. **preconceit** (-sēt'), **preconception** (-sep'-), *n.*

preconcert (prēkənsœt'), *v.t.* to contrive or agree on by previous arrangement. *n.* (-kon'sət), an arrangement previously made. **preconcertedly,** *adv.* **preconcertedness,** *n.*

precondition (prēkəndish'ən), *n.* a necessary preliminary condition. *v.t.* to prepare beforehand, to put into a desired condition or frame of mind beforehand.

preconize, -ise, (prē'kəniz), *v.t.* to proclaim publicly; to cite or summon publicly; of the Pope, to confirm publicly (an appointment or one nominated). **preconization,** *n.*

pre-Conquest (prēkon'kwest), *n.* before the time of the Norman Conquest (1066).

preconscious (prēkon'shəs), *a.* pertaining to a state antecedent to consciousness.

precontract (prēkon'trakt), *n.* a previous contract. *v.i.*, *v.t.* (-trakt'), to contract beforehand.

precook (prēkuk'), *v.t.* to cook beforehand.

precordia (prikaw'diə), FRAECORDIA.

precostal (prikos'təl), *a.* in front of the ribs.

precritical (prēkrit'ikəl), *a.* preceding the critical treatment of a subject, esp. preceding the critical philosophy of Kant.

precurrent (prikŭ'rənt), *a.* occurring beforehand, precursory. **precursive,** *a.* precursory. **precursor,** *n.* a forerunner, a harbinger; one who or that which precedes the approach of anything; a predecessor in office etc. **precursory,** *a.* preceding and indicating ¬s a forerunner or harbinger; preliminary, introductory.

pred., (*abbr.*) predicate.

predaceous, -acious (pridā'shəs), *a.* living by prey, predatory; pertaining to animals living by prey. **predacean,** *n.* **predacity** (-das'-), *n.* **predation,** *n.* the way of life of a predator, the relationship between a predator and its prey; depredation. **predator** (pred'ə-), *n.* a predatory animal. **predatory,** *a.* habitually hunting and killing other animals for food; living by plunder; pertaining to or characterized by plunder or pillage; rapacious, exploitive.

pre-Darwinian (prēdahwin'iən), *a.* preceding the doctrines of evolution etc. propounded by Charles Darwin in 1859.

predate (pridāt'), *v.t.* to antedate.

predecease (prēdisēs'), *n.* the death of a person before some other. *v.t.* to die before (a particular person).

predecessor (prē'disesə), *n.* one who precedes another in any position, office etc.; a thing preceding another thing;

predefine   632   prefix

a forefather, an ancestor.
predefine (prēdifīn'), *v.t.* to define, limit or settle beforehand. **predefinition** (-defi-), *n.*
predella (pridel'ə), *n.* (*pl.* **predelle** (-le)) the platform on which an altar stands or the highest of a series of altar-steps; a painting on the face of this; a painting on a step- or shelf-like appendage, usu. at the back of the altar. [It.]
predesignate (prēdez'ignāt), *v.t.* to designate or indicate beforehand. **predesignation,** *n.*
predestinate (prides'tināt), *v.t.* to predestine. *a.* (-nət), ordained or appointed beforehand. **predestinarian** (-neə'ri-), *a.* pertaining to predestination. *n.* a believer in predestination. **predestination,** *n.* the act of predestining, esp. the act of God in foreordaining some to salvation and some to perdition. **predestinator,** *n.* **predestine,** *v.t.* to appoint beforehand by irreversible decree; to pre-ordain (to salvation, to do a certain deed etc.); to predetermine.
predetermine (prēditœ'min), *v.t.* to determine or settle beforehand; to foreordain; to predestine. *v.i.* to determine beforehand. **predeterminable,** *a.* **predeterminate** (-nət), *a.* **predetermination,** *n.*
predial (prē'diəl), *a.* consisting of lands or farms; attached to lands or farms; arising from or produced by land.
predicable (pred'ikəbl), *a.* capable of being predicated or affirmed of something. *n.* anything that may be predicated of something; (*Log.*) one of Aristotle's five classes of predicates – genus, species, difference, property, accident. **predicability** (-bil'-), *n.*
predicament (pridik'əmənt), *n.* that which is predicted, a category; a particular situation, condition or position, esp. a critical, dangerous or unpleasant one. **predicamental** (predikəmen'-), *a.*
predicant (pred'ikənt), *n.* a preaching friar, esp. a Dominican; a predikant. *a.* engaged in preaching.
predicate (pred'ikāt), *v.t.* to affirm, to assert as a property etc.; (*Log.*) to assert about the subject of a proposition; (*N Am.*) to found, to base (an argument etc. on). *v.i.* to make an affirmation. *n.* (-kət), (*Log.*) that which is predicated, that which is affirmed or denied of the subject; (*Gram.*) the entire statement made about the subject, including the copula as well as the logical predicate; an inherent quality. **predication,** *n.* **predicative** (-dik'-), *a.* **predicatively,** *adv.*
predicatory (pred'ikətəri), *a.* of or pertaining to a preacher or to preaching.
predict (pridikt'), *v.t.* to forecast; to foretell, to prophesy.
predictable, *a.* able to be forecast or foretold; (occurring or apt to behave in a manner which is) easily foreseen. **predictability** (-bil'-), *n.* **prediction** (-dik'-), *n.* **predictive,** *a.* **predictively,** *adv.* **predictor,** *n.* one who predicts; a range-finding and radar device for anti-aircraft use.
predigest (prēdijest', -di-), *v.t.* to digest partially before introducing into the stomach; to render more easily digestible. **predigested,** *a.* **predigestion,** *n.*
predikant (pred'ikənt), *n.* a minister of the Dutch Reformed Church, esp. in S Africa. [Dut.]
predilection (prēdilek'shən, pred-), *n.* a prepossession in favour of something, a preference, a partiality.
predispose (prēdispōz'), *v.t.* to dispose or incline beforehand; to make susceptible or liable to. **predisponent** (-nənt), *n., a.* **predisposition** (-zish'ən), *n.*
predominate (pridom'ināt), *v.i.* to be superior in strength, influence or authority; to prevail, to have the ascendancy (over); to have control (over); to preponderate. **predominance, -ancy,** *n.* **predominant,** *a.* predominating (over); superior, overruling, controlling. **predominantly,** *adv.* **predominatingly,** *adv.*
pre-eclampsia (prēiklamp'siə), *n.* a serious toxic condition occurring in late pregnancy, characterized by high blood pressure and oedema.
pre-elect (prēilekt'), *v.t.* to elect beforehand. **pre-election,** *n.*

pre-embryo (prēem'briō), *n.* name given to the developing embryo during the first 14 days after fertilization of the ovum.
pre-eminent (prēem'inənt), *a.* eminent beyond others; superior to or surpassing all others, outstanding. **pre-eminence,** *n.* **pre-eminently,** *adv.*
pre-empt (priempt'), *v.t.* to secure by pre-emption; to seize on to the exclusion of others; to act before another (in order to thwart), to anticipate. *v.i.* (*Bridge*) to make a pre-emptive bid. **pre-emption,** *n.* the act or right of buying before others; a government's right to seize the property of subjects of another state while in transit, esp. in wartime. **pre-emptive,** *a.* anticipating, forestalling. **pre-emptive bid,** *n.* (*Bridge*) an unusually high bid intended to shut out opposition. **pre-emptive strike,** *n.* an attack on enemy installations intended to forestall a suspected attack on oneself. **pre-emptor,** *n.*
preen (prēn), *v.t., v.i.* to clean and arrange (feathers) using the beak; to take great trouble with; an excessive interest in (one's appearance); to pride or congratulate oneself (on).
pre-engage (prēingāj'), *v.t.* to engage by previous contract or pledge; to preoccupy; to engage in conflict beforehand.
pre-engagement, *n.*
pre-establish (prēistab'lish), *v.t.* to establish beforehand.
pre-established harmony HARMONY.
pre-exilian (prēigzil'iən), *a.* before the period of exile, esp. of the Jewish exile in Babylon. **pre-exilic,** *a.*
pre-exist (prēigzist'), *v.i.* to exist previously. **pre-existence,** *n.* **pre-existent,** *a.*
pref., (*abbr.*) preface; preference; preferred; prefix.
prefab (prē'fab), *n.* (*coll.*) a prefabricated building, esp. a small house.
prefabricate (prēfab'rikāt), *v.t.* to manufacture the component parts of (a structure) in advance for rapid on-site assembly. **prefabrication,** *n.*
preface (pref'əs), *n.* something spoken or written as introductory to a discourse or book; an introduction; a preamble, a prelude; the thanksgiving etc. forming the prelude to the consecration of the Eucharist. *v.t.* to furnish with a preface; to introduce (with preliminary remarks etc.). **prefacer,** *n.* **prefatorial** (-taw'ri-), **prefatory** (pref'-), *a.* **prefatorily,** *adv.*
prefect (prē'fekt), *n.* (*Rom. Ant.*) a commander, a governor, a chief magistrate; the civil governor of a department in France, or of a province in Italy; in some schools, a senior pupil with limited disciplinary powers over others.
prefectoral (-fek'-), **-torial** (-taw'ri-), *a.* **prefecture** (-chə), *n.* the office, jurisdiction, official residence, or the term of office of a prefect. **prefectural** (-fek'chə-), *a.*
prefer (prifœ'), *v.t.* (*past, p.p.* **preferred**) to set before, to hold in higher estimation, to like better; to bring forward, to submit; to promote; to recommend, to favour. **preferred debt,** *n.* one having priority of payment. **preferred shares, stock** etc., *n.pl.* preference shares etc. **preferable** (pref'ə-), *a.* **preferability** (-bil'-), *n.* **preferably,** *adv.* **preference** (pref'ə-), *n.* the act of preferring; liking for one thing over another; right or liberty of choice; that which is preferred; favour displayed towards a person or country before others, esp. in commercial relations; (*Law*) priority of right to payment, esp. of debts. **preference-bonds, -shares, -stock,** *n.pl.* those entitled to a dividend before ordinary shares. **preferential** (prefəren'shəl), *n.* giving, receiving or constituting preference; favouring certain countries in commercial relations. **preferentialism,** *n.* **preferentialist,** *n.* **preferentially,** *adv.* **preferment,** *n.* advancement; promotion; a superior office or dignity, esp. in the church.
prefigure (prēfig'ə), *v.t.* to represent by antecedent figures, types or similitudes; to foreshadow; to picture to oneself beforehand. **prefiguration,** *n.* **prefigurative,** *a.*
prefix (prēfiks'), *v.t.* to put, place or set in front of; to

prefloration 633 **premunire**

attach at the beginning (as an introduction, prefix etc.).
n. (pré'-), a letter, syllable or word put at the beginning of a word to modify the meaning; a title prefixed to a name. **prefixture** (-chə), *n.*
prefloration (prēflərā'shən), *n.* (*Bot.*) aestivation.
preform (prēfawm'), *v.t.* to form beforehand. **preformation**, *n.* the act of preforming. **preformative**, *a.* forming beforehand. *n.* a formative letter or other element prefixed to a word.
prefrontal (prēfrŭn'təl), *a.* situated in front of the frontal bone or the frontal region of the skull. *n.* a prefrontal bone, esp. in reptiles and fishes.
preggers (preg'əz), *a.* (*coll.*) pregnant.
pre-glacial (prēglā'shəl), *a.* belonging to the period before the glacial epoch.
pregnable (preg'nəbl), *a.* capable of being taken by force.
pregnant (preg'nənt), *a.* being with child or young, gravid; fruitful, big (with consequences etc.); inventive, imaginative; full of meaning or suggestion, significant; portentous, fraught; (*Gram. etc.*) implying more than is expressed. **pregnancy**, *n.* **pregnantly**, *adv.*
prehallux (prihal'əks), *n.* a rudimentary digit or toe found in certain mammals, reptiles etc.
preheat (prēhēt'), *v.t.* to heat beforehand.
prehensile (prihen'sīl), *a.* seizing, grasping; adapted to seizing or grasping, as the tails of monkeys. **prehensility** (prēhensil'-), *n.* **prehension** (-shən), *n.* the act of taking hold of or seizing; apprehension. **prehensive**, *a.* **prehensor**, *n.*
prehistoric (prēhisto'rik), *a.* of or pertaining to the time prior to that known to written history. **prehistorically**, *adv.* **prehistory** (-his'-), *n.*
pre-ignition (prēignish'ən), *n.* premature ignition of the explosive mixture in the cylinder of an internal-combustion engine.
prejudge (prējŭj'), *v.t.* to judge before a case has been fully heard, to condemn in advance; to form a premature opinion about. **prejudgment**, **prejudication** (-joodikā'-), *n.*
prejudice (prej'ədis), *n.* opinion, bias or judgment formed without due consideration of facts or arguments; intolerance or hostility toward a particular group, race etc.; mischief, damage or detriment arising from unfair judgment or action. *v.t.* to prepossess with prejudice, to bias; to affect injuriously, esp. to impair the validity of a right etc. **without prejudice**, (*Law*) without impairing any pre-existing right, detracting from any subsequent claim, or admitting any liability. **prejudiced**, *a.* prepossessed, biased. **prejudicial** (-dish'əl), *a.* causing prejudice or injury; mischievous, detrimental. **prejudicially**, *adv.*
prelapsarian (prēlapseə'riən), *a.* of the time before the Fall and the consequent loss of innocence.
prelate (prel'ət), *n.* an ecclesiastical dignitary of the highest order, as an archbishop, bishop etc., formerly including abbot and prior. **prelacy**, *n.* the office, dignity or see of a prelate; (*collect.*) prelates; episcopacy (in a hostile sense). **prelateship**, *n.* **prelatess** (-tis), *n.* an abbess or prioress; (*facet.*) the wife of a prelate. **prelatic, -ical** (-lat'-), *a.* **prelatically**, *adv.* **prelatize, -ise**, *v.i.* to support or encourage prelacy. *v.t.* to bring under the influence of the prelacy.
prelature (-chə), *n.* prelacy.
prelect (prilekt'), *v.i.* to read a lecture or discourse in public. **prelection**, *n.* **prelector**, *n.*
prelibation (prēlibā'shən), *n.* a foretaste.
preliminary (prilim'inəri), *a.* introductory; previous to the main business or discourse. *n.* something introductory; (*pl.*) introductory or preparatory arrangements etc. **preliminarily**, *adv.* **prelims** (prē'limz), *n.pl.* preliminary matter of a book; preliminary examinations at university.
prelingual (prēling'gwəl), *a.* preceding the acquisition or development of language.
prelude (prel'ūd), *n.* something done, happening etc., introductory or preparatory to that which follows; a harbinger, a precursor; (*Mus.*) a short introductory strain preceding the principal movement, a piece played as introduction to a suite. *v.t.* to perform or serve as a prelude to; to introduce with a prelude; to usher in, to foreshadow. *v.i.* to serve as a prelude (to); to begin with a prelude; (*Mus.*) to play a prelude. **preludial** (-loo'-), *a.* **preludize, -ise**, *v.i.* **prelusive** (-loo'siv), *a.* **prelusively**, *adv.* **prelusory**, *a.*
premarital (prēma'ritəl), *a.* occurring before marriage.
premature (premətū'-, prem'-), *a.* ripe or mature too soon; happening, arriving, existing or performed before the proper time; born after a gestation period of less than 37 weeks. *n.* (*Mil.*) the premature explosion of a shell. **prematurely** (-tū'-), *adv.* **prematureness, prematurity** (-tū'-), *n.*
premaxillary (prēmaksil'əri), *n., a.* (the bone) situated in front of the maxilla or upper jaw.
premed (prēmed'), *n., a.* short for PREMEDICAL, PREMEDICATION.
premedical (prēmed'ikəl), *a.* of or pertaining to a course of study undertaken before medical studies. *n.* a premedical student; premedical studies. **premedicate**, *v.t.* **premedication**, *n.* drugs administered to sedate and to prepare a patient for general anaesthesia.
premeditate (primed'itāt), *v.t.* to meditate on beforehand; to plan and contrive beforehand. *v.i.* to deliberate previously. **premeditatedly**, *adv.* **premeditation**, *n.* **premeditator**, *n.*
premenstrual (prēmen'strual), *a.* preceding menstruation. **premenstrual tension**, *n.* nervous tension caused by the hormonal changes which precede menstruation.
premier (prem'iə), *a.* first, chief, principal. *n.* a prime or chief minister. **premiership**, *n.*
première (prem'ieə, -iə), *n.* a first performance of a play or film. *v.t.* to give a first performance of. [F]
premillennial (prēmilen'iəl), *a.* previous to the millennium. **premillenarian** (-neə'ri-), *n.* one believing that the Second Advent will precede the millennium. **premillenarianism, premillennialism**, *n.*
premise (prem'is), *n.* a proposition laid down, assumed or proved from which another is inferred; (*Log.*) one of the two propositions of a syllogism from which the conclusion is drawn (see MAJOR, MINOR). *v.t.* (primīz', prem'is), to put forward as preparatory to what is to follow; to lay down as an antecedent proposition or condition. *v.i.* to lay down antecedent propositions.
premises (prem'isiz), *n.pl.* (*Law*) matters previously set forth (in a deed or conveyance), esp. the aforesaid house or lands etc.; hence a piece of land and the buildings upon it, esp. considered as a place of business.
premiss (prem'is), PREMISE.
premium (prē'miəm), *n.* a reward, a recompense, a prize; a sum paid in addition to interest, wages etc., a bonus; a fee for instruction in craft, profession etc.; a payment (usu. periodical) made for insurance; the rate at which shares, money etc. are selling above their nominal value. **at a premium**, above the nominal value, above par; in great demand. **to put a premium on**, to render more than usually valuable or advantageous. **Premium Bond**, *n.* British government bond bearing no interest but subject to a monthly draw for money prizes.
premolar (prēmō'lə), *n., a.* (one of the teeth) situated in front of the molars.
premonition (premənish'ən), *n.* previous warning or notice; a foreboding, a presentiment. **premonitor** (-mon'-), *n.* **premonitory**, *a.*
Premonstratensian (primonstrəten'siən), *n.* a member of an order of regular canons, founded at Prémontré, near Laon, France, in 1119, or of the corresponding order of nuns. *a.* belonging to the Premonstratensians. **Premonstrant** (-mon'-), *n.*
premunire (prēmūni'ri), PRAEMUNIRE.

prenatal (prēnā'təl), *a.* anterior to birth. **prenatally,** *adv.*
prenti (pren'ti), *n.* the lizard of Central Australia.
prentice etc. APPRENTICE.
preoccupy (priok'ūpī), *v.t.* to take possession of beforehand or before another; to prepossess; to pre-engage, to engross (the mind etc.). **preoccupancy,** *n.* the act or right of taking possession before others. **preoccupation,** *n.* prepossession, prejudice; prior occupation; the state of being preoccupied or engrossed (with); that which preoccupies, as a business affair etc. **preoccupiedly,** *adv.*
preordain (prēawdān'), *v.t.* to ordain beforehand. **preordainment, preordination** (-di-), *n.*
pre-owned (prēōnd'), *a.* second-hand.
prep (prep), *n.* (*School sl.*) preparation or private study; (*chiefly N Am.*) a preparatory school. **prep school,** *n.* a preparatory school. **preppy,** *a.* (*N Am.*) denoting a young but classic look in clothes, clean-cut, conventionally smart. *n.* (*N Am.*) a student at a preparatory school; a person who dresses in a preppy fashion.
prep., (*abbr.*) preparation; preparatory; preposition.
prepack (prēpak'), *v.t.* to wrap or pack (food etc.) before sale.
prepaid (prēpād'), *a.* paid in advance (as postage etc.).
prepare (pripeə'), *v.t.* to make ready; to bring into a suitable condition, to fit for a certain purpose; to make ready or fit (to do, to receive etc.); to produce; to construct, to put together, to draw up; to get (work, a speech, a part etc.) ready by practice, study etc.; (*Mus.*) to lead up to (a discord) by sounding the dominant note in a consonance. *v.i.* to get everything ready; to take the measures necessary (for); to make oneself ready. **preparation** (prepə-), *n.* the act of preparing; the state of being prepared; (*often pl.*) a preparatory act or measure; anything prepared by a special process, as food, a medicine, a part of a body for anatomical study etc.; the preparing of lessons or schoolwork; (*Mus.*) the introduction of a note to be continued in a subsequent discord. **preparative** (-pa'rə-), *a.* preparatory. *n.* that which tends or serves to prepare; an act of preparation; (*Mil., Nav.*) a signal to make ready. **preparatively,** *adv.* **preparatory,** *a.* tending or serving to prepare; introductory (to). **preparatory school,** *n.* a private school for pupils usu. aged 6–13, which prepares them for entry to a public school; (*N Am.*) a private secondary school which prepares students for college. **preparatorily,** *adv.* **prepared,** *a.* to be prepared, to be ready; to be willing (to). **preparedly** (-rid-), *adv.* **preparedness** (-rid-), *n.* **preparer,** *n.*
prepay (prēpā'), *v.t.* to pay beforehand; to pay in advance, esp. by affixing a postage stamp to (a telegram etc.). **prepayable,** *a.* **prepayment,** *n.*
prepense (pripens'), *a.* premeditated, deliberate. **malice prepense,** intentional malice.
preperception (prēpəsep'shən), *n.* previous perception; (*Psychol.*) an impress forming the material of a percept.
preponderate (pripon'dərāt), *v.i.* to be heavier; to be superior or to outweigh in number, power, influence etc.; to sink (as the scale of a balance). **preponderance,** *n.* **preponderant,** *a.* **preponderantly, preponderatingly,** *adv.* **preponderation,** *n.*
preposition (prepəzish'ən), *n.* a word or group of words, e.g. *at, by, in front of*, used to relate the noun or pronoun it is placed in front of to other constituent parts of the sentence. **prepositional,** *a.* **prepositionally,** *adv.* **prepositive** (pripoz'-), *a.* prefixed, intended to be placed before (a word). *n.* a prepositive word or particle.
prepossess (prēpəzes'), *v.t.* to occupy beforehand; to imbue (with an idea, feeling etc.); to bias (esp. favourably); (of an idea etc., to preoccupy. **prepossessing,** *a.* biasing; tending to win favour, attractive. **prepossessingly,** *adv.* **prepossession** (-shən), *n.*
preposterous (pripos'tərəs), *a.* contrary to nature, reason or common sense; obviously wrong, foolish, absurd. **preposterously,** *adv.* **preposterousness,** *n.*
prepotent (pripō'tənt), *a.* very powerful; possessing superior force or influence; overbearing; (*Biol.*) possessing superior fertilizing influence. **prepotence, -ency,** *n.* **prepotential** (prēpəten'-), *a.* **prepotently,** *adv.*
preppy PREP.
pre-prandial (prēpran'diəl), *a.* done, happening etc. before dinner.
pre-preference (prēpref'ərəns), *a.* ranking before preference shares etc.
prepuce (prē'pūs), *n.* the foreskin, the loose covering of the glans penis; a similar fold of skin over the clitoris. **preputial** (-pū'-), *a.*
prequel (prē'kwel), *n.* a book or film written or made later, but dealing with events happening before those in the original.
Pre-Raphaelite (prēraf'iəlīt), *n.* an artist who aimed at reviving the spirit and technique that characterized painting before the time of Raphael. *a.* having the characteristics of Pre-Raphaelitism. **Pre-Raphaelite Brotherhood,** a small group of painters formed in London in 1848, including Holman Hunt, Millais and D. G. Rossetti, to cultivate the spirit and methods of the early Italian painters, esp. in respect to truth to nature and vividness of colour. **Pre-Raphaelitism,** *n.*
prerecord (prērikawd'), *v.t.* to record (a radio etc. programme) in advance, for broadcasting later.
pre-Reformation (prērefəmā'shən), *a.* happening, or dating from, before the Reformation.
prerelease (prērilēs'), *n., v.t.* to release (or the release of) a record or film before the official date.
preremote (prērimōt'), *a.* occurring, done etc., still more remotely in the past.
prerequisite (prērek'wizit), *a.* required beforehand. *n.* a requirement that must be satisfied in advance, a precondition.
prerevolution (prērevəloo'shən), **-tionary,** *a.* of or pertaining to the time before the (esp. French or Russian) revolution.
prerogative (prirog'ətiv), *n.* an exclusive right or privilege vested in a particular person or body of persons, esp. a sovereign, in virtue of his or her position or relationship; any peculiar right, option, privilege, natural advantage etc. *a.* of, pertaining to or having a prerogative.
Pres., (*abbr.*) President.
pres., (*abbr.*) present.
presage (pres'ij), *n.* something that foretells a future event, an omen; foreboding, presentiment. *v.t.* (*also* prisāj') to foreshadow, to betoken; to give warning by natural signs etc.; to forebode, to foretell. **presager,** *n.*
presbyopia (prezbiō'piə), *n.* a form of long-sightedness with indistinct vision of near objects, caused by alteration in the refractive power of the eyes with age. **presbyope** (prez'-), *n.* one affected with this. **presbyopic** (-op'-), *a.*
presbyter (prez'bitə), *n.* an elder who had authority in the early Church; in the Episcopal Church, a minister of the second order, a priest; in the Presbyterian Church, a minister of a presbytery, an elder. **presbyteral** (-bit'-), **-terial** (-tiə'ri-), *a.* **presbyterate** (-bit'ərət), **presbytership,** *n.* **Presbyterian** (-tiə'ri-), *n.* any adherent of Presbyterianism; a member of a Presbyterian Church. *a.* pertaining to Church government by presbyters; governed by presbyters. **Presbyterian Church,** *n.* a Church governed by elders, including ministers, all equal in rank. **United Presbyterian Church,** the Church formed by the union of the United Secession and Relief Churches in 1847, united in 1900 with the Free Church of Scotland. **Presbyterianism,** *n.* **presbytery** (-ri), *n.* a body of elders in the early Church; a court consisting of the pastors and ruling elders of the Presbyterian churches of a given district, ranking above the kirk-session and below the synod; the district

represented by a presbytery; the eastern portion of a chancel beyond the choir in a cathedral or other large church, the sanctuary; in the Roman Catholic Church, a priest's residence.
preschool (prē'skool), *a.* (for children who are) under school age.
prescient (pres'iənt), *a.* foreknowing, far-seeing. **prescience**, *n.* **presciently**, *adv.*
prescientific (prēsiəntif'ik), *a.* pertaining to the period before the rise of science or of scientific method.
prescind (prisind'), *v.t.* to cut off; to abstract, to consider independently. *v.i.* to separate one's consideration (from).
prescribe (priskrīb'), *v.t.* to lay down with authority; to appoint (a rule of conduct etc.); to direct to be used as a remedy. *v.i.* to write directions for medical treatment; (*Law*) to assert a prescriptive title (to or for). **prescriber**, *n.* **prescript** (prē'skript), *n.* a direction, a command, a law. *a.* prescribed, directed. **prescription** (-skrip'-), *n.* the act of prescribing; that which is prescribed, esp. a written direction for the preparation of medical remedies, and the manner of using them; the medication etc. prescribed; (*Law*) long-continued or immemorial use or possession without interruption, as giving right or title; right or title founded on this; ancient or long-continued custom, esp. when regarded as authoritative; a claim based on long use. **prescriptive** (-skrip-), *a.* acquired or authorized by long use; based on long use or prescription; laying down rules. **prescriptively**, *adv.* **prescriptiveness**, *n.*
preselector (prēsilek'tə), *n.* a system whereby a gear can be selected before it is engaged. **preselect**, *v.t.*
presence (prez'əns), *n.* the quality or state of being present; the immediate vicinity of a person; (a person with) an imposing or dignified bearing; personal magnetism; the ability to grasp and hold an audience's attention; a military or diplomatic delegation in a place in order to exercise influence; an influence as of a being invisibly present. **presence of mind**, a calm, collected state of mind. **real presence**, the actual existence of the body and blood of Christ in the Eucharist. **presence-chamber, -room**, *n.* the room in which a great person receives company.
presensation (prēsensā'shən), *n.* sensation, feeling, or consciousness of something before it exists.
present[1] (prez'nt), *a.* being in a place referred to; being in view or at hand; found or existing in the thing referred to; being under discussion, consideration etc.; now existing, occurring, going on etc.; (*Gram.*) expressing what is actually going on; ready at hand, assisting in emergency. *n.* the present time; the present tense; (*pl.*) these writings, a term used in documents to express the document itself. **at present**, at the present time, now. **for the present**, for the time being; just now; so far as the time being is concerned. **present-day**, *a.* contemporary, of the current time. **present tense**, *n.* the form of the verb expressing being or action at the present time.
present[2] (prizent'), *v.t.* to introduce to the acquaintance or presence of, esp. to introduce formally; to submit (oneself) as a candidate, applicant etc.; to offer (a clergyman) to a bishop for institution (to a benefice); to exhibit, to show; to hold in position or point (a gun etc.); to offer or suggest (itself) to the attention; to offer for consideration, to submit; to exhibit (an actor, a play etc.) on the stage; to act as the presenter of (as a television programme); to portray, to depict, to represent; to offer, to give, to bestow, esp. in a ceremonious way; to invest or endow (with a gift); to tender, to deliver. *v.i.* to come forward as a patient (with); of a foetus, to be in a specified position during labour with respect to the mouth of the uterus. *n.* (prez'nt), that which is presented, a gift. **to present arms**, to hold a rifle etc. in a perpendicular position in front of the body to salute a superior officer. **presentable**, *a.* fit to be presented; of suitable appearance for company etc.; fit to be shown or exhibited. **presentably**, *adv.* **presentability** (-bil'-), *n.*
presentation (prezəntā'shən), *n.* the act of presenting; a formal offering or proffering; a present, a gift; an exhibition, a theatrical representation; a verbal report on, or exposé of, a subject, often with illustrative material; the manner of presenting, esp. the appearance, arrangement, neatness etc. of material submitted; an introduction, esp. a formal introduction to a superior personage; (*Law*) the act or right of presenting to a benefice; (*Obstetrics*) the particular position of the foetus at birth; (*Psych.*) the process by which an object becomes present to consciousness, or the modification of consciousness involved in the perception of an object. **presentation copy**, *n.* a book presented gratis by an author or publisher. **presentational**, *a.* **presentationism**, *n.* the doctrine that the mind has immediate cognition of objects of perception, or of elemental categories such as space, time etc. **presentationalist, -tionist**, *n.* **presentative** (prizen'-), *a.* pertaining to or of the nature of mental presentation; subserving mental presentation; of a benefice, admitting of the presentation of an incumbent.
presentee (prezntē'), *n.* one presented to a benefice; one recommended for office; one receiving a present.
presenter (prizen'tə), *n.* one who presents; a broadcaster who introduces and comperes, or provides a linking commentary for, a radio or television programme.
presentient (prizen'shənt, -sen'-), *a.* feeling or perceiving beforehand.
presentiment (prizen'timənt), *n.* apprehension or anticipation, more or less vague, of an impending event, esp. of evil, a foreboding.
presentive (prizen'tiv), *a.* presenting an object or conception directly to the mind, opp. to symbolic; presentative. **presentiveness**, *n.*
presently (prez'ntli), *adv.* soon, shortly; (*chiefly N Am., Sc.*) at the present time.
presentment (prizent'mənt), *n.* the act of presenting; a theatrical representation, a portrait, a likeness; a statement, an account, a description; the act or mode of presentation to the mind; a report by a grand jury respecting an offence, from their own knowledge.
preserve (prizœv'), *v.t.* to keep safe, to guard, to protect; to save, to rescue; to maintain in a good or the same condition; to retain, to keep intact; to keep from decay or decomposition by chemical treatment, boiling, pickling etc.; to keep (a stream, covert, game etc.) for private use by preventing poaching etc. *v.i.* to make preserves; to maintain protection for game in preserves. *n.* fruit prepared with sugar or preservative substances, jam; a place where game is preserved; water where fish are preserved; a special domain, something reserved for certain people only. **preservable**, *a.* **preservation** (prezə-), *n.* **preservationist**, *n.* one who is interested in preserving traditional and historic things. **preservative**, *a.* having the power of preserving; tending to preserve. *n.* that which preserves, esp. a chemical substance used to prevent decomposition in foodstuffs. **preserver**, *n.* **preserving pan**, *n.* a large pan used for making jams and preserves.
preset (prēset'), *v.t.* (*pres.p.* **-setting**, *past, p.p.* **-set**) to set (controls etc.) in advance for later automatic operation.
preshrunk (prēshrŭngk'), *a.* of a fabric, shrunk before being made up into a garment etc.
preside (prizīd'), *v.i.* to be set in authority over others; to sit at the head of a table; to act as director, controller, chairman or president; to lead, to superintend; to officiate (at the organ, piano etc.). **presidency** (prez'idənsi), *n.* the office, jurisdiction or term of office of a president; the territory administered by a president. **president** (prez'idənt), *n.* one (usu. elected) presiding over a temporary or permanent body of persons; the chief magistrate or elective head of the government in a modern republic; one presiding over the meetings of a society; the

presidial — **pretax**

chief officer of certain colleges and universities, esp. in the US; (*N Am.*) the permanent chairman and chief executive officer of a corporation, board of trustees, government department etc. **Lord President of the Council**, a member of the House of Lords who acts as president of the Privy Council. **presidential** (-den'-), *a.* **presidentially**, *adv.* **presidentship**, *n.* **presider**, *n.*
presidial (prisid'iəl), **-ary**, *a.* pertaining to a garrison; having or serving as a garrison.
presidium (prisid'iəm), *n.* (*pl.* **-diums**, **-dia** (-diə)) a permanent executive committee in a Communist country.
presignify (prēsig'nifī), *v.t.* to signify or intimate beforehand.
press[1] (pres), *v.t.* to act steadily upon with a force or weight; to push (something up, down, against etc.) with steady force; to put or hold (upon etc.) with force; to squeeze, compress; to extract juice from; to make by pressing in a mould; esp. to make (a gramophone record from a matrix; to clasp, to embrace, to hug; to crowd upon; to urge, to ply hard, to bear heavily on; to invite with persistent warmth; to put forward vigorously and persistently; to weigh down, to distress; to constrain; to enforce strictly, to impress; to force (upon); to make smooth by pressure (as cloth or paper). *v.i.* to exert pressure; to bear or weigh heavily; to be urgent; to throng, to crowd, to encroach; to strive eagerly, to hasten, to strain, to push one's way. *n.* the act of pressing, urging or crowding; a crowd; urgency, hurry; an upright case, cupboard or closet, for storing things, esp. linen; a book-case; an instrument or machine for compressing any body or substance, forcing it into a more compact form, shaping, extracting juice etc.; a machine for printing; a printing-establishment; the process or practice of printing; the reaction of newspapers etc. to a person, event etc. **freedom of the press**, the right to print and publish statements, opinions etc. without censorship. **in the press**, being printed; on the eve of publication. **the press**, the news media collectively, esp. newspapers and periodicals; journalists. **to go to press**, to start printing, to begin to be printed. **to press on, ahead, forward**, to continue (determinedly) on one's way. **press agent**, *n.* a person employed to handle relations with the press, esp. to ensure good publicity for an actor, organization etc. **press-box**, *n.* a shelter for reporters on a cricket-field etc. **press conference**, *n.* a meeting of a statesman etc. with journalists to announce a policy or answer questions. **press-cutting**, *n.* a clipping from a newspaper. **press-gallery**, *n.* a gallery set aside for reporters, esp. in the Houses of Parliament. **pressman**, *n.* one who manages a printing-press; a journalist. **press-mark**, *n.* a number, symbol or other mark indicating the place of a book on the shelves of a library. **press release**, *n.* an official statement or report given to the press. **press-room**, *n.* the room in a printing-office where the presses are. **press-stud**, *n.* a fastener consisting of two small round buttons, one of which has a small raised knob which snaps into a hole in the other. **press-up**, *n.* a gymnastic exercise in which the body is held rigid in a prone position and raised and lowered by bending and straightening the arms. **press-work**, *n.* the work or management of a printing-press; journalistic work. **presser**, *n.* **pressing**, *a.* urgent, importunate, insistent. *n.* the gramophone records made from a single matrix at one time. **pressingly**, *adv.*
pression (-shən), *n.* the act of pressing; pressure.
press[2] (pres), *v.t.*, *v.i.* to force (men) into naval or military service. *n.* a compulsory enlisting of men into naval or military service; a commission to force men into service. **press-gang**, *n.* a detachment of men employed to impress men, usu. into the navy. **press-money**, *n.* money paid to men who enlist.
pressiroster (presiros'tə), *n.* any individual of the Pressirostres, a group of wading birds with a compressed beak.

pressirostral, *a.*
pressure (presh'ə), *n.* the act of pressing; the state of being pressed; a force steadily exerted upon or against a body by another in contact with it; the amount of this, usu. measured in units of weight upon a unit of area; constraining force, compulsion; moral force; persistent attack; stress, urgency; trouble, affliction, oppression. *v.t.* to apply pressure to; to constrain, to subject to compelling moral force against one's will. **high pressure** HIGH. **pressure-cabin**, *n.* (*Aviat.*) a pressurized cabin in an aircraft. **pressure-cooker**, *n.* an apparatus for cooking at a high temperature under high pressure. **pressure-cook**, *v.t.* **pressure group**, *n.* a group or small party exerting pressure on government etc. for its own ends. **pressure point**, *n.* any of various points on the body where a blood vessel may be pressed against a bone to check bleeding. **pressure-suit**, *n.* an airman's suit that inflates automatically if there is a failure in the pressure-cabin. **pressurize**, **-ise**, *v.t.* to fit an aircraft cabin with a device that maintains normal atmospheric pressure at high altitudes; (to seek to) coerce. **pressurized-water reactor**, a type of nuclear reactor that uses water under pressure as coolant and moderator.
Prestel® (pres'tel), *n.* the British Telecom viewdata system.
Prester John (pres'tə), *n.* a mythical Christian sovereign and priest, supposed in the Middle Ages to rule in Abyssinia or somewhere in the interior of Asia.
presternum (prēstœ'nəm), *n.* the front part of the sternum; (*Ent.*) the prosternum.
prestidigitation (prestidijitā'shən), *n.* sleight of hand, conjuring. **prestidigitator** (-dij'-), *n.*
prestige (prestēzh'), *n.* influence or weight derived from fame, excellence, achievements etc. *a.* having or conferring prestige; superior, very high-quality, very stylish etc. **prestigious** (-tij'əs), *a.* having or conferring prestige.
presto[1] (pres'tō), *adv.* (*Mus.*) quickly. *a.* quick. *n.* a quick movement. **prestissimo** (-tis'imō), *adv.* very fast indeed. *a.* very fast. *n.* a very fast movement. [It.]
presto[2] (pres'tō), **hey presto**, *adv.*, *int.* immediately (to indicate the speed with which e.g. a conjuring trick is performed).
prestressed (prēstrest'), *a.* of concrete, reinforced with stretched steel wires or rods.
presume (prizūm'), *v.t.* to venture on without leave; to take for granted or assume without previous inquiry or examination. *v.i.* to venture without previous inquiry; to form over-confident or arrogant opinions; to behave with assurance or arrogance. **to presume on**, to rely on; to take unfair advantage of. **presumable**, *a.* **presumably**, **presumedly** (-mid-), *adv.* **presumer**, *n.* **presuming**, *a.* presumptuous. **presumingly**, *adv.* **presumption** (-zŭmp'-), *n.* the act of presuming; assumption of the truth or existence of something without direct proof; that which is taken for granted; the ground for presuming; over-confidence, arrogance, effrontery. **presumption of fact**, an inference as to a fact from facts actually known. **presumption of law**, assumption of the truth of a proposition until the contrary is proved; an inference established by law as universally applicable to particular circumstances. **presumptive**, *a.* giving grounds for or based on presumption. **heir presumptive**, an heir whose actual succession may be prevented by the birth of one nearer akin to the present holder of a title, estate etc. **presumptive evidence**, *n.* evidence derived from circumstances which necessarily or naturally attend a fact. **presumptively**, *adv.* **presumptuous** (-tūəs), *a.* full of presumption; arrogant; rash, venturesome. **presumptuously**, *adv.* **presumptuousness**, *n.*
presuppose (prēsəpōz'), *v.t.* to assume beforehand; to imply as a necessary antecedent. **presupposition** (-sŭpəzi'-), *n.*
pretax (prētaks'), *a.* of income or profit before tax has

been deducted.
pretend (pritend'), *v.t.* to assume the appearance of; to feign to be; to simulate; to allege or put forward falsely; to put forward, to assert, to claim. *v.i.* to feign, to make believe; to put forward a claim (to). *a.* make-believe. **pretence**, (*esp. N Am.*) **pretense** (-tens'), *n.* a claim (true or false); a false profession; a pretext; display, ostentation; (an act of) pretending or feigning; (a) semblance. **pretendedly**, *adv.* **pretender**, *n.* one who makes a claim, esp. one that cannot be substantiated; a claimant, esp. to a throne held by another branch of the same family. **Old Pretender**, James Stuart, 1688-1766, son of James II. **Young Pretender**, Charles Edward Stuart, 1720-1788 son of the Old Pretender. **pretendership**, *n.* **pretendingly**, *adv.* **pretension** (-shən), *n.* (*often pl.*) a claim; (*often pl.*) an aspiration; a pretext; pretentiousness. **pretentious** (-shəs), *a.* full of pretension; making specious claims to excellence etc.; ostentatious, arrogant, conceited. **pretentiously**, *adv.* **pretentiousness**, *n.*
preter-, *pref.* beyond; beyond the range of; more than.
preterhuman (prētəhū'mən), *a.* more than human, superhuman.
preterist (pret'ərist), *n.* one whose chief interest is in the past.
preterite, (*esp. N Am.*) **preterit** (pret'ərit), *a.* (*Gram.*) denoting completed action or existence in past time; past, gone by. *n.* the preterite tense. **preteriteness**, *n.* **preterition** (prētərish'ən), *n.* the act of passing over or omitting; the state of being passed over; (*Theol.*) the passing over of the non-elect, opp. to *election*. **preteritive** (prīte'-), *a.*
pretermit (prētəmit'), *v.t.* to pass by or over, to neglect, to omit (to mention, to do etc.); to discontinue. **pretermission** (-shən), *n.*
preternatural (prētənat'ūrəl), *a.* beyond what is natural; out of the regular course of nature. **preternaturalism**, *n.* the state of being preternatural; a preternatural occurrence, thing etc. **preternaturalist**, *n.* **prenaturally**, *adv.* **preternaturalness**, *n.*
pretext (prē'tekst), *n.* an excuse; an ostensible reason or motive. *v.t.* (pritekst'), to allege as pretext or motive.
pretone (prē'tōn), *n.* the vowel or syllable preceding the accented syllable. **pretonic** (-ton'-), *a.*
pretor etc. **PRAETOR**.
pretty (prit'i), *a.* good-looking, attractive, appealing (though without the striking qualities or perfect proportions of beauty); aesthetically pleasing (with the same qualification); superficially or conventionally attractive; (*coll., derog.*) of a man, effeminate-looking; (*chiefly iron.*) nice, fine; considerable, large. *adv.* moderately, fairly; very. *n.* a pretty thing or person. **a pretty penny** PENNY. **pretty much, well**, nearly, almost. **to be sitting pretty** SIT. **to pretty up**, to prettify, to adorn. **pretty-pretty**, *a.* affectedly pretty, over-pretty. *n.pl.* knick-knacks, gewgaws. **pretty-spoken**, *a.* speaking in a pleasing manner. **prettify** (-fī), *v.t.* to make pretty; to put or depict in a pretty way. **prettification**, *n.* **prettily**, *adv.* **prettiness**, *n.* **prettyish**, *a.* **prettyism**, *n.*
pretypify (prētip'ifī), *v.t.* to prefigure.
pretzel (pret'sl), *n.* a crisp biscuit of wheaten flour flavoured with salt, usu. in the shape of a stick or a knot. [G]
prevail (prīvāl'), *v.i.* to have the mastery or victory (over, against etc.); to predominate; to be in force, to be current or in vogue); to be customary. **to prevail on, upon**, to succeed in persuading, to induce. **prevailing**, *a.* predominant, most frequent; current, generally accepted. **prevailingly**, *adv.* **prevalence** (prev'ə-), *n.* the act of prevailing; superiority, predominance; frequency, vogue, currency. **prevalent**, *a.* **prevalently**, *adv.*
prevaricate (privā'rikāt), *v.i.* to shuffle, to quibble; to act or speak evasively; to equivocate. **prevarication**, *n.* **prevaricator**, *n.*

prevenance, (prev'ənəns), *n.* anticipation of the wants or wishes of others.
prevenient (privēn'yənt), *a.* going before, preceding, previous; preventive (of). **prevenient grace**, *n.* grace preceding repentance and predisposing to conversion. **prevenience**, *n.*
prevent (privent'), *v.t.* to keep from happening; to hinder, to thwart, to stop; to go before, to precede. **preventable, -ible**, *a.* capable of prevention. **preventative** PREVENTIVE. **preventer**, *n.* one who or that which prevents or hinders; (*Naut.*) a supplementary rope, chain, spar, stay etc., to support another. **preventingly**, *adv.* **prevention**, *n.* the act of preventing; hindrance, obstruction. **preventive**, (*loosely*) **preventative**, *a.* tending to hinder or prevent; prophylactic; of or belonging to the coastguard or customs and excise service. *n.* that which prevents; a medicine or precaution to ward off disease; a contraceptive. **preventive detention**, *n.* a system for dealing with a habitual criminal by detention for a definite period after the completion of the sentence for a specific crime. **preventive service**, *n.* the coastguard or customs and excise service. **preventively**, *adv.*
preview (prē'vū), *n.* an advance view, a foretaste; an advance showing of a play, film, art exhibition etc. before its general presentation to the public; (also, *esp. N Am.*) **prevue**) a television or cinema trailer. *v.t.* to view or show in advance.
previous (prē'viəs), *a.* going before, antecedent; prior (to); (*sl.*) premature, hasty. *adv.* before, previously (to). **Previous Examination**, *n.* (*Camb. Univ.*) *n.* the first examination for the degree of BA. **previous question**, *n.* in the House of Commons, a motion 'that the question be not now put', which, if carried, has the effect of delaying a vote; in the House of Lords and US legislature, a motion to proceed to a vote immediately; in public meetings, a motion to proceed with the next business. **previously**, *adv.* **previousness**, *n.*
previse (privīz'), *v.t.* to know beforehand, to foresee; to forewarn. **prevision** (-vizh'ən), *n.* **previsional**, *a.* **previsionally**, *adv.*
prewar (prēwaw'), *a.* of or pertaining to a period before a war, esp. World War II.
prewash (prē'wosh), *n.* an additional wash before the main one in an automatic washing machine.
prey (prā), *n.* that which is or may be seized to be devoured by carnivorous animals; booty, spoil, plunder; a victim. *v.i.* to take booty or plunder; to take food by violence. **beast, bird of prey**, a carnivorous beast or bird. **to prey on**, to rob, to plunder; to chase and seize as food; to make a victim of, to subject to robbery, extortion etc.; to have a depressing or obsessive effect on. **preyer**, *n.*
Priapus (prīā'pəs), *n.* (*Class. Myth.*) the god of procreation; (**priapus**) a phallus; (**priapus**) the penis. **priapean** (-pē'ən), **priapic** (-ap'-), *a.* **priapism** (prī'ə-), *n.* lasciviousness; continuous erection of the penis without sexual excitement. [L]
pribble (prib'l), *n.* empty chatter.
price (prīs), *n.* the amount asked for a thing or for which it is sold; the cost of a thing; the amount needed to bribe somebody; that which must be expended, sacrificed, done etc. to secure a thing; (*Betting sl.*) the odds; estimation, value, preciousness. *v.t.* to fix the price of, to value, to appraise. **above, beyond, without price**, priceless. **a price on one's head**, the reward offered for one's killing or capture. **at a price**, for a lot of money etc. **not at any price**, under no circumstances. **to price oneself out of the market**, to lose trade by charging too high prices. **what price?**, (*iron.*) what about? **price control**, *n.* the fixing by government of maximum prices for goods and services. **price-fixing**, *n.* the setting of prices by agreement between producers and distributors; price control. **price list**, *n.* a table of the current prices of merchandise, stocks etc.

price ring, *n.* a group of manufacturers or traders who co-operate to maintain prices at an agreed, high level. **price tag,** *n.* the label attached to an object showing its price; price, cost. **priceless,** *a.* invaluable, inestimable; (*sl.*) very funny. **pricelessness,** *n.* **pricey,** *a.* (*comp.* **pricier,** *superl.* **priciest**) expensive.
prick (prik), *n.* the act of pricking; the state or the sensation of being pricked; a puncture; a dot, point or small mark made by or as by pricking; a pointed instrument, a goad; a sharp, stinging pain; (*sl.*) the penis; (*sl.*) an obnoxious or inept man. *v.t.* to pierce slightly, to puncture; to make by puncturing; to mark off (names etc.) with a prick, hence, to select; to goad, to rouse, to incite. *v.i.* to ride rapidly, to spur; to point upward; to feel as if pricked. **to kick against the pricks,** to hurt oneself in unavailing struggle against something. **to prick off, out,** to mark a pattern out with dots; to plant seedlings more widely apart with a view to transplanting later to their permanent quarters. **to prick (up) the ears,** of dogs etc., to raise the ears as if listening; to become very attentive. **prick-eared,** *a.* of dogs etc., having erect or pointed ears. **prick-ears,** *n.pl.* **prick-song,** *n.* written music. **pricker,** *n.* a sharp-pointed instrument, a bradawl.
pricket (prik'it), *n.* a buck in its second year; a sharp point for sticking a candle on. **pricket's sister,** *n.* the female fallow deer in the second year.
prickle[1] (prik'l), *n.* a small, sharp point; a thorn-like growth capable of being peeled off with the skin or bark, opp. to thorn or spine; (*loosely*) a small thorn, spine etc. *v.t.* to prick slightly; to give a pricking or tingling sensation to. *v.i.* to have such a sensation. **prickle-back,** *n.* the stickleback. **prickly,** *a.* full of or armed with prickles. **prickly heat,** *n.* a skin condition characterized by itching and stinging sensations, prevalent in hot countries. **prickly pear,** *n.* (the pear-shaped fruit of) any cactus of the genus *Opuntia,* usu. covered with prickles. **prickliness,** *n.*
prickle[2] (prik'l), *n.* a variety of wicker basket.
pride (prīd), *n.* inordinate self-esteem, unreasonable conceit of one's own superiority; arrogance; sense of dignity, self-respect; generous elation or satisfaction arising out of some accomplishment, possession or relationship; a source of such elation; the highest point, the best condition; a collection of lions. **pride of place,** the highest or most important position. **to pride oneself on,** to be proud of oneself for. **prideful,** *a.* (*chiefly Sc.*). **pridefully,** *adv.* **pridefulness,** *n.* **prideless** (-lis), *a.*
prie-Dieu (prēdyœ'), *n.* a kneeling desk for prayers. **prie-Dieu chair,** a chair with a tall sloping back, esp. for praying. [F]
prier (prī'ə), *n.* one who pries.
priest (prēst), *n.* one who officiates in sacred rites, esp. by offering sacrifice; a minister of the second order, below a bishop and above a deacon; (*coll.*) a clergyman (esp. in a hostile sense); a small mallet or club for killing fish when caught. *v.t.* to ordain as a priest. **priest-craft,** *n.* priestly policy based on material interests. **priest-in-the-pulpit** PRIEST'S HOOD. **priest-ridden,** *a.* (*derog.*) dominated by priests. **priest's hole,** *n.* a hiding-place for fugitive priests, esp. in England under the penal laws. **priest's hood,** *n.* the wild arum. **priest-vicar,** *n.* a minor canon in certain cathedrals. **priestess** (-tis), *n. fem.* **priest-hood** (-hud), *n.* **priestless** (-lis), *a.* **priestlike,** *a.* **priestling** (-ling), *n.* **priestly,** *a.* of, pertaining to or befitting a priest or the priesthood; sacerdotal. **priestliness,** *n.*
prig (prig), *n.* a conceited, formal or moralistic person. **priggery,** *n.* **priggish,** *a.* conceited, affectedly precise, formal, moralistic. **priggishly,** *adv.* **priggishness,** *n.*
prim (prim), *a.* formal, affectedly proper, demure. *v.t.* (*past, p.p.* **primmed**) to put (the lips, mouth etc.) into a prim expression; to deck with great nicety or preciseness. *v.i.* to make oneself look prim. **primly,** *adv.* **primness,** *n.*
prima (prē'mə), *a.* first, chief, principal. **prima ballerina,** *n.* the leading female ballet dancer. **prima donna,** *n.* (*pl.* **prime donne** (-mā don'ā) a chief female singer in an opera; a person who is temperamental, hard to please, and given to histrionics. [It.]
primacy (prī'məsi), *n.* the dignity or office of a primate; pre-eminence.
primaeval (prīmē'vəl), PRIMEVAL.
prima facie (prī'mə fā'shi), *adv.* at first sight, on the first impression. **prima facie case,** (*Law*) a case apparently established by the evidence. [L]
primal (prī'məl), *a.* primary, original, primitive; fundamental, chief. **primally,** *adv.*
primaquine (prī'məkwēn), *n.* a synthetic drug used to treat malaria.
primary (prī'məri), *a.* first in time, order or origin; original, radical, firsthand; primitive, fundamental; first in rank or importance, chief; first or lowest in development, elementary; of or being an industry that produces raw materials; of or being the inducing current or its circuit in an induction coil or transformer; pertaining to the lowest series of strata, Palaeozoic. *n.* that which stands first in order, rank or importance; a celestial body round which other members of a system orbit; a meeting or election for the selection of party candidates by voters of a state or region, esp. in the US; one of the large quill-feathers of a bird's wing; a primary colour; a primary school. **primary cell,** *n.* a battery in which chemical energy is converted into electrical energy. **primary colours** COLOUR. **primary consumer,** *n.* a herbivore. **primary education,** *n.* education in primary, junior and infant schools. **primary feather,** *n.* **primary school,** *n.* a school for children aged under 11 (England and Wales) or under 12 (Scotland). **primary winding,** *n.* the winding of a transformer which is on the input side. **primarily** (prī'-, -ma'ri-), *adv.* **primariness,** *n.*
primate (prī'mət), *n.* the chief prelate in a national episcopal church, an archbishop; a member of the order Primates. **Primate of all England,** the archbishop of Canterbury. **Primate of England,** the archbishop of York. **primates** (prīmā'tēz), *n.pl.* the highest group of mammals, comprising humans, apes, monkeys and lemurs. **primateship,** *n.* **primatial** (-mā'shəl), *a.* **primatology** (-tol'-), *n.* the study of primates. **primatologist,** *n.*
prime[1] (prīm), *a.* first in time, rank, excellence or importance; first-rate, excellent; original, primary, fundamental; in the vigour of maturity; divisible by no integral factors except itself and unity (as 2, 3, 5, 7, 11, 13). *n.* the period or state of highest perfection; the best part (of anything); the first canonical hour of the day, beginning at 6 a.m. or at sunrise; in the Roman Catholic Church, the office for this hour; the first stage, the beginning (of anything); dawn, spring, youth; a prime number; (*Fencing*) the first of the eight parries or a thrust in this position. **prime cost,** *n.* the cost of material and labour in the production of an article. **prime-meridian** MERIDIAN. **prime minister,** *n.* the first Minister of State, esp. in the UK. **prime mover,** *n.* one who or that which originates a movement or an action, esp. the force putting a machine in motion; God. **prime rate,** *n.* the lowest commercial rate of interest charged by a bank at a particular time. **prime time,** *n.* peak viewing or listening time for television or radio audiences, for which advertising rates are highest. **prime vertical,** *n.* a great circle of the heavens passing through the east and west points of the horizon and the zenith. **primely,** *adv.* **primeness,** *n.*
prime[2] (prīm), *v.t.* to prepare (a gun) for firing; to supply (with information); to coach; to fill (with liquor); to fill (a pump) with fluid to expel the air before starting; to inject fuel into the float chamber of (a carburettor); to lay the first coat of paint, plaster or oil on. *v.i.* of a boiler, to carry over water with the steam to the cylinder; of a tide, to come before the mean time. **primer**[1], *n.* a person or

thing that primes; a priming-wire; a detonator; (a type of) paint used as a sealant and a base for subsequent coats.
priming, *n.* the act of preparing a firearm for discharge; a train of powder connecting a blasting-charge with the fuse; fluid introduced before starting a pump; a first layer of paint etc.; water carried from the boiler into the cylinder of a steam-engine; hasty instruction, cramming; the shortening of the interval between tides (from neap to spring tides), opp. to *lagging.* **priming-iron, -wire,** *n.* a wire for piercing a cartridge when home, or for clearing the vent of a gun etc.
primer[2] (prī'mə), *n.* an elementary reading-book for children; a short introductory book; (prim'ə), one of two sizes of type, great primer and long primer.
primero (primeə'rō), *n.* a game of cards fashionable in the 16th and 17th cents., the original of poker.
primeval, primaeval (primē'vəl), *a.* belonging to the earliest ages, ancient, original, primitive. **primevally,** *adv.*
priming PRIME[2].
primiparous (primip'ərəs), *a.* bringing forth a child for the first time.
primitiae (primish'iē), *n.pl.* first fruits, annates; the discharge of fluid from the uterus before parturition.
primitive (prim'itiv), *a.* pertaining to the beginning or the earliest periods; early, ancient, original, primordial; (*Gram.*) radical, not derivative; simple, plain, old-fashioned; crude, uncivilized; of a culture or society, not advanced, lacking a written language and all but basic technical skills; (*Geol.*) belonging to the lowest strata or the earliest period; (*Biol.*) pertaining to an early stage of development; (*Art*) belonging to the period before the Renaissance; not conforming to the traditional standards of Western painting; painting in a naive, childlike or apparently untaught manner. *n.* (a picture by) a primitive painter; a primitive word; a Primitive Methodist. **primitive colours,** *n.pl.* primary colours. **Primitive Methodism,** *n.* a sect aiming at a preponderance of lay control in church government, established in 1810 by secession from the Methodist Church. **Primitive Methodist,** *n.* **primitive rocks,** *n.pl.* primary rocks. **primitively,** *adv.* **primitiveness,** *n.*
primo (prē'mō), *n.* (*pl.* **-mos**) the first part (in a duet etc.).
primo basso, *n.* the chief bass singer. [It.]
primogeniture (primōjen'ichə), *n.* seniority by birth amongst children of the same parents; the right, system or rule under which, in cases of intestacy, the eldest son succeeds to the real estate of his father. **primogenital, -ary,** *a.* **primogenitive,** *a.,* *n.* **primogenitor,** *n.* the first father or ancestor; an ancestor.
primordial (primaw'diəl), *a.* first in order, original, primitive; existing at or from the beginning; first-formed. *n.* an origin; a first principle or element. **primordiality** (-al'-), *n.* **primordialism,** *n.* **primordially,** *adv.*
primp (primp), *v.t.* to make prim; to prink. *v.i.* to prink oneself; to preen; to put on affected airs.
primrose (prim'rōz), *n.* a common British wild plant, with pale yellow flowers in early spring; a pale yellow colour. *a.* like a primrose; of a pale yellow colour. **Primrose Day,** *n.* the anniversary of the death of Benjamin Disraeli, 19 Apr. 1881, commemorated by the wearing of primroses (said to have been his favourite flower). **Primrose League,** *n.* a conservative league formed in memory of Benjamin Disraeli, having for its objects 'the maintenance of religion, of the estates of the realm, and of the imperial ascendancy of the British Empire'. **primrose path,** *n.* the path of ease and pleasure, esp. as leading to perdition.
Primula (prim'ūlə), *n.* a genus of herbaceous plants comprising the primrose, cowslip etc.; (**primula**) a plant of this genus.
primum mobile (prī'məm mō'bili), *n.* in the Ptolemaic system, an imaginary sphere believed to revolve from east to west in 24 hours, carrying the heavenly bodies with it;

the first source of motion, the mainspring of any action. [L]
primus (prī'məs), *a.* first, eldest (of the name, among boys in a school). *n.* the presiding bishop in the Scottish Episcopal Church. **Primus®,** *n.* a portable paraffin cooking stove used esp. by campers. [L]
prince (prins), *n.* (*now rhet.*) a sovereign, a monarch; the ruler of a principality or small state, usu. feudatory to a king or emperor; a male member of a royal family, esp. the son or grandson of a monarch; a member of a foreign order of nobility usu. ranking next below a duke; a chief, leader or foremost representative. **Prince of Darkness,** the Devil. **Prince of Peace,** the Messiah, Christ. **prince of the Church,** a cardinal. **Prince of Wales,** the title customarily conferred on the heir-apparent to the British throne. **Prince Rupert's drop,** a pear-shaped lump of glass formed by falling in a molten state into water, bursting to dust when the thin end is nipped off. **prince-bishop,** *n.* a bishop whose see is a principality. **Prince Charming,** *n.* an ideal suitor. **Prince Consort,** *n.* a prince who is the husband of a reigning female sovereign. **Prince Regent,** *n.* a prince acting as regent. **prince royal,** *n.* the eldest son of a sovereign. **prince's feather,** *n.* a popular name for several plants, esp. the Mexican *Amaranthus hypochondriacus.* **prince's metal,** *n.* an alloy of copper and zinc. **princedom** (-dəm), *n.* **princelet** (-lit), **princeling** (-ling), *n.* **princelike,** *a.* **princely,** *a.* pertaining to or befitting a prince; having the rank of a prince; stately, dignified; generous, lavish. *adv.* as becomes a prince. **princeliness,** *n.* **princeship,** *n.* **princess** (-ses, -ses'), *n.* the daughter or granddaughter of a sovereign; the wife of a prince; a princesse. **Princess Regent,** *n.* a princess acting as regent; the wife of a Prince Regent. **princess royal,** *n.* title conferrable for life on the eldest daughter of a reigning sovereign.
princeps (prin'seps), *a.* (*pl.* **-cipes** (-sipēz)) first. *n.* a chief or head man. **editio princeps,** (*pl.* **editiones** (-ō'nēz), **principes**) the original edition of a book. **facile** (fas'ilē) **princeps,** easily first, beyond question the chief or most important. [L]
princesse, princess (prinses'), *n., a.* (a dress) having a close-fitting bodice and flared skirt cut in one piece. [F]
principal (prin'sipəl), *a.* chief, leading, main; first in rank, authority, importance, influence or degree; constituting the capital sum invested. *n.* a chief or head; a president, a governor, the head of a college etc.; a leader or chief actor in any transaction, the person ultimately liable; a person employing another as agent; the actual perpetrator of a crime, or one aiding and abetting; a performer who takes a leading role; a civil servant in charge of a section; a capital sum invested or lent, as distinguished from income; a main rafter, esp. one extending to the ridge-pole; an organ-stop of the open diapason family. **principal boy,** *n.* the leading male role in a pantomime, usu. taken by a woman. **principal parts,** *n.pl.* those inflected forms of a verb from which all other inflections can be derived. **principally,** *adv.* chiefly, mainly, for the most part. **principalship,** *n.*
principality (prinsipal'iti), *n.* the territory or jurisdiction of a prince; the country from which a prince derives his title. **the Principality,** Wales.
principate (prin'sipāt), *n.* (*Rom. Hist.*) the form of government under the early emperors when some republican features were retained; a principality.
principia (prinsip'iə), *n.pl.* origins, elements, first principles. [L]
principle (prin'sipl), *n.* a source, an origin; a fundamental truth or proposition from which others are derived; a general truth forming a basis for reasoning or action; a rule of action or conduct deliberately adopted; the habitual regulation of conduct by moral law; a law of nature by virtue of which a given mechanism etc. brings about certain results; the mechanical contrivance, combination

of parts, or mode of operation, forming the basis of a machine, instrument, process etc.; (*Chem.*) the constituent that gives specific character to a substance. **in principle**, as far as the basic idea or theory is concerned. **on principle**, because of the fundamental (moral) issue involved; in order to assert a principle. **principled**, *a.* guided by principle; based on a principle.
prink (pringk), *v.i.* to make oneself smart, esp. excessively so. *v.t.* to dress up. **prinker**, *n.*
print (print), *n.* an indentation or other mark made by pressure, an imprint, an impression; an impression from type, an engraved plate etc.; printed lettering; printed matter; a printed publication, esp. a newspaper; an engraving, a lithograph etc.; a reproduction of a work of art made by a photographic process; printed cotton cloth; an article made of such material; a fingerprint; (*Phot.*) a positive image produced from a negative. *v.t.* to impress, to mark by pressure; to take an impression of, to stamp; to impress or make copies of by pressure, as from inked types, plates or blocks, on paper, cloth etc.; to cause (a book etc.) to be so impressed or copied, to publish; to reproduce a design, writing etc. by any transfer process; to mark with a design etc. by stamping; to imprint, to form (letters etc.) in imitation of printing; to impress (on the memory etc.) as if by printing; (*Phot.*) to produce (a positive image) from a negative. *v.i.* to practise the art of printing; to publish books etc.; to form letters etc. in imitation of printing. **in print**, in a printed form; on sale (of a printed book etc.). **out of print**, no longer obtainable from the publisher. **to print out**, to produce a print-out (of). **print-maker**, *n.* one who makes engravings etc. **print-out**, *n.* (a) printed copy automatically produced by a computer. **print run**, *n.* (the number of copies produced in) a single printing of a book etc. **print-seller**, *n.* one who deals in engravings. **print-shop**, *n.* a place where printing is carried on; a place where engravings etc. are sold. **print-works**, *n.* an establishment for printing cotton fabrics. **printable**, *a.* able to be printed, or printed on or from; fit to appear in print. **printed circuit**, *n.* an electronic circuit consisting of conductive material etched or deposited by electrolysis on to an insulating base. **printless**, *a.* **printer**, *n.* one engaged in printing books, newspapers etc.; a typesetter; one who carries on a printing business; a machine or instrument for printing copies, designs etc.; a device for producing print-out. **printer's devil**, *n.* a boy of all work in a printing-office. **printer's ink**, *n.* a viscous mixture of black pigment and oil or varnish used as ink in printing. **printer's mark**, *n.* an engraved design used as a trade-mark by a printer or publisher. **printer's pie** PIE [3]. **printing**, *n.* the act, process, or practice of impressing letters, characters or figures on paper, cloth or other material; the business of a printer; typography. **printing-ink**, *n.* printer's ink. **printing machine**, *n.* a machine for taking impressions from type etc., esp. a power-operated one. **printing-office**, *n.* an establishment where printing is carried on. **printing-paper**, *n.* paper suitable for printing. **printing-press**, *n.* a printing-machine; a hand-press for printing.
prior[1] (prī'ə), *a.* former, preceding; earlier; taking precedence. *adv.* previously, antecedently (to).
prior[2] (prī'ə), *n.* a superior of a monastic house or order next in rank below an abbot. **claustral prior**, one acting as assistant to an abbot. **priorate**, **priorship**, *n.* **prioress** (-ris), *n. fem.* **priory**, *n.* a religious house governed by a prior or prioress.
priority (prīo'riti), *n.* going before, antecedence; precedence, a superior claim or entitlement; something given or meriting special attention; the right to proceed while other vehicles wait. *a.* having or entitling to priority. **prioritize, -ise**, *v.t.* to give (something) priority; to arrange (tasks etc.) according to their importance.
prise (prīz), *n.* leverage. *v.t.* to wrench; to force open with or as with a lever; to extract with difficulty.

prism (priz'm), *n.* a solid having similar, equal and parallel plane bases or ends, its sides forming similar parallelograms; a transparent solid of this form, usu. triangular, with two refracting surfaces set at an acute angle to each other, used as an optical instrument; a spectrum produced by refraction through this; any medium acting on light etc. in a similar manner. **prismal, prismatic** (-mat'-), *a.* pertaining to or resembling a prism; formed, refracted or distributed by a prism; (*Cryst.*) orthorhombic. **prismatic binoculars**, *n.pl.* binoculars shortened by the insertion of prisms. **prismatic colours** COLOUR. **prismatic compass**, *n.* a hand-compass with an attached prism by which the dial can be read while taking the sight. **prismatically**, *adv.* **prismatoid** (-matoid), *n.* a solid with parallel polygonal bases connected by triangular faces. **prismatoidal** (-toi'-), *a.* **prismoid** (-moid), *n.* **prismoidal**, *a.* **prismy** (-mi), *a.*
prison (priz'n), *n.* a place of confinement, esp. a public building for the confinement of criminals, persons awaiting trial etc.; confinement, captivity. **prison-bird** JAILBIRD. **prison-breaker**, *n.* one who escapes from legal imprisonment. **prison-breaking**, *n.* **prison-house**, *n.* (*poet.*) a prison. **prison officer**, *n.* a person who guards and supervises prisoners in a prison. **prisoner**, *n.* one confined in a prison; one under arrest; a captive. **prisoner at the bar**, a person in custody or on trial upon a criminal charge. **prisoner of conscience**, one whose political, religious etc. beliefs have led to imprisonment. **prisoner of war**, a person captured in war. **to take prisoner**, to capture; to arrest and hold in custody. **prisoner's-base**, *n.* a game played by two sides occupying opposite goals or bases, the object being to touch and capture a player away from his base.
prissy (pris'i), *a.* (*coll.*) prim, fussy. **prissily**, *adv.* **prissiness**, *n.*
pristine (pris'tēn, -tīn), *a.* pertaining to an early state or time; pure, uncorrupted; as new.
prithee (pridh'i), *int.* (*old-fashioned*) pray, please.
privacy PRIVATE.
private (prī'vət), *a.* not public; retired, secluded; secret, confidential; not holding a public position, not administered or provided by the state; not part of, or being treated under, the National Health Service; not official; personal, not pertaining to the community; one's own; secretive, reticent. *n.* a soldier of the lowest rank; (*pl.*) the private parts. **in private**, privately, confidentially; in private life. **private member's bill**, one introduced by and sponsored by a member of Parliament who is not a government minister. **private act or bill**, *n.* one affecting a private person or persons and not the general public. **private company**, *n.* a company with a restricted number of shareholders, whose shares are not offered for sale to the general public. **private detective** DETECTIVE. **private enterprise**, *n.* economic activity undertaken by private individuals or organizations. **private eye**, *n.* (*coll.*) a private detective. **private investigator**, *n.* a private detective. **private judgment**, *n.* one's individual judgment, esp. as applied to a religious doctrine or passage of Scripture. **private parts**, *n.pl.* the genitals. **private school**, *n.* one run independently by an individual or group esp. for profit. **private secretary**, *n.* one entrusted with personal and confidential matters; a civil servant acting as an aide to a minister or senior government official. **private sector**, *n.* the part of the economy which is not state owned or controlled. **private soldier**, *n.* a private. **private view**, *n.* occasion when only those invited to an exhibition are admitted. **privacy** (prī'vəsi, priv'-), *n.* **privately**, *adv.* **privateness**, *n.* **privatize, -ise**, *v.t.* to take back into the private sector, to return to private ownership. **privatization**, *n.*
privateer (prīvətiə'), *n.* an armed ship owned and officered

by private persons commissioned by Government to engage in war against a hostile nation, esp. to capture merchant shipping; an officer or one of the crew of such a ship. *v.i.* to serve in a privateer. **privateering**, *n.* **privateersman**, *n.*
privation (privā'shən), *n.* deprivation or lack of what is necessary; destitution; absence, loss, negation (of). **privative** (priv'ə-), *a.* causing privation; consisting in the absence of something; expressing privation or absence of a quality etc., negative. *n.* a prefix or suffix (as *un-* or *-less*) giving a negative meaning to a word. **privatively**, *adv.*
privet (priv'it), *n.* an evergreen, white-flowered shrub largely used for hedges.
privilege (priv'ilij), *n.* a benefit, right, advantage or immunity pertaining to a person, class, office etc.; favoured status, the possession of privileges, a special advantage; (*Law*) a particular right or power conferred by a special law; an exemption pertaining to an office; a right of priority or precedence in any respect. *v.t.* to invest with a privilege; to license, to authorize (to do); to exempt (from). **breach of privilege**, infringement of rights belonging to Parliament. **privilege of clergy**, benefit of clergy. **privileged**, *a.* **privileged communication**, *n.* (*Law*) a communication which there is no compulsion to disclose in evidence.
privy (priv'i), *a.* secluded, hidden, private; cognizant of something secret with another, privately knowing (with *to*). *n.* a latrine; (*Law*) a person having an interest in any action or thing. **privy chamber**, *n.* a private apartment in a royal residence. **Privy Council**, *n.* the private council of the British sovereign (the functions of which are now largely exercised by the Cabinet and committees). **Privy Councillor**, *n.* **privy purse**, *n.* an allowance of money for the personal use of the sovereign; the officer in charge of this. **Privy Seal**, *n.* the seal appended to grants etc. which have not to pass the Great Seal. **Lord Privy Seal**, the officer of State entrusted with the Privy Seal. **privily**, *adv.* secretly, privately. **privity**, *n.* the state of being privy to (certain facts, intentions etc.); (*Law*) any relationship to another party involving participation in interest, reciprocal liabilities etc.
prize[1] (prīz), *n.* that which is offered or won as the reward of merit or superiority in any competition, exhibition etc.; a sum of money or other object offered for competition in a lottery etc.; a fortune, or other desirable object of perseverance, enterprise etc. *a.* offered or gained as a prize; first-class, of superlative merit. *v.t.* to value highly, to esteem. **prize-fellow**, *n.* one who holds a fellowship awarded for pre-eminence in an examination. **prize-fellowship**, *n.* **prize-fight**, *n.* a boxing-match for stakes. **prize-fighter**, *n.* **prize-fighting**, *n.* **prizeman**, *n.* the winner of a prize. **prize-money**, *n.* **prize-ring**, *n.* the roped space (now usu. square) for a prize-fight; prize-fighting. **prize-winner**, *n.* **prize-winning**, *a.* **prizeless**, *a.*
prize[2] (prīz), *n.* that which is taken from an enemy in war, esp. a ship or other property captured at sea. *v.t.* to make a prize of. **prize-court**, *n.* a court adjudicating on cases of prizes captured at sea, in England and US, a department of the courts of Admiralty. **prize-money**, *n.* the proceeds of the sale of a captured vessel etc.
prize[3] (prīz), PRISE.
PRO (*abbr.*) Public Records Office; public relations officer.
pro[1] (prō), *prep.* for; in favour of. *adv.* in favour. **pro and con**, for and against; on both sides. **pros and cons**, reasons or arguments for and against. **pro bono publico** (bō'nō pub'likō), for the public good. **pro forma** (faw'mə), as a matter of form; of an invoice etc. made out to show the market price of goods. **pro rata** (rah'tə), in proportion, proportionally; proportional. **pro tempore** (tem'pəri), for the time being; temporary.
pro[2] (prō), *n.* (*coll.*) a professional football player, cricketer, actor etc.; (*sl.*) a prostitute.

pro-[1] (prō), *pref.* in favour of; replacing, substituting for; (in existing latinate compounds) onward, forward, before.
pro-[2] (prō'), *pref.* before (in time or position); earlier than; projecting, forward; rudimentary.
proa (prō'ə), *n.* a long, narrow, swift Malayan canoe, usu. equipped with both sails and oars.
proactive (prōak'tiv), *a.* energetic, enterprising, taking the initiative.
pro-am (prō'am), *a.* involving both professionals and amateurs.
prob., (*abbr.*) probably.
probabilism (prob'əbilizm), *n.* in Roman Catholic theology, the doctrine that, in matters of conscience about which there is disagreement or doubt, it is lawful to adopt any course, at any rate if this has the support of any recognized authority. **probabilist**, *n.*
probability (probəbil'iti), *n.* the quality of being probable; that which is or appears probable; (*Math.*) likelihood of an event measured by the ratio of the favourable chances to the whole number of chances. **in all probability**, most likely.
probable (prob'əbl), *a.* likely to prove true, having more evidence for than against, likely. *n.* a person likely to be chosen for a team, post etc. **probably**, *adv.*
probang (prō'bāng), *n.* a slender whalebone rod with a piece of sponge etc. at the end, for introducing into or removing obstructions in the throat.
probate (prō'bāt), *n.* the official proving of a will; a certified copy of a proved will. **probate duty**, *n.* a tax charged upon the personal property of deceased persons, now merged in estate duty.
probation (prəbā'shən), *n.* a trial period of employment for a candidate for a job; a method of dealing with criminals by allowing them to go at large under supervision during their good behaviour; the act of proving; evidence, proof. **on probation**, being tested for suitability etc.; under the supervision of a probation officer. **probation officer**, *n.* a court official whose duty it is to supervise and assist offenders who are on probation. **probational**, *a.* serving for or pertaining to probation or trial. **probationary**, *a.* probational; undergoing probation. *n.* a probationer. **probationer**, *n.* one on probation or trial. **probationership**, *n.*
probative (prō'bətiv), *a.* proving or tending to prove; serving as proof, evidential. **probator** (-bā'-), *n.* an examiner; an approver.
probe (prōb), *n.* a surgical instrument, usu. a silver rod with a blunt end, for exploring cavities of the body, wounds etc.; (*Elec.*) a lead containing or connected to a monitoring circuit; a docking device, esp. a projecting pipe which connects with the drogue of a tanker aircraft to permit in-flight refuelling; an unmanned vehicle used to send back information from exploratory missions in space; an exploratory survey; a thorough investigation, as by a newspaper of e.g. alleged corruption. *v.t.* to search or examine (a wound, ulcer etc.) with, or as with, a probe; to scrutinize thoroughly. *v.i.* to make a tentative or exploratory investigation. **probe-scissors**, *n.pl.* scissors with the points tipped with buttons, used to open wounds. **probeable**, *a.* **prober**, *n.*
probity (prob'iti, prō'-), *n.* tried honesty, sincerity or integrity; high principle, rectitude.
problem (prob'ləm), *n.* a question proposed for solution; a matter, situation or person that is difficult to deal with or understand; a source of perplexity or distress; (*Geom.*) a proposition requiring something to be done; (*Phys. etc.*) an investigation starting from certain conditions for the determination or illustration of a law etc.; an arrangement of pieces on a chess-board from which a certain result has to be attained, usu. in a specified number of moves. **no problem**, it's all right; it doesn't matter. **to have a problem, problems**, to have difficulty (with); to be in trouble. **problem child**, *n.* one whose character presents

parents, teachers etc. with exceptional difficulties. **problem play, picture,** *n.* a play or picture dealing with a social problem or with tricky moral questions. **problematic, -ical** (-mat'-), *a.* doubtful, questionable, uncertain. **problematically,** *adv.* **problemist, problematist,** *n.* one who studies or composes (chess) problems.
pro-Boer (prōbooǝ'), *n.* one who favoured the Boers in the S African war of 1899–1902. *a.* favouring the Boers.
proboscis (prǝbos'is), *n.* (*pl.* **-ides** (-idēz)), the trunk of an elephant or the elongated snout of a tapir etc.; the elongated mouth of some insects; the suctorial organ of some worms etc.; (*facet.*) the human nose. **proboscis monkey,** *n.* a monkey of Borneo with a long, flexible nose. **proboscidean** (probǝsid'-), *a.* having a proboscis; pertaining to the Proboscidea, an order of mammals containing the elephants, the extinct mastodon etc. *n.* any individual of the Proboscidea. **proboscidiferous** (-dif'-), *a.*
proc., (*abbr.*) proceedings.
procacious (prǝkā'shǝs), *a.* forward, pert, petulant.
procaine (prō'kān), *n.* a crystalline substance used as a local anaesthetic.
pro-cathedral (prōkǝthē'drǝl), *n.* a church or other building used as a substitute for a cathedral.
proceed (prǝsēd'), *v.i.* to go (in a specified direction or to a specified place); to go on; to go forward, to advance, to continue to progress; to carry on a series of actions, to go on (with or in); to take steps; to act in accordance with a method or procedure; to be carried on; to issue or come forth, to originate; to take a degree; to take or carry on legal proceedings. **procedure** (-dyǝ), *n.* the act or manner of proceeding; the (customary or established) mode of conducting business etc., esp. in a court or at a meeting; a course of action; an action, a step in a sequence of actions. **procedural,** *a.* **proceeder,** *n.* one who proceeds, esp. to a university degree. **proceeding,** *n.* progress, advancement; a line of conduct; a transaction; (*pl.*) events, what is going on; (*pl.*) steps in the prosecution of an action at law; (*pl.*) the records of a learned society. **proceeds** (prō'-), *n.pl.* material results, profits, as the amount realized by the sale of goods.
proceleusmatic (prosǝlūsmat'ik), *a.* (*Pros.*) denoting a metrical foot of four short syllables. *n.* such a foot; (*pl.*) verse in this metre.
procellarian (prosǝleǝr'iǝn), *a.* (*Zool.*) belonging to the genus *Procellaria* or the Procellaridae, a family of Tubinares containing the petrels. *n.* a bird of this genus.
procephalic (prōsifal'ik), *a.* of or pertaining to the anterior part of the head, esp. in invertebrates.
procerebrum (prǝse'rǝbrǝm), *n.* the prosencephalon. **procerebral,** *a.*
process[1] (prō'ses), (*esp. N Am.*) (pros'-), *n.* a course or method of proceeding or doing, esp. a method of operation in manufacture, scientific research etc.; a natural series of continuous actions, changes etc.; a progressive movement or state of activity, progress, course; the course of proceedings in an action at law; a writ or order commencing this; a method of producing a printing surface by photography and mechanical or chemical means; (*Anat., Zool. etc.*) an out-growth, a protuberance of a bone etc. *v.t.* to institute legal proceedings against; to serve a writ on; to treat (food etc.) by a preservative or other process; to subject to routine procedure, to deal with; (*Compu.*) to perform operations on (data); to reproduce by a photomechanical process. **in the process,** during the carrying-out (of a specified operation). **in the process of,** engaged in; undergoing. **process block,** *n.* a printing block produced by photomechanical means. **process-server,** *n.* a sheriff's officer who serves processes or summonses. **processor,** *n.* a person or thing that processes; a device or program that processes data; a central processing unit.
process[2] (prǝses'), *v.i.* to go in procession.

procession (prǝsesh'ǝn), *n.* a train of persons, vehicles etc. proceeding in regular order for a ceremony, display, demonstration etc.; the movement of such a train; the act or state of issuing forth, emanation (as of the Holy Ghost from the Father). *v.i.* to go in procession. **processional,** *a.* pertaining to or used in processions. *n.* a service-book giving the ritual of or the hymns sung in religious processions; a processional hymn. **processionary, processive** (-siv), *a.* **processionist,** *n.* one who takes part in a procession. **processionize, -ise,** *v.i.*
procès-verbal (proseverbal'), *n.* (*pl.* **-baux** (-bō)), an official record of proceedings, minutes. [F]
prochain (proshen'), *a.* (*Law*) nearest, next. **prochain ami** or **amy** (amē'), *n.* the nearest friend, who is entitled to sue etc. on behalf of an infant. [F]
pro-choice (prōchois'), *a.* in favour of a woman's right to choose whether or not to have an abortion.
prochronism (prō'krǝnizm), *n.* an error in chronology dating an event before its actual occurrence.
procidence (prō'sidǝns), *n.* a slipping from the normal position, a prolapsus. **procident,** *a.*
procinct (prǝsingkt'), *a.* prepared, ready.
proclaim (prǝklām'), *v.t.* to announce publicly; to declare openly, to publish; to announce the accession of; to outlaw by public proclamation. **proclaimer,** *n.* **proclamation** (proklǝ-), *n.* **proclamatory** (-klam'-), *a.*
proclitic (prǝklit'ik), *a.* (*Gr. Gram.*) attached to and depending in accent upon a following word. *n.* a monosyllable attached to a following word and having no separate accent.
proclivity (prǝkliv'iti), *n.* tendency, propensity. **proclivitous,** *a.* steep, abrupt. **proclivous** (-klī'-), *a.* inclined or sloping forward (of teeth).
preoconsul (prōkon'sǝl), *n.* a Roman magistrate, usu. an ex-consul, governing a province; a governor or viceroy of a modern dependency etc. **proconsular** (-sū-), *a.* **proconsulate** (-sūlǝt), **proconsulship,** *n.*
procrastinate (prǝkras'tināt), *v.i.* to put off action; to be dilatory. **procrastinatingly,** *adv.* **procrastination,** *n.* **procrastinative, -tory,** *a.* **procrastinator,** *n.*
procreate (prō'kriāt), *v.t.* to generate, to beget. **procreant,** *a.* **procreation,** *n.* **procreative,** *a.* **procreativeness,** *n.* **procreator,** *n.*
Procrustean (prǝkrŭs'tiǝn), *a.* reducing to strict conformity by violent measures. [Gr. *Prokroustēs*, a mythical robber of Attica, who stretched or mutilated his victims till their length was exactly that of his couch]
proctalgia (proktal'jiǝ), *n.* pain in the anus.
proctectomy (proktek'tǝmi), *n.* excision of the rectum or anus.
proctitis (proktī'tis), *n.* inflammation of the anus or rectum.
proct(o)-, *comb. form.* pertaining to the anus.
proctor (prok'tǝ), *n.* a university official charged with the maintenance of order and discipline; a person employed to manage another's cause, esp. in an ecclesiastical court. **Queen's** or **King's Proctor,** an officer of the Crown who intervenes in probate, divorce or nullity cases when collusion or other irregularity is alleged. **proctorage** (-rij), *n.* management by a proctor. **proctorial** (-taw'ri-), *a.* **proctorship,** *n.* **proctorize, -ise,** *v.t.* to deal with (an under-graduate) in the capacity of proctor. **proctorization,** *n.*
procumbent (prǝkŭm'bǝnt), *a.* lying down on the face; (*Bot.*) lying or trailing along the surface of the ground.
procuration (prokūrā'shǝn), *n.* the act of procuring or obtaining; action on behalf of another; a document authorizing one person to act for another; the fee paid by the clergy for the bishop, archdeacon etc. at their visitations; procuring of girls for unlawful purposes.
procurator (prok'ūrātǝ), *n.* one who manages another's affairs, esp. those of a legal nature, an agent, a proxy, an attorney; (*Mediaeval and Sc. Univ.*) an elective officer

having financial, electoral and disciplinary functions, a proctor; (*Rom. Hist.*) a fiscal officer in an imperial province having certain administrative powers. **procurator fiscal**, *n*. (*Sc.*) the public prosecutor in a county or district. **procuratorship**, *n*. **procuratorial** (-taw'ri-), *a*. **procuratory**, *n*. a power of attorney. **procuratrix** (-triks), *n. fem.* one of the superiors managing the temporal affairs in a nunnery.

procure (prəkūə'), *v.t.* to obtain, to get by some means or effort; to acquire, to gain; to obtain as a pimp. *v.i.* to act as procurer or procuress, to pimp. **procurable**, *a*. **procural, procurance, procurement**, *n*. **procurer**, *n*. one who procures or obtains, esp. one who procures a woman to gratify a person's lust, a pimp, a pander. **procuress** (-ris), *n. fem.*

procureur (prokürœ'), *n*. a procurator. **procureur général** (zhänäral'), *n*. a public prosecutor acting in a court of appeal or of cassation. [F]

Procyon (prō'siən), *n*. the lesser dog-star; a genus of mammals containing the racoons, typical of the Procyonidae.

prod (prod), *n*. a pointed instrument, a goad, a poke with or as with this. *v.t.* (*past, p.p.* **prodded**) to poke with or as with such an instrument; to goad, to irritate, to incite. **prodder**, *n*.

prod-, *pref.* form of PRO- before vowels.

prodelision (prōdəlizh'ən), *n.* (*Pros.*) elision of the initial vowel (of a word etc.).

prodigal (prod'igəl), *a.* given to extravagant expenditure; wasteful, lavish (of). *n.* a prodigal person, a spendthrift. **prodigality** (-gal'-), *n.* extravagance, profusion; lavishness, waste. **prodigalize, -ise**, *v.t.* to spend prodigally. **prodigally**, *adv.*

prodigy (prod'iji), *n.* something wonderful or extraordinary; a wonderful example (of); a person, esp. a child, or thing with extraordinary gifts or qualities; something out of the ordinary course of nature, a monstrosity. **prodigious** (-dij'-), *a.* wonderful, astounding; enormous in size, quality, extent etc. **prodigiously**, *adv.* **prodigiousness**, *n.*

prodrome (prod'rōm), **prodromus** (-rəməs), *n. (pl.)* (**dromes** (-rəmēz), **-dromi** (-rəmī)) an introductory book or treatise; (*Med.*) a symptom of aproaching disease. **prodromal, prodromatic** (-drom'-), *a.*

produce (prədūs'), *v.t.* to bring into view; to publish, to exhibit; to bring into existence; to bear, to yield, to manufacture; to make; to bring about, to cause; to extend, to continue (a line) in the same direction; (*Theat., Cinema, TV*) to act as producer of. *n.* (prod'-), that which is produced or yielded; the result (of labour, skill etc.); (*collect.*) natural or agricultural products of a country etc. **producer**, *n.* one who produces, esp. a cultivator, manufacturer etc. as distinguished from a consumer; a furnace used for the manufacture of carbon monoxide gas; (*Theat., TV*) (*dated*) a director; (*Theat., Cinema, TV*) a person who exercises general administrative and financial control over a play, film or broadcast. **producer gas**, *n.* gas produced in a producer. **producible**, *a.* **producibility** (-bil'-), *n.* **product** (prod'ŭkt), *n.* that which is produced by natural processes, labour, art or mental application; effect, result; the quantity obtained by multiplying two or more quantities together; (*Chem.*) a compound not previously existing in a substance but produced by its decomposition. **product placement**, *n*. an oblique form of advertising in which a product is used or displayed during a film or TV programme. **productile** (-dŭk'til), *a.* capable of being produced or extended. **production** (-dŭk'-), *n.* the act of producing, esp. as opposed to consumption; a thing produced; the amount produced, the ouput; (*Econ.*) the creation of goods and services with exchange value; the work of a film etc. producer; (preparation for) the public presentation of a stage work. **production line**, *n.* a system of stage-by-stage manufacture in which a product undergoes various processes or operations as it passes along a conveyor belt. **production model**, *n.* a standard mass-produced model of esp. a car. **productive** (-dŭk'-), *a.* producing or tending to produce; yielding in abundance, fertile; (*Econ.*) producing commodities having exchangeable value. **productively**, *adv.* **productiveness, productivity** (prodŭktiv'-), *n.* yield in abundance; efficiency of production. **productivity deal**, *n.* an agreement making wage increases dependent on increased efficiency and output. **productor** (-dŭk'-), *n.* **productress** (-tris), *n. fem.*

pro-educational (prōedūkā'shənəl), *a.* in favour of education.

proem (prō'em), *n.* a preface, a preamble, an introduction, a prelude. **proemial** (-ē'-), *a.*

proembryo (prōem'briō), *n.* a cellular structure of various forms in plants from which the embryo is developed. **proembryonic** (-on'-), *a.*

proenzyme (prōen'zīm), *n.* the inactive form in which some enzymes are produced, to be activated after secretion.

prof (prof), (*coll.*) short for PROFESSOR.

Prof., (*abbr.*) Professor.

profane (prəfān'), *a.* not sacred, not initiated into sacred or esoteric rites or knowledge; secular; irreverent towards holy things; irreverent, impious, blasphemous; heathenish; common, vulgar. *v.t.* to treat with irreverence; to desecrate, to violate, to pollute. **profanation** (profə-), *n.* **profanely**, *adv.* **profaneness, profanity** (-fan'-), *n.* **profaner**, *n.*

profess (prəfes'), *v.t.* to make open or public declaration of; to affirm one's belief in or allegiance to; to affirm one's skill or proficiency in; to undertake the teaching or practice of (an art, science etc.); to teach (a subject) as a professor; to lay claim to, to make a show of, to pretend (to be or do). *v.i.* to act as a professor; to make protestations or show of. **professed**, *a.* avowed, declared, acknowledged; alleged; in the Roman Catholic Church, of a religious person who has taken vows. **professedly** (-sid-), *adv.* by profession; avowedly; pretendedly, ostensibly. **profession** (-shən), *n.* the act of professing; a declaration, an avowal; a protestation, a pretence; an open acknowledgment of sentiments, religious belief etc.; a vow binding oneself to, or the state of being a member of a religious order; a calling, a vocation, esp. an occupation involving high educational or technical qualifications; the body of persons engaged in such a vocation. **the profession**, (*coll.*) actors. **the three professions**, divinity, law, medicine. **professional**, *a.* of or pertaining to a profession; engaging in an activity as a means of livelihood, esp. as opposed to amateur; characterized by or conforming to the technical or ethical standards of a profession; competent, conscientious. *n.* a member of a profession; one who makes his living by some art, sport etc. as distinguished from one who engages in it for pleasure; a person who shows great skill and competence in any activity. **professional foul**, *n.* (*Sport*) a deliberate foul, to prevent the other team from scoring. **professionalism**, *n.* the qualities, stamp or spirit of a professional; participation by professionals, esp. in sports. **professionalize, -ise**, *v.t.* **professionally**, *adv.* **professionless**, *a.* **professor**, *n.* one who makes profession, esp. of a religious faith; a public teacher of the highest rank, esp. in a university. **professoriate** (-saw'riət), *n.* **professoress** (-ris), *a.* **professorship**, *n.* **professorial** (-saw'ri-), *a.* **professorially**, *adv.*

proffer (prof'ə), *v.t.* to offer or tender for acceptance. *n.* an offer, a tender. **profferer**, *n.*

proficient (prəfish'ənt), *a.* well versed or skilled in any art, science etc., expert, competent. *n.* one who is proficient, an expert. **proficiency**, *n.* **proficiently**, *adv.*

profile (prō'fīl), *n.* an outline; a side view, esp. of the hu-

profit — man face; a drawing, silhouette, or other representation of this; the outline of a vertical section of a building etc.; a vertical section of soil or rock showing the various layers; a wooden framework used as a guide in forming an earthwork; a short biographical sketch. *v.t.* to draw in profile or in vertical section; to shape (stone, wood, metal etc.) to a given profile; to write a profile of. **high, low profile**, a position, attitude or behaviour calculated (not) to call attention to oneself; extensive or limited involvement. **to keep a low profile**, to avoid calling attention to oneself. **high-profile**, *a.* **low-profile**, *a.* **profiler, profilist**, *n.* one who draws profiles.

profit (prof'it), *n.* any advantage or benefit, esp. one resulting from labour or exertion; excess of receipts or returns over outlay, gain *(often in pl.)*; (*Econ.*) the portion of the gains of an industry received by the capitalist or the investors. *v.t.* to benefit, to be of advantage to. *v.i.* to be of advantage (to); to receive benefit or advantage (by or from). **profit and loss**, (*Accountancy*) gains credited and losses debited in an account so as to show the net loss or profit. **profit motive**, *n.* (*Econ.*) the incentive of private profit for the production and distribution of goods. **profit-sharing**, *n.* a system of remuneration by which the workers in an industrial concern are apportioned a percentage of the profits in order to give them an interest in the business. **profitable**, *a.* yielding or bringing profit or gain, lucrative; advantageous, beneficial, useful. **profitableness**, *n.* **profitably**, *adv.* **profiteer** (-tiə'), *v.i.* to make excessive profits at the expense of the public, esp. in a time of national stress. *n.* one guilty of this social crime. **profiteering**, *n.* **profitless**, *a.* **profitlessly**, *adv.* **profitlessness**, *n.*

profiterole (prəfē'tərōl, -fit'-), *n.* a small, hollow ball of choux pastry with a sweet or savoury filling.

profligate (prof'ligət), *a.* licentious, dissolute; wildly extravagant. *n.* a profligate person. **profligacy, profligateness**, *n.* **profligately**, *adv.*

pro forma PRO¹.

profound (prəfownd'), *a.* having great intellectual penetration or insight; having great knowledge; requiring great study or research, abstruse, recondite; deep, intense; deep-seated, far below the surface; reaching to or extending from a great depth; coming from a great depth; thorough-going, extensive; very low (as an obeisance). *n.* a vast depth, an abyss. **profoundly**, *adv.* **profoundness, profundity** (-fun'di-), *n.*

profunda (prəfŭn'də), *n.* one of various deep-seated veins or arteries.

profundity PROFOUND.

profuse (prəfūs'), *a.* poured forth lavishly, copious, superabundant; prodigal, extravagant. **profusely**, *adv.* **profuseness, profusion** (-zhən), *n.*

Prog, (*abbr.*) (*coll.; sometimes offensive*) progressive.

prog¹ (prog), *v.i.* to poke about (esp. for food); to forage, to beg. *n.* (*sl.*) victuals, food.

prog² (prog), *n.* (*Oxf. and Camb. sl.*) a proctor. *v.t.* to proctorize. **proggins** (-inz), *n.* a prog.

prog³ (prog), *n.* short for (radio or TV) PROGRAMME.

progenitor (prəjen'itə), *n.* an ancestor in the direct line, a parent; a predecessor, an original. **progenitive**, *a.* **progenitiveness**, *n.* **progenitorial** (-taw'ri-), *a.* **progenitress** (-tris), **-trix** (-triks), *n. fem.* **progenitorship**, *n.* **progeniture** (-chə), *n.* begetting, generation; offspring.

progeny (proj'əni), *n.* offspring of human beings, animals or plants; children, descendants; results, consequences.

progeria (prōjiə'riə), *n.* premature old age.

progesterone (prəjes'tərōn), *n.* a female steroid hormone that prepares and maintains the uterus for pregnancy.

progestogen (prəjes'təjən), *n.* any of a range of hormones of the progesterone type, synthetic progestogens being used in oral contraceptives.

proggins PROG².

proglottis (prōglot'is), *n.* (*pl.* **-ides** (-idēz)) a segment of a tapeworm forming a distinct animal with genital organs. **proglottic**, *a.*

prognathic (prognath'ik), **prognathous** (prog'-), *a.* having the jaws projecting; projecting (of the jaws). **prognathism** (prog'-), *n.*

prognosis (prognō'sis), *n.* (*pl.* **-noses** (-sēz)) an opinion as to the probable course or result of an illness; a forecast, a prediction.

prognostic (prognos'tik), *n.* a sign or indication of a future event; a forecast; a symptom. *a.* foreshowing; indicative of something future by signs or symptoms. **prognosticable**, *a.* **prognosticate**, *v.t.* to foretell from present signs; to foreshadow, to presage, to betoken. **prognostication**, *n.* **prognosticative**, *a.* **prognosticator**, *n.*

programme, (*esp. N Am., Comput.*) **program** (prō'gram), *n.* (a paper, booklet etc. giving) a list of the successive items of any entertainment, public ceremony, conference, course of study etc. plus other relevant information; the items on such a list; a broadcast presented at a scheduled time; (*Comput.*) a sequence of instructions which, when fed into a computer, enable it to process data in specified ways; material for programmed instruction; a plan or outline of proceedings or actions to be carried out. *v.t.* to arrange a programme for; to enter in a programme; to cause to conform to a certain pattern, esp. of thought, behaviour; (*usu.* **program, -gramming, -grammed**) to prepare as a program for, to feed a program into, a computer. **programme music**, *n.* music intended to suggest a definite series of scenes, incidents etc. **programmable** (-gram'-), *a.* (*Comput.*). **programmability** (-bil'-), *n.* **programmatic** (-mat'-), *a.* of or having a programme; of, or of the nature of, programme music. **programmed**, *a.* **programmed instruction, learning**, *n.* a teaching method involving the breaking down of the subject matter into small items in a logical sequence on which students can check themselves as they proceed. **programmer**, *n.* (*Comput.*). **programming**, *n.* **programming language**, *n.* any of various code systems used in writing computer programs and giving instructions to computers.

progress (prō'gres, *esp. N Am.* prog'-), *n.* a moving or going forward; movement onward, advance; advance towards completion, a higher state; increased proficiency; growth, development. *v.i.* (prəgres'), to move forward, to advance; to advance, to develop; to make improvement. **in progress**, going on, proceeding. **progression** (-gresh'ən), *n.* progress, motion onward; movement in successive stages; a regular succession of notes or chords in melody or harmony; regular or proportional advance by increase or decrease of numbers. **arithmetical progression** ARITHMETIC. **geometrical progression** GEOMETRY. **progressional**, *a.* **progressionist**, a believer in or advocate of social and political progress. **progressionism**, *n.* **progressist** (prō'-, prog'-, -gres'-), *a.* advocating progress, esp. in politics. *n.* a progressive, a reformer. **progressive** (-gres'-), *a.* moving forward or onward; improving; (of a disease) increasing in extent or severity; in a state of progression, proceeding step by step, successive; continuously increasing; believing in or advocating social and political reform; denoting an educational system which allows flexibility and takes the needs and abilities of the individual child as its determinant; denoting a verb form which expresses action in progress. *n.* a progressive person; (**Progressive**) an adherent of a party called progressive; the progressive form of a verb. **progressive whist** or **bridge**, *n.* whist or bridge played by a number of sets of players at different tables, each winning player moving to another table at the end of each hand or series of hands. **progressively**, *adv.* **progressiveness**, *n.* **progressivism**, *n.* the principles of a progressive or reformer.

prohibit (prəhib'it), *v.t.* to forbid authoritatively; to hinder, to prevent. **prohibiter, prohibitor**, *n.* **prohibition**

(prōibish'-), *n.* the act of prohibiting; an order or edict prohibiting; the forbidding by law of the manufacture and the sale of intoxicating liquors for consumption as beverages. **prohibitionist,** *n.* one in favour of prohibiting the sale of intoxicating liquors. **prohibitive, -itory,** *a.* tending to prohibit or preclude; (of costs, prices etc.) such as to debar purchase, use etc. **prohibitively,** *adv.* **prohibitiveness,** *n.*

project (proj'ekt, prō'-), *n.* a plan, a scheme, a design; an (esp. large-scale) undertaking; a piece of work undertaken by a pupil or group of pupils to supplement and apply classroom studies. *v.t.* (-jekt'), to throw or shoot forward; to cause to extend forward or jut out; to cast (light, shadow, an image) on to a surface or into space; to enable (one's voice) to be heard at a distance; to transport in the imagination; to make a prediction based on known data; to impute (something in one's own mind) to another person, group or entity; to contrive, to plan; to make (an idea etc.) objective; (*Geom.*) to draw straight lines from a given centre through every point of (a figure) so as to form a corresponding figure on a surface; to produce (such a projection); to make a projection of. *v.i.* to jut out, to protrude; to make oneself audible at a distance; to communicate effectively. **projectile** (prəjek'til), *n.* a body projected or thrown forward with force; a self-propelling missile, esp. one adapted for discharge from a heavy gun. *a.* impelling forward; adapted to be forcibly projected, esp. from a gun. **projection** (prəjek'shən), *n.* the act or state of projecting, protruding, throwing or impelling; a part or thing that projects; the act of planning; the process of externalizing an idea or making it objective; a mental image viewed as an external object; the showing of films or slides by projecting images from them on to a screen; a prediction based on known data; the geometrical projecting of a figure; the representation of the terrestrial or celestial sphere, or a part of it, on a plane surface; the process whereby one ascribes to others mental factors and attributes really in ourselves. **Mercator's projection** (məkā'təz), a projection of the surface of the earth upon a plane so that the lines of latitude are represented by horizontal lines and the meridians by parallel lines at right angles to them. **projectionist,** *n.* one who operates a film projector. **projective,** *a.* pertaining to or derived by projection; (*Geom.*) such that they may be derived from one another by projection (of two plane figures); externalizing or making objective. **projective property,** a property that remains unchanged after projection. **projectively,** *adv.* **projectment** (-jekt'-), *n.* a scheme, a design, a contrivance. **projector** (-jek'-), *n.* one who forms schemes; an instrument or apparatus for projecting rays of light, images etc.

prokaryote (prōka'riot), *n.* an organism whose cells have no distinct nucleus, their genetic material being carried in a single filament of DNA.

prolapse (prō'laps, -laps'), *n.* prolapsus. *v.i.* to fall down or out. **prolapsus** (-lap'səs), *n.* a falling down or slipping out of place of an organ or part, as the uterus or rectum.

prolate (prō'lāt), *a.* extended in the direction of the longer axis, elongated in the polar diameter, opposed to oblate. **prolately,** *adv.* **prolateness,** *n.* **prolative** (-lā'-, prō'-), *a.* (*Gram.*) extending or completing the predicate.

prole (prōl), *n., a.* (*derog., offensive*) short for PROLETARIAN.

proleg (prō'leg), *n.* one of the soft, fleshy appendages or limbs of caterpillars etc., distinct from the true legs.

prolegomenon (prōligom'inən), *n.* (*usu. in pl.* **-ena** (-inə)) an introductory or preliminary discourse prefixed to a book etc. **prolegomenary, -enous,** *a.*

prolepsis (prəlep'sis), *n.* anticipation; (*Rhet.*) a figure by which objections are anticipated or prevented; the anticipatory use of a word as attributive instead of a predicate, as in *their murdered man* for *the man they intended to murder;* a prochronism. **proleptic, -ical,** *a.* **proleptically,** *adv.*

proletarian (prōlətea'riən), *a.* of or pertaining to the common people. *n.* a member of the proletariat. **proletarianism,** *n.* **proletariat** (-riət, -at), *n.* the class of the community without property, the wage-earners, the unprivileged classes. **proletary** (prō'-), *n., a.*

prolicide (prō'lisīd), *n.* the crime of killing one's offspring, esp. before or immediately after birth. **prolicidal** (-sī'-), *a.*

pro-life (prōlīf'), *a.* favouring greater restrictions on the availability of legal abortions and/or a ban on the use of human embryos for experimental purposes. **pro-lifer,** *n.*

proliferate (prəlif'ərāt), *v.i.* to grow or reproduce itself by budding or multiplication of parts; to grow or increase rapidly and abundantly; to become more widespread. *v.t.* to produce by proliferation. **proliferation,** *n.* **proliferative, proliferous,** *a.* **proliferously,** *adv.*

prolific (prəlif'ik), *a.* bearing offspring, esp. abundantly; fruitful, productive, fertile; abounding (in); very productive (of); (*Bot.*) bearing fertile seed. **prolificacy, prolificity** (prolifis'-), **prolificness,** *n.* **prolifically,** *adv.* **prolification,** *n.* the generation of animals or plants; (*Bot.*) the production of buds from leaves etc., the development of an abnormal number of parts.

prolix (prō'liks), *a.* long and wordy; tedious, tiresome, diffuse. **prolixity** (-lik'si-), **prolixness,** *n.* **prolixly,** *adv.*

prolocutor (prōlok'ūtə), *n.* a chairman or speaker, esp. of the Lower House of Convocation in the Church of England. **prolocutorship,** *n.*

prologue (*esp. N Am.*) **prolog** (prō'log), *n.* an introductory discourse, esp. lines introducing a play; an act or event forming an introduction to some proceeding or occurrence; the speaker of a prologue. **prologize, -ise,** *v.t.* to introduce, to preface.

prolong (prəlong'), *v.t.* to extend in duration; to lengthen, to extend in space or distance; to lengthen the pronunciation of. **prolongable,** *a.* **prolongation** (prō-), *n.* the act of lengthening or extending; a lengthening in time or space; the part by which anything is lengthened. **prolonger,** *n.*

prolonge (prəlonj'), *n.* a rope in three pieces connected by rings with a hook at one end and a toggle at the other, used for moving a gun etc.

prolusion (prəloo'zhən), *n.* a prelude; a preliminary essay, exercise or attempt. **prolusory** (-səri), *a.*

PROM (prom), (*Comput.*) acronym for P*rogrammable* R*ead* O*nly* M*emory.*

prom (prom), *n.* short for (sea-front) PROMENADE, PROMENADE CONCERT, PERFORMANCE.

promenade (promənahd'), *n.* a walk, drive or ride for pleasure, exercise or display; a place for promenading, esp. a paved terrace on a sea-front; a processional sequence in a square or country dance; (*N Am.*) a dance at college, school or association. *v.i.* to take a walk etc. for pleasure, exercise or show; (*Dancing*) to perform a promenade. *v.t.* to take a promenade along, about or through; to lead (a person) about, esp. for display. **promenade concert, performance** (prom'-), *n.* a concert or performance at which the floor of the hall is left bare for the audience to stand or walk about. **promenade deck,** *n.* an upper deck on a ship where passengers may stroll. **promenader,** *n.* a standing member of the audience at a promenade concert or performance.

Promerops (prom'ərops), *n.* a S African genus of birds, comprising the Cape promerops with a slender, curved bill and a long tail.

Promethean (prəmē'thiən), *a.* of, pertaining to, or like Prometheus; original creative. **promethium** (-thiəm), *n.* a metallic element, at. no. 61; chem. symbol Pm, obtained as a fission product of uranium. [Gr. *Promētheus,* said to have stolen fire from Olympus, and to have bestowed it on mortals, teaching them the use of it and the arts of civilization]

prominent (prom'inənt), *a.* standing out, jutting, projecting, protuberant; conspicuous; distinguished. **prominence, -nency,** *n.* the state of being prominent; a prominent point or thing, a projection; an eruption of incandescent gas from the surface of the sun. **prominently,** *adv.*
promiscuous (prəmis'kūəs), *a.* mixed together in a disorderly manner; of different kinds mingled confusedly together; not restricted to a particular person, kind etc.; indulging in indiscriminate sexual intercourse; fortuitous, accidental, casual. **promiscuity** (promiskū'-), *n.* sexual promiscuousness; communal marriage. **promiscuously,** *adv.* **promiscuousness,** *n.*
promise (prom'is), *n.* a verbal or written engagement to do or forbear from doing some specific act; that which is promised; ground or basis of expectation, esp. of success, improvement or excellence. *v.t.* to engage to do or not do; to engage to give or procure; to make a promise of something to; to give good grounds for expecting. *v.i.* to bind oneself by a promise; to give grounds for favourable expectations. **breach of promise** BREACH. **land of promise** or **promised land,** the land of Canaan promised to Abraham and his seed; heaven; any place of expected happiness or prosperity. **to promise well or ill,** to hold out favourable or unfavourable prospects. **promise-breaker,** *n.* **promisee** (-sē'), *n.* (*Law*) one to whom a promise is made. **promiser,** *n.* **promising,** *a.* giving grounds for expectation or hope; hopeful, favourable. **promisingly,** *adv.* **promisor,** *n.* one who enters into a covenant. **promissory,** *a.* containing or of the nature of a promise, esp. a promise to pay money. **promissory note,** *n.* a signed engagement to pay a sum of money to a specified person or the bearer at a stated date or on demand.
promo (prō'mō), *n.* (*pl.* **-mos**), (*coll.*) something used to promote a product, esp. a pop video.
promontory (prom'əntəri), *n.* a headland; a point of high land projecting into the sea; a rounded protuberance. **promontoried,** *a.*
promote (prəmōt'), *v.t.* to contribute to the growth, increase or advancement of; to foster, to encourage; to raise to a higher rank or position, to exalt; to bring to the notice of the public, to encourage the sale of by advertising; (*Chess*) to raise (a pawn) to the rank of queen. **promoter,** *n.* one who or that which promotes or furthers; one who organizes or finances a joint-stock company; one who organizes a sporting event, esp. a boxing match. **promoterism,** *n.* the practice of floating joint-stock companies. **promotion,** *n.* advancement in position; furtherance, encouragement, a venture, esp. in show business; (an advertising campaign, special offer etc. intended as a means of) bringing a product, person to public notice. **on promotion,** awaiting, expecting or preparing oneself for promotion; on one's good behaviour. **promotional,** *a.* **promotive,** *a.*
prompt (prompt), *a.* acting quickly or ready to act as occasion demands; done, made or said with alacrity. *n.* time allowed for payment of a debt as stated in a prompt-note; the act of prompting, or the thing said to prompt an actor etc. *v.t.* to urge or incite (to action or to do); to instigate; to suggest to the mind, to inspire, to excite (thoughts, feelings etc.); to assist (a speaker, actor etc.) when at a loss, by suggesting the words forgotten. **prompt-book,** *n.* a copy of the play for the use of the prompter at a theatre. **prompt-note,** *n.* a note reminding a purchaser of a sum due and the date of payment. **prompt-side,** *n.* the side of a stage on which the prompter stands, usu. to the left of the actor. **prompter,** *n.* one who prompts, esp. one employed at a theatre to prompt actors. **promptitude** (-titūd), **promptness,** *n.* **promptly,** *adv.*
promulgate (prom'əlgāt), *v.t.* to make known to the public; to publish abroad, to disseminate; to announce publicly. **promulgation,** *n.* **promulgator,** *n.*

promuscis (prəmŭs'is), *n.* the proboscis of the Hemiptera and some other insects.
pron., (*abbr.*) pronoun; pronounced; pronunciation.
pronaos (pronā'os), *n.* (*Gr. and Roman Ant.*) the area immediately before a temple enclosed by the portico; the vestibule.
pronate (prō'nāt), *v.t.* to lay (a hand or forelimb) prone so as to have the palm downwards. **pronation** (-nā'shən), *n.* **pronator** (-nā'-), *n.* a muscle that brings about pronation.
prone (prōn), *a.* lying with the face downward, opp. to supine; lying flat, prostrate; sloping downwards; disposed, inclined. **pronely,** *adv.* **proneness,** *n.*
prong (prong), *n.* a spike of a fork or sharp-pointed instrument; a spike-like projection. *v.t.* to pierce, stab or prick with a prong. **prongbuck, pronghorn** or **prong-horned antelope,** *n.* a N American ruminant, *Antilocapra americana*. **pronged,** *a.*
pronominal (prənom'inəl), *a.* pertaining to or of the nature of a pronoun. **pronominally,** *adv.*
pronoun (prō'nown), *n.* a word used in place of a noun to denote a person or thing already mentioned or implied.
pronounce (prənowns'), *v.t.* to utter articulately, to say correctly; to utter formally, officially or rhetorically; to declare, to affirm. *v.i.* to articulate; to declare one's opinion or judgment (on, for, against etc.). **pronounceable,** *a.* **pronounced,** *a.* strongly marked, emphatic, decided. **pronouncedly** (-sid-), *adv.* **pronouncement,** *n.* a statement. **pronouncer,** *n.* **pronouncing,** *a.* pertaining to, indicating or teaching pronunciation. **pronunciation** (-nūnsi-), *n.* the act or mode of pronouncing sounds or words; the correct pronouncing of words etc.
pronto (pron'tō), *adv.* (*coll.*) without delay, quickly.
pronunciation PRONOUNCE.
proof (proof), *n.* the act of proving, a test, a trial; testing, assaying, experiment; demonstration; a sequence of steps establishing the correctness of a mathematical or logical proposition; convincing evidence of the truth or falsity of a statement, charge etc. esp. oral or written evidence submitted in the trial of a cause; (*Sc. Law*) evidence taken before a judge, or a trial before a judge instead of by jury; the state or quality of having been proved or tested; proved impenetrability, esp. of armour; a standard degree of strength in proof spirit; a trial impression from type for correction; the alcoholic strength of a beverage based on this; an impression of an engraving taken with special care before the ordinary issue is printed; a first or early impression of a photograph, coin, medal etc. *a.* proved or tested as to strength, firmness etc.; impenetrable; able to resist; used in testing, verifying etc.; of a certain degree of alcoholic strength. *v.t.* to make proof, 'esp. waterproof. **proofread,** *v.t.* to read and correct (printer's proofs). **proof-reader,** *n.* **proof-reading,** *n.* **proof spirit,** *n.* a mixture of alcohol and water containing a standard amount of alcohol, in Britain 57.1% by volume.
-proof, *comb. form.* (to make) resistant, impervious.
prop[1] (prop), *n.* a support, esp. a prop a loose or temporary one; a person supporting a cause etc.; a prop forward. *v.t.* (*past, p.p.*) **propped**) to support or hold (up) with or as with a prop; of a prop, to support, to hold up. **to prop up,** to support in an upright position; to keep going with financial etc. help. **prop forward,** *n.* (*Rugby*) either of the two forwards supporting the hooker in the front row of the scrum.
prop[2] (prop), *n.* a stage property. **props,** *n.pl.* stage properties; (*sing. in constr.*) (also **props man, -mistress** etc.) PROPERTY-MAN
prop[3] (prop), *n.* an aeroplane propeller. **propfan,** *n.* (an aircraft with) a jet engine having turbine-driven, rear-mounted propellors. **propjet,** TURBOPROP under TURBO.
prop., (*abbr.*) proper, property; proposition; proprietor.
propaedeutic (prōpēdu'tik), *a.* pertaining to or of the nature of introductory or preparatory study. *n.pl.* prelimi-

nary learning or instruction introductory to any art or science.
propaganda (propəgan'də), *n.* information, ideas, opinions etc. propagated as a means of winning support for, or fomenting opposition to, a government, cause, institution etc.; an organization, scheme or other means of propagating such information etc. **Propaganda**, *n.* in the Roman Catholic Church, a congregation of cardinals charged with all matters connected with foreign missions. **propagandism**, *n.* **propagandist**, *n.* one devoted to or engaged in propaganda. *a.* propagandistic. **propagandistic** (-dis'-), *a.* **propagandize, -ise**, *v.t., v.i.* to spread (by) propaganda; to subject to propaganda.
propagate (prop'əgāt), *v.t.* to cause to multiply by natural generation or other means; to reproduce; to pass down to the next generation; to cause to spread or extend, to disseminate; to impel forward, to transmit, to cause to extend in space. *v.i.* to be reproduced or multiplied by natural generation or other means; to have offspring, to spread or extend. **propagable**, *a.* **propagation**, *n.* the act of propagating; dissemination, diffusing; extension or transmission through space. **propagative**, *a.* **propagator**, *n.* a person or thing that propagates; a heated, covered box for growing plants from seed or cuttings.
propane (prō'pān), *n.* a flammable, gaseous alkane used as fuel.
propel (prəpel'), *v.t.* (*past, p.p.* **propelled**) to drive or cause to move forward or onward. **propellant**, *a.* that propels. *n.* that which propels, esp. the fuel mixture used by a rocket engine, or the gas in an aerosol. **propellent**, *a.* **propeller**, *n.* one who or that which propels; (also **screw-propeller**) a rotating device, usu. consisting of two to four blades set at an angle and twisted like the thread of a screw, at the end of a shaft driven by steam, electricity etc. for propelling a vessel through the water or an aeroplane or airship through air. **propelling pencil**, *n.* one having a metal or plastic casing and a replaceable lead which can be extended or retracted. **propelment**, *n.*
propene (prō'pēn), PROPYLENE under PROPYL.
propensity (prəpen'siti), *n.* bent, natural tendency, inclination.
proper (prop'ə), *a.* belonging or pertaining exclusively or peculiarly (to); correct, just; suitable, appropriate; fit, becoming; decent, respectable; strictly decorous; real, genuine, according to strict definition (*usu. following its noun*); (*coll.*) thorough, complete (*Her.*) in the natural colours. *n.* that part of the mass which varies according to the liturgical calendar. **proper fraction**, *n.* a fraction less than unity. **proper name, noun**, *n.* one designating an individual person, animal, place etc. **properly**, *adv.* in a proper manner, fitly, suitably; correctly, accurately; (*coll.*) thoroughly, quite. **properness**, *n.*
property (prop'əti), *n.* an inherent quality or attribute; a characteristic; that which is owned; a possession, possessions; a piece of real estate; exclusive right of possession, ownership; (*pl.*) articles required for the production of a play on the stage. **property-man, -master, -mistress, -woman**, *n.* the man or woman in charge of theatrical properties. **property-tax**, *n.* a direct tax on property. **propertied**, *a.*
propfan, PROP³.
prop-forward, PROP¹.
prophecy (prof'əsi), *n.* a prediction, esp. one divinely inspired; the prediction of future events; the gift or faculty of prophesying.
prophesy (prof'əsī), *v.t.* to predict, to foretell. *v.i.* to utter prophecies. **prophesier** (-siə), *n.*
prophet (prof'it), *n.* one who foretells future events, esp. under divine inspiration; a revealer or interpreter of the divine will; a religious leader, a founder of a religion; a preacher or teacher of a cause etc. **major prophets**, Isaiah, Jeremiah, Ezekiel, Daniel. **minor prophets**, the prophets in the Old Testament from Hosea to Malachi. **prophet of doom**, a person who is continually predicting ruin and disaster. **the Prophet**, Mohammed. **the Prophets**, (the books written by) the prophetic writers of the Old Testament. **prophetess** (-tis), *n. fem.* **prophethood** (-hud), **prophetship**, *n.* **prophetic, -al** (-fet'-), *a.* of, pertaining to, or containing prophecy; predictive, anticipative. **prophetically**, *adv.* **propheticism** (-fet'isizm), *n.*
prophylactic (profilak'tik), *a.* protecting against disease; preventive. *n.* a preventive medicine; (*esp. N Am.*) a condom. **prophylaxis** (-lak'sis), *n.*
propinquity (prəping'kwiti), *n.* nearness in time, space, or relationship; similarity.
propionic acid (prōpion'ik), *n.* a fatty acid used esp. in making flavourings and perfumes.
propitiate (prəpish'iāt), *v.t.* to appease, to conciliate; to render favourable. **propitiable**, *a.* **propitiation**, *n.* **propitiator**, *n.* **propitiatory**, *a.* intended or serving to propitiate.
propitious (-shəs), *a.* favourable; disposed to be kind or gracious; auspicious, suitable (for etc.). **propitiously**, *adv.* **propitiousness**, *n.*
propjet, PROP³.
propolis (prop'əlis), *n.* a resinous substance obtained by bees from buds etc. and used to cement their combs, stop up crevices etc., bee-glue.
proponent (prəpō'nənt), *n.* one who makes a proposal or proposition; one who argues for, an advocate.
proportion (prəpaw'shən), *n.* the comparative relation of one part or thing to another with respect to magnitude, number or degree; ratio; due relation, suitable adaptation of one part or thing to others; a proportional part, a share; (*pl.*) dimensions; equality of ratios between pairs of quantities; a series of such quantities; the rule by which from three given quantities a fourth may be found bearing the same ratio to the third as the second bears to the first, the rule of three. *v.t.* to adjust in suitable proportion; to make proportionate (to). **geometrical, harmonic proportion** GEOMETRY, HARMONIC. **in, out of proportion (to)**, (not) in due relation as to magnitude, number etc.; (not) consistent with the real importance of the matter in hand. **proportionable**, *a.* being in proportion, corresponding, proportional. **proportionableness**, *n.* **proportionably**, *adv.* **proportional**, *a.* having due proportion; pertaining to proportion; (*Math.*) having a constant ratio. *n.* a quantity in proportion with others, one of the terms of a ratio. **proportional representation**, *n.* an electoral system in which the representation of parties in an elected body is as nearly as possible proportional to their voting strength. **proportionalism**, *n.* **proportionalist**, *n.* an advocate of proportional representation. **proportionality** (-nal'-), *n.* **proportionally**, *adv.* **proportionate** (-nət), *a.* in due or a certain proportion (to). *v.t.* (-nāt), to make proportionate. **proportionately**, *adv.* **proportionateness**, *n.* **proportioned**, *a.* (*usu. in comb., as well-proportioned*). **proportionment**, *n.*
propose (prəpōz'), *v.t.* to put forward, to present for consideration etc.; to nominate; to offer (a toast); to intend. *v.i.* to put forward a plan or intention; to make an offer, esp. of marriage. **proposal**, *n.* the act of proposing; something proposed; an offer of marriage; an application for insurance. **proposer**, *n.*
proposition (propəzish'ən), *n.* that which is propounded or proposed; (*Log.*) a sentence in which something is affirmed or denied; (*Math.*) a formal statement of a theorem or problem, sometimes including the demonstration; an invitation to have sexual intercourse; (*coll.*) a person or thing that has to be dealt with. *v.t.* to make a proposition to, esp. to invite to have sexual intercourse. **propositional**, *a.*
propound (prəpownd'), *v.t.* to state or set forth for consideration, to propose; to bring forward (a will etc.) for probate. **propounder**, *n.*
proprietor (prəprī'ətə), *n.* an owner, one who has the

propriety 648 **protactinium**

exclusive legal right or title to anything. **proprietary**, *n.* a body of proprietors collectively; proprietorship. *a.* of or pertaining to a proprietor or proprietorship; owned as property; made and marketed under a patent, trademark etc. **proprietorial** (-taw'ri-), *a.* **proprietorially**, *adv.* **proprietorship**, *n.* **proprietress** (-tris), **-trix** (-triks), *n. fem.*
propriety (prəprī'əti), *n.* the quality of being conformable to an acknowledged or correct standard or rule; fitness, correctness, rightness; correctness of behaviour, becomingness.
proprioception (prōpriəsep'shən), *n.* reception of or activation by, stimuli from within the organism. **proprioceptive**, *a.* **proprioceptor**, *n.* any reeptor which receives such stimuli.
props PROP [2].
propulsion (prəpŭl'shən), *n.* the act of propelling, a driving forward; that which propels, a driving force. **propulsive**, *a.*
propyl (prō'pil), *n.* a hydrocarbon radical derived from propane. **propylene** (prop'ilēn), *n.* a colourless, gaseous alkene obtained from petroleum.
pro rata PRO [1].
prorate (prōrāt', prō'-), *v.t.* (*chiefly N Am.*) to distribute proportionally. **pro-rateable**, *a.*
prorogue (prərōg'), *v.t., v.i.* to put an end to the meetings of (as Parliament) without dissolving it. **prorogation**, *n.*
pros- *pref.* to, towards; before, in addition.
prosaic (prəzā'ik), *a.* pertaining to or resembling prose; unpoetic, unimaginative; dull, commonplace. **prosaically**, *adv.* **prosaicness**, **prosaism** (prō'-), **prosaicism** (-zā'isizm), *n.* **prosaist** (prō'-), *n.* a writer of prose; a prosaic person.
proscenium (prəsē'niəm), *n.* (*pl.* **-nia** (-ə)) the part of a stage between the curtain or drop-scene and the orchestra; in ancient Greek and Roman theatres, the space in front of the scenery, the stage. **proscenium (arch),** *n.* the frame through which the audience views the traditional type of stage.
prosciutto (prəshoo'tō), *n.* cured, usu. smoked, Italian ham.
proscribe (prəskrīb'), *v.t.* to publish the name of, as doomed to death, forfeiture of property etc., to outlaw; to banish, to exile; to denounce as dangerous; to interdict, to forbid. **proscriber**, *n.* **proscription** (-skrip'-), *n.* **proscriptive** (-skrip'-), *a.*
prose (prōz), *n.* ordinary written or spoken language not in metre, as opposed to verse; a passage for translation into a foreign language; commonplaceness; a tedious or unimaginative discourse. *a.* written in or consisting of prose; dull, commonplace, prosaic. *v.i.* to write or talk in a dull, tedious manner. *v.t.* to write or utter in prose; to turn into prose. **prose poem**, *n.* a piece of prose that has some of the characteristics of poetry. **prosy**, *a.* dull, tedious, long-winded. **prosily**, *adv.* **prosiness**, *n.*
prosecute (pros'ikūt), *v.t.* to pursue or follow up with a view to attain or accomplish; to carry on (work, trade etc.); to take legal proceedings against; to seek to obtain by legal process. *v.i.* to act as a prosecutor; to institute a prosecution. **prosecutable**, *a.* **prosecution** (-kū'-), *n.* the act of prosecuting; the exhibition of a charge against an accused person before a court; the instituting and carrying on of a civil or criminal suit; the prosecutor or prosecutors collectively. **prosecutor**, *n.* **public prosecutor**, an officer conducting criminal proceedings on behalf of the public. **prosecutrix** (-triks), *n. fem.*
proselyte (prosiəlīt), *n.* a new convert to some religion, party or system, esp. a gentile convert to Judaism. **proselytism** (-li-), *n.* **proselytize, -ise** (-li-), *v.t., v.i.* to convert or make converts. **proselytizer, -iser**, *n.*
prosencephalon (prosənsef'əlon), *n.* (*pl.* **-la** (-lə)) the anterior part of the brain comprising the cerebral hemispheres etc. **prosencephalic** (-ensəfal'-), *a.*

prosenchyma (prəsen'kimə), *n.* plant tissue composed of elongated thick-walled cells, found esp. in woody tissue. **prosenchymatous** (prosənkim'-), *a.*
prosimian (prōsim'iən), *a., n.* (of or being) one of a primitive suborder of primates which includes lemurs, lorises and tarsiers.
prosit (prō'sit), *int.* may it benefit you, success, used in (German) drinking toasts.
prosody (pros'ədi), *n.* the science of versification, formerly a branch of grammar. **prosodiacal** (dī'ə-), **prosodial** (-sō'-), **prosodic** (-sod'-), *a.* **prosodist**, *n.*
prosopopoeia (prosōpəpe'yə), *n.* a rhetorical figure by which abstract things are represented as persons, or absent persons as speaking.
prospect [1] (pros'pekt), *n.* an extensive view; the way a house etc. fronts or looks; a scene; a pictorial representation of a view; a mental picture of what is to come; expectation, ground of expectation; an indication of the presence of ore; a place likely to yield an ore; a sample of ore for testing; the mineral obtained by testing; (*pl.*) expectation of money to come or of an advancement in career; (*pl.*) chances of success; a prospective customer. **prospect** [2] (prəspekt'), *v.i.* to search, to explore, esp. for minerals. *v.t.* to search or explore (a region) for minerals. **prospective**, *a.* pertaining to the future; anticipated, expected, probable. **prospectively**, *adv.* **prospectiveness**, *n.* **prospectless**, *a.* **prospector**, *n.* one who searches for minerals or mining sites.
prospectus (prəspek'təs), *n.* a descriptive circular announcing the main objects and plans of a commercial scheme, institution, literary work etc.
prosper (pros'pə), *v.t.* to make successful or fortunate. *v.i.* to succeed; to thrive. **prosperity** (-spe'-), *n.* the condition of being prosperous; success, wealth. **prosperous**, *a.* successful, thriving, making progress, or advancement; wealthy. **prosperously**, *adv.*
prost (prōst), PROSIT.
prostaglandin (prostəglan'din), *n.* any of a group of hormone-like substances which have wide-ranging effects on body processes, e.g. muscle contraction.
prostate (pros'tāt), *n., a.* (of) the prostate gland. **prostate gland**, *n.* a gland situated before the neck of the bladder in male mammals, that secretes the major liquid component of semen. **prostatectomy** (-stətek'təmi), *n.* surgical removal of the prostate gland. **prostatic** (-tat'-), *a.* **prostatitis** (-təti'tis), *n.* inflammation of the prostate.
prosthesis (pros'thəsis, -thē'-), *n.* (*pl.* **-ses** (-sēs)) (*Gram.*) the addition of one or more letters to the beginning of a word; (*Surg.*) the addition of an artificial part to supply a defect; an artificial part thus supplied. **prosthetic** (-thet'-), *a.* **prosthetically**, *adv.* **prosthetics**, *n.sing.* the branch of surgery or dentistry concerned with prosthesis. **prosthetist**, *n.*
prostitute (pros'titūt), *v.t.* to hire (oneself, another) out for sexual purposes; to offer or sell for base or unworthy purposes; to devote to base uses. *n.* a person (esp. a woman or homosexual man) who engages in sexual intercourse for money. **prostitution**, *n.* **prostitutor**, *n.*
prostrate (pros'trāt), *a.* lying flat or prone, face down; of a plant, extending, trailing or growing in a horizontal position; lying in a posture of humility or at mercy; overcome, exhausted. *v.t.* (-trāt'), to lay flat; to cast (oneself) down, esp. in reverence or adoration (before); to throw down, to overthrow, to overcome, to demolish; to reduce to physical exhaustion. **prostration**, *n.*
prostyle (prō'stīl), *a.* having a row of columns, usu. four, entirely in front of the building. *n.* a portico supported on columns entirely in front of the building.
prosy PROSE.
Prot., (*abbr.*) Protectorate; Protestant.
prot- PROT(O)-.
protactinium, *n.* a radioactive element, at. no. 91; chem.

protagonist 649 protract

symbol Pa, yielding actinium on disintegration.
protagonist (prətăg'ənist), *n.* the leading character or actor in a Greek play; a leading character, advocate, champion etc.
protasis (prot'əsis), *n. (pl.* **-ases**) a clause containing the antecedent, esp. in a conditional sentence; the first part of a classic drama, in which the characters are introduced and the argument explained. **protatic** (-tat'-), *a.*
Protea (prō'tiə), *n.* a S African genus of evergreen shrubs cultivated for their showy flower heads.
protean (prō'tiən), *a.* readily assuming different shapes or aspects; variable, changeable.
protect (prətekt'), *v.t.* to shield, defend or keep safe (from or against injury, danger etc.); to support (industries) against foreign competition by imposing duties on imports; to provide funds so as to guarantee payment of (bills etc.). **protectingly**, *adv.* **protection**, *n.* the act of protecting; the state of being protected; that which protects, a covering, shield or defence; protection money; freedom from injury, molestation etc. purchased by protection money; the promotion of home industries by bounties or by duties on imports. **protection money,** *n.* a bribe extorted by gangsters from shopkeepers etc. by threats of damage to property, personal assault etc. **protectionism,** *n.* the doctrine or system of protecting home industries against foreign competition. **protectionist,** *n., a.*
protective, *a.* affording protection; intended to protect; of a person, desirous of shielding another from harm or distress. *n.* something that protects; a condom. **protective coloration, colouring,** *n.* colouring that enables animals to escape detection by blending with their surroundings, camouflage. **protective custody,** *n.* detention before trial in order to ensure an accused's personal safety. **protectively,** *adv.* **protectiveness,** *n.*
protector (prətek'tə), *n.* one who protects against injury or evil etc.; a protective device, a guard; one in charge of the kingdom during the minority, incapacity etc. of the sovereign; a title of Oliver Cromwell, Lord Protector of the Commonwealth (1653–8), and his son Richard Cromwell (1658–9). **protectoral,** *a.* **protectorate** (-rət), *n.* protection, usu. combined with partial control, of a weak State by a more powerful one; territory under such protection; the office of protector of a kingdom; the period of this, esp. that of Oliver and Richard Cromwell. **protectorless,** *a.* **protectorship,** *n.* **protectress** (-tris), **-trix** (-triks), *n. fem.*
protégé (prot'ə'zhā, prō'-), *n.* one under the protection, care, or patronage of another. **protégée,** *n. fem.*
protein (prō'tēn), *n.* any of a large group of complex organic compound, containing carbon, oxygen, hydrogen and nitrogen, usu. with some sulphur, found in all organic bodies and forming an essential constituent of animal foods. **protein engineering,** *n.* the production of new proteins using the techniques of genetic engineering to cause mutations in existing proteins. **proteinaceous** (-nā'shəs), *a.* **proteinic** (-tē'-), **proteinous** (-tē'-), *a.*
pro tem (prō tem'), *(abbr.)* PRO TEMPORE under PRO.
proteolysis (prōtiol'isis), *n.* the splitting up of proteins by the process of digestion or the application of enzymes. **proteolytic** (-lit'-), *a.*
Proterozoic (protərəzō'ik), *a., n.* (pertaining to) an era of geological history between the Archaeozoic and the Palaeozoic.
protest[1] (prətest'), *v.i.* to make a solemn affirmation; to express dissent or objection. *v.t.* to affirm or declare formally or earnestly; to object (that); *(esp. N Am.)* to express one's disapproval of or objection to; to make a formal declaration, usu. by a notary public, that payment (of a bill) has been demanded and refused. **protester, -tor,** *n.* **protestingly,** *adv.*
protest[2] (prō'-), *n.* the act of protesting; a solemn or formal declaration of opinion, usu. of dissent or remonstrance; an expression or demonstration of dissent, disapproval etc.; a formal declaration by the holder of the non-payment of a bill.
Protestant (prot'istənt), *n.* one of the party adhering to Luther at the Reformation, who protested against the decree of the majority involving submission to the authority of the Roman Catholic Church; a member of a Church upholding the principles of the Reformation, or (loosely) of any Church not within the Roman communion; **(protestant** *(also* -tes'-)) one who makes a protest. *a.* pertaining to the Protestants, or to Protestantism. **Protestantism,** *n.* **Protestantize, -ise,** *v.t., v.i.* (to cause) to become Protestant.
protestation (protəs-), *n.* a solemn affirmation or declaration; a protest.
prothalamion, (prōthəlā'miən), **-mium** (-əm), *n. (pl.* **-mia)** a song in honour of the bride and bridegroom before the wedding.
prothallium (prəthal'iəm), **-thallus** (-thal'əs), *n. (pl.* **-llia** (-liə), **-lli** (-li)) *(Bot.)* a cellular structure bearing the sexual organs in vascular cryptogams.
prothesis (proth'isis), *n.* the placing of the elements in readiness for use in the Eucharist; hence, a credencetable, or the part of a church in which this stands; *(Gram.)* prosthesis. **prothetic** (-thet'-), *a.*
prothonotary (prəthon'ətəri), *n.* a chief clerk or notary; the chief clerk or registrar of a court, now chiefly in some American and foreign courts, and formerly of the Courts of Chancery, Common Pleas and King's Bench; a member of the Roman Catholic College of Prothonotaries Apostolic who register the papal acts. **prothonotariat** (-teə'riət, -at), *n.* **prothonotarial,** *a.* **prothonotaryship,** *n.*
prothrombin (prōthrom'bin), *n.* a protein in blood plasma, converted to thrombin in the blood-clotting process.
Protista (prətis'tə), *n.pl.* a large group or kingdom of organisms, e.g. protozoans and single-celled algae, whose position (as animals or plants) is doubtful. **protist** (prō'-), *n.* any individual of the Protista.
protium (prō'tiəm), *n.* ordinary hydrogen of atomic weight 1 (as opposed to deuterium, tritium).
prot(o)- *comb. form.* chief; earliest, original, primitive; denoting that chemical compound in a series in which the distinctive element or radical combines in the lowest proportion with another element.
protocol (prō'təkol), *n.* the original draft of an official document or transaction, esp. minutes, or a rough draft of a diplomatic instrument or treaty, signed by the parties to a negotiation; the formal etiquette and procedure governing diplomatic and ceremonial functions; the official formulas used in diplomatic instruments, charters, wills etc.; a set of rules for transmitting data between computers in a communications system.
protomartyr (prōtəmah'tə), *n.* a first martyr (applied esp. to St Stephen); the first who suffers in any cause.
proton (prō'ton), *n.* a particle occurring in atomic nuclei and identical with the nucleus of the hydrogen atom, having an electric charge equal and opposite to that of the electron, and a mass 1840 times as great.
protonotary (protənō'təri), PROTHONOTARY.
protoplasm (prō'təplazm), *n.* the viscid, semifluid substance of a living cell. **protoplasmatic** (-mat'-), **protoplasmic** (-plaz'-), *a.*
prototype (prō'tətip), *n.* an original or primary type or model, an exemplar, an archetype; a pre-production model on which tests can be carried out to trace design faults, indicate possible improvements etc. **prototypal** (-tī'-), **prototypic, -al** (-tip'-), *a.*
Protozoa (prōtəzō'ə), *n.pl. (sing.* **-zoon** (-on)) the division of the animal kingdom, comprising the simplest forms consisting of a single cell or a colony of cells. **protozoal,** *a.* **protozoan,** *a., n.* **protozoic,** *a.* **protozoology** (-zōol'-), *n.* the branch of zoology dealing with the Protozoa.
protract (prətrakt'), *v.t.* to extend in duration, to prolong;

to draw (a map, plan etc.) to scale, esp. with a scale and protractor; to extend (a body part). **protractedly**, *adv*.
protractile (-til), *a*. of the organ etc. of an animal, capable of extension. **protraction**, *n*. **protractor**, *n*. an instrument, usu. in the form of a graduated arc, for laying down angles on paper etc.; a muscle that protracts or extends a limb.
protrude (prətrood'), *v.t.* to thrust forward or out; to cause to project or issue. *v.i.* to project, to be thrust forward. **protrudent**, *a*. **protrusible** (-si-), *a*. **protrusile** (-sil), *a*. **protrusion** (-zhən), *n*. **protrusive**, *a*. **protrusively**, *adv*.
protuberant (prətū'bərənt), *a*. swelling, bulging out, prominent. **protuberance**, *n*. a swelling, a prominence, a knob, a bump. **protuberantly**, *adv*.
proud (prowd), *a*. having high or inordinate self esteem; haughty, arrogant; having a due sense of dignity; feeling honoured, pleased, gratified; grand, imposing; of words, looks etc., stately, inspired by pride; of deeds etc., inspiring pride, noble, grand; vigorous, spirited, fearless; projecting, standing out above a plane surface. **proud flesh**, *n*. swollen flesh growing about a healing wound. **proudly**, *adv*. **proudness**, *n*.
Prov., (*abbr.*) Provençal, Provence; Proverbs; Province; Provost.
prove (proov), *v.t.* (*p.p.* **proved, proven** (proo'vən, prō'-)) to put to a test, to try by a standard; to show to be true; to establish or demonstrate by argument, reasoning or testimony; to establish the authenticity or validity of, esp. to obtain probate of (a will); to show or ascertain the correctness of, as by a further calculation. *v.i.* to turn out (to be); of dough, to rise and become aerated before baking. **not proven** (prō'-), (*Sc. Law*) not proved (a verdict given when there is not sufficient evidence to convict). **proving ground**, *n*. a place where something, esp. a vehicle is subjected to trials and scientific tests; any testing experience or situation. **provable**, *a*. **provableness**, *n*. **provably**, *adv*. **prover**, *n*.
provenance (prov'inəns), *n*. origin, source.
Provençal (provēsahl'), *n*. a native or inhabitant of Provence, France; the language of Provence. *a*. pertaining to Provence, its language or inhabitants.
provender (prov'əndə), *n*. dry food for beasts, fodder; (*facet.*) provisions, food.
provenience (prəvēn'yəns), PROVENANCE.
proverb (prov'əb), *n*. a short, pithy sentence, containing some truth or wise reflection proved by experience or observation; a maxim, a saw, an adage; a typical example, a byword. **Proverbs**, *n.sing*. a collection of maxims forming a book of the Old Testament. **proverbial** (-vœ'-), *a*. of the nature of a proverb; noted as an example (of a specified characteristic etc.). **proverbialist**, *n*. a writer, composer or collector of proverbs. **proverbially**, *adv*.
provide (prəvid'), *v.t.* to furnish, to supply; to equip (with); to lay down as a preliminary condition, to stipulate. *v.i.* to make preparation or provision (for or against); to furnish means of subsistence (for). **provided**, *conj*. on the understanding or condition (that). **provider**, *n*. **providing**, *n. conj*. provided.
providence (prov'idəns), *n*. foresight, timely care or preparation; frugality, economy, prudence; the beneficent care or control of God over His creatures; God or nature regarded as exercising such care; a manifestation of such care. **provident**, *a*. making provision for the future, thrifty; showing foresight, prudent. **provident society** FRIENDLY SOCIETY. **providential** (-den'-), *a*. due to or effected by divine providence; lucky, fortunate, opportune. **providentially**, *adv*. **providently**, *adv*.
province (prov'ins), *n*. (*Rom. Hist.*) a country or territory beyond the confines of Italy under a Roman governor; a large administrative division of a kingdom, country or state; the territory under the authority of an archbishop or metropolitan; (*pl.*) all parts of a country except the metropolis; proper sphere of action, business, knowledge etc. **provincial** (-vin'shəl), *a*. pertaining to a province; constituting a province; of, pertaining to, or characteristic of the provinces; narrow, rustic, unpolished. *n*. one who belongs to or comes from a province or the provinces; (*Eccles.*) the superior of an order etc., in a province; a person who is unpolished or narrow in outlook. **provincialism**, *n*. the quality of being provincial; a mode of speech, thought, behaviour etc. or a word or expression, peculiar to a province or the provinces; restriction of outlook or interests to local matters. **provincialist**, *n*. **provinciality** (-shial'-), *n*. **provincialize, -ise,** *v.t.* **provincially**, *adv*.
proving PROVE.
provision (prəvizh'ən), *n* the act of providing; previous preparation; a precautionary measure; a stipulation or condition providing for something; (*pl.*) victuals, eatables etc. *v.t.* to provide with provisions. **provisional**, *a*. temporary, not permanent; requiring future confirmation. **Provisional**, *n., a*. (a member) of the militant breakaway faction of the IRA or Sinn Fein. **provisionally**, *adv*.
proviso (prəvī'zō), *n*. (*pl.* **-sos, -soes**) a provisional condition, a stipulation; a clause in a covenant or other document rendering its operation conditional.
provisory (prəvī'zəri), *a*. conditional; provisional. **provisorily**, *adv*.
Provo (prō'vō), *n*. short for PROVISIONAL.
provocation, provocative PROVOKE.
provoke (prəvōk'), *v.t.* to rouse; to incite or stimulate to action, anger etc.; to irritate, to incense, to exasperate; to instigate, to call forth, to cause. **provocation** (provə-), *n*. **provocative** (-vok'-), *a*. tending to provoke; irritating, annoying, esp. with the intention to excite anger or rouse to action; intended or serving to incite sexual desire. **provocatively**, *adv*. **provocativeness**, *n*. **provoking**, *a*. annoying, exasperating. **provokingly**, *adv*.
provost (prov'əst), *n*. one appointed to superintend or hold authority; the head of a college; the head of a chapter, a prior, a dignitary in a cathedral corresponding to a dean; (*Sc.*) the chief magistrate in a municipal corporation or burgh; (*Hist.*) an officer in charge of a body of men, establishment etc., a steward. **Lord Provost**, *n*. the chief magistrate of Edinburgh, Glasgow, Aberdeen, Perth and Dundee. **provost-marshal** (prəvō'-), *n*. a commissioned officer, the head of the military police in a camp or in the field. **provostship**, *n*.
prow (prow), *n*. the fore part of a vessel, the bow.
prowess (prow'is), *n*. valour, bravery, gallantry; outstanding ability or skill.
prowl (prowl), *v.i.* to rove (about) stealthily as if in search of prey. *v.t.* to go through or about in this way. *n*. the act or an instance of prowling. **prowler**, *n*. **prowlingly**, *adv*.
proximal (prok'siməl), *a*. nearest the centre of the body or the point of attachment, opposed to distal. **proximally**, *adv*.
proximate (prok'simət), *a*. nearest, next; immediately preceding or following; very near, imminent; approximate. **proximate cause**, that which immediately precedes and produces the effect. **proximately**, *adv*. **proximity** (-sim'-), *n*. immediate nearness in place, time, relation, etc. **proximo** (-mō), *a*. in or of the month succeeding the present.
proxy (prok'si), *n*. the agency of a substitute for a principal; one deputed to act for another, esp. in voting; a document authorizing one person to act or vote for another; *a*. done, made etc. by proxy. **proxyship**, *n*.
prude (prood), *n*. a person who affects great modesty or propriety, esp. in regard to sexual matters **prudery** (-əri), *n*. **prudishness**, *n*. **prudish**, *a*. **prudishly**, *adv*.
prudent (proo'dənt), *a*. cautious, discreet, circumspect; worldly-wise, careful of consequences; showing good judgement or foresight; **prudence**, *n*. the quality of being prudent; caution, circumspection; care in the use of re-

sources; consideration for one's best interests. **prudential** (-den'shəl), *a.* actuated or characterized by prudence; worldly-wise, mercenary. **prudentialism, prudentiality** (-shial'-), *n.* **prudentially,** *adv.* **prudently,** *adv.*
prudery PRUDE.
pruinose (proo'inōs), *a.* (*Bot.*) covered with a powdery substance or bloom, frosted.
prune[1] (proon), *n.* the dried fruit of various kinds of *Prunus domestica*, the common plum; a plum; a dark purple colour; (*coll.*) a stupid or uninteresting person.
prune[2] (proon), *v.t.* to cut or lop off the superfluous branches etc. from; to cut or lop (off, away etc.); to free from anything superfluous; to reduce as if by pruning. *v.i.* to cut away superfluous branches, material etc. **pruner,** *n.* **pruning-hook, -knife, -shears,** *n.* instruments of various forms for pruning trees and shrubs.
prurient (pruər'iənt), *a.* disposed to, characterized by or arousing an unhealthy interest in sexual matters; characterized by a morbid curiosity. **prurience, -ency,** *n.* **pruriently,** *adv.*
prurigo (proori'gō), *n.* a papular disease of the skin attended with intolerable itching. **pruriginous** (-rij'-), *a.* **pruritus** (-təs), *n.* itching.
Prussian (prŭsh'ən), *a.* of or pertaining to Prussia. *n.* a native or inhabitant of Prussia. **Prussian blue,** *n.* a deep-blue pigment obtained from ferrocyanide or iron. **Prussianism,** *n.* practices or policies (e.g. the imposition of rigid discipline, militaristic organization) held to be typically Prussian. **Prussianize, -ise,** *v.t.* **Prussianizer, -iser,** *n.* **prussic** (prŭs'-), *a.* of or derived from Prussian blue. **prussic acid,** hydrocyanic acid, first obtained from Prussian blue.
pry[1] (prī), *v.i.* to look closely or inquisitively; to search or inquire curiously or impertinently (into). *n.* the act of prying. **prying,** *a.* **pryingly,** *adv.*
pry[2] (prī), *v.t.* PRISE.
Przewalski's horse (pəzhəval'skiz), *n.* a primitive, wild horse of central Asia, having an erect mane and no forelock. [after Nikolai *Przewalski*, 1839–88, the Russian explorer who discovered it]
PS, (*abbr.*) Police Sergeant; postscript; private secretary; (*Theat.*) prompt side.
Ps., Psa., (*abbr.*) Psalm; Psalms.
PSA, (*abbr.*) Property Services Agency; Public Services Authority.
psalm (sahm), *n.* a sacred song or hymn. **the Psalms,** a book of the Old Testament consisting of sacred songs, many of which are ascribed to David. **psalmist,** *n.* **the psalmist,** David or the composer of any of the Psalms. **psalmody** (sah'mədi, sal'-), *n.* the act, art or practice of singing psalms, esp. in divine worship; psalms collectively. **psalmodic** (-mod'-), *a.* **psalmodist,** *n.* a composer or singer of psalms. **Psalter** (sawl'tə), *n.* The Book of Psalms; a book containing the Psalms for use in divine service, esp. the version of the Psalms in the Prayer Book or the Latin collection used in the Roman Catholic Church.
psalterium (sawltiə'riəm), *n.* (*pl.* **-ria** (-riə)) the third stomach of a ruminant.
psaltery (sawl'təri), *n.* a mediaeval stringed instrument somewhat resembling the dulcimer, but played by plucking the strings.
PSBR, (*abbr.*) public sector borrowing requirement.
PSDR, (*abbr.*) public sector debt repayment.
psephology (sefol'əji), *n.* the statistical and sociological study of elections. **psephological,** (-loj'-), *a.* **psephologist,** *n.*
pseud (sūd), *n.* (*coll.*) an affected or pretentious person; a sham. *a.* pseudo. **pseudery,** *n.* (*coll.*) pretentiousness; falseness. **pseudo** (-dō), *a.* false, sham, spurious; affected, pretentious.
pseud(o)-, *comb. form.* false, counterfeit, spurious; closely resembling, as in *pseudoclassical, pseudo-Gothic, pseudohistorical.*
pseudepigrapha (sūdəpig'rəfə), *n.pl.* spurious writings, esp. uncanonical writings ascribed to Scriptural authors etc. **pseudepigraphal, pseudepigraphical** (-depigraf'-), *a.* **pseudepigraphy,** *n.* the ascription of false names of authors to books.
pseudocarp (sū'dōkahp), *n.* a fruit composed of other parts besides the ovary.
pseudograph (sū'dəgraf), *n.* a spurious writing, a literary forgery. **pseudography** (-dog'-), *n.*
pseudomorph (sū'dəmawf), *n.* a mineral having the crystalline form of another. **pseudomorphic, -ous** (-maw'-), *a.* **pseudomorphism** (-maw'-), *n.*
pseudonym (sū'dənim), *n.* a fictitious name, esp. a nom de plume. **pseudonymity** (-nim'-), *n.* **pseudonymous** (-don'-), *a.* **pseudonymously,** *adv.*
pseudopodium (sūdōpō'diəm), *n.* (*pl.* **-podia** (-diə)) a process formed by the protrusion of the protoplasm of a cell or a unicellular animal, serving for locomotion, ingestion of food etc.; a false pedicel in mosses etc. **pseudopodial,** *a.*
pshaw (pshaw), *int.* an exclamation of contempt, impatience, disdain or dislike.
psi[1] (si), *n.* the twenty-third letter of the Greek alphabet, equivalent to *ps*; paranormal or psychic phenomena collectively. **psi particle,** *n.* an elementary particle formed by electron-positron collision.
psi[2] (*abbr.*) pounds per square inch.
psilocybin (silōsī'bin), *n.* a hallucogenic drug obtained from Mexican mushrooms.
psittaceous (sitā'shəs), **psittacine** (sit'əsin), *a.* belonging or allied to the parrots; parrot-like. **psittacosis** (sitəkō'sis), *n.* a disease of parrots communicable to man, with a high mortality.
psoas (sō'əs), *n.* either of the two large hip-muscles. **psoatic** (-at'-), *a.*
psora (saw'rə), *n.* (*Path.*) the itch or an analogous skin-disease. **psoriasis** (-rī'əsis), *n.* a dry, scaly skin disease. **psoriatic** (-at'-), *a.* **psoric,** *a.*
PST, (*abbr.*) Pacific Standard Time.
PSV, (*abbr.*) public service vehicle.
psych, psyche (sīk), *v.t.* (*coll.*) to psychoanalyse. **to psych (out),** (*N Am.*) to work out, to divine, to anticipate correctly; to intimidate or defeat by psychological means. **to psych up,** to prepare or stimulate psychologically as a preliminary to action.
psych- *comb. form.* PSYCH(O)-.
psyche (sī'ki), *n.* the soul, the spirit, the mind; the principles of emotional and mental life.
psychedelic (sīkədel'ik), *a.* pertaining to new, altered or heightened states of consciousness and sensory awareness as induced by the use of certain hallucinatory drugs; of drugs, capable of producing such states; having an effect on the mind similar to that of psychedelic drugs; resembling the phenomena of psychedelic experience; of colours, unnaturally vivid, fluorescent. **psychedelically,** *adv.*
psychiatry (sīkī'ətri), *n.* the study and treatment of mental disorders. **psychiatric** (-kiat'-), *a.* **psychiatrically,** *adv.* **psychiatrist,** *n.*
psychic (sī'kik), *a.* pertaining to the human soul, spirit, or mind; of or pertaining to phenomena that appear to be outside the domain of physical law, paranormal, extrasensory; sensitive to non-physical or paranormal forces and influences. *n.* one having psychic powers; a medium. **psychical,** *a.* psychic. **psychically,** *adv.*
psycho (sī'kō), *n., a.* (*coll.*) short for PSYCHOPATH(IC), PSYCHOTIC.
psych(o)- *comb. form.* mental; psychical.
psychoactive (sīkōak'tiv), *a.* of drugs, capable of affecting the mind or behaviour.

psychoanalysis (sīkōənal'isis), n. a method devised by Sigmund Freud for exploring and bringing to light concepts, experience etc. hidden in the unconscious mind, as a form of treatment for functional nervous diseases or mental illness. **psychoanalyse** (-an'əlīz), v.t. to subject to, or treat by, psychoanalysis. **psychoanalyst** (-an'ə), n. **psychoanalytic, -ical** (-anəlit'-), a.
psychobabble (sīkōbabl), n. (excessive or unnecessary use of) psychological jargon.
psychodrama (sīkōdrahmə), n. an improvised dramatization of events from a patient's past life, used as a form of mental theory. **psychodramatic** (-drəmat'-), a.
psychodynamics (sīkōdīnam'iks), n. sing. the study of mental and emotional forces and their effect on behaviour. **psychodynamic,** a.
psychogenesis (sīkōjen'əsis), n. (the study of) the origin or development of the mind: orgination in the mind. **psychogenetic** (-net'-), a.
psychogenic (sīkōjen'ik), a. of an illness etc., of mental, as opposed to physical, origin.
psychographics (sīkōgraf'iks), n.sing. the classification of people according to psychological characteristics such as attitudes and motivation, used esp. in marketing.
psychokinesis (sīkōkinē'sis), n. apparent movement of, or alteration in physical objects produced by mind power. **psychokinetic** (-net'-), a.
psycholinguistics (sīkōlinggwis'tiks), n.sing. the study of the psychology of language, its acquisition, development, use etc. **psycholinguist** (-ling'-), n. **psycholinguistic,** a.
psychology (sīkol'əji), n. the science of the human mind; a system or theory of mental laws and phenomena within this science; (characteristic) mentality or motivation; (coll.) skill in understanding or motivating people. **psychological** (-loj'-), a. pertaining to psychology; pertaining to or affecting the mind; existing only in the mind. **psychological moment,** n. the critical moment, the exact time for action etc. **psychological warfare,** n. the use of propaganda to reduce enemy morale. **psychologically,** adv. **psychologist,** n. **psychologize, -ise,** v.t., v.i. to explain or interpret in psychological terms.
psychometrics (sikōmet'riks), n.sing. the branch of psychology dealing with the measurement of mental capacities and attributes, esp. by the use of psychological tests and statistical methods. **psychometry** (-kom'-), n. psychometrics; the occult faculty of divining by touching a physical object, the character, surroundings, experiences etc. of persons who have touched it. **psychometric, -ical,** a. **psychometrically,** adv. **psychometrist,** n.
psychomotor (sīkōmō'tə), a. pertaining to muscular action proceeding from mental activity.
psychoneuroimmunology (sīkōnūrōimūnol'əji), n. immunology concerned with the effect of the mind on the body's immune response.
psychoneurosis (sīkōnūrō'sis), n. (pl. **-roses** (-sēz)), a neurosis, esp. one due to emotional conflict. **psychoneurotic** (-rot'-), a.
psychopath (sīkəpath), n. one suffering from a severe personality disorder characterized by anti-social behaviour and a tendency to commit such acts of violence. **psychopathic** (-path'-), a. **psychopathically,** adv.
psychopathology (sīkōpəthol'əji), n. (the branch of psychology dealing with) mental and behavioural aberrance. **psychopathological** (-pathəloj'-), a. **psychopathologist,** n.
psychophysics (sīkōfiz'iks), n.sing. the science of the relations between mind and body, esp. between physical stimuli and psychological sensation. **psychophysicist** (-sist), n.
psychophysiology (sīkōfiziol'əji), n. the branch of physiology treating of mental phenomena.
psychosexual (sīkōsek'shəl), a. pertaining to the psychological aspects of sex. **psychosexually,** adv.

psychosis (sīkō'sis), n. (pl. **-ses** (-sēz)), n. a mental derangement, not due to organic lesion, characterized by a severe distortion of the sufferer's concept of reality. **psychotic,** (-kot'-), a., n.
psychosomatic (sīkōsəmat'ik), a. denoting a physical disorder caused by or influenced by the patient's emotional condition.
psychotherapeutic (sīkōtherəpū'tik), a. treating disease by psychological methods **psychotherapeutics,** n. sing. **psychotherapy** (-the'rəpi), n. the treatment of disease by psychological or hypnotic means. **psychotherapist,** n.
psychotic PSYCHOSIS.
psychrometer (sīkromitə), n. the wet-and-dry bulb hygrometer for measuring the humidity of the atmosphere.
PT, (abbr.) Pacific time; physical training.
Pt¹, (abbr.) point, port.
Pt², (chem. symbol) platinum.
pt, (abbr.) part; pint(s); (Math.) point.
PTA, (abbr.) Parent-Teacher Association; Public Transport Authority.
ptarmigan (tah'migən), n. a bird allied to the grouse, having grey or brown plumage in the summer and white in the winter.
Pte, (abbr.) (Mil.) Private.
pter-, pteri-, ptero-, comb. form. winged; having processes resembling wings.
-ptera comb. form. organisms having a certain number or type of wings. **-pteran, -pterous,** a. comb. form.
pteridology (teridol'əji), n. the science of ferns. **pteridological** (-loj'-), a. **pteridologist,** n.
pteridophyte (tĕridōfīt), n. a member of the Pteridophyta. **Pteridophyta,** n.pl. a division of vascular plants with roots, stems and leaves, that reproduce by spores, the ferns, horsetails and club mosses.
pterodactyl (terədak'til), n. an extinct winged reptile from the Mesozoic strata.
pteropod [te'rəpod), n. (Zool.) any individual of the Pteropoda. **Pteropoda** (tərop'ədə), n.pl. a sub-class of Mollusca in which the foot is expanded into wing-like lobes or paddles.
pterosaur (te'rsaw), n. any individual of the Pterosauria. **Pterosauria** (-saw'riə), n.pl. an order of flying reptiles of the Mesozoic age.
pterygoid (te'rigoid), a. (Anat.) wing-shaped; of or connected with the pterygoid processes n. a pterygoid bone or process. **pterygoid process,** n. either of the wing-like processes descending from the great wings of the sphenoid bone of the skull.
ptisan (tiz'n, -zan'), n. Barley-water or other mucilaginous decoction used as a nourishing beverage.
PTO, (abbr.) please turn over.
Ptolemaic (toləmā'ik), a. pertaining to Ptolemy, Alexandrian astronomer (2nd cent. AD) who maintained that the earth was a fixed body in the centre of the universe, the sun and moon revolving round it; pertaining to the Ptolemies, kings of Egypt, 323–30 BC. **Ptolemaic system,** n. Ptolemy's conception of the universe.
ptomaine (tō'mān), n. one of a class of sometimes poisonous amines derived from decaying animal and vegetable matter. **ptomaine poisoning,** n. food poisoning, formerly, and erroneously, thought to be due to ptomaines.
ptosis (tō'sis), n. a drooping of the upper eyelid from paralysis of the muscle raising it.
ptyalin (tī'əlin), n. an enzyme contained in saliva, which converts starch into dextrin.
Pu, (chem. symbol) plutonium.
pub (pub), n. a public house. v.i. (past, p.p. **pubbed)** to visit public houses (esp), n. **go pubbing, pub-crawl,** n. (coll.) a drinking tour of a number of pubs.
pub., (abbr.) public; published, publisher, publishing.
puberty (pū'bərti), n. the period of life at which persons become capable of begetting or bearing children. **pubertal,**

a. **pubes** (-bēz), *n. (pl.* **pubes**) the hypogastric region which in the adult becomes covered with hair; the hair of the pubic region. **pubescence** (-bes'ns), *n.* the state or age of puberty; soft, hairy down on plants or parts of animals. **pubescent**, *a.* arrived at the age of puberty; covered with soft hairy down. **pubic**, *a.* of or pertaining to the pubes or pubis. **pubis** (-bis), *n.* a bone forming the anterior part of the pelvis.

public (pŭblik), *a.* pertaining to or affecting the people as a whole, opp. to personal or private; open to the use or enjoyment of all, not restricted to any class; done, existing or such as may be observed by all, not concealed or clandestine; open, notorious; well-known, prominent; of or pertaining to the affairs or service of the people. *n.* the people in general; any particular section of the people. **in public**, openly, publicly. **to go public**, to become a public company; to make publicly known. **public-address system**, a system of microphones, amplifiers, loudspeakers etc. used for addressing a large audience. **public lending right**, the right of authors to royalties when their books are borrowed from public libraries. **public bar**, *n.* a bar in a public house, usu. less well appointed and serving drinks at cheaper prices than a saloon bar. **public company**, *n.* one whose shares can be purchased on the stock exchange by members of the public. **public convenience**, *n.* a public lavatory. **public corporation**, *n.* an organization set up by government to run a nationalized service or industry. **public domain**, *n.* the status in law of a published work on which the copyright has expired. **public enemy**, *n.* somebody, esp. a notorious criminal, considered to be a menace to the community. **public health**, *n.* the field of responsibility for the general health of the community covering e.g., sanitation, food-handling in shops and restaurants, hygiene in public places. **public house**, *n.* a house licensed for the retail of intoxicating liquors, an inn, a tavern. **public law**, *n.* international law. **public nuisance**, *n. (Law)* an illegal act affecting the whole community rather than an individual; *(coll.)* a generally objectionable person. **public opinion**, *n.* the views of the general public or the electorate on political and social issues. **public opinion poll**, an assessment of public opinion on an issue, based on the responses of a sample of the community to certain questions. **public orator**, *n.* the official spokesperson for a university. **public prosecutor**, *n.* an official who conducts criminal prosecutions on behalf of the state. **public relations**, *n.sing. or pl.* the relationship between an organization and the public; (a department entrusted with) the maintenance of goodwill towards, and a favourable image of, an organization in the mind of the public. **public relations officer, public school**, *n.* a school under the control of a publicly elected body; a school whose headmaster is a member of the Headmasters' Conference, usu. an endowed school providing a liberal education for such as can afford it. **public sector**, *n.* the state-owned part of the economy. **public servant**, *n.* a government employee. **public spirit**, *n.* interest in or devotion to the community. **public-spirited**, *a.* **public-spiritedly**, *adv.* **public-spiritedness**, *n.* **public utility**, *n.* an enterprise concerned with the provision of an essential service, e.g. gas, water, electricity, to the public. **public works**, *n.pl.* roads, buildings etc. constructed for public use by or on behalf of the government. **publicness**, *n.* **publicly**, *adv.*

publican (pŭb'likən), *n. (Rom. Hist.)* a collector or farmer of the revenues, taxes etc.; a keeper of a public house.

publication (pŭblikā'shən), *n.* the act of making publicly known; the act of publishing a book, periodical, musical composition etc.; a work printed and published.

publicist (pŭb'lisist), *n.* a writer or authority on international law; a writer on current social or political topics, esp. a journalist; one who publicizes, esp. a press or publicity agent. **publicism**, *n.* **publicistic** (-sis'-), *a.*

publicity (pəblis'iti), *n.* the quality of being public; public attention or interest; the process of attracting public attention to a product, person etc; anything calculated to arouse public interest, as a newsworthy event or information, advertising etc. **publicity agent**, *n.* a person employed to keep before the public the name of a product, film etc. **publicity stunt**, *n.* an unusual or attention-grabbing event engineered specifically for purposes of publicity.

publicize, -ise (pŭb'lisīz), *v.t.* to make known to the public; to advertise.

publish (pŭb'lish), *v.t.* to make public, to promulgate, to announce publicly; to issue or print and offer for sale to the public; to issue the works of (an author); to put into circulation. *v.i.* to print and offer for sale; to have one's work published. **publishable**, *a.* **publisher**, *n.* one who publishes esp. books and other literary productions. **publishing**, *n.*

puce (pūs), *a.* brownish purple.

puck[1] (pŭk), *n.* a mischievous sprite, elf or fairy. **puckish**, *a.*

puck[2] (pŭk), *n.* a vulcanized rubber disc used instead of a ball in ice hockey.

pucka, pukka (pŭk'ə), *a. (Ang.-Ind.)* durable, substantial; genuine; superior.

pucker (pŭk'ə), *v.t.* to gather into small folds or wrinkles. *v.i.* to become wrinkled or gathered into small folds etc. *n.* a fold, a wrinkle, a bulge. **puckery**, *a.*

puckish PUCK[1].

pud (pud), *n. (coll.)* short for PUDDING.

puddening (pud'əning), *n. (Naut.)* a pad of rope etc. used as a fender.

pudding (pud'ing), *n.* a mixture of animal or vegetable ingredients, usu. with flour or other farinaceous basis, of a soft or moderately hard consistency, baked or boiled, and eaten either as a main dish or as a sweet; meat or fruit cooked in a flour-based casing; dessert; a skin or intestine stuffed with minced meat etc., a large sausage; *(coll.)* a fat or podgy person; *(Naut.)* a puddening. **pudding-ball**, *n. (Austral.)* a fish resembling the mullet. **pudding-face**, *n.* a fat, round, smooth face. **pudding-faced**, *a.* **pudding-head**, *n. (coll.)* a stupid person. **pudding-stone**, *n.* a conglomerate of pebbles in a siliceous matrix. **puddingy**, *a.*

puddle (pŭd'l), *n.* a small pool of liquid, esp. of rainwater; clay and sand worked together to form a watertight lining for a pond, canal etc.; a muddle; a bungler, an awkward person. *v.i.* to dabble (in mud, water etc.); to mess, to muddle (about). *v.t.* to make dirty or muddy; to work (clay etc.) into puddle; to line or render watertight with puddle; to stir up (molten iron) in a furnace so as to convert it into wrought-iron. **puddler**, *n.* one who puddles, esp. a worker employed in puddling iron. **puddling**, *n.* **puddly**, *a.*

pudency (pū'dənsi), *n.* modesty, shamefacedness. **pudendum** (-den'dəm), *n. (often pl.* **pudenda** (-də)), the external genitals, esp. of a woman. **pudendal, pudic** (pū'-), *a.* pertaining to the pudenda.

pudge (pŭj), *n.* a short, thick or fat person or figure. **pudgy**, *a.*

pudic PUDENCY.

pudsy (pŭd'zi), *a. (coll., dial.)* plump.

pueblo (pweb'lō), *n.* a village or settlement, of the Indians of New Mexico etc.; a village or town in Spanish America. **pueblan**, *a.*

puerile (pū'əril), *a.* childish, silly, inane. **puerilely**, *adv.* **puerility** (-ril'-), *n.*

puerperal (pūœ'pərəl), *a.* pertaining to or resulting from childbirth. **puerperal fever**, *n.* a fever, caused by infection of the genital tract, attacking women after childbirth. **puerperalism**, *n. (Path.)*

puff (pŭf), *v.i.* to breathe, to blow, to emit or expel air, steam etc. in short, sudden blasts; to move or go while

puffing; to breathe hard; to come (out) in a short, sudden blast; to become inflated or distended. *v.t.* to emit, to blow out, with a short sudden blast or blasts; to blow, disperse or drive thus; to draw at (a cigarette, pipe etc.); to utter pantingly; to inflate, to blow (up or out); to blow (away etc.); to cause to be out of breath; to praise or advertise in an exaggerated or misleading way. *n.* a short, sudden blast of breath, smoke, steam etc., a whiff, a gust; the sound made by this; a small amount of breath, smoke etc., emitted at one puff; an act of drawing in and exhaling whilst smoking a cigarette, pipe etc.; a light, puffy thing or small mass of any material; a cake, tart etc. of light or spongy consistency; a light wad, pad or tuft for applying powder to the skin; an exaggerated or misleading advertisement, review etc. **puffed (out)**, (*coll.*) out of breath. **puffed up**, inflated; swollen up with conceit or self-importance. **puff-adder**, *n.* a highly venomous African snake, which inflates part of its body when aroused. **puff-ball**, *n.* a fungus of the genus *Lycoperdon*, the roundish spore-case of which emits dry, dust-like spores. **puff-bird**, *n.* a bird of the family Bucconidae, so called from their habits of puffing out their plumage. **puff-box**, *n.* a toilet-box for holding powder and puff. **puff-paste, -pastry**, *n.* (a rich dough used to make) a light, flaky pastry etc. **puffer**, *n.* a person or thing that puffs, esp. a steamboat, steam engine etc.; a globefish. **puffily**, *adv.* **puffing**, *a.*, *n.* **puffingly**, *adv.* **puffy**, *a.* puffing, blowing or breathing in puffs; short-winded; swollen, distended; tumid, turgid, bombastic. **puffiness**, *n.*
puffin (pŭf'in), *n.* a sea-bird of the genus *Fratercula*, esp. the N Atlantic *F. arctica.*
puffing, puffly, puffy, etc. PUFF.
pug[1] (pŭg), *n.* a pug-dog; a proper name for a fox; a pug-engine. **pug-dog**, *n.* a small, short-haired dog with wrinkled face, up-turned nose and tightly curled tail. **pug-engine**, *n.* a small locomotive for shunting etc. **pug-faced**, *a.* **pug-nose**, *n.* a short squat nose. **pug-nosed**, *a.*
pug[2] (pŭg), *n.* clay and other material mixed and prepared for making into bricks. *v.t.* (*past, p.p.* **pugged**) to grind (clay etc.) and render plastic for brick-making; to puddle with clay; to pack (a wall, floor etc.) with sawdust etc. to deaden sound. **pug-mill**, *n.* a mill in which clay is made into pug. **pugging**, *n.*
pug[3] (pŭg), *n.* (*Ang.-Ind.*) the footprint or trail of an animal. *v.i.* (*past, p.p.* **pugged**) to track game etc.
pugaree, puggree (pŭg'əri), *n.* an Indian light turban; a long piece of muslin wound round a hat or helmet in hot climates to protect from the sun. **pugareed**, *a.*
pugging PUG[2].
pugilist (pū'jilist), *n.* a boxer, a prize-fighter; (*fig.*) a fighter, a pugnacious controversialist etc. **pugilism**, *n.* **pugilistic** (-lis'-), *a.*
pugnacious (pŭgnā'shəs), *a.* inclined to fight; quarrelsome. **pugnaciously**, *adv.* **pugnacity** (-nas'-), **pugnaciousness**, *n.*
puisne (pū'ni), *a.* junior or inferior in rank (applied to judges); (*Law*) later, more recent. *n.* a puisne judge.
puissant (pū'isənt, pwis'-), *a.* (*poet.*) powerful, strong, mighty. **puissance**, *n.* power, strength; (pwĕ'säs), a showjumping event that tests a horse's power to jump high obstacles. **puissantly**, *adv.*
puke (pūk), *v.t.*, *v.i.* (*sl.*) to vomit. *n.* vomit; the act of vomiting. **puker**, *n.*
pukka PUCKA.
puku (poo'koo), *n.* a red African antelope, *Cobuis vardoni.*
pulchritude (pŭl'kritūd), *n.* beauty. **pulchritudinous**, (-tū'din-), *a.*
pule (pūl), *v.i.* to cry plaintively or querulously, to whine, to whimper; to pipe, to chirp. *v.t.* to utter in a querulous, whining tone. **puling**, *a.*, *n.* **pulingly**, *adv.*
pulka (pŭl'ka), *n.* a travelling sleigh with a prow like a canoe, used by Laplanders.

pull (pul), *v.t.* to draw towards one by force; to drag, to haul, to tug; to draw (up, along, nearer etc.); to move (a vehicle) in a particular direction; to pluck; to remove by plucking; to strip of feathers; to draw the entrails from (a fowl); to bring out (a weapon); to strain (a muscle or tendon); to row (a boat); to take (a person in a boat) by rowing; (*coll.*) to attract (a crowd); (*sl.*) to carry out esp. with daring and imagination or with deceptive intent; (*coll.*) to withdraw; (*sl.*) to seduce; (*sl.*) to make a raid upon (a gambling-house); (*sl.*) to arrest; (*Print.*) to take (an impression) by a hand-press, to take (a proof); (*Cricket*) to strike (a ball) from the off to the on side; (*Golf*) to strike (a ball) to the left; (*Racing*) to rein in (a horse), esp. so as to lose a race. *v.i.* to give a pull; to tug, to haul; to move in a motor vehicle in a particular direction; to strain against the bit (of a horse); to draw, to suck (at a pipe); to pluck, to tear (at). *n.* the act of pulling, a tug; that which is pulled; force of pulling or drawing towards; a handle by which beer is drawn, a door opened, a bell rung etc.; a quantity of beer etc. drawn; a draught, a swig; a draw on a pipe; an impression from a hand-press, a proof; (*Cricket*) a stroke by which a ball is sent from the off to the on side; (*Golf*) a stroke sending a ball to the left; the checking of a horse by its rider, esp. to secure defeat; a spell of rowing; a stroke of an oar; (*coll.*) a hold, unfair or illegitimate influence; (*coll.*) a spell of hard exertion; (*coll.*) power to attract, influence etc. **to pull about**, to pull to and fro, to handle roughly. **to pull a fast one**, FAST. **to pull apart**, to pull asunder or into pieces; to become separated or severed; to pull to pieces. **to pull back**, to retreat; to withdraw. **to pull down**, to demolish; to degrade, to humble; to weaken, to cause (prices etc.) to be reduced. **to pull faces**, FACE; **to pull in**, to retract, to make tighter; of a train, to enter a station; of a vehicle or driver, to stop (at), to pull over; (*coll.*) to attract (audiences etc.); (*sl.*) to arrest; to earn. **to pull off**, to accomplish (something difficult or risky). **to pull oneself together**, to regain one's composure or self-control. **to pull one's, somebody's leg**, LEG. **to pulll one's punches** PUNCH[1]. **to pull one's weight**, WEIGHT. **to pull out**, to leave, to depart; to withdraw; to cease to participate in; to move out from the side of the road or from behind another vehicle; of aircraft, to level off after a dive. **to pull over**, to draw in to the side of the road (and stop). **to pull round**, to (cause to) recover. **to pull through**, to (cause to) survive, recover or not fail against the odds. **to pull together**, to cooperate. **to pull to pieces**, to tear (a thing) up; to criticize, to abuse. **to pull up**, to drag up forcibly; to pluck up; to cause to stop; to come to a stop; to rebuke; to gain on, to draw level with. **pull-back**, *n.* a drawback, a restraint, hindrance; a device for holding back and keeping in parts of a woman's skirt; a retreat, a withdrawal. **pull-down**, *a.* (*Comput.*) of or being a menu which can be accessed during the running of the program and which brings a list of options down over the screen. **pull-in**, *n.* a stopping place; a transport café. **pull-on**, *a.* (a garment) without fastenings, requiring simply to be pulled on. **pull-out**, *n.* a removable section of a magazine; a large fold-out leaf in a book. **pullover**, *n.* a jersey which is pulled over the head. **pull-through**, *n.* a cord with a rag attached, used for cleaning the barrel of a firearm. **pulled**, *a.* plucked, stripped, as fowls, skins etc.; depressed in health, spirits etc., dragged (down). **puller**, *n.* one who or that which pulls; an implement, machine etc. for pulling; a horse that pulls against the bit, a hard-mouthed or high-spirited horse.
pullet (pul'it), *n.* a young fowl, esp. a hen before the first moult.
pulley (pul'i), *n.* a wheel with a grooved rim, or a combination of such wheels, mounted in a block for changing the direction or for increasing the effect of a force; a wheel used to transmit power or motion by means of a

belt, chain etc. passing over its rim. *v.t.* to lift or hoist with a pulley; to furnish or fit with pulleys. **fast and loose pulley**, a pair of pulleys on a shaft, one fixed and revolving with the shaft, the other loose, for throwing the shaft into or out of gear by means of a belt running round the one or the other.
Pullman (pul'mən), *n.* a Pullman car; a train made up of these. **Pullman car**, *n.* a luxurious railway saloon or sleeping-car originally built at the Pullman works, Illinois. [George M. *Pullman* 1831-97, US inventor]
pullulate (pŭl'ūlāt), *v.i.* to shoot, to bud, to germinate; to breed; to swarm; to develop, to spring up. **pullulant**, *a.* **pullulation**, *n.*
pulmo-, *comb. form.* pertaining to the lungs.
pulmonary (pŭl'mənəri, pul-), *a.* pertaining to the lungs. **pulmonary artery**, *n.* the artery carrying blood from the heart to the lungs. **pulmonary disease**, *n.* lung disease, esp. consumption. **pulmonic** (-mon'-), *a.* pulmonary; affected with or subject to disease of the lungs. *n.* one having diseased lungs; a medicine for lung-diseases.
pulmonate (pŭl'mənāt, pul-), *a.* (*Zool.*) furnished with lungs. *n.* a pulmonate mollusc.
pulp (pŭlp), *n.* any soft, moist, coherent mass; the fleshy or succulent portion of a fruit; the soft tissue of an animal body or in an organ or part, as in the internal cavity of a tooth; the soft mixture of rags, wood etc. from which paper is made; a magazine or book printed on cheap paper and sentimental or sensational in content; (*Mining*) pulverized ore mixed with water. *v.t.* to convert into pulp; to extract the pulp from. *v.i.* to become pulpy. **pulper**, *n.* **pulpify** (-fī), *v.t.* **pulpless**, *a.* **pulplike**, *a.* **pulpy**, *a.* **pulpiness**, *n.*
pulpit (pul'pit), *n.* an elevated enclosed stand from which a preacher delivers a sermon; the medium through which a message or opinion is expressed. *a.* pertaining to the pulpit or to preaching. **the pulpit**, preachers generally; preaching. **pulpitarian** (-teə'ri-), *a.*, *n.* **pulpiteer** (-tiə'), *n.* (*contempt.*) a preacher. **pulpiteering**, *n.*
pulpike, pulpy etc. PULP.
pulque (pul'ki), *n.* a Mexican vinous beverage made by fermenting the sap of species of agave. **pulque brandy**, a liquor distilled from this.
pulsar (pŭl'sah), *n.* an interstellar source of regularly pulsating radio waves, prob. a swiftly rotating neutron star. [*puls*ating st*ar*]
pulsate (pŭl'sāt), *v.i.* to move, esp. to expand and contract, with rhythmical alternation, to beat, to throb; to vibrate, to thrill; (*Phys.*) to vary periodically in force, magnitude, intensity etc. **pulsatile** (pŭl'sətil), *a.* pulsatory; (*Mus.*) played by beating, percussive. **pulsation**, *n.* the action of pulsating; the movement of the pulse. **pulsatory** (pŭl'sə-), *a.* of or pertaining to pulsation; actuated by or having the property of pulsation. **pulsator**, *n.* a machine for separating diamonds from earth, a jigging-machine; part of a milking-machine; a pulsometer.
pulsatilla (pŭlsətil'ə), *n.* the pasque-flower.
pulse[1] (pŭls), *n.* the rhythmic beating of the arteries caused by the propulsion of blood along them from the heart; a beat of the arteries or the heart; a pulsation, a vibration; a short-lived variation in some normally constant value in a system, as in voltage etc.; an electromagnetic or sound wave of brief duration; a quick, regular stroke or recurrence of strokes (as of oars); a throb, a thrill. *v.i.* to pulsate. *v.t.* to send (forth, out etc.) by or as by rhythmic beats. **to feel one's pulse**, to gauge the rate or regularity of one's pulse as a sign of health etc.; to sound one's intentions, views etc. **to keep one's finger on the pulse of**, to keep up to date with developments in. **pulse-rate**, *n.* the number of pulse beats per minute. **pulseless**, *a.* **pulselessness**, *n.* **pulsimeter** (-sim'itə), *n.* an instrument for measuring the rate, force, regularity etc. of the pulse.
pulsometer (-som'itə), *n.* a pumping device operated by the admission and condensation of steam in alternate chambers; a pulsimeter.
pulse[2] (pŭls), *n.* the seeds of leguminous plants; a plant producing such seeds.
pulverize, -ise (pŭl'vəriz), *v.t.* to reduce to fine powder or dust; to demolish, to smash, to defeat utterly. *v.i.* to be reduced to powder. **pulverate**, *v.t.* **pulverable**, *a.* **pulverizable**, *a.* **pulverization**, *n.* **pulverizer**, *n.* one who or that which pulverizes; a machine for reducing a liquid to fine spray; a machine for pulverizing earth. **pulverous**, *a.* **pulverulent** (-vœ'ū-), *a.* consisting of fine powder; covered with powder, powdery; liable to disintegrate into fine powder. **pulverulence**, *n.*
pulvillus (pŭlvil'əs), *n.* (*pl.* **-lli** (-li)) the pad or cushion of an insect's foot. **pulvillar, pulvilliform** (-fawm), *a.*
pulvinate (pŭl'vināt), *a.* (*Nat. Hist.*) cushion shaped, pad-like. **pulvinated** (-nātid), *a.* having a convex face (as a frieze).
puma (pū'mə), *n.* the cougar, *Felis concolor*, a large feline carnivore of the Americas.
pumice (pŭm'is), *n.* a light, porous or cellular kind of lava, used as a cleansing and polishing material. *v.t.* to rub, polish or clean with this. **pumice-stone**, *n.* pumice. **pumicate** (-pū'-), *v.t.* **pumiceous** (pūmish'əs), *a.*
pummace (pŭm'əs), POMACE.
pummel (pŭm'l), *v.t.* (*past, p.p.* **pummelled**) to strike or pound repeatedly, esp. with the fists. *n.* POMMEL.
pump[1] (pŭmp), *n.* a device or engine, usu. in the form of a cylinder and piston, for raising water or other liquid; a machine for exhausting, transferring a gas or compressing; the act of pumping, a stroke of a pump; an attempt at extracting information from a person. *v.t.* to raise, remove or drive with a pump; to free from water or make dry with a pump; to propel, to pour, with or as with a pump; to supply liberally; to move up and down as if working a pump-handle; to put out of breath (*usu. in p.p.*); to elicit information from by artful interrogations. *v.i.* to work a pump; to raise water etc., with a pump; to move up and down in the manner of a pump-handle; to spurt. **to pump iron**, to do weight-lifting exercises. **to pump up**, to inflate (a pneumatic tyre); to inflate the tyres of (a cycle etc.). **pump-action**, *a.* of a shotgun, requiring a pump-like movement to bring a shell into the chamber. **pump-brake**, *n.* the handle of a ship's pump. **pump-handle**, *n.* the handle by which a pump is worked; (*coll.*) the hand or arm. **pump-head**, *n.* the casing at the head of a chain pump for directing the water into the discharge-spout. **pump-priming**, *n.* introducing fluid into a pump to expel the air before operation; investing money to stimulate commercial activity, esp. in stagnant or depressed areas. **pump-room**, *n.* a room where a pump is worked; a room at a spa where the waters from the medicinal spring are dispensed. **pumpage** (-ij), *n.* **pumper**, *n.*
pump[2] (pŭmp), *n.* a light low-heeled, slipper-like shoe, usu. of patent leather, worn with evening dress and for dancing; a plimsoll.
pumpernickel (pŭm'pənikl), *n.* German whole-meal rye bread.
pumpkin (pŭmp'kin), *n.* the large globular fruit of *Cucurbita pepo*; the trailing, annual plant bearing this fruit.
pun[1] (pŭn), *n.* the playful use of a word in two different senses, or of words similar in sound but different in meaning. *v.i.* (*past, p.p.* **punned**) to make a pun. **punnage** (-ij), punning, *n.* **punningly**, *adv.* **punster**, *n.* one who makes puns; one addicted to pun-making.
pun[2] (pŭn), *v.t.* (*past, p.p.* **punned**) to pound, to crush, to consolidate by ramming; to work (clay etc.) with a punner.
puna (poo'nə), *n.* a cold high plateau between the two ranges of the Cordilleras, the cold wind prevalent there; mountain-sickness.
Punch (pŭnch), *n.* the chief character in the popular

puppet-show of Punch and Judy, represented as a grotesque humped-backed man. **as pleased as Punch**, highly delighted. [short for PUNCHINELLO]
punch[1] (pŭnch), *n.* a tool, usu. consisting of a short cylindrical piece of steel tapering to a sharp or blunt end, for making holes, indenting, forcing bolts out of holes etc.; a machine in which a similar tool is used, esp. one for making holes in paper or cardboard; a tool or machine for stamping a die or impressing a design; a blow with the fist; vigour, forcefulness; striking power. *v.t.* to stamp or perforate with a punch; to make (a hole or indentation) thus; to drive (out etc.) with a punch; to strike, esp. with the fist; to press in vigorously (as a key or button); to record by pressing a key. **to pull one's punches** (*usu. in neg.*) to strike or criticize with less than full force. **punchbag**, *n.* a heavy, stuffed bag struck with the fists as exercise or by boxers in training; an unresisting victim. **punchball**, *n.* a ball usu. suspended or on an elastic stand used for punching practice. **punch-card, punched card**, *n.* a card in which data are represented by perforations, used in computers. **punch-drunk**, *a.* suffering a form of cerebral concussion from having taken repeated blows to the head; dazed. **punchline**, *n.* the conclusion of a joke, story that shows the point of it, produces the laugh or reveals an unexpected twist. **punch-up**, *n.* (*coll.*) a brawl, a fistfight. **puncher**, *n.* **punchy**, *a.* (*coll.*) forceful, incisive; punch-drunk.
punch[2] (pŭnch), *n.* a beverage compounded of wine or spirit, water or milk, lemons, sugar, spice etc. **punch-bowl**, *n.*
punch[3] (pŭnch), *n.* a short, fat fellow; a stout-built carthorse. **punchy**, *a.*
puncheon[1] (pŭn'chən), *n.* a short upright timber, used for supporting the roof in a mine or as an upright in the framework of a roof; a perforating or stamping tool, a punch.
puncheon[2] (pŭn'chən), *n.* a large cask holding from 72 to 120 gallons (324–540 l).
punchinello (pŭnchinel'ō), *n.* a clown or buffoon in Italian puppet shows; a buffoon, a Punch, a grotesque person. [It. *Pulcinello*, a character in Neapolitan low comedy]
punchline, punch-up, punchy PUNCH [1].
punctate (pŭngk'tāt), *a.* covered with points, dots, spots etc. **punctation**, *n.* **punctiform** (-tifawm), *a.* like a point or dot; punctate.
punctilio (pŭngktil'iō), *n.* (*pl.* **-ios**) a nice point in conduct, ceremony or honour; precision in form or etiquette. **punctilious**, *a.* precise or exacting in punctilio; strictly observant of ceremony or etiquette. **punctiliously**, *adv.* **punctiliousness**, *n.*
punctual (pŭngk'chuəl), *a.* observant and exact in matters of time; done, made, or occurring exactly at the proper time; (*Geom.*) of or pertaining to a point. **punctualist**, *n.* one who is very exact in observing forms and ceremonies. **punctuality** (-al'-), *n.* **punctually**, *adv.*
punctuate (pŭngk'chuāt), *v.t.* to mark with stops, to divide into sentences, clauses etc. with stops; to interrupt or intersperse; (*coll.*) to emphasize, to accentuate; to enforce (with). **punctuation**, *n.* **punctuation mark**, *n.* any of the symbols used in punctuating written material, as the comma, question mark and full stop. **punctuative**, *a.*
punctum (pŭngk'təm), *n.* (*pl.* **-ta** (-tə)) a point, a speck, a dot, a minute spot of colour etc. **punctule** (-tūl), *n.* a minute point, speck or dot. **punctulate** (-lət), *a.* **punctulation**, *n.* a point.
puncture (pŭngk'chə), *n.* a small hole made with something pointed, a prick; a small hole in a pneumatic tyre; the act of pricking or perforating. *v.t.* to make a puncture in; to pierce or prick with something pointed. *v.i.* of a tyre, balloon etc., to sustain a puncture.
pundit (pŭn'dit), *n.* a Hindu learned in the Sanskrit language and the science, laws and religion of India; a learned person; a pretender to learning.
punga (pŭng'ə), *n.* a New Zealand tree-fern, the pith of which is edible.
pungent (pŭn'jənt), *a.* sharply affecting the senses, esp. those of small or taste; pricking or stinging; acrid, keen, caustic, biting; piquant, stimulating; (*Nat. Hist.*) sharp-pointed, adapted for pricking or piercing. **pungency**, *n.* **pungently**, *adv.*
Punic (pū'nik), *a.* pertaining to the Carthaginians, Carthaginian; treacherous, faithless. *n.* the language of the Carthaginians.
punier, puniness etc. PUNY.
punish (pŭn'ish), *v.t.* to inflict a penalty on for an offence; to visit judicially with pain, loss, confinement or other penalty, to chastise; to inflict a penalty for (an offence); to inflict pain or injury on, to handle severely, to maul; to give great trouble to (opponents in a game, race etc.); (*coll.*) to consume large quantities of (food etc.). **punishable**, *a.* **punishability** (-bil'-), **punishableness**, *n.* **punishably**, *adv.* **punisher**, *n.* **punishing**, *a.* severe, wearing. **punishment**, *n.* **punitive** (pū'-), **punitory** (pū'-), *a.* awarding or inflicting punishment; retributive.
Punjabi (pŭnjah'bi), *n.* (*pl.* **-bis**) a native or inhabitant of the Punjab, a state in NW India; the language of the Punjab. *a.* of or belonging to the Punjab or Punjabis.
punk[1] (pŭngk), *n.* wood decayed through the growth of a fungus, touch-wood; amadou, a composition for igniting fireworks.
punk[2] (pŭngk), *n.* worthless articles; (a follower of) a youth movement of the late 1970s and 1980s, characterized by a violent rejection of established society, outlandish (often multi-coloured) hairstyles, and the use of worthless articles such as safety pins, razor blades, as decoration; punk rock; (*esp. N Am.*) a novice; (*esp. N Am.*) a petty criminal; (*formerly*) a prostitute. *a.* associated with the punk movement or punk rock; (*N Am.*) inferior. **punk rock**, *n.* a style of popular music associated with the punk movement and characterized by a driving beat often crude or obscene lyrics and an aggressive performing style. **punk rocker**, *n.*
punkah (pŭng'kə), *n.* a large portable fan; a large screen-like fan suspended from the ceiling and worked by a cord.
punner (pŭn'ə), *n.* a tool used for ramming earth, in a hole etc.
punnet (pŭn'it), *n.* a small, shallow basket for fruit, flowers etc.
punster PUN[1].
punt[1] (pŭnt), *n.* a shallow, flat-bottomed, square-ended boat, usu. propelled by pushing against the bottom of the stream with pole. *v.t.* to propel (a punt etc.) thus; to convey in a punt. *v.i.* to propel about thus; to go (about) in a punt. **punter, puntist, puntsman**, *n.*
punt[2] (pŭnt), *v.i.* in the games of basset, faro, ombre etc. to stake against the bank; (*sl.*) to bet on a horse etc. to a point in the game of faro; the act of playing basset, faro etc.; a punter. **punter**, *n.* a petty backer of horses; a small gambler to the Stock Exchange; a prostitute's client; (*coll.*) any customer or client.
punt[3] (pŭnt), *v.t.* (*Football*) to kick (the ball) after dropping it from the hand and before it touches the ground. *n.* such a kick.
punt[4] (pŭnt), *n.* the Irish pound.
punter PUNT[1, 2].
punto (pŭn'tō), *n.* a thrust or pass in fencing.
punty (pŭn'ti), *n.* (*pl.* **-ties**) a round ornamental mark on a glass article, like the hollow left by the end of a pontil.
puny (pū'ni), *a.* (*comp.* **punier**, *superl.* **puniest**) small and feeble, tiny, undersized, weak, poorly developed; petty, trivial. **puniness**, *n.*
pup (pŭp), *n.* a puppy. *v.t.* (*past, p.p.* **pupped**) to bring forth (pups). *v.i.* to bring forth pups; to whelp, to litter.

in pup, pregnant. **to sell a pup to**, (*sl.*) to trick into buying something worthless; to swindle. **pup tent**, *n*. a very small and basic shelter tent.

pupa (pū'pə), *n*. (*pl*. **-pae** (-pē), **-pas**) an insect at the immobile, metamorphic stage between larva and imago. **pupal**, *a*. **pupate** (-pāt'), *v.i.* to become a pupa. **pupation**, *n*. **Pupipara** (-pip'ərə), *n.pl.* a division of Diptera in which the young are developed as pupae within the body of the mother. **pupiparous**, *a*. **pupivorous** (-piv'ərəs), *a*. feeding on the pupae of other insects. **pupoid** (-poid), *a*.

pupil[1] (pū'pl), *n*. a young person of either sex under the care of a teacher; one who is being, or has been, taught by a particular person; (*Law*) a boy or girl under the age of puberty and under the care of a guardian, a ward. **pupil-teacher**, *n*. formerly, one in apprenticeship as a teacher and receiving general education at the same time. **pupillage** (-ij), **pupilship**, *n*. the state or period of being a pupil. **pupillarity** (-la'ri-), *n*. (*Sc. Law*) the period before puberty. **pupillary**, *a*.

pupil[2] (pū'pl), *n*. the circular opening of the iris through which rays of light pass to the retina. **pupillary**, *a*. **pupillate** (-lət), **pupilled**, *a*. of ocelli, having a central spot like a pupil. **pupillometer** (-lom'itə), *n*. an instrument for measuring the pupil of the eye or the distance between the eyes. **pupillometry**, *n*.

Pupipara, etc. PUPA.

puppet (pŭp'it), *n*. an articulated toy figure moved by strings, wires or rods, a marionette; a small figure with a hollow head and cloth body into which the operator's hand is inserted; one whose actions are under another's control; a mere tool. **puppet-clack** PUPPET-VALVE. **puppet-play, -show**, *n*. a play with puppets as dramatis personae. **puppet-theatre**, *n*. **puppet-valve**, *n*. a disc on a stem with vertical motion to and from its seat. **puppeteer** (-tiə'), *n*. one who manipulates puppets. **puppetry**, *n*. the art of making and manipulating puppets and presenting puppet shows.

puppy (pŭp'i), *n*. a young dog; a silly young fellow, a coxcomb, a fop. **puppy-dog**, *n*. a puppy. **puppy fat**, *n*. temporary plumpness in children or adolescents. **puppy-headed**, *a*. **puppy love**, *n*. temporary infatuation in adolescence. **puppydom** (-dəm), **puppyhood** (-hud), *n*. the state of being a puppy. **puppyish**, *a*. **puppyism**, *n*.

purana (poorah'nə), *n*. any of a great division of Sanskrit poems comprising the whole body of Hindu mythology. **puranic**, *a*.

Purbeck (pœ'bek), *n*. Purbeck stone. **Purbeck marble**, *n*. one of the finer varieties of Purbeck stone, used for shafts etc. in architecture. **Purbeck limestone**, *n*. a hard limestone from Purbeck. [Isle of *Purbeck*, a peninsula in Dorset]

purblind (pœ'blīnd), *a*. partially blind, near-sighted, dim-sighted; lacking understanding or insight. **purblindly**, *adv*. **purblindness**, *n*.

purchase (pœ'chəs), *v.t.* to obtain by payment of an equivalent; to buy; to acquire at the expense of some sacrifice, exertion, danger etc.; to haul up, hoist, or draw in by means of a pulley, lever, capstan etc. *n*. the act of purchasing or buying; that which is purchased; annual value, annual return, esp. from land; the acquisition of property by payment of a price or value, any mode of acquiring property other than by inheritance; advantage gained by the application of any mechanical power, leverage; an appliance furnishing this, as a rope, pulley etc.; an effective hold or position for leverage, a grasp, a foothold; the system of buying commissions in the army, abolished in 1871. **purchase-money**, *n*. the price paid or contracted to be paid for anything purchased. **purchase tax**, *n*. formerly, a differential tax on certain goods sold to the public. **purchasable**, *a*. **purchaser**, *n*.

purdah (pœ'də), *n*. a curtain or screen, esp. one keeping women from the view of strangers; the custom in India and elsewhere of secluding women.

pure (pūə), *a*. unmixed, unadulterated; free from anything foul or polluting, clear, clean; of unmixed descent, free from admixture with any other breed; mere, sheer, absolute; free from moral defilement, innocent, guiltless; unsullied, chaste; free from discordance, harshness etc., perfectly correct in tone-intervals; having a single sound or tone, not combined with another; of sciences, entirely theoretical, not applied. **purebred**, *a*. of a pure strain through many generations of controlled breeding. **pure science**, *n*. science based on self-evident truths, as logic, mathematics etc. **purely**, *adv*. **pureness**, *n*.

purée (pū'rā), *n*. a smooth thick pulp of fruit, vegetables etc. obtained by liquidizing, sieving etc; a thick soup made by boiling meat or vegetables to a pulp and straining it. *v.t.* to reduce to a purée.

purfle (pœfl), *v.t.* to decorate with an ornamental border, to border; to ornament the edge of (a canopy etc.) with embroidery. *n*. a border or edging of embroidered work. **purfling**, *n*. ornamental bordering; the ornamental border on the backs and bellies of stringed instruments.

purgation (pəgā'shən), *n*. the act of purging, purification; cleansing of the bowels by the use of purgatives; (*Hist.*) the act of clearing oneself from an imputed crime by oath or ordeal; in the Roman Catholic Church, the process of spiritual purification of souls in purgatory. **purgative** (pœ'gə-), *a*. having the quality of cleansing, esp. evacuating the intestines, aperient. *n*. an aperient or cathartic. **purgatively**, *adv*.

purgatory (pœ'gətri), *n*. a place or state of spiritual purging, esp. a place or state succeeding the present life in which, according to the Roman Catholic Church, the souls of the faithful are purified from venial sins by suffering; any place of temporary suffering or tribulation; (*coll.*) an acutely uncomfortable experience. *a*. cleansing, purifying. **purgatorial** (-taw'ri-), *a*. **purgatorian**, *a*., *n*.

purge (pœj), *v.t.* to cleanse or purify; to free (of or from impurity, sin etc.); to remove (off or away) by cleansing; to clear (of an accusation, suspicion etc.); to get rid of (persons actively in opposition); to rid (a nation etc.) of such people; to atone for, expiate or annul (guilt, spiritual defilement etc.); to clear of guilt; to cleanse (the bowels) by cathartic action; *v.i.* to become or grow pure; to become purged; to cause purging. *n*. a purgative medicine; an act of purging. **purger**, *n*. **purging**, *a*., *n*.

puri (poo'ri), *n*. an Indian food of unleavened whole-wheat bread, deep-fried and sometimes containing a spicy vegetable etc. mixture.

purification etc. PURIFY.

puriform (pū'rifawm), *a*. in the form of pus; like pus.

purify (pū'rifī), *v.t.* to make pure, to cleanse; to free from sin, guilt, pollution etc.; to make ceremonially clean; to clear of or from foreign elements, corruptions etc. **purification** (-fi-), *n*. the act of physical or spiritual purifying; the act or process of cleansing ceremonially. **purificator** (-fi), *n*. a piece of linen used to wipe the chalice and paten at the Eucharist. **purificatory**, *a*. having power to purify; tending to purify. **purifier**, *n*.

Purim (puə'rim), *n*. a Jewish festival instituted in commemoration of the deliverance of the Jews from the destruction threatened by Haman's plot (Esther ix.20–32).

purin, purine (pū'rin), *n*. a crystalline solid derivable from uric acid, of which caffeine, xanthine etc. are derivatives; a derivative of this that is a constituent of a nucleic acid.

puriri (poorē'rē), *n*. the New Zealand oak or teak.

purist (pū'rist), *n*. one advocating or affecting purity, esp. in the choice of words; a rigorous critic of literary style. **purism**, *n*. **puristic, -al** (-ris'-), *a*.

Puritan (pū'ritən), *n*. one of a party or school of English Protestants of the 16th and 17th cents., who aimed at purifying religious worship from all ceremonies etc. not authorized by Scripture, and advocating the strictest pur-

ity of conduct; (*usu. derog.*) any person practising or advocating extreme strictness in conduct or religion. *a.* pertaining to the Puritans; excessively strict in religion or morals. **puritanic, -ical** (-tan'-), *a.* **puritanically,** *adv.* **Puritanism,** *n.* **puritanize, -ise,** *v.t., v.i.*
purity (pū'riti), *n.* the state of being pure, cleanness; freedom from pollution, adulteration or admixture of foreign elements; moral cleanness, innocence, chastity.
purl[1] (pœl), *n.* an edging or fringe of twisted gold or silver wire; the thread or cord of which this is made; a small loop on the edges of pillow lace; a series of such loops as an ornamental hem or edging; an inversion of the stitches in knitting. *v.t.* to border or decorate with purl or purls; to knit with an inverted stitch. *v.i.* to knit in purl stitch.
purl[2] (pœl), *v.i.* to flow with a soft, bubbling, gurgling or murmuring sound. *n.* a gentle bubbling, gurgling or murmuring sound.
purl[3] (pœl), *v.t., v.i.* to upset, to overturn. *n.* a heavy fall, an overturn. **purler,** *n.* (*coll.*) a heavy fall or throw, a cropper, a spill; a knockdown blow.
purlieu (pœ'lū), *n.* (*usu. pl.*) the bounds or limits within which one ranges; (*pl.*) outlying parts, outskirts, environs; (*Hist.*) the borders or outskirts of a forest, esp. a tract of land, once included in forest but entirely or partially disafforested.
purlin (pœ'lin), *n.* a horizontal timber resting on the principal rafters and supporting the common rafters or boards on which the roof is laid.
purloin (pœloin'), *v.t.* to steal, to take by theft. *v.i.* to practise theft; to pilfer. **purloiner,** *n.*
purple (pœ'pl), *a.* of the colour of red and blue blended, the former predominating; (*Rom. Ant.*) of the colour obtained from the mollusca *Purpura* and *Murex,* prob. crimson; dyed with or as with blood; imperial, regal; of literary style, florid, highly rhetorical. *n.* this colour; a purple pigment or dye; (cloth for) a purple dress or robe, esp. of an emperor, king, Roman consul or a bishop; imperial or regal power; the robe of a cardinal; the cardinalate. *v.t.* to make or dye purple. *v.i.* to become purple. **born in the purple,** of high and wealthy, esp. royal or imperial, family (see PORPHYROGENITE). **royal purple,** a deep violet tending to blue. **purple emperor,** *n.* a variety of butterfly. **purple heart,** *n.* a mauve, heart-shaped, amphetamine tablet taken as a stimulant; (*Purple Heart*) a US decoration for wounds received on active service. **purple passage, patch,** *n.* a passage of obtrusively elevated or ornate writing. **purplish, purply,** *a.*
purport (pəpawt'), *v.t.* to convey as the meaning, to imply, to signify; to profess, to be meant to appear (to). *n.* (pœ'-), meaning, tenor, import; object, purpose. **purportless** (pœ'-), *a.*
purpose (pœ'pəs), *n.* an end in view, an object, an aim; the reason why something exists; effect, result, consequence; determination, resolution. *v.t.* to intend, to design. *v.i.* to have an intention or design. **on purpose,** intentionally, designedly, not by accident; in order (that). **to the purpose,** with close relation to the matter in hand, relevantly; usefully. **purpose-built,** *a.* constructed to serve a specific purpose. **purposeful,** *a.* having a definite end in view; determined. **purposefully,** *adv.* **purposefulness,** *n.* **purposeless,** *a.* **purposelessly,** *adv.* **purposelessness,** *n.* **purposelike,** *a.* **purposely,** *adv.* of set purpose, intentionally, not by accident. **purposive,** *a.* having or serving a purpose; purposeful. **purposiveness,** *n.*
Purpura (pə'pūrə), *n.* a genus of gasteropods, many species of which secrete a fluid from which the ancients obtained their purple dye; (**purpura**) a morbid condition of the blood or blood-vessels characterized by livid spots on the skin. **purpure,** *n.* (*Her.*) purple, represented in engraving by diagonal lines from left to right. **purpureal** (-pū'riəl), *a.* (*poet.*) purple. **purpurescent** (-res'nt), *a.* purplish. **purpuric** (-pū'-), *a.* of or pertaining to the disease purpura; of or pertaining to a purple colour.
purr (pœ), *n.* a soft vibratory murmuring as of a cat when pleased. *v.i.* to make this sound. *v.t.* to signify, express or utter thus. **purring,** *a.*, *n.* **purringly,** *adv.*
purse (pœs), *n.* a small bag or pouch for money, usu. carried in the pocket; (*N Am.*) a woman's handbag; money, funds, resources, a treasury; a sum of money subscribed or collected or offered as a gift, prize etc.; a definite sum (varying in different Eastern countries); a baglike receptacle, a pouch, a cyst. *v.t.* to wrinkle, to pucker. *v.i.* to become wrinkled or puckered. **a light purse, an empty purse,** poverty, want of resources. **a long purse, a heavy purse,** wealth, riches. **privy purse** PRIVY. **public purse,** *n.* the national treasury. **purse-bearer,** *n.* one who has charge of the purse of another person or of a company etc.; an officer who carried the Great Seal in a purse before the Lord Chancellor. **purse-net,** *n.* a net the mouth of which can be drawn together with cords like an old-fashioned purse. **purse-proud,** *a.* proud of one's wealth. **purse-seine,** *n.* a large purse-net for sea-fishing. **purse-strings,** *n.* strings for drawing together the mouth of an old-fashioned purse; control of expenditure. **purseful,** *n.* **purseless,** *a.*
purser (pœ'sə), *n.* an officer on board ship in charge of the provisions, clothing, pay and general business. **pursership,** *n.*
purslane (pœ'slin), *n.* a succulent herb, *Portulaca oleracea,* used as a salad and pot-herb.
pursue (pəsū'), *v.t.* to follow with intent to seize, kill etc.; to try persistently to gain or obtain, to seek; to proceed in accordance with; to apply oneself to, to practise continuously; of consequences etc., to attend persistently, to haunt; to continue to discuss, to follow up; to attend, to accompany. *v.i.* to follow, to seek (after); to go in pursuit; to go on, to proceed, to continue. **pursuable,** *a.* **pursuance,** *n.* carrying out, performance; implementation. **pursuant,** *a.* in accordance, consonant, conformable (to). *adv.* in accordance or conformably (to). **pursuantly,** *adv.* **pursuer,** *n.* One who pursues; (*Sc. Law*) a plaintiff, a prosecutor. **pursuit** (-sūt), *n.* the act of pursuing a following; a prosecution; an endeavour to attain some end; an employment, occupation, business or recreation that one follows persistently.
pursuivant (pœ'sivant, -swi-), *n.* (*Her.*) an attendant on a herald, an officer of the College of Arms of lower rank than a herald; (*poet.*) a follower, an attendant.
pursy[1] (pœ'si), *a.* short-winded, asthmatical; fat, corpulent. **pursiness,** *n.*
pursy[2] (pœ'si), *a.* like a purse, puckered up like a purse-mouth; moneyed, purse-proud.
purtenance (pœ'tənəns), APPURTENANCE.
purulent (pūə'ūlənt), *a.* consisting of or discharging pus or matter. **purulence, -lency,** *n.* **purulently,** *adv.*
purvey (pəvā'), *v.t.* to provide, to supply; to supply (provisions etc.); *v.i.* to make provision; to act as purveyor. **purveyance,** *n.* the purveying or providing of provisions; provisions supplied; (*Hist.*) the old royal prerogative of buying up provisions, impressing horses etc. **purveyor,** *n.* one who purveys; one who supplies provisions etc., a caterer, esp. on a large scale.
purview (pœ'vū), *n.* extent, range, scope, intention; range of vision, knowledge etc.; (*Law*) the body of a statute consisting of the enacting clauses.
pus (pūs), *n.* the matter secreted from inflamed tissues, the produce of suppuration.
Puseyism (pū'ziizm), *n.* the High Church tenets of the Oxford School of which Dr Edward Pusey (1800–82) was a prominent member. Tractarianism. **Puseyite** (-īt), *a., n.*
push (push), *v.t.* to press against with force, tending to urge forward; to move (a body along, up, down etc.) thus; to make (one's way) vigorously, to impel, to drive; to put pressure on (a person); to develop and carry (as a

point, an argument), esp. to extremes; (with a number) to approach; to seek to promote, esp. to promote the sale of; (*sl.*) to peddle (drugs). *v.i.* to exert pressure (against, upon etc.); to press forward, to make one's way vigorously, to hasten forward energetically; to thrust or butt (against); to be urgent and persistent; (*Billiards*) to make a push-stroke. *n.* the act of pushing, a thrust, a shove; a vigorous effort, an attempt, an onset; pressure; an exigency, a crisis, an extremity; persevering energy, self-assertion; (*Mil.*) an offensive; (*Billiards*) a stroke in which the ball is pushed, not struck; (*Austral.*) a gang, a clique, a party. **at a push, if it comes to the push**, if really necessary. **to get, give the push**, (*sl.*) to dismiss, be dismissed, esp. from a job. **to push around**, (*coll.*) to bully, to treat with contempt. **to push for**, (*coll.*) to advocate vigorously, to make strenuous efforts to achieve. **to push in**, (*coll.*) to force one's way into (esp. a queue) ahead of others. **to push off**, to push against the bank with an oar so as to move a boat off; (*coll.*) to go away. **to push on**, to press forward; to hasten; to urge or drive on. **to push one's luck**, (*coll.*) to take risks, esp. by overplaying an existing advantage. **to push through**, to secure the acceptance of speedily or by compulsion. **to push up the daisies**, (*coll.*) to be dead and buried. **push-cart**, *n.* (N Am.) a barrow. **push-bike, -bicycle**, *n.* one worked by the rider as distinguished from a motor-bicycle. **push button**, *n.* a device for opening or closing an electric circuit by the pressure of the finger on a button. **push-button**, *a.* operated by means of a push button. **pushchair**, *n.* a light, folding chair on wheels for a child. **pushover**, *n.* (*coll.*) something easy; a person, team etc. easy to defeat. **push-pull**, *a.* of or being any piece of apparatus in which electrical or electronic devices, e.g. two transistors in an amplifier, act in opposition to each other. **push-start**, *v.t.* to set a vehicle in motion by pushing, then engage a gear thus starting the engine. *n.* an act of push-starting a vehicle. **push-stroke**, *n.* (*Billiards*) a push. **pusher**, *n.* a person or thing that pushes, esp. a device used in conjunction with a spoon for feeding very young children; a pushy person; (*sl.*) a drug peddler. **pusher aeroplane**, *n.* an aeroplane with its propeller at the rear. **pushful**, *a.* (*coll.*) self-assertive, energetic, vigorous or persistent in advancing oneself. **pushfulness**, *n.* **pushing**, *n.* enterprising, energetic. **pushingly**, *adv.* **pushy**, *a.* (*coll.*) pushful.
Pushtoo, Pushtu, Pushto (-tō), PASHTO.
pusillanimous (pūsilan'iməs), *a.* destitute of courage, firmness or strength of mind, fainthearted. **pusillanimity** (-nim'-), *n.* **pusillanimousness, pusillanimously**, *adv.*
puss (pus), *n.* a pet name for a cat, esp. in calling; a hare; (*coll.*) a child, a girl. **puss-moth**, *n.* a large bombycid moth, *Cerura vinula*. **pussy**, *n.* a pet name for a cat, puss; (*taboo sl.*) the female pudenda. **pussy-cat**, *n.* a cat; anything woolly or fuzzy, as a willow catkin. **pussyfoot**, *v.i.* to move stealthily or warily; to avoid committing oneself. **pussy-willow**, *n.* a willow with silky grey catkins, as a small American willow, *Salix discolor*.
pustule (pŭs'tūl), *n.* a small vesicle containing pus, a pimple; a small excrescence, a wart, a blister. **pustular**, *a.* **pustulate**, *v.t.*, *v.i.* to form into pustules. *a.* (-lət), covered with pustules or excrescences. **pustulation**, *n.* **pustulous**, *a.*
put[1] (put), *v.t.* (*pres. p.* **putting**, *past, p.p.* **put**) to move so as to place in some position; to set, lay, place or deposit; to bring into some particular state or condition; to append, to affix; to connect, to add; to assign; to express, to state; to render, to translate (into); to apply, to set, to impose; to stake (money on); to invest; to inflict; to subject, to commit (to or upon); to advance, to propose (for consideration etc.); to submit (to a vote); to constrain, to incite, to force, to make (a person do etc.); to make (one) appear in the right, wrong etc.; to repose (as trust, confidence etc.); to estimate; to hurl, to cast, to throw; to thrust, to stab with. *v.i.* (*Naut.*) to go, to proceed, to steer one's course (in a specified direction). *n.* the act of putting; a cast, a throw (of a weight etc.); an agreement to sell or deliver (stock, goods etc.) at a stipulated price within a specified time; a thrust. **not to put it past somebody**, to consider a person capable of. **to put about**, to inconvenience; (*Naut.*) to go about, to change the course of to the opposite tack; (*coll.*) to make public, to spread abroad. **to put across**, to communicate effectively. **to put away**, to remove; to return to its proper place; to lay by; to divorce; to imprison; (*coll.*) to consume. **to put back**, to retard, to check the forward motion of; to postpone; to move the hands of (a clock) back; to replace; (*Naut.*) to return (to land etc.). **to put by**, to put, set or lay aside; to evade; to put off with evasion; to desist from. **to put down**, to suppress, to crush; to take down, to snub, to degrade; to confute, to silence; to reduce, to diminish; to write down, to enter, to subscribe; to reckon, to consider, to attribute; to put (a baby) to bed; to kill (esp. an old or ill animal); to pay (as a deposit); of an aircraft, to land. **to put forth**, to present to notice; to publish, to put into circulation; to extend; to shoot out; to exert; to sprout, to bud. **to put forward**, to set forth, to advance, to propose; to thrust (oneself) into prominence; to move the hands of (a clock) onwards. **to put in**, to introduce, to interject, to interpose; to insert, to enter; to install in office etc.; to present, to submit (as an application, request etc.); to enter a harbour; (*coll.*) to spend, to pass (time). **to put in mind**, to remind. **to put it across someone**, (*coll.*) to defeat someone by ingenuity; to put it on, (*coll.*) to pretend (to be ill etc.); to exaggerate. **to put off**, to lay aside, to discard, to take off; to postpone; to disappoint, to evade; to hinder, to distract the attention of; to dissuade (from); to cause aversion to; to foist, to palm off (with). **to put on**, to take on; to clothe oneself with; to assume; to add, to affix, to apply; to come to have an increased amount of; to bring into play, to exert; to cause to operate; to stage, to produce; to appoint; to move the hands of (a clock) forward; to impose, put upon; (*coll.*) to intentionally deceive or mislead. **to put one over on**, (*coll.*) to deceive into believing or accepting. **to put out**, to invest, to place (at interest); to eject; to extinguish; to disconcert; to annoy, to irritate; to inconvenience; to exert; to dislocate; to publish, to broadcast; to give out (work) to be done at different premises; to render unconscious. **to put over**, to put across. **to put through**, to carry out, accomplish; to process; to connect by telephone; to cause suffering to. **to put to it**, to distress; to press hard. **to put up**, to raise; to offer, to present, as for sale, acution; to give, to show (a fight, resistance etc.); to provide (money, a prize); to offer (oneself) as a candidate; to present as a candidate; to publish (banns etc.); to pack up; to place in a safe place; to lay aside; to sheathe; to erect, to build; to lodge and entertain; to take lodgings. **to put upon**, to impose upon; to take undue advantage of. **to put up to**, to incite to; to make conversant with. **to put up with**, to tolerate, to submit to. **to stay put**, STAY. **put-down**, *n.* a snub; an action or remark intended to humiliate. **put-off**, *n.* an evasion, an excuse. **put-on**, *n.* an attempt to deceive or mislead. **put-up**, *a.* something secretly pre-arranged for purposes of deception. **putter**[1], *n.* one who puts; a shot-putter. **putting (the shot)**, *n.* the act or sport of throwing a heavy weight from the shoulder by an outward thrust of the arm.
put[2] (pŭt), PUTT.
putative (pū'ətiv), *a.* reputed, supposed; commonly regarded as. **putatively**, *adv.*
putlog (pŭt'log), *n.* a short horizontal piece of timber for the floor of a scaffold to rest on.
putrefy (pū'trifī), *v.t.* to make putrid, to cause to rot or decay; to make carious or gangrenous; to corrupt. *v.i.* to become putrid, to rot, to decay; to fester, to suppurate.

putrefaction (-fak'-), *n.* **putrefactive,** *a.* **putrescent** (-tres'nt), *a.* **putrescence,** *n.* **putrescible,** *a.* **putrescin** (-in), *n.* (*Chem.*) a poisonous alkaloid contained in decaying animal matter. **putrid** (-trid), *a.* in a state of putrefaction, decomposition, or decay; tainted, foul, noxious; corrupt; (*sl.*) very bad, worthless. **putridity** (-trid'-), **putridness,** *n.* **putridly,** *adv.*

putsch (puch), *n.* a rising, revolt.

putt (pŭt), *v.t.*, *v.i.* (*Golf*) to strike (the ball) gently with a putter so as to get it into the hole on the putting-green. *n.* this stroke. **putter**[2] (pŭt'ėr), *n.* a short, stiff golf-club, used for striking the ball on the putting-green; a person who putts. **putting-green,** *n.* the piece of ground around each hole on a golf-course, usu. kept rolled, closely mown and clear of obstacles; an area of smooth grass with several holes for putting games.

puttee (pŭt'i, -ē), *n.* a long strip of cloth wound spirally round the leg, usu. from ankle to knee, as a form of gaiter, worn esp. as part of former military uniform.

putter[1] PUT[1].
putter[2] PUTT.
puttier PUTTY.
putting PUT[1].
putting-green PUTT.

putto (put'ō), *n.* (*pl.* **putti** (-ē)) a figure of a small boy, cherub or cupid, in Renaissance and Baroque art.

putty (pŭt'i), *n.* calcined tin or lead used by jewellers as polishing-powder for glass, metal etc.; whiting and linseed-oil beaten up into a tenacious cement, used in glazing; fine lime-mortar used by plasterers for filling cracks etc.; yellowish- or brownish-grey colour; one who is easily manipulated. *v.t.* to fix, cement, fill up, or cover with putty. **up to putty,** (*Austral. sl.*) no good, valueless, of bad quality. **putty-faced,** *a.* having a smooth, colourless face like putty. **putty-powder,** *n.* jewellers' putty in the form of powder, used for polishing. **puttier,** *n.* a worker with putty, a glazier.

puy (pwē), *n.* a conical hill of volcanic origin, esp. in Auvergne, France.

puzzle (pŭz'l), *n.* a state of bewilderment or perplexity; a perplexing problem, question or enigma; a toy, riddle or other contrivance for exercising ingenuity or patience. *v.t.* to perplex, to mystify. *v.i.* to be bewildered or perplexed. **to puzzle out,** to discover, or work out by mental labour. **puzzledom** (-dəm), **puzzlement, puzzler,** *n.* **puzzlingly,** *adv.*

puzzolana (putsəlah'nə), POZZOLANA.
PVC, (*abbr.*) polyvinyl chloride.
PW, (*abbr.*) policewoman.
p.w., (*abbr.*) per week.
PWA, (*abbr.*) person with AIDS.
PWARC, (*abbr.*) person with AIDS-related complex.
PWR, (*abbr.*) pressurized water reactor.

pyaemia, pyemia (pīē'miə), *n.* (*Path.*) blood-poisoning, due to the absorption of putrid matter into the system causing multiple abscesses. **pyaemic,** *a.*

pycnidium (piknid'iəm), *n.* (*pl.* **-dia** (-diə)) a receptacle bearing pycnidiospores or stylospores in certain fungi. **pycnid** (pik'-), *n.* **pycnidiospore** (-əspaw), *n.* a stylospore developed in a pycnidium.

pycno-, *comb. form.* thick, dense.

pycnogonid (piknog'ənid), *n.* (*Zool.*) a marine arthropod belonging to the group Pycnogonida, comprising the sea-spiders. **pycnogonoid** (-noid), *a.*, *n.*

pycnometer (piknom'itər), *n.* a bottle or flask used in measuring the specific gravity of fluids.

pycnospore (pik'nəspaw), PYCNIDIOSPORE.

pye (pī), PIE[1].

pye-dog (pī'dog), *n.* a pariah dog, a cur.

pyelitis (pīəlī'tis), *n.* (*Path.*) inflammation of the pelvis of the kidney. **pyelitic** (-lit'-), *a.*

pyel(o)-, *comb. form.* pertaining to the kidneys.

pyelonephritis (pīəlōnəfrī'tis), *n.* inflammation of the kidney and of the renal pelvis. **pyelonephritic** (-frit'-), *a.*

pyemia (pīē'miə), PYAEMIA.

pygal (pī'gəl), *a.* of, pertaining to, or near the rump or hind quarters. *n.* the pygal shield or plate of the carapace of a turtle.

pygmy, pigmy (pig'mi), *n.* one of a race of dwarfish people mentioned by Herodotus and other ancient historians as living in Africa and India; one of various dwarf races living in Malaysia and Central Africa, esp. the Akka, Batwa and Obongo of equatorial Africa; an (abnormally) small person; anything very diminutive; a pixy, a fairy; one having a certain faculty or quality in relatively a very small degree. *a.* diminutive, dwarf; small and insignificant. **pygmaean** (-mē'ən), *a.*

pygostyle (pī'gəstil), *n.* the vomer or ploughshare bone forming the end of the vertebral column in most birds. **pygostyled,** *a.*

pyjamas, (*esp. N Am.*) **pajamas** (pəjah'məz), *n.pl.* loose trousers of silk, cotton etc. worn by both sexes among Muslims in India and Pakistan; a sleeping-suit consisting of a loose jacket and trousers.

pylon (pī'lən), *n.* a gateway of imposing form or dimensions, esp. the monumental gateway of an Egyptian temple; a stake marking out the course in an aerodrome; a structure, usu. of steel, supporting an electric cable; a rigid, streamlined support for an engine etc., on the outside of an aircraft.

pylorus (pīlawr'əs), *n.* the contracted end of the stomach leading into the small intestine; the adjoining part of the stomach. **pyloric** (-lor'ik), *a.*

pyo-, *comb. form.* pus.

pyogenesis (pīōjen'əsis), *n.* the formation of pus, suppuration. **pyogenetic** (-net'-), **pyogenic,** *a.*

pyoid (pī'oid), *a.* of the nature of pus.

pyorrhoea, (*esp. N Am.*) **pyorrhea** (pīərē'ə), *n.* discharge of pus; inflammation of the teeth sockets.

pyosis (pīō'sis), *n.* suppuration.

pyr- PYR(O)-.

pyracanth (pī'rəkanth), *n.* an evergreen thorny shrub, *Crataegus pyracantha,* with white flowers and coral-red berries, also called the evergreen thorn, commonly trained against walls as an ornamental climber.

pyramid (pi'rəmid), *n.* a monumental structure of masonry, with a square base and triangular sloping sides meeting at the apex; a similar solid body or figure, with a triangular or polygonal but usu. square base; a structure, object or formation of this shape; a tree trained in this form; a hierarchical structure or system visualized as having this shape; a game of pool played with 15 coloured balls and a cue-ball. **the Pyramids,** the great pyramids of ancient Egypt. **pyramid selling,** *n.* a fraudulent system of selling whereby batches of goods are sold to agents who sell smaller batches at increased prices to sub-agents and so on down. **pyramidal** (-ram'-), **pyramidic, -ical** (-mid'-), *a.* **pyramidally, pyramidically,** *adv.* **pyramidist,** *n.* a student or investigator of the origin, structure etc. of ancient pyramids, esp. those of Egypt. **pyramidwise,** *adv.* **pyramidon** (-ram'idon), *n.* an organ stop having stopped pipes like inverted pyramids, producing very deep tones.

pyrargyrite (pirah'jirīt), *n.* (*Min.*) a native sulphide of silver and antimony.

pyre (pīə), *n.* a funeral pile for burning a dead body; any pile of combustibles.

pyrene (pī'rēn), *n.* the stone of a drupe.

Pyrethrum (pīrēth'rəm), *n.* a genus of compositous plants (usu. regarded as a sub-division of *Chrysanthemum*), comprising the feverfew; an insecticide made from the dried heads of these. **pyrethrin** (-rin), *n.* either of two oily, insecticidal compounds found in pyrethrum flowers.

pyretic (pīret'ik), *a.* of, relating to, or producing fever; remedial in fever. *n.* a pyretic medicine.

pyrexia (pīrek'siə), n. fever, feverish condition. **pyrexial, -ical**, a.
Pyrex, (pī'reks), a., n. (made of) heat-resistant glass containing oxide of boron.
pyrheliometer (pəhēliom'itə), n. an instrument for measuring the amount of solar radiation. **pyrheliometric** (-met'-), a.
pyridine (pī'ridēn, pī'-), n. a liquid alkaloid obtained from bone-oil, coal-tar naphtha etc., used as a remedy for asthma. **pyridoxine** (piridok'sēn), n. a derivative of pyridine, vitamin B₆, important esp. in amino acid metabolism.
pyriform (pī'rifawm), a. pear-shaped.
pyrimidine (pirim'idēn), n. a nitrogen-containing organic base with a strong odour; a derivative of this base that is a constituent of a nucleic acid.
pyrites (pīrī'tēz), n. a native sulphide of iron, one of two common sulphides, chalcopyrite, yellow or copper pyrites, or marcasite, usu. called iron pyrites. **pyritic, -ical** (-rit'-), **pyritous** (pī'ri-), a. **pyritiferous** (-tif'-), a.
pyro (pī'rō), n. pyrogallic acid.
pyr(o)-, comb. form. fire, heat; obtained (as if) by heating.
pyroclastic (pirəklas'tik), a. formed from or consisting of the fragments broken up or ejected by volcanic action.
pyroelectric (pīr'ōlek'trik), a. of a mineral, becoming electropolar on heating. **pyro-electricity**, n.
pyrogallic (pirəgal'ik), a. produced from gallic acid by heat. **pyrogallol** (-ol), n. pyrogallic acid, used as a developing agent in photography.
pyrogen (pī'rəjen), n. a substance, such as ptomaine, that produces fever on being introduced into the body. **pyrogenetic** (-net'-), **pyrogenic** (-jen'-), a. producing heat; producing feverishness; pyrogenous. **pyrogenous** (-roj'-), a. produced by fire, igneous.
pyrognostic (pirəgnos'tik), a. of or pertaining to those properties of a mineral that are determinable by heat.
pyrography (pirog'rəfi), n. the art of making designs in wood by means of fire, poker-work. **pyrograph** (pī'rəgraf), v.i. **pyrographer, -phist**, n. **pyrographic** (-graf'-), a. **pyrogravure** (-rōgrəvūə'), n. pyrography; a picture produced by this.
pyrolatry (pirol'ətri), n. fire-worship.
pyroligneous (pirōlig'niəs), a. derived from wood by heat. **pyrolignite** (-nit), n. a salt of pyroligneous acid.
pyrology (pirol'əji), n. the science of fire or heat, esp. the branch of chemistry dealing with the application of heat, blow-pipe analysis etc. **pyrological** (-loj'-), a. **pyrologist**, n.
pyrolusite (pirəloo'sit), n. (Min.) native manganese dioxide, one of the most important of the ores of manganese.
pyrolysis (pirol'isis), n. the decomposition of a substance by heat. **pyrolyse**, (esp. N Am.) **-yze** (pī'rəliz), v.t. to subject to this process. **pyrolytic** (-lit'-), a.
pyromagnetic (pirōmagnet'ik), a. of or pertaining to the alterations of magnetic intensity due to changes in temperature. **pyromagnetic generator**, n. a dynamo for generating electricity by induction through changes in the temperature of the field-magnets.
pyromancy (pīr'əmansi), n. divination by fire. **pyromantic** (-man'-), a.
pyromania (pirəmā'niə), n. a compulsive desire to destroy by fire. **pyromaniac** (-ak), n. **pyromaniacal** (-mənī'ə-), a.
pyrometer (pirom'itə), n. an instrument for measuring high temperatures; an instrument for measuring the expansion of bodies by heat. **pyrometric, -ical** (-met'-), a. **pyrometrically**, adv. **pyrometry**, n.
pyrope (pī'rōp), n. a deep-red garnet.
Pyrophane (pī'rōfūn), n. a variety of opal that absorbs melted wax and becomes translucent when heated and opaque again on cooling. **Pyrophanous** (-rof'-), a.
pyrophone (pī'rəfōn), n. a musical instrument the notes of which are produced in glass tubes each containing two hydrogen flames.
pyrophoric (pirəfo'rik), a. igniting spontaneously on contact with air; of an alloy, emitting sparks when struck.
pyrophosphoric (pirōfosfo'rik), a. derived by heat from phosphoric acid.
pyrophotograph (pirəfō'təgraf), n. a photographic picture fixed on glass or porcelain by firing. **pyrophotographic** (-graf'-), a. **pyro-photography** (-tog'-), n.
pyrophysalite (pirōfis'əlit), n. a coarse, nearly opaque variety of topaz which swells on being heated.
pyroscope (pīr'əskōp), n. an instrument for measuring the intensity of radiant heat.
pyrosis (pirō'sis), n. heartburn, water-brash.
pyrosome (pī'rəsōm), n. an animal of the genus *Pyrosoma*, consisting of highly phosphorescent compound ascidians united in free-swimming cylindrical colonies, mostly belonging to tropical seas.
pyrotechnic (pirətek'nik), a. pertaining to fireworks or their manufacture; of the nature of fireworks; (fig.) resembling a firework show, brilliant, dazzling. **pyrotechnics**, n. sing. the art of making fireworks; (sing. or pl.) a display of fireworks; (sing. or pl.) a dazzling or virtuoso display. **pyrotechnical**, a. **pyrotechnically**, adv. **pyrotechnist**, n. **pyrotechny** (pī'rətekni), n.
pyrotic (pirot'ik), a. caustic. n. a caustic substance.
pyroxene (pī'roksēn, -rok'-), n. a name used for a group of silicates of lime, magnesium or manganese, of various forms and origin. **pyroxenic** (-sen'-), a.
pyroxyle (pirok'sil), **pyroxylin** (-lin), n. any explosive, including gun-cotton, obtained by immersing vegetable fibre in nitric or nitrosulphuric acid, and then drying it.
pyroxylic (-sil'-), a. denoting the crude spirit obtained by the distillation of wood in closed vessels.
Pyrrhic (pi'rik), a. of or pertaining to Pyrrhus. **Pyrrhic victory**, n. a victory that is as costly as a defeat, like that of Pyrrhus, king of Epirus, over the Romans at Asculum (279 BC)
pyrrhic (pi'rik), n. a metrical foot of two short syllables. a. consisting of two short syllables.
Pyrrhonism (pi'rənizm), n. the sceptical philosophy taught by Pyrrho of Ellis, the Greek sceptical philosopher of the 4th cent. BC; universal doubt, philosophic nescience. **Pyrrhonian** (-rō'-), **Pyrrhonic** (-ron'-), a. **Pyrrhonist**, n.
Pyrus (pī'rəs), n. a genus of Rosaceae comprising the apple and pear. **Pyrus Japonica** (jəpon'ikə), n. a small tree or shrub of this genus bearing bright scarlet flowers.
Pythagoras's theorem (pithag'ərəs), n. the theorem that the square on the hypotenuse of a right-angled triangle is equal to the sum of the squares on the other two sides.
Pythagorean (-rē'ən), n. a follower of Pythagoras of Samos (6th cent. BC), philosopher and mathematician. a. pertaining to Pythagoras or his philosophy. **Pythagoreanism, Pythagorism**, n.
Pythian (pith'iən), a. pertaining to Delphi, to Apollo, or to his priestess who delivered oracles at Delphi. n. Apollo or his priestess at Delphi. **Pythic**, a. **Pythian** or **Pythic games**, n.pl. one of the four great Panhellenic festivals, celebrated once every four years near Delphi.
pythogenic (pithəjen'ik), a. produced by filth or putrid matter. **pythogenesis** (-əsis), n. generation from or through filth.
python¹ (pī'thən), n. a large non-venomous serpent that crushes its prey.
python² (pī'thən), n. a familiar spirit or demon; one possessed by this, a soothsayer, a diviner. **pythoness** (-nis), n. a woman possessed by a familiar spirit or having the gift of prophecy, a witch; applied esp. to the priestess of the temple of Apollo at Delphi who delivered the oracles. **pythonic** (-thon'-), a. inspired, oracular, prophetic. **pythonism**, n.
pyuria (piū'riə), n. the presence of pus in the urine.

pyx (piks), *n.* (*Eccles.*) the covered vessel, usu. of precious metal, in which the host is kept; a box at the Royal Mint in which sample coins are placed for testing at the annual trial by a jury of the Goldsmiths' Company. *v.t.* to test (a coin) by weighing and assaying.

pyxidium (piksid'iəm), *n.* (*pl.* **-dia** (diə)) (*Bot.*) a capsule or seed-vessel dehiscing by a transverse suture, as in the pimpernel.

pyxis (pik'sis), *n.* a box, a casket; (*Bot.*) a pyxidium.

pzazz (pizaz'), PIZZAZZ.

Q

Q, q, the 17th letter and the 13th consonant (*pl.* **Ques, Q's, Qs**), is normally followed by *u*, the combination *qu* having the sound of *kw.* **to mind one's p's and q's** P[1]. **Q-boat,** *n.* an armed vessel disguised as a merchantman, employed to lure and surprise hostile submarines. **Q factor,** *n.* the difference between stored energy and the rate at which energy is being expended; the heat released in a nuclear explosion. [*Quality factor*] **Q fever,** *n.* an acute fever whose symptoms include fever and pneumonia. [*Query fever*, the cause being originally unknown] **Q-ship,** *n.* a ship with concealed guns which acts as a decoy. [*Query -ship*]

Q., (*abbr.*) heat; queen; question.

q., (*abbr.*) quart; quarter; quarterly; query; quire; question.

QANTAS (kwon'təs), (*acronym*) Queensland and Northern Territory Air Services, the national airline of Australia.

QARANC, (*abbr.*) Queen Alexandra's Royal Nursing Corps.

Qawwali (kawal'i), *n.* the traditional devotional music of the Sufis.

QB, (*abbr.*) Queen's Bench.

QC, (*abbr.*) Queen's Counsel.

QED, (*abbr.*) quod erat demonstrandum. [L, which was to be proved]

qindar (kindah'), *n.* (*pl.* **-darka** (-kə)) a monetary unit of Albania, one-hundredth of a lek.

QM, (*abbr.*) quartermaster.

QSO, (*abbr.*) quasi-stellar object, a quasar.

qt, (*abbr.*) quantity; quart, quarts; quiet. **on the qt,** (*coll.*) secretly, on the sly.

qua (kwā, kwah), *conj.* in the character of, by virtue of being, as. [L]

quack[1] (kwak), *v.i.* to make a harsh cry like that of a duck; to chatter loudly, to brag. *n.* the cry of a duck; a noisy outcry.

quack[2] (kwak), *n.* a mere pretender to knowledge or skill, esp. one in medicine offering pretentious remedies and nostrums; an ignorant practitioner, an empiric, a charlatan; (*coll.*) a doctor. *a.* pertaining to quacks or quackery. **quackery, quackism,** *n.* **quackish,** *a.*

quacksalver (kwak'salver), *n.* one who brags of his medicines or salves; a quack.

quad[1] (kwod), *n.* a quadrangle or court, as of a college etc.

quad[2] (kwod), *n.* (*Print.*) a quadrat. *v.t.* to insert quadrats (in a line of type).

quad[3] (kwod), *n.* (*coll.*) short for QUADRUPLET.

quad[4] (kwod), *a.* short for QUADRAPHONIC.

quadr- *comb.form.* QUADR(I)-.

quadragenarian (kwodrəjənēə'riən), *a.* 40 years old. *n.* one 40 years old.

Quadragesima (kwodrəjes'imə), *n.* the first Sunday in Lent, also called Quadragesima Sunday. **quadragesimal,** *a.* pertaining to or used in Lent, Lenten.

quadrangle (kwod'rang-gl), *n.* a plane figure having four angles and four sides, esp. a square or rectangle; an open square or four-sided court surrounded by buildings; such a court together with the surrounding buildings. **quadrangular** (-rang-gū-), *a.* **quadrangularly,** *adv.*

quadrant (kwod'rənt), *n.* the fourth part of the circumference of a circle, an arc of 90°; a plane figure contained by two radii of a circle at right angles to each other and the arc between them; a quarter of a sphere; any of the four parts into which a plane is divided by two axes at right angles to one another; an instrument shaped like a quarter-circle graduated for taking angular measurements; (*Naut.*) such an instrument formerly used for taking the altitude of the sun, now superseded by the sextant. **quadrantal** (-ran'-), *a.*

quadraphonics, quadrophonics (kwodrəfon'iks), *n.sing.* a system of recording and reproducing sound using four independent sound signals or speakers. **quadraphonic,** *a.* **quadraphony** (-raf'-), *n.*

quadrat (kwod'rət), *n.* a block of type-metal lower than the type, used for spacing out lines etc.; a square of vegetation taped off for intensive study.

quadrate (kwod'rət), *a.* square, rectangular; of the quadrate bone. *n.* the quadrate bone; a square, cubical or rectangular object. *v.t., v.i.* (-rāt', kwod'-), to make square; to agree, to match, to correspond. **quadrate bone,** *n.* in birds and reptiles, a bone by means of which the jaws are articulated with the skull. **quadratic** (rat'-), *a.* involving the second and no higher power of the variable or unknown quantity. *n.* a quadratic equation; (*pl.*) the part of algebra dealing with quadratic equations. **quadratrix** (-rā'triks), *n.* (*pl.* **-trices** -trisēz) a curve by means of which straight lines can be found equal to the circumference of circles or other curves and their several parts. **quadrature** (-chə), *n.* the act of squaring or finding a square equal in area to a given curved figure; the position of a heavenly body with respect to another 90° distant; (*Elec.*) the position when there is a phase difference of one quarter of a cycle or 90° between two waves.

quadrella (kwodrel'ə), *n.* (*Austral.*) a form of betting where the person making the bet must pick the winners of four races.

quadrennial (kwodren'iəl), *a.* comprising or lasting four years; recurring every four years. **quadrennially,** *adv.* **quadrennium** (-əm), *n.* a period of four years.

quadr(i)-, *comb.form.* four.

quadric (kwod'rik), *a.* of the second degree; quadratic. *n.* a quantic, curve or surface of the second degree.

quadricentennial (kwodrisenten'iəl), *n.* the 400th anniversary of an event. *a.* pertaining to a period of 400 years.

quadriceps (kwod'riseps), *n.* a four-headed muscle acting as extensor to the leg. **quadricipital,** *a.*

quadridentate (kwodriden'tāt), *a.* having four indentations or serrations.

quadrifid (kwod'rifid), *a.* cleft into four parts, segments or lobes.

quadrifoliate (kwodrifō'liət), *a.* four-leaved; having four leaflets.

quadriga (kwodrē'gə), *n.* (*pl.* **-gae** -jē) an ancient Roman two-wheeled chariot drawn by four horses abreast.

quadrilateral (kwodrilat'ərəl), *a.* having four sides and four angles. *n.* a quadrilateral figure or area. **the Quadrilateral,** the district in N Italy defended by the fortresses of Mantua, Verona, Peschiera and Legnano. **quadrilaterality, quadrilateralness,** *n.*

quadrilingual (kwodriling'gwəl), *a.* speaking or written in four languages.

quadrille (kwədril'), *n.* a dance consisting of five figures executed by four sets of couples; a piece of music for

such a dance; a game of cards played by four persons with 40 cards, fashionable in the 18th cent. *v.i.* to dance a quadrille; to play music for a quadrille.
quadrillion (kwədril'yən), *n.* in Britain, the number produced by raising a million to its fourth power, represented by 1 followed by 24 ciphers; in N America, the fifth power of 1000, 1 followed by 15 ciphers.
quadrilobate (kwodrilō'bāt), *a.* having four lobes.
quadrilocular (kwodrilok'ūlə), *a.* having four cells or chambers.
quadrinomial (kwodrinō'miəl), *a.* consisting of four terms. *n.* a quantity consisting of four algebraic terms.
quadripartite (kwodripah'tīt), *a.* divided into or consisting of four parts; affecting or shared by four parties. **quadripartitely**, *adv.* **quadripartition** (-tish'ən), *n.* division by four or into four parts.
quadriplegia (kwodriplēj'iə), *n.* paralysis of all four limbs. **quadriplegic**, *n., a.*
quadrisyllabic (kwodrisilab'ik), *a.* consisting of four syllables. **quadrisyllable** (-sil'-), *n.*
quadrivalent (kwodrivā'lənt), *a.* having a valency or combining power of four. **quadrivalency, -valence,** *n.*
quadrivium (kwodriv'iəm), *n.* in the Middle Ages, an educational course consisting of arithmetic, music, geometry, and astronomy. **quadrivial,** *a.* having four ways meeting in a point; pertaining to the quadrivium.
quadroon (kwədroon'), *n.* the offspring of a mulatto and a white; a person of quarter Negro and three-quarters white blood; applied to similarly proportioned hybrids in human, animal and vegetable stocks.
quadrophonics QUADRAPHONICS.
Quadrumana (kwodroo'mənə), *n.pl.* an order of mammals in which the hind as well as the fore feet have an opposable digit and are used as hands, containing the monkeys, apes, baboons and lemurs. **quadrumane** (kwod'rəmān), *n.* **quadrumanous,** *a.*
quadruped (kwod'rəped), *n.* a four-footed animal, esp. a mammal. *a.* having four legs and feet. **quadrupedal** (-roo'pidəl), *a.*
quadruple (kwod'rupl, -roo'-), *a.* fourfold; consisting of four parts; involving four members, units etc.; multiplied by four; equal to four times the number or quantity of. *n.* a number or quantity four times as great as another; four times as much as many. *v.i.* to become fourfold as much, to increase fourfold. *v.t.* to make four times as much, to multiply fourfold. **quadruplet** (kwod'ruplit, -roo'-), *n.* a compound or combination of four things working together; one of four children born of the same mother at one birth; four notes to be played in a time value of three. **quadruplex** (kwod'rupleks), *a.* fourfold; used four times over (of a telegraphic wire). *n.* an electrical apparatus by means of which four messages may be sent simultaneously over one telegraphic wire. *v.t.* to arrange (a wire etc.) for quadruplex working. **quadruplicate** (-roo'plikət), *a.* fourfold; four times as many or as much; four times copied. *n.* one of four copies or similar things; quadruplicity. *v.t.* (-kāt), to make fourfold, to quadruple. **quadruplication,** *n.* **quadruplicity** (-plis'-), *n.* **quadruply,** *adv.*
quaere (kwiə'rē), *v. imper.* ask, inquire, it is a question. *n.* a question, a query. [L]
quaestor (kwēs'tə), *n.* in ancient Rome, a magistrate having charge of public funds, a public treasurer, paymaster etc. **quaestorial** (-taw'ri), *a.* **quaestorship,** *n.*
quaff (kwof), *v.t.* to drink in large draughts. *v.i.* to drink copiously. *n.* a copious draught. **quaffer,** *n.*
quag (kwag), *n.* a piece of marshy or boggy ground.
quagmire, *n.* a quaking bog, a marsh, a slough; a difficult or awkward situation. **quaggy,** *a.* **quagginess,** *n.*
quagga (kwag'ə), *n.* a S African quadruped, *Equus quagga*, intermediate between the ass and the zebra, now extinct; Burchell's zebra, *Equus burchellii.*

quaggy, quagmire QUAG.
quahaug (kwəhawg'), **quahog** (-hog), *n.* the common round or hard clam, *Venus mercenaria*, of the Atlantic coast of N America.
quaich, quaigh (kwäkh), *n.* (*Sc.*) a shallow drinking-vessel, usu. of wood.
Quai d'Orsay (käydawsä'), *n.* term for the French Foreign Office, from its location on the Quai d'Orsay, on the S bank of the Seine in Paris.
quail[1] (kwāl), *v.i.* to shrink, to be cowed, to lose heart; to give way (before or to).
quail[2] (kwāl), *n.* a small migratory bird of the genus *Coturnix*, allied to the partridge, esp. *C. coturnix;* one of various alllied gallinaceous birds. **quail-hawk,** *n.* the New Zealand sparrow-hawk.
quaint (kwānt), *a.* old-fashioned and pleasantly odd; pleasing by virtue of strangeness, oddity or fancifulness; odd, whimsical, singular. **quaintish,** *a.* **quaintly,** *adv.* **quaintness,** *n.*
quake (kwāk), *v.i.* to shake, to tremble; to quiver, to rock, to vibrate. *n.* a tremulous motion, a shudder; (*coll.*) an earthquake. **quaking,** *a.* trembling; unstable. **quaking ash,** *n.* the aspen. **quaking-grass,** *n.* grass of the genus *Briza*, the spikelets of which have a tremulous motion. **quaky,** *a.* **quakiness,** *n.*
Quaker (kwākə), *n.* a member of the Christian sect of the Society of Friends, founded by George Fox (1624–91). *a.* pertaining to Quakers or Quakerism. **quaker-bird,** *n.* the sooty albatross. **quaker-gun,** *n.* a wooden gun mounted to deceive the enemy. **Quakerdom** (-dəm), *n.* **Quakeress** (-ris), *n. fem.* **Quakerish,** *a.* **Quakerism,** *n.* **Quakerly,** *a.* like a Quaker.
qualify (kwol'ifī), *v.t.* to invest or furnish with the requisite qualities; to make competent, fit, or legally capable (to be or do, or for any action, place, office or occupation); to modify, to limit, to narrow the scope, force etc. of (a statement, opinion or word); to moderate, to mitigate, to temper; to reduce the strength of flavour of (spirit etc.) with water, to dilute; to attribute a quality to, to describe or characterize as. *v.i.* to be or become qualified or fit; to make oneself competent, suitable or eligible (for). **qualifiable,** *a.* **qualification** (-fi-), *n.* the act of qualifying or the state of being qualified; modification, restriction or limitation of meaning, exception or partial negation restricting completeness or absoluteness; any natural or acquired quality fitting a person or thing (for an office, employment etc.); a condition that must be fulfilled for the exercise of a privilege etc. **qualificative, -tory,** *n., a.* **qualified,** *a.* **qualifier,** *n.* **qualifying,** *n., a.* **qualifying round,** *n.* a preliminary round in a competition. **qualifyingly,** *adv.*
qualitative QUALITY.
quality (kwol'iti), *n.* relative nature or kind, distinguishing character; a distinctive property or attribute, that which gives individuality; a mental or moral trait or characteristic; particular efficacy, degree of excellence, relative goodness; high social status; the affirmative or negative nature of a proposition; that which distinguishes sounds of the same pitch and intensity, timbre; quality newspaper. *a.* of high quality or grade, excellent, superior; designed to appeal to an educated reader or viewer. **the quality,** (*formerly*) persons of high rank, the upper classes. **quality circle,** *n.* a periodic meeting of employees and management to discuss problems, complaints etc. with a view to improving performance. **quality control,** *n.* the testing of manufactured products to ensure they are up to standard. **quality time,** *n.* (leisure) time spent constructively in improving oneself or one's lifestyle, family relationships etc. **qualitative** (-tə-), *a.* of or pertaining to quality, opp. to *quantitative.* **qualitative analysis,** *n.* the detection of the constituents of a compound body. **qualitatively,** *adv.* **qualitied,** *a.*

qualm (kwahm), *n.* a sensation of nausea, a feeling of sickness; a sensation of fear or uneasiness; a misgiving, a scruple, compunction. **qualmish**, *a.* **qualmishly**, *adv.* **qualmishness**, *n.* **qualmy**, *a.*

qualy *n.* (kwol'i), a method of measuring the quality of length of the life given to a patient and medical treatment, used to assess the cost-effectiveness of treatment, and to compare different, expensive, treatments. [acronym for *quality-adjusted life year*]

quamash (kwom'ash, -mash'), *n.* the bulb of a liliaceous plant, *Camassia esculenta*, eaten by various N American peoples.

quandary (kwon'dəri), *n.* a state of difficulty or perplexity; an awkward predicament, a dilemma.

quandong (kwan'dong), *n.* a small Australian tree, *Fusanus acuminatus*, with edible drupaceous fruit.

quango (kwang'gō), *n.* a board set up by central government to supervise activity in a specific field, e.g. the Race Relations Board. [acronym for *q*uasi-*a*utonomous *n*on-*g*overnmental *o*rganization]

quant (kwont), *n.* a punting-pole with a flange at the end to prevent its sinking in the mud. *v.t.* to propel with this. *v.i.* to propel a boat with this.

quanta (kwon'tə), *pl.* QUANTUM.

quantic (kwon'tik), *n.* a rational integral homogeneous function of two or more variables. **quantical**, *a.*

quantify (kwon'tifī), *v.t.* to determine the quantity of, to measure as to quantity; to express the quantity of; to define the application of as regards quantity. **quantifiable**, *a.* **quantification** (-fī), *n.* **quantifier**, *n.* that which indicates quantity.

quantitative (kwon'titātiv), **quantitive**, *a.* pertaining to or concerned with quantity, opp. to *qualitative*; esp. of verse, relating to or based on the quantity of vowels. **quantitative analysis**, the determination of the amounts and proportions of the constituents of a compound body. **quantitatively, quantitively**, *adv.*

quantity (kwon'titi), *n.* that property in virtue of which anything may be measured; extent, measure, size, greatness, volume, amount or number; a sum, a number; a certain or a large number, amount or portion; *(pl.)* large quantities, abundance; the duration of a syllable; the extent to which a predicate is asserted of the subject of a proposition; a thing having such relations, of number or extension, as can be expressed by symbols, a symbol representing this. **unknown quantity**, a person, thing or number whose importance or value is unknown. **quantity-surveyor**, *n.* one employed to estimate the quantities of materials used in erecting a building.

quantize, -ise (kwon'tīz), *v.t.* to restrict or limit to a set of fixed values; to express in terms of quantum theory. **quantization, -isation**, *n.*

quantum (kwon'təm), *n.* (*pl.* -ta) a quantity, an amount; a portion, a proportion, a share; an amount required, allowed or sufficient; the smallest indivisible quantity by which some physical properties (as energy) can change or be transferred. **quantum jump, leap**, *n.* (*coll.*) a sudden transition; unexpected and spectacular progress. **quantum mechanics**, *n.sing.* a branch of mechanics based on quantum theory, applied to elementary particles and atoms which do not behave according to Newtonian mechanics. **quantum mechanical**, *a.* **quantum mechanically**, *adv.* **quantum number**, *n.* a set of integers or half-integers which serve to describe the energy states of a particle or system of particles. **quantum theory**, *n.* the theory that energy transferences occur in bursts of a minimum quantity.

quaquaversal (kwahkwəvœ'səl), *a.* pointing in every direction; of dip, inclined outwards and downwards in all directions. **quaquaversally**, *adv.*

quarantine (kwo'rəntēn), *n.* the prescribed period of isolation imposed on persons, animals or ships coming from places infected with contagious disease; the enforced isolation of such persons, animals, ships, goods etc. or of persons or houses so infected; a place where quarantine is enforced; any period of enforced isolation. *v.t.* to isolate or put in quarantine. **quarantine flag**, *n.* the yellow flag flown from ships to show infectious disease aboard.

quarant' ore (kwa'rəntaw'rā), *n.* in the Roman Catholic Church, 40 hours' exposition of the blessed Sacrament. [It., 40 hours]

quarenden (kwo'rəndən), **-der** (-də), *n.* a large red variety of apple, esp. grown in Devon and Somerset.

quark (kwahk), *n.* any of several hypothetical particles thought to be the fundamental units of other subatomic particles. [a word coined by James Joyce in *Finnegan's Wake*, 1939]

quarrel[1] (kwo'rəl), *n.* a short, heavy bolt or arrow with a square head, formerly used for shooting from cross-bows or arbalests; a square or diamond-shaped pane of glass used in lattice-windows.

quarrel[2] (kwo'rəl), *n.* a falling-out or breach of friendship; a noisy or violent contention or dispute, an altercation, a brawl, a petty fight; a ground or cause of complaint or dispute, a reason for strife or contention. *v.i.* (*past, p.p,* **quarrelled**) to fall out, to break off friendly relations (with); to dispute violently, to wrangle, to squabble; to cavil, to take exception, to find fault (with). **quarreller**, *n.* **quarrelling**, *a.* **quarrelsome** (-səm), *a.* inclined or apt to quarrel, contentious; irascible, choleric, easily provoked. **quarrelsomely**, *adv.* **quarrelsomeness**, *n.*

quarrian, -rrion (kwo'riən), *n.* a cockatiel found in inland Australia.

quarry[1] (kwo'ri), *n.* a place whence building-stone, slates etc. are dug, cut, blasted etc.; a source from which information is extracted. *v.t.* to dig or take from or as from a quarry; to make a quarry in. *v.i.* to dig in or as in a quarry. **quarryman, -woman**, *n.* a person employed in a quarry. **quarrymaster**, *n.* the owner of a quarry. **quarriable**, *a.* **quarrier**, *n.*

quarry[2] (kwo'ri), *n.* orig., a part of the entrails etc. of a deer placed on a skin and given to the hounds; any animal pursued by hounds, hunters, a bird of prey etc.; game, prey; any object of pursuit.

quarry[3] (kwo'ri), *n.* a square or diamond-shaped pane of glass, a quarrel; a square stone or tile. *v.t.* to glaze with quarries; to pave with quarries. **quarry-tile**, *n.* an unglazed floor tile.

quart[1] (kwawt), *n.* a measure of capacity, the fourth part of a gallon, two pints (1·136 l); a measure, bottle or other vessel containing such quantity.

quart[2] (kaht), *n.* a sequence of four cards in piquet etc.

quartan (kwaw'tən), *a.* occurring or recurring every fourth day. *n.* quartan ague or fever. **quartan ague, fever**, one recurring every third or, inclusively, every fourth day.

quarte CARTE[2].

quarter (kwaw'tə), *n.* a fourth part, one of four equal parts; the fourth part of a year, three calendar months; the fourth part of a cwt. (28 lb., 12·7 kg); a grain measure of 8 bushels (2·91 hl); the fourth part of a pound weight; the fourth part of a dollar, 25 cents; one of four parts, each comprising a limb, into which the carcase of an animal or bird may be divided; *(pl.)* the similar parts into which the body of a criminal or traitor was formerly divided after execution; a haunch; one of the divisions of a shield when this is divided by horizontal and perpendicular lines meeting in the fesse point; either side of a ship between the main chains and the stern; the fourth part of a period of the moon; one of the four phases of increase or decrease of the moon's face during a lunation; a point of time 15 minutes before or after the hour; one of four periods into which a game is divided; one of the four chief points of the compass; one of the main divisions of the globe corresponding to this; a particular direction, region, or locality; place of origin or supply, source; a division of a town, esp. one assigned to or occupied by a particular class; (*usu. pl.*) allotted position, proper place or station, esp. for troops; (*pl.*) place of lodging or abode, esp. a station or encampment occupied by troops; exemption from death allowed in war to a surrendered enemy; mercy, clemency. *v.t.* to divide into four equal parts; to cut the body of (a traitor) into quarters; to bear or arrange (charges or coats of arms) quarterly on a shield etc., to add (other arms) to those of one's family, to divide (a shield) into quarters by vertical and horizontal lines; to put into quarters, to assign quarters to, to provide (esp. soldiers) with lodgings and food; of a hound, to range over (a field) in all directions. *v.i.* to be stationed or lodged; to range in search of game; of the wind, to blow on a ship's quarter. **at close quarters**, close at hand. **quarter of an hour**, a period of 15 minutes. **a bad quarter of an hour**, a short disagreeable experience. **quarter-back**, *n.* a player in American football who directs the attacking play of his team. **quarter-bell**, *n.* a bell sounding the quarter-hours. **quarter-binding**, *n.* leather or cloth on the back only of a book, with none at the corners. **quarter-bound**, *a.* **quarter-bred**, *a.* of horses or cattle, having one-fourth pure blood. **quarter-butt**, *n.* (*Billiards*) a long cue, shorter than a half-butt. **quarter-day**, *n.* the day beginning each quarter of the year (Lady Day, 25 Mar., Midsummer Day, 24 June, Michaelmas Day, 29 Sept. and Christmas Day, 25 Dec.) on which tenancies etc. begin and end, payments are due etc. **quarter-deck**, *n.* the upper deck extending from the stern to the mainmast, usu. assigned for the use of officers and cabin passengers. **quarter-final**, *n.* the round before the semi-final, in a knockout competition. **quarter-finalist**, *n.* **quarter-horse**, *n.* (*N Am.*) a horse capable of running short distances at great speed. **quarter-hour**, *n.* a quarter of an hour; the point of time 15, 30 or 45 minutes before or after the hour. **quarter-hourly**, *adv.* **quarter light**, *n.* the small window in the front door of a car, often for ventilation. **quartermaster**, *n.* a regimental officer appointed to provide and assign quarters, lay out camps, and issue rations, clothing, ammunition etc.; a petty officer, having charge of the steering, signals, stowage etc. **quartermaster-general**, *n.* a staff-officer in charge of the department dealing with quartering, encamping, moving, or embarking troops. **quartermaster-sergeant**, *n.* a sergeant assisting the quartermaster. **quarter-miler**, *n.* an athlete who specializes in the quarter-mile race. **quarter note**, *n.* (*N Am.*) a crotchet. **quarter-plate**, *n.* a photographic plate measuring 4¼ × 3¼ in. (10·8 × 8·3 cm); a picture produced from this. **quarter-sessions**, *n.pl.* a general court of limited criminal and civil jurisdiction formerly held by the Justices of the Peace in every county (and in boroughs where there is a Recorder). **quarterstaff**, *n.* an iron-shod pole about 6½ ft. (2 m) long, formerly used as a weapon of offence or defence, usu. grasped by one hand in the middle and by the other between the middle and one end. **quarter-tone**, *n.* an interval of half a semitone. **quarterage** (-rij'), *n.* a quarterly payment, wages, allowance etc. **quartered**, *a.* **quartering**, *n.* a dividing into quarters or fourth parts; the assignment of quarters or lodgings; the grouping of several coats of arms on a shield; one of the coats so quartered. **quarterly**, *a.* containing a quarter; occurring or done every quarter of a year. *adv.* once in each quarter of the year; in quarters, arranged in the four quarters of the shield. *n.* a periodical published every quarter.

quartern (kwaw'tən), *n.* a quarter or fourth part of various measures, esp. of a pint, peck or pound. **quartern-loaf**, *n.* a loaf of the weight of 4 lb. (1·8 kg).

quartet, quartette (kwawtet'), *n.* a musical composition for four voices or four instruments; a group or set of four similar things.

quartic (kwaw'tik), *a.* pertaining to the fourth degree; *n.* a curve of the fourth degree.

quartile (kwaw'tīl), *a.* denoting the aspect of two heavenly bodies when distant from each other a quarter of a circle or 90°. *n.* a quartile aspect; a quarter of the individuals studied in a statistical survey, whose characteristics lie within stated limits; one of the three values of a variable that divides its distribution into four such groups.

quarto (kwaw'tō), *n.* (*pl.* **-tos**) a size obtained by folding a sheet twice, making four leaves or eight pages (usu. written 4to); a book, pamphlet etc., having pages of this size. *a.* having the sheet folded into four leaves.

quartz (kwawts), *n.* a mineral consisting of pure silica or silicon dioxide, either massive or crystallizing hexagonally. **quartz clock**, *n.* a synchronous electric clock of high accuracy in which the alternating current frequency is determined by the mechanical resonance of a quartz crystal. The mechanical strain of the crystal is translated into an electrical signal by the piezoelectric effect. **quartz crystal**, *n.* a piece of piezoelectric quartz cut and ground so that it vibrates at a natural frequency. **quartz glass**, *n.* glass made of almost pure silica, transparent to ultraviolet radiation and resistant to high temperatures. **quartz (iodine) lamp**, *n.* a light source, based on iodine vapour, used for high-intensity lighting in car-lamps and cine projectors. **quartziferous** (-if'-), *a.* **quartzite** (-it), *n.* a massive or schistose metamorphic rock consisting of sandstone with a deposition of quartz about each grain. **quartzitic** (-it'-), *a.* **quartzose** (-ōs), **quartzy**, *a.*

quas KVASS.

quasar (kwā'sah), *n.* any of a group of unusually bright, star-like objects outside our galaxy, with large red-shifts. They are a powerful source of radio waves and other energy sources. [from *quasi*-stell*ar* radio source]

quash (kwosh), *v.t.* to anul or make void; to put an end to, esp. by legal procedure; to suppress, to extinguish.

quasi (kwā'si), *conj.* as if. **quasi-**, *comb. form.* apparent, seeming, not real; practical, half, not quite, as in quasi-crime, quasi-historical, quasi-public, quasi-sovereign.

quasi-stellar object (kwā'sistel'ə), *n.* any of various classes of very distant celestial bodies, including quasars.

quassia (kwosh'ə), *n.* a genus of S American and W Indian (esp. Surinam) trees, the bitter wood, bark and oil of which yield a tonic. **quassic** (kwăs'-, kwos'-), *a.* [named by Linnaeus after *Quassi*, a W African slave who discovered its curative properties]

quater-centenary (kwatəsəntē'nəri, -ten'-), *n.* a 400th anniversary. **quater-centennial**, *n.*, *a.*

quaternary (kwətœ'nəri), *a.* consisting of four, having four parts, esp. composed of four elements or radicals arranged in fours; fourth in order; (**Quaternary**) applied to the post-Tertiary geological period or most recent strata or those above the Tertiary. *n.* a set of four; the number four; (**Quaternary**) the Quaternary geological period or system of rocks. **quaternity**, *n.* a set of four.

quaternion (kwətœ'niən), *n.* a set, group or system of four; a quire of four sheets once folded; (*Maths.*) an operator that changes one vector into another, so called as depending upon four irreducible geometrical elements; (*pl.*) the form of the calculus of vectors employing this. *v.t.* to divide into or arrange in quaternions, files or companies.

quatrain (kwot'rān), *n.* a stanza of four lines, usu. rhyming alternately.

quatrefoil (kat'rəfoil), *n.* an opening, panel or other figure in ornamental tracery, divided by cusps into four foils; a leaf or flower composed of four divisions or lobes.

Quattrocento (kwahtrōchen'tō), *n.* the 15th cent., regarded as a distinctive period in Italian art and literature. **quattrocentism**, *n.* **quattrocentist**, *n.*

quaver (kwā'və), *v.i.* to quiver, to tremble, to vibrate; to sing or play with tremulous modulations or trills. *v.t.* to sing or utter with a tremulous sound. *n.* a shake or rapid vibration of the voice, a trill; a quiver or shakiness in speaking; a note equal in duration to half a crotchet or one-eighth of a semibreve. **quaverer**, *n*. **quavering**, *a*. **quaveringly**, *adv*. **quavery**, *a*.

quay[1] (kē), *n*. a landing-place or wharf, usu. of masonry and stretching along the side of or projecting into a harbour, for loading or unloading ships. *v.t.* to furnish with a quay or quays. **quayage** (-ij), *n*. a system of quays; a charge imposed for the use of a quay. **quayside**, *n*. the edge of a quay.

quean (kwēn), *n*. a slut, a hussy, a jade, a strumpet; (*Sc*.) a young or unmarried woman, a lass.

queasy (kwē'zi), *a*. sick at the stomach, affected with nausea; causing or tending to cause nausea; unsettling the stomach; easily nauseated; fastidious, squeamish. **queasily**, *adv*. **queasiness**, *n*.

Quebecker, Quebecer (kwibek'ə), *n*. a native or inhabitant of Quebec. **Québecois** (kābekwah'), *n*. a French-speaking inhabitant of Quebec.

quebracho (kibrah'chō), *n*. (*pl*. **-chos**) one of several N American trees producing a medicinal bark, used esp. in cases of fever.

Quechua (kech'wə), *n*. a member of any of various groups of S Am Indian peoples, including the Incas; their language. **Quechuan**, *a*., *n*.

queen (kwēn), *n*. the wife of a king; a queen-dowager; a female sovereign of a kingdom; a court-card bearing a conventional figure of a queen; the most powerful piece in chess; a queen-bee; a fully developed fertile female ant or termite; a woman of majestic presence; one masquerading as a sovereign or presiding at some festivity; a city, nation or other thing regarded as the supreme example of its class; a female cat; (*sl*., *derog*.) an effeminate male homosexual, often an aging one. *v.t.* to make (a woman) queen; (*Chess*) to make (a pawn) into a queen. *v.i.* to act the queen; to act in a superior or arrogant way; to become a queen. **Queen Anne is dead**, stale news. **Queen Anne's bounty** BOUNTY. **Queen Anne's lace**, the wild carrot. **Queen Anne's style**, the architectural style prevalent in the reign of Queen Anne (*c*. 1700–20), characterized by plain and unpretentious design with classic details; also applied to a style of decorative art typified by Chippendale furniture. **queen-bee**, *n*. a fully-developed female bee; a woman in a dominating position, socially or in business. **queen-cake**, *n*. a small, soft, usu. heart-shaped currant cake. **queen-consort** CONSORT[1]. **queen-dowager**, *n*. the widow of a king. **queen-mother**, *n*. a queen-dowager who is also the mother of the reigning sovereign. **queen olive**, *n*. a type of large, fleshy olive which can be used for pickling. **queen-post**, *n*. one of two suspending or supporting posts between the tie-beam and rafters in a roof. **Queen regent**, *n*. a queen who reigns as regent. **Queen regnant**, *n*. a reigning queen. **Queen's Bench** BENCH. **Queen's Counsel** COUNSEL. **Queen's English**, *n*. southern British English when taken as the standard. **Queen's flight**, a unit of the RAF reserved for the use of the royal family, established as King's Flight in 1936. **Queen's Guide, Scout**, *n*. a guide or scout who has passed the highest tests of proficiency and ability. **queen substance**, *n*. a secretion of the queen bee fed to worker bees to stop the development of their ovaries. **queen truss**, *n*. a truss in a roof, framed with queen posts. **queen's-ware**, *n*. glazed Wedgewood earthenware of a creamy colour. **queencraft**, *n*. **queendom** (-dəm), **queenhood** (-hud), **queenship**, *n*. **queenless** (-lis), **queenlike**, **queenly**, *a*. **queenliness**, *n*.

Queensberry Rules (kwēnz'bəri), *n.pl*. standard rules of boxing drawn up by the 8th Marquess of Queensberry in 1867.

Queensland nut (kwēnz'lənd), *n*. a proteaceous tree of Queensland and New South Wales; its edible nut.

queer (kwiə), *a*. strange, odd; singular, droll; eccentric, touched, slightly insane; curious, questionable, suspicious; out of sorts; (*coll*.) bad, worthless, counterfeit; (*sl*., *derog*.) homosexual. *n*. (*sl*., *derog*.) a homosexual. *v.t.* (*coll*.) to spoil, to put out of order. **in Queer Street**, (*coll*.) in trouble, esp. financial. **to queer one's pitch**, to spoil one's chances. **queer fish**, *n*. (*coll*.) a strange person. **queerish**, *a*. **queerly**, *adv*. **queerness**, *n*.

quell (kwel), *v.t.* to suppress, to put down, to subdue; to crush; to cause to subside; to calm, to allay, to quiet. **queller**, *n*.

quench (kwench), *v.t.* to extinguish, to put out, esp. with water; to cool (heat or a heated thing) with water; to allay, to slake; to suppress, to subdue. **quenchable**, *a*. **quencher**, *n*. one who or that which quenches; (*coll*.) a draught that allays thirst. **quenching**, *a*. **quenchless** (-lis), *a*. that cannot be quenched; inextinguishable. **quenchlessly**, *adv*. **quenchlessness**, *n*.

quenelle (kənel'), *n*. a ball of savoury paste made of meat or fish, usu. served as an entrée.

querimonious (kwerimō'niəs), *a*. complaining, querulous, discontented. **querimoniously**, *adv*. **querimoniousness**, *n*.

querist (kwiə'rist), *n*. one who asks questions, an inquirer.

quern (kwœn), *n*. a simple hand-mill for grinding corn, usu. consisting of two stones resting one on the other; a small hand-mill for grinding spices. **quernstone**, *n*.

querulous (kwe'rələs, -ū-), *a*. complaining; discontented, peevish, fretful; of the nature of complaint. **querulously**, *adv*. **querulousness**, *n*.

query (kwiə'ri), *n*. a question (often used absolutely as preface to a question); a point or objection to be answered; a mark of interrogation (?), question mark. *v.t.* to put a question; to express a doubt or question. *v.t.* to question, to call in question; to express doubt concerning; to mark with a query. **querying**, *n*., *a*. **queryingly**, *adv*.

quesadilla (kāsədē'yə), *n*. a tortilla filled, fried and topped with cheese.

quest (kwest] *n*. the act of seeking, a search; an expedition or venture in search or pursuit of some object, esp. in the days of chivalry; the object of such an enterprise; an official inquiry; a jury or inquest. *v.t.* to seek for or after. *v.i.* to make quest or search; to go (about) in search of something. **quester**, *n*. **questful**, *a*. **questing**, *a*. **questingly**, *adv*.

question (kwes'chən), *n*. the act of asking or inquiring, interrogation, inquiry; a sentence requiring an answer, an interrogative sentence; a subject for inquiry, a problem requiring solution; a subject under discussion; a proposition or subject to be debated and voted on, esp. in a deliberative assembly; a subject of dispute, a difference, doubt, uncertainty, objection. *v.t.* to ask a question or questions of, to interrogate, to examine by asking questions; to study (phenomena etc.) with a view to acquiring information; to call in question, to treat as doubtful or unreliable, to raise objections to. *v.i.* to ask a question or questions; to doubt, to be uncertain. **a burning question**, a subject causing intense interest. **beyond all, past question**, undoubtedly, unquestionably. **in question**, referred to, under discussion, **leading question**. LEAD[2]. **open question**, a question that remains in doubt or unsettled. **out of the question**, not worth discussing, impossible. **previous question** PREVIOUS. **to beg the question** BEG. **to call in question** CALL[1]. **to pop the question** POP[1]. **to put the question**, to put to the vote, to divide the meeting or House upon. **question-mark, -stop**, *n*. a mark of interrogation (?). **question master**, *n*. a person who puts questions, e.g. the person who asks the questions in a quiz or game. **question time**, *n*. time set aside each day in Parliament where ordinary members may question ministers. **questionable**, *a*. open to doubt or suspicion; disputable. **questionability** (-bil'-), **questionableness**, *n*. **questionably**, *adv*. **questioner**, **-ist**, *n*. **questioning**, *n*., *a*. **questioningly**, *adv*. **questionless**, (-lis), *adv*. beyond all ques-

quetzal — quinoline

tion or doubt. **questionnaire** (-neə'), *n*. a series of questions designed to collect information.
quetzal (kwet'səl), *n*. a brilliant Guatemalan trogon, *Pheromacrus mocinno*; the monetary unit of Guatemala.
queue (kū), *n*. a plaited tail hanging at the back of the head, a pigtail; a file of persons, vehicles etc. waiting their turn. *v.i.* to form into a waiting queue. **queue-jumping**, *n*. pushing into a queue ahead of people already waiting in it. **queue-jumper**, *n*.
quibble (kwib'l), *n*. an evasion of the point, an equivocation; a trivial or sophistical argument or objection, esp. one exploiting a verbal ambiguity. *v.i.* to evade the point in question; to employ quibbles. **quibbler**, *n*. **quibbling**, *a*. **quibblingly**, *adv*.
quiche (kēsh), *n*. a savoury pastry shell filled with egg custard, and usually cheese, bacon, onion or other vegetables.
quick (kwik), *a*. lively, vigorous, ready, alert; acutely sensitive or responsive, prompt to feel or act; intelligent; rapid in movement, acting swiftly; done or happening in a short time, speedy; quickset. *adv*. in a short space, at a rapid rate. *n*. living flesh, esp. the sensitive flesh under the nails; (*fig*.) the feelings, the seat of the feelings. **the quick and the dead**, the living and the dead. **to have a quick one**, (*coll*.) to have a quick (alcoholic) drink. **quick-change**, *a*. making rapid changes of costume or appearance (of actors etc.). **quick-change artist**, *n*. a performer who executes quick changes; someone who frequently changes their opinions. **quick-firer**, *n*. a gun with a mechanism for firing shots in rapid succession. **quick-firing**, *a*. **quick-freeze**, *n*. very rapid freezing to retain the natural qualities of food; a receptacle in which such food is kept frozen. **quick march**, *n*. a march in quick time; the music for such a march. *interj*. a command to begin such a march or to make haste. **quick step**, *n*. the step used in marching at quick time; a fast foxtrot. **quick-tempered**, *a*. easily irritated, irascible. **quickthorn**, *n*. the hawthorn, esp. when planted as a hedge. **quick time**, *n*. the ordinary rate of marching in the British Army, usu. reckoned at 128 paces of 33 in. to the minute or 4 miles an hour. **quick-trick**, *n*. a card that should win a trick during the opening rounds of play. **quick-witted**, *a*. having a keen and alert mind; having a ready wit. **quick-wittedly**, *adv*. **quick-wittedness**, *n*. **quickie** (-i), *n*. (*coll*.) something that is done rapidly; a swiftly consumed (alcoholic) drink; a swift act of sexual intercourse. **quickly**, *adv*. **quickness**, *n*.
quicken (kwik'ən), *v.t.* to give or restore life or animation to; to stimulate, to inspire, to kindle; to accelerate. *v.i.* to come to life; to move with increased rapidity; to be in that state of pregnancy in which the child gives signs of life; to give signs of life in the womb. **quickener**, *n*. **quickening**, *a*.
quicklime (kwik'līm), *n*. burned lime not yet slaked.
quicksand (kwik'sand), *n*. loose wet sand easily yielding to pressure and engulfing persons, animals etc.; a bed of such sand.
quickset (kwik'set), *a*. of a hedge, composed of living plants, esp. hawthorn bushes. *n*. slips of plants, esp. hawthorn, put in the ground to form (a quickset hedge).
quicksilver (kwik'silvə), *n*. mercury. *a*. mercurial, unpredictable. *v.t.* to coat with an amalgam of quicksilver and tin-foil. *a*. **quicksilvered**, *a*. **quicksilvering**, *n*. **quicksilverish**, *a*. **quicksilvery**, *a*.
quid[1] (kwid), *n*. a piece of tobacco for chewing.
quid[2] (kwid), *n*. (*pl.* **quid**) (*sl.*) a pound (sterling). **quids in** (*sl.*) in a profitable position.
quiddity (kwid'iti), *n*. the essence of a thing; a quibble. **quiddative, quidditative**, *a*.
quidnunc (kwid'nŭngk), *n*. a newsmonger, a gossip.
quid pro quo (kwid prō kwō), something in return (for something), an equivalent; the substitution of one thing for another, or a mistake or blunder consisting in this.

quiescent (kwies'ənt), *a*. at rest, still, inert, dormant; tranquil, calm; not sounded. **quiescence, -cy**, *n*. **quiescently**, *adv*.
quiet (kwī'ət), *a*. in a state of rest; calm, unruffled, placid, tranquil, peaceful, undisturbed; making no noise, silent, hushed; gentle, mild, peaceable; unobtrusive, not glaring or showy; not overt, private; retired, secluded. *n*. a state of rest or repose; freedom from disturbance, tranquillity; silence, stillness, peace, calmness; peace of mind, calm, patience, placidness. *v.t.* to bring to a state of rest; to soothe, to calm, to appease. *v.i.* to become quiet. **on the quiet**, secretly. **quieten**, *v.t., v.i.* to quiet. to make, or become, calm, quiet. **quieter**, *n*. **quietly**, *adv*. **quietness, quietude**, *n*.
Quietism (kwī'ətizm), *n*. a form of religious mysticism based on the withdrawal of the soul from external objects and fixing it upon the contemplation of God; a state of calmness and placidity. **Quietist**, *a*. quietistic. *n*. an adherent of Quietism. **quietistic** (-tis'-), *a*.
quietus (kwīē'təs), *n*. a final discharge or settlement; death.
quiff (kwif), *n*. a tuft of hair brushed back to curve up over the forehead.
quill (kwil), *n*. the hollow stem or barrel of a feather; one of the large strong feathers of a bird's wing or tail; a pen made from such a feather, a pen; also, a plectrum, toothpick, angler's float etc. made from this; a spine of a porcupine; a tube; or hollow stem on which weavers wind their thread, a bobbin, a spool; a musical pipe made from a hollow cane, reed etc.; a strip of cinnamon or cinchona bark rolled into a tube; a fluted fold. **quill-driver**, *n*. (*contemp*.) a writer, an author, a clerk. **quill-feather**, *n*. a large wing or tail feather. **quilled**, *a*. (*usu. in comb*., as *longquilled*). **quilling**, *n*. lace, tulle or ribbon, gathered into small round plaits resembling quills.
quilt (kwilt), *n*. a bed-cover or coverlet made by stitching one cloth over another with some soft warm material as padding between them, a counterpane. *v.t.* to pad or cover with padded material; to stitch together, esp. with crossing lines of stitching, (two pieces of cloth) with soft material between them; to stitch in crossing lines or ornamental figures, like the stitching in a quilt; to sew up, as in a quilt. **continental quilt**, a quilt or duvet stuffed with down. **quilted**, *a*. **quilter**, *n*. **quilting**, *n*. the process of making quilted work; material for making quilts; quilted work.
quin, *n*. one child (of quintuplets).
quinary (kwī'nəri), *a*. consisting of or arranged in fives.
quinate (-nət), *a*. composed of five leaflets (of a leaf).
quince (kwins), *n*. the hard, acid, yellowish fruit of a shrub or small tree, *Pyrus cydonia*, used in cookery for flavouring and for preserves etc.
quincentenary (kwinsəntē'nəri, -ten'-), QUINGENTENARY.
quincunx (kwin'kŭngks), *n*. an arrangement of five things in a square or rectangle, one at each corner and one in the middle, esp. such an arrangement of trees in a plantation. **quincuncial** (-kŭn'shəl), *a*. **quincuncially**, *adv*.
quinella (kwinel'ə), *n*. (*Austral*.) a form of betting where the person placing the bet must pick the first- and second-placed winners.
quingentenary (kwinjəntē'nəri, -ten'-), *n*. a 500th anniversary; its celebration.
quinine (kwinēn', kwin'-), *n*. a bitter alkaloid obtained from cinchona barks, used as a febrifuge, tonic etc.; sulphate of quinine (the form in which it is usually employed as a medicine). **quinic** (kwin'-), *a*. **quinicine** (kwin'isīn), *n*. a yellow resinous amorphous alkaloid compound obtained from quinidine or quinine. **quinidine** (kwin'idēn), *n*. an alkaloid, isomeric with quinine, contained in some cinchona barks.
quinol (kwin'ol), *n*. hydroquinone. **quinotic** (-not'-), *a*.
quinoline (kwin'əlēn), *n*. a colourless, pungent, liquid

compound, obtained by the dry distillation of bones, coal and various alkaloids, forming the basis of many dyes and medicinal compounds. **quinology**, etc.
quinone (kwin'ōn, -nōn'), *n.* (*Chem.*) a yellow crystalline compound, usu. produced by the oxidation of quinic acid; any of a series of similar compounds derived from the benzene hydrocarbons by the substitution of two oxygen atoms for two of hydrogen.
quinqu-, quinque-, quinqui-, comb.form. relating to five.
quinquagenarian (kwinkwəjineə'riən), *n.* a person 50 years old. *a.* 50 years old. **quinquagenary** (-jē'-), *a.* quinquagenarian. *n.* a quinquagenarian; a 50th anniversary.
quinquagesima (kwinkwəjes'imə), *n.* Quinquagesima Sunday. **Quinquagesima Sunday**, the Sunday next before Lent, about 50 days before Easter. **quinquagesimal**, *a.* pertaining to the number 50; pertaining to 50 days.
quinquenniad (kwinkwen'iad), *n.* a quinquennium. **quiquennium** (-əm), *n.* (*pl.* **-nia** -niə) a period of five years. **quinquennial**, *a.* recurring once in five years; lasting five years. **quinquennially**, *adv.*
quinquereme (kwin'kwirēm), *n.* a galley having five banks of rowers.
quinquina (kinkē'nə, kwinkwi'-), *n.* Peruvian bark, cinchona.
quinsy (kwin'zi), *n.* inflammatory sore throat, esp. with suppuration of one tonsil or of both. **quinsied**, *a.*
quint[1] (kwint), *n.* in piquet, a sequence of five cards of the same suit; a fifth; a stop giving tones a fifth above the normal. **quint major**, the cards from ten to ace. **quint minor**, those from seven to knave.
quint[2] (kwint), *n.* (*N Am., Can.*) a quin.
quinta (kwin'tə), *n.* a country-house or villa, in Portugal, Madeira and Spain.
quintain (kwin'tən), *n.* a post, or a figure or other object set up on a post, in the Middle Ages, to be tilted at, often fitted with a sandbag, sword or other weapon that swung round and struck a tilter who was too slow; the exercise of tilting at this.
quintal (kwin'təl), *n.* a weight of 100 or 112 lb. (45.36 or 50.8 kg); 100 kg or 220½ lb.
quintan (kwin'tən), *a.* recurring every fourth (or inclusively fifth) day. *n.* an intermittent fever or ague the paroxysms of which return every fourth day.
quinte (kĩt), *n.* the fifth of the thrusts or parries.
quintessence (kwintes'əns), *n.* the fifth, last or highest essence, apart from the four elements of earth, air, fire and water, forming the substance of the heavenly bodies and latent in all things; the pure and concentrated essence of any substance, a refined extract; the essential principle or pure embodiment (of a quality, class of things etc.). **quintessential** (-sen'shəl), *a.* **quintessentially**, *adv.*
quintet, quintette (kwin tet'), *n.* a musical composition for five voices or instruments; a party, set or group of five.
quintillion (kwintil'yən), *n.* the fifth power of a million, represented by 1 followed by 30 ciphers; (*F, N Am.*) the sixth power of a thousand, 1 followed by 18 ciphers. **quintillionth**, *n.*, *a.*
quintuple (kwin'tūpl, -tū'-), *a.* fivefold. *n.* a fivefold thing, group or amount. *v.t.* to multiply fivefold. *v.i.* to increase fivefold. **quintuplet** (-plit), *n.* a set of five things; (*pl.*) five children at a birth; five notes played in the time of four. **quintuplicate** (-tū'plikət), *a.* consisting of five things (parts) etc. *n.* a set of five; one of five similar things. *v.t.* (-kāt), to multiply by five. **quintuplication**, *n.*
quip (kwip), *n.* a sarcastic jest or sally; a witty retort; a smart saying. *v.t.* (*past, p.p.* **quipped**) to remark wittily or sarcastically. *v.i.* to make a quip. **quippish, quipsome** (-səm), *a.* **quipster** (-stə) *n.* someone who makes witty remarks.
quipu (kē'poo, kwip'oo), *n.* a contrivance of coloured threads and knots used by the ancient Peruvians in place of writing.

quire[1] (kwīə), *n.* 24 sheets of paper; orig. a set of four sheets of paper or parchment folded into 8 leaves, as in mediaeval manuscripts.
quire[2] (kwīə), CHOIR.
quirk (kwœk), *n.* an odd or idiosyncratic character trait; a mannerism; an unexpected twist or vagary; as *a quirk of fate*; a flourish in drawing, writing or music; an acute recess between the moulding proper and the fillet or soffit. **quirk-moulding**, *n.* **quirkish, quirksome** (-səm), *a.* **quirky**, *a.* **quirkiness**, *n.*
quirt (kwœt), *n.* a riding-whip with a short handle and a long, braided leather lash. *v.t.* to strike with a quirt.
quisling (kwiz'ling), *n.* a traitor; one who openly allies himself with his nation's enemy. [Vidkun *Quisling*, 1887–1945, Norwegian collaborator]
quit (kwit), *v.t.* (*past, p.p.* **quitted, quit**) to rid (oneself) of: to give up, to renounce, to abandon; to leave, to depart from; to cease, to desist from; to free, to liberate; to acquit, to behave, to conduct (one, them etc., usu. without 'self'); *v.i.* to leave, to depart; to give up, to stop doing something; to admit defeat. *a.* clear, absolved; rid (of). **quits**, even, left on even terms, so that neither has the advantage. **double or quits** DOUBLE[2]. **to be, to cry quits**, to declare things to be even, to agree not to go on with a contest, quarrel etc., to make it a draw. **quitclaim**, *n.* a renunciation of right or claim. *v.t.* to renounce claim or title (to). **quit-rent**, *n.* a rent (usu. small) paid by a freeholder or copyholder in discharge of other services. **quittance**, *n.* a discharge or release from a debt or obligation; a receipt, an acquittance. **quitter**, *n.* one who quits; a shirker, a coward.
quitch (kwich), *n.* couch-grass, *Triticum repens.* **quitch-grass**, *n.*
quite[1] (kwīt), *adv.* completely, entirely, altogether, to the fullest extent; somewhat, fairly; rather; certainly, definitely. *int.* (*also* **quite so**) certainly, decidedly (a form of affirmation). **quite something**, *n.* someone or something remarkable. **quite the thing**, quite proper or fashionable.
quittance QUIT.
quiver[1] (kwiv'ə), *n.* a portable case for arrows. **quivered**, *a.* **quiverful**, *n.* **to have one's quiver full**, to have many children.
quiver[2] (kwiv'ə), *v.i.* to tremble or be agitated with a rapid tremulous motion; to shake, to shiver. *v.t.* to cause (wings etc.) to quiver. *n.* a quivering motion. **quivering**, *a.* **quiveringly**, *adv.* **quiverish**, *a.* **quivery**, *a.*
qui vive (kē vēv), **on the qui vive**, on the look-out, alert, expectant.
quixotic (kwiksot'ik), *a.* extravagantly romantic, visionary; aiming at lofty but impracticable ideals. **quixotically**, *adv.* **quixotism -try** (kwik'-, -tri), *n.* **quixotize, -ise** (kwik'-), *v.t., v.i.*
quiz (kwiz), *n.* a test of knowledge; a radio or television game based on this; a quizzer; an odd-looking or eccentric person. *v.t.* (*past, p.p.* **quizzed**) to banter, to chaff, to make fun of; to look at in a mocking or offensively curious way; to question closely, interrogate; (*N Am.*) to test the knowledge of. *v.i.* to behave in a bantering or mocking way. **quizzable**, *a.* **quizzer**, *n.* one given to quizzing. **quizzical**, *a.* questioning; gently mocking. **quizzically**, *adv.*
quod (kwod), *n.* (*sl.*) prison, jail.
quod erat demonstrandum (kwod irat' demənstran'dəm), which was to be proved. [L]
quod erat faciendum (kwod irat' fāshien'dəm), which was to be done. [L]
quodlibet (kwod'libet), *n.* (*Mus.*) a fantasia, a medley; a scholastic discussion or argument; a knotty point, a subtlety.
quoin (koin), *n.* a large stone, brick etc. at the external angle of a wall, a corner-stone; the external angle of a building; an internal angle, a corner; a wedge-shaped

block of wood used by printers etc. for various purposes, as locking up type in a form, raising the level of a gun etc. *v.t.* to raise or secure with a quoin or wedge. **quoining**, *n.*
quoit (koit, kwoit), *n.* a flattish circular ring of iron, rubber etc. for throwing at a mark; (*pl.*) a game of throwing such rings.
quokka (kwok'ə), *n.* a variety of bandicoot with short ears.
quondam (kwon'dam), *a.* having formerly been, sometime, former.
Quonset hut® (kwon'sit), *n.* (*N Am.*) a hut similar to a Nissen hut.
Quorn (kwawn), *n.* a high-protein vegetable foodstuff, based on a tiny plant of the mushroom family.
quorum (kwaw'rəm), *n.* (*pl.* **-ums**) the minimum number of officers or members of a society, committee etc. that must be present to transact business. **quorate** (-rət), *a.* being or consisting of a quorum.
quota (kwō'tə), *n.* a proportional share, part, or contribution; a prescribed number, e.g. of students to be admitted to a given college at the beginning of each year.
quote (kwōt), *v.t.* to adduce or cite from (an author, book etc.); to repeat or copy out the words of (a passage in a book etc.); to name the current price of; to make or give an estimate of the cost of. *v.i.* to cite or adduce a passage (from); to estimate costs. *n.* a quotation; (*pl.*) quotation marks. **quote-unquote**, an expression used to show the beginning and end of a quotation. **quotable**, *a.* worth quoting. **quotability** (-bil'-), **quotableness**, *n.* **quotably**, *adv.* **quotation**, *n.* the act of quoting; a passage quoted; a price quoted or current; an estimate; (*Print.*) a quadrat for filling up blanks etc. **quotation marks**, *n.pl.* punctuation marks (in Eng. usu. double or single inverted commas) at the beginning and end of a passage quoted. **quoted**, *a.* **quoted company**, *n.* a company whose shares are quoted on the Stock Exchange. **quoter**, *n.* **quoteworthy**, *a.*
quoth (kwōth), *v.t.* (*1st and 3rd pers.*) said, spoke.
quotidian (kwətid'iən), *a.* daily; (*Path.*) recurring every day; (*fig.*) commonplace, everyday. *n.* a fever or ague of which the paroxysms return every day.
quotient (kwō'shənt), *n.* the result obtained by dividing one quantity by another; a ratio, usu. multiplied by a hundred, between e.g. a test score and a standard measurement. **quotiety** (-tī'əti), *n.* relative frequency.
quotum (kwō'təm), QUOTA.
quo vadis? (kwo vah'dis), whither goest thou? [L]
quo warranto (kwō woran'tō), *n.* (*Law*) a writ requiring a person or body to show the authority by which some office or franchise is claimed or exercised. [med. L, by what warrant?]
Qurán, Qur'an (kərahn'), KORAN.
qv (*abbr.*) quod vide, which see (*imp.*), an instruction to look up a cross-reference. [L]
qwerty (kwœ'ti), *n.* the standard English typewriter or keyboard layout.

R

R, r, the 18th letter, and the 14th consonant of the English alphabet (*pl.* **Ars, R's** or **Rs**), has two sounds: the first when it precedes a vowel, as in *ran, morose;* the second, at the end of syllables and when it is followed by a consonant, as in *her, martyr, heard.* **the three Rs,** reading, writing and arithmetic, the fundamental elements of primary education.
Ra, (*chem. symbol*) radium.
rabbet (rab′it), *v.t.* to cut a groove or slot along the edge of (a board) so that it may receive the edge of another piece cut to fit it; to unite or fix in this way. *n.* a groove or slot made in the edge of a board that it may join with another; a joint so made; a rabbet-plane. **rabbet-plane,** *n.* a plane for cutting rabbets.
rabbi (rab′i), *n.* (*pl.* **-bbis**) a Jewish doctor or teacher of the law, esp. one ordained and having certain juridical and ritual functions. **rabbinate** (-nət), *n.* the office of rabbi; rabbis collectively. **rabbinic** (-bin′-), *n.* the language or dialect of the rabbins, later Hebrew. *a.* rabbinical. **rabbinical,** *a.* pertaining to the rabbins, their opinions, learning or language. **rabbinically,** *adv.* **rabbinism,** *n.* **rabbinist,** *n.* **rabbinistic** (-nis′-), *a.* **rabbinite** (-nit), *n.* a person who follows the traditions of the rabbis and the Talmud.
rabbit (rab′it), *n.* a burrowing rodent, *Lepus cuniculus,* allied to the hare, killed for its flesh and fur; (*sl.*) a bungling player at an outdoor game. *v.i.* to hunt rabbits; (*often with* **on**) to talk at length, often aimlessly. **rabbit fever,** *n.* tularaemia. **rabbit-hutch,** *n.* a cage for rearing tame rabbits in. **rabbit punch,** *n.* a sharp blow to the back of the neck that can cause unconsciousness or death. **rabbit-warren, rabbitry,** *n.* a piece of ground where rabbits are allowed to live and breed. **rabbiter,** *n.* **rabbity,** *a.*
rabble[1] (rab′l), *n.* a noisy crowd of people, a mob; the common people, the lower orders. **rabble-rouser,** *n.* someone who stirs up the common people, who manipulates mass anger or violence; a demagogue. **rabble-rousing,** *n., a.* **rabblement,** *n.*
rabble[2] (rab′l), *n.* an iron tool used for stirring molten metal.
Rabelaisian (rabəlā′ziən), *a.* of, pertaining to or characteristic of the French satirical humorist François Rabelais (1483–1553); extravagant, grotesque, coarsely and boisterously satirical. *n.* a student or admirer of Rabelais. **Rabelaisianism,** *n.*
rabi (rūb′i), *n.* the grain crop reaped in the spring, the chief of the three Indian crops.
rabic RABIES.
rabid (rab′id), *a.* furious, violent; fanatical, headstrong, excessively zealous or enthusiastic; affected with rabies. **rabidity** (bid′-), **rabidness,** *n.* **rabidly,** *adv.*
rabies (rā′biz), *n.* a disease of the nervous system arising from the bite of a rabid animal, characterized by hydrophobia. **rabic** (rab′-), **rabietic** (-et′-), **rabific** (-bif′-), *a.*
RAC, (*abbr.*) Royal Armoured Corps; Royal Automobile Club.
raccoon (rəkoon′), *n.* a furry ring-tailed N American carnivore of the genus *Procyon,* allied to the bears, esp. *P. lotor.*
race[1] (rās), *n.* a rapid movement, a swift rush; a rapid current of water, esp. in the sea or a tidal river; a channel of a stream, esp. an artificial one; a contest of speed between horses, runners, ships, motor-vehicles etc.; (*fig.*) any competitive contest depending chiefly on speed; a course or career; a channel or groove along which a piece of mechanism, as a shuttle, glides to and fro; (*Austral.*) a fenced passage in a sheep-fold; (*pl.*) a series of racing contests for horses. *v.i.* to run or move swiftly; to go at full speed; to go at a violent pace owing to diminished resistance (as a propeller when lifted out of the water); to contend in speed or in a race (with); to attend races. *v.t.* to cause to contend in a race; to contend against in speed; to cause (a horse) to run in a race; to cause (an engine) to run at high speed. **race-card,** *n.* a programme of a race-meeting with particulars of the horses, prizes etc. **race-course, -track,** *n.* a piece of ground on which horse-races are run; a mill-race. **race-goer,** *n.* someone who frequently goes to race-meetings. **race-going,** *n.* **raceground,** *n.* a racecourse. **race-horse,** *n.* a blood-horse bred for racing. **race-meeting,** *n.* a meeting for horse-racing. **raceway,** *a.* (*N Am.*) a channel or passage for water, as a mill-race; (*Mach.*) a groove for the passage of a shuttle etc.; (*Elec.*) a conduit or subway for wires or a cable. **racer,** *n.* one who races or contends in a race; a race-horse; a yacht, cycle, motor-car etc. built for racing. **racing,** *n.* **racing-car,** *n.* a car specially built to go at high speeds in competition. **racy,** *a.* RACY.
race[2] (rās), *n.* a group or division of persons, animals or plants sprung from a common stock; a particular ethnic stock; a subdivision of this, a tribe, nation or group of peoples, distinguished by less important differences; a clan, a family, a house; a genus, species, stock, strain or variety, of plants or animals, persisting through several generations; (*fig.*) lineage, pedigree, descent; a class of persons or animals differentiated from others by some common characteristic. **race-hatred,** *n.* hatred of other people on grounds of race. **race relations,** *n.pl.* the relations between people of different races within a single community; the study of such relations. **race riot,** *n.* a riot caused by a feeling of being discriminated against on grounds of race. **racial** (-shəl), *a.* pertaining to race or lineage. **racially,** *adv.* **racism, racialism,** *n.* antagonism between different races; a tendency towards this; a belief in the superiority of one race over another; discrimination based on this belief. **racist, racialist,** *n.*
raceme (rəsēm′), *n.* a centripetal inflorescence in which the flowers are attached separately by nearly equal stalks along a common axis. **racemate** (-āt), *n.* a racemic compound. **racemed,** *a.* **racemic,** *a.* pertaining to or obtained from grape-juice; composed of equal amounts of dextrorotatory and laevorotatory isomers. **racemiferous** (rasəmif′-), *a.* **racemism** (ras′ə-), *n.* the quality of being racemic. **racemize, -ise** (ras′ə-), *v.t., v.i.* to change into a racemic form. **racemization,** *n.* **racemose** (ras′əmōs), **-mous,** *a.*
racer, raceway etc. RACE[1].
rachi-, rachio-, *comb. form.* pertaining to the spine.
rachis (rak′is), *n.* (*pl.* **-ides** (-idēz)) the axis of an inflorescence; the axis of a pinnate leaf or frond; the spinal column; the shaft of a feather, esp. the part bearing the barbs.
rachitis (rəki′tis), *n.* rickets. **rachitic** (-kit′-), *a.*

Rachmanism (rakh'mənizm), *n.* the conduct of an unscrupulous landlord who exploits his tenants and charges extortionate rents for slum property.
racial RACE².
racily etc. RACY.
rack¹ (rak), *v.t.* to stretch or strain, esp. on the rack; to torture, to cause intense pain or anguish to; to strain, tear, shake violently or injure; (*fig.*) to strain, to puzzle (one's brains etc.); to wrest, to exaggerate (a meaning etc.); to extort or exact (rent) in excess or to the utmost possible extent; to harass (tenants) by such exaction of rent. *n.* an apparatus for torture consisting of a framework on which the victim was laid, his wrists and ankles being tied to rollers which were turned so as to stretch him, to the extent sometimes of dislocating the joints. **on the rack**, under torture; under great stress. **to rack one's brains**, to use great mental effort. **rack-rent**, *n.* an exorbitant rent, approaching the value of the land. *v.t.* to extort such a rent from (a tenant, land etc.). **rack-renter**, *n.* a landlord extorting such a rent; a tenant paying it. **racking**¹, *a.*
rack² (rak), *n.* an open framework or set of rails, bars, woven wire etc. for placing articles on; a grating or framework of metal or wooden rails or bars for holding fodder for cattle etc.; a bar or rail with teeth or cogs for engaging with a gear-wheel, pinion or worm. *v.t.* to place on or in a rack. **rack and pinion**, *n.* a device for converting rotary motion into linear motion and vice versa, with a gearwheel which engages in a rack; a type of steering gear found in some vehicles. **rack-railway**, *n.* a railway (usu. on a steep incline) with a cogged rail between the bearing rails. **rack-wheel**, *n.* a cog-wheel.
rack³ (rak), *n.* light vapoury clouds, cloud-drift; (perh. var. of WRACK) destruction, wreck. *v.i.* to fly, as cloud or vapour before the wind. **to go to rack and ruin**, to fall completely into ruin.
rack⁴ (rak), ARRACK.
rack⁵ (rak), *v.t.* to draw off (wine etc.) from the lees.
racking-can, **-cock**, **-engine**, **-faucet**, **-pump**, *n.* kinds of vessel, tap, pump etc. used in racking off wine.
rack⁶ (rak), *n.* a horse's mode of going in which both hooves of one side are lifted from the ground almost or quite simultaneously, all four legs being off the ground entirely at times. *v.i.* to go in this manner (of a horse). **racker**, *n.* a horse that goes at a racking pace. **racking**², *a.*
racket¹, **racquet** (rak'it), *n.* a kind of bat, with a network of catgut instead of a blade, with which players at tennis, squash, badminton or rackets strike the ball; a snow-shoe resembling this; (*pl.*) a game of ball resembling tennis, played against a wall in a four-walled court. *v.t.* to strike with or as with a racket. **racket-court**, **-ground**, *n.* a four-walled court where rackets is played. **racket-press**, *n.* a press for keeping the strings of a racket taut. **racket tail**, *n.* a type of humming-bird which has two long, racket-shaped tail feathers. **racket-tailed**, *a.*
racket² (rak'it), *n.* a clamour, a confused noise, a din; a commotion, a disturbance, a fuss; (*sl.*) a scheme, a dodge, an underhand plan; an underhand combination; an organized illegal or unethical activity; (*sl.*) business; a mediaeval instrument of the woodwind family, with a deep bass pitch, like a bassoon. *v.i.* to make a noise or din; to frolic, to revel, to live a gay life, to knock about. **to stand the racket**, to stand the expenses, to pay the score; to put up with the consequences; to get through without mishap. **racketer**, **racketing**, *n.* confused, tumultuous mirth. **rackety**, *a.* **racketeer** (-tiə'), *n.* a member of a gang engaged in systematic blackmail, extortion or other illegal activities for profit. *v.t.* to operate an illegal business or enterprise for profit. **racketeering**, *n.*
racking¹,², RACK ¹,⁶.
racking-can etc. RACK ⁵.

rack-rent RACK ¹.
racon (rā'kon), *n.* a radar beacon. [acronym for *ra*dar bea*con*]
raconteur (rakôntœ'), *n.* a (good, skilful etc.) storyteller.
raconteuse (-tœz'), *n. fem.* **raconteuring**, *n.*
racoon RACCOON.
racquet RACKET ¹.
racy (rā'si), *a.* having a distinctive flavour; lively, pungent, piquant, spirited; (*coll.*) suggestive, bordering on the indecent, risqué. **racily**, *adv.* **raciness**, *n.*
rad ¹, (*abbr.*) radian; radical (in politics); radius.
rad ², (rad), *n.* a unit measuring the dosage of ionized radiation absorbed, equivalent to 100 ergs of energy per gram of mass of irradiated material. [*rad*iation]
RADA (rah'də), (*abbr.*) Royal Academy of Dramatic Art.
radar (rā'dah), *n.* the employment of reflected or retransmitted radio waves to locate the presence of objects and to determine their angular position and range; the equipment used for this. **radar beacon**, *n.* a fixed radio transmitter which sends out signals which allow an aircraft or ship to determine its own position. **radar gun**, *n.* a device like a gun, used by the police, which, when 'fired' at a moving car, uses radar to record the car's speed. **radarscope**, *n.* a cathode-ray oscilloscope capable of showing radar signals. **radar trap**, *n.* a device which uses radar to allow the police to identify vehicles exceeding the speed limit. [acronym for *ra*dio *d*etection *a*nd *r*anging]
raddle (rad'l), *n.* ruddle. *v.t.* to paint or colour with red ochre; to apply rouge (to the face) excessively or badly. **raddled**, *a.* dilapidated; unkempt; haggard-looking due to age or debauchery.
radial (rā'diəl), *a.* of, pertaining to or resembling a ray, rays or radii; extending or directed from a centre as rays or radii, divergent; having radiating parts, lines etc.; of or pertaining to the radius of the forearm. *n.* a radiating part, bone, nerve, artery etc.; a radial-ply tyre. **radial artery**, *n.* artery of the forearm, felt at the wrist when taking the pulse. **radial engine**, *n.* an internal-combustion engine which has its cylinders arranged radially. **radial-ply**, *a.* of a tyre having the fabric in the outer casing placed radially to the centre for increased flexibility. **radial symmetry**, *n.* the state of having several planes arranged symmetrically around a common axis. **radially symmetrical**, *a.* **radial velocity**, *n.* the component of velocity of an object along the line of sight between the observer and the object. **radiality** (-al'-), *n.* radial symmetry. **radialize**, **-ise**, *v.t.* to cause to radiate as from a centre. **radialization**, *n.* **radially**, *adv.* **radian** (-ən), *n.* an arc equal in length to the radius of its circle; the angle subtending such an arc, 57·296°.
radiant (rā'diənt), *a.* emitting rays of light or heat; issuing in rays; (*fig.*) shining, beaming (with joy, love etc.); splendid, brilliant; radiating, radiate. *n.* the point from which a star-shower seems to proceed; the point from which light or heat radiates; a straight line proceeding from a fixed pole about which it is conceived as revolving. **radiant energy**, *n.* energy given out in the form of electromagnetic waves. **radiant flux**, *n.* the rate at which radiant energy is emitted or transmitted. **radiant heat**, *n.* heat by radiation, employed therapeutically in rheumatism by the use of electric lamps. **radiance**, **-ancy**, *n.* **radiantly**, *adv.*
radiata pine (rādiah'tə), *n.* a pine tree grown in Australia and New Zealand for timber.
radiate (rā'diāt), *v.i.* to emit rays of light or heat; to send out rays from or as from a centre; to issue and proceed in rays from a central point. *v.t.* to send out rays or from a central point; to send forth in all directions; to show clearly. *a.* (-ət) having rays or parts diverging from a centre, radiating; radially arranged, marked etc., radially symmetrical. **radiately**, *adv.* **radiation**, *n.* the act of radiating or emitting rays; the transmission of heat, light etc.

in the form of electromagnetic waves, from one body to another without raising the temperature of the intervening medium; a travelling outwards, as radii, to the periphery; a group of rays of the same wave-length; the gamma rays emitted in nuclear decay. **radiation sickness,** *n.* illness caused by too great absorption of radiation in the body, whose symptoms include fatigue, nausea, vomiting, internal bleeding, loss of hair and teeth, and in extreme cases, leukaemia. **radiational,** *a.* **radiative,** *a.* **radiato-,** *comb. form.* **radiator,** *n.* that which radiates; a device charged with hot air, water, steam etc. for radiating heat in a building; a device for dissipating the heat absorbed by the cooling-water of an engine jacket; the part of an aerial which radiates electromagnetic waves. **radiatory,** *a.*
radical (rad'ikəl), *a.* pertaining to the root, source or origin; inherent, fundamental; going to the root, thoroughgoing, extreme; favouring or producing extreme changes in political, social etc. conditions; arising from or close to the root; of or pertaining to the root of a number or quantity. *n.* one promoting extreme measures or holding advanced views on either side of the political spectrum; a root; a quantity that is, or is expressed as, the root of another; the radical sign ($\sqrt{}$, $\sqrt[3]{}$ etc.); an element, atom or group of atoms forming the base of a compound and not decomposed by the reactions that normally alter the compound. **radical chic,** *n.* (*derog.*) superficial, dilettantish left-wing radicalism. **radical mastectomy** MASTECTOMY. **radical sign,** *n.* the symbol $\sqrt{}$ placed before a number to show that the square root, or some higher root as shown by a superscript number (e.g. $\sqrt[3]{}$), is to be calculated. **radicalism,** *n.* the principles of radical politics. **radicalistic** (-lis'-), *a.* **radicalistically,** *adv.* **radicality** (-kal'-), *n.* **radicalness,** *n.* **radicalize, -ise,** *v.t., v.i.* **radicalization,** *n.* **radically,** *adv.* thoroughly, fundamentally, essentially.
radicand (rad'ikand), *n.* a number from which a root is to be extracted, usually preceded by a radical sign, e.g. three is the radicand of $\sqrt{}$ 3.
radicchio (rədē'kiō), *n.* (*pl.* **-chios**) a type of chicory from Italy with purple and white leaves eaten raw in salads. [It.]
radices RADIX.
radicle (rad'ikl), *n.* the part of an embryo that develops into the primary root; a small root, a rootlet; a root-like part of a nerve, vein etc.; a radical. **radicular** (-dik'ū-), *a.*
radii RADIUS.
radio (rā'diō), *n.* electromagnetic waves used in two-way broadcasting; any device which can send signals through space using electromagnetic waves; a wireless receiving set; radio broadcasting; the programmes broadcast on the radio. *v.t., v.i.* to communicate by wireless.
radio-[1] (rā'diō), *comb. form.* pertaining to radio, radio frequency or broadcasting; pertaining to radiation or radioactivity.
radio-[2] (rā'diō), *comb. form.* radiate; pertaining to the outer bone of the forearm.
radioactive (rādiōak'tiv), *a.* having the property of emitting invisible rays that penetrate bodies opaque to light, affecting the electrometer, photographic plates etc. **radioactive decay,** *n.* the disintegration of a nucleus as a result of electron capture. **radioactive series,** *n.* a series of nuclides which each undergo radioactive decay, finally creating a stable element, usually lead. **radioactively,** *adv.* **radioactivity** (-tiv'-), *n.*
radio-astronomy (rādiōəstron'əmi), *n.* the study of radio waves received from celestial objects.
radio beacon (rā'diō), *n.* a transmitting station which sends out signals to aid navigators.
radiobiology (rādiōbiol'əji), *n.* the study of the effects of radiation on the body using radioactive tracers. **radiobiological** (-loj'-), *a.* **radiobiologically,** *adv.* **radiobiologist,** *n.*
radiocarbon (rādiōkah'bən), *n.* carbon-14, a radioactive carbon isotope. **radiocarbon dating,** *n.* a method of dating organic material by measuring the carbon-14 levels.
radiochemistry (rādiōkem'istri), *n.* the chemistry of radioactive elements.
radiocompass (rādiōkŭm'pəs), *n.* a device for navigation which can determine the direction of incoming radio waves from a beacon.
radio control (rā'diō), *n.* remote control using radio signals. **radio-controlled,** *a.*
radio data system (rādiō dātə sis'təm), *n.* a system of incorporating digital data, e.g. travel information, automatic tuning to the strongest signal, in ordinary FM transmissions in the form of an inaudible signal to which suitably equipped receivers can respond.
radioelement (rādiōel'əmənt), *n.* a chemical element with radioactive powers.
radio frequency (rādiō frē'kwənsi), *n.* frequency which is within the range for radio transmission. **radio-frequency amplifier,** a high-frequency amplifier.
radiogram (rā'diōgram), *n.* a radio and record player; an X-ray; a radiotelegraphic message.
radiograph (rā'diəgrahf), *n.* an actinograph; a negative produced by X-rays; a print from this. *v.t.* to obtain a negative of by means of such rays. **radiographer** (-og'-), *n.* one who takes X-ray pictures of parts of the body. **radiographic** (-graf'-), *a.* **radiographically,** *adv.* **radiography** (-og'-), *n.*
radio-immuno-assay (rādiōimūnōas'ā), *n.* an immunological assay which uses radioactive labelling of various levels, such as hormone levels.
radioisotope (rādiōī'sətōp), *n.* a radioactive isotope, produced in an atomic pile or in an atomic bomb explosion. **radioisotopic** (-top'-), *a.* **radioisotopically,** *adv.*
radiolarian (rādiōleə'riən), *n., a.* (a member) of a large order of marine protozoans with radiating filamentous pseudopodia.
radiology (rādiol'əji), *n.* the branch of medical science concerned with radioactivity, X-rays and other diagnostic or therapeutic radiations. **radiologic, radiological** (-loj'-), *a.* **radiologically,** *adv.* **radiologist,** *n.* a medical practitioner trained in radiology, such as one who interprets X-ray pictures of parts of the body.
radiometer (rādiom'itə), *n.* an instrument for illustrating the conversion of radiant light and heat into mechanical energy. **radiometric** (-met'-), *a.*
radiomimetic (rādiōmimet'ik), *a.* pertaining to a chemical or substance which affects living tissue in a similar way to ionizing radiation.
radio-opaque (rādiōōpāk'), *a.* not allowing X-rays or other radiation to pass through.
radiopaging (rā'diōpājing), *n.* a system for alerting a person, using a small device which emits a sound in response to a signal at a distance.
radiophonic (rādiōfon'ik), *a.* pertaining to music produced electronically. **radiophonics,** *n.* the art of producing music by electronic means. **radiophonically,** *adv.* **radiophony** (-of'əni), *n.*
radioscopy (rādios'kəpi), *n.* examination of bodies by means of X-rays.
radiosonde (rā'diōsond), *n.* a miniature radio transmitter sent up in a balloon and dropped by parachute, for sending information on pressures, temperatures etc.
radio source (rā'diō), *n.* any celestial object, e.g. a quasar, which emits radio waves.
radio spectrum (rā'diō), *n.* that range of electromagnetic frequencies, between 10 kHz and 300,000 MHz, used in radio transmissions.
radiotelegraphy (rādiōteleg'rəfi), *n.* telegraphy which transmits messages using radio waves. **radio telegraph,** *n.* **radiotelegraphic** (-graf'-), *a.*
radiotelephone (rādiōtel'ifōn), *n.* apparatus for sending telephone messages using radio waves. *v.t.* to telephone using radiotelephone. **radiotelephonic** (-fon'-), *a.* **radio-**

telephony (-lef'-), *n.*
radio telescope (rā'diō), *n.* an apparatus for collecting radio waves from outer space.
radioteletype (rādiōtel'itīp), *n.* a teleprinter which can transmit or receive messages using radio waves; a network of teleprinters used to communicate news and messages.
radiotherapy (rādiothe'rəpi), *n.* the treatment of disease by means of radiation; actinotherapy. **radiotherapist,** *n.*
radio wave (rā'diō), *n.* an electromagnetic wave of radio frequency.
radish (rad'ish), *n.* a cruciferous plant, *Raphanus sativus*, cultivated for its root, which is eaten as a salad.
radium (rā'diəm), *n.* a highly radioactive metallic element resembling barium, at. no. 88; chem. symbol Ra, obtained from pitchblende. **radium therapy,** *n.* treatment of disease, esp. cancer, using radiation from radium.
radius (rā'diəs), *n.* (*pl.* **-dii** (-diī)) the shorter of the two long bones of the forearm; the corresponding bone in animals and birds; the straight line from the centre of a circle or sphere to any point in the circumference; the length of this, half the diameter; a radiating line, part, object etc., as a spoke; a circular area measured by its radius; the outer zone of a composite flower; a floret in this; a branch of an umbel. **radius vector,** *n.* (*pl.* **radii vectores** (-taw'rēz)) the distance from a fixed point to a curve; a line drawn from the centre of a heavenly body to that of another revolving round it.
radix (rā'diks), *n.* (*pl.* **radices** (-disēz)) a quantity or symbol taken as the base of a system of enumeration, logarithms etc.; a source or origin; a root. **radix point,** *n.* any point which separates the integral part from the fractional part of a number, such as a decimal point.
radome (rā'dōm), *n.* a protective covering for radar antennae, through which radio waves can pass. [*radar, dome*]
radon (rā'don), *n.* a gaseous radioactive element emitted by radium, at. no. 86; chem. symbol Rn.
radula (rad'ūlə), *n.* (*pl.* **-lae** (-lē)) the odontophore or lingual ribbon of some molluscs. **radular,** *a.*
RAF (raf), (*abbr.*) Royal Air Force.
raffia, raphia (raf'iə), *n.* a Madagascar palm with a short stem and gigantic pinnate leaves; fibre prepared from these used for tying, ornamental work etc.
raffish (raf'ish), *a.* disreputable, dissipated-looking; tawdry, flashy. **raffishly,** *adv.* **raffishness,** *n.*
raffle (raf'l), *n.* a kind of lottery in which the prizes are goods rather than money. *v.t.* to dispose of by means of a raffle. **raffler,** *n.*
Rafflesia (rəflē'ziə), *n.* a genus of very large stemless parasitic plants from Java and Sumatra.
raft[1] (rahft), *n.* a number of logs, planks etc. fastened together for transport by floating; a flat floating framework of planks or other material used for supporting or carrying persons, goods etc.; a slab of thick concrete providing a foundation for a building on soft ground. *v.t.* to transport on or as on a raft; to fasten together as a raft. *v.i.* to travel on a raft; to work on a raft; **rafter**[2], **raftsman,** *n.* one who manages or works on a raft.
raft[2] (rahft), *n.* (*N Am., coll.*) a large number, a crowd, a lot.
rafter[1] (rahf'tə), *n.* a sloping piece of timber supporting a roof, or the framework on which the tiles etc. of the roof are laid.
rafter[2], **raftsman** RAFT[1].
rag[1] (rag), *n.* a fragment of cloth, esp. an irregular piece detached from a fabric by wear and tear; (*pl.*) tattered or shabby clothes; (*fig.*) a remnant, a scrap, the smallest piece (of anything); (*collect.*) torn fragments of cloth, linen etc., used as material for paper, stuffing etc.; a handkerchief; (*contemptuous*) a flag, sail, drop-curtain, a newspaper etc.; ragtime music. **ragamuffin** (-əmŭfin), *n.* a ragged, beggarly person. **ragamuffinly,** *adv.* **rag-and-bone man,** an itinerant dealer in household refuse. **ragbag,** *n.* a bag for scraps of cloth; (*coll.*) a carelessly-dressed woman. **rag-bolt,** *n.* a bolt with jags on the shank to prevent its being easily withdrawn. **rag-book,** *n.* a book for a child made out of cloth instead of paper. **rag-doll,** *n.* a doll made from cloth. **ragman**[1], **-woman,** *n.* one who collects or deals in rags. **rag-paper,** *n.* paper made from rags. **rag-picker,** *n.* someone who collects rags from rubbish bins etc. **ragtag or ragtag and bobtail,** the riff-raff, the rabble. **rag-time,** *n.* irregular syncopated time in music, played esp. on the piano. **rag-trade,** *n.* (*coll.*) the clothing industry. **rag-weed,** *n.* ragwort. **rag worm,** *n.* any of several burrowing marine worms used as bait in fishing. **ragwort,** *n.* a yellow-flowered plant of the genus *Senecio*. **ragged** (-id), *a.* rough, jagged, or uneven in outline or surface; disjointed, irregular; lacking in uniformity or cohesion; worn into rags, tattered; wearing tattered clothes; shabby, poor, miserable in appearance. **ragged robin,** *n.* a crimson-flowered plant, *Lychnis floscuculi*, the petals of which have a tattered appearance. **raggedly,** *adv.* **raggedness,** *n.* **raggedy,** *a.* tattered.
rag[2] (rag), *n.* a hard, coarse, rough stone, usu. breaking up into thick slabs; a large roofing-slate with a rough surface on one side (also called **ragstone**). **rag work,** *n.* thick slabs of masonry.
rag[3] (rag), *v.t.* (*past, p.p.* **ragged**) to tease, irritate or play rough practical jokes on. *n.* an act of ragging; in British universities, a series of light-hearted events, processions etc. staged to raise money for charity. **to lose one's rag,** (*sl.*) to lose one's temper. **rag-day, -week,** *n.* a day or week devoted to university rag. **ragging,** *n.*
raga (rah'gə), *n.* in traditional Hindi music, a form or a mode which forms the basis for improvisation; a composition composed following such a pattern.
ragamuffin, rag-bolt RAG[1].
rage (rāj), *n.* violent anger, fury; a fit of passionate anger; (*fig.*) extreme violence, vehemence or intensity (of); a violent desire or enthusiasm (for); intense emotion, passion or ardour; (*coll.*) an object of temporary pursuit, enthusiasm or devotion. *v.i.* to be furious with anger; to be violently incensed or agitated; to express anger or passion violently; to be violent, to be at the highest state of vehemence, intensity or activity. **all the rage,** an object of general desire, quite the fashion. **rageful,** *a.* **ragefully,** *adv.* **raging,** *a.* angry, furious; violent, intense. *n.* violence; fury. **ragingly,** *adv.*
ragged RAG[1].
raggle-taggle (ragltag'l), *a.* unkempt, untidy.
ragi (rah'gi), **raggee** (rag'i), **raggy,** *n.* an Indian foodgrain, *Eleusine coracana*.
Raglan (rag'lən), *n.* a loose overcoat with no seams on the shoulders, the sleeves going up to the neck. *a.* cut in this way. **raglan sleeve,** *n.* a sleeve which continues to the collar, with no seams on the shoulders.
ragman RAG[1].
ragout (ragoo'), *n.* a stewed dish of small pieces of meat and vegetables, highly seasoned. *v.t.* to make into a ragout.
ragstone RAG[2].
rag-time, ragwort RAG[1].
rah (rah), *int.* a short form of HURRAH.
Raï (rī), *n.* a form of dance music, originating in Algeria, which combines a traditional vocal line with a modern, pop-style backing.
raid (rād), *n.* a hostile or predatory incursion, a foray; a sudden invasion or attack, incursion, esp. of police or customhouse officers; an air raid. *v.i., v.t.* to make a raid (upon). **to raid the market,** artificially to upset stock market prices for future gain. **raider,** *n.*
rail[1] (rāl), *n.* a bar of wood or metal or series of such bars resting on posts or other supports, forming part of a fence, banisters etc.; one of a continuous line of iron or steel bars, resting on sleepers etc. laid on the ground,

rail

usu. forming one of a pair of such lines constituting the track of a railway or tramway; one of a similar pair of lines serving as track for part of a machine; the railway as a means of travel or transportation; a bar fixed on a wall on which to hang things; a horizontal structural support in a door. *v.t.* to enclose with rails; to furnish with rails; to send by rail. **to go off the rails**, to go awry; to go mad. **rail-car**, *n.* a motor-driven vehicle on railway lines. **rail card**, *n.* an identity card issued to certain people (e.g. pensioners and students) allowing the holder cheap rail fares. **railhead**, *n.* a terminus; the farthest point to which rails have been laid. **railman, -woman**, *n.* a railway worker. **railroad**, *n.* (*chiefly N Am.*) a railway. *v.t.* to force to take action or a decision hurriedly; to push through with undue haste. **railroader**, *n.* **railer**, *n.* one who makes or fits rails. **railing**, *n.* a fence made of wooden or other rails; materials for railings; the laying of rails. **railless** (-lis), *a.*
rail[2] (rāl), *v.i.* to use abusive or derisive language; to scoff (at or against). **railer**, *n.* one who rails or scoffs. **railing**, *n., a.* **railingly**, *adv.* **raillery**, *n.* good-humoured ridicule or pleasantry, banter.
rail[3] (rāl), *n.* a bird of the family Rallidae, esp. of the genus *Rallus*, comprising *R. aquaticus*, the water-rail, and the corncrake or landrail, *Crex pratensis*.
railer, railing etc. RAIL[1, 2].
raillery RAIL[2].
railway (rāl'wā), *n.* a track formed of rails of iron or steel along which trains and vehicles are driven, usu. by locomotives; a track laid with rails for the passage of heavy horse-vehicles, travelling-cranes, trucks etc.; a system of tracks, stations, rolling-stock and other apparatus worked by one company or organization. *v.i.* to make railways; to travel by rail. **railway-carriage**, *n.* a railway vehicle for passengers. **railway-crossing**, *n.* a crossing of two railway lines, or a road and a railway. **railwayman, -woman**, *n.* a railway worker.
raiment (rā'mənt), *n.* (*poet.*) dress, apparel, clothes.
rain (rān), *n.* the condensed moisture of the atmosphere falling in drops; a fall of such drops; a large quantity of anything falling quickly; (*pl.*) the rainy season in a tropical country. *v.i.* (*usu. impers.*); to fall in drops of water from the clouds to fall in showers like rain. *v.t.* to pour down (rain); to send down in showers like rain. **come rain, come shine**, whatever the weather, whatever the circumstances. **right as rain**, perfectly all right. **to be rained off, out**, to be cancelled or postponed due to rain or bad weather. **to rain cats and dogs** CAT[1]. **rain-bird**, *n.* one of various birds supposed to foretell rain, esp. the green woodpecker. **rainbow** (-bō), *n.* a luminous arc showing the prismatic colours, caused by the reflection, double refraction and dispersion of the sun's rays passing through raindrops. *a.* coloured like the rainbow; many-coloured; consisting of, or bringing together, (esp. minority) groups from different races or of different political views. **rainbow-coloured**, *a.* **rainbow-tinted**, *a.* **rainbow trout**, *n.* a brightly-coloured Californian trout, *Salmo irideus*. **rainbowy**, *a.* **raincheck**, *n.* (*esp. N Am.*) a ticket for a sports event which allows readmission on another day if rain stops play; the postponing of a decision. **to take a raincheck**, to postpone accepting an invitation till a later date. **rain-cloud**, *n.* a cloud producing rain, a nimbus. **raincoat**, *n.* a waterproof coat or cloak for wearing in wet weather; a mackintosh. **rain-doctor**, *n.* a wizard professing to cause rain by incantations. **raindrop**, *n.* a particle of rain. **rainfall**, *n.* the amount of rain which falls in a particular district in a given period; a shower of rain. **rain forest**, *n.* a dense tropical forest of mostly evergreen trees with a very heavy rainfall. **rain-gauge**, *n.* an instrument for measuring the amount of rain falling on a given surface. **rain-maker** RAIN-DOCTOR. **rainproof, -tight**, *a.* impervious to rain. *v.t.* to make something impervious to rain. **rain shadow**, *n.* the leeward side of hills, which has a relatively light rainfall compared to the windward side. **rainstorm**, *n.* a storm with very heavy rain. **rain-wash**, *n.* the movement of soil and stones effected by rain. **rainwater**, *n.* (pure) water that has fallen in the form of rain. **rainless** (-lis), *a.* **rainy**, *a.* characterized by much rain; showery, wet. **rainy day**, a time of misfortune or distress, esp. pecuniary need. **rainily**, *adv.* **raininess**, *n.*

Rais (rās), REIS[1].
raise (rāz), *v.t.* to cause to rise, to lift; to cause to stand up, to set upright; to restore to life; to arouse; to stir up (against, upon etc.); to erect, to build, to construct; to rear, to cause to grow, to breed; to produce, to create, to cause; to assemble (e.g. an army); to collect, to procure, to levy (money etc.); to bid more money at cards; to suggest (a point etc.); to advance, to promote; to make higher or nobler; to cause (a bump, blister etc.) to form; to increase the amount of; to come in sight of (land etc.); to establish radio links with; to multiply a number by itself a specified number of times; to bring to an end (a blockade, siege). *n.* (*N Am.*) a rise in salary. **to raise a hand to**, to hit. **to raise Cain** CAIN. **to raise cloth**, to put a nap on cloth. **to raise hell**, (*coll.*) to make a lot of trouble. **to raise money on**, to sell or pawn something to make money. **to raise one's eyebrows**, to look surprised or bemused; to look disapproving. **to raise one's eyes**, to look upwards (to). **to raise one's glass (to)**, to drink a toast (to). **to raise the hat to** HAT. **to raise the wind** WIND. **raised beach** BEACH. **raiser**, *n.* **raising agent**, *n.* a natural or chemical substance which causes dough or cakes to rise.
raisin (rā'zin), *n.* a dried grape, the partially dried fruit of various species of vine.
raison d'être (rāzôn detr''), *n.* the reason for a thing's existence. [F]
raisonné (rā'zənā), *a.* arranged systematically (of a catalogue). [F]
raj (rahj), *n.* the British rule of India before 1947.
Rajah, -ja (rah'jə), *n.* an Indian king, prince or tribal chief, a title of a dignitary or noble; a Malayan or Javanese chief. **Rajahship**, *n.*
Rajpoot, -put (rahj'put), *n.* one of an Indian warrior caste who claim descent from the Kshatriyas; one of a Hindu aristocratic class.
rake[1] (rāk), *n.* an implement having a long handle with a cross-bar set with teeth, used for drawing loose material together, smoothing soil etc.; any similarly shaped implement, e.g. as used by a croupier for gathering in money or chips; a two-wheeled implement for gathering hay together etc. *v.t.* to collect or gather (up or together) with a rake; to scrape, scratch, smooth, clean etc. (soil) with a rake; to search through; to sweep (a ship, deck, line of soldiers etc.) from end to end with gunfire; to scan along the length of; to scrape or graze. *v.i.* to use or work with a rake; to search (about etc.) with or as with a rake. **to rake in**, (*sl.*) to accumulate, usu. money. **to rake off**, (*coll.*) to receive a share of the profits from an illegal job. **to rake up**, to collect together; to revive something, such as a quarrel, the past etc. **rake-hell**, (*Hist.*) RAKE[2]. **rake-off**, *n.* (*sl.*) commission on a job; more or less illicit profits from a job. **raker**, *n.* **raking**, *n.*
rake[2] (rāk), *n.* a dissolute or immoral man, a debauchee, a libertine. **rakery** (-əri), *n.* dissoluteness. **rakish**, *a.* **rakishly**, *adv.* **rakishness**, *n.*
rake[3] (rāk), *n.* inclination, slope, esp. backward slope; projection of the stem or stern of a vessel beyond the extremities of the keel; the slope of the stage, an auditorium. *v.i.* to slope backwards from the perpendicular. *v.t.* to give such an inclination to. **raker**, *n.* a sloping shore or support. **rakish**, *a.* with masts sharply inclined and apparently built for speed; dashing, jaunty; smart-looking but slightly piratical. **rakishly**[2], *adv.*
raki, -kee (rahkē', rak'i), *n.* an aromatic liquor made from

spirit or grape-juice, usu. flavoured with mastic, used in the E Mediterranean region.

rakish etc. RAKE [2,3].

râle (rahl), *n.* a rattling sound in addition to that of respiration, heard with the stethoscope in lungs affected by disease.

rallentando (raləntan'dō), *adv.* (*Mus.*) gradually slower.

rallier RALLY [2].

rally [1] (ral'i), *v.t.* to reunite, to bring (disordered troops) together again; to restore, to reanimate, to revive; to organize for a common purpose. *v.i.* to reassemble, to come together again after a reverse or rout; to regain strength, to return to a state of health, vigour or courage. *n.* the act of rallying or recovering order, strength, health, energy etc.; an assembly, a reunion; (*Lawn tennis etc.*) rapid exchange of strokes; a sharp increase in trade on the Stock Exchange after a period of decline; a motor race, usu. over public roads, which tests the driver's skill or the quality of the vehicle. **to rally round**, to come to (someone's) aid morally or financially. **to rally to**, to join together to support. **rally-cross**, *n.* a motor race in which specially adapted saloon cars race over a course with a rough, uneven surface. **rallying-point**, *n.* a spot or moment for making a rally.

rally [2] (ral'i), *v.t.* to banter, to chaff. *n.* banter, raillery.

rallier, *n.* **rallyingly**, *adv.*

RAM (ram), *n.* a temporary storage space in a computer which is lost when the computer is switched off. [acronym for *r*andom-*a*ccess *m*emory]

ram (ram), *n.* an uncastrated male sheep; a battering-ram; a beak of steel at the bow of a warship for cutting into a hostile vessel; the drop-weight of a pile-driver or steamhammer; a hydraulic engine for raising water, lifting etc.; a rammer; the compressing piston of a hydrostatic press; the plunger of a force-pump; a spar for driving planks etc. by impact. *v.t.* (*past, p.p.* **rammed**) to beat, drive, press or force (down, in, into etc.) by heavy blows; to stuff, to compress, to force (into) with pressure; to make firm by ramming; to strike (a ship) with a ram; to drive or impel (a thing against, into etc.) with violence. **the Ram**, the constellation or zodiacal sign Aries. **to ram something down someone's throat**, to force someone to accept an idea, for example, by arguing aggressively or forcefully. **ram-jet, ram-jet engine**, *n.* a form of aeroengine where the compressed air produced by the forward movement of the aircraft is used to burn the fuel; an aircraft powered by such an engine. **ramrod**, *n.* a rod for forcing down the charge of a muzzle-loading gun. **rammer**, *n.* one who or that which rams; an instrument for pounding, driving etc.; a ramrod.

Ramadan (ram'ədan, -dahn'), *n.* the ninth and holiest month of the Islamic year, the time of the great annual fast.

Raman effect (rah'mən), *n.* the change in wavelength which light undergoes when it passes through a transparent medium. **Raman spectroscopy**, *n.* the study of the properties of molecules using the Raman effect. [Sir Chandrasekhara *Raman*, 1888–1970, Indian physicist]

ramble (ram'bl), *v.i.* to walk or move about freely or aimlessly, to rove; to wander to be incoherent in speech, writing etc. *n.* a roaming about; a walk for pleasure or without a definite object, a stroll. **rambler**, *n.* one who rambles about; especially a person who takes long walks in the countryside; a variety of climbing-rose. **rambling**, *a.* wandering about; desultory, disconnected, irregular, straggling. **ramblingly**, *adv.*

Ramboesque (rambœsk'), Rambo *a.* pertaining to the fictional film character Rambo, looking or behaving like him, i.e. with mindless violence and aggression. **Ramboism** (ram'-), *n.* [*Rambo, First Blood II*, film released in Britain in 1985]

rambunctious (rambŭngk'shəs), *a.* unruly, boisterous, exuberant. **rambunctiously**, *adv.*

rambutan (rambooʹtən), *n.* the red, hairy, pulpy fruit of a Malaysian tree, *Nephelium lappaceum*.

RAMC, (*abbr.*) Royal Army Medical Corps.

ramekin (ram'ikin), *n.* a dish of cheese, eggs, breadcrumbs etc., baked in a small dish or mould; the mould itself.

ramie (ram'i), *n.* a bushy Chinese and E Indian plant, *Boehmeria nivea*, of the nettle family; the fine fibre of this woven as a substitute for cotton.

ramify (ram'ifī), *v.i.* to divide into branches or subdivisions, to branch out, to send out offshoots; to develop a usually complicated consequence. *v.t.* to cause to divide into branches etc. **ramification** (-fi-), *n.* the act of ramifying; a subdivision in a complex system, structure etc.; the production of figures like branches; the arrangement of branches; a consequence.

rammer RAM [1].

ramose (rā'mōs, rəmōs'), **ramous** (rā'məs), *a.* branching; branched; full of branches. **ramosely, -mously**, *adv.* **ramosity** (-mos'-), *n.*

ramp [1] (ramp), *v.i.* to dash about wildly; to act in a violent or aggressive manner; to ascend or descend to another level (of a wall). *v.t.* to build or provide with ramps. *n.* a slope or inclined plane or way leading from one level to another; a moveable stairway for boarding a plane; a hump in the road designed to slow traffic down; a difference in level between the abutments of a rampart arch; a sloping part in the top of a hand-rail, wall, coping etc.; the act of ramping; a vulgar, badly-behaved woman. **to ramp and rage**, to act in a violent manner.

ramp [2] (ramp), *v.t.* (*sl.*) to force (one) to pay a bet etc.; to swindle. *n.* a swindle, con. one involving exorbitant price increases.

rampage (rampāj'), *v.i.* to dash about, to storm, to rage, to behave violently or boisterously. *n.* (ram'-), boisterous, excited or violent behaviour. **on the rampage**, violently excited; on a drunken spree. **rampageous** (-pā'-), *a.* **rampageously**, *adv.* **rampageousness**, *n.* **rampager**, *n.*

rampant (ram'pənt), *a.* of a heraldic lion, standing upright on the hind legs; unrestrained, aggressive, wild, violent; rank, luxuriant (of weeds etc.); springing from different levels (of an arch). **rampancy**, *n.* **rampantly**, *adv.*

rampart [1] (ram'paht), *n.* an embankment, usu. surmounted by a parapet, round a fortified place, or such an embankment together with the parapet; a defence. *v.t.* to fortify or defend with or as with a rampart.

rampion (ram'piən), *n.* a bell-flower, *Campanula rapunculus*, with red, purple or blue blossoms.

ramrod etc. RAM [1].

ramshackle (ram'shakl), *a.* shaky, tumble-down, rickety. **ramshackly**, *a.* in bad repair.

ramson (ram'zən, -sən), *n.* (*usu. in pl.*) the broad-leaved garlic, *Allium ursinum*, or its bulbous root, eaten as a relish.

ran (ran), (*past*) RUN.

Rana (rā'nə), *n.* a genus of batrachians comprising the frogs and toads.

ranch (rahnch), **rancho** (-ō), *n.* (*N Am.*) a farm for rearing livestock esp. cattle and horses; a house belonging to such a farm; any large farm devoted to a particular crop. *v.t.* to manage or work on a ranch. **ranchman, rancher, ranchero** (-cheə'rō), *n.*

rancherie (rahn'chəri), *n.* any settlement of N American Indians in British Columbia, Canada.

rancid (ran'sid), *a.* having the taste or smell of stale oil or fat; rank; stale. **rancidify** (ransid'ifī), *v.t., v.i.* to become rancid. **rancidness**, *n.* **rancidly**, *adv.*

rancour (rang'kə), *n.* inveterate spite, resentment or enmity, malignancy, deep-seated malice. **rancorous**, *a.* **rancorously**, *adv.*

rand (rand), *n.* orig. a border, edge or margin; a strip of leather between the sole and heel-piece of a boot or shoe;

a thin inner sole; (*S Afr.*) the highlands bordering a river-valley; the standard monetary unit of S Africa. **the Rand**, abbr. for Witwatersrand, the gold and diamond country in S Africa of which Johannesburg is the centre.
R and B, (*abbr.*) rhythm and blues.
R and D, (*abbr.*) research and development.
random (ran'dəm), *a.* done, made etc. without calculation or method; left to chance; of a statistical value which cannot be determined, only defined in terms of probability. **at random**, at haphazard; without direction or definite purpose. **random access**, *n.* direct access to specific data in a larger store of computer data. **random shot**, *n.* a shot discharged without direct aim; *orig.* a shot fired at the extreme range attainable by elevating the muzzle of a gun. **randomize, -ise**, *v.t.* to set up (e.g. a survey) in a deliberately random way to make any results more viable statistically. **randomization**, *n.* **randomizer**, *n.* **randomly**, *adv.* **randomwise**, *adv.*
R and R, (*abbr.*) rock and roll.
randy (ran'di), *a.* sexually aroused or eager, lustful; (*chiefly Sc.*) disorderly, reckless. **randily**, *adv.* **randiness**, *n.*
ranee, -ni (rah'ni), *n.* a Hindu queen; the consort of a rajah.
rang (rang), (*past*) RING [2].
rangatira (rŭng·gatiə'rə), *n.* a Maori chief of either sex.
range (rānj), *v.t.* to set in a row or rows; to arrange in definite order, place, company etc., to classify; to rank; to roam or pass over, along or through; to set a gun to fire a certain distance. *v.i.* to lie, extend or reach; to vary (from one specified point to another); to roam, to wander, to rove, to sail (along etc.); of a gun, to have a specified range. *n.* a row, rank, line, chain or series; a stretch, a tract, esp. of grazing or hunting-ground; the area, extent, scope, compass or sphere of power, activity, variation, voice-pitch etc.; a variety, a cross-section; a number of things forming a distinct class or series, esp. the products of a manufacturer or dealer; the extreme horizontal distance attainable by a gun; the distance between a weapon and its target; a piece of ground with targets etc. for firing practice; the distance a vehicle, ship, plane etc. can travel without refuelling; a cooking-stove or fireplace, usu. containing a boiler, oven or ovens, hotplate etc.; the set of values of a dependent variable in statistics. **to range oneself**, to align, classify oneself (with). **range-finder**, *n.* an instrument for measuring the distance of an object from the observer, used in shooting a gun or focusing a camera. **ranger**, *n.* one who ranges, a rover, a wanderer; the superintendent of a royal forest or park; (*chiefly N Am.*) a person who patrols a national park or forest; a Girl Guide of 16 and upwards; (*N Am.*) a commando in the US army; (*N Am.*) a member of a body of armed troops used to police a district or State. **rangership**, *n.* **ranging rod**, *n.* a usually red and white striped rod used in surveying. **rangy**, *a.* tall, wiry, strong; (*Austral.*) mountainous. **rangily**, *adv.* **ranginess**, *n.*
rangiora (rang·giaw'rə), *n.* a broad-leaved shrub found in New Zealand.
rani RANEE.
rank[1] (rangk), *n.* a row, a line, a row of soldiers ranged side by side; a row of taxis for hire; relative position, degree, standing, station, class; high station, dignity, eminence; relative degree of excellence etc.; a line of squares stretching across a board from side to side. *v.t.* to draw up or marshal in rank; to classify, to estimate, to give a (specified) rank to. *v.i.* to hold a (specified) rank; to have a place or position (among, with etc.); to have a place on the list of claims on a bankrupt's estate; (*N Am.*) to take precedence (over). **rank and fashion**, people of high society. **rank and file**, common soldiers; (*fig.*) ordinary people. **the ranks**, private soldiers collectively. **to close ranks**, to maintain solidarity. **to pull rank**, to take precedence by virtue of higher rank, sometimes unfairly. **to take rank with**, to be placed on a level or be ranked with. **ranker**, *n.* a soldier in the ranks; a commissioned officer promoted from the ranks. **ranking**, *a.* (*N Am., Can.*) highly placed; prominent; (*Carib. sl.*) stylish; exciting. *n.* a position on a scale of excellence.
rank[2] (rangk), *a.* luxuriant in growth; over-fertile, over-abundant; coarse, gross; rancid, offensive, strong, evil-smelling; flagrant, arrant, utter; complete, total (e.g. a rank outsider). **rankly**, *adv.* **rankness**, *n.*
ranker RANK[1].
rankle (rang'kl), *v.i.* to irritate, to cause pain, anger or bitterness.
rankly RANK[2].
rankness RANK[2].
ransack (ran'sak), *v.t.* to search thoroughly, to rummage; to pillage, to plunder. **ransacker**, *n.*
ransom (ran'səm), *n.* release from captivity in return for a payment; a sum of money paid for such release or for goods captured by an enemy. *v.t.* to redeem from captivity or obtain the restoration of (property) by paying a sum of money; to demand or exact a ransom for; to hold to ransom; to release in return for a ransom; to redeem from sin, to atone for. **a king's ransom**, a very large sum of money or amount of valuables. **to hold to ransom**, to keep in confinement until a ransom is paid. **ransomable**, *a.* **ransomer**, *n.* **ransomless** (-lis), *a.*
rant (rant), *v.i.* to use loud, bombastic or violent language; to declaim or preach in a theatrical or noisy fashion. *n.* bombastic or violent declamation; a tirade, a noisy declamation; inflated talk. **ranter**, *n.* one who rants; a declamatory preacher; (*pl.*) a nickname given to the Primitive Methodists. **ranterism**, *n.* **ranting**, *a.* **rantingly**, *adv.*
ranunculus (rənŭng'kūləs), *n.* (*pl.* **-luses**, **-li** (-lī)) a genus of plants including the buttercup. **ranunculaceous** (-lā'shəs), *a.* pertaining to or belonging to this genus of plants.
rap[1] (rap), *v.t.* (*past, p.p.* **rapped**) to strike with a slight, sharp blow; to strike smartly; (*fig.*) to utter in a quick, abrupt way; to rebuke. *v.i.* to strike a sharp, quick blow, esp. at a door; to make a sharp, quick sound like a light blow; (*N Am., coll.*) to talk freely and informally; to perform a rap. *n.* a slight, sharp blow; a sound like the blow from a knocker, the knuckles etc. on a door; a similar sound made by some agency as a means of communicating messages at a spiritualistic séance; a sharp rebuke; a fast monologue spoken, often impromptu, over music. **to beat the rap**, (*N Am., Can. sl.*) to be acquitted of a crime, to escape punishment. **to rap on the knuckles**, to reprove, reprimand. **to take the rap**, (*coll.*) to take the blame for another. **rapper**, *n.* one who raps; a spirit-rapper; a door-knocker. **rapping**, *n.* the art of performing rhythmic monologues to music.
rap[2] (rap), *n.* a counterfeit Irish coin, passing for a half-penny in the time of George I; a thing of no value. **not worth a rap**, worthless.
rapacious (rəpā'shəs), *a.* grasping, extortionate; given to plundering or seizing by force, predatory; living on food seized by force (of animals). **rapaciously**, *adv.* **rapaciousness**, **rapacity** (-pas'-), *n.*
rape[1] (rāp), *v.t.* to ravish, to force (a woman) to have sexual intercourse against her will; to despoil, to violate. *n.* penetrative sexual intercourse of someone (usu. a woman) against the person's will; seizing or carrying off by force; violation, despoiling (e.g. of the countryside). **rapist**, *n.*
rape[2] (rāp), *n.* a plant, *Brassica napus*, allied to the turnip, grown as food for sheep; a plant, *B. campestris oleifera*, grown for its seed which yields oil, cole-seed. **wild rape**, charlock. **rape-cake**, *n.* the compressed seeds and husks of rape after the oil has been expressed, used for feeding cattle and as manure. **rape-oil**, *n.* oil obtained from the seed of *B. napus*. **rape-seed**, *n.* the seed of *B. napus*.
rape[3] (rāp), *n.* the refuse stems and skins of grapes after

the wine has been expressed, used to make vinegar.
Raphaelesque (rafāelesk'), *a.* after the style of the Italian painter Raphael (1483–1520). **Raphaelism,** *n.* the idealistic principles of Raphael in painting. **Raphaelite** (-līt), *n.*
raphe (rā'fē), *n.* a seam-like suture or line of union; a suture or line of junction, a median line or rib, a fibrovascular cord connecting the hilum of an ovule with the base of the nucleus.
raphia RAFFIA.
raphis (rā'fis), *n.* (*usu. in pl.* **raphides** (-dēz)) needle-shaped transparent crystals, usu. of calcium oxalate, found in the cells of plants.
rapid (rap'id), *a.* very swift, quick, speedy; done, acting, moving or completed in a very short time; descending steeply. *n.* (*usu. pl.*) a sudden descent in a stream, with a swift current. **rapid eye movement** (*abbr.* **REM**), *n.* the rapid eye movement which usu. occurs during the dreaming phase of sleep. **rapid-fire,** *a.* quick-firing (of guns). **rapid-firer,** *n.* **rapid transit,** *n.* (*N Am.*) fast passenger transport, usually by underground, in urban areas. **rapidity** (-pid'-), **rapidness,** *n.* **rapidly,** *adv.*
rapier (rā'piə), *n.* a light, narrow sword, used only in thrusting, a small-sword. **rapier-fish,** *n.* a sword-fish.
rapine (rap'īn), *n.* the act of plundering or carrying off by force; plunder, spoliation, robbery.
rappee (rəpē'), *n.* a coarse kind of snuff.
rappel (rəpel'), *n.* the beat of a drum calling soldiers to arms; abseiling. *v.i.* to abseil.
rapport (rəpaw'), *n.* correspondence, sympathetic relationship, agreement, harmony. **rapporteur** (rapawtœ'), *n.* a person responsible for carrying out an investigation and presenting a report on it to a higher committee.
rapprochement (raprosh'mē), *n.* reconciliation, the re-establishment of friendly relations, esp. between two nations. [F]
rapscallion (rapskal'iən), *n.* a rascal, a scamp, a good-for-nothing. *a.* rascally.
rapt (rapt), *a.* carried away by one's thoughts or emotions, absorbed, engrossed; enraptured, entranced. **raptly,** *adv.* **raptness,** *n.* **raptor,** *n.* one of the Raptores, birds of prey.
Raptores (raptaw'rēz), *n.* an order of birds of prey containing the eagles, hawks and owls. **raptorial** (-taw'ri-), *n.*, *a.* (pertaining to a) bird of prey; (of) its talons, adapted for seizing. **raptorious,** *a.*
rapture (rap'chə), *n.* ecstasy, transport, ecstatic joy; (*pl.*) a fit or transport of delight. **raptured, rapturous,** *a.* **rapturously,** *adv.*
rara avis (rah'rə ah'vis, reə'rə ā'vis), *n.* a rarity, something very rarely met with. [L, rare bird]
rare[1] (reə), *a.* of sparse, tenuous, thin or porous substance, not dense; exceptional, seldom existing or occurring, not often met with, unusual, scarce, uncommon; especially excellent, singularly good, choice, first-rate. **rare earth metals,** a group of rare metals (in many of their properties resembling aluminium) which occur in some rare minerals. **rare gas,** *n.* an inert gas. **rarely,** *adv.* seldom; exceptionally; remarkably well. **rareness,** *n.*
rare[2] (reə), *a.* of meat, half-cooked, underdone. **rarebit** (-bit), WELSH RABBIT.
raree-show (reə'rēshō), *n.* a show carried about in a box, a peep-show. [corr. of RARE[1], SHOW]
rarefy (reə'rifī), *v.t.* to make rare, thin, porous or less dense and solid; to expand without adding to the substance of; to purify, to refine, to make less gross. *v.i.* to become less dense. **rarefaction** (-fak'-), **rarefication** (-fi-), *n.* **rarefactive** (-fak'-), *a.* **rarefied,** *a.* thin; exalted.
raring (reə'ring), *a.* ready, eager. **raring to go,** eager to get started.
rarity (reə'riti), *n.* rareness; tenuity; unusual excellence; a rare thing; a thing of exceptional value through being rare.
rasbora (razbaw'rə), *n.* any of the small, brightly coloured cyprinid fishes from tropical Asia and E Africa, popular for aquariums.
rascal (ras'kəl, rahs'-), *n.* a rogue, a tricky, dishonest or contemptible person, a knave, a scamp; applied playfully to a child or animal etc. *a.* worthless, low, mean. **rascaldom** (-dəm), **rascalism, rascality** (-kal'-), *n.* **rascallion** (-kal'yən), *n.* a rascal. **rascally,** *a.* dishonest, contemptible. *adv.* in a dishonest manner.
rase RAZE.
rash[1] (rash), *a.* hasty, impetuous, venturesome; reckless, thoughtless, acting or done without reflection. **rashly,** *adv.* **rashness,** *n.*
rash[2] (rash), *n.* an eruption of spots or patches on the skin; a series of unwelcome, unexpected events.
rasher (rash'ə), *n.* a thin slice of bacon or ham for frying.
rashly, rashness RASH[1].
rasp (rahsp), *v.t.* to rub down, scrape or grate with a coarse, rough implement; to file with a rasp; to irritate; to utter in a grating tone. *v.i.* to rub, to grate; to make a grating sound; to grate (upon feelings etc.). *n.* an instrument like a coarse file with projections or raised teeth for scraping away surface material; a harsh, grating noise. **raspatory,** *n.* a rasp for scraping the outer membrane from bones etc. **rasper,** *n.* a rasp, scraper, a rasping-machine.
raspberry (rahz'bəri), *n.* the fruit of various species of *Rubus*, esp. *R. idaeus*, consisting of red or sometimes white or yellow drupes set on a conical receptacle; (*coll.*) a rude derisive sound with the lips. **raspberry-cane,** *n.* a long woody shoot of the raspberry plant.
rasse (ras'i, ras), *n.* a feline carnivore allied to the civet, inhabiting the E Indies and S China.
Rastafarian (rastəfeə'riən), *n.* a member of the religious and political, largely Jamaican, cult, which believes Ras Tafari, the former Emperor of Ethiopia, Haile Selassie, to be God. *a.* pertaining to Rastafarians or Rastafarianism. **Rasta** (ras'-), *n.*, *a.* Rastafarian. **Rastafarianism,** *n.* **Rastaman** (ras'-), *n.* a Rastafarian.
raster (ras'tə), *n.* the scanning lines which appear as a patch of light on a television screen and which reproduce the image.
rat (rat), *n.* one of the large rodents of the mouse family, esp. the black rat, *Mus rattus*, and *M. decumanus*, the grey, brown or Norway rat; one who deserts his party or his friends, a turncoat; (*coll.*) a worker who works for less than the trade-union rate of wages or who stands aloof from or works during a strike, a blackleg; a despicable person. *v.i.* (*past*, *p.p.* **ratted**) to hunt or kill rats (esp. of dogs); (*coll.*) to play the rat in politics, in a strike etc. **like a drowned rat,** soaked to the skin. **rats!** (*sl.*) an exclamation of annoyance or derision. **to rat on,** to betray, to divulge secret information, to inform against. **to smell a rat,** to be suspicious. **ratbag,** *n.* (*sl.*) a despicable or unpleasant person. **rat-catcher,** *n.* one who gets his living by catching rats. **rat kangaroo,** *n.* a kangaroo-like marsupial about the size of a rabbit. **rat race,** *n.* the continual competitive scramble of everyday life. **rat-run,** *n.* a minor road regularly used as a short-cut by drivers too impatient to follow the signposted route. **ratsbane,** *n.* poison for rats. **rat-snake,** *n.* an Indian snake, *Zamenis mucosus*, which preys on rats. **rat's-tail, rat-tail,** *n.* an excrescence growing from the pastern to the middle of the shank of a horse; a disease in horses in which the hair of the tail is lost; a tail like a rat's; a type of spoon with a ridge along the back of the bowl. **rat-tailed,** *a.* **rat-trap,** *n.* a trap for catching rats; a rat-trap pedal; a one-way system or arrangement of road-blocks designed to prevent drivers using minor roads as rat runs. *a.* applied to a cycle-pedal consisting of two parallel notched or toothed steel plates. *v.t.* to frustrate drivers' attempts to take short-cuts by means of a rat-trap. **ratter,** *n.* person or animal who or which catches rats. **ratting,** *n.*, *a.* **ratty,** *a.* infested with or characteristic of rats; (*sl.*) annoyed, ill-tempered.

rata (rah'tə), *n.* a large New Zealand forest tree of the myrtle family, having beautiful crimson flowers and yielding hard red timber.

ratable (rā'təbl), RATEABLE under RATE¹.

ratafia (ratəfē'ə), *n.* a liqueur or cordial flavoured with the kernels of cherry, peach, almond or other kinds of stone fruit; (also **ratafia biscuit**) a small almond-flavoured macaroon.

ratan (rətan', rat'-), RATTAN¹.

ratany (rat'əni), RHATANY.

rat-a-tat (ratətat'), RAT-TAT.

ratatouille (ratətwē'), *n.* a vegetable casserole from Provence, France, made with aubergines, tomatoes, peppers etc., stewed slowly in olive oil. [F]

ratchet (rach'it), *n.* a wheel or bar with inclined angular teeth, between which a pawl drops, permitting motion in one direction only; the pawl or detent that drops between the teeth of a ratchet-wheel. **ratchet-bar**, *n.* a bar with teeth into which a pawl drops. **ratchet-brace, -coupling, -drill, -jack, -lever, -punch, -wrench**, *n.* various tools or mechanical appliances working on the principle of the ratchet-bar or wheel with a pawl. **ratchet-wheel**, *n.* a wheel with toothed edge.

rate¹ (rāt), *n.* the proportional measure of something in relation to some other thing, ratio, comparative amount, degree etc.; a standard by which any quantity or value is fixed; valuation, price, value, relative speed etc.; a sum levied upon property for local purposes, distinguished from taxes which are for national purposes. *v.t.* to estimate the value, relative worth, rank etc. of; to fix the rank of (a seaman etc.); to assess for local rates; to subject to payment of local rates; (*coll.*) to consider, to regard as. *v.i.* to be rated or ranked (as). **at any rate**, in any case; even so. **at that rate**, if that is so. **rate-book**, *n.* a book of rates or prices; a record of local valuations for assessment of rates. **rate-cap**, *v.t.* to restrict the amount of money a local authority may levy in rates. **rate-capping**, *n.* **ratepayer**, *n.* one who is liable to pay rates. **rate support grant**, *n.* the money given to local authorities by central government to supplement the amount it raises in rates. **rateable**, *a.* liable to be rated, subject to assessment for municipal rates; capable of being rated or valued; proportional, estimated proportionally. **rateable value**, *n.* the estimated value of a property, used annually to assess the rates chargeable on the property. **rateability** (-bil'-), *n.* **rateably**, *adv.* **rater**, *n.* one who rates or assesses.

rate² (rāt), *v.t.* to chide angrily, to scold.

ratel (rā'təl), *n.* a nocturnal carnivore of the genus *Mellivora*, allied to the badger, with two species, *M. indicus*, from India, and *M. capensis*, the honey-badger of W and S Africa.

ratepayer, rater etc. RATE¹.

ratfink (rat'fingk), *n.* (*derog.*) a despicable person. *a.* mean, despicable.

rath (rahth), *n.* a prehistoric Irish hill-fort or earthwork.

rather (rah'dhə), *adv.* sooner, more readily or willingly, preferably, for choice; with more reason, more properly, rightly, or truly, more accurately; in a greater degree, to a greater extent; to a certain extent; slightly, somewhat. *int.* (*coll.*) very much, assuredly; yes, certainly. **the rather**, by so much the more. **ratherish**, *adv.* (*N Am.*). **ratherly**, *adv.* (*Sc.*).

ratify (rat'ifī), *v.t.* to confirm, to establish or make valid (by formal consent or approval). **ratifiable**, *a.* **ratification** (-fi-), *n.* **ratifier**, *n.*

rating¹ (rā'ting), *n.* the act of assessing, judging, renting etc.; a classification according to grade; a classification according to the conditions under which an appliance will operate satisfactorily; an ordinary seaman; (*pl.*) an evaluation of the popularity of radio or television programmes.

rating² (rā'ting), *n.* a scolding, a harsh reprimand.

ratio (rā'shiō), *n.* the relation of one quantity or magnitude to another of the same kind, measured by the number of times one is contained by the other, either integrally or fractionally; the relation existing between speeds of driving and driven gears, pulleys etc. **turns ratio**, the ratio of the number of turns of wire in the primary winding of a transformer to the number in the secondary.

ratiocinate (rashios'ināt), *v.i.* to reason or argue; to deduce consequences from premises or by means of syllogisms. **ratiocination**, *n.* **ratiocinative**, *a.* **ratiocinator**, *n.* **ratiocinatory**, *a.*

ration (rash'ən), *n.* a fixed allowance of food served out for a given time; a portion of provisions etc. allowed to one individual; (*pl.*) provisions, esp. food. *v.t.* to supply with rations; to put on fixed rations. **ration book**, *n.* a book issued periodically containing coupons etc. authorizing the holder to draw rations.

rational¹ (rash'ənəl), *a.* having the faculty of reasoning, endowed with mental faculties; agreeable to reasoning, reasonable, sensible, not foolish, not extravagant; based on or conforming to what can be tested by reason; of a number which can be expressed as the ratio of two integers, quantities. **rational number**, *n.* a number expressed as the ratio of two integers. **rationally**, *adv.* **rationale** (rashənahl'), *n.* a statement or exposition of reasons or principles; the logical basis or fundamental reason (of anything). **rationalism** (rash'ənəlizm), *n.* the determination of all questions of belief, esp. in religious matters, by the reason, rejecting supernatural revelation; the doctrine that reason supplies certain principles for the interpretation of phenomena that cannot be derived from experience alone. **rationalist**, *n.* **rationalistic** (-lis'-), *a.* **rationalistically**, *adv.* **rationality** (-nal'-), **rationalness**, *n.* the quality of being rational; the power of reasoning; reasonableness. **rationalize, -ise**, *v.t.* to render rational or reasonable; to clear (an equation etc.) of radical signs; to reorganize so as to make more efficient and economic; to find plausible (and usu. creditable) reasons to justify (one's conduct). *v.i.* to provide plausible reasons for one's conduct. **rationalization**, *n.* the act of rationalizing; the attempt to supply a conscious reason for an unconscious motivation in the explanation of behaviour. **rationalizer**, *n.*

ratite (rat'īt), *n.* of or belonging to the group Ratitae or birds with a keelless sternum and abortive wings, comprising the ostrich, emu, cassowary, kiwi, moa etc. **ratitous**, *a.*

ratline (rat'lin), **ratling** (-ling), *n.* one of the small ropes extended across the shrouds on each side of a mast, forming steps or rungs.

ratoon, rattoon (rətoon'), *n.* a sprout from the root of a sugar-cane that has been cut down. *v.t.* to cut down so as to encourage growth.

ratsbane RAT.

rattan (rətan', rat'-), *n.* the long, thin, pliable stem of various species of E Indian climbing palms of the genus *Calamus*; a switch or walking-stick of this material; (*collect.*) such stems used for wickerwork etc.

rat-tat (rat-tat'), *n.* a rapid knocking sound as of a knocker on a door.

ratter RAT¹.

rattle (rat'l), *v.i.* to make a rapid succession of sharp noises, as of things clattered together or shaken in a hollow vessel; to talk rapidly, noisily or foolishly; to move, go or act with a rattling noise; to run, ride or drive rapidly. *v.t.* to cause to make a rattling noise, to make (a window, door etc.) rattle; to utter, recite, play etc. (off, away etc.) rapidly; (*coll.*) to disconcert, to fluster. *n.* a rapid succession of sharp noises; an instrument, esp. a child's toy, with which such sounds are made; a rattling noise in the throat; rapid, noisy or empty talk, chatter; an incessant chatterer; noise, bustle, racket, boisterous gaie-

ty; the horny articulated rings in the tail of the rattlesnake, which make a rattling noise; a plant (**red rattle, yellow rattle**) having seeds that rattle in their cases.
rattle-brain, -head, -pate, *n.* **rattle-brained, -headed, -pated,** *a.* giddy, wild, empty-headed. **rattlesnake,** *n.* a snake of the American genus *Crotalus,* the tail of which is furnished with a rattle. **rattletrap,** *n.* a rickety object, esp. a vehicle. **rattlewort,** *n.* a plant of the genus *Crotalaria.*
rattler, *n.* one who or that which rattles; (*N Am., coll.*) a rattlesnake. **rattling,** *a.* making a rapid succession of sharp noises; (*coll.*) brisk, vigorous; (*sl.*) first-rate, excellent.
raucous (raw'kəs), *a.* hoarse, rough or harsh in sound. **raucity** (-si-), *n.* **raucously,** *adv.* **raucousness,** *n.*
raunchy (rawn'chi), *a.* (*sl.*) earthy; smutty; slovenly. **raunchily,** *a.* **raunchiness,** *n.*
raupo (row'pō), *n.* the giant bulrush of New Zealand.
rauwolfia (row·wool'fiə), *n.* a tropical flowering shrub from SE Asia; the root of this plant, which yields a drug.
ravage (rav'ij), *n.* devastation, ruin, havoc, waste; (*pl.*) devastating effects. *v.t.* to devastate; to spoil, to pillage. *v.i.* to make havoc. **ravager,** *n.*
rave (rāv), *v.i.* to wander in mind, to be delirious, to talk wildly, incoherently or irrationally; to speak in a furious way (against, at etc.); to act, move or dash furiously, to rage; to be excited, to go into raptures (about etc.); (*sl.*) to enjoy oneself wildly. *v.t.* to utter in a wild, incoherent or furious manner. *n.* (*coll.*) (an expression of) extravagant enthusiasm; (*sl.*) a wild party, esp. a very large party with rock music held e.g. outdoors or in a warehouse and for which tickets may be sold. **rave-up,** *n.* (*sl.*) a very lively party. **raver,** *n.* (*sl.*) a person with a wild, uninhibited social life; (*sl.*) a person who attends a rave. **raving,** *a.* frenzied; marked. *n.pl.* extravagant, irrational utterances. **ravingly,** *adv.*
ravel (rav'əl), *v.t.* (*past, p.p.* **ravelled**) to entangle, to confuse, to complicate, to involve; to untwist, to disentangle, to separate the component threads of; to fray. *v.i.* to become tangled. *n.* a tangle. **raveller,** *n.* **ravelling,** *n.* the act of entangling, confusing etc.; the act of unravelling; anything, as a thread, separated in the process of unravelling.
ravelin (rav'əlin), *n.* (*Fort.*) a detached work with a parapet and ditch forming a salient angle in front of the curtain of a larger work.
raven[1] (rā'vən), *n.* a large, black, omnivorous bird, *Corvus corax,* of the crow family. *a.* resembling a raven in colour, glossy black.
raven[2] (rav'ən), *v.t.* to devour with voracity; to ravage, to plunder. *v.i.* to plunder; to go about ravaging; to prowl after prey; to be ravenous. **ravening,** *n.,* a **raveningly,** *adv.*
ravenous (rav'ənəs), *a.* voracious, hungry, famished; furiously rapacious, eager for gratification. **ravenously,** *adv.* **ravenousness,** *n.*
ravine (rəvēn'), *n.* a long, deep hollow caused by a torrent, a gorge, a narrow gully or cleft. **ravined**[2], *a.*
ravingly RAVE [1].
ravioli (raviō'li), *n.* small pasta cases with a savoury filling.
ravish (rav'ish), *v.t.* to enrapture, to transport (with pleasure etc.); to violate, to rape. **ravisher,** *n.* **ravishing,** *a.* enchanting, charming, entrancing, transporting, filling one with rapture. **ravishingly,** *adv.* **ravishment,** *n.*
raw (raw), *a.* uncooked; in the natural state; not manufactured, requiring further industrial treatment; unhemmed; untrained, unskilled, inexperienced, undisciplined, immature, fresh; crude, untempered; having the skin off, having the flesh exposed, galled, inflamed, sore; cold and damp, bleak (of weather). *n.* a raw place on the body, a sore, a gall. **in the raw,** in its natural state; naked. **to touch on the raw,** to wound in a sensitive spot. **rawboned,** *a.* having bones scarcely covered with flesh, gaunt. **raw deal,** *n.* (*coll.*) unfair treatment. **rawhide,** *n.* untanned leather; a whip made of this. *v.t.* to whip with this. **raw material,** *n.* the material used for any manufacturing process; anything or any person regarded as usable for a purpose. **raw silk,** *n.* natural and untreated silk fibre; material made from untreated silk fibres. **raw umber,** *n.* umber that has not been calcined; the colour of this. **rawish,** *a* **rawly,** *adv..* **rawness,** *n.*
rawinsonde (rā'winsond), *n.* a hydrogen balloon which carries meteorological instruments to measure wind velocity. [*radar, wind, radiosonde*]
ray[1] (rā), *n.* a line or beam of light proceeding from a radiant point; a straight line along which radiant energy, esp. light or heat, is propagated; a gleam, a vestige or slight manifestation (of hope, enlightenment etc.); one of a series of radiating lines or parts; the outer whorl of florets in a composite flower; one of the bony rods supporting the fin of a fish, one of the radial parts of a starfish or other radials. *v.t.* to shoot out (rays), to radiate; to adorn with rays. *v.i.* to issue or shine forth in rays. **Becquerel rays** BECQUEREL. **Röntgen rays** RÖNTGEN. **ray flower, floret,** *n.* any of the small strap-shaped flowers in the flower head of some composite plants, such as the daisy. **ray fungus,** *n.* the Actinomyces bacterium which forms radiating threads. **ray-gun,** *n.* in science fiction, a gun which sends out rays to kill or stun. **rayed,** *a.* **rayless** (-lis), *a.* **raylet** (-lit), *n.* a small ray.
ray[2] (rā), *n.* any of several large cartilaginous fish allied to the sharks, with a flat disk-like body and a long, slender tail.
ray[3] (rā), *n.* the second note in the tonic sol-fa notation.
rayed, rayless RAY [1].
rayon (rā'on), *n.* (a fabric woven from) artificial silk made from cellulose.
raze, rase (rāz), *v.t.* to scratch (out), to erase; to obliterate; to demolish, to level with the ground, to destroy. **razed** *a.*
razer *n.*
razoo (rəzoo'), *n.* (*Austral. sl.*) a farthing. **not a brass razoo,** not a farthing.
razor (rā'zə), *n.* a cutting instrument for shaving off the hair of the beard or head. *v.t.* to shave with a razor. **razor-back,** *n.* a sharp back like a razor; (*N Am.*) a species of wild pig with such a back. *a.* having a sharp back or ridge like a razor. **razor-backed,** *a.* **razor-bill,** *n.* a bird with a bill like a razor, esp. the razor-billed auk, *Alca torda.* **razor-billed,** *a.* **razor-blade,** *n.* a blade used in a razor, for cutting or shaving. **razor-edge,** *n.* the edge of a razor; a keen edge; a sharp crest or ridge, as of a mountain; a critical situation, a crisis; a sharp line of demarcation, esp. between parties or opinions. **razor-shell,** *n.* a bivalve mollusc with a shell like a razor. **razor wire,** *n.* strong wire set across with pieces of sharp metal.
razz (raz), *n.* (*sl.*) a sound of contempt, a raspberry. *v.t.* (*N Am., Can.*) to jeer at, to heckle.
razzamatazz (razəmətaz'), *n.* colourful, noisy, lively atmosphere or activities.
razzle-dazzle (razldaz'l), *n.* (*coll.*) bewilderment, excitement, stir, bustle; intoxication. *v.t.* to dazzle, to daze; to bamboozle. **on the razzle-dazzle,** on the spree.
Rb, (*chem. symbol*) rubidium.
RC, (*abbr.*) Red Cross; Roman Catholic.
RCA, (*abbr.*) Radio Corporation of America; Royal College of Art.
RCN, (*abbr.*) Royal Canadian Navy; Royal College of Nursing.
RCP, (*abbr.*) Royal College of Physicians.
RCS, (*abbr.*) Royal College of Surgeons; Royal Corps of Signals.
RCVS, (*abbr.*) Royal College of Veterinary Surgeons.
rd, (*abbr.*) road; rod.
RE, (*abbr.*) religious education; Royal Engineers.
Re, (*chem. symbol*) rhenium.
re[1] (rā), *n.* the second note of a major scale; the second note of the scale of C major, D.

re² (rē), *prep.* in the matter of; (*coll.*) as regards, about. **reabsorb**, *v.t.* to absorb anew or again. **reabsorption**, *n.* **reaccommodate**, *v.t.* to accommodate or adjust afresh or again. **reaccuse**, *v.t.* to accuse again.
reach (rēch), *v.t.* to stretch out; to extend; to extend towards so as to touch, to extend as far as; to amount to; to attain to, to arrive at; to hit, to affect; to hand, to deliver, to pass; to make contact with, to communicate with. *v.i.* to reach out, to extend; to reach or stretch out the hand; to make a reaching effort, to be extended so as to touch, to have extent in time, space etc.; to attain (to). *n.* the act or power of reaching; extent, range, compass, power, attainment; an unbroken stretch of water, as between two bends; the direction travelled by a vessel on a tack. **reach-me-down**, *a.* of clothes, cheap ready-made or second-hand. *n.pl.* ready-made or second-hand clothes. **reachable**, *a.* **reacher**, *n.* **reaching**, *a.*
react (riakt'), *v.i.* to act in response (to a stimulus etc.); to have a reciprocal effect, to act upon the agent; to act or tend in an opposite manner, direction etc.; to exert an equal and opposite force to that exerted by another body; to exert chemical action (upon). **reactance**, *n.* the impedance to the flow of an alternating current due to inductance or capacitance. **reactant**, *n.* a substance that takes part in a chemical reaction. **reaction**, *n.* reciprocal action; the response of an organ etc. to stimulation; an emotional, intellectual etc. response to events; an effect produced by a drug or by a substance to which a person is allergic; the chemical action of one substance upon another; the equal and opposite force exerted upon the agent by a body acted upon; contrary action or condition following the first effects of an action; action in an opposite direction, esp. in politics after a reform movement, revolution etc. **reaction engine, -motor**, *n.* an engine or motor which develops thrust by expelling gas at high speed. **reaction turbine**, *n.* a turbine where the working fluid is accelerated through expansion in the static nozzles and the rotor blades. **reactionary**, *a.* involving or tending towards reaction, esp. in politics, retrograde, conservative. *n.* a reactionary person. **reactionism**, *n.* **reactionist**, *n.* **reactivate**, *v.t.* to restore to a state of activity. **reactivation**, *n.* **reactive**, *a.* **reactively**, *adv.* **reactivity** (-tiv'-), **reactiveness**, *n.* **reactor**, *n.* a substance which undergoes a reaction; a vessel in which reaction takes place; a nuclear reactor; a person sensitive to a given drug or medication.
read (rēd), *v.t.* (*past, p.p.* **read** (red)), to perceive and understand the meaning of (printed, written or other characters, signs, symbols, significant features etc.), to peruse; to reproduce mentally or vocally or instrumentally (words, notes etc. conveyed by symbols etc.); to receive and understand radio communications from; to discover by observation, to interpret, to explain; to see through; to learn or ascertain by reading; to study by reading; to bring into a specified condition by reading; to study for an examination; to indicate or register (of a meteorological instrument etc.); (*Comput*) to obtain data from a storage device. *v.i.* to follow or interpret the meaning of a book etc.; to pronounce (written or printed matter) aloud; to render written music vocally or instrumentally (well, easily etc.); to acquire information (about); to study by reading; to mean or be capable of interpretation (in a certain way etc.); to sound or affect (well, ill etc.) when perused or uttered. *n.* an act of reading, a perusal. **to read a lesson, lecture, to someone**, to scold, to reprimand someone. **to read between the lines** LINE¹. **to read into**, to extract a meaning not explicit. **to read off**, to take a reading, or information, from an instrument, e.g. a thermometer. **to read out**, to read aloud. **to read someone's mind**, to make an accurate guess as to what someone is thinking. **to read up**, to acquire knowledge (of) by intensive reading. **to take as read** TAKE. **read-out**, *n.* the data retrieved from a computer; the act of retrieving data from computer storage facilities for display on screen or as print-out. **read-write head**, *n.* the electromagnetic head in a computer disk-drive which reads or writes data on magnetic tape or disk. **readable**, *a.* worth reading, interesting; legible. **readableness, readability** (-bil'-), *n.* **readably**, *adv.* **reader**, *n.* one who reads; one who reads much; a person employed by a publisher to read and report upon manuscripts etc. offered for publication; a proofreader; a lay reader; a lecturer, usu. ranking next below a professor, in some universities, Inns of Court etc.; a textbook, a book of selections for translation, a reading-book for schools. **readership**, *n.* the post of university reader; a body of readers, especially of a particular author or publication. **reading**, *n.* the act, practice or art of reading; the study or knowledge of books, literary research, scholarship; a public recital or entertainment at which selections etc. are read to the audience; the form of a passage given by a text, editor etc.; the way in which a passage reads, an interpretation, a rendering; an observation made by examining an instrument; the recital of the whole or part of a Bill as a formal introduction or measure of approval in a legislative assembly. **first reading**, the formal introduction of a Bill. **second reading**, a general approval of the principles of a Bill. **third reading**, the final acceptance of a Bill together with the amendments passed in committee. **reading-book**, *n.* a book of selections to be used as exercises in reading. **reading-desk**, *n.* a stand for books etc. for the use of a reader, esp. in church, a lectern. **reading-lamp**, *n.* a lamp for reading by. **reading-room**, *n.* a room in a library, club etc. furnished with books, papers etc.
readapt, *v.t.* **readaptation**, *n.* [RE-ADAPT]
readdress (rēədres'), *v.t.* to put a new (esp. a corrected) address upon.
reader, reading READ.
readily, readiness READY.
readjust, *v.t.* to arrange or adjust afresh. **readjustment**, *n.*
readmit, *v.t.* to admit again. **readmission, readmittance**, *n.* **readopt**, *v.t.* to adopt again. **readoption**, *n.* **readorn**, *v.t.* to adorn afresh. **readvertise**, *v.t., v.i.* to advertise again.
ready (red'i), *a.* in a state of preparedness, fit for use or action; willing; about (to); quick, prompt; able, expert; within reach, handy, quickly available; held in the position preparatory to presenting and aiming (of a firearm). *adv.* (*usu. in comb. with p.p.*) in a state of preparedness. *n.* (*sl.*) ready money. *v.t.* to make ready, to prepare. **at the ready**, of a firearm, held in the ready position; poised ready for use. **ready, steady, go!** words used to start a race. **ready to hand**, nearby. **ready-to-wear**, *a.* off-the-peg. *n.* off-the-peg clothing. **to make ready**, to prepare; to prepare a forme before printing. **ready-made**, *a.* made beforehand, not made to order (esp. of clothing in standard sizes). **ready-mix**, *n.* a food or concrete mix which only needs liquid to be added to make it ready for use. **ready money**, *n.* actual cash, ready to be paid down. **ready-reckoner**, *n.* a book with tables of interest etc. for facilitating business calculations. **readily**, *adv.* without trouble or difficulty, easily; willingly, without reluctance. **readiness**, *n.* the state of being ready, preparedness; willingness, prompt compliance; facility, ease.
reaffirm, *v.t.* to affirm again. **reaffirmation**, *n.* **reafforest**, *v.t.* to convert again into forest. **reafforestation**, *n.*
Reaganism (rā'gənizm, rē'-), *n.* the political, economic etc. policy and philosophy associated with Ronald *Reagan*, US President, 1980–88. **Reaganomics** (-om'iks), *n.* the economic policies of Reagan's administration, e.g. major tax-cutting.
reagent (riā'jənt), *n.* a substance used to detect the presence of other substances by means of their reaction; a force etc. that reacts. **reagency**, *n.* reciprocal action.
real¹ (riəl), *a.* actually existing; not fictitious, imaginary,

theoretical etc.; true, genuine; not counterfeit, not spurious; having substantial existence, objective; consisting of fixed or permanent things, as lands or houses; having an absolute and independent existence, opp. to nominal or phenomenal; of incomes, prices etc., reckoned in terms of purchasing power rather than nominal value. **for real,** (*sl.*) in reality; in earnest. **the real McCoy,** the genuine article; the best. **the real thing,** the genuine article and not a substitute. **real ale,** *n.* beer which is allowed to ferment and mature in the cask and is not pumped up from the keg with carbon dioxide. **real estate,** *n.* landed property. **real life,** *n.* actual human life, as opposed to fictional representation of human life. **real number,** *n.* any rational or irrational number. **realpolitik** (räahl′politĕk), *n.* politics based on practical reality rather than moral or intellectual ideals. **real presence,** *n.* the actual presence of the body and blood of Christ in the Eucharist. **real property,** *n.* immovable property such as freehold land. **real-time,** *a.* pertaining to the processing by computer of data as it is generated. **really,** *adv.* in fact, in reality; (*coll.*) positively, I assure you; is that so?

real² (rā′al), *n.* (*pl.* **reales** (-lēz)) a Spanish silver coin or money of account.

real³ (riəl), *a.* royal. **real tennis,** *n.* royal tennis, played in a walled indoor court.

realgar (rial′gah), *n.* orange-red native disulphide of arsenic, used as a pigment and in the manufacture of fireworks.

realign (rēəlīn′), *v.t.* to align again; to group together on a new basis. **realignment,** *n.*

realism (riə′lizm), *n.* the scholastic doctrine that every universal or general idea has objective existence, opp. to *nominalism* and *conceptualism;* the doctrine that the objects of perception have real existence, opp. to *idealism;* the doctrine that in perception there is an immediate cognition of the external object; the practice of representing objects, persons, scenes etc. as they are or as they appear to the painter, novelist etc., opp. to *idealism* and *romanticism;* a practical or down-to-earth attitude towards or acceptance of the facts of life. **realist,** *n.* a believer in realism; a practical person. **realistic** (-lis′-), *a.* pertaining to realism; matter-of-fact, common-sense. **realistically,** *adv.*

reality (rial′iti), *n.* the quality of being real, actuality, actual existence, being, that which underlies appearances; truth, fact; that which is real and not counterfeit, imaginary, suppositious etc.; the real nature (of); (*Law*) the permanent quality of real estate. **in reality,** in fact.

realize, ise (riə′līz), *v.t.* to become aware of; to apprehend clearly and vividly; to bring into actual existence, to give reality to, to accomplish; to cause to seem real; to convert into money; to sell; to bring in, as a price. **realizable,** *a.* **realization,** *n.* **realizer,** *n.*

really etc. REAL ¹.

realm (relm), *n.* a kingdom; sphere, field of interest.

realtor (riəl′taw), *n.* (*N Am.*) estate agent, esp. a member of the National Association of Real Estate Boards; a dealer in land for development. **realty,** *n.* real property.

ream¹ (rēm), *n.* 480 sheets or 20 quires of paper (often 500 or more sheets to allow for waste); (*pl.*) large quantities (of written matter). **printer's ream,** *n.* 516 sheets.

ream² (rēm), *v.t.* to enlarge (a hole in metal). **reamer,** *n.* an instrument or tool used in reaming.

reanimate (rēan′imāt), *v.t.* to restore to life; to revive, to encourage, to give new spirit to. **reanimation,** *n.*

reap (rēp), *v.t.* to cut with a scythe, sickle or reaping-machine; to gather in (a harvest etc.); to cut the harvest off (ground etc.); (*fig.*) to obtain as return for labour, deeds etc. *v.i.* to perform the act of reaping; to receive the consequences of labour, deeds etc. **reaper,** *n.* one who reaps; a reaping-machine. **the grim reaper,** death. **reaping-hook,** *n.* a sickle. **reaping-machine,** *n.*

reapparel, *v.t.* to clothe again. **reappear,** *v.i.* to appear again. **reappearance,** *n.* **reapply,** *v.t.* to apply again. **reapplier,** *n.* **reapplication,** *n.* **reappoint,** *v.t.* to appoint again. **reappointment,** *n.* **reapportion,** *v.t.* to share out again. **reapportionment,** *n.* **reappraise,** *v.t.* to revalue. **reappraisal,** *n.* **reapproach,** *v.t.* to approach again.

rear¹ (riə), *v.t.* to raise, to set up, to elevate to an upright position; to build, to erect, to uplift; to bring up, to breed, to educate; to raise, to cultivate, to grow. *v.i.* of a horse, to stand on the hind legs. **rearer,** *n.*

rear² (riə), *n.* the back or hindmost part, esp. the hindmost division of a military or naval force; the back (of); a place or position at the back; (*euphem.*) the buttocks. *a.* pertaining to the rear. **in the rear,** at the back. **to bring up the rear,** to come last. **rear-admiral,** *n.* a naval officer next below the rank of vice-admiral. **rear-end,** *n.* the back part of anything; (*coll.*) the buttocks. *v.t.* to crash into the rear of (a vehicle). **rear-guard,** *n.* a body of troops protecting the rear of an army. **rear-lamp, -light,** *n.* a red light at the back of a bicycle or motor vehicle. **rear-view mirror,** *n.* a small mirror in a motor vehicle which allows the driver to observe the traffic behind him. **rearmost,** *a.* coming or situated last of all. **rearward** (-wəd), *n.* the rear, esp. of an army. *a.* situated in or towards the rear. *adv.* rearwards.

rearwards, *adv.* towards the rear.

reargue, *v.t.* to argue or debate afresh. **rearguement,** *n.*

rearm, *v.t.* to arm afresh, esp. with more modern weapons. **rearmament,** *n.*

rearrange (rēəranj′), *v.t.* to arrange in a new order; to put back in the original order; to fix a new time for. **rearrangement,** *n.* **rearrest,** *v.t.* to arrest again. *n.* a second arrest. **reascend,** *v.t., v.i.* to ascend again. **reascension,** *n.* **reascent,** *n.*

reason (rē′zən), *n.* that which is adduced to support or justify, or serves as a ground or motive for an act, opinion etc.; that which accounts for anything, a final cause; the premise of an argument, esp. the minor premise when stated after the conclusion; the intellectual faculties, esp. the group of faculties distinguishing man from animals; the intuitive faculty which furnishes a priori principles, categories etc.; the power of consecutive thinking, the logical faculty; good sense, judgment, sanity; sensible conduct; moderation; the exercise of the rational powers. *v.i.* to use the faculty of reason; to argue, esp. to employ argument (with) as a means of persuasion; to reach conclusions by way of inferences from premises. *v.t.* to debate, discuss or examine by means of the reason or reasons and inferences; to assume, conclude or prove by way of argument; to persuade or dissuade by argument; to set forth in orderly argumentative form. **by reason of,** because, on account of, in consequence of. **in, within reason** within moderation; according to good sense. **pure reason,** reason without the benefit of experience. **reasons of state,** politics or state security used to justify immoral acts. **to listen to reason,** to allow oneself to be persuaded. **to stand to reason,** to follow logically; to be obvious. **reasonable,** *a.* endowed with reason; rational, reasoning, governed by reason; conformable to reason, sensible, proper; not extravagant, moderate, esp. in price, fair, not extortionate. **reasonableness,** *n.* **reasonably,** *adv.* **reasoned,** *a.* well-thought-out or well-argued. **reasoner,** *n.* **reasoning,** *n.* the act of drawing conclusions from premises or using the reason; argumentation; a statement of the reasons justifying a course, opinion, conclusion etc. **reasonless** (-lis), *a.*

reassemble, *v.t., v.i.* **reassembly,** *n.* **reassert,** *v.t.* **reassertion,** *n.* **reassess,** *v.t.* **reassessment,** *n.* **reassign,** *v.t.* **reassignment,** *n.* **reassume,** *v.t.* **reassumption,** *n.* **reassure,** *v.t.* to assure or confirm again; to restore to confidence, to give fresh courage to; to reinsure. **reassurance,** *n.* **reassurer,** *n.* **reassuring,** *a.* **reassuringly,** *adv.*

reattach, *v.t.* to attach afresh. **reattachment,** *n.* **reattain,** *v.t.* to attain again. **reattainment,** *n.* **reattempt,** *v.t.* to

Réaumur (rä'ōmuə), *a.* applied to the thermometer invented by the French physicist R. A. F. de *Réaumur* (1683-1757), or to his thermometric scale, the zero of which corresponds to freezing-point and 80° to boiling-point.

rebarbative (ribah'bətiv), *a.* repellent, grim, forbidding; surly; uncooperative.

rebate[1] (rē'bāt, -bāt'), *v.t.* to make a deduction from; to refund a part of the amount payable; to make blunt, to dull; to remove a portion of (a charge). *n.* (rē'-) a deduction, a discount; a partial repayment. **rebatable, rebateable,** *a.* **rebatement,** *n.* **rebater,** *n.*

rebate[2] (ribāt'), RABBET.

rebeck, rebec (rē'bek), *n.* a mediaeval three-stringed musical instrument played with a bow.

rebel (reb'l), *a.* rebellious; pertaining to rebellion or to rebels. *n.* one who forcibly resists the established government or renounces allegiance thereto; one who resists authority or control. *v.i.* (ri bel') (*past, p.p.* **rebelled**) to engage in rebellion (against); to revolt (against any authority or control); to feel or show repugnance (against). **rebel-like,** *a.* **rebellion** (-bel'yən), *n.* organized resistance by force of arms to the established government; opposition to any authority. **rebellious** (-bel'yəs), *a.* engaged in rebellion; defying or opposing lawful authority; disposed to rebel, refractory, insubordinate, difficult to manage or control. **rebelliously,** *adv.* **rebelliousness,** *n.*

rebid, *v.t., v.i.* to bid again, usu. in bridge, to bid on the same suit as the previous bid. **rebind,** *v.t.* (*past, p.p.* **rebound**[1]) to bind again; to give a new binding to. **rebirth,** *n.* a second birth, esp. an entrance into a new sphere of existence, as in reincarnation.

reborn (rēbawn'), *a.* born again (esp. of spiritual life). **rebore,** (rēbaw') *v.t.* to bore again, e.g. a cylinder, so as to clear it.

rebound[1] (rēbownd'), *past, p.p.* REBIND.

rebound[2] (ribownd'), *v.i.* to bound back, to recoil (from a blow etc.); to react, to recoil (upon). *n.* (rē'-) the act of rebounding, a recoil; reaction (of feeling etc.). **on the rebound,** in the act of bouncing back; as a reaction to a disappointment. **rebounder,** *n.* a small trampoline used for performing jumping exercises on. **rebounding,** *n.* the act of rebounding; a form of exercise involving jumping up and down on a rebounder. **rebroadcast,** *v.t.* to broadcast again. *a.* broadcast again. *n.* a second broadcast.

rebuff (ribūf'), *n.* a rejection, a check (to an offer or to one who makes advances etc.); a curt denial, a snub; a defeat, a sudden or unexpected repulse. *v.t.* (*past, p.p.* **rebuffed**) to give a rebuff to, to repel.

rebuild, *v.t.* (*past, p.p.* **rebuilt**) to build again, to reconstruct.

rebuke (ribūk'), *v.t.* to reprove, to reprimand, to chide; to censure, to reprehend (a fault etc.). *n.* the act of rebuking; a reproof. **rebukable,** *a.* **rebuker,** *n.* **rebukingly,** *adv.*

rebury, *v.t.* **reburial,** *n.*

rebus (rē'bəs), *n.* a picture or figure representing enigmatically a word, name or phrase, usu. by objects suggesting words or syllables; (*Her.*) a device representing a proper name or motto in this way.

rebut (ribūt'), *v.t.* (*past, p.p.* **rebutted**) to contradict or refute by plea, argument or countervailing proof. **rebuttable,** *a.* **rebutment, rebuttal,** *n.* **rebutter,** *n.* one who rebuts; the answer of a defendant to a plaintiff's surrejoinder.

rec., (*abbr.*) receipt; recipe; record, recorded, recorder.

recalcitrant (rikal'sitrənt), *a.* refractory, obstinately refusing submission. *n.* a recalcitrant person. **recalcitrance, -cy,** *n.*

recalculate, *v.t.* **recalculation,** *n.*

recalesce (rēkales'), *v.i.* to grow hot again (esp. of iron or steel in the process of cooling). **recalescence,** *n.* **recalescent,** *a.*

recall (rikawl'), *v.t.* to call back; to summon to return; to bring back to mind, to recollect; to revoke, to annul, to take back. *n.* (rē'-), a calling back; a summons to return; a signal calling back soldiers, a ship etc.; the power of recalling, remembering, revoking or annulling; (*N Am.*) the right of electors to dismiss an elected official by popular vote. **total recall,** the ability to remember in great detail. **recallable,** *a.* **recallment,** *n.*

recant (rikant'), *v.t.* to retract, to renounce, to abjure; to disavow. *v.i.* to disavow or abjure opinions or beliefs formerly avowed, esp. with a formal acknowledgment of error. **recantation** (rē-), *n.* **recanter,** *n.*

recap (rē'kap. -kap'), *v.t.* to recapitulate.

recapitulate (rēkəpit'ūlāt), *v.t.* to repeat in brief (as the principal heads of a discourse or dissertation), to sum up, to summarize. **recapitulation,** *n.* the act of recapitulating, e.g. at the end of a speech; the apparent repetition of the evolutionary stages of a species in the embryonic development of a member of that species; the repeating of earlier themes in a piece of music. **recapitulative, recapitulatory,** *a.*

recapture (rēkap'chə), *n.* the act of recapturing; that which is recaptured. *v.t.* to capture again, to recover (a prize from the captor); to recover (a former ability); to experience again. **recast** (rēkahst'), *v.t.* to cast, found or mould again; to fashion again, to remodel; to compute or add up again; (*Theat*) to reassign parts in (a play). *n.* that which has been recast; the process or result of recasting. **recaster,** *n.*

recce (rek'i), *n.* (*pl.* **recces**) reconnaissance. *v.t., v.i.* to reconnoitre.

recd, (*abbr.*) received.

recede[1] (rīsēd'), *v.i.* to go back or away (from); to be gradually lost to view by distance; to incline, slope or trend backwards or away; to retreat, to withdraw (from); to diminish; of hair, to cease to grow at the temples.

recede[2] (rēsēd'), *v.t.* to cede again, to restore to a former possessor.

receipt (risēt'), *n.* the act or fact of receiving or being received; (*usu. in pl.*); that which is received, esp. money; a written acknowledgment of money or goods received; a recipe. *v.t.* to give a receipt for; to write an acknowledgment of receipt on (a bill etc.).

receive (risēv'), *v.t.* to obtain, get or take as a thing due, offered, sent, paid or given; to be given, to be furnished or supplied with; to acquire; to accept with approval or consent, to admit, as proper or true; to admit to one's presence, to welcome, to entertain as guest; to encounter, to take or stand the onset of; to be a receptacle for; to understand, to regard (in a particular light); to accept (stolen goods) from a thief; to convert incoming electrical signals into sounds or pictures by means of a receiver; to return the service in tennis or squash. *v.i.* to hold a reception of visitors or callers. **receivable,** *a.* **receivability** (-bil'-), **receivableness,** *n.* **received,** *a.* generally accepted or believed. **Received (Standard) English,** *n.* English spoken by educated British people, taken as the standard of the language. **Received Pronunciation,** *n.* the non-localized pronunciation of British English, taken as the standard. **receiver** (risē'və), *n.* one who receives; one who receives stolen goods, a fence; a vessel for receiving the products of distillation or collecting gas; an officer appointed to administer property under litigation, esp. that of bankrupts; an apparatus for the reception of radio or television signals; the part of a telephone which contains the earpiece and mouthpiece. **official receiver,** a person appointed by a bankruptcy court to receive the sums due to and administer the property of a bankrupt. **receiving-house, -office, -room** etc., *n.* places for the receipt of parcels, money, recruits etc. **receiving-order,** *n.* an order from a bankruptcy court staying separate action against a debtor and placing his affairs in the hands of an official receiver. **receivership**, *n.* the office of receiver; the state of being administered by the receiver.

recency RECENT.

recension (risen'shən), *n.* a critical revision of a text; a revised edition. **recensor**, *n.*

recent (rē'sənt), *a.* of or pertaining to the present or time not long past; that happened, existed or came into existence lately; late (of existence); modern, fresh, newly begun or established; pertaining to the existing epoch, Post-Pliocene, Quaternary. **recency** (-si), **recentness**, *n.* **recently**, *adv.*

receptacle (risep'təkl), *n.* that which receives, holds or contains; a vessel, space or place of deposit; a part forming a support, as the portion of a flower on which the sexual organs are set, the axis of a flower cluster etc. **receptacular** (resiptak'ū-), *a.*

reception (risep'shən), *n.* the act of receiving; the state of being received; receipt, acceptance, admission; the receiving, admitting or accommodating of persons, esp. guests, new members of a society etc.; a formal party for guests, esp. after a wedding; an area, e.g. in a hotel or offices, where visitors, clients etc. are received, appointments and enquiries dealt with etc.; the act or process of receiving (ideas or impressions) into the mind; mental acceptance, admission or recognition (of a theory etc.); the quality of received radio or television signals. **reception centre**, *n.* a place where people can receive immediate assistance for problems, such as drugs or homelessness. **reception order**, *n.* the official order required for detention in a mental hospital. **reception-room**, *n.* a room for receptions; (*coll.*) a room to which visitors are admitted, opp. to bedrooms, kitchen etc. **receptionist**, *n.* person at a hotel or elsewhere, whose duty it is to receive and look after visitors. **receptor**, *n.* any of various devices which receive signals or information; an organ adapted for receiving stimuli; a sensory nerve ending which changes stimuli into nerve impulses.

receptive (risep'tiv), *a.* having ability or capacity to receive; quick to receive impressions, ideas etc. **receptively**, *adv.* **receptiveness**, **receptivity** (rēseptiv'-), *n.*

recess (rises'), *n.* cessation or suspension of public or other business, a vacation; a part that recedes, a depression, indentation, hollow, niche or alcove; a secluded or secret place; (*N Am.*) a break between classes at school. *v.t.* to put into a recess; to build a recess in a wall. *v.i.* to adjourn. **recessed**, *a.*

recession[1] (risesh'ən), *n.* the act of receding; a receding part or object; a slump, esp. in trade or economic activity; the withdrawal of the clergy and choir after a church service. **recessional**, *a.* pertaining to the recession of the clergy and choir from the chancel. *n.* a hymn sung during this ceremony. **recessive**, *a.* tending to recede; of a stress accent, tending to move towards the beginning of a word. **recessive gene**, *n.* one that must be inherited from both mother and father in order to show its effect in the individual. **recessively**, *adv.* **recessiveness**, **recessivity** (-siv'-), *n.*

recession[2] (risesh'ən), *n.* the act of giving back to a former owner.

recharge (rēchahj'), *v.t.* to make a new charge against; to charge or attack again or in return; to restore the electrical charge in e.g. a battery. *n.* a new charge or a charge in return. **to recharge one's batteries**, to renew one's vigour. **rechargeable**, *a.*

réchauffé (rāshōfā'), *n.* a dish warmed up again; (*fig.*) a rehash. [F]

recheck, *v.t.* to check again. *n.* the act of checking something again.

recherché (rəshœ'shā), *a.* (*fem.* **-chée** (-shā)) out of the common; precious, affected. [F]

rechristen, *v.t.* to christen again; to give a new name to.

recidivist (risid'ivist), *n.* a relapsed or inveterate criminal, usu. one serving or who has served a second term of imprisonment. **recidivation**, **recidivism**, *n.* a habitual relapse into crime. **recidivistic** (-vis'-), *a.*

recipe (res'ipi), *n.* a formula or prescription for compounding medical or other mixtures; directions for preparing a dish; a remedy, expedient, device or means for effecting some result.

recipient (risip'iənt), *a.* receiving; receptive. *n.* one who receives, a receiver. **recipience**, **recipiency**, *n.*

reciprocal (risip'rəkəl), *a.* acting, done or given in return, mutual; mutually interchangeable, inversely correspondent, complementary; expressing mutual action or relation. *n.* the quotient resulting from dividing unity by a quantity. **reciprocal pronoun**, *n.* a pronoun which expresses a mutual action or relationship. **reciprocal ratio**, *n.* the ratio between the reciprocals of two quantities. **reciprocal terms**, *n.* terms having the same signification and therefore interchangeable. **reciprocality** (-kal'-), *n.* reciprocity. **reciprocally**, *adv.* **reciprocant**, *n.* a different invariant.

reciprocate (risip'rəkāt), *v.i.* to alternate, to move backwards and forwards; to return an equivalent, to make a return in kind. *v.t.* to give alternating or backward-and-forward motion to; to give and take mutually, to interchange; to give in return. **reciprocating engine**, *n.* an engine performing work with a part having **reciprocating motion**, i.e. backward-and-forward or up-and-down motion, as of a piston. **reciprocation**, *n.* the act of reciprocating; giving and returning; reciprocal motion. **reciprocative**, *a.* **reciprocator**, *n.* **reciprocatory**, *a.* reciprocating, opp. to rotatory.

reciprocity (resipros'iti), *n.* the state of being reciprocal, reciprocation of rights or obligations; mutual action or the principle of give-and-take, esp. interchange of commercial privileges between two nations.

recirculate, *v.t.* to pass or go round again. **recirculation**, *n.*

recital (risī'təl), *n.* the act of reciting; an enumeration or narrative of facts or particulars, a story; the part of a document formally stating facts, reasons, grounds etc.; a public entertainment consisting of recitations; a musical performance, esp. by one person or of the works of one person. **recitalist**, *n.*

recitation (resitā'shən), *n.* the recital of prose or poetry, esp. the delivery of a composition committed to memory; a composition intended for recital.

recitative (resitətēv'), *n.* a style of rendering vocal passages intermediate between singing and ordinary speaking, as in oratorio and opera; a piece or part to be sung in recitative. *a.* (risī'tativ), pertaining or suitable for recitative; pertaining to a recital. **recitatively**, *adv.*

recite (resīt'), *v.t.* to repeat aloud or declaim from memory, esp. before an audience; to narrate, to rehearse (esp. facts etc. in a legal document); to quote, to cite; to enumerate. *v.i.* to give a recitation. **recitable**, *a.* **reciter**, *n.*

reck (rek), *v.t.* (*chiefly poet.*) to care, to heed. **reckless** (-lis), *a.* careless, heedless; rash, venturesome; regardless, indifferent, neglectful, heedless of the consequences. **recklessly**, *adv.* **recklessness**, *n.*

reckon (rek'ən), *v.t.* to count, to add (up), calculate or compute; to count or include (in or among), to regard (as), to account, to esteem, to consider (to be); to be of the opinion. to calculate, to guess (that). *v.i.* to compute, to calculate, to settle accounts with; to suppose, to believe, to guess, to calculate. **to reckon on**, to rely upon, to expect. **to reckon with**, **without**, to take, or fail to take, into account. **reckoner**, *n.* any of several devices or tables for quick calculations. **reckoning**, *n.* the act of calculating or counting; a statement of accounts or charges, a bill, esp. for liquor; a settling of accounts; an estimate or calculation of a ship's position or course. **day of reckoning**, the day of settling accounts or scores. **dead reckoning** DEAD.

reclaim (riklām'), *v.t.* to bring back from error, vice, wildness etc., to reform; to make available for cultivation; to demand back, to claim the restoration of; to recover usable substances from waste products. *n.* the act of reclaiming or being reclaimed, reclamation. **reclaimable,** *a.* **reclaimably,** *adv.* **reclaimer,** *n.* **reclamation** (reklə-), *n.* the act of reclaiming; the state of being reclaimed; the recovery for cultivation of waste land; the recovery of usable substances from waste products.

réclame (rāklahm'), *n.* notoriety; puffing, self-advertisement. [F]

reclassify, *v.t.* **reclassification,** *n.*

recline (riklīn'), *v.t.* to lay or lean (one's body, head, limbs etc.) back, esp. in a horizontal or nearly horizontal position. *v.i.* to assume or be in a leaning or recumbent posture, to lie down or lean back upon cushions or other supports. **reclinable,** *a.* **reclinate** (rek'linət), *a.* of plants, inclined from an erect position, bending downwards. **reclination,** *n.* **recliner,** *n.* someone who or something which, reclines; a type of armchair which can be adjusted to recline backwards.

reclose, *v.t.,* *v.i.* **reclothe,** *v.t.*

recluse (riklōōs'), *a.* retired from the world; solitary, secluded, retired, sequestered. *n.* one who lives retired from the world, esp. a religious devotee who lives in a solitary cell and practises austerity and self-discipline.

recognition (rekəgnish'ən), *n.* act of recognizing; state of being recognized; acknowledgment, notice taken; a perceiving as being known. **recognitive** (-kog'-), **recognitory** (-kog'-), *a.*

recognizance (rəkog'nizəns), *n.* (*Law*) a bond or obligation entered into a court or before a magistrate to perform a specified act, fulfil a condition etc. (as to keep the peace or appear when called upon); a sum deposited as pledge for the fulfilment of this.

recognize, -ise (rek'əgnīz), *v.t.* to know again; to recall the identity of; to acknowledge, to admit the truth, validity, existence etc. of; to acknowledge the legality of (a government); to reward, to thank; to show appreciation of. *v.i.* (*N Am.*) to enter into recognizances. **recognizable,** *a.* **recognizability** (-bil'-), *n.* **recognizably,** *adv.* **recognizant** (rikog'ni-), *a.* **recognizer, -iser,** *n.*

recoil (rikoil'), *v.i.* to start or spring back; to rebound; to shrink back, as in fear or disgust; to be driven back; to retreat; to go wrong and harm the perpetrator. *n.* the act of recoiling; a rebound; the act or feeling of shrinking back, as in fear or disgust; the backward kick of a gun when fired; the change in motion of an atom caused by emission of a particle. **recoiler,** *n.* **recoilingly,** *adv.* **recoilment,** *n.*

recollect[1] (rēkəlekt'), *v.t.* to gather together again; to collect or compose (one's ideas, thoughts or feelings); to summon up, to rally, to recover (one's strength, spirit etc.). *v.i.* to come together again.

recollect[2] (rekəlekt'), *v.t.* to recall to memory, to remember, to succeed in recalling the memory of. *v.i.* to succeed in remembering. **recollection,** *n.* the act or power of recollecting; a memory, a reminiscence; the period of past time over which one's memory extends. **recollective,** *a.* **recollectively,** *adv.*

recolonize, -ise, *v.t.* to colonize afresh. **recolonization, -isation,** *n.* **recolour,** *v.t.* to colour again. **recombine,** *v.t., v.i.* to combine again. **recombinant** (rikom'bi-), *n., a.* **recombinant DNA,** *n.* DNA prepared in the laboratory by combining DNA molecules from different individuals or species. **recombination,** *n.* the process of combining genetic material from different sources. **recomfort,** *v.t.* to comfort or console again; to give new strength to. **recomforture,** *n.* **recommence,** *v.t., v.i.* to begin again. **recommencement,** *n.*

recommend (rekəmend'), *v.t.* to commend to another's notice, use or favour, esp. to represent as suitable for an office or employment; to advise (a certain course of action etc.), to counsel; to render acceptable or serviceable (of qualities etc.); to give or commend (one's soul, a person etc.) in charge (to God etc.). **recommendable,** *a.* **recommendableness, recommendability** (-bil'-), *n.* **recommendably,** *adv.* **recommendation,** *n.* the act of recommending; a quality or feature that tends to procure a favourable reception, a ground of approbation; a letter recommending a person for an appointment etc. **recommendatory,** *a.* **recommender,** *n.*

recommission, *v.t.* **recommit,** *v.t.* **recommitment, recommittal,** *n.* **recommunicate,** *v.t.*

recompense (rek'əmpens), *v.t.* to make a return or give an equivalent for, to repay (a person, a service, an injury etc.); to compensate (for), to make up (for). *n.* that which is given as a reward, compensation, requital or satisfaction (for a service, injury etc.). **recompensable,** *a.* **recompenser,** *n.* **recompensive,** *a.*

reconcile (rek'ənsīl), *v.t.* to restore to friendship after an estrangement; to make content, acquiescent or submissive (to); to harmonize, to make consistent or compatible (with); to adjust, to settle (differences etc.); in the Roman Catholic Church, to reconsecrate (a desecrated church etc.). **reconcilable,** *a.* **reconcilability** (-bil'-), *n.* **reconcilableness,** *n.* **reconcilably,** *adv.* **reconcilement, reconciliation** (-sili-), *n.* **reconciler,** *n.* **reconciliatory** (-sil'i-), *a.*

recondite (rek'əndīt), *a.* out of the way, abstruse, little known, obscure; pertaining to abstruse or special knowledge, profound. **reconditely,** *adv.* **reconditeness,** *n.*

recondition (rēkəndish'ən), *v.t.* to repair, to make as new. **reconditioned,** *a.*

reconduct, *v.t.* **reconfirm,** *v.t.*

reconnaissance (rikon'əsəns), *n.* the act of reconnoitring, a preliminary examination or survey, esp. of a tract of country or a coast-line in war-time to ascertain the position of the enemy, the strategic features etc.; a detachment of soldiers or sailors performing this duty.

reconnoitre (rekənoi'tə), *v.t.* to make a reconnaissance of; to make a preliminary examination or survey of. *v.i.* to make a reconnaissance. *n.* a reconnaissance. **reconnoitrer,** *n.* [F]

reconquer, *v.t.* **reconquest,** *n.* **reconsecrate,** *v.t.* **reconsecration,** *n.* **reconsider,** *v.t., v.i.* to consider again (esp. with a view to rescinding); to review, to revise. **reconsideration,** *n.* **reconsolidate,** *v.t., v.i.* **reconsolidation,** *n.* **reconstitute,** *v.t.* **reconstituent,** *n. a.* **reconstitution,** *n.* **reconstruct,** *v.t.* to construct again; to rebuild; to build up a picture of something from the available evidence, e.g. of a crime. **reconstructible,** *a.* **reconstruction,** *n.* the act or process of reconstruction; (*N Am. Hist.*) the process by which the southern States were restored to Federal rights and privileges after the Civil War of 1861–5. **reconstructional,** *a.* **reconstructionary,** *a.* **reconstructive,** *a.* **reconstructively,** *adv.* **reconstructor,** *n.*

reconvene, *v.t., v.i.* **reconvention,** *n.* **reconvert,** *v.t.* **reconversion,** *n.* **reconvey,** *v.t.* **reconveyance,** *n.*

record[1] (rikawd'), *v.t.* to write an account of, to set down permanent evidence of, to imprint deeply on the mind; to make a recording (on disc, tape, video tape etc.) of for subsequent reproduction; to bear witness to; to indicate, to register. *v.i.* to make a record or recording. **recordable,** *a.* **recorded,** *a.* **recorded delivery,** *n.* a postal service where an official record of posting and delivery is kept. **recording,** *a.* registering waveforms arising from sound sources, or the readings of meteorological and other instruments. *n.* a record of sound or image on record, tape or film; the record, tape or film so produced; a radio or television programme which has been recorded. **Recording Angel,** *n.* an angel supposed to keep a record of every person's good and bad deeds.

record[2] (rek'awd, -əd), *n.* a written or other permanent account or statement of a fact or facts; a register, a report, a minute or minutes of proceedings, a series of marks made by a recording instrument; a thin plastic disc

on to which sound is recorded; an official report of proceedings, judgment etc. to be kept as authentic legal evidence, or an official memorial of particulars, pleadings etc. to be submitted as a case for decision by a court; the state of being recorded, testimony, attestation; the past history of a person's career, esp. as an index of character and abilities; a list of crimes which a person has committed; the authentic register of performances in any sport; the best performance or the most striking event of its kind recorded; a portrait, monument or other memento of a person, event etc. **court of record**, a court whose proceedings are officially recorded and preserved as evidence. **for the record**, for the sake of accuracy. **off the record**, in confidence, not said officially. **on record**, recorded, esp. with legal authentication. **to beat, break, the record**, to surpass all former achievements or events of the kind. **to go on record**, to state one's beliefs publicly. **to have a record**, to be a known, previously convicted criminal. **to put, set, the record straight**, to correct an error or false impression. **record-breaker**, n. **record-breaking**, n., a. **Record Office**, n. an official repository for state papers. **record-player**, n. a machine for playing and reproducing sounds on a record. **recordable**, a. **records**, n.pl. **public records**, official statements of public deeds or acts.
recorder (rikaw'də), n. one who or that which records; a magistrate having a limited criminal and civil jurisdiction in a city or borough and presiding over quarter-sessions; a machine for recording sound onto tape; a vertical form of flute with a fipple. **recordership**, n.
recount[1] (rikownt'), v.t. to relate in detail, to narrate. **recountal**, n.
recount[2] (rēkownt'), v.t. to count over again, esp. of votes at an election. n. a new count.
recoup (rikoop'), v.t. to reimburse, to indemnify (oneself) for a loss or expenditure; to compensate, to make up for (a loss, expenditure etc.); (*Law*) to keep back (a part of something due). v.i. (*Law*) to make such a deduction. **recoupable**, a. **recouper**, n. **recoupment**, n.
recourse (rikaws'), n. resorting or applying (to) as for help; a source of help, that which is resorted to; the right to demand payment. **to have recourse to**, to go to for advice, help etc., esp. in emergency. **without recourse**, a qualified endorsement of a bill or negotiable instrument which shows that the endorser takes no responsibility for non-payment.
recover[1] (rikŭv'ə), v.t. to regain, to repossess oneself of, to win back; to make up for, to retrieve; to save (the by-products of an industrial process); to bring (a weapon) back after a thrust etc.; to obtain by legal process. v.i. to regain a former state, esp. after sickness, misfortune etc.; to come back to consciousness, life, health etc.; to be successful in a suit; to make a recovery. n. the position of a weapon or the body after a thrust etc.; the act of coming back to this. **recoverable**, a. **recoverableness**, n. **recoverability**, n. **recoverer**, n. **recovery**, n. the act of recovering or the state of having recovered; restoration to health after sickness etc.; the obtaining of the right to something by the judgment of a court; a golf stroke played on to the fairway or green from a bunker or the rough; the action of bringing the arm, an oar etc. back ready for another stroke; the retrieval of by-products from an industrial process.
re-cover[2] (rēkŭv'ə), v.t. to cover again, to put a new covering on.
recreant (rek'riənt), a. craven, cowardly; disloyal. n. a coward, a mean-spirited wretch, an apostate, a deserter. **recreance**, n. **recreancy**, n. **recreantly**, adv.
recreate[1] (rek'riāt), v.t. to entertain, to amuse. v.i. to take recreation. **recreation**[1], n. the act of refreshing oneself or renewing one's strength after toil; amusement, diversion; an amusing or entertaining exercise or employment. **recreation ground**, n. a communal open space in an urban area. **recreational, recreative**, a. refreshing, reinvigorating. **recreatively**, adv. **recreativeness**, n.
re-create[2] (rēkriāt'), v.t. to create anew. a. re-created. **re-creation**[2], n. **re-creator**, n. **re-creative** etc.
recriminate (rikrim'ināt), v.i. to retort an accusation, to bring counter-charges against. v.t. to accuse in return. **recrimination**, n. the act of bringing a counter-charge, of accusing in return; a counter-charge. **recriminative, recriminatory**, a. **recriminator**, n.
recrudesce (rēkrudes'), v.i. to open, break out or become raw or sore again. **recrudescence**, n. the state of becoming sore again; a relapse, a breaking-out again; the production of a young shoot from a ripened spike etc.; (*loosely*) a renewal, a reappearance. **recrudescent**, a.
recruit (rikroot'), v.t. to enlist (soldiers, sailors or airmen); to bring into an organization as an employee, member etc.; to supply (an army, regiment, crew etc.) with recruits; to replenish with fresh supplies, to fill up gaps etc.; to restore to health, to refresh, to reinvigorate. v.i. to gain new supplies; to seek to recover health; to (seek to) obtain recruits. n. a service man or woman newly enlisted; (*fig.*) one who has newly joined an organization. **recruitable**, a. **recruital, recruitment**, n. **recruiter**, n. **recruiting ground**, n. any source of, or place from which, recruits may be gained. **recruiting-officer, -party, -sergeant**, n. persons engaged in enlisting recruits.
rectal RECTUM.
rectangle (rek'tang·gl), n. a plane rectilinear quadrilateral figure with four right-angles. **rectangled**, a. having an angle or angles of 90°. **rectangular** (-tang'gū-), a. shaped like a rectangle; rectangled; placed or having parts placed at right-angles. **rectangular coordinates**, n.pl. in a Cartesian system, coordinates which have axes perpendicular to each other. **rectangular hyperbola**, n. a hyperbola with asymptotes at right-angles. **rectangularity** (-la'-), n. **rectangularly**, adv.
rect(i)-, *comb. form.* straight; right.
rectify (rek'tifī), v.t. to set right, to correct, to amend, to adjust; to remedy, to redress; to refine or purify (spirit etc.) by repeated distillations and other processes; to determine the length of (an arc etc.); to transform (an alternating current) into a continuous one. **rectifiable**, a. **rectification** (-fi-), n. rectifying in all its senses; (*Radio*) the conversion of an alternating current into a direct current. **rectifier**, n.
rectilineal (rektilin'iəl), **rectilinear**, a. consisting of, lying or proceeding in a straight line; straight; bounded by straight lines. **rectilineally, rectilinearly**, adv. **rectilinearity** (-nia'ri-), n.
rectitude (rek'titūd), n. uprightness, rightness of moral principle, conformity to truth and justice.
recto (rek'tō), n. the right-hand page of an open book (usu. that odd numbered), opp to *verso*.
recto- RECTUM.
rector (rek'tə), n. a parson or incumbent of a Church of England parish who was formerly entitled to the whole of the tithes; the head of a religious institution, university, incorporated school etc.; a clergyman in charge of a parish in the Episcopalian Church. **Lord Rector** LORD. **rectorate** (-rət), n. a rector's term of office. **rectorial** (-taw'ri), a. pertaining to rector; the election of a Lord Rector. **rectorship**, n. **rectory**, n. the benefice or living of a rector with all its rights, property etc.; the house of a rector.
rectress (-tris), n. *fem.*
rectoscope, recto-uterine etc. RECTUM.
rectrix (rek'triks), n. (*pl.* **rectrices** (-sēz)) any of the quill-feathers in a bird's tail which guide its flight.
rectum (rek'təm), n. (*pl.* **-ta** (-tə)) the lowest portion of the large intestine extending to the anus. **rectal**, a. **rectally**, adv. **recto-**, comb. form. **rectocele** (-təsēl), n. prolapse of the anus with protrusion.
rectus (rek'təs), n. (*pl.* **-ti** (-tī)) one of various straight

muscles, esp. of the abdomen, thigh, neck and eyes.
recumbent (rikŭm′bənt), *a.* lying down, reclining. **recumbence, recumbency,** *n.* **recumbently,** *adv.*
recuperate (rikū′pərāt, -koo′-), *v.t.* to recover, to regain (financial losses etc.). *v.i.* to recover (from sickness, loss of power etc.). **recuperable,** *a.* **recuperation,** *n.* **recuperative, recuperatory,** *a.* **recuperator,** *n.*
recur (rikœ′), *v.i.* (*past, p.p.* **recurred**) to return, to go back to in thought etc.; to come back to one's mind; to happen again, to happen repeatedly; to be repeated indefinitely. **recurrence** (-kŭ′rəns), *n.* **recurring,** *a.* happening or being repeated. **recurring fever,** *n.* a relapsing fever. **recurring decimals,** *n.* figures in a decimal fraction that recur over and over again in the same order. **recursion** (-shən), *n.* the act of returning; the computation of a sequence from a preceding mathematical value. **recursive,** *a.* **recursively,** *adv.*
recurrent (rikŭ′rənt), *a.* returning, recurring, esp. at regular intervals; turning in the opposite direction (of veins, nerves etc.), running in an opposite course to those from which they branch. *n.* a recurrent nerve or artery, esp. one of the laryngeal nerves. **recurrently,** *adv.*
recurve (rikœv′), *v.t., v.i.* to bend backwards. **recurvate** (-vət), *a.* recurved, reflexed. **recurvation** (rē-), **recurvature** (-vəchə), *n.*
recusant (rek′ūzənt), *n. a.* (one) obstinately refusing to conform, esp. (*Eng. Hist.*) to attend the services of the Established Church. **recusance, -cy,** *n.*
recycle (rēsī′kl), *v.t.* to pass again through a system of treatment or series of changes, esp. a waste product (e.g. paper, glass), so as to make it reusable. *v.i.* to return to the original position so the operation can begin again, esp. of electronic devices. *n.* the repetition of a sequence of events. **recyclable,** *a.* **recycler,** *n.*
red[1] (red), *a.* of a bright warm colour, as blood, usually including crimson, scarlet, vermilion etc., of the colour at the least refracted end of the spectrum or that farthest from the violet; flushed, stained with blood; revolutionary, anarchistic; a sign of danger. *n.* a red colour or a shade of this; the red colour in roulette etc.; the red ball at billiards; a red pigment; red clothes; red wine; a revolutionary, an extreme radical, an anarchist. **to be in the red,** to be overdrawn at the bank. **to be on red alert,** to be in a state of readiness for a crisis or disaster. **to paint the town red,** to have a riotous time. **to put out the red carpet,** to give an impressive welcome. **to see red,** to become enraged. **red admiral** ADMIRAL. **red algae,** *n.pl.* one family of seaweeds, the Rhodophyceae, with red pigment as well as chlorophyll. **red-backed,** *a.* **red-backed shrike,** *n.* the butcher-bird, *Lanius collurio.* **red-backed spider,** *n.* (*Austral.*) a venomous spider with red spots on its back. **red bark,** *n.* a variety of cinchona. **red-bearded, -bellied, -berried, -billed,** *a.* having a red beard, belly, berries etc. **red biddy,** *n.* red wine mixed with methylated spirits. **red-blind,** *a.* colour-blind with regard to red. **red-blindness,** *n.* **red blood cell,** *n.* any blood cell containing haemoglobin, which carries oxygen to the tissues. **red-blooded,** *a.* vigorous, virile, crude. **red-bloodedness,** *n.* **redbreast,** *n.* the robin, *Erythacus rubecula.* **red-breasted,** *a.* **redbrick university,** *n.* one of the pre-1939 provincial universities in Britain. **redbud,** *n.* the Judas-tree, *Cercis canadensis.* **red cabbage,** *n.* a reddish-purple cabbage, used for pickling. **red-cap,** *n.* a popular name for any small bird with a red head, esp. the goldfinch; (*sl.*) a military policeman; (*N Am., coll.*) a railway porter. **red card,** *n.* a piece of red cardboard shown by a soccer referee to a player to indicate that he has been sent off. **red carpet,** *n.* a strip of red carpet put out for a celebrity or important person to walk on; deferential treatment. **red cedar,** *n.* any of various species of cedar, esp. a juniper with fragrant, red wood; the timber from such a tree. **red cent,** *n.* (*N Am. coll.*) a trifle of money. **red-cheeked,** *a.* **red-coat,**
n. a British soldier, so called from the scarlet tunics worn by line regiments. **red-coated,** *a.* **red coral,** *n.* any of several pinkish red corals used to make ornaments and jewellery. **red corpuscle,** *n.* a red blood cell. **Red Crescent,** *n.* the Red Cross Society in Muslim countries. **red cross,** *n.* St George's Cross, the English national emblem. **Red Cross Society,** *n.* an international society or organization for the provision of ambulance and hospital service for the wounded in time of war, and to assist in severe epidemics and national disasters in peace time, in accordance with the Geneva Convention of 1864. **redcurrant,** *n.* the small, red, edible berry from a shrub of the gooseberry family. **red deer,** *n.* a large species of deer with reddish coat and branching antlers *Cervus elaphus,* still wild in the Scottish Highlands, Exmoor etc. **red duster,** *n.* (*sl.*) the red ensign. **red dwarf,** *n.* a star with a relatively small mass and low luminosity. **red earth,** *n.* soil coloured red by iron compounds, found in tropical savanna. **red ensign,** *n.* a red flag with the Union Jack in one corner, used as the ensign of the British Merchant Navy. **red-eye,** *n.* the name of several American fishes, also of the European *Leuciscus erythrophthalmus,* with scarlet lower fins; low quality whisky. **red-eyed,** *a.* having bloodshot eyes, with lids red and inflamed with weeping. **red-faced,** *a.* flushed with embarrassment; with a red, florid complexion. **redfin,** *n.* any of several small fish with red fins. **red-fish,** *n.* the name of various American fishes, including the red-drum, the blue-back salmon, *Oncorhynchus nerka* etc.; a male salmon in the spawning season. **red flag,** *n.* the symbol of revolution or of communism; a danger signal. **red fox,** *n.* the common European fox, *Vulpes vulpes.* **red giant,** *n.* a giant red star with high luminosity. **red grouse** GROUSE[1]. **Red Guard,** *n.* a member of the militant Maoist youth movement in China, formed to preserve popular support for the regime and active during the cultural revolution in the 1960s. **red-haired,** *a.* having red hair. **red-handed,** *a.* having hands red with blood; (caught) in the very act (originally of homicide). **red hat,** *n.* a cardinal's hat; a staff officer. **redhead,** *n.* any person with red hair. **redheaded,** *a.* **red-heat,** *n.* the temperature at which a thing is red-hot; the state of being red-hot. **red herring,** *n.* herring dried and smoked; (*fig.*) anything which diverts attention from the real issue or line of enquiry. **to draw a red herring across the track,** (*fig.*) to distract attention by starting an irrelevant discussion. **red-hot,** *a.* heated to redness; (*fig.*) excited, furious, wildly enthusiastic. **red-hot poker,** *n.* the flame-flower. **Red Indian,** *n.* (*offensive*) a N American Indian. **red lead,** *n.* red oxide of lead used as a pigment. **red-letter day,** *n.* an auspicious or memorable day, because saints' days were marked with red letters in the calendar. **red light,** *n.* any signal to stop; a danger signal. **red-light area, district,** *n.* an area or district in a town where there is a collection of brothels etc. **red man** REDSKIN. **red meat,** *n.* beef and mutton. **red mullet,** *n.* an edible fish found in European waters. **redneck,** *n.* (*N Am., derog.*) a poor white farm labourer in the South; a reactionary person or institution. *a.* reactionary. **red ochre,** *n.* any of several red earths, used as pigments. **red pepper,** *n.* any of various pepper plants cultivated for their hot red fruits; the fruit of such a plant; cayenne pepper; the red fruit of the sweet pepper. **Red Planet,** *n.* Mars. **redpole, -poll,** *n.* a popular name for two species of birds of the Fringillidae family, from the red hue of their heads, esp. the greater redpole, the male linnet; (*pl.*) red-haired polled cattle. **red rag,** *n.* anything that excites rage as a red object is supposed to enrage a bull. **red rattle** RATTLE. **Red River cart,** *n.* (*Can.*) a strong, two-wheeled, horse- or ox-drawn cart from W Canada. **red rose,** *n.* the emblem of the House of Lancaster during the Wars of the Roses (1455–85). **red salmon,** *n.* any salmon with red flesh, especially the sockeye salmon. **red sanders,** *n.* red sandalwood. **redshank,** *n.*

the red-legged sand-piper, *Tringa totanus*. **red shift**, *n.* the shift of lines in the spectrum towards the red, caused by a receding light source. **red-shirt**, *n.* a follower of Garibaldi (1807–82); a revolutionary. **Redskin**, *n.* (*offensive*) a N American Indian. **red snapper**, *n.* any of several edible fish of the snapper family. **red snow**, *n.* snow reddened by a minute alga, *Protococcus nivalis*, frequent in Arctic and Alpine regions. **red spider**, *n.* a mite infesting vines and other hot-house plants. **Red Spot**, *n.* a reddish spot, oval in shape and about 48,000 km long which drifts around the southern hemisphere of Jupiter. **red squirrel**, *n.* a reddish squirrel found in Europe and some parts of Asia. **redstart**, *n.* a red-tailed migratory song-bird, *Phoenicurus phoenicurus*. **red-tape**, *n.* extreme adherence to official routine and formality (from the red tape used in tying up official documents). *a.* characterized by this. **red-tapery, redtapism**, *n.* **red-tapist**, *n.* **red tide**, *n.* the sea, when discoloured and made toxic by red protozoans. **red underwing**, *n.* a large moth with red and black hind wings. **red-water**, *n.* haematuria in cattle and sheep, the most marked symptom of which is the red urine. **red weed**, *n.* the corn poppy; herb Robert and other plants. **red wine**, *n.* wine coloured by grape skins. **redwing**, *n.* a variety of thrush, *Turdus musicus*, with red on the wings. **redwood**, *n.* a name of various trees and their timber, esp. the gigantic Californian *Sequoia sempervirens*. **redden**, *v.t.* to make red. *v.i.* to become red, esp. to blush. **reddish, reddy**, *a.* **reddishness**, *n.* **redly**, *adv.* **redness**, *n.*

red² (red), REDD.

redaction (ridak'shən), *n.* reduction to order, esp. revising, rearranging and editing a literary work; a revised or re-arranged edition. **redact**, *v.t.* to reduce to a certain form, esp. a literary form, to edit, to prepare for publication. **redacteur** (-tœ'), *n.* **redactional**, *a.* **redactor**, *n.* an editor. **redactorial** (redaktaw'ri-), *a.*

redan (ridan'), *n.* (*Fort.*) a work having two faces forming a salient towards the enemy.

redbreast etc. RED¹.

redd (red), *v.t.* (*Sc.*) to clean out, to get rid of; to adjust, to clear up, to put in order, to tidy, to make ready. **to redd up**, to put in order. **redder**, *n.*

redden etc. RED¹.

reddle, *var. of* RUDDLE².

rede (rēd), *n.* (*old use*) counsel, advice; resolve, intention. *v.t.* to counsel, to advise; to read or interpret (a riddle etc.).

redecorate, *v.t.* **redecoration**, *n.* **rededicate**, *v.t.* **rededication**, *n.* **rededicatory**, *a.*

redeem (ridēm'), *v.t.* to buy back, to recover by paying a price; to recover (mortgaged property), to discharge (a mortgage), to buy off (an obligation etc.); to perform (a promise); to recover from captivity by purchase, to ransom; to deliver, to save, to rescue, to reclaim; to deliver from sin and its penalty; to atone for, to make amends for; to make good. **redeemability** (-bil'-), **redeemableness**, *n.* **redeemable**, *a.* **redeemer**, *n.* one who redeems, esp. Christ, the Saviour of the world. **redeeming**, *a.* compensating for faults.

redefine, *v.t.* to define again or afresh.

redeliver, *v.t.* to deliver back, to restore, to free again; to repeat, to report. **redeliverance, redelivery**, *n.* **redemand**, *v.t.* to demand again or back. **redemise** (rēdimīz'), *v.t.* (*Law*) to transfer an estate etc.) back. *n.* a retransfer.

redemption (ridemp'shən), *n.* the act of redeeming or the state of being redeemed, esp. salvation from sin and damnation by the atonement of Christ; release by purchase, ransom; reclamation (of land etc.); purchase (of admission to a society etc.); that which redeems. **redemptive, redemptory**, *a.*

redeploy (rēdiploi'), *v.t.* to transfer (troops, labour force) from one area to another; to assign a new task to. **redeployment**, *n.* improved internal arrangements in a factory etc. as a means to improving output.

redescend, *v.t., v.i.* **redescribe**, *v.t.* **redescription**, *n.* **redesign**, *v.t.* **redetermine**, *v.t., v.i.* **redetermination**, *n.* **redevelop**, *v.t.* to develop again; to renovate and build in a depressed urban area. **redevelopment**, *n.* **redevelopment area**, *n.* an urban area where existing buildings are either demolished and rebuilt or renovated.

redid REDO.

redingote (red'ing-gōt), *n.* a woman's long double-breasted coat; orig. a similar coat worn by men. [F]

redintegrate (ridin'tigrāt), *v.t.* to restore to completeness, make united or perfect again; to renew, to re-establish. **redintegration**, *n.* **redintegrative**, *a.*

redirect, *v.t.* to direct onto a new route; to re-address. **redirection**, *n.* **rediscover**, *v.t.* **rediscovery**, *n.* **redissolve**, *v.t., v.i.* **redissoluble, redissolvable**, *a.* **redissolution**, *n.* **redistribute**, *v.t.* to distribute again. **redistribution**, *n.* the act of distributing again; the reallocation of seats in the Canadian House of Commons to each province according to population, carried out every 10 years from a census.

redivivus (redivī'vəs) *a.* come to life again. [L]

redo (rēdoo'), *v.t.* (*past* **redid**, *p.p.* **redone** (-dŭn')) to do again; to redecorate.

redolent (red'ələnt), *a.* giving out a strong smell; suggestive, reminding one (of). **redolence, -cy**, *n.* **redolently**, *adv.*

redouble, *v.t.* to double again; to increase by repeated additions, to intensify; to multiply; to fold back; to double an opponent's double, in bridge. *v.i.* to become increased by repeated additions, to grow more intense, numerous etc.; to be repeated, to re-echo. *n.* the act of redoubling. **redoublement**, *n.*

redoubt (ridowt'), *n.* a detached outwork or field-work enclosed by a parapet without flanking defences; a last retreat.

redoubtable (ridow'təbl), *a.* formidable; valiant. **redoubtableness**, *n.* **redoubtably**, *a.*

redound (ridownd'), *v.i.* to have effect, to conduce or contribute (to one's credit etc.); to result (to), to act in return or recoil (upon).

redox (rē'doks), *a.* pertaining to a chemical reaction where one agent is reduced and another oxidized. [*reduction oxidation*]

redpole etc. RED¹.

redraft, *v.t.* to draft or draw up a second time. *n.* a second draft. **redraw**, *v.t.* to draw again. *v.i.* to draw a fresh bill of exchange to cover a protested one. **redrawer**, *n.*

redress¹ (rēdres'), *v.t., v.i.* to dress again.

redress² (ridres'), *v.t.* to set straight or right again, to readjust, to rectify; to remedy, to amend, to make reparation for. *n.* redressing of wrongs or oppression; reparation; rectification. **redressable**, *a.* **redresser**, *n.* **redressment**, *n.*

redshank etc. RED¹.

reduce (ridūs'), *v.t.* to bring to a specified condition; to bring back (to); to modify (so as to bring into another form, a certain class etc.), to make conformable (to a rule, formula etc.); to bring by force (to a specified condition, action etc.), to subdue, to conquer; to bring down, to lower, degrade, to diminish; to weaken; to change from one denomination to another; to cause a chemical reaction with hydrogen; to remove oxygen atoms; to bring about an increase in the number of electrons; to lessen the density of a photographic print or negative. *v.i.* to resolve itself; to lessen. **to reduce to the ranks**, to degrade to the rank of private soldier. **reduced**, *a.* **in reduced circumstances**, poor, hard-up. **reducement**, *n.* **reducent**, *n., a.* **reducer**, *n.* one who or that which reduces; a piece of pipe for connecting two other pieces of different diameter. **reducible**, *a.* **reducibility** (-bil'-), **reducibleness**, *n.* **reducibly**, *adv.* **reducing**, *a.* **reducing agent**, *n.* a substance which reduces another in a chemical process. **redu-**

cing glass, *n.* a lens or mirror which reflects an image smaller than the actual object observed. **reductase** (-ās), *n.* any enzyme which reduces organic compounds. **reductive,** *a.*

reduction (ridŭk'shən), *n.* the act of reducing; the state of being reduced; a conquest; a decrease, a diminution; a reduced copy of anything; the process of making this; the process of finding an equivalent expression in terms of a different denomination; the process of reducing the opacity of a negative etc.; a term applied to any process whereby an electron is added to an atom. **reductio ad absurdum** (-tiō ad əbsœ'dəm), a method of disproving a proposition by showing that its logical consequences would be absurd; proof of the truth of a proposition by showing that its contrary has absurd consequences; (*coll.*) carrying something to the point of absurdity. **reductionism,** *n.* the explaining of complex data or phenomena in simpler terms. **reductionist,** *n., a.* **reductionistic** (-nis'-), *adv.*

redundant (ridŭn'dənt), *a.* superfluous, excessive, superabundant; deprived of one's job as one is no longer necessary; using more words than are necessary, pleonastic, tautological. **redundance, -cy,** *n.* **redundantly,** *adv.*

reduplicate (rēdū'plikāt), *v.t.* to redouble, to repeat; to repeat a letter or syllable to form a tense. *a.* doubled, repeated; of petals or sepals, with edges turned out. *n.* a duplicate. **reduplication,** *n.* **reduplicative,** *a.*

redwing, redwood etc. RED[1].

ree[1] (rē), *n.* the female ruff.

ree[2] REIS[1].

reebok (rē'bok), *n.* (*S Afr.*) a small S African antelope, *Pelea capreola.*

re-echo (rēck'ō), *v.t.* to echo again; to return the sound, to resound. *v.i.* to echo again; to reverberate.

reed (rēd), *n.* (the long straight stem of) any of a number of water or marsh plants belonging to the genera *Phragmites, Arundo* or *Ammophila;* (*collect.*) these as material for thatching etc.; a musical pipe made of this; a thin strip of metal or wood inserted in an opening in a musical instrument, set in vibration by a current of air to produce the sound; hence, a musical instrument or organ-pipe constructed with this (*usu. in pl.*); an implement or part of a loom for separating the threads of the warp and beating up the weft; a semicircular moulding, usu. in parallel series (*usu. in pl.*). *v.t.* to thatch with reed; to fit (an organ-pipe etc.) with a reed; to decorate with reeds. **broken reed,** *n.* an unreliable person. **reedband,** *n.* a consort of reed instruments. **reed-bird,** *n.* the bobolink. **reedbuck,** *n.* an antelope with a buff-coloured coat, found south of the Sahara in Africa. **reed-bunting,** *n.* a common European bunting with a black head, *Emberiza schoeniclus.* **reed-grass,** *n.* any of the reeds or any grasses of similar habit. **reed-instrument,** *n.* a woodwind instrument with a reed, such as an oboe. **reed-mace,** *n.* the bulrush. **reed-organ,** *n.* a musical instrument with a keyboard, the sounds of which are produced by reeds of the organ type. **reed-pheasant,** *n.* the bearded titmouse. **reed-pipe,** *n.* a reeded organ-pipe; a musical pipe made of a reed. **reed-stop,** *n.* an organ-stop controlling a set of reed-pipes. **reed warbler,** *n.* a common European bird, *Acrocephalus scirpaceus.* **reeding,** *n.* a semi-cylindrical moulding or series of these; milling on the edge of a coin. **reedless** (-lis), *a.* **reedling** (-ling), *n.* the bearded titmouse. **reedy,** *a.* abounding in reeds; like a reed; sounding like a reed, thin, sharp in tone; thin, frail in form. **reediness,** *n.*

re-edit, *v.t.* to edit afresh.

re-educate (rēed'ūkāt), *v.t.* to educate once more; to teach new skills to; to rehabilitate through education. **re-education,** *n.*

reef[1] (rēf), *n.* a ridge of rock, coral sand etc. in the sea at or near the surface of the water; a lode or vein of gold-bearing quartz. **reefy,** *a.*

reef[2] (rēf), *n.* one of the horizontal portions across the top of a square sail or the bottom of a fore-and-aft sail, which can be rolled up or wrapped and secured in order to shorten sail. *v.t.* to reduce the extent of a sail by taking in a reef or reefs; to take in a part of (a bowsprit, top-mast etc.) in order to shorten it. **reef-knot,** *n.* a square or symmetrical double knot. **reef-line,** *n.* a small rope passing through eyelet-holes for reefing a sail. **reef-point,** *n.* a short length of rope stitched to a sail, for attaching a reef. **reefer**[1], *n.* one who reefs; a reef-knot; a reefing-jacket; (*Naut. sl.*) a midshipman. **reefing-jacket,** *n.* a stout, close-fitting double-breasted jacket.

reefer[2] (rē'fə), *n.* a marijuana cigarette.

reek (rēk), *n.* smoke; vapour, steam, fume; a foul, stale or disagreeable odour, a foul atmosphere. *v.i.* to emit smoke, vapour or steam; to give off a disagreeable odour; to give a strong impression of. **Auld Reekie** (-i), *n.* Edinburgh. **reeking, reeky,** *a.* smoky; filthy, dirty.

reel[1] (rēl), *n.* a cylindrical frame or other device on which thread, cord, wire, paper etc. can be wound, either in the process of manufacture or for winding and unwinding as required; a bobbin; the spool on which a film is wound; a quantity of (material) wound on a reel. *v.t.* to wind out a reel; to unwind or take (off) a reel. **to reel in, up,** to wind (thread, a line etc.) on a reel; to draw (a fish etc.) towards one by winding line onto a reel. **to reel off,** to unwind or pay out from a reel; (*fig.*) to tell (a story), recite (a list) etc. fluently and without a hitch. **reel-fed,** *a.* pertaining to printing on a reel or web of paper. **reel-line,** *n.* an angler's line wound on a reel, esp. the back part as distinguished from the casting-line. **reel-plate,** *n.* a metal plate on an angler's reel fitting into a groove etc. on a fishing-rod. **reel-to-reel,** *a.* of magnetic tape, wound from one reel to another; using such tape. **reelable,** *n.* **reeler,** *n.* **reeling,** *n., a.* **reelingly,** *adv.*

reel[2] (rēl), *v.i.* to stagger, to sway; to go (along) unsteadily; to have a whirling sensation, to be dizzy; to be staggered, to rock. *n.* a staggering or swaying motion or sensation. **reeling,** *n., a.* **reelingly,** *adv.*

reel[3] (rēl), *n.* a lively Scottish or Irish dance in which the couples face each other and move in figures-of-eight; a piece of music for this. *v.i.* to dance a reel. **Virginia reel,** an American country dance.

re-elect, *v.t.* to elect again. **re-election,** *n.* **re-elective,** *v.t.* to elevate again. **re-elevation,** *n.* **re-eligible,** *a.* capable of being re-elected to the same position. **re-eligibility,** *n.* **re-embark,** *v.t., v.i.* to embark again. **re-embarkation,** *n.* **re-embattle,** *v.t.* to array again for battle. **re-embody,** *v.t.* to embody again. **re-embodiment,** *n.* **re-embrace,** *v.t.* to embrace again. *n.* a second embrace. **re-emerge,** *v.i.* to emerge again. **re-emergence,** *n.* **re-emergent,** *a.* **re-enable,** *v.t.* to make able again. **re-emphasize, -ise,** *v.t.* to emphasize again. **re-enact,** *v.t.* to enact again. **re-enactment,** *n.* **re-endow,** *v.t.* to endow again. **re-enforce,** *v.t.* to enforce again. **re-engage,** *v.t., v.i.* to engage again. **re-engagement,** *n.* **re-engine,** *v.t.* to furnish with new engines. **re-enlist,** *v.t., v.i.* to enlist again. **re-enter,** *v.t., v.i.* to enter again. **re-entrance,** *n.* **re-entrant,** *a.* re-entering, pointing inward. *n.* an inward-pointing angle, esp. in fortification, opp. to *salient.* **re-entry,** *n.* the act of re-entering; a new entry in a book etc.; the return of a spacecraft into the earth's atmosphere. **re-equip,** *v.t.* to equip again. **re-erect,** *v.t.* to erect again. **re-erection,** *n.* **re-establish,** *v.t.* to establish anew, to restore. **re-establisher,** *n.* **re-establishment,** *n.* **re-evaluate,** *v.t.* to evaluate again. **re-evaluation,** *n.*

reeve[1] (rēv), *n.* historically, a chief officer or magistrate of a town or district, holding office usually under the king but sometimes by election.

reeve[2] (rēv), *v.t.* (*past, p.p.* **rove** (rōv), **reeved**) to pass (the end of a rope, a rod etc.) through a ring, a hole in a

block etc.; to fasten (a rope etc.) by this means.
reeve[3] (rēv), *n.* the female of the ruff.
re-examine, *v.t.* to examine again. **re-examination**, *n.* **re-exchange**, *v.t.* to exchange again. **re-exhibit**, *v.t.* to exhibit again. **re-exist**, *v.i.* to exist again. **re-existence**, *n.* **re-existent**, *a.* **re-export**, *v.t.* to export again; to export after having been imported; to export (imported goods) after processing. *n.* a commodity re-exported. **re-exportation**, *n.* **re-exporter**, *n.* **reface**, *v.t.* to put a new face or surface on. **refacing**, *n.* **refashion**, *v.t.* to fashion anew. **refashioner**, *n.* **refasten**, *v.t.* to fasten again.
ref, (*abbr.*) referee; reference.
ref (ref), *n.* referee. *v.t., v.i* (*past, p.p.* **reffed**) to referee.
refectory (*rifek'təri*), *n.* a room or hall where meals are taken in religious houses, colleges etc. **refectory table**, *n.* a long narrow dining table, esp. on trestles.
refer (rifœ'), *v.t.* (*past, p.p.* **referred**) to trace back, to assign (to a certain cause, source, place etc.); to hand over (for consideration and decision); to send or direct (a person) for information etc.; to fail (an examinee). *v.i.* to apply for information; to appeal; to cite, to allude, to direct attention (to); to be concerned (with), to have relation (to). **referable** (ref'ə-, rifœ'-), *a.* **referrable**, *a.* **referent** (ref'ə-), *n.* that to which a word or phrase refers. **referential** (refərən'-), *a.* **referentially**, *adv.* **referral**, *n.* the act of referring or being referred. **referred**, *a.* **referred sensation**, *n.* pain or other sensation localized at a different point from the part actually causing it. **referrible** REFERABLE.
referee (refərē'), *n.* person to whom a point or question is referred; a person to whom a matter in dispute is referred for settlement or decision, an arbitrator, an umpire; a person who is prepared to testify to the abilities and character of someone, and who furnishes testimonials. *v.t., v.i.* to act as a referee at (a football match etc.).
reference (ref'ərəns), *n.* the act of referring; relation, respect; allusion, directing of attention (to); a note or mark referring from a book to another work or from the text to a commentary, diagram etc.; that which is referred to; a person referred to for information, evidence of character etc., a referee; a testimonial. **cross-reference** CROSS[1]. **in, with reference to**, with regard to, as regards, concerning. **terms of reference**, the specific limits of the scope (of an investigation or piece of work). **without reference to**, irrespective of, regardless of. **reference Bible**, *n.* a Bible with cross-references in the margin. **reference book, work**, *n.* an encyclopaedia, dictionary or the like, consulted when occasion requires, not for continuous reading. **reference library**, *n.* a library where books may be consulted but not borrowed.
referendum (refərən'dəm), *n.* (*pl.* **-da** (-də), **-dums**) the submission of a political question to the whole electorate for a direct decision by general vote.
refill, *v.t.* to fill again. *n.* (rē'-), that which is used to refill; a fresh fill (as of lead for a pocket-pencil, tobacco for a pipe etc.). **refillable**, *a.* **refind**, *v.t.* to find again.
refine (rifīn'), *v.t.* to clear from impurities, defects etc., to purify, to clarify; to free from coarseness, to educate, to cultivate the taste, manners etc. of; to make (a statement, idea etc.) more sophisticated, subtle, complex or abstract; to transform or modify into a more sophisticated, subtler or more abstract form. *v.i.* to become pure or clear; to become polished or more highly cultivated in talk, manners etc.; to affect subtlety of thought or language; to draw subtle distinctions (upon). **refinable**, *a.* **refined**, *a.* freed from impurities; highly cultivated, elegant. **refinedly** (-nid-), *adv.* **refinedness**, *n.* **refinement** (rifīn'mənt), *n.* the act or process of refining, the state of being refined; elegance of taste, manners, language etc.; high culture or sophistication; subtlety; a (more) subtle distinction or piece of reasoning. **refiner** (rifī'nə), *n.* one who refines, esp. a person whose business it is to refine metals etc.; an apparatus for purifying coal-gas etc.; one who invents superfluous subtleties or distinctions. **refinery**, *n.* a place for refining raw materials, such as sugar or oil.
refit, *v.t.* to make fit for use again, to repair, to fit out anew (esp. a ship). *v.i.* to be repaired, be made fit for use again. *n.* (rē'-), the repairing or renewing of what is damaged or worn out, esp. the repairing of a ship. **refitment**, *n.*
refl., (*abbr.*) reflection, reflective, reflectively; reflex, reflexive, reflexively.
reflate (rēflāt'), *v.t.* to inflate again; to be subject to reflation. **reflation**, *n.* an increase in economic activity, esp. through an increase in the supply of money and credit, after deflation. **reflationary**, *a.*
reflect (riflekt'), *v.t.* to turn or throw (light, heat, sound etc.) back, esp. in accordance with certain physical laws; to mirror, to throw back an image of; (*fig.*) to reproduce accurately, to correspond to in features or effects; to cause to accrue or to cast (honour, disgrace etc.) upon; to show, to give an idea of. *v.i.* to throw back light, heat, sound etc.; to turn the thoughts back, to think, to ponder, to meditate (on); to remind oneself (that); to bring shame or discredit (on or upon). **reflectance**, *n.* the amount of light or radiation reflected by a surface. **reflectible**, *a.* **reflecting**, *a.* **reflecting factor**, *n.* reflectance. **reflecting microscope**, *n.* a microscope with a series of mirrors instead of lenses. **reflecting telescope**, *n.* a telescope in which the object glass is replaced by a polished reflector, from which the image is magnified by an eyepiece. **reflectingly**, *adv.* casting censure or reflections (upon), reproachfully. **reflection**, (*dated*) **reflexion** (-flek'shən), *n.* the act of reflecting; the state of being reflected; that which is reflected; rays of light, heat etc. or an image thrown back from a reflecting surface; the act or process by which the mind takes cognizance of its own operations; further or continued consideration, thought, meditation; a thought, idea, comment or opinion resulting from deliberation; censure, reproach (brought or cast upon etc.); reflex action. **reflectional**, *a.* **reflectionless** (-lis), *a.* **reflective**, *a.* throwing back an image, rays of light, heat etc.; pertaining to or concerned with thought or reflection; meditative, thoughtful; taking cognizance of mental operations. **reflectively**, *adv.* **reflectiveness**, *n.* **reflectivity** (rēflektiv'-), *n.* the ability to reflect radiation. **reflector**, *n.* one who or that which reflects, esp. a reflecting surface that throws back rays of light, heat etc., usu. a polished, concave surface, as in a lamp, lighthouse, telescope, surgical or other instrument etc. **reflet** (rə-flā'), *n.* a metallic lustre or glow.
reflex (rē'fleks), *a.* turned backward; introspective; turned back upon itself or the source, agent etc.; bent back, recurved; reflected, lighted by reflected light; involuntary, produced independently of the will under stimulus from impressions of the sensory nerves. *n.* a reflection; a reflected image, reproduction or secondary manifestation; an involuntary nervous action; reflected light, colour etc. *v.t.* (-fleks') to bend or fold back, to recurve. **conditioned reflex**, a behaviouristic mechanism or reaction in which the previous experience of the responder produces the same reaction to a given stimulus at each presentation. **reflex anal dilatation**, involuntary widening of the anus on physical examination, used somewhat controversially as a diagnostic procedure to detect repeated anal penetration, esp. in cases involving suspected sexual abuse in children. **reflex action**, *n.* the involuntary contraction of a muscle in response to stimulus from without the body. **reflex arc**, *n.* the nervous pathway which nerve impulses travel along to produce a reflex action. **reflex camera**, *n.* a camera in which the main lens is used as a viewfinder. **reflexed**, *a.* bent backwards or downwards. **reflexible**, *a.* **reflexibility** (-bil'-), *n.* **reflexive**, *a.* denoting action upon the agent; implying action by the subject upon itself/himself/herself

or referring back to the grammatical subject. **reflexive verb,** *n.* a verb that has for its direct object a pronoun which is also for the agent or subject. **reflexively,** *adv.* **reflexiveness,** *n.* **reflexivity** (rēfleksiv'-), *n.* **reflexly,** *adv.*
reflexology (rēfleksol'-), *n.* a form of alternative medical therapy where certain areas of the soles of the feet are massaged to stimulate supposedly corresponding areas of the body's circulation and nerves, and so release tension. **reflexologist,** *n.*
reflexion REFLECT.
refloat, *v.t., v.i.* to float again. **reflorescence,** *n.* a second florescence. **reflorescent,** *a.* **reflourish,** *v.i.* to flourish anew. **reflow,** *v.i.* to flow back; to ebb. *n.* a reflowing, a reflux; the ebb (of the tide). **reflower,** *v.i.* to flower again.
refluent (ref'luənt), *a.* flowing back; ebbing. **refluence,** *n.*
reflux (rē'flŭks), *n.* a flowing back; a return, an ebb; the boiling of liquid in a flask fitted with a condenser, so that the vapour condenses and flows back into the flask.
refold, *v.t.* to fold again. **refoot,** *v.t.* to put a new foot to (a stocking, etc.). **reforest,** *v.t.* to reafforest. **reforestation,** *n.* **reforge,** *v.t.* to forge over again; to refashion.
re-form (rēfawm'), *v.t.* to form again or anew. **re-formation,** *n.* **re-former,** *n.*
reform (rifawm'), *v.t.* to change from worse to better by removing faults, imperfections, abuses etc.; to improve, to amend, to redress, to remedy. *v.i.* to amend one's habits, morals, conduct etc.; to abandon evil habits etc. *n.* the act of reforming; an alteration for the better, amendment, improvement, etc. **Reform Acts,** *n.pl.* Acts passed in 1832, 1867, and 1884 for enlarging the electorate and reforming the constitution of the House of Commons. **Reform Judaism,** *n.* a form of Judaism which adapts Jewish Law to contemporary life. **reform school,** *n.* a reformatory. **reformable,** *a.* **reformability** (-bil'-), *n.*
reformation (refəmā'shən), *n.* the act of reforming; the state of being reformed; redress of grievances or abuses, esp. a thorough change or reconstruction in politics, society or religion. **the Reformation,** *n.* the great religious revolution in the 16th cent. which resulted in the establishment of the Protestant Churches. **reformational,** *a.* **reformationist,** *n.* **reformative** (-faw'mə-), *a.* tending to produce reformation. **reformatory** (-faw'mə-), *a.* reformative. *n.* formerly, an institution for the detention and reformation of juvenile offenders.
reformed (rifawmd'), *a.* corrected, amended, purged of errors and abuses. **Reformed Church,** *n.* one of the Protestant Churches that adopted Calvinistic doctrines, distinguished from Lutheran Churches.
reformer (rifaw'mə), *n.* one who effects a reformation; one who favours (esp. political) reform; one who took a leading part in the Reformation of the 16th cent. **reformism,** *n.* any policy advocating religious or political reform.
reformulate, *v.t.* to formulate again. **reformulation,** *n.* **refortify,** *v.t.* to fortify anew. **refound,** *v.t.* to found anew; to recast.
refract (rifrakt'), *v.t.* to deflect or turn (a ray of light etc.) from its direct course (of water, glass or other medium differing in density from that through which the ray has passed). **refractable,** *a.* **refracted,** *a.* deflected from a direct course, as a ray of light or heat; bent back at an acute angle. **refracting,** *a.* **refracting telescope,** *n.* the earliest form of telescope, in which the image of an object is received direct through a converging lens and magnified by an eye-piece. **refraction,** *n.* the deflection that takes place when a ray of light, heat etc. passes at any other angle than a right angle from one medium into another medium of different density; the amount of deflection which takes place. **astronomical refraction,** the deflection of a ray of light proceeding from a heavenly body to the eye of a spectator on the earth, due to the refracting power of the atmosphere. **double refraction,** the splitting of a ray of light, heat etc. into two polarized rays which may be deflected differently on entering certain materials, e.g. crystals. **refractional, refractive,** *a.* **refractive index,** *n.* the amount by which a medium refracts light. **refractivity** (rēfraktiv'-), *n.* **refractometer** (rēfraktom'itə), *n.* any instrument which measures the refractive index of a medium. **refractometric** (-met'-), *a.* **refractometry** (-tri), *n.* **refractor,** *n.* a refracting medium, lens or telescope.
refractory (rifrak'təri), *a.* perverse, obstinate in opposition or disobedience, unmanageable; not amenable to ordinary treatment; not easily fused or reduced, not easily worked; fire-resistant; slowly, or not, responsive to stimulus (of nerves etc.). *n.* a fire-resistant material used for lining kilns etc. **refractorily,** *adv.* **refractoriness,** *n.*
refracture, *v.t.* to fracture again.
refrain[1] (rifrăn'), *n.* the burden or chorus of a song, a phrase, line or group of lines usu. repeated at the end of every stanza.
refrain[2] (rifrăn'), *v.i.* to hold back, to restrain, to curb, to abstain (from an act or doing). **refrainer,** *n.*
reframe, *v.t.* to frame again, to fashion anew.
refrangible (rifran'jibl), *a.* capable of being refracted. **refrangibility** (-bil'-), *n.*
refreeze, *v.t.* to freeze again.
refresh (rifresh'), *v.t.* to make fresh again; to reanimate, to reinvigorate; to revive or restore after depression, fatigue etc.; to stimulate (one's own memory); to restore; (*coll.*) to give (esp. liquid) refreshments to. *v.i.* (*coll.*) to take (esp. liquid) refreshment. **refresher,** *n.* one who or that which refreshes; an extra fee paid to counsel when a case is adjourned or continued from one term or sitting to another; (*coll.*) a drink. **refresher course,** *n.* a course to bring up to date one's knowledge of a particular subject. **refreshing,** *a.* (re)invigorating, (re)animating. **refreshingly,** *adv.* **refreshingness,** *n.* **refreshment,** *n.* the act of refreshing; the state of being refreshed; that which refreshes, esp. (*pl.*) food or drink. **refreshment room,** *n.* a room at a railway station etc. for the supply of refreshments.
refrigerate (rifrij'ərāt), *v.t.* to make cool or cold; to freeze or keep at a very low temperature in a refrigerator so as to preserve in a fresh condition. **refrigerant,** *a.* cooling, allaying heat. *n.* a liquid or other substance used for cooling; a medicine for allaying fever or inflammation. **refrigeration,** *n.* **refrigerative,** *n., a.*
refrigerator (rifrij'ərātə), *n.* an apparatus for keeping meat and other provisions in a frozen state or at a very low temperature, in order to preserve their freshness. **refrigeratory,** *a.* cooling. *n.* a vessel attached to a still for condensing vapour; a refrigerator.
refringent (rifrin'jənt), *a.* refractive. **refringence, refringency** (-si), *n.*
refuel, *v.t.* to provide with fresh fuel. *v.i.* to take on fresh fuel.
refuge (ref'ūj), *n.* (a) shelter, retreat, sanctuary or protection from danger or distress; a place, thing, person or course of action that shelters or protects from danger, distress or calamity; an expedient, a subterfuge; a raised area in the middle of a road forming a safe place for pedestrians to halt at. **house of refuge,** a charitable institution for the destitute and homeless. **refugee** (-jē', *n.* one who flees to a place of refuge, esp. to a foreign country in time of war or persecution or political commotion. **refugium** (rifū'jiəm), *n.* (*pl.* **-gia** (-jiə)) a geographical region which has not been changed by geographical or climatic conditions and so becomes a haven for varieties of flora and fauna rare elsewhere.
refulgent (rifūl'jənt), *a.* shining brightly, brilliant, radiant, splendid. **refulgence,** *n.* **refulgently,** *adv.*
refund (rifŭnd'), *v.t.* to pay back, to repay, to restore; to reimburse. *n.* (rē'-) the money etc. reimbursed. **refundable,** *a.* **refunder,** *n.* **refundment,** *n.*
re-fund (rēfŭnd'), *v.t.* to fund anew; to pay off an old debt

by borrowing more money.
refurbish, *v.t.* to fit out anew.
refurnish, *v.t.* to furnish anew; to supply with new furniture.
refusal (rifū'zəl), *n.* (an) act of refusing; denial of anything solicited, demanded or offered. **first refusal** the choice or option of taking or refusing something before it is offered to others.
refuse[1] (rifūz'), *v.t.* to decline to do, yield, grant etc.; to deny the request, demand or marriage proposal of; to decline to jump over (a fence etc.). *v.i.* to decline (to comply, yield, grant, do etc.); to fail to jump (of a horse); to be unable to follow suit. **refusenik, refusnik** (-nik), *n.* a Soviet Jew who has been refused permission to emigrate; (*coll.*) a person who refuses to cooperate in some way. **refuser**, *n.*
refuse[2] (ref'ūs), *a.* rejected; valueless. *n.* that which is refused or rejected as worthless; waste or useless matter. **refuse tip**, *n.* a place where refuse is heaped or disposed of.
re-fuse (rēfūz'), *v.t.* to fuse or melt again. **re-fusion** (-zhən), *n.*
refusenik, refusnik REFUSE[1].
refute (rifūt'), *v.t.* to prove (a statement, argument etc.) false or erroneous, to disprove. **refutable**, *a.* **refutably**, *adv.* **refutal, refutation** (ref-), *n.* **refutatory, refuter**, *n.*
reg., (*abbr.*) regent; regiment; register, registrar; registry; regular, regularly; regulation.
regain (rigān'), *v.t.* to recover possession of; to reach again; to gain again, to recover. **regainable**, *a.* **regainer**, *n.* **regainment**, *n.*
regal[1] (rē'gəl), *a.* pertaining to or fit for a sovereign or sovereigns. *n.* kingly, royal, magnificent. **regally**, *adv.*
regal[2] (rē'gəl), *n.* a small portable reed-organ held in the hands, in use in the 16th and 17th cents. (*often in pl.* form, as *a pair of regals*).
regale (rigāl'), *v.t.* to entertain sumptuously; to delight, to entertain (with). **regalement**, *n.* **regaler**, *n.*
regalia (rigā'liə), *n.pl.* the insignia of royalty, esp. the emblems worn or displayed in coronation ceremonies etc.; ceremonial insignia or finery in general.
regalism (rē'gəlizm), *n.* the doctrine of the royal supremacy in ecclesiastical affairs. **regality** (-gal'-), *n.* royalty, kingship; sovereign jurisdiction.
regally REGAL[1].
regalvanize, -ise, *v.t.* to galvanize again.
regard (rigahd'), *v.t.* to look at, to observe, to notice; to give heed to, to pay attention to, to take into account; to value, to pay honour to, to esteem; to look upon or view (in a specified way or with fear, reverence etc.), to consider (as); to pertain to, to concern. *n.* a look, a gaze; observant attention, heed, care, consideration; esteem, kindly or respectful feeling; reference, relation; (*pl.*) compliments, good wishes. **as regards**, with respect to, concerning. **in, with regard to**, concerning. **in this regard**, on this point. **with kind regards**, with good wishes. **regardant**, *a.* (*Her.*) looking backward; observant, watchful. **regarder**, *n.* **regardful**, *a.* showing regard, respect or consideration. **regardfully**, *adv.* **regardfulness**, *n.* **regarding**, *prep.* respecting, concerning. **regardless** (-lis), *a.* heedless, careless, negligent (of). *adv.* in spite of anything, heedlessly. **regardlessly**, *adv.* **regardlessness**, *n.*
regather, *v.t., v.i.* to gather or collect again.
regatta (rigat'ə), *n.* a race-meeting at which yachts or boats contend for prizes.
regency (rē'jənsi), *n.* the post commission or government of a regent; a body entrusted with the office or duties of a ruler; the period of office of a regent or a body so acting. *a.* (*often with cap.*) of the style of architecture, art etc. obtaining during the early 19th cent. in Britain. **the Regency**, *n.* the period (1810–20) when George, Prince of Wales, was regent for George III. **regent** (rē'jənt), *a.* (*pred.*) exercising the authority of regent. *n.* a person appointed to govern a kingdom during the minority, absence or disability of a sovereign. **regentess** (-tis), *n. fem.* **regentship**, *n.*
regenerate (rijen'ərāt), *v.t.* to change fundamentally and reform the moral and spiritual nature of; to impart fresh vigour to; to generate anew, to give new existence to. *a.* (-rət) regenerated, renewed. **regeneracy, regenerateness, regeneration**, *n.* **regenerative** (-tiv), *a.* **regenerator**, *n.* **regeneratory**, *a.* **regeneratively**, *adv.* **regeneratrix** (-triks), *n. fem.*
regenesis, *n.* the state of being born again or reproduced.
regerminate, *v.i.* to germinate anew. **regermination**, *n.*
reggae (reg'ā), *n.* a form of rhythmical W Indian rock music in 4/4 time.
regicide (rej'isid), *n.* (one who takes part in) the killing of a king. **regicidal** (-sī'-), *a.*
regild, *v.t.* to gild again.
regime, régime (rāzhēm'), *n.* mode, conduct or prevailing system of government or management; the prevailing social system or general state of things. **ancien régime** (äsyī'), the system of government and society prevailing in France before the Revolution of 1789.
regimen (rej'imən), *n.* the systematic management of food, drink, exercise etc. for the preservation or restoration of health; a course of food etc. governed by such management; the syntactical dependence of one word on another. **regiminal** (rijim'inəl), *a.*
regiment (rej'imənt), *n.* a body of soldiers forming the largest permanent unit of the army, usu. divided into two battalions comprising several companies or troops, and commanded by a colonel. *v.t.* (-ment) to form into a regiment or regiments; to organize into a system of bodies or groups; to discipline esp. (over-) strictly or harshly. **regimental** (-men'-), *a.* of or pertaining to a regiment. *n.pl.* military uniform. **regimentally**, *adv.* **regimentation**, *n.* (esp. excessive) organization into a regiment or a system of groups etc.
regina (rijī'nə), *n. fem.* (esp. *Law*) a reigning queen.
region (rē'jən), *n.* a tract of land, sea, space etc. of large but indefinite extent having certain prevailing characteristics, as of fauna or flora; a part of the world or the physical or spiritual universe (*often in pl.*); a district, a sphere, a realm; a civil (local government) division of a town or district; one of the strata into which the atmosphere or the sea may be divided; an area of the body surrounding an organ etc. **in the region of**, near; approximately. **the infernal, lower, nether regions**, hell, Hades. **upper regions**, the higher strata of the atmosphere or the sea; the sky; heaven. **regional**, *a.* sectionalism on a regional basis; loyalty to one's region. **regionalist**, *n.* **regionalistic** (-lis'-), *a.* **regionalize, ise**, *v.t.* to organize into administrative regions. **regionalization**, *n.* **regionally**, *adv.* **regioned**, *a.*
régisseur (rāzhēsœ'), *n.* an official in a dance company whose responsibilities include directing. **régisseuse** (-œz'), *n. fem.*
register (rej'istə), *n.* an official written record; a book, roll or other document in which such a record is kept; an official or authoritative list of names, facts etc., as of births, marriages, deaths, persons entitled to vote at elections, shipping etc.; an entry in such a record or list; a cash register; a contrivance for regulating the admission of air or heat to a room, ventilator, fireplace etc.; the range or compass of a voice or instrument; a particular portion of this; (*a*) level of colloquialism or formality of language used in a particular situation; a sliding device in an organ for controlling a set of pipes; (*Print.*) precise correspondence of lines etc. on one side of the paper to those on the other; (*Colour print.*) exact overlaying of the different colours used; (*Phot.*) correspondence in position of the surface of a sensitized film to that of the focusing-screen; a computer device which can store small amounts

of data. *v.t.* to enter in a register; to record as in a register; to cause to be entered in a register, esp. (a letter etc.) at a post office for special care in transmission and delivery; to record, to indicate (of a recording or measuring instrument; to show visibly (an emotion etc.)); (*Print., Phot. etc.*) to cause to correspond precisely in position. *v.i.* to enter one's name in or as in a register; to make an impression; (*Print. etc.*) to be in register; (*Artill.*) to carry out experimental shots in order to ascertain the exact range of a target. **in register**, exactly corresponding in position (of printed matter, photographic and colour plates etc.). **parish register** PARISH. **ship's register** SHIP. **register office**, *n*. a registry office. **register ton**, *n*. a unit used to measure the internal capacity of a ship, of 100 cu. ft. (about 3 m³). **registered**, *a*. **Registered General Nurse**, *n*. a nurse who has passed the General Nursing Council for Scotland's examination. **registered post**, *n*. a postal service where a registration fee is paid for mail and compensation paid in case of loss; mail sent by this service. **registered trademark**, *n*. a trademark which is legally registered and protected. **registrable**, *a*. **registrant**, *n*. a person registering (esp. a trademark etc.).

registrar (rejistrah', rej'-), *n*. an official keeper of a register or record; an official charged with keeping registers of births, deaths and marriages; a hospital doctor between the grades of houseman and consultant. **Registrar-General**, *n*. a public officer who superintends nationally the registration of births, deaths and marriages. **registrarship**, *n*. **registration**, *n*. the act of registering; the state of being registered; an entry in a register. **registration document**, *n*. a document which shows the official details of a motor vehicle. **registration number**, *n*. a combination of letters and numbers, displayed by every motor-vehicle, showing place and year of registration.

registry (rej'istri), *n*. an office or other place where a register is kept; registration; a register; the place where a ship is registered. **registry office**, *n*. an employment agency for domestic servants; a registrar's office where civil marriages etc. are performed.

regius (rē'jiəs), *a*. royal; appointed by the sovereign. **Regius Professor**, *n*. one of several professors at Oxford and Cambridge Univs. whose chairs were founded by Henry VIII, or in Scottish Univs. whose chairs were founded by the Crown.

reglaze, *v.t.* to glaze again. **reglazing**, *n*.

regnal (reg'nəl), *a*. of or pertaining to a reign. **regnal day**, *n*. the anniversary of a sovereign's accession. **regnal year**, *n*. the year of a reign dating from the sovereign's accession (used in dating some documents). **regnant**, *a*. reigning, ruling; predominant, prevalent. **regnancy**, *n*.

regorge, *v.t.* to disgorge, to vomit up. *v.i.* to gush or flow back (from a river etc.). **regraft**, *v.t.* to graft again. **regrant**, *v.t.* to grant anew. *n*. a renewed or fresh grant.

regress (rē'gres), *n*. passage back; return, regression. *v.i.* (rigres') to move back, to return. **regression** (-shən), *n*. going backwards, esp. in terms of progress etc.; reversion to type; return to an earlier form of behaviour; the turning back of a curve upon itself. **regressive**, *a*. **regressively**, *adv*. **regressiveness**, *n*. **regressivity** (rēgresiv'-), *n*. **regressor**, *n*.

regret (rigret'), *n*. distress or sorrow for a disappointment, loss or want; (sometimes *pl.*) grief, repentance or remorse for a wrong-doing, fault or omission (esp. in offering an apology); disappointment. *v.t.* (*past, p.p.* **regretted**) to be distressed or sorry for (a disappointment, loss etc.); to regard (a fact, action etc.) with sorrow or remorse. **regretful**, *a*. sorry for past action. **regretfully**, *adv*. **regretfulness**, *n*. **regrettable**, *a*. to be regretted. **regrettably**, *adv*.

regroup, *v.t., v.i.* to group together again. **regrow**, *v.t., v.i.* to grow again. **regrowth**, *n*.

Regt, (*abbr.*) regent; regiment.

regulable REGULATE.

regular (reg'ūlə), *a*. conforming to or governed by rule, law, type or principle; systematic, methodical, consistent, symmetrical, unvarying; harmonious, normal; acting, done or happening in an orderly, uniform, constant or habitual manner, not casual, fortuitous or capricious; conforming to custom, etiquette etc., not infringing conventions; complete, thorough, out-and-out, unmistakable; (*N Am.*) popular, likeable; conforming to the normal type of inflection; governed throughout by the same law, following consistently the same process; having the sides and angles equal; of a soldier, belonging to the standing army, opp. to territorials, yeomanry etc. *n*. a soldier belonging to a permanent army; (*coll.*) a person permanently employed or constantly attending (as a customer etc.). **regularity** (-la'-), *n*. **regularize, -ise**, *v.t.* **regularization**, *n*. **regularly**, *adv*.

regulate (reg'ūlāt), *v.t.* to adjust, control or order, esp. by rule; to subject to restrictions; to adjust to requirements, to put or keep in good order; to reduce to order. **regulable**, *a*. **regulation**, *n*. the act of regulating; the state of being regulated; a prescribed rule, order or direction. *a*. (*coll.*) prescribed by regulation; formal, normal, accepted, ordinary, usual. **regulative**, *a*. tending to regulate. **regulator** (reg'ūlātə), *n*. one who or that which regulates; a clock keeping accurate time, used for regulating other timepieces; the lever of a watch or other contrivance for regulating or equalizing motion, pressure etc. **regulator valve**, *n*. a valve in a locomotive which controls the supply of steam to the cylinders. **regulatory**, *a*.

regulo (reg'ūlō), *n*. the temperature of a gas oven, given by numbers on a scale. [from Regulo®, trademark for a type of thermostat on gas ovens]

regulus (reg'ūləs), *n*. (*pl.* **-luses** or **-li** (-lī)), an intermediate product of smelting retaining to a greater or lesser extent the impurities of the ore.

regurgitate (rigœ'jitāt), *v.t.* to throw or pour back again; to bring back (partially digested food) into the mouth after swallowing. *v.i.* to gush or be poured back. **regurgitant**, *a*. **regurgitation**, *n*.

rehabilitate (rēhəbil'itāt), *v.t.* to restore to a former rank, position, office or privilege, to reinstate; to re-establish (one's character or reputation); to make fit (after disablement, imprisonment etc.) for making a living or playing a part in the life of society; to restore (a building) to good condition. **rehabilitation**, *n*. re-establishment of character or reputation; the branch of occupational therapy which deals with the restoration of the maimed or unfit to a place in society. **rehabilitative**, *a*.

rehandle, *v.t.* to handle or deal with again. **rehang**, *v.t.* to hang (e.g. curtains) again. **reharness**, *v.t.* to harness again. **rehash**, *v.t.* to work over again; to remodel (a design etc.), esp. in a perfunctory or ineffective manner. *n*. (rē'-), something stated or presented again under a new form.

rehear, *v.t.* to hear a second time; to try (a legal case) over again. **rehearing**, *n*. a second hearing; a retrial.

rehearse (rihœs'), *v.t.* to repeat, to recite; to relate, to recount, to enumerate; to recite or practise (a play, musical performance, part etc.) before public performance. *v.i.* to take part in a rehearsal. **rehearsal**, *n*. the act of rehearsing; a preparatory performance of a play etc. **rehearser**, *n*. **rehearsing**, *n*.

reheat, *v.t.* to heat again; to inject fuel into a jet aircraft's exhaust gases, to produce more thrust. *n*. the process by which thrust is produced in an aircraft by the ignition of fuel added to exhaust gases. **reheater**, *n*. an apparatus for reheating, esp. in an industrial process. **reheel**, *v.t.* to heel (a shoe etc.) again. **rehire**, *v.t.* to hire again (usu. after dismissal). **rehouse**, *v.t.* to house anew. **rehumanize, -ise**, *v.t.* to humanize again. **rehypothecate**, *v.t.* to hypothecate again; to pledge again. **rehypothecation**, *n*.

rehoboam (rēəbō'əm), *n*. a wine bottle (especially a

champagne bottle) which holds six times the amount of a standard bottle, approximately 156 fl. oz. (about 4·6 l).
Reich (rīkh), *n*. the German realm considered as an empire made up of subsidiary states. **First Reich**, the Holy Roman Empire (962–1806). **Second Reich**, Germany under the Hohenzollern emperors (1871–1918). **Third Reich**, Germany under the Nazi regime (1933–45). **Reichsmark** (rīkhs′mahk), *n*. the standard monetary unit of Germany between 1924 and 1948. **Reichsrat** (rīkhs′raht), *n*. the old Austrian parliament. **Reichstag** (-tahg), *n*. the parliament of the German Reich (1867–1933); the building this parliament met in. [G, kingdom]
reify (rē′ifī), *v.t.* to make (an abstract idea) concrete; to treat as real. **reification** (-fi-), *n*. **reificatory**, *a*. **reifier**, *n*.
reign (rān), *n*. supreme power, sovereignty, dominion; rule, sway, control, influence; the period during which a sovereign reigns; a kingdom, realm, sphere. *v.i.* to exercise sovereign authority, to be a king or queen; to predominate, to prevail.
reignite, *v.t.* to ignite again. **reillume, reillumine**, *v.t.* to light up again, to illumine again. **reillumination**, *n*. **reimburse**, *v.t.* to compensate (one who has spent money); to refund (expenses etc.). **reimbursable**, *a*. **reimbursement**, *n*. **reimburser**, *n*. **reimplant**, *v.t.* to implant again. **reimplantation**, *n*. **reimport**, *v.t.* to import again after exportation; to import goods made from exported raw materials. **reimportation**, *n*. **reimpose**, *v.t.* to impose again. **reimposition**, *n*. **reimpress**, *v.t.* to impress anew. **reimpression**, *n*. **reimprint**, *v.t.* to imprint again; to reprint. **reimprison**, *v.t.* to imprison again. **reimprisonment**, *n*.
rein (rān), *n*. a long narrow strip, usu. of leather, attached at each end to a bit, for guiding and controlling a horse or other animal in riding or driving; (*in pl.*) a similar device for controlling a young child; (*fig.*) means of restraint or control (*often in pl.*). *v.t.* to check, to control, to manage with reins; to pull (in or up) with reins; (*fig.*) to govern, to curb, to restrain. **to draw rein**, to pull up. **to give rein, the reins to**, to leave unrestrained; to allow (a horse) to go its own way. **to keep on a tight rein**, to control carefully. **to rein in**, to cause a horse to stop by pulling on the reins. **to take the reins**, to assume guidance, direction, office etc.
reinaugurate, *v.t.* to inaugurate again. **reincarnate**, *v.t.* to incarnate anew; to cause to be born again. *a*. born again in a new body. **reincarnation**, *n*. passage of the soul from one body to another after physical death. **reincarnationism**, *n*. (the) belief in reincarnation. **reincarnationist**, *n*.
reincense, *v.t.* to incense anew. **reincite**, *v.t.* to incite anew. **reincorporate**, *v.t.* to incorporate again. **reincorporation**, *n*. **reincur**, *v.t.* to incur again.
reindeer (rān′diə), *n*. a deer now inhabiting the sub-arctic parts of the northern hemisphere, domesticated for the sake of its milk and as a draught animal.
reinduce, *v.t.* to induce again. **reinduction**, *n*. **reinfect**, *v.t.* to infect again. **reinfection**, *n*.
reinforce (rēinfaws′), *v.t.* to add (new) strength to; to strengthen or support (a military force etc.); to strengthen by adding to the size, thickness etc. of, to add a strengthening part to; to enforce again. **reinforceable**, *a*. **reinforced concrete**, *n*. concrete given great tensile strength by the incorporation of rods of iron etc., ferro-concrete. **reinforcement**, *n*. the act of reinforcing; the state of being reinforced; anything that reinforces; additional troops, ships etc. (*usu. in pl.*). **reinforcer**, *n*.
reinform, *v.t.* to inform again. **reinfuse**, *v.t.* to infuse again. **reingratiate**, *v.t.* to ingratiate (oneself) again. **reinhabit**, *v.t.* to inhabit again. **re-ink**, *v.t.* to ink again. **reinoculate**, *v.t.* to inoculate again. **reinoculation**, *n*.
reins (rānz), *n.pl.* the kidneys; the loins (formerly supposed to be the seat of the affections and passions).

reinscribe, *v.t.* to inscribe again. **reinsert**, *v.t.* to insert again. **reinsertion**, *n*.
reinspect, *v.t.* to inspect again. **reinspection**, *n*. **reinspire**, *v.t.* to inspire again. **reinstall**, *v.t.* to install again. **reinstalment**, *n*. **reinstate**, *v.t.* to restore, to replace (in a former job, post, position, state etc.); to replace, to repair (property damaged by fire etc.). **reinstatement**, *n*. **reinstruct**, *v.t.* to instruct again or in turn. **reinstruction**, *n*. **reinsure**, *v.t.* to insure against insurance risks. **reinsurance**, *n*. **reinsurer**, *n*. **reintegrate**, *v.t.*, *v.i.* to integrate again. **reintegration**, *n*. **reinter**, *v.t.* to inter or bury again. **reinterment**, *n*. **reinterpret**, *v.t.* to interpret again, or differently. **reinterpretation**, *n*. **reinterrogate**, *v.t.* to interrogate again. **reinterrogation**, *n*. **reintroduce**, *v.t.* to introduce or bring back into again. **reintroduction**, *n*. **reinvade**, *v.t.* to invade again. **reinvasion**, *n*. **reinvent**, *v.t.* to invent again. **reinvention**, *n*. **reinvest** (rēinvest′), *v.t.*, *v.i.* to invest again. **reinvestment**, *n*. **reinvestigate**, *v.t.* to investigate again. **reinvestigation**, *n*. **reinvigorate**, *v.t.* to reanimate; to give fresh vigour to. **reinvigoration**, *n*. **reinvite**, *v.t.* to invite again. **reinvolve**, *v.t.* to involve again. **reissue**, *v.t.*, *v.i.* to issue again. *n*. a second issue. **reissuable**, *a*. **reiterate**, *v.t.* to repeat (again and again). **reiteratedly**, *adv*. **reiteration**, *n*. **reiterative**, *a*. expressing or characterized by reiteration. *n*. a word or part of a word repeated so as to form a reduplicated word. **reiteratively**, *adv*.
reject (rijekt′), *v.t.* to put aside, to discard, to cast off; to refuse to accept (e.g. an implanted organ); to refuse to receive, grant etc., to deny (a request etc.); to repel, to cast up again, to vomit. *n*. (rē′-) something that has been rejected; something which is not perfect, substandard and offered for sale at a discount. **rejectable**, *a*. **rejectamenta** (-təmen′tə), *n.pl.* matter rejected, refuse, excrement. **rejecter, -tor**, *n*. **rejection**, *n*. **rejective**, *a*.
rejig (rējig′), *v.t.* to rearrange. *n*. (rē′-) the act of rejigging. **rejigger**, *n*.
rejoice (rijois′), *v.t.* to make joyful, to gladden. *v.i.* to feel joy or gladness in a high degree; to be glad (that or to); to delight or exult (in); to express joy or gladness; to celebrate, to make merry. **to rejoice in**, to be glad because of; (*often facet.*) to have. **rejoicer**, *n*. **rejoicing**, *n*. joyfulness; the expression of joyfulness, making merry, celebrating a joyful event (*usu. in pl.*). **rejoicingly**, *adv*.
rejoin (rijoin′), *v.t.* to join again; to join together again; to reunite after separation; to answer to a reply, to retort. *v.i.* to come together again; to answer a charge or pleading, esp. as the defendant. **rejoinder** (-də), *n*. an answer to a reply, a retort; a reply or answer in general; the answer of a defendant to the plaintiff's replication.
rejoint, *v.t.* to reunite the joints of; to fill up the joints of (stone-, brickwork etc.) with new mortar, to point. **rejudge**, *v.t.* to judge again; to re-examine (a legal case).
rejuvenate (rijoo′vənāt), *v.t.* to make young again; to restore to vitality. *v.i.* to become young or restored to vitality again. **rejuvenation**, *n*. **rejuvenator**, *n*. **rejuvenesce** (-nes′), *v.i.* to grow young again; (*Biol.*) to acquire fresh vitality (of cells). *v.t.* to rejuvenate. **rejuvenescence**, *n*. **rejuvenescent**, *a*. **rejuvenize, -ise**, *v.t.*, *v.i.*
rekindle, *v.t.* to kindle again; (*fig.*) to inflame or rouse anew. *v.i.* (*lit.*, *fig.*) to be kindled or aroused again.
rel., (*abbr.*) relative(ly); religion, religious.
relabel, *v.t.* to label again. **reland**, *v.t.*, *v.i.* to land again.
relapse (rilaps′), *v.i.* to fall or slip back (into a former bad or vicious state or practice), esp. into illness after partial recovery; to lapse again after moral improvement, conversion or recantation. *n*. a falling or sliding back into a former bad state, esp. of (poor) health after partial recovery. **relapser**, *n*.
relate (rilāt′), *v.t.* to tell, to narrate, to recount; to bring into relation or connection (with); to ascribe to as source or cause, to show a relation (with). *v.i.* to have relation or

regard (to); to refer (to); to get on well with. **relatable**, *a.* **related**, *a.* narrated; causally etc. connected; connected or allied by blood or marriage. **relatedness**, *n.* **relater**, *n.*
relation (rilā'shən), *n.* the act of relating; that which is related; a narrative, an account, a story; (in) respect (to); the condition of being related or connected; the way in which a thing stands or is conceived in regard to another as dependence, independence, similarity, difference, correspondence, contrast etc.; connection by blood or marriage, kinship; a person so connected, a relative, a kinsman or kinswoman; (*pl.*) business, dealings, affairs (with); (*euphem.*) sexual intercourse. **relational**, *a.* having, pertaining to or indicating relation. **relationally**, *adv.* **relationless** (-lis), *a.* **relationship**, *n.* the state of being related; connection by blood etc., kinship; (a) mutual connection between people or things; an emotional or sexual affair.
relative (rel'ətiv), *a.* being in relation to, association with, something, involving or implying relation, correlative; resulting from relation, in proportion (to something else), comparative; not absolute but depending on comparison with or relation to something else; corresponding, related; relevant, pertinent, closely related (to); (in correspondence) having reference, relating; (*Gram.*) referring or relating to another word, sentence or clause, called the antecedent; (*Mus.*) having the same key signature. *n.* a person connected by blood or marriage, a kinsman or kinswoman, a relation; a relative word, esp. a pronoun; something relating to or considered in relation to another thing, a relative term. **relative density**, *n.* the density of a substance as compared to the density of a standard substance under the same, or special, conditions. **relative frequency**, *n.* in statistics, the actual number of favourable events as compared to the total number of possible events. **relative humidity**, *n.* the amount of water vapour present in the air as compared to the same amount of saturated air at the same temperature. **relative majority**, *n.* the majority held by the winner of an election where no candidate has won more than 50% of the vote. **relatival** (-tī'-), *a.* **relatively**, *adv.* in relation to something else; comparatively. **relativeness**, *n.* **relativism**, *n.* the doctrine that truth, morals etc. are not absolute but relative to the specific culture concerned. **relativist**, *n.* **relativistic** (-vis'-), *a.* **relativity** (-tiv'-), *n.* **relativity theory**, *n.* a theory enunciated by Albert Einstein founded on the postulate that motion is relative, and developing the Newtonian conception of space, time, motion and gravitation.
relax (rilaks'), *v.t.* to slacken, to loosen; to allow or cause to become less tense or rigid; to make less strict or severe, to abate, to mitigate; to relieve from strain or effort. *v.i.* to become less tense, rigid, stern or severe; to grow less energetic; to take relaxation, to follow a leisure pursuit. **relaxant**, *a.* relaxing. *n.* a relaxing medicine. **relaxation** (rilək-), *n.* the act of relaxing; the state of being relaxed; (a) diminution (of tension, severity, application or attention); cessation from work, indulgence in recreation, amusement. **relaxative**, *a.* **relaxed**, *a.* informal. **relaxedly** (-sid-), *adv.* **relaxer**, *n.* **relaxing**, *a.*
relay (rē'lā), *n.* a supply of fresh horses, workers, hounds etc. to relieve others when tired; a supply of anything to be ready when required; a contrivance for strengthening a current over an unusually long distance; (a component used in) a type of electronic switch; the passing along of something by stages; a race run in stages by a succession of participants. *v.t.* (rilā'), to spread (information) by relays; to convey (information from one person to another); to broadcast (signals or a programme received from another station).
re-lay (rēlā'), *v.t.* to lay again.
release (rilēs'), *v.t.* to set free (from restraint), to liberate; to deliver (from pain, care, trouble, grief or other evil); to free (from obligation or penalty); to make (information) public; to emit; (*Law*) to surrender (a right, debt, claim etc.). *n.* liberation from restraint, pain, care, obligation or penalty; a discharge from liability, responsibility etc.; (*Law*) surrender or conveyance of property or right to another; the instrument or document in accordance with which this is carried out; a handle, catch or other device by which a piece of mechanism is released; anything newly issued for sale or to the public; a news item available for broadcasting. **releasee** (-sē'), *n.* (*Law*) a person to whom property is released. **releaser**, *n.* **releasor**, *n.* (*Law*) one releasing property or a claim to another.
re-lease (rēlēs'), *v.t.* to lease again.
relegate (rel'əgāt), *v.t.* to send away, to banish, to demote; to consign or dismiss (usu. to some inferior position etc., such as a football team to a lower division). **relegable**, *a.* **relegation**, *n.*
relent (rilent'), *v.i.* to become less harsh, severe or obdurate; to give way, to yield. **relenting**, *n.*, *a.* **relentingly**, *adv.* **relentless** (-lis), *a.* merciless, pitiless; unrelenting, constant. **relentlessly**, *adv.* **relentlessness**, *n.*
relet, *v.t.* to let again.
relevant (rel'əvənt), *a.* pertinent, applicable to the matter in hand, apposite. **relevance, -cy**, *n.* **relevantly**, *adv.*
reliable, reliant RELY.
relic (rel'ik), *n.* some part or thing remaining after loss or decay of the rest, a remnant, a fragment, a scrap, a survival, a trace; any ancient object of historical interest; something remaining or kept in remembrance of a person, esp. a part of the body or other object religiously cherished because of having belonged to some saint; a keepsake, a souvenir, a memento. **relic-monger**, *n.* one who trades in relics.
relict (rel'ikt), *a.* of a geological or geographical feature, such as a mountain, which is a remnant of an earlier, now eroded formation.
relief[1] (rilēf'), *n.* alleviation (of pain, grief, discomfort etc.); that which alleviates; assistance given to people in poverty or distress, esp. under the Poor Law; redress of a grievance etc., esp. by legal remedy or compensation; release from a post or duty by a person or persons acting as replacement or substitute; such a substitute; the raising of the siege of a besieged town; (*fig.*) anything that breaks monotony or relaxes tension. **comic relief**, dialogue, incidents or scenes of a comic nature alleviating the stress in a tragic play or story. **relief work**, *n.* charity organized for the unemployed, refugees etc.
relief[2] (rilēf'), *n.* the projection of carved or moulded figures or designs from a surface in approximate proportion to the objects represented; a piece of sculpture, moulding etc. with the figures etc. projecting; apparent projection of forms and figures due to drawing, colouring and shading; distinctness of contour, clearness, vividness. **relief map**, *n.* one in which hills and valleys are shown by prominences and depressions (usu. in exaggerated proportion) instead of contour lines.
relieve (rilēv'), *v.t.* to alleviate, to mitigate, to relax, to lighten; to free wholly or partially (from pain, grief, discomfort, legal burden etc.); to release (from a post, duty, responsibility etc.), esp. to take turn with (someone) on guard; to raise the siege of; (*coll.*) to take away from, to deprive of; to break or mitigate (monotony, dullness etc.); to give relief or prominence to, to bring out or make conspicuous by contrast. **to relieve oneself**, (*euphem.*) to defecate or urinate. **to relieve someone of something**, to take something without the owner's approval, to steal it. **relievable**, *a.* **relieved**, *a.* experiencing relief, especially from worry or emotion. **relievedly** (-vid-), *adv.* **reliever**, *n.* **relieving**, *n.*, *a.* **relieving arch**, *n.* one constructed in a wall to take the weight off a part underneath. **relieving-officer**, *n.* an officer appointed to superintend the relief of the poor in a parish or union.
relight, *v.t.*, *v.i.* to light, kindle or illumine again.

religieuse (rilēzhœz'), *n.* a nun. **religieux** (-zhœ'), *n.* a monk. [F]
religion (rilij'ən), *n.* belief in a superhuman being or beings, esp. a personal god, controlling the universe and entitled to worship and obedience; the feelings, effects on conduct and the dogma and practices resulting from such belief; a system of faith, doctrine and worship; the monastic state, the state of being bound by religious vows; anything of great personal importance. **to get religion**, to be converted. **religionism**, *n.* profession or affectation of religion; excessive or exaggerated religious zeal. **religionist**, *n.*, *a.* **religionize**, **-ise**, *v.t.* to make religious, to imbue with religion. *v.i.* to profess or display religion. **religionless** (-lis), *a.* **religiose** (-iōs), *a.* morbidly; pious, sanctimonious. **religiosely**, *adv.* **religiosity** (-ios'-), *n.* **religious**, *a.* pertaining to, or imbued with, religion; pious, devout, godly; bound by vows to a monastic life, belonging to a monastic order; (*fig.*) extremely conscientious, rigid, strict. *n.* one bound by monastic vows. **religious house**, *n.* a house for monks or nuns, a monastery, a convent. **religiously**, *adv.* in a religious manner; (*coll.*) scrupulously. **religiousness**, *n.*
reline, *v.t.* to line again, to give a new lining to.
relinquish (riling'kwish), *v.t.* to forsake, to abandon, to resign; to quit, to desist from; to give up a claim to, to surrender; to let go. **relinquisher**, *n.* **relinquishment**, *n.*
reliquary (rel'ikwəri), *n.* a depository for relics, a casket for keeping a relic or relics in. **reliquaire**, (relikweə'), *n.* a reliquary.
relish (rel'ish), *n.* taste, distinctive flavour, esp. pleasing; something taken with food to give a flavour, a pickled condiment; enjoyment (of food etc.), gusto, appetite, zest, fondness, liking; a slight admixture or flavouring, a smack, a trace (of); pleasing anticipation. *v.t.* to partake of with pleasure, to like, to be gratified by, to enjoy; to look forward to with pleasure. *v.i.* to have a flavour, to taste or smack (of). **relishable**, *a.*
relisten, *v.i.* to listen again. **relive**, *v.i.* to live again, to revive. *v.t.* to live over again, esp. in the imagination. **relivable**, *a.* **reload**, *v.t.*, *v.i.* to load (a firearm) again. **relocate**, *v.t.*, *v.i.* to locate again; to move (e.g. a factory, workers, business) to a new location. **relocation**, *n.*
reluctant (rilŭk'tənt), *a.* struggling or resisting, unwilling, averse, disinclined (to); doing, done or granted unwillingly. **reluctance**, *n.* unwillingness; the ratio of the opposition between a magnetic substance and the magnetic flux. **reluctantly**, *adv.*
relume, relumine, *v.t.* to light again, to rekindle; to make bright or light up again.
rely (rilī'), *v.i.* to trust or depend (upon) with confidence. **reliable** (-lī'-), *a.* that may be relied on; trustworthy. **reliability** (-bil'-), **reliableness**, *n.* **reliably**, *adv.* **reliance**, *n.* confident dependence (upon), trust; a person or thing relied on. **reliant**, *a.* **reliantly**, *adv.*
REM, (*abbr.*) rapid eye movement.
rem, (*abbr.*) remark(s).
rem (rem), *n.* a unit of radiation dosage which has the same biological effects as 1 rad of X-ray or gamma radiation. [röntgen equivalent *man* or *mammal*]
remain (rimān'), *v.i.* to stay behind or be left over after use, separation, destruction etc.; to survive; to continue in a place or state; to last, to abide, to continue, to endure; to continue to be; to be yet to be done or dealt with. *n.* (*usu. pl.*) that which remains behind after destruction, consumption etc.; a dead body, a corpse; literary productions published after one's death; relics. **remainder** (-də), *n.* anything left over after a part has been taken away, the rest, the residue; the quantity left over after subtraction, the fractional excess remaining after division into whole numbers; (*Law*) an interest in an estate which takes effect and is enjoyed after a prior estate is determined; copies of an edition left unsold after the demand has ceased and offered at a reduced price. *v.t.* to offer such copies at a reduced price. **remaining**, *a.*
remake, *v.t.* to make again or anew. *n.* (rē'-), anything made again from the original materials; a new version (of an old film, record etc.). **reman**, *v.t.* to man (a ship, gun etc.) again; to equip with a new complement of men.
remand (rimahnd'), *v.t.* to send back (to); to recommit in custody after a partial hearing. *n.* the act of remanding; the state of being remanded. **on remand**, in custody awaiting trial. **remand centre**, *n.* a place of detention for people awaiting trial. **remand home**, *n.* (*formerly*) a place where children aged 8–14 years were detained as punishment for criminal offences. **remandment**, *n.*
remanent (rem'ənənt), *a.* remaining, left behind, surviving. **remanence**, *n.* the ability of any magnetized substance to remain magnetic when the magnetizing force becomes zero. **remanet** (-nit), *n.* a remainder; a case postponed to another term; a Bill deferred to another parliamentary session.
remargin, *v.t.* to give a fresh margin to (a page etc.).
remark (rimahk'), *v.t.* to take notice of, to observe with particular attention, to perceive; to utter by way of comment, to comment (upon). *v.i.* to make a comment or observation (on). *n.* the act of noticing, observation; an observation, a comment; (*usu. pl.*) anything said, conversation; (*Engraving,* also **remarque**) a distinguishing mark, usu. a marginal sketch, indicating the particular state of an engraved plate. **remarkable**, *a.* worthy of special observation or notice, notable; unusual, extraordinary, striking. **remarkableness**, *n.* **remarkably**, *adv.* **remarked**, *a.* conspicuous; with a remark etched on. **remarker**, *n.*
re-mark, *v.t.* to mark again.
remarque REMARK.
remarry, *v.t.*, *v.i.* to marry again. **remarriage**, *n.* **remast**, *v.t.* to furnish with a new mast or masts. **remaster**, *v.t.* to make a new, master recording from an older original to improve the sound quality. **remasticate**, *v.t.* to chew over again. **remastication**, *n.*
Rembrandtesque (rembrantesk'), *a.* in the style or resembling the effects of the Dutch painter Rembrandt van Rijn (1609–69), esp. in chiaroscuro. **Rembrandtish** (rem'-), *a.*
REME (rē'mi), (*abbr.*) Royal Electrical and Mechanical Engineers.
remeasure, *v.t.* to measure again. **remeasurement**, *n.*
remedy (rem'ədi), *n.* that which cures a disease; medicine, healing treatment; that which serves to remove, counteract or repair any evil; redress, reparation; the tolerated variation in the standard weight of coins. *v.t.* to cure, to heal; to repair, to rectify, to redress. **remediable** (rimē'-), *a.* **remediableness**, *n.* **remediably**, *adv.* **remedial** (rimē'-), *a.* affording, containing or designed to provide for a remedy; pertaining to the teaching of slow learners, and people with special needs, such as the disabled, their books and materials, classes etc. **remedially**, *adv.* **remediless** (-lis), *a.*
remelt, *v.t.* to melt again.
remember (rimem'bə), *v.t.* to bear or keep in mind, not to forget; to know by heart; to recall to mind, to recollect; to keep in mind with gratitude, reverence or respect; (*coll.*) to convey a greeting from; to be good to, to make a present to, to tip; to commemorate (e.g. the dead). *v.i.* to have the power of, and exercise the, memory. **rememberable**, *a.* **rememberability** (-bil'-), *n.* **rememberably**, *adv.* **remembrance** (-brəns), *n.* the act of remembering; memory; the time over which memory extends; the state of being remembered; a recollection, a memory; that which serves to recall to or preserve in memory; a keepsake, a memento, a memorial; (*pl.*) regards, greetings. **Remembrance Day, Sunday,** *n.* the Sunday nearest to 11 Nov., when the dead of the two World Wars are remembered, also called Armistice Day. **remembrancer**, *n.*

remerge | 697 | **renascent**

one who or that which puts one in mind of something; a reminder, a memento.
remerge, *v.t.* to merge again.
remex (rē'meks), *n.* (*pl.* **remiges** (rem'ijēz)) one of the quill feathers of a bird's wings. **remiform** (rem'ifawm), *a.* oar-shaped. **remigate,** *v.i.* to row. **remigation,** *n.* **remigial** (rimij'iəl), *a.*
remigrate (rem'igrāt, rēmi'grāt), *v.i.* to migrate back again, to return to a former place or state. **remigration,** *n.*
remind (rimīnd'), *v.t.* to put in mind (of); to cause to remember (to do etc.). **reminder,** *n.* a person who, or thing which, reminds.
reminiscence (reminis'əns), *n.* the act or power of remembering or recalling past knowledge; that which is remembered; (*pl.*) a collection of personal recollections of past events; (*fig.*) something reminding or suggestive (of); the philosophical doctrine that the human mind has seen archetypes of everything before in an earlier, disembodied, existence. **reminisce,** *v.i.* to talk, think or write about past experiences. **reminiscent,** *a.* recalling past events to mind; of the nature of or pertaining to reminiscence; reminding or suggestive (of). *n.* one who records reminiscences. **reminiscential** (-sen'shəl), **reminiscitory,** *a.* **reminiscently,** *adv.*
remint, *v.t.* to mint over again.
remise (rimīz'), *n.* (*Law*) a release of property; a surrender of a claim; (*Fencing*) (-mēz'), a thrust following up one that misses before the opponent has time to recover. *v.t.* (-*Law*, mīz'), to surrender, to release or grant back; (-*Fencing*, mēz'), to make a remise.
remiss (rimis'), *a.* careless or lax in the performance of duty or business; heedless, negligent; slow, slack, languid. **remissible,** *a.* that may be remitted, admitting of remission. **remissibility** (-bil'-), *n.* **remissly,** *adv.* **remissness,** *n.*
remission (rimish'ən), *n.* the act of remitting; the remitting or discharge of a debt, penalty etc.; forgiveness, pardon; abatement (e.g. in the symptoms of a disease); diminution, reduction (e.g. of a prison sentence); remittance (of money etc.). **remissive,** *a.* **remissively,** *adv.* **remissly, remissness** REMISS. **remissory,** *a.* remitting, relieving, abating; forgiving.
remit (rimit'), *v.t.* (*past, p.p.* **remitted**) to send or put back; to transmit (cash, bills etc.); to refer or submit, to send back for consideration, to refer (to a lower court); to relax, to slacken, to mitigate, to desist from partially or entirely; to refrain from exacting etc., to forgo, to discharge from (a fine, penalty etc.). *v.i.* to become less intense, to abate. **remitment,** *n.* remittance (of money). **remittal,** *n.* a giving up, a surrender; remission from one court to another; remission (of offences etc.). **remittance,** *n.* the act of remitting money, bills or the like, in payment for goods etc.; a sum so remitted. **remittance-man,** *n.* an emigrant depending on remittances from home for his living. **remittee** (-ē'), *n.* one receiving a remittance. **remittent,** *a.* of an illness, having alternate increase and decrease of intensity. **remittently,** *adv.*
remix, *v.t.* to change the balance of (a recording). *n.* (rē'-) a remixed version of a recording.
remnant (rem'nənt), *n.* that which is left after a larger part has been separated, lost or destroyed; the remainder; the part surviving after destruction; the last part of a piece of cloth etc., esp. a portion offered at a reduced price; a scrap, a fragment, a surviving trace. *a.* surviving; remaining.
remodel, *v.t.* (*past, p.p.* **remodelled**) to model again; to refashion in a differente style. **remodify,** *v.t.* to modify again. **remodification,** *n.* **remonetize, -ise,** *v.t.* to reinstate (a metal etc.) as legal currency. **remonetization,** *n.*
remonstrance (rimon'strəns), *n.* the act of remonstrating; an expostulation, a protest; a formal representation or protest against public grievances etc. **remonstrant,** *a.* con-

taining or of the nature of remonstrance, expostulatory. *n.* one who remonstrates. **remonstrantly,** *adv.* **remonstrate** (rem'ənstrāt), *v.t.* to say or state in remonstrance. *v.i.* to make a remonstrance (with). **remonstratingly,** *adv.* **remonstration,** *n.* **remonstrative** (-mon'-), **remonstratory** (rem'-), *a.* **remonstrator,** *n.*
remontant (rimon'tənt), *a.* blooming more than once in the season (of roses). *n.* a rose blooming more than once in the season.
remora (rem'ərə), *n.* a sucking-fish having a disk for attaching itself to sharks, sword-fishes etc. and believed by the ancients to have the power of stopping ships in this way.
remorse (rimaws'), *n.* the pain caused by a sense of guilt, bitter repentance; compunction, reluctance to commit a wrong or to act cruelly. **remorseful,** *a.* feeling remorse; penitent. **remorsefully,** *adv.* **remorsefulness,** *n.* **remorseless** (-lis), *a.* without remorse; cruel, unstoppable, unceasing. **remorselessly,** *adv.* **remorselessness,** *n.*
remote (rimōt'), *a.* far off, distant in time or space; not at all closely connected or related; separated, different, alien, foreign; out-of-the-way, sequestered; slight, inconsiderable, little (*usu. in superl.*). *n.* a device for remote control of an apparatus etc. **remote control,** *n.* electric or radio control of apparatus etc. from a distance. **remote-control(led),** *a.* **remotely,** *adv.* **remoteness,** *n.*
rémoulade (rāmulahd'), *n.* a sauce, often made with mayonnaise, flavoured with herbs, mustard and capers and served with fish, cold meat, salads etc.
remould, *v.t.* to mould, fashion or shape anew. *n.* (rē'-) a used tyre which has had a new tread bonded into the casing and the walls coated with rubber. **remount,** *v.t.* to mount again, to re-ascend; to mount or set up (a gun, jewellery, a picture etc.) again; to supply (a regiment etc.) with fresh horses. *v.i.* to mount a horse again; to make a fresh ascent. *n.* (rē'-) a fresh horse for riding on; a fresh mount or setting.
remove (rimoov'), *v.t.* to move from a place; to move to another place; to take away, to get rid of; to transfer to another job, post or office; to dismiss. *v.i.* to go away (from), esp. to change one's place of abode. *n.* a degree of difference or gradation; a class or form (in some public schools); removal, change of place or position, departure. **removable,** *a.* able to be moved; liable to removal. **removability** (-bil'-), *n.* **removal,** *n.* the act of removing or displacing; a change of place, site or abode; dismissal; (*euphem.*) murder. **removed,** *a.* distant in space or time; distant in relationship. **once, twice removed,** separated by one or two degrees of closeness of relationship. **remover,** *n.* one who removes, esp. one whose business is to remove furniture from one house to another.
remunerate (rimū'nərāt), *v.t.* to reward, to recompense, to pay for a service etc.; to serve as recompense or equivalent (for or to). **remuneration,** *n.* **remunerative, remuneratory,** *a.* producing a due return for outlay; paying, profitable. **remuneratively,** *adv.* **remunerativeness,** *n.*
remurmur, *v.t.* to utter back in murmurs. *v.i.* to return a murmuring echo.
REN (ren), *abbr.* Ringer Equivalence Number, a unit by which the maximum number of appliances which may be connected to a single telephone line may be calculated.
Renaissance (rinā'səns, *esp. N Am.* ren'əsăs), *n.* the revival of art and letters in the 14th–16th cents.; the period of this; the style of architecture, painting, literature and science that was developed during this period; any revival of a similar nature, a rebirth. **Renaissance man, woman,** *n.* a person who has wide expertise and learning.
renal (rē'nəl), *a.* pertaining to the kidneys. **renal pelvis,** *n.* the cavity joining the kidney and the ureter.
rename, *v.i.* to name anew, to give a new name to.
renascent (rinas'ənt, -nā'-), *a.* coming into being again; pertaining to the Renaissance. **renascence,** *n.* rebirth,

renewal, a springing into fresh life; the Renaissance.
rencounter (rənkown'tə), **rencontre** (rĕkön'tr'), *n.* a hostile meeting or collision, an encounter, a combat, a duel, a skirmish; an unexpected meeting or encounter. *v.i.* to come together, to clash; to meet an enemy unexpectedly.
rend (rend), *v.t.* (*past, p.p.* **rent**[1] (rent)) to tear, pull or wrench (off, away, apart, asunder etc.); to split or separate with violence; (*fig.*) to cause anguish to; to pierce, disturb, with sound. *v.i.* to become torn or pulled apart. **render**[1], *n.*
render[2] (ren'də), *v.t.* to give in return; to pay or give back; to give up, to surrender; to bestow, to give, to pay, to furnish; to present, to submit, to hand in; to reproduce, to express, to represent, to interpret, to translate, to perform, to execute; to make, to cause to be; to boil down, to melt and clarify (fat); to give the first coat of plaster to. *n.* a return, a payment in return; the first coat of plaster on a wall etc. **renderable**, *a.* **renderer**, *n.*
rendering, *n.* a return; a translation, a version; an interpretation, execution (of a piece of music, a dramatic part etc.); the first coat of plaster on brickwork etc. **rendition**, *n.* surrender, giving up; translation, interpretation; an execution, performance, rendering (of music etc.).
rendezvous (ron'dāvoo, rē'-), *n.* (*pl. unchanged*, -vooz) a place appointed for assembling, esp. of troops, warships etc.; a (time and) place agreed upon for meeting; a general meeting-place. *v.i.* to meet or assemble at a rendezvous.
rendition RENDER[2].
renegade (ren'əgād), *n.* an apostate, esp. from Christianity; a deserter; a turncoat. *v.i.* to turn renegade. **renegation**, *n.* **renegue, renege** (rinĕg', -nāg'), *v.i.* to fail to follow suit at cards, to revoke; to make denial; to go back (on one's commitments or promises). **reneguer, reneger**, *n.*
renerve, *v.t.* to put fresh nerve or vigour into.
renew (rinū'), *v.t.* to make new again or as good as new, to renovate; to restore to the original or a sound condition; to make fresh or vigorous again, to reanimate, to revivify, to regenerate; to repair, to patch up, to replace (old or worn-out things with new); to make, do or say over again, to recommence, to repeat; to grant a further period of (a lease, patent, mortgage etc.); to obtain such a grant. *v.i.* to become young or new again; to grow again; to begin again. **renewable**, *a.* **renewability** (-bil'-), *n.* **renewal**, *n.* the act of renewing; the state of being renewed; revival, regeneration; a fee paid for continuance of anything. **renewer**, *n.* **renewing**, *n.*
renidify, *v.i.* to build another nest. **renidification**, *n.*
reniform (rē'nifawm), *a.* kidney-shaped.
renin (rē'nin), *n.* a protein enzyme secreted by the kidneys, which helps to maintain blood pressure.
rennet[1] (ren'it), *n.* curdled milk from the stomach of an unweaned calf etc. or an aqueous infusion from the stomach-membrane of the calf, used to coagulate milk; a similar preparation from seeds or other vegetable sources.
rennet[2] (ren'it), *n.* a name for several varieties of apple, esp. pippins.
renominate, *v.t.* to nominate again. **renomination**, *n.*
renounce (rinowns'), *v.t.* to reject or cast off formally, to repudiate, to disclaim, to disown; to forsake, to abandon; to forswear, to abjure; to give up, to withdraw from; (*Law*) to decline or resign a right or trust; to fail to follow suit at cards through having none left of that suit. *v.i.* to fail to follow suit. *n.* a failure to follow suit. **renouncement, renouncer**, *n.*
renovate (ren'əvāt), *v.t.* to make new again; to restore to a state of soundness or vigour; to repair. **renovation**, *n.* **renovative**, *a.* **renovator**, *n.*
renown (rinown'), *n.* exalted reputation, fame, celebrity. **renowned**, *a.* famous, celebrated. **renownedly** (-nid-), *adv.*

rent[1] (rent), (*past, p.p.*) REND.
rent[2] (rent), *n.* a tear, slit or breach, an opening made by rending or tearing asunder; a cleft, a fissure, a chasm; (*fig.*) a schism, a separation brought about by violent means.
rent[3] (rent), *n.* a sum of money payable periodically for the use of lands, tenements etc.; payment for the use of any kind of property; the return from cultivated land after production costs have been subtracted. *v.t.* to occupy, hold in tenancy or use in return for rent; to let for rent; to hire. *v.i.* to be let (at a certain rent). **for rent**, available for use on payment of rent. **rent-a-**, *comb. form* (*usu. facet.*) rented or hired, e.g. *rent-a-crowd, rent-a-mob*. **rent boy**, *n.* a young male homosexual prostitute. **rent-charge**, *n.* a periodical charge on land etc. granted to some person other than the owner. **rent collector**, *n.* **rent-day**, *n.* the day on which rent is due. **rent-free**, *a.* exempted from the payment of rent. *adv.* without payment of rent. **rent-restriction**, *n.* restrictions on a landlord's powers to charge rent. **rent-roll**, *n.* a schedule of a person's property and rents; a person's total income from this source. **rent strike**, *n.* a refusal by tenants to pay their rent. **rentable**, *a.* **rental**, *n.* the total income from rents of an estate; a rent-roll; property available for rent; (*esp. N Am.*) a hired item, usu. a car. **renter**, *n.* one who holds an estate or tenement by paying rent; one who lets property for rent; a tenant; a hirer. **rentless** (-lis), *a.*
renter RENT[3].
renumber, *v.t.* to number again.
renunciation (rinūnsiā'shən), *n.* the act of renouncing; a declaration or document expressing this; self-denial, self-sacrifice, self-resignation. **renunciant**, *n.*, *a.* **renunciative, -tory**, *a.*
renverse (renvœs'), *v.t.* to reverse, to turn the other way; to overthrow, to upset. **renversé** (rĕversā'), **renverse** (-vers'), *a.* inverted, reversed.
reobtain, *v.t.* to obtain again. **reobtainable**, *a.* **reoccupy**, *v.t.* to occupy again. **reoccupation**, *n.* **reoffer**, *v.t.* to offer again for public sale. **reopen**, *v.t., v.i.* to open again. **reopening clause**, *n.* in collective bargaining, a clause in a contract which allows an issue to be reconsidered before the expiry date of the contract. **reordain**, *v.t.* to ordain again; to appoint or enact again. **reordination**, *n.* **reorder**, *v.t.* to put in order again, to rearrange; to order or command again. **reorganize, -ise**, *v.t.* to organize anew. **reorganizer**, *n.* **reorganization**, *n.* **reorient**, *a.* (*poet.*) rising again. *v.t.* to orient again. **reorientate**, *v.t.* to restore the normal outlook of; to orientate (oneself) again.
rep.[1], (*abbr.*) report; reporter; republic, republican.
rep[2] (rep), (*n.*) (*coll.*) a travelling salesperson (short for REPRESENTATIVE). *v.i.* (*past, p.p.* **repped**[1]) to work as a travelling salesperson.
rep[3] (rep), *n.* a textile fabric of wool, cotton or silk, with a finely-corded surface. **repped**[2], *a.* having a surface like rep.
rep[4] (rep), short for REPERTORY THEATRE.
repacify, *v.t.* to pacify again. **repack**, *v.t.* to pack again. **repacker**, *n.* **repaganize, -ise**, *v.t., v.i.* to make or become pagan again. **repaid**, (*past, p.p.*) REPAY. **repaint**, *v.i.* to paint again.
repair[1] (ripeə'), *v.i.* (*dated or facet.*) to go, to betake oneself, to resort (to). *n.* a place to which one goes often or which is frequented by many people.
repair[2] (ripeə'), *v.t.* to restore to a good or sound state after dilapidation or wear; to make good the damaged or dilapidated parts of, to renovate, to mend; to remedy, to set right, to make amends for. *n.* restoration to a sound state; good or comparative condition. **in repair** or **in good repair**, in sound working condition. **in bad repair** or **out of repair**, in a dilapidated condition, needing repair. **repairable** REPARABLE. **repairer**, *n.*
repaper, *v.t.* to paper (walls etc.) again.

reparable (rep'ərəbl), *a.* capable of being made good, put in a sound state, or repaired. **reparation**, *n.* the act of repairing or restoring; the state of being repaired; satisfaction for wrong or damage, amends, compensation; (*pl.*) repairs. **reparative**, *a.* (-pa'-) **reparatory**, *a.*

repartee (repahtē'), *n.* (skill at) smart or witty conversation.

repartition, *n.* a fresh distribution or allotment.

repass, *v.t.* to pass again; to go past again; to recross. *v.i.* to pass in the opposite direction; to pass again (into, through etc.). **repassage**, *n.*

repast (ripahst'), *n.* a meal; food, victuals; the act of taking food.

repatriate, *v.t.* to restore (someone) to his or her native country. *v.i.* to return to one's country. *n.* (-ət) a person who has been repatriated. **repatriation**, *n.*

repay (ripā'), *v.t.* (*past, p.p.* **repaid**) to pay back, to refund; to return, to deal (a blow etc.) in retaliation or recompense; to pay (a creditor etc.), to make recompense for, to requite. *v.i.* to make a repayment or requital. **repayable**, *a.* **repayment**, *n.*

repeal (ripēl'), *v.t.* to revoke, to rescind, to annul; to recall or retract. *n.* abrogation, revocation, annulment. **repealable**, *a.* **repealer**, *n.* one who repeals; one who advocates repeal. **repealist**, *n.*

repeat (ripēt'), *v.t.* to do, make or say over again; to reiterate; to rehearse; to reproduce, to imitate. *v.i.* to do something over again; to recur, to happen again; to strike over again the last hour or quarter-hour struck (of a watch etc.); to rise to the mouth, to be tasted again (of food once eaten). *n.* repetition, esp. of a song or other item on a programme; (*Mus.*) a passage to be repeated, a sign indicating this; another supply of goods corresponding to the last; the order for this; a radio or television programme broadcast for the second or subsequent time. **repeatable**, *a.* **repeatedly**, *adv.* **repeater**, *n.* one who repeats; a repeating firearm; a watch or clock striking the hours and parts of hours when required; a repeating signal etc.; an indeterminate decimal in which the same figures continually recur in the same order. **repeating decimal** RECURRING. **repeating rifle**, *n.* one constructed usu. with a magazine, so as to fire several shots without reloading.

repechage (rep'əshahzh), *n.* a heat, esp. in rowing or fencing, where contestants beaten in earlier rounds get another chance to qualify for the final.

repel (ripel'), *v.t.* (*past, p.p.* **repelled**) to drive or force back or (magnetically) apart; to check the advance of; to repulse, to ward off; to keep at a distance; (of fluids etc.) to refuse to mix with (each other); to tend to drive back, to be repulsive, hateful or antagonistic to. **repellence, -ency**, *n.* **repellent**, *a.* repelling or tending to repel; repulsive. *n.* that which repels. **repellently**, *adv.* **repeller**, *n.*

repent (ripent'), *v.i.* to feel sorrow, regret, or pain for something done or left undone, esp. to feel such sorrow for sin, to be penitent or contrite; to be sorry. *v.t.* to feel contrition or remorse for, to regret. **repentance**, *n.* **repentant**, *a.* **repentantly, repentingly**, *adv.* **repenter**, *n.*

repercussion (rēpəkŭsh'ən), *n.* the act of driving or forcing back; recoil; echo, reverberation; consequence, usu. unfavourable, dangerous etc.; (*Mus.*) frequent repetition of the same subject, note, chord etc. **repercussive**, *a.* driving back; causing reverberation or consequences; driven back; reverberated.

repertoire (rep'ətwah), *n.* a stock of musical or theatrical pieces, songs etc., that a person or company is prepared and ready to perform.

repertory (rep'ətəri), *n.* a storehouse, a collection, a magazine, esp. of information, statistics etc.; a repertoire; (*also* **repertory theatre**) a theatre served by a stock or permanent company prepared to present a number of different plays, called a **repertory company**; this type of theatre company or production generally.

reperuse, *v.t.* to peruse again. **reperusal**, *n.*

repetend (rep'ətend, -tend'), *n.* something repeated, a recurring word or phrase, a refrain; that part of a repeating decimal which keeps recurring.

répétiteur (rāpātētœ'), *n.* a person who coaches opera singers.

repetition (repitish'ən), *n.* (an instance of) the act of repeating, (an) iteration; recital from memory; that which is repeated, a piece set to be learnt by heart; a copy, a reproduction, a replica; (*Mus.*) the ability of a musical instrument to repeat a note in rapid succession. **repetitional, -ary, repetitious, repetitive** (-pet'-), *a.* (esp. boringly or annoyingly) repeating. **repetitive strain injury**, a condition in which the joints and tendons of usu. the hands become inflamed, typically as the result of repeated use of (usu. industrial) apparatus or machinery. **repetitiously**, *adv.* **repetitiousness, repetitiveness**, *n.*

repiece, *v.t.* to piece together again.

repine (ripīn'), *v.i.* to fret, to be discontented (at); to murmur; to complain, to grumble. **repiner**, *n.* **repiningly**, *adv.*

replace (riplās'), *v.t.* to put back again in place; to take the place of, to succeed; to be a substitute for; to supersede, to displace; to put a substitute in place of, to fill the place of (with or by); to put in a different place. **replaceable**, *a.* **replacer**, *n.* **replacement**, *n.* the act of replacing; one that replaces; a substitute.

replant, *v.t.* to plant (a tree etc.) again; to re-establish, to resettle; to put plants in (ground) again. **replantation**, *n.*

replay, *v.t.* to play again (a record, game etc.). *n.* (rē'-) a second game between two contestants; (also **action replay**) the playing again of part of a broadcast match or game, often in slow motion.

replenish (riplen'ish), *v.t.* to fill up again; to fill completely; to stock abundantly. **replenisher**, *n.* **replenishment**, *n.*

replete (riplēt'), *a.* completely filled; abundantly supplied or stocked (with); filled to excess, sated, gorged (with). **repletion**, *n.* the state of being replete; eating and drinking to the point of being full; surfeit.

replevy (riplev'i), *v.t.* (*Law*) to recover possession of (distrained goods) upon giving security to submit the matter to a court and to surrender the goods if required. *n.* a replevin. **replevisable, replevisable** (-vis-), *a.* **replevin** (-in), *n.* an action for replevying; the writ by which goods are replevied.

replica (rep'likə), *n.* a duplicate (of a picture, sculpture etc.) by the artist who executed the original; an exact copy, a facsimile (sometimes on a smaller scale).

replicate (rep'likət), *a.* folded back on itself. *n.* a tone one or more octaves above or below a given tone. *v.t.* (-kāt) to fold back on itself; to add a replicate to (a tone); to reproduce, to make a replica of. **replication**, *n.* a reply, a rejoinder; (*Law*) the reply of a plaintiff to the defendant's plea; an echo, a repetition; a copy, an imitation; (*Mus.*) a replicate. **replicative**, *a.*

replunge, *v.t., v.i.* to plunge again. *n.* the act of plunging again. **replunger**, *n.*

reply (riplī'), *v.i.* (*pres.p.* **replying**, *past, p.p.* **replied**) to answer, to respond, to make answer orally, in writing, or by action; (*Law*) to plead in answer to a defendant's plea. *v.t.* to return (as) in answer; to answer (that etc.). *n.* the act of replying; that which is said, written, or done in answer, a response. **replier**, *n.*

repoint, *v.t.* to repair the joints of brickwork etc. with new cement or mortar. **repopulate** (rēpop'ūlāt), *v.t.* to populate again.

report (ripawt'), *v.t.* to bring back as an answer; to give an account of, to describe or to narrate, esp. as an eyewitness; to state as a fact or as news; to prepare a description or record of, esp. for official use or for publication; to announce (that), to make a formal or official statement about; to give information against, to make a complaint about the wrongdoing or misbehaviour of (a person). *v.i.* to make or tender a report; to act as a reporter; to present oneself (at a certain place etc.); to be responsible (to an employer or supervisor). *n.* that which is reported, esp. the formal statement of the result of an investigation, trial etc.; a detailed account (of a speech, meeting etc.), esp. for publication in a newspaper; common talk, popular rumour; periodic statement of a pupil's work and behaviour at school; a loud noise, esp. of an explosive kind. **report stage**, *n.* the stage of progress with a Bill in the House of Commons when a committee has reported. **reportable**, *a.* **reportage** (-tij. repawtahzh'), *n.* the art of reporting news; writing in a factual or journalistic style; a type of film journalism using no words. **reported speech**, *n.* indirect speech. **reporter**, *n.* one who reports; one who draws up official statements of law proceedings and decisions of legislative debates; one who gathers news etc., for a newspaper or broadcasting company. **reportorial** (-taw'ri-), *a.*

repose[1] (ripōz'), *v.t.* to place, to put (confidence etc. in). **reposal**, *n.*

repose[2] (ripōz'), *v.t.* (lay to) rest; to refresh with rest; to place at rest or recline. *v.i.* to (lie at) rest; to be (laid) in a recumbent position, esp. in sleep or death; (*poet.*) to rest or be supported (on). *n.* the act of resting or being at rest; rest, cessation of activity, excitement, toil etc.; sleep, quiet, tranquillity, calmness, composure, ease of manner etc. **reposedness**, (-zid-) *n.* **reposeful**, *a.* **reposefully**, *adv.* **reposefulness**, *n.*

reposit (ripoz'it), *v.t.* to lay up; to deposit, as in a place of safety. **reposition** (repəzish'-), *n.* **repositor**, *n.* (*Surg.*) an instrument for replacing anything out of place. **repository**, *n.* a place where things are deposited, esp. for safety or preservation; a depository, a museum, a store, a magazine, a shop, a warehouse, a vault, a sepulchre; a person to whom a secret etc. is confided.

repossess, *v.t.* to possess again; to legally take possession again of goods not paid for. **repossession**, *n.*

repot, *v.t.* to put (a plant) into a fresh pot.

repoussé (ripoo'sā), *a.* formed in relief by hammering from behind (of ornamental metal work). *n.* metal work ornamented in this way. **repoussage** (-sahzh'), *n.*

repp (rep), same as **rep**[3].

repped REP[2], REP[3].

reprehend (reprihend'), *v.t.* to find fault with; to censure, to blame, to chastise. **reprehender**, *n.* **reprehensible**, *a.* open to censure or blame. **reprehensibleness**, *n.* **reprehensibly**, *adv.* **reprehension** (-shən), *n.*

represent (reprizent'), *v.t.* to present to or bring before the mind by describing, portraying, imitating etc.; to serve as a likeness of, to depict (of a picture etc.); to set forth, to state (that); to describe (as), to make out (to be); to enact (a play etc.) on the stage; to play the part of; to serve as symbol for, to stand for, to be an example or specimen of; to take the place of temporarily, as deputy, substitute etc.; to act as agent or spokesman for, esp. in a representative chamber; to bring a mental image of (an event, object etc.) before the mind. **representable**, *a.* **representation**, *n.* the act of representing; a depiction of any kind; a dramatic performance; a statement of arguments etc.; the system of representing bodies of people in a legislative assembly; representatives collectively. **proportional representation**, *n.* an electoral system by which parties are represented in parliament in proportion to the total numbers of votes cast for each. **representational**, *a.* **representationism**, *n.* the doctrine that the immediate object in perception is only an idea, image, or other representation of the external thing. **representationist**, *n.* **representative**, *a.* serving to represent or symbolize, able or fitted to represent, typical; acting as agent, delegate, deputy etc.; consisting of delegates etc.; based on representation by delegates. *n.* one who or that which represents; an example, a specimen, a typical instance or embodiment; an agent, deputy, or substitute, esp. a person chosen by a body of electors; a travelling salesperson, a sales representative; (*Law*) one who stands in the place of another as heir etc. **House of Representatives**, the lower house of the US Congress. **representative government**, *n.* system of government by representatives elected by the people. **representatively**, *adv.* **representativeness**, *n.* **representer**, *n.*

re-present (rēprizent'), *v.t.* to present again. **re-presentation** (-prezən-), *n.*

repress (ripres'), *v.t.* to restrain, to keep under restraint; to put down, to suppress, to quell; to prevent from breaking out etc.; to banish (unpleasant thoughts etc.) to the subconscious. **repressible**, *a.* **repression** (-shən), *n.* the act of repressing; unconscious exclusion from the conscious mind of thoughts and memories which are in conflict with conventional behaviour. **repressive**, *a.* tending to repress; characteristic of psychological repression. **repressively**, *adv.*

re-press, *v.t.* to press again.

reprieve (riprēv'), *v.t.* to suspend the execution of for a time; to grant a respite to; to rescue, to save (from). *n.* the temporary suspension of a sentence on a prisoner; the warrant authorizing this; a respite.

reprimand (rep'rimahnd), *n.* a severe rebuke, esp. a public or official one. *v.t.* (-mahnd') to rebuke severely, esp. publicly or officially.

reprint, *v.t.* to print (a book etc.) again; to print a further impression of. *n.* (rē'-) a new impression of a printed work without considerable alteration of the contents.

reprisal (riprī'zəl), *n.* the act of seizing from an enemy by way of retaliation; that which is so taken; any act of retaliation.

reprise (riprēz'), *n.* (*Mus.*) a refrain, a repeated phrase etc.

reprize, *v.t.* to prize anew.

repro (rē'prō), *n.* (*pl.* **-pros**) short for REPRODUCTION.

reproach (riprōch'), *v.t.* to censure, to upbraid; to find fault with (something done). *n.* censure mingled with regret or grief; a rebuke; shame, infamy, disgrace; an object or cause of shame, scorn or derision. **above, beyond reproach**, blameless, pure. **reproachable**, *a.* **reproachably**, *adv.* **reproacher**, *n.* **reproachful**, *a.* containing or expressing reproach; shameful, infamous, base. **reproachfully**, *adv.* **reproachfulness**, *n.* **reproachingly**, *adv.* **reproachless** (-lis), *a.* **reproachlessness**, *n.*

reprobate (rep'rəbāt), *a.* abandoned to sin, lost to virtue or grace; depraved. *n.* one who is abandoned to sin; a wicked, depraved wretch; (*coll.*) a rascal, a slightly roguish person.

reprocess, *v.t.* to process again; to treat a substance or material in order to make it reusable in a new form. **reprocessing**, *n.*, *a.*

reproduce (rēprədūs'), *v.t.* to produce again; to copy or simulate; to produce new life through sexual or asexual processes. *v.i.* to produce offspring; to duplicate (well, badly etc.) as a copy. **reproducer**, *n.* **reproducible**, *a.* **reproduction** (-dŭk'-), *n.* the act of reproducing; any of the sexual or asexual processes by which animals or plants produce offspring; a copy, imitation. **reproductive, -tory**, *a.* **reproductively**, *adv.* **reproductiveness**, **reproductivity** (-tiv'-), *n.*

reprography (riprog'rəfi), *n.* the art or process of reproducing printed matter (e.g. by photocopying). **reprographic** (reprəgraf'-), *a.* **reprographically**, *adv.*

reproof (riproof'), *n.* (an expression of) censure, blame, reprehension. **reprove** (-proov'), *v.t.* to rebuke, to censure, esp. to one's face, to chide. **reprovable**, *a.* **reprover**, *n.* **reprovingly**, *adv.*

re-prove, *v.t.* to prove again.

reprovision, *v.t.* to provision (a ship etc.) afresh. **reprune**,

reps *v.t.* to prune again.
reps (reps), var. of REP³.
reptile (rep'til), *a.* creeping, crawling, moving on the belly or on small, short legs; grovelling, servile, mean, base. *n.* a crawling animal, one of a class of animals comprising the snakes, lizards, turtles, crocodiles etc.; a grovelling, mean, base person. **reptilian** (-til'-), *n., a.* **reptiliferous** (-tilif'-), *a.* containing fossil reptiles. **reptiliform** (-til'ifawm), **reptilious** (-til'-), **reptiloid** (-loid), *a.* **reptilivorous** (-liv'ərəs), *a.* devouring reptiles.
republic (ripūb'lik), *n.* a state or a form of political constitution in which the supreme power is vested not in a sovereign but in the people or their elected representatives, a commonwealth. **republican**, *a.* pertaining to or consisting of a republic; characteristic of the principles of a republic; believing in or advocating these. *n.* one who favours or advocates a republican form of government; (*also* **Republican**) a member or supporter of the Republican party in the US; (*also* **Republican**) a supporter of republicanism in N Ireland. **Republican Party**, *n.* the alternative political party to the Democratic in the government of the US. **republicanism**, *n.* **republicanize, -ise**, *v.t.*
republish, *v.t.* to publish again; to print a new edition or impression of. **republication**, *n.*
repudiate (ripū'diāt), *v.t.* to refuse to acknowledge, to disown, to disclaim (a debt etc.); to disavow, to reject, to refuse to admit, accept, recognize (e.g. a claim) etc.; to cast off, to put away, to divorce (one's wife). **repudiation**, *n.* **repudiator**, *n.*
repugnance, repugnancy (ripŭg'nəns(i)), *n.* antipathy, dislike, distaste, disgust, aversion; incompatibility, or opposition, of mind, disposition, statements, ideas etc. **repugnant**, *a.* **repugnantly**, *adv.*
repullulate, *v.i.* to sprout, shoot, or bud again; to break out again, to recur, to reappear (of a disease or morbid growth). **repullulation**, *n.* **repullulescent**, *a.*
repulse (ripŭls'), *v.t.* to repel, to beat or drive back, esp. by force of arms; to reject, esp. in a rude manner, to rebuff, to snub; (*fig.*) to defeat in argument. *n.* an act of repulsing; the state of being repulsed; a rebuff, a refusal, a failure, a disappointment. **repulser**, *n.* **repulsion** (-shən), *n.* the act of repulsing; the state of being repulsed; (*Phys.*) the tendency of certain bodies to repel each other, opp. to *attraction*; dislike, repugnance, aversion. **repulsive**, *a.* acting so as to repel; unsympathetic, forbidding; repellent, loathsome, disgusting; (*Phys.*) acting by repulsion. **repulsively**, *adv.* **repulsiveness**, *n.*
repurchase, *v.t.* to purchase back or again. *n.* the act of buying again; that which is so bought.
reputable (rep'ūtəbl), *a.* being in good repute; respectable, trustworthy, creditable. **reputableness**, *n.* **reputably**, *adv.*
reputation (repūtā'shən), *n.* the estimation in which one is generally held, repute; good estimation, good fame, credit, esteem, respectability; the repute, honour, or credit derived from favourable public opinion or esteem.
repute (ripūt'), *v.t.* (*chiefly in p.p.*) to consider, to reckon, to regard (as). *n.* reputation, fame; character attributed by public report. **reputed**, *a.* generally regarded (sometimes with implication of doubt etc.). **reputedly**, *adv.*
request (rikwest'), *n.* an expression of desire, or the act of asking, for something to be granted or done; a petition; that which is asked for; the state of being demanded or sought after. *v.t.* to ask (that); to address a request to. **on request**, if or when asked for. **request stop**, *n.* a stop on a route where a bus etc. will stop only if signalled to do so. **requester**, *n.*
requicken, *v.t.* to quicken again, to reanimate.
requiem (rek'wiem), *n.* a mass for the repose of the soul of a person deceased; the musical setting of this. [L (the first word of the introit *Requiem aeternam dona eis, Domine*)]
requiescat (rekwies'kat), *n.* a wish or prayer for the repose of the dead. [L *requiescat in pace*, let him (or her) rest in peace]
require (rikwiə'), *v.t.* to ask for or claim as a right or by authority, to order; to demand (something of a person), to insist (that); to have need of imperatively, to depend upon for completion etc. **requirable**, *a.* **requirement**, *n.* that which is required; an essential condition.
requisite (rek'wizit), *a.* required by the nature of things, necessary (for completion etc.), indispensable. *n.* that which is required; a necessary part or quality. **requisiteness**, *n.*
requisition (rekwizish'ən), *n.* the act of requiring or demanding; application made as of right; a formal and usu. written demand or request for the performance of a duty etc.; an authoritative order for the supply of provisions, business materials etc.; the state of being called upon or put in use. *v.t.* to make a formal or authoritative demand for, esp. for commercial or military purposes; to make such a demand upon (a town etc.); to make a requisition for, to call in for use. **requisitionist**, *n.* one who makes a requisition.
requite (rikwit'), *v.t.* to repay, to make return to, to recompense; to give or deal (esp. love) in return; to reward, to avenge. **requital**, *n.* **requiter**, *n.*
rerail, *v.t.* to put (rolling stock) on the rails again. **reread**, *v.t.* to read or peruse again.
re-record, *v.t., v.i.* to record again.
reredos (riə'rədōs, riə'dos), *n.* (*pl.* **-doses**) the ornamental screen at the back of an altar; the back of an open hearth, a fire-back.
rerun, *v.t.* to run (a race etc.) again; to show (a film or television programme) again. *n.* (rē'-) a repeated film etc.; a race run a second time.
res (rās), *n.* (*pl.* **res**; *Law*) a thing, property; subject matter of a court action. [L]
res., (*abbr.*) research; reserve; residence; resigned; resolution.
resaddle (rēsad'l), *v.t., v.i.* to saddle again. **resail** (rēsāl'), *v.t.* to sail (a race etc.) again. *v.i.* to sail back again. **resale** (rēsāl'), *n.* a second sale; a sale at second hand. (rē'-).
rescind (risind'), *v.t.* to annul, to cancel, to revoke, to abrogate. **rescission** (-sizh'ən), *n.* **rescissory**, *a.*
rescript (rē'skript), *n.* the answer or decision of a Roman emperor to a question or appeal, esp. on a point of jurisprudence; a Pope's written reply to a question of canon law, morality etc.; an edict, a decree, an order, or official announcement; something rewritten, the act of rewriting.
rescue (res'kū), *v.t.* to deliver from confinement, danger, evil, or injury; (*Law*) to liberate by unlawful means from custody, to recover (property etc.) by force. *n.* deliverance from confinement, danger, evil or injury; forcible seizure (of a person, property etc.) from the custody of the law. **rescuable**, *a.* **rescuer**, *n.*
research (risœch'), *n.* diligent and careful inquiry or investigation; systematic study of phenomena etc., a course of critical, usu. scientific, investigation. *v.i.* to make researches. *v.t.* to make careful and systematic investigation into. **researcher**, *n.*
reseat, *v.t.* to seat again; to furnish (a church etc.) with new seats; to provide (a chair, pair of trousers etc.) with a new seat.
resect (risekt'), *v.t.* (*Surg.*) to excise a section of an organ or part; to cut or pare down, esp. the articular extremity of a bone. **resection**, *n.*
reseek, *v.t.* to seek again; **reseize**, *v.t.* to seize again; to take possession of. **resell**, *v.t.* to sell again.
resemble (rizem'bl), *v.t.* to be like, to be similar to, to have features, nature etc., like those of. **resemblance**, *n.* similarity in nature or appearance. **resemblant**, *a.*
resend, *v.t.* to send back or again.
resent (rizent'), *v.t.* to regard as an injury or insult; to feel

or show displeasure or indignation at; to cherish bitter feelings about or towards. **resenter**, *n.* **resentful**, *a.* **resentfully**, *adv.* **resentfulness**, *n.* **resentingly**, *adv.* **resentment**, *n.*

reservation (rezəvā'shən), *n.* the act of reserving; that which is reserved; the booking of accommodation in a hotel, train, ship etc.; in N America, a tract of land reserved for native Indian tribes or for public use; in the Roman Catholic church, the reserving of the right of nomination to benefices, of the power of absolution, or of a portion of the consecrated elements of the Eucharist; an expressed or implicit limitation, exception, or qualification; *(also* **central reservation)** a strip of land separating the two roads of a dual carriageway. **mental reservation**, an unexpressed qualification, exception or point of view radically affecting or altering the meaning of a statement, oath etc.

reserve (rizœv'), *v.t.* to keep back (for future use, enjoyment, treatment etc.), to hold over, to postpone, to keep in store; to retain (for oneself or another), esp. as an exception from something granted; to book, keep or set apart; to retain the right of nomination to a benefice for the Pope; to set apart (a case) for absolution by the Pope, a bishop etc.; to retain a portion of the consecrated elements of the Eucharist; *(in p.p.)* to set apart for a certain fate, to destine. *n.* that which is reserved; a sum of money or a fund reserved, esp. by a company or by bankers, to meet any unforeseen or emergency demand; a piece of land set aside for a special use; troops kept for any emergency, such as to act as reinforcements or cover a retreat; a part of the military or naval forces not embodied in the regular army and navy, but liable to be called up in case of emergency; a member of these forces, a reservist; the state of being reserved or kept back for a special purpose; *(Sport)* a substitute; a lower limitation attached to a prize at auction; an award to an exhibit entitling it to a prize if a winner should be disqualified; reticence, self-restraint, caution in speaking or action; *(usu. in pl.)* normally hidden resources or stocks (of a quality, virtue etc.). **in reserve**, reserved and ready for use in emergency. **without reserve**, fully, without qualification or hesitation; (offered for sale) to the highest bidder without the condition of a reserve price. **reserve currency**, *n.* a foreign currency acceptable as a medium of international banking transactions and held in reserve by many countries. **reserve price**, *n.* a price below which no offer will be accepted at auction. **reserved**, *a.* reticent, backward in communicating one's thoughts or feelings, undemonstrative, distant; retained for a particular use, person etc., booked. **reserved list**, *n.* a list of naval officers not on active service but liable to be called up in emergency. **reserved occupation**, *n.* a vital type of employment which exempts one from military service in the event of war. **reserved sacrament**, *n.* portion of the consecrated elements reserved after communion, for adoration. **reservedly** (-vid-), *adv.* **reservedness** (-vid-), *n.* **reservist**, *n.* a member of the military or naval reserve.

reservoir (rez'əvwah), *n.* a receptacle in which a quantity of anything, esp. fluid, may be kept in store; an earthwork or masonry structure for the storage of water in large quantity; a part of an implement, machine, animal or vegetable organ etc., acting as a receptacle for fluid; a reserve supply or store of anything, e.g. strength. *v.t.* to collect or store in a reservoir.

reset (rĕset'), *v.t.* to set (type, a jewel etc.) again. **resettable**, *a.*

resettle, *v.t.* and *i.* to settle again. **resettlement**, *n.* **res gestae** (rāz ges'tī, rĕz jes'tē) *n.pl.* achievements; exploits; *(Law)* relevant facts or circumstances admissible in evidence.

reshape, *v.t.* to shape again. **reshuffle**, *v.t.* to shuffle again; to rearrange or reorganize, esp. cabinet or a government department. *n.* (rē'-) the act of reshuffling.

reside (rizīd'), *v.i.* to dwell permanently or for a considerable length of time, to have one's home (at or in); to be in official residence; (of qualities, rights etc.) to be vested (in). **residence** (rez'i-), *n.* the act or state of residing in a place; the act of living or remaining where one's duties lie or one's job is pursued; the place where one dwells, one's abode; a house of some size or pretensions. **in residence**, actually resident; of an artist, writer etc., acting in a regular capacity for a limited period at a gallery, university or other community etc. **residency**, *n.* formerly, the official residence of a resident or governor of a British protectorate in India; (the post held during) a period of specialized training undertaken by a doctor following internship. **resident** (rez'i-), *a.* residing; having a residence, esp. official quarters in connection with one's duties; non-migratory; inherent, vested (in). *n.* one who dwells permanently in a place as dist. from a visitor; a guest staying overnight at a hotel; a representative of the British government in a British protectorate; a junior doctor who lives and works in a hospital to gain specialized experience; a non-migratory bird or animal. **residential** (rezidən'shəl), *a.* suitable for residence or for residences; pertaining to residence. **residentiary** (rezidən'-), *a.* maintaining or bound to an official residence. *n.* an ecclesiastic bound to an official residence.

residue (rez'idū), *n.* that which is left or remains over, the rest, remainder, esp. that which remains of an estate after payment of all charges, debts, and particular bequests; that which is left after any process of separation or purification, esp. after combustion, evaporation etc.; the remainder left by subtraction or division. **residual** (-zid'-), *a.* of the nature of a residue or residuum; remaining after a part has been taken away; *(Math.)* left by a process of subtraction; remaining unexplained or uneliminated. *n.* a residual quantity, a remainder; a payment to an artist for reuse of a film, recording etc. **residuary** (-zid'-), *a.* pertaining to or forming a residue, residual, remaining, esp. pertaining to the residue of an estate. **residuum** (-zid'ūəm), *n. (pl.* **-dua** (-ə)) a residue.

re-sign (rēsīn'), *v.t.* to sign again.

resign (rizīn'), *v.t.* to give up, to surrender, to relinquish (to); to renounce, to abandon; to yield, to submit, to reconcile (oneself, one's mind etc. to). *v.i.* to give up office, to leave a job, to retire (from). **resignation** (rezig-), *n.* the act of resigning; an office; a document announcing this; the state of being reconciled, patience, acquiescence, submission. **resigned**, *a.* submissive, acquiescent or enduring; surrendered, given up. **resignedly** (-nid-), *adv.* **resigner**, *n.*

resile (rizīl'), *v.i.* to spring back, to rebound, to recoil; to resume the original shape after compression, stretching etc.; to show elasticity *(Law)* to back out (from an agreement etc.). **resilience** (-zil'yəns), **-cy**, *n.* elasticity; (of a person) an ability to recover quickly from physical illness, misfortune etc. **resilient**, *a.*

resin (rez'in), *n.* an amorphous inflammable vegetable substance secreted by plants and usu. obtained by exudation, esp. from the fir and pine; a similar substance obtained by the chemical synthesis of various organic materials, used esp. in making plastics. *v.t.* to treat with resin. **resiniferous** (-nif'-), **resiniform** (-ifawm), *a.* **resinify** (-ifi), *v.t., v.i.* **resinification** (-fi-), *n.* **resinoid** (-noid), *n., a.* **resinous**, *a.* pertaining to or resembling resin; obtained from resin. **resinously**, *adv.* **resinousness**, *n.* **resiny**, *a.*

resist (rizist'), *v.t.* to stand or strive against, to act in opposition to; to endeavour to frustrate; to oppose successfully, to withstand, to stop, to repel, to frustrate; to be proof against. *v.i.* to offer resistance. *n.* a substance applied to a surface etc. to prevent the action of a chemical agent, such as the acid-resistant coating on etched printed circuit boards. **resistant**, *a.* offering resistance.

resistance, *n.* the act or power of resisting; opposition, refusal to comply; that which hinders or retards, esp. the opposition exerted by a fluid to the passage of a body or the opposition exerted by a substance to the passage of electric current, heat etc. through it; the body's natural power to withstand disease; an electronic resistor; a resistance movement. **the line of least resistance,** the easiest course of action. **resistance movement,** *n.* an underground organization of civilians and others in an enemy-occupied country directed to sabotaging the invaders' plans and rendering their position as difficult as possible. **resistance thermometer,** *n.* a type of thermometer which accurately measures high temperatures from the change in resistance of a wire coil or semiconductor as the temperature varies. **passive resistance** PASSIVE. **resistor,** *n.* an electronic component with a specified resistance. **resistible,** *a.* **resistibility** (-bil'-), **resistibleness,** *n.* **resistibly, resistingly,** *adv.* **resistive,** *a.* **resistivity** (rēsistiv'-), *n.* (formerly called **specific resistance**) the electrical resistance of a conducting material of a given cross-sectional area and length. **resistless** (-lis), *a.* **resistlessly,** *adv.* **resistlessness,** *n.*

resit, *v.t.*, *v.i.* to sit (an examination) again after failing. *n.* (rē'-) an examination which one must sit again.

res judicata (rās joodikah'tə), *n.* (*Law*) an issue that has already been settled in court and cannot be raised again.

resoluble, *a.* capable of being dissolved again, resolved or analysed.

resolute (rez'əloot), *a.* having a fixed purpose, determined, constant in pursuing an object, firm, decided, unflinching. **resolutely,** *adv.* **resoluteness,** *n.* **resolution,** *n.* (*Chem.*) the act or process of resolving or separating anything into its component parts, decomposition, analysis; the disappearance of inflammation without production of pus; (*Mus.*) the conversion of a discord into a concord; (*Prov.*) the substitution of two short syllables for a long one; the sharpness, definition of a television or film image; mental analysis, solution of a problem etc.; a formal proposition, statement of opinion, or decision by a legislative or corporate body or public meeting; a proposition put forward for discussion and approval; resoluteness, determination, firmness in adhering to one's purpose. **resolutionist,** *n.* one who makes or supports a resolution. **resolutive,** *a.* having the power of or tending to resolve or dissolve.

resolve (rizolv'), *v.t.* (*Chem.*) to separate into component parts; to dissolve, to analyse, to disintegrate, to dissipate; to analyse mentally, to solve, to explain, to clear up, to answer; to convert (into) by analysis; (*Med.*) to cause to disperse or pass away without suppuration; (*Mus.*) to convert (a discord) into concord; to pass by vote a resolution (that); to cause (a person) to decide on a particular course of action. *v.i.* to separate into component parts, to dissolve, to break up, to be analysed; (*Med.*) to pass away without suppuration; (*Mus.*) to be converted from discord into concord; to make one's mind up, to decide or determine (upon); to pass a resolution. *n.* a resolution, a firm decision or determination; resoluteness, firmness of purpose. **resolvable,** *a.* **resolvability** (-bil'-), **resolved,** *a.* determined, resolute. **resolvedly** (-vid-), *adv.* **resolvent,** *a.* having the power of resolving. *n.* that which has the power of resolving, esp. a chemical substance, drug, or medical application. **resolver,** *n.* **resolving power,** *n.* (the measure of) the ability of a microscope or telescope to distinguish or produce separable images of small adjacent objects; the ability of a photographic emulsion to produce fine detailed images.

resonant (rez'ənənt), *a.* (re-)echoing, resounding; having the property of prolonging or reinforcing sound, esp. by vibration; (of sounds) prolonged or reinforced by vibration or reverberation. **resonance,** *n.* the quality or state of being resonant; sympathetic vibration; (*Eng.*, *Elec.*) the specially large vibration of a body or system when subjected to a small periodic force of the same frequency as the natural frequency of the system; the amplification of human speech by sympathetic vibration in the bone structure of the head and chest resounding in the vocal tract, often used in singing; (*Chem.*) the description of the electronic structure of a molecule in certain compounds in terms of different arrangements of two or more bonds. **resonantly,** *adv.* **resonate,** *v.i.* to resound; to reverberate. **resonator,** *n.* a body or system that detects and responds to certain frequencies; a device for enriching or amplifying sound by resonance.

resorb (risawb'), *v.t.* to absorb again. **resorbence,** *n.* **resorbent,** *a.* **resorption** (-sawp'-), *n.* **resorptive,** *a.*

resort[1] (rizawt'), *v.i.* to go, to repair, to betake oneself; to have recourse, to apply, to turn to (for aid etc.). *n.* the act of frequenting a place; the state of being frequented; the place frequented, esp. a place popular with holiday makers; recourse; that to which one has recourse, an expedient. **last resort,** thing or person to which one comes for aid or relief when all else has failed; a final attempt. **resorter,** *n.*

re-sort[2] (rēsawt'), *v.t.* to sort again.

resound (rizownd'), *v.i.* to ring, to (re-)echo, to reverberate (with); to be filled with sound; to be re-echoed, to be repeated, reinforced, or prolonged (of sounds, instruments etc.); to make a sensation, become widely significant (of news, events etc.). *v.t.* to sound again; to echo the sound of; to spread the fame of. **resounding,** *a.* (of a victory, defeat etc.) clear, decisive; ringing, resonant. **resoundingly,** *adv.*

resource (risaws'), *n.* a means of assistance, aid, support, or safety; an expedient, a device; (*pl.*) means of (esp. financial) support and defence, e.g. of a country; capacity for finding or devising means; practical ingenuity. **resourceful,** *a.* ingenious, clever. **resourcefully,** *adv.* **resourcefulness,** *n.* **resourceless** (-lis), *a.* **resourcelessness,** *n.*

re-speak (rēspēk'), *v.t.* to speak again; to echo back.

respect (rispekt'), *n.* relation, regard, reference; attention, heed (to); particular, aspect, point; esteem, deferential regard, demeanour, or attention; (*pl.*) expressions of esteem sent as a complimentary message. *v.t.* to esteem, to regard with deference; to treat with consideration, to spare from insult, injury, interference, invasion of privacy etc.; to relate or have reference to. **in respect of, to,** with regard to. **to pay one's respects,** to express or send a message of esteem or compliment. **respectable,** *a.* worthy of respect, of good repute; of good social standing, honest, decent, not disreputable; fairly good, tolerable, passable; not inconsiderable, above the average (in number, quantity, merit etc.). **respectability** (-bil'-), **respectableness,** *n.* the quality or character of being respectable. **respectably,** *adv.* **respecter,** *n.* (no) **respecter of persons,** one who pays (or does not pay) undue consideration to and is biased by others' wealth and station. **respectful,** *a.* showing respect. **respectfully,** *adv.* **respectfulness,** *n.* **respecting,** *prep.* with regard to, in respect of. **respective,** *a.* relating severally to each of those in question, comparative, relative. **respectively,** *adv.*

respell, *v.t.* to spell again.

respire (rispīə'), *v.i.* to breathe; to inhale or take air into and exhale it from the lungs. *v.t.* to inhale and exhale, to breathe out, to emit (perfume etc.). **respirable** (res'pi-), *a.* capable of being respired; fit to be breathed. **respirability** (-bil'-), **respirableness,** *n.* **respiration** (respi-), *n.* the act or process of breathing; one cycle of inhaling and exhaling; the absorption of oxygen and emission of carbon dioxide by living organisms. **respirator** (res'pi-), *n.* an appliance worn over the mouth or mouth and nose to exclude poisonous gases, fumes etc., or to protect the lungs from the sudden inspiration of cold air; a machine for aiding breathing; a gas-mask. **respiratory,** *a.* **respiratory disease,** *n.* any disease involving an organ concerned in respiration. **respiratory quotient,** *n.* (*Biol.*) the ratio of carbon dioxide expired to the volume of oxygen consumed by an organism or tissue in a given period. **respirometer** (respirom'itə), *n.* an instrument for measuring respiration; an apparatus for supplying a diver with air for breathing.

respite (res'pit), *n.* a temporary intermission (from labour, effort, suffering etc.), esp. a delay in the execution of a sentence; an interval of rest or relief, a reprieve. *v.t.* to relieve by a temporary cessation of labour, suffering etc.; to grant a respite to, to reprieve; to suspend the execution of (a sentence); to postpone, to defer, to delay. **respiteless,** *a.*

resplendent (risplen'dənt), *a.* shining with brilliant lustre; vividly or gloriously splendid. **resplendence, -cy,** *n.* **resplendently,** *adv.*

respond (rispond'), *v.i.* to answer, to make reply (esp. of a congregation returning set answers to a priest); to do something or show an effect in answer or correspondence (to); to react (to an external stimulus); to show sympathy or sensitiveness (to). *v.t.* to answer, to say in response. **respondent,** *a.* giving response, answering; responsive (to); in the position of defendant. *n.* one who answers; one who maintains a thesis in reply; one who answers in a suit at law, a defendant, esp. in a divorce case.

response (rispons'), *n.* the act of answering; that which is answered in word or act, an answer, a reply; a portion of a liturgy said or sung in answer to the priest; the ratio of the output to the input level on an electrical transmission system at any particular frequency; the reaction of an organism to stimulation. **responsible,** *a.* answerable, liable, accountable (to or for); to blame or to be thanked (for); morally accountable for one's actions, able to discriminate between right and wrong; respectable, trustworthy; of a job etc., involving responsibility. **responsibility** (-bil'-), *n.* the state of being responsible, as for a person, trust etc.; ability to act according to the laws of right and wrong; that for which one is responsible. **responsibly,** *adv.* **responsive,** *a.* answering or inclined to answer; of the nature of an answer; reacting to stimulus; responding readily, sympathetic. **responsively,** *adv.* **responsiveness,** *n.*

rest[1] (rest), *n.* cessation from bodily or mental exertion or activity, repose, sleep; freedom from care, disturbance, or molestation, peace, tranquillity; a period of such cessation or freedom, esp. a brief pause or interval; a stopping-place, a place for lodging, a shelter for cabmen, sailors etc.; that on which anything stands or is supported, a prop, a support, a device for steadying a rifle on when taking aim, for supporting the cutting-tool in a lathe etc.; a long pole with a cross-piece at one end used as a support for a billiard cue in playing; (*Mus.*) (the sign indicating) an interval of silence; a pause in a verse, a caesura; (*poet.*) death. *v.i.* to cease from exertion, motion, or activity; to be relieved from toil or exertion; to repose, to lie in sleep or death, to lie buried; to be still, to be without motion; to be free from care, disturbance, or molestation, to be tranquil, to be at peace; to be allowed to lie fallow (of land); to be positioned, spread out, supported or fixed, to be based, to lean, to recline, to stand (on); to depend, to rely (upon); (of eyes) to be fixed, to be directed steadily (upon); to remain; (*US*) (of an attorney) to call no more evidence; (*sl.*) of actors, to be unemployed. *v.t.* to cause to cease from exertion; to give repose to, to lay at rest; (*in p.p.*) to refresh by resting; to place for support, to base, to establish, to lean, to lay, to support. **at rest,** reposing; not in motion; still; not disturbed, agitated, or troubled; (*euphem.*) dead, in the grave. **to lay to rest,** to bury; (*fig.*) to quell (e.g. rumours). **to rest with,** to be (left) in the hands of. **rest-cure,** *n.* a period of seclusion and repose (usu. in bed) as a method of treatment for illness or nervous disorders. **rest-day,** *n.* a day of rest, inactivity; Sunday. **rest mass,** *n.* the mass of an object at rest. **rest room,** *n.* (*N Am., euphem.*) a room with toilet facilities etc. in a public building. **restful,** *a.* inducing rest, soothing, free from disturbance; at rest, quiet. **restfully,** *adv.* **restfulness,** *n.* **resting-place,** *n.* a place for rest; the grave. **restless** (-lis), *a.* not resting, never still, agitated, uneasy, fidgety, unsettled, turbulent; not affording sleep, sleepless. **restlessly,** *adv.* **restlessness,** *n.*

rest[2] (rest), *n.* that which is left, the remaining part or parts, the residue, the remainder, the others; a reserve fund, a balance or surplus fund for contingencies. *v.i.* to remain, to stay, to continue (in a specified state). **all the rest,** all that remains, all the others. **(as) for the rest,** as regards the remaining persons, matters, or things, as regards anything else.

restamp, *v.t.* to stamp again.

restart, *v.t.,* *v.i.* to start afresh.

restate, *v.t.* to state again or express differently.

restaurant (res'tərənt, -trənt, -trä), *n.* a place for refreshment; an eating house. **restaurant car,** *n.* a railway dining-car. **restaurateur** (-tərətœ'), *n.* the keeper or owner of a restaurant.

restful etc. REST[1].

restiform (res'tifawm), *a.* (*Anat.*) rope- or cord-like (applied to two bundles of fibrous matter connecting the medulla oblongata with the cerebellum). **restiform body,** *n.* either of these bundles.

resting-place REST[1].

restitution (restitū'shən), *n.* the act of restoring something taken away or lost; making good, reparation, indemnification (after causing a loss etc.). **restitutive, -ory** (res'-), *a.*

restive (res'tiv), *a.* unwilling to go forward, standing still, halting; restless, fidgety, impatient of control, unmanageable. **restively,** *adv.* **restiveness,** *n.*

restock, *v.t.* to stock again.

restoration, restorative RESTORE.

restore (ristaw'), *v.t.* to bring back to a former state, to repair, to reconstruct; to put back in place, to replace, to return; to bring back (to health), to cure; to bring back (to a former position), to reinstate; to bring into existence, law or use again, to re-establish, to renew; to represent (an extinct animal, mutilated picture, ruin etc.) as it is supposed to have been originally; to give back, to make restitution of. **restorable,** *a.* **restoration** (restə-), *n.* the act of restoring; a building etc., restored to its supposed original state; a drawing, model, or other representation of a building, extinct animal etc., in its supposed original form. **the Restoration,** the return of Charles II in 1660 and the re-establishment of the monarchy after the Commonwealth; the return of the Bourbons to France in 1814. **restorationism,** *n.* the doctrine of the final return of all men to happiness and sinlessness in the afterlife. **restorationist,** *n.* **restorative,** *a.* tending to restore health, strength etc. *n.* food, drink, a medicine etc., for restoring strength, vigour etc., a stimulant, a tonic. **restoratively,** *adv.* **restorer,** *n.*

re-strain (rēstrān'), *v.t.* to strain again.

restrain (ristrān'), *v.t.* to hold back, to check, to curb; to keep under control, to repress, to hold in check, to restrict; to confine, to imprison. **restrainable,** *a.* **restrainedly** (-nid-), *adv.* **restrainer,** *n.* **restraint,** *n.* the act of restraining; the state of being restrained; check, control, (self-)repression, avoidance of excess; (a) constraint, reserve; (a) restriction, limitation; abridgment of liberty, confinement. **restraint of trade,** interference with free competition in business.

restrengthen, *v.t.* to strengthen anew.

restrict (ristrikt'), *v.t.* to limit, to confine, to keep within certain bounds. **restricted,** *a.* limited, confined; out of bounds to the general public; denoting a zone where a

speed limit or waiting restriction applies for vehicles. **restriction**, *n.* something that restricts; a limiting law or regulation; the state of restricting or being restricted. **restrictive**, *a.* restricting or tending to restrict; (*Gram.*) describing a relative clause or phrase which restricts the application of the verb to the subject. **restrictive practice**, *n.* (*usu. in pl.*) a trading agreement contrary to the public interest, e.g. a cartel; a practice by a trade union, e.g. the closed shop, regarded as limiting managerial flexibility. **restrictedly, restrictively**, *adv.*
restrike, *v.t.* to strike again. **restuff**, *v.t.* to stuff anew.
result (rizült′), *v.i.* to be the actual, or follow as the logical consequence, to ensue; to have a given issue, to terminate or end (in). *n.* consequence, issue, outcome, effect; a quantity, value, or formula obtained from a calculation; a final score in a game or contest. **resultance**, *n.* **resultant**, *a.* resulting; following as a result; consequent on the combination of two factors, agents etc. *n.* that which results; the force resulting from the combination of two or more forces acting in different directions at the same point, ascertained by a parallelogram of forces. **resultful**, *a.* **resulting**, *a.* **resultless** (-lis), *a.*
resume (rizūm′), *v.t.* to take back, to take again, to reoccupy, to recover (command etc.); to begin again, to recommence, to go on with after interruption. *v.i.* to continue after interruption, to recommence. **resumable**, *a.*
résumé (rez′ümā), *n.* a summary, a recapitulation, a condensed statement, an abstract; (*esp. N Am.*) a CURRICULUM VITAE.
resummon, *v.t.* to summon again; to convene again. **resummons**, *n.*
resumption (rizümp′shən), *n.* (the) act of resuming. **resumptive**, *a.*
resupinate, *a.* (*Bot.*) inverted, apparently upside-down. **resupination**, *n.*
resurge (risœj′), *v.i.* to rise again; to return to popularity, vogue. **resurgence**, *n.* **resurgent**, *a.* rising again, rising from the dead; returning to popularity, vogue. *n.* one who rises from the dead.
resurrect (rezərekt′), *v.t.* to bring back to life; to bring again into vogue or currency, to revive; to exhume. **resurrection**, *n.* a rising again from the dead, esp. the rising of Christ from the dead, and the rising of all the dead at the Last Day; the state of being risen again; a springing again into life, vigour, vogue, or prosperity; exhumation, body-snatching. **resurrectionism**, *n.* belief in the Christian doctrine of resurrection. **resurrection-man**, **resurrectionist**, *n.* a body-snatcher. **resurrection plant**, *n.* a desert plant which curls into a tight ball in drought and unfolds when moistened. **resurrectional**, *a.*
resurvey, *v.t.* to survey again; to read and examine again. *n.* a renewed survey (-sœ′).
resuscitate (risūs′itāt), *v.t.* to revive, to restore from apparent death; to revivify, to restore to vigour, animation, usage, currency etc. *v.i.* to revive, to come to life again. **resuscitant**, *n.*, *a.* **resuscitation**, *n.* **resuscitative**, *a.* **resuscitator**, *n.* a machine for causing recommencement of breathing after it has ceased.
ret., (*abbr.*) retain; retired; return(ed).
retable (ritā′bl), *n.* a shelf, ledge or panelled frame above the back of an altar for supporting ornaments.
retail (rē′tāl), *n.* the sale of commodities in small quantities to the end-user; a dealing out in small portions. *a.* pertaining to selling by retail. *v.t.* to sell in small quantities; (-tāl′) to tell (a story etc.) in detail, to recount, to retell, to spread about. *v.i.* to be sold by retail (at or for a specified price). **retail price index**, an index of the cost of living, based on average retail prices of selected goods, usu. updated monthly. **retailer**, *n.*
retain (ritān′), *v.t.* to hold or keep possession of, to keep; to continue to have, to maintain, to preserve; to hire, to engage the services of (esp. legal counsel) by paying a preliminary fee; of walls etc., to hold back, to keep in place; to remember. **retainable**, *a.* **retainer**, *n.* one who or that which retains; an attendant, a follower, esp. of a feudal chieftain; (*Law*) an agreement by which an attorney acts in a case; a preliminary fee paid (esp. to a legal counsel) to secure his services. **retaining fee**, *n.* a retainer. **retaining wall**, *n.* a massive wall built to support and hold back the earth of an embankment, a mass of water etc.
retake, *v.t.* (*pres.p.* **-taking**, *past* **-took**, *p.p.* **-taken**) to take again; to recapture; to shoot a film (scene) again, to record (a performance) again. *n.* (rē-) a second photographing (of a scene), a re-recording.
retaliate (rital′iāt), *v.i.* to return like for like, esp. evil for evil, to return or retort in kind, to make reprisals. **retaliation**, *n.* the act of retaliating; the imposition of import duties by one country against another which imposes duties on imports from the first country. **retaliative, -tory**, *a.*
retard (ritahd′), *v.t.* to cause to move more slowly; to hinder, to impede, to check, restrain or delay the growth, advance, arrival or occurrence of. *v.i.* to be delayed; to happen or arrive later or abnormally late. **retardant**, *n.* a substance that slows down a chemical reaction. *a.* serving to delay or slow down. **retardation** (rē-), *n.* **retardative, -tory**, *a.* **retarded**, *a.* underdeveloped intellectually or emotionally; subnormal in learning ability. **retarder**, *n.* someone or something that retards; a retardant, esp. an additive that delays cement setting. **retardment**, *n.*
retch (rech), *v.i.* to make an effort to vomit; to strain, as in vomiting. *n.* the act or sound of retching.
retd, (*abbr.*) retained; retired; returned.
rete (rē′tē), *n.* (*pl.* **retia** (-tiə, -shiə)) a network of nerves or blood-vessels. **retial** (-shi-), *a.*
retell, *v.t.* (*past*, *p.p.* **retold**) to tell again.
retention (riten′shən), *n.* the act of retaining; the state of being retained; continued confinement; the power of retaining, esp. ideas in the mind; (*Med.*) failure to evacuate urine, faeces etc., power of retaining food etc. **retentive**, *a.* capable of, good at, retaining. **retentively**, *adv.* **retentiveness**, *n.*
rethink, *v.t.*, *v.i.* to think again, to reconsider (a plan, decision etc.) and take an alternative view.
retiarius (rētiəˈriəs, -shi-), *n.* (*pl.* **-arii** (-riī)) a gladiator armed with a net and trident.
reticent (retˈisənt), *a.* reserved in speech; not disposed to communicate one's thoughts or feelings; inclined to keep one's own counsel; silent, taciturn. **reticence**, *n.* **reticently**, *adv.*
reticle (retˈikl), *n.* a network of fine lines, etc., drawn across the focal plane of an optical instrument. **reticular** (-tik′ū-), *a.* having the form of a net or network; formed with interstices. **reticularly**, *adv.* **reticulate**, *v.t.* to make or divide into or arrange in a network, to mark with fine intersecting lines. *v.i.* to be divided into or arranged in a network. *a.* (-lət), formed of or resembling a network. **reticulately**, *adv.* **reticulation**, *n.* **reticule** (-kūl), *n.* a kind of lady's handbag, orig. of network; a reticle. **reticulum** (-tikˈūləm), *n.* (*pl.* **reticula** (-lə)) (*Anat.*) the second stomach of ruminants; a net-like or reticulated structure, membrane etc.
retiform (rē′tifawm, ret′-), *a.* reticular.
retina (retˈinə), *n.* (*pl.* **-nas**, **-nae** (-nē)) a net-like layer of sensitive nerve-fibres and cells behind the eyeball in which the optic nerve terminates. **retinal**, *a.*
retinitis (retinī′tis), *n.* inflammation of the retina.
retin(o)-, *comb. form.* pertaining to the retina.
retinoscopy (retinos′kəpi), *n.* examination of the eye using an instrument that throws a shadow on to the retina. **retinoscopic** (-skop′-), *a.* **retinoscopically**, *adv.*
retinue (retˈinū), *n. sing.* the attendants on a distinguished person.
retire (ritīə′), *v.i.* to withdraw, to go away, to fall back, to retreat, to recede; to withdraw from business or working

life; to resign one's office or appointment, to cease from or withdraw from active service; to go to or as to bed; to go into privacy or seclusion. *v.t.* to cause to retire or resign; to order (troops) to retire; to withdraw (a bill or note) from circulation or currency. **retiral**, *n.* the act of retiring. **retired**, *a.* private, withdrawn from society, given to privacy or seclusion; secluded, sequestered; having given up business, work etc. **retired list**, *n.* a list of retired officers, etc., usually on half-pay. **retiredness**, *n.* **retirement**, *n.* **retiring**, *a.* unobtrusive, not forward, shy, unsociable. **retiringly**, *adv.* **retiringness**, *n.*
retold (rētōld'), *past, p.p.* RETELL.
retool, *v.t., v.i.* to replace or re-equip (a factory etc.) with new tools. *v.t.* to remake, esp. by means of tools.
retort[1] (ritawt'), *n.* a vessel with a bulb-like receptacle and a long neck bent downwards used for distillation of liquids etc.; a large receptacle of fire-clay, iron etc. of analogous shape, used for the production of coal-gas. *v.t.* to purify (mercury etc.) by treatment in a retort.
retort[2] (ritawt'), *v.t.* to turn or throw back; to turn (an argument, accusation, etc.) on or against the author; to pay back (an attack, injury etc.) in kind; to say, make, or do, by way of repartee etc.. *v.i.* to turn an argument or accusation against the originator or aggressor. *n.* the turning of an accusation, taunt, attack etc. against the author or aggressor; a sharp rejoinder. **retorter**, *n.* **retortion**, *n.* bending, turning, or twisting back; the act of retorting; retaliation.
retouch, *v.t.* to touch again; to improve (a photograph, picture etc.) by making good any blemishes etc. *n.* an act of retouching; (rē'-) a photograph, painting etc. that has been retouched. **retoucher**, *n.* **retrace**, *v.t.* to trace back to the beginning, source etc.; to go back over (one's course or track) again; (*fig.*) to go over again in memory, to try to recollect; to trace (an outline) again. **retraceable**, *a.*
retract (ritrakt'), *v.t.* to draw or pull back or in; to take back, to revoke, to recall, to recant, to disavow, to acknowledge to be false or erroneous. *v.i.* to draw, or be capable of drawing, back or in; to withdraw, to shrink back; to withdraw or recall a declaration, statement, accusation, promise, concession etc. **retractable, retractile** (-tīl), *a.* capable of being retracted. **retractability** (-bil'-), **retractility** (-til'-), *n.* **retractation** (rē-), *n.* the act of retracting, disavowing, or recanting. **retraction**, *n.* the act or process of drawing back or in; the act of retracting. **retractive**, *a.* serving to retract or draw in. **retractor**, *n.* a muscle used for drawing back; an instrument or bandage for holding back parts in the way of a surgeon.
retral (rē'trəl, ret'-), *a.* situated at the back, posterior, hinder; bending backward.
retransfer, *v.t.* to transfer again. *n.* an act of retransferring. **retransform**, *v.t.* to transform anew. **retransformation**, *n.* **retranslate**, *v.t.* to translate again; to translate back again to the original language. **retranslation**, *n.* **retread**, *v.t.* (*past* **retrod**, *p.p.* **-trodden**) to tread again; to remould a tyre. *n.* (rē'-) a used tyre which has had its worn tread replaced.
retreat (ritrēt'), *n.* the act of withdrawing, esp. of an army before an enemy; a signal for such withdrawal; a drumbeat at sunset; a state of retirement or seclusion; a period of retirement from society at large; a place of retirement, safety, privacy, or seclusion; a refuge, as an asylum for lunatics, inebriates, aged persons etc.; retirement for meditation, prayer etc. *v.i.* to move back, to withdraw, esp. before an enemy or from an advanced position; to withdraw to a place of privacy, seclusion, or safety; to recede. *v.t.* to cause to withdraw; (*Chess*) to move (a piece) back.
retrench (ritrench'), *v.t.* to cut down, to reduce, to curtail, to diminish, to shorten, to abridge; to cut out or pare down; (*Mil.*) to furnish with a retrenchment. *v.i.* to curtail expenses, to make economies. **retrenchment**, *n.* the act of retrenching; a fortification constructed with or behind another to prolong a defence.
retrial (rē'trīəl), *n.* a new trial.
retribution (retribū'shən), *n.* recompense, a suitable return, esp. for evil; requital, vengeance; the distribution of rewards and (*esp.*) punishments in the future life. **retributive, -tory** (-trib'-), *a.* **retributor**, *n.*
retrieve (ritrēv'), *v.t.* to find and bring in (esp. of a dog bringing in game or a stick, ball etc.); to recover by searching or recollecting, to recall to mind; to regain (that which has been lost, impaired etc.); to rescue (from); to restore, to re-establish (one's fortunes etc.); to remedy, to make good, to repair (e.g. a difficult situation); in tennis etc., to return (a difficult ball) successfully; to recover (data stored in a computer system). *v.i.* to fetch (of a dog). **retrievable**, *a.* **retrievably**, *adv.* **retrieval**, *n.* **retriever**, *n.* person or thing that retrieves; a breed of dog trained to fetch game that has been shot.
retrim, *v.t.* to trim again.
retro-, *pref.* backwards, back; in return; (*Anat. etc.*) behind.
retroact (retrōakt'), *v.i.* to act backwards or in return; to act retrospectively. **retroaction, retroactivity** (-tiv'-), *n.* **retroactive**, *a.* (of laws etc.) applying to the past; operating backward. **retroactively**, *adv.*
retrochoir (ret'rōkwīə), *n.* a part of a cathedral or other large church east of or beyond the high altar.
retrofit (retrōfit'), *v.t.* (*pres.p.* **-fitting**, *past, p.p.* **-fitted**) to equip or modify (an aircraft, car etc.) with new parts or safety equipment after manufacture.
retroflected (retrəflek'tid), **retroflex** (ret'rəfleks), **-flexed**, *a.* turned or curved backward; (of vowels or consonants) articulated with the tip of the tongue bent upwards and backwards. **retroflexion** (-flek'shən), **retroflection**, *n.*
retrograde (ret'rəgrād), *a.* going, moving, bending or directed backwards; inverted, reversed; declining, degenerating, deteriorating, reversing progress; (*Astron.*) applied to the motion of a planet relative to the fixed stars when it is apparently from east to west. *n.* a backward movement or tendency, deterioration, decline; a degenerate person. *v.i.* to move backward; to decline, to deteriorate, to revert, to recede; (*Astron.*) to (appear to) move from east to west relative to the fixed stars. **retrogradation**, *n.*
retrogress (retrəgres'), *v.i.* to go backward, to retrograde; to degenerate. **retrogression** (-shən), *n.* **retrogressive**, *a.* **retrogressively**, *adv.*
retrorocket (ret'rōrokit), *n.* a small rocket on a spacecraft, satellite etc. which produces thrust in the opposite direction to flight for deceleration or manoeuvring.
retrorse (ritraws'), *a.* turned or bent backwards, reverted. **retrorsely**, *adv.*
retrospect (ret'rəspekt), *n.* a looking back on things past; view of, regard to, or consideration of previous conditions etc.; a review of past events. *v.i.* to look or refer back (to). *v.t.* to view or consider retrospectively. **retrospection** (-spek'-), *n.* **retrospective** (-spek'-), *a.* in retrospection; viewing the past; applicable to what has happened in the past. *n.* an exhibition of an artist's life's work. **retrospectively**, *adv.*
retroussé (ritrōō'sā), *a.* (of the nose) turned up at the end.
retrovert (retrəvœt'), *v.t.* to turn back. **retroversion** (-shən), *n.*
retrovirus (ret'rōvīərəs), *n.* any of a group of viruses that uses RNA to synthesize DNA, reversing the normal process of cellular transcription of DNA into RNA. Many cause cancer in animals and one is the cause of Aids in humans.
retry, *v.t.* (*past, p.p.* **retried**) to try again.
retsina (retsē'nə), *n.* a resin-flavoured white wine from Greece.
returf, *v.t.* to turf again.
return (ritœn'), *v.i.* to come or go back, esp. to the same

retuse 707 **reverse**

place or state; to revert, to happen again, to recur. *v.t.* to bring, carry, or convey back; to give, render, or send back, to requite; to repay, to put or send back or in return; to say in reply, to retort; to make an official report, statement, count etc. of; to elect; to play a card of the same suit as another player has led. *n.* the act of coming or going back; the act of giving, paying, putting, or sending back; that which is returned; an unwanted ticket previously bought for an event; an official account, statement or report; a returning officer's announcement of a candidate's election; the act of electing or returning; the state of being elected; the proceeds or profits on labour, investments etc. (*often in pl.*); a return bend in a pipe etc.; (*Arch.*) a receding part of a façade etc., a part of a wall etc. turning in another direction, esp. at right angles; (*Games, Fencing etc.*) a stroke, thrust etc. in return; a return match or game; (*Law*) the rendering back or delivery of a writ, precept, or execution to the proper officer or court. *a.* (for) coming back or going back to the point of origin, as a journey. **by return (of post),** by the next post back to the sender. **in return,** in reply or response; in requital. **many happy returns,** a birthday greeting. **to return thanks,** to offer thanks; to answer a toast. **return game** or **match,** *n.* a second meeting of the same competitors, participants, clubs or teams. **return ticket,** *n.* a ticket for a journey to a place and back again; the return half of such a ticket. **returnable,** *a.* **returnee,** *a.* **returner,** *n.* **returning officer,** *n.* the presiding officer at an election. **returnable, returnless** (-lis), *a.*
retuse (ritūs'), *a.* having a round end with a depression in the centre.
reunion (rēūn'yən), *n.* the act of reuniting; the state of being reunited; a periodic meeting or social gathering, esp. of friends, former associates, or partisans. **reunite** (-nit'), *v.t.* to join again after separation; to reconcile after variance. *v.i.* to become united again.
reuse (rēūz'), *v.t.* to use again. *n.* (-ūs') the act of using again.
rev (rev), *n.* a revolution in an engine. *v.t., v.i.* (*pres.p.* **revving,** *past, p.p.* **revved**) to run an engine quickly by increasing the speed of revolution. **rev counter,** *n.* (*coll.*) TACHOMETER.
Rev., (*abbr.*) Revelation (of St John the Divine); Reverend.
rev., (*abbr.*) revenue; reverse(d); review; revise(d), revision; revolution; revolving.
revalorization (rēvalərīzā'shən), *n.* (*Econ.*) restoration of the value of currency.
revalue, *v.t.* to adjust the exchange rate of a currency, usu. upwards; to reappraise. **revaluation,** *n.*
revamp (rēvamp'), *v.t.* to renovate, to restore or alter the appearance of. *n.* something renovated or revamped; the act or process of revamping.
revanche (rivash', -vahnch'), *n.* a policy directed towards restoring lost territory or possessions. **revanchism,** *n.* **revanchist,** *n., a.*
Revd, (*abbr.*) Reverend.
reveal[1] (rivēl'), *v.t.* to make known by supernatural or divine means; to disclose, to divulge (something secret, private or unknown), to betray; to cause or allow to become visible. **revealable,** *a.* **revealer,** *n.* **revealing,** *a.* significant, telling; exposing more of the body than is usual (of a dress etc.). **revealingly,** *adv.* **revealment,** *n.*
reveal[2] (rivēl'), *n.* the depth of a wall as seen in the side of an aperture, doorway or window.
réveillé (rival'i), *n.* a morning signal by drum or bugle to awaken soldiers or sailors.
revel (rev'əl), *v.i.* (*pres.p.* **revelling,** *past, p.p.* **revelled**) to make merry, to feast, to carouse, esp. boisterously; to take unrestrained enjoyment (in). *v.t.* to spend or waste in revelry. *n.* an act of revelling, a carouse, a merry-making. **reveller,** *n.* **revelry** (-ri), *n.*
revelation (revəlā'shən), *n.* the act of revealing, a disclo-

sure of knowledge or information; that which is revealed, esp. by God to man; the title of the last book of the New Testament, the Apocalypse; an astonishing disclosure. **revelational,** *a.* **revelationist,** *n.* one who believes in divine revelation; the author of the Apocalypse. **revelative, -tory** (rev'-), *a.*
reveller revelry REVEL.
revenant (rev'ənənt), *n.* one who returns from the grave or from exile, esp. a ghost.
revendication (rivendikā'shən), *n.* (*International Law*) a formal claim for the surrender of rights, esp. to territory. **revendicate,** *v.t.*
revenge (rivenj'), *v.t.* to exact satisfaction or retribution for, to requite, to retaliate against; to avenge or satisfy (oneself) with such retribution or retaliation. *n.* retaliation, requital, retribution or spiteful return for an injury; a means, mode or act of revenging; the desire to inflict revenge, vindictiveness. **revengeful,** *a.* **revengefully,** *adv.* **revengefulness,** *n.* **revengeless** (-lis), *a.* **revenger,** *n.* **revengingly,** *adv.*
revenue (rev'ənū), *n.* income, esp. from a considerable amount of many forms of property (*often in pl.*); the annual income of a state, derived from taxation, customs, excise etc.; the department of the Civil Service collecting this. **inland revenue** INLAND. **revenue-cutter,** *n.* a vessel employed to prevent smuggling. **revenue-officer,** *n.* a customs officer.
reverberate (rivœ'bərāt), *v.t.* to send back, to re-echo, to reflect (sound, light, or heat); *v.i.* to be driven back or to be reflected (of sound, light, heat); to resound, to re-echo; to rebound, to recoil. **reverb,** *n.* an electronic device which creates an artificial echo in recorded music. **reverberation,** *n.* **reverberator,** *n.* **reverberatory,** *a.* producing or acting by reverberation.
revere (riviə'), *v.t.* to regard with awe mingled with affection, to venerate. **reverence** (rev'ə-), *n.* the act of revering, veneration; a feeling of or the capacity for feeling, awe mingled with affection; title given to the clergy (in *his reverence* etc.). *v.t.* to regard or treat with reverence, to venerate. **saving your reverence,** (*facet.*) with all respect to you. **reverencer,** *n.* **reverend** (rev'ə-), *a.* worthy or entitled to reverence or respect, esp. as a title of respect given to clergy (a dean is addressed as **Very Reverend,** a bishop as **Right Reverend,** and an archbishop as **Most Reverend**). **reverent** (rev'ə-), *a.* feeling or expressing reverence; submissive, humble. **reverential** (revəren'-), *a.* **reverentially,** *adv.* **reverently** (rev'ə-), *adv.*
reverie (rev'əri), *n.* listless musing; a day-dream, a loose or irregular train of thought; a dreamy musical composition.
revers (riviə'), *n.* (*pl.* **revers**) a part of a coat, esp. a lapel, turned back so as to show the lining.
reverse (rivœs'), *a.* turned backward, inverted, reversed, upside down; having an opposite direction, contrary. *n.* the contrary, the opposite; the back surface (of a coin etc.), the opposite of *obverse;* a complete change of affairs for the worse, a check, a defeat. *v.t.* to turn in the opposite direction, to turn the other way round, upside down, or inside out; to invert, to transpose; to cause to have a contrary motion or effect; to revoke, to annul, to nullify. *v.i.* to change to a contrary condition, direction etc.; to put a car into reverse gear; in dancing, to (begin to) turn in the opposite direction. **to reverse the charges,** to make a telephone call for which the recipient pays. **reverse video,** *n.* (*Comput.*) a technique for highlighting information on a VDU in which the normal colours for text and background are reversed. **reversal,** *n.* act of reversing. **reversal film,** *n.* positive film, slide film as opp. to negatives. **reversed,** *a.* turned in a reverse direction; changed to the contrary; made or declared void; applied to the spire of a shell turning from right to left. **reversedly** (-sid-), *adv.* **reversely,** *adv.* **reverser,** *n.* **reversible,** *a.* **reversibility** (-bil'-), *n.* **reversing,** *n., a.* **reversing light,** *n.* a

light on the rear of a motor vehicle which is illuminated when reverse gear is engaged. **reversion** (-shən), *n.* return to a former condition, habit etc.; the tendency of an animal or a plant to revert to ancestral type or characters; the returning of an estate to the grantor or his heirs after a particular period is ended; the right of succeeding to an estate after the death of the grantee etc.; a sum payable upon some event, as a death, esp. an annuity or life assurance; the right or expectation of succeeding to an office etc., on relinquishment by the present holder. **reversional, reversionary,** *a.* **reversionally,** *adv.* **reversioner,** *n.* one who holds the reversion to an estate etc. **reverso** (-sō), *n.* the left-hand page of an open book, usu. even-numbered, a verso.

revert (rivœt'), *v.t.* to turn (esp. the eyes) back. *v.i.* to return, to go back, to fall back, to return (to a previous condition, habits, type etc., esp. to a wild state); to turn the attention again (to); (*Law*) to come back by reversion to the possession of the former proprietor. **revertant,** *a.* (*Her.*) bent back, esp. like the letter S. **reverter,** *a.* one who or that which reverts. **revertible,** *a.*

revet (rivet'), *v.t.* (*past, p.p.* **revetted**) to face (a wall, scarp, parapet etc.) with masonry. **revetment,** *n.* a facing of stones, concrete etc. to protect a wall or embankment; a retaining wall.

revictual, *v.t.* to victual again.

review (rivū'), *v.t.* to view again; to look back on, to go over in memory, to revise; to survey, to look over carefully and critically; to write a critical essay or description of; to hold an inspection of. *v.i.* to write reviews. *n.* a repeated examination, a reconsideration, a second view; a retrospective survey; a revision, esp. by a superior court of law; a critical account (of a book etc.); a periodical publication containing essays and criticisms; a display or a formal or official inspection of military or naval forces. **court of review,** one to which sentences and decisions are submitted for judicial revision. **review copy,** *n.* a copy of a new book sent to an individual or periodical for review. **review order,** *n.* (*Mil.*) parade uniform and arrangement; full dress, full rig. **reviewable,** *a.* **reviewal,** *n.* **reviewer,** *n.* one who reviews, esp. books.

revile (rivīl'), *v.t.* to address or describe with insulting or scandalous language, to abuse, to vilify. *v.i.* to be abusive, to rail. **revilement,** *n.* **reviler,** *n.* **revilingly,** *adv.*

revise (rivīz'), *v.t.* to examine for correction or emendation; to correct, alter, or amend; to reread (course notes etc.) for an examination. *n.* a revision; a proof-sheet in which corrections made in rough proof have been embodied; a revised form or version. **revisable,** *a.* **revisal,** *n.* **Revised Version,** *n.* the revision of the Authorized Version of the Bible. **reviser, revisor,** *n.* **revisership,** *n.* **revision** (-vizh'ən), *n.* the act or process of revising; the process of rereading course notes etc. before an exam; a revised version. **revisional, -ary, revisory,** *a.* **revisionism,** *n.* **revisionist,** *n.* a reviser of the Bible; one who believes in the broadening and evolution of the theories of Marxism; (*derog.*) one who departs from the strict principles of orthodox Communism.

revisit, *v.t.* to visit again. *n.* a further visit. **revisitation,** *n.*

revisor, revisory REVISE.

revitalize, -ise, *v.t.* to vitalize again. **revitalization,** *n.*

revival REVIVE.

revive (rivīv'), *v.i.* to return to life, consciousness, health, vigour activity, vogue, the stage etc.; to gain new life or vigour; to recover from a state of obscurity, neglect, or depression; to come back to the mind again, to reawaken. *v.t.* to bring back to life, consciousness, vigour, etc.; to produce again on stage; to reanimate; to resuscitate, to renew, to renovate, to reawaken, to re-establish, to reencourage; (*Chem.*) to restore or reduce to its natural or metallic state. **revivable,** *a.* **revivably,** *adv.* **revival,** *n.* the act of reviving; the state of being revived; return or re-covery of life, consciousness, vigour, activity or vogue; a renaissance; a new production of a dramatic work previously neglected or forgotten; a religious reawakening. **revivalism,** *n.* **revivalist,** *n.*, *a.* **reviver,** *n.* one who or that which revives; a preparation for renovating leather, cloth etc.; (*sl.*) a drink, a stimulant. **revivingly,** *adv.*

revivify, *v.t.* to restore to life; to reanimate, to reinvigorate, to put new life into; to revive. **revivification** (-fi-) *n.* **revivingly** REVIVE.

reviviscent, *a.* recovering life and strength, reviving. **reviviscence, -cy,** *n.*

revoke (rivōk'), *v.t.* to annul, to cancel, to repeal, to rescind. *v.i.* (*Cards*) to fail to follow suit when this is possible and mandatory. *n.* act of revoking at cards. **revocable,** *a.* **revocability** (-bil'-), **revocableness,** *n.* **revocably,** *adv.* **revocation** (revə-), *n.* **revocatory** (rev'ə-), *a.*

revolt (rivōlt'), *v.i.* to renounce allegiance, to rise in rebellion or insurrection, to turn away (from); to be repelled (by), to feel disgust (at), to turn away in loathing (from), to feel repugnance (at). *v.t.* to repel, to nauseate, to disgust. *n.* a renunciation of allegiance and subjection; a rebellion, a rising, an insurrection; a change of sides. **revolter,** *n.* **revolting,** *a.* causing disgust, repulsion, or abhorrence. **revoltingly,** *adv.*

revolution (revəloo'shən), *n.* the act or state of revolving; the circular motion of a body on its axis, rotation; (a complete) rotation or movement round a centre; the period of this; a round or cycle or regular recurrence or succession; a radical change or reversal of circumstances, conditions, relations or things; a fundamental change in government, esp. by the forcible overthrow of the existing system and substitution of a new ruler or political system. **American Revolution,** the successful revolt of the 13 American colonies from Great Britain in the War of Independence (1775–81). **Cultural Revolution,** the period of upheaval in the People's Republic of China in the 1960s, when Chairman Mao sought to destroy the power of the bureaucracy. **French Revolution,** the overthrow of the French monarchy in 1789 and the succeeding years. **Glorious Revolution,** that by which James II was driven from the English throne in 1688. **Russian Revolution,** the overthrow of the Czarist regime by the Bolshevists and establishment of a socialist republic, in 1917. **revolutionary,** *a.* pertaining to or tending to produce a revolution in government; radical. *n.* an advocate of revolution; one who takes an active part in a revolution. **revolutionism,** *n.* **revolutionize, -ise,** *v.t.* to bring about a revolution in; to cause radical change in.

revolve (rivolv'), *v.i.* to turn round; to move round a centre, to rotate; to move in a circle, orbit, or cycle; to roll along. *v.t.* to cause to revolve or rotate; to turn over and over in the mind, to meditate on, to ponder over. **revolver,** *n.* one who or that which revolves; a pistol having a revolving breech cylinder by which it can be fired several times without reloading. **revolving,** *a.* **revolving credit,** *n.* credit which allows repeated borrowing of a fixed sum as long as that sum is never exceeded. **revolving door,** *n.* a door, usu. with four leaves at right angles, that rotates about a vertical axis.

revue (rivū'), *n.* a light entertainment with songs, dances etc., representing topical characters, events, fashions etc.

revulsion (rivŭl'shən), *n.* a sudden or violent change or reaction, esp. of feeling; reduction of a disease in one part of the body by treatment of another part, as by counter-irritation; disgust. **revulsive,** *a.* causing or tending to cause revulsion. *n.* an application causing revulsion, a counter-irritant. **revulsively,** *adv.*

reward (riwawd'), *v.t.* to repay, to requite, to recompense (a service or offence, a doer or offender). *n.* that which is given in return for good or evil done or received; a recompense, a requital; a sum of money offered for the detection of a criminal, for the restoration of anything lost,

etc. **rewardable**, *a.* **rewardableness**, *n.* **rewardably**, *adv.* **rewarder**, *n.* **rewardfulness**, *n.* **rewarding**, *a.* profitable; personally satisfying. **rewardingly**, *adv.* **rewardless** (-lis), *a.*
rewind, *v.t.* (*past* **-wound**) to wind (film or tape) on to the original spool, reel etc. *n.* something rewound. **rewire**, *v.t.* to install new electrical wiring in a house etc. **reword**, *v.t.* to put into new words. **rework**, *v.t.* to treat or use again; to revise, to reprocess for renewed use. **rewrite**, *v.t.* to write over again; to revise. *n.* (rē'-) something rewritten or revised.
rex (reks), *n.* a reigning king; the official title used by a king, esp. on documents, coins etc.
reynard (ren'əd, -ahd, rā'-), *n.* a proper or poetic name for the fox.
RF, (*abbr.*) radio frequency.
RFC, (*abbr.*) Royal Flying Corps; Rugby Football Club.
RFU, (*abbr.*) Rugby Football Union.
RGN, (*abbr.*) Registered General Nurse.
RGS, (*abbr.*) Royal Geographical Society.
rh, RH, (*abbr.*) right hand.
Rh, (*chem. symbol*) rhodium.
Rh, (*abbr.*) rhesus.
rhabdomancy (rab'dəmansi), *n.* divination by a rod, esp. the discovery of minerals, underground streams etc. with the divining-rod. **rhabdomancer**, *n.* **rhabdomantic** (-man'-), *a.*
rhachis (rak'is), **rhachitis** (rəki'tis), etc. RACHIS.
rhamphoid (ram'foid), *a.* beak-shaped.
rhapsody (rap'sədi), *n.* a high-flown, enthusiastic composition or utterance; (*Mus.*) an irregular and emotional composition, esp. of the nature of an improvisation; in ancient Greece, an epic poem, or a portion of this for recitation at one time by a rhapsodist. **rhapsode** (-sōd, -*a.* an ancient Greek reciter of epic poems, esp. one of a professional school who recited the Homeric poems. **rhapsodic** (-sod'-), *a.* of or pertaining to rhapsody; emotionally enthusiastic; high-flown, extravagant. **rhapsodically**, *adv.* **rhapsodize, -ise**, *v.t.* to speak or write with emotion or enthusiasm; *v.i.* to recite or write rhapsodies. **rhapsodist**, *n.* a rhapsode; any professional reciter or improviser of verses; one who writes or speaks rhapsodically.
rhatany (rat'əni), *n.* a Peruvian shrub or its root, from which an extract is obtained used in medicine and for adulterating port wine.
rhea[1] (rē'ə), *n.* a genus of birds containing the S American three-toed ostriches; a bird of this genus.
rhea[2] (rē'ə), *n.* the ramie plant.
rhematic (rēmat'ik), *a.* pertaining to the formation of words, esp. verbs.
Rhemish (rē'mish), *a.* of or pertaining to Rheims (applied esp. to an English translation of the New Testament by Roman Catholic students in 1582).
Rhenish (ren'ish), *a.* pertaining to the Rhine or Rhineland. *n.* Rhine wine, hock.
rhenium (rē'niəm), *n.* a metallic element, at. no. 75; chem. symbol Re, occurring in certain platinum and molybdenum ores.
rheo-, *comb. form.* pertaining to a current; flow. **rheology** (rēol'əji), *n.* the science dealing with the flow and deformation of matter. **rheologic, -ical** (-loj'-), *a.* **rheologist**, *n.* **rheostat** (rē'əstat), *n.* a variable resistance for adjusting and regulating an electric current, a potentiometer. **rheostatic** (-stat'-), *a.* **rheotropism** (rēo'trəpizm), *n.* the tendency in growing plant-organs exposed to running water to dispose their longer axes either in the direction of or against the current.
rhesus (rē'səs), *n.* one of the macaques, an Indian monkey, held sacred in some parts of India. **rhesus (Rh) factor**, *n.* an antigen substance occurring in the red blood corpuscles of most human beings and many mammals (e.g. the rhesus monkey). **Rh positive** or **Rh negative** indicate whether this substance is present or absent.

rhetor (rē'tə), *n.* in ancient Greece, a teacher or professor of rhetoric; a professional orator. **rhetoric** (ret'ərik), *n.* the art of effective speaking or writing, the rules of eloquence; a treatise on this; the use of language for effect or display, esp. affected or exaggerated oratory or declamation. **rhetorical** (-to'-), *a.* pertaining to or of the nature of rhetoric; designed for effect or display, florid, showy, affected, declamatory. **rhetorical question**, one put merely for the sake of emphasis and requiring no answer. **rhetorically**, *adv.* **rhetorician** (retərish'ən), *n.* a teacher of rhetoric; a skilled orator; a flamboyant or affected speaker.
rheum[1] (room), *n.* the thin serous fluid secreted by the mucous glands as tears, saliva, or mucus; (*poet.*) tears; mucous discharge, catarrh. **rheumatic** (-mat'-), *a.* pertaining to, of the nature of, suffering from, or subject to, rheumatism. **rheumatically**, *adv.* **rheumatic fever**, *n.* a disease characterized by fever, acute pain in the joints and potential inflammation of the heart and pericardium. **rheumatics**, *n.pl.* (*coll.*) rheumatism. **rheumatism**, *n.* an inflammatory disease affecting muscles and joints of the human body, and attended by swelling and pain. **rheumat(o)-**, *comb. form.* rheumatic.
rheum[2] (rē'əm), *n.* a genus of plants comprising the rhubarbs.
rheumatoid (roo'mətoid), *a.* rheumatic. **rheumatoid arthritis**, *n.* disease of the tissues of the joints.
rheumatology (roomətol'əji), *n.* the study of rheumatism. **rheumatological** (-loj'-), *a.* **rheumatologist**, *n.*
rheumophthalmia (rooməfthal'miə), *n.* rheumatic inflammation of the sclerotic membrane.
rhinal (rī'nəl), *n.* of or pertaining to the nose or nostrils.
rhinalgia (rinal'jiə), *n.* nasal neuralgia.
Rhine (rīn), *a.* pertaining to or derived from the Rhine or its bordering regions. **rhinestone**, *n.* a species of rock crystal; a colourless artificial gem cut to look like a diamond. **Rhine wine**, *n.* a wine made from grapes grown in the neighbourhood of the Rhine.
rhinencephalon (rīnensef'əlon), *n.* the olfactory lobe of the brain. **rhinencephalic** (-fal'-), **rhinencephalous**, *a.*
rhinestone RHINE.
rhinitis (rīni'tis), *n.* inflammation of the nose.
rhino (rī'nō), *n.* (*pl.* **-nos, -no**) a rhinoceros.
rhin(o)-, *comb. form.* pertaining to the nose or nostrils.
rhinoceros (rīnos'ərəs), *n.* (*pl.* **-oses**) a large pachydermatous quadruped, now found only in Africa and S Asia, with one or two horns on the nose. **rhinoceros-bird**, *n.* the African beef-eater or ox-pecker. **rhinoceroid** (-roid), **rhinocerotic** (-rot'-), **rhinocerotoid** (-se'rətoid), *n., a.*
rhinology (rīnol'əji), *n.* the branch of science dealing with the nose and nasal diseases. **rhinological** (-loj'-), *a.* **rhinologist**, *n.*
rhinopharyngeal (rīnōfərin'jiəl), *a.* pertaining to the nose and the pharynx. **rhinopharyngitis** (rīnōfarinji'tis), *n.* inflammation of the nose and pharynx.
rhinoplasty (rī'nōplasti), *n.* an operation for restoring an injured nose or changing the nose by plastic surgery. **rhinoplastic** (-plas'-), *a.*
rhinorrhoea (rīnərē'ə), *n.* discharge of blood from the nose.
rhinoscleroma (rīnōsklirō'mə), *n.* a disease affecting the nose, lips etc. with a tuberculous growth.
rhinoscope (rī'nəskōp), *n.* an instrument for examining the nasal passages. **rhinoscopic** (-skop'-), *a.* **rhinoscopy** (-nos'-), *n.*
rhipid(o)-, *comb. form.* having fan-shaped processes.
rhipipteran (ripip'tərən), *n.* a fan-winged. **rhipipterous**, *a.*
rhiz(a)-, rhiz(o)-, *comb. form.* pertaining to a root; having roots or root-like processes.
rhizanth (rī'zanth), *n.* a plant flowering or seeming to flower from the root. **rhizanthous** (-zan'-), *a.*
rhizic (rī'zik), *a.* of or pertaining to the root of an equation.

rhizocarp (rī'zōkahp), *n.* a plant having a perennial root but a stem that withers annually. **rhizocarpean** (-kah'-), *a.* **rhizocarpic, -carpous,** *a.*
rhizodont (rī'zōdont), *a.* having teeth rooted and fused with the jaw (as crocodiles). *n.* a rhizodont reptile.
rhizogen (rī'zəjən), *n.* a plant parasitic on the roots of another plant. **rhizogenic** (-jen'-), *a.* root-producing.
rhizoid (rī'zoid), *a.* root-like. *n.* a filiform or hair-like organ serving for attachment, in mosses etc.
rhizome (rī'zōm), **rhizoma** (-zō'mə), *n.* a prostrate, thickened, root-like stem, sending roots downwards and yearly producing aerial shoots etc. **rhizomatous** (-zom'-), *a.*
rhizomorph (rī'zōmawf), *n.* a root-like mycelial growth by which some fungi attach themselves to higher plants. **rhizomorphoid** (-maw'foid), **-phous,** *a.*
rhizophagous (rīzof'əgəs), *a.* feeding on roots. **rhizophagan** (-gən), *n., a.*
rhizophore (rī'zəfaw), *n.* a structure bearing the roots in certain plant species. **rhizophorous** (-zof'-), *a.* root-bearing.
rho (rō), *n.* (*pl.* **rhos**) the 17th letter of the Greek alphabet (P, ρ).
rhodamine (rō'dəmēn), *n.* any of a group of fluorescent, usu. red dyestuffs.
Rhode Island Red (rōd), *n.* an American breed of domestic fowl with reddish-brown plumage.
Rhodes Scholar (rōdz), *n.* a student holding one of the Rhodes scholarships at Oxford University founded under the will of Cecil Rhodes (1853–1902) for students from the British Commonwealth and the US. [Cecil *Rhodes*, 1853–1902]
Rhodian[1] (rō'diən), *a.* pertaining to Rhodes, an island in the Aegean Sea. *n.* a native or inhabitant of Rhodes.
rhodium[1] (rō'diəm), *n.* the Jamaica rosewood; the hard, white, scented wood of either of two shrubby convolvuluses growing in the Canary Islands, also called rhodium- or rhodian-wood. **oil of rhodium,** an oil obtained from this. **rhodian**[2], *a.*
rhodium[2] (rō'diəm), *n.* a greyish-white metallic element, at. no. 45; chem. symbol Rh, belonging to the platinum group. **rhodic, rhodous,** *a.*
rhod(o)-, *comb. form.* rose; rose-like; rose-coloured; scented like a rose.
rhododendron (rōdədən'drən), *n.* a genus of evergreen shrubs akin to the azaleas, with brilliant flowers; a shrub of this genus.
rhodolite (rō'dəlīt), *n.* a pale pink or purple garnet used as a gemstone.
rhodonite (rō'dənīt), *n.* a rose-pink silicate of manganese.
rhodophyl (rō'dəfil), *n.* the compound pigment giving red algae their colour. **rhodophyllous** (-fil'-), *a.*
rhodopsin (rədop'sin), *n.* a purplish pigment found in the retina, visual purple.
rhoeadic (rēad'ik), *a.* derived from the red poppy. **rhoeadic acid,** *n.* a compound found in the flowers of this, the principle of their colouring-matter. **rhoeadine** (rē'ədin), *n.* an alkaloid obtained from the red poppy.
rhomb (rom, romb), *n.* an oblique parallelogram, with equal sides; (*Cryst.*) a rhombohedron. **rhomb-spar,** *n.* a perfectly crystallized variety of dolomite. **rhombic** (rom'bik), *a.* **rhombiform** (-fawm), *a.*
rhomb(o)-, *comb. form.* pertaining to a rhomb.
rhombohedron (rombōhē'drən), *n.* (*pl.* **-dra** (-drə)) a solid figure bounded by six equal rhombs. **rhombohedral,** *a.*
rhomboid (rom'boid), *n.* a parallelogram the adjoining sides of which are not equal and which contains no right angle; a rhomboid muscle. *a.* having the shape or nearly the shape of a rhomboid. **rhomboid ligament,** *n.* a ligament connecting the first rib and the end of the clavicle. **rhomboid muscle,** *n.* either of two muscles connecting the vertebral border of the scapula with the spine. **rhomboidal** (-boi'-), *a.* **rhomboidally,** *adv.*

rhomboideum (romboi'diəm), *n.* (*pl.* **-dea** (-diə)) a rhomboid ligament.
rhombus (rom'bəs), *n.* (*pl.* **-buses, -bi** (-bī)) a rhomb.
rhotacism (rō'təsizm), *n.* exaggerated or erroneous pronunciation of the letter *r*, burring; (*Philol.*) the change of *s* into *r*, as in Indo-European languages. **rhotacize, -ise,** *v.i.*
RHS, (*abbr.*) Royal Historical Society; Royal Horticultural Society; Royal Humane Society.
rhubarb (roo'bahb), *n.* any herbaceous plant of the genus *Rheum*, esp. the English, French, common or garden rhubarb, the fleshy and juicy stalks of which are cooked when young as a substitute for fruit; the sound made by actors to simulate background conversation; (*coll.*) nonsense. **rhubarby,** *a.*
rhumb (rŭm), *n.* a line cutting all the meridians at the same angle, such as a ship would follow sailing continuously on one course. **rhumbline,** *n.*
rhumba RUMBA.
rhyme, rime[1] (rīm), *n.* a correspondence of sound in the final accented syllable or group of syllables of a line of verse with that of another line, consisting of identity of the vowel sounds and of all the consonantal sounds but the first; verse characterized by rhyme (*sing.* or *pl.*); poetry, verse; a word rhyming with another. *v.i.* to make rhymes, to versify; to make a rhyme with another word or verse; to be in accord, to harmonize (with). *v.t.* to cause to rhyme; to to put into rhyme; to pass or waste (time etc.) in rhyming. **without rhyme or reason,** inconsiderately, thoughtlessly; pointless. **rhyme royal,** *n.* a seven-lined decasyllabic stanza rhyming *a b a b b c c,* so called because employed by James I of Scotland in the *King's Quhair.* **rhyme scheme,** *n.* the pattern of rhymes in a stanza, poem etc. **rhymeless** (-lis), *a.* **rhymelessness,** *n.* **rhymer, rhymester** (-stə), *n.* a poet, esp. of inferior talent; a minstrel. **rhyming slang,** *n.* a form of slang originating among Cockneys in London in which the word to be disguised (often an indecent one) is replaced by a rhyming phrase of which often only the first element is used, so that the rhyme itself disappears (e.g. *Barnet fair* becomes *Barnet* meaning *hair*).
rhynch(o)-, *comb. form.* having a snout or snout-like process.
rhynchophore (ring'kəfaw), *n.* one of the division of beetles containing the weevils or snout-beetles. **rhyncophoran** (-kof'-), *n., a.* **rhynchophorous** (-kof'-), *a.*
rhynchosaur (ring'kəsaw), **rhynchosauros** (-saw'rəs), *n.* a genus of toothless saurians from the Devonian strata.
rhyolite (rī'əlīt), *n.* an igneous rock of structure showing the effect of lava-flow, composed of quartz and feldspar with other minerals. **rhyolitic** (-lit'-), *a.*
rhyparographer (ripərog'rəfə), *n.* a painter of squalid subjects. **rhyparographic** (-graf'-), *a.* **rhyparography,** *n.*
rhysimeter (rīsim'itə), *n.* an instrument for measuring the velocity of a fluid or the speed of a ship.
rhythm (ridh'm), *n.* movement characterized by action and reaction or regular alternation of strong and weak impulse, stress, accent, motion, sound etc.; metrical movement determined by the regular recurrence of harmonious succession of groups of long and short or stressed and unstressed syllables called feet; the flow of words in verse or prose characterized by such movement; the regulated succession of musical notes according to duration; structural system based on this; (*Art*) correlation of parts in a harmonious whole; any alternation of strong and weak states or movements. **rhythm and blues,** a style of popular music integrating elements of folk, rock and roll and blues. **rhythm method,** *n.* an unreliable method of contraception requiring sexual abstinence during the period when ovulation is most likely to occur. **rhythm section,** *n.* (the players of) the section of instruments (usu. piano, double-bass and drums) in a band whose main task is to provide the rhythm. **rhythmic, -al,** *a.*

rhythmically, *adv.* **rhythmist**, *n.* **rhythmless** (-lis), *a.*
RI, (*abbr.*) refractive index; religious instruction; Royal Institution.
ria (rē'ə), *n.* a long, narrow inlet into the sea-coast.
rial (riahl', rī'əl), *n.* the unit of currency in Iran, Oman, Saudi Arabia and the Yemen Arab Republic (see also RIYAL).
riant (rī'ənt), *a.* (*fem.* **riante** (-ənt)) smiling, cheerful.
rib¹ (rib), *n.* one of the bones extending outwards and forwards from the spine, and in man forming the walls of the thorax; a ridge, strip, line etc., analogous in form or function to this; a cut of meat including one or more ribs; a curved timber extending from the keel for supporting the side of a ship etc.; a raised moulding or groin on a ceiling or vaulted roof; a timber or iron beam helping to support a bridge; a hinged rod forming part of an umbrella-frame; a purlin; a main vein in a leaf; a raised row in a knitted or woven fabric; (*facet.*) a wife, in allusion to Eve (**Adam's rib**). *v.t.* (*past, p.p.* **ribbed**) to furnish with ribs; to mark with ribs or ridges; to enclose, strengthen etc. with ribs; to knit alternate plain and purl stitches to form a raised row. **false ribs**, the lower five pairs of ribs. **floating ribs**, the two lowest pairs of ribs. **ribcage**, *n.* the structure of ribs and tissue which forms the enclosing wall of the chest. **rib-vaulting**, *n.* **ribbed**, *a.* **ribbing**, *n.* a system or arrangement of ribs, as in a vaulted roof etc. a method of ploughing, raftering or half-ploughing. **ribless** (-lis), *a.*
rib² (rib), *v.t.* (*past, p.p.* **ribbed**) (*coll.*) to tease, make fun of. **ribbing**, *n.*
RIBA, (*abbr.*) Royal Institute of British Architects.
ribald (rib'əld), *n.* one using scurrilous language. *a.* scurrilous, coarse, licentious, lewd (of language). **ribaldry** (-ri), *n.*
riband (rib'ənd), RIBBON.
ribble-rabble (rib'lrab'l), *n.* a rabble, a mob.
ribbon (rib'ən), *n.* a narrow woven strip or band of silk, satin etc., used for ornamenting dress etc.; such a strip or band worn as a distinctive mark of an order, college, club etc. or in place of a medal awarded for military service etc.; a narrow strip of anything; an ink-impregnated cloth strip used in typewriters etc.; torn shreds, ragged strips. **Blue Ribbon**, BLUE. **ribbon development**, *n.* urban extension in the form of a single depth of houses along roads radiating from the town. **ribbon-fish**, *n.* a long, narrow, flattish fish of various species. **ribbon-grass**, *n.* an American grass grown for ornamental purposes. **ribboned**, *a.* wearing or adorned with ribbons (*usu. in comb.*).
Ribes (rī'bēz), *n.* the genus comprising the currant and gooseberry plants.
riboflavin, riboflavine (rībōflā'vin), *n.* a yellow vitamin of the B complex found esp. in milk and liver, which promotes growth in children.
ribonuclease (rībōnū'kliās), *n.* any of several enzymes that catalyse the hydrolysis of RNA.
ribonucleic acid (rībōnūklē'ik), **RNA**, *n.* any of a group of nucleic acids present in all living cells and playing an essential role in the synthesis of proteins.
ribose (rī'bōs), *n.* a sugar occurring in RNA and riboflavin.
ribosome (rī'bəsōm), *n.* any of numerous minute granules containing RNA and protein in a cell, which are the site for protein synthesis.
Ricardian (rikah'diən), *a.* of or pertaining to the economist David Ricardo (1772-1823) or his opinions. *n.* a follower of Ricardo.
rice (rīs), *n.* the white grain or seeds of an E Indian aquatic grass extensively cultivated in warm climates for food. **rice-biscuit**, *n.* **rice milk**, *n.* milk boiled and thickened with rice. **rice-paper**, *n.* a paper made from the pith of a Taiwanese tree, and used by Chinese artists for painting on; a thin edible paper made from rice straw, used in baking.

rich (rich), *a.* wealthy, having many possessions, abounding (in resources, productions etc.); abundantly supplied; producing ample supplies; fertile, abundant, well-filled; valuable, precious, costly; elaborate, splendid; abounding in qualities pleasing to the senses, sweet, luscious, high-flavoured, containing much fat, oil, eggs, sugar, spices etc.; vivid, bright; mellow, deep, full, musical (of sounds); comical, funny, full of humorous suggestion. **riches**, *n. pl.* abundant possessions, wealth, opulence, affluence. **richly**, *adv.* in a rich manner, abundantly, thoroughly. **richness**, *n.*
Richter scale (rikh'tə), *n.* a logarithmic scale for registering the magnitude of earthquakes. [from Charles *Richter*, 1900-85, US seismologist]
ricinic (risin'ik), **ricinoleic** (-nəlē'-), **ricinolic** (-nol'-), *a.* derived from castor-oil.
rick¹ (rik), *n.* a stack of corn, hay etc., regularly built and thatched. *v.t.* to make or pile into a rick. **rick-barton, -yard**, *n.* space on a farm reserved for ricks. **rick-stand**, *n.* a platform of short pillars and joists for keeping a rick above the ground.
rick² (rik), *v.t.* to wrench or sprain. *n.* a wrench or sprain.
rickets (rik'its), *n.* a disease of children consisting in the softening of the bones (esp. the spine), bow-legs, emaciation etc., owing to lack of vitamin D in the bones. **rickety**, *a.* affected with or of the nature of rickets; feeble in the joints; shaky, tumble-down, fragile, unsafe.
rickettsia (riket'siə), *n.* (*pl.* **-siae** (-ē), **-sias**) any of a group of microorganisms found in lice, ticks etc. which when transmitted to man cause serious diseases, e.g. typhus. **rickettsial**, *a.* [after Howard T. *Ricketts*, 1871-1910, US pathologist]
rickshaw (rik'shaw), **ricksha** (-shah), *n.* a light two-wheeled shaded carriage drawn by one or two men, or attached to a bicycle etc. See TRISHAW.
ricochet (rik'əshā), *n.* a bounding or skipping of a stone, projectile, or bullet off a hard or flat surface; a hit so made. *v.i.* (*past, p.p.* **ricocheted** (-shād)) to skip or bound in this manner. *v.t.* to aim at or hit thus. [F, etym. doubtful]
ricotta (rikot'ə), *n.* a soft white Italian cheese made from sheep's milk, and widely used in cooking both sweet and savoury dishes.
RICS, (*abbr.*) Royal Institution of Chartered Surveyors.
rictus (rik'təs), *n.* (*pl.* **rictuses, rictus**) the expanse of a person's or animal's open mouth, gape; grimace, esp. as in death; the opening of a two-lipped corolla.
rid (rid), *v.t.* (*past* **ridded, rid**, *p.p.* **rid**, †**ridded**) to free, to clear, to disencumber (of). **to be** or **get rid of**, to free oneself or become free from. **riddance**, *n.* clearance; deliverance; relief. (a) **good riddance**, a welcome relief from someone or something undesirable.
ridden, *p.p.* RIDE.
riddle¹ (rid'l), *n.* a question or proposition put in ambiguous language to exercise the ingenuity; a puzzle, conundrum or enigma; any person, thing, or fact of an ambiguous, mysterious, or puzzling nature. *v.i.* to speak in riddles. *v.t.* to solve, to explain (a riddle, problem etc.). **riddler**, *n.*
riddle² (rid'l), *n.* a coarse sieve for sifting gravel, cinders etc., or washing ore. *v.t.* to pass through a riddle, to sift; to perforate with holes, as with shot.
ride (rīd), *v.i.* (*past* **rode** (rōd), *p.p.* **ridden**, †**rid**) to sit and be carried along, as on a horse, cycle, public conveyance etc., esp. to go on horseback; to practise horsemanship; to float, to seem to float; to lie at anchor; to be supported, to be on something, esp. in motion; to work (up); to be in a (specified) condition for riding. *v.t.* to sit on and be carried along by (a horse etc.); to traverse on a horse, cycle etc.; to execute or accomplish this; to cause to ride, to give a ride to; to be upborne by, to float over; to oppress, to tyrannize, to domineer (over); (*sl.*) to copu-

rider 712 **right**

late with. *n.* the act of riding; a journey on horseback, in a public conveyance or other vehicle; a road for riding on, esp. through a wood; (*sl.*) an act of copulation; (*sl.*) a sexual partner. **to let (something) ride,** to let (something) alone, let (something) continue without interference. **to ride out,** to come safely through (a storm etc.); to endure successfully. **to ride to death,** to overdo. **to ride to hounds,** to hunt. **to ride up,** of a skirt etc., to work up out of normal position. **to take for a ride,** to kidnap and murder; to play a trick on. **ride and tie,** *n.* a cross-country race between teams consisting of two people with one horse who ride and run alternately. **ridable, rideable,** *a.* **-ridden,** *comb. form* oppressed, dominated by, or excessively concerned with something in particular. **rider,** *n.* one who rides, a horse; an additional clause to a document, act etc.; an opinion, recommendation etc. added to a verdict; a subsidiary problem, a corollary, an obvious supplement; an additional timber, plate or other device used for strengthening. **riderless** (-lis), *a.* **riding**[1], *n.* the act or state of one who rides; a road for riding on, esp. a grassed track through or beside a wood. **riding-crop,** *n.* a whip with a short lash used by riders on horseback. **riding-school,** *n.* a place where riding is taught.
rider etc. RIDE.
ridge (rij), *n.* the long horizontal angle formed by the junction of two slopes; an elevation of the earth's surface long in comparison with its breadth; a long and narrow hill-top or mountain-crest; a continuous range of hills or mountains; a strip of ground thrown up by a plough or other implement; a tongue of high pressure on a weather map. *v.t., v.i.* to form into ridges; **ridge-pole,** *n.* a horizontal timber along the ridge of a roof; the horizontal pole of a long tent. **ridgeway,** *n.* a road or way along a ridge. **ridgy,** *a.*
ridicule (rid'ikūl), *n.* words or actions intended to express contempt and excite laughter; derision, mockery. *v.t.* to laugh at; to make fun of; to expose to derision. **ridiculous** (-dik'-), *a.* meriting or exciting ridicule. **ridiculously,** *adv.* **ridiculousness,** *n.*
riding[1] RIDE.
riding[2] (rī'ding), *n.* one of the three former administrative divisions of Yorkshire.
riesling (rēz'ling, rīz'-), *n.* a dry white wine, or the grape that produces it.
rieve (rēv), **river** REAVE.
rife (rīf), *a.* occurring in great quantity, number etc.; current, prevalent, abundant.
riff (rif), *n.* a phrase or figure played repeatedly in jazz or rock music, usu as background to an instrument solo.
riffle (rif'l), *v.t.* to flick through rapidly (the pages of a book etc.); to shuffle playing cards by halving the deck and flicking the corners together using both thumbs. *v.i.* to flick cursorily (with *through*).
riffraff (rif'raf), *n.* worthless people; rabble.
rifle[1] (rī'fl), *v.t.* to search and rob; to plunder, to pillage, to strip; to snatch and carry off. **rifler,** *n.*
rifler[2] (rī'fl), *n.* a firearm having the barrel spirally grooved so as to give a rotary motion to the projectile; (*pl.*) troops armed with rifles. *v.t.* to furnish (a firearm or the bore or barrel of a firearm) with spiral grooves in order to give a rotary motion to the projectile. **rifle-bird,** *n.* an Australian bird with velvety black plumage. **rifleman,** *n.* **rifle range,** *n.* an area for target practice using rifles. **rifling,** *n.* the spiral grooves in the bore of a firearm which cause the rotation of the projectile fired.
rift (rift), *n.* a cleft, a fissure; a wide crack, rent, or opening, made by riving or splitting; a break in cloud; a break in friendly relations, a disagreement. *v.t. v.i.* to break open, to split. **rift valley,** *n.* a narrow valley formed by the subsidence of the earth's crust between two faults.
rig[1] (rig), *v.t.* (*pres.p.* **rigging,** *past, p.p.* **rigged**) to furnish or fit (a ship) with spars, gear, tackle, etc.; to dress,

clothe, or fit (up or out); to put together or fit (up) in a hasty or make-shift way. *n.* the way in which the masts and sails of a ship are arranged; (*coll.*) an outfit, a turnout; (*coll.*) an articulated lorry; an oil rig. **rig-out, -up,** *n.* dress, outfit. **rigger,** *n.* **rigging,** *n.* the system of tackle, ropes etc. supporting the masts, and controlling the sails etc. of a ship; the adjustment or alignment of the components of an aeroplane.
rig[2] (rig), *v.t.* to manipulate fraudulently; to hoax, to trick. **to rig the market,** to manipulate the market so as to raise or lower prices for underhand purposes.
rigging etc. RIG[1].
right (rīt), *a.* required by or acting, being, or done in accordance with truth and justice; equitable, just, good; proper, correct, true; fit, suitable; most suitable, the preferable, the more convenient; sound, sane, well; properly done, placed etc., not mistaken, satisfactory; on or towards the side of the body which is to the south when the face is to the east; straight; direct; formed by lines meeting perpendicularly; not oblique, involving or based on a right-angle or angles (of cones, pyramids, cylinders etc.); politically conservative, right-wing. *adv.* in accordance with truth and justice, justly, equitably; exactly, correctly, properly, satisfactorily, well; very, quite, to the full; to or towards the right hand; straight; at once; all the way to, completely. *n.* that which is right or just; fair or equitable treatment; the cause or party having justice on its side; just claim or title, esp. a claim enforceable at law, justification; that which one is entitled to; (*pl.*) proper condition, correct or satisfactory state; (*coll.*) the right hand; a punch with this; the right-hand side, part or surface of anything; a thing, part etc. pertaining or corresponding to this; the more conservative party or division within a party. *v.t.* to set in or restore to an upright, straight, correct, or proper position; to correct, make right, to rectify; to do justice to, to vindicate, to rehabilitate; to relieve from wrong or injustice. *v.i.* to resume a vertical position. *int.* expressing approval, compliance, enthusiasm. **all right,** (*coll.*) correct, satisfactory, in good condition, safe etc.; yes. **by right(s),** properly; with justice. **in one's own right,** by right independent of another person. **in one's right mind,** sane, lucid. **in the right,** correct; in accordance with reason or justice. **right and left,** in all directions; on both sides; with both hands etc. **right away, right off,** *adv.* at once, immediately. **right, left and centre,** on every side. **right of way,** the right established by usage or by dedication to the public to use a track, path, road etc. across a person's land; the right of a vehicle or vessel to take precedence in passing according to law or custom; (*N Am.*) permanent way of a railway. **righto** (-ō), **right oh! right you are,** (*coll.*) forms of assent, approval etc. **to put** or **set to rights,** to arrange, to put in order. **to serve one right,** to be thoroughly well deserved. **right-about,** *n.* the opposite direction, the reverse to the opposite direction. **right-angle,** *n. adv.* one formed by two lines meeting perpendicularly. **right-angled,** *a.* having a right-angle or angles. **at right-angles,** placed at or forming a right-angle. **right hand,** *n.* the hand on the right side, esp. as the better hand; position on or direction to this side; one's best or most efficient assistant, aid, or support. **right hand,** *a.* situated on or towards the right hand; denoting one whose help is most useful or necessary. **right-hand man,** *n.* one's best assistant, aid etc. **right-handed,** *a.* using the right hand more readily than the left; used by or fitted for use by the right hand (of tools etc.); turning to the right (of the thread of a screw etc.). **right-handedness,** *n.* **Right Honourable,** a title given to peers, peeresses, privy councillors etc. **right-minded,** *a.* properly, justly, or equitably disposed. **rights issue,** *n.* (*Stock Exch.*) an issue of new shares by a company to its existing shareholders on more favourable terms. **right-thinking,** *a.* holding acceptable opinions. **rightable,** *a.*

righter, *n*. **rightful**, *a*. just, equitable, fair; entitled, holding, or held by legitimate claim. **rightfully**, *adv*. **rightist**, *n*. a conservative, an adherent of the right in politics. **rightless** (-lis), *a*. **rightly**, *adv*. justly, fairly, equitably; honestly, uprightly; correctly, accurately, properly. **rightness**, *n*. **rightward** (-wəd), *a.*, *adv*. **right wing**, *n*. the conservative section of a party or grouping; the right side of an army, football ground etc. **right-wing**, *a*. of or on the right wing; having or relating to conservative political views. **right-winger**, *n*.
righteous (rī'chəs), *a*. just, upright, morally good; equitable, deserved, justifiable, fitting. **righteously**, *adv*. **righteousness**, *n*.
rightful, rightly etc. RIGHT.
rigid (rij'id), *a*. stiff, not easily bent, not pliant, unyielding; rigorous, strict, punctilious, inflexible, harsh, stern, austere. **rigidify** (-jid'ifī), *v.i.* (*pres.p.* **rigidifying**, *past*, *p.p.* **rigidified**) to make or become rigid. **rigidity** (-jid'-), **rigidly**, *adv*.
rigmarole (rig'mərōl), *n*. long, unintelligible, rambling talk; a long, complicated or pointless procedure, set of instructions, etc. *a.* incoherent.
rigor (rī'gaw, rig'ə), *n*. a feeling of chill, a shivering attended with stiffening etc., premonitory of fever etc.; a state of rigidity (of the muscles). **rigor mortis** (maw'tis), *n*. the stiffening of the body following death.
rigour (rig'ə), *n*. strictness, exactness in enforcing rules; stiffness or inflexibility of opinion, doctrine, observance etc., austerity of life; sternness, harshness, asperity; inclemency of the weather etc., hardship, distress; (*pl.*) harsh proceeding, severities. **rigorism**, *n*. **rigorist**, *a*. and *n*. **rigorous**, *a*. strict, precise, severe, stern, inflexible; logically accurate, precise, stringent; inclement, harsh. **rigorously**, *adv*.
Rig-Veda (rigvā'də), *n*. the oldest and most original of the Vedas.
Riksdag (riks'dahg), *n*. the Swedish parliament.
rile (rīl), *v.t.* to make angry, to vex, to irritate.
rill (ril), *n*. a small brook, a rivulet; a trench or furrow; a rille. **rille**, *n*. a furrow, trench or narrow valley on Mars or the moon.
rim (rim), *n*. an outer edge, border or margin, esp. of a vessel or other circular object; a ring or frame; the peripheral part of the framework of a wheel, between the spokes or hub and the tyre. *v.t.* (*past*, *p.p.* **rimmed**) to form or furnish with a rim. **rimless** (-lis), *a*. **rimmed**, *a*. having a rim.
rime (rīm), *n*. hoar-frost; a deposit of ice caused by freezing fog or low temperatures. **rimy**, *a*.
rimose (rī'mōs), **-mous**, *a*. full of chinks or cracks, as the bark of trees.
rimu (rē'moo), *n*. the red pine of New Zealand.
rind (rīnd), *n*. the outer coating of cheese, fruits etc.; bark, peel, husk, skin.
rinderpest (rin'dəpest), *n*. a malignant contagious disease attacking cattle.
ring[1] (ring), *n*. a circlet; a circlet of gold etc., worn usu. on a finger or in the ear as an ornament, token etc.; anything in the form of a circle; a line, mark, moulding, space or band round or the rim of a circular or cylindrical object or sphere; a concentric band of wood formed by the annual growth of a tree; a group or concourse of people, things etc. arranged in a circle; a circular space, enclosure or arena for boxing, circus performances etc.; a combination of persons acting in concert; a closed chain of atoms; in mathematics, a closed set. *v.t.* (*past*, *p.p.* **ringed**) to put a ring round; to encircle, to enclose, to hem in; to fit with a ring; to cut a ring of bark from (a tree); to put a ring on or in. **the ring**, boxing. **to run rings round**, to outstrip easily. **to throw one's hat into the ring**, to challenge; to present oneself as a candidate or contestant. **ring-bark**, *v.t.* to cut a ring of bark from a tree in order to check growth, kill it or induce it to fruit. **ring binder**, *n*. a binder consisting of metal rings which hold loose-leaf pages by means of perforations in the paper. **ring-bolt**, *n*. a bolt with a ring or eye at one end. **ring-dove**, *n*. a wood-pigeon. **ring-fence** FENCE. **ring-finger**, *n*. the third finger, esp. of the left hand, on which the wedding-ring is worn. **ring-leader**, *n*. the leader of a riot, mutiny, piece of mischief etc. **ring main**, *n*. an electrical system in which power points are connected to the supply through a closed circuit. **ringmaster**. *n*. the manager and master of ceremonies of a circus performance. **ring-necked**, *a*. having a band or bands of colour round the neck. **ring-ouzel**, *n*. a thrush-like bird allied to the blackbird, having a white band on the breast. **ring pull**, *n*. a metal ring attached to a can of soft drink, beer etc. which opens it when pulled. **ring road**, *n*. a road circumnavigating an urban centre. **ringside**, *n.*, *a*. (of) the area or seats immediately beside a boxing or wrestling ring or any sporting arena; (of) any position affording a close and unobstructed view. **ringworm**, *n*. a contagious skin-disease caused by a white fungus. **ringed**, *a*. having, encircled by or marked with a ring or rings (*often in comb.*); annular. **ringer**[1], *n*. a quoit falling round the pin; a quoit so thrown or a throw resulting in this; one who rings; (*Austral.*) the fastest shearer in a shearing-shed. **ringless** (-lis), *a*. **ringlet** (-lit), *n*. a curly lock of hair; a satyrid butterfly. **ringleted**, *a*.
ring[2] (ring), *v.i.* (*past* **rang** (rang), *p.p.* **rung** (rŭng)) to give a clear vibrating sound, as a sonorous metallic body when struck; to re-echo, to resound, to reverberate; to have a sensation as of vibrating metal, to tingle (of the ears); to give a summons or signal by ringing. *v.t.* to cause to ring; to sound (a knell, peal etc.) on a bell or bells; to telephone. *n*. the sound of a bell or other resonant body; the quality of resonance; the characteristic sound of a voice, statement etc.; (*coll.*) a telephone call. **to ring a bell**, to sound familiar, to cause to remember. **to ring down, up**, to lower or raise the curtain in a theatre. **to ring false** or **true**, to appear genuine or insincere. **to ring in**, to report in by telephone. **to ring off**, to end a telephone call; to hang up (the receiver). **to ring the changes** CHANGE. **to ring up**, to call on the telephone; **ringer**[2], *n*. one who rings, e.g. church bells; a horse, athlete etc. racing under the name of another; (*coll.*) a person or thing almost identical to another, as in a **dead ringer**. **ringing**, *a*. sounding like a bell; sonorous, resonant. *n*. a sound of or as a bell. **ringing-tone**, *n*. the tone heard on a telephone after an unengaged number is dialled. **ringingly**, *adv*.
ringgit (ring'git), *n*. the unit of currency of Malaysia, the Malaysian dollar, equal to 100 cents.
ringhals (ring'hals), **rinkhals** (-kals), *n*. (*pl.* **-hals, -halses**) a venom-spitting snake of southern Africa.
rink (ringk), *n*. a strip of ice or of a green marked off for playing bowls or curling; a division of a side so playing; a prepared floor for roller-skating or an area of usu. artificially-formed ice for ice-skating; the building or structure housing a skating-rink.
rinse (rins), *v.t.* to wash, to cleanse with an application of clean water; to remove soap by rinsing. *n*. the act of rinsing; a hair tint. **rinser**, *n*.
Rioja (riō'khə), *n*. a type of Spanish table wine. [a region in N Spain]
riot (rī'ət), *n*. a disturbance, an outbreak of lawlessness, a tumult, an uproar; unrestrained conduct, revelry; unrestrained indulgence in something; luxuriant growth, lavish display; in law, a tumultuous disturbance of the peace by three or more persons; (*coll.*) a person or thing which is hilariously funny. *v.i.* to take part in a riot; to revel. **to run riot**, to act without control or restraint; to grow luxuriantly. **Riot Act**, *n*. an Act of 1715 enjoining riotous persons to disperse within an hour of the Act

being read by a magistrate. **to read the riot act**, to give a severe warning that something must cease; to reprimand severely. **riot gun**, *n*. a type of gun used to disperse rioters. **riot police**, *n*. police specially trained and equipped to deal with rioters. **riot shield**, *n*. a large transparent shield used as protection by riot police. **rioter**, *n*. **riotous**, *a*. **riotously**, *adv*. **riotousness**, *n*.

RIP, (*abbr*.) *requiescat, requiescant in pace*, may he, she, they rest in peace. [L]

rip[1] (rip), *v.t*. (*past, p.p.* **ripped**) to tear or cut forcibly (out, off, up etc.); to rend, to split; to saw (wood) along the grain; to take out or away by cutting or tearing; to make a long tear or rent in; to utter (an oath etc.) with violence. *v.i*. to come or be torn forcibly apart, to tear; to go (along) at a great pace. *n*. a rent made by ripping, a tear. **to let it rip**, (*coll*.) to allow to proceed without restraint. **to let rip**, (*coll*.) to speak, act or proceed without restraint. **to rip off**, to steal (from); to cheat. **rip cord**, *n*. a cord which when pulled, releases a parachute from its pack or opens the gas-bag of a balloon allowing it to descend. **rip-off**, *n*. (*sl*.) a theft; a cheat; an exploitative imitation. **rip-roaring**, *a*. (*coll*.) noisy, unrestrained, exuberant. **ripsaw**, *n*. one for sawing along the grain. **ripper**, *n*. (*sl*.) an excellent person or thing; a murderer who mutilates the victim's body. **ripping**, *a*. (*sl*.) excellent, fine, splendid. **rippingly**, *adv*. (*sl*.)

rip[2] (rip), *n*. an eddy, an overfall, a stretch of broken water; a riptide. **rip current, riptide**, *n*. a rip caused by tidal currents flowing away from the land.

riparian (rīpeə'riən), *a*. pertaining to or dwelling on the banks of a river. *n*. an owner of property on the banks of a river.

ripe (rīp), *a*. ready for reaping, gathering or eating; mature, fully developed, fit for use, ready or in a fit state (for); resembling ripe fruit, rosy, rounded, luscious. **ripely**, *adv*. **ripen**, *v.t., v.i*. **ripeness**, *n*.

riposte (ripost'), *n*. in fencing, a quick lunge or return thrust; a quick reply, a retort, a repartee; a counterstroke. *v.i*. to reply with a riposte.

ripper, ripping etc. RIP[1].

ripple (rip'l), *v.i*. to run in small waves or undulations; to sound as water running over a rough surface. *v.t*. to agitate or cover with small waves or undulations. *n*. the ruffling of the surface of water; a wavelet; an undulation (of water, hair etc.); a sound as of rippling water. **ripplemark**, *n*. a mark as of ripples or wavelets on sand, mud, rock etc. **ripple-marked**, *a*. **rippling**, *n*., *a*. **ripply**, *a*.

ripsaw RIP[1].

rise (rīz), *v.i*. (*past* **rose** (rōz), *p.p.* **risen** (riz'n)) to move upwards, to ascend, to leave the ground, to mount, to soar; to get up from a lying, kneeling, or sitting position, or out of bed, to become erect, to stand up; to adjourn, to end a session; to come to life again; to swell or project upwards; to increase; to be promoted, to thrive; to increase in confidence, cheerfulness, energy, force, intensity, value, price etc.; to slope up; to arise, to come into existence, to originate; to come to the surface, to come into sight; to become audible; to become higher in pitch; to respond esp. with annoyance; to become equal to; to break into insurrection, to revolt, to rebel (against). *n*. the act of rising; ascent, elevation; an upward slope, the degree of this; source, origin, start; increase or advance in price, value, power, rank, age, prosperity, height, amount, salary etc. **to give rise to**, to cause. **to rise and shine**, to rise from bed in the morning. **to rise to the occasion**, to become equal to the demands of an occurrence, event etc. **to take a rise out of someone**, to provoke a person to anger, by teasing. **riser**, *n*. one who or that which rises; the vertical part of a step etc. **rising**, *n*. a mounting up or ascending; a revolt, an insurrection; the agent causing dough to rise. *a*. increasing; growing. *adv*. (*coll*.) approaching, nearing. **rising damp**, *n*. the absorption of ground moisture into the fabric of a building.

rishi (rish'i), *n*. a seer, a saint, an inspired poet, esp. one of the seven sages said to have communicated the Vedas to mankind.

risible (riz'ibl), *a*. inclined to laugh; exciting laughter; pertaining to laughter. **risibility** (-bil'-), *n*.

rising RISE.

risk (risk), *n*. hazard, chance of harm, injury, loss etc.; a person or thing liable to cause a hazard or loss. *v.t*. to expose to risk or hazard; to venture on, to take the chances of. **at risk**, in danger (of); vulnerable (to). **to run a risk**, to incur a hazard; to encounter danger. **riskily**, *adv*. **riskiness**, *n*. **risky**, *a*. dangerous, hazardous; venturesome, daring; **risqué** (-kā), *a*. suggestive of indecency, indelicate.

risotto (rizot'ō), *n*. an Italian dish of rice cooked in stock, with onions, cheese, chicken, ham etc.

risqué RISK.

rissole (ris'ōl), *n*. a ball or flat cake of minced meat, fish etc., fried with breadcrumbs etc.

rite (rīt), *n*. a religious or solemn prescribed act, ceremony or observance; the prescribed acts, ceremonies or forms of worship of any religion. **rite of passage**, a ceremony marking an individual's change of status, esp. into adulthood or matrimony.

ritual (rit'ūəl), *a*. pertaining to, consisting of or involving rites. *n*. a prescribed manner of performing divine service; performance of rites and ceremonies, esp. in an elaborate or excessive way; a book setting forth the rites and ceremonies of a particular Church; any formal or customary act or series of acts consistently followed. **ritualism**, *n*. punctilious or exaggerated observance of ritual. **ritualist**, *n*. **ritualistic** (-lis'-), *a*. **ritualistically**, *adv*. **ritualize, -ise**, *v.t., v.i*. **ritually**, *adv*.

ritzy (rit'si), *a*. (*coll*.) elegant, showy, luxurious, rich. **ritzily**, *adv*. **ritziness**, *n*. [after the *Ritz* hotels estab. by César Ritz, 1850-1918]

rival (rī'vəl), *n*. a person or thing considered as equal to another; a person, team etc. striving to surpass another in a quality, pursuit etc., a competitor. *a*. being a rival, having the same claims or pretensions, emulous; in competition. *v.t*. (*past, p.p.* **rivalled**) to vie with, to emulate, to strive to equal or surpass; to be, or almost be, the equal of. **rivalry** (-ri), *n*.

rive (rīv), *v.t*. (*p.p.* **riven** (rivn), to tear split or asunder; to wrench or rend (away, from, off etc.). *v.i*. to split, to cause to split.

river (riv'ə), *n*. a large stream of water flowing in a channel over a portion of the earth's surface and discharging itself into the sea, a lake, a marsh or another river; a large and abundant stream, a copious flow. *a*. of, or dwelling in or beside, a river or rivers. **to sell down the river**, to let down, to betray. **river-bed**, *n*. the channel in which a river flows. **river-craft**, *n. pl*. small craft plying only on rivers. **river-god**, *n*. a deity presiding over or personifying a river. **river-side**, *n*. the ground along the bank of a river. *a*. built on or pertaining to this. **riverine** (-rīn), *a*. of, pertaining to, resembling or produced by a river; riparian.

rivet (riv'it), *n*. a short bolt, pin or nail, usu. with a flat head at one end, the other end being flattened out and clinched by hammering, used for fastening metal plates together, ornamenting denim jeans etc. *v.t*. (*past, p.p.* **riveted**) to join or fasten together (as) with a rivet or rivets; to fix (attention, eyes etc. upon); to engross the attention of. **riveter**, *n*. **riveting**, *n*., *a*.

riviera (riviə'rə), *n*. a coastal strip reminiscent of the French and Italian Rivieras on the Mediterranean.

rivulet (riv'ūlit), *n*. a small stream.

riyal (riyahl'), *n*. the unit of currency of Dubai, Qatar, Saudi Arabia and the Yemen (also **rial**).

RL, (*abbr.*) Rugby League.
Rly, (*abbr.*) railway.
RM, (*abbr.*) Royal Mail; Royal Marines.
rm, (*abbr.*) ream; room.
RMA, (*abbr.*) Royal Marine Artillery; Royal Military Academy.
RN, (*abbr.*) Registered Nurse; Royal Navy.
Rn, (*chem. symbol*) radon.
RNA, *n.* RIBONUCLEIC ACID [ribonucleic acid].
RNIB, (*abbr.*) Royal National Institute for the Blind.
RNLI, (*abbr.*) Royal National Lifeboat Institution.
roach[1] (rōch), *n.* a freshwater fish allied to the carp.
roach[2] (rōch), *n.* the upward curve in the foot of a square sail.
roach[3] (rōch), *n.* (*N Am.*) short for COCKROACH; (*sl.*) the butt or filter of a cannabis cigarette.
road (rōd), *n.* a track or way for travelling on, esp. a broad, tarmacked strip of ground suitable for motor-vehicles, forming a public line of communication between places, a highway; a street; route, course; a roadstead. **one for the road**, (*coll.*) a last drink before leaving. **on the road**, passing through, travelling, touring (often as a way of life). **to get in, out (of) someone's, the road**, (*esp. Sc.*) to get in or out of the way. **to hit the road**, to leave, to begin travelling. **to take (to) the road**, to set out on a journey; to become an itinerant. **roadblock**, *n.* an obstruction placed across a road by the army or police checking for escaped criminals, terrorists etc. **road hog**, *n.* a selfish motorist or cyclist paying no regard to the convenience of other people using the road. **road movie**, *n.* a film or genre of film that has a central plot involving a journey. **roadhouse**, *n.* a public house, restaurant etc. on a highway, which caters for motorists. **road roller**, *n.* a vehicle with a large metal roller for compacting the surface of a newly-laid (section of) road. **road-sense**, *n.* the instinct of a road-user which enables him or her to cope with a traffic emergency, avoid an accident etc. **road-show**, *n.* a touring group of performers; a musical or theatrical performance by such a group; a live or prerecorded outside broadcast by a touring radio or television unit. **road-side**, *n., a.* **roadstead**, *n.* an anchorage for ships some distance from the shore. **road test**, *n.* a test for roadworthiness; a working test. *v.t.* **roadway**, *n.* the part of a highway used by vehicles. **roadwork**, *n.* physical training comprising running or jogging along roads; (*pl.*) repairs to or under a section of road. **roadworthy**, *a.* **roadworthiness**, *n.* fit for use or travel. **roadie** (-i), *n.* (*sl.*) a person employed to transport, set up and dismantle the instruments and equipment of a touring band or pop group. **roadster** (-stə), *n.* a cycle suitable for the road; (*N Am. dated*) a two-seater automobile.
roadway, roadworthy ROAD.
roam (rōm), *v.i.* to wander about without any definite purpose, to rove, to ramble. *v.t.* to range, to wander over. **roamer**, *n.*
roan (rōn), *n., a.* of a bay, sorrel or dark colour, with spots of grey or white thickly interspersed.
roar (raw), *v.i.* to make a loud, deep, hoarse, continued sound, as a lion; to make a confused din like this (of a person in rage, distress or loud laughter, the sea, thunder, guns, fire etc.). *v.t.* to shout, say, sing or utter with a roaring voice. *n.* a loud, deep, hoarse, continued sound as of a lion etc.; a confused din resembling this; a burst of mirth or laughter; a loud engine noise. **roarer**, *n.* **roaring**, *a.* shouting, noisy, boisterous, stormy; brisk, active. *n.* a loud, continued or confused noise. *adv.* extremely, boisterously. **roaring drunk**, extremely and noisily drunk. **roaring trade**, *n.* thriving and profitable business. **the roaring forties** FORTY. **roaringly**, *adv.*
roast (rōst), *v.t.* to cook by exposure to the direct action of radiant heat, esp. at an open fire or in an oven; to dry and parch (coffee beans etc.) by exposure to heat; to heat excessively or violently; (*coll.*) to criticize strongly. *v.i.* to dress meat by roasting; to be roasted. *a.* roasted. *n.* roast meat or a dish of this, a roast joint. **roaster**, *n.* one who or that which roasts; an oven or other device for roasting; a chicken or other animal or vegetable etc. suitable for roasting. **roasting**, *n., a.*
rob (rob), *v.t.* (*past, p.p.* **robbed**) to steal from; to defraud, cheat, deprive of. **robber**, *n.* **robbery**, *n.* an instance of stealing. **daylight robbery**, flagrant extortion or overpricing.
robe (rōb), *n.* a dress, gown or vestment of state, rank or office (*often in pl.*); a dressing-gown, bathrobe or other loose, long garment. *v.t.* to clothe, to dress. *v.i.* to put on a robe or vestments.
robin (rob'in), *n.* a small warbler, the redbreast, also called robin redbreast.
robot (rō'bot), *n.* a machine capable of performing various functions in a human manner; a humanoid; an automaton; a brutal, mechanically efficient person who is devoid of sensibility; (*S Afr.*) traffic lights. **robot-like**, *a.* **robotic** (-bot'-), *a.* **robotic dancing**, *n.* a style of dancing with robot-like mechanical movements. **robotics**, *n.pl.* the branch of technology concerned with the design, construction, maintenance and application of robots.
robust (rəbŭst'), *a.* strong, hardy, vigorous, capable of endurance, having excellent health and physique; sturdy, hardy (of plants); full-bodied (of wine). **robustly**, *adv.* **robustness**, *n.*
ROC, (*abbr.*) Royal Observer Corps.
roc (rok), *n.* a mythical bird of immense size and strength.
rocaille (rokī'), *n.* a decorative work of rock, shell or a similar material.
rocambole (rok'əmbōl), *n.* a plant related to the leek, Spanish garlic.
rochet (roch'it), *n.* an open-sided vestment with tight sleeves, worn by bishops and abbots.
rock[1] (rok), *n.* a mass of stone forming a hill, promontory, islet, cliff etc.; a boulder, a stone, a pebble; (*sl.*) a diamond or other precious or large gem; a hard sweet often in the form of a stick; a person or thing providing refuge, stability, supportiveness etc. **on the rocks**, poor, hard up; at an end, destroyed (e.g. of a marriage); of a drink, served with ice. **the Rock**, Gibraltar. **the Rockies**, the Rocky Mts. **rock-bottom**, *n.* the lowest point (e.g. of despair). *a.* lowest possible (of prices). **rock-cake**, *n.* a bun with a hard rough surface. **rock-climber**, *n.* a mountaineer who scales rock-faces. **rock-climbing**, *n.* **rock crystal**, *n.* the finest and most transparent kind of quartz, usu. found in hexagonal prisms. **rock-face**, *n.* the surface of a vertical or nearly vertical cliff or mountain-side. **rock garden**, *n.* a rockery; a garden containing a rockery or rockeries. **rock-goat**, *n.* the ibex. **rock-plant**, *n.* any of various plants dwelling among rocks, esp. an alpine. **rock-rabbit**, *n.* a hyrax. **rock-rose**, *n.* the cistus. **rock-salt**, *n.* salt found in stratified beds, halite. **rockery**, *n.* a mound or slope of rocks, stones and earth, for growing alpine and other plants. **rocky**[1], *a.* full of or abounding with rocks; consisting of or resembling rock; solid; rugged, hard, obdurate, difficult. **rockiness**, *n.*
rock[2] (rok), *v.t., v.i.* (to cause) to move backwards and forwards or to and fro; to cause to sway or reel. *v.i.* to dance to rock or rock and roll music. *n.* rocking motion; rock music. *a.* pertaining to rock music. **off one's rocker**, crazy. **rockabilly** (-əbili), *n.* a quick-paced type of Southern American rock and country music originating in the 1950s. **rock and roll, rock'n'roll**, *n.* a type of music popular from the 1950s which combines jazz and country and western music; the type of dancing executed to this music. *a.* pertaining to this type of music or style of dancing. *v.i.* to dance to the rock-and-roll style. **rock and roller, rock'n'roller**, *n.* **rock music**, *n.* a type of popular music characterized by a strong persistent beat which de-

veloped out of rock and roll. **Rocker**, *n.* one of a teenage band of leather-clad motorcyclists of the 1960s. **rocker**, *n.* a rocking-chair; a curved piece of wood on which a cradle, rocking-chair etc., rocks; a low skate with a curved blade; the curve of this blade; of machinery, applied to various devices and fittings having a rocking motion. **rockily**, *adv.* unsteadily. **rockiness**, *n.* **rocking**, *n.*, *a.* **rocking-chair**, *n.* a chair mounted on rockers. **rocking-horse**, *n.* a large toy horse mounted on rockers. **rocky**², *a.* (*coll.*) unsteady, tottering, fragile. **rockiness**, *n.*

rock-bottom, ROCK ¹.

rocket (rok'it), *n.* a firework consisting of a cylindrical case of metal or paper filled with a mixture of explosives and combustibles, used for display, signalling, conveying a line to stranded vessels and in warfare; a device with a warhead containing high explosive and propelled by the mechanical thrust developed by gases generated through the use of chemical fuels; (*coll.*) a severe scolding, a telling off. *v.t.* to propel by means of a rocket. *v.i.* to rise rapidly (e.g. of prices); to advance to a high position speedily (e.g. of a promoted person). **rocket range**, *n.* a place for testing rocket projectiles. **rocketry** (-ri), *n.* the scientific study of rockets.

rococo (rəkō'kō), *n.* a florid style of ornamentation (in architecture, furniture etc.) flourishing under Louis XV in the 18th cent.; design or ornament of an eccentric and over-elaborate kind. *a.* in this style.

rod (rod), *n.* a straight, slender piece of wood, a stick, a wand; this or a bundle of twigs etc. as an instrument of punishment; fishing-rod; a slender bar of metal, esp. forming part of machinery etc.; a unit of lineal measure, equal to 5½ yards (about 5 m); (*N Am. sl.*) a revolver; a rod-like body or structure in the retina of the eye.

rode (rōd), *past* RIDE.

rodent (rō'dənt), *n.* any animal of an order of small mammals having two (or sometimes four) strong incisors and no canine teeth, comprising the squirrel, beaver, rat etc.

rodeo (rōdā'ō, rō'diō), *n.* a driving together or rounding-up of cattle; an outdoor entertainment or contest exhibiting the skills involved in this (extended to other contests suggestive of this).

roe¹ (rō), *n.* a small species of deer. **roebuck**, *n.* the male roe. **roedeer**, *n.* the roe.

roe² (rō), *n.* the mass of eggs forming the spawn of fishes, amphibians etc., called the hard roe; the sperm or milt, called the soft roe.

roentgen (ront'gən, -jən, rœnt'-), **röntgen**, *n.* the international unit of quantity of X- or gamma-rays. **röntgen rays**, *n.* former name for an X-RAY. a form of radiant energy penetrating most substances opaque to ordinary light, employed for photographing hidden objects and for therapeutic treatment of lupus, cancer etc., also known as X-rays. [W.K. von *Röntgen*, German physicist, 1845–1923]

rogation (rəgā'shən), *n.* (*usu. in pl.*) a solemn supplication, esp. that chanted in procession on Rogation Days. **Rogation Days**, *n.pl.* the Monday, Tuesday and Wednesday preceding Ascension Day marked by prayers, processions, supplications.

roger (roj'ə), in radio communications etc., an expression meaning 'received and understood'; an expression of agreement or acquiescence. *v.t.* (*sl.*) used of a man, to have sexual intercourse with. **jolly Roger** JOLLY.

rogue (rōg), *n.* a rascal, a scamp, a trickster, a swindler; a playful term of endearment for a child or mischievous person; a vicious wild animal cast out or separate from the herd, esp. an elephant; an inferior or intrusive plant among seedlings; a variation from the standard type or variety. **rogues' gallery**, *n.* a collection of photographic portraits taken in police records for identification of criminals. **rogue-buffalo**, **elephant** etc., *n.* a solitary savage animal. **roguery**, *n.* **roguish**, *a.* mischievous, high-spirited,

saucy. **roguishly**, *adv.* **roguishness**, *n.*

roister (rois'tə), *v.i.* to behave uproariously, to revel boisterously; to swagger. **roisterer**, *n.* a swaggering, noisy reveller. **roisterous**, *adv.*

role, rôle (rōl), *n.* a part or character taken by an actor; any part or function one is called upon to perform. **role-play, -playing**, *n.* an enactment of a possible situation or playing of an imaginary role as therapy, training etc.

roll (rōl), *n.* anything rolled up, a cylinder of any flexible material formed by or as by rolling or folding over on itself; a small loaf of bread; a pastry or cake rolled round a filling; an official record, a register, a list, esp. of names; a cylindrical or semi-cylindrical mass of anything; a rolling motion or gait; a resounding peal of thunder etc.; a continuous beating of a drum with rapid strokes; (*N Am.*) a wad of money. *v.t.* to send, push or cause to move along by turning over and over; to cause to rotate; to cause to revolve between two surfaces; to knead, press, flatten or level with or as with a roller or rollers; to en-wrap (in), to wrap (up in); to form into a cylindrical shape by wrapping round and round or turning over and over; to carry or impel forward with a sweeping motion; to carry (oneself along) with a swinging gait; to utter with a prolonged, deep, vibrating sound. *v.i.* to move along by turning over and over and round and round; to revolve; to operate or cause to operate; to move along on wheels; (*coll.*) to progress; to move or slip about with a rotary motion (of eyes etc.); to wallow about; to sway, to reel, to go from side to side; to move along with such a motion; of a ship, to turn back and forth on her longitudinal axis; of an aircraft, to make a full corkscrew revolution about the longitudinal axis; to undulate or sweep along; to be formed into a cylindrical shape by turning over upon itself; to grow into a cylindrical or spherical shape by turning over and over; to spread (out) under a roller. **all rolled into one**, combined together. **a roll in the hay**, (*coll.*) sexual intercourse; a period of love-play. **heads will roll**, persons will be severely punished. **Master of the Rolls**, the head of the Record Office, an ex-officio judge of the Court of Appeal and member of the Judicial Committee. **to be on a roll**, (*coll.*) to be in a period of good fortune, success or luck. **to roll along**, to move or push along by rolling; to walk in a casual manner or with an undulating gait; to have a casual or unambitious approach to life. **to roll in**, to come in quantities or numbers; to arrive in a casual manner; to wind in; to push in by rolling. **to roll up**, to wind up (e.g. a car window); to make a cigarette by hand; to wind into a cylinder; to assemble, to come up. **to strike off the roll(s)**, to remove from the official list of qualified solicitors; to debar, expel. **rollbar**, *n.* a metal strengthening bar which reinforces the frame of a (racing) vehicle which may overturn. **roll-call**, *n.* the act of calling a list of names to check attendance. **rollmop**, *n.* a rolled-up fillet of herring pickled in vinegar and usu. garnished with onion. **roll-neck**, *n.*, *a.* (of) an upper garment usu. a jumper with a high neck folded over. **roll on!** *int.* hurry along, come quickly (of a day, date, event). **roll-on**, *n.* a step-in elastic corset that fits by stretching; a deodorant applied by a plastic rolling ball in the neck of its container. **roll-on-roll-off**, *n.*, *a.* (of) a vessel carrying motor vehicles which drive on and off when embarking and disembarking. **roll-top**, *a.* **roll-top desk**, *n.* a desk with a flexible cover sliding in grooves. **roll up**, *n.* (*coll.*) a hand-made cigarette made with tobacco and a cigarette paper; (*Austral.*) an assemblage. **rollable**, *a.* suitable for rolling; capable of being rolled. **rolled**, *a.* **rolled gold**, *n.* metal covered by a thin coating of gold. **roller**, *n.* one who or that which rolls; a cylindrical body turning on its axis, employed alone or forming part of a machine, used for inking, printing, smoothing, spreading out, crushing etc.; a long, heavy, swelling wave; a long, broad bandage, rolled up for convenience; any of various birds remarkable for their habit of turning somersaults in the air; a tumbler pigeon; a small cylinder for curling the hair, a curler. **rollerball**, *n.* a type of pen with a nib consisting of a rolling ball which controls the flow of ink. **roller-bearing**, *n.* a bearing comprised of strong steel rollers for giving a point of contact. **roller-coaster**, *n.* a switchback railway at an amusement park, carnival, fair etc. **roller derby**, *n.* a (often boisterous) roller-skating race. **roller-skate**, *n.* a skate mounted on wheels or rollers for skating on asphalt etc. *v.i.* to skate on these. **roller-skater**, *n.* **roller-skating**, *n.* **roller-towel**, *n.* a continuous towel hung on a roller. **rolling**, *a.*, *n.*, *adv.* r semi-cylindrical mass of anything; a **rolling stone**, a person who cannot settle down in one place. **to be rolling in it**, to be extremely wealthy. **rolling-mill**, *n.* a factory in which metal is rolled out by machinery into plates, sheets, bars etc. **rolling-pin**, *n.* a hard cylinder for rolling out dough, pastry etc. **rolling-stock**, *n.* the carriages, vans, locomotives etc. of a railway.

rollick (rol'ik), *v.i.* to behave in a careless, merry fashion; to frolic, to revel, to be merry or enjoy life in a boisterous fashion. *n.* a frolic, a spree, an escapade. **rollicking**, *a.* boisterous, carefree. *n.* (*coll.*) a scolding.

rollock ROWLOCK.

roly-poly (rō'lipō'li), *a.* round, plump, podgy. *n.* a pudding made of a sheet of suet paste, spread over with jam, rolled up and baked or boiled; a plump or dumpy person, esp. a child.

ROM (rom), *n.* a data-storage device in computers which retains information permanently in an unalterable state. [read *only* memory]

Rom., (*abbr.*) Roman; Romance (language); Rumania(n); Romans.

rom., (*abbr.*) roman (type).

Romaic (rəmā'ik), *n.* the vernacular language of modern Greece. *a.* of, pertaining to or expressed in modern Greek.

Roman (rō'mən), *a.* pertaining to the modern or ancient city of Rome or its territory or people; denoting numerals expressed in letters, not in figures; belonging to the Roman alphabet; of or pertaining to the Roman Catholic Church, papal. *n.* an inhabitant or citizen of Rome; a Roman Catholic; a letter of the Roman alphabet. *n.pl.* an epistolary book in the New Testament written by St Paul to the Christians of Rome. **roman**, *a.* denoting ordinary upright characters used in print as distinct from italic or gothic. *n.* roman type. **Roman architecture**, *n.* a style of architecture in which the Greek orders are combined with the use of the arch, distinguished by its massive character and abundance of ornament. **Roman candle** CANDLE. **Roman Catholic**, *a.* of or pertaining to the Church of Rome. *n.* a member of this Church. **Roman Catholicism**, *n.* **Roman Empire**, *n.* the empire established by Augustus, 27 BC, divided in AD 395 into the Western or Latin and Eastern or Greek Empires. **Roman holiday**, *n.* an entertainment or enjoyment which depends on others suffering. **Roman law**, *n.* the system of law evolved by the ancient Romans which forms the basis of many modern legal codes. **Roman nose**, *n.* one with a high bridge, an aquiline nose. **Roman numerals**, *n. pl.* the ancient Roman system of numbering consisting of letters representing cardinal numbers, occasionally still in use. **Romanic** (-man'-), *a.* derived from Latin; Romance (of languages or dialects), derived or descended from the Romans. **Romanish**, *a.* of, pertaining to or characteristic of the Church of Rome. **Romanism**, *n.* **Romanist**, *n.* **Romanistic** (-nis'-), *a.* **romanize, -ise**, *v.t.* to subject to the authority of ancient Rome; to Latinize; to convert to the Roman Catholic religion. *v.i.* to use Latin words or idioms; to conform to Roman Catholic opinions. **romanization**, *n.* **Romano-**, *comb.form* Roman.

roman à clef (romā ā klā'), *n.* 'novel with a key', a novel in which a knowing reader is expected to identify real people under fictitious names or actual events disguised as fictitious. [F]

Romance (rəmans'), *n.* one of a group of languages derived from Latin, e.g. French, Spanish, Rumanian. *a.* of or pertaining to this group of languages. **romance**, *n.* a mediaeval tale, usu. in verse, orig. in early French or Provençal, describing the adventures of a hero of chivalry; a story, usu. in prose, rarely in verse, with characters, scenery and incidents more or less remote from ordinary life; fiction of this character; a modern literary genre of sentimental love-stories, romantic fiction; the spirit or atmosphere of imaginary adventure, chivalrous or idealized love, strangeness and mystery; a love-affair; a fabrication; a short musical composition of simple character, usu. suggestive of a love-song. *v.i.* to imagine or tell romantic or extravagant stories; to make false, exaggerated or imaginary statements. **romancer**, *n.* **romancing**, *n.*, *a.*

Romanesque (rōmənesk'), *a.* of the styles of architecture that succeeded the Roman and lasted till the introduction of Gothic. *n.* Romanesque art, architecture etc.

Romanian RUMANIAN.

Romansch, Romansh (rəmansh'), *n.*, *a.* (of) the Rhaeto-Romanic language or dialects of part of E Switzerland.

romantic (rəman'tik), *a.* pertaining to, of the nature of or given to romance; imaginative, visionary, poetic, extravagant, fanciful; given to, inspiring or expressing love, esp. of a sentimental nature; wild, picturesque, suggestive of romance (of scenery etc.); pertaining to the movement in literature and art reacting against classicism and tending towards the unfettered expression of ideal beauty and grandeur. *n.* a romantic poet, novelist etc., a romanticist; a romantic person; a person given to sentimental thoughts or acts of love. **romantically**, *adv.* **romanticism** (-sizm), *n.* the quality or state of being romantic; the reaction from classical to mediaeval forms and to the unfettered expression of romantic ideals which originated in Germany about the middle of the 18th cent., and reached its culmination in England and France in the first half of the 19th cent., the **Romantic Movement** or **Romantic Revival**. **romanticist**, *n.* **romanticize, -ise**, *v.t.*, *v.i.* **romanticization**, *n.*

Romany (rō'məni), *n.* a gipsy; the gipsy language. *a.* gipsy.

Rome (rōm), *n.* capital city and ancient State of Italy, capital of the Roman Empire; the Church of Rome. **Rome was not built in a day**, accomplishments of any lasting worth require time and patience. **Romish**, *a.* (*derog.*) belonging to or tending towards Roman Catholicism.

Romeo (rō'miō), *n.* a man who is an ardent lover, from the hero of Shakespeare's *Romeo and Juliet*.

romp (romp), *v.i.* to play or frolic roughly or boisterously; to go rapidly (along, past etc.) with ease. *n.* one who romps; rough or boisterous play; a swift run; an easy win. **to romp home**, to win easily. **rompers**, *n. pl.* a one-piece play-suit for infants, a romper suit.

rondavel (ron'dəvel, -dah'-), *n.* a round hut or building in S Africa.

rondeau (ron'dō), *n.* (*pl.* **-deaux** (-dōz)) a poem in iambic verse of eight or ten syllables and ten or thirteen lines, with only two rhymes, the opening words coming twice as a refrain; a rondo. **rondel** (-dəl), *n.* a particular form of rondeau, usu. of thirteen or fourteen lines with only two rhymes throughout; a circular piece, disc, pane of glass etc. **rondo** (-dō), *n.* (*pl.* **-dos**) a musical composition having a principal theme which is repeated after each subordinate theme, often forming part of a symphony etc.

rone, rone-pipe (rōn), *n.* (*Sc.*) a gutter, a pipe for channelling rainwater from a roof.

röntgen ROENTGEN.

roo (roo), *n.* (*Austral. coll.*) a kangaroo.

rood (rood), *n.* the cross of Christ, a crucifix, esp. one set

on a rood-beam or screen; a measure of land, usu. the fourth part of an acre (about 0·1 ha). **rood-screen,** *n.* a stone or wood screen between the nave and choir, usu. elaborately designed and decorated with carving etc., orig. supporting the rood.
roof (roof), *n.* (*pl.* **roofs**) the upper covering of a house or other building; the covering or top of a vehicle etc.; any analogous part, as of a furnace, oven etc.; the palate; (*fig.*) the top of a mountain or plateau; a covering, a canopy; a house, shelter etc.; an upper limit, a ceiling. *v.t.* to cover with or as with a roof; to shelter. **roof garden,** *n.* a garden of plants and shrubs growing in soil-filled receptacles on a flat roof. **roof rack,** *n.* a detachable rack on the roof of a motor vehicle for holding luggage etc. **rooftop,** *n.* **rooftree,** *n.* the ridge-pole of a roof. **roofage** (-ij), *n.* **roofed,** *a.* **roofer,** *n.* **roofing,** *n.*, *a.* **roofless** (-lis), *a.*
rooinek (rō'inek), *n.* (*S Afr.*) a nickname for an Englishman.
rook[1] (ruk), *n.* a gregarious bird of the crow family with glossy black plumage; a cheat, a swindler, a sharper, esp. at cards, dice etc. *v.t.* to cheat, to swindle; to charge extortionately. **rookery,** *n.* breeding place of rooks, seals, or seabirds.
rook[2] (ruk), *n.* the castle in chess.
rookie, rooky (ruk'i), *n.* (*sl.*) a raw recruit or beginner.
room (room), *n.* space regarded as occupied or available for occupation, accommodation; capacity, vacant space or standing-ground; opportunity, scope; a portion of space in a building enclosed by walls, floor and ceiling; those present in a room; (*pl.*) apartments, lodgings, accommodation. *v.i.* to occupy rooms, to lodge. **to give, leave, make room,** to withdraw so as to leave space for other people. **room mate,** *n.* one with whom a person shares a room or lodgings. **room service,** *n.* in a hotel, the serving of food and drink to guests in their rooms. **roomed,** *a.* having rooms (*usu. in comb.* as *six-roomed*). **roomer,** *n.* (*N Am.*) a lodger. **roomette** (-et'), *n.* (*N Am.*) a sleeping compartment in a train. **roomful,** *n.* (*pl.* **-fuls**). **rooming,** *a.*, *n.* **rooming-house,** *n.* (*N Am.*) a lodging-house. **roomy,** *a.* having ample room; spacious, extensive. **roomily,** *adv.* **roominess,** *n.*
roost (roost), *n.* a pole or perch for birds to rest on; a place for fowls to sleep in at night; a resting-place. *v.i.* to perch on or occupy a roost. **come home to roost,** to recoil on one. **rooster,** *n.* the domestic cock.
root[1] (root), *n.* the descending part of a plant which fixes itself in the earth and draws nourishment therefrom; (*pl.*) the ramifying parts, rootlets or fibres into which this divides, or the analogous part of an epiphyte etc.; an esculent root; the embedded part of a tooth, hair, etc.; the basis, the bottom, the fundamental part or that which supplies origin, sustenance, means of development etc.; (*pl.*) one's ancestry, origins, place of origin or belonging; the elementary, unanalysable part of a word as distinguished from its inflexional forms and derivatives; the fundamental note of a chord; in mathematics, the quantity or number that, multiplied by itself a specified number of times, yields a given quantity. *v.i.* to take root. *v.t.* to cause to take root; to fix or implant firmly (to the spot); to pull or dig (up) by the roots. **root and branch,** utterly, radically. **to root out,** to uproot; to extirpate. **to take, strike root,** to become planted and send out living roots or rootlets; to become immovable or established. **root beer,** *n.* (*N Am.*) a fizzy soft drink made from the roots of sassafras and other plants. **root-cap,** *n.* a protective covering of cells on the tip of a root. **root-crop,** *n.* a crop of plants with esculent roots. **rootstock,** *n.* a rhizome; the original source or primary form of anything. **root vegetable,** *n.* a vegetable that is or has an esculent root. **rooted,** *a.* **rooter,** *n.* **rootless** (-lis), *a.* **rootlet** (-lit), *n.* a small root, a radicle. **rootlike,** *a.* **rooty,** *a.* **rootiness,** *n.*
root[2] (root), *v.t.* to dig, turn or grub (up) with the snout, beak, etc. *v.i.* to turn up the ground in this manner in search of food; to hunt (up or out), to rummage (about, in etc.); **rooter,** *n.* **rootle** (-l), *v.t.*, *v.i.*
root[3] (root), *v.i.* (*coll.*) to cheer, to shout encouragements to, to support. **rooter,** *n.* one who roots, cheers, supports.
rooti (roo'ti), **roti** (rō'-), **ruti,** *n.* Indian unleavened bread, food.
rope (rōp), *n.* a stout cord of twisted fibres of hemp, flax, cotton etc., or wire; a series of things strung together in a line e.g. of garlic, onions, pearls; a slimy or gelatinous formation in beer etc. *v.t.* to tie, fasten or secure with a rope; to enclose or close (in) with rope. **the rope,** a hangman's noose; death by hanging. **to give (someone) enough (or plenty of) rope to hang himself, herself,** to allow someone enough freedom of speech or action to commit a blunder or cause his or her own downfall. **to know the ropes,** to be well acquainted with the circumstances, methods and opportunities in any sphere. **to rope in,** to capture or pull in a steer, horse etc. with a rope, to lasso; (*coll.*) to enlist or persuade someone to join a group or enter into an activity. **rope-ladder,** *n.* a ladder made of two ropes connected by rungs usu. of wood. **rope's-end,** *n.* a short piece of rope formerly used for flogging, esp. on shipboard. **rope-walk,** *n.* a long piece of usu. covered ground where ropes are twisted. **rop(e)able,** *a.* (*Austral. coll.*) wild, intractable; angry, out of temper, irascible; capable of being roped. **roped,** *a.* **rop(e)y,** *a.* resembling a rope; glutinous, viscid; (*coll.*) inferior, shoddy; (*coll.*) unwell. **ropily,** *adv.* **ropiness,** *n.*
Roquefort (rok'faw), *n.* French cheese made from goats' and ewes' milk.
roquet (rō'kā), *v.t.* in croquet, to make one's ball strike another; to strike another ball (of one's ball). *v.i.* to make this stroke. *n.* this stroke or a hit with it.
ro-ro (rō'rō), (*abbr.*) roll-on-roll-off.
rorqual (raw'kwəl), *n.* a whale with dorsal fins, the finback.
Rorschach test (raw'shahkh, -shahk), *n.* a test for personality traits and disorders based on the interpretation of random ink-blots. [Hermann *Rorschach*, 1884–1922, Swiss psychiatrist]
rort (rawt), *n.* (*Austral. sl.*) a party or boisterous celebration; a deception. **rorter,** *n.* **rorty,** *a.* rowdy, noisy.
rosace (rō'zas), *n.* a rose-shaped centre-piece or other ornament, a rosette; a rose-window. **rosacea** (-zā'siə), *n.* a chronic skin-disease characterized by redness of the skin. **rosaceous** (-zā'shəs), *a.* pertaining to the Rosaceae family of plants to which the rose belongs; rose-like, rose-coloured. **rosarian** (-zeə'ri-), *n.* a rose-fancier, a cultivator of roses. **rosarium** (-əm), *n.* a rose-garden. **rosary,** *n.* a rose-garden, a rose-plot; a form of prayer in the Roman Catholic Church in which three sets of five decades of aves, each decade preceded by a paternoster and followed by a gloria, are repeated; this series of prayers; a string of beads by means of which account is kept of the prayers uttered.
rose[1] (rōz), *n.* any plant or flower of the genus *Rosa*, consisting of prickly bushes or climbing and trailing shrubs bearing single or double flowers, usu. scented, of all shades of colour from white and yellow to dark crimson; one of various other flowers or plants (with distinctive adjective or phrase) having some resemblance to the rose; a light crimson or pink colour; a complexion of this colour (*often in pl.*); a device, rosette, knot, ornament or other object shaped like a rose; a perforated nozzle for a hose or watering-pot; a rose-window; a rose-shaped ornament or a fitting for an electrical wire on a ceiling; a circular card, disk or diagram with radiating lines, used in a mariner's compass etc. *a.* coloured like a rose, pink or pale red. *v.t.* (*chiefly in p.p.*) to make rosy. **a bed of roses,** a luxurious situation, ease. **all roses, roses all the way,** completely pleasant, unproblematic or easy. **every-**

thing is coming up roses, everything is turning out successfully. **rose of Jericho,** a small annual cruciferous plant of N Africa and Syria, etc., having fronds that expand with moisture, also called the resurrection plant. **rose of Sharon,** an Eastern plant sometimes identified with the meadow saffron, the cistus and the polyanthus narcissus; a species of St John's wort. **under the rose,** in secret; privately, confidentially, sub-rosa. **Wars of the Roses,** the civil wars (1455-85) between the Houses of Lancaster and York, who respectively took a red and a white rose as their emblems. **rose-apple,** *n.* a tropical tree cultivated for its foliage, flowers and fruit. **rose-bay,** *n.* the great willow-herb; the oleander; the rhododendron. **rose bowl,** *n.* a bowl-shaped ornamental vase for roses. **rose-bud,** *n.* a flower-bud of a rose; a young girl. *a.* like a rose-bud (of a mouth). **rose-campion,** *n.* a garden plant with crimson flowers. **rose-chafer, beetle,** *n.* a European beetle infesting roses. **rose-coloured,** *a.* sanguine, optimistic; of a rose-colour. **to see through rose-coloured or tinted spectacles or glasses,** to view matters in an extremely optimistic light. **rose-cut,** *a.* cut with a flat surface below and a hemispherical or pyramidal part above covered with facets (of diamonds etc.). **rosehip,** *n.* a red berry, the fruit of the rose plant. **rose-water,** *n.* scented water distilled from rose leaves. *a.* affectedly delicate, fine or sentimental. **rose-window,** *n.* a circular window filled with tracery branching from the centre, usu. with mullions arranged like the spokes of a wheel. **rosewood,** *n.* a hard close-grained fragrant wood of a dark-red colour obtained chiefly from various species of *Dalbergia.* **rosewood oil,** *n.* oil obtained from a species of rosewood. **roseate,** *a.* (-ət), rose-coloured, rosy; smiling, promising, optimistic. **roselike,** *a.* **rosery** (-zəri), *n.* a place where roses grow, a rose-plot, a rosarium. **rosy,** *a.* resembling a rose; blooming; flourishing; favourable, auspicious. **rosy-cheeked,** *a.* having a healthy bloom, pink cheeks. **rosiness,** *n.*

rose[2] (rōz), *past* RISE.

rosé (rō'sā), *n., a.* (of) a pink-coloured wine made from red grapes with their skins removed or combined with red and white wines.

rose-bud, rose-water etc. ROSE[1].

rosella (rōzel'ə), *n.* (*Austral.*) a variety of brightly-coloured parakeet. [from *Rose Hill,* a district near Sydney where it was first observed]

rosemary (rōz'məri), *n.* an evergreen fragrant shrub of the mint family, leaves of which yield a perfume and oil and are used in cooking etc.

roseola (rōzē'ələ), *n.* a non-contagious febrile disease with rose-coloured spots, German measles; a rash occurring in measles etc. **roseolar,** *a.*

Rosetta stone (rōzet'ə), *n.* a basalt stele with an inscription in hieroglyphics, demotic characters and Greek, discovered at Rosetta, in Egypt (1799) from which Egyptian hieroglyphics were deciphered.

rosette (rəzet'), *n.* a bunch of ribbons arranged concentrically more or less as the petals of a rose (usu. worn as a badge or given as a prize); an architectural ornament or a decoration in the form of a rose; a rose-window; a circular group of leaves usu. round the base of a stem. **rosetted,** *a.*

rosewood etc. ROSE[1].

Rosh Hashanah (rosh həshah'nə), *n.* the Jewish New Year or the festival celebrating it.

Rosicrucian (rozikroo'shən), *n.* a member of a secret religious society devoted to the study of occult science, which became known to the public early in the 17th cent., and was alleged to have been founded by a German noble, Christian Rosenkreuz, in 1484. *a.* of or pertaining to Rosenkreuz, this society or its members.

rosin (roz'in), *n.* resin, esp. the solid residue left after the oil has been distilled from crude turpentine, colophony. *v.t.* to rub, smear etc. with rosin, esp. to apply it to a violin etc. bow. **rosined,** *a.* **rosiny,** *a.*

rosiness ROSE[1].

ROSPA (ros'pə), (*abbr.*) Royal Society for the Prevention of Accidents.

roster (ros'tə), *n.* a list showing the order of rotation in which employees, officers, members etc. are to perform their turns of duty. *v.t.* to put on a roster.

rostrum (ros'trəm), *n.* (*pl.* **-stra** (-strə)) the beak or prow of a Roman galley; a platform from which public orations etc. are delivered; a beak, bill, beak-like snout, part or process. **rostral,** *a.* pertaining to, situated on or resembling a rostrum or beak; (*Rom. Ant.*) decorated with the beaks of war-galleys or representations of these (of columns etc.).

rot (rot), *v.i.* (*past, p.p.* **rotted**) to decay, to decompose by natural change, to putrefy; to be affected with a decaying disease; to become morally corrupt; to pine away. *v.t.* to cause to rot, to decompose, to make putrid. *n.* putrefaction, rottenness; dry-rot, wet-rot; a malignant liver-disease in sheep etc.; (*coll.*) nonsense, rubbish. *int.* expressing disbelief or disagreement. **rotgut,** *n.* (*sl.*) an alcoholic drink of inferior quality.

rota (rō'tə), *n.* a list of names, duties etc., a roster; (**Rota**) in the Roman Catholic Church, the supreme court deciding on ecclesiastical and secular causes. **rotary,** *a.* rotating on its axis; acting or characterized by rotation. *n.* a rotary machine or part of a machine. **Rotary Club, the Rotary,** *n.* a local business club for mutual benefit and service. **rotary cultivator,** *n.* a horticultural machine with revolving blades or claws for tilling. **rotary press,** *n.* a printing press in which the printing surface is a revolving cylinder. **Rotarian** (-teə'ri-), *n., a.* (a member) of a Rotary Club.

rotate (rətāt'), *v.i.* to revolve round an axis or centre; to act in rotation. *v.t.* to cause (a wheel etc.) to revolve; to arrange (crops etc.) in rotation. *a.* (rō'tət) wheel-shaped (of a calyx, corolla etc.). **rotatable,** *a.* **rotation,** *n.* the act of rotating, rotary motion; alternation, recurrence, regular succession. **rotational,** *a.* **rotative** (rō'-), *a.* **rotator,** *n.* that which moves in or gives a circular motion; a muscle imparting rotatory motion. **rotatory,** *a.* **rotovate,** *v.i.* **Rotovator**® (rō'təvātə), *n.* a rotary cultivator.

rote (rōt), *n.* mere repetition of words, phrases etc. without understanding; mechanical, routine memory or knowledge.

roti ROOTI.

rotifer (rō'tifə), *n.* (*pl.* **-fers**) one of the wheel-animalcules, a phylum of minute aquatic animals with swimming organs appearing to have a rotary movement. **rotiferal, rotiferous** (-tif'-), *a.*

rotisserie (rōtē'səri, -tis'-), *n.* a device with a spit on which food esp. meat is roasted or barbecued; a restaurant specializing in meat cooked in this way.

rotogravure (rōtəgrəvūə'), *n.* a process of photogravure-printing on a rotary machine; a print produced by this process.

rotor (rō'tə), *n.* name given to any system of revolving blades that produce lift in aircraft; the rotating part of a machine.

rotten (rot'n), *a.* decomposed, decayed, decaying, tainted, putrid, fetid; unsound, liable to break, tear etc.; morally corrupt, unhealthy, untrustworthy, defective; (*coll.*) poor or contemptible in quality; disagreeable, annoying, unpleasant; (*coll.*) unwell. **rotten borough** BOROUGH. **rottenstone,** *n.* a friable siliceous limestone used for polishing. **rottenly,** *adv.* **rottenness,** *n.* **rotter,** *n.* (*sl.*) a good-for-nothing or detestable person.

Rottweiler (rot'wilə), *n.* a large, heavily built German breed of dog with a smooth black coat, noted for its aggressive tendencies.

rotund (rətūnd'), *a.* rounded, circular or spherical; orotund, sonorous, magniloquent (of speech or language);

rouble, ruble 720 **round**

plump, well-rounded. **rotunda** (-də), *n.* a circular building, hall etc., esp. with a dome. **rotundity**, *n.* **rotundly**, *adv.*
rouble, ruble (roo'bl), *n.* the Russian monetary unit, equal to 100 kopecks.
roué (roo'ā), *n.* a rake, a debauchee.
rouge (roozh), *n.* a cosmetic used to colour the cheeks red; red oxide of iron used for polishing metal, glass etc. *v.t.* to colour with rouge. **rouge et noir** (ä nwah'), a gambling card-game played by a 'banker' and a number of persons on a table marked with four diamonds, two red and two black.
rough (rŭf), *a.* having an uneven, broken or irregular surface, having prominences or inequalities, not smooth, level or polished; shaggy, hairy, of coarse texture; rugged, hilly, hummocky; harsh to the senses, astringent, discordant, severe; violent, boisterous, tempestuous; turbulent, disorderly; harsh or rugged in temper or manners; cruel, unfeeling; rude, unpolished; lacking finish or completeness, not completely wrought, crude; approximate, not precise or exact, general; difficult, hard (to bear). *adv.* roughly, in a rough manner. *n.* a rough or unfinished state; rough ground; the ground to right and left of a golf fairway; a rough person, a rowdy; a draft, a rough drawing; (*collect.*) rough or harsh experiences, hardships. *v.t.* to make rough, to roughen; to plan or shape (out) roughly or broadly. **the rough side of one's tongue**, (*coll.*) a scolding, a rebuke. **to cut up rough**, to be upset, to grow quarrelsome. **to rough in**, to outline, to draw roughly. **to rough it**, to put up with hardships; to live without the ordinary conveniences. **to rough up**, (*sl.*) to beat up, to injure during a beating. **to sleep rough**, to sleep out-of-doors. **to take the rough with the smooth**, to be subject to unpleasantness or difficulty as well as ease, happiness etc. **rough-and-ready**, *a.* hastily prepared, without finish or elaboration; provisional, makeshift. **rough-and-tumble**, *a.* disorderly, irregular, boisterous, haphazard. *n.* an irregular fight, contest, scuffle etc. *adv.*
rough-cast, *v.t.* to form or compose roughly; to coat (a wall) with coarse plaster. *n.* a rough model or outline; a coarse plastering, usu. containing gravel, for outside walls etc. *a.* formed roughly, without revision or polish; coated with rough-cast. *n.* **rough-cut**, *n.* the first assembly of a film by an editor from the selected takes which are joined in scripted order. **rough diamond**, *n.* a person of rough exterior or manners but with a genuine or warm character. **rough draft**, *v.t.*, *n.* (to make) a rough sketch, a preliminary version. **rough-dry**, *v.t.* to dry without smoothing or ironing. *a.* (of clothes, etc.) dried and ready for ironing. **rough-hew**, *v.t.* to hew out roughly; to give the first crude form to. **rough-hewn**, *a.* rugged, rough, unpolished. **rough house**, *n.* (*coll.*) horse-play, brawling. **rough justice**, *n.* justice appropriate to a crime but not strictly legal; a sentence or verdict hastily reached and executed. **rough-neck**, *n.* (*N Am. coll.*) a rowdy, a hooligan; an oil-worker employed to handle drilling equipment on a rig. **rough-rider**, *n.* a horse-breaker; a bold skilful horseman able to ride unbroken horses. **rough-shod**, *a.* shod with roughened shoes. **to ride rough-shod over**, to treat in a domineering and inconsiderate way. **rough stuff**, *n.* (*sl.*) violence, violent behaviour. **rough trade**, *n.* (*sl.*) a usu. casual homosexual partner who is uncultivated or aggressive. **roughage** (-ij), *n.* food materials containing a considerable quantity of cellulose, which resist digestion and promote peristalsis. **roughed-up**, *a.* beaten-up. **roughen**, *v.t.*, *v.i.* **roughish**, *a.* **roughly**, *adv.* **roughness**, *n.*
roulade (roolahd'), *n.* (*Mus.*) a run of notes on one syllable, a flourish; in cookery, a rolled piece usu. of meat, and usu. filled with a stuffing.
roulette (roolet'), *n.* a game of chance played with a ball on a table with a revolving disk; a wheel with points for making dotted lines, used in engraving, for perforating

etc.; in geometry, a curve that is the locus of a point rolling on a curve.
Roumanian RUMANIAN.
round (rownd), *a.* spherical, circular, cylindrical or approximately so; convexly curved in contour or surface, full, plump, not hollow, corpulent; going and returning to the same point, with circular or roughly circular course or motion; continuous, unbroken; plain, open, frank, candid, fair; quick, smart, brisk (of pace etc.); full-toned, resonant; articulated with lips formed into a circle (of sounds); liberal, ample, large, considerable; evenly divisible, approximate, without fractions. *n.* a round object, piece, slice etc.; a thick cut from a joint (of beef); that which goes round, circumference, extent; a circular course, a circuit, a heat, a cycle, a recurrent series, a bout, a session, a spell, an allowance, a series of actions etc.; an order of drinks for several people, each of whom is buying drinks for the group in turn; a burst of applause; a single shot or volley fired from a firearm or gun; ammunition for this; the state of being completely carved out in the solid, opp. to relief; a circuit of inspection, the circuit so made; a piece of music sung by several voices each taking it up in succession. *adv.* on all sides so as to encircle; so as to come back to the same point; to or at all points on the circumference or all members of a party etc.; by a circuitous route; with rotating motion. *prep.* on all sides of; so as to encircle; to or at all parts of the circumference of; in all directions from (in the relation of a body to its axis or centre); revolving round. *v.t.* to make round or curved; to pass, go or travel round; to collect together, to gather (up); to fill out, to complete; to pronounce fully and smoothly. *v.i.* to grow or become round; to go the rounds, as a guard; (*chiefly Naut.*) to turn round. **in round numbers**, approximately; to the nearest large number. **in the round**, able to be viewed from every side; in full. **round the bend**, **twist**, mad, crazy. **to bring someone round**, to resuscitate someone; to persuade someone to accept an idea, a situation etc. **to come round**, to revive; to begin to accept an idea, situation etc. **to get round someone**, to take advantage of by flattery or deception. **to round down** or **up**, to lower or raise a number to avoid fractions or reach a convenient figure. **to round off**, to shape (angles etc.) to a round or less sharp form; to finish off, complete, perfect. **to round on**, to turn upon, to attack. **to round out**, to fill out, become more plump. **to round up**, to gather together. **round about**, *prep.*, *adv.* in or as in a circle round, all round; circuitously, indirectly; approximately. **roundabout**, *a.* circuitous, indirect. *n.* a merry-go-round; a device at a cross-roads whereby traffic circulates in one direction only. **round dance**, *n.* a dance in which the performers are ranged or move in a circle, esp. a waltz. **round-eyed**, *a.* **round-faced**, *a.* **round-hand**, *n.* writing in which the letters are round and full. **Roundhead**, *n.* a term applied by the Cavaliers during the Civil War to the Parliamentarians, from their wearing their hair short. **round-house**, *n.* (*N Am.*) a circular building containing a turntable for servicing railway locomotives. **round robin**, *n.* (*coll.*) a petition with the signatures placed in a circle so that no name heads the list; a tournament in which each contestant plays every other contestant. **round-shouldered**, *a.* bent forward so that the back is rounded. **roundsman**, *n.* one who makes calls to collect orders, deliver goods etc.; (*N Am.*) a policeman making a round of inspection. **round table**, *n.* a conference or meeting at which all parties are on an equal footing. **round-the-clock**, *a.* continuous; lasting 24 hours a day. **round trip**, *n.* a return journey to a place and back. *a.* (*N Am.*) a return. **round-up**, *v.t.* to gather (horses, cattle etc.) together. *n.* a gathering together of cattle etc.; a similar gathering of people, objects, news, facts etc. (e.g. a news round-up).
roundworm, *n.* a parasitic elongated worm, a nematode.

rounded, *a.* **roundel** (-dəl), *n.* anything of a round shape; a round disc, panel, heraldic circular charge etc.; a rondel or rondeau; a roundelay. **roundelay** (-dəlā), *n.* a simple song, usu. with a refrain; a mediaeval round dance. **rounders**, *n. pl.* a game with a short bat and a ball, between two sides, with four bases to which a player hitting the ball has to run without being hit by it. **roundish**, *a.* **roundly**, *adv.* in a round or roundish form; bluntly, straightforwardly, plainly, emphatically. **roundness**, *n.*
roundel, roundelay ROUND.
roup (rowp), *v.t.* (*Sc., North.*) to sell by auction. *n.* a sale by auction.
rouse (rowz), *v.t.* to raise or startle (game) from a covert; to wake; to excite to thought or action; to provoke, to stir (up), to agitate. *v.i.* to wake or be wakened; to start up; to be excited or stirred (up) to activity etc. *n.* reveille. **to rouse on**, (*Austral.*) to scold, to tell off. **rouseabout**, *n.* (*Austral.*) an odd-job man in a shearing-shed or on a station. **rouser**, *n.* **rousing**, *a.* having power to awaken, excite or rouse. **rousingly**, *adv.* **roust** (rowst), *v.t.* to rouse, to rout (out). **roustabout**, *n.* (*N Am.*) a labourer on wharves; (*N Am., Austral.*) a casual labourer; a rouseabout; an unskilled worker on an oil rig.
roust, roustabout ROUSE.
rout[1] (rowt), *n.* a crowd, a miscellaneous or disorderly concourse; (*Law*) an assembly and attempt of three or more people to do an unlawful act; an utter defeat and overthrow; a disorderly and confused retreat of a defeated army etc. *v.t.* to defeat utterly and put to flight.
rout[2] (rowt), *v.t.* to root (up or out); to turn, fetch, drive out etc. (of bed, house etc.); to gouge, to scoop, to tear (out etc.). *v.i.* to root (about), to search. **router**, *n.* a plane, a saw or any of various other tools for hollowing out or cutting grooves.
route (root, *Mil.* rowt), *n.* the course, way or road(s) travelled or to be travelled. *v.t.* to send by a certain route; to arrange or plan the route of. **en route** (ē), on the way. **route-march**, *n.* an arduous military-training march; (*coll.*) a long tiring walk.
routine (rootēn'), *n.* a course of procedure, business or official duties etc., regularly pursued; any regular or mechanical habit or practice; a sequence of jokes, movements, steps etc. regularly performed by a comedian, dancer, skater, stripper etc.; (*coll.*) tiresome or insincere speech or behaviour; a computer program or part of one which performs a particular task. *a.* tiresome, repetitive, commonplace; of or pertaining to a set procedure. **routinely**, *a.*
roux (roo), *n.* a sauce base, the thickening element in a sauce made from fat and flour cooked together.
rove[1] (rōv), *past* REEVE[2].
rove[2] (rōv), *v.i.* to wander, to ramble, to roam. *v.t.* to wander over, through etc. *n.* the act of roving. **Rover, Rover Scout**, *n.* (*formerly*) a member of the Rovers or Rover Scouts, a branch of the Scouts for boys over 16 years old, now a *Venture Scout*. **rover**, *n.* a pirate, a buccaneer; a wanderer; in archery, a mark chosen at random, a mark for long-distance shooting. **roving**, *n., a.* **roving commission**, *n.* a commission without a rigidly defined area of authority. **roving eye**, *n.* a promiscuous sexual interest.
rove[3] (rōv), *v.t.* to draw out and slightly twist slivers of wool, cotton etc., before spinning into thread. *n.* fibres prepared in this way.
row[1] (rō), *n.* a series of persons or things in a straight or nearly straight line; a line, a rank (of seats, vegetables etc.); a street usu. of identical houses. **in a row**, (placed) one after the other, (ordered) in succession.
row[2] (rō), *v.t.* to propel by oars; to convey by rowing. *v.i.* to row a boat; to labour with an oar; to be impelled by oars. *n.* a spell at rowing; an excursion in a rowing-boat. **row-, rowing-boat**, *n.* a boat propelled by rowing. **rowlock, rollock** (rol'ək), *n.* a crotch, notch or other device on the gunwale of a boat serving as a fulcrum for an oar. **rower**, *n.* **rowing**, *n., a.* **rowing-machine**, *n.* an exercise machine fitted with oars and a sliding seat.
row[3] (row), *n.* a noisy disturbance, a noise, a din, a commotion; a quarrel; a scolding. *v.i.* to make a row, to quarrel.
rowan (row'ən, rō'-), *n.* the mountain-ash. **rowan-berry**, *n.* the small red fruit of the rowan. **rowan-tree**, *n.*
rowdy (row'di), *n.* (*pl.* **-dies**) a noisy, rough or disorderly person. *a.* rough, riotous. **rowdily**, *adv.* **rowdiness, rowdyism**, *n.*
rowel (row'əl), *n.* a spiked disk or wheel on a spur; a roll or disk of various materials with a hole in the centre for placing under a horse's skin to discharge purulent matter. *v.t.* to insert a rowel in (a horse etc.); to prick or goad with a rowel.
rower, rowlock ROW[2].
royal (roi'əl), *a.* of, pertaining to, suitable to or befitting a king or queen; under the patronage of, in the service of, chartered or founded by royalty; regal, kingly, princely; noble, magnificent, majestic; on a great scale, splendid, first-rate. *n.* a royal stag; a royal mast or sail next above the topgallant; a royal personage; royal paper. **Burgh Royal** BURGH. **rhyme royal** RHYME. **the Royals**, the royal family; (*formerly*) the first regiment of foot in the British service, the Royal Scots. **the royal we**, the customary use of the 1st person plural by a sovereign referring to himor herself. **Royal Academy**, *n.* an academy of fine arts in London (est. 1768). **Royal Air Force**, *n.* the airforce of Great Britain. **royal assent** ASSENT. **royal blue**, *n., a.* (of) a deep blue. **Royal Commission** COMMISSION. **royal fern**, *n.* the flowering fern, *Osmunda regalis*. **royal flush** FLUSH. **royal icing**, *n.* a hard icing on wedding cakes, fruit cakes etc. **royal jelly**, *n.* the food secreted and fed by workerbees to developing queen-bees; a health preparation of this substance. **Royal Marines**, *n. pl.* corps specializing in commando and amphibious operations. **royal mast**, *n.* the topmost part of a mast above the topgallant. **Royal Navy**, *n.* the navy of Great Britain. **royal palm**, *n.* a tall palm of tropical America. **royal paper**, *n.* a size of paper 20×25 in. (about 50×63 cm) for printing, 19×24 in. (about 48×61 cm) for writing. **royal stag**, *n.* a stag with antlers having 12 or more points. **royal standard**, *n.* flag with the royal arms. **royal tennis**, *n.* real tennis, court tennis. **royal warrant**, *n.* one authorizing the supply of goods to a royal household. **royalism**, *n.* **royalist**, *n.* an adherent or supporter of royalism or of monarchical government, esp. a supporter of the royal cause in the Civil War. *a.* supporting monarchical government; belonging to the Royalists. **royally**, *adv.* **royalty**, *n.* (*pl.* **-ties**) the office or dignity of a king or queen, sovereignty; royal rank birth or lineage; kingliness, queenliness; a royal person or persons; a member of a reigning family; a right or prerogative of a sovereign; (*usu. pl.*) a share of profits paid to a landowner for the right to work a mine, to a patentee for the use of an invention, to an author on copies of books sold etc.
rozzer (roz'ə), *n.* (*sl.*) a policeman.
RP, (*abbr.*) Reformed Presbyterian; Regius Professor; Received Pronunciation.
RPI, (*abbr.*) retail price index.
rpm, (*abbr.*) resale price maintenance; revolutions per minute.
RR, (*abbr.*) Right Reverend.
-rrhagia, *comb. form* abnormal discharge, excessive flow, as in *menorrhagia*.
-rrhoea, -rrhea, *comb. form* a discharge, a flow, as in *diarrhoea*.
RS, (*abbr.*) Royal Society.
RSA, (*abbr.*) Republic of South Africa; Royal Scottish Academy or Academician; Royal Society of Arts.

RSFSR, *(abbr.)* Russian Soviet Federated Socialist Republic.
RSM, *(abbr.)* Regimental Sergeant-Major; Royal School of Music; Royal Society of Medicine.
RSPB, *(abbr.)* Royal Society for the Protection of Birds.
RSPCA, *(abbr.)* Royal Society for the Prevention of Cruelty to Animals.
RSV, *(abbr.)* Revised Standard Version (of the Bible).
RSVP, *(abbr.)* *répondez s'il vous plaît*, reply, if you please.
RTE, *(abbr.)* *Radio Telefís Eireann*, Irish radio and television.
Rt. Hon., *(abbr.)* Right Honourable.
RU, *(abbr.)* Rugby Union.
Ru, *(chem. symbol)* ruthenium.
rub[1] (rŭb), *v.t. (past, p.p.* **rubbed**) to apply friction to, to move one's hand or other object over the surface of; to polish, to clean, to scrape, to graze; to slide or pass (a hand or other object) along, over or against something; to take an impression of (a design) with chalk and graphite on paper laid over it; to remove by rubbing; to affect (a person or feelings etc.) as by rubbing; to spread on or mix into by rubbing. *v.i.* to move or slide along the surface of, to grate, to graze, to chafe (against, on etc.); to get (along, on, through etc.) with difficulty; to meet with a hindrance (of bowls). *n.* the act or a spell of rubbing; a hindrance, an obstruction, a difficulty. **to rub along,** to manage, just to succeed; to cope despite difficulties; to keep on friendly terms. **to rub down,** to bring to smaller dimensions or a lower level by rubbing; to clean, smooth or dry by rubbing. **to rub in,** to force in by friction; to enforce or emphasize (a grievance etc.). **to rub noses,** to do this as an Eskimo greeting. **to rub off onto (someone),** to pass on by example or close association. **to rub one's hands,** to express expectation, glee, satisfaction etc. in this manner. **to rub shoulders,** to associate or mix (with). **to rub someone's nose in it,** to refer to or remind someone of an error, indiscretion or misfortune. **to rub (up) the wrong way,** to irritate. **to rub out,** to remove or erase by friction; to kill. **to rub up,** to polish, to burnish; to freshen (one's recollection of something). **rubdown,** *n.* the act of rubbing down. **rubbed,** *a.* **rubber**[1], *n.* one who or that which rubs; an instrument, cloth, etc. used for rubbing; india-rubber or caoutchouc; a piece of india-rubber for erasing pencil marks etc.; (*N Am.*) a condom; (*pl.*) galoshes, rubber over-shoes. *a.* made of, yielding or relating to india-rubber. **rubber band,** *n.* a continuous band of rubber of varying widths and thicknesses, for securing the hair, packages etc. **rubber cement,** *n.* an adhesive containing rubber. **rubberneck, -necker,** *n.* (*esp. N Am.*) a sightseer; one who gapes out of curiosity. *v.i.* (*esp. N Am.*) to sightsee; to gape foolishly. **rubber plant,** *n.* a plant common to Asia and related to the fig, with large shiny leaves, in dwarf form grown as a popular houseplant. **rubber stamp,** *n.* a routine seal of approval, an automatic endorsement; a device with a rubber pad for marking or imprinting; a person who makes routine authorizations, a cipher. *v.t.* to imprint with a rubber stamp; to approve or endorse as a matter of routine. **rubber tree,** *n.* a tropical tree native to S America from which latex (the chief constituent of rubber) is obtained. **rubberize, -ise,** *v.t.* **rubbery,** *a.* **rubbing,** *n.* an impression made on paper laid over an image and rubbed with chalk, wax etc.
rubaiyat (roo'bīyat), *n.* (*Pers.*) a verse form consisting of quatrains.
rubato (rubah'tō), *n.* (*pl.* **-ti** (-tē), **-tos**) flexibility of rhythm, fluctuation of tempo within a musical piece. *a.*, *adv.* (to be) performed in this manner.
rubber[2] (rŭb'ə), *n.* a series of three games at whist, bridge, back-gammon etc.; two games out of three or the game that decides the contest.
rubbish (rŭb'ish), *n.* waste, broken or rejected matter; refuse, junk, garbage, litter, trash; nonsense. *a. (coll.)* bad, useless, distasteful etc. *v.t. (coll.)* to criticize, to reject as rubbish. **on the rubbish heap**, *(coll.)* discarded as ineffective or worthless. **rubbishy,** *a.*
rubble (rŭb'l), *n.* rough, broken fragments of stone, brick etc.; rubble-work. **rubble-work,** *n.* masonry composed of irregular fragments of stone, or in which these are used for filling in. **rubbly,** *a.*
rube (roob), *n.* (*N Am. sl.*) an unsophisticated country-dweller, a country bumpkin.
rubella (rubel'ə), *n.* German measles. **rubellite** (roo'bəlīt, -bel'-), *n.* a pinky-red tourmaline. **rubeola** (-bē'ələ), *n.* measles.
Rubicon (roo'bikən), *n.* a small stream in Italy, bounding the province of Caesar, which he crossed it before the war with Pompey, exclaiming, 'The die is cast!'; hence, an irrevocable step, a point of no return. **rubicon,** *n.* in piquet, the winning of the game before one's opponent has scored 100 points. **to cross the Rubicon,** to take a decisive step.
rubicund (roo'bikənd), *a.* ruddy, rosy. **rubicundity** (-kŭn'-), *n.*
rubidium (rubid'iəm), *n.* a silvery-white metallic element, at. no. 37; chem. symbol Rb, belonging to the potassium group. **rubidic,** *a.*
rubiginous (rubij'inəs), *a.* rusty or brownish-red in colour.
Rubik('s) cube® (roo'bik), *n.* a puzzle invented by the Hungarian designer Ernö Rubik (b. 1944) consisting of a cube each face of which is divided into nine coloured segments which can be revolved to obtain the same colour on each face.
ruble ROUBLE.
rubric (roo'brik), *n.* a title, initial letter, chapter-heading, entry, set of rules, commentary or direction, orig. printed in red or distinctive lettering, esp. a liturgical direction in the Prayer Book etc. **rubrical,** *a.* **rubrically,** *adv.*
ruby (roo'bi), *n.* a precious stone of a red colour, a variety of corundum; the colour of ruby, esp. a purplish red; something resembling, containing or made of a ruby. *a.* of the colour of a ruby; made of, containing or resembling a ruby or rubies; marking a 40th anniversary. **ruby-coloured, ruby-red,** *a.* of the deep red colour of a ruby. **ruby wedding,** *n.* a 40th wedding anniversary.
RUC, *(abbr.)* Royal Ulster Constabulary.
ruche (roosh), *n.* a quilled or pleated strip of gauze, lace, silk or the like used as a frill or trimming. *v.t.* to trim with ruche. **ruching,** *n.*
ruck[1] (rŭk), *n.* a heap, a mass; a multitude, a crowd, esp. the mass of horses left behind by the leaders in a race; the common run of people or things; in Rugby, a gathering of players round the ball when it is on the ground. *v.i.* to form a ruck in Rugby.
ruck[2] (rŭk), *n.* a crease, a wrinkle, a fold. *v.i., v.t.* to wrinkle, to crease.
rucksack (rŭk'sak), *n.* a bag carried on the back by means of straps by campers, hikers, climbers etc, a back pack.
ruckus (rŭk'əs), *n.* (*chiefly N Am.*) a row, a disturbance, an uproar.
ruction (rŭk'shən), *n. (coll.)* a commotion, a disturbance, a row.
rudbeckia *n.* a plant of a genus of N American plants of the aster family, also called the cone-flowers. [Olaus Rudbeck, 1630–1702, Swedish botanist]
rudd (rŭd), *n.* a fish akin to the roach, also called the red-eye.
rudder (rŭd'ə), *n.* a flat wooden or metal framework or solid piece hinged to the stern of a boat or ship and serving as a means of steering; a vertical moving surface in the tail of an aeroplane for providing directional control and stability; any steering device; a principle etc., which guides, governs or directs the course of anything. **rudderless** (-lis), *a.*

ruddily, ruddiness RUDDY.
ruddle (rŭd'l), **raddle** (rad'l), **reddle** (red'l), *n.* a variety of red ochre used for marking sheep. *v.t.* to colour or mark with ruddle.
ruddy (rŭd'i), *a.* of a red or reddish colour; of a healthy complexion, fresh-coloured; (*euphem.*) bloody. **ruddily,** *adv.* **ruddiness,** *n.*
rude (rood), *a.* simple, primitive, crude, uncultivated, uncivilized, unsophisticated, unrefined; coarse, rough, rugged; unformed, unfinished; coarse in manners, uncouth; impolite, uncivil, insolent, offensive, insulting; violent, boisterous, abrupt, ungentle, tempestuous; hearty, robust, strong. **rudely,** *adv.* **rudeness,** *n.* **rudery,** *n.*
rudiment (roo'dimənt), *n.* (*often pl.*) an elementary or first principle of knowledge etc.; (*often pl.*) the undeveloped or imperfect form of something, a beginning, a germ; a partially-developed, aborted or stunted organ, structure etc., a vestige. **rudimental, rudimentary** (-men'-), *a.* **rudimentarily,** *adv.*
rue[1] (roo), *n.* a shrubby evergreen plant, of rank smell and acrid taste, formerly used as a stimulant etc. in medicine.
rue[2] (roo), *v.t.* (*pres. p.* **rueing, ruing**) to grieve or be sorry for, to regret, to repent of. **rueful,** *a.* **ruefully,** *adv.* **ruefulness,** *n.*
ruff[1] (rŭf), *n.* an act of trumping when one cannot follow suit.
ruff[2] (rŭf), *n.* a broad plaited or fluted collar or frill of linen or muslin worn by both sexes, esp. in the 16th cent.; anything similarly puckered or plaited; a growth like a ruff, as the ring of feathers round the necks of some birds; a bird of the sandpiper family (perh. from the conspicuous ruff in the male in the breeding season); a breed of pigeons related to the jacobin. **ruff-like,** *a.*
ruff[3], **ruffe** (rŭf), *n.* a small freshwater fish related to and resembling the perch.
ruffian (rŭf'iən), *n.* a lawless, brutal person, a bully, a violent hoodlum. **ruffianism,** *n.* **ruffianly,** *a.*
ruffle (rŭf'l), *v.t.* to disorder, to disturb the smoothness or order of, to rumple, to disarrange; to annoy, to disturb, to upset, to discompose; to turn the pages (of a book) quickly. *v.i.* to grow rough or turbulent. *n.* a strip or frill of fine plaited or goffered lace etc., attached to some part of a garment, esp. at the neck or wrist; a ruff; a low, vibrating beat of the drum; a wrinkle, a surface irregularity. **ruffled,** *a.* **ruffling,** *n.*, *a.*
rufous (roo'fəs), *a.* of a brownish or yellowish red.
rug (rŭg), *n.* a thick, heavy wrap, coverlet, blanket etc. a carpet or floor-mat, usu. woollen or of skin with the hair or wool left on; of similar material; (*N Am.*) a carpet; (*coll.*) a false hairpiece, a wig. **to pull the rug (out) from under,** to put (someone) in a defenceless or discomposed state, to undermine someone.
ruga (roo'gə), *n.* (*pl.* **-gae** (-gē)) (*Anat.*) a wrinkle, crease, fold or ridge. **rugose** (-gōs), *a.* wrinkled, ridged, corrugated. **rugosely,** *adv.* **rugosity** (-gos'-), *n.*
Rugby, rugby, Rugby football (rŭg'bi), *n.* a game of football in which players are allowed to use their hands in carrying and passing an oval ball and tackling their opponents. **Rugby League,** *n.* a form of rugby played by two teams each consisting of 13 players of amateur or professional status. **Rugby Union,** *n.* a form of rugby played by teams of 15 players of amateur status. **rugger,** *n.* (*coll.*) rugby.
rugged (rŭg'id), *a.* having a surface full of inequalities, extremely uneven, broken and irregular; rocky, craggy, of abrupt contour; strongly marked (of features); rough in temper, stern, unbending, severe; rude, unpolished; strenuous, hard; hardy, sturdy. **ruggedly,** *adv.* **ruggedness,** *n.*
rugose, rugosity RUGA.
ruin (roo'in), *n.* a disastrous change or state of wreck or disaster, overthrow, downfall; bankruptcy; a cause of destruction, downfall or disaster, havoc, bane; the state of being ruined or destroyed; the remains of a structure, building, city etc. that has become demolished or decayed (*often in pl.*); a person who has suffered a downfall, e.g. a bankrupt. **ruination,** *n.* **ruined,** *a.* **ruiner,** *n.* **ruining,** *a.*, *n.* **ruinous,** *a.* fallen into ruin, dilapidated; causing ruin, baneful, destructive, pernicious. **ruinously,** *adv.* **ruinousness,** *n.*
rule (rool), *n.* the act of ruling or the state or period of being ruled, government, authority, sway, direction, control; that which is established as a principle, standard or guide of action or procedure; a line of conduct, a regular practice, an established custom, canon or maxim; method, regularity; an authoritative form, direction or regulation, a body of laws or regulations, to be observed by an association, religious order etc. and its individual members; a strip of wood, plastic, metal etc. usu. graduated in inches or centimetres and fractions of an inch or millimetres, used for linear measurement or guidance; a prescribed formula, method etc. for solving a mathematical problem of a given kind; an order, direction or decision by a judge or court, usu. with reference to a particular case only; in printing, a thin metal strip for separating columns, headings etc.; the line printed with this; the general way of things; (*pl.*) Australian football. *v.t.* to govern, to manage, to control; to curb, to restrain; to be the rulers, governors or sovereign of; to lay down as a rule or as an authoritative decision; to mark (paper etc.) with straight lines. *v.i.* to exercise supreme power (usu. over); to decide, make a decision; to dominate, to be prevalent. **as a rule,** usually, generally. **rule of three,** (*Arith.*) simple proportion. **rule of thumb,** practical experience, as dist. from theory, as a guide in doing anything. **to rule out,** to exclude, to eliminate (as a possibility). **to rule the roast,** to be the leader, to be dominant. **rulable,** *a.* capable of being ruled; (*N Am.*) permissible, allowable. **ruler,** *n.* one who rules or governs; an instrument with straight edges or sides, used as a guide in drawing straight lines, a rule. **ruling,** *n.* a authoritative decision, esp. with regard to a special legal case; a ruled line or lines. *a.* having or exercising authority or control; predominant, pre-eminent.
rum[1] (rŭm), *n.* a spirit distilled from fermented molasses or cane-juice. **rum baba** BABA. **rum-butter,** *n.* butter mixed with sugar and flavoured with rum. **rum-punch, -toddy,** *n.* a punch or toddy made with rum.
rum[2] (rŭm), *a.* (*sl.*) strange, singular, odd, queer. **rumly,** *adv.* **rumness,** *n.*
Rumanian, Romanian, Roumanian (rumā'miən), *a.* of or pertaining to the country of Rumania, its people or language. *n.* the language of Rumania; a native or inhabitant of Rumania.
rumba, rhumba (rŭm'bə), *n.* a complex and rhythmic Cuban dance; a ballroom dance developed from this; a piece of music for this dance.
rumble (rŭm'bl), *v.i.* to make a low, heavy, continuous sound, as of thunder, heavy vehicles etc.; to move (along) with such a sound; (*N Am.*, *sl.*) to be involved in a gang fight. *v.t.* to cause to move with a rumbling noise; to utter with such a sound; (*sl.*) to be undeceived about; to see through. *n.* a rumbling sound; (*N. Am.*, *sl.*) a gang fight. *n.* (*N Am.*) a dicky; an outside folding-seat on some early motor vehicles. **rumbler,** *n.* **rumbling,** *a.* **rumbly,** *a.*
rumbustious (rŭmbŭs'chəs), *a.* (*coll.*) boisterous, turbulent. **rumbustiously,** *adv.* **rumbustiousness,** *n.*
rumen (roo'mən), *n.* (*pl.* **-mens, -mina** (-minə)) the first cavity of the complex stomach of a ruminant.
ruminant (roo'minənt), *n.* any member of the division of cud-chewing animals with a complex stomach comprising the ox, sheep, deer etc.; any other cud-chewing animal (e.g. the camel). **ruminate,** *v.i.* to chew over the cud; to

muse, to meditate. *v.t.* to chew over again (what has been regurgitated); to ponder over. **rumination**, *n.* **ruminative**, *a.* **ruminatively**, *adv.* **ruminator**, *n.*
rummage (rŭm'ij), *v.t.*, *v.i.* to make a careful search (in or through), esp. by throwing the contents about; to disarrange or throw into disorder by searching. *n.* the act of rummaging, a search; miscellaneous articles, odds and ends (got by rummaging) **rummage sale**, *n.* (*chiefly N Am.*) a sale of miscellaneous articles, esp. in aid of charity, a jumble sale. **rummager**, *n.*
rummer (rŭm'ə), *n.* a large drinking-glass.
rummy (rŭm'i), *n.* any of several card-games in which the object is to collect combinations and sequences of cards.
rumness RUM².
rumour, (*N Am.*) **rumor** (roo'mə), *n.* popular report, hearsay, common talk, gossip; a current story without any known authority. *v.t.* to report or circulate as a rumour.
rumourmonger, *n.* (*coll.*) one who spreads rumours.
rump (rŭmp), *n.* the end of the backbone with the adjacent parts, the posterior, the buttocks (usu. of beasts or contemptuously of human beings); in birds, the uropygium; the fag- or tail-end of anything. **the Rump**, the remnant of the Long Parliament, after the expulsion of those favourable to Charles I by Pride's Purge in 1648, or after its restoration in 1659. **rump steak**, *n.* beefsteak cut from the rump. **rumpless** (-lis), *a.*
rumple (rŭm'pl), *v.t.* to wrinkle, to make uneven, to crease, to disorder clothes etc. *n.* a fold, a crease, a wrinkle. **rumply**, *a.*
rumpus (rŭm'pəs), *n.* a disturbance, an uproar, a row. **rumpus room**, *n.* (*chiefly N Am.*) a play or games room esp. for children.
run (rŭn), *v.i.* (*past* **ran** (ran), *p.p.* **run**) to move or pass over the ground by using the legs more quickly than in walking, esp. with a springing motion, so that both feet are never on the ground at once; to hasten; to amble, trot, or canter (of horses etc.); to flee, to try to escape; to make a run at cricket; to compete in a race; to complete a race in a specific position; to seek election etc.; to move or travel rapidly; to make a quick or casual trip or visit; to be carried along violently; to move along on or as on wheels; to revolve; to be in continuous motion, to be in action or operation; to go smoothly; to glide, to elapse; to flow; to fuse, to melt, to dissolve and spread; to flow (with), to be wet, to drip, to emit liquid, mucus etc.; to go, to ply; to spread or circulate rapidly or in profusion; of a shoal of fish, to migrate, esp. upstream for spawning; to range; to move from point to point; to rove; to extend, to take a certain course, to proceed, to go on, to continue (for a certain distance or duration); to pass or develop (into etc.); to play, feature, print or publish; to tend, to incline; to be current, valid, in force or effect; to occur inherently, persistently or repeatedly; to pass freely or casually; to occur in sequence; to perform, execute quickly or in sequence; to be allowed to wander unrestrainedly or grow (wild); to pass into a certain condition or reach a specific state; to elapse; of a loan, debt etc. to accumulate; to ladder, to unravel; to sail before the wind. *v.t.* to cause to run or go; to cause or allow to pass, penetrate etc., to thrust with; to drive, to propel; to track; to pursue, to chase, to hunt; to press (hard) in a race, competition etc.; to accomplish (as if) by running, to perform or execute (a race, an errand etc.), to follow or pursue (a course etc.); to cause to ply; to bring to a specific state (as if) by running; to keep going, to manage, to conduct, to carry on, to work, to operate; to enter or enrol (as a contender); to introduce or promote the election of (a candidate); to get past or through (e.g. a blockade); to cross, to traverse; to cause to extend or continue; to discharge; to flow with; to cause to pour or flow from; to fill (a bath) from a flowing tap; to convey in a motor vehicle, to give a lift; to pass through a process, routine or treatment; to be affected by or subjected to; to sail with the wind; to graze animals (in open pasture); in billiards, cricket etc., to hit or score a successful sequence of shots, runs etc.; to sew quickly; to have or keep current; to publish; to cast, to found, to mould; to deal in; to smuggle; to incur, to expose oneself to, to hazard; to allow a bill etc. to accumulate before paying. *n.* an act or spell of running; the distance or duration of a run or journey; a trip, a short excursion; in a vehicle the running of two batsmen from one wicket to the other in cricket without either's being put out; a unit of score in cricket; the distance a golf ball rolls along the ground; a complete circuit of the bases by a player in baseball etc.; a continuous course, a sustained period of operation or performance; a sequence series, stretch or succession (e.g. of cards, luck etc.); a succession of demands (on a bank etc.); a pipe or course for flowing liquid; the ordinary succession, trend or general direction, the way things tend to move; a ladder or rip in a stocking, jumper, pair of tights etc.; general nature, character, class or type; a batch, flock, drove or shoal of animals, fish etc. in natural migration; a periodical passage or migration; an inclined course esp. for winter sports; a habitual course or circuit; a regular track (of certain animals), a burrow; a grazing-ground; an enclosure for fowls; free use or access, unrestricted enjoyment; an attempt; a mission involving travel (e.g. a smuggling operation, a bombing run); (*Mus.*) a roulade. **at a run,** running, in haste. **in the long run**, eventually. **in the short run**, in the short term. **on the run**, in flight, fugitive. **the runs**, (*coll.*) diarrhoea. **to run across**, to traverse at a run; to encounter by chance, to discover by accident. **to run after**, to pursue with attentions; to chase. **to run along**, to leave, to go away. **to run at**, to rush at, to attack. **to run a temperature**, to have an abnormally high body temperature. **to run away**, to flee, to abscond, to elope. **to run away with**, to win an easy victory. **to run down**, to stop through not being wound up, recharged etc.; to become enfeebled by overwork etc.; to pursue and overtake; to search for and discover; to disparage, to abuse; to run against or over and sink or collide with. **to run dry**, to stop flowing; to end (of a supply). **to run for it**, to make an escape attempt, to run away. **to run foul of** FOUL. **to run in**, to call, to drop in; to arrest, to take into custody; to break in (e.g. a motor vehicle, machine) by running or operating; to insert (e.g. printed matter); to approach. **to run in the family**, to be hereditary. **to run into**, to incur, to fall into; to collide with; to reach (a specified number, amount etc.); to meet by chance. **to run into the ground**, to exhaust or wear out with overwork. **to run off**, to print; to cause to pour or flow out. **to run off with**, to elope with; to steal, remove. **to run on**, to talk volubly or incessantly; to be absorbed by (of the mind); to continue without a break. **to run out**, to come to an end; to leak. **to run out on**, to abandon. **to run over**, to review or examine cursorily; to recapitulate; to overflow; to pass, ride or drive over. **to run rings around** RING. **to run riot** RIOT. **to run short** SHORT. **to run the show**, to manage; to have control of something in one's own hands. **to run through**, to go through or examine rapidly; to squander; to pervade; to transfix; to pierce with a weapon; to strike out by drawing a line through. **to run to**, to extend to. **to run to earth, ground**, to track down, to find after hard or prolonged searching. **to run together**, to fuse, blend, mix. **to run to seed** SEED. **to run up**, to increase quickly, to accumulate (a debt etc.); to build, make or sew in a hasty manner; to raise or hoist. **runabout**, *n.* a light motor-car or aeroplane; **runaround**, *n.* evasive and deceitful treatment. **runaway**, *n.* one who runs away, a deserter, a fugitive; a bolting horse; an escape, a flight. *a.* breaking from restraint; out of control, rising quickly (e.g. of prices); prodigious, decisive (e.g. of a success); easily won; fleeing as a runaway. **run-

down, *a*. exhausted, worn out; dilapidated. **run-down**, *a*. exhausted, worn out; dilapidated. *n*. a brief or rapid resumé; a reduction in number, speed, power etc. **run-in**, *n*. (*coll*.) an argument, a row, a contention; an approach; an insertion of printed matter within a paragraph. **run-off**, *n*. overflow or liquid drained off; an additional tie-breaking contest, race etc. **run-of-the-mill**, *a*. undistinguished, ordinary, mediocre. **run-on**, *n*. continuous printed matter; an additional word, quantity, expense etc. **run-through**, *n*. a quick examination, perusal or rehearsal. **run-up**, *n*. an approach; a period preceding an event etc., e.g. a general election. **runway**, *n*. a landing-strip for aircraft; a ramp, passageway or chute. **runner**, *n*. one who runs; a racer, a messenger, a scout, a spy; one who solicits custom etc., an agent, a collector, a tout; a smuggler; that on which anything runs, revolves, slides etc.; the blade of a skate; a piece of wood or metal on which a sleigh runs; a groove, rod, roller etc. on which a part slides or runs, esp. in machinery; a creeping stem thrown out by a plant, such as a strawberry, tending to take root; a twining or climbing plant, esp. a kidney bean; a longer strip of carpet for a passage etc., or cloth for a table etc. **to do a runner**, (*sl*.) to abscond, run away, leave clandestinely or quickly. **runner bean**, *n*. a trailing bean also called scarlet runner. **runner-up**, *n*. (*pl*. **runners-up**) the unsuccessful competitor in a final who takes second place. **running**, *n*. the act of one who or that which runs; smuggling; management, control, operation; maintenance, working order; competition, chance of winning a race etc. *a*. moving at a run; cursive (of handwriting); flowing, continuous, uninterrupted; discharging matter; following in succession, repeated; trailing (of plants); current, done at, or accomplished with, a run. **take a running jump**, under no circumstances; go away. **in** or **out of the running**, having or not having a chance of winning. **to make the running**, to set the pace. **running battle**, *n*. a battle between pursuers and pursued; a continuous or long-running argument. **running-board**, *n*. the footboard of an (early) motor-car. **running commentary**, *n*. an oral description, usu. by broadcasting, of an event in progress, e.g. a race. **running head, title**, *n*. the title of a book used as a headline throughout. **running knot**, *n*. a knot which slips along the rope, line etc. **running lights**, *n.pl*. lights visible on moving vehicles, vessels or aircraft at night. **running mate**, *n*. a horse teamed or paired with another; a subordinate candidate, one standing for the less important of two linked offices in a US election, esp. the vice-presidency. **running repairs**, *n.pl*. repairs carried out while a machine etc. is in operation, minor repairs. **running stitch**, *n*. a simple continuous stitch used for gathering or tacking. **runningly**, *adv*. **runny**, *a*.
runcible spoon (rŭn'sibl), *n*. a three-pronged fork hollowed out like a spoon and with one of the prongs having a cutting edge. [nonsense word invented by Edward Lear]
rune (roon), *n*. a letter or character of the earliest Teutonic alphabet or futhorc, formed from the Greek alphabet by modifying the shape to suit carving, used chiefly by the Scandinavians and Anglo-Saxons; any mysterious mark or symbol; a canto or division in Finnish poetry. **runic**, *a*. of, pertaining to, consisting of, written or cut in runes.
rung[1] (rŭng), *n*. a stick or bar forming a step in a ladder; a rail or spoke in a chair etc.; a level, a stage, a position. **rungless** (-lis), *a*.
rung[2] (rŭng), *past, p.p.* RING[2].
runic RUNE.
runnel (rŭn'l), *n*. a rivulet, a little brook; a gutter.
runt (rŭnt), *n*. the smallest or feeblest animal in a litter esp. a piglet; a large variety of domestic pigeon; (*derog*.) any animal or person who is stunted in growth, deficient or inferior. **runtish, runty**, *a*. **runtiness**, *n*.
rupee (rupē'), *n*. the standard monetary unit of various Asian countries including India, Pakistan, Sri Lanka, Nepal, Bhutan, the Maldives, Mauritius and the Seychelles.
rupiah (roo'piə), *n*. (*pl*. **-ah, -ahs**) the standard monetary unit of Indonesia.
rupture (rŭp'chə), *n*. the act of breaking or the state of being broken or violently parted, a break, a breach; a breach or interruption of concord or friendly relations; hernia. *v.t.* to burst, to break, to separate by violence; to sever (a friendship etc.); to affect with hernia. *v.i.* to suffer a breach or disruption. **rupturable**, *a*.
rural (roo'rəl), *a*. pertaining to the country as distinguished from town; pastoral, agricultural; suiting or resembling the country, rustic. **rural dean**, *n*. a clergyman, ranking below an archdeacon, charged with the inspection of a district. **ruralism**, *n*. **ruralist**, *n*. **rurality** (-ral'-), **ruralize, -ise**, *v.i., v.t.* **ruralization**, *n*. **rurally**, *adv*.
Ruritania (rooritā'niə), *n*. a fictitious state in SE Europe, scene of great adventures invented by Anthony Hope in *The Prisoner of Zenda;* an imaginary kingdom. **Ruritanian**, *n., a*.
ruse (rooz), *n*. a stratagem, artifice, trick or wile.
rush[1] (rŭsh), *n*. a plant with long thin stems or leaves, growing mostly on wet ground, used for making baskets, mats, seats for chairs etc.; a stem of this plant; applied to various other similar plants, e.g. the bulrush; a rushlight; something of little or no worth. *a*. (made) of rush or rushes. **rush-candle, -light**, *n*. a small candle made of the pith of a rush dipped in tallow; any weak flickering light. **rushlike**, *a*. **rushy**, *a*.
rush[2] (rŭsh), *v.t.* to drive, urge, force, move or push with violence and haste, to hurry; to perform or complete quickly; to take by sudden assault; to surmount, to pass, to seize and occupy, with dash or suddenness; (*coll*.) to cheat, to swindle; (*coll*.) to overcharge. *v.i.* to move or run impetuously or precipitately; to enter or go (into) with undue eagerness or lack of consideration; to run, flow or roll with violence and impetuosity. *n*. the act of rushing; a violent or impetuous movement, advance, dash or onslaught; a sudden onset of activity, movement or thronging of people (to a gold-field etc.); (*usu. pl*.) the first print from a film; a violent demand (for) or run (on) a commodity etc.; (*sl*.) a surge of euphoria induced by a drug. *a*. characterized by or requiring much activity, speed or urgency. **to rush one's fences**, to act too hastily or precipitously. **rush hour**, *n*. a period when traffic is very congested owing to people going to or leaving work. **rusher**, *n*.
rusk (rŭsk), *n*. a piece of bread or cake crisped and browned in the oven, often given as baby food.
Russ., (*abbr*.) Russia; Russian.
russet (rŭs'it), *a*. of a reddish-brown colour. *n*. a reddish-brown colour; a rough-skinned reddish-brown variety of apple. **russety**, *a*.
Russian (rŭsh'ən), *a*. of or pertaining to Russia, its people or their language. *n*. a native or inhabitant of Russia; the Russian language. **russia leather**, *n*. a soft leather made from hides prepared with birch-bark oil. **Russian roulette**, *n*. a test of mettle or act of bravado involving firing a revolver loaded with a single bullet at one's own head after spinning the chamber; any dangerous or foolish undertaking. **Russian salad**, *n*. a salad of pickles and diced vegetables in mayonnaise dressing. **Russianization, -isation**, *n*. **Russianize**, *v.t.* **Russki, Russky** (rŭs'ki), *n., a*. (*offensive*) Russian. **Russo-**, *comb. form*.
rust (rŭst), *n*. the red incrustation on iron or steel caused by its oxidation when exposed to air and moisture; any similar incrustation on metals; any corrosive or injurious accretion or influence; a dull or impaired condition due to idleness etc.; a plant disease caused by parasitic fungi of the order Uredinales, blight; any of these fungi; the colour of rust, an orangey-red shade of brown. *a*. rust-coloured. *v.i.* to contract rust; to be oxidated; to be

attacked by blight; to degenerate through idleness or disuse. *v.t.* to affect with rust, to corrode; to impair by idleness, disuse etc. **rust-coloured**, *a.* **rust-proof**, *a.* impervious to corrosion. **rust-proofing**, *n.* treatment against rusting. **rusted**, *a.* **rustily**, *adv.* **rustiness**, *n.* **rustless** (-lis), *a.* **rusty**, *a.* covered or affected with or as with rust; rust-coloured; faded, discoloured by age; antiquated in appearance; harsh, husky (of the voice); impaired by disuse, inaction, neglect etc. **rustily**, *adv.* **rustiness**, *n.*
rustic (rŭs'tik), *a.* pertaining to the country, rural; like or characteristic of country life or people, unsophisticated, simple, artless; rude, unpolished; awkward, uncouth, clownish; of rough workmanship, coarse, plain; rusticated. *n.* a country person or dweller; an artless, unsophisticated, uncouth or clownish person; rustic work. **rustic work**, *n.* woodwork or masonry with a characteristically rough surface. **rustically**, *adv.* **rusticate**, *v.i.* to retire to or to dwell in the country; to become rustic. *v.t.* to suspend for a time from residence at a university or exile to the country, as a punishment; to make rustic in style, finish etc. **rusticated**, *a.* **rustication**, *n.* **rusticator**, *n.* **rusticity** (-tis'-), *n.*
rustle (rŭs'l), *v.i.* to make a quick succession of small sounds like the rubbing of silk or dry leaves; to move or go along with this sound; (*N Am.*) to bustle, to move quickly and energetically; (*N Am.*) to steal cattle. *v.t.* to cause to make this sound; to acquire by rustling. *n.* a rustling sound. **to rustle up**, to gather up, to put together; to prepare or make quickly, or without preparation or

prior notice. **rustler**, *n.* one who or that which rustles; (*N Am.*) a pushing, bustling person; (*N Am.*) a cattle thief. **rustling**, *n.*, *a.* **rustlingly**, *adv.*
rut[1] (rŭt), *n.* the sexual excitement or heat of deer and some other animals. *v.i.* (*past*, *p.p.* **rutted**) to be in a period of this. **rutting**, *n.*, *a.* **ruttish**, *a.* lustful, libidinous, lewd. **ruttishness**, *n.*
rut[2] (rŭt), *n.* a sunken track made by wheels or vehicles; a hollow, a groove; a settled habit or course. *v.t.* (*past*, *p.p.* **rutted**) to make ruts in. **to be in a rut**, to be stuck in tedious routine.
rutabaga (roo'təbā'gə), *n.* (*N Am.*) the Swedish turnip, the swede.
ruth (rooth), *n.* pity, compassion; remorse, sorrow. **ruthless** (-lis), *a.* pitiless, merciless, cruel, barbarous. **ruthlessly**, *adv.* **ruthlessness**, *n.*
ruthenium (ruthē'niəm), *n.* a white, spongy metallic element of the platinum group, at. no. 44; chem. symbol Ru.
rutile (roo'tīl), *n.* red dioxide of titanium.
RV, (*abbr.*) Revised Version (of the Bible).
-ry (-ri), *suf.* shortened form of -ERY, as in *Englishry, poultry, yeomanry.*
rye (rī), *n.* the seeds or grain of a cereal allied to wheat, used to make (black) bread, whisky etc.; the plant bearing this; rye whisky; (*N Am.*) rye bread. *a.* of rye. **rye bread**, *n.* bread (white or dark) made from rye flour. **rye flour**, *n.* flour made from rye grain. **rye-grass**, *n.* one of various grasses, cultivated for fodder grass. **rye whisky**, *n.* whisky distilled from rye grain.

S

S, s, the 19th letter of the English alphabet (*pl.* **Ss, S's, Esses**) in an S-shaped object or curve. **collar of SS** COLLAR.
S¹, (*abbr.*) Sabbath; Saint; Saxon; siemens; Signor; society; South, Southern; sun.
S², (*chem. symbol*) sulphur.
s, (*abbr.*) second; shilling; singular; snow; son; succeeded.
-s¹ (-s, -z), *suf.* forming plurals of most nouns.
-s² (-s, -z), *suf.* forming third pers. sing. pres. tense of most verbs.
-'s¹ (-s, -z), *suf.* forming possessives of sing. nouns and pl. nouns not ending in *s*.
-'s² (-z), *suf.* short for is, has; us.
SA, (*abbr.*) Salvation Army; sex appeal; limited liability company [F *société anonyme*]; South Africa; South America; South Australia.
Sabaoth (sab'āoth, -bā'-), *n.* (*Bibl.*) hosts, armies (in the title 'Lord God of Sabaoth').
sabbat (sab'ət), SABBATH.
Sabbatarian (sabəteə'riən), *n.* a Jew who strictly observes the seventh day of the week; a Christian who observes Sunday as a Sabbath, or who is specially strict in its observance. *a.* observing or inculcating the observance of the Sabbath or Sunday. **Sabbatarianism**, *n.*
Sabbath (sab'əth), *n.* the seventh day of the week, Saturday, set apart, esp. by the Jews, for rest and divine worship; the Christian Sunday observed as a day of rest and worship; a time of rest. (**sabbath, witches' Sabbath, sabbat**) midnight assembly of witches, wizards and demons, supposed to be convoked by the devil. **Sabbath-breaker**, *n.* one who profanes the Sabbath. **Sabbath-breaking**, *n.* **Sabbath Day**, *n.* the Jewish Sabbath (Saturday); Sunday. **Sabbatic, -ical** (-bat'-), *a.* pertaining to or befitting the Sabbath. **sabbatical**, *n.* an extended period of leave from one's work. **Sabbatical year**, *n.* every seventh year, during which the Hebrews were not to sow their fields or prune their vineyards, and were to liberate slaves and debtors; a year's leave of absence orig. granted every seven years esp. to university teachers. **Sabbatism**, *n.* **Sabbatismal** (-tiz'-), *a.*
sabelline (səbel'in), *a.* pertaining to the sable; coloured like its fur.
saber SABRE.
sabin (sab'in, sā'-), *n.* a unit of acoustic absorption. [Wallace C *Sabine*, 1868–1919, US physicist]
Sabine (sab'īn), *n.* one of an ancient Italian race inhabiting the central Apennines. *a.* of or pertaining to this people.
sable (sā'bl), *n.* a small Arctic and sub-Arctic carnivorous quadruped, *Mustela zibellina*, allied to the marten, the brown fur of which is very highly valued; its skin or fur; a painter's brush made of its hair; (*Her.*) black; (*poet.*) black, esp. as the colour of mourning; (*pl.*) mourning garments. *a.* black; (*poet.*) dark, gloomy. *v.t.* to make dark or dismal. **sable-coloured**, *a.* black. **sable-stoled, sable-vested**, *a.* clothed in sables. **sabled**, *a.*
sabot (sab'ō), *n.* a wooden shoe, usu. made in one piece, worn by peasantry etc. in France, Belgium etc.; a wooden-soled shoe.
sabotage (sab'ətahzh), *n.* malicious damage to a railway, industrial plant, machinery etc., as a protest by discontented workers, or as a non-military act of warfare; any action designed to hinder or undermine. **saboteur** (-tœ'), *n.* one who commits sabotage.
Sabra (sah'brə), *n.* an Israeli born in Israel. [Heb.]
sabre, (*esp. N Am.*) **saber** (sā'bə), *n.* a cavalry sword having a curved blade; (*pl.*) cavalry; a light fencing sword with a tapering blade. *v.t.* to cut or strike down or kill with the sabre. **sabre-bill, -wing**, *n.* S American birds. **sabre-fish**, *n.* the silver eel, *Trichiurus lepturus*. **sabre-rattling**, *n.* a display of military power or aggression. **sabre-toothed tiger**, *n.* a large extinct feline mammal with long upper canines.
sac (sak), *n.* a pouch, a cavity or receptacle in an animal or vegetable; a pouch forming the envelope of a tumour, cyst etc. **saccate** (-āt), **sacciform** (sak'sifawm), *a.* having the form of a pouch.
saccate SAC².
saccharine, saccharin (sak'ərin), *n.* an intensely sweet compound obtained from toluene, a product of coal-tar, used as a sugar substitute in food. *a.* pertaining to sugar; having the qualities of sugar; sickly sweet, sugary; ingratiatingly pleasant or polite. **saccharic** (-ka'-), *a.* pertaining to or obtained from sugar. **saccharide** (-rīd), *n.* a carbohydrate, esp. a sugar. **sacchariferous** (-rif'-), *a.* producing or containing sugar. **saccharify** (-ka'rifī), *v.t.* to break down into simple sugars. **saccharimeter** (-rim'itə), *n.* an instrument for determining the quantity of sugar in solutions, esp. by means of a polarized light. **saccharimetry** (-tri), *n.* **saccharite** (-rīt), *n.* a white or whitish granular variety of feldspar. **saccharoid** (-roid), *a.* (*Geol.*) having a granular structure. *n.* a sugar-like substance. **saccharoidal** (-roi'-), *a.* **saccharometer** (-rom'itə), *n.* a saccharimeter, esp. a hydrometer for measuring sugar concentration. **saccharose** (-rōs), *n.* sucrose. **saccharous**, *a.* **Saccharum** (-rəm), *n.* an invert sugar obtained from cane sugar; a genus of grasses comprising the sugar-cane.
sacchar(o)-, *comb. form.* sugar.
saccharose, Saccharum SACCHARINE.
sacciform SAC².
saccule (sak'ūl), **-ulus** (-ləs), *n.* (*pl.* **-les, -li** (-lē)) a small sac, esp. the smaller of two cavities in the labyrinth of the inner ear. **saccular, sacculate, -ated**, *a.* **sacculation**, *n.*
sacerdotal (sasədō'təl), *a.* pertaining to priests or the priesthood; priestly; attributing sacrificial power and supernatural or sacred character to priests; claiming or suggesting excessive emphasis on the authority of the priesthood. **sacerdotalism**, *n.* **sacerdotalist**, *n.* **sacerdotalize, -ise**, *v.t.* **sacerdotally**, *adv.*
sachem (sā'chəm), *n.* a chief of certain tribes of N American Indians; a magnate, a prominent person; (*US*) one of the governing officers of the Tammany Society in New York City. **sachemship**, *n.*
sachet (sash'ā), *n.* a small ornamental bag or other receptacle containing perfumed powder for scenting clothes etc.; a small packet of shampoo etc.
sack¹ (sak), *n.* a large, usu. oblong bag of strong coarse material, for holding corn, raw cotton, wool etc.; the quantity a sack contains, as a unit of capacity and weight; a sack together with its contents; a loose garment, gown or appendage to a dress, of various kinds, a sacque; a loose-fitting waistless dress; (*coll.*) dismissal from employment; (*sl.*) bed. *v.t.* to put into a sack; to give the

sack to. **to give** or **to get the sack**, to dismiss or be dismissed. **to hit the sack**, (*sl.*) to go to bed. **sackcloth**, *n.* sacking; this worn formerly in token of mourning or penitence. **sack-race**, *n.* a race in which the competitors jump along with their legs in sacks. **sackful**, *n.* **sacking**, *n.* coarse stuff of which sacks, bags etc. are made.

sack² (sak), *v.t.* to plunder or pillage (a place taken by storm); *n.* the pillaging of a captured place.

sack³ (sak), *n.* an old name for various white wines, esp. those from Spain and the Canaries.

sackbut (sak'būt), *n.* a mediaeval bass trumpet with a slide like the modern trombone; (*Bibl.*) an Aramaic musical stringed instrument.

sacking SACK¹.

sacque (sak), *n.* a loose-fitting woman's gown; a loose-fitting coat hanging from the shoulders.

sacra, sacral SACRUM.

sacrament (sak'rəmənt), *n.* a religious rite instituted as an outward and visible sign of an inward and spiritual grace (applied by the Eastern and Roman Catholic Churches to baptism, the Eucharist, confirmation, matrimony, penance, holy orders and anointing of the sick, and by most Protestants to the first two of these); the Lord's Supper, the Eucharist; the consecrated elements of the Eucharist; a sacred token, symbol, influence etc.; **sacramental** (-men'-), *a.* pertaining to or constituting a sacrament; *n.* a rite or observance ancillary or analogous to the sacraments. **sacramentalism**, *n.* the doctrine of the spiritual efficacy of the sacraments. **sacramentalist**, *n.* **sacramentality** (-tal'-), *n.* sacramental nature. **sacramentally**, *adv.* **Sacramentarian** (-teə'ri-), *a.* (*also* **sacramentarian**) relating to the sacraments or the Sacramentarians. *n.* one holding extreme or 'high' doctrines regarding the spiritual efficacy of the sacraments. **Sacramentarianism**, *n.* **sacramentary** (-men'-), *a.* pertaining to a sacrament or to the Sacramentarians. *n.* an ancient book of ritual in the Western Church, containing the rites for Mass and for the administration of the sacraments generally etc.

sacrarium² (səkreə'riəm), *n.* the complex sacrum of a bird.

sacred (sā'krid), *a.* dedicated to religious use, consecrated; dedicated or dear to a divinity; set apart, reserved or specially appropriated (to); pertaining to or hallowed by religion or religious service, holy; sanctified by religion, reverence etc., not to be profaned, inviolable. **sacred beetle**, *n.* a scarab. **Sacred College**, *n.* the collegiate body of cardinals in the Roman Catholic Church. **sacred cow**, *n.* an institution, custom, etc. regarded with excessive reverence and as beyond criticism. **sacredly**, *adv.* **sacredness**, *n.*

sacrifice (sak'rifis), *n.* the act of offering an animal, person etc., esp. by ritual slaughter, or the surrender of a valued possession to a deity, as an act of propitiation, atonement or thanksgiving; that which is so offered, a victim, an offering; the Crucifixion as Christ's offering of himself; the Eucharist as a renewal of this or as a thanksgiving; the giving up of anything for the sake of another person, object or interest; the sale of goods at a loss; a great loss or destruction (of life etc.). *v.t.* to offer to a deity as a sacrifice; to surrender for the sake of another person, object etc., to devote; (*coll.*) to sell at a much reduced price. **sacrificial** (-fish'əl), *a.* **sacrificially**, *adv.*

sacrilege (sak'rilij), *n.* the violation or profanation of sacred things, esp. larceny from a consecrated building; irreverence towards something or someone (considered) sacred. **sacrilegious** (-li'jəs), *a.* **sacrilegiously**, *adv.* **sacrilegiousness**, *n.*

sacrist (sā'krist), *n.* an officer in charge of the sacristy of a church or religious house with its contents. **sacristan** (sak'ristən), *n.* a sacrist; a sexton. **sacristy** (sak'risti), *n.* an apartment in a church in which the vestments, sacred vessels, books etc. are kept.

sacro-, *comb. form.* sacrum, sacral. **sacro-costal** (sakrōkos'təl), *a.* pertaining to the sacrum and of the nature of a rib. *n.* a sacro-costal part. **sacroiliac** (-il'iak), *a.* pertaining to the sacrum and the ilium. **sacro-pubic** (-pū'bik), *a.* pertaining to the sacrum and the pubis.

sacrosanct (sak'rəsangkt), *a.* inviolable by reason of sanctity; regarded with extreme respect, revered. **sacrosanctity** (-sangk'titi), *n.*

sacrum (sā'krəm, sak'-), *n.* (*pl.* **-cra**) a composite bone formed by the union of vertebrae at the base of the spinal column, and constituting the dorsal part of the pelvis. **sacral**, *a.*

sad (sad), *a.* (*comp.* **sadder**, *superl.* **saddest**) sorrowful, mournful; expressing sorrow; causing sorrow, unfortunate; lamentable, bad, shocking; of bread, cake etc., heavy, not well raised; shabby. **sad-eyed, -faced, -hearted**, *a.* (*poet.*) sorrowful or looking sorrowful. **sad-iron**, *n.* a solid smoothing iron. **sadden**, *v.t.* to make sad; *v.i.* to become sad. **saddish**, *a.* **sadly**, *adv.* **sadness**, *n.*

saddle (sad'l), *n.* a seat placed on an animal's back, to support a rider; a similar part of the harness of a draught animal; a seat on a cycle, agricultural machine etc.; an object resembling a saddle; a saddle-shaped marking on an animal's back; the rear part of a male fowl's back; a joint of mutton, venison etc., including the loins; a supporting piece in various machines, suspension-bridges, gun-mountings, tackle etc.; a depressed part of a ridge between two summits, a col. **in the saddle**, in full control; in occupation (of a post). *v.t.* to put a saddle on; to load or burden with a duty etc. **saddleback**, *n.* a roof or coping sloping up at both ends or with a gable at each end; a saddlebacked hill; an animal with a marking suggestive of a saddle; the hooded crow; a black pig with a white band across the back. *a.* saddlebacked. **saddlebacked**, *a.* of a horse, having a low back with an elevated neck and head; curving up at each end. **saddle-bag**, *n.* one of a pair of bags connected by straps slung across a horse etc. from the saddle; a bag attached to the back of the saddle of a bicycle etc. **saddle-blanket**, *n.* a saddle-cloth. **saddle-bow**, *n.* the pommel. **saddle-cloth**, *n.* a cloth laid on a horse under the saddle. **saddle-corporal, -sergeant**, *n.* a regimental saddler. **saddle horse**, *n.* a horse for riding. **saddle-pillar**, *n.* the saddle support of a cycle. **saddle soap**, *n.* a special type of oily soap used for cleaning and preserving leather. **saddle-sore**, *adj.* sore from riding a horse or bicycle. **saddle-tree**, *n.* the frame of a saddle; the tulip-tree. **saddleless**, *a.* **saddler**, *n.* a maker or dealer in saddles and harness. **saddlery**, *n.* the trade or shop of a saddler; saddles and harnesses collectively.

Sadducee (sad'ūsē), *n.* one of a sect among the Jews, arising in the 2nd cent. BC, who adhered to the written law to the exclusion of tradition. **Sadducean** (-sē'ən), *a.* **Sadduceeism**, *n.*

sadhu, saddhu (sah'doo), *n.* a Hindu usu. mendicant holy man.

sad-iron SAD.

sadism (sā'dizm), *n.* sexual perversion characterized by a passion for cruelty; pleasure derived from inflicting pain. **sadist**, *n.* **sadistic** (sədis'-), *a.* **sadistically**, *n.* **sadomasochism** (-dōmas'əkizm), *n.* sadism and masochism combined in one person. **sadomasochist**, *n.* **sadomasochistic** (-kis'-), *a.*

sae, (*abbr.*) stamped addressed envelope.

safari (səfah'ri), *n.* a hunting or scientific expedition, esp. in E Africa. **safari jacket**, *n.* a light, usu. cotton jacket with breast pockets and a belt. **safari park**, *n.* a park containing uncaged wild animals, such as lions and monkeys. **safari suit**, *n.* a suit having a safari jacket.

safe (sāf), *a.* free or secure from danger, damage or evil; uninjured, unharmed, sound; affording security; not dangerous, hazardous or risky; cautious, prudent, trusty;

unfailing, certain, sure; no longer dangerous, secure from escape or from doing harm. *n.* a receptacle for keeping things safe, a steel fire-proof and burglar-proof receptacle for valuables, a strong-box; a cupboard or other receptacle for keeping meat and other provisions. (*sl.*) a condom. **safe-blower, -breaker, -cracker,** *n.* one who opens safes to steal. **safe-breaking, -cracking,** *n.* **safe-conduct,** *n.* an official document or passport ensuring a safe passage, esp. in a foreign country or in time of hostilities. **safe-deposit, safety deposit,** *n.* a specially-constructed building or basement with safes for renting. **safeguard,** *n.* one who or that which protects; a proviso, precaution, circumstance etc. that tends to save loss, trouble, danger etc.; a safe-conduct, a passport. *v.t.* to make safe or secure by precaution, stipulation etc. **safe-house,** *n.* a place that can be used as a refuge. **safe-keeping,** *n.* the act of keeping or preserving in safety; secure guardianship; custody. **safe period,** *n.* the part of the menstrual cycle when conception is least likely to occur. **safe seat,** *n.* a Parliamentary seat that is certain to be held by the same party as previously. **to be on the safe side,** as a precaution. **safely,** *adv.* **safeness,** *n.*

safety (säf'ti), *n.* the state of being safe, freedom from injury, danger or risk; safe-keeping or custody; a safety-catch; **safety belt,** *n.* a seat-belt; a belt fastening a person working high up to a fixed object to prevent falling. **safety-catch,** *n.* a device to prevent the accidental operation of e.g. a lift or a firearm. **safety curtain,** *n.* a fireproof curtain in a theatre that cuts off the stage from the audience. **safety deposit,** *n.* a safe-deposit. **safety-fuse,** *n.* a fuse that allows an explosive to be fired without danger to the person igniting it. **safety glass,** *n.* glass layered with a sheet of plastic to resist shattering; glass treated to prevent splintering when broken. **safety-lamp,** *n.* a miner's lamp protected by wire or gauze so as not to ignite combustible gas. **safety-match,** *n.* a match that ignites only on a surface treated with a special ingredient. **safety net,** *n.* a net to catch tight-rope and trapeze performers if they should fall; a safeguard, precaution. **safety-pin,** *n.* a pin with a part for keeping it secure and guarding the point. **safety-razor,** *n.* one mounted on a handle with a guard to prevent cutting the skin. **safety-valve,** *n.* a valve on a boiler automatically opening to let steam escape to relieve pressure and prevent explosion; any harmless means of relieving anger, excitement etc.

safflower (saf'lowə), *n.* a thistle-like plant, *Carthamus tinctorius*, with orange flowers yielding a red dye, and seeds rich in oil.

saffron (saf'rən), *n.* the dried deep orange stigmas of a crocus, *Crocus sativus*, used for colouring and flavouring food; this plant; the colour deep orange; the meadow saffron, *Colchicum autumnale*; the bastard saffron or safflower. *a.* saffron-coloured, deep yellow. **saffrony,** *a.* **safranin, -ine** (-nin, -nēn), *n.* any of a series of basic compounds used in dyeing.

sag (sag), *v.i.* (*past, p.p.* **sagged**) to droop, to sink, to yield or give way esp. in the middle, under weight or pressure; to bend, to hang sideways; to lose vigour, to weaken; of prices, esp. of stocks, to decline. *v.t.* to cause to give way, bend, or curve sideways. *n.* the act or state of sagging or giving way; the amount of this. **saggy,** *a.*

saga (sah'gə), *n.* a mediaeval prose narrative recounting family or public events in Iceland or Scandinavia; a story of heroic adventure; a series of books relating the history of a family; (*coll.*) a long involved story or account.

sagacious (səgā'shəs), *a.* intellectually keen or quick to understand or discern, intelligent, perspicacious, shrewd, wise; of policy etc., characterized by wisdom and discernment. **sagaciously,** *adv.* **sagaciousness, sagacity** (-gas'-), *n.*

sagamore (sag'əmaw), *n.* a N American Indian chief, a sachem.

sage[1] (sāj), *n.* a grey-leaved aromatic plant of the genus *Salvia*, esp. *S. officinalis*, formerly much used in medicine, now employed in cookery. **sage-brush,** *n.* a shrubby plant of the various species of *Artemisia*, abounding in the plains in the W US. **sage-cheese,** *n.* cheese flavoured and coloured with layers of or an infusion of sage. **sage-cock, -grouse,** *n.* the largest of the American grouse, *Centrocercus urophasianus*, frequenting the sage-brush regions. **sage-green,** *n.* a greyish green. **sagy,** *a.*

sage[2] (sāj), *a.* wise, prudent; judicious, well-considered; grave, serious- or solemn-looking. *n.* a person of great wisdom, esp. one of past times with a traditional reputation for wisdom, and consulted accordingly for advice and guidance. **sagely,** *adv.* **sageness,** *n.*

saggar (sag'ə), *n.* a vessel of fireproof pottery in which delicate porcelain is enclosed for protection while in a kiln. **saggar-house,** *n.*

saggy SAG.

Sagitta (səjit'ə), *n.* (*Geom.*) the versed sine of an arc; a genus of small transparent pelagic worms; a northern constellation. **sagittal,** *a.* of or pertaining to the join between the two parietal bones forming the sides and top of the skull; in or parallel to the mid-plane of the body. **Sagittarius** (sajitεə'riəs), *n.* the Archer, a zodiacal constellation and the ninth sign of the zodiac, which the sun enters on 22 Nov. **sagittate** (saj'ität), *a.* esp. of a leaf, shaped like an arrow-head.

sago (sā'gō), *n.* (*pl.* **-gos**) a starchy substance obtained from the soft inner portion of the trunk of several palms or cycads and used as food. **sago-palm,** *n.*

sagy SAGE[1].

Saharan (səhah'rən), *adj.* of or relating to the Sahara Desert in N. Africa.

sahib (sah'ib), *n.* the title used in colonial India in addressing a European man or a man of rank.

said (sed), *past, p.p.* SAY[1].

sail (sāl), *n.* a piece of canvas or other fabric spread on rigging to catch the wind, and cause a ship or boat to move in the water; some or all of a ship's sails; a ship or vessel with sails; a specified number of ships in a squadron etc.; an excursion by sail or (*loosely*) by water; anything like a sail in form or function; the arm of a windmill; the dorsal fin of some fish; a wing. *v.i.* to move or be driven forward by the action of the wind upon sails; to be conveyed in a vessel by water; to set sail; to handle or make journeys in a vessel equipped with sails as a sport or hobby; to pass gently (along), to float (as a bird), to glide; to go along in a stately manner. *v.t.* to pass over in a ship, to navigate; to perform by sailing; to manage the navigation of (a ship); to cause to sail, to set afloat. **full sail,** with all sails set and a fair wind. **to make sail,** to set sail; to extend an additional quantity of sail. **to sail close to the wind** WIND[1]. **to sail into,** to attack vigorously. **to set sail,** to begin a voyage. **to shorten sail,** to reduce the amount of sail spread. **to strike sail,** to lower sails suddenly. **under sail,** with sails spread. **sail-arm,** *n.* an arm of a windmill. **sailboard,** *n.* a moulded board with a single mast and sail, used in windsurfing. **sailboarding,** *n.* **sailcloth,** *n.* canvas etc. for making sails; a kind of dress-material. **sail-fish,** *n.* a fish with a large dorsal fin, as the basking shark. **sail-loft,** *n.* a large apartment where sails are cut out and made. **sailplane,** *n.* a glider that rises in an upward air current. **sailplane,** *v.i.* **sail-room,** *n.* an apartment on board ship where spare sails are stowed. **sail-yard,** *n.* a horizontal spar on which sails are extended. **sailer,** *n.* a ship (with reference to her power or manner of sailing). **sailing,** *n.* **sailing-boat, -ship,** *n.* a boat or ship with sails. **sailing-master,** *n.* an officer whose duty it is to navigate a yacht etc. **sailless,** *a.*

sailor (sā'lə), *n.* a seaman, a mariner, esp. one of the crew as distinguished from an officer. **good, bad sailor,** *n.* one who is not, or who is, liable to be seasick. **sailor-hat,** *n.* a

sainfoin 730 **saliva**

flat-crowned narrow-brimmed straw hat worn by women, or one with a turned-up brim for children. **sailor-man**, *n.* (*coll.*) a seaman. **sailor's-knot**, *n.* a kind of reef-knot used in tying a neck-tie. **sailor-like, sailorly,** *a.* **sailoring,** *n.* **sailorless,** *a.*
sainfoin (sān'foin), *n.* a leguminous herb, *Onobrychis sativa*, resembling clover, grown for fodder.
saint (sānt, (*as pref. also*) sənt), *n.* a person eminent for piety and virtue, a holy person; one of the blessed in heaven; one canonized or recognized by the Church as preeminently holy and deserving of veneration; (*pl.*) the name used by the Mormons and members of some other sects in speaking of themselves. *v.t.* to canonize; to regard or address as a saint. **cross of St Anthony, St Andrew's cross, St George's cross** CROSS. **St Anthony's fire,** *n.* erysipelas. **St Bernard (dog),** *n.* a large and powerful breed of dog orig. kept by the monks of the Hospice in the Great St Bernard Pass to rescue travellers. **St Elmo's fire** (-el'mōz), *n.* the corposant. **St-John's-wort,** *n.* any plant of the genus *Hypericum*. **St Martin's summer,** *n.* a spell of mild weather in late autumn. **St Monday,** *n.* Monday turned into a holiday by workers. **saint's day,** *n.* a day dedicated to the commemoration of a particular saint, esp. the patron saint of a church, school etc. **St Stephen's,** *n.* the British parliament (so named from the chapel within the precincts of the Houses of Parliament). **St Valentine's day** VALENTINE. **St Vitus's dance** (vī'təs), *n.* DANCE. **sainthood** (-hud), *n.* **sainted,** *a.* canonized; gone to heaven; holy, pious. **saintlike, saintly,** *a.* **saintliness,** *n.* **saintship,** *n.*
St-Simonian (sāntsimō'niən), *n.* an adherent of the Comte de St-Simon (1760-1825), who advocated the establishment of State ownership and distribution of earnings according to capacity and labour. *a.* of or pertaining to his doctrines. **St-Simonianism, -Simonism** (-si'mə-), *n.*
saith (seth), *3rd sing.* SAY [1].
sake [1] (sāk), *n.* end, purpose; desire of obtaining; benefit or interest. **for God's sake,** a solemn adjuration. **for old time's sake,** in memory of days gone by. **for the sake of, for someone's** or **something's sake,** because of, out of consideration for.
sake [2], **saké, saki** (sak'i), *n.* a Japanese fermented liquor made from rice.
saki (sak'i), SAKE [2].
sal (sal), *n.* (*Chem., Pharm.*) salt (used only with qualifying word). **sal-ammoniac** (-əmō'niak), *n.* ammonium chloride. **sal volatile** (vəlat'ili), *n.* an aromatic solution of ammonium carbonate.
salaam (səlahm'), *n.* a ceremonious salutation or obeisance in Eastern countries. *v.i.* to make a salaam.
salable SALE.
salacious (səlā'shəs), *a.* lustful, lecherous; arousing lust, erotic, lewd. **salaciously,** *adv.* **salaciousness, salacity** (-las'-), *n.*
salad (sal'əd), *n.* a dish of (mixed) raw vegetables; a cold dish of precooked vegetables, or of fruit, often mixed with a dressing; any herb or other vegetable suitable for eating raw. **salad-cream,** *n.* a kind of mayonnaise. **salad days,** *n.pl.* the time of youth and inexperience. **salad-dressing,** *n.* a mixture of oil, vinegar, mustard etc., for dressing salads. **salad-oil,** *n.* a vegetable oil suitable for use in salad-dressings.
salal (sal'əl), *n.* an evergreen shrub, *Gaultheria shallon*, of California etc., bearing grape-like edible berries.
salamander (sal'əmandə), *n.* a lizard-like animal anciently believed to be able to live in fire; a spirit or genie fabled to live in fire; anyone who can stand great heat, under fire; an amphibian of the family *Urodela*. **salamandrine** (-man'drin), *a.* **salamandroid** (-man'droid), *n., a.*
salami, salame (səlah'mi), *n.* (*pl.* **-mis, -mes**) a highly-seasoned Italian sausage.
salary (sal'əri), *n.* fixed pay given periodically, usu. monthly, for work not of a manual or mechanical kind. *v.t.* to pay a salary to. **salaried,** *a.*
salchow (sal'kov), *n.* an ice-skating jump with turns in the air. [Ulrich *Salchow*, 1877-1949, Swed. skater]
sale (sāl), *n.* the act of selling; the exchange of a commodity for money or other equivalent; an event at which goods are sold; an auction; a disposal of goods at reduced prices; demand, market; (*pl.*) quantity of goods sold; (*pl.*) the activities involved in selling goods collectively. **bill of sale** BILL [3]. **sale and lease back,** *n.* a method of raising capital for a business by selling property and then renting it from the new owner. **sale of work,** a sale of home-made goods for charitable purposes. **sale or return,** an arrangement by which a retailer may return unsold goods to the wholesaler. **saleroom,** *n.* (chiefly *Brit.*) a room in which goods are displayed for sale, an auction-room. **sales-clerk,** *n.* (*chiefly N Am.*) a shop assistant. **salesman, -woman, -person,** *n.* a person employed to sell goods, esp. in a shop; a sales representative. **salesmanship,** *n.* the art of selling, skill in persuading prospective purchasers. **sales representative,** *n.* a person employed to secure orders for a company's products, usu. in an assigned geographical area. **sales resistance,** *n.* opposition of a prospective customer to purchasing a product. **sales talk, sales pitch,** *n.* persuasive or attractive arguments to influence a possible purchaser. **sales tax,** *n.* a tax levied on retail sales receipts and added to selling prices by retailers. **saleable, salable,** *a.* **saleableness, saleability** (-bil'-), *n.*
saleratus (salərā'təs), *n.* (*N Am.*) an impure bicarbonate of potash or soda, much used as baking powder.
Salian (sā'liən), *a.* of or pertaining to a Frankish tribe on the lower Rhine to which the ancestors of the Merovingians belonged. *n.* one of this tribe. **Salic** (sal'-), *a.* Salian. **Salic law** or **code,** *n.* a Frankish law-book current during the Merovingian and Carolingian periods. **Salic law, Salique** (səlēk'), **law,** *n.* a law derived from this excluding females from succession to the throne, esp. as the fundamental law of the French monarchy.
salicet (sal'iset), **salicional** (-lish'ə-), *n.* organ-stops with notes of a reedy quality. **salicetum** (-sē'təm), *n.* (*pl.* **-tums, -ta** (-tə)) a garden or arboretum of willows. [as foll.]
salicin (sal'isin), *n.* a bitter crystalline compound obtained from the bark of willows and poplars, used medicinally. **salicylate** (-lis'ilāt), *n.* a salt of salicylic acid. **salicylic** (-sil'-), *a.* derived from the willow; belonging to a series of benzene derivatives of salicin; derived from salicylic acid. **salicylic acid,** *n.* an acid whose derivatives, including aspirin, are used to relieve pain and to treat rheumatism. **silicylize, -ise** (-lis'-), *v.t.* to impregnate with salicylic acid.
salient (sā'liənt), *a.* leaping, jumping, springing; pointing or projecting outwards; conspicuous, prominent, noticeable; (*Her.*) represented in a leaping posture. *n.* a salient angle; a portion of defensive works or of a line of defence projecting towards the enemy. **salience, -ency,** *n.* **saliently,** *adv.*
saliferous (səlif'ərəs), *a.* of rock strata, bearing or producing salt. **saliferous system,** *n.* the Triassic rocks, from the deposits of salt.
saline (sā'līn), *a.* consisting of or having the characteristics of salt; containing or impregnated with salt or salts. *n.* a salina; a saline substance, esp. a purgative; a saline solution, esp. with the same concentration as body fluids. **salina** (səlī'nə), *n.* a salt-marsh, -lake, -spring etc.; saltworks. **salinlferous** (salinif'-), SALIFEROUS. **salinity** (səlin'-), *n.* **salineness,** *n.* **salinometer** (salinom'itə), *n.* an instrument for ascertaining the density of brine in the boilers of marine steam-engines.
Salique SALIC under SALIAN [2].
saliva (səlī'və), *n.* an odourless, colourless, somewhat viscid liquid secreted by glands into the mouth where it

lubricates ingested food, spittle. **salivary** (sal'i-), *a.* of or producing saliva. **salivate** (sal'i-), *v.t.* to excite an unusual secretion and discharge of saliva in, as by the use of mercury. *v.i.* to secrete or discharge saliva in excess. **salivation** (sali-), *n.*
Salk vaccine (sawlk), *n.* a vaccine against poliomyelitis. [Jonas *Salk*, b. 1914, US doctor]
sallee (sal'i), *n.* name given to several kinds of acacia; a species of eucalyptus.
sallow[1] (sal'ō), *n.* a willow-tree, esp. one of the low shrubby varieties; a willow-shoot, an osier; any of various moths feeding on willows. **sallowy**, *a.*
sallow[2] (sal'ō), *a.* of a sickly-yellowish or pale-brown colour. *v.t.* to make sallow. **sallowish**, *a.* **sallowness**, *n.*
sally[1] (sal'i), *n.* a sudden rushing out or sortie of troops from a besieged place against besiegers; an excursion; a sudden or brief outbreak of spirits etc., an outburst; a flight of fancy or wit, a bantering remark etc. *v.i.* of troops, to rush out suddenly; to go (out) on a journey, excursion etc.
sally[2] (sal'i), *n.* the part of a bell-ringer's rope covered with wool for holding; the first movement of a bell when set for ringing.
Sally-lunn (salilŭn'), *n.* a sweet tea-cake eaten hot and buttered. [*Sally Lunn*, who hawked them at Bath, *c.* 1800]
salmagundi (salməgŭn'di), *n.* a dish of chopped meat, anchovies, eggs, oil, vinegar etc.; a medley, a miscellany.
salmi, salmis (sal'mē), *n.* (*pl.* **salmis**) a ragout, esp. of game-birds stewed with wine.
salmis SALMI.
salmon (sam'ən), *n.* (*pl.* in general **salmon**; in particular **salmons**) a larger silvery, pink-fleshed fish of the genus *Salmo*, esp. *S. salar*, an anadromous fish; extended to various fish resembling the salmon. *a.* salmon-coloured. **salmon-colour**, *n.* the colour of salmon flesh, orangey-pink. **salmon-coloured**, *a.* **salmon-ladder, -leap, -pass, -stair, -weir**, *n.* a series of steps, zigzags, or other contrivances to enable salmon to get past a dam or waterfall. **salmon-parr** PARR. **salmon-trout**, *n.* an anadromous fish, *Salmo trutta*, resembling the salmon but smaller. **salmonoid** (sal'mənoid), *n.*, *a.*
salmonella (salmənel'ə), *n.* a genus of bacteria, many of which cause food poisoning. **salmonellosis** (-ō'sis), *n.* infection with bacteria of the genus *Salmonella*.
salmonet SALMON.
salon (sal'on), *n.* a reception-room, esp. in a great house in France; a periodical reunion of eminent people in the house of someone socially fashionable, esp. a lady; (*pl.*) fashionable circles; a hall for exhibiting paintings etc.; the business premises of a hairdresser, beautician etc.
saloon (səloon'), *n.* a large room or hall, esp. one suitable for social receptions, public entertainments etc., or used for a specified purpose; a large room for passengers on board ship; a closed motor-car with no internal partitions; (*esp. N Am.*) a drinking-bar, a public-house. **saloon bar**, *n.* the more reserved bar in a public-house. **saloon-pistol, -rifle**, *n.* firearms suitable for short-range practice in a shooting saloon.
salopettes (saləpets'), *n.pl.* thick usu. quilted trousers with shoulder straps, used for skiing.
Salopian (səlō'piən), *n.* a native of or inhabitant of Shropshire. *a.* pertaining to Shropshire.
salpicon (sal'pikon), *n.* a stuffing or thick sauce made with chopped meat and vegetables.
salpiglossis (salpiglos'is), *n.* a genus of S American herbaceous plants with handsome flowers.
salpinx (sal'pingks), *n.* (*pl.* **salpinges**, (-pin'jēz)) the Eustachian tube; the Fallopian tube. **salpingitis** (-ji'tis), *n.* inflammation of the Eustachian or the Fallopian tubes. **salpingitic** (-jit'-), *a.*
salsa (sal'sə), *n.* a Puerto Rican dance or the music for

this; a spicy sauce of tomatoes, onions and hot peppers.
salsify (sal'sifi), *n.* a composite plant, *Tragopogon porrifolius*, the long whitish root of which is eaten; this root.
SALT (sawlt), (*abbr.*) Strategic Arms Limitation Talks.
salt (sawlt), *n.* chloride of sodium, used for seasoning and preserving food, obtained from sea-water or brine by evaporation or in crystalline form in beds of various geological age; that which gives flavour; relish, piquancy, pungency, wit, repartee, brilliance in talk etc.; a salt-cellar; (*coll.*) a sailor; a salt-marsh or salting; a compound formed by the union of basic and acid radicals, an acid the hydrogen of which is wholly or partially replaced by a metal; (*pl.*) smelling salts; (*pl.*) any of various mineral salts used as a medicine, esp. as a purgative. *a.* impregnated or flavoured with or tasting of salt, saline; cured with salt; living or growing in salt water; of wit etc., pungent. *v.t.* to sprinkle or cover with salt; to season with salt; to cure or preserve with salt; (*Phot.*) to treat (paper etc.) with a solution of a salt; to add liveliness to (a story, etc.); to misrepresent as valuable by the addition of material, esp. to add pieces of ore etc. to (a mine) so as to represent it as profitable to work. *v.i.* to deposit salt from a saline substance. **in salt**, sprinkled with salt or steeped in brine for curing. **(not) worth one's salt**, (not) worth keeping, (not) useful. **salt of lemon**, acid oxalate of potassium. **salt of the earth**, person or persons of very great worth. **to salt an account** etc., to put down excessively high prices. **to salt away**, to save or hoard (money etc.). **with a grain or pinch of salt**, with reservations or caution. **salt bath**, *n.* a bath of molten salts used in the hardening or tempering of steel. **salt-bush**, *n.* a shrubby plant of the goosefoot family on which stock feed. **salt-cake**, *n.* crude sulphate of soda, prepared for the use of glass- and soap-makers. **salt flat**, *n.* a salt-covered flat area formed by the total evaporation of a body of water. **salt-glaze**, *n.* a glaze produced on pottery by putting salt into the kiln after firing. **salt-lake**, *n.* an inland body of salt water. **salt-lick**, *n.* a place to which cattle go to lick ground impregnated with salt. **salt-marsh**, *n.* land liable to be overflowed by the sea, esp. used for pasturage or for collecting salt. **salt-mine**, *n.* a mine for rocksalt. **salt-pan**, *n.* a shallow depression in the land in which salt water evaporates to leave salt; a vessel in which brine is evaporated at a salt-works. **salt-pit**, *n.* a pit where salt is obtained. **salt-water**, *a.* living in or pertaining to salt water, esp. the sea. **salt-works**, *n.* a factory for making salt. **salt-wort**, *n.* any of various plants of the genus *Salsola* or *Salicornia*, growing in salt-marshes and on seashores. **salter**, *n.* one who salts (fish etc.); a worker at a salt-works. **saltern** (-tən), *n.* a salt manufactory; a series of pools for evaporating sea-water. **salting**, *n.* the application of salt for preservation etc.; (*pl.*) salt-lands, a salt-marsh. **saltish**, *a.* **saltishly**, *adv.* **saltishness**, *n.* **saltless**, *a.* **saltness**, *n.* **salty**, *a.* of or containing salt; tasting (strongly) of salt; of the sea or life at sea; witty, earthy, coarse. **saltily**, *adv.* **saltiness**, *n.*
saltarello (saltərel'ō), *n.* an Italian or Spanish dance characterized by sudden skips; the music for such a dance.
saltation (saltā'shən), *n.* a leaping or bounding; an abrupt transition or variation in the form of an organism. **saltatorial** (-tətaw'ri-), **saltatorian, saltatorious, saltatory** (sal'-), *a.*
salt-cellar (sawlt'selə), *n.* a container for holding salt at table.
saltigrade (sal'tigrād), *a.* formed for leaping.
saltpetre, (*esp. N Am.*) **saltpeter** (sawltpē'tə), *n.* potassium nitrate. **Chile saltpetre**, impure sodium nitrate. **saltpetrous**, *a.*
salubrious (səlū'briəs), *a.* of climate etc., promoting health, wholesome; spiritually wholesome, respectable. **salubriously**, *adv.* **salubriousness, salubrity**, *n.*
Saluki (səloo'ki), *n.* a Persian greyhound.

salutary (sal'ūtəri), *a.* promoting good effects, beneficial, corrective, profitable; salubrious, wholesome. **salutarily**, *adv.* **salutariness**, *n.*

salute (səloot'), *v.t.* to greet with a gesture or words of welcome, respect or recognition; to accost or welcome (as with a bow, kiss, oath, volley etc.); to honour by the discharge of ordnance etc.; to show respect to (a military superior) by a salute; to praise, acknowledge. *v.i.* to perform a salute. *n.* gesture of welcome, homage, recognition etc., a salutation; a prescribed method of doing honour or paying a compliment or respect, as discharge of ordnance, dipping colours, presenting arms etc. **salutation** (salū-), *n.* the act of saluting; that which is said or done in the act of greeting; words of greeting or communicating good wishes or courteous inquiries. **salutational**, *a.* **salutatorian** (salūtətaw'ri-), *n.* a student at a N American college who pronounces the salutatory. **salutatory**, *a.* pertaining to or of the nature of a salutation; pertaining to a salutatory. *n.* an oration delivered by a graduating student at the degree-giving ceremony in N American colleges.

salvable (sal'vəbl), *a.* capable of being saved. **salvableness, salvability** (-bil'-), *n.* **salvably**, *adv.*

Salvadorian, Salvadorean (salvədaw'riən), *n.* a native or inhabitant of El Salvador. *adj.* of or relating to El Salvador.

salvage (sal'vij), *n.* the act of saving a ship, goods etc. from shipwreck, capture, fire etc.; compensation allowed for such saving; property so saved; the saving and recycling of waste or scrap material; material saved for re-use. *v.t.* to save or recover from wreck, capture, fire etc.; to save from ruin or destruction; to gain (something of value) from a failure. **salvage-money**, *n.* **salvageable**, *a.* **salvager**, *n.*

salvation (salvā'shən), *n.* the act of saving from destruction; deliverance, preservation from danger, evil etc.; deliverance of the soul, or of believers from sin and its consequences; that which delivers, preserves etc. **Salvation Army**, *n.* a religious organization on a military pattern working among the poor. **Salvationism**, *n.* the principles and practices of the Salvation Army. **Salvationist**, *n.* a member of the Salvation Army.

salve (salv), *n.* a healing ointment; anything that soothes or palliates. *v.t.* to dress or anoint with a salve; to soothe, to ease, to palliate.

salve (sal'vä, -vi), *n.* a Roman Catholic antiphon beginning with the words *Salve Regina*, 'Hail, holy Queen', addressed to the Virgin; music for this.

salver (sal'və), *n.* a tray, usu. of silver, brass, electro-plate etc., on which refreshments, visiting-cards etc. are presented.

Salvia (sal'viə), *n.* a genus of labiate plants comprising the common sage and many cultivated species with brilliant flowers.

salvo (sal'vō), *n.* (*pl.* **-voes, -vos**) a discharge of guns etc. as a salute; a volley of cheers etc.; a concentrated fire of artillery, release of missiles etc.

sal volatile SAL [1].

salvor (sal'və), *n.* a person or ship effecting salvage.

SAM (sam), (*abbr.*) surface-to-air missile.

Sam., (*Bibl. abbr.*) Samuel.

samara (səmah'rə), *n.* a one-seeded indehiscent dry fruit with wing-like extensions, produced by the sycamore, ash etc.

Samaritan (səma'ritən), *a.* pertaining to Samaria. *n.* a native or inhabitant of Samaria; the language of Samaria; one adhering to the Samaritan religious system; a kind, charitable person, in allusion to the 'good Samaritan' of the parable (Luke x.30-37); a member of a voluntary organization formed to give help to people in despair.

samarium (səmeə'riəm), *n.* a silvery-grey metallic chemical element, at. no. 62; chem. symbol Sm, one of the rare-earth metals.

samba (sam'bə), *n.* a Brazilian dance; a ballroom dance in imitation of this; music for this.

sambo (sam'bō), *n.* a person of three-quarters black African descent; (*offensive*) a black person.

Sam Browne (sam brown), *n.* a military officer's belt with a light strap over the right shoulder; a belt of similar design made from a fluorescent material and worn by motor-cyclists, cyclists etc. [Sir *Samuel Browne*, 1824–1901]

same (sām), *a.* identical; not other, not different; identical or similar in kind, quality, degree etc.; exactly alike; just mentioned, aforesaid; unchanged, unchanging, uniform, monotonous. *pron.* the same thing; the aforesaid. **all the same**, nevertheless; notwithstanding what is said, done, altered etc. **at the same time**, nevertheless, still. **sameness**, *n.* **samey** (sā'mi), *adj.* (*coll.*) monotonous, insufficiently varied.

Samian (sā'miən), *a.* pertaining to Samos. *n.* a native of Samos. **Samian ware**, *n.* red or black earthenware pottery found on Roman sites.

samisen (sam'isen), *n.* a Japanese three-stringed guitar-like instrument played with a plectrum.

samizdat (sam'izdat), *n.* the clandestine publishing of banned literature in the USSR; also, such literature.

Samoan (səmō'ən), *n.* a native or inhabitant or the language of Samoa. *a.* pertaining to Samoa.

samosa (səmō'sə), *n.* (*pl.* **samosas, samosa**) an Indian savoury of spiced meat or vegetables in a triangular pastry case.

samovar (sam'əvah), *n.* a Russian tea-urn heated by burning charcoal in an inner tube.

Samoyed (sam'əyed), *n.* a member of a Mongolian people inhabiting middle Siberia; their language; a breed of white sledge-dog. **Samoyedic** (-yed'-), *a.* pertaining to such a people. *n.* their language.

samp (samp), *n.* (*N Am.*) maize coarsely ground or made into porridge.

sampan (sam'pan), *n.* a Chinese flat-bottomed river boat, frequently used for habitation.

samphire (sam'fīə), *n.* a herb, *Crithmum maritimum*, growing on sea-cliffs, the aromatic leaves of which are pickled as a condiment; glasswort.

sample (sam'pl, sahm'-), *n.* a part taken, offered or used as illustrating the whole, a specimen, an example, a pattern, a model. *v.t.* to take samples of, to test, to try; to have an experience of. **sampler**, *n.* one who or that which takes samples; a piece of embroidered work done as a specimen of skill. **sampling**, *n.* the act of sampling; the taking of extracts from existing popular songs and putting them together to form a new one.

Samson (sam'sən), *n.* a man of abnormal strength (Judges xiv.6 *passim*).

samurai (sam'urī), *n.* (*pl.* **samurai**) a member of the military caste under the Japanese feudal regime, or a military retainer; now, an army officer.

sanatorium (sanətaw'riəm), *n.* (*esp. N Am.*) **sanitarium** (-teə'ri-), *n.* (*pl.* **-riums, -ria**, (-riə)) an institution for the treatment of chronic diseases; a place to which people resort for the sake of their health; an institution for invalids, esp. convalescents; a sick room, esp. in a boarding school. **sanatory** (san'-), *a.*

sanbenito (sanbənē'tō), *n.* a penitential garment worn by heretics who recanted, or at an auto-da-fé by persons condemned by the Inquisition.

sanctify (sangk'tifī), *v.t.* to make holy, to consecrate; to set apart or observe as holy; to purify from sin; to give a sacred character to, to sanction, to make inviolable; to render productive of holiness. **sanctification** (-fi-), *n.* **sanctifier**, *n.*

sanctimony (sangk'timəni), *n.* affectation of piety. **sanctimonious** (-mō'-), *a.* making a show of piety or saintliness. **sanctimoniously**, *adv.* **sanctimoniousness**, *n.*

sanction (sangk'shən), *n.* the act of ratifying, ratification, confirmation by superior authority; a provision for enforcing obedience, a penalty or reward; anything that gives binding force to a law, oath etc.; countenance, support, encouragement conferred by usage etc.; that which makes any rule of conduct binding; (*usu. in pl.*) a coercive measure taken by one state against another in order to enforce compliance with international law or a change in policy etc. *v.t.* to give sanction to, to authorize, to ratify; to countenance, to approve. **sanctionable**, *a.* **sanctionless**, *a.*
sanctity (sangk'titi), *n.* the state of being holy, holiness; spiritual purity, saintliness; sacredness, inviolability; (*pl.*) sacred things, feelings etc. **sanctitude**, *n.* holiness, saintliness; sacredness.
sanctuary (sangk'chuəri), *n.* a holy place; a church, temple or other building or enclosure devoted to sacred uses, esp. an inner shrine or most sacred part of a church etc., as the part of a church where the altar is placed; a church or other consecrated place in which debtors and malefactors were free from arrest; any similar place of immunity, an asylum, a refuge; immunity, protection; a place where deer, birds etc. are left undisturbed.
sanctum (sangk'təm), *n.* (*pl.* **-tums, -ta,** (tə)) a sacred or private place; (*coll.*) a private room, den or retreat. **sanctum sanctorum,** *n.* the holy of holies in the Jewish temple; (*coll.*) one's sanctum.
Sanctus (sangk'tus), *n.* the liturgical phrase 'Holy, holy, holy', in Latin or English; the music for this. **Sanctus-bell,** *n.* a bell, usu. in a turret or bell-cote over the junction of nave and chancel, rung at the Sanctus before the Canon of the Mass.
sand (sand), *n.* comminuted fragments of rock, esp. of chert, flint and other quartz rocks, reduced almost to powder; a particle of this; (*pl.*) tracts of sand, stretches of beach or shoals or submarine banks of sand; (*pl.*) particles of sand in an hour-glass; (*N Am. coll.*) grit, endurance, pluck. *v.t.* to sprinkle or treat with sand; to mix sand with to adulterate; to cover or overlay with or bury under sand; to smooth or rub with sandpaper or a similar abrasive. **the sands are running out,** the end is approaching. **sandbag,** *n.* a bag or sack filled with sand, used in fortification for making defensive walls, as ballast, for stopping crevices, draughts etc. as a weapon for stunning a person etc. *v.t.* to fortify or stop up with sandbags; to strike or fell with a sandbag. **sandbagger,** *n.* **sand-bank,** *n.* a bank or shoal of sand, esp. in the sea, a river etc. **sand-bar,** *n.* a ridge of sand built up by currents in a sea or river. **sand-bath,** *n.* a vessel containing hot sand used for heating, tempering etc. **sand-blast,** *n.* a jet of sand used for engraving and cutting glass, cleaning stone surfaces etc. *v.t.* to cut, clean etc. with a sand-blast. **sand-blight,** *n.* (*Austral.*) an eye inflammation caused by sand. **sand-box,** *n.* a box containing sand carried by a locomotive etc., for sprinkling the rails when slippery; (*Golf*) a box for sand used in teeing; a large open box containing sand for children to play in. **happy as a sand-boy,** happily engrossed. **sand-castle,** *n.* a model of a castle in sand. **sand-crack,** *n.* a fissure in the hoof of a horse, liable to cause lameness; a crack or flaw in a brick due to defective mixing. **sand-eel,** *n.* an eel-like fish of the genus *Ammodytes*. **sand-flea,** *n.* a sand-hopper. **sand-fly,** *n.* a species of midge; an angler's fly. **sand-glass,** *n.* an hour-glass. **Sandgroper,** *n.* (*Austral., derog.*) a Western Australian. **sand-heat,** *n.* heat imparted by warmed sand in chemical operations. **sand-hopper,** *n.* a crustacean, *Talitrus locusta*. **sand-iron,** *n.* a golf club used for lifting the ball from sand. **sandman,** *n.* a being in fairy-lore who makes children sleepy by casting sand in their eyes. **sand-martin,** *n.* a small swallow, *Hirundo riparia*, which makes its nest in sand-banks etc. **sandpaper,** *n.* a paper or thin cloth coated with sand, used for smoothing wood etc. *v.t.* to rub or smooth with this. **sand-pipe,** *n.* (*Geol.*) a deep cylindrical hollow, filled with sand and gravel, penetrating chalk. **sandpiper,** *n.* a popular name for several birds haunting sandy places, chiefly of the genera *Tringa* and *Totanus;* (*Austral.*) the rainbow bird. **sand-pit,** *n.* a container of sand for children to play in, a sand-box. **sand-pump,** *n.* a pump used for extracting wet sand from a drill-hole, caisson etc. **sand-shoe,** *n.* a light shoe, usu. of canvas with a rubber sole, for walking on sands. **sandstone,** *n.* stone composed of an agglutination of grains of sand. **sand-storm,** *n.* a storm of wind carrying along volumes of sand in a desert. **sand-trap,** *n.* (*chiefly N Am., Golf*) a bunker. **sand-worm,** *n.* the lug-worm. **sandwort,** *n.* any plant of the genus *Arenaria*, low herbs growing in sandy soil. **sand-yacht,** *n.* a yacht-like vehicle with wheels and sails for use on sand. **sanded,** *a.* sprinkled with sand; filled, covered or dusted with sand. **sander,** *n.* one who or that which sands, esp. a power tool for smoothing etc. by means of an abrasive belt or disk. **sandy,** *a.* consisting of or abounding in sand; of the colour of sand; of hair, yellowish-red; having hair of this colour; (*N Am. coll.*) plucky, brave, having plenty of grit or sand; shifting, unstable. **sandiness,** *n.* **sandyish,** *a.*
sandal[1] (san'dəl), *n.* a kind of shoe consisting of a sole secured by straps passing over the foot and often round the ankle; a strap for fastening a low shoe. **sandalled,** *a.* wearing sandals; fitted or fastened with a sandal.
sandal[2] (san'dəl), *n.* sandalwood. **sandalwood,** *n.* the fragrant wood of various trees of the genus *Santalum*, esp. *S. album*, much used for cabinet work; a tree that yields sandalwood; a similar wood or a tree that yields it. **red sandalwood** RED.
sanderling (san'dəling), *n.* a small wading bird, *Crocethia alba*.
sandwich (san'wich), *n.* two thin slices of bread and butter with meat etc. between them; anything resembling a sandwich in layered arrangement. *v.t.* to put, lay or insert between two things of a dissimilar kind. **sandwich course,** *n.* an educational course containing one or more periods of industrial work. **sandwich-man,** *n.* a man carrying two advertisement boards (**sandwich-boards**) hung from his shoulders, one in front and one behind.
sandy SAND.
sane (sān), *a.* sound in mind, not deranged; of views etc., sensible, reasonable. **sanely,** *adv.* **saneness,** *n.*
sang (sang), *past* SING[1].
sangaree (sang·gərē'), *n.* wine and water sweetened, spiced and usu. iced.
sang-de-boeuf (sädəbœf'), *n.* a dark-red colour such as that of some old Chinese porcelain. *a.* of this colour.
sangfroid (säfrwah'), *n.* coolness, calmness, composure in danger etc.
sangria (sang·grē'ə), *n.* a Spanish drink of diluted (red) wine and fruit juices.
sanguinary (sang'gwinəri), *a.* accompanied by bloodshed or carnage; delighting in bloodshed, bloodthirsty, murderous. **sanguinarily,** *adv.* **sanguinariness,** *n.*
sanguine (sang'gwin), *a.* having the colour of blood; of the complexion, ruddy, florid; hopeful, cheerful, confident, optimistic, ardent, enthusiastic; full of blood, plethoric. *n.* blood colour, deep red; a crayon of this colour prepared from iron oxide. **sanguinely,** *adv.* **sanguineness,** *n.* **sanguineous** (-gwin'-), *a.* pertaining to, forming or containing blood; sanguinary; of a blood colour; full-blooded, plethoric. **sanguinity** (-gwin'-), *n.* sanguineness; consanguinity. **sanguinivorous** (-niv'-), **sanguivorous** (-gwiv'-), *a.* feeding on blood.
Sanhedrin (san'idrin, -hē'-, -hed'-), *n.* the supreme court of justice and council of the Jewish nation, down to AD 425.
sanicle (san'ikl), *n.* a small woodland plant of the umbelliferous genus *Sanicula*, allied to the parsley.

sanies (sā'niĕz), *n.* a thin fetid discharge, usu. stained with blood, from sores or wounds. **sanious**, *a.*
sanious SANIES.
sanitary (san'itəri), *a.* relating to or concerned with the preservation of health, pertaining to hygiene; free from dirt, disease-causing organisms etc., hygienic. **sanitary towel** or (*esp. N Am.*) **napkin**, *n.* an absorbent pad used to absorb menstrual flow. **sanitary ware**, *n.* ceramic plumbing fixtures such as sinks, toilet bowls etc. **sanitarian** (-teə'ri-), *n.*, *a.* **sanitarily**, *adv.* **sanitariness**, *n.* **sanitarist**, *n.* **sanitarium** SANATORIUM. **sanitation**, *n.* **sanitationist**, *n.*
sanitize, -ise, *v.t.* to make sanitary; to remove offensive language etc. from, make respectable. **sanitization**, *n.*
sanity (san'iti), *n.* saneness, mental soundness; reasonableness, moderation.
sank (sangk), *past* SINK.
sans (sanz, sã), *prep.* without. **sans-culotte** (sãkulot', sanzkü-), *n.* a republican in the French Revolution; a radical extremist, a revolutionary. *a.* republican, revolutionary. **sans-culottism**, *n.* **sans serif, sanserif** (sanse'rif), *n.* a printing type without serifs.
Sanskrit (san'skrit), *n.* the ancient language of the Hindu sacred writings, the oldest of the Indo-European group. **Sanskritic** (-skrit'-), *a.* **Sanskritist**, *n.*
Santa Claus (san'təklawz, -klawz'), *n.* a mythical whitebearded old man bringing presents at Christmas and putting them in children's stockings.
santolina (santəlē'nə), *n.* a genus of fragrant shrubby composite plants allied to the camomile.
santonica (santon'ikə), *n.* the unexpanded flower-heads of an Oriental species of *Artemisia* or wormwood, containing santonin. **santonin** (san'tənin), *n.* the bitter principle of santonica, used as an anthelmintic. **santoninic** (-nin'-), *a.*
sap[1] (sap), *n.* the watery juice or circulating fluid of living plants; the sapwood of a tree; vital fluid, strength, vigour; (*sl.*) a gullible person, a saphead. *v.t.* (*past*, *p.p.* **sapped**) to draw off sap; to exhaust the strength or vitality of. **sap-green**, *n.* a green pigment obtained from the juice of blackthorn berries; the colour of this. *a.* of this colour.
saphead, *n.* (*coll.*) a stupid person. **sap-rot**, *n.* dry-rot. **sap-tube**, *n.* a plant vessel conducting sap. **sapwood**, *n.* the soft new wood next to the bark, alburnum. **sapless**, *a.* **sapling** (-ling), *n.* a young tree; a youth. **sappy**, *a.* **sappiness**, *n.*
sap[2] (sap), *v.t.* (*past*, *p.p.* **sapped**) to undermine; to approach by mines, trenches etc.; to render unstable by wearing away the foundation; to subvert. *v.i.* to make an attack or approach by digging trenches or undermining. *n.* a deep ditch, trench or mine for approach to or attack on a fortification; insidious undermining or subversion of faith etc. **sapper**, *n.* one who saps; (*coll.*) an officer or private of the Royal Engineers.
sapan-wood, sappan-wood (sap'ənwud), *n.* a brownish-red dyewood obtained from trees of the genus *Caesalpinia*, esp. *C. sappan*, from S Asia and Malaysia.
sapele (səpē'li), *n.* a reddish-brown wood resembling mahogany obtained from W African trees of the genus *Entandrophragma*; a tree that yields sapele.
saphena (səfē'nə), *n.* either of two prominent veins of the leg. **saphenal, saphenous**, *a.*
sapid (sap'id), *a.* possessing flavour that can be relished, savoury; not insipid, vapid or uninteresting. **sapidity** (-pid'-), **sapidness**, *n.*
sapient (sā'piənt), *a.* wise, sagacious, discerning. **sapiently**, *adv.* **sapience**, *n.* **sapiential** (-en'shəl), *a.* of or conveying wisdom.
sapling SAP[1].
sapodilla (sapədil'ə), *n.* the edible fruit of a large evergreen tree, *Achras sapota*, growing in the W Indies and Central America; the tree itself; its durable wood.
saponaceous (sapənā'shəs), *a.* soapy; resembling, containing or having the qualities of soap.

Saponaria (sapəneə'riə), *n.* a genus of plants comprising the soapworts.
saponify (səpon'ifi), *v.t.* to convert into soap by combination with an alkali. *v.i.* of an oil, fat etc., to become converted into soap. **saponifiable**, *a.* **saponification** (-fi-), *n.*
saponin (sap'ənin), *n.* any of various glucosides obtained from the soapwort, horse-chestnut etc. that produce a soapy foam and are used in detergents.
sappan-wood SAPAN-WOOD.
sapper SAP[2].
sapphic (saf'ik), *a.* pertaining to Sappho, a poetess (*c.* 600 BC) from the Greek island of Lesbos; applied to a stanza or metre used by her, esp. a combination of three pentameters followed by a dipody. *n.pl.* sapphic verses or stanzas. **sapphism**, *n.* lesbianism. **sapphist**, *n.*
sapphire (saf'iə), *n.* any transparent blue variety of corundum; an intense and lustrous blue, azure; *a.* sapphireblue. **sapphirine** (-rin), *a.* having the qualities, esp. the colour, of sapphire. *n.* a mineral of a pale blue colour, esp. a silicate of alumina and magnesia or a blue spinel.
sapraemia (səprē'miə), *n.* septic poisoning. **sapraemic**, *a.*
sapr(o)-, *comb. form.* indicating rotting or dead matter.
saprogenic (saprəjen'ik), *a.* producing or produced by putrefaction.
saprophagous (səprof'əgəs), *a.* feeding on decomposing matter.
saprophyte (sap'rəfit), *n.* a plant, bacterium or fungus that grows on decaying organic matter. **saprophytic** (-fit'-), *a.* **saprophytically**, *adv.* **saprophytism**, *n.*
saprozoic (saprəzō'ik), *a.* saprophagous.
sapsucker SAP[1].
sapwood SAP[1].
saraband, sarabande (sa'rəband), *n.* a slow and stately Spanish dance; a piece of music for this in strongly accented triple time.
Saracen (sa'rəsən), *n.* a nomad Arab of the Syrian-Arabian desert in the times of the later Greeks and Romans; a Muslim or Arab at the time of the Crusades. **Saracenic** (-sen'-), *a.*
Saratoga, Saratoga trunk (sarətō'gə), *n.* a variety of lady's large travelling trunk.
sarcasm (sah'kazm), *n.* a bitter, taunting, ironical or wounding remark; bitter or contemptuous irony or invective. **sarcastic** (-kas'-), *a.* containing or characterized by sarcasm; given to using sarcasm. **sarcastically**, *adv.*
sarcenet (sah'snit), *n.* a thin, fine soft-textured silk used chiefly for linings, ribbons etc.
Sarcina (sah'sinə), *n.* a genus of bacteria or schizomycetous fungi in which the cocci break up into cuboidal masses.
sarcine (sah'sin), *n.* a nitrogenous compound existing in the juice of flesh.
sarco-, *comb. form.* flesh.
sarcoblast (sah'kōblahst), *n.* a germinating particle of protoplasm. **sarcoblastic** (-blas'-), *a.*
sarcocarp (sah'kōkahp), *n.* the fleshy part of a drupaceous fruit.
sarcocolla (sahkōkol'ə), *n.* sarcocol; a genus of S African shrubs of the family Penaeaceae.
sarcode (sah'kōd), *n.* animal protoplasm. **sarcodal, sarcodic** (-kod'-), *a.*
sarcoderm (sah'kōdœm), *n.* an intermediate fleshy layer in certain seeds.
sarcoid (sah'koid), *a.* resembling flesh. *n.* a particle of sponge tissue; a swelling, nodule.
sarcolemma (sahkōlem'ə), *n.* the tubular membrane sheathing muscular tissue. **sarcolemmic**, *a.*
sarcology (sahkol'əji), *n.* the branch of anatomy concerned with the soft parts of the body. **sarcological** (-loj'-), *a.* **sarcologist**, *n.*
sarcoma (sahkō'mə), *n.* (*pl.* **-mas, -mata** -tə) a tumour of connective tissue. **sarcomatosis** (-tō'sis), *n.* the formation

and spread of sarcomas. **sarcomatous**, *a*.
sarcophagous (sahkof'əgəs), **sarcophagic** (-faj'-), *a*. feeding on flesh. **sarcophagon** (-gon), *n*. an insect of the order Sarcophaga, a flesh-fly. **sarcophagy** (-ji), *n*. the practice of eating flesh.
sarcophagus (sahkof'əgəs), *n*. (*pl*. **-gi** (-jī)) **-guses**) a stone coffin, esp. one of architectural or decorated design.
sarcoplasm (sah'kōplazm), *n*. the substance between the columns of muscle-fibre. **sarcoplasmic**, *a*.
sarcosis (sahkō'sis), *n*. (*pl*. **-ses**) a fleshy tumour, a sarcoma.
sarcous (sah'kəs), *a*. composed of flesh or muscle tissue.
sard (sahd), *n*. a precious stone, a variety of cornelian.
sardelle, sardel (sahdel'), *n*. a small Mediterranean clupeoid fish like the sardine.
sardine (sahdēn'), *n*. a fish, *Clupea pilchardus*, caught off Brittany and Sardinia, and cured and preserved in oil; any of various other small fish preserved in the same way.
Sardinian (sahdin'iən), *a*. pertaining to the island or the former kingdom of Sardinia. *n*. a native or inhabitant of Sardinia.
sardonic (sahdon'ik), *a*. unnatural, forced, affected, insincere; of laughter etc., sneering, malignant, bitterly ironical. **sardonically**, *adv*.
sargasso (sahgas'ō), *n*. (*pl*. **-sos**) the gulfweed, *Sargassum bacciferum*; a floating mass of this or similar vegetation. **Sargasso Sea**, *n*. a part of the Atlantic abounding with this. **sargassum** (-əm), *n*. a plant of the genus *Sargassum*, sargasso.
sarge (sahj), *n*. short for SERGEANT.
sari, saree (sah'ri), *n*. a Hindu woman's dress.
sarky (sah'ki), *a*. (*coll*.) sarcastic; bad-tempered.
Sarmatian (sahmā'shən), *a*. pertaining to ancient Sarmatia, now Poland and part of Russia, or its people. *n*. a native or inhabitant of Sarmatia.
sarmentose (sahmen'tōs), **-tous**, *a*. (*Bot*.) having or producing runners. **sarmentum** (-təm), *n*. (*pl*. **-ta** (-tə), **-tums**) a prostrate shoot rooting at the nodes, a runner.
sarong (sərong'), *n*. a loose, skirt-like garment traditionally worn by men and women in the Malay Archipelago.
saros (seə'ros), *n*. a cycle of 6585¼ days in which solar and lunar eclipses repeat themselves.
sarothrum (sərō'thrəm), *n*. the pollen brush on the leg of a honey-bee.
Sarracenia (sarəsē'niə), *n*. a genus of insectivorous plants with pitcher-shaped leaves.
sarrusophone (sərooz'əfōn), *n*. a brass musical instrument resembling an oboe with a metal tube.
sarsaparilla (sahspəril'ə), *n*. the dried roots of various species of *Smilax*, used as a flavouring and formerly in medicine as an alterative and tonic; a plant of this genus; a carbonated drink flavoured with sassafras.
sarsen (sah'sən), *n*. a sandstone boulder such as those scattered over the chalk downs of Wiltshire. **sarsen-boulder, -stone**, *n*.
sarsenet (sahs'nit), SARCENET.
sartorial (sahtaw'riəl), *a*. pertaining to a tailor or tailored clothing.
sartorius (sahtaw'riəs), *n*. a muscle of the thigh that helps to flex the knee.
SAS, (*abbr*.) Special Air Service.
sash[1] (sash), *n*. an ornamental band or scarf worn round the waist or over the shoulder, frequently as a badge or part of a uniform. **sashed**, *a*.
sash[2] (sash), *n*. a frame of wood or metal holding the glass of a window; a sliding light in a greenhouse etc. *v.t.* to furnish with sashes. **sash-cord, -line**, *n*. a stout cord attached to a sash and the sash-weights. **sash-frame**, *n*. the frame in which a sash slides up and down. **sash-pocket**, *n*. the space in which the sash-weights are hung. **sash-weight**, *n*. a weight used to balance a sash and hold it in an open position. **sash-window**, *n*. a window having a movable sash or sashes. **sashed**, *a*. **sashless**, *a*.
sashay (sashā'), *v.i*. (*chiefly N Am*.) to walk or move in a nonchalant or sauntering manner; to strut, swagger.
sashimi (sashim'i), *n*. a Japanese dish of thin slices of raw fish. [Jap.]
Sask., (*abbr*.) Saskatchewan.
Sasquatch (sas'kwach), *n*. a hairy humanoid creature reputedly living in W Canada.
sass (sas), *n*. (*N Am., coll.*) impudence, cheek, sauce. *v.t.* to talk impudently to. **sassy**, *a*. (*N Am., coll.*) cheeky, saucy.
sassaby (səsā'bi), *n*. a large S African antelope, *Alcelaphus lunatus*, the bastard hartebeest.
sassafras (sas'əfras), *n*. a tall N American tree, *Sassafras albidum*, of the laurel family; the dried bark of its root used as an aromatic stimulant and flavouring.
Sassanid (sas'ənid), *n*. one of the descendants of Sasan, ancestor of the last pre-Islamic dynasty of Persia (AD 226–642).
Sassenach (sas'ənakh), *n*. (*Sc. and Ir., chiefly derog.*) a Saxon, an English person. *a*. English.
sassoline (sas'əlēn), *n*. a mineral composed of a native triclinic form of boric acid.
sassy SASS.
sastra (sahs'trə), SHASTER under SHASTRA.
Sat., (*abbr*.) Saturday.
sat (sat), *past, p.p*. SIT.
Satan (sā'tən), **Satanas** (sat'ənas), *n*. the arch-fiend, the devil. **satanic, -ical** (sətan'-), *a*. pertaining to, emanating from or having the qualities of Satan; devilish, infernal. **satanically**, *adv*. **satanism**, *n*. a diabolical disposition, doctrine or conduct; the deliberate pursuit of wickedness; Satan-worship. **satanist**, *n*.
satanology (sātənolə'ji), *n*. the study of or a treatise on doctrines relating to the devil.
satara (sat'ərə, -tah'-), *n*. a heavy, horizontally-ribbed woollen or broad-cloth.
satay (sat'ā), *n*. a Malaysian and Indonesian dish of cubed meat served with a spicy peanut sauce.
SATB, (*abbr*.) soprano, alto, tenor, bass.
satchel (sach'əl), *n*. a small rectangular bag, often suspended by a strap passing over one shoulder, esp. for schoolchildren to carry books etc. in. **satchelled**, *a*.
sate (sāt), *v.t*. to satisfy (an appetite or desire); to satiate, to surfeit, to glut, to cloy.
sateen (sətēn'), *n*. a glossy woollen or cotton fabric made in imitation of satin. [from SATIN]
satellite (sat'əlit), *n*. a secondary planet revolving round a primary one; a man-made device projected into space to orbit the earth, moon etc., used for communications, broadcasting, weather forecasting, surveillance etc.; something dependent on or subordinate to another; an obsequious follower, dependant or henchman. **satellite broadcasting**, *n*. the transmission of broadcast programmes via an artificial satellite for reception in the home. **satellite dish**, *n*. a parabolic aerial for reception of programmes, transmitted via an artificial satellite. **satellite programme**, *n*. such programmes. **satellite state**, *n*. a country subservient to a greater power. **satellite town**, *n*. a small town dependent upon a larger town in the vicinity.
satiate (sā'shiət), *v.t*. to satisfy (as a desire or appetite) fully; to sate, to glut, to surfeit. **satiable**, *a*. **satiation**, *n*.
satiety (səti'əti), *n*. the state of being sated or glutted; excess of gratification producing disgust.
satin (sat'in), *n*. a silken fabric with an overshot weft and a highly-finished glossy surface on one side only. *a*. made of or resembling this, esp. in smoothness. *v.t*. to give (paper etc.) a glossy surface like satin. **white satin** WHITE. **satin-bird**, *n*. an Australian bower-bird, *Ptilonorhyncus violoceus*. **satin-finish**, *n*. a lustrous polish given to silverware with a metallic brush. **satin-flower**, *n*. honesty; the greater stitchwort. **satin-paper**, *n*. a fine, glossy writing-paper.

satin-spar, *n.* a finely fibrous variety of aragonite, calcite, or gypsum. **satin-stitch**, *n.* a stitch in parallel lines giving the appearance of satin. **satin-wood**, *n.* an ornamental cabinet wood of various species from the E and W Indies. **satinet, satinette** (-net'), *n.* a thin satin; a glossy fabric made to imitate satin. **satining-machine**, *n.* a machine for giving paper etc. a satiny surface. **satinize, -ise**, *v.t.* to satin. **satiny**, *a.*

satire (sat'iə), *n.* a composition in verse or prose in which wickedness or folly or individual persons are held up to ridicule; ridicule, sarcasm or the use of ridicule, irony and invective ostensibly for the discrediting of vice or folly. **satiric, -ical** (-ti'-), *a.* **satirically**, *adv.* **satiricalness**, *n.* **satirist** (-i-), *n.* one who writes or employs satire. **satirize, -ise** (-i-), *v.t.* to ridicule by means of satire. *v.i.* to use or write satire.

satisfy (sat'isfī), *v.t.* to supply or gratify to the full; to content, to gratify, to please; to pay (a debt etc.); to fulfil, to comply with; to be sufficient for, to meet the desires, expectations or requirements of; to make sexually fulfilled; (*Math., Log.*) to fulfil the conditions of; to free from doubt; to convince; to meet (a doubt, objection etc.) adequately. *v.i.* to give satisfaction; to make payment, compensation or reparation, to atone. **satisfaction** (-fak'-), *n.* the act of satisfying; the state of being satisfied; gratification, contentment; payment of a debt, fulfilment of an obligation; a source of satisfaction; reparation, compensation, amends; atonement, esp. the atonement for sin achieved by Christ's death; the performance of penance. **satisfactory** (-fak'-), *a.* giving satisfaction, sufficient, adequate, meeting all needs, desires or expectations; relieving the mind from doubt; atoning, making amends. **satisfactorily**, *adv.* **satisfactoriness**, *n.* **satisfiable**, *a.* **satisfier**, *n.* **satisfying**, *a.* **satisfyingly**, *adv.*

satori (sətaw'ri), *n.* in Zen Buddhism, an intuitive enlightenment.

satrap (sat'rəp), *n.* a governor of a province under the ancient Persian empire, a viceroy; a governor, a ruler of a dependency etc., esp. one who affects despotic ways. **satrapy**, *n.* the territory, office or period of office of a satrap.

satsuma (sat'sumə), (-soo'-), *n.* a seedless type of mandarin orange; a tree that bears such fruit. **Satsuma ware**, *n.* a cream-coloured variety of Japanese pottery.

saturate (sach'ərāt), *v.t.* to soak, impregnate, or imbue thoroughly; to fill or charge (a body, substance, gas, fluid etc.) with another substance, fluid, electricity etc. to the point where no more can be held; to cause (a chemical compound) to combine until no further addition is possible; to overwhelm (a target) with bombs or projectiles; to swamp (an area) (with police, salesmen, etc.); to swamp (a market) with merchandise so that no further expansion of business is possible. **saturable**, *a.* **saturant**, *a.* saturating. *n.* a substance neutralizing acidity or alkalinity. **saturated**, *a.* of a solution, containing as much dissolved material as possible at a given temperature; full of water, soaked; of an organic compound, containing only single bonds between carbon atoms and not reacting to add further groups to the molecule. **saturated fat**, *n.* a fat containing mostly saturated fatty acids. **saturater**, *n.* **saturation**, *n.* the state of being saturated; the presence in the atmosphere of the maximum amount of water vapour at any particular temperature; the point at which increasing magnetizing force fails to increase any further the flux-density of the magnet; the purity of a colour, freedom from mixture with white. **saturation bombing**, *n.* bombing that completely covers a target area. **saturation current**, *n.* the maximum value of electric current that can be carried. **saturation point**, *n.* the point at which no more can be taken in, held etc.

Saturday (sat'ədi), *n.* the seventh day of the week.

Saturn (sat'ən), *n.* an ancient Roman god of agriculture, usu. identified with the Greek Kronos, father of Zeus; the sixth of the major planets in distance from the sun. **Saturnalia** (-nā'liə), *n.pl.* an ancient Roman annual festival held in December in honour of Saturn, regarded as a time of unrestrained licence and merriment; (*often sing.*) a season or occasion of unrestrained revelry. **saturnalian**, *a.* **saturnalian** *a.* of or pertaining to the planet Saturn; denoting the accentual metre of early Latin poetry. *n.* an inhabitant of Saturn; (*pl.*) saturnian verses. **saturnine** (-nin), *a.* born under the influence of the planet Saturn; dull, phlegmatic, gloomy, morose. **saturninely**, *adv.* **saturnite** (-nit), *n.* a mineral substance containing lead.

satyagraha (sah'tyəgrah·hə, -tyah'-), *n.* non-violent resistance to authority as practised orig. by Mahatma Gandhi.

satyr (sat'ə), *n.* an ancient sylvan Greek deity represented with the legs of a goat, budding horns, and goat-like ears, identified by the Romans with the fauns; a lascivious man. **satyriasis** (-rī'əsis), *n.* unrestrained sexual appetite in men. **satyric** (-ti'-), *a.*

sauce (saws), *n.* a preparation, usu. liquid, taken with foods as an accompaniment or to enhance the taste; (*fig.*) anything that gives piquancy or makes palatable; (*coll.*) sauciness, impertinence, impudence, cheek. *v.t.* (*fig.*) to flavour, to make piquant or pungent; (*coll.*) to be saucy or impudent towards; to treat with sauce, to season. **sauce-boat**, *n.* a table-vessel for holding sauce. **sauce-box**, *n.* (*coll.*) an impudent person, esp. a child. **saucepan**, *n.* a metal pan or pot, usu. cylindrical with a long handle, for boiling or stewing; orig. a pan for cooking sauces. **saucy**, *a.* pert, impudent, insolent to superiors, cheeky; smart, sprightly. **saucily**, *adv.* **sauciness**, *n.* **saucer** (saw'sə), *n.* a shallow china vessel for placing a cup on and catching spillings; any small flattish vessel, dish or receptacle of similar use. **saucer-eyes**, *n. pl.* large, round, staring eyes. **saucer-eyed**, *a.* **saucerful**, *n.* (*pl.* -fuls). **saucerless**, *a.*

Saudi (sawdi), *adj.* of or relating to Saudi Arabia. *n.* a native or inhabitant of Saudi Arabia.

sauerkraut (sowə'krowt), *n.* finely chopped cabbage compressed with salt until it ferments.

sauger (saw'gə), *n.* the smaller N American pike-perch, *Stizostedion canadense.*

sauna (saw'nə), *n.* a Finnish-style steam bath; a building or room used for saunas.

saunter (sawn'tə), *v.i.* to wander about idly and leisurely; to walk leisurely (along). *n.* a leisurely ramble or stroll; a sauntering gait. **saunterer**, *n.*

saury (saw'ri), *n.* a sea-fish, *Scomberesox saurus,* with elongated body ending in a beak.

sausage (sos'ij), *n.* an article of food consisting of pork or other meat minced, seasoned and stuffed into a length of animal's gut or a similar receptacle; anything of similar cylindrical shape. **sausage balloon**, *n.* an observation balloon shaped like an inflated sausage. **sausage-dog**, *n.* (*coll.*) a dachshund. **sausage-meat**, *n.* meat used for stuffing sausages, esp. cooked separately as stuffing etc. **sausage-roll**, *n.* sausage-meat enclosed in pastry and baked.

sauté (sō'tā), *a.* lightly fried. *v.t.* (*past, p.p.* **sautéed, sautéd**) to fry lightly. *n.* a dish of this. *adj.* sautéed until slightly brown.

Sauternes (sōtœn'), *n.* a sweet white Bordeaux wine.

sauve qui peut (sōv kē pœ), *n.* a state of panic or chaos.

savable etc. SAVE.

savage (sav'ij), *a.* uncultivated, untamed, wild; uncivilized, in a primitive condition; fierce, brutal, cruel, violent, ferocious; (*coll.*) extremely angry, enraged; *n.* a human being in a primitive state, esp. a member of a nomadic tribe living by hunting and fishing; a person of extreme brutality or ferocity. *v.t.* to attack violently, esp. of an animal, to bite, tear or trample; to subject to fero-

savanna 737 **sayyid**

cious criticism. **savagely**, *adv.* **savagedom** (-dəm), **savageness, savagery** (-jəri), *n.*
savanna, savannah (səvan'ə), *n.* an extensive treeless plain covered with low vegetation, esp. in tropical America. **savanna (black)bird**, *n.* the W Indian bird *Crotophaga ani.* **savanna flower**, *n.* an evergreen shrub of various species of *Echites.*
savant (sav'ənt), *n.* a person of learning.
savate (savaht'), *n.* a style of boxing in which the feet are used as well as the hands.
save (sāv), *v.t.* to preserve, rescue or deliver as from danger, destruction or harm of any kind; to deliver from sin, to preserve from damnation; to keep undamaged or untouched; to keep from being spent or lost; to reserve and lay by, to husband, to refrain from spending or using; to spare, to exempt (*with double object*); to obviate, to prevent; to prevent or obviate the need for; (*Soccer, hockey, etc.*) to prevent a score by stopping (a struck ball or puck). *v.i.* to be economical, to avoid waste or undue expenditure; (*also with up*) to set aside money for future use. *prep.* except, saving; leaving out, not including. *conj.* except. *n.* the act of preventing an opponent from scoring a goal; something saved, an economy. **save-all**, *n.* anything that prevents things from being wasted. **save-as-you-earn**, *n.* a government savings scheme in which regular contributions are deducted from earnings. **savable**, *a.* **saver**, *n.* (*usu. in comb.*, as *life-saver*). **saving**, *a.* preserving from danger, loss, waste etc.; economical, frugal; reserving or expressing a reservation, stipulation etc. *n.* the act of economizing; (*usu. pl.*) that which is saved, an economy; (*pl.*) money saved, esp. regularly or over a period of time; an exception, a reservation. *prep.* save, except; with due respect to. **savings account**, *n.* a bank account intended for longer-term deposits and with a higher rate of interest. **savings bank**, *n.* a bank receiving small deposits and usu. devoting any profits to the payment of interest. **savingly**, *adv.*
saveloy (sav'əloi), *n.* a highly-seasoned dried sausage of salted pork (orig. of brains).
savin, savine (sav'in), *n.* an evergreen bush or low tree, *Juniperus sabina*, with bluish-green fruit, yielding an oil formerly used medicinally.
saviour (sāv'yə), *n.* one who preserves, rescues, or redeems. **our** or **the Saviour**, Christ, regarded as the Redeemer of mankind.
savoir faire (savwah feə'), *n.* quickness to do the right thing, esp. in social situations, tact, presence of mind.
savory (sā'vəri), *n.* a plant of the aromatic genus *Satureia*, esp. *S. hortensis*, used in cookery.
savour, (*esp. N Am.*) **savor** (sā'və), *n.* (characteristic) flavour, taste, relish; a particular taste or smell; characteristic quality; suggestive quality, smack or admixture (of); *v.t.* to give a flavour to; to relish, to enjoy the savour of. *v.i.* to have a particular smell or flavour, to smack (of); **savourless**, *a.* **savoury**, *a.* having a pleasant savour; palatable, appetizing; free from offensive smells; salty, spicy etc. (as opp. to sweet); respectable, wholesome. *n.* a savoury dish, esp. as served as an appetizer or digestive. **savourily**, *adv.* **savouriness**, *n.*
savoy (səvoi'), *n.* a hardy variety of cabbage with wrinkled leaves.
Savoyard (səvoi'ahd), *n.* a native of Savoy; a devotee, producer, or performer of the comic operas of W.S. Gilbert and A.S. Sullivan. *a.* of Savoy.
savvy (sav'i), *v.t., v.i.* (*sl.*) to know, to understand. *n.* understanding, knowingness, cleverness.
saw[1] (saw), *n.* a cutting-instrument, usu. of steel, with a toothed edge, worked by hand, or power-driven, as in circular or ribbon form; a tool or implement used as a saw. *v.t.* (*past* **sawed**, *p.p.* **sawn**) to cut with a saw; to form or make with a saw; to make motions as if sawing. *v.i.* to use a saw; to undergo cutting with a saw; to make motions of one sawing. **saw-bill**, *n.* a tropical or sub-tropical American bird, the motmot, with serrated mandibles; a duck with a serrated beak. **saw-bones**, *n.* (*sl.*) a surgeon. **sawdust**, *n.* small fragments of wood produced in sawing. **saw-fish**, *n.* a fish of the genus *Pristis*, with an elongated, saw-like snout. **saw-fly**, *n.* any of various hymenopterous insects, as of the genus *Tenthredo*, furnished with a saw-like ovipositor. **saw-horse**, *n.* a rack on which wood is laid for sawing. **sawmill**, *n.* a mill with machinery for sawing timber. **sawn-off**, *a.* of a shotgun, having the end of the barrel cut off with or as with a saw. **sawpit**, *n.* a pit over which timber is sawed, one person standing above and the other below the log. **saw-set**, **-wrest**, *n.* a tool for slanting the teeth of a saw alternately outward. **saw-toothed**, *a.* serrated. **saw-whet**, *n.* a small N American owl, *Nyctale acadica*, with a harsh cry. **sawwort**, *n.* any plant of the genus *Serratula*, having serrated leaves yielding a yellow dye. **sawyer** (-yə), *n.* one employed in sawing timber into planks, or wood for fuel; a wood-boring larva; (*N Am.*) a tree fallen into a river and swept along, sawing up and down in the water; (*New Zealand*) a kind of grasshopper, the weta.
saw[2] (saw), *n.* a saying, a proverb, a familiar maxim.
saw[3] (saw), *past* SEE[1].
sawder (saw'də), *n.* blarney, flattery.
sawwort, sawyer etc. SAW[1].
sax[1] (saks), *n.* a slate-cutter's chopping and trimming tool with a point for making holes.
sax[2] (saks), *n.* short for SAXOPHONE.
saxatile (sak'sətil, -til), *a.* pertaining to or living among rocks.
saxhorn (saks'hawn), *n.* a brass musical wind-instrument with a long winding tube, a wide opening and several valves.
saxicolous (saksik'ələs), **saxicoline** (-lin), *a.* inhabiting or growing among rocks, saxatile.
saxifrage (sak'sifrāj), *n.* any plant of the genus *Saxifraga* consisting largely of Alpine or rock plants with tufted, mossy or encrusted foliage and small flowers. **saxifragaceous** (-frəgā'shəs), *a.*
Saxon (sak'sən), *n.* one of a Teutonic people from N Germany who conquered England in the 5th and 6th cents.; an Anglo-Saxon; the old Saxon or the Anglo-Saxon language. *a.* pertaining to the Saxons, their country or language; Anglo-Saxon. **Saxon-blue**, *n.* indigo dissolved in sulphuric acid, used by dyers; saxe-blue.
Saxony (sak'səni), *n.* a fine wool or woollen material produced in Saxony.
saxophone (sak'səfōn), *n.* a brass musical wind-instrument with a single reed used as a powerful substitute for the clarinet. **saxophonist** (-sof'ə-), *n.*
saxtuba (saks'tūbə), *n.* a bass saxhorn.
say (sā), *v.t.* (*past, p.p.* **said** sed, †*3rd sing. pres.* **saith** seth) to utter in or as words, to speak, to pronounce; to recite, to rehearse, to repeat; to tell, to affirm, to assert, to state; to allege, to report; to promise; to suppose, to assume; to give as an opinion or answer, to decide; to convey by means of artistic expression. *v.i.* to speak, to talk, to answer. *n.* what one says or has to say, an affirmation, a statement; (*coll.*) one's turn to speak; authority, influence. *adv.* approximately, about; for example. **I say**, an exclamation of mild surprise, protest etc. or calling for attention. **it is said** or **they say**, it is generally reported or rumoured. **not to say**, indeed one might say, perhaps even. **that is to say**, in other words. **to say nothing of**, as well as. **to say the least**, without the least exaggeration. **say-so**, *n.* a dictum; an unfounded assertion; right of decision, authority. **says, sez, you!** *int.* (*sl.*) an expression of incredulity. **saying**, *n.* that which is said; a maxim, an adage, a saw.
SAYE, (*abbr.*) save-as-you-earn.
sayyid (sī'id), *n.* a Muslim title of respect; a descendant of

certain members of Mohammed's family.
Sb, (*chem. symbol*) antimony.
SC, (*abbr.*) South Carolina; Special Constable; Supreme Court.
Sc, (*chem. symbol*) scandium.
sc, (*abbr.*) scene; *scilicet* (namely); *sculpsit* (he/she sculptured it); scruple; small capitals.
s/c, (*abbr.*) self-contained.
scab (skab), *n.* an incrustation formed over a sore etc., in healing; a highly-contagious skin-disease resembling mange, attacking horses, cattle and esp. sheep; one of various fungoid plant-diseases; a despicable scoundrel; a worker who refuses to join in a strike or who takes the place of a striker, a blackleg. *v.i.* (*past, p.p.* **scabbed**) to form a scab; to work as a scab or blackleg. **scabmite**, *n.* the itch-mite. **scabbed, scabby**, *a.* **scabbily**, *adv.* **scabbiness**, *n.*
scabbard (skab'əd), *n.* the sheath of a sword or similar weapon. **scabbard fish**, *n.* a small silver sea-fish (*Lepidopus candatus*) with a blade-like body; any of various related fishes.
scabies (skā'biz, -biēz), *n.* the itch, a contagious skin-disease.
scabious (skā'biəs), *a.* consisting of or covered with scabs; affected with itch. *n.* a plant of the herbaceous genus *Scabiosa*, having involucrate heads of blue, pink and white flowers.
scabrous (skā'brəs), *a.* rough, rugged or uneven; scaly, scurfy; difficult, thorny, awkward to handle; approaching the indecent, indelicate. **scabridity** (skəbrid'-), **scabrousness**, *n.*
scad (skad), *n.* the horse-mackerel.
scaffold (skaf'əld), *n.* a temporary structure of poles and ties supporting a platform for the use of workers building or repairing a house or other building; a temporary raised platform for the execution of criminals; a platform, or stage for shows or spectators; the bony framework of a structure, esp. one to be covered by developed parts. *v.t.* to furnish with a scaffold; to uphold, to support. **scaffolder**, *n.* **scaffolding**, *n.* a scaffold or system of scaffolds for builders, shows, pageants etc.; a framework; materials for scaffolds.
scaglia (skal'yə), *n.* a red, white or grey Italian limestone corresponding to chalk. **scagliola** (-yō'lə), *n.* a hard, polished plaster, coloured in imitation of marble.
scalable (skā'ləbl), *a.* that may be scaled.
scalar (skā'lə), *a.* scalariform; of the nature of a scalar. *n.* a pure number, esp. the term in a quaternion that is not a vector; a quantity having magnitude but no direction (e.g. time). **scalariform** (skəla'rifawm), *a.* of the structure of cells, vessels, veins etc., ladder-shaped.
scald[1] (skawld), *v.t.* to burn with or as with a hot liquid or vapour; to clean (out) with boiling water; to cook briefly in hot water or steam; to raise (milk) nearly to boiling point. *n.* an injury to the skin from hot liquid or vapour. **scalder**, *n.* **scalding**, *n.* **scalding-hot**, hot enough to scald.
scald[2], **skald** (skawld), *n.* an ancient Norse poet or reciter of poems, a bard. **scaldic**, *a.*
scalder, scalding SCALD[1].
scaldic SCALD[2].
scale[1] (skāl), *n.* one of the thin horny plates forming a protective covering on the skin of fishes, reptiles etc.; a modified leaf, bract, hair, feather, disk, husk or other structure resembling this; a thin flake of dry skin; a scab; a carious coating; an incrustation; a coating deposited on the insides of pipes, kettles etc. by hard water; a small plate or flake of metal etc. *v.t.* to strip the scales off; to remove in scales or layers; to deposit scale on. *v.i.* to form scales; to come off in scales; to become coated with scale; (*Austral.*) to ride on a tram or bus without paying the fare. **scale-board**, *n.* a thin board for the back of a picture etc. **scale insect**, *n.* an insect, esp. of the family Coccidae, whose female secretes a protective waxy shell and lives attached to a host plant. **scale-winged**, *a.* having the wings covered with scales, lepidopterous. **scale-work**, *n.* an arrangement of overlapping scales, imbricated work. **scaled**, *a.* having scales (*usu. in comb.* as *thick-scaled*). **scaleless**, *a.* **scaly**, *a.* **scaliness**, *n.*
scale[2] (skāl), *n.* the dish of a balance; (*usu. pl.*) a simple balance; (*usu. pl.*) a machine for weighing. *v.t.* to amount to in weight.
scale[3] (skāl), *n.* anything graduated or marked with lines or degrees at regular intervals, as a scheme for classification, gradation etc.; a basis for a numerical system in which the value of a figure depends on its place in the order; a system of correspondence between different magnitudes, relative dimensions etc.; a line, numerical ratio etc. for showing this; a set of marks or a rule or other instrument marked with these showing exact distances, proportions, values etc., used for measuring, calculating etc.; (*Mus.*) all the tones of a key arranged in ascending or descending order according to pitch. *v.t.* to climb by or as by a ladder; to clamber up; to draw or otherwise represent to scale or proper proportions; to alter the scale of; to arrange, estimate or fix according to a scale; to adjust according to a standard. *v.i.* to have a common scale, to be commensurable. **scaling-ladder**, *n.* a ladder used in storming fortified places.
scalene (skā'lēn), *a.* of a triangle, having no two sides equal; of a cone or cylinder, having the axis inclined to the base; pertaining to the scalenus muscles. *n.* a scalene triangle; a scalenus muscle. **scalenum** (skəlē'nəm), *n.* a scalene triangle. **scalenus** (skəlē'nəs), *n.* (*pl.* **-ni**, (-nī)) one of a series of irregularly triangular muscles at the neck.
scaliness SCALE[1].
scaling-ladder SCALE[3].
scallion (skal'yən), *n.* a variety of onion or shallot.
scallop (skol'əp, skal'-), *n.* a bivalve mollusc of the genus *Pecten* or a related genus, with ridges and flutings radiating from the middle of the hinge and an undulating margin; the large adductor muscle of a scallop eaten as food; a single shell of a scallop worn as a pilgrim's badge; such a shell or a small shallow dish or pan used for cooking and serving oysters etc. in; (*pl.*) an ornamental undulating edging cut like that of a scallop-shell. *v.t.* to cut or indent the edge of thus; to cook in a scallop. **scallop-shell**, *n.* **scalloping**, *n.* **scalloping-tool**, *n.*
scallywag (skal'iwag), *n.* a scamp, a rascal.
scalp (skalp), *n.* the top of the head; the skin of this with the hair belonging to it, formerly torn off by N American Indians as a trophy of victory; a trophy or symbol of conquest. *v.t.* to tear or take the scalp from; to cut the top part, layer etc. off (anything); to flay, to lay bare; to criticize or abuse savagely; (*chiefly N Am.*) to buy (cheaply) and resell so as to make a large profit; (*N Am.*) to buy and sell so as to take small quick profits on (stocks etc.). *v.i.* (*N Am.*) to take small profits to minimize risk. **scalp-lock**, *n.* a solitary tuft of hair left on the shaven crown of the head by the warriors of some American tribes. **scalper**, *n.* one who scalps; (*chiefly N Am.*) a ticket tout. **scalping**, *n., a.* **scalping-knife**, *n.* **scalpless**, *a.*
scalpel (skal'pəl), *n.* a small knife used in surgical operations and anatomical dissections.
scalper, etc. SCALP.
scaly SCALE[1].
scam (skəm), *n.* (*coll.*) a fraudulent trick or operation.
scammony (skam'əni), *n.* the Asiatic *Convolvulus scammonia;* a purgative gum-resin from the root of this.
scamp[1] (skamp), *n.* a worthless person, a rogue; a mischievous child. **scampish**, *a.*
scamp[2] (skamp), *v.t.* to do or execute (work etc.) in a careless manner or with bad material.

scamper (skam'pə), *v.i.* to run rapidly, playfully, hastily, or impulsively. *n.* a hasty or playful run; a hurried excursion, a hurried tour.
scampi (skam'pi), *n.* (*pl.* **scampi**) large prawns, esp. when fried in breadcrumbs or batter.
scampish SCAMP[1].
scan (skan), *v.t.* (*past, p.p.* **scanned**) to count, mark or test the metrical feet or the syllables of (a line of verse); to examine closely or intently, to scrutinize; to examine sequentially or systematically; to glance at or read through hastily; to continuously traverse (an area or object) with a beam of laser light, electrons etc. in order to examine or to produce or transmit an image; to observe with a radar beam; to examine and produce an image of (a body part) using ultrasound, X rays etc. *v.i.* to be metrically correct, to agree with the rules of scansion. *n.* an act of scanning; an image or display produced by scanning. **scanning,** *n., a.* **scanning beam,** *n.* the beam of light or electrons with which an image is scanned for television. **scanning disk,** *n.* a disk with a spiral of holes with or without lenses, used for dividing a transmitted picture into a series of narrow strips. **scanning electron microscope,** *n.* an electron microscope in which a beam of electrons scan an object to produce a three-dimensional image. **scanner,** *n.* one who or that which scans; the aerial of a radar device; an instrument used in scanning the human body, esp. one that takes radiographic photographs from various angles and combines them into a three-dimensional image.
scandal (skan'dl), *n.* indignation, offence or censure at some act or conduct, esp. as expressed in common talk; damage to reputation, reproach, shame, disgrace; malicious gossip, aspersion of character; (*Law*) a defamatory statement, esp. of an irrelevant nature; a disgraceful action, person etc., an affront. **scandal-monger,** *n.* one who disseminates scandal. **scandalize, -ise,** *v.t.* to offend by improper or outrageous conduct, to shock. **scandalous,** *a.* **scandalously,** *adv.* **scandalousness,** *n.* **scandal sheet,** *n.* a newspaper or periodical dealing in scandal and gossip.
scandent (skan'dənt), *a.* climbing, as ivy.
Scandinavian (skandinā'viən), *a.* pertaining to Scandinavia (Norway, Sweden, Denmark and Iceland), its language or literature. *n.* a native or inhabitant of Scandinavia; the languages of Scandinavia collectively.
scandium (skan'diəm), *n.* a rare metallic element, at. no. 21; chem. symbol Sc, discovered in certain Swedish yttrium ores.
scanner, scanner, scanning beam etc. SCAN.
scansion (skan'shən), *n.* the act of scanning verse; a system of scanning.
scant (skant), *a.* not full, large or plentiful; scarcely sufficient, not enough, deficient; short (of). *v.t.* to limit, to skimp, to stint; to dole out grudgingly. **scanty,** *a.* scant, deficient, insufficient; limited or scarcely adequate in extent, size or quantity. **scantily,** *adv.* **scantiness,** *n.*
scantling (skant'ling), *n.* a small quantity or portion; a rough draft or sketch; a beam less than 5 in. (12·7 cm) in breadth and thickness; the sectional measurement of timber; the measurement of stone in all three dimensions; a set of fixed dimensions, esp. in shipbuilding; a trestle for a cask.
scantiness, scanty SCANT.
scape (skāp), *n.* a leafless radical stem bearing the flower; the basal part of an insect's antenna; the shaft of a feather. **scapeless,** *a.*
-scape (-skāp), *comb. form.* scene, view, as in *seascape, townscape.*
scapegoat *n.* (*Bibl.*) a goat on whose head the high priest laid the sins of the people and then sent it away into the wilderness; one made to bear blame due to another.
scapegrace, *n.* a graceless, worthless person, esp. a child.

scapeless SCAPE.
scapolite (skap'əlīt), *n.* one of a group of tetragonal silicate minerals of calcium, aluminium and sodium.
scapula (skap'ūlə), *n.* (*pl.* **-lae, -las**) the shoulder-blade. **scapular,** *a.* pertaining to the scapula or shoulder. *n.* in the Roman Catholic Church, a vestment worn by certain monastic orders usu. consisting of two strips or panels of cloth across the shoulders and hanging down the breast and back; an adaptation of this worn as a badge of affiliation to a religious order; any of a series of feathers springing from the base of the humerus in birds, and lying along the side of the back. **scapulo-,** *comb. form.* **scapulo-humeral,** *a.* pertaining to the scapula and the humerus. **scapulo-ulnar,** *a.* pertaining to the scapula and the ulnus.
scar[1] (skah), *n.* a mark left by a wound, burn, ulcer etc., a cicatrice; the mark left on a plant by the fall of a leaf, stem, seed, deciduous part etc.; the after-effects of emotional distress, a psychological trauma etc. *v.t.* (*past, p.p.* **scarred**) to mark with a scar or scars; to leave with lasting adverse effects. *v.i.* to form a scar, to cicatrize. **scarless,** *a.* **scarry,** *a.*
scar[2] (skah), *n.* a crag, a cliff, a precipitous escarpment.
scarab (ska'rəb), *n.* an ancient Egyptian sacred beetle; a seal or gem cut in the shape of a beetle, worn as an amulet by the Egyptians; a scarabaeid. **Scarabaeus** (-bē'əs), *n.* a genus of beetles typical of the Scarabaeidae. **scarabaeid** (-bē'id), *a.* of or pertaining to the Scarabaeidae, a family of beetles containing the dung beetles. *n.* a beetle of this family. **scarabaeoid** (-oid), *n., a.*
scaramouch (ska'rəmowch, -mooch), *n.* a coward and braggart.
scarce (skeəs), *a.* infrequent, seldom met with, rare, uncommon; insufficient, not plentiful, scantily supplied; *adv.* hardly, scarcely. **to make oneself scarce,** to keep out of the way; to be off, to decamp. **scarcely,** *adv.* hardly, barely, only just; only with difficulty; not quite (used as a polite negative). **scarceness, scarcity,** *n.* deficiency; rareness; a dearth (of); a famine.
scare (skeə), *v.t.* to frighten, to alarm, to strike with sudden fear; to drive (away) through fear. *v.i.* to become frightened. *n.* a sudden fright, a panic; a widespread terror of e.g. invasion, epidemic etc. **scarecrow,** *n.* a figure set up to frighten birds away from crops etc.; a bugbear; a shabby or absurd-looking person. **scaredy-cat,** *n.* (*coll.*) person who is easily frightened. **scaremonger,** *n.* one who causes scares, esp. by circulating unfounded reports etc. **scary,** *a.*
scarf[1] (skahf), *n.* (*pl.* **scarfs, scarves** (-vz)) a long strip or square of some material worn round the neck and shoulders or over the head for warmth or decoration. **scarf-pin, -ring,** *n.* a pin or ring, usu. of gold, used to fasten a neck-tie. **scarfskin,** *n.* the outer layer of skin, the cuticle.
scarf[2] (skahf), *v.t.* to join the ends of (timber) by means of a scarf-joint; to cut a scarf in or on; to flench (a whale). *n.* (*pl.* **scarfs**) (**scarf, scarf-joint**) a joint made by bevelling or notching so that the thickness is not increased, and then bolting or strapping together; a bevelled or notched end that forms such a joint; an incision or groove cut along the body of a whale before stripping off the blubber.
scarify (skeə'rifī), *v.t.* (*Surg.*) to scratch or make slight incisions in; to loosen the surface of (soil); to criticize mercilessly. **scarification,** *n.* **scarificator** (-fi-), *n.* a surgical instrument with lancet-points used in scarifying. **scarifier,** *n.* one who scarifies; a scarificator; an implement or machine for breaking up soil etc.
scarlatina (skahlətē'nə), *n.* (a mild form of) scarlet fever.
scarless SCAR[1].
scarlet (skah'lit), *n.* a bright red colour tending towards orange; cloth or dress of this colour esp. official robes or

uniform. *a.* of a scarlet colour; dressed in scarlet. **scarlet-bean** SCARLET RUNNER. **scarlet fever,** *n.* an infectious fever characterized by the eruption of red patches on the skin. **scarlet hat,** *n.* a cardinal's hat; the rank of cardinal. **scarlet rash,** *n.* roseola. **scarlet runner,** *n.* a trailing bean, *Phaseolus multiflorus,* with scarlet flowers. **scarlet woman,** *n.* worldliness or sensuality; pagan or papal Rome (see Rev. xvii.4–5); a prostitute.

scarp (skahp), *n.* a steep or nearly perpendicular slope; the interior slope of the ditch at the foot of the parapet of a fortification. *v.t.* to cut down so to be steep or nearly perpendicular.

scarper (skah'pə), *v.i.* (*sl.*) to leave in a hurry; to go away without notice or warning.

scarves, *pl.* SCARF[1].

scary SCARE.

scat[1] (skat), *int.* go away!, be off! *v.i.* (*past, p.p.* **scatted**) (*coll.*) to depart hastily; (*chiefly N Am., coll.*) to move quickly.

scat[2] (skat), *n.* jazz singing in meaningless syllables. *v.i.* (*past, p.p.* **scatted**) to sing in this way.

scathing, *a.* hurtful, harmful; of sarcasm etc., very bitter or severe, withering. **scathingly,** *adv.*

scatology (skatol'əji), *n.* the study of fossil excrement or coprolites; the biological study of excrement, esp. to determine diet; interest in or literature characterized by obscenity. **scatological** (skatəloj'-), *a.* **scatophagous** (-tof'əgəs), *a.* feeding on dung. **scatoscopy** (-tos'kəpi), *n.* diagnosis by means of faeces.

scatter (skat'ə), *v.t.* to throw loosely about, to fling in all directions; to strew, to bestrew; to cause to separate in various directions, to disperse; to dissipate; to diffuse (radiation) or cause to spread out. *v.i.* to disperse; to be dissipated or diffused. *n.* the act of scattering; a small number scattered about; the extent of scattering. **scatterbrain,** *n.* a giddy, heedless person. **scatter-brained,** *a.* **scatter-cushion, -rug,** *n.* a small cushion or rug which can be moved to any position in a room. **scattered,** *a.* irregularly situated, not together; widely apart. **scattering,** *n.* the act of dispersing or strewing something; a small amount or number irregularly strewn; the deflecting or spreading out of a beam of radiation in passing through matter.

scatty (skat'i), *a.* (*coll.*) incapable of prolonged concentration, empty-headed, giddy. **scattily,** *adv.* **scattiness,** *n.*

scaup, scaup-duck (skawp), *n.* a sea-duck found in the northern regions.

scauper (skaw'pə), *n.* a wood-engraver's gouge-like tool.

scavenger (skav'ənjə), *n.* a person employed to clean the streets by sweeping, scraping, and carrying away refuse; one who collects waste or discarded objects; an organism feeding on refuse, carrion etc.; a child employed in a spinning-mill to collect loose cotton; anyone willing to do 'dirty work' or delighting in filthy subjects. **scavenger-beetle, -crab,** *n.* a beetle or crab feeding on carrion. **scavenge,** *v.t.* to clean (streets etc.); to search for or salvage (something usable) from among waste or discarded material; to remove impurities from (molten metal) by causing their chemical combination; to remove (impurities etc.). *v.i.* to act as a scavenger; to search for usable material. **scavengery,** *n.*

ScD, (*abbr.*) Doctor of Science.

SCE, (*abbr.*) Scottish Certificate of Education.

scena (shā'nə), *n.* (*pl.* **scene,** (-nā) a long elaborate solo piece or scene in opera.

scenario (sinah'riō), *n.* (*pl.* **-rios**) a sketch or outline of the scenes and main points of a play etc. the script of a film with dialogue and directions for the producer during the actual shooting; an account or outline of projected, expected or imagined future events.

scene (sēn), *n.* the stage in a Greek or Roman theatre; hence, the stage, the theatre; the place where anything occurs or is exhibited as on a stage; the place in which the action of a play or story is supposed to take place; one of the painted frames, hangings or other devices used to give an appearance of reality to the action of a play; a division of a play comprising so much as passes without change of locality or break of time, or, in French drama, without intermediate entrances or exits; a single event, situation or incident in a play or film; a film or television sequence; a description of an incident, situation etc. from life; a striking incident, esp. an exhibition of feeling or passion; a landscape, a view, regarded as a piece of scenery; (*coll.*) one's usual or preferred social environment, area of interest etc.; (*coll.*) an area of activity or business. **behind the scenes,** at the back of the stage; in possession of facts etc., not generally known. **change of scene,** change of surroundings by travel. **scene-dock,** *n.* a place near the stage in a theatre for storing scenery. **scene-painter,** *n.* one who paints scenery for theatres. **scene-painting,** *n.* **scene-shifter,** *n.* a person employed in a theatre to move scenery. **scenery** (-əri), *n.* the various parts or accessories used on the stage to represent the actual scene of the action; the views presented by natural features, esp. when picturesque. **scenic,** *a.* of or pertaining to the stage; of or pertaining to natural scenery; characterized by beautiful natural scenery, picturesque; arranged for effect, dramatic, theatrical; of a painting etc., depicting a scene or incident. **scenic railway,** *n.* a switchback railway at a fun-fair.

scent (sent), *v.t.* to perceive by smell; to recognize the odour of; to begin to suspect; to trace or hunt (out) by or as by smelling; to perfume. *v.i.* to exercise sense of smell. *n.* odour, esp. of a pleasant kind; the odour left by an animal forming a trail by which it can be followed (as by hounds); pieces of paper left as a trail in a paper-chase; a trail to be pursued; a clue; a liquid essence containing fragrant extracts from flowers etc., a perfume; the sense of smell, esp. the power of recognizing or tracing things by smelling. **scent-bag,** *n.* an external pouch-like scent-gland, as in the musk-deer; a bag containing aniseed etc., used to leave a track of scent for hounds to follow. **scent-bottle,** *n.* a bottle for holding perfume. **scent-gland,** *n.* a gland secreting an odorous substance, as in the musk-deer, civet etc. **scent-organ,** *n.* **scented,** *a.* having a scent (*usu. in comb.*, as *keen-scented*).

sceptic, (*esp. N Am.*) **skeptic** (skep'tik), *n.* one who doubts the truth of a revealed religion; an agnostic; an atheist; a person of a questioning, doubting or incredulous habit of mind; one who casts doubt on any statement, theory etc., esp. in a cynical manner; one who questions or denies the possibility of attaining knowledge of truth; an adherent of philosophical scepticism, a Pyrrhonist. *a.* sceptical. **sceptical,** *a.* pertaining to or characteristic of a sceptic; doubting or denying the truth of revelation, or the possibility of knowledge; given to doubting or questioning, incredulous. **sceptically,** *adv.* **scepticism** (-sizm), *n.*

sceptre, (*esp. N Am.*) **scepter** (sep'tə), *n.* a staff or baton borne by a sovereign as a symbol of authority; royal authority. *v.t.* (*in p.p.*) to invest with a sceptre or with royal authority. **sceptred,** *a.*

sch., (*abbr.*) school.

schadenfreude (shah'dənfroidə), *n.* pleasure in others' misfortunes.

schedule (shed'ūl, *esp. N Am.* sked'-), *n.* a written or printed table, list, catalogue or inventory (appended to a document); a timetable; a planned programme of events, tasks etc. *v.t.* to enter in a schedule; to make a schedule or list of; to arrange for a particular time.

scheelite (shē'līt), *n.* a vitreous variously-coloured mineral, a tungstate of calcium.

schema (skē'mə), *n.* (*pl.* **-mata** (-tə)) a scheme, summary, outline or conspectus; a chart or diagram; the abstract fig-

scheme ure of a syllogism; a figure of speech; in Kant's philosophy, the form, type or rule under which the mind applies the categories to the material of knowledge furnished by sense-perception. **schematic** (-mat'-), *a.* having, or in the nature of, a plan or schema. **schematically,** *adv.* **schematize, -ise,** *v.t.* to formulate or express by means of a scheme; to apply the Kantian categories in philosophy to. **schematism,** *n.*

scheme (skēm), *n.* a plan, a project, a proposed method of doing something; a contrivance, an underhand design; a table or schedule of proposed acts, events etc., a syllabus; a systematic statement, representation, diagram or arrangement of facts, objects, principles etc.; a table of classification. *v.t.* to plan, to design to contrive, to plot. *v.i.* to form plans; to plot. **schemer,** *n.* **scheming,** *a.* given to forming schemes.

scherzo (skœt'sō), *n.* (*pl.* **-zi** (-sē), **-zos**) a light playful movement in music, usu. following a slow one, in a symphony or sonata. **scherzando** (-san'dō), *adv.* (*Mus.*) playfully.

schiavone (skyavō'nā), *n.* a 17th-cent. basket-hilted broadsword, so called because the Schiavoni or Slav bodyguards of the Doge were armed with it.

Schick test (shik), *n.* a test to determine susceptibility to diphtheria by injecting diluted diphtheria toxin into the skin.

schilling (shil'ing), *n.* the standard monetary unit of Austria.

schindylesis (skindilē'sis), *n.* (*pl.* **-ses** (-sēz)) an articulation in which a thin part of one bone fits into a groove in another. **schindyletic** (-let'-), *a.*

schipperke (skip'əki), *n.* a small black variety of lapdog.

schism (siz'm, skiz'm), *n.* a split or division in a community; division in a Church, esp. secession of a part or separation into two Churches; the sin of causing such division. **schismatic** (-mat'-), *n., a.* **schismatical,** *a.* **schismatically,** *adv.* **schismatist,** *n.* **schismatize, -ise,** *v.t., v.i.*

schist (shist), *n.* a rock of a more or less foliated or laminar structure, tending to split easily. **schistaceous** (-tā'shəs), *a.* slate-grey. **schistose** (-tōs), *a.* of the nature or structure of schist.

schistosoma (shistəsō'mə), *n.* the *Bilharzia* genus of worms. **schistosomiasis** (-səmi'əsis), *n.* a disease caused by infestation with worms of the genus *Schistosoma*.

Schizanthus (skizan'thəs), *n.* a genus of annual plants from Chile with much-divided leaves and showy flowers.

schizo (skit'sō), *n.* (*coll.*) short for SCHIZOPHRENIC.

schiz(o)-, *comb. form.* marked by a cleft or clefts; tending to split.

schizocarp (skiz'əkahp), *n.* a fruit splitting into several one-seeded portions without dehiscing. **schizocarpous** (-kah'-), *a.*

schizogenesis (skitsəjen'əsis), *n.* reproduction by fission. **schizogenic, -genetic** (-net'-), *a.* **schizogenically, -genetically,** *adv.* **schizogony** (-sog'əni), *n.* schizogenesis.

schizognathous (skitsog'nəthəs), *a.* having the bones of the palate cleft from the vomer and each other, as in the gulls, plovers etc. **schizognathism,** *n.*

schizomycete (skitsəmi'sēt), *n.* (*Bot.*) one of the Schizomycetes, a class of microscopic organisms comprising bacteria. **schizomycetous** (-sē'-), *a.*

schizophrenia (skitsəfrē'niə), *n.* a severe psychological disorder characterized by loss of contact with reality, personality disintegration, hallucinations, delusions etc. **schizophrenic** (-fren'-), *n., a.*

schizopod (skit'səpod), *n.* one of the Schizopoda, a suborder of podophthalmate crustaceans with the feet apparently cleft. **schizopodous** (-sop'-), *a.*

schizothymia (skitsəthī'miə), *n.* introversion exhibiting elements of schizophrenia but within normal limits. **schizothymic,** *a.*

schläger (shlā'gə), *n.* a German student's duelling sword, pointless, but with sharpened edges towards the end.

schlemiel, schlemihl (shləmēl'), *n.* (*chiefly N Am., coll.*) a bungling clumsy person who is easily victimized.

schlepp, (shlep), *v.t.* (*chiefly N Am., coll.*) to drag, pull. *n.* an unlucky or incompetent person.

schlieren (shlē'rən), *n.* small streaks of different composition in igneous rock; streaks in a transparent fluid caused by regions of differing density and refractive index.

schloss (shlos), *n.* a castle (in Germany).

schmaltz, schmalz (shmawlts), *n.* over-sentimentality, esp. in music. **schmaltzy,** *a.*

schmelze (shmɛlt'sə), *n.* one of various kinds of coloured glass, esp. that coloured red and used to flash white glass.

schnapps (shnaps), *n.* a type of Dutch gin, (esp. in Germany) any strong spirit.

schnauzer (shnow'zə), *n.* a wire-haired German terrier.

schnitzel (shnit'səl), *n.* a veal cutlet.

schnorkel (shnaw'kl), SNORKEL.

schnozzle (shnoz'l), *n.* (*chiefly N Am., coll.*) a nose.

scholar (skol'ə), *n.* a learned person, esp. one with a profound knowledge of literature; an undergraduate on the foundation of a college and receiving assistance from its funds, usu. after a competitive examination; a person acquiring knowledge, a (good or apt) learner; a disciple; a pupil, a student, one attending school. **scholarly,** *a.* befitting a scholar; learned. **scholarship,** *n.* high attainments in literature or science; education, instruction; education, usu. with maintenance, free or at reduced fees, granted to a successful candidate after a competitive examination; the emoluments so granted to a scholar.

scholastic (skəlas'tik), *a.* pertaining to school, schools, universities etc.; educational, academic; pedagogic, pedantic; pertaining to or characteristic of the schoolmen of the Middle Ages; given to precise definitions and logical subtleties. *n.* a schoolman of the Middle Ages; one characterized by the method and subtlety of the schoolman; a mere scholar, an academic person; a Jesuit of the third grade. **scholastically,** *adv.* **scholasticism** (-sizm), *n.*

scholiast (skō'liast), *n.* a commentator, esp. an ancient grammarian who annotated the classics. **scholiastic** (-as'-), *a.* **scholium** (-əm), *n.* (*pl.* **-lia** (-ə)) a marginal note, esp. an explanatory comment on the Greek and Latin authors by an early grammarian.

school[1] (skool), *n.* a shoal of fish, porpoises etc. *v.i.* to form a school, swim in a school.

school[2] (skool), *n.* an institution for education or instruction, esp. one for instruction of a more elementary kind than that given at universities; a faculty of a university; an establishment offering specialized teaching; the building or buildings of a school; the body of pupils of a school; a session or time during which teaching is carried on; a lecture-room; a seminary in the Middle Ages for teaching logic, metaphysics and theology; (*pl.*) the mediaeval universities, professors, teaching etc.; scholasticism; any of the branches of study with separate examinations taken by candidates for honours; the hall where such examinations are held; (*pl.*) the final BA examination at Oxford Univ.; the body of disciples or followers of a philosopher, artist etc., or of adherents of a cause, principle, system of thought etc.; (*Mus.*) a book of instruction, a manual; any sphere or circumstances serving to discipline or instruct; (*coll.*) a group of people assembled for a common purpose, as playing poker. *v.t.* to instruct, to educate; to train, to drill; to discipline, to bring under control; to send to school. **school board,** *n.* a public body (1870-1902) elected to provide for the elementary instruction of children in their district. **schoolbook,** *n.* a book for use in schools. **schoolboy, -girl,** *n.* a boy or girl attending a school. *a.* pertaining to schoolboys or schoolgirls. **schoolfellow,** *n.* one who attends the same school. **schoolhouse,** *n.* a building used as a school; the dwelling-house of a schoolmaster or schoolmistress; the

head-teacher's house or the chief boarding-house at a public school. **school-ma'am, -marm**, *n.* (*N Am. coll.*) a schoolmistress. **schoolman**, *n.* a teacher or professor in a mediaeval university; one versed in the theology, logic, or metaphysics of the mediaeval schools or the niceties of academic disputation. **schoolmaster, -mistress**, *n.* a head or assistant teacher in a school; a pedagogue; one who or that which trains or disciplines. **schoolmate**, *n.* one attending the same school. **schoolroom**, *n.* a room where teaching is given, in a school, house etc. **schoolteacher**, *n.* one who teaches in a school. **schooling**, *n.* instruction or education at school; training, tuition, coaching, guidance; school fees; discipline; the training of a horse for riding, or in dressage, jumping etc.
schooner (skoo'nə), *n.* a vessel with two or more masts with fore-and-aft rigging; (*N Am.*) a large emigrant-wagon or van; a tall glass for beer or ale; a tall glass for sherry.
schottische (shətesh'), *n.* a dance resembling a polka; the music for it.
schuss (shus), *n.* a straight fast ski slope; a run made on this. *v.i.* to make such a run.
schwa (shwah. shvah), *n.* a neutral unstressed vowel sound; the symbol (ə) used to represent this.
sci, (*abbr.*) science, scientific. **sci fi** (sī'fī), (*abbr.*) science fiction.
sciatic (siat'ik), *a.* pertaining to the hip; of or affecting the sciatic nerve; of the nature of or affected by sciatica. **sciatic nerve**, *n.* the nerve that extends from the pelvis down the back of the thigh. **sciatica** (-kə), *n.* neuralgia of the hip and thigh; pain in the great sciatic nerve. **sciatically**, *adv.*
science (sī'əns), *n.* systematized knowledge; a department of systematized knowledge, a system of facts and principles concerning any subject; a natural science; the pursuit of such knowledge or the principles governing its acquirement. **science fiction**, *n.* fiction dealing with space travel, life on one of the planets etc. **science park**, *n.* a place where academic scientific research is applied to commercial developments. **sciential** (-en'shəl), *a.* of or producing science; having knowledge. **scientially**, *adv.*
scientific (-tif'-), *a.* pertaining to, used or engaged in science; treating of or devoted to science; made or done according to the principles of science, systematic, exact; of boxing etc., skilful, expert. **scientifically**, *adv.* **scientism**, *n.* scientific methods or attitudes; (belief in) the application of scientific methods to investigate and explain social and psychological phenomena. **scientist**, *n.* one who studies or is expert in a (natural) science. **Scientology**®, *n.* a religious movement advocating self-improvement of one's physical and mental condition through psychological and scientific means. **scientologist**, *n.*
scienter (sien'tə), *adv.* (*Law.*) with knowledge, wittingly, deliberately.
sci fi SCI.
scilicet (sī'liset), *adv.* to wit, namely.
scilla (sil'ə), *n.* a genus of bulbous liliaceous plants containing the squills.
Scillonian (silō'niən), *n., a.* (a native) of the Scilly Isles.
scimitar (sim'itə), *n.* a short oriental sword, single-edged, curved and broadest towards the point.
scintigraphy (sintig'rəfi), *n.* a diagnostic technique that uses the radiation emitted following administration of a radioactive isotope to produce a picture of an internal body organ. **scintigram** (sin'tigram), *n.* a picture produced by scintigraphy.
scintilla (sintil'ə), *n.* a spark; a trace, hint. **scintillate** (sin'-), *v.i.* to emit sparks; to emit flashes of light when bombarded by electrons, photons etc.; to sparkle, to twinkle; to be brilliantly witty or interesting. **scintillant**, *a.* **scintillation**, *n.* **scintillation counter**, *n.* an instrument for measuring radiation from a source by electronically counting the flashes of light produced by the absorption of radioactive particles by a phosphor.
sciolist (sī'əlist), *n.* one who knows many things superficially, a pretender to knowledge. **sciolism**, *n.* **sciolistic** (-lis'-), *a.*
sciolto (shol'tō), *adv.* (*Mus.*) freely, to one's taste; staccato.
scion (sī'ən), *n.* a shoot, esp. for grafting or planting; a descendant, a child.
sciotheism (sīōthē'izm), *n.* ghost-worship, esp. of departed ancestors.
scire facias (sīə'ri fā'shias), *n.* (*Law*) a writ to enforce the execution of or annul judgments etc.
scirocco (shirok'ō), SIROCCO.
scirrhus (si'rəs, ski'-), *n.* (*pl.* **-rrhi** (-rī)) a hard (cancerous) tumour. **scirrhoid** (-roid), **scirrhous**, *a.* **scirrhosity** (-ros'-), *n.*
scissel (sis'l), *n.* metal clippings; the remainder of plates after discs have been punched out in coining.
scissile (sis'il), *a.* that may be cut. **scission** (sish'ən), *n.* the act of cutting or dividing; a division, separation or split.
scissors (siz'əz), *n.pl.* (**scissors, pair of scissors**) a cutting instrument consisting of two blades pivoted together that cut objects placed between them; a gymnastic movement in which the legs open and close with a scissor-like action. **scissor**, *v.t.* to cut with scissors; to clip or cut (out) with scissors. **scissors-and-paste**, *n., a.* (of) compilation, as distinguished from original literary work. **scissors hold**, *n.* a wrestling hold in which the legs lock round the opponent's head or body. **scissors kick**, *n.* a swimming kick in which the legs move in a scissor-like action. **scissor-tooth**, *n.* a tooth working against another like a scissor-blade, in certain carnivores.
sciurine (sī'ūrin. -rīn), *a.* pertaining to or resembling the squirrel family. *n.* a squirrel. **sciuroid** (-roid), *a.*
sclera (sklíə'rə), *n.* the sclerotic. **scleritis** (sklərī'tis), *n.* sclerotitis.
sclerenchyma (sklāreng'kimə), *n.* the strong tissue forming the hard or fibrous parts of plants, such as the walls of nuts and fruit-stones, leaf midribs etc.; the calcareous tissue in coral. **sclerenchymatous** (-kim'ətəs), *a.*
scleriasis (sklāri'əsis), *n.* hardening or induration of tissue.
sclerite (sklíə'rīt), *n.* one of the definite component parts of the hard integument of various invertebrates. **scleritic** (-rit'-), *a.*
scler(o)-, *comb. form.* hard, dry; sclerotic.
scleroderm (skle'rədœm), *n.* a hardened integument or exoskeleton, esp. of corals; a fish of the family Sclerodermi, having hard scales. **scleroderma** (-dœ'-), **-mia** (-miə), *n.* a chronic induration of the skin. **sclerodermatous** (-dœ'-), **sclerodermic** (-dœ'-), *a.* **sclerodermite** (-dœ'mīt), *n.* one of the hard segments of the body in crustaceans. **sclerodermitic** (-mit'-), *a.*
sclerogen (sklíə'rəjən), *n.* the hard matter deposited in the cells of certain plants, as the ivory-nut. **sclerogenous** (sklərōj'-), *a.*
scleroid (sklíə'roid), *a.* (*Bot., Zool.*) hard in texture.
scleroma (sklərō'mə), *n.* (*pl.* **-mata** (-tə)) hardening of cellular tissue, scleriasis.
scleroprotein (sklíərōprō'tēn), *n.* an insoluble protein, as keratin, forming the skeletal tissues of the body.
sclerosis (sklərō'sis), *n.* hardening of a plant cell-wall by the deposit of sclerogen; thickening or hardening of a body tissue. **sclerosed** (sklíə'rōzd), *a.*
sclerotal (sklərō'təl), *n.* one of the bony plates of the sclerotic coat in some birds and reptiles; the sclerotic. *a.* pertaining to the sclerotal; sclerotic. **sclerotic** (-rot'-), *a.* of the outer coat or tunic of the eye, hard, indurated; of or affected with sclerosis. *n.* the firm white membrane forming the outer coat of the eye, the white of the eye; a medicine hardening the parts to which it is applied. **scleritis** (-rī'tis), **sclerotitis** (sklərətī'tis), *n.* inflammation of

the sclerotic.
sclerotium (sklərō′tiəm), *n.* a compact tuberous mass formed on the mycelium of certain higher fungi, as ergot; a cyst-like part of a plasmodium in the Mycetozoa. **sclerotioid** (-oid), *a.* resembling a sclerotium.
sclerous (sklīə′rəs), *a.* hard, indurated, ossified.
scoff[1] (skof), *n.* an expression of contempt, derision, or mockery; a gibe, a taunt; an object of derision, a laughing-stock. *v.i.* to speak in derision or mockery, to mock or jeer (at). **scoffer**, *n.* **scoffingly**, *adv.*
scoff[2] (skof), *v.t.* (*coll.*) to eat ravenously. *n.* food.
scold (skōld), *v.i.* to find fault noisily or angrily; to rail (at). *v.t.* to chide or find fault with noisily or angrily; to chide, to rate, to rail at. *n.* a noisy, railing, nagging woman; a scolding. **scolder**, *n.* **scolding**, *n.*, *a.* **scoldingly**, *adv.*
scoliosis (skoliō′sis), *n.* lateral curvature of the spine. **scoliotic** (-ot′-), *a.*
scollop (skol′əp), *n.* SCALLOP.
Scolopax (skol′əpaks), *n.* a genus of birds containing the woodcock and formerly the snipe and redshank. **scolopaceous** (-pā′shəs), *a.*
Scolopendra (skoləpen′drə), *n.* a genus of myriapods containing the larger centipedes; a millipede or centipede. **scolopendriform** (-drifawm), **scolopendrine** (-drin, -drīn), *a.*
Scolopendrium (skoləpen′driəm), *n.* a genus of ferns containing the hart's tongue, Phyllitis.
Scolytus (skol′itəs), *n.* a genus of bark-boring beetles. **scolytid** (-tid), *n.* **scolytoid** (-toid), *a.*
Scomber (skom′bə), *n.* a genus of fish containing the mackerel. **scombrid** (-brid), *n.* **scombroid** (-broid), *a.* of or belonging to the Scombroidea, a suborder of fishes including the mackerels, tunas and swordfishes. *n.* one of the Scombroidea.
scon (skon) var. of SCONE.
sconce[1] (skons), *n.* a candle-holder fixed to a wall; the socket of a candlestick into which the candle is inserted.
sconce[2] (skons), *n.* a block-house, a bulwark, a small detached fort; a shelter, a covering, a shed. *v.t.* to fortify with a sconce.
scone (skon, *Eng. also* skōn), *n.* a soft thin plain cake, usu. in small round or triangular pieces, cooked on a griddle or in an oven.
scoop (skoop), *n.* a short-handled shovel-like implement for drawing together, lifting and moving loose material such as coal, grain, sugar, potatoes etc.; a large ladle or dipping-vessel; a gouge-like implement used by grocers, surgeons etc. or for spooning out shaped pieces of ice-cream or other soft food; the bucket of a dredging-machine; a coal-scuttle; the act or movement of scooping; the amount scooped at once; (*coll.*) a large profit made in a speculation or competitive transaction; the publication or broadcasting of a piece of sensational news in advance of rival newspapers etc.; a news item so published. *v.t.* to ladle or dip (out) or to hollow (out) with a scoop; to lift (up) with a scoop; to scrape, gouge or hollow (out); (*coll.*) to gain (a large profit) by a deal etc.; to forestall (rival newspapers etc.) with a piece of sensational news. **scoopwheel**, *n.* a wheel with buckets round it used to raise water or for dredging. **scooper**, *n.* one who or that which scoops; a tool used by engravers. **scoopful**, *n.*
scoot (skoot), *v.i.* (*coll.*) to dart off, bolt, to scurry away. **scooter**, *n.* a two-wheeled toy vehicle on which a child can ride with one foot, propelling with the other; a larger, motorized two-wheeled vehicle with a seat.
scopa (skō′pə), *n.* (*pl.* **-pae** (-pē)) a brush-like tuft of bristly hairs as on the legs of bees. **scopate** (-pāt), *a.* brush-shaped; covered with brush-like hairs. **scopula** (skop′ūlə), *n.* a small brush-like tuft on the legs of bees and spiders, a scopa. **scopulate** (skop′ūlət), *a.*
scope[1] (skōp), *n.* range of action or observation, outlook, reach, sphere; extent of or room for activity, development etc.; outlet, opportunity, vent; (*Naut.*) length of cable at which a vessel rides.
scope[2] (skōp), *n.* short for OSCILLOSCOPE under OSCILLATE, PERISCOPE, TELESCOPE etc.
-scope (-skōp), *suf.* denoting an instrument of observation etc., as in *microscope, spectroscope.* **-scopic** (-skopik), *suf.* pertaining to this or to observation etc. as in *microscopic, spectroscopic.* **-scopy** (-skəpi), *suf.* observation by the instrument etc., specified, as in *microscopy, spectroscopy.*
scopolamine (skəpol′əmēn), *n.* hyoscine hydrobromide, used, among other purposes, as a sedative and a truth drug.
scorbutic (skawbū′tik), *a.* pertaining to, like or affected with scurvy. *n.* a person affected with scurvy. **scorbutically**, *adv.*
scorch (skawch), *v.t.* to burn the outside of so as to injure or discolour without consuming, to singe to parch, to dry or shrivel (up); to affect harmfully with or as with heat; to criticize or censure severely. *v.i.* to be parched, singed or dried up with or as with heat; (*coll.*) to go at an excessive rate of speed. *n.* a burn or mark caused by scorching; (*coll.*) an act or spell of scorching. **scorched**, *a.* **scorched earth**, a descriptive term for the destruction of everything in a country that might be of service to an invading army. **scorching**, *a.* **scorchingly**, *adv.* **scorcher**, *n.* one who or that which scorches; (*coll.*) an extremely hot day; (*sl.*) a striking or staggering example, a stunner.
scordato (skawdah′tō), *a.* (*Mus.*) put out of tune. **scordatura** (-too′rə), *n.* an intentional departure from normal tuning to secure special effects.
score (skaw), *n.* a notch or mark on a tally; an account, a bill, a debt; anything laid up or recorded against one, a grudge; the points made by a player or side at any moment in, or in total in certain games and contests; the record of this; the act of gaining a point in a game or contest; a mark from which a race starts, competitors fire in a shooting-match etc.; a line drawn or scratched through writing etc.; a scratch, incision; a copy of a musical work in which all the component parts are shown, either fully or in a compressed form, so called from the line orig. drawn through all the staves; the music for a film, play etc.; the notation for a choreographed work; twenty, a set of twenty; (*pl.*) large numbers; account, category, reason; (*sl.*) a remark etc. in which one scores off another person; (*coll.*) the situation, the facts. *v.t.* to mark with notches, cuts, scratches, lines etc.; to gash, to groove, to furrow; to make or mark (lines etc.); to mark (out) with lines; to mark (up) or enter in a score; to gain (a point, a win etc.) in a game or contest; to arrange in score; to orchestrate; to arrange for an instrument; to prepare the sound-script for (a film). *v.i.* to keep a score; to win points, advantages etc.; (*sl.*) to obtain illegal drugs; (*sl.*) of a man, to successfully seduce someone into having sexual intercourse. **to pay off old scores**, to pay someone out or have revenge for an offence of old standing. **to score off**, (*coll.*) to get the better of; to triumph over in argument, repartee etc. **scoreboard**, *n.* a board on which the score at any point in a game or contest is displayed. **scorecard**, *n.* a card for recording the score in a game, esp. golf. **scorer**, *n.* **scoring**, *n.*
scoria (skaw′riə), *n.* (*pl.* **-riae** (-riē)) cellular lava or ashes; the refuse of fused metals, dross. **scoriaceous** (-ā′shəs), *a.* **scoriform** (-fawm), *a.* **scorify** (-rifī), *v.t.* to reduce to dross; to assay (metal) by fusing its ore in a scorifier with lead and borax. **scorification** (-fi-), *n.* **scorifier**, *n.*
scorn (skawn), *n.* contempt, disdain; mockery, derision; a subject or object of extreme contempt. *v.t.* to hold in extreme contempt or disdain; to regard as unworthy, paltry or mean. **scorner**, *n.* **scornful**, *a.* **scornfully**, *adv.* **scornfulness**, *n.*
scorper (skaw′pə), *n.* a gouging-tool for working in con-

Scorpio 744 scrap

cave surfaces in wood, metal or jewellery.
Scorpio (skaw'piō), *n.* a zodiacal constellation, and the eighth sign of the zodiac. **scorpioid** (-oid), *a.* (*Bot.*) curled up like the end of a scorpion's tail and uncurling as the flowers develop. *n.* a scorpioid inflorescence.
scorpion (skaw'piən), *n.* one of an order of arachnids, with claws like a lobster and a sting in the jointed tail; (*Bibl.*) a whip armed with points of iron; a form of ballista; the constellation Scorpio. **scorpion-fish**, *n.* the sea-scorpion. **scorpion-fly**, *n.* a fly of the family Panorpidae, named from the forceps-like point of the abdomen. **scorpiongrass**, *n.* the myosotis or forget-me-not.
Scot (skot), *n.* a native of Scotland; (*pl.*) *orig.*, a Gaelic people migrating to Scotland from Ireland in the 5th or 6th cent.
scot (skot), *n.* a payment, an assessment, a tax. **scot and lot**, a town or parish tax levied according to ability to pay. **to pay scot and lot**, to settle outstanding accounts, obligations etc. **scot-free**, *a.* free from payment, untaxed; unpunished; unhurt, safe.
Scot., (*abbr.*) Scotland, Scottish.
Scotch (skoch), *a.* Scottish. *n.* (a glass of) (Scotch) whisky; the Scots. **Scotch-barley**, *n.* pot or husked barley. **Scotch broth**, *n.* a clear broth containing barley and chopped vegetables. **Scotch cap**, *n.* a brimless woollen cap, either a Balmoral or a Glengarry. **Scotch catch** or **snap**, *n.* a short note followed by a long note in two played to the same beat. **Scotch egg**, *n.* a hard-boiled egg encased in sausage-meat and breadcrumbs. **Scotch fir**, *n.* the Scots pine. **Scotch mist**, *n.* a wet dense mist; fine drizzle. **Scotch terrier**, *n.* a breed of dog characterized by short legs and a rough coat. **Scotch thistle**, *n.* one of various thistles regarded as the Scottish national emblem, esp. *Carduus lanceolatus* or *C. nutans*. **Scotch whisky**, *n.* whisky made from malted barley and distilled in Scotland. **Scotchman, -woman**, *n.* a Scotsman, -woman.
Scotchman grass, *n.* (*New Zealand*) a variety of grass with sharp points.
scotch (skoch), *v.t.* (*dated*) to cut with narrow incisions; to wound slightly, to cripple, to disable. *n.* (*dated*) a slight cut or incision; a mark for hopping from, as in the game of hopscotch.
scotch[2] (skoch), *n.* a block for a wheel or other round object. *v.t.* to block, wedge or prop (a wheel, barrel etc.) to prevent rolling; to frustrate (a plan etc.).
scoter (skō'tə), *n.* a large sea-duck of the genus *Melanitta*.
scotia (skō'shə), *n.* a hollow moulding in the base of a column.
Scotism (skō'tizm), *n.* the scholastic philosophy of Johannes Duns Scotus (d. 1308). **Scotist**, *n., a.*
Scotland Yard (skot'lənd), *n.* the headquarters of the London Metropolitan Police; the Criminal Investigation Department of the police; police detectives.
scoto-, *comb. form.* dark, dullness.
scotodinia (skotōdin'iə), *n.* dizziness, vertigo, with dimness of vision.
scotoma (skotō'mə), *n.* (*pl.* **-mas, -mata** (-tə)) a blind spot in the field of vision; dizziness or swimming of the head with dimness of sight. **scotomatous**, *a.*
Scots (skots), *a.* Scottish (applied to the people, language and law). *n.* the language of Scotland; (*pl.* of SCOT) the people of Scotland. **Scots pine**, *n.* a European pine, *Pinus sylvestris*, prob. indigenous in N Britain. **Scotsman, -woman**, *n.*
Scottish (skot'ish), *a.* pertaining to Scotland, its people, language, dialect or literature. *n.* the Scots language. **Scottish terrier**, *n.* a Scotch terrier. **Scotticism** (-sizm), *n.* a Scottish idiom. **Scotticize, -ise**, *v.t.* to make Scottish.
Scottie, Scotty (-i), *n.* a nickname for a Scotsman; (*coll.*) a Scotch terrier. **Scottishness**, *n.*
scoundrel (skown'drəl), *n.* an unprincipled person, a rogue, a rascal, a villain.

scour[1] (skowə), *v.t.* to clean, polish or brighten by friction; to remove or clean (away, off etc.) by rubbing; to flush or clear out; of water etc., to pass swiftly through or over; to purge violently. *v.i.* to clean; to be scoured or cleaned (well, easily etc.); to be purged to excess. *n.* scouring; a swift, deep current; a rapid; the clearing action of this; dysentery in cattle; a cleanser for various fabrics. **scourer**, *n.*
scour[2] (skowə), *v.i.* to rove, to range; to skim, to scurry; to search about. *v.t.* to move rapidly over, esp. in search; to search thoroughly.
scourge (skœj), *n.* a whip with thongs used as an instrument of punishment; any means of inflicting punishment, vengeance or suffering; a pestilence or plague. *v.t.* to whip with or as with a scourge; to afflict, to harass, to chastise. **scourger**, *n.*
Scouse (skows), *n.* a native or inhabitant of Liverpool; the dialect of Liverpool.
scout[1] (skowt), *n.* one sent out to bring in information, esp. one employed to watch the movements etc. of an enemy; one employed to search for people with talent in a particular field, new sales markets etc.; the act of watching or bringing in such information, a scouting expedition; a member of an organization, established in Great Britain by Lord Baden-Powell in 1908 and now worldwide, intended to train and develop qualities of leadership, responsibility etc. in boys; a college servant at Oxford Univ. *v.t.* to act as a scout. *v.i.* to search for. **Boy Scout**, *n.* the former name for a Scout. **scoutmaster**, *n.* formerly, the leader of a group of Boy Scouts; a person in charge of a troop of scouts. **Scouting**, *n.*
scout[2] (skowt), *v.t.* to treat with contempt and disdain, to reject contemptuously.
scow (skow), *n.* a large flat-bottomed, square-ended boat. *v.t.* to transport in a scow.
scowl (skowl), *v.i.* to frown, to look sullen or ill-tempered; to have a threatening aspect. *n.* an angry frown; a look of sullenness, ill-temper or discontent. **scowlingly**, *adv.*
scrabble (skrab'l), *v.i.* to make irregular or unmeaning marks; to scrawl, to scribble; to scramble; to scrape, scratch or grope (about) as if to obtain something. *v.t.* to scribble on or over. *n.* a scribble, scrawl; a scratching or scraping; a scramble, struggle; (**Scrabble**®) a word-building board game. **scrabbler**, *n.*
scrag (skrag), *n.* anything thin, lean or shrivelled; a lean or bony person or animal; a lean or bony piece of meat, esp. the lean end of neck of mutton. *v.t.* (*past, p.p.* **scragged**) (*sl.*) to wring the neck of, to throttle; to kill by hanging. **scragged, scraggy**, *a.* **scraggedness** (-gid-), **scragginess**, *n.* **scraggily**, *adv.*
scram[1] (skram), *v.i.* (often *imp.*) (*sl.*) to go away quickly.
scram[2], *n.* an emergency shutdown of a nuclear reactor. *v.t.* to shutdown (a nuclear reactor) in an emergency.
scramble (skram'bl), *v.i.* to climb or move along by clambering, crawling, wriggling etc., esp. with the hands and knees; to move with urgent or disorderly haste; to seek or struggle (for, after etc.) in a rough-and-tumble or eager manner; to climb or spread irregularly; of an aircraft or its crew, to take-off immediately. *v.t.* to put or collect together hurriedly or haphazardly; to mix or jumble up; to prepare (eggs) by breaking into a pan and stirring up during cooking; to order (an aircraft or crew) to scramble; to make (a transmission) unintelligible with an electronic scrambler. *n.* the act of scrambling; a climb or walk over rocks etc., or in a rough-and-tumble manner; a rough or unceremonious struggle for something; an emergency take-off of fighter aircraft; a motor-cycle race over rough ground. **scrambler**, *n.* one who scrambles; an electronic device for scrambling speech transmitted by radio or telephone by altering frequencies.
scrap[1] (skrap), *n.* a small detached piece, a bit, a fragment; a picture, paragraph etc., cut from a newspaper

scrap etc., for preservation; refuse, waste, esp. old pieces of discarded metal collected for melting down etc.; (*pl.*) bits, odds-and-ends, leavings; (*pl.*) leftover fragments of food; (*usu. pl.*) refuse of fat from which the oil has been expressed. *v.t.* (*past, p.p.* **scrapped**) to make scrap of, to consign to the scrap-heap; to condemn and discard as worn out, obsolete etc. **scrapbook**, *n.* a blank book into which pictures, cuttings from newspapers etc. are pasted for preservation. **scrap-cake**, *n.* fish-scrap compressed into cakes. **scrap-heap**, *n.* a heap of scrap metal; a rubbish-heap. **scrap iron** or **metal**, *n.* discarded metal for reprocessing. **scrapyard**, *n.* a place where scrap, esp. scrap metal, is collected or stored. **scrappy**, *a.* consisting or made up of scraps; disconnected. **scrappily**, *adv.* **scrappiness**, *n.*

scrap[2] (skrap), *n.* (*coll.*) a fight, a scuffle, a dispute. *v.i.* (*past, p.p.* **scrapped**) to engage in a fight. **scrapper**, *n.* **scrapping-match**, *n.*

scrape (skrāp), *v.t.* to rub the surface of with something rough or sharp; to abrade, smooth or shave (a surface) thus; to remove, to clean (off, out etc.) thus; to erase; to rub or scratch (out); to excavate or hollow (out) by scraping; to rub against with a rasping or grating noise; to draw or rub along something with a scraping noise; to damage or graze by rubbing on a rough surface; to collect or get together by scraping; to save or amass with difficulty or by small amounts. *v.i.* to rub the surface of something with a rough or sharp instrument; to abrade, to smooth, to clean something thus; to rub (against something) with a scraping or rasping noise; to make such a noise; to get through with difficulty or by a close shave; to be saving or parsimonious; to play awkwardly on a violin etc.; to make an awkward bow with a drawing back of the foot. *n.* the act, sound or effect of scraping; an awkward bow with a drawing back of the foot; (*coll.*) an awkward predicament, esp. one due to one's own conduct. **to scrape acquaintance with**, to contrive to make the acquaintance of. **to scrape along** or **by**, (*coll.*) to keep going somehow. **to scrape away**, to abrade, to reduce by scraping. **to scrape down**, to scrape away; to scrape from head to foot or top to bottom; to silence or put down by scraping the feet. **to scrape the barrel** BARREL. **scraper**, *n.* one who scrapes; an instrument for scraping, esp. for cleaning the dirt off one's boots before entering a house; an awkward fiddler; a miser; a prehistoric flint implement used for scraping skins etc. **scraperboard**, *n.* a board with a surface that can be scraped off to form a design; this method of producing designs. **scraping**, *n.*

scratch[1] (skrach), *v.t.* to tear or mark the surface of lightly with something sharp; to wound slightly; to rub or scrape with the nails; to hollow out with the nails or claws; to chafe the surface of; to erase, to obliterate, to score (out, through etc.); to expunge (esp. the name of a horse in a test of entries for a race); to withdraw from a contest; to cancel (a match, game etc.); to form by scratching; to scrape (up or together). *v.i.* to use the nails or claws in tearing, scraping, marking, hollowing out etc.; to rub or scrape one's skin with the nails; to chafe, rub; to scrape the ground as in searching; to make a grating noise; to withdraw one's entry from a contest; to get by or manage with difficulty. *n.* a mark made by scratching; a slight wound; a sound of scratching; an act or spell of scratching; a scratch-wig; a mark from which competitors start in a race, or a line across a prize-ring at which boxers begin; (*pl.*) a horse-disease characterized by scabs or chaps between the heel and pastern-joint. *a.* improvised; put together hastily or haphazardly, multifarious, nondescript; (*Sport*) without handicap. **to be** or **come up to (the) scratch**, to be satisfactory, to fulfil the desired standard or requirements. **to scratch along**, to scrape along. **to start from scratch**, to start from the very beginning, with no advantage. **scratch pad**, *n.* (*chiefly N Am.*) a notebook, a scribbling block. **scratch video**, *n.* a collage on video of previously existing pieces of television and cinema film; the technique or genre of making scratch videos. **scratcher**, *n.* one who or that which scratches; a bird that scratches for food, one of the Rasores. **scratching**, *n.* a scratchy sound effect produced by manually rotating a (pop) record backwards and forwards, used in some styles of pop music; (*pl.*) refuse strained out of melted lard. **scratchy**, *a.* consisting of or characterized by scratches; tending to scratch or rub, rough; making a noise like scratching; uneven, irregular, heterogeneous. **scratchily**, *adv.* **scratchiness**, *n.*

scrawl (skrawl), *v.t.* to draw, write or mark clumsily, hurriedly or illegibly, to scribble. *v.i.* to scribble, to mark with illegible writing, etc. *n.* a piece of hasty, clumsy or illegible writing. **scrawler**, *n.* **scrawly**, *a.*

scrawny (skraw'ni), *a.* excessively lean, thin, bony; meagre.

scray (skrā), *n.* the tern or sea-swallow.

scream (skrēm), *v.i.* to make a shrill, piercing, prolonged cry as if in extreme pain or terror; to give out a shrill sound, to whistle, hoot or laugh loudly; to speak or write excitedly or violently; to be over-conspicuous or vivid. *v.t.* to utter or say in a screaming tone. *n.* a loud, shrill, prolonged cry, as of one in extreme pain or terror; (*coll.*) something or someone excruciatingly funny. **screamer**, *n.* one who or that which screams, esp. the swift; any bird of the S American semi-aquatic family Palamedeidae, from their harsh cry; (*coll.*) a sensational headline. **screamingly**, *adv.* extremely.

scree (skrē), *n.* loose fragments or debris of rock on a steep slope; a slope covered with this.

screech (skrēch), *v.i.* to scream out with a sharp, harsh, shrill voice; to make a shrill, strident noise. *v.t.* to utter or say with such a voice. *n.* a shrill, harsh cry as of terror or pain. **screech-owl**, *n.* an owl, *Tyto alba*, that screeches instead of hooting. **screecher**, *n.* **screechy**, *a.*

screed (skrēd), *n.* a long harangue or tirade; a strip of mortar, wood etc. put on a wall etc. that is to be plastered, as a guide to evenness of surface etc.; a screeding; a piece, a fragment, a strip; a long and tedious piece of writing. **screeding**, *n.* the final rendering of concrete to get a smooth surface.

screen (skrēn), *n.* a partition separating a portion of a room or of a church from the remainder, esp. one between the choir and the nave or ambulatory; a movable piece of furniture, usu. consisting of a light framework covered with paper, cloth etc., used to shelter from excess of heat, draught etc.; anything serving to shelter, protect or conceal; a surface on which images can be projected; a board or structure on which notices etc. can be posted; a frame containing a mesh placed over a window, door etc. to keep out flies; the part of a television set, VDU etc. on which the image appears; the film industry, moving pictures collectively. *v.t.* to shelter or protect from inconvenience, injury, hurt or pain, to shield; to hide, to conceal wholly or partly; to separate with a screen; to test for the presence of disease, weapons etc.; to examine or check thoroughly in order to assess suitability, sort into categories etc.; to project (a film) on a screen; to portray in film. **screenplay**, *n.* a film script including stage directions and details of characters and sets. **screen printing** or **process** SILK-SCREEN PRINTING under SILK. **screen writer**, *n.* a writer of screenplays. **screenings**, *n.pl.* small stuff or refuse separated by screening.

screw (skroo), *n.* a cylinder with a spiral ridge or groove round its outer surface (called a male or exterior screw) or round its inner surface (called a female or internal screw), esp. a male screw used for fastening boards etc. together; a male or female screw forming part of a tool, mechanical appliance or machine and conveying motion to another part or bringing pressure to bear; something resembling a screw in spiral form; a screw-propeller; a screw steamer; a turn of a screw; a sideways motion or tendency like that of a screw, a twist; backspin given to a ball in snooker, billiards etc.; a twisted-up paper (of tobacco etc.); (*sl.*) a stingy person; (*coll.*) salary; (*sl.*) a prison warder; (*sl.*) an act of sexual intercourse. *v.t.* to fasten, secure, tighten, join etc. with a screw or screws; to turn (a screw); to turn round or twist as a screw; to give a spiral thread or groove to; to press hard, to oppress, esp. by exactions, to grind; to extort, to squeeze (money etc.) out of; (*sl.*) to cheat; to twist, to contort, to distort (as the face); (*sl.*) to have sexual intercourse with. *v.i.* to turn as a screw; to twist, to move obliquely or spirally, to swerve; (*sl.*) to have sexual intercourse. **to have a screw loose**, to be slightly crazy. **to put the screws on,** (*sl.*) to put pressure on. **to screw up,** to tighten up with or as with a screw; to fasten with a screw or screws; to shut (a person) in thus; to twist; (*sl.*) to bungle, mess up; (*sl.*) to make confused or neurotic, to disturb. **to screw up courage,** to summon up resolution. **screwball,** *a.* (*chiefly N Am. coll.*) eccentric, crazy, zany. *n.* an eccentric person. **screw-cutter,** *n.* a tool for cutting screws. **screw driver,** *n.* a tool like a blunt chisel for turning screws. **screw eye,** *n.* a screw with a loop instead of a slotted head, for attaching cords to picture-frames etc. **screw-gear,** *n.* an endless screw or worm for working a cogwheel etc. **screw-jack,** *n.* a lifting-jack with a screw rotating in a nut; a dentist's implement for pressing teeth apart etc. **screw-pile,** *n.* a pile armed with a screw-point, sunk by turning instead of hammering. **screw-pine,** *n.* any tree of the E Indian genus *Pandanus*, with leaves clustered spirally. **screw-press,** *n.* a press worked by means of a screw. **screw-propeller** PROPELLER. **screwtop,** *n.* (a bottle or jar with) a top that opens and closes with a screwing motion. **screw-wrench,** *n.* a tool for gripping the head of a large screw or nut; a wrench with jaws worked by a screw. **screwed,** *a.* (*sl.*) drunk, tipsy. **screwer,** *n.* **screwy,** *a.* (*sl.*) eccentric, absurd, zany; (*sl.*) mad, crazy.

scribal, SCRIBE.

scribble (skrib'l), *v.i.* to write hastily, illegibly or without regard to correctness of handwriting or composition; to make random or meaningless marks with a pen, crayon etc.; (*derog.*) to be a journalist or author. *v.t.* to write hastily, carelessly or without regard to correctness. *n.* hasty or careless writing; a scrawl; something written hastily or carelessly. **scribbler,** *n.* a minor author. **scribblingly,** *adv.* **scribbly,** *a.*

scribe (skrīb), *n.* a writer, a penman; a secretary, a copyist; an ancient Jewish writer or keeper of official records, one of a class of commentators, interpreters and teachers of the sacred law; a pointed instrument for marking lines on wood, bricks etc., a scriber. *v.t.* to mark with a scriber; to mark and fit one piece to the edge of another. **scribal,** *a.* **scriber, scribing-awl, -iron, -tool,** *n.* a tool used for scoring or marking lines etc.

scrim (skrim), *n.* strong cotton or linen cloth used for lining in upholstery and for cleaning.

scrimmage (skrim'ij), *n.* a tussle, a confused or rough-and-tumble struggle, a skirmish; in Rugby football, a scrummage; in American football, the period or activity between the ball coming into play and the time it is dead.

scrimp (skrimp), *v.t.* to make small, scant or short; to limit or straiten, to skimp. *v.i.* to skimp, to be niggardly. *a.* scanty, narrow. *adv.* scarcely, barely. *n.* (*N Am.*) a niggard, a pinching miser. **scrimpy,** *a.* **scrimpily, scrimply,** *adv.* **scrimpiness, scrimpness,** *n.*

scrimshank (skrim'shangk), *v.i.* (*coll.*) to avoid work, to get out of doing one's duty.

scrimshaw (skrim'shaw), *v.t.* to decorate (ivory, shells etc.) with carvings and coloured designs. *v.i.* to produce decorated work of this kind. *n.* a piece of such work.

scrip (skrip), *n.* orig. a writing, a list, as of names, a schedule; a provisional certificate given to a subscriber for stock of a bank or company; such certificates collectively.

script (skript), *n.* a piece of writing; handwriting as dist. from print; printed cursive characters, type in imitation of writing; handwriting in imitation of type; the written text or draft of a film, play or radio or television broadcast as used by the actors or performers; (*Law*) a writing, an original document. *v.t.* to write the script for. **scriptwriter,** *n.* one who writes scripts, esp. for broadcasting or for the cinema. **scriptorium** (-taw'ri-), *n.* (*pl.* **-riums, -ria** (-ə)) a writing-room, esp. in a monastery.

Script., (*abbr.*) Scripture.

scripture (skrip'chə), *n.* a sacred writing or book; the Bible, esp. the books of the Old and New Testament without the Apocrypha; a passage from the Scriptures. **Holy Scripture,** the Bible. **the Scriptures,** the Bible. **scriptural,** *a.* pertaining to, derived from, based upon, or contained in the Scriptures. **scripturally,** *adv.*

scrivener (skriv'ənə), *n.* one whose business was to draw up contracts or other documents, a notary; formerly, a financial agent, a broker, a money-lender.

scrobe (skrōb), *n.* (*Ent.*) a groove, as that receiving the base of the antenna in a weevil. **scrobicule** (-ikūl), *n.* (*Biol.*) a small pit or depression. **scrobicular** (-bik'-), **scrobiculate** (-bik'ūlət), **-lated** (-lātid), **scrobiculous** (-bik'-), *a.*

scrofula (skrof'ūlə), *n.* (*dated*) a form of tuberculosis affecting esp. the lymph glands of the neck, also called king's evil. **scrofulous,** *a.* **scrofulously,** *adv.*

scroll (skrōl), *n.* a roll of paper or parchment; an ancient book or volume in this form; a convoluted or spiral ornament more or less resembling a scroll of parchment, as a volute, the curved head of a violin etc., a band or ribbon bearing an inscription, a flourish, or tracery consisting of spiral lines; (*Her.*) the ribbon upon which a motto is inscribed. *v.t.* to roll up like a scroll; to decorate with scrolls; to enter in a scroll; (*Comput.*) to move (text) across a screen. *v.i.* to curl up like a scroll; (*Comput.*) to move text upwards, sideways etc. on a screen so as to display the next line or section. **scroll saw,** *n.* a fret-saw for cutting scrolls. **scroll work,** *n.* ornamental work in spiral lines, esp. cut out with a scroll-saw.

Scrooge (skrooj), *n.* a miserly person, named after Ebenezer *Scrooge*, a character in Dickens's *A Christmas Carol*.

scrophularia (skrofūleə'riə), *n.* a genus of plants typical of the family Scrophulariaceae, containing the figwort. **scrophulariaceous** (-ā'shəs), *a.*

scrotum (skrō'təm), *n.* (*pl.* **-ta** (-tə) **-tums**) the pouch enclosing the testes in the higher mammals. **scrotal,** *a.*

scrounge (skrownj), *v.t.* (*coll.*) to pilfer; to cadge. *v.i.* to forage or hunt around; to cadge things. **scrounger,** *n.*

scrub[1] (skrŭb), *v.t.* (*past, p.p.* **scrubbed**) to rub hard with something coarse and rough, esp. with soap and water used with a scrubbing-brush for the purpose of cleaning or scouring; to purify (a gas) with a scrubber; (*coll.*) to get rid of, cancel, delete, erase. *v.i.* to clean, scour or brighten things by rubbing hard; to work hard and penuriously, to drudge; to scrub the hands and arms before carrying out surgery. *n.* the act of scrubbing; a worn-out brush or broom; a lotion containing abrasive granules for cleansing the skin. **scrubber,** *n.* one who or that which scrubs; a scrubbing-brush; a gas-purifier for removing tar and ammonia by spraying with water; (*sl.*) a prostitute or promiscuous woman. **scrubbing,** *n., a.* **scrubbing-board,** *n.* a ribbed board used in washing for rubbing clothes on. **scrubbing-brush,** *n.* a stiff brush for scrubbing floors etc.

scrub[2] (skrŭb), *n.* (a tract of) brushwood, undergrowth or stunted trees; a stunted tree, bush etc.; a paltry, stingy person; *n.* an inferior animal; something mean or despicable; (*N Am.*) a player not of the first team. *a.* mean, paltry, petty, niggardly, contemptible. **scrubland,** *n.* land covered with scrub vegetation. **scrub turkey,** *n.* (*Austral.*)

the lowan or mallee mound bird. **scrubber**, *n*. (*Austral.*) a bullock that has run wild. **scrubby**, *a*. mean, stunted, insignificant; covered with brushwood; rough, unshaven. **scrubbiness**, *n*.

scruff¹ (skrŭf), *n*. the nape or back of the neck, esp. as grasped by a person dragging another.

scruff² (skrŭf), *n*. (*coll.*) an unkempt or scruffy person. **scruffy**, *a*. scurvy; untidy, dirty, shabby, down-at-heel. **scruffiness**, *n*.

scrum (skrŭm), **scrummage** (-ij), *n*. a set struggle in rugby between the forwards of both sides grappling in a compact mass with the ball on the ground in the middle; a scuffle. *v.i.* (*past, p.p.* **scrummed**) to form a scrum. **to scrum down**, to scrum. **scrum half**, *n*. the half-back who puts the ball into the scrum.

scrump (skrŭmp), *v.t., v.i.* (*dial.*) to steal (apples) from an orchard. **scrumpy**, *n*. rough cider.

scrumptious (skrŭmp'shəs), *a*. (*coll.*) first-class, stylish; of food, delicious.

scrunch (skrŭnch), *v.t.* to crunch; to crush, to crumple; to hunch up. *v.i.* to make or move with a crunching sound. *n.* a crunch. **scrunch-dry**, *v.t.* to dry (the hair) with a hair-dryer whilst crushing in the hand, to give body.

scruple (skroo'pl), *n*. a weight of 20 grains (1·296 g), the third part of a dram (apothecaries' weight); (*dated*) a small quantity, a tiny fraction, a particle; a doubt, objection or hesitation from conscientious or moral motives. *v.i.* to have scruples, to doubt, to hesitate, to be reluctant (to do etc.). **scrupulous** (-pū-), *a*. influenced by scruples; careful, cautious, extremely conscientious, punctilious, precise, exact. **scrupulously**, *adv*. **scrupulousness**, *n*.

scrutiny (skroo'tini), *n*. close observation or investigation; minute inquiry; critical examination; an official examination of votes given at an election to verify the correctness of a declared result. **scrutineer** (-niə'), *n*. one who acts as examiner in a scrutiny of votes. **scrutinize, -ise,**, *v.t.* to examine narrowly or minutely. **scrutinizer**, *n*. **scrutinizingly**, *adv*.

scruto (skroo'tō), *n*. a trapdoor with springs, made flush with a theatre stage, for rapid disappearances etc.

scry (skrī), *v.t.* to crystal-gaze; to descry.

scuba (skoo'bə, skū'-), *n*. an aqualung. **scuba-diving**, *n*. underwater swimming with an aqualung. **scuba-dive**, *v.i.* **scuba-diver**, *n*. [acronym for *self-contained underwater breathing apparatus*]

scud (skŭd), *v.i.* (*past, p.p.* **scudded**) to run or fly swiftly; (*Naut.*) to run fast before a gale with little or no sail spread. *v.t.* to move swiftly over. *n*. the act or a spell of scudding; loose, vapoury clouds driven swiftly by the wind; a light passing shower. **scudder**, *n*.

scudo (skoo'dō), *n*. (*pl.* **-di** (-dē)) an old Italian silver coin and money of account.

scuff (skŭf), *v.i.* to drag or scrape with the feet in walking, to shuffle; to become abraded or roughened, esp. by use. *v.t.* to scrape or shuffle (the feet); to abrade, scratch or roughen the surface of. *n*. the act or noise of scuffing; a mark or roughened place caused by scuffing. **scuffed**, *a*. **scuffy**, *a*. worn, shabby.

scuffle (skŭf'l), *v.i.* to fight or struggle in a rough-and-tumble way. *n*. a confused and disorderly fight or struggle; a soft, shuffling sound. **scuffler**, *n*.

sculduggery (skŭldŭg'əri), SKULDUGGERY.

scull (skŭl), *n*. one of a pair of short oars used by one person for propelling a boat; an oar used with twisting strokes over the stern; one who sculls a boat. *v.t.* to propel (a boat) by a scull or sculls. *v.i.* to propel a boat thus. **sculler**, *n*. one who sculls; a boat rowed thus.

scullery (skŭl'əri), *n*. a place where dishes and utensils are washed up, vegetables prepared etc.

scullion (skŭl'yən), *n*. a servant who cleans pots, dishes etc., a kitchen drudge.

sculp (skŭlp), *v.t.* (*coll.*) to carve, to sculpture.

sculpture (skŭlp'chə), *n*. the art of cutting, carving, modelling or casting wood, stone, clay, metal etc. into representations of natural objects or designs in round or in relief; carved or sculptured work collectively; a piece of this; raised or sunk markings on a shell, elytrum etc. *v.t.* to represent in or by sculpture; to ornament with sculpture; to shape by or as by carving, moulding etc. **sculpt**, *v.t., v.i.* to sculpture. **sculptor, -tress**, *n*. one who sculptures. **sculptural, sculpturesque** (-resk'), *a*. **sculpturally**, *adv*.

scum (skŭm), *n*. impurities that rise to the surface of liquid, esp. in fermentation or boiling; the scoria of molten metal; froth, foam or any film of floating matter; (*fig.*) refuse, dregs, the vile and worthless part. *v.t.* to clear of scum, to skim. *v.i.* (*past, p.p.* **scummed**) to rise as scum, to form a scum; to become covered with scum. **scummer**, *n*. **scummy**, *a*.

scumble (skŭm'bl), *v.t.* to cover (an oil-painting) lightly with opaque or semi-opaque colours so as to soften the outlines or colours; to produce a similar effect on (a drawing) by lightly rubbing; to soften (a colour) thus; to prepare (a painted wall) for repainting. *n*. a material for scumbling; the effect produced.

scuncheon (skŭn'chən), *n*. a bevelling, splay or elbow in a window-opening etc.; arching etc., across the angles of a square tower supporting a spire.

scunner (skŭn'ə), *v.t.* (*Sc.*) to disgust to nauseate. *v.i.* to feel loathing, to be sickened. *n*. loathing, disgust an object of loathing.

scupper (skŭp'ə), *n*. a hole or tube through a ship's side to carry off water from the deck. *v.t.* to sink (a ship); (*coll.*) to ruin, to do for. **scupper-hole**, *n*. **scupper-hose, -shoot**, *n*. a spout hanging from a scupper to carry the water clear of the side.

scurf (skœf), *n*. flakes or scales thrown off by the skin, esp. of the head; any loose scaly matter adhering to a surface. **scurfy**, *a*. **scurfiness**, *n*.

scurrilous (skŭ'riləs), *a*. using or expressed in low, vulgar, grossly abusive or indecent language. **scurrilously**, *adv*. **scurrilousness, scurrility** (-ril'-), *n*.

scurry (skŭ'ri), *v.i.* to go with great haste, to hurry, to scamper. *n*. an act or the noise of scurrying.

scurvy (skœ'vi), *a*. mean, paltry, base, shabby, contemptible. *n*. a disease caused by lack of vitamin C and characterized by swollen gums, extravasation of blood and general debility, arising orig. esp. among those on shipboard from a deficiency of vegetables. **scurvied**, *a*. **scurvily**, *adv*. **scurviness**, *n*.

scurvy grass (skœ'vigrahs), *n*. a plant, *Cochlearia officinalis*, formerly used as a remedy for scurvy.

scut (skŭt), *n*. a short tail, as of a hare, rabbit or deer.

scuta SCUTUM.

scutage (skū'tij), *n*. money paid by a feudal tenant in lieu of personal attendance on his lord in war.

scutal, scutate SCUTUM.

scutch (skŭch), *v.t.* to dress (cotton, flax etc.) by beating. *n*. a scutcher; coarse tow separated from flax by scutching. **scutch-blade, -rake, scutcher, scutching-sword**, *n*. an implement of various kinds used in scutching flax.

scutcheon (skŭch'ən), *n*. an escutcheon; a cover or frame for a keyhole; a name-plate. **scutcheoned**, *a*.

scute SCUTUM.

scutellum (skūtel'əm), *n*. (*pl.* **-lla**) a small shield, plate, scale or horny segment in or on a plant or animal. **scutellar, scutellate** (skū'tələt), **scutellated**, (-lātid), *a*. **scutellation**, *n*. **scutelliform** (-tel'ifawm), *a*. shield-shaped.

scuttle¹ (skŭt'l), *n*. a metal or other receptacle for carrying or holding coals, esp. for a fire-place, usu. called a coal-scuttle. **scuttleful**, *n*.

scuttle² (skŭt'l), *n*. a hole with a movable lid or hatch in a wall or roof on the deck or side of a ship; the lid or hatch covering this. *v.t.* to cut holes through the bottom or

sides of (a ship); to sink by cutting such holes. **scuttlebutt, -cask,** *n.* a cask of drinking-water, usu. pierced with a hole for dipping through, kept on the deck of a ship. **scuttler,** *n.*
scuttle[3] (skŭt'l), *v.i.* to hurry along, to scurry; to make off, to bolt. *n.* a hurried run or gait, a hasty flight, a bolt. **scuttler,** *n.*
scutum (skū'təm), *n. (pl.* **-ta** (-tə)) the shield of the heavyarmed Roman legionaries; a scute; the kneepan. **scutal, scutate** (-tət), *a.* covered with scutes or bony plates; shield-shaped. **scute,** *n.* a shield-like plate, scale or bony or horny segment as of the armour of a crocodile, turtle etc. **scutiform** (-tifawm), *a.* **scutulum** (-tūləm), *n. (pl.* **-la** (-lə)) a shield-shaped scale or scab, esp. in ringworm of the scalp. **scutulate** (-lət), *a.*
Scylla (sil'ə), *n.* a rock on the Italian shore of the Straits of Messina, facing Charybdis, described by Homer as a monster devouring sailors. **between Scylla and Charybdis,** caught between alternative risks, escape from one of which entails danger from the other.
scyphus (sī'fəs), *n. (pl.* **-phi** (-fī)) a bowl-shaped footless Greek cup with two handles; a cup-shaped plant part or organ. **scyphiform** (-fifawm), *a.* **scyphose** (-fōs), *a.*
scythe (sīdh), *n.* a long curved blade with a crooked handle used for mowing or reaping; a curved blade projecting from the axle of an ancient war-chariot. *v.t.* to cut with a scythe. **scythe-stone,** *n.* a whetstone for sharpening scythes. **scythed,** *a.*
Scythian (sidh'iən), *a.* pertaining to ancient Scythia, the region north of the Black Sea, or the ancient race inhabiting it. *n.* one of this race; the Scythian language.
SDI, *(abbr.)* Strategic Defence Initiative.
SDLP, *(abbr.)* Social and Democratic Labour Party.
SDR, *(abbr.)* special discretion required; special drawing rights.
SE, *(abbr.)* south-east, south-eastern.
Se, *(chem. symbol)* selenium.
sea (sē), *n.* the body of salt water covering the greater part of the earth's surface, the ocean; a definite part of this, or a very large enclosed body of (usu. salt) water; the swell or motion of the sea; a great wave, a billow; the set or direction of the waves; a vast quantity or expanse, an ocean, a flood (of people, troubles etc.). *a.* of, pertaining to, living, growing or used in, on or near the sea, marine, maritime. **at full sea,** at high tide; at the acme of culmination. **at sea,** on the open sea; out of sight of land; perplexed, uncertain, wide of the mark. **four seas** FOUR. **high seas** HIGH. **over** or **beyond seas,** to or in countries separated by sea. **Seven Seas** SEVEN. **to go to sea, to follow the sea,** to become or to be a sailor. **to put to sea,** to leave port or land. **sea-anchor** DRAG-ANCHOR under DRAG. **sea-anemone,** *n.* a solitary tentacled polyp of the order Actinaria; an actinia. **sea-ape** SEA-FOX; SEA-OTTER. **sea-bass** BASS[2]. **seabed,** *n.* the floor of the sea. **sea-bird,** *n.* **sea-board,** *n.* land bordering on the sea; the sea-coast; the seashore. *a.* bordering on the sea. **seaborne,** *a.* conveyed by sea. **sea-breach,** *n.* irruption of the sea through an embankment. **sea-bream,** *n.* a marine food-fish of the family Sparidae. **sea breeze,** *n.* a breeze blowing from the sea, usu. by day, in alternation with a land-breeze at night. **sea-calf,** *n.* the common seal. **sea captain,** *n.* the captain of a vessel, as dist. from a military officer; a great commander by sea. **sea card,** *n.* the card of the mariner's compass; a map or chart. **sea change,** *n.* a transformation or transmutation (produced by the sea). **seacoast,** *n.* a coast. **seacock,** *n.* a valve through which the sea can be admitted into the hull of a ship. **sea cow,** *n.* a sirenian; a walrus. **sea-cucumber,** *n.* a holothurian such as the trepang. **sea dog,** *n.* the common seal; the dog-fish; an old sailor, esp. of the Elizabethan era. **sea eagle,** *n.* the osprey; any of various fishing-eagles and other large sea-birds. **sea fan,** *n.* a coral of the genus *Gorgonia* or a related genus, having fan-like branches. **seafarer,** *n.* a sailor. **seafaring,** *a.* travelling by sea; following the occupation of a sailor. *n.* travel by sea; the occupation of a sailor. **seafight,** *n.* a naval engagement. **sea-flower,** *n.* the sea-anemone. **sea food,** *n.* edible saltwater fish and crustaceans, esp. shellfish. **sea-fox,** *n.* the long-tailed shark, *Alopias vulpes.* **seafront,** *n.* the part of a town that faces the sea. **seagoing,** *a.* making foreign voyages, as opp. to coasting; seafaring. **sea green,** *n., a.* (of) a faint bluish-green. **seagull,** *n.* a gull. **sea holly,** *n.* an umbelliferous plant, *Eryngium maritimum*, with spiny leaves. **sea horse,** *n.* the hippocampus or a similar fish; a fabulous animal, half horse and half fish. **sea-island cotton,** *n.* a fine variety of cotton originally grown on the islands off the coasts of Georgia, S Carolina and Florida. **sea-kale,** *n.* a cruciferous plant, *Crambe maritima*, grown as a culinary vegetable for its young shoots. **sea lane,** *n.* a route for ships at sea. **sea lavender,** *n.* any species of *Statice*, esp. *S. limonium.* **sea legs,** *n.pl.* ability to walk on the deck of a vessel at sea on a stormy day. **sea level,** *n.* a level continuous with that of the surface of the sea at mean tide, taken as a basis for surveying etc. **sea lion,** *n.* a large-eared seal, esp. of the genus *Otariai;* (*Her.*) a fabulous animal, half lion and half fish. **Sea Lord,** *n.* one of four naval Lords of the Admiralty. **seaman,** *n. (pl.* **-men**) a mariner, a sailor, esp. one below the rank of officer; a person able to navigate a ship, a navigator. **seamanship,** *n.* **sea-mile,** GEOGRAPHICAL MILE under MILE. **sea monster,** *n.* a huge sea-creature, natural or mythical. **sea mouse,** *n.* an iridescent sea-worm, *Aphrodite aculeata.* **sea otter,** *n.* a marine otter, *Enhydra marina*, of the shores of the N Pacific. **sea-pink,** *n.* thrift, *Armeria maritima.* **sea plane,** *n.* an aeroplane fitted with floats to enable it to take off from and alight on the water. **seaport,** *n.* a town with a harbour on the coast. **sea salt,** *n.* salt obtained from seawater by evaporation. **seascape,** *n.* a sea-piece. **sea scorpion,** *n.* any fish of the genus *Scorpaena;* the sculpin, *Cottus scorpius.* **Sea Scouts,** *n.pl.* a branch of the Scouts specializing in sailing etc. **sea serpent,** *n.* a sea-snake; a creature of immense size and serpentine form, believed by mariners to inhabit the depths of the ocean. **seashell,** *n.* the shell of a marine mollusc. **seashore,** *n.* the shore, coast or margin of the sea; (*Law*) the space between highand low-water mark; land adjacent to the sea. **seasick,** *a.* suffering from sea-sickness. **seasickness,** *n.* nausea and vomiting brought on by the motion of a ship. **seaside,** *n.* a place or district close to the sea, esp. a holiday resort. *a.* bordering on the sea. **sea snail,** *n.* a snail-like marine gasteropod; a slimy fish of the family Liparididae, the unctuous sucker. **seasnake,** *n.* a venomous marine snake of the family Hydrophidae inhabiting the Indian Ocean and other tropical seas; the sea-serpent. **sea squirt,** *n.* an ascidian. **sea swallow,** *n.* the tern. **sea trout,** *n.* the salmon-trout, bull-trout and some other fishes. **sea urchin,** *n.* an echinus. **seawall,** *n.* a wall or embankment for protecting land against encroachment by the sea. **seaway,** *n.* a ship's progress; a clear way for a ship at sea. **seaweed,** *n.* any alga or other plant growing in the sea. **seaworthy,** *a.* of a ship, in a fit state to go to sea. **seaworthiness,** *n.* **seaward,** *a.* directed or situated towards the sea. *adv.* towards the sea. *n.* a seaward side or aspect. **seawards,** *adv.*
seal[1] (sēl), *n.* a carnivorous amphibious marine mammal of various species of the family Phocidae, having flipperlike limbs adapted for swimming and thick fur; applied to allied mammals belonging to the family Otariidae, distinguished by having visible external ears, comprising the sea-lions and fur-seals; sealskin. *v.i.* to hunt seals. **sealrookery,** *n.* a breeding-place of seals. **sealskin,** *n.* the under-fur of the fur-seal, esp. prepared for use as material for jackets etc.; a sealskin garment. **sealer,** *n.* a ship or person engaged in seal-hunting.

seal² (sēl), *n.* a die or stamp having a device, usu. in intaglio, for making an impression on wax or other plastic substance; a piece of wax, lead or other material stamped with this and attached to a document as a mark of authenticity etc., or to an envelope, package, box etc. to prevent its being opened without detection etc.; the impression made thus on wax, lead etc.; a stamped wafer- or other mark affixed to a document in lieu of this; any device that must be broken to give access; any act, gift or event regarded as authenticating, ratifying or guaranteeing; a symbolic, significant or characteristic mark or impress; anything used to close a gap, prevent the escape of gas etc.; water in the trap of a drain-pipe preventing the ascent of foul air. *v.t.* to affix a seal to; to stamp with a seal or stamp, esp. as a mark of correctness or authenticity; to fasten with a seal; to close hermetically, to shut up; to close (the lips etc.) lightly; to confine securely; to secure against leaks, draughts etc.; to make (as wood) impermeable to rain, etc. by applying a coating; to fix or fill with plaster etc.; to confirm; to ratify, to certify; to set a mark on, to designate or destine irrevocably. **Great Seal**, the official seal of the United Kingdom used to seal treaties, writs summoning Parliament, and other state documents of great importance. **Privy Seal** PRIVY. **seal ring**, *n.* a finger-ring with a seal. **sealed-beam**, *a.* pertaining to electric lights, as car headlights, in which the reflector and bulb are in one sealed unit. **sealing-wax**, *n.* a composition of shellac and turpentine with a pigment used for sealing letters etc. **sealable**, *a.* **sealant**, *n.* a substance for sealing wood, stopping up gaps etc.

sealskin SEAL.

sealyham (sē′lihəm), *n.* a breed of Welsh terrier.

seam (sēm), *n.* a ridge or other visible line of junction between two parts or things, esp. two pieces of cloth etc. sewn together, planks fitted edge to edge, or sheet-metal lapped over at the edges; (*Anat.*) a suture; a mark of separation, a crack, a fissure; a line on the surface of anything, esp. the face, a wrinkle, a cicatrix, a scar; a thin layer separating two strata of rock; a thin stratum of coal; (*N Am.*) a piece of sewing. *v.t.* to join together with or as with a seam; to mark with a seam, furrow, scar etc. **seamless**, *a.* **sempstress**, (semp′-), *n.* a woman whose occupation is sewing. **seamy**, *a.* showing the seams; disreputable, sordid, unpleasant.

Seanad Eireann (shan′ədh eə′rən), *n.* the upper house, or senate, of the parliament of Eire.

seance, séance (sā′ons, -ōs), *n.* a meeting for exhibiting, receiving or investigating spiritualistic manifestations.

SEAQ, *n.* a computerized system for recording trade and price changes in shares, used by the London Stock Exchange. [*Stock Exchange Automated Quotations*]

sear (-siə), *v.t.* to burn or scorch the surface of to dryness and hardness; to cauterize; (*fig.*) to brand; (*dated*) to make callous or insensible; (*poet.*) to wither up, to blast. **seared**, *a.* hardened, insensible, callous. **searing**, *a.* **searingly**, *adv.*

search (sœch), *v.t.* to go over and examine for what may be found or to find something; to examine (esp. a person) for concealed weapons etc.; to explore, to probe; to look for, to seek (out). *v.i.* to make a search, inquiry or investigation. *n.* the act of seeking, looking or inquiring; investigation, exploration, inquiry, quest, examination. **right of search**, the right claimed by a belligerent nation to board neutral vessels and examine their papers and cargo for contraband. **search me!** *int.* how should I know? I have no idea. **searchlight**, *n.* an electric arc-light or other powerful illuminant concentrated into a beam that can be turned in any direction for lighting channels, discovering an enemy etc. **search party**, *n.* one going out to search for a lost, concealed or abducted person or thing. **search warrant**, *n.* a warrant granted by a justice of the peace, authorizing entry into a house etc. to search for stolen property etc. **searchable**, *a.* **searcher**, *n.* **searching**, *a.* making search or inquiry; penetrating, thorough, minute, close. **searchingly**, *adv.*

season (sē′zən), *n.* one of the four divisions of the year, spring, summer, autumn, winter; a period of time of a specified or indefinite length; the period of the greatest activity of something, or when it is in vogue, plentiful, at its best etc.; a favourable, opportune, fit, suitable or convenient time; a period when a mammal is on heat; seasoning; a season ticket. *v.t.* to make sound or fit for use by preparation, esp. by tempering, maturing, acclimatizing, inuring, habituating or hardening; to make mature or experienced; to render palatable or give a higher relish to by the addition of condiments etc.; to make more piquant or pleasant, to add zest to; to mitigate, to moderate, to qualify (justice with mercy etc.). *v.i.* to become inured, habituated, accustomed etc.; of timber, to become hard and dry. **in season**, in vogue; in condition for shooting, hatching, use, mating, eating etc.; of a mammal, on heat; at a fit or opportune time. **in season and out of season**, at all times, continuously or indiscriminately. **season ticket**, *n.* a railway or other ticket, usu. issued at a reduced rate, valid for any number of journeys etc., for the period specified. **seasonable**, *a.* occurring or done at the proper time, opportune; suitable to the season. **seasonableness**, *n.* **seasonably**, *adv.* **seasonal**, *a.* of or occurring at a particular season; required, done, etc. according to the season. **seasonally**, *adv.* **seasoner**, *n.* **seasoning**, *n.* anything added to food to make it more palatable; anything that increases enjoyment.

seat (sēt), *n.* that on which one sits or may sit, a chair, bench, stool etc.; the part of a chair etc. on which a person's weight rests in sitting; the part of a machine or other structure on which another part or thing is supported; the buttocks or the part of trousers etc. covering them; a place for sitting or where one may sit; the place where anything is, location, site, situation; a place in which authority is vested; a country residence, a mansion; the right of sitting, esp. in a legislative body; manner or posture of sitting. *v.t.* to cause to sit down, to place or set on a seat; to assign seats to; to provide (a church etc.) with seats; to provide (a chair, trousers etc.) with a seat; to settle, to locate, to install, to establish, to fix in place. *v.i.* of a garment, to become baggy from sitting. **in the hot seat**, having ultimate responsibility for decisions taken. **seat belt**, *n.* a strap to hold a person in a seat in a car, aeroplane etc. **seated**, *a.* sitting. **seater**, *n.* (*usu. in comb.* as *two-seater*). **seating**, *n.* the provision of seats; the seats provided or their arrangement; material for seats; a support on which something rests.

SEATO (sē′tō), (*abbr.*) South East Asia Treaty Organization.

sebaceous (sibā′shəs), *a.* fatty; made of fatty or oily matter; of glands, ducts, follicles etc., containing, conveying, or secreting fatty or oily matter. **sebacic** (-bas′-), *a.* **seborrhoea**, (*N Am.*) **-rrhea**, (sebərē′ə)), *n.* excessive secretion of sebum. **seborrhoeic**, *a.* **sebum** (sē′bəm), *n.* the fatty matter secreted by the sebaceous glands, which lubricates the hair and skin.

Sebat (sē′bat), *n.* the fifth month of the Jewish civil year and the 11th of the ecclesiastical year, corresponding to part of Jan. and Feb.

sec¹ (sek), *a.* of wine, dry.

sec² (sek), *n.* short for SECOND¹.

sec³ (sek), (*abbr.*) secant.

sec., (*abbr.*) second, secondary; secretary; section; according to.

secant (sē′kənt), *a.* cutting. *n.* a straight line intersecting a curve, esp. a radius of a circle drawn through the second extremity of an arc of this and terminating in a tangent to the first extremity; the ratio of this to the radius; the ratio of the hypotenuse to the base of a right-angled triangle

formed by drawing a perpendicular to either side of the angle.
secateurs (sekətœz'), *n.pl.* pruning-scissors.
secco (sek'ō), *a.* (*Mus.*) plain, unadorned. *n.* tempera-painting.
secede (sisēd'), *v.i.* to withdraw from fellowship, association or communion, as with a political or religious body. **seceder**, *n.*
secern (sisœn'), *v.t.* to separate, to distinguish; to secrete or excrete. **secernent**, *a.* secretory. *n.* a secretory organ; a drug etc. promoting secretion. **secernment**, *n.*
secession (sisesh'ən), *n.* the act of seceding. **secessionism**, *n.* **secessionist**, *n.* a seceder or advocate of secessionism, esp. one who took part with the Southern States in the American Civil War of 1861-5.
sech (sesh), (*abbr.*) hyperbolic secant.
seckel (sek'l), *n.* a small, pulpy variety of pear.
seclude (siklood'), *v.t.* to shut up or keep (a person, place etc.) apart or away from society; to cause to be solitary or retired. **secluded**, *a.* hidden from view, private; away from others, solitary. **secludedly**, *adv.* **secludedness**, *n.* **seclusion** (-zhən), *n.* **seclusive**, *a.*
second[1] (sek'ənd), *a.* immediately following the first in time, place or position; next in value, authority, rank or position; secondary, inferior; other, alternate; additional, supplementary; subordinate, derivative; (*Mus.*) lower in pitch. *n.* the next after the first in rank, importance etc.; a second class in an examination etc.; a person taking this; another or an additional person or thing; one who supports another, esp. one who attends on the principal in a duel, boxing match etc.; the 60th part of a minute of time or angular measurement; (*coll.*) a very short time; (*pl.*) goods that have a slight flaw or are of second quality; (*pl.*) coarse, inferior flour, or bread made from this; (*Mus.*) the interval of one tone between two notes, either a whole tone or a semi-tone; the next tone above or below; two tones so separated combined together; a lower part added to a melody when arranged for two voices or instruments; (*coll.*) an alto; second gear. *v.t.* to forward, to promote, to support; to support (a resolution) formally to show that the proposer is not isolated. **second-best**, *a.* of second quality. **second chamber**, *n.* the upper house in a legislative body having two chambers. **second childhood**, *n.* senile dotage. **second-class**, *a.* of second or inferior quality, rank etc., second-rate; treated as inferior or second-rate; of the second class. **second class**, *n.* the category next to the first or highest; the second level of an honours degree. **Second Coming**, *n.* the return of Christ to earth as prophesied in the Bible. **second floor**, *n.* the second from the ground-floor. (In the US the term is applied to the first storey.) **second gear**, *n.* the forward gear next above first gear in a car etc. **second-hand**, *a.* not primary or original; not new, sold or for sale after having been used or worn; dealing in second-hand goods. **at second hand**, indirectly, from or through another. **second nature**, *n.* something that has become effortless or instinctual through constant practice. **second-rate**, *a.* of inferior quality, size, value etc. **second sight**, *n.* the power of seeing things at a distance in space or time as if they were present, clairvoyance. **second thoughts**, *n.pl.* reconsideration of a previous opinion or decision. **second wind**, *n.* a renewed burst of energy, stamina etc. after a concentrated effort. **secondly**, *adv.* in the second place; as the second item.
second[2] (sikond'), *v.t.* to retire (a military officer) temporarily without pay in order that he or she may take a civil or other appointment; to transfer temporarily or release for temporary transfer to another position, branch of an organization etc. **secondment**, *n.*
secondary (sek'əndəri), *a.* coming next in order of place or time to the first; not primary, not original, derivative, supplementary, subordinate; of the second or of inferior rank, importance etc.; revolving round a primary planet; between the tertiary geological formation above and the primary below, Mesozoic; of or being a feather on the second joint of a bird's wing; pertaining to or carrying an induced current. *n.* a delegate or deputy; a cathedral dignitary of secondary rank; a secondary planet, a satellite; the secondary geological epoch or formation; a secondary feather; a hind wing in an insect; a secondary coil, circuit etc. **secondary cell**, *n.* a rechargeable cell or battery using reversible chemical reactions to convert chemical into electrical energy. **secondary coil** or **winding**, *n.* a coil in which the current in the primary winding induces the electric current. **secondary colours** COLOUR[1]. **secondary education** or **school**, *n.* that provided for children who have received an elementary education. **secondary electrons**, *n.pl.* the electrons emitted by secondary emission. **secondary emission**, *n.* the emission of electrons from a surface or particle bombarded by primary electrons at high velocity. **secondary picketing**, *n.* picketing of an organization by workers with whom there is no direct dispute. **secondary sex characteristics**, *n.pl.* attributes related to the sex of an individual that develop from puberty. **secondary tumour**, **growth**, *n.* a tumour occurring somewhere other than at the site of the original cancer. **secondarily**, *adv.* **secondariness**, *n.*
seconde (sikond', səgōd'), *n.* (*Fencing*) a position in parrying or lungeing.
secondo (sikkon'dō), *n.* the second part or the second performer in a musical duet.
secrecy (sē'krəsi), *n.* the state of being secret, concealment; the quality of being secretive, secretiveness; solitude, retirement, seclusion.
secret (sē'krit), *a.* concealed from notice, kept private, hidden, not to be revealed or exposed, privy; unseen, occult, mysterious; given to secrecy, secretive, close, reserved, reticent; secluded, private. *n.* something to be kept back or concealed; a thing kept back from general knowledge; a mystery, something that cannot be explained; the explanation or key to this; secrecy; in the Roman Catholic Church, a prayer in a low tone recited by the celebrant at Mass. **in secret**, secretly, privately. **open secret**, something known generally. **secret agent**, *n.* an agent of the secret service. **secret police**, *n.* a police force operating in secret, usu. dealing with political rather than criminal matters. **secret service**, *n.* a government service for obtaining information or carrying out other work of which no account is given to the public. **secretly**, *adv.* **secretness**, *n.*
secretaire (sekrətea'), *n.* an escritoire, a bureau.
secretary (sek'rətəri), *n.* an officer appointed by a company, firm, society etc. to conduct its correspondence, keep its records and represent it in business transactions etc.; (also **private secretary**) a person employed by another to assist in literary work, correspondence etc.; Secretary of State; an escritoire. **Secretary of State**, a minister in charge of a government department; the Foreign Secretary of the US. **secretary-bird**, *n.* a S African bird preying on snakes etc. (named from its pen-like tufts in the ear). **secretary-general**, *n.* the person in charge of the administration of an organization. **secretary hand**, *n.* (*Hist.*) a style of handwriting used for legal documents until the 17th cent. **secretarial** (-teə'ri-), *a.* **secretariat** (-teə'riat), *n.* the post of a secretary; an administrative office headed by a Secretary; the administrative workers of an organization. **secretaryship**, *n.*
secrete (sikrēt'), *v.t.* to conceal, to hide; to keep secret; to separate from the blood, sap etc. by the process of secretion. **secretion**, *n.* the act of secreting or concealing; the process of separating materials from the blood, sap etc. for the service of the body or for rejection as excreta; any matter thus secreted, as mucus, gastric juice, urine etc. **secretional**, **-nary**, *a.* **secretor**, *n.* **secretory**, *a.*

secretive (sē'krətiv), *a.* given to secrecy, reserved, uncommunicative; (-krē'-), promoting or causing secretion. **secretively**, *adv.* **secretiveness**, *n.*
secretly, secretness SECRET.
secretor, secretory SECRETE.
sect (sekt), *n.* a body of persons who have separated from a larger body, esp. an established church, on account of philosophical or religious differences; a religious denomination, a nonconformist church (as regarded by opponents); the body of adherents of a particular philosopher, school of thought etc. **sectarian**, *n., a.* **sectarianism**, *n.* **sectarianize, -ise**, *v.t.* **sectary**, *n.* a member of a sect; (*Hist.*) a Dissenter, esp. an Independent or Presbyterian in the epoch of the Civil War.
sectant (sek'tənt), *n.* (*Geom.*) a portion of space separated by three intersecting planes but extending to infinity.
sectile (sek'tīl), *a.* capable of being cut.
section (sek'shən), *n.* separation by cutting; that which is cut off or separated, a part, a portion; one of a series of parts into which anything naturally separates or is constructed so as to separate for convenience in handling etc.; a division or subdivision of a book, chapter, statute etc.; a section-mark; a distinct part of a country, people, community, class etc.; a thin slice of any substance prepared for microscopic examination; a cutting of a solid figure by a plane, or the figure so produced; a vertical plan of a building etc. as it would appear upon an upright plane cutting through it; a part of an orchestra consisting of all the instruments of one class. *v.t.* to divide or arrange in sections; to represent in sections. **section-mark**, *n.* the sign §, marking a reference or the beginning of a section of a book, chapter etc. **sectional**, *a.* **sectionalism**, *n.* **sectionalize, -ise**, *v.t.* **sectionally**, *adv.*
sector (sek'tə), *n.* a portion of a circle or other curved figure included between two radii and an arc; a mathematical rule consisting of two hinged arms marked with sines, tangents etc.; a section of a battle front; a distinct part, a section. **sectoral**, *a.* **sectorial** (-taw'ri-), *a.* denoting a tooth on each side of either jaw, adapted for cutting like scissors with the corresponding one, as in many Carnivora; sectoral. *n.* a sectorial tooth.
secular (sek'ūlə), *a.* pertaining to the present world or to things not spiritual or sacred, not ecclesiastical or monastic; worldly, temporal, profane; lasting, extending over, occurring in or accomplished during a century, an age or a very long period of time; pertaining to secularism. *n.* a layman; a Roman Catholic priest bound only by the vow of chastity and belonging to no regular order; a church official who is not ordained. **secularism**, *n.* the state of being secular; applied to an ethical system founded on natural morality and opposed to religious education or ecclesiasticism. **secularist**, *n., a.* **secularize, -ise**, *v.t.* **secularization**, *n.* **secularly**, *adv.*
secund (sikŭnd', sek'-, sē'-), *a.* of flowers etc., arranged all on one side of the rachis. **secundly**, *adv.*
secundine (sek'əndīn, -dīn, -kŭn'-), *n.* (*often pl.*) the placenta and other parts connected with the foetus, ejected after parturition, the after-birth; (*Bot.*) the membrane immediately surrounding the nucleus.
secundogeniture (sikŭndōjen'ichə), *n.* the right of inheritance belonging to a second son.
secure (sikūə'), *a.* free from danger, risk or apprehension; safe from attack, impregnable; reliable, confident, certain, sure (of); in safe keeping, safe not to escape. *v.t.* to make safe or secure; to put into a state of safety from danger; to fasten; to close securely, to enclose or confine securely; to make safe against loss, to guarantee payment of; to get, to obtain, to gain possession of. **to secure arms**, to hold rifles muzzle downwards with the lock under the armpit as a protection from rain. **securable**, *a.* **securely**, *adv.* **securement**, *n.* **secureness**, *n.* security. **securer**, *n.*
security (sikū'riti), *n.* (*pl.* **-ties**) the state of being or feeling secure; freedom from danger or risk, safety; certainty, assurance, over-confidence; that which guards or secures; (an organization which sees to) the protection of premises etc. against burglary, espionage etc.; a pledge, a guarantee; something given or deposited as a pledge for payment of a loan, fulfilment of obligation etc., to be forfeited in case of non-performance; one who becomes surety for another; a document constituting evidence of debt or of property, a certificate of stock, a bond etc. **Security Council**, *n.* a body of the United Nations charged with the maintenance of international security and peace. **security guard**, *n.* one employed to guard buildings, money in transit etc. **security risk**, *n.* a person or thing considered to be a threat to (national) security. **securitization, -isation**, *n.* the putting together of a number of stocks, mortgages etc. into a single bond which is traded like a security. **securitize**, *v.t.*
sedan (sidan'), *n.* (also **sedan-chair**) a covered chair for one person, carried by two men by means of a pole on each side; (*N Am.*) a closed car with a single compartment for driver and passengers, a saloon car.
sedate (sidāt'), *a.* composed, calm, tranquil, staid, not impulsive. *v.t.* to administer a sedative to. **sedately**, *adv.* **sedateness**, *n.* **sedation**, *n.* a state of calmness or relaxation; the administration of a sedative. **sedative** (sed'ə-), *a.* allaying nervous irritability, soothing, assuaging pain. *n.* a sedative medicine, influence etc.
sedentary (sed'əntəri), *a.* sitting; accustomed or inclined, or obliged by occupation, to sit a great deal; involving or requiring much sitting; caused by sitting much; not migratory, attached to one place, not free-moving. **sedentarily**, *adv.* **sedentariness**, *n.*
Seder (sā'də), *n.* a ceremonial meal eaten on the first night (or the first two nights) of Passover.
sedge (sej), *n.* a coarse grass-like plant of the genus *Carex*, usu. growing in marshes or beside water; any coarse grass growing in such spots; a sedge-fly. **sedgy**, *a.*
sedilia (sidil'iə), *n.pl.* a series of (usu. three) stone seats, usu. canopied and decorated, on the south side of the chancel in churches, for the priest, deacon and subdeacon.
sediment (sed'imənt), *n.* the matter which subsides to the bottom of a liquid; lees, dregs, settlings. **sedimentary** (-men'-), *a.* **sedimentary rocks**, *n.pl.* rocks or strata laid down as sediment from water. **sedimentation**, *n.*
sedition (sidish'ən), *n.* agitation, disorder or commotion in a state, not amounting to insurrection; conduct tending to promote treason or rebellion. **seditionary**, *n., a.* **seditious**, *a.* **seditiously**, *adv.* **seditiousness**, *n.*
seduce (sidūs'), *v.t.* to lead astray, to entice from rectitude or duty, esp. to induce a person to have sexual intercourse; to entice or lure, esp. by offering rewards. **seducer**, *n.* **seducible**, *a.* **seducing**, *a.* **seducingly**, *adv.* **seduction** (sidŭk'shən), *n.* the act of seducing, esp. of persuading a woman to have sexual intercourse; the state of being seduced; that which seduces, an enticement, an attraction, a tempting or attractive quality, a charm. **seductive**, *a.* **seductively**, *adv.* **seductiveness**, *n.* **seductress** (-tris), *fem. n.*
sedulous (sed'ūləs), *a.* assiduous, constant, steady and persevering in business or endeavour; industrious, diligent. **sedulity** (-dū'-), **sedulousness**, *n.* **sedulously**, *adv.*
Sedum (sē'dəm), *n.* a genus of fleshy-leaved plants including the stonecrop, orpine etc.
see[1] (sē), *v.t.* (*past* **saw**, (saw) *p.p.* **seen**) to perceive by the eye; to discern, to descry, to observe, to look at; to perceive mentally, to understand, to apprehend, to have an idea of; to witness, to experience, to go through, to have knowledge of; to imagine, to picture to oneself; to call on, to pay a visit to, to grant an interview to, to receive; to escort, to attend, to conduct (a person home etc.); (*Poker* etc.) to accept (a challenge, bet etc., or

person offering this). *v.i.* to have or exercise the power of sight; to discern, to comprehend; to inquire, to make an investigation (into); to reflect, to consider carefully; to ascertain by reading; to take heed; to give attention; to make provision for; to look out; to take care (that); (*imper.*) to refer to. **let me see**, a formula asking for time to consider or reflect. **see you (later)**, goodbye for the present. **to see about**, to give attention to; to make preparations etc.; a polite form of refusal. **to see daylight**, (*coll.*) to begin to comprehend. **to see fit, good**, to think advisable. **to see life**, to gain experience of the world, esp. by dissipation. **to see off**, to escort on departure; (*coll.*) to get rid of. **to see out**, to escort out of a house etc.; to outlive, outlast; to last to the end of. **to see over**, to inspect. **to see the light**, to be born; to realize the truth; to be converted to a religion or to any other belief. **to see through**, to penetrate, not to be deceived by; to persist (in a task etc.) until it is finished; to help through a difficulty, danger etc. **to see to**, to look after. **to see to it that**, to take care that. **see-through**, *a.* (semi-)transparent, esp. of clothing. **seeing**, *n.* sight; (*Astron.*) atmospheric conditions for observation. *conj.* inasmuch as, since, considering (that). **seeing eye**, *n.* a guide dog for the blind. **seer**, *n.* one who sees; one who foresees, a prophet.

see[2] (sē), *n.* the diocese or jurisdiction of a bishop or archbishop. **Holy See**, the Papacy, the papal Court.

seed (sēd), *n.* the mature fertilized ovule of a flowering plant, consisting of the embryo germ or reproductive body and its covering; (*collect.*) seeds, esp. in quantity for sowing; (*dated*) the male fertilizing fluid, semen; the germ from which anything springs, first principle, beginning or source; (*Bibl.*) progeny, offspring, descendants. *v.t.* to sow or sprinkle with seed; to put a small crystal into (a solution) to start crystallization; to scatter solid particles in (a cloud) to bring on rain; to remove the seeds from (fruit etc.); in sport, to arrange the draw in (a tournament) so that the best players do not meet in the early rounds; to classify (a good player) in this way. *v.i.* to sow seed; to run to seed. **to run to seed**, to become shabby; to lose self-respect. **seedbed**, *n.* a piece of ground where seedlings are grown; a place where anything develops. **seed cake**, *n.* a sweet cake containing aromatic seeds, esp. caraway. **seed coral**, *n.* coral in small seed-like pieces. **seed corn, -grain**, *n.* corn set aside for sowing. **seed money**, *n.* the money with which a project is set up. **seed oyster**, *n.* oyster-spat. **seed pearl**, *n.* a small seed-like pearl. **seed plot**, *n.* a piece of ground on which seeds are sown; a nursery or hotbed (of seditions etc.). **seed potato**, *n.* a potato tuber used for planting. **seed vessel**, *n.* the pericarp. **seed wool**, *n.* raw cotton from which the seeds have not yet been removed. **seeded**, *a.* **seeder**, *n.* a seed-drill or other device for planting seeds; a device for removing the seeds from raisins etc. **seedless**, *a.* **seedling** (-ling), *a.* raised from seed. *n.* a plant reared from seed; a very young plant. **seedy**, *a.* abounding in seeds; run to seed; (*coll.*) shabby, down-at-heel; off colour, as after a debauch; out of sorts. **seedily**, *adv.* **seediness**, *n.* shabbiness, near poverty; a state of poor health.

seek (sēk), *v.t.* (*past, p.p.* **sought**, (sawt)) to go in search of; to try to find, to look for; to ask, to solicit (a thing of a person); to aim at, to try to gain, to pursue as an object; to search (a place etc. through), to resort to. *v.i.* to make search or inquiry (after or for); to endeavour, to try (to do). **to seek out**, to search for; to cultivate the friendship of. **sought-after**, *a.* in demand, much desired or courted. **seeker**, *n.*

seem (sēm), *v.i.* to give the impression of being, to be apparently though not in reality; to appear (to do, to have done, to be true or the fact that); to be evident or apparent. **I can't seem to**, (*coll.*) I am unable to. **it seems**, it appears, it is reported (that). **it would seem**, it appears, it seems to one. **seeming**, *a.* appearing, apparent, but not real; apparent and perhaps real. *n.* appearance, semblance, esp. when false. **seemly** (sēm'li), *a.* proper, decent; suited to the occasion, purpose etc.

seen (sēn), *p.p.* SEE[1].

seep (sēp), *v.i.* to soak, to percolate, to ooze. **seepage** (-ij), *n.*

seer SEE[1].

seersucker (siə'sŭkə), *n.* a thin striped linen or cotton fabric with a puckered appearance.

seesaw (sē'saw), *n.* a game in which two persons sit one on each end of a board balanced on a support in the middle and move alternately up and down; the board so used; alternate or reciprocating motion. *a.* moving up and down or to and fro; vacillating. *v.i.* to cause to move in a seesaw fashion. *v.i.* to play at seesaw; to move up and down or backwards and forwards; to act in a vacillating manner.

seethe (sēdh), *v.t.* (*dated*) to cook by boiling; to soak in liquid. *v.i.* to boil; to be in a state of ebullition; to be agitated, to bubble over. **seething**, *a.* **seethingly**, *adv.*

segment (seg'mənt), *n.* a portion cut or marked off as separable, a section, a division, esp. one of a natural series (as of a limb between the joints, the body of an articulate animal, a fruit or plant organ divided by clefts); (*Geom.*) a part cut off from any figure by a line or plane. *v.i.* to divide or be divided into segments; to undergo cleavage. *v.t.* to divide into segments. **segmental, -ary, -ate** (-men'-, -tət), *a.* **segmentally**, *adv.* **segmentation**, *n.* **segmented**, *a.* composed of segments; divided into segments.

segregate (seg'rigāt), *v.t.* to separate from others, to set apart, to isolate; to place in a separate class; to split (a community) into separate parts on the basis of race. *v.i.* of a pair of alleles, to become separated during meiosis; (*Cryst.*) to separate from a mass and collect about nuclei and lines of fracture. **segregation**, *n.* the act of segregating; separation of a community on racial grounds. **segregational**, *a.* **segregationist**, *n.* a believer in racial segregation. **segregative**, *a.* **segregator**, *n.*

seguidilla (segidēl'yə), *n.* a lively Spanish dance in triple time; the music for this.

seicento (sāchen'tō), *n.* the 17th cent. in Italian art, architecture or literature.

seiche (sāsh), *n.* an undulation in a body of water usu. due to disturbance of atmospheric pressure or to subterranean movements.

Seidlitz powder (sed'lits), *n.* a mild laxative, composed of a mixture of Rochelle salt, bicarbonate of soda and finely powdered tartaric acid, mixed separately in water to form an effervescing drink.

seigneur (senyœ'), **seignior** (sān'-), *n.* (*F Hist.*) a feudal lord. **seigneurial**, *a.* **seigneury** (sān'-), *n.* (*F Hist.*) the territory or lordship of a seigneur. **seigniorage** (sān'yərij), *n.* something claimed by the sovereign or by a feudal superior as a prerogative, esp. an ancient right of the Crown to a percentage on bullion brought to the mint to be coined; the profit derived from issuing coins at a rate above their intrinsic value; a royalty. **seigniorial** (-yaw'ri-), SEIGNEURIAL. **seigniory**, *n.* feudal lordship; power as sovereign lord; the territory or domain of a feudal lord; (*It. Hist.*) the municipal council of an Italian republic.

seine (sān), *n.* a large fishing-net with floats at the top and weights at the bottom for encircling. *v.t.* to catch with this. *v.i.* to fish with it.

seise (sēz), *v.t.* (*usu. in p.p.*) (*Law*) to put in possession of. **to be, stand seised of**, to have in legal possession. **seisable**, *a.* **seisin** (-zin), *n.* possession of land under a freehold; the act of taking possession; the thing possessed.

seisor, *n.*

seismic (sīz'mik), **seismal** (-məl), *a.* of, pertaining to or produced by an earthquake.

seismo- 753 **self-image**

seismo-, *comb. form.* pertaining to an earthquake.
seismogram (sīz'məgram), *n.* a record given by a seismograph. **seismograph** (sīz'məgraf), *n.* an instrument for recording the period, extent and direction of the vibrations of an earthquake. **seismographer** (-mog'-), *n.* **seismographic, -ical** (-graf'-), *a.* **seismography** (-mog'-), *n.*
seismology (sīzmol'əji), *n.* the study or science of earthquakes. **seismological** (-loj'-), *a.* **seismologically**, *adv.* **seismologist**, *n.*
seismometer (sīzmom'itə), *n.* a seismograph; a seismoscope. **seismometric, -ical** (-met'-), *a.* **seismometry**, *n.*
seize (sēz), *v.t.* to grasp or lay hold of suddenly, to snatch, to take possession of by force; to grasp mentally, to comprehend; to come upon, to affect suddenly and forcibly; (*Naut.*) to fasten, to lash with cord etc.; (*Law*) to seise: to take possession of; to impound, to confiscate. *v.i.* to lay hold (upon); to jam, to become stuck. **seizable**, *a.* **seizer**, *n.* **seizin** (-zin), etc. SEISIN, under SEISE. **seizing-up**, *n.* the locking or partial welding together of sliding surfaces from lack of lubrication. **seizure** (-zhə), *n.* the act of seizing; a sudden attack, as of a disease.
sejant (sē'jənt), *a.* (*Her.*) sitting with the forelegs erect.
selachian (silā'kiən), *n.* a fish of the group Selachii comprising the sharks, dog-fishes etc. *a.* pertaining to this group.
Selaginella (seləjinel'ə), *n.* a genus of evergreen moss-like cryptogamic plants many of which are cultivated for ornamental purposes.
selah (sē'lə), *n.* a word occurring in the Psalms and in Habakkuk, always at the end of a verse, variously interpreted as indicating a pause, a repetition, the end of a strophe etc.
seldom (sel'dəm), *adv.* rarely, not often. *a.* rare.
select (silekt'), *a.* chosen, picked out, choice; taken as superior to or more suitable than the rest; strict in selecting new members etc., exclusive, more valuable. *v.t.* to choose, to pick out (the best etc.). **select committee**, *n.* members of parliament specially chosen to examine a particular question and to report on it. **selection**, *n.* the act of selecting; the right or opportunity of selecting, choice; that which is selected; a natural or artificial process of sorting out organisms suitable for survival; a range of goods (as in a shop) from which to choose; (*Austral.*) FREE SELECTION under FREE. **selective**, *a.* selecting, exercising a power or ability to choose. **selectively**, *adv.* **selectivity** (sēlektiv'-), *n.* the quality of being selective; the efficiency of a radio receiver in separating the different broadcasting stations. **selectness**, *n.* **selector**, *n.* one who or that which selects; (*Austral.*) a settler who takes up a piece of select land.
selenium (silē'niəm), *n.* a non-metallic element, at. no. 34; chem. symbol Sc; obtained especially as a by-product in the manufacture of sulphuric acid, similar in chemical properties to sulphur and tellurium, utilized for its varying electrical resistance in light and darkness. **selenium cell**, *n.* a type of photoelectric cell using a strip of selenium. **selenate** (sel'ināt), *n.* a salt of selenic acid. **selenic** (silen'ik, -lē'-), *a.* containing or derived from (high valency) selenium; of or derived from the moon. **selenide** (se'līnid), *n.* a compound of selenium with an element or radical. **selenious**, *a.* containing or derived from (low valency) selenium. **selenite** (se'līnīt), *n.* a transparent variety of gypsum or sulphate of lime; a salt of selenious acid; an inhabitant of the moon. **selenitic** (-nit'-), *a.*
selen(o)-, seleni-, *comb. form.* pertaining to or containing selenium; pertaining to the moon.
selenodont (silē'nədont), *a.* of molar teeth, having crescent-shaped ridges; pertaining to the Selenodonta. *n.* a selenodont mammal.
selenography (selinog'rəfi), *n.* a description of the moon and its phenomena; the art of delineating the face of the moon. **selenograph** (-lē'nəgraf), *n.* **selenographer**, *n.* **selenographic, -ical** (-graf'-), *a.*
selenology (selinol'əji), *n.* the branch of astronomical science treating of the moon. **selenological** (-lēnəloj'-), *a.* **selenologically**, *adv.* **selenologist**, *n.*
self (self), *n.* (*pl.* **selves**) the individuality of a person or thing, as the object of reflexive consciousness or action; one's individual person; one's private interests etc.; furtherance of these; a flower of a uniform or of the original wild colour. *a.* uniform, pure, unmixed; same-coloured; of one piece or the same material throughout. *pron.* (*coll., facet.*) myself, yourself etc. **selfhood** (-hud), *n.* **selfish**, *a.* attentive only to one's own interests; not regarding the interests or feelings of others; actuated by or proceeding from self-interest. **selfishly**, *adv.* **selfishness**, *n.* **selfless**, *a.* having no regard for self, unselfish. **selflessly**, *adv.* **selflessness**, *n.*
self- (self-), *comb. form.* by, of, in, to, for, etc. the self; of oneself or itself; automatic. **self-abnegation**, *n.* denial of one's own interests. **self-absorption**, *n.* preoccupation with oneself to the exclusion of others. **self-abuse**, *n.* denigration of one's abilities etc.; masturbation. **self-acting**, *n.* automatic. **self-addressed**, *a.* of an envelope, addressed to return to sender; directed to oneself. **self-aggrandizement**, *n.* the act of enhancing one's own power or importance. **self-aggrandizing**, *a.* **self-appointed**, *a.* having power or authority without the consent of others. **self-assertion**, *n.* the act of stating one's views etc., esp. in an aggressive manner. **self-asserting**, *a.* **self-assertive**, *a.* **self-assurance**, *n.* certain of the worth and validity of one's own ideas, opinions etc. **self-assured**, *a.* **self-catering**, *a.* of holiday accommodation, not providing meals, cleaning etc. **self-centred**, *a.* interested only in oneself and one's own affairs. **self-coloured**, *a.* of a uniform colour throughout. **self-confessed**, *a.* according to one's own admission. **self-confidence**, *n.* confidence in one's own abilities, judgement etc. **self-confident**, *a.* **self-confidently**, *adv.* **self-conscious**, *a.* aware of one's actions, behaviour, situation etc., esp. as observed by others. **self-consciously**, *adv.* **self-consciousness**, *n.* **self-contained**, *a.* complete in itself; of a flat, room etc., having its own kitchen and bathroom; reserved, uncommunicative. **self-contradiction**, *n.* contradiction of oneself; *a.* statement containing two contradictory elements. **self-contradictory**, *a.* **self-control**, *n.* power of controlling one's feelings, impulses etc. **self-controlled**, *a.* **self-deception, -deceit**, *n.* the act of deceiving oneself. **self-deceptive**, *a.* **self-defence**, *n.* the act of defending one's own person, property, ideas etc. **the noble art of self-defence**, boxing. **self-defensive**, *a.* **self-denial**, *n.* refusal to gratify one's own appetites or desires. **self-denying**, *a.* **self-determination**, *n.* freedom of an individual to act without external influence; the right of a people or nation to choose its own political status or government. **self-discipline**, *n.* the act of disciplining one's own emotions, actions etc. **self-disciplined**, *a.* **self-doubt**, *n.* lack of confidence in one's abilities. **self-effacement**, *n.* the act of making oneself, one's actions etc. inconspicuous, esp. through modesty or timidity. **self-effacing**, *a.* **self-employed**, *a.* running one's own business, or working freelance. **self-employment**, *n.* **self-esteem**, *n.* confidence in oneself; vanity. **self-evident**, *a.* obvious of itself, not requiring proof or demonstration. **self-evidently**, *adv.* **self-explanatory**, *a.* obvious, not requiring further explanation. **self-expression**, *n.* the expression of one's own personality (through art etc.). **self-expressive**, *a.* **self-government**, *n.* control by a group, nation etc. of its own political affairs. **self-governing**, *a.* **self-governed**, *a.* **self-heal**, *n.* a small plant *prunella vulgaris*, supposedly having healing properties. **self-help**, *n.* the act or practice of attaining one's ends without the aid of others.
selfhood SELF.
self-image, *n.* one's own idea of oneself. **self-important**, *a.*

having an exaggerated belief in one's own worth, abilities etc. **self-importance**, *n.* **self-improvement**, *n.* improvement of one's own social or economic position by one's own efforts. **self-induced**, *a.* brought on by one's own actions. **self-induction**, *n.* production of an electromotive force in a circuit by variation of a current in the same circuit. **self-indulgent**, *a.* gratifying one's own desires etc. **self-indulgence**, *n.* **self-interest**, *n.* one's personal advantage; absorption in selfish aims.
selfish etc. SELF.
self-justification, *n.* the act of excusing one's own behaviour.
selfless etc. SELF.
self-loading, *a.* of a firearm, reloading automatically. **self-love**, *n.* undue regard for oneself or one's own interests; selfishness; conceit. **self-made**, *a.* successful, wealthy etc. through one's own efforts. **self-opinionated**, *a.* overvaluing oneself or one's own opinions; obstinately adhering to one's own views. **self-pity**, *n.* excessive preoccupation with one's own misfortunes. **self-pitying**, *a.* **self-pityingly**, *adv.* **self-possessed**, *a.* calm, imperturbable, in control of one's emotions etc. **self-possession**, *n.* **self-preservation**, *n.* preservation of oneself from injury. **self-propelled**, *a.* of a vehicle, provided with its own means of propulsion. **self-propelling**, *a.* **self-raising**, *a.* of flour, having the raising agent already added. **self-realization**, *n.* the full development of one's potential. **self-regard**, *n.* consideration or respect for oneself or one's own interests. **self-reliance**, *n.* reliance on one's own abilities. **self-reliant**, *a.* **self-reproach**, *n.* the act of blaming oneself. **self-reproachful**, *a.* **self-respect**, *n.* due regard for one's character and integrity. **self-respecting**, *a.* **self-restraint**, *n.* restraint imposed on one's feelings, impulses etc. **self-righteous**, *a.* possessing undue regard for one's own virtuousness. **self-righteously**, *adv.* **self-righteousness**, *n.* **self-sacrifice**, *n.* surrender or subordination of one's own interests and desires to those of others. **self-sacrificing**, *a.* **selfsame**, *a.* exactly the same, absolutely identical. **self-satisfied**, *a.* showing smug or complacent satisfaction with one's own actions. **self-satisfaction**, *n.* **self-sealing**, *a.* of an envelope, able to seal itself. **self-seeking**, *a.* exclusively pursuing one's own interests; selfish. *n.* the act of selfishly pursuing one's own interests, profit etc. **self-seeker**, *n.* **self-service**, *a.*, *n.* (a restaurant, shop etc.) where the customer serves himself. **self-serving**, *a.* giving priority to one's own interests at the expense of others. **self-starter**, *n.* an electric motor for starting a motor car; a strongly motivated person. **self-styled**, *a.* assuming a name or title oneself without authorization, claimed, pretended. **self-sufficient**, *a.* capable of fulfilling one's own requirements, needs etc. without aid. **self-sufficiency**, *n.* **self-sufficiently**, *adv.* **self-will**, *n.* obstinacy. **self-willed**, *a.* **self-winding**, *a.* of a clock, wristwatch etc., winding itself automatically.
Seljuk (seljook'), *n.* a member of the Turkish dynasties who ruled in Central and W Asia during 11th–13th cents. *a.* of or pertaining to the Seljuks.
sell (sel), *v.t.* (*past, p.p.* **sold**) to transfer or dispose of (property) to another for an equivalent in money; to yield or give up (one's life etc.) exacting some return; to be a regular dealer in; to give up, surrender or betray for a price, reward or bribe; (*sl.*) to disappoint, to cheat, to play a trick upon; to inspire others with a desire to possess. *v.i.* to be a shopkeeper or dealer; to be purchased, to find purchasers. *n.* (*sl.*) a disappointment, a fraud; a manner of selling (*hard*, *aggressive marketing*; or *soft*, *gentle persuasion*). **sold on**, enthusiastic about. **to sell off**, to sell the remainder of (goods); to clear out (stock), esp. at reduced prices. **to sell one a pup**, (*sl.*) to swindle. **to sell out**, to sell off (one's stock etc.); to sell completely; to dispose of (one's shares in a company etc.); to betray. **to sell up**, to sell the goods of (a debtor) to pay his debt; to sell one's business, one's house and possessions etc. **sell-out**, *n.* a betrayal; a performance etc. for which all the tickets have been sold. **seller**, *n.* **seller's market**, *n.* one in which demand exceeds supply and sellers make the price. **selling race**, *n.* a horse-race, the winner of which is sold by auction.
Sellotape® (sel'ətāp), *n.* a cellulose or plastic adhesive tape for mending, binding etc. **sellotape**, *v.t.* to fix or fasten with Sellotape.
seltzer (selt'sə), *n.* a natural effervescing mineral water or synthetic imitation, used as a refreshing drink.
selva (sel'və), *n.* tropical rain forest in the Amazon basin.
selvage, selvedge (sel'vij), *n.* the edge of cloth woven so as not to unravel; a narrow border. **selvaged**, *a.*
selves (selvz), *n.pl.* SELF.
Sem., (*abbr.*) Seminary; Semitic.
semantic (siman'tik), *a.* of or pertaining to semantics; concerned with the meaning of words and symbols. **semantically**, *adv.* **semantics**, *n.sing.* the area of linguistics concerned with meaning. **semanticist** (-sist), *n.*
semaphore (sem'əfaw), *n.* an apparatus or system for signalling using oscillating arms or flags etc. **semaphoric** (-fo'), *a.*
semasiology (simāsiol'əji), *n.* semantics. **semasiological** (-loj'-), *a.* **semasiologically**, *adv.* **semasiologist**, *n.*
sematic (simat'ik), *a.* of the nature of a sign, significant, esp. pertaining to markings on animals serving to attract, to warn off enemies etc.
semblance (sem'bləns), *n.* external appearance, seeming; a mere show; a likeness, an image.
semé (sem'ā), *a.* (*Her.*) applied to a field, or charge, strewn over with figures, as stars, crosses etc.
sememe (sem'ēm, sē'-), *n.* the meaning of a morpheme.
semen (sē'mən), *n.* the fertilizing fluid containing spermatozoa, produced by the generative organs of a male animal.
semester (simes'tə), *n.* a college half-year in German, some American and other universities.
semi (sem'i), *n.* (*coll.*) a semidetached house; a semifinal.
semi- (semi-), *pref.* half; partially, imperfectly.
semiannual, *a.* occurring every six months; half-yearly. **semiannually**, *adv.*
semiarid, *a.* of a climate, less dry than a desert climate, supporting scrub vegetation.
semiautomatic, *a.* partly automatic; of a firearm, self-loading. *n.* such a firearm.
semibreve (sem'-), *n.* a musical note equal to half a breve, or two minims.
semicircle (sem'-), *n.* a half circle. **semicircular**, *a.* **semicircular canals**, *n.pl.* three fluid-filled tubes in the inner ear, concerned with the maintenance of balance.
semicolon, *n.* a mark (;) used in punctuation, now usu. intermediate in value between the period and the comma.
semiconductor, *n.* a substance (as silicon) whose electrical conductivity lies between those of metals and insulators and increases as its temperature rises; a device using such a substance.
semiconscious, *a.* half-conscious. **semiconsciously**, *adv.* **semiconsciousness**, *n.*
semidetached, *a.* (of a building) joined to another by a common wall.
semifinal, *n.* (*Sport*) the match or round before the final. **semifinalist**, *n.*
semifluid, *a.* imperfectly fluid. *n.* a semifluid substance.
semiliquid, *n.*, *a.* semifluid.
semilunar, *a.* resembling or shaped like a half-moon or crescent. **semilunar bone**, *n.* a semi-lunar bone. **semilunar valve**, one of two half-moon-shaped valves in the heart.
semimetal, *n.* an element having metallic properties but non-malleable. **semimetallic**, *a.*
semimonthly, *a.* occurring twice a month; issued at half-monthly intervals.

seminal (sem'inəl), *a.* of or pertaining to semen or reproduction; germinal, propagative; important to the future development of anything; containing new ideas, original. **seminally**, *adv.*

seminar (sem'inah), *n.* a group of students undertaking an advanced course of study or research together, usu. under the guidance of a professor; such a course; a discussion group, or a meeting of it.

seminary (sem'inəri), *n.* (*pl.* **-naries**) a place of education, a school, academy or college, esp. a (foreign) Roman Catholic school for training priests. **seminarian** (-neə'ri-), *n.*, *a.* **seminarist**, *n.*

seminiferous (seminif'erəs), *a.* bearing or producing seed; conveying semen.

semiology (semiol'əji), *n.* the study of the symptoms of disease; the study of signs and symbols generally. **semiological** (-loj'-), *a.*

semiotics (semiot'iks), *n.sing.* (*Linguistics*) the study of signs and symbols and their relationships in language. **semiotic**, *a.*

semipalmate, *a.* half-webbed, as the toes of many shore-birds.

semipermeable, *a.* permeable by small molecules but not by large ones.

semiprecious, *a.* valuable, but not regarded as a precious stone.

semiprofessional, *a.* (of a person) engaged in a sport or some activity part-time but for pay; (of a sport etc.) engaged in by semiprofessionals; relating to someone who is partly professional. *n.* a semiprofessional person.

semiquaver (sem'-), *n.* (*Mus.*) a note of half the duration of a quaver.

semirigid, *a.* of an airship, having a flexible gas container and a rigid keel.

semi-Saxon, *a.* intermediate between Anglo-Saxon and English, pertaining to the early period of Middle English, c. 1150–1250. *n.* the semi-Saxon language.

semiskilled, *a.* of a worker, having some basic skills but not highly trained.

semisolid, *a.* so viscous as to be almost solid.

Semite (sem'it, se'-), *n.* a member of one of the peoples (including Hebrews, Phoenicians, Aramaeans, Assyrians, Arabs and Abyssinians) reputed to be descended from Shem, or who speak a Semitic language. *a.* Semitic. **Semitic** (simit'-), *a.* pertaining to the Semites or their languages. *n.* one of the Semitic group of languages.

semitone (sem'-), *n.* (*Mus.*) an interval equal to half a major tone on the scale. **semitonic**, *a.* **semitonically**, *adv.*

semitrailer, *n.* a trailer which has back wheels but is supported in front by the towing vehicle.

semitropical, *a.* partly within or bordering on the tropics.

semivowel (sem'-), *n.* a sound having the character of both vowel and consonant as *w* and *y*; sometimes applied to consonants such as *l*, *m*, *r* and *z*, that are not mute; a character representing such.

semolina (seməlē'nə), *n.* the hard grains of wheat left after sifting floor, used for puddings etc.

sempervivum (sempəvē'vəm), *n.* a genus of fleshy plants of the family Crassulaceae containing the houseleeks.

semipiternal (sempitœ'nəl), *a.* everlasting, eternal, endless. **sempiternally**, *adv.*

semplice (sem'plichi), *adv.* (*Mus.*) simply, plainly, without embellishment.

sempre (sem'pri), *adv.* (*Mus.*) in the same manner throughout.

Semtex® (sem'teks), *n.* a malleable plastic explosive.

SEN, (*abbr.*) State Enrolled Nurse.

Sen., (*abbr.*) senate; senator; senior.

sen (sen), *n.* a Japanese monetary unit, one-hundredth of a yen. [Jap.]

senate (sen'ət), *n.* an assembly or council performing legislative or administrative functions; the state council of the ancient Roman republic and empire of ancient Athens, Lacedaemon etc., of the free cities of the Middle Ages etc.; the upper legislative house in various bicameral parliaments, as of the US and France; the governing body of the various universities; any venerable deliberative or legislative body. **senate-house**, *n.* a building in which a senate meets. **senator**, *n.* a member of a senate. **senatorial** (-taw'ri-), *a.* **senatorially**, *adv.*

send (send), *v.t.* (*past*, *p.p.* **sent**) to cause or bid to go or pass or to be conveyed or transmitted to some destination; to dispatch; to cause to go (in, up, off, away etc.); to propel, to hurl, to cast; to cause to come or befall, to grant, to bestow, to inflict; to cause to be, to bring about. *v.i.* (*Naut.*) to pitch (of a vessel). **to send down**, to expel from a university; (*coll.*) to send to prison. **to send for**, to require the attendance of a person or the bringing of a thing; to summon; to order. **to send forth**, to put forth; to emit. **to send in**, to submit (as a competition entry). **to send off**, to dispatch; to give a send-off to one departing; in sport, to order (a player) off the field because of an infringement of the rules. **to send on**, to forward (mail); to send (luggage) in advance. **to send out**, to send forth. **to send up**, to parody; to ridicule. **send-off**, *n.* a start, as in a race; a leave-taking, a friendly demonstration to one departing on a journey. **send-up**, *n.* **sender**, *n.*

senescent (sines'ənt), *a.* growing old. **senescence**, *n.*

seneschal (sen'ishəl), *n.* an officer in the houses of princes and high dignitaries in the Middle Ages having the superintendence of feasts and domestic ceremonies, sometimes dispensing justice; a steward or major-domo. **seneschalship**, *n.*

sengreen (sen'grēn), *n.* the houseleek or sempervivum.

senhor (senyaw'), *n.* a man, in a Portuguese-speaking country; the Portuguese or Brazilian title corresponding to the English Mr or sir. **senhora** (-rə), *n.* a lady; Mrs, madam. **senhorita** (-rē'tə), *n.* a young unmarried girl; Miss.

senile (sē'nīl), *a.* pertaining to or proceeding from the infirmities etc. of old age; suffering from the physical and mental infirmities of old age. **senility** (sinil'-), *n.*

senior (sēn'yə), *a.* older, elder (appended to names to denote the elder of two persons with identical names, esp. father and son); older or higher in rank or service. *n.* one older than another; one older or higher in rank, service etc.; (*N Am.*) a student in his or her third or fourth year. **senior common room**, a common room for the use of staff at a college. **senior citizen**, *n.* an old-age pensioner. **senior service**, *n.* the Royal Navy. **senior wrangler**, **optime**, *n.* (*Camb. Univ.*) first in first class of mathematical tripos. **seniority** (-nio'-), *n.*

senna (sen'ə), *n.* the dried, purgative leaflets or pods of several species of cassia.

señor (senyaw'), *n.* (*pl.* **-ñors**, **-ñores**) a man, in a Spanish-speaking country; the Spanish form of address equivalent to Mr or sir. **señora** (-rə), *n.* (*pl.* **-ras**) a lady; Mrs, madam. **señorita** (-rē'tə), *n.* (*pl.* **-tas**) a young girl; Miss.

sensation (sensā'shən), *n.* the mental state of affection resulting from the excitation of an organ of sense, the primary element in perception or cognition of an external object; the content of such a mental state or affection, a state of excited feeling or interest, esp. affecting a number of people; the thing or event exciting this. **sensational**, *a.* causing, or pertaining to, sensation; (*coll.*) very good. **sensationalism**, *n.* the employment of sensational methods in literary composition, political agitation etc.; (*Phil.*) the theory that all knowledge is derived from sensation. **sensationalist**, *n.* **sensationalistic** (-lis'-), *a.* **sensationalize**, **-ise**, *v.t.* **sensationally**, *adv.*

sense (sens), *n.* one of the five faculties by which sensation is received through special bodily organs (sight, hearing, touch, taste, smell); also, the muscular sense giving a sensation of physical effort; the faculty of sensation, perception or ability to perceive through the senses, sensi-

tiveness; bodily feeling, sensuousness; intuitive perception, comprehension, appreciation; consciousness, conviction (of); sound judgment, sagacity, common sense, good mental capacity; meaning, signification; general feeling or judgment, consensus of opinion; (*pl.*) normal command or possession of the senses, sanity. *v.t.* to be aware of, to perceive by the senses; to understand. **in one's senses**, sane. **out of one's senses**, insane. **to make sense**, to be intelligible. **to make sense of**, to understand. **sense datum**, *n.* an item of experience received directly through a sense organ. **sense-organ**, *n.* a bodily part or organ concerned in the production of sensation. **sense perception**, *n.* **senseless**, *a.* incapable of sensation, insensible; contrary to reason, foolish, nonsensical. **senselessly**, *adv.* **senselessness**, *n.*
sensible (sen'sibl), *a.* perceptible by the senses; appreciable; acting with or characterized by good sense or judgment, judicious, reasonable; having perception (of). *n.* that which is sensible or perceptible. **sensibility** (-bil'-), *n.* (*pl.* **-ties**) capacity to see or feel; sensitivity; awareness; (*often pl.*) emotional feelings; the capacity to respond to emotional, aesthetic or moral stimuli. **sensibleness**, *n.* **sensibly**, *adv.*
sensitive (sen'sitiv), *a.* of or depending on the senses, sensory; readily or acutely affected by external influences; delicate, susceptible, excitable or responsive; (*Phot.*) susceptible to the action of light; of information, secret, classified. **sensitive plant**, *n.* a plant, *Mimosa pudica* or *M. sensitiva*, the leaves of which shrink from the touch. **sensitively**, *adv.* **sensitiveness**, *n.* **sensitivity** (-tiv'-), *n.* **sensitize, -ise**, *v.t.* to make sensitive; to render (paper etc.) sensitive to light; to render (a person) sensitive (to an allergen, drug etc.). **sensitization**, *n.* **sensitizer**, *n.* **sensitometer** (-tom'itǝ), *n.* an apparatus for determining the sensitiveness of photographic films etc.
sensor (sen'sǝ), *n.* anything which responds to, and signals, a change in a physical stimulus, for information or control purposes.
sensorium (sensaw'riǝm), *n.* (*pl.* **-ria**) the seat or organ of sensation, the brain; the nervous system, comprising the brain, spinal cord etc.; the grey matter of these. **sensory** (sen'sǝri), *a.* of the senses or of sensation.
sensual (sen'shuǝl, -sū-), *a.* pertaining to or affecting the senses, carnal as dist. from spiritual or intellectual; pertaining or devoted to the indulgence of the appetites or passions, esp. those of sex, voluptuous. **sensuality** (-al'-), *n.* **sensualist**, *n.* **sensually**, *adv.*
sensuous (sen'shuǝs, -sū-), *a.* pertaining to or derived from the senses; abounding in or suggesting sensible images; readily affected through the senses. **sensuously**, *adv.* **sensuousness**, *n.*
sent (sent), *past, p.p.* SEND.
sentence (sen'tǝns), *n.* a series of words, containing a subject, predicate etc., expressing a complete thought; a penalty or declaration of penalty upon a condemned person; a judicial decision, verdict; (*dated*) a pithy saying, a maxim, a proverb; two or more musical phrases forming a unit. *v.t.* to pronounce judgment on; to condemn to punishment. **sentential** (-ten'shǝl), *a.*
sententious (senten'shǝs), *a.* abounding in pithy sentences, axioms or maxims; terse, brief and energetic; pompous in tone. **sententiously**, *adv.* **sententiousness**, *n.*
sentient (sen'shiǝnt, -tiěnt), *a.* having the power of sense perception; having sense of feeling; conscious. **sentience**, *n.* **sentiently**, *adv.*
sentiment (sen'timǝnt), *n.* mental feeling excited by aesthetic, moral or spiritual ideas; a thought, view or mental tendency derived from or characterized by emotion; susceptibility to emotion; (*often pl.*) an opinion or attitude.
sentimental (sentimen'tǝl), *a.* characterized by sentiment; swayed by emotion; mawkish; displaying excessive emo-

tion. **sentimental value**, *n.* the value of an object in terms not of money but of associations, memories etc. **sentimentalism, sentimentality** (-tal'-), *n.* unreasonable or uncontrolled emotion; mawkishness. **sentimentalist**, *n.* **sentimentalize, -ise**, *v.i.* to affect sentimentality. *v.t.* to render sentimental. **sentimentally**, *adv.*
sentinel (sen'tinǝl), *n.* one who keeps watch to prevent surprise, esp. a soldier on guard. *v.t.* to watch over, to guard; to set sentinels at or over.
sentry (sen'tri), *n.* (*pl.* **-tries**) a soldier who guards the entrance to a military base etc. to prevent unauthorized access. **sentry-box**, *n.* a shelter for a sentry.
senza (sent'sǝ), *prep.* (*Mus.*) without. **senza tempo**, without strict time.
Sep., (*abbr.*) September; Septuagint.
sepal (sep'ǝl), *n.* one of the segments, divisions or leaves of the calyx of a flower.
separate (sep'ǝrāt), *v.t.* to disunite, to set or keep apart; to break up into distinct parts, to disperse; to come or be between, to be the boundary of. *v.i.* to part, to be disconnected, to withdraw (from); to disperse; of a married couple, to agree to live apart. *a.* (-rǝt), disconnected, considered apart; distinct, individual; disunited from the main body. **separate maintenance**, *n.* an allowance made by a husband to a wife from whom he is separated by consent. **separability** (-bil'-), *n.* **separable**, *a.* **separably**, *adv.* **separately**, *adv.* **separateness**, *n.* **separates**, *n.pl.* women's clothes that cover part of the body and are worn together, e.g. skirts and jackets. **separation**, *n.* the act of separating or the state of being separated, esp. partial divorce, consisting of cessation of cohabitation between married persons. **separatism**, *n.* **separatist**, *n.* one who advocates secession, from a church, political party, federation etc. **separator**, *n.* one who separates; a machine that separates the cream from milk.
Sephardi (sifah'di), *n.* (*pl.* **-dim** (-dim)) a Spanish, Portuguese or N African Jew. **Sephardic**, *a.*
sepia (sē'piǝ), *n.* a dark brown pigment; this pigment prepared from the black secretion of the cuttlefish; a cuttlefish; (**Sepia**) a genus of cephalopodous molluscs containing this; a photograph or drawing in sepia. *a.* made in sepia; of the colour sepia.
sepoy (sē'poi), *n.* an Indian soldier disciplined in the European manner, esp. one in the former British Indian army.
sepsis (sep'sis), *n.* (*pl.* **sepses** (-sēz)) putrefaction; infection by disease-causing bacteria, e.g. from a wound, blood-poisoning.
sept (sept), *n.* a clan or branch of a clan, esp. in Scotland or Ireland.
Sept., (*abbr.*) September; Septuagint.
septa (sep'tǝ), (*pl.*) SEPTUM.
septal (sep'tǝl), *a.* of or pertaining to a septum.
September (septem'bǝ), *n.* the ninth month of the year (the seventh after March, first month of the ancient Roman year).
septenary (sep'tinǝri), *a.* consisting of or relating to seven; by sevens; lasting seven years. *n.* a set of seven years, things etc.
septennium (septen'iǝm), *n.* a period of seven years. **septennial**, *a.* **septennially**, *adv.*
septet (septet'), *n.* a group of seven, esp. singers, voices, instruments etc.; a musical composition for seven performers.
septic (sep'tik), *a.* causing or tending to promote sepsis or putrefaction. **septic tank**, *n.* a tank in which sewage is partially purified by the action of bacteria. **septicaemia** (-sē'miǝ), *n.* a disease of the blood caused by the absorption of septic matter. **septicaemic**, *a.* **septically**, *adv.*
septilateral (septilat'ǝrǝl), *a.* seven-sided.
septillion (septil'yǝn), *n.* the seventh power of a million; (*N Am.*) the eighth power of a thousand. **septillionth**, *n., a.*

septimal (sep'timəl), *a.* of, relating to or based on the number seven. **septime** (sep'tēm), *n.* the seventh parry in fencing.
septimole (sep'timōl), *n.* (*Mus.*) a group of seven notes to be played in the time of four or six.
septuagenarian (septūəjineə'riən), *n.* a person of 70 years of age, or between 69 and 80. *a.* of such an age.
Septuagesima (-jes'imə), *n.* the third Sunday before Lent, so called because about 70 days before Easter.
Septuagint (sep'tūəjint), *n.* a Greek version of the Old Testament including the Apocrypha (*c.* 3rd cent. BC), so called because, according to tradition, about 70 persons were employed on the translation.
septum (sep'təm), *n.* (*pl.* **-ta**) a partition as in a chambered cell, the cell of an ovary, between the nostrils etc. **septal**, *a.* **septulum** (-tūlum), *n.* a small septum. **septulate** (-lət), *a.*
septuple (sep'tūpl), *a.* sevenfold. *n.* a set of seven things. *v.t., v.i.* to multiply by seven. **septuplet** (-tū'plit), *n.* a septimole; one of seven children born in a single multiple birth.
sepulchre (sep'əlkə), *n.* a tomb, esp. one hewn in the rock or built in a solid and permanent manner; a burial-vault. *v.t.* to place in a sepulchre, to entomb. **sepulchral** (-pŭl'krəl), *a.* pertaining to burial, the grave or to monuments raised over the dead; suggestive of a sepulchre, grave, dismal, funereal. **sepulchrally**, *adv.*
sepulture (sep'əlchə), *n.* interment, burial.
sequel (sē'kwəl), *n.* that which follows; a succeeding part, a continuation (of a story etc.); the upshot, consequence or result (of an event etc.).
sequela (sikwē'lə), *n.* (*pl.* **lae** -lē) an abnormal condition occurring as the consequence of some disease; an inference, a consequence.
sequence (sē'kwəns), *n.* succession, the process of coming after in space, time etc.; a series of things following one another consecutively or according to a definite principle; a set of consecutive cards; (*Mus.*) a succession of similar harmonious formations or melodic phrases at different pitches; a scene in a film. **sequent**, *a.* **sequential** (sikwen'shəl), *a.* **sequentiality** (-shial'-), *n.* **sequentially**, *adv.*
sequester (sikwes'tə), *v.t.* (*esp. in p.p.*) to set apart, to isolate, to seclude; (*Law*) to separate (property etc.) from the owner temporarily; to take possession of (property in dispute) until some case is decided or claim is paid; to confiscate, to appropriate. **sequestral** SEQUESTRUM. **sequestrate** (sē'kwistrāt), *v.t.* (*Law*) to sequester. **sequestration**, *n.* **sequestrator**, *n.*
sequestrum (sikwes'trəm), *n.* (*pl.* **-ra** (-trə)) a piece of dead and separated bone remaining in place. **sequestral** (-kwes'-), *a.*
sequin (sē'kwin), *n.* (*Hist.*) a venetian gold coin; a small disc of shiny metal, jet etc., used as trimming for dresses etc.
Sequoia (sikwoi'ə), *n.* a Californian genus of gigantic conifers, with two species.
sera SERUM.
sérac (sirak', se'-), *n.* one of the large angular or tower-shaped masses into which a glacier breaks up at an ice-fall.
seraglio (siral'yō), *n.* (*pl.* **lios**) a walled palace, esp. the old palace of the Turkish Sultan, with its mosques, government offices etc. at Istanbul; a harem.
serai SERE [3].
serang (sərang'), *n.* a boatman; leader of a lascar crew.
serape (sərah'pā), *n.* a Mexican blanket or shawl.
seraph (se'rəf), *n.* (*pl.* **seraphs**, **seraphim** (-fim)) an angel of the highest order. **seraphic** (-raf'-), *a.* of or like a seraph; ecstatic, rapturous; (*coll.*) very good, well-behaved. **seraphically**, *adv.*
Serb (sœb), **Serbian**, *a.* of or pertaining to Serbia, one of the federated republics of Yugoslavia, its people or its language. *n.* a native of Serbia; the Slav language of Serbia. **Serbo-Croat** (-krō'at), *n.* the Slav language which has Serbian and Croat as its main dialects, the official language of Yugoslavia.
serdab (sədab', sœ'-), *n.* a secret passage or cell in an ancient Egyptian tomb.
sere [1] SEAR [1].
sere [2] (siə), *n.* a series of ecological communities following one another in one area. **seral**, *a.*
serein (sərān'), *n.* a fine rain falling from a clear sky after sunset, esp. in tropical regions.
serenade (serənād'), *n.* a song or piece of music played or sung in the open air at night, esp. by a lover beneath his lady's window; a nocturne, a serenata. *v.t.* to sing or play a serenade to or in honour of. *v.i.* to perform a serenade. **serenader**, *n.* **serenata** (-nah'tə), *n.* a cantata or simple form of symphony, usu. with a pastoral subject, for the open air.
serendipity (serəndip'iti), *n.* the happy knack of making unexpected and delightful discoveries by accident. **serendipitous**, *a.* [coined by Horace Walpole, after the fairy tale *The Three Princes of Serendip* (an old name for Sri Lanka)]
serene (sərēn'), *a.* of the sky, atmosphere etc., calm, fair and clear; placid, tranquil, undisturbed; applied as a title to certain continental princes. **Serene Highness**, title accorded to certain European princelings. **serenely**, *adv.* calmly, quietly, deliberately. **serenity** (-ren'-), *n.*
serf (sœf), *n.* a feudal labourer or peasant attached to an estate, a villein; a slave, a drudge. **serfdom** (-dəm), **-hood** (-hud), *n.*
serge (sœj), *n.* a strong and durable twilled cloth, of worsted, cotton, rayon etc.
sergeant, (*Law*) **serjeant** (sah'jənt), *n.* a non-commissioned military officer ranking next above corporal, teaching drill, commanding small detachments etc.; a police-officer ranking next below an inspector. **colour-sergeant** COLOUR. **Sergeant-**, **Serjeant-at-Arms**, *n.* an officer of the Houses of Parliament attending the Lord Chancellor or the Speaker, and carrying out arrests etc.; an officer with corresponding duties attached to other legislative bodies; one of several court and city officers with ceremonial duties. **sergeant-fish**, *n.* a fish with lateral stripes resembling a chevron. **sergeant-major**, *n.* the chief sergeant of a regiment, of a squadron of cavalry or of a battery of artillery. **Sergt.**, (*abbr.*) Sergeant.
serial (siə'riəl), *a.* pertaining to, consisting of or having the nature of a series; of a novel, published in instalments in a periodical; occurring as part of a series of a set of repeated occurrences as in *serial murder*; pertaining to the computer processing of tasks one after another; of music, based on a fixed, arbitrary series of notes, not on a traditional scale. *n.* a serial story; a serial publication, a periodical. **serial number**, *n.* a number stamped on an item which identifies it in a large series of identical items. **serialism**, *n.* (*Mus.*). **serialist**, *n.* **seriality** (-al'-), *n.* **serialize**, **-ise**, *v.t.* to publish (a novel) in instalments. **serially**, *adv.* **seriate** (-ət), **-ated** (-ātid), *a.* arranged in a series or regular sequence. *v.t.* to arrange thus. **seriately**, *adv.* **seriatim** (-ā'tim), *adv.* in regular order; one point etc. after the other. **seriation**, *n.*
sericeous (sirish'iəs), *a.* pertaining to or consisting of silk; silky, downy, soft and lustrous.
sericterium (seriktiə'riəm), *n.* (*pl.* **-teria**) the silk-spinning gland in silkworms.
sericulture (se'rikulchə), *n.* the breeding of silkworms and the production of raw silk. **sericultural** (-kŭl'-), *a.* **sericulturist**, *n.*
seriema (serie'mə), *n.* a long-legged Brazilian and Paraguayan bird, the crested screamer.
series (siə'riz), *n.* (*pl.* **series**), a number, set or continued

succession of things similar to each other or each bearing a definite relation to that preceding it; a sequence, a row, a set; a set of volumes, parts, articles, periodicals etc., consecutively numbered or dated or issued in the same format under one general title; (*Math.*) a number of terms, the successive pairs of which are related to each other according to a common law or mode of derivation, a progression; the connection of two or more electric circuits so that the same current traverses all the circuits; a group of allied strata forming a subdivision of a geological system.

serif (se′rif), *n.* in printing, one of the fine cross-lines at the top and bottom of letters.

serigraph (se′rigraf), *n.* a silk-screen print. **serigrapher** (-rig′-), *n.* **serigraphic** (-graf′-), *a.* **serigraphy** (-rig′-), *n.*

serin (se′rin), *n.* a small green finch allied to the canary.

seringa (səring′gə), *n.* a Brazilian rubber-tree of various species.

serious (siə′riəs), *a.* grave, sober, sedate, thoughtful, earnest, not frivolous; of great importance, momentous; in earnest, not ironical or pretended, sincere; having serious consequences, dangerous; (*coll.*) in large or excessive amounts, *serious money*; (*coll.*) high quality. **seriocomic, -comical**, *a.* mingling the serious and the comic; serious in meaning with the appearance of comedy, or comic with a grave appearance. **serioso** (seriō′sō), *adv.* (*Mus.*) with gravity, solemnly. **seriously**, *adv.* **seriousness**, *n.*

serjeant etc. SERGEANT.

sermon (sœ′mən), *n.* a discourse founded on a text from the Bible delivered in church in exposition of doctrine or instruction in religion or morality; a similar discourse delivered elsewhere; a moral reflection; a serious exhortation or reproof. **sermonize, -ise**, *v.i., v.t.* **sermonizer**, *n.*

sero-, *comb. form.* serum.

serology (sərol′əji), *n.* the study of blood serum, its composition and properties. **serological** (-loj′-), *a.* **serologist**, *n.*

serosity SEROUS.

sérotine (se′rətin), *n.* a small reddish bat, *Vesperugo serotinus*, flying in the evening.

serotinous (sirot′inəs), *a.* (*Bot.*) appearing late in the season.

serotonin (sirot′ənin), *n.* a compound found in many body tissues which acts as a vasoconstrictor.

serous (siə′rəs), *a.* pertaining to or resembling serum. **serous membrane**, *n.* a thin, transparent membrane lining certain large body cavities, and secreting a thin fluid which allows movement of the organs in the cavities. **serosity** (-ros′-), *n.*

serpent (sœ′pənt), *n.* (*dated*) a snake; a northern constellation; a treacherous, insinuating person; an old-fashioned wind-instrument of serpentine form; the Devil. **serpentine** (-tīn), *a.* pertaining to, resembling or having the qualities of a serpent; coiling, winding, twisting, sinuous; subtle, wily, treacherous. *n.* a massive or fibrous rock consisting of hydrated silicate of magnesia richly coloured and variegated and susceptible of a high polish, used for making various ornamental articles. **serpentinely**, *adv.*

serpigo (səpī′gō), *n.* a skin-disease, esp. a form of herpes or spreading ring-worm. **serpiginous** (-pij′i-), *a.*

Serps (sœps), (*acronym*) State earnings-related pensions scheme.

serpula (sœ′pūlə), *n.* (*pl.* **lae**) a brilliantly coloured marine worm living in a contorted or spiral shell.

serrate¹ (se′rāt), *a.* notched on the edge, like a saw, serrated.

serrate² (sərāt′), *v.t.* (*usu. in p.p.*) to cut into notches and teeth, to give a saw-like edge to. **serration**, *n.*

serrato-, serri-, serro-, *comb. forms.* serrated.

serricorn (se′rikawn), *n.* (*Ent.*) having serrated antennae. *n.* a serricorn beetle.

serried (se′rid), *a.* close-packed, in compact order (esp. of soldiers).

serriped (se′riped), *a.* (*Ent.*) having serrated feet.

serrirostrate (seriros′trət), *a.* (*Ornith.*) having a serrated bill.

serrulate (se′rələt), *a.* finely serrate; having minute notches. **serrulation**, *n.*

serum (siə′rəm), *n.* (*pl.* **serums, sera** (-rə)) the thin transparent part that separates from the blood in coagulation; a constituent of milk and other animal fluids, lymph; animal serum used as an antitoxin etc. **serum hepatitis**, *n.* an acute viral infection of the liver, marked by inflammation and jaundice, spread by contact with infected blood.

serval (sœ′vəl), *n.* an African wild cat with long legs and a black-spotted tawny coat.

servant (sœ′vənt), *n.* a person employed by another person or body of persons to work under direction for wages, an employee, esp. in a household, a domestic. **civil servant** CIVIL. **public servant** PUBLIC.

serve (sœv), *v.t.* to act as servant to, to be in the employment of; to be useful to, to render service to; to be subservient or subsidiary to; to satisfy, to avail, to suffice; to supply, to perform (a purpose, function etc.); to carry out the duties of, to do the work of (an office etc.); to behave towards, to treat (well, ill etc.); to dish (up), to bring to and set on table; to distribute to those at table; to furnish, to supply (a person with); to deliver (a summons, writ etc.) in the manner prescribed by law; to throw or send (a ball etc.); of a male animal, to mate with. *v.i.* to be employed, to perform the duties of or to hold an office etc.; to perform a function, to take the place of be used as, to be a satisfactory substitute (for); to suffice, to avail; to be satisfactory, favourable or suitable; to be in subjection; to deliver the ball in certain games; to attend a celebrant at the altar. *n.* the act of or turn for serving at tennis etc. a service. **serves you right**, (*coll.*) you've got your deserts. **to serve a rope**, (*Naut.*) to lash or whip a rope with thin cord to prevent fraying. **to serve a sentence**, to undergo the punishment prescribed. **to serve at table**, to act as waiter or waitress. **to serve one's time**, to serve one's sentence; to go through an apprenticeship; to hold an office etc. for the full period. **to serve up**, to serve out (food). **serving-maid**, *n.* a female servant. **server**, *n.* one who serves at table; a utensil (as a tray or spoon) used to serve food; in tennis etc., one who serves; one who assists the celebrant at mass. **serving**, *n.* a portion of food, a helping.

Servian (sœ′viən), SERB.

service¹ (sœ′vis), *n.* the act of serving; work done for an employer or for the benefit of another; the organized supply, installation or maintenance of goods or provision of a public need; a benefit or advantage conferred on someone; the state of being a servant, esp. the place or position of a domestic servant; a department of state or public work or duty, the organization performing this or the persons employed in it; use, assistance; a liturgical form for worship, an office; a performance of this; formal legal delivery, posting up or publication (of a writ, summons etc.); a set of dishes, plates etc. required for serving a meal; the act or manner of serving food or drink; the act of serving the ball at tennis etc.; maintenance work undertaken by the vendor after a sale; (*pl.*) the armed forces; (*pl.*) the service area of a motorway; (*pl.*) provision of water, electricity etc. to a property. *v.t.* to repair or maintain a car etc. after sale; to meet interest on (a debt); of a male animal, to serve. **in active service**, engaged in actual duty in the army, navy etc. **service area**, *n.* a place beside a motorway where petrol, food etc. are available. **service car**, *n.* (*New Zealand*) a long-distance bus. **service charge**, *n.* a percentage of a bill, charged in addition to the total to pay for service. **service flat**, *n.* a flat for which

an inclusive sum is charged for rent and full hotel service. **service industry**, *n.* one concerned with providing a service rather than with manufacturing. **serviceman**, *n.* a member of the armed forces. **service road**, *n.* a minor road running alongside a main road and carrying local traffic only. **service station**, *n.* a roadside establishment providing petrol etc. to motorists. **servicewoman**, *n., fem.* **serviceable**, *a.* able or willing to render service; useful, benificial, advantageous; durable, fit for service. **serviceability** (-bil'-), *n.* **serviceably**, *adv.*
service[2] (sœ'vis), *n.* **service-tree**, *n.* a European tree with small pear-like fruit; the wild service-tree.
serviette (sœviet'), *n.* a table-napkin.
servile (sœ'vīl), *a.* of, pertaining to or befitting a slave or slaves; slavish, abject, fawning, menial, servile, obedient. **servility** (-vil'-), *n.*
servitude (sœ'vitūd), *n.* the condition of a slave, slavery, bondage; subjection to or as to a master; (*Law*) the subjection of property to an easement for the benefit of a person other than the owner or of another estate.
servo (sœ'vō), *n., a.* (*pl.* **-vos**) (of or pertaining to) a servomechanism or servomotor. **servomechanism**, *n.* an automatic device using a small amount of power which controls the performance of a much more powerful system. **servomotor**, *n.*, a motor which powers a servomechanism.
sesame (ses'əmi), *n.* an E Indian annual herb with oily seeds used as food. **open sesame**, a magic formula for opening a door, mentioned in the *Arabian Nights;* an easy means of entry to a profession etc. **sesamoid** (-moid), *a.* of or relating to one of several small bones developed in tendons as in the knee-cap, the sole of the foot etc.
sesqui-, *comb. form.* denoting a proportion of 1½ to 1 or 3 to 2; denoting combinations of three atoms of one element with two of another.
sesquicentennial (seskwisəntenʹ-), *n., a.* (of) a 150th anniversary or its celebration.
sessile (ses'īl), *a.* (*Bot., Zool.*) attached by the base, destitute of a stalk or peduncle. **sessile oak**, *n.* the durmast. **sessility** (-sil'-), *n.*
session (sesh'ən), *n.* a sitting or meeting of a court, council, legislature, academic body etc. for the transaction of business; the period during which such meetings are held at short intervals; the time of such meeting; a university or school year; any period devoted to an activity, meeting, etc. the lowest court of the Presbyterian Church, called the Kirk-Session. **Court of Session** COURT. **sessional**, *a.*
sesterce (ses'təs), **sestertius** (sistœ'shəs), *n.* (*pl.* **-ces, -tii** (-shii)) an ancient Roman silver (afterwards bronze) coin worth 2½ asses or ¼ denarius.
sestet (sestet'), *n.* a SEXTET; the last six lines of a sonnet.
sestina (sestē'nə), *n.* a form of verse consisting of six sixlined stanzas with a final triplet, each stanza having the same terminal words to the lines but in different order.
set[1] (set), *v.t.* (*pres.p.* **setting**, *past, p.p.* **set**) to place, to put, to stand; to fix; to plant (usu. *out*); to bring, put, place or station in a specified or right position, posture, direction or state; to place in a particular time or place (of a novel, play, etc.); to adjust; to prescribe; to arrange or dispose for use, action, display etc.; to apply (a thing to something else); to attach, to fasten, to join; to determine, to appoint, to settle, to establish; to cause to sit; to apply (oneself, one's energies etc., to), to cause (to work etc.); to present, to offer (an example, task etc.); to fix (the hair) in waves etc.; to adapt or fit (words etc.) to music usu. composed for the purpose; (*Naut.*) to hoist, to spread (sail). *v.i.* to become solid, hard or firm from a fluid condition, to congeal, to solidify; to take shape, to become fixed; to move, tend or incline in a definite or specified direction; of seeds or fruit, to mature, to develop; of a dog, to point; to dance face to face; to pass below the horizon; to decline, to pass away. *a.* fixed, unyielding, immovable; determined, intent (on or upon); rigid, motionless; stationary; established, prescribed; regular, in due form; stereotyped, conventional. **to set about**, to begin; to prepare or take steps (to do etc.); to attack. **to set against**, to balance (one thing) against another; to make (a person) unfriendly to or prejudiced against. **to set aside**, to reserve; to reject; to annul, to quash. **to set at ease**, to relieve of anxiety, fear, bashfulness etc.; to make comfortable. **to set at naught** NAUGHT. **to set back**, to turn backwards, to reverse the movement of; to hinder the progress of, to impede; (*coll.*) to cost. **to set by the ears** EAR[1]. **to set down**, to let (a passenger) alight from a vehicle; to put in writing, to note; to attribute; to explain (as); to snub, to rebuke. **to set eyes on** EYE[1]. **to set fire to** FIRE[1]. **to set foot (on)**, to tread (on). **to set forth**, to expound, to make known, to state; to start (on a journey etc.). **to set free**, to release. **to set in**, of the tide, to move steadily shoreward; of the weather, to become settled. **to set in order**, to arrange, to adjust. **to set little, much by**, to value little or highly. **to set off**, to make more attractive or brilliant by contrast; to act as a foil to; to place over, against, as an equivalent; to start or cause to start (laughing etc.); to set out; to detonate. **to set on**, to incite, to instigate, to urge (to attack); to make an attack on. **to set oneself**, to apply oneself, to undertake. **to set one's hand to**, to begin (a task). **to set on foot** FOOT. **to set out**, to mark off; to display, to arrange; to expound, to state at length, to publish; to plant out; to start (upon a journey etc.). **to set right**, to correct. **to set sail** SAIL. **to set store by** BY. **to set the heart on** HEART. **to set the teeth on edge** EDGE. **to set to**, to apply oneself vigorously; to begin to fight. **to set to work**, to begin; to cause to begin working. **to set up**, to erect, to build; to raise, to exalt, to establish; (to enable to) start a business (as); to begin to utter (e.g. a wail); (*sl.*) to arrange for (someone else) to be blamed, to frame; to restore to health. **setter**, *n.* a large dog trained to point at game by standing rigid. **setting**, *n.* the framing etc. in which something (as a jewel) is set; the framing, surroundings or environment of a thing, event etc.; the scenery, background and other accessories of a play, film, etc.; the music to which words, a song etc. are fitted.
set[2] (set), *n.* a number of similar, related or complementary things or persons, a collection, a group, a company, a clique; a number of things intended to be used together or required to form a whole; a collection of mathematical objects, numbers etc.; a group of games played together, counting as a unit, esp. in tennis; a predisposition to respond in a certain way to a psychological stimulus; posture, pose, carriage; permanent inclination, bend, displacement, bias; an apparatus for transmitting or receiving radio or television signals; the repetoire of a performer, musical group, etc. during a session; the act of pointing at game etc. (by a setter); a young plant for setting out, a shoot, a slip for planting; the amount of margin in type determining the distance between letters; the props, scenery, etc. forming the setting of a play or film. **dead set** DEAD. **set-back**, *n.* a check, an arrest; an overflow, a counter-current; a relapse. **set-down**, *n.* a rebuke, a snub, a rebuff. **set-in**, *a.* of a part of a garment, made up separately and then sewn in. **set-line**, *n.* a long fishing-line with shorter lines attached to it. **set-off**, *n.* a thing set off against another, an offset, a counterpoise, a counterclaim; a decorative contrast, an embellishment; (*Print.*) an accidental transference of ink from one printed sheet to another. **set-piece**, *n.* a carefully prepared and usually elaborate performance; an elaborate, formalized piece of writing, painting etc.; a carefully arranged display of fireworks or a large firework built up with scaffolding; in sport, a formal movement to put the ball back into play. **set point**, *n.* in tennis etc., a point which,

if won by one of the players, will win the set. **set-screw**, *n.* one which secures parts of machinery together and prevents relative movement. **set-square**, *n.* a right-angled triangular piece of wood etc. used in mechanical drawing. **set theory**, *n.* a branch of mathematics which studies the properties and relationships of sets. **set-to**, *n.* a fight, esp. with the fists; a heated argument. **set-up**, *n.* an arrangement; a situation; (*N Am., sl.*) a situation which has a predetermined outcome.
seta (sē'tə), *n.* (*pl.* -**tae**) a bristle or bristle-like plant or animal part. **setaceous** (sitā'shəs), *a.* bristly; set with, consisting of or resembling bristles.
set-out, set-square SET².
sett, set³ (set), *n.* a small rectangular block of stone used for road paving; the burrow of a badger.
settee (sitē'), *n.* a long seat for several persons, with a back; a sofa.
setter, setting SET¹.
settle¹ (set'l), *n.* a long, high-backed seat or bench for several persons.
settle² (set'l), *v.t.* to place firmly, to put in a permanent or fixed position, to establish; to put in order; to cause to sit down or to become fixed; to determine, to decide; to colonize; to settle in as colonists; to install in or take up residence; to cause to sink or subside, to precipitate; to clear of dregs; to deal with, to dispose of, to finish with, to do for; to adjust and liquidate (a disputed account); to pay (an account); to secure (property, an income etc., on); to adjust, to accommodate, end (a quarrel, dispute etc.). *v.i.* to sit down, to alight; to cease from movement, agitation etc.; to become motionless, fixed or permanent; to take up a permanent abode, mode of life etc.; to become established, to become a colonist (in); to subside, to sink to the bottom; to become clarified; to determine, to resolve (upon); to adjust differences, claims or accounts. **to settle down**, to become regular in one's mode of life, to become established; to begin to apply oneself (to a task etc.). **to settle for**, to accept, to be content with. **to settle in**, to make or become comfortably established. **to settle up**, to pay what is owing. **settlement**, *n.* the act of settling; the state of being settled; a subsidence; a place or region newly settled, a colony; a community or group of persons living together, esp. in order to carry out social work among the poor; (*Law*) the conveyance of property or creation of an estate to make provision for the support of a person or persons or for some other object; the property so settled. **settler**, *n.* one who settles, esp. a colonist. **settlings**, *n.pl.* sediment, lees, dregs.
set-to SET².
set-up SET¹.
seven (sev'n), *n.* the number or figure 7 or vii; the age of seven; the seventh hour after midnight or midday; (*pl.*) a rugby game or tournament played with teams of seven players; a set of seven persons or things, esp. a card with seven pips. *a.* seven in number; aged seven. **seven deadly sins**, pride, covetousness, lust, gluttony, anger, envy, sloth. **seven wonders of the world**, the Pyramids, the Hanging Gardens of Babylon, the Temple of Diana at Ephesus, the tomb of Mausolus of Caria, the Colossus of Rhodes, the statue of Zeus by Phidias and the Pharos of Alexandria. **seven-year itch**, the supposed onset of boredom, leading to infidelity, after seven years of marriage. **seven seas**, *n.pl.* the N and S Atlantic, N and S Pacific, Arctic, Antarctic and Indian oceans. **sevenfold**, *a., adv.* **seventh** (-th), *n.* one of seven equal parts; (*Mus.*) the interval between a given tone and the seventh above it (inclusively) on the diatonic scale; a combination of these two. *n., a.* (the) last of seven (people, things etc.); the next after the sixth. **Seventh-Day Adventists**, a sect that believes in the imminent second advent of Christ and observes Saturday as the sabbath. **seventh heaven**, *n.* (*coll.*) a state of perfect bliss. **seventhly**, *adv.*

seventeen (sevntēn'), *n.* the number or figure 17 or xvii; the age of 17. *a.* 17 in number; aged 17. **seventeenth**, *n.* one of 17 equal parts. *n., a.* (the) last of 17 (people, things etc.); the next after the 16th.
seventy (sev'nti), *n.* (*pl.* -**ties**) the number or figure 70 or lxx; the age of 70. *a.* 70 in number; aged 70; (*pl.*) the period of time between one's 70th and 80th birthdays; the range of temperature between 70° and 80°; the period of time between the 70th and 80th years of a century. **seventy-eight**, *n.* (*coll.*) a gramophone record playing at 78 revolutions per minute. **seventieth** (-əth), *n.* one of 70 equal parts. *n., a.* (the) last of 70 (people, things etc.); the next after the 69th.
sever (sev'ə), *v.t.* to part, to separate, to divide, to cleave; to cut or break off (apart from the whole); to conduct or carry on independently. *v.i.* to separate, to part. **severable**, *a.* **severance**, *n.* **severance pay**, *n.* a sum of money paid to a worker as compensation for loss of employment.
several (sev'ərəl), *a.* separate, distinct, individual; various; not common, not shared with others, pertaining to individuals; consisting of a number, more than two but not many. *n.* a few, an indefinite number, more than two but not many. **severally**, *adv.* **severalty** (-ti), *n.* (*Law*) exclusive tenure or ownership.
severance SEVER.
severe (siviə'), *a.* rigorous, strict, austere, harsh, merciless; trying, hard to endure or sustain; distressing, bitter, painful; grave, serious; rigidly conforming to rule, unadorned, restrained. **severely**, *adv.* **severity** (-ve'ri-), *n.*
Sèvres (porcelain) (sev'r), *n.* porcelain made at Sèvres in Seine-et-Oise, France.
sew (sō), *v.t.* (*p.p.* **sewn**, (sōn), **sewed**) to fasten together by thread worked through and through with a needle; to make, mend, close up, attach, fasten on or in etc. by sewing. *v.i.* to work with a needle and thread. **to sew up**, to mend, join etc. by sewing; (*sl.*) to exhaust, nonplus; (*sl.*) to complete satisfactorily. **sewer**¹, *n.* sewing, *n., a.* **sewing-machine**, *n.* a machine for stitching etc. driven electrically or by a treadle or a crank turned by hand.
sewage SEWER².
sewer¹ SEW.
sewer² (soo'ə, sū'ə), *n.* a channel, underground conduit or tunnel for carrying off the drainage and liquid refuse of a town etc. **sewer-gas**, *n.* foul air from a sewer. **sewer-rat**, *n.* the common brown rat. **sewage** (-ij), *n.* the waste matter carried off through the sewers. **sewage-farm**, *n.* a place where sewage is treated esp. for use as manure. **sewage-works**, *n.sing.* a place where sewage is treated before being discharged. **sewerage** (-rij), *n.* the system of draining by means of sewers; sewers, drains etc. collectively; sewage.
sewin (sū'in), *n.* a variety of sea or salmon-trout.
sewing-machine SEW.
sewn SEW¹.
sex (seks), *n.* the sum total of the physiological, anatomical and functional characteristics which distinguish male and female; the quality of being male or female; (*collect.*) male or females, men or women; sexual intercourse. *a.* of or relating to sexual matters or differences between sexes. *v.t.* to determine the sex of. **sex appeal**, *n.* what makes a person attractive to the opposite sex. **sex chromosome**, *n.* the chromosome responsible for the initial determination of sex. **sex-linked**, *a.* of a gene, located on a sex chromosome; of a character, determined by a sex-linked gene. **sex object**, *n.* a person perceived solely as an object of sexual desires and fantasies. **sex shop**, *n.* a shop selling sexual aids, pornography, etc. **sexed**, *a.* **sexism**, *n.* discrimination (esp. against women) on the grounds of sex. **sexist**, *n., a.* **sexless**, *a.* **sexology** (-ol'-), *n.* the science dealing with human sexual behaviour and relationships. **sexological** (-loj'-), *a.* **sexologist**, *n.* **sexy**, *a.* sexually stimulating; sexually aroused; exciting, trendy, desirable.

sexily, *adv*. **sexiness**, *n*.
sex(a)-, *comb. form.* containing six; sixfold.
sexagenarian (seksəjineə'riən), *a*. 60 years of age or between 59 and 70. *n*. a sexagenarian person.
Sexagesima, Sexagesima Sunday (seksəjes'imə), *n*. the second Sunday before Lent. **sexagesimal**, *a*. pertaining to, proceeding by or based on 60. *n*. a sexagesimal fraction.
sexcentenary (seksəntē'nəri), *a*. pertaining to or consisting of 600 years. *n*. a 600th anniversary or its celebration.
sexennial (seksen'iəl), *a*. occurring once every six years; lasting six years. **sexennially**, *adv*.
sexism, sexless, sexology SEX.
sexploitation (seksploitā'shən), *n*. the portrayal or manipulation of sex for financial profit in films, magazines etc.
sext (sekst), *n*. in the Roman Catholic Church, the office for the sixth hour or noon.
sextant (sek'stənt), *n*. the sixth part of a circle; an instrument used in navigation and surveying for measuring angular distances or altitudes.
sextet (sekstet'), *n*. a group of six; a group of six singers or musicians; a composition for such a group.
sextillion (sekstil'yən), *n*. the sixth power of a million, represented by 1 followed by 36 ciphers; (*N Am., Fr.*) the seventh power of a thousand, 1 followed by 21 ciphers.
sexton (sek'stən), *n*. an officer having the care of a church, its vessels, vestments etc., and frequently acting as parish-clerk and a grave-digger; a sexton-beetle. **sexton-beetle**, *n*. a beetle that buries carrion to serve as a nidus for its eggs, a burying-beetle.
sextuple (seks'tūpl), *a*. six times as many or much. *n*. a sextuple amount. **sextuplet** (-tū'plit), *n*. one of six born at one birth; (*Mus.*) a group of six notes played in the time of four.
sexual (sek'sūəl, -shəl), *a*. of, pertaining to or based on sex or the sexes or on the distinction of sexes; pertaining to generation or copulation, venereal. **sexual harassment**, *n*. persistent unwelcome sexual attentions esp. towards a woman in her place of work. **sexual intercourse**, *n*. a sexual act in which the male's erect penis is inserted into the female's vagina. **sexual selection**, *n*. a method of selection based on the struggle for mating which, according to one school of thought, accounts for the origin of secondary sexual characteristics. **sexuality** (-al'-), *n*. **sexually**, *adv*. **sexually transmitted disease**, *n*. a venereal disease.
sexy, SEX.
SF, (*abbr.*) San Francisco; science fiction; Society of Friends.
sf., (*abbr.*) sforzando; sforzato.
SFA, (*abbr.*) Scottish Football Association; (*sl.*) Sweet Fanny Adams.
sforzando (sfawtzan'dō), **sforzato**, *adv*. (*Mus.*) emphatically, with sudden vigour.
SG, (*abbr.*) Solicitor General.
sg, (*abbr.*) specific gravity.
sgraffito (skrafē'tō), GRAFFITO.
sgt, (*abbr.*) sergeant.
sh (sh), *int*. calling for silence.
shabby (shab'i), *a*. ragged, threadbare; in ragged or threadbare clothes; mean, paltry, despicable. **shabbily**, *adv*. **shabbiness**, *n*.
shack (shak), *n*. a rude cabin or shanty, esp. one built of logs. **to shack up (with)**, (*sl.*) to live (with), usu. having a sexual relationship.
shackle (shak'l), *n*. a fetter, gyve or handcuff; the bow of a padlock; a coupling link; an insulating spool or support for a telegraph wire; (*pl.*) fetters, restraints, impediments. *v.t.* to chain, to fetter; to restrain, to impede, to hamper. **shackler**, *n*.
shad (shad), *n*. a name for several anadromous deep-bodied food-fish, esp. the American or white shad.
shaddock (shad'ək), *n*. the large orange-like fruit of a Malaysian and Polynesian tree, the pomelo.
shade (shād), *n*. obscurity or partial darkness caused by the interception of the rays of light; gloom, darkness; a place sheltered from the sun, a secluded retreat; the dark or darker part of a picture; a screen for protecting from or moderating light, esp. a covering for a lamp, or a shield worn over the eyes; (*N Am.*) a window blind; a glass cover for protecting an object; a colour; gradation of colour, esp. with regard to its depth or its luminosity; a scarcely perceptible degree, a small amount; something unsubstantial, unreal or delusive; a spectre, a ghost; (*pl.*) Hades; (*pl. coll.*) sunglasses; (*pl.*) suggestions (of), undertones. *v.t.* to shelter or screen from light or heat; to cover, to obscure, to darken (an object in a picture) so as to show gradations of colour or effects of light and shade. *v.i.* to pass off by degrees or blend (with another colour). **shaded**, *a*. **shadeless**, *a*. **shading**, *n*. **shady** (shā'di), *a*. sheltered from the light and heat of the sun; casting shade; (*coll.*) disreputable, of equivocal honesty. **shadily**, *adv*. **shadiness**, *n*.
shadoof (shədoof'), *n*. a water-raising contrivance consisting of a long pole with bucket and counterpoise, used on the Nile etc.
shadow (shad'ō), *n*. shade; a patch of shade; the dark figure of a body projected on the ground etc. by the interception of light; an inseparable companion; one who follows another closely and unobtrusively; darkness, obscurity, privacy; the dark part of a picture, X-ray etc.; a reflected image; an imperfect or faint representation, an adumbration; something causing distress or sorrow; a faint trace, the slightest degree; something unsubstantial or unreal; a phantom, a ghost. *v.t.* to darken, to cloud; to set (forth) dimly or in outline, to adumbrate; to watch secretly, to spy upon, to dog. **shadow boxing**, *n*. boxing against an imaginary opponent when training. **shadow cabinet**, *n*. a group of leading members of a party out of office, who would probably constitute the cabinet if in power. **shadower**, *n*. **shadowy**, *a*. **shadowiness**, *n*.
shady SHADE.
shaft (shahft), *n*. the slender stem or stock of a spear, arrow etc.; anything more or less resembling this, as a ray (of light), a bolt or dart (of lightning, ridicule etc.); a column between the base and the capital; a small column in a cluster or in a window-joint; a stem, a stalk, a trunk; the scape of a feather; any long, straight and more or less slender part; a penis; the handle of a tool; one of the bars between a pair of which a horse is harnessed; a large axle, arbor or long cylindrical bar, esp. rotating and transferring motion; a well-like excavation, usu. vertical, giving access to a mine; an upward vent in a building for a lift, etc. *a*. feigned; denoting a member or members of the main Parliamentary opposition party. *v.t.* (*N Am., sl.*) to cheat, treat unfairly; (*sl.*) to have sexual intercourse with. **shafted**, *a*.
shag (shag), *n*. (*past, p.p.* **shagged**) a rough coat of hair, a bushy mass; cloth having a long coarse nap; strong tobacco cut into fine shreds; the crested or green cormorant. *a*. shaggy. *v.t., v.i.* (*taboo*) to have sexual intercourse with. **to shag out**, to exhaust. **shaggy**, *a*. rough-haired, hairy, hirsute; coarse, tangled, unkempt. **shaggy dog story**, a long, inconsequential story, funny but lacking a punch-line. **shaggily**, *adv*. **shagginess**, *n*.
shagreen (shəgrēn'), *n*. a kind of leather with a granular surface prepared from the skins of horses, asses, etc.; the skins of various sharks, rays etc. covered with hard papillae.
shah (shah), *n*. (*formerly*) a sovereign of Iran.
shake (shāk), *v.t.* (*past* **shook**, (shuk), *p.p.* **shaken**) to move forcibly or rapidly to and fro or up and down; to cause to tremble or quiver; to shock, to convulse, to agitate, to disturb; to brandish; to weaken the stability of, to impair, to undermine; to trill; to upset another's compo-

sure; to cause another to doubt; (*Austral., sl.*) to steal; (*N Am., coll.*) to get rid of. *v.i.* to move quickly to and fro or up and down, to tremble, to totter, to shiver; to quiver, to rock; to change the pitch or power of the voice, to make trills; (*N Am.*) to shake hands. *n.* the act or an act of shaking; a jerk, a jolt, a shock; the state of being shaken, agitation, vibration, trembling; a trill; a milk-shake; a crack in timber or rock; (*N Am., Austral.*) an earthquake. **no great shakes,** (*sl.*) of no great account. **the shakes,** a fit of trembling, caused by fever, withdrawal from alcohol etc. **to shake down,** to bring down (fruit etc.) by shaking; to cause (grain etc.) to settle into a compact mass; to become compact; (*N Am., sl.*) to swindle, extort (from); (*coll.*) to go to bed. **to shake hands** HAND¹. **to shake in one's shoes,** to be very frightened. **to shake off,** to get rid of (as if by) shaking, to cast off. **to shake one's head,** to move the head from side to side in token of refusal, dissent, disapproval etc. **to shake out,** to open out or empty by shaking. **to shake up,** to mix, disturb etc. by shaking; (*coll.*) to reorganize drastically; (*coll.*) to shock, perturb. **shake-down,** *n.* a makeshift bed; (*N Am., sl.*) a swindle, an act of extortion. **shake-up,** *n.* **shakeable,** *a.* **shaker,** *n.* a container for mixing or sprinkling by shaking. **the Shakers,** an American religious sect who hold that Christ's second advent has already taken place (named from their religious dances). **shaky,** *a.* liable to shake, unsteady, rickety, unstable, tottering; of doubtful integrity, solvency, ability etc. **shakily,** *adv.* **shakiness,** *n.*

Shakespearean, Shakespearian (shăkspiə'riən), *a.* pertaining to or resembling Shakespeare or his works. *n.* a student of Shakespeare's works.

shako (shak'ō), *n.* (*pl.* **-kos**) a military cylindrical hat, usu. flat-topped, with a peak in front, usu. tilting forward, and decorated with a pompom, plume or tuft.

shaky, SHAKE.

shale (shāl), *n.* a laminated argillaceous rock resembling soft slate, often containing much bitumen. **shale oil,** *n.* oil obtained from bitumen shale. **shaly,** *a.*

shall (shal), *v.aux.* (*past, subj.* **should** (shud)) in the 1st pers., used to express simple futurity or a conditional statement; in the 2nd and 3rd pers., to express a command, intention, promise, permission etc., to express future or conditional obligation, duty etc., or to form a conditional protasis etc.

shallop (shal'əp), *n.* a light open boat.

shallot (shəlot'), *n.* a plant allied to garlic with similar but milder bulbs.

shallow (shal'ō), *a.* not having much depth; superficial, trivial, silly. *n.* a shallow place. *v.i.* to become shallow or shallower. *v.t.* to make shallow. **shallowly,** *adv.* **shallowness,** *n.*

shaly SHALE.

sham (sham), *v.t.* (*past, p.p.* **shammed**) to feign, to make a pretence of. *v.i.* to feign, to pretend. *n.* an imposture, a false pretence, a fraud, one who or that which pretends to be someone or something else. *a.* feigned, pretended, counterfeit.

shaman, *n.* a priest of shamanism; an exorcist or medicine man among shamanist tribes, esp. of N Am. Indians. **shamanism** (shah'mənizm, shā'-), *n.* a form of religion based on the belief in good and evil spirits which can be influenced by shamans, prevailing esp. in N Asia. **shamanist,** *n., a.*

shamateur (sham'ətə), *n.* a person classed as an amateur in sport, but who accepts payment.

shamble (sham'bl), *v.i.* to walk in an awkward, shuffling or unsteady manner. *n.* a shambling walk or gait. **shambling,** *a.*

shambles (sham'blz), *n.sing.* or *pl.* a butcher's slaughterhouse; a place of carnage or execution; utter confusion.

shambolic (shambol'ik), *a.* (*coll.*) chaotic, very confused.

shame (shām), *n.* a painful feeling due to consciousness of guilt, degradation, humiliation etc.; the instinct to avoid this, the restraining sense of pride, modesty, decency, decorum; a state of disgrace, discredit or ignominy; anything that brings reproach, a disgrace; (*coll.*) an unfairness. *v.t.* to make ashamed; to bring shame on, to cause to blush or feel disgraced; to disgrace. **shame!** *int.* that is unfair! disgraceful! what a pity! **shame on you,** you should be ashamed. **to put to shame,** to humiliate by exhibiting better qualities. **shameful,** *a.* **shamefully,** *adv.* **shamefulness,** *n.* **shameless,** *a.* immodest. **shamelessly,** *adv.* **shamelessness,** *n.*

shamefaced, *a.* bashful, shy; showing a sense of shame. **shamefacedly** (-fāsth, -fâsid-), *adv.*

shammy, (sham'i), **shammy leather,** *n.* (*coll.*) CHAMOIS.

shampoo (shampoo'), *v.t.* to lather, wash and rub (the hair of); to wash (carpets, upholstery) with shampoo. *n.* the act of shampooing; a liquid soap or detergent used for this.

shamrock (sham'rok), *n.* a species of trefoil forming the national emblem of Ireland.

shandy (shan'di), *n.* (*pl.* **-dies**) a mixture of beer and ginger-beer or lemonade.

shanghai (shanghī'), *v.t.* to drug and ship as a sailor while stupefied; to kidnap; to trick into doing something; (*Austral., New Zealand*) to shoot with a catapult. *n.* (*Austral., New Zealand*) a catapult.

shank (shangk), *n.* the leg, esp. the part from the knee to the ankle, the shin; the corresponding part in other vertebrates; a cut of meat from this part of an animal; the straight part of an instrument, tool etc. connecting the acting part with the handle. *v.i.* of plants, to be affected or fall (off) with decay in the footstalks. **Shanks's mare,** **pony,** on foot.

shanny (shan'i), *n.* the smooth blenny.

shan't (shahnt), *contr. shall not.*

shantung (shantŭng'), *n.* a plain fabric woven in coarse silk yarns.

shanty¹ (shan'ti), *n.* a rude hut or cabin; a hastily built or rickety building; (*Austral.*) a low public-house, a grogshop. **shanty-town,** *n.* a poor part of a town, city etc. consisting of shanties.

shanty² (shant'ti), *n.* a song sung by sailors etc. one with a strong rhythm sung while working.

SHAPE (shāp), (*acronym*) *S*upreme *H*eadquarters *A*llied *P*owers *E*urope.

shape (shāp), *v.t.* (*p.p.* **shaped**) to form, to create; to make into a particular form, to fashion; to adapt, to fit, to adjust, to make conform (to); to regulate, to direct; to conceive, to conjure up. *v.i.* to take shape, to develop (well, badly etc.); to become fit or adapted (to). *n.* the outward form, figure, configuration or contour; outward aspect, form, appearance; concrete form, embodiment, realization; definite, fit or orderly form or condition; an image, an appearance, an apparition; a pattern, a mould. **to shape up,** to develop a shape; to develop satisfactorily. **shap(e)able,** *a.* **shaped,** *a.* having a shape (*usu. in comb.* as *square-shaped*). **shapeless,** *a.* having no regular form; lacking in symmetry. **shapelessness,** *n.* **shapely,** *a.* well-formed, well-proportioned; having beauty or regularity. **shapeliness,** *n.* **shaper,** *n.*

shard (shahd), **sherd** (shœd), *n.* a broken piece, esp. a fragment of pottery, glass, etc.; the wing-case of a beetle.

share¹ (sheə), *n.* a part or portion detached from a common amount or stock; a part to which one has a right or which one is obliged to contribute, a fair or just portion; a lot, an allotted part, esp. one of the equal parts into which the capital of a company is divided. *v.t.* to divide into portions, to distribute among a number, to apportion; to give away a portion of; to partake of; to have or endure with others, to participate in. *v.i.* to have a share or shares (in), to be a sharer or sharers (with), to participate. **deferred shares**, those on which a reduced or no dividend is paid until a fixed date or contingent event. **preference, preferred shares** PREFERENCE. **to go shares**, to divide equally with others. **to share out**, to divide into equal shares and distribute. **share-crop**, *v.i.* (*N Am.*) to farm as a sharecropper. **sharecropper**, *n.* (*N Am.*) a tenant farmer who pays over part of the crop as rent. **shareholder**, *n.* one who holds a share or shares in a company etc. **sharer**, *n.*

share² (sheǝ), *n.* a ploughshare.

sharia(h) (shǝrē'ǝ), *n.* the body of Islamic religious law.

shark (shahk), *n.* a selachoid sea-fish of various species with lateral gill openings and an inferior mouth, mostly large and voracious and armed with formidable teeth; a grasping, rapacious person; a rogue, a swindler. **sharkskin**, *n.* a smooth woven fabric of rayon etc.

sharon fruit (sha'rǝn), *n.* a kind of persimmon.

sharp (shahp), *a.* having a keen edge or fine point; terminating in a point or edge; peaked, pointed, edged; angular, abrupt; clean-cut, clearly outlined or defined; pungent, acid, sour; shrill, biting, piercing; harsh, sarcastic, acrimonious; severe, painful, intense; acute, keen-witted; vigilant, attentive, alert, penetrating; alive to one's interests; unscrupulous, dishonest, underhand; keen, penetrating (of the wind); (*coll.*) stylish, flashy. (*Phon.*) surd, voiceless; above the true pitch, esp. a semi-tone higher. *adv.* punctually, exactly; above the true pitch; sharply. *n.* a long and slender sewing-needle; a note a semitone above the true pitch; the sign (♯) indicating this. **at the sharp end**, taking the most important or difficult part in any enterprise. **sharp practice**, *n.* (*coll.*) underhand or questionable dealings. **sharp-shooter**, *n.* a skilled marksman. **sharp-shooting**, *n.* **sharp-witted**, *a.* having a keen wit, judgment or discernment. **sharpen**, *v.t., v.i.* to make sharp; to raise the pitch of (a note). **sharpener**, *n.* **sharper**, *n.* (*coll.*) a swindler, a rogue. **sharpish**, *a.* rather sharp. *adv.* (*coll.*) rather quickly. **sharply**, *adv.* **sharpness**, *n.*

shastra (shas'trǝ), **shaster** (-tǝ), *n.* any of the Vedas and other sacred scriptures of Hinduism.

shat (shat), *past, p.p.* SHIT.

shatter (shat'ǝ), *v.t.* to break up at once into many pieces; to smash, to shiver; to destroy, to dissipate, to overthrow, to ruin; to upset, distress; (*sl.*) to tire out. *v.i.* to break into fragments. **shatter-proof**, *a.* made so as to be resistant to shattering. **shattered**, *a.* **shattering**, *a.* **shatteringly**, *adv.*

shave (shāv), *v.t.* to remove hair from (the face, a person etc.) with a razor; to remove (usu. off) from a surface with a razor; to pare or cut thin slices off the surface of (leather, wood etc.); to pass by closely with or without touching, to brush past, to graze. *v.i.* to shave oneself. *n.* the act of shaving or the process of being shaved; an implement for shaving, paring or scraping; a thin slice; a narrow escape or miss. **shaven**, *a.* (*often in comb.*). **shaver**, *n.* an electric razor; (*coll.*) a young boy. **shaving**, *n.* the act of one who shaves; a thin slice pared off. *a.* used while shaving. **shaving cream**, **foam**, *n.* soapy cream or lather applied to the face before shaving.

Shavian (shā'viǝn), *a.* of or in the manner of George Bernard Shaw (1856–1950). *n.* a follower of Shaw.

shawl (shawl), *n.* a square or oblong garment worn chiefly by women as a loose wrap for the upper part of the person. **shawl collar**, *n.* on a coat etc., a collar of a rolled shape that tapers down the front of the garment.

shawm (shawm), *n.* an ancient wind instrument similar to the oboe.

shay (shā), *n.* (*dial., facet.*) a chaise.

she (shē), *pron.* (*obj.* **her**, (hœ), *poss.* **her**, **hers** (hœz)) the female person, animal or personified thing mentioned or referred to. *n.* a female. *a.* female (*esp. in comb.*, as **she-cat**, **she-devil**, **she-goat** etc.).

shea (shē, shē'ǝ), *n.* a tropical African tree, yielding a kind of butter.

sheading (shē'ding), *n.* any one of the six divisions of the Isle of Man.

sheaf (shēf), *n.* (*pl.* **sheaves** (-vz)) a quantity of things bound or held together lengthwise, esp. a bundle of wheat, oats, barley etc. *v.t.* to collect and bind into sheaves, to sheave.

shear (shiǝ), *v.t.* (*past* **sheared**, *p.p.* **shorn** shawn, **sheared**) to cut or clip esp. with shears; to reduce or remove nap from (cloth etc.) by clipping; to remove (wool etc.) thus; to fleece, to strip. *v.i.* to use shears; to cut, to penetrate, to separate. *n.* the act or an instance of shearing; a strain caused by pressure upon a solid body in which the layers of its substance move in parallel planes; (*Geol.*) alteration of structure by transverse pressure; (*pl.*) a cutting-instrument with two large blades crossing each other like scissors and joined together by a spring; (*pl.*) SHEERS. **shear-legs** SHEERS. **shearer**, *n.* one who shears sheep. **shearling** (-ling), *n.* a sheep that has been shorn once.

shearwater (shiǝ'wawtǝ), *n.* a bird of the genus *Procellaria*, esp. the Manx shearwater allied to the petrels.

sheat-fish (shēt'fish), *n.* a large catfish, the largest European freshwater fish.

sheath (shēth), *n.* a case for a blade, weapon or tool, a scabbard; (*Nat. Hist.*) an envelope, a case, a cell-covering, investing tissue, membrane etc.; any similar protective covering or case; a condom. **sheath-knife**, *n.* a large knife enclosed in a sheath. **sheathe** (shēdh), *v.t.* to put into a sheath; to protect by a casing or covering. **sheathing** (-dh-), *n.* that which sheathes, esp. a metal covering for a ship's bottom.

sheave¹ (shēv), *n.* the grooved wheel in a block or pulley over which the rope runs.

sheave² (shēv), *v.t.* to gather into sheaves, to sheaf.

sheaves (shēvz), (*pl.*) SHEAF.

shebang (shibang'), *n.* (*chiefly N Am.*, *sl.*) a concern, affair, matter, etc.

shebeen (shibēn'), *n.* (*chiefly Ir.*) an unlicenced place where liquors are sold.

shed¹ (shed), *v.t.* (*past, p.p.* **shed**) to pour out, to let fall, to drop, to spill; to effuse; to throw off, to emit; to diffuse, to spread around. *v.i.* to cast off seed, a covering, clothing etc.; of an animal, to moult. *n.* a watershed. **to shed light on**, to clarify (a situation etc.). **shedder**, *n.*

shed² (shed), *n.* a slight simple building, usu. a roofed structure with the ends or ends and sides open; a hut, an outhouse.

she'd (shid, shēd), contr. of **she had**, **she would**.

sheen (shēn), *n.* brightness, splendour, lustre, glitter. **sheeny**, *a.*

sheep (shēp), *n.* (*pl.* **sheep**) a gregarious ruminant animal reared for its flesh and wool; a timid, subservient, unoriginal person who follows the crowd; a bashful or embarrassed person. **black sheep**, BLACK. **sheep-cote** SHEEP-FOLD. **sheep-dip**, *n.* a preparation for killing vermin or preserving the wool on sheep. **sheepdog**, *n.* a breed of heavy, rough-coated, short-tailed dogs employed by shepherds; a collie. **sheepdog trial**, *n.* (*usu. pl.*) a competition in which working sheepdogs are tested. **sheepfold**, *n.* a pen or enclosure for sheep. **sheep's eyes**, *n.pl.* (*dated*) a wishful or amorose glance. **sheepshank**, *n.* a knot used to shorten a rope temporarily. **sheep-shearer**, *n.* one who shears sheep. **sheep-shearing**, *n.* **sheepskin**, *n.* the skin of a sheep, esp. used as a coat or rug. **sheep-walk**, *n.* land for pasturing sheep. **sheepish**, *a.* bashful, diffident, timid; ashamed. **sheepishly**, *adv.* **sheepishness**, *n.*

sheer¹ (shiǝ), *a.* pure, unmixed, simple, mere, absolute, downright; perpendicular, unbroken by a ledge or slope;

of a fabric, very thin, diaphanous. *adv.* vertically, plumb; entirely, outright. **to shear off**, to break off vertically.

sheer² (shiə), *v.i.* (*Naut.*) to deviate from a course; of a horse, to start aside, to shy. *n.* the upward curvature of a vessel towards the bow and stern; the position of a ship riding at single anchor; a swerving or curving course. **to sheer off**, to move off, to go away.

sheers (shiəz), **sheer-legs**, *n.pl.* an apparatus consisting of two masts, or legs, secured at the top, for hoisting heavy weights.

sheet (shēt), *n.* a thin, flat, broad piece of anything, esp. a rectangular piece of linen, cotton or nylon used in a bed to keep the blankets etc. from a sleeper's body; a piece of metal etc., rolled out, hammered, fused etc. into a thin sheet; a thin piece of paper or glass; a newspaper; a broad expanse or surface; a rope attached to the clew of a sail for moving, extending it etc. *v.t.* to cover, wrap or shroud in a sheet or sheets. **in sheets**, of rain, very heavy. **sheet-anchor**, *n.* a large anchor, usu. one of two carried outside the waist of a ship for use in emergencies; a chief support, a last refuge. **sheet bend**, *n.* a kind of knot used for joining ropes of different thicknesses. **sheet-lightning**, *n.* lightning in wide extended flashes. **sheet-metal**, *n.* metal rolled out, hammered or fused into thin sheets. **sheet music**, *n.* music printed on unbound sheets of paper. **sheeting**, *n.* fabric used for making sheets.

sheik, sheikh (shāk, shēk), *n.* the head of an Arab family, clan or tribe. **sheik(h)dom** (-dəm), *n.*

sheila (shē'lə), *n.* (*Austral., sl.*) a girl, a woman.

shekel (shek'l), *n.* a Jewish weight of 1/60 of a mina; (*Hist.*) a silver coin of this weight; the main unit of currency of Israel; (*pl. dated sl.*) money, riches.

sheldrake, shellduck (shel'drāk), *n.* a large wild duck with vivid plumage, breeding on sandy coasts. **sheldduck, shellduck**, *n.fem.*

shelf (shelf), *n.* (*pl.* **shelves**) a horizontal board or slab set in a wall or forming one of a series in a bookcase, cupboard etc., for standing books etc. on; a projecting layer of rock, a ledge; a reef, a sandbank. **off the shelf**, available from stock. **on the shelf**, put aside, discarded; of a woman, considered too old to marry. **shelf-life**, *n.* the length of time a foodstuff or manufactured item can be stored before deteriorating.

shell (shel), *n.* a hard outside covering, as of a nut, egg, testaceous animal etc.; a husk, a pod, a wing-case or elytron, a pupa case, an exoskeleton, a carapace etc.; the framework or walls of a house, ship etc., with the interior removed or not yet built; a light, long and narrow racing-boat; a hollow explosive projectile; a case containing the explosive in fireworks, cartridges etc.; mere outer form or semblance; a spherical area outside the nucleus of an atom occupied by electrons of almost equal energy. *v.t.* to strip or break off the shell from; to remove from the shell; to throw shells at; to bombard. *v.i.* to come away or fall (off) in scales; to cast the husk or shell. **to come out of one's shell**, to stop being shy or reserved. **to shell out**, (*sl.*) to pay up, to pay the required sum. **shellback**, *n.* an old sailor. **shelldrake** SHELDRAKE. **shellfish**, *n.* any aquatic mollusc or crustacean having a shell. **shellproof**, *a.* impenetrable to shells, bomb-proof. **shell-shock** COMBAT FATIGUE. **shell-shocked**, *a.* **shelled**, *a.* (*usu. in comb.*, as *hard-shelled*). **shell-less**, *a.* **shell-like**, *a.* **shelly**, *a.*

she'll (shel), contr. of *she shall, she will.*

shellac (shəlak'), *n.* a thermoplastic resin obtained by purifying the resinous excreta of certain jungle insects, used in the manufacture of varnishes. *v.t.* (*past, p.p.* **shellacked**) to varnish with this.

Shelta (shel'tə), *n.* a secret jargon made up largely of Gaelic or Irish words, used by tinkers, etc.

shelter (shel'tə), *n.* anything that covers or shields from injury, danger, heat, wind etc.; protection, security; a place of safety. *v.t.* to shield from injury, danger etc.; to protect, to cover; to conceal, to screen. *v.i.* to take shelter (under). **sheltered**, *a.* protected from weather or from outside influence; of housing, providing a safe, supervised environment for the disabled or elderly. **shelterer**, *n.*

shelty, -tie (shel'ti), *n.* a Shetland pony or sheepdog.

shelve¹ (shelv), *v.t.* to place on a shelf or shelves; to put aside, to defer indefinitely; to fit with shelves. **shelver**, *n.* **shelving**, *n.* shelves collectively; material for making shelves.

shelve² (shelv), *v.i.* to slope gradually.

shelves (*pl.*) SHELF.

shemozzle (shimoz'l), *n.* (*sl.*) an uproar, a violent row; a confused situation.

shenanigan (shinan'igən), *n.* (*often pl., sl.*) trickery, deception; noisy, boisterous behaviour.

she-oak (shē'ōk), *n.* an Australian tree of the genus *Casuarina*.

Sheol (shē'ōl, -əl), *n.* the place of the dead, often translated 'hell' in the Authorized Version of the Bible.

shepherd (shep'əd), *n.* one employed to tend sheep; a pastor, a Christian minister. *v.t.* to tend, as a shepherd; to drive or gather together. **Good Shepherd**, Jesus Christ. **shepherd's pie**, *n.* cooked minced meat, covered with mashed potatoes and baked in an oven. **shepherd's purse**, *n.* a common cruciferous weed with triangular flat seed pods. **shepherdess** (-dis), *n.fem.*

Sheraton (she'rətən), *a.* applied to furniture of a severe style designed and introduced into England by Thomas Sheraton (*1751–1806*) towards the end of the 18th cent.

sherbet (shœ'bit), *n.* an effervescent powder used in sweets or to make fizzy drinks; a water-ice.

sherd SHARD.

sherif, shereef (shərēf'), *n.* a descendant of Mohammed through his daughter Fatima and Hassan Ibn Ali; the chief magistrate of Mecca.

sheriff (she'rif), *n.* (also **high sheriff**) the chief Crown officer of a county or shire; in London, Bristol, Norwich and Nottingham the sheriffs are civic authorities; (*N Am.*) an elected county official responsible for keeping the peace etc.; (*Sc.*) a judge (sheriff-principal or sheriff-substitute). **sheriff-clerk**, *n.* (*Sc.*) the registrar of the sheriff's court. **sheriff-court**, *n.* (*Sc.*) a sheriff's court, hearing civil and criminal cases. **sheriff-principal**, *n.* (*Sc.*) the chief judge of a county or city. **sheriffdom** (-dəm), *n.* SHRIEVALTY.

Sherpa (shœ'pə), *n.* one of a mountaineering people living on the southern slopes of the Himalayas.

sherry (she'ri), *n.* a fortified Spanish white wine orig. from Xeres. **sherry-cobbler** COBBLER.

she's (shiz, shēz), contr. of *she is* or *she has.*

Shetland (shet'lənd), *n.* a Shetland pony. **Shetland lace**, *n.* an ornamental openwork trimming made of woollen yarn. **Shetland pony**, *n.* a very small variety of the horse with flowing mane and tail, peculiar to Shetland.

shew (shō), etc. (*Sc., Bibl.*) SHOW.

Shia (shē'ə), *n.* one of the two main branches of Islam (see also SUNNA), which regards Ali (Mohammad's cousin and son-in-law) as the first rightful imam or caliph and rejects the three Sunni caliphs. *a.* belonging to, or characteristic of, the Shia sect. **Shiism**, *n.* **Shiite** (-it), *n.*, *a.* (a member) of the Shia sect. **Shiitic** (-it'-), *a.*

shibboleth (shib'əlth), *n.* a criterion, test or watchword of a party etc.; an old-fashioned or discredited doctrine, slogan, etc.

shickered, *a.* (*Austral., sl.*) drunk.

shiel SHIELING.

shield (shēld), *n.* a broad piece of defensive armour made of wood, leather or metal, usu. carried on the left arm to protect the body, usu. straight across the top and tapering to a point at the bottom; a shield-shaped trophy in, e.g., a sporting competition; a screen or framework or other device used as a protection; a shield-like part in an animal

or a plant; (*Her.*) an escutcheon or field bearing a coat of arms; defence, a protection, a defender; (*N Am.*) a sheriff's or detective's badge; a structure of lead, concrete etc., round something highly radioactive to protect against radiation; a mass of very ancient rock at the centre of a continent. *v.t.* to screen or protect with or as with a shield. **shield-like**, *a.*
shieling (shē'ling), **shiel**, *n.* (*Sc.*) a hut used by shepherds, sportsmen etc.; a piece of summer pasturage.
shier, shiest SHYER, SHYEST under SHY [1], [2].
shift (sift), *v.t.* to move from one position to another; to change the position of; to change (one thing) for another; (*sl.*) to dispose of, sell; to remove, dislodge. *v.i.* to move or be moved about; to change place or position; to change into a different place, form, state etc.; to resort to expedients, to do the best one can, to manage, to contrive; (*sl.*) to move quickly. *n.* a change of place, form or character; a substitution of one thing for another, a vicissitude; a relay of workers; the period of time for which a shift works; a woman's loose, unshaped dress; a contrivance, an expedient; an artifice, an evasion. **to make shift**, to manage, to contrive (to do, to get on etc.). **shifter**, *n.* **shiftless**, *a.* incompetent, incapable, without ambition. **shiftlessness**, *n.* **shifty**, *a.* furtive, sly, unreliable. **shiftily**, *adv.* **shiftiness**, *n.*
Shigella (shigel'ə), *n.* a genus of rod-shaped bacteria which cause dysentery in human beings.
Shiitake mushroom, *n.* a dark brown mushroom with pale-beige gills used in Oriental cookery, usu. only available abroad in dried form (**dried Chinese mushroom**).
Shiite etc. SHIA.
shillelagh (shilā'lə, -lə), *n.* (*Ir.*) an oak or blackthorn sapling used as a cudgel.
shilling (shil'ing), *n.* a former British silver (or, later, cupronickel), coin equal in value to 12 old pence (5 new pence); the basic monetary unit of several E African countries.
shilly-shally (shil'ishali), *v.i.* to act irresolutely, to hesitate; to be undecided. *n.* irresolution, hesitation.
shily (shī'li), SHYLY under SHY [1].
shim (shim), *n.* a wedge, piece of metal etc., used to tighten up joints, fill in spaces etc. *v.t.* (*past, p.p.* **shimmed**) to fill in, wedge or fit with this.
shimmer (shim'ə), *v.i.* to glimmer, beam or glisten faintly. *n.* a faint or tremulous light. **shimmering, shimmery**, *a.*
shimmy (shim'i), *n.* an orig. N American dance in which the body is shaken rapidly; abnormal vibration in an aircraft or motor vehicle. *v.i.* to dance a shimmy; of a car or aircraft, to vibrate.
shin (shin), *n.* the forepart of the human leg between the ankle and the knee; a cut of beef, the lower foreleg. *v.i.* (*past, p.p.* **shinned**) to climb up (a tree etc.) by means of the hands and legs alone. *v.t.* to kick on the shins. **shinbone**, *n.* the tibia.
shindig (shin'dig), **shindy** (shin'di), *n.* a noisy or rowdy party or dance; a row, a quarrel, a commotion.
shine (shin), *v.i.* (*past, p.p.* **shone**, (shon)) to emit or reflect rays of light; to be bright, to beam, to glow; to be brilliant, eminent or conspicuous; to be lively or animated. *v.t.* (*past, p.p.* **shined**) to cause to shine, to make bright, to polish; (*N Am.*) to clean shoes etc. *n.* brightness, lustre; (*sl.*) a fancy, a liking. **to take a shine to**, to like at first sight. **shiner**, *n.* that which shines; a popular name for several silvery fishes; (*sl.*) a black eye. **shiny**, *a.* **shininess**, *n.*
shingle[1] (shing'gl), *n.* a thin piece esp. of wood laid in overlapping rows as a roof-covering; a woman's haircut in which the hair is layered, showing the shape of the head. *v.t.* to roof with shingles; to cut (hair) in a shingle. **shingler**, *n.*
shingle[2] (shing'gl), *n.* coarse rounded gravel on the seashore. **shingly**, *a.*

shingles (shing'glz), *n. sing.* a viral infection, *Herpes zoster*, marked by pain and inflammation of the skin along the path of an affected nerve (usu. on the chest or abdomen).
shingly SHINGLE [2].
Shinto (shin'tō), *n.* the indigenous religion of the people of Japan existing along with Buddhism. **Shintoism**, *n.* **Shintoist**, *n.*
shinty (shin'ti), *n.* a game somewhat resembling hockey, played by teams of 12 people.
shiny SHINE.
ship (ship), *n.* a large sea-going vessel, esp. one with three or more square-rigged masts; a ship's crew; (*coll.*) an aircraft, a spacecraft. *v.t.* (*past, p.p.* **shipped**) to put on board a ship; to send, take or carry in a ship; to engage for service on board a ship; to send (goods) by any recognized means of conveyance; to send off, away; to bring, take aboard. *v.i.* to embark on a ship; to engage for service as a sailor. **ship of the desert**, a camel. **when one's ship comes in**, when one becomes rich. **shipboard**, *a.* used, intended for or occurring on board ship. **shipbuilder**, *n.* **shipbuilding**, *n.* **ship-canal**, *n.* a canal along which ocean-going vessels can pass. **ship-chandler**, *n.* one who deals in commodities for fitting out ships. **ship-chandlery**, *n.* **ship-load**, *n.* the quantity of cargo, passengers etc. that a ship carries. **shipmaster**, *n.* the master, captain or commander of a vessel. **shipmate**, *n.* a fellow-sailor. **ship-money**, *n.* (*Hist.*) a tax formerly charged on the ports, towns, cities, boroughs and counties of England for providing certain ships for the navy. **shipowner**, *n.* one who owns a ship or ships or shares therein. **ship('s)-biscuit**, *n.* a hard coarse kind of bread or biscuit used on board ship, hard-tack. **shipshape**, *adv.* in good order. *a.* well arranged, neat, trim. **ship's papers**, *n.pl.* documents carried by a ship containing details of ownership, nationality, destination and cargo. **shipworm**, *n.* a bivalve that bores into ship's timbers, piles etc. **shipwreck**, *n.* the destruction or loss of a ship, by foundering, striking a rock or other cause; the wreck of a ship or part of a ship; destruction, ruin. *v.t.* to cause to suffer shipwreck; to ruin. **shipwright**, *n.* a craftsman employed in shipbuilding. **shipyard**, *n.* a yard etc. where ships are built or repaired. **shipment**, *n.* the act of shipping; goods or commodities shipped, a consignment. **shipper**, *n.* one who or a company that ships or sends goods by a common carrier. **shipping**, *a.* pertaining to ships. *n.* the act or business of sending goods etc. by ship; ships collectively, esp. the ships of a country or port.
-ship (-ship), *suf.* denoting state, condition; status, office; tenure of office; skill in the particular capacity specified; the whole group of people of a specified type; as in *fellowship, friendship, judgeship, ladyship, marksmanship, scholarship.*
shippo (ship'ō), *n.* Japanese cloisonné enamel ware.
shipshape, shipwreck, shipwright, shipyard SHIP.
shiralee (shirəlē'), *n.* (*Austral.*) a swag, a tramp's bundle.
shire (shīə), *n.* a county, esp. one whose name ends in 'shire'. (*Austral.*) a rural district with its own elected local council. **the Shires**, the predominantly rural midland counties of England, esp. Leicestershire and Northamptonshire, noted for foxhunting. **shire-horse**, *n.* a large breed of draught horse, orig. raised in the midland shires.
shirk (shœk), *v.t.* to avoid or get out of unfairly. *v.i.* to avoid the performance of work or duty. *n.* one who shirks. **shirker**, *n.*
shirr (shœ), *n.* an elastic cord or thread inserted in cloth etc. to make it elastic; a gathering or fulling. *v.t.* to draw (a sleeve, dress etc.) into gathers by means of elastic threads; (*N Am.*) to bake eggs in a buttered dish. **shirring**, *n.*
shirt (shœt), *n.* a loose garment of linen, cotton, wool, silk or other material, extending from the neck to the thighs,

and usu. showing at the collar and wristbands, worn by men and boys under the outer clothes; a woman's blouse with collar and cuffs. **boiled shirt**, BOIL[1]. **Blackshirt**, BLACK. **brown-shirt**, BROWN. **in one's shirt sleeves**, with one's jacket off. **red-shirt**, RED. **to keep one's shirt on**, (*coll.*) to keep calm. **to put one's shirt on**, (*coll.*) to bet all one has. **shirt-front**, *n.* the part of a shirt covering the breast, esp. if stiffened and starched; a dicky. **shirtwaist**, **shirtwaister**, *n.* a woman's dress with the bodice tailored to resemble a shirt. **shirted**, *a.* **shirting**, *n.* **shirtless**, *a.* **shirty**, *a.* (*sl.*) cross, ill-tempered.
shish kebab KEBAB.
shit (shit), **shite** (shit), *v.i.* (*past, p.p.* **shit, shat**, (shat)) (*taboo*) to empty the bowels. *n.* (*taboo*) ordure, excrement; nonsense; a worthless or despicable person or thing. *int.* (*taboo*) expressing anger, disappointment etc. **the shits**, (*taboo*) diarrhoea. **shitty**, *a.* soiled with excrement; very bad or inferior; despicable.
Shiva, SIVA.
shiver[1] (shiv'ə), *n.* a tiny fragment, a sliver. *v.t., v.i.* to break into shivers.
shiver[2] (shiv'ə), *v.i.* to tremble or shake, as with fear, cold or excitement. *n.* the act of shivering, a shivering movement. **the shivers**, a feeling or movement of horror; a chill, ague. **shivering**, *n.*, *a.* **shivery**, *a.*
shivoo (shivoo'), *n.* (*Austral. sl.*) a (noisy) party; an entertainment.
shoal[1] (shōl), *n.* a shallow, a submerged sand-bank. *a.* of water, shallow, of little depth. *v.i.* to become shallower. **shoaly**, *a.*
shoal[2] (shōl), *n.* a large number, a multitude, a crowd, esp. of fish moving together. *v.i.* to form a shoal or shoals.
shock[1] (shok), *n.* a violent collision of bodies, a concussion, an impact, a blow, a violent onset; a sudden and violent sensation, as that produced on the nerves by a discharge of electricity; prostration brought about by a violent and sudden disturbance of the system; a sudden mental agitation, a violent disturbance (of belief, trust etc.). *v.t.* to give a violent sensation of disgust, horror or indignation to; to shake or jar by a sudden collision. *v.i.* to behave or appear in an improper or scandalous fashion; (*poet.*) to collide. **shock-absorber**, *n.* an apparatus to neutralize mechanical shock, esp. the shock of axle-springs on recoil. **shock-proof**, *a.* resistant to damage from shock. **shock tactics**, *n.pl.* any action relying on sudden and violent action for its success. **shock therapy, treatment**, *n.* the treatment of certain mental and other disorders by administering an electric shock. **shock troops**, *n.pl.* selected soldiers employed on tasks requiring exceptional endurance and courage. **shock wave**, *n.* a very strong sound wave, accompanied by a rise in pressure and temperature, caused by an explosion or by something travelling supersonically. **shockable**, *a.* **shocker**, *n.* (*coll.*) something or someone that shocks, esp. a sensational story, film, play, etc. **shocking**, *a.* causing a shock; disgraceful; dreadful. **shocking pink**, *n.* a garish, intense shade of pink. **shockingly**, *adv.*
shock[2] (shok), *n.* a collection of sheaves of grain, usu. 12 but varying in number. *v.i.* to collect sheaves into shocks.
shock[3] (shok), *n.* a thick, bushy mass or head of hair.
shocker, shocking SHOCK[1].
shod (shod), *past, p.p.* SHOE.
shoddy (shod'i), *n.* fibre obtained from old shredded cloth; inferior cloth made from a mixture of this with new wool etc.; anything of an inferior, sham or adulterated kind. *a.* inferior, not genuine, sham.
shoe (shoo), *n.* an outer covering for the foot, esp. one distinguished from a boot by not coming up to the ankles; (*N Am.*) a boot; a horseshoe; anything resembling a shoe in form or function, as a socket, ferrule, wheel-drag or parts fitted to implements, machinery etc.

to take friction, thrust etc. *v.t.* (*pres.p.* **shoeing**, *past, p.p.* **shod**, (shod)) to furnish (esp. a horse) with shoes; to cover at the bottom or tip. **dead man's shoes** DEAD. **to be in another's shoes**, to be in another's place or plight. **shoe-black**, *n.* a person earning a living by cleaning the shoes of passers-by. **shoe-horn**, *n.* a device to assist one in putting on a shoe. **shoe-lace**, *n.* a string of cotton etc. for fastening on a shoe; *v.t.* to cram, squeeze into a tight space. **shoe-leather**, *n.* leather for making shoes; shoes. **shoemaker**, *n.* **shoestring**, *n.* (*N Am.*) a shoe-lace; (*coll.*) an inadequate or barely adequate sum of money. **on a shoestring**, (done) on a tight budget. **shoetree**, *n.* a support inserted in a shoe to keep its shape. **shoeless**, *a.* **shoer**, *n.* one who makes or puts on shoes; a farrier.
shogun (shō'gən), *n.* the hereditary commander-in-chief of the army and virtual ruler of Japan under the feudal regime, abolished in 1868. **shogunate** (-ət), *n.*
shone (shon), *past, p.p.* SHINE.
shoo (shoo), *int.* begone, be off. *v.t.* to drive (fowls etc. away) by crying 'shoo'. *v.i.* to cry 'shoo'.
shook[1] (shuk), *past* SHAKE.
shook[2] (shuk), *n.* a set of staves and headings for a cask ready for setting up; a set of boards for a box etc.
shoot (shoot), *v.i.* (*past, p.p.* **shot** (shot)) to dart, rush or come (out, along, up etc.) swiftly; to sprout, to put out buds etc. to extend in growth; to protrude, to project, to jut out; to discharge a missile, esp. from a firearm; to hunt game etc. thus. *v.t.* to propel, let fly, discharge, eject or send with sudden force; to cause (a bow, firearm etc.) to discharge a missile; to hit, wound or kill with a missile from a bow, firearm etc.; to hunt thus over (ground, an estate etc.); to pass swiftly through, over or down; to protrude, to push out; to put forth; in various games, to hit or kick at a goal; to take (photographs) or record (on cine-film). *n.* a young branch, sprout or sucker; an inclined plane or trough down which water, goods etc. can slide, a chute, a rapid; a place where rubbish can be shot; a shooting-party, match or expedition, a hunt. **shoot!** *int.* (*esp. N Am.*) speak out! say it! **the whole shoot**, (*coll.*) the whole amount, everything. **to shoot ahead**, to get swiftly to the front in running, swimming etc. **to shoot a line**, (*sl.*) to boast, to exaggerate. **to shoot down**, to destroy, kill, by shooting; to defeat the argument of. **to shoot down in flames**, to criticize severely; to defeat soundly. **to shoot home**, to hit the target or mark. **to shoot one's mouth off**, (*sl.*) to brag, exaggerate. **to shoot through**, (*Austral., coll.*) to leave, to depart hastily. **to shoot up**, to grow rapidly; (*sl.*) to inject a drug into a vein. **shoot-out**, *n.* a fight, esp. to the death, using guns; a direct confrontation. **shooter**, *n.* (*usu. in comb.*) one who or that which shoots, as *six-shooter*; (*Cricket*) a ball that darts along the ground without bouncing. **shooting**, *n.* the act of discharging firearms or arrows; a piece of land rented for shooting game; the right to shoot over an estate etc. **shooting-box, -lodge**, *n.* a small house or lodge for use during the shooting season. **shooting-brake**, *n.* an estate car. **shooting-gallery, -range**, *n.* a piece of ground or an enclosed space with targets and measured ranges for practice with firearms. **shooting-star**, *n.* an incandescent meteor shooting across the sky. **shooting-stick**, *n.* a walking-stick that may be adapted to form a seat.
shop (shop), *n.* a building in which goods are sold by retail; a building in which a manufacture, craft or repairing is carried on; (*coll.*) one's business, profession etc. or talk about this. *v.i.* (*past, p.p.* **shopped**) to visit shops for the purpose of purchasing goods. *v.t.* (*sl.*) to inform against, esp. to the police. **all over the shop**, (*coll.*) scattered around. **to shop around**, to try several shops to find the best value. **to shut up shop**, to give up doing something. **to talk shop**, to talk about one's occupation. **shop assistant**, *n.* one who serves in a retail shop. **shop-floor**, *n.* the part of a workshop where the machinery is si-

tuated; the work-force as opposed to the management. **shopkeeper**, *n.* the owner of a shop, a trader who sells goods by retail. **shopkeeping**, *n.* **shoplifter**, *n.* one who steals from a shop under pretence of purchasing. **shoplifting**, *n.* **shop-soiled**, *a.* dirty or faded from being displayed in a shop; tarnished; hackneyed. **shop steward**, *n.* a trade union member elected by the work-force to represent them. **shop-walker**, *n.* a person employed in a large shop to direct customers etc. **shopworn** SHOP-SOILED. **shopper**, *n.* one who shops; a bag for carrying shopping. **shopping**, *n.* the act or an instance of buying goods from shops; goods purchased from shops. **shopping centre**, *n.* an area or complex where there are many shops. **shopping mall**, *n.* a shopping centre with covered walkways.

shore[1] (shaw), *n.* the land on the borders of a large body of water, the sea, a lake etc.; (*Law*) the land between high- and low-water marks. **shore-leave**, *n.* of a sailor, naval officer, etc., permitted absence from duty spent on land. **shoreline**, *n.* the line where the land and a body of water meet. **shoreless**, *a.* **shoreward(s)** (-wəd), *a.*, *adv.*

shore[2] (shaw), *n.* a prop, a stay, a support for a building, wall etc. *v.t.* to support or hold (up) with shores. **shoring**, *n.*

shore[3] (shaw), *past* SHEAR.

shorn (shawn), *p.p.* SHEAR.

short (shawt), *a.* measuring little in linear extension, not long; not extended in time or duration, brief; below the average in stature, not tall; not coming up to a certain standard; deficient, scanty, defective, in want (of); breaking off abruptly; brief, concise; abrupt, curt; brittle, friable, crumbling or breaking easily; (*coll.*) neat, undiluted; of vowels and syllables, not prolonged, unaccented; (*Comm., Stock Exchange etc.*) not having goods, stocks etc. in hand at the time of selling; of stocks etc., not in hand, sold. *adv.* abruptly, at once; so as to be short or deficient; briefly. *n.* a short syllable or vowel, or a mark (˘) indicating that a vowel is short; a short circuit; a single-reel film; (*pl.*) knee- or thigh-length trousers; (*pl., chiefly N Am.*) underpants; a drink of, or containing, spirits; (*pl.*) short-dated bonds. *v.t.* to make of no effect; to short-circuit. **for short**, as an abbreviation. **in short**, briefly, in few words. **in short supply**, scarce. **short for**, a shortened form of. **the long and the short of it** LONG[1]. **to be taken, caught short**, to feel a sudden need to urinate or defecate. **to cut, bring, pull up short**, to check or pause abruptly. **to fall short** FALL. **to make short work of**, to deal with quickly and expeditiously. **to run short**, to exhaust the stock in hand (of a commodity). **to sell short**, to cheat; to disparage. **to stop short**, to come to a sudden stop; to fail to reach the point aimed at. **shortbread**, **shortcake**, *n.* a brittle, dry cake like a biscuit made with much butter and sugar. **short-change**, *v.t.* to give too little change to; (*sl.*) to cheat. **short circuit**, *n.* (*Elec.*) an accidental crossing of two conductors carrying a current by another conductor of negligible resistance, which shortens the route of the current. **short-circuit**, *v.t.* to form or introduce a short circuit; to dispense with intermediaries; to take a short cut. **short-coming**, *n.* a failure of performance of duty etc., a defect. **short-cut**, *n.* a shorter route than the usual. **short-dated**, *a.* of a security etc., having only a little time to run. **shortfall**, *n.* the amount by which something falls short, deficit. **shorthand**, *n.* a system of contracted writing used for reporting etc., stenography. **short-handed**, *a.* short of workers, helpers etc. **short haul**, *n.* transport etc. over a short distance. **shorthorn**, *n.* one of a breed of cattle with short horns. **short list**, *n.* a selected list of candidates from whom a final choice will be made. **short-list**, *v.t.* **short-lived** (-livd), *a.* not living or lasting long, brief. **short odds**, *n.pl.* in betting, a nearly equal chance. **short order**, *n.* an order in a restaurant for food that can be prepared quickly. *a.* producing or relating to such food. **short-range**, *a.* having a small range in time or distance. **short-sight**, *n.* inability to see clearly at a distance, myopia; lack of foresight. **short-sighted**, *a.* **short-sightedly**, *adv.* **short-sightedness**, *n.* **short-sleeved**, *a.* having sleeves reaching not below the elbow. **short-spoken**, *a.* curt and abrupt in speech. **short-staffed**, *a.* short-handed. **short supply**, *n.* general shortage of a commodity. **short-tempered**, *a.* having little self-control, irascible. **short-term**, *a.* of or covering a short period of time. **short time**, *n.* the condition of working fewer than the normal number of hours per week. **short-waisted**, *a.* unusually short from the shoulders to the waist. **short wave**, *n.* a radio wave of between 10 and 100 metres wavelength. **short-winded** (-win'-), *a.* easily put out of breath; brief, concise. **shortage** (-tij), *n.* a deficiency; the amount of this. **shorten**, *v.t.* to make short in time, extent etc.; to curtail; to reduce the amount (of sail spread). *v.i.* to become short, to contract. **shortening**, *n.* butter or another fat added to dough to make short pastry. **shortie**, **shorty** (-ti-), *n.* (*coll.*) a shorter than average person, garment etc. **shortish**, *a.* **shortly**, *adv.* **shortness**, *n.*

shot (shot), *n.* a missile for a firearm, esp. a solid or nonexplosive projectile; the act of shooting; the discharge of a missile from a firearm or other weapon; an attempt to hit an object with such a missile; a photographic exposure; the film taken between the starting and stopping of a camera; (*coll.*) an injection by hypodermic needle; (*coll.*) a glass of an alcoholic drink, esp. a spirit, a short; a stroke at various games; an attempt to guess etc.; the distance reached by a missile, the range of a firearm, bow etc.; a marksman or woman; (*pl.* **shot**) a small lead pellet, a quantity of which is used in a charge or cartridge for shooting game. *a.* having a changeable colour, as shot silk. **a shot in the arm**, something which encourages or invigorates. **a shot in the dark**, a random guess. **big shot**, BIG[1]. **like a shot**, immediately, eagerly. **moonshot** MOON. **to call the shots**, (*sl.*) to be in charge. **to get shot of**, (*coll.*) to get rid of. **to give it one's best shot**, (*coll.*) to try one's very best. **to have a shot at**, to attempt. **to have shot one's bolt**, to be unable to take further action. **shotgun**, *n.* a light gun for firing small-shot. *a.* enforced. **shot putting** PUTTING THE SHOT under PUT[1]. **shot silk**, *n.* silk with warp and weft of different colours, chatoyant silk.

should (shud), *past* SHALL.

shoulder (shōl'də), *n.* the part of the body at which the arm, foreleg or wing is attached to the trunk; (*pl.*) the upper part of the back; the part of a garment covering the shoulder; the fore-quarter of an animal cut up as meat; anything resembling a shoulder, such as a projecting part of a mountain, tool etc.; the verge of a road. *v.t.* to push with the shoulder; to jostle, to make (one's way) thus; to take on one's shoulders; to accept a responsibility; to carry vertically (a rifle etc.) at the side of the body. **cold shoulder**, COLD. **shoulder to shoulder**, (standing in rank) with shoulders nearly touching; with hearty cooperation, with mutual effort. **to rub shoulders** RUB. **shoulder-blade**, *n.* the scapula. **shoulder-strap**, *n.* a strap over the shoulder supporting a bag, a garment, etc. **shouldered**, *a.* (*usu. in comb.*) having shoulders, as *broad-shouldered*.

shout (showt), *n.* a loud, vehement and sudden call or outcry of joy, triumph or the like; (*Austral. coll.*) a round of drinks. *v.i.* to utter a loud cry or call; to speak at the top of one's voice; (*Austral. coll.*) to buy a round of drinks. *v.t.* to utter with a shout; to say at the top of one's voice. **to shout down**, to silence or render inaudible by shouting. **shouter**, *n.*

shove (shŭv), *v.t.*, *v.i.* to push, to move forcibly along; to push against, to jostle. *n.* a strong or hard push. **to shove off**, to push off from the shore etc.; (*sl.*) to go away. **shove-halfpenny**, *n.* a game in which coins are slid over a flat board which is marked off into sections. **shover**, *n.*

shovel (shŭv'l), *n.* an implement consisting of a wide blade or scoop with a handle, used for shifting loose material. *v.t.* (*past, p.p.* **shovelled**) to shift, gather together or take up and throw with, or as with, a shovel. *v.i.* to use a shovel. **shovel-head, -nose,** *n.* a popular name for kinds of sturgeon, sharks etc. **shovelful,** *n.* **shoveller,** *n.* one who shovels; (also **shoveler**) the spoon-bill duck.

show (shō), *v.t.* (*past*, **showed**, *p.p.* **shown,** †**showed**) to cause or allow to be seen, to disclose, to offer to view, to exhibit, to present, to expose, to reveal; to give, to bestow, to offer; to make clear, to point out, to explain, to demonstrate, to prove; to cause (a person) to see or understand; to conduct (round or over a house etc.). *v.i.* to become visible or noticeable, to appear; to have a specific appearance. *n.* the act of showing; outward appearance, semblance, pretence; display, ostentation, parade, pomp; a spectacle, a pageant, a display, an exhibition; a theatrical performance or other entertainment; (*sl.*) an opportunity, a chance, a concern, a business; an indication, a trace. **to give the show away,** to let out the real nature of something; to blab. **to show off,** to set off, to show to advantage; to make a display of oneself, one's talents etc. **to show one's hand** HAND [1]. **to show up,** to expose; to be clearly visible; to be present, to arrive; (*coll.*) to humiliate. **show biz,** *n.* (*coll.*) show business. **show-boat,** *n.* a steamboat fitted as a theatre. **showbread, shewbread,** *n.* 12 loaves (one for each tribe) displayed by the Jewish priests in the Temple, and renewed every Sabbath. **show business,** *n.* the entertainment industry, theatre, television, cinema. **showcase,** *n.* a glass case for exhibiting specimens, articles on sale etc.; anything providing a similar function. *v.t.* to exhibit. **showdown,** *n.* an open or final confrontation. **showgirl,** *n.* a girl who appears in variety entertainments, esp. a dancer or singer. **show house,** *n.* one of a group of new houses, open to the public as an example of the type. **showjumper,** *n.* **showjumping,** *n.* competitive riding over a set course containing obstacles. **showman,** *n.* the presenter, proprietor or producer of a show; one skilled in presentation. **showmanship,** *n.* **show-off,** *n.* one who shows off, an exhibitionist. **showpiece,** *n.* a particularly fine specimen, used for display. **showplace,** *n.* a place tourists etc. go to see. **showroom,** *n.* a room where goods for sale are set out for inspection. **shower**[1], *n.* **showing,** *n.* **showy,** *a.* ostentatious, gaudy. **showily,** *adv.* **showiness,** *n.*

shower[2] (show'ə), *n.* a fall of rain, hail or snow of short duration; a brief fall of sparks, bullets etc.; a copious supply (of); a shower-bath; (*chiefly N Am.*) a party (e.g. for a bride-to-be or expectant mother) at which gifts are given; (*coll., derog.*) a collection of (inferior etc.) people. *v.t.* to sprinkle or wet with a shower; to discharge or deliver in a shower. *v.i.* to take a shower. **shower-bath,** *n.* a bath in which a stream of water is sprayed over the body from an overhead nozzle. **showery,** *a.*

shrank (shrangk), *past* SHRINK.

shrapnel (shrap'nəl), *n.* bullets enclosed in a shell with a small charge for bursting and spreading in a shower; shell-splinters from a high-explosive shell. [Gen. Henry *Shrapnel*, 1761–1842, inventor]

shred (shred), *n.* a piece torn off; a strip, a rag, a fragment, a tiny piece. *v.t.* (*past, p.p.* **shredded**) to tear or cut into shreds. **shredder,** *n.*

shrew (shroo), *n.* a shrewmouse; a bad-tempered, scolding woman, a virago. **shrewmouse,** *n.* a small nocturnal insectivorous mammal. **shrewish,** *a.* **shrewishly,** *adv.* **shrewishness,** *n.*

shrewd (shrood), *a.* astute, sagacious, discerning. **shrewdly,** *adv.* **shrewdness,** *n.*

shriek (shrēk), *v.i.* to utter a sharp, shrill, inarticulate cry; to scream, to screech, as in a sudden fright; to laugh wildly. *v.t.* to utter with a shriek. *n.* a sharp, shrill, inarticulate cry. **shrieker,** *n.*

shrieval (shrē'vəl), *a.* of or pertaining to a sheriff. **shrievalty** (-ti), *n.* the office or jurisdiction of a sheriff; the tenure of this.

shrift (shrift), *n.* confession; absolution. **short shrift,** *n.* summary treatment.

shrike (shrīk), *n.* the butcher-bird, feeding on insects and small birds and having the habit of impaling them on thorns for future use.

shrill (shril), *a.* high-pitched and piercing in tone, sharp, acute; noisy, importunate. *v.i., v.t.* to utter a piercing sound; to sound shrilly. **shrillness,** *n.* **shrilly,** *adv.*

shrimp (shrimp), *n.* a slender long-tailed edible crustacean, allied to the prawn; (*coll.*) a very small person. *v.i.* to fish for shrimps. **shrimper,** *n.*

shrine (shrīn), *n.* a chest or casket in which sacred relics were deposited; a tomb, altar, chapel etc. of special sanctity; a place hallowed by its associations. *v.t.* to enshrine.

shrink (shringk), *v.i.* (*past* **shrank** (shrangk), *p.p.* **shrunk** (shrŭngk), *part.a.* **shrunken** (-ən)) to grow smaller, to contract, to shrivel; to give way, to recoil; to flinch. *v.t.* to cause to shrink, to make smaller. *n.* (*sl.*) a psychiatrist. **shrink-wrap,** *v.t.* to wrap in plastic film, which is then shrunk, e.g. by heating, to make a tight-fitting, sealed package. *n.* **shrinkable,** *a.* **shrinkage** (-ij), *n.* **shrinker,** *n.* **shrinking,** *a.*

shrive (shrīv), *v.t.* (*past* **shrove,** shrōv, *p.p.* **shriven,** (shriv'n)) to receive the confession of, impose penance on and absolve; to confess (oneself) and receive absolution. *v.i.* to confess, impose penance and administer absolution. **shriver,** *n.* a confessor.

shrivel (shriv'l), *v.i.* (*past, p.p.* **shrivelled**) to contract, to wither, to become wrinkled. *v.t.* to cause to contract or become wrinkled.

shriven, shriver SHRIVE.

shroud (shrowd), *n.* a winding sheet; anything that covers or conceals; (*pl.*) ropes extending from the lower mastheads to the sides of the ship, serving to steady the masts. *v.t.* to dress in a shroud for burial; to cover, disguise or conceal. **shroudless,** *a.*

shrove (shrōv), *past* SHRIVE.

Shrovetide, *n.* the period before Lent, when people formerly went to confession and afterwards celebrated. **Shrove Tuesday,** *n.* the day before Ash Wednesday.

shrub[1] (shrŭb), *n.* a drink composed of the sweetened juice of lemons or other fruit with spirit.

shrub[2] (shrŭb), *n.* a woody plant smaller than a tree, with branches proceeding directly from the ground without any supporting trunk. **shrubbery,** *n.* a plantation of shrubs; shrubs collectively. **shrubby,** *a.* **shrubbiness,** *n.* **shrublike,** *a.*

shrug (shrŭg), *v.t., v.i.* (*past, p.p.* **shrugged**) to draw up (the shoulders) to express dislike, doubt etc. *n.* this gesture. **to shrug off,** to disregard, ignore; to throw off, get rid of.

shrunk (shrŭngk), *p.p.,* **shrunken** (-ən), *part.a.* SHRINK.

shuck (shŭk), *n.* (*chiefly N Am.*) a shell, husk or pod; (*pl.*) something utterly valueless. *v.t.* to remove the shell etc. from. **shucker,** *n.* **shucks,** *int.* expressive of contempt, annoyance, embarrassment etc.

shudder (shŭd'ə), *v.i.* to shiver suddenly as with fear; to tremble, to quake. *n.* a sudden shiver or trembling. **shuddering,** *n., a.* **shudderingly,** *adv.* **shuddery,** *a.*

shuffle (shŭf'l), *v.t.* to shift or shove to and fro or from one to another; to move (cards) over each other so as to mix them up; to mix (up), to throw into disorder. *v.i.* to change the relative positions of cards in a pack; to shift ground; to prevaricate; to move (along) with a dragging gait. *n.* the act of shuffling; a shuffling movement of the feet etc.; the shuffling of cards; a mix-up, a general change of position. **shuffler,** *n.*

shuffleboard, *n.* a game played (now usu. on a ship's deck) by shoving wooden disks with the hand or a mace towards marked compartments.

shufti, -ty (shŭf'ti, shŭf'-), *n*. (*sl.*) a (quick) look (at something).
shun (shŭn), *v.t.* (*past, p.p.* **shunned**) to avoid, to eschew, to keep clear of.
'shun (shŭn), *int.* short for ATTENTION.
shunt (shŭnt), *v.t.* to turn (a train etc.) on to a side track; to get rid of, suppress, evade or divert. *v.i.* of a train etc., to turn off on to a side track. *n.* the act of shunting; a conductor joining two points of a circuit through which part of an electric current may be diverted; a passage connecting two blood-vessels, diverting blood from one to the other; (*sl.*) a car crash. **shunter**, *n*.
shush (shush, shŭsh), *int*. calling for silence. *v.i., v.t.* to be or make quiet.
shut (shŭt), *v.t.* (*pres.p.* **shutting**, *past, p.p.* **shut**) to close by means of a door, lid, cover etc.; to cause (a door, lid, cover etc.) to close an aperture; to keep (in or out) by closing a door; to bar (out), to exclude, to keep from entering or participating in; to bring the parts or outer parts together. *v.i.* to become closed; of teeth, scissor-blades etc., to come together. **to shut down**, to pull or push down (a window-sash etc.); of a factory, business etc., to stop working. **to shut in**, to confine; to encircle. **to shut off**, to stop the inflow or escape of (gas etc.) by closing a tap etc.; to separate; to block off. **to shut out**, to exclude, to bar out; to prevent the possibility of. **to shut to**, to close (a door); of a door, to shut. **to shut up**, to close all the doors, windows etc. of a house etc.; to close and fasten up (a box etc.); to put away in a box etc.; to confine; (*coll*.) to confute, to silence. **shut-down**, *n*. a closure (e.g. of a factory); a stoppage. **shut-eye**, *n*. (*coll*.) sleep. **shut-off**, *n*. a cessation, a stoppage; a device for shutting something off. **shut up!** *int*. be quiet! stop talking!
shutter (shŭt'ə), *n*. one who or that which shuts; a cover of wooden battens or panels or metal slats for sliding, folding, rolling or otherwise fastening over a window; a device in a camera which allows exposure of the film for a predetermined period; a contrivance for closing the swell-box of an organ. **shuttered**, *a*.
shuttle (shŭt'l), *n*. a boat-shaped contrivance enclosing a bobbin used by weavers for passing the thread of the weft between the threads of the warp; the holder carrying the lower thread for making lock-stitches in a sewing machine; a shuttle service; a vehicle used on a shuttle service or one that goes between two points. *v.i.* to move or travel regularly between two points or places. **shuttle service,** *n*. transport service running to and fro between two points.
shuttlecock (shŭt'lkok), *n*. a light cone-shaped object with feathered flights, used in the games of battledore and badminton; anything repeatedly passed to and fro.
shy[1] (shi), *a*. (*comp*. **shyer, shier,** *superl*. **shyest, shiest**) easily frightened, fearful, timid; bashful, coy, shrinking from approach or familiarity; wary, cautious, suspicious; circumspect, careful, watchful (of); elusive; (*chiefly N Am.*) short (of); (*in comb*.) showing reluctance or aversion, as *workshy*. *v.i.* of a horse, to start or turn aside suddenly; to recoil, draw back. *n.* the act of shying. **shyly,** *adv*. **shyness,** *n*.
shy[2] (shī), *v.t., v.i.* (*coll*.) to fling, to throw. *n.* the act of shying; a try, an attempt. **shyer,** *n*.
shyster (shistə), *n*. (*coll*.) a tricky or disreputable person, esp. a lawyer.
SI, *n*. the now universally used system of scientific units, the basic units of which are the metre, second, kilogram, ampere, kelvin, candela and mole. [F *Système International d'Unités*]
Si, (*chem. symbol*) silicon.
si (sē), *n*. (*Mus*.) te.
sial (sī'əl), *n*. the outer layer of the earth's crust, rock rich in silicon and aluminium.
sialogogue (sī'ələgog), *n*. (*Med*.) a medicine promoting salivary discharge.
siamang (sī'əmang), *n*. a large gibbon from the Malay peninsula and Sumatra.
Siamese (siəmēz'), *n., a*. Thai. **Siamese cat,** *n*. a breed of cat with blue eyes and dark-coloured ears, face, tail and paws. **Siamese twins,** *n.pl*. identical twins born joined together at some part of the body.
SIB, (*abbr.*) Securities and Investments Board.
sib (sib), *n*. a brother or sister; a blood relative. **sibling** (-ling), *n*. one of two or more children having one or both parents in common.
sibilant (sib'ilənt), *a*. hissing; having a hissing sound. *n.* a letter which is pronounced with a hissing sound, as *s* or *z*. **sibilance, -ancy,** *n*. **sibilate,** *v.t., v.i.* **sibilation,** *n*.
sibling, SIB.
sibyl (sib'l), *n*. one of a number of women who prophesied in ancient times under the supposed inspiration of a deity; a prophetess, a sorceress, a fortune-teller. **sibylline** (-bil'in), *a*. pertaining to or composed or uttered by a sibyl; prophetic, oracular, cryptic, mysterious.
sic[1] (sĕk, sik), *adv*. thus, so (usu. printed after a doubtful word or phrase to indicate that it is quoted exactly as in the original).
sic[2] (sik), SICK[2].
siccative (sik'ətiv), *n*. a substance that accelerates drying, esp. one used with oil-paint.
Sicilian (sisil'yən), *a*. of or pertaining to Sicily or its inhabitants. *n.* a native of Sicily. **siciliana** (-siliah'nə), *n*. a traditional, rustic dance of Sicily; the music (in 6/8 or 12/8 time) for it.
sick[1] (sik), *a*. ill, affected by some disease, in bad health; affected with nausea, inclined to vomit; tending to cause sickness; disgusted, feeling disturbed, upset, pining (for etc.); tired (of); of a room, quarters etc., set apart for sick persons; of humour, macabre, cruel, using subjects not usu. considered suitable for jokes. *n.* (*coll*.) vomit. **sick-bay,** *n*. a room for the sick or injured, esp. on board a ship. **sick building syndrome,** *n*. (*Med.*) a syndrome thought to be caused by working in a fully airconditioned building. **sick to one's stomach**, (*chiefly N Am.*) affected with nausea, vomiting. **to be, feel sick**, to vomit or be inclined to vomit. **sick-bed,** *n*. a bed occupied by one who is ill. **sick headache,** *n*. migraine. **sick-leave,** *n*. leave of absence on account of illness. **sick-list,** *n*. a list of persons, esp. in a regiment, ship etc., laid up by sickness. **on the sick-list,** laid up by illness. **sick-making,** *a*. (*coll*.) galling, annoying. **sick pay,** *n*. a benefit paid to a worker on sick-leave. **sicken,** *v.i.* to grow ill, to show symptoms of illness; to feel disgust (at). *v.t.* to make sick; to affect with nausea; to disgust. **sickener,** *n*. **sickening,** *a*. disgusting, offensive; (*coll.*) annoying. **sickeningly,** *adv*. **sickish,** *a*. **sickishly,** *adv*. **sickishness,** *n*. **sickly,** *a*. habitually indisposed, weak in health, invalid, marked by sickness; languid, faint, weakly-looking; nauseating, mawkish. *adv*. in a sick manner. **sickliness,** *n*. sickness, *n*.
sick[2] (sik), *v.t.* to incite to chase or attack, to urge to set upon.
sickle (sik'l), *n*. an implement with a long curved blade set on a short handle, used for reaping, lopping etc. **sicklebill,** *n*. a bird of various species with a sickle-shaped bill. **sickle-cell anaemia,** *n*. a severe form of anaemia marked by the presence of sickle-shaped red blood-cells.
sickly, sickness SICK[1].
side (sīd), *n*. one of the bounding surfaces (or lines) of a material object, esp. a more or less vertical inner or outer surface (as of a building, a room, a natural object etc.); such a surface as dist. from the top and bottom, back and front, or the two ends; the part of an object, region etc. to left or right of its main axis or part facing one; either surface of a plate, sheet, layer etc.; the part of a person or animal on the right hand or left, esp. that between the

hip and shoulder; direction or position, esp. to right or left, in relation to a person or thing; an aspect or partial view of a thing; one of two opposing bodies, parties or sects; one of the opposing views or causes represented by them; line of descent through father or mother; a twist or spin given to a billiard ball; (*sl.*) swagger, bumptiousness, pretentiousness. *v.i.* to take part with, to put oneself on the side of. *a.* situated at or on the side, lateral; being from or towards the side, oblique, indirect. **on the side**, in addition to the main aim, occupation; (*N Am.*) as a side-dish. **to choose sides**, to select parties for competition in a game. **to take sides**, to support one side in an argument etc. **sidearms**, *n.pl.* weapons, as swords or pistols, carried at the side. **side-band**, *n.* the band of radio frequencies on either side of the carrier frequency, caused by modulation. **sideboard**, *n.* a flat-topped table or cabinet placed at the side of a room to support decanters, dining utensils etc.; (*pl.*) side-whiskers. **sideburns**, *n.pl.* side-whiskers. **sidecar**, *n.* a car with seats, attached to the side of a motor-cycle; a kind of cocktail. **side-dish**, *n.* a supplementary dish at dinner etc. **side-drum**, *n.* a small double-headed drum with snares, carried at the drummer's side. **side effect**, *n.* a secondary effect (e.g. of a drug), often adverse. **sidekick**, *n.* a close associate or assistant, often in a shady enterprise. **sidelight**, *n.* light admitted into a building etc. from the side; an incidental illustration (of a subject etc.); a small light at the side of a vehicle; one of the two navigational lights carried by a ship at night. **sideline**, *n.* an incidental or subsidiary branch of business; a line marking the side of a sports pitch, tennis-court etc. **on the sidelines**, watching a game etc. from the side of the pitch etc.; not participating directly in an activity. **side-saddle**, *n.* a saddle for sitting sideways on a horse. **sideshow**, *n.* a subordinate show, business affair etc. **side-slip**, *n.* a skid; a slip or shoot from a plant; a movement of an aeroplane downwards and outwards from its true course. *v.i.* esp. of aircraft, motor vehicles etc., to skid, to slip sideways. **sidesman**, *n.* a church officer assisting the churchwarden. **side-splitting**, *a.* of laughter, a joke etc., convulsing. **sidestep**, *n.* a step or movement to one side. *v.i.* to step aside. *v.t.* to avoid, dodge. **side-stroke**, *n.* a swimming stroke performed sideways. **sideswipe**, *n.* a glancing blow; an incidental criticism. **sidetrack**, *n.* a diversion or digression; (*N Am.*) a siding. *v.t.* to divert or distract from one's main purpose; to defer indefinitely. **side-view**, *n.* a view from the side, a profile. **sidewalk**, *n.* (*N Am.*) a pavement. **side-whiskers**, *n.pl.* hair grown by a man on the side of the face in front of the ears, sideboards. **sidewinder** (-winˈdə), *n.* a N American rattlesnake that moves by a kind of sideways looping movement; (*N Am.*) a heavy punch from the side. **sided**, *a.* (*usu. in comb.*, as *many-sided*). **sideless**, *a.* **sidelong**, *adv.* obliquely; laterally. *a.* oblique. **sider**, *n.* one who sides with a particular party etc. **sideward** (-wəd), *adv.*, *a.* **sidewards**, *adv.* **sideways**, *adv.* **siding**, *n.* the act of taking sides; a short line of metals beside a railway line, used for shunting and joining this at one end.

sidereal (sīdiəˈriəl), *a.* of or pertaining to the fixed stars or the constellations; measured or determined by the apparent motion of the stars. **sidereal day**, *n.* the time between two successive upper culminations of a fixed star or of the vernal equinox, about four minutes shorter than the solar day. **sidereal month**, *n.* the mean period required by the moon to make a circuit among the stars, amounting to 27·32166 days. **sidereal time**, *n.* time as measured by the apparent diurnal motion of the stars. **sidereal year**, *n.* the time occupied by a complete revolution of the earth round the sun, longer than the tropical year.

siderite (sidˈərīt), *n.* a rhombohedral carbonate of iron; an iron meteorite; a blue variety of quartz.

sidero-, *comb. form.* iron.

siderolite (sidˈərəlīt), *n.* a meteorite consisting partly of stone and partly of iron.

siderosis (sidərōˈsis), *n.* a lung disease caused by breathing iron or other metal dust.

siderostat (siˈdərōstat), *n.* an astronomical instrument by which a star under observation is kept within the field of the telescope.

siding SIDE.

sidle (sīˈdl), *v.i.* to go or move sideways; to move furtively.

siege (sēj), *n.* the process of besieging or the state of being besieged; the offensive operations of an army etc. before or round a fortified place to compel surrender.

siemens (sēˈmənz), *n.* the SI unit of electrical conductance.

Sienese (sēənēzˈ), *a.* of or pertaining to Siena, a city of Italy, or the Sienese school of painters (13th and 14th cents.). *n.* a native of Siena.

sienna (sienˈə), *n.* a pigment composed of a native clay coloured with iron and manganese, known as raw (yellowish-brown) or burnt (reddish-brown) sienna according to the mode of preparation.

sierra (sicˈrə), *n.* a long serrated mountain-chain.

siesta (siesˈtə), *n.* a short midday sleep, esp. in hot countries.

sieve (siv), *n.* an instrument for separating the finer particles of substances from the coarser by means of meshes or holes through which the former pass and the others are retained. *v.t.* to sift.

sievert (sēˈvət), *n.* the SI unit of ionizing radiation, equal to 100 rems.

sift (sift), *v.t.* to separate into finer and coarser particles by means of a sieve; to separate (from, out etc.); to sprinkle (sugar, flour etc.) as with a sieve; to examine minutely, to scrutinize, to analyse critically. *v.i.* to fall or be sprinkled as from a sieve. **sifter**, *n.*

sig., (*abbr.*) signature.

sigh (sī), *v.i.* to draw a deep, long respiration, as an involuntary expression of grief, fatigue, relief etc.; to yearn (for); to make a sound like sighing. *v.t.* to utter with sighs. *n.* the act or sound of sighing. **sigher**, *n.*

sight (sīt), *n.* the faculty of seeing; the act of seeing; vision, view; range of vision; point of view, estimation; visibility; that which is seen, a spectacle, a display, a show, esp. something interesting to see; a device on a firearm, optical instrument etc. for enabling one to direct it accurately to any point; (*coll.*) a great quantity (of); a strange object, a fright. *v.t.* to get sight of, to espy, to discover by seeing; to adjust the sights of; to aim by means of sights. **at, on sight**, as soon as seen; immediately; on presentation for payment. **in sight**, visible. **out of sight**, where it cannot be seen; disappeared; forgotten. **sight unseen**, without prevous inspection (of the object to be bought etc.). **to lose sight of**, to cease to see; to overlook; to forget. **sight-read**, *v.t.* **sight-reader**, *n.* one who reads (music etc.) at sight. **sight-reading**, *n.* **sight screen**, *n.* a white screen set on the boundary of a cricket field to help the batsman see the ball. **sightsee**, *v.i.* to visit the sights, notable buildings, etc. of a place. **sightseeing**, *n.* **sightseer** (-sēə), *n.* **sighted**, *a.* having sight (*in comb.*, of a specified kind, as *short-sighted*). **sightless**, *a.* wanting sight, blind; invisible. **sightlessly**, *adv.* **sightlessness**, *n.* **sightly**, *a.* pleasing to the eye. **sightliness**, *n.*

sigma (sigˈmə), *n.* the name of the Greek letter Σ σ, or s, equivalent to *s*. **sigmoid** (-moid), *a.* (*chiefly Anat.*) curved like the sigma or the letter *S*; having a double or reflexed curve; *n.* such a curve. **sigmoidal** (-moiˈ-), *a.* sigmoid.

sign (sīn), *n.* a mark expressing a particular meaning; a conventional mark used for a word or phrase to represent a mathematical process (as + or −); a symbol, a token, a symptom or proof (of), esp. a miracle as evidence of a supernatural power; a trace, an indication; a password, a secret formula, motion or gesture by which confederates

signal etc. recognize each other; a motion, action or gesture used instead of words to convey information, commands etc.; a device, usu. painted on a board, displayed as a token or advertisement of a trade, esp. by innkeepers; one of 12 ancient divisions of the zodiac named after the constellations formerly in them. *v.t.* to mark with a sign, esp. with one's signature, initials or an accepted mark as an acknowledgment, guarantee, ratification etc.; to convey (away) by putting one's signature to a deed etc.; to engage or to be taken (on) as an employee, etc. by signature; to order, request or make known by a gesture; to write (one's name) as signature. *v.i.* to write one's name as signature. **to sign away**, to transfer or dispose of (as if) by signing. **to sign for**, to acknowledge receipt of by signing. **to sign in**, to record arrival by signing. **to sign off**, to stop work for the time; to discharge from employment; to stop broadcasting; to end a letter by signing. **to sign on**, to commit (oneself or another) to an undertaking or employment; to register as unemployed. **to sign out**, to record departure by signing. **to sign up**, to enlist or cause to enlist. **signboard**, *n.* a board on which a tradesman's sign or advertisement is painted, or on which a notice is posted. **sign language**, *n.* a system of communication that uses visual signals rather than the spoken word. **signpainter**, *n.* one who paints signboards etc. **signpost**, *n.* a post supporting a sign, esp. as a mark of direction at crossroads etc. **signable**, *a.* **signer**, *n.*
signal (sig'nəl), *n.* a sign agreed upon or understood as conveying information, esp. to a person or persons at a distance; an event that is the occasion for some action; an electronic transmission conveying a warning, message, etc.; the apparatus by which this is conveyed. *v.t.* (*past, p.p.* **signalled**) to make signals to; to convey, announce, order etc. by signals. *v.i.* to make signals. *a.* distinguished from the rest, conspicuous, noteworthy, extraordinary. **signal-box**, *n.* the cabin from which railway signals and points are worked. **signalman**, *n.* one who works railway signals. **signalize, -ise** (-ise), *v.t.* to make signal or remarkable; to point out or indicate particularly. **signaller**, *n.* **signally**, *adv.*
signature (sig'nəchə), *n.* the name, initials or mark of a person written or impressed with his or her own hand; a distinguishing letter or number at the bottom of the first page of each sheet of a book; such a sheet after folding; (*Mus.*) the signs of the key and rhythm placed at the beginning of a staff to the right of the clef; all such signs including the clef; a significant mark, sign or characteristic. **signature tune**, *n.* a distinctive piece of music used to identify a programme, performer etc. **signatory**, *a.* having signed, bound by signature. *n.* one who signs, esp. as representing a state.
signet (sig'nit), *n.* a small seal, esp. for use in lieu of or with a signature as a mark of authentication. **signet-ring**, *n.* a finger ring set with a signet.
signify (sig'nifī), *v.t.* to make known by signs or words; to communicate, to announce; to be a sign of, to mean, to denote. *v.i.* to be of consequence, to matter. **significance** (-nif'ikəns), *n.* the quality of being significant, expressiveness; meaning, real import; importance, moment, consequence. **significant**, *a.* meaning something; expressing or suggesting something more than appears on the surface; meaning something important, weighty, noteworthy, not insignificant. **significantly**, *adv.* **signification**, *n.* the act of signifying; that which is signified, the precise meaning, sense or implication (of a term etc.). **significative**, *a.* conveying a meaning or signification; serving as a sign or evidence (of). **signifier**, *n.*
signor, signior (sēn'yaw), *n.* an Italian man; the Italian form of address corresponding to sir or Mr. **signora** (-yaw'rə), *n.* a lady; Mrs, madam. **signorina** (-yərē'nə), *n.* a girl; Miss.
sika (sē'kə), *n.* a small Japanese deer, now introduced into Britain.
Sikh (sēk), *n.* one of a monotheistic religious community, founded in the 16th cent. in the Punjab. *a.* of or relating to the Sikhs or their religion. **Sikhism**, *n.*
silage (sī'lij), ENSILAGE.
sild (sild), *n.* a young herring, esp. canned in Norway.
silence (sī'ləns), *n.* the state of being silent, taciturnity, absence of noise, stillness; the fact of not mentioning a thing, secrecy; absence of mention, oblivion. *v.t.* to make silent, esp. by refuting with unanswerable arguments; to stop from sounding; to compel to cease firing. **silencer**, *n.* one who or that which silences; a device for reducing or muffling noise, fitted to firearms; a muffling device fitted to the exhaust of a motor on a vehicle etc.
Silene (sīlē'nī), *n.* a genus of caryophyllaceous plants comprising the catch-fly etc.
silent (sīlənt), *a.* not speaking, not making any sound, noiseless, still; of a letter, not pronounced; of a film, having no soundtrack; not loquacious, taciturn, making no mention, saying nothing (about). **silent majority**, *n.* the large majority of a population who are presumed to have moderate views but do not bother to express them. **silent partner**, *n.* one having no voice in the management of a business. **silently**, *adv.* **silentness**, *n.*
silhouette (siluet'), *n.* a portrait in profile or outline, usu. black on a white ground or cut out in paper etc.; the outline of a figure as seen against the light or cast as a shadow. *v.t.* to represent or cause to be visible in silhouette.
silica (sil'ikə), *n.* a hard, crystalline silicon dioxide, occurring in various mineral forms, esp. as sand, flint, quartz etc. **silica gel**, *n.* a granular form of hydrated silica, used to absorb water and other vapours. **silicate** (-kət), *n.* a salt of silicic acid. **siliceous** (-lish'əs), **silicic** (-lis'-), *a.* **silicify** (-lis'ifī), *v.t.* to convert into or impregnate with silica, to petrify. *v.i.* to become or be impregnated with silica. **silicification** (-fi-), *n.* **silicon** (-kən), *n.* a non-metallic semi-conducting element, at. no. 14; chem. symbol Si, usu. occurring in combination with oxygen as quartz or silica, and next to oxygen the most abundant of the elements. **silicon chip** CHIP[1]. **silicones** (-kōnz), *n.pl.* water-repellent oils of low melting-point, the viscosity of which changes little with temperature, used as lubricants, constituents of polish etc. **silicone rubber**, *n.* a synthetic rubber stable up to comparatively high temperatures. **silicosis** (-kō'sis), *n.* an occupational disease of the lungs occasioned by the inhalation of silica dust.
siliqua (sil'ikwə), *n.* (*pl.* **-quae** (-kwē)) a dry, elongated pericarp or pod containing the seeds, as in plants of the mustard family. **silique** (silēk'), *n.* (*Bot.*) a siliqua. **siliquose** (-kwōs), **siliquous**, *a.*
silk (silk), *n.* a fine, soft, glossy fibre spun by the larvae of certain moths, esp. the common silkworm; similar thread spun by the silk-spider and other arachnids; thread or cloth made of silk; (*pl.*) varieties of this or garments made of it; the soft fibres surrounding an ear of maize. *a.* made of silk. **to take silk**, to become a KC or QC. **silk-cotton**, *n.* the silky covering of the seed-pods of the bombax and other trees, kapok. **silk-gland**, *n.* a gland in the silkworm, certain spiders etc., secreting silk. **silk hat**, *n.* a top hat. **silk-screen**, *a.* of a stencil method of printing in which paint or ink is forced through a screen of silk or other fine-meshed fabric. **silk-screen printing**, *n.* **silkworm**, *n.* the larva of *Bombyx mori* or allied moths which enclose their chrysalis in a cocoon of silk. **silken**, *a.* **silky**, *a.* like silk, glossy, soft, silken; smooth, ingratiating, suave. **silky oak**, *n.* an Australian tree yielding wood suitable for furniture, fittings etc. **silkiness**, *n.*
sill (sil), *n.* a block or timber forming a basis or foundation in a structure, esp. a slab of timber or stone at the foot of a door or window; a sheet of intrusive igneous rock between other strata.
sillabub (sil'əbŭb), SYLLABUB.

silly (sil'i), *a.* foolish, fatuous, weak-minded; showing want of judgment, unwise, imprudent; mentally weak, imbecile; trifling, frivolous; dazed, stunned; (*Cricket*) close to the batsman's wicket. *n.* a silly person. **silly-billy,** *n.* (*coll.*) a silly person. **silly season,** *n.* the late summer, when newspapers are traditionally full of trivial stories, for want of anything serious to print. **silliness,** *n.*
silo (silō), *n.* (*pl.* **-los**) a store-pit or air-tight chamber for pressing and preserving green fodder; a tall construction in which grain etc. can be stored; an underground store and launch pad for a guided missile.
silt (silt), *n.* fine sediment deposited by water. *v.t.* to choke or fill (up) with silt. *v.i.* to be choked (up) with silt. **silty,** *a.*
Silurian (siloo'riən, si-), *a.* of or pertaining to the Silurian system. *n.* the lowest sub-division of the Palaeozoic strata, next above the Cambrian (well developed in S Wales where these strata were first examined).
Silurus (siloo'rəs), *n.* a genus of fishes typical of the Siluridae, containing the sheat-fish. **silurid** (-rid), *n., a.*
silva, sylva (sil'və), *n.* a group of trees, a forest, a wood.
silvan, sylvan, *a.* wooded; pertaining to a wood or forest; growing in woods; rural, rustic. *n.* a deity of the woods, a satyr; a rustic, a forest-dweller. **silvicultural** (-vikŭl'-), *a.* **silviculture** (sil'-), *n.* the cultivation of trees, forestry. **silviculturist** (sil'-), *n.*
silver (sil'və), *n.* a precious ductile and malleable metallic element of a white colour, at. no. 47; chem. symbol Ag; domestic utensils, ornaments etc. made of this; silver or cupronickel coin; the colour or lustre of or as of silver; a silver medal. *a.* made of silver; resembling silver, white or lustrous like silver; esp. of bells or speech, soft and clear in tone. *v.t.* to coat or plate with silver or a silver substitute; to give a silvery lustre to. **the silver lining (in every cloud),** the bright or compensating side of any misfortune, trouble etc. **the silver screen,** the cinema screen; cinematography. **Silver Age,** *n.* in Greek and Roman mythology, the age preceding the Golden Age; the period of Latin literature following the classical period. **silver birch,** *n.* a common variety of birch with a silvery-white trunk. **silver bromide,** *n.* a pale yellow salt used in the production of photographic emulsions. **silver chloride,** *n.* a white salt used in the production of photographic paper and emulsions, and antiseptics. **silver fir,** *n.* a tall species of fir with silvery bark and two white lines on the underside of the leaves. **silver-fish,** *n.* a silvery fish of various species, esp. a white variety of goldfish; any of various small wingless insects which occur in buildings and can be destructive to books, cloth etc. **silver-foil,** *n.* silver-leaf.
silver-fox, *n.* a variety of common red fox in a phase during which its coat becomes black mixed with silver; the pelt of this animal. **silver-gilt,** *n.* silver or silverware gilded. **silver grey,** *n., a.* pale luminous grey. **silver-haired,** *a.* **silver iodide,** *n.* a yellow salt used in medicine, photography, and in seeding clouds to make artificial rain. **silver-leaf,** *n.* silver beaten out into thin leaves or plates; a disease of plum trees. **silver medal,** *n.* the medal awarded to the competitor who comes second in a contest, etc. **silver plate,** *n.* silver-ware; (metal articles coated with) a thin layer of silver, electroplate. *v.t.* to coat with this, to electroplate. **silver-plated,** *a.* **silver-plating,** *n.* the process of coating metal articles with a layer of silver, esp. by electroplating. **silver service,** *n.* a set of silver dining utensils; a manner of serving food in restaurants using a fork and spoon in one hand. **silverside,** *n.* the upper and choicer part of a round of beef. **silversmith,** *n.* a maker of or worker in silver articles. **silversmithing,** *n.* **silver-tongued,** *a.* eloquent. **silverware,** *n.* articles of silver, esp. table utensils; silver plate. **silver wedding,** *n.* a 25th wedding anniversary. **silverweed,** *n.* any of various silvery-leaved plants, esp. *Potentilla anserina.* **silvering,** *n.* **silvery,** *a.* having the appearance of silver; having a soft

clear sound. **silveriness,** *n.*
sima (sī'mə), *n.* the inner part of the earth's crust.
simian, *n., a.* (of or like) a monkey or ape.
similar (sim'ilə), *a.* like, having a resemblance (to); resembling (each other); alike; (*Geom.*) made up of the same number of parts arranged in the same manner, corresponding. **similarity** (-la'-), *n.* **similarly,** *adv.*
simile (sim'ili), *n.* a comparison of two things which have some strong point of resemblance, esp. as an illustration or poetical figure. **similitude** (-mil'itūd), *n.* likeness, resemblance; counterpart.
simmer (sim'ə), *v.i.* to boil gently; to be just below boiling-point; to be on the point of bursting into laughter, anger etc. *v.t.* to boil gently; to keep just below boiling-point. *n.* a state of simmering; the point of breaking out. **to simmer down,** to become less agitated or excited, to calm down.
simnel cake (sim'nəl), *n.* a rich fruit cake, formerly eaten on Mid-Lent Sunday, Easter and Christmas Day.
simony (sim'əni, sī'-), *n.* the buying or selling of ecclesiastical preferment. **simonist** (sī'-), *n.*
simoom (simoom'), **simoon** (-moon), *n.* a hot dry wind blowing over the desert, esp. of Arabia, raising great quantities of sand and causing intense thirst.
simp (simp), *n.* (*coll.*) a simpleton.
simpatico (simpat'ikō), *a.* (*coll.*) sympathetic; congenial.
simper (sim'pə), *v.i.* to smile in an affected manner, to smirk. *v.t.* to utter with a simper. *n.* an affected smile or smirk. **simperer,** *n.* **simpering,** *a.* **simperingly,** *adv.*
simple (sim'pl), *a.* consisting of only one thing; uncompounded, unmingled, all of one kind, not analysable, not subdivided, elementary; not complicated, not complex, straightforward; not elaborate, not adorned, not sumptuous; plain, homely, humble, of low degree; insignificant, trifling, unaffected, unsophisticated, natural, artless, sincere; credulous; clear, intelligible; weak in intellect, silly, inexperienced, ignorant; absolute, mere. *n.* a medicinal herb or a medicine made from it. **simple fraction,** *n.* a fraction having whole numbers for the denominator and numerator, a common or vulgar fraction. **simple fracture** FRACTURE. **simple-hearted,** *a.* genuine, sincere; uncomplicated, guileless. **simple interest,** *n.* interest upon the principal only. **simple machine,** *n.* any of various simple mechanisms, including the pulley, wedge, lever, wheel and axle, screw, and inclined plane, which overcome resistance at one point by applying force usu. at another point. **simple-minded,** *a.* foolish, stupid; unsophisticated; mentally deficient, feeble-minded. **simple-mindedness,** *n.* **simple sentence,** *n.* one with only one main clause. **simpleton** (-tən), *n.* a silly, gullible or feeble-minded person. **simplex** (-pleks), *a.* simple, not compound; in telecommunications, allowing the transmission of a signal in only one direction at a time. *n.* the most rudimentary geometric figure of a given dimension (e.g. a line in one-dimensional space, a triangle in two-dimensional space). **simplicity** (-plis'-), *n.* **simplify,** *v.t.* to make simple; to make simpler or easier to understand, to reduce to essentials. **simplification** (-plifi-), *n.* **simplifier** (-fiə), *n.* **simplism,** *n.* affectation of simplicity; oversimplification. **simplistic** (-plis'-), *a.* oversimplified, naive, superficial, unrealistically limited, shallow etc; oversimplifying. **simplistically,** *adv.* **simply,** *adv.*
simulacrum (simūlā'krəm), *n.* (*pl.* **-cra** (-krə), **-crums**) a misleading representation, a semblance.
simulate (sim'ūlāt), *v.t.* to assume the likeness or mere appearance of; to counterfeit, to feign, to imitate, to put on, to mimic; to reproduce the structure, movement or conditions of (e.g. in an experiment, by computer etc.). **simulated,** *a.* pretended, false, feigned; of leather, fur etc., imitation; of the flight of an aircraft, spaceship etc., reproduced or represented (by a model). **simulation,** *n.* **simulative,** *a.* **simulator,** *n.*

simulcast (sim'əlkahst), *n.* (the transmission of) a simultaneous broadcast on radio and television. *v.t.* to broadcast a programme in this way.
simultaneous (siməltā'niəs), *a.* happening, done or acting at the same time. **simultaneity** (-nē'-), **simultaneousness**, *n.* **simultaneously**, *adv.*
sin¹ (sin), *n.* transgression of duty, morality, or the law of God; wickedness, moral depravity; a transgression, an offence; a breach of etiquette, social standards etc. *v.i.* (*past, p.p.* **sinned**) to commit a sin; to offend (against). **mortal sin**, deliberate sin that deprives the soul of divine grace. **original sin** ORIGINAL. **seven deadly sins** SEVEN. **to live in sin**, of a couple, to cohabit without being married. **sin bin**, *n.* (*sl.*) an area to the side of an ice-hockey pitch where players who have committed fouls are temporarily sent; (*coll.*) a special unit for unruly children. **sinful**, *a.* **sinfully**, *adv.* **sinfulness**, *n.* **sinless**, *a.* **sinlessly**, *adv.* **sinlessness**, *n.* **sinner**, *n.*
sin² (sin), (*abbr.*) sine.
Sinanthropus (sin'an'thrəpəs), *n.* an ape-like human, the remains of whom have been discovered in China, Peking man.
since (sins), *adv.* after or from a time specified or implied till now; at some time after such a time and before now; from that time before this, before now, ago. *prep.* from the time of; throughout or during the time after; after and before now. *conj.* from the time that or when, during the time, after that; inasmuch as; because.
sincere (sinsiə'), *a.* not feigned or put on, genuine, honest, undissembling, frank. **sincerely**, *adv.* **sincereness**, *n.* **sincerity** (-se'ri-), *n.*
sinciput (sin'sipŭt), *n.* the upper part of the head, especially from the forehead to the coronal suture. **sincipital** (-sip'-), *a.*
sine¹ (sin), *n.* in trigonometry, of an angle, the ratio of the length of the line opposite the angle to the length of the hypotenuse in a right-angled triangle.
sine² (sin'i, sī'-), *prep.* without, lacking. **sine die** (dī, dē'ā), without any day (being fixed). **sine qua non** (sin'i kwah non'), an indispensable condition. [L]
sinecure (sin'ikūə, sī'-), *n.* an ecclesiastical benefice without cure of souls; any paid office with few or no duties attached. **sinecurism**, *n.* **sinecurist**, *n.*
sinew (sin'ū), *n.* a tendon, a fibrous cord connecting muscle and bone; (*pl.*) muscles; that which gives strength or power. **the sinews of war**, money. **sinewless**, *a.* **sinewy**, *a.* **sinewiness**, *n.*
sinfonia (sinfō'niə), *n.* (*pl.* **-nias, -nie**) a symphony; a symphony orchestra. **sinfonietta** (-fonyet'ə), *n.* a short or light symphony; a small symphony orchestra.
sing (sing), *v.i.* (*past* **sang** (sang), **sung** (sŭng), *p.p.* **sung**) to utter words in a tuneful manner, to render a song vocally, to make vocal melody; to emit sweet or melodious sounds; of a kettle etc., to make a whistling sound; to ring, to buzz; (*sl.*) to confess, to inform, to grass; to utter loudly and clearly. *v.t.* to utter (words, a song, tune etc.) in a tuneful or melodious manner; to relate, proclaim or celebrate in verse or poetry; to celebrate; to accompany with singing; to greet, acclaim, lull, usher (in or out) etc., with singing; to chant. **to sing along**, of an audience, to accompany a performer in singing popular songs. **to sing out**, to call out loudly, to shout. **singsong**, *a.* of a voice, accent etc., having a rising and falling inflection; having a monotonous rhythm. *n.* a monotonous rising and falling (of a voice etc.); an informal session of singing usu. well-known songs. **singable**, *a.* **singer** (sing'ér), *n.* **singing**, *n.*, *a.* **singing telegram**, *n.* a (usu. congratulatory) telegram with a message which is delivered in song (often by someone in costume). **singing-voice**, *n.* the voice as used in singing. **singingly**, *adv.*
sing., (*abbr.*) singular.
singe (sinj), *v.t.* (*pres.p.* **singeing**, *past, p.p.* **singed**) to burn slightly; to burn the surface of or the tips of (hair etc.); to scorch; to burn bristles, nap etc. off (an animal carcase, fabric). *n.* a slight or superficial burn.
Singhalese SINHALESE.
single (sing'gl), *a.* consisting of one only, sole; particular, individual, separate, solitary, alone, unaided, unaccompanied; unmarried; simple, not compound, not complicated, not combined with others; involving or performed by one or by one on each side; designed for use by or with one person, thing etc.; of petals or blooms, not double, not clustered. *n.* a hit for one run in cricket etc.; a single measure, amount or thing; an unmarried person; a flower with the usual number of petals or blooms, not a double or cluster; a gramophone record with only one track recorded on each side; a rail, bus etc ticket for a journey in one direction; (*pl.*) unmarried people; (*pl.*) a game, esp. of tennis, consisting of a single player on either side. *v.t.* to pick out from among others. **single-acting**, *a.* of an engine or pump, working by means of pressure on one side of the piston or pistons only. **single-action**, *a.* of a gun that must be cocked before firing. **single bed**, *n.* a bed intended to be used by one person. **single-breasted**, *a.* of a jacket, coat etc., having only one thickness of cloth over the chest when closed, with one central set of buttons, holes etc. **single cream**, *n.* a pouring cream of a less fatty consistency than double cream. **single-decker**, *n.* (*coll.*) a bus with one level of passenger seating only. **single-end**, *n.* (*Sc.*) a one-room dwelling. **single-entry** ENTRY. **single figures**, *n.pl.* a number etc. under 10. **single file**, *n.* a line (of people, etc.) ranged one behind the other. **single-handed**, *a.* done without assistance; unassisted, alone. *adv.* without assistance. **single-handedly**, *adv.* **single-handedness**, *n.* **single-minded**, *a.* intent on one purpose only; dedicated. **single-mindedly**, *adv.* **single-mindedness**, *n.* **single parent**, *n.* one parent raising a child or children alone. **single-parent family**, *n.* a one-parent family. **singles bar, club**, *n.* a bar or club where unmarried people meet. **single-sex**, *a.* of a school or other institution, admitting members of one sex only. **single-track**, *a.* of a road or railway, having only one lane or track. **singletree**, *n.* (*N Am., Austral.*) a swingletree. **singleness**, *n.* **singlet** (-glit), *n.* a vest; a garment resembling this worn by athletes. **singleton** (-tən), *n.* a single-card of any particular suit in a player's hand at whist, bridge etc.; a mathematical set of one; a single object, person etc. as opposed to a group or pair. **singly**, *adv.*
singsong SING.
singular (sing'gūlə), *a.* standing alone, out of the usual course, remarkable, extraordinary, unique, distinguished; peculiar, odd, eccentric; of a word or inflected form of a word, denoting or referring to one person or thing, not plural; of a logical proposition, referring to a specific thing or person, not general. *n.* the singular number; a word denoting this. **singularity** (-la'-), *n.* **singularize, -ise**, *v.t.* to alter a word that looks like a plural (as *pease*) to the singular form (as *pea*); to make striking or eye-catching. **singularization**, *n.* **singularly**, *adv.*
singultus (singgŭl'təs), *n.* hiccups, hiccuping.
sinh (shin. sīnch'), *n.* a hyperbolic sine.
Sinhalese (sinəlēz), **Singhalese**, *a.* of or pertaining to Sri Lanka (formerly Ceylon), or to its majority people or their language, Sri Lankan. *n.* a native or citizen of Sri Lanka; a member of the Sinhalese people largely inhabiting Sri Lanka; the official language of Sri Lanka and the Sinhalese people.
sinister (sin'istə), *a.* (*Her.*) on the left side (of a shield etc.); ominous, threatening evil; malevolent, villainous. **sinisterly**, *adv.* **sinisterness**, *n.* **sinistral**, *a.* being on, pertaining to, directed towards, or inclined to the left; left-handed. **sinistrally**, *adv.* **sinistrorse** (-traws), *a.* (*Bot.*) twining to the left; sinistral. **sinistrorsal** (-traw-), *a.*

sink (singk), *v.i.* (*past* **sank** (sangk), †**sunk** (sŭngk), *p.p.* **sunk**, *part.a.* **sunken** (-kən)) to go downwards, to descend, to fall gradually; to disappear below the surface or the horizon; to fall or descend by force of gravity; to decline to a lower level of health, morals etc.; to deteriorate; to subside, to droop, to despond; to expire or come to an end by degrees; to become lower in intensity, pitch, value, price etc.; to become shrunken or hollow, to slope downwards, to recede; to go deep or deeper into, to penetrate, to be impressed into, to be absorbed. *v.t.* to cause to sink; to submerge (as) in a fluid, to send below the surface; to excavate, to make by excavating; to cause to disappear; to conceal, to suppress; to allow to fall or droop; to invest; to invest unprofitably, to lose, to squander; (*coll.*) to drink, to quaff; to cause to penetrate, to insert; (of a ball, to send into a pocket, hole, etc. *n.* a plastic, porcelain or metal basin, usu. fitted to a water supply and drainage system in a kitchen; a place of iniquity; a depression or area of ground in which water collects, a sink-hole, a swallow-hole; in physics, a device, body or process which absorbs or dissipates energy, as *heat sink.* **to sink in**, to become absorbed, to penetrate; to become understood. **to sink or swim**, to either succeed or fail (in a venture etc.). **sink-hole**, *n.* a hole for the discharge of foul waste; a hole or series of holes in limestone strata through which water sinks below the surface or a stream disappears underground, a swallow-hole. **sink-tidy**, *n.* a receptacle with a perforated bottom for holding washing-up utensils; a small sieve placed over a plug-hole for catching refuse. **sink unit**, *n.* a sink and draining board set in a structure with a drawer and cupboards. **sinkable**, *a.* **sinkage** (-kij), *n.* the act, operation or process of sinking; the amount of sinking; a depression. **sinker**, *n.* one who or that which sinks; a weight used to sink a fishing-line, net etc. **sinking**, *n., a.* **sinking-fund**, *n.* a fund set aside for the reduction of a debt.
sinless, sinner etc. SIN [1].
Sinn Fein (shinfān'), *n.* an Irish republican party which was formed in 1905 by the coalescence of all the Irish separatist organizations, which is the political wing of the Irish Republican Army. **Sinn Feiner**, *n.* a member of Sinn Fein.
Sino-, *comb. form.* Chinese.
sinology (sinol'əji), *n.* the study of Chinese languages, culture, literature etc. **sinological** (-loj'-), *a.* **sinologist**, **sinologue** (sin'əlog), *n.*
Sino-Tibetan (sinōtibet'ən), *n.* a family of languages comprising most Chinese languages, Tibetan, Burmese and usu. Thai. *a.* of or pertaining to this family of languages.
sinsemilla, *n.* a specially potent type of marijuana; the (variety of) cannabis plant from which it is obtained.
sinter (sin'tə), *n.* a calcareous or siliceous rock precipitated from (hot) mineral waters. *v.t.* to form (metal powder, ceramics, glass etc.) into a solid mass by pressure or heating at a temperature below melting-point. *v.i.* to be formed into such a mass.
sinuate (sin'ūət), **-ated** (-ātid), *a.* esp. of the edges of leaves etc., bending, curving or winding in and out, sinuous. **sinuately**, *adv.* **sinuosity** (-os'-), *n.* **sinuous**, *a.* bending in and out, winding; serpentine, tortuous; supple, lithe. **sinuously**, *adv.* **sinuousness**, *n.*
sinus (sī'nəs), *n.* (*pl.* **sinuses**) a cavity or pouch-like hollow, esp. in bone or tissue; the cavity in the skull which connects with the nose; a fistula; a rounded recess or curve, as in the margin of a leaf. **sinusitis** (-sī'tis), *n.* (painful) inflammation of a nasal sinus.
Sion (sī'ən), ZION.
Sioux (soo), *n.* (*pl.* **Sioux**) a member of a N American Indian people comprising the Dakotas; any of various languages spoken by this group. **Siouan**, *n.* a family of central and eastern N American languages.
sip (sip), *v.t., v.i.* (*past, p.p.* **sipped**) to drink or imbibe in small quantities using the lips. *n.* a very small draught of liquid; the act of sipping. **sippet** (-it), *n.* a small piece of toast or fried bread used as a garnish or sop.
siphon, syphon (sī'fən), *n.* a curved tube having one branch longer than the other, used for conveying liquid over the edge of a cask, tank etc., through the force of atmospheric pressure; a soda siphon; a suctorial or other tubular organ, esp. in cephalopods, gastropods etc. *v.t.* to convey or draw off by a siphon. **siphon bottle**, *n.* (*chiefly N Am.*) a soda siphon. **siphonal, siphonic** (-fon'-), *a.* **siphonophore** (-əfaw), *n.* one of the Siphonophoridae, variously regarded as a colony of medusoid zooids or as a single individual composed of a cluster of tubular organs.
sir (sœ), *n.* a term of courteous or formal address to a man; (**Sir**) a title prefixed to the names of baronets and knights.
sirdar (sœ'dah), *n.* a military leader, or commander in India; the former British commander-in-chief of the Egyptian army; a leader, chief, foreman etc.
sire (sīə), *n.* a title used in addressing a king or a sovereign prince; the male parent of an animal, esp. a stallion. *v.t.* esp. of stallions or male domestic animals, to father.
siren (sī'rən), *n.* in Greek mythology, a sea-nymph, half-woman and half-bird, one of several dwelling on a rocky isle and luring sailors to shipwreck by their singing; a charming or seductive woman, esp. a dangerous temptress; an apparatus for producing a loud warning sound by means of a rotating perforated disc through which steam or compressed air is emitted; an electrical warning device emitting a similarly piercing sound; one of a family of American eel-like amphibians, with two anterior feet and permanent branchiae. **sirenian** (-rē'-), *n.* one of an order of marine herbivorous mammals, allied to the whales, but having the fore limbs developed into paddles, comprising the manatees and dugongs. *a.* of or pertaining to this order.
Sirius (si'riəs), *n.* the dog-star, the brightest star in the sky.
sirloin (sœ'loin), *n.* the loin or upper part of the loin of beef.
sirocco (sirok'ō), *n.* (*pl.* **-ccos**) a hot oppressive wind blowing from N Africa across to the northern Mediterranean coast; applied generally to a sultry southerly wind.
sirrah (si'rə), *n.* fellow, sir (an obsolete term of address used in anger or contempt).
sirree (sirē'), *n.* (*N Am. coll.*) sir, sirrah (used for emphasis often with *yes* or *no*).
sirup (si'rəp), SYRUP.
sis, siss (sis), *n.* (*coll.*) short for SISTER. **sissy, cissy** (sis'i), *n.* an effeminate, feeble or cowardly fellow. *a.* effeminate, feeble, cowardly.
-sis (-sis), *suf.* (*pl.* **-ses** (-sēz)) denoting a process or action of, or condition caused by.
sisal (sis'əl), **sisal-grass, -hemp**, *n.* the fibre of the American aloe used for cordage etc.
siskin (sis'kin), *n.* a small migratory song-bird, allied to the goldfinch, the aberdevine.
sister (sis'tə), *n.* a female born of the same parents as another; applied to a half-sister or a foster-sister; a senior nurse, one in charge of a hospital ward; a female fellow-member of the same group, society, religious community etc.; a nun in the Roman Catholic Church; (*chiefly N Am. coll.*) a woman (usu. as a form of address). *a.* closely related, similar, of the same design, type, origins, as *sister ships*. **half-sister** HALF. **sisterhood** (-hud), *n.* the state of being a sister, the relation of sisters; a community of women bound together by religious vows, common interests etc. **sister-in-law**, *n.* (*pl.* **sisters-in-law**), a husband's or wife's sister; a brother's wife. **sisterliness**, *n.* **sisterly**, *a.*
sistrum (sis'trəm), *n.* (*pl.* **-trums, -tra** (-trə)) a jingling instrument used by the ancient Egyptians in the worship of Isis.

Sisyphean (sisifē'ən), *a.* unceasingly or fruitlessly laborious.

sit (sit), *v.i.* (*pres.p.* **sitting**, *past, p.p.* **sat** (sat)) to set oneself or be in a resting posture with the body nearly vertical supported on the buttocks; of birds and various animals, to be in a resting posture; to perch, to roost; to cover eggs in order to hatch, to brood; to be in a specified position, quarter etc.; to be situated; of clothes etc., to suit, to fit, to hang; to rest or weigh (on); to take a position, to pose (for one's portrait etc.); to meet, to hold a session; to hold or occupy a seat (on) a deliberative body or in a specified capacity (as in judgment); to take up a position, to encamp (before) so as to besiege; to rest, weigh; to remain unused or inactive. *v.t.* to cause to sit, to set; to place (oneself) in a seat; to hold or keep a sitting position on (a horse etc.); to be a candidate for an examination; to baby-sit. *n.* an act or time of sitting; a sit-down. **to sit back**, to withdraw from active participation. **to sit by**, to observe without taking an active part. **to sit down**, to place oneself on a seat after standing. **to sit for**, to take an examination; to pose for a portrait. **to sit in**, to observe, be present at, or participate in a discussion, meeting, lecture etc., as a visitor; to take part in a sit-in. **to sit on**, to be a member of a committee, discussion group etc.; (*coll.*) to repress severely, to snub; (*coll.*) to suppress. **to sit out**, to sit out of doors; to sit apart from (a dance, meeting etc.); to remain till the end of (a concert etc.). **to sit tight**, to hold firm and do nothing. **to sit up**, to rise from a recumbent position; to sit with the body erect; suddenly to pay attention, take notice, become alert; not to go to bed. **sit-down**, *n.* (*coll.*) a rest, a break; a type of passive resistance offered by participants in a demonstration; a sit-down strike. *a.* of a meal, eaten while seated at a table. **sit-down strike**, *n.* a strike in which employees occupy their place of work. **sit-in**, *n.* the occupation of premises (e.g. at a university) as a form of protest. **sit-up**, *n.* a physical exercise in which the upper torso is raised from a reclining into a sitting position using the abdominal muscles. **sitter**, *n.* one who sits, esp. for a portrait; a hen that sits on a clutch of eggs to incubate them; a baby-sitter. **sitting**, *n.* a period of continuous sitting (as for a meal, a portrait); a session, a meeting for business; brooding; a clutch of eggs for hatching. *a.* seated; of a hen, brooding; holding office; occupying or in possession of; in session. **sitting duck** DUCK[2]. **to be sitting pretty**, to be in an advantageous position. **sitting-room**, *n.* a room for sitting in, a lounge; room or space for persons sitting. **sitting target**, *n.* a sitting duck. **sitting tenant**, *n.* one occupying a flat, house etc.

sitar (sitah', sit'-), *n.* an Indian stringed musical instrument with a long neck.

sitcom (sit'kom), *n.* (*coll.*) a situation comedy.

site (sīt), *n.* local position, situation; ground on which anything, esp. a building, stands, has stood, or will stand. *v.t.* to position, locate.

sitter, sitting SIT.

situate (sit'ūāt), *v.t.* to place; to locate. *a.* situated. **situated**, *a.* placed or being in a specified situation, condition or relation. **situation**, *n.* the place in which something is situated, position, locality; position of affairs or circumstances, esp. a critical juncture in a story or play; a paid office, post or place. **situation comedy**, *n.* a serialized comedy on radio or esp. television involving the same set of characters in a different comic situation in each episode. **situational**, *a.*

sitz-bath (sits'bahth), *n.* a bath in which one sits, a hip-bath.

Siva (sē'və), **Shiva** (shē-), *n.* the god which is associated with Brahma and Vishnu in the Hindu triad, known as the destroyer and reproducer of life. **Sivaism**, *n.*

six (siks), *n.* the number or figure 6 or vi; the age of six; the sixth in a series (e.g. a playing card); a division of a Cub Scout or Brownie pack; that which represents, amounts to or is worth six, esp. a hit for six runs in cricket; a set of six; the sixth hour after midday or midnight. *a.* six in number; aged six. **at sixes and sevens**, in disorder or confusion. **six (of one) and half a dozen (of the other)**, having alternatives of equal acceptability, merit etc. **six of the best**, a severe beating, esp. with a cane. **the Six Counties**, the former counties of Northern Ireland. **to knock for six**, to overcome completely, to defeat; to astonish; to stagger. **six-footer**, *n.* (*coll.*) a person 6 ft. tall. **six-gun**, *n.* a six-shooter. **six-pack**, *n.* a pack of six cans or bottles, esp. of beer. **sixpence**, *n.* (*formerly*) a cupronickel coin equivalent to six old pennies. **six-shooter**, *n.* (*coll.*) a six-chambered revolver. **sixer**, *n.* the leader of a Cub Scout or Brownie six. **sixfold**, *a., adv.* **sixmo** (-mō), *n.* sexto. **sixth**, *n.* one of six equal parts; a sixth form; a musical interval between a tone and the sixth (inclusively) above or below it on the diatonic scale; a note separated from another by this interval; a tone and its sixth sounded together. *n., a.* (the) last of six (people, things etc.); the next after the fifth. **sixth form**, *n.* the highest form in a secondary school. **sixth-form college**, *n.* one where subjects are taught at sixth-form level. **sixth-former**, *n.* sixth form. **sixth sense**, *n.* the power of intuition. **sixthly**, *adv.*

sixte (sikst), *n.* a parry in fencing in which the hand is opposite the right breast and the point of the sword raised and a little to the right.

sixteen (sikstēn'), *n.* the number or figure 16 or xvi; the age of 16. *a.* 16 in number; aged 16. **sixteenmo** (-mō), *n.* sextodecimo. **sixteenth**, *n.* one of 16 equal parts. *n., a.* (the) last of 16 (people, things etc.); the next after the 15th. **sixteenth note**, *n.* (*N Am.*) a semiquaver.

sixty (siks'ti), *n.* the number or figure 60 or lx; the age of 60; 60°F. *a.* 60 in number; aged 60. **sixties**, *n.pl.* the period of time between one's 60th and 70th birthdays; the range of temperature (fahrenheit) between 60 and 70 degrees; the period of time between the 60th and 70th years of a century. **sixtieth**, *n.* one of 60 equal parts. *n., a.* (the) last of 60 (people, things etc.); the next after the 59th.

size[1] (sīz), *n.* measurement, extent, dimensions, magnitude; one of a series of standard grades or classes with which garments and other things are divided according to their relative dimensions; (*coll.*) quality, character, condition (of any one or anything). *v.t.* to sort or arrange according to size; to cut or shape to a required size. **to size up**, to form a rough estimate of the size of; to judge the capacity (of a person). **sizable, sizeable**, *a.* of considerable size. **sizably**, *adv.* **sized**, *a.* having a particular size (*usu. in comb.*, as *small-sized*). **sizer**, *n.*

size[2] (sīz), **sizing**, *n.* a gelatinous solution used to glaze surfaces (e.g. of paper), stiffen fabrics etc. *v.t.* to coat, glaze or prepare with size.

sizer SIZE[1].

sizzle (siz'l), *v.i.* to make a hissing noise as of frying; (*coll.*) to be extremely hot; (*coll.*) to be in a rage. *n.* a hissing noise. **sizzler**, *n.* (*coll.*) a hot day; (*coll.*) anything which is striking or racy (e.g. a dress, a novel). **sizzling**, *a.*

SJ, (*abbr.*) Society of Jesus.

SJA, (*abbr.*) Saint John's Ambulance (Association or Brigade).

sjambok (sham'bok), *n.* a short heavy whip, usu. of rhinoceros hide.

ska (skah), *n.* an early form of reggae music originating in Jamaica.

skald (skawld), SCALD[2].

skat (skat), *n.* a three-handed card-game reminiscent of piquet.

skate[1] (skāt), *n.* a fish of the genus *Raia*, distinguished by having a long pointed snout.

skate[2] (skāt), *n.* (a boot or shoe equipped with) a metal

device fitted with a steel blade for gliding on ice, an ice-skate; a similar device with four wheels for gliding on a smooth surface, a roller-skate; the blade or runner on a skate; a period of skating. *v.i.* to move over ice or a smooth surface on skates. **to get one's skates on**, to hurry up. **to skate around**, to avoid talking about or confronting an issue, subject etc. **to skate on thin ice**, to be in a precarious or dangerous situation. **to skate over**, to avoid talking about, dealing with, or confronting an issue, subject etc., to gloss over. **skateboard**, *n.* a board mounted on roller-skate wheels on which both feet can be placed when momentum is achieved. *v.i.* to ride on a skateboard. **skateboarder**, *n.* **skateboarding**, *n.* **skater**, *n.* **skating**, *n.*, *a.* **skating rink**, *n.* a place with an artificial floor or sheet of ice for skating.
skean-dhu (skē'ɔndoo), *n.* a knife or dagger worn with Scottish Highland dress (in the stocking).
skedaddle (skidad'l), *v.i.* to run away, as in haste or panic. *int.* go away, beat it! *n.* a hasty flight, retreat or dispersal.
skeet (skēt), *n.* a type of clay-pigeon shooting in which targets are hurled in different directions, angles etc. from two traps.
skein (skān), *n.* a quantity of yarn, silk, wool, cotton etc., wound in a coil which is folded over and knotted; something resembling this; a flock of wild geese, swans etc. in flight.
skeleton (skel'itən), *n.* the bones of a person or animal dried, preserved and fastened together in the posture of the living creature; (*coll.*) a very lean person or emaciated animal; the hard supporting or protective framework of an animal or vegetable body, comprising bones, cartilage, shell and other rigid parts; the supporting framework of any structure; an outline or rough draft containing the essential portions (of a plot, etc.). *a.* reduced to the essential parts or a minimum. **skeleton in the cupboard**, an unpleasant or shameful secret from the past. **skeleton-key**, *n.* a key with most of the inner bits removed, used for picking locks, a pass key. **skeleton staff**, *n.* a staff reduced to the minimum number able to run a factory, office etc. **skeletal**, *a.* pertaining to the skeleton; thin, emaciated. **skeletonize, -ise**, *v.t.* to reduce to or as to a skeleton framework or outline.
skeptic (skep'tik), (*N Am.*) SCEPTIC.
skerrick (ske'rik), *n.* (*chiefly Austral., N. Zealand*) a tiny amount. **not a skerrick left**, nothing left at all.
skerry (ske'ri), *n.* (*chiefly Sc.*) a rocky islet in the sea; a reef.
sketch (skech), *n.* a rough, hasty, unfinished or tentative delineation; a preliminary study, a rough draft, an outline, a short account without details; a play, comic routine, descriptive essay, musical composition etc., of a brief, unelaborated or slight character. *v.t.* to make a sketch of; to present in rough draft or outline without details. *v.i.* to make a sketch or sketches. **sketch-book**, *n.* a book of drawing paper for sketching in; a collection of descriptive essays etc. **sketcher**, *n.* **sketchy**, *a.* **sketchily**, *adv.* **sketchiness**, *n.*
skew (skū), *v.i.* to move sideways, to turn aside, to swerve; to squint, to look askance. *v.t.* to cause to skew; to distort. *a.* oblique, twisted, turned askew; distorted, unsymmetrical. *n.* an oblique course, position or movement. **skewback**, *n.* a stone, plate or course of masonry at the top of an abutment taking the spring of an arch. **skew-backed**, *a.* **skew-whiff**, *a.* (*coll.*) askew, to one side. **skewed**, *a.* **skewness**, *n.*
skewbald (skū'bawld), *a.* piebald with spots of white and a colour other than black. *n.* an animal of this colour.
skewer (skū'ə), *n.* a long pin of wood or metal for holding meat together; a similar implement used for various other purposes. *v.t.* to fasten with a skewer; to pierce with or as with a skewer.

ski (skē), *n.* (*pl.* **ski, skis**) a long narrow runner of waxed wood, plastic etc. fastened one to each foot and used for sliding over snow or water-skiing. *v.i.* (*pres.p.* **skiing**, *past, p.p.* **skied, ski'd** (skēd)) to move on skis. **skibob** (-bob), *n.* a snow vehicle with a low seat and steering device supported on two skis. **skibobber**, *n.* **skibobbing**, *n.* **skijorer** (-jawrə), *n.* **skijoring**, *n.* a sport in which a skier is towed by a horse or vehicle. **ski jump**, *n.* a ski-slope or run surmounted by a ramp from which skiers jump. *v.i.* to execute a jump from this. **ski jumper**, *n.* **ski-jumping**, *n.* **ski lift**, *n.* any of various forms of lifting apparatus for transporting skiers up a slope (e.g. a chair lift). **ski pants**, *n.pl.* stretch trousers with stirrups which fit under the feet. **ski-run**, *n.* a slope for skiing on. **ski stick**, *n.* one of a pair of pointed sticks used in skiing to balance or propel. **skier**, *n.*
skid (skid), *n.* a support, prop, or other device used for sliding heavy things on; a shoe or other device acting as a brake; the act of skidding, a slip on muddy ground, an icy road etc. *v.t.* (*past, p.p.* **skidded**) to cause to skid; to check or brake with a skid. *v.i.* of wheels or vehicles, to slip sideways; to revolve rapidly without progressing. **the skids**, a downward path into ruin. **skid row**, *n.* (*chiefly N Am., sl.*) an area of a city inhabited by vagrants, down-and-outs, etc.
skiff (skif), *n.* a small light boat.
skiffle (skif'l), *n.* a type of music popular in the 1950s played on unconventional percussion instruments and guitars. **skiffle band, group**, *n.* a band composed of players who perform this music.
skill (skil), *n.* familiar knowledge of any art or science combined with dexterity; expertness, ability, practical mastery of a craft, trade, sport etc., often attained by training. **skilful**, *a.* having, showing or requiring skill; expert, adept, clever, adroit, dexterous. **skilfully**, *adv.* **skilled**, *a.* having skill, skilful; involving or requiring skill or specialized training.
skillet (skil'it), *n.* a metal pan with a long handle used for frying, as a saucepan etc.
skilly (skil'i), *n.* thin broth, soup or gruel.
skim (skim), *v.t.* (*past, p.p.* **skimmed**) to clear the scum etc. from the surface of; to take (cream etc.) from the surface of a liquid; to touch lightly or nearly touch the surface of, to graze; to throw so as to cause to graze or pass lightly over a surface; to glance over or read superficially. *v.i.* to pass lightly and rapidly over or along a surface; to glance (over) rapidly and superficially. *n.* the act or process of skimming; the thick matter which forms on, or is removed from, the surface of a liquid; a thin layer or coating. **skimmed milk**, *n.* milk from which the cream has been skimmed. **skimmer**, *n.* one who or that which skims; a perforated ladle for skimming; a sea-bird which skims small fishes from the water with its lower mandible.
Skimmia (skim'iə), *n.* an Asiatic genus of evergreen shrub with red berries.
skimp (skimp), *v.t.* to supply in a niggardly manner, to stint (a person, provisions etc.); to perform with insufficient attention or inadequate effort. *v.i.* to be stingy or parsimonious. **skimpy**, *a.* **skimpily**, *adv.* **skimpiness**, *n.*
skin (skin), *n.* the natural membranous outer covering of an animal body; the hide or integument of an animal removed from the body, with or without the hair; a vessel made of the skin of an animal for holding liquids (as wine); the outer layer or covering of a plant, fruit etc.; a film (e.g. a liquid); a membrane; the outer cladding of a vessel, rocket etc.; (*coll.*) a skinhead. *v.t.* (*past, p.p.* **skinned**) to strip the skin from, to flay, to peel; to graze (as the knee); to cover (over) with or as with skin; (*sl.*) to cheat, to swindle. *v.i.* to become covered (over) with skin, to cicatrize; made of, intended for, or used on the skin. **by the skin of one's teeth**, very narrowly, by a close shave. **no skin off one's nose**, making no difference to

one, not perturbing to one. **skin and bone**, extremely thin, emaciated. **to get under one's skin**, to interest or annoy one intensely. **to save one's skin**, to escape injury. **skin-deep**, *a.* superficial, not deep. **skin-diver**, *n.* one who dives deep wearing no protective clothing. **skin-diving**, *n.* **skin flick**, *n.* (*coll.*) a film which features nudity and sex scenes. **skinflint**, *n.* a niggardly person, a miser. **skin graft**, *n.* the transfer of skin from a sound to a disfigured or injured part. **skinhead**, *n.* a member of a gang of aggressive and often racist youths characterized by their cropped hair, heavy-duty boots and braces; (*coll.*) someone with cropped hair. **skin test**, *n.* a test performed on the skin to determine its resistance to disease or substances liable to cause an allergic reaction. **skintight**, *a.* of garments, tight, clinging. **skinful**, *n.* (*sl.*) a large quantity of liquor, enough to make one drunk. **skinless**, *a.* **skinned**, *a.* (*usu. in comb.*, as *thin-skinned*). **skinny**, *a.* of or resembling a skin; very lean or thin. **skinny-dip**, *v.i.* (*coll.*) to swim in the nude.
skink (skingk), *n.* a small lizard of Africa and SW Asia.
skint (skint), *a.* (*sl.*) hard-up for money, penniless.
skip¹ (skip), *v.i.* (*past, p.p.* **skipped**) to move about with light bounds, hops or capers, esp. by shifting rapidly from one foot to another; to jump repeatedly over a skipping-rope; to pass rapidly from one thing to another; to pass over quickly or make omissions; (*sl.*) to make off hurriedly, to abscond. *v.t.* to cause to skim over a surface, e.g. of a stone, to skim; to omit, to miss deliberately, to absent oneself from (a meal, a class, a church service etc.); (*chiefly N Am.*) to leave (town) quickly and quietly, to abscond. *n.* a light leap or spring, esp. from one foot to the other; an act of omitting, leaving out or passing over. **skip it!** forget about it! it does not matter! **skip distance**, *n.* the minimum distance around a radio transmitter at which it is possible to receive an ionospheric wave. **skipjack**, *n.* any of various kinds of fish, beetles etc., that move with skips or springs, esp. the clickbeetle and two varieties of tuna, the skipjack tuna and black skipjack. **skip zone**, *n.* an area around a broadcasting station where it is impossible to receive a transmission. **skipper**¹, *n.* one who, or that which, skips; a saury; one of the lepidopteran family Hesperidae, from their short, jerky flight. **skipping**, *n.* the act, recreation or exercise of jumping over a rope repeatedly. **skipping-rope**, *n.* a rope or cord used for skipping (over) as a game or form of physical exercise.
skip² (skip), *n.* a container for collecting and moving refuse, building materials etc.; a cage or bucket lift in a mine.
skip³ (skip), *n.* (*coll.*) short for SKIPPER².
skipper¹ SKIP¹.
skipper² (skip'ə), *n.* a sea captain, the master of a vessel; the captain of a team or side; an aircraft captain. *v.t.* to act as skipper (of).
skipping-rope SKIP¹.
skirl (skœl), *v.i.* (*Sc.*) to make a shrill noise like that of the bagpipes. *n.* the sound of the bagpipes.
skirmish (skœ'mish), *n.* a slight or irregular fight, esp. between small parties or scattered troops; a contest, clash, struggle, esp. of a preliminary, brief or minor nature. *v.i.* to engage in a skirmish. **skirmisher**, *n.*
skirt (skœt), *n.* the part of a coat or other garment hanging below the waist; a woman's outer garment hanging from the waist; (*sl., offensive*) a woman, a girl; the edge of anything, a border, a margin; the outer flap surrounding the base of a hovercraft; (*pl.*) the extremities or outer parts. *v.t.* to lie or go along or by the edge of; to pass round (the edge of), to avoid; to border; to edge or border (with). *v.i.* to lie or move (along, round, on) the border or outskirts. **skirted**, *a.* **skirting**, *n.* material suitable for skirts; a skirting-board; (*pl.*) (*Austral., N. Zealand*) the inferior parts of wool trimmed from a fleece.
skirting-board, *n.* a board running round the bottom of the wall of a room.
skit (skit), *n.* a satirical piece, lampoon or humorous theatrical sketch. **skite** (skīt), *v.i.* (*Sc.*) to dart aside, to slip, to slide; (*Austral., N. Zealand coll.*) to boast. *v.t.* to hit with a darting blow. *n.* an act or instance of darting aside or slipping; a sharp blow, esp. in a slanting direction; (*Austral., N. Zealand coll.*) one who boasts; boastful talk.
skitter (skit'ə), *v.i.* to glide, skim over or skip rapidly, esp. along a surface; to fish by drawing a bait etc. along the surface; (*Sc.*) to have diarrhoea. *v.t.* to cause to skitter.
skittish (skit'ish), *a.* of horses, excitable, nervous, easily frightened; capricious, uncertain, coquettish, too lively. **skittishly**, *adv.* **skittishness**, *n.*
skittle (skit'l), *n.* one of the blocks or pins set up to be thrown at in ninepins; (*pl.*) ninepin bowling.
skive¹ (skīv), *v.t.* to split (leather) into thin layers; to shave or pare (hides). **skiver**, *n.* a thin leather split from a sheep-skin.
skive² (skīv), *v.t., v.i.* (*coll.*) to avoid performing (a duty, task etc.). *n.* (*coll.*) a period of shirking or an evasion of duty etc.; (*coll.*) work etc. which is far from onerous. **skiver**, *n.*
skivvy (skiv'i), *n.* (*sl.*) a maid or general servant. *v.i.* to work as a skivvy; to drudge.
skoal, skol (skōl), *int.* cheers! good health! (usu. as a toast).
Skt, Skr, (*abbr.*) Sanskrit.
skua (skū'ə), *n.* a dark-coloured predatory sea-bird allied to the gulls.
skulduggery, skullduggery (skŭldŭg'əri), *n.* (*coll.*) underhand trickery.
skulk (skŭlk), *v.i.* to lurk, to withdraw and conceal oneself; to lie concealed, to move about furtively; to sneak away, esp. from duty, work, danger etc. *n.* one who skulks, a skulker. **skulker**, *n.*
skull (skŭl), *n.* the bony case enclosing the brain, the skeleton of the head, the cranium; (*usu. derog.*) the brain, the intelligence. **skullcap**, *n.* a light, brimless cap fitting closely to the head; the sinciput; a plant with blue, helmet-shaped flowers. **skull and crossbones**, a representation of a human skull surmounting two crossed thigh-bones, used as an emblem of death or danger. **skulled**, *a.* (*usu. in comb.*, as *thick-skulled*).
skunk (skŭngk), *n.* a N American carnivorous quadruped, with bushy tail and white stripes down the back, which when irritated ejects a fetid secretion from the anal glands; (*coll.*) a base or obnoxious person.
sky (skī), *n.* the apparent vault of heaven, the firmament; the upper region of the atmosphere, the region of clouds; climate, the weather; (*pl.*) the heavens. *v.t.* to hit (a ball) high into the air. **the sky is the limit**, the possibilities for achievement are unbounded. **to the skies**, lavishly, extravagantly. **sky-blue**, *n., a.* (a) pale blue. **skydive**, *v.i.* to jump from an aircraft and delay opening the parachute, esp. in order to execute manoeuvres. *n.* an instance of this. **skydiver**, *n.* **skydiving**, *n.* **sky-high**, *adv.* high as the sky. *a.* very high. **to blow skyhigh**, to blow up, to destroy completely. **skyjack** (-jak), *v.t.* to hijack (an aircraft). **skylark**, *n.* a lark, that flies singing high into the air. *v.i.* (*coll.*) to lark, to frolic, to play practical jokes etc. **skylight**, *n.* a window set in a roof or ceiling. **skyline**, *n.* outline against the sky of the configuration of the land or buildings; the horizon. **skypilot**, *n.* (*sl.*) a clergyman, a chaplain. **skyrocket**, *n.* a rocket. *v.i.* to rise rapidly to a high level. **skysail**, *n.* a light sail set above the royal in a square-rigged ship. **skyscraper**, *n.* a very high multi-storeyed building. **sky-writing**, *n.* (the formation of) writing, traced in the sky by smoke discharged from an aeroplane. **skywriter**, *n.* **skyward**, *a., adv.* **skywards**, *adv.*
Skye terrier (skī), *n.* a small rough-haired variety of Scotch terrier with long body and short legs.

skylight etc. SKY.
skyr (skiə), *n.* curd.
slab (slab), *n.* a thick, flat, regularly shaped piece of anything, esp. of stone, concrete etc.; a large slice of bread, cake etc.; the outside piece sawn from a log in squaring the side; (*Austral., N. Zealand*) a rough plank; (*coll.*) a mortuary table. *a.* (*Austral., N. Zealand*) made of rough planks. *v.t.* (*past, p.p.* **slabbed**) to saw slabs from (a log etc.); to cut or form into a slab or slabs; to cover or line with slabs.
slack (slak), *a.* not drawn tight, loose; limp, relaxed; careless, remiss, not zealous, eager or active; lax; sluggish, dull, slow. *adv.* in a slack manner; insufficiently. *n.* the part of a rope etc. that hangs loose; a slack period in trade etc.; a cessation of flow (of water); small coal, screenings; (*pl.*) casual trousers. *v.i.* to become loose or looser; to become remiss or lazy. *v.t.* to slow; to lessen; to cause to abate; to loosen, to relax; to slake (lime). **to slack off**, to loosen, to reduce the tension on (a rope etc.); to shirk work. **slackwater,** *n.* the interval between the flux and the reflux of the tide. **slacken,** *v.i., v.t.* to (become) slack. **slacker,** *n.* a shirker, a lazy or remiss person. **slackly,** *adv.* **slackness,** *n.*
slag (slag), *n.* the fused refuse or dross separated in the reduction of ores, cinder; volcanic scoria; (*sl.*) a slovenly or immoral woman; a mixture of mineral dross and dust produced in coal mining. *v.i.* to form slag. *v.t.* to convert into slag; (*sl.*) to criticize, to disparage. **to slag off,** to make disparaging remarks about. **slagheap,** *n.* a hill or heap of waste material produced in coal mining. **slagger,** *n.* (*coll.*) one who disparages another. **slaggy,** *a.*
slain (slān), *p.p.* SLAY¹.
slake (slāk), *v.t.* to quench, to assuage, to satisfy; to mix (lime) with water so as to form a chemical combination. *v.i.* of lime, to become slaked. **slaked lime,** *n.* calcium hydroxide. **slakeable, slakable,** *a.*
slalom (slah'ləm), *n.* a ski or canoe race on a zig-zagged course marked with artificial obstacles.
slam (slam), *v.t.* (*past, p.p.* **slammed**) to shut suddenly with a loud noise; to put (a thing down) thus; (*coll.*) to hit, to thrash, to defeat completely; in whist etc., to beat by winning every (a grand slam) or all but one (a little or small slam) trick; (*coll.*) to criticize severely; to put into action suddenly or violently. *v.i.* of a door, to shut violently or noisily; (*coll.*) to move, esp. to enter or leave, angrily or violently. *n.* a noise or act of slamming; in whist or bridge, a grand slam, a small or little slam. **slammer,** *n.* (*sl.*) prison.
slander (slahn'də), *n.* a false report maliciously uttered to injure a person; defamation, calumny; (*Law*) false defamatory language or statements. *v.t.* to injure by the malicious utterance of a false report, to defame falsely. **slanderer,** *n.* **slanderous,** *a.*
slang (slang), *n.* words or language used colloquially but not regarded as correct English; the special language or dialect of a particular class, group etc., jargon. *v.i.* to abuse with slang. **slanging,** *n., a.* **slanging match,** *n.* a quarrel in which strong insults are exchanged. **slangy,** *a.* **slangily,** *adv.* **slanginess,** *n.*
slant (slahnt), *v.i.* to slope; to incline from or be oblique to a vertical or horizontal line; to be biased (towards). *v.t.* to cause to slant; to present with a bias. *a.* sloping, oblique. *n.* a slope; inclination from the vertical or horizontal; a solidus; an angle of approach, information concerning; a bias. **slanted,** *a.* **slanting,** *a.* **slantways, -wise,** *adv.*
slap (slap), *v.t.* (*past, p.p.* **slapped**) to strike with the open hand, to smack; to bring down or throw forcefully or quickly. *n.* a blow, esp. with the open hand. *adv.* as with a sudden blow, plump, bang. **slap and tickle,** sex play. **slap in the face,** a rebuff. **slap on the wrist,** a reprimand. **to slap down,** to quash. **to slap on,** to apply (paint etc.) hurriedly, carelessly, forcefully or thickly. **to slap on the back,** to congratulate. **slap-bang,** *adv.* suddenly, violently, headlong; exactly, precisely. **slap-dash,** *adv.* in a careless, rash, impetuous manner. *a.* hasty, impetuous, careless, happy-go-lucky. *n.* rough and haphazard work; roughcast. *v.t.* to roughcast; to do hastily or carelessly. **slaphappy,** *a.* careless, irresponsible; punch-drunk; happy-go-lucky, carefree. **slapstick,** *n.* broad comedy or knockabout farce. *a.* knockabout. **slap-up,** *a.* (*coll.*) first-rate; lavish.
slash (slash), *v.t.* to cut by striking violently; to reduce drastically; to make long incisions or narrow gashes in, to slit; (*usu. in p.p.*) to make slits in (sleeves etc.) to show the lining; to criticize severely; to lash (with a whip etc.). *v.i.* to strike (at etc.) violently and at random with a knife, sword etc.; to lash. *n.* a long cut, slit or incision; a slashing cut or stroke; a slit in a garment designed to reveal the lining as a decorative feature; a solidus; (*sl.*) the act of urinating. **slashed,** *a.* **slasher,** *n.*
slat (slat), *n.* a thin narrow strip, usu. of wood or metal. **slatted** equipped with slats. **slatted,** *a.*
slate¹ (slāt), *n.* a fine-grained laminated rock easily splitting into thin, smooth, even slabs; a slab or trimmed piece of this, esp. for use as a roofing-tile; (*chiefly N Am.*) a preliminary list of candidates; the colour of slate, a dull blue-grey. *v.t.* to cover or roof with slates; (*chiefly N Am.*) to place (a candidate) on a list; (*in passive*) to propose; (*in passive*) to plan. *a.* made or consisting of slate; slate-coloured. **a clean slate,** an unblemished record. **on the slate,** on credit, on the tab. **to wipe the slate clean,** to start afresh, to erase past crimes, errors etc. **slate-colour,** *n.* **slate-coloured,** *a.* **slater,** *n.* one who manufactures slates; one who slates roofs; a wood-louse. **slaty,** *a.* **slatiness,** *n.* **slating,** *n.*
slate² (slāt), *v.t.* to criticize savagely, to abuse, to berate. **slating**², *n.* severe reprimand or critical onslaught.
slattern (slat'ən), *n.* an untidy or sluttish woman. **slatternly,** *a.* **slatternliness,** *n.*
slaty SLATE¹.
slaughter (slaw'tə), *n.* wholesale or indiscriminate killing, carnage; the killing of animals for market. *v.t.* to kill wantonly or ruthlessly, to massacre; to kill animals for market; (*coll.*) to defeat resoundingly. **slaughterhouse,** *n.* a place where beasts are slaughtered. **slaughterman,** *n.* one who kills livestock for market. **slaughterer,** *n.* **slaughterous,** *a.* **slaughterously,** *adv.*
Slav (slahv), *n.* one of any of various peoples inhabiting eastern Europe who speak a Slavonic language. *a.* Slavonic. **Slavic,** *n., a.* Slavonic. **Slavism,** *n.* **Slavonian** (sləvō'ni-), *n., a.* Slovene. **Slavonic** (-von'-), *a.* pertaining to a group of languages belonging to the Indo-European family including Russian, Bulgarian, Polish etc.; pertaining to the peoples who speak these. *n.* the Slavonic language(s). **Slavophil(e)** (-əfīl), *n.*
slave (slāv), *n.* one who is the property of another; one who is entirely under the influence (of) or a helpless victim (to); a drudge; a mean, abject person; one device which is controlled by another, or imitates the action of a similar device. *v.i.* **slave-driver,** *n.* an overseer of slaves; an exacting taskmaster. **slave labour,** *n.* **slave-ship,** *n.* a vessel engaged in the slave-trade. **slave state,** *n.* one of the southern States of N America in which slavery flourished prior to the Civil War. **slave-trade,** *n.* the trade of procuring, buying and transporting slaves, esp. from Africa to America in the 16th-18th cents. **slave-trader,** *n.* **slaver**¹, *n.* one who deals in slaves; a slave-ship. **slavery**¹ (-vəri), *n.* **slavey,** *n.* (*coll.*) a maid-servant, a household drudge. **slavish,** *a.* characteristic of a slave; subservient, servile, abject, ignoble; entirely imitative, devoid of originality; consisting in drudgery. **slavishly,** *adv.* **slavishness,** *n.*
slaver² (slav'ə), *v.i.* to let saliva flow from the mouth, to dribble; to fawn, to drool. *v.t.* to let saliva dribble upon

or over. n. saliva dribbling from the mouth. **slaverer**, n. **slavering**, a.
slavery SLAVE.
slavey SLAVE.
Slavic etc. SLAV.
Slavonian, Slavonic etc. SLAV.
slay (slā), v.t. (past **slayed**, **slew** (sloo), p.p. **slayed**, **slain** (slān)) to put to death, to kill; (coll.) to impress powerfully; (coll.) to amuse to an overwhelming degree. **slayer**, n.
sleaze (slēz), n. (coll.) that which is squalid, distasteful, disreputable; sleaziness. **sleazy**, a. slatternly; squalid. **sleazily**, adv. **sleaziness**, n.
sled (sled), **sledge**[1] (slej), n. a vehicle on runners used on snow or ice; a sleigh; a toboggan. v.t. to convey on a sled. **sledding, sledging**, n. **sledger**, n.
sledge, sledgehammer[2] (slej), n. a heavy hammer wielded by both hands. a. hardhitting.
sleek (slēk), a. of fur, skin etc., smooth, glossy; oily, unctuous, smooth-spoken; well groomed, prosperous-looking. **sleekly**, adv. **sleekness**, n. **sleeky**, a.
sleep (slēp), v.i. (past, p.p. **slept** (slept)) to take rest in sleep, to be asleep; to be or lie dormant, inactive or in abeyance. v.t. to rest in (sleep); to lodge (a certain number). n. a state of rest in which consciousness is almost entirely suspended; a spell of this; torpor, rest, quiet, death. **to sleep around**, (coll.) to be sexually promiscuous. **to sleep in**, to sleep on the premises; to oversleep. **to sleep off**, to rid or recover from (e.g. the effects of alcohol) by sleeping. **to sleep on it**, to postpone making a decision until the next day. **to sleep rough**, to sleep out of doors. **to sleep together, with**, to have sexual intercourse (with). **sleep-walk**, v.i. to walk while asleep. **sleep-walker**, n. a somnambulist. **sleep-walking**, n. **sleeper**, n. one who sleeps; a support for the rails on a railway track; a sleeping compartment or carriage of a train; a train with these; (coll.) (e.g. a secret agent) who lies dormant before coming into action; (coll.) something (e.g. a film, a book) which becomes valuable or popular after a period of being neither. **sleeping**, n., a. **sleeping-bag**, n. a bag of some warm material in which one can sleep. **sleeping-car**, n. a railway carriage fitted with berths. **sleeping partner**, n. a partner having no share in the management of a business. **sleeping pill**, n. a sedative in tablet form for inducing sleep. **sleeping policeman**, n. (coll.) a hump in the road for slowing traffic. **sleeping sickness**, n. a disease characterized by fever and lethargy, and caused by a parasite *Trypanosoma gambiense*. **sleepless**, a. **sleeplessly**, adv. **sleeplessness**, n. **sleepy**, a. inclined to sleep, drowsy; dull, lazy, indolent, habitually inactive; tending to induce sleep; **sleepy-head**, n. a lazy or sleepy person. **sleepily**, adv. **sleepiness**, n.
sleet (slēt), n. hail or snow mingled with rain. v.i. to snow or hail with a mixture of rain. **sleety**, a. **sleetiness**, n.
sleeve (slēv), n. the part of a garment that covers the arm; the cardboard cover for a gramophone record. **to have up one's sleeve**, to hold secretly in reserve or in readiness. **to laugh in, up one's sleeve** LAUGH. **to roll up one's sleeves**, to get ready for hard work, a fight etc. **sleeve notes**, n.pl. the printed information on a record cover. **sleeved**, a. having sleeves (of a stated type). **sleeveless**, a. of a dress, blouse etc. without sleeves.
sleigh (slā), n. a vehicle mounted on runners for driving over snow or ice, a sledge. **sleigh-bell**, n. a small bell hung on a sleigh or its harness. **sleighing**, n.
sleight (slīt), n. dexterity, skill in manipulating things; trickery, cunning. **sleight of hand**, manual dexterity as used esp. in conjuring tricks.
slender (slen′də), a. small in circumference or width as compared with length; thin, slim; slight, scanty, meagre, inadequate, small, poor; feeble, not strong. **slenderly**, adv. **slenderness**, n.
slept (slept), past, p.p. SLEEP.

sleuth (slooth), n. (coll.) a detective. v.i. to act as a detective.
slew[1] (sloo), past SLAY[1].
slew[2], **slue** (sloo), v.t., v.i. to swing (round, about etc.) as on a pivot. n. such a turn. **slewed, slued**, a. (sl.) tipsy, drunk.
slice (slīs), n. a broad thin piece cut off; a part, share etc.; an implement used for slicing, a broad thin knife for lifting fish etc. from a frying-pan or for serving it; a slicing stroke in tennis or golf; v.t. to cut (usu. up) into broad, thin pieces; to cut (off) slices from; to cut, to divide; to strike a ball with a drawing motion. **sliceable**, a. **slicer**, n. a machine for slicing bread etc. **slicing**, n., a.
slick (slik), a. smooth, sleek; oily, smooth of speech etc.; polished, glossy; (coll.) dexterous, adroit; neatly or deftly performed; clever, smart, specious. n. a smooth or slippery surface patch, esp. of oil spilled on water. v.t. to make smooth or sleek. **slicker**, n. (coll.) a plausible, cunning person, a swindler. **slickly**, adv. **slickness**, n.
slide (slīd), v.i. (past, p.p. **slid** (slid)), to move smoothly along a surface with continuous contact, to glide, to slip, esp. to glide over ice, snow or other slippery surface, without skates; to pass (away, into etc.) smoothly, gradually or imperceptibly to drift, to take its own course. v.t. to cause to move smoothly along with a slippery motion; to pass secretly or unobtrusively, to slip. n. an act of sliding; a slip, a thing, piece or part that slides (e.g. on a machine); a glass carrying an object to be viewed in a microscope; a photographic transparency mounted in card or plastic; a surface, series of grooves, guide-bars etc., on which a part slides; an inclined channel, shute etc., esp. which children slide down in play; a polished track on ice on which persons slide; a prepared slope for coasting or tobogganing; a landslip; a hairslide, a clasp; a downward turn (e.g. in value). **to let slide**, to allow to deteriorate. **slide rule**, n. a device, consisting of one rule sliding within another, whereby several mathematical processes can be performed mechanically. **slidable**, a. **slider**, n. **sliding**, n., a. **sliding scale**, n. a scale of duties, prices etc., varying directly or inversely according to fluctuations of value or other conditions.
slight (slīt), a. inconsiderable, insignificant; small in amount, intensity etc.; inadequate, paltry, superficial, negligible; slender, slim; frail, flimsy, weak. n. an act of disrespect or neglect. v.t. to treat disrespectfully, to snub. **slighter**, n. **slighting**, a. disparaging. **slightingly**, adv. **slightish**, a. **slightly**, adv. **slightness**, n.
slily SLYLY.
slim (slim), a. (comp. **slimmer**, superl. **slimmest**) slender, thin, of slight shape or build; poor, slight; v.i. (past, p.p. **slimmed**) to adopt devices such as dieting and exercises in order to keep the body slim. v.t. to make slim. **to slim down**, to reduce in size or scale. **slimline**, a. slim in shape. **slimmer**, n. one who loses or attempts to lose weight through dieting or exercise. **slimmers' disease**, n. anorexia nervosa. **slimming**, n. **slimmish**, a. **slimness**, n.
slime (slīm), n. a soft, glutinous or viscous substance. **slimy**, a. consisting of or of the nature of slime; covered with or abounding with slime; slippery; repulsively mean, dishonest, cringing or obsequious. **slimily**, adv. **sliminess**, n.
sling[1] (sling), v.t. (past, p.p. **slung** (slŭng)) to throw, to hurl, esp. from a sling; (coll.) to cast out; to suspend in or as in a swing, to hang so as to swing; to hoist by means of a sling; (coll.) to throw, to hurl; (coll.) to pass. n. a throw; a short leather strap for hurling a small missile; a band, loop, or other arrangement of rope, chains, straps etc., for suspending, hoisting or transferring anything; a band of cloth for supporting an injured limb.
sling[2] (sling), n. a sweetened drink of water mixed with spirits, esp. gin.
slink[1] (slingk), v.i. (past, p.p. **slunk** (slŭngk)), to steal or

sneak away in a furtive, ashamed or cowardly manner; to move sinuously and provocatively. **slinky**, *a.* sinuous, slender; clinging, figure-hugging. **slinkily**, *adv.* **slinkiness**, *n.*

slip[1] (slip), *v.i.* (*past, p.p.* **slipped**) to slide, to glide; to slide unintentionally or out of place, to miss one's footing; to move, go or pass unnoticed, furtively or quickly; to get away, become free, or escape thus; to commit a small mistake or oversight; to go (along) swiftly; to decline; to elapse; (*coll.*) to lose control (of a situation); of a clutch, to fail to engage; to pass from the memory; *v.t.* to cause to move in a sliding manner; to put (on or off) or to insert (into) with a sliding, stealthy, hasty or careless motion; to let loose, to unleash, to undo; to put on or remove (a garment) speedily or easily; to escape or free oneself from; *n.* the act or state of slipping; an unintentional error, a small offence; a woman's sleeveless undergarment; an inclined ramp; a landslide; in cricket, any of three off-side positions or the fielders playing these positions. **slip of the pen, tongue**, a written or spoken error. **to give the slip**, to escape from, to evade. **to let slip** LET[1]. **to slip away, off**, to leave quickly or unobtrusively. **to slip up**, to make a mistake. **slipcase**, *n.* an open-ended cover for one or more books which reveals the spine(s). **slipknot**, *n.* a knot that slips up and down the string etc. on which it is made, a running knot. **slip-on**, *n., a.* (a garment or item of footwear) which can be put on or removed easily and quickly. **slipover**, *n., a.* (a garment) easily put on over the head, a pullover. **slipped disc**, *n.* a protrusion of an intervertebral disc causing painful pressure on spinal nerves. **slipshod**, *a.* careless, slovenly. **slip road**, *n.* an access or exit road from or onto a motorway. **slipstream**, *n.* the stream of air behind a vehicle etc. **slip up**, *n.* an error, a blunder. **slipway**, *n.* a slip for the repair, laying up or launch of vessels. **slippage** (-ij), *n.* an act, instance, amount or degree of slipping or failure to meet a target. **slipper**, *n.* a loose shoe easily slipped on or off, esp. for wearing indoors. *v.t.* (*coll.*) to beat with a slipper. **slipper bath**, *n.* a bath with covered end, roughly resembling a slipper. **slippery**, *a.* so smooth, wet, muddy etc. as to cause slipping, not allowing a firm footing or hold; difficult to hold, elusive, not to be depended on, shifty, artful, cunning; unstable. **slipperiness**, *n.* **slippy**, *a.* (*coll.*) slippery; (*coll.*) quick, sharp. **slippiness**, *n.*

slip[2] (slip), *n.* a long narrow strip of paper, wood or other material; a cutting for grafting or planting; a scion, a descendant; a young, thin person, a stripling.

slit (slit), *v.t.* (*past, p.p.* **slit**) to cut lengthways; to cut into long pieces or strips; to make a long cut in. *n.* a long cut or narrow opening. **slit-trench**, *n.* a narrow trench for one or two soldiers. **slitter**, *n.*

slither (slidh'ə), *v.i.* to slip, to slide unsteadily (along etc.); to move with a sliding motion. **slithery**, *a.*

sliver (sliv'ə), *n.* a piece of wood or similar material torn off, a splinter. *v.i.* to break to slivers.

Sloane (Ranger) (slōn rān'jə), *n.* a wealthy young (upper-) middle class person with a home in London and in the country, characteristically wearing expensive informal country clothes.

slob (slob), *n.* (*coll.*) a messy, slovenly or boorish person. **slobbish, slobby**, *a.*

slobber (slob'ə), *v.i.* to let saliva run from the mouth, to dribble; to talk or behave in a maudlin manner. *v.t.* to wet with saliva, to dribble over. *n.* saliva or spittle running from the mouth; maudlin talk or behaviour. **slobberer**, *n.* **slobbery**, *a.* **slobberiness**, *n.*

sloe (slō), *n.* the fruit of the blackthorn, *Prunus spinosa*, or the shrub bearing it. **sloe-eyed**, *a.* having dark, slanted or almond-shaped eyes. **sloe gin**, *n.* gin flavoured with sloes.

slog (slog), *v.t., v.i.* (*past, p.p.* **slogged**) to hit vigorously and at random, esp. in batting or with the fists; to work hard; to move slowly or cumbersomely. *n.* a spell of hard work; a heavy blow; an exhausting walk. **slogger**, *n.*

slogan (slō'gən), *n.* a political catchword; a catchy advertising phrase or word.

sloop (sloop), *n.* a fore-and-aft rigged vessel with one mast.

slop (slop), *n.* water or other liquid carelessly thrown about; (*pl.*) liquid food refuse fed to animals, esp. pigs; dirty water, liquid refuse; (*pl.*) poor-quality food. *v.t.* (*past, p.p.* **slopped**) to spill or allow to overflow. *v.i.* to tramp through slush or mud; to become spilled, to overflow the side of a vessel. **to slop out**, of prisoners, to clean out slops from a chamber pot. **sloppy**, *a.* wet, splashed, covered with spilt water or puddles; slovenly, done carelessly; weakly sentimental, maudlin or effusive. **sloppily**, *adv.* **sloppiness**, *n.*

slope (slōp), *n.* an inclined surface, line or direction, an incline, ground whose surface makes an angle with the horizon; the degree of such inclination. *v.i.* to be inclined at an angle to the horizon, to lie obliquely, to slant. **to slope off**, to leave, esp. furtively, to sneak away. **sloping**, *a.* **slopingly**, *adv.*

slosh (slosh), *v.t.* (*coll.*) to strike hard; (*coll.*) to splash, spread or pour (liquid); to move (something) about in liquid; to wet by splashing. *v.i.* to move or splash through slush, mud, water etc.; *n.* (*coll.*) a heavy blow; slush, mud etc.; the slapping or splashing sound of liquid. **sloshed**, *a.* (*coll.*) drunk.

slot (slot), *n.* a groove, channel, depression or opening, esp. in timber or a machine for some part to fit into; the aperture into which coins are put in a slot-machine; a place or niche (e.g. in an organization); a position in a sequence or schedule. *v.t.* (*past, p.p.* **slotted**) to fit or place (as) into a slot. *v.i.* to fit together or into by means of a slot or slots. **to slot in**, (*coll.*) to fit or settle in easily. **slot-machine**, *n.* a machine (for gambling, dispensing sweets, drinks etc.) operated by means of coins or tokens pushed or dropped through a narrow aperture. **slotted**, *a.* **slotter**, *n.*

sloth (slōth), *n.* laziness, indolence, sluggishness; a S American arboreal mammal characterized by its slow and awkward movements on the ground. **slothful**, *a.* **slothfully**, *adv.* **slothfulness**, *n.*

slouch (slowch), *n.* an ungainly or negligent drooping attitude, gait, or movement; (*sl.*) an awkward, slovenly or incapable person. *v.i.* to droop or hang down carelessly; to stand or move in a loose, negligent or ungainly attitude. **sloucher**, *n.* **slouching**, *a.* **slouchy**, *a.* **slouchiness**, *n.*

slough[1] (slow), *n.* a place full of mud, a bog, a quagmire; a marsh, a swamp; a state of abject depression. **slough of despond**, extreme despondency.

slough[2] (slŭf), *n.* the cast skin of a snake. *v.t.* to cast off (a skin, dead tissue etc.).

Slovak (slō'vak), *n.* an inhabitant of Slovakia; one of a Slavonic people inhabiting E Czechoslovakia; the language of this people. *a.* of or pertaining to this people, their language or the region they inhabit. **Slovakian** (-vak'-), *n., a.*

sloven (slŭv'n), *n.* one who is careless of dress or negligent of cleanliness; an untidy, careless, lazy person. **slovenly**, *a.*, **slovenliness**, *n.*

Slovene (slōvēn'. slō'-), *n.* an inhabitant of Slovenia; one of or S Slavonic people inhabiting Yugoslavia; the language of this people. *a.* of or pertaining to this people, their language or the region they inhabit. **Slovenian** (-vēn'-), *n., a.*

slow (slō), *a.* not quick, moving at a low speed; taking a long time; not prompt or willing; tardy, backward; slack; behind the right time; stupid, dull; tedious, lifeless; preventing or designed to prevent fast movement. *adv.* slowly. *v.i.* to slacken speed, to go slower. *v.t.* to reduce the speed of. **slowcoach**, *n.* one who is slow in moving, acting, deciding etc. **slowdown**, *n.* a slackening of pace.

slow handclap, *n.* a slow regular clapping expressing audience discontent. **slow-fuse, -match,** *n.* a fuse or match burning slowly for igniting explosives. **slow-motion,** *n.* in film and video, a technique which allows action to appear slower than normal. *a.* of or pertaining to this; operating or moving at a slower speed than is normal. **slow-witted,** *a.* **slowish,** *a.* **slowly,** *adv.* **slowness,** *n.*

slow-worm (slō'wœm), *n.* a small limbless snake-like lizard, *Anguis fragilis,* the blind-worm.

sludge (slŭj), *n.* mud, mire, slush; an oozy or slimy sediment, as of ore and water; a hard precipitate produced in the treatment of sewage. **sludgy,** *a.*

slue SLEW [2].

slug (slŭg), *n.* a shell-less air-breathing gastropod; a seaslug; a bullet; a small, roughly rounded lump of metal; (*coll.*) a quantity of liquor which can be gulped at one go. *v.t.* (*past, p.p.* **slugged**) (*coll.*) to hit hard. **sluggard** (-əd), *n.* a person habitually lazy. **sluggardly,** *a.* **slugger,** *n.* **sluggish,** *a.* habitually lazy, dull, inactive; slow in movement or response, inert, torpid. **sluggishly,** *adv.* **sluggishness,** *n.*

sluice (sloos), *n.* a waterway with a valve or hatch by which the level of a body of water is controlled; a sluicegate or flood-gate; the stream above, below, or passing through a flood-gate; an inclined trough or channel. *v.t.* to drench, to wash thoroughly, to rinse; **sluice-gate,** *n.* a floodgate. **sluice-way,** *n.* a channel into which water passes from a sluice.

slum (slŭm), *n.* a squalid neighbourhood in a town; a house, flat etc. which is overcrowded, in a deteriorated condition etc. *v.i.* (*past, p.p.* **slummed**) to live in squalid or poverty-stricken conditions; to visit a place or affect a life style inferior to what one is accustomed to out of curiosity or for amusement. **slummer,** *n.* **slummy,** *a.*

slumber (slŭm'bə), *v.i.* to sleep, esp. lightly; to be inactive or dormant. *n.* light sleep; a state of dormancy, inactivity. **slumberer,** *n.*

slump (slŭmp), *v.i.* to fall or sink suddenly; of prices, prosperity etc., to come down, to collapse; to decline quickly or drastically. *n.* the act or an instance of slumping; a heavy fall or decline, a collapse (of prices etc.).

slung (slŭng), *past, p.p.* SLING [1].

slunk (slŭngk), *past, p.p.* SLINK.

slur (slœ), *v.t.* (*past, p.p.* **slurred**) to calumniate; to speak slightingly of; to pass lightly over; to pronounce indistinctly. *v.i.* to speak or articulate indistinctly; to pass lightly or slightingly (over). *n.* a stain, a stigma, a reproach or disparagement; a blurred impression in printing; a slurring in pronunciation or singing; **slurred,** *a.*

slurry (slŭ'ri), *n.* a thin, fluid paste made by mixing certain materials (esp. cement) with water.

slurp (slœp), *n.* a sucking sound produced when eating or drinking noisily. *v.i.* to eat or drink noisily.

slush (slŭsh), *n.* liquid mud, sludge; half-melted snow; (*sl.*) mawkishly sentimental talk or writing, gush. *v.t.* to throw slush over, to soak or bedaub with slush. **slush fund,** (*coll.*) *n.* a fund of money used to finance corrupt business or political practices. **slushiness,** *n.* **slushy,** *a.*

slut (slŭt), *n.* a dirty, slovenly or sexually promiscuous woman. **sluttery,** *n.* **sluttish,** *a.* **sluttishly,** *adv.* **sluttishness,** *n.*

sly (slī), *a.* (*comp.* **slyer, slier,** *superl.* **slyest, sliest**) crafty, cunning, underhand, furtive, not open or frank; playfully roguish, knowing, arch. **on the sly,** slyly, in secret, on the quiet. **slyly, slily,** *adv.* **slyness,** *n.*

SM, (*abbr.*) sergeant major.

Sm, (*chem. symbol*) samarium.

smack[1] (smak), *n.* a slight taste or flavour; a suggestion, trace, tincture or dash (of); a smattering; (*sl.*) heroin. *v.i.* to have a taste, flavour or suggestion (of).

smack[2] (smak), *n.* a one-masted vessel, like a sloop or cutter, used in fishing etc.

smack[3] (smak), *n.* a quick, smart report as of a blow with something flat, a crack of a whip etc.; a blow with the flat of the hand, a slap; a loud kiss. *v.t.* to strike with the flat of the hand, to slap; to separate (the lips) with a sharp noise; *v.i.* of the lips, to make a sharp noise as of opening quickly; *adv.* suddenly, plump, directly. **to smack one's lips,** to gloat, to relish, to anticipate. **smacker,** *n.* a noisy kiss; a resounding blow; (*sl.*) a pound or dollar note.

small (smawl), *a.* deficient or relatively little in size, age, stature, degree, power, amount, number, weight etc.; of less dimensions than the standard kind; of minor importance, trifling, petty; concerned or dealing with business etc., of a restricted or minor kind; of low degree, poor, humble, plebeian; unpretentious; paltry, mean, narrow-minded. *adv.* into small pieces. *n.* the slender part of the back; (*pl.*) undergarments; **the small screen,** television. **to feel small,** to feel humiliated or insignificant. **small ads,** *n.pl.* classified advertisements. **small-arms,** *n.pl.* portable firearms, as rifles, pistols etc. **small beer,** *n.* a trivial matter; an insignificant person. **small-bore,** *a.* of a low-calibre gun, having a chamber with a small bore. **small change,** *n.* coins of low denominations. **small fry,** *n.* (*coll.*) an insignificant person or thing; (*pl.*) (*coll.*) children. **small holder,** *n.* the tenant of a small holding. **small holding,** *n.* a portion of land of limited area and rental let for cultivation. **small hours,** *n.pl.* the time from midnight till 3 or 4 AM. **small intestine,** *n.* the part of the intestine comprising the duodenum, jejunum and the ileum. **small letters,** *n.pl.* lowercase letters. **small-minded,** *a.* restricted in outlook, petty. **small-mindedness,** *n.* **smallpox,** *n.* variola, a contagious, feverish disease, characterized by eruptions on the skin. **small-scale,** *a.* of limited scope or extent. **small talk,** *n.* light or superficial conversation, gossip. **small-time,** *a.* (*coll.*) insignificant, unimportant; amateurish. **small-timer,** *n.* **smallish,** *a.* **smallness,** *n.*

small-arms, smallpox etc. SMALL.

smarm (smahm), *n.* (*coll.*) gush, fawning behaviour. *v.t.* to plaster, to flatten, (with hair oil etc.); to ingratiate oneself (with). *v.i.* to fawn, to ingratiate. **smarmy,** *a.* (*coll.*) sleek and smooth; having a wheedling manner. **smarmily,** *adv.* **smarminess,** *n.*

smart (smaht), *v.i.* to feel or give or cause a sharp pain or mental distress; to rankle; to feel wounded. *adv.* smartly. *n.* a sharp, lively pain, a stinging sensation; a feeling of irritation; distress, anguish. *a.* stinging, pungent, severe, vigorous, lively, brisk; astute, clever, intelligent, ingenious; quick at repartee, impertinently witty; shrewd, wide-awake, sharp; spruce, well groomed, stylish, fashionable. **to look smart,** to hurry up, be quick. **smart alec(k)** (-alik), *n.* (*coll.*) a know-it-all. **smart-alecky,** *a.* **smart ass,** *n.* (*chiefly N Am. sl.*) a smart alec. **smartmoney,** *n.* money bet or invested by experienced gamblers or businessmen. **smart set,** *n.* a fashionable social group of people. **smarten,** *v.t.,* *v.i.* **smartish,** *adv.* (*coll.*) **smartly,** *adv.* **smartness,** *n.* **smarty, smarty pants,** *n.* (*coll.*) a smart alec.

smash (smash), *v.t.* to break to pieces by violence, to shatter, to dash, to wreck, to crash; to hit with a crushing blow; to overthrow completely, to rout, to crush; to hit (a shuttlecock, tennis ball etc.) with a forceful over-head stroke. *v.i.* to break to pieces; to collide or crash (into); to perform a smash (in badminton, tennis etc.); to come to pieces under force. *n.* an act or instance of smashing; a breaking to pieces; the sound this makes; a smash-up; in badminton, tennis etc., a forceful overhead stroke; (*coll.*) a smash-hit; a break up, a collapse; a disaster. *adv.* with a smash. **smash-and-grab,** *a.* (*coll.*) descriptive of a theft where a shop window is broken and goods within hurriedly removed. **smash hit,** *n.* (*coll.*) a great success. **smash-up,** *n.* a violent collision between vehicles, a car

crash. **smashable**, *a*. **smashed**, *a*. broken; (*sl.*) very drunk. **smasher**, *n*. one who or that which smashes; (*sl.*) something of staggering size, quality, effectiveness etc.; (*sl.*) an outstandingly attractive or amiable person. **smashing**, *a*. (*coll.*) very fine, wonderful. **smashing**, *adv*.
smatter (smat'ə), *n*. a smattering. **smattering**, *n*. a slight superficial knowledge; a small quantity.
smear (smiə), *v.t*. to rub or daub with anything greasy; to rub (a lens etc.) so as to blur; to apply thickly; to soil, to pollute; to slander. *v.i.* to make a smear; to become smudged. *n*. a stain or mark made by smearing; an attack on a person's reputation; a substance (e.g. vaginal secretion) smeared on a glass slide for examination under a microscope. **smear campaign**, *n*. a series of orchestrated attacks on the reputation of a politician, institution etc. **smear test**, *n*. a microscopic examination of a smear, e.g. for cervical cancer. **smearer**, *n*. **smeary**, *a*. **smearily**, *adv*. **smeariness**, *n*.
smell (smel), *n*. the sense by which odours are perceived; the sensation or the act of smelling; that which affects the organs of smell, scent, odour; a bad odour, a stench; a characteristic quality, a trace, an aura. *v.t. (past, p.p.* **smelt**[3] (smelt), **smelled**) to perceive the odour of; to inhale the odour of anything with the nose; to detect by means of scent; to hunt, trace or find (out) by or as by the scent. *v.i.* to give out an odour (of etc.); to have a specified smell; to smack (of); to have or exercise the sense of smell; to stink. **to smell out**, to detect by instinct or prying; to pollute (e.g. a room with smoke). **smellingbottle**, *n*. a small bottle or phial for holding smelling-salts. **smelling-salts**, *n.pl*. an aromatic preparation of ammonium carbonate used in cases of faintness etc. **smelly**, *a*. malodorous. **smelliness**, *n*.
smelt[1] (smelt), *v.t.* to fuse (an ore) so as to extract the metal; to extract (metal) from ore thus. **smelter**, *n*. **smelting**, *n*. **smelting-furnace**, *n*.
smelt[2] (smelt), *n*. (*pl.* **smelt, smelts**) a small food-fish, *Osmerus eperlanus*, allied to the salmon.
smelt[3] (smelt), *past, p.p.* SMELL.
smew (smū), *n*. a small merganser or diving-duck, *Mergus albellus*.
smidgen, smidgeon, smidgin (smij'in), *n*. (*coll.*) a tiny amount.
smile (smīl), *v.i*. to express kindness, love, pleasure, amusement or contempt by an instinctive lateral and upward movement of the lips and cheeks; of the weather, fortune etc., to look bright or favourable. *v.t.* to express by or as by a smile. *n*. the act of smiling; a cheerful expression. **to come up smiling**, to end up in a favourable state, esp. after misfortune. **smiler**, *n*. **smiley**, *a*. **smilingly**, *adv*.
smirch (smœch), *v.t.* to soil, to smear, to defame. *n*. a stain, a smear.
smirk (smœk), *v.i.* to smile affectedly or scornfully, to simper. *n*. an affected or scornful smile, a simper. **smirker**, *n*. **smirky**, *a*.
smite (smīt), *v.t.* (*past* **smote** (smōt), *p.p.* **smitten** (-ən)) to strike, to deal a severe blow to; to inflict damage or disaster upon; (*usu. in p.p.*) to strike or affect (with a feeling, disease etc.).
smith (smith), *n*. one who works in metals; a blacksmith.
smithy (smidh'i), *n*. a blacksmith's workshop.
smithereens (smidhərēnz'), *n.pl*. little bits, tiny fragments.
smithy SMITH.
smitten, *p.p.* (*coll.*) enamoured. [SMITE]
smock (smok), *n*. a woman's loose frock, artist's overall etc. resembling this.
smog (smog), *n*. a smoky fog containing chemical fumes. **smoggy**, *a*. **smogless** (-lis), *a*.
smoke (smōk), *n*. volatile products of combustion in the form of visible vapour or fine particles escaping from a burning substance; a suspension of particles in gas; anything ephemeral or unsubstantial; the act of smoking a pipe, cigar etc.; (*sl.*) a cigarette. *v.i.* to emit smoke; to emit vapour, fumes etc.; of a chimney etc., to send smoke into a room; to draw into the mouth or inhale and exhale the smoke of tobacco etc.; *v.t.* to apply smoke to; to blacken, colour, cure, flavour, suffocate, drive out, destroy, cleanse etc., with smoke; to draw with the mouth or inhale and exhale the smoke of (tobacco etc.). **no-smoking zone, area**, part of a building, vehicle etc. where it is forbidden to smoke. **the smoke**, a (big) city; London. **to go, end up in smoke**, of a scheme, a desire, to come to nothing; to be burned to nothing. **to smoke out**, to drive out with smoke; to discover, to bring into the open. **smoke bomb**, *n*. a projectile containing a composition that emits a dense smoke. **smoke box**, *n*. in a steam locomotive, the chamber through which smoke and gases pass from the boiler tubes to the funnel. **smoke-screen**, *n*. a dense volume of smoke produced by chemicals used to conceal the movements of ships, troops etc., from the enemy; something used to obscure or deceive. **smoke signals**, *n.pl*. a method of conveying messages by a series of puffs of smoke. **smoke-stack**, *n*. a funnel, esp. on a steamer. **smokable**, *a*. **smoked**, *a*. **smoked herring**, *n*. a kipper. **smokeless** (-lis), *a*. **smokeless zone**, *n*. an area in which it is forbidden to emit smoke from chimneys. **smoker**, *n*. one who smokes tobacco; one who dries, cures, fumigates etc., with smoke; a smoking compartment. *n.*, *a*. **smoking jacket**, *n*. a (usu. velvet) jacket, formerly worn by men after smoking. **smoking carriage, compartment, room**, *n*. a railway-carriage etc., reserved for smokers. **smoky, smokey**, *a*. resembling smoke in colour, smell, flavour etc.; filled with smoke; emitting smoke; dirtied by smoke. *n*. **smokily**, *adv*. **smokiness**, *n*.
smolder (smōl'də), SMOULDER.
smooch (smouch), *v.i.* (*coll.*) to kiss and cuddle.
smooth (smoodh), *a*. having a continuously even surface, free from roughness, projections or indentations; not hairy; of water, unruffled; free from obstructions or impediments; offering no resistance; of sound, taste etc., not harsh; equable, calm, pleasant, bland, suave, polite, flattering. *v.t.* to make smooth; to free from harshness, discomforts, obstructions, irregularities etc.; to extenuate, to soften, to alleviate, to dispel. *v.i.* to become smooth. *n*. the act of smoothing; a smooth place or part; (*coll.*) that which is pleasant or easy. **to smooth over**, to gloss over (a difficulty). **smooth-spoken, -tongued**, *a*. polite, plausible, flattering. **smoother**, *n*. **smoothie, -thy** (-i), *n*. (*coll., usu. derog.*) an excessively sauve or plausible person, esp. a man. **smoothing**, *n.*, *a*. **smoothing-iron**, *n*. a polished iron implement formerly used for smoothing linen etc. **smoothly**, *adv*. **smoothness**, *n*.
smorgasbord (smaw'gəsbawd), *n*. a buffet comprising an assortment of hors d'oeuvres and other dishes.
smorzando (smawtsan'dō), **smorzato** (-sah'tō), *a., adv*. (*Mus.*) with a gradual fading or dying away.
smote (smōt), (*past*) SMITE.
smother (smŭdh'ə), *v.t.* to suffocate, to stifle; to kill by suffocation etc.; to keep (a fire) down by covering it with ashes etc.; to suppress, to keep from being divulged, to conceal; to overcome, to overwhelm; to cover thickly, to enclose, to envelop. **smotheringly**, *adv*. **smothery**, *a*. **smotheriness**, *n*.
smoulder, (*esp. N Am.*) **smolder** (smōl'də), *v.i.* to burn in a smothered way without flame; to exist in a suppressed or latent condition; to feel or show strong repressed emotions (as anger, jealousy).
smudge (smŭj), *v.t.* to smear or blur (writing, drawing etc.); to make a dirty smear, blot or stain on; *v.i.* to become smeared or blurred. *n*. a dirty mark, a smear, a blur. **smudgy**, *a*. **smudginess**, *n*.
smug (smŭg), *a*. (*comp.* **smugger**, *superl*. **smuggest**) self-satisfied, complacent. **smugly**, *adv*. **smugness**, *n*.
smuggle (smŭg'l), *v.t.* to import or export illegally; to con-

smut (smŭt), *n.* a particle of soot or other dirt, a mark or smudge made by this; obscene or ribald talk, language, stories etc. *v.t.* (*past, p.p.* **smutted**) to stain or mark with smut. *v.i.* **smutty,** *a.* **smuttily,** *adv.* **smuttiness,** *n.*

Sn, (*chem. symbol*) tin.

snack (snak), *n.* a slight, hasty meal. *v.i.* to have a snack. **snack-bar,** *n.* a cafe, self-service restaurant or other place offering light meals or refreshments.

snaffle (snaf'l), *v.t.* (*coll.*) to steal, to appropriate for oneself.

snag (snag), *n.* a jagged projection; a tear, a pull, a flaw in fabric; an unexpected or concealed difficulty, an obstacle. *v.t.* (*past, p.p.* **snagged**) to tear or catch (fabric) on a snag. *v.i.* to become snagged. **snagged, snaggy,** *a.*

snail (snāl), *n.* a gasteropodous mollusc of various species, usu. bearing a shell; the sea snail; a sluggish person or thing. **snail's pace,** *n.* a slow rate of progress. **snaillike,** *a.* **snaily,** *a.*

snake (snāk), *n.* a limbless reptile of a venomous or non-venomous type having a forked tongue and the ability to swallow prey whole; a snake-like lizard or amphibian; a sneaking, treacherous person; anything resembling a snake in appearance or movement; in the EEC, a system which allows the currencies of member countries to fluctuate within narrow limits. *v.i.* to wind, to move quietly or snakily. **snake in the grass,** a treacherous or underhand person. **snakes and ladders,** a board game in which counters can advance more speedily up ladders or move backwards down ladders. **snake bite,** *n.* the venomous bite of a snake; (*coll.*) a drink of beer and cider. **snake-charmer,** *n.* an entertainer who appears to mesmerize snakes by playing music. **snake-charming,** *n.* **snakeskin,** *n., a.* (made of) the skin of a snake. **snakelike, snakish, snaky,** *a.* **snakily,** *adv.* **snakiness,** *n.*

snap (snap), *v.t.* (*past, p.p.* **snapped**) to bite (at); to grasp, seize or snatch (at); to make a sharp, quick sound, like a crack or slight explosion; to break, part, close or fit into place suddenly with such a noise; to collapse (with pressure, strain of work etc.), to speak sharply or irritably. *v.t.* to seize suddenly, to snatch; to take advantage of eagerly; to utter abruptly or irritably; to cause (a whip, the fingers etc.) to make a sharp crack or report; to break with such a noise; to shut (to) or bring (together) thus; to photograph casually. *n.* the act or an instance of snapping; the sound produced by this; a sudden spell of severe weather; a crisp gingerbread cake; a children's game of cards; a snapshot; vigour, briskness, (*coll.*) something that is easy or profitable, a cinch. *a.* closing or fastening with a snap; *adv.* with (the sound of) a snap. *int.* uttered when playing the game of snap; hence, indicating similarity, identicalness or synchronicity. **to snap one's fingers (at),** to show contempt or defiance. **to snap out of it,** to change one's mood abruptly (for the better). **to snap someone's head off,** to retort abruptly, irritably or rudely. **to snap up,** to purchase eagerly. **snap decision,** *n.* one taken without deliberation. **snapdragon,** *n.* a plant of the genus *Antirrhinum,* with a flower opening like a dragon's mouth. **snap fastener,** *n.* a press-stud. **snap-on,** *a.* designed to be attached by a snap fastening or spring clip. **snapshot,** *n.* a photograph taken without preparation or posing. **snapper,** *n.* **snapping,** *n., a.* **snappingly,** *adv.* **snappish,** *a.* **snappishly,** *adv.* **snappishness,** *n.* **snappy,** *a.* brisk, sharp, lively; smart, up-to-date, stylish. **to make it snappy,** to hurry up. **snappily,** *adv.*

snare (sneə), *n.* a trap for catching animals; a trick, trap or stratagem by which one is brought into difficulty, defeat, disgrace, sin etc. *v.t.* to catch in a snare; to ensnare, entrap or inveigle. **snarer,** *n.* **snary,** *a.*

snarl[1] (snahl), *v.i.* to growl in a sharp tone with teeth bared, as an angry dog; to speak in a harsh, surly or savage manner. *v.t.* to express or say (out) with a snarl. *n.* a sharp-toned growl. **snarler,** *n.* **snarling,** *n., a.* **snarlingly,** *adv.* **snarly,** *a.*

snarl[2] (snahl), *n.* a tangle, a knot of hair, thread etc.; (*fig.*) an entanglement, embarrassing difficulty. *v.t.* to entangle; to cause to become confused or complicated. *v.i.* to become entangled, muddled, complicated. **to snarl up,** to (cause to) become confused, disordered, inoperable, immobile etc. **snarl-up,** *n.* an instance or state of confusion, obstruction, disorder etc. (esp. a traffic jam). **snarled,** *a.* **snarler,** *n.*

snatch (snach), *v.t.* to seize suddenly, eagerly or without permission; to remove or rescue (from, away etc.) suddenly or hurriedly; to win or gain narrowly. *v.i.* to try to seize, to make a sudden motion (at) as if to seize. *n.* an act of snatching, a grab (at); that which is snatched; a short spell of rest, work etc.; a fragment of talk, song etc.; in weightlifting, a kind of lift in which the weight is raised overhead in one motion; (*coll.*) a kidnapping. **snatcher,** *n.* **snatchily, snatchingly,** *adv.* **snatchy,** *a.*

snazzy (snaz'i), *a.* (*sl.*) up-to-date, showy, smart, attractive (e.g. of clothes). **snazzily,** *adv.* **snazziness,** *n.*

sneak (snēk), *v.i.* (*past, p.p.* **sneaked,** (*N Am.*) **snuck**) to creep, slink or steal (about, away, off etc.), as if afraid or ashamed to be seen; to behave in a mean, cringing, cowardly or underhand way; to tell tales. *v.t.* (*sl.*) to steal; to place or remove stealthily. *n.* one who sneaks; a talebearer. **to sneak away, off,** to leave unobtrusively. **sneak preview,** *n.* a privileged advance look at a film, product etc. before it is shown to the public. **sneak-thief,** *n.* a pilferer, one who steals from open windows or doors. **sneaker,** *n.* **sneakers,** *n.pl.* (*chiefly N Am.*) rubber-soled shoes. **sneakingly,** *adv.* **sneaky,** *a.* **sneakily,** *adv.* **sneakiness,** *n.*

sneer (sniə), *v.i.* to show contempt by a smile or grin; to scoff, to jibe. *v.t.* to utter or express with a sneer. *n.* a grimace or verbal expression of contempt or derision. **sneerer,** *n.* **sneering,** *n., a.* **sneeringly,** *adv.* **sneery,** *a.*

sneeze (snēz), *v.i.* to eject air etc. through the nostrils audibly and convulsively, owing to irritation of the inner membrane of the nose. *n.* the act of sneezing. **not to be sneezed at,** not to be despised, worth consideration. **sneezer,** *n.* **sneezy,** *a.*

snib (snib), *n.* a bolt or catch. *v.t.* (*past, p.p.* **snibbed**) to fasten with this.

snick (snik), *v.t.* to cut, to nick to snip; in cricket, to hit (the ball) lightly with a glancing stroke. *n.* a slight cut, nick or notch; a light glancing hit, in cricket.

snicker (snik'ə), *v.i.* to snigger. *v.t.* to utter with a snigger. *n.* a snigger. **snickerer,** *n.* **snickery,** *a.*

snide (snīd), *a.* malicious, sneering, disparaging, sly, mean. **snidely,** *adv.* **snideness,** *n.*

sniff (snif), *v.i.* to draw air audibly up the nose in order to smell, clear the nasal passages, inhale a drug, express contempt etc. *v.t.* to draw (*usu.* up) with the breath; to smell, to perceive by sniffing. *n.* the act or sound of sniffing. **to sniff at,** to express contempt or disdain for. **to sniff out,** to discover (as if) by sniffing. **sniffer,** *n.* **sniffer dog,** *n.* a dog trained to smell out drugs or explosives. **sniffle** (-l), *v.i.* to sniff (as with a cold, when weeping etc.), to snuffle. *n.* an act or sound of sniffling; a snuffle. **the sniffles,** a slight head-cold. **sniffler,** *n.* **sniffly,** *a.* **sniffily,** *adv.* **sniffiness,** *n.* **sniffy,** *a.* (*coll.*) given to sniffing, disdainful. **snifter,** *n.* (*coll.*) a small quantity of alcoholic drink.

snigger (snig'ə), *v.i.* to laugh in a half-suppressed or discourteous manner. *n.* a suppressed laugh.

snip (snip), *v.t.* (*past, p.p.* **snipped**) to cut or clip off sharply or quickly with shears or scissors. *v.i.* to make such a cutting movement. *n.* the act, movement or sound of snipping; a cut with scissors or shears; a small piece snipped off; (*coll.*) a certainty, a cinch, a bargain.

snipper, *n.* **snippet** (-it), **snipping,** *n.* a small bit snipped off; (*pl.*) scraps, fragments (of news etc.). **snippety,** *a.* **snippetiness,** *n.* **snippy,** *a.*
snipe (snīp), *n.* a long-billed marsh- and shore-bird of the genus *Gallinago*; *v.i.* to criticize, to find fault, to carp; to pick off members of the enemy, usu. from cover. **sniper,** *n.* a marksman firing from a concealed position. **sniping,** *n.*
snipper, snippet etc. SNIP.
snitch (snich), *v.t.* (*sl.*) to steal, to pilfer.
snivel (sniv'l), *v.i.* (*past, p.p.* **snivelled**) to run at the nose; to cry or fret with snuffling; to be tearful. *n.* audible or affected weeping. **sniveller,** *n.* **snivelly,** *a.*
SNO, (*abbr.*) Scottish National Orchestra.
snob (snob), *n.* one who regards the claims of wealth and position with an exaggerated respect; one who condescends to, patronizes, or avoids those felt to be of lower standing. **snobbery, snobbishness,** *n.* **snobbish, snobby,** *a.* **snobbishly,** *adv.*
SNOBOL (snō'bol), *n.* a language used in computer programming for handling strings of symbols. [acronym for String Orientated Symbolic Language]
Sno-Cat® (snō'kat), *n.* a type of vehicle designed to travel on snow.
snog (snog), *v.i.* (*coll.*) (*past, p.p.* **snogged**) to kiss and cuddle. *n.* (*coll.*) the act or an instance of this.
snook (snook), *n.* **to cock a snook,** to put the thumb to the nose and spread the fingers; to defy.
snooker (snoo'kə), *n.* a game resembling pool or pyramids played on a billiard-table; a shot or situation in this game in which the cue ball is blocked by another ball making a direct stroke impossible. *v.t.* to put (one's opponent) in this position; to place in difficulty (by presenting an obstacle); to thwart; to defeat.
snoop (snoop), *v.i.* to go about in an inquisitive or sneaking manner, to pry. *n.* an act or instance of snooping; a snooper. **snooper,** *n.* a prying busybody. **snoopy,** *a.*
snooty (snoo'ti), *a.* (*coll.*) supercilious, snobbish. **snootily,** *adv.* **snootiness,** *n.*
snooze (snooz), *v.i.* to take a short sleep, esp. in the day. *n.* a short sleep, a nap. **snoozer,** *n.* **snoozy,** *a.*
snore (snaw), *v.i.* to breathe through the mouth and nostrils with a hoarse noise in sleep. *n.* the act or sound of snoring. **snorer,** *n.* **snoring,** *n.*
snorkel (snaw'kl), *n.* a breathing apparatus used in diving and underwater swimming consisting of a tube which extends from the mouth to above the surface of the water. *v.i.* (*past, p.p.* **snorkelled**) to swim with a snorkel. **snorkelling,** *n.*
snort (snawt), *v.i.* to force air violently and loudly through the nostrils like a frightened or excited horse (e.g. as an expression of contempt); to make a noise like this; to inhale drugs, esp. habitually. *v.t.* to utter or throw (out) with a snort; (*sl.*) to inhale (a drug). *n.* the act or sound of snorting; (*sl.*) an instance of inhaling a drug, or the amount inhaled in one snort; (*coll.*) a snifter. **snorter,** *n.* one who or an animal that snorts; (*coll.*) anything of extraordinary size, violence etc. (as a strong wind). **snortingly,** *adv.*
snot (snot), *n.* mucus from the nose; (*sl.*) a low or contemptible person. **snotty,** *a.* (*coll.*) soiled with nasal mucus; (*sl.*) snobbish, snooty; (*sl.*) contemptible, low. **snotty-nosed,** *a.* **snottily,** *adv.* **snottiness,** *n.*
snout (snowt), *n.* the projecting nose or muzzle of an animal; (*sl.*) the nose; a nozzle. **snouted,** *a.* **snoutish,** *a.* **snoutless** (-lis), *a.* **snoutlike,** *a.* **snouty,** *a.*
snow (snō), *n.* watery vapour frozen and falling to the ground in flakes; a fall of this; anything resembling snow, esp. in whiteness; (*sl.*) cocaine; a mass of white dots on a television or radar screen caused by interference. *v.i.* to fall in or as snow. *v.t.* to cover, sprinkle or block (up) with snow; to send or scatter down as snow; **to snow under,** to overwhelm (with work etc.). **snowball,** *n.* a round mass of snow pressed together in the hands and flung as a missile; a drink of advocaat and lemonade. *v.i.* to throw snowballs; to accumulate with increasing rapidity, to accelerate. **snowbird,** *n.* a small finch, bunting or sparrow, esp. the snow-bunting. **snow-blind,** *a.* partially or totally blinded, usu. temporarily, through the glare of reflected light from the surface of snow. **snow-blindness,** *n.* **snowblower,** *n.* a machine which clears snow from a road by blowing it to the side. **snow-bound,** *a.* imprisoned or kept from travelling by snow. **snow-bunting,** *n.* a northern finch, *Plectrophanex nivalis*, visiting Britain in winter. **snowcap,** *n.* the cap of snow on top of a mountain. **snow-capped,** *a.* crowned with snow. **snowdrift,** *n.* a mass of snow accumulated by the wind. **snowdrop,** *n.* a bulbous plant, *Galanthus nivalis*, with a white flower appearing in early spring. **snow-fall,** *n.* a fall of snow; the amount of snow falling in a given place during a given time. **snow-field,** *n.* an expanse of snow, esp. in polar or lofty mountain regions. **snow-flake,** *n.* a fleecy cluster of ice crystals falling as snow. **snow-goose,** *n.* a white Arctic goose with black wing-tips. **snow-leopard,** *n.* the ounce. **snow-line,** *n.* the lowest limit of perpetual snow on mountains etc. **snowman,** *n.* a large snowball shaped roughly like a human figure. **snowmobile,** *n.* a motor vehicle with runners or caterpillar tracks enabling it to travel over snow. **snow-plough,** *n.* an implement used to clear a road or railway track of snow. **snow-shoe,** *n.* a long, light, racket- or ski-shaped frame worn to prevent sinking when walking on snow. **snow-slip,** *n.* an avalanche. **snow tyre,** *n.* a heavy tyre with deep treads for use on snow. **snow-white,** *a.* as white or pure as snow. **snowless** (-lis), *a.* **snow-like,** *a.*, *adv.* **snowy,** *a.* white like snow; abounding with snow; covered with snow; spotless, unblemished. **snowy owl,** *n.* a white, black-barred northern owl, *Nyctea scandiaca.* **snowily,** *adv.* **snowiness,** *n.*
SNP, (*abbr.*) Scottish National Party.
Snr, snr, (*abbr.*) senior.
snub (snŭb), *v.t.* (*past, p.p.* **snubbed**) to slight in a pointed or offensive manner. *n.* an act of snubbing, a rebuff, a slight. *a.* short, stubby. **snub-nose,** *n.* **snub-nosed,** *a.* having a short upturned nose; of a gun, having a short barrel. **snubber,** *n.* **snubbing,** *n.*, *a.* **snubbingly,** *adv.* **snubby,** *a.*
snuck SNEAK.
snuff[1] (snŭf), *n.* a sniff; powdered tobacco inhaled through the nose; a pinch of this; a state of resentment, a huff. **up to snuff,** knowing, sharp, not easily imposed upon; in good condition, up to scratch. **snuff-box,** *n.* a small container for carrying snuff. **snuffer,** *n.* **snuffing,** *n.*, *a.* **snuffy,** *a.* **snuffiness,** *n.*
snuff[2] (snŭf), *n.* the charred part of the wick in a candle or lamp. *v.t.* to extinguish (a flame) by or as by snuffing. **to snuff it,** (*sl.*) to die. **snuffers,** *n.pl.*
snuffle (snŭf'l), *v.i.* to breathe noisily or make a sniffing noise as when the nose is obstructed; to talk through the nose; to snivel, to whine. *v.t.* to utter through the nose; to sniff. *n.* the act or sound of snuffling; a nasal tone or voice. **the snuffles,** obstruction of the nostrils by mucous nasal catarrh. **snuffler,** *n.* **snuffly,** *a.*
snuffy, snuffiness SNUFF[1].
snug (snŭg), *a.* lying close, sheltered, and comfortable; cosy, comfortable; compact, trim, well secured; not exposed to view. **snuggle** (-l), *v.i.* to move or lie close (up to) for warmth. *v.t.* to draw close to one, to cuddle. *n.* the act of snuggling. **snugly,** *adv.* **snugness,** *n.*
so[1] (sō), *adv.* in such a manner or to such an extent, degree etc. (with *as* expressed or understood); in the manner or to the extent, degree, intent, result etc. (with *that* or *but*); on condition or provided (that); also, in addition; indeed, certainly; as, compared; extremely, very; for this

reason, therefore, consequently, accordingly; thus, this, that, then, as follows; in such a case, or state. *conj.* in order that; well; therefore; with the result that. *int.* expressing surprise, dawning awareness or dissent etc. *a.* true, corresponding; put in a set order, right. *pron.* the same. **and so forth, and so on** FORTH. **or so**, or thereabouts. **so be it**, let it be thus (in affirmation, resignation etc.). **so help me God** HELP. **so long!** *int.* good-bye. **so much**, (of, in or to) a particular amount, degree or extent. **so much as**, however much, to whatever extent. **so much for**, there is nothing more to be said; expressing contempt at a failure. **so much so**, to such a degree, extent (that). **so what?** what about it? **so-and-so**, *n.* an indefinite person or thing (*euphem.*) an unpleasant person or disliked thing. **so-called**, *a.* usually called thus (with implication of doubt). **so-so**, *a.* middling, mediocre. *adv.* indifferently.
so[2] SOH.
**So., **(*abbr.*) south, southern.
soak (sōk), *v.t.* to suck (in or up), to absorb (liquid); to put in liquid to become permeated, to steep, to wet thoroughly, to drench; to extract or remove by steeping in a liquid; (*sl.*) to overcharge, to make (a person) pay. *v.i.* to lie in liquid so as to become permeated, to penetrate, to permeate (into, through etc.); to drink excessively, to tipple. *n.* the process of soaking or being soaked; the period for which something is immersed; (*sl.*) a heavy drinker. **to soak in**, to become fully understood, appreciated, felt etc. **soakaway**, *n.* a hole or depression dug in the ground to allow drainage to percolate into the soil. **soakage** (-ij), *n.* **soaker**, *n.* **soaking**, *n., a.*
soap (sōp), *n.* an unctuous compound of a fatty acid and a base, usu. of sodium or potassium, used for washing and cleansing; (*sl.*) flattery, glib or persuasive talk; a soap opera. *v.t.* to rub or wash with soap; to flatter. **soft soap**, SOFT. **soapbox**, *n.* a box or improvised stand used as a platform by an orator. **soap-bubble**, *n.* a thin inflated film of soapy water. **soap opera**, *n.* a serialized television or radio drama usu. following a regular set of characters through various domestic or sentimental situations. **soapsuds**, *n.pl.* water impregnated with soap to form a foam. **soapy**, *a.* of the nature of or resembling soap; smeared or combined with soap; unctuous, flattering, smooth. **soapily**, *adv.* **soapiness**, *n.*
soar (saw), *v.i.* to fly aloft, to rise; of a bird, aircraft etc, to sail, float at a great height; to rise or mount intellectually or in spirit, status, position etc.; to increase or rise rapidly in amount, degree etc.; to tower. **soarer**, *n.* **soaringly**, *adv.*
sob (sob), *v.i.* (*past, p.p.* **sobbed**) to catch the breath in a convulsive manner, as in violent weeping. *v.t.* to utter with a sob or sobs; to bring on (a certain state) by sobbing. *n.* a convulsive catching of the breath, as in sobbing. **sob-story**, *n.* a hard-luck story intended to elicit pity. **sobstuff**, *n.* sentimental speech, writing, film etc. intended to arouse tears. **sobbingly**, *adv.*
sober (sō'bə), *a.* not drunk; temperate in the use of alcoholic liquors etc.; moderate, well-balanced, sane; self-possessed, dispassionate; serious, solemn, sedate; of colours etc., subdued, quiet. *v.t.* to make sober. *v.i.* to become calm, quiet or grave. **sober-minded**, *a.* **sober-mindedness**, *n.* **sobering**, *a.* **soberly**, *adv.* **soberness**, *n.* **sobriety** (-brī'ə-)
sobriety SOBER.
sobriquet (sō'brikā), **soubriquet** (soo'-), *n.* a nickname; an assumed name.
Soc., soc., (*abbr.*) socialist; society.
so-called SO[1].
soccer (sok'ə), *n.* (*coll.*) Association Football.
sociable (sō'shəbl), *a.* fit or inclined to associate or be friendly, companionable, affable. **sociability** (-bil'-), **sociableness**, *n.* **sociably**, *adv.*

social (sō'shəl), *a.* of or pertaining to society or its divisions, or to the intercourse, behaviour or mutual relations of humans; living in communities, gregarious, not solitary, tending to associate with others, fitted for existence in an organized, cooperative system of society; of, relating to, or conducive to shared activities or companionship; pertaining to the social services. *n.* a social gathering. **social anthropology**, *n.* a discipline within the social sciences concerned with systems of belief and cultural organization in a society. **social climber**, *n.* one who seeks membership of a higher social class, esp. by ingratiating him- or herself. **social climbing**, *n.* **social contract**, *n.* a collective agreement between members of a society and a government that secures the rights and liberties of each individual. **social democracy**, *n.* the theories and practices of socialists who believe in transforming a capitalist society into a socialist one by democratic means. **social democrat**, *n.* a supporter of social democracy; (**Social Democrat**) a member of the Social Democratic party. **social insurance**, *n.* insurance against unemployment, sickness or old age provided by the state out of contributions from employers and wage-earners. **social science**, *n.* the study of society and the interaction and behaviour of its members. **social scientist**, *n.* **social security**, *n.* state provision for the unemployed, aged or sick through a system of pensions or benefits. **social services**, *n.pl.* welfare services provided by the state or a local authority. **social work**, *n.* any of various types of welfare service provided by the social services. **social worker**, *n.* **Socialism**, *n., a.* involving the collective ownership of the sources and instruments of production, democratic control of industries, cooperation and state distribution of the products, free education etc. **socialist**, *n., a.* **socialistic** (-is'-), *a.* **socialistically**, *adv.* **socialite** (-līt), a member of fashionable society. **sociality** (-shial'-), *n.* **socialize, -ise**, *v.t.* to prepare, make fit for social life. *v.i.* to behave in a convivial or sociable manner. **socialization, -isation**, *n.* **socially**, *adv.*
socialite, sociality, socialize, socially SOCIABLE.
society (səsī'əti), *n.* a social community; the general body of persons, communities or nations constituting mankind regarded as a community; social organization; the privileged and fashionable classes of a community; a body of persons associated for some common object, an association; companionship, fellowship. *a.* of or pertaining to fashionable society. **Society of Friends**, the Quakers. **Society of Jesus**, the Jesuits. **societal**, *a.* of or pertaining to (human) society. **societally**, *adv.*
socio-, *comb. form.* social; society.
sociobiology (sōsiōbīol'əji), *n.* the study of human or animal behaviour from a genetic or evolutionary basis.
sociocultural (sōsiōkŭl'chərəl), *a.* of, pertaining to or involving social and cultural factors. **socioculturally**, *adv.*
socioeconomic (sōsiōekənom'ik, -ek-), *a.* of, pertaining to or involving social and economic factors. **socioeconomically**, *adv.*
sociolinguistics (sōsiōling-gwis'tiks), *n. sing.* study of language as it functions in society. **sociolinguist** (-ling'-), *n.*
sociology (sōsiol'əji), *n.* the science of the organization and dynamics of human society. **sociological** (-loj'-), *a.* **sociologically**, *adv.* **sociologist**, *n.* [F *sociologie*]
sociometry (sōsiom'ətri), *n.* the study of social relationships within a group. **sociometric** (-met'-), *a.* **sociometrist**, *n.*
sociopolitical (sōsiōpəlit'ikəl), *a.* of, pertaining to or involving social and political factors.
sock[1] (sok), *n.* a short stocking. **put a sock in it**, be quiet, shut up. **to pull up one's socks**, to make a vigorous effort to do better.
sock[2] (sok), *v.t.* (*sl.*) to hit with a blow. *n.* a hit, a blow. **to sock it to**, to act, speak etc. with great vigour or force.
socket (sok'it), *n.* a natural or artificial hollow place or

fitting adapted for receiving and holding another part or thing; an electric power point. **socket-joint** BALL AND SOCKET JOINT under BALL¹. **socketed**, *a*.
Socratic (səkrat′ik), *a*. of, pertaining to or according to Socrates, Greek philosopher, 469–399 BC. *n*. an adherent of Socrates or his philosophy. **Socratic method**, *n*. the dialectical method of procedure by question and answer introduced by Socrates. **Socratically**, *adv*.
sod¹ (sod), *n*. a piece of surface soil filled with the roots of grass etc., turf.
sod² (sod), *n*. (*sl*.) a despicable person, esp. male; a fellow, chap. *int*. used like a swear word to express annoyance. **sod all**, nothing at all. **sod off**, go away, get lost. **Sod's law** (sodz), *n*. a wry maxim that anything which can possibly go wrong will do so.
soda (sō′də), *n*. any of various compounds of sodium, e.g. sodium carbonate, sodium hydroxide, sodium bicarbonate; soda-water; (*N Am*.) a fizzy soft drink. **soda bread**, a type of bread made with baking soda. **soda-fountain**, *n*. a device for dispensing soda-water; (*N Am*.) a counter serving soft drinks, ice creams etc. **soda siphon**, *n*. a pressurized bottle for dispensing soda-water. **soda-water**, *n*. an effervescent drink composed of water charged with carbon dioxide. **sodaic** (-dā′-), *a*. **sodium** (-diəm), *n*. a silver-white metallic element, the base of soda. **sodium bicarbonate**, *n*. a white powder used in baking powder and antacid preparations, sodium hydrogencarbonate, baking soda. **sodium carbonate**, *n*. a crystalline salt used in the manufacture of cleaning agents, glass etc., washing soda. **sodium chloride**, *n*. common salt. **sodium lamp** SODIUM(-VAPOUR) LAMP. **sodium nitrate**, *n*. a white crystalline salt occurring naturally as Chile saltpetre, used in fertilizers, explosives etc. **sodium(-vapour) lamp**, *n*. an electric lamp used esp. in street lighting, consisting of a glass tube containing sodium vapour and neon which emits an orange light when current is passed through it.
sodality (sədal′iti), *n*. a fellowship, a confraternity, esp. a charitable association in the Roman Catholic Church.
sodden (sod′n), *a*. soaked, saturated; bloated and stupid, esp. with drink. **soddenness**, *n*.
sodium SODA.
sodomite (-mit), *n*. one who practises sodomy. **sodomitic, -ical** (-mit′-), *a*. **sodomy** (-i), *n*. anal intercourse with a man or woman, or sexual relations with an animal.
soever (sōev′ə), *adv*. appended, sometimes as a suffix, and sometimes after an interval, to pronouns, adverbs or adjectives to give an indefinite or universal meaning.
sofa (sō′fə), *n*. a long stuffed couch or seat with raised back and ends. **sofa-bed**, *n*. a sofa that can be extended so as to serve as a bed, a bed-settee.
soft (soft), *a*. yielding easily to pressure, easily moulded, cut or worked, malleable, pliable, plastic, opp. to *hard*; affecting the senses in a mild or gentle manner; of a day, a breeze etc., balmy, gentle; low-key, non-insistent; of an image, blurred; smooth to the touch; of colours, outlines etc., not brilliant, glaring, or abrupt; not loud or harsh, low-toned; of water, free from mineral salts that prevent lathering; of a drug, relatively harmless or nonaddictive; gentle or mild in disposition, conciliatory; impressionable, sympathetic, compassionate; flaccid, out of condition, pampered; easily imposed on, lenient; weak, timorous, effeminate; sentimental, tender-hearted; (*Phon*.) not guttural or explosive, sibilant (as *c* in *cede* or *g* in *gem*), voiced (as *b, d,* and *g*); easy. *n*. a soft part, object or material. *adv*. softly, gently, quietly. **soft in the head**, feeble-minded, foolish. **to be soft on**, to be lenient towards. **softball**, *n*. a game resembling baseball played with a larger and softer ball. **soft-boiled**, *a*. (of an egg) boiled for a brief time so that the yoke remains soft. **soft-core**, *a*. of pornography, relatively inexplicit and mild. **soft-cover**, *n., a*. (a) paperback. **soft currency**, *n*. a currency that is unstable owing to the uncertainty of its gold back-

ing. **soft drinks**, *n.pl*. non-intoxicant beverages. **soft-focus**, *n., a*. (having, designed to produce) a slightly out-of-focus image with blurred edges. **soft-furnishings**, *n.pl*. textile furnishings such as carpets, curtains, chair covers etc. **soft goods**, *n.pl*. textiles. **soft-headed**, *a*. silly, stupid. **soft-headedness**, *n*. **soft-hearted**, *a*. tender-hearted, compassionate. **soft-heartedly**, *adv*. **soft-heartedness**, *n*. **soft-land**, *v.i., v.t*. (to cause a spacecraft) to land gently on the moon, a planet, without incurring damage. **soft landing**, *n*. **soft option**, *n*. an option offering least difficulty. **soft palate**, *n*. the posterior part of the palate terminating in the uvula. **soft-pedal**, *n*. a foot pedal for subduing the tone of notes played on the piano. *v.i., v.t*. to play down, avoid the issue of. **soft porn(ography)**, *n*. soft-core pornography. **soft sell**, *n*. selling by means of gentle persuasiveness or suggestion. **soft soap**, *n*. semiliquid soap made with potash; (*coll*.) flattery, blarney. **soft-soap**, *v.t*. (*coll*.) to flatter for some ulterior object. **soft-spoken**, *a*. speaking softly; mild, affable, conciliatory. **soft spot**, *n*. tenderness, fondness. **soft touch**, *n*. (*sl*.) someone easily influenced or imposed upon. **software**, *n*. computer programs. **soft water**, *n*. water free from mineral salts that prevent lathering. **softwood**, *n*. the wood of a coniferous tree. **soften** (sof′n), *v.t*. to make soft or softer; to palliate, to mitigate, tone down. *v.i*. to become soft or softer. **to soften up**, to make more sympathetic to; to break down the resistance of. **softener**, *n*. **softening**, *n*. **softening of the brain**, a softening of cerebral tissue resulting in mental deterioration; (*coll*.) stupidity, excessive credibility. **softly**, *adv*. **softness**, *n*. **softy, softie** (-ti), *n*. a silly, weak-minded person; one who is physically unfit or flaccid; a tender-hearted person.
soften, softly, softy etc. SOFT.
SOGAT (sō′gat), (*acronym*) *S*ociety *o*f *G*raphical and *A*llied *T*rades.
soggy (sog′i), *a*. soaked, sodden, thoroughly wet; heavy with moisture; (*coll*.) dull, heavy, spiritless. **soggily**, *adv*. **sogginess**, *n*.
soh, so (sō), **sol** (sol), *n*. the fifth tone of the diatonic scale; the syllable denoting it.
soi-disant (swahdēzä′), *a*. self-styled, pretended, so-called.
soigné, (*fem*.) **soignée** (swahn′yä), *a*. well-turned-out, elegant, exquisite in taste.
soil¹ (soil), *n*. the ground, esp. the top stratum of the earth's crust whence plants derive their mineral food; land, country; that which nourishes or promotes development.
soil² (soil), *v.t*. to make dirty; to sully, to tarnish, to pollute. *v.i*. to become sullied or dirty, to tarnish. *n*. any foul matter, filth, refuse, dung, compost. **soil-pipe**, *n*. a pipe carrying waste material and water from a toilet. **soiled**, *a*.
soirée (swah′rä), *n*. an evening party or gathering for conversation and social intercourse etc.
sojourn (soj′œn. sō′-), *v.i*. to stay or reside (in, among etc.) temporarily. *n*. a temporary stay or residence. **sojourner**, *n*.
Sol¹, (*abbr*.) Solomon.
sol², (*abbr*.) soluble; solution.
solace (sol′əs), *n*. comfort in grief, trouble etc., consolation, compensation; *v.t*. to comfort or console, in trouble etc.
solar (sō′lə), *a*. pertaining to, proceeding from, measured by or powered by the sun. **solar battery**, *n*. a battery powered by solar cells. **solar cell**, *n*. a cell that converts solar energy into electricity. **solar energy**, *n*. energy derived from the sun. **solar panel**, *n*. a panel of solar cells functioning as a power source. **solar plexus**, *n*. the epigastric plexus, a network of nerves behind the stomach. **solar power**, *n*. solar energy. **solar system**, *n*. the sun and the various heavenly bodies revolving about it. **solarium** (-leə′riəm), *n*. (*pl*. **-ria, riums**) a room or building con-

sold (sōld), *past, p.p.* SELL [1].

structed for the enjoyment of, or therapeutical exposure of the body to, the rays of the sun.

solder (sŏl'də), *n.* a fusible alloy for uniting the edges etc. of less fusible metals; anything that cements or unites. *v.t.* to unite or mend with or as with or as with solder. *v.i.* to become united or mended (as) with solder. **solderer**, *n.* **soldering**, *n., a.* **soldering-iron**, *n.* a tool used hot for melting and applying solder.

soldier (sōl'jə), *n.* a person engaged in military service; a person of military skill or experience; one who works diligently for a cause. **soldier of fortune**, a military adventurer, a mercenary; one who lives on his or her wits. **to soldier on**, to persevere doggedly in the face of difficulty. **soldier-like, -ly,** *a., adv.* **soldiering,** *n.* **soldiership,** *n.* **soldiery,** *n.* soldiers collectively; a body of soldiers.

sole[1] (sōl), *n.* the flat under side or bottom of the foot; the part of a boot or shoe under the foot, esp. the part in front of the heel. *v.t.* to furnish (a boot etc.) with a sole.

sole[2] (sōl), *n.* a flat-fish of various species highly esteemed as food.

sole[3] (sōl), *a.* single, only, unique, alone in its kind. **solely,** *adv.*

solecism (sol'isizm), *n.* a deviation from correct idiom or grammar; any incongruity, error or absurdity; a breach of good manners, an impropriety. **solecist,** *n.* **solecistic, -ical** (-sis'-), *a.* **solecistically,** *adv.*

solely SOLE [3].

solemn (sol'əm), *a.* performed with or accompanied by rites, ceremonies or due formality; awe-inspiring, impressive; grave, serious, momentous; formal, affectedly grave, self-important, pompous. **solemnify** (-lem'nifi), *v.t.* to make solemn. **solemnification** (-fi-), *n.* **solemnity** (-lem'ni-), *n.* solemnness, impressiveness; affected gravity or formality; (*often pl.*) a rite or ceremony, esp. one performed with religious reverence. **solemnize, -ise** (-niz), *v.t.* to dignify or to celebrate with solemn formalities or ceremonies. **solemnization, -isation,** *n.* **solemnly,** *adv.* **solemnness, solemness,** *n.*

solenoid (sol'ənoid, sō'-), *n.* a magnet consisting of a cylindrical coil traversed by an electric current. **solenoidal** (-noi'-), *a.*

sol-fa (solfah'), *n.* solmization; tonic sol-fa.

solicit (səlis'it), *v.t.* to make earnest or importunate request for; to make earnest or persistent requests or appeals to; to entice or incite (someone) to do something illegal or immoral; of a prostitute, openly to offer sexual relations in exchange for money. *v.i.* to make earnest or importunate appeals; of a prostitute, to proposition someone as a potential client. **solicitant,** *n., a.* **solicitation,** *n.* **solicitor,** *n.* a legal practitioner authorized to advise clients and prepare causes for barristers. **Solicitor General,** *n.* a law officer of the British Crown ranking next to the Attorney General, appointed by the government in power to advise and represent it in legal matters. **solicitorship,** *n.* **solicitous,** *a.* anxious, concerned, apprehensive, disturbed (about, for etc.); eager (to). **solicitously,** *adv.* **solicitousness, solicitude** (-tūd), *n.*

solid (sol'id), *a.* composed of particles closely cohering, dense, compact; not hollow, devoid of cavities, interstices or crevices, not porous; uniform, uninterrupted; firm, unyielding, stable; sound, substantial, not flimsy; real, genuine, reliable, well-grounded; the same throughout, homogeneous; thinking, feeling or acting unanimously; (*Geom.*) of three dimensions, cubic; *adv.* in a solid manner. *n.* a rigid, compact body; a body or magnitude possessing length, breadth, and thickness. **solid fuel,** *n.* fuel composed of solid matter (e.g. coal) rather than gas or liquid; solid propellant. **solid state,** *n., a.* **solid-state physics,** *n.* a branch of physics dealing with the properties and nature of solid matter. **solidify** (-lid'ifi), *v.t., v.i.* to make or become solid. **solidifiable,** *a.* **solidification** (-fi-), *n.* **solidifier,** *n.* **solidity** (-lid'-), **solidness,** *n.* **solidly,** *adv.*

solidarity (solida'riti-), *n.* cohesion, mutual dependence; community of interests, feelings, responsibilities etc.

solidus (sol'idəs), *n.* (*pl.* **-di** (-dī)), the stroke (/) formerly denoting a shilling (as in 2/6); also used in writing fractions (e.g. 1/4), separating numbers (e.g. in dates) or alternative words (as in him/her) etc.

soliloquy (səlil'əkwi), *n.* a talking to oneself; a speech or discourse, esp. in a play, uttered to oneself, a monologue. **soliloquist,** *n.* **soliloquize, -ise,** *v.i.* **soliloquizer, -iser,** *n.*

solipsism (sol'ipsizm), *n.* the theory that the only knowledge possible is that of oneself, absolute egoism. **solipsist,** *n., a.* **solipsistic** (-sis'-), *a.*

solitaire (sol'iteə), *n.* a gem, esp. a diamond, set singly, in a ring etc.; a game played by one person on a board with hollows and marbles, holes and pegs etc.; a card game for one player, patience.

solitary (sol'itəri), *a.* living or being alone, lonely, not gregarious; of plants, growing singly; passed or spent alone; unfrequented, sequestered, secluded; single, individual, sole. *n.* one who lives in solitude a recluse. **solitary confinement,** *n.* in a prison, incarceration without the company of others, isolation. **solitarily,** *adv.* **solitariness,** *n.* **solitude,** (-tūd), *n.* loneliness, solitariness, seclusion.

solmization, -isation (solmizā'shən), *n.* the association of certain syllables with the notes of the musical scale, a recital of the notes of the gamut, sol-faing.

solo (sō'lō), *n.* (*pl.* **solos, soli** (-lē)) a composition or passage played by a single instrument or sung by a single voice, usu. with an accompaniment; solo whist, a call in this game; a solo flight. *a., adv.* unaccompanied, alone. **solo flight,** *n.* a flight in an aircraft by a single person. **soloist,** *n.*

Solomon (sol'əmən), *n.* a very wise man (after King *Solomon* of Israel, d. *c.* 930 BC). **Solomonic** (-mon'-), *a.*

so long, *int.* (*coll.*) good-bye.

solstice (sol'stis), *n.* the time (about 21 June and 22 Dec.) and point at which the sun is farthest from the celestial equator (north in summer and south in winter). **solstitial** (-stish'əl), *a.*

soluble (sol'ūbl), *a.* capable of being dissolved in a fluid; capable of being solved. **solubility** (-bil'-), *n.* the quality or state of being soluble; the number of grams of substance required to saturate 100 grams of solvent. **solubilize, -ise,** *v.t.* **solution** (səloo'shən), *n.* the liquefaction of a solid or gaseous body by mixture with a liquid; the liquid combination so produced; the condition of being dissolved; the resolution or act or process of solving a problem, difficulty etc.; the correct answer to a problem etc.

solus, (sō'ləs), *a.* (*fem.* **sola** (-lə), alone (used esp. in stage directions).

solute, solution SOLUBLE.

solve (solv), *v.t.* to resolve or find an answer to (a problem etc.); to clear up, to settle, to put an end to. **solvable,** *a.* **solvability** (-bil'-), *n.* **solvent** (-vənt), *a.* having the power to dissolve; able to pay all just debts or claims. *n.* a liquid that can dissolve a substance. **solvent abuse,** *n.* the use of solvents (such as glue or petrol) as drugs, by inhaling their fumes. **solvency,** *n.* **solver,** *n.*

solver SOLVE.

Som., (*abbr.*) Somerset.

soma (sō'mə), *n.* (*pl.* **-mata** (-tə), **-mas**) the axial part of the body, i.e. without the limbs; the body as distinct from the germ cells.

Somali (səmah'li), *n.* (*pl.* **-lis, -li**) a member of a people inhabiting Somalia in NE Africa; the language of this people. *a.* of or pertaining to this people, their language or their country.

somatic (səmat'ik), *a.* pertaining to the body or the body

wall, corporeal, physical. *n.pl.* somatology. **somatically**, *adv.*
somato-, *comb. form.* body.
somatology (sōmətol'əji), *n.* the science of organic bodies, esp. human anatomy and physiology. **somatological** (-loj'-), *a.* **somatologist**, *n.*
sombre (som'bə), *a.* dark, gloomy; solemn, melancholy. **sombrely**, *adv.* **sombreness**, *n.*
sombrero (sombreə'rō), *n. (pl.* **-ros**) a wide-brimmed hat worn largely in Mexico.
some (sŭm), *a.* an indeterminate quantity, number etc. of; an appreciable if limited amount etc. of; several; a few, a little; a considerable quantity, amount etc. of; a certain, a particular but not definitely known or specified (person or thing); (*chiefly N Am.*) striking, outstanding. *adv.* about, approximately; (*coll.*) to some extent. *pron.* a particular but undetermined part or quantity; certain not definitely known or unspecified ones. **somebody**, *pron.* some person. *n.* a person of consequence. **someday**, *adv.* at some unspecified time in the future. **somehow**, *adv.* in some indeterminate way; in some way or other; by some indeterminate means. **someone**, *pron.* somebody. **someplace**, *adv.* somewhere. **something**, *n.* some indeterminate or unspecified thing; some quantity or portion if not much; a thing of consequence or importance. *adv.* in some degree. **something else**, (*coll.*) a person or thing inspiring wonder, awe, disbelief etc. **something like** LIKE [1]. **sometime**, *adv.* once, formerly, at one time; at some unspecified time. *a.* former, late. **sometimes**, *adv.* at some times, now and then. **someway**, *adv.* in some unspecified way. **somewhat**, *adv.* to some extent, rather. *n.* a certain amount or degree; something. **somewhere**, *adv.*, *adv.* (in, at or to) some unknown or unspecified place; (in) some place or other. **to get somewhere**, to make headway, to progress.
-some [1] (-səm), *suf.* forming adjectives, full of, as in *gladsome, troublesome, winsome;* forming nouns, denoting a group with a specified number of members, as in *threesome, foursome.*
-some [2] (-sōm), *comb. form.* a body, as in *chromosome.*
somebody, -how etc. SOME.
somersault, summersault (sŭm'əsawlt), *n.* a leap in which one turns heels over head and lands on one's feet. *v.i.* to execute a somersault.
something, -time, -what etc. SOME.
somnambulism *n.* the act or condition of walking or performing other actions in sleep or a condition resembling sleep. **somnambulist**, *n.* **somnambulistic** (-lis'-), *a.*
somnolent (som'nələnt), *a.* sleepy, drowsy; inducing sleep. **somnolence, -ency**, *n.* **somnolently**, *adv.*
son (sŭn), *n.* a male child in relation to a parent or parents; a male descendant; a form of address used by an old person to a youth, a priest or teacher to a disciple etc.; a native of a country; an inheritor, exponent or product of (a quality, art, occupation etc.). **son of a bitch**, a despicable or unpleasant man. **son of a gun**, (*coll.*) fellow (used as a jocular mode of address). **the Son (of Man)**, the second person in the Trinity, Christ, the Messiah. **son-in-law**, *n. (pl.* **sons-in-law**) the husband of a daughter. **sonless** (-lis), *a.* **sonny**, *n.* a familiar, often patronizing or derogatory term of address to a boy or man. **sonship**, *n.* the state of being a son.
sonar (sō'nah), *n.* a device which detects the presence and position of underwater objects by means of echo-soundings.
sonata (sənah'tə), *n.* an instrumental composition, esp. for the piano, usu. of three or four movements in different rhythms. **sonatina** (sonətē'nə), *n.* a short or simple sonata.
sonde (sond), *n.* a scientific device for gathering information about atmospheric conditions at high altitudes.
son et lumière (son à loomieə'), *n.* an outdoor entertainment at a historic location which recreates past events associated with it using sound effects, a spoken narration, music, and special lighting.
song (song), *n.* musical or modulated utterance with the voice, singing; a melodious utterance, as the musical cry of a bird; a musical composition accompanied by words for singing; an instrumental piece of a similar character; a short poem intended or suitable for singing, esp. one set to music; poetry, verse; a trifle, a small sum. **going for a song**, selling for a trifle. **song and dance**, a fuss. **Song of Songs** or **of Solomon**, an Old Testament book attributed to Solomon, containing love songs. **song-bird, -sparrow, -thrush**, *n.* a bird that sings. **song cycle**, *n.* a sequence of songs concerned with the same subject or theme. **songwriter**, *n.* one who composes (esp. popular) songs. **songwriting**, *n.* **songless** (-lis), *a.* **songster** (-stə), *n.* one skilled in singing; a song-bird. **songstress** (-stris), *n. fem.*
sonic (son'ik), *a.* of, pertaining to or producing soundwaves; travelling at about the speed of sound. **sonic boom**, *n.* the loud noise caused by a shock-wave produced by an aircraft or projectile travelling at supersonic speed. **sonically**, *adv.*
sonnet (son'it), *n.* a poem of 14 iambic pentameter lines, octave rhyming *a b b a a b b a*, and a sestet with three rhymes variously arranged. **sonneteer** (-tiə'), *n.* a writer of sonnets. *v.i.* to compose sonnets.
sonny SON.
sonorous (son'ərəs), *a.* giving out sound, resonant; loud sounding, sounding rich or full; high sounding, impressive. **sonority** (-no'-), **sonorousness**, *n.* **sonorously**, *adv.*
soon (soon), *adv.* in a short time from now or after a specified time, early; quickly, readily, willingly. **as soon as, so soon as**, at the moment that; immediately after; not later than. **no sooner ... than**, immediately. **sooner or later**, sometime or other; inevitably, eventually. **sooner rather than later**, fairly soon.
soot (sut), *n.* a black substance composed of carbonaceous particles rising from fuel in a state of combustion and deposited in a chimney etc. *v.t.* to cover, manure or soil with soot. **sooty**, *a.* **sootily**, *adv.* **sootiness**, *n.*
sooth (sooth), *n.* truth, reality. **soothsayer**, *n.* a seer, a diviner.
soothe (soodh), *v.t.* to calm, to tranquillize; to soften, to mitigate, to assuage; to humour, to flatter, to gratify. **soother**, *n.* **soothingly**, *adv.*
sop (sop), *n.* something given to pacify. **sopping**, *a.* wet through. **soppy**, *a.* maudlin, sentimental, weak-minded. **soppily**, *adv.* **soppiness**, *n.*
sophism (sof'izm), *n.* a specious but fallacious argument. **sophist**, *n.* one of a class of men in ancient Athens who taught philosophy, dialectic, rhetoric etc., for pay; a fallacious reasoner, a quibbler. **sophistic, -ical** (-fis'-), *a.* **sophistically**, *adv.* **sophisticate** (-fis'tikàt), *v.t.* to deprive of simplicity, to make perverted, affected or artificial; to make more complicated or refined. *v.i.* to be sophistical. *n.* (-kət), a sophisticated person. **sophisticated**, *a.* worldly-wise, superficially; self-assured; complex, highly developed; subtle; refined; cultured. **sophistication**, *n.* **sophisticator**, *n.* **sophistry**, *n.* a specious but fallacious argument; the art of reasoning using such argument(s); this kind of reasoning.
Sophoclean (sofəklē'ən), *a.* pertaining to or characteristic of Sophocles, the Greek tragic poet.
sophomore (sof'əmaw), *n. (chiefly N Am.)* a second-year student. **sophomoric, -ical** (-mo'-), *a.*
-sophy, *comb. form.* denoting (a branch of) knowledge.
soporific (sopərif'ik), *a.* causing or tending to cause sleep; drowsy, sleepy. *n.* a soporific medicine or agent.
soprano (səprah'nō), *n. (pl.* **-nos, -ni** (-nē)) a female or boy's voice of the highest kind; a singer having such a voice; a musical part for such voices; an instrument which

has the highest range within a family of instruments. *a.* of or having a treble part, voice or pitch. **sopranino** (sopranē'nō), *n.*, *a.* (*pl.* **-nos**) (an instrument) possessing the highest pitch in a family of instruments. **sopranist**, *n.*
sorb (sawb), *n.* the service-tree (also **sorb-apple**); its fruit.
sorbefacient (sawbifā'shənt), *a.* promoting absorption. *n.* a substance or preparation promoting absorption.
sorbet (saw'bā), *n.* an ice flavoured with fruit juice, spirit etc.; sherbet.
sorcerer (saw'sərə), *n.* one who uses magic, witchcraft or enchantments, a wizard. **sorceress** (-ris), *n.* **sorcerous**, *a.* **sorcery**, *n.*
sordid (saw'did), *a.* mean, ignoble, vile. **sordidly**, *adv.* **sordidness**, *n.*
sordine (saw'dēn), **-dino** (-dē'nō), *n.* (*pl.* **-dini** (-nē)) a contrivance for deadening the sound of a musical instrument, a mute, a damper. **con** or **senza sordini, -dino**, to be played with or without mute.
sore (saw), *a.* tender and painful to the touch, esp. through disease or irritation; mentally distressed, aggrieved, vexed; easily annoyed; touchy; causing annoyance, irritating, exasperating; (*coll.*) annoyed. *adv.* (*obs.*) sorely, grievously, severely, intensely. *n.* a sore place on the body where the surface is bruised, broken or inflamed by a boil, ulcer etc.; that which excites resentment, remorse, grief etc. **to stand out like a sore thumb**, to be highly conspicuous. **sore point**, *n.* a subject etc. which arouses irritation, annoyance, hurt feelings etc. **sorely**, *adv.* **soreness**, *n.*
sorel(l) SORREL 2.
sorghum (saw'gəm), *n.* any member of a genus of plants containing the Indian millet, durra etc., much cultivated in the US for fodder etc.
sorites (səri'tēz), *n.* a series of syllogisms so connected that the predicate of one forms the subject of that which follows, the subject of the first being ultimately united with the predicate of the last; a sophistical argument in this form. **soritic, -ical** (-rit'-), *a.*
soroptimist (sərop'timist), *n.* a member of an international organization of women's clubs, Soroptimist International.
sororicide (-sīd), *n.* the murder of a sister; the murderer of a sister.
sorority (-ro'ri-), *n.* a body or association of women, a sisterhood; (*N Am.*) a society of women students. **sorority house**, *n.* (*N Am.*) the residence of members of an academic sorority, usu. on the campus of a college or university.
sorosis (sərō'sis), *n.* a fleshy fruit formed by the cohesion of numerous flowers etc., as the pineapple.
sorrel[1] (so'rəl), *n.* a herb with acid leaves, *Rumex acetosa,* allied to the dock, wood sorrel.
sorrel[2] (so'rəl), *a.* of a reddish or yellowish-brown. *n.* this colour; a horse (**sorrel horse**) or other animal of this colour.
sorrily, sorriness SORRY.
sorrow (so'rō), *n.* mental pain or distress from loss, disappointment etc., grief, sadness; an event, thing or person causing this, an affliction; a misfortune; mourning, lamentation. *v.i.* to grieve; to lament. **sorrower**, *n.* **sorrowful**, *a.* **sorrowfully**, *adv.* **sorrowfulness**, *n.* **sorrowing**, *n., a.*
sorry (so'ri), *a.* feeling or showing grief or pity for some loss etc., regretful; poor, paltry, pitiful, despicable; apologetic. *int.* expressing apology. **sorrily**, *adv.* **sorriness**, *n.*
sort (sawt), *n.* a number (of things etc.) having the same or similar qualities, a class, kind, type or species; an example or instance of a kind; an example or instance of something or someone sharing inadequate or remote characteristics with a kind; (*coll.*) a person, a type (of person); fashion, way, manner. *v.t.* to separate into sorts, classes etc.; to select from a number; to arrange. **a good sort**, an attractive, companionable person; a decent type. **of a sort, of sorts**, of an inferior or inadequate kind. **out of sorts**, irritable, moody; slightly unwell. **sort of**, rather, to a degree, as it were. **to sort out**, to solve or resolve; to clear out; tidy up; to separate; to arrange; (*coll.*) to beat, to punish. **sortable**, *a.* **sorter**, *n.* one who or that which sorts (e.g. postal material). **sortment**, *n.*
sortie (saw'ti), *n.* a sally, esp. of troops from a besieged place in order to attack or raid; a mission or attack by a single aircraft.
sorus (saw'rəs), *n.* (*pl.* **sori** (sō'rī)) a heap, group or cluster, esp. of spore-cases on the fronds of ferns.
S O S (es ō es'), *n.* an internationally recognized distress call in Morse code; any distress call or plea for help (e.g. an emergency broadcast on television or radio). *v.i.* to call for help or rescue.
so-so so.
sostenuto (sostinū'tō), *adv., a.* in a steadily sustained manner.
sot (sot), *n.* an habitual drunkard, one habitually muddled (as if) with excessive drinking. **sottish**, *a.*
soteriology, *n.* the doctrine of salvation. **soteriological**, *a.*
sotto voce (so'tō vō'chā), *adv.* in an undertone. [It]
sou (soo), *n.* a French copper coin, formerly worth 1/12 of a livre; the 5-centime piece; a very small amount of money.
soubrette (soobret'), *n.* an intriguing, mischievous coquettish female character in a comedy, esp. the role of a lady's maid; one who displays similar characteristics, a flirt, a coquette.
soubriquet SOBRIQUET.
souchong (soo'shong. -chong'), *n.* a black tea made from the youngest leaves.
souffle (soo'fl), *n.* a low whispering or murmur heard in the auscultation of an organ etc.
soufflé (soo'flā), *n.* a light dish made of beaten whites of eggs etc. *a.* made light and frothy. **souffléed**, *a.*
sough (sow, sŭf. *Sc.* sookh) *v.i.* to make a murmuring, sighing sound, as the wind. *n.* such a sound.
sought (sawt), *past, p.p.* SEEK.
souk (sook), *n.* an outside, often covered market in a Muslim country (esp. in N Africa and the Middle East).
soul (sōl), *n.* the spiritual part of a person; a spiritual being; the moral and emotional part of a person; the rational part of a person, consciousness; the vital principle and mental powers possessed by humans in common with lower animals; the essential or animating or inspiring force or principle, the life, the energy in anything; one who inspires, a leader, a moving spirit; the heart; spirit, courage, nobility; a disembodied spirit; a human being, a person; an epitome, embodiment or exemplification; soul music. *a.* of or pertaining to soul music; of, relating to or characteristic of Black Americans or their culture, food etc. **the life and soul**, the liveliest or most entertaining person at a party, in company etc. **soul brother, sister**, *n.* a fellow Black person. **soul-destroying**, *a.* unrewarding, frustrating, boring. **soul food**, *n.* (*coll.*) the traditional foods of American Blacks in the south (e.g. yams, chitterlings). **soul mate**, *n.* one with whom a person feels a close affinity. **soul music**, *n.* a popular type of Black music combining elements of blues, gospel, jazz and pop. **soul-searching**, *n.* a critical and close examination of one's motives, actions etc. **souled**, *a.* (*usu. in comb.*), as *high-souled*) **soulful**, *a.* having or showing deep or sad feeling. **soulfully**, *adv.* **soulfulness**, *n.* **soulless**, *a.* **soullessly**, *adv.* **soullessness**, *n.*
sound[1] (sownd), *a.* whole, unimpaired, free from injury, defect or decay; (of sleep) deep, unbroken; not diseased or impaired, healthy; well-grounded, wise, well established; orthodox; trustworthy, honest; solid, stable, firm; based on truth or reason, valid, correct; solvent; thorough, complete. *adv.* soundly, fast (asleep). **soundly**, *adv.* **soundness**, *n.*

sound² (sownd), *n.* the sensation produced through the organs of hearing; that which causes this sensation, the vibrations affecting the ear, esp. those of a regular and continuous nature as opp. to noise; a specific tone or note; an articulate utterance corresponding to a particular vowel or consonant; (*usu. pl.*) (*sl.*) music, esp. popular music. *a.* of or pertaining to radio as opposed to television. *v.i.* to make or give out sound; to convey a particular impression by or as by sound; to summons, to call; to resonate. *v.t.* to cause to sound, to utter audibly; to give a signal for by sound; to cause to resound, to make known, to proclaim; to test by sound. **to sound off**, to boast; to speak loudly, volubly, angrily etc. **sound barrier**, *n.* the shock wave produced when a moving body attains the speed of sound. **sound board**, *n.* a board for enhancing the sounds made by various musical instruments; a sounding board. **sound-box**, *n.* a hollow cavity in the belly of some musical instruments. **sound effect**, *n.* (*often pl.*) an imitation or reproduction of a sound used in the performance of a play or on the soundtrack of a film or broadcast. **sound-proof**, *a.* impenetrable to sound. *v.t.* to make impenetrable to sound, to insulate against sound. **sound-track**, *n.* the portion along the side of a film which bears the continuous recording of the accompanying sound; the synchronized sound-recording accompanying a film etc. **soundwave**, **sounder**, *n.* that which causes or emits a sound. **sounding**, *a.* plausible, pompous, high-flown. **sounding board**, *n.* a person, institution, group etc., used to test reaction to a new idea or plan. **soundless**, *a.* without a sound, silent. **soundlessly**, *adv.*
sound³ (sownd), *n.* a narrow passage of water, as a strait connecting two seas.
sound⁴ (sownd), *v.t.* to measure the depth of (a sea, channel, water in a ship's hold etc.) with a sounding-line or rod; to test or examine by means of a probe etc. *v.i.* to take soundings, to ascertain the depth of water. *n.* an instrument for exploring cavities of the body, a probe; **to sound out**, to test, to examine, to endeavour to discover (intentions, feelings etc.); **sounder**, *n.* a device for taking soundings. **sounding**, *n.* the act of measuring the depth of water; (*usu. pl.*) a measurement of depth taken thus; a test, an examination, a probe.
sounding SOUND ², ⁴.
soup (soop), *n.* a liquid food made from meat, fish or vegetables and stock. (*coll.*) anything resembling soup in consistency etc. (e.g. a thick fog). **in the soup**, (*sl.*) in difficulties, in trouble. **to soup-up**, to modify (the engine of a car or motorcycle) in order to increase its power. **soup-kitchen**, *n.* a public establishment for supplying soup to the poor; a mobile army kitchen. **soup-plate**, **-spoon**, *n.* a deep plate or spoon for holding or drinking soup. **soupy**, *a.*
soupçon (soop'sŏ), *n.* a mere trace, taste or flavour (of). [F]
sour (sowǝ), *a.* sharp or acid to the taste, tart; tasting thus through fermentation, rancid; harsh of temper, crabbed, morose, peevish; disagreeable, jarring, inharmonious; of soil, excessively acidic or infertile. *v.t.* to make sour. *v.i.* to become sour. **sour cream**, *n.* fresh cream soured by the introduction of bacteria, used in salads, cooking etc. **sourdough**, *a.* of bread, made with fermenting yeast. **sour grapes**, GRAPE. **sourpuss**, *n.* (*coll.*) a habitually morose person. **soured**, *n.*, *a.* **souring**, *n.* the process of becoming or turning sour. **sourish**, *a.* **sourly**, *adv.* **sourness**, *n.*
source (saws), *n.* the spring or fountain-head from which a stream of water proceeds, a first cause, a generating force etc.; an origin, a beginning; one who or that which gives out, initiates or creates something; a person or thing that provides inspiration or information. **source-book**, *n.* a book containing original documents for study. **source program**, *n.* an original computer program which has been converted into machine language.
sourdine (suǝdēn'), SORDINE.
sousaphone (soo'zǝfōn), *n.* a brass wind-instrument resembling the tuba.
souse (sows), *n.* a dip or plunging into water or another liquid; (*sl.*) a drunkard. *v.t.* to pickle; to plunge into or drench thoroughly with water etc.; (*sl.*) to inebriate. **soused**, *a.* (*sl.*) drunk; pickled; drenched.
soutane (sootan'), *n.* a cassock.
south (sowth), *n.* that one of the four cardinal points of the compass directly opposite to the north; the South. *a.* situated in the south; facing in the southern direction; of the wind, coming from the south. *adv.* towards the south; of the wind, from the south. **the South**, a southern part or region, esp. the area of England south of the Wash or the American states south of the Mason–Dixon Line; the less developed countries of the world, the Third World. **the South Sea(s)**, the Pacific Ocean; the seas south of the equator. **South African**, *a.* of or pertaining to the Republic of South Africa, its inhabitants or any of their languages. *n.* a native, citizen or inhabitant of South Africa. **southbound**, *a.* going or leading south. **Southdown**, *a.* of or pertaining to the South Downs, Sussex; *n.* a breed of hornless sheep originating here. **south-east**, *n.* the point of the compass equally distant from the south and the east; *a.* pertaining to or coming from the SE. *adv.* towards the SE. **the South-east**, the southeastern area of Britain including London. **south-easter**, *n.* a SE wind. **south-easterly**, *a.*, *adv.*, *n.* **south-eastern**, *a.* **south-easternmost**, *a.* **southeastwards**, *a.* in the direction of, or coming from, the SE. *n.* a southeasterly direction or southeastern area. *adv.* towards the SE. **southeastwardly**, *a.*, *adv.* **south-eastwards**, *adv.* **southpaw**, *n.*, *a.* (of or pertaining to) a left-hand person, esp. a left-handed boxer. **South Pole**, *n.* the most southerly point on the earth's axis or the celestial sphere. **south south-east**, **south-west**, *n.* a compass point midway between south and south-east, south and south-west. **south-west**, *n.*, *a.* **the South-west**, the south-western part of Britain. **southwester**, *n.* a wind from the SW. **sou'wester**, *n.* a waterproof hat with a wide brim hanging down behind, worn by sailors etc. **southwesterly**, *a.*, *adv.*, *n.* **south-western**, *a.* **southwesternmost**, *a.* **south-westward**, *a.*, *adv.*, *n.* **south-westwards**, *adv.* **souther**, *n.* a south wind. **southerly** (sŭdh'ǝli), *a.*, *adv.* tending towards the south. *n.* a south wind. **southerliness**, *n.* **southern** (sŭdh'ǝn), *a.* of or pertaining to or situated in or towards the south; coming from the south. **Southern Cross** CROSS ¹. **southern lights**, *n.pl.* aurora australis. **Southerner**, *n.* an inhabitant or native of the South, esp. of Southern England or the Southern States of the US. **southernmost**, *a.* **southmost**, *a.* **southward** *adv.*, *a.*, *n.* **southwardly**, *adv.* **southwards**, *adv.*
souvenir (soo'vǝniǝ), *n.* a keepsake, a memento.
sovereign, **sovran** (sov'rin), *a.* supreme; possessing supreme power, dominion or jurisdiction; royal; efficacious, effectual, as a remedy. *n.* a supreme ruler, a king, an emperor, a monarch; a former English gold coin, worth one pound. *adv.* **sovereignty** (-ti), *n.*
Soviet (sō'viǝt, sov'-), *a.* of or pertaining to the USSR, its government or people. *n.* (*pl.*) the government or people of the USSR; (**soviet**) a local council elected by workers and inhabitants of a district in the USSR; a regional council selected by a number of these; the national congress consisting of delegates from regional councils. *a.* of or pertaining to a soviet. **sovietic**, *a.* **sovietism**, *n.* **sovietize**, **-ise**, *v.t.* to transform (a country etc.) to the Soviet model of economic, social and political activity. **sovietization**, **-isation**, *n.*
sow¹ (sō), *v.t.* (*past* **sowed**, *p.p.* **sown** (sōn), **sowed**) to scatter (seed) for growth; to scatter seed over (ground etc.); to scatter over, to cover thickly with; to dissemi-

nate, to spread; to implant, to initiate. *v.i.* to scatter seed for growth. **to sow the seeds of,** to introduce, initiate or implant (a doubt, a suspicion etc.). **sower,** *n.* **sowing,** *n.*
sow² (sow), *n.* a female pig.
soya bean (soi'ə), *n.* a leguminous herb, *Glycine soja,* orig. cultivated in Japan as a principal ingredient of soy sauce, in more recent years grown as a source of oil or flour. **soy** (soi), **soya sauce,** *n.* a thin brown sauce with a salty meaty flavour made from fermented soya beans and used extensively in Chinese cookery.
sozzled (soz'ld), *a.* (*coll.*) drunk.
Sp., (*abbr.*) Spain; Spaniard; Spanish.
sp., (*abbr.*) special; (*pl.* **spp**) species; specific; spelling.
spa (spah), *n.* a mineral spring; a resort or place where there is such a spring.
space (spās), *n.* continuous extension in three dimensions or any quantity or portion of this; the universe beyond the earth's atmosphere, outer space; an interval between points etc.; an interval of time; room, an unoccupied seat, an empty place; *v.t.* to set so that there will be spaces between; to put the proper spaces between (words, lines etc.). **the space age,** the era in which space travel and exploration have become possible. **space-age,** *a.* of or pertaining to the space age; modern. **space-bar,** *n.* a bar on a typewriter or computer keyboard for making spaces (between words etc.). **spacecraft,** *n.* a manned or unmanned craft for travelling through outer space. **space flight,** *n.* a flight beyond the earth's atmosphere. **Space Invaders**®, *n.* a video game in which the object is to shoot down images of alien invaders from outer space. **spaceman, -woman,** (*pl.* **-men**) a space traveller. **spaceplatform, -station,** *n.* a platform planned in outer space to serve as a landing-stage in space travel and as a base for scientific investigations. **space probe,** *n.* a spacecraft carrying equipment for collecting scientific measurements of conditions in space. **spaceship,** *n.* a manned spacecraft. **space shuttle,** *n.* a spacecraft designed to carry people and materials to and from a space-station; (**space shuttle**) a reusable rocket-launched manned spacecraft that returns to earth. **space-suit,** *n.* clothing specially adapted for space travel. **space-time (continuum),** *n.* the four-dimensional manifold for continuum which in accordance with Einstein's theory of relativity, is the result of fusing time with three-dimensional space. **space travel,** *n.* travel beyond the earth's atmosphere. **spacewalk,** *n.* a trip by an astronaut outside a spacecraft when it is in space. *v.i.* to float or move in space while attached by a line to a spacecraft. **spacer,** *n.* **spacial** SPATIAL. **spacing,** *n.* **spacious** (-shəs), *a.* having ample room; capacious, roomy, wide, extensive. **spaciously,** *adv.* **spaciousness,** *n.*
spade¹ (spād), *n.* an implement for digging, having a broad blade fitted into a long handle, and worked with both hands and one foot; a tool of similar form employed for various purposes. **to call a spade a spade,** to be outspoken, not to mince matters. **spadework,** *n.* tiresome preliminary work. **spadeful,** *n.* (*pl.* **-fuls**)
spade² (spād), *n.* a playing-card with a black figure or figures shaped like a heart with a small triangular handle; *pl.* this suit of cards; (*offensive*) a Negro.
spaghetti (spəget'i), *n.* a long, thin variety of pasta. **spaghetti junction,** *n.* a road junction, esp. at a motorway, at which there are many intersecting roads and/or flyovers. **spaghetti western,** *n.* a (type of) cowboy film (pop. in the 1960s) filmed in Italy or Spain often with a violent or melodramatic content.
spalpeen (spal'pēn'), *n.* (*chiefly Ir.*) a scamp, a rascal.
Spam® (spam), *n.* a tinned luncheon meat of chopped and spiced ham.
Span., (*abbr.*) Spanish.
span¹ (span), *v.t.* (*past, p.p.* **spanned**) to extend from side to side of (a river etc.); to measure with one's hand expanded; to encompass, to cover. *n.* a brief space of distance or time; an entire stretch of distance or time (e.g. a lifespan, attention span); the space from end to end of a bridge etc.; the horizontal distance between the supports of an arch; a wingspan. **span-roof,** *n.* an ordinary roof with two sloping sides.
span² (span), *n.* a pair of horses, usu. matched in colour etc., harnessed side by side; a yoke or team of oxen etc.
span³ (span), *past* SPIN.
spandrel, -dril (span'drəl), *n.* the space between the shoulder of an arch and the rectangular moulding etc. enclosing it, or between the shoulders of adjoining arches and the moulding etc.
spangle (spang'gl), *n.* a small disc of glittering metal or other material, used for ornamenting dresses etc.; any small sparkling object. *v.t.* to set or adorn with spangles. **spangled,** *a.* **spangler,** *n.* **spangly,** *a.*
Spaniard (span'yəd), *n.* a native or inhabitant of Spain.
spaniel (span'yəl), *n.* a popular name for a class of dogs, distinguished chiefly by large drooping ears, long silky or curly coat and a gentle disposition; a servile, cringing person.
Spanish (span'ish), *a.* of or pertaining to Spain, its people or their language. *n.* the Spaniards; the official language of Spain and its former colonies. **Spanish America,** *n.* the predominantly Spanish-speaking parts of Central and S America, and the West Indies (former Spanish colonies). **Spanish-American,** *n.,* *a.* **Spanish-fly,** *n.* a cantharis; a (supposedly aphrodisiac) preparation made from this. **Spanish guitar,** *n.* classical guitar music; the type of guitar this is played on. **Spanish main,** *n.* the NE coast of S America and the adjacent part of the Caribbean Sea.
spank¹ (spangk), *v.t.* to strike with the open hand, to slap, esp. on the buttocks. *n.* a resounding blow with the open hand, a slap, esp. on the buttocks. **spanking,** *n.* a series of such slaps.
spank² (spangk), *v.i.* to move briskly along, esp. at a pace between a trot and a gallop.
spanker *n.* one who spanks; (*coll.*) an exceptionally fine specimen; a stunner. **spanking,** *a.* (*coll.*) dashing, brisk, stunning; of a breeze, strong.
spanner (span'ə), *n.* an instrument for tightening up or loosening the nuts on screws, a wrench. **to throw a spanner in the works,** to cause an impediment, to cause confusion or difficulty.
spar¹ (spah), *n.* a round timber, a pole, esp. used as a mast, yard, boom, shears etc.
spar² (spah), *n.* a name for various lustrous minerals occurring in crystalline or vitreous form. **sparry,** *a.*
spar³ (spah), *v.i.* (*past, p.p.* **sparred**), to move the arms about in defence of offence as in boxing; to engage in a contest of words etc. **sparrer,** *n.* **sparring,** *n.* **sparringmatch,** *n.* **sparring-partner,** *n.* a boxer with whom one in training practises; a person with whom one engages in lively repartee.
spare (speə), *a.* meagre, scanty, frugal; thin, lean, wiry; concise (of style); that can be spared, kept in reserve, available for use in emergency etc. *v.t.* to use frugally, to be chary of using; to refrain from using; to dispense with; to refrain from inflicting upon; to refrain from punishing, injuring, destroying etc.; to relieve, to release; to be able to afford. *n.* that which is surplus to immediate requirements and available for use. **to go spare,** (*sl.*) to become excessively angry, agitated or distraught. **to spare,** extra, surplus, more than required. **spare part,** *n.* a replacement for a machine part which may break, wear out etc. **spare-part surgery,** *n.* surgery including the implanting of artificial organs or parts. **sparerib,** *n.* a piece of pork consisting of the ribs with only a little meat. **spare room,** *n.* a guest bedroom. **spare tyre,** *n.* a tyre carried in a vehicle as a replacement in case of a puncture; (*coll.*) a bulge of fat around the midriff. **sparely,** *adv.* **spareness,** *n.* **sparer,** *n.* **sparing,** *a.* **sparingly,** *n.* **sparingness,** *n.*

sparing, SPARE.

spark (spahk), *n.* an incandescent particle thrown off from a burning substance; a brilliant point, facet, gleam etc.; a flash of wit, a particle of life or energy; a trace, a hint (of kindled interest etc); a fashionable young man; a vivacious and witty person; the luminous effect of a disruptive electrical discharge. *v.i.* to give out sparks; to produce sparks at the point of broken continuity in an electrical circuit. **to make sparks fly**, to start a violent quarrel, to cause a row. **to spark off**, to start, to kindle, to enliven. **spark-plug, sparking-plug**, *n.* a device for igniting the explosive mixture in the cylinder of an internal combustion engine. **sparkish**, *a.* **sparklet** (-lit), *n.* **sparks**, *n. sing.* (*sl.*) the wireless operator on board ship; an electrician.

sparkle (spah'kl), *n.* a gleam, a glittering, glitter, brilliance; vivacity, wit; effervescence. *v.i.* to emit sparks; to glisten, to glitter, to twinkle; to be vivacious, witty, scintillating. **sparkler**, *n.* (*sl.*) a diamond; a hand-held firework that emits fizzling sparks. **sparkling wine**, *n.* a wine made effervescent by carbon monoxide gas. **sparklingly**.

sparrer, sparring-match SPAR [3].

sparrow (spa'rō), *n.* a small brownish-grey bird of the genus *Passer*, esp. *P. domesticus*, the house-sparrow; any of various other small birds resembling this, e.g. the hedge-sparrow. **sparrow-hawk**, *n.* a small hawk, *Accipiter nisus*, preying on small birds etc.

sparry SPAR [2].

sparse (spahs), *a.* thinly scattered, set or occurring at considerable intervals, not dense. **sparsely**, *adv.* **sparseness, sparsity**, *n.*

Spartan (spah'tən), *n.* a native or inhabitant of Sparta; one bearing pain, enforcing discipline etc., like a Spartan. *a.* of or pertaining to Sparta or the Spartans; like a Spartan, hardy, strict etc.; austere, rigorous, frugal.

spasm (spaz'm), *n.* a convulsive and involuntary muscular contraction; a sudden or convulsive act, movement etc.; a violent burst of emotion or effort. **spasmodic, -ical** (-mod'-), *a.* **spasmodically**, *adv.* **spastic** (spas'tik), *a.* of, affected by, resembling or characterized by spasms; (*derog. sl.*) ineffectual, incapable. *n.* a sufferer from cerebral palsy; (*derog. sl., often considered offensive*) an ineffectual, clumsy person. **spastically**, *adv.* **spasticity** (-tis'-), *n.*

spat[1] (spat), *n.* (*usu. pl.*) a short gaiter fastening over and under the shoe.

spat[2] (spat), *n.* (*chiefly N Am.*) a petty quarrel; (*chiefly N Am.*) a splash, a drop, a smattering (e.g. of rain).

spat[3] (spat), *past* SPIT[2].

spate (spāt), *n.* a heavy flood, esp. in mountain stream or river; a sudden downpour; a sudden onrush or outburst.

spatial, spacial (spā'shəl), *a.* of, relating to, existing or occurring in space. **spatiality** (-al'-), *n.* **spatially**, *adv.* **spatiotemporal** (-shiō-), *a.* of space-time; of, concerned with or existing in both space and time. **spatiotemporally**, *adv.*

spatter (spat'ə), *v.t.* to scatter or splash (water etc.) about; to sprinkle or splash with water, mud etc. *v.i.* to sprinkle drops of saliva etc. about; to be scattered about thus. *n.* a shower, a sprinkling, a pattering.

spatula (spat'ūlə), *n.* a broad knife or trowel-shaped tool used for spreading plasters, working pigments, mixing foods etc. **spatular, -late** (-lət), *a.* **spatule**, *n.* a broad, spatuliform part, as in the tail of many birds; a spatula.

spawn (spawn), *v.t.* of fish, amphibians etc., to deposit or produce (eggs, young etc.); (*derog.*) of human beings, to bring forth. *v.i.* of fish etc., to deposit eggs; (*derog.*) to issue, to be brought forth, esp. in abundance. *n.* the eggs of fish, frogs and molluscs; white fibrous matter from which fungi are produced; (*derog.*) offspring. **spawner**, *n.*

spay (spā), *v.t.* to destroy or remove the ovaries of female animals.

speak (spēk), *v.i.* (*past* **spoke** (spōk), *p.p.* **spoken**) to utter articulate sounds or words in the ordinary tone as dist. from singing; to talk, to converse; to deliver a speech or address; to communicate or intimate by other means; of a picture etc., to be highly expressive or lifelike; to be on speaking terms; to be a spokesman (for); to produce a (characteristic) sound. *v.t.* to utter articulately; to make known, to tell, to declare; to talk or converse in (a language). **nothing to speak of**, insignificant, unimportant. **so to speak**, as it were. **to speak for**, to act as an advocate for, to represent, to witness. **to speak of**, to mention. **to speak one's mind**, to speak freely and frankly. **to speak out, up**, to speak loudly; to speak without constraint, to express one's opinion freely. **to speak to**, to address; to speak in support or confirmation of. **to speak volumes**, to be of great or peculiar significance (for etc.). **to speak well of**, to furnish favourable evidence of. **speakeasy**, *n.* a premises where illicit liquor was sold during the time of Prohibition. **speakable**, *a.* **speaker**, *n.* one who speaks, esp. one who delivers a speech; an officer presiding over a deliberative assembly (esp. the House of Commons); a loudspeaker. **speakership**, *n.* **speaking**, *a.* able to speak; (*in comb.*) able to speak a specific language; transmitting speech. **on speaking terms**, amicable. **speaking in tongues** THE GIFT OF TONGUES under TONGUE. **strictly speaking**, in the strict sense of the words. **speaking clock**, *n.* a telephone service that tells the caller the time by means of a recorded message.

spear (spiə), *n.* a weapon with a pointed head on a long shaft; a spearman; a sharp-pointed instrument with barbs, for stabbing fish etc.; a blade or stalk of grass. *v.t.* to pierce, kill or capture with a spear. **spear-grass**, *n.* grass of various species having long, sharp leaves. **speargun**, *n.* a gun for firing spears under water. **spearhead**, *n.* the pointed end of a spear; the person or group leading a campaign, thrust or attack. *v.t.* to lead a campaign, an assault etc. **spearmint**, *n.* the garden mint, *Mentha viridis*.

spec[1] (spek), *n.* (*coll.*) short for SPECULATION, SPECIFICATION. *a.* **on spec**, (*coll.*) on the chance of, in the hope that, as a gamble.

spec., (*abbr.*) special.

special (spesh'əl), *a.* particular, peculiar, not ordinary or general; additional; close, intimate; designed for a particular purpose, environment or occasion. *n.* a person or thing designed for a special purpose etc.; a special train, constable, edition of a newspaper, item on a menu etc. **Special Branch**, *n.* a branch of the British police force dealing with political security. **special constable** [CONSTABLE]. **special delivery**, *n.* express delivery. **special effect**, *n.* an extraordinary visual or sound effect, esp. one created on a film, video tape, or television or radio broadcast. **special licence**, LICENCE. **special pleading**, *n.* specious or unfair argument. **special school**, *n.* a school established to meet the educational needs of handicapped children. **specialism**, *n.* a special area of expertise etc., a speciality. **specialist**, *n.* one who devotes him- or herself to a particular branch of a profession etc. **specialistic**, *a.* **speciality**, (-shial'-), *n.* a special characteristic or feature, a peculiarity; a special area of expertise, pursuit, occupation, service, commodity etc. **specialize, -ise**, *v.i.* to differentiate, limit, adapt or apply to a specific use, function, environment, purpose or meaning. *v.i.* to become differentiated, adapted or applied thus; to employ oneself as or train oneself for a specialist. **specialization, -isation** (-zā'shùn), *n.* **specially**, *adv.* **specialness**, *n.* **specialty**, *n.* (*chiefly N Am.*) a speciality.

species (spē'shiz), *n.* (*pl.* **species**) a group of organisms (subordinate to a genus) generally resembling each other and capable of reproduction; (*Log.*) a group of individuals having certain common attributes and designated by a common name (subordinate to a genus); a kind, a sort, a variety; the form or shape given to any material. **speciation**, *n.* the development of a biological species. **specie** (-shē), *n.* coin as dist. from paper money. **in specie**, in

coin; in kind.
specify (spes'ifī), *v.t.* to mention expressly, to name distinctively; to include in a specification. **specifiable**, *a.*
specific (-sif'-), *a.* clearly specified or particularized, explicit, definite, precise; constituting, pertaining to, characterizing or particularizing a species; distinctive, peculiar, special. *n.* a medicine, remedy, agent etc. for a particular part of the body. **specific gravity** *n.* the relative weight or density of a solid or fluid expressed by the ratio of its weight to that of an equal volume of a substance taken as a standard, water in the case of liquids and solids, air for gases. **specific heat (capacity)** *n.* the heat required to raise the temperature of one unit of a given substance by one degree. **specifically**, *adv.* **specification** (spesifi-), *n.* the act of specifying; an article or particular specified; a detailed statement of particulars, esp. of materials, work, workmanship to be undertaken or supplied by an architect, builder, manufacturer etc.; a detailed description of an invention by an applicant for a patent. **specificative** (-sif'i-), *a.* **specificity** (-fis'-), **specificness**, *n.* **specified**, *a.*
specimen (spes'imən), *n.* a part or an individual intended to illustrate or typify the nature of a whole or a class, an example, an illustration, an instance; a sample of blood, urine etc. taken for medical analysis; (*coll., usu. derog.*) a person.
specious (spē'shəs), *a.* apparently right or fair, plausible; deceptively pleasing to the eye, showy. **speciosity** (-os'-), **speciousness**, *n.* **speciously**, *adv.*
speck (spek), *n.* a small spot, fleck, stain or blemish; a minute particle. *v.t.* to mark with a speck or specks. **speckless**, *a.* **specky**[1], *a.*
speckle (spek'l), *n.* a small spot, stain or patch of colour, light etc. *v.t.* to mark (as) with speckles. **speckled**, *a.*
specks SPECS.
specky SPECK.
specs, specks (speks), *n.pl.* (*coll.*) short for SPECTACLES.
spectacle (spek'təkəl), *n.* a show, something exhibited to the view; a pageant, a sight; (*coll.*) a sight attracting ridicule, laughter etc.; *pl.* an optical instrument, consisting of a lens for each eye mounted in a light frame, used to assist the sight. **spectacled**, *a.* **spectacular** (-tak'ū-), *a.* of the nature of a spectacle; marked by great display; dramatic; thrilling; stunning, striking. *n.* an elaborate show in a theatre, on television etc. **spectacularly**, *adv.*
spectator (spektā'tə), *n.* one who looks on, esp. at a show or spectacle. **spectator sport**, *n.* a sport that attracts a large number of spectators. **spectatorial** (-tətaw'ri-), *a.* **spectatorship**, *n.* **spectatress** (-tris), **-trix** (-triks), *n. fem.*
spectre, (*esp. N Am.*) **specter** (spek'tə), *n.* an apparition, a ghost; a person, thought, event etc. causing alarm or threat. **spectral**, *a.* ghostlike, of or pertaining to ghosts. **spectrality** (-tral'-), *n.* **spectrally**, *adv.*
spectro-, *comb. form* spectrum.
spectrometer (spektrom'itə), *n.* an instrument for measuring the angular deviation of a ray of light passing through a prism. **spectrometric** (-met'-), *a.* **spectrometry**, *n.*
spectroscope (spek'trəskōp), *n.* an instrument for forming and analysing the spectra of rays emitted by bodies. **spectroscopic, -ical** (-skop'-), *a.* **spectroscopically**, *adv.* **spectroscopist** (-tros'-), *n.* **spectroscopy** (-tros'-), *n.*
spectrum (spek'trəm), *n.* (*pl.* **-tra** (-trə)) an image produced by the decomposition of rays of light or other radiant energy by means of a prism, in which the parts are arranged according to their refrangibility; an image persisting on the retina after stimulation has ceased, an after image; a range, a series (of interests, activities etc.) **spectrum analysis**, *n.* chemical analysis with the spectroscope. **spectral**, *a.* of, pertaining to, or like a spectrum.
speculate (spek'ūlāt), *v.i.* to pursue an inquiry or form conjectures or views by consideration in the mind; to make purchases, investments etc. on the chance of profit. **speculation**, *n.* the act or practice of speculating; a mental inquiry, train of thought or series of conjectures about a subject; a speculative business transaction, investment or undertaking. **speculative**, *a.* **speculatively**, *adv.* **speculativeness**, *n.* **speculator**.
speculum (spek'ūləm), *n.* (*pl.* **-la** (-lə)) a surgical instrument for dilating passages of the body, to facilitate inspection; a mirror, esp. one of polished metal used as a reflector in a telescope. **specular**, *a.* of or pertaining to a mirror; mirror-like; of or pertaining to a speculum.
sped (sped), *past, p.p.* SPEED.
speech (spēch), *n.* the faculty or act of uttering articulate sounds or words; that which is spoken, an utterance, a remark; a public address, an oration; the language or dialect of a nation, region etc.; an individual's characteristic manner of speech. (*Linguistics*) parole. **speech community**, *n.* a community sharing a common dialect or language. **speech-day**, *n.* the annual day for presenting prizes in schools etc. **speechify**, *v.i.* (*often derog.*) to make a (pompous or lengthy) speech or speeches, to harangue. **speechifier**, *n.* **speechless** (-lis), *a.* unable to speak, silent, esp. through emotion; dumb, dumb-founded. **speechlessly**, *adv.* **speechlessness**, *n.*
speed (spēd), *n.* rapidity, swiftness, rate of motion; the ratio of the distance covered to the time taken by a moving body; (*sl.*) amphetamine; the numerical expression of the sensitivity of a photographic plate, film or paper to light; a measure of the power of a lens to take in light; a ratio of gears in a motor vehicle, on a bicycle etc. *v.i.* (*past, p.p.* **sped** (sped)) to move rapidly, indicate its speed. **speed trap**, *n.* a stretch of road monitored by police using to hasten; to drive, to travel at an excessively high, dangerous or illegal speed. *v.t.* to cause to go fast, to urge to send at great speed. **at speed**, quickly. **speedball**, *n.* (*sl.*) a amphetamine. **speedboat**, *n.* a light boat driven at great speed by a motor-engine. **speed limit**, *n.* the legal limit of speed for a vehicle on a particular road in particular conditions. **speedometer** (-dom'itə), *n.* a device attached to a vehicle to measure and indicate its speed. **speed trap**, *n.* a stretch of road monitored by police using radar devices to catch speeding drivers. **speed-way** *n.* a racecourse for motorcycles; a dirt track; the sport of motorcycle racing on a track. **speeder**, *n.* **speedy**, *a.* **speedily**, *adv.* **speediness**, *n.* **speeding**, *a.* an excessive, dangerous or illegal speed. *a.* travelling at such a speed. **speedo** (-ō), *n.* (*pl.* **-dos**) (*coll.*) a speedometer.
spell[1] (spel), *n.* a series of words used as a charm, an incantation; occult power, fascination, enchantment; a powerful attraction. **under a spell**, held under or as if under an occult power. **spellbind**, *v.t.* to put a spell on; to entrance. **spellbinder**, *n.* one or that which entrances, esp. an eloquent speaker, film, book etc. **spellbinding**, *a.* **spellbound**, *a.* under the influence of a spell; enchanted, fascinated.
spell[2] (spel), *v.t.* (*past, p.p.* **spelled, spelt**) to say or write the letters forming (a word); to read or decipher with difficulty; of letters, to form a word; to mean, to import, to portend. *v.i.* to put letters together in such a way as to (correctly) form a word. **to spell (it) out**, *n.* to utter or write letter by letter; to make clear, easy to understand. **speller**, *n.* one who spells. **spelling**, *n.* the act or ability of one who spells; orthography; the particular formation of letters making up a word. **spelling-bee**, *n.* a competition in spelling. **spelling-book**, *n.* a book for teaching to spell.
spell[3] (spel), *n.* a turn of work or rest; a (usu. short) period of time.
spelt[1] (spelt), *n.* a variety of wheat formerly much cultivated in S Europe etc.
spelt[2] (spelt), *past, p.p.* SPELL[2].
spelunker (spilŭng'kə), *n.* (*N Am.*) one who explores or studies caves as a sport or hobby. **spelunking**, *n.*

spencer (spen'sə), *n.* a short overcoat or jacket, for men or women; a woman's undergarment, a vest. [Earl *Spencer*, 1758–1834]
spend (spend), *v.t.* (*past, p.p.* **spent** (spent)) to pay out (money etc.); to consume, to use up; to pass (time); to squander, to waste; to expand; to wear out, to exhaust; *v.i.* to expend money; to waste away, to be consumed. **spendthrift,** *a.* prodigal, wasteful. *n.* a prodigal or wasteful person. **spendable,** *a.* **spender,** *n.* **spending,** *n.* **spending money,** *n.* pocket-money, money for spending. **spent,** *a.* exhausted, burnt out, used up. **a spent force,** one who or that which is used up, exhausted, useless etc.
Spenserian (spensə'riən), *a.* of, pertaining to or in the style of the poet Edmund Spenser (1552–99) or his verse.
sperm (spœm), *n.* the male seminal fluid of animals; a male gamete; a sperm whale or cachalot. **sperm bank,** *n.* a storage place for semen required for artificial insemination. **sperm whale,** *n.* a large toothed whale, *Physeter catodon.*
sperm- SPERMAT(O)-.
-sperm, *comb. form.* a seed.
spemaceti (spœməsē'ti), *n.* a white, fatty, brittle substance, existing in solution in the oily matter in the head of the sperm-whale, used for candles, ointments etc.
spermary, spermarium (-meə'riəm), *n.* (*pl.* **-ries, -ria** (riə)) the male spermatic gland, testicle or other organ.
spermat- SPERMAT(O)-.
spermatic (spœrmat'ik), **ical, spermic,** *a.* consisting of, pertaining to or conveying sperm or semen; of or pertaining to the spermary. **spermatically,** *adv.*
spermat(o)-, sperm(o)-, *comb. form.* sperm, seed. [SPERM]
spermatoblast (spœ'mətəblast), *n.* a cell from which a spermatozoon develops.
spermatocyte (spœma'tōsīt), *n.* a cell which develops into a sperm cell.
spermatogenesis (spœmătōjen'sis), *n.* the development of spermatozoa. **spermatogenic** (-net'-), **-genic, spermatogenous,** *a.* **spermatogeny,** *n.*
spermatophyte (spœmat'ōfīt), *n.* a seed-bearing plant. **spermatophytic** (-fit'-), *a.*
spermatozoon (spœmatōzō'on), *n.* (*pl.* **-zoa** (-zō'ə)) one of the minute living bodies in the seminal fluid essential to fecundation by the male; a male germ cell. **spermatozoal, -zoan, -zoic** (-zō'ik), *a.*
spermic (spœ'mik), SPERMATIC.
spermicide (spœ'misīd), *n.* a substance that kills spermatozoa. **spermicidal** (-sī'-), *a.*
sperm(o)- SPERMAT(O)-.
spew (spū), *v.t.* to vomit; to cast out with abhorrence; to spit out; to emit or eject violently or in great quantity. *v.i.* to vomit; to stream, gush or flood out. *n.* vomit, that which is ejected with great force or in great quantity. **spewy,** *a.*
sp. gr., (*abbr.*) specific gravity.
sphagnum (sfăg'núm), *n.* a genus of crytograms containing the bog- or peat-mosses. **sphagnous,** *a.*
sphen- SPHEN(O)-.
sphen(o)-, *comb. form.* pertaining to or resembling a wedge.
Sphenodon (sfē'nədon), *n.* a genus of nocturnal lizard-like reptiles, now confined to New Zealand.
sphenoid (sfen'oid), *a.* wedge-shaped. *n.* a sphenoid bone; *n.* a wedge-shaped crystal enclosed by four equal isosceles triangles. **sphenoid bone,** *n.* a wedge-shaped bone lying across the base of the skull. **sphenoidal** (-noi'-), *a.*
sphere (sfiə), *n.* a solid bounded by a surface every part of which is equally distant from a point within called the centre; a solid figure generated by the revolution of a semicircle about its diameter; a figure approximately spherical, a ball, a globe, esp. one of the heavenly bodies; a globe representing the earth or the apparent heavens; one of the spherical shells revolving round the earth as centre in which, according to ancient astronomy, the heavenly bodies were set; the sky, the heavens; area of knowledge, discipline; field of action, influence etc., scope, range, province, place, position; social class. **spheral,** *a.* spherical; of or pertaining to the celestial spheres or the music of the spheres. **spherical** (sfe'rikəl), **spheric** (sfer'-), *a.* sphere-shaped, globular; relating to spheres. **spherical angle,** *n.* the angle between two intersecting great circles of a sphere. **spherical coordinates,** *n.pl.* coordinates used to locate a point in space comprising a radius vector and two angles measured from a vertical and a horizontal line. **spherically,** *adv.* **sphericity** (sfiris'-), *n.* **spherics** (sfe'riks), *n.pl.* spherical geometry and trigonometry. **spheroid** (sfe'roid), *n.* a body nearly spherical; a solid generated by the revolution of an ellipse about its minor axis (called an **oblate spheroid**) or its major axis (called a **prolate spheroid**). **spheroidal, -roidic, -ical** (-roi'-), *a.* **spheroidally,** *adv.* **spheroidity,** (-roi'-), **-dicity** (-dis'-), *n.*
spherograph (sfe'rəgraf), *n.* a stereographic projection of the earth with meridians and lines of latitude, used for the mechanical solution of problems in navigation, etc.
spherometer (sfirom'itə), *n.* an instrument for measuring the radii and curvature of spherical surfaces.
spherule (sfer'ūl), *n.* a small sphere. **spherular,** *a.*
spherulite (sfer'ūlīt), *n.* a rounded concretion occurring in various rocks; a radiolite.
sphincter (sfingk'tə), *n.* a ring muscle that contracts or shuts any orifice or tube. **sphincteral, -terial** (-tiə'ri-), **-teric** (-te'-), *a.*
sphinx (sfingks), *n.* in Greek mythology, a winged monster, half woman and half lion, said to have devoured the inhabitants of Thebes till a riddle she had proposed was solved; an ancient Egyptian figure with the body of a lion and a human or animal head; a taciturn or enigmatic person. **sphinxlike,** *a.*
sphygm(o)-, *comb. form.* pertaining to a pulse.
sphygmus (sfig'məs), *n.* a pulse, a pulsation. **sphygmic,** *a.*
spica (spī'kə), *n.* (*Bot.*) a spike; a spiral surgical bandage with the turns reversed. **spicate** (-kāt), **-ated, spiciform** (-sifawm), *a.* (*Bot.*) pointed, having spikes. [L]
spiccato (spikah'tō), *n.*, *a.* (a musical passage) played so that the bow rebounds lightly from the strings.
spice (spīs), *n.* any aromatic and pungent vegetable substance used for seasoning food; a flavour, a touch, a trace. something which adds zest or interest. *v.t.* to season with spice; to add interest to. **spicy,** *a.* flavoured with spice; abounding in spices; pungent, piquant; suggestive, indelicate; showy, smart. **spicily,** *adv.* **spiciness,** *n.*
spick-and-span (spikənspan'), *a.* new and fresh, clean and smart.
spicule (spik'ūl), *n.* a small sharp needle-shaped body, such as the calcareous or siliceous spikes in sponges etc.; (*Bot.*) a small or subsidiary spike. a spiked flare of hot gas ejected from the surface of the sun. **spicular, -late** (-lət), **spiculiform** (-fawm), *a.* **spiculiferous** (-lif'-), **-ligerous** (-lij'-), *a.*
spicy SPICE.
spider (spī'də), *n.* an eight-legged arachnid of the order Araneida, usu. furnished with a spinning apparatus utilized by most species for making webs to catch their prey; an arachnid resembling this. **spider-crab,** *n.* a crab with long thin legs. **spider-monkey,** *n.* a monkey belonging to the American genus *Ateles* or *Eriodes* with long limbs and slender bodies. **spider plant,** *n.* a house plant having streamers of long narrow leaves with central white or yellow stripes. **spider's web, spider-web,** *n.* **spider-like,** *a.*, *adv.* **spidery,** *a.*
spiel (shpēl, spēl), *n.* a speech, sales patter. *v.i.* to reel off patter. **spieler,** *n.* a card-sharper, a trickster.
spiffing (spif'ing), *a.* (*dated coll.*) excellent. **spiffy,** *a.* (*coll.*) smartly dressed, spruce.
spifflicate (spif'likăt), *v.t.* to smash, to crush, to beat up.

spigot / spirit

spiflication, *n.*
spigot (spig'ət), *n.* a peg or plug for stopping the vent-hole in a cask; the turning-plug in a faucet; a faucet, a tap.
spike (spīk), *n.* a pointed piece of metal, as one of a number fixed on the top of a railing, fence, or wall, or worn on boots to prevent slipping; any pointed object, a sharp point; a large nail or pin, used in structures built of large timbers, on railways etc. *v.t.* to fasten with spikes; to furnish with spikes; to sharpen to a point; to pierce or impale; to fasten on with a spike or spikes; to lace a drink with spirits; to render useless. **to spike someone's guns**, to foil someone's plans. **spiky**, *a.*
spikenard (spīk'nahd), *n.* a herb, *Nardostachys atamansi*, related to the valerian; an ancient and costly aromatic ointment prepared chiefly from the root of this.
spiky SPIKE.
spile (spīl), *n.* a small wooden plug, a spigot; a large timber driven into the ground to protect a bank etc., a pile.
spill[1] (spil), *n.* a slip of paper or wood used to light a candle, pipe etc.
spill[2] (spil), *v.t.* (*past, p.p.* **spilt** (spilt), **spilled**) to suffer to fall or run out of a vessel; to shed; (*coll.*) to throw out of a vehicle or from a saddle. *v.i.* of liquid, to run or fall out. *n.* a tumble, a fall, esp. from a vehicle or saddle. **to spill the beans**, to divulge a secret. **spillway**, *n.* a passage for the overflow of water from a reservoir etc.
spillikin (spil'ikin), *n.* a small strip or pin of bone, wood etc., used in spillikins. **spillikins**, *n. sing.* a game in which players attempt to remove spillikins from a pile one at a time without disturbing the others.
spilt, SPILL[2].
spin (spin), *v.t.* (*pres. p.* **spinning** *past* **spun** (spŭn), *p.p.* **spun**) to draw out and twist (wool, cotton etc.) into threads; to make (yarn etc.) thus; of spiders etc., to produce (a web, cocoon etc.) by drawing out a thread of viscous substance; to tell, compose etc., at great length; to make (a top etc.) rotate rapidly. *v.i.* to draw out and twist cotton etc., into threads; to make yarn etc., thus; to whirl round; to turn round quickly; to go along with great swiftness. *n.* the act or motion of spinning, a whirl; a brief run in a motor car etc.; a rapid diving descent accompanied by a continued gyration of the aeroplane. **in a flat spin**, extremely agitated or confused. **to spin a yarn**, to tell a story. **to spin out**, to compose or tell (a yarn etc.) at great length; to prolong, to protract; to spend (time) in tedious discussion etc. **spin-bowling**, *n.* in cricket, a style of bowling in which the ball is delivered slowly with an imparted spin to make it bounce unpredictably. **spin-bowler**, *n.* **spin-drier**, *n.* a machine that dries washing to the point of being ready for ironing by forcing out the water by centrifugal force. **spin-dry**, *v.t.* **spin-off**, *n.* a by-product, something derived from an existing idea or product. **spinner**, *n.* one who spins; a machine for spinning thread; (*Cricket*) a ball bowled with a sharp spin; a spinbowler. **spinneret** (-əret), *n.* the spinning organ of a spider through which the silk issues; the orifice through which liquid cellulose is projected to form the threads of rayon or artificial silk. **spinning-jenny**, *n.* a mechanism invented by Hargreaves in 1767 for spinning several strands at once. **spinning-mill**, *n.* a factory where spinning is carried on. **spinning-wheel**, *n.* a wheel driven by the foot or hand, formerly used for spinning wool, cotton, or flax.
spinabifida SPINE.
spinach (spin'ich, -ij), *n.* an annual herb of the genus *Spinacia*, esp. *S. oleracea*, with succulent leaves cooked as food; other herbs used in the same fashion. **spinaceous** (-nā'shəs), *a.*
spinal SPINE.
spindle (spin'dl), *n.* a pin or rod in a spinning-wheel for twisting and winding the thread; a rod used for the same purpose in hand-spinning; a pin bearing the bobbin in a spinning-machine; a rod or axis which revolves or on which anything revolves; a slender object or person. **spindle-legged, -shanked**, *a.* having long, thin legs. **spindle-legs, -shanks**, *n.pl.* **spindle-shaped**, *a.* tapering from the middle towards both ends, fusiform. **spindly**, *a.* tall and thin; elongated.
spindrift (spin'drift), *n.* fine spray blown up from the surface of water.
spine (spīn), *n.* the spinal column; a sharp, stiff woody process; a sharp ridge, projection, out-growth etc.; the back of a book, usu. bearing the title and the author's name. **spina bifida**, *n.* a congenital condition in which malformation of the spine causes the meninges to protrude, producing enlargement of the head and paralysis of the lower body. **spinal**, *a.* pertaining to the spine. **spinal canal**, *n.* a passage that contains the spinal cord. **spinal column**, *n.* the interconnected series of vertebrae in the skeleton which runs the length of the trunk and encloses the spinal cord. **spinal cord**, *n.* a cylindrical structure of nerve-fibres and cells within the vertebral canal and forming part of the central nervous system. **spine-chiller**, *n.* a book, film, event etc. that causes terror. **spine-chilling**, *a.* **spined**, *a.* **spineless** (-lis), *a.* without a spine; of weak character, lacking decision. **spine-tingling**, *a.* causing a sensation of terror. *a.* **spinoid** (-noid), *a.* **spinose** (-nōs'), **spinous**, *a.* **spinosity** (-nos'-), *n.* **spiny**, *a.* **spininess**, *n.*
spinet (spinet'), *n.* an obsolete musical instrument, similar in construction to but smaller than the harpsichord.
spini- *comb. form.* pertaining to the spine.
spinnaker (spin'əkə), *n.* a large jib-shaped sail carried opposite the mainsail on the mainmast of a racing-yacht.
spinner, spinneret SPIN.
spinney (spin'i), *n.* a small wood with undergrowth, a copse.
spinning-, jenny, -mill etc. SPIN.
spinode (spī'nōd), *n.* a stationary point on a curve, a cusp.
spinoid, spinose etc. SPINE.
Spinozism (spinō'zizm), *n.* the monistic system of Baruch de Spinoza (1632–77), a Spanish Jew, who resolved all being into extension and thought, which he considered as attributes of the Sole Substance, God. **Spinozist**, *n.* **Spinozistic** (-zis'-), *a.*
spinster (spin'stə), *n.* an unmarried woman. **spinsterhood** (-hud), *n.* **spinsterish**, *a.* [SPIN, -STER]
spinthariscope (spintha'riskōp), *n.* an instrument for showing the rays emitted by radium by the scintillations caused by their impact against a fluorescent screen. **spinthariscopic** (-skop'-), *a.*
spiny SPINE.
spiracle (spī'rəkl), *n.* a breathing-hole, a vent-hole for lava etc. **spiracular** (-rak'ū-), **-late** (-lət), **spiraculiform** (-kū'lifawm), *a.*
spiral, *a.* forming a spire, spiral, or coil; continually winding about and receding from a centre; continually winding, as the thread of a screw. *n.* a spiral curve, formation, spring, or other object; a plane curve formed by a point revolving round a central point while continuously advancing on or receding from it; a helix; a continuous upward or downward movement, e.g. of prices; flight in a spiral motion. **spirality** (-ral'-), *n.* **spirally**, *adv.* **spiraled**, *a.*
spirant (spī'rənt), *n.* a fricative consonant. *a.* fricative.
spire (spīə), *n.* a tapering, conical, or pyramidal structure, esp. the tapering portion of a steeple.
spirillum (spīril'əm), *n.* (*pl.* **-lla** (-lə)) a bacterium of a genus, *Spirillum*, of bacteria having a spiral structure. **spirillar, spirilliform** (-ifawm), *a.*
spirit (spi'rit), *n.* the immaterial part of a person, the soul; this as not connected with a physical body, a disembodied soul, a ghost; an incorporeal or supernatural being, a sprite, an elf, a fairy; a person considered with regard to his or her peculiar qualities of mind, temper, etc.; a person of great mental or moral force; vigour of mind or

intellect; vivacity, energy, ardour, enthusiasm; temper, disposition; (*often pl.*) mental attitude, mood, humour; real meaning or intention; actuating principle, pervading influence, peculiar quality or tendency; (*usu. pl.*) distilled alcoholic liquors, as brandy, whisky etc.; a solution (of a volatile principle) in alcohol. *v.t.* to convey (away, off etc.) secretly and rapidly. **spirit, spirits of wine,** pure alcohol. **the Spirit, the Holy Spirit,** the Third Person of the Trinity. **spirit-lamp,** *n.* a lamp burning methylated or other spirit. **spirit-level,** *n.* an instrument used for determining the horizontal by an air-bubble in a tube containing alcohol. **spirit-rapper,** *n.* one professing to communicate with spirits by means of raps on a table etc. **spirit-rapping,** *n.* **spirited,** *a.* full of spirit, fire, or life, animated, lively, courageous; (*in comb.*), having a particular mental attitude, as *high-spirited*. **spiritedly,** *adv.* **spiritedness,** *n.* **spiritless** (-lis), *a.* **spiritlessly,** *adv.* **spiritlessness,** *n.* **spiritoso** (-tō'sō), *adv.* (*Mus.*) in a spirited manner. **spiritual** (-chəl), *a.* pertaining to or consisting of spirit; immaterial; incorporeal; pertaining to the soul or the inner nature; derived from or pertaining to God, pure, holy, scared, divine, inspired; pertaining to sacred things, not lay or temporal. *n.* a type of hymn sung by Negroes of the southern US. **spirituality** (-al'-), *n.* immateriality, incorporeity; the quality of being spiritual or unworldly; that which belongs to the church, or to an ecclesiastic on account of a spiritual office. **spiritualism,** *n.* a system of professed communication with departed spirits, chiefly through persons called mediums; (*Phil.*) the doctrine that the spirit exists as distinct from matter or as the only reality, opp. to materialism. **spiritualist,** *n.* **spiritualistic** (-lis'-), *a.* **spiritualize, -ise,** *v.t.* **spiritualization, -isation,** *n.* **spiritualizer, -iser,** *n.* **spiritually,** *adv.* **spiritualness,** *n.* **spirituous** (-tūəs), *a.* containing spirit, alcoholic, distilled as distinguished from fermented.
spiro-[1], *comb. form.* pertaining to a coil.
spiro-[2], *comb. form.* pertaining to breathing.
Spirochaeta (spīrōkā'tə), *n.* a genus of spiral-shaped bacteria which includes the causative agents of syphilis, relapsing fever, and epidemic jaundice.
spirograph (spī'rōgraf), *n.* an instrument for recording the movement in breathing.
spirometer (spīrom'itə), *n.* an instrument for measuring the capacity of the lungs. **spirometric** (-met'-), *a.* **spirometry** (-rom'itri), *n.*
spiroscope (spī'rōskōp), *n.* a spirometer.
spirt (spoet), SPURT[1,2].
Spirula (spī'rələ), *n.* a genus of tropical cephalopods having a flat spiral shell.
spiry SPIRE[1].
spit[1] (spit), *n.* a long pointed rod on which meat for roasting is rotated before a fire; a point of land or a narrow shoal extending into the sea.
spit[2] (spit), *v.t.* (*pres. p.* **spitting,** *past, p.p.* **spat,** (spat)) to eject (saliva etc.) from the mouth; to utter in a violent or spiteful way. *v.i.* to eject saliva from the mouth; of an angry cat, to make a spitting noise; of rain, to drizzle. *n.* spittle, saliva; spitting; likeness, counterpart. **spit and polish,** (*coll.*) (obsessive) cleanliness, attention to details, as in the army. **to spit it out,** (*coll.*) to speak, tell immediately. **spitfire,** *n.* an irascible person. **spitter,** *n.* **spitting,** *n.* **within spitting distance,** (*coll.*) near. **spitting image,** *n.* (*coll.*) an exact likeness, one who or that which resembles another. **spittle**[1] (-l), *n.* saliva, esp. ejected from the mouth. **spittoon** (-toon'), *n.* a receptacle for spittle.
spite (spīt), *n.* ill will, malice, malevolence; rancour, a grudge, *v.t.* to thwart maliciously; to vex or annoy. **in spite of,** notwithstanding. **spiteful,** *a.* **spitefully,** *adv.* **spitefulness,** *n.*
spitfire SPIT[2].
spitter, spittle[1] SPIT[2].
spittoon SPIT[2].

spitz (spits), *n.* a sharp-muzzled breed of dog, called also *Pomeranian*.
spiv (spiv), *n.* a hanger-on in dubious circles; a man cheaply over-dressed without apparent occupation; one who dresses flashily; a petty black-market dealer. **spivvy,** *a.*
splanchnic (splangk'nik), *a.* pertaining to the bowels.
splanchno-, *comb. form.* pertaining to the bowels or viscera. [see SPLANCHNIC]
splash (splash), *v.t.* to bespatter with water, mud etc.; to dash (liquid etc., about, over, etc.); to make (one's way) thus; to spend recklessly; to display prominently in a newspaper. *v.i.* to dash water or other liquid about; to be dashed about in drops, to dabble, to plunge; to move or to make one's way (along etc.) thus. *n.* the act of splashing; water or mud splashed about; a noise as of splashing; a spot or patch of liquid, colour etc.; a vivid display; a dash; a small amount of soda-water etc. mixed with an alcoholic drink. **to make a splash,** (*sl.*) to make a sensation, display, etc. **splash-board,** *n.* a guard in front of a vehicle to protect the occupants from mud. **splashdown,** *n.* the landing of a spacecraft on the ocean. **splash down,** *v.i.* **splashy** *a.*
splat (splat), *n.* the slapping sound made by a soft or wet object striking a surface.
splatter (splat'ə), *v.i.* to spatter; to make a continuous splash or splashing noise. *v.t.* to utter thus, to sputter. *n.* a spatterer; a splash.
splay (splā), *v.t.* to spread out. *v.i.* to become splayed. *a.* turned outwards. **splay-foot,** *n.* a broad, flat foot turned outwards. **splay-footed,** *a.*
spleen (splēn), *n.* a soft vascular organ situated to the left of the stomach in most vertebrates which produces lymphocytes, antibodies, and filters the blood; spitefulness, ill temper; low spirits, melancholy. **spleen-wort,** *n.* a fern of the genus *Asplenium*, formerly supposed to be a specific for spleen. **spleenful, spleenish, spleeny,** *a.* **spleenfully, spleenishly,** *adv.* **spleenless** (-lis), *a.*
splen-, *comb. form.* pertaining to the spleen.
splendid (splen'did), *a.* magnificent, gorgeous, sumptuous; glorious, illustrious; brilliant, lustrous, dazzling; fine, excellent, first-rate. **splendidly,** *adv.* **splendiferous** (-dif'-), *a.* (*facet.*) splendid. **splendour,** *n.*
splenetic (splənet'ik), *a.* of or pertaining to the spleen; affected with spleen; peevish, ill-tempered. *n.* a person affected with spleen; a medicine for disease of the spleen. **splenetically,** *adv.*
splenial SPLENIUS.
splenic (splen'ik), *a.* pertaining to or affecting the spleen.
splenitis (-nī'tis), inflammation of the spleen. **splenitic** (-nit'-), *a.*
splenius (splē'niəs), *n.* a muscle extending in two parts on either side of the neck serving to bend the head backwards. **splenial,** *a.* of or pertaining to this; splint-like.
splenology (splənol'əji), *n.* scientific study of the spleen. **splenological** (splənəloj'-), *a.*
splenotomy (splənot'əmi), *n.* the dissection of or an incision into the spleen.
splice (splīs), *v.t.* to unite (two ropes etc.) by interweaving the strands of the ends; to unite (timbers etc.) by bevelling, overlapping, and fitting the ends together; to join the trimmed ends of (film, tape, etc.) with adhesive material; (*coll.*) to unite in marriage. *n.* a union of ropes, timbers etc., by splicing; the point of juncture between two pieces of film; the joint on the handle of a cricket bat which fits into the blade. **to splice the main-brace** BRACE. **splicer,** *n.*
spline (splīn), *n.* a flexible strip of wood or rubber used in laying down large curves in mechanical drawing; a key fitting into a slot in a shaft and wheel to make these revolve together; the slot itself.
splint (splint), *n.* a thin piece of wood or other material

splinter   **spoon**

used to keep the parts of a broken bone together; a thin strip of wood used in basketmaking etc.; *v.t.* to secure or support with splints.
splinter (splin'tə), *n.* a thin piece broken, split, or shivered off. *v.t.* to split, shiver, or rend into splinters. *v.i.* to split or shiver into splinters. **splintery**, *a.* **splinter group, party**, *n.* a small group that has broken away from its parent political etc. organization. **splinter-proof**, *a.* proof against the splinters of bursting shells or bombs.
split (split), *v.t.* (*pres. p.* **splitting**, past, *p.p.* **split**) to break, cleave, tear, or divide, esp. longitudinally or with the grain; to divide into two or more parts, thicknesses etc.; to divide into opposed parties; to divide (one's vote or votes) between different candidates. *v.i.* to be broken or divided, esp. longitudinally or with the grain; to break up, to go to pieces; to divide into opposed parties; (*sl.*) to betray the secrets of, to inform (on); (*coll.*) to burst with laughter; (*sl.*) to depart. *n.* the act or result of splitting; a crack, rent, or fissure; a separation into opposed parties, a rupture, a schism; something split; a dessert dish of sliced fruit (e.g. banana) and ice cream etc. (*pl.*) an acrobat's feat of sitting down with the legs stretched out right and left; (*sl.*) a half bottle of soda water; a half glass of liquor. *a.* having been split; fractured; having splits. **to split hairs** HAIR. **to split one's sides**, to laugh heartily. **to split the difference**, to compromise by showing the average of two amounts. **to split the infinitive**, to insert a word between *to* and the verb, as *to completely defeat*. **to split up**, *v.t.* to separate into sections, parts or units. *v.i.* to be thus separated; to cease to live as a couple, to divorce. **split-level**, *a.* of a house, etc., divided into more than one level. **split pea**, *n.* a dried pea split in half and used in soups etc. **split personality**, *n.* a personality comprising two or more dissociated groups of attitudes and behaviour. **split ring**, *n.* a metal ring so constructed that keys can be put on it or taken off. **split screen**, *n.* a cinematic technique in which different images are displayed simultaneously on separate sections of the screen. **split-screen**, *a.* **split second**, *n.* an instant, a fraction of a second. **split-second**, *a.* **split shift**, *n.* a work period divided into two parts separated by a long interval. **splitter**, *n.* one who splits. **splitting**, *a.* of a headache, acute, severe.
splodge (sploj), **splotch** (sploch), *n.* a daub, a blotch, an irregular stain. **splodgy, splotchy**, *a.*
splosh (splosh), (*coll.*) SPLASH.
splotch, splotchy SPLODGE.
splurge (splœj), *v.i.* to spend a lot of money (on). *n.* the act or an instance of splurging.
splutter (splút'ə), *v.t., v.i.* to sputter. *n.* a sputter, a noise, a fuss. **splutterer**, *n.* **spluttery**, *a.*
spode (spōd), *n.* porcelain made by the potter Josiah Spode (1754–1827). **spode-ware**, *n.*
spoil (spoil), *v.t.* (*past & p.p.* **spoilt** (spoilt), **spoiled**) to mar, to vitiate, to impair the goodness, usefulness etc., of; to impair the character of by over-indulgence. *v.i.* of perishable food, to decay, to deteriorate through keeping; to be eager or over-ripe (for a fight). *n.* waste material obtained in mining, quarrying, excavating, etc.; (*usu. pl. or collect.*) plunder, booty; offices, honours, or emoluments acquired as the result of a party victory, esp. in the US. **spoilsport**, *n.* a person who interferes with another's pleasure. **spoilage** (-ij), *n.* an amount wasted or spoiled; the act of spoiling or the state of being spoiled. **spoiler**, *n.* one who spoils; an aerodynamic device fitted to an aircraft wing to increase drag and reduce lift; a similar device fitted to the front or rear of a motor vehicle to maintain stability at high speeds.
spoke¹ (spōk), *n.* one of the members connecting the hub with the rim of a wheel; a rung of a ladder; one of the radial handles of a wheel, e.g. as in a ship's wheel. *v.t.* to furnish with spokes. **to put a spoke in someone's wheel**,

to thwart someone. **spokeshave**, *n.* a plane with a handle at each end for dressing spokes, curved work etc.
spoke², (spōk), *past*, **spoken**, *p.p.* SPEAK.
spokesman (spōks'mən), **spokeswoman, spokesperson**, *n.* one who speaks for another or others.
spoliation (spōliā'shən), *n.* robbery, pillage, the act or practice of plundering, esp. of neutral commerce, in time of war; (*Law*) destruction, mutilation, or alteration of a document to prevent its use as evidence; taking the emoluments of a benefice under an illegal title. **spoliatory**, *a.*
spondee (spon'dē), *n.* a metrical foot of two long syllables. **spondaic** (-dā'ik), *a.*
spondylitis (spondili'tis), *n.* inflammation of the vertebrae.
sponge (spŭnj), *n.* a marine animal with pores in the body-wall; the skeleton or part of the skeleton of a sponge or colony of sponges, esp. of a soft, elastic kind used as an absorbent in bathing, cleansing etc.; any sponge-like substance or implement; dough for baking before it is kneaded; sponge-cake; an application of a sponge; (*coll.*) a parasite, a sponger; (*coll.*) a heavy drinker. *v.t.* to wipe, wet, or cleanse with a sponge; to obliterate, to wipe (out) with or as with a sponge; to absorb, to take (up) with a sponge; to extort or obtain by parasitic means. *v.i.* to suck in, as a sponge; to live parasitically or by practising mean arts (on). **to throw in** or **up the sponge** or **towel** to acknowledge oneself beaten; to give up the contest. **sponge bag**, *n.* a small waterproof bag for carrying toiletries. **sponge bath**, *n.* a cleansing of the body using a wet sponge or cloth, as for bedridden persons. **sponge-cake**, *n.* a light, spongy cake. **sponger**, *n.* one who or that which sponges; (*coll.*) a mean parasite. **spongy**, *a.* **sponginess**, *n.*
sponsion (spon'shən), *n.* the act of becoming surety for another; (*International Law*) an act or engagement on behalf of a state by an agent not specially authorized.
sponson (spon'sən), *n.* a projection from the sides of a vessel, for a gun on a warship, or to support a bearing, a paddle wheel etc.; a device attached to the wings of a seaplane to give it steadiness when resting on the water.
sponsor (spon'sə), *n.* a surety, one who undertakes to be responsible for another; a godfather or godmother; (*chiefly N Am.*) a person or firm that pays the costs of mounting a radio or TV programme in exchange for advertising time; a person or organization that provides esp. financial support for another person or group or for some activity; a person who promises to pay a sum of money usu. to charity, the amount of which is determined by the performance of an entrant in a fund-raising event; a member who introduces a bill into a legislative assembly. *v.t.* to act as a sponsor for. **sponsored**, *a.* **sponsorial** (-saw'ri-), *a.* **sponsorship**, *n.*
spontaneous (spontā'niəs), *a.* arising, occurring, done, or acting without external cause; not due to external constraint or suggestion, voluntary; not due to conscious volition or motive; instinctive, automatic, involuntary; self-originated, self-generated. **spontaneity** (-tənē'-, -nā'-), **spontaneousness**, *n.* **spontaneously**, *adv.*
spoof (spoof), *v.t.* to hoax, to fool; to parody. *n.* a deception, a hoax; a parody, humorous take-off (of a play, poem etc).
spook (spook), *n.* a ghost; (*chiefly N Am.*) a spy *v.t.* (*chiefly N Am.*) to startle or frighten. *v.i.* (*chiefly N Am.*) to become frightened. **spooky**, *a.* ghostly; frightening.
spool (spool), *n.* a small cylinder for winding thread, photographic film etc., on; the central bar of an angler's reel; a reel (of cotton etc.). *v.t.* to wind on a spool.
spoon¹ (spoon), *n.* a domestic utensil consisting of a shallow bowl on a stem or handle, used for conveying liquids or liquid food to the mouth etc.; an implement or other thing shaped like a spoon, as an oar with the blade curved lengthwise, a golf-club with a lofted face, a spoon-bait etc. *v.t.* to take (up etc.) with a spoon; (*Cricket*

etc.) to hit a ball (usu. up) with little force. *v.i.*(*dated*) to indulge in demonstrative love-making. **spoon-bait**, *n.* a spoon-shaped piece of bright metal with hooks attached used as a revolving lure in fishing. **spoon-bill**, *n.* a bird with a broad, flat bill. **spoon-fed**, *a.* pampered; provided with information, opinions etc. in a ready-made form which precludes the need for independent thought or effort. **spoon-feed**, *v.t.* **spoonful**, *n.* (*pl.* **-fuls**). **spoony**, **spooney**, *a.* (*dated*) mawkishly amorous.
spoonerism (spoo'nərizm), *n.* accidental or facetious transposition of the initial letters or syllables of words, e.g. 'I have in my breast a half-warmed fish.' [Rev. W.A. Spooner 1844–1930, Warden of New College, Oxford]
spoor (spoor), *n.* the track of a wild animal. *v.i.* to follow a spoor.
sporadic (spərad'ik), *a.* separate, scattered, occurring here and there or irregularly, intermittent. **sporadically**, *adv.*
sporangium (spəranj'iəm), *n.* (*Bot.*) (*pl.* **-ges, -gia** (-jiə)) a spore-case. **sporangial**, *a.*
spore (spaw), *n.* the reproductive body in a cryptogram, usu. composed of a single cell not containing an embryo; a minute organic body that develops into a new individual, as in protozoa etc., a germ.
spor(o)- *comb. form.* pertaining to spores.
sporogenesis (sporōjen'əsis), *n.* spore formation.
sporophyte (spaw'rəfit), *n.* the nonsexual phase in certain plants exhibiting alternation of generations. **sporophytic** (-fit'-), *a.*
sporozoan (spawrōzō'ən), *n.* any of a group of spore-producing parasitic protozoans, that includes the malaria parasite.
sporran (spo'rən), *n.* a pouch, usu. covered with fur, worn by Scottish Highlanders in front of the kilt.
sport (spawt), *n.* diversion, amusement; fun, jest; mockery; game, pastime, esp. athletic or outdoor pastime, as hunting, shooting, fishing, racing, running etc.; (*coll.*) a good loser; (*Austral.*) a form of address used esp. between males; (*coll.*) one who likes to womanize, gamble or otherwise enjoy himself; an animal or plant deviating remarkably from the normal type; (*pl.*) a meeting for outdoor games etc. *v.i.* to play, to divert oneself; to vary remarkably from the normal type. *v.t.* to wear or display in an ostentatious manner; *a.* (*pl.*) suitable for sport. **to make sport of**, to jeer at, to ridicule. **sporter**, *n.* **sporting**, *a.* relating to, used in, or fond of sports; calling for sportsmanship; involving a risk, as in sports competition. **sportingly**, *adv.* **sportive**, *a.* frolicsome, playful. **sportively**, *adv.* **sportiveness**, *n.* **sports car**, *n.* a low usu. two-seater car built for high speed performance. **sportscast**, *n.* (*N Am.*) a broadcast of sports news. **sportscaster**, *n.* **sports jacket**, *n.* a casual jacket for men, usu. made of tweed. **sportsman**, *n.* one skilled in or devoted to sports, esp. hunting, shooting, fishing etc.; one who acts fairly towards opponents or who faces good or bad luck with equanimity. **sportsmanlike**, *a.* **sportsmanship**, *n.* **sports medicine**, *n.* the medical supervision of athletes in training and in competition and the treatment of their injuries. **sportswear**, *n.* clothing intended for wear during sporting or leisure activities. **sportswoman**, *n. fem.* **sporty**, *a.* taking pleasure in sports; vulgar, showy; dissipated. **sportiness**, *n.*
sporule (spaw'rool), *n.* a spore, esp. a small or secondary spore.
spot (spot), *n.* a small mark or stain, a speck, a blot; a stain on character or reputation; a small part of a surface of distinctive colour or texture; a small extent of space; a particular place, a definite locality; (*Billiards*) a mark near the top of a billiard-table on which the red ball is placed; a spot-stroke; (*coll.*) a small amount of anything; a spotlight; a place on a television or radio programme for an entertainer; an opportunity in the interval between programmes for advertisers; a pimple; (*coll.*) a night spot.

v.t. (*past, p.p.* **spotted**) to mark, stain, or discolour with a spot or spots; (*coll.*) to pick out, to notice, to detect; to place on the spot at billiards. *v.i.* to become or be liable to be marked with spots. **in a spot**, in an awkward situation. **high spot** HIGH[1]. **on the spot**, at once, without change of place; in danger or difficulty; in the immediate locality. **soft spot** SOFT. **tight spot**, a dangerous or complicated situation. **to change one's spots**, to reform one's ways. **to knock spots off**, to outdo easily. **weak spot**, a flaw in one's character; a gap in a person's knowledge. **spot-ball**, *n.* (*Billiards*) a white ball marked with a black spot. **spot cash**, *n.* (*coll.*) money down. **spot check**, *n.* a random examination or check without prior warning. **spotlight**, *n.* an apparatus for throwing a concentrated beam of light on an actor on the stage; the patch of light thus thrown. **in the spotlight**, the focus of attention. **spot-on**, *a.* (*coll.*) absolutely accurate. **spot-stroke**, *n.* (*Billiards*) a winning-hazard off the red ball when on the spot. **spot-weld**, *v.t.* to join two pieces of metal with a circular weld. *n.* a weld of this type. **spotless** (-lis), *a.* **spotlessly**, *adv.* **spotlessness**, *n.* **spotted**, *a.* **spotted dick**, *n.* a steamed suet pudding with currants. **spotted fever**, *n.* cerebrospinal meningitis, characterized by spots on the skin. **spotter**, *n.* observer trained to detect the approach of enemy aircraft; one whose hobby is taking note of the registration numbers of trains, aircraft etc. **spotty**, *a.* **spottiness**, *n.*
spouse (spows), *n.* a husband or wife. **spousal** (-zəl), *a.* pertaining to marriage.
spout (spowt), *v.t.* to pour out or discharge with force or in large volume; to utter or recite in a declamatory manner. *v.i.* to pour out or issue forcibly or copiously; to declaim. *n.* a short pipe, tube, or channelled projection for carrying off water from a gutter, conducting liquid from a vessel, shooting things into a receptacle etc.; a continuous stream, jet, or column of water etc.; a water-spout. **up the spout**, (*sl.*) ruined, failed; pregnant. **spouter**, *n.*
sprag (sprag), *n.* a billet of wood, esp. a prop for the roof of a mine; a chock or bar for locking the wheel of a vehicle.
sprain (sprān), *v.t.* to twist or wrench the muscles or ligaments of (a joint) so as to injure without dislocation. *n.* such a twist or wrench or the injury due to it.
sprang (sprang), *past* SPRING.
sprat (sprat), *n.* a small food-fish of the herring tribe; applied to the young of the herring and to other small fish.
sprawl (sprawl), *v.i.* to lie or stretch out the body and limbs in a careless or awkward posture; to straggle, to be spread out in an irregular or ungraceful form. *n.* a careless or awkward posture; an irregular arrangement; an act of sprawling. **sprawling, sprawly**, *a.*
spray[1] (sprā), *n.* water or other liquid flying in small, fine drops; a perfume or other liquid applied in fine particles with an atomizer; an appliance for spraying. *v.t.* to throw or apply in the form of spray; to treat with a spray. **spray gun**, *n.* an appliance which sprays paint etc. **sprayer**, *n.*
spray[2] (sprā), *n.* a small branch or sprig, esp. with branchlets, leaves, flowers etc., used as a decoration; an ornament resembling a sprig of leaves, flowers etc.
spread (spred), *v.t.* (*past, p.p.* **spread**) to extend in length and breadth by opening, unrolling, unfolding, flattening out etc.; to scatter, to diffuse; to disseminate, to publish; to cover the surface of; to display to the eye or mind. *v.i.* to be extended in length and breadth; to be scattered, diffused, or disseminated. *n.* the act of spreading; breadth, extent, compass, expansion; diffusion, dissemination; (*coll.*) a meal set out, a feast; two facing pages in a book, magazine etc. **spreadeagle**, *v.t.* to cause to stand or lie with arms and legs stretched out. *v.i.* to stand or lie thus. *a.* (also **spreadeagled**) lying or standing with the arms and legs stretched out. **spreadsheet**, *n.* a computer program which can perform rapid calculations on figures

displayed on a VDU in rows and columns, used for business accounting and financial planning. **spreader**, *n*.
spree (sprē), *n*. a lively frolic, esp. with drinking; a bout of extravagance or excess.
sprig (sprig), *n*. a small branch, twig, or shoot; an ornament resembling this; a small headless nail or brad. *v.t.* (*past, p.p,* **sprigged**) to ornament with sprigs; to drive small brads into. **spriggy**, *a*.
sprightly (sprīt'li), *a*. lively, spirited, gay, vivacious.
spring (spring), *v.i.* (*past* **sprang** (sprang), *p.p.* **sprung** (sprŭng)) to leap, to bound, to jump; to move suddenly by or as by the action of a spring; to rise, to come (up) from or as from a source, to arise, to originate, to appear, esp unexpectedly; of wood etc., to warp, to split. *v.t.* to cause to move, fly, act etc., suddenly by or as by releasing a spring; (*sl.*) to bring about the escape from prison of; to cause to explode; to cause (timber etc.) to warp, crack, or become loose; of a vessel, to develop (a leak) thus; to cause to happen unexpectedly; to make known suddenly. *n*. a leap; a backward movement as from release from tension, a recoil, a rebound; the starting of a plank, seam, leak etc.; elastic force; an elastic body or structure, usu. of bent or coiled metal used to prevent jar, to convey motive power etc.; a source of energy, a cause of action, a motive; a natural issue of water from the earth, a fountain; a source, an origin; the first of the four seasons of the year, that preceding summer roughly March, April, and May in the N hemisphere; the early part, youth. **spring balance**, *n*. a balance weighing objects by the tension of a spring. **spring-board**, *n*. a springy board giving impetus in leaping, diving etc.; anything that provides a starting point or initial impetus. **spring chicken**, *n*. (*N Am.*) a tender young chicken, usu. from 2 to 10 months old; (*coll.*) a young, active, inexperienced person. **spring-clean**, *v.t.* to clean (a house) thoroughly in preparation for summer. **spring clean**, *n*. **spring loaded**, *a*. having or secured by means of a spring. **spring lock**, *n*. a lock with a spring-loaded bolt. **spring-mattress**, *n*. a mattress containing a series of spiral springs set in a frame. **spring onion**, *n*. an onion with a tiny thin-skinned bulb and long leaves, eaten in salads. **spring roll**, *n*. a Chinese dish comprising a thin pancake filled with a savoury mixture and deep fried. **springtail**, *n*. an insect having bristles on its under side enabling it to leap. **spring tide**, *n*. a high tide occurring a day or two after the new or the full moon; (*poet.*) springtime. **springtime**, *n*. the season of spring. **springer**, *n*. one who or that which springs; breeds of spaniel orig. used to rouse game; (*Arch.*) the part or stone where the curve of an arch begins; the rib of a groined roof; the lowest stone of a gable-coping. **springless** (-lis), *a*. **springlike**, *a*. **springy**, *a*. elastic, like a spring. **springiness**, *n*.
springbok (-bok), *n*. a southern African gazelle that leaps in play and when alarmed; (**Springbok**) a sportsman or sportswoman representing South Africa in international competitions.
springe (sprinj), *n*. a noose, a snare, usu. for small game. *v.t.* to catch in this.
springer, springless, springy etc. SPRING.
sprinkle (spring'kl), *v.t., v.i.* to scatter in small drops or particles. *n*. a sprinkling, a light shower. **sprinkler**, *n*. that which sprinkles. **sprinkler system**, *n*. a system of fire-extinction in which a sudden rise in temperature triggers the release of water from overhead nozzles. **sprinkling**, *n*. a small quantity or number.
sprint (sprint), *v.i.* to run, cycle, etc. at top speed. *n*. a short burst of running, cycling etc. at top speed; a race run or cycled thus. **sprinter**, *n*.
sprit (sprit), *n*. a small spar set diagonally from the mast to the top outer corner of a sail. **spritsail** (-sl), *n*.
sprite (sprīt), *n*. a fairy, an elf; a computer generated display shape that can be manipulated by a programmer to create fast and complex animation sequences.
spritzer (sprit'sə), *n*. a drink made from white wine and soda water.
sprocket (sprok'it), *n*. one of a set of teeth on a wheel etc., engaging with the links of a chain; a sprocket-wheel; a wheel with teeth for advancing film in a camera or projector. **sprocket-wheel**, *n*. a wheel set with sprockets.
sprout (sprowt), *v.i.* to shoot forth, to develop shoots, to germinate; to grow, like the shoots of plants. *v.t.* to cause to put forth sprouts or to grow. *n*. a new shoot on a plant; (*pl.*) brussels sprouts.
spruce[1] (sproos), *a*. neat, trim, smart. **to spruce up**, to smarten (oneself) up. **sprucely**, *adv*. **spruceness**, *n*.
spruce[2], **spruce fir** (sproos), *n*. a pine of the genus *Picea*. **spruce-beer**, *n*. a fermented liquor made from the leaves and small branches of the spruce.
sprue[1] (sproo), *n*. a hole or channel through which molten metal or plastic is poured into a mould; the corresponding projection in a casting.
sprue[2] (sproo), *n*. a tropical disease characterized by diarrhoea, anaemia, and wasting.
spruit (sproo'it), *n*. (*S Afr.*) a small tributary stream, esp. one dry in summer.
sprung (sprŭng), *past, p.p.* SPRING.
spry (sprī), *a*. (*comp.* **sprier, spryer,** *superl.* **spriest, spryest**) active, lively, agile, nimble. **spryly**, *adv*. **spryness**, *n*.
spud (spŭd), *n*. a short spade-like tool for cutting up weeds by the root etc.; (*coll.*) a potato. *v.t.* to dig (up) or clear (out) with a spud. **to spud in**, to begin drilling an oil well.
spume (spūm), *n*. froth, foam. *v.i.* to froth, to foam. **spumous, spumy**, *a*.
spun (spŭn), (*past, p.p.*) SPIN *a*. formed or produced by spinning. **spun glass**, *n*. fibreglass. **spun gold**, *n*. gold thread. **spun silk**, *n*. yarn made from silk waste and spun like woollen yarn.
spunk (spŭngk), *n*. mettle, spirit, pluck; touchwood, tinder; (*taboo*) semen. **spunky**, *a*. plucky, spirited.
spur (spœ), *n*. an instrument worn on a horseman's heel having a sharp or blunt point or a rowel; instigation, incentive, stimulus, impulse; a spur-shaped projection, attachment, or part, as the pointed projection on a cock's leg, or a steel point or sheath fastened on this in cockfighting; the largest root of a tree; a ridge or buttress projecting from a mountain range; a tubular projection on the columbine and other flowers; a railway siding or branch line. *v.t.* (*past, p.p.* **spurred**) to prick with spurs; to urge, to incite; to furnish with spurs. *v.i.* to ride hard. **on the spur of the moment**, on impulse. **to win one's spurs**, to achieve distinction, to make oneself famous. **spur wheel**, *n*. a gear-wheel with radial teeth projecting from the rim.
spurge (spœj), *n*. a plant of the genus *Euphorbia* with milky and usu. acrid juice. **spurge-laurel**, *n*. a bushy evergreen shrub with poisonous berries.
spurious (spū'riəs), *a*. not genuine, not proceeding from the true or pretended source, false, counterfeit; like an organ in form or function but physiologically or morphologically different. **spuriously**, *adv*. **spuriousness**, *n*.
spurn (spœn), *v.t.* to reject with disdain; to treat with scorn, *v.i.* to show contempt (at). *n*. scornful rejection. **spurner**, *n*.
spurt (spœt), *v.i.* to gush out in a jet or sudden stream; to make a sudden intense effort. *v.t.* to send or force out thus. *n*. a forcible gush or jet of liquid; a short burst of intense effort or speed.
sputa (spū'tə), *pl.* SPUTUM.
sputnik (spŭt'nik, sput'-), *n*. the name given by the Russians to a series of artificial earth satellites, the first of which was launched in 1957.
sputter (spŭt'ə), *v.i.* to emit saliva in scattered particles, to splutter; to speak in a jerky, incoherent, or excited way.

sputum — **squeeze**

v.t. to emit with a spluttering noise; to utter rapidly and indistinctly; to remove atoms from (a surface) by bombardment with high energy ions; to coat a surface with (a metallic film) by such a process. *n.* the process or act of sputtering; confused, incoherent speech. **sputterer**, *n.*
sputum (spū'təm), *n.* (*pl.* **-ta** (-tə)) spittle, saliva; matter expectorated in various diseases.
spy (spī), *v.t.* to see, to detect, to discover, esp. by close observation; to explore or search (out) secretly; to discover thus. *v.i.* to act as a spy; to search narrowly, to pry. *n.* one sent secretly into an enemy's territory, a rival firm, etc., to obtain information that may be useful in the conduct of hostilities etc.; one who keeps a constant watch on the actions, movements etc., of others. **spyglass**, *n.* a small telescope.
sq, (*abbr.*) square; (*pl* **sqq**) sequens (the following).
Sqd, (*abbr.*) squadron.
squab (skwob), *a.* fat, short, squat. *n.* a short, fat person; a young pigeon, esp. unfledged; a stuffed cushion, a sofa padded throughout, an ottoman. **squabby**, *a.*
squabble (skwob'l), *v.i.* to engage in a petty or noisy quarrel, to wrangle. *n.* a petty or noisy quarrel, a wrangle. **squabbler**, *n.*
squad (skwod), *n.* a small number of soldiers assembled for drill or inspection; a small party of people or players. **squad car**, *n.* a police car. **squaddy**, *n.* (*coll.*) a private soldier.
squadron (skwod'rən), *n.* a main division of a cavalry regiment, usu. consisting of two troops containing 120–200 men; a detachment of several warships employed on some particular service; an air force formation of two or more flights. **squadron-leader**, *n.* a commissioned officer in an air force equivalent in rank to a major in the army.
squalid (skwol'id), *a.* dirty, mean, poverty-stricken; sordid. **squalidity** (-lid'-), **squalidness**, **squalor**, *n.* **squalidly**, *adv.*
squall (skwawl), *v.i., v.t.* to cry out; to scream discordantly. *n.* a harsh, discordant scream, esp. of a child; a sudden, violent gust or succession of gusts of wind, esp. accompanied by rain, hail, snow, sleet etc. **squaller**, *n.* **squally**, *a.*
squalor SQUALID.
squama (skwā'mə), *n.* (*pl.* **-mae** (-mē)) a scale or scale-like structure, feather, part of bone etc. **squamose** (-ōs), **squamous**, **squamulose** (-mūlōs), *a.*
squander (skwon'də), *v.t.* to spend wastefully; to dissipate by foolish prodigality. **squanderer**, *n.*
square (skweə), *n.* a rectangle with equal sides; any surface, area, object, part etc., of this shape; a rectangular division of a chess- or draught-board, window-pane etc.; an open quadrilateral area surrounded by buildings, usu. laid out with trees, flower-beds, lawns etc.; an L- or T-shaped instrument for laying out and testing right angles; the product of a quantity multiplied by itself; (*sl.*) a conventional, old-fashioned person, one out of keeping with modern ways of thought. *a.* having four equal sides and four right angles; rectangular; at right angles (to); broad with straight sides or outlines; (*Football etc.*) in a straight line across the pitch; just, fair, honest; evenly balanced, even, settled, complete, thorough, absolute; full, satisfactory; (*coll.*) dull, conventional. *adv.* at right angles; honestly; fairly; evenly. *v.t.* to make square or rectangular; to adjust, to bring into conformity (with or to); to make even, to settle, to pay; (*coll.*) to bribe, to gain over thus; to multiply (a number or quantity) by itself; *v.i.* to be at right angles (with); to conform precisely, to agree, to harmonize; to level a score, make an account even, etc.; to put oneself in an attitude for boxing. **back to square one**, back to where one started without having made any progress. **on the square**, at right angles; fairly, honestly. **square dinkum**, (*Austral. coll.*) absolutely honestly. **to square away**, (*N Am. coll.*) to put in order, tidy up. **to square off**, to assume a posture of defence or attack. **to square up**, to settle an account. **to square the circle**, to construct geometrically a square equal in area to a given circle; hence, to attempt impossibilities. **square-bashing**, *n.* (*sl.*) military drill. **square bracket**, *n.* either of a pair of written or printed characters, [ ], used to enclose a section of writing or printing, or used as a sign of aggregation in a mathematical formula. **square dance**, *n.* a dance in which the couples form squares. **square dancer**, *n.* **square dancing**, *n.* **square knot**, *n.* a reef knot. **square leg**, *n.* (*Cricket*) a fielder standing about 20 yd. directly behind a batsman as he receives the bowling. **square meal**, *n.* a meal which is full and satisfying. **square measure**, *n.* a system of measures expressed in square feet, metres etc. **square number**, *n.* the product of a number multiplied by itself. **square-rigged**, *a.* having the principal sails extended by horizontal yards suspended from the middle. **square root**, *n.* the quantity that, mulipied by itself, will produce the given quantity. **square-sail** (-sl), *n.* a four-cornered sail set on a yard, esp. on a fore-and-aft rigged vessel. **squarely**, *adv.* **squareness**, *n.* **squarer**, *n.* one who squares. **squarish**, *a.*
squarrose (skwo'rōs), *a.* rough with projecting scale-like processes.
squash¹ (skwosh), *v.t.* to crush, to press flat or into a pulp; to suppress, to overcome; to crush, to snub, to humiliate. *v.i.* to be crushed or beaten to pulp as by a fall; to squeeze (into). *n.* a thing or mass crushed or squeezed to pulp; the fall of a soft body; the sound of this; a throng, a squeeze; a game with rackets and balls played in a court; a drink made from usu. concentrated fruit juice diluted with water. **squasher**, *n.* **squashy**, *a.* **squashiness**, *n.*
squash² (skwosh), *n.* a fleshy, edible, gourd-like fruit; the trailing plant of the genus *Curcurbita* which bears it.
squat (skwot), *v.i.* (*past, p.p.* **squatted**) to sit down or crouch on the haunches; chiefly of animals, to crouch, to cower; to settle on land or occupy a building without any title. *a.* short, thick, dumpy. *n.* a squatting posture; a building occupied by squatters. **squatly**, *adv.* **squatness**, *n.* **squatter**, *n.* one who occupies property or land without title; (*Austral.*) one who owns extensive land for pasturage. **squattocracy** (-ok'rəsi), *n.* (*Austral.*) squatters as a corporate and influential body.
squaw (skwaw), *n.* (*offensive*) a N American Indian woman or wife. **squaw man**, *n.* (*derog.*) a White man married to a N American Indian.
squawk (skwawk), *v.i.* to utter a loud, harsh cry; (*coll.*) to protest loudly. *n.* such a cry or protest.
squeak (skwēk), *v.i.* to utter a sharp, shrill, usu. short cry; to inform; to confess. *v.t.* to utter with a squeak. *n.* a sharp, shrill sound; (*coll.*) a narrow escape or margin, a close shave. **squeaky**, *a.* **squeaky-clean**, *a.* spotless; above reproach. **squeakily**, *adv.*
squeal (skwēl), *v.i.* to utter a more or less prolonged shrill cry as in pain, etc.; (*sl.*) to turn informer; (*coll.*) to complain. *n.* a more or less prolonged shrill cry. **squealer**, *n.*
squeamish (skwē'mish), *a.* easily nauseated, disgusted or offended; fastidious, finicky, hypercritical, prudish, unduly scrupulous. **squeamishly**, *adv.* **squeamishness**, *n.*
squeegee (skwē'jē), *n.* an implement, composed of a strip of rubber fixed to a handle for cleaning windows; a similar implement, usu. with a rubber roller, used by photographers for squeeezing and flattening. *v.t.* to sweep, smooth etc., with a squeegee.
squeeze (skwēz), *v.t.* to press closely, esp. between two bodies or with the hand, so as to force juice etc., out; to extract (juice etc.) thus; to force (oneself etc., into, out of etc.); to extort money etc., from, to harass by exactions; to exact (money etc.) by extortion etc.; to put pressure on, to oppress, to constrain by arbitrary or illegitimate

means. *v.i.* to press, to push, to force one's way (into, through etc.). *n.* the act of squeezing; pressure; a close embrace; a throng, a crush; a tight clasp; restriction; a period of (e.g. economic) restriction; an amount got by squeezing; in bridge, whist, play that forces an opponent to discard a potentially winning card, squeeze play; (*chiefly N Am., coll.*) a current boyfriend or girlfriend, the love of one's life. **squeezable,** *a.* **squeezer,** *n.*
squelch (skwelch), *v.t.* to crush; to silence, to extinguish, to discomfit. *v.i.* to make a noise as of treading in wet snow; to walk making this sound. *n.* a crushing retort; a squelching noise. **squelcher,** *n.* **squelchy,** *a.*
squib (skwib), *n.* a firework emitting sparks and exploding with a bang, usu. thrown by the hand; lampoon. **damp squib,** something which fails to make the intended impact or impression.
squid (skwid), *n.* (*pl.* **squid, squids**) a small kind of cuttlefish.
squiffy (skwif'i), *a.* (*coll.*) slightly drunk.
squiggle (skwig'l), *v.i.* to squirm, to wriggle; to make wriggly lines. *n.* a wriggly line. **squiggler,** *n.* **squiggly,** *a.*
squill (skwil), *n.* a liliaceous plant resembling the bluebell; the sliced bulb of this used as an expectorant, diuretic etc.
squinch (skwinch), *n.* an arch across the internal angle of a square tower to support the side of an octagonal spire etc.
squint (skwint), *v.i.* to look with the eyes differently directed; to be affected with strabismus; to look obliquely; to look with eyes half shut. *v.t.* to shut or contract (the eyes) quickly. *a.* having a squint; (*coll.*) crooked. *n.* an affection of the eyes causing the axes to be differently directed, strabismus; a stealthy look, a sidelong glance; (*coll.*) a look, a glance. **squinter,** *n.* **squinty,** *a.*
squire (skwīə), *n.* a country gentleman, esp. the chief landowner in a place; (*Hist.*) an attendant on a knight; (*coll.*) a term of address between men. *v.t.* to escort (a woman). **squir(e)archy** (-ahki), *n.* landed proprietors collectively; the political influence of, or government by these. **squirearchal, -archical** (-ah'-), *a.*
squirm (skwœm), *v.i.* to wriggle, to writhe about; to display discomfort, embarrassment etc. *n.* a wriggling movement.
squirrel (skwi'rəl), *n.* a brown or grey bushy-tailed rodent quadruped living chiefly in trees; the fur of a squirrel; (also **barking squirrel**) a prairie-dog; (*coll.*) a person who hoards things. *v.t.* to hoard (away). **squirrel cage,** *n.* a small cylindrical cage with a treadmill; the rotor of an induction motor with cylindrically arranged copper bars. **squirrel-monkey,** *n.* a small S American monkey with soft golden fur.
squirt (skwœt), *v.t.* to eject in a jet or stream from a narrow orifice. *v.i.* of liquid, to be so ejected. *n.* a jet (of liquid); (*coll.*) a pert, conceited, small or insignificant person. **squirter,** *n.*
squish (skwish), *v.t.* to crush so as to make a squelching or sucking noise. *v.i.* to make a squelching or sucking sound. *n.* the sound of squishing. **squishy,** *a.*
squit (skwit), *n.* (*sl.*) an insignificant person; nonsense.
Sr[1], (*abbr.*) Señor; Sir.
Sr[2], (*chem. symb.*) strontium.
sr, (*abbr.*) steradian.
Sra, (*abbr.*) Señora.
SRC, (*abbr.*) Science Research Council; Student Representative Council.
SRN, (*abbr.*) State Registered Nurse.
Srta, (*abbr.*) Señorita.
SS, (*abbr.*) Saints; Hitler's bodyguard, used as security police, concentration camp guards etc. [G *Schutzstaffel*, elite guard]; steamship.
SSE, (*abbr.*) south-southeast.
SSR, (*abbr.*) Soviet Socialist Republic.
SSW, (*abbr.*) south-southwest.

St, (*abbr.*) Saint; statute; Strait; Street.
st., (*abbr.*) stanza; stone.
stab (stab), *v.t.* (*past, p.p.* **stabbed**) to pierce or wound with a pointed weapon; to plunge (a weapon into); to inflict pain upon or to injure by slander etc. *v.i.* to aim a blow with or as with a pointed weapon (at). *n.* a blow or thrust with a pointed weapon; a wound inflicted thus; a sudden sharp sensation, emotion, pang, etc.; a secret malicious injury. **to make a stab at,** (*coll.*) to attempt. **to stab in the back,** to betray; to injure the reputation of someone esp. a colleague, friend etc. **stabber,** *n.*
Stabat Mater (stah'batmah'tə), *n.* a Latin hymn reciting the seven dolours of the Virgin at the Cross, beginning with these words; a musical setting of this.
stable[1] (stā'bl), *a.* firmly fixed, established; not to be moved, shaken or destroyed easily; firm, resolute, constant, not changeable, unwavering; (*Chem.*) durable, not readily decomposed; not radioactive. **stabile** (-bil), *a.* fixed; stable. *n.* an abstract art form similar to a mobile but stationary. **stability** (stəbil'-), *n.* the quality of being stable; the property of mechanical, electrical or aerodynamic systems that makes them return to a state of equilibrium after disturbance. **stabilize, -ise,** *v.t.* to make stable. **stabilization,** *n.* the act of stabilizing. **stabilizer,** *n.* any thing or person that stabilizes; a device that gives extra stability to an aircraft, vessel, children's bicycle etc., an additive which retards chemical action. **stableness,** *n.* **stably,** *adv.*
stable[2] (stā'bl), *n.* a building or part of a building for horses or (sometimes) cattle; the race-horses belonging to a particular stable; a group of people with particular skills or from the same place of training, e.g. athletes under single management, actors from the same drama school; any collection or group. *v.t.* to put or keep in a stable. *v.i.* of horses, etc., to lodge in a stable. **stable-boy, -girl, -man,** *n.* one employed in a stable. **stable lad,** *n.* a groom in a racing stable. **stabling,** *n.* accommodation in a stable or stables.
staccato (stəkah'tō), *n., a., adv.* (*Mus.*) (a piece of music) played with each note sharply distinct and detached, opp. to *legato.*
stachys (stăk'is), *n.* a labiate plant with white or reddish spikes of flowers, also called the wound-wort.
stack (stak), *n.* a round or rectangular pile of corn in the sheaf, hay, straw etc.; a pile, a heap, esp. of an orderly kind; a pyramidal pile of rifles standing on their butts with the muzzles together; a measure of wood or coal, 108 cu. ft. (3·05 m^3); (*coll.*) a great quantity; a chimney, a funnel, a smoke-stack; a towering isolated mass of rock; (*usu. pl.*) compact bookshelves, in a library, usu. with restricted public access; aircraft circling an airport at different altitudes waiting for instructions to land; a temporary storage area in a computer memory. *v.t.* to pile in a stack or stacks; to load with stacks; to assign (waiting aircraft) to a particular altitude in preparation for landing at an airport. **to stack the cards,** to interfere with a deck of cards secretly for the purpose of cheating; to arrange (matters) to the disadvantage or advantage, of someone. **stacked,** *a.* (*sl.*) of a woman, having generous proportions, esp. large breasts.
stadium (stā'diəm), *n.* (*pl.* **-diums, -dia** (-diə)) an ancient Greek measure of about 607 ft. (184 m), the course for foot-races at Olympia; an enclosure, usu. an amphitheatre, where games, races, outdoor concerts, etc. can be watched by a large number of spectators.
Stadtholder, Stadholder (staht'hōldə, stat'-), *n.* (*Hist.*) a viceroy, governor, or deputy-governor of a province or town in the Netherlands; the chief magistrate of the United Provinces.
staff[1] (stahf), *n.* (*pl.* **staffs, staves** (stāvz)) a stick carried for help in walking etc., or as a weapon; support; a stick, rod, pole etc., borne as an emblem of office or authority;

staff a shaft, pole etc., forming a support or handle, as a flagstaff; a rod used in surveying a rod-like appliance, instrument, part, fitting etc.; (*Mil.*) a body of officers assisting an officer in command whose duties are concerned with a regiment or an army as a whole; a body of persons working under a manager, editor etc.; (*Mus.*) a set of five parallel lines and spaces on or between which notes are written representing the pitch of tones. *v.t.* to provide with a staff. **(the) staff of life**, staple foodstuff, esp. bread. **staff-notation**, *n.* (*Mus.*) notation by the staff as dist. from sol-fa. **staff nurse**, *n.* a qualified nurse next in rank below a sister. **staff-officer, -sergeant** etc., *n.* one serving on a staff.
staff[2] (stahf), *n.* a composition of plaster, cement etc., used as building material etc., esp. in temporary structures.
stag (stag), *n.* the adult male of a deer; (*sl.*) a male unaccompanied by a woman at a social function; (*Stock Exch.*) one who applies for or purchases stock or shares in a new issue solely with the object of selling at a profit immediately on allotment. **stag-beetle**, *n.* a beetle with large mandibles, in the male branching like a stag's horns. **stag party**, *n.* (*coll.*) a party for men only, esp. one given for a man about to be married.
stage (stāj), *n.* an elevated platform; a raised platform on which theatrical performances take place; the theatre, the drama, the profession of an actor, actors collectively; a scene of action; one of a series of regular stopping-places on a route; the distance between two such stations; a definite portion of a journey; a point in a progressive movement, a definite period in development; a stagecoach; a detachable propulsion unit of a rocket; the small platform on a microscope where the slide is mounted for examination; part of a complex electronic circuit. *v.t.* to put on the stage; to plan and execute an event. **to go on the stage**, to become a professional actor or actress. **stage-coach**, *n.* a horse-drawn coach that ran regularly by stages for conveyance of parcels, passengers etc. **stage-craft**, *n.* the art of writing or staging plays. **stage-direction**, *n.* an instruction directing the movements etc., of actors in a play. **stage-door**, *n.* a door to a theatre for the use of actors, workmen, etc. **stagefright**, *n.* a fit of nervousness in facing an audience. **stage-hand**, *n.* a worker who moves scenery etc. in a theatrical production. **stage-manage**, *v.t.* to direct or supervise (from behind the scenes). **stage-manager**, *n.* one who superintends the scenic effects etc., of a play. **stage-struck**, *a.* smitten with the theatre. **stage-whisper**, *n.* an audible aside; something meant for the ears of others than the person ostensibly addressed. **stager**, *n.* a person of long experience in anything, esp. in the compound *old-stager*. **staging**, *n.* a scaffolding; the act of putting a play on the stage. **staging area**, *n.* an assembly point for troops in transit. **staging post**, *n.* a regular stopover point esp. on an air route. **stagy**, *a.* theatrical, unreal. **staginess**, *n.*
stagflation (stagflā'shən), *n.* a combination of high inflation and falling industrial output and employment.
stagger (stag'ə), *v.i.* to move unsteadily in standing or walking, to totter, to reel; to begin to give way, to waver, to hesitate. *v.t.* to cause to reel; to cause to hesitate; to shock with surprise etc.; to overlap, to place zig-zag; of working hours etc., to arrange so as not to coincide with others. *n.* a staggering movement; (*pl.*) a disease affecting the brain and spinal cord in horses and cattle, characterized by vertigo etc.; giddiness, vertigo. **staggerer**, *n.* one who staggers; a staggering blow, argument etc. **staggering**, *a.* **staggeringly**, *adv.*
stagnant (stag'nənt), *a.* still; without current, motionless; dull, sluggish, inert. **stagnancy**, *n.* **stagnate** (-nāt', stag'-), *v.i.* to be or become stagnant. **stagnation**, *n.*
staid (stād), *a.* sober, steady, sedate. **staidly**, *adv.* **staidness**, *n.*

stain (stān), *v.t.* to discolour, to soil, to sully; to tarnish, to blemish (a reputation etc.); to colour by means of dye or other agent acting chemically or by absorption, opp. to painting; to impregnate (an object for microscopic examination) with a colouring matter affecting certain parts more powerfully than others. *v.i.* to cause discoloration; to take stains. *n.* a discoloration; a spot of a distinct colour; a blot, a blemish; a solution or dye for colouring wood, leather, cloth, etc. **stained glass**, *n.* glass coloured for use in windows. **stainable**, *a.* **stainer**, *n.* **stainless** (-lis), *a.* without a stain, immaculate; resistant to rust or tarnish. **stainless steel**, *n.* a rustless alloy steel used for cutlery etc. **stainlessly**, *adv.*
stair (steə), *n.* one of a series of steps, esp. for ascending from one storey of a building to another; (*usu. pl.*), a flight of stairs. **backstairs** BACK. **below stairs**, in the servants' quarters or relating to their affairs. **flight of stairs**, a set of stairs, as from one landing to another. **staircase**, *n.* a flight of stairs with banisters, supporting structure etc. **moving staircase** MOVE. **stairway**, *n.* a staircase. **stairwell**, *n.* the vertical shaft which contains the staircase.
stake (stāk), *n.* a stick or post pointed at one end and set in the ground, as a support, part of a railing etc.; a post to which persons condemned to death by burning were bound; a prop or upright part of fitting for supporting a machine etc., anything, esp. money, wagered on a competition or contingent event; (*pl.*) money competed for in a race etc.; (*pl.*) a race for this; (*fig.*) anything contended for; an interest held e.g. in a company. *v.t.* to fasten, support, or protect with a stake or stakes; to mark (out or off), lay claim to with or as with stakes; to wager, to venture (on an event etc.). **at stake**, in hazard, at issue, in question. **to pull up stakes**, to move home, to move on. **to stake one's claim**, to assert one's right to possess (something). **to stake out**, *v.t.* to place under surveillance. **stakeout**, *n.* a place, person etc. under surveillance; a (police) surveillance operation covering a particular building or area.
Stakhanovism (stəkhan'əvizm), *n.* USSR system for increasing production by utilizing each worker's initiative. **Stakhanovite** (-vit), *n.*, *a.* [after A G *Stakhanov*, 1906–77, Soviet miner]
stalactite (stal'əktīt), *n.* a deposit of carbonate of lime, hanging from the roof of a cave etc., in the form of a thin tube or a large icicle, produced by the evaporation of percolating water. **stalactiform** (-fawm), **stalactitic** (-tit'-), *a.*
stalag (stah'lag), *n.* in World War II, a German prisoner-of-war camp, esp. for men from the ranks and non-commissioned officers.
stalagmite (stal'əgmīt), *n.* a deposit of the same material as in a stalactite which projects upwards from the floor of a cave etc. **stalagmitic** (-mit'-), *a.*
stale (stāl), *a.* not fresh, dry, musty; vapid or tasteless from being kept too long; trite, over used; in poor condition from overtraining. *n.* urine of horses etc. *v.i.* of horses etc., to urinate. **staleness**, *n.*
stalemate (stāl'māt), *n.* (*Chess*) the position when the king, not actually in check, is unable to move without placing himself in check, and there is no other piece that can be moved; a situation of deadlock. *v.t.* to subject to a stalemate; to bring to a standstill.
Stalinism (stah'linizm), *n.* the brutal authoritarian regime associated with the Russian dictator Joseph Stalin (1879–1953), developed from the ideology of Marxism–Leninism. **Stalinist**, *n.*
stalk[1] (stawk), *v.i.* to walk with high, pompous steps; to go stealthily, to steal (up to game or prey) under cover. *v.t.* to pursue stealthily by the use of cover. *n.* the act of stalking game or prey; a pompous gait. **stalker**, *n.* **stalking-horse**, *n.* a horse or figure like a horse behind

which a person hides when stalking game; a mask, a pretence, a pretext.
stalk[2] (stawk), *n.* the stem or axis of a plant; the peduncle of a flower; the supporting peduncle of a barnacle etc.; the stem of a wineglass etc. **stalk-eyed**, *a.* having the eyes set on peduncles (as certain crustaceans). **stalked**, *a.* (*usu. in comb.* as *thin-stalked*). **stalkless** (-lis), *a.* **stalky**, *a.*
stall (stawl), *n.* a division or compartment for a horse, ox etc., in a stable or byre; a booth or shed in a market, street etc., or a bench, table etc., in a building for the sale of goods; a finger-stall; a seat in the choir of a large church, enclosed at the back and sides and usu. canopied, for a clergyman, chorister etc.; one of a set of seats in a theatre, usu. in the front part of the pit; (*pl.*) the area containing these; an instance of an aircraft or motor stalling; a diversion, an evasion, a delaying tactic. *v.t.* to put or keep in a stall (esp. cattle for fattening). *v.i.* to stick fast (in mud etc.); of a car etc. engine, to cease working suddenly; to allow an aeroplane to lose its forward impetus and thus cause a downward fall; to play for time; to be evasive. **starting stalls**, a group of stalls from which horses emerge at the start of a race. **to stall for time**, to postpone or hold off as long as possible. **stall-feed**, *v.t.* to fatten in a stall.
stallion (stal'yən), *n.* an uncastrated male horse, esp. one kept for breeding purposes.
stalwart (stawl'wət), *a.* strong in build, sturdy; stout, resolute. *n.* a strong, resolute dependable person. **stalwartly**, *adv.* **stalwartness**, *n.*
stamen (stā'mən), *n.* (*pl.* **stamens, stamina** (stam'inə)), the pollen-bearing male organ of a flower. **staminate** (-nāt), *a.* having stamens (but no pistils). **staminiferous** (-nif'-), *a.*
stamina (stam'inə), *n.* strength, vigour, power of endurance.
stammer (stam'ə), *v.i., v.t.* to speak with halting articulation, nervous hesitation, or repetitions of the same sound; to stutter. *n.* a stammering utterance or vocal disorder characterized by this. **stammerer**, *n.* **stammering**, *n., a.* **stammeringly**, *adv.*
stamp (stamp), *v.t.* to make a mark or impression upon with a pattern etc.; to affix a stamp to; to impress (a device etc.) upon something; to impress deeply (on the memory etc.); to bring (the foot etc.) down heavily; to extinguish thus, to put (out); to crush by downward force or pressure; to destroy; to repress; to show, to distinguish. *v.i.* to strike the foot forcibly on the ground; to walk in this manner. *n.* the act of stamping; an instrument for stamping marks, designs etc.; the mark made by this; an official mark set on things chargeable with some duty or tax, to show that such is paid; a small piece of paper officially stamped for affixing to letters, receipts etc.; a label, imprint, or other mark certifying ownership, quality, genuineness etc.; distinguishing mark, impress, kind, sort; a downward blow with the foot; a blow with a stamping-machine; an instrument for crushing ore; (*coll.*) a national insurance contribution. **to stamp out**, to extinguish (a fire) by stamping; to suppress, extirpate. **stamp album**, *n.* a book to hold a postage-stamp collection. **stamp duty**, *n.* a tax imposed on certain legal documents. **stamping-ground**, *n.* a habitual meeting place, a favourite resort. **stamper**, *n.*
stampede (stampēd'), *n.* a sudden fright causing horses or cattle to scatter and run; any impulsive movement or action on the part of a large number of persons. *v.i.* to take part in a stampede. *v.t.* to cause to do this.
stance (stans, stahns), *n.* the position adopted by a person when standing; the position taken for a stroke in golf, cricket etc.; a personal attitude, political position etc.; (*chiefly Sc.*) place, site, station.
stanch (stahnch) *v.t.* to prevent or stop the flow of (blood etc., from a wound).

stanchion (stan'shən), *n.* a prop, post, pillar etc., forming a support or part of a structure; a vertical bar or pair of bars for confining cattle in a stall. *v.t.* to support with a stanchion.
stand (stănd), *v.i.* (*past, p.p.* **stood**, (stud)) to be upon the feet; to be or become or remain erect; to be in a specified state, attitude, position, situation, rank etc.; to have a specified height or stature; to be or remain in a stationary position, to cease from motion, to stop, to be or remain immovable, not to give way; to remain firm or constant, to abide, to endure, to persist; to hold good, to remain valid or unimpaired; to be motionless, to lie stagnant; to move into a specified position and remain in it; to hold a specified course, to steer; of a setter, to point; to become a candidate. *v.t.* to set in an erect or a specified position; to endure, to sustain, without giving way or complaining; to undergo (a trial etc.); to sustain the expense of (a drink etc.). *n.* a cessation of motion or progress, a halt, a state of inactivity, a standstill; the act of standing, esp. with firmness, in a fixed or stationary position, place, or station; resistance, opposition, defensive effort etc.; a small frame or piece of furniture for supporting anything; a place in a town where cabs etc., stand for hire; an erection for spectators to stand or sit on; in cricket, a lengthy partnership between two batsmen at the wicket. **it stands to reason**, it is logically manifest (that). **one-night stand**, a performance given by a musical group, theatrical company etc. in one spot for one night only before moving on; (*coll.*) a sexual relationship that lasts one night only. **to stand a chance** CHANCE. **to stand by**, to be present, to be a bystander; to look on passively; to uphold, to support firmly; to abide by; to stand near in readiness to act promptly as directed. **to stand down**, to withdraw; of a committee, to be dissolved; to leave the witness box in a law court; to come off duty. **to stand fast**, to stay firm, to be unmoved. **to stand for**, to support the cause of; to represent, to imply; to offer oneself as a candidate for; to endure. **to stand good**, to remain valid. **to stand in**, to take an actor's place in a scene until the cameras are ready; to deputize for. **to stand off**, to keep at a distance; to move away; to suspend (an employee); to come to a stalemate. **to stand on**, to insist on (ceremony etc.); (*Naut.*) to keep on the same course. **to stand one's ground**, to remain resolute, to stay fixed in position. **to stand on one's own feet**, to manage without the help of others. **to stand out**, to project; to be conspicuous; to persist (in opposition against); to endure without giving way. **to stand over**, to supervise closely. **to stand to**, (*Mil.*) to take up positions in preparation for possible attack. **to stand up**, to rise to one's feet; to fail to keep an appointment with. **to stand up for**, to maintain, to support, to take the part of. **to stand up to**, to resist, endure, confront, withstand (criticism, opposition, etc.). **stand-by**, *n.* a thing or person to be confidently relied upon; a substitute or replacement kept esp. for use in an emergency. *a.* of a ticket, not booked in advance, subject to availability. **on stand-by**, held in readiness for use in an emergency etc; of an airline passenger, awaiting an empty seat, not having booked in advance. **stand-in**, *n.* a minor actor who takes the place of a star in a scene until the cameras are ready; a substitute. **stand-off, stand-off half** FLY-HALF. **stand-offish**, *a.* distant, reserved. **stand-offishly**, *adv.* **stand-offishness**, *n.* **standpipe**, *n.* an upright pipe serving as a hydrant, to provide a head of water for pressure etc. **stand-point**, *n.* a point of view. **standstill**, *n.* a stoppage, a cessation of progress. **stand-up**, *a.* of a collar, upright; of a comedian, telling jokes etc. directly to the audience in a solo performance; taken, done, etc. in a standing position. **stand-up fight** FIGHT. **stander**, *n.* **standing**, *a.* erect; fixed, established, permanent, not temporary or for a special occasion; performed or begun from a standing position; stagnant. *n.* the act of one that

stands; relative place or position; repute, estimation, esp. good estimation, duration, existence. **standing army,** *n.* a peacetime army of professional soldiers. **standing order,** *n.* an instruction to a bank by a customer to pay fixed sums at regular intervals in payment of bills etc.; an order made by an organization as to the manner in which its business shall be conducted. **standing rigging,** *n.* the fixed ropes and chains by which the masts etc. on a ship, are secured. **standing room,** *n.* room for standing, esp. after all seats are filled. **standing stone,** *n.* a large erect stone set in the ground in prehistoric times. **standing wave,** *n.* a wave that has a fixed amplitude at any given point along its axis.

standard (stan'dəd), *n.* a flag as the distinctive emblem of an army, government etc.; a measure of extent, quantity, value etc., established by law or custom as an example or criterion for others; any type, fact, thing etc., serving as a criterion; the degree of excellence required for a particular purpose; comparative degree of excellence; in coinage the proportion of gold or silver and alloy fixed by authority; a grade of classification in elementary schools; an upright pillar, post, or other support; a tree or shrub growing on a single upright stem, or supported on its own stem. *a.* recognized as a standard for imitation, comparison etc. **standard of living,** a level of subsistence or material welfare of an individual, group or community. **standardbearer,** *n.* a soldier carrying a standard; a leader of a movement or cause. **standard deviation,** *n.* a measure of the scatter of the value of a variable about a mean in a frequency distribution. **standard lamp,** *n.* a movable lamp on a tall pedestal. **standard time,** *n.* the official time of a locality reckoned from a conventionally-adopted meridian (for most purposes this is the meridian of Greenwich). **standardize, -ise,** *v.t.* **standardization,** *n.*

standing, standpoint, standstill etc. STAND.

stank (stangk), *past* STINK.

stannary (stan'əri), *n.* a tin-mine, tin-works; a tin-mining district.

stannic (stan'ik), *a.* of or containing (tetravalent) tin.

stannous, *a.* of or containing (bivalent) tin.

stanza (stan'zə), *n.* a group of rhymed lines adjusted to each other in a definite scheme. **stanzaed** (-zəd), **stanzaic** (-zā'-), *a.*

stapes (stā'pēz), *n.* the innermost of the three ossicles of the middle ear, the stirrup bone.

staphyl(o)-, *comb. form.* uvula.

Staphylococcus (stafilōkok'əs), *n.* a genus of microorganisms (*cocci*) forming the bacteria most frequently found in cutaneous affections of a suppurative kind.

staple[1] (stā'pl), *n.* a U-shaped piece of metal driven into a post, wall etc., to receive part of a fastening or to hold wire etc.; a similarly shaped piece of thin wire used to fasten papers etc.; a bent wire used in wire-stitching. *v.t.* to fasten or attach with staples. **staple gun,** *n.* a device for fixing staples on to a surface. **stapler,** *n.* a device for inserting staples.

staple[2] (stā'pl), *n.* the principal commodity sold or produced in any place, country etc.; the main element of diet etc.; the chief material or substance of anything; raw material; the length, strength etc., of the fibre of wool, cotton etc., as a criterion of quality. *a.* chief, principal, main. *v.t.* to sort or classify (wool etc.) according to staple.

star (stah), *n.* a celestial body appearing as a fixed point, esp. one of the fixed stars or those so distant that their relative position in the heavens appears constant; an object, figure, or device resembling a star, esp. one with radiating points used as an emblem or ornament; an asterisk (*); a white spot on the forehead of a horse etc.; a brilliant or prominent person, esp. an actor or singer; a heavenly body regarded as having influence over a person's life. *v.t.* (*past, p.p.* **starred**) to set, spangle, or decorate with stars; to put an asterisk against (a name etc.). *v.i.* of an actor, singer, etc., to appear as a star. *v.t.* to mark with a star or stars. **giant star** GIANT. **star-of-Bethlehem,** a bulbous plant of the lily family with star-shaped white flowers striped outside with green. **Star of David,** the emblem of Judaism and the State of Israel consisting of a six-pointed star made from two superimposed equilateral triangles. **Stars and Stripes,** the national flag of the US. **to see stars,** to see small points of light as a result of e.g. a bump on the head; to be dazed. **Star-Chamber,** *n.* a court of civil and criminal jurisdiction at Westminster (abolished 1641). **stardust,** *n.* a large concentration of distant stars appearing as dust; a romantic or magical feeling. **starfish,** *n.* an echinoderm with five or more rays or arms. **star fruit,** *n.* the yellow, edible fruit, star-shaped in section, of a SE Asian tree the carambola. **star gaze,** *v.i.* to gaze at the stars; to daydream. **stargazer,** *n.* **stargazing,** *n.* **starlight,** *n.* the light of the stars. **starlit,** *a.* **star sapphire,** *n.* a sapphire reflecting light in the figure of a star. **star-spangled,** *a.* covered with stars. **Star-spangled Banner,** the Stars and Stripes; the national anthem of the US. **star-studded,** *a.* of a film, play etc., having a large proportion of famous performers. **Star Wars,** *n.* (*funct. as sing.*) (*coll.*) the Strategic Defence Initiative, a proposed defence plan of the US involving laser-armed satellites deployed in space for destroying enemy missiles. **starwort,** *n.* a plant of the genus *Stellaria* or *Aster.* **stardom** (-dəm), *n.* the state or status of being a star in films etc. **starless** (-lis), *a.* **starlet** (-lit), *n.* a young actress who is being trained and promoted as a future star performer. **starlike,** *a.* **starry,** *a.* filled, adorned with stars; shining like, or illuminated by stars. **starry-eyed,** *a.* acting or thinking in a dreamy, overoptimistic manner. **starriness,** *n.*

starboard (stah'bəd), *n.* the right-hand side of a vessel or aircraft looking forward. *a.* pertaining to or located on the starboard. *v.t.* to put or turn to starboard.

starch (stahch), *n.* a white, tasteless, odourless, amorphous compound, found in all plants except fungi, but esp. in cereals, potatoes, beans etc., an important constituent of vegetable foods, and used as a soluble powder to stiffen linen etc.; food, e.g. potatoes, which contains a lot of starch; stiffness, preciseness, formality. *v.t.* to stiffen with starch. **starch-reduced,** *a.* having the starch content reduced, as in bread etc. eaten by slimmers. **starcher,** *n.* **starchy,** *a.* pertaining to or containing starch; stiff, unyielding. **starchiness,** *n.*

stare (steə), *v.i.* to look with eyes fixed and wide open, as in admiration, surprise, horror etc.; to stand out, to be prominent. *n.* a staring gaze. **to stare one in the face,** to be obvious to one. **starer,** *n.*

stark (stahk), *a.* downright, sheer; blunt, bare; unadorned, severe; desolate. *adv.* wholly, absolutely. **stark-naked** (stahk-), *a.* quite naked. **starkers** (-kəz), *a.* (*coll.*) stark-naked. **starkly,** *adv.* **starkness,** *n.*

starling (stah'ling), *n.* a small black and brown speckled bird of the genus *Sturnus*, esp. the common starling.

start (staht), *v.i.* to make a sudden involuntary movement, as from fear, surprise etc.; to move abruptly (aside, etc.); of timber, rivets etc., to give way, to become loose etc.; to set out, to begin a journey, a task, etc.; to make a beginning (on a journey etc.); to commence. *v.t.* to cause to start, to rouse; to originate, to set going; to set (people) working; to give the signal to (persons) to start in a race; to begin (work etc.); to cause (timbers etc.) to start. *n.* a sudden involuntary movement, as of fear, surprise etc.; the beginning of a journey, enterprise etc., a setting-out; a starting-place; the amount of lead originally given to a competitor in a race etc.; advantage gained in a race, business etc. **by fits and starts** FIT[1]. **for a start,** in the first place. **to start in,** to begin. **to start off,** to begin a journey; to commence an activity; to cause someone or something

startle   805   stave

to begin. **to start on**, (*coll.*) to pick a fight with; to reprimand. **to start out**, to begin a journey; to take the first steps in a particular activity. **to start up**, to rise suddenly; of an engine, to start; to originate. **starter**, *n.* one who starts; one who gives the signal for starting a race etc.; a horse or other competitor starting in a race; (also **self starter**) a device for starting an internal-combustion engine; anything that initiates a process; (*often pl.*) the first course of a meal. **for starters**, (*coll.*) in the first place, to begin with. **starting**, *a.* **starting block**, *n.* (*usu. pl.*) a device consisting of angled wooden blocks or metal pads used by sprinters to brace their feet in crouch starts. **starting gate** GATE. **starting-point**, *n.* a point of departure. **starting-post**, *n.* a post from which competitors start in a race. **starting price**, *n.* the odds on a horse at the beginning of a race.

startle (stah'tl), *v.t.* to cause to start; to alarm, to shock. **startler**, *n.* **startling**, *a.* surprising, alarming.

starve (stahv), *v.i.* to perish or suffer severely from hunger; to be in want or penury; to suffer from the lack of mental or spiritual nutriment. *v.t.* to cause to perish or be extremely distressed by lack of food; to force (into surrender etc.) thus; to deprive of physical or mental nutriment. **starvation**, *n.*

stash (stash), *v.t.* (*coll.*) to store, (money etc.) in a secret place (usu. with *away*). *n.* a secret store, a hideaway.

stasis (stā'sis), *n.* stagnation of the blood, esp. in the small vessels or capillaries; a state of equilibrium or inaction.

-stat, *comb. form.* designating a device that causes something to remain stationary or constant, as in *thermostat*.

state (stāt), *n.* condition, mode of existence, situation, relation to circumstances; a political community organized under a government, a commonwealth, a nation, the body politic; such a community forming part of a federal republic; civil government; dignity, pomp, splendour; (*coll.*) a nervous or excited condition. *a.* of or pertaining to the state or body politic; run or financed by the State; used, reserved for or pertaining to ceremonial occasions. *v.t.* to utter esp. with explicitness and formality. **state of affairs**, a certain situation, set of circumstances. **State Registered Nurse**, a fully qualified nurse. **the States**, (*coll.*) the US. **to lie in state**, of an important dead person, to lie in a coffin in some place where the public may come to visit as a token of respect. **statecraft**, *n.* statesmanship. **State Department**, *n.* that part of the US government responsible for foreign affairs. **statehouse**, *n.* the building which houses a US state legislature. **state-of-the-art**, *a.* using the most advanced technology available at the time. **stateroom**, *n.* a room reserved for ceremonial occasions; a private sleeping apartment on a liner etc. **state school**, *n.* a goverment-financed school for the provision of free education. **stateside**, *a.*, *adv.*, of, in, or towards the US. **statesman**, *n.* one skilled in the art of government; one taking a leading part in the administration of the state. **statesmanlike**, **statesmanly**, *a.* **statesmanship**, *n.* **stateswoman**, *n.fem.* **state socialism**, *n.* government ownership of the leading industries, financial institutions etc. in the public interest. **statable**, **stateable**, *a.* **statehood**, *n.* **stateless** (-lis), *a.* without nationality. **stately**, *a.* grand, lofty, dignified, elevated, imposing. **stately home**, *n.* a large country mansion, usu. of historic interest and open to public view. **stateliness**, *n.* **statement**, *n.* the act of stating; that which is stated; a formal account, recital, or narration; a formal presentation of accounts. **statism**, *n.* belief in the control of economic and social affairs by the state. **statist**, *n.*

static (stat'ik), *a.* pertaining to bodies at rest or in equilibrium; acting as weight without producing motion; pertaining to or causing stationary electric charges; relating to interference of radio or television signals. *n.* static electricity; atmospherics; electrical interference of radio or television signals causing crackling, hissing, and a speckled picture. **static electricity**, *n.* electrical effects caused by stationary charges, as opposed to charged particles flowing in a current. **statics**, *n. sing.* the branch of dynamics which treats the relations between forces in equilibrium. **statically**, *adv.*

Statice (stat'isē), *n.* a genus of plants containing the sea-lavender.

station (stā'shən), *n.* the place where a person or thing stands, esp. an appointed or established place; a place where police, fire services, coastguards, naval or military forces etc., have their headquarters, a military post; a place or building at which railway-trains or buses stop for setting down or taking up passengers or goods; position, occupation, standing, rank, esp. high rank; (*Austral.*) the ranch-house or homestead of a sheep-farmer; in the Roman Catholic Church, a church to which a procession resorts for devotion; any of a series of 14 images or pictures (in a church) representing successive scenes in Christ's passion; the area inhabited by a particular organism, a habitat; a radio or television channel; a place reserved for a particular activity, service, etc., as a power station, petrol station, etc. *v.t.* to assign to or place in a particular station, to post. **station house**, *n.* (*chiefly N Am.*) a police station; a fire station. **stationmaster**, *n.* the official in charge of a railway station. **station wagon**, *n.* (*chiefly N Am.*) an estate car. **stationary**, *a.* remaining in one place, not moving; intended to remain in one place, fixed, not portable; of planets, having no apparent movement in longitude; not changing in character, condition, magnitude etc. **stationary wave** STANDING WAVE.

stationer (stā'shənə), *n.* one who sells papers, pens, ink, and writing-materials. **stationery**, *n.*

statism STATE.

statistics (stətis'tiks), *n.pl.* numerical facts, arranged and classified, esp. respecting social conditions; (*sing. in constr.*) the science of collecting, organizing, and applying statistics. **statistical**, *a.* **statistically**, *adv.* **statistician** (statistish'ən), *n.*

stator (stā'tə), *n.* the fixed part of an electrical generator.

statoscope (stat'əskōp), *n.* a sensitive aneroid barometer for showing minute fluctuations of pressure.

statue (stat'ū), *n.* a representation of a person or animal sculptured or cast, e.g. in marble or bronze, esp. about life-size. **statuary**, *a.* of or for statues. *n.* statues collectively; the art of making statues. **statuesque** (-esk'), *a.* having the dignity or beauty of a statue. **statuesquely**, *adv.* **statuesqueness**, *n.* **statuette** (-et'), *n.* a small statue.

stature (stach'ə), *n.* the natural height of a body, esp. of a person; eminence.

status (stā'təs, *esp. N Am.* stat'-), *n.* (*pl.* **-toses**) relative standing, rank, or position in society; (*Law*) legal position or relation to others; situation, state of affairs. **status symbol**, *n.* a possession regarded as indicative of a person's elevated social rank or wealth.

status quo (stā'təs kwō, *esp N Am.* stat'-), *n.* the existing state of affairs.

statute (stat'ūt), *n.* a law enacted by a legislative body; an ordinance of a corporation or its founder intended as a permanent law. **statute of limitations**, a statute prescribing a period of time within which proceedings must be taken to enforce a right or bring an action at law. **statute-book**, *n.* a book in which statutes are published. **statute law**, *n.* law enacted by a legislative body. **statutory**, *a.* enacted, regulated, enforced, or recognized by statute.

staunch[1] (stawnch), STANCH[1].

staunch[2] (stawnch), *a.* (*rarely*) watertight; loyal, constant, trustworthy; firm, solid. **staunchly**, *adv.* **staunchness**, *n.*

stave (stāv), *n.* one of the curved strips forming the side of a cask etc.; a strip of wood or other material used for a similar purpose; a stanza, a verse; (*Mus.*) a staff. *v.t.* (*past, p.p.* **staved**, **stove** (stōv)) to break a hole in (a cask,

boat etc.); to make (a hole) thus; to furnish or fit with staves; to stop, avert, or ward (off).
staves, *pl.* STAFF ¹, STAVE.
stavesacre (stăv'zăkə), *n.* a species of larkspur, the seeds of which were formerly used as a poison for lice etc.
stay ¹ (stā), *v.i.* to continue in a specified place or state; to remain; to dwell or have one's abode temporarily (at, with etc.); (*Sc.*, *S Afr.*) to live (at); to tarry, to wait; to keep going or last out (in a race etc.). *v.t.* to hinder, to stop (the progress etc., of); to postpone, to suspend; to satisfy temporarily. *n.* the act of staying or dwelling; continuance in a place etc.; a check, a restraint or deterrent; suspension of judicial proceedings. **to stay over,** (*coll.*) to remain overnight. **to stay put,** to remain in one's place. **stay-at-home,** *n.*, *a.* (one who is) unenterprising. **staying power,** *n.* stamina. **stayer,** *n.*
stay ² (stā), *n.* a support, a prop; (*pl.*) a corset. *v.t.* to prop (usu. *up*), to support. **stay-bar, -rod,** *n.* one used as a stay or support in a building etc.
stay ³ (stā), *n.* a rope, chain etc. supporting or bracing a mast, spar, chimney etc.; (*pl.*) the position of a sailing vessel relative to the wind. **to miss or refuse stays,** of a sailing vessel, to fail in tacking. **staysail** (-sl), *n.* a sail extended by a stay.
stayer STAY ¹.
stead (sted), *n.* (*rarely*) place or room which another had or might have had. **in one's stead,** instead of one. **to stand in good stead,** to be of service to.
steadfast, stedfast (sted'fəst, -fahst), *a.* firm, resolute, unwavering. **steadfastly,** *adv.* **steadfastness,** *n.*
steady (sted'i), *a.* firmly fixed, not wavering; moving or acting in a regular way, uniform, constant, continuous; free from intemperance, irregularity, constant in mind, or conduct. *n.* (*coll.*) a regular boy friend or girl friend. *int.* a warning to keep callm, etc.; an order to a helmsman to keep on course; part of a command to start, e.g. a race, as in *ready, steady, go! v.t.* to make steady. *v.i.* to become steady. **steady-state theory,** in cosmology, the theory that the Universe has always existed in a steady state, matter being created continuously as it expands. cp. BIG-BANG THEORY. **to go steady** GO ². **steadily,** *adv.* **steadiness,** *n.* **steadying,** *a.*
steak (stāk), *n.* any of several cuts of beef such as *stewing steak, braising steak*; a slice of beef, cut for grilling etc.; a slice or cut of fish, pork, veal, minced meat, etc. **steakhouse,** *n.* a restaurant that specializes in serving steaks.
steal (stēl), *v.t.* (*past* **stole** (stōl), *p.p.* **stolen** (stō'lən)) to take away without right or permission, to take feloniously; to secure covertly or by surprise; to secure insidiously. *v.i.* to take anything feloniously; to go or come furtively or silently. *n.* (*coll.*) the act of stealing; something stolen; (*coll.*) a bargain. **to steal a march on,** to be beforehand with, to get the start of. **to steal someone's thunder,** to take the credit due to another. **stealer,** *n.*
stealth (stelth), *n.* furtiveness, secrecy; secret procedure. **stealthily,** *adv.* **stealthy** (-thi), *a.*
steam (stēm), *n.* water in the form of vapour or the gaseous form to which it is changed by boiling; the visible mass of particles of water into which this condenses; any vaporous exhalation; (*coll.*) energy, force, go. *v.i.* to give off steam; to rise in steam or vapour; to move by the agency of steam; (*coll.*) to move or proceed rapidly or with force. *v.t.* to treat with steam for the purpose of softening, melting etc., esp. to cook by steam. **steamed up,** of windows etc., clouded by steam; (*coll.*) angry, indignant. **to go under one's own steam,** to go by one's own efforts, to go without help. **to let off steam,** to relieve one's feelings. **steamboat,** *n.* a vessel propelled by steam. **steam engine,** *n.* a boiler in a steam engine. **steam engine,** *n.* an engine worked by the pressure of steam on a piston moving in a cylinder etc. **steam iron,** *n.* an electric iron with a compartment in which water is heated and then emitted as steam to aid pressing and ironing. **steam jacket,** *n.* a hollow casing round a cylinder etc., for receiving steam to heat the latter. **steamroller,** *n.* a heavy roller propelled by steam, used in road-making and repairing; any crushing force. *v.t.* to crush (opposition etc.) by overwhelming pressure. **steamship,** *n.* a ship propelled by steam. **steam turbine,** *n.* a machine in which steam acts on moving blades attached to a drum. **steamer,** *n.* a vessel propelled by steam; a receptacle for steaming articles, esp. for cooking food. **steamy,** *a.* of, like, full of, emitting, or covered with steam; (*sl.*) erotic. **steaminess,** *n.*
stearin, -rine (stiə'rin), *n.* a fatty compound contained in the more solid animal and vegetable fats tristearin; stearic acid as used for candles. **stearic** (stia'-), *a.* of or pertaining to fat or stearic acid. **stearic acid,** *n.* a fatty acid obtained from solid fats and used in making candles and soap.
steatite (stē'ətīt), *n.* massive talc, soapstone. **steatitic** (-tit'-), *a.*
steat(o)-, *comb. form.* fat.
steatopygia, steatopyga (stēəto'pijiə-pigə), *n.* excessive fatness of the buttocks.
steatopygous (stēətop'igəs), *a.* characterized by fat buttocks.
steed (stēd), *n.* a horse, esp. a war-horse.
steel (stēl), *n.* iron combined with carbon in various proportions, remaining malleable at high temperatures and capable of being hardened by cooling; ; a steel rod with roughened surface for sharpening knives; a steel strip for stiffening; a quality of hardness, toughness etc. in a person. *v.t.* to cover, point, or face with steel; to harden (the heart etc.). **cold steel** COLD. **steel band,** *n.* a type of band (orig. from the Caribbean islands) which plays percussion instruments made from oil drums. **steel-engraving,** *n.* the art of engraving upon steel plates; an engraving on a steel plate; an impression from this. **steel grey,** *n.*, *a.* bluish-grey like steel. **steel wool,** *n.* fine steel shavings bunched together for cleaning and polishing. **steelworker,** *n.* **steelworks,** *n. sing.* or *pl.* a plant where steel is made. **steely,** *a.* **steeliness,** *n.*
steelyard (stēl'yahd), *n.* a balance with unequal arms, the article weighed being hung from the shorter arm and a weight moved along the other till they balance.
steenbok (stēn'bok, stän'-), *n.* a small S. African antelope.
steep ¹ (stēp), *a.* sharply inclined, sloping at a high angle; (*coll.*) of prices etc., excessive, exorbitant (*coll.*) far-fetched. **steepen,** *v.t.*, *v.i.* **steeply,** *adv.* **steepness,** *n.*
steep ² (stēp), *v.t.* to soak in liquid; to wet thoroughly; to imbue, to saturate. *n.* the process of steeping; a liquid for steeping. **to steep in,** to impregnate or imbue with. **steeper,** *n.*
steepen etc. STEEP ¹.
steeper STEEP ².
steeple (stē'pl), *n.* a lofty structure rising above the roof of a building, esp. a church tower with a spire. **steeplechase** *n.* a horse race over a course in which hedges etc., have to be jumped; a track race over obstacles including hurdles and water jumps. **steeplechaser,** *n.* **steeplechasing,** *n.* **steeplejack,** *n.* one who climbs steeples etc., to do repairs etc. **steepled,** *a.*
steer ¹ (stiə), *v.t.* to guide (a ship, aeroplane, car etc.) by a rudder, wheel, handle etc.; to direct (one's course) thus. *v.i.* to guide a ship etc., or direct one's course by or as by this means; to be steered (easily etc.). **to steer clear of,** to avoid. **steersman,** *n.* one who steers. **steerable,** *a.* **steerage** (-rij), *n.* the part of a ship, usu. forward and on or below the main deck, allotted to passengers travelling at the lowest rate; the effect of the helm on a ship; **steerage-way,** *n.* sufficient motion of a vessel to enable her to answer the helm. **steerer,** *n.* **steering,** *n.* **steering committee,** *n.* a committee which determines the order of business for a legislative assembly or other body. **steering**

steer — **wheel**, *n*. the wheel which controls the rudder of a ship, or the stub axles of the front wheels of a motor vehicle, etc.
steer² (stiə), *n*. a young male of the ox kind, esp. a castrated bullock.
steerage etc. STEER¹.
stegosaur (steg'əsaw), **-saurus** (-saw'rəs), *n*. any of several quadrupedal herbivorous dinosaurs of the Jurassic period, with armour-like bony plates.
stein (stīn), *n*. a large, usu. earthenware beer mug, often with a hinged lid.
steinbock (stīn'bok), STEENBOK.
stele (stē'lē), (*pl*. **-lae**) an upright slab usu. with inscriptions and sculpture, for sepulchral or other purposes; the cylindrical vascular portion in the stems and roots of plants. **stelar,** *a*.
stellar (stel'ə), *a*. of, pertaining to or resembling stars. **stellate** (-āt), **-ated,** *a*. star-shaped, radiating. **stellately,** *adv*. **stellular** (-ū-), *a*. set with or shaped like small stars.
stem¹ (stem), *n*. the stock, stalk, or ascending axis of a tree, shrub, or other plant; the slender stalk or peduncle of a flower, leaf etc.; an analogous part, as the slender part between the body and foot of a wine glass, the tube of a tobacco-pipe etc., the part of a noun, verb etc., to which case-endings etc., are affixed; the stock of a family, a branch of a family; the upright piece of timber or iron at the fore end of a ship to which the sides are joined. *v.t.* (*past, p.p*. **stemmed**) to remove the stem or stems of. **from stem to stern,** from one end of the ship to the other. **to stem from,** to originate in. **stemlike,** *a*. **stemmed,** *a*.
stem² (stem), *v.t.* (*past, p.p*. **stemmed**) to draw up, to check, to hold back; to make headway against; in skiing, to slow down by pushing the heel of one or both skis outward from the direction of travel. *n*. in skiing, the process of stemming, used to turn or slow down.
stemma (stem'ə), *n*. (*pl*. **stemmata** (-tə)) pedigree, a family tree.
stench (stench), *n*. a foul or offensive smell.
stencil (sten'sil), *n*. a thin plate of metal or other material out of which patterns have been cut for painting through the spaces on to a surface; a decoration, etc., produced thus. *v.t.* (*past, p.p*. **stencilled**) to paint (letters, designs etc.) by means of a stencil; to decorate (a wall etc.) thus.
Sten gun (sten), *n*. a light sub-machine gun. [Sheperd and Turpin, the designers, and *E*nfield, as in BREN GUN]
steno- *comb. form*. contracted.
stenograph (sten'əgraf), *n*. a character used in shorthand; a form of typewriter using stenographic characters. **stenographer** (-nog'-), *n*. a shorthand writer. **stenography** (-nog'-), *n*.
stenosis (stənō'sis), *n*. constriction of a bodily passage or orifice; constipation. **stenotic** (-not'-), *a*.
stentor (sten'taw), *n*. a person with a loud, strong voice; a howling monkey, esp. the ursine howler. **stentorian** (-taw'ri-), *a*.
step (step), *v.i.* (*past, p.p*. **stepped**) to lift and set down a foot or the feet alternately; to walk a short distance in a specified direction; to walk or dance slowly or with dignity; to enter (into); to tread, press. *v.t.* to go through, perform, or measure by stepping; to insert the foot of (a mast etc.) in a step. *n*. a single complete movement of one leg in the act of walking, dancing etc.; the distance traversed in this; a short distance; an action or measure taken in a series directed to some end; that on which the foot is placed in ascending or descending, a single stair or a tread in a flight of stairs; a rung of a ladder, a support for the foot in stepping in or out of a vehicle, a doorstep etc.; a footprint; a break in the outline at the bottom of a float or hull of a seaplane which assists in lifting it from the surface of the water; (*pl*.) a stepladder; a degree or grade in progress, rank, or precedence; a socket supporting a frame, etc., for the end of a mast, shaft, etc. **in step,** in marching, dancing etc., in conformity or unison with others; (*coll.*) in agreement (with). **out of step,** not in step; (*coll.*) not in agreement or harmony (with others). **step by step,** *adv*. gradually, with deliberation, taking one step at a time. **to break step,** to cease marching in unison. **to step down,** to resign, retire, relinquish one's position etc.; to decrease the voltage of. **to step in,** to intervene; to visit briefly. **to step on it,** to hurry, to increase speed. **to step out,** to leave (a room etc.) briefly; to take longer, faster strides. **to step out of line,** to depart from normal or acceptable behaviour. **to step up,** to advance by one or more stages; to increase the voltage of; to come forward. **step-down,** *a*. of a transformer, reducing voltage. **stepladder,** *n*. a self-supporting, portable ladder with fixed or hinged prop. **step-up,** *a*. of a transformer, increasing the voltage. *n*. an increase in size, quantity, position, etc. **stepped,** *a*. **stepper,** *n*. **stepping,** *a*. **stepping stone,** *n*. a raised stone in a stream or swampy place on which one steps in crossing; a means to an end.
step- (step-), (*pref.*) used to express relation only by the marriage of a parent. **stepbrother, -sister,** *n*. a stepfather's or stepmother's child by a former marriage. **stepchild, -daughter, -son,** *n*. the child of one's husband or wife by a former marriage. **stepfather, -mother, -parent,** *n*. the later husband or wife of one's parent.
stephanotis (stefənō'tis), *n*. a tropical climbing plant with fragrant waxy flowers.
stepmother etc. STEP-.
steppe (step), *n*. a vast plain devoid of forest, esp. in Russia and Siberia.
stepping stone etc. STEP.
-ster (-stə), *suf.* denoting an agent, as in *gangster, songster*.
stercoraceous (stœrkərā'shəs), *a*. pertaining to, composed of, or like dung.
stere (stiə), *n*. a cubic metre (35·147589 cu. ft.) used to measure timber.
stereo¹ (ste'riō), *n*. (*pl*. **stereos**) stereophonic music reproduction; a piece of stereophonic music equipment such as a record player, tape deck etc. *a*. stereophonic.
stereo² (ste'riō), short for STEREOTYPE, STEREOSCOPE, STEREOSCOPIC.
stereo-, *comb. form*. solid, three-dimensional.
stereochemistry (steriōkem'istri), *n*. chemistry concerned with the composition of matter as exhibited in the relations of atoms in space.
stereogram (ste'riəgram), *n*. a three-dimensional picture or image, a stereograph. **stereograph** (stē'riəgrahf, -graf), *n*. a pair of almost identical images which when viewed together through a stereoscope give a three-dimensional effect.
stereoisomer (steriōī'sōmə), *n*. an isomer of a molecule in which the atoms are linked in the same order but have a different spatial arrangement. **stereoisomerism,** *n*.
stereophonic (steriōfon'ik), *a*. of a sound recording or reproduction system involving the use of two or more separate microphones and loudspeakers to split the sound into separate channels to create a spatial effect. **stereophonically,** *adv*. **stereophony** (-of'əni), *n*.
stereoscope (ste'riəskōp), *n*. a binocular instrument for blending into one two pictures taken from slightly different positions, thus giving an effect of three dimensions. **steroscopic** (-skop'-), *a*. giving the effect of solidity. **stereoscopy** (-os'-), *n*.
stereotype (ste'riōtīp), *n*. a printing-plate cast from a mould taken from movable type; a hackneyed convention, idea etc.; one who or that which conforms to a standardized image. *v.t.* to make a stereotype of; to fix or establish in a standard form. **stereotyper, -typist,** *n*. **stereotyped,** *a*. hackneyed, unoriginal. **stereotypy** (ste'riōtipi), *n*. the process of making stereotype plates; meaninglessly, repetitive action or thought.
steric (ste'rik), **-ical,** *a*. pertaining or caused by to the spa-

tial arrangement of atoms in a molecule.
sterile (ste'rīl), *a.* barren, unfruitful; not producing crops, fruit, young etc.; containing no living bacteria, microbes etc., sterilized; destitute of ideas or sentiment. **sterility** (-ril'-), *n.* **sterilize, -ise** (-ri-), *v.t.* to rid of living bacteria; to make sterile; to render incapable of procreation. **sterilization,** *n.* **sterilizer,** *n.*
sterling (stœ'ling), *a.* of coins and precious metals, of standard value, genuine, pure; sound, of intrinsic worth, not showy. *n.* British (as distinct from foreign) money; genuine British money. **sterling area,** *n.* a group of countries that keep their reserves in sterling rather than in gold or dollars.
stern [1] (stœn), *a.* severe, grim, forbidding, austere; harsh, rigid, strict; ruthless, unyielding, resolute. **sternly,** *adv.* **sternness,** *n.*
stern [2] (stœn), *n.* the hind part of a ship or boat; the rump or tail of an animal. **stern-chaser** CHASE [1]. **sternforemost,** *adv.* (moving) with the stern in front. **stern-post,** *n.* a timber or iron post forming the central upright of the stern and usu. carrying the rudder. **stern-sheets,** *n.pl.* the space in a boat between the stern and the aftermost thwart. **sternway,** *n.* the movement of a ship backwards. **stern-wheel,** *n.* a paddle-wheel at the stern of a riversteamer. **stern-wheeler,** *n.* **sternmost,** *a.* **sternward,** *a.*, *adv.* **-wards,** *adv.*
sternum (stœ'nəm), *n.* the breast-bone. **sternal,** *a.* pertaining to the sternum.
sternutation (stœnūtā'shən), *n.* the act of sneezing, a sneeze. **sternutator,** *n.* a substance that causes sneezing, tears, wheezing etc. (used in chemical warfare). **sternutatory,** *a.* causing (one) to sneeze. *n.* a sternutative substance, as snuff.
sternward etc. STERN [2].
steroid (ste'roid), *n.* any of a group of compounds of similar chemical structure, including sterols, bile acids and various hormones.
sterol (ste'rol), *n.* any of various solid alcohols, such as cholesterol, ergosterol.
stertorous (stœ'tərəs), *a.* characterized by deep snoring or snore-like sounds. **stertorously,** *adv.* **stertorousness,** *n.*
stet (stet), *v.t.* (*Print.*) let it stand (cancelling a previous correction); to write 'stet' against. *n.* this word or a mark indicating it.
stethoscope (steth'əskōp), *n.* an instrument used in auscultation of the chest etc. **stethoscopic** (-skop'-), *a.* **stethoscopy** (-thos'kəpi), *n.*
stetson (stet'sən), *n.* a broad-brimmed slouch hat. [from John Stetson, 1830–1906, hatmaker]
stevedore (stē'vədaw), *n.* one whose occupation is to load or unload ships.
stew [1] (stū), *v.t.* to cook by boiling slowly or simmering. *v.i.* to be cooked thus; to be stifled or oppressed by a close atmosphere; (*coll.*) to be anxious, agitated. *n.* meat etc., cooked by stewing; (*coll.*) a state of mental agitation or worry. **to stew in one's own juice,** to suffer alone the consequences of one's actions. **stewed,** *a.*
stew [2] (stū), *n.* a fish-pond or tank for keeping fish alive for the table; an artificial oyster-bed.
steward (stū'əd), *n.* a person employed to manage the property or affairs of another or other persons (esp. the paid manager of a large estate or household); one in the service of provisions, etc., in a college, club etc.; an attendant on a ship, aircraft etc. in charge of provisions, cabins etc.; one of the officials superintending a ball, show, public meeting etc. **Lord High Steward,** an officer of State regulating precedence at coronations etc. **stewardess** (-dis), *n. fem.* **stewardship,** *n.*
sthenic (sthen'ik), *a.* exhibiting an extreme degree of energy or vital action.
stibnite (-nīt), *n.* a grey mineral consisting of antimony sulphide.

-stichous, *comb. form* having a certain number of rows.
stick [1] (stik), *n.* a shoot or branch of a tree or shrub broken or cut off, or a slender piece of wood or other material used as a rod, staff, baton, walking-cane etc., or as part of something; anything resembling this in shape; a drumstick, gear stick, fiddle-stick etc.; the control-rod of an aircraft; a number of bombs dropped in succession; (*Naut.*) a mast, a spar; (*coll.*) an awkward, incompetent or stupid person; (*coll.*) blame, hostile criticism. *v.t.* (*past, p.p.* **sticked**) to provide (a plant) with sticks for support. **in a cleft stick,** in a difficult situation. **the sticks,** (*often derog.*) remote rural areas, the backwoods; the far-out suburbs of a town or city. **to give someone stick,** to blame or criticize someone. **wrong end of the stick,** a complete misunderstanding of a situation. **stick insect,** *n.* an insect which resembles dry twigs.
stick [2] (stik), *v.t.* (*past, p.p.* **stuck** (stŭk)) to thrust the point of (in, through etc.); to fix or insert (into); to thrust (out or up); to protrude to fix upright; to fix on or as on a point; to pierce, to stab; to set with something pointed; to cause to adhere to; to set or compose (type); (*coll.*) to tolerate, endure (it); (*sl.*) to force something unpleasant or illegal on (one); to baffle, to puzzle; (*coll.*) to place, to put (somewhere). *v.i.* to be inserted or thrust (into); to protrude project, or stand (up, out etc.); to become fixed, to adhere; to remain attached (to); to be inseparable, to be constant (to); to persist, to persevere; to be stopped, hindered, or checked; to be perplexed or embarrassed; to have scruples or misgivings, to hesitate (at). **to get stuck in(to)** GET. **to stick around** or **about,** (*coll.*) to remain in the vicinity. **to stick at nothing,** not to be deterred or feel scruples. **to stick by,** to stay close to; to remain faithful to, to support. **to stick one's neck out,** (*coll.*) to invite trouble; to take a risk. **to stick out,** to protrude; to hold out, to resist. **to stick out for,** to demand, to insist upon. **to stick to,** to adhere to; to persevere. **to stick up,** to put up, to erect; to stand up, to be prominent; to paste or post up; (*chiefly N Am., sl.*) to rob at gun-point. **to stick up for,** to take the part of, to defend. **to stick up to,** to stand up against, to resist. **sticking plaster,** *n.* an adhesive plaster for wounds etc. **stick-in-the-mud,** *a.* dull, slow, unprogressive; *n.* such a person. **stick-up,** *n.* (*sl.*) an armed robbery. **stuck-up,** *a.* puffed up, conceited, giving oneself airs. **sticker,** *n.* one who or that which sticks; (*coll.*) a knife; an adhesive label or poster; (*Cricket*) a batsman who stays in long, making few runs; a hard-working or persevering person. **sticky,** *a.* tending to stick, adhesive; viscous, glutinous; (*coll.*) difficult, painful; humid, warm. **sticky end,** *n.* (*coll.*) a disagreeable end or death. **sticky-fingered,** *a.* (*coll.*) prone to stealing. **sticky wicket,** *n.* a damp cricket pitch which is difficult to bat on; (*coll.*) a difficult situation. **stickily,** *adv.* **stickiness,** *n.*
stickle (stik'l), *v.i.* to contend pertinaciously for some trifle. **stickler,** *n.* one who stands out for trifles.
stickleback (stik'lbak), *n.* a small spiny-backed, freshwater fish.
stickler STICKLE.
sticky etc. STICK [2].
stiff (stif), *a.* rigid, not easily bent or moved; not pliant, not flexible, not yielding, not working freely; obstinate, stubborn, firm, persistent; constrained, not easy, not graceful, awkward, formal, precise, affected; hard to deal with or accomplish; difficult; strong, e.g. of liquor, a breeze; of prices, high; not fluid, thick and tenacious, viscous. *adv.* (*coll.*) utterly, extremely, as in *bored stiff, frozen stiff. n.* (*sl.*) a corpse. **stiff with,** (*coll.*) packed with, full of. **stiff-necked,** *a.* stubborn, self-willed. **stiffen,** *v.t., v.i.* to make or become stiff. **stiffener,** *n.* something which stiffens; (*coll.*) a strong alcoholic drink. **stiffening,** *n.* **stiffish,** *a.* **stiffly,** *adv.* **stiffness,** *n.*
stifle (stī'fl), *v.t., v.i.* to smother, to suffocate; to suppress; to stamp out.

stigma (stig'mə), *n.* (*pl.* **-mas, -mata** (-mətə)) a mark or indication of infamy, disgrace etc.; a natural mark or spot on the skin; the part of the pistil which receives the pollen; (*Path.*) a small red spot on the skin from which blood oozes in excitement etc. (*pl.*) in the Roman Catholic Church, marks miraculously developed on the body, corresponding to the wounds of Christ. **stigmatic** (-mat'-), *a.* pertaining to, like, or having stigmas or stigmata; anastigmatic. **stigmatism,** *n.* (*Phys.*) anastigmatism, the condition characterized by stigmata. **stigmatist,** *n.* one on whom stigmata are said to be impressed. **stigmatize, -ise,** *v.t.* to mark with a brand of disgrace etc.; to distinguish as different, bad etc.; to cause stigmata to appear on. **stigmatization,** *n.*

stile (stīl), *n.* a series of steps or other contrivance by which one may get over or through a fence etc.; a turnstile.

stiletto (stilet'ō), *n.* (*pl.* **-ttos**) a small dagger; a pointed instrument for making eyelet-holes etc., a stiletto heel. *v.t.* to stab with a stiletto. **stiletto heel,** *n.* an excessively tapered, high heel for a woman's shoe.

still[1] (stil), *a.* at rest, motionless; quiet, calm; silent, noiseless, hushed; not effervescent or sparkling. *n.* stillness, calm, quiet; (*Cinema.*) an enlargement of a frame or a single photograph for record or publicity purposes. *adv.* now, then, or for the future, as previously; even till now or then, yet; nevertheless, all the same; quietly, without moving. *v.t.* to quiet, to calm; to silence; to appease. **still birth,** *n.* the bringing forth of a dead child; a child born dead. **stillborn,** *a.* **still life,** *n.* the representation of fruit, flowers, and other inanimate objects. **stillness,** *n.*

still[2] (stil), *n.* a vessel or apparatus employed in distillation, esp. of spirits, consisting of a boiler, a tubular condenser or worm enclosed in a refrigerator, and a receiver. **still-room,** *n.* a room for distilling; a pantry or store-room for liquors, preserves etc.

stillness STILL[1].

stilt (stilt), *n.* a pole having a rest for the foot, used in pairs, to raise a person above the ground in walking; any of a number of tall supports or columns for raising a building above the ground; a long-legged, three-toed, shore-bird related to the plover. **stilted,** *a.* raised on or as on stilts; of literary style etc., bombastic, inflated, unnatural. **stiltedly,** *adv.* **stiltedness,** *n.*

Stilton (stil'tən), *n.* a rich, white, veined cheese, orig. made at *Stilton,* in Cambridgeshire.

stimulus (stim'ūləs), *n.* (*pl.* **stimuli** (-lī)), that which stimulates; an incitement, a spur; that which excites reaction in a living organism; (*Med.*) a stimulant. **stimulant,** *a.* serving to stimulate; producing a quickly diffused and transient increase in physiological activity. *n.* anything that stimulates, such as drugs or alcohol. **stimulate,** *v.t.* to rouse to action or greater exertion; to spur on, to incite; to excite organic action. *v.i.* to act as a stimulus. **stimulating,** *a.* **stimulation,** *n.* **stimulative,** *a.* **stimulator,** *n.*

sting (sting), *v.t.* (*past, p.p.* **stung** (stŭng)) to pierce or wound with a sting; to cause acute physical or mental pain to; (*coll.*) to cheat, to overcharge. *v.i.* to have or use a sting; to have an acute and smarting pain. *n.* a sharp-pointed defensive or offensive organ, often conveying poison, with which certain insects, scorpions and plants are armed; the act of stinging; the wound or pain so caused; any acute pain, ache, smart, stimulus etc.; (*coll.*) a deception, a stratagem for extracting money, etc. **stingray,** *n.* a tropical ray with a venomous spine on its tail. **stinger,** *n.* one who, or that which, stings; a smarting blow. **stinging,** *a.* **stinging-nettle,** *n.* **stingy**[1], *n.*

stingy[2] (stin'ji), *a.* tight-fisted, meanly parsimonious, niggardly. **stingily,** *adv.* **stinginess,** *n.*

stink (stingk), *v.i.* (*past* **stank** (stangk), **stunk** (stüngk), *p.p.* **stunk**) to emit a strong, offensive smell; (*coll.*) to have an evil reputation; to be corrupt, bad or unacceptable. *v.t.* to annoy with an offensive smell. *n.* a strong, offensive smell; (*sl.*) a disagreeable exposure, a row. **to raise a stink,** (*sl.*) to complain; to stir up trouble, esp. adverse publicity. **to stink out,** to drive out by creating an offensive smell; to cause to stink. **stink bomb,** *n.* a small glass sphere which releases a foul-smelling liquid when broken. **stinkhorn,** *n.* an evil-smelling fungus. **stinker,** *n.* a stinking person, animal etc.; (*sl.*) an unpleasant person or thing. **stinking,** *a.* emitting an offensive smell; (*coll.*) offensive, repulsive, objectionable; (*coll.*) extremely drunk. *adv.* (*coll.*) extremely, very. **stinkingly,** *adv.* **stinko,** *a.* (*sl.*) drunk.

stint (stint), *v.t.* to give or allow scantily or grudgingly; to supply scantily or grudgingly (with food etc.). *n.* limit, bound, restriction; an allotted amount, quantity, turn of work etc.; a small sandpiper, esp. the dunlin. **stinter,** *n.*

stipe (stīp), *n.* a stalk, stem or stem-like support, also **stipes** (-pēz), *pl.* **-pites** (stip'itēz)). **stipel** (-pl), *n.* a secondary stipule at the base of a leaflet. **stipellate** (-āt, stipel'-), **stipitiform** (-fawm), **stipitiform** (-tifawm), *a.*

stipend (stī'pend), *n.* a periodical payment for services rendered, a salary, esp. of a clergyman. **stipendiary** (-pen'-), *a.* performing services for or receiving a stipend. *n.* one receiving a stipend.

stipes etc. STIPE.

stipple (stip'l), *v.t., v.i.* to engrave, paint or draw by means of dots or light dabs instead of lines etc. *n.* this method; work produced thus. **stippler,** *n.* **stippling,** *n.*

stipular etc. STIPULE.

stipulate (stip'ūlāt), *v.t.* to lay down or specify as essential to an agreement; (*rarely*) to guarantee. *v.i.* to make as a condition of an agreement. **stipulation,** *n.* **stipulator,** *n.*

stipule (stip'ūl), *n.* a small leaf-like appendage, usu. in pairs at the base of a petiole. **stipular,** *a.*

stir (stœ), *v.t.* (*past, p.p..* **stirred**) to cause to move, to agitate, to disturb; to move vigorously, to bestir (oneself etc.); to rouse (up), to excite, to animate, to inflame, to awaken. *v.i.* to move, to be in motion, not to be still. *n.* agitation, commotion, bustle, excitement; a movement; the act of stirring; (*sl.*) prison. **to stir up,** to agitate; to incite. **stir-crazy,** *a.* (*esp. N Am., sl.*) mentally unbalanced by a term in prison. **stir-fry,** *v.t.* to cook food rapidly in the Chinese manner by stirring in hot oil over a high heat in a wok or frying pan. *n.* a dish cooked in this way. **stirrer,** *n.* one who or that which stirs; (*coll.*) a person who deliberately causes trouble. **stirring,** *a.* moving; animating, rousing, exciting, stimulating. **stirringly,** *adv.*

stirk (stœk), *n.* (*dial.*) a yearling ox or cow.

stirps (stœps), *n.* (*pl.* **-pes** (-pēz)), (*Law*) stock, family, progenitor; (*Zool.*) a classificatory group.

stirrup (sti'rəp), *n.* a horse rider's foot-rest consisting of an iron loop suspended from the saddle by a strap; (*Naut.*) a rope with an eye for carrying a foot-rope. **stirrup-cup,** *n.* a parting cup, esp. orig. on horseback. **stirrup-iron,** *n.* a stirrup. **stirrup-strap,** *n.* **stirrup-pump,** *n.* a portable hand-pump with a length of hose, to be worked by one or two persons.

stitch (stich), *n.* a sharp intense pain in the side; a single pass of the needle in sewing; a single turn of the wool or thread round a needle in knitting; the link of thread, wool etc., thus inserted; (*coll.*) the least bit of clothing. *v.t., v.i.* to sew. **in stitches,** helpless with laughter. **to stitch up,** to sew together or mend; to suture; (*sl.*) to incriminate by informing on. **stitch-bird,** *n.* the New Zealand honey-eater. **stitchwort,** *n.* a plant with starry white flowers, which is common in hedges. **stitcher,** *n.* **stitchery,** *n.*

stoa (stō'ə), *n.* (*pl.* **stoae** (-ē), **stoas**) a portico.

stoat (stōt), *n.* the ermine, esp. in its summer coat; applied also to the weasel, ferret etc.

stochastic (stəkas'tik), *a.* random; involving chance or probability.

stock¹ (stok), *n.* the trunk or main stem of a tree or other plant; a family, a breed, a line of descent, a distinct group of languages; (*Biol.*) a colony, an aggregate organism; a post, a butt, a stump; the principal supporting or holding part of anything, the handle, block, base, body etc.; liquor from boiled meat, bones etc., used as a basis for soup; the aggregate of goods, raw material etc., kept on hand for trade, manufacture etc., or as a reserve store; a repertoire of plays; the beasts on a farm (called livestock), or implements of husbandry and produce; money lent to a government represented by certificates entitling the holders to fixed interest; the capital of a corporate company divided into shares entitling the holders to a proportion of the profits; (*pl.*) the shares of such capital; a gillyflower; the related plant, Virginian stock; (*pl.*) a frame of timber with holes in which the ankles, and sometimes also the wrists, of petty offenders were formerly confined; (*pl.*) a timber framework on which a vessel rests during building; a band of silk, leather etc., worn as a cravat in the 18th cent., now part of riding dress. *a.* kept in stock; habitually used, standing, permanent. *v.t.* to provide with goods, live stock, or other requisites; to keep in stock; to furnish with a handle, butt etc. *v.i.* to take in supplies; to tiller. **in, out of stock,** available or not available to be sold immediately. **on the stocks,** in preparation. **to take stock,** to make an inventory of goods etc. on hand; to survey one's position, prospects etc.; to examine, to form an estimate (of a person, etc.). **to take stock in,** to attach importance to. **stockbreeder,** *n.* one who raises livestock. **stockbroker,** *n.* one engaged in the purchase and sale of stocks on commission. **stockbroker belt,** *n.* (*coll.*, *sometimes derog.*) the prosperous commuter area around London. **stockbroking,** *n.* **stock car,** *n.* a production (saloon) car modified for racing. **stockdove,** *n.* the European wild pigeon, smaller and darker than the ring-dove. **stock exchange,** *n.* the place where stocks or shares are publicly bought and sold. **stockholder,** *n.* a proprietor of stock in the public funds or shares in a stock company; (*Austral.*) a grazier. **stock-in-trade,** *n.* goods, tools and other requisites of a trade etc.; resources, capabilities. **stockjobber,** *n.* formerly, a dealer who speculated in stocks so as to profit by fluctuations of price and acted as an intermediary between buying and selling stockbrokers. **stockjobbing, -jobbery,** *n.* **stockman,** *a.* one in charge of livestock, also called a **stock-keeper. stock-market,** *n.* a stock exchange or the business transacted there. **stockpile,** *v.t.* to accumulate commodities, esp. reserves of raw materials. **stockpot,** *n.* a pot for making or storing stock for soup. **stock-still,** *adv.* motionless. **stocktaking,** *n.* **stock whip,** *n.* a short-handled whip with a long lash for herding cattle. **stockyard,** *n.* an enclosure with pens etc., for cattle at market etc. **stockily,** *adv.* **stockiness,** *n.* **stockist,** *n.* one who keeps certain goods in stock. **stocky,** *a.* thick-set, short and stout, stumpy.

stockade (stəkād'), *n.* a line or enclosure of posts or stakes. *v.t.* to surround or fortify with a stockade.

stockfish (stok'fish), *n.* cod, ling etc. split open and dried in the sun without salting.

stocking (stok'ing), *n.* (*usu. in pl.*) a close-fitting covering for the foot and leg; an elastic covering used as a support for the leg in cases of varicose veins etc. **stocking filler,** *n.* a gift suitable for inclusion in a Christmas stocking. **stocking mask,** *n.* a nylon stocking pulled over the head as a disguise, e.g. as worn by burglars. **stocking stitch,** *n.* in knitting, alternate rows of plain and purl stitches. **stockinged** (-lis), *a.* **stockinet** (-net'), *n.* an elastic knitted material for undergarments etc.

stocky etc. STOCK ¹.

stodgy (stoj'i), *a.* of food, heavy, starchy, indigestible; dull, heavy, matter-of-fact. **stodge,** *n.* (*coll.*) stodgy food; *v.i.* to stuff with food. **stodginess,** *n.*

stoep (stoop), *n.* (*S. Afr.*) an open, roofed platform in front of a house.

Stoic (stō'ik), *n.* a philosopher of the school founded by Zeno, *c.* 308 BC, teaching that virtue is the highest good, and that the passions and appetites should be rigidly subdued. **stoic,** *n.* a person who displays stoical qualities. *a.* stoical. **stoical,** *a.* resigned, impassive. **stoically,** *adv.* **stoicism** (-sizm), *n.* the philosophy of the Stoics; indifference to pleasure or pain.

stoichiometry (-om'ətri), *n.* the branch of chemistry treating of chemical combination in definite proportions, the mathematics of chemistry. **stoichiometric** (-met'-), *a.*

stoke (stōk), *v.t.* to tend (a furnace, esp. of a steam-engine); to stir a fire. *v.i.* to act as stoker. **to stoke up,** to feed a fire or furnace with fuel; to fill oneself with food. **stokehold,** *n.* the compartment on a ship where the furnaces are tended. **stoke-hole,** *n.* an aperture in a furnace etc. for a stirring tool and adding fuel; a stokehold. **stoker,** *n.* a person who stokes a furnace.

STOL (stol), *n.* a system by which aircraft take off and land over a short distance; an aircraft using this system. cp. VTOL. [acronym for short *t*ake-*o*ff and *l*anding]

stole¹ (stōl), *n.* a narrow band of silk etc. worn over both shoulders by priests, and by deacons over the left shoulder; a band of fur etc. worn round the neck or shoulders by women.

stole² (stōl), *past* **stolen** (-lən), *p.p.* STEAL.

stolid (stol'id), *a.* dull, impassive, phlegmatic, stupid. **stolidity** (-lid'-), **stolidness,** *n.* **stolidly,** *adv.*

stolon (stō'lən), *n.* a trailing or prostrate shoot that takes root and develops a new plant; an underground shoot in mosses developing leaves. **stoloniferous** (-nif'-), *a.*

stoma (stō'mə), *n.* (*pl.* **stomata** (-tə)) a minute orifice, a pore; an aperture for respiration in a leaf.

stomach (stŭm'ək), *n.* a digestive cavity formed by a dilatation of the alimentary canal, or (in certain animals) one of several such cavities; (*loosely*) the belly, the abdomen; appetite, inclination, liking. *v.t.* to accept as palatable; to put up with, to brook. **stomach-ache,** *n.* an abdominal pain, indigestion device. **stomach-pump,** *n.* a suction for withdrawing the contents of the stomach. **stomacher,** *n.* an ornamental covering for the breast and upper abdomen worn by women in the 15th–17th cents. **stomachful,** *n.* **stomachic** (-mak'-), *a.* pertaining to the stomach; exciting the action of the stomach or aiding digestion. *n.* a stomachic medicine.

stomatitis (stōməti'tis), *n.* inflammation of the mouth. **stomatology** (-tol'əji), *n.* the science of diseases of the mouth.

stomp (stomp), *v.t.*, *v.i.* to stamp with the feet. *n.* an early jazz composition with a bold rhythm; a lively jazz dance involving heavy stamping of the feet. **stomper,** *n.*

stone (stōn), *n.* a piece of rock, esp. a small one, a pebble, cobble, or piece used in road-making etc.; rock as material for building, paving etc.; a piece of this shaped and prepared for a special purpose, as a millstone, tombstone, curling stone etc.; a gem, usu. called a precious stone; a calculus, the disease calculus; the seed of a grape etc., the hard case of the kernel in a drupe or stonefruit; a hailstone; (*pl.* **stone**) a measure of weight 14 lb. (6·35 kg). *a.* made of stone or a hard material like stone. *v.t.* to pelt with stones; to face, wall, or pave with stone; to free (fruit) from stones. **to leave no stone unturned,** to use all available means to effect an object. **Stone Age,** *n.* the period in which primitive man used implements of stone, not metal. **stone-blind,** *a.* completely blind. **stonechat,** *n.* a small black and reddish-brown songbird. **stone-cold,** *a.* completely cold. **stone-cold sober,** completely sober. **stonecrop,** *n.* any species of *Sedum,* esp. *S. acre.* **stonecurlew,** *n.* the thick-knee curlew or any bird of the family Burhinidae. **stonecutter,** *n.* one whose occupation is to cut stones for building, etc. **stonecutting,** *n.* **stone-deaf,**

stood *a.* completely deaf. **stonefish**, *n.* a poisonous tropical fish resembling a stone on the seabed. **stonefly**, *n.* an insect with aquatic larvae harbouring under stones, used as bait for trout. **stonefruit**, *n.* a fruit with seeds covered by a hard shell, as peaches, plums etc., a drupe. **stone-ground**, *a.* (of flour) ground between millstones. **stonemason**, *n.* one who dresses stones or builds with stone. **stonepine**, *n.* a Mediterranean pine with a spreading top. **stone's throw**, *n.* a short distance. **stonewall**, *v.i.* to obstruct parliamentary business by making long speeches; (*Cricket*) to stay in batting without trying to make runs. **stoneware**, *n., a.* (pottery) made from clay and flint or a hard siliceous clay. **stone-washed**, *a.* of clothes, denim etc., given a faded surface by the abrasive action of small pieces of pumice. **stonework**, *n.* masonry. **stoned**, *a.* (*sl.*) under the influence of drugs or alcohol. **stony**, *a.* pertaining to, made or consisting of, abounding in or resembling stone; hard, cruel, pitiless; impassible; obdurate. **stony-broke**, *a.* (*sl.*) destitute or nearly destitute of money. **stony-hearted**, *a.* unfeeling. **stonily**, *adv.* **stoniness**, *n.*

stood (stud), *past, p.p.* STAND.

stooge (stooj), *n.* a butt, a confederate, a decoy; a subordinate; a scapegoat. *v.i.* to act as a stooge.

stook (stuk), *n.* (*chiefly Sc.*) a bundle of sheaves set up to dry. *v.t.* to set up in stooks.

stool (stool), *n.* a seat without a back, for one person, usu. with three or four legs; a low bench for kneeling or resting the feet on; the seat used in evacuating the bowels; an evacuation; the stump of a timber-tree from which shoots are thrown up; a plant or stock from which young plants are produced by layering etc. *v.i.* to shoot out stems from the root. **to fall between two stools** FALL. **stool ball**, *n.* a game like cricket, played in S England. **stool-pigeon**, *n.* a pigeon used as a decoy; a decoy; an informer for the police.

stoop[1] (stoop), *v.i.* to bend the body downward and forward; to have an habitual forward inclination of the head and shoulders; to condescend, to lower, to bring oneself down (*to*). *v.t.* to incline (the head, shoulders etc.) downward and forward. *n.* the act of stooping; an habitual inclination of the shoulders etc. **stoopingly**, *a.*

stoop[2] (stoop), *n.* (*N Am.*) a small platform with steps and a roof in front of a building, or an open porch.

stop (stop), *v.t.* (*past, p.p.* **stopped**) to close by filling or obstructing, to stanch, to plug (up); to fill a crack, a cavity etc.; to impede; to cause to cease moving, going, working, or acting (or from moving etc.); to prevent the doing or performance of; to keep back, to cut off, to suspend; (*Mus.*) (of an instrument) to press a string, close an aperture etc. so as to alter the pitch. *v.i.* to come to an end, to come to rest; to discontinue, to cease or desist (*from*); (*coll.*) to stay, to remain temporarily, to sojourn; to punctuate. *n.* the act of stopping or the state of being stopped, a cessation, a pause, an interruption; a punctuation mark indicating a pause; a block, peg, pin etc. used to stop the movement of something at a particular point; (*Mus.*) the pressing down of a string, closing of an aperture etc., effecting a change of pitch; a key, lever or other device employed in this; a set of pipes in an organ having tones of a distinct quality; a knob bringing these into play; a perforated diaphragm for regulating the passage of light; a sound produced by closure of the mouth, a mute consonant. **to pull out all the stops**, to play at maximum volume; to make the utmost effort. **to stop a gap** GAP. **to stop at nothing**, to be ruthless, to be ready to do anything to achieve one's ends. **to stop off, over**, to break one's journey. **stop bath**, *n.* an acidic solution used to halt the action of a developer or a photographic negative or print. **stopcock**, *n.* a small valve used to stop the flow of fluid in a pipe. **stopgap**, *n.* a temporary substitute or expedient. **stop-go**, *a.* of a policy etc., alternately active and inactive. **stopoff, stopover**, *n.* a break in a journey. **stoppress**, *n., a.* (applied to) news inserted in a paper after the printing has commenced. **stop-watch**, *n.* a watch with an additional hand which can be stopped by a special device at any second or fraction of a second, used for timing races etc. **stoppage** (-ij), *n.* the act or state of being stopped; a deduction from pay; a cessation of work, as in a strike. **stopper**, *n.* one who or that which stops; a plug, a stopple; (*Naut.*) a rope, plug, clamp etc. for checking the motion of a cable etc. *v.t.* to close or secure with a stopper. **stopping**, *n.* (*coll.*) material for filling a cavity in a tooth. **stopple** (stop'l), *n.* that which stops or closes the mouth of a vessel, a stopper, plug, bung etc.; *v.t.* to close with a stopple.

storage STORE.

storax (staw'raks), *n.* a balsamic vanilla-scented resin obtained from *Styrax officinalis*, formerly used in medicine etc.; the tree itself.

store (staw), *n.* a stock laid up for drawing upon; an abundant supply, plenty, abundance (*often in pl.*); a place where things are laid up or kept for sale, a storehouse, a warehouse; a large establishment where articles of various kinds are sold; (*N Am.*) a shop; (*pl.*) articles kept on hand for special use, esp. ammunition, arms, military and naval provisions etc., a supply of such articles. *v.t.* to accumulate or lay (usu. up or away) for future use; to stock or supply (*with*); to deposit in a warehouse etc. for safe keeping; to hold or keep in (as water etc.); (*Comput.*) to enter data into a computer memory or in a storage device. **in store**, in reserve; ready for use. **to set store by**, to value highly. **storehouse**, *n.* a place where things are stored up, a warehouse, a granary, repository, etc. **storekeeper**, *n.* one who has the charge of stores; (*N Am.*) a shopkeeper. **storeroom**, *n.* **storable**, *a.* **storage** (-rij), *n.* the act of storing, warehousing etc.; the price paid for or the space reserved for this; (*Comput.*) the action of storing date in computer memory or on disk etc. **storage battery**, *n.* an accumulator. **storage capacity**, *n.* the maximum amount of data that can be held in a computer memory. **storage device**, *n.* a piece of computer hardware such as a magnetic tape, optical disk etc. that can store data. **storage heater**, *n.* a type of radiator which stores heat during periods of off-peak electricity.

storey, story (staw'ri), *n.* (*pl.* **-reys, -ries**) *n.* a horizontal division of a building, a floor or level; a set of rooms on the same floor. **storeyed, storied**, *a.* having storeys.

stork (stawk), *n.* a long-necked, long-legged wading-bird allied to the heron, esp. the white or house-stork. **stork's-bill**, *n.* a plant allied to the geranium.

storm (stawm), *n.* a violent disturbance of the atmosphere attended by wind, rain, snow, hail, or thunder and lightning, a tempest; a violent disturbance or agitation of society lite, the mind etc.; a tumult, commotion etc.; a violent outburst (of cheers etc.); a direct assault on a fortified place. *v.i.* to rage (of wind, rain etc.); to bluster, to fume, to behave or move violently. *v.t.* to take by storm. **a storm in a teacup**, a fuss about nothing. **to take by storm**, to capture by means of a violent assault; to captivate, overwhelm. **storm-bird**, *n.* the stormy petrel. **stormbound**, *a.* stopped or delayed by storms. **storm-centre**, *n.* the place of lowest pressure in a cyclonic storm; a place etc., liable to violent disturbance. **storm-cock**, *n.* the mistle-thrush, fieldfare, or green woodpecker. **storm-cone**, *n.* cone of canvas, hoisted as warning of an approaching storm. **storm door**, *n.* an additional stout, outer door. **storm lantern** HURRICANE LAMP. **storm-trooper**, *n.* a semi-military member of the Nazi party; one of a force of shock troops. **stormy**, *a.* characterized by storms; tempestuous; violent, vehement, passionate. **stormy petrel**, *n.* any of various small petrels. **stormily**, *adv.* **storminess**, *n.*

Storting, Storthing (staw'ting), *n.* the Norwegian parliament.

story¹ (staw'ri), *n.* a written narrative or verbal recital in prose or verse, of actual or fictitious events, a tale, short novel, romance, anecdote, legend or myth; the plot or incidents of a novel, epic or play; a series of facts of special interest connected with a person, place etc.; an account of an incident, experience etc.; a descriptive article in a newspaper; (*coll.*) a falsehood, a fib. **the same old story**, (*coll.*) familiar sequence of events. **the story goes**, it is commonly said. **to cut a long story short**, to be brief (in speaking), to tell omitting details. **story-book**, *n.* a book containing a story or stories. *a.* fairy-tale. **story line**, *n.* the main plot of a book, film etc. **story-teller**, *n.* **storytelling**, *n.*
story² STOREY.
stoup, stoop (stoop), *n.* a basin for holy water.
stoush (stowsh), *n.* (*Austral. coll.*) a fight, a brawl.
stout (stowt), *a.* strong, sound, sturdy, stanch, well-built; brave, resolute, intrepid; corpulent, bulky, fleshy. *n.* a malt liquor, very strong porter. **stout-hearted**, *a.* **stout-heartedly**, *adv.* **stout-heartedness**, *n.* **stoutish**, *a.* **stoutly**, *adv.* **stoutness**, *n.*
stove¹ (stōv), *n.* an apparatus, wholly or partially closed, in which fuel is burned for heating, etc.; a cooker. **stovepipe**, *n.* a pipe for conducting smoke etc., from a stove to a chimney.
stove² (stōv), *past* STAVE.
stow (stō), *v.t.* to put or pack (often away) in a suitable or convenient place or position; to pack or fill compactly with things. **stow it**, (*sl.*) drop it! stop (joking etc.). **stow-away**, *n.* one who conceals himself on a ship, aircraft etc. in order to get a free passage. **stowage** (-ij), *n.* an area or place for stowing goods or the charge for this; the act or state of being stowed; things for stowing.
strabismus (strəbiz'məs), *n.* squinting, a squint, produced by a muscular defect of the eye. **strabismal, -mic**, *a.*
strad (strad), *n.* short for STRADIVARIUS.
straddle (strad'l), *v.i.* to stand, walk or sit with the legs wide apart; to sit on the fence. *v.t.* to stand or sit astride of this; to shoot beyond and short of a target to determine the range. *n.* the act of straddling; the distance between the legs of one straddling; (*Stock Exch.*) a contract securing the right of either a put or call; a high-jumping technique in which the legs straddle the bar while the body is parallel to it. **straddler**, *n.*
Stradivarius (stradivəə'riəs), *n.* a violin made by Antonio *Stradivari* of Cremona (1644–1737). Often shortened to (*coll.*) **Strad**.
strafe (sträf, strahf), *v.t.* to rake with machine-gun fire from the air; to punish severely. *n.* an attack from the air; a severe punishment.
straggle (strag'l), *v.i.* to wander away from the main body or direct course; to get dispersed; to spread irregularly (of plants etc.). **straggler**, *n.* **straggly**, *a.*
straight (strāt), *a.* extending uniformly in one direction, not bent, curved or crooked; upright, honest, not deviating from truth or fairness, correct, accurate, right; level, even; unobstructed, uninterrupted; reliable, trustworthy, authoritative; (of a drink) undiluted; (*coll.*) equal; (*sl.*) not using drugs; (*coll.*) conventional; (*sl.*) heterosexual. *n.* a straight part, piece or stretch of anything; in poker, five cards in sequence irrespective of suit; (*coll.*) a conventional person; (*sl.*) a heterosexual person; (*sl.*) a cigarette containing only tobacco (and not marijuana, etc.); a sequence of cards; the straight part of a racetrack. *adv.* in a straight line; directly, without deviation; immediately, at once; without cheating, lying, etc.; candidly; continuously. **straight away, straight off**, at once, without delay. **the straight and narrow**, the honest and virtuous way of life. **to go straight**, to abandon criminal activities and become honest. **to keep a straight face**, to refrain from smiling. **straight angle**, *n.* an angle of 180°. **straightedge**, *n.* a strip of metal or wood having one edge straight, used as a ruler etc. **straight fight**, *n.* a contest between two candidates or sides only. **straight flush**, *n.* (in poker) a hand with five cards of the same suit in sequence. **straightforward**, *a.* straight; upright, honest, frank, open; (of a task) simple, presenting no difficulties. **straightforwardly**, *adv.* **straightforwardness**, *n.* **straight man**, *n.* one who acts as a stooge to a comedian. **straight-out**, *a.* (*N Am. coll.*) outright, complete; blunt, honest. **straighten**, *v.t.* to make straight. **to straighten out**, to resolve, unscramble; (*chiefly N Am.*) to reform or become reformed. **straightener**, *n.* **straightly**, *adv.* **straightness**, *n.*
strain¹ (strān), *v.t.* to stretch tight; to exert to the utmost; to weaken, injure, or distort by excessive effort or overexertion; to force beyond due limits; to apply (rules etc.) beyond the proper scope or intent; to press closely, to embrace; to constrain, to make unnatural, artificial, or uneasy; to purify from extraneous matter by passing through a colander or other strainer; to remove (solid matter) by straining (out). *v.i.* to exert oneself, to make violent efforts (after etc.); to pull or tug (at); to be filtered, to percolate. *n.* the act of straining, a violent effort, a pull, tension; an injury, distortion, or change of structure, caused by excessive effort, exertion, or tension; mental tension, fatigue from overwork etc.; a song, a tune, a melody, a piece of poetry; tone, spirit, manner, style, pitch. **strained**, *a.* unnatural, forced; tense, stressful. **strainer**, *n.* a filter; a sieve, colander.
strain² (strān), *n.* race, stock, family, breed; natural tendency or disposition.
strait (strāt), *n.* a narrow passage of water between two seas (*usu. in pl.*); a trying position, distress, difficulty (*usu. in pl.*). **strait-jacket**, *n.* a garment for confining the arms of the violently insane etc. **strait-laced**, *a.* puritanically strict in morals or manners. **straiten**, *v.t.* to distress; place in difficulty. **straitly**, *adv.* **straitness**, *n.*
strake (strāk), *n.* a continuous line of planking or plates from stem to stern of a vessel; part of the metal rim on a cart-wheel.
stramonium (strəmō'niəm), *n.* a drug prepared from the thorn-apple used for nervous complaints.
strand¹ (strand), *n.* a shore or beach of the sea, lake, or large river. *v.t.* to run or force aground; (*in p.p.*) to bring to a standstill or into straits, esp. from lack of funds. *v.i.* to run aground. **stranded**, *a.* in difficulties; without resources.
strand² (strand), *n.* one of the fibres, threads, wires etc. of which a rope etc. is composed; a length of hair; a string of pearls or beads; an element forming part of a complex whole. *v.t.* to make (a rope etc.) by twisting strands together.
strange (strānj), *a.* alien; not one's own; not well known, unfamiliar, new; unusual, singular, extraordinary, queer, surprising, unaccountable; fresh or unused (to), unacquainted, awkward. **strange particle**, *n.* an elementary particle (e.g. hyperon) which possesses a quantum strangeness number different from zero. **strangely**, *adv.* **strangeness**, *n.* the quality of being strange; the quantum number, conserved in strong but not in weak interactions, introduced to explain the paradoxically long lifetimes of certain elementary particles.
stranger (strān'jə), *n.* one from another place; a foreigner; a guest, a visitor; a person unknown (to one); one ignorant or unaccustomed (to).
strangle (strang'gl), *v.t.* to kill by compressing the windpipe, to choke, to throttle; to suppress, to stifle. **strangler**, *n.* **stranglehold**, *n.* a choking grip used in wrestling; a restrictive force or influence. **strangles**, *n.pl.* an infectious disease affecting horses etc.
strangulate (strang'gūlāt), *v.t.* to strangle; to compress a blood-vessel, intestine etc. **strangulation**, *n.*
strangury (strang'gūri), *n.* a disease characterized by pain in passing the urine, which is excreted in drops.

strap (strap), *n.* a long, narrow strip of leather, or similar material for fastening about things, carrying or lifting things, or holding on to; a strip, band or plate for holding parts together; a shoulder-strap; a strop; a strap-shaped blade or part, a ligula. *v.t.* (*past, p.p.* **strapped**) to fasten (*often* down, up etc.) with a strap; to beat with a strap; to sharpen, to strop. **the strap**, chastisement with a strap. **strap-hanger**, *n.* (*coll.*) a standing passenger in a bus or train. **strap-work**, *n.* ornamentation in the form of crossed or interlacing bands. **strapping**, *a.* tall, lusty, strong, muscular.
strata (strah'tə), *pl.* STRATUM.
stratagem (strat'əjəm), *n.* an artifice, trick or manoeuvre esp. for deceiving an enemy.
strategy (strat'əji), *n.* the art or science of war, esp. the art of directing military movements so as to secure the most advantageous positions and combinations of forces; a long-term plan aimed at achieving a specific goal; a stratagem. **strategic, -al** (-tē'-), *a.* pertaining to, used in or of the nature of strategy; (of missiles etc.) for use against an enemy's homeland rather than on the battlefield. **strategically**, *adv.* **strategics**, *n.* **strategist**, *n.* an expert in strategy.
strath (strath), *n.* a wide valley through which a river runs.
strathspey (-spā'), *n.* a Scottish dance slower than a reel; music in 4/4 time for this.
strati-, *comb. form.* layer.
straticulate (-tːkˈūlət), *a.* (*Geol.*) arranged in numerous thin strata. **stratification** (-fi-), *n.*
stratify (strat'ifī), *v.t.* (*past, p.p.* **-fied**) to form or arrange in strata. **stratification** (-fi-), *n.* **stratified**, *a.*
stratigraphy (strətigˈrəfi), *n.* the branch of geology dealing with the succession, classification, nomenclature etc. of stratified rocks; the analysis of layers in archaeology. **stratigraphic** (-graf'-), *a.*
stratocumulus (strahtōkūˈmūləs), *n.* a layer of cloud in dark round masses. **stratopause** (strat'əpawz), *n.* the upper boundary of the stratosphere. **stratosphere** (strat'əsfiə), *n.* the upper layer of atmosphere extending upwards from about 6 to 50 miles (10 to 80 km) above the earth's surface in which temperature does not decrease with the height. **stratospheric** (-sfe'-), *a.*
stratum (strah'təm), *n.* (*pl.* **-ta** (-tə), **-tums**) a horizontal layer of any material; a bed of sedimentary rock; a layer of tissue or cells; a layer of sea or atmosphere; a social level. **stratal**, *a.*
stratus (strah'təs), *n.* (*pl.* **-ti** (-tī)) a continuous horizontal sheet of cloud.
straw (straw), *n.* the dry, ripened stalk or stalks of certain species of grain, esp. wheat, rye, oats etc.; such a stalk or a piece of one; anything proverbially worthless; a long thin plastic or paper tube for sucking up a drink; the colour of straw, pale yellow; that which one resorts to in desperation. *a.* made of straw; resembling straw. **last straw** LAST[1]. **man of straw** MAN. **straw in the wind**, a hint or indication of future events. **to clutch, grasp at a straw (or straws)**, to resort to desperate remedies. **to draw the short straw**, to be the one selected for a difficult or unpleasant task. **strawboard**, *n.* a thick cardboard made from straw. **straw-coloured**, *a.* pale yellow. **straw poll**, *n.* an unofficial ballot test of opinion. **strawy**, *a.*
strawberry (straw'bəri), *n.* a low, stemless perennial plant bearing a fleshy red fruit with small achenes on the surface; the fruit of this. *a.* of the colour (purplish-red) or flavour of strawberries. **strawberry blonde**, *a.* a woman with reddish-blonde hair. **strawberry-mark**, *n.* a soft reddish birthmark. **strawberry tomato**, *n.* a Cape gooseberry.
strawberry-tree, *n.* an evergreen arbutus bearing a strawberry-like fruit.
stray (strā), *v.i.* to wander from the direct or proper course, to go wrong, to lose one's way; to wander from the path of rectitude; to digress; to lose concentration. *n.* any domestic animal that has gone astray; a waif; a homeless person. *a.* gone astray; straggling, occasional, sporadic. **strayer**, *n.*
streak (strēk), *n.* an irregular line or long narrow mark of a distinct colour from the ground; a vein, or strip; a course or stretch, esp. of good or bad luck; a flash of lightning; a trace; an act of running naked in public. *v.t.* to mark with streaks. *v.i.* (*coll.*) to run naked through a public place as a prank; to move in a straight line at speed; to become streaked. **yellow streak** YELLOW. **streaked**, *a.* **streaker**, *n.* **streaky**, *a.* marked with streaks; striped; (of bacon) having alternate layers of meat and fat. **streakily**, *adv.* **streakiness**, *n.*
stream (strēm), *n.* a body of flowing water or other fluid; a river, a brook; a steady flow, a current, a drift; anything in a state of continuous progressive movement, a moving throng etc., (*often in pl.*) a group of schoolchildren of the same general academic ability. *v.i.* to flow, move, or issue in or as a stream; to pour out or emit liquid abundantly; to float, hang, or wave in the wind etc. *v.t.* to pour out or flow with liquid abundantly; to group (schoolchildren) into streams. **streamline**, *n.* the shape given to aircraft, vehicles etc., in order to cause the minimum of resistance. *v.t.* to give this shape (to). **streamlined**, *a.* having a contoured shape to offer minimum resistance to air or liquid; effectively organized, efficient, simplified; graceful. **stream of consciousness**, *n.* the flow of thoughts and feelings forming an individual's conscious experience; a literary technique used to express the unspoken thoughts and emotions of a fictional character, without using conventional narrative or dialogue. **streamer**, *n.* a long, narrow flag, strip of coloured paper or ribbon, a pennon; a column of light shooting across the sky; (*Comput.*) a device which copies data from a hard disk onto magnetic tape as a backup against accidental erasure or loss. **streamlet** (-lit), *n.* **streamy**, *n.*
street (strēt), *n.* a road in a city or town usu. with houses on one side or on both; the part of the road used by vehicles; the people living in a street. *a.* of or relating to life in urban centres. **on the streets**, living by prostitution; homeless, destitute. **streets ahead of**, far better than, more developed or advanced than. **streets apart**, completely different. **(right) up one's street**, ideally suited to one's talents, inclinations etc. **street arab** ARAB[1]. **streetcar**, *n.* (*N Am.*) a tram. **street credibility**, *n.* knowledge of the customs, language etc. associated with the urban counterculture (also **street cred**). **street value**, *n.* the monetary value of a commodity, esp. drugs, in terms of the price paid by the ultimate user. **streetwalker**, *n.* a prostitute who solicits on the streets. **streetwise**, *a.* familiar with life among the poor, criminals etc. in an urban environment.
strength (strength), *n.* the quality of being strong; muscular force; firmness, solidity; power, potency; intensity; amount or proportion of the whole number (of an army, ships etc.). **from strength to strength**, with continually increasing success. **in strength**, in considerable numbers. **on the strength of**, in reliance on. **strengthen**, *v.t.* to make strong or stronger. *v.i.* to increase in strength. **strengthener**, *n.*
strenuous (stren'ūəs), *a.* involving, requiring or characterized by effort, vigour, endeavour. **strenuously**, *adv.* **strenuousness**, *n.*
strepitoso (strepitō'sō), *adv.* (*Mus.*) in a noisy, impetuous manner.
strepto-, *comb. form.* twisted chain; flexible.
streptococcus (-kokˈəs), *n.* (*pl.* **-i** (-kok'sī)) a genus of bacteria consisting of spherical organisms in chains of varying length.
streptomycin (-mī'sin), *n.* an antibiotic obtained from soil bacterium and used in the treatment of tuberculosis and other bacterial infections.

stress (stres), *n.* constraining or impelling force; physical, mental or emotional strain; tension, pressure, violence; weight, importance, or influence; emphasis accent; force exerted upon or between the parts of a body. *v.t.* to lay the stress or accent on; to subject to stress or force. **stressful,** *a.*
stretch (strech), *v.t.* to draw out, to extend, to extend in any direction or to full length; to tighten, to draw tight; to extend lengthwise, to straighten; to cause to extend, to hit so as to prostrate; to distend, to strain; to do violence to; to exaggerate; (*sl.*) to hang by the neck. *v.i.* to be extended in length or breadth; to have a specified extension, to reach; to be drawn out or admit of being drawn out; to extend or straighten one's body or limbs. *n.* the act of stretching or state of being stretched; extent or reach; a reach, sweep, or tract (of land, water etc.); the distance covered on a section of racetrack; period of a prison sentence. *a.* having the capacity to stretch, e.g. stretch pants. **at a stretch,** at one go, continuously; with difficulty, involving special effort; if necessary. **to stretch a point,** to go beyond what might be expected. **stretchmarks,** *n.pl.* marks left on the abdomen after pregnancy. **stretcher,** *n.* that which stretches; a litter or other appliance for carrying a sick, wounded or disabled person in a recumbent position; a brick or stone laid lengthwise in a course in a wall; a beam used as a tie in a structural framework; a cross-piece in a boat for a rower to press his feet against; a cross-piece between the legs of a chair or table. **stretcher-bearer,** *n.* one who helps to carry a stretcher with the wounded etc. **stretchy,** *a.* **stretchiness,** *n.*
strew (stroo), *v.t.* (*p.p.* **strewn, strewed**) to scatter, to spread thus; to cover by scattering or by being scattered over.
strewth (strooth), *int.* an exclamation of surprise or alarm etc.
stria (strī'ə), *n.* (*pl.* **striae** (-ē)) a superficial furrow, a thin line or groove, mark or ridge. **striate** (-ət), **-ted,** *a.* marked with striae. *v.t.* (-āt'), to mark with striae. **striation,** *n.*
stricken (strik'n), *a.* afflicted by disease, illness, grief, love, etc.
strict (strikt), *a.* enforcing or observing rules precisely, not lax; rigorous, severe, stringent; defined or applied exactly or absolutely, accurate, precise. **strictly,** *adv.* **strictness,** *n.*
stricture (strik'chə), *n.* a censure, a sharp criticism; (*Path.*) a contraction of duct or channel, as of the urethra. **strictured,** *a.*
stride (strīd), *v.i.* (*past* **strode** (strōd), *p.p.* **stridden** (strid'n), **strid**) to walk with long steps. *v.t.* to pass over in one step; to bestride. *n.* a long or measured step or the distance covered by this; progress or rate of progress; a stride piano; (*pl.*) (*chiefly Austral. coll.*) men's trousers. **to make great strides,** to progress or develop rapidly. **to take in one's stride,** to achieve without difficulty or effort. **stride piano,** *n.* a style of jazz piano in which the right hand plays the melody, while the left alternates in a swinging rhythm between single bass notes (on strong beats) and chords.
stridence (strī'dəns), **-cy** (-si), *n.* loudness or harshness of tone. **strident,** *a.* sounding harsh, grating. **stridently,** *adv.*
stridor (-daw), *n.* a harsh, whistling noise made during respiration and caused by blockage of the air passages; a harsh high-pitched sound. **stridulate** (strid'ūlāt), *v.i.* to make a shrill creaking noise (esp. of cicadas and grasshoppers by rubbing hard parts of their body together). **stridulant, stridulous,** *a.* **stridulation,** *n.* **stridulator,** *n.*
strife (strīf), *n.* contention, conflict, hostile struggle.
strigose (-gōs), *a.* having short stiff hair, bristle or hair-like scales.
strike (strīk), *v.t.* (*past* **struck** (strŭk), *p.p.* **struck, stricken** (strik'n)) to hit, to deliver a blow or blows upon; to deliver, to deal, to inflict (a blow etc.); to afflict (*usu. in p.p.*) to drive, to send (a ball etc.) with force; to attack an enemy craft, location etc.; to produce, make, form, effect, or bring into a particular state by a stroke, as to ignite (a match), to stamp or mint (a coin); to afflict suddenly, to blind, to deafen etc.; to make (a bargain); to cause (a bell etc.) to sound; to notify by sound; to cause to penetrate, to thrust (into); to hook (a fish) by jerking the tackle upwards; to effect forcibly, to impress strongly; to occur suddenly to the mind of; to cause (a cutting etc.) to take root; to lower (sails, a flag, tent etc.); to surrender by lowering (a flag etc.); to leave off (work), esp. to enforce a demand for higher wages etc.; to level corn etc. in (a measure) by scraping off the surplus; to determine (a balance, average etc.); to assume (an attitude); to discover, to come across. *v.i.* to hit, to deliver a blow or blows (upon); to collide, to dash (against, upon etc.); to be driven on shore, a rock etc.; to sound (the time) by a stroke (of a bell etc.); to lower sails, flag etc. in token of surrender etc.; to take root; to leave off work to enforce a demand for higher wages etc.; to arrive suddenly, to happen (upon); to enter or turn (into a track etc.); (*Geol.*) to extend in a particular direction (of strata). *n.* the act of striking for an increase of wages etc.; an attack upon an enemy location, craft etc.; a straight-edge for levelling something, as a measure of grain; (*Geol.*) the horizontal direction of an outcrop; a discovery (as of oil); a luck find, unexpected success; in tenpin bowling, the knocking down of all ten pins with the first bowl, or the score in doing this; (*Baseball*) a good pitched ball missed by the batter and counting against him; (*Cricket*) in a position to receive the bowling; an attack on a target from the air. **to strike back,** to return a blow, retaliate. **to strike down,** to make ill or cause to die, esp. suddenly. **to strike home,** to hit the intended target; to achieve the desired effect. **to strike it rich,** to find a deposit of oil, minerals etc.; to make an unexpected large financial gain. **to strike lucky,** to be fortunate. **to strike off,** to remove, separate, dislodge etc. by or as if by a blow; to erase, to strike out. **to strike out,** to blot out, to efface, to expunge; to make vigorous strokes (in skating, swimming etc.); to hit from the shoulder (in boxing); to start out; (*N Am.*) to fail. **to strike up,** to begin to play or sing; to enter into, to start (a conversation etc.). **strikebound,** *a.* of a factory etc., closed or disrupted because of a strike. **strikebreaker,** *n.* a blackleg, worker brought in to replace one out on strike. **strike-pay,** *n.* an allowance for subsistence from a trade union to workers on strike. **striker,** *n.* one who or that which strikes, esp. a worker on strike; in soccer, an attacking player, a forward. **striking,** *a.* surprising, forcible, impressive, noticeable. **striking circle,** *n.* in hockey, the semi-circular area in front of the goal from within which the ball must be struck to score. **strikingly,** *adv.* **strikingness,** *n.*
strine (strīn), *n.* used humorously, Australian English. [a rendering of *Australian* in an Australian accent]
string (string)), *n.* twine, a fine line, usu. thicker than thread and thinner than cord; a length of this or strip of leather, tape, or other material, used for tying, fastening, binding together, connecting etc.; a string-like fibre, tendon, nerve etc.; a piece of wire, catgut etc., yielding musical sounds or notes when caused to vibrate in a piano, violin etc.; (*pl.*) the stringed instruments in an orchestra; a cord or thread upon which anything is strung, hence a series of things or persons connected together or following in close succession; (*Racing*) the horses under training at a particular stable; (*pl.*) conditions, complications; a sequence of alphabetic or numeric characters in a computer program. *v.t.* (*past, p.p.* **strung** (strŭng)) to furnish with a string or strings; to fasten the string on (a bow); to make (nerves etc.) tense (*usu. in p.p.*); to thread

stringendo 815 **structure**

on a string; to strip (beans etc.) of strings or fibres. *v.i.* to become stringy. **on a string**, totally dependent, e.g. emotionally; held in suspense. **no strings attached**, (*coll.*) with no conditions or restrictions. **to have two strings to one's bow** BOW [1]. **to pull strings**, to exert influence unobtrusively. **to string along**, *v.i.* (*coll.*) to accompany; to agree with, go along with. *v.t.* (*coll.*) to fool, deceive. **to string up**, (*coll.*) to hang. **string-band**, *n.* a band of stringed instruments. **string-bean**, *n.* (*N Am.*) a runner bean, a french bean. **stringboard**, *n.* a timber receiving the ends of stairs in a staircase. **string-course**, *n.* a projecting horizontal band or moulding running along a building. **string-halt** SPRING-HALT. **string-piece**, *n.* a supporting timber forming the edge of a framework, esp. of a floor; a stringboard. **string quartet**, *n.* a combination of four string instruments, viz. two violins, a viola and a cello; music written for this combination. **string tie**, *n.* a narrow necktie. **stringed**, *a.* **stringer**, *n.* a stringboard; a long horizontal member in a structural framework; (*coll.*) a journalist who works part-time for a newspaper or news agency in a particular area. **stringlike**, *a.* **stringy**, *a.* consisting of strings or small threads, fibrous, ropy, viscous; wiry, sinewy (of a person). **stringy-bark**, *a.* a name for many of the Australian gum-trees, from their fibrous bark. **stringiness**, *n.*
stringendo (strinjen'dō), *adv.* (*Mus.*) in accelerated time.
stringent (strin'jənt), *a.* strict, precise, binding, rigid; hampered, tight, unaccommodating (of the money-market etc.). **stringency**, *n.* **stringently**, *adv.*
stringer, stringy, etc. STRING.
strip (strip), *v.t.* (*past, p.p.* **stripped**) to pull the covering from, to denude, to skin, to peel, to husk, to clean; to deprive (of), to despoil, to plunder; to remove (clothes, paint, bark, branches etc.); to milk (a cow) to the last drop. *v.i.* to take off one's clothes, to undress; to come away in strips; to have the thread torn off (of a screw), to be discharged without spin (of a projectile). *n.* a long, narrow piece; an airstrip; the clothes worn by a football team etc.; a striptease. **comic strip** (also called **strip cartoon**) COMIC. **to strip down**, to dismantle. **to tear (someone) off a strip**, (*coll.*) to criticise or rebuke severely. **strip club**, *n.* a club in which striptease artists perform. **strip lighting**, *n.* lighting by long fluorescent tubes. **strip mine**, *n.* an opencast mine. **strip mining**, *n.* **striptease**, *n.* a cabaret turn in which an actress partially or wholly undresses herself. **stripper**, *n.*
stripe (strip), *n.* a long, narrow band of a distinctive colour or texture; a chevron on the sleeve of a uniform indicating rank; a stroke with a whip, scourge etc. *v.t.* to mark with stripes. **striped**, *a.* **stripy**, *a.*
stripling (strip'ling), *n.* a youth, a lad.
strive (strīv), *v.i.* (*past* **strove** (strōv), *p.p.* **striven** (striv'ən)), to make efforts, to endeavour earnestly, to struggle; to contend, to vie, to emulate. **striver**, *n.*
strobe (strōb), *n.* a stroboscope. **strobe lighting**, *n.* high-intensity flashing light; the apparatus that produces it.
stroboscope (-əskōp), [-SCOPE], *n.* an instrument for observing periodic motion by making the moving body visible at certain points through the use of synchronized flashing light. **stroboscopic, -al** (-skop'-), *a.* **stroboscopically**, *adv.*
strobile (strob'il, -il), **strobilus** (-bī'ləs), *n.* a multiple fruit such as a pine-cone.
stroboscope, etc. STROBE.
strode (strōd), *past* STRIDE.
stroganoff (strog'ənof), *n.* a dish of meat, usu. beef in strips cooked with onions and mushrooms in a sour-cream sauce (also called **beef stroganoff**).
stroke[1] (strōk), *n.* the act of striking, a blow; the impact, shock, noise etc, of this; a sudden attack (of disease, affliction etc.), a sudden onset of paralysis; a single movement of something, esp. one of a series of recurring movements, as of the heart, an oar, wing, swimmer, piston, striking clock etc.; the length, manner, rate etc. of such a movement; a mark made by a single movement of a pen, pencil etc., a solidus; the oarsman nearest the stern who sets the rate of rowing. *v.t.* to act as stroke for (a boat or crew). **at a stroke**, by a single action. **off one's stroke**, not at one's best. **on the stroke**, punctually. **stroke play**, *n.* in golf, scoring by counting the number of strokes played as opposed to the number of holes won.
stroke[2] (strōk), *v.t.* to pass the hand over the surface of caressingly. *n.* the act of stroking.
stroll (strōl), *v.i.* to walk leisurely or idly, to saunter. *v.t.* to saunter or ramble on foot. *n.* a leisurely ramble. **stroller**, *n.* (*N Am.*) a pushchair.
stroma (strō'mə), *n.* (*pl.* **-mata** (-tə)) the framework of tissue of an organ or cell; a dense mass of hyphae produced by some fungi, in which fructification may develop; the dense framework of a chloroplast etc.
strong (strong), *a.* (*comp.* **stronger** (-gə); *super.* **strongest** (-gəst)) able to exert great force, powerful, muscular, able, capable; acting with great force, vigorous, forcible, energetic; having great powers of resistance or endurance; healthy, robust, hale; firm, tough, solid; having great numbers, resources etc.; having a specified number of personnel etc.; concentrate; having a powerful effect on the senses, loud and penetrating, glaring, pungent, ill-smelling, intoxicating, heady; (*Gram.*) forming inflexions by internal vowel-change, and not by addition of a syllable (*as* **strike, struck, stride, strode**). **going strong**, prospering, getting on famously, in good form or spirits. **to come on strong**, to act or behave in a forceful, reckless, or defiant way. **strong-arm**, *a.* using or involving physical force. *v.t.* to show violence towards. **strongbox**, *n.* a safe or robust trunk for storing valuables. **strong drink**, *n.* alcoholic liquors. **stronghold**, *n.* a fortress or fortified place; a place of refuge for, or domination by, a certain group of people. **strong interaction**, *n.* an interaction between elementary particles responsible for the forces that bind nucleons together in an atomic nucleus. **strong language**, *n.* swearing. **strong-minded**, *a.* having a vigorous mind; resolute, determined. **strong-mindedly**, *adv.* **strong-mindedness**, *n.* **strong point**, *n.* something at which one excels. **strongroom**, *n.* a specially reinforced room for storing valuables. **strong-willed**, *a.* **strongly**, *adv.*
strontium (stron'tiəm), *n.* a yellowish metallic element, at. no. 38; chem. symbol Sr resembling calcium. **strontium-90**, *n.* strontium with atomic weight of 90, a radioactive product of nuclear fission which tends to accumulate in bones.
strop (strop), *n.* a strip of leather etc., for sharpening razors etc., on. *v.t.* (*past, p.p.* **stropped**) to sharpen with or on a strop.
strophanthus (strəfan'thəs), *n.* any of various tropical gamopetalous small trees or shrubs; the seeds of these. **strophanthin** (-thin), *n.* a poisonous drug made from strophanthus seeds.
strophe (strō'fi), *n.* the turning of the chorus from right to left in an ancient Greek drama; a part of the ode (consisting of strophe, antistrophe, and epode) sung whilst so turning, esp. the first part, the strophe proper. **strophic**, *a.*
stroppy (strop'i), *a.* (*coll.*) rowdy, angry; awkward, quarrelsome. **stroppily**, *adv.* **stroppiness**, *n.*
strove (strōv), *past* STRIVE.
struck (strŭk), *past, p.p.* STRIKE.
structure (strŭk'chə), *n.* a combination of parts, as a building, machine, organism etc., esp. the supporting or essential framework; the manner in which a complex whole is constructed, put together, or organically formed; the arrangement of parts, organs, atoms etc., in a complex whole. *v.t.* to create a structure. **structural**, *a.*

structural formula, *n.* a chemical formula showing the arrangement of atoms and bonds in a molecule. **structuralism**, *n.* an approach to the human sciences, literature, linguistics etc. as coded systems comprising self-sufficient and self-determining structures of interrelationships and rules of combination through which meaning is generated and communicated. **structuralist**, *n.* **structurally**, *adv.*

structured, *a.* (*usu. in comb.*, as *loose-structured*) **structureless** (-lis), *a.*

strudel (stroo'dl), *n.* a thin pastry rolled up with a filling (e.g. apple) and baked.

struggle (strŭg'l), *v.i.* to make violent movements; to put forth great efforts, esp. against difficulties or opposition; to strive (to); to contend (with or against); to make one's way (along etc.) against difficulties, opposition etc. *n.* an act or spell of struggling; a strenuous effort; a fight or contest. **struggling**, *a.*

strum (strŭm), *v.t.*, *v.i.* to play noisily or carelessly, to thrum on a stringed instrument. *n.* **strumming**.

struma (stroo'mə), *n.* (*pl.* **-mae**) (*dated*) scrofula; goitre; (*Bot.*) a cushion-like swelling on a petiole etc. **strumose** (-mōs), **strumous**, *a.*

strumpet (strŭm'pit), *n.* a prostitute, a harlot.

strung (strŭng), *past, p.p.* STRING.

strut[1] (strŭt), *v.i.* to walk with a pompous, conceited gait. *n.* such a gait. **strutter**, *n.* **strutting**, *a.* **struttingly**, *adv.*

strut[2] (strŭt), *n.* a timber or iron beam inserted in a framework so as to keep other members apart, a brace. *v.t.* to brace with a strut or struts.

struthious, *a.* of or resembling the ostriches; flightless.

styrchnine (strik'nēn), *n.* a highly poisonous alkaloid obtained from species of *Strychnos*, esp. *S. nux vomica*, used in medicine as a stimulant etc.

stub (stŭb), *n.* the stump of a tree, tooth etc.; a stump, end or remnant of anything, e.g. of a cigarette; (*N Am.*) a cheque counterfoil. *v.t.* (*past, p.p.* **stubbed**) to grub up by the roots; to clear of stubs; to strike one's toe against something; to extinguish a cigarette etc. (foll. by *out*).

stubby, *a.* short and thickset. *n.* (*coll. Austral.*) a small squat beer bottle. **stubbiness**, *n.*

stubble (stŭb'l), *n.* the stumps of wheat, barley etc. covering the ground after harvest; short, bristly hair, whiskers etc. **designer-stubble** DESIGN. **stubbled**, **stubbly**, *a.*

stubborn (stŭb'ən), *a.* unreasonably obstinate, not to be persuaded; obdurate, inflexible, intractable, refractory. **stubbornly**, *adv.* **stubbornness**, *n.*

stucco (stŭk'ō), *n.* (*pl.* **-coes, -ccos**) fine plaster for coating walls or moulding into decorations in relief; any of various types of plaster used for coating the outside of buildings. *v.t.* (*past, p.p.* **-ccoed**) to coat with stucco.

stuck (stŭk), *past, p.p.* STICK.

stud[1] (stŭd), *n.* a large-headed nail, knob, head of a bolt etc., esp. fixed as an ornament; an ornamental button for wearing in a shirt-front etc.; a cross-piece in a link of chain-cable; a post or scantling to which laths are nailed in a partition. a protrusion on the sole of a football boot, etc. *v.t.* (*past, p.p.* **studded**) to set with studs or ornamental knobs; to set thickly, to bestrew.

stud[2] (stŭd), *n.* a number of horses kept for riding, racing, breeding etc.; any male animal used for breeding; an animal-breeding establishment; (*sl.*) a sexually potent man. *a.* of or relating to an animal-breeding establishment or an animal kept for breeding. **stud-book**, *n.* a register of pedigrees of horses or cattle. **stud-farm**, *n.* a farm where horses are bred. **studhorse**, *n.* a stallion. **stud poker**, *n.* a variety of poker.

studdingsail (stŭn'sl), *n.* an additional sail set beyond the sides of a square sail in light winds.

student (stū'dənt), *n.* a person engaged in study, esp. one receiving instruction at a university, college or other institution for higher education or technical training; (*esp. N Am.*) a schoolboy or girl; a studious person; a person receiving an annual grant for study or research from a foundation etc. **studentship**, *n.* a grant for study at a university.

studiedly STUDY.

studio (stū'diō), *n.* the working-room of a sculptor, painter, photographer etc.; the room in which records, radio and television programmes are recorded, or films made; the place from which television and radio programmes are broadcast; (*pl.*) the buildings used for making films by a television or film company. **studio flat**, *n.* a flat consisting of one room.

studious (stū'diəs), *a.* devoted to study; careful, observant (of); studied, deliberate, intended. **studiously**, *adv.* **studiousness**, *n.*

study (stŭd'i), *n.* (*pl.* **studies**) mental application to books, art, science etc., the pursuit of knowledge; something that is studied or worth studying; a sketch or other piece of work done for practice or as a preliminary design for a picture etc.; (*Mus.*) a composition designed to test or develop technical skill; (*Theat.*) one who learns a part; a room devoted to study, literary work etc.; a reverie, a fit of musing; earnest endeavour, watchful attention; the object of this. *v.t.* (*past, p.p.* **studied**) to apply the mind to for the purpose of learning; to inquire into, to investigate; to contemplate, to consider attentively; to commit to memory; to apply thought and pains to, to be zealous for; (*in p.p.*) deliberate, premeditated, intentional. *v.i.* to apply oneself to study, esp. to reading; to meditate, to cogitate, to muse; to be assiduous, diligent, or anxious (to do). **studied** (-id), *a.* deliberate, intentional. **studiedly**, *adv.* **studiedness**, *n.*

stuff (stŭf), *n.* the material of which anything is made or may be made; the fundamental substance, essence, or elements of anything; household goods, furniture, utensils etc.; a textile fabric, esp. woollen, as opp. to silk or linen; worthless matter, nonsense, trash. *v.t.* to cram, to pack, to fill or stop (up); to fill (a fowl etc.) with stuffing or seasoning for cooking; to fill the skin of (a dead animal) so as to restore its natural form; to fill with food; to cram, press, ram, or crowd into a receptacle, confined space etc.; to fill with ideas, notions, nonsense etc.; (*coll.*) to impose on, to hoax. *v.t.* (*sl.*) to have sexual intercourse with a woman. *v.i.* to cram oneself with food. **bit of stuff**, (*sl. derog.*) a girl or woman. **hot stuff**, (*coll.*) an attractive or potent person or thing. **that's the stuff!** just what is needed. **to do one's stuff**, (*coll.*) to act as one is expected. **to know one's stuff**, (*coll.*) to be competent in one's chosen field. **stuffed**, *a.* (of poultry etc.) filled with stuffing; having blocked nasal passages (foll. by *up*). **get stuffed!**, *int.* (*taboo sl.*) expressing anger, contempt etc. against another person. **stuffed shirt**, *n.* (*coll.*) a pompous person. **stuffer**, *n.* **stuffing**, *n.* material used to stuff something; a mixture of ingredients used to stuff poultry etc. before it is cooked. **to knock the stuffing out of**, to beat (an opponent thoroughly). **stuffing-box**, *n.* a chamber packed with stuffing so as to be air-tight or water-tight, in which a piston-rod etc. can work freely. **stuffy**, *a.* ill-ventilated, close, fusty; strait-laced; stuffed up. **stuffiness**, *n.*

stultify (stŭl'tifi), *v.t.* (*past, p.p.* **stultified**) to render absurd, to cause to appear self contradictory, inconsistent, or ridiculous. **stultification** (-fi-), *n.* **stultifier**, *n.*

stum (stŭm), *n.* unfermented grape-juice, must. *v.t.* to prevent (wine) from fermenting by adding stum.

stumble (stŭm'bl), *v.i.* to trip in walking or to strike the foot against something without falling, to have a partial fall; to act, move, or speak blunderingly; to come (upon) by chance. *n.* an act of stumbling. **stumbler**, *n.* **stumbling**, *a.* **stumbling-block**, *n.* an obstacle, an impediment, a cause of difficulty, hesitation etc. **stumblingly**, *adv.*

stumer (stū'mə), *n.* (*sl.*) a cheque that has no money to back it; a returned cheque; a forgery.

stump (stŭmp), *n.* the part left in the earth after a tree has fallen or been cut down; any part left when the rest of a branch, limb, tooth etc., has been cut away, amputated, destroyed, or worn out, a stub, a butt; (*Cricket*) one of the three posts of a wicket; a platform from which speeches are made. (*pl.*) the legs; a pointed roll of leather or paper used to rub down the strong lines of a crayon or pencil drawing etc. *v.i.* to walk stiffly, awkwardly, or noisily, as on wooden legs; to make speeches. *v.t.* to work upon (a drawing etc.) with a stump; to go about (a district) making speeches; to put out (the batsman) at cricket by touching the wicket while he is out of the crease; (*coll.*) to pose, to put at a loss; (*sl.*) to pay (up) at once. **on the stump**, going about making political speeches. **to stump up**, to pay up; to produce the money required. **stumper**, *n.* **stumpy**, *a.* short, thick-set, stocky; full of stumps, stubby. **stumpiness**, *n.*

stun (stŭn), *v.t.* (*past, p.p.* **stunned**) to daze or deafen with noise; to render senseless with a blow; to stupefy, to overpower. **stunner**, *n.* one who or that which stuns; (*sl.*) something astonishing or first-rate. **stunning**, *a.* stupefying; (*sl.*) impressive, fine etc. **stunningly**, *adv.*

stung (stŭng), *past, p.p.* STING.

stunk (stŭngk), *past, p.p.* STINK.

stunsail, stuns'l (stŭn'sl), STUDDINGSAIL.

stunt[1] (stŭnt), *v.t.* to check in growth or development, to dwarf, to cramp. *n.* a check in growth; a stunted animal or thing.

stunt[2] (stŭnt), *n.* a performance serving as a display of strength, skill, or the like; a feat; a thing done to attract attention; a feat of aerobatics. **stuntman, stuntwoman,** *n.* one who performs dangerous feats (esp. as a stand-in for an actor).

stupa (stoo'pə), *n.* a Buddhist domed shrine; a tope.

stupe (stūp), *n.* a medicated compress of soft material used to relieve pain.

stupefy (stū'pəfī), *v.t.* (*pres.p.* **stupefying**, *past, p.p.* **stupefied**) to make stupid or senseless; to deprive of sensibility. **stupefacient** (-fā'shənt), *n., a.* (of) a stupefying drug. **stupefaction** (-fak'-), *n.* astonishment; the act of stupefying or state of being stupefied. **stupefier**, *n.*

stupendous (stūpen'dəs), *a.* astounding in magnitude, force, degree etc., marvellous, amazing, astonishing. **stupendously**, *adv.* **stupendousness**, *n.*

stupid (stū'pid), *a.* in a state of stupor, stupefied; dull of apprehension, wit or understanding, obtuse; senseless, nonsensical. **stupidity** (-pid'-), **stupidness**, *n.* **stupidly**, *adv.*

stupor (stū'pə), *n.* a dazed condition, torpor, deadened sensibility.

sturdy (stœ'di), *a.* robust, vigorous, strong. **sturdily**, *adv.* **sturdiness**, *n.*

sturgeon (stœ'jən), *n.* a large anadromous fish of the genus *Acipenser*, characterized by bony scales, esp. *A. sturio*, which yields caviare and isinglass.

stutter (stŭt'ə), *v.i.* to keep hesitating or repeating sounds spasmodically in the articulation of words; to make a sudden repetitive sound. *n.* this act or habit. **stutterer**, *n.* **stutteringly**, *adv.*

sty[1] (stī), *n.* (*pl.* **sties**) a pen or enclosure for pigs; a mean or filthy habitation; a place of debauchery. *v.t.* to shut up in or as in a sty. *v.i.* to live in or as in a sty.

sty[2], **stye** (stī), *n.* (*pl.* **sties, styes**) a small inflamed swelling on the edge of the eyelid.

Stygian (stij'iən), *a.* pertaining to the river Styx; gloomy, impenetrable (of darkness).

style (stīl), *n.* a pointed instrument for writing on wax-covered tablets, a stylus; the gnomon of a sun-dial; (*Bot.*) the prolongation of an ovary, bearing the stigma; manner of writing, expressing ideas, speaking, behaving, doing etc., as dist. from the matter expressed or done; sort, kind, make, pattern; the general characteristics of literary diction, artistic expression, or mode of decoration, distinguishing a particular people, person, school, period etc.; the proper expression of thought in language; manner or form of a superior or fashionable character, fashion, distinction; mode of designation or address, title, description. *v.t.* to designate, to describe formally by name and title; to design or shape. **stylebook,** *n.* a book containing rules of grammar, typography etc. for printers and editors. **stylar**, *a.* of or pertaining to a style for writing etc. *a.* **styling mousse,** *n.* a light foam which holds the hair in different styles. **stylish,** *a.* fashionable in style, smart, showy. **stylishly,** *adv.* **stylishness,** *n.* **stylist,** *n.* a writer having or cultivating a good style; a clothes designer; a hairdresser who styles hair. **stylistic** (-lis'-), *a.* **stylistically,** *adv.* **stylize, -ise,** *v.t.* to give a conventional form to. **stylization,** *n.*

stylet (stī'lit), *n.* (*Surg.*) the stiffening wire of a catheter; a probe.

stylite (stī'līt), *n.* a religious recluse in ancient and mediaeval times who lived on the top of a pillar.

stylobate (stī'ləbāt), *n.* a continuous base for a range of columns.

stylograph (stī'ləgraf), *n.* a pen with a tubular point fed with ink from a reservoir in the shaft. **stylographic** (-graf'-), *a.* **stylographically,** *adv.* **stylography** (-log'-), *n.* the art, process etc. of using a style or stylograph.

styloid (stī'loid), *a.* resembling a stylus. *n.* the styloid process, a spine projecting from the base of the temporal bone.

stylus (stī'ləs), *n.* (*pl.* **-li, -luses**) a pointed instrument for writing, drawing or engraving, a style; a device attached to the cartridge in the arm of a record player that follows the groove in a record.

stymie (stī'mi), *n.* (*Golf*) (*formerly*) the position when an opponent's ball lies between the player's ball and the hole. *v.t.* to hinder or obstruct.

styptic (stip'tik), *a.* that stops bleeding. *n.* a drug that arrests bleeding. **stypticity** (-tis'-), *n.*

styrene (stī'rēn), *n.* a colourless volatile liquid derived from benzene used in the manufacture of plastics and synthetic rubber.

suable (sū'əbl), *a.* capable of being sued. **suability** (-bil'-), *n.*

suasion (swā'zhən), *n.* (*rare*) persuasion as opp. to compulsion. **suasive** (-siv), *a.* **suasively,** *adv.*

suave (swahv), *a.* agreeable, bland, gracious, polite. **suavely,** *adv.* **suavity, suaveness** *n.*

sub (sŭb), *n.* short for SUBALTERN, SUBEDITOR, SUBMARINE, SUBORDINATE, SUBSTITUTE; a small loan or advance payment of wages etc. *v.i.* (*past, p.p.* **subbed**) to act as a substitute or as a subeditor; to receive pay in advance on account of wages due later. *v.t.* to grant (a small loan or advance) to; to subedit.

sub- (sŭb-), *pref.* under, situated below; from below, upward; denoting inferior or subordinate position; subdivision of, part of; slightly, rather; approximately, bordering on; (*Chem.*) less than normal; containing in small proportion; (*Math.*) denoting the inverse of a ratio.

subacid (sŭbas'id), *a.* slightly acid or sour. *n.* a subacid substance. **subacidity** (-sid'-), *n.*

subalpine (sŭbal'pīn), *a.* pertaining to elevated regions not above the timber-line.

subaltern (sŭb'əltən), *a.* subordinate; of inferior rank; (*Log.*) particular, ranking below universal. *n.* a junior army officer, one below the rank of captain. **subalternation** (-awl'tə-), *n.* (*Log.*) the relation between a particular and a universal proposition of the same quality.

subantarctic (sŭbantahk'tik), *a.* pertaining to the region bordering on the Antarctic.

subaqua (sŭbak'wə), *a.* pertaining to underwater sports.

subarctic (sŭbahk'tik), *a.* pertaining to the region border-

ing on the Arctic.
subassembly (sŭb'əsembli), *n.* an assembled unit forming part of a larger product.
subatomic (sŭb'ətom'ik), *a.* of or occurring inside an atom; making up an atom; smaller than an atom.
subbasement, *n.* the area below the basement in a building.
subclinical (sŭbklin'ikəl), *a.* having symptoms sufficiently slight as to be undetectable clinically. **subclinically**, *adv.*
subcommittee (sŭb'kəmiti), *n.* a small committee appointed from among its members by a larger committee to consider and report on a particular matter.
subconscious (sŭbkon'shəs), *a.* slightly or partially conscious; existing in the mind but measurable to one's full awareness. *n.* that part of the field of consciousness which at any given moment is outside the range of one's attention; the accumulation of past conscious experiences which are forgotten or for the moment are out of one's thoughts. **subconsciously**, *adv.* **subconsciousness**, *n.*
subcontinent (sŭbkon'tinənt), *n.* a region large enough to be a continent though itself forming part of a yet larger continent. **subcontinental**, *a.*
subcontract[1] (sŭbkon'trakt), *n.* a contract sublet from another.
subcontract[2] (sŭbkəntrakt'), *v.t., v.i.* to make a subcontract. **subcontractor**, *n.*
subcritical (sŭbkrit'ikəl), *a.* (pertaining to nuclear fuel) of insufficient mass to sustain a chain reaction.
subculture (sŭb'kŭlchə), *n.* a social or ethnic group with a characteristic culture differing from that of the national culture. **subcultural**, *a.*
subcutaneous, *a.* (*Med.*) situated or used beneath the skin. **subcutaneously**, *adv.*
subdeacon, *n.* in the Roman Catholic Church, a cleric who assists at High Mass. **subdeaconry, subdeaconship**, *n.*
subdivide (sŭbdivīd'), *v.t., v.i.* to divide again or into smaller parts. **subdivisible** (-viz'-), *a.* **subdivision** (-vizh'ən), *n.*
subdominant (sŭbdom'inənt), *n., a.* (*Mus.*) (pertaining to) the tone next below the dominant, the fourth of the scale.
subdue (səbdū'), *v.t.* to conquer, to reduce to subjection, to vanquish, to overcome; to check, to curb; to tame, to render gentle or mild; to tone down, to soften, to make less glaring. **subduable**, *a.* **subdual**, *n.* **subdued**, *a.* quiet, passive, cowed; toned down, not harsh or glaring.
subedit (sŭb'edit), *v.t.* to edit written or printed material.
subeditor, *n.* an assistant editor; someone who checks copy for printing, esp. on a newspaper.
subequatorial, *a.* characteristic of regions above or below the equator.
suberose, subereous (sūbiə'riəs), **suberic** (-be'-), *a.* of the nature or texture of, pertaining to or derived from cork.
subfusc (sŭb'fŭsk), *a.* dusky, sombre; drab, dingy. *n.* formal academic dress at Oxford Univ. **subfuscous** (-fūs'-), *a.* dusky.
subgelatinous, *a.*
subhead (sŭb'hed), **-heading**, *n.* a heading, often explanatory, beneath the main heading of a book, article etc.
subhuman (sŭbhū'mən), *a.* less than human or that which is normal to humans; pertaining to animals lower than humans.
subimago (sŭbimā'gō), *n.* a stage in the metamorphosis of certain insects.
subintrant (sŭbin'trənt), *a.* characterized by paroxysms that succeed each other so rapidly as to be almost continuous.
subitamente (soobētamen'tā), *adv.* (*Mus.*) suddenly. **subito** (-tō), *adv.* (*Mus.*) suddenly, immediately.
subj., (*abbr.*) subject; subjunctive.
subjacent (sŭbjā'sənt), *a.* underlying; lower in position.
subject[1] (sŭb'jikt), *a.* being under the power, control or authority of another; exposed, liable, prone, disposed (to); dependent, conditional; submissive. *n.* one under the dominion or political rule of a person or state, one owing allegiance to a sovereign, a member of a state as related to the sovereign or government; that which is treated or to be treated in any specified way; the topic under consideration; the theme of discussion or description, or artistic expression or representation; the leading phrase or motif in music; a branch of learning or study; a dead body for dissection etc.; the cause or occasion (for); a person regarded as subject to any specific disease, mental tendency, psychic influence etc.; (*Log.*) that member of a proposition about which something is predicated; the noun or its equivalent about which something is affirmed, the nominative of a sentence; the ego, as distinguished from the object or non-ego, the mind, the conscious self; the substance or substratum to which attributes must be referred. **subject to**, conditional upon (ratification etc.); conditionally upon. **subject-heading**, *n.* a heading in an index, catalogue etc. under which references are given. **subject-matter**, *n.* the object of consideration, discussion etc. **subjection** (-jek'-), *n.* **subjective** (jek'-), *a.* concerned with or proceeding from the consciousness or the mind, as opp. to objective or external things; due to or proceeding from the individual mind, personal; lacking reality, fanciful, imaginary; (*Art*) characterized by the prominence given to the individuality of the author or artist; denoting the case of the subject of a verb, nominative. *n.* the subjective case. **subjectively**, *adv.* **subjectiveness, subjectivity** (-tiv'-), *n.* **subjectivism**, *n.* the doctrine that human knowledge is purely subjective, and therefore relative. **subjectivist** (sŭb'-), *n.*
subject[2] (səbjekt'), *v.t.* to subdue, to reduce to subjection (to); to expose, to make liable; to cause to undergo.
subjoin (sŭbjoin'), *v.t.* to add at the end, to append, to affix. **subjoinder** (-də), *n.*
sub judice (sŭb joo'disi), *a.* under consideration, esp. by a court or judge.
subjugate (sŭb'jəgāt), *v.t.* to subdue, to conquer, to bring into subjection, to enslave. **subjugable**, *a.* **subjugation**, *n.* **subjugator**, *n.*
subjunctive (səbjŭngk'tiv), *a.* denoting the mood of a verb expressing condition, hypothesis or contingency. *n.* the subjunctive mood. **subjunctively**, *adv.*
sublease (sŭb'lēs), *n.* a lease of property by a tenant or lessee. *v.t.* to grant or obtain a sublease of (property). **sublessee** (-ē'), *n.* **sublessor** (-les'ə), *n.*
sublet (sŭblet'), *v.t.* (*pres.p.* **subletting**, *past, p.p.* **sublet**) to let property already rented or held on lease. *n.* a subletting.
sublieutenant *n.* the lowest ranked Commissioned officer in the Royal Navy. **sublieutenancy**, *n.*
sublimate (sŭb'limāt), *v.t.* to convert (a solid substance) by heat directly to vapour without passing through the liquid state (followed by an equivalent return to solidity by cooling); to refine, to purify, to etherealize; to divert by sublimation. *n.* the product of sublimation. **sublimation**, *n.* the result of sublimating; (*Psych.*) the diversion by the subject of certain instinctive impulses, esp. sexual, into altruistic or socially acceptable channels. **sublimatory**, *n., a.*
sublime (səblīm'), *a.* of the most lofty or exalted nature; characterized by grandeur, nobility or majesty; inspiring awe; unparalleled, outstanding. *v.t.* to sublimate; to elevate, to purify; to make sublime. *v.i.* to pass directly from solid to vapour, to be sublimated; to be elevated or purified; to become sublime. **sublimely**, *adv.* **sublimity** (-blim'-), *n.*
subliminal (səblim'inəl), *a.* not reaching the threshold of consciousness, hardly perceived; pertaining to subconsciousness. **subliminal advertising**, *n.* advertising directed to and acting on the unconscious. **subliminally**, *adv.*
sublunary (sŭbloo'nəri), **sublunar**, *a.* situated beneath the

earth and the moon; pertaining to the earth.
submachine gun (sŭbməshēn'), *n.* a light automatic or semiautomatic rapid-firing gun fired from the hip or shoulder.
submarine (sŭbmərēn'), *a.* situated, acting or growing beneath the surface of the sea. *n.* (sŭb'-) a vessel, esp. a warship, that may be submerged. **submariner** (-ma'rinə), *n.* a sailor in a submarine.
submaxilla (sŭbmaksil'ə), *n.* (*pl.* **-llae** (-ē), **-llas**), the lower jaw. **submaxillary** (-mak'-), *a.*
submediant (sŭbmē'diənt), *n.*, *a.* (*Mus.*) (of) the sixth note of the diatonic scale.
submerge (səbmœj'), *v.t.* to put under water etc., to flood; to inundate, to overwhelm. *v.i.* to sink under water etc.
submergence, *n.* **submerse** (-mœs'), *v.t.* to submerge. **submersed**, *a.* being or growing under water. **submersible**, *n.*, *a.* (a vessel) capable of being submersed. **submersion** (-shən), *n.*
submission (-shən), *n.* the act of submitting; the state of being submissive, compliance, obedience, resignation, meekness. **submissive**, *a.* **submissively**, *adv.* **submissiveness**, *n.*
submit (səbmit'), *v.t.* (*past*, *p.p.* **submitted**) to yield or surrender (oneself); to subject to a process, treatment etc.; to present or refer for consideration, decision etc.; to put forward deferentially. *v.i.* to yield, to surrender, to give in; to be submissive. **submitter**, *n.*
subnormal (sŭbnaw'məl), *a.* less than normal, below the normal standard; having lower intelligence than is normal. **subnormality** (-mal'-), *n.*
subnuclear, *a.* referring to particles within or smaller than the nucleus of an atom.
suborbital (sŭbaw'bitəl), *a.* beneath the orbit of the eye; less than a complete orbit of the earth, moon etc.
suborder (sŭb'awdə), *n.* a subdivision of a taxonomic order. **subordinal** (-aw'-), *a.*
subordinate[1] (səbaw'dinət), *a.* inferior in order, rank, importance, power etc.; subject, subservient, subsidiary (to). *n.* a person working under another or inferior in official standing. **subordinate clause**, *n.* a clause that functions as a noun, adjective or adverb rather than as a sentence. **subordinately**, *adv.*
subordinate[2] (səbaw'dināt), *v.t.* to make subordinate; to treat or consider as of secondary importance; to make subject or subservient (to). **subordination**, *n.* **subordinative**, *a.*
suborn (səbawn'), *v.t.* to procure by underhand means, esp. bribery, to commit perjury or other criminal act. **subornation**, *n.* **suborner**, *n.*
subplot (sŭb'plot), *n.* a secondary or subordinate plot in a novel, play etc.
subpoena (səpē'nə), *n.* a writ commanding a person's attendance in a court of justice under a penalty. *v.t.* (*pres.p.* **subpoenaing**, *past*, *p.p.* **subpoenaed** (-nəd)) to serve with such a writ.
subrogation (sŭbrəgā'shən), *n.* the substitution of one person in the place of another with succession to his or her rights to a debt etc. **subrogate** (sŭb'-), *v.t.*
sub rosa (sŭb rō'zə), *adv.* secretly; in confidence.
subroutine (sŭb'rootēn), *n.* a sequence of computer instructions for a particular task that can be used at any point in a program.
subscribe (səbskrīb'), *v.t.* to write (one's name etc.) at the end of a document etc.; to sign (a document, promise etc.); to contribute or pledge to contribute (an annual or other specified sum) to or for a fund, object etc.; to publish by securing subscribers beforehand. *v.i.* to write one's name at the end of a document; to assent or give support (to an opinion etc.); to engage to pay a contribution, to allow one's name to be entered in a list of contributors; to undertake to receive and pay for shares or a periodical, service etc. **subscriber**, *n.* **subscriber trunk dialling**, a telephone dialling system allowing subscribers to dial direct to any number in the system. **subscript** (sŭb'skript), *n.*, *a.* (a character) written or printed underneath another or below the base line. **subscription** (-skrip'-), *n.* the act of subscribing; a signature; a contribution to a fund etc.; a membership fee; a raising of money from subscribers; an advance payment for several issues of a periodical; an application to purchase shares.
subsection (sŭb'seksh ən), *n.* a subdivision of a section.
subsequent (sŭb'sikwənt), *a.* coming immediately after in time or order; following, succeeding, posterior (to). **subsequence**, *n.* **subsequently**, *adv.*
subserve (səbsœv'), *v.t.* to serve as a means or instrument in promoting (an end etc.). **subservient**, *a.* useful as an instrument or means; obsequious, servile. **subservience**, **-ency**, *n.* **subserviently**, *adv.*
subset (sŭb'set), *n.* (*esp. Math.*) a set contained within a larger set.
subsextuple (sŭbseks'tŭpl), *a.* in the ratio of one to six.
subshrub (sŭb'shrŭb), *n.* a small woody plant with nonwoody tips.
subside (səbsīd'), *v.i.* to sink, to fall in level; to settle; to sink in, to collapse; to settle down, to abate, to become tranquil. **subsidence** (-sī'-, sŭb'si-), *n.*
subsidiary (səbsid'iəri), *a.* aiding, auxiliary, supplemental; subordinate or secondary in importance. *n.* a subsidiary person or thing, an auxiliary, an accessory; a company whose shares are mostly owned by another. **subsidiarily**, *adv.*
subsidy (sŭb'sidi), *n.* (*Hist.*) pecuniary aid granted by parliament to the sovereign for purposes of state, a tax to defray special expenses; a sum paid by one government to another, usu. to meet the expenses of a war; a contribution by the state, a public corporation etc., to a commercial or charitable undertaking of benefit to the public; financial aid. **subsidize, -ise**, *v.t.*
subsist (səbsist'), *v.i.* to exist, to remain in existence; to live, to have means of living, to find sustenance, to be sustained (on). **subsistence**, *n.* the state or means of subsisting; the minimum required to support life. **subsistence farming**, *n.* farming in which most of the yield is consumed by the farmer with little over for sale. **subsistence level**, *n.* an income or living standard that provides the basic necessities for life. **subsistent**, *a.*
subsoil (sŭb'soil), *n.* the stratum of earth immediately below the surface-soil.
subsonic (sŭbson'ik), *a.* pertaining to, using or travelling at speeds below that of sound.
substance (sŭb'stəns), *n.* that of which a thing consists; matter, material, as opp. to form; matter of a definite or identifiable chemical composition; the essence, the essential part, pith, gist or main purport; that which is real, solidity, firmness, solid foundation; material possessions, property, wealth, resources; (*Phil.*) the permanent substratum in which qualities and accidents are conceived to inhere, the self-existent ground of attributes and phenomena.
substandard (sŭbstan'dəd), *a.* below an accepted or acceptable standard.
substantial (səbstan'shəl), *a.* having physical substance; real, actually existing, not illusory; solid, stout, strongly constructed, durable; possessed of substance, having sufficient means, well-to-do, financially sound; of considerable importance, value, extent, amount etc.; material, practical, virtual. *n.* (*usu. pl.*) the essential parts, reality. **substantialism**, *n.* the doctrine that there are substantial realities underlying phenomena. **substantialist**, *n.* **substantiality** (-shial'-), *n.* **substantialize, -ise**, *v.t.*, *v.i.* **substantially**, *adv.*
substantiate (səbstan'shiāt), *v.t.* to make real or actual; to establish, to prove, to make good (a statement etc.). **substantiation**, *n.*

substantive (sŭb'stəntiv), *a.* expressing real existence; having substance or reality, having or pertaining to the essence or substance of anything; independently existent, not merely implied, inferential or subsidiary; denoting or functioning as a noun; of a dye or dyeing process, not requiring a mordant; (*Mil.*) permanent (rank). *n.* a noun or part of a sentence used as a noun. **substantival** (-tī'-), *a.* **substantivally, substantively,** *adv.*
substation (sŭb'stāshən), *n.* a subsidiary station, esp. one in which electric current from a generating station is modified before distribution.
substitute (sŭb'stitūt), *n.* a person or thing put in the place of or serving for another. *v.t.* to put or use in the place of another person or thing; to replace (an atom or group in a molecule) with another; to introduce a substitute for. *v.i.* to act as a substitute. **substitutable,** *a.* **substitution,** *n.* **substitutive,** *a.*
substratum (sŭb'strahtəm), *n.* (*pl.* **-ta** (-tə)) that which underlies anything; a layer or stratum lying underneath; the subsoil; the ground or basis (of phenomena etc.), foundation. **substrate** (-strāt), *n.* a substratum; the substance on which an enzyme acts; a base on which something lives or is formed.
substructure (sŭb'strŭkchə), *n.* an under-structure or foundation. **substructural** (-strŭk'-), *a.*
substyle (sŭb'stīl), *n.* the line on which the style or gnomon of a dial stands. **substylar,** *a.*
subsume (səbsūm'), *v.t.* to include under a more general class or category. **subsumable,** *a.* **subsumption** (-sŭmp'-), *n.*
subtangent (sŭbtan'jənt), *n.* the portion of the axis of a curve intercepted between an ordinate and a tangent both drawn from the same point.
subtemperate (sŭbtem'pərət), *a.* pertaining to slightly colder than temperate regions.
subtenant (sŭb'tenənt), *n.* a tenant holding property from one who is also a tenant. **subtenancy,** *n.*
subtend (səbtend'), *v.t.* to extend under or be opposite to (of a chord relatively to an arc, or the side of a triangle to an angle); (*Bot.*) to be lower than and enclose.
subterfuge (sŭb'təfūj), *n.* a deception, prevarication etc. employed to avoid an inference, censure etc., or to evade or conceal something.
subterposition (sŭbtəpəzish'ən), *n.* position under something else, esp. of strata.
subterranean (sŭbtərā'niən), **-aneous,** *a.* underground; hidden, concealed. **subterraneously,** *adv.* **subterrestrial** (-res'tri-), *a.*
subtitle (sŭb'tītl), *n.* an additional or subsidiary title to a book etc.; a half-title, usu. placed before the title page in books; a printed explanatory caption to a silent film or a printed translation of the dialogue in a foreign film. *v.t.* to provide a subtitle for.
subtle (sŭt'l), *a.* rarefied, attenuated, delicate, hard to seize, elusive; difficult to comprehend, not obvious, abstruse; making fine distinctions, acute, discerning; ingenious, skilful, clever; artful, cunning, crafty, insidious. **subtleness,** *n.* **subtlety** (-ti), *n.* **subtly** (-li), *adv.*
subtonic (sŭbton'ik), *n.* (*Mus.*) the note next below the tonic.
subtract (səbtrakt'), *v.t.* to take away (a part, quantity etc.) from the rest, to deduct. **subtracter,** *n.* **subtraction,** *n.* a subtracting; an arithmetical operation to find the difference in amount between two numbers. **subtractive,** *a.* **subtrahend** (sŭb'trəhend), *n.* the number or quantity to be subtracted from another.
subtropical (sŭbtrop'ikəl), *a.* characterized by features common to both the temperate and tropical zones; pertaining to the regions near the tropics. **subtropics,** *n.pl.*
subulate (sŭb'ūlət), *a.* (*chiefly Bot.*) awl-shaped.
suburb (sŭb'œb), *n.* an outlying part of a city or town; (*pl.*) the residential outskirts of a city or large town. **suburban** (-œ'-), *a.* pertaining to a suburb or the suburbs; (*fig.*) descriptive of an outlook on life limited by certain narrow conventions. **suburbanite** (-nīt), *n.* one who lives in the suburbs. **suburbanize, -ise,** *v.t.* **suburbia** (-biə), *n.* (the inhabitants of) residential suburbs collectively; the lifestyle, culture etc. held to be characteristic of these.
subvention (səbven'shən), *n.* a grant in aid, a subsidy.
subvert (səbvœt'), *v.t.* to overthrow, to destroy, to overturn; to corrupt, to pervert. **subversion** (-shən), *n.* the act of subverting a legal government, authority etc.; destruction, ruin. **subversive,** *n.,* *a.* **subversively,** *adv.* **subversiveness,** *n.* **subverter,** *n.*
subway (sŭb'wā), *n.* an underground passage, tunnel, conduit etc.; (*N Am.*) an underground railway.
subzero (sŭbziə'rō), *a.* below zero (degrees Celsius).
succedaneum (sŭksidā'niəm), *n.* (*pl.* **-nea** (-niə)) that which is used instead of something else, a substitute (esp. a medical drug). **succedaneous,** *a.*
succeed (səksēd'), *v.t.* to follow, to come after (in time or order), to be subsequent to; to take the place previously occupied by, to be heir or successor to. *v.i.* to follow in time or order, to be subsequent (to); to be the heir or successor (to an office, estate etc.); to be successful, to attain a desired object, to end well or prosperously. **succeeder,** *n.* **succeedingly,** *adv.*
success (səkses'), *n.* the act of succeeding, favourable result, attainment of what is desired or intended; attainment of worldly prosperity; someone or something that is successful. **successful,** *a.* **successfully,** *adv.* **successfulness,** *n.*
succession (səksesh'ən), *n.* a following in order; a series of things following in order; the act or right of succeeding to an office or inheritance; the order in which persons so succeed; the line of persons so succeeding; the order of descent in the development of species. **successional,** *a.* **successionally,** *adv.* **successor,** *n.* one who or that which follows or succeeds another.
successive (səkses'iv), *a.* following in order or uninterrupted succession, consecutive. **successively,** *adv.* **successiveness,** *n.*
succinct (səksingkt'), *a.* compressed into few words, brief, concise. **succinctly,** *adv.* **succinctness,** *n.*
succotash (sŭk'ətash), *n.* (*N Am.*) a dish composed of green maize and beans cooked together.
succour, (*esp. N Am.*) **succor** (sŭk'ə), *v.t.* to come to the aid of; to help or relieve in difficulty or distress. *n.* aid in time of difficulty or distress.
succuba (sŭk'ūbə), **-bus** (-bəs), *n.* (*pl.* **-bae** (-bē), **-bi** (-bī)) a demon believed to assume the shape of a woman and have sexual intercourse with men in their sleep.
succulent (sŭk'ūlənt), *a.* juicy; (of a plant, stem etc., thick and fleshy. *n.* a plant which survives in arid conditions by storing moisture in its tissues. **succulence,** *n.* **succulently,** *adv.*
succumb (səkŭm'), *v.i.* to cease to resist etc., to give way; to yield, to submit (to force etc.); to die.
succursal (səkœ'səl), *a.* auxiliary (used esp. of an ecclesiastical building, as a chapel of ease).
such (sŭch), *a.* of that, or the same, or the like kind or degree (as); of the kind or degree mentioned or implied; being the same in quality, degree etc.; so great, intense etc. (*usu.* as *or* that). *adv.* so (in *such a nice day* etc.). *pron.* such a person, persons or things (as); suchlike. **such and such,** not known or specified, some. **such as,** of a kind like; for example. **suchlike,** *a.* of such a kind. *pron.* things of that sort.
suck (sŭk), *v.t.* to draw (milk etc.) into the mouth by the action of the lips or lips and lungs; to imbibe, to drink in, to absorb (up or in), to acquire, to gain; to engulf, to draw (in); to draw liquid from with or as with the mouth; to dissolve or eat thus; to hold and take in the mouth

suckle   821   **suit**

with a sucking action. *v.i.* to draw liquid etc. in by suction; to draw milk, nourishment etc. in thus; to make the sound of sucking. *n.* an act or spell of sucking, suction; force of suction; a small draught or drink. **to suck up (to)**, to act in an obsequious manner (towards), toady. **sucker**, *n.* one who or that which sucks; a newly-born animal not yet weaned; a fish that sucks in food or has a suctorial mouth, esp. one of the N American Catostomidae; the piston of a suction-pump; a pipe or tube through which anything is drawn by suction; (*Biol.*) an organ, such as an acetabulum, acting on the same principle, a suctorial organ; a shoot from a root or a subterranean part of a stem; (*coll.*) a ready dupe, a gullible person; (*coll.*) a person who is very fond of or unable to resist a specified thing. *v.t.* to strip suckers from (a plant). *v.i.* to send out suckers.
suckle (sŭk'l), *v.t.* to give milk from the breast or udder to. **suckling** (-ling), *n.* a child or animal not yet weaned.
sucre (soo'krä), *n.* the monetary unit of Ecuador.
sucrose (sū'krōs), *n.* sugar as obtained from sugar cane or sugar beet.
suction (sŭk'shən), *n.* the act or process of sucking; the production of a vacuum in a confined space causing fluid to enter, or a body to adhere to something, under atmospheric pressure. **suction-pump**, *n.* pump in which liquid is forced up by atmospheric pressure. **suctorial** (-taw'ri-), *a.* adapted for sucking or for adhering by suction.
sudamina (sūdam'inə), *n.pl.* minute transparent vesicles arising from a disorder of the sweatglands. **sudaminal**, *a.*
Sudanese (soodənēz'), *a.* of or pertaining to the Republic of the Sudan, the region of Central Africa south of Egypt. *n.* (*pl.* **Sudanese**) a native or inhabitant of the Sudan.
sudatorium (-dətaw'riəm), *n.* (*pl.* **-ria** (-iə)) a hot-air bath. **sudatory** (-də-), *a.* exciting perspiration. *n.* a sudatorium.
sudd (sŭd), *n.* a floating mass of vegetation, trees etc. obstructing navigation in the White Nile.
sudden (sŭd'n), *a.* happening unexpectedly, without warning; instantaneous, abrupt, swift, rapid. **on, of a sudden**, suddenly; unexpectedly. **sudden infant death syndrome**, cot death. **sudden death**, *n.* an extended period of play to decide a tie in a game or contest, ending when one side scores. **suddenly**, *adv.* **suddenness**, *n.*
sudoriferous (sūdərif'ərəs), *a.* producing or secreting perspiration. **sudorific** (-rif'-), *a.* causing perspiration. *n.* a sudorific drug.
Sudra (soo'drə), *n.* a member of the lowest of the four great Hindu castes.
suds (sŭdz), *n.pl.* soapy water forming a frothy mass. **sudsy**, *a.*
sue (sū), *v.t.* to prosecute or to pursue a claim (for) by legal process; to entreat, to petition. *v.i.* to take legal proceedings (for); to make entreaty or petition (to or for).
suede (swād), *n.* leather given a nap surface by rubbing.
suet (sū'it), *n.* the hard fat about the kidneys and loins of oxen, sheep etc. **suet pudding**, *n.* a boiled or steamed pudding made with suet. **suety**, *a.*
suffer (sŭf'ə), *v.t.* to experience, to undergo (something painful, disagreeable or unjust); to endure, to sustain, to support (unflinchingly etc.); to tolerate, to put up with; to permit, to allow. *v.i.* to undergo or endure pain, grief, injury, loss etc.; to undergo punishment, esp. to be executed; to experience damage; to be at a disadvantage. **sufferable**, *a.* **sufferably**, *adv.* **sufferance**, *n.* negative consent, toleration, allowance, tacit or passive permission; suffering; endurance, patience, submissiveness. **on sufferance**, merely tolerated. **sufferer**, *n.* **suffering**, *n.* **sufferingly**, *adv.*
suffice (səfīs'), *v.i.* to be enough, to be adequate or sufficient (for or to do etc.). *v.t.* to be enough for, to content, to satisfy. **sufficiency** (-fish'ənsi), *n.* the quality of being sufficient; an adequate supply (of). **sufficient**, *a.* enough, adequate, sufficing (for); self-sufficient, *n.* (*coll.*) enough,

a sufficiency. **sufficiently**, *adv.*
suffix[1] (sŭf'iks), *n.* a letter or syllable appended to the end of a word.
suffix[2] (sŭf'iks, -fiks'), *v.t.* to add as a suffix, to append.
suffocate (sŭf'əkāt), *v.t.* to choke, to kill by stopping respiration; to smother, to stifle; to cause difficulty of respiration to. *v.i.* to be or feel suffocated. **suffocatingly**, *adv.* **suffocation**, *n.*
suffragan (sŭf'rəgən), *a.* assisting (said of a bishop consecrated to assist another bishop or of any bishop in relation to the metropolitan). *n.* a suffragan or auxiliary bishop. **suffraganship**, *n.*
suffrage (sŭf'rij), *n.* a vote in support of an opinion etc., or of a candidate for office; approval, consent; the right to vote, esp. in parliamentary elections; a short intercessory prayer by the congregation, esp. one of the responses in the Litany. **suffragette** (-jet'), *n.* a female agitator for women's right to vote. **suffragist**, *n.* an advocate of extension of the right to vote, esp. to women. **suffragism**, *n.*
suffruticose (sŭffroo'tikōs), *a.* having a woody perennial base with nonwoody branches.
suffuse (səfūz'), *v.t.* of a blush, fluid etc., to overspread, as from within. **suffusion** (-zhən), *n.* **suffusive**, *a.*
sufi (soo'fi), *n.* a Muslim pantheistic philosopher and mystic. **sufic**, *a.* **sufism**, *n.*
sugar (shug'ə), *n.* a sweet, crystalline substance obtained from the expressed juice of various plants, esp. the sugar cane and the sugar beet; any substance resembling sugar, esp. in taste; (*Chem.*) one of various sweet or sweetish soluble carbohydrates, such as glucose, sucrose, lactose etc.; flattering or seductive words, esp. used to mitigate or disguise something distasteful; a term of affection, dear. *v.t.* to sweeten, cover or sprinkle with sugar; to mitigate, disguise or render palatable. **sugar bean**, *n.* a variety of kidney-bean. **sugar beet**, *n.* a variety of common beet from which sugar is extracted. **sugar candy**, *n.* candy. **sugar cane**, *n.* a very tall grass with jointed stems from the juice of which sugar is made. **sugar-coated**, *a.* covered with sugar; made superficially attractive, esp. to hide something less pleasant. **sugar daddy**, *n.* (*coll.*) a well-to-do usu. elderly man who spends money on a young girlfriend. **sugar loaf** LOAF[1]. **sugar maple**, *n.* a N American tree, the sap of which yields sugar. **sugarplum**, *n.* a sweetmeat, esp. boiled sugar formed into a ball etc. **sugar refiner**, *n.* **sugar refinery**, *n.* **sugar tongs**, *n.pl.* a pair of small tongs for lifting lumps of sugar at table. **sugared**, *a.* **sugary**, *a.* containing or resembling sugar; (excessively) sweet-tasting; exaggeratedly charming or flattering. **sugariness**, *n.*
suggest (səjest'), *v.t.* to cause (an idea etc.) to arise in the mind; to propose (a plan, idea etc.) for consideration; to hint, indicate. **suggester**, *n.* **suggestible**, *a.* that may be suggested; readily yielding to (hypnotic) suggestion. **suggestion** (-chən), *n.* the act of suggesting; that which is suggested; a hint, a prompting, an insinuation; insinuation of an idea or impulse to a receptive mind or the mind of a hypnotized person; the spontaneous calling up of an associated idea in the mind. **suggestive**, *a.* containing or conveying (a) suggestion; tending to suggest thoughts etc., esp. of a prurient nature. **suggestively**, *adv.* **suggestiveness**, *n.*
suicide (sū'isīd), *n.* the act of intentionally taking one's own life; any self-inflicted action of a disastrous nature; a person who takes his or her own life intentionally. **suicide pact**, *n.* an agreement between people to commit suicide at the same time. **suicidal** (-sī'-), *a.* **suicidally**, *adv.*
sui generis (sū'ī jen'əris, sū'i), *a.* unique, of its own kind.
suint (sū'int, swint), *n.* the natural grease of wool.
suit (soot, sūt), *n.* the act of suing, petition, request; courtship; a legal prosecution or action for the recovery of a right etc.; one of the four sets in a pack of cards; those

cards in a hand belonging to one of these; a set of outer clothes (now usu. jacket and trousers or a skirt), esp. when made of the same cloth; a set (of sails or other articles used together). *v.t.* to adapt, to accommodate, to make fitting (to); to satisfy, to please, to meet the desires etc. of; to agree with, to befit, to be appropriate to. *v.i.* to agree, to accord, to correspond (with); to be convenient. **to follow suit**, (*Cards*) to play a card of the suit led; (*fig.*) to follow an example. **suitcase**, *n.* a small travelling case. **suitable**, *a.* suited, fitting, convenient, proper, becoming. **suitability** (-bil'-), **suitableness**, *n.* **suitably**, *adv.* **suiting**, *n.* cloth for suits.
suite (swēt), *n.* a company, a retinue; a set (of connecting rooms, matching furniture etc.); a series of instrumental compositions, orig. of dance-tunes.
suitor (soo'tə, sū'-), *n.* a petitioner, an applicant; a wooer, a lover; a party to a lawsuit.
suivez (swē'vä), *v.i.* (*Mus.*) follow (direction to the accompanist to adapt his or her time to the soloist).
sukiyaki (sookiyak'i), *n.* a Japanese dish of thin slices of meat and vegetables cooked together with soy sauce, saké etc.
sulcate (sŭl'kāt), *a.* having longitudinal furrows, grooves or channels. **sulcus** (-kəs), *n.* (*pl.* **-ci** (-sī)) a groove, furrow; a furrow separating convolutions of the brain.
sulf(o)-, sulfate, sulfur etc. SULPH(O)-, SULPHATE, SULPHUR.
sulk (sŭlk), *v.i.* to be sulky. *n.* (*often pl.*) a fit of sulkiness. **sulky**, *a.* sullen, morose, ill-humoured, resentful. *n.* a light, two-wheeled vehicle for a single person. **sulkily**, *adv.* **sulkiness**, *n.*
sullage (sŭl'ij), *n.* filth, refuse; sewage; silt.
sullen (sŭl'ən), *a.* persistently ill-humoured, morose, sour-tempered, cross; dismal, forbidding, unpropitious, baleful. *n.pl.* (*dated*) a fit of sullenness, the sulks. **sullenly**, *adv.* **sullenness**, *n.*
sully (sŭl'i), *v.t.* to soil, to tarnish; to defile, to disgrace. *v.i.* to be soiled or tarnished.
sulph- SULPH(O)-.
sulphadiazine (sŭlfədī'əzin), *n.* a sulpha drug used to treat pneumonia and meningitis.
sulpha drugs (sŭl'fə), *n.pl.* a group of sulphonamide drugs with a powerful antibacterial action.
sulphanilamide (sŭlfənil'əmid), *n.* a drug formerly administered for combating streptococcal and other bacterial diseases.
sulphate (sŭl'fāt), *n.* a salt of sulphuric acid. *v.t.* to treat with a sulphate or cause to turn into a sulphate; to cause the formation of lead sulphate on the plates of an accumulator. **sulphation**, *n.*
sulphide (sŭl'fīd), *n.* a compound of sulphur, with an element or radical.
sulphite (sŭl'fīt), *n.* a salt of sulphurous acid. **sulphitic**, *a.*
sulph(o)-, (*esp. N Am.*) **sulf(o)-**, *comb. form.* sulphur.
sulphonamide (sŭlfon'əmid), *n.* an amide of a sulphonic acid; a sulpha drug.
sulphonic acid (sŭlfon'ik), *n.* any of a class of strong organic acids used in making dyes, drugs and detergents.
sulphur, (*esp. N Am.*) **sulfur** (sŭl'fə), *n.* a pale-yellow nonmetallic element, at. no. 16; chem. symbol S, insoluble in water, occurring in crystalline or amorphous forms, used in the manufacture of chemicals, matches etc., one of various pale-yellow butterflies. *a.* of the colour of sulphur, pale-yellow. **sulphur bottom (whale)**, *n.* the blue whale, *Sibbaldus musculus.* **sulphur dioxide**, *n.* a pungent gas used industrially and as a bleach and food preservative, that is a major source of air pollution. **sulphur ore**, *n.* iron pyrites. **sulphur spring**, *n.* a spring of water impregnated with sulphur or sulphide etc. **sulphureous** (-fū'ri-), *a.* consisting of or having the qualities of sulphur; sulphur-coloured. **sulphureously**, *adv.* **sulphureousness**, *n.* **sulphuric** (-fū'-), *a.* derived from or containing sulphur, esp. in its highest valency. **sulphuric acid**, *n.* a corrosive oily liquid acid, oil of vitriol. **sulphurize, -ise** (-fū-), *v.t.* to impregnate with or subject to the action of sulphur, esp. in bleaching. **sulphurization**, *n.* **sulphurous** (-fū-), *a.* containing sulphur in its lower valency; sulphureous.
sultan (sŭl'tən), *n.* a Muslim sovereign, esp. a former ruler of Turkey; a bird of the water-hen family with splendid blue and purple plumage; a white-crested variety of domestic fowl, orig. from Turkey. **sultana** (-tah'nə), *n.* the wife, mother or daughter of a sultan; the mistress of a king, prince etc.; a yellow, seedless raisin grown esp. in the Mediterranean; an American purple sultan-bird. **sultanate** (-nət), *n.* **sultaness** (-nis), *n. fem.*
sultry (sŭl'tri), *a.* very hot, close and heavy, oppressive; passionate, sensual. **sultrily**, *adv.* **sultriness**, *n.*
sum (sŭm), *n.* the aggregate of two or more numbers, magnitudes, quantities or particulars, the total; substance, essence, summary; a particular amount of money; an arithmetical problem or the process of working it out. *v.t.* (*past, p.p.* **summed**) to add, collect or combine into one total or whole. **to sum up**, to put in a few words, to condense; to recapitulate; to form a rapid opinion or estimate of.
sumach, sumac (soo'mak, shoo'-), *n.* a tree or shrub of the genus *Rhus*, the dried and powdered leaves of which are used in tanning, dyeing etc.; a preparation of the dried leaves.
Sumerian (səmiə'riən), *n., a.* (a native or the language) of Sumer, an ancient region of SW Asia.
summary (sŭm'əri), *a.* condensed into narrow compass or few words, abridged, concise, compendious; done briefly or unceremoniously. *n.* an abridged or condensed statement, an epitome. **summary offence**, *n.* an offence tried in a magistrate's court. **summarily**, *adv.* **summariness**, *n.* **summarize, -ise**, *v.t.* to make or be a summary of.
summation (səmā'shən), *n.* the act or process of making a sum, addition; a summing-up; a summary. **summational**, *a.*
summer (sŭm'ə), *n.* that season of the year when the sun shines most directly upon a region, the warmest season of the year; (*pl.*) years of age. *a.* pertaining to or used in summer. *v.i.* to pass the summer. *v.t.* to feed or keep (cattle etc.) during the summer. **summerhouse**, *n.* a light building in a garden, for shade etc. in summer. **summer lightning**, *n.* sheet lightning seen too far off for the thunder to be heard. **summer pudding**, *n.* a pudding of soft fruit in a bread casing. **summer school**, *n.* a course of study held during the summer vacation. **summertime**, *n.* the season of summer. **summer time**, *n.* the official time of one hour in advance of Greenwich mean time that comes into force between stated dates in the summer in Britain.
summersault (sŭm'əsawlt), SOMERSAULT.
summit (sŭm'it), *n.* the highest point, the top, the vertex; utmost elevation, degree etc.; a summit conference. **summit conference**, *n.* a conference between heads of states.
summon (sŭm'ən), *v.t.* to call, cite or command to meet or attend; to order by a summons to appear in court; to call upon to do something; to rouse, call up (courage etc.). **summoner**, *n.*
summons (sŭm'ənz), *n.* (*pl.* **summonses**) the act of summoning; an authoritative call or citation, esp. to appear before a court or judge. *v.t.* to serve with a summons, to summon.
sumo (soo'mō), *n.* traditional Japanese wrestling in which a contestant attempts to force his opponent out of the designated area or to touch the ground with a part of the body other than the feet.
sump (sŭmp), *n.* a well in the floor of a mine, to collect water for pumping; a receptacle for lubricating oil in the crank-case of an internal-combustion engine; a pit to collect metal at its first fusion.

sumptuary (sŭmp′tūəri), *a.* pertaining to or regulating expenditure. **sumptuary law, edict,** *n.* one restraining private excess in dress, luxury etc.

sumptuous (sŭmp′tūəs), *a.* costly, expensive; showing lavish expenditure; splendid, magnificent. **sumptuously,** *adv.* **sumptuousness,** *n.*

sun (sŭn), *n.* the star round which the earth revolves and which gives light and heat to the earth and other planets of the solar system; the light or warmth of this, sunshine, a sunny place; a fixed star that has satellites and is the centre of a system; (*poet.*) a day, a sunrise; anything splendid or luminous, or a chief source of light, honour etc. *v.t.* (*past, p.p.* **sunned**) to expose to the rays of the sun. *v.i.* to sun oneself. **a place in the sun,** a favourable situation, scope for action etc. **to take,** (*sl.*) **shoot the sun,** (*Naut.*) to ascertain the sun's altitude in order to determine the latitude. **under the sun,** in the world, on earth. **sunbath,** *n.* exposure of the body to the sun or a sunlamp; insolation. **sunbathe,** *v.i.* to expose the body to the sun. **sunbather,** *n.* **sunbeam,** *n.* a ray of sunlight. **sunbed,** *n.* an array of ultraviolet-emitting light tubes under which one lies to tan the skin; a portable folding bed used for sunbathing. **sunbird,** *n.* any of the Nectariniidae, small birds of brilliant metallic plumage with a striking resemblance to humming-birds. **sunblind,** *n.* a window-shade or awning. **sunblock,** *n.* a cream, lotion etc. for the skin that blocks out the sun's ultraviolet rays; a sunscreen. **sunbonnet,** *n.* a large bonnet of light material with projections at the front and sides and a pendant at the back. **sunburn,** *n.* tanning or inflammation of the skin due to exposure to the sun. **sunburned, -burnt,** *a.* **sunburst,** *n.* a strong or sudden burst of sunlight; a design or ornament resembling the pattern of radiating sun rays. **sundeck,** *n.* the upper deck of a passenger ship. **sundew,** *n.* a low, hairy, insectivorous bog-plant of the genus *Drosera*. **sundial,** *n.* an instrument for telling the time of day by means of the shadow of a gnomon cast on a dial etc. **sundog,** *n.* a parhelion. **sundown,** *n.* sunset. **sundowner,** *n.* (*Austral. sl.*) a tramp who times his arrival at sundown in order to get a night's lodging. **sundress,** *n.* a lightweight, lowcut, sleeveless dress for wearing in the sun. **sundried,** *a.* dried in the sun. **sun filter,** *n.* a sunscreen. **sunfish,** *n.* a large fish of various species with a body like a sphere truncated behind. **sunflower,** *n.* a plant of the genus *Helianthus,* esp. *H. annuus,* with yellow-rayed flowers. **sunglasses,** *n.pl.* darkened glasses for protecting the eyes from glare. **sun-god,** *n.* the sun worshipped as a deity. **sun-hat, -helmet,** *n.* a light hat with a broad brim etc., to protect from the sun. **sun lamp,** *n.* a lamp that gives out ultraviolet rays for curative purposes or tanning the skin. **sunlight,** *n.* **sunlit,** *a.* **sun lounge,** (*N Am.*) **parlour,** *n.* a room with large windows to admit sunlight. **sunproof,** *a.* **sunrise,** *n.* the first appearance of the sun above the horizon; the time of this. **sunrise industry,** *n.* a high-technology industry (with good prospects for the future). **sunroof,** *n.* a car roof with a panel that slides open. **sunscreen,** *n.* a substance included in suntan preparations to protect the skin by screening out some of the ultraviolet radiation from the sun. **sunset,** *n.* the disappearance of the sun below the horizon; the time of this; the decline (of life etc.). **sunshade,** *n.* a parasol, awning, blind etc. used as a protection against the sun. **sunshine,** *n.* the light of the sun; the space illuminated by this; warmth, brightness, cheerfulness, favourable influence. **sunshine roof,** *n.* a sunroof. **sunshiny,** *a.* **sunspot,** *n.* a dark patch sometimes seen on the surface of the sun. **sunstroke,** *n.* heatstroke due to exposure to the sun in hot weather. **suntan,** *n.* a browning of the skin caused by the formation of pigment induced by exposure to the sun or a sunlamp. **suntanned,** *a.* **suntrap,** *n.* a sheltered sunny place, as in a garden. **sunup,** *n.* (*chiefly N Am.*) sunrise. **sun-worship,** *n.* **sun-worshipper,** *n.* **sunless** (-lis), *a.* **sun-like,** *a.* **sunny,** *a.* bright with or warmed by sunlight; bright, cheerful, cheery, genial; proceeding from the sun. **sunnily,** *adv.* **sunniness,** *n.*

Sun., (*abbr.*) Sunday.

sundae (sŭn′dā), *n.* an ice-cream containing fragments of nuts and various fruits.

Sunday (sŭn′dā. -di), *n.* the 1st day of the week, the Christian Sabbath. **month of Sundays** MONTH. **Sunday best,** *n.* (*coll.*) best clothes for use on Sundays. **Sunday school,** *n.* a school held on Sundays for religious instruction.

sunder (sŭn′də), *v.t.* (*dated*) to part, to separate; to keep apart.

sundew, sundown, sundowner etc. SUN.

sundry (sŭn′dri), *a.* several, various, miscellaneous. *n.pl.* matters, items or miscellaneous articles, too trifling or numerous to specify. **all and sundry** ALL.

sung (sŭng), *past, p.p.* SING.

sunk (sŭngk), **sunken,** *p.p.* SINK.

Sunna (sŭn′ə), *n.* the traditional part of the Muslim law, based on the sayings or acts of Mohammed, accepted as of equal authority to the Koran by orthodox Muslims or the Sunni but rejected by the Shiites. **Sunni** (-i), *n.* **Sunnite** (-īt), *n.*, *a.*

sunny, sunrise, sunshine etc. SUN.

sup (sŭp), *v.t.* (*past, p.p.* **supped**) to take (soup etc.) in successive sips or spoonfuls. *v.i.* to take in liquid or liquid food by sips or spoonfuls; (*dated*) to take supper. *n.* a mouthful (of liquor, soup etc.).

sup., (*abbr.*) superior; superlative; supine; supplement, supplementary; supra (above).

super (soo′pə, sū′-), *a.* (*coll.*) excellent, very good, enjoyable etc.

super- (soopə-, sū-), *pref.* over, above; above in position, on the top of; over in degree or amount, excessive, exceeding, more than, transcending; besides, in addition; of a higher kind.

superable (sū′pərəbl), *a.* that may be overcome, conquerable. **superability, superableness,** *n.* **superably,** *adv.*

superabound (soopərəbownd′), *v.i.* to be more than enough. **superabundance** (-bŭn′-), *n.* **superabundant,** *a.* **superabundantly,** *adv.*

superannuate (soopəran′ūāt), *v.t.* to dismiss, discard, disqualify or incapacitate on account of age; to pension off on account of age. **superannuable,** *a.* **superannuated,** *a.* **superannuation,** *n.* the act of superannuating; the state of being superannuated; a regular payment made by an employee to a pension scheme; the pension paid.

superb (sūpœb′, sū-), *a.* grand, majestic, imposing, magnificent, splendid, stately; excellent, first-rate. **superbly,** *adv.* **superbness,** *n.*

super-calendered (soopəkal′əndəd), *a.* of paper, highly finished.

supercargo (soopəkah′gō), *n.* (*pl.* **-goes**) an officer in a merchant-ship who superintends sales etc. and has charge of the cargo.

supercentre (soo′pəsentə), *n.* a very large or important centre for a particular activity; a very large self-service store, a hypermarket, superstore.

supercharge (soo′pəchahj), *v.t.* to charge or fill greatly or to excess with emotion, vigour etc.; to fit a supercharger to. *n.* one charge borne upon another. **supercharger,** *n.* a mechanism in an internal-combustion engine which provides for the complete filling of the cylinder with explosive material when going at high speed.

superciliary (soopəsil′iəri), *a.* pertaining to or situated above the eyebrows.

supercilious (soopəsil′iəs), *a.* contemptuous, overbearing, haughtily indifferent, arrogant, disdainful. **superciliously,** *adv.* **superciliousness,** *n.*

superclass (soo′pəklahs), *n.* a taxonomic category between a phylum or division and a class.

supercolumnar (soopəkəlum'nə), *a.* having one order of columns placed over another. **supercolumniation** (-ni-), *n.*
supercomputer (soo'pəkəmpūtə), *n.* a very powerful computer capable of over 100 million arithmetic operations per second. **supercomputing,** *n.*
superconductivity (soopəkondəktiv'iti), *n.* the total loss of electrical resistance exhibited by some metals and alloys at very low temperatures. **superconduction,** *n.* **superconduction,** *n.* **superconducting** (soo'-), **superconductive** (-dŭk'-), *a.* **superconductor** (soo'-), *n.*
supercool (soo'pəkool), *v.t.* to cool (a liquid) below its freezing-point without solidification.
super-duper (soo'pə-doo'pə), *a.* (*coll.*) very pleasing, wonderful.
superego (soo'pərēgō), *n.* (*Psych.*) the unconscious inhibitory morality in the mind which criticizes the ego and condemns the unworthy impulses of the id.
superelevation (soopərelivā'shən), *n.* the difference in height between the opposite sides of a curved section of road, railway track etc.
supererogation (soopərerəgā'shən), *n.* performance of more than duty requires. **works of supererogation,** voluntary works, besides, over and above God's Commandments. **supererogate** (-e'rə-), *v.i.* **supererogatory** (-irog'ə-), *a.*
superfamily (soo'pəfamili), *n.* a taxonomic category between a suborder and a family; an analogous category of languages.
superfecundation (soopəfekəndā'shən), *n.* the conception of two embryos from ova produced at one time, by separate acts of sexual intercourse.
superficial (soopəfish'əl), *a.* pertaining to or lying on the surface; not penetrating deep; not deep or profound, shallow. **superficiality** (-shial'-), **superficialness,** *n.* **superficially,** *adv.*
superficies (soopəfish'iēz), *n.* (*pl.* **superficies**) a surface; its area; external appearance or form. [L]
superfine (soopərfin'), *a.* exceedingly fine, surpassing in fineness, of extra quality; extremely fine in size; overrefined. **superfineness,** *n.*
superfluous (supœ'fluəs), *a.* more than is necessary or sufficient, excessive, superabundant, redundant. **superfluity** (soopəfloo'-), *n.* the state of being superfluous; something unnecessary; an excess, a superabundance. **superfluously,** *adv.* **superfluousness,** *n.*
supergiant (soo'pəjiənt), *n.* a very large, very bright star of low density.
superglue (soo'pəgloo), *n.* an adhesive that gives an extremely strong bond on contact.
supergrass (soo'pəgrahs), *n.* (*coll.*) a police informer whose information implicates many people or concerns major criminals or criminal activities.
superheat (soopəhēt'), *v.t.* to heat to excess, to heat (steam) above the boiling-point of water so no condensation occurs; to heat (a liquid) above boiling point without vaporization. **superheater** (soo'-), *n.*
superheterodyne (soopəhet'ərədin), *n.* a radio receiver with a high degree of selectivity.
superhuman, *a.* having supernatural powers; beyond normal human powers or capacity. **superhumanly,** *adv.*
superimpose (soopərimpōz'), *v.t.* to lay upon something else. **superimposition** (-pəzish'-), *n.*
superinduce (soopərindūs'), *v.t.* to bring in as an addition, to superadd. **superinduction** (-dŭk'-), *n.*
superintend (soopərintend'), *v.t.* to have or exercise the management or oversight of, to direct, to control. **superintendence,** *n.* superintending, supervision. **superintendency,** *n.* the office or district of a superintendent. **superintendent,** *n.* one who superintends; a police officer ranking above an inspector.
superior (supiə'riə), *a.* upper, of higher position, class, grade, rank, excellence, degree etc.; better or greater relatively (to); of a quality above the average; of wider application; situated near the top, or in the higher part; (*Bot.*) growing above another, as the calyx or the ovary; above being influenced by or amenable (to); supercilious. *n.* a person superior to one or to others, one's better; the head of a monastery, convent or other religious house. **superior planet,** *n.* one further from the sun than the earth is. **superiority** (-o'ri-), *n.* **superiority complex,** *n.* an inflated opinion of one's worth.
superl., (*abbr.*) superlative.
superlative (supœ'lətiv), *a.* raised to the highest degree, consummate, supreme; (*Gram.*) expressing the highest or utmost degree. *n.* the superlative degree; a word or phrase in the superlative degree; an exaggeration. **superlatively,** *adv.* **superlativeness,** *n.*
superlunar (soopəloo'nə), **-nary,** *a.* above the moon, celestial, not mundane.
superman (soo'pəman), *n.* a hypothetical superior being, esp. one advanced in intellect and morals; a person of outstanding ability or achievements.
supermarket (soo'pəmahkit), *n.* a large, self-service shop where food and domestic goods are sold.
supermundane (soopəmŭn'dān), *a.* above or superior to worldly things.
supernal (supœ'nəl), *a.* of a loftier kind, nature or region; celestial, heavenly, divine, lofty.
supernatant (soopənā'tənt), *a.* floating on the surface. **supernatation,** *n.*
supernatural (soopənach'ərəl), *a.* existing by, due to, or exercising powers above the usual forces of nature, outside the sphere of natural law. **supernaturalism,** *n.* belief in the supernatural. **supernaturalist,** *n.* **supernaturalistic** (-lis'-), *a.* vastly beyond the normal. **supernaturally,** *adv.* **supernaturalness,** *n.*
supernormal, *a.* **supernormality,** *n.* **supernormally,** *a.*
supernova (soo'pənōvə), *n.* (*pl.* **-vae** (-vē), **-vas**) a nova up to 100 million times brighter than the sun, produced by the eruption of a star following its implosion.
supernumerary (soopənū'mərəri), *a.* being in excess of a prescribed or customary number. *n.* a supernumerary person or thing, esp. a person appearing on the stage without a speaking part.
superorder (soo'pərawdə), *n.* a taxonomic category between an order and a subclass or a class. **superordinal,** *a.*
superordinate (soopəraw'dinət), *a.* superior in rank or status; having the relation of superordination. **superordination,** *n.* the ordination of a person to fill an office not yet vacant; (*Log.*) the relation of a universal proposition to a particular proposition that it includes.
superphosphate (soopəfos'fāt), *n.* a phosphate containing the greatest amount of phosphoric acid that can combine with the base; a mixture of phosphates used as a fertilizer.
superpose (soopəpōz'), *v.t.* to lay over or upon something. **superposable,** *a.* **superposition,** *n.*
superpower (soo'pəpowə), *n.* a very powerful nation, such as the US.
supersaturated (soopəsat'ūrātid), *a.* containing more material than a saturated solution or vapour. **supersaturate,** *v.t.* **supersaturation,** *n.*
superscribe (soo'pəskrīb), *v.t.* to write on the top or outside of something or above; to write a name, inscription, address etc. on the outside or top of. **superscript** (-skript), *a.* written at the top or outside; set above the line, superior. *n.* a superior character. **superscription,** *n.*
supersede (soopəsēd'), *v.t.* to put a person or thing in the place of, to set aside, to annul; to take the place of, to displace, to supplant. **supersedence, supersedure** (-dyə), **supersession** (-sesh'ən), *n.*
supersonic (soopəson'ik), *a.* pertaining to sound waves with such a high frequency that they are inaudible; above

the speed of sound; travelling at or using such speeds. **supersonically**, *adv.*

superstar (soo'pəstah), *n.* a very popular film, music, sports etc. star.

superstition (soopəstish'ən), *n.* credulity regarding the supernatural, the occult or the mysterious; ignorant or unreasoning dread of the unknown; a religion, particular belief or practice originating in this, esp. a belief in omens, charms etc. **superstitious**, *a.* **superstitiously**, *adv.* **superstitiousness**, *n.*

superstore (soo'pəstaw), *n.* a very large supermarket; a very large store selling goods other than food.

superstratum (soopəstrah'təm), *n.* (*pl.* **-ta** (-tə)) a stratum resting on another.

superstructure (soo'pəstrŭkchə), *n.* the part of a building above the ground; an upper part of a structure; a concept etc. based on another. **superstructural**, *a.*

supertanker (soo'pətangkə), *n.* a very large tanker ship.

supertax (soo'pətaks), *n.* a tax in addition to the basic income tax, levied on incomes above a certain level.

supertonic (soopəton'ik), *n.* the note next above the tonic in the diatonic scale.

supervene (soopəvēn'), *v.i.* to come or happen as something extraneous or additional **supervenient**, *a.* **supervention** (-ven'-), *n.*

supervise (soo'pəviz), *v.t.* to have oversight of, to oversee, to superintend. **supervision** (-vizh'ən), *n.* **supervisor**, *n.* **supervisory**, *a.*

supinate (soo'pināt, sū'-), *v.t.* to turn the palm of (the hand) upwards or forwards. **supinator**, *n.* either of two muscles which do this. **supination**, *n.* the placing or holding of the palm of the hand upwards or forwards.

supine (soo'pīn, sū'-), *a.* lying on the back with the face upwards; negligent, indolent, listless, careless. *n.* a Latin verbal noun formed from the p.p. stem and ending in *-um* (1st supine) or *-u* (2nd supine). **supinely**, *adv.* **supineness**, *n.*

supp., suppl., (*abbr.*) supplement(ary).

supper (sŭp'ə), *n.* the last meal of the day, esp. a light one; an evening social affair including supper. **the Lord's Supper**, Holy Communion. **supperless** (-lis), *a.*

supplant (səplahnt'), *v.t.* to take the place of or oust, esp. by craft or treachery. **supplanter**, *n.*

supple (sŭp'l), *a.* pliant, flexible, easily bent; lithe, able to move and bend easily; yielding, compliant, soft, submissive, obsequious, servile. **supplejack**, *n.* a tough climbing shrub, from which walking-sticks are made. **suppleness**, *n.* **supply** (-li), *adv.*

supplement[1] (sŭp'limənt), *n.* an addition, esp. one that supplies a deficiency; an addition or update to a book, newspaper or periodical; the angle that added to another will make the sum two right angles. **supplementary** (-men'-), *a.* **supplementary benefit**, *n.* a weekly grant of money paid by the state to people whose income falls below a certain minimum level.

supplement[2] (sŭpliment'), *v.t.* to make additions to; to complete by additions. **supplementation**, *n.* [as prec.]

suppliant (sŭp'liənt), *a.* entreating, supplicating; expressing entreaty or supplication. *n.* a humble petitioner. **suppliance**, *n.* **suppliantly**, *adv.*

supplicate (sŭp'likāt), *v.t.* to beg or ask for earnestly and humbly; to address in earnest prayer; to beg humbly (to grant etc.). *v.i.* to petition earnestly, to beseech. **supplicant**, *a.* suppliant. *n.* a suppliant. **supplication**, *n.* **supplicatory**, *a.*

supply[1] SUPPLE.

supply[2] (səpli'), *v.t.* to furnish with what is wanted, to provide (with); to furnish, to provide, to satisfy; to serve instead of; to fill (the place of), to make up for (a deficiency etc.). *n.* the act of supplying things needed; that which is supplied; a sufficiency of things required or available for use; (*often pl.*) necessary stores or provisions; the quantity of goods or services offered for sale at a particular time; (*pl.*) a grant of money by Parliament to meet the expenses of government, an allowance; one who supplies a position temporarily, a substitute. **supply and demand** DEMAND. **supplier**, *n.*

support (səpawt'), *v.t.* to bear the weight of, to hold up, to sustain; to keep from yielding or giving way, to give strength or endurance to; to furnish with necessaries, to provide for; to give assistance to; to advocate, to defend, to back up, to second; to promote, to encourage; to bear out, to substantiate, to corroborate; to bear; to endure, to put up with; to keep up, to be able to carry on; to maintain; to act as, to represent (a character etc.); to play a secondary role to (the main character) in a film or play; to accompany (a pop group, feature film etc.) in a subordinate role. *n.* the act of supporting or the state of being supported; one who or that which supports; aid, countenance, assistance; subsistence, livelihood. **supportable**, *a.* **supportableness**, *n.* **supportably**, *adv.* **supporter**, *n.* one who or that which supports or maintains. **supporting**, *a.* giving support; playing or having a secondary or subordinate role. **supportive**, *a.* providing support, esp. moral or emotional encouragement.

suppose (səpōz'), *v.t.* to lay down without proof, to assume by way of argument or illustration; to imagine, to believe; to take to be the case, to accept as probable, to surmise; to believe (to exist); to involve or require as a condition, to imply; (*usu. pass.*) to require or expect, to be obliged to. **supposable**, *a.* **supposed** (-pōzd', -pō'zid), *a.* believed to be so. **supposedly** (-zid-), *adv.* **supposer**, *n.* **supposition** (sŭpəzish'-), *n.* **suppositional**, *a.* **suppositionally**, *adv.*

supposititious (səpozitish'əs), **suppositious**, *a.* substituted for something else, not genuine, spurious. **supposititiously**, *adv.* **supposititiousness**, *n.*

suppositive (səpoz'itiv), *a.* including or implying supposition. *n.* a conjunction implying supposition. **suppositively**, *adv.*

suppository (səpoz'itəri), *n.* a medicinal body introduced into an internal passage, as the vagina or rectum, and left to dissolve.

suppress (səpres'), *v.t.* to put down, to overpower, to subdue, to quell; to keep in or back, to withhold, to stifle, to repress; to keep back from disclosure or circulation, to conceal. **suppressant**, *n.* **suppressible**, *a.* **suppression** (-shən), *n.* **suppressionist**, *n.* **suppressive**, *a.* **suppressor**, *n.* one who suppresses; a device for preventing electrical interference in a circuit.

suppurate (sŭp'ūrāt), *v.i.* to generate pus, to fester. **suppuration**, *n.* **suppurative**, *a.*

supra- (soopra-, sū-), *pref.*, above, over, super-; beyond.

supraciliary SUPERCILIARY.

supraclavicular (sooprəklavik'ūlə), *a.* situated above the clavicle.

supradorsal (sooprədaw'sl), *a.* on the back; above the dorsal surface.

supramaxillary (soopromak'silori), *a.* of or pertaining to the upper jaw. *n.* the upper maxillary bone.

supramundane (soopromūn'dān), *a.* above the world.

supranational (soopronash'ənəl), *a.* overriding national sovereignty.

supraorbital (sooproaw'bitəl), *a.* being above the eyesocket.

supraposition SUPERPOSITION.

suprarenal (soopərē'nəl), *a.* situated above the kidneys. **suprarenal glands**, *n.pl.* the adrenal glands.

suprascapular (sooprəskap'ūlə), **-lary**, *a.* situated above the shoulder-blade.

supreme (suprēm', sū-), *a.* highest in authority or power, highest in degree or importance, utmost, extreme, greatest possible; last, final. **Supreme Being**, *n.* God. **Supreme Court**, *n.* the highest judicial court in a state. **Supreme**

Court of Judicature JUDICATURE. **supremacist** (-prem'-), *n.* one who believes or promotes the supremacy of a particular group. **supremacy** (-prem'-), *n.* the quality or state of being supreme; the highest authority or power. **supremely**, *adv.*
supremo (suprē'mō, sū-), *n.* (*pl.* **-mos**) (*coll.*) a supreme leader or head.
supt., (*abbr.*) superintendent.
sur-, *pref.* super-, as in *surcharge, surface, surfeit.*
sura (soo'rə), *n.* a chapter of the Koran.
surah (sū'rə), *n.* a soft, twilled, silk material.
sural (sū'rəl), *a.* pertaining to the calf of the leg.
surat (surat'), *n.* coarse, short cotton grown near Surat, India; cloth made from this.
surbase (sœ'bās), *n.* the cornice or moulding at the top of a pedestal or base. **surbased**, *a.*
surcease (sœsēs'), *n.* (*dated*) cessation. *v.i., v.t.* to cease, desist (from).
surcharge (sœ'chahj), *v.t.* to overload, to overburden, to overfill; to put an extra charge on, to overcharge; to show an omission in (an account) for which credit should be allowed; to impose payment of (a sum) or on (a person) for amounts in official accounts disallowed by an auditor; to overprint (as a stamp) with a surcharge. *n.* an excessive load, burden or charge; an extra charge or cost; an overcharge; an amount surcharged on official accounts; another valuation or other matter printed on a postage- or revenue-stamp; a stamp so treated; an additional charge imposed as a penalty for false returns of income or other taxable property.
surcingle (sœ'sing-gl), *n.* a belt or girth put round the body of a horse etc., for holding a saddle or blanket on its back; the girdle of a cassock. *v.t.* to put a surcingle on; to fasten with this.
surcoat (sœ'kōt), *n.* an outer coat, esp. a loose robe worn over armour; an outer jacket worn by women (14th–16th cents.).
surculus (sœ'kūləs), *n.* (*pl.* **-li** (-lī)) a shoot rising from a root-stock, a sucker. **surculose** (-lōs), *a.*
surd (sœd), *a.* not capable of being expressed in rational numbers; uttered with the breath and not with the voice. *n.* (*Math.*) an irrational quantity; a surd consonant, as *p, f, s*, opp. to the vocals *b, v, z*.
sure (shuə, shaw), *a.* certain, confident, undoubting; free from doubts (of); positive, believing, confidently trusting (that); infallible, certain (to); safe, reliable, trustworthy, unfailing; unquestionably true; certain (of finding, gaining etc.). *adv.* (*chiefly N Am.*) surely, certainly; yes. **for sure**, surely, certainly. **sure enough**, in reality, not merely expectation. **to be sure**, (*coll.*) without doubt, certainly, of course. **to make sure**, to make certain, to ascertain; to make secure. **to make sure of**, to consider as certain. **well, I'm sure**, an exclamation of surprise. **sure-fire**, *a.* (*coll.*) bound to succeed, assured. **sure-footed**, *a.* not liable to stumble or fall. **sure thing**, *n.* something certain of success. *int.* (*coll.*) certainly, yes. **surely**, *adv.* securely, safely; certainly (frequently used by way of asseveration or to deprecate doubt); undoubtedly. **sureness**, *n.*
surety (shuə'rəti, shaw'-), *n.* a person undertaking responsibility for payment of a sum, discharge of an engagement or attendance in court by another, a guarantor; a pledge deposited as security against loss or damage or for payment or discharge of an engagement etc.; certainty. **suretyship**, *n.*
surf (sœf), *n.* the swell of the sea breaking on the shore, rocks etc.; the foam of this. *v.i.* to ride on the surf, engage in surfing. **surfboard**, *n.* a long narrow board used in surfing. **surfboat**, *n.* a strong and buoyant boat for use in surf. **surfboatman**, *n.* **surfer**, *n.* one who engages in surfing. **surfie** (-i), *n.* (*Austral., coll.*) one whose life centres round surfing. **surfing**, *n.* the sport of riding on a board on the surf of an incoming wave. **surfy**, *a.*

surface (sœ'fəs), *n.* the exterior part of anything, the outside, the superficies; (*Geom.*) that which has length and breadth but not thickness; that which is apparent at first view or on slight consideration. *v.t.* to put a surface on; to smooth, to polish. *v.i.* to rise to the surface; (*coll. facet.*) to wake up or get out of bed. **surface-to-air**, *a.* pertaining to missiles launched from land to an airborne target. **surface-active**, *a.* capable of lessening the surface-tension of a liquid. **surface mail**, *n.* mail sent by land or sea. **surface-tension**, *n.* the tension of a liquid causing it to act as an elastic enveloping membrane tending to contract to the minimum area, as seen in the bubble, the drop etc. **surface-water**, *n.* water collecting on the surface of the ground. **surfaced**, *a.* **surfacer**, *n.*
surfactant (sœfak'tənt), *n.* a surface-active substance, as a detergent. [*surface-active agent*]
surfeit (sœ'fit), *n.* excess, esp. in eating and drinking; oppression resulting from this, satiety, nausea; an excessive supply or amount. *v.t.* to fill or feed to excess, to overload, to cloy.
surg., (*abbr.*) surgeon, surgery, surgical.
surge (sœj), *v.i.* of waves, to swell, to heave, to move up and down; to well up, to move with a sudden rushing or swelling motion; to rise suddenly. *n.* a large wave, a billow, a swell; a heaving and rolling motion; a sudden increase or rise. **surger**, *n.*
surgeon (sœ'jən), *n.* a medical practitioner treating injuries, deformities and diseases by manual procedure, often involving operations; a practitioner holding the diploma of the Royal College of Surgeons; a medical officer in the army, navy or a military hospital; a surgeon fish. **surgeonfish**, *n.* a sea-fish of the genus *Teuthis*, with lance-like spines at the tail. **surgeon general**, *n.* the chief medical officer in the army or navy; the head of the public health service in the US.
surgery (sœ'jəri), *n.* (the branch of medicine dealing with) the treatment of injuries, deformities or diseases by manual procedure, often operations; the office or consulting-room of a doctor, dentist etc., or its hours of opening; the time during which an MP is available for consultation. **surgical**, *a.* **surgical spirit**, *n.* methylated spirits used for sterilizing, cleaning the skin etc. **surgically**, *adv.*
suricate (sū'rikāt), *n.* a small S African burrowing carnivore *Suricata tetradactyla* allied to the weasel, and often domesticated as a mouser.
surly (sœ'li), *a.* sullen, rude, gruff, uncivil. **surlily**, *adv.* **surliness**, *n.*
surmise (səmīz'), *n.* a supposition on slight evidence, a guess, a conjecture. *v.t.* to guess, to imagine, with but little evidence; to conjecture, to suspect. *v.i.* to conjecture, to guess, to suppose. **surmisable**, *a.* **surmiser**, *n.*
surmount (səmownt'), *v.t.* to overcome, to vanquish, to rise above; to get or climb to the top of and beyond; to overtop, to cap; to surpass. **surmountable**, *a.*
surname (sœ'nām), *n.* a name added to the first or Christian name; a family name; orig. an appellation signifying occupation etc., or a nickname ultimately becoming hereditary. *v.t.* to call by a surname; to give a surname to.
surpass (sœpahs'), *v.t.* to excel, to go beyond in amount, degree etc.; to go beyond the range or capacity of, to transcend. **surpassable**, *a.* **surpassing**, *a.* excellent in an eminent degree. **surpassingly**, *adv.*
surplice (sœ'plis), *n.* a loose, flowing vestment of white linen, with full sleeves, worn by clergy and choristers. **surpliced**, *a.*
surplus (sœ'pləs), *n.* that which remains over, excess beyond what is used or required; the balance in hand after all liabilities are paid.
surprise (səpriz'), *n.* a taking unawares or unprepared; emotion excited by something sudden or unexpected; astonishment; an event exciting this, something unexpected. *v.t.* to come or fall upon suddenly and un-

expectedly, esp. to attack unawares; to strike with astonishment, to be contrary to or different from expectation; (*usu. p.p.*) to shock, to scandalize; to disconcert; to lead or drive unawares (into an act etc.). **surprisal,** *n.* **surprisedly** (-zid-), *adv.* **surpriser,** *n.* **surprising,** *a.* causing surprise. **surprisingly,** *adv.*
surrealism (sərē'əlizm), *n.* an artistic and literary movement of the 20th cent. which aimed at expressing the subconscious activities of the mind by presenting images with the chaotic incoherency of a dream. **surrealist,** *n.*, *a.* **surrealistic** (-lis'-), *a.*
surrebut (sœribüt'), *v.i.* to reply to a defendant's rebutter. **surrebutter,** *n.* the plaintiff's reply to the defendant's rebutter.
surrejoin (sœrijoin'), *v.i.* to reply to a defendant's rejoinder. **surrejoinder** (-də), *n.* the plaintiff's reply to the defendant's rejoinder.
surrender (səren'də), *v.t.* to yield up to the power or control of another; to give up possession of, esp. upon compulsion or demand; to yield (oneself) to any influence, habit, emotion etc. *v.i.* to yield something or to give oneself up into the power of another, esp. to an enemy in war; to give in, to yield, to submit; to appear in court in discharge of bail etc. *n.* the act of surrendering or the state of being surrendered; the voluntary relinquishing of a (life) insurance policy by its holder, usu. in return for a payment (the policy's **surrender value**). **surrenderee** (-ē'), *n.* (*Law*) one to whom an estate is surrendered. **surrenderer,** *n.*
surreptitious (sŭriptish'əs), *a.* done by stealth or fraud; secret, clandestine. **surreptitiously,** *adv.*
surrogate (sŭ'rəgət), *n.* a deputy; a deputy of a bishop or his chancellor appointed to grant marriage licences and probates; a substitute. **surrogate mother,** *n.* a woman who bears a child for a (childless) couple, often after (artificial) insemination or embryo implantation. **surrogacy** (-si), **surrogateship,** *n.* **surrogation,** *n.*
surround (sərownd'), *v.t.* to lie or be situated all round, to encompass, to environ, to encircle, to invest, to enclose; to cause to be surrounded in this way. *n.* an edging, a border; the floor-covering, or staining of floorboards, between the skirting and the carpet. **surroundings,** *n.pl.* things around a person or thing, environment, circumstances.
surtax (sœ'taks), *n.* an additional tax; an additional graduated income tax formerly imposed in the UK in place of the supertax on all incomes above a certain amount. *v.i.* to impose a surtax.
surtitles (sœ'titlz), *n.pl.* a translation of the text of an opera etc. projected on a screen above the stage.
surtout (sœtoo'), *n.* a man's overcoat, esp. one like a frock-coat.
surveillance (səvā'ləns), *n.* observation, close watch, supervision.
survey[1] (səvā'), *v.t.* to look over, to take a general view of; to view with a scrutinizing eye; to examine closely; to examine and ascertain the condition, value etc. of (esp. a building); to determine by accurate observation and measurement the boundaries, extent, position, contours etc. of (a tract of country, coast, estate etc.). *v.i.* to carry out a survey. **surveyable,** *a.* **surveying,** *n.* **surveyor,** *n.* one who surveys, esp. one who measures land; an inspector (of customs, weights and measures etc.). **surveyorship,** *n.*
survey[2] (sœ'vā), *n.* the act or process of surveying; a general view; a careful examination, investigation, inspection or scrutiny; an account based on this; the operation of surveying land etc.; a department carrying this on; a map, plan etc. recording the results of this; (*N Am.*) a district for the collection of customs.
survive (səviv'), *v.t.* to live longer than, to outlive, to outlast (an event, period etc.). *v.i.* to be still alive or in existence.

survival, *n.* the act or condition of surviving; a person, thing, custom, opinion etc. surviving into a new state of things. **survival of the fittest,** the preservation of forms of life that have proved themselves best adapted to their environment, the process or result of natural selection. **survivor,** *n.*
sus[1] (sŭs), *n.* (*sl.*) suspicion of loitering with criminal intent.
sus[2] suss.
susceptible (səsep'tibl), *a.* admitting (of); capable of being influenced or affected, accessible, liable (to); impressionable, sensitive. **susceptibility** (-bil'-), *n.* the condition or quality of being susceptible; (*pl.*) sensitive feelings, sensibilities. **susceptibly,** *adv.* **susceptive,** *a.* readily receiving impressions etc., susceptible; receiving emotional impressions. **susceptiveness, susceptivity** (sŭseptiv'-), **susceptor,** *n.*
sushi (soo'shi), *n.* a Japanese dish of cold rice cakes with a vinegar dressing and garnishes of raw fish etc.
suspect[1] (səspekt'), *v.t.* to imagine to exist, to have an impression of the existence of without proof, to surmise; to be inclined to believe to be guilty but upon slight evidence, to doubt the innocence of; to hold to be uncertain, to doubt, to mistrust. *v.i.* to be suspicious.
suspect[2] (sŭs'pekt), *a.* suspected, under suspicion, suspicious; doubtful, uncertain. *n.* a person suspected of crime etc.
suspend (səspend'), *v.t.* to hang up, to hang from something above; to sustain from falling or sinking; to hold (particles) in a suspension; to render temporarily inoperative or cause to cease for a time, to intermit; to defer; to debar temporarily from a privilege, office etc. **to suspend payment,** to be unable to meet one's financial engagements. **suspended animation,** *n.* temporary ceasing of the body's vital functions. **suspended sentence,** *n.* a prison sentence that is not served unless a further crime is committed. **suspender,** *n.* one who or that which suspends; (*pl.*) attachments to hold up socks or stockings; (*N Am.*) braces. **suspender belt,** *n.* a belt with stocking suspenders attached. **suspendible, suspensible** (-si-), *a.* **suspendibility** (-bil'-), *n.*
suspense (səspens'), *n.* a state of uncertainty, doubt or apprehensive expectation or waiting; (*Law*) a temporary cessation of a right etc. **suspension** (-shən), *n.* the act of suspending; the state of being suspended; a dispersion of solid particles in a fluid; a system of springs etc. that supports the body of a vehicle on the axles. **suspension-bridge,** *n.* a bridge sustained by flexible supports passing over a tower or elevated pier and secured at each extremity. **suspensive,** *a.* having power to suspend; uncertain, doubtful. **suspensively,** *adv.* **suspensor,** *n.* **suspensory,** *n.* a supporting ligament, part etc., esp. the bone or bones by which the lower jaw is suspended from the cranium in vertebrates.
suspicion (səspish'ən), *n.* the act or feeling of one who suspects; belief in the existence of wrong or guilt on inadequate proof, doubt, mistrust; a very slight amount; a trace. **suspicious,** *a.* inclined to suspect; entertaining suspicion; expressing or showing suspicion; exciting or likely to excite suspicion. **suspiciously,** *adv.* **suspiciousness,** *n.*
suss, sus (sŭs), *v.t.* (*past, p.p.* **sussed**) to work out or discover the true facts of. **to suss out,** to investigate, find out about.
sustain (səstān'), *v.t.* to bear the weight of, to hold up, to keep from falling; to bear up against or under; to stand, to undergo without yielding; to experience, to suffer; to nourish, provide sustenance for; to enable to bear something, to keep from failing, to strengthen, to encourage, to keep up; to prolong; to maintain, to uphold; to establish by evidence; to support, to confirm, to bear out, to substantiate. **sustainable,** *a.* **sustainer,** *n.*

sustenance (sūs'tənəns), *n.* that which sustains, the means of support or maintenance; the nourishing element in food; food, subsistence; the act of sustaining.
sustentation (sŭstəntā'shən), *n.* support, maintenance. **sustentator, sustentor** (sŭs'-), *n.*
susurrant (sūsŭ'rənt), **-rous**, *a.* whispering, rustling, murmuring. **susurration** (sūsə-), **susurrus** (sūsŭ'rəs), *n.*
sutler (sŭt'lə), *n.* a person who follows an army and sells provisions, liquor etc. **sutlership**, *n.* **sutlery**, *n.*
sutor (soo'tə, sū'-), *n.* a cobbler. **sutorial** (-taw'ri-), *a.*
sutra (soo'trə), *n.* a rule, a precept, an aphorism; (*pl.*) Brahminical books of rules, doctrine etc.
suttee (sŭ'tē, -tē'), *n.* a Hindu custom by which the widow was burnt on the funeral pyre with her dead husband; a widow so burnt. **sutteeism**, *n.*
suture (soo'chə), *n.* the junction of two parts by their margins as if by sewing, esp. of the bones of the skull; the uniting of two body surfaces, esp. the edges of a wound, by stitching; catgut, silk etc. used in uniting body surfaces; a stitch or seam made in this way. *v.t.* to unite by a suture. **sutural, sutured,** *a.* **suturally,** *adv.* **suturation,** *n.*
suzerain (soo'zərän, -rin), *n.* a feudal lord, a lord paramount; a state having sovereignty or control over another. **suzerainty,** *n.*
svelte (svelt), *a.* esp. of a woman's figure, slender, lissom.
SW, (*abbr.*) southwest, southwestern; short wave.
swab (swob), *n.* a mop for cleaning floors, decks, the bore of a gun etc.; a small piece of cotton-wool or gauze used for removing blood, dressing wounds, taking specimens etc.; (*sl.*) a clumsy, useless person. *v.t.* (*past, p.p.* **swabbed**) to rub, wipe or clean with a swab or mop. **swabber,** *n.*
swaddle (swod'l), *v.t.* to wind or swathe in or as in a bandage, wrap, or wraps; to wrap in swaddling-clothes to restrict movement. **swaddling-clothes**, *n.pl.* cloth bands used for swaddling an infant.
swag (swag), *v.i.* (*past, p.p.* **swagged**) to hang loose and heavy; to sag. *v.t.* to hang or arrange in swags. *n.* an ornamental festoon; a heavy, loosely hanging fold of fabric; booty obtained by robbery, esp. burglary; a pack or bundle of personal effects, baggage. **swag-bellied,** *a.* having a large prominent belly. **swag-belly,** *n.* **swagman,** *n.* (*Austral.*) a man who carries his swag about with him in search of work. **swagshop,** *n.* (*sl.*) a shop where cheap and trashy goods are sold.
swage (swāj), *n.* a tool for shaping wrought-iron etc. by hammering or pressure. *v.t.* to shape with a swage. **swage-block,** *n.* a heavy iron block or anvil with grooves etc. used for shaping metal.
swagger (swag'ə), *v.i.* to walk, strut or go (about etc.) with an air of defiance, self-confidence or superiority; to talk or behave in a blustering, boastful or hectoring manner. *v.t.* to bluster or bluff (a person into, out of etc.). *n.* a swaggering walk, gait or behaviour; bluster, dash, self-conceit. *a.* (*dated coll.*) smart, fashionable, swell. **swagger-cane, -stick,** *n.* a short cane with metal head carried by soldiers. **swagger-coat,** *n.* a loose coat made on full lines that sways when the wearer walks. **swaggerer,** *n.* **swaggeringly,** *adv.*
swagman, -shop SWAG.
Swahili (swəhē'li), *n.* a people of Zanzibar and the adjoining coast; their language, spoken in Kenya, Tanzania and several other parts of E Africa.
swain (swān), *n.* (*poet.*) a young rustic; a country gallant; a male lover.
swallow[1] (swol'ō), *n.* a small, swift, migratory bird of the genus *Hirundo*, with long, pointed wings and forked tail; a swift or other bird resembling the swallow. **one swallow does not make summer,** a warning against jumping to conclusions. **swallow dive,** *n.* a dive with the arms outstretched. **swallowtail,** *n.* a deeply-forked tail; a butterfly with such a tail, also a humming-bird; the points of a burgee; a dove-tail; (*often pl.*) a swallow-tailed coat, a dress-coat. **swallow-tailed,** *a.* with deeply-forked tail.
swallow[2] (swol'ō), *v.t.* to take through the mouth and throat into the stomach; to absorb, to engulf, to overwhelm, to consume (up); to accept with credulity; to accept without resentment, to put up with; to refrain from showing or expressing; to retract, to recant; to say indistinctly. *v.i.* to perform the action of swallowing. *n.* (*rare*) the gullet or oesophagus; the amount swallowed at once; a swallow-hole. **swallow-hole,** (*dial.*) *n.* an opening in limestone into which a stream or streamlet runs. **swallowable,** *a.* **swallower,** *n.*
swam (swam), *past* SWIM.
swami (swah'mi), *n.* a Hindu religious teacher.
swamp (swomp), *n.* a tract of wet, spongy land, a bog, a marsh. *v.t.* to cause (a boat etc.) to be filled with or to sink in water; to plunge or sink into a bog; to overwhelm, to render helpless with difficulties, numbers etc. *v.i.* to fill with water, to sink, to founder. **swamp gum,** *n.* (*Austral.*) a variety of eucalyptus growing in swamps. **swamp oak,** *n.* the casuarina. **swampy,** *a.*
swan (swon), *n.* a large, web-footed aquatic bird of the genus *Cygnus*, with a long neck and usu. white plumage, noted for its grace in the water; the constellation Cygnus; a poet, a singer (with alln. to the swan-song). *v.i.* to wander aimlessly (about, around etc.). **swan dive,** *n.* (*N Am.*) a swallow dive. **swan herd,** *n.* one who tends swans, esp. a royal officer superintending swan-marks. **swan-hopping** SWAN-UPPING. **swan maiden,** *n.* in German folklore, a maiden able to take the shape of a swan. **swan mark,** *n.* a mark on a swan showing ownership, usu. a notch on the upper mandible. **swan marker,** *n.* **swan neck,** *n.* a pipe, tube, rail etc. curved like a swan's neck, esp. the end of a discharge-pipe. **swansdown,** *n.* down obtained from a swan; a thick cotton cloth with a downy nap on one side. **swan shot,** *n.* a large size of shot. **swan-skin,** *n.* a swan's skin with the feathers on; a soft, fine-twilled flannel. **swan-song,** *n.* the song traditionally believed to be sung by a dying swan; the last or dying work, esp. of a poet; any final work, performance etc. **swan-upping** (-ŭp'ing), *n.* the annual inspection and marking of Thames swans. **swanlike,** *a.* **swannery,** *n.* a place where swans are kept or bred.
swank (swangk), *v.i.* (*sl.*) to swagger, to show off, to bluster. *n.* swagger, bluster. **swanky,** *a.* showing off, showy; stylish, elegant.
swap (swop), **swop,** *v.t., v.i.* (*past, p.p.* **swapped**) to exchange, to barter. *n.* an exchange, a barter; something exchanged in a swap. **swapper,** *n.* **swapping,** *a.* large, strapping.
SWAPO (swah'pō), (*abbr.*) South-West Africa People's Organization.
swaraj (swərahj'), *n.* sellf-government for India; agitation to secure it.
sward (swawd), *n.* a surface of land covered with thick short grass; turf. **swarded, swardy,** *a.*
swarf (swawf), *n.* grit, metal filings, chips, grindings.
swarm[1] (swawm), *n.* a large number of small animals, insects, people etc., esp. when moving in a confused mass; (*pl.*) great numbers; a cluster of honey-bees issuing from a hive with a queen-bee and seeking a new home. *v.i.* of bees, to collect together in readiness for emigrating, to leave (or go out of) a hive in a swarm; to congregate, to throng, to be exceedingly numerous; to move (about etc.) in a swarm; of places, to be thronged or overcrowded (with).
swarm[2] (swawm), *v.t., v.i.* to climb (up a tree, rope, pole etc.) by embracing it with the arms and legs.
swarthy (swaw'dhi), *a.* dark or dusky in complexion. **swarthily,** *adv.* **swarthiness,** *n.*

swash (swosh), *v.i.* to make a noise as of splashing water; of liquid, to wash or splash about; to strike noisily or violently. *v.t.* to strike noisily or violently. *n.* a washing, dashing or splashing of water; a blustering noise. **swashbuckler**, *n.* a bully, a bravo; an adventurer, a daredevil. **swash-buckling**, *a.* **swash letter**, *n.* an ornamental italic capital with tails and flourishes. **swasher**, *n.*
swastika (swos'tikə), *n.* a cross with arms bent at a right angle, used as a symbol of anti-semitism or Nazism.
swat (swot), *v.t.* (*past*, *p.p.* **swatted**) to hit sharply; to crush (a fly) thus. *n.* a sharp blow. **swatter**, *n.*
swatch (swoch), *n.* a sample of cloth.
swath (swoth), **swathe** (swādh), *n.* a row or ridge of grass, corn etc. cut and left lying on the ground; the space cut by a scythe, machine etc. in one course; a broad strip or band; a space left as if by a scythe.
swathe[1] (swādh), *v.t.* to bind or wrap in or as in a bandage, cloth etc. *n.* a bandage, a wrapping.
swathe[2] SWATH.
sway (swā), *v.i.* to move backwards and forwards, to swing, to oscillate irregularly; to be unsteady, to waver, to vacillate; to lean or incline to one side or in different directions; (*dated*) to bear rule, to govern. *v.t.* to cause to oscillate, waver, or vacillate; to cause to incline to one side; to bias; to influence, to control, to rule. *n.* (*dated*) rule, dominion, control; the act of swaying, a swing. **sway-back**, *n.* a hollowed or sagging back, esp. in horses. **sway-backed**, **swayed**, *a.* having the back hollowed, strained or weakened.
swear (sweə), *v.i.* (*past* **swore** (swaw), *p.p.* **sworn** (swawn)) to affirm solemnly invoking God or some other sacred person or object as witness or pledge, to take an oath; to appeal (to) as witness of an oath; to use profane or obscene language; to give evidence on oath; to promise on oath. *v.t.* to utter or affirm with an oath, to take oath (that); to cause to take oath, to administer an oath to, to bind by an oath; to declare, to vow, to promise or testify upon oath; to utter profanely or obscenely. *n.* an act or spell of swearing; a profane oath. **to swear by**, (*coll.*) to have or profess great confidence in. **to swear in**, to induct into office with the administration of an oath. **to swear off**, to renounce solemnly. **swear-word**, *n.* an obscene or taboo word.
sweat (swet), *n.* the moisture exuded from the skin of an animal, perspiration; moisture exuded from or deposited in drops on any surface; the act or state of sweating (*coll.*) drudgery, toil, hard labour, exertion; (*coll.*) a state of anxiety, a flurry; (*coll.*) an old soldier. *v.i.* (*past*, *p.p.* **sweated**, **sweat**) to exude sweat, to perspire; to emit moisture; of moisture, to exude; to collect surface moisture; to be in a flurry or state of anxiety, panic etc., to smart; to toil, to labour, to drudge; to be employed; to carry on business on the sweating-system. *v.t.* to emit as sweat; to make (an animal etc.) sweat by exertion; to employ at starvation wages, to exact the largest possible amount of labour from at the lowest pay, by utilizing competition; to bleed, to subject to extortion; to subject (hides, tobacco etc.) to fermentation; to wear away (coins) by friction etc.; to remove sweat from (horses etc.) with a scraper; to melt (solder etc.) by heating; to unite (metal pieces) in this way; to heat (esp. vegetables) in fat until the juices exude. **no sweat**, (*sl.*) no difficulty or problem, without trouble. **to sweat blood**, (*sl.*) to work or worry to an extreme degree. **to sweat out**, to remove or get rid of by sweating; (*coll.*) to endure, live through. **sweatband**, *n.* a band of absorbent material round the forehead or wrist, as worn in some sports to keep sweat out of the eyes or from the hands. **sweatshirt**, *n.* a loose, long-sleeved sweater made from cotton jersey. **sweatshop**, *n.* a factory or other work-place that employs the sweating system. **sweated**, *a.* pertaining to or produced by the sweating system. **sweater**, *n.* a (thick) jersey, jumper or pull-over; one who or that which causes to sweat. **sweating**, *a.* causing or enduring sweat. **sweating sickness**, *n.* a form of malaria epidemic in the 15th and 16th cents. **sweating system**, *n.* the practice of employing operatives at starvation wages in unhealthy conditions and for long hours. **sweaty**, *a.* **sweatily**, *adv.* **sweatiness**, *n.*
Swede (swēd), *n.* a native or inhabitant of Sweden. **swede**, *n.* a Swedish turnip, *Brassica rutabaga*.
Swedish (swē'dish), *a.* pertaining to Sweden or its inhabitants. *n.* the language of the Swedes.
sweeny (swē'ni), *n.* atrophy of a muscle, esp. of the shoulder in horses.
sweep (swēp), *v.i.* (*past*, *p.p.* **swept** (swept)) to glide, move or pass along with a strong, swift continuous motion; of the eye, to range unchecked; of land, a curve etc., to extend continuously; to go with a stately motion. *v.t.* to carry (along, away etc.) with powerful or unchecked force; to move swiftly and powerfully over, across or along, to range, to scour; esp. of the eyes, to pass over in swift survey; to pass over destructively; to rake, to enfilade, to clear; to gain an overwhelming victory in; to dredge (the bottom of a river etc.); to wipe out, remove, destroy; to clear dirt etc. from or clean with or as with a broom etc.; to collect or gather (up) with or as with a broom; to propel with sweeps; to cause to move with a sweeping motion. *n.* the act of sweeping; a clearance, a riddance; a sweeping motion; a sweeping curve, direction, piece of road etc.; a broad expanse; the range, reach or compass of a sweeping motion or of an instrument, weapon, implement etc. having this motion; a long oar used to propel barges or sailing-vessels in a calm; a swape; a chimney-sweeper; (*dated*) a blackguard; a sweepstake. **to make a clean sweep**, to get rid of entirely. **to sweep the board**, to win everything. **sweepback**, *n.* the angle relatively to the axis at which an aircraft wing is set back. **sweepnet**, *n.* a sweep-seine; a butterfly net. **sweep (second) hand**, *n.* a watch or clock hand that registers seconds. **sweep-seine**, *n.* a long seine used for sweeping a large area. **sweeper**, *n.* one who sweeps; a carpet-sweeper; (*Austral.*) a worker in the woolsheds, a 'broomie'; (*Austral.*, *coll.*) a slow train; a defensive player in soccer positioned behind the main defensive line. **sweeping**, *a.* that sweeps; covering a wide area; wide-ranging, comprehensive; without discrimination or qualification. *n.pl.* things collected by sweeping; (*fig.*) rubbish, refuse, litter. **sweepingly**, *adv.* **sweepingness**, *n.*
sweepstake (swēp'stāk), **sweepstakes**, *n.* a lottery in which a number of persons stake sums on an event, esp. on a horse-race, the total amount staked being divided among the winning betters.
sweet (swēt), *a.* having a taste like that of honey or sugar; containing sugar or a sweetening ingredient; pleasing to the senses; fragrant; pleasant or melodious in sound; refreshing, restful; fresh, not salt or salted, not sour, bitter, stale or rancid; of butter, fresh, unsalted; free from acids or other corrosive substances; pleasant to the mind, agreeable, delightful; charming, amiable, gracious, lovable, dear, beloved. *n.* a sweet thing; an article of confectionery, a sweetmeat; a sweet dish, as a tart, pudding, ice etc.; the course at a meal after the meat; (*pl.*) pleasures, delights, pleasant experiences; dear one, darling. *adv.* sweetly. **to be sweet on**, to be in love with; to be very fond of. **to have a sweet tooth**, to be fond of sweet-tasting things. **sweet-and-sour**, *a.* cooked with sugar and vinegar or lemon juice. **sweet bay**, *n.* the laurel, bay-tree. **sweetbread**, *n.* the pancreas or thymus-gland, esp. of a calf or sheep, used as food. **sweet-brier** BRIER. **sweet chestnut** CHESTNUT. **sweetcorn**, *n.* a variety of maize with kernels rich in sugar; the kernels eaten as a vegetable when young. **sweet-flag** *n.* a plant with an aromatic root-stock used in medicine, confectionery etc. **sweet-gale** GALE[3]. **sweetheart**, *n.* a lover. *v.i.* to be love-making. **sweet-john**,

swell

n. the narrow-leaved variety of sweet-william. **sweetmeat**, *n.* an article of confectionery, usu. consisting wholly or principally of sugar, a sugar-plum, a bonbon; a fruit candied with sugar. **sweet pea**, *n.* an annual leguminous climbing plant, *Lathyrus odoratus*, with showy flowers. **sweet pepper**, *n.* a mild-flavoured capsicum fruit with a thick fleshy wall. **sweet potato**, *n.* a tropical climbing plant, *Batatas edulis*, with an edible root. **sweet root**, *n.* liquorice-root. **sweet shop**, *n.* a shop where sweets are sold. **sweetsop**, *n.* a tropical American tree, *Anona squamosa*, allied to the custard-apple, with sweet, pulpy fruit. **sweet talk**, *n.* (*coll.*) flattery, blandishment. **sweet-talk**, *v.t.* to flatter, esp. in order to coax or persuade. **sweet violet**, *n.* the scented or wood-violet, *Viola odorata*. **sweet william**, *n.* a biennial species of pink, *Dianthus barbatus*, with dense clusters of showy and fragrant flowers. **sweet-willow**, *n.* the sweet-gale. **sweet wood**, *n.* the true laurel, *Laurus nobilis*; applied to other trees and shrubs of the family Lauraceae. **sweeten**, *v.t.* to make sweet or sweeter; to make more agreeable or less unpleasant; to mollify, pacify. **sweetener**, *n.* a (sugar-free) sweetening agent; (*sl.*) a bribe. **sweetening**, *n.* **sweetie** (-i), **sweety**, *n.* a sweet; a term of endearment. **sweeting**, *n.* a sweet variety of apple; (*dated*) a term of endearment. **sweetish**, *a.* **sweetishness**, *n.* **sweetly**, *adv.* **sweetness**, *n.*

swell (swel), *v.i.* (*p.p.* **swollen** (swō'lən), **swelled**) to dilate or increase in bulk or extent, to expand; to rise up from the surrounding surface, to bulge, to belly (out); to become greater in volume, strength or intensity; to rise in altitude; to be puffed up, to be elated, to strut; to be inflated with anger etc. *v.t.* to increase the size, bulk, volume or dimensions of; to inflate, to puff up. *n.* the act or effect of swelling; rise, increase, augmentation; a succession of long, unbroken waves in one direction, as after a storm; a bulge, a bulging part; (*Mus.*) an increase followed by a decrease in the volume of sound; a combined crescendo and diminuendo; a contrivance for gradually increasing and diminishing sound in an organ etc.; a swell-organ; (*coll.*) a person of high standing or importance, a showy, dashing or fashionable person. *a.* (*coll.*) characterized by showiness or display, smart, foppish, dandified; of distinction; (*N Am.*, *coll.*) excellent, fine. **swell-blind**, *n.* one of the movable slats forming the front of a swell box. **swell box**, *n.* a chamber containing the pipes of a swell organ, which is opened and closed to change the volume. **swell-organ**, *n.* an organ or partial organ with the pipes enclosed in a swell box. **swelled**, **swollen head**, *n.* conceit. **swelling**, *n.* the act of expanding etc., or the state of being swollen or augmented; an unnatural enlargement of a body part.

swelter (swel'tə), *v.i.* of the weather etc., to be hot, moist and oppressive, to cause faintness, languor or oppression; to be overcome and faint with heat; to sweat profusely. *n.* (*coll.*) a sweltering condition. **sweltering**, *a.* oppressively hot. **swelteringly**, *adv.*

swept (swept), *past, p.p.* SWEEP. **sweptback**, *a.* of an aircraft wing, slanting backwards, having sweepback. **sweptwing**, *a.* having sweptback wings.

swerve (swœv), *v.i.* to turn to one side, to deviate, to diverge from the direct or regular course. *v.t.* to cause to diverge, to deflect. *n.* the act of swerving, a sudden divergence or deflection. **swervable**, **swerving**, *a.* **server**, *n.*

SWG, (*abbr.*) standard wire gauge.

swift (swift), *a.* moving or able to move with great rapidity, fleet, rapid, quick, speedy; ready, prompt, expeditious; passing rapidly, soon over, brief, unexpected, sudden. *adv.* swiftly. *n.* a small, long-winged insectivorous bird of the family Apodidae, esp. *Apus apus*, closely resembling the swallow; the common newt; a ghost-moth; the sail of a windmill. **swift-footed**, **-winged**, *a.* running, flying with swiftness. **swiftly**, *adv.* **swiftness**, *n.*

swig (swig), *v.t.*, *v.i.* (*past, p.p.* **swigged**) (*coll.*) to drink in large draughts. *n.* (*coll.*) a large or deep draught of liquor.

swill (swil), *v.t.* to wash, to rinse; to drink greedily. *v.i.* to drink to excess. *n.* a rinsing; liquid food for animals, esp. pigs, hog-wash; (liquid) rubbish, slops; a swig. **swiller**, *n.* **swillings**, *n.pl.* hog-wash.

swim (swim), *v.i.* (*pres.p.* **swimming**, *past* **swam** (swam), *p.p.* **swum** (swŭm)) to float or be supported on water or other liquid; to move progressively in the water by the motion of the hands and feet, or fins, tail etc.; to glide along; to be drenched or flooded (with water etc.); to seem to reel or whirl round one; to have a feeling of dizziness. *v.t.* to pass, traverse or accomplish by swimming; to compete in (a race); to perform (a particular swimming stroke); to cause (a horse, boat etc.) to swim or float; to bear up, to float (a ship etc.). *n.* the act or a spell of swimming; a pool or run frequented by fish in a river; the swimming-bladder; the main current of life, business etc. **swim bladder** SWIMMING BLADDER. **swimsuit** SWIMMING COSTUME. **swimmable**, *a.* **swimmer**, *n.* **swimmeret** (-ret), *n.* one of the appendages of a crustacean serving as a swimming-organ. **swimming-bath**, **-pool**, *n.* a bath or artificial pool for swimming in. **swimming bladder**, *n.* the air-bladder or sound of a fish. **swimming costume**, *n.* a woman's one-piece garment for swimming. **swimmingly**, *adv.* smoothly, easily, without impediment.

swindle (swin'dl), *v.t.*, *v.i.* to cheat; to defraud grossly or deliberately. *n.* the act or process of swindling; a gross fraud or imposition, a fraudulent scheme; (*coll.*) a thing that is not what it pretends to be, a deception, a fraud. **swindler**, *n.*

swine (swīn), *n.* (*pl.* **swine**) an ungulate omnivorous mammal of the family Suidae, esp. the genus *Sus*, a pig, a hog; a greedy, vicious or debased person; something difficult or unpleasant. **swine bread**, *n.* the truffle; the sow-bread. **swine fever**, **-plague**, *n.* an infectious lung-disease affecting the pig. **swineherd**, *n.* one who tends swine. **swine pox**, *n.* a form of chicken-pox affecting swine. **swine's-snout**, *n.* the dandelion. **swinish**, *a.* **swinishly**, *adv.* **swinishness**, *n.*

swing (swing), *v.i.* (*past p.p.* **swung** (swŭng)) to move to and fro, as a body suspended by a point or one side, to sway, hang freely as a pendulum, to oscillate, to rock; to turn on or as on a pivot, to move or wheel (round etc.) through an arc; to go with a swaying, undulating or rhythmical gait or motion; to go to and fro in a swing; (*coll.*) to hit out (at) with a swinging arm movement; to be hanged; to play swing music; to have the rhythmical quality of swing music; to fluctuate between emotions, decisions etc.; (*coll.*) to be lively or up-to-date. *v.t.* to cause to move to and fro, to sway, to oscillate; to wave to and fro, to brandish; to cause to turn or move around, as on a pivot or through an arc; to cause to go to and fro in a swing; to play or perform in the style of swing music; (*coll.*) to manipulate, influence; (*coll.*) to cause to happen, bring about. *n.* the act or state of swinging; a swinging or oscillating motion; a swinging gait or rhythm; the compass or sweep of a moving body; a curving or sweeping movement; a blow delivered with a sweeping arm movement; free course, unrestrained liberty; regular course of activity; a seat suspended by ropes etc., in which a person or thing may swing to and fro; a spell of swinging in this; swing music; a shift in opinion, condition etc. **in full swing**, in full activity or operation. **to swing the lead**, to trump up an excuse for evading a duty. **swingboat**, *n.* a boat-shaped carriage for swinging in at fairs etc. **swing bridge**, *n.* a drawbridge opening by turning horizontally. **swing music**, **swing**, *n.* a style of playing jazz in which the basic melody and rhythm persist through individual interpretations of the theme, impromptu variations etc. **swing-plough**, *n.* a plough without wheels. **swing wheel,** *n.* the wheel driving a clock-pendulum, corresponding to the balance-wheel of a watch. **swing-wing,** *a.* of an aircraft, having movable wings allowing varying degrees of sweep-back at different speeds. **swinger,** *n.* **swinging,** *a.* that swings; (*coll.*) lively or up-to-date. **swingingly,** *adv.*

swingeing (swinj'ing), *a.* severe, great, huge.

swingle (swing'gl), *v.t.* to clean (flax) by beating with a swingle. *n.* a wooden instrument for beating flax to separate the woody parts from the fibre. **swinglebar, -tree,** *n.* the cross-bar pivoted in the middle to which the ends of a horse's traces are attached.

swinish, etc. SWINE.

swipe (swip), *v.t.* to hit with great force, in cricket, golf etc.; to drink off, to gulp down; (*sl.*) to pilfer. *v.i.* to hit out with a swipe; to drink the contents of a glass at one go. *n.* a hard, swiping blow, esp. at cricket; (*pl.*, *coll.*) thin, washy or inferior beer. **swiper,** *n.*

swirl (swœl), *v.i.* to form eddies, to whirl about. *v.t.* to carry (along, down etc.) with an eddying motion. *n.* a whirling motion, an eddy; the furious rush of a fish through water, or the disturbance so caused; a winding or curling pattern or figure. **swirly,** *a.*

swish (swish), *v.i.* to make a whistling sound in cutting through the air; to move with such a sound. *v.t.* to make such a whistling movement with; to strike or cut (off) with such a sound; to flog, to thrash, esp. with a birch. *n.* a whistling sound, movement or blow; a stroke with a birch etc. *a.* (*coll.*) smart, elegant.

Swiss (swis), *a.* of or pertaining to Switzerland or its inhabitants. *n.* (*pl.* **Swiss**) a native or inhabitant of Switzerland. **Swiss chard** CHARD. **Swiss Guards,** *n.pl.* mercenaries formerly employed as bodyguards in France, Naples etc., and still at the Vatican. **swiss roll,** *n.* a thin sponge cake, rolled up around a filling, esp. of jam.

switch (swich), *n.* a small flexible twig or rod; a (false) tress of hair; a mechanism for diverting railway trains or vehicles from one line to another, or for completing or interrupting an electric circuit, transferring current from one wire to another etc.; a shift, change; an exchange. *v.t.* to lash or beat with a switch; to move, whisk or snatch (away etc.) with a jerk; to shift (a train etc.) from one line to another; to turn (on or off) with a switch; to connect or disconnect (a user of a telephone) thus; to change, divert. *v.i.* to move or swing with a careless or jerking movement, to whisk; to cut (off) connection on a telephone etc.; to make a change, to shift. **to switch off,** (*coll.*) to stop listening or paying attention, to lose interest. **to switch on,** (*coll.*) to become alive or responsive to. **switchblade,** *n.* a flick knife. **switchboard,** *n.* a board on which switches are fixed controlling electric or telephonic circuits. **switchman,** *n.* a man in charge of railway switches, a shunter.

switchback (swich'bak), *n.* a zigzag railway for ascending or descending steep inclines; a steeply ascending and descending road, track etc.; a railway on which the vehicles are carried over a series of ascending inclines by the momentum of previous descents, used for amusement at fairs etc.

swither (swidh'ə), *v.i.* (*Sc.*) to hesitate.

Switzer (swit'sə), *n.* (*rarely*) a Swiss.

swivel (swiv'l), *n.* a link or connection comprising a ring and pivot or other mechanism allowing the two parts to revolve independently; a support allowing free horizontal rotation; a swivel-gun. *v.i.*, *v.t.* (*past*, *p.p.* **swivelled**) to turn on a swivel or pivot. **swivel chair,** *n.* a chair that revolves on its base. **swivel gun,** *n.* a gun mounted on a pivot.

swizzle (swiz'l), *n.* a mixed drink of various kinds; (*coll.*) a cheat, a fraud (also **swizz**). **swizzle stick,** *n.* a stick for frothing drinks.

swob (swob), etc. SWAB.

swollen (swōln), *p.p.* SWELL.

swoon (swoon), *v.i.* to fall into a fainting fit; of music etc., to sink or die away. *n.* a faint. **swooningly,** *adv.*

swoop (swoop), *v.i.* to descend upon prey etc. suddenly, as a hawk, to come (down) upon, to attack suddenly. *v.t.* to fall on suddenly and seize, to snatch (up). *n.* a sudden plunge of or as of a bird of prey on its quarry; a sudden descent, attack, seizing or snatching; (*coll.*) a snatching up of all at once.

swoosh (swoosh), *v.i.* to move with or make a rushing sound. *n.* this sound.

swop SWAP.

sword (sawd), *n.* a weapon, usu. consisting of a long blade fixed in a hilt with a guard for the hand, used for cutting or thrusting; a swordlike (body) part or object; the power of the sword, military power, sovereignty; war, destruction in war, death. **sword of Damocles** DAMOCLEAN. **sword of justice,** judicial authority. **sword of State,** a sword carried before the sovereign etc. on ceremonial occasions. **to put to the sword,** to kill (esp. those captured or defeated in war). **swordbearer,** *n.* an officer who carries a sword of State. **sword belt,** *n.* a belt from which a sword is slung. **sword bill,** *n.* a S American hummingbird with a long sword-shaped bill. **sword blade,** *n.* **sword cane,** *n.* a hollow walking-stick enclosing a long, pointed blade. **sword cut,** *n.* a cut or scar inflicted by a sword. **sword dance,** *n.* a dance in which swords are brandished or clashed together or in which women pass under crossed swords; a Highland dance performed over two swords laid crosswise on the floor. **swordfish,** *n.* a sea-fish of the genus *Xiphias*, allied to the mackerel, having the upper jaw prolonged into a formidable swordlike weapon. **sword grass,** *n.* a species of sedge with sword-like leaves. **sword-guard,** *n.* the part of a sword-hilt protecting the hand. **sword knot,** *n.* a ribbon or tassel tied to the hilt of a sword, orig. used for securing it to the wrist. **sword lily,** *n.* the gladiolus. **swordplay,** *n.* a combat between gladiators, fencing; repartee. **sword-shaped,** *a.* **swordsman,** *n.* one who carries a sword; one skilled in the use of the sword. **swordsmanship,** *n.* **swordstick,** *n.* a swordcane. **sworded,** *a.* wearing or armed with a sword. **swordlike,** *a.*

swore (swaw), *past*, **sworn** (swawn), *p.p.* SWEAR.

swot (swot), *v.i.*, *v.t.* (*past*, *p.p.* **swotted**) (*coll.*) to study hard. *n.* hard study; a piece of hard work; one who studies hard.

swum (swŭm), *p.p.* SWIM.

swung (swŭng), *past*, *p.p.* SWING.

swy (swi), *n.* (*Austral.*, *coll.*) the game of two-up.

sybarite (sib'ərit), *n.* a native or inhabitant of Sybaris, an ancient Greek colony in S Italy, noted for effeminacy, voluptuousness and luxury; a sensual and luxurious person. *a.* sybaritic. **sybaritic** (-rit'-), *a.* **sybaritically,** *adv.* **sybaritism,** *n.*

sybil (sib'l), SIBYL.

sycamine (sik'əmin), *n.* the black mulberry-tree.

sycamore (sik'əmaw), *n.* a medium-sized Eurasian tree, *Acer pseudoplatanus*, allied to the maple and plane; the sycamore-fig; (*N Am.*) a plane tree. **sycamore-fig,** *n.* a Syrian and Egyptian tree with a fig-like fruit.

syconium (sikō'niəm), *n.* (*pl.* **-nia** (-niə)) a multiple fruit developed from a fleshy receptacle having numerous flowers, as in the fig.

sycophant (sik'əfant), *n.* a servile flatterer, a parasite. **sycophancy,** *n.* **sycophantic** (-fan'-), *a.* **sycophantism,** *n.*

sycosis (sikō'sis), *n.* a pustular eruption or inflammation of the scalp or bearded part of the face.

syenite (sī'ənīt), *n.* a granular igneous rock consisting of orthoclase and hornblende, with or without quartz. **syenitic** (-nit'-), *a.*

syllable (sil'əbl), *n.* a sound forming a word or part of a word, containing one vowel sound, with or without a con-

syllabub sonant or consonants, and uttered at a single effort or vocal impulse; the least expression or particle of speech. *v.i.* to pronounce by syllables, to articulate. **syllabary**, *n.* a catalogue of characters representing syllables; such characters collectively, serving the purpose of an alphabet in certain languages. **syllabic** (-ab'-), *a.* pertaining to, consisting of or based on a syllable or syllables; having each syllable distinctly articulated; representing the sound of a whole syllable, as distinct from *alphabetic*. **syllabically**, *adv.* **syllabicate, syllabify** (-lab'ifī), **syllabize, -ise**, *v.t.* to separate into or pronounce by syllables. **syllabication, syllabification** (-fi-), *n.* **syllabled**, *a.* (*usu. in comb.*, as *two-syllabled*).

syllabub (sil'əbŭb), *n.* a cold dessert made of cream mixed with sugar, wine and lemon juice.

syllabus (sil'əbəs), *n.* (*pl.* **-buses, -bi** (-bī)) a list, outline, summary, abstract etc., giving the principal points or subjects of a course of lectures, teaching or study, examination requirements, hours of attendance etc.

syllepsis (silep'sis), *n.* (*pl.* **-ses** (-sēz)) the connection of a verb or adjective with two nouns, with only one of which it is in syntactical agreement, as in 'Neither he nor I am there.' **sylleptic**, *a.* **sylleptically**, *adv.*

syllogism (sil'əjizm), *n.* a form of argument consisting of three propositions, a major premise or general statement, a minor premise or instance, and a third deduced from these called the conclusion. **syllogistic** (-jis'-), *a.* **syllogistically**, *adv.* **syllogize, -ise**, *v.i.*, *v.t.* to reason or deduce by syllogisms. **syllogization**, *n.* **syllogizer**, *n.*

sylph (silf), *n.* an elementary being inhabiting the air, intermediate between material and immaterial beings; a graceful and slender girl; a S American humming-bird with a long, brilliantly-coloured tail. **sylphlike**, *a.*

sylvan (sil'vən), SILVAN.

sylvanite (sil'vənīt), *n.* a gold or silver telluride mineral.

sylviculture SILVICULTURE under SILVA.

sym-, *pref.* SYN- (before *b*, *m*, or *p*).

symbiont (sim'biont), *n.* an organism living in a state of symbiosis. **symbiontic** (-on'-), *a.* **symbiosis** (-ō'sis), *n.* the vital union or partnership of certain organisms, such as the fungus and alga in lichens. **symbiotic** (-ot'-), *a.* **symbiotically**, *adv.*

symbol (sim'bl), *n.* an object typifying or representing something by resemblance, association etc., a type, an emblem; a mark, character or letter accepted as representing or signifying some thing, idea, relation, process etc., as the letters of the alphabet, those representing chemical elements, the signs of mathematical relations etc. *v.t.* (*past*, *p.p.* **symbolled**) to symbolize. **symbolic, -ical** (-bol'-), *a.* pertaining to, serving as or using symbols. **symbolic logic**, *n.* logic that uses symbols to represent and clarify principles etc. **symbolically**, *adv.* **symbolics**, *n. sing.* **symbolism**, *n.* representation by symbols or signs; a system of symbols; symbolic significance; the use of symbols, esp. in art and literature; a late 19th-cent. movement among (French) artists and writers using symbolic images to express or suggest the essential nature of things, mystical ideas, emotions etc. **symbolist**, *n.* **symbolistic** (-lis'-), *a.* **symbolistically**, *adv.* **symbolize, -ise**, *v.t.* to be the symbol of, to typify; to represent by symbols; to treat as symbolic, not literal; to make symbolic or representative of something. *v.i.* to use symbols. **symbolization**, *n.* **symbolizer**, *n.* **symbology** (-bol'-), *n.* the use of symbols as a means of expression; the study or analysis of symbols. **symbological** (-loj'-), *a.* **symbologist**, *n.*

symmetry (sim'itri), *n.* due proportion of the several parts of a body or any whole to each other, congruity, parity, regularity, harmony; beauty of form arising from this; arrangement of parts on either side of a dividing line or point so that the opposite parts are exactly similar in shape and size; regularity of structure so that opposite halves exactly correspond; regularity of number in sepals, petals, stamens etc., each whorl comprising the same number or multiples of this. **symmetric** (-met'-), *a.* **symmetrically**, *adv.*

sympathy (sim'pəthi), *n.* the quality of being affected with the same feelings as another, or of sharing emotions, affections, inclinations etc. with another person, animal etc.; fellow-feeling, agreement, harmony; (*often pl.*) a feeling of accord (with); loyalty or support; compassion (for); unity or correlation of action; response of an organ or part to an affection in another without actual transmission of the cause; the relation between inanimate bodies by which the vibration of one sets up a corresponding vibration in another. **sympathetic** (-thet'-), *a.* pertaining to, expressive of, or due to sympathy; having sympathy or common feeling with another, sympathizing; being or acting in sympathy or agreement, concordant; in accord with one's mood or disposition, congenial; proceeding from or due to pain or injury in another organ or part; pertaining to or mediated by the sympathetic nervous system; of acoustic, electrical, and other vibrations, produced by impulses from other vibrations. **sympathetic nervous system**, the part of the autonomic nervous system in which nerve impulses are transmitted chiefly by adrenalin and related substances. **sympathetically**, *adv.* **sympathize, -ise**, *v.i.* to have or express sympathy with another, as in pain, pleasure etc.; to be of the same disposition, opinion etc. **sympathizer**, *n.* **sympathomimetic** (-thōmimet'-), *a.* having or causing physiological effects like those produced by the sympathetic nervous system.

symphony (sim'fəni), *n.* a complex and elaborate composition for an orchestra, usu. consisting of four varied movements; an instrumental passage or composition occurring as an interlude in or introduction to a vocal work; a symphony orchestra; a harmonious composition. **symphony orchestra**, *n.* a large orchestra containing wind, string and percussion sections. **symphonic** (-fon'-), *a.* **symphonic poem**, *n.* a tone poem. **symphonist**, *n.* a composer or performer of symphonies.

symphysis (sim'fisis), *n.* (*pl.* **-ses** (-sēz)) the joint formed by the union of two parts of the skeleton by growing together or the intervention of cartilage; the growing together or union of two plant parts. **symphyseal** (-fiz'iəl), *a.*

symploce (sim'pləsi), *n.* the repetition of a word or phrase at the beginning and of another at the end of successive clauses.

sympodium (simpō'diəm), *n.* (*pl.* **-dia** (-diə)) a false plant axis or stem composed of superimposed branches. **sympodial**, *a.* **sympodially**, *adv.*

symposium (simpō'ziəm), *n.* (*pl.* **-sia** (-ziə), **-siums**) in ancient Greece, a drinking together, a convivial party usu. following a banquet, with music, dancing etc.; a series of brief articles expressing the views of different writers, in a magazine etc.; a conference or formal meeting at which several speakers give addresses on a particular topic.

symptom (simp'təm), *n.* a perceptible change in the appearance or functions of the body indicating disease; a sign, a token, an indication. **symptomatic, -ical** (-mat'-), *a.* **symptomatically**, *adv.* **symptomatology** (-mətol'-), *n.* a branch of medicine concerned with disease symptoms; the symptoms associated with a disease.

syn., (*abbr.*) synonym, synonymous.

syn-, *pref.* with; together; alike.

synaeresis, syneresis (siniə'rəsis), *n.* the contraction of two vowels or syllables into one; the expulsion of liquid from a gel by contraction.

synaesthesia, (*N Am.*) **synesthesia** (sinəsthē'ziə), *n.* sensation experienced at a point distinct from the point of stimulation. **synaesthetic**, (*N Am.*) **synesthetic**, *a.*

synagogue (sin'əgog), *n.* a Jewish congregation for religious instruction and observances; a building or place of meeting for this. **synagogal, synagogical** (-goj'-), *a.*

synalepha (sinəlē'fə), *n.* a blending of two syllables into one, esp. by the suppression of a final vowel before an initial vowel.

synallagmatic (sinəlagmat'ik), *a.* of a contract or treaty, imposing reciprocal obligations.

synangium (sinan'jiəm), *n.* (*pl.* **-gia** (-jiə)) the boat-shaped sorus composed of sporangia in some ferns; an arterial trunk.

synapse (sī'naps), *n.* the point at which a nerve impulse is transmitted from one neuron to another. **synapsis**, *n.* (*pl.* **-ses**) the pairing of homologous chromosomes occurring at the start of cell division by meiosis; a synapse. **synaptic**, *a.* **synaptically**, *adv.*

synarthrosis (sinahthrō'sis), *n.* (*pl.* **-ses** (-sēz)) an articulation not permitting motion, as in sutures, symphysis etc. **synarthrodial** (sinahthrō'diəl), *a.*

sync, synch (singk), *v.t., v.i., n.* short for SYNCHRONIZE, SYNCHRONIZATION under SYNCHRONISM.

syncarp (sin'kahp), *n.* an aggregate fruit, as the blackberry. **syncarpous** (-kah'-), *a.*

synch SYNC.

synchondrosis (sinkəndrō'sis), *n.* the almost immovable articulation of bones by means of cartilage, as in the vertebrae.

synchromesh (sing'krəmesh), *a.* pertaining to a system of gearing in which the drive and driving members are automatically synchronized before engagement, thus avoiding shock and noise in changing gear.

synchronism (sing'krənizm), *n.* concurrence of two or more events in time, coincidence, simultaneousness; a tabular arrangement of historical events or personages according to their dates. **synchronistic** (-nis'-), *a.* **synchronistically**, *adv.* **synchronize, -ise**, *v.i.* to concur in time, to happen at the same time. *v.t.* to cause to occur in unison or at the same time; to cause to agree in time or indicate the same time; to match (the sound-track of a film) exactly with the picture. **synchronized swimming**, *n.* a sport in which one or a team of swimmers perform a series of dance-like movements to music. **synchronization**, *n.* **synchronizer**, *n.* **synchronous**, *a.* occurring simultaneously; operating or recurring together at the same rate. **synchronous motor**, *n.* an electric motor whose speed is proportional to the frequency of the supply current. **synchronously**, *adv.*

synclinal (singklī'nəl), *a.* (*Geol.*) sloping downward towards a common point or line, opp. to *anticlinal*. **syncline** (sing'-), *n.* a synclinal flexure or axis.

syncopate (sing'kəpāt), *v.t.* to contract (a word) by omitting one or more letters or syllables from the middle; to modify (a musical note, rhythm etc.) by beginning on an unaccented and continuing with an accented beat. **syncopation**, *n.*

syncope (sing'kəpi), *n.* the elision of a letter or syllable from the middle of a word; (*Med.*) a faint. **syncopal, -copic** (-kop'-), *a.* **syncopist**, *n.*

syncretism (sing'krətizm), *n.* the attempted reconciliation of various philosophic or religious schools or systems of thought, as against a common opponent; in language, the fusion of different inflectional forms. **syncretic** (-kret'-), *n., a.* **syncretist**, *n.* **syncretistic** (-tis'-), *a.* **syncretize, -ise**, *v.t., v.i.*

syndactyl (sindak'til), *a.* having the digits united, as in webbed feet. **syndactylism**, *n.*

syndesmosis (sindesmō'sis), *n.* (*pl.* **-ses** (-sēz)) an articulation of bones by ligaments. **syndesmotic** (-mot'-), *a.*

syndetic (sindet'ik), *a.* (*Gram.*) serving to connect, copulative.

syndic (sin'dik), *n.* an officer or magistrate invested with varying powers in different places and times; a business agent of a university, corporation etc. **syndicate** (-kət), *n.* an association of persons or firms formed to promote some special interest or undertake a joint project; an agency that supplies material for simultaneous publication in several newspapers or periodicals. *v.t.* (-kāt), to combine in a syndicate; to manage by means of a syndicate; to sell for simultaneous publication in several newspapers or periodicals; to sell (a television programme) for broadcasting by several different stations. **syndication**, *n.*

syndicalism (sin'dikəlizm), *n.* the economic doctrine that all the workers in any trade or industry should participate in the management and control and in the division of the profits (and that in order to bring about this condition the workers in different trades should federate together and enforce their demands by sympathetic strikes). **syndicalist**, *n.*

syndrome (sin'drōm), *n.* concurrence; the aggregate of symptoms characteristic of any disease or disorder; a pattern or set of feelings, actions etc. characteristic of a condition or problem.

syne (sīn), *adv.* (*Sc.*) long ago, since.

synecdoche (sinek'dəki), *n.* a rhetorical figure by which a part is put for the whole or the whole for a part. **synecdochical** (-dok'-), *a.*

synechia (sinikī'ə), *n.* morbid adhesion of the iris to the cornea or to the capsule of the crystalline lens.

synecology (sinikol'əji), *n.* the ecology of plant and animal communities. **synecologic, -ical** (-loj'-), *a.* **synecologically**, *adv.*

syneresis SYNAERESIS.

synergism (sin'əjizm), *n.* the working together of two drugs, muscles etc. such that their combined action exceeds the sum of their individual actions. **synergic, -getic**, *a.* of muscles etc., working together, cooperative. **synergist**, *n.* something that acts with, or increases the effect of, another. **synergistic** (-jis'-), *a.* **synergy** (-ji), *n.* combined action between different organs etc., synergism.

synesis (sin'əsis), *n.* grammatical construction according to the sense rather than syntax.

synesthesia SYNAESTHESIA.

syngamy (sing'gəmi), *n.* sexual reproduction by union of gametes, syngenesis. **syngamic** (-gam'-), *a.*

syngenesis (sinjen'əsis), *n.* reproduction by the union of the ovum and the spermatozoon. **syngenetic** (-net'-), *a.*

synod (sin'əd), *n.* an ecclesiastical council; a deliberative assembly, a meeting for discussion. **synodal** (-nod'-), *a.*

synoecious (sinē'shəs), *a.* having male and female organs in the same flower or receptacle.

synonym (sin'ənim), *n.* a word having the same meaning as another of the same language; a word denoting the same thing but differing in some senses, or in range of application. **synonymic** (-nim'-), **-ical**, *a.* of or pertaining to synonymy. **synonymity** (-nim'-), *n.* the quality of being synonymous. **synonymize, -ise** (-non'-), *v.t.* to express by synonyms or a synonym. **synonymous** (-non'iməs), *a.* expressing the same thing by a different word or words; having the same meaning, conveying the same idea. **synonymously**, *adv.* **synonymy** (-mi), *n.* a system of synonyms; a treatise on synonyms; synonymity.

synopsis (sinop'sis), *n.* (*pl.* **-ses** (-sēz)) a general view, a conspectus, a summary. **synoptic** (-tik), *a.* of the nature of a synopsis, affording a general view. *n.* one of the synoptic gospels. **synoptic gospels**, *n.pl.* those of Matthew, Mark and Luke. **synoptical**, *a.* **synoptically**, *adv.* **synoptist**, *n.* one of the writers of the synoptic gospels.

synovia (sinō'viə), *n.* an albuminous lubricating fluid secreted by the synovial membranes lining joints and tendon sheaths. **synovial**, *a.* pertaining to or secreting synovia. **synovitis** (-əvī'tis), *n.* inflammation of a synovial membrane.

syntax (sin'taks), *n.* (the part of grammar that deals with) the due arrangement of words or the construction of sentences. **syntactic** (-tak'-), *a.* of, pertaining to or according to the rules of syntax. **syntactically**, *adv.* **syntactics**, *n.sing.* semiology dealing with the formal relations and properties of signs.

synthesis (sin'thəsis), *n.* (*pl.* **-ses** (-sēz)) the putting of two or more things together, combination, composition; the building up of a complex whole by the union of elements, esp. the process of forming concepts, general ideas, theories etc.; the production of a substance by chemical reaction; the formation of compound words by means of composition and inflexion, as opp. to analysis which employs prepositions etc. **synthesist**, *n.* **synthesize, -tize** (-tīz), **-ise**, *v.t.* **synthesizer**, *n.* one who, or that which, synthesizes; a usu. keyboard-operated electronic instrument that can produce and manipulate a wide variety of sounds, imitate conventional musical instruments etc. **synthetic, -ical** (-thet'-), *a.* pertaining to or consisting in synthesis; artificially produced, man-made, false, sham. **synthetically**, *adv.*
syphilis (sif'ilis), *n.* an infectious venereal disease caused by a microorganism introduced into the system by direct contact or due to heredity, having three stages: primary syphilis, affecting the genitals etc.; secondary syphilis, attacking the skin and mucous membranes; and tertiary syphilis, spreading to the muscles, bones and brain. **syphilitic** (-lit'-), *a.* **syphiloid** (-loid), *a.*
syphon (sī'fən), SIPHON.
syren (sī'rən), SIREN.
Syriac (si'riak), *a.* pertaining to Syria or its language. *n.* the language of the ancient Syrians, western Aramaic. **Syrian**, *n.*, *a.* (a native) of Syria.
syringa (siring'gə), *n.* the mock-orange, *Philadelphus;* a genus of plants containing the lilacs.
syringe (sirinj'), *n.* a cylindrical instrument with a piston used to draw in a quantity of liquid by suction and eject or inject it in a stream, spray or jet, a squirt. *v.t.* to water, spray or cleanse with a syringe.
syrinx (si'ringks), *n.* (*pl.* **syringes** (-rin'jēz), **syrinxes** the Eustachian tube; the organ of song in birds, the inferior larynx, a modification of the trachea where it joins the bronchi; a surgically-made passage or fistula; a Pan-pipe; a narrow gallery cut in the rock in ancient Egyptian tombs. **syringeal** (-rin'ji-), *a.*
syrup (si'rəp), *n.* a saturated solution of sugar in water, usu. combined with fruit-juice etc. for use in cookery, as a beverage etc., or with a medicinal substance; the uncrystallizable fluid separated from sugar-cane juice in the process of refining molasses, treacle; excessive sweetness or sentimentality. **syrupy**, *a.*
syssarcosis (sisahkō'sis), *n.* (*pl.* **-ses** (-sēz)) a connection of parts of the skeleton by intervening muscle.
systaltic (sistal'tik), *a.* of the heart, alternately contracting and dilating, pulsatory.
system (sis'təm), *n.* coordinated arrangement, organized combination, organization, method; an established method or procedure; a coordinated body of principles, facts, theories, doctrines etc.; a logical grouping, a method or plan of classification; a coordinated arrangement or organized combination or assembly of things or parts, for working together, performing a particular function etc. a group of related or linked natural objects, as mountains, the rocks of a geological period etc.; any complex and coordinated whole; any organic structure taken as a whole, as the animal body, the universe etc. **The System**, *n.* (*Austral. Hist.*) the whole question of transportation, including the treatment of convicts; **(the system)** (*coll., derog.*) the establishment, bureaucracy or society generally, esp. when regarded as a destroyer of individualism. **systems analysis**, *n.* the analysis of an industrial, medical, business etc. procedure or task in order to identify its requirements and devise a (computer) system to fulfil these. **systems analyst**, *n.* **systematic, -ical** (-mat'-), *a.* methodical; done, formed or arranged on a regular plan, not haphazard; taxonomic. **systematically**, *adv.* **systematics**, *n.sing.* (the study of) classification or taxonomy. **systematist**, *n.* **systematize, -ise**, *v.t.* **systematization**, *n.* **systematizer**, *n.* **systemic** (-stem'-, -stē'-), *a.* pertaining to or affecting the bodily system as a whole; of an insecticide etc., absorbed by the tissues of a plant etc., thus making it toxic. **systemically**, *adv.*
systole (sis'təli), *n.* the contraction of the heart forcing the blood outwards, alternating with diastole. **systolic** (-tol'-), *a.*
syzygy (siz'iji), *n.* the conjunction or opposition of any two of the heavenly bodies, esp. of a planet with the sun; (*Biol. etc.*) conjunction or union. **syzygetic** (-jet'-), *a.* **syzygetically**, *adv.*

T

T, t, the 20th letter and the 16th consonant (*pl.* **Ts, T's, Tees**), is a hard voiceless dental mute; followed by *h* it has two distinct sounds, surd or breathed, as in *think, thank, thought* (shown in this dictionary by 'th'), and sonant or vocal, as in *this that, though* (shown here by 'dh'). *n.* a T-shaped thing or part. *a.* T-shaped (*usu. in comb.*, as *T-bar, -square*). **to a T** TEE. **T-bandage,** *n.* a bandage in the shape of a 'T'. **T-bar,** *n.* a metal etc. bar in the shape of a 'T'. **T-bar lift,** a ski lift with a T-bar. **T-bone,** *n.* a bone in the shape of a 'T', as in a sirloin steak. **T-shirt,** *n.* an informal light-weight, (short-sleeved) garment for the upper body. **T-square,** *n.* a 'T' shaped ruler.
T, (*chem. symbol*) tritium.
t, (*abbr.*) temperature; tempo; tense; tenor; ton(s); town; transitive.
t' (tə), (*dial.*) the.
't (t), it.
TA, (*abbr.*) territorial army.
Ta, (*chem. symbol*) tantalum.
ta (tah), *int.* thank you.
Taal (tahl), *n.* S African Dutch, Afrikaans.
tab[1] (tab), *n.* a small flap, tag, tongue etc., as the flap of a shoe, the tag or tip of a lace etc.; a small paper flap attached to a file for identification purposes; a strap, a loop; military insignia; (*N Am.*) the bill; a check, close surveillance. *v.t.* to put tabs on. **to keep tabs on,** to keep a watch on.
tab[2] (tab), *n.* (short for) tablet, tabulator. *v.t.* to tabulate.
tabard (tab'əd), *n.* an outer garment worn over armour; a herald's sleeveless coat blazoned with the arms of the sovereign.
tabaret (tab'əret), *n.* a fabric of alternate satin and watered-silk stripes used for upholstery.
Tabasco® (təbas'kō), *n.* a hot, capsicum sauce.
tabasheer (təbəshiə'), *n.* a hydrated opaline silica deposited in the joints of the bamboo, used in the E Indies as a medicine.
tabbinet TABINET.
tabby (tab'i), *n.* silk or other stuff with a watered surface; a tabby-cat; a cat, esp. a female cat; (*fig.*) a gossipy old maid or old woman. *v.t.* to give a wavy or watered appearance to. *a.* wavy, watered. **tabby-cat,** *n.* a grey or brownish cat with dark stripes.
tabefaction, tabefy TABES.
taberdar (tab'ədah), *n.* a scholar of Queen's College, Oxford.
tabernacle (tab'ənakl), *n.* a tent-like structure used by the Jews as a sanctuary before settlement in Palestine; (*fig.*) the human body as the temporary abode of the soul; a non-conformist place of worship; an ornamental receptacle for the consecrated Elements or the pyx; a canopy, canopied stall or niche, a canopy-like structure over a tomb etc.; a socket or hinged post for unstepping the mast on a river-boat. *v.i.* to sojourn. *v.t.* to give shelter to. **Feast of Tabernacles,** an autumn feast of the Jews in memory of the sojourn in the wilderness. **tabernacle-work,** *n.* carved canopies and tracery over a pulpit, stall etc. **tabernacled, tabernacular** (-nak'ū-), *a.*
tabes (tā'bēz), *n.* wasting away, emaciation; a wasting disease. **tabefaction** (tabifak'shən), *n.* wasting away from disease, emaciation. **tabefy** (tab'ifī), *v.i., v.t.* **tabes dorsalis** (dawsah'lis), *n.* an advanced form of syphilis which attacks the spinal cord. **tabescence** (təbes'əns), *n.* **tabescent, tabetic** (təbet'-), **tabic** (tab'-), **tabid** (tab'id), *a.* **tabidly,** *adv.* **tabidness, tabitude** (tab'itūd), *n.*
tabinet, tabbinet (tab'inet), *n.* a watered fabric of silk and wool, used for window-curtains etc.
tabla (tah'blə), *n.* a pair of small Indian drums with variable pitch, played with the hands.
tablature (tab'lətyə, -chə), *n.* a painting on a wall or ceiling; a picture; (*fig.*) a vivid description, mental image etc.; a system of notation for instruments of the lute and violin class, showing string and fret position, and indicating rhythm and fingering.
table (tā'bl), *n.* an article of furniture consisting of a flat surface resting on one or more supports, used for serving meals upon, working, writing, playing games etc.; the food served upon a table, cuisine; the company sitting at a table; a table or board adapted for a particular game (*usu. in comb.*, as *billiard-table*); either half of a backgammon-table; a part of a machine or machine-tool on which the work is put to be operated on; any apparatus consisting of a plane surface; a slab of wood or other material; such a slab with writing or an inscription; hence, the contents of such writing etc.; a list of numbers, references, or other items arranged systematically, esp. in columns; a flat surface, a plateau, the flat face of a gem; a horizontal band of moulding; the sound board of a guitar, cello etc. *v.t.* to lay (a bill etc.) on the table in front of the Speaker in the House of Commons, i.e. to submit for discussion; (*N Am.*) to suspend indefinitely discussion of (a bill etc.). **at table,** taking a meal. **to lay** or **lie on the table,** (*Parl.*) to defer or be deferred indefinitely. **to turn the tables,** to reverse the conditions or relations. **under the table,** illicit, secret; (*coll.*) drunk. **table-beer,** *n.* beer for drinking at meals. **table-cloth,** *n.* a cloth, usu. of white linen, for covering a table, esp. at mealtimes. **table-cover,** *n.* a cloth, usu. coloured, for covering a table at other times. **table-cut,** *a.* cut with a flat face (of gems). **table d'hôte** (tahbl`dōt'), *n.* (*pl.* **tables-**) a hotel or restaurant meal at fixed price, limited to certain dishes arranged by the proprietor. **table-knife,** *n.* a knife for use at meals. **table-land,** *n.* a plateau. **table licence,** *n.* one which permits the holder to serve alcohol with food. **table-linen,** *n.* (*collect.*) table-cloths, napkins etc. **table-lifting, -moving, -rapping, -turning** etc., *n.* making a table rise, move or turn over without apparent cause, as by spiritualistic agency. **table manners,** *n.pl.* accepted behaviour during meals. **table mat,** *n.* a mat placed on a table to protect the surface from hot dishes. **table-money,** *n.* an allowance to general and flag officers for hospitality; a charge to members of clubs for use of the dining-room. **table salt,** *n.* fine, free-flowing salt used at table. **table-skittles,** *n.pl.* a game of skittles set up on a board, and knocked down by a ball suspended above the board. **tablespoon,** *n.* a large spoon, four times the size of a teaspoon and holding half a fluid ounce. **tablespoonful,** *n.* the amount contained in a tablespoon. **table-talk,** *n.* talk at table or meals; familiar conversation, miscellaneous chat. **table tennis,** *n.* a game like lawn tennis played on a table with small bats and hollow balls. **table top,** *n.* the flat top of a table; any flat top. **table-topped,** *a.* **table-**

turning TABLE-LIFTING. **table-ware**, *n*. dishes, plates, knives, forks etc., for use at meals. **table wine**, *n*. an unfortified wine drunk with meals. **tabled**, *a*. **tableful**, *n*. **tabling**, *n*.
tableau (tab'lō), *n*. (*pl*. **-leaux** (-lōz)) a picture; a striking or vivid representation or effect. **tableau vivant** (vē'vă), *n*. (*pl*. **tableaux vivants** (-lō vē'vă)) a motionless group of performers dressed and arranged to represent some scene or event. [F]
tablet (tab'lit), *n*. a thin flat piece of wood etc. for writing on; (*pl*.) a set of these; a small table or slab, esp. used as a memorial; a small flat piece or cake of medicinal or other substance; (*Sc*.) a sweetmeat made from sugar and condensed milk.
tabling TABLE.
tabloid (tab'loid), *n*. proprietary name for a compressed dose of a drug; a popular newspaper measuring about 12 in. (30 cm) by 16 in. (40 cm), informal in style with lots of photographs; (*coll. derog*.) a cheap, sensational newspaper. *a*. in the tabloid style of journalism.
taboo, tabu (təboo'), *n*. a Polynesian custom prohibiting the use of certain persons, places or things; (*fig*.) ban, prohibition; any ritual restriction, usu. of something considered to be unclean or unholy. *a*. banned, prohibited, by social, religious or moral convention. *v.t*. to put under taboo.
tabor, tabour (tā'bə), *n*. (*Hist*.) a small drum used to accompany the pipe.
tabouret (tab'ərit), *n*. a small stool; an embroidery frame; a needle-case.
tabular (tab'ūlə), *a*. in the form of a table, having a broad flat surface; formed in laminae or thin plates; set forth, arranged in, or computed from tables. **tabula** (-lə), (*l*.) a flat surface; a writing tablet. **tabula rasa** (rah'zə), *n*. the mind in its original state, before any impressions have been made on it; a fresh start. **tabularly**, *adv*. **tabulate**, *v.t*. to reduce to or arrange (figures etc.) in tabular form; to shape with a flat surface. *a*. (-lət), table-shaped, broad and flat; arranged in laminae. **tabulation**, *n*. **tabulator**, *n*. an attachment to a typewriter to facilitate tabulation work; a machine which prints data from punched cards, producing tables etc. **tabulatory**, *a*.
tacamahac (tak'əməhak), *n*. a resinous exudation from various S American trees; the balsam poplar.
tace (tā'sē, tah'kā), be silent! **tacet** (tā'set, tas'-, tak'-), *imper*. a direction on a musical score indicating a certain instrument or singer is silent. [L]
tache (tash), *n*. a freckle, a blotch on the skin; a spot, stain or blemish; a catch, a fastening.
tach-, *comb. form* speed, speedy.
tacheometer TACHYMETER.
tachism(e) (tash'izm), *n*. a form of action painting with haphazard blobs of colour. **tachist(e)**, *n., a*.
tachistoscope (təkis'təskōp), *n*. an instrument which flashes images on to a screen for very brief spaces of time, usually a fraction of a second, used in the study of learning and perception. **tachistoscopic** (-skop'-), *a*.
tachogram (tak'əgram), *n*. a visual record produced by a tachograph.
tachograph (tak'əgraf), *n*. a tachometer in a motor vehicle which records its speed, and the distance travelled.
tachometer (təkom'itə), *n*. an instrument for indicating the speed of rotation of a revolving shaft. **tachometry**, *n*.
tachycardia (takikah'diə), *n*. abnormally rapid heart beat.
tachygraphy (təkig'rəfi), *n*. shorthand, esp. one of the ancient Greek or Roman systems. **tachygrapher**, *n*. **tachygraphic, -al** (-graf'-), *a*.
tachylyte (tak'ilit), *n*. a black, vitreous basalt. **tachylytic** (-lit'-), *a*.
tachymeter (təkim'itə), *n*. a surveying-instrument for measuring distances rapidly. **tachymetrical, tacheometrical** (-met'-), *a*. **tachymetry, tacheometry**, *n*.

tachyon (tak'ion), *n*. a theoretical elementary particle which travels faster than the speed of light.
tachyphylaxis (takifilak'sis), *n*. the rapid development of tolerance or immunity to the effects of a specific drug.
tacit (tas'it), *a*. implied but not expressed, understood, existing though not stated. **tacitly**, *adv*. **tacitness**, *n*.
taciturn (tas'itœn), *a*. habitually silent, reserved. **taciturnity** (-tœ'-), *n*. **taciturnly**, *adv*.
tack[1] (tak), *n*. a small, sharp, flat-headed nail; a stitch, esp. one of a series of long, rapid stitches for fastening temporarily; a rope by which the forward lower corner of certain sails is fastened; the part of a sail to which such rope is fastened; the course of a ship as determined by the position of her sails; the act of tacking or changing direction to take advantage of a side-wind etc.; (*fig*.) course of action, policy; stickiness; shoddiness; vulgar ostentation; cheapness; seediness; (*coll*.) food, fare. *v.t*. to fasten with tacks; to stitch together in a hasty manner; to annex, to append (to or on to); to change the course of a ship to the opposite tack. *v.i*. to change the course of a ship by shifting the tacks and position of the sails; to zigzag; (*fig*.) to alter one's conduct or policy. **hard tack**, ship's biscuit. **on the right (wrong) tack**, on the right (wrong) lines. **to come down to brass tacks**, to face realities, to state facts. **tack hammer**, *n*. a small hammer for driving in tacks. **tacker**, *n*. one who tacks; one who makes additions. **tacky**, *a*. slightly sticky; shoddy; vulgar and ostentatious; seedy. **tackily**, *adv*. **tackiness**, *n*.
tack[2] (tak), *n*. saddles, bridles, harness etc.
tacking (tak'ing), *n*. the act of one who tacks; attaching a clause with a different object to a Bill in order to enable this to pass the House of Lords; the right of a mortgagee to priority of a subsequent mortgage over an intermediate one of which he had no notice.
tackle (tak'l), *n*. apparatus, esp. of ropes, pulleys etc. for lifting, hoisting etc., or for working spars, sails etc.; the implements, gear or outfit for carrying on any particular work or sport; an instance of tackling. *v.t*. to grapple with; to seize hold of, stop and challenge (an opponent); (*coll*.) to set to work vigorously upon; to secure or make fast with tackle. **tackler**, *n*. **tackling**, *n*. (*collect*.) tackle.
taco (tah'kō), *n*. (*pl*. **-cos**) a thin pancake from Mexico, usually with a meat or spicy vegetable filling.
tact (takt), *n*. an intuitive sense of what is fitting or right, or adroitness in doing or saying the proper thing; the stroke in beating time. **tactful**, *a*. **tactfully**, *adv*. **tactfulness**, *n*. **tactless**, *a*. **tactlessly**, *adv*. **tactlessness**, *n*.
tactics (tak'tiks), *n*. (*sing. or pl*.) the art of manoeuvring military or naval forces, esp. in actual contact with the enemy; (*pl*.) procedure or devices to attain some end. **tactic**, *n*. a manoeuvre or procedure, a piece of tactics. **tactical**, *a*. skilful, diplomatic; or pertaining to tactics. **tactical voting**, *n*. the practice of voting for the candidate most likely to defeat the favourite candidate, rather than one's preferred candidate. **tactically**, *adv*. **tactician** (-tish'ən), *n*.
tactile (tak'tīl), *a*. of, pertaining to or perceived by the sense of touch. **tactility** (-til'-), *n*. **tactual**, *a*. **tactually**, *adv*.
tadpole (tad'pōl), *n*. the larva of an amphibian, esp. of a frog or toad, before the gills and tail disappear.
taedium vitae (tē'diəm vē'tī, vī'tē), *n*. weariness of life. [L]
taekwondo (tīkwondō'), *n*. a form of Korean self-defence involving kicks and punches.
tael (tāl), *n*. a Chinese weight of 1½ oz. (42·5 g), and a silver monetary unit.
ta'en (tān), *contr*. TAKEN.
taenia (tē'niə), *n*. (*pl*. **-niae**) a band or fillet separating the Doric frieze from the architrave; a band or ribbon-like part; a genus of internal parasites containing the tapeworm. **taeniacide** (-sīd), *n*. a chemical or substance which destroys tapeworms. **taeniasis** (-ni'əsis), *n*. infestation with tapeworms. **taeniate** (-āt), **taenioid** (-oid), *a*.

tafferel (taf'rəl), *n.* the upper part of a ship's stern.
taffeta (taf'itə), *n.* a light, thin, glossy silk fabric; applied also to silk and linen or silk and wool fabrics.
taffrail (taf'rāl), *n.* the rail round a ship's stern.
Taffy (taf'i), *n.* (*coll.*) a Welshman. [Welsh pron. of *Davy*, short for *David*]
taffy (taf'i), TOFFEE.
tafia, taffia (taf'iə), *n.* a variety of rum distilled from molasses.
tag (tag), *n.* any small appendage, as a metal point at the end of a lace; a loop for pulling a boot on; a label, esp. one tied on; a loose or ragged end or edge; a loose tuft of wool on a sheep; the tail or tip of the tail of an animal; anything tacked on at the end; the refrain of a song; a well-worn phrase or quotation; a children's game in which the players try to escape being touched by one; the act of tagging in wrestling. *v.t.* (*past, p.p.* **tagged**) to fit, furnish or mark with a tag; to furnish with tags or trite phrases; to attach (to, on to or together); to touch in the game of tag; in wrestling, to touch a team-mate's hand as a signal that he may take his turn in the ring; (*coll.*) to follow closely or persistently (after); to call or name; to remove tags from (a sheep). **to tag along with**, to go along with someone, to follow. **tag day**, *n.* a flag day. **tag end**, *n.* the final part of something. **tag-rag** RAGTAG under RAG [1]. **tag-tail**, *n.* a hanger-on, a sycophant. **tagged**, *a.* **tagged atom**, *n.* the radioactive isotope of a tracer element. **tagger**, *n.* one who tags, esp. the pursuer in the game of tag; (*pl.*) thin tin-plate or sheet iron. **tagging**, *n.*
Tagetes (təjē'tēz), *n.* a genus of showy American plants of the aster family comprising the French and African marigolds.
tagliatelle (talyətel'i), *n.* pasta in the form of thin strips. [It.]
tahini (təhē'ni), *n.* a thick paste made from ground sesame seeds. [Arab.]
tahr (teə), **thar** (tah), *n.* a beardless Himalayan goat.
tahsil (tahsēl'), *n.* a division for revenue and other administrative purposes in some Indian states. **tahsildar** (-dah'), *n.* a tahsil officer.
taiaha (tī'əhah), *n.* (*New Zealand*) a chieftain's walking-stick, a wand of office.
Taic THAI.
t'ai chi ch'uan (tī chē chwahn'), *n.* a Chinese form of exercise and self-defence which requires good coordination and balance.
taiga (tī'gə), *n.* the coniferous forests found in subarctic North America and Eurasia. [Rus.]
taihoa (tīhō'ə), *n.* (*New Zealand*) a phrase meaning 'Wait!'
tail[1] (tāl), *n.* the hindmost part of an animal, esp. when it extends beyond the rest of the body; anything resembling this in shape or position, as a prolongation of the body of or a pendant or appendage to anything, a bird's end feathers, a fish's caudal fin, the slender end or luminous train of a comet, the stem of a note in music; (*usu. pl.*) the skirt of a coat; the horizontal unit at the rear of an aeroplane; the hind or lower or inferior part of anything; the lower end of a stream or pool; a retinue, a queue; a person employed to follow another; (*Turkey*) a horse-tail formerly carried before a pasha; something coming from behind, e.g. a tail wind; (*sl.*) the buttocks; (*sl.*) female genitalia; (*sl.*) a woman. *v.t.* to furnish with a tail; (*coll.*) to remove the tails or ends from; to join (on to another thing); to insert one end of (a timber etc.) into a wall etc. *v.i.* to follow closely (after); (*Austral.*) to herd sheep or cattle; to fall behind or drop (away or off) in a scattered line; to swing (up and down stream) with the tide (of a vessel). **bit (piece) of tail**, (*sl. derog.*) a woman. **on someone's tail**, very close behind someone. **to tail away, off**, to dwindle. **to turn tail**, to run away. **with one's tail between one's legs**, beaten, in a state of defeat. **tailback**, *n.* a queue of traffic stretching back from an obstruction or traffic problem. **tail-board**, *n.* the hinged or sliding board at the back of a cart, wagon etc. **tail-coat**, *n.* a coat with tails or the skirt divided at the back, a morning or evening coat. **tail-coated**, *a.* **tail covert**, *n.* the covert feathers around a bird's tail. **tail-end**, *n.* the fag-end. **tail-end Charlie, tail-ender**, *n.* someone bringing up the rear. **tail-feather**, *n.* a rudder feather in a bird's tail. **tail-gate**, *n.* the lower gate of a canal-lock; a tailboard. *v.i.* to drive very closely behind another vehicle. **tail-light**, *n.* a red warning light at the rear of a motor-car. **tail-piece**, *n.* an ornamental design at the end of a chapter or section of a book; a triangular block on a violin etc. to which the strings are attached. **tail-pipe**, *n.* the suction-pipe in a pump. **tailplane**, *n.* the fixed horizontal portion of the tail of an aeroplane. **tail-race**, *n.* the part of a mill-race below a water-wheel. **tail-skid**, *n.* a device to take the weight at the rear end of an aeroplane's fuselage while taxiing. **tail-spin**, *n.* a vertical, nose-foremost dive by an aeroplane, during which it describes a spiral. **tail-stock**, *n.* an adjustable casting on a lathe which supports the free-end of a workpiece. **tail wind**, *n.* a wind blowing in the same direction as one is travelling in. **tailed**, *a.* (*usu. in comb.*, as *long-tailed*). **tailing**, *n.* the action of one that tails; the part of a stone or brick inserted into a wall; (*pl.*) the refuse part of ore, grain etc. **tailless** (-lis), *a.*
tail[2] (tāl), *n.* limitation of ownership, limited ownership; an estate of inheritance limited to a person and the heirs of his body.
tailed, tailless TAIL [1].
tailor (tā'lə), *n.* one whose occupation is to cut out and make clothes, esp. for men. *v.i.* to work as a tailor. *v.t.* to make clothes for (*usu. in p.p.*, as *well-tailored*); to fashion to a particular purpose or need. **tailor-bird**, *n.* an oriental bird that sews together leaves to form its nest. **tailor-made**, *a.* made by a tailor, well cut and close-fitting; exactly fitted to its purpose. **a tailored article of clothing**. **tailor's chalk**, *n.* pipeclay used by tailors and dressmakers to mark material. **tailor's tack**, *n.* loose tacking stitches used to transfer marks from the pattern to the material. **tailored**, *a.* **tailoress** (-ris), *n.fem.* **tailoring**, *n.* **tailorize, -ise**, *v.t.*
Taino (tī'nō), *n.* a member of an extinct American Indian race of the W Indies; their language.
taint (tānt), *n.* a trace of decay, disease etc.; a corrupting influence, infection; a stain, a disgrace. *v.t.* to infect with a poisonous or corrupting element; to sully, to tarnish. *v.i.* to be infected or affected with incipient putrefaction; to weaken. **tainted**, *a.* **taintless** (-lis), *a.* **taintlessly**, *adv.*
taipan (tī'pan), *n.* a large and extremely venomous Australian snake.
taipo (tī'pō), *n.* a devil, an evil spirit; a surveyor's instrument, a theodolite. [Maori]
tait (tāt), *n.* a long-snouted phalanger of Western Australia.
taj (tahj), *n.* a crown, a head-dress of distinction, esp. a tall cap worn by Muslim dervishes.
takahe (tah'kəhē), *n.* the notornis. [Maori]
takapu (takah'poo), *n.* the New Zealand gannet.
take (tāk), *v.t.* (*past* **took** (tuk), *p.p.* **taken**) to lay hold of, to grasp, seize, capture, steal, catch, arrest, gain possession of, win, captivate, transport, escort, charm etc.; to carry off, to remove, carry away, carry with one, convey, conduct, extract, exact, withdraw, extort etc.; to go by means of; to receive, obtain, procure, acquire, consume, eat, appropriate, to assume; to accept, endure, hold, adopt, select, receive and retain, submit to, put up with; to ascertain by inquiry, weighing, measuring etc.; to follow a course of study or action; to understand, detect, apprehend, grasp, suppose, consider, infer, conclude, interpret; to be infected with, to contract, to be affected with, to feel, to experience; to bear in a specified way, to regard (as); to perform (an action etc.); to undertake the duties of; to photograph. *v.i.* to deduct something from,

to derogate, to detract; to have a desired effect, to work, to operate; to come out well (in a photograph); to please, to be popular (with); to be attracted or inclined (to); to betake oneself (to); to be attracted by a bait; to make an acquisition; to fall ill. *n.* the act of taking; that which is taken; the amount (of fish etc.) taken at one catch or in one season; takings; the amount of copy taken at one time; a scene that has been filmed. **on the take**, making money dishonestly. **to take account of**, to consider. **to take advantage of**, to make use of circumstances to the prejudice of; to use to advantage. **to take a fancy to** FANCY. **to take after**, to resemble physically, mentally etc. **to take against**, to take a dislike to. **to take aim**, to direct a missile etc. **to take apart**, to separate; (*coll.*) to criticize severely. **to take as read**, to assume. **to take away**, to subtract; to remove. **to take back**, to withdraw, to retract. **to take care**, to be careful, cautious, or vigilant. **to take care of**, to look after, to provide for. **to take down**, to write down; to swallow; to pull to pieces; to humiliate. **to take effect** EFFECT. **to take fire**, to ignite; (*fig.*) to become excited. **to take five**, (*coll.*) to take a few minutes' break. **to take for**, to mistake for. **to take for a ride**, to deceive. **to take for granted**, to accept or assume without question; to fail to appreciate. **to take from**, to deduct from; to diminish, to derogate. **to take heed** HEED. **to take hold of**, to seize. **to take in**, to admit, to receive; to undertake, to do (washing, typewriting etc.); to include, to comprise; to contract, to furl (sails); to understand, to accept as true; to deceive. **to take in hand** HAND [1]. **to take into one's head**, to seize the idea or belief (that), to resolve (to). **to take it**, to accept misfortune or punishment. **to take it or leave it**, to accept with its problems, or not at all. **to take it out of**, (*coll.*) to exhaust the strength or freshness of. **to take it out on**, to vent one's anger on. **to take leave** LEAVE [1]. **to take notice**, to observe; to show alertness. **to take oath**, to swear (that). **to take off**, to remove; to deduct (from); to drink off, to swallow; to mimic; to jump (from); to begin flight. **to take on**, to engage (workmen etc.); to undertake (work etc.); (*coll.*) to be very upset. **to take one up on**, to accept a person's challenge. **to take out**, to remove (a stain etc.); to bring, lead or convey out; to procure; to copy. **to take over**, to assume the management, ownership etc. of. **to take place**, to happen. **to take root**, to strike root. **to take someone out of himself, herself**, to distract someone from their problems or shyness. **to take the air** AIR. **to take the field** FIELD. **to take to**, to resort to; to form a habit or liking for. **to take to heart** HEART. **to take to pieces**, to separate something into its various components; (*coll.*) to criticize severely. **to take to task**, to reprove. **to take up**, to lift (up); to receive into a vehicle; to enter upon, to begin; to pursue; to occupy, to engage; to take into custody; to accept; to pick up and secure; to take possession of; to criticize. **to take up on**, to argue (with someone). **to take upon**, to assume. **to take upon oneself**, to take responsibility for. **to take up with**, to associate with. **takeaway**, *n., a.* (food) bought from a restaurant for consumption at home; the restaurant or shop where such food is bought; a takeaway meal. **takedown**, *a.* made to be disassembled; a humiliation. **take-home pay**, *n.* the amount of salary left after deductions (income tax, national insurance etc.). **take-in**, *n.* a deception, a fraud, an imposition. **take-off**, *n.* caricature; the spot from which one's feet leave the ground in leaping; a stroke in croquet by which a player sends his own ball forward and touches another ball without shifting it; the rising of an aircraft into the air. **take-out**, *n.* a takeaway. **take-over**, *n.* the act of seizing control. **take-over bid**, *n.* an offer to purchase enough shares to obtain control of a company. **take-up**, *n.* the act of claiming something, especially of services or state benefit. **tak(e)able**, *a.* **taker**, *n.* one who takes, esp. one who accepts a bet. **taking**, *a.* that takes; pleasing, alluring, attractive; infectious. *n.* the act of one that takes; capture, arrest; a state of agitation; (*pl.*) money taken; receipts. **takingly**, *adv.* **takingness**, *n.*

takin (tah′kin, -kēn), *n.* a hollow-horned, goat-like antelope inhabiting the Mishmi Hills of SE Tibet.

talapoin (tal′əpoin), *n.* a Buddhist priest or monk in Burma, Ceylon etc.; an African monkey, *Cercopithecus talapoin.*

talaria (təleə′riə), *n.pl.* the winged boots or sandals of Hermes, Iris etc. [L]

talbot (tawl′bət), *n.* a large variety of hound, formerly used for tracking and hunting.

talbotype (tawl′bətīp), *n.* a process invented by Fox Talbot in 1840 of producing a latent image upon sensitized paper, the basis of the photographic process. [W.H. Fox *Talbot*, 1800–1877]

talc (talk), *n.* a fibrous, greasy magnesium silicate occurring in prisms and plates, used as a lubricator etc.; (*coll.*) mica; talcum powder. **talcite** (-īt), *n.* a massive variety of talc. **talcky, talcoid** (-oid), **-cose** (-ōs), **-cous**, *a.* **talcum powder** (-kəm), *n.* powdered magnesium silicate; a powder made from purified talc, used to absorb excess body moisture, usually perfumed.

tale (tāl), *n.* a narrative, an account, a story, true or fictitious, esp. an imaginative or legendary story; an idle or malicious report. **an old wive's tale** WIFE. **to tell one's (it's) own tale**, to speak for oneself (itself). **to tell tales**, to tell lies; to report malicious stories to someone in authority. **to tell tales out of school**, to break confidences. **talebearer**, *n.* one who spreads malicious reports. **talebearing**, *n., a.* **tale-teller**, *n.*

Talegalla (taligal′ə), *n.* a genus of birds comprising the brush-turkey and allied megapods of Australia and New Guinea; a bird of this genus, esp. the brush-turkey.

talent (tal′ənt), *n.* a weight and denomination of money in ancient Greece, Rome, etc. differing in various countries at different times; a particular aptitude, gift, or faculty; mental capacity of a superior order; persons of talent; (*coll.*) attractive members of the opposite sex, collectively. **talent scout, spotter**, *n.* a person who is employed to discover talented people, e.g. for sports' clubs or the entertainment industry. **talent show**, *n.* a show which gives amateur entertainers the opportunity to display their ability. **talented**, *a.* endowed with talents or ability. **talentless** (-lis), *a.*

tales (tā′lēz), *n.* a writ for summoning jurors to make up a deficiency; a list of such as may be thus summoned. **talesman** (tā′lēzmən, tālz′-), *n.* a person thus summoned.

taliacotian (taliəkō′shən), *a.* pertaining to the Italian anatomist Tagliacozzi (*d.* 1599). **taliacotian operation**, *n.* the operation of forming a new nose by taking a graft from the arm or forehead, dissevered only after union has taken place.

taligrade TALIPED.

talion (tal′iən), *n.* the law of retaliation. **talionic** (-on′-), *a.*

taliped (tal′iped), *a.* club-footed; having the feet twisted into a peculiar position (of the sloth). **taligrade** (-grād), *a.* walking on the outer side of the foot. **talipes** (-pēz), *n.* club-foot; the sloth-like formation of the feet.

talipot (tal′ipot), *n.* an E Indian fan-palm.

talisman (tal′ismən), *n.* (*pl.* **-mans**) a charm, an amulet, a magical figure to which wonderful effects were ascribed; (*fig.*) something producing wonderful effects. **talismanic** (-man′-), *a.* **talismanically**, *adv.*

talk (tawk), *v.i.* to speak; to converse, to communicate ideas or exchange thoughts in spoken words or through other means; to have the power of speech; to make sounds as in speech. *v.t.* to express in speech; to converse about, to discuss; to speak (a specified language); to persuade or otherwise affect by talking. *n.* conversation, chat; a subject of conversation; gossip, rumour; a short speech or address. **tall talk**, *n.* (*coll.*) exaggeration. **to talk**

about, to discuss; to gossip about. **to talk at him/her**, to address remarks to indirectly or incessantly; to talk, esp. offensively, about (a person) in his presence. **to talk away**, to spend or use up (time) in talking. **to talk back**, to answer impudently. **to talk big**, to boast. **to talk down**, to silence by loud or persistent talking. **to talk into**, to persuade by argument. **to talk of**, to discuss; to mention; (*coll.*) to suggest. **to talk out**, to kill a motion by discussing it until the time of adjournment. **to talk out of**, to dissuade from doing something by talking. **to talk over**, to discuss at length; to persuade or convince by talking. **to talk round**, to discuss without coming to a decision; to persuade. **to talk shop**, to talk about one's job out of work hours. **to talk tall**, to boast. **to talk through one's hat** HAT. **to talk to**, to speak to; (*coll.*) to remonstrate with, to reprove. **to talk up**, to speak loudly, boldly; to praise. **talkback**, *n.* a two-way radio system. **talk-show**, *n.* a chat-show. **talkative**, *a.* given to talking. **talkatively**, *adv.* **talkativeness**, *n.* **talker**, *n.* **talkie** (-i), *n.* an early film with sound. **talking**, *a.* that talks; able to talk. **talking of**, concerning. **talking book**, *n.* a recording of a book for the blind. **talking head**, *n.* (*TV*) a person shown from the shoulders up only, without any action or illustrative material. **talking picture**, *n.* a talkie. **talking point**, *n.* a matter to be talked about. **talking shop**, *n.* (*often derog.*) a meeting for discussion rather than action. **talking-to**, *n.* a telling-off, a reproof.
tall (tawl), *a.* high in stature, above the average height; having a specified height; (*sl.*) extravagant, boastful, excessive. **tall order**, *n.* an exacting or unreasonable demand. **tall ship**, *n.* a square-rigged sailing ship. **tall story**, *n.* an exaggerated account. **tallish**, *a.* **tallness**, *n.*
tallboy (tawl'boi), *n.* a high chest of drawers, often on legs.
tallier TALLY.
tallith (tal'ith), *n.* a scarf worn by Jews during prayer.
tallow (tal'ō), *n.* a substance composed of the harder or less fusible fats, chiefly of animals, esp. beef- or muttonfat, used for making candles, soap etc. *v.t.* to grease or smear with tallow. **tallow-candle**, *n.* **tallow-chandler** *n.* one who makes or deals in tallow-candles. **tallow-face**, *n.* a person with a pale complexion. **tallow-faced**, *a.* **tallow-tree**, *n.* one of various trees yielding vegetable tallow. **tallower**, *n.* **tallowish**, **tallowy**, *a.*
tally (tal'i), *n.* a stick in which notches are cut as a means of keeping accounts; such a notch or mark, a score; a reckoning; anything made to correspond with something else, a counterpart, a duplicate (of); a mark registering number (of things received, delivered etc.); such a number used as a unit of reckoning; a label or tag for identification. *v.t.* to score as on a tally, to record, to register; to put (a sheet etc.) aft. *v.i.* to agree, to correspond (with). **tally-clerk**, *n.* the person on the wharf who checks a ship's cargo against its cargo list. **tally-man, -woman**, *n.* one who keeps a tally; one who keeps a tallyshop; one who collects hire purchase payments. **tallyshop**, *n.* a shop at which goods are sold on the tally system. **tally system**, *n.* the system of giving and receiving goods on credit, to be paid for by regular instalments. **tallier**, *n.*
tally-ho (tal'ihō'), *n.* the huntsman's cry to hounds; a four-in-hand coach. *v.i.* to utter this cry.
talma (tal'mə), *n.* a long cape or cloak, worn by men or women early in the 19th cent.
Talmud (tal'mud), *n.* the body of Jewish civil and religious law not comprised in the Pentateuch, containing the Mishna and the Gemara. **Talmudic, -ical** (-mu'-), *a.* **Talmudism**, *n.* **Talmudist**, *n.* a person learned in the Talmud. **Talmudistic** (-dis'-), *a.*
talon (tal'ən), *n.* a claw, esp. of a bird of prey; anything hooked or claw-like; the projection on a lock-bolt against which the key presses; the heel of a sword-blade; the cards left in the pack after dealing; an ogee moulding. **taloned**, *a.*
talpa (tal'pə), *n.* an encysted tumour, a wen; the genus of insectivorous animals typified by the common mole.
talus[1] (tā'ləs), *n.* (*pl.* **-li** (-lī)) the ankle bone; talipes; the slope or inclination of a wall etc., tapering towards the top. [L]
talus[2] (tā'ləs), *n.* (*pl.* **-luses**) a mass or sloping heap of fragments accumulated at the base of a cliff, scree.
tam TAM-O'-SHANTER.
tamable, tamability, tamableness TAME.
tamale (təmah'li), *n.* a Mexican dish of maize and meat highly seasoned.
tamandua (təman'dūə), **tamanoir** (təmənwah'), *n.* a genus or subgenus of tropical American ant-eaters.
tamanu (tam'ənoo), *n.* a large E Indian and Polynesian tree yielding tacamahac.
tamara (tam'ərə), *n.* a condiment used largely in Italy, consisting of powdered cinnamon, cloves, coriander, aniseed, and fennel-seeds.
tamarack (tam'ərak), *n.* the American or black larch; a N American pine, *Pinus murrayana*.
tamari (təmah'ri), *n.* a concentrated sauce made from soya beans.
tamarillo (təməril'ō), *n.* (a S American shrub bearing) an edible red fruit.
tamarin (tam'ərin), *n.* a S American marmoset, esp. *Midas rosalia*.
tamarind (tam'ərind), *n.* a tropical tree, *Tamarindus indica*; its pulpy leguminous fruit, used in making cooling beverages, as a food flavouring and as a laxative; its wood.
tamarisk (tam'ərisk), *n.* an evergreen shrub with slender feathery branches and white and pink flowers.
tambour (tam'buə), *n.* a drum, esp. a bass drum; a circular frame on which silk etc. is embroidered; silk or other stuff embroidered thus; a cylindrical stone, as one of the courses of the shaft of a column; a ceiled vestibule in a porch etc., for preventing draughts; a sliding door, or rolling top, on cabinets and desks etc. *v.t., v.i.* to embroider with or on a tambour. **tamboura** (-boo'rə), *n.* an eastern stringed instrument, plucked like a guitar.
tambourin (-rin), *n.* a Provençal tabor or drum; a dance accompanied by this and the pipe; the music for such a dance.
tambourine (tambərēn'), *n.* a small drum-like instrument composed of a hoop with parchment stretched across one head and loose jingles in the sides, played by striking with the hand etc. **tambourinist**, *n.*
tame (tām), *a.* having lost its native wildness; domesticated, not wild; tractable, docile; subdued, spiritless; insipid; (*coll.*) cultivated, produced by cultivation. *v.t.* to make tame; to domesticate, to make docile; to subdue, to humble. **tam(e)able**, *a.* capable of being tamed. **tam(e)ability** (-bil'-), **tam(e)ableness**, *n.* **tameless** (-lis), *a.* **tamely**, *adv.* **tameness**, *n.* **tamer**, *n.*
Tamil (tam'il), *n.* a Dravidian language spoken in S India and Sri Lanka, one of the people speaking this language. *a.* pertaining to this language or people.
tamma (tam'ə), *n.* (*Austral.*) a variety of wallaby.
Tammany (tam'əni), *n.* a political organization in New York affiliated to the Democratic party, also called *Tammany Hall* or *Tammany Society*; (*fig.*) political corruption. **Tammanyism**, *n.* [*Tammany* Hall, meeting-place]
tammuz (tam'uz), *n.* the fourth month in the Jewish calendar according to biblical reckoning, the tenth in the civil year, usually falling in June and July.
tammy (tam'i), *n.* a tam-o'-shanter.
tam-o'-shanter (taməshan'tə), **tam** (tam), *n.* a cap fitting closely round the brows but wide and full above. [Burns's poem *Tam o' Shanter*]
tamp (tamp), *v.t.* to fill up (a blast-hole) with rammed clay above the charge; to ram down (railway ballast, road-

metal etc.). **tamper**[1], *n*. one who or that which tamps; a reflective casing around the core of a nuclear weapon which increases its efficiency. **tamping**, *n*. the act of filling up a hole; the material used. **tampion** (-piən), **tompion** (tom'-), *n*. a stopper for the mouth of a gun; a stopper for the top of an organ-pipe. **tampon** (-pon), *n*. a plug of lint etc. used for stopping haemorrhage and to absorb bodily secretions such as menstrual blood. *v.t.* to plug with a tampon. **tamponade** (-nād'), **tamponage** (-nij), *n*. the surgical use of a tampon.

tampan (tam'pan), *n*. a S African tick with venomous bite.
tamper[1] TAMP.
tamper[2] (tam'pə), *v.i.* to meddle (with); to interfere illegitimately, esp. to alter documents etc., to adulterate, or employ bribery. **tamperer**, *n*. **tampering**, *n*.
tampon TAMP.
tam-tam (tam'tam), TOM-TOM.
tan[1] (tan), *n*. the bark of the oak or other trees, used for tanning hides; the colour of this, yellowish brown; bronzing of the complexion. *a.* tan-coloured. *v.t.* (*past, p.p.* **tanned**) to convert (raw hide) into leather by treating with tannin etc.; to make brown by exposure to the sun; to subject to a hardening process; (*coll.*) to flog, to thrash. *v.i.* to become brown by exposure to the sun. **to tan someone's hide**, to beat somebody very badly, to thrash. **tan-balls**, *n.pl.* spent tan compressed into balls for fuel. **tanbark**, *n*. the bark of some trees, such as the oak, a source of tannin. **tan-coloured**, *a.* **tan-liquor, -ooze**, *n*. an infusion used in tanning. **tan-pit, -vat**, *n*. a vat for steeping hides in tannin. **tanstove**, *n*. a hot-house with a bark-bed. **tanyard**, *n*. a tannery. **tannable**, *a.* **tannage** (-ij), *n*. tanning; that which is tanned. **tannate** (-ət), *n*. a tannic acid salt. **tanned**, *a.* **tanner**[1], *n*. **tannery**, *n*. a place where tanning is done. **tannic**, *a.* pertaining to or derived from tan. **tannic acid**, *n*. tannin. **tanniferous** (-if'-), *a.* **tannin** (-in), *n*. an astringent substance obtained from oak-bark etc., used in tanning leather, making writing ink etc., and in medicine. **tanning**, *n*.
tan[2] (tan), abbrev. for TANGENT.
tanager (tan'əjə), *n*. an American bird related to the finches, usu. with brilliant plumage. **tanagrine** (-grin), **tanagroid** (-groid), *a.*
tandem (tan'dəm), *adv.* one behind the other. *a.* arranged thus. *n*. a vehicle with two horses so harnessed; a cycle for two riders one behind the other; an arrangement of two things one behind the other. **in tandem**, with one thing behind another. **tandemwise**, *adv.*
tandoori (tandoo'ri), *n*. an Indian method of cooking meat, vegetables and bread in a clay oven.
tang[1] (tang), *n*. a strong taste or flavour; a distinctive quality. **tangy**, *a.*
tang[2] (tang), *n*. a projecting piece, tongue etc., as the shank of a knife, chisel etc. inserted into the haft. *v.t.* to furnish with a tang. **tanged**, *a.*
tang[3] (tang), *v.t.* to make a ringing, twanging noise. *n*. such a noise.
tang[4] (tang), *n*. one of various seaweeds.
tanga (tang'gə), *n*. a very brief bikini.
tangelo (tan'jəlō), *n*. (*pl.* -**los**) a tangerine and pomelo hybrid.
tangent (tan'jənt), *a.* meeting at a single point. *n*. a straight line meeting a circle or curve without intersecting it (even if produced); in trigonometry, the ratio of the sine to the cosine; a small piece of metal in a clavichord, that strikes the string. **tangent of an angle**, (*Trig.*) the ratio of a perpendicular subtending the angle in a right-angled triangle to the base. **to go or fly off at a tangent**, to diverge suddenly from a course of thought or action. **tangency**, *n*. **tangential** (-jen'shəl), *a.* of a tangent; along the line of a tangent; digressive, irrelevant. **tangentiality** (-shial'-), *n*. **tangentially**, *adv.*
Tangerine (tanjərēn'), *a.* of or pertaining to Tangiers. *n*. a native of Tangiers; (**tangerine**) a small, loose-skinned orange from Tangiers.

tanghin (tang'gin), *n*. a Madagascan tree, the fruit of which has a poisonous kernel; this poison.
tangible (tan'jibl), *a.* perceptible by touch; definite, capable of realization, not visionary; corporeal. *n*. a tangible thing, property as opposed to goodwill. **tangibility** (-bil'-), **tangibleness**, *n*. **tangibly**, *adv.*
tangle (tang'gl), *v.t.* to knot together or intertwine in a confused mass; to entangle, to ensnare, to entrap; to complicate. *v.i.* to become thus knotted together or intertwined; to come into conflict with; to embrace. *n*. a confused mass of threads, hairs etc. intertwined; (*fig.*) a state of confusion; a complicated situation or problem; various kinds of seaweed. **tangle-foot**, *n*. (*N Am. sl.*) intoxicant, bad whisky. **tangled**, *a.* **tanglement**, *n*. **tangler**, *n*. **tanglesome**, *a.* **tangling** *n.*, *a.* **tanglingly**, *adv.* **tangly**, *a.*
tango (tang'gō), *n*. (*pl.* -**gos**) a dance of a complicated kind for couples, in 4-4 time. *v.i.* to dance the tango.
tangram (tang'gram), *n*. a Chinese puzzle consisting of a square cut into several differently shaped pieces which have to be fitted together.
tangy TANG[1].
tanh (than, tansh), *n*. hyperbolic tangent.
tanist (tan'ist), *n*. (*Anc. Ir.*) the heir presumptive to a chief, elected from his family. **tanistry** (-ri), *n*. this system of succession.
tank (tangk), *n*. a cistern or vessel of large size for holding liquid, gas etc.; a reservoir for water; the part of a locomotive-tender containing the supply of water for the boiler; a heavily-armoured motor vehicle running on caterpillar tractors and carrying guns. *v.t.* to store or treat in a tank; (*sl.*) to defeat. **to tank up**, to fill a vehicle with fuel; to drink, or cause to drink, a large quantity of alcohol. **tank-car**, *n*. a railway wagon carrying a tank or tanks. **tank-engine**, *n*. a locomotive with a water-tank over the boiler, and without a tender. **tank farm**, *n*. an area with oil storage tanks. **tank farmer**, *n*. **tank-farming**, *n*. hydroponics. **tank top**, *n*. a sleeveless top with low neck, usually worn over a shirt or blouse. **tankage** (-ij), *n*. storage in tanks; a charge for this; the cubic capacity of a tank or tanks; the residuum from rendering refuse fats etc., used as a fertilizer. **tanked**, *a.* **tanked-up**, *a.* (*sl.*) drunk. **tanker**, *n*. a specially-built steamer or motor vessel fitted with tanks for carrying a cargo of oil; an aircraft for refuelling other aircraft in the air. **tankful**, *n*. **tanking**, *n*. (*sl.*) a defeat.
tanka[1] (tang'kə), *n*. (*collect.*) the descendants of an aboriginal tribe now living in boats or by the waterside at Canton, China. **tanka-boat**, *n*.
tanka[2] (tang'kə), *n*. a Japanese verse form with five lines.
tankard (tang'kəd), *n*. a large drinking-vessel, usu. of metal and often with a cover.
tanner[1], **tannic** TAN[1].
tanner[2] (tan'ə), *n*. (*old sl.*) sixpence.
Tannoy® (tan'oi), *n*. a public announcement and loudspeaker system.
tanrec, tenrec (ten'rek, tan'-), *n*. a small insectivorous mammal from Madagascar, allied to the hedgehog.
tansy (tan'zi), *n*. a yellow-flowered perennial herb with much-divided, bitter, aromatic leaves.
tantalite, tantalize, tantalum TANTALUS.
Tantalus (tan'tələs), *n*. (*Gr. Myth.*) a son of Zeus, condemned to stand up to his chin in water, which perpetually shrank away when he attempted to quench his thirst; a genus of wading-birds allied to the ibis; (**tantalus**) a spirit-stand in which the decanters remain in sight but are secured by a lock. **tantalus-cup**, *n*. a scientific toy consisting of a figure of a man in a cup, illustrating the principle of the siphon. **tantalate** (-lāt), *n*. salt of tantalic acid. **Tantalean, Tantalian** (-tā'-), **Tantalic** (-tal'-), *a.* of Tantalus. **tantalic**, *a.* of tantalum. **tantalism**, *n*. the punishment

of Tantalus; torment. **tantalite** (-līt), *n.* a black mineral found in granite, an ore of tantalum. **tantalize, -ise,** *v.t.* to tease or torment by holding out some desirable object and keeping it out of reach. **tantalization,** *n.* **tantalizer,** *n.* **tantalizing,** *a.* **tantalizingly,** *adv.* **tantalum** (-ləm), *n.* a metallic element, unable to absorb water and used in surgical instruments, chem. symbol Ta, at. no. 73.
tantamount (tan'təmownt), *a.* equivalent (to) in value or effect.
tantara (tantah'rə), *n.* a quick succession of notes on a trumpet, hunting-horn etc.
tantivy (tantiv'i), *n.* a hunting-cry. *adv.* swiftly, speedily. *v.i.* to hasten, to rush.
tant mieux (tä myœ'), so much the better. [F]
tanto (tan'tō), *a.* (*Mus.*) too much.
tantony (tan'təni), *n.* the smallest pig in a litter, usu. **tantony pig.**
tant pis (tä pē'), so much the worse. [F]
tantra (tan'trə), *n.* one of a class of later Sanskrit religious text-books dealing chiefly with magical powers. **tantric,** *a.*
tantrism, *n.* the teaching of the tantras; a mystical movement based on tantra texts in Hinduism and Buddhism. **tantrist,** *n.*
tantrum (tan'trəm), *n.* a burst of ill-temper, a fit of passion.
Taoiseach (tē'shəkh), *n.* the Prime Minister of Eire. [Ir.]
Taoism (tah'ōizm), *n.* the Chinese religious system based on the teachings of Lao-tze (b. 604 BC). **Tao,** *n.* in Taoism, the principle of creative harmony in the universe; the rational basis of human conduct; the relation between human life and eternal truth. **Taoist,** *n.* **Taoistic** (-is'-), *a.* [Chin. *tao*, way, -ISM]
tap [1] (tap), *v.t.* to strike lightly or gently; to strike lightly with; to apply a protective piece to the heel of (a shoe). *v.i.* to strike a gentle blow. *n.* a light or gentle blow, a rap; the sound of this; a protective piece put on the heel of a shoe; TAPS. **tap-dance,** *n.* a step dance where the performers wear shoes with metal studs in the heels to make a rhythmic sound as they dance. *v.i.* to perform such a dance. **tap-dancer,** *n.* **tap-dancing,** *n.* **tap-shoe,** *n.* a shoe with specially fitted taps, for tap-dancing. **tapper** [1], *n.* **tapping,** *n.*, *a.*
tap [2] (tap), *n.* a valve for drawing water or other fluid through; a faucet, a spigot; a plug for closing a cask etc.; liquor of a particular brew or quality; (*coll.*) a tap-room; a tool for cutting female or internal screw-threads; a device for listening, connected secretly to a telephone. *v.t.* (*past, p.p.* **tapped**) to pierce (a cask etc.) so as to let out a liquid; to let out or draw off (a liquid) thus; to furnish with a tap; to draw (fluid) from; to draw upon a source of supply, usually for the first time; to divert current from (a wire); to attach a receiver to a telephone so as to overhear private conversations; to make an internal screw in. **on tap,** ready to be drawn (of liquor); (*coll.*) readily available. **tap-bolt,** *n.* a bolt with a head on one end and a thread on the other for screwing into some fixed part. **taphole,** *n.* a hole in a furnace through which molten metal can be run off. **taphouse,** *n.* an inn. **tap-room,** *a.* a bar where liquor on tap is sold. **tap-root,** *n.* the main root of a plant penetrating straight downwards for some depth. **tap-water,** *n.* water from a tap, rather than from a bottle. **tappable,** *a.* that may be tapped (of rubber-trees etc.). **tapper** [2], *n.* **tapping,** *n.* **tapster** (-stə), *n.* one who serves liquor in a bar.
tapa (tah'pə), *n.* a kind of tough cloth-like paper made from the bark of a tree, used in Polynesia for clothes, nets etc.
tapadero (tahpədea'rō), *n.* a leather guard worn in front of the stirrup in California. [Sp.]
tape (tāp), *n.* a narrow strip of woven linen, cotton etc., used for tying things together, in dressmaking, bookbinding etc.; such a strip stretched across a race-course at the winning-post; a tape-line or tape-measure; a continuous strip of paper or magnetic tape on which messages are recorded by a recording machine; a strong flexible band rotating on pulleys in printing and other machines. *v.t.* to furnish, fasten or tie up with tapes; to label (sections of a book) with tape bands; to get a measure of; to record (sound) on magnetic tape. **magnetic tape,** plastic tape coated with magnetic powder which can be magnetized in patterns corresponding to recorded music, speech etc. **red tape** RED. **to breast the tape,** in a race on foot, to touch or break the tape across the course to win. **to have someone or something taped,** (*coll.*) to have a complete understanding of. **tape deck,** *n.* a machine for recording sound on to magnetic tape and which replays this sound through an independent amplifier. **tape-line, -measure,** *n.* a tape or strip of metal, marked with inches etc., for measuring, usu. coiled in a round flat case. **tape-machine,** *n.* a telegraphic instrument that records news, stock prices etc. **tape recorder,** *n.* an electronic apparatus for recording music etc., on magnetic tape. **tape record,** *v.t.* **tape recording,** *n.* **tape-script,** *n.* a recording of written text. **tapeworm,** *n.* a cestoid worm infesting the alimentary canal of man and other vertebrates. **tapeless,** *a.* **taper** [1], *n.*
taper [2] (tā'pə), *n.* a small wax-candle; anything giving a very feeble light; tapering form. *v.t., v.i.* to (cause to) become gradually smaller towards one end, or smaller or less important. **tapered,** *a.* tapering in form; lighted by tapers. **taperer,** *n.* a person who bears a taper. **tapering,** *a.* **taperingly,** *adv.* **taperwise,** *adv.*
tapestry (tap'istri), *n.* a textile fabric in which the wool is supplied by a spindle instead of a shuttle, with designs applied by stitches across the warp; any ornamental fabric with designs applied in this manner. *v.t.* to hang with or as with tapestry. *a.* of tapestry. **tapestried,** *a.*
tapetum (təpē'təm), *n.* (*pl.* **-ta** (-tə)) a layer of cells lining the cavity of anthers in flowering plants or of the sporangia in ferns; a portion of the choroid membrane of the eye in certain vertebrates.
tapeworm TAPE.
tapioca (tapiō'kə), *n.* a starchy, granular substance produced by beating cassava, forming a light farinaceous food.
tapir (tā'pə), *n.* an ungulate herbivorous, swine-like mammal of the family Tapiridae, allied to the rhinoceros, with a short, flexible proboscis. **tapiroid** (-roid), *n.*, *a.*
tapis (tap'ē), *n.* tapestry (formerly used as a table-covering). **to be** or **come on the tapis,** to be or come under consideration. [F]
tappable TAP [2].
tapper TAP [1], TAP [2].
tappet (tap'it), *n.* a projecting arm or lever imparting intermittent motion to some part in machinery. **tappet-loom,** *n.* one in which the heddles are worked by tappets. **tappet-motion, -rod, -wheel,** *n.*
tapping TAP [2].
taps (taps), *n.* the last bugle call at night, as a signal for lights-out; a similar call at a military funeral; a song signalling the close of a meeting in the Guide movement.
tapster TAP [2].
tapu (tap'oo), TABOO.
tar [1] (tah), *n.* a thick, dark, viscid oily liquid produced by the dry distillation of organic bodies and bituminous minerals; coal tar. *v.t.* (*past, p.p.* **tarred**) to cover with tar. **tarred with the same brush,** having the same bad characteristics. **to tar and feather,** to smear with tar and then cover with feathers, a punishment inflicted usually by rioters. **tarbrush,** *n.* a brush used to apply tar. **tar macadam** (məkad'əm), **tarmac, Tarmac**® (-mak), *n.* a mixture of tar and road metal giving a smooth, dustless surface. **tar-paper,** *n.* paper treated with tar, used in the building trade. **tar-seal,** *v.t.* to cover the surface of a road with tarmacadam. *n.* the bitumen surface of a road. **tar-**

sealed, *a.* **tar-water**, *n.* a cold infusion of tar, formerly used as a medicine. **tarry**[1], *a.*
tar[2] (tah), *n.* (*coll.*) a sailor.
tara (tah'rə), *n.* the tara-fern. **tara-fern**, *n.* the New Zealand and Tasmanian edible fern.
taradiddle TARRADIDDLE.
taraire (təri'rē), *n.* a white-wood New Zealand tree.
tarakihi (tahrəkē'hē), *n.* an edible fish from New Zealand waters.
taramasalata (tarəməsəlah'tə), *n.* a pale pink creamy Greek pâté, made from grey mullet or smoked cod roe, olive oil and garlic.
tarantass (tarəntas'), *n.* a large Russian four-wheeled carriage without springs.
tarantella (tarəntel'ə), *n.* a rapid Neapolitan dance in triplets; the music for such a dance.
tarantula (təran'tūlə), *n.* a large, venomous spider of S Europe, whose bite was formerly supposed to produce **tarantism** (ta'rən-), an epidemic dancing mania; a large hairy spider of tropical America.
tarata (tərah'tə), *n.* the lemon-wood tree of New Zealand.
taratantara (tarətan'tərə), *n.* the sound of a trumpet, bugle etc.
taraxacum (tərak'səkəm), *n.* (a plant of) a genus containing the dandelion; a plant of this family; a drug prepared from this. **taraxacin** (-səsin), *n.* a bitter principle believed to be the basis of this drug.
tarboosh, tarboush, tarbush (tahboosh'), *n.* a brimless cap or fez, usu. red.
tardamente (tahdəmen'tā), *adv.* (*Mus.*) slowly. [It.]
tardigrade (tah'digrād), *a.* slow-moving. *n.* one of the Tardigrada, a division of edentates containing the sloths.
tardy (tah'di), *a.* moving slowly, sluggish; late, dilatory; reluctant. **tardily**, *adv.* **tardiness**, *n.*
tare[1] (teə), *n.* a vetch, esp. *Vicia sativa*, the common vetch; a weed, perh. darnel.
tare[2] (teə), *n.* an allowance for the weight of boxes, wrapping etc. in which goods are packed; the weight of a motor vehicle without fuel, load, passengers or equipment; the weight of the vessel in which a substance is weighed. *v.t.* to ascertain the amount of tare of.
targe (tahj), *n.* a light shield.
target (tah'git), *n.* an object set up as a mark to be fired at in archery, musketry etc., painted with concentric bands surrounding a bull's eye; the objective of an air-raid; (*coll.*) the aim, sum of money etc. to be reached by a combined effort; the specific objective or aim of any (concerted) effort; any person or thing made the object of attack, criticism etc.; the anti-cathode used in a discharge-tube to set up X-rays; (*Railway*) a small signal at a switch etc.; the neck and breast of a lamb, as a joint of meat. *v.t.* to make a target of; to aim at. **on target**, on the right course; on schedule. **target area**, *n.* an area with a target located in it; an area which is a target. **target language**, *n.* the language into which a text etc. is to be translated. **target practice**, *n.* shooting practice to improve one's aim. **targetable**, *a.* which can be aimed at. *n.* **targeted**, *a.* **targeteer** (-tiə), *n.* a soldier armed with a target. **targeting**, *n.* the art or practice of targeting, esp. in Britain the practice of directing the resources of the social services to those in most need.
Targum (tah'gəm), *n.* one of various ancient Aramaic versions or paraphrases of the Old Testament scriptures. **Targumic** (-goo'-), **Targumistic** (-mis'-), *a.* **Targumist**, *n.*
tariff (ta'rif), *n.* a list or table of duties or customs payable on the importation or export of goods; a duty on any particular kind of goods; a table of charges; a method of charging for gas and electricity; the charges imposed on these. *v.t.* to draw up a list of duties on (goods); to price, to put a valuation on. **tariff reform**, *n.* the removal of defects or abuses in the tariff, free trade or approximation to this.

tarlatan (tah'lətən), *n.* a fine, transparent muslin.
tarn (tahn), *n.* a small mountain lake.
tarnation (tahnā'shən), (*N Am. coll.*), *int.* expressing annoyance etc.
tarnish (tah'nish), *v.t.* to diminish or destroy the lustre of; to sully, to stain. *v.i.* to lose lustre. *n.* loss of lustre, a stain, a blemish; the film of discoloration forming on the exposed face of a mineral. **tarnishable**, *a.* **tarnished**, *a.* **tarnisher**, *n.*
taro (tah'rō), *n.* a tropical plant of the arum family, the roots of which are used as food by Pacific islanders.
taroc (ta'rok), **tarot** (-rō), *n.* a figured playing-card, one of a pack of 78, used in an old (orig. Italian) card-game; a pack of such cards, now widely used for fortune-telling; any game played with tarot cards.
tarpan (tah'pan), *n.* a small wild horse of the steppes of Russia and Tartary.
tarpaulin (tahpaw'lin), *n.* a canvas-cloth coated with tar or other waterproof compound; a sheet of this; a sailor's broad-brimmed tarred or oiled hat; (*coll.*) a sailor.
Tarpeian (tahpē'ən), *a.* relating to *Tarpeia*, said to have been buried at the foot of the Tarpeian rock. **Tarpeian rock**, *n.* a cliff in ancient Rome from which state criminals were hurled.
tarpon (tah'pon), *n.* a large and powerful game-fish of the herring family common in W Indian and western Atlantic waters.
tarradiddle, taradiddle (ta'rədidl), *n.* (*coll.*) a lie, a fib; nonsense.
tarragon (ta'rəgən), *n.* a perennial herb allied to wormwood, used in cookery etc. **tarragon vinegar**, *n.* vinegar flavoured with tarragon.
tarrock (ta'rək), *n.* the young kittiwake; the tern; the guillemot.
tarry[1] TAR[1].
tarry[2] (ta'ri), *v.i.* to stay, to remain behind; to wait; to linger, to delay, to be late. **tarrier**, *n.*
tarsal, tarsi TARSUS.
tarsia (tah'siə), *n.* an Italian mosaic or inlaid woodwork.
tarsus (tah'səs), *n.* (*pl.* **-si** (-sī)) the set of bones (seven in man) between the lower leg and the metatarsus, the ankle; the shank of a bird's leg; the terminal segment in the leg of an insect or crustacean; a plate of connective tissue in the eyelid. **tarsal**, *a.* pertaining to the tarsus or the ankle. *n.* a tarsal bone. **tarsier** (-siə), *n.* a small arboreal lemur with very large eyes and ears, and long tarsal bones. **tarsioid** (-sioid), *a.* pertaining to the tarsier. **tarso-**, *comb. form.* **tarsometatarsal** (-sōmetətah'səl), *a.* pertaining to the tarsus and the metatarsus. **tarsometatarsus** (-səs), *n.*
tart[1] (taht), *a.* sharp to the taste, acid; (*fig.*) biting, cutting, piercing. **tartish**, *a.* **tartly**, *adv.* **tartness**, *n.*
tart[2] (taht), *n.* a pie containing fruit; a piece of pastry with jam etc.; (*coll.*) a girl, esp. one of doubtful character; a prostitute. **to tart up**, to make more showy; to dress cheaply, in a vulgar way. **tarted-up**, *a.* **tartine** (-ēn), *n.* a slice of bread and butter. **tartlet** (-lit), *n.* a small savoury or sweet tart. **tarty**, *a.* (*sl.*) cheap; promiscuous.
tartan[1] (tah'tən), *n.* a woollen fabric cross-barred with stripes of various colours forming patterns distinguishing each of the various Highland clans; such a pattern; a garment, esp. a plaid, made of it; (*fig.*) a Highlander or a Highland regiment. *a.* consisting, made of, or like tartan. **tartaned**, *a.*
tartan[2], **tartane** (tah'tən), *n.* a small Mediterranean one-masted vessel with bowsprit and lateen sail.
tartar[1] (tah'tə), *n.* partially purified argol, the impure tartrate of potassium deposited from wines; cream of tartar; a yellowish incrustation of calcium phosphate deposited on the teeth. **cream of tartar** CREAM. **tartar emetic**, *n.* a tartrate of potassium and antimony used as an emetic and purgative. **tartar sauce**, *n.* TARTAR(E) SAUCE. **tartareous**

(-teə′ri-), *a.* of or like tartar. **tartaric** (-ta′-), *a.* pertaining to, or containing, tartar or tartaric acid. **tartaric acid,** *n.* a crystalline acid from plants, used as a food acid (E334) and in medicines. **tartarize, -ise,** *v.t.* **tartarization,** *n.* **tartarous,** *a.* **tartrate** (-trāt), *n.* a salt of tartaric acid. **tartrated,** *a.* in the form of tartrate. **tartrazine** (-trəzēn), *n.* a yellow dye used in textiles, medicines and food (E102).
Tartar², **Tatar** (tah′tə), *a.* of or pertaining to Tartary or the races comprising Turks, Cossacks and Kirghis Tartars. *n.* a native of Tartary or a member of this group of races; their language; (*fig.*) a person of an intractable, irritable temper or more than one's match. **to catch a Tartar,** to find an opponent stronger than was expected. **Tartarian** (-teə′ri-), *n.*, *a.* **Tartaric** (-ta′-), *a.* **Tartarly,** *adv.*
Tartarean TARTARUS.
tartareous, -ric, tartarize etc. TARTAR¹.
tartar(e) sauce (tah′tah), *n.* a relish served with fish etc. made of mayonnaise, chopped capers, herbs etc.
Tartarus (tah′tərəs), *n.* (*Gr. Myth.*) a deep abyss below Hades where the Titans were confined; the abode of the wicked in Hades. **Tartarean** (-teə′ri-), *a.*
tartish, tartly etc. TART¹.
tartlet TART².
tartrate, tartrazine TARTAR¹.
Tartuffe (tahtuf′, -toof′), *n.* a hypocritical pretender. **Tartuffish,** *a.* **Tartuffism,** *n.* [F, a character in Molière's *Tartuffe*]
tarwhine (tah′win), *n.* any of several edible Australian seafish, esp. the sea bream.
Tarzan (tah′zən), *n.* a man of great physical strength. [from Edgar Rice Burroughs' stories]
tash (tash), *n.* (*coll.*) short for MOUSTACHE.
tasimeter (təsim′itə), *n.* an instrument for measuring small changes in atmospheric pressure or temperature. **tasimetry,** *n.*
task (tahsk), *n.* a definite amount of work imposed; a lesson to be learned at school; a piece of work undertaken voluntarily. *v.t.* to impose a task upon; to overtax. **to take to task,** to reprove, to reprimand. **task force,** *n.* a group formed to carry out a specific task; a military or police group formed to undertake a specific mission. **taskmaster, taskmistress,** *n.* one who imposes a task or demands hard or continuous work. **task-work,** *n.* work imposed or performed as a task.
taslet (tas′lit), *n.* a tassel.
Tasmanian (tazmā′niən), *a.* of or pertaining to Tasmania. *n.* a native or inhabitant of Tasmania. **Tasmanian devil,** *n.* the dasyure, a fierce, cat-like marsupial. **Tasmanian tiger, wolf,** *n.* a wolf-like, carnivorous marsupial, the thylacine, now almost extinct.
tasmanite (taz′mənit), *n.* a resinous mineral found in some Tasmanian shales.
tass (tas), *n.* a cup, a goblet; a small draught.
Tass (tas), (*acronym*) the main Soviet news agency. [Telegrafnoye agentstvo Sovetskovo Soyuza, Telegraphic Agency of the Soviet Union]
tassel¹ (tas′l), *n.* a pendent ornament, usu. composed of a tuft of threads, cords, silk etc. attached to the corners of cushions, curtains etc.; the staminate inflorescence on Indian corn. *v.t.* to furnish or adorn with tassels. *v.i.* to form tassels. **tasselled,** *a.* **tasselling,** *n.* **tasselly,** *adv.*
tassel² (tas′l), TERCEL.
taste (tāst), *v.t.* to try the flavour of by taking into the mouth; to perceive the flavour of; to experience; (*coll.*) to eat a little of. *v.i.* to take or eat a small portion of food etc., to partake (of); to have experience (of); to have a smack or flavour (of). *n.* the sensation excited by the contact of various soluble substances with certain organs in the mouth, flavour; the sense by which this is perceived; the act of tasting; a small quantity tasted, drunk or eaten, a bit taken as a sample; the mental faculty or power of apprehending and enjoying the beautiful and the sublime in nature and art, or of appreciating and discerning between the degrees of artistic excellence; manner, style, execution, as directed or controlled by this; an inclination, a predilection (for). **good taste,** an intuitive feeling for what is aesthetically correct. **to one's taste,** to one's liking. **taste bud,** *n.* any of the tiny organs on the tongue sensitive to taste. **tastable,** *a.* **tasteful,** *a.* having, characterized by, or done with good taste; having or showing aesthetic taste. **tastefully,** *adv.* **tastefulness,** *n.* **tasteless** (-lis), *a.* having no flavour, insipid; vapid, dull; lacking aesthetic taste. **tastelessly,** *adv.* **tastelessness,** *n.* **taster,** *n.* one who tastes, esp. one employed to test the quality of teas, liquors etc., by tasting, orig. one employed to taste food and drink before it was served; an implement for cutting a small cylindrical sample from cheese, a small cup used by a wine-taster etc. **tasty,** *a.* savoury, toothsome, pleasant to the taste; (*coll.*) in good taste. **tastily,** *adv.* **tastiness,** *n.*
tat¹ (tat), *v.t.* (*past, p.p.* **tatted**) to make by knotting. *v.i.* to make tatting. *n.* knotted work or lace used for edging etc., also called **tatting.**
tat² (tat), *n.* a coarse E Indian canvas or matting, esp. gunny.
tat³, tatt TATTY.
tat⁴ (tat), TATTOO³.
ta-ta (tatah′), *int.* (*coll.*) goodbye.
Tatar TARTAR².
tater (tā′tə), (*vulg.*) POTATO.
tatler (tat′lə), TATTLE.
tatou, tatu (tah′too), *n.* an armadillo.
tatter (tat′ə), *n.* a rag; a torn and hanging piece or shred. *v.i.* to fall into tatters. **tatterdemalion** (-dəmā′liən), *n.* a ragged fellow. **tattered, tattering, tattery,** *a.*
Tattersall's (tat′əsawlz), *n.* a lottery based in Melbourne; a sportsman's club. **tattersall,** *n.* material with stripes in a checked pattern. [Richard *Tattersall,* 1724–95, English horseman]
tatting TAT¹.
tattle (tat′l), *v.i.* to chatter, to gossip; to tell tales or secrets. *n.* prattle, gossip, idle talk; a gossip. **tattletale,** *n.*, *a.* tell-tale. **tattler,** *n.* one who tattles, a gossip; a sandpiper. **tattlingly,** *adv.*
tattoo¹ (tatoo′), *n.* the beat of drum recalling soldiers to their quarters; a military pageant, esp. by night. *v.i.* to beat the tattoo.
tattoo² (tatoo′), *v.t.* to mark (the skin) by pricking and inserting pigments. *n.* a mark or pattern so produced. **tattooage** (-ij), *n.* **tattooer, tattooist,** *n.*
tattoo³ (tŭt′oo), *n.* a native-bred pony.
tatty (tat′i), *a.* (*sl.*) untidy, unkempt; shabby, of poor quality. **tat(t),** *n.* rubbish, rags; something which is pretentious but of little real value. **tattily,** *adv.* **tattiness,** *n.*
tatu TATOU.
tau (taw), *n.* the Greek letter τ; a tau cross; the American toad-fish *Batrachus tau.* **tau cross,** *n.* a cross shaped like a T, a St Anthony's cross.
taught (tawt), *past, p.p.* TEACH.
taunt¹ (tawnt), *v.t.* to reproach or upbraid sarcastically or contemptuously. *n.* a bitter or sarcastic reproach. **taunter,** *n.* **taunting,** *n.*, *a.* **tauntingly,** *adv.*
taunt² (tawnt), *a.* tall (of masts).
taupe (tōp), *n.* a brownish-grey colour. *a.* of this colour.
Taurus (taw′rəs), *n.* the Bull, the second zodiacal constellation; the second sign of the zodiac. **Taurean** (-ri′ən), *n.*, *a.* (a person) born under the sign of Taurus. **tauric,** *a.* **tauriform** (-fawm), *a.* having the form of a bull. **taurine** (-rīn), *a.* bull-like; bovine; of or pertaining to Taurus. **tauromachy** (-rom′əki), *n.* bull-fighting; a bull-fight.
taut (tawt), *a.* tight, tense, not slack; in good order, trim. **tauten,** *v.t.*, *v.i.* to make or become taut. **tautly,** *adv.* **tautness,** *n.*
tauto-, *comb. form.* same, identical.

tautochrone (taw'tōkrōn), *n.* a curve such that a heavy body rolling down it from a state of rest will always reach the same point in the same time from whatever point it starts. **tautochronism** (-tok'rə), *n.* **tautochronous** (-tok'rə-), *a.*
tautog (tawtog'), *n.* a food-fish common on the Atlantic coast of the US, the N American black-fish.
tautology (tawtol'əji), *n.* repetition of the same thing in different words; in logic, a statement that is always true. **tautologic, -ical** (-loj'-), *a.* **tautologically,** *adv.* **tautologist,** *n.* **tautologize, -ise,** *v.i.*
tautomerism (tawtom'ərizm), *n.* the ability of two isomers to change into one another so that they may co-exist in equilibrium. **tautomer** (taw'-), *n.* a readily changing isomer. **tautomeric** (-me'-), *a.*
tautonym (taw'tənim), *n.* a two-part name in which the specific name repeats or reflects the generic name, e.g. *Rattus rattus* (black rat). **tautonymic** (-nim'-), **tautonymous** (-ton'-), *a.* **tautonymy** (-tonimi), *n.*
tautophony (tawtof'əni), *n.* the repetition of sounds. **tautophonical** (-fon'-), *a.*
tavern (tav'ən), *n.* a public-house, an inn. **taverna** (-vœ'nə), *n.* a Greek hotel with its own bar; a Greek restaurant.
TAVR (*abbr.*) (*formerly*) Territorial and Army Volunteer Reserve.
taw[1] (taw), *v.t.* to dress or make (skins) into leather with mineral agents, as alum, instead of tannin. **tawer,** *n.* **tawery,** *n.* a place where skins are dressed in this way. **tawing,** *n.*
taw[2] (taw), *n.* a game at marbles; the line from which to play in this; a marble.
tawa (tah'wə), *n.* a New Zealand tree whose wood is used for box-making.
tawahi (təwah'hē), *n.* the New Zealand penguin.
tawdry (taw'dri), *a.* showy without taste or elegance; gaudy. *n.* tasteless or worthless finery. **tawdrily,** *adv.* **tawdriness,** *n.*
tawer, tawery TAW[1].
tawhai (tah'wī), *n.* one of the New Zealand beeches.
tawny (taw'ni), *a.* brownish-yellow, tan-coloured. **tawny eagle,** *n.* a tawny-coloured eagle found in Africa and Asia. **tawny owl,** *n.* a European owl with reddish-brown plumage. **tawniness,** *a.*
taws, tawse (tawz), *n.* (*chiefly Sc.*) a leather strap, usually with the end cut into thin strips, used as an instrument of punishment; a lash.
tax (taks), *n.* a compulsory contribution levied on persons, property or businesses to meet the expenses of government or other public services; (*fig.*) a heavy demand, requirement, strain etc. *v.t.* to impose a tax on; (*fig.*) to lay a heavy burden or strain upon, to make demands upon; to fix amounts of (costs etc.); to register for payment of tribute; to accuse. **tax avoidance,** *n.* legal avoidance of tax. **tax-collector, -gatherer,** *n.* **tax deductible,** *a.* of expenses which can be legally deducted before assessment for tax. **tax disk,** *n.* a paper disk on a car's windscreen showing payment of road tax. **tax evasion,** *n.* illegal avoidance of tax. **tax exile,** *n.* a person who lives abroad to avoid paying (high) taxes. **tax-free,** *a.* exempt from taxation. **tax haven,** *n.* a country where taxes are low, and which attracts tax exiles. **tax-payer,** *n.* **tax return,** *n.* a yearly statement of income used to calculate liability for tax. **tax shelter,** *n.* a financial arrangement to lessen tax payable. **taxability** (-bil'-), *n.* **taxable,** *a.* **taxableness,** *n.* **taxably,** *adv.* **taxation,** *n.* **taxational,** *a.* **taxer,** *n.* **taxing,** *a.* demanding, burdensome, tiring. **taxing master,** *n.* (*Law*) the official who assesses costs of actions.
taxi (tak'si), *n.* (*coll.*) a motor-cab fitted with a taximeter, also called **taxi-cab.** *v.i.* (*pres.p.* **taxiing,** *past, p.p.* **taxied**) (of an aircraft) to travel along the ground. **taxi-rank, taxi-stand,** *n.* a cab rank. **taxiway,** *n.* a marked path from an airport terminal to a runway.
taxidermy (tak'sidœmi), *n.* the art of preparing and mounting the skins of animals so that they resemble the living forms. **taxidermal, taxidermic** (-dœ'-), *a.* **taxidermist,** *n.*
taximeter (tak'simētə), *n.* an automatic instrument fitted in a cab for registering distances and indicating fares.
taxin (tak'sin), *n.* a resinous substance extracted from yew leaves.
taxis (tak'sis), *n.* (*Gram. and Rhet.*) order, arrangement; (*Zool. etc.*) classification; (*Biol.*) movement in response to a stimulus; (*Surg.*) methodical application of manual pressure to restore parts to their places.
taxonomy (takson'əmi), *n.* the department of biology treating of the principles of classification; classification; also called **taxology. taxon** (tak'-), *n.* (*pl.* **-xa** (-sə)) any taxonomical category or group. **taxonomic, -ical** (-nom'-), *a.* **taxonomically,** *adv.* **taxonomist,** *n.*
tazza (tat'sə), *n.* (*pl.* **-ze** (-sā)) a flattish or saucer-shaped cup, esp. one on a high foot. [It.]
TB, (*abbr.*) tuberculosis.
Tb, (*chem. symbol*) terbium.
tbs., tbsp., (*abbr.*) tablespoon, tablespoonful.
Tc, (*chem. symbol*) technetium.
tchick (chik), *n.* a sound made by pressing the tongue against the palate and withdrawing it quickly. *v.i.* to make this sound, as in urging a horse.
Te, (*chem. symbol*) tellurium.
te, ti (tē), *n.* in music, the seventh note of a scale.
tea (tē), *n.* the dried and prepared leaves of a small evergreen tree or shrub of the camellia family; a decoction or infusion of tea-leaves for drinking; a light afternoon or a more substantial evening meal at which tea is served; an infusion or decoction of other vegetable or animal substances for drinking, esp. for medicinal purposes. *v.i.* to take tea. *v.t.* to supply with tea. **black tea,** tea which has been allowed to ferment between the rolling and firing processes; tea without milk. **green tea,** tea roasted while fresh. **high tea, meat tea,** a cooked evening meal at which tea is drunk. **Russian tea,** tea drunk with lemon instead of milk. **tea-bag,** *n.* a small perforated bag containing tea. **tea-bread,** *n.* light, spongy fruit bread. **tea-caddy** CADDY[1]. **tea-cake,** *n.* a light cake, often toasted for eating at tea. **tea-canister,** *n.* **tea-chest,** *n.* a box lined with thin sheet-lead, in which tea is imported. **tea-cloth,** *n.* a dish-cloth. **tea-cosy,** *n.* a cover for a tea-pot to keep the contents hot. **tea-cup,** *n.* a small cup for drinking tea from. **teacupful,** *n.* **tea-dealer,** *n.* **tea-drinker,** *n.* **tea-fight,** *n.* (*coll.*) a tea-party. **tea-garden,** *n.* a garden where tea and other refreshments are served to the public. **tea-gown,** *n.* a woman's loose gown for wearing at afternoon tea. **tea-kettle,** *n.* a kettle for boiling water to make tea. **tea-leaf,** *n.* a leaf of tea or the tea-plant; (*pl.* **-leaves**) such leaves after infusion. **tea-party,** *n.* **tea-plant,** *n.* **teapot,** *n.* a vessel in which tea is infused. **tea-room,** *n.* a restaurant etc. where afternoon teas are provided. **tea rose,** *n.* a rose with scent supposed to resemble tea. **tea-saucer, -service, -set, -spoon,** *n.* utensils used in serving tea. **teaspoonful,** *n.* the quantity contained in a teaspoon; one-eighth of a fluid oz. **tea-table,** *n.* **tea-taster,** *n.* one whose business it is to test and sample teas by the taste. **tea-things,** *n.pl.* (*coll.*) cups, saucers etc. for tea. **tea-time,** *n.* the time of the day when tea is had. **tea-tray,** *n.* **tea-tree,** *n.* the tea-plant or shrub; one of the Australian myrtaceous plants that furnished a tea substitute for early settlers. **tea-trolley,** *n.* **tea-urn,** *n.* a vessel for supplying hot water for tea, or tea in large quantities.
teach (tēch), *v.t.* (*past, p.p.* **taught** (tawt)) to cause (a person etc.) to learn (to do) or acquire knowledge or skill in, to instruct or train in; to impart knowledge or information concerning (a subject etc.), to give lessons in; to educate; to explain, to show, to make known. *v.i.* to perform the duties of a teacher; to give instruction.

teach-in, *n.* an informal conference on a specific subject involving specialists and students. **teachable,** *a.* that may be taught (of a subject etc.); apt to learn, docile. **teachableness,** *n.* **teacher,** *n.* **teaching,** *n.* the act of one who teaches; that which is taught, doctrine. *a.* which teaches; instructive. **teaching aid,** *n.* any device which helps in teaching. **teaching hospital,** *n.* one where medical students are trained. **teaching machine,** *n.* any machine which gives information to the user and corrects the user's answers to questions set.

teak (tēk), *n.* a large E Indian tree yielding a hard, heavy timber used largely for furniture, shipbuilding etc.; this timber.

teal (tēl), *n.* a small freshwater duck related to the mallard.

team (tēm), *n.* two or more horses, oxen etc. harnessed together; a number of persons working together, forming a side in a game etc. *v.t.* to harness or join together in a team; to haul, convey, etc.; with a team; to match. **teammate,** *n.* a fellow team-member. **team-spirit,** *n.* the willingness to act as a team, for the good of the team. **teamwork,** *n.* cooperation with other members of a team or group. **teamster** (-stə), *n.* one who drives a team; (*N Am.*) a lorry-driver. **teamwise,** *adv.*

teapoy (tē'poi), *n.* a small three- or four-legged table for holding a tea-service etc.

tear[1] (teə), *v.t.* (*past* **tore** (taw), *p.p.* **torn** (tawn)) to pull forcibly apart; to lacerate; to make (a rent, tear, wound etc.) thus; to pull violently (away, out etc.); to drag, remove or sever thus. *v.i.* to pull violently (at); to part or separate on being pulled; to rush, move or act with violence. *n.* a rent. **that's torn it,** (*sl.*) that's spoiled it. **to tear a strip off,** (*sl.*) to reprimand. **to tear into,** to attack furiously. **to tear oneself away,** to leave reluctantly. **to tear one's hair,** to be overcome with grief; to be very frustrated. **tear-away,** *n.* (*coll.*) a reckless, sometimes violent, young person. *a.* reckless, impetuous. **tear sheet,** *n.* a page in a publication that is designed to be torn out. **tear-strip,** *n.* a perforated strip along the edge of a paper envelope or other wrapper, which is torn in order to open the envelope. **tearer,** *n.* **tearing,** *a.* (*coll.*) violent, furious, tremendous, as *tearing hurry.*

tear[2] (tiə), *n.* a drop of the saline liquid secreted by the lachrymal glands, moistening the eyes or flowing down in strong emotion etc.; a drop of liquid; a solid, transparent drop or drop-like object. **tear-drop,** *n.* **tear-duct,** *n.* the nasal duct. **tear gas,** *n.* a poison gas that affects the lachrymal ducts and causes violent watering of the eyes. **tear-jerker,** *n.* a book, film or song which is excessively sentimental. **tear-shell,** *n.* a shell that on explosion liberates tear gas. **tear-stained,** *a.* **tearful,** *a.* shedding tears; about to shed tears; sad. **tearfully,** *adv.* **tearfulness,** *n.* **tearless,** *n.* **teary,** *a.*

tease (tēz), *v.t.* to pull apart or separate the fibres of; to comb or card (wool or flax); to annoy, to irritate, to vex with petty requests, importunity, jesting or raillery; to importune (to do something). *n.* one who teases or irritates thus. **teaser,** *n.* one who or that which teases; (*coll.*) an awkward question, problem or situation. **teasing,** *a.* **teasingly,** *adv.*

teasel, teazel, teazle (tē'zl), *n.* a plant with large burs or heads covered with stiff, hooked awns, which are used for raising a nap on cloth; this bur or head; any implement used as a substitute for this. *v.t.* to dress with teasels. **teaseller,** *n.*

teat (tēt), *n.* the nipple of the female breast or udder, through which milk is drawn; a projection or appliance resembling this, such as the attachment on a baby's feeding bottle through which milk etc. is sucked. **teated,** *a.* **teatlike,** *a.*

Tebeth (teb'əth), *n.* the 10th month of the Jewish ecclesiastical year, comprising parts of December and January.

tec (tek), (*sl.*) short for DETECTIVE.

tech.[1] (tek), short for TECHNICAL COLLEGE.
tech.[2] (*abbr.*) technical, technically; technology.
technetium (teknē'shiəm), *n.* a chemical element, at. no. 43; chem. symbol Tc, whose radioisotope is used in radiotherapy.
technic (tek'nik), *n.* technique, technics. *a.* technical.
technical (tek'nikl), *a.* of or pertaining to the mechanical arts; of or pertaining to any particular art, science, business etc.; strictly interpretated within a specific field; pertaining to technique. **technical college,** *n.* a further education college which specializes in technical, secretarial and industrial skills. **technical drawing,** *n.* the study and practice of draughtsmanship. **technical hitch,** *n.* a temporary breakdown caused by a mechanical fault. **technical knockout,** *n.* in boxing, the referee's decision to end the fight because one boxer is too badly injured to continue. **technicality** (-kal'-), *n.* technicalness; a technical term, expression etc.; a petty, formal detail. **technically,** *adv.* **technicalness,** *n.*
technician (teknish'ən), **technicist** (tek'nisist), *n.* one skilled in the technical side of a subject, a technical expert.
Technicolor® (tek'nikülə), *n.* proprietary name for a colour cinematograph process.
technicon (tek'nikən), *n.* a gymnastic apparatus for training the hands of organists, pianists etc.
technics (tek'niks), *n.sing.* technical rules, terms, methods etc.; technology.
technique (teknēk'), *n.* a mode of artistic performance or execution; mechanical skill in art, craft etc.; proficiency in some skill; a particular way of carrying out a scientific, medical etc. operation.
technocracy (teknok'rəsi), *n.* government by technical experts such as scientists. **technocrat** (tek'nōkrat), *n.* **technocratic** (-krat'-), *a.*
technology (teknol'əji), *n.* the science of the industrial arts; the practical application of science to industry and other fields; the total technical means and skills available to a particular human society. **technologic, -ical** (-loj'-), *a.* **technologist,** *n.*
techy (tech'i), etc. TETCHY.
tecnology (teknol'əji), *n.* the scientific study of children; a treatise on children, their diseases etc. **tecnonymy** (-non'imi), *n.* the custom of naming the parent from the child. **tecnonymous,** *a.*
tectology (tektol'əji), *n.* morphology dealing with the organism as a group of organic individuals, structural morphology. **tectological** (-loj'ikl), *a.*
tectonic (tekton'ik), *a.* of or pertaining to building or construction; structural; (*Geol.*) of or resulting from movements of the earth's crust. **tectonics,** *n.sing.* the art of constructing or building; structural geology. **tectonically,** *adv.*
tectorial (tektaw'riəl), *a.* forming a covering (esp. of a membrane of the ear). **tectorium** (-əm), *n.* (*pl.* **-ia** (-iə)).
tectrices (tek'trisēz, -trī'-), *n.pl.* the feathers covering the wing or tail. **tectricial** (-trish'əl), *a.*
Ted (ted), TEDDY BOY.
ted (ted), *v.t.* (*past, p.p.* **tedded**) to turn over and spread (hay) so as to expose to the sun and air. **tedder,** *n.* an implement to do this.
teddy[1] (ted'i), TEDDY-BEAR.
teddy[2] (ted'i), *n.* a woman's one-piece undergarment.
teddy-bear (ted'i), *n.* a stuffed toy bear. [Theodore (*Teddy*) Roosevelt, 1858-1919]
Teddy boy (ted'i), *n.* an adolescent seeking self-expression by affecting clothes reminiscent of the late Edwardian period, and frequently by unruly behaviour. **Teddy girl,** *n. fem.*
Tedesco (tides'kō), *n., a.* (*pl.* **-chi** (-kē)) German (used in connection with painting etc.).
Te Deum (tādā'əm), *n.* a hymn of praise sung at morning

service or as a special thanksgiving; a musical setting for this; a thanksgiving service at which it is sung. [L]
tedious (tē'diəs), *a.* tiresome, wearisome; monotonous, fatiguing. **tediously**, *adv.* **tediousness, tedium** (-əm), *n.*
tee[1] (tē), *n.* the 20th letter of the alphabet, T, t; anything shaped like this letter; a T-shaped pipe, joint etc.; a mark for quoits, curling-stones etc. **to a tee, T,** exactly (right); to a nicety. **tee-shirt** T-SHIRT.
tee[2] (tē), *n.* in golf, a small pile of sand or a plastic cone from which the ball is played at the commencement of each hole. *v.t.* (*past, p.p.* **teed**) to put the ball on this. **to tee off**, to play from this; (*fig.*) to begin.
tee-hee (tēhē'), *int.* an exclamation of laughter. *n.* laughter, a chuckle. *v.i.* to laugh, to titter.
teem[1] (tēm), *v.i.* to be prolific; to be stocked to overflowing. **teemer**, *n.* **teeming**, *a.*
teem[2] (tēm), *v.t.* to pour out (esp. molten metal); to empty. *v.i.* (*dial., coll.*) to pour (down) as rain etc.
teen (tēn), *n.* a teenager. *a.* teenaged.
-teen (-tēn), *suf.* denoting the addition of 10 (in numbers 13–19). **-teenth** (-th), *suf.* forming ordinal numbers from the cardinals 13–19.
teens (tēnz), *n.pl.* the years of one's age from 13 to 19. **teenage, teenaged**, *a.* in the teens; pertaining to teenagers. **teenager**, *n.* an adolescent in his or her teens. **teeny-bopper**, *n.* (*coll.*) a young, usually girl, teenager, who follows the latest trends in clothes and pop-music with great enthusiasm.
teeny (tē'ni), (*childish*) TINY.
teetee[1] (tē'tē), *n.* (*New Zealand*) the diving petrel.
teetee[2] (tē'tē), *n.* a small S American monkey.
teeter (tē'tə), *v.i.* to see-saw; to move to and fro unsteadily, to sway. *v.t.* to move to and fro, to tip up, to tilt. *n.* a see-saw.
teeth (tēth), *pl.* TOOTH.
teethe (tēdh), *v.i.* to cut or develop teeth. **teething**, *n., a.* **teething ring**, *n.* a ring for a teething baby to chew on. **teething troubles**, *n.pl.* the soreness and irritation caused when cutting the first teeth; problems encountered in the early stages of any project.
teetotal, TT (tētō'təl), *a.* of, pertaining to, pledged to, or advocating total abstinence from intoxicants. **teetotalism**, *n.* **teetotaller**, *n.* **teetotally**, *adv.*
teetotum (tētō'təm), *n.* a toy, orig. four-sided, turning like a top, used in a game of chance.
teff (tef), *n.* an Ethiopian cereal, yielding flour used there for bread, elsewhere used as a fodder-plant.
TEFL (tef'l), (*acronym*) the teaching of *E*nglish as a *f*oreign *l*anguage.
Teflon® (tef'lon), *n.* polytetrafluoroethylene, non-stick coating for saucepans etc.
teg (teg), *n.* a female fallow-deer; a doe or a sheep in the second year; its fleece.
tegmen (teg'mən), *n.* (*pl.* **-mina** (-minə)) a covering of an organ or part in an animal or plant; the leathery forewing in orthopteran insects. **tegmental** (-men'-), *a.* **tegminal**, *a.*
tegument (-ūmənt), *n.* a protective covering, envelope or membrane in animals. **tegumental** (-men'-), **tegumentary** (-men'-), *a.* **tegmentum, tegumentum** (-men'-), *n.* (*pl.* **-ta** (-tə)).
tegular (teg'ūlə), *a.* pertaining to, resembling, or consisting of tiles. **tegularly**, *adv.* **tegulated** (-lātid), *a.*
tegument TEGMEN.
tehee (tēhē'), *n.* TEE-HEE.
Te igitur (tēij'itə, tā ig'ituə), *n.* the first two words of the canon of the Mass; the book containing this. [L]
teil (tēl), *n.* the lime-tree or linden. **teil-tree**, *n.*
teinoscope (tī'nəskōp), *n.* an optical instrument, consisting of two prisms so combined that the chromatic aberration of light is corrected, and the linear dimensions of objects are increased or diminished.
tektite (tek'tīt), *n.* a small, dark, glassy stone, thought to be of meteoric origin.
teknology (teknol'əji), **teknonymy** (-non'imi), etc. TECNOLOGY.
tel., (*abbr.*) telegram; telegraph(ic); telephone.
tela (tē'lə), *n.* (*Anat.*) a web, a web-like membrane, structure etc. **telar, telary**, *a.*
telaesthesia, (*esp. N Am.*) **telesthesia** (telǝsthē'ziǝ, -'zhǝ), *n.* the perception of events beyond the normal range of sense perceptions. **telaesthetic, telesthetic** (-thet'-), *a.*
telamon (tel'əmən), *n.* (*pl.* **-mones** (-mō'nēz)) a male figure serving as a column or pilaster.
telangiectasis (telanjiek'təsis), *n.* abnormal dilatation of the small arteries or capillaries. **telangiectatic** (-tat'-), *a.*
tele-, *comb. form* far, distant; television.
tele-ad (tel'iad), *n.* a classified advertisement sent to a newspaper etc. by telephone.
telebarometer (telibərom'itə), *n.* an instrument showing atmospheric pressure at a distance. **telebarograph** (-ba'rəgraf), *n.*
telecast (tel'ikahst), *n.* a television programme. *v.t.* to broadcast by television. **telecaster**, *n.*
telecom (tel'ikom), *n.* short for TELECOMMUNICATIONS.
telecommunication (telikəmūnikā'shən), *n.* communication at a distance, by cable, telephone, radio etc.; (*pl.*) the science of telecommunication.
teledu (tel'ədoo), *n.* the stinking badger of Java and Sumatra.
telefilm (tel'ifilm), *n.* a film made specifically to be shown on television.
telega (tilā'gə), *n.* a four-wheeled springless Russian cart.
telegenic (telijen'ik), *a.* suitable for television.
telegnosis (telinō'sis, -gnō'-), *n.* knowledge of distant events not obtained through normal sense perceptions.
telegony (tileg'əni), *n.* the supposed influence that a female's first mate has on her offspring by subsequent mates. **telegonic** (-gon'-), **telegonous**, *a.*
telegram (tel'igram), *n.* a communication sent by telegraph, now superseded by the telemessage.
telegraph (tel'igraf), *n.* an apparatus or device for transmitting messages or signals to a distance, esp. by electrical agency. *v.t.* to transmit (a message etc.) by telegraph; to signal in any way; to give advance warning of something. *v.i.* to send a message by telegraph; to signal (to etc.). **telegraph-cable, -line, -pole, -post, -wire**, *n.* a cable etc., used in establishing telegraphic connection. **telegraph-plant**, *n.* an E Indian plant of the bean family the leaves of which have a spontaneous jerking movement. **telegraph-table**, *n.* a board on which the names of horses in a race, cricket-scores etc. are displayed. **telegrapher, telegraphist** (-leg'-), *n.* **telegraphist's cramp** MORSE-KEY PARALYSIS. **telegraphese**, *n.* jargon used in telegrams; contracted language. **telegraphic** (-graf'-), *a.* pertaining to the telegraph, sent by telegraph; suitable for the telegraph, brief, concisely worded. **telegraphically**, *adv.* **telegraphophone** (-graf'əfōn), *n.* an instrument for reproducing phonographic sounds or records at a distance. **telegraphy** (-leg'rəfi), *n.* the art or practice of communicating by telegraph or of constructing or managing telegraphs.
telekinesis (telikīnē'sis), *n.* the movement of bodies at a distance and without the interposition of a material cause. **telekinetic** (-net'-), *a.*
telemark (tel'əmahk), *n.* a swinging turn in skiing.
telemessage (tel'imesij), *n.* a message sent by telex or telephone, superseding the telegram.
telemeter (tilem'itə), *n.* an instrument for determining distances, used in surveying, artillery practice etc. *v.t.* to obtain and transmit data from a distance. **telemetric** (telimet'-), *a.* **telemetry** (-tri), *n.* the use of radio waves to transmit data.
telencephalon (telənsef'əlon), *n.* the front part of the brain, made up of the cerebrum, parts of the hypothala-

mus and the third ventricle. **telencephalic** (-fal'-), *a.*
teleology (teliol'əji, tē-), *n.* the doctrine of final causes. **teleologic, -ical** (-loj'-), *a.* **teleologically,** *adv.* **teleologist,** *n.*
teleostean (-os'tiən), *a.* of or belonging to the Teleostei, an order of osseous fishes.
telepathy (təlep'əthi), *n.* communication between minds at a distance without the agency of the senses, mindreading. **telepathic** (telipath'-), *a.* **telepathically,** *adv.* **telepathist,** *n.* **telepathize, -ise,** *v.t., v.i.*
telephone (tel'ifōn), *n.* an instrument for transmitting sounds to distances by a wire or cord, esp. by electrical agency. *v.t.* to transmit by means of a telephone. *v.t.* to speak thus (to). **telephone-box, -booth, -kiosk,** *n.* a callbox. **telephone directory,** *n.* a book listing names, addresses and telephone numbers in a given area. **telephonic** (-fon'-), *a.* **telephonically,** *adv.* **telephonist** (-lef'-), *n.* a person who operates a telephone switchboard. **telephony** (-lef'əni), *n.*
telephotograph (telifō'təgraf), *n.* a picture reproduced at a distance; a picture obtained by telephotography. *v.t.* to photograph thus. **telephotographic** (-graf'-), **telephoto** (-ō), *a.* **telephoto lens,** *n.* a lens of long focal length, for obtaining photographs of very distant objects. **telephotography** (-tog'rəfi), *n.*
teleprinter (tel'iprintə), *n.* a telegraphic apparatus with a keyboard transmitter and a typeprinting receiver, whereby messages are received in printed form.
teleprompter (tel'iprɒmptə), *n.* an apparatus which enables a speaker on television to see his/her text without this being visible to the viewers.
telerecording (telirikaw'ding), *n.* a recording for broadcasting on television.
telesales (tel'isālz), *n.pl.* the selling of goods or services by telephone.
telescope (tel'iskōp), *n.* an optical instrument for increasing the apparent size of distant objects. *v.t.* to drive or force (sections, trains etc.) into each other, like the sliding sections of a telescope. *v.i.* to move or be forced into each other thus. **telescopic** (-skop'-), *a.* performed by, pertaining to, a telescope; capable of retraction and protraction. **telescopic sight,** *n.* a small telescope mounted on a rifle, used as a sight. **telescopically,** *adv.* **telescopist** (-les'-), *n.* **telescopy** (-les'-), *n.*
teleseme (tel'isēm), *n.* a system of electric transmitters with an annunciator used for signalling.
telespectroscope (telispek'trəskōp), *n.* an instrument for spectroscopic examination of the heavenly bodies. **telestereoscope** (-ste'riō-), *n.* an optical instrument presenting distant objects in relief.
telestich (tiles'tik, tel'-), *n.* a poem in which the final letters of each line make up a word or words.
teletext (tel'itekst), *n.* written data transmitted by television companies and viewable on a television screen supplied with a special adaptor.
telethermograph (telithœ'məgraf), *n.* a self-registering telethermometer; a record made by this. **telethermometer** (-thəmom'itə), *n.* a thermometer registering at a distance by electrical means.
telethon (tel'ithon), *n.* a very long television programme, usu. to raise funds for charities.
Teletype® (tel'itīp), *n.* a kind of teleprinter. **teletypewriter** (-tīp'-), *n.* a teleprinter.
televangelist (telivan'jəlist), **tele-evangelist** (teli-ivan'-), *n.* an evangelical preacher who hosts a television show in order to reach a wide audience. **televangelism,** *n.*
teleview (tel'ivū), *v.t., v.i.* to view with a television receiver. **televiewer,** *n.*
televise (tel'ivīz), *v.t.* to transmit by television.
television (tel'ivizhən), *n.* the transmission by radio or other means of visual images so that they are displayed on a cathode-ray tube screen. A rapid succession of such images gives a visual impression of an event as it actually occurs; a television set. **high-definition television,** a television system using a larger number of lines and a wider screen to give a clearer picture. **television set,** *n.* a device designed to receive and decode incoming electrical, television signals. **televisual** (-vizh'əl), *a.* pertaining to television. **televisor** (-vīzə), *n.* a television receiver.
telex (tel'eks), *n.* a teleprinter-hiring service run by the Post Office; a teleprinter used for this service, the message sent. *v.t.* to send a message by telex.
telic (tel'ik), *a.* expressing end or purpose; purposive.
tell (tel), *v.t. (past, p.p.* **told** (tōld)) to relate, to recount; to make known, to express in words, to communicate, to divulge; to inform, to assure; to order, to direct; to distinguish, to ascertain. *v.i.* to give information or an account (of); *(coll.)* to inform, to tattle; to produce a marked effect. **all told,** all included. **to tell apart,** to distinguish between. **to tell off,** to count off; *(coll.)* to scold. **to tell on,** to give away secrets. **to tell one's beads,** to recite the rosary. **to tell the time,** to read the time from a clock. **you're telling me,** *(coll.)* you are telling me something I know all about. **telltale,** *a.* telling tales; conveying information. *n.* one who tells tales, esp. about the private affairs of others; *(fig.)* a sign, an indication, a token; any automatic device for giving information as to condition, position etc.; an index in front of the wheel or in the cabin to show the position of the tiller. **tellable,** *a.* **teller,** *n.* one who tells; one who numbers or counts, esp. one of four appointed to count votes in the House of Commons; an officer in a bank etc. appointed to receive or pay out money. **tellership,** *n.* **telling,** *a.* **telling-off,** *n.* a rebuke. **tellingly,** *adv.*
tellural (tilūə'rəl), *a.* of or pertaining to the earth. **tellurian** (-loo'-), *n., a.* (an inhabitant) of the earth. **telluric** (-loo'-), *a.* **tellurion** (-loo'riən), *n.* an apparatus for illustrating the real and apparent movements of the earth, the phenomena of eclipses, day and night, the seasons etc.
tellurium (tiloo'riəm), *n.* a rare silvery-white non-metallic element, at. no. 52, chem. symbol Te, found in association with gold, silver and bismuth. **tellurate** (tel'ūrāt), *n.* a salt of telluric acid. **telluride** (tel'ūrid), *n.* **telluric, tellurous** (tel'ūrəs), *a.* **telluriferous** (telūrif'-), *a.* **tellurite** (tel'ūrīt), *n.* native oxide of tellurium; a salt of tellurous acid. **tellurometer** (telūrom'itə), *n.* in surveying, an electronic instrument which measures distances using radio waves.
telly (tel'i), *n. (coll.)* television.
telotype (tel'ətīp), *n.* a printing electric telegraph; a telegram printed by this.
telpher (tel'fə), *n.* a form of suspended monorail on which a truck runs, carrying its load hanging below the level of the truck and rail. **telpherline, -way,** *n.* **telpherage** (-rij), *n.* transportation of this nature, operated usually by electricity.
telson (tel'sən), *n.* the last somite or joint in the abdomen of Crustacea.
Telugu (tel'ugoo), *n.* the most extensive of the Dravidian languages spoken on the Coromandel coast of India.
temenos (tem'ənos), *n. (Gr. Ant.)* a sacred enclosure, esp. the precinct of a temple.
temerarious (temərea'riəs), *a.* rash, reckless, headstrong; careless, done at random. **temerariously,** *adv.* **temerity** (time'riti), *n.* excessive rashness, recklessness.
temp., *(abbr.)* temperature.
temp (temp), *n.* short for TEMPORARY. *n.* a temporary, usu. secretarial or clerical, worker.
Tempean (tempē'ən), *a.* of or like Tempe, a beautiful vale in Thessaly which was much praised by classic poets; delightful, lovely.
temper (tem'pə), *v.t.* to mix in due proportion; to bring (steel etc.) to a proper degree of hardness by heating and cooling; *(fig.)* to qualify by admixture, to modify, to mod-

erate, to tone down; to adjust the tones of (an instrument) according to a particular temperament. *v.i.* to be tempered. *n.* disposition of mind, esp. as regards the passions or emotions; self-command; anger, irritation, passion; the state of a metal as regards hardness and elasticity; condition or consistency (of a plastic mixture as mortar). **bad temper**, an angry mood, a tendency to anger. **good temper**, good-nature, calmness of mood. **out of temper**, irritable, in a bad temper. **to keep one's temper**, to remain calm and rational. **to lose one's temper**, to become angry. **temperable**, *a.* **temperative**, *a.* **tempered**, *a.* adjusted according to equal temperament; having a temper. **temperedly**, *adv.* (*usu. in comb. as hot-tempered, hot-temperedly*). **temperer**, *n.*
tempera (tem'pərə), *n.* painting in distemper. [It.]
temperament (tem'pərəmənt), *n.* individual character as determined by the reaction of the physical upon the mental constitution, natural disposition; the adjustment of the tones of an instrument to fit the scale in any key, esp. by a compromise in the case of instruments of fixed intonation, as an organ or piano. **equal temperament**, a system of tuning where the octave is divided into twelve equal intervals, or semitones. **temperamental** (-men'-), *a.* resulting from or connected with temperament; having an erratic or neurotic temperament. **temperamentally**, *adv.*
temperance (tem'pərəns), *n.* moderation, self-restraint, esp. in the indulgence of the appetites and passions; moderation in the use of intoxicants; (*incorr.*) total abstinence. *a.* advocating, promoting moderation, esp. in alcoholic drinks. **temperance hotel**, *n.* one in which alcoholic liquors are not supplied.
temperate (tem'pərət), *a.* moderate, self-restrained; abstemious; not liable to excess of heat or cold, mild (of climate). **temperate zone**, *n.* that part of the earth which, between the tropics and the polar circles, has a moderate climate. **temperately**, *adv.* **temperateness**, *n.*
temperative TEMPER.
temperature (tem'prəchə), *n.* degree of heat or cold in a body or the atmosphere, esp. as registered by a thermometer; (*coll.*) body temperature above normal. **temperature-humidity index**, one which measures temperature and humidity and the effect of these on human comfort.
tempered, etc. TEMPER.
tempest (tem'pəst), *n.* a violent storm of wind, esp. with heavy rain, hail or snow; (*fig.*) violent tumult or agitation. **tempestuous** (-pes'tū-), *a.* **tempestuously**, *adv.* **tempestuousness**, *n.*
Templar (tem'plə), *n.* a member of a religious and military order (the Knights Templars), founded in the 12th cent., for the protection of pilgrims to the Holy Land; a lawyer or a law-student having chambers in the Temple, in London.
template TEMPLET.
temple[1] (tem'pl), *n.* an edifice dedicated to the service of some deity or deities, esp. of the ancient Egyptians, Greeks or Romans; one of the three successive buildings that were the seat of Jewish worship at Jerusalem; a place of public Christian worship; (*London*) two Inns of Court, on the ancient site of the Temple, the establishment of the Knights Templars; (*Bibl. etc.*) a place in which the divine presence specially resides. **templar**, *a.* pertaining to a temple.
temple[2] (tem'pl), *n.* the flat portion of the head between the forehead and ear. **temporal**[2] (-pərəl), *a.* positioned at the temples. **temporal bone**, *n.* one of the two compound bones at the sides of the skull. **temporal lobe**, *n.* one of the large lobes on either side of the cerebral hemisphere, associated with hearing and speech.
temple[3] (tem'pl), *n.* an attachment in a loom for keeping the fabric stretched.
templet, template (tem'plət), *n.* a pattern, gauge or mould,

usu. of thin wood or metal, used as a guide in shaping, turning or drilling; a short timber or stout stone placed in a wall to distribute the pressure of beams etc.
tempo (tem'pō), *n.* (*pl.* **-pi** (-pē)) quickness or rate of movement, time. [It.]
temporal[1] (tem'pərəl), *a.* pertaining to this life; civil, secular; pertaining to or expressing time. **temporal lords**, *n.pl.* the peers of the realm, as distinguished from the archbishops and bishops. **temporal power**, *n.* that of the Pope or the Church in temporal as distinguished from ecclesiastical affairs. **temporally**, *adv.* **temporalness**, *n.* **temporality** (-ral'-), *n.* the laity; a secular possession, esp. (*pl.*) the revenues of a religious corporation or an ecclesiastic; temporalness.
temporal[2] TEMPLE[2].
temporary (tem'pərəri), *a.* lasting or intended only for a time or a special occasion; transient. *n.* a person working on a short-term contract. **temporarily** (tem'-, ra'ri-, -re'ri-), *adv.* **temporariness**, *n.*
temporize, -ise (tem'pəriz), *v.i.* to pursue an indecisive, procrastinating or time-serving policy; to comply with or yield to the requirements of time and occasion; to delay. **temporization**, *n.* **temporizer**, *n.* **temporizingly**, *adv.*
tempt (tempt), *v.t.* to incite or entice (to something or to do); to attract, to allure; to invite. **temptable**, *a.* **temptability** (-bil'-), *n.* **temptation**, *n.* **tempter**, *n.* one who tempts; the devil. **tempting**, *a.* enticing, inviting. **temptingly**, *adv.* **temptress** (-tris), *n. fem.*
tempus fugit (tem'pəs fū'jit), time flies. [L]
ten (ten), *n.* the number or figure 10 or X; the age of 10; the 10th hour after midnight or midday; a group of 10 people or things; a playing-card with 10 pips; a size of shoe or article of clothing designated by the number 10. *a.* 10 in number; aged 10. **tenfold**, *a., adv.* made up of 10 parts; 10 times as much. **ten-gallon hat**, *n.* a wide-brimmed hat worn by American cowboys. **ten minute rule**, *n.* in Parliament, a procedure where a member may make a short, ten-minute speech, introducing a bill. **tenpence**, *n.* **tenpenny**, *a.* priced or sold at tenpence. **ten-pin bowling, ten-pins**, *n.* a game played with ten pins in a skittle-alley. **tenth** (-th), *n.* one of 10 equal parts. *n., a.* (the) last of 10 (people, things etc.); the next letter after the 9th. **tenthly**, *adv.*
tenable (ten'əbl), *a.* capable of being held, retained or maintained against attack. **tenability** (-bil'-), *n.* **tenableness**, *n.* **tenably**, *adv.*
tenace (ten'ās), *n.* (*Whist etc.*) the best and third best cards of a suit held in the same hand. **minor tenace**, the second and fourth best cards thus held. [F]
tenacious (tənā'shəs), *a.* holding fast; inclined to hold fast, obstinate, unyielding; retentive, sticky; highly cohesive, tough. **tenaciously**, *adv.* **tenaciousness**, *n.* **tenacity** (-nas'-), *n.*
tenaculum (tənak'ūləm), *n.* (*pl.* **tenacula** (-lə)) a surgeon's finely-hooked instrument for seizing blood-vessels etc.
tenail, (tənāl'), *n.* (*Fort.*) a low outwork in the enceinte ditch in front of the curtain between two bastions.
tena koe (tē'nakō'ē), *n.* a Maori greeting.
tenant (ten'ənt), *n.* a person holding a land or tenement from a landlord; (*Law*) one holding lands or tenements by any kind of title; a defendant in a real action; (*loosely*) an occupant, a dweller, an inhabitant. *v.t.* to hold as tenant; to occupy. **tenant at will**, (*Law*) one who holds possession of lands at the will of the owner or lessor. **tenant-farmer**, *n.* one cultivating land leased from the owner. **tenantright**, *n.* the right allowed by custom to a well-behaved tenant not to be liable to injurious increase of rent or to be deprived of tenancy without compensation. **tenancy** (-si), *n.* the holding of lands etc.; the period of such property, or office. **tenantable**, *a.* fit for occupation by a tenant. **tenantableness**, *n.* **tenantless** (-lis), *a.* **tenantry** (-ri), *n.* (*collect.*) tenants; the state of being a

tenant. **tenantship**, *n.*
tench (tench), *n.* a freshwater fish of the carp family.
tend[1] (tend), *v.i.* to move, hold a course, or be directed (in a certain direction etc.); to have a bent, inclination or attitude to aim, to conduce (to). **tendency** (-dənsi), *n.* bent, drift, inclination, disposition. **tendentious** (-den'shəs), *a.* with an underlying purpose, intended to further a cause. **tendentiously**, *adv.* **tendentiousness**, *n.*
tend[2] (tend), *v.t.* to attend, to watch, to look after, to take charge of. *v.i.* to attend, to wait (upon). **tended**, *a.*
tendency, tendentious TEND[1].
tender[1] (ten'də), *n.* one who tends; a carriage attached to a locomotive carrying the supply of fuel, water etc.; a vessel attending a larger one, to supply provisions, carry dispatches etc.
tender[2] (ten'də), *v.t.* to offer, to present for acceptance; to offer in payment. *v.i.* to make a tender (to do certain work etc.). *n.* an offer for acceptance; an offer in writing to do certain work or supply certain articles, at a certain sum or rate; (*Law*) a formal offer of money or other things in satisfaction of a debt or liability; (*N Am.*) a bid. **legal tender** LEGAL. **tenderer**, *n.* **tendering**, *n.*
tender[3] (ten'də), *a.* easily impressed, broken, bruised etc., soft, delicate, fragile, weakly, frail; sensitive, easily pained or hurt, susceptible to pain, grief etc., impressible, sympathetic; loving, affectionate; careful, solicitous, considerate (of), requiring to be treated delicately or cautiously; easily chewed (of food). **tender-eyed**, *a.* having gentle eyes. **tenderfoot**, *n.* (*N Am., Austral. sl.*) a newcomer, a novice; one of the lowest grade of Scouts or Girl Guides. **tender-hearted**, *a.* having great sensibility or susceptibility. **tender-heartedly**, *adv.* **tender-heartedness**, *n.* **tender-loin**, *n.* the tenderest part of the loin in beef or pork; (*N Am.*) the undercut, fillet. **tender-minded**, *a.* **tenderize, -ise**, *v.t.* to make tender (e.g. meat) e.g. by pounding and so breaking down the fibres. **tenderization**, *n.* **tenderizer**, *n.* an instrument for pounding meat; a substance which makes (meat) tender. *a.* **tenderling** (-ling), *n.* **tenderly**, *adv.* **tenderness**, *n.*
tendon (ten'dən), *n.* one of the strong bands or cords of connective tissue forming the termination or connection of the fleshy part of a muscle. **tendinous, tendonous**, *a.*
tendril (ten'dril), *n.* a leafless organ by which a plant clings to another body for support. **tendrilled**, *a.*
tenebrae (ten'ibrē), *n.pl.* in the Roman Catholic Church, the offices of matins and lauds for the last three days in Holy Week. **tenebrific** (-brif'-), *a.* causing or producing darkness. **tenebrism**, *n.* a 17th cent. Spanish and Neapolitan school of painting, characterized by areas of dark colour. **tenebrist**, *n.* **tenebrous, tenebrious** (-neb'ri-), *a.* dark, gloomy.
tenement (ten'əmənt), *n.* an apartment or set of apartments used by one family; a dwelling-house; (*esp. Sc.*) a large building divided into rooms and flats; (*Law*) any kind of permanent property that may be held, as lands, houses etc. **tenement-house**, *n.* a house let out in tenements, esp. in a poor district. **tenemental, tenementary** (-men'-), *a.*
tenendum (tinen'dəm), *n.* (*pl.* **tenenda** (-də)) the clause in a deed in which the tenure is defined. [L]
tenesmus (tinez'məs), *n.* an impotent desire, accompanied by painful effort and straining, to evacuate the bowels. **tenesmic**, *a.*
tenet (ten'it, tē'-), *n.* an opinion, principle, doctrine or dogma.
tenfold TEN.
tenner (ten'ə), *n.* (*coll.*) a ten-pound note.
tennis (ten'is), *n.* a game for two or four persons played by striking a ball to and fro with rackets over a net stretched across a walled court (now usu. called **real tennis**). **lawn tennis**, a game for two (singles) or four (doubles) simpler than tennis and omitting the wall. **table tennis**, an indoor game resembling lawn tennis but played on a table; ping-pong. **tennis-arm, -elbow, -knee**, *n.* an arm etc. strained or sprained in tennis-playing, or through other exercise. **tennis-ball**, *n.* **tennis-court**, *n.* **tennis-player, -racket, -shoe**, *n.*
Tennysonian (tenisō'niən), *a.* pertaining to, or in the style of, Alfred, Lord Tennyson (1809–92), poet.
tenoid (tē'noid), TAENIOID under TAENIA.
tenon (ten'ən), *n.* the projecting end of a piece of timber fitted for insertion into a mortise etc. *v.t.* to cut a tenon on; to join by a tenon. **tenon-saw**, *n.* a thin saw with a strong brass or steel back used for cutting tenons etc. **tenon-machine**, *n.* **tenoner**, *n.*
tenor (ten'ə), *n.* a settled course, tendency or direction; general purport or drift (of thought etc.); (*Law*) an exact transcript or copy; the highest of male chest voices between baritone and alto; the part for this; one with a tenor voice; an instrument, esp. the viola, playing a part between bass and alto. *a.* pertaining to or adapted for singing or playing the tenor part. **tenor-clef**, *n.* the C clef placed upon the fourth line of the stave. **tenorist**, *n.*
tenosynovitis (tēnōsinōvī'tis, ten-), *n.* swelling and inflammation in the tendons, usually in joints, caused by repetitive use of the joint concerned.
tenotomy (tinot'əmi), *n.* the cutting of a tendon.
tenpence, tenpenny, etc. TEN.
tenrec TANREC.
tense[1] (tens), *n.* a form taken by a verb to indicate the time, and also the continuance or completedness, of an action. **tenseless**, *a.*
tense[2] (tens), *a.* stretched tight, strained to stiffness (*lit. and fig.*); under or producing emotional stress. *v.t., v.i.* to make or become tense. **tensely**, *adv.* **tenseness, tensity**, *n.* **tensible**, *a.* **tensibility** (-bil'-), *n.* **tensile** (-sīl), *a.* of or pertaining to tension; capable of extension. **tensile strength**, *n.* a measure of the greatest stress a given substance can withstand before breaking. **tensility** (-sil'-), *n.*
tension (-shən), *n.* the act of stretching or the state of being stretched; strain, stress, effort; mental strain, stress or excitement; a state of hostility or anxiety; stress tending to draw asunder the particles of a body, as in a belt, sheet etc. that is being pulled; the expansive force of a gas or vapour. **tension-rod**, *n.* a rod in a structure preventing the spreading of opposite members. **tensiometer**, *n.* any of various instruments which measure tensile strength, comparative vapour pressure, surface tension of liquid, moisture content of soil. **tensiometry** (-siom'itri), *n.* the branch of physics which has to do with tension and tensile strength. **tensional**, *a.* **tensionless** (-lis), *a.* **tensor** (-sə, -saw), *n.* a muscle that stretches or tightens a part.
tent[1] (tent), *n.* a portable shelter consisting of canvas or other flexible material stretched over and supported on poles. *v.t.* to cover with or lodge in a tent. *v.i.* to encamp in a tent. **bell tent**, a circular tent supported on a central pole. **tent bed**, *n.* a bed with curtains which hang from a central point, in the style of a tent. **tent-fly**, *n.* a loose piece of canvas etc. fastened over the ridge-pole to shelter a tent from sun and rain. **tent-maker**, *n.* **tent-peg, -pin**, *n.* a strong peg or pin driven into the ground to secure a tent to the ground. **tent-pole**, *n.* a pole supporting a tent. **tented**, *a.* **tenter**[1], *n.* **tentful**, *n.* **tent-wise**, *adv.*
tent[2] (tent), *n.* a small roll of lint, sponge etc. inserted in a wound, ulcer etc. to keep it open. *v.t.* to keep open with a tent.
tent[3] (tent), *n.* a Spanish wine of a deep red colour, used for sacramental purposes.
tentacle (ten'təkl), *n.* a long slender organ of touch, prehension or locomotion, a feeler, as an arm of a cuttle-fish; a sensitive hair. **tentacled**, *a.* **tentacular** (-tak'ū-), **tentaculate** (-tak'ūlət), **-lated** (-lātid), **-loid** (-loid), *a.* **tentaculiferous** (-lif'-), **-ligerous** (-lij'-), *a.* bearing, or producing, tentacles. **tentaculiform** (-tak'ūlifawm), *a.* **tentaculum**

(-tak'ūləm), n. (pl. -ula (-lə)).
tentative (ten'tətiv), a. consisting or done as a trial or essay, experimental; hesitant, uncertain. n. an experiment, a trial. **tentatively**, adv.
tenter[1] **and** [2] TENT [1] and [4].
tenter[3] (ten'tə), n. a frame or machine for stretching cloth to dry to make it set even and square; a tenterhook. **tenterhook**, n. one of a set of hooks used in stretching cloth on the tenter. **on tenterhooks**, in a state of suspense and anxiety.
tenth, etc. TEN.
tentorium (tentaw'riəm), n. (Anat.) a membranous partition stretched across the cranium between the cerebrum and the cerebellum.
tenui-, comb. form slender, thin.
tenuity (tinū'iti), n. thinness, slenderness; rarity; (fig.) meagreness. **tenuis** (ten'ūis), n. (Gr. Gram.) one of the hard or surd mutes, k, p, t.
tenuifolious (tenūifō'liəs), a. having thin or narrow leaves.
tenuiroster (tenūiros'tə), n. one of the Tenuirostres, a group of insessorial birds with long, slender bills. **tenuirostral**, a.
tenuous (ten'ūəs), a. thin, slender, small, minute; rare, rarefied, subtle, over-refined; insignificant. **tenuously**, adv. **tenuousness**, n.
tenure (ten'yə), n. the act, manner or right of holding property, etc. real estate or office; the manner or conditions of holding; the period or term of holding; the holding of a university or college post for an assured period of time. **tenurial**, a.
tenuto (tinoo'tō), a. sustained, held on for the full time, opp. to staccato. [It.]
teocalli (tēōkal'i, tă-), n. a pyramidal mound or structure, usu. surmounted by a temple, used for worship by the ancient peoples of Mexico, Central America etc.
tepee, teepee (tē'pē), n. a N American Indian tent, cone-shaped, and made of animal skins.
tepefy (tep'ifī), v.t. to make tepid. v.i. to become tepid. **tepefaction** (-fak'-), n.
tephrite (tef'rīt), n. a volcanic rock allied to basalt. **tephritic** (-frit'-), **tephritoid** (-ritoid), a. **tephromancy** (-rōmansi), n. divination by the inspection of sacrificial ashes.
tepid (tep'id), a. moderately warm; lukewarm. **tepidarium** (-deə'riəm), n. (pl. **-ria** (-riə)) (Rom. Ant.) the room between the frigidarium and the caldarium in a Roman bath; a boiler in which the water was heated. **tepidity** (-pid'-), **tepidness**, n. **tepidly**, adv.
tequila (tikē'lə), n. a Mexican spirit distilled from agave which forms the basis of many drinks; the plant from which this spirit is distilled.
ter., (abbr.) terrace; territory.
ter-, comb. form thrice, three times.
tera-, comb. form 10 to the power of 12.
teraphim (te'rəfim), n.pl. household gods or idols among the Jews, consulted as oracles.
terato-, comb. form pertaining to a monster.
teratogeny (terətoj'əni), n. the production of monsters or abnormal growths. **teratogen**, n. a substance that induces abnormalities in a foetus. **teratogenic** (-jen'-), a. **teratism** (te'-), n. a monster, a malformed person or animal, esp. at the foetal stage.
teratology (terətol'əji), n. the branch of biology dealing with monsters and malformations; a work on the marvellous, a marvellous tale etc. **teratological** (-loj'-), a. **teratologist**, n.
teratoma (terətō'mə), n. a tumour consisting of a type of cell not normally found in that part of the body.
teratosis (terətō'sis), n. monstrosity.
terbium (tœ'biəm), n. a rare metallic element, at. no. 65, chem. symbol Tb, found in association with erbium and yttrium. **terbic**, a.

terce (tœs), TIERCE.
tercel (tœ'səl), TIERCEL.
tercentenary (tœsəntē'nəri), a. comprising 300 years. n. a 300th anniversary. **tercentennial** (-ten'-), a. of 300 years. n. a 300th anniversary.
tercet (tœ'sit), n. a verse triplet.
terebene (te'ribēn), n. a liquid hydrocarbon obtained by treating oil of turpentine with sulphuric acid, used as an antiseptic, disinfectant etc. **terebic** (-reb'-), a.
terebinth (te'rəbinth), n. the turpentine-tree, from which turpentine is obtained; its resin. **terebinthine** (-bin'thin), a. pertaining to or partaking of the qualities of terebinth or turpentine.
terebra (te'rəbrə), n. (pl. **-brae** (-brē)) (Ent.) an ovipositor adapted for boring. **terebrate**, v.t. to bore. **terebrant**, n., a. **Terebratula** (-brat'ūlə), n. (pl. **-lae** (-lē)) (Zool.) a genus of brachiopods, largely extinct. **terebratular**, a.
teredo (tərē'dō), n. (pl. **-dos**, **-dines** (-dinēz)) a mollusc that bores into submerged timber, the ship-worm. **teredine** (te'rədin, -din), n.
terek (te'rek), n. a species of sandpiper with the bill curved slightly upward, frequenting E Asia.
Terentian (tiren'shiən), a. of, pertaining to, or in the style of the Roman dramatist Terence.
terete (tərēt'), a. rounded, cylindrical and smooth.
tergal (tœ'gəl), a. of or pertaining to the back or a tergite.
tergeminate (təjem'inət), a. having a pair of leaflets on each of two secondary petioles and at the base.
tergiferous (təjif'ərəs), a. bearing or carrying on the back, as ferns their seeds.
tergite (tœ'jit), n. the upper or dorsal plate of a somite or segment of an articulate animal, also called **tergum** (-gəm) (pl. **-ga** (-gə)). **tergal**, a.
tergiversate (tœ'jivəsāt), v.i. to practise evasions or subterfuges, to equivocate; to change sides. **tergiversation**, n. **tergiversator**, n.
tergum TERGIFEROUS.
term (tœm), n. a limited period; the period during which instruction is regularly given or the courts are in session; an appointed day or date; (Law) an estate to be enjoyed for a fixed period; the period during which childbirth is due; a word having a definite and specific meaning; (pl.) language or expressions used; (pl.) conditions, stipulations, price, charge, rate of payment; relative position, relation, footing; a word or group of words that may be the subject or predicate of a proposition; (Math.) the antecedent or consequent of a ratio; one of the parts of an expression connected by the plus or minus signs. v.t. to designate, to call, to denominate. **terms of reference**, the specific points which a committee or other body is charged to decide. **terms of trade**, the ratio of export prices to import prices. **to be on speaking terms**, to be well enough acquainted to speak to each other; to be friends with. **to bring to terms**, to force or induce to accept conditions. **to come to terms**, to conclude an agreement (with); to find a way of living (with); to give way. **termer, -or**, (Law) one who has an estate for a term of years or for life. **term insurance**, n. insurance of a specific period only. **termtime**, n. **termless** (-lis), a. unlimited, boundless. **termly**, a. occurring every term. adv. term by term; every term; periodically.
terma (tœ'mə), n. (pl. **-mata** (-tə)) a thin layer of grey matter at the front of the 3rd ventricle of the brain. **termatic** (-mat'-), n., a.
termagant (tœ'məgənt), n. a shrewish, abusive, violent woman. a. violent, boisterous, turbulent, shrewish. **termagancy**, n.
termatic TERMA.
termer TERM.
termes (tœ'mēz), n. (pl. **-mites** (-mitēz)) a termite.
terminable, etc. TERMINATE.
terminal (tœ'minəl), a. pertaining to or forming a

terminate 851 **territory**

boundary, limit or terminus; forming or situated at the end of a series or part; ending in death; occurring every term; (*sl.*) extreme, very great. *n.* that which terminates; a limit, an extremity, an end, esp. one of the free ends of an electrical conductor from a battery etc. a rail or air terminus; a device (usu. keyboard and visual display unit) with input and output links with a computer at a distance; a site where raw materials are unloaded, processed and distributed. **terminal illness,** *n.* a fatal disease or disorder. **terminally ill,** *a.* **terminal velocity,** *n.* the speed of an object when it reaches its target; the maximum speed attained by a rocket, missile etc. in a parabolic flight path, or by an aircraft; the maximum speed attained by an object falling through a fluid under gravity. **terminally,** *adv.*
terminate (tœ'mināt), *v.t.* to bound, to limit; to form the extreme point or end of; to put an end to. *v.i.* to stop, to end (in etc.). *a.* limitable, limited, bounded; (*Math.*) finite. **terminable,** *a.* capable of being terminated; having a given term or period. **terminableness,** *n.* **termination,** *n.* **terminational,** *a.* **terminative, terminatory,** *a.* **terminatively,** *adv.* **terminator,** *n.* one who or that which terminates; the dividing-line between the illuminated and the dark part of a heavenly body.
terminer OYER.
terminism (tœ'minizm), *n.* the doctrine that there is a limited period in each man's life for repentance and grace; nominalism. **terminist,** *n.*
terminology (tœminol'əji), *n.* the science or study of the (correct) use of terms; the terms used in any subject. **terminological** (-loj'-), *a.* **terminological inexactitude,** *n.* (*facet.*) a lie. **terminologically,** *adv.*
terminus (tœ'minəs), *n.* (*pl.* **-ni** (-nē)) a boundary, a boundary-mark; the station at the end of a railway, bus route etc.; (**Terminus**) the Roman god of boundaries. **terminus ad quem,** the limit to which; the terminal point; destination. **terminus a quo,** the limit from which, the starting point, beginning. [L]
termite (tœ'mit), *n.* a white ant found in warm regions. **termitarium** (-teə'riəm), **termitary** (-təri), *n.* a nest of or cage for termites. **termitic,** *a.*
termless, termor etc. TERM.
tern[1] (tœn), *n.* a gull-like sea-bird, slenderly-built, with narrow, sharp-pointed wings. **ternery,** *n.*
tern[2] (tœn), *a.* ternate. *n.* a set of three, esp. three lottery numbers winning a large prize if won together; the prize thus won. **ternal, ternary,** *a.* proceeding by or consisting of three. **ternate** (-nət, -nāt), *a.* arranged in threes or whorls of three. **ternately,** *adv.*
terne (tœn), *n.* sheet-iron coated with an alloy of tin and lead; inferior tin-plate. **terne-plate,** *n.*
ternery TERN[1].
terotechnology (terōteknol'əji), *n.* the application of managerial, financial and engineering skills to the installation and efficient operation of equipment and machinery.
terpene (tœ'pēn), *n.* one of various isomeric oily hydrocarbons derived chiefly from coniferous plants. **terpin** (-pin), *n.* a derivative of oil of turpentine and other terpenes. **terpineol** (-pin'iol), *n.* a terpene alcohol used in perfumes.
Terpsichorean (tœpsikərē'ən, -kaw'ri-), *a.* pertaining to Terpsichore (tœpsik'əri), the Muse of dancing; dancing.
terra (te'rə), *n.* (*pl.* **-rrae** (-rē)) earth. **terra alba** (al'bə), *n.* any of various white, earthy substances, e.g. gypsum, kaolin, magnesia etc. **terra-cotta** (-kot'ə), *n.* a hard, unglazed pottery used as a decorative building-material, for statuary etc.; a statue or figure in this; the brownish-orange colour of terra-cotta. **terra firma** (fœ'mə), *n.* dry land. **terra incognita** (inkognē'tə), *n.* unknown country. **terra japonica** (jəpon'ikə), *n.* gambier. [It. and L]
terrace (te'rəs), *n.* a raised level space or platform, either artificially constructed or natural; a balcony; a paved patio; a row of houses joined by party-walls; a street lined with such houses; (*Geol.*) an old shore-line or raised beach; the open tiers around a football stadium where spectators stand. *v.t.* to form into or furnish with terraces. **terraced,** *a.* **terraced house,** *n.* one which forms part of a terrace.
terrain (tərān'), *n.* a region, a tract, an extent of land of a definite geological character; a tract of country which is the scene of operations.
terramara (tərəmah'rə), *n.* (*pl.* **-re** (-rā)) a deposit in parts of S Europe containing prehistoric remains, analogous to that of the kitchen-middens.
Terramycin® (terəmī'sin), *n.* an antibiotic used to treat a wide range of bacterial infections.
terraneous (tərā'niəs), *a.* (*Bot.*) growing on land.
terrapin (te'rəpin), *n.* a freshwater tortoise, esp. the N American saltmarsh or diamond-back terrapin, highly esteemed for food.
terraqueous (terā'kwiəs, -rak'-), *a.* consisting of land and water, as the globe.
terrarium (tereə'riəm), *n.* a vivarium for land animals; a rounded glass container for growing plants.
terrazzo (terat'sō), *n.* (*pl.* **-zzos**) a mosaic floor-covering made by setting marble or other chips into cement, which is then polished.
terrene (terēn'), *a.* pertaining to the earth, earthy; terrestrial. *n.* a region. **terrenely,** *adv.*
terreplein (teə'plān), *n.* the upper surface of a rampart where guns are mounted. [F]
terrestrial (təres'triəl), *a.* pertaining to or existing on the earth, not celestial; consisting of land, not water; living on the ground, not aquatic, arboreal etc.; pertaining to this world, worldly. *n.* an inhabitant of the earth. **terrestrial magnetism,** *n.* the magnetic properties possessed by the earth as a whole, which actuate the magnetic compass. **terrestrially,** *adv.*
terret (te'rit), *n.* one of the rings or loops on harness through which the driving-reins pass; the ring on a dog's collar for attaching the lead.
terrible (ter'ibl), *a.* causing terror or dread; awful, formidable, terrifying, appalling, shocking; very bad, of very poor quality; (*coll.*) excessive, extreme. **terribleness,** *n.* **terribly,** *adv.*
terricolous (tərik'ələs), *a.* living on or in the earth.
terrier[1] (te'riə), *n.* a small active dog of various breeds with an instinct for pursuing its quarry underground; (*coll.*) a member of the Territorial Army.
terrier[2] (te'riə), *n.* (*Hist.*) a book or roll in which lands are described by site, boundaries, acreage etc.
terrific (tərif'ik), *a.* causing terror; frightful, terrible; (*coll.*) very good, excellent. **terrifically,** *adv.* frighteningly; (*coll.*) exceedingly, surprisingly. **terrify** (te'rifī), *v.t.* to strike with terror, to frighten. **terrifying,** *a.* **terrifyingly,** *adv.*
terrigenous (tərij'inəs), *a.* produced by or derived from the earth; of geological deposits, formed in the sea from debris from land erosion.
terrine (tərēn'), *n.* an earthenware jar containing some table-delicacy, sold with its contents; an earthenware pot for cooking; the food, esp. pâté, cooked in such a pot.
territory (te'ritəri), *n.* the extent of land within the jurisdiction of a particular sovereign, state or other power; a large tract of land; (*US*) (**Territory**) a division of the country not yet granted full State rights or admitted into the Union; a field of action; the area defended by an animal or bird. **territorial** (-taw'ri-), *a.* pertaining to territory; limited to a given district; of or pertaining to the Territorial Army; (*US*) pertaining to a Territory or the Territories. *n.* (*coll.*) a member of the Territorial Army. **Territorial Army,** *n.* a military force established in 1907 for home defence to supersede the militia, yeomanry and volunteers (full name: **Territorial and Volunteer Reserve**). **territorial waters,** *n.pl.* the area of sea, usu. three miles out, adjoin-

ing the coast and adjudged to be under the jurisdiction of the country occupying that coast. **territorialize, -ise,** *v.t.* **territoriality,** *n.* **territorially,** *adv.* **territoried,** *a.*
terror (te'rə), *n.* extreme fear; an object of fear; government or revolution by terrorism; (*coll.*) an exasperating nuisance, bore, troublesome child etc. **king of terrors,** death. **Reign of Terror,** the bloodiest period of the French Revolution (April 1793–July 1794). **terror-stricken, -struck,** *a.* terrified, paralysed with fear. **terrorism,** *n.* organized violence and intimidation, usu. for political ends; the act of terrorizing. **terrorist,** *n.* **terroristic** (-ris'-), *a.* **terrorize, -ise,** *v.t.* to terrify; to coerce with threats of violence etc. **terrorization,** *n.*
terry (te'ri), *n.* a pile fabric in which the loops are not cut. **terry-towelling,** *n.* **terry-velvet,** *n.*
Tersanctus (tœsangk'təs), *n.* (*Eccles.*) the Trisagion.
terse (tœs), *a.* concise, pithy, abrupt; of style, neat and compact. **tersely,** *adv.* **terseness,** *n.*
tertial (tœ'shəl), *a.* pertaining to the tertiary feathers. *n.* one of the tertiary feathers.
tertian (tœ'shən), *a.* occurring or recurring every other day. *n.* a fever or ague, the paroxysms of which recur every other day.
tertiary (tœ'shəri), *a.* of the third order, rank or formation; pertaining to the Tertiary; pertaining to higher education. *n.* one of the feathers attached to the proximal joint of a bird's wing; the third geological period, following the Secondary or Mesozoic; in the Roman Catholic Church, a member of the 3rd order of a monastic body. **tertiary college,** *n.* a 6th-form college which teaches vocational courses.
tertius (tœ'shəs), *a.* 3rd (of the name). **tertium quid** (-shəm kwid'), *n.* a 3rd (or intermediate) something.
teru-tero (ter'ute'rō), *n.* the Cayenne lapwing.
tervalent (tœvā'lənt), TRIVALENT.
Terylene® (te'rilēn), *n.* a synthetic polyester textile material.
terza rima (tœt'sərē'mə), *n.* (*pl.* **-ze** (-sā), **-me** (-mā)) a form of triplet in iambic decasyllables or hendecasyllables rhyming *aba bcb.* [It.]
terzetto (tœtset'ō), *n.* a short composition for three performers or singers.
TESL (tes'l), (*abbr.*) teaching of English as a second language.
tesla (tes'lə), *n.* the unit of magnetic flux density equal to a flux of one weber per square metre. **tesla coil,** *n.* a transformer which produces high voltages at high frequencies. [Nikola *Tesla*, 1857–1943, US inventor]
tessellated (tes'əlātid), *a.* composed of tesserae, inlaid; (*Nat. Hist.*) coloured or marked in checkered squares. **tessellar,** *a.* **tessellation,** *n.*
tessera (tes'ərə), *n.* (*pl.* **-sserae** (-rē)) a small cubical piece of marble, earthenware etc., used in mosaics. **tesseral,** *a.* of or composed of tesserae; (*Cryst.*) isometric. **tessular** (-ülə), *a.* (*Cryst.*) tesseral.
tessitura (tesituə'rə), *n.* the natural pitch of a voice or piece of vocal music. [It.]
test[1] (test), *n.* a vessel used in refining gold and silver, a cupel; a critical trial or examination; a means of trial, a standard, a criterion; judgment, discrimination; (*Chem.*) a substance employed to detect one or more of the constituents of a compound; a removable hearth in a reverberatory furnace; an oath or declaration of loyalty or beliefs. *v.t.* to put to the test, to try, to prove by experiment; to try severely, to tax (one's endurance etc.); (*Chem.*) to examine by the application of some reagent; (*Metal.*) to refine in a cupel. **test ban,** *n.* the banning, by agreement, of the testing of nuclear weapons. **test-bed,** *n.* an area for testing machinery etc. **test case,** *n.* a case taken to trial in order that the court shall decide some question that affects other cases. **test-drive,** *n.*, *v.t.* a trial drive of (to drive) a car or other motor vehicle to assess its performance, before purchase. **test-flight,** *n.* a trial flight of a new aircraft. **test-fly,** *v.t.* **test match,** *n.* a cricket match forming one of a series of international matches. **test-paper,** *n.* bibulous paper saturated with a chemical solution that changes colour when exposed to the action of certain chemicals. **test pilot,** *n.* a pilot who test-flies new aircraft. **test-tube,** *n.* a narrow glass tube closed at one end, used in chemical tests. **test-tube baby,** a baby born from an ovum fertilized in an artificial womb in a laboratory, then implanted into the mother's womb; a baby conceived by artificial insemination. **testable**[1], *a.* **tester**[1], *n.* **testing,** *n.*
test[2] (test), *v.t.* to attest, to verify.
test[3] (test), *n.* a shell, a hard covering or exoskeleton.
testa (tes'tə), *n.* (*pl.* **testae** (-tē)) the outer integument of a seed; a test. [L]
testable[1] TEST[1].
testable[2] (tes'təbl), *a.* (*Law*) that may be given in evidence; that may be devised or bequeathed.
Testacea (testā'shiə), *n.pl.* an order of protozoans having shells, shell-bearing invertebrates excluding crustaceans. **testacean,** *n.*, *a.* **testaceous** (-shəs), *a.* **testacel** (tes'təsel), *n.* any species of the Testacella, a group of carnivorous slugs; a member of the Testacella. **testaceology** (-ol'-), *n.*
testacy (tes'təsi), *n.* the state of being testate.
testament (tes'təmənt), *n.* that which testifies proof, attestation; a solemn instrument in writing by which a person disposes of his or her personal estate after death, a will; one of the two main divisions of the Scriptures; (*coll.*) a copy of the New Testament. **New Testament,** the portion of the Bible dealing with the Christian dispensation, composed after the birth of Christ. **Old Testament,** the portion treating of the old or Mosaic dispensation. **testamentary** (-men'-), *a.* **testamentarily,** *adv.* **testamur** (-tā'muə), *n.* a certificate that a student has passed an examination.
testate (tes'tāt), *a.* having made and left a will. *n.* one who has left a will in force. **testation,** *n.* **testator** (-tā-), *n.* someone who dies testate. **testatrix** (-tā'triks), *n. fem.*
tester[1] TEST[1].
tester[2] (tes'tə), *n.* a canopy, esp. over a four-post bedstead.
tester[3] (tes'tə), *n.* a shilling of Henry VIII; (*old sl.*) a sixpence.
testes (tes'tēz), *pl.* TESTIS under TESTICLE.
testicle (tes'tikl), *n.* one of the two glands which secrete the seminal fluid in males. **testicular** (-tik'ū-), **testiculate** (-tik'ūlāt), *a.* **testis** (tes'tis), *n.* (*pl.* **-tes** (-tēz)) a testicle; a round organ or part resembling this.
testify (tes'tifī), *v.i.* to bear witness (to, against, concerning etc.); (*Law*) to give evidence; to make a solemn declaration. *v.t.* to bear witness to; to attest; to affirm or declare; to be evidence or serve as proof of. **testifier,** *n.*
testily TESTY.
testimony (tes'timəni), *n.* a solemn declaration or statement; (*Law*) a statement under oath or affirmation; evidence, proof, confirmation; a solemn declaration of approval or protest; (*Bibl.*) the law as set forth in the two tables, the decalogue. **testimonial** (-mō'-), *n.* a certificate of character, services, qualifications etc. of a person; a formal statement of fact; a gift formally (and usu. publicly) presented to a person as a token of esteem and acknowledgment of services etc. **testimonialize, -ise,** *v.t.* to present with a testimony.
testiness TESTY.
testing TEST[1].
testosterone (testos'tərōn), *n.* a steroid hormone secreted by the testes.
testudo (testū'dō), *n.* (*pl.* **-dos, -dines** (-dinēz)) (*Rom. Ant.*) a screen formed by shields held above their heads and overlapping by soldiers advancing to the attack of a

testy (tes'ti), *a.* irritable, peevish, pettish, petulant. **testily**, *adv.* **testiness**, *n.*

fortress; any similar screen, esp. one used by miners working in places liable to cave in; a genus of tortoises.
testudinal, *a.* pertaining to or resembling the tortoise.
testudinarious (-neə'ri-), *a.* mottled like tortoiseshell.
testudinary, *a.* **testudinated** (-nātid), **-dinate** (-nət), *a.* shaped or arched like the back of a tortoise. **testudineous** (-din'iəs), *a.* resembling the shell of a tortoise.
testy (tes'ti), *a.* irritable, peevish, pettish, petulant. **testily**, *adv.* **testiness**, *n.*
tetanus (tet'ənəs), *n.* a disease marked by long-continued spasms of voluntary muscles, esp. those of the jaws, as in lock-jaw. **tetanal**, *a.* **tetanic** (-tan'-), *a.* pertaining to or characteristic of tetanus. *n.* a medicine acting on the muscles through the nerves, as strychnine. **tetanize, -ise**, *v.t.* **tetanization** *n.* **tetanoid** (-noid), *a.* **tetany** (-ni), *n.* an intermittent spasm of the muscles.
tetchy (tech'i), *a.* fretful, irritable, touchy. **tetchily**, *adv.* **tetchiness**, *n.*
tête-à-tête (tātahtāt', tetahtet'), *a.* private, confidential. *adv.* in private or close confabulation. *n.* (*pl.* **têtes-à-têtes** or **tête-à-têtes** (-tet')) a private interview, a close or confidential conversation; a sofa for two persons, esp. with seats facing in opposite directions so that the occupants face one another.
tether (tedh'ə), *n.* a rope or halter by which a grazing animal is prevented from moving too far; (*fig.*) prescribed range, scope. *v.t.* to confine with or as with a tether. **at the end of one's tether**, at the limit of one's strength, endurance or patience.
tetr(a)-, *comb. form* four.
tetrabasic (tetrəbā'sik), *a.* of an acid, having four replaceable hydrogen atoms.
tetrabranchiate (tetrəbrang'kiət), *a.* having four gills.
tetrachord (tet'rəkawd), *n.* a scale series of half an octave, esp. as used in ancient music. **tetrachordal** (-kaw'-), *a.*
tetract (tet'rakt), *a.* having four rays or branches, as a sponge-spicule. *n.* a four-rayed sponge-spicule. **tetractinal** (-trak'ti-), **-nose** (-nōs), *a.*
tetracyclic (tetrəsiklik), *a.* having four whorls or rings.
tetracycline (tetrəsi'klēn, -klin), *n.* any of several antibiotics, some of which are derived from a bacterium, used to treat a wide range of infections.
tetrad (tet'rad), *n.* the number four; a collection, group, or set of four things; (*Chem.*) an atom or element that can unite with or replace four atoms of hydrogen. **tetradic** (-trad'-), *a.*
tetradactyl (tetrədak'til), *n.*, *a.* (an animal) having four digits on each limb. **tetradactylous**, *a.*
tetradecapod (tetrədek'əpod), *a.* having 14 feet. **tetradecapodon** (-kap'ədon), *n.*, *a.* **tetradecapodous** (-kap'-), *a.*
tetraethyl (tetrəē'thil, -eth'il), *a.* having four ethyl groups. **tetraethyl lead**, *n.* an anti-knock, insoluble liquid used in petrol.
tetragon (tet'rəgon), *n.* a plane figure having four angles. **tetragonal** (-trag'ə-), *a.* having the form of a tetragon; pertaining to the crystal system characterized by three axes at right angles, of which only two are equal. **tetragonally**, *adv.*
tetragram (tet'rəgram), *n.* a word of four letters; a quadrilateral figure. **tetragrammaton** (-gram'əton), *n.* the group of four letters representing the name Jehovah or some other sacred word.
tetragynian (tetrəjin'iən), **tetragynous** (-traj'i-), *a.* having four pistils.
tetrahedron (tetrəhē'drən), *n.* a solid figure bounded by four planes, esp. equilateral, triangular faces. **tetrahedral**, *a.* **tetrahedroid** (-droid), *n.*
tetrahexahedron (tetrəheksəhē'drən), *n.* a solid bounded by 24 equal faces, four corresponding to each face of the cube. **tetrahexahedral**, *a.*
tetralogy (titral'əji), *n.* a collection of four dramatic works, esp. (*Gr. Ant.*) a trilogy or three tragedies, followed by a satyric piece.
tetrameral (titram'ərəl), **tetramerous**, *a.* consisting of four parts.
tetrameter (titram'itə), *n.* a verse consisting of four measures.
tetraplegia (tetrəplē'jə), *n.* quadriplegia, paralysis of both arms and legs. **tetraplegic**, *n.*, *a.*
tetraploid (tet'rəploid), *a.* having four times the haploid number of chromosomes; a tetraploid nucleus or cell.
tetrapod (tet'rəpod), *a.* having four feet or limbs; belonging to the Tetrapoda, a division of butterflies with only four perfect legs. *n.* a four-footed animal, esp. one of the Tetrapoda. **tetrapodous** (-trap'-), *a.*
tetrapterous (titrap'tərəs), *a.* having four wings or wing-like appendages, as certain fruits. **tetrapteran**, *a.* tetrapterous; *n.* a tetrapterous insect.
tetrarch (tet'rahk), *n.* a governor of the fourth part of a province under the Roman empire, also a tributary prince; the commander of a subdivision of the ancient Greek phalanx. **tetrarchate** (-kət), **tetrarchy** (-ki), *n.* **tetrarchical** (-trah'-), *a.*
tetrastich (tet'rəstik), *n.* a stanza, poem, or epigram consisting of four lines of verse.
tetrastichal (-tras'-), **tetrastichic** (-stik'-), *a.* **tetrastichous** (-tras'-), *a.* in four rows.
tetratomic (tetrətom'ik), *a.* having four atoms to a molecule.
tetravalent (tetrəvā'lənt), *a.* having a valency of four. **tetravalency**, *n.*
tetrode (tet'rōd), *n.* a thermionic valve containing four electrodes.
tetroxide (tetrok'sīd), *n.* any oxide containing four oxygen atoms per molecule.
tetryl (tet'ril), *n.* a yellow crystalline explosive solid, used as a detonator.
tetter (tet'ə), *n.* a name applied to several cutaneous diseases.
Teuton (tū'ton), *n.* orig. one of a German tribe, first mentioned as dwelling near the Elbe, *c.* 300 BC; a member of any Teutonic race. **Teuto-**, *comb. form.* **Teutonic** (-ton'-), *a.* pertaining to the Teutons; pertaining to the Germanic peoples, including Scandinavians, Anglo-Saxons etc., as well as the German races; characteristic of the Germans. *n.* the language or languages of the Teutons collectively; Germanic. **Teutonic languages**, *n.pl.* a group of Aryan or Indo-European languages including High and Low German and the Scandinavian languages. **Teutonicism** (-ton'-), **Teutonism**, *n.* **Teutonize, -ise**, *v.t.* **Teutonization**, *n.*
Tex., (*abbr.*) Texas, Texan.
Tex-Mex (teksmeks'), *a.* pertaining to or denoting the Texan version of something Mexican, such as food etc.
text (tekst), *n.* the original words of an author, esp. as opp. to a translation, commentary, or revision etc.; the actual words of a book or poem; the words of something as printed, written, or displayed on a video display unit; a verse or passage of Scripture, esp. one selected as the theme of a discourse; a subject, a topic; text-hand; any book or novel which is studied as part of an educational course. **textbook**, *n.* a standard book for a particular branch of study; a manual of instruction. *a.* conforming to textbook descriptions; ideal; typical. **textbookish**, *a.* **textual** (-tū-), *a.* pertaining to or contained in the text. **textual criticism**, *n.* the study of texts, esp. the Bible, to establish the original text; a close reading and analysis of any literary text. **textualist**, *n.* one who adheres strictly to the text. **textualism**, *n.* **textually**, *adv.* **textuary**, *a.*, *n.*
textile (teks'tīl), *a.* woven; suitable for weaving; pertaining to weaving. *n.* a woven fabric; raw material suitable to be made into cloth. **textorial** (-taw'ri-), *a.* pertaining to weaving.
textual TEXT.

texture (teks'chə), *n.* the particular arrangement or disposition of threads, filaments etc., in a woven fabric; the disposition of the constituent parts of any body, structure, or material; the structure of tissues, tissue; (*Art.*) the representation of the surface of objects in works of art; the quality of something as perceived by touch; the distinct character or quality. *v.t.* to give texture to. **textural**, *a.* **textured**, *a.* **textureless**, *a.* **texturize**, **-ise**, *v.t.* **texturized vegetable protein, TVP**, *n.* a substitute made from soya beans which resembles meat in texture and taste.
TGWU, (*abbr.*) Transport and General Workers Union.
Th., (*abbr.*) Thomas; Thursday.
Th, (*chem. symbol*) Thorium.
-th, (-th), *suf.* forming abstract names [cp. -NESS], as *filth*, *wealth*; forming ordinal numbers, as *fifth*, *fiftieth*.
Thai (tī), *a.* of or pertaining to Thailand, formerly known as Siam. *n.* the language of Thailand; a native or citizen of Thailand.
thalamus (thal'əməs), *n.* (*pl.* **-mi** (-mī)) (*Gr. Ant.*) an inner room, the women's apartment; a part of the mid-brain which has a co-ordinating function in directing nerve impulses to different parts of the brain cortex; the receptacle of a flower. **thalamic** (-lām'-), *a.* **thalamifloral** (-miflaw'rəl), *a.* having the petals, stamens etc., inserted on the thalamus. **thalamium** (-lā'miəm), *n.* (*pl.* **-mia** (-miə),), a spore-case in algae; a form of hymenium in some fungi.
thalassaemia, thalassemia (thaləsē'miə), *n.* a hereditary disorder of the blood due to defects in the synthesis on haemoglobin, sometimes fatal in children.
thalassic (thəlas'ik), *a.* of or pertaining to the sea, marine.
thaler (tah'lə), *n.* an old German silver coin.
Thalia (thəlī'ə), *n.* the Muse of comedy and pastoral poetry. **Thalian**, *a.*
thalidomide (thəlid'əmid), *n.* a drug formerly used as a sedative, withdrawn from use in 1961, as it was shown to be associated with malformation of the foetus when taken by pregnant women. **thalidomide baby**, *n.* a baby born showing the effects of thalidomide. [ph*thalic* acid + *id* (from *imide*)]
thallium (thal'iəm), *n.* a rare soft, white, metallic element, at. no. 81, chem. symbol Tl, the spectrum of which contains a bright-green line (whence the name), used in alloys and glass-making. **thallic, thallous**, *a.*
thallus (thal'əs), *n.* (*pl.* **-lusses, -lli** (-ē)) a plant-body without true root, stem, or leaves. **thalliferous** (-lif'-), *a.* **thalloid** (-oid), *a.* **thallophyte** (-ləfīt), *n.* any member of the lowest division of the vegetable kingdom, lacking stems, leaves and roots, and including algae, fungi, lichens and bacteria. **thallophytic** (-fit'-), *a.*
thalweg, talweg (tahl'veg), *n.* the longitudinal outline of a riverbed; the line of steepest descent from a point on the land surface.
than (dhən, dhan), *conj.* used after adjectives and adverbs expressing comparison, such as *more, better, worse, rather* etc., to introduce the second member of a comparison.
thanage THANE.
thanat(o)-, *comb. form* death.
thanatism (than'ətizm), *n.* the doctrine of annihilation at death. **thanatist**, *n.*
thanatoid (than'ətoid), *a.* resembling death; apparently dead; (*Zool.*) poisonous, deadly.
thanatology (thanətol'əji), *n.* the scientific study of death.
thanatopsis (thanətop'sis), *n.* a view, or contemplation, of death.
thane, thegn (thān), *n.* (*OE Hist.*) a freeman holding land by military service and ranking between ordinary freemen and the nobles; (*Scot Hist.*) a hereditary tenant of the king. **thanage** (-ij), *n.* thaneship; the land held by a thane; the tenure of this. **thanedom** (-dəm), **thanehood** (-hud), **thaneship**, *n.*
thank (thangk), *n.* (*now pl.*) an expression of gratitude; a formula of acknowledgment of a favour, kindness, benefit etc. *v.t.* to express gratitude (to or for); to make acknowledgment to for a gift, offer etc. (often used ironically, esp. as a contemp. refusal). **thanks to**, because of, owing to. **thank you**, a formula expressing thanks, polite refusal etc. **thank-offering**, *n.* an offering made as an expression of gratitude, esp. a Jewish sacrifice of thanksgiving. **thanksgiver**, *n.* **thanksgiving**, *n.* the act of returning thanks or expressing gratitude, esp. to God; a form of words expressive of this; (*Bibl.*) a thank-offering. **Thanksgiving Day**, *n.* (*N Am.*) a day set apart annually for thanksgiving to God for blessings enjoyed individually and nationally (usu. the last Thursday in Nov.). **thankful**, *a.* grateful; expressive of thanks. **thankfully**, *adv.* **thankfulness**, *n.* **thankless** (-lis), *a.* insensible to kindness, ungrateful; not deserving thanks, unprofitable. **thanklessly**, *adv.* **thanklessness**, *n.*
that (dhat, dhət), *a.* (*pl.* **those** (dhōz)) the (person or thing) specifically designated, pointed out, implied, or understood; (correlated with *this*) the more remote or less obvious of two things; such (usu. followed by *as*). *pron.* the person or thing specifically designated, pointed out, implied, or understood; who or which (now usu. demonstratively and introducing a restrictive or defining clause). *adv.* in such a manner, to such a degree. *conj.* introducing a clause, stating a fact or supposition; implying purpose, so that, in order that; implying result, consequence etc.; implying reason or cause, on the ground that, because, since. **and (all) that**, and everything of that sort. **at that**, at that point; moreover. **(just) like that**, effortlessly, straight off. **that away**, (*coll.*) that way, in that direction. **that is**, to be more precise, precisely. **that's that**, there is no more to be done, said etc. **thatness**, the state of being a definite thing.
thatch (thach), *n.* a roof-covering of straw, rushes, reeds etc. *v.t.* to cover with this. *v.i.* to do thatching. **thatched**, *a.* **thatcher**, *n.* **thatching**, *n.* the act of thatching; the materials used in thatching. **thatch palm**, *n.* any of several palms used in thatching.
Thatcherism (thach'ərizm), *n.* the political, economic etc. philosophy and policies of Margaret Thatcher, British Prime Minister, (1979-), **Thatcherite**, *n.* a supporter of Margaret Thatcher or her policies. *a.* pertaining to Margaret Thatcher or her policies.
thauma(t)-, *comb. form* pertaining to wonder or miracles.
thaumaturge (thaw'mətœj), *n.* a worker of miracles; a wonder-worker, a magician or conjurer. **thaumaturgic, -al** (-tœ'-), *a.* **thaumaturgist**, *n.* **thaumaturgy**, *n.*
thaw (thaw), *v.i.* of ice, snow etc., to melt, dissolve, or become liquid; of weather, to become so warm as to melt ice or snow; (*fig.*) to relax one's stiffness, to unbend, to become genial. *v.t.* to melt, to dissolve; to infuse warmth or into. *n.* the act of thawing or the state of being thawed; warm weather that thaws; a relaxation of tension, an increase in friendliness. **to thaw out**, to return to normal from a frozen condition; to become more relaxed or more friendly. **thawless**, *a.* **thawy**, *a.*
the (dhə, dhi, dhē), *a.* applied to a person or thing or persons or things already mentioned, implied, or definitely understood; used before a singular noun to denote a species; prefixed to adjectives used absolutely, giving them the force of a substantive; before nouns expressing a unit to give distributive force (as '90p. the pint'); emphatically (dhē), to express uniqueness (as '*the* famous Duke of Wellington'), *adv.* used before adjectives and adverbs in the comparative degree, to that extent, to that amount, by so much.
the(o)-, *comb. form*, pertaining to God or a god.
theanthropic (thēənthrop'ik), **-al**, *a.* being both human and divine; tending to embody deity in human forms. **theanthropism** (-an'-), *n.*
thearchy (thē'ahki), *n.* Government by God or gods; a

body, class, or order of gods or deities.
theater THEATRE.
theatre, (*esp. N Am.* **theater**) (thē'ətə), *n.* a building for dramatic spectacles, a play-house; a cinema; a room, hall etc.; with a platform at one end, and seats arranged in ascending tiers, used for lectures, demonstrations etc.; the room in a hospital etc. used for operations; the drama, the stage; the place or scene of an action, event etc.; matter suitable to be staged. **theatre-in-the-round,** a theatre where the seats are arranged around a central acting area; the style of producing plays in such a theatre. **theatre of cruelty,** a branch of the theatre which seeks to express pain and suffering and an awareness of evil. **theatre of the absurd,** a branch of the theatre which juxtaposes the fantastic and the bizarre with the irrationality and tragedy of human existence. **the theatre,** the world of actors, producers, theatre companies etc. **theatre-goer,** *n.* a person who goes to the theatre regularly. **theatre organ,** *n.* a type of organ usu. electrically wind-controlled, with effects of most instruments of an orchestra, employed for entertainment purposes in cinemas and theatres. **theatric** (thiat'-).
theatrical (thiat'rikl), *n.* of or pertaining to the theatre; befitting the stage, dramatic; suitable or calculated for display, pompous, showy; befitting or characteristic of actors, stagy, affected. **theatricalism, theatricality** (-kal'-), *n.* **theatricalize, -ise,** *v.t.* **theatrically,** *adv.* **theatricals,** *n.pl.* dramatic performances, esp. private.
thebaine (thē'bāēn), *n.* (*Chem.*) a poisonous crystalline alkaloid obtained from opium.
Theban (thē'bən), *a.* pertaining to ancient Thebes (in Greece or in Egypt). *n.* a native or inhabitant of Thebes. **Theban year,** *n.* the Egyptian year of 365¼ days. **Thebaid** (-bāid), *n.* the territory of Egyptian Thebes.
theca (thē'kə), *n.* (*pl.* **-cae** (-kē)) (*Bot., Zool., etc.*) a sheath, a case. **thecal, thecate** (-kət, -kāt), *a.* **theciferous** (-sif'-), *a.* **theciform** (-sifawm), *a.*
thé dansant (tā dāsã'), *n.* (*pl.* **thés dansants** (tā dāsã')) a dance held during afternoon tea, popular in the 1920s and 1930s. [F literally dancing tea]
thee (thē), *obj.* THOU.
theft (theft), *n.* the act of thieving or stealing; larceny; that which is stolen.
thegn, etc. THANE.
their, theirs, *poss.* THEY.
theism (thē'izm), *n.* belief in a God, as opp. to atheism; belief in a righteous God supernaturally revealed, as opp. to Deism. **theist,** *n.* **theistic, -al** (-is'-), *a.*
them (dhem, dhəm), *obj.* THEY.
theme (thēm), *n.* a subject on which a person writes or speaks; short dissertation or essay by a student, school pupil etc., on a certain subject; the part of a noun or verb remaining unchanged by inflexions; a melodic subject usu. developed with variations; (*Log.*) the subject of thought; an underlying unifying principle. **theme park,** *n.* a park designed for leisure, where all the activities are based on a single subject. **theme song,** *n.* a recurring melody in a film, musical etc. which is associated with the production or a specific character; a signature tune; (*coll.*) a person's characteristic complaint, repeated phrase etc. **thematic** (thimat'-), *a.* **thematic catalogue,** *n.* a catalogue giving the opening theme of each piece of music. **thematically,** *adv.*
Themis (them'is, thē'-), *n.* the Greek goddess of Justice or Law; one of the asteroids.
themselves (dhəmselvz'), *pron.* the emphatic and reflexive form of the third plural personal pronoun.
then (dhen), *adv.* at that time; afterwards, soon after, after that, next; at another time. *conj.* in that case; therefore; consequently; this being so, accordingly. *a.* (*coll.*) of or existing at that time. *n.* that time, the time mentioned or understood. **by then,** by that time. **then and there,** on the spot, immediately. **then or thenabouts,** about that time.
thenar (thē'nah), *n.* the palm, the sole. *a.* of or pertaining to the palm of the hand or the sole of the foot.
thence (dhens), *adv.* from that place; for that reason, from that source; from that time. **thenceforth, thenceforward,** *adv.* from that time onward.
theo-, *comb. form* concerning God or gods.
theobroma (thēōbrō'mə), *n.* a genus of tropical trees, one of which, *Theobroma cacao* yields cocoa and chocolate. **theobromic,** *a.* **theobromine** (-mēn, -min), *n.* a bitter alkaloid resembling caffeine contained in the seeds of *T. cacao.*
theocracy (thiok'rəsi), *n.* Government by the immediate direction of God or through a sacerdotal class; a state so governed. **theocrat** (thē'əkrat), *n.* **theocratic, theocratical** (-krat'-), *a.* **theocratist,** *n.*
theocrasy (thiok'rəsi), *n.* mixed worship of different gods, polytheism; the union of the soul with God in contemplation.
Theocritean (thiokritē'ən), *a.* of, pertaining to, or in the style of the Greek pastoral poet Theocritus; pastoral, idyllic, Arcadian.
theodicy (thiod'isi), *n.* a vindication of divine justice in respect to the existence of evil. **theodicean** (-sē'ən), *n.*
theodolite (thiod'əlīt), *n.* a portable surveying instrument for measuring horizontal and vertical angles. **theodolitic** (-lit'-), *a.*
theogony (thiog'əni), *n.* the genealogy of the gods; a poem treating of this. **theogonic** (-gon'-), *a.* **theogonist,** *n.*
theol., (*abbr.*) theologian, theological, theology.
theology (thiol'əji), *n.* the science of God and His attributes and relations to the universe; the science of religion, esp. Christianity. **natural theology,** the science dealing with the knowledge of God as derived from His works. **theologian** (thēəlō'-), *n.* one versed in theology; a teacher or student of theology. **theological** (-loj'-), *a.* **theologically,** *adv.* **theologize, -ise,** *v.t.* to make theological. *v.i.* to speculate on theology.
theomachy (thiom'əki), *n.* a combat against or among the gods. **theomachist,** *n.*
theomancy (thē'əmansi), *n.* divination by oracle or by people inspired by god.
theomania (thēəmā'niə), *n.* religious insanity; a delusion that one is God. **theomaniac** (-ak), *n.*
theomorphic (thēəmaw'fik), *n.* having the form or semblance of God, opp. to anthropomorphic. **theomorphism,** *n.*
theophany (thiof'əni), *n.* the manifestation or appearance of God to man. **theophanic** (-fan'-), *a.*
theorbo (thiaw'bō), *n.* a stringed instrument resembling a two-necked lute used in the 16th–17th cents. **theorbist,** *n.*
theorem (thē'ərəm), *n.* a proposition to be proved; a principle to be demonstrated by reasoning; (*Math.*) a rule or law, esp. one expressed by symbols, etc. **theorematic, -al** (-mat'-), *a.* **theorematist** (-rem'ətist), *n.*
theoretic (thēəret'ik), **-al,** *a.* pertaining to or founded on theory not facts or knowledge, not practical, speculative. **theoretically,** *adv.* **theoretics,** *n.* the speculative parts of a science. **theoretician** (-tish'ən), *n.* a person interested in the theory rather than the practical application of a given subject.
theory (thē'əri), *n.* supposition explaining something, esp. a generalization explaining phenomena as the results of assumed natural causes; a speculative idea of something; mere hypothesis; speculation, abstract knowledge; an exposition of the general principles of a science etc.; a body of theorems illustrating a particular subject. **theorist,** *n.* one who theorizes; one given to forming theories. **theorize, -ise,** *v.i.* **theorization,** *n.* **theorizer,** *n.*
theosophy (thios'əfi), *n.* a form of speculation, mysticism, or philosophy aiming at the knowledge of God by means of intuition and contemplative illumination or by direct

communion; a term commonly applied to a system founded in the USA, in 1875, which claims to show the unity of all religions in their esoteric teaching, manifested by occult phenomena. **theosoph** (thē'əsof), **theosopher**, **-phist**, *n*. **theosophic**, **-al** (-sof'-), *a*. **Theosophical Society**, *n*. a religious society founded in 1875 by Madame Blavatsky and others, and derived from Brahmanism and Buddhism. **theosophism**, *n*. **theosophist**, *n*. **theosophistical** (-fis'-), *a*. **theosophize**, **-ise**, *v.i*.
therapeutic (therəpū'tik), *a*. pertaining to the healing art; curative. *n.pl.* the branch of medical science dealing with the treatment of disease and the action of remedial agents in both health and disease. **therapeutical**, *a*. **therapeutically**, *adv*. **therapeutist**, *n*. **therapy** (the'rəpi), *n*. therapeutics, the treatment of disease or physical and mental disorders from a curative and preventive point of view; physiotherapy; psychiatric or psychological therapy. **therapist**, *n*. a practitioner of therapy, esp. a psychologist or psychiatrist.
there (dheə, dhə), *adv*. in or at that place, point, or stage; to that place, thither; frequently used before the verb, especially the verb *to be*, when the true subject follows it. *n*. that place. *int*. expressing direction, confirmation, triumph, alarm etc. **all there**, (*sl*.) wide awake; fully competent; knowing all about it; of normal intelligence. **not all there**, (*coll*.) not fully competent; mentally deficient. **here and there** HERE. **so there**, an expression of derision or triumph. **thereabout**, **-bouts**, *adv*. near that place, number, degree etc. **thereafter**, *adv*. after that; according to that. **thereat**, *adv*. at that place; thereupon; on that account. **thereby**, *adv*. by that means; in consequence of that; thereabouts. **therefore**, *adv*. for that reason, consequently, accordingly. **therefrom**, *adv*. from this or that time, place etc. **therein**, *adv*. in that or this time, place, respect etc. **thereinafter**, *adv*. later in the same (document etc.). **thereinbefore**, *adv*. earlier in the same (document etc.). **thereof**, *adv*. of that or it. **thereon**, *adv*. on that or it. **thereto**, *adv*. to that or this; besides, over and above. **thereunder**, *adv*. under that or this. **thereupon**, *adv*. in consequence of that; immediately after or following that; upon that. **therewith**, *adv*. with that; thereupon. **therewithal**, *adv*. with all this, besides.
therianthropic (thiərianthrop'ik), *a*. of or pertaining to deities represented as half man and half beast or to their worship. **therianthropism** (-an'-), *n*. **theriomorphic** (-maw'fik), **-phous**, *a*. having the form of a beast.
therm (thœm), *n*. a British unit of heat, equal to 100,000 British Thermal Units. **British Thermal Unit**, *n*. 1/180th part of the quantity of heat required to raise the temperature of 1 lb (0.45 kg) of water from 32° F to 212° F (0–100°C).
thermae (thœ'mē), *n.pl.* hot springs or baths, esp. the public baths of the ancient Romans. [L]
thermal (thœ'məl), *a*. of or pertaining to heat or thermae; of clothing, insulating the body against very low temperatures. *n*. a rising current of warm air; (*pl*.) thermal (under)clothes. **thermal barrier**, *n*. the heating effect of air friction, making flight at high speeds difficult. **thermal reactor**, *n*. a nuclear reactor in which fission is induced using mainly thermal neutrons in thermal equilibrium with their surroundings. **thermal springs**, *n.pl.* hot springs. **thermally**, *adv*. **thermic**, *a*.
Thermidor (thœ'midaw), *n*. the 11th month of the French Republican year, 19 July–17 Aug.
thermion (thœ'mion), *n*. an electrically charged particle emitted by an incandescent body. **thermionic** (-on'-), *a*. **thermionic valve**, *n*. a vacuum tube in which a stream of electrons flows from one electrode to another and is controlled by one or more other electrodes. **thermionics**, *n*. the branch of electronics dealing with the emission of electrons from hot bodies; the study of the behaviour of these electrons in a vacuum.

thermistor (thœmis'tə), *n*. a semi-conducting device whose resistance decreases with rising temperature. [*therm*al re*sistor*]
thermite, (thœ'mīt), **Thermit**® (-mit), *n*. a mixture of finely-divided aluminium and a metallic oxide, esp. of iron, producing intense heat on combustion.
thermo-, *comb. form.* heat.
thermobarometer (thœmōbərom'itə), *n*. an apparatus for measuring atmospheric pressure by the boiling-point of water.
thermochemistry (thœmōkem'istri), *n*. the branch of chemistry dealing with the relations between chemical reactions and the heat liberated or absorbed.
thermocline (thœ'mōklīn), *n*. a layer of water in a lake etc., in which the water temperature decreases rapidly between the epilimnion and hypolimnion.
thermodynamics (thœmōdīnam'iks), *n*. the branch of physics dealing with the relations between heat and other forms of energy. **thermodynamic**, *a*.
thermoelectricity (thœmōeliktris'əti), *n*. electricity generated by differences of temperature.
thermogenesis (thœmōjen'əsis), *n*. the production of heat, esp. by physiological processes. **thermogenetic** (-net'-), **thermogenic**, *a*.
thermograph (thœ'məgraf), *n*. an instrument for automatically recording variations of temperature. **thermogram** (-gram), *n*.
thermoluminescence (thœmōloomines'əns), *n*. phosphorescence produced by heating an irradiated substance. **thermoluminescent**, *a*.
thermomagnetism (thœmōmag'nətizm), *n*. magnetism as modified or produced by the action of heat. **thermomagnetic** (-net'-), *a*.
thermometer (thərmom'itə), *n*. an instrument for measuring temperature, usu. by the expansion or contraction of a column of mercury or alcohol in a graduated tube of small bore with a bulb at one end. **thermometric, -al,** (thœmōmet'-), *a*. **thermometrically**, *adv*. **thermometry**, (-tri), *n*.
thermonuclear (thœmōnū'kliə), *a*. used of the fusion of nuclei, as in **thermonuclear reaction**, which is the fusion of nuclei at very high temperatures, as in the hydrogen bomb.
thermophile (thœ'məfil), **-philic** (-fil'-), *a*. (*Biol*.) thriving in a high temperature.
thermopile (thœr'mōpīl), *n*. a thermoelectric battery, esp. one employed to measure small quantities of radiant heat.
thermoplastic (thœmōplas'tik), *n*., *a*. (a substance) which softens under heat without undergoing any chemical change, and can, therefore, be heated repeatedly.
Thermos® (thœr'məs), *n*. a type of vacuum flask.
thermosetting (thœmōset'ing), *a*. of a substance which softens initially under heat but subsequently hardens and becomes infusible and insoluble.
thermosphere (thœ'məsfiə), *n*. the part of the earth's atmosphere above the mesosphere, from about 50 miles (80 km), in which the temperature rises steadily with height.
thermostat (thœ'məstat), *n*. a self-acting apparatus for regulating temperatures. **thermostatic** (-stat'-), *a*.
thermotaxis (thœmōtak'sis), *n*. the movement of an organism in reaction to heat stimulus. **thermotactic**, (-tak'tik), **-taxic**, *a*.
thermotropism (thœmōtrō'pizm), *n*. the orientation of a plant in response to temperature difference. **thermotropic** (-trop'-), *a*.
theroid (thiə'roid), *a*. resembling a beast.
THES, (*abbr*.) Times Higher Education Supplement.
thesaurus (thisaw'rəs), *n*. (*pl*. **-i**, (-ī)) a cyclopaedia or lexicon; a collection of words, phrases etc., esp. arranged as groups of synonyms.
these (dhēz), *pl*. THIS.

thesis (thē'sis), *n.* (*pl.* **-ses** (-sēz)) a proposition advanced or maintained; an essay or dissertation, esp. one based on original research and submitted by a candidate for a degree, etc.; (*Log.*) an affirmation, as opp. to an hypothesis.
Thespian (thes'piən), *a.* pertaining to Thespis, traditional Greek dramatic poet; (**thespian**) relating to tragedy or the drama. *n.* (**thespian**) an actor.
Thess. (*abbr.*) Thessalonians.
theta (thē'tə), *n.* the eighth letter of the Greek alphabet (θ, ϑ), transliterated by *th*.
theurgy (thē'œji), *n.* divine or supernatural agency, esp. in human affairs; supernatural as distinguished from natural magic. **theurgic, -al** (-oe'-), *a.* **theurgist,** *n.*
thew (thū), *n.* (*usu. in pl.*) muscles, sinews; strength, vigour. **thewed, thewy,** *a.* **thewless,** (-lis), *a.*
they (dhā), *pron.* (*obj.* **them,** (dhem, dhəm), *poss.* **their,** (dheə), *absol.* **theirs** (dheəz)) the plural of the third personal pronoun (*he, she* or *it*); people in general; (*coll.*) those in authority. **they'd** (dhād), they had; they would. **they'll,** (dhāl), they shall; they will. **they're,** (theə), they are. **they've,** (dhāv), they have.
thiamine (thī'əmēn), *n.* vitamin B, important for metabolism and nerve function.
thick (thik), *a.* having great or specified extent or depth from one surface to the opposite; arranged, set or planted closely, crowded together, close packed or abounding (with); following in quick succession; dense, viscous; opaque; cloudy, foggy; dull, stupid; of articulation etc., indistinct, muffled; (*coll.*) very friendly, familiar. *adv.* thickly, in close succession; indistinctly. *n.* the thickest part; the middle, esp. of a fight. **a bit thick,** unreasonable. **through thick and thin,** under any conditions, undauntedly, resolutely. **thick ear,** (*coll.*) a swollen ear as a result of a blow. **thickhead,** *n.* a blockhead; (*Austral.*) a bird of the Pachycephalidae family, akin to the flycatchers. **thick-headed,** *a.* **thick-knee,** *n.* the stone-curlew. **thick-lipped,** *a.* **thick-set,** *a.* planted, set or growing close together; solidly built, stout, stumpy. *n.* a thick-set hedge. **thick-skinned,** *a.* not sensible to taunts, reproaches etc, insensitive. **thick-skin,** *n.* **thick-skull,** *n.* **thick-skulled, -witted,** *a.* **thicken,** *v.t., v.i.* **thickening,** *n.* **thicket,** *n.* a thick growth of small trees, bushes, etc. **thickish,** *a.* **thickly,** *adv.* **thickness,** *n.* the state of being thick; extent from upper surface to lower, the dimension that is neither length nor breadth; a sheet or layer of cardboard etc. **thicky** (-i), *n.* (*sl.*) a stupid person.
thief (thēf), *n.* (*pl.* **thieves** (thēvz)) one who steals, esp. furtively and without violence. **thieve** (thēv), *v.i.* to practise theft; to be a thief. *v.t.* to take by theft. **thieves' Latin** LATIN. **thievery,** (-vəri), **thievishness,** *n.* **thievish,** *a.* **thievishly,** *adv.*
thigh (thī), *n.* the thick, fleshy portion of the leg between the hip and knee in man; the corresponding part in other animals. **thigh-bone,** *n.* the principal bone in the thigh, the femur.
thimble (thim'bl), *n.* a cap of metal etc., worn to protect the end of the finger in sewing; a sleeve or short metal tube; a ferrule; an iron ring having an exterior groove worked into a rope or sail to receive another rope or lanyard. **thimble-case,** *n.* **thimbleful,** *n.* as much as a thimble holds; a very small quantity. **thimblerig,** (-rig), *n.* a sleight-of-hand trick with three thimbles and a pea, persons being challenged to bet which cover the pea is under. **thimblerigger,** *n.*
thin (thin), *a.* having the opposite surfaces close together, of little thickness, slender; not close-packed, not dense; sparse, scanty, meagre; lean, not plump; not full, scant, bare; flimsy, easily seen through. *adv.* thinly. *v.t.* (*past, p.p.* **thinned**) to make thin; to make less crowded; to remove fruit, flowers etc., from (a tree or plant) to improve the rest. (also **thin out**) *v.i.* to become thin or thinner; to waste away. **thin on the ground,** present in small numbers, scarce. **thin on top,** balding. **thin-skinned,** *a.* sensitive, easily offended. **thinly,** *adv.* **thinness,** *n.* **thinner,** *n.* a solvent used to thin, e.g. paint. **thinnish,** *a.*
thine (dhīn), THY.
thing [1] (thing), *n.* any object or thought; whatever exists or is conceived to exist as a separate entity esp. an inanimate object as distinguished from a living being; an act, a fact, affair, circumstance etc.; (*coll.*) a person or other animate object regarded with commiseration, disparagement etc.; (*pl.*) clothes, belongings, luggage etc. **one of those things,** a happening that one cannot do anything about. **the thing,** the proper thing (to do etc.). **to be all things to all men,** to adjust one's behaviour, opinions etc. to suit the circumstances. **to be hearing, seeing things,** to hear or see imaginary things. **to be on to a good thing,** (*coll.*) to have found something which can be turned to one's advantage. **to do one's own thing,** (*coll.*) to do what one likes or what one pleases. **to have a good thing going,** (*coll.*) to be in a position one can profit from. **to have a thing about,** to have an unaccountable prejudice or fear about. **to know a thing or two,** to be well-informed or shrewd. **to make a thing of,** to make an issue or argument out of. **thingamajig, thingummygig** (-əmijig), **thingumabob,** (-bob), **thingummy, thingy** (-i), *n.* (*coll.*) a thing, what d'you call it.
think (thingk), *v.t.* (*past, p.p.* **thought** [1], (thawt)) to regard or examine in the mind, to reflect, to ponder (over etc.); to consider, to be of opinion, to believe; to design, to intend; to effect by thinking; (*coll.*) to remember, to recollect. *v.i.* to exercise the mind actively, to reason; to meditate, to cogitate, to consider (on, about etc.). *n.* (*coll.*) an act of thinking; a thought. **to have another think coming,** (*coll.*) to be wrong about what one assumes will happen. **to think again, to think better of it,** to change one's mind, to decide not to pursue (a course of action). **to think of,** to have in mind, to conceive, to imagine; to call to mind, to remember; to have a particular opinion or feeling about, to esteem. **to think out,** to devise; to solve by long thought. **to think over,** to ponder, to reflect on. **to think through,** to consider all the possible consequences of, problems that might occur in etc. **to think twice,** to hesitate and consider very carefully; to think again. **to think up,** to invent, concoct by thinking. **think-tank,** *n.* a group of experts in any field who meet to solve problems and produce new ideas in that field. **thinkable,** *a.* **thinker,** *n.* **thinking,** *a.,* *n.* **thinkingly,** *adv.*
thinly, etc. THIN.
third (thœd), *n.* one of three equal parts; (*Mus.*) an interval between a tone and the next but one on the diatonic scale; a tone separated by this interval; the consonance of two such tones; a third-class honours degree; the third gear in a motor vehicle. *n., a.* (the) last of three (people, things etc); the next after the second. **third-class, -rate,** *a.* of the class coming next to the second; inferior, worthless. **third degree,** *n.* (*sl.*) intimidation or torture, esp. to extract information. **third party,** *n., a.* (a person) other than the principals (in a contract etc.). **third rail,** *n.* an extra rail through which electricity is supplied to an electric locomotive. **Third World,** the countries not aligned with either of the superpowers; the developing countries, esp. in Africa, Asia and S America. **thirdly,** *adv.*
thirst (thœst), *n.* the uneasiness or suffering caused by want of drink; desire for drink; eager longing or desire. *v.i.* to feel thirst (for or after). **thirstless,** (-lis), *a.* **thirsty,** *a.* feeling thirst; dry, parched; (*coll.*) exciting thirst. **thirstily,** *adv.* **thirstiness,** *n.*
thirteen (thœtēn'), *n.* the number or figure 13 or xiii; the age of 13. *a.* 13 in number; aged 13. **thirteenth,** (-th), *n.* one of 13 equal parts. *n., a.* (the) last of 13 (people, things etc.); the next after the 12th.
thirty (thœ'ti), *n.* three times ten; the number or figure 30 or xxx; the age of 30. *a.* 30 in number; aged 30. **thirties,**

n.pl. the period of time between one's 30th and 40th birthdays; the range of temperature between 30 and 40 degrees; the period of time between the 30th and 40th years of a century. **thirtieth**, (-əth), *n.* one of 30 equal parts. *n.*, *a.* (the) last of 30 (people, things etc.); the next after the 29th. **thirty-something**, *a.* referring to the lifestyle etc. of affluent people born in the 1950s and 1960s.
this (dhis), *a.*, *pron.* (*pl.* **these**, (dhēz)) used to denote the person or thing that is present or near in place or time, or already mentioned, implied, or familiar. *adv.* to this extent. **this and that**, (*coll.*) random and usu. unimportant subjects of conversation. **thisness**, *n.* haecceity.
thistle (this'l), *n.* a plant of several genera of the aster family with prickly stems, leaves and involucres. **thistledown**, *n.* the feathery seeds produced by the thistle. **thistly**, *a.*
thither (dhidh'ə), *adv.* to that place; to that end, point or result.
thixotropic (thiksətrop'ik), *a.* of certain gels (e.g. non-drip paints), becoming fluid when shaken or stirred. **thixotropy** (-sot'rəpi), *n.*
tho' (dhō), THOUGH.
thole (thōl), *n.* a pin in the gunwale of a boat serving as fulcrum for the oar, also called **thole pin**.
Thomism (tō'mizm), *n.* the scholastic philosophy and theology of St Thomas Aquinas (1227–74). **Thomist**, *a.*, *n.* **Thomistic** (-mis'-), *a.*
-thon (-thon), *comb. form.* a large-scale event or related series of events lasting a long time or demanding endurance of the participants (as *telethon*).
thong (thong), *n.* a strip of leather used as a whip-lash, for reins, or for fastening anything; (*pl.*) (*N Am.*) flip-flop sandals.
Thor (thaw), *n.* the ancient Scandinavian god of thunder, war, and agriculture. **Thor's hammer**, *n.* a flint implement.
thoracic, etc. THORAX.
thorax (thaw'raks), *n.* (*pl.* **thoraces**, (-rəsēz), **thoraxes**,) the part of the trunk between the neck and the abdomen; the middle division of the body of insects. **thoracic** (thəras'-), *a.* **thoraci(co)-, thoraco-**, *comb. form.*
thorium (thaw'riəm), *n.* a radioactive metallic element, at. no. 90; chem. symbol Th; found chiefly in thorite. **thoria**, (-riə), *n.* oxide of thorium, used in the manufacture of heat-resistant materials. **thoric**, **thorinic** (-rin'-), *a.* **thorite**, (-rit), *n.* a massive dark hydrous silicate of thorium, found in Norway.
thorn (thawn), *n.* a spine, a sharp-pointed process, a prickle; a thorny shrub, tree, or herb (*usu. in comb.* as *blackthorn, whitehorn*); the OE letter þ (th). **a thorn in one's side**, (or **flesh**), a constant source of trouble. **thorn-apple**, *n.* a plant with prickly seed-capsules, *Datura stramonium*. **thornback**, *n.* the British ray or skate, *Raja clavata*, the back and tail of which are covered with spines. **thorn-bill, -tail**, *n.* a name for various hummingbirds. **thorn-bush**, *n.* **thornless**, (-lis), *a.* **thorny**, *a.*
thorough (thŭ'rə), *a.* complete, perfect, not superficial; painstaking, meticulous; utter, out-and-out. **thoroughbass** (-bās), *n.* (*Mus.*) a bass part accompanied by shorthand marks, usu. figures, written below the stave, to indicate the harmony; this method of indicating harmonies; the science of harmony. **thoroughbred** *a.* of pure breed; high-spirited, mettlesome. *n.* a thoroughbred animal, esp. a horse. **thoroughfare** *n.* a passage through from one street, etc., to another, an unobstructed road or street; a road or street for public traffic. **thoroughgoing** *a.* going or ready to go to any lengths; thorough, uncompromising; utter, complete. **thoroughly**, *adv.* **thoroughness**, *n.*
thorp, thorpe (thawp), *n.* a village, a hamlet (esp. in placenames).
those, (dhōz), *pl.* THAT.

thou[1] (dhow), *pron.* (*obj.* **thee**, *thē*) the second personal pronoun singular, denoting the person spoken to (now used only in addresses to the Deity and in poetry).
thou[2] (thow), *n.* (*pl.* **thou(s)**) (*coll.*) short for THOUSAND; a thousandth of an inch (0.0254 mm).
though (dhō), *conj.* notwithstanding that; even if; granting or supposing that; (*ellipt.*) and yet; however. **as though** AS.
thought[1], (thawt), *past*, *p.p.* THINK.
thought[2] (thawt), *n.* the act or process of thinking; reflection, serious consideration, meditation; deep concern or solicitude; considerateness; the faculty of thinking or reasoning; that which is thought; a conception, an idea, a reflection, a judgment, conclusion etc.; (*pl.*) one's views, ideas, opinions, etc., **a thought**, (*coll.*) a very small degree, etc., a shade, somewhat. **happy thought**, an apposite or timely suggestion, idea etc. **thought-reader**, *n.* one who perceives by telepathy what is passing in another person's mind. **thought-reading**, *n.* **thought-transference**, *n.* telepathy. **thought-wave**, *n.* a telepathic undulation or vibration. **thoughtful**, *n.* **thoughtfully**, *n.* **thoughtfully**, *adv.* **thoughtfulness**, *n.* **thoughtless**, (-lis), *a.* **thoughtlessly**, *adv.* **thoughtlessness**, *n.*
thousand (thow'zənd), *a.*, *n.* ten hundred, 1000; a great many. **thousand-fold**, *a.* and *adv.* **thousandth** (-th), *a.*, *n.*
thrall (thrawl), *n.* a slave, a serf; bondage, thraldom. *a.* in thrall. *v.t.* to enthral, to enslave. **thraldom**, (-dəm), *n.*
thrash (thrash), to beat soundly, esp. with a stick or whip; to overcome, to defeat, to conquer; to thresh. *v.i.* to strike out wildly and repeatedly. *n.* a thrashing; (*sl.*) a party. **to thrash out**, to discuss thoroughly in order to find a solution. **thrasher**[1], *n.* one who thrashes; a thresher-shark. **thrashing**, *n.*
thrasher[2] (thrash'ə), *n.* (*N Am.*) a N American songbird of the genus *Harporhyncus*, resembling the thrush, esp. the brown thrasher, *H. rufus*, common in the eastern States.
thrasonical (thrəson'ikəl), *a.* bragging, boastful. **thrasonically**, *adv.*
thrawn (thrawn), *a.* (*Sc.*) twisted; perverse, stubborn.
thread (thred), *n.* a slender cord consisting of two or more varns doubled or twisted; a single filament of cotton, silk, wool etc., esp. lisle thread; anything resembling this; a fine line of colour etc.; a thin seam or vein; the spiral on a screw; a continuous course (of life etc.); the continuing theme or linking element in an argument or story. *v.t.* to pass a thread through the eye or aperture of; to string (beads etc.) on a thread; to pick (one's way) or to go through an intricate or crowded place etc.; to streak (the hair) with grey etc.; to cut a thread on (a screw). **threadbare**, *a.* worn so that the thread is visible, having the nap worn off; worn, trite, hackneyed. **threadbareness**, *n.* **thread-mark**, *n.* a mark produced by coloured silk fibres in banknotes to prevent counterfeiting. **threadpaper**, *n.* soft paper for wrapping up thread. **threadworm**, *n.* a thread-like nematode worm, esp. one infesting the rectum of children. **threader**, *n.* **threadlike**, *a.*, *adv.* **thready**, *a.* **threadiness**, *n.*
threat (thret), *n.* a declaration of an intention to inflict punishment, loss, injury etc.; an indication of something unpleasant or harmful to come; a source of danger. **threaten**, *v.t.* to use threats to; to announce intention (to inflict injury etc.); to announce one's intention to inflict (injury etc.). *v.i.* to use threats; to have a threatening appearance. **threatener**, *n.* **threateningly**, *adv.*
three (thrē), *n.* the number or figure 3 or iii; the age of three; the third hour after midnight or midday; a group of three. *a.* three in number; aged three. **rule of three** RULE. **three R's**, reading, writing, arithmetic. **three-colour process**, the printing of coloured illustrations by the superposition of the three primary colours. **three-cornered**, *a.* **three-decker**, *n.* a vessel carrying guns on

three decks; a pulpit in three stories; anything having three layers or levels. **three-dimensional**, *a.* giving the effect of being seen or heard in three dimensions. **three-handed**, *a.* having three hands; of some card-games, for three players. **three-headed**, *a.* **three-legged race**, *n.* race in which runners compete in pairs, each having one leg tied to his or her partner's. **three-master**, *n.* a vessel, esp. a schooner, with three masts. **three-pence** (threp'əns, thrip'-, thrŭp'-), *n.* the sum of threepence. **threepenny**, (threp'ni, thrip'-, thrŭp'-), *a.* **threepenny bit**, *n.* formerly a small coin value threepence. **three-phase**, *a.* a term applied to an alternating-current system in which the currents flow in three separate circuits. **three-piece**, *a.* consisting of three matching pieces, as a suit of clothes, a suite of furniture etc. **three-ply**, *a.* having three strands, thicknesses etc. *n.* plywood of three layers. **three-point landing**, *n.* one in which an aeroplane touches all three wheels down simultaneously. **three-point turn**, *n.* an about-turn in a narrow space made by a vehicle moving obliquely forwards, backwards and forwards again. **three-quarter**, *a.* of three-fourths the usual size or number; of portraits, showing three-fourths of the face, or going down to the hips. *n.* (*Rugby*) player positioned between the half backs and the full back. **three-ring circus**, *n.* a circus with simultaneous performances in three rings; a situation in which a bewildering variety of events are taking place. **threescore**, *a.* 60. *n.* the age of 60. **three-fold**, *a.*, *adv.* **threesome**, (-səm), *a.* threefold, triple. *n.* a party of three; a game for three.
thremmatology (thremətol'əji), *n.* the branch of biology dealing with the breeding of animals and plants.
threnody (thren'ədi), **threnode** (thrē'nōd, thren'-), *n.* a song of lamentation; a poem on the death of a person. **threnetic, -al** (thrinet'-), **threnodial** (thrinō'-), **threnodic** (-nod'-), *a.* **threnodist** (thren'-), *n.*
thresh (thresh), *v.t.*, *v.i.* to beat out or separate the grain (from corn etc.); to thrash. **thresher**, *n.* one who threshes; a threshing machine; a large, long-tailed shark. **threshing**, *n.* **threshing floor**, *n.* a floor or area on which grain is threshed. **threshing machine**, *n.*
threshold (thresh'ōld), *n.* the stone or plank at the bottom of a doorway; an entrance, a doorway; a beginning; the minimum strength of a stimulus, etc., that will produce a response as *threshold of pain.*
threw, (throo), *past* THROW.
thrice (thris), *adv.* three times; (*fig.*) very much.
thrift (thrift), *n.* frugality; good husbandry, economical management; the sea-pink, *Armeria maritima.* **thriftless**, (-lis), *a.* **thriftlessly**, *adv.* **thriftlessness**, *n.* **thrifty**, *a.* frugal, careful, economical. **thriftily**, *adv.* **thriftiness**, *n.*
thrill (thrill), *v.t.* to penetrate; to affect with emotion so as to give a sense as of vibrating or tingling; of emotion to go through one. *v.i.* to penetrate, vibrate, or quiver (through, along etc., of emotion); to have a vibrating, shivering or tingling sense of emotion. *n.* an intense vibration, shiver or wave of emotion; a vibratory or tremulous resonance observed in auscultation; (*coll.*) anything exciting. **thriller**, *n.* a sensational or exciting novel, film etc. **thrillingly**, *adv.* **thrillingness**, *n.*
thrips (thrips), *n.* a minute insect of the genus *Thrips* or allied genus injurious to plants, esp. grain.
thrive (thriv), *v.i.* (*past*, **throve** (thrōv), **thrived**, *p.p.* **thriven** (-thriv'ən), **thrived**) to prosper, to be fortunate, to be successful; to grow vigorously. **thriver**, *n.* **thrivingly**, *adv.* **thrivingness**, *n.*
thro' (throo), THROUGH.
throat, *n.* the front part of the neck, containing the gullet and windpipe; the gullet, the pharynx, the windpipe, the larynx; a throat-shaped inlet, opening, or entrance, a narrow passage, strait etc. **sore throat**, an inflamed condition of the membranous lining of the gullet etc., usu. due to a cold. **to be at one another's throats**, fighting or quarrelling violently. **to cut one another's throats**, to engage in a ruinous competition. **to cut one's own throat**, (*fig.*) to adopt a policy that will harm or ruin one. **to lie in one's throat**, to lie outrageously. **to ram something down someone's throat** RAM. **throated**, *a.* **throaty**, *a.* guttural; hoarse; having a large or prominent throat. **throatily**, *adv.* **throatiness**, *n.*
throb (throb), *v.i.* (*past*, *p.p.* **throbbed**) to beat rapidly or forcibly (of the heart or pulse); to vibrate, to quiver. *n.* a strong pulsation, a palpitation. **throbbingly**, *adv.*
throe (thrō), *n.* a violent pain, a pang, esp. (*pl.*) the pains of childbirth. **in the throes of**, struggling with (a task etc.).
thrombosis (thrombō'sis), *n.* (*pl.* **-oses** (-sēz)) local coagulation of the blood in the heart or a blood-vessel; (*coll.*) a coronary thrombosis. **thrombotic** (-bot'-), *a.* **thrombin** (-bin), *n.* an enzyme concerned in the clotting of blood. **thrombocyte** (-bəsit), *n.* a blood platelet. **thrombose** (throm'-), *v.t.*, *v.i.* to affect with or undergo thrombosis. **thrombosed**, *a.* **thrombus**, (-bəs), *n.* the clot of blood closing a vessel in thrombosis.
throne (thrōn), *n.* a royal seat, a chair or seat of state for a sovereign, bishop etc.; sovereign power; one of the third order of angels. **thronal**, *a.* **throneless**, (-lis), *a.*
throng (throng), *n.* a multitude of persons or living things pressed close together, a crowd. *v.i.* to crowd or press together; to come in multitudes. *v.t.* to crowd, to fill to excess; to fill with a crowd; to press or impede by crowding upon.
throstle (thros'l), *n.* the songthrush, a machine for continuously twisting and winding wool, cotton etc.
throttle (:hrot'l), *n.* the wind pipe, the gullet, the throat; a throttle-valve. *v.t.* to choke, to strangle; to shut off, reduce or control (the flow of steam in a steam-engine or of explosive mixture to an internal-combustion engine). **throttle-valve**, *n.* a valve regulating such flow.
through (throo), *prep.* from end to end of, from side to side of, between the sides or walls of; over the whole extent of, throughout; from place to place within; by means, agency, or fault of, on account of; (*N Am.*) up to and including; during; at or to the end of. *adv.* from end to end or side to side, from beginning to end; to a final (successful) issue; completely. *a.* going through or to the end, proceeding right to the end or destination, esp. (of railway or steamboat tickets etc.) over several companies' lines; direct; completed. **all through**, all the time, throughout. **through and through**, through again and again; searchingly; completely; in every way. **to be through**, (*coll.*) to have finished. **to carry through** CARRY. **to fail through** FALL. **to go through, to go through with** GO [1]. **throughout**, *adv.* right through, in every part; from beginning to end. *prep.* right through, from beginning to end of. **throughput**, *n.* the amount of raw material put through or processed in e.g. a factory, computer.
throve (thrōv), *past* THRIVE.
throw (thrō), *v.t.* (*past* **threw**, (throo), *p.p.* **thrown** (thrōn)) to fling, to hurl, to cast, esp. to a distance with some force; to cast down, to cause to fall; to drive, to impel; to make (a cast) with dice; to turn or direct quickly or suddenly (the eyes etc.); to move (a switch) so as to engage or disengage; to cause to be in a certain condition, esp. suddenly; to be subjected to (a fit); to deliver (a punch); to put on (clothes etc.) hastily or carelessly; to cast off (the skin, as a snake); of rabbits etc., to bring forth (young); to twist, to wind into threads; to shape on a potter's wheel; (*coll.*) to hold (a party); to puzzle or astonish; (*sl.*) to lose (a contest) deliberately. *v.i.* to hurl or fling a missile (at etc.); to cast dice. *n.* the act of throwing, a cast; a cast of the dice; the distance to which a missile is thrown; the extent of motion (of a crank etc.); (*Geol.*) a faulting, a dislocation; the extent of dislocation; (*N Am.*) a rug or decorative cloth put over a piece of

thru 860 **thyroid**

furniture. **to throw away**, to cast from one; to reject carelessly; to spend recklessly, to squander; to lose through carelessness or neglect; to fail to take advantage of. **to throw back**, to reflect, as light etc.; to revert (to ancestral traits). **to throw down**, to overturn; to lay (oneself) down prostrate. **to throw in**, to interject, to interpolate; to put in without extra charge, to add as a contribution or extra. **to throw in one's hand**, to give up a job etc., as hopeless. **to throw off**, to cast off, to get rid of, to abandon, to discard; to produce without effort; to evade (pursuit). **to throw oneself on**, to commit oneself to the protection, favour etc., of. **to throw open**, to open suddenly and completely; to make freely accessible. **to throw out**, to reject; to emit; to give utterance to, to suggest; to cause (a building etc.) to stand out or project; to utter (a suggestion etc.) casually; to confuse. **to throw over**, to abandon, to desert. **to throw together**, to put together hurriedly or carelessly; to bring into casual contact. **to throw up**, to raise or lift quickly; to abandon, to resign; (*coll.*) to vomit. **throwaway**, *a.* disposable; of something written or said, deliberately casually. **throwback** *n.* a reversion to an earlier type. **throw-in**, *n.* a method in soccer of putting the ball back in play from touch. **thrower**, *n.* **throwing**, *n.*
thru (throo), *prep.*, *adv.* (*N Am.*) THROUGH.
thrum[1] (thrŭm), *v.i.* (*past, p.p.* **thrummed**) to play carelessly or unskilfully (on a stringed instrument); to tap, to drum monotonously (on a table etc.). *v.t.* to play (an instrument) thus; to tap or drum on. *n.* the act or sound of such drumming or playing.
thrum[2] (thrŭm), *n.* the fringe of warp-threads left when the web has been cut off, or one of such threads; loose thread, fringe etc., a tassel; (*pl.*) coarse or waste yarn. **thrummy**, *a.*
thrush[1] (thrŭsh), *n.* a bird of the family Turdidae, esp. the song-thrush or throstle, *Turdus philomelos*.
thrush[2] (thrŭsh), *n.* a vesicular disease of the mouth and throat, usu. affecting children, caused by the fungus *Candida albicans;* a similar infection of the vagina, caused by the same fungus; an inflammatory affection of the frog in the feet of horses.
thrusher (thrŭsh'ə), THRASHER[2].
thrust (thrŭst), *v.t.* (*past, p.p.* **thrust**) to push suddenly or forcibly; to stab. *v.i.* to make a sudden push (at); to stab (at); to force or squeeze (in etc.). *n.* a sudden or violent push; an attack as with a pointed weapon, a stab; force exerted by a jet engine, propeller etc. To give forward movement; force exerted by one body against another, esp. horizontal outward pressure, as of an arch against its abutments; the forceful part, or gist, of an argument etc. **to thrust oneself in**, to intrude; to interfere. **to thrust through**, to pierce. **thruster**, *n.* **thrusting**, *a.* dynamic and ambitious.
thud (thŭd), *n.* a dull sound as of a blow on something soft. *v.i.* (*past, p.p.* **thudded**) to make a thud; to fall with a thud.
thug (thŭg), *n.* one of a fraternity of religious assassins in India (suppressed 1828–35); a violent, brutal man, esp. a criminal. **thuggee** (-ē), **thuggery, thuggism**, *n.*
Thuja THUYA.
Thule (thū'lē), *n.* the name given by the voyager Pytheas of Massilia to the northernmost land he reached, variously identified with the Shetlands, Iceland, Norway etc. **ultima Thule**, (ŭl'timə), a very remote place. **thulium** (-liəm), *n.* a rare silver-grey malleable metallic element at. no. 69; chem. symbol Tm; a member of the rare-earth group.
thumb (thŭm), *n.* the short thick digit of the human hand; the corresponding digit in animals; the part of a glove which covers the thumb. *v.t.* to handle, perform or play awkwardly; to soil or mark with the thumb; to turn (the pages of a book) with the thumb. *v.i.* to hitch-hike. **one's fingers all thumbs**, fumbling, clumsily. **rule of thumb**, a rough, practical method. **thumbs down**, an indication of failure or disapproval. **thumbs up** an indication of success or approval. **to thumb a lift**, to get a lift from a passing car by signalling with up-raised thumb. **under one's thumb**, completely under one's power or influence. **thumb index**, *n.* an index in a book in which the letters are printed on the fore-edge, spaces being cut away from preceding pages to expose them to sight. **thumb-mark**, *n.* a mark made with a (dirty) thumb. **thumb-nail sketch**, *n.* a brief, vivid description. **thumb-nut**, *n.* a nut with wings for screwing up with the thumb. **thumbscrew**, *n.* a screw adapted to be turned with the finger and thumb; an old instrument of torture for compressing the thumb. **thumbstall**, *n.* a case, sheath, or covering for an injured or sore thumb. **thumb-tack**, *n.* (*N Am.*) a drawing-pin. **thumbed**, *a.* **thumbless**, (-lis), *a.*
thummim URIM.
thump (thŭmp), *v.t.* to strike with something giving a dull sound, esp. with the fist. *v.i.* to beat, to knock, to hammer (on, at etc.). *n.* a blow giving a dull sound; the sound of this. **thumper**, *n.* **thumping**, *a.* (*coll.*) very large.
thunder (thŭn'də), *n.* the sound following a flash of lightning, due to the disturbance of the air by the electric discharge; a thunderbolt; a loud noise; a vehement denunciation or threat. *v.i.* to make the noise of thunder; to make a loud noise, esp. while moving; to make loud denunciations etc. *v.t.* to emit or utter as with the sound of thunder. **thunderbolt**, *n.* an electric discharge with lightning and thunder; a supposed missile or mass of heated matter formerly believed to be discharged in this; an irresistible force; something that leaves one thunder-struck. **thunder-clap, -crack, -peal**, *n.* a sudden burst of thunder. **thunder-cloud**, *n.* a cloud from which lightning and thunder are produced. **thunder-shower, -storm**, *n.* a storm with thunder. **thunder-struck**, *a.* struck by lightning; astounded. **thunderer**, *n.* **thundering**, *a.* producing thunder or loud sound like thunder; (*sl.*) extreme, remarkable, tremendous, out-and-out. *adv.* unusually, remarkably, tremendously. **thunderingly**, *adv.* **thunderless, thunderous**, very loud; angry, threatening. **thundery**, *a.* characterized by or giving a warning of thunder. **thunderously**, *adv.*
thurible (thū'ribl), *n.* a censer. **thurifer**, (-fə), *n.* one who carries a censer. **thuriferous** (rif'-), *a.* producing frankincense. **thurification**, (-fi-), *n.* the act of burning incense.
Thursday (thœz'dā, -di), *n.* the fifth day of the week.
thus (dhŭs), *adv.* in this manner; in the way indicated or about to be indicated; accordingly; to this extent. **thusness**, *n.* (*facet.*).
thuya, thuja (thū'yə), *n.* a genus of coniferous trees or shrubs, also called arbor-vitae.
thwack (thwak), WHACK.
thwart (thwawt), *a.* transverse, oblique. *n.* a transverse plank in a boat serving as seat for a rower. *v.t.* to foil, to frustrate. **thwarter**, *n.* **thwartingly**,
thy (dhi), *pron.*, *a.* (*before vowels usu. and absolutely* **thine** (dhīn)) of or pertaining to thee (poss. corresponding to THOU).
thylacine (thī'ləsēn, -sīn, -sin), *n.* the Tasmanian wolf, *Thylacinus cynocephalus*, the largest predatory marsupial, poss. extinct.
thyme (tīm), *n.* any plant of the genus *Thymus*, esp. the garden thyme, *T. vulgaris*, a pungent aromatic herb used in cookery. **thymy** (-i), *a.*
-thymia (-thīmiə), *comb. form.* forming nouns indicating a mental or emotional condition.
thymine (thī'mēn), *n.* one of the bases in DNA and RNA, containing nitrogen.
thymol (thī'mol), *n.* a phenol obtained from oil of thyme, used as an antiseptic.
thymus (thī'məs), *n.* (*pl.* **-mi** (-mī)) a gland situated in the lower region of the neck, usu. degenerating after puberty.
thyroid (thī'roid), *a.* of or connected with the thyroid gland or cartilages. *n.* the thyroid body or gland; the

thyroid cartilage; a thyroid artery. **thyroid body** or **gland**, *n.* a large ductless gland consisting of two lobes situated on each side of the larynx and the upper part of the windpipe, which regulates metabolism and hence growth and development. **thyroid cartilage**, a large cartilage in the larynx, called in man the Adam's apple. **thyroid extract**, *n.* an extract prepared from the thyroid glands of oxen, sheep and pigs, and employed therapeutically. **thyro-**, *comb. form.* **thyroxin** (thīrok'sin), *n.* the main hormone produced by the thyroid gland, an amino acid containing iodine.

thyrsus (thœ'səs), *n.* (*pl.* **-si** (-sī)) (*Gr. Ant.*) a spear or shaft wrapped with ivy or vine branches and tipped with a fir-cone, an attribute of Bacchus; (also **thyrse**) (*Bot.*) an inflorescence consisting of a panicle with the longest branches in the middle. **thyrsoid**, (-soid), *a.*

thyself (dhīself'), *pron.* a reflexive and emphatic form used after or instead of 'thou' or 'thee'.

Ti, (*chem. symbol*) titanium.

ti (tē), *n.* (*Mus.*) TE.

tiara (tiah'rə), *n.* the head-dress of the ancient Persian kings, resembling a lofty turban; the triple crown worn by the Pope as a symbol of his temporal, spiritual, and purgatorial power; hence, the papal dignity; a jewelled coronet or headband worn as an ornament by women. **tiara'd, tiaraed**, (-rəd),

Tibetan (tibet'n), *a.* of or pertaining to the country of Tibet or its language; *n.* an inhabitant or the language of that country.

tibia (tib'iə), *n.* (*pl.* **-biae**, (-biē), **-bias**) the shinbone, the anterior and inner of the two bones of the leg; the fourth joint of the leg in an arthropod. **tibial**, *a.* **tibio-**, *comb. form.*

tic (tik), *n.* a habitual convulsive twitching of muscles, esp. of the face, tic douloureux. **tic douloureux**, (doolərœ', dolərōō'), *n.* facial neuralgia characterized by spasmodic twitching. [F]

tick[1] (tik), *n.* a name for various parasitic acarids infesting some animals and occasionally man; (*sl., dated*) an unpleasant or despicable person. **tick fever**, *n.* a disease transmitted by ticks.

tick[2] (tik), *n.* a cover or case for the filling of mattresses and beds; the material for this, usu. strong striped cotton or linen cloth, also called **ticking**.

tick[3] (tik), *n.* (*coll.*) credit.

tick[4] (tik), *v.i.* to make a small regularly recurring sound like that of a watch or clock. *v.t.* to mark (off) with a tick. *n.* the sound made by a going watch or clock; (*coll.*) a moment; a small mark used in checking items. **to tick off**, to mark off (a series) by ticks; (*coll.*) to scold. **to tick over**, of an engine, to run slowly with gear disconnected; to operate smoothly, at a low level of activity. **what makes one tick**, (*coll.*) one's main interest or most striking characteristic. **tick-tack**, *n.* a recurring, pulsating sound; a code of signalling employed by bookmakers whereby their agents can keep them informed of the betting odds. **tick-tock**, (-tok), *n.* the noise of a clock ticking. **ticker**, *n.* (*coll.*) a watch; the heart.

ticker tape *n.* a ribbon of paper on which some telegraphic machines print out information.

ticket (tik'it), *n.* a card or paper with written or printed contents entitling the holder to admission to a concert etc., conveyance by train etc., or other privilege; a tag or label giving the price etc. of a thing it is attached to; (*sl.*) a visiting-card; (*coll.*) a parking ticket; (*Mil. coll.*) discharge from the Army; (*Naut. coll.*) a master's certificate; (*Aviat. coll.*) a pilot's certificate; (*N Am.*) the list of candidates put up by a party, hence the principles or programme of a party. *v.t.* to put a ticket on. **the ticket, just the ticket**, (*coll.*) the right, desirable or appropriate thing. **ticket of leave**, formerly a licence to a prisoner to be at large under certain restrictions before expiration of the sentence. **ticket-of-leave man**, a person holding this. **ticket-day**, *n.* the day before settling-day on the Stock Exchange when the brokers and jobbers learn the amount of stocks and shares that are passing between them and are due for settlement, and the names of the actual purchasers. **ticket-punch**, *n.* a punch for cancelling or marking tickets.

ticking TICK[2].

tickle (tik'l), *v.t.* to touch lightly so as to cause a thrilling sensation, resembling tingling or prickling, and usually producing laughter; to please, to gratify, to amuse. *v.i.* to feel the sensation of tickling. *n.* the act or sensation of tickling. *a.* (*prov.*) ticklish, uncertain. **tickled pink** (or **to death**), very amused, very pleased. **tickler**, *n.* **ticklish**, *a.* sensible to the feeling of tickling; difficult, critical, precarious, needing tact or caution. **ticklishly**, *adv.* **ticklishness**, *n.*

tidal TIDE.

tidbit (tid'bit), TITBIT.

tiddle (tid'l) TITTLE[2].

tiddler (tid'lə), *n.* a stickleback or other very small fish; (*coll.*) anything very small. **tiddling**, *n.* fishing for these. *a.* very small or insignificant. **tiddly**,[1] *a.*

tiddly[2] (tid'li), *a.* (*coll.*) slightly drunk, drunk.

tiddlywinks (tid'liwingks), *n.* a game in which the players snap small plastic or ivory disks into a tray.

tide (tīd), *n.* time, season; the alternative rise and fall of the sea, due to the attraction of the sun and moon; a rush of water, a flood, a torrent, a stream; the course or tendency of events. *v.i.* (*Naut.*) to work in or out of a river or harbour by the help of the tide. **to tide over**, (to help) to surmount difficulties in a small way or temporarily. **tidemark**, *n.* a line along a shore showing the highest level of the tide; (*coll.*) a dirty line round a bath indicating the level of the bath water; (*coll.*) a line on the body showing the limit of washing. **tide-waiter**, *n.* (formerly) a custom-house officer who boards ships entering port in order to enforce customs regulations. **tidewater**, *n.* water affected by the movement of the tide; (*N Am.*) low-lying coastal land. **tideway**, *n.* the channel in which the tide runs; the ebb or flow of the tide in this. **tidal** *a.* pertaining or relating to the tides; periodically rising and falling or ebbing and flowing, as the tides. **tidal basin, dock**, or **harbour**, *n.* one in which the level of the water rises or falls with the tide. **tidal river**, *n.* one in which the tides act a long way inland. **tidal wave**, *n.* a wave following the sun and moon from east to west and causing the tides; (*loosely*) a large wave due to an earthquake etc.; a great movement of popular feeling. **tideless**, (-lis), *a.*

tidings (-tī'dingz), *n.pl.* news, a report.

tidy (tī'di), *a.* orderly, in becoming order, neat, trim; (*coll.*) considerable, pretty large. *n.* (*N Am.*) an ornamental covering for a chair-back etc.; a receptacle for odds and ends. *v.t.* to make tidy, to put in order. **tidily**, *adv.* **tidiness**, *n.*

tie (-tī), *v.t.* (*pres.p.* **tying** (tī'ing)) to fasten with a cord etc., to secure, to attach, to bind; to arrange together and draw into a knot, bow etc.; to bind together, to unite; to confine, to restrict, to bind (down etc.); (*Mus.*) to unite (notes) by a tie. *v.i.* to be exactly equal (with) in a score. *n.* something used to tie things together; a neck-tie; a bond, a link; a restraint; an obligation; a beam or rod holding parts of a structure together; (*Mus.*) a curved line placed over two or more notes to be played as one; an equality of votes, score etc., among candidates, competitors etc.; a match between any pair of a number of players or teams; (*N Am.*) a railway sleeper. **to tie in**, to agree or coordinate (with); to be associated or linked (with). **to tie up**, to fasten securely to a post etc.; to restrict, to bind by restrictive conditions; to be compatible or coordinated (with); to keep occupied to the exclusion of other activities. **tie-beam**, *n.* a horizontal beam connecting rafters.

tie-break(er), n. a contest to decide the winner after a tied game etc. **tie-clip**, n. metal clip for fastening a tie to a shirt. **tie-dyeing**, **tie and dye**, n. a method of dyeing in which parts of the fabric are knotted or tied tightly to avoid being coloured. **tie-dyed**, a. **tie-in**, n. a connection; something linked to something else, esp. a book to a film. **tie-line**, n. a telephone line between two branch exchanges. **tie-pin**, n. ornamental pin used to fasten a neck-tie to a shirt. **tie-up**, n. (N Am.) a deadlock, a stand-still, esp. in business or industry, through a strike etc.
tied, a. of a public house, bound to obtain its supplies from one brewer etc.; of a dwelling-house, owned by an employer and rented to a current employee; of a game etc., ending with an equal score on each side. **tier**[1] n.
tier[2] (tiə), n. a row, a rank, esp. one of several rows placed one above another. v.t. to pile in tiers.
tierce (tiəs), n. a cask of 42 gallons; a sequence of three cards of the same suit; in fencing, the third position for guard, parry, or thrust; (Eccles.) the office for the third hour.
tiercel (tiə'səl), n. a male falcon.
tiercet (tiə'sit), TERCET.
tiff (tif), n. a fit of peevishness; a slight quarrel.
tiffany (tif'əni), n. a kind of thin silk-like gauze.
tiffin (tif'in), n. in British India, a light meal between breakfast and dinner. v.i. to take this.
tig (tig), v.t. to touch in the game of tig. v.i. (Sc.) to give light touches. n. the game of tag.
tiger (tii'gə), n. a large Asiatic carnivorous feline mammal, Felis tigris, tawny with black stripes; applied to other large feline animals as the American tiger or jaguar, the red tiger or cougar, etc; a fierce, relentless, very energetic and forceful or cruel person. **tiger-beetle**, n. a predaceous beetle with striped or spotted wing-cases. **tiger-cat**, n. a wild cat of various species; (Austral.) the dasyure. **tiger-flower**, n. a plant of the genus Tigridia spotted with orange and yellow. **tiger-lily**, n. a lily, Lilium tigrinum, with orange-spotted flowers. **tiger-moth**, n. one of the Arctiidae, with streaked hairy wings. **tiger's-eye**, n. a gem with brilliant chatoyant lustre. **tiger shark**, n. a voracious Indian Ocean shark with a striped or spotted body. **ti-gersnake**, n. a deadly Australian snake which is brown with black cross bands. **tigerish**, a. **tigress** (-gris), n. fem.
tigrine, (-grin), a.
tight (tiit), a. compactly built or put together, not leaky; impervious, impermeable (often in comb. as water-tight); drawn, fastened, held, or fitting closely; tense, stretched to the full, taut; neat, trim, compact; close-fisted, parsimonious; under strict control; not easily obtainable (of money); (coll.) awkward, difficult; (sl.) drunk. adv. tightly. **tight-fisted**, a. mean, stingy. **tight-knit**, a. tightly integrated or organized. **tight-lipped**, a. having the lips pressed tightly together, in anger etc.; taciturn. **tight-rope**, n. a rope stretched between two points upon which an acrobat walks, dances etc. **tighten**, v.t., v.i. **tightener**, n. **tightly**, adv. **tightness**, n. **tights**, n.pl. a close-fitting garment made of nylon or wool etc. covering the legs and the body below the waist and worn by women, male acrobats, ballet dancers etc.
tigon (ti'gon), n. the offspring of a tiger and a lioness. [tiger, lion]
tigress, **tigrine** TIGER.
tike (tik), TYKE.
tiki (tē'ki), n. a Maori neck ornament or figurine, a stylized representation of an ancestor etc.
tilde (til'də, tild), n. a diacritical sign (˜) in Spanish put over n to indicate the sound ny; in Portuguese put over vowels to indicate nasalization.
tile (til), n. a thin slab of baked clay, used for covering roofs, paving floors, constructing drains etc.; a similar slab of porcelain or other material used for ornamental paving. v.t. to cover with or as with tiles. **on the tiles**, enjoying oneself wildly, usu. drunkenly. **to have a tile loose**, to be eccentric, half-crazy. **tilery**, (-ləri), n. **tiling**, n.
tiliaceous (tiliā'shəs), a. allied to or resembling the linden or lime-tree.
till[1] (til), v.t. to cultivate. **tillable**, a. **tillage**, (-ij), n. **tiller**[1], n.
till[2] (til), prep. up to, up to the time of, until. conj. up to the time when. **till now**, up to the present time. **till then**, up to that time.
till[3] (til), n. a money-drawer in or on a counter; a cash-register.
till[4] (til), n. an unstratified clay containing boulder, pebbles, sand etc., deposited by glaciers. **tilly**, a.
tiller[1] TILL[1].
tiller[2] (til'ə), n. the lever on the head of a rudder by which this is turned. **tiller-chain**, **-rope** n. one connecting the tiller with the steering-wheel.
tiller[3] (til'ə), n. the shoot of a plant springing from the base of the original stalk; a sucker; a sapling.
tilt[1] (tilt), n. a covering for a cart or wagon; an awning over the stern-sheets of a boat etc. v.t. to cover with a tilt.
tilt[2] (tilt), v.i. to heel over, to tip, to be in a slanting position; to charge with a lance, to joust, as in a tournament. v.t. to raise at one end, to cause to heel over, to tip, to incline; to thrust or aim (a lance); to hammer or forge with a tilt-hammer. n. an inclination from the vertical, a slanting position; a tilting, a tournament, a charge with the lance; a tilt-hammer. **at full tilt**, at full speed or full charge, with full force **tilt-hammer**, n. a large hammer on a pivoted lever. **tilt-yard**, n. a place for tilting. **tilter**, n.
tilth (tilth), n. tillage, cultivation; the depth of soil tilled.
Tim., (abbr.) Timothy.
timbal, **tymbal** (tim'bl), n. a kettledrum.
timbale (tibahl'), n. a dish of meat or fish pounded and mixed with white of egg, cream etc., and cooked in a mould. [F]
timber (tim'bə), n. wood suitable for building, carpentry etc.; trees yielding wood suitable for constructive purposes, trees generally; a piece of wood prepared for building, esp. one of the curved pieces forming the ribs of a ship; (Hunting) fences, hurdles etc. v.t. to furnish or construct with timber. int. a warning when a tree is about to fall. **timber line** TREE. **timber-toes**, n. (coll.) a wooden-legged person. **timberwolf**, n. a grey-coloured type of wolf once common in N America. **timber-yard**, n. a yard where timber is stored, etc. **timbered**, a. wooded (usu. in comb., as well-timbered). **timbering**, n. the using of timber; temporary timber supports for the sides of an excavation.
timbre (tim'bə, tībr'), n. the quality of tone distinguishing particular voices, instruments etc., due to the individual character of the sound-waves.
timbrel (tim'brl), n. an ancient instrument like the tambourine.
time (tim), n. the general relation of sequence or continuous or successive existence; duration or continuous existence regarded as divisible into portions or periods, a particular portion of this; a way of measuring or expressing time; a clock-setting in a particular time-zone; a period characterized by certain events, persons, manners etc., an epoch, an era (sometimes in pl.); a portion of time allotted to one or to a specified purpose, the time available or at one's disposal; the period of an apprenticeship, of gestation, of a round at boxing etc.; a portion of time as characterized by circumstances, conditions of existence etc.; a point in time, a particular moment, instant, or hour; the moment of childbirth or death; the end of an allotted period, esp. for a sporting contest or drinking in a public bar; a date, a season, an occasion, an opportunity; the relative duration of a note or rest; rate of movement, tempo; style of movement, rhythm; (sl.) a term of imprisonment. v.t. to do, begin, or perform at the proper

timid

moment; to regulate as to time; to ascertain or mark the time, duration, or rate of; to measure, as in music. *v.i.* to keep time (with). **apparent time, solar time,** time as reckoned by the apparent motion of the sun. **at the same time** SAME. **at times,** at intervals, now and then. **for the time being,** for the present. **from time to time** FROM. **Greenwich time** GREENWICH. **in good time,** early, with time to spare. **in no time,** very quickly. **in time,** not too late; early enough; in course of time; sometime or other, eventually; in accordance with the time, rhythm, etc. **mean time,** an average of apparent time. **on time,** punctually. **quick time** QUICK. **sidereal time,** time shown by the apparent diurnal revolutions of the stars. **time and motion study,** investigation into working methods with a view to increasing efficiency. **time and (time) again,** repeatedly. **time enough,** soon enough. **time of day,** the hour by the clock; a greeting appropriate to this; (*sl.*) the latest aspect of affairs. **time off,** time away from work. **time out of mind, time immemorial,** time beyond legal memory. **to beat time** BEAT. **to do time,** (*coll.*) to serve a prison sentence. **to make time,** to find an opportunity. **to lose time,** to delay. **to pass the time of day,** to greet, to say 'good-day' to. **time-ball,** *n.* a ball dropped from the top of a staff at an observatory at a prescribed instant of time, usu. 1 P.M. **time-bomb,** *n.* a bomb set to explode at some prearranged time. **time-book, -card,** *n.* one specifying or recording hours of work for workmen etc. **time capsule,** *n.* a container filled with objects characteristic of the present day, which is buried to be dug up in the future. **time-clock,** *n.* an apparatus which stamps time-cards as workers arrive and leave work. **time-consuming,** *a.* taking up a lot of time. **time exposure,** *n.* (a photograph taken by) exposure of a film for a relatively long time. **time-honoured,** *a.* sanctioned by long tradition. **time-keeper,** *n.* a clock, watch, or chronometer; a person who records time, esp. of workmen; a person considered in terms of punctuality (as a *good timekeeper*). **time lag,** *n.* the interval that elapses between cause and result. **time-lapse,** *a.* of a method of filming a slow process by taking still photographs at regular intervals and showing them as a normal-speed film. **time-limit,** *n.* the period within which a task must be completed. **time-piece,** *n.* a clock or watch. **time-server,** *n.* one who suits his conduct, opinions, and manners to those in power. **time-serving,** *a.*, *n.* **time-sharing,** *a.* simultaneous access to a computer by several users on different terminals; the purchase of the use of holiday accommodation for the same period every year (also **time-share**). **time-sheet,** *n.* a sheet of paper on which hours of work are recorded. **time-signal,** *n.* a signal issued by an observatory or broadcasting station to indicate the exact time. **time signature,** *n.* an indication of time at the beginning of a piece of music. **timetable,** *n.* a printed list of the times of departure and arrival or trains etc.; a record of times of employees, school lessons etc.; a table containing the relative value of every note in music. *v.t.* to put on a timetable; to arrange in a timetable. **time-work,** *n.* work paid for by time, opp. to piece-work. **time-worn,** *a.* antiquated, dilapidated. **time zone,** *n.* a geographical region in which the same standard time is used. **timeful,** *a.* seasonable, timely, early. **timeless,** (-lis), *a.* untimely, premature; without end; ageless; not restricted to a particular period. **timely** *a.* seasonable, opportune. **timeliness,** *n.* **timeous, timous** (timəs), *a.* (*Sc.*) **timeously, timously,** *adv.* **timer,** *n.* an instrument which measures or records time; one which operates a machine etc., at a preset time. **times** *prep.* multiplied by. **timing,** *n.* reckoning the time taken; the precise instant at which ignition occurs in an internal-combustion engine, and at which the valves open and close; the controlling mechanism for this; choosing of the best time (to do something).

timid (tim'id), *a.* easily frightened, shy. **timidity** (-id'-), **timidness,** *n.* habitual shyness or cowardice. **timidly,** *adv.*

tinker

timing, timist TIME.
timocracy (timok'rəsi), *n.* a form of government in which a certain amount of property is a necessary qualification for office. **timocratic** (-krat'-), *a.*
timorous (tim'ərəs), *a.* fearful, timid. **timoroso** (-rō'sō) [It.], *adv.* (*Mus.*) with hesitation. **timorously,** *adv.* **timorousness,** *n.*
Timothy grass (tim'əthi), *n.* a valuable fodder-grass, *Phleum pratense.* [Timothy Hanson, an American through whom it first came into use, c 1720]
timous, etc. TIME.
timpano (tim'pənō), *n.* (*pl.* **-ni** (-ni)) an orchestral kettledrum. **timpanist,** *n.*
tin (tin), *n.* a lustrous white metallic element, at. no. 50; chem. symbol Sn; easily beaten into thin plates, much used for cooking utensils etc., esp. in the form of thin plates of iron coated with tin; a pot or other utensil made of this; a tinplate container that can be hermetically sealed to preserve food or drink; the contents of such a container; an area at the bottom of the front wall of a squash court; (*sl.*) money. *v.t.* (*past, p.p.* **tinned**) to coat or overlay with tin; to preserve (meat, fruit etc.) in tins. **(little) tin god,** a person of local, undeserved importance; a self-important person. **tin pan alley,** the world of popular music; the writers and publishers of such music. **tin can,** *n.* **tin fish,** *n.* (*Nav. sl.*) a torpedo. **tinfoil,** *n.* tin, tin alloy or aluminium beaten into foil for wrapping foodstuffs etc. *v.t.* to coat or cover with this. **tin hat,** *n.* a steel shrapnel helmet. **tin lizzie** (liz'i), *n.* (*sl.*) an old or dilapidated motor car. **tinman, tinsmith,** *n.* one who makes articles of tin or tinplate. **tin-opener,** *n.* an implement for opening airtight tins of preserved meat, fruit etc. **tin-plate,** *n.* iron-plate coated with tin. *v.t.* to coat with tin. **tin-pot,** *a.* (*sl.*) worthless, rubbishy. **tin roof,** *n.* one made of corrugated iron. **tin-stone,** *n.* cassiterite, the commonest form of tin ore. **tintack,** *n.* a carpet tack, tack coated with tin. **tin-type,** *n.* ferrotype. **tinware,** *n.* vessels or utensils of tin or tin-plate. **tinner,** *n.* a tin-miner or tinsmith. **tinny,** *a.* of or like tin; making a thin, metallic sound; cheap, made of flimsy materials. **tinnily,** *adv.*
tinamou (tin'əmoo), *n.* a S American quail-like gallinaceous game-bird.
tinct, tinction, tinctorial TINCTURE.
tincture (tingk'chə), *n.* an alcoholic or other solution of some principle, usu. vegetable, used in medicine; a tinge or shade (of colour); a tint; a slight taste or flavour, a spice (of); (*Her.*) one of the colours, metals, or furs used in emblazoning. *v.t.* to imbue with a colour or tint, to tinge; to flavour; to give a flavour or tinge (of some quality etc.). **tinct,** *n.* a stain, colour, or tint. **tinction,** *n.* colouring-material; the act or process of colouring; (*Med.*) a modification of a remedy by admixture etc. **tinctorial** (-taw'ri-), *a.* pertaining to colour or dyes; colouring.
tinder (tin'də), *n.* any dry, very combustible substance, esp. charred linen, used to kindle fire from a spark. **tinder-box,** *n.* a box furnished with tinder, flint and steel, for this purpose. **tinder-like, tindery,** *a.*
tine (tin), *n.* the prong, point, or spike of an antler, fork, harrow etc. **tined,** *a.*
tinea (tin'iə), *n.* a fungal disease of the skin, ringworm.
tinfoil TIN.
ting (ting), *n.* a tinkling sound, as of a small bell. *v.i.* to make this sound. **ting-a-ling,** (-əling'), *n.*
tinge (tinj), *v.t.* to colour slightly, to stain (with); to modify slightly the character or qualities of. *n.* a slight admixture of colour, a tint; a smack, flavour. **tinger,** *n.* **tingible, ting(e)ing,** *a.*
tingle (ting'gl), *v.i.* to feel a stinging, prickly sensation; to give this sensation.
tinker (ting'kə), *n.* an itinerant mender of pots, kettles, pans etc.; a rough-and-ready worker or repairer; the act of tinkering, patching, botching. *v.t.* to mend pots,

tinkle   864   tithe

kettles etc.; to mend, alter, or patch up in a rough-and-ready way, or in a clumsy, makeshift, or ineffective manner. *v.i.* to work thus (at or with); to interfere, to meddle; to experiment (with). **not give a tinker's cuss**, not care at all. **tinkerly**, *a*. **tinkler** (-klə), *n*. (*Sc.*)
tinkle (ting'kl), *v.i.* to make a succession of sharp, metallic sounds as of a bell; (*sl.*) to urinate. *v.t.* to cause to tinkle, to ring. *n*. such a sound; (*sl.*) a telephone call. **tinkler**, *n*.
tinman, tinner TIN.
tinnitus (tinii'təs), *n*. (*Med.*) ringing in the ears.
tinny, etc. TIN.
tinsel (tin'sl), *n*. brass, tin or other lustrous metallic substances beaten into thin sheets and used in strips, disks, or spangles to give a sparkling effect in decoration; a fabric adorned with this; a cloth composed of silk and silver; superficial brilliancy or display. *a*. gaudy, showy, superficially fine. *v.t.* to adorn with tinsel. **tinselly**, *a*.
tin-stone TIN.
tint (tint), *n*. a variety of colour, esp. one produced by admixture with another colour, esp. white; a pale colour; a slight tinge (of another colour); in engraving, an effect of shading texture obtained by a closed series of parallel lines. *v.t.* to give a tint or tints to; to tinge. **tinter**, *n*. one who or that which tints; an engraving-tool or machine for tinting; a plain lantern-slide of one colour. **tintless**, (-lis), *a*. **tinty**, *a*. inharmoniously tinted.
tintack TIN.
tintinnabulum (tintinab'ūləm), *n*. (*pl*. **-la** (-lə)) a bell, esp. a small tinkling one for signalling, fitting to harness etc.; a ringing, tinkling or jingling of bells, plates etc. **tintinnabular, -lary**, *a*. **tintinnabulation** *n*.
tinware TIN.
tiny (tī'ni), *a*. (*comp*. **tinier**, *superl*. **tiniest**) very small.
tip[1] (tip), *n*. the point, end, or extremity, esp. of a small or tapering thing; a small piece or part attached to anything to form a point or end, as a ferrule or shoe-tip; a brush used in laying on goldleaf. *v.t.* (*past, p.p.* **tipped**) to put a tip on; to form the tip of. **on the tip of one's tongue**, about to be uttered. **tipstaff** *n*. a metal-tipped staff carried by a sheriff's officer; a sheriff's officer. **tiptoe**, *adv*. on the tips of the toes. *v.i.* to walk or stand on tiptoe. **tip-top**, *n*. the highest point, the very best; *a*. of the very best. *adv*. in a first-rate way. **tip-topper**, *n*.
tip[2] (tip), *v.t.* to cause to lean, to tilt (up, over etc.); to overturn, to upset; to discharge (the contents of a cart, vessel etc.) thus; to strike lightly, to tap, to touch; to give a small gratuity to; to toss or throw lightly, to give; (*coll.*) to give private information to about a horse, an investment etc. *v.i.* to lean over, to tilt; to upset. *n*. a small present in money, a gratuity; private information, esp. for betting or investment purposes; a slight touch, push or hit; a place where rubbish is dumped. **to tip in**, (*Print.*) to insert a loose plate by pasting the back margin to the page following. **to tip off**, to give a warning hint; in basketball, to start play by throwing the ball high between players of the two sides. **to tip the balance**, to be the deciding factor in a result; to make more or less favourable to. **to tip the scales**, to weigh; to tip the balance. **to tip the wink**, (*coll.*) to hint, to inform furtively. **tip-cat**, *n*. a game with a piece of wood pointed at both ends which is hit with a stick; the tapering piece of wood. **tip-off**, *n*. **tip-up**, *a*. of a (theatre) seat, able to be tilted up on a hinge or pivot. **tipper**, *n*. one who or that which tips; a lorry or truck whose platform can be tilted towards the rear to empty out the load. **tipster**, (-stə), *n*. one who supplies tips about races etc.
tippet (tip'it), *n*. a fur covering for the neck and shoulders, worn by women; an ecclesiastical vestment; part of the official costume of judges etc.
tipple (tip'l), *v.i.*, *v.t.* to drink (alcoholic liquors) habitually. *n*. strong drink; one's favourite (alcoholic) drink. **tippler**, *n*.

tipsy (tip'si), *a*. fuddled, partially intoxicated, proceeding from or inducing intoxication. **tipsy-cake**, *n*. a sponge cake soaked in wine served with custard. **tipsily**, *adv*. **tipsiness**, *n*.
tiptoe, tip-top TIP[1].
TIR, (*abbr.*) International Road Transport (F, Transports Internationaux Routiers).
tirade (tīrād'), *n*. a long, vehement speech, declamation, or harangue, esp. of censure or reproof.
tire[1] (tīə), *v.t.* to exhaust the strength of by toil or labour; to fatigue, to weary; to exhaust the patience or attention of. *v.i.* to become weary or exhausted. **tired**, *a*. fatigued; bored, impatient, irritated; stale, hackneyed. **tiredness**, *n*. **tireless** (-lis), *a*. unwearied, untirable. **tirelessly**, *adv*. **tiresome**, (-səm), *a*. fatiguing, tiring; wearisome, tedious, annoying. **tiresomely**, *adv*. **tiresomeness**, *n*. **tiring**, *n*.
tire[2], **tyre** (tīə), *n*. a band of iron, steel, etc., placed round the rim of a wheel; (*N Am.*) a tyre. **tired**, *a*. (*usu*. *in comb.*, as *rubber-tired*). **tireing**, *n*. **tireless** (-lis), *a*.
tireless, etc. TIRE[1,2].
tiro (tī'rō), TYRO.
Tirolese (tiraléz'), *a*. pertaining to the Tirol, in Austria. *n*. a native of Tirol.
tirwit (tœ'wit), *n*. the lapwing.
'tis (tiz), short for IT IS.
tisane (tizan'), *n*. a ptisan; a medicinal infusion of dried leaves or flowers.
Tishri (tish'ri), **Tisri** (tiz'-), *n*. the first month of the Hebrew civil and the seventh of the ecclesiastical year, corresponding to parts of September and October.
tissue (tish'ōō), *n*. any fine, gauzy, or transparent woven fabric; a fabric of cells and their products, forming the elementary substance of plant and animal organs; a fabrication, a connected series (of lies, accidents etc.); tissue paper; a paper handkerchief. *v.t.* to form into tissue; to interweave, to variegate. **tissue culture**, *n*. the growing of pieces of biological tissue in a nutritive medium in a laboratory. **tissue-paper**, *n*. a thin, gauzy, unsized paper, used for wrapping articles, protecting engravings etc. **tissue typing**, *n*. the ascertaining of types of body tissue, e.g. in order to match organs for transplant. **tissued**, *a*.
Tit. (*abbr.*) Titus.
tit[1] (tit), *n*. a titmouse; a titlark. **titbit**, *n*. a delicate or dainty morsel of food or gossip. **titlark**, *n*. a small bird of the genus *Anthus*, esp. *A. pratensis*, the meadow-pipit. **titling**, (-ling), *n*. a titmouse; a titlark.
tit[2] (tit), *n*. a tap, a slight blow. **tit for tat**: (-tat), blow for blow, retaliation.
tit[3] (tit), *n*. a teat or nipple; (*sl.*) a woman's breast; (*sl.*) a stupid, objectionable or contemptible person.
Titan (tī'tən), *n*. (*Gr. Myth.*) one of the 12 children of Uranus and Ge of gigantic size and strength; the sun-god as the offspring of Hyperion, one of the Titans; a person of superhuman strength or genius. *a*. Titanic. **titanesque** (-nesk'), **titanic** (-tan'-), *a*. **Titaness**, (-nis), *n. fem*.
titanium (titā'niəm, tī-), *n*. a dark-grey metallic element, at. no. 22; chem symbol Ti; found in small quantities in various minerals. **titanium dioxide**, *n*. a white pigment. **titanate**, (-nāt), *n*. a salt of titanic acid. **titanic** (titan'ik), *a*. of quadrivalent titanium. **titaniferous** (tītənif'-), *a*. **titanite** (tī'tənit), *n*. an intensely hard titanosilicate of calcium, sphene. **titano-**, *comb. form*. **titanous**, (tī'tə-), *a*.
titbit TIT[1].
titch (tich), *n*. (*coll.*) a very small person; a very small amount. **titchy**, *a*.
titfer (tit'fə), *n*. (*sl.*) a hat. [rhyming sl. *tit for tat*]
tithe (tīdh), *n*. the 10th part of anything; a tax of one-tenth, esp. of the yearly proceeds from land and personal industry, payable for the support of the clergy and Church. *v.t.* to impose tithes upon. **tithe barn**, *n*. a barn in which the parson stored his corn and other tithes. **tithe-pig**, *n*. one pig out of 10 set apart for tithe. **tithable**,

a. **tither**, *n.* **tithing**, *n.* the taking or levying of tithes.
titi[1] (tē'tē), *n.* the New Zealand diving petrel.
titi[2] (tē'tē), *n.* a small brightly-coloured S American monkey.
titian (tish'ən), *a.* reddish-brown in colour.
titillate (tit'ilāt), *v.t.* to tickle; to excite or stimulate pleasurably. **titillation** *n.*
titivate (tit'ivāt), *v.t.*, *v.i.* to dress up, to adorn, to make smart. **titivation**, *n.*
titlark TIT[1].
title (tī'tl), *n.* an inscription serving as a name or designation, esp. of a book, chapter, poem etc.; the entire contents of the title page of a book; a book or publication; a brief part of this containing the essentials; a title page; the distinguishing formula at the head of a legal document, statute etc.; a division of a document, treatise etc., including caption and text, as arranged for reference; a personal appellation denoting office, nobility, distinction, or other qualification; (*Law*) the right to ownership of property; the legal evidence of this; a title-deed; an acknowledged claim; the grounds of this; fineness, esp. of gold, expressed in carats; (*Eccles.*) a source of income and a fixed sphere of duty required as a condition precedent to ordination; in the Roman Catholic Church, a church or parish; a subtitle in a film; (*pl*) the credits in a film; in a sport, a championship. *v.t.* to give a title to. **title-deed**, *n.* a legal instrument giving the evidence of a person's right to property. **title-holder**, *n.* one holding a title in sport. **title page**, *n.* the page at the beginning of a book giving the subject, author's name etc. **title-rôle**, *n.* the character or part from whose name the title of a play is taken. **titled**, *a.* bearing a title of nobility. **titleless**, (-lis), *a.*
titling, *n.* the act of impressing the title on the back of a book.
titling TIT[1]; TITLE.
titmouse (tit'mows), *n.* (*pl.* -**mice** (-mīs)) a small insectivorous bird of the subfamily Parinae, usu. nesting in holes in tree-trunks.
Titoism (tē'tōizm), *n.* the kind of Communism introduced by Pres. Tito (1892–1980) in Yugoslavia as opposed to that of Russia. **Titoist**, *n.*
titrate (tī'trāt),), *v.t.* to determine the amount of a particular constituent in a solution by adding a known quantity of another chemical capable of reacting upon it. **titration** *n.* **titre** (-tə), *n.* the concentration of a substance in a solution, as ascertained by titration.
titter (tit'ə), *v.i.* to laugh in a restrained manner, to snigger, to giggle. *n.* a restrained laugh. **titterer**, *n.*
tittie[1] (tit'i), *n.* (*Sc. coll.*) a sister.
tittie[2] (tit'i), *n.* childish dim. of TIT[3].
tittle (tit'l), *n.* any small diacritic or punctuation mark; a particle, an iota.
tittlebat (tit'lbat), *n.*, STICKLEBACK.
tittle-tattle (tit'ltatl), *n.* gossip. *v.i.* to gossip.
tittup (tit'əp), *v.i.* (*coll.*) to go, act, or behave in a lively manner, to prance, to frisk. *n.* a tittuping action or movement. **tittupy**, *a.*
titubation (titūbā'shən), *n.* fidgeting or stumbling caused by nervous disorder.
titular (tit'ūlə), *a.* existing in name or in title only, or holding a title without the office or duties attached, nominal; of, pertaining to, or held in virtue of a title; conferring a title. *n.* one who holds the title of an office or benefice without the authority or duties pertaining to it. **titularly**, *adv.* **titulary**, *a.*, *n.*
tizzy (tiz'i), *n.* (*coll.*) a state of extreme agitation (also **tizz**).
Tl, (*chem. symbol*) thallium.
TLS, (*abbr.*) Times Literary Supplement.
TM, (*abbr.*) trademark; transcendental meditation.
Tm, (*chem. symbol*) thulium.
TN, (*abbr.*) Tennessee; trade name.
TNT, (*abbr.*) trinitrotoluene.

to (tu, tə, too), *prep.* in a direction towards (a place, person, thing, state or quality); as far as; no less than, in comparison with, in respect of, in correspondence with; concerning; in the relation of, for, as; against, adjoining; before; accompanied by (music); preceding the indirect object or the person or thing affected by the action etc.; the sign of the infinitive mood, expressing futurity, purpose, consequence etc., limiting the meaning of adjectives, or forming verbal nouns; (*ellipt.*) denoting the infinitive of a verb mentioned or understood. *adv.* towards the condition or end required; into the normal condition, esp. to a standstill or a closed state. **as to** AS. **to and fro** FRO. **to-be**, *a.* about to be (always after the noun, as *mother-to-be*). **to-do** (tədoo'), *n.* fuss, commotion.
toad (tōd), *n.* a tailless amphibian like a frog, usu. with a warty body, terrestrial except during breeding; a repulsive or detestable person. **toad-eater**, *n.* an obsequious parasite, a sycophant. **toad-eating**, *a.*, *n.* **toad-fish**, *n.* a batrachoid fish of the Atlantic coast of N America. **toadflax**, *n.* a perennial herb of the genus *Linaria*, usu. with yellow or bluish personate flowers. **toad-in-the-hole**, *n.* a piece of beef, sausage, or the like, baked in batter. **toadstool** *n.* an umbrella-shaped fungus, esp. a poisonous mushroom.
toadstone (tōd'stōn), *n.* an igneous rock of Carboniferous age, occurring in veins and sheets in limestone.
toady (tō'di), *n.* a toad-eater. *v.i.*, *v.t.* to fawn upon, to play the toady (to). **toadyish**, *a.* **toadyism**, *n.*
toast (tōst), *n.* a slice of bread browned at the fire or under a grill etc., eaten dry, buttered or with some other dish; a drinking or a call for drinking to the health of some person, cause, sentiment etc., (from the old custom of putting toast in liquor perh. through an incident recorded in the *Tatler*); the person or other object of this; (*dated*) a woman often toasted. *v.t.* to brown (bread etc.) under a grill or over a fire; to warm (the feet etc.) at an open fire; to drink to the health or in honour of. *v.i.* to be toasted. **on toast**, of food, served on a piece of toast; at one's mercy. **toast and water** or **toast-water**, *n.* a cooling drink made by pouring boiling water on toast. **toastmaster, -mistress** *n.* an official who announces the toasts at public dinners etc. **toast-rack**, *n.* a table-utensil for holding slices of toast. **toaster**, *n.* an electrical device for toasting bread; a toasting-fork. **toasting-fork**, *n.* a fork to hold bread, etc., for toasting.
tobacco (təbak'ō), *n.* a plant of American origin of the genus *Nicotiana*, with narcotic leaves which are used, after drying and preparing, for smoking, chewing, snuff etc.; the leaves of this, esp. prepared for smoking. **tobacco-pipe**, *n.* a pipe used in smoking tobacco. **tobacco-plant**, *n.* **tobacco-pouch**, *n.* a pouch for carrying tobacco in. **tobacconist**, *n.* a dealer in tobacco.
toboggan (təbog'ən), *n.* a long low sled used for sliding down snow- or ice-covered slopes. *v.i.* to slide on a toboggan. **toboggan-shoot, -slide**, *n.* a prepared course for tobogganing, on a hillside or a timber structure. **tobogganer**, *n.* **tobogganing**, *n.*
toby (tō'bi), *n.* a mug or jug shaped like an old man wearing a three cornered hat.
toccata (təkah'tə), *n.* a composition orig. designed to exercise the player's touch. **toccatella** (tokətel'ə), **toccatina** (tokətē'nə), *n.* a short or easy toccata.
tocher (tokh'ə), *n.* (*Sc.*) a woman's dowry. *v.t.* to give a dowry to. **tocherless** (-lis), *a.*
tocopherol (təkof'ərol), *n.* vitamin E.
tocsin (tok'sin), *n.* an alarm-bell; the ringing of an alarm-bell, an alarm-signal.
tod (tod), *n.* **on one's tod**, (*sl.*) on one's own. [rhyming slang, *on one's Tod Sloan* (name of an American jockey)]
today, to-day (tədā'), *adv.* on or during this or the present day; at the present day. *n.* this day.
toddle (tod'l), *v.i.* to walk with short unsteady steps, as a

child; to walk in a careless or leisurely way, to saunter. *v.t.* to walk (a certain distance etc.) thus. *n.* a toddling walk; a stroll. **toddler,** *n.* (*coll.*) a toddling child.

toddy (tod'i), *n.* the fermented juice of various palm trees; a beverage of spirit and hot water sweetened.

toe (tō), *n.* one of the five digits of the foot, the part of a boot, stocking etc., covering the toes; the fore part of the hoof of a horse etc.; the calk in the front of a horse-shoe; a projection from the foot of a buttress etc., to give it greater stability; the end of the head of a golf-club; the lower end or a projecting part in a shaft, spindle, rod, lever, organ pipe etc. *v.t.* to touch (a line, mark etc.) with the toes; to furnish (socks, shoes etc.) with toes; (*Golf*) to strike (a ball) with the toe of a club. **on one's toes,** alert, ready to act. **to toe in or out,** to turn the toes in or out in walking etc. **to toe the line,** to conform, to bow to discipline. **to tread on someone's toes,** to offend someone. **to turn up one's toes,** (*sl.*) to die. **toecap,** *n.* a stiffened part of a boot or shoe covering the toes. **toehold,** *n.* in climbing, a small foothold; any slight or precarious means or access or progress. **toe-in,** *n.* a slight convergence towards the front given to the front wheels of motor vehicles. **toenail,** *n.* a nail on a toe; a nail driven in obliquely. **toe-rag,** (*sl.*) a mean or despicable person. **toed,** *a.* (*usu. in comb.* as *three-toed*). **toeless,** (-lis), *a.*

toff (tof), *n.* (*sl.*) a swell, a dandy, a person of consequence.

toffee, toffy (tof'i), *n.* a sweetmeat made of boiled sugar or molasses and butter. **toffee-apple,** *n.* a toffee-coated apple on a stick. **toffee-nosed,** *a.* (*sl.*) conceited, arrogant.

to-fore (təfaw'), *prep., adv.* before.

toft (toft), *n.* a homestead; a homestead and the adjoining land; (*dial.*) a hillock or knoll.

tofu (tō'foo), *n.* unfermented soya bean curd.

tog[1] (tog), *n.* (*sl.*) (*usu. in pl.*) clothes. *v.t.* (*past, p.p.* **togged**) to dress (up or out), esp. in one's best.

tog[2] (tog), *n.* a unit of measurement of the heat insulation of clothing, fabrics etc.

toga (tō'gə), *n.* a loose flowing robe, the principal outer garment of an ancient Roman citizen. **toga praetexta** PRAETEXTA. **toga virilis** (viri'lis), the toga assumed by the ancient Roman at the age of 14. **togaed** (-gəd), *a.*

together (təgedh'ə), *adv.* in company or union, conjointly, unitedly; in the same place or at the same time; into union, so as to unite or be joined; without cessation or intermission. *a.* (*coll.*) competent, assured, composed, well-organized. **to get it together,** (*coll.*) to succeed in (doing something); to become well-organized. **togetherness,** *n.* a friendly feeling of being together as a group.

toggle (tog'l), *n.* a pin put through a loop or eye at the end of a rope for securing this; a cross-piece for securing a watch-chain; a toggle-joint; (*Comput.*) a switch which is pressed to turn a feature on or off. **toggle-joint,** *n.* a knee-joint formed by two plates hinged together so as to change the direction of pressure from vertical to horizontal. **toggle-press,** *n.* a press acting by means of toggle-joints. **toggle-switch,** *n.* an electric switch with a projecting lever which is pushed, usu. up or down.

toheroa (tōərō'ə), *n.* an edible mollusc on the New Zealand shores.

toil[1] (toil), *v.i.* to labour with pain and fatigue of body or mind; to move or progress painfully or laboriously. *v.t.* to fatigue or wear out with toil. *n.* hard and unremitting work, labour, drudgery. **toil-worn,** *a.* worn with toil. **toiler,** *n.* **toilful, toilsome,** (-səm), *a.* **toilfully, toilsomely,** *adv.* **toilless,** (-lis), *a.* **toilsomeness,** *n.*

toil (toil),[2] *n.* (*now in pl.*) a net or snare.

toile (twahl), *n.* cloth; a model of a garment made up in cheap cloth. [F]

toiler TOIL[1].

toilet (toi'lit), *n.* the act or process of dressing etc.; style or fashion of dress; dress, costume (also **toilette** (twalet')); a toilet-table; a cover for this; a water-closet; (*Med.*) the cleansing of a part after an operation etc. **to make one's toilet,** to dress, arrange one's hair etc. **toilet-cover,** *n.* a cloth for a toilet-table. **toilet-paper,** *n.* soluble paper for use in a water-closet. **toilet-service, -set,** *n.* a set of utensils for a toilet-table. **toilet-soap,** *n.* **toilet-table,** *n.* a dressing-table with looking-glass etc. **toilet-training,** *n.* training a child to control its bowels and bladder and to use a lavatory. **toilet water,** *n.* a form of perfume lighter than an essence. **toiletry,** (-ri), *n.* (*often pl.*) an article or preparation used in the process of washing or beautifying oneself.

toilful, etc. TOIL[1].

toilinet(te), (twahlinet'), *n.* a fabric of silk and cotton with woollen filling.

Tokay (tōkā', tō'-), *n.* a rich aromatic wine made at Tokaj in Hungary; a white grape from which it is made.

token (tō'kən), *n.* something representing or recalling another thing, event etc.; a sign, a symbol; an evidence, an indication, a symptom; a memorial of love or friendship, a keepsake; a sign proving authenticity; a piece of metal like a coin, formerly issued by tradesmen, banks etc., representing money of greater intrinsic value; a metal or plastic disk used instead of a coin, e.g. in a slot machine; a voucher that can be used as payment for goods to a certain value. *a.* serving as a token; nominal, perfunctory, done, given, invited etc. for form's sake only. **by the same token,** in corroboration. **more by token** MORE[1]. **token payment,** *n.* a small payment made to indicate that the debt or obligation is not repudiated. **tokenism,** *n.* the practice of making only a token effort. **tokenless,** (-lis), *a.*

toko TOCO.

tolbooth TOLL[1].

told, *past, p.p.* TELL.

Toledo (təlē'dō), *n.* a sword or sword-blade made at Toledo in Spain. **Toledan,** *a.*, *n.*

tolerate (tol'ərāt), *v.t.* to suffer, to endure, to permit by not preventing or forbidding; to abstain from judging harshly or condemning (persons, religions, votes, opinions etc.); to sustain, to endure (pain, toil etc.); to sustain (a drug etc.) with impunity. **tolerable,** *a.* endurable, supportable; passable, fairly good. **tolerableness,** *n.* **tolerably,** *adv.* **tolerance,** *n.* the act or state of toleration; permissible variation in weight, dimension, fitting etc. **tolerant,** *a.* showing toleration. **tolerantly,** *adv.* **toleration** *n.* the act of tolerating; the spirit of tolerance; recognition of the right of private judgment in religious matters and of freedom to exercise any forms of worship. **tolerationist,** *n.* **tolerator,** *n.*

toll[1] (tōl), *n.* a tax or duty charged for some privilege, service etc., esp. for the use of a road, bridge, market etc.; damage, deaths, etc., suffered in an accident, natural disaster etc. *v.i.* to pay toll; to take toll. **toll-bar, -gate,** *n.* a gate or bar placed across a road to stop passengers or vehicles till toll is paid. **tolbooth,** (tol'-), **tollbooth** *n.* (*Sc.*) a town jail; *orig.* a temporary structure for the collection of market-tolls. **toll-bridge,** *n.* a bridge where toll is charged for passing over it. **toll call,** *n.* (*N Am.*) a long-distance telephone call. **toll-gatherer, -man,** *n.* **toll-house,** *n.* the house at a toll-gate occupied by a toll-collector. **tollable,** *a.* **tollage,** *n.* **toller,** *n.*

toll[2] (tōl), *v.t.* to cause (a bell) to sound with strokes slowly and uniformly repeated; to give out (a knell etc.) with a slow, measured sound (of a bell, clock etc.); to ring on account of. *v.i.* to sound or ring (of a bell) with slow, regular strokes. *n.* a tolling or a stroke of a bell. **toller,** *n.*

tollable, tollage, etc. TOLL[1].

toller[1, 2] TOLL[1, 2].

Toltec (tol'tek), *n.* one of a people which ruled in Mexico during the 7th–11th cents., before the Aztecs. *a.* of or

pertaining to this people. **Toltecan**, *a*.
tolu (toloo'), *n*. a balsam derived from a S American tree, *Myroxylon balsamum*. **toluate** (tol'ūāt), *n*. a salt of toluic acid. **toluene** (tol'ūēn), *n*. a liquid compound belonging to the aromatic series. **toluic** (-loo'-), *a*.
tom (tom), *n*. a male animal, esp. a tom-cat. **Tom and Jerry**, a hot drink of rum and water with eggs beaten up etc. **Tom, Dick and Harry**, average commonplace people, any taken at random. **tom-boy**, *n*. a girl who likes the rough outdoor activities traditionally associated with boys. **tom-cat**, *n*. a male cat. **Tom Collins** (-kol'inz), a collins made with gin. **tom-fool**, *n*. a ridiculous fool, a trifler. *a*. very foolish. *v.i*. to play the fool, to act nonsensically. **tomfoolery**, *n*. **tom-noddy**, *n*. a blockhead, a dolt; the puffin. **Tom Thumb**, *n*. a midget. **tom-tit**, *n*. a small bird, a tit, esp. a titmouse.
tomahawk (tom'əhawk), *n*. a N American Indian battle-axe or hatchet with a stone, horn, or steel head. *v.t*. to strike or kill with a tomahawk.
tomato (təmah'tō), *n*. (*pl*. **-toes**) the red or yellow pulpy edible fruit (used as a vegetable) of a trailing plant, *Lycopersicon esculentum*, of the nightshade family or Solanaceae; the plant itself.
tomb (toom), *n*. a grave; a vault for the dead; a sepulchral monument. **tombless**, (-lis), *a*. **tombstone**, *n*. a stone placed as a memorial over a grave.
tombac, tomback (tom'bak), *n*. one of various copper and zinc alloys.
tombola (tombō'lə), *n*. an instant lottery at a fête etc. in which tickets are drawn from a revolving drum.
tomboy, tom-cat, etc. TOM.
tome (tōm), *n*. a volume, esp. a ponderous one.
-tome (-tōm), *suf*. used to form nouns, meaning a cutting instrument of a specified kind, as *microtome*.
tomentum (təmen'təm), *n*. a pubescence consisting of matted woolly hairs; (*Anat*.) the inner surface of the pia mater, flocculent with tiny vessels. **tomentose** (-tōs), **-ous**, *a*.
tom-fool, etc. TOM.
tommy (tom'i), *n*. a British private soldier (from *Tommy Atkins*, of disputed orig.); (*sl*.) bread, food, provisions, esp. carried by workmen or given to them in lieu of wages; this method of payment, the truck system; a form of wrench; a rod inserted in a box-spanner. **tommy-gun**, *n*. a short-barrelled, quick-firing firearm. **tommy rot**, *n*. (*coll*.) nonsense. **tommy-shop**, *n*. a shop or other place where the truck system is in force.
tomography (təmog'rəfi), *n*. diagnostic radiography of plane sections of the human body.
tomorrow, to-morrow (təmo'rō), *n*. the next day after today, the morrow. *adv*. on or during this. **like there's no tomorrow**, (*coll*.) recklessly, extravagantly.
tompion (tom'piən), *n*. a lithographic inking-pad; a tampion.
tom-tit TOM.
tom-tom (tom'tom), *n*. a long, narrow, hand-beaten drum used in India, Africa, etc.
-tomy, *suf*. indicating cutting (as *anatomy*) or surgical incision (as *phlebotomy*).
ton[1] (tŭn), *n*. a measure of weight, 20 cwt or 2240 lb. av. (1016.05 kg), also called **long ton**; (*N Am*.) 2000 lb. av. (907.18 kg) also called **short ton**; 1000 kg (2205 lb. av.), also called **metric ton**, **tonne** (tŭn, tŭn'i); a measure of capacity (for timber or cargo on shipboard, 40 cubic ft. (1.132 cu. m) a measure of the displacement of a ship, 2240 lb. av. (1016.05 kg) 35 cu. ft. (0.991 cu. m) of sea water, also called **displacement ton**, a measure of the cargo space on a ship, 100 cu. ft. (2.83 cu. m), also called **register ton**; (*coll*.) an unspecified great weight; (*usu. pl., coll*.) a large quantity; (*sl*.) a score or total of 100; (*sl*.) 100 mph. **ton-up**, *adj*. (*sl*.) having achieved a score, speed etc. of 100. **-tonner**, *comb. form* a ship of a specified tonnage, as a 3000-*tonner*.
ton[2] (tō), *n*. the prevailing fashion or mode. **tonish**, (tō'-), *a*. **tonishness**, *n*. **tony**[1] (tō'-), *a*. (*sl*.)
tonal (tō'nəl), *a*. pertaining to tone or tonality; having tonality. **tonality** (-nal'-), *n*. (*Mus*.) the character or quality of a tone or tonal system; a system of tones, a key; (*Painting*) the general colour-scheme of a picture. **tonally**, *adv*.
tondo (ton'dō), *n*. a circular painting or relief. [It.]
tone (tōn), *n*. sound, with reference to pitch, quality, and volume; a musical sound; modulation or inflexion of the voice to express emotion etc.; general disposition, temper, mood, prevailing sentiment, spirit; timbre; an interval of a major second; an ancient psalm-tune, esp. one of the Gregorian tones; (*Gram*.) syllabic stress; degree of luminosity of a colour; the general effect of a picture, esp. as regards colour and luminosity, the tint or shade of colour; the shade or colour of a photographic print; healthy general condition of the bodily organs, tissues etc. *v.t*. to give tone or quality to; to tune; to modify the colour of a photographic picture by a chemical bath. *v.i*. to harmonize in colour, tint etc.; to receive a particular tone or tint. **to tone down**, to subdue, to soften (the tint, tone, pitch, intensity etc., of); to modify, to reduce, to soften (a statement, demands etc.); to become softer, less emphatic etc. **to tone up**, to become firmer, more vigorous; to heighten, intensify. **tone arm**, *n*. the pick-up arm of a record player. **tone-deaf**, *a*. unable to distinguish accurately between musical sounds of different pitch. **tone language**, *n*. a language in which variation of tone serves to distinguish between words otherwise pronounced in the same way. **tone poem**, *n*. an orchestral composition in one movement which illustrates a train of thought external to the music. **tone row**, *n*. the basic series of notes in serial music. **tone-wheel**, *n*. a high-speed commutator used for the reception of continuous radio waves. **toned**, *a*. **toneless**, (-lis), *a*. **toner**, *n*. one who, or that which, tones; a lotion applied to the face to tighten the pores. **tonometer** (-nom'itə), *n*. a tuning-fork or other instrument for determining the pitch of a tone; an instrument for measuring strains in liquids.
tong (tong), *n*. a Chinese secret society.
tonga (tong'gə), *n*. in India, a light two-wheel cart for four persons.
tongs (tongz), *n.pl*. an implement consisting of two limbs, usu. connected near one end by a pivot, used for grasping coals etc. (also **a pair of tongs**).
tongue (tŭng), *n*. a fleshy muscular organ in the mouth, used in tasting, swallowing and (in man) speech; the tongue of an ox, sheep etc., as food; a tongue-shaped thing or part; the clapper of a bell; the pin in a buckle; a piece of leather closing the gap in the front of a laced shoe; the index of a scale or balance; a vibrating slip in the reed of a flageolet and other instruments; a pointed rail in a railway-switch; a projecting edge for fitting into a groove in match-board; a long low promontory, a long narrow inlet; speech, utterance, the voice; manner of speech; a language; hence a nation, a race. *v.t*. to modify (the sounds of a flute etc.) with the tongue; to put a tongue on (matchboard etc.). *v.i*. to use the tongue in playing some wind instruments. **the gift of tongues**, the power of speaking in unknown tongues, esp. as miraculously conferred on the Apostles on the day of Pentecost. **to give tongue** GIVE[1]. **to hold one's tongue** HOLD[1]. **with one's tongue in one's cheek**, ironically. **tongue-lashing**, *n*. a severe scolding. **tongue-tie**, *n*. shortness of fraenum impeding movement of the tongue. **tongue-tied**, *a*. impeded in speech by this; afraid of or prevented from speaking freely. **tongue-twister**, *n*. a series of words difficult to articulate without stumbling. **tongued**, *a*. (*usu. in comb*., as *loud-tongued*). **tongueless**, (-lis), *a*. **tonguelet**, (-lit), *n*.
tonic (ton'ik), *a*. invigorating, bracing; of or pertaining to

tones; (*Mus.*) pertaining to or founded on the key-note; (*Phonet.*) denoting a voiced sound; stressed; (*Path.*) pertaining to tension, unrelaxing (of spasms). *n.* a tonic medicine; tonic water; (*Mus.*) the key-note. **tonic sol-fa,** *n.* a system of musical notation in which diatonic scales are written always in one way (the key-note being indicated), the tones being represented by syllables or initials, and time and accents by dashes and colons. **tonic solfaist,** *n.* one versed in or advocating this system. **tonic water,** *n.* a carbonated drink flavoured with quinine often used as a mixer with alcoholic drinks. **tonically,** *adv.* **tonicity** (-nis'-), *n.* the state of being tonic; tone; elasticity or contractility of the muscles.
tonight, to-night (tənīt'), *n.* the present night; the night of today. *adv.* on or during this.
tonish, etc. TON[2].
tonka bean (tong'kə), *n.* the fruit of a S American tree, *Dipterix odorata,* the fragrant seeds of which are used in perfumery.
tonnage (tŭn'ij), *n.* the carrying capacity or internal cubic capacity of a vessel expressed in tons; the aggregate freightage of a number of vessels, esp. of a country's merchant marine; a duty on ships, formerly assessed on tonnage, now on dimensions.
tonne (tŭn, tŭn'i), *n.* the metric ton (see TON[1]).
tonneau (ton'ō), *n.* the rear part of a motor-car containing the back seats. [F]
-tonner TON[1].
tonometer TONE.
tonsil (ton'sil), *n.* either of two organs situated in the hinder part of the mouth on each side of the fauces. **tonsillar, tonsillitic** (-lit'ik), *a.* **tonsillectomy,** (-lek'təmi), *n.* surgical removal of the tonsils. **tonsillitis** (-lī'tis), *n.* inflammation of the tonsils.
tonsorial (tonsaw'riəl), *a.* pertaining to a barber or his art.
tonsure (ton'shə), *n.* the shaving of the crown (as in the Roman Catholic Church before 1972) or of the whole head (as in the Greek Church) on admission to the priesthood or a monastic order; the part of the head thus shaved; admission into holy orders. *v.t.* to confer the tonsure on.
tontine (ton'tēn, -tēn'), *n.* a form of annuity in which the shares of subscribers who die are added to the profits shared by the survivors, the last of whom receives the whole amount.
ton-up TON.
tonus (tō'nəs), *n.* tonicity; (*Path.*) a tonic spasm.
Tony (tō'ni), *n.* (*pl.* **Tonys**) an annual American award for work in the theatre. [Antoinette (*Tony*) Perry, d. 1946, US actress]
tony TON[2].
too (too), *adv.* in excessive quantity, degree etc.; more than enough; as well, also, in addition, at the same time; moreover; (*coll.*) extremely, superlatively.
took (tuk), *past* TAKE.
tool (tool), *n.* a simple implement, esp. one used in manual work; a machine used in the making of machines; (*fig.*) anything used as an instrument or apparatus in one's occupation or profession; a person employed as an instrument or agent, a cat's paw; (*bookbinding*) a handstamp or design used in tooling; (*sl.*) the penis. *v.t.* to impress designs on (a bookcover); to work or shape with a tool; (also **tool up**) to equip with (machine) tools for a particular purpose. *v.i.* to work with a tool; (*sl.*) to drive, to ride. **tooled-up,** (*sl.*) carrying firearms. **tool-holder,** *n.* a device for pressing the tool against the work in a lathe; a handle for use with various tools. **toolkit,** *n.* a set of tools. **toolmaker,** *n.* a worker who makes and repairs machine tools in a workshop, etc. **toolmaking,** *n.* **toolroom,** *n.* the part of a workshop where tools are made or repaired. **toolshed,** *n.* a shed for storing esp. tools. garden tools. **tooler,** *n.* one who or that which tools; a stone-mason's broad chisel. **tooling,** *n.*
toon (toon), *n.* a large E Indian tree, *Toona ciliata,* with close-grained red wood.
toot (toot), *v.i.* to make a noise with an instrument or the mouth like that of a horn; to give out such a sound; to call (of grouse). *v.t.* to sound (a horn etc.) thus; to give out (a blast etc.) on a horn. *n.* a tooting sound or blast. **tooter,** *n.*
tooth (tooth), *n.* (*pl.* **teeth** (tēth)) one of the hard dense structures, originating in the epidermis, growing in the mouth or pharynx of vertebrates, and used for mastication; a false or artificial tooth made by a dentist; a toothlike projection on the margin of a leaf etc.; a projecting pin, point, cog etc.; (*pl.*) powers, esp. to compel compliance. *v.t.* to furnish with teeth; to indent. *v.i.* to interlock. **in the teeth of,** in spite of; in direct opposition to; in the face of (the wind). **long in the tooth,** elderly, old (as in horses). **sweet tooth,** a liking for sweet things. **to cast in one's teeth** CAST[1]. **to get one's teeth into,** to become actively and concentratedly involved with. **to one's teeth,** to one's face; in open opposition. **tooth and nail,** with all one's power. **to set the teeth on edge** EDGE. **to show one's teeth,** to adopt a threatening attitude. **toothache,** *n.* pain in the teeth. **tooth-brush,** *n.* a small brush for the teeth. **tooth-comb,** *n.* a fine-toothed comb. **tooth ornament** (*Arch.*) DOG'S-TOOTH. **tooth-paste, -powder,** *n.* paste or powder for cleaning the teeth. **toothpick,** *n.* a pointed instrument of bone, quill etc., for removing particles of food etc., from between the teeth. **toothwort** *n.* a herb, *Lathraea squamaria,* allied to the broom-rape, with toothlike scales on the root-stock; the shepherd's purse, *Capsella bursa-pastoris,* and other plants. **toothed,** *a.* **toothed whale,** *n.* any of a number of whales having simple teeth, as porpoises etc. **toothless** (-lis), *a.* **toothlet,** (-lit), *n.* **toothlike,** *a.* **toothsome** (-səm), *a.* palatable, pleasing to the taste. **toothsomely,** *adv.* **toothsomeness,** *n.* **toothy,** *a.* having prominent teeth.
tootle (too'tl), *v.i.* to toot gently or continuously, as on a flute; (*coll.*) to amble, to trot.
tootsie (tut'si), *n.* a child's word for a foot.
top[1] (top), *n.* the highest part or point of anything, the summit; the upper side or surface; a cover for the upper part of anything, a lid, a cap; the head of a page in a book; the part of a plant above ground; the crown of the head; the upper end or head of a table; the highest position, place, rank etc.; the highest degree, the apex, the culmination, the height; (*Naut.*) a platform round the head of a lower mast, forming an extended base for securing the topmast shrouds; a garment, esp. for women, covering the upper part of the body; top gear; top spin. *v.t.* (*past, p.p.* **topped**) to remove the top or extremity of (a plant, etc.); to put a top or cap on; to cover the top of; to rise to the top of, to surmount; to excel, to surpass, to be higher than; to be (of a specified height); to hit (a ball) on the upper half; (*sl.*) to kill, esp. by hanging. *a.* being on or at the top or summit; highest in position, degree etc. **big top,** a big circus tent. **off the top of one's head,** without preparation, impromptu. **on top,** in the lead; in control. **on top of,** added to; in control of; very close to. **over the top,** on the attack; too far, to excess. **to top oneself,** (*sl.*) to commit suicide esp. by hanging oneself. **to top off,** to complete by putting the top or uppermost part to; to finish, to complete. **to top out,** to put the last or highest stone etc. on, (on a building). **to top up,** to fill up (with petrol, oil, etc.). **top-boot,** *n.* a boot having high tops, usu. of distinctive material and colour. **top brass,** *n.* (*sl.*) the highest-ranking officials or officers. **top-coat,** *n.* an overcoat. **top dog,** *n.* (*coll.*) the uppermost fellow, the boss. **top-dress,** *v.t.* to manure on the surface, as distinguished from digging or ploughing in. **top-dressing,** *n.* **top-flight,** *a.* of the highest rank or quality. **topgallant** (təgal'ənt, top-), *a.* (*Naut.*) applied to the mast, rigging

and sail, next above the topmast. **top gear,** *n.* the highest forward gear in a motor vehicle; maximum effort, output etc. **top-hat,** *n.* a tall silk cylindrical hat. **top-heavy,** *a.* having the top or upper part too heavy for the lower; (*coll.*) intoxicated. **top-hole,** *a.* (*sl, dated*) excellent, first rate. **top-knot,** *n.* an ornamental knot or bow worn on the top of the head; a tuft or crest growing on the head. **top-level,** *a.* at the highest level. **topmast,** *n.* the mast next above the lower mast. **top-notch,** *a.* excellent. **topsail** (topsl), *n.* a square sail next above the lowest sail on a mast; a fore-and-aft sail above the gaff. **top secret,** *a.* requiring the highest level of secrecy. **top side,** *n.* a cut of beef from the thigh; (*pl.*) the sides of a vessel above the water-line. **topsman,** *n.* (*Sc. or dial.*) a head servant, bailiff or overseer; a chief drover; (*sl.*) a hangman. **top-soil,** *n.* the upper layer of soil. *v.t.* to remove this from (a piece of ground). **top-soiling,** *n.* **top spin,** *n.* spin, imparted to a ball by hitting the upper half of it, which makes it travel further and faster. **topless,** (-lis), *a.* without a top; of women's clothing, leaving the breasts bare; of an entertainment etc., featuring women with topless clothing. **topmost,** *a.* highest, uppermost. **topper,** *n.* one who or that which tops; fruit etc., of better quality put at the top in a basket etc.; (*coll.*) a top-hat. **topping,** *a.* (*sl.*) very fine, excellent. *n.* something which forms a top layer, esp. a sauce for food. **topping-up,** *n.* the addition of distilled water to an accumulator cell to compensate for loss by evaporation. **toppingly,** *adv.* **tops,** *n.* the best. *a.* excellent, best.

top², (top), *n.* a wooden or metal toy, usu. conical- or pear-shaped, made to rotate with great velocity on a metal point underneath, by the rapid unwinding of a string or spring or with the hand. **to sleep like a top,** to sleep very soundly.

topaz (tō'paz), *n.* a transparent or translucent fluosilicate of aluminium, usu. white or yellow, but sometimes green, blue, red, or colourless, valued as a gem; a large and brilliant humming-bird. **topazolite** (-paz'əlit), *n.* a yellow or green variety of garnet resembling topaz.

top-boot, -coat, -dress, etc TOP[1].
tope[1] (tōp), *v.i.* to drink alcoholic liquors excessively or habitually, to tipple. **toper,** *n.* a tippler, a heavy drinker.
tope[2] (tōp), *n.* a small shark of the genus *Galeus*, the dog-fish.
topgallant TOP[1].
toph (tōf), **tophus** (-fəs), *n.* calcareous matter deposited round the teeth and at the surface of the joints in gout. **tophaceous** (-fā'shəs), *a.*
Tophet (tō'fet), *n.* a place in the valley of Hinnom, SE of Jerusalem, once used for idolatrous worship, and afterwards for the deposit of the city refuse, to consume which fires were continually kept burning; (*fig.*) hell.
topi, topee (tō'pi), *n.* a sun-hat, a pith helmet. **sola topi,** *n.* a helmet made of sola pith.
topiary (tō'piəri), *a.* shaped by cutting or clipping. *n.* the art of cutting and clipping trees or shrubs etc., into fanciful shapes. **topiarian** (-eə'ri), *n.* **topiarist,** *n.*
topic (top'ik), *n.* the subject of a discourse, argument, literary composition, or conversation. **topical,** *a.* pertaining to or of the nature of a topic comprising or consisting of allusions, esp. to current or local topics. **topically,** *adv.*
topless, topman, etc. TOP[1].
topography (təpog'rəfi), *n.* the detailed description of particular places; representation of local features on maps etc.; the artificial or natural features of a place or district; the mapping of the surface or the anatomy of particular regions of the body. **topographer,** *n.* **topographic, -al** (-graf'ik), *a.* **topographically,** *adv.*
topology (təpol'əji), *n.* topography; the study of geometrical properties and relationships which are not affected by distortion of a figure. **topological** (-loj'-), *a.*
toponym (top'ənim), *n.* a place-name. **toponomy, topony-my** (-pon'əmi), *n.* the science of place-names; a register of place-names of a district etc.; *n.* the naming of regions of the body. (also **toponymics** (-nim'-)) **toponymic, -ical,** *a.*
topper, topping, etc. TOP[1].
topple (top'l), *v.i.* to totter and fall; to project as if about to fall. *v.t.* to cause to topple, to overturn.
topsail, top-sawyer, etc. TOP[1]
topsy-turvy (topsitœ'vi), *adv.*, *a.* upside down; in an upset or disordered condition. *adv.* in a confused manner. *n.* a topsy-turvy state. *v.t.* to turn topsy-turvy; to throw into confusion. **topsy-turviness, topsy-turvydom,** (-dəm), **topsy-turvyism,** *n.*
toque (tōk), *n.* a small, brimless, close-fitting bonnet; a cap or head-dress, usu. small and close-fitting, worn at various periods by men and women. [F]
tor (taw), *n.* a prominent hill or rocky peak, esp. on Dartmoor and in Derbyshire.
torah (taw'rə), *n.* the Mosaic law; the Pentateuch.
torc TORQUE.
torch (tawch), *n.* a light made of resinous wood, twisted flax, hemp etc., soaked in oil or tallow, for carrying in the hand; a hand-lamp containing an electric battery and bulb; an apparatus producing a very hot flame for welding etc.; a source of enlightenment. **to carry a torch for,** to suffer from unrequited love for. **torch-bearer,** *n.* **torch-dance,** *n.* a dance in which each performer carries a torch. **torch-fishing** or **torching,** *n.* fishing at night by torch-light. **torch-light,** *n.* **torch-race,** *n.* a race among the ancient Greeks, in which the runners carried lighted torches. **torch-singer,** *n.* **torch-song,** *n.* a sad song about unrequited love. **torchère** (-sheə'), *n.* (F.) an ornamental stand for a lamp.
tore[1] (taw), *past* TEAR[1].
tore[2] (taw), TORUS.
toreador (toriədaw'), *n.* a bull-fighter, esp. one who fights on horse-back.
torero (təreə'rō), *n.* (*pl.* **-ros**) a bullfighter, esp. one who fights on foot.
toreutic (tərōō'tik), *a.* pertaining to carved, chased, or embossed work, esp. in metal. **toreutics,** *n.pl.* the art of this. **toreumatography** (-mətog'rəfi), *n.* a description of or treatise on toreutics.
tori, toric TORUS.
torii (tō'riē), *n.* (*pl.* **torii**) a gateless gateway composed of two up-rights with (usu.) three superimposed cross-pieces, at the approach to a Shinto temple.
torment[1] (taw'ment), *n.* extreme pain or anguish of body or mind; a source or cause of this.
torment[2] (tawment'), *v.t.* to subject to torment; to afflict, to vex, to irritate. **tormentingly,** *adv.* **tormentor,** *n.* **tormentress,** (-tris), *n. fem.*
tormentil (taw'məntil), *n.* a low herb, *Potentilla tormentilla*, with four-petalled yellow flowers, the astringent root-stock of which is used for medicine.
torn, (tawn), *p.p.* TEAR[1]. **that's torn it,** (*coll.*) exclamation expressing annoyance at one's plans, activities etc., having been ruined.
tornado (tawnā'dō), *n.* (*pl.* **does**) a storm of extreme violence covering a very small area at once, but progressing rapidly, usu. having a rotary motion with electric discharges; (loosely) a very strong wind, a hurricane; a person or thing of great, usu. violent, energy. **tornadic,** *a.*
toroid (to'roid), *n.* a figure shaped like a torus. **toroidal** (-roi'-), *a.* of or like a torus.
torpedo (tawpē'dō), *n.* (*pl.* **-does**) a long, cigar-shaped apparatus charged with explosive, used for attacking a hostile ship below the water-line; a detonating fog-signal placed on a railway track to be exploded by the wheels of a train; a cartridge for exploding in an oil-well etc.; a mine or shell buried in the way of a storming-party; a mixture of fulminate and grit exploded on the ground as a toy; an electric ray, a sea-fish having an electrical appara-

tus for disabling or killing its prey. *v.t.* to attack, blow up, or sink with a torpedo; to destroy or wreck suddenly.
aerial torpedo, a torpedo launched from an aircraft.
torpedo-boat, *n.* a small swift vessel fitted for firing torpedoes. **torpedo-net**, *n.* a wire net hung round a ship to intercept torpedoes. **torpedo-tube**, *n.* (*Nav.*) a tube for the discharge of torpedoes. **torpedoist** *n.*
torpid (taw'pid), *a.* having lost the power of motion or feeling; benumbed; dormant (of a hibernating animal); apathetic, sluggish, inactive. **torpefy** (-pifī), *v.t.* **torpescent** (-pes'ənt), *a.* **torpescence**, *n.* **torpidity** (-pid'-), **torpidness, torpor,** (-pə), *n.* **torpidly,** *adv.* **torporific** (-pərif'-), *a.*
torque (tawk), *n.* (also **torc**) a twisted necklace of gold or other metal, worn by the ancient Gauls etc.; the movement of a system of forces causing rotation. **torque converter**, *n.* a device for transmitting and amplifying torque, esp. through a hydraulic medium. **torque wrench**, *n.* a type of wrench fitted with a gauge for measuring torque. **torquate** (-kwāt), **-ated**, *a.* (*Zool.*) having a ring of distinctive colour about the neck. **torqued**, *a.* twisted; (*Her.*) wreathed.
torr (taw), *n.* a unit of pressure, equal to 1/760 of a standard atmosphere.
torrefy (to'rifī), *v.t.* to dry or parch; to roast (ores etc.). **torrefaction** (-fak'-), *n.*
torrent (to'rənt), *n.* a violent rushing stream (of water, lava etc.); a flood (of abuse, passion etc.). **torrential** (-ren'shəl), *a.* **torrentially,** *adv.*
Torricellian (torichel'iən), *a.* pertaining to the Italian physicist and mathematician E. *Torricelli* 1608–47. **torricellian tube,** *n.* the barometer. **torricellian vacuum,** *n.* the vacuum above the mercury in this.
torrid (to'rid), *a.* dried up with heat, parched, scorching, very hot; intense, passionate. **torrid zone,** *n.* the broad belt of the earth's surface included between the tropics. **torridity** (-rid'-), **torridness,** *n.* **torridly,** *adv.*
torse (taws), *n.* (*Her.*) a wreath. **torsade** (-sād'), *n.* an ornamental twisted cord, ribbon, etc.
torsion (taw'shən), *n.* the act of twisting or the state of being twisted; (*Mech.*) the force with which a body tends to return to its original state after being twisted; the stress produced by twisting. **torsion balance**, *n.* an instrument for estimating very minute forces by the action of a twisted wire. **torsibility** (-sibil'-), *n.* **torsional,** *a.* **torsionally,** *adv.* **torsionless,** (-lis), *a.*
torsk (tawsk), *n.* a food-fish, *Brosmius brosme*, allied to the cod.
torso (taw'sō), *n.* (*pl.* **-sos**) the trunk of a statue or body without the head and limbs.
tort (tawt), *n.* (*Law*) a private or civil wrong; not arising out of a contract, for which damages can be sought.
tortious (-shəs), *a.* **tortiously,** *adv.*
torte (taw'tə, tawt), *n.* a rich gateau or tart, with fruit, cream etc.
torticollis (tawtikol'is), *n.* a spasmodic affection of the neck-muscles, stiff-neck.
tortilla (tawtē'yə), *n.* in Mexican cooking, a thin flat maize cake baked on an iron plate. [Sp.]
tortious, tortiously TORT.
tortoise (taw'təs), *n.* a terrestrial or freshwater turtle; (*Rom. Ant.*) a testudo; a very slow person. **tortoiseshell,** *n.* the mottled horny plates of the carapace of some sea-turtles, used for combs, ornaments, inlaying etc. *a.* made of this; resembling this in marking and colour. **tortoiseshell butterfly,** *n.* any of the genus *Nymphalis*, with mottled yellow, orange and black wings. **tortoiseshell cat,** *n.* a female domestic cat with mottled yellow, brown and black coat.
tortrix (taw'triks), *n.* a genus of British moths typical of the family Tortricidae, called the leaf-rollers. **tortricid,** *n.* a moth of this genus.

tortuous (taw'tūəs), *a.* twisting, winding, crooked; roundabout, devious, not open and straightforward. **tortuose,** (-ōs), *a.* (*Bot.*) **tortuosity** (-os'-), *n.* **tortuousness,** *n.* **tortuously,** *adv.*
torture (taw'chə), *n.* the infliction of extreme physical pain as a punishment or to extort confession etc.; excruciating pain or anguish of mind or body. *v.t.* to subject to torture; to wrest from the normal position; to distort; to pervert the meaning of (a statement etc.). **torturable,** *a.* **torturer,** *n.* **torturingly,** *adv.* **torturous,** *a.*
torula (taw'ūlə), *n.* a chain of spherical bacteria; (*Bot.*) a genus of microscopic yeast-like fungi. **toruliform,** (-fawm), *a.* **torulose,** (-lōs), **-lous,** *a.* (*Bot.*) having alternate swells and contractions like the growth of torula.
torus (taw'əs), *n.* (*pl.* **-ri -rī**)) a semi-circular projecting moulding, esp. in the base of a column; the receptacle or thalamus of a flower, the modified end of a stem supporting the floral organs; (*Anat.*) a rounded ridge; (*Geom.*) a ring-shaped surface generated by a circle rotated about a line which does not intersect the circle. **toric,** *a.*
Tory (taw'ri), *n.* one of the party opposed to the exclusion of the Duke of York (James II) from the throne and to the Revolution of 1688; a member of the Conservative party; (*pl.*) this party. *a.* pertaining to the Tories. **Toryism,** *n.*
tosh (tosh), *n.* (*sl.*) rubbish, nonsense.
toss (tos), *v.t.* (*past, p.p.* **tossed,** *poet.* **tost**) to throw up with the hand, esp. palm upward; to throw, to pitch, to fling, with an easy or careless motion; of an animal, to throw into the air with the horns; to throw back (the head) with a jerk; to jerk; to throw about or from side to side, to cause to rise and fall, to agitate; to throw (up) a coin into the air to decide a wager etc., by seeing which way it falls; hence, to settle a wager or dispute with (a person) thus. *v.i.* to roll and tumble about, to be agitated; to throw oneself from side to side. *n.* the act of tossing; the state of being tossed. **to take a toss,** to be thrown by a horse. **to toss off,** to swallow at a draught; to produce or do quickly or perfunctorily; (*sl.*) to masturbate. **to toss up,** to toss a coin. **to win the toss,** to have something decided in one's favour by tossing up a coin. **toss-pot,** *n.* a toper, a drunkard. **toss-up,** *n.* the tossing up of a coin; a doubtful point, an even chance. **tosser,** *n.* **tossily,** *adv.* (*coll.*) pertly, indifferently. **tossy,** *a.*
tot[1] (tot), *n.* anything small or insignificant, esp. a small child; (*coll.*) a dram of liquor. **tottie, totty** (-i), *n.*, *a.*
tot[2] (tot), *n.* a sum in simple or compound addition. *v.t.* (*past, p.p.* **totted**) to add (up) *v.i.* to mount (up). **totting-up,** *n.* adding together to make a total, esp. driving offences until there are sufficient to cause disqualification.
tot[3] (tot), *n.* (*sl.*) something re-usable salvaged from a dustbin etc. **totter,** *n.* one who scavenges from dustbins etc.; a scrap-dealer. **totting,** *n.*
total (tō'təl), *a.* complete, comprising everything or constituting the whole; comprising everything; absolute, entire, thorough. *n.* the total sum or amount; the aggregate. *v.t.* (*past, p.p.* **totalled**) to ascertain the total of; to amount to as a total; (*N Am., sl.*) to wreck (a car) completely in a crash. *v.i.* to amount (to) as a total. **total abstinence,** complete abstention from intoxicating liquors. **total recall,** *n.* the ability to remember the past in great detail. **total war,** *n.* warfare in which all available resources, military and civil, are employed. **totalitarian** (-talitea'ri-), *a.* permitting no rival parties or policies; controlling the entire national resources of trade, natural wealth, and manpower. **totalitarianism,** *n.* **totality** (-tāl'-), *n.* **totalize, -ise,** *v.t.* to total. *v.i.* to use a totalizator. **totalization,** *n.* **totalizator,** *n.* a machine for showing the total amount of bets staked on a race in order to divide the whole among those betting on the winner. **totally,** *adv.*
totara (tō'tərə), *n.* the New Zealand red pine.
tote[1] (tōt), *v.t.* to carry, to lead, to haul. **to tote bag,** *n.* a

large bag for shopping etc.
tote[2] (tōt), *n.* short for TOTALIZATOR.
totem (tō'təm), *n.* a natural object, usu. an animal, taken as a badge or emblem of an individual or clan on account of a supposed relationship; an image of this. **totem-pole**, *n.* a post on which totems are carved or hung. **totemic** (-tem'-), **totemistic** (-mis'-), *a.* **totemism**, *n.* **totemist**, *n.*
tother, t'other (tŭdh'ə), *a.*, *pron.* the other.
totipalmate (tōtipal'mət, -māt), *a.* wholly webbed. **totipalmation**, *n.*
totitive TOTIENT.
totter[1] (tot'ə), *v.i.* to walk or stand unsteadily, to stagger; to be weak, to be on the point of falling. **totterer**, *n.* **totteringly**, *adv.* **tottery**, *a.*
totter[2] TOT[3].
tottie, totty TOT[1].
toucan (too'kən), *n.* a brilliantly-coloured tropical American bird with an enormous beak.
touch (tŭch), *v.t.* to meet the surface of, to have no intervening space between at one or more points, to be in, or come into, contact with; to bring or put the hand or other part of the body or a stick etc., into contact with; to cause (two objects) to come into contact; to put the hand to (the hat etc.); to reach, to attain; to meddle, to interfere with; to injure slightly; to approach, to compare with; to impair; to concern, to relate to; to treat of hastily or lightly; to strike lightly, to tap, to play upon lightly, to mark or delineate lightly, to put (in) fine strokes with a brush etc.; to be tangent to; to produce a mental impression on; to affect with tender feeling, to soften; (*sl.*) to beg or borrow money. *v.i.* to come into contact (of two or more objects); to come to land, to call (at a port etc.) *n.* the act of touching; the state of touching or being touched, contact; the sense by which contact, pressure etc., are perceived; a slight effort, a light stroke with brush or pencil; a stroke, a twinge; a trace, a minute quantity, a tinge; characteristic manner or method of handling, working, executing, playing on the keys or strings of a musical instrument etc.; the manner in which the keys of a piano etc., respond to this; characteristic impress; intimate correspondence, intercourse, or communication, accord, sympathy; magnetization of a steel bar by contact with magnets; a test, a proof, a touchstone; (*Med.*) the exploring of organs etc., by touch; (*Football*) the part of the field outside the touch-lines and between the goal-lines. **in touch**, outside the touch-lines; in contact, communication or correspondence; having up-to-date knowledge. **out of touch**, no longer in contact, communication or correspondence; not having up-to-date knowledge. **to touch down**, (*Rugby*) to touch the ground with the ball behind the opponent's goal; of an aircraft etc., to land. **to touch off**, to cause to begin, to trigger. **to touch on** or **upon**, to allude to; to deal with (a subject etc.) briefly; to have a bearing on. **to touch up**, to correct or improve by slight touches, as paint or make-up, to retouch; to strike or stimulate (a horse, etc.) gently; (*sl.*) to try to arouse sexually by touching. **touch-and-go**, *n.* a state of uncertainty. *a.* highly uncertain, very risky or hazardous. **touch-down**, *n.* a touching down. **touch-hole**, *n.* the priming hole or vent of a gun. **touch judge**, *n.* official who indicates whether and where the ball has gone into touch in rugby. **touch-lines**, *n.pl.* (*Football*) the two longer or side boundaries of the field. **touch-me-not**, *n.* the plant noli-me-tangere. **touch-needle**, *n.* a needle of gold alloy of known composition employed in assaying other alloys by comparison of the marks made on the touchstone. **touchpaper**, *n.* paper saturated with nitrate of potash for igniting gunpowder etc. **touchstone**, *n.* a dark stone, usu. jasper, schist, or basanite used in conjunction with touch-needles for testing the purity of gold and other alloys; a standard, a criterion. **touch-type**, *v.i.* to type without looking at the typewriter keyboard. **touch-typing**,

n. **touchwood**, *n.* wood decayed by the action of fungi, easily ignited and burning like tinder. **touchable**, *a.* **touched**, *a.* moved by some emotion, e.g. pity or gratitude; (*coll.*) slightly insane. **toucher**, *n.* **touching**, *a.* affecting, moving, pathetic. *prep.* concerning, with regard to. **touchingly**, *adv.* **touchingness**, *n.*
touché (tooshā'), *int.* acknowledging a hit in fencing, or a point scored in argument. [F]
touchy (tŭch'i), *a.* apt to take offence, irascible, irritable. **touchily**, *adv.* **touchiness**, *n.*
tough (tŭf), *a.* firm, strong, not easily broken; resilient, not brittle; able to endure hardship; stubborn, unyielding; aggressive, violent; laborious; difficult; (*coll.*) hard, severe (of luck etc.) *n.* (*N Am.*) a rough, a bully. **toughen**, *v.t.*, *v.i.* **toughish**, *a.* **toughly**, *adv.* **toughness**, *n.*
toupee, toupet (too'pā), *n.* a small wig to cover a bald spot.
tour (tuər), *n.* a journeying round from place to place in a district, country etc.; an extended excursion or ramble; a circuit; a shift or turn of work or duty, esp. a period of duty on a foreign station; a trip made by a theatrical company or solo performer, sports team etc., stopping at various places to play. *v.i.* to make a tour. *v.t.* to make a tour through. **touring car, tourer**, *n.* a large, long car with room for a lot of luggage. **tour operator**, *n.* a travel agency which organizes package tours. **tourism**, *n.* organized touring, esp. from or to a foreign country. **tourist**, *n.* a person making a tour, esp. a holidaymaker or sportsman. **tourist class**, *n.* the lowest category of passenger accommodation in a ship or aircraft. **tourist trap**, *n.* a place tawdrily got up to appeal to the ignorant tourist. **touristy**, *a.* full or tourists; designed to attract tourists.
touraco (too'rəkō), *n.* a brilliantly-coloured African bird of the genus *Turacus corythaix*.
tour de force (tuə də faws'), *n.* a brilliant feat or performance. [F.]
tourist TOUR.
tourmaline (tuə'məlēn), *n.* a black or coloured transparent or translucent silicate with electrical properties, some varieties of which are used as gems.
tournament (tuə'nəmənt), *n.* a contest, exercise, or pageant in which mounted knights contested, usu. with blunted lances etc.; any contest of skill in which a number of persons take part.
tournedos (tuə'nədō), *n.* (*pl.* **-dos** (-z)) a thick round fillet steak.
tourney (tuə'ni), *n.* a tournament.
tourniquet (tuə'nikā), *n.* a bandage for compressing an artery and checking haemorrhage.
tousle (tow'zəl), *v.t.* to pull about; to disarrange, to rumple, to dishevel, to put into disorder. *v.i.* to toss about, to rummage. *n.* a tousled mass (of hair etc.) **tously, tousy**, *a.*
tout[1] (towt), *v.i.* to solicit custom in an obtrusive way; to observe secretly, to spy (esp. on horses in training for a race). *v.t.* to publicize or praise loudly. *n.* one employed to tout; one who watches horses in training and supplies information; one who sells tickets (esp. for over-booked theatrical or sporting events) at very high prices (also **ticket tout**). **touter**, *n.*
tout[2] (too), *a.* all, whole. *adv.* entirely. **tout à fait** (tootafe'), entirely. **tout de suite** (tootswēt'), immediately. **tout ensemble** (tootèsē'bl'), all together. **tout le monde** (toolamōd'), everybody. [F.]
tovaris(c)h (təvah'rish), *n.* comrade. [Rus.]
tow[1] (tō), *v.t.* to pull (a boat, ship etc.) through the water by a rope etc.; to pull a vehicle behind another; to drag (a net) over the surface of water to obtain specimens; to pull, to drag behind one. *n.* the act of towing; the state of being towed. **in tow**, being towed; following; under control or guidance. **on tow**, of a vehicle, being towed. **tow-bar**, *n.* a strong bar on the back of a vehicle for

attaching a trailer. **towboat**, *n*. a tug; a boat, barge etc., that is being towed. **tow(ing)-line, -rope**, *n*. a hawser or rope used in towing. **tow(ing)-net**, *n*. one for towing along the surface of water to collect specimens. **tow(ing)-path**, *n*. a track beside a canal or river for animals towing barges etc. **towable**, *a*. **towage** (-ij), *n*.

tow² (tō), *n*. the coarse broken fibres of hemp or flax after heckling etc. **tow-headed**, *a*. having very pale hair. **towy**, *a*.

toward (təwawd', twawd, tawd), **towards**, *prep*. in the direction of; as regards, with respect to; for, for the purpose of; near, about. *adv*. in preparation, at hand.

towel (tow'əl), *n*. an absorbent cloth for wiping and drying after washing etc. *v.t.* (*past, p.p.* **towelled**) to wipe with a towel; (*sl.*) to thrash. **to throw in the towel**, to concede defeat, esp. in boxing. **towelling**, *n*. material for making towels; (*sl.*) a thrashing.

tower (tow'ə), *n*. a structure lofty in proportion to the area of its base, and circular, square, or polygonal in plan, frequently forming part of a church, castle, or other large building; a place of defence, a protection. *v.i.* to rise to a great height, to soar; to be relatively high, to reach high (above). **tower of strength**, person or thing that can be counted on for esp. moral support. **tower block**, *n*. a very tall residential or office building. **towered**, *a*. **towering**, *a*. very high, lofty; of passion etc., violent, outrageous. **towery**, *a*.

town (town), *n*. a collection of dwelling-houses larger than a village, esp. one not constituted a city; this as contrasted with the country; the people of a town; the chief town of a district or neighbourhood, esp. London. **on the town**, in pursuit of pleasure and entertainment. **to go to town**, (*coll.*) to let oneself go, to drop all reserve. **town and gown** [GOWN]. **town-clerk**, *n*. formerly, the clerk to a municipal corporation; the keeper of the records of a town. **town-council**, *n*. the governing body in a town. **town-councillor**, *n*. **town-crier** CRIER. **town gas**, *n*. manufactured coal gas, opp. to *natural gas*. **town hall**, *n*. a large public building for the transaction of municipal business, public meetings, and entertainments etc. **town house**, *n*. a private residence in town, opp. to country house; a town hall; a modern urban terraced house, esp. a fashionable one. **town-planning**, *n*. the regulating of the laying out or extension of a town. **town-planner**, *n*. **townscape**, *n*. a picture of an urban scene; the visual design of an urban development. **townsfolk**, *n.pl.* (*collect.*) the people of a town or city. **town-talk**, *n*. the subject of general conversation. **townee** (-'nē), *n*. (*sl.*) one who habitiuallly or for preference lives in town. **townish**, *a*. **townless** (-lis), *a*. **townlet** (-lit), *n*. **township**, *n*. (*Hist.*) a division of a large parish, comprising a village or town; (*N Am.*) a territorial district subordinate to a county invested with certain administrative powers; (*Austral.*) any town or settlement, however small; in S Africa, an urban area designated for non-white people. **township jazz, jive**, style of popular music mingling traditional African elements and rock and roll, originating from the S African townships. **townsman**, *n*. an inhabitant of a town; one's fellow citizen. **townspeople**, *n.pl.* (*collect.*) **townward**, *a.*, *adv*. **townwards**, *adv*.

towy TOW².

tox(i)-, toxico- *comb. form* poisonous.

toxaemia, (*esp. N Am.*) **toxemia** (toksē'miə), *n*. blood-poisoning.

toxanaemia, (*esp. N Am.*) **toxanemia** (toksənē'miə), *n*. anaemia due to blood-poisoning.

toxic (tok'sik), *a*. of or pertaining to poison; poisonous. **toxic shock syndrome**, a group of symptoms including vomiting, fever and diarrhoea, attributed to the use of tampons by menstruating women, thought to be caused by a toxin arising from staphylococcal infection. **toxically**, *adv*. **toxicant**, *a*. poisonous. *n*. a poison. **toxication**, *n*.

toxicity (-sis'-), *n*. **toxicology** (-kol'əji), *n*. the branch of medicine treating of poisons and their antibodies. **toxicologist**, *n*. **toxicological** (-loj'-), *a*. **toxicologically**, *adv*.

toxicomania (-mā'niə), *n*. a morbid desire for poison.

toxicosis (-kō'sis), *n*. a morbid state due to the action of toxic matter. **toxigenic** (-jen'-), *a*. producing poison. **toxiphobia** (-fō'biə), *n*. unreasonable fear of being poisoned.

toxin (tok'sin), *n*. a poisonous compound causing a particular disease; any poisonous ptomaine.

toxocariasis (toksōkəri'əsis), *n*. a disease in humans caused by the larvae of a parasitic worm (*Toxocara*) found in cats and dogs, causing damage to the liver and eyes.

toxophilite (toksof'ilit), *n*. one skilled in or devoted to archery. *a*. pertaining to archery. **toxophilitic** (-lit'-), *a*.

toy (toi), *n*. a plaything, esp. for a child; something of an amusing or trifling kind, not serious or for actual use. *v.i.* to trifle, to amuse oneself, to flirt; to fiddle; to consider idly. **toy-boy**, *n*. (*coll.*) a (woman's) very young lover. **toy dog, spaniel**, or **terrier** a pigmy variety of dog kept as a curiosity or pet. **toyshop**, *n*. a shop where toys are sold. **toyer**, *n*. **toyingly**, *adv*.

trabeate (trā'biət), **trabeated**, *a*. (*Arch.*) built with horizontal beams, not with arches or vaults. **trabeation**, *n*.

trabecula (trəbek'ūlə), *n*. a band or bar of connective tissue, esp. one forming the framework of an organ. **trabecular, trabeculate** (-lət), **trabeculated** (-lātid), *a*.

trace¹ (trās), *n*. one of the two straps, chains, or ropes by which a vehicle is drawn by horses etc. **in the traces**, in harness. **to kick over the traces** KICK.

trace² (trās), *n*. a mark left by a person or animal walking or thing moving, a track, a trail, a footprint, a rut etc. (*usu. in pl.*); a line made by a recording instrument; a token, vestige, or sign of something that has existed or taken place; a minute quanity. *v.t.* to follow the traces or track of; to note the marks and vestiges of; to ascertain the position or course of; to pursue one's way along; to delineate, to mark out; to sketch out (a plan, scheme etc.); to copy (a drawing etc.) by marking the lines on transparent paper or linen laid upon it. *v.i.* to be followed back to the origins, date back. **trace element**, *n*. a chemical element present in small quantities, esp. one that is valuable for an organism's physiological processes. **traceable**, *a*. **traceability** (-bil'-), **traceableness**, *n*. **traceably**, *adv*. **tracer**, *n*. one who traces; a trace-horse; an artificially produced radio-active isotope introduced into the human body where its course can be followed by its radiations. **tracer bullet, shell**, *n*. a bullet or shell whose course is marked by a smoke trail or a phosphorescent glow. **tracery** (-əri), *n*. ornamental open-work in Gothic windows etc.; any decorative work or natural markings resembling this. **traceried**, *a*. **tracing**, *n*. (the act of making) a traced copy. **tracing-paper, -cloth, -linen**, *n*. a thin transparent paper or linen used for copying drawings etc., by tracing.

trachea (trəkē'ə), *n*. (*pl.* -**cheae** (-kē'ē) the windpipe, the air-passage from the larynx to the bronchi and lungs; one of the tubes by which air is conveyed from the exterior in insects and arachnids; (*Bot.*) a duct, a vessel. **tracheal, trachean, tracheate** (trā'kiət), *a*. **trachearian** (trākiəɔ'ri-), **tracheary** (trā'-), *a*. belonging to the Trachearia, a division of arachnids having tracheae. *n*. one of this division.

tracheo-, *comb. form*. of or in the trachea.

tracheotomy (trakiot'əmi), *n*. the operation of making an opening into the windpipe.

trachoma (trəkō'mə), *n*. a disease of the eye characterized by papillary or granular excrescences on the inner surface of the lids.

trachyte (trak'īt), *n*. a gritty-surfaced volcanic rock containing glassy feldspar crystals. **trachytic** (-kit'-), *a*.

tracing, etc. TRACE².

track (trak), *n*. a series of marks left by the passage of a person, animal, or thing, a trail; a series of footprints (*usu. in pl.*); a path, esp. one not constructed but beaten

tract 873 **traffic**

by use; a course, the route followed by ships etc.; a racecourse; a set of rails, a monorail, or a line of railway with single or double tracks; the distance between the points where a pair of wheels are in contact with the ground, or rails etc.; (*N Am.*) a railway line; the groove in a gramophone record in which the needle travels; anything, e.g. a song, recorded on a gramophone record; one of several paths on a magnetic recording device on which esp. sound from a single input channel is recorded; the endless band on which a tractor propels itself; the conveyor which carries the items being assembled in a factory. *v.t.* to follow the track or traces of; to trace, to follow out (the course of anything); to follow the flight of (a spacecraft etc.) by receiving signals emitted by or reflected from it; to film (a subject) by moving the camera along a fixed path; to tow; of the stylus of a pickup arm, to follow the groove on a record; of a camera, to move along a fixed path while shooting. **beaten track**, *n.* the usual method; the ordinary way. **in one's tracks**, where one stands. **to keep, lose, track of,** to keep, fail to keep, oneself informed of the whereabouts or progress of somebody or something. **to make tracks,** to leave, to run away, to decamp; to head (for). **to track down,** to discover by tracking. **track event,** in athletics, a race of any kind. **tracklayer**, *n.* (*N Am.*) a plate-layer. **track record,** *n.* the past achievements, performance, experience etc. of a person or thing. **track shoe,** *n.* a light running shoe with spikes on the sole to improve grip. **tracksuit,** *n.* a light, loose-fitting suit for wearing before and after vigorous exercise, or as a leisure garment. **tracker,** *n.* **tracker dog,** *n.* a dog that uses its sense of smell to find e.g. drugs, persons or smuggled goods. **trackless** (-lis), *a.* pathless, unmarked by feet; untrodden, untravelled. **tracklessly,** *adv.* **tracklessness,** *n.*
tract[1] (trakt), *n.* a region or area of land or water of a considerable but undefined extent; (*Anat.*) the region of an organ or system; a period (of time).
tract[2] (trăkt), *n.* a short treatise or pamphlet, esp. on religion or morals.
tractable (trak'təbl), *a.* that may be easily led, managed, or controlled; docile, manageable. **tractability** (-bil'-), *n.* **tractableness,** *n.* **tractably,** *adv.*
Tractarian (traktea'riən), *n.* one of the authors of *Tracts for the Times* (1833–41) a series enunciating the principles of **Tractarianism,** reaction towards primitive Catholicism and against rationalism; an adherent of this, a High Churchman. *a.* pertaining to Tractarianism.
tractate (trak'tāt), *n.* a treatise.
tractile (trak'tīl), *a.* capable of being drawn out.
traction (trak'shən), *n.* the act of drawing something along a surface, esp. by motive power; the state of being so drawn; adhesive friction on a surface, esp. that of a wheel; a pulling force exerted on a part of the body to cure an abnormal condition. **traction-engine,** *n.* a locomotive for drawing heavy loads on ordinary roads.
tractor (trak'tə), *n.* a self-propelling vehicle capable of drawing other vehicles, farm implements etc.; the front section of an articulated lorry, consisting of a chassis, engine and driver's cab, which pulls the trailer; an aircraft with its propellor or propellors mounted in front of the engine.
trad (trad), *a* (*coll.*) short for TRADITIONAL. *n.* traditional jazz.
trade (trād), *n.* a business, handicraft, or mechanical or mercantile occupation carried on for subsistence or profit, distinguished from agriculture, unskilled labour, the professions etc.; the exchange of commodities, buying and selling, commerce; an exchange of one thing for another; the amount of business done in a particular year, place etc.; (*collect.*) persons engaged in a particular trade; (*coll.*) a deal, a bargain (in business or politics); (*pl.*) the tradewinds; (*sl.*) a homosexual sexual partner. *v.i.* to buy and sell, to barter, to exchange, to traffic, to deal (in); to carry on commerce or business (with); to carry merchandise (between etc.); to buy and sell (political influence, patronage etc.) corruptly. *v.t.* to sell or exchange in commerce, to barter; to swap. **to trade in,** to give in part payment. **to trade on,** to take advantage of. **domestic** or **home trade,** that carried on within a country. **foreign trade,** interchange of commodities by importation or exportation with other countries. **trade cycle,** *n.* the recurrent alternation of prosperity and depression in trade. **trade gap,** *n.* the amount by which a country's visible imports exceeds its visible exports. **trade-in,** *n.* a transaction in which an item is given in part payment for another; the item thus given. **trademark,** *n.* a registered symbol or name used by a manufacturer or merchant to guarantee the genuineness of goods; a distinguishing feature of a person or thing. *v.t.* to provide with a trademark. **trade name,** *n.* the name by which an article is called in the trade; the name of a proprietary article. **trade-off,** *n.* the exchange of one thing for another, esp. as a compromise. **trade-price,** *n.* the price charged to dealers for articles to be sold again. **trade secret,** *n.* a process, formula etc. used to make a commercial product, known to only one manufacturer. **tradesman,** *n.* a retail dealer, a shopkeeper; a craftsman. **tradespeople,** *n.* (*collect.*) people engaged in trades, tradesmen and their families. **tradeswoman,** *n. fem.* **trade union, trades union,** *n.* an organized body of workers in any trade, formed for the promotion and protection of their common interests. **Trades Union Congress,** the association of which all major British trade unions are members. **trade-unionism,** *n.* **trade-unionist,** *n.* **trade-wind,** *n.* a wind blowing from the north or south toward the thermal equator and deflected in a westerly direction by the easterly rotation of the earth; (*pl.*) these and the anti-trades. **tradeless** (-lis), *a.* **trader,** *n.* a person engaged in trade; a merchant, a tradesman; a vessel employed in trade. **trading estate,** *n.* an area of buildings intended for commercial or light industrial use. **trading stamp,** *n.* a stamp given free with a purchase, which can be saved and later exchanged for goods.
tradition (trədish'ən), *n.* the handing down of opinions, practices, customs etc., from ancestors to posterity, esp. by oral communication; a belief, custom etc., so handed down; a doctrine believed to have divine authority but not found in Scripture, as the oral law said to have been given by God to Moses on Mt Sinai, the oral teaching of Christ not recorded in the New Testament; the acts and sayings of Mohammed not recorded in the Koran; the principles, maxims etc., derived from the usage and experience of artists, dramatists, actors etc.; (*Law*) formal delivery (of property). **traditional,** *a.* of tradition; of or concerning a type of jazz which began in New Orleans in the 1900s. **traditionalism,** *n.* adherence to tradition, esp. superstitious regard to tradition in religious matters; a philosophic system attributing human knowledge, esp. of religion and ethics, to revelation and tradition. **traditionalist, traditionist,** *n.* **traditionalistic** (-lis'-), *a.* **traditionally,** *adv.*
traduce (trədūs'), *v.t.* to defame, to calumniate, to misrepresent. **traducement,** *n.* **traducer,** *n.* **traducible,** *a.* **traducingly,** *adv.*
traffic (traf'ik), *n.* the exchange of goods by barter or by the medium of money; trade, commerce, esp. of an illegal kind; the trade (in a particular commodity etc.); the transportation of persons, animals, or goods by road, rail, sea or air; the signals transmitted over a communications system; the passing to and fro of persons, vehicles etc., on a road etc.; the person, vehicles etc. passing thus; amount of goods or number of persons conveyed; intercourse, dealings (with). *v.i.* (*p.p.* **trafficked,** *pres.p.* **trafficking**) to trade, to buy and sell goods, to have business (with). *v.t.* to barter. **traffic circle,** *n.* (*N Am.*) a

roundabout. **traffic island**, *n.* a raised section in the middle of a road, which acts as a lane-divider, refuge for pedestrians etc. **traffic jam**, *n.* vehicles that are stationary or slow-moving because of the large volume of traffic. **traffic lights**, *n.pl.* coloured lights at street intersections to control the flow and direction of traffic. **traffic warden**, *n.* a person employed to enforce observance of parking restrictions, esp. by issuing parking tickets. **trafficker**, *n.* **trafficless**, *a.*
trafficator (traf'ikātə), *n.* formerly, a movable arm on a car that indicates the driver's intention to turn to right or left.
tragacanth (trag'əkanth), *n.* a whitish or reddish demulcent gum obtained from species of *Astragalus*, used in pharmacy, calico-printing etc.; a low, spiny, leguminous shrub of this genus growing in SW Asia.
tragedian (trəjē'diən), *n.* a writer of tragedies; an actor in tragedy. **tragedienne** (-en'), *n.* an actress in tragedy.
tragedy (traj'ədi), *n.* a drama in verse or elevated prose dealing with a lofty theme of a sad, pathetic, or terrible kind, usu. with an unhappy ending; the branch of drama to which such plays belong; a fatal or calamitous event, esp. a murder or fatal accident with dramatic accompaniments. **tragic**, *a.* of the nature or in the style of tragedy; characterized by loss of life; lamentable, sad, calamitous. **tragically**, *adv.* **tragicality** (-kal'-), **tragicalness**, *n.* **tragicomedy**, *n.* a drama in which tragic and comic scenes or features are mingled. **tragi-comic**, **-ical**, *a.* **tragicomicality**, *n.* **tragi-comically**, *adv.*
tragic, **tragi-comedy**, etc TRAGEDY.
tragus (trā'gəs), *n.* (*pl.* **-gi** (-jī)) a small process on the front of the orifice in the external ear.
trail (trāl), *v.t.* to drag along behind, esp. along the ground; to follow by the track or trail; to carry (a rifle, etc.) in a horizontal or oblique position in the right hand with the arm extended; to tread down (grass) to make a path; to lag behind (sb, e.g. a runner in a race). *v.i.* to be dragged along behind, to hang down loosely or grow to some length along the ground, over a wall etc.; to lag behind; to be losing in a contest etc. *n.* anything trailing behind a moving thing, a train, a floating appendage etc.; the end of a gun-carriage resting on the ground when the gun is unlimbered; a track left by an animal etc.; the scent followed in hunting; a beaten track through forest or wild country. **to trail away, off**, to become fainter, weaker. **trail blazer**, *n.* one who blazes a trail; a pioneer in a field of endeavour. **trailing edge**, *n.* (*Aviat.*) the rear edge of a streamlined body, or of a control surface.
trailer (trā'lə), *n.* one who or that which trails; a trailing plant; any vehicle, sled etc., drawn behind another; (*N Am.*) a caravan; a short film giving advance publicity to a forthcoming production.
train (trān), *n.* that which is drawn or dragged along behind; an extended part of a gown, robe etc., trailing behind the wearer; the tail of a comet; a long trailing tail or tail-feathers of a bird; the trail of a gun-carriage; a retinue, a suite; a line or long series or succession of persons or things; a series of railway carriages or trucks drawn by an engine; a line of combustible material leading fire to a charge or mine; a set of wheels, pinions etc., transmitting motion; process, orderly succession, progressive condition. *v.t.* to bring to a state of proficiency by prolonged instruction, practice etc.; to instruct, to drill, to accustom (to perform certain acts or feats); to prepare by diet and exercise (for a race etc.); to bring (a plant etc.) by pruning, manipulation etc., into a desired shape, position etc.; to bring to bear, to aim (a gun, telescope etc.) at. *v.i.* to prepare oneself or come into a state of efficiency for (a race, match etc.); to go by train. **train-**, **trained-band**, *n.* a company of citizen soldiers organized at various dates during the 16th-18th cents. **train-bearer**, *n.* an attendant employed to hold up the train of a robe etc. **train ferry**, *n.*

a ferry on to which a train is run to be conveyed across water to a track on the farther side. **train-mile**, *n.* a mile travelled by a train, the unit of work in railway statistics. **train-spotter**, *n.* one whose hobby is to collect train numbers. **trainable**, *a.* **trainee** (-nē'), *n.* a person undergoing training. **trainer**, *n.* one who trains; esp. one who prepares men, horses etc., for races etc. **training**, *n.* the preparation of a person or animal for a particular activity, occupation etc; state of being trained or physically fit. **training-college**, **school**, *n.* one for training teachers. **training-ship**, *n.* a ship for instructing boys in navigation, seamanship, etc.
train-oil (trā'noil), *n.* oil obtained from the blubber or fat of whales.
traipse (trāps), *v.i.* to trudge, to drag along wearily.
trait (trāt, trā), *n.* a distinguishing or peculiar feature; a stroke, a touch (of).
traitor (trā'tə), *n.* one who violates his allegiance; one guilty of disloyalty, treason, or treachery. **traitorous, traitorously**, *adv.* **traitorousness**, *n.* **traitress** (-tris), *n. fem.*
trajectory (trəjek'təri), *n.* the path described by a body, comet, projectile etc., under the action of given forces; a curve or surface cutting the curves or surfaces of a given system at a constant angle.
tram[1] (tram), *n.* a tramway, a tram-car. **tram-car**, *n.* **tram-line**, *n.* a tramway; (*pl.*) the lines at the side of a tennis-court which mark the boundaries of the singles and doubles court. **tram-way**, *n.* a street railway on which passenger-cars are drawn by horses, or by electricity, steam, or other mechanical power.
tram[2] (tram), *n.* silk thread made up of two or more strands twisted together, used for the weft of the finer kinds of silk goods.
tram-line etc. TRAM[1].
trammel (tram'l), *n.* a net of various forms for catching fish, esp. a trammel-net; a shackle or fetter, esp. one used in teaching a horse to amble; anything restraining freedom or activity (usu. in *pl.*). *v.t.* (*past, p.p.* **trammelled**) to confine, to hamper, to restrict. **trammel-net**, *n.* a net formed by a combination of three seines, in which fish become entangled. **trammelled**, *a.* confined, hampered.
tramontane (trəmon'tan), *a.* lying, situated or coming from beyond the Alps (as seen from Italy); hence, foreign, barbarous.
tramp (tramp), *v.i.* to walk or tread heavily, to walk, to go on foot. *v.t.* to tread heavily on, to trample; to go over or traverse, or to perform (a journey etc.) on foot; to hike. *n.* an act of tramping, the tread of persons etc., walking or marching; the sound of this; a walk, a journey on foot; an itinerant beggar, a vagrant; (*sl.*) a promiscuous girl or woman; a freight-vessel having no regular line.
trample (tram'pl), *v.t.* to tread under foot, esp. in scorn, triumph etc.; to tread down, to crush thus; to treat with arrogance or contemptuous indifference. *v.i.* to tread heavily (on); to tread (on) in contempt. *n.* the act or sound of trampling. **trampler**, *n.*
trampoline (tram'pəlēn), *n.* a sheet of canvas suspended by springs or elasticated cords from a frame, used for bouncing on or for assisting jumps in gymnastics.
trance (trahns), *n.* a state in which the soul seems to have passed into another state of being; ecstasy, rapture; a state of insensibility to external surroundings with suspension of some of the vital functions, catalepsy; the hypnotic state.
tranche (trahnsh), *n.* a portion, esp. of a larger sum of money.
trannie, tranny (tran'i), *n.* (*coll.*) short for TRANSISTOR RADIO.
tranquil (trang'kwil), *a.* calm, peaceful, serene, quiet, undisturbed. **tranquillity** (-kwil'-), **tranquilness**, *n.* **tranquillize, -ise**, *v.t.* to make calm, esp. with a sedative

drug. **tranquillization,** *n.* **tranquillizer,** *n.* that which makes tranquil; a sedative drug. **tranquillizingly,** *adv.* **tranquilly,** *adv.*
trans-, *comb. form.* across, over; beyond; through; into another state or place.
trans., *(abbr.)* transitive; translated, translation, translator.
transact (tranzakt', trən-), *v.t.* to do, to perform, to manage, to carry out. *v.i.* to do business, to conduct matters (with). **transaction,** *n.* the management or carrying out of a piece of business etc.; that which has been transacted, a piece of business, an affair, a proceeding; *(pl.)* the reports of the proceedings of learned societies; adjustment of a dispute by mutual concessions etc. **transactor,** *n.*
transalpine (tranzal'pīn), *a.* lying or situated beyond the Alps (usu. as seen from Italy).
transatlantic (tranzətlan'tik), *a.* lying or being beyond the Atlantic; crossing the Atlantic.
transceiver (transē'və), *n.* a device for transmitting and receiving radio signals.
transcend (transend'), *v.t., v.i.* to rise above, to surpass, to excel, to exceed; to pass or be beyond the range, sphere, or power (of human understanding etc.). **transcendence, -dency,** *n.* **transcendent,** *a.* excelling, surpassing, supremely excellent; in Scholastic philosophy, applied to concepts higher or of wider signification than the categories of Aristotle; in Kantian philosophy, beyond the sphere of knowledge or experience; above and independent of the material universe. *n.* that which is transcendent. **transcendental** (-den'-), *a.* in Kantian philosophy, transcendent, beyond the sphere of experience; belonging to the a priori elements of experience, implied in and necessary to experience; explaining matter and the universe as products of mental conception; transcending ordinary ideas; abstruse, speculative, vague, obscure; *(Math.)* not capable of being produced by the fundamental operations of algebra, addition, multiplication etc. *n.* a transcendent concept. **transcendental meditation,** *n.* a form of meditation intended to induce spiritual balance and harmony through silent repetition of a mantra. **transcendentally,** *adv.*
transcontinental (tranzkontinen'təl), *a.* extending or travelling across a continent.
transcribe (transkrīb'), *v.t.* to copy in writing, to write out in full (shorthand notes etc.); to translate, transliterate; to transfer (data) from one recording medium to another; to record (spoken sounds) in the form of phonetic symbols. **transcriber,** *n.* **transcript** (tran'skript), *n.* a written copy. **transcription** (-skrip'-), *n.* transcribing or being transcribed; what is transcribed; *(Mus.)* the arrangement of a vocal composition for an instrument, or the readjustment of a composition for another instrument. **transcriptional, transcriptive** (-skrip'-), *a.*
transducer (tranzdū'sə), *n.* a power-transforming device for which the input and output are of different kinds, electrical, acoustic, optical etc., e.g. loudspeaker, microphone, photoelectric cell, etc.
transect (tran'sekt), *v.t.* to cut across; *(Anat.)* to dissect transversely. **transection** (-sek'-), *n.* [as SECT²]
transept (tran'sept), *n.* either of the transverse arms extending north and south in a cruciform church.
transfer (transfœ'), *v.t. (past, p.p.* **transferred)** to convey, remove, or shift from one place or person to another; to make over the possession of; *(Football)* to sell or release (a player) to another club; to convey (a design etc.) from one surface to another, esp. in lithography; to remove (a picture etc.) from a wall etc., to canvas or other surface. *v.i.* to move from one place to another; to change from one bus, train, etc to another. *n.* (trans'-), the removal or conveyance of a thing from one person or place to another; *(Football)* the sale or release of a player to another club; the act of conveying a right, property etc., from one person to another; the deed by which this is effected; that which is transferred; a design conveyed or to be conveyed from paper etc., to some other surface; *(N Am.)* a ticket which allows a passenger on public transport to change routes. **transfer-book,** *n.* a register of transfers of stocks, shares, etc. **transfer list,** *n.* a list of footballers available for transfer to other clubs. **transferable** (-foe'-), *a.* **transferable vote,** *n.* in a system of proportional representation, a vote that can be transferred to a second candidate if the first loses a preliminary ballot. **transferability** (-bil'-), *n.* **transferee** (-rē'), *n.* **transference** (trans'fə-), *n.* **transferrer** (-fœ'-), *n.* **transferential** (-ren'shəl), *a.*
transfiguration (transfigūrā'shən), *n.* a change of form or appearance, esp. that of Christ on the Mount (Matt. xvii.1-9); a festival on 6 Aug. in commemoration of this. **transfigure** (-fig'ə), *v.t.* to change the outward appearance of, esp. so as to elevate and glorify.
transfix (transfiks'), *v.t.* to pierce through, to impale; to render motionless with shock, fear etc., **transfixion** (-fik'shən), *n.* the act of transfixing; amputation by piercing and cutting outwards.
transform (transfawm'), *v.t.* to change the form, shape, or appearance of, to metamorphose; to change in disposition, character etc. *n.* (trans'-), the result of a mathematical or linguistic transformation. **transformable,** *a.* **transformation,** *n.* the act of transforming; the state of being transformed, a metamorphosis, a transmutation; a change from solid to liquid or liquid to gaseous form or the reverse; *(Math.)* the change of a figure or expression with another equivalent to it; *(Physiol.)* the change in the blood in its passage through the capillaries of the vascular system; a morbid change of tissue into a form not proper to that particular part; a rule for the transforming of the underlying structures of a language into actual sentences. **transformational,** *a.* **transformational grammar,** *n.* a grammar which describes the structure of a language in terms of a set of rules for transforming the underlying structures of the language into an infinite number of actual sentences. **transformative,** *a.* **transformer,** *n.* one or that which transforms; a device which changes the current and voltage of an alternating electrical supply. **transformism** (trans'-), *n.* the theory of the development of one species from another; the theory that complex animals were developed from organisms originally free, united into a colony and then into organs of a differentiated whole. **transformist,** *n.* **transformistic** (-mis'-), *a.*
transfuse (transfūz'), *v.t.* to cause to pass from one vessel etc., into another; to transfer (blood) from the veins of one person or animal to those of another; to inject (a liquid) into a blood-vessel or cavity to replace loss or wastage. **transfusion** (-zhən), *n.* **transfusionist,** *n.* **transfusive,** *a.*
transgress (tranzgres'), *v.t.* to break, to violate, to infringe, *v.i.* to offend by violating a law or rule, to sin. **transgression** (-shən), *n.* **transgressive,** *a.* **transgressively,** *adv.* **transgressor,** *n.*
tranship (tranship'), *v.t.* to transfer from one ship, vehicle etc., to another. **transhipment,** *n.*
transient (tran'siənt, -ziənt), *a.* not lasting or durable; transitory, momentary, hasty, brief; *(Mus.)* passing, serving merely to connect or introduce; *(N Am.)* (of a hotel guest) staying one night only. *n.* a transient person or thing; a transient fluctuation in the amount of current flowing through an electrical circuit. **transience, -ency, transientness,** *n.* **transiently,** *adv.*
transistor (tranzis'tə), *n.* a device made primarily of a semi-conductor capable of giving current and power amplification; *(coll.)* a transistor radio. **transistor radio,** *n.* a small portable radio. **transistorize,** *v.t.* to equip with transistors. **transistorization,** *n.*
transit (tran'zit, -sit), *n.* the act of passing, conveying, or

being conveyed, across, over, or through; conveyance; a line of passage, a route; the apparent passage of a heavenly body over the meridian of a place; the passage of a heavenly body across the disk of another, esp. of Venus or Mercury across the sun's disk. **in transit**, being conveyed. **transit camp**, *n.* a camp where people stay temporarily before moving on to another place. **transit-duty**, *n.* duty paid upon goods passing through a country.

transition (tranzish'ən), *n.* passage or change from one place, state, or action to another; a change in architecture, painting, literature etc.; a change from one musical key to another or from the major to the relative minor. **transition stage** or **period**, *n.* the stage or period of transition in art etc. **transitional, -ary**, *a.* **transitionally**, *adv.*

transitive (tran'sitiv), *a.* of verbs, expressing an action having a direct object. **transitively**, *adv.* **transitiveness**, *n.*

transitory (tran'sitəri), *a.* lasting a short time, transient, not durable, short-lived. **transitorily**, *adv.* **transitoriness**, *n.*

translate (translāt'), *v.t.* to render or express the sense of (a word, passage, or work) into or in another language; to interpret, to express in clearer terms; to express, paraphrase, or convey (an idea etc.) from one art or style into another; to transform, to change; to remove from one office to another (esp. a bishop to another see); (*Mech.*) to move (a body) so that all parts follow the same direction, to give motion without rotation. *v.i.* to be engaged in translation. **translatable**, *a.* **translation**, *n.* **translational**, *a.* **translator**, *n.* **translatory**, *a.* **translatress** (-tris), *n. fem.*

transliterate (tranzlit'ərāt), *v.t.* to represent (words, sounds etc.) in the corresponding or approximately corresponding characters of another language. **transliteration**, *n.* **transliterator**, *n.*

translucent (transloo'sənt), *a.* allowing light to pass through but not transparent; (*loosely.*) transparent. **translucence, -cency**, *n.*

translunary (tranzloo'nəri), *a.* situated beyond the moon, opp. to sublunary; ethereal, visionary.

transmarine (tranzmərēn'), *a.* situated beyond the sea.

transmigrate (tranz'mīgrāt, -grāt'), *v.i.* of the soul, to pass from one body into another, to undergo metempsychosis; to pass from one place, country, or jurisdiction to another, to migrate. **transmigrant** (tranz'-, -mī'-), *n.* one who transmigrates, a migrant; an alien passing through one country on the way to another. **transmigrant**, *a.* **transmigration**, *n.* **transmigrationism**, *n.* the doctrine of metempsychosis. **transmigrator** (tranz'-), *n.* **transmigratory** (-mī'grə-), *a.*

transmit (tranzmit'), *v.t.* (*past, p.p.* **transmitted**) to send, transfer, convey, or communicate from one person or place to another; to suffer to pass through, to act as a medium for, to conduct; to broadcast (a TV or radio programme). **transmissible** (-mis'-), *a.* **transmissibility** (-bil'-), *n.* **transmission** (-shən), *n.* the act of transmitting; the conveying of electrical energy from place to place; the radiation of ether waves; signals sent out by a transmitter; the gear by which power is conveyed from the engine to the live axle; a radio or TV broadcast. **transmissive** (-mis'-), *a.* **transmitter**, *n.* a person or thing that transmits: any form of machine that transmits telegraphic messages; the apparatus required for radiating a signal.

transmogrify (tranzmog'rifī), *v.t.* (*coll.*) to transform, esp. as if by magical means. **transmogrification** (-fi-), *n.*

transmute (tranzmūt'), *v.t.* to change from one form, nature, or substance into another; to transform (into). **transmutable**, *a.* **transmutability** (-bil'-), *n.* **transmutably**, *adv.* **transmutative**, *a.* **transmuter**, *n.* **transmutation**, *n.* the act of transmuting; the state of being transmuted; the change of base metals into gold or silver; the change of one species into another; (*Geom.*) the reduction of one figure or body into another of the same area or content; the conversion of one element or nuclide into another either naturally or artificially.

transoceanic (tranzōshian'ik), *a.* situated or coming from beyond the ocean; crossing the ocean.

transom (tran'səm), *n.* a horizontal bar of wood or stone across a window or other opening; a horizontal bar across the top of a doorway separating it from the fan-light; (*N Am.*) a fanlight; one of the beams bolted across the sternpost of a ship, supporting the after-end of the deck. **transom-window**, *n.* a window divided by a transom; a window over the transom of a door. **transomed**, *a.*

transonic (transon'ik), *a.* relating to or being a speed near the speed of sound.

transparent (transpa'rənt, -peə'-), *a.* allowing the uninterrupted pasage of light; easily seen through; plain, evident, clear; frank, sincere. **transparence**, **transparentness**, *n.* **transparently**, *adv.* **transparency**, *n.* transparentness; a thing that is transparent; a positive photograph on a transparent base mounted on a frame for viewing by means of a projector.

transpire (transpīə'), *v.t.* to emit through the excretory organs (of the skin or lungs), to emit as vapour, to exhale. *v.i.* of perspiration etc. to be emitted through the the excretory organs, to pass off as vapour; to leak out, become known; to happen. **transpirable**, *a.* **transpiration** (-pi-), *n.* **transpiratory**, *a.*

transplant (transplahnt'), *v.t.* to remove and plant in another place; to remove from one place and establish in another; to transfer (living tissue) from one part or person to another. *n.* (trans'-), the surgical procedure for transplanting an organ; the organ thus transplanted. **transplantable**, *a.* **transplantation**, *n.* **transplanter**, *n.* one who or that which transplants; a machine for removing trees with earth and replanting; a tool for taking up plants thus.

transponder (transpon'də), *n.* a radio or radar device which automatically transmits a signal in response to a signal received.

transport¹ (transpawt'), *v.t.* to carry or convey from one place to another; to remove (a criminal) to a penal colony; (*chiefly in p.p.*) to carry away by powerful emotion, to entrance, to ravish. **transportable**, *a.* that may be transported. **transportability** (-bil'-), *n.* **transporter**, *n.* a large vehicle for transporting goods. **transporter bridge**, *n.* a device for carrying road traffic across a river on a moving platform. **transportation**, *n.* the act of transporting or conveying; the state of being transported; (means of) conveyance; carriage of persons or things from one place to another; banishment to a penal colony.

transport² (trans'pawt), *n.* transportation, conveyance from one place to another; a transport ship or aircraft; a vehicle, aircraft etc. used for transporting people or goods; ecstasy. **transport café**, *n.* a roadside café used predominantly by lorry drivers. **transport ship** or **vessel**, *n.* one used to carry troops, munitions of war, stores, etc. **transport-worker**, *n.* a worker on any system of transport.

transpose (transpōz'), *v.t.* to cause to change places; to change the natural order or position of (words or a word) in a sentence; to transfer a mathematical term from one side of an equation to the other, changing the sign; (*Mus.*) to write or play in a different key. **transposal**, **transposition**, *n.* the act of transposing; the state of being transposed. **transpositional**, **transpositive** (-poz'-), *a.*

transputer (transpū'tə), *n.* (*Comput.*) fast, powerful microchip equivalent to a 32-bit microprocessor with RAM facility.

transsexual (transek'shuəl, -shəl), *a., n.* (of) a person who dresses and lives for all or most of the time as a member of the opposite sex; (of) a person who undergoes surgery and medical treatment to adopt the physical characteristics of the opposite sex.

trans-ship (tranship'), TRANSHIP.

transubstantiation, *n.* a conversion of the whole substance of the bread and wine in the Eucharist into the body and blood of Christ.

transuranic, *a.* of an atomic element, having an atomic number higher than uranium.

transverse (tranzvəs', tranz'-), *a.* lying, being, or acting across or in a cross direction. *n.* that which is transverse, esp. a transverse muscle. *v.t.* to lie or pass across.

transversal, *a.* transverse; running or lying across. *n.* a straight line cutting a system of lines. **transversally, transversely,** *adv.*

transvestism (tranzvest'izm), *n.* the adoption of clothing and manners properly belonging to the opposite sex. **transvestite** (-tīt), *n.* one who practices this.

Transylvanian (transilvā'niən), *a.* of or belonging to Transylvania, in Romania.

trap (trap), *n.* a contrivance for catching game, vermin, and other animals, consisting of a pitfall, enclosure, or mechanical arrangement, esp. with a door or lid closing with a spring, often baited; a trick or artifice for misleading or betraying a person, an ambush, a stratagem; a device for suddenly releasing a bird or propelling an object into the air to be shot at; a device for hurling clay-pigeons into the air; a U-shaped bend or other contrivance in a soil-pipe etc., for sealing this with a body of liquid and preventing the return flow of foul gas; a trap-door; (*sl.*) the mouth. *v.t.* (*past, p.p.* **trapped**) to catch in or as in a trap; to retain, hold back; to stop or hold (gas etc.) in a trap. *v.i.* to catch animals in traps. **trap-door,** *n.* a door in a floor or roof. **trapshooting,** *n.* clay-pigeon shooting. **trapper,** *n.* one who traps animals, esp. for furs.

trapeze (trəpēz'), *n.* an apparatus consisting of a suspended bar on which gymnasts perform swinging, balancing, and other feats. **trapezial,** *a.* trapeziform. **trapezian,** *a.* (*Cryst.*). **trapeziform** (-ifawm), *a.* **trapezium,** (-iəm), *n.* (*pl.* **-zia** (-ziə), **-ziums**) a quadrilateral figure no two or only two sides of which are parallel. **trapezoid** (trap'əzoid, -pē'-), *a.* trapeziform. *n.* a quadrilateral only two or no two of whose sides are parallel. **trapezoidal** (-zoi'-), *a.*

trapper TRAP[1].

trappings, *n.pl.* decorations, adornments, esp. those pertaining to an office etc., finery.

Trappist (trap'ist), *n.* a member of a Cistercian order, following the strict rule of La Trappe, a monastery founded at Soligny-la-Trappe, France, in 1140.

trapshooting TRAP[1].

trash (trash), *n.* any waste or worthless matter, refuse, rubbish; (*N Am.*) domestic refuse; a rubbishy article or production of any kind; a poor or worthless person or group of people. *v.t.* to subject to criticism, to denigrate. **trash-can,** (*N Am.*) a dust-bin. **trashman,** *n.* (*N Am.*) a refuse collector. **trashery, trashiness,** *n.* **trashily,** *adv.* **trashy,** *a.* **trashiness,** *n.*

trattoria (tratərē'ə), *n.* (*pl.* **-rias, -rie** (-rē'ā)) an Italian restaurant.

trauma (traw'mə), *n.* a wound or external injury; the morbid condition produced by this; a psychological shock having a lasting effect on the subconscious. **traumatic** (-mat'-), *a.* pertaining to or adapted to the cure of wounds. *n.* a medicine for wounds. **traumatism** *n.* **traumatize,** *v.t.* to wound or injure; to subject to mental shock. **traumato-,** *comb. form.*

travail (trav'āl), *n.* painful toil, painful exertion or effort; the pangs of childbirth.

travel (trav'l), *v.i.* (*past, p.p.* **travelled**) to make a journey, esp. to foreign lands; of a machine or part to move (along, in, up and down etc.); to move, to go, to pass through space; to make journeys as a commercial traveller for securing orders etc.; of food or drink, to survive transportation in a specified way; (*coll.*) to move quickly. *v.t.* to journey over. *n.* the act of travelling; (*pl.*) an account of travelling, usu. in distant countries; the length of stroke, the range or scope, of a piston etc. **travel agent,** *n.* one who sells holidays, air, train, bus tickets etc. **travel agency,** *n.* **travel-soiled, -stained, -worn,** *n.* soiled or worn with travel. **travelled,** *a.* having travelled; experienced in travelling. **traveller,** *n.* one who travels; a commercial traveller; a gipsy; (*Naut.*) an iron ring etc., sliding on a spar, rope etc. **traveller's cheque,** *n.* a cheque available in various denominations, sold by a financial institution for use abroad by a traveller, who signs it on receipt and countersigns it in order to cash it. **traveller's-joy,** *n.* the wild clematis, *C. vitalba.* **travelling expenses,** *n.pl.* expenses incurred by a commercial traveller etc., and paid by the employers. **travelling salesman, salesperson,** *n.* one who travels from place to place promoting and selling the products or services of his or her company. **travelogue** (-log), *n.* a lecture or talk on travel illustrated by cinematograph films.

traverse (trav'œs), *n.* anything, esp. a part of a building or mechanical structure, crossing something else; (*Naut.*) a zigzag line described by a ship owing to contrary winds etc.; the act of traversing or travelling across; the sideways travel of part of a machine; a sideways movement of climbers on a mountain-side or precipice to avoid obstacles. *v.t.* (trəvœs', trav'-), to travel across; to make a traverse along (a cliff etc.); to lie across or through; to examine, consider, or discuss thoroughly. *v.i.* to make a traverse. **traversable,** *a.* **traverser,** *n.* one who or that which traverses. **traversing,** *n.* a method of plane-table surveying by measured connected lines.

travesty (trav'əsti), *n.* a burlesque imitation; a ridiculous misrepresentation; a parody. *v.t.* to make a travesty of, to burlesque.

travolator (tra'vəlātə), *n.* a moving pavement for conveying pedestrians, e.g. at an airport.

trawl (trawl), *n.* a net, shaped like a flattened bag, for dragging along the sea-bottom; a trawl-line; the act of trawling. *v.i.* to fish with a trawl-net; to gather data etc. from a great number of different sources. **trawl-boat,** *n.* **trawl-line,** *n.* a line of great length, with short lines carrying baited hooks, buoyed up at intervals, for deep-sea fishing. **trawl-net,** *n.* **trawler,** *n.* one who trawls; a fishing-vessel using a trawl-net. **trawling,** *n.*

tray (trā), *n.* a flat shallow vessel, used for holding or carrying small articles on; a shallow coverless box, esp. one forming a compartment in a trunk etc. **trayful,** *n.*

treacherous (trech'ərəs), *a.* violating allegiance, disloyal, perfidious; deceptive, illusory; unreliable, unsafe. **treacherously,** *n.* **treacherousness, treachery,** *n.*

treacle (trē'kl), *n.* a syrup drained from sugar in refining; molasses. **treacly,** *a.*

tread (tred), *v.i.* (*past* **trod** (trod), *p.p.* **trodden** (trod'n)) to set the foot on the ground; to walk, to step, to go; to deal (cautiously etc.); to follow (in a person's footsteps). *v.t.* to step or walk on; to crush with the feet; to trample on; to walk (a distance, journey etc.); to dance (a measure etc.). *n.* the act or manner of walking; the sound of walking, a footstep; the flat part of a stair or step; a piece of rubber, metal etc., placed on this to reduce wear or noise; the outer face of a tyre that is in contact with the road; the part of a sole that rests on the ground. **to tread down,** to press down or crush with the feet; to trample on; to destroy. **to tread in,** to press in or into with the feet. **to tread on,** to trample on; to set the foot on; to follow closely. **to tread on someone's toes,** to offend someone's susceptibilities. **to tread upon one's heels** HEEL[1]. **to tread out,** to press out (wine, etc.) with the feet; to extinguish by stamping on. **to tread water,** to remain upright and afloat by making walking motions with the legs; to undergo period of relative inactivity. **treadmill,** *n.* a revolving cylinder driven by the weight of a person or persons, horses etc., treading on movable steps on the periphery,

formerly used as a punishment in prisons; wearisome monotony or routine. **treader**, *n.*
treadle (tred'l), *n.* a lever worked by the foot giving motion to a lathe, sewing-machine, bicycle etc.
treadmill TREAD.
Treas., *(abbr.)* treasurer; treasury.
treason (trē'zən), *n.* a violation of allegiance by a subject against his sovereign or government, esp. an overt attempt to subvert the government; an act of treachery, a breach of faith. **high treason,** violation of allegiance to the sovereign or the state. **treasonable,** *a.* consisting of or involving treason. **treasonableness,** *n.* **treasonably,** *adv.*
treasure (trezh'ə), *n.* precious metals in any form, or gems; a quantity of these hidden away or kept for future use, a hoard; accumulated wealth; anything highly valued, a precious or highly-prized thing, esp. if portable; a person greatly valued, a beloved person. *v.t.* to lay (up) as valuable, to hoard, to store (up); to prize, to lay (up) in the memory as valuable. **treasure-house,** *n.* a building in which treasures or highly-valued things are kept. **treasure hunt,** *n.* a game in which people compete to be the first to find something hidden. **treasure trove** (trōv), *n.* money, gold, silver, plate, or bullion found hidden in the earth or private place.
treasurer (trezh'ərə), *n.* one who has the charge of the funds of a company, society, club etc. **treasurership,** *n.*
treasury (trezh'əri), *n.* a place or building in which treasure is stored; a place where the public revenues are kept; a government department in charge of the public revenue; the offices of this; a repository, a book etc., full of information on any subject. **treasury bench,** *n.* the front bench on the right hand of the Speaker in the House of Commons, appropriated to the First Lord of the Treasury, the Chancellor of the Exchequer, and other members of the ministry. **Treasury bill,** *n.* an instrument of credit issued by the government as an acknowledgment of money lent by a private person for three, six or twelve months. **Treasury bond,** *n.* a government promissory note running for a definite period not exceeding six years, bearing interest at a fixed rate, and redeemable at par; an Exchequer bond. **Treasury note,** *n.* a demand note issued by the Treasury; a currency note.
treat (trēt), *v.t.* to act or behave to or towards; to deal with or manipulate for a particular result, to apply a particular process to, to subject to the action of a chemical agent etc.; to handle or present or express (a subject etc.) in a particular way; to supply with food, drink, or entertainment at one's expense. *v.i.* to arrange terms (with), to negotiate; to discuss, to discourse (of). *n.* an entertainment given to school-children etc.; an unusual pleasure or gratification. **a treat,** *(coll.)* excellently, very well. **to stand treat,** *(coll.)* to pay for drinks, etc. **treatable,** *a.* **treater,** *n.*
treatise (trē'tiz), *n.* a literary composition expounding, discussing, and illustrating a subject in a thorough way.
treatment (trēt'mənt), *n.* any medical procedure intended to bring about a cure; the act or manner of treating. **the treatment,** *(coll.)* the usual way of dealing with something in a particular situation.
treaty (trē'ti), *n.* an agreement formally concluded and ratified between different states; an agreement between persons etc.; negotiation, the act of treating for the adjustment of differences etc. **treaty port,** *n.* a seaport kept open by treaty to foreign commerce.
treble (treb'l), *a.* triple, threefold; soprano. *n.* a soprano voice, singer, or part; a high-pitched musical instrument; the higher part of the frequency range, esp. in electronic sound reproduction. *v.t.* to multiply by three. *v.i.* to become threefold. **treble chance,** *n.* a type of bet in football pools in which one wins by accurately predicting the number of draws, and home and away wins. **treble clef,** *n.* the clef that places G above middle C on the second line of the staff. **trebly,** *adv.*

trecento (trāchen'tō), *n.* the 14th cent. as characterized by a distinctive style of Italian literature and art. [It.]
treddle (tred'l), TREADLE.
tree (trē), *n.* a perennial woody plant rising from the ground with a single supporting trunk or stem; a thing resembling a tree, esp. in having a stem and branches; a family or genealogical tree; a gibbet; a cross of crucifixion; a diagram with branching lines; a timber beam or framework, as an axle-tree, swingle-tree etc. **at the top of the tree,** having attained the highest position in a profession etc. **tree of knowledge,** a tree in the Garden of Eden, the fruit of which gave knowledge of good and evil (Gen. iii). **tree of life,** a tree in the Garden of Eden of which Adam and Eve were forbidden to eat (Gen. ii.9); the arborvitae. **up a tree,** in a fix, cornered. **tree-frog,** *n.* a frog with arboreal habits. **tree surgeon,** *n.* an expert in the treatment of diseased trees. **tree surgery,** *n.* **treeless** (-lis), *a.*
treenail (trē'nāl, tren'əl), *n.* a pin or peg of hard wood used in fastening timbers, esp. in shipbuilding.
trefoil (tref'oil, trē'-), *n.* a plant with three leaflets or three-lobed leaves, esp. of the genus *Trifolium,* as the clover, the black medick etc.; a three-lobed or three-cusped ornament in window-tracery etc.; any object in this shape. **trefoiled,** *a.*
treillage (trā'lij), *n.* a light frame of posts and rails to support espaliers; a trellis.
trek (trek), *v.i.* of oxen, to draw a vehicle or load; to travel by ox-wagon; to journey, esp. with difficulty on foot. *n.* a journey with a wagon; a stage or day's march; any long, arduous journey, esp. on foot. **trekker,** *n.*
trellis (trel'is), *n.* open-work of strips of wood crossing each other and nailed together, used for verandas, summer-houses etc.; a lattice, a grating; a screen or other structure made of this. *v.t.* to interlace into a trellis; to furnish with trellis.
trematode (trem'ətōd), *a.* pertaining to the Trematoda, an order of parasitic worms containing the fluke-worms. **trematoid** (-toid), *a.*, *n.*
tremble (trem'bl), *v.i.* to shake involuntarily, as with fear, cold, weakness etc.; to be in a state of fear or agitation; to be alarmed (for); to totter, to oscillate, to quaver. *n.* the act or state of trembling; fear. **trembler,** *n.* one who trembles; an automatic vibrator for making or breaking an electrical circuit. **tremblingly,** *adv.* **trembly,** *a.*
tremendous (trimen'dəs), *a.* terrible, dreadful; of overpowering magnitude, violence etc.; *(coll.)* extraordinary, considerable. **tremendously,** *adv.* **tremendousness,** *n.*
tremolo (trem'əlō), *n.* a tremulous or quavering effect in singing, playing etc.; an organ or harmonium stop producing a vibrating tone. [It.]
tremor (trem'ə), *n.* a trembling, shaking, or quivering; a thrill. **tremorless,** *a.*
tremulous (trem'ūləs), *a.* trembling, shaking, quivering; timid, irresolute, wavering. **tremulously,** *adv.* **tremulousness,** *n.* **tremulant,** *a.* tremulous. *n.* a tremolo; an organstop or similar device on an electronic instrument for producing this.
trench (trench), *n.* a long, narrow cut or deep furrow in the earth, a ditch, esp. a long narrow ditch, usu. with a parapet formed by the excavated earth, to cover besieging troops etc. *v.t.* to cut a trench or trenches in (ground etc.); to turn over (ground) by cutting a successive series of trenches and filling in with the excavated soil; to ditch. *v.i.* to encroach (on). **to open the trenches,** to begin to dig or to form trenches or lines of approach. **trench-coat,** *n.* a heavy, lined macintosh crossing over in front and furnished with belt and storm sleeves; a similar raincoat, worn by men or women. **trench-fever,** *n.* a remittent or relapsing fever affecting men living in trenches, etc., and transmitted by the excrement of lice. **trench mortar,** *n.* a mortar used for throwing bombs. **trench-,** **trenching-**

plough, *n.* a plough for cutting deep furrows. **trench warfare**, *n.* a type of warfare in which soldiers take up positions in trenches facing the enemy.
trenchant (tren'chənt), *a.* sharp, keen; cutting, biting, incisive. **trenchancy,** *n.* **trenchantly,** *adv.*
trencher (tren'chə), *n.* a wooden plate, now used for cutting bread upon. **trencher-man,** *n.* a (good or poor) feeder or eater.
trend (trend), *v.i.* to extend or lie along in a particular direction; to incline; to bend (away etc.); to have a general tendency or direction. *n.* general tendency, bent, or inclination; mode, fashion. **trendsetter,** *n.* one who originates, dictates fashions. **trendy,** *a.* (*sometimes derog.*) following the latest trends; fashionable. *n.* a trendy person.
trental (tren'təl), *n.* in the Roman Catholic Church, a series of thirty masses for the dead.
trepan (tripan'), *n.* a surgeon's cylindrical saw for removing portions of the skull. *v.t.* to perforate with a trepan. **trepanation** (trepə-), **trepanning,** *n.*
trephine (trifēn', -fīn'), *n.* an improved trepan with a centre-pin. *v.t.* to operate on with this.
trepidation (trepidā'shən), *n.* a state of alarm, excitement, or agitation; a trembling of the limbs, as in paralysis.
treponema (trepənē'mə), *n.* a member of a genus of spirochetes (*Treponema*) that cause syphilis and other diseases.
trespass (tres'pəs), *n.* a transgression against law, duty etc., an offence, a sin; a wrongful act involving injury to the person or property of another, any transgression other than treason, misprision of treason, or felony, esp. unauthorized entry into another's land. *v.i.* to commit an illegal intrusion (upon the personal or property rights of another); to intrude, encroach, or make undue claims (upon). **trespasser,** *n.*
tress (tres), *n.* a lock or plait of hair, esp. from the head of a girl or woman; (*pl.*) hair. *v.t.* to arrange in tresses. **tressed, tressy,** *a.*
trestle (tres'l), *n.* a movable frame for supporting a table, platform etc., usu. consisting of a pair of divergent legs, fixed or hinged; an open braced framework of timber or iron for supporting the horizontal portion of a bridge etc.; (*pl.*) the props or shores of a ship in process of building etc. **trestle-bridge,** *n.* **trestle-table,** *n.* a table formed of boards supported on movable trestles.
trevally (trival'i), *n.* the silver bream.
trevet TRIVET.
trews (trooz), *n.pl.* trousers, esp. made of tartan.
trey (trā), *n.* the three at cards or dice. Cp. TRAY ².
TRH, (*abbr.*) Their Royal Highnesses.
tri-, *comb. form.* three; three times; triple.
triable (trī'əbl), *a.* that may be tried or tested.
triad (trī'ad), *n.* a collection of three; (*Welsh Lit.*) a composition in which statements etc., are grouped in threes; (*Chem.*) an element or radical with a combining power of three; (*Mus.*) a chord of three notes; a common chord; a Chinese secret society, often engaging in illegal activities. **triadic** (-ad'-), *a.*
triadelphous (trīədel'fəs), *a.* of a plant having the stamens in three bundles.
trial (trī'əl), *n.* the act or process of trying or testing; experimental treatment; a test, an examination, an experiment; that which tries or tests strength, endurance, and other qualities; hardship, trouble, suffering etc.; (*Law.*) the judicial examination and determination of the issues in a cause between parties before a judge, judge and jury, or a referee. **on trial,** undergoing a test; being tried in a lawcourt. **trial and error,** a method of solving problems by trying several solutions and choosing the most successful. **trial balance,** *n.* a comparison of the debit and credit totals in double-entry book-keeping. **trial run,** *n.* a preliminary test of a new procedure etc. **trial-trip,** *n.* a test trip by a new vessel to show her capabilities.
Triandria (trīan'driə), *n.pl.* a Linnæan class consisting of plants with hermaphrodite flowers having three stamens. **triandrian, -drous,** *a.*
triangle (trī'ang-gl), *n.* a figure, esp. a plane figure, bounded by three lines, esp. straight lines; a drawing-implement or other thing or ornament of this shape; a combination of three spars lashed together at the top for shifting weights; a steel rod bent into a triangle and sounded by striking with a steel rod; a northern constellation; any situation involving three people or elements. **triangular** (-ang'gū-), *a.* having the shape of a triangle; three-cornered; involving three people or elements. **triangular compasses,** *n.pl.* compasses with three legs. **triangularity** (-la'-), *n.* **triangularly,** *adv.* **triangulate** (-lāt), *v.t.* to make triangular; to divide into triangles. esp. (an area) in surveying; to ascertain by this means. *a.* (-lət), (*Zool.*) marked with triangles. **triangulation,** *n.* **triangulately,** *adv.*
Trias (trī'əs), *n.* the division of rock strata between the Carboniferous and the Jurassic (divided in Germany into three groups, whence the name). **Triassic** (-as'-), *a.* and *n.*
triathlon (trīath'lən), *n.* athletic contest involving three different sports (usu. swimming, cycling, running). **triathlete,** *n.* a competitor in this.
triatomic (trīətom'ik), *a.* having three atoms in the molecule.
triaxal (trīak'səl), **-axial** (-siəl), *a.* having three axes.
tribadism (tri'bədizm), *n.* simulated heterosexual intercourse between women, lesbianism.
tribal (trī'bəl), *a.* belonging to or pertaining to a tribe. **tribally,** *adv.* **tribalism,** *n.* loyalty to a tribe or group.
tribasic (trībā'sik), *a.* having three atoms of hydrogen replaceable by a base or basic radical; of a molecule, having three monovalent basic atoms or groups.
tribe (trīb), *n.* a group of people ethnologically related and forming a community or a political division; a group claiming common descent or affinity, a clan or group of clans, esp. a group of savage clans under a chief; a number of persons of the same character, profession etc. (*usu. contemp.*); a family, esp. a large one; (*Bot. and Zool.*) a more or less indefinite group of plants or animals, usu. above a genus and below an order. **tribesman, tribeswoman,** *n.*
triblet (trib'lit), *n.* a mandrel used in forging tubes, nuts, and rings etc.
tribology (trībol'əji), *n.* the study of friction, lubrication and wear between moving surfaces.
triboluminescence (trībōloomines'əns), *n.* luminescence produced by friction.
tribometer (trībom'itə, trib-), *n.* a sled-like apparatus for measuring sliding friction.
tribrach (trībrak, trib'-), *n.* a metrical foot of three short syllables. **tribrachic** (-brak'-), *a.*
tribrachial (trībrā'kiəl) *n.* a three-armed tool or implement.
tribulation (trībūlā'shən), *n.* severe affliction, suffering, distress.
tribunal (trībū'nəl, trib-), *n.* a court of justice; a board of arbitrators etc.; a seat or bench for judges, magistrates etc., a judgment-seat.
tribune (trib'ūn), *n.* one of two (later ten) representatives elected by the people of ancient Rome to protect their rights and liberties against the patricians, also, one of various civil, fiscal, and military officers; a champion of popular rights and liberties. **tribunate** (-nət, -nāt), **tribuneship,** *n.* **tribunicial** (-nish'əl), *a.*
tributary (trib'ūtəri), *a.* paying or subject to tribute; subsidiary, contributory; serving to increase a larger stream. *n.* a tributary person or state; a tributary stream. **tributarily,** *adv.* **tributariness,** *n.*
tribute (trib'ūt), *n.* a sum of money or other valuable thing paid by one prince or state to another in token of submission, for peace or protection, or by virtue of a treaty; the state of being under obligation to pay this; a contribution,

gift, or offering (of praise etc.); a share of ore paid to a miner under the system of tribute-work. **tribute-money**, *n*. **tribute-work**, *n*. (*Mining*). **tributer**, *n*. one doing tribute-work. [L *tribūtum*, neut. of *tribūtus*, p.p. of *tribuere*, to give, to pay]
tricapsular (trīkap'sūlə), *a*. having three capsules.
tricarpous (trīkah'pəs), *a*. having three carpels.
tricaudate (trīkaw'dāt), *a*. having three tail-like processes.
trice (trīs), *n*. an instant. **in a trice**, in a moment.
Tricel®(trī'sel), *n*. a partly synthetic textile fibre used in dress fabrics.
tricentenary (trīsəntē'nəri), TERCENTENARY.
tricephalous (trīsef'ələs), *a*. three-headed.
triceps (trī'seps), *a*. of muscles, three-headed. *n*. a three-headed muscle, esp. the large muscle at the back of the upper arm.
triceratops (trīse'rətops), *n*. a large herbivorous dinosaur of the Cretaceous period with three horns and a bony crest on the head.
trichiasis (triki'əsis), *n*. entropion or inversion of the eyelashes; a disease of the kidneys in which filamentous matter is passed in the urine; a swelling of the breasts due to obstruction of milk-excretion in child-bearing women.
trichina (triki'nə), *n*. (*pl*. **-nae** (-nē)) a hair-like nematode parasitic worm, infesting the intestine or muscles of pigs, man, etc. **trichiniasis** (trikini'əsis), **trichinosis** (-nō'-), *n*. a disease due to the presence of trichinæ in the system.
trichinize, -ise, (trik'-), *v.t*. **trichinization, -isation**, *n*.
trichinozed (trik'inōzd), **trichinotic** (trikinot'-), **trichinous**, *a*.
trichite (trik'īt), *n*. a minute hair-like form occurring in certain vitreous volcanic rocks; a minute fibril found in some sponge-spicules, a spicule composed of these.
trichogenous (trikoj'inəs), *a*. promoting growth of the hair. **trichogen** (trī'kəjen, trik'-), *n*.
trichology (trikol'əji), *n*. the study of the human hair. **trichological** (-loj'-), *a*. **trichologist**, *n*.
trichoma (trikō'mə), *n*. one of the threads composing the thallus in filamentous algae; a disease of the hair, *plica*.
trichomatose (-tōs), *a*. affected with this. **trichome** (trī'kōm), *n*. (*Bot*.) a hair, filament, scale, prickle, or an outgrowth.
tricopathy (trikop'əthi), *n*. any disease of the hair. **tricopathic** (-path'-), *a*.
trichosis (trikō'sis), *n*. any disease of the hair.
trichotomy (trikot'əmi), *n*. division into three, esp. of the human being into body, soul, and spirit. **trichotomize, -ise**, *v.t*. **trichotomous**, *a*. **trichotomously**, *adv*.
trichroism (trikroizm), *n*. the property of exhibiting different colours in three different directions when viewed by transmitted light. **trichroic** (-krō'-), *a*.
trichromatic (trīkrəmat'ik), *a*. three-coloured, having the normal three fundamental colour-sensations (of red, green, and purple). **trichromatism** (-krō'-), *n*.
trick (trik), *n*. an artful device or stratagem; a foolish or malicious act, a prank, a practical joke; a feat of dexterity, esp. of sleight of hand; an ingenious or peculiar way of doing something, a knack; a particular habit or practice, a mannerism, a personal peculiarity; (*Cards*.) the whole number of cards played in one round; a round; a point gained as the result of a round. *v.t*. to cheat, to deceive, to delude, to inveigle (into, out of etc.); to dress, to deck (out or up). *v.i*. to practise trickery. **how's tricks?** how are you? **to do the trick**, to achieve the required effect. **to know a trick or two**, to know of some expedient. **to trick out**, to decorate, to dress up. **trick-track** TRICTRAC. **trick cyclist**, *n*. (*sl*.) a psychiatrist. **trick or treat**, children's Hallowe'en custom of asking for treats from householders under threat of playing tricks on them. **tricker, trickster** (-stə), *n*. **trickery**, *n*. **tricky**, *a*. (*coll*.) difficult, awkward; requiring tactful or skilful handling; deceitful. **trickishly**, *adv*. **trickishness**, *n*. **trickily**, *adv*.

trickiness, *n*. **tricksy** (-si), *a*. playful, sportive; excessively elaborate.
trickle (trik'l), *v.i*. to flow in drops or in a small stream. *v.t*. to cause to flow thus. *n*. a trickling; a small stream, a rill. **trickle-down theory**, *n*. the idea that the prosperity of the wealthy in a society will always eventually benefit its poorer members. **tricklet** (-lit), *n*. **trickly**, *a*.
triclinium (triklin'iəm), *n*. (*pl*. **-nia**) (*Rom. Ant*) a set of couches arranged round three sides of a dining-table; a dining-table furnished with this; a dining-room with this.
tricolour (trī'kūlə, trik'ələ), *n*. a flag or banner having three colours, esp. arranged in equal stripes, as the national standard of France of blue, white and red, divided vertically. *a*. three-coloured. **tricoloured**, *a*.
tricorn (trī'körn), *n*. a three-cornered hat.
tricot (trē'kō), *n*. a hand-knitted woollen fabric or a machine-made imitation; a soft, ribbed cloth.
tricotyledonous (trīkotilē'dənəs), *a*. having three cotyledons. [TRI-, COTYLEDONOUS]
tric-trac (trik'trak), *n*. a complicated form of backgammon.
tricuspid (trī'kŭs'pid), *a*. of molar teeth, a valve of the heart etc. having three cusps or points. **tricuspidate** (-dāt), **-dated**, *a*.
tricycle (trī'sikl), *n*. a three-wheeled cycle. *v.i*. to ride on this. **tricyclist**, *n*.
tricyclic (trīsī'klik), *a*. of a compound, having three rings in its molecule.
tridactyl (trīdak'til), **-tylous**, *a*. having three fingers or toes.
trident (trī'dənt), *n*. a three-pronged implement or weapon, esp. a fish-spear; a three-pronged sceptre or spear, the emblem of Poseidon or Neptune as god of the sea. **tridental** (-den'-), *a*. **tridentate** (-den'tāt), *a*. having three teeth or prongs.
Tridentine (trīden'tīn), *a*. of or pertaining to Trent, or the Council held there 1545–63. *n*. one who accepts the decrees of the Council of Trent, or prefers these to the decrees of the Second Vatican Council.
tridimensional (trīdimen'shənəl), *a*. having three dimensions.
triduo (trē'duō), **triduum** (trīd'ūəm), *n*. in the Roman Catholic Church, a three days' service of prayer etc.
tried (trīd), *p.p*. TRY. *a*. shown to be effective, durable etc. by testing or use, proven.
triennial (trīen'iəl), *a*. lasting for three years; happening every three years. *n*. a triennial plant, publication, etc.; every third anniversary of an event. **triennially**, *adv*.
triennium (-iəm), *n*. (*pl*. **-nnia** (-iə)) a period of three years.
trier (trī'ə), *n*. one who keeps on endeavouring or persisting.
trifacial (trīfā'shəl), *a*. three-fold and pertaining to the face (as the trigeminus). *n*. the trigeminus.
trifarious (trīfeə'riəs), *a*. (*Bot*.) arranged in three rows; facing three ways.
trifid (trī'fid), *a*. (*Bot. and Zool*.) divided wholly or partially into three, three-cleft.
trifle (trī'fl), *n*. a thing, matter, fact etc., of no value or importance; a small amount of money etc.; a light confection of whipped cream or white of egg, with cake, jam, wine etc. *v.i*. to act or talk with levity; to sport, to jest, to fool. *v.t*. to waste, fritter, or fool away (time) in trifling. **to trifle with**, to treat with disrespect or lack of proper seriousness; to toy (with). **trifler**, *n*. **trifling**, *a*. insignificant, trivial. **triflingly**, *adv*. **triflingness**, *n*.
trifloral (trīflaw'rəl), *a*. (*Bot*.) bearing three flowers. **triflorous**, *a*. **trifoliate** (-fō'liət), **-ated** (-ātid), *a*. (*Bot*.) three-leaved, consisting of three leaflets. **trifoliolate** (-fō'liələt), *a*. having three leaflets. **Trifolium** (-iəm), *n*. a genus of low herbs containing the trefoils or clovers.
trifocal (trīfō'kəl), *a*. having three focusses or focal

triforium (trīfaw'riəm), *n.* (*pl.* **-ria** (-iə)) a gallery or arcade in the wall over the arches of the nave or choir, or sometimes the transepts, in a large church.

trifurcate (trīfœ'kət), **-cated** (-kātid), *a.* having three branches or forks; trichotomous. *v.t., v.i.* to divide into three.

trig (trig), **trigon**, (*abbr.*) trigonometry.

trigamous (trig'əməs), *a.* married three times; having three wives or three husbands at once; having male, female, and hermaphrodite flowers on the same head. **trigamist**, *n.* **trigamy**, *n.*

trigeminal (trījem'inəl), *a.* threefold; (*Anat.*) of or pertaining to the trigeminus. *n.* the trigeminus. **trigeminus** (-nəs), *n.* the fifth cranial or trifacial nerve dividing into the superior and inferior maxillary and the ophthalmic nerves.

trigger (trig'ə), *n.* a catch or lever for releasing the hammer of a gun-lock; any similar device for releasing a spring, etc., in various forms of mechanism; anything that initiates a process, sequence of events etc. *v.t.* to cause to happen, to set off; to activate, to put into operation. **trigger-happy**, *a.* over eager to fire (a gun etc); over eager to take action.

triglot (trī'glot), *a.* written in three languages.

trigon (trī'gon), *n.* a triangle; a set of three signs of the zodiac arranged at the angles of an equilateral triangle. **trigonal** (trig'-), *a.* triangular, three-cornered; (*Math.*) denoting a system of trilinear coordinates. **trigonally**, *adv.* **trigonous** (trig'-), *a.*

trigoneutic (trigənū'tik), *a.* (*Ent.*) producing three broods in a year.

trigonometry (trigənom'itri), *n.* the branch of mathematics treating of the relations of the sides and angles of triangles, and applying these to astronomy, navigation, surveying etc. **trigonometer**, *n.* an instrument for the mechanical solution of plane right-angled triangles. **trigonometric**, **-ical** (-met'-), *a.* **trigonometric function**, *n.* any of a group of functions of an angle expressed in terms of the ratios of the sides of a right-angled triangle; the inverse of trigonometric function. **trigonometrically**, *adv.*

trigram (trī'gram), *n.* a trigraph; a set of three straight lines in one plane not all intersecting in the same point. **trigrammatic** (-mat'-), **trigrammic** (-gram'-), *a.*

trigraph (trī'graf), *n.* a group of three letters representing a single sound.

trigynous (trij'inəs), *a.* having three pistils.

trihedron (trīhē'drən), *n.* a three-sided figure. **trihedral**, *a.*

trike (trīk), (*coll.*) short for TRICYCLE.

trilabe (trī'lāb), *n.* a three-pronged grasping instrument used in surgery.

trilaminar (trīlam'inə), *a.* having or consisting of three layers.

trilateral (trīlat'ərəl), *a.* having three sides. **trilaterally**, *adv.*

trilby (tril'bi), *n.* a man's soft felt hat with a dent in the middle.

trilemma (trīlem'ə), *n.* a syllogism involving three alternatives.

trilinear (trīlin'iə), *a.* consisting of three lines.

trilingual (trīling'gwəl), *a.* pertaining to or expressed in three languages.

triliteral (trīlit'ərəl), *a.* consisting of or using three letters (*esp.* of Semitic roots). *n.* a triliteral word or root. **triliteralism**, **triliterality** (-ral'-), *n.*

trill (tril), *v.i.* to sing or give forth a sound with a pronounced tremulous vibration. *v.t.* to sing or utter with a quavering or shake. *n.* a tremulous or quavering sound; a consonant pronounced with a trilling sound, as r; a shake, a rapid alternation of two notes a tone or semitone apart.

trillion (tril'yən), *n.* the product of a million raised to the third power; (*esp.* *N Am.*) a million million; (*pl.*) (*coll.*) an indefinite large number. **trillionth** (-th), *a.*

trilogy (tril'əji), *n.* a series of three tragedies, each complete in itself, but connected by the story or theme, and adapted for performance in immediate succession; a group of three plays, operas, novels etc., each complete in itself, but similarly connected.

trim (trim), *v.t.* (*past, p.p.* **trimmed**) to put in good order, to make neat and tidy; to remove irregularities, excrescences, or superfluous or unsightly parts from; to cut, lop, or clip (those) away or off; to dress, to smooth, to plane (wood, boards etc.); to put (a lamp etc.) in order by clipping or renewing a wick, carbons etc.; to reduce (e.g. costs); to adjust (sails, yards etc.) to the wind; to adjust (a ship) by arranging the cargo, ballast etc.; (*coll.*) to reprove sharply, to chastise. *v.i.* to adopt a middle course, between parties, opinions etc. *a.* properly adjusted, in good order; well-equipped, neat, tidy, smart. *n.* state of preparation or fitness, order, condition, esp. of a ship or her cargo, ballast, masts etc.; the angle at which an aeroplane flies in given conditions; the interior panels, decorative fascia, etc. of a vehicle; an act of trimming, esp. hair; material used to trim clothes etc. **trimly**, *adv.* **trimmer**, *n.* one who or that which trims; an implement or machine for clipping timber etc.; one who trims between parties, esp. in politics, a time-server. **trimming**, *n.* the act of one who trims; material sewn on a garment, upholstery etc. for ornament; (*coll.*, *pl.*) accessories to a dish; (*pl.*) anything additional to the main item. **trimness**, *n.*

trimaran (trī'məran), *n.* a sailing vessel with three hulls.

trimer (trī'mə), *n.* a polymer whose molecule is formed from three molecules of a monomer. **trimerous**, *a.* having three parts, joints, members etc.

trimeter (trim'itə), *n.* a verse consisting of three measures of two feet each. *a.* consisting of three measures. **trimetric**, **-ical** (trimet'-), *a.*

trimethyl (trīmeth'il, -mēthīl), *a.* containing three methyl groups. **trime*thylamine**, *n.* the tertiary amine of methyl, a frequent constituent of stale herringbrine.

trimonthly (trīmŭnth'li), *a.* occurring every three months; lasting three months.

trimorphism (trīmaw'fizm), *n.* the existence in certain species of plants and animals of three distinct forms, colours etc., esp. having flowers with pistils or stamens of three different relative lengths; the property of crystallizing in three distinct forms. **trimorphic**, **-morphous**, *a.*

trine (trīn), *a.* threefold, triple; (*Astrol.*) pertaining to or in trine. *n.* a triad, a set of three; (*Astrol.*) the aspect of planets distant from each other 120°. **trinal**, **-ary**, *a.*

trinervate (trīnœ'vət), *a.* three-nerved -veined, or -ribbed.

trinitarian TRINITY.

trinitrotoluene (trīnītrōtol'uēn), *n.* a chemical compound, usually known as TNT, largely used as a high explosive.

trinity (trin'iti), *n.* a group or union of three individuals, a triad; the state of being three or threefold; the union of three persons (the Father, the Son, and the Holy Ghost) in one Godhead; the doctrine of the Trinity; a symbolical representation of the Trinity frequent in art, as the triangle or three interlacing circles. **Trinity House**, *n.* an association for licensing pilots, managing lighthouses, beacons, buoys etc., in British waters. **Trinity Sunday**, *n.* the Sunday next after Whit Sunday. **Trinity term**, *n.* the term beginning after Easter at some universities. **Trinitarian** (-teə'ri-), *a.* of or pertaining to the doctrine of the Trinity; *n.* one who believes in this. **Trinitarianism**, *n.*

trinket (tring'kit), *n.* a small personal ornament of no great value as a jewel, esp. a ring; any small ornament or fancy article.

trinodal (trīnōdəl), *a.* (*Bot.*, *Anat.*, *etc.*) having three nodes or joints.

trinomial (trīnō'miəl), *a.* consisting of three terms, esp. (*Alg.*) connected by the signs + or −; *n.* a trinomial name or expression. **trinomialism**, *n.* trinomial nomenclature, esp. in biology. **trinomially**, *adv.*

trio (trē'ō), *n.* a group of three people or things; a musical composition for three voices or three instruments; a set of three singers or players; the second part of a minuet, march etc.; three aces, kings, queens, knaves, or tens.
triode (trī'ōd), *n.* a thermionic valve with three electrodes.
trioecious (trīē'shəs), *a.* (*Bot.*) having male, female and hermaphrodite flowers, each on different plants of the same species.
triole (trē'ōl), *n.* (*Mus.*) a triplet.
trional (trī'ōnəl), *n.* a hypnotic drug prescribed in cases of mental disease and neurasthenia.
trip (trip), *v.i.* (*past, p.p.* **tripped**) to move, step, walk, or run lightly or nimbly; of rhythm etc. to go lightly or evenly; to make a false step, to stumble; to make a short journey; to be under the influence of a hallucinogenic drug; to catch the foot (over something) so as nearly to fall; to err, to go wrong; to be activated. *v.t.* to cause to fall by catching or obstructing the feet etc.; to catch or detect in a fault, mistake, or offence; to release (a part of a machine) by unfastening; to activate, to set off. *n.* a light nimble step; a leaping movement of the feet; a short excursion, voyage, or journey; (*sl.*) a period spent under the influence of a hallucinogenic drug; (*sl.*) an unpleasant experience; (*sl.*) a pleasurable or engrossing experience; a sudden stroke or catch by which a wrestler trips up his antagonist; a stumble; a false step; a failure, a mistake; any device for activating a mechanism. **trip-hammer**, *n.* a tilt-hammer. **trip-wire**, *n.* a wire that trips a mechanism when pulled. **tripper**, *n.* one who goes on a trip, an excursionist; a device that trips a mechanism.
tripartite (trīpah'tīt), *a.* divided into three corresponding parts or copies; made or concluded between three parties. **tripartitely**, *adv.* **tripartition** (-tish'ən), *n.*
tripe (trīp), *n.* a part of the stomach of ruminating animals prepared for food; (*coll.*) silly stuff; rubbish, nonsense.
tripennate (trīpen'ət), TRIPINNATE.
tripetalous (trīpet'ələs), *a.* having three petals.
triphane (trī'fān), *n.* spodumene.
triphthong (trif'thong), *n.* a combination of three vowels forming one sound. **triphthongal** (-thong'gəl), *a.*
triphyllous (trīfil'əs), *a.* three-leaved.
tripinnate (trīpin'ət), *a.* triply pinnate. **tripinnately**, *adv.*
triplane (trī'plān), *n.* an aeroplane with three supporting planes.
triple (trip'l), *a.* consisting of three parts or three things united, threefold; multiplied by three; of musical rhythm, having three beats to the bar. *n.* a threefold quantity; three of anything. *v.t.* to treble, to make threefold. *v.i.* to become three times as large or as many. **triple crown**, the crown or tiara worn by the Pope. **triple-crowned**, *a.* **triple-expansion engine**, *n.* (*Mach.*) an engine in which the steam expands successively in high, intermediate, and low pressure cylinders, all of which work on the same shaft. **triple-headed**, *a.* **triple jump**, *n.* an athletic event in which the competitor performs a hop, a step and a jump in succession. **triple point**, *n.* the temperature and pressure at which the solid, liquid and vapour phases of a substance are in equilibrium. **triplet** (-lit), *n.* a set or group of three; (*coll.*) each of three children at a birth; three verses rhyming together; (*Mus.*) three notes performed in the time of two. **triplex** (-leks), *n.* triple-time; a composition in three parts; (**Triplex**®) a type of laminated glass. **triplicate** (-likət), *a.* threefold. *n.* a copy, document, or other thing corresponding to two others of the same kind. *v.t.* (-kāt), to make triplicate, to treble. **in triplicate**, written out or copied three times. **triplicate ratio**, the ratio of the cubes (of two quantities). **triplication**, **triplicature** (-kəchə), *n.* **triplicity** (-plis'-), *n.* the state of being triple. **triply**, *adv.*
tripod (trī'pod), *n.* a three-legged stand, stool, utensil, seat, table etc.; a three-legged support for a camera etc.
tripodal (trip'-), *a.*

tripos (trī'pos), *n.* (*pl.* **-ses**) either part of the examination for an honours BA at Cambridge Univ.
tripper TRIP.
triptane (trip'tān), *n.* (*Aviat.*) a very powerful fuel, trimethyl butane.
triptych (trip'tik), *n.* a picture, carving, or other representation, on three panels side by side, frequently used for altar-pieces; a group of three associated pictures etc.
triptyque (trip'tēk), *n.* customs pass, made out in triplicate, for importing or exporting a motor vehicle. [F]
tripwire TRIP.
triquetra (trikwet'rə, -kwet'-), *n.* an ornament composed of three interlacing arcs.
triradial (trīrā'diəl), **triradiate** (-ət), **-ated** (-ātid), *a.* having three rays or radiating branches.
trireme (trī'rēm), *n.* a war-galley with three benches of oars.
trisagion (trisag'iən), *n.* a hymn with a threefold invocation of God as holy, in the liturgies of the Greek and Eastern Churches.
trisect (trīsekt'), *v.t.* to divide into three (esp. equal) parts. **trisection**, *n.*
triserial (trīsiə'riəl), **-ate** (-iət), *a.* (*Anat.*, *Bot.*, *etc.*) arranged in three rows.
trishaw (trīshaw), *n.* a three-wheeled rickshaw.
triskelion (triskel'iən), *n.* a form of fylfot, usu. consisting of three human legs, bent, and joined at the thigh, as in the arms of the Isle of Man.
trismegistus (trismijis'təs), *a.* thrice great (epithet of Hermes).
trismus (triz'məs), *n.* lock-jaw.
trispermous (trīspœ'məs), *a.* three-seeded.
trisplanchnic (trīsplangk'nik), *a.* of or pertaining to the three great viscera of the body, cranial, thoracic, and abdominal.
trisporous (trīspaw'rəs), *a.* having three spores. **trisporic**, *a.*
tristich (tris'tik), *n.* a strophe or set of three lines. **tristichous**, *a.* (*Bot.*) arranged in three vertical rows.
tristigmatic (tristigmat'ik), *a.* (*Bot.*) having three stigmas.
tristylous (tristī'ləs), *a.* (*Bot.*) having three styles.
trisulcate (trīsŭl'kāt), *a.* (*Bot.*) having three furrows or grooves; (*Zool.*) having three digits or hoofs.
trisyllable (trīsil'əbl), *n.* a word of three syllables. **trisyllabic** (-lab'-), *a.* **trisyllabically**, *adv.*
tritanopia (trītənō'piə), *n.* a reduced ability to distinguish the colour blue.
trite (trīt), *a.* worn out; commonplace, hackneyed, stale. **tritely**, *adv.* **triteness**, *n.*
tritheism (trī'thēizm), *n.* the doctrine that the three persons of the Trinity are each distinct Gods. **tritheist**, *n.* **tritheistic** (-is'-), *a.*
triticum (trit'ikəm), *n.* a genus of grasses including wheat.
tritium (trit'iəm), *n.* an isotope of hydrogen with a mass three times that of ordinary hydrogen.
Triton (trī'tən), *n.* in Greek mythology, a son of Poseidon (Neptune) by Amphitrite, or one of a race of minor seagods, represented as half man and half fish, and blowing a spiral shell.
tritone (trī'tōn), *n.* (*Mus.*) an augmented fourth, containing three whole tones.
triturate (trit'ūrāt), *v.t.* to rub or grind down to a fine powder; to masticate with the molar teeth. **triturable**, *a.* **trituration**, *n.* **triturator**, *n.* **triturium** (-tū'riəm), **-torium** (-taw'-), *n.* (*pl.* **-ia** (-riə)), a vessel for separating liquids of different densities.
triumph (trī'əmf), *n.* in ancient Rome, a pageant in honour of a victorious general; the state of being victorious; victory, success; joy or exultation for success. *v.i.* to enjoy a triumph; to gain a victory, to prevail (over); to boast or exult (over); to exult. **triumphal** (-ŭm'-), *a.* of or pertaining to a triumph. **triumphal arch**, *n.* an arch built to celebrate a victory or other notable event. **triumphal-**

ism, *n.* an arrogant pride in one's own success.
triumphant (-ŭm'-), *a.* victorious, successful; exultant.
triumphantly, triumphingly (-ŭm'-), *adv.*
triumvir (trīūm'viə), *n.* (*pl.* **triumvirs, -viri** (-rī), any one of three men united in office, esp. a member of the first or second triumvirate in ancient Rome. **triumviral**, *a.*
triumvirate (-rət), *n.* the office of a triumvir; a group of triumvirs; a coalition of three men in office or authority, esp. the first triumvirate, of Pompey, Julius Caesar and Crassus in 60 BC, or the second, of Mark Antony, Octavian, and Lepidus, in 43 BC.
triune (trī'ūn), *a.* three in one. **triunity**, *n.*
trivalent (trīvā'lənt), *a.* having a valency or combining power of three. **trivalence, trivalency**, *n.*
trivalvular (trīval'vūlə), *a.* having three valves. **trivalve** (trī'-), *a.*, *n.*
trivet (triv'it), *n.* a three-legged stand, esp. a metal tripod or movable bracket for supporting cooking vessels at a fire. **right as a trivet**, (*coll.*) firm, stable; hence in first-rate health, circumstances, position etc.
trivia (triv'iə), *n.pl.* trifles, inessentials. [see foll.]
trivial (triv'iəl), *a.* of little value or importance; trifling; inconsiderable; commonplace, ordinary. **triviality** (-al'-), **trivialness**, *n.* **trivialize**, *v.t.* to cause to seem trivial, to minimize. **trivially**, *adv.* **trivium** (-əm), *n.* in Mediaeval schools, the first three liberal arts: grammar, rhetoric, and logic.
tri-weekly (trīwēk'li), *a.* happening, issued or done three times a week or once every three weeks.
-trix (-triks), *suf.* denoting a feminine agent, as in *executrix*, *testatrix*.
trizone (trī'zōn), *n.* the British, American and French zones of occupation in Germany after World War II. **trizonal** (-zō'-), *a.*
troat (trōt), *n.* the cry of a buck in rutting time. *v.i.* to cry thus.
trocar (trō'kah), *n.* an instrument for draining an internal part of fluid, used in dropsy, hydrocele etc.
trochaic TROCHEE.
trochal TROCHE.
troche (trōsh, trōk), *n.* a lozenge, usu. circular, of medicinal substance. **trochal** (-kəl), *a.* wheelshaped, rotiform.
trochee (trō'kē), *n.* a metrical foot of two syllables, long and short. **trochaic** (-kā'-), *a.*, *n.*
trochiter (trok'itə), *n.* the greater tuberosity of the humerus for the insertion of several muscles.
trochlea (trok'liə), *n.* (*pl.* **-leae** (-iē)) a pulley-like anatomical part or surface, esp. that of the humerus articulating with the ulna. **trochlear**, *a.* (*Anat.*, *Bot.*). **trochleate** (-iət), *a.* (*Bot.*).
trochoid (trō'koid), *a.* (*Anat.*) rotating on its own axis, pivotal; of a trochoidal. *n.* a curve generated by a point in the plane of one curve rolling upon another; (*Anat.*) a trochoid joint. **trochoidal** (-koi'-), *a.* (*Geom.*).
trochometer (trōkom'itə), *n.* an hodometer.
trochophore (trō'kōfaw), *n.* a free-swimming ciliate larva of many invertebrates.
trod, trodden TREAD.
troglodyte (trog'lədit), *n.* a cave-dweller, **troglodytic, -ical** (-dit'-), *a.* **troglodytism**, *n.*
trogon (trō'gon), *n.* one of a family of tropical American insectivorous birds, with brilliant plumage.
troika (troi'kə), *n.* a team of three horses harnessed abreast; a travelling-carriage drawn by this.
troilism (troi'lizm), *n.* sexual activity involving three people of both sexes.
Trojan (trō'jən), *a.* pertaining to ancient Troy. *n.* an inhabitant of ancient Troy; a person of pluck or determination. **Trojan horse**, *n.* the huge wooden horse in which the Greeks secretly entered Troy; any subterfuge intended to undermine an organization etc. from within.
troll [1] (trōl), *v.t.* to roll or reel out (a song) in a careless manner; to fish (water) by trailing or spinning a revolving bait, esp. behind a boat. *v.i.* to fish thus; to walk, to stroll. *n.* a catch; a reel on a fishing-rod; a spinning bait, a spoon-bait etc.
troll [2] (trōl), *n.* a giant or giantess in Scandinavian mythology, endowed with supernatural powers; later, a familiar but impish dwarf.
trolley, trolly [1] (trol'i), *n.* a four-wheeled truck or low car, esp. one the body of which can be tilted over; a costermonger's cart; a set of shelves with wheels, used for moving things, e.g. trays of food, around; a basket on wheels used for containing goods to be purchased in a grocery shop, supermarket etc.; a grooved wheel on a pole used for conveying current to the motor on electric railways, tramways etc. **off one's trolley**, (*sl.*) crazy, mad. **trolley bus**, *n.* an omnibus deriving its motive power through a trolley from overhead wires. **trolley-car**, *n.* (*N Am.*) a tramcar. **trolley-pole**, *n.* **trolley-system**, *n.* the system of working electric railways, tramways etc., by means of trolleys.
trollop (trol'əp), *n.* a careless, slovenly woman, a slattern; a woman of bad character. **trollopy**, *a.*
trolly [1] TROLLEY.
trolly [2] (trol'i), *n.* a kind of lace with the pattern outlined by thick thread or a number of threads combined. **trolly-lace**, *n.*
trombone (trombōn'), *n.* a large and powerful wind-instrument of the trumpet kind usu. played by means of a sliding tube. **trombonist**, *n.*
trommel (trom'l), *n.* a rotating cylindrical sieve for cleaning and sizing ore.
tromometer (trəmom'itə), *n.* an instrument for measuring earth tremors. **tromometric** (trəmōmet'-), *a.*
trompe (tromp), *n.* an apparatus worked by a descending column of water for producing a blast in a furnace.
trompe l'oeil (trŏmp lœy'), *n.* (a painting etc. giving) a very deceptive appearance of reality. [F]
-tron (-tron), *suf.* elementary particle, e.g. *plectron;* particle accelerator, e.g. *cyclotron*.
trona (trō'nə), *n.* a native hydrous carbonate of soda.
troop (troop), *n.* an assemblage of persons or animals, a crowd, a company; (*pl.*) soldiers; a band or company of performers, a troupe; the unit of cavalry formation, usu. consisting of 60 troopers, commanded by a captain. *v.i.* to come together, to assemble, to come thronging (up, together etc.); to move (along a way etc.) in a troop; to hurry (off etc.). **troop carrier**, *n.* a vehicle, aircraft, or ship designed for conveying soldiers. **troop-ship**, *n.* a transport for soldiers. **trooper**, *n.* a cavalry-soldier; a private in a cavalry regiment; a troop-ship; (*Austral.*) a mounted policeman. **trooping the colour**, a ceremonial parade at which the colour is carried between the files of troops.
trop., (*abbr.*) tropical.
trop- TROPO(-).
trope (trōp), *n.* a figurative use of a word.
troph- TROPH(O)-.
trophesy, etc. TROPHIC.
trophic (trof'ik), *a.* pertaining to nutrition. **trophesy** (-əsi), *n.* (*Path.*) deranged nutrition due to nervous disorder. **trophesial** (-fē'-), *a.*
-trophic, *comb. form* relating to nutrition.
troph(o)-, *comb. form.*
trophotropism (trəfot'rəpizm), *n.* the movement of the organs of a growing plant toward or away from nutrient substances, induced by the chemical nature of its surroundings. **trophotropic** (trōfətrop'-), *a.*
trophy (trō'fi), *n.* anything preserved as a memorial of victory or success; an ornamental group of typical or symbolical objects placed on a wall etc. **trophied**, *a.*
-trophy, *comb. form* a specified form of nourishment of growth.

tropic (trop'ik), *n.* either of the two parallels of latitude situated at 23° 27' from the equator, the northern called the **tropic of Cancer**, and the southern the **tropic of Capricorn**, (*pl.*) the regions of the torrid zone between these; either of the corresponding parallels of declination on the celestial sphere. *a.* of or pertaining to the tropics, tropical. **tropical**, *a.* pertaining to, lying within, or characteristic of the tropics; of the weather, very hot, passionate, fervent. **tropical month**, *n.* the mean period of the moon's passing through 360° of longitude, i.e. 27 days, 7 hours, 43 min., 4.7 secs. **tropical year**, *n.* a solar year. **tropically**, *adv.* **tropicopolitan** (-kəpol'itən), *a.* inhabiting and confined to the tropics; *n.* a tropicopolitan animal or plant.
tropism (trō'pizm), *n.* the direction of growth in a plant or other organism that is due to an external stimulus.
trop(o)-, *comb. form* turn(ing); tropism.
tropology (trəpol'əji), *n.* the use of tropical or figurative language; interpretation of the Scriptures in a figurative sense. **tropological** (-loj'-), *a.* **tropologically**, *adv.*
tropopause (trop'ōpawz), *n.* the boundary between the troposphere and the stratosphere.
troposphere (trop'əsfiə), *n.* the hollow sphere of atmosphere surrounding the earth, bounded by the stratosphere, in which temperature varies and the weather functions.
troppo (trop'ō), *adv.* (*Mus.*) too much, excessively. [It., too much]
trot (trot), *v.i.* (*past, p.p.* **trotted**) of a horse or other quadruped, to move at a steady rapid pace by simultaneously lifting one fore-foot and the hind-foot of the opposite side alternately with the other pair, the body being unsupported at intervals; to run with short brisk strides. *v.t.* to cause to trot; to cover (a distance etc.) by trotting. *n.* the pace, motion, or act of a horse etc., in trotting; a brisk steady pace. **on the trot**, one after the other, successively. **the trots**, (*coll.*) diarrhoea. **to trot out**, (*coll.*) to utter (esp. something familiar or trite). **trotter**, *n.* one who or that which trots, esp. a horse trained for fast trotting; (*pl.*) sheep's or other animals' feet used as food.
Trot (trot), *n.* (*coll., often derog.*) a Trotskyite or other leftwinger.
troth (trōth), *n.* faith, fidelity, truth. **to plight one's troth** PLIGHT[1].
Trotskyism (trot'skiizm), *n.* the political theories of Trotsky, esp. that of worldwide proletarian revolution.
troubadour (troo'bədua), *n.* one of a class of lyric poets who flourished in Provence in the 11th cent.
trouble (trŭb'l), *v.t.* to agitate, to disturb; to annoy, to molest; to distress, to afflict; to inconvenience, to put to some exertion or pains. *n.* affliction, distress, worry, perplexity, annoyance, misfortune; labour, exertion, inconvenience. **to ask for trouble**, (*sl.*) to lack caution. **to get into trouble**, to incur censure or punishment; to become pregnant. **troublemaker**, *n.* a person who stirs up discontent, strife, etc. **troubleshooter**, *n.* a person who finds the causes of problems and solves them. **trouble spot**, *n.* a place where there is frequent disturbance, e.g. strikes or fights. **troubler**, *n.* **troublesome** (-səm), *a.* giving trouble; annoying, vexatious; tiresome, wearisome, importunate. **troublesomely**, *adv.* **troublesomeness**, *n.*
trough (trof), *n.* a long, narrow, open receptacle of wood, iron etc., for holding water, fodder etc., for domestic animals; a deep narrow channel, furrow, or depression (in land, the sea etc.); an area of low atmospheric pressure; a hollow between the crests of a wave of radiation; a low point, e.g. in economic activity, in demand, etc.; a state of low spirits.
trounce (trowns), *v.t.* to beat severely; to inflict a decisive defeat upon. **trouncing**, *n.*
troupe (troop), *n.* a company of actors, performers etc. **trouper**, *n.* a member of such a company; a reliable person.
trousers (trow'zəz), *n. pl.* a two legged outer garment reaching from the waist to the ankles. **caught with one's trousers, pants down**, in a situation where one is unprepared. **to wear the trousers**, to be in the position of authority, esp. in a family. **trouser suit**, *n.* a suit of a jacket and a pair of trousers, often when worn by a woman. **trousered**, *a.* **trousering**, *n.* cloth for making trousers.
trousseau (troo'sō), *n.* (*pl.* **-sseaux, -sseaus** (-sōz)) the clothes and general outfit of a bride.
trout (trowt), *n.* a freshwater game-fish, *Salmo fario*, allied to but smaller than the salmon. **trout-stream**, *n.* **troutlet** (-lit), **-ling** (-ling), *n.* **trouty**, *a.*
trouvere (troovea'), *n.* one of the mediaeval poets of N France.
trove SEE TREASURE.
trowel (trow'əl), *n.* a flat-bladed, usu. pointed, tool used by masons etc., for spreading mortar etc.; a scoop-shaped tool used in digging up plants etc. *v.t.* to apply or dress with a trowel. **to lay it on with a trowel**, to flatter grossly.
troy (troi), *n.* a system of weights (12 oz. av. to 1 lb. (340-454 g)) used chiefly in weighing gold, silver and gems, also called **troy weight**.
truant (troo'ənt), *n.* one who shirks or neglects duty; a child who stays away from school without leave. *v.i.* to play truant. **to play truant**, to stay away from school without leave. **truancy**, *n.* **truantly**, *adv.*
Trubenise[®] (troo'bəniz), *v.t.* a method to stiffen fabrics with cellulose acetate.
truce (troos), *n.* a temporary cessation of hostilities; an agreement to cease hostilities; an armistice; a temporary intermission, alleviation, or respite. **truce-breaker**, *n.*
truck[1] (trŭk), *v.t., v.i.* to exchange; to barter. *n.* exchange of commodities; barter; intercourse, dealings; the truck system. **to have no truck with**, to have no dealings with. **truck farmer**, (*N Am.*) a market-gardener. **truck shop**, a shop where the truck system is carried on, a tommy shop. **truck system**, the practice of paying wages in goods instead of money.
truck[2] (trŭk), *n.* a strong, usu. four-wheeled vehicle for conveying heavy goods; an open railway wagon; a low barrow with two small wheels used by porters etc., for moving luggage etc.; a framework and set of wheels for supporting the whole or part of a railway carriage etc.; (*N Am.*) a lorry. *v.t.* to convey on a truck. *v.i.* to work as a lorry-driver. **truckload**, *n.* the amount conveyed by a truck. **trucker**, *n.* a lorry-driver; one who transports goods by lorry.
truckle (trŭk'l), *v.i.* to give way obsequiously (to the will of another); to cringe, to be servile (to). **truckle-bed**, *n.* a low bed on castors or wheels for rolling under another. **truckler**, *n.*
truculent (trŭk'ūlənt), *a.* savage, ferocious, barbarous, violent. **truculence, -lency**, *n.* **truculently**, *adv.* in a truculent manner.
trudge (trŭj), *v.i., v.t.* to travel on foot esp. with labour and fatigue. *n.* a walk of this kind.
true (troo), *a.* conformable to fact or reality, not false or erroneous; in accordance with appearance, not deceptive, counterfeit, or spurious; genuine; in accordance with right or law, legitimate, rightful; corresponding to type or standard; of a voice etc. in perfect tune; faithful, loyal, constant; of a compass bearing, determined in relation to the earth's geographical, rather than its magnetic pole. *v.t.* to make true, exact, or accurate. *adv.* truly. **in/out of true**, correctly/not correctly aligned. **not true**, (*coll.*) amazing, incredible. **true to type**, normal, what might be expected. **true-blue**, staunch, faithful, genuine; (*Brit.*) loyal to the Conservative Party. **true-born**, *a.* of legitimate birth; such by birth or blood. **true-bred**, *a.* of genuine or right breed. **true-hearted**, *a.* **true-heartedness**, *n.* **true-love**, *n.* one truly loved or loving; one's sweetheart.

truffle (trŭf'l), *n.* a fleshy fungus of the genus *Tuber*, used for seasoning etc.; a sweet flavoured with rum or chocolate, resembling a truffle in shape. **truffle-dog,** *n.* a dog trained to find truffles.

trug (trŭg), *n.* a wooden basket used by gardeners, greengrocers etc.; a hod for mortar.

truism (troo'izm), *n.* a self-evident or unquestionable truth; an obvious statement, a platitude. **truistic** (-is'-), *a.*

truly (troo'li), *adv.* sincerely, in accordance with truth, accurately; genuinely; in reality; faithfully, honestly, loyally; really, indeed. **yours truly,** conventional formal ending to a letter; I, myself, me.

trump (trŭmp), *n.* any card of a suit ranking for the time being above the others; (*coll.*) a good fellow; a generous or reliable person. *v.t.* to take with a trump. *v.i.* to play a trump-card. **to come up trumps,** to be useful or helpful at an opportune moment. **trump-card,** *n.* the card turned up to determine which suit is to be trumps; any card of this suit; an infallible expedient.

trumpery (trŭm'pəri), *a.* showy but worthless, delusive, rubbishy.

trumpet (trŭm'pit), *n.* a musical wind instrument, usu. consisting of a long, straight, curved, or coiled tube with a wide termination, usu. of brass; a thing resembling this in shape, as a funnel; a reed-stop in an organ; a sound of or as of a trumpet, e.g. that made by an elephant. *v.t.* to proclaim by or as by sound of trumpet. *v.i.* to make a loud sound as of a trumpet (esp. of the elephant). **to blow one's own trumpet,** to boast. **trumpet-call,** *n.* a call by sound of trumpet; an imperative call to action. **trumpet-flower,** *n.* a plant with large tubular flowers. **trumpet-major,** *n.* the head trumpeter in a cavalry regiment. **trumpeter,** *n.* one who sounds a trumpet, esp. a soldier giving signals on the trumpet in a cavalry regiment.

trump up, to fabricate, to concoct.

truncate (trŭng'kāt), *v.t.* to cut the top or end from. **truncation, truncature** (trŭng'kəchə), *n.*

truncheon (trŭn'shən, -chən), *n.* a short staff, club, or cudgel, esp. one carried by a police officer in Britain; a baton, a staff of authority.

trundle (trŭn'dl), *v.t., v.i.* to move heavily (as if) on wheels.

trunk (trŭngk), *n.* the main stem of a tree, opp. to the branches or roots; the body of an animal apart from the limbs, head, and tail; the main body of anything; a trunk-line; the shaft of a column; a box or chest with a hinged lid for packing clothes etc., in for travel; (*N Am.*) the boot of a motorcar; a ventilating shaft, conduit, chute, flume etc.; proboscis of an elephant or any analogous organ; (*pl.*) men's shorts for swimming. **trunk-call,** *n.* a long-distance telephone call. **trunk exchange,** *n.* a telephone exchange connected by trunk lines or other trunk exchanges. **trunk-line,** *n.* the main line of a railway, canal, telephone etc. **trunk road,** *n.* any major road for long-distance travel. **trunkful,** *n.* **trunkless** (-lis), *a.*

trunnion (trŭn'yən), *n.* one of the cylindrical projections from the sides of a cannon or mortar.

truss (trŭs), *v.t.* to support or brace with a truss; to fasten (a fowl or the wings of a fowl etc.) with a skewer or twine before cooking; to tie up securely, to bind; *n.* a timber or iron supporting and strengthening structure in a roof, bridge etc.; a large corbel; a padded belt or other apparatus worn round the body for preventing or compressing a hernia; a bundle (56 lb., 25·4 kg) of old, (60 lb., 27·2 kg) of new hay, or (36 lb., 16·3 kg) of straw; a compact terminal cluster of flowers. **to truss up,** to make up in to a bundle; to bind or tie up; to hang. **truss-beam,** *n.* **trussbridge,** *n.*

trust (trŭst), *n.* confident reliance on or belief in the integrity, veracity, justice, friendship, power, protection etc., of a person or thing; confidence, firm expectation (that); the person or thing on which reliance is placed; reliance on (assumed honesty etc.) without examination; commercial credit; (*Law*) confidence reposed in a person to whom property is conveyed for the benefit of another; the right to or title in such property as distinct from its legal ownership; the property or thing held in trust; the legal relation between such property and the holder; something committed to one's charge or care; the obligation of one who has received such a charge; (*Comm.*) a combination of a number of businesses or companies under one general control for the purpose of defeating competition, creating a monopoly etc. *v.t.* to place confidence in, to believe in, to rely upon; to believe, to have a confident hope of expectation; to commit to the care of a person, to entrust; to entrust (a person with a thing). *v.i.* to have trust or confidence. **trust deed,** an instrument of conveyance that creates a trust. **trust fund,** *n.* money etc. held in trust. **trust-house,** a public house owned by a trust company and not by a brewer. **trust territory,** *n.* a territory governed by another country by the authority of the United Nations. **trustable,** *a.* **trustee** (-tē'), *n.* one to whom property is committed in trust for the benefit of another; one of a body of people, often elective, managing the affairs of an institution. **trusteeship,** *n.* the office of a trustee; a trust territory. **truster,** *n.* **trustful,** *a.* full of trust; trusting, confiding. **trustfully,** *adv.* **trustfulness,** *n.* **trusting,** *a.* **trustingly,** *adv.* **trustworthy,** *a.* deserving of trust or confidence. **trustworthiness,** *n.* **trusty,** *a.* trustworthy, reliable; not liable to fail in time of need. *n.* a prisoner trusted with a certain amount of liberty to do jobs, etc. **trustily,** *adv.* **trustiness,** *n.*

truth (trooth), *n.* the state or quality of being true; conformity to fact or reality; that which is true, a fact, a verity; honesty, veracity, sincerity; fidelity, constancy; true religion. **in truth,** in reality, in fact, truly. **truth drug,** (*coll.*) any drug used to render a person more liable to tell the truth when being interrogated. **truth-teller,** *n.* **truth-value,** *n.* the truth or falsity of a statement. **truthful,** *a.* habitually speaking the truth, veracious, reliable, conformable to truth. **truthfully,** *adv.* **truthfulness,** *n.* **truthlessness,** *n.*

try (trī), *v.t.* (*past, p.p.* **tried** (trīd)) to test, to examine by experiment; to determine the qualities etc., of by reference to a standard; to find out by experiment or experience; to attempt, to endeavour (to do etc.); to subject to a severe or undue test, to strain; to subject to hardship, suffering etc., as if for a test, to afflict; to investigate (a charge, issue etc.); judicially, to subject (a person) to judicial trial; to prove or settle by a test or experiment; *v.i.* to endeavour, to make an attempt, to put forth efforts. *n.* (*coll.*) an attempt; in rugby, the right to carry the ball and try to kick a goal from in front, earned by touching the ball down behind the opponents' goal line. **to try for,** to aim at; to attempt to secure; to apply for. **to try on,** to put (clothes) on to see if they fit; (*sl.*) to see how much a person will tolerate. **try-on,** *n.* (*Brit., coll.*) something done or said to test a person's patience or gullibility. **to try out,** to test. **try-sail** (trī'sl), *n.* a fore-and-aft sail set on a gaff abaft the foremast and mainmast. **triable,** *a.* **trying,** *a.* irritating, annoying. **try-, trying-square,** *n.* a carpenter's square with a wooden stock and steel limb. **tryout,** *n.* a trial, e.g. of a new method.

tryout TRY.

trypanosome (trip'ənəsōm), *n.* one of the Trypanosomata, an order of flagellate infusorians infesting the blood of man and pathogenic to him. **trypanosomiasis** (-mī'əsis, -pan-), *n.* a disease caused by an infection with a trypanosome.

trypsin (trip'sin), *n.* a ferment contained in the pancreatic juice etc. **tryptic** (-tik), *a.* **tryptone** (-tōn), *n.* a peptone formed during digestion by the action of trypsin on proteins. **tryptophan** (-tōfan), *n.* an amino acid widely distributed in proteins and essential for life.

tryst (trist, trīst), *n.* an appointed meeting, an appointment; a rendezvous.
Tsar CZAR.
TSB, (*abbr.*) Trustee Savings Bank.
tsetse (tsetˊsi), *n.* a S African fly, *Glossina morsitans*, the bite of which is often fatal to cattle, horses, dogs etc., and transmits to man the trypanosomes of sleeping-sickness.
T-shirt, T-square T.
tsunami (tsoonahˊmi), *n.* a very large wave at sea caused by a submarine earthquake, volcanic eruption etc. [Jap.]
TT, (*abbr.*) teetotaller; tuberculin tested.
TUC, (*abbr.*) Trades Union Congress.
Tuareg (twahˊreg), *n.* a member of a nomadic Berber tribe of the Sahara; their language.
tub (tŭb), *n.* an open wooden (usu. round) vessel constructed of staves held together by hoops, used for washing, holding butter etc.; the amount (of butter etc.) that a packing-tub holds; a small cask; a small, usu. plastic, container for ice-cream, margarine etc.; a bath-tub, a sponge-bath, a bath in a tub; a bucket, box, or truck for bringing up ore etc. from a mine; a short clumsy boat; a boat for practising rowing in. *v.i.* (*past, p.p.* **tubbed**) to take a bath in a tub; to row in a tub. **tub-thumper,** *n.* (*coll.*) a ranting preacher. **tubbing,** *n.* **tubbish,** *a.* **tubful,** *n.* **tubby,** *a.* tub-shaped, corpulent.
tuba (tūˊbə), *n.* (*pl.* **-bas, -bae** (-bē)) a brass wind-instrument with a low pitch; a powerful reed-stop in an organ.
tube (tūb), *n.* a long hollow cylinder for the conveyance of fluids and various other purposes, a pipe; a cylindrical vessel of thin flexible metal for holding pigment, toothpaste etc.; the main body of a wind-instrument; the central portion of a heavy gun round which the jackets are fixed by shrinking; (*coll.*) a tubular electric railway; (*N Am.*) a radio valve; a tubular vessel in an animal or plant for conveying air, fluids etc.; (*esp. Austral. sl.*) a can of beer. *v.t.* to furnish with or enclose in a tube or tubes.
tube-railway, *n.* an underground electric railway running in a tubular tunnel. **tubewell,** *n.* a pipe with a sharp point and perforations just above this for driving into the ground to obtain water from a depth. **tubal, tubar,** *a.*
tubeless tyre, *n.* a type of tyre designed to be airtight without an inner tube. **tubing,** *n.*
tuber (tūˊbə), *n.* a short, thick portion of an underground stem, set with eyes or modified buds, as in the potato; a genus of subterranean fungi, containing the truffle; (*Anat.*) a swelling or prominence. **tuberiferous** (-rifˊ-), *a.* **tuberiform** (-ifawm), *a.* **tuberosity** (-rosˊ-), **tuberousness,** *n.* **tuberous,** *a.* having prominent knobs or excrescences; like or bearing tubers.
tubercle (tūˊbəkl), *n.* a small prominence, esp. in bone; a small granular non-vascular tumour or nodule formed within the substance of an organ as the result of morbid action; a warty excrescence. **tubercled, tubercular** (-bœˊkū-), **tuberculate** (-lət), **-lated** (-lātid), **tuberculoid** (-loid), **tuberculose** (-lōs), **-lous,** *a.* **tuberculation,** *n.* formation of tubercles; a system of tubercles; the state of being tuberculous. **tubercularize, tuberculize,** *v.t.* to infect with tuberculosis. **tuberculization,** *n.* **tuberculin** (-lin), *n.* a ptomaine produced by the action of the tubercle-bacillus; a fluid used hypodermically in the diagnosis of tuberculosis. **tuberculin-tested,** *a.* of milk, produced by cows tested and found free of infection and tuberculosis. **tuberculosis** (-lōˊsis), *n.* a diseased condition characterized by the presence of tubercles in the tissues, esp. pulmonary tuberculosis or consumption. **tuberculosed,** (-bœˊ-), *a.*
tuberiferous etc. TUBER.
tuberosity, tuberous TUBER.
tubiform (tūˊbifawm), *a.* having the shape of a tube.
tubing TUBE.
tubular (tūˊbŭlə), *a.* tube-shaped; having or consisting of a tube or tubes. **tubular bells,** *n.pl.* an orchestral percussion instrument consisting of metal tubes suspended vertically and struck to produce a bell-like sound. **tubular boiler,** *n.* one in which the water circulates in a number of pipes in contact with a fire. **tubular bridge,** *n.* one consisting of a large rectangular tube through which a roadway or railway passes. **tubulate** (-lət), **-lated** (-lātid), *a.* [L]
tuck (tŭk), *v.t.* to press close together or press, fold, or roll the loose ends or parts of compactly (up, in etc.); to wrap or cover (up or in) closely or snugly; to gather up, to fold or draw together or into small compass; to push or press, to cram, to stuff, to stow (away, into, etc.); to gather or stitch (a dress etc.) in folds. *v.i.* to make tucks; to be got rid of by tucking away (of loose cloth etc.). *n.* a horizontal fold in a dress etc., esp. one of a series made for ornament or to dispose of loose material; a tuck-net; (*sl.*) food, esp. sweets, pastry etc.; a type of dive in which the knees are bent and held close to the chest by embracing the shins. **tucker,** *n.* esp. (*Austral.*) food. **to tuck in,** to eat greedily, **to tuck away,** to eat heartily; to place somewhere hidden or isolated. **tuck-in, -out,** *n.* a hearty meal, a spread. **tuck-shop,** *n.* (*sl.*) a shop, esp. in a school, where sweets and pastry are sold.
tucker[1] TUCK [1].
tucker[2] (tŭkˊə), *n.* an ornamental frilling of lace or muslin round the top of a woman's dress.
tucker[3] (tŭkˊə), *v.t.* (*esp. N Am. coll.*) to exhaust (often with *out*).
-tude (-tūd), *suf.* forming abstract nouns, as *altitude, beatitude, fortitude.*
Tudor (tūˊdə), *a.* pertaining to the English royal line (from Henry VII to Elizabeth), founded by Owen Tudor of Wales, who married the widow of Henry V, or to their period. **Tudor rose,** *n.* a five-lobed flower adopted as badge by Henry VII. **Tudor style,** *n.* the late Perpendicular style in Gothic architecture.
Tues., (*abbr.*) Tuesday.
Tuesday (tūzˊdā, -di), *n.* the third day of the week.
tuff (tŭf), *n.* an earthy, sometimes fragmentary, deposit of volcanic materials of the most heterogeneous kind. **tuffaceous** (-āˊshəs), *a.*
tuffet (tŭfˊit), *n.* a low mound or seat. [var. of TUFT]
tuft (tŭft), *n.* a cluster, a bunch, a collection of hairs, threads, feathers etc., held or fastened together at one end; (*coll.*) a goatee, an imperial; *v.t.* to adorn with or as with tufts; to pass thread through (a mattress etc,) at regular intervals and fasten a button or tuft in the depression thus made. **tufted, tufty,** *a.*
tug (tŭg), *v.t.* to pull or draw with great effort or with violence; to haul, to tow. *v.i.* (*past, p.p.* **tugged**) to pull violently (at). *n.* the act or a spell of tugging; a vigorous or violent pull; a violent effort, a severe struggle; a small powerful steam-vessel for towing others. **tug of love,** *n.* a dispute between parents or guardians over custody of a child. **tug of war,** *n.* a contest between two sets of persons pulling a rope from opposite ends across a line marked on the ground; a struggle between two sides.
tug(h)rik (tooˊgrēk), *n.* the standard unit of currency in Mongolia. [Mongolian]
tuition (tūishˊən), *n.* teaching, instruction, esp. in a particular subject or group of subjects as dist. from education; a fee for this. **tuitional, tuitionary,** *a.*
tularaemia (toolərēˊmiə), (*esp. N Am.*) **-emia,** *n.* an acute infectious bacterial disease of rodents, sometimes communicated to humans.
tulip (tūˊlip), *n.* any plant of the genus *Tulipa*, bulbous plants of the lily family, with gorgeous bell-shaped flowers of various colours. **tulip-tree,** *n.* a large N American tree, *Liriodendron tulipifera*, of the magnolia family, bearing greenish-yellow, tulip-like flowers. **tulip-wood,** *n.* the wood of this tree.
tulle (tūl), *n.* a fine silk net, used for veils etc., orig. man-

ufactured in the French city of Tulle.

tum[1] (tŭm), **tummy** (-i), *n.* (*coll.*) short for STOMACH.

tum[2] (tŭm), **tum-tum** (-tŭm), *n.* the sound of a stringed musical instrument like the banjo.

tumble (tŭm'bl), *v.i.* to fall (down etc.) suddenly or violently; to roll or toss about; to walk, run, or move about, in a careless or headlong manner; to perform acrobatic feats, esp. without special apparatus; to decrease quickly. *v.t.* to toss or fling forcibly; to throw or push (down etc.); to cause to tumble or fall; to throw into disorder, to rumple; to dry (clothes) in a tumble-dryer. *n.* a fall; a state of disorder; an acrobatic feat, esp. a somersault. **to tumble to**, (*sl.*) to understand, to comprehend. **tumbledown**, *a.* dilapidated. **tumble-dry**, *v.t.* to dry (clothes) in a tumble-dryer. **tumble-dryer**, *n.* a domestic appliance with a revolving cylinder into which damp clothes are placed and dried by having warm air blown through them as they turn. **tumbleweed**, *n.* a plant that breaks away from its roofs in autumn, e.g. an amaranth and is blown around by the wind. **tumbling**, *n.* **tumbling-barrel, -box**, *n.* a revolving box etc., in which castings are cleaned by friction. **tumbly**, *a.*

tumbler (tŭm'blə), *n.* one who or that which tumbles; one who performs somersaults, an acrobat; a variety of pigeon, from its habit of turning over in flight; a toy that turns somersaults; a stemless drinking-glass; a springlatch in a lock, that engages a bolt unless lifted by the key; a part of the lock in a fire-arm attached to the hammer and engaging with the trigger. **tumbler switch**, *n.* a simple form of switch used for electric light connections. **tumblerful**, *n.*

tumbril (tŭm'bril), *n.* a cart of the type used to convey prisoners to the guillotine during the French Revolution.

tumid (tū'mid), *a.* swollen, enlarged, distended; pompous, bombastic, turgid. **tumescent** (-mes'ənt), *a.* swollen, enlarged; becoming swollen or enlarged. **tumescence**, *n.* **tumidity** (-mid'-), **tumidness**, *n.* **tumidly**, *adv.* **tumefy** (-mifi), *v.t.* to cause to swell; to inflate. *v.i.* to swell; to rise in or as in a tumour. **tumefacient** (-fā'shənt), *a.* **tumefaction** (-fak'-), *n.*

tummy TUM[1].

tumour, (*N Am.*) **tumor** (tū'mə), *n.* a swelling on some part of the body, esp. if due to a morbid growth.

tum-tum TUM[1].

tumult (tū'mŭlt), *n.* the commotion, disturbance, or agitation of a crowd, esp. with a confusion of sounds; a confused outbreak or insurrection; uproar, stir, riot; excitement, agitation, or confusion of mind. **tumultuous** (-mŭl'-), *a.* **tumultuously**, *adv.* **tumultuousness**, *n.*

tumulus (tū'mūləs), *n.* (*pl.* **-li**) a mound of earth, sometimes combined with masonry, usually sepulchral, a barrow. **tumular, tumulous**, *a.*

tun (tŭn), *n.* a large cask, esp. for alcoholic liquors; a wine measure, 252 galls. (11·46 hl); a brewer's fermenting-vat. **tunnage** (-ij), *n.* a tax on imported wine levied on each cask or tun, usu. coupled with POUNDAGE.

tuna (tū'nə), *n.* (*pl.*) **tuna, tunas**, (*Zool.*) any of a genus of large scombroid sea-fish found in warmer waters; (also **tuna-fish**) its flesh as food.

tunable, etc. TUNE.

tundra (tŭn'drə), *n.* a marshy treeless plain in the arctic and subarctic regions, with permanently frozen subsoil and covered largely with mosses and lichens.

tune (tūn), *n.* a melodious succession of musical tones forming a coherent whole, an air, a melody; correct intonation in singing or playing; proper adjustment of an instrument for this. *v.t.* to put in tune; to adjust, to adapt, to attune; to adjust (an engine) for optimum performance; to adjust (a radio, TV set) for optimum reception of an incoming signal. **in, out of tune**, at, not at the correct pitch; correctly, incorrectly adjusted for pitch; (not) in harmony, sympathy, agreement (with). **to call the tune**, to give orders, to say what is to be done. **to change one's tune**, CHANGE. **to the tune of**, (*coll.*) to the sum or amount of. **to tune in**, (*Radio.*) to adjust a circuit to obtain resonance at a required frequency; to switch on a radio, TV set and start listening, watching. **to tune up**, (of a group of musicians) to adjust (instruments) to a common pitch before playing; to improve the performance of an engine by tuning. **tunable**, *a.* **tunableness**, *n.* **tunably**, *adv.* **tuneful**, *a.* melodious, musical. **tunefully**, *adv.* **tunefulness**, *n.* **tuneless** (-lis), *a.* unmusical, inharmonious. **tuned circuit**, *n.* (*Radio.*) an oscillatory circuit adjusted to yield resonance at a required wave-length. **tuner**, *n.* one who tunes, esp. one whose occupation is to tune musical instruments; a knob, dial etc. by which a radio or TV set is tuned to different wavelengths. **tuning**, *n.* the act of tuning; (*Mus.*) a set of pitches to which (the strings of) stringed instruments are tuned; the state of adjustment of an engine, radio receiver etc. **tuning condenser**, *n.* (*Radio.*) a variable condenser embodied in a tuning circuit. **tuning-fork**, *n.* a two pronged steel instrument giving a fixed note when struck, used to measure the pitch of musical tones, etc. **tuning-hammer**, *n.* a hammer-shaped wrench for tuning pianofortes, harps, etc.

tungsten (tŭng'stən), *n.* a heavy, greyish-white metallic element, at. no. 74; chem. symbol W (also known as **wolfram**) of unusually high melting point. **tungstic**, *a.* **tungstate** (-stāt), *n.* a salt of tungstic acid.

tunic (tū'nik), *n.* a short-sleeved body-garment reaching nearly to the knees, worn by the ancient Greeks and Romans; a mediaeval surcoat worn over armour; a modern loose coat or short overskirt gathered in or belted at the waist, now worn only by women and children; a military or policeman's jacket; (*Anat.*) a membrane or envelope covering some part or organ; (*Bot.*) a membranous skin. **tunicate** (-kət), *n.* any individual of the order Tunicata, a division of Metazoa, forming a connecting-link between the Vertebrata and the Invertebrata.

tuning-crook, etc. TUNE.

tunnage TUN.

tunnel (tŭn'l), *n.* an artificial underground passage or gallery, esp. one under a hill, river etc., for a railway, road, or canal; a passage dug by a burrowing animal; a mining level. *v.t.* (*past*, *p.p.* **tunnelled**) to make a tunnel through (a hill etc.); *v.i.* to cut or make a tunnel. **tunnel diode**, *n.* a semi-conductor diode capable of giving äa amplification. **tunnel-net**, *n.* a net with a wide mouth narrowing towards the other end. **tunnel vision**, *n.* a medical condition in which one can only see objects directly in front of one; extreme narrowness of viewpoint.

tunny (tŭn'i), TUNA.

tupelo (too'pəlō), *n.* a N American tree of the genus *Nyssa*, esp. the black- or sour-gum; the wood of this.

Tupi (toopē'), *n.* (*pl.* **Tupis, Tupi**) a member of a S American people dwelling in the Amazon region; their language.

tuppence (tŭp'ns), **tuppenny** (-ni), (*coll.*) TWOPENCE, TWOPENNY.

turban (tœ'bən), *n.* an Oriental head-dress consisting of a sash or scarf wound round the cap; a woman's head-dress imitating this; a narrow-brimmed or brimless hat worn by women and children.

turbid (tœ'bid), *a.* muddy, discoloured, thick; disordered, unquiet, disturbed. **turbidity** (-bid'-), **turbidness**, *n.* **turbidly**, *adv.*

turbine (tœ'bīn), *n.* a water-wheel or motor enclosed in a case or tube in which a flowing stream acts by direct impact or reaction upon a series of vanes or buckets; a similar wheel or motor driven by steam or air; a vessel propelled by a turbine.

turbo, *n.*, *a.* (a model of, car etc.) incorporating a turbocharger. **turbo-**, *comb. form* having or driven by a turbine. **turbocharger** (tœ'bōchahjə), *n.* a supercharger,

esp. for motor car engines, driven by exhaust gas turbines. **turbofan** (-fan), *n.* (an aircraft powered by) a gas-turbine aero-engine with a large fan which forces air out with the exhaust gases, thus increasing thrust. **turbojet** (-jet), *n.* (an aircraft powered by) a turbojet engine. **turbojet engine**, *n.* an engine with a turbine-driven compressor for supplying compressed air to the combustion chamber. **turboprop** (-prop), *n.* (an aircraft powered by) an engine with a turbine-driven propeller.
turbot (tœ'bət), *n.* a large European flat-fish, *Psetta maxima*, with bony tubercles, highly valued as food.
turbulent (tœ'būlənt), *a.* disturbed, tumultuous; insubordinate, disorderly. **turbulence**, *n.* **turbulently**, *adv.*
Turcophil (-fil), *n.* a lover of Turkey and the Turks. **Turcophilism**, *n.* **Turcophobe** (-fōb), *n.*
turd (tœd), *n.* (*taboo*) a lump of excrement or dung; a contemptible person.
tureen (tūrēn', tə-), *n.* a deep covered dish or vessel for holding soup.
turf (tœf), *n.* (*pl.* **turfs**, **turves** (-tœvz)) surface earth filled with the matted roots of grass and other small plants; a piece of this, a sod; greensward, growing grass; peat. *v.t.* to cover or line with turfs or sods. **the turf**, the racecourse; the occupation or profession of horse-racing. **to turf out**, (*coll.*) to throw out, to eject forcibly. **turf accountant**, *n.* a bookmaker. **turf-clad**, *a.* covered with turf. **turfiness**, *n.* **turfy**, *a.*
turgid (tœ'jid), *a.* swollen, bloated, morbidly distended; tumid; pompous, inflated, bombastic. **turgescent** (-jes'ənt), *a.* **turgescence**, *n.* **turgidity** (-jid'-), **turgidness**, *n.* **turgidly**, *adv.*
Turk (tœk), *n.* a native, inhabitant or citizen of Turkey; a native speaker of a Turkic language; a troublesome person, esp. a boy. **Turk's-cap**, *n.* a martagon lily; the melon-cactus.
Turkey[1] (tœ'ki), *n.* the country of the Turks. **Turkey carpet**, *n.* a soft velvety woollen carpet, orig. made in Turkey. **Turkey red**, *n.* a brilliant red dye orig. obtained from madder; cotton cloth dyed with this. **Turkey-stone**, *n.* novaculite; turquoise.
turkey[2] (tœ'ki), *n.* a large gallinaceous bird of the genus *Meleagris*, allied to the pheasant; (*Austral.*) the wild turkey, the Callegalla or brush turkey, and the mallee-bird or sand turkey; (*esp. N Am., sl.*) a flop. **to talk turkey**, (*esp. N Am.*) to come to the point, to talk business. **turkey-buzzard**, **-vulture**, *n.* an American vulture, *Cathartes*. **turkey-cock**, *n.* a male turkey; a conceited, pompous person. **turkey-corn**, *n.* maize. **turkey-trot**, *n.* a round dance with little or no bending of the knees and a swing of the body.
Turki (tœ'ki), *a.* of the Turkish, as distinct from the Tatar, branch of the Turko-Tatar languages. *n.* a Turki language or speaker. **Turkic**, **Turko-tatar** (-kō-), *a.*, *n.* (of) that branch of the Altaic languages to which Turkish belongs.
Turkish (tœ'kish), *a.* pertaining to Turkey or the Turks. *n.* the language of the Turks. **Turkish bath**, *n.* a hot-air bath in which one is sweated, washed, rubbed, massaged etc., and conducted through a series of cooling-rooms. **Turkish carpet** TURKEY CARPET. **Turkish delight**, *n.* a gelatinous sweetmeat.
Turkoman (tœ'kōmən), *n.* (*pl.* **-mans**) a member of any of the Turkish or Tatar peoples living in Turkestan or the adjoining regions of Persia, Afghanistan and Russia.
turmalin TOURMALINE.
turmeric (tœ'mərik), *n.* an E Indian plant, *Curcuma longa*, of the ginger family; the powdered rhizome of this used as dye-stuff, a stimulant, or a condiment, esp. in curry. **turmeric-paper**, *n.* unsized white paper saturated with turmeric used as a test for alkalis, which change the colour from yellow to red.
turmoil (tœ'moil), *n.* commotion, disturbance, tumult.
turn (tœn), *v.t.* to cause to move round on or as on an axis, to give a rotary motion to; to cause to go, move, aim, point, look etc., in a different direction; to expose the other side of, to invert, to reverse; to renew (a cuff, collar etc.) by reversing; to bring lower soil to the surface by digging or ploughing; to revolve in the mind; to perform (a somersault); to apply or devote to a different purpose or object, to give a new direction to; to bend, to adapt, to change in form, condition, nature etc.; to cause to become, to convert, to transform, to transmute; to translate, to paraphrase; to pass, go, or move to the other side of, to go round; to pass round the flank of (an army) so as to attack it from the flank or rear; to reach or pass beyond (a certain age, time); to cause to ferment, to make sour; to nauseate; to infatuate, to unsettle, to make giddy; to cause to go, to send, to put by turning; to shape in a lathe or on a potter's wheel; to give a shapely form to, to mould, to round (a sentence etc.); to cause an enemy agent to become a double agent. *v.i.* to have a circular or revolving motion, to rotate, to revolve, to move round or about; to move the body, face, or head in a different direction, to change front from right to left, etc.; to change in posture, attitude, or position; to take a particular direction; to be changed in nature, form, condition etc.; to change colour; to become sour or spoiled; to become unsettled, infatuated, or giddy; to become nauseated; to result, to terminate; to undergo the process of turning on the lathe; (*Cricket*) to spin, to deviate from line; (*Cricket*) (of wicket) to assist spin bowling. *n.* the act of turning, rotary motion; a revolution; the state of being turned; a change of direction, position or tendency, a deflection; a bend, a curve, a winding, a corner; a single round or coil of a rope etc.; a change, a vicissitude; a turning-point; a point of change in time; a short walk, a stroll, a promenade; a performance, bout or spell (of doing something); an occasion, opportunity, or time (for doing something); coming in succession to each of a number of persons; succession, alternation, rotation; (*coll.*) a nervous shock; shape, form, mould, character, disposition, temper; (*Theat.*) (the performer of) a short, theatrical act. **a good, bad turn**, a helpful service, a disservice. **at every turn**, constantly; everywhere. **by turns**, alternately; at intervals. **done to a turn**, cooked exactly right. **ill turn** ILL. **in turn**, in order of succession, in rotation. **on the turn**, just turning (of the tide); beginning to go sour; the point of changing. **out of turn**, out of the proper order of succession; at an inappropriate time. **to serve one's turn**, to serve one's purpose; to help or suit one. **to take a turn for the better, worse**, to improve, deteriorate. **to take turns**, to alternate, to perform or participate in rotation or succession. **to turn a blind eye to**, to pretend not to see, to overlook. **to turn about**, to turn the face in another direction; to turn round. **to turn a deaf ear to**, to refuse to listen to. **to turn adrift**, to unmoor (a boat) and allow to float away; (*fig.*) to cast off. **to turn again**, to return. **to turn against**, (cause to) become hostile to; to use against. **to turn aside**, to deviate; to divert, to avert. **to turn down**, to fold or double down; to lower (a light, the volume on a radio etc.); to lay (a card) face downwards; (*coll.*) to reject. **to turn in**, to direct or incline inwards; to fold or double in; to send, put, or drive in; to hand over, to surrender; to give, to execute (a performance etc.); (*coll.*) to go to bed. **to turn off**, to deflect; to deviate; to dismiss; to shut or switch off; to cause to lose interest in, esp. sexually. **turn-off**, (*sl.*) *n.* a person or thing that excites distaste. **to turn on**, to open a way to (gas etc.) by turning the tap; to switch on; to direct, to aim; to hinge or depend upon; to attack suddenly and unexpectedly; (*coll.*) to excite, to arouse the interest of, esp. sexually; (*sl.*) to introduce to drugs; (*sl.*) to have got high on drugs. **turn-on**, (*sl.*) *n.* a person or thing that causes emotional or sexual arousal. **to turn one's hand**, to apply oneself. **to turn out**, to drive out, to expel; to point or to cause to point outwards; to

turn (pockets etc.) inside out; (of a room) to clean thoroughly; to bring to view; to produce, as the result of labour; to prove to be; to switch off; to dress, to groom, to look after the appearance of; to gather, to assemble; to go out; (*coll.*) to get out of bed. **to turn over**, to change the position of, to invert, to reverse; (of an engine) to (cause to) start or run at low revolutions; to surrender, to hand over; to transfer (to), to put under other control; to cause to turn over, to upset; to do business to the amount of; to consider, to ponder; (*sl.*) to rob. **to turn round**, to face about; to adopt new views, attitude, policy etc.; to complete the processing of; to complete the unloading and reloading of (a ship, aircraft); to restore to profitability. **to turn tail**, TAIL. **to turn to**, to have recourse to; to change or be changed into; to direct towards; to find (a page) in a book; to set to work. **to turn turtle**, TURTLE[2]. **to turn up**, to bring to the surface; to unearth, to bring to light; to place (a card etc.) with the face upwards; to tilt up; to find and refer to (a passage) in a book; to point upwards; to come to light; to happen; to make one's appearance, **to turn upon**, to hinge on; to attack; to direct or aim at. **turn and turn about**, alternately, successively. **turnabout**, *n.* the act of facing in an opposite direction; a complete reversal (of opinion, policy etc.). **turnaround**, turnabout; turnround. **turn-buckle**, *n.* a coupling for metal rods etc. allowing adjustment of length. **turncap**, *n.* a chimney cowl turning round with the wind. **turncoat**, *n.* one who turns his coat; one who deserts his party or principles. **turn-down**, *a.* folded or doubled down. **turn indicator**, *n.* (*Aviat.*) a gyroscopic instrument which indicates any deviation in the course of an aircraft. **turnkey**, *n.* one who has the charge of the keys of a prison, a warder. **turn-out**, *n.* a turning out for duty; an assembly, a large party; a showy or well-appointed equipage; dress, get-up; a quantity of articles or products manufactured in a given time. **turn-over**, *n.* an upset; a semicircular pie or tart made by turning over half the crust; the amount of money turned over in a business in a given time; the rate at which stock in trade is sold and replenished; the rate at which employees leave and have to be replaced. **turnround**, *n.* (the time taken by) the process of unloading a ship, aircraft and reloading it ready for its next trip; (the time taken by) the complete processing of anything; a change to an opposite and usu. better state. **turnstone**, *n.* a bird, *Arenaria interpres*, allied to the plover. **turntable**, *n.* a platform rotating in a horizontal plane used for shifting rolling-stock from one line of rails to another; the rotating table which supports a gramophone record while being played. **turn-up**, *n.* a turned-up fold at the bottom of a trouser leg; (also **a turn-up for the book**) a sudden and unexpected (fortunate) occurrence. **turner**, *n.* one who turns, esp. one who turns articles in a lathe; a variety of tumbler-pigeon. **turnery**, *n.* **turning**, *n.* the act of one who or of that which turns; a bend, a corner, the point where a road meets another; such a road. **turning circle**, *n.* the smallest circle in which a vehicle can turn round. **turning-point**, *n.* the point in place, time etc., on or at which a change takes place, the decisive point.

turnip (tœr'nip), *n.* a plant of the genus *Brassica*, with a fleshy globular root used as a vegetable and for feeding sheep.

turnpike (tœn'pīk), *n.* a gate set across a road to stop carriages etc., from passing till the toll is paid; a turnpike road; (*N Am.*) a motorway on which a toll is payable.

turnstile (tœn'stīl), *n.* a post with four horizontal revolving arms, set at the entrance to an enclosure, building etc., allowing persons to pass through one at a time often after a toll or fee is paid.

turpentine (tœ'pəntīn), *n.* an oleoresin exuding naturally or from incisions in several coniferous trees, esp. the terebinth; oil or spirit of turpentine, popularly called **turps** (tœps), used for mixing paint, varnishes etc. and in medicine; white spirit, also called **turpentine substitute**.

turpitude (tœ'pitūd), *n.* baseness, depravity.

turps TURPENTINE.

turquoise (tœ'koiz, -kwoiz, -kwahz), *n.* a sky-blue or bluish-green translucent or opaque precious stone; a pale greenish-blue. *a.* of turquoise colour.

turret (tŭ'rit), *n.* a small tower attached to a building, and rising above it; a low flat cylindrical or conical armoured tower, usu. revolving, so that the guns command a wide radius on a warship, tank or fort; a similar structure on an aircraft. **turreted**, *a.*

turtle (tœ'tl), *n.* a marine reptile encased in a carapace, like a tortoise, with flippers used in swimming; a chelonian, esp. the green turtle, *Chelonia mydes*, used for soup; turtle-soup. **to turn turtle**, to turn completely over, to capsize. **turtle-neck**, *n.* (a sweater with) a round, high, close-fitting neck. **turtle-necked**, *a.* **turtle-soup**, *n.* rich soup made from fatty parts of the turtle.

turtle-dove, *n.* the common wild dove, esp. *Turtur communis*, noted for its soft cooing.

turves, (tœvz), *pl.* TURF.

Tuscan (tŭs'kən), *a.* pertaining to Tuscany. *n.* a native or the language of Tuscany; the Tuscan order. **Tuscan order**, *n.* (*Arch.*) the simplest of the five classic orders, a Roman modification of Doric.

tusche (tush), *n.* a substance used in lithography for drawing in the design which resists the printing medium.

tush (tŭsh), *int.* an expression of contempt or impatience.

tusk (tŭsk), *n.* a long pointed tooth, esp. one protruding from the mouth as in the elephant, narwhal etc. **tusked**, **tusky**, *a.* **tusker**, *n.* an elephant or wild boar with well-developed tusks.

tussle (tŭs'l), *v.i.* to struggle, to scuffle (with or for). *n.* a struggle, a scuffle.

tussock (tŭs'ək), *n.* a clump, tuft, or hillock of growing grass. **tussock-grass**, *n.* a grass, *Dactylis caespitosa*, forming tufts 5–6 ft. (1·7-2·0 m) high, growing in Patagonia and the Falkland Islands.

tussore, tussur, tusser (tŭs'aw. tŭs'ə), *n.* an Indian silkworm moth, *Antherea mylitta*, feeding on the jujube tree etc., or a Chinese oak-feeding silkworm moth, *A. pernyi*; a strong, coarse silk obtained from these.

tut (tŭt), *int.*, *n.* an exclamation of impatience, rebuke, or contempt. *v.i.* to make this exclamation.

tutee (tūtē'), *n.* person who is tutored.

tutelage (tū'təlij), *n.* Guardianship; the state of being under a guardian; the period of this. **tutelar, -lary**, *a.* having the care or protection of a person or thing, protective; pertaining to a guardian.

tutor (tū'tə), *n.* a private teacher, esp. one having the general care and instruction of a pupil in preparation for a university etc.; (*Eng. Univ.*) an officer directing the studies of undergraduates in a college and charged with discipline etc.; a college or university teacher who teaches and holds discussions with students in small groups; an instruction book; (*Law*) a guardian of a minor. *v.t.* to act as a tutor to; to instruct, to teach. **tutorage** (-ij), *n.* **tutoress**, (-ris), *n. fem.* **tutorial** (-taw'ri-), *a.* of a tutor. *n.* a teaching session or conference with a tutor. **tutorially**, *adv.* **tutorship**, *n.*

tutti (tut'i), *adv.* (*Mus. direction*) all together. *n.* a composition or passage for singing or performing thus. [It.]

tutti-frutti (toot'ifroo'ti), *n.* a confection, as ice-cream, made of or flavoured with different fruits. [It.]

tutu (too'too), *n.* a ballet-dancer's short, stiff skirt that spreads outwards. [F]

tuum (tū'əm), *n.* thine, yours; thy or your property. [L]

tu-whit tu-whoo (təwit təwoo'), *int.* an imitation of the cry of an owl.

tuxedo (tŭksē'dō), *n.* (*pl.* **-dos, -does**) (*N Am.*) a dinner jacket.

tuyère (twēyeə', tooyeə'), *n.* the blast-pipe or nozzle in a

furnace, forge etc.
TV (*abbr.*) television. **TV dinner**, *n.* a complete, ready-packaged and frozen dinner that only needs reheating before being eaten.
TVP (*abbr.*) textured vegetable protein.
TWA (*abbr.*) Trans-World Airlines.
twa (twaw), (*Sc.*) TWO.
twaddle (twod'l), *v.i.* to talk unmeaningly; to chatter. *n.* unmeaning talk, silly chatter, nonsense. **twaddler**, *n.* **twaddly**, *adv.*
twain (twān), *a.* two. *n.* a pair, a couple. **in twain**, in two, asunder.
twang (twăng), *v.i.* to make a ringing metallic sound as by plucking the string of a musical instrument; to play (on) thus; to speak or be uttered with a nasal sound. *v.t.* to cause to sound with a twang; to play (an instrument) thus; to utter or pronounce with a nasal sound. *n.* such a ringing metallic sound; a nasal tone (in speaking etc.).
'twas (twoz). [short for IT WAS]
twat (twat, twot), *n.* (*taboo*) the female genitals; (*sl.*) a stupid or contemptible person.
tweak (twēk), *v.t.* to pinch and twist or pull with a sudden jerk, to twitch. *n.* a sharp pinch or pull, a twitch.
twee (twē), *a.* excessively dainty and prettified; sentimentally sweet.
tweed (twēd), *n.* a twilled woollen or wool-and-cotton fabric with unfinished surface, used chiefly for outer garments.
tweet (twēt), **tweet-tweet** (-twēt'), *int.* imitation of the sound made by a small bird. *v.i.* make this sound. **tweeter**, *n.* loudspeaker used to produce higher frequencies.
tweezers (twē'zǝz), *n.pl.* small pincers for picking up minute things, plucking out hairs etc., usually called a pair of tweezers.
twelfth (twelfth), *a.* (*Mus.*) an interval of an octave and a fifth. *n.* one of twelve equal parts. *n.*, *a.* the last of 12 (people, things etc.); the next after the 11th. **the glorious twelfth**, 12 Aug., when grouse-shooting begins. **Twelfth Day**, *n.* the 12th day after Christmas, the festival of the Epiphany, 6 Jan. **Twelfth Night**, *n.* the eve of this, 5 Jan.
twelve (twelv), *n.* the number or figure 12 or XII; the age of 12; midnight or midday. *a.* 12 in number; aged 12. **the Twelve**, the twelve Apostles. **twelvemonth**, *n.* a year.
twenty (twen'ti), *n.* the number or figure 20 or XX; the age of 20. *a.* 20 in number; aged 20. **twentieth** (-tiǝth), *n.* one of 20 equal parts *n.*, *a.* the last of 20 (people, things etc.); the next after the 19th. **twentyfold**, *a.*, *adv.* **twenty-pence (piece)**, *n.* a British coin worth 20p. **twenty-twenty**, *a.* of vision, normal.
'twere (twœ). [short for IT WERE]
twerp, twirp (twœp), *n.* (*sl.*) a contemptible or silly person. [etym. unknown]
twi-, *comb. form.* two; double.
twice (twīs), *adv.* two times; doubly. **twice-told**, *a.* related twice; well-known, hackneyed.
twiddle (twid'l), *v.t.* to rotate; to twirl idly; to fiddle with. *v.i.* to twirl; to fiddle or trifle (with). **to twiddle one's thumbs**, to sit idle.
twig[1] (twig), *n.* a small shoot or branch of a tree, bush, etc., a branchlet; a divining rod; (*Anat.*) a small branch of an artery or other vessel; (*Elec.*) a small distributing conductor. **twigged**, *a.* **twiggy**, *a.* **twigless** (-lis), *a.*
twig[2] (twig), *v.t.* (*past, p.p.* **twigged**) (*coll.*) to understand, to comprehend, to catch the drift of; to see, to notice.
twilight (twī'līt), *n.* the diffused light from the sky appearing a little before sunrise and after sunset; a faint light, shade, obscurity; indistinct or imperfect perception, revelation, or knowledge. *a.* pertaining to, happening, or done in the twilight; dim, shady, obscure. **twilight of the gods**, (*Norse Myth.*) a conflict in which the gods were overcome and the world destroyed. **twilight sleep**, *n.* (*Med.*) a state of semi-consciousness produced by administering scopolamine and morphine in which labour pains are mitigated and forgotten when over. **twilight zone**, *n.* a transitional or intermediate zone; a decaying urban area esp. between the commercial centre and the residential suburbs.
twill (twil), *n.* a fabric in which the weft-threads pass alternately over one warp-thread and then under two or more, producing diagonal ribs or lines. *v.t.* to weave thus.
'twill (twil), short for IT WILL.
twin (twin), *a.* being one of two born at a birth; being one of a similar or closely related pair of things, parts etc.; double, twofold; (*Bot.*) growing in pairs or divided into two equal parts. *n.* one of two children or young produced at a birth; a person or thing very closely resembling or related to another; an exact counterpart. *v.t.* to couple, to pair (with). **dissimilar, binovular twins**, twins proceeding from the fertilization of two oocytes. **identical, uniovular twins**, twins that have developed from a single oocyte. **the Twins**, (*Astron.*) the zodiacal sign of the Gemini. **twin bed**, *n.* one of a matching pair of single beds. **twinscrew**, *n.* a steamer with two propellers twisted in opposite directions. **twin set**, *n.* a jumper and cardigan made to match. **twin town**, *n.* a town which has forged close civic and cultural links with a town in a foreign country. **twin-tub**, *n.* washing machine with two separate drums, one for washing, the other for spin-drying. **twinship**, *n.*
twine (twīn), *v.t.* to twist; to form (thread etc.) by twisting together; to wind or coil round, to embrace; to form by interweaving. *v.i.* to be interwoven; to entwine, to coil (about, round etc.); to wind, to meander. *n.* a twist, a convolution, a coil; the act of twining or entwining; an interlacing, a tangle; strong string made of two or three strands twisted together. **twiner**, *n.* **twiningly**, *adv.*
twinge (twinj), *v.t.* to affect with a sharp, sudden pain. *n.* a sharp, sudden, shooting pain; a pang, as of remorse or sorrow.
twinkle (twing'kl), *v.i.* to shine with a broken quivering light, to gleam fitfully, to sparkle; to appear and disappear in rapid alternation, to move tremulously. *v.t.* to flash or emit (light) intermittently in rapid gleams. *n.* a tremulous gleam, a sparkle; a glimmer; a rapid tremulous movement. **twinkling**, *n.* a twinkle; the time of this, an instant.
twirl (twœl), *v.t.* to cause to rotate rapidly, esp. with the fingers, to spin; to whirl (round); to twiddle, to twist, to curl (the moustache etc.). *v.i.* to revolve or rotate rapidly, to whirl (round). *n.* a rapid circular motion; a quick rotation; a twist, a curl, a flourish.
twirp TWERP.
twist (twist), *v.t.* to wind a thread, filament, strand etc., round another; to form (a rope or threads etc., into) a rope etc.) thus; to intertwine (with or in with); to give a spiral form to by turning the ends in opposite directions; to wrench, to distort; to pervert, to misrepresent; to twine, to wreathe; to cause (a ball) to rotate while following a curved path; to make (one's way) in a winding manner. *v.i.* to be turned or bent round and round upon itself; to be or grow in a spiral form; to move in a curving, winding, or irregular path; to writhe, to squirm; to dance the twist. *n.* the act or manner of twisting or the state of being twisted; a quick or vigorous turn, a whirling motion given to a ball etc.; a dance, popular in the 1960's in which the dancer gyrates his or her hips in time to the music while remaining more or less on the same spot; a sharp bend; a peculiar tendency, a bent, an idiosyncrasy; an unexpected development in, or conclusion to, the plot of a story; (*Phys.*) a twisting strain; the angle or degree of torsion of a rod etc.; forward motion combined with rotation; thread, cord, string, rope etc., made from twisted strands, esp. strong silk thread or cotton yarn; twisted tobacco; a small piece

of lemon rind. **round the twist** (*coll.*) crazy. **to twist somebody's arm**, to use force or psychological pressure to persuade someone. **twistable**, *a.* **twister**, *n.* one who or that which twists; a ball delivered with a twist at cricket, billiards etc.; (*esp. N Am. coll.*) a tornado, a waterspout; (*coll.*) a cheat, a rogue.

twit[1] (twit), *v.t.* (*past, p.p.* **twitted**) to reproach, taunt, or upbraid (with some fault etc.). **twittingly**, *adv.*

twit[2] (twit), *n.* (*coll.*) a fool.

twitch (twich), *v.t.* to pull with a sudden or sharp jerk; to snatch. *v.i.* to pull or jerk (at); to move with a spasmodic jerk or contraction. *n.* a sudden pull or jerk; a sudden involuntary contraction of a muscle etc. **twitcher**, *n.* one who or that which twitches. **twitchy**, *a.* nervous.

twitter (twit'ə), *v.i.* to utter a succession of short, tremulous, intermittent notes; to chirp; to have a tremulous motion of the nerves, to be agitated. *v.t.* to utter with tremulous, intermittent sounds. *n.* such a succession of sounds, a chirping; (*coll.*) a state of excitement or nervous agitation.

twittingly TWIT.

'twixt (twikst), [short for BETWIXT]

two (too), *n.* the number or figure 2 or II; the age of 2; the second hour after midnight or midday. *a.* two in number; aged 2. **in two**, into two parts; as under. **one or two, two or three**, a few. **to put two and two together**, to draw inferences. **two-bit**, *a.* (*N Am., coll.*) insignificant, small-time. **two-dimensional**, *a.* having two dimensions; lacking (the appearance of) depth. **two-dimensionality**, *n.* **two-edged**, *a.* having an edge on both sides (of a knife etc.); cutting both ways. **two-faced**, *a.* having two faces; deceitful, insincere. **twofold**, *a.* double; *adv.* doubly. **two-foot**, *a.* (*coll.*) measuring two feet. **two-handed**, *a.* having two hands; having to be used with both hands; played, worked etc., by two persons; using both hands with equal dexterity, ambidextrous. **two-headed**, *a.* **twopence** (tŭp'əns), *n.* the sum of two pence. **two-pence piece**, *n.* a coin worth two-pence. **twopenny** (tŭp'ni), *a.* worth twopence; cheap, worthless, common, vulgar. **twopenny-halfpenny**, *a.* worth or costing twopence-halfpenny; paltry, insignificant. **two-piece**, *n., a.* (a garment) consisting of two usu. matching parts. **two-ply**, *a.* having two strands (as cord) or two thicknesses (as carpets, cloth, plywood etc.). **two-sided**, *a.* having two sides or aspects. **twosome** (-səm), *n.* a couple; a dance, game of golf, etc. involving two people. **two-speed**, *n.* giving or adapted to two rates of speed. **two-step**, *n.* (*Dancing*) a kind of round dance to march or polka time. **two-stroke**, *a., n.* (being, having) an internal-combustion engine with a cycle of two strokes. **two-tier**, *a.* having an upper and a lower level, as a legislature with an upper and a lower house. **two-time**, *v.t.* (*coll.*) to be unfaithful to; to double-cross. **two-timing**, *a.* **two-timer**, *n.* **two-tone**, *a.* having two colours or shades. **two-way**, *a.* arranged to allow movement in either of two directions; of a radio, able to send and receive; reciprocal; (*Math.*) having a double mode of variation. **two-way mirror**, *n.* a half-silvered piece of glass that acts as a mirror viewed from one side but is translucent from the other.

TX, (*abbr.*) Texas.

-ty (-ti), *suf.* forming abstract nouns as *bounty, cruelty, fealty*; as in *fifty, twenty*.

Tyburn tree, *n.* the gallows.

tycoon (tīkoon'), *n.* a financial or political magnate.

tyke (tīk), *n.* a dog; a cur; (*dial.*) an ill-mannered fellow; a Yorkshireman.

tylosis (tīlō'sis), *n.* inflammation of the eyelids with thickening and roughening of the margins. **tylotic** (-lot'-), *a.*

tympan (tim'pən), *n.* a frame stretched with paper cloth or parchment, used for equalizing the pressure in some printing-presses; any thin sheet or membrane tightly stretched; a tympanum.

tympanic TYMPANUM.

tympanites (timpənī'tēz), *n.* (*Path.*) distension of the abdomen, due to the accumulation of air in the intestine, etc. **tympanitic** (-nit'-), *a.*

tympanitis (timpənī'tis), *n.* (*Path.*) inflammation of the lining membrane of the middle ear.

tympanum (tim'pənəm), *n.* (*pl.* **-na** (-nə)) the middle ear; the tympanic membrane or ear-drum; (*Arch.*) a triangular area, usu. recessed, in a pediment, the space between the lintel of a doorway and the arch enclosing it. **tympanic** (-pan'-), *a.* like a drum; acting like a drum-head; (*Anat.*) pertaining to the tympanum.

Tynewald, Tynwald (tin'wawld), *n.* the legislature of the Isle of Man.

typ(o), (*abbr.*) TYPOGRAPHER; TYPOGRAPHIC; TYPOGRAPHY.

type (tīp), *n.* a distinguishing mark, a symbol, an emblem, an image; any person or thing that stands as an illustration, pattern, characteristic example, or representative specimen of another thing or class of things; a kind, a class, a category; (*coll.*) a person (of a specified kind); a prophetic similitude; (*Biol.*) a general form or structure common to a number of individuals; an organism exhibiting the essential characteristics of its group; (*Chem.*) a compound, such as hydrochloric acid, water, ammonia or methane, illustrating other compounds by analogy; an original conception, object, or work of art, serving as a model or guide to later artists; any of a class of objects embodying the characteristics of a group or class, esp. as a model, pattern, or exponent (of beauty or other qualities); a piece of metal or hard wood bearing a letter or character usu. in relief, for printing with; (*collect.*) a set or quantity or kind of these. *v.t.* to typewrite. **-type**, *comb. form* of the kind specified, resembling. **in type**, set in type. **type-cast**, *v.t.* (*past, p.p.* **typecast**) to cast continually in the same kind of part. **typeface**, *n.* the printing surface of type; a design of printing type. **type-founder**, *n.* one who casts types. **type-foundry**, *n.* **type-metal**, *n.* an alloy of lead, antimony, and tin, used for making printing-type. **typescript**, *n.* typewritten matter. **typesetter**, *n.* a compositor; a machine for setting type. **typesetting**, *n., a.* **typewrite**, *v.i.* to write with a typewriter. **typewriter** *n.* a machine for producing printed characters as a substitute for handwriting; **typewriting**, *n.* **typewritten**, *a.*

typhlitis (tiflī'tis), *n.* (*Path.*) inflammation of the caecum. **typhlitic** (-lit'-), *a.* **typhlo-**, *comb. form.*

typhoid (tī'foid), *a.* pertaining to or resembling typhus. *n.* typhoid fever, an infectious fever characterized by an eruption of red spots on the chest and abdomen, severe intestinal irritation, inflammation, diarrhoea etc., enteric. **typhoidal** (-foi'-), *a.* **typhomalarial** (-fō-), *a.* malarial with typhoidal symptoms. **typhomania** (-mā'niə), *n.* the low muttering delirium characteristic of typhus and typhoid fever. **typhonia** (-fō'niə), *n.* a form of sleepless and delirious stupor characteristic of typhus.

typhoon (tīfoon'), *n.* a violent cyclonic hurricane occurring in the China Seas and the West Pacific. **typhonic** (-fon'-), *a.*

typhus (tī'fəs), *n.* a contagious fever marked by an eruption of dark purple spots, great prostration, stupor and delirium. **typhous**, *a.*

typical (tip'ikəl), *a.* of the nature of or serving as a type; representative, emblematic, symbolical (of); embodying the characters of a group, class etc.; characteristic (of). **typically**, *adv.* **typicalness**, *n.* **typify** (-fī), *v.t.* to be a type of, to exemplify. **typification** (-fi-), *n.* **typifier**, *n.*

typist (tī'pist), *n.* one who types letters etc.

typo (tī'pō), *n.* (*pl.* **typos**) (*coll.*) a typographical error.

typography (tīpog'rəfi), *n.* the art of printing; the arrangement, character, or appearance of printed matter. **typographer**, *n.* **typographic, -al** (-graf'-), *a.* **typographically**, *adv.*

typolite (tī'pəlit), *n.* a stone impressed with the figure of a

typolithography

plant or animal, a fossil.
typolithography (tipōlithog'rəfi), *n.* the process of printing from lithographic stones which have previously received transferred impressions from type. **typolithographic** (-graf'-), *a.*
typology (tīpol'əji), *n.* the doctrine of interpretation of types, esp. those of the Scriptures. **typological** (-loj'-), *a.* **typologist**, *n.*
typonym (tī'pənim), *n.* (*Biol.*) the name based on a type. **typonimal** (-pon'-), *a.* **typonimic** (-nim'-), *a.*
tyrannosaurus (tīranōsaw'rəs), *n.* (*Palaeont.*) a genus of carnivorous dinosaurs, about 40 ft. (12 m) in length.
tyranny TYRANT.
tyrant (tī'rənt), *n.* an oppressive or cruel ruler or master; an oppressor, a despot, an autocrat, esp. (*Hist.*) one obtaining power by usurpation; an arbitrary or despotic ruler. **tyrannical** (tiran'-), *a.* acting like or characteristic of a tyrant; despotic, arbitrary, imperious. **tyrannically**, *adv.* **tyrannicalness**, *n.* **tyrannicide** (tiran'isīd), *n.* the act of killing a tyrant; one who kills a tyrant. **tyrannicidal** (-sī'-), *a.* **tyrannize** (ti'rə-), *v.i.* to act the tyrant; to rule despotically or oppressively (over). *v.t.* to rule (a person etc.) despotically. **tyrannous** (ti'rə-), *a.* **tyrannously**, *adv.*
tyranny (ti'rəni), *n.* arbitrary, or oppressive exercise of power; an arbitrary, despotic, or oppressive act.
tyre (tīə), *n.* an air-filled rubber casing, a strip of solid rubber or a band of metal surrounding a wheel of a vehicle.
tyriasis (tirī'əsis), *n.* (*Path.*) a form of elephantiasis; tyroma. **tyroma** (tīrō'mə), *n.* falling off of the hair through a fungoid growth at the roots.
tyro (tī'rō), *n.* (*sometimes derog.*) a beginner, a novice.
Tyrolean (tirəlē'ən), **Tyrolese** (-lēz'), TIROLESE.
tyroma TYRIASIS.
tyrosine (tī'rəsēn, -sin), *n.* an amino acid formed by the decomposition of proteins.
Tzar, (zah), etc. CZAR.
tzetze (tset'si), TSETSE.
Tzigane (tsigahn), *a.* of or pertaining to the Hungarian gipsies or their music. *n.* an Hungarian gipsy. [Hung.]

U

U¹, u, the twenty-first letter and the fifth vowel (*pl.* **Us, U's, Ues**,), has five principal sounds; (1) as in *rule*, (rool); (2) as in *bull* (bul); (3) as in *but* (būt); (4) as in *bur* (bœ); (5) as in *due* dū. **U**, *a.* (*coll.*) of words, phrases, behaviour etc., associated with the so-called Upper Classes. **U-boat** *n.* a German submarine. [G *Unterseeboot*] **U-turn**, *n.* a turn made by a motor vehicle which takes it back along the direction from which it has come, without reversing; any complete reversal of policy etc.
U², (*abbr.*) Unionist; university; universal (used of a film certified suitable for viewing without age limit).
U³, (*chem. symbol*) uranium.
UAE, (*abbr.*) United Arab Emirates.
UAR, (*abbr.*) United Arab Republic.
UB40, *n.* a card issued to a person registered as unemployed; (*coll.*) an unemployed person.
ubiety (ūbī'iti), *n.* the state of being in a particular place; the relation of locality, whereness.
ubiquity (ūbik'witi), *n.* the quality or state of being everywhere or in an indefinite number of places at the same time, omnipresence. **ubiquitarian** (-teə'ri-), *n.* (*Theol.*) a believer in the omnipresence of Christ's body, esp. with reference to the Eucharist. *a.* of or pertaining to this belief. **ubiquitarianism**, *n.* **ubiquitous**, *a.* **ubiquitously**, *adv.* **ubiquitousness**, *n.*
uc, (*abbr.*) upper case.
UCATT, (*abbr.*) Union of Construction, Allied Trades and Technicians.
UCCA (ŭk'ə), (*abbr.*) Universities Central Council on Admissions.
UCW, (*abbr.*) Union of Communication Workers.
UDA, (*abbr.*) Ulster Defence Association.
udal (ū'dəl), *n.* freehold tenure based on uninterrupted possession, as in N Europe before feudalism and in Orkney and Shetland at the present day. **udaller**, *n.* the holder of such tenure.
UDC, (*abbr.*) Urban District Council.
udder (ŭd'ə), *n.* the milk-secreting organ of a cow, ewe etc. **uddered**, *a.* **udderless**, *a.*
UDF, (*abbr.*) United Democratic Front, a South African anti-apartheid association.
UDI, (*abbr.*) unilateral declaration of independence.
udometer (ūdom'itə), *n.* a rain-gauge. **udometric** (-met'-), *a.*
UDR, (*abbr.*) Ulster Defence Regiment.
UEFA (ūä'fä), (*abbr.*) Union of European Football Associations.
UFO, (*abbr.*) unidentified flying object. **ufology** (ūfol'əji), *n.* the study of flying saucers and other UFOs. **ufologist**, *n.*
UGC, (*abbr.*) University Grants Committee.
ugh (ükh. üh. uh), *int.* an exclamation of disgust or horror.
ugli (ŭg'li), *n.* a cross between a grapefruit and a tangerine.
ugly (ŭg'li), *a.* unpleasing to the sight, not beautiful; unsightly, ungraceful, not comely; morally repulsive, unpleasant; suggesting evil; awkward, cantankerous; threatening, formidable. **ugly duckling**, *n.* an unpromising person or thing which turns out surprisingly successful etc. **uglify** (-fī), *v.t.* **uglily**, *adv.* **ugliness**, *n.*
Ugrian (oog'riən, ŭg'-), **Ugric**, *n.* a member of the Eastern branch of the Finno-Ugrian peoples, esp. the Magyars; their group of languages. *a.* of or relating to these peoples or their languages.
UHF, (*abbr.*) ultra-high frequency.
Uhlan (oo'lən, ū'-), *n.* a cavalryman armed with a lance, in the old German and some other Continental armies.
UHT, (*abbr.*) ultra-heat treated (milk).
uitlander (āt'landə), *n.* an immigrant into the Transvaal.
UK, (*abbr.*) United Kingdom.
UKAEA, (*abbr.*) United Kingdom Atomic Energy Authority.
ukase (ūkāz'), *n.* an edict or decree of the Imperial Russian Government; any arbitrary decree.
Ukrainian (ūkrān'iən), *n.* a native or inhabitant of the Ukraine or Ukrainian Soviet Socialist Republic; the language of the Ukraine. *a.* of the Ukraine, its people or its language.
ukulele (ūkəlā'li), *n.* a small four-stringed instrument resembling a guitar.
ulcer (ŭl'sə), *n.* an open sore on the outer or inner surface of the body accompanied by a secretion of pus or other discharge; a source of corruption or moral pollution. **ulcerable**, *a.* **ulcerate**, *v.t.* to affect with or as with an ulcer *v.i.* to form an ulcer; to become ulcerous. **ulceration**, *n.* **ulcerative**, *a.* **ulcered**, **ulcerous**, *a.* **ulcerously**, *adv.* **ulcerousness**, *n.*
-ule, (-ūl), *dim. suf.* as in *globule, pustule.*
ulema (oo'limə), *n.* the body of Muslim doctors of law and interpreters of the Koran in a country, esp. in Turkey; a member of such a body.
-ulent -LENT.
ulitis (ūli'tis), *n.* inflammation of the gums.
ullage (ŭl'ij), *n.* the quantity that a cask falls short of being full.
ulna (ŭl'nə), *n.* (*pl.* **-nae** (-nē)) the larger and longer of the two bones of the fore-arm. **ulnar**, *a.* **ulno-**, *comb. form.*
ulosis (ūlō'sis), *n.* cicatrization.
ulotrichous (ūlot'rikəs), *a.* having woolly or curly hair. **ulotrichan**, *a.* **ulotrichy** (-i), *n.*
ulster (ŭl'stə), *n.* a long, loose overcoat for men or women, usu. with a belt. [*Ulster*, province of Ireland]
ult. (ŭlt), (*abbr.*) ultimo.
ulterior (ŭltiə'riə), *a.* lying beyond or on the other side of any line or boundary; more remote or distant; not at present in view, under consideration, or pertinent; not yet disclosed, unavowed, esp. deliberately not revealed. **ulteriorly**, *adv.*
ultima (ŭl'timə), *n.* the last syllable of a word. **ultima Thule**, THULE.
ultimate (ŭl'timət), *a.* last, final, beyond which there is nothing existing or possible; incapable of further analysis; fundamental, elementary, primary; greatest, most important. *n.* something final, fundamental or of greatest moment. **ultimately**, *adv.* **ultimateness**, *n.*
ultimatum (ŭltimä'təm), *n.* (*pl.* **-tums**, **-ta** (-tə)), a final proposal, statement of conditions, or concession, the rejection of which may involve rupture of diplomatic relations and a declaration or war; anything final, essential, or fundamental.
ultimo (ŭl'timō), *adv.* last month.
ultimogeniture (ŭltimōjen'ichə), *n.* inheritance by the

youngest son.
ultra (ŭl'trə), *a.* extreme, advocating extreme views or measures; uncompromising, extravagant. *n.* an extremist. **ultraism,** *n.* **ultraist,** *n.*
ultra- (ŭltrə-), *pref.* beyond, on the other side of; beyond the ordinary limit or range of; beyond the reasonable, excessive(ly). **ultracentrifuge,** *n.* a centrifuge that spins at very high speeds. *v.t.* to spin in an ultracentrifuge. **ultraclassical,** *a.* extravagantly classical in style etc. **ultraconservative,** *a.* extravagantly conservative. **ultrafiche,** *n.* a sheet of microfilm, like a microfiche but holding more microcopies. **ultrahigh,** *a.* of radio frequencies, between 300 and 3000 megahertz. **ultra-microscope,** *n.* one with a light source at the side, for examining particles too small to be seen with the ordinary microscope. **ultramicroscopic,** *a.* too small to be seen with an ordinary optical microscope; pertaining to or using an ultramicroscope. **ultramicroscopy,** *n.* **ultramodern,** *a.* **ultramontane,** *a.* being or lying beyond the mountains, esp. the Alps, esp. on the Italian side; hence, supporting the absolute power and infallibility of the pope. *n.* one who resides south of the Alps; a supporter of ultramontanism. **ultramontanism** (-mŏn'tə-), *n.* in the Roman Catholic Church, the principle that all ecclesiastical power should be concentrated in the hands of the pope, in contradistinction to the independent development of national Churches. **ultramontanist** (-mŏn'tə-), *n.* **ultramundane** (-mŭn'-), *a.* external to the world or the solar system; pertaining to the supernatural or another life. **ultra-orthodox,** *a.* **ultra-Protestant,** *a.* **ultra-religious,** *a.* **ultra-sensual,** *a.* **ultrashort waves,** *n.pl.* electromagnetic waves below 10 metres in wavelength. **ultrasonic,** *a.* pertaining to or using sound waves of higher than audible frequency. **ultrasonically,** *adv.* **ultrasonics,** *n.sing.* the science of ultrasonic waves. **ultrasound** (ŭl'-), *n.* ultrasonic waves, used esp. for medical diagnosis. **ultrastructure** (ŭl'-), *n.* the structure of an organism or cell as made visible by a microscope, esp. an electron microscope. **ultraviolet,** *n., a.* (electromagnetic radiation) having a wavelength shorter than the violet end of the visible spectrum but longer than X-rays. **ultravirus,** *n.* a very small virus that can pass through the finest of filters.
ultramarine (ŭtrəmərēn'), *a.* situated, being, or lying beyond the sea. *n.* a deep-blue pigment formerly obtained from lapis lazuli; the colour of this.
ultra vires (ŭl'trə vīə'rēz, vē'rāz), *a., adv.,* beyond one's legal power or authority.
ululate (ū'ūlāt), *v.i.* to howl, as a dog or wolf, to hoot. **ululant,** *a.* **ululation,** *n.*
umbel (ŭm'bl), *n.* an inflorescence in which the flower stalks spring from one point and spread like the ribs of an umbrella forming a flattish surface, as in the parsley family. **umbellar, umbellate** (-lət), **-lated** (-lātid), *a.* **umbelliferous** (-lif'-), *a.* belonging to the Umbelliferae, a family of plants whose flower heads are umbels. **umbellule** (-lūl), *n.* **umbellifer** (-bel'ifə), *n.* an umbelliferous plant.
umber (ŭm'bə), *n.* a dark yellowish-brown pigment derived from a mineral ferric oxide containing manganese; the colour of umber. *a.* of the colour of umber, dark, dusky. *v.t.* to colour with or as with umber. **burnt umber,** umber heated so as to produce a much redder brown. **raw umber,** umber in the natural state.
umbilical (ŭmbil'ikl), *a.* of, or pertaining to, or situated near the navel; of the umbilical cord. **umbilical cord,** *n.* the rope-like structure of vessels and connective tissue connecting the foetus with the placenta; a cable carrying electricity, air etc., from a servicing point to a spacecraft, astronaut, diver etc. **umbilicate** (-kət), **-ed,** *a.* having or shaped like an umbilicus; have a central depression. **umbilication,** *n.* **umbilicus** (-kəs, -lī'-), *n.* the navel; (*Nat. Hist.*) a navel-shaped depression or other formation; the hilum; a depression at the axial base of some univalve shells.
umbles (ŭm'blz), *n.pl.* the entrails of a deer (cp. HUMBLE-PIE.)
umbo (ŭm'bō), *n.* (*pl.* **-bos, -bones,** (-bō'nēz)) the boss or projecting point in the centre of a shield; (*Nat. Hist.*) a boss, knob, prominence, or elevation. **umbonal, umbonate** (-nət), **umbonic** (-bon'-), *a.*
umbra (ŭm'brə), *n.* (*pl.* **-brae** (-brē), **-bras**) the part of the shadow of a planet etc., esp. the earth or moon, in which the light of the sun is entirely cut off; the dark central portion of a sun-spot.
umbrage (ŭm'brij), *n.* a sense of injury, offence; (that which affords) shade. **umbrageous** (-brā'jəs), *a.* shady, shaded. **umbrageously,** *adv.* **umbrageousness,** *n.* shadiness.
umbral, umbrated UMBRA.
umbrella (ŭmbrel'ə), *n.* a light screen of nylon etc., stretched on a folding frame of radiating ribs on a stick, for holding above the head as a protection against rain; the umbrella-shaped disc of a medusa used as a swimming organ; a protection, a cover; a screen of aircraft or of gunfire covering a military movement; an organization which protects or coordinates the activities of a number of separate groups; a general heading etc. encompassing several individual ones. **umbrella-bird,** *n.* a S American bird of the genus *Cephalopterus,* with a large erectile spreading crest. **umbrella-stand,** *n.* a stand for holding umbrellas, in an entrance hall etc. **umbrella-tree,** *n.* a small magnolia with flowers and leaves in an umbrella-like whorl at the ends of the branches. **umbrellaed** (-ləd), *a.*
Umbrian (ŭm'briən), *a.* of or pertaining to Umbria, in Central Italy, esp. of the school of painting to which Raphael and Perugino belonged; the language of Umbria, one of the principal Italic dialects. *n.* a native of ancient Umbria.
umiak (oo'miak), *n.* an Eskimo boat made of skins stretched on a framework.
umlaut (ŭm'lowt), *n.* change of the vowel in a syllable through the influence of an *i* or *u* (usu. lost or modified) in the following syllable; (*Print.*) the diaeresis mark used over German vowels.
umpire (ŭm'pīə), *n.* a person chosen to enforce the rules and settle disputes in a game, as cricket; a person chosen to decide a question of controversy; (*Law*) a third person called in to settle a disagreement between arbitrators. *v.t., v.i.* to act as umpire (in). **umpirage** (-rij), **umpireship,** *n.*
umpteen (ŭmptēn'), *a.* an indefinitely large number. **umpteenth,** *a.*
UN, (*abbr.*) United Nations.
'un (ŭn), *pron.* (*coll.*) ONE.
un- (ŭn-), *pref.* giving a negative sense to adjectives, adverbs, and nouns; used with verbs to denote reversal or annulment of the action of the simple verb (sometimes ambiguous, thus *unrolled* may mean 'not rolled up', or 'opened out after having been rolled up'). Since there is no limit to the use of this prefix the meaning of words not given in the following selection can be ascertained by reference to the simple verb, adjective etc.
UNA, (*abbr.*) United Nations Association.
un- (cont.) **unabashed,** *a.* **unabated,** *a.* **unable,** *a.* not able (to); not having sufficient power or ability; incapable, incompetent. **unabridged,** *a.* **unacademic,** *a.* **unaccented,** *a.* **unacceptable,** *a.* **unacceptableness, unacceptability,** *n.* **unaccommodating,** *a.* **unaccompanied,** *a.* unattended; (*Mus.*) without accompaniment. **unaccomplished,** *a.* unfinished, not carried out or effected; lacking accomplishments. **unaccountable,** *a.* not accountable or responsible; inexplicable. **unaccountability, unaccountableness,** *n.* **unaccountably,** *adv.* **unaccounted (for),** *a.* not explained; not included in an account or list. **unaccredited,** *a.* **unaccustomed,** *a.* not usual or familiar; not used (to). **unachievable,** *a.* **unachieved,** *a.* **unacknowledged,** *a.* not

acknowledged, not recognized. **una corda** (oō'nə kaw'də), *a.*, *adv.* (*Mus.*) using the soft pedal. [It., one string]
un- (cont.) **unacquainted**, *a.* **unactable**, *a.* not capable of being acted; unfit for representation. **unacted**, *a.* **unadaptable**, *a.* **unadapted**, *a.* unfitted (for). **unaddressed**, *a.* **unadjusted**, *a.* **unadministered**, *a.* **unadmired**, *a.* **unadmonished**, *a.* **unadopted**, *a.* not adopted; of a road etc. not taken over by the local authority. **unadorned**, *a.* not adorned, without decoration. **unadulterate, -ated**, *a.* not adulterated, unmixed; pure, genuine. **unadventurous**, *a.* **unadvertised**, *a.* **unadvisable**, *a.* **unadvisability, unadvisableness**, *n.* **unadvised**, *a.* not advised; not prudent or discreet, rash. **unadvisedly**, *adv.* **unadvisedness**, *n.* **unaffected**, *a.* not influenced or affected; without affectation, sincere, genuine. **unaffectedly**, *adv.* **unaffectedness**, *n.* **unaffiliated**, *a.* **unafflicted**, *a.* **unafraid**, *a.* **unaggressive**, *a.* **unaided**, *a.* **unalarmed**, *a.* **unalienable**, etc. INALIENABLE. **unaligned**, *a.* **unalleviated**, *a.* **unallowable**, *a.* **unalloyed**, *a.* **unalterable**, *a.* **unalterability, unalterableness**, *n.* **unalterably**, *adv.* **unaltered**, *a.* **unambiguous**, *a.* plain, clear. **unambiguously**, *adv.* **unambiguousness**, *n.* **unambitious**, *a.* **unambivalent**, *a.* **unamenable**, *a.* **unamended**, *a.* **un-American**, *a.* not American; alien to or incompatible with American ideas or principles. **unamiable**, *a.* **unamused**, *a.* **unamusingly**, *adv.* **unanalysable**, *a.* **unanalysed**, *a.* **unanimated**, *a.*
unanimous (ūnan'iməs), *a.* being all of one mind, agreeing in opinion; formed, held, or expressed with one accord. **unanimity** (-nim'-), **unanimousness**, *n.* **unanimously**, *adv.*
un- (cont.) **unannounced**, *a.* **unanswerable**, *a.* that cannot be satisfactorily answered or refuted. **unanswerability, unanswerableness**, *n.* **unanswerably**, *adv.* **unanswered**, *a.* **unanticipated**, *a.* **unapologetic**, *a.* **unapparent**, *a.* **unappealing**, *a.* **unappeasable**, *a.* **unappeased**, *a.* **unappetizing, -ising**, *a.* **unappetizingly**, *adv.* **unapplied**, *a.* **unappreciated**, *a.* **unappreciative**, *a.* **unapprehended**, *a.* **unapprehensible**, *a.* **unapprised**, *a.* **unapproachable**, *a.* that cannot be approached, inaccessible; reserved, distant in manner; that cannot be rivalled. **unapproachability, unapproachableness**, *n.* **unapproachably**, *adv.* **unappropriated**, *a.* **unapproved**, *a.* **unapproving**, *a.* **unapprovingly**, *adv.* **unapt**, *a.* **unaptly**, *adv.* **unaptness**, *n.* **unarguable**, *a.* **unarm**, *v.t.*, *v.i.* to disarm. **unarmed**, *a.* **unarmoured**, *a.* **unarranged**, *a.* **unarrayed**, *a.* **unarticulated**, *a.* **unartificial**, *a.* not artificial; natural. **unartificially**, *adv.* **unartistic**, *a.* **unascertainable**, *a.* **unascertained**, *a.* **unashamed**, *a.* **unasked**, *a.* **unaspirated**, *a.* **unaspiring**, *a.* **unaspiringly**, *adv.* **unassailable**, *a.* incapable of being assailed; incontestable. **unassailed**, *a.* **unassayed**, *a.* **unassignable**, *a.* **unassigned**, *a.* **unassimilated**, *a.* **unassisted**, *a.* **unassuming**, *a.* not arrogant or presuming, modest. **unassured**, *a.* **unatoned**, *a.* **unattached**, *a.* not attached; (*Law*) not seized for debt; not belonging to any particular club, regiment etc.; not married or engaged. **unattainable**, *a.* **unattainableness**, *n.* **unattempted**, *a.* **unattended**, *a.* **unattested**, *a.* **unattractive**, *a.* **unattractively**, *adv.* **unattractiveness**, *n.* **unaugmented**, *a.* **unauthentic**, *a.* **unauthenticated**, *a.* **unauthorized, -ised**, *a.* **unavailable**, *a.* **unavailability, unavailableness**, *n.* **unavailing**, *a.* ineffectual; vain, useless. **unavailingly**, *adv.* **unavenged**, *a.* **unavoidable**, *a.* inevitable; that cannot be made null or void. **unavoidableness**, *n.* **unavoidably**, *adv.* **unavowed**, *a.* **unaware**, *a.* not aware, ignorant (of); careless, inattentive. *adv.* (*loosely*) unawares. **unawares**, *adv.* without warning; by surprise, unexpectedly; undesignedly. **unbacked**, *a.* of a horse, not taught to bear a rider, unbroken; unsupported, having no backers; of a chair etc., without a back. **unbaked**, *a.* **unbalance**, *v.t.* to throw off one's balance. **unbalanced**, *a.* not balanced, not in equipoise; not brought to an equality of debit and credit; without mental balance, unsteady, erratic. **unban**, *v.t.* (*past, p.p.*

-**banned**) to lift the ban on. **unbaptized, -ised**, *a.* **unbar**, *v.t.* to remove a bar or bars from; to unfasten, to open. **unbathed**, *a.* **unbearable**, *a.* not to be borne, intolerable. **unbearably**, *adv.* **unbearded**, *a.* **unbeatable**, *a.* that cannot be beaten, unsurpassable. **unbeaten**, *a.* not beaten; not conquered or surpassed; untrodden. **unbeautiful**, *a.* not beautiful; ugly. **unbecoming**, *a.* not becoming, not suited (to); not befitting; improper, indecorous, indecent. **unbecomingly**, *adv.* **unbecomingness**, *n.* **unbefitting**, *a.* **unbefriended**, *a.* **unbegot, unbegotten**, *a.* not begotten; self-existent. **unbeknown(st)**, *a.* (*coll.*) not known; unknown (to); *adv.* without the knowledge. **unbelief**, *n.* the withholding of belief; incredulity; scepticism; disbelief (in, esp. divine revelation). **unbelievable**, *a.* **unbeliever**, *n.* one who does not believe, esp. in a religion. **unbelieving**, *a.* **unbeloved**, *a.* **unbelt**, *v.t.* **unbend**, *v.t.* (*past, p.p.* -**bent**) to change or free from a bent position; to straighten; to relax from exertion, tension, constraint etc.; (*Naut.*) to unfasten (sails) from the yards and stays; to cast loose or untie (a cable or rope). *v.i.* to become straightened; to relax from constraint, formality etc.; to be affable. **unbendable**, *a.* **unbending**, *a.* unyielding, resolute, inflexible. **unbendingly**, *adv.* **unbendingness**, *n.* **unbeneficed**, *a.* **unbespoken**, *a.* **unbestowed**, *a.* **unbias**, *v.t.* to set free from bias. **unbiased**, *a.* **unbiblical**, *a.* not in or according to the Bible. **unbidden**, *a.* not commanded; not called for, spontaneous; uninvited. **unbigoted**, *a.* **unbind**, *v.t.* (*past, p.p.* -**bound**) to untie, to unfasten; to release from a binding; to free from bonds, to release. **unbirthday**, *a.* (*coll.*) of a present, given on an occasion other than a birthday. **unblamable**, *a.* **unblamableness**, *n.* **unblamably**, *adv.* **unblamed**, *a.* **unbleached**, *a.* **unblemished**, *a.* **unblinking**, *a.* showing no surprise or other emotion. **unblock**, *v.t.* **unblooded**, *a.* **unbloody**, *a.* not stained with blood; not accompanied with bloodshed; not bloodthirsty. **unblotted**, *a.* **unblushing**, *a.* shameless, barefaced, impudent. **unblushingly**, *adv.* **unboiled**, *a.* **unbolt**, *v.t.* to undo the bolts of; to unfasten, to open. **unbolted**, *a.* not fastened by a bolt. **unbolted**, *a.* of flour etc., not bolted or sifted. **unborn**, *a.* not yet born; still to have existence **unbosom**, *v.t.* to disclose (one's feelings etc.). *v.i.* to disclose one's secret feelings, opinions, or intentions; to open one's heart. **unbound**, *past, p.p.* UNBIND. **unbounded**, *a.* boundless, not bounded (by); infinite, not subject to check or control. **unboundedly**, *adv.* **unboundedness**, *n.* **unbowed** (-bowd'), *a.* not bowed; unconquered. **unbox**, *v.t.* **unbrace**, *v.t.* to remove or relax the braces of; to free from tension, to loosen, to relax. **unbraid**, *v.t.* to separate the strands of; to unweave, to disentangle. **unbranched**, *a.* **unbranching**, *a.* **unbreakable**, *a.* **unbribable**, *a.* **unbridle**, *v.t.* to remove the bridle from; to set free from restraint. **unbridled**, *a.* freed from the bridle; unrestrained, unruly, ungovernable. **unbroken**, *a.* not broken; not subdued; uninterrupted, regular; not violated; not broken in, not accustomed to the saddle etc.; not opened up by the plough; of a record, not bettered. **unbrotherly**, *a.* **unbrotherliness**, *n.* **unbruised**, *a.* **unbuckle**, *v.t.* to unfasten the buckle of. **unbuilt**, *a.* not built; of land, not yet built upon. **unburden**, *v.t.* to free from a load or burden; to relieve (the mind etc.) by disclosing or confession. **unburdened**, *a.* **unburied**, *a.* **unburned, -burnt**, *a.* **unbusinesslike**, *a.* **unbutton**, *v.t.* to unfasten the buttons of. *v.t.* to undo one's buttons; (*coll.*) to talk without restraint. **uncage**, *v.t.* **uncalled**, *a.* **uncalled for**, not necessary; not asked for, gratuitous, impertinent. **uncanny**, *a.* not canny, weird, mysterious; incautious, rash, dangerous. **uncanonized, -ised**, *a.* **uncap**, *v.t.* to remove the cap or cover from; to remove an upper limit from. *v.i.* to remove one's cap or hat (in salutation). **uncapped**, *a.* **uncared-for**, *a.* not cared for, neglected. **uncaring**, *a.* **uncarpeted**, *a.* **uncase**, *v.t.* **uncashed**, *a.* **uncastrated**, *a.* **uncatalogued**, *a.* **uncaught**, *a.* **uncauterized, -ised**, *a.* **unceasing, *a.* not ceasing, incessant, continual. **unceasingly**, *adv.* **uncensored**, *a.* **uncensured**, *a.* **unceremonious**, *a.* without ceremony, formality, or courtesy; familiar, brusque, abrupt. **unceremoniously**, *adv.* **unceremoniousness**, *n.* **uncertain**, *a.* not certain; not sure; doubtful; not certainly or precisely known; not to be relied on; undecided, changeable, fickle, capricious. **in no uncertain terms**, forcefully, unambiguously. **uncertainly**, *adv.* **uncertainty**, *n.* **uncertainty principle**, *n.* the principle that the position and velocity of a subatomic particle cannot both be ascertained at the same time. **uncertificated**, *a.* **uncertified**, *a.* **unchain**, *v.t.* **unchallengeable**, *a.* **unchallenged**, *a.* **unchangeable**, *a.* **unchangeableness**, *n.* **unchangeably**, *adv.* **unchanged**, *a.* **unchanging**, *a.* **unchangingly**, *adv.* **uncharacteristic**, *a.* **uncharitable**, *a.* not harmonizing with Christian feeling; harsh, censorious. **uncharitableness**, *n.* **uncharitably**, *adv.* **uncharted**, *a.* not marked on a chart; unmapped. **unchartered**, *a.* **unchaste**, *a.* **unchastely**, *adv.* **unchastened**, *a.* **unchecked**, *a.* not checked or repressed; unrestrained, uncontrolled; not examined. **unchivalrous**, *a.* **unchivalrously**, *adv.* **unchosen**, *a.* **unchristian**, *a.* not Christian, heathen; not according to or befitting the spirit of Christianity; (*coll.*) outrageous. **unchristianly**, *a.* **unchurch**, *v.t.* to expel from a Church; to excommunicate; to deprive of the character or standing of a Church.

uncial (ŭn'shəl), *a.* denoting a kind of majuscule writing somewhat resembling modern capitals used in manuscripts of the 4-8th cents. *n.* an uncial letter or manuscript.

uncinate (ŭn'sinət), *a.* (*Bot.*) hooked at the end; (*Anat.* etc.) having a hooked appendage. **uncinal**, *a.* **unciform** (-fawm), *a.*

un- (*cont.*) **uncircumcised**, *a.* not circumcised; not Jewish; heathen, unholy, profane. **uncivil**, *a.* not civil, discourteous, ill-mannered; (*poet.*) rude, boisterous. **uncivilly**, *adv.* **uncivilized**, **-ised** (-siv'ilizd), *a.* **unclad**, *a.* **unclaimed** (-klāmd'), *a.* **unclasp**, *v.t.* to unfasten the clasp of; to release from a grip. **unclassifiable**, *a.* **unclassified**, *a.* not divided into categories; of information, not restricted.

uncle (-ŭng'kl), *n.* the brother of one's father or mother; the husband of one's aunt; (*N Am.*) an elderly man (a friendly mode of address); (*sl.*) a pawnbroker. **Uncle Sam**, *n.* the government or a typical representative of the US. **Uncle Tom**, *n.* (*derog or offensive*) an American Negro considered to be servile in his manner towards white people. **uncleship**, *n.*

un- (*cont.*) **unclean**, *a.* not clean; foul, dirty; lewd, unchaste; (*Jewish Law*) not ceremonially clean. **uncleanness**, *n.* **uncleanly** (-klen'-), *a.* **uncleanliness**, *n.* **unclear**, *a.* **unclench**, *v.t.*, *v.i.* **unclerical**, **unclinch**, *v.t.* **uncloak**, *v.t.*, *v.i.* **unclog**, *v.t.* **unclose**, *v.t.*, *v.i.* to open. **unclothe**, *v.t.* **unclouded**, *a.* not obscured by clouds; clear, bright. **unclubbable**, *a.* **uncluttered**, *a.* **uncock**, *v.t.* to let down the hammer of (a gun, etc.) without exploding the charge. **uncoil**, *v.t.*, *v.i.* to unwind. **uncoined**, *a.* genuine. **uncoloured**, *a.* not coloured; told with simplicity or without exaggeration, unvarnished. **uncombed**, *a.* **uncomeatable**, *a.* (*coll.*) that cannot be come at; not attainable, not obtainable. **uncomely**, *a.* **uncomfortable**, *a.* feeling or causing discomfort. **uncomfortably**, *adv.* **uncommercial**, *a.* **uncommitted**, *a.* not pledged to support any particular policy, party etc. **uncommon**, *a.* not common, unusual, remarkable, extraordinary. **uncommonly**, *adv.* remarkably, to an uncommon degree. **uncommonness**, *n.* **uncommunicative**, *a.* reserved, taciturn. **uncommunicatively**, *adv.* **uncommunicativeness**, *n.* **uncompensated**, *a.* **uncompetitive**, *a.* **uncomplaining**, *a.* **uncomplainingly**, *adv.* **uncomplaisant**, *a.* **uncomplaisantly**, *adv.* **uncompleted**, *a.* **uncomplicated**, *a.* **uncomplimentary**, *a.* **uncompounded**, *a.* **uncomprehending**, *a.* **uncomprehensive**, *a.* **uncompromising**, *a.* not compromising or admitting of compromise; determined, rigid, inflexible, strict. **uncompromisingly**, *adv.* **unconcealed**, *a.* **unconcern**, *n.* absence of concern or anxiety; indifference, apathy. **unconcerned**, *a.* not concerned (in or with); free from anxiety. **unconcernedly**, *adv.* **unconcluded**, *a.* **uncondemned**, *a.* **uncondensable**, *a.* **uncondensed**, *a.* **unconditional**, *a.* not conditional; absolute. **unconditionality**, **unconditionalness**, *n.* **unconditionally**, *adv.* **unconditioned**, *a.* not learned or conditioned, innate. **unconfinable**, *a.* that cannot be confined; unbounded. **unconfined**, *a.* **unconfinedly**, *adv.* **unconfirmed**, *a.* **unconformable**, *a.* **unconformability** (-bil'iti), **unconformableness**, *n.* **unconformably**, *adv.* **unconformity**, *n.* **uncongenial**, *a.* **uncongenially**, *adv.* **unconnected**, *a.* **unconquerable**, *a.* **unconquerably**, *adv.* **unconquered**, *a.* **unconscientious**, *a.* **unconscientiously**, *adv.* **unconscientiousness**, *n.* **unconscionable**, *a.* not reasonable, inordinate; not influenced or restrained by conscience; (*Law*) grossly unfair, inequitable. **unconscionableness**, *n.* **unconscionably**, *adv.* **unconscious**, *a.* not conscious, ignorant, unaware (of); temporarily deprived of consciousness; not perceived by the mind. *n.* (*Psych.*) a term which includes all processes which cannot be made conscious by an effort of the will. **unconsciously**, *adv.* **unconsciousness**, *n.* **unconsecrated**, *a.* **unconsenting**, *a.* **unconsidered**, *a.* not taken into consideration. **unconstitutional**, *a.* not authorized by or contrary to the principles of the constitution. **unconstitutionality**, *n.* **unconstitutionally**, *adv.* **unconstrained**, *a.* **unconstrainedly** (-nid), *adv.* **unconstricted**, *a.* **unconsumed**, *a.* **unconsummated**, *a.* **uncontainable**, *a.* **uncontaminated**, *a.* **uncontemplated**, *a.* not contemplated or expected. **uncontested**, *a.* **uncontradicted**, *a.* **uncontrollable**, *a.* unmanageable. **uncontrollableness**, *n.* **uncontrollably**, *adv.* **uncontrolled**, *a.* **uncontrolledly** (-lid), *adv.* **uncontroversial**, *a.* **uncontroversially**, *adv.* **uncontroverted**, *a.* **unconventional**, *a.* not fettered by convention or usage; informal, free and easy, bohemian. **unconventionality**, *n.* **unconventionally**, *adv.* **unconversant**, *a.* **unconverted**, *a.* **unconvertible**, *a.* **unconvicted**, *a.* **unconvinced**, *a.* **unconvincing**, *a.* **uncooked**, *a.* **uncool**, *a.* (*sl.*) not cool; unfashionable, unsophisticated. **uncooperative**, *a.* **uncoordinated**, *a.* **uncork**, *v.t.* to take the cork out of; to give vent to (one's feelings etc.). **uncorrected**, *a.* **uncorroborated**, *a.* **uncorroded**, *a.* **uncorrupted**, *a.* **uncorruptible**, *a.* **uncountable**, *a.* **uncounted**, *a.* not counted; innumerable. **uncountenanced**, *a.* **uncouple**, *v.t.* to disconnect; to let loose, to release. **uncourtly**, *a.* **uncourtliness**, *n.*

uncouth (ŭnkooth'), *a.* awkward, clumsy; outlandish, odd, ungainly. **uncouthly**, *adv.* **uncouthness**, *n.*

un- (*cont.*) **uncovenanted**, *a.* not bound by a covenant; not promised or secured by a covenant. **uncover**, *v.t.* to remove a covering from; to divest of covering; to make known, to disclose. *v.i.* to take off the hat in salutation. **uncovered**, *a.* **uncoveted**, *a.* **uncowl**, *v.t.* **uncreated**, *a.* not yet created; existing independently of creation. **uncreative**, *a.* **uncritical**, *a.* not critical, not inclined to criticize; not according to the rules of criticism. **uncritically**, *adv.* **uncross**, *v.t.* to change from a crossed position. **uncrossed**, *a.* not crossed; as a cheque; not opposed. **uncrowded**, *a.* **uncrown**, *v.t.* to discrown, to depose, to dethrone. **uncrowned**, *a.* discrowned; not yet crowned; having the power without the title of king. **uncrushable**, *a.* **uncrystallized**, **-ised**, *a.*

UNCTAD (ŭngk'tad), (*acronym*) United Nations Commission for Trade and Development.

unction (ŭngk'shən), *n.* the act of anointing with oil or an unguent, as a symbol of consecration or for medical purposes; that which is used in anointing, an unguent or ointment; anything soothing or ingratiating; a quality in speech conveying deep religious or other fervour; effusive or affected emotion, gush; relish, gusto; (*Theol.*) grace. **unctuous**, *a.* greasy, oily, soapy to the touch; full of

unction; oily, effusive, hypocritically or affectedly fervid. **extreme unction**, EXTREME. **unctuously**, *adv.* **unctuousness**, *n.*
uncus (ŭng'kəs), *n.* (*pl.* **unci** (ŭn'sī)) (*Nat. Hist.*) a hook, claw, or hook-like part or appendage.
un- (*cont.*) **uncushioned**, *a.* not cushioned or padded. **uncut**, *a.* not cut; of leaves of a book, having the margins untrimmed; not shortened or abridged. **undam**, *v.t.* **undamaged**, *a.* **undated**, *a.* **undaunted**, *a.* not daunted; fearless. **undauntedly**, *adv.* **undauntedness**, *n.* **undead**, *a.* **undebated**, *a.* **undebauched**, *a.*
undecagon (ŭndĕk'əgon), *n.* a plane figure having 11 angles and 11 sides.
un- (*cont.*) **undeceive**, *v.t.* to free from deception or error; to open the eyes of. **undeceived**, *a.* **undecided**, *a.* not decided or settled; irresolute, wavering. **undecidedly**, *adv.* **undecipherable**, *a.* **undeclared**, *a.* **undecorated**, *a.* **undefeated**, *a.* **undefended**, *a.* **undefiled**, *a.* not defiled; pure. **undefinable**, *a.* **undefined**, *a.* not defined; indefinite, vague. **undelegated**, *a.* **undelivered**, *a.* **undemanded**, *a.* **undemanding**, *a.* **undemocratic**, *a.* **undemonstrated**, *a.* **undemonstrative**, *a.* not demonstrative; not exhibiting strong feelings; reserved. **undeniable**, *a.* not capable of being denied; indisputable; decidedly good, excellent. **undeniably**, *adv.* **undenounced**, *a.* **undependable**, *a.* not to be depended on. **undepressed**, *a.* **undeprived**, *a.*
under (ŭn'də), *prep.* in or to a place or position lower than, below; at the foot or bottom of; covered by, on the inside of, beneath the surface of; beneath the appearance or disguise of; inferior to or less than in quality, rank, degree, number, amount etc.; subject to, subordinate or subservient to; governed, controlled, or directed by; liable to, on condition or pain of, in accordance with; by virtue of; in the time of; attested by; planted or sown with; because of; in the process of; in a group consisting of. *adv.* in or into a lower or subordinate place, condition, or degree; into a covered or sheltered position; into a state of unconsciousness. *a.* lower, inferior, subordinate; inadequate, falling short. **under age**, not of full age. **under arms**, ARM ². **under a cloud**, out of favour. **under fire**, FIRE ¹. **under sail**, SAIL. **under sentence**, having received sentence or judgment. **under the breath**, in a low voice; very softly. **under way**, WAY. **underling** (-ling), *n.* an inferior agent or assistant. **undermost**, *a.* lowest in place, position, rank etc.
under-, (ŭndə-), *pref.* under, below (the substantive to which it is prefixed); underneath, beneath, lower than, in position, rank etc., subordinate; insufficiently, incompletely, immaturely. Only a selection of compounds with this prefix is given; others can be explained by reference to the simple adjective, noun, or verb. **underachieve**, *v.i.* to fail to achieve as much (esp. academically) as expected. **underachiever**, *n.* **underact**, *v.t., v.i.* to act or to play inadequately, or in a restrained way. **underactive**, *a.* **underarm** (ŭn'-), *a., adv.* (made or done) with the arm below shoulder level. **underbelly** (ŭn'-), *n.* the underside of an animal, nearest to the ground and consequently less protected; any soft or vulnerable part or aspect (of an organization etc.). **underbid**, *v.t.* to bid less than (as at auction). **underblanket**, *n.* **underbody**, *n.* the underpart of the body of an animal, car etc. **underbred**, *a.* not thoroughbred; ill-bred. **underbrush** (ŭn'-), *n.* undergrowth, underwood. **underbuy**, *v.t.* to buy at a lower price than that paid by others; to buy for less than the proper value. **undercapitalized, -ised**, *a.* of a business, having less capital than that needed to operate efficiently. **undercarriage** (ŭn'-), *n.* the main alighting gear of an aircraft. **undercharge**, *v.t.* to charge less than the fair price for; to put an insufficient charge in (a gun, etc.). **underclass** (ŭn'-), *n.* a social class falling outside the standard classification, very deprived in economic, educational etc., terms. **under-clerk**, *n.* **under-clerkship**, *n.* **underclothes** (ŭn'-), *n.pl.* clothes worn under others, esp. next to the skin. **underclothing**, *n.* **undercoat** (ŭn'-), *n.* a layer of fine fur underneath an animal's main coat; a coat of paint serving as a base for the main coat. **undercoat**, *v.t.* **undercover**, *a.* done in secret. **undercook**, *v.t.* **undercroft** (un'-), *n.* a vault, esp. under a church or large building, a crypt. **undercurrent**, *n.* a current running below the surface; a secret or unapparent tendency or influence. **undercut**, *v.t.* to cut under (coal etc.) so as to remove it easily; to cut away the material beneath (a carved design) to give greater relief; to make a price lower than that of a competitor; (*Golf*) to hit (a ball) so as to make it rise high. *n.* (ŭn'-), the act or effect of undercutting; a blow upwards; the underside of a sirloin, the tenderloin. **underdeveloped**, *a.* not sufficiently or adequately developed; (of a country, economically backward. **underdo**, *v.t.* to do inadequately; to cook insufficiently. **underdog** (un'-), *n.* an oppressed person, one in an inferior position. **underdone**, *a.* insufficiently cooked. **underdress**, *v.t., v.i.* to dress insufficiently or too plainly. **underemployed**, *a.* **underemployment**, *n.* **underestimate**, *v.t.* to estimate at too low a rate. *n.* an inadequate estimate. **underestimation**, *n.* **underexpose**, *v.t.* (*Phot.*). **underexposure**, *n.* **underfeed**, *v.t., v.i.* **underfelt** (ŭn'-), *n.* a felt underlay. **underfloor**, *a.* of a method of central heating using hot air piped under the floor. **underflow** (ŭn'-), *n.* an undercurrent. **underfoot**, *adv.* under the feet; beneath; in the way. **underfund**, *v.t.* **underfunding**, *n.* **undergarment** (ŭn'-), *n.* one worn under others. **undergo**, *v.t.* (*past* **-went**, *p.p.* **-gone**) to experience, to pass through; to suffer; to bear up against, to endure with firmness. **undergraduate**, *n.* a member of a university who has not yet taken a degree. **undergraduateship**, *n.* **underground** (-ŭn'-), *a.* situated below the surface of the earth; obscure, secret, unperceived by those in authority; ignoring, or subversive of, established trends, avant-garde. *n.* that which is underground; an underground railway; a secret or subversive group or organization. *adv.* below the surface of the earth. **undergrown**, *a.* **undergrowth** (ŭn'-), *n.* small trees or shrubs, growing under larger ones. **underhand**, *adv.* secretly, not openly, clandestinely; slyly, unfairly, by fraud; underarm *a.* (ŭn'-). clandestine, secret; sly, unfair, fraudulent; of bowling, with the hand underneath both the elbow and the ball, underarm. **underhanded**, *a.* underhand. **underhandedly**, *adv.* **underhandedness**, *n.* **underhold** (ŭn'-), *n.* (*Wrestling*) a hold round the body with the arms underneath one's opponent's. **underhung** (-hŭng', *attributively* ŭn'-), *a.* of the lower jaw, projecting beyond the upper jaw; having the lower jaw projecting before the upper. **underinsured**, *a.* **underlap**, *v.t.* to be folded or extend under the edge of (cp. OVERLAP). **underlay** ¹, *v.t.* (*past, p.p.* **-laid**,) to lay something under; to support by something laid under. *n.* (-ŭn'-), a piece of paper etc., placed beneath type etc., to bring it to the proper level for printing; a thick felt or rubber sheet laid under a carpet. **underlay** ², *past* UNDERLIE. **underlet**, *v.t.* to let below the proper value; to sublet. **underletter**, *n.* **underletting**, *n.* **underlie**, *v.t.* (*past* **-lay**, *p.p.* **-lain**,) to lie under or beneath; to be the basis or foundation of. **underline**, *v.t.* to mark with a line underneath, esp. for emphasis; to stress, emphasize forcefully. *n.* (-ŭn'-), an announcement of a subsequent theatrical performance at the foot of a playbill. **underlinen** (ŭn'-), *n.* linen underclothing. **underman**, *v.t.* to furnish (as a ship) with less than the proper complement of men. **undermanned**, *a.* **undermentioned**, *a.* mentioned below or later. **undermine**, *v.t.* to dig a mine or excavation under; to render unstable by digging away the foundation of; to injure by clandestine or underhand means; to wear away (one's strength etc.) by imperceptible degrees. **underminer**, *n.* **underneath**, *adv., prep.* beneath, below. *n.* an underside. **undernote** (ŭn'-), *n.* a subdued note, an undertone. **undernourish**, *v.t.* under-

underpaid   898   **undiscerning**

nourishment, *n.* **underpaid,** *a.* **underpants** (ŭn'-), *n.pl.* a man's undergarment covering the body from the waist to the thighs. **underpart** (ŭn'-), *n.* a lower part; a part lying underneath. **underpass** (ŭn'-), *n.* a road passing under a railway or another road. **underpay,** *v.t.* (*past, p.p.* **-paid**) to pay inadequately. **underperform,** *v.i.* to perform less well than expected. **underpin,** *v.t.* to support (a wall etc.) by propping up with timber, masonry, etc.; to strengthen the foundations of (a building). **underpinning,** *n.* **underplay,** *v.t.* to play (a part) inadequately; to play with restraint or underemphasis for effect. *v.i.* to play a low card whilst one holds a higher one of the same suit; to underact. *n.* (ŭn'-), the act of underplaying. **underplot** (ŭn'-), *n.* a subordinate plot in a play, novel, etc. **underpopulated,** *a.* **underpraise,** *v.t.* to praise less than is deserved. **underprice,** *v.t.* **underprivileged,** *a.* lacking the economic and social privileges enjoyed by most members of society. **underprize,** *v.t.* to value below one's merits. **underproduce,** *v.t., v.i.* **underproduction,** *n.* lower or less production than the normal or the demand. **underproof** (ŭn'-), *a.* containing less alcohol than proof spirit. **underquote,** *v.t.* to offer (goods etc.) at lower prices than (others). **underrate,** *v.t.* to rate or estimate too low. **under-reckon,** *v.t.* **underripe,** *a.* **underrun,** *v.t.* to run beneath, to pass under. *v.i.* to fall short of the desired amount. **underscore,** *v.t.* to underline. **undersea,** *a.* **underseal,** *v.t.* to coat the exposed underparts of (a vehicle) with a corrosion-resistant substance. *n.* (ŭn'-), such a substance. **undersecretary,** *n.* a secretary immediately under the principal secretary, a senior civil servant. **undersecretaryship,** *n.* **undersell,** *v.t.* to sell cheaper than. **underseller,** *n.* **under-servant,** *n.* **undersexed,** *a.* having less than the normal sexual drive. **undershirt,** *n.* (*N Am.*) a vest or singlet. **undershorts** (ŭn'-), *n. pl.* **undershoot** (ŭn'-), *n.* (*Aviat.*) falling short of the mark in landing. **undershot,** *a.* of a water-wheel, driven by water passing under it; of the lower jaw, underhung. **undershrub** (ŭn'-), *n.* a plant of shrubby habit, but smaller than a shrub. **underside** (ŭn'-), *n.* a lower side or surface. **undersign,** *v.t.* to sign under or at the foot of. **the undersigned,** the person or persons signing a document etc. **undersized,** *a.* below the normal or average size. **underskirt** (ŭn'-), *n.* a skirt worn under another. **underslung,** *a.* (*Motor.*) descriptive of a chassis with the frame below the axles. **undersoil** (ŭn'-), *n.* subsoil. **underspend,** *v.i.* to spend less than expected or allowed for. *v.t.* to spend less than (an allowed budget). **understaffed,** *a.*
understand (-ŭndəstand'), *v.t.* (*past, p.p.* **-stood** (-stud)) to take in, know, or perceive the meaning of; to comprehend fully, to have complete apprehension of; to perceive the force or significance of; to suppose to mean, to take as meant or implied, to gather, assume, or infer from information received; to supply (a word, explanation etc.) mentally; to be informed or told, to hear (that); to feel sympathetic towards, to identify with. *v.i.* to have or exercise the power of comprehension; to have a sympathetic attitude towards someone or something. **understandable,** *a.* **understandably,** *adv.* **understanding,** *a.* intelligent; sensible; sympathetic, tolerant. *n.* the act of one who understands; comprehension; the power or faculty of apprehension; the faculty of thinking or of apprehending relations and drawing inferences; discernment; clear insight and intelligence in practical matters; union of minds or sentiments, accord; an informal agreement or compact. **on the understanding that,** provided that. **understandingly,** *adv.*
under- (*cont.*) **understate,** *v.t.* to represent as less, inferior, etc., than the truth. **understated,** *a.* restrained, simple. **understatement,** *n.* **understeer** (ŭn'-), *n.* a tendency in a motor vehicle to turn less sharply than expected. **understock,** *v.t.* to furnish (a shop etc.) with insufficient stock. **understood,** (-stud), *past, p.p.* UNDER-STAND. **understudy,** *v.t.* to study (a part) in order to play it if the usual actor is unable; to study the acting of (an actor or actress) thus. *n.* (ŭn'-), one who studies a part or actor thus. **undersubscribe,** *v.t.*
undertake (-ŭndətāk'-), *v.t.* (*past* **-took,** (-tuk'), *p.p.* **-taken,** (-kən)) to take upon oneself, to assume, to engage in, to enter upon (a task, enterprise, responsibility etc.); to engage oneself, to promise (to do); to guarantee, to affirm, to answer for it (that). **undertaker** (ŭn'-), *n.* one who undertakes something; one who manages funerals. **undertaking** (ŭn'-), *n.* the act of one who undertakes any business; that which is undertaken, a task, an enterprise, an agreement, a promise, a stipulation.
under- (*cont.*) **undertenant** (-ŭn'-), *n.* a tenant under another tenant. **undertenancy,** *n.* **underthrust** (ŭn'-), *n.* (*Geol.*) a fault in which rocks on a lower plane have moved underneath a more stable upper layer. **undertint** (ŭn'-), *n.* a subdued tint. **undertone** (ŭn'-), *n.* a low or subdued tone, esp. in speaking; an unstated meaning or emotional tone; a subdued colour, an undertint. **undertook** (-ŭndətuk'), *past* UNDERTAKE. **undertow** (-ŭn'-), *n.* a backward current opposite to that on the surface, an underset, esp. the backward flow under waves breaking on a shore. **undertrump,** *v.t.* to play a lower trump than (another person or another trump played). **underuse,** *n., v.t.* **underutilize, -ise,** *v.t.* **undervalue,** *v.t.* to value too low; to despise. **undervaluation,** *n.* **undervaluer,** *n.* **undervest** (ŭn'-), *n.* **underwater,** *a., adv.* **underwear** (ŭn'-), *n.* clothes worn underneath others, underclothing; the wearing of these. **underweight,** *n., a.* (being) less than the average or expected weight. **underwent,** *past* UNDERGO. **underwhelm,** *v.t.* (*coll.*) to fail to impress, to disappoint (formed from *overwhelm*). **underwing,** (ŭn'-), *n.* a nocturnal moth with conspicuous markings on the hind or under wings. **underwood** (ŭn'-), *n.* undergrowth. **underwork,** *v.t.* to work for a lower price than, to undercut. *v.i.* to work inadequately. *n.* (ŭn'-), subordinate or inferior work. **underworld,** (ŭn'-), *n.* the nether world, the infernal regions; the antipodes; the earth as the sublunary sphere; the criminal class of society. **underwrite,** *v.t.* (*past* **-wrote,** *p.p.* **-written**) to execute and deliver a policy of insurance); to engage to buy all the stock in (a new company etc.) not subscribed for by the public; to agree to purchase (shares) at a fixed price; to write beneath, to subscribe. *v.i.* to act as an underwriter, to practise insurance. **underwriter,** (ŭn'-), *n.* **underwriting,** (ŭn'-), *n.* **underwrought,** *a.* insufficiently wrought.
un- (*cont.*) **undescended,** *a.* not having descended; of a testis, not having descended into the scrotum. **undeserved,** *a.* **undeservedly** (-vid-), *adv.* **undeserving,** *a.* **undeservingly,** *adv.* **undesignated,** *a.* **undesigned,** *a.* not designed, unintentional. **undesignedly** (-nid-), *adv.* **undesignedness,** *n.* **undesigning,** *a.* **undesirable,** *a.* not desirable; unpleasant, inconvenient. *n.* an undesirable person. **undesirability,** *n.* **undesirableness,** *n.* **undesirably,** *adv.* **undesired,** *a.* not desired; not asked for. **undesirous,** *a.* **undetachable,** *a.* **undetectable,** *a.* **undetected,** *a.* **undetermined,** *a.* not determined, not decided, not fixed; irresolute; indeterminate. **undeterred,** *a.* **undeveloped,** *a.* **undeviating,** *a.* **undeviatingly,** *adv.* **undiagnosed,** *a.* **undid,** *past* UNDO.
undies (ŭn'diz), *n.pl.* (*coll.*) women's underwear.
un- (*cont.*) **undifferentiated,** *a.* **undiffused,** *a.* **undigested,** *a.* **undignified** (-dig'nifid), *a.* not dignified; not consistent with one's dignity. **undiluted,** *a.* **undiminished,** *a.* **undimmed,** *a.*
undine (ŭn'dēn), *n.* a female water sprite without a soul, but capable of obtaining one by marrying a mortal and bearing a child; (*Med.*) a form of eye-irrigator. **undinal** (-dē'-), *a.*
un- (*cont.*) **undiplomatic,** *a.* not diplomatic. **undirected,** *a.* **undiscerned,** *a.* **undiscernible,** *a.* **undiscerning,** *a.*

undiscerningly   899   **unfortified**

undiscerningly, *adv.* **undischarged,** *a.* **undisciplined,** *a.* **undisclosed,** *a.* **undisconcerted,** *a.* **undiscouraged,** *a.* **undiscoverable,** *a.* **undiscovered,** *a.* **undiscriminating,** *a.* **undiscriminatingly,** *adv.* **undiscussed,** *a.* **undisguised,** *a.* not disguised; open, frank, plain. **undisguisedly** (-zid-), *adv.* **undisheartened,** *a.* **undismayed,** *a.* **undispelled,** *a.* **undisplayed,** *a.* **undisputed,** *a.* **undissected,** *a.* **undissolved,** *a.* **undistinguishable,** *a.* **undistinguished,** *a.* **undistorted,** *a.* **undistracted,** *a.* **undistressed,** *a.* **undistributed,** *a.* (*chiefly Log.*) not distributed. **undisturbed,** *a.* **undisturbedly** (-bid-), *adv.* **undiversified,** *a.* **undiverted,** *a.* **undivided,** *a.* **undividedly,** *adv.* **undivorced,** *a.* **undivulged,** *a.* **undo,** *v.t.* (*past* **-did,** *p.p.* **-done**) to reverse (something that has been done), to annul; to unfasten, to untie; to unfasten the buttons, garments etc., of (a person); to bring ruin on, to destroy, to corrupt. **undoer,** *n.* **undoing,** *n.* **undocumented,** *a.* **undomesticated,** *a.* **undone,** *a.* not done; unfastened; ruined, destroyed. **undoubted,** *a.* not called in question, not doubted or disputed, genuine. **undoubtedly,** *adv.* without doubt. **undrained,** *a.* **undramatic,** *a.* **undrape,** *v.t.* to remove drapery from, to uncover. **undreamed, undreamt,** *a.* **undreamed-of,** *a.* not thought of. **undress,** *v.t.* to divest of clothes, to strip; to take the dressing, bandages, etc., from (a wound etc.). *v.i.* to undress oneself. *n.* the state of being partly or completely undressed; ordinary dress, opp. to full dress or uniform; negligent attire. *a.* pertaining to everyday dress; commonplace. **undressed,** *a.* (partially) naked; improperly clothed; not treated or processed; without a dressing. **undrinkable,** *a.* **undue,** *a.* excessive, disproportionate; not yet due; improper; illegal. **undulate** (ūn′dūlət), *v.i.* wavy, bending in and out or up and down. *v.i.* (-lāt), to have a wavy motion; of water, to have a wavy margin, form or markings; to rise and fall. *v.t.* to cause to move in waves. **undulant,** *a.* undulating. **undulant fever,** *n.* brucellosis in human beings, so called because the fever is intermittent. **undulately,** *adv.* **undulatingly,** *adv.* **undulation,** *n.* the act of undulating; a wavy or sinuous form or motion, a gentle rise and fall, a wavelet. **undulatory,** *a.* having an undulating character; rising and falling like waves; pertaining or due to undulation. **un-** (cont.) **unduly,** *adv.* excessively. **undutiful,** *a.* **undutifully,** *adv.* **undutifulness,** *n.* **undying,** *a.* unceasing, immortal. **undyingly,** *adv.* **unearned,** *a.* not earned. **unearned income,** *n.* income from rents, investments etc. **unearned increment,** *n.* increase in the value of land due to increased population etc., not to labour or expenditure on the part of the owner. **unearth,** *v.t.* to pull or bring out of the earth; to cause (a fox etc.) to leave his earth; to dig up; to bring to light, to find out. **unearthly,** *a.* not earthly; not of this world, supernatural; weird, ghostly. **unearthliness,** *n.* uneasy, *a.* restless, troubled, anxious, uncomfortable, ill at ease; difficult; awkward, stiff, constrained. **unease,** *n.* **uneasily,** *adv.* **uneasiness,** *n.* **uneatable,** *a.* **uneaten,** *a.* **unecclesiastical,** *a.* **uneclipsed,** *a.* **uneconomic,** *a.* not economic, not financially viable. **uneconomical,** *a.* not economical, profligate. **unedifying,** *a.* **unedited,** *a.* **uneducated,** *a.* **unelaborated,** *a.* **unelated,** *a.* **unelected,** *a.* **unelucidated,** *a.* **unemancipated,** *a.* **unembarrassed,** *a.* **unemotional,** *a.* **unemotionally,** *adv.* **unemphatic,** *a.* **unemphatically,** *adv.* **unemployed,** *a.* not in use; having no paid work. *n.* a person out of work; workless people generally. **unemployable,** *a.*, *n.* **unemployment,** *n.* **unemployment benefit,** *n.* a regular payment by the State to an unemployed worker. **unempowered,** *a.* **unenclosed,** *a.* **unencumbered,** *a.* **unending,** *a.* having no end, endless. **unendingly,** *adv.* **unendowed,** *a.* **unendurable,** *a.* **unendurably,** *adv.* **unenforced,** *a.* **un-English,** *a.* not English; not characteristic or worthy of English people. **unenjoyable,** *a.* **unenlightened,** *a.* **unenterprising,** *a.* **unenterprisingly,** *adv.* **unenterprisingness,** *n.* **unentertaining,** *a.* **unentertainingly,** *adv.* **unenthusiastic,** *a.*

unenviable, *a.* **unenviably,** *adv.* **unenvied,** *a.* **unequal,** *a.* not equal; not equal to; uneven; of a contest, etc., not evenly balanced. **unequalize, -ise,** *v.t.* **unequalled,** *a.* **unequally,** *adv.* **unequipped,** *a.* **unequivocal,** *a.* not equivocal, not ambiguous; plain, manifest. **unequivocally,** *adv.* **unequivocalness,** *n.* **unerased,** *a.* **unerring,** *a.* committing no mistake; not missing the mark, certain, sure. **unerringly,** *adv.* **unescapable,** *a.* **UNESCO, Unesco** (ūnes′kō), (*acronym*) United Nations Educational, Scientific and Cultural Organization. **un-** (cont.) **unescorted,** *a.* **unessayed,** *a.* **unessential,** *a.* not essential, not absolutely necessary; not of prime importance. *n.* some thing or part not absolutely necessary or indispensable. **unestablished,** *a.* **unestimated,** *a.* **unethical,** *a.* **unevaporated,** *a.* **uneven,** *a.* not even, level, or smooth; not uniform, regular, or equable. **unevenly,** *adv.* **unevenness,** *n.* **uneventful,** *a.* **unexaggerated,** *a.* **unexamined,** *a.* **unexampled,** *a.* not exampled; having no parallel; unprecedented. **unexcelled,** *a.* **unexceptionable,** *a.* not exceptionable; to which no exception can be taken; unobjectionable, faultless. **unexceptionableness,** *n.* **unexceptionably,** *adv.* **unexceptional,** *a.* not exceptional, ordinary. **unexciting,** *a.* **unexclusive,** *a.* **unexecuted,** *a.* **unexpected,** *a.* **unexpectedly,** *adv.* **unexpectedness,** *n.* **unexpiated,** *a.* **unexpired,** *a.* **unexplainable,** *a.* **unexplained,** *a.* **unexploded,** *a.* **unexploited,** *a.* **unexplored,** *a.* **unexposed** (-spōzd′), *a.* **unexpounded,** *a.* **unexpressed,** *a.* **unexpressive,** *a.* **unexpurgated,** *a.* **unextended,** *a.* **unextinguishable,** *a.* **unfadable,** *a.* **unfaded,** *a.* **unfading,** *a.* **unfadingly,** *adv.* **unfailing,** *a.* not liable to fail or run short; unerring, infallible; reliable, certain. **unfailingly,** *adv.* **unfailingness,** *n.* **unfair,** *a.* not fair; not equitable, not impartial; dishonourable, fraudulent. **unfairly,** *adv.* **unfairness,** *n.* **unfaithful,** *a.* not faithful; adulterous. **unfaithfully,** *adv.* **unfaithfulness,** *n.* **unfaltering,** *a.* **unfalteringly,** *adv.* **unfamiliar,** *a.* strange, unknown; not familiar or acquainted with). **unfamiliarity,** *n.* **unfamiliarly,** *adv.* **unfashionable,** *a.* **unfashionableness,** *n.* **unfashionably,** *adv.* **unfashioned,** *a.* not fashioned by art; shapeless. **unfasten,** *v.t.* **unfathered,** *a.* not acknowledged by its father or author; (*poet.*) fatherless. **unfatherly,** *a.* **unfathomable,** *a.* **unfathomableness,** *n.* **unfathomably,** *adv.* **unfathomed,** *a.* **unfatigued,** *a.* **unfavourable,** *a.* **unfavourableness,** *n.* **unfavourably,** *adv.* **unfeasible,** *a.* **unfeathered,** *a.* not feathered; unfledged. **unfed,** *a.* **unfeed,** *a.* not retained by a fee. **unfeeling,** *a.* insensible; hard-hearted, cruel. **unfeelingly,** *adv.* **unfeelingness,** *n.* **unfeigned,** *a.* **unfeignedly** (-nid-), *adv.* **unfelt,** *a.* not felt, not perceived. **unfeminine,** *a.* **unfenced,** *a.* not enclosed by a fence; not fortified. **unfermented,** *a.* **unfertilized, -ised,** *a.* **unfetter,** *v.t.* to free from fetters or restraint. **unfettered,** *a.* **unfilial,** (-fil′iəl), *a.* **unfilially,** *adv.* **unfilled,** *a.* **unfiltered,** *a.* **unfinished,** *a.* not finished, incomplete; not having been through a finishing process. **unfit,** *a.* not fit (to do, to be, for etc.); improper, unsuitable. *v.t.* to make unfit or unsuitable; disqualify. **unfitly,** *adv.* **unfitness,** *n.* **unfitted,** *a.* not fitted; unfit; not fitted up, not furnished with fittings. **unfitting,** *a.* **unfittingly,** *adv.* **unfix,** *v.t.* **unflagging,** *a.* **unflappable,** *a.* (*coll.*) not readily upset or agitated, imperturbable. **unflattering,** *a.* **unflatteringly,** *adv.* **unflavoured,** *a.* **unfledged,** *a.* not yet fledged; underdeveloped, immature. **unflinching,** *a.* **unflinchingly,** *adv.* **unflustered,** *a.* **unfocus(s)ed,** *a.* **unfold,** *v.t.* to open the folds of; to spread out; to discover, to reveal; to display. *v.i.* to spread open, to expand, to develop. **unforced,** *a.* not forced, not constrained; natural, easy. **unfordable,** *a.* **unforeseeable,** *a.* **unforeseen,** *a.* **unforgettable,** *a.* **unforgivable,** *a.* **unforgiven,** *a.* **unforgiving,** *a.* **unforgivingly,** *adv.* **unforgivingness,** *n.* **unforgotten,** *a.* **unformed,** *a.* devoid of form, shapeless, amorphous, structureless; not yet fully developed, immature. **unformulated,** *a.* **unforthcoming,** *a.* **unfortified,** *a.* **unfortu-**

nate, *a.* not fortunate, unlucky, unhappy. *n.* one who is unfortunate. **unfortunately**, *adv.* **unfounded**, *a.* having no foundation of fact or reason, groundless; not yet established. **unfranked**, *a.* **unfreeze**, *v.t., v.i.* (*past, -froze, p.p. -frozen*). **unfrequent**, *a.* **unfrequented**, *a.* **unfriended**, *a.* without a friend or friends. **unfriendly**, *a.* not friendly; not user-friendly. **unfriendliness**, *n.* **unfrock**, *v.t.* to take the frock or gown from; hence, to deprive of the character and privileges of a priest. **unfruitful**, *a.* **unfruitfully**, *adv.* **unfruitfulness**, *n.* **unfulfilled**, *a.* **unfunded**, *a.* not funded; of a debt etc., floating. **unfunny**, *a.* **unfurl**, *v.t., v.i.* to open or spread out (a sail, banner etc.). **unfurnished**, *a.* not furnished (with); without furniture. **unfused**, *a.* not fused, not melted.
ungainly (ŭngān'li), *a.* clumsy, awkward. **ungainliness**, *n.*
un- (cont.) **ungallant**, *a.* not gallant, not courteous to women. **ungarnished**, *a.* **ungenerous**, *a.* **ungenerously**, *adv.* **ungenial**, *a.* **ungenteel**, *a.* **ungenteelly**, *adv.* **ungentle**, *a.* not gentle, harsh, rude, unkind; ill-bred. **ungentlemanly**, *a.* not becoming a gentleman; rude, ill-bred. **ungentlemanliness**, *n.* **unget-at-able**, *a., adv.* difficult of access. **ungifted**, *a.* **ungird**, *v.t.* (*past, p.p. -girt*) to undo or remove a girdle from; to unbind. **unglazed**, *a.* **unglue**, *v.t.* to separate by or as if by removing glue. **ungodly**, *a.* not godly, heathen; wicked; (*coll.*) outrageous. **ungodlily**, *adv.* **ungodliness**, *n.* **ungovernable** *a.* not governable; unruly, wild, passionate, licentious. **ungovernably**, *adv.* **ungraceful**, *a.* not graceful; clumsy, inelegant. **ungracefully**, *adv.* **ungracefulness**, *n.* **ungracious**, *a.* wanting in graciousness; discourteous, rude, unmannerly, offensive. **ungraciously**, *adv.* **ungraded**, *a.* **ungraduated**, *a.* **ungrammatical**, *a.* not according to the rules of grammar. **ungrammatically**, *adv.* **ungrateful**, *a.* **ungratefully**, *adv.* **ungratefulness**, *n.* **ungratified**, *a.* **ungroomed**, *a.* **ungrounded**, *a.* unfounded, baseless. **ungrudging**, *a.* **ungrudgingly**, *adv.* **unguarded**, *a.* not guarded; careless, incautious; incautiously said or done. **unguardedly**, *adv.*
ungual UNGUIS.
unguent (ŭng'gwənt), *n.* any soft composition used as an ointment or for lubrication. **unguentary** (-gwen'-), *a.*
unguis (ŭng'gwis), *n.* (*pl.* **gues** (-gwēz)), a nail, claw or hoof; the narrow base of a petal. **ungual**, *a.* of, pertaining to, or having a nail, claw, or hoof. **unguicular** (-gwik'ū-), *a.* **unguiculate, -lated**, *a.*
ungulate (ŭng'gūlət), *a.* hoofed; hoof-shaped; belonging to the Ungulata; *n.* an ungulate animal. **Ungulata** (-lā'tə), *n.pl.* a division of mammals comprising those with hoofs.
un- (cont.) **ungum**, *v.t.* to unglue. **unhallowed**, *a.* not consecrated, not holy; profane; sinful, ungodly. **unhampered**, *a.* **unhand**, *v.t.* to take the hand or hands off; to let go from one's grasp. **unhandsome**, *a.* not handsome; not generous, petty, ungracious. **unhandsomely**, *adv.* **unhandy**, *a.* not handy; clumsy, awkward, inconvenient. **unhandily**, *adv.* **unhandiness**, *n.* **unhang**, *v.t.* **unhappy**, *a.* not happy, miserable, wretched; unlucky, unfortunate; inappropriate. **unhappily**, *adv.* **unhappiness**, *n.* **unharmed**, *a.* **unharness**, *v.t.* to remove harness from. **unhatched**, *a.* of eggs, not hatched. **unhealthful**, *a.* **unhealthfully**, *adv.* **unhealthfulness**, *n.* **unhealthy**, *a.* not enjoying or promoting good health; (*coll.*) dangerous. **unhealthily**, *adv.* **unhealthiness**, *n.* **unheard**, *a.* not heard. **unheard of**, not heard of; unprecedented. **unheated**, *a.* **unheeded**, *a.* not heeded; disregarded, neglected. **unheedful**, *a.* **unheedfully**, *adv.* **unheeding**, *a.* **unhelpful**, *a.* **unhelpfully**, *adv.* **unhemmed**, *a.* **unheralded**, *a.* **unheroic**, *a.* **unhesitating**, *a.* **unhesitatingly**, *adv.* **unhidden**, *a.* **unhindered**, *a.* **unhinge**, *v.t.* to take (a door) off the hinges; to unsettle (the mind etc.). **unhinged**, *a.* **unhistoric, -ical**, *a.* **unhitch**, *v.t.* to unfasten or release from a hitch. **unholy**, *a.* not holy, not hallowed; impious, wicked; (*coll.*) hideous, frightful. **unholily**, *adv.* **unholiness**, *n.* **unhonoured**, *a.* **unhook**, *v.t.* to remove from a hook; to open or undo by disengaging the hooks of. **unhoped**, *a.* not hoped (for); unexpected, beyond hope. **unhorse**, *v.t.* to remove from horseback; to dislodge. **unhouse**, *v.t.* to drive from a house; to deprive of shelter. **unhuman**, *a.* not human. **unhung**, *a.* not hung; not hanged. **unhurried**, *a.* **unhurt**, *a.* **unhusk**, *v.t.* **unhygienic**, *a.*
uni- (ūni-), *comb. form.* one; single. **uniarticulate** *a.* single-jointed.
Uniat, (ū'niət), **Uniate** (-ât), *n.* a member of any community of Oriental Christians acknowledging the supremacy of the Pope but retaining its own liturgy, rites and ceremonies. *a.* of or pertaining to the Uniats.
uni- (cont.) **uniaxial**, *a.* having a single axis. **uniaxially**, *adv.* **unicameral**, *a.* of a legislative body, consisting of a single chamber. **unicameralism**, *n.* **unicameralist**, *n.* **unicapsular**, *a.* (*Bot.*) having a single capsule.
UNICEF (ū'nisef), **Unicef**, (*acronym*) United Nations International Children's Emergency Fund (now called United Nations Children's Fund).
uni- (cont.) **unicellular** *a.* consisting of a single cell. **unicolour, -ed**, *a.* of one colour.
unicorn (ū'nikawn), *n.* a fabulous animal like a horse, with a long, straight, tapering horn; (*Bibl.*) a two-horned animal, perh. the urus (a mistranslation of Heb. *re'em*). **unicorn-beetle**, *n.* a large beetle with a single horn on the prothorax. **unicornous** (-kaw'-), *a.* one-horned.
uni- (cont.) **unicostate**, *a.* having one principal rib or nerve; (*Bot.*) having a midrib. **unicycle** (ū'nisīkl), *n.* a monocycle.
un- (cont.) **unidentifiable**, *a.* **unidentified**, *a.*
uni- (cont.) **unidimensional**, *a.* **unidirectional**, *a.*
UNIDO (ūnē'dō), (*acronym*) United Nations Industrial Development Organization.
unification, unifier UNIFY.
uni- (cont.) **uniflorous**, *a.* bearing but a single flower. **unifoliar, unifoliate, unifoliolate**, *a.* consisting of one leaf or leaflet.
uniform (ū'nifawm), *a.* having always one and the same form, appearance, quality, character etc., always the same, not varying, not changing, homogeneous; conforming to one rule or standard, applying or operating without variation. *n.* dress of the same kind and appearance worn by members of the same body; a set of such clothes. **uniformed**, *a.* dressed in uniform. **uniformity** (-faw'-), *n.* the quality or state of being uniform; consistency, sameness. **uniformitarian** (-teə'ri-), *n.* one who believes that there has been essential uniformity of cause and effect throughout the physical history of the world, opp. to catastrophism. **uniformitarianism**, *n.* **uniformly**, *adv.*
unify (ū'nifī), *v.t.* to make a unit of; to regard as one; to reduce to uniformity. **unification** (-fī-), *n.* **unifier**, *n.*
uni- (cont.) **unilateral**, *a.* affecting or occurring on one side; arranged on or turned towards one side only; applied or undertaken by one side or party only. **unilateralism**, *n.* **unilateralist**, *n.* **unilaterally**, *adv.* **uniliteral**, *a.* consisting of only one letter.
un- (cont.) **unilluminated**, *a.* not illuminated; dark; ignorant. **unillumined**, *a.* **unillustrated**, *a.*
uni- (cont.) **unilocular, -loculate**, *a.* having or consisting of a single cell or chamber.
un- (cont.) **unimaginable**, *a.* that cannot be imagined; inconceivable. **unimaginably**, *adv.* **unimaginative**, *a.* **unimaginatively**, *adv.* **unimaginativeness**, *n.* **unimpaired**, *a.* **unimpassioned**, *a.* **unimpeachable**, *a.* **unimpeachability**, *n.* **unimpeachableness**, *n.* **unimpeded**, *a.* **unimportance**, *n.* **unimportant**, *a.* **unimposing**, *a.* **unimpressionable**, *a.* **unimpressed**, *a.* **unimpressive**, *a.* **unimpressively**, *adv.* **unimpressiveness**, *n.* **unimproved**, *a.* not improved; of land, not tilled, drained etc., not used to advantage. **unimpugned**, *a.* **unindexed**, *a.* **unindicted**, *a.* **uninfected**, *a.* **uninflammable**, *a.* **uninflated**, *a.* **uninfluenced**, *a.* **uninfluential**, *a.* **uninforma-**

tive, *a.* **uninformed**, *a.* not informed (about); ignorant generally. **uninhabitable**, *a.* **uninhabited**, *a.* **uninhibited**, *a.* **uninitiated**, *a.* **uninjured**, *a.* **uninspired**, *a.* **uninspiring**, *a.* **uninstructive**, *a.* **uninstructively**, *adv.* **uninsulated**, *a.* **uninsurable**, *a.* **uninsured**, *a.* **unintelligent**, *a.* **unintelligently**, *adv.* **unintelligible**, *a.* **unintelligibility**, *n.* **unintelligibly**, *adv.* **unintended**, *a.* **unintentional**, *a.* **unintentionally**, *adv.* **uninterested**, *a.* not taking any interest (in). **uninteresting**, *a.* **uninterestingly**, *adv.* **uninterpretable**, *a.* **uninterrupted**, *a.* **uninterruptedly**, *adv.*
uni- (cont.) **uninuclear, uninucleate**, *a.*
un- (cont.) **uninventive**, *a.* **uninventively**, *adv.* **uninvestigated**, *a.* **uninvited**, *a.* **uninviting**, *a.* not inviting, not attractive, repellent. **uninvitingly**, *adv.* **uninvoked**, *a.* **uninvolved**, *a.*
union (ūn'yən), *n.* the act of uniting; the state of being united; marriage; junction, coalition; a trade-union; agreement or concord of mind, will, affection, or interests; a combination of parts or members forming a whole, an amalgamation, a confederation, a league; the unit formed by such a political combination, esp. the UK, US, USSR; a students' club; (*Med.*) the growing together of parts separated by injury; (*Hist.*) two or more parishes consolidated for administration of the Poor Laws; a workhouse established by this; in plumbing, a device for connecting pipes; a fabric made of two different yarns, as linen and cotton; (*UK*) a device emblematic of union borne in the upper corner next the staff of a flag; this used as a flag, called a **union jack**, or **union flag**. **Union of Soviet Socialist Republics**, English form of the official title of the government of Russia. **unionism**, *n.* the principle of combining; esp. the system of combination among workers engaged in the same occupation or trade, trade-unionism; (**Unionism**) the principles of the Unionist party. **unionist**, *n.* a member of a trade-union; a promoter or advocate of trade-unionism; (**Unionist**) a member of a political party formed to uphold the legislative union between Great Britain and Ireland and to oppose Home Rule; (**Unionist**) (*N Am.*) an opponent of secession before and during the American Civil War. **unionistic** (-nis'-), *a.* **unionize, -ise**, *v.t.*, *v.i.* to organize into a trade-union. **unionization**, *n.*
uni- (cont.) **uniparous** (ūnip'ərəs), *a.* bringing forth normally but one at a birth; (*Bot.*) having one axis or stem. **unipartite**, *a.* not divided. **uniped** (ū'-), *a.* having only one foot. *n.* a one-footed animal. **unipersonal**, *a.* (*Gram.*) used only in one person. **uniplanar**, *a.* lying or occurring in one plane. **unipolar**, *a.* of a nerve cell, having but one process; having or using a single electric or magnetic pole; (*Elec.*) exhibiting but one kind of polarity. **unipolarity**, *n.*
unique (ūnēk'), *a.* having no like or equal; unmatched, unparalleled. *n.* a unique person or thing. **uniquely**, *adv.* **uniqueness**, *n.*
un- (cont.) **unironed**, *a.*
uni- (cont.) **unisex** (ū'-), *a.* that can be used, worn, etc., by both sexes. **unisexual** (-sek'-), *a.* of one sex only; not hermaphrodite, having only one kind of sexual organs. **unisexually**, *adv.*
unison ('nisən), *n.* (*Mus.*) coincidence of sounds proceeding from equality in rate of vibrations, unity of pitch; an interval of one or more octaves; the act or state of sounding together at the same pitch; concord, agreement, harmony. *a.* sounding together; coinciding in pitch; sounding alone. **unisonal, -nant, -nous** (-nis'-), *a.* **unisonance** (-nis'-), *n.*
unit (ū'nit), *n.* a single person, thing, or group, regarded as one and individual for the purposes of calculation; each one of a number of things, persons etc., forming a plurality; a quantity represented by the number one; a quantity adopted as the standard of measurement or calculation; a part of a machine which performs a particular function; a piece of furniture which forms part of a set, designed for a particular use in a kitchen etc.; having a specific function; a group within a larger organization; a part of a larger military formation; a quantity of a drug, vitamin etc., which produces a specific effect. **unit price**, *n.* the price of a commodity expressed per unit of weight, volume etc. **unit pricing**, *n.* the practice of showing unit price on goods for sale. **unit trust**, *n.* an investment company which purchases holdings in a range of different enterprises and allocates proportions of these holdings according to the amount invested. **Unitarian** (-teə'ri-), *n.* a member of a Christian body which rejects the doctrine of the Trinity; a monotheist; one who advocates unity or unification, esp. in politics. *a.* pertaining to the Unitarians. **Unitarianism**, *n.* **Unitarianize, -ise**, *v.t.* **unitary**, *a.* of or pertaining to a unit or units; of the nature of a unit, whole, integral. **unitize, -ise**, *v.t.*
UNITA (ūnē'tə), (*acronym*) National Union for the Total Liberation of Angola. [Port. *União Nacional por Independência Total de Angola*]
unite (ūnīt'), *v.t.* to join together so as to make one; to combine, to conjoin, to amalgamate; to cause to adhere, to attach together. *v.i.* to become one; to become consolidated, to combine, to coalesce, to agree, to co-operate. **United Brethren**, *n.pl.* the Moravians. **United Kingdom** KINGDOM. **United Nations**, *n. sing.* or *pl.* an international organization of sovereign states, founded in 1945 'to save succeeding generations from the scourge of war'. **United Reformed**, *a.* of a church formed in 1972 from the union of the Presbyterian and Congregational churches in England and Wales. **United States**, *n. sing.* or *pl.* a federal union or republic of sovereign States, esp. that of N America. **unitedly**, *adv.* **uniter**, *n.* **unitive** (ū'ni-), *a.* uniting; producing or characterized by unity.
unity (ū'niti), *n.* the state or condition of being one or individual, oneness, as opp. to plurality or division; the state of being united, union; an agreement of parts or elements, harmonious interconnection, structural coherence; concord, agreement, harmony; a thing forming a coherent whole; (*Math.*) the number one, a factor that leaves unchanged the quantity on which it operates; (*Drama etc.*) the condition that the action of a play should be limited to the development of a single plot, that the supposed time should coincide with the actual duration of the play or to a single day, and that there should be no change of scene (called the three dramatic unities of action, time and place).
uni- (cont.) **univalent**, *a.* (*Chem.*) having a valency or combining power of one. **univalency, -ence**, *n.* **univalve** (ū'-), *a.* having only one valve. *n.* a univalve mollusc. **univalvular** (-val'-), *a.*
universal (ūnivœ'səl), *a.* of or pertaining to the whole world or all persons or things in the world or in the class under consideration; common to all cases, unlimited, all-embracing, general; applicable to all purposes, conditions or cases; (*Log.*) predicable of all the individuals of a class, opp. to *particular*; occurring or existing everywhere or throughout the world; capable of many uses. *n.* (*Log.*) a universal proposition; (*Phil.*) a universal concept; a thing or nature predicable of many. **universal coupling**, or **joint**, *n.* a device for connecting two parts or things, allowing freedom of movement in any direction. **universal time**, *n.* Greenwich Mean Time. **universalism**, *n.* the quality of being universal; (*Theol.*) the doctrine that all men will eventually be saved. **universalist**, *a.*, *n.* **universalistic** (-lis'-), *a.* **universality** (-sal'-), *n.* the state of being universal. **universalize, -ise**, *v.t.* **universalization**, *n.* **universally**, *adv.*
universe (ū'nivœs), *n.* the aggregate of existing things; all created things viewed as constituting one system or whole, the cosmos, including or excluding the Creator; all human beings; (*Log.*) all the objects that are the subjects

of consideration.
university (ūnivœ'siti), *n.* an educational institution for both instruction and examination in the higher branches of knowledge with the power to confer degrees, usu. comprising subordinate colleges, schools etc.; the members of this collectively; (*coll.*) a team or crew representing a university, as distinguished from a college team etc. **University Extension,** EXTENSION.
uni- (cont.) **univocal,** *a.* of a word, having only one meaning; (*Mus.*) having unison of sounds. **univocally,** *adv.* **univocation,** *n.* agreement of name and meaning.
un- (cont.) **unjoin,** *v.t.* **unjoint,** *v.t.* to disjoint, to separate the joints of. **unjust,** *a.* not just; not conformable to justice. **unjustly,** *adv.* **unjustifiable,** *a.* **unjustifiableness,** *n.* **unjustifiably,** *adv.* **unjustified,** *a.*
unkempt (ŭnkempt'), *a.* uncombed; rough, unpolished.
un- (cont.) **unkennel,** *v.t.* to release or drive out from a kennel; to let loose. **unkept,** *a.* **unkind,** *a.* not kind, harsh, hard, cruel. **unkindly,** *adv.* **unkindness,** *n.* **unkink,** *v.t., v.i.* **unkneaded,** *a.* **unknit,** *v.t.* **unknot,** *v.t.* **unknowable,** *a.* **unknowably,** *adv.* **unknowing,** *a.* not knowing; ignorant or unaware (of). **unknowingly,** *adv.* **unknown,** *a.* not known; untold, incalculable, inexpressible; of quantities in equations etc., unascertained. *n.* an unknown person, thing or quantity. **Unknown Soldier,** *n.* an unidentified soldier whose body is buried in a memorial as a symbol of all soldiers killed in war. **unlabelled,** *a.* **unlaboured,** *a.* not produced by labour, untilled, unworked; of style etc., spontaneous, natural, easy. **unlace,** *v.t.* to loose or unfasten by undoing the lace or laces of. **unladylike,** *a.* **unlaid,** *a.* **unlamented,** *a.* **unlatch,** *v.t.* to unfasten the latch of (a door etc.). **unlawful,** *a.* **unlawfully,** *adv.* **unlawfulness,** *n.* **unlay,** *v.t.* (*Naut.*) to untwist (rope etc.). **unleaded,** *a.* of petrol, without added lead compounds. **unlearn,** *v.t.* to forget the knowledge of; to expel from the mind (that which has been learned), to get rid of (a vice etc.). **unlearned¹, -learnt,** *a.* not learnt. **unlearned²** (-nid), *a.* not learned. **unlearnedly,** *adv.* **unlearnedness,** *n.* **unleash,** *v.t.* **unleavened,** *a.*
unless (ŭnles'), *conj.* if it be not the case that; except when.
un- (cont.) **unlettered,** *a.* illiterate. **unliberated,** *a.* **unlicensed,** *a.* **unlike,** *a.* not like; dissimilar. *prep.* not like; not characteristic of. **unlikeness,** *n.* **unlik(e)able,** *a.* difficult to like, not likeable. **unlikely,** *a.* improbable; unpromising. *adv.* improbably. **unlikelihood,** **unlikeliness,** *n.* **unlimber,** *v.t.* **unlimited,** *a.* not limited; having no bounds, indefinite, unmeasured, unnumbered; unconfined, unrestrained. **unlimitedly,** *adv.* **unlimitedness,** *n.* **unlined,** *a.* without lines; without a lining. **unlink,** *v.t.* **unliquidated,** *a.* **unlisted,** *a.* not on a list; of securities, not listed on the Stock Exchange; (*N Am.*) of a telephone number, ex-directory. **unlit,** *a.* **unlived-in,** *a.* not lived in; over-tidy. **unload,** *v.t.* to discharge the load from; to discharge (a load); to withdraw the charge from (a gun etc.); *v.i.* to discharge a load or freight; **unloader,** *n.* **unlock,** *v.t.* to unfasten the lock of (a door, box etc.); to disclose. **unlodge,** *v.t.* to dislodge. **unlooked-for,** *a.* not looked for, unexpected. **unloose,** *v.t.* to unfasten, to loose; to set at liberty. **unloosen,** *v.t.* **unlovable,** *a.* **unloved,** *a.* **unlovely,** *a.* not lovely; not beautiful or attractive. **unloveliness,** *n.* **unloving,** *a.* **unlovingly,** *adv.* **unlucky,** *a.* not lucky or fortunate; unsuccessful, unfortunate; disastrous; inauspicious, ill-omened. **unluckily,** *adv.* **unluckiness,** *n.* **unmade,** **unmaidenly,** *a.* **unmailable,** *a.* incapable of being sent by post. **unmaimed,** *a.* **unmaintainable,** *a.* **unmake,** *v.t.* (*past, p.p.* **-made**) to undo, to destroy; to annihilate; to depose; to alter fundamentally. **unmalleable,** *a.* **unman,** *v.t.* (*past, p.p.* **-manned**) to deprive of courage or fortitude; to deprive of men; to deprive of maleness or manly qualities. **unmanageable,** *a.* not manageable; not easily controlled. **unmanful,** *a.* **unmanfully,** *adv.* **unmanly,** *a.* not

like a man; effeminate; childish; cowardly. **unmanliness,** *n.* **unmanned,** *a.* not manned, having no crew. **unmannered,** *a.* without mannerism; lacking good manners. **unmannerly,** *a.* not mannerly; rude, ill-bred. **unmannerliness,** *n.* **unmarked,** *a.* not marked; not noticed, unobserved. **unmarketable,** *a.* **unmarriageable,** *a.* **unmarried,** *a.* **unmasculine,** *a.* **unmask,** *v.t.* to remove the mask from; to expose. *v.i.* to take one's mask off; to reveal oneself. **unmatchable,** *a.* **unmatched,** *a.* **unmated,** *a.* **unmatured,** *a.* **unmeaning,** *a.* having no meaning; senseless; expressionless, vacant. **unmeaningly,** *adv.* **unmeaningness,** *n.* **unmeant,** *a.* not meant, not intended. **unmeasured,** *a.* not measured; indefinite, unlimited, unmeasurable. **unmechanical,** *a.* **unmechanized, -ised,** *a.* **unmelodious,** *a.* **unmelodiously,** *adv.* **unmelodiousness,** *n.* **unmelted,** *a.* **unmemorable,** *a.* **unmendable,** *a.* **unmentionable,** *a.* not mentionable, not fit to be mentioned. *n.pl.* (*facet.*) trousers; underwear. **unmentionableness,** *n.* **unmerchantable,** *a.* **unmerciful,** *a.* **unmercifully,** *adv.* **unmercifulness,** *n.* **unmerited,** *a.* **unmethodical,** *a.* **unmetrical,** *a.* not metrical; not according to the rules or requirements of metre. **unmetrically,** *adv.* **unmilitary,** *a.* **unmindful,** *a.* not mindful, heedless (of). **unmindfully,** *adv.* **unmindfulness,** *n.* **unmirthful,** *a.* **unmirthfully,** *adv.* **unmistakable,** *a.* that cannot be mistaken; manifest, plain. **unmistakably,** *adv.* **unmistaken,** *a.* **unmitigated,** *a.* not mitigated; unqualified, unconscionable. **unmixed,** *a.* **unmodernized, -ised,** *a.* **unmodified,** *a.* **unmodulated,** *a.* **unmolested,** *a.* **unmoor,** *v.t.* to loose the moorings of; to release partially by weighing one of two or more anchors. *v.i.* to weigh anchor. **unmoral,** *a.* not moral. **unmorality,** *n.* **unmorally,** *adv.* **unmortgaged,** *a.* **unmotherly,** *a.* **unmotivated,** *a.* lacking in motive or incentive. **unmould,** *v.t.* to take from a mould; to change the form of. **unmounted,** *a.* **unmourned,** *a.* **unmoved,** *a.* not moved; not changed in purpose, unshaken, firm; not affected, not having the feelings excited. **unmoving,** *a.* motionless; unaffecting. **unmown,** *a.* **unmurmuring,** *a.* not complaining. **unmurmuringly,** *adv.* **unmusical,** *a.* not pleasing to the ear, discordant; not interested or skilled in music. **unmusicality,** *n.* **unmusically,** *adv.* **unmutilated,** *a.* **unmuzzle,** *v.t.* **unnail,** *v.t.* **unnameable,** *a.* **unnamed,** *a.* **unnatural,** *a.* not natural; contrary to nature; not in accordance with accepted standards of behaviour; monstrous, inhuman; artificial, forced, strained, affected. **unnaturalize, -ise,** *v.t.* to make unnatural. **unnaturalized,** *a.* **unnaturally,** *adv.* alien. **unnaturalness,** *n.* **unnavigable,** *a.* **unnecessary,** *a.* not necessary; needless, superfluous. *n.* (*usu. pl.*) that which is unnecessary. **unnecessarily,** *adv.* **unneeded,** *a.* **unneedful,** *a.* **unnegotiable,** *a.* **unneighbourly,** *a.* **unneighbourliness,** *n.* **unnerve,** *v.t.* to deprive of nerve, strength or resolution. **unnerved,** *a.* **unnoted,** *a.* not heeded. **unnoticed,** *a.* **unnourished,** *a.* **unnumbered,** *a.* not marked with numbers; countless.
UNO, (*abbr.*) United Nations Organization.
un- (cont.) **unobjectionable,** *a.* **unobjectionably,** *adv.* **unobliging,** *a.* **unobscured,** *a.* **unobservant,** *a.* **unobserved,** *a.* **unobserving,** *a.* **unobstructed,** *a.* **unobtainable,** *a.* **unobtrusive,** *a.* **unobtrusively,** *adv.* **unobtrusiveness,** *n.* **unoccupied,** *a.* **unoffending,** *a.* not offending; harmless, innocent. **unoffered,** *a.* **unofficial,** *a.* not having official character or authorization. **unofficially,** *adv.* **unopened,** *a.* **unopposed,** *a.* **unordained,** *a.* **unorganized, -ised,** *a.* not organized or arranged; not unionized. **unoriginal,** *a.* not original, derived; not possessed of originality. **unornamental,** *a.* not ornamental; plain, ugly. **unornamented,** *a.* **unorthodox,** *a.* **unostentatious,** *a.* **unostentatiously,** *adv.* **unostentatiousness,** *n.* **unowned,** *a.* **unpacified,** *a.* **unpack,** *v.t.* to open and take out the contents of; to take (things) out of a package etc. **unpaged,** *a.* not having the pages numbered. **unpaid,** *a.* of a debt etc. not paid, not discharged; not having received the payment due; acting

unpaid for   903   **unreproachful**

gratuitously. **unpaid for,** not paid for; taken on credit. **unpainted,** *a.* **unpaired,** *a.* **unpalatable,** *a.* **unpalatability,** *adv.* **unparalleled,** *a.* not paralleled; unequalled, unprecedented. **unpardonable,** *a.* **unpardonableness,** *n.* **unpardonably,** *adv.* **unpared,** *a.* **unparental,** *a.* **unparented,** *a.* **unparliamentary,** *a.* esp. of language, contrary to the rules or usages of Parliament. **unparliamentarily,** *adv.* **unparliamentariness,** *n.* **unpasteurized, -ised,** *a.* **unpatented,** *a.* **unpatriotic,** *a.* **unpatriotically,** *adv.* **unpaved,** *a.* **unpeaceful,** *a.* **unpedantic,** *a.* **unpedigreed,** *a.* **unpeeled,** *a.* **unpeg,** *v.t.* to take out the pegs from; to open or unfasten thus. **unpen,** *v.t.* **unpensioned,** *a.* **unperceived,** *a.* **unperforated,** *a.* **unperformed,** *a.* **unperfumed,** *a.* **unperson,** *n.* a person whose existence is officially ignored or denied. **unpersuadable,** *a.* **unpersuaded,** *a.* **unpersuasive,** *a.* **unperturbed,** *a.* **unperused,** *a.* **unperverted,** *a.* **unphilosophical,** *a.* not in a philosophic way; lacking philosophy. **unphilosophically,** *adv.* **unpick,** *v.t.* to loosen, take out, or open, by picking. **unpicked,** *a.* not picked; not picked out or selected. **unpicturesque,** *a.* **unpiloted,** *a.* **unpin,** *v.t.* (*past, p.p.* **-pinned**) to remove the pins from; to unfasten (something held together by pins). **unpitied,** *a.* **unpitying,** *a.* **unpityingly,** *adv.* **unplaced,** *a.* not placed; not holding a place; not among the first three at the finish of a race. **unplagued,** *a.* **unplait,** *v.t.* **unplaned,** *a.* **unplanned,** *a.* **unplanted,** *a.* **unplastered,** *a.* **unplated,** *a.* **unplausible,** *a.* **unplausibly,** *adv.* **unplayable,** *a.* **unpleasant,** *a.* not pleasant; disagreeable. **unpleasantly,** *adv.* **unpleasantness,** *n.* the quality of being unpleasant; a slight disagreement. **unpleased,** *a.* **unpleasing,** *a.* **unpleasingly,** *adv.* **unpledged,** *a.* **unploughed** (-ploud'), *a.* **unplucked,** *a.* **unplug,** *v.t.* (*past, p.p.* **-plugged**) to remove a plug or obstruction from; to disconnect (an electrical appliance) from a source of electricity. **unplumbed,** *a.* **unpoetic, -ical,** *a.* **unpoetically,** *adv.* **unpointed,** *a.* not having a point; not punctuated; not having the vowel-points or diacritical marks; of masonry; not pointed. **unpolished,** *a.* **unpolitical,** *a.* not related to or interested in politics. **unpolled,** *a.* not polled, not having registered one's vote. **unpolluted,** *a.* **unpopular,** *a.* not popular; not enjoying the public favour. **unpopularity,** *n.* **unpopularly,** *adv.* **unposed,** *a.* **unpossessed,** *a.* not possessed; not in possession (of). **unposted,** *a.* of a letter etc. not posted; not posted up; without information. **unpractical,** *a.* **unpracticality,** *n.* **unpractically,** *adv.* **unpractised,** *a.* not put in practice; unskilful, inexperienced. **unpraised,** *a.* **unprecedented,** *a.* being without precedent, unparalleled; new. **unpredictable,** *a.* that cannot be predicted; whose behaviour cannot be predicted or relied on. **unpredicted,** *a.* **unprefaced,** *a.* **unprejudiced,** *a.* **unpremeditated,** *a.* not premeditated, not planned beforehand; unintentional. **unpremeditatedly,** *adv.* **unprepared,** *a.* not prepared, impromptu; not ready (for etc.). **unpreparedness,** *n.* **unprepossessing,** *a.* **unprescribed,** *a.* **unpresentable,** *a.* not presentable; not fit to be seen. **unpresuming,** *a.* **unpresumptuous,** *a.* **unpretending,** *a.* **unpretendingly,** *adv.* **unpretentious,** *a.* **unpretentiously,** *adv.* **unpretentiousness,** *n.* **unpreventable,** *a.* **unpriced,** *a.* having the price or prices not fixed, quoted or marked up; priceless. **unprincipled,** *a.* not dictated by moral principles; destitute of principle, immoral. **unprintable,** *a.* that cannot be printed (because obscene or libellous). **unprinted,** *a.* **unprivileged,** *a.* **unproclaimed,** *a.* **unprocurable,** *a.* **unproductive,** *a.* **unproductively,** *adv.* **unproductiveness,** *n.* **unprofessed,** *a.* **unprofessional,** *a.* not pertaining to one's profession; contrary to the rules or etiquette of a profession; not belonging to a profession. **unprofitable,** *a.* **unprofitableness,** *n.* **unprofitably,** *adv.* **unprogrammable,** *a.* **unprogrammed,** *a.* **unprogressive,** *a.* not progressive, conservative. **unprogressiveness,** *n.* **unprohibited,** *a.* **unpromising,** *a.* not promising success. **unprompted,** *a.* of one's own free will or initiative. **unpronounceable,** *a.* **unpropagated,** *a.* **unpropitious,** *a.* **unpropitiously,** *adv.* **unpropitiousness,** *n.* **unproportional,** *a.* not in proportion, disproportionate. **unproposed,** *a.* **unprosperous,** *a.* **unprotected,** *a.* **unprotecting,** *a.* **unprotested,** *a.* **unprotesting,** *a.* **unprovable,** *a.* **unproved, -proven,** *a.* **unprovided,** *a.* not provided; not furnished (with supplies etc.). **unprovoked,** *a.* having received no provocation; not instigated. **unpruned,** *a.* not pruned. **unpublished,** *a.* not made public; of books etc., not published. **unpunctual,** *a.* **unpunctuality,** *n.* **unpunctuated,** *a.* **unpunishable,** *a.* **unpunished,** *a.* **unpurchased,** *a.* **unpurified,** *a.* **unputdownable,** *a.* (*coll.*) of a book, too exciting to put down before it is finished. **unquailing,** *a.* **unquailingly,** *adv.* **unqualified,** *a.* not qualified; not fit, not competent; not having passed the necessary examination etc.; not qualified legally; not limited by conditions or exceptions, absolute. **unqualifiedly** (-fiid-), *adv.* **unquarried,** *a.* **unquelled,** *a.* **unquenchable,** *a.* **unquenchably,** *adv.* **unquenched,** *a.* **unquestionable,** *a.* not to be questioned or doubted, indisputable. **unquestionably,** *adv.* **unquestioned,** *a.* not called in question, not doubted; having no questions asked, not interrogated. **unquestioning,** *a.* not questioning, not doubting; implicit. **unquestioningly,** *adv.* **unquiet,** *a.* restless, uneasy, agitated. **unquietly,** *adv.* **unquotable,** *a.* **unquote,** *v.i.* to close a quotation. *int.* indicating the end of a (spoken) quotation. **unquoted,** *a.* **unransomed,** *a.* **unravaged,** *a.* **unravel,** *v.t.* (*past, p.p.* **-ravelled**) to separate the threads of; to disentangle, to untwist; to solve, to clear up (the plot of a play etc.). *v.i.* to be disentangled; to be opened up or revealed. **unravelment,** *n.* **unreachable,** *a.* **unread** (-red), *a.* not read; not well-read, unlearned, illiterate. **unreadable,** *a.* **unreadableness,** *n.* **unready,** *a.* not ready; not prompt to act, etc. **unreadily,** *adv.* **unreadiness,** *n.* **unreal,** *a.* not real; unsubstantial, visionary, imaginary. **unreality,** *n.* **unreally,** *adv.* **unrealistic,** *a.* **unrealizable, -isable,** *a.* **unrealized,** *a.* **unreaped,** *a.* **unreason,** *n.* want of reason; folly, absurdity. **unreasonable,** *a.* not reasonable; exorbitant, extravagant, absurd; not listening to reason. **unreasonableness,** *n.* **unreasonably,** *adv.* **unreasoned,** *a.* not reasoned or thought out rationally. **unreasoning,** *a.* not reasoning; foolish; not having reasoning faculties. **unreasoningly,** *adv.* **unrebuked,** *a.* **unrecallable,** *a.* **unrecanted,** *a.* **unreceived,** *a.* **unreciprocated,** *a.* **unreceptive,** *a.* **unreckoned,** *a.* **unreclaimed,** *a.* not reclaimed; unregenerate. **unrecognizable, -isable,** *a.* **unrecognizably,** *adv.* **unrecognized,** *a.* not recognized; not acknowledged. **unrecompensed,** *a.* **unreconciled,** *a.* **unreconstructed,** *a.* clinging to old-fashioned social or political notions. **unrecorded,** *a.* **unrectified,** *a.* not corrected. **unredeemed,** *a.* not redeemed; not fulfilled; not taken out of pawn; not recalled by payment of the value; not counterbalanced by any redeeming quality, unmitigated. **unredressed,** *a.* **unreel,** *v.t.* to unwind. *v.i.* to become unwound. **unrefined,** *a.* not refined; not purified; of unpolished manners, taste etc. **unreflecting,** *a.* **unreformable,** *a.* **unreformed,** *a.* **unrefreshed,** *a.* **unrefuted,** *a.* **unregarded,** *a.* **unregardful,** *a.* **unregenerate,** *a.* **unregistered,** *a.* **unregretted,** *a.* **unregulated,** *a.* **unrehearsed,** *a.* **unreined,** *a.* not held in check by the rein; unrestrained, unbridled. **unrelated,** *a.* **unrelaxed,** *a.* **unrelaxing,** *a.* **unrelenting,** *a.* **unrelentingly,** *adv.* **unrelentingness,** *n.* **unreliable,** *a.* **unreliability,** **unreliableness,** *n.* **unreliably,** *adv.* **unrelieved,** *a.* **unreligious,** *a.* irreligious; not connected with religion, secular. **unremarkable,** *a.* **unremembered,** *a.* **unremitting,** *a.* not relaxing; incessant, continued. **unremittingly,** *adv.* **unremunerative,** *a.* not profitable. **unrenewed,** *a.* **unrenounced,** *a.* **unrepairable,** *a.* **unrepaired,** *a.* **unrepealed,** *a.* **unrepeatable,** *a.* that cannot be done or said again; of language, too foul to repeat. **unrepentant,** *a.* **unrepented,** *a.* **unreplenished,** *a.* **unreported,** *a.* **unrepresentative,** *a.* **unrepresented,** *a.* **unrepressed,** *a.* **unrequested,** *a.* **unreproachful,** *a.* **unre-**

proved, *a.* **unrequested**, *a.* **unrequited**, *a.* not requited; not recompensed. **unrescinded**, *a.* **unresented**, *a.* **unresenting**, *a.* **unresentingly**, *adv.* **unreserve**, *n.* lack of reserve, frankness, candour. **unreserved**, *a.* not reserved; open, frank; given, offered, or done without reservation. **unreservedly** (-vid-), *adv.* **unreservedness**, *n.* **unresisted**, *a.* **unresisting**, *a.* **unresistingly**, *adv.* **unresolved**, *a.* not resolved, undecided, irresolute; unsolved, not cleared up. **unresponsive**, *a.* **unrest**, *n.* restlessness, agitation, disquiet, uneasiness, unhappiness. **unrestful**, *a.* **unrestfully**, *adv.* **unrestfulness**, *n.* **unrestored**, *a.* **unrestrainable**, *a.* **unrestrainably**, *adv.* **unrestrained**, *a.* **unrestrainedly**, (-nid-), *adv.* **unrestraint**, *n.* **unrestricted**, *a.* **unrestrictedly**, *adv.* **unretarded**, *a.* **unretentive**, *a.* **unretracted**, *a.* **unrevealed**, *a.* **unrevenged**, *a.* **unreversed**, *a.* **unrevised**, *a.* **unrevoked**, *a.* **unrewarded**, *a.* **unrewarding**, *a.* **unrhymed**, *a.* **unrhythmic**, **-ical**, *a.* **unridable**, *a.* **unridden**, *a.* **unrighted**, *a.* **unrighteous**, *a.* not righteous, not just; contrary to justice or equity; evil, wicked, sinful. **unrighteously**, *adv.* **unrighteousness**, *n.* **unrip**, *v.t.* (*past*, *p.p.* **-ripped**) to rip open, to undo or unfasten by ripping. **unripe**, *a.* not ripe; not mature; premature. **unripened**, *a.* **unripeness**, *n.* **unrivalled**, *a.* having no rival; unequalled, peerless. **unrobe**, *v.t.*, *v.i.* **unroll**, *v.t.* to unfold (a roll of cloth etc.); to display, to lay open. *v.i.* to be unrolled; to be displayed. **unromantic**, *a.* **unromantically**, *adv.* **unroot**, *v.t.* to tear up by the roots; to extirpate, to eradicate. **unrounded**, *a.*

UNRRA (ūn′rə), (*acronym*) United Nations Relief and Rehabilitation Administration.

un- (cont.) **unruffled**, *a.* not ruffled, unperturbed. **unruled**, *a.* not governed; of paper etc. not ruled with lines. **unruly**, *a.* not submitting to restraint; lawless, turbulent, ungovernable. **unruliness**, *n.*

UNRWA (ūn′rə), (*acronym*) for *United Nations Relief and Works Agency*.

un- (cont.) **unsaddle**, *v.t.* to remove the saddle from; to unseat. *v.i.* to unsaddle a horse. **unsafe**, *a.* dangerous, perilous, risky; not to be trusted. **unsafely**, *adv.* **unsaid**, *a.* not said, unspoken. **unsalaried**, *a.* **unsaleable**, *a.* **unsaleability**, **unsaleableness**, *n.* **unsalted**, *a.* **unsanctified**, *a.* **unsanctioned**, *a.* **unsanitary**, *a.* unhealthy. **unsated**, *a.* **unsatisfactorily**, *adv.* **unsatisfactoriness**, *n.* **unsatisfied**, *a.* **unsatisfying**, *a.* **unsatisfyingly**, *adv.* **unsaturated**, *a.* not saturated; of fats, having a high proportion of fatty acids with double bonds in the molecular structure. **unsaved**, *a.* **unsavoury**, *a.* unattractive, repellent, disgusting; morally offensive; unpleasant to taste or smell. **unsavourily**, *adv.* **unsavouriness**, *n.* **unsay**, *v.t.* (*past*, *p.p.* **-said**) to retract or withdraw (what has been said). **unsayable**, *a.* **unscalable**, *a.* that cannot be climbed. **unscarred**, *a.* **unscathed**, *a.* not scathed, uninjured. **unscented**, *a.* **unscheduled**, *a.* **unscholarly**, *a.* **unschooled**, *a.* **unscientific**, *a.* not in accordance with scientific principles or methods; lacking scientific knowledge. **unscientifically**, *adv.* **unscorched**, *a.* **unscoured**, *a.* **unscourged**, *a.* **unscramble**, *v.t.* to restore to order from a scrambled state; to make (a scrambled message) intelligible. **unscrambler**, *n.* **unscreened**, *a.* **unscrew**, *v.t.* to withdraw or loosen (a screw); to unfasten thus. **unscripted**, *a.* not using a script; unplanned unrehearsed. **unscriptural**, *a.* not in conformity with the Scriptures. **unscripturally**, *adv.* **unscrupulous**, *a.* having no scruples of conscience; unprincipled. **unscrupulously**, *adv.* **unscrupulousness**, *n.* **unseal**, *v.t.* to break or remove the seal of; to open. **unsealed**, *a.* **unsearchable**, *a.* incapable of being searched out, inscrutable. **unsearched**, *a.* **unseasonable**, *a.* **unseasonableness**, *n.* **unseasonably**, *adv.* **unseasoned**, *a.* **unseat**, *v.t.* to remove from one's seat; to throw from one's seat on horseback; to deprive of a parliamentary seat or political office. **unseated**, *a.* thrown from or deprived of a seat; not furnished with seats; having no seat. **unseaworthy**, *a.* **unsecluded**, *a.* **unseconded**, *a.* **unsectarian**, *a.* **unsectarianism**, *n.* **unsecured**, *a.* **unseduced**, *a.* **unseductive**, *a.* **unseeded**, *a.* in a sporting tournament, not put with the best players in the competition draw. **unseeing**, *a.* blind; unobservant, unsuspecting. **unseemly**, *a.* not seemly; unbefitting, unbecoming. **unseemliness**, *n.* **unseen**, *a.* not seen; invisible; not seen previously (as a piece to be translated). *n.* an unseen passage for translation. **unsegregated**, *a.* **unselected**, *a.* **unselfconscious**, *a.* **unselfish**, *a.* regarding or prompted by the interests of others rather than one's own. **unselfishly**, *adv.* **unselfishness**, *n.* **unsensational**, *a.* **unsent**, *a.* **unsentenced**, *a.* **unsentimental**, *a.* **unseparated**, *a.* **unserviceable**, *a.* **unserviceableness**, *n.* **unset**, *a.* of a gem, trap, the sun etc., not set. **unsettle**, *v.t.* to change from a settled state or position; to make uncertain or fluctuating; to derange, to disturb. **unsettled**, *a.* not settled, fixed or determined; undecided, hesitating; changeable; having no fixed abode; not occupied, uncolonized; of debts etc., unpaid. **unsex**, *v.t.* to deprive of the qualities of one's sex, esp. to make (a woman) less feminine. **unshackle**, *v.t.* **unshaded**, *a.* **unshadowed**, *a.* **unshak(e)able**, *a.* **unshaken**, *a.* **unshapely**, *a.* misshapen. **unshared**, *a.* **unshaven**, *a.* **unsheathe**, *v.t.* to draw from its sheath. **unshed**, *a.* **unsheltered**, *a.* **unship**, *v.t.* (*past*, *p.p.* **-shipped**) to unload from a ship; to disembark; (*Naut.*) to remove from a regular or fixed place. *v.i.* to become unshipped. **unshipped**, *a.* **unshockable**, *a.* **unshocked**, *a.* **unshod**, *a.* **unshorn**, *a.* not shorn, clipped, or shaven. **unshown**, *a.* **unshrinkable**, *a.* **unshrinking**, *a.* not recoiling, undaunted, unhesitating. **unshrinkingly**, *adv.* **unshrunk**, *a.* **unshut**, *a.* **unshuttered**, *a.* **unsifted**, *a.* **unsighted**, *a.* not sighted, not seen; invisible; not seeing; of a gun etc., not furnished with sights. **unsightly**, *a.* unpleasing to the sight, ugly. **unsightliness**, *n.* **unsigned**, *a.* **unsilvered**, *a.* **unsinkable**, *a.* **unsinning**, *a.* **unsisterly**, *a.* **unsisterliness**, *n.* **unsized**[1], *a.* not arranged by size. **unsized**[2], *a.* not sized, not stiffened. **unskilful**, *a.* **unskilfully**, *adv.* **unskilfulness**, *n.* **unskilled**, *a.* destitute of skill or special knowledge or training; produced without or not requiring special skill or training. **unslaked**, *a.* **unsleeping**, *a.* **unsmiling**, *a.* **unsmoked**, *a.* **unsnarl**, *v.t.* to remove a snarl or tangle from. **unsociable**, *a.* not sociable, solitary; reserved. **unsociability**, **unsociableness**, *n.* **unsociably**, *adv.* **unsocial**, *a.* not social, solitary; antisocial; of hours of work, falling outside the usual working day. **unsoiled**, *a.* **unsold**, *a.* **unsoldierly**, *a.* **unsolicited**, *a.* **unsolicitous**, *a.* **unsolvable**, *a.* **unsolved**, *a.* **unsoothed**, *a.* **unsophisticated**, *a.* simple, artless, free from artificiality, inexperienced; not corrupted or adulterated, pure, genuine. **unsophisticatedly**, *adv.* **unsophisticatedness**, *n.* **unsorted**, *a.* **unsought**, *a.* **unsound**, *a.* not sound; weak, decayed; unreliable; diseased; ill-founded; not valid, fallacious. **unsoundly**, *adv.* **unsoundness**, *n.* **unsoured**, *a.* **unsown**, *a.* **unsparing**, *a.* liberal, profuse, lavish; unmerciful. **unsparingly**, *adv.* **unsparingness**, *n.* **unspeak**, *v.t.* to retract, to unsay. **unspeakable**, *a.* unutterable, inexpressible, beyond expression; inexpressibly bad or evil. **unspeakably**, *adv.* **unspeakableness**, *n.* **unspecified**, *a.* **unspeculative**, *a.* **unspent**, *a.* **unspilt**, *a.* **unspiritual**, *a.* **unspoiled**, **-spoilt**, *a.* **unspoken**, *a.* **unspontaneous**, *a.* **unsporting**, *a.* **unsportsmanlike**, *a.* unbecoming a sportsman. **unspotted**, *a.* free from spots; unblemished, uncontaminated; faultless, perfect. **unsprung**, *a.* not equipped with springs. **unstable**, *a.* not stable, not firm; liable to sudden shifts of moods; of a chemical compound, atom etc., decaying or decomposing rapidly or easily. **unstained**, *a.* not stained; unblemished, unsullied. **unstamped**, *a.* not having a stamp affixed. **unstarched**, *a.* **unstartling**, *adv.* **unstated**, *a.* **unstatesmanlike**, *a.* **unsteady**, *a.* not steady, not firm; changeable, variable; unstable, precarious; (*coll.*) irregular in habits or conduct. **unsteadily**, *adv.* **unsteadiness**, *n.*

unstep, v.t. (past, p.p. **-stepped**) (Naut.) to take out of a step or socket. **unsterilized**, **-ised**, a. **unstick**, v.t. (past, p.p. **-stuck**). **unstigmatized**, **-ised**, a. **unstimulated**, a. **unstinted**, a. **unstirred**, a. **unstitch**, v.t. to open by unpicking the stitches of. **unstop**, v.t. (past, p.p. **-stopped**) to free from obstruction; to remove the stopper from, to open. **unstoppable**, a. **unstopped**, a. **unstopper**, v.t. **unstored**, a. **unstrained**, a. not strained, not filtered; not subjected to strain; not forced; easy, natural. **unstrap**, v.t. **unstratified**, a. **unstreamed**, a. **unstressed**, a. not subjected to stress; unaccented. **unstring**, v.t. (past, p.p. **-strung**) to take away the string or strings of; to loosen; to loosen the string or strings of, to relax the tension of (nerves etc.); to remove (pearls, etc.) from a string. **unstructured**, a. not having a formal or rigid structure; loose; relaxed, unceremonious. **unstuck**, a. (sl.) disarranged, disorganized. **to come unstuck**, (sl.) esp. of a plan or course of action, to go wrong or fail. **unstudied**, a. not studied; easy, natural. **unstuffed**, a. **unsubdued**, a. **unsubjugated**, a. **unsubmissive**, a. **unsubmissively**, adv. **unsubmissiveness**, n. **unsubscribed**, a. **unsubstantial**, a. not substantial; having little solidity or validity; unreal. **unsubstantiality**, n. **unsubstantially**, adv. **unsubstantiated**, a. **unsubstantiation**, n. **unsuccessful**, a. **unsuccessfully**, adv. **unsugared**, a. **unsuitable**, a. **unsuitability**, **unsuitableness**, n. **unsuitably**, adv. **unsuited**, a. not suited, not fit or adapted (for or to). **unsullied**, a. **unsummoned**, a. **unsung**, a. not sung; (poet.) not celebrated in verse. **unsupervised**, a. **unsupplied**, a. **unsupported**, a. **unsuppressed**, a. **unsure**, a. **unsurmised**, a. **unsurmountable**, a. **unsurmounted**, a. **unsurpassable**, a. **unsurpassably**, adv. **unsurpassed**, a. **unsurveyed**, a. **unsusceptible**, a. **unsuspected**, a. **unsuspectedly**, adv. **unsuspecting**, a. **unsuspectingly**, adv. **unsuspicious**, a. **unsuspiciously**, adv. **unsuspiciousness**, n. **unsustainable**, a. **unsustained**, a. **unswallowed**, a. **unswathe**, v.t. **unswayed**, a. not swayed, biased, or influenced. **unsweetened**, a. **unswept**, a. **unswerving**, a. **unswervingly**, adv. **unsworn**, a. not sworn; not bound by an oath. **unsymmetrical**, a. out of symmetry; lacking in symmetry. **unsymmetrically**, adv. **unsympathetic**, a. **unsystematic**, a. **untack**, v.t. to undo (something that has been tacked). **untainted**, a. **untalented**, a. **untam(e)able**, a. **untam(e)ableness**, n. **untamed**, a. **untangle**, v.t. to disentangle. **untanned**, a. **untapped**, a. **untarnished**, a. **untasted**, a. **untaught**, a. not instructed, illiterate; ignorant; natural, spontaneous. **untaxed**, a. **unteachable**, a. **untechnical**, a. **untempered**, a. not moderated or controlled. **untempted**, a. **untenable**, a. **untenability**, **untenableness**, n. **untenanted**, a. **untended**, a. **untendered**, a. not offered. **untested**, a. **untether**, v.t. **unthanked**, a. **unthankful**, a. **unthankfully**, adv. **unthankfulness**, n. **unthatched**, a. **unthinkable**, a. incapable of being thought or conceived; (coll.) highly improbable. **unthinking**, a. heedless, careless; done without thought or care. **unthinkingly**, adv. **unthought**, a. not remembered or thought (of). **unthoughtful**, a. **unthoughtfulness**, n. **unthread**, v.t. to take a thread out of (a needle etc.); to find one's way out of (a maze etc.). **unthreaded**, a. not threaded. **unthrifty**, a. **unthriftily**, adv. **unthriftiness**, n. **unthrone**, v.t. to dethrone. **unthwarted**, a. **untidy**, a. **untidily**, adv. **untidiness**, n. **untie**, v.t. to undo (a knot); to unfasten; to free from restrictions. v.i. to become untied. **untied**, a.

until (ŭntīl'), prep. up to the time of; as far as. conj. up to the time when.

un- (cont.) **untiled**, a. not covered with tiles. **untillable**, a. **untilled**, a. **untimbered**, a. **untimely**, a. unseasonable, inopportune; premature. adv. unseasonably, prematurely. **untimeliness**, n. **untinged**, a. **untiring**, a. **untiringly**, adv. **untitled**, a.

unto (ŭn'tu), prep. to.

un- (cont.) **untold**, a. not told, revealed, or communicated; not counted, innumerable. **untormented**, a. **untorn**, a. **untortured**, a. **untouchable**, n. a Hindu belonging to one of the lowest castes or to no caste (nowadays the preferred term is HARIJAN). **untouched**, a. **untoward**, a. unlucky, unfortunate, awkward. **untowardly**, adv. **untraceable**, a. **untraced**, a. **untracked**, a. **untraded**, a. **untrained**, a. **untrammelled**, a. **untransferable**, a. **untranslatable**, a. **untranslated**, a. **untransportable**, a. **untravelled**, a. not having travelled; not travelled over. **untraversed**, a. **untreated**, a. **untried**, a. **untrimmed**, a. **untrod**, **-trodden**, a. **untroubled**, a. not disturbed by care, sorrow, etc.; calm, unruffled. **untrue**, a. not in accordance with facts, false; not faithful, disloyal, inconstant; not conforming to a correct standard. **untruly**, adv. **untruss**, v.t. **untrussed**, a. **untrustworthy**, a. **untrustworthiness**, n. **untruth**, n. contrariety to truth; a falsehood, a lie; want of veracity. **untruthful**, a. **untruthfully**, adv. **untruthfulness**, n. **untuck**, v.t. to unfold or undo, as a tuck. **untunable**, a. **untuneful**, a. **untunefully**, adv. **untutored**, a. uninstructed; raw, crude. **untwine**, v.t., v.i. **untwist**, v.t., v.i. **unurged**, a. **unusable**, a. **unused**, a. not having been used; not accustomed (to). **unusual**, a. not usual; uncommon, strange, remarkable. **unusually**, adv. **unusualness**, n. **unutilized**, **-ised**, a. **unutterable**, a. unspeakable, inexpressible, indescribable, ineffable. **unutterably**, adv. **unuttered**, a. **unvaccinated**, a. **unvalued**, a. not esteemed; not appraised, not estimated; invaluable, inestimable. **unvanquished**, a. **unvaried**, a. **unvarnished**, a. not covered with varnish; not embellished, plain, simple. **unvarying**, a. **unvaryingly**, adv. **unveil**, v.t. to remove a veil or covering from, esp. with public ceremony from (a statue etc.); to reveal, to disclose. v.i. to take one's veil off; to be revealed. **unveiling**, n. **unvenomous**, a. **unventilated**, a. **unverifiable**, a. **unverified**, a. **unversed**, a. not versed or skilled (in). **unvexed**, a. **unviable**, a. **unviolated**, a. **unvisited**, a. **unvoiced**, a. not spoken, not uttered; (Phon.) not voiced. **unwaged**, a. not paid a wage; unemployed or not doing paid work. **unwaked**, **unwakened**, a. **unwalled**, a. **unwanted**, a. **unwarlike**, a. **unwarmed**, a. **unwarned**, a. **unwarrantable**, a. not defensible or justifiable, inexcusable; improper, illegitimate. **unwarrantableness**, n. **unwarrantably**, adv. **unwarranted**, a. not authorized; not guaranteed; not justified. **unwary**, a. **unwarily**, adv. **unwariness**, n. **unwashed**, a. not washed. **the great unwashed**, the mob, the rabble. **unwasted**, a. **unwatched**, a. **unwatchful**, a. **unwatchfulness**, n. **unwatered**, a. not watered, not furnished with water, not diluted, not irrigated. **unwavering**, a. steady, steadfast, firm. **unwaveringly**, adv. **unweaned**, a. **unwearable**, a. **unwearied**, a. **unweariedly**, adv. **unweary**, a. **unwearying**, a. **unwearyingly**, adv. **unweave**, v.t. to undo (something that has been woven); to separate the threads of. **unwed**, **-wedded**, a. **unweeded**, a. **unweighed**, a. **unwelcome**, a. **unwelcoming**, a. **unwell**, a. not well; sick, indisposed. **unwept**, a. not lamented, not mourned. **unwhispered**, a. **unwholesome**, a. **unwholesomely**, adv. **unwholesomeness**, n. **unwieldy**, a. that cannot be easily wielded; bulky, ponderous, clumsy. **unwieldily**, adv. **unwieldiness**, n. **unwifely**, a. **unwill**, v.t. to will the reverse of. **unwilled**, a. involuntary. **unwilling**, a. not willing; averse, reluctant, undesirous (of, to, for, etc.); involuntary. **unwillingly**, adv. **unwillingness**, n. **unwind** (-wīnd'), v.t. (past, p.p. **-wound** (-wownd')) to pull out (something that has been wound); to free from entanglement. v.i. to become unwound; (coll.) to relax. **unwinking**, a. watchful, vigilant. **unwisdom**, n. lack of wisdom, folly. **unwise**, a. not wise, without judgment; foolish. **unwisely**, adv. **unwished**, a. not desired; not sought (for). **unwithered**, a. **unwitnessed**, a. **unwitting** (ŭnwit'ing), a. unconscious, unintentional, inadvertent. **unwittingly**, adv.

un- (cont.) **unwomanly**, a. **unwon**, a. **unwonted**, a. not accustomed; unusual. **unwontedly**, adv. **unwontedness**, n.

unwooded, *a.* unwooed, *a.* **unworkable**, *a.* **unworkmanlike**, *a.* **unworldly**, *a.* not worldly, spiritually minded; pertaining to spiritual things. **unworldliness**, *n.* **unworn**, *a.* never worn, new; not impaired by use. **unworried**, *a.* **unworshipped**, *a.* **unworthy**, *a.* not worthy, not deserving (of); not becoming, not seemly, discreditable. **unworthily**, *adv.* **unworthiness**, *n.* **unwound** (-wownd'), *past, p.p.* UNWIND. **unwounded** (-woon'-), *a.* **unwoven**, *a.* **unwrap**, *v.t.* **unwrinkled**, *a.* **unwritable**, *a.* **unwritten**, *a.* not written; traditional; not distinctly expressed; not written upon, blank. **unwritten law**, *n.* one not formulated in statutes etc., esp. one based on custom and judicial decisions. **unwrought**, *a.* **unyielding**, *a.* unbending, stiff; firm, obstinate. **unyieldingly**, *adv.* **unyieldingness**, *n.* **unyoke**, *v.t.* to loose from or as from a yoke. **unzip**, *v.t. (past, p.p.* -zipped) to undo the zip of.
UP, *(abbr.)* United Presbyterian; Uttar Pradesh.
up (ŭp), *adv.* to a higher place, position, degree, amount, rank, price, musical pitch etc.; to London, to a capital, university, a place farther north, or other place regarded as higher; at or to the time or place referred to; to a standstill; off the ground; to or in an erect or standing posture or a position or condition for action, out of bed, on one's legs, in the saddle; in a state of proficiency; above the horizon; so as to be level with, as high or as far as, equal (to); completely, entirely, effectually; together; into existence, prominence, operation; ready for consideration or attention; appearing in court as a defendant. *prep.* from a lower to a higher place or point of; in an ascending direction on or along, towards the higher part of; towards the interior of; at or in a higher part of. *a.* moving, sloping, or directed towards a higher or more central part; towards the capital; out of bed; finished; ready; knowledgeable (in); ahead (in a competition). *n.* that which is up; a high or higher position; a state or period of success, prosperity etc. *v.t. (past, p.p.* **upped**) to raise; to increase. *v.i.* to do something suddenly and unexpectedly. **on the up and up**, becoming steadily more successful; straight, honest. **time is up**, the allotted time is past; the appointed moment has arrived. **to come up with**, to overtake. **up against**, confronting, having to deal with. **up-and-coming**, *a.* enterprising, alert, keen; promising. **up and doing**, active and busy. **up and down**, (moving) alternately upwards and downwards; undulating; marked by alternate success and failure, good health and poor health etc.; in one place and another; from one place to another; in every direction. **up-and-over**, *a.* of a door, opened by pulling it upwards to a horizontal position. **up and running**, under way, in operation. **up front**, at the front; at the forefront; of money etc., (paid) in advance; honest, straightforward, confident, positive, enthusiastic. **ups and downs**, rises and falls, undulations; vicissitudes, changes of fortune. **up the pole**, crazy; pregnant. **up to**, to an equal height with; equal to; *(sl.)* incumbent upon. **up to anything**, *(coll.)* ready for any devilment, sport, etc. **up to date**, *(coll.)* recent, abreast of the times. **up top**, *(sl.)* in one's head or brain. **up to snuff**, *(coll.)* knowing, cunning, acute, sharp. **up yours**, *(sl. offensive)* expressing contempt, defiance etc. **what's up**, what is going on? **upper**[1], *n. (sl.)* a stimulant drug.
up- *pref.* up, upwards, upper.
Upanishad (oopan'ishad), *n.* one of the philosophical treatises forming the third division of the Vedas.
upas (ū'pəs), *n.* the upas-tree; the poisonous sap of this and other Malaysian trees; corrupting or pernicious influence. **upas-tree**, *n.* a Javanese tree, *Antiaris toxicaria*, the acrid milky juice of which contains a virulent poison, used for poisoning arrows, and formerly believed to destroy animal or vegetable life in its immediate neighbourhood.
up- *(cont.)* **upbeat** (ŭp'-), *n. (Mus.)* an unaccented beat, on which the conductor raises his baton. *a. (coll.)* cheerful, optimistic.
upbraid (ŭpbrād'), *v.t.* to charge; to reproach (with); to reprove with severity. *v.i.* to chide. **upbraider**, *n.* **upbraidingly**, *adv.*
up- *(cont.)* **upbringing** (-ŭp'-), *n.* bringing up, education. **upcast**, *v.t.* to cast or throw up. *a.* (-ŭp'-, *predicatively* -kast'), directed upwards, cast up. *n.* a casting or throwing upwards; the shaft by which air ascends after ventilating a mine. **upcoming**, *a.* about to appear, happen etc. **up-country** (ŭp'-), *adv., a.* towards the interior of a country, inland. **update**, *v.t.* to bring up to date. *n.* a bringing up to date; that which has been updated. **up-end**, *v.t.* to turn over on its end; to transform completely. **upgrade**, *v.t.* to raise (a worker or a job) to a higher grade or status; to raise in quality, value, esteem etc. **upgrowth** (ŭp'-), *n.* the act or process of growing up; that which grows up. **upheaval**, *n.* the act or process of heaving up; *(Geol.)* an elevation of part of the crust of the earth; a violent disturbance, revolution etc. **upheave**, *v.t.* to lift up from beneath. *v.i.* to heave up. **upheld**, *past, p.p.* UPHOLD. **uphill** (ŭp'-), *a.* leading or going up a hill; difficult, arduous, severe. *adv.* (-hil'), in an ascending direction, upwards; against difficulties. *n.* ascending ground. **uphold**, *v.t.* to hold up, to keep erect; to support, to sustain, to maintain; to defend; to approve, to countenance. **upholder**, *n.*
upholster (ŭphōl'stə), *v.t.* to furnish with curtains, carpets, furniture etc.; to furnish or adorn (chairs etc.) with stuffing, cushions, coverings etc.; to cover (with etc.) **upholsterer**, *n.* **upholstery**, *n.*
up- *(cont.)* **upkeep** (ŭp'-), *n.* (cost of) maintenance. **upland** (ŭp'-), *n.* the higher part of a district. *a.* situated on or pertaining to the uplands. **uplands**, *n.pl.* high land, upland. **uplift**, *v.t.* to lift up, to raise; to raise morally or spiritually, edify. *n.* (ŭp'-), an uplifting or upheaval; *(coll.)* spiritual improvement, edification. *a.* uplifted. **uplighting** (ŭp'-), *n.* spot lighting or other room lighting directed upwards. **uplighter**, *n.* **uplying** (ŭp'-), *a.* **upmarket**, *a.* pertaining to buying or dealing in goods of relatively high quality or price. *adv.* towards or into this sector of the market.
upon (ŭpon'), *prep., adv.* on.
upper[1], UP.
upper[2] (ŭp'ə), *a.* higher in place; superior in rank, dignity etc.; *(Geol.)* being the later division of period or system. *n.* the part of a boot or shoe above the sole. **on one's uppers**, *(sl.)* destitute. **upper case**, *n. (Print.)* the case holding capitals, reference marks etc. **upper class**, *n.* the economically and socially most powerful class in a society. **upper cut**, *n.* in boxing, a punch delivered in an upwards direction with a bent arm. **upper deck**, *n. (Naut.)* the full-length deck of a ship above the water-level. **upper hand**, *n.* superiority, mastery. **Upper House**, *n.* The House of Lords. **upper works**, *n. (Naut.)* the parts above the water when a ship is in proper trim for a voyage. **uppermost**, *a.* highest in place, rank, authority etc.; predominant. *adv.* in the highest place; on, at, or to the top. **uppish** (ŭp'ish), *a.* self-assertive, pretentious, putting on airs, snobbish. **uppishly**, *adv.* **uppishness**, *n.*
uppity (ŭp'iti), *(coll.)* uppish.
up- *(cont.)* **upraise**, *v.t.* to raise up; to uplift. **uprate**, *v.t.* to raise to a higher rank, rate or power. **upright** (ŭp'-), *a.* erect, perpendicular; righteous, honest, not deviating from moral rectitude. *adv.* erect, vertically. *n.* an upright timber, pillar, post, or other part of a structure; an upright piano etc. **upright piano**, *n.* one with a vertical case for the strings. **uprightly**, *adv.* **uprightness**, *n.* **uprising** (ŭp'-), *n.* the act of rising up, esp. from bed; an insurrection, a rising, a riot.
uproar (ŭp'raw), *n.* a noisy tumult, a violent disturbance, bustle and clamour. **uproarious** (-raw'-), *a.* noisy and disorderly; extremely funny. **uproariously**, *adv.* noisily; hi-

lariously. **uproariousness**, *n*.
up- (cont.) **uproot**, *v.t.* to tear up by or as by the roots.
uprush (ŭp'-), *n*. an upward rush.
upsadaisy (ŭp'sədāzi), UPSYDAISY.
up- (cont.) **upscale**, (*chiefly N Am. coll.*) *a., adv.* upmarket; trendy.
upset (ŭpset'), *v.t.* (*pres.* **upsetting**, *past, p.p.* **upset**) to overturn; to put out of one's normal state, to disconcert, to distress; to make slightly ill, to put out of sorts; to shorten and thicken (a tire or other metal object) by hammering or pressure. *v.i.* to be overturned. *n*. (-ŭp'-), the act of upsetting; the state of being upset. **upset price**, the lowest price at which property is offered for sale by auction, a reserve price.
upshot (ŭp'shŏt), *n*. the final issue, result, or conclusion (of a matter).
upside-down (ŭpsid·down'), *adv., a*. with the upper part under; in complete disorder and confusion.
up- (cont.) **upstage**, *adv*. at the rear of a stage; away from a film or television camera. *a*. situated upstage; standoffish. *v.t.* to force (an actor) to face away from the audience by taking a position upstage; to draw attention away from. **upstair** (ŭp'-), *a*. pertaining to or in an upper storey. **upstairs** (-steəz), *adv., n*. (in or to) the upper storey or storeys. **to kick upstairs**, KICK. **upstanding**, *a*. erect; honest, upright. **upstart** (ŭp'-), *n*. one who rises suddenly from humble origins to wealth, power or consequence; one who assumes an arrogant bearing. **upstate**, *adv., a*. to or in that part of a state of the US which is away from, and usu. to the north of, the principal city. **upstream**, *a., adv*. against the current; (situated) higher up a river. **upstroke** (ŭp'-), *n*. an upward line in writing. **upsurge** (ŭp'-), *n*. a sudden, rapid rise. *v.i.* (-sœj'), to surge up. **upswept**, *a*. swept or brushed upwards. **upswing** (ŭp'-), *n*. an upward rise; an increase or improvement, esp. in economic terms.
upsydaisy (ŭp'sidāzi), *int*. a reassuring expression to accompany the lifting up of someone, esp. a child, who has stumbled or fallen.
up- (cont.) **uptake** (ŭp'-), *n*. the act of taking or lifting up; the process of taking, absorbing or accepting what is on offer; a pipe, shaft or flue with an upward current. **quick on the uptake**, quick to understand or learn.
up- (cont.) **up-tempo**, *a*. played or sung at a fast tempo.
upthrow (ŭp'-), *n*. a throwing up, an upheaval; (*Geol.*) the upward displacement on one side of a fault. **upthrust** (ŭp'-), *n*. an upward thrust, esp. a geological upheaval.
uptight, *a*. tense, nervy; irritated, indignant; conventional, strait-laced. **uptime** (ŭp'-), *n*. the time during which a machine, esp. a computer, is actually working.
up-to-date, up-to-the-minute UP.
up- (cont.) **uptown** (ŭp'-), *a., adv., n*. (*chiefly N Am*.) (in or towards) the upper, or residential, part of town. **upturn** (ŭp'-), *n*. an upward trend or turn towards improved conditions, higher prices etc.; an upheaval. *v.t.* to turn up or over; to direct upwards.
UPU, (*abbr.*) Universal Postal Union.
upward (ŭp'wəd), *a*. directed, turned, or moving towards a higher place or state. **upwardly**, *adv*. upwards. **upwardly mobile**, aspiring to improve one's lifestyle, social status etc. **upwards**, *adv*. towards a higher place, in an upward direction; towards the source or spring; more. **upwards of**, more than.
up- (cont.) **upwind**, *adv., a*. against the wind; (to or) on the windward side of.
uracil (ū'rəsil), *n*. one of the pyrimidine bases occurring in ribonucleic acid.
uraemia, (*esp. N Am.*) **uremia** (ūrē'miə), *n*. a condition caused by the retention of urea and other noxious substances in the kidneys and bladder. **uraemic**, *a*.
uraeus (ūrē'əs), *n*. the serpent emblem worn on the headdress of ancient Egyptian divinities and kings.

Ural-Altaic (ūrəlaltā'ik), *a*. of or pertaining to the Ural and Altaic mountain ranges or the people inhabiting them; (*Philol.*) denoting a family of Mongoloid, Finnic and allied languages of agglutinative structure spoken in N Europe and Asia.
Uralic (ūral'ik), *a., n*. (relating to) a language group comprising the Finno-Ugric and Samoyed languages.
uranium (ūrā'niəm), *n*. a rare, heavy, white, metallic element found in pitchblende etc. It is radioactive and fissionable, as in the first atom bomb. **uranium bomb**, *n*. an atom bomb using uranium as explosive. **uranic** (-ran'-), **uranous** (ū'-), *a*.
urano-, *comb. form*. sky, the heavens.
uranography (ūrənog'rəfi), *n*. descriptive astronomy. **uranographic, -al** (-graf'-), *a*. **uranographist** (-nog'-), *n*.
uranometry (ūrənom'itri), *n*. the measurement of the heavens or of stellar distances; a map of the heavens showing the relative positions and apparent magnitudes of the stars.
Uranus (ūrā'nəs, ū'rə-), *n*. (*Gr. Myth.*) the most ancient of all the Greek gods, son of Ge and father of Kronos or Saturn and the Titans; a planet situated between Saturn and Neptune, discovered by Sir William Herschel in 1781.
urate (ū'rāt), *n*. (*Chem.*) a salt of uric acid.
urban (œ'bən), *a*. of or pertaining to, situated or living in a city or town. **urban district**, *n*. a district comprising a small town or towns with a small aggregate population or not yet incorporated as a borough. **urban guerrilla**, *n*. a terrorist active in cities and towns. **urban renewal**, *n*. slum clearance or redevelopment. **urbanist**, *n*. (*N Am.*) a town-planner. **urbanite** (-nīt), *n*. a town-dweller. **urbanize, -ise**, *v.t.*
urbane (œbān'), *a*. courteous, polite, suave, refined, polished. **urbanely**, *adv*. **urbanity** (-ban'-), *n*.
urceolus (œsē'ələs), *n*. (*pl.* **-li** (-lī)) (*Nat. Hist.*) a pitcher- or urn-shaped organ. **urceolar** (œ'-), **-late** (-lāt), *a*. pitcher-shaped, with a swelling body and contracted orifice.
urchin (œ'chin), *n*. a roguish, mischievous child; a sea-urchin.
urd (œd), *n*. a bean plant grown for its edible seed, eaten esp. in India; this seed.
Urdu (uə'doo), *n*. Hindustani; the official language of Pakistan, also widely used in India esp. by Muslims.
-ure (-ūə, -yə, -ə), *suf*. forming abstract nouns, as *censure, portraiture, seizure*.
urea (ūrē'ə), *n*. a soluble crystalline compound contained in urine, esp. of mammals. **ureal**, *a*.
ureter (ūrē'tə), *n*. the duct conveying the urine from the kidneys into the bladder. **ureteral, ureteric** (-ter'ik), *a*. **ureteritis** (-ī'tis), *n*. inflammation of the ureter.
urethane (ū'rəthān), *n*. a chemical compound used esp. as a solvent or anaesthetic; polyurethane.
urethra (ūrē'thrə), *n*. (*pl.* **-thrae** (-rē)) the duct by which the urine is discharged from the bladder. **urethral**, *a*. **urethritis** (-thrī'tis), *n*. inflammation of the urethra.
urethroscope (ūrē'thrəskōp), *n*. an instrument for examining the interior of the urethra.
uretic (ūret'ik), *a*. pertaining to urine.
urge (œj), *v.t.* to drive; to impel; to force onwards; to press earnestly with argument, entreaty etc.; to press the acceptance or adoption of, to insist on. *n*. a strong impulse, an inner drive or compulsion. **urgency** (-jənsi), *n*. the quality or state of being urgent; pressure of necessity. **urgent**, *a*. pressing, demanding early attention; demanding or soliciting with importunity. **urgently**, *adv*. **urger**, *n*.
-urgy (-əji, -œji), *comb. form*. technology; technique.
-uria (ūriə), *comb. form*. diseased condition of the urine.
uric URINE.
urinal (ūri'nəl), *a*. a toilet-vessel or fixed receptacle for the use of persons passing urine; a public or private room,

building, enclosure etc. containing urinals; a glass receptacle for holding urine for medical inspection.
urine (ū'rin), *n.* a pale-yellow fluid with an acid reaction secreted from the blood by the kidneys, stored in the bladder, and discharged through the urethra, the chief means for the removal of nitrogenous matter resulting from the decay of tissue. **uric**, *a.* **uric acid**, a white, tasteless and odourless, almost insoluble compound found chiefly in excrement of birds and reptiles, and in small quantities in the urine of mammals. **urinary**, *a.* pertaining to urine. *n.* a reservoir for urine etc. **urinate**, *v.i.* to pass urine. **urination**, *n.* **urinative**, *a.* provoking the discharge of urine; diuretic. **urinogenital** (ūrinōjen'itəl), GENITO-URINARY. **urinology** (-nol'-), UROLOGY. **urinometer** (-nom'itə), *n.* an instrument for ascertaining the specific gravity of urine. **urinometric** (-met'-), *a.* **urinometry** (-nom'-), *n.* **urinoscopy** (-nos'kəpi), -SCOPY. UROSCOPY. **urinous**, *a.*
urite (ū'rit), *n.* the ventral portion of an abdominal segment in arthropods.
urn (œn), *n.* a vase with a foot and usually a rounded body used for preserving the ashes of the dead, for holding water, as a measure, and other purposes; (*fig.*) something in which the remains of the dead are preserved, a grave; a vase-shaped vessel with a tap, and usually a spirit-lamp or other heater, for keeping tea, coffee, bouillon etc., hot. **urn-shaped**, *a.*
uro-[1] *comb. form.* tail, hind part.
uro-[2] *comb. form.* urine.
urochord (ū'rōkawd), *n.* the notochord of larval ascidians and some tunicates; an individual of the Urochordata or Tunicata.
urogenital (ūrōjen'itəl), GENITO-URINARY.
urolith (ū'rəlith), *n.* a calculus in the urinary tract.
urology (ūrol'əji), *n.* the branch of medicine concerned with the genito-urinary tract. **urological, -logic**, *a.* **urologist**, *n.*
uropygium (ūrəpij'iəm), *n.* (*Ornith.*) the terminal part of the body or the rump. **uropygial**, *a.* **uropygial gland**, *n.* a gland at the base of a bird's tail that secretes an oily liquid used in preening the feathers.
uroscopy (ūros'kəpi), *n.* the diagnostic examination of the urine. **uroscopic**, *a.*
urostyle (ū'rəstil), *n.* a bone forming the posterior extremity of the vertebral column in the tailless amphibians. **urostylar** (-stī'-), *a.*
Ursa Major (œ'sə), *n.* (*Astron.*) the constellation, the Great Bear. **Ursa Minor**, *n.* the Little Bear. **ursiform** (-fawm), *a.* like a bear. **ursine** (-sin), *a.* pertaining to or resembling a bear.
Ursuline (œ'sūlin), *n.* one of an order of nuns founded in 1537, devoted chiefly to nursing and the education of girls. *a.* belonging to this. [St *Ursula*]
urticaceous (œtikā'shəs), *a.* of or having the character of nettles; belonging to the plant family that includes the nettles. **urticaria** (-keə'riə), *n.* nettle-rash.
urubu (ooruboo'), *n.* the Central American black vulture.
urus (ū'rəs), *n.* an extinct wild ox, *Bos urus* or *primigenius*, the aurochs.
us (ŭs), *pron.* the objective form of the first person plural pronoun 'we'; used for the singular 'me' in formal statements by the sovereign or a newspaper editor, or in very colloquial spoken use.
US, (*abbr.*) United States.
USA, (*abbr.*) United States Army; United States of America.
usable USE[2].
USAF, (*abbr.*) United States Air Force.
usage (ū'sij, ū'zij), *n.* the manner of using or treating, treatment; customary or habitual practice, esp. as authorizing a right etc.; (an instance of) the way a language is actually used; (*Law*) a uniform and recognized practice.

usance (ū'zəns), *n.* a period of time allowed for payment of a foreign bill of exchange.
use[1] (ūs), *n.* the act of using; the state of being used; employment in or application to a purpose; occasion, need, or liberty to use; the quality of being useful or serving a purpose; utility, serviceableness; custom, practice, wont, usage; a form of ritual, etc., peculiar to a church, diocese, or country; (*Law*) enjoyment of the benefit or profit of lands and tenements held by another in trust for the beneficiary. **in use**, being employed; in customary practice. **to have no use for**, to dislike, to disapprove of. **to make use of**, to use, to employ; to exploit (a person). **useful**, *a.* of use, serving a purpose; good, beneficial, profitable, advantageous; (*coll.*) clever, competent, highly satisfactory. **usefully**, *adv.* **usefulness**, *n.* **useless**, *a.* not of use, serving no useful end or purpose; unavailing, ineffectual; (*coll.*) inept, stupid, unfit. **uselessly**, *adv.* **uselessness**, *n.*
use[2] (ūz), *v.t.* to employ, to apply to a purpose, to put into operation; to turn to account, to avail oneself of; to treat in a specified way; to take advantage of, exploit; to consume, expend, use up; to make a practice of employing; to take (as drugs) regularly; (*usu. in p.p.*) to accustom, to habituate, to inure. *v.i.* (*usu. in past*) to be accustomed, to be wont. **to use up**, to consume, to exhaust. **usable**, *a.* capable of being used. **used**, *a.* already made use of; secondhand; exploited. **used-up**, *a.* exhausted, finished. **user**, *n.* one who uses; (*Law*) continued use or enjoyment of a thing. **user-friendly**, *a.* of computers or software, easy to operate; easy to use, understand etc. **user-friendliness**, *n.*
usher (ŭsh'ə), *n.* an officer or servant acting as doorkeeper (esp. in a court or public hall), or whose business it is to introduce strangers or to walk before a person of rank; a seat-attendant at a cinema, theatre etc. *v.t.* to act as usher to; to introduce, as a forerunner or harbinger, bring or show (in etc.). **usherette** (-ret'), *n.* a female usher at a cinema or theatre. **ushership**, *n.*
USM, (*abbr.*) unlisted securities market.
USN, (*abbr.*) United States Navy.
usquebaugh (ŭs'kwibah. -baw), *n.* whisky; an Irish liqueur made of brandy, spices etc.
USS, (*abbr.*) United States Senate; United States ship.
USSR, (*abbr.*) Union of Soviet Socialist Republics.
usu., (*abbr.*) usual, usually.
usual (ū'zhəl), *a.* such as ordinarily occurs, customary, habitual, common, ordinary, frequent. **usually**, *adv.* **usualness**, *n.*
usufruct (ū'zūfrŭkt), *n.* right to the use and enjoyment of property belonging to another without waste or destruction of its substance. *v.t.* to hold in or subject to usufruct. **usufructuary** (-frŭk'chu-), *n.* one who has usufruct. *a.* relating to or of the nature of a usufruct.
usurer (ū'zhərə), *n.* one who lends money at exorbitant interest. **usurious** (ūzhuə'riəs), *a.* practising usury, exacting exorbitant interest; pertaining to or of the nature of usury. **usuriously**, *adv.* **usuriousness**, *n.* **usury** (-ri), *n.* the practice of lending money at exorbitant interest, esp. higher than that allowed by law; exorbitant interest.
usurp (ūzœp'), *v.t.* to seize or take possession of without right. *v.i.* to take possession without right; to encroach (upon). **usurpation**, *n.* **usurper**, *n.* **usurping**, *adv.*
usury USURER.
USW, (*abbr.*) ultrashort wave.
UT, (*abbr.*) universal time.
ut (ŭt), *n.* (*Mus.*) the first or key note in Guido's musical scale, now usu. superseded by do (see DO[2])).
utensil (ūten'sil), *n.* an implement, an instrument, esp. one used in cookery or domestic work.
uterus (ū'tərəs), *n.* (*pl.* -**ri** (-rī)) the womb. **uterine**, (-rin), *a.* pertaining to the womb; born of the same mother but not the same father. **uteritis** (-rī'tis), *n.* inflammation of

the womb. **uter(o)-**, *comb. form.*
utilitarian (ūtilitēə'riən), *a.* of or pertaining to utility or to utilitarianism; concerned with, or made for, practical use rather than beauty. *n.* an advocate of utilitarianism. **utilitarianism**, *n.* the ethical doctrine that actions are right in proportion to their usefulness or as they tend to promote happiness; the doctrine that the end and criterion of public action is the greatest happiness of the greatest number.
utility (ūtil'iti), *n.* usefulness, serviceableness; that which is useful; the ability to satisfy human wants; a public service, as the supply of water or electricity. *a.* designed or adapted for general use; practical, utilitarian; pertaining to goods mass-produced to standard designs. **utility room**, *n.* a room (in a private house) used for storage, laundry etc.
utilize, -ise, (ū'tiliz), *v.t.* to make use of, to turn to account. **utilizable,** *a.* **utilization,** *n.*
utmost (ŭt'mōst), *a.* being or situated at the farthest point or extremity; farthest, extreme, greatest, ultimate. *n.* the utmost extent or degree.
Utopia (ūtō'piə), *n.* a place or state of ideal perfection. **Utopian**, *a.* pertaining to or resembling Utopia; ideal, perfect or highly desirable but impracticable. *n.* an inhabitant of Utopia; an ardent but visionary political or social reformer. **Utopianism**, *n.* [lit. nowhere, coined by Sir Thomas More as the title of his book (published 1516) describing an imaginary island with a perfect social and political system]

utricle (ū'trikl), *n.* a pouch or sac of an animal or plant; a sac-like cavity, esp. one in the labyrinth of the inner ear. **utricular** (-trik'ū-), *a.*
utter[1] (ŭt'ə), *a.* complete, total, perfect, entire; absolute, unconditional. **utter barrister,** *n.* a junior barrister not allowed to plead within the bar. **utterly,** *adv.* **uttermost,** *a.* utmost. **utterness,** *n.*
utter[2] (ŭt'ə), *v.t.* to give forth audibly; to give expression to; to put (notes, base coin etc.) into circulation. **utterable,** *a.* **utterance,** *n.* the act of uttering; vocal expression; speech, words; power of speaking. **utterer,** *n.*
UU, (*abbr.*) Ulster Unionist.
UV, (*abbr.*) ultraviolet.
uva (ū'və), *n.* (*Bot.*) a succulent indehiscent fruit with a central placenta, as a grape.
uvea (ū'viə), *n.* (*Anat.*) the inner coloured layer of the iris of the eye. **uveal,** *a.*
UVF, (*abbr.*) Ulster Volunteer Force.
uvula (ū'vūlə), *n.* (*pl.* **-lae** (-lē)) a fleshy body hanging from the posterior margin of the soft palate. **uvular,** *a.* pertaining to the uvula; produced with vibration of the uvula.
uxorious (ŭksaw'riəs), *a.* excessively or foolishly fond of one's wife, doting. **uxorial,** *a.* of or pertaining to a wife; uxorious. **uxoricide** (-sid), *n.* wife-murder; a wife-murderer. **uxoriously,** *adv.* **uxoriousness,** *n.*
Uzbeg (ŭz'beg), **Uzbek** (-bek), *n.* a member of one of the Turkish races of Turkestan.

V

V¹, **v**, the 22nd letter, and the 17th consonant (*pl.* **Vs, V's, Vees**), is a voiced labiodental spirant or fricative, produced by the junction of the lower lip and upper teeth, corresponding to the voiceless *f*, which is similarly produced; (*Roman numeral*) 5. **V-bomb**, *n.* a self-propelled rocket or bomb launched by Germany in World War II mainly against Britain, typically **V-1's** or **V-2's**. **V-Day**, *n.* Victory Day, esp. VE Day. **V-neck**, *n.* the neck of any garment when it is shaped like the letter V. **V-necked**, *a.* **V-sign**, *n.* a sign made with index and middle fingers in the form of the letter V, palm outwards as a victory salute, palm inwards as a sign of contempt or derision.
V², (*chem. symbol*) vanadium.
vac (vak), *n.* (*coll.*) short for VACATION, esp. when applied to university holidays.
vacant (vä'kənt), *a.* unfilled, empty, unoccupied; unemployed, at leisure; unintelligent, empty-headed, silly, inane. **vacant possession**, *n.* availability of a house or other property for immediate occupation. **vacancy**, *n.* the state of being vacant, emptiness; mental vacuity, idleness, inanity; empty space, a gap, a chasm; an unfilled or vacant post or office; an unoccupied room in a guest house, hotel etc. **vacantly**, *adv.*
vacate (vəkāt'), *v.t.* to make vacant, to give up occupation or possession of; to annul, to make void.
vacation (vəkā'shən), *n.* the act of vacating; a period of cessation of legal or other business, or of studies at university etc.; a holiday.
vaccinate (vak'sināt), *v.t., v.i.* to inoculate with vaccine to procure immunity from smallpox, or with the modified virus of any disease so as to produce a mild form of it and prevent a serious attack. **vaccination**, *n.* **vaccinationist**, *n.* **vaccinator**, *n.* **vaccine** (-sēn), *a.* of, pertaining to, or obtained from cows; of or pertaining to vaccination. *n.* the virus of cowpox prepared for use in vaccination; any agent used for inoculation and immunization; software designed to halt the progress or neutralize the effect of a computer virus. **vaccinal, vaccinic** (-sin'-), *a.* **vaccinia** (-sin'iə), *n.* cowpox, esp. as produced by inoculation.
vacherin (vash'ri), *n.* a dessert of meringue filled with cream, fruit, nuts etc. (F)
vacillate (vas'ilāt), *v.i.* to sway to and fro, to waver; to oscillate from one opinion or resolution to another, to be irresolute. **vacillatingly**, *adv.* **vacillation**, *n.* **vacillator**, *n.*
vacuole (vak'ūōl), *n.* (*Biol.*) a minute cavity in an organ, tissue etc., containing air, fluid etc. **vacuolar**, *a.* **vacuolate** (-lət), *a.*
vacuous (vak'ūəs), *a.* empty, unfilled, void; unintelligent, blank, expressionless. **vacuousness**, *n.* **vacuity** (-kū'-), *n.*
vacuum (vak'ūəm, -yəm), *n.* (*pl.* **vacuums, vacua** (-ūə)), a space completely devoid of matter; a space or vessel from which the air has been exhausted to the furthest possible extent by an air-pump or analogous means; a partial diminution of pressure, as in a suction-pump, below the normal atmospheric pressure, a state or feeling of emptiness or of deprivation, a void. **vacuum-brake**, *n.* a continuous train-brake in which the pressure applying the brakes is caused by the exhaustion of the air from a bellows pulling the brake-rod as it collapses. **vacuum-cleaner**, *n.* a machine for removing dirt by suction. **vacuum flask**, *n.* a flask constructed with two walls between which is a vacuum, for the purpose of keeping the contents hot or cold. **vacuum-gauge**, *n.* a gauge indicating the pressure consequent on the production of a vacuum. **vacuum packed**, *a.* sealed in a container from which most of the air has been removed. **vacuum pump**, *n.* an airpump used to remove air or other gas, and so create a vacuum. **vacuum tube**, *n.* (*N. Am.*) an electronic valve.
vade-mecum (vahdimē'kəm, -mā'-), *n.* a pocket companion or manual for ready reference.
vagabond (vag'əbond), *a.* wandering about, having no settled habitation, nomadic; driven or drifting to and fro, aimless. *n.* one who wanders about without any settled home, a wanderer, esp. an idle or disreputable one, a vagrant; a scamp, a rogue. **vagabondage** (-dij), **vagabondism**, *n.* **vagabondish**, *a.*
vagal (vā'gəl), VAGUS.
vagary (vā'gəri), *n.* a whimsical idea, an extravagant notion, a freak.
vagina (vəji'nə), *n.* (*pl.* **-nas, -nae** (-nē)) a sheath, a sheath-like envelope or organ; the genital passage of a female from the vulva to the uterus. **vaginal** (-jī'-, vaj'i-), **vaginate** (vaj'ināt, -nət), **-nated**, *a.* sheathed. **vaginismus** (vajiniz'məs), *n.* spasmodic contraction of the vaginal sphincters. **vaginitis** (vajini'tis), *n.* inflammation of the vagina. **vaginotomy** (vajinot'əmi), *n.* incision of the vagina.
vagrant (vā'grənt), *a.* wandering about without a settled home; itinerant, strolling; roving, unrestrained. erratic, random. *n.* a wanderer, an idle person, a vagabond, a tramp; (*Law*) a person wandering about begging or without visible means of subsistence. **vagrancy**, *n.* **vagrantly**, *adv.*
vague (vāg), *a.* indistinct, of doubtful meaning or application, ambiguous, indefinite, ill-defined; absentminded, inclined to or expressive of imprecision of thought. **vaguely**, *adv.* **vagueness**, *n.*
vagus (vā'gəs), *n.* the tenth cranial nerve which regulates the heart beat, rhythm of breathing etc. **vagal**, *a.*
vain (vān), *a.* empty, unsubstantial, unreal, worthless; fruitless, ineffectual, unavailing, unproductive; proud of petty things or of trifling attainments, conceited, self-admiring; foolish, silly. **in vain**, to no purpose; ineffectually. **to take someone's name in vain**, to use someone's name in a pejorative, insulting or blasphemous way. **vainglory**, *n.* excessive vanity; vain pomp or show; pride, boastfulness. **vainglorious**, *a.* **vaingloriously**, *adv.* **vaingloriousness**, *n.* **vainly**, *adv.*
vair (veə), *n.* (*Her.*) a fur represented by shield-shaped figures of argent and azure alternately.
Vaishnava (vīsh'navə), *n.* one of the great sects of reformed Brahmins who worship Vishnu as supreme among the Hindu gods.
Vaisya (vīs'yə), *n.* the third of the four chief Hindu castes; a member of this.
valance (val'əns), *n.* a short curtain; the hanging round the frame or tester of a bedstead. [perh. from *Valence* in France]
vale (vāl), *n.* a valley; a little trough or channel. **Vale of tears**, human life, existence, the world.
valediction (validik'shən), *n.* a bidding farewell; a farewell, an adieu. **valedictorian** (-taw'ri-), *n.* (*N Am.*) a student

valence 911 **vampire**

who delivers a valedictory. **valedictory,** *a.* bidding farewell; pertaining to or of the nature of a farewell. *n.* a parting address or oration, esp. at graduation in an American university.
valence (vā'ləns), VALENCY.
Valenciennes (valēsyen'), *n.* a fine variety of lace the design of which is made with and of the same thread as the ground. [*Valenciennes* in France]
valency (vālənsi), *n.* the combining or replacing power of an element or radical reckoned as the number of monovalent elements it can replace or combine with; a unit of combining capacity.
valentine (val'əntīn), *n.* a sweetheart chosen on St Valentine's day; a letter or card of an amatory or satirical kind sent to a person of the opposite sex on St Valentine's day. **St Valentine's day**, 14 Feb., commemorating the day when St Valentine was beheaded by the Romans and when birds were supposed to begin to mate.
valerian (vəliə'riən), *n.* a herbaceous plant of the genus *Valeriana* with clusters of pink or white flowers; a preparation from the root of *V. officinalis* used as a mild stimulant etc. **valeric acid,** *n.* a fatty acid of disagreeable smell obtained from valerian. **valerate** (val'ərət), *n.* a salt of valeric acid. **valeric** (-le'rik), *a.*
valet (val'it, val'ā), *n.* a manservant who attends on his master's person. *v.t.* to act as valet to. **valet de chambre** (val'ädəshăbr''), a valet.
valeta VELETA.
valetudinarian (valitūdineə'riən), *a.* sickly, infirm, delicate; morbidly anxious about one's state of health. *n.* an invalid; a valetudinarian person; a hypochondriac. **valetudinarianism,** *n.* **valetudinary** (-tū'-), *n., a.* valetudinarian.
valgus (val'gəs), *n.* the condition of a bone or body part of being twisted away from the line of the body.
Valhalla (valhal'ə), *n.* in Norse mythology, the palace of immortality where the souls of heroes slain in battle were carried by the valkyries.
valiant (val'iənt), *a.* brave, courageous, intrepid. **valiantly,** *adv.*
valid (val'id), *a.* well-grounded, sound, cogent, logical, incontestable; (*Law*) legally sound, sufficient, and effective, binding. **validate,** *v.t.* to make valid, to ratify, to confirm, to make binding. **validation,** *n.* **validity** (-lid'-), **validness,** *n.* **validly,** *adv.*
valine (vā'lēn, val'-), *n.* an amino acid that is essential to health and growth in humans and other vertebrates.
valise (vəlēs'), *n.* a small travelling bag or case; (*N Am.*) a suit-case.
Valium® (val'iəm), *n.* a brand name for the tranquillizer diazepam.
Valkyrie (val'kiri, -kiə'ri), *n.* in Norse mythology, one of 12 maidens who were sent by Odin to select those destined to be slain in battle and to conduct their souls to Valhalla. **Valkyrian** (-ki'ri-), *a.*
vallecula (vəlek'ūlə), *n.* (*pl.* **lae** (-lē)) (*Anat., Bot. etc.*) a groove or furrow.
valley (val'i), *n.* a depression in the earth's surface bounded by hills or mountains, and usu. with a stream flowing through it; any hollow or depression between higher ground or elevations of a surface; the area drained by a river; the internal angle formed by two inclined sides of a roof.
vallonia, valonia (vəlō'niə), *n.* the large acorn-cup of the vallonia oak, used for dyeing, tanning, ink-making etc. **vallonia oak,** an evergreen oak of the Greek archipelago etc.
vallum (val'əm), *n.* a Roman rampart, an agger.
valonia VALLONIA.
valorize, -ise (val'əriz), *v.t.* to increase or stabilize the price by an officially organized scheme.
valour (val'ə), (*esp. N. Am.*) **valor,** *n.* personal bravery, courage, esp. as displayed in fighting; prowess. **valorous,** *a.* **valorously,** *adv.*
valse (vals), *n.* a waltz. [F]
valuable, valuation VALUE.
value (val'ū), *n.* worth, the desirability of a thing, esp. as compared with other things; the qualities that are the basis of this; worth estimated in money or other equivalent, the market price; the equivalent of a thing; valuation, estimation, appreciation of worth; (*pl.*) moral principles, standards, priorities; meaning, signification, import; the relative duration of a tone as indicated by the note; the relation of the parts of a picture to each other with regard to light and shade, apart from colour; the amount or quantity denoted by a symbol or expression. *v.t.* to estimate the value of, to appraise; to esteem, to rate highly, to prize; to reckon the monetary worth of. **commercial, economic, exchange,** or **exchangeable value, value in exchange,** the value in terms of other commodities, the purchasing power of a commodity in the open market; the market price as determined by economic laws. **value-added tax,** in Britain, a tax levied at each stage of production and distribution of a commodity or service and paid by the buyer as a purchase tax. **value judgement,** *n.* a subjective and personal estimate of merit in a particular respect. **valuable,** *a.* having great value, worth, or price, costly, precious; capable of being valued or appraised; worthy, estimable. **valuableness,** *n.* **valuably,** *adv.* **valuate,** *v.t.* to appraise the value of. **valuation,** *n.* the act of valuing or appraising; estimation of the value of a thing; estimated value or worth, the price placed on a thing. **valuator,** *n.* an appraiser. **valueless** (-lis), *a.* of no value, worthless, futile. **valuelessness,** *n.* **valuer,** *n.* one who values, an appraiser, esp. of property, jewellery etc.
valuta (vəloo'tə), *n.* the value of one currency in terms of another.
valve (valv), *n.* an automatic or other contrivance for opening or closing a passage or aperture so as to permit or prevent passage of a fluid, as water, gas, or steam; (*Anat.*) a membranous part of a vessel or other organ preventing the flow of liquids in one direction and allowing it in the other; a device on a brass instrument for varying the effective length of the tube; (*Bot.*) one of the segments into which a capsule dehisces, either half of an anther after its opening; a shortened form of electronic or thermionic valve; a vacuum tube or bulb containing electrodes and exhibiting sensitive control by one or more electrodes of the current flowing between the others; one of the parts or divisions of a shell. **valval,** *a.* **valvate** (-vāt), *a.* like or having a valve; descriptive of petals which meet at the margins only. **valved,** *a.* (*usu. in comb.* as *three-valved*). **valveless** (-lis), *a.* **valvelet** (-lit), **valvule** (-vūl), *n.* a little valve. **valvular** (-vū-), *a.* pertaining to, operating by or having a valve; resembling a valve. **valvular disease,** *n.* disordered action of the heart owing to defects in the cardiac valves.
vamoose (vəmoos'), **vamose** (vəmōs'), *v.i.* (*chiefly N Am. sl.*) to decamp, to be gone, to be off. *v.t.* to decamp from.
vamp¹ (vamp), *n.* the part of a boot or shoe upper in front of the ankle seams; a patch intended to give a new appearance to an old thing; (*Mus.*) an improvised accompaniment. *v.t.* to put a new vamp on (a boot etc.); to give a new appearance to, to make more modern; (*Mus.*) to improvise an accompaniment to. *v.i.* to improvise accompaniments. **vamper,** *n.*
vamp² (vamp), *n.* an adventuress, a woman who exploits her charms to take advantage of men. *v.t.* to fascinate, to ensnare. *v.i.* to act the role of vamp.
vampire (vam'pīə), *n.* a ghost of a heretic, criminal, or other outcast, supposed to leave the grave at night and suck the blood of sleeping persons; one who preys upon other, a blood-sucker; a bat, e.g. of the genus *Desmodus*, which feeds on the blood of animals; (*Theat.*) a small double spring-door used for sudden entrances and exits.

van 912 **variance**

vampiric (-pi'rik), *a.* **vampirism** (-pi-), *n.* belief in vampires; blood-sucking; (*fig.*) extortion.
van[1] (van), *n.* the vanguard.
van[2] (van), *n.* a motor vehicle, usu. covered, for conveying goods, furniture etc.; a closed railway-carriage for luggage or for the guard. *v.t.* (*past, p.p.* **vanned**) to convey in a van.
van[3] (van), *n.* a fan or machine for winnowing grain; (*Mining*) a test of the quality of ore by washing on a shovel etc. *v.t.* to test (ore) thus. **vanner,** *n.*
vanadium (vənā'diəm), *n.* a rare, silver-white metallic element, at. no. 23; chem. symbol V, used to give tensile strength to steel and, in the form of its salts, to produce an intense permanent black colour. **vanadate** (van'ədāt), *n.* a salt of vanadic acid. **vanadic** (-nad'-), **vanadous** (van'ə-), *a.*
Van Allen belt (van al'ən), *n.* a belt of intense particle radiation in the earth's outer atmosphere. [James A. *Van Allen,* b. 1914, US physicist]
V and A, (*abbr.*) Victoria and Albert Museum.
Vandal (van'dəl), *n.* one of a Teutonic people from the shores of the Baltic that overran Gaul, Spain, and N Africa and Rome in the 5th cent., destroying works of art etc.; (**vandal**) one who wilfully or ignorantly destroys or damages anything. **Vandalic** (-dal'-), *a.* **vandalize, -ise,** *v.t.* to destroy or damage deliberately and senselessly. **vandalism,** *n.*
Van de Graff generator (van'dəgrahf), *n.* an electro-static generator. [Robert *Van de Graff*, 1901–1967, US physicist]
Vandyke (vandīk'), *n.* a picture by the Flemish painter Sir Anthony Van Dyck (1599–1641); any one of the series of points forming an ornamental border to lace, linen etc.; a Vandyke collar or cape. *a.* applied to the style of dress, esp. ornamented with vandykes, worn by the figures in Van Dyck's portraits. *v.t.* to cut the edge of (linen etc.) into Vandykes. **Vandyke beard,** *n.* a pointed beard. **Vandyke brown,** *n.* a reddish-brown colour or pigment. **Vandyke cape, collar,** *n.* one ornamented with vandykes.
vane (vān), *n.* a weathercock, flag or arrow pointing in the direction of the wind; a similar device on an axis turned by a current of water etc., as in a meter; a fin on a bomb to ensure its falling on its war-head; the arm of a windmill; the blade of a propeller etc.; a horizontal part on a surveyor's levelling-staff for moving up and down to the line of sight of the telescope; the sight on a quadrant, compass etc.; the broad part of a feather; (*Naut.*) a slender streamer used to show the direction of the wind, a dog-vane. **vaned,** *a.* **vaneless** (-lis), *a.*
vang (vang), *n.* either of a pair of guy-ropes running from the peak of a gaff to the deck to steady it.
vanguard (van'gahd), *n.* the troops who march in the front of an army, an advance-guard; the leaders or leading position in a movement etc.
vanilla (vanil'ə), *n.* any of a genus (*Vanilla*) of tall, epiphytal orchids, natives of tropical Asia and America, bearing fragrant flowers; the fruit of *Vanilla planifolia* and other species used for flavouring; an extract from this used for flavouring ices, syrups etc. **vanillic,** *a.* pertaining to or derived from vanilla. **vanillin,** *n.* the aromatic principle of vanilla.
vanish (van'ish), *v.i.* to disappear suddenly; to become imperceptible, to be lost to sight, to fade away, to dissolve; to pass away, to pass out of existence; (*Math.*) to become zero. **vanishing cream,** *n.* a cosmetic which is rapidly absorbed into the pores leaving no trace of grease. **vanishing point,** *n.* the point at which all parallel lines in the same plane tend to meet.
vanity (van'iti), *n.* the quality or state of being vain; empty pride, conceit of one's personal attainments or attractions; ostentation, show; emptiness, futility, unreality, worthlessness; that which is vain, unreal or deceptive.

vanity bag, case, *n.* a small case used to carry a woman's make-up and toiletries.
vanner VAN [3].
vanquish (vang'kwish), *v.t.* to conquer, to overcome, to subdue, to refute. **vanquishable,** *a.* **vanquisher,** *n.*
vantage (vahn'tij), *n.* advantage; a situation, condition, or opportunity favourable to success; (*Lawn Tennis*) the point scored by either side after deuce or five all. **vantage-ground,** *n.* superiority of position or place.
vapid (vap'id), *a.* insipid, flat, spiritless. **vapidity** (-pid'-), **vapidness,** *n.* **vapidly,** *adv.*
vaporize etc. VAPOUR.
vaporetto (văpəret'ō), *n.* (*pl.* **-ttos**) a small steamship that travels the canals of Venice.
vapour (vā'pə), *n.* moisture in the air, light mist; (*loosely*) any visible diffused substance floating in the atmosphere; the gaseous form of a substance that is normally liquid or solid; an unreal or unsubstantial thing; or foolish imagining, a fanciful notion; a remedial preparation applied by inhaling; (*pl.*) depression of spirits. *v.i.* to give out vapour; to boast, to brag, to bluster. **vapour density,** *n.* the density of a gas or vapour relative to hydrogen at the same temperature and pressure. **vapour pressure,** *n.* the pressure exerted by a vapour that is in equilibrium with its solid or liquid form. **vapour trail,** *n.* a white trail of condensed vapour left in the sky after the passage of an aircraft. **vapour ware,** *n.* (*coll.*) computer hardware or software that is planned or promised but does not materialize. **vaporiferous** (-rif'-), **vaporific** (-rif'-), *a.* **vaporimeter** (-rim'itə), *n.* an instrument for measuring the pressure of vapour. **vaporize, -ise,** *v.t.* to convert into vapour; to destroy by causing to become vapour; to cause to vanish suddenly as if by conversion into vapour. *v.i.* to become vaporized. **vaporizable,** *a.* **vaporization,** *n.* **vaporizer,** *n.* one that vaporizes; an atomizer. **vaporous,** *a.* resembling, in the form of, containing of, full of vapour; unsubstantial, ethereal, flimsy; fanciful, foolish. **vaporosity** (-ros'-), **vaporousness,** *n.* **vaporously,** *adv.* vapoury. **vapourish,** *a.* **vapourishness,** *n.* **vapoury,** *a.* like or full of vapour; given to the vapours.
vapourer moth, *n.* a tussock moth, the female of which has vestigial wings and cannot fly.
vaquero (vəkeə'rō), *n.* (*pl.* **-ros**) (*Mexico, US*) a herdsman, a cowherd.
varactor (vərak'tə), *n.* a two-electrode semi-conductor device in which capacitance varies with voltage.
Varangian (vəran'jiən), *n.* one of the Norse sea-rovers in the 8th to 12 cents. who ravaged the coasts of the Baltic and conquered part of Russia. **Varangian Guard,** *n.* the body-guard of the Byzantine emperors, formed partly of Varangians.
varec (va'rik), *n.* an impure carbonate of soda made in Brittany.
vari-, *comb. form.* various, variegated.
variable (veə'riəbl), *a.* capable of varying, liable to change; changeable, unsteady, fickle, inconstant; able to be varied, adapted or adjusted; quantitatively indeterminate, susceptible of continuous change of value, esp. assuming different values while others remain constant; (*Biol.*) tending to variations of structure, function etc. *n.* that which is variable; (*Math.*) a variable quantity; a symbol representing this; a variable star; (*Naut.*) a shifting wind, (*pl.*) the region between northerly and southerly tradewinds. **variable cost,** *n.* a cost which varies with level of output. **variable star,** *n.* a star whose apparent magnitude is not constant. **variability** (-bil'-), **variableness,** *n.* **variably,** *adv.*
variance (veə'riəns), *n.* the state of being variant, disagreement, difference of opinion, dissension, discord; (*Law*) disagreement between the allegations and proof or between the writ and the declaration; (*Statistics*) a measure of the dispersion of a set of observations. **variant,** *a.*

showing variation, differing in form, character, or details; varying slightly from the standard; tending to vary, changeable. *n.* a variant form, reading, type etc.

variation (veəriā'shən), *n.* the act, process or state of varying; alteration, change, modification, deviation, mutation; the extent to which a thing varies; deviation of a heavenly body from the mean orbit or motion; the angle of deviation from true north or of declination of the magnetic needle; the deviation in structure or function from the type or parent form; (*Math.*) the relation between the changes of quantities that vary as each other; permutation; (*Mus.*) a repetition of a theme with fanciful elaborations and changes of form. **variational**, *a.* **variative** (veə'ri-), *a.*

varicella (varisel'ə), *n.* chicken-pox. **varicellar, varicelloid** (-loid), *a.*

varices, *n.pl.* VARIX.

varicoloured (veə'rikŭləd), *a.* variously coloured, variegated.

varicose (va'rikōs), *a.* esp. of veins, abnormally and permanently dilated, affected with varix; pertaining to varices. **varicocele** (-sēl), *n.* a tumour formed by varicose veins of the spermatic cord. **varicosed**, *a.* **varicosity** (-kos'-), *n.* the state of being varicose; a varicose part or vessel.

varied VARY.

variegate (veə'rigāt), *v.t.* to diversify in colour, to mark with patches of different hues; to dapple, to chequer. **variegated**, *a.* **variegation**, *n.*

variety (vərī'əti), *n.* the quality or state of being various; diversity, absence of sameness or monotony, many-sidedness, versatility; a collection of diverse things; a minor class or group of things differing in some common peculiarities from the class they belong to; a kind, a sort, a thing of such a sort or kind; an individual or group differing from the type of its species in some transmittable quality but usually fertile with others of the species, a sub-species. **variety entertainment, show**, *n.* an entertainment consisting of singing, dancing, acrobatic turns, conjuring etc. **variety theatre**, *n.* one for variety shows, a music-hall. **varietal**, *a.* pertaining to or constituting a variety, esp. a biological variety; of a wine, made from a single grape variety. *n.* a varietal wine. **varietally**, *adv.* **variform** (veə'rifawm), *a.* varying in form, of different shapes.

variola (vərī'ələ), *n.* smallpox. **variolar, variolic** (veəriol'-), **variolous**, *a.* **variole** (va'riōl), *n.* a shallow pit-like depression. **variolate** (-lāt), **-lated**, *a.* **variolite** (va'riəlīt), *n.* a variety of spherulitic basalt with a surface resembling skin marked with smallpox. **variolitic**, *a.* **varioloid** (va'riəloid), *a.* resembling or of the nature of smallpox. *n.* a mild form of smallpox, esp. as modified by previous inoculation.

variorum (veəriaw'rəm), *a.* of an edition of a work, with notes of various commentators inserted. **variorum edition**, *n.* an edition of a classic etc. with comparisons of texts and notes by various editors and commentators.

various (veə'riəs), *a.* differing from each other, diverse; unlike, dissimilar; sundry, divers, several; variable; separate, distinct. **variously**, *adv.* **variousness**, *n.*

varistor (vəris'tə), *n.* a semiconductor device whose resistance depends on the applied voltage. [*variable resistor*]

varix (va'riks), *n.* (*pl.* **-ices** (-risēz)) a permanent dilatation of a vein or other vessel; a varicose vessel; one of the ridges traversing the whorls of a univalve shell.

varlet (vah'lit), *n.* a page, an attendant preparing to be a squire; a menial; a knave, a rascal.

varmint (vah'mint), *n.* (*prov.*) a troublesome or mischievous person or animal.

varna (vah'nə), *n.* any of the four great Hindu castes.

varnish (vah'nish), *n.* a thin resinous solution for applying to the surface of wood, metal etc., to give it a hard, transparent, shiny coating; any lustrous or glossy appearance on a surface; the lustrous surface or glaze of pottery etc.; superficial polish, gloss, palliation, whitewash. *v.t.* to cover with varnish; to give an improved appearance to, to gloss over, to whitewash. **varnish-tree**, *n.* any tree from which the material for varnish is obtained. **varnisher**, *n.*

varsity (vah'siti), *n.* (*coll.*) university.

varus (veə'rəs), *n.* the condition of a bone or body part of being twisted inwards towards the midline of the body; a variety of club-foot in which the foot is bent inwards, talipes varus.

varve (vahv), *n.* a seasonal layer of clay deposited in still water, used to fix Ice Age chronology.

vary (veə'ri), *v.t.* (*past, p.p.* **varied** (-rid)) to change, to alter in appearance, form or substance; to modify, to diversify; to make variations of (a melody etc.). *v.i.* to be altered in any way; to undergo change; to be different or diverse, to differ, to be of different kinds; to increase or decrease proportionately with or inversely to the increase or decrease of another quantity.

vas (vas), *n.* (*pl.* **vasa** (vā'sə)) a vessel or duct. **vas deferens** (-def'ərənz), *n.* (*pl.* **vasa deferentia** (-defərən'shiə)) the spermatic duct. **vasal** (vā'-), *a.*

vascular (vas'kūlə), *a.* of, consisting of, or containing vessels or ducts for the conveyance of blood, chyle, sap etc.; containing or rich in blood-vessels. **vascularity** (-la'ri-), *n.* **vascularize, -ise**, *v.t.* **vascularization**, *n.* **vascularly**, *adv.*

vasculum (vas'kūləm), *n.* (*pl.* **-la** (-lə),) a botanist's collecting-case, usu. of tin.

vase (vahz), *n.* an ornamental vessel of glass, pottery etc., used esp. for holding cut flowers, and in ancient times for various domestic purposes. **vaseful**, *n.*

vasectomy (vəsek'təmi), *n.* excision of the vas deferens or part of it to produce sterility.

Vaseline® (vas'əlēn), *n.* a yellow, soft, medicated paraffin jelly employed as a lubricant etc.

vasi-, vaso-, *comb. form.* pertaining to a vas, vessel or duct.

vasiform (vas'ifawm), *a.* having the form of a vas.

vaso- VASI-.

vasoconstrictor (văzōkənstrik'tə), *n.* a nerve fibre, drug etc. causing constriction of a blood-vessel. **vasoconstriction**, *n.*

vasodilator (văzōdilā'tə), *n.* a nerve, drug etc. causing dilatation of a blood vessel. **vasodilatation** (-dilə-), *n.*

vasomotor (văzōmō'tə), *a.* causing constriction or dilatation in a vessel. *n.* a vasomotor agent or drug. **vasomotorial** (-taw'ri-), *a.*

vasopressin (văzōpres'in), *n.* a pituitary hormone that raises blood pressure and decreases urine production. **vasopressor**, *n.*, *a.* (an agent) causing a rise in blood pressure by constricting the arteries.

vassal (vas'əl), *n.* one holding land under a superior lord by feudal tenure, a feudatory; a slave, a humble dependant. *a.* pertaining to a vassal; servile. **vassalage** (-lij), *n.* the state or condition of a vassal; the obligation of a vassal to feudal service; servitude, dependence; a fief; vassals collectively.

vast (vahst), *a.* of great extent, immense, huge, boundless; very great in numbers, amount, degree etc. *n.* a boundless expanse. **vastly**, *adv.* **vastness**, *n.* **vasty**, *a.* (*poet.*) vast.

vastus (vas'təs), *n.* a large muscular mass on the outer or inner surface of the thigh.

VAT (vat), (*abbr.*) value-added tax.

vat (vat), *n.* a large tub, tank, or other vessel used for holding mash or hop-liquor in brewing and in many manufacturing operations in which substances are boiled or steeped. *v.t.* (*past, p.p.* **vatted**) to put into or treat in a vat.

vatic (vat'ik), *a.* prophetic, oracular.

Vatican (vat'ikən), *n.* the palace of the Pope on the Vatican

vaticinate 914 velarize

hill in Rome; the papal government. **Vatican City,** *n.* a small area on the Vatican Hill set up as an independent state in 1929. **Vatican Council,** *n.* the 20th Ecumenical Council (1869–70) at which the infallibility of the Pope when speaking ex cathedra was affirmed; (also **Vatican II**) a similar council held between 1962 and 1965.
vaticinate (vətis'ināt), *v.t.*, *v.i.* to prophesy. **vaticinal,** *a.* prophetic. **vaticination,** *n.* **vaticinator,** *n.* a prophet.
vaudeville (vawd'əvil), *n.* a slight dramatic sketch or pantomime interspersed with songs and dances; a miscellaneous series of sketches, songs etc., a variety entertainment. **vaudevillian** (-vil'-), *n.* one who performs in vaudeville.
Vaudois[1] (vōdwah'), *a.* of or pertaining to the canton of Vaud, Switzerland. *n. (pl.* **Vaudois**) a native or inhabitant of Vaud; the Vaudois dialect.
Vaudois[2] (vōdwah'), *a.* of or pertaining to the Waldenses. *n. (pl.* **Vaudois**) one of the Waldenses.
vault[1] (vawlt), *n.* an arched roof, a continuous arch or semi-cylindrical roof, a series of arches connected by radiating joints; an arched chamber, esp. underground; a cellar; a place of interment built of masonry under a church or in a cemetery; any vault-like covering or canopy, as the sky; (*Anat.*) an arched roof of a cavity. *v.t.* to cover with, or as with, a vault or vaults; to construct a vault or vaults; a vaulted structure or framework.
vault[2] (vawlt), *v.i.* to leap, to spring, esp. with the hands resting on something or with the help of a pole. *v.t.* to leap over thus. *n.* such a leap. **vaulting-horse,** *n.* a wooden horse or frame for vaulting over in a gymnasium. **vaulter,** *n.*
vaunt (vawnt), *v.i.* to boast, to brag. *v.t.* to boast of; to praise or display boastfully. *n.* a boast. **vaunter,** *n.* **vauntingly,** *adv.*
vavasour (vav'əsuə), *n.* a vassal holding land from a great vassal and having other vassals under him. **vavasory** (-səri), *n.* the tenure or lands of a vavasour.
VC, (*abbr.*) Victoria Cross; Vice-Chancellor; Vice-Consul; Vice-Chairman.
vc, (*abbr.*) violoncello.
VCR, (*abbr.*) video cassette recorder.
VD, (*abbr.*) venereal disease.
VDT, (*abbr.*) visual display terminal.
VDU, (*abbr.*) visual display unit.
VE, (*abbr.*) Victory in Europe. **VE Day,** *n.* the day, 8 May 1945, on which hostilities in Europe in World War II officially ceased.
've, contr. form of *have.*
veal (vēl), *n.* the flesh of a calf as food. **vealy,** *a.*
vector (vek'tə), *n.* a vector quantity; a line in space or in a diagram representing the magnitude and direction of a vector quantity; an organism (such as an insect) that carries a disease-causing agent from one host to another. *v.t.* to direct (aircraft) to a particular point. **vector quantity,** *n.* a quantity having both magnitude and direction (e.g. velocity but not temperature). **vectorial** (-taw'ri-), *a.*
Veda (vā'də), *n.* the ancient Hindu scriptures, divided into four portions or books (the *Rig-, Yajur-, Sāma-,* and *Artharva-Veda*). **Vedanga** (-dang'gə), *n.* a work supplementary or auxiliary to the Veda. **Vedanta** (-dan'tə), *n.* a system of philosophy founded on the Veda. **Vedantic,** *a.* **Vedantist,** *n.*, *a.* **Vedic,** *a.*
vedette (videt'), *n.* a sentinel (usu. mounted) stationed in advance of an outpost; a small vessel used for scouting purposes etc.
vee (vē), *n.* the 22nd letter of the alphabet, V, v; anything in the shape of this letter.
veer (viə), *v.i.* of the wind, to change direction, esp. in the direction of the sun; to shift, to change about, esp. in opinion, conduct etc. *v.t.* to let out or slacken (a rope etc.); to wear (a ship); to shift, to change. *n.* a shift in direction or course.

veg (vej), *n.* (*coll.*) short for VEGETABLE.
vegan (vē'gən), *n.* one who believes in the use for food, clothing etc. of vegetable products only, thus excluding dairy products, leather etc.; one who uses no animal products.
vegetable (vej'təbl), *n.* a plant; a herbaceous plant used for culinary purposes; one whose mental or physical capabilities are minimal as a result of injury, disease etc.; a dull, idle or passive person. *a.* pertaining to, of the nature of, or resembling, a plant; made of or pertaining to culinary vegetables. **vegetable-ivory** IVORY. **vegetable kingdom,** *n.* the division of organic nature comprising plants. **vegetable marrow,** *n.* the fruit of a species of gourd used as a culinary vegetable. **vegetable oil,** *n.* an oil obtained from seeds or plants, used in cooking etc. **vegetable oyster,** *n.* salsify. **vegetal** (vej'ə-), *a.* pertaining to, or of the nature of plants; vegetative. **vegetarian** (-əteə'ri-), *n.* one who abstains from animal food, and lives on vegetables, fruits, cereals etc. and sometimes eggs and dairy products. *a.* of vegetarians; consisting of vegetables, fruit and cereals. **vegetarianism,** *n.* **vegetate** (-itāt), *v.i.* to grow in the manner of a plant, to exercise the functions of a vegetable; to live an idle, passive, monotonous life. **vegetation,** *n.* the act or process of vegetating; vegetables or plants collectively, plant-life; all the plants in a specified area; an excrescence on the body. **vegetative,** *a.* growing; pertaining to or involved with growth; concerned with or promoting plant life or growth; of reproduction or propagation, not involving sexual processes; of the involuntary functions of growth, circulation, secretions, digestion etc. common to plants and animals; lacking intellectual activity or stimulation, passive, dull. **vegetatively,** *adv.* **vegetativeness,** *n.*
veggie (vej'i), *n.* (*coll.*) a vegetarian; a vegetable.
vehement (vē'əmənt), *a.* proceeding from or exhibiting intense fervour or passion, ardent, passionate, impetuous; acting with great force, energy or violence. **vehemently,** *adv.* **vehemence,** *n.*
vehicle (vē'ikl), *n.* any kind of carriage or conveyance for use on land, having wheels or runners; any liquid etc. serving as a medium for pigments, medicinal substances etc.; any person or thing employed as a medium for the transmission of thought, feeling etc. **vehicular, -lary** (-hik'ū-), *a.*
veil (vāl), *n.* a more or less transparent piece of cloth, muslin etc., usu. attached to the head-dress, worn to conceal, shade, or protect the face; a nun's head-dress; a curtain or other drapery for concealing or protecting an object; a mask, a disguise, a pretext; a velum. *v.t.* to cover with a veil; to hide, to conceal, to disguise. **veiling,** *n.* a veil; material for veils. **to draw a veil over,** to conceal discretely; to refrain from mentioning. **to take the veil,** to assume the veil according to the custom of a woman on becoming a nun; to retire to a convent. **veilless** (-lis), *a.*
vein (vān), *n.* one of the tubular vessels in animal bodies conveying blood to the heart; (*loosely*) any blood-vessel; a rib or nervure in an insect's wing or a leaf; a fissure in rock filled with material deposited by water; a seam of any substance; a streak or wavy stripe of different colour, in wood, marble, or stone; a distinctive trait, quality, tendency or cast of mind; particular mood or humour. *v.t.* to fill or cover with, or as with veins. **veining,** *n.* **veinless** (-lis), *a.* **veinlet** (-lit), *n.* **veinlike, veiny,** *a.*
velamen (vilā'mən), **velamentum** (veləmen'təm), *n.* (*pl.* **-mina, -menta** (-ə)), (*Anat.*) a membraneous covering or envelope; the corky outer layer of the aerial roots of some orchids. **velamentous,** *a.*
velar (vē'lə), VELUM. **velarium** (-leə'riəm), *n.* (*pl.* **-ia** (-iə),)) the great awning stretched over the seats in a theatre or amphitheatre as a protection against rain or sun; a velum.
velarize, -ise (vē'ləriz), *v.t.* to sound a guttural further back than the hard palate.

velatura (velətoo'rə), *n.* the glazing of pictures by rubbing on a thin coating of colour with the hand.

Velcro® (vel'krō), *n.* a fastening for clothes etc. which consists of two nylon strips, one consisting of hooks the other of loops, which stick together when pressed.

veld, veldt (velt, felt), *n.* open country in southern Africa, suitable for pasturage, esp. the high treeless plains in N Transvaal and NW Natal.

veld-schoen, veld-skoen (velt'skoon, felt'-), *n.* (*S Afr.*) a shoe made of raw hide.

veldt VELD.

veleta, valeta (vəlē'tə), *n.* a dance or dance tune in slow waltz time.

vell (vel), *n.* the fourth stomach of a calf used in making rennet.

velleity (vilē'iti), *n.* a low degree of desire or volition unaccompanied by effort.

vellicate (vel'ikāt), *v.i.* to twitch spasmodically. **vellication,** *n.* **vellicative,** *a.*

vellum (vel'əm), *n.* a fine parchment orig. made of calf-skin; a manuscript written on this; a heavy cream-coloured paper. **vellumy,** *a.*

veloce (vilō'chä), *adv.* (*Mus.*) with great quickness. [It.]

velocipede (vilos'ipēd), *n.* any kind of carriage propelled by the feet; an early form of cycle. **velocipedist,** *n.*

velocity (vilos'iti), *n.* swiftness, rapidity, rate of motion; (*Phys.*) rate of motion in a particular direction. **velocimeter** (velǝsim'itǝ), *n.* an apparatus for measuring velocity.

velodrome (vel'ǝdrōm), *n.* a building containing a cycle-racing track.

velour, velours (vilooǝ'), *n.* a fabric with a velvet-like surface or finish.

velouté (vǝloo'tā), *n.* a thick creamy sauce or soup. [F]

velum (vē'lǝm), *n.* (*pl.* **-la** (-lǝ)) (*Anat. etc.*) a membrane, a membranous covering, envelope or part; the soft palate. **velar,** *a.* of or forming a velum; produced by the back of the tongue in contact with the soft palate.

velvet (vel'vit), *n.* a closely-woven fabric, usu. of silk, with a short, soft nap or cut pile on one side; a fabric with a velvet-like pile; the furry skin covering the growing antlers of a deer; (*sl.*) money won by gambling or speculation. *a.* velvety; as soft as velvet. **cotton velvet,** velvet made with cotton back and silk face. **on velvet,** (*coll.*) in a position of comfort, luxury, wealth etc. **velvet glove,** *n.* gentleness concealing strength. **velvet-paper,** *n.* flock wall-paper. **velvet-pile,** *n.* a pile like that of velvet; a fabric with such a pile. **velvety,** *a.* resembling velvet; soft, smooth. **velveteen** (-tēn'), *n.* a cotton velvet or cotton fabric with a velvet-pile.

vena (vē'nə), *n.* (*pl.* **-nae** (-nē)) a vein. **venal**[1], *a.* **venation,** *n.* (the arrangement of) the veins on leaves, insects' wings etc. **venational,** *a.*

venal[2] (vē'nəl), *a.* ready to be bought over for financial gain or to sacrifice honour or principle for sordid considerations; mercenary, hireling. **venality** (-nal'-), *n.* **venally,** *adv.*

venatic (vinat'ik), **-al,** *a.* pertaining to or used in hunting; fond of the chase. **venatically,** *adv.* **venatorial** (venətaw'ri-), *a.*

venation VENA.

vend (vend), *v.t.* to sell; to offer (small wares) for sale. *v.i.* to be sold; to engage in selling. **vending machine,** *n.* a slot machine dispensing goods, e.g. cigarettes, drinks, sweets. **vendee** (-dē'), *n.* a buyer. **vendor, vender,** *n.* a seller, esp. the seller of a house or other property; a vending machine. **vendible,** *a.* **vendibility** (-bil'-), *n.*

vendace (ven'dās), *n.* a small and delicate white-fish found in some British lakes.

vendee etc. VEND.

vendetta (vendet'ə), *n.* a blood-feud, often carried on for generations, in which the family of a murdered or injured person seeks vengeance on the offender or any member of his family, prevalent, esp. in Corsica, Sardinia and Sicily; this practice; a feud, private warfare or animosity.

vendible, vendor VEND.

veneer (viniə'), *v.t.* to cover with a thin layer of fine or superior wood; to face with a veneer; to put a superficial polish on, to disguise, to gloss over. *n.* a thin layer of superior wood for veneering; a decorative or protective surface or facing applied to a backing; a superficial appearance. **veneering,** *n.*

venepuncture, venipuncture (vē'nipŭngkchə, ven'-), *n.* the piercing of a vein, esp. with a hypodermic needle.

venerable (ven'ərəbl), *a.* worthy of veneration; rendered sacred by religious or other associations; applied as a title to archdeacons in the Church of England, and to a person who has attained the first of three degrees in canonization in the Roman Catholic Church. **venerability** (-bil'-), **venerableness,** *n.* **venerably,** *adv.*

venerate (ven'ərāt), *v.t.* to regard or treat with profound deference and respect, to revere. **veneration,** *n.* **venerative,** *a.* **venerator,** *n.*

venereal (viniə'riəl), *a.* pertaining to sexual desire or sexual intercourse; produced by sexual intercourse; pertaining to or affected with venereal disease; involving the genitals. **venereal disease,** *n.* disease conveyed by sexual intercourse, as gonorrhoea, syphilis and chancroid. **venereology** (-ol'-), *n.* the study of venereal diseases. **venery**[1] (ven'-), *n.* sexual indulgence.

venery[2] (ven'əri), *n.* hunting, the chase.

venesect, venisect (ven'isekt), *v.t., i.* to make an incision in a vein. **venesection** (-sek'-), *n.*

Venetian (vinē'shən), *a.* pertaining to the city or province of Venice, in N Italy. *n.* a native or inhabitant of Venice; (*coll.*) a venetian blind. **Venetian blind,** *n.* a blind made of thin slats on braid or webbing arranged to turn so as to admit or exclude light. **Venetian glass,** *n.* a delicate ornamental glass-ware made at or near Venice. **Venetian mast,** *n.* a pole painted spirally in two or more colours, used for street decorations.

vengeance (ven'jəns), *n.* punishment inflicted in return for an injury or wrong, retribution. **with a vengeance,** forcibly, emphatically, undoubtedly, extremely. **vengeful,** *a.* vindictive, revengeful. **vengefully,** *adv.* **vengefulness,** *n.*

venial (vē'niəl), *a.* that may be pardoned or excused. **venial sin,** *n.* in the Roman Catholic Church, a sin that is not a mortal sin. **veniality** (-al'-), **venialness,** *n.* **venially,** *adv.*

Venice (ven'is), *a.* Venetian. **Venice glass,** *n.* Venetian glass.

venipuncture VENEPUNCTURE.

venisect VENESECT.

venison (ven'isən, ven'zən), *n.* the flesh of the deer as food.

Venite (vinī'tē), *n.* Psalm xcv, 'O come let us sing', used as a canticle; a musical setting of the same.

Venn diagram (ven), *n.* a diagram in which sets and their relationships are represented by circles or other figures. [John *Venn,* 1834–1923, British mathematician]

venom (ven'əm), *n.* a poisonous fluid secreted by snakes, scorpions etc., and injected by biting or stinging; spite, malignity, virulence; poison. **venomous,** *a.* full of venom; spiteful; poisonous; having venom-secreting glands. **venomously,** *adv.* **venomousness,** *n.*

venose (vē'nōs), *a.* having veins; covered with veins. **venous** (vē'nəs), *a.* pertaining to, circulating in or contained in the veins; having or consisting of veins. **venosity** (-nos'-), *n.* local excess of veins or of venous blood. **venously,** *adv.*

vent (vent), *n.* a hole or aperture, esp. for the passage of air, water etc. out of a confined place; the flue of a chimney, a touch-hole, a finger-hole in a wind-instrument etc.; an outlet of a volcano; the opening of the cloaca, the

anus in animals below mammals; a means or place of passage, escape etc., an outlet, free play, utterance, expression etc.; a split in a garment as in the back of a coat or jacket. *v.t.* to make a vent in; to give vent to; to release through a vent; to utter, to pour forth. **to give vent to**, to express freely. **vent-hole**, *n.* **ventless** (-lis), *a.*
venter (ven'tə), *n.* the belly, the abdomen, any large cavity containing viscera; (*Nat. Hist.*) an expanded or hollowed part or surface; the fleshy part of a muscle, the belly; (*Law*) the womb, hence, a mother. **ventral** *a.* pertaining to the venter; pertaining to or situated on the anterior or lower surface or point; (*Bot.*) situated on the side turned towards the axis. **ventrally**, *adv.* **ventricose** (-trikōs), *a.* having a protruding belly; (*Nat. Hist.*) distended, inflated, esp. on one side.
ventilate (ven'tilāt), *v.t.* to supply with fresh air, to cause a circulation of air in (a room etc.); to oxygenate (the blood); to give publicity to, to throw open for discussion etc. **ventilation**, *n.* **ventilative**, *a.* **ventilator**, *n.*
ventr- VENTR(O)-.
ventral VENTER.
ventricle (ven'trikl), *n.* a cavity or hollow part in an animal body; a cavity of the vertebrate brain containing cerebrospinal fluid; either of the two chambers of the heart that receive blood from the atria. **ventricular, -lous** (-trik'ū-), *a.*
ventricose VENTER.
ventriloquism (ventril'əkwizm), *n.* the act or art of speaking or producing sounds so that the sound appears to come not from the person speaking but from a different source. **ventriloquist**, *n.* **ventriloquy** (-kwi), *n.* **ventriloquize, -ise**, *v.i.* **ventriloquial** (-lō'-), **ventriloquistic** (-kwis'-), **ventriloquous**, *a.*
ventr(o)-, *comb. form.* pertaining to the abdomen or a venter.
venture (ven'chə), *n.* the undertaking of a risk, a hazard; an undertaking of a risky nature; a commercial speculation; a stake, that which is risked. *v.t.* to expose to hazard or risk, to hazard, to stake; to dare; to brave the hazards of; to dare to express. *v.i.* to dare; to have the courage or presumption (to do etc.); to undertake a risk. **venture capital**, *n.* capital invested in new, esp. speculative, business enterprises by people or organizations other than the owners of the enterprise. **Venture Scout**, *n.* a senior member of the Scouts organization usu. over 15 years old. **at a venture**, at random. **to venture on** or **upon**, to dare to enter upon or engage in etc. **venturer**, *n.* **venturesome** (-səm), *a.* ready to take risks, daring; risky. **venturesomely**, *adv.* **venturesomeness**, *n.* **venturous**, *a.* venturesome. **venturously**, *adv.* **venturousness**, *n.*
venturi (tube) (ventū'ri), *n.* a tube or duct, wasp-waisted and expanding at the ends, used in measuring the flow rates of fluids, as a means of accelerating air flow, or to provide a suction source for vacuum-operated instruments. [G.B. *Venturi*, 1746–1822, Italian physicist]
venue (ven'ū), *n.* (*Law*) the place or country where a crime is alleged to have been committed and where the jury must be empanelled and the trial held; the clause in an indictment indicating this; the scene of an activity or event; a meeting place.
Venus (vē'nəs), *n.* the goddess of love, esp. sensual love; a planet between the earth and Mercury, the brightest heavenly body after the sun and moon. **Mount of Venus**, the female pubes, the mons veneris; (*Palmistry*) the elevation at the base of the thumb. **Venus's comb**, *n.* an annual herb of the parsley family. **Venus's flytrap**, *n.* an insectivorous herb of the sundew family. **Venus's looking-glass**, *n.* a plant of the genus *Specularia*, esp. *S. speculum.* **Venusian** (vinū'zhən, -ziən), *n.*, *a.* (an inhabitant) of the planet Venus.
veracious (virā'shəs), *a.* habitually speaking or disposed to speak the truth; characterized by truth and accuracy;

true. **veraciously**, *adv.* **veracity** (-ras'-), *n.*
veranda, verandah (viran'də), *n.* a light external gallery or portico with a roof on pillars, along the front or side of a house.
veratrine (ve'rətrin, -trēn), *n.* a highly poisonous amorphous compound obtained from hellebore and other plants, formerly used as a local counter irritant in neuralgia, rheumatism etc.
veratrum (virā'trəm), *n.* the hellebore; (**Veratrum**) a genus of plants containing the hellebore.
verb (vœb), *n.* that part of speech which predicates, a word that asserts something in regard to something else (the subject). **verbal**, *a.* of, pertaining to or using words; respecting words only, not ideas etc.; literal, word for word; pertaining to or derived from a verb; oral, spoken, not written. *n.* a verbal noun; an oral statement; (*sl.*) an admission of guilt made by a suspect when arrested. *v.t.* (*past, p.p.* **verballed**) (*sl.*) to attribute such an admission to (someone). **verbal noun**, *n.* a noun derived from a verb, esp. an Eng. word ending in -ING. **verbalist**, *n.* one who deals in words only; a literal adherent to or a minute critic of words. **verbalism**, *n.* **verbalize, -ise**, *v.t.* to convert or change into a verb; to express in words. *v.i.* to use many words, to be verbose. **verbalization**, *n.* **verbally**, *adv.*
verbatim (vəbā'tim), *adv.* word for word. **verbatim et literatim** (et litərah'tim), word for word and letter for letter. [L]
Verbena (vəbē'nə), *n.* a large genus of plants of which *Verbena officinalis*, the common vervain, is the type. **verbenaceous** (-nā'shəs), *a.*
verbiage (vœ'biij), *n.* the use of many words without necessity, verbosity, wordiness. **verbicide** (-sid), *n.* (*facet.*) word-slaughter; a word-slaughterer.
verbose (vəbōs'), *a.* using or containing more words than are necessary. **verbosely**, *adv.* **verboseness**, **verbosity** (-bos'-), *n.*
verboten (vəbō'tən, fə-), *a.* forbidden by authority. [G]
verdant (vœ'dənt), *a.* green; covered with growing plants or grass; fresh, flourishing; inexperienced, unsophisticated, easily taken in. **verdancy**, *n.* **verde antico** (vœ'di antē'kō), **verd antique**, *n.* an ornamental stone composed chiefly of serpentine, usu. green and mottled or veined; a green incrustation on ancient bronze. **verdantly**, *adv.*
verderer (vœ'dərə), *n.* historically, a judicial officer who has charge of the royal forests.
verdict (vœ'dikt), *n.* the decision of a jury on an issue of fact submitted to them in the trial of any cause, civil or criminal; a decision, judgment. **open verdict**, one reporting the commission of a crime without specifying the guilty person; of an inquest, one failing to state the cause of death. **special verdict**, one in which specific facts are placed on record but the court is left to form conclusions on the legal aspects.
verdigris (vœ'digrēs), *n.* a green crystalline substance formed on copper by the action of dilute acetic acid, used as a pigment and in medicine; greenish rust on copper etc.
verdure (vœ'dyə), *n.* greenness of vegetation, fresh vegetation or foliage. **verdured, verdurous**, *a.* **verdureless** (-lis), *a.*
verge[1] (vœj), *n.* the extreme edge, brink, border or margin; the threshold, limit; the grass-edging of a bed or border or alongside a road; a rod, wand or staff, carried as an emblem of authority, esp. before a bishop or other dignitary; the edge of the tiles projecting over a gable etc.; a spindle, shaft etc., in the mechanism of a watch, loom, and other machines. **on the verge of**, on the brink of.
verge[2] (vœj), *v.i.* to approach, to come near, to border (on); to serve as an edge. *v.t.* to edge, border.
verger (vœ'jə), *n.* an officer carrying the verge or staff of

verglas office before a bishop or other dignitary; an official in a church acting as usher or as pew-opener. **vergership**, *n.*
verglas (vɛə'glah), *n.* a film of ice on rock.
veridical (virid'ikəl), *a.* truthful, veracious. **veridicality**, *n.* **veridically**, *adv.*
verify (vɛ'rifi), *v.t.* to confirm the truth of; to inquire into the truth of, to authenticate; to fulfil; to affirm under oath, to append an affidavit to (pleadings). **verifiable**, *a.* **verifiability** (-bil'-), *n.* **verification** (-fi-), *n.* **verifier**, *n.*
verily (vɛ'rili), *adv.* (*dated, poet.*) in very truth, assuredly.
verisimilitude (verisimil'itūd), *n.* the appearance of or resemblance to truth; probability, likelihood; something apparently true or a fact. **verisimilar** (-sim'-), *a.*
verism (vɛ'rizm), *n.* extreme naturalism in art or literature.
verity (vɛ'riti), *n.* truth, correspondence (of a statement) with fact; a true statement, a truth; a thing really existent, a fact. **veritable**, *a.* real, genuine; actual, true. **veritably**, *adv.*
verjuice (vœ'joos), *n.* an acid liquid expressed from crab-apples, unripe grapes etc. and used in cooking and for other purposes. **verjuiced**, *a.*
verkrampte (fəkramp'tə), *n.*, *a.* in S Africa, (pertaining to) a person of Afrikaner Nationalist opinions who opposes any liberalization of government policy, esp. in matters of race.
verligte (fəlikh'tə), *n.*, *a.* in S Africa, (pertaining to) a person of liberal political attributes, esp. towards black and coloured people.
vermeil (vœ'mil), *n.* silver-gilt; a transparent varnish for giving a lustre to gilt; vermilion.
verm(i)-, *comb. form.* pertaining to worms.
vermicelli (væmisel'i, -chel'i), *n.* pasta in the form of long slender threads. **chocolate vermicelli**, *n.* small thin pieces of chocolate used for cake decoration.
vermicide (vœ'misid), *n.* a medicine or drug that kills worms, an anthelmintic. **vermicidal** (-sī'-), *a.*
vermicular (vəmik'ūlə), *a.* of or pertaining to a worm; resembling the motion or track of a worm; of reticulated work etc., tortuous, marked with intricate wavy lines; worm-eaten in appearance; vermiform.
vermiculate (vəmik'ūlət), *a.* worm-eaten; vermicular. *v.t.* (-lāt), to decorate with vermicular lines or tracery. **vermiculation**, *n.* motion after the manner of a worm, as in the peristaltic motion of the intestines; the art of vermiculating; vermiculated work; the state of being worm-eaten.
vermiculite (vəmik'ūlīt), *n.* an altered mica that expands on heating to form a lightweight water-absorbent material used as insulation and in seed beds.
vermiform (vœ'mifawm), *a.* worm-shaped; having the form or structure of a worm. **vermiform appendix**, *n.* a small worm-like organ of no known function situated at the extremity of the caecum, the appendix.
vermifuge (vœ'mifūj), *n.* a medicine or drug that destroys or expels intestinal worms, an anthelmintic. **vermifugal** (-gəl), *a.*
vermivorous (vœmiv'ərəs), *a.* feeding on worms.
vermilion (vəmil'yən), *n.* a brilliant red pigment consisting of mercuric sulphide obtained by grinding cinnabar or by the chemical treatment of mercury and sulphur; the colour of this. *a.* of a beautiful red colour. *v.t.* to colour with or as with vermilion.
vermin (vœ'min), *n.pl.* a collective name for certain offensive animals, as the smaller mammals or birds injurious to crops or game, noxious or offensive insects, grubs or worms, esp. lice, fleas etc.; low, despicable or repulsive persons. **verminous**, *a.* **verminously**, *adv.*
vermis (vœ'mis), *n.* the middle lobe connecting the two halves of the cerebellum.
vermouth (vœ'məth), *n.* a liqueur made of white wine flavoured with aromatic herbs, orig. wormwood.
vernacular (vənak'ūlə), *a.* of architecture etc., native, indigenous; of language, idiom etc., belonging to the country of one's birth; of or being the common name of a plant or animal. *n.* one's native tongue; the native idiom or dialect of a place or country; the jargon of a particular group of people. **vernacularism**, *n.* **vernacularize, -ise**, *v.t.* **vernacularization, -isation**, *n.* **vernacularly**, *adv.*
vernal (vœ'nəl), *a.* pertaining to, prevailing, done, or appearing in spring; pertaining to youth. **vernal equinox** EQUINOX. **vernal grass**, *n.* a fragrant grass, *Anthoxanthum odoratum*, sown among hay. **vernally**, *adv.* **vernalization, -isation**, *n.* the beating of seeds before sowing, as by chilling, in order to hasten flowering. **vernalize**, *v.t.* **vernation**, *n.* the arrangement of the young leaves within the leaf-bud.
vernicle (vœ'nikl), *n.* a cloth for wiping sweat, held to have Christ's face miraculously impressed on it when St Veronica wiped his face; any representation of this; a medal or badge bearing it worn by pilgrims who have visited Rome.
vernier (vœ'niə), *n.* a movable scale for measuring fractional portions of the divisions of the scale on a measuring instrument, as a barometer, theodolite etc. [Pierre *Vernier*, *c.* 1580–1637, French mathematician]
Veronal® (ve'rənəl), *n.* a hypnotic drug, diethylbarbituric acid, also called barbitone.
veronica (viron'ikə), *n.* a herb or shrub of the fig-wort family, with blue, purple or white flowers, the speedwell; a handkerchief or cloth bearing a portrait of Christ, esp. that of St Veronica said to have been miraculously so impressed when she wiped the sweat from Christ's face on the way to Calvary.
veronique (verənēk'), *a.* (used after the noun) served with white grapes, e.g. *sole veronique*. [F]
verruca (vəroo'kə), *n.* (*pl.* **-cae** (-sē), **-cas**) a wart; (*Nat. Hist.*) a wart-like elevation. **verruciform** (-sifawm), *a.* **verrucose** (-kōs), **-cous**, *a.* covered in warts or wart-like projections.
vers (veə), *n.* verse. **vers libre** (lēbr'), *n.* free verse. **verslibrist**, *n.* a writer of free verse. [F]
versant (vœ'sənt), *n.* an area of land sloping in one direction; general lie or slope.
versatile (vœ'sətīl), *a.* turning easily, readily applying oneself to new tasks, occupations, subjects etc., many-sided; having many uses; changeable, variable, inconstant; (*Bot.*, *Zool.*) of anthers, antennae etc., moving freely round or to and fro on its support. **versatilely**, *adv.* **versatility** (-til'-), *n.*
verse (vœs), *n.* a metrical line consisting of a certain number of feet; a group of metrical lines, a stanza; metrical composition as distinguished from prose; a particular type of metrical composition; a poem; one of the short divisions of a chapter of the Bible; a short sentence in a liturgy etc. *v.t.*, *v.i.* to versify. **verset** (-sit), *n.* a short organ interlude or prelude; a short verse; a verside.
versicle (-sikl), *n.* a short verse, esp. one of a series recited in divine service by the minister alternately with the people. **versicular** (-sik'ū-), *a.* pertaining to verses; relating to division into verses. **versify** (-sifī), *v.t.* to turn (prose) into verse; to narrate or express in verse. *v.i.* to make verses. **versification** (-fi-), *n.* **versifier**, *n.*
versed (vœst), *a.* skilled, familiar, experienced, proficient (in); (*Trig.*) turned about, reversed. **versed sine**, *n.* one minus the cosine.
verset, versicle etc. VERSE.
versicolour (vœ'sikūlə), **-coloured**, *a.* having various colours, variegated; changeable from one colour to another, with differences of light.
versify etc. VERSE.
version (vœ'shən), *n.* that which is translated from one language into another, a translation; a statement, account, or description of something from one's particular point of view; a particular form, a variant; the turning of a child in the womb to facilitate delivery; the adaptation of a

vers libre 918 **vestry**

work of art into another medium. **versional**, *a.*
vers libre vers.
verso (vœ'sō), *n.* a left-hand page of a book, sheet etc.; the other side of a coin or medal to that on which the head appears.
verst (vœst), *n.* a Russian measure of length, 3500.64 ft., nearly two-thirds of a mile (about 1 km).
versus (vœ'səs), *prep.* against.
vert (vœt), *n.* (*Law*) everything in a forest that grows and bears green leaves; the former right to cut green or growing wood; (*Her.*) the tincture green.
vertebra (vœ'tibrə), *n.* (*pl.* **-brae**) one of the bony segments of which the spine or backbone consists. **vertebral**, *a.* **vertebral column**, *n.* the spinal column. **vertebrally**, *adv.* **Vertebrata** (-brā'tə), *n.pl.* a division of animals comprising those with a backbone, including mammals, birds, reptiles, amphibians and fishes. **vertebrate** (-brət), *n.*, *a.* (a member) of the Vertebrata.
vertex (vœ'teks), *n.* (*pl.* **-tices** (-tisēz),) the highest point, the top, summit, or apex; the point on the limb (of sun, moon or planet) furthest above the observer's horizon; the point of an angle, cone, pyramid etc.; the top of the arch of the skull. **vertical** (-ti), *a.* of, pertaining to, or situated at the vertex or highest point; situated at or passing through the zenith; perpendicular to the plane of the horizon, upright; of or pertaining to the vertex of the head; pertaining to a hierarchical structure or arrangement in strata; comprising or combining different or consecutive stages or levels. *n.* a vertical position or plane. **vertical angles**, *n.* either pair of opposite angles made by two intersecting lines. **vertical circle**, *n.* an azimuth-circle. **vertical plane**, *n.* a plane passing through the zenith perpendicular to the horizon. **vertical take-off**, *n.* the take-off of an aeroplane without a preliminary run or taxiing. **verticality** (-kal'-), **verticalness**, *n.* **vertically**, *adv.*
verticil (vœ'tisil), *n.* (*Biol.*) a whorl, an arrangement of parts in a circle round a stem etc. **verticillate** (-tis'ilət), **-lated**, *a.* **verticillately**, *adv.*
vertigo (vœ'tigō), *n.* giddiness, dizziness; a feeling as if one were whirling round. **vertiginous** (-tij'i-), *a.* pertaining to, having or causing vertigo; whirling; changing, unstable. **vertiginously**, *adv.* **vertiginousness**, *n.*
vertu (vœtoo'), virtu.
vervain (vœ'vān), *n.* a wild plant or weed, with small purplish flowers, of the genus *Verbena*, esp. *V. officinalis*, formerly credited with medical and other virtues.
verve (vœv), *n.* spirit, enthusiasm, energy, esp. in literary or artistic creation.
vervet (vœ'vit), *n.* a small S African monkey, usu. black-speckled greyish-green, with reddish-white face and abdomen.
very (ve'ri), *a.* (now chiefly used intensively), real, true, actual, genuine, being what it seems or is stated to be, selfsame. *adv.* in a high degree; to a great extent; greatly, extremely, exceedingly.
Very light (veə'ri), *n.* a firework to produce a flare for lighting up the countryside. [Edward W. *Very*, 1852-1910, US naval officer]
Very pistol, *n.* a pistol for firing Very lights.
Very Rev., (*abbr.*) Very Reverend.
vesica (ve'sikə), *n.* (*pl.* **-cae** (-sē)) a bladder, cyst etc. esp. the urinary bladder. **vesica piscis** (pis'is, -kis), *n.* the elliptic aureole in which the Saviour and the saints were often depicted by early painters. **vesical**, *a.* **vesicant**, *n.* a blister-producing counter-irritant; a poison-gas that causes blisters. *a.* causing blistering. **vesicate**, *v.t.* to raise vesicles or blisters on. **vesication**, *n.* **vesicatory**, *n.*, *a.* vesicant. **vesicle** (-kl), *n.* a small bladder or cavity, sac, cyst, bubble, or hollow structure. **vesicular** (-sik'ū-), **-late** (-lət), *a.* **vesiculation**, *n.* the formation of residues.
vesper (ves'pə), *n.* the evening star, Venus, appearing just after sunset; (*fig.*) evening; (*pl.*) in the Roman Catholic and Greek Churches, the sixth of the seven canonical hours; (*pl.*) the evening service. *a.* pertaining to the evening or to vespers. **Sicilian Vespers** sicilian. **vesperbell**, *n.* the bell that summons to vespers. **vesperal**, *n.* the part of the antiphonary containing the chants for vespers.
vespertine (ves'pətin, -tin), *a.* of, pertaining to, or done in the evening; (*Zool.*) flying in the evening; (*Bot.*) opening in the evening; (*Astrol.*) descending towards the horizon at sunset.
vespiary (ves'piəri), *n.* a nest of wasps, hornets etc. **vespiform** (-fawm), *a.* resembling a wasp. **vespine** (-pīn), *a.*
vessel (ves'l), *n.* a hollow receptacle, esp. for holding liquids, as a jug, cup, dish, bowl, barrel etc.; a ship or craft of any kind, esp. one of some size; a tube, a duct, or canal in which the blood or other fluids are conveyed; (*Bot.*) a canal or duct formed by the breaking down of the partitions between cells; a person regarded as receiving or containing (grace, wrath etc.). **the weaker vessel**, (*now usu. facet.*) woman (I Pe'er iii.7). **vesselful**, *n.*
vest (vest), *n.* (*esp. N Am.*) a waistcoat; an undergarment for the upper part of the body, a singlet; a close jacket formerly worn by women, now a (usu. V-shaped) piece on the front of the body or waist of a gown. *v.t.* (*poet.*) to clothe with or as with a garment; to invest or endow (with authority, etc.); to confer an immediate fixed right of present or future possession of (property in a person). *v.i.* of property, a right etc., to come or take effect (in a person). **vest pocket**, *a.* (*esp. N Am*) small enough to fit into a waistcoat pocket. **vested**, *a.* wearing vestments, robed; (*Law*) held by or fixed in a person, not subject to contingency. **vested interest**, *n.* a source of gain to which the owner considers himself entitled by custom and right; a strong interest in a political, social etc. state of affairs or system held for personal reasons; a person or group holding such an interest.
Vesta (ves'tə), *n.* the goddess of the hearth and the hearth-fire; the fourth asteroid; (**vesta**) a wax match igniting by friction. **vestal**, *a.* pertaining to the goddess Vesta or the vestal virgins; pure, chaste. *n.* a vestal virgin; a woman of spotless chastity; a nun. **vestal virgin**, *n.* one of the virgin priestesses, vowed to perpetual chastity, who had charge of the temple of Vesta at Rome, and of the sacred fire which burned perpetually on her altar.
vestiary (ves'tiəri), *a.* pertaining to dress. *n.* a wardrobe, a robing-room.
vestibule (ves'tibūl), *n.* a small hall, lobby, or antechamber next to the outer door of a house, from which doors open into the various inner rooms; a porch; a covered passage between the cars in a corridor train; (*Anat.*) a chamber, cavity or channel communicating with others, as the central chamber of the labyrinth of the ear. **vestibular** (-tib'-), **-late** (-lət), *a.* **vestibuled**, *a.*
vestige (ves'tij), *n.* the mark of a foot made in passing, a footprint; a sign, a mark or trace of something no longer present or in existence; (*coll.*) a minute amount, a particle; (*Biol.*) an organ or part that has degenerated and become nearly or entirely useless. **vestigial, vestigiary** (-tij'-), *a.*
vestment (vest'mənt), *n.* a garment, esp. a robe of state or office; any of the ritual garments of the clergy, choristers, etc., esp. a chasuble; an altar-cloth.
vestry (ves'tri), *n.* a room or place attached to a church in which the vestments are kept and in which the clergy, choristers etc., robe; a chapel or room attached to a non-liturgical church; a meeting of the ratepayers of a parish (called a common, general or ordinary vestry) or of their elected representatives (called a select vestry) for dealing with parochial business, formerly exercising sanitary and other powers of local government, as such now superseded by the parish council. **vestry-clerk**, *n.* an officer appointed by a vestry to keep the accounts etc. **vestryman**, *n.* a member of a vestry. **vestral**, *a.* **vestrydom**

(-dəm), n. government by a vestry, esp. if corrupt or incompetent.
vesture (ves'chə), n. (poet.) dress, clothes, apparel; a covering. v.t. to clothe, to dress. **vestural**, a.
Vesuvian (vizoo'viən, -soo'-), a. pertaining to Vesuvius, a volcano near Naples, Italy. n. vesuvianite. **vesuvianite** (-nit), n. a vitreous brown or green silicate first found among the ejections of Vesuvius.
vet[1] (vet), n. short for VETERINARY SURGEON. v.t. (past, p.p. **vetted**) to subject to careful scrutiny and appraisal; to examine or treat (esp. an animal) medically.
vet[2] (vet), n. (chiefly N Am.) short for VETERAN.
vetch (vech), n. a plant of the genus Vicia of the bean family, including several wild and cultivated species used for forage, esp. the common vetch or tare. **vetchling** (-ling), n. a plant of the genus Lathyrus, allied to the vetches. **vetchy**, a.
veteran (vet'ərən), a. grown old or experienced, esp. in the military service; of or pertaining to veterans. n. one who has had long experience in any service, occupation or art, esp. as a soldier; (N Am.) an ex-service man. **veteran car**, n. a motor car built before 1916, esp. one built before 1905.
veterinary (vet'ərinəri), a. pertaining to treatment of the diseases of animals, esp. domestic animals, as cows, horses, dogs etc. n. a veterinary surgeon. **veterinary surgeon**, n. a person qualified to diagnose and treat diseases and injuries in animals. **veterinarian** (-neə'ri-), n. (chiefly N Am.) a veterinary surgeon.
veto (vē'tō), n. (pl. **vetoes**) the power or right of a sovereign, president, or branch of a legislature to prevent or reject the enactments of another branch; the act of exercising such right; any authoritative prohibition, refusal, negative, or interdict. v.t. to refuse approval to (a Bill etc.); to prohibit, to forbid. **vetoer, vetoist**, n.
vex (veks), v.t. to cause trouble or annoyance to, to irritate; to distress, to worry; (poet.) to agitate, to throw (the sea etc.) into commotion. **vexation**, n. the act of vexing or the state of being vexed, irritation, annoyance, trouble; that which causes irritation, an annoyance; a harassing by process of or under the cover of law. **vexatious**, a. **vexatiously**, adv. **vexatiousness**, n. **vexed**, a. annoyed, worried, troubled, filled with vexation; of a question or doctrine much debated or contested. **vexedly** (-sid-), adv. **vexing**, a. **vexingly**, adv. **vexer**, n.
vexillum (veksil'əm), n. (pl. **vexilla** (lə)) a square flag forming the standard of a body of cavalry troops in ancient Rome; a body of troops under a separate vexillum; the large upper petal of a papilionaceous flower; the web of a feather. **vexil** (vek'-), n. (Bot.) **vexillar, -ary** (vek'-), **-llate** (-lət), a. **vexillology**, n. the study of flags. **vexillologist**, n.
VF, (abbr.) video frequency; voice frequency.
VG, (abbr.) Vicar-General.
VHF, (abbr.) very high frequency.
via (vī'ə, vē'ə), prep. by way of, through. **via dolorosa** (vē'ə dolərō'sə), n. the way to Calvary. **via media** (vī'ə mē'diə, vē'əmā'diə), n. a middle way, a mean between extremes.
viable (vī'əbl), a. of a foetus etc. capable of maintaining independent existence, able to survive; (Bot.) able to live in a particular climate; likely to become actual or to succeed. **viability** (bil'-), n.
viaduct (vī'ədŭkt), n. a bridge-like structure, esp. one composed of masonry and a considerable number of arches carrying a road or railway over a valley etc.
vial (vī'əl), n. a phial.
viand (vī'ənd), n. (usu. pl.) an article of food, esp. meat.
viaticum (vīat'ikəm), n. (pl. **cums, -ca** (-kə)) a supply of provisions or an allowance of money for a journey; the Eucharist as given to a person at the point of death. **viator** (-ā'-), n. a traveller, a wayfarer.

vibes (vībz), n.pl. (sl.) feelings, intuitions or sensations experienced or communicated, vibrations.
vibex (vī'beks), n. (pl. **vibices** (-bisēz),) a purple spot appearing on the skin in certain fevers.
vibraculum (vībrak'ūləm), n. (pl. **-la** (-lə)) one of the filamentous whip-like appendages of many polyzoa, bringing particles of food within reach by their lashing movements. **vibracular**, a.
vibrant (vī'brənt), a. vibrating, tremulous; resonant; full of life and vigour. **vibrancy**, n. **vibrantly**, adv.
vibraphone (vī'brəfōn), n. a percussion instrument similar to a xylophone but with metal bars placed over electronic resonators.
vibrate (vībrāt'), v.i. to move to and fro rapidly, to swing, move to and fro, to oscillate; to thrill, to quiver, to throb; to move to and fro ceaselessly, esp. with great rapidity. v.t. to cause to swing, move to and fro, oscillate or quiver; to measure (seconds etc.) by vibrations or oscillations, to emit with or through vibrations. **vibratile** (vī'brətil), a. **vibratility** (-til'-), n. **vibration**, n. the act of vibrating; oscillation; rapid motion backward and forward, esp. of the parts of an elastic solid or of a liquid the equilibrium of which has been disturbed; one such complete movement; (pl.) a characteristic aura or atmosphere held to emanate from a person or place; (pl.) (coll.) feelings communicated from one person to another, vibes. **vibrational, vibrative** (vī'-), a. **vibrator**, n. one who or that which vibrates; a vibrating reed used to chop a continuous current and thus produce an alternating current; a vibrating electrical apparatus used in massage and to provide sexual stimulation. **vibratory** (vī'-), a.
vibrato (vēbra'tō), n. a pulsating effect, esp. in singing or string-playing, produced by the rapid variation of emphasis on the same tone. [It.]
vibrio (vib'riō), n. a form of bacterium more or less screw-shaped with a filament at each end, as that causing Asiatic cholera.
vibrissa (vibris'ə), n. (pl. **-sae** (-ē)) a stiff coarse hair or bristle in the nostrils of man and about the mouths of most mammals; one of the bristle-like feathers about the mouths of some birds, as the flycatchers; one of the bristles about the mouths of some flies.
viburnum (vībœ'nəm), n. a shrub or small tree of a genus containing the guelder rose and the laurustinus etc., of the honeysuckle family.
vicar (vik'ə), n. the priest of a parish the greater tithes of which belong to a chapter or a layman, he himself receiving the smaller tithe or a stipend. **lay vicar**, in the Church of England, a cathedral officer who sings some portion of the service. **Vicar of Bray**, a turncoat. **Vicar of Christ**, one of the Pope's titles. **vicar apostolic**, n. in the Roman Catholic Church, a titular bishop appointed where no episcopate has been established etc. **vicar choral**, n. in the Church of England, a clerical or lay assistant in the choral part of a cathedral service. **vicar forane**, n. in the Roman Catholic Church, a functionary appointed by a bishop with limited (chiefly disciplinary) jurisdiction over clergy etc. **vicar-general**, n. in the Roman Catholic Church, an officer appointed by a bishop as his assistant, esp. in matters of jurisdiction; in the Church of England, an officer assisting a bishop or archbishop in ecclesiastical causes and visitations. **vicarage** (-rij), n. the benefice of a vicar; the house or residence of a vicar. **vicariate** (-keə'riət), a. having delegated power, vicarious. n. delegated office or power; a vicarship, esp. the jurisdiction of a vicar apostolic.
vicarious (vikeə'riəs, vī-), a. deputed, delegated; acting on behalf of another; performed, executed or suffered for or instead of another; experienced at second-hand through imagining the experiences of another. **vicariously**, adv.
vice[1] (vīs), n. an evil or immoral practice or habit; evil

vice conduct, gross immorality, depravity; a fault, a blemish, a defect; a bad habit or trick in a domestic animal, esp. a horse; sexual immorality, esp. prostitution. **vice squad**, *n.* a police squad dealing with enforcement of prostitution, pornography and gambling laws.
vice² (vīs), *n.* an instrument with two jaws, brought together by a screw or lever, between which an object may be clamped securely. *v.t.* to secure in or as in a vice. **vicelike**, *a.*
vice³ (vī'si, vīs), *prep.* in place of.
vice⁴ (vīs), *n.* (*coll.*) short for VICE-PRESIDENT, VICE-CHAIRMAN etc.
vice- (vīs), *pref.* denoting one acting or qualified to act in place of or as deputy of another, or one next in rank. **vice-admiral**, *n.* a naval officer next in rank below an admiral, and next above a rear-admiral. **vice-admiralty**, *n.* **vice-chairman**, *n.* **vice-chairmanship**, *n.* **vice-chamberlain**, *n.* the deputy of the Lord Chamberlain. **vice-chancellor**, *n.* a deputy-chancellor; an officer who discharges most of the administrative duties of a university; in the Roman Catholic Church, the head cardinal of the branch of Chancery dealing with bulls and briefs; formerly a subordinate judge in Chancery. **vice-chancellorship**, *n.* **vice-consul**, *n.* **vice-consulship**, *n.* **vice-governor**, *n.* **vice-president**, *n.* a deputy-president. **vice-presidentship**, **-presidency**, *n.* **vice-principal**, *n.* **viceregal**, *a.* of a viceroy. **viceregally**, *adv.* **vice-reine** (-rān), *n.fem.* the wife of a viceroy; a female viceroy. **viceroy** (vīs'roi), *n.* a ruler acting with royal authority in a colony, dependency etc. **viceroyal**, *a.* **viceroyalty**, **viceroyship**, *n.*
vicegerent (vīsje'rənt), *n.*, *a.* having or exercising delegated power. *n.* an officer exercising delegated authority, a deputy. **vicegerency**, *n.*
vicenary (vis'inəri), *a.* consisting of or pertaining to 20.
vicennial (visen'iəl), *a.* happening every 20 years; lasting 20 years.
vice versa (vīsi vœ'sə, vīs), *adv.* the order or relation being inverted, the other way round.
Vichy (vē'shi), *n.* a town in the Allier Department, France. **Vichy Water**, *n.* an effervescent mineral water found at Vichy. **vichyssoise** (vēshiswahz'), *n.* a cream soup of leeks and potatoes, often served chilled.
vicinage (vis'inij), *n.* neighbourhood, vicinity, surrounding places, environs; the state of being neighbours, neighbourliness. **vicinal**, *a.* near, neighbouring. **vicinity** (-sin'-), *n.* the neighbourhood, the adjoining or surrounding district; the state of being near, proximity; near relationship (to).
vicious (vish'əs), *a.* characterized by some vice, fault or defect; faulty, imperfect, defective, incorrect; contrary to moral principles or to rectitude; addicted to vice, depraved, wicked; cruel, violent, spiteful, malignant. **vicious circle**, *n.* a set of circumstances or events in which the solution to one problem creates further difficulties which in turn aggravate the initial problem; reasoning in which a conclusion is drawn from premises which themselves depend on the truth of the conclusion. **viciously**, *adv.* **viciousness**, *n.*
vicissitude (visis'itūd), *n.* a change of condition, circumstances or fortune, a mutation, a revolution; (*poet.*) regular change or mutation of fortune. **vicissitudinary, -dinous** (-tū'), *a.*
victim (vik'tim), *n.* a living creature sacrificed to some deity or in the performance of some religious rite; a person or thing destroyed, injured or ill-treated in the pursuit of some object; a person killed or injured in an accident, or in an epidemic; a dupe, a gull. **victimize, -ise**, *v.t.* to make a victim of; to dupe, to swindle; to treat or discriminate against unfairly. **victimization**, *n.* **victimizer**, *n.*
victor (vik'tə), *n.* a person, nation, team etc. that conquers in battle or wins in a contest. **victory**, *n.* the defeat of an enemy in battle, or of an opponent in a contest; success in a struggle or undertaking. **Victory sign**, *n.* the first and second fingers extended in the form of a V. (see V-SIGN under v). **victorious** (-taw'ri-), *a.* having conquered in battle or any contest; triumphant; associated or connected with victory. **victoriously**, *adv.* **victoriousness**, *n.*
victoria (viktaw'riə), *n.* a four-wheeled carriage with a folding top, a raised seat for the driver, seats for two persons over the back axle and a low seat for two persons over the front axle; a gigantic variety of water-lily; a variety of domestic pigeon; (also **victoria plum**) a large red plum. **Victoria Cross**, *n.* a British naval and military decoration in the shape of a Maltese cross, instituted by Queen Victoria (1856), bestowed for conspicuous bravery or devotion in the presence of the enemy. **Victorian**, *a.* of, pertaining to, or flourishing or living in the reign of Victoria; characteristic of the standards and moral values held to be prevalent in Victorian times, esp. inclined to stuffy formality and prudishness. *n.* a person, esp. a writer, living or flourishing then; a native of Victoria, Australia. **victoriana** (-ah'nə), *n.* objects, ornaments etc. of the Victorian period. [Queen *Victoria*, 1819–1901]
victress VICTOR.
victual (vit'l), *n.* (*usu. in pl.*) food, provisions. *v.t.* to supply or store with provisions. *v.i.* to lay in provisions; to eat. **victualler**, *n.* one who supplies victuals; a victuallingship. **licensed victualler** LICENSE. **victualless** (-lis), *a.* **victualling-bill**, *n.* a custom-house warrant for the shipment of provisions for a voyage. **victualling-department, -office**, *n.* the office managing the supply of provisions to the navy. **victualling-ship**, *n.* one conveying provisions to other ships or to a fleet. **victualling-yard**, *n.* one, usu. adjoining a dockyard, where warships are provisioned.
vicuña, vicugna, vicunia (vikoon'yə), *n.* a S American animal, allied to the camel, native to the Andes; a fine cloth made of its wool.
vidame (vēdahm'), *n.* (*F Hist.*) a minor noble holding lands under a bishop.
vide (vī'dē, vē'dā), *v. imper.* see (in reference to a passage in a book etc.). [L]
videlicet (vidē'liset, -dä'liket), *adv.* namely, that is to say, to wit (usu. abbrev. to VIZ.). [L]
video (vid'iō), *n.* a video recorder; a video cassette; a video recording. *a.* relating to or employed in the transmission or reception of a televised image; concerned with, or operating at video frequencies. *v.t.*, *v.i.* to make a video recording (of). **video camera**, *n.* a camera which records its film on video tape. **video cassette**, *n.* one containing video tape. **video conferencing**, *n.* meeting in conference by means of a video. **video disk**, *n.* one on which television picture and sound can be played back. **video frequency**, *n.* that in the range required for a video signal. **video game**, *n.* an electronically operated game played by means of a visual display unit. **video jockey**, *n.* one who introduces videos of popular music or similar. **video nasty**, *n.* a video film which includes horrific or gruesome scenes of violence, sexual outrage or other atrocities. **videophone, video telephone**, *n.* a telephone which can also transmit a picture of each speaker. **video (tape) recorder**, *n.* a machine for recording and playing back television broadcasts or for playing films made on videotape. **videotape**, *n.* magnetic tape used for recording a television programme or similar for subsequent transmission. *v.t.* to record on videotape. **video (tape) recording**, *n.* recording both television picture and sound on magnetic tape. **video-record**, *v.t.* to record on a video recorder. **videotex** VIEW-DATA. **video tube**, *n.* a television tube.
vidette (videt'), VEDETTE.
vidimus (vī'diməs), *n.* (*pl.* **-muses**) an examination or inspection of accounts etc.; an abstract or summary. [L]
vie (vī), *v.i.* to strive for superiority; to contend; to be equal or superior (with or in). **vier**, *n.* **vying**, *a.*
Vienna (vien'ə), *n.* the capital of Austria. **Vienna loaf**, *n.* a

Viennese — vindictive

long round-ended loaf of white bread. **Vienna steak**, *n*. a meat rissole.
Viennese (vēənēz'), *a*. pertaining to Vienna or its inhabitants. *n*. a native or the inhabitants of Vienna.
Vietnamese (vietnəmēz'), *a*. of or pertaining to Vietnam in SE Asia; *n*. a native or the language of Vietnam.
vieux jeu (viœ zhœ), *a*. old-fashioned. [F]
view (vū), *n*. survey or examination by the eye; range of vision; power of seeing; that which is seen, a scene, a prospect; a picture or drawing of this; an intellectual or mental survey; the manner or mode of looking at things, considering a matter etc.; judgment, opinion, theory; intention, purpose, design; inspection by a jury etc. *v.t.* to examine with the eye; to survey mentally or intellectually; to consider, to form a mental impression or judgment of; to watch television. **in view**, in sight. **in view of**, considering, having regard to. **on view**, open to public inspection. **to take a dim view of**, to regard unfavourably. **with a view to**, for the purpose of; with an eye to. **view-data**, *n*. a communications system by which data can be transferred through a telephone line and displayed on TV or video. **view-finder**, *n*. a device of mirrors in a camera which shows the view to be taken. **view-halloo**, *n*. a huntsman's shout on seeing the fox break cover. **viewphone** VIDEOPHONE. **view-point**, *n*. a point of view, an aspect. **viewable**, *a*. **viewer**, *n*. **viewing**, *n*. **viewless** (-lis), *a*. (*poet.*) invisible. **viewy**, *a*. having peculiar or impracticable views, faddy, visionary. **viewiness**, *n*.
vigesimal (vījes'iməl), *a*. 20th. **vigesimo** (-mō), *n*., *a*. 20mo. (of books). **vigesimo-quarto**, *n*. a 24mo.
vigia (vijē'ə), *n*. a warning of a rock, shoal etc. on a hydrographical chart.
vigil (vij'il), *n*. keeping awake during the customary hours of rest, watchfulness; the eve of a church festival. **vigilance**, *n*. the state of being vigilant; (*Path*.) insomnia. **vigilance committee**, *n*. a self-organized committee for maintaining order or inflicting summary justice in an ill-ordered community or district. **vigilant**, *a*. awake and on the alert; watchful, wary, circumspect. **vigilantly**, *adv*. **vigilante** (-lan'ti), *n*. a member of a vigilance committee; a self-appointed keeper of public order. **vigilantism**, *n*.
vigneron (vē'nyərō), *n*. a wine-grower. [F]
vignette (vinyet'), *n*. an ornamental flourish round a capital letter in a manuscript; an engraving not enclosed within a definite border, esp. on the title page of a book; a photograph, drawing or other portrait showing the head and shoulders with a background shading off gradually; a brief description or character-sketch. *v.t.* to shade off (a portrait, drawing etc.); thus; to make a photograph or portrait of in this style. **vignetter**, **vignettist**, *n*.
vigoroso (vigərō'sō), *adv*. (*Mus*.) with energy. [It.]
vigour (vig'ə), *n*. active physical or mental strength or energy; abounding vitality, vital force, robustness; exertion of strength, force, activity; forcibleness, trenchancy. **vigourless** (-lis), *a*. **vigorous**, *a*. **vigorously**, *adv*. **vigorousness**, *n*.
viking (vī'king), *n*. (*often* Viking) a rover, freebooter or pirate, esp. one of the Scandinavian warriors of the 8th–10th cents. **vikingism**, *n*.
vilayet (vilah'yit), *n*. a province of the old Turkish empire.
vile (vīl), *a*. worthless, morally base, depraved, despicable, abject, villainous, odious; (*coll*.) disagreeable, abominable. **vilely**, *adv*. **vileness**, *n*.
vilify (vil'ifi), *v.t.* to traduce, to defame. **vilification** (-fi-), *n*. **vilifier**, *n*.
vilipend (vil'ipend), *v.t.* to speak of disparagingly or contemptuously, to depreciate. **vilipender**, *n*.
villa (vil'ə), *n*. a country house; a detached suburban house. **villadom** (-dəm), *n*. villas collectively; (*fig*.) the middle classes.
village (vil'ij), *n*. a small assemblage of houses, smaller than a town or city and larger than a hamlet; the inhabitants of a village. *a*. pertaining to a village; rustic, countrified. **villager**, *n*. an inhabitant of a village. **villagization**, *n*. the resettlement and rehousing of a population in new villages outside their own area, often achieved by force. **villagize, -ise,** *v.t.*
villain (vil'ən), *n*. a person guilty or capable of crime or great wickedness; a scoundrel, a wretch; (*coll*.) a rogue, a rascal; (*often* villein) a feudal serf, a bondsman attached to a feudal lord or to an estate. *a*. pertaining to, composed of, or performed by a villain or villains. **villainage** (-ij), *n*. **villainess**, *n.fem*. **villainous**, *a*. worthy or characteristic of a villain; depraved, vile; very bad. **villainously**, *adv*. **villainousness**, *n*. **villainy**, *n*.
villanelle (vilənel'), *n*. a poem in five tercets and a final quatrain on two rhymes.
Villarsia (vilah'siə), *n*. a genus of marsh or aquatic plants of the order Gentianaceae with yellow flowers. [Dominique Villars, 1745–1814, French botanist]
-ville, *comb. form* (*sl.*) a place, condition or quality with a character as specified, e.g. *dullsville, dragsville, squaresville*. [from the suffix *-ville* in names of towns]
villeggiatura (vilejətoo'rə), *n*. retirement or a stay in the country. [It.]
villein VILLAIN.
villus (vil'əs), *n*. (*pl*. -li (-lī)) one of the short hair-like processes on certain membranes, as those on the inner surface of the small intestine; (**villi**) long, close, soft hairs. **villiform** (-lifawm), *a*. **villoid** (-oid), *a*. **villose** (-ōs), **-lous**, *a*. **villosity** (-los'-), *n*.
vim (vim), *n*. (*coll*.) energy, vigour. **(full of) vim and vigour**, (abounding in) energy and vitality.
vimana (vimah'nə), *n*. the central shrine of an Indian temple with pyramidal roof; a temple gate.
viminal (vim'inəl), *a*. pertaining to, producing or consisting of twigs or shoots. **vimineous** (-min'i-), *a*.
vin (vĩ), *n*. wine. **vin du pays** (dü pāē), local wine. **vin blanc**, *n*. white wine. **vin ordinaire** (awdineə'), *n*. inexpensive table wine for daily use. **vin rosé** (rō'zā), *n*. rosé wine. **vin rouge** (roozh), *n*. red wine. [F]
vina (vē'nə), *n*. an Indian stringed instrument with a fretted fingerboard over two gourds.
vinaceous (vīnā'shəs), *a*. pertaining to wine or grapes; of the nature or colour of wine.
vinaigrette (vinigret'), *n*. an ornamental bottle or perforated case for holding aromatic vinegar etc., a smelling-bottle; a salad dressing consisting of oil, vinegar and seasoning. **vinaigrous** (-nā'-), *a*. sour, acid; (*fig*.) cross, crabbed.
vinasse (vinas'), *n*. a residual product containing potassium salts from the wine-press or beets from which sugar has been extracted.
Vincentian (vinsen'shən), *a*. pertaining to or founded by St Vincent de Paul (1577–1660). *n*. a member of a religious and charitable order founded by him, a Lazarist.
vincible (vin'sibl), *a*. capable of being conquered, not invincible. **vincibility** (-bil'-), **vincibleness**, *n*.
vincristine (vinkris'tēn), *n*. an alkaloid substance derived from the tropical periwinkle, used in the treatment of some types of leukaemia.
vinculum (ving'kūləm), *n*. (*pl*. -la (-lə)) (*Math*.) a straight line drawn over several terms to show that they are all alike, to be added to or deducted from those preceding or following; (*Anat*.) a fraenum; (*Print*.) a brace.
vindaloo (vindəloo'), *n*. a type of hot Indian curry.
vindicate (vin'dikāt), *v.t.* to maintain (a claim, statement etc.) against attack or denial; to defend (a person) against reproach, accusation etc.; to prove to be true or valid, to establish, to justify, to uphold. **vindicable**, *a*. **vindicability** (-bil'-), *n*. the state of being vindicable. **vindication**, *n*. **vindicative**, *a*. **vindicator**, *n*. **vindicatress** (-tris), *n.fem*.
vindicatory, *a*. tending to vindicate or justify; punitive.
vindictive (vindik'tiv), *a*. revengeful; characterized or

prompted by revenge. **vindictive damages**, *n.pl.* damages given to punish the defendant. **vindictively**, *adv.* **vindictiveness**, *n.*
vine (vīn), *n.* a slender climbing plant of the genus *Vitis*, esp. the common or grape-vine; any plant with a slender climbing or trailing stem. **vine-borer**, *n.* one of various insects that injure vines by boring into the stems, twigs etc. **vine-clad**, *a.* covered with vines. **vine-disease**, *n.* any disease attacking the grape-vine, esp. that caused by the phylloxera. **vine-dresser**, *n.* one who dresses, trims or prunes vines. **vine-fretter**, *n.* a small insect infesting vines. **vineyard** (vin'yəd), *n.* a plantation of grape-vines. **vined** *a.* **vinery** (-əri), *n.* a greenhouse for vines. **viny**, *a.* pertaining to vines. **vinic** (vin'-), *a.* pertaining to or derived from wine. **vinous**, *a.* of, pertaining to, or having the qualities of wine. **vinosity** (-nos'-), *n.*
vinegar (vin'igə), *n.* an acid liquid obtained by oxidation of acetous fermentation from wine, cider etc., used as a condiment and as a preservative in pickling; anything sour or soured, as a disposition etc. *v.t.* to put vinegar on or into; (*fig.*) to make sour. **vinegar-eel**, *n.* a minute worm infesting vinegar, sour paste etc. **vinegar-plant**, *n.* a microscopic fungus producing acetous fermentation. **vinegarish**, **vinegary**, *a.*
vinery, vineyard VINE.
vingt-et-un (vītaü'), *n.* a card game in which the object is to make the aggregate number of the pips on the cards as nearly as possible 21 without exceeding this. [F]
vini-, *comb. form.* pertaining to wine or vines.
viniculture (vin'ikŭlchə), *n.* the cultivation of grape-vines. **vinicultural**, *a.* **viniculturist**, *n.*
vinifacteur, *n.* any apparatus used for wine-making.
vinificator (vin'ifikātə), *n.* an apparatus for condensing the alcoholic vapours from the fermenting must in wine-making.
vino (vē'nō), *n.* (*coll.*) wine. [It.]
vinometer (vīnom'itə), *n.* an instrument for measuring the percentage of alcohol in wine.
vinous VINE.
vintage (vin'tij), *n.* the yield of a vineyard or vine-district for a particular season; the season of gathering grapes; the product of a particular year. *a.* produced at some particular time, esp. of a past season; produced in a good year; of good or lasting quality, classic; of the best examples of, or most typical of. **vint**, *v.t.* to make (wine). **vintage car**, *n.* an old-fashioned car (esp. one built between 1919 and 1930). **vintage wine**, *n.* wine of a good vintage year. **vintager**, *n.* a grape-gatherer.
vintner (vint'nə), *n.* a wine-merchant. **vintnery**, *n.*
vinyl (vī'nil), *n.* an organic radical CH$_2$CH; any vinyl resin or plastic, esp. PVC. *a.* of or made of a vinyl resin, esp. conventional gramophone records. **vinyl resins, plastics**, *n.pl.* thermoplastic resins, polymers or co-polymers of vinyl compounds.
viol (vī'əl), *n.* a mediaeval stringed musical instrument, the predecessor of the violin; a violoncello or bass-viol. **viol class**, *n.* instruments like the violin, violoncello etc., played with a bow and having no frets, thus being capable of continuous gradation. **violist**, *n.* a player on the viol or viola.
viola[1] (viō'lə), *n.* an instrument like a large violin, the alto or tenor violin; a viol.
viola[2] (vī'ələ), *n.* a plant or flower of the genus containing the violet and pansy. **violaceous** (-lā'shəs), *a.* of a violet colour; of the violet family.
viola da gamba (viō'lə da gam'bə), an early form of bass-viol. [It.]
violate (vī'əlāt), *v.t.* to infringe or transgress, to break, to disobey (a law, obligation, duty etc.); to treat irreverently, to desecrate; to harm, do violence to, outrage; to ravish, to rape. **violable**, *a.* **violation**, *n.* **violative**, *a.* **violator**, *n.*

violence (vī'ələns), *n.* the state or quality of being violent; violent exercise of power; violent treatment; injury, outrage; vehemence, intensity or impetuosity of feeling, action etc.; the illegal exercise of physical force, an act of intimidation by the show or threat of force. **to do violence to**, to do a physical injury to, to outrage, to violate. **violent**, *a.* acting with or characterized by the exertion of great physical force; vehement, impetuous, furious; intense, abrupt, immoderate; produced by or resulting from extraneous force or poison, not natural (of death etc.). **violently**, *adv.*
violet (vī'ələt), *n.* a plant or flower of the genus *Viola*, esp. the sweet violet, the dog violet and some other species with small blue, purple or white flowers; a colour seen at the opposite end of the spectrum to red, produced by a slight mixture of red with blue; a small violet-coloured butterfly of various species. *a.* of the colour of violet. **shrinking violet**, (*coll.*) a shy, hesitant person. **violet-wood**, *n.* one of several kinds of wood, esp. king-wood and myall-wood. **violescent** (-les'ənt), *a.* tending to a violet colour.
violin[1] (vīəlin'), *n.* a musical instrument of the viol class with four strings, played with a bow; a player on this. **violinist**, *n.*
violin[2] (vī'əlin), *n.* an emetic substance contained in the common violet.
violinist VIOLIN[1].
violist VIOL.
violoncello (vīələnchel'ō), *n.* a four-stringed musical instrument of the viol class rested on the ground between the legs, a bass-viol. **violoncellist**, *n.*
violone (viəlō'nā), *n.* a mediaeval double-bass viol; an organ-stop of string-like tone.
VIP, (*abbr.*) very important person.
viper (vī'pə), *n.* a venomous snake of the family Viperidae, esp. the European viper or adder, the only poisonous British snake; a mischievous or malignant person. **viper's bugloss**, *n.* the blue weed or blue thistle, *Echium vulgare*. **viper's-grass**, *n.* a perennial plant of the aster family. **viperiform** (-rifawm), **viperine** (-rin), **viperish**, **viperoid** (-roid), **viperous**, *a.*
virago (virah'gō), *n.* an impudent, turbulent woman; a termagant; a woman of masculine strength and courage. **viragoish, virago-like**, *a.*
viral (vī'rəl), *a.* pertaining to or caused by a virus.
virelay (vi'rilā), *n.* an old form of French verse with two rhymes to a stanza and usu. a refrain.
virement (vīə'mənt), *n.* authorized transference of a surplus to balance a deficit.
vireo (vi'riō), *n.* an American passerine insectivorous singing-bird.
virescent (vires'ənt), *a.* (*Bot.*) green, tending to become green, viridescent; abnormally green (of petals etc.). **virescence**, *n.*
virgate (vœ'gət), *a.* long, straight and erect, rod-like.
Virgilian (vəjil'iən), *a.* pertaining to or in the style of Virgil, Latin poet (70-19 BC).
virgin (vœ'jin), *n.* a woman who has had no sexual relations with a man; a member of an order of women under vows of chastity; a madonna; a female insect that produces eggs without fertilization; the constellation Virgo; a man who has had no sexual relationship. *a.* being a virgin; pure, chaste, undefiled; befitting a virgin; maidenly, modest; unworked, untried, not brought into cultivation; producing eggs without impregnation (of insects). **the Virgin**, the mother of Christ. **virgin-born**, *a.* born of a virgin. **Virgin Queen**, *n.* name applied to Queen Elizabeth I (1533-1603). **virgin's bower**, *n.* the plant traveller's-joy. **virginal**, *a.* pertaining to or befitting a virgin; pure, chaste, maidenly. *n.* a keyed musical instrument, shaped like a box, used in the 16th-17th cents., also called a pair of virginals. **virginally**, *adv.* **virginhood** (-hud), **virginity**

(-jin'-), n. the state of being a virgin, purity, innocence.
Virginia (vəjin'yə), n. tobacco from Virginia. **Virginia creeper**, n. a woody vine with ornamental foliage. **Virginian**, n., a.
Virgo (vœ'gō), n. one of the 12 ancient zodiacal constellations; the sixth sign of the zodiac.
virgule (vœ'gūl), n. a small rod, a twig; (*Print.*) a solidus. **virgulate** (-lət), a.
viridescent (viridés'ənt), a. greenish; becoming slightly green. **viridescence**, n. **viridian** (-rid'-), n. a bright green pigment, a form of chromic oxide. **viridigenous** (-dij'inəs), a. imparting greenness (esp. to oysters). **viridity** (-rid'-), n. greenness, the colour of fresh vegetation; greenness in oysters, due to feeding on green organisms.
virile (vi'ril), a. of or pertaining to man or the male sex, procreative; characteristic of a man, masculine, manly; of a male, sexually potent. **virilism**, n. the development in the female of masculine characteristics, mental and physical. **virility** (-ril'-), n.
virole (virōl'), n. a ferrule; (*Her.*) a hoop or ring encircling a horn.
virology etc. VIRUS.
virose (vī'rōs), **virous**, a. poisonous; (*Bot.*) emitting a fetid odour.
virtu (vœtoo'), n. love of or taste for the fine arts. **articles** or **objects of virtu**, rare, old or beautiful works of decorative art. [It.]
virtuoso (vœtūō'sō, -zō), n. (*pl.* **-sos**) a connoisseur of articles of virtu; a skilled performer in some fine art, esp. music. **virtuosic**, a. **virtuosity** (-os'-), **virtuosoship**, n.
virtue (vœ'choo), n. moral excellence, goodness, uprightness, rectitude; conformity with or practice of morality or duty; a particular moral excellence; sexual purity, chastity, esp. in women; inherent power, goodness or efficacy. **cardinal virtues** CARDINAL. **by** or **in virtue of**, by or through the efficacy or authority of, on the strength of. **virtueless** (-lis), a. **virtual**, a. being such in essence or effect though not in name or appearance, equivalent so far as effect is concerned, in effect. **virtual memory**, n. (*Comput.*) a facility which allows more data to be stored than the capacity of the main memory by transferring data to and from disks. **virtuality** (-al'-), n. **virtually**, adv. **virtuous**, a. characterized by virtue, morally good; chaste. **virtuously**, adv. **virtuousness**, n.
virulent (vi'rulənt), a. extremely poisonous; highly infective; having a severe effect; extremely bitter, acrimonious or malignant. **virulence**, n. **virulently**, adv.
virus (vī'rəs), n. a very small infective agent capable of self-propagation only in living matter, the causative agent of many diseases; a disease caused by this; moral taint or corrupting influence; virulence, malignity; (*Comput.*) a short program inserted into a main program, which can damage data and the operation of a computer, and can spread to other computers via electronic links. **virus infection, disease**, n. **virology** (virol'-), n. the study of viruses and virus diseases. **virological** (-loj'-), a. **virologist**, n.
vis (vis), n. (*pl.* **vires** (vī'rēz)) (*Mech.*) force, energy, potency. **vis inertiae** INERT. **vis mortua** (maw'tūə), n. dead force; force doing no work. **vis viva** (vī'və, vē'-), n. living force, measured by the mass of a moving body multiplied by the square of its velocity. [L]
visa (vē'zə), **visé** (-zā), n. an official endorsement on a passport showing that it has been examined and found correct. v.t. (*pres.p.* **visaing**, *past*, *p.p.* **visaed**) to certify (a passport).
visage (viz'ij), n. the face, the countenance. **visaged**, a. having a look of a particular type.
visagiste (vēzəzhēst'), n. one who specializes in facial make-up.
visard VISOR.
vis-à-vis (vēzahvē'), adv. face to face, opposite to; in relation to. n. a person facing another as in certain dances,

e.g. a quadrille; a carriage or couch for two persons sitting vis-à-vis. [F]
viscacha (viskah'chə), n. a S American burrowing rodent like a chinchilla.
viscera (vis'ərə), n.pl. the internal organs of the great cavities of the body, as the skull, thorax and abdomen, esp. those of the abdomen, the intestines. **visceral**, a. **viscerally**, adv. **viscerate**, v.t. to disembowel. **visceri-, viscero-**, comb.form.
viscid (vis'id), a. sticky, adhesive; semifluid in consistency. **viscidity** (-sid'-), n. **viscin** (-in), n. a viscid liquid obtained from mistletoe etc., the chief constituent of bird-lime.
visco-, comb. form. pertaining to viscosity.
viscometer (viskom'itə), **viscosimeter** (-kəsim'-), n. an apparatus for determining the viscosity of liquids. **viscometry** (-kom'itri), n.
viscose (vis'kōs), n. (the cellulose sodium salt used in the manufacture of) artificial silk. **viscosity** (-kos'-), n. stickiness; thickness of a fluid etc.; the property of fluids, semifluids and gases which expresses their resistance to flow, change of shape or re-arrangement of molecules. **viscous** (-kəs), **viscose**, a. **viscousness**, n.
viscount (vī'kownt), n. a British peer ranking next below an earl, and above a baron. **viscountcy** (-si), n. **viscountess** (-tis), n.fem. **viscountship, viscounty**, n.
viscous VISCOSE.
Viscum (vis'kəm), n. a genus of parasitic shrubs comprising the mistletoe.
viscus (vis'kəs), n. viscera.
Vishnu (vish'noo), n. the preserver god of the Hindu sacred Triad.
visible (viz'ibl), a. capable of being seen, perceptible by the eye; in view, apparent, open, conspicuous. **visible Church**, n. the body of professing Christians. **visible horizon**, n. the apparent limit bounding the view. **visible means**, n.pl. means or resources which are apparent to or ascertainable by others. **visible radiation**, n. electromagnetic radiation which can be detected by the eye; light. **visible speech**, n. phonetic symbols representing every possible articulate utterance. **visibility** (-bil'-), n. **high, low visibility**, n. (*Meteor.*) clear or indistinct visibility. **visibleness**, n. **visibly**, adv.
Visigoth (viz'igoth), n. one of the western Goths who settled in S Gaul and Spain in the 4th and 5th cents. **Visigothic** (-goth'-), a.
vision (vizh'ən), n. the act or faculty of seeing, sight; that which is seen, an object of sight; a mental representation of a visual object, esp. in a dream or trance; a supernatural or prophetic apparition; a creation of the imagination or fancy, foresight, an appreciation of what the future may hold. v.t. to see in or as in a vision; to imagine; to present as in a vision. **vision mixer**, n. one who blends or combines different camera shots in television or films. **visional**, a. **visionally**, adv. **visionary**, a. existing in a vision or in the imagination only; imaginary, unreal, unsubstantial, unpractical; given to day-dreaming, fanciful theories etc. n. a visionary person. **visionariness**, n. **visionist**, n. **visionless** (-lis), a.
visit (viz'it), v.t. to go or come to see a person or place; to come or go to for the purpose of inspection, supervision etc.; to afflict (of diseases etc.); (*Bibl.*) to chastise; to comfort, to bless. v.i. to call on or visit people; to keep up friendly intercourse. n. the act of visiting, or going to see a person, place or thing; a stay (with or at); a formal or official call or inspection. **visitable, visitatorial** (-tətaw'ri-), a. **visiting-book**, n. one in which calls received or intended are entered. **visiting-card**, n. a small card, bearing one's name etc. to be left in making a call. **visiting professor**, n. a professor invited to join academic staff for a limited time. **visitant**, n. a migratory bird that visits a country at certain seasons; (*poet.*) a visitor, a

guest. **visitation**, *n.* the act of visiting; a formal or official visit for the purpose of inspection etc.; (*Internat. Law*) the boarding of a foreign vessel in time of war to ascertain her character etc.; the right to do this; a divine dispensation, esp. a chastisement or affliction; (*Her.*) the official visit of a herald to a district for the examination and verification of arms, pedigrees etc.; in the Roman Catholic Church, a festival held on 2 July in honour of the visit of the Virgin Mary to Elizabeth (Luke i.39); (*Zool.*) an abnormal and extensive irruption of animals into a region. **visitation of the sick**, an Anglican office for the comfort and consolation of sick persons. **visitor**, *n.* one who makes a call; one who visits a place; an officer appointed to make a visitation of any institution. **visitors' book**, *n.* one in which visitors' names are entered, esp. in which visitors to a hotel, boarding-house, museum etc. write remarks. **visitor's passport**, *n.* a simplified short-term passport.
visiting etc. VISIT.
visor (vīz'ə), *n.* the movable perforated part of a helmet defending the face; a projecting part on the front of a cap, for shielding the eyes. **visored**, *a.* **visorless** (-lis), *a.*
vista (vis'tə), *n.* a long view shut in at the sides, as between rows of trees; (*fig.*) a mental view far into the past or future. **vistaed**, *a.* [It.]
visual (vizh'ūəl), *a.* of, pertaining to or used in sight or seeing; serving as an organ or instrument of seeing. *n.pl.* photographs, a piece of film etc., as distinct from the accompanying sound-track. **visual aid**, *n.* a picture, film, photograph, diagram etc. used as an aid to teaching or imparting information. **visual arts**, *n.pl.* painting, sculpture, film etc. as opposed to music, literature etc. **visual display unit**, (*abbr.* VDU) *n.* a device, usu. with a keyboard, which displays characters etc. representing data stored in a computer memory. **visuality** (-al'-), *n.* **visualism**, *n.* **visualist**, *n.* **visualize, -ise**, *v.t.* to make visual or visible; to make visible to the eye; to form a visual image of in the mind. **visualization**, *n.* **visualizer**, *n.* **visually**, *adv.* **visually challenged**, *a.* blind or having severely impaired sight.
vital (vī'təl), *a.* pertaining to, necessary to or supporting organic life; containing life; affecting life; indispensable, essential; very important; lively, energetic. *n.pl.* the parts or organs of animals essential to life, as the heart, brain etc. **vital centre**, *n.* the point in the body at which a wound appears to be instantly fatal, esp. the respiratory nerve-centre in the medulla oblongata. **vital functions**, *n.pl.* the bodily functions that are essential to life such as the circulation of blood. **vital force** or **principle**, *n.* one assumed as accounting for organic life etc. **vital spark**, *n.* the principle of life, hence life or trace of life; (*coll.*) the moving spirit behind an enterprise etc. **vital statistics**, *n.pl.* those relating to birth, marriage, mortality; (*coll.*) the measurements of a woman's bust, waist and hips. **vitalism**, *n.* the doctrine that life is derived from something distinct from physical forces. **vitalist**, *n., a.* **vitalistic** (-lis'-), *a.* **vitality** (-tal'-), *n.* **vitalize, -ise**, *v.t.* to give life to; to animate, to make more lively. **vitalization**, *n.* **vitally**, *adv.*
vitamin (vit'əmin), *n.* (*esp. N Am.* (vī'-)) one of a number of naturally occurring substances (the main vitamins are A, B group, C, D, E and K) which are necessary, though in minute quantities, for normal growth and metabolism.
vitellus (vitel'əs), *n.* (*pl.* -**li** (-lī)) yolk of egg, the protoplasmic contents of the ovum. **vitellary** (vit'-), **vitelline** (-īn), *a.* **vitelli-, vitello-**, *comb. forms.* **vitellicle** (-ikl), *n.* a yolk-sac.
vitiate (vish'iāt), *v.t.* to impair the quality of; to corrupt; to render faulty or imperfect; to render invalid or ineffectual. **vitiation**, *n.* **vitiator**, *n.* **vitiosity** (-os'-), *n.* the state of being vicious; depravity, corruption.
viticide (vit'isid), *n.* an insect or other vermin injurious to vines.
viticolous (vitik'ələs), *a.* living on or infesting the vine.
viticulture (vit'ikŭlchə), *n.* the cultivation of the grape-vine. **viticultural**, *a.* **viticulturist**, *n.*
vitiosity VITIATE.
Vitis (vī'tis), *n.* a genus of plants comprising the grape-vine.
vitreous (vit'riəs), *a.* consisting of or resembling glass; obtained from glass. **vitreous electricity**, *n.* electricity generated by friction on glass, formerly regarded as positive. **vitreous humour**, *n.* the jelly-like substance filling the posterior chamber of the eye, between the lens and the retina. **vitreosity** (-os'-), **vitreousness**, *n.* **vitrescent** (-tres'ənt), *a.* **vitrescence**, *n.* **vitrescible**, *a.* vitrifiable.
vitric, *a.* of or like glass; pertaining to vitrics. *n.pl.* fused siliceous compounds, glass and glassy materials, opp. to ceramics; the science or history of glass-manufacture. **vitriform** (-rifawm), *a.*
vitric etc. VITREOUS.
vitrify (vit'rifī), *v.t.* to convert into glass or a glassy substance by heat and fusion. *v.i.* to be converted into glass. **vitrification** (-fi-), **vitrifaction** (-fak'-), *n.* **vitrifiable**, *a.* **vitrifiability** (-bil'-), *n.*
vitrine (vit'rēn, -rin), *n.* a glass show-case.
vitriol (vit'riəl), *n.* sulphuric acid as made from green vitriol; any salt of this, a sulphate; malignancy, caustic criticism etc. **black vitriol**, an impure copper sulphate. **blue** or **copper vitriol**, copper sulphate. **green vitriol**, ferrous sulphate. **oil of vitriol**, sulphuric acid. **red, rose vitriol**, cobalt sulphate. **vitriolation, -lization**, *n.* **vitriolic** (-ol'-), *a.* pertaining to, obtained from, or having the qualities of vitriol; caustic, bitter, malignant. **vitrioline** (-lin), *a.* **vitriolizable**, *a.* **vitriolize, -ise, vitriolate**, *v.t.* to convert into a sulphate.
vitro-di-trina (vitrōditrē'nə), *n.* white Venetian glass in which fine threads of cane form a lace-like pattern. [It.]
vitrophyre (vit'rəfīə), *n.* a porphyritic-volcanic rock of a vitreous structure. **vitrophyric** (-fī'rik), *a.*
Vitruvian (vitroo'viən), *a.* (*Arch.*) of or in the style of Marcus Vitruvius Pollio, a Roman architect of the Augustan age. **Vitruvian scroll**, *n.* a pattern consisting of convoluted undulations, used in friezes etc.
vitta (vit'ə), *n.* (*pl.* -**tae** (-tē)) the lappet of a mitre; an oil-tube in the fruit of the parsley family etc.; a band or stripe of colour. **vittate** (-ət), *a.*
vittles (vit'lz), *n.pl.* an alternative or dialect form of VICTUALS.
vitular (vit'ūlə), -**lary**, **vituline** (-lin, -līn), *a.* of or pertaining to a calf or calving; calf-like.
vituperate (vitū'pərāt), *v.t.* to upbraid, to abuse, to rail at. **vituperable**, *a.* **vituperation**, *n.* **vituperative**, *a.* **vituperatively**, *adv.* **vituperator**, *n.*
viva[1] (vē'və), *n., int.* an exclamation of applause or joy. [It.]
viva[2] (vī'və), VIVA VOCE.
vivace (vivah'chā), *adv.* (*Mus.*) in a brisk, lively manner. [It.]
vivacious (vivā'shəs), *a.* lively, bright, animated, sprightly, gay; (*Bot.*) tenacious of life, living through the winter, perennial. **vivaciously**, *adv.* **vivacity** (-vas'-), **vivaciousness**, *n.*
vivandière (vēvãdyeə'), *n.fem.* a female sutler attached to a continental, esp. French, regiment. [F]
vivarium (vīveər'iəm), *n.* (*pl.* -**ria** (-riə)) a park, garden or other place artificially prepared in which animals etc. are kept alive as nearly as possible in their natural state. **vivary** (vī'-), *n.*
vivat (vē'vat), *n., int.* the cry 'long live'. **vivat rex** or **regina**, *int.* long live the king or queen. [L]
viva voce (vī'və vō'chi), *adv., a.* by word of mouth, orally. *n.* a viva voce or oral examination. [L]
viverriform (vive'rifawm), *a.* having the shape or structure

vives (vīvz), *n.* a disease of the ear-glands in horses.
vivi-, *comb. form.* alive, living.
vivid (viv'id), *a.* vigorous, lively; very bright, intense, brilliant; clear, strongly marked, highly coloured. **vividly**, *adv.* **vividness**, *n.*
vivify (viv'ifī), *v.t.* to give life to, to quicken, to animate, to enliven. **vivification** (-fi-), *n.* **vivifier**, *n.*
viviparous (vīvip'ərəs), *a.* bringing forth young alive, opp. to *oviparous* and *ovoviviparous*; (*Bot.*) producing bulbs or seeds that germinate while still attached to the parent plant. **viviparously**, *adv.* **viviparity** (vivipa'ri-), **viviparousness**, *n.*
vivisection (vivisek'shən), *n.* the dissection of or performance of inoculative or other experiments on living animals. **vivisect**, *v.t.* to dissect (a living animal). **vivisectional**, *a.* **vivisectionist**, *n.* **vivisector** (viv'-), *n.* **vivisepulture** (-sep'əlchə), *n.* burial alive.
vivo (vē'vō), *adv.* (*Mus.*) with life and animation. [It.]
vixen (vik'sən), *n.* a she-fox; a shrewish, quarrelsome woman; a scold. **vixenish, vixenly**, *a.*
viz (viz), VIDELICET.
vizcacha (viskah'chə), VISCACHA.
vizier (viziə'), *n.* a high officer or minister of state in Muslim countries. **grand vizier**, the prime minister in the Turkish empire etc. **vizierate** (-rət), **viziership**, *n.* **vizierial**, *a.*
vizor (vī'zə), VISOR.
Vizsla (vizh'lə), *n.* a Hungarian breed of hunting dog with a smooth red or rust-coloured coat. [*Vizsla*, a town in Hungary]
VJ, (*abbr.*) video jockey.
vla, (*abbr.*) viola.
Vlach (vlahk), *n.* a Wallachian.
vlei (vlī, flī), *n.* (*S Afr.*) a swampy tract, a place where water lies in rainy seasons. [Afrikaans]
VLF, (*abbr.*) very low frequency.
vln, (*abbr.*) violin.
vocable (vō'kəbl), *n.* a word, esp. as considered phonologically.
vocabulary (vəkab'ūləri), *n.* a list or collection of words used in a language, science, book etc., usu. arranged in alphabetical order, and explained; a word-book; the stock of words at one's command.
vocal (vō'kəl), *a.* of or pertaining to the voice or oral utterance; having a voice; uttered or produced by the voice; resounding with or as with voices; voiced, sonant, not surd; having the character of a vowel. *n.* a vocal sound, a vowel; (*pl.*) the sung part of a (jazz or pop) song, as distinct from the instrumental part; in the Roman Catholic Church, a person authorized to vote in certain elections. **vocal cords**, *n.pl.* the elastic folds of the lining membrane of the larynx about the opening of the glottis. **vocal music**, *n.* music composed for or produced by the voice as distinct from instrumental music. **vocal score**, *n.* a musical score showing the singing parts in full. **vocalic** (-kal'-), *a.* pertaining to or consisting of vowel sounds. **vocalism**, *n.* the exercise of the vocal organs; a vowel sound. **vocalist**, *n.* a singer, opp. to an instrumental performer. **vocality** (-kal'-), **vocalness**, *n.* **vocalize, -ise**, *v.t.* to form or utter with the voice, esp. to make sonant; to insert the vowel-points in (Hebrew etc.). *v.i.* to exercise the voice, to speak, to sing etc. **vocalization**, *n.* **vocally**, *adv.*
vocation (vəkā'shən), *n.* a call or sense of fitness for a particular career; a divine call or guidance to undertake a duty, occupation etc.; one's calling or occupation. **vocational**, *a.* **vocationally**, *adv.*
vocative (vok'ətiv), *a.* pertaining to or used in addressing a person or thing. *n.* the case of a noun used in addressing a person or thing.
vociferate (vəsif'ərāt), *v.t.* to cry loudly, to shout; to make one's views known loudly and strongly. **vociferance, vociferation**, *n.* **vociferant**, *n.*, *a.* **vociferator**, *n.* **vociferous**, *a.* **vociferously**, *adv.* **vociferousness**, *n.*
vocule (vok'ūl), *n.* the faint sound made after articulating final *k*, *p* or *t*.
vodka (vod'kə), *n.* a strong spirituous liquor distilled from rye, orig. used in Russia.
voe (vō), *n.* (*Orkney, Shetland*) a small inlet, bay or creek.
vogue (vōg), *n.* fashion prevalent at any particular time; popular acceptance or usage. **vogue word**, *n.* a word much used at a particular time or period. **voguish**, *a.*
voice (vois), *n.* the sound uttered by the mouth, esp. by a human being, in speaking, singing etc.; the faculty or power of vocal utterance, speech, language; (*fig.*) expression of the mind or will in words whether spoken or written etc.; one's opinion or judgment, one's right to express this, one's choice, vote or suffrage; one expressing the will or judgment of others; a sound suggestive of human speech; sound produced by the breath acting on the vocal cords, sonancy; the verb-form expressing the relation of the subject to the action, as active, passive or middle. *v.t.* to give utterance to, to express; to regulate the tones of, to tune; to write the voice-parts for; to give voice or sonancy to. **with one voice**, unanimously. **voice-box**, (*coll.*) the larynx. **voicemail**, *n.* a form of electronic mail in which sounds are converted to a digital form and stored in a computer file. **voice-over**, *n.* the voice of an unseen narrator, actor etc. in a film etc. **voice-print**, *n.* an electronically recorded graphic representation of a person's voice. **voiced**, *a.* sonant; having a voice (*usu. in comb.* as *loud-voiced*). **voiceful**, *a.* vocal, sonorous. **voiceless** (-lis), *a.* having no voice or vote; speechless, mute; (*Phon.*) not voiced. **voicelessness**, *n.* **voicer**, *n.*
void (void), *a.* empty, unfilled; having no holder, occupant or incumbent; free from, destitute (of); useless, vain, ineffectual; having no legal force, null, invalid. *n.* an empty space; a vacuum. *v.t.* to invalidate, to nullify; to discharge, to emit from the bowels; to evacuate. **voidable**, *a.* **voidance**, *n.* the act of voiding or ejecting from a benefice; the state of being vacant. **voided**, *a.* made void; (*Her.*) of a charge, having the inner part cut away so that the field shows through. **voider**, *n.* **voidly**, *adv.* **voidness**, *n.*
voile (voil), *n.* a thin, semi-transparent dress material.
voir dire (vwah diə'), *n.* (*Law*) an oath administered to a witness.
voivode (voi'vōd), **vaivode** (vā'-), *n.* orig. a military commander in Slavonic countries; formerly, a liege prince in Romania, Wallachia etc.; the chief of an administrative division in Poland; an inferior administrative officer in Turkey. **voivodeship**, *n.*
vol (vol), *n.* (*Her.*) two outspread wings united at the base.
vol., (*abbr.*) volcano; volume; volunteer.
vola (vō'lə), *n.* (*pl.* **-lae** (-lē)) (*Anat.*) the palm of the hand; the sole of the foot. **volar**, *a.*
volant (vol'ənt), *a.* passing through the air; flying, able to fly; current; (*poet.*) nimble, active, rapid; (*Her.*) represented as flying.
volante (volan'tā), *n.* a two-wheeled covered horse-drawn vehicle with very long shafts and a chaise-body slung in front of the axle.
Volapük (vol'əpuk), *n.* a universal language invented (1879) by Johann Maria Schleyer. **Volapükist**, *n.*
volar VOLA.
volatile (vol'ətil), *a.* readily evaporating; (*fig.*) lively, sprightly, brisk, gay; fickle, changeable; (*Comput.*) of or pertaining to a type of memory which loses its stored data when the power is switched off. **volatility** (-til'-), *n.* **volatilize, -ise** (-lat'-), *v.t.* to cause to pass off in vapour. *v.i.* to

evaporate. **volatilizable**, *a.* **volatilization**, *n.*
vol au vent (vol'əvē), *n.* a raised or puff case filled with a variety of filling, often savoury. [F]
volcano (vəlkā'nō), *n.* (*pl.* **-noes**) an opening in the earth's surface through which lava, cinders, gases etc. are ejected from the interior, esp. at the top of a hill or mountain formed by the successive accumulations of ejected matter; (*fig.*) any situation where danger, upheaval etc. seems likely. **volcanic** (-kan'-), *a.* pertaining to, produced by, or of the nature of a volcano. **volcanic glass**, *n.* rock without a crystalline structure, as obsidian, pumice etc. produced by the rapid cooling of molten lava. **volcanic mud, sand**, *n.* volcanic ash which has been deposited under water and sorted and stratified. **volcanic rocks**, *n.pl.* those formed by volcanic agency. **volcanically**, *adv.* **volcanicity** (volkənis'-), **volcanism** (vol'-). *n.* **volcanist, volcanologist** (-nol'-), *n.* **volcanize, -ise** (vol'-), *v.t.* **volcanization**, *n.* **volcanology** (volkənol'-), *n.* the study of volcanoes. **volcanological** (-loj'-), *a.*
vole[1] (vōl), *v.t.* (*Cards*) to win all the tricks. *n.* the act of winning all the tricks in a deal.
vole[2] (vōl), *n.* a mouse-like rodent of the sub-family Arvicolinae, often called a water-rat.
volente Deo (vəlen'ti dā'ō), God willing. [L]
volery (vol'əri), *n.* an aerodrome; a flight of birds.
volet (vol'ā), *n.* a wing or panel of a triptych.
volitant (vol'itənt), *a.* flying, volant.
volition (vəlish'ən), *n.* exercise of the will; the power of willing. **volitient**, *a.* **volitional, -ary,** *a.* **volitionally**, *adv.* **volitionless** (-lis), *a.* **volitive** (-vol'-), *a.*
Volksraad (folks'raht), *n.* the former legislative assemblies of the Transvaal and the Orange Free State.
Volkswagen® (folks'vahgən, völks'wagən), *n.* a German make of car designed for popular use. [G]
volley (vol'i), *n.* a flight or simultaneous discharge of missiles; the missiles thus discharged; (*fig.*) a noisy outburst or emission of many things at once; a return of the ball at tennis and similar games before it touches the ground; (*Cricket*) a ball that flies straight at the head of the wicket after once hitting the ground. *v.t.* to discharge in or as in a volley; to return or bowl in a volley. *v.i.* to discharge or be discharged in a volley; to fire together (of guns); to return a ball before it touches the ground. **half-volley**, *n.* a return immediately after the ball has touched the ground. **volley-ball**, *n.* a game in which a large ball is hit back and forward over a high net by hand, played by two teams; the ball used in this game. **volleyer**, *n.*
volplane (vol'plān), *v.i.* (*Aviat.*) to fly downwards at a considerably higher angle than that of a glide. *n.* such a descending flight.
volt[1] (volt), *n.* a circular tread, the gait of a horse going sideways round a centre; a sudden leap to avoid a thrust in fencing.
volt[2] (vōlt), *n.* the SI unit of electric potential or potential difference, the difference of potential between two points in a conductor carrying a current of 1 ampere when the power dissipated between them is 1 watt. **voltmeter**, *n.* an instrument for measuring electromotive force directly, calibrated in volts. **volta-**, *comb. form* voltaic. *a.* **voltage** (-tij), *n.* electromotive force or potential difference as measured or expressed in volts. **voltaic** (-tā'ik), *a.* pertaining to electricity produced by chemical action or contact. **voltaic cell**, *n.* a primary cell. **voltaic pile**, *n.* a galvanic pile. **voltaism**, *n.* **voltameter** (-tam'itə), *n.* a device which measures electric charge. **voltite** (-tīt'), *n.* an insulating material for electric wires. [Alessandro *Volta*, 1745–1827, Italian physicist]
volta (vol'tə), *n.* (*pl.* **-te** (-tā)) (*Mus.*) time, turn. [It.]
Voltairism (voltea'rizm), **Voltaireanism** (-riən-), *n.* the principles or practices of Voltaire; scoffing scepticism.
Voltairean, *n.*, *a.* [François-Marie Arouet, 1694–1778, better known as *Voltaire*]

voltaism etc. VOLT.
volte-face (voltfas'), *n.* a turn round; an entire change of front in opinions, attitudes etc.
voltmeter VOLT.
volubilate etc. VOLUBLE.
voluble (vol'ūbl), *a.* characterized by a flow of words, fluent, glib, garrulous; (*Bot.*) volubilate. **volubilate** (-lū'bilət), **volubile** (-bil), *a.* (*Bot.*) twining, climbing by winding round a support; turning or rotating readily. **volubility** (-bil'-), **volubleness**, *n.* **volubly**, *adv.*
volucrine (vol'ūkrin, -krīn), *a.* of or pertaining to birds.
volume (vol'ūm), *n.* a collection of (usu. printed) sheets of paper, parchment etc. bound together, forming a book or work or part of one; (*loosely*) a book, a tome; (*Ant.*) a roll or scroll of papyrus, vellum etc. constituting a book; cubical content; mass, bulk; (*Mus.*) fullness or roundness of tone; loudness, or the control for adjusting it on a radio, television etc. **to speak, express volumes**, to mean much; to be very significant. **volumed**, *a.* (*usu. in comb.* as *three-volumed*). **volumenometer** (-inom'itə), *n.* an apparatus for measuring the volume of a solid body by the quantity of fluid that it displaces. **volumenometry** (-tri), *n.* **volumeter** (-lū'mitə), *n.* an instrument for measuring the volume of a gas; a hydrometer. **volumetric, -ical** (-met'-), *a.* **volumetrically**, *adv.* **voluminal** (-lū'mi-), *a.* pertaining to volume.
voluminous (vəloo'minəs), *a.* consisting of many volumes; producing many or bulky books (of a writer); of great volume, bulk or size. **voluminosity** (-nos'-), **voluminousness**, *n.* **voluminously**, *adv.*
voluntary (vol'əntəri), *a.* proceeding from or determined by one's own free will or choice, not under external constraint; acting or done willingly, spontaneous, intentional, purposive, designed; endowed with or exercising the power of willing; subject to or controlled by the will (of muscles, movement etc.); brought about, established or supported by voluntary action (of a church, school etc.); (*Law*) done without constraint or by consent, without valuable consideration; gratuitous. *n.* an organ solo played in a church etc. before, during or after service; a supporter of the principle that the Church (and usu. education) should be independent of the State and maintained by voluntary effort. **voluntarily**, *adv.* **voluntariness**, *n.* **voluntarism**, *n.* the reliance on voluntary subscriptions rather than on state aid for the upkeep of schools, churches etc. **voluntarist**, *n.* **Voluntary Aid Detachment**, *n.* an official organization of men and women to render first aid in time of war and assist in hospital work etc.
volunteer (voləntiə'), *n.* one who undertakes a job etc. voluntarily; one who enters into any service of his/her own free will. *a.* voluntary. *v.t.* to offer or undertake voluntarily. *v.i.* to offer one's services voluntarily, esp. orig. to offer to serve (for a campaign etc.) as a volunteer.
voluptuary (vəlŭp'chuəri), *n.* one given to luxury or sensual pleasures. *a.* pertaining to, promoting or devoted to sensual pleasure. **voluptuous**, *a.* pertaining to, contributing to, or producing sensuous or sensual gratification; of a woman, sexually alluring because of shapeliness or fullness of figure. **voluptuously**, *adv.* **voluptuousness**, *n.*
volute (vol'ūt, -lūt'), *n.* a spiral scroll used in Ionic, Corinthian and Composite capitals; a volutoid gastropod, usu. of tropical seas and having a beautiful shell. *a.* (*Bot.*) rolled up. **voluted**, *a.* **volution**, *n.* a spiral turn, a convolution; a whorl of a spiral shell. **volutoid** (-toid), *n.*, *a.*
volvox (vol'voks), *n.* a genus of simple, freshwater, greenish organisms united in spherical colonies, composed of minute flagellate cells which set up a revolving motion.
volvulus (-vūləs), *n.* a twisting of an intestine causing obstruction of the intestinal canal.
vomer (vō'mə), *n.* a small thin bone forming the chief portion of the partition between the nostrils in man. **vomerine** (-rin), *a.*
vomica VOMIT.

vomit (vom'it), *v.t.* to eject from the stomach by the mouth; to eject or discharge violently, to belch out. *v.i.* to eject the contents of the stomach by the mouth; to be sick. *n.* matter ejected from the stomach by the mouth; an emetic. **vomit-nut,** *n.* nux vomica. **vomica** (-kə), *n.* (*pl.* **-cae** (-sē)) an encysted collection of pus, esp. in the lung. **vomiter,** *n.* **vomito** (-tō), *n.* the yellow fever in its worst form. **vomitory,** *a.* emetic. *n.* an emetic; (*Rom. Ant.*) one of the openings for entrance or exit in an ancient theatre or amphitheatre. **vomiturition** (-tūrish'ən), *n.* an ineffectual attempt to vomit; retching.

voodoo (voo'doo), *n.* a system of magic orig. including snake-worship practised by Creoles and Negroes in Haiti and other parts of the W Indies and in the southern US; a sorcerer or conjurer skilled in this. *v.t.* to put a spell on or bewitch with voodoo. **voodooism,** *n.* **voodooish,** *a.*

Voortrekker (vuə'trekə, fuə'-, -trek'ə), *n.* one of the Dutch farmers from Cape Colony who took part in the Great Trek into the Transvaal in 1836 and following years; (**voortrekker**) a pioneer.

voracious (vərā'shəs), *a.* greedy in eating; ravenous, gluttonous; insatiable, very eager. **voraciously,** *adv.* **voraciousness, voracity** (-ras'-), *n.* **vorant** (vaw'rənt), *a.* (*Her.*) devouring.

-vore (-vaw), *comb. form.* an animal which feeds on, as *carnivore, herbivore*.

-vorous (-vərəs), *comb. form.* feeding on, living on, as *carnivorous, herbivorous.*

vortex (vaw'teks), *n.* (*pl.* **vortices** (-tisēz)) a whirling or rotating mass of fluid, esp. a whirlpool; a situation or activity that takes up all one's attention or energy. **vortex-ring,** *n.* one the axis of which is a closed curve. **vortical, vorticose** (-kōs), **vorticular** (-tik'ū-), *a.* **vortically,** *adv.* **vorticel** (-tisel), *n.* a bell-animalcule. **vorticelloid** (-oid), *a.* **vorticity** (-tis'-), *n.* **vortiginous** (-tij'i-), *a.* vortical, whirling. [L]

vorticism (vaw'tisizm), *n.* a school of early 20th-cent. painting which seeks to represent nature in formal designs of straight and angular patterns. **vorticist,** *n.*

votary (vō'təri), *n.* one who is devoted or consecrated by a vow or promise; one who is devoted to some particular service, study, pursuit etc. **votaress** (-ris), *n. fem.*

vote (vōt), *n.* a formal expression of opinion, will or choice, in regard to the election of a candidate, the passing or rejection of a resolution, law etc., usu. signified by voice, gesture or ballot; anything by which this is expressed, as a ballot, ticket etc.; that which is voted, as a grant of money; the aggregate votes of a party etc.; the right to vote, the suffrage. *v.i.* to give one's vote; to express one's approval (for). *v.t.* to give one's vote for; to enact, resolve, ratify or grant by a majority of votes; (*coll.*) to declare by general consent. **to vote down,** to defeat or suppress by vote. **to vote in,** to elect. **vote of no confidence,** the legal method of forcing the resignation of a government or governing body or person. **to vote with one's feet,** to indicate one's dissatisfaction by leaving. **votable,** *a.* **voteless** (-lis), *a.* **voter,** *n.* **voting-paper,** *n.* a paper by means of which one votes, esp. by ballot in a parliamentary election.

votive (vō'tiv), *a.* given, paid or dedicated in fulfilment of a vow. **votively,** *adv.*

vouch (vowch), *v.t.* to uphold or guarantee by assertion, proof etc., to confirm, to substantiate. *v.i.* to give testimony, to answer (for). **vouchee** (-chē'), *n.* the person vouched or summoned in a writ of right. **voucher,** *n.* one who or that which vouches for or attests; a document etc. serving to confirm or establish something, as a payment, the correctness of an account etc.

vouchsafe (vowchsāf'), *v.t.* to condescend to grant; to concede. *v.i.* to deign, to condescend (to).

voussoir (vooswah'), *n.* one of the wedge-shaped stones forming an arch.

vow (vow), *n.* a solemn promise or pledge, esp. made to God or to a saint etc., undertaking an act, sacrifice, obligation etc. *v.t.* to promise solemnly; to dedicate by a vow; to affirm solemnly. *v.i.* to make a vow. **to take vows,** to enter a religious order and commit oneself to the vows of chastity, poverty and obedience.

vowel (vow'əl), *n.* a sound able to make a syllable or to be sounded alone; an open and unimpeded sound as opp. to a closed, stopped or mute sound or consonant; a letter representing this, esp. the simple vowels, *a, e, i, o, u.* **vowel-gradation,** *n.* ablaut. **vowel-mutation,** *n.* umlaut. **vowel-point,** *n.* one of the marks indicating the vowels in Hebrew etc. **vowelize, -ise,** *v.t.* to insert vowel-points. **vowelled,** *a.* (*usu. in comb.* as *open-vowelled*). **vowelless** (-lis), *a.* **vowelly,** *a.*

vox (voks), *n.* voice. **vox humana** (hūmah'nə), *n.* an organ-stop producing tones approximating to those of the human voice. **vox populi** (pop'ūlī), **vox pop,** *n.* the voice of the people, popular feeling. [L]

voyage (voi'ij), *n.* a journey by water, esp. by sea to a distant place. *v.i.* to make a voyage. *v.t.* to travel over by water. **voyageable,** *a.* **voyager,** *n.* **voyageur** (vwayazhœ'), *n.* a Canadian boatman or trapper.

voyeur (vwayœ'), *n.* one who derives gratification from watching sexual acts, people undressing etc. **voyeurism,** *n.* **voyeuristic** (-ris'-), *n.* [F]

VP, (*abbr.*) Vice-President.

VR, (*abbr.*) variant reading; Victoria Regina.

vraic (vrāk), *n.* a seaweed used for fuel and manure, found in the Channel Islands.

vroom (vroom), *n.* (*coll.*) power, drive, energy (*also interj.*). *v.i.* to drive fast.

vraisemblance (vrāsēblās'), *n.* an appearance of truth, verisimilitude.

vs., (*abbr.*) versus.

VSO, (*abbr.*) Very Superior Old, used to indicate that port or brandy is between 12 and 17 years old; in Britain, Voluntary Service Overseas, an organization which sends volunteers overseas to use and teach their skills.

VSOP, (*abbr.*) Very Special Old Pale, used to indicate that brandy or port is between 20 and 25 years old.

VTOL (vē'tol), (*abbr.*) vertical take-off and landing, a system by which aircraft take off and land without taxiing.

VTR, (*abbr.*) video tape recorder.

Vulcan (vŭl'kən), *n.* the Roman god of fire and metalworking. **Vulcanian** (-kā'ni-), *a.* **vulcanic** (-kan'-), **vulcanism** etc. VOLCANIC, -ISM etc.

vulcanite (vŭl'kənit), *n.* vulcanized rubber, ebonite. **vulcanize, -ise,** *v.t.* to treat rubber with sulphur at a high temperature so as to increase its strength and elasticity, producing vulcanite (the hard form) or soft and flexible rubber. **vulcanization,** *n.*

vulgar (vŭl'gə), *a.* pertaining to or characteristic of the common people, plebeian, common, coarse, unrefined; ordinary, in common use. **the vulgar,** the common people, the uneducated. **vulgar era,** *n.* the Christian era. **vulgar fraction,** *n.* a fraction having the numerator less than the denominator. **vulgar tongue,** *n.* the vernacular. **vulgarian** (-geə'ri-), *n.* a vulgar person, esp. a rich person with vulgar manners etc. **vulgarism,** *n.* **vulgarity** (ga'ri-), *n.* **vulgarize, -ise,** *v.t.* **vulgarization,** *n.* **vulgarly,** *adv.*

Vulgate (vŭl'gət), *n.* the Latin translation of the Bible made by St Jerome, 383–405.

vulnerable (vŭl'nərəbl), *a.* capable of being wounded; susceptible of or liable to injury, attack etc. **vulnerability** (-bil'-), **vulnerableness,** *n.* **vulnerary,** *a.* useful in healing wounds or for the cure of external injuries. *n.* a vulnerary plant, drug or composition.

Vulpes (vŭl'pēz), *n.* the genus which includes the common fox. **vulpine** (vŭl'pīn), *a.* pertaining to or characteristic of a fox; crafty, cunning. **vulpicide** (-pisid), *n.* the killing of

a fox, esp. otherwise than by hunting; a fox-killer. **vulpinism** (-pin-), *n.*
vulsella (vŭlsel'ə), *n.* a forceps with hooked teeth or claws.
vulture (vŭl'chə), *n.* a large falconoid bird with head and neck almost naked, feeding chiefly on carrion; a rapacious person. **vulturine** (-rin), **vulturish, vulturous,** *a.*

vulva (vŭl'və), *n.* an opening, an entrance, esp. the external opening of the female genitals. **vulvar, vulvate** (-vət), **vulviform** (-vifawm), *a.* **vulvitis** (-vī'tis), *n.* inflammation of the vulva. **vulvo-,** *comb. form.*
vv, (*abbr.*) vice versa.
vying (vī'ing), *pres.p.* VIE.

W¹, w¹, the 23rd letter of the English alphabet, taking its form and name from the union of two V's, V formerly having the name and force of U. W (*pl.* **ws, w's**) has the sound of a semi-vowel, as in *was, will, forward*.
W², (*abbr.*) watt; West, Western; women('s size).
W³, (*chem. symbol*) tungsten.
w², (*abbr.*) week; weight; white; wicket; wide; width; wife; with.
WA, (*abbr.*) West Africa; Western Australia.
WAAC (wak), *n.* (a member of) the Women's Auxiliary Army Corps.
WAAF (waf), *n.* (a member of) the Women's Auxiliary Air Force Service (later Women's Royal Air Force).
wabble (wob'l), WOBBLE.
wacke (wak'ə), *n.* an earthy or clayey rock produced by the decomposition of igneous rocks.
wacky (wak'i), *a.* (*sl.*) crazy, eccentric, absurd. **wackily**, *adv.* **wackiness**, *n.*
wad¹ (wod), *n.* a small, compact mass of some soft material, used for stopping an opening, stuffing between things etc.; a felt or paper disk used to keep the charge in place in a gun, cartridge etc.; (*coll.*) a number of currency notes, a lot of money. *v.t.* (*past, p.p.* **wadded**) to compress into a wad; to stuff or line with wadding; to pack, stop up or secure with a wad. **wadding**, *n.* a spongy material, usu. composed of cotton or wool, used for stuffing garments, cushions etc.; material for gun-wads.
wad² (wod), *n.* an earthy ore of manganese; (*dial.*) plumbago.
wadable WADE.
wadding WAD¹.
waddle (wod'l), *v.i.* to walk with an ungainly rocking or swaying motion and with short, quick steps, as a duck or goose. *n.* a waddling gait. **waddler**, *n.* **waddlingly**, *adv.*
waddy (wod'i), *n.* an Australian war-club, usu. bent like a boomerang or with a thick head. *v.t.* to hit with a waddy. **waddy wood**, *n.* a Tasmanian tree from which this is made.
wade (wād), *v.i.* to walk through water or a semi-fluid medium, as snow, mud etc.; to make one's way with difficulty and labour. *v.t.* to pass through or across by wading; to ford (a stream) on foot. **to wade in(to)**, (*coll.*) to tackle or attack vigorously. **wadable**, *a.* **wader**, *n.* one who wades; a high, waterproof boot, worn by anglers etc. for wading; a wading bird. **wading bird**, *n.* a long-legged bird that wades.
wadi, wady (wod'i), *n.* the valley or channel of a stream that is dry except in the rainy season. [Arab.]
wafer (wā'fə), *n.* a small, thin, sweet cake or biscuit; a thin disk of unleavened bread used in the Eucharist, the Host; a thin adhesive disk of dried paste for sealing letters, fastening documents etc.; a thin disk of silicon or other semiconductor material on which integrated electrical circuits are formed before being cut into individual chips. *v.t.* to seal or attach with a wafer. **wafer-cake**, *n.* **wafery**, *a.*
waffle¹ (wof'l), *n.* a thin batter cake baked on a waffle-iron. **waffle-iron**, *n.* a utensil with hinged plates for baking waffles.
waffle² (wof'l), *v.i.* (*dial.*) to wave, to fluctuate; (*coll.*) to chatter or write aimlessly and at length. *n.* vague or inconsequential talk or writing. **waffling**, *n.*, *a.*
waft (wahft. woft), *v.t.* to carry or convey through the air; to carry lightly or gently along. *v.i.* to float or be borne on the air. *n.* an act of wafting, as a sweep of a bird's wing; a breath or whiff of odour etc.
wag¹ (wag), *v.t.* (*past, p.p.* **wagged**) to shake up and down or backwards and forwards lightly and quickly, esp. in playfulness, reproof etc.; to move (the tongue) in chatter or gossip. *v.i.* to move up and down or to and fro, to oscillate; of the tongue, to move in chatter or gossip; to move on, to keep going, to proceed. *n.* an act or motion of wagging, a shake.
wag² (wag), *n.* a facetious person, a wit, a joker. **waggery**, **waggishness**, *n.* jocularity, playful merriment, practical joking. **waggish**, *a.* **waggishly**, *adv.*
wage (wāj), *n.* (*often pl.*) payment for work done or services rendered, esp. fixed periodical pay for labour of a manual kind; (*usul pl.*) recompense, reward, requital. *v.t.* to engage in, to carry on (a battle, war etc.). **wage-earner**, *n.* the person who earns the (most substantial) wage in a household. **wage-freeze**, *n.* the fixing of a wage-level for a prolonged period. **wage-, wages-fund**, *n.* the portion of the capital of a community expended in paying the wages of labour. **wage-slave**, *n.* (*coll.*) a person dependent on a wage or salary. **wagedom, wagery** (-əri), *n.* **wageless** (-lis), *a.*
wager (wā'jə), *n.* something staked or hazarded on the outcome of a contest etc.; a bet; something on which bets are laid. *v.t., v.i.* to stake, to bet. **wager of battle** BATTLE. **wager of law**, a compurgation. **wagerer**, *n.*
wagga (blanket) (wog'ə), *n.* a rug made from corn sacks cut open and sewn together.
waggery, waggish etc. WAG².
waggle (wag'l), *v.t., v.i.* to wag quickly and frequently. *n.* a short, quick wagging. **waggly**, *a.*
waggon WAGON.
Wagnerian (vahgniə'riən), *a.* pertaining to or in the style of Wagner's music or musical dramas. **Wagnerianism, Wagnerism** (vahg'nərizm), *n.* **Wagnerist** (vahg'-), **Wagnerite**, *n.* [Richard *Wagner*, 1813-83]
wagon, waggon (wag'ən), *n.* a strong four-wheeled vehicle for the transport of heavy loads, usu. with a rectangular body, often with a removable cover; an open railway truck. **on (off) the wagon**, (*coll.*) abstaining (no longer abstaining) from alcohol. **wagon-ceiling, -roof, -vault**, *n.* a barrel-vault. **wagon-load**, *n.* **wagon-train**, *n.* a column of horse-drawn wagons carrying supplies, pioneer settlers etc. **wagonage** (-ij), *n.* wagons collectively. **wagoner**, *n.* one who drives or leads a wagon, a charioteer; the constellation Auriga. **wagonette** (-net'), *n.* a four-wheeled pleasure carriage of light construction, for six or eight persons on seats facing each other, often with a removable cover, drawn by one or more horses. **wagonful**, *n.* **wagonless**, *a.*
wagon-lit (vagōlē'), *n.* (*pl.* **wagons-lits, wagon-lits**, -lē') a sleeping-car. [F]
wagtail (wag'tāl), *n.* any of various small, black and grey or white or yellow birds, named from the wagging of their tails.
Wahabi, Wahhabi (wəhah'bi), *n.* (*pl.* **-bis**) one of a sect founded about the middle of the 18th cent. cultivating a

strict form of Islam. **Wahabiism**, *n.*
wahine (wǝhē′ni), *n.* a Maori or Polynesian woman. [Maori]
waif (wāf), *n.* a person or thing found astray, ownerless, or cast up by or adrift on the sea; a homeless wanderer, esp. a forsaken or unowned child.
wail (wāl), *v.t.* to lament loudly over, to bewail. *v.i.* to lament, to utter wails; to make a plaintive sound (as the wind). *n.* a loud, high-pitched lamentation, a plaintive cry; a sound like this. **wailer**, *n.* **wailful**, *a.* **wailing**, *n., a.* **Wailing Wall**, *n.* a wall in Jerusalem, a remnant of an ancient temple, held sacred by the Jews as a place of worship and lamentation. **wailingly**, *adv.*
wain (wān), *n.* (*poet.*) a four-wheeled vehicle for the transportation of goods, a wagon; Charles's Wain. **wainwright**, *n.* one who makes wagons.
wainscot (wāns′kǝt), *n.* a wooden, usu. panelled, lining or casing of the walls of a room; the lower part of the walls of a room when lined or finished differently from the upper part; fine-grade oak for wainscot panelling etc. *v.t.* (*past, p.p.* **wainscoted, -tted**) to line with this. **wainscoting, -tting**, *n.* (material for) a wainscot or wainscots.
wairepo (wīrē′pō), *n.* (*New Zealand*) grog, spirits. [Maori]
waist (wāst), *n.* that part of the human body below the ribs and thorax and above the hips; this part as normally more contracted than the rest of the trunk; the constriction between the abdomen and thorax of a wasp etc.; the middle part of an object, as an aircraft fuselage, esp. if more contracted than the other parts; the part of a ship between the quarter-deck and the forecastle; the part of a garment encircling the waist; (*N Am.*) a blouse, a shirtwaist. **waist-band**, *n.* a band or belt worn round the waist, esp. a band forming the upper part of a skirt, trousers etc. **waist-belt**, *n.* **waist-cloth**, *n.* a loin-cloth. **waistcoat** (wās′kǝt), *n.* a short, close-fitting garment, usu. without sleeves, extending from the neck to the waist. **waist-deep, -high**, *a., adv.* as deep, as high, or in (water etc.) as far as the waist. **waistline**, *n.* the waist of a dress etc., not necessarily corresponding with the wearer's natural waist.
wait (wāt), *v.i.* to remain inactive or in the same place until some event or time for action, to stay; to be in a state of expectation or readiness; to be on the watch (for); to be ready or in a fit state for use etc.; to act as a waiter, to attend (on persons) at table. *v.t.* to wait for; to postpone, to delay. *n.* the act of waiting; time taken in waiting, delay; watching, ambush; (*pl.*) a band of singers and players performing carols in the streets etc. at Christmas-time. **to lie in wait**, to wait for in secret, to waylay. **to wait on, upon**, to attend upon as a servant; to pay a visit to deferentially; to await; of consequences etc., to follow. **to wait up**, to remain out of bed waiting (for). **waiter**, *n.* one who waits; an attendant on the guests at a hotel, restaurant etc.; a dumb-waiter. **waiting**, *n., a.* **in waiting**, in attendance, esp. on the sovereign. **waiting game**, *n.* a holding back of action in the hope of more advantageous circumstances later. **waiting list**, *n.* a list of people waiting for a vacancy, treatment etc. **waiting-maid, -woman**, *n.* a female attendant. **waiting-room**, *n.* a room at a railway-station etc. where persons can rest while waiting. **waitress** (-ris), *n.fem.*
waive (wāv), *v.t.* to forgo, to relinquish; to refrain from using, insisting on etc.; to defer, postpone. **waiver**, *n.* (*Law*) the act of waiving a claim, right etc.; a written statement of this.
waiwode (wā′wōd), VOIVODE.
waka (wok′ǝ), *n.* a Maori canoe.
wake[1] (wāk), *v.i.* (*past* **woke** (wōk), **waked**, *p.p.* **waked, woken**) to be aroused from sleep, to cease to sleep; to revive from a trance, death etc.; to be awake, to be unable to sleep; to be roused or to rouse oneself from inaction, inattention etc. *v.t.* to rouse from sleep; to revive, to raise from the dead; to arouse, to stir (up) to excite, to alert; to break the silence of, to disturb. *n.* the act of waking or being awake; a vigil. **wake-robin**, *n.* the wild arum, 'lords and ladies'. **wakeful**, *a.* not disposed or unable to sleep, restless; passed without sleep, disturbed; watchful, alert. **wakefully**, *adv.* **wakefulness**, *n.* **waker**, *n.*
wake[2] (wāk), *n.* the feast of the dedication of a church, formerly kept by watching all night; a merry-making held in connection with this; the watching of a dead body, prior to burial, by friends and neighbours of the deceased, with lamentations often followed by a merry-making. *v.t.* to hold a wake over.
wake[3] (wāk), *n.* the track left by a vessel passing through water; the track or path left after something has passed. **in the wake of**, following.
waken (wā′kǝn), *v.t.* to rouse from sleep; to rouse to action etc.; to call forth. *v.i.* to wake, to cease from sleeping. **wakener**, *n.*
waker WAKE[1].
Waldenses (wolden′sēz), *n.pl.* a religious sect founded about 1170 by Peter Waldo, a merchant of Lyons, in a reform movement leading to persecution by the Church. **Waldensian**, *n., a.*
waldgrave (wawld′grāv), *n.* a German title of nobility, orig. a head forester.
waldhorn (wawld′hawn), *n.* a hunting-horn; a french horn without valves.
wale (wāl), *n.* a ridge on the skin, a weal; a ridge on the surface of cloth; a wide plank extending along a ship's side. *v.t.* to weal.
waler (wā′lǝ), *n.* (*Austral.*) a riding-horse (orig. as supplied by military authorities in New South Wales).
Walhalla (valhal′ǝ), VALHALLA.
walk (wawk), *v.i.* to go along by raising, advancing and setting down each foot alternately, never having both off the ground at once; of a quadruped, to go along with a slow gait keeping at least two feet on the ground at any time; to go at the ordinary pace, not to run, not to go or proceed rapidly; to go or travel on foot; of a ghost, to move about or become visible; (*coll.*) to depart, to be off, to be dismissed. *v.t.* to walk over, on or through; to tread; to cause to walk; to lead, drive or ride at a walking pace; to accompany on foot; to move (an object) by alternately lifting one side then the other. *n.* the act of walking; the pace, gait or step of a person or animal that walks; a distance walked; a stroll, a promenade; the route chosen for this; a piece of ground laid out for walking, a foot-path etc.; the district or round of a hawker, postman etc.; a sheep-walk; one's profession, occupation, sphere of action etc. **to walk away with**, to win or gain easily. **to walk into**, (*sl.*) to thrash; to abuse; to meet with unexpectedly; to eat heartily of. **to walk off with**, (*coll.*) to carry off, to steal; to walk away with. **to walk one's chalks**, the chalk CHALK. **to walk out**, to go on strike; to leave suddenly, to abandon. **to walk out with**, (*dated*) to go courting with. **to walk the plank** PLANK. **to walk the streets** STREET. **walk-about**, *n.* a wandering journey by Australian Aborigines; an informal walk to meet the public by a politician, member of royalty etc. *adv.* moving from place to place. **walk-in**, *a.* of a wardrobe etc., large enough to walk and move around in. **Walkman**®, *n.* a small portable cassette recorder with headphones. **walk-on**, *n.* a small (non-speaking) part in a play etc. **walk-out**, *n.* a strike; the act of leaving (a meeting etc.) as a protest. **walk-over**, *n.* an easy victory. **walkway**, *n.* a path etc. for pedestrian use only; a place for walking, a walk. **walkable**, *a.* **walker**, *n.* one who walks; a shop-walker; a bird that steps instead of hopping; a gallinaceous bird; a frame for supporting a baby, disabled person etc. when walking. **walking**, *n., a.* **walking-dress**, *n.* a dress for wearing out of doors. **walking-leaf**, *n.* an insect mimicking a leaf. **walking-papers**, *n.pl.* (*coll.*) notice of dismissal. **walking-

stick, *n.* a stick carried in walking; a stick insect.
walkie-talkie (wawkitaw'ki), *n.* a portable combined transmitter and receiver.
walkyrie (valkiə'ri, val'-), VALKYRIE.
wall (wawl), *n.* a continuous structure of stone, brick etc., narrow relatively to its height, forming an enclosure, fence, or the front, back or side, or an internal partition of a building; (*usu. pl.*) a rampart, a fortification; anything resembling a wall, as a cliff, a mountain-range etc.; the enclosing sides of a vessel, cavity etc.; a defence. *v.t.* to furnish, enclose, surround or defend with a wall; to separate or divide with a wall; to block (up) with a wall. **off the wall**, (*sl.*) crazy, unreliable. **to give the wall to**, to allow as a courtesy to walk or pass by on the side of a pavement etc. away from the gutter. **to go to the wall**, to get the worst in a contest; to be ruined. **to have one's back to the wall**, to be in a desperate position. **up the wall**, (*coll.*) in or into a state of distraction or exasperation. **wall bars**, *n.pl.* horizontal bars fixed to a wall, used for gymnastics. **wall-creeper**, *n.* a Eurasian songbird frequenting walls and cliffs. **wall-cress**, *n.* a plant of the genus *Arabis* growing in crevices. **wallflower**, *n.* a sweet-smelling plant of the genus *Cheiranthus*, with yellow, brown, and crimson flowers; (*coll.*) one excluded from the main social activity, esp. a lady without a partner at a dance. **wall-fruit**, *n.* fruit grown on trees trained against walls. **wall game**, *n.* a kind of football played only at Eton. **wall-painting**, *n.* one painted on a wall, a fresco. **wallpaper**, *n.* paper, usu. with decorative patterns or texture, for pasting on the walls of rooms. **wall-pellitory** PELLITORY. **wall-pepper**, *n.* stone-crop. **wall-plate**, *n.* a piece of timber let into a wall as a bearing for the ends of the joists etc. **wall-rue**, *n.* a small evergreen fern growing on old walls, cliffs etc. **Wall Street**, *n.* the New York stock exchange and money market. **wall-tie**, *n.* a metal bond between the sides of a cavity wall. **wall-to-wall**, *a.* of carpet etc., covering all the floor; (*coll.*) continuous, non-stop. **walled**, *a.* **walling**, *n.*
wallaba (wol'əbə), *n.* a leguminous tree from Guyana, used in carpentry and building.
wallaby (wol'əbi), *n.* one of the small species of kangaroo. **on the wallaby**, (*Austral., sl.*) tramping about looking for work etc.
Wallach (wol'ək), *n.* a Wallachian or Vlach, a Romance-speaking inhabitant of Romania. **Wallachian** (-lā'-), *a.* of or pertaining to Wallachia. *n.* a native or the language of Wallachia.
wallah, walla (wol'ə), *n.* (*often in comb.*) an agent, worker or any one employed in a specific type of work; (*coll., sometimes derog.*) a person, a fellow.
wallaroo (woləroo'), *n.* (*Austral.*) one of the large species of kangaroo.
waller etc. WALL.
wallet (wol'it), *n.* a bag or sack for carrying necessaries for a journey or march; a small bag or case for carrying paper money, credit cards etc.; a folder for papers, documents etc.
wall-eye (waw'li), *n.* a condition of the eye due to opacity of the cornea or to a divergent squint; an eye with a very light-coloured iris, esp. due to this; a large, glaring eye, as in fish. **wall-eyed**, *a.*
wallflower WALL.
Walloon (wəloon'), *n.* one of a French-speaking people in SE Belgium and the adjoining parts of France; their language. *a.* pertaining to the Walloons or their language.
wallop (wol'əp), *v.i.* to boil with a noisy bubbling and rolling motion; to move along in a clumsy tumbling fashion, to waddle. *v.t.* (*coll.*) to thrash; to defeat or beat decisively. *n.* (*coll.*) a blow, a punch; (*coll.*) forceful impact, power; (*sl.*) beer. **walloper**, *n.* **walloping**, *n.* (*coll.*) a thrashing. *a.* (*coll.*) big, thumping, whopping.
wallow (wol'ō), *v.i.* to roll or tumble about in mire, water etc.; to revel grossly or self-indulgently (in vice etc.). *v.t.* to roll (oneself) about in mire etc. *n.* the act of wallowing; a mud-hole or other place in which animals wallow. **wallower**, *n.*
wally (wol'i), *n.* (*sl.*) an incompetent or stupid person.
walnut (wawl'nŭt), *n.* a tree of the genus *Juglans*, bearing a nut enclosed in a green fleshy covering; the unripe fruit of this used for pickling; the ripe nut; the timber of this or other species of the same genus used in cabinet-making and for gun-stocks; a yellowish-brown colour.
Walpurgis night (valpuə'gis), *n.* the eve of 1 May, when witches are supposed to hold revel and dance with the devil, esp. on the Brocken, in Germany.
walrus (wawl'rəs), *n.* a large, amphibious, long-tusked, seal-like mammal of the Arctic seas. **walrus moustache**, *n.* a thick moustache with long drooping ends.
waltz (wawlts), *n.* a dance in triple time in which the partners pass round each other smoothly as they progress; the music for such a dance. *v.i.* to dance a waltz; to move quickly. **to waltz into**, (*coll.*) to rebuke severely. **waltzer**, *n.* one who waltzes; a type of fairground roundabout.
waly (wā'li), *a.* beautiful, fine, excellent; strong, robust.
wamble (wom'bl), *v.i.* of the stomach, to rumble, to heave; to be affected with nausea; to move unsteadily. *n.* a heaving; a feeling of nausea; an unsteady gait.
wampee (wompē'), *n.* a tree of the rice family cultivated in China and the E Indies bearing a grape-like, pulpy berry.
wampum (wom'pəm), *n.* small beads made of shells, used by the N American Indians formerly as money, or for decorating belts, bracelets etc.
wan (won), *a.* (*comp.* **wanner**, *superl.* **wannest**) pale or sickly in hue; lacking vigour or liveliness; of light etc., dim, faint. **wanly**, *adv.* **wanness**, *n.* **wannish**, *a.*
wand (wond), *n.* a long, slender rod, esp. one used by conjurers or as a staff of office; a conductor's baton; a light-pen used for reading bar codes. **wandy**, *a.*
wander (won'də), *v.i.* to travel or go here and there without any definite route or object, to rove, ramble or roam; to follow an irregular or winding course; to lose one's way; to deviate from the right or proper course; to talk or think incoherently or senselessly, to be delirious; to digress from the subject in hand. *v.t.* to wander over, to traverse in a random way. **wanderer**, *n.* **wandering**, *n., a.* (*usu. pl.*) **Wandering Jew**, *n.* a legendary character condemned, for an insult to Christ, to wander from place to place until the Day of Judgment; any of several trailing or climbing plants. **wanderingly**, *adv.* **wanderment**, *n.*
wanderlust (won'dəlŭst), *n.* the desire to travel.
wanderoo, wanderu (wondəroo'), *n.* the lion-tailed macaque, with a large greyish beard, of W India; a monkey from Sri Lanka.
wandoo (won'doo), *n.* the white gum-tree of W Australia.
wane (wān), *v.i.* to diminish in size and brilliance, as the illuminated portion of the moon; to decrease in power, strength etc. *n.* the act or process of waning, decrease, diminution; the period when the moon wanes; a defective edge or corner on timber.
wangle (wang'gl), *v.t.* (*coll.*) to manipulate, to employ cunningly; to falsify (accounts etc.); to achieve or gain by devious means. *n.* an act or instance of wangling. **wangler**, *n.*
wank (wangk), *v.i.* (*taboo*) to masturbate. *n.* (*taboo*) an instance of masturbating. **wanker**, *n.* (*taboo*) one who masturbates; (*sl.*) a worthless, incompetent or contemptible person.
wanly etc. WAN.
wannabee (won'əbē), *n., a.* (*sl.*) (a person) anxious to become somebody or something.
want (wont), *n.* the state or condition of not having, lack, deficiency, absence, need (of); need, privation, poverty; a longing or desire for something that is necessary or required for happiness etc.; that which is not possessed but

is so desired. *v.t.* to be without, to lack, to be deficient in; to need, to require; to be short by, to require in order to be complete; to feel a desire for, to crave; to desire or request the presence or assistance of; ought (to). *v.i.* to be in need, to be in want (for); to be deficient (in), to fall short (in); to be lacking, to have need (for). **want-ad**, *n.* a classified advertisement in a newspaper etc. specifying an item or service wanted. **wanter**, *n.* **wanting**, *a.* absent, missing, lacking; not meeting the required or expected standard; lacking (in), deficient (in); (*coll.*) witless, daft, deficient in intelligence. *prep.* without, less, save. **wantless** (-lis), *a.*
wanton (won'tən), *a.* frolicsome, playful; unrestrained, wild, unruly; extravagant, luxuriant; licentious, lascivious; random, reckless, purposeless. *n.* an unchaste person, esp. a woman; a trifler; a playful, idle creature. *v.i.* to sport, to frolic; to move, act or grow at random or unrestrainedly. **wantonly**, *adv.* **wantonness**, *n.*
wapacut (wop'əkŭt), *n.* a large white N American owl.
wapens(c)haw WAPINSHAW.
wapentake (wop'əntāk), *n.* a name formerly given in certain English counties to a division corresponding to a hundred.
wapinshaw, wapinschaw, wap(p)ens(c)haw (wop'nshaw, wap'-), *n.* (*Sc., formerly*) a review of persons under arms; a meeting for rifle-shooting, curling-matches etc.
wapiti (wop'iti), *n.* a N American stag related to the red deer, erroneously called the elk.
wappens(c)haw WAPINSHAW.
war (waw), *n.* a contest carried on by force of arms between nations, or between parties in the same state; the state of things brought about by this, a state of hostilities with suspension of ordinary international relations; hostile operations, military or naval attack, invasion; the military art, strategy; hostility, active enmity, strife; (a) conflict, feud, struggle. *v.i.* (*past, p.p.* **warred**) to make or carry on war; to contend, to strive, to compete; to be in opposition, to be inconsistent. **art of war**, strategy and tactics. **at war**, engaged in hostilities (with). **civil war** CIVIL. **cold war** COLD. **holy war** HOLY. **in the wars**, (*coll.*) (bruised or injured as from) fighting or quarrelling. **man-of-war** MAN. **to be, go on the war-path**, to be ready for or engaged in conflict; to be thoroughly roused or incensed. **war-bond** *n.* a government bond issued as a means of raising a war loan. **war correspondent**, *n.* a journalist who reports on current events from the scene of a war or battle. **war crime**, *n.* one committed in connection with a war. **war criminal**, *n.* **war-cry**, *n.* a name or phrase formerly shouted in charging etc.; a watchword; a party cry. **war-dance**, *n.* a dance practised, as by some N American Indian tribes, as a preparation for battle. **war footing**, *n.* a condition (of the military or naval establishments) of readiness for active hostilities. **war-game**, *n.* a simulated military battle or campaign; an enactment of a battle using models. **war-god**, *n.* a deity worshipped as giving victory, as Mars or the Greek Ares. **warhead**, *n.* the head of a torpedo, aerial bomb, rocket etc., charged with explosive, removable in peace practice. **war-horse**, *n.* a charger; (*fig.*) a veteran, a person full of warlike memories etc.; a standard, (overly-)familiar piece of music etc. **war-loan**, *n.* a loan raised to meet the cost of a war. **warlord**, *n.* a military leader or commander of (a part of) a nation. **war machine**, *n.* the economic, industrial, administrative apparatus used by a state in order to prepare for and engage in war. **war-marked**, *a.* bearing the marks or traces of war. **warmonger**, *n.* one who traffics in war, or promotes it by every means possible. **war neurosis** SHELL-SHOCK. **War Office**, *n.* a government department administering the affairs of the army. **war-paint**, *n.* paint put on the face and body, esp. by N American Indians, before going into battle; full dress; (*coll.*) make-up. **war-path**, *n.* the path taken by an attacking party of N American Indians; hence a warlike expedition. **warplane**, *n.* a military aircraft for use in war. **warship**, *n.* an armed ship for use in war. **war-song**, *n.* a song sung at a war-dance or before battle; a song on a martial theme. **wartime**, *n.* **war-wearied**, *a.* **war-whoop**, *n.* a shout or yell raised by N American Indians in attacking. **war-widow**, *n.* a woman whose husband has been killed in war. **warworn**, *a.* exhausted by or experienced in war. **warless** (-lis), *a.* **warlike**, *a.* fit or ready for war; fond of war, martial, soldier-like, military; threatening war, hostile.
War., (*abbr.*) Warwickshire.
waratah (wo'rətə), *n.* one of a genus of Australian proteaceous shrubs with a large, brilliant crimson flower.
warble[1] (waw'bl), *n.* a small hard tumour on a horse's back caused by the rubbing of the saddle; a small tumour produced by the larva of the bot-fly. **warble-fly**, *n.* the bot-fly.
warble[2] (waw'bl), *v.i.* of birds, to sing in a continuous quavering or trilling manner; of streams etc., to make a continuous melodious sound; to sing (with trills and variations etc.). *v.t.* to sing or utter thus. *n.* the act or sound of warbling; a song; a trill. **warbler**, *n.* one who, or that which, warbles; one of the Sylviidae, a family of small birds comprising the nightingale, black-cap, hedge-sparrow, robin etc. **warbling**, *a.* **warblingly**, *adv.*
ward (wawd), *n.* watch, guard, the act of guarding, protection; an inner court of a castle; a parrying or guard in fencing; confinement, custody; guardianship, control; a minor or person under guardianship; an administrative or electoral division of a town or city; a separate division of a hospital, prison or workhouse; a projection inside a lock preventing the turning of any but the right key. *v.t.* to parry, to turn aside, to keep (off). **watch and ward** WATCH. **ward in Chancery**, *n.* a minor under the guardianship of the Court of Chancery. **ward-room**, *n.* a room on a warship for commissioned officers below the rank of commander. **wardship**, *n.* guardianship, tutelage.
-ward(s) (-wəd(z)), *suf.* expressing direction as in *backward, forward, homeward, inwards, outwards* etc.
warden[1] (waw'dn), *n.* a keeper, a guardian; a governor; the head of some colleges, schools and hostels; (*N Am.*) a prison governor; one who keeps ward, a watchman; one of the officials in a Civil Defence organization; (*Austral.*) a government official in charge of a goldfield. **Warden of the Cinque Ports**, the governor of the Cinque Ports. **wardenship**, *n.*
warden[2] (waw'dn), *n.* a variety of cooking pear.
warder (waw'dr), *n.* a keeper, a jailer, a prison guard. **wardress** (-dris), *n.fem.*
wardian (waw'diən), *a.* applied to a close-fitting case with glass sides and top, retaining moisture, for transporting delicate plants, esp. ferns.
Wardour Street (waw'də), *n.* the British film industry; the bogus antique language etc. of costume films. [locality in London where the cinema industry is centred]
wardrobe (waw'drōb), *n.* a cabinet, cupboard or other place, where clothes are hung up; a person's stock of wearing apparel; the stage costumes of a theatre or company. **wardrobe dealer**, *n.* one who deals in used or second-hand clothing.
-wards -WARD.
wardship etc. WARD.
ware[1] (weə), *n.* manufactured articles of a specified kind, esp. pottery, as *table-ware, stone-ware*; (*pl.*) articles of merchandise, articles for sale, goods.
ware[2] (weə), *v.t.* (*imper.*) beware! look out for, guard against, keep clear of.
warehouse (weə'hows), *n.* a building in which goods are stored, kept for sale or in bond; a wholesale or large retail store. *v.t.* (-howz), to deposit or secure (furniture, bonded goods etc.) in a warehouse. **warehouseman**, *n.*

warfare 933 **wash**

one who keeps or is employed in a warehouse. **warehousing**, *n.* (*coll.*) the practice of anonymously building up a shareholding in a company, using nominees etc. to purchase the shares.
warfare (waw'feə), *n.* a state of war, hostilities; conflict, strife. *v.i.* to carry on war; to engage in war; to contend.
warfarin (waw'fərin), *n.* a compound used as a rodent poison and to prevent blood clotting.
warily etc. WARY.
warlike etc. WAR.
warlock (waw'lok), *n.* a wizard, a sorcerer.
warm (wawm), *a.* being at a rather high temperature; having heat in a moderate degree; promoting, emitting or conveying heat; having the sensation of rather more than ordinary heat, esp. with the temperature of the skin raised by exercise etc.; ardent, zealous, enthusiastic, cordial; sympathetic, emotional, affectionate; amorous, erotic, indelicate; animated, heated, excited, vehement, passionate, excitable; of a skirmish etc., violent, vigorous, brisk, strenuous, lively; of colours, being predominantly red or yellow; of a scent, fresh, strong; near the object sought (in children's games); (*coll.*) well off, in comfortable circumstances; (*coll.*) unpleasant, uncomfortable. *v.t.* to make warm; to make ardent or enthusiastic, to excite; (*sl.*) to thrash. *v.i.* to become warm; to become animated, zealous, sympathetic or enthusiastic (to or towards). **to warm up**, to make or become warm; to reheat (cooked food); to prepare for a contest, performance etc., esp. to prepare for an athletic contest etc. by exercising and stretching; to make (an audience) more receptive to a main act by a preliminary entertainment. **warm-up**, *n.* **warm-blooded**, *a.* having warm blood, esp. between 98° and 112° F (36·6 and 44·4° C), as mammals and birds; emotional, passionate, excitable; amorous, erotic. **warm-bloodedness**, *n.* **warm front**, *n.* the advancing edge of a mass of warm air. **warm-hearted**, *a.* having warm, affectionate, kindly or susceptible feelings. **warm-heartedly**, *adv.* **warm-heartedness**, *n.* **warmer**, *n.* (*usu. in comb.* as *foot-warmer*). **warming**, *n.* (*sl.*) a thrashing. **warming-pan**, *n.* a closed pan, usu. of brass with a long handle, for holding live coals, formerly used to warm a bed; a person who holds a post temporarily till another is qualified to fill it. **warmish**, *a.* **warmly**, *adv.* **warmth**, **warmness**, *n.*
warn (wawn), *v.t.* to give notice to, to inform beforehand; to caution or put on one's guard against; to expostulate with, to admonish; to order to go or stay (away, off etc.). *v.i.* to give a warning. **warner**, *n.* **warning**, *n.* the act of cautioning or admonishing against danger etc.; previous notice, esp. to quit one's service etc.; that which serves to warn. *a.* serving to warn. **warningly**, *adv.*
warp (wawp), *n.* the threads running the long way of a woven fabric, crossed by the weft or woof; a rope, usu. smaller than a cable, used in towing a vessel; the state of being twisted, a twist or distortion in timber etc.; a perversity or aberration of mind or disposition; an alluvial deposit of water artificially introduced into low lands. *v.t.* to turn or twist out of shape, to make crooked, to distort; to pervert, to bias, to turn awry; to fertilize by means of artificial inundations; to tow or move with a line attached to a buoy, anchor or other fixed point etc.; to run (yarn) off for weaving. *v.i.* to become twisted, crooked or distorted; to turn aside; to become perverted. **warper**, *n.* **warping-bank**, *n.* a bank for retaining the water let on to ground for fertilizing purposes. **warping-hook**, *n.* a ropemaker's hook used in warping. **warping mill**, *n.* a revolving wooden frame upon which fabric threads are wound when being made into a warp. **warping-post**, *n.* a post used in warping rope-yarn.
warragal (wo'rəgl), WARRIGAL.
warrant (wo'rənt), *v.t.* to answer or give an assurance for, to guarantee; to give authority to, to sanction; to serve as guarantee for; to attest the truth of; to serve as grounds or justification for. *n.* anything that authorizes a person to do something, authorization; sanction; reason, grounds, justification; anything that attests or bears out a statement etc.; one who or that which vouches, a voucher; a document authorizing a person to receive money etc.; an instrument giving power to arrest a person, levy a distress etc.; a certificate of office held by a warrant-officer. **warrant of attorney** ATTORNEY[2]. **warrant-officer**, *n.* an officer next below a commissioned officer, acting under a warrant from the Admiralty or War Office as a gunner, boatswain or sergeant-major. **warrantable**, *a.* justifiable, defensible; of deer, old enough to be hunted. **warrantableness**, *n.* **warrantably**, *adv.* **warrantee** (-tē'), *n.* **warranter**, *n.* **warrantor**, *n.* (*Law*). **warranty**, *n.* a warrant, an authorization; a promise or undertaking from a vendor to a purchaser that the thing sold is the vendor's to sell and is good and fit for use etc.; an express or implied undertaking in a contract that a fact is as stated.
warren (wo'rən), *n.* a piece of ground where rabbits live and breed; a place for keeping and breeding small game animals; an overcrowded district; a maze of interconnecting streets or passages. **warrener**, *n.*
warrigal (wo'rigl), *n.* (*Austral.*) the dingo; a wild native; an outlaw, a rascal. *a.* untamed, wild.
warrior (wo'riə), *n.* a person experienced or distinguished in war, a distinguished soldier.
wart (wawt), *n.* a small hard excrescence on the skin of the hands etc. due to irregular growth of the papillae, caused by a virus; a spongy excrescence on the hind pastern of a horse; a small protuberance on the surface of a plant. **warts and all**, without concealing any blemishes or imperfections. **wart-hog**, *n.* an African large-headed wild pig with warty excrescences on the face. **warted**, **warty**, *a.* **wartless** (-lis), *a.*
wary (weə'ri), *a.* cautious, watchful against deception, dangers etc.; circumspect; done with or characterized by caution. **warily**, *adv.* **wariness**, *n.*
was (woz), **wast** (-wost), **were** (wœ), **wert** (-wœt), *past tense* BE.
wase-goose WAYZGOOSE.
wash (wosh), *v.t.* to cleanse with water or other liquid; to remove or take out, off, away etc. thus; to pass water or other liquid through or over; to purify; of dew, waves, the sea etc., to fall upon, cover, moisten or dash against; to carry along, to sweep away etc., to scoop (out) by or as by the action of moving liquid; to separate the earthy and lighter parts from (ore); to cover with a thin coat of colour; to overlay with a thin coat of metal. *v.i.* to cleanse oneself with water etc.; to stand washing without fading or being injured in any way; (*coll.*) of a story etc., to stand examination; of water etc., to move or splash or sweep along; to drift or be carried along on water. *n.* the act or operation of washing; the state of being washed; a quantity of clothes etc. washed at one time; the motion of a body of water esp. the swirling and foaming caused by the passage of a vessel; soil removed and accumulated by water; land washed by the sea or a river; waste liquor from the kitchen often used as food for pigs; thin liquid food, slops; a liquid used for toilet purposes, a lotion; a thin coating of colour spread over broad masses of a painting, pen-and-ink drawing etc.; a thin liquid for coating a wall etc.; a thin coat of metal; the blade of an oar; fermented wort from which spirit has been extracted; a disturbance in the air caused by the passage of an aircraft. **the wash**, the washing of clothes, linen etc. **to come right in the wash**, (*coll.*) to come right in the end. **to wash down**, to wash the whole of; to accompany (food) with a drink. **to wash one's hands of**, to disclaim any responsibility for. **to wash out**, to remove by washing; to wash free of something unwanted; (*coll.*) to cancel, to annul. **to wash up**, to wash dishes etc.; (*esp. N Am.*) to wash one's

hands and face. **wash-basin,** *n.* a wash-hand basin. **wash-board,** *n.* a board with a ribbed surface for scrubbing clothes on; a board to keep the water from washing over a gunwale or through a port etc. of a ship. **wash-boiler,** *n.* one for boiling clothes in the process of washing. **wash-bottle,** *n.* an apparatus for washing gases, precipitates etc. by passing them through a liquid. **wash-bowl,** *n.* a wash-hand basin. **wash-cloth,** *n.* a piece of cloth used in washing dishes etc. **wash-day,** *n.* the day on which domestic washing is done or sent to the laundry. **washed out,** *a.* limp, exhausted, worn-out; faded, colourless. **washed-up,** *a.* (*coll.*) no longer successful or effective, finished, failed. **wash-gilding,** *n.* water-gilding. **wash-hand basin,** a basin for washing the hands etc., forming part of the furnishings of a room. **wash-house,** *n.* a building for washing clothes etc., a laundry; a scullery. **wash-leather,** *n.* chamois leather or an imitation of this. **wash-out,** *n.* a scooping out or sweeping away of rock, earth etc. by a rush of water; a cleansing by washing out; (*coll.*) a failure, a muddle; (*coll.*) a muddler. **wash-rag,** *n.* (*N Am.*) a face-cloth, a flannel. **wash-room,** *n.* (*N Am.*) a lavatory. **wash-stand,** *n.* a piece of furniture for holding a ewer or pitcher, basin etc. for washing one's face and hands etc. **wash-tub,** *n.* a tub in which clothes etc. are washed. **washable,** *a.* **washability,** *n.* **washer,** *n.* a ring or perforated disk of metal, rubber etc. for placing beneath a nut etc. to tighten the joint; one who or that which washes; a washing machine. **washerman,** *n.* a laundryman. **washerwoman,** *n. fem.* **washing,** *n.* the act of cleansing by water etc., ablution; clothes etc. washed together or for washing. **washing-machine,** *n.* an electrical machine in which clothes are washed automatically. **washing-powder,** *n.* a preparation used in washing clothes. **washing soda,** *n.* crystalline sodium carbonate. **washing up,** *n.* the washing of dishes etc., esp. after a meal; dishes, cutlery etc. to be washed. **washy,** *a.* watery, too much diluted, weak, thin; wanting in solidity, stamina or vigour, feeble. **washily,** *adv.* **washiness,** *n.*
Wash., (*abbr.*) Washington.
Washingtonia (woshingtō'niə), *n.* a gigantic Californian sequoia. [George *Washington*, 1st Pres., US, 1732–99]
wasn't (woz'ənt), contr. form of *was not.*
WASP (wosp), *n.* an American of N European descent, considered in N America as belonging to a privileged class. [*W*hite *A*nglo-*S*axon *P*rotestant]
wasp (wosp), *n.* a predatory hymenopterous insect of solitary or social habits, esp. the common wasp, a European insect with a slender waist, black and yellow stripes and a powerful sting; a spiteful or irritable person. **wasp-bee, -beetle, -fly,** *n.* one somewhat resembling a wasp, but without a sting. **wasp-waisted,** *a.* having a very thin waist. **waspish,** *a.* snappish, petulant, irritable. **waspishly,** *adv.* **waspishness,** *n.*
wassail (wos'l, -āl), *n.* a festive occasion, a drinking-bout, a carouse; spiced ale or other liquor prepared for a wassail; formerly, a toast to a person's health. *v.i.* to carouse, to make merry; to go from house to house singing carols at Christmas. **wassail-bowl, -cup, -horn,** *n.* one from which wassail was drunk. **wassailer,** *n.*
Wassermann test (vas'əmən, was'-), *n.* a diagnostic test for the presence of syphilis. [A. von *Wassermann*, 1866–1925]
wast (wost), 2nd pers. sing. of past tense of BE.
waste (wāst), *a.* desolate, empty, untilled, devastated; barren, unproductive, dreary, dismal, cheerless; superfluous, left over as useless or valueless. *v.t.* to devastate, to lay waste, to wear away gradually; to cause to lose weight, strength and health; to consume, to spend, to use up unnecessarily, carelessly or lavishly; to fail to use to advantage; (*Law*) to injure or impair (an estate) by neglect; (*sl.*) to kill. *v.i.* to wear away gradually, to dwindle, to wither; to lose weight, strength and health. *n.* the act of wasting, squandering or throwing away to no purpose; the state or process of being wasted or used up, gradual diminution of substance, strength, value etc.; material, food etc. rejected as superfluous, useless or valueless, refuse; waste-products; a desolate or desert region, a wilderness; a dreary scene, an empty space, a void; (*Law*) damage or injury to an estate etc. caused by the act or neglect of a life-tenant etc. **to lay waste,** to render desolate; to devastate, to ruin. **waste-book,** *n.* an account book for entering transactions as they take place before carrying them over to the ledger. **waste-land,** *n.* a barren uninhabited place. **waste paper,** *n.* spoiled, used or discarded paper. **waste (paper) basket,** a receptacle for waste paper. **waste-pipe,** *n.* a discharge-pipe for used or superfluous water. **waste-product,** *n.* material produced by a process as a useless by-product; an unusable product of metabolism. **wastage** (-ij), *n.* loss by use, decay, leakage etc.; avoidable loss of something useful; reduction in numbers of employees etc. by retirement, voluntary resignation etc. **wasteful,** *a.* extravagant, spending or using recklessly, unnecessarily or too lavishly. **wastefully,** *adv.* **wastefulness,** *n.* **wasteless** (-lis), *a.* inexhaustible. **waster,** *n.* one who wastes; a prodigal, a spendthrift; a good-for-nothing, a wastrel; an article spoilt and rendered unmarketable in manufacture. **wasting,** *n., a.* **wasting asset,** *n.* a non-renewable asset that is gradually used up, as a mine.
wastrel (wās'trəl), *n.* a waif; a profligate; a wasteful person.
wat (waht), *n.* a Thai Buddhist temple or monastery.
watch (woch), *n.* the act or state of watching; a state of alertness, vigilance, close observation or attention; vigil, look-out, waiting in a state of expectancy, dread etc.; a watchman or body of watchmen, a guard; a division of the night; a small timepiece actuated by a spring or battery for carrying on the person; a period of or of keeping guard; the period of time during which each division of a ship's crew is alternately on duty (four hours except during the dog-watches of two hours); either half (starboard or port watch) into which the officers and crew are divided, taking duty alternately. *v.i.* to be on the watch, to be vigilant, observant or expectant; to look out (for); to act as a protector or guard (over); to keep awake at night, to keep vigil. *v.t.* to guard; to observe closely, to keep one's eye or eyes on; to observe with a view to detecting etc.; to look at, view; to tend, look after; to be careful of; to look out for, to await, to bide (one's time etc.). **on the watch,** vigilant, on the look-out. **to watch out,** to be on the look-out (for); to take care. **watch and ward,** continuous watch; orig. watch by night and day. **watch-box,** *n.* a sentry-box. **watch-case,** *n.* the case enclosing the works of a watch. **watch-chain,** *n.* a metal watch-guard or watch-strap. **Watch Committee,** *n.* a local committee supervising the policing of the district. **watch-dog,** *n.* a dog kept to guard premises etc. and give notice of burglars etc.; a person or group that monitors the activities of an organization or in some sphere to guard against illegal or undesirable practices etc. **watch-fire,** *n.* a fire in a camp etc. at night or used as a signal. **watch-glass,** *n.* a glass covering the face of a watch; an hour- or half-hourglass for measuring the period of a nautical watch; a curved disk of glass used in a laboratory to hold small samples etc. **watch-guard,** *n.* a chain, cord, ribbon etc. for securing a watch to the person. **watch-house,** *n.* a house occupied by a watch or guard; a lock-up. **watching brief,** *n.* a brief issued to a barrister instructed to watch a case on behalf of a client not directly concerned in the action. **watch-key,** *n.* a key for winding up a watch. **watchmaker,** *n.* **watchmaker's oil,** *n.* a fine thin oil for lubricating the works of watches etc. **watchmaking,** *n.* **watchman,** *n.* a guard, a sentinel; one who guards the streets of a town at night; a man guarding a large build-

ing etc. **watch-night,** *n.* the last night of the year when services are held in Protestant churches. **watch-oil,** *n.* watchmaker's oil. **watch-spring,** *n.* the mainspring of a watch. **watch-strap,** *n.* a strap for securing a watch round the wrist. **watch-tower,** *n.* a tower of observation or one on which sentinels are placed. **watchword,** *n.* a word given to sentinels etc. as a signal that one has the right of admission etc.; a motto, word or phrase symbolizing or epitomizing the principles of a party etc. **watcher,** *n.* **watchful,** *a.* vigilant, observant, cautious. **watchfully,** *adv.* **watchfulness,** *n.*

water (waw'tə), *n.* a colourless, transparent liquid, tasteless and odourless possessing a neutral reaction, a compound of hydrogen with oxygen; (*often pl.*) a (natural) body of water, as a sea, a lake, a river; the surface of a body of water; a liquid consisting chiefly or partly of water, as various solutions or products of distillation; tears, sweat, urine or another secretion of animal bodies; (*usu. pl.*) the amniotic fluid surrounding a foetus; the transparency or lustre of a diamond, pearl etc.; a wavy lustrous finish on silk etc.; stock issued without any corresponding increase of paid-up capital. *v.t.* to apply water to, to moisten, sprinkle, dilute, adulterate, irrigate or supply with water; to furnish with water for drinking; to increase (nominal capital etc.) by the issue of stock without corresponding increase of assets; (*in p.p.*) to give an undulating sheen to the surface of (silk etc.) by moistening, pressing and heating in manufacture. *v.i.* of the mouth, eyes etc., to secrete, shed or run with water; to get or take in water; of cattle etc., to drink. **high water** HIGH. **in deep water, waters,** in difficulties, troubles or distress. **in hot water,** in trouble, difficulty or disgrace. **in smooth water,** out of one's troubles or difficulties. **low water** LOW [1]. **of the first water,** of the purest quality; of the highest excellence. **strong waters** STRONG. **table water** TABLE. **to hold water,** to be sound or valid, to stand scrutiny. **to keep one's head above water,** to avoid financial ruin. **to make, pass water** MAKE [2]. **to make one's mouth water,** to make one very desirous. **to throw cold water on** COLD. **troubled waters** TROUBLE. **water of crystallization,** the water that unites with salts in crystallization. **water on the brain,** hydrocephalus. **water-bailiff,** *n.* an officer employed to watch a river or other fishery to prevent poaching. **water-bed,** *n.* a bed with a rubber mattress filled with water. **water-beetle,** *n.* **water-bellows,** *n.* a valved vessel suspended mouth downwards in water for producing an air-current by alternate raising and lowering. **water-bird,** *n.* **water-biscuit,** *n.* a thin plain biscuit of flour and water. **water-blister,** *n.* one containing watery fluid without pus or blood. **water-boatman,** *n.* a water-bug with paddle-like hind legs. **water-borne,** *a.* conveyed by water. **water-brash,** *n.* a form of indigestion, with water eructations. **water-buffalo,** *n.* the common domesticated Asian buffalo. **water-bug,** *n.* an aquatic insect. **water bus,** *n.* a river craft carrying passengers on a regular service. **water-butt,** *n.* a large open-headed barrel for catching and preserving rain-water. **water cannon,** *n.* a high-pressure water hose used for quelling riots etc. **water-carriage,** *n.* conveyance by water. **water-cart,** *n.* a wheeled tank etc. for carrying a supply of water or for watering the streets. **water-cement,** *n.* hydraulic cement. **water-chestnut,** *n.* (an Asian sedge with) an edible tuber used in Chinese cooking; an aquatic plant with an edible nutlike fruit. **water-chute,** *n.* a structure with a slide down which water is kept running, for tobogganing down in a boat-like sled. **water-clock,** *n.* a clepsydra. **water-closet,** *n.* (a room containing) a toilet with a water-supply for flushing the basin and preventing the rise of sewer-gas. **water-colour,** *n.* a pigment ground up with water and gum instead of oil; a water-colour painting; (*often pl.*) the art of painting in water-colours. **water-colourist,** *n.* **water-cool,** *v.t.* to cool by (circulating) water. **water-cooled,** *a.* **water-cooler,** *n.* a device for cooling drinking water. **watercourse,** *n.* a stream, a river; a channel in which a natural stream or canal flows. **water craft,** *n.* ships, vessels, boats etc. **water-cress,** *n.* a creeping aquatic plant eaten as salad. **water-cure,** *n.* hydropathy. **water-diviner,** *n.* a dowser. **water-dog,** *n.* one accustomed to the water. **water-drain,** *n.* **water-drainage,** *n.* **water-drop,** *n.* a drop of water, a tear etc. **water-engine,** *n.* an engine driven by water; an engine to raise water. **waterfall,** *n.* a steep or perpendicular descent of a river etc. **water-finder,** *n.* a dowser. **water-flag,** *n.* the yellow iris. **water-flea,** *n.* a minute freshwater crustacean. **water-flood,** *n.* **water-fly,** *n.* a fly whose larvae lurk under stones in streams. **waterfowl,** *n.* (*pl.* **waterfowl**) a bird that frequents rivers, lakes etc. **waterfront,** *n.* the part of a town facing or bordering a sea, lake etc. **water-gas,** *n.* a gas obtained by the decomposition of water and treatment with carbon, used as a fuel. **water-gate,** *n.* a flood-gate; a gate giving access to a river etc. **water-gauge,** *n.* an instrument for indicating the height of the water inside a boiler etc. **water-glass,** *n.* a tube with a glass end for enabling one to see objects under water; soluble glass, esp. as used for fixing a water-colour drawing on dry plaster; a water-clock; a viscous solution of sodium or potassium silicate in water, used in industry and as a preservative for eggs. **water-gruel,** *n.* gruel made with water instead of milk. **water-hammer,** *n.* the concussion of water in a pipe when a tap is turned off or steam admitted. **water-hammering,** *n.* **water-hen,** *n.* the moor-hen. **water-hole,** *n.* a hole where water collects, a water pool. **water-ice,** *n.* a frozen confection made from water, sugar etc. **water-jacket,** *n.* a casing filled with water surrounding a part of a machine that is to be kept cool. **water-joint,** *n.* a watertight joint. **water-jump,** *n.* a ditch, stream etc. to be jumped, esp. in a steeplechase. **water-kelpie,** *n.* a malignant water-sprite. **water-laid,** *a.* of rope, cable-laid. **water-lens,** *n.* a magnifying lens formed by a glass-bottomed brass cell containing water. **water-level,** *n.* the level of the water in the sea etc., esp. used as datum; a levelling instrument in which water is employed instead of spirit. **water-lily,** *n.* a plant with large floating leaves and white or coloured flowers. **water-line,** *n.* the line up to which the hull of a vessel is submerged in the water. **waterlogged,** *a.* of a vessel, flooded with water so as to lie like a log; esp. of ground, saturated with water. **water-main,** *n.* a main pipe in a system of water-supply. **water-man,** *n.* a boatman plying for hire on rivers etc.; a (good or bad) oarsman. **watermanship,** *n.* **watermark,** *n.* a mark indicating the level to which water rises in a well etc.; the limits of the rise and fall of the tide etc.; a translucent design stamped in paper in the process of manufacture to show the maker, size etc. *v.t.* to stamp with this. **water-meadow,** *n.* a meadow fertilized by being flooded at certain seasons from an adjoining stream. **water-melon,** *n.* an African melon with a hard dark-green skin and red watery flesh. **water-meter,** *n.* a contrivance for measuring a water-supply. **water-mill,** *n.* a mill driven by water. **water-monkey,** *n.* an earthenware long-necked jar for drinking-water, used in hot countries. **water-motor,** *n.* a motor driven by water under pressure, a turbine, a water-wheel. **water-nymph,** *n.* a naiad. **water-ouzel,** *n.* the dipper. **water-pipe,** *n.* a pipe for conveying water. **water-pistol,** *n.* a toy pistol that shoots a jet of water. **water-plane,** *n.* the plane in which the water-line of a vessel lies; a hydro-aeroplane. **water-plant,** *n.* **water-plate,** *n.* a double plate containing hot water for keeping food warm. **water-polo,** *n.* a game like polo in which swimmers hit a ball with the hand. **water-pot,** *n.* **water-power,** *n.* the power of water employed or capable of being employed as a prime mover. **water-pox,** *n.* chicken-pox. **waterproof,** *a.* impervious to water. *n.* a waterproof coat or other garment. *v.t.* to render waterproof. **waterproofer,** *n.* **waterproofing,** *n.* **water-rail,** *n.* the common European rail.

water-ram, *n.* a hydraulic ram. **water-rat** WATER-VOLE. **water-rate**, *n.* a charge for the supply of water. **water-repellent, -resistant**, *a.* resistant but not impervious to water. **water-sail**, *n.* a sail set in very light airs, next to the water. **water-seal**, *n.* a small body of water in a bend etc., used to prevent the escape of gas from a pipe etc. **watershed**, *n.* a ridge or other line of separation between two river-basins or drainage-systems; a dividing point between two distinct periods etc. **water-shoot**, *n.* a discharge pipe or trough for rain-water etc. **waterside**, *n.* the margin of a river, stream, lake or the sea. **water-ski**, *n.* a type of ski used for planing over water towed by a speedboat. **water-skiing**, *n.* **water-skin**, *n.* a bag or bottle of skin for carrying water. **water-snake**, *n.* **water-softener**, *n.* a device or chemical used to remove or chemically alter the substances that cause hardness in water. **water-soldier**, *n.* an aquatic plant with long narrow leaves rising above the water. **water-spaniel**, *n.* a spaniel used in hunting waterfowl. **water-splash**, *n.* part of a road etc. always submerged by a crossing stream. **watersports**, *n.pl.* **waterspout**, *n.* a phenomenon which occurs during a tornado over the sea, in which water appears to be drawn up from the sea in a whirling column. **water-sprite**, *n.* **water-supply**, *n.* a system for storing and supplying water for the service of a town etc.; the amount of water supplied. **water-table**, *n.* a projecting ledge or string-course for throwing off the water on a building; a level below which ground is saturated with water. **water-tank**, *n.* **water-tiger**, *n.* the predatory larva of some water-beetles. **watertight**, *a.* so tightly fastened or fitted as to retain or not to admit water. **water-tower**, *n.* an elevated building carrying a large tank or reservoir for giving pressure to a water-supply. **water-tube**, *n.* a tube for containing water, esp. one of a series in a boiler in which water circulates exposed to the gases of combustion. **water-vapour**, *n.* water in gaseous form, esp. when evaporated below boiling temperature. **water-violet**, *n.* any plant of the aquatic genus *Hottonia*. **water-vole**, *n.* a large aquatic vole, the water-rat. **water-wagtail**, *n.* the pied wagtail. **waterway**, *n.* a navigable channel; a fairway; the thick planks along the edge of a deck in which a channel is hollowed for conducting water to the scuppers. **water-weed**, *n.* **water-wheel**, *n.* a wheel moved by water and employed to turn machinery. **water-wings**, *n.pl.* floats used in teaching swimming. **water-witch**, *n.* a dowser. **water-works**, *n.sing.* an establishment for the collection, preservation and distribution of water for use of communities, working of machinery etc.; an artificial fountain; (*euphem.*) the urinary system; (*coll.*) crying, tears. **water-worn**, *a.* worn away by the action of water. **watered**, *a.* **watered capital**, *n.* an increase in the nominal value of stock without a corresponding increase in assets or paid-up capital. **watered-down**, *a.* diluted or weakened by or as by the addition of water; reduced in force, effect etc. **waterer**, *n.* **watering**, *n., a.* **watering-can**, *n.* a vessel with a (perforated) nozzle for sprinkling water on plants etc. **watering-hole**, *n.* a water-filled hollow where animals can drink; (*coll.*) a pub, bar etc. **watering-place**, *n.* a place where water may be procured for cattle etc.; a place to which people resort to drink mineral waters or bathe, a spa. **watering-trough**, *n.* a drinking-trough for animals. **waterless** (-lis), *a.* **watery**, *a.* containing much water; moist, sodden; suffused or running with water; thin, transparent or pale, like water; rainy-looking; consisting of water; tasteless, insipid. **wateriness**, *n.*
waterloo (wawtəloo'), *n.* a downfall, a decisive defeat. [*Waterloo*, Belgian town where Napoleon was finally defeated, 1815]
watt (wot), *n.* a unit of power equal to a rate of working of one joule per second or the power available when the electromotive force is one volt and the current is one ampere. **watt-hour**, *n.* a unit of (electrical) energy equal to a power of one watt operating for one hour. **watt-hour meter, wattmeter**, *n.* **wattage** (-ij), *n.* amount of power in watts. [James *Watt*, 1736–1819, Scottish engineer]
wattle (wot'l), *n.* a hurdle of interwoven twigs or wickerwork; the fleshy lobe under the throat of the domestic fowl, turkey etc.; a barbel of a fish; one of various Australian and Tasmanian species of acacia; the national flower of Australia. *v.t.* to interweave, to interlace, to plait; to form or construct by plaiting etc. **wattle and daub**, *n.* a method of constructing walls of wicker-work covered with mud or clay. **wattle-work**, *n.* wicker-work. **wattled**, *a.* **wattling**, *n.*
waul (wawl), *v.i.* to cry as a cat, to squall. [onomat.]
wave (wāv), *v.i.* to move to and fro with a sinuous or sweeping motion as a flag in the wind, to flutter, to undulate; to have an undulating shape or conformation; to beckon or signal (to) by waving the hand, a handkerchief etc. *v.t.* to cause to move to and fro, to give an undulating motion to, to brandish; to give an undulating surface, conformation or appearance to; to indicate, direct or command by a waving signal. *n.* a moving ridge or long curved body of water or other liquid, esp. one formed on the surface of the sea, rising into an arch and breaking on the shore; (*poet., often pl.*) the sea, water; a disturbance of the equilibrium of a fluid medium continuously propagated from point to point without a corresponding advance of the particles in the same direction, by which motion, heat, light, sound, electricity etc. are transmitted; a single curve or cycle in such a motion; an undulation; a waviness of the hair; a wave-like stripe or streak; the act or gesture of waving, as a signal etc.; a heightened volume or intensity of some force, influence, emotion, activity etc.; a movement like that of a wave on the sea; a widespread advance or influx; a prolonged spell of cold or esp. hot weather; a waveform; rhythmical electromagnetic disturbance propagated through space. **permanent wave** PERMANENT. **wave-band**, *n.* a range of frequencies or wave lengths for radio transmissions of a particular type. **waveform**, *n.* (the shape of) the graph of a wave, showing the variation in a varying quantity against time. **waveguide**, *n.* a metal tube used for carrying and guiding electromagnetic waves, esp. microwaves. **wavelength**, *n.* the distance measured between the crests of two adjacent waves; the space intervening between corresponding points, as the maximum positive points of two successive waves. **wave mechanics**, *n.sing.* quantum mechanics based on the wave-like properties and behaviour of particles. **wave-meter**, *n.* an instrument used for measuring the wavelength or frequency of an electromagnetic wave. **wave-motion**, *n.* **waveson** (-sən), *n.* (*Law*) goods left floating on the sea after shipwreck. **wave-worn**, *a.* **waveless**, *a.* **wavelet** (-lit), *n.* **wave-like**, *a.* **wavy**, *a.* rising or swelling in waves; undulating. **Wavy Navy**, *n.* (*coll.*) the Royal Naval Volunteer Reserve. **wavily**, *adv.* **waviness**, *n.*
waver (wā'və), *v.t.* to play or move to and fro; to flicker, to quiver; to begin to give way, to falter; to be in a state of indecision, to vacillate. **waverer**, *n.* **waveringly**, *adv.* **waveringness**, *n.*
waveson WAVE.
wavey (wā'vi), *n.* the snow-goose.
wavily etc., **wavy** WAVE.
wax[1] (waks), *n.* a yellow, plastic, fatty substance excreted by bees and used for the cells of honeycombs, beeswax; this purified and bleached, used for candles, modelling and pharmaceutical and other purposes; any one of various plant or animal substances that are principally esters of fatty acids or alcohols; a mineral substance, as ozocerite, composed of hydrocarbons; any one of various substances resembling beeswax, as cobbler's wax, sealing-wax etc.; cerumen; (*coll.*) a person who is compliant or easily influenced; (*coll.*) a gramophone record; (*sl.*) a rage. *a.* waxen. *v.t.* to smear, rub, polish, treat or join with wax.

wax-berry, *n.* the fruit of the wax-myrtle. **waxbill,** *n.* a small bird with a bill resembling red sealing-wax in colour. **wax-chandler,** *n.* a maker or seller of wax candles. **wax-cloth,** *n.* a floor-cloth. **wax doll,** *n.* a doll with a face made of wax. *a.* having a face like this, pretty but devoid of expression. **wax-insect,** *n.* an insect producing wax. **wax-light,** *n.* a taper, match etc. made of wax. **wax-moth,** *n.* a bee-moth. **wax-myrtle** CANDLEBERRY-MYRTLE. **wax-painting,** *n.* encaustic painting. **wax-palm,** *n.* a S American palm the trunk or leaves of which yield wax. **wax-paper, waxed-paper,** *n.* paper waterproofed with wax. **waxpod bean,** *n.* a French bean with waxy, yellow pods. **wax-tree,** *n.* a tree yielding wax which exudes from it or is deposited by insects. **waxwing,** *n.* a bird whose wing-tips resemble pieces of red sealing-wax. **wax-work,** *n.* anatomical and other figures, models of fruit, flowers etc. in wax; *(pl.)* an exhibition of wax figures. **wax-worker,** *n.* **waxen,** *a.* made or consisting of wax; with a surface resembling wax; like wax, impressible, plastic; very pale. **waxer,** *n.* **waxy,** *a.* resembling wax; pliable, easily moulded; containing or covered with wax, like wax in consistency; pallid, wan; *(sl.)* angry, cross. **waxily,** *adv.* **waxiness,** *n.*
wax[2] (waks), *v.i.* to increase gradually in size and brilliance, as the illuminated portion of the moon between new and full; to become larger, to grow in numbers, strength, intensity etc.; to become gradually.
waxen, waxy etc. WAX[1].
way (wā), *n.* a road, path, track or other place of passage; length of space passed over, distance to be traversed; the route followed or to be followed between two places or to reach a place; direction in which a thing or place lies or in which motion etc. takes place; the method or manner of doing something; a line or course of action; a usual or habitual mode of action or conduct, a personal peculiarity; one's line of business or occupation, range, scope; one's course of life; that which one wants; relation, respect, point; condition, state; room for passage or advance, ground over which one would proceed; onward movement, progress, advance, headway; esp. of a ship, motion, impetus; *(pl.)* the framework of timbers over which a ship is launched. *adv.* far, away. **by the way,** in passing, parenthetically; during the journey. **by way of,** by the route or for the purpose of; as a form of, to serve as. **each way,** *adv. (Racing)* for win and for place. **in a way,** to some degree; from one point of view. **in the family way** FAMILY. **in the way,** in a position or of a nature to obstruct or hinder. **in the way of,** so as to fall in with or obtain; as regards, by way of. **on the way,** in progress. **out of the way** OUT. **right of way** RIGHT. **the way of all flesh,** death. **to give way** GIVE[1]. **to go, take one's own way,** to follow one's own plan, to act independently. **to go one's way,** to depart. **to have one's way,** to get what one wants. **to lead the way** LEAD[2]. **to make one's way,** to proceed; to prosper, esp. by one's own exertions. **to make way** MAKE[2]. **to pave the way for,** to prepare a way of attaining some object. **under way,** of a ship etc. in motion; in progress. **Way of the Cross,** a series of pictures in a church representing the successive stages of Christ's progress to Calvary; a series of devotions suited to each of these. **ways and means,** means of doing, esp. of providing money. **Committee of Ways and Means,** a committee of the House of Commons for considering proposed taxes etc. **way-bill,** *n.* a list of passengers in a public conveyance or of goods sent by a common carrier. **way-board,** *n.* a thin layer between strata of some thickness. **wayfarer,** *n.* a traveller, esp. on foot. **wayfaring,** *n., a.* **wayfaring-tree,** *n.* a large shrub with white flowers and black berries, common by roadsides. **waylay,** *v.t.* to wait in the way of with a view to rob etc., to lie in wait for. **waylayer,** *n.* **wayleave,** *n.* a right of way over the land of another, esp. rented by a company etc. **way-out,** *a. (coll.)* out-of-the-ordinary, unconventional, experimental; *(sl.)* excellent. **wayside,** *n.* the side of the road. *a.* situated or growing by the wayside. **way-station,** *n. (N Am.)* a railway halt. **way-train,** *n. (N Am.)* a local train. **way-worn,** *a.* wearied with travel.
-ways (-wāz), *suf.* forming adverbs of position, direction, manner etc., as *always, lengthways*.
wayward (wā'wəd), *a.* perverse, wilful, freakish, capricious, obstinate. **waywardly,** *adv.* **waywardness,** *n.*
waywode (wā'wōd), VOIVODE.
wayzgoose, wase-goose (wāz'goos), *n. (pl.* **-gooses**) an annual dinner, picnic or other entertainment given to or held by the persons employed in a printing-office.
Wb, *(abbr.)* weber.
WC, *(abbr.)* water closet; West Central.
WCC, *(abbr.)* World Council of Churches.
WD, *(abbr.)* War Department; Works Department.
we (wē), *nom. pl. of* first *(pers.)* pron. the plural of I, denoting the person speaking and others associated with or represented by him or her; used by a sovereign, the editor of a newspaper, the writer of an unsigned article etc.; people in general.
WEA, *(abbr.)* Workers' Educational Association.
weak (wēk), *a.* deficient in physical strength, not robust, vigorous or powerful; feeble, sickly, easily exhausted or fatigued; deficient in mental or moral strength, of defective intelligence, lacking strength of will or resisting power; yielding readily to temptation; characterized by or showing lack of resolution or will-power; deficient in durability, force or efficiency; fragile, brittle, pliant; unreliable, ineffective, inefficacious; deficient in number, quantity, weight etc.; lacking in flavour, dilute; poor, inadequate, trivial; unsustained, unconvincing, controvertible; of verbs, inflected by the addition of *-ed, -d* or *-t* to the stem in forming the past tense and p.p., not by internal vowel-change; denoting the verse-ending in which the stress falls on a normally unaccented or proclitic word; showing a downward trend in price, characterized by falling prices. **weak-eyed,** *a.* having eyes easily fatigued or not seeing well. **weak-headed,** *a.* weak in intellect. **weak interaction,** *n.* an interaction between subatomic particles responsible for certain decay processes (cp. *strong interaction*). **weak-kneed,** *a.* giving way easily; lacking in resolution. **weak-minded,** *a.* feeble in intelligence or in resolution. **weak-mindedness,** *n.* **weak side,** *n.* those traits of a person's character by which he or she is most easily influenced. **weak-sighted,** *a.* **weak-spirited,** *a.* timid, pusillanimous. **weaken,** *v.t., v.i.* **weakener,** *n.* **weaker sex,** *(derog.* or *facet.) n.* women. **weakish,** *a.* **weakling** (-ling), *n.* a feeble person. **weakly,** *adv.* in a weak manner. *a.* not strong in constitution; feeble, infirm, sickly. **weakness,** *n.* the quality or state of being weak; a particular defect, failing or fault, one's weak point; lack of resisting power.
weal[1] (wēl), *n.* a sound, healthy or prosperous state of persons or things. **the public, general** or **common weal,** *n.* the welfare or prosperity of the community.
weal[2] (wēl), *n.* a ridge or raised streak made by a rod or whip on the flesh. *v.t.* to mark with weals by flogging.
weald (wēld), *n.* a tract of open forest land, esp. the portion of Kent, Surrey, Sussex and Hants between the N and S Downs. **weald-clay,** *n.* the upper part of the Wealden strata, comprising beds of clay, iron-stone etc., rich in fossils. **Wealden,** *a.* pertaining to the Weald of Kent and Sussex, esp. geologically. **Wealden strata,** *n.* the series of lower Cretaceous freshwater strata best displayed in the Weald.
wealth (welth), *n.* riches, large possessions of money, goods or lands; affluence; abundance, a profusion, great plenty (of). **wealth-tax,** *n.* **wealthy,** *a.* rich, affluent, having many possessions. **wealthily,** *adv.* **wealthiness,** *n.*
wean (wēn), *v.t.* to accustom (a child or animal) to nourishment other than mother's milk; to detach or estrange from a habit, indulgence, desire etc. *n.* (wän), *(Sc.)* a

child. **to be weaned on**, to be familiar or grow up with from an early age. **weaner**, *n*. a young animal newly weaned. **weanling**, *n*. a child or animal newly weaned. *a*. newly weaned.
weapon (wep'n), *n*. an instrument of offence or defence, a thing used to inflict bodily harm; any means used for attack or defence; a claw, sting, thorn, prickle etc. **weapon-schaw** (-shaw), WAPINSHAW. **weaponed**, *a*. **weaponless**, *a*. **weaponry** (-ri), *n*. weapons collectively.
wear[1] (weə), *v.t.* (*past* **wore** (waw), *p.p.* **worn** (wawn)) to carry on the person, to have on as clothing or ornament; to be dressed in, esp. habitually; to arrange (hair or clothes) in a specified manner; to bear, to carry, to maintain, to exhibit, to display; to consume, diminish, waste, impair, efface or alter by rubbing or use; to exhaust, fatigue or weary; to produce (a hole, channel etc.) by attrition; (*coll.*) to stand for, to tolerate, accept. *v.i.* to be consumed, diminished, effaced, altered etc. by rubbing or use; to be exhausted, to be tired (out); to stand continual use (well, badly etc.); to become by use, age, attrition etc.; to resist the effects of use, age, attrition etc., to endure; to pass gradually (away etc.). *n*. the act of wearing; the state of being worn; that which is worn or to be worn, clothing; damage or diminution by attrition, use etc.; durability, fitness for use. **to wear down**, to overcome gradually by persistent pressure. **to wear off**, to remove, efface or diminish, or to be effaced or diminished by attrition, to rub off; to decline or pass away gradually. **to wear out**, to use until no longer of use, to consume, waste or render worthless by use; to exhaust, to tire out; to be used up, consumed or gradually exhausted by attrition and use. **to wear the breeches** BREECH. **wear and tear**, waste, diminution or injury caused by ordinary use. **wearable**, *a*. **wearability** (-bil'-), *n*. **wearer**, *n*. **wearing**, *n.*, *a*.
wear[2] (weə), *v.t.* (*past, p.p.* **wore**) to bring (a ship) about tack by putting the helm up. *v.i.* of a ship, to come round thus.
wear[3] (wiə), WEIR.
weary (wiə'ri), *a*. tired, fatigued, exhausted; expressing weariness or exhaustion; dispirited, impatient or sick (of); tiresome, tedious, exhausting, irksome. *v.t.* to tire, to fatigue; to make weary or impatient (of). *v.i.* to become tired or fatigued; to become weary (of). **wearily**, *adv*. **weariness**, *n*. **wearisome** (-səm), *a*. tedious, tiresome, causing weariness. **wearisomely**, *adv*. **wearisomeness**, *n*.
weasel (wē'zl), *n*. a small British reddish-brown, whitebellied quadruped related to the stoat, ferret etc., with a long lithe body and short legs, preying on small birds, mice etc. *v.i.* (*past, p.p.* **weaselled**) (*coll.*) to evade or extricate oneself from a responsibility, obligation etc.; (*chiefly N Am. coll.*) to equivocate. **weasel-faced**, *a*. having a sharp, thin face. **weasel word**, *n*. a word designed to mislead or to be evasive. **weaselly**, *a*.
weather (wedh'ə), *n*. the state of the atmosphere with reference to cold or heat, humidity, rain, pressure, wind, electrical conditions etc., esp. the state of the sky at any given time with reference to clouds and rain; (*usu. pl.*) change, vicissitude. *v.t.* to encounter and pass through (storms or bad weather) in safety; to endure and come through in safety; (*Naut.*) to get to windward of (a cape etc.) in spite of inclement weather; to expose to the action of the weather; (*usu. p.p.*) to wear, disintegrate or discolour (rock, cliffs, masonry etc.) by this; to slope (tiles etc.) down so as to overlap. *v.i.* to stand the effects of weather; to disintegrate or discolour by exposure to weather. *a*. situated towards the wind; windward. **stress of weather**, storms, winds etc. **to make good, bad weather**, of a vessel, to behave well or ill in a storm. **to make heavy weather of**, to exaggerate the difficulty of. **under the weather**, poorly, unwell. **weather-beaten**, *a*. seasoned or tanned by exposure to weather, storms etc. **weatherboard**, *v.t.* to furnish with weather-boarding. *n*. a board used for weather-boarding. **weather-boarding**, *n*. boards fastened together so as to overlap and to throw off rain, snow etc. from roofs, walls etc. **weather-bound**, *a*. detained by bad weather. **weather-box**, **-house**, *n*. a toy weather-indicator, the figures of a man and woman emerging at the sides of a toy house indicating wet or dry weather respectively. **weather-bureau**, *n*. a meteorological department or office. **weather-chart**, **-map**, *n*. a chart of a wide area indicating the state of the weather in different parts. **weathercock**, *n*. a revolving vane, often in the shape of a cock, mounted on the top of a steeple or other high point to show the direction of the wind; an inconstant person. **weather-eye**, *n*. **to keep one's weather-eye open**, (*coll.*) to be on the alert. **weather-gauge** GAUGE. **weather-glass**, *n*. a barometer. **weather-house** WEATHER-BOX. **weatherman**, *n*. a person who forecasts and reports on the weather. **weather-map** WEATHER-CHART. **weather-moulding**, *n*. a dripstone over a door, window etc. **weatherproof**, *a*. proof against the weather. *v.t.* to render weatherproof. **weather-prophet**, *n*. **weather-report**, *n*. an official daily report of meteorological observations and probable changes in the weather. **weather-service**, *n*. an organization carrying out meteorological observations. **weather-ship**, *n*. a ship engaged on meteorological work. **weather-stain**, *n*. discoloration by exposure to the atmosphere. **weather-stained**, *a*. **weather-station**, *n*. a place where meteorological observations are taken or recorded. **weather-strip**, *n*. a piece of board, rubber or the like fastened across a door, window etc. to keep out draught. **weather-tiling**, *n*. tiles hung on outside walls to protect against damp. **weather-vane** WEATHER-COCK. **weather window**, *n*. a limited period of time when the weather conditions are suitable for a particular activity or project. **weather-wise**, *a*. skilful in forecasting the weather. **weathering**, *n*. disintegration etc. through exposure to the weather. **weatherly**, *a*. of a ship, presenting such lateral resistance to the water as to make little leeway. **weatherliness**, *n*. **weathermost**, *a*. farthest to windward.
weave (wēv), *v.t.* (*past* **wove** (wōv), *p.p.* **woven**, **wove**) to form (threads, yarns etc.) into a fabric by interlacing; to produce (cloth or a cloth article) thus; to construct by intertwining canes, rushes etc.; of a spider, to form (a web); to interweave (facts, details etc.) into a story, theory etc.; to construct (a scheme, plot etc.) thus. *v.i.* to make fabrics by interlacing threads etc.; to work at a loom. *n*. a manner of weaving; the texture of a woven fabric. **to get weaving** GET. **weavable**, *a*. **weaver**, *n*. one who weaves, esp. one whose occupation is to weave cloth etc.; a weaver-bird. **weaver-bird**, *n*. a finch-like bird of the warmer parts of Asia, Africa and Australia, that constructs elaborate nests of woven grass.
web (web), *n*. (a piece of) woven fabric; a cobweb or a similar structure woven by spiders, caterpillars etc.; an artfully contrived plot etc.; a large roll of paper for printing etc. as it comes from the mill; connective tissue; the membrane between the toes of swimming-birds etc.; the vane of a feather; the thin part of the plate in a girder connecting the upper and lower plates; the part of a railway-carriage wheel between the nave and rim; the blade of a saw etc. *v.t.* (*past, p.p.* **webbed**) to connect, furnish or cover with or as with a web. **web-eye**, *n*. a disease of the eye caused by a film. **web-eyed**, *a*. **web-fingered**, **-footed**, **-toed**, *a*. **web-fingers**, **-feet**, **-toes**, *n. pl.* those with the digits connected by a web. **web offset**, *n*. offset printing using a continuous roll of paper. **web-worm**, *n*. the gregarious larva of any insect weaving a web or tent as a shelter. **webbed**, *a*. **webbing**, *n*. a strong woven band of fibre etc., used for girths, the bottoms of seats, beds etc.; any strong woven tape or edging; a woven structure. **webby**, *a*.
weber (vā'bə, web'ə), *n*. a unit of magnetic flux. [Wilhelm

Weber, 1804-91, German physicist]
wed (wed), *v.t.* (*past*, **wedded**, *p.p.* **wedded**, **wed**) to marry; to give in marriage; to unite, to attach firmly. *v.i.* to marry. **wedded**, *a.* married, pertaining to matrimony; intimately united; strongly attached (to). **wedding**, *n.* a marriage ceremony, usu. with the accompanying festivities. **penny-wedding** PENNY. **silver, golden, diamond wedding**, the 25th, 50th or 60th anniversaries of a wedding. **wedding-breakfast**, *n.* a celebratory meal given after a wedding ceremony. **wedding-cake**, *n.* an iced cake distributed to the guests at a wedding. **wedding-day**, *n.* the day of a marriage or its anniversary. **wedding-favour**, *n.* a knot of white ribbons or a rosette worn at a wedding. **wedding-garment**, *n.* **wedding-ring**, *n.* a plain ring given by one partner to the other, esp. by the bridegroom to the bride, during the marriage ceremony, and worn thereafter.
Wed., (*abbr.*) Wednesday.
we'd (wēd), contr. form of *we had*, *we would*.
wedge (wej), *n.* a piece of wood or metal thick at one end and tapering to a thin edge at the other, used for splitting wood, rocks etc., for exerting great pressure, raising great weights etc., forming one of the mechanical powers; an object or portion of anything in the shape of a wedge; a shoe without an instep, having the heel and sole together forming a wedge; something that causes a separation or divide. *v.t.* to cleave or split with or as with a wedge; to crowd or push (in), as a wedge forces its way; to fix or fasten with a wedge or wedges. **the thin end of the wedge**, a first step, measure or change likely to have important ulterior results. **wedge-shaped**, *a.* **wedgewise**, *adv.* **wedge-tailed**, *a.* of a bird, having a wedge-shaped tail owing to the greater length of the middle feathers. **wedgy**, *a.*
Wedgwood® (wej'wud), *n.* a type of fine pottery, made by Josiah Wedgwood (1730-95) and his successors, often bearing a white cameo-like design in relief. **Wedgwood blue**, *n.* a light greyish-blue.
wedlock (wed'lok), *n.* matrimony, the married state.
Wednesday (wenz'di, wed'nz-), *n.* the fourth day of the week.
wee[1] (wē), *a.* (*Sc.*) very small, tiny, little.
wee[2] (wē), WEE-WEE.
weed (wēd), *n.* a useless or troublesome plant in cultivated land, a plant springing up where not wanted in a garden etc.; any useless or troublesome intrusive thing; a leggy, loose-bodied horse; (*coll.*) a cigar; (*coll.*) a weak or weedy person. *v.t.* to clear (ground) of weeds; to pull up (a noxious or intrusive plant); to clear of anything hurtful or offensive; to sort (out) (useless or inferior elements, members etc.); to rid of these. *v.i.* to pull up weeds from a garden etc. **the weed**, *n.* tobacco. **weed-grown**, *a.* overgrown with weeds. **weed-killer**, *n.* a chemical or other product (usu. poisonous) for destroying weeds. **weedicide** (-disīd), *n.* a chemical weed-killer. **weeding-chisel, -fork, -hook, -tongs**, *n.* a tool used in weeding. **weeder**, *n.* one who weeds; a weeding-tool. **weedy**, *a.* containing weeds; (*coll.*) thin, weak, lacking stamina. **weediness**, *n.*
weeds (wēdz), *n.pl.* mourning worn by a widow.
week (wēk), *n.* a period of seven days, esp. from Sunday to Saturday inclusively; the five or six working days, excluding Sunday, or Saturday and Sunday. **a week of Sundays**, (*coll.*) seven weeks; a long time. **today, tomorrow, yesterday week**, the day later or earlier by a week than the one specified. **weekday**, *n.* any day of the week except Sunday and usu. also Saturday. **weekend**, *n.* the days ending one and beginning the following week (usu. Saturday and Sunday), as a time for holiday etc. *v.i.* to make a holiday etc. on these. **weekender**, *n.* **week-night**, *n.* a night of a weekday. **weekly**, *a.* happening, issued or done once a week or every week; lasting a week; pertaining to or reckoned by the week. *adv.* once a week;

week by week. *n.* a weekly periodical.
weem (wēm), *n.* a subterranean chamber, dwelling or passage, usu. lined with rough stones.
ween (wēn), *v.i.* (*old-fashioned*) to be of opinion; to think, to fancy.
weeny (wē'ni), *a.* (*coll.*) very small, tiny. **weeny-bopper**, *n.* a pre-adolescent fan of pop music and pop stars.
weep (wēp), *v.i.* (*past*, *p.p.* **wept** (wept)) to shed tears; to lament, mourn (for); to let fall or to be emitted; to drip, to exude, to run or be suffused with drops of moisture. *v.t.* to shed tears over; to lament, to bewail; to shed (tears); to exude. **weeper**, *n.* one who or that which weeps; a hired mourner; a widow's white cuff or black crape veil or a man's sash-like hatband worn as a token of mourning. **weeping-ash, -birch, -willow**, *n.* an ash, birch or willow with delicate pendulous branches. **weepy**, *a.* liable to weep. *n.* (*coll.*) a film or book of a very sentimental nature.
weever (wē'və), *n.* either of two British fishes, inflicting painful wounds with their dorsal and opercular spines.
weevil (wē'vl), *n.* a small beetle with the head elongated, feeding on grain, nuts, roots, leaves etc., esp. one infesting corn. **weevilled, weevilly**, *a.*
wee-wee (wē'wē), *v.i.* a child's word for urinate. *n.* an act of urinating; urine.
w.e.f., (*abbr.*) with effect from.
weft (weft), *n.* the threads passing through the warp from selvedge to selvedge, the woof; a web.
weigh (wā), *v.t.* to find the weight of by means of a balance etc.; to be equivalent to in weight; to weigh out (a particular amount); to consider carefully, to estimate the relative value or advantages of; to raise (an anchor). *v.i.* to have a specified weight; to be considered as important, to have weight or influence; to be burdensome or oppressive (on or upon); (*Naut.*) to weigh anchor, to start on a voyage. *n.* the act or process of weighing. **to weigh anchor** ANCHOR. **to weigh down**, to cause to sink by weight, to force down; to oppress. **to weigh in**, of a jockey, to be weighed before a race; (*coll.*) to intervene. **to weigh out**, to take (a particular weight of something) from a quantity; to distribute or apportion according to weight; of a jockey, to be weighed after a race; (*sl.*) to pay (money) out. **under weigh** WAY. **weigh-beam**, *n.* a portable steelyard suspended in a frame. **weighbridge**, *n.* a machine with an iron platform, on which lorries etc. are weighed. **weigh-house**, *n.* a public building at which goods are weighed. **weighable**, *a.* **weighage** (-ij), *n.* a duty paid for weighing goods. **weigher**, *n.* **weighing-cage**, *n.* a cage in which live animals may be weighed. **weighing-machine**, *n.* a machine for weighing persons, objects etc.
weight (wāt), *n.* the force with which bodies tend towards a centre of attraction, esp. the centre of the earth, the downward tendency caused by gravity less the centrifugal tendency due to the earth's rotation; heaviness, esp. as expressed in terms of some standard unit; the standard amount that something or someone, as a boxer, should weigh; a scale or graduated system of units of weight; a unit of weight used in such a system; a piece of metal etc. of known weight used with scales for weighing goods etc.; a heavy object or mass used for mechanical purposes, as in a clock etc.; a heavy load, a burden; pressure, oppressiveness; importance, consequence, impressiveness, efficacy, preponderance; a value given to an item in a frequency distribution to represent its relative importance. *v.t.* to attach weights to; to add weight to; to burden; to treat with minerals etc., to adulterate; to assign a statistical weight to; to bias. **to pull one's weight**, to take one's due share of work or responsibility. **weighting (allowance)**, *n.* an allowance paid in addition to basic salary to offset the higher living costs of a particular area. **weight-lifter**, *n.* **weight-lifting**, *n.* the

sport of lifting barbells of increasing weight using standard lifting techniques. **weight-training,** *n.* physical training using weights to strengthen and tone muscles. **weight-watcher,** *n.* a person who is attempting to lose weight by dieting. **weight-watching,** *n.* **weightless,** *a.* **weightlessness,** *n.* the condition of having no apparent weight which exists when there are no forces opposing gravity, as in an orbiting spacecraft. **weighty,** *a.* having great weight, heavy; important, momentous; convincing, influential. **weightily,** *adv.* **weightiness,** *n.*
weir (wiə), *n.* a dam across a stream for raising the level of the water above it; a fence or enclosure of stakes, nets etc. set in a stream to catch fish.
weird (wiəd), *n.* (*chiefly Sc.*) fate, destiny; (*pl.*) the Fates. *a.* pertaining to fate or destiny; supernatural, uncanny; strange, peculiar. **Weird Sisters,** *n.pl.* the Fates. **weirdly,** *adv.* **weirdness,** *n.* **weirdo, weirdie** (-dō, -di), *n.* (*pl.* **-dos, -dies**) (*coll.*) a strange or eccentric person.
Weismannism (vīs'mənizm), *n.* (*Biol.*) the doctrines of August *Weismann* (1834–1915) with regard to the continuity of the germ-plasm, and the impossibility of transmitting acquired characters.
weka (wē'kə, wä'-, wek'ə), *n.* the New Zealand woodhen.
Welch (welsh), *n.*, *a.* Welsh; spelling used for Welch Fusiliers, Welch Regiment.
welch (welsh), WELSH [2].
welcome (wel'kəm), *a.* admitted or received with pleasure and cordiality (often used ellipt. as an int. addressed to a guest etc.); producing satisfaction or gladness; gladly permitted (to do etc.). *n.* a salutation to a newcomer; a kind or cordial reception or entertainment of a guest etc.; a willing acceptance of an offer etc. *v.t.* to greet cordially; to receive or entertain with kindness or cordiality; to receive (news etc.) with pleasure; to greet or receive in a particular way. **to outstay, overstay one's welcome,** to stay too long. **welcomeness,** *n.* **welcomer,** *n.*
weld [1] (weld), *n.* dyer's-weed, a branched mignonette yielding luteolin, a yellow dye-stuff.
weld [2] (weld), *v.t.* to unite or join (pieces of metal or plastic) together by heat or by compressing, esp. after they have been softened by heat; to make, produce or repair thus; to unite into a coherent mass, body etc. *v.i.* to unite (well or ill) by this process. *n.* a joint or junction made by welding. **weldable,** *a.* **weldability** (-bil'-), *n.* **welder,** *n.*
welfare (wel'feə), *n.* prosperity, success; health, well-being; welfare work; financial and other aid given to those in need. **on Welfare,** (*N Am.*) receiving financial benefits from the state. **Welfare State,** *n.* (a state operating) a system in which the government promotes and assumes responsibility for the general welfare, usu. by introducing social security measures. **welfare work,** *n.* efforts to improve living conditions for the very poor, elderly etc. **welfare worker,** *n.*
welkin (wel'kin), *n.* (*poet.*) the sky, the vault of heaven.
well [1] (wel), *adv.* (*comp.* **better** (bet'ə), *superl.* **best** (best)) in a good or right manner, properly, satisfactorily; happily, fortunately; skilfully; prosperously, successfully; adequately, fully, perfectly, abundantly, sufficiently; intimately; to a considerable extent; heartily, cordially, gratifyingly; with kindness, with approval, on good terms; justly, fairly, reasonably, wisely, befittingly; very possibly, indeed. *a.* (*pred. only*) in good health; in a satisfactory state, position or circumstances; sensible, advisable; fortunate. *n.* that which is well. *int.* expressing astonishment, expectation, resignation, concession etc.; often used as an expletive in resuming one's discourse. **as well,** in addition; equally, as much (as), not less than; just as reasonably, with no worse results; proper, right, not unadvisable (to). **well away,** making rapid progress; (*coll.*) drunk. **well done,** an expression of congratulation. **well-acquainted,** *a.* **well-advised,** *a.* prudent, judicious, wise.

well-appointed, *a.* fully furnished or equipped. **well-balanced,** *a.* sensible, sane. **well-behaved,** *a.* **well-being,** *n.* welfare. **well-born,** *a.* of good birth. **well-bred,** *a.* of good breeding or manners; of good or pure stock. **well-built,** *a.* sturdy, robust, muscular. **well chosen,** *n.* selected with judgment. **well-conditioned,** *a.* of good temper; in good condition. **well-conducted,** *a.* well-behaved. **well-connected,** *a.* related to good families. **well-developed,** *a.* **well-disposed,** *a.* of favourable and kindly feeling (to or towards). **well done,** *a.* of food, cooked thoroughly. **well-dressed,** *a.* **well earned,** *a.* **well enough** ENOUGH. **well-favoured,** *a.* handsome, good-looking. **well-fed,** *a.* **well-found,** *a.* well-appointed. **well-founded,** *a.* based on certain or well-authenticated grounds. **well-groomed,** *a.* neat and elegant in dress and appearance. **well-grounded,** *a.* well-founded; having all the basic knowledge of a subject etc. **well-heeled,** *a.* wealthy. **well-informed,** *a.* having ample information; having a knowledge of numerous subjects. **well-intentioned,** *a.* having good intentions (usu. with alln. to unsatisfactory results). **well-judged,** *a.* skilfully, tactfully or accurately done. **well-knit,** *a.* esp. of a person's body, compact, firmly built. **well known,** *a.* known to many people, familiar, notorious. **well-liked,** *a.* **well-looking,** *a.* of pleasing appearance. **well-mannered,** *a.* well-bred, polite. **well-meaning,** *a.* having good intentions. **well-meant,** *a.* well met, *int.* hail! welcome! **well-nigh,** *adv.* almost, nearly. **well off,** *a.* prosperous. **well-oiled,** *a.* (*sl.*) drunk. **well-preserved,** *a.* young-looking for one's age. **well-proportioned,** *a.* **well-read,** *a.* having read extensively, having wide knowledge gained from books. **well-reputed,** *a.* of good reputation. **well-respected,** *a.* **well-rounded,** *a.* pleasantly curved or rounded; symmetrical, complete; broad in scope, full, varied. **well set,** *a.* firmly set, well-knit, muscular. **well-spoken,** *a.* speaking well, eloquent; well-mannered. **well-thought-of,** *a.* respected, esteemed. **well-thumbed,** *a.* of a book, marked from much handling. **well-timed,** *a.* **well-to-do,** *a.* well off. **well-tried,** *a.* often tried or tested with satisfactory results. **well-trod, -trodden,** *a.* much used or frequented. **well-turned,** *a.* shapely; aptly expressed. **well-upholstered,** *a.* (*facet.*) plump. **well-wisher,** *n.* a person who wishes well to one. **well-woman,** *a.* pertaining to or designed for the health and well-being of women, esp. through preventive and educative measures. **well-worn,** *a.* worn out, trite, hackneyed. **wellness,** *n.*
well [2] (wel), *n.* a shaft bored in the ground to obtain water, oil, brine, gas etc.; a hole, space or cavity more or less resembling this; a space in the middle of a building enclosing the stairs or a lift or left open for light and ventilation; a space occupied by counsel etc. in a law-court; the boxed-in space enclosing the pumps of a vessel; a compartment in a fishing-vessel with a perforated bottom where fish are kept alive; the receptacle holding the ink in an inkstand; a spring, a fountain; a natural pool fed by this; a source. *v.i.* to spring or issue (forth etc.) as a fountain. **well-boat,** *n.* a fishing-boat having a well for conveying fish alive. **well-deck,** *n.* the space enclosed between the forecastle and poop on some ships. **well-dish,** *n.* one with a hollow for gravy to collect in. **well-head,** *n.* the source or fountain-head of a river etc. **well-hole,** *n.* the pit or shaft of a well; the well of a staircase etc. **well-room,** *n.* a room at a spa where the waters are served to visitors. **well-sinker,** *n.* one who digs or sinks wells. **well-sinking,** *n.* **well-spring,** *n.* a source of continual supply. **well-water,** *n.*
we'll (wēl), contr. form of *we will, we shall*.
wellington (boot) (wel'ingtən), *n.* a waterproof boot, usu. rubber, coming up to the mid-calf or knee. [after the first Duke of *Wellington* 1769–1852]
Wellingtonia (welingtō'niə), *n.* a sequoia.
welly, wellie (wel'i), *n.* (*coll.*) a wellington boot.

wels (wɛlz), *n.* the sheat-fish.
Welsh[1] (welsh), *a.* pertaining to Wales or its inhabitants. *n.* the language of the Welsh; (*pl.*) the Welsh people. **Welsh dresser**, *n.* a dresser with open shelves above drawers and cupboards. **Welsh harp**, *n.* one with three rows of strings. **Welshman, -woman**, *n.* **welsh rabbit, rarebit** (rɛə'bit), *n.* cheese melted and spread over toasted bread.
welsh[2], **welch** (welsh), *v.i.* to evade an obligation, esp. to fail to pay a (gambling) debt. **welsher**, *n.*
welt (welt), *n.* a strip of leather sewn round a boot or shoe between the upper and the sole for sewing them together; the border or trimming of a garment; a weal. *v.t.* to furnish with a welt; to weal, to flog.
weltanschauung (velt'anshowung), *n.* (*Phil.*) a survey of the world as an entity. [G]
welter (wel'tə), *v.i.* to roll, to tumble about, to wallow; of waves etc., to heave and roll about confusedly. *n.* a weltering movement, a turmoil, a confusion.
welterweight (wel'təwāt), *n.* a boxer weighing 10st–10st 7lb (63·5–66·7 kg) if professional, 10st–10st 8lb (63·5–67·1 kg) if amateur.
weltpolitik (velt'politik), *n.* a policy aiming at the predominance of a country, specifically Germany, in the affairs of the whole world. [G]
welwitschia (welwich'iə), *n.* a plant from SW tropical Africa, with a trunk several feet wide and only a foot high (over 1 m wide and 30 cm high), and no leaves except the two cotyledonous ones, which grow to 6 ft (1·8 m) or more.
wen (wen), *n.* a sebaceous cyst, frequently occurring on the scalp or neck; (*fig.*) an abnormal growth; an overgrown city, esp. London.
wench (wench), *n.* (*now chiefly facet.*) a girl or young woman; a female servant. **wench-like**, *a.* **wenchless** (-lis), *a.*
Wend (wend), *n.* one of a Slavonic people inhabiting Saxony and Prussia (now N and E Germany). **Wendic**, *a.* **Wendish**, *n.*, *a.*
wend (wend), *v.t.* to go or direct (one's way). *v.i.* to go.
wendigo (wen'digō), *n.* in Amer. Indian mythology, an evil spirit which eats human flesh.
Wendy house® (wen'di), *n.* a small toy house for children to play in.
Wensleydale (wenz'lidāl), *n.* a breed of long-haired sheep; a type of crumbly white cheese. [*Wensleydale* in Yorkshire]
went (went), *past* GO[1].
wentletrap (wen'tltrap), *n.* a univalve marine mollusc with a long, many-whorled shell.
wept (wept), *past, p.p.* WEEP.
were (wœ), WAS.
we're (wiə), contr. form of *we are*.
weren't (wœnt), contr. form of *were not*.
werewolf (wœ'wulf), *n.* (*pl.* **-wolves** (-wulvz)) a person turned or supposed to have the power of turning into a wolf.
wergild (wœ'gild), *n.* (*OE Law*) a fine or monetary compensation for manslaughter and other offences against the person, paid by the kindred of the offender to the kindred of the injured person to avoid blood-feud.
wernerite (wœ'nərīt), *n.* scapolite.
wert (wœt), (*old-fashioned*) 2nd pers. sing. past BE.
Wesleyan (wez'liən), *a.* of or belonging to the Church or sect founded by John *Wesley* (1703-91). *n.* a member of this, a Wesleyan Methodist. **Wesleyanism**, *n.*
west (west), *adv.* at, in or towards the quarter opposite the east, or where the sun sets at the equinox. *n.* that one of the four cardinal points exactly opposite the east; the region or part of a country or of the world lying opposite to the east, esp. the western part of England, Europe or the US; the non-Communist countries of Europe and N America; a wind blowing from the west. *a.* being, lying or living in or towards the west; (blowing) from the west. **to go west**, to die; to be destroyed. **west country**, *n.* the SW part of England. **west-countryman, -countrywoman**, *n.* **West End**, *n.* the fashionable part of London, immediately west of Charing Cross, where the main shops, theatres etc. are located. **West-end**, *a.* **westering**, *a.* passing to the west (of the sun). **westerly**, *a.* in, situated or directed towards the west; (blowing) from the west. *n.* a wind from the west. *adv.* towards the west. **western**, *a.* in, facing or directed towards the west; belonging to or to do with the west; (blowing) from the west. *n.* a film, play or novel dealing with the western states of the US in the wilder periods of their history. **the Western Empire**, *n.* the western division of the Roman Empire having Rome as capital, after AD 395. **Western Church**, *n.* the Latin Church which continued to acknowledge the pope after the schism of the Greek and Latin Churches in the 9th cent. **Western European Union**, *n.* a political and military association, formed in 1955, of Belgium, France, Italy, Luxemburg, Netherlands, UK and West Germany. **Western Powers**, *n.pl.* a loose term for the European powers (and the US) contrasted with the USSR and her satellite powers. **westerner**, *n.* **westernize, -ise**, *v.t.* **westernization**, *n.* **westernmost**, *a.* **westing**, *n.* distance travelled or amount of deviation towards the west. **westward** (-wəd), *a.*, *adv.* **westwards**, *adv.*
Westinghouse brake (wes'tinghows), *n.* a brake worked by compressed air for use on railway trains and motor cars. [G *Westinghouse*, 1846–1914]
Westminster (west'minstə), *n.* the British Parliament. [London borough in which the Houses of Parliament are situated]
wet (wet), *a.* moistened, soaked, saturated, covered with or containing water or other liquid; rainy; not yet dry or hard; using a liquid; (*coll.*, *dated.*) drunk; (*coll.*) feeble, characterless, sentimental; (*N Am.*) allowing or favouring the sale of alcoholic beverages, opp. to prohibitionist. *n.* wetness, moisture; anything that wets, esp. rain; (*sl.*) a drink; (*coll.*) a feeble person; (*coll.*) a moderate conservative politician. *v.t.* (*past, p.p.* **wetted**) to make wet; to moisten, drench or soak with liquid; to urinate on; (*sl.*) to celebrate (a bargain etc.) with drink. **The Wet**, *n.* (*Austral.*) the monsoon season. **to wet one's whistle**, to drink. **wetback**, *n.* an illegal immigrant to the US from Mexico. **wet blanket**, *n.* (*coll.*) a person who damps enthusiasm, zeal etc. **wet bob** BOB[2]. **wet bulb** DRY-BULB THERMOMETER under DRY. **wet dock**, *n.* a dock in which vessels can float. **wet dream**, *n.* an erotic dream with emission of semen. **wetland**, *n.* (*often pl.*) swamp, marsh land. **wet-look**, *a.* (made from a material) having a shiny finish. **wet-nurse**, *n.* a woman employed to suckle a child not her own. *v.t.* to act as wet-nurse to; to coddle. **wet-pack**, *n.* (*Med.*) a wet sheet in which a patient is wrapped. **wet-rot**, *n.* decay in wood caused by fungal attack in timber subjected to prolonged damp. **wet-shod**, *a.* having the shoes wet. **wet suit**, *n.* a tight-fitting usu. rubber garment for divers etc. that allows water in whilst retaining body heat. **wetly**, *adv.* **wetness**, *n.* **wetting**, *n.* **wettish**, *a.*
wether (wedh'ə), *n.* a castrated ram.
we've (wēv), contr. form of *we have*.
wey (wā), *n.* a weight or measure (for wood, grain etc.) varying with different articles.
wf., (*abbr.*) (*Print.*) wrong fount.
whack (wak), *v.t.* to strike heavily; (*sl.*) to share out (plunder etc.). *n.* a heavy blow; (*sl.*) a share, a portion; (*coll.*) an attempt. **whacked**, *a.* (*coll.*) exhausted. **whacking**, *n.* a beating, a thrashing. *a.*, *adv.* (*coll.*) large, whopping.
whacky WACKY.
whale[1] (wāl), *n.* a large marine fish-like mammal, of various species of *Cetacea*, several of which are hunted chiefly

for their oil and whalebone; (*coll.*) something very big, exciting etc. *v.i.* to engage in whale-fishing. **whale-back**, *n.* a vessel with the main decks covered in and rounded over as a protection against rough seas. **whale-boat**, *n.* a boat sharp at both ends, such as those used in whaling. **whalebone**, *n.* a horny, elastic substance occurring in long, thin plates, found in the palate of certain whales. **whalebone whale**, *n.* **whale-calf**, *n.* a young whale. **whale-fin**, *n.* whalebone. **whale-fishery**, **-fishing**, *n.* **whale-line**, *n.* rope of great strength used in whaling. **whale-man**, *n.* (*chiefly N Am.*) a whaler. **whale-oil**, *n.* oil obtained from the blubber of whales; spermaceti. **whaler**, *n.* a person employed in whaling; a ship employed in whaling. **whaling**, *n.* the catching and processing of whales. **whaling-gun**, *n.* a gun for firing harpoons at whales. **whaling-master**, *n.* the captain of a whaler.
whale² (wāl), WEAL².
wham (wam), *n.* a forceful blow; the noise of this. *v.t.*, *v.i.* (*past, p.p.* **whammed**) to strike or crash, or cause to strike or crash with a loud, forceful blow.
whang¹ (wang), *v.t.* to beat noisily, to bang. *v.i.* of a drum etc., to make a noise as if whanged. *n.* a whanging blow, a bang.
whang² (wang), *n.* a tough leather strap or thong.
whangee (-ē'), *n.* a flexible bamboo cane.
wharf (wawf), *n.* (*pl.* **wharfs**, **wharves** (wawvz)) a platform built alongside a river, harbour, canal etc. for loading and unloading ships. *v.t.* to moor at a wharf; to deposit or store goods on a wharf. **wharf-rat**, *n.* the brown or Norway rat. **wharfage** (-ij), *n.* the charge for using a wharf; the use of a wharf. **wharfing**, *n.* **wharfinger** (-finjə), *n.* a person who owns or has charge of a wharf.
what (wot), *pron.* (*interrog.*) which thing or things (*often used ellipt.*); (*rel.*) that which, those which, the things that; which things, how much! (as an exclamation); (*dial.*) that or which. *a.* which thing, kind, amount, number etc. (in asking questions); how great, remarkable, ridiculous etc. (used in an exclamatory sense); (*rel.*) such as, as many or as many as, any that. *adv.* (*interrog.*) to what extent, in what respect? **what for**, for what reason, purpose? etc. **to give what for** GIVE¹. **what have you**, (*coll.*) anything else of the kind. **what ho!** an exclamation of greeting or accosting. **what next?** (*int.*) monstrous! absurd! **what not** WHAT HAVE YOU. **what of that**, no matter, never mind. **what's what**, (*coll.*) the real thing or situation. **what's up?** UP. **what though**, what does it matter if? admitting that. **what time** TIME. **what-d'ye-call-it**, *n.* (*coll.*) a phrase put for something that has slipped one's memory. **whate'er**, *pron.* (*poet.*) whatever. **whatever**, *pron.* anything soever that; all that which. *a.* no matter what (thing or things). **whatnot**, *n.* a piece of furniture with shelves for ornaments, books etc. **whatsit** (-sit), *n.* (*coll.*) a person or thing whose name is unknown or temporarily forgotten. **whatsoever**, **whatsoe'er**, *pron.*, *a.* WHATEVER.
whau (wow), *n.* a New Zealand tree with very light-weight cork-like wood.
whaup (hwawp), *n.* (*chiefly Sc.*) the curlew.
wheal (wēl), *n.* (*Cornwall*) a mine (usu. a tin-mine).
wheat (wēt), *n.* an annual cereal grass cultivated for its grain which is ground into flour for bread. **wheat belt**, *n.* the area east of the Rocky Mountains in Canada and the US where wheat is extensively cultivated. **wheat-ear**, *n.* an ear of wheat. **wheat-fly**, *n.* any of various flies that injure wheat, esp. the Hessian fly. **wheat germ**, *n.* the embryo of the wheat grain, rich in vitamins. **wheat-grass**, *n.* couch-grass. **wheat-meal**, *n.*, *a.* (made from) flour containing much of the original wheat grain. **wheat-moth**, *n.* one of various moths, the larvae of which destroy wheat. **wheaten**, *a.*
wheatear (wē'tiə), *n.* a small white-rumped songbird.
Wheatstone automatic (wēt'stən), *n.* a mechanical system of telegraphy for transmitting and receiving high-speed signals automatically. **Wheatstone bridge**, *n.* a device for measuring an unknown electrical resistance by means of a known resistance. [Sir C. Wheatstone, 1802–75]
whee (wē), *int.* an exclamation of delight or excitement.
wheedle (wē'dl), *v.t.* to win over or persuade by coaxing or flattery; to cajole, to humour; to obtain from or get (out of) by coaxing and flattery. **wheedler**, *n.* **wheedling**, *a.* **wheedlingly**, *adv.*
wheel (wēl), *n.* a circular frame or solid disk turning on its axis, used in vehicles, machinery etc. to reduce friction and facilitate motion; a machine, implement, device etc. consisting principally of a wheel, esp. a spinning-wheel, potter's wheel, steering-wheel etc.; (*N Am.*, *coll.*) a cycle; (*pl.*, *coll.*) a car; an object resembling a wheel, a disk; a catherine-wheel; an instrument of torture formerly used for breaking the limbs of criminals; torture with this; the act of wheeling, circular motion, rotation; a turn, a revolution; the turning or swinging round of a body of troops or a line of warships as on a pivot; (*pl.*) the forces controlling or activating an organization etc. *v.t.* to move or push (a wheeled vehicle etc.) in some direction; to cause to turn or swing round as on a pivot. *v.i.* to turn or swing round thus; (*lit. or fig.*) to change direction, to face another way; to gyrate; to ride a cycle etc. **to break upon the wheel** BREAK. **wheels within wheels**, concealed reasons or interdependent circumstances. **wheel and axle**, one of the mechanical powers, consisting of a cylindrical axle on which a wheel is fastened concentrically, the difference between their respective diameters supplying leverage. **wheel-animalcule**, *n.* a rotifer. **wheelbarrow**, *n.* a barrow usu. supported on a single wheel, with two handles by which it is wheeled. **wheel-base**, *n.* the distance between the front and rear hubs of a vehicle. **wheel-brace**, *n.* a brace-shaped spanner for adjusting bolts on a wheel. **wheel-chair**, *n.* a chair on wheels, esp. for invalids. **wheel clamp**, *n.* a clamp fixed on to the wheel of an illegally parked car to prevent it from being driven away before a fine is paid. **wheel-clamp**, *v.t.* **wheel-horse**, *n.* a wheeler. **wheel-house**, *n.* a shelter for the steersman. **wheelman**, *n.* (*N Am.*) a cyclist. **wheel-seat**, *n.* the part of an axle carrying a fixed wheel and fastened into its hub. **wheel-shaped**, *a.* **wheel-spin**, *n.* the revolution of wheels without a grip of the road. **wheelstone**, *n.* an entrochite. **wheel-tread**, *n.* the part of a rim or tyre that touches the ground. **wheel-window**, *n.* a rose-window. **wheelwright**, *n.* a person whose occupation is to make wheels etc. **wheeled**, *a.* (*usu. in comb.*, as *four-wheeled*). **wheeler**, *n.* one who wheels; a wheelwright; a horse next to the wheels in a tandem etc. **wheeler-dealer**, *n.* one who operates shrewdly and often ruthlessly in business, politics etc. **wheeler-dealing**, *n.* **wheelie** (-i), *n.* (*coll.*) a manoeuvre in which a bicycle or motorcycle is briefly supported on the rear wheel alone. **wheelless** (-lis), *a.* **wheely**, *a.*
wheeze (wēz), *v.i.* to breathe hard and with an audible sound, as in asthma. *v.t.* to utter thus. *n.* a wheezing sound; (*coll.*, *esp. dated*) a joke, a trick, a scheme. **wheezer**, *n.* **wheezy**, *a.* **wheezily**, *adv.* **wheeziness**, *n.*
whelk¹ (welk), *n.* a marine spiral-shelled gasteropod used for food.
whelk² (welk), *n.* a small pustule or pimple. **whelked**, *a.*
whelm (welm), *v.t.* (*old-fashioned*) to overwhelm, to engulf, to submerge.
whelp (welp), *n.* the young of a dog, a pup; the young of a beast of prey, a cub; an offensive or ill-bred child or youth. *v.i.* of a bitch etc., to bring forth young. *v.t.* to bring forth (a pup or cub); (*derog.*) to give birth to or produce. **whelpless**, *a.*
when (wen), *adv.*, *conj.* (*interrog.*) at what or which time? (*rel.*) at which (time), at the time that, at any time that, at whatever time; as soon as; at or just after the time that;

whence ... **whip**

after which, and then; although; considering that; while (*often ellipt. with pres.p.*). *pron.* what or which time. **whenas**, *adv.* (*old-fashioned*) when; whereas, while. **whenever**, *adv.* at whatever time. **whensoever**, *adv.* at what time soever.

whence (wens), *adv., conj.* (*interrog.*) from what place or which? where from? how?; (*rel.*) from which place, origin, source etc.; for which reason, wherefore; (*ellipt.*) to or at the place from which. *pron.* what or which place or starting-point. **whencesoever**, *adv., conj.* from whatsoever place or source.

whenever etc. WHEN.

where (weə), *adv., conj.* (*interrog.*) at or in what place, situation, case, circumstances? etc.; to what place? whither? in what direction?; (*rel.*) in which (place or places), in or to the place, direction etc. in which; whereas. *pron.* what or which place. **whereabout**, *adv., conj.* about which, in regard to which; whereabouts. **whereabouts**, *adv., conj.* near what or which place roughly. *n.* the approximate locality or the locality in or near which a person or thing is. **whereas**, *conj.* the fact or case being that, considering that (in legal preambles etc.); the fact on the contrary being that, when in reality. **whereat**, *adv., conj.* at which. **whereby**, *adv.* by which (means). **wherefore**, *adv.* for what reason? why? for which reason, on which account. *n.* the reason why. **wherefrom**, *adv.* from which, whence. **wherein**, *adv., conj.* in what place, respect? etc.; in which thing, place, respect etc. **whereinsoever**, *adv.* **whereinto**, *adv.* **whereness**, *n.* **whereof**, *adv., conj.* of what? of which or whom. **whereon**, *adv., conj.* **wheresoever**, *adv., conj.* in or to what place soever. **whereto** *adv., conj.* to which or what place or end. **whereunder**, *adv., conj.* **whereunto**, *adv., conj.* to what end or purpose? to which. **whereupon**, *adv., conj.* upon which; in consequence of or immediately after which. **wherever**, *adv., conj.* at, in or to whatever place. **wherewith**, *adv., conj.* with what? with which. **wherewithal**, *adv., conj.* wherewith. *n.* the necessary means or resources, esp. money.

wherry (we'ri), *n.* a light shallow rowing-boat for plying on rivers. **wherryman**, *n.*

whet (wet), *v.t.* (*past, p.p.* **whetted**) to sharpen by rubbing on a stone or similar substance; to excite, to stimulate. *n.* the act of whetting; anything taken to whet or stimulate the appetite. **whetstone**, *n.* a piece of stone used for sharpening cutlery etc.; anything that sharpens or stimulates. **whetter**, *n.*

whether (wedh'ə), *conj.* introducing (an indirect question in the form of) an alternative clause followed by an alternative *or, or not*, or *or whether*, or with the alternative unexpressed.

whetstone, whetter WHET.

whew ((h)woo, fū), *int.* an exclamation of astonishment or consternation.

whey (wā), *n.* the watery part of milk that remains after the casein etc. have formed curds and been separated. **whey-face**, *n.* a pale-faced person. **whey-faced**, *a.* **whey-tub**, *n.* **wheyey** (-i), **wheyish**, *a.*

which (wich), *pron.* (*interrog.*) what person, thing, persons or things of a definite number; (*rel.*) representing in a subordinate clause a noun expressed or understood in the principal sentence or previous clause. *a.* (*interrog.*) what (person, thing etc.) of a definite number; (*rel.*) used with a noun defining an indefinite antecedent. **whichever**, *a., pron.* which (person or thing) of two or more.

whicker (wik'ə), *v.i.* to neigh softly.

whidah (-bird), whydah (wid'ə), *n.* a small W African weaver-bird, the male of which has four tail-feathers of enormous length.

whiff[1] (wif), *n.* a sudden expulsion of smoke etc., a puff, esp. one carrying an odour; a small amount, a trace; a small cigar; a light outrigged sculling boat. *v.t., v.i.* to puff or blow lightly; (*coll.*) to smell (unpleasant). **whiffy**, *a.* (*coll.*) smelly.

whiff[2] (wif), *v.i.* to fish with a hand-line, usu. from a boat, towing the bait near the surface.

whiff[3] (wif), *n.* a European or W Indian flat-fish.

whiffle (wif'l), *v.i.* of the wind etc., to veer about; to change from one opinion or course to another, to prevaricate, to equivocate. **whiffler**, *n.*

Whig (wig), *n.* a member of the British political party that contended for the rights and privileges of Parliament in opp. to the Court party or Tories, supported the Revolution of 1688 and the principles it represented, and was succeeded by the Liberals; orig. a Scottish Covenanter; an American colonist who supported the cause of independence in the American Revolution; a member of an American political party from about 1834–54, representing commercial and financial interests. **Whiggery, Whiggism**, *n.* **Whiggish**, *a.* **Whiggishly**, *adv.* **Whiggishness**, *n.*

whigmaleerie (wigməliə'ri), *n.* a trinket, a gewgaw; a whim.

while (wīl), *n.* a space of time, esp. the time during which something happens or is done. *conj.* during the time that, as long as, at the same time as (*often used ellipt. with pres.p.*); at the same time that, whereas (followed by a correlative sentence bringing out a contrast). **once in a while**, occasionally. **to be worthwhile**, to be worth the time, labour or expense involved. **to while away**, to pass (time etc.) pleasantly or without weariness. **whiles**, *adv.* while; (*Sc.*) sometimes. **whilst** (-st), *conj.* while.

whilom (wī'ləm), *adv.* (*old-fashioned*) formerly, once, of old. *a.* former.

whilst WHILE.

whim (wim), *n.* a sudden fancy, a freak, a caprice; a hoisting device, usu. consisting of a vertical winch worked by a horse, formerly used in mines for raising ore. **whimmy**, *a.* whimsical. **whimsical** (-zikl), *a.* full of whims; oddly humorous; odd-looking, curious, fantastic. **whimsicality** (-kal'-), **whimsicalness**, *n.* **whimsically**, *adv.* **whimsy** (-zi), *n.* a whim, a crotchet. **whimwham** (-wam), *n.* a plaything, a whim, a fancy.

whimbrel (wim'brəl), *n.* a small European curlew.

whimper (wim'pə), *v.i.* to cry with a low, broken, whining voice; to whine. *v.t.* to utter in such a tone. *n.* a low, querulous or whining cry. **whimperer**, *n.* **whimperingly**, *adv.*

whimsical, whimwham etc. WHIM.

whin[1] (win), *n.* furze, gorse. **whinberry**, *n.* the bilberry. **whinchat**, *n.* a small thrush-like songbird. **whinny**[1], *a.* abounding in whin.

whin[2] (win), **whinsill, whinstone**, *n.* a very hard, resistant rock, esp. basalt, chert or quartzose sandstone.

whine (wīn), *v.i.* to make a plaintive, long-drawn cry; to complain or find fault in a peevish way. *v.t.* to utter with a whine or in a peevish way. *n.* a whining cry, sound or tone; a peevish complaint. **whiner**, *n.* **whiningly**, *adv.* **whiny, whiney** (-i), *a.*

whinge (winj), *v.i.* to cry fretfully; to complain, to whine. *n.* a complaint. **whingeing**, *n.* **whinger**[1], *n.*

whinger[2] (wing'ə), *n.* a dirk, a short sword or hanger.

whinny[1] WHIN[1].

whinny[2] (win'i), *v.i.* to neigh, esp. in a gentle or delighted way. *n.* the act or sound of whinnying.

whinsill, whinstone WHIN[2].

whip (wip), *v.t.* (*past, p.p.* **whipped**) to move suddenly and quickly, to snatch, to dart, to jerk (out, away etc.); to drive or urge (on) with a whip; to beat (out of etc.); to strike forcefully with or as if with a whip; to beat (eggs, cream etc.) into a froth; to fish (a stream) by casting a line over the water; (*coll.*) to beat, to overcome; to manage or discipline (the members of a political party); to bind with a close wrapping of twine, thread etc.; to bind (twine) round a joint etc.; to oversew (a seam) with close stitches; to twist (goods etc.) with a rope passed

through a pulley; (*sl.*) to steal. *v.i.* to move or start suddenly, to start, to dart (out, in etc.). *n.* an instrument for driving horses etc., or for punishing persons, consisting of a lash tied to a handle or rod; a coachman or driver; a whipper-in; a member of Parliament appointed to enforce discipline and to summon the members of the party to divisions etc.; a summons sent out by a whip to ensure such attendance; a hoisting apparatus consisting of a single rope and pulley; a whipping motion; a dessert made with whipped eggs, cream etc. **to whip up,** to excite, arouse, stimulate; to produce hurriedly. **whip-cord,** *n.* a hard twisted cord for making a whip; a very durable corded cloth made from worsted yarns. **whip-crane,** *n.* a crane used with a whip for rapid hoisting. **whip-gin,** *n.* a block for use in hoisting. **whip-graft,** *n.* a graft made by inserting a tongue in a scion into a slit cut in the stock. *v.t.* to graft by this method. **whip-hand,** *n.* the hand holding the whip; the advantage or control. **whip-handle,** *n.* **whiplash,** *n.* **whiplash injury,** *n.* an injury to the neck caused by a sudden uncontrolled forwards and backwards movement of the unsupported head. **whip-ray,** *n.* a stingray. **whip-round,** *n.* (*coll.*) a collection of money. *v.i.* to make a collection. **whip-saw,** *n.* a narrow saw-blade with the ends fastened in a frame. *v.t.* to saw with this; (*N Am. sl.*) to beat (a person) at every point in a game or in betting. **whip-snake,** *n.* a slender whip-like snake. **whip-stitch,** *n.* a small stitch used for oversewing. **whip-stock,** *n.* the rod or handle of a whip. **whip-tail,** *n.* a Tasmanian fish; a small kangaroo. **whip-top,** *n.* a top kept spinning with a whip. **whip-like,** *a.* **whipper,** *n.* **whipper-in,** *n.* (*pl.* **whippers-in**) a person employed to assist the huntsman by looking after the hounds; a parliamentary whip. **whippersnapper,** *n.* a noisy, presuming, insignificant person. **whipping,** *n.* **whipping-boy,** *n.* a boy formerly educated with a young prince and taking his punishments for him; a scapegoat. **whipping-post,** *n.* a post to which offenders were tied to be whipped (usu. attached to stocks). **whipping-top** WHIP-TOP. **whippy,** *a.* flexible, springy.
whippet (wip'it), *n.* a breed of racing-dogs, similar to but smaller than the greyhound.
whipple-tree (wip'l), *n.* a swingle-tree.
whip-poor-will (wip'əwil), *n.* a small N American nocturnal bird allied to the goat-suckers.
whir, whirr (wœ), *v.i.* (*past, p.p.* **whirred**) to revolve, move or fly quickly with a buzzing or whizzing sound. *n.* a whirring sound.
whirl (wœl), *v.t.* to swing round and round rapidly; to cause to revolve or fly round with great velocity; to carry (away or along) rapidly; to hurl or fling. *v.i.* to turn round and round rapidly, to rotate, to gyrate, to spin; to be carried or to travel rapidly in a circular course; to move along swiftly; of the brain etc., to be giddy, to seem to spin round. *n.* a whirling motion; a confused state, giddiness; commotion, bustle; (*coll.*) an attempt, a trial. **whirl-bone,** *n.* the bone of a ball-and-socket joint, esp. the patella or knee-cap. **whirligig** (-ligig), *n.* a child's spinning or rotating toy; a merry-go-round; a water-beetle that darts about in a circular manner over the surface of pools etc.; something that continually moves or changes; a revolving or rotating course. **whirlpool,** *n.* an eddy or vortex. **whirlwind,** *n.* a funnel-shaped column of air moving spirally round an axis, which at the same time has a progressive motion. **whirler,** *n.* **whirling,** *n.,* *a.* **whirling-table,** *n.* a machine for exhibiting the effects of centrifugal and centripetal forces; a potter's wheel.
whirr WHIR.
whish[1] (wish), **whisht** (wisht), *int.* hush! silence!
whish[2] (wish), *v.i.* to move through the air or water with a whistling sound. *n.* a whistling sound.
whisk (wisk), *v.t.* to sweep, brush or flap (away or off); to carry off or take (away) swiftly or suddenly; to shake, flourish or wave about with a quick movement; to beat up (eggs etc.). *v.i.* to move or go swiftly or suddenly. *n.* a whisking movement; a small bunch of grass, straw, feathers, hair etc. used as a brush or for flapping away flies, dust etc.; an instrument for beating up cream, eggs etc.
whisker (wis'kə), *n.* (*usu. pl.*) hair growing on the cheeks of a man; one of the bristly hairs growing round the mouth of a cat or other animal; a narrow margin; a very fine and strong hairlike crystal. **whiskered,** *a.*
whisky[1], (*Ir., US*) **whiskey** (wis'ki), *n.* a spirit distilled usu. from barley, sometimes from wheat, rye etc. **whisky-liver,** *n.* cirrhosis of the liver caused by alcoholic poisoning. **whisky-mac,** *n.* a drink made from whisky and ginger wine. **whisky-toddy,** *n.* **whiskified** (-fid), *n.* (*coll.*) intoxicated by or as by whisky.
whisky[2] (wis'ki), *n.* a light one-horse chaise or gig for fast travelling.
whisky-jack (wis'kijak), **-john** (-jon), *n.* a grey jay, common in northern N America.
whisper (wis'pə), *v.i.* to speak with articulation but without vocal vibration; to speak in a low voice so as not to be overheard; to converse privately or in a whisper; to devise mischief, to plot, to talk slander; to rustle. *v.t.* to tell or bid in a whisper or privately; to utter or disseminate thus; to hint or suggest privately. *n.* a whispering tone or voice; a whispered remark or speech; a hint, an insinuation, a rumour. **whisperer,** *n.* **whispering,** *n.* **whispering campaign,** *n.* the organized spread of injurious rumours. **whispering-gallery,** *n.* a gallery, corridor etc. in which the faintest sounds made at particular points are audible at other distant points though inaudible elsewhere. **whisperingly,** *adv.*
whist (wist), *n.* a card game, usu. for four persons, played with the entire pack of 52 cards. **dummy whist** DUMMY. **whist drive,** *n.* a competitive series of games of whist.
whistle (wis'l), *v.i.* to make a shrill musical sound by forcing the breath through a small opening of the lips or with an instrument; an appliance on a steam-engine etc.; of an instrument, engine etc., to emit this sound; of birds etc., to make a similar sound; to make such a sound by swift motion of a missile, the wind etc. *v.t.* to emit or utter (a tune etc.) by whistling; to call or give a signal to thus. *n.* a whistling sound, note or cry; an instrument for producing such a sound; (*sl.*) the throat. **to wet one's whistle** WET. **to whistle for,** to stand little or no chance of getting. **to whistle for a wind,** the superstitious practice of old sailors in a calm. **whistle-stop,** *n.* (*N Am.*) a small station where trains stop only on signal; (*orig. N Am.*) a brief visit to a town, as by a political candidate. **whistler,** *n.* one who or that which whistles; (*N Am.*) the whistling or hoary marmot; a whistling duck or other bird; a broken-winded horse. **whistling duck,** *n.* the American widgeon.
whit (wit), *n.* a jot, the least particle, an iota.
white (wit), *a.* being of the colour produced by reflection of all the visible rays in sunlight, as of pure snow, common salt, foam of clear water etc.; approaching this colour, pale, pallid, bloodless, transparent, colourless; silvery, whitish-grey; belonging to a light-complexioned group or race; pure, clean, stainless; spotless, innocent; grey, silvery or hoary as from age etc.; of coffee, containing milk or cream; having snow; clothed in white; having white or pale fur, hair etc.; not malicious or malevolent; fair, happy, propitious. *n.* a white colour, paint or pigment; a white person esp. one of European descent; a white animal; a white part of anything, having the colour of snow; the sclerotic coat of the eye surrounding the iris; the albuminous material surrounding the yolk of an egg; (*pl.*) white clothes; (*pl.*) leucorrhoea. **white ant** TERMITE.
whitebait, *n.* the fry of several clupeoid fish eaten when about 2 in. (5 cm) long. **whitebeam,** *n.* a shrub or small tree with silvery undersides to the leaves. **white bear,** *n.*

whitening 945 **whole**

the polar bear. **white-beard,** *n.* an old man. **white blood cell,** a leucocyte. **white-cap,** *n.* the redstart and other birds; a white-crested wave. **white-collar,** *a.* pertaining to non-manual employees, as office and clerical workers. **white corpuscle, cell,** *n.* a leucocyte. **white-crested, -crowned,** *a.* **white crops,** *n.pl.* wheat, barley etc., which whiten as they ripen. **white damp,** *n.* (*Mining*) carbon monoxide. **white dwarf,** *n.* a type of small, very faint, dense star. **white-eared,** *a.* **white elephant** ELEPHANT. **white ensign** NAVAL ENSIGN under ENSIGN. **white-faced,** *a.* pale-faced; having a white front or surface; of animals, having a white spot or streak on the front of the head. **white feather,** *n.* **to show the white feather** FEATHER. **white fish,** *n.* a general term for food-fish other than salmon, esp. whitings and haddocks. **whitefish,** *n.* a N American salmonoid food-fish; the menhaden and other fish. **white flag** FLAG. **white flour,** *n.* flour made of wheat from which most of the bran and germ have been removed. **whitefly,** *n.* (*pl.* **-flies, -fly**) a small insect which is a pest of plants. **White Friar,** *n.* a Carmelite (from the white cloak). **white frost,** *n.* hoar-frost. **white gold,** *n.* a whitish alloy of gold with palladium, nickel etc. **white goods,** *n.pl.* household linen; large kitchen appliances, as freezers and cookers. **white-handed,** *a.* **white heat,** *n.* the degree of heat at which objects become incandescent and appear white; a high pitch of excitement, passion etc. **white horses,** *n.pl.* foam-crested waves. **white-hot,** *a.* **White House,** *n.* the official residence of the President of the US at Washington. **white-iron,** *n.* thin sheet iron with a coating of tin. **white knight,** *n.* one who gives (financial) support to a person or organization in a difficult situation. **white-land,** *n.* a tough, clayey soil, of a whitish hue when dry. **white lead,** *n.* carbonate of lead, esp. used as a basis of white oil-paint. **white lie** LIE [1]. **white light,** *n.* light containing more or less equal intensities of all wavelengths in the visible spectrum. **white-limed,** *a.* whitewashed. **white-lipped,** *a.* pale to the lips, esp. with fear. **white-livered,** *a.* cowardly. **white magic,** *n.* magic not involving the devil, sorcery; magic used for good. **white man,** *n.* (*dated, coll.*) an honourable, upright man. **white man's burden,** *n.* the white man's supposed obligation to promote the welfare of the less-developed non-white races. **White Man's Grave,** the lands along the Guinea Coast of W Africa where the atmosphere was thought peculiarly unhealthy to Europeans. **white meat,** *n.* meat that appears white after cooking, as poultry, veal, pork. **white metal,** *n.* a tin- or sometimes lead-based alloy for bearings, domestic utensils etc. **white night,** *n.* a sleepless night. **white noise,** *n.* noise, esp. electrical, consisting of sound waves of a wide range of frequencies. **white-out,** *n.* a condition of uniform whiteness occurring in polar or similar snow-covered regions in heavy cloud. **white paper,** *n.* a government report on a matter recently investigated. **white pepper** PEPPER. **white sale,** *n.* a sale of household linen at reduced prices. **white sauce,** *n.* a thick sauce made with flour and milk or a fish or white-meat stock. **white slave,** *n.* a woman or child procured, and usu. exported, for immoral purposes. **white slaver,** *n.* **white slavery,** *n.* **whitesmith,** *n.* a tinsmith; one who finishes or galvanizes iron-work. **white spirit,** *n.* a distillate of petroleum used as a paint solvent and thinner. **white squall,** *n.* a squall not preceded by clouds, as in tropic seas. **whitestone,** *n.* a fine, white granite. **white thorn,** *n.* the hawthorn. **whitethroat,** *n.* a small warbler of the genus *Sylvia*. **white tie,** *a.* (men's formal evening wear, with) a white bow tie. **whitewash,** *n.* a mixture of quicklime and water or of whiting and size used for whitening walls, ceilings etc.; a false colouring given to a person's character or memory to counteract disreputable allegations. *v.t.* to cover with whitewash; to cover up or conceal (a misdemeanour etc.); to clear (a person's name) thus; (*coll.*) to defeat decisively. **white-washer,** *n.* **white water,** *n.* foaming water in breakers,

rapids etc. **white wedding,** *n.* one in which the bride wears white. **white whale,** *n.* the beluga. **white wine,** *n.* any wine of a pale yellow colour. **white witch,** *n.* one using her power for beneficent purposes. **white-wood,** *n.* (any one of various trees yielding) light-coloured timber. **whited,** *a.* **whited sepulchre,** *n.* a hypocrite (from Christ's allusion to the scribes and Pharisees, Matt. xxiii.27). **whitely,** *adv.* **whiten,** *v.t., v.i.* **whitener,** *n.* **whiteness,** *n.* **whitening,** *n.* the act of making white; the state of becoming white; whiting [1]. **whitey** [1] (-i), *n.* (*derog.*) a white person; white people collectively. **whitish,** *a.* **whitishness,** *n.*
whitening, whitesmith, whitethroat, whitewash etc. WHITE.
whitey [1] WHITE.
whitey [2] (wi'ti), *n.* (*Austral.*) a flour-and-water scone cooked in wood ashes.
whither (widh'ə), *adv., conj.* (*interrog.*) to what or which place, where; (*rel.*) to which; whithersoever, wheresoever. **whithersoever,** *adv., conj.* to what place soever.
whiting [1] (wi'ting), *n.* fine chalk pulverized, washed and prepared for use in whitewashing, polishing etc.
whiting [2] (wi'ting), *n.* a sea-fish used for food. **whiting-pout,** *n.* a gadoid fish resembling this with an inflatable membrane over the eyes.
whitish etc. WHITE.
whitleather (wit'ledhə), *n.* leather dressed with alum, white leather; the paxwax of the ox.
Whitley Council (wit'li), *n.* an industrial council comprising employers and employees to settle disputes and promote welfare in the industry. [J.H. *Whitley*, 1866–1935, first chairman]
whitlow (wit'lō), *n.* a pus-filled inflammation, esp. round the nail of a finger or toe. **whitlow-grass,** *n.* a minute white-flowered, grass-like herb.
Whitsun (wit'sn), *a.* pertaining to Whit-Sunday or Whitsuntide. **Whit-Sunday,** *n.* the seventh Sunday after Easter, a festival commemorating the day of Pentecost. **Whit-Monday** etc., *n.* **Whitsuntide,** *n.* Whit-Sunday and the following days. **Whit-week,** *n.*
whittle (wit'l), *v.t.* to trim, shave or cut pieces or slices off with a knife; to shape thus; to thin down; to reduce, pare away or bring (down) in amount etc., gradually or by degrees. *v.i.* to keep on paring, shaving or cutting away (at a stick etc.) with a knife.
whity (wi'ti), *a.* whitish, inclining to white (*usu. in comb.* as *whity-brown*, between white and brown).
whiz, whizz (wiz), *v.i.* (*past, p.p.* **whizzed**) to make or move with a hissing sound, like an arrow or ball flying through the air; to move about rapidly. *n.* a whizzing sound; a person who is very good at some activity. **whizz-bang,** *n.* a small high-velocity shell. **whiz(z)-kid,** *n.* (*coll.*) one who is outstandingly successful or clever, esp. at a relatively young age. **whizzingly,** *adv.*
who (hoo), *pron.* (*obj.* **whom** (hoom), *poss.* **whose** (hooz)) (*interrog.*) what or which person or persons?; (*rel.*) that (identifying the subject or object in a relative clause with that of the principal clause). **whodunit** (-dŭn'it), *n.* (*coll.*) a detective or mystery story. **whoever, whoso, whosoever,** *pron.* (*obj.* **whomever, whomsoever** etc.) any one without exception who, no matter who.
WHO, (*abbr.*) World Health Organization.
whoa (wō), *int.* stop! (used chiefly by drivers to horses).
whole (hōl), *a.* sound, in good health; unimpaired, uninjured, not broken, intact; restored to health; complete or entire; containing the total number of parts, undivided, undiminished; integral, composed of units, not fractional; from which no constituents have been removed, entire, as *whole blood. n.* a thing complete in all its parts, units etc.; all that there is of a thing, the entirety; a complete system, a complete combination of parts, an organic unity. **on the whole,** all things considered; in most cases. **whole-bound,** *a.* bound entirely in leather,

opp. to half- or quarter-bound. **whole-coloured**, *a.* having the same colour throughout. **wholefood**, *n.* food that has undergone little or no processing or refining. **wholehearted**, *a.* done or intended with all one's heart, hearty, generous, cordial, sincere. **wholeheartedly**, *adv.* **wholeheartedness**, *n.* **whole-hogger**, *n.* a thorough-paced supporter. **whole-hoofed**, *a.* having undivided hoofs. **whole-length**, *a.* of a portrait etc., exhibiting the whole figure. **wholemeal**, *n., a.* (made from) flour ground from the entire wheat grain. **whole number**, *n.* an integer. **wholeness**, *n.*
wholesale (hōl'sāl), *n.* the sale of goods in large quantities as dist. from retail. *a.* buying or selling thus; done etc. in the mass, on the large scale, indiscriminate. *adv.* by wholesale, in large quantities; by the mass, on the large scale. **wholesaler**, *n.*
wholesome (hōl'səm), *a.* tending to promote health, salutary, salubrious; promoting moral or mental health, not morbid. **wholesomely**, *adv.* **wholesomeness**, *n.*
who'll (hool), contr. form of *who will, who shall.*
wholly (hō'li), *adv.* entirely, completely; totally, exclusively.
whom, whomsoever etc. WHO.
whoop (woop), *v.i.* to utter the cry 'whoop'; to shout or cry out loudly in excitement, encouragement, exultation etc.; to halloo. *v.t.* to utter with a whoop; to urge (on) with whoops; to mock at with loud cries. *n.* the cry 'whoop'; a loud shout of excitement, encouragement etc.; the sound made in whooping-cough. **to whoop it up, to make whoopee**, to celebrate noisily or riotously. **whoopee** (-ē'), *int.* an exclamation of excitement or delight. *n.* riotous enjoyment; a noisy, jolly time. **whoopee cushion**, *n.* a cushion that when sat upon emits a sound as of someone breaking wind. **whooper**, *n.* **whooper swan**, *n.* a large swan with a whooping call. **whooping-cough** (hoo'-), *n.* an infectious disease, pertussis, esp. of children, characterized by a violent cough followed by a loud convulsive respiration. **whoops**, *int.* an exclamation of surprise or apology.
whoosh (wush), *n.* a rushing or hissing sound as of something moving swiftly through the air. *v.i.* to make or move with such a sound.
whop (wop), *v.t.* (*past, p.p.* **whopped**) (*coll.*) to beat, to thrash; to defeat; to cause to drop with a loud noise. *v.i.* to fall with a loud noise. **whopper**, *n.* (*coll.*) anything uncommonly large etc.; a monstrous lie. **whopping**, *a.* (*coll.*) uncommonly large.
whore (haw), *n.* a prostitute, a courtesan. *v.i.* to fornicate; (*Bibl.*) to practise idolatry. **whoredom** (-dəm), *n.* fornication; idolatry. **whorehouse**, *n.* a brothel. **whoremaster**, *n.* a pimp; a whoremonger. **whoremonger**, *n.* a fornicator. **whoreson** (-sən), *n.* (*old-fashioned*) a bastard. *a.* bastardlike, mean, scurvy. **whoring**, *n.* **whorish**, *a.* **whorishly**, *adv.* **whorishness**, *n.*
whorl (wœl, wawl), *n.* a circular set or ring of leaves, sepals or other organs on a plant; one convolution or turn of a spiral, as in a univalve shell; a coil, spiral, convolution. **whorled**, *a.*
whortleberry (wœ'tlberi), *n.* the bilberry.
who's (hooz), contr. form of *who is, who has.*
whose, whoso, whosoever WHO.
why (wī), *adv., conj.* (*interrog.*) for what reason or purpose?; (*rel.*) on account of which. *n.* the reason, explanation or purpose of anything. *int.* expressing surprise etc.
whydah WHIDAH.
WI, (*abbr.*) West Indies; Women's Institute.
wick[1] (wik), *n.* a piece or bundle of fibrous or spongy material used in a candle or lamp to convey the melted grease or oil by capillary attraction to the flame. **to get on someone's wick**, (*coll.*) to annoy or irritate someone.
wick[2] (wik), *v.t.* to strike (a stone) obliquely in the game of curling. *n.* such a hit.
wick[3] (wik), *n.* a town, village or municipal district (chiefly in place-names).
wicked (wik'id), *a.* sinful, addicted to evil or vice, wilfully transgressing against the divine or moral law, immoral, depraved; mischievous, roguish; very bad, harmful, injurious; (*sl.*) very good, excellent. **wickedly**, *adv.* **wickedness**, *n.*
wicken (wik'ən), *n.* the rowan or mountain ash.
wicker (wik'ə), *n.* flexible twigs, esp. of willow, plaited into a material for baskets, chairs etc. *a.* made of this material. **wickerwork**, *n.* **wickered**, *a.*
wicket (wik'it), *n.* a small gate, door or other entrance, esp. one close beside or forming part of a larger one; a small aperture in a door or wall, having a grille, or opened and closed by means of a sliding panel; a set of three stumps surmounted by two bails at which the bowler directs the ball in cricket; the ground on which this is set up; the innings or turn of each batsman at the wicket; the pitch between the wickets, esp. as regards condition for bowling; (*coll.*) situation, circumstances. **to keep wicket**, in cricket, to be wicket-keeper. **wicket-door, -gate**, *n.* **wicket-keeper**, *n.* the fielder who stands behind the batsman's wicket in cricket.
widdershins (wid'əshinz), WITHERSHINS.
wide (wid), *a.* having a great relative extent from side to side, broad, opp. to narrow; having a specified degree of breadth; far-extending; vast, spacious, extensive; not limited or restricted, large, free, liberal, comprehensive; distant or deviating by a considerable extent or amount from a mark, point, purpose etc.; fully open or expanded; (*sl.*) crafty, shrewd. *adv.* widely; to a great distance, extensively; far from the mark or purpose. *n.* in cricket, a wide ball, one bowled too far to the side and out of the batsman's reach. **broke to the wide**, (*sl.*) absolutely penniless. **wide-angle lens**, *n.* a camera lens with an angle up to 100° used for photographing buildings etc. **wideawake**, *a.* having one's eyes open; alert, wary; keen, sharp, knowing. *n.* a soft felt hat with a broad brim. **wideawakeness**, *n.* **wide-bodied**, *a.* of an aircraft having a wide fuselage. **wide boy**, *n.* a crafty, shrewd fellow, inclined to sharp practice. **wide-eyed**, *a.* surprised, astonished; naive. **wide open**, *a.* open to attack; of indeterminate or unpredictable outcome; (*N Am.*) lawless, disorderly. **widespread**, *a.* widely disseminated. **widely**, *adv.* **widen**, *v.t., v.i.* **widener**, *n.* **wideness**, *n.* **widish**, *a.*
-wide (-wīd), *comb. form.* extending throughout, as *nationwide*.
widgeon, wigeon (wij'ən), *n.* a wild duck living in marshes.
widget (wij'it), *n.* a gadget; a thingumajig, a whatsit.
widow (wid'ō), *n.* a woman who has lost her husband by death and remains unmarried; a woman whose husband devotes much time to a (sporting) activity that takes him away from home; a short final line of a paragraph etc. at the top of a printed column or page. *v.t.* to bereave of a husband; to make a widow or widower; to deprive (of). **grass-widow** GRASS. **widow-hunter**, *n.* one who courts a widow for her fortune. **widow's curse**, *n.* an unfailing source of supply (I Kings xvii.16). **widow's mite**, *n.* a small but ill-afforded contribution (Mark xii.42). **widow's peak**, *n.* the natural growth of hair to a point in the middle of the forehead. **widow's weeds**, *n.pl.* deep mourning worn with a flowing veil of black crepe. **widower**, *n.* a man who has lost his wife by death and remains unmarried. **widowhood** (-hud), *n.*
width (width), *n.* extent of a thing from side to side, breadth, wideness; a piece of material cut from the full width of a roll etc.; comprehensiveness of mind, liberality, catholicity. **widthways, -wise**, *adv.* in a crosswise direction.
wield (wēld), *v.t.* to have the management or control of; to

wiener · · · wilt

sway; to handle, to use or employ. **to wield the sceptre**, to rule with supreme command. **wieldable**, *a*. **wielder**, *n*.
wiener (wē'nə), *n*. (*N Am.*) a type of frankfurter. **Wiener schnitzel** (shnit'sl), *n*. a cutlet of veal or pork, coated with a breadcrumb mixture. [G]
wife (wīf), *n*. (*pl*. **wives** (wīvz)) a married woman, esp. in relation to her husband; (*dial. or in comb.*) a woman. **old wives' tale**, a legend, a foolish story. **wife-swapping**, *n*. the temporary exchange of spouses for sex. **wifehood** (-hud), *n*. **wifeless** (-lis), *a*. **wifelike, wifely**, *a*. **wifie** (-i), *n*.
wig[1] (wig), *n*. a covering for the head composed of false hair, worn to conceal baldness, as a disguise, for ornament or as part of an official costume, esp. by judges, lawyers, servants in livery etc. **wigmaker**, *n*. **wigged**, *a*. **wigless** (-lis), *a*.
wig[2] (wig), *v.t.* (*past, p.p.* **wigged**) to rate, to reprimand, to scold. **wigging**, *n*. a scolding.
wigan (wig'ən), *n*. an open canvas-like fabric used for stiffening.
wigeon WIDGEON.
wiggle (wig'l), *v.t., v.i.* to move jerkily, esp. from side to side. *n*. an act of wiggling.
wight (wīt), *n*. a person.
wigwag (wig'wag), *v.t.* to wag to and fro. *v.i.* to move to and fro, to wag; to signal by waving flags.
wigwam (wig'wam), *n*. a N American Indian hut or cabin, usu. consisting of a framework covered with bark, matting, hides etc.
wilco (wil'kō), *int.* used in radio communications etc. to indicate that a message received will be complied with.
wild (wīld), *a*. living in a state of nature, esp. inhabiting or growing in the forest or open country; esp. of animals and plants, not tamed, domesticated or cultivated; not civilized, savage; unsettled, uncultivated, irregular, uninhabited; wayward, loose or disorderly in conduct, lawless, reckless, incautious, rash; ill-considered, imprudent, extravagant, inordinate; ungoverned, unrestrained; turbulent, stormy, furious; anxiously eager, passionate, mad (with etc.); excited, enthusiastic (about etc.); of horses etc., shy, easily startled, given to shying; (*Bot.*) growing in a state of nature; of a playing card, able to represent any card the holder chooses. *n*. a desert or uninhabited and uncultivated tract. *adv.* in a wild manner. **to run wild**, to be unconfined or unrestrained. **wild and woolly**, (*coll.*) rough-mannered; fantastic, ill-thoughtout. **wild-boar**, *n*. **wild-born**, *a*. born in a wild state. **wildcat**, *n*. an undomesticated species of cat native to Europe; a quick-tempered, fierce person; an exploratory drilling for oil or natural gas. **wildcat scheme**, *n*. a rash and risky speculation or other scheme. **wildcat strike**, *n*. a strike not approved by the relevant union, or undertaken in breach of a contract. **wild-duck** DUCK[2]. **wild-fire**, Greek fire; will-o'-the-wisp. **like wild-fire**, very quickly. **wildfowl**, *n*. (*collect.*) birds of various species pursued as game, esp. waterfowl. **wild-fowling**, *n*. **wild-goose chase**, a foolish or hopeless enterprise. **wildlife**, *n*. wild animals. **wild oats**, *n.pl.* youthful excesses, esp. sexual ones. **wild rice**, *n*. the edible black seeds of a N American marsh grass. **Wild West**, *n*. the N American West during the lawless period of its early settlement. **wild-wood**, *n*. a tract of natural wood or forest. *a*. consisting of or pertaining to this. **wilding**, *n*. a plant that springs up by natural agency, esp. a wild fruit-tree; the fruit of such a plant. **wildish**, *a*. **wildly** *adv.* **wildness**, *n*.
wildebeest (wil'dibēst), *n*. a gnu.
wilder (wil'də), *v.t.* (*old-fashioned*) to bewilder.
wilderness (wil'dənis), *n*. an uninhabited or uncultivated land, a desert; a waste, a scene of disorder or confusion; a portion of a garden left to run wild; a confused mass or quantity (of). **in the wilderness**, out of office; not wielding power.
wildgrave (wīld'grāv), *n*. a German title of nobility; orig.

the head keeper of a forest.
wilding, wildish etc. WILD.
wile (wīl), *n*. a trick, an artifice, a stratagem or deception. *v.t.* to entice, to cajole (into, away etc.).
wilful (wil'fəl), *a*. intentional, voluntary, deliberate; done of one's own free will, without compulsion, not accidental; due to malice or evil intent; obstinate, self-willed, headstrong, perverse. **wilfully**, *adv.* **wilfulness**, *n*.
wilga (wil'gə), *n*. (*Austral.*) the dogwood tree.
wilily, wiliness WILY.
will[1] (wil), *v.t.* (*past, cond.* **would** (wud), *coll. neg.* **won't, wouldn't** (wŏnt, wud'nt)), *v.t.* to desire, to wish, to choose, to want (a thing, that etc.); to be induced, to consent, to agree (to etc.); to be in the habit or accustomed (to); to be able (to). *v.aux* (*in second and third pers., or in first pers. in reported statement*) to be about or going to (expressing simple futurity or conditional action); (*in first pers.*) to intend, desire or have a mind to; to be certain or probable as a natural consequence, must. **willer**, *n*. **willing**, *a*. inclined, ready, not averse or reluctant (to); cheerfully acting, done, given etc. **to show willing**, to indicate a readiness to help, comply etc. **willingly**, *adv.* **willingness**, *n*. **would-be**, *pref.* desirous, vainly aspiring to be.
will[2] (wil), *n*. the mental power or faculty by which one initiates or controls one's activities, opp. to external causation and to impulse or instinct; the exercise of this power, an act of willing, a choice of volition, an intention, a fixed or authoritative purpose; determination, energy of character, power of carrying out one's specific intentions or dominating others; that which is willed, resolved or determined upon; arbitrary disposal, discretion or sufferance; inclination or disposition towards others; the legal declaration of one's intentions as to the disposal of one's property (esp. freehold or landed) after one's death, embodied in a written, witnessed instrument. **at will**, at one's pleasure or discretion. **with a will**, heartily, zealously. **willed**, *a*. (*usu. in comb.* as *strong-willed*). **willless** (-lis), *a*.
will[3] (wil), *v.t.* to intend or bring about by the exercise of one's will, to resolve, to determine; to direct, control or cause to act in a specified way by the exercise of one's will-power; to bequeath or devise by will. *v.i.* to exercise will-power. **will-power**, *n*. control exercised deliberately over impulse or inclinations.
willet (wil'it), *n*. a N American sandpiper allied to the snipe.
willie (wil'i), *n*. (*coll.*) a childish or facetious word for PENIS.
willies (wil'iz), *n.pl.* (*coll.*) nervousness, apprehensiveness.
willing etc. WILL[1].
will-o'-the-wisp (wiladhawisp'), *n*. an ignis fatuus; an illusory hope, goal etc.
willow[1] (wil'ō), *n*. a tree or shrub usu. growing near water, characterized by long, slender, pliant branches, largely yielding osiers and timber used for cricket-bats etc.; hence, a cricket-bat. **willow-herb**, *n*. a plant of the genus *Epilobium*, esp. the rose-bay. **willow-pattern**, *n*. a decorative pattern of Chinese style in blue on a white ground for china, introduced in 1780. **willow-warbler, -wren**, *n*. the chiff-chaff. **willowed, willowy**, *a*. abounding with willows; lithe, slender or graceful, like a willow.
willow[2], **willy**[1] (wil'ō, -i), *n*. a machine for the preliminary process of beating, picking and cleaning wool. *v.t.* to treat (wool) thus.
willy[2] WILLIE.
willy-nilly (wilinil'i), *adv.* willingly or unwillingly; randomly, haphazardly. *a*. happening whether it is desired or not; random, haphazard.
willy-willy (wiliwil'i), *n*. (*Austral.*) the tropical cyclone that sweeps over NW Australia in the late summer.
wilt (wilt), *v.i.* to wither, to droop; to lose freshness or

Wilton 948 **wind**

vigour. *v.t.* to cause to wilt. *n.* the act of wilting or the state of being wilted; a plant disease characterized by wilting.
Wilton (wil'tən), *n.* a carpet with the loops cut open into an elastic velvet pile, orig. manufactured at Wilton, in Wiltshire.
Wiltshire (wilt'shə), *n.* a breed of pigs; a kind of mild-cured bacon; a kind of cheese. [Eng. county]
wily (wī'li), *a.* using or full of wiles; cunning, crafty. **wilily**, *adv.* **wiliness**, *n.*
wimble (wim'bl), *n.* a boring-instrument, a gimlet, brace-and-bit etc. *v.t.* to bore with this.
WIMP, (*abbr.*) (*Comput.*) windows, icons, mouse, pull-down menus.
wimp (wimp), *n.* (*coll.*) a feeble, ineffectual person.
wimple (wim'pl), *n.* a covering of silk, linen etc., worn over the head, neck and sides of the face formerly by women and still by some nuns. *v.t.* to cover with a wimple.
Wimpy[®] (wim'pi), *n.* a type of hamburger inside a bread roll.
Wimshurst machine (wimz'hœst), *n.* a friction machine by which static electricity can be generated and stored. [British engineer James *Wimshurst*, 1832–1903]
win (win), *v.t.* (*past, p.p.* **won**[1] (wŭn)) to gain, obtain, achieve or attain by fighting, struggling or superiority in a contest, competition, wager etc.; to gain by toil etc., to earn; to be victorious in; to reach; to persuade, to secure the support, favour or assent of; to get or extract (ore etc.) by mining, smelting etc.; (*sl.*) to steal. *v.i.* to be successful or victorious in a fight, contest, wager etc.; to make one's way by struggle or effort (to etc.); to produce an attractive effect (upon). *n.* a success, a victory. **to win one's spurs** SPUR. **to win out**, to be successful, to prevail. **you can't win**, *int.* used to express resignation when one has failed, been defeated etc. **winnable**, *a.* **winner**, *n.* a person or thing that wins; a person or thing that is bound to succeed. **winning**, *a.* that wins; attractive, charming. *n.pl.* the amount won at racing, in a game of cards etc. **winning hazard** HAZARD. **winning post**, *n.* a post marking the end of a race. **winningly**, *adv.*
wince (wins), *v.i.* to shrink, start back, recoil or flinch, as from pain, trouble or a blow. *n.* the act of wincing.
wincer, *n.*
wincey (win'si), *n.* a cotton cloth with wool filling. **winceyette** (-et'), *n.* a light-weight cotton cloth raised on both sides.
winch (winch), *n.* a windlass, a hoisting-machine; a crank or handle for turning an axle etc. *v.t.* to raise using a winch.
wind[1] (wind, *poet.* wīnd), *n.* air in motion, a natural air-current, a breeze, a gale; air set in motion artificially; air used or stored for use in a musical instrument, machine etc.; (*collect.*) (those playing the) wind-instruments in an orchestra etc.; breath as acquired by the body in exertion; power of breathing in exertion, lung power; meaningless talk or rhetoric; the gas produced in the stomach during digestion etc., flatulence; scent or odour carried on the wind; hence, a hint, suggestion or indication (of); the windward position (of). *v.t.* (*past, p.p.* **winded**) to perceive the presence of by scent; to cause to be out of breath; to bring the wind up from the stomach of (a baby), e.g. by patting his/her back; to expose to the wind, to ventilate; (wind) to sound (a horn) by blowing. **how the wind blows**, the position or state of affairs. **in the wind's eye**, towards the precise point from which the wind blows. **it's an ill wind**, few situations are so bad that nobody at all benefits from them. **to get the wind up**, to become nervous, frightened. **to get wind of**, to find out about. **the four winds**, the four cardinal points. **to break wind**, to discharge wind from the anus. **to raise the wind**, to pro-cure the necessary amount of cash. **to sail close to the wind**, to take risks. **to take the wind out of someone's sails**, to frustrate someone's plans, to disconcert. **windbag**, *n.* a bag inflated with wind; a person of mere words, a long-winded speaker. **wind band**, *n.* a musical ensemble composed of wind instruments. **wind-blown**, *a.* blown by the wind, said esp. of trees deformed by a prevailing wind. **wind-borne**, *a.* **wind-bound**, *a.* prevented from sailing by contrary winds. **windbreak**, *n.* a wall, line of trees etc. serving to break the force of the wind. **windburn**, *n.* skin irritation caused by the wind. **wind-cheater**, *n.* a close-textured garment to keep out the wind; an anorak. **wind-chest**, *n.* the box or reservoir for compressed air in an organ. **windchill**, *n.* a measure of the combined chilling effect of low-temperature and wind, as in *windchill factor*. **wind-colic**, *n.* pain in the abdomen caused by flatulence. **wind cone, sock**, *n.* an open-ended fabric sleeve flying from a mast, serving as an indicator of the strength and direction of the wind. **wind-egg**, *n.* an imperfect or shell-less egg. **windfall**, *n.* something blown down by the wind; unexpected good fortune. **windfallen**, *a.* blown down by the wind. **wind-fanner**, *n.* a windhover. **wind farm**, *n.* a groupp of windmills for generating electricity. **wind-flower**, *n.* the wood-anemone. **wind-gall**, *n.* a soft tumour on the fetlock joint of a horse. **wind-gauge**, *n.* an anemometer; an instrument for showing the pressure in the wind-chest of an organ; a contrivance attached to the sight of a gun to show the allowance necessary for deflection due to the wind. **windhover**, *n.* the kestrel. **wind-instrument**, *n.* a musical instrument in which the tones are produced by air from a bellows or the mouth. **wind-jammer**, *n.* a merchant sailing-ship. **wind machine**, *n.* a machine used for producing an air-stream. **windmill**, *n.* a mill driven by the action of the wind on sails; a similar device for generating electricity, pumping water etc. (*pl.*) imaginary adversaries, chimeras (with alln. to Don Quixote). *v.t., v.i.* to (cause to) move like a windmill. **wind-pipe**, *n.* the breathing passage, the trachea. **wind power**, *n.* electrical power produced by harnessing wind energy, e.g. by a windmill. **wind-pump**, *n.* a pump operated by the force of the wind on a propeller. **windrose**, *n.* a diagram with radiating lines indicating the velocity and direction of winds affecting a place. **windrow** (-rō), *n.* a row of hay raked together, corn-sheaves, peats etc. set up for drying. **windscreen**, *n.* a glass screen forming the front window of a car. **windscreen wiper**, *n.* a rubber-edged blade which sweeps across a windscreen to wipe off water. **windshaken**, *a.* **windshield**, *n.* (*N Am.*) a windscreen. **wind sock** WIND CONE. **wind-surfing**, *n.* the sport of sailing standing upright on a surfboard fitted with a sail. **wind-swept**, *a.* exposed to the wind. **wind-tight**, *a.* airtight, excluding the wind. **wind-tunnel**, *n.* a tunnel-like device for producing an airstream of known velocity for testing the effect of wind on the structure of model vehicles, aircraft etc. **windage** (-ij), *n.* the difference between the diameter of the bore of a muzzle-loading rifled gun and that of the projectile; the influence of wind deflecting a projectile; allowance for this. **winded**, *a.* **windless** (-lis), *a.* **windward** (-wəd), *n.* the direction from which the wind blows. *a.* lying in or directed towards this. *adv.* in the direction from which the wind blows. **to get to the windward of**, to get to this side of; to get the advantage over. **windy**, *a.* characterized by wind, stormy, boisterous; exposed to the wind; flatulent, caused by flatulence; verbose, loquacious, empty; (*coll.*) frightened. **windily**, *adv.* **windiness**, *n.*
wind[2] (wīnd), *v.i.* (*past, p.p.* **wound** (wownd)) to turn, move, go or be twisted or coiled in a spiral, curved or tortuous course or shape; to be circular, spiral, tortuous or crooked; to meander; to proceed circuitously, to twist one's way or insinuate oneself (into etc.); to be wrapped spirally (round, into etc.). *v.t.* to cause to turn spirally, to

windage wrap, twine or coil; to encircle, to coil round, to entwine; to pursue (one's course) in a spiral, sinuous or circuitous way; to hoist or move by means of a windlass, capstan etc. **to wind down**, to reduce gradually; to relax. **winding-down**, *n.* **to wind off**, to unwind; to stop talking. **to wind up**, to coil up; to coil or tighten up the spring of (a watch etc.); to put into a state of tension or readiness for activity; (*coll.*) to irritate, to annoy; (*coll.*) to find oneself in a certain state or situation; to bring or come to a conclusion; to arrange the final settlement of the affairs of (a business etc.); to go into liquidation. **winding-up**, *n.* **winder**, *n.* **winding**, *n.*, *a.* (material) wound or coiled round something, e.g. wire in an electric motor. **winding-engine**, *n.* a hoisting engine. **winding-sheet**, *n.* the sheet in which a corpse is wrapped. **winding-stair**, *n.* a stair built around a newel. **winding-tackle**, *n.* **windingly**, *adv.*
windage, windhover etc. WIND¹.
windlass (wind'ləs), *n.* a machine consisting of a cylinder on an axle turned by a crank, used for hoisting or hauling. *v.t.* to hoist or haul with this.
windless, windmill WIND¹.
windlestraw (win'dlstraw), *n.* the old stalks of various grasses.
window (win'dō), *n.* an opening in the wall or roof of a building, vehicle or other structure, usu. with the wooden or metal glazed framework filling it, for the admission of light or of light and air; the sash of a window-frame; a brief period of time when the conditions allow a particular activity; (*Comput.*) a rectangular area on a VDU where information can be displayed. **window-bar**, *n.* the bar of a sash or window-frame. **window-blind**, *n.* **window-box**, *n.* a flower-box for a window-sill. **window-curtain**, *n.* **window-dressing**, *n.* the arrangement of goods for display in a shop window; deceptive display, insincere argument. **window-envelope**, *n.* one with an open or transparent panel through which the address can be seen. **window-frame**, *n.* the framework in a window holding the sashes. **window-glass**, *n.* **window-sash**, *n.* a frame in which panes of glass for windows are set. **window-seat**, *n.* a seat in the recess of a window. **window-shop**, *v.i.* to gaze idly at the displays in shop-windows. **window-shopper**, *n.* **window shopping**, *n.* **window-sill**, *n.* **windowed**, *a.* (*usu. in comb.* as *many-windowed*). **windowless** (-lis), *a.*
windpipe, windrow WIND¹.
Windsor (win'zə), *n.* a town in Berks, England. **brown Windsor**, Windsor soup; a common, tasteless soup. **Windsor chair**, *n.* a strong, plain wooden chair with a back curved into supports for the arms. **Windsor soap**, *n.* a brown scented soap formerly made at Windsor.
windward, windy etc. WIND¹.
wine (win), *n.* the fermented juice of grapes; the juice of certain fruits etc. prepared in imitation of this; intoxication; a dark-red colour. **spirit of wine** SPIRIT. **to wine and dine**, to entertain with food and alcohol. **winebag**, *n.* a wineskin; a wine-bibber. **wine bar**, *n.* a bar that serves mostly wine, esp. with food. **wine-bibber**, *n.* a wine-drinker, a tippler. **wine-bibbing**, *n.* **wine-bottle**, *n.* **wine-bowl**, *n.* **wine-box**, *n.* a cardboard box with a plastic lining, fitted with a tap for dispensing wine. **wine-carriage**, *n.* a wheeled receptacle for circulating a wine-bottle at table. **wine-cask**, *n.* **wine-cellar**, *n.* **wine-cooler**, *n.* a vessel for cooling wine in bottles with ice. **wine-cup**, *n.* **wine-glass**, *n.* a small glass for drinking wine from. **wineglassful**, *n.* about 2 fl oz (6 cl). **winegrower**, *n.* **wine-merchant**, *n.* **wine-palm**, *n.* a palm tree from which palm-wine is obtained. **winepress**, *n.* an apparatus in which grapes are pressed; the place in which this is done. **wineskin**, *n.* a skin, usu. of a goat, sewn into a bag for holding wine. **wine-stone**, *n.* a deposit of crude tartar or argal in wine-casks. **wine-tasting**, *n.* an occasion when people can sample various wines. **wine-vault**, *n.* a vault in which wine is stored; a bar or tap-room where wine is retailed. **wineless**, *a.* **winery**, *n.* an establishment where wine is made. **winy**, *a.*
wing (wing), *n.* one of the limbs or organs of flight in birds, insects etc.; one of the supporting parts of a flying-machine; motion by means of wings, flight, power of flight; (*coll.*) an arm; a part of a building, fortification, army, bone, implement etc. projecting laterally; an RAF unit of three squadrons; in football and similar games, a player on one or other extreme flank; the position in which such a player plays; one of the extreme factions of a party, group etc.; one of the front-wheel mudguards of a car; (*pl.*) the sides of a stage or pieces of scenery placed there; two lateral petals of a papilionaceous flower which stand opposite each other; (*pl.*) the mark of proficiency a pilot qualified in the RAF is entitled to wear on his uniform. *v.t.* to furnish with wings; to enable to fly or move with swiftness; to traverse or travel on wings; to wound in the wing or (*coll.*) the arm. *v.i.* to fly. **in the wings**, waiting in readiness. **on the wing**, flying; in motion. **to take under one's wing**, to take under one's protection. **to take wing**, to begin flying, to fly away; to disappear. **wingbeat**, *n.* a complete stroke of the wing in flying. **wing-case**, *n.* the horny cover or case, consisting of a modified wing, protecting the flying wings of coleopterous insects. **wing-chair**, *n.* an armchair having a high back with side-pieces stretching forward. **wing-collar**, *n.* a stiff upright shirt-collar with the points turned down. **wing-commander**, *n.* a commissioned officer in the RAF equivalent to a lieutenant-colonel. **wing-covert**, *n.* one of the small feathers covering the insertion of a bird's flight-feathers. **wing-footed**, *a.* as if having wings on the feet; swift. **wing nut**, *n.* a nut that is tightened by two flat wings on its sides. **wing-sheath**, *n.* a wing-case. **wing-span**, *n.* the distance from one wing-tip of a bird, aircraft etc. to the other. **wing-stroke**, *n.* a wing-beat. **wing-tip**, *n.* **winged**, *a.* furnished with wings; (*poet.*, wing'id) going straight to the mark, powerful (of words etc.). **winger**, *n.* a football-player etc. positioned on the wing. **wingless** (-lis), *a.* **winglet** (-lit), *n.*
winge WHINGE.
wink (wingk), *v.i.* to close and open the eyes quickly, to blink; of an eye, to close and open; to give a sign or signal by such a motion of the eye; to twinkle, to flicker. *v.t.* to close and open (an eye or the eyes). *n.* the act of winking, esp. as a signal; a hint, a private intimation; a moment, an instant. **forty winks**, (*coll.*) a nap. **to tip the wink**, to give a hint privately. **to wink at**, to affect not to see; to connive at. **winker**, *n.* **winking**, *n.* **like winking**, very rapidly; with great vigour. **winkingly**, *adv.*
winkle (wing'kl), *n.* an edible sea-snail, a periwinkle. **to winkle out**, (*coll.*) to extract with difficulty; to elicit (information etc.) with difficulty. **winkle-pickers**, *n.pl.* shoes with pointed toes.
winning etc. WIN.
winnow (win'ō), *v.t.* to separate and drive the chaff from (grain); to fan chaff (away, out etc.); to sift, to sort, to examine or analyse thoroughly; to blow on, to stir (hair etc.). **winnower**, *n.* **winnowing**, *n.*
wino (wī'nō), *n.* (*pl.* **winos**) (*coll.*) an alcoholic, esp. one who drinks mainly wine.
winsey (win'si), WINCEY.
winsome (win'səm), *a.* engaging, winning, charming, attractive; graceful, lovely. **winsomely**, *adv.* **winsomeness**, *n.*
winter (win'tə), *n.* the cold season of the year, astronomically in northern latitudes from the December solstice to the March equinox. usu. regarded as including December, January, February; a period of inactivity, a cheerless or depressing state of things; (*poet.*) a year of life. *a.* pertaining, suitable to or lasting for the winter. *v.i.* to pass the winter; to hibernate. *v.t.* to keep, manage

or maintain through the winter. **winter-apple**, *n*. an apple that keeps well or ripens in winter. **winter-barley, -wheat**, *n*. varieties of cereal sown in autumn. **winterberry**, *n*. a N American shrub of the genus *Ilex*, bearing bright red berry-like drupes. **winter-cress**, *n*. a herb of the mustard family grown in the winter as a salad. **winter-crop**, *n*. **winter-garden**, *n*. a large conservatory or glass-house for plants not hardy enough to withstand the climate outside during winter. **wintergreen**, *n*. any of several herbs keeping green throughout the winter. **winter-quarters**, *n.pl.* the quarters occupied by an army etc. during the winter. **winter sport**, *n*. sport practised on snow and ice, usu. outdoors, e.g. skiing, skating. **winterless** (-lis), *a*. wintry (-tri), *a*. of or like winter, e.g. of a smile, look etc., cold and cheerless. **wintriness**, *n*.
winy WINE.
winze (winz), *n*. (*Mining*) a shaft sunk from one level to another for communication or ventilation.
wipe (wīp), *v.t.* to rub with something soft in order to clean or dry; to apply solder to with something soft; to clear (a magnetic tape or videotape) of recorded material; to apply (grease etc.) by wiping. *v.i.* to strike (at). *n*. the act of wiping; a sweeping blow. **to wipe away**, to remove by wiping; to get rid of. **to wipe one's eyes**, to cease weeping. **to wipe out**, to clean out by wiping; to efface, to obliterate; to destroy, to annihilate. **to wipe the floor with**, (*sl.*) to defeat utterly. **wipe-out**, *n*. an act or instance of wiping out; interference that renders impossible the reception of other signals. **wipe-out area**, *n*. the vicinity of a transmitting station where wipe-out occurs. **wiper**, *n*. cloth etc. used for wiping; a windscreen wiper.
wire (wiə), *n*. metal drawn out into a slender and flexible rod or thread of uniform diameter; such a slender rod, thread or strand of metal; the electric telegraph, a telegram; a wire barrier or fence. *v.t.* to apply wire to, to fasten, secure, bind or stiffen with wire; to string (beads) on a wire; to snare with wire; (*coll.*) to telegraph to. *v.i.* to send a telegram. **to pull wires**, to control politics etc. by clandestine means. **to wire in**, (*sl.*) to apply oneself vigorously. **wire brush**, *n*. a brush with wire bristles used, e.g. for scraping rust off metal. **wire-cloth, -gauze, -netting**, *n*. a fabric of woven wire. **wire-cutter**, *n*. an implement for cutting wire. **wire-dancer**, *n*. an acrobat performing on a tight wire. **wiredraw**, *v.t.* (*p.p.* **-drawn**) to form (metal) into wire by forcibly drawing through a series of gradually diminishing holes; to overstrain or overrefine (an argument etc.). **wiredrawer**, *n*. **wire-edge**, *n*. an edge turned back like wire, on a knife etc., by oversharpening. **wire gauge**, *n*. an instrument for measuring the diameter of wire. **wire-gun**, *n*. a heavy gun constructed of steel wire of rectangular section coiled round a tube. **wire-haired**, *a*. having stiff, wiry hair (esp. of terriers). **wire-heel**, *n*. a disease of the foot in horses. **wire-netting**, *n*. wire woven into a net, used for fencing. **wire-puller**, *n*. **wire-pulling**, *n*. **wire-rope**, *n*. **wire service**, *n*. (*N Am.*) a news-agency. **wire-tap**, *v.t.* to tap (a telephone). **wire-wool**, *n*. abrasive material consisting of a mass of very fine wires, used for cleaning etc. **wireworm**, *n*. a vermiform larva of a click-beetle, destructive to roots of vegetables, cereals etc. **wire-like**, *a*. **wirer**, *n*. **wiring**, *n*. a system of wires, esp. one carrying electric current. **wiry**, *a*. made of or resembling wire; tough and flexible; lean but sinewy; stiff (of hair etc.). **wirily**, *adv*. **wiriness**, *n*.
wireless (wiə'lis), *n*. wireless telegraphy, radio; any process or method whereby sound can be transmitted by electromagnetic waves without the intervention of wires; an instrument for receiving such messages etc.; the programmes of entertainment etc. thus transmitted. *v.t.*, *v.i.* (*coll.*) to communicate with or inform by this.
wis (wis), *v.i. first sing.* (*old-fashioned*) (I) know.
wisdom (wiz'dəm), *n*. the quality or state of being wise; knowledge and experience together with ability to make use of them rightly, practical discernment, sagacity, judgment, common sense. **wisdom-tooth**, *n*. the third molar appearing about the age of 20.
wise[1] (wīz), *a*. having or characterized by the power or faculty of discerning or judging rightly, or by knowledge and experience together with ability to apply them rightly, sagacious, sensible, discreet, prudent, judicious; experienced, understanding; informed, aware; (*N Am. coll.*) insolent, cocksure. **to wise up**, (*esp. N Am. coll.*) to (cause to) be aware or informed. **wisecrack**, *n*. (*coll.*) a smart but not profound epigram; a witty comment. **wisecracker**, *n*. **wisecracking**, *a*. **wise guy**, *n*. (*esp. N Am. coll.*) an insolent or cocksure person. **wise woman**, *n*. a witch, a fortune-teller. **wisely**, *adv*.
wise[2] (wīz), *n*. manner, way, mode of acting, behaving etc.
-wise (-wīz), *suf*. forming adverbs of manner, as *anywise, lengthwise, likewise, otherwise*; with regard to, concerning, as in *jobwise, weatherwise*.
wiseacre (wī'zākə), *n*. one pretending to learning or wisdom.
wish (wish), *v.t.* to have a strong desire, aspiration or craving (that etc.), to crave, to covet, to want; to frame or express a desire or wish concerning, to invoke, to bid. *v.i.* to have a strong desire (for); to make a wish. *n*. a desire, a longing, an aspiration; an expression of this, a request, a petition, an invocation; that which is desired. **to wish someone, something on someone**, to foist someone or something on someone. **wishbone**, **wishing-bone**, *n*. the merrythought, the longer part of which when broken by two persons is supposed to entitle the holder to the fulfilment of some wish. **wisher**, *n*. (*usu. in comb*. as *well-wisher*). **wishful**, *a*. wishful thinking, *n*. belief based on desires rather than facts. **wishfully**, *adv*. **wishfulness**, *n*. **wishing-cap**, *n*. a magic cap conferring the power of realizing one's wishes.
wishtonwish (wish'tɔnwish), *n*. the N American prairie-dog.
wishy-washy (wish'iwoshi), *a*. vague, ill-defined; lacking strength, forcefulness etc.
wisp (wisp), *n*. a small bunch or handful of straw, hay etc.; a tuft; a thin band or streak; a slim or delicate person, esp. a girl. **wispy**, *a*.
wist (wist), *past* WIT[1].
wistaria (wistə'riə), **wisteria** (-tiə'-), *n*. a leguminous climbing shrub with racemes of lilac-coloured flowers. [US anatomist Caspar *Wistar* 1761–1818]
wistful (wist'fəl), *a*. full of vague yearnings, esp. for unattainable things, sadly longing; thoughtful in a melancholy way, pensive. **wistfully**, *adv*. **wistfulness**, *n*.
wistiti, ouistiti (wis'titi), *n*. the marmoset.
wit[1] (wit), *v.t.*, *v.i.* (*old-fashioned*) (*first sing*. **wot** (wot), *second sing*. **wottest** (-ist), *past* **wist** (wist); *no other parts used*) to know (esp. in the infinitive 'to wit', namely). **witting**, *a*. **wittingly**, *adv*. consciously, knowingly, intentionally.
wit[2] (wit), *n*. (*often pl.*) intelligence, understanding, sense, sagacity; (*pl.*) sanity; the power of perceiving analogies and other relations between apparently incongruous ideas or of forming unexpected, striking or ludicrous combinations of them; a person distinguished for this power, a witty person. **at one's wits' end**, at a complete loss what further steps to take. **to have one's wits about one**, to be alert. **witless** (-lis), *a*. **witlessly**, *adv*. **witlessness**, *n*. **witted**, *a*. **witticism** (-sizm), *n*. a witty phrase or saying, a jest. **witty**, *a*. **wittily**, *adv*. **wittiness**, *n*. (*usu. in comb*. as *slow-witted*).
witan (wit'ən), WITENAGEMOT.
witch[1] (wich), *n*. a woman having dealings with evil spirits or practising magic or sorcery; a bewitching or fascinating woman; an old and ugly woman. *v.t.* to bewitch, fasci-

nate, to enchant. **witchcraft,** *n.* the practices of witches; sorcery, magic. **witch-doctor,** *n.* in some tribal societies, a man who invokes supernatural powers, esp. to cure people. **witch-finder,** *n.* (*Hist.*) one whose business was to discover witches. **witch hunt,** *n.* the searching out and public exposure of opponents accused of disloyalty to a state, political party etc. **witchery,** *n.* **witching,** *a.* **witching hour,** *n.* midnight. **witchingly,** *adv.*
witch [2], **witch-elm, witch-hazel** (wich), WYCH.
witchetty (wich'əti), *n.* (*Austral.*) the edible grub of a longicorn beetle.
witenagemot (wit'ənəgəmōt, -mōt'), *n.* the Anglo-Saxon national assembly or parliament. [OE]
with (widh), *prep.* in or into company of or the relation of accompaniment, association, simultaneousness, cooperation, harmoniousness etc.; having, possessed of, marked or characterized by; in the possession, care or guardianship of; by the means, instrumentality, use or aid of; by the addition or supply of; because of, owing to, in consequence of; in regard to, in respect of, concerning, in the case of; in separation from; in opposition to, against; in spite of, notwithstanding. **with child,** pregnant. **with it,** *a.* (*coll.*) up-to-date, fashionable; alert to what is being done or said. **with young,** (of a mammal) pregnant.
withal (widhawl'), *adv.* with the rest, in addition, at the same time, further, moreover.
withdraw (widhdraw'), *v.t.* (*past* **-drew** (-droo'), *p.p.* **-drawn** (-drawn')) to draw back, aside or apart; to take away, to remove, to retract. *v.i.* to retire from a presence or place; to go apart or aside; to retract a statement, accusation etc.; to isolate oneself socially, emotionally etc.
withdrawal (-əl), *n.* the act of withdrawing; the process of stopping the use of drugs and breaking an addiction. **withdrawer,** *n.* **withdrawing-room,** *n.* a drawing-room. **withdrawn,** *a.* very reserved, socially isolated etc.
withe (widh, with, widh), *n.* a tough, flexible branch, esp. of willow or osier, used in binding things together; a band or tie made of osiers, twigs, straw etc.
wither (widh'ə), *v.t.* to cause to fade, shrivel or dry, to shrivel and dry (up); to cause to lose freshness, soundness, vitality or vigour; (*fig.*) to blight, to blast; to make abashed. *v.i.* to become dry and wrinkled; to dry and shrivel (up); to lose freshness, soundness, vigour etc.; to fade away, to languish, to droop. **witheredness,** *n.* **withering,** *a.* **witheringly,** *adv.*
withers (widh'əz), *n.pl.* the ridge between the shoulderblades of a horse. **wither-wrung,** *a.* injured or hurt in the withers.
withershins (widh'əshinz), *adv.* anti-clockwise, to the left or opposite to the direction of the sun.
withhold (widhhōld'), *v.t.* (*past, p.p.* **-held** (-held')) to keep from action, to hold back, to refuse to grant, to refrain. **withholder,** *n.* **withholdment,** *n.*
within (widhin'), *adv.* inside, in or to the inside, in the inner part or parts, internally, indoors, in the mind, heart or spirit. *n.* the inside. *prep.* in or to the inner or interior part or parts of, inside; in the limits, range, scope or compass of; not beyond, not outside of, not farther off than; in no longer a time than.
without (widhowt'), *adv.* in, at or to the outside, outside, outwardly, externally, out of doors. *n.* the outside. *prep.* not having, not with, having no, destitute of, lacking, free from; outside of; out of the limits, compass or range of, beyond.
withstand (widhstand'), *v.t.* (*past, p.p.* **-stood** (-stud')) to stand up against, to resist, to oppose. *v.i.* (*poet.*) to make a stand or resistance (against). **withstander,** *n.*
withy (widh'i), *n.* a withe; a willow.
witless etc. WIT [2].
witness (wit'nis), *n.* attestation of a fact etc., testimony, evidence; a thing that constitutes evidence or proof, confirmation; a thing or person serving as testimony to or proof of; a spectator, a person present at an event; one who gives evidence in a law-court or for judicial purposes, esp. on oath; one who affixes his name to a document to testify to the genuineness of the signature. *v.t.* to see or know by personal presence, to be a spectator of; to attest, to sign as witness; to indicate, to show, to prove. *v.i.* to bear testimony, to testify, to give evidence; to serve as evidence (against, for etc.). **to bear witness,** to give testimony; to be a sign (of). **witness-box,** *n.* an enclosure in a law-court for witnesses (*esp. N Am.* **witness-stand**). **witnessable,** *a.* **witnesser,** *n.*
witter (wit'ə), *v.i.* to talk, chatter (on) at great length.
witticism etc. WIT [2].
witting etc. WIT [1].
witty WIT [2].
wive (wīv), *v.t.* (*old-fashioned*) to take for a wife; to provide with a wife. *v.i.* to marry a wife.
wiver, wivern (wī'vən), WYVERN.
wives WIFE.
wiz (wiz), *n., a.* (*coll.*) short for WIZARD.
wizard (wiz'əd), *n.* a sorcerer, an enchanter, a magician; one who works wonders. *a.* magic, enchanting, enchanted; (*sl.*) wonderful, marvellous. **wizardly,** *a.* **wizardry** (-ri), *n.*
wizen (wiz'n), *v.t., v.i.* to wither, to dry up, to shrivel. *a.* wizened.
wk, (*abbr.*) week.
wkly, (*abbr.*) weekly.
wkt, (*abbr.*) wicket.
WNP, (*abbr.*) Welsh National Party.
WO, (*abbr.*) Warrant Officer.
wo (wō), WHOA.
woad (wōd), *n.* a cruciferous plant yielding a blue dye; this dye formerly in use for staining the body, esp. by the ancient Britons. **woaded,** *a.*
wobbegong (wob'əgong), *n.* (*Austral.*) a mottle-skin shark also known as the carpet shark.
wobble (wob'l), *v.i.* to incline to one side and then the other alternately, as a rotating body when not properly balanced; to oscillate; to stagger; to waver, to be inconsistent or inconstant. *v.t.* to cause to wobble. *n.* a rocking, uneven motion, a stagger, a swerve; an act of hesitation, inconsistency or vacillation. **wobbler,** *n.* **wobbly,** *a.* inclined to wobble; unsteady. *n.* (*coll.*) a tantrum.
wobbles (wob'lz), *n.* (*Austral.*) a W Australian horse- and cattle-disease caused by eating poisonous palm-leaves.
wodge (woj), *n.* (*sl.*) a thick slice or chunk.
woe (wō), *n.* sorrow, affliction, distress, calamity, overwhelming grief. **woe worth the day** WORTH [2]. **woebegone** (-bigon), *a.* overcome with woe, sorrowful-looking, dismal. **woeful,** *a.* sorrowful, miserable; pitiful, inadequate. **woefully,** *adv.* **woefulness,** *n.*
wog (wog), *n.* (*offensive*) any dark-skinned person.
woggle (wog'l), *n.* a leather or plastic ring used to tie a Boy Scout's kerchief at the front.
wok (wok), *n.* a large metal bowl with curved sides and handles used in Chinese cooking.
woke, woken WAKE [1].
wold (wōld), *n.* a tract of open country, esp. downland or moorland.
wolf (wulf), *n.* (*pl.* **wolves** (wulvz)) a usu. grey carnivorous quadruped, closely allied to the dog, preying on farm animals and hunting larger animals in packs; a rapacious, ravenous, greedy or cruel person; (*coll.*) a man who is rapacious in the pursuit of women for sexual purposes; (*Mus.*) a discordant sound in certain chords of a keyboard instrument, esp. an organ, due to unequal temperament. *v.t.* to devour ravenously, to gulp or swallow (down) greedily. **to cry wolf,** to raise a false alarm. **to keep the wolf from the door,** to keep off starvation. **wolf in sheep's clothing,** a person who disguises malicious intentions behind a pre-

tence of innocence. **wolf-cub**, *n.* (*formerly*) a cub scout. **wolf-dog**, *n.* a large dog used for guarding sheep against wolves; a cross between a wolf and a dog. **wolf-fish**, *n.* a large voracious fish, also called a sea-wolf. **wolf-hound**, *n.* a large powerful dog of Russian or Irish breed. **wolf's-bane**, *n.* a species of aconite or monk's-hood. **wolf's-claw, -foot**, *n.* club-moss. **wolf's-fist**, *n.* a puff-ball. **wolf-tooth**, *n.* a small additional pre-molar in horses. **wolf-whistle**, *n.* a whistle made by a male at the sight of an attractive girl. **wolfish**, *a.* **wolfishly**, *adv.* **wolfishness**, *n.*
wolfram (wul'frəm), *n.* a native tungsten ore composed of tungstate of iron and manganese; tungsten.
wolverine (wul'vərēn), *n.* a small N American carnivorous animal also called the glutton or carcajou.
wolves *pl.* WOLF.
woman (wum'ən), *n.* (*pl.* **women** (wim'in)), an adult human female; womankind, the female sex; womanly feeling, womanliness; an effeminate or timid and tender man; (*coll.*, *often offensive*) a mistress or girlfriend; a female attendant or servant. *a.* female. *v.t.* to address or speak of as 'woman'. **to make an honest woman of** HONEST. **woman of the world**, a woman knowledgeable in the ways of the world; a society woman. **woman-born**, *a.* born of a woman. **woman-hater**, *n.* a misogynist. **womankind** (-kind), *n.* women collectively, the female sex; the women of a household. **woman suffrage**, *n.* the exercise of the electoral franchise by or its extension to women. **womanhood** (-hud), *n.* womanish, *a.* having the character or qualities of a woman, effeminate. **womanishly**, *adv.* **womanishness**, *n.* **womanize, -ise**, *v.t.* to make effeminate, to unman. *v.i.* of a man, to have casual sexual relationships with many women. **womanizer**, *n.* **womanless** (-lis), *a.* **womanlike**, *a.*, *adv.* **womanly**, *a.* having the qualities becoming a woman, truly feminine. *adv.* in the manner of a woman. **womanliness**, *n.* **women**, *n.pl.* **womenfolk**, *n.* women collectively; a group of women. **Women's Institute**, *n.* an organization of women in Britain, for mutual training and improvement in domestic and social life. **women's liberation, women's lib**, *n.* a movement (women's movement) which began in the 1960s and advocated the social, sexual and psychological emancipation of women from the dominance of men.
womb (woom), *n.* the organ in a female mammal in which the young is developed before birth, the uterus; the place where anything is engendered or brought into existence. **wombed**, *a.* having a womb; capacious. **womb-like**, *a.*
wombat (wom'bat), *n.* an Australian nocturnal marsupial mammal resembling a small bear.
women (wim'in), *pl.* WOMAN.
womerah (wom'ərə), *n.* (*Austral.*) a throwing-stick.
won[1] (wŭn), *past, p.p.* WIN.
won[2] (won), *n.* the standard monetary unit in N and S Korea.
wonder (wŭn'də), *n.* a strange, remarkable or marvellous thing, event, action, incident etc.; a miracle, a prodigy; the emotion excited by that which is unexpected, strange, extraordinary or inexplicable, or which arrests by its grandeur; surprise mingled with admiration. *v.i.* to be struck with wonder or surprise; to look with or feel wonder; to feel doubt or curiosity (about etc.). *v.t.* to speculate about. **nine days' wonder** NINE. **no (or small) wonder**, it is not surprising (that etc.). **seven wonders of the world** SEVEN. **wonderland**, *n.* a land of marvels, fairyland. **wonder-struck, -stricken**, *a.* **wonder-worker**, *n.* one who performs wonders. **wonder-working**, *a.* **wonderer**, *n.* **wonderful** *a.* astonishing, strange, admirable; exciting wonder or astonishment. **wonderfully**, *adv.* admirably; strangely; greatly. **wonderfulness**, *n.* **wondering**, *n.*, *a.* **wonderingly**, *adv.* **wonderment**, *n.* amazement, awe; curiosity. **wondrous** (-drəs), *a.* wonderful, marvellous, strange. *adv.* wonderfully, exceedingly. **wondrously**, *adv.*
wonky (wong'ki), *a.* (*sl.*) unsteady, shaky.

wont (wōnt), *a.* used, accustomed (to); using or doing habitually. *n.* custom, habit, use. *v.aux.* to be accustomed or used (to). **wonted**, *a.* customary, habitual, usual.
won't (wōnt), *contr.* form of *will not*.
woo (woo), *v.t.* to court, to solicit in marriage; to seek to gain or attain; to solicit, to coax, to importune. *v.i.* to go courting. **wooer**, *n.* **wooing**, *n.*
wood (wud), *n.* a large and thick collection of growing trees, a forest (*often pl.*); the fibrous substance of a tree between the bark and the pith; trees, timber; (*Bowls*) a bowl; a golf-club with a wooden head; (*Mus.*) the woodwind. **from the wood**, from the cask. **not to see the wood for the trees**, to be prevented by excessive details from getting an overall view. **out of the wood**, out of danger. **wood-agate**, *n.* an agate derived from wood by silicification and still showing the woody structure. **wood alcohol**, *n.* methyl alcohol, formerly produced by the distillation of wood, now synthesized. **wood-anemone**, *n.* the wild anemone. **wood-ashes**, *n.pl.* the ashes of burnt wood or plants. **woodbine**, *n.* the wild honeysuckle. **woodblock**, *n.* a die cut in wood for striking impressions from; a wood-cut. **wood-carving**, *n.* **woodchat**, *n.* a type of shrike of Europe and N Africa. **woodchuck**, *n.* a N American marmot. **wood-coal**, *n.* charcoal; lignite. **woodcock**, *n.* a game-bird related to the snipe. **woodcraft**, *n.* skill in anything pertaining to life in the woods or forest, esp. in hunting. **woodcut**, *n.* an engraving on wood; a print or impression from this. **woodcutter**, *n.* one who cuts wood or timber; an engraver on wood. **wood-engraver**, *n.* an engraver on wood; a beetle that bores under the bark of trees. **wood-engraving**, *n.* **wood-fibre**, *n.* fibre obtained from wood, used for papermaking etc. **wood-fretter**, *n.* an insect that eats into wood. **wood-gas**, *n.* illuminating gas produced by dry distillation of wood. **wood-grouse**, *n.* the capercailzie. **wood-hole**, *n.* a place where wood is stored. **wood-house**, *n.* **wood-ibis**, *n.* a variety of stork from the southern US. **woodland**, *n.* land covered with woods, wooded country. *a.* pertaining to this, sylvan. **wood-lark**, *n.* a European lark smaller than the skylark. **wood-leopard**, *n.* a moth the caterpillars of which live in the wood of fruit trees. **wood-louse**, *n.* a wingless isopod insect infesting decayed wood etc. **woodman**, *n.* a forester; one who fells timber; a wood-cutter. **wood-note**, *n.* a wild or natural note or song. **woodnymph**, *n.* a dryad; a brilliantly-coloured moth; a variety of humming-bird. **wood-opal**, *n.* silicified wood. **woodpaper**, *n.* paper made from wood-fibre. **wood-pavement**, *n.* paving composed of blocks of wood. **woodpecker**, *n.* a bird living in woods and tapping trees to discover insects. **wood-pie**, *n.* the great spotted woodpecker. **wood-pigeon**, *n.* the ringdove, a European pigeon whose neck is nearly encircled by a ring of whitish-coloured feathers. **woodpile** *n.* a pile of wood. **a nigger in the woodpile** NIGGER. **wood-pulp**, *n.* wood-fibre pulped in the process of manufacturing paper. **woodruff**, *n.* a woodland plant with fragrant flowers. **wood-screw**, *n.* a metal screw for fastening pieces of wood together. **woodshed**, *n.* a shed for storing wood, esp. firewood. **woodsman**, *n.* one who lives in the woods; a woodman. **woodsorrel**, *n.* a creeping woodland plant with acid juice and small white flowers. **wood-tar**, *n.* tar obtained from wood. **wood-vetch**, *n.* a climbing vetch. **wood-warbler**, *n.* an American warbler; a woodwren. **wood-wasp**, *n.* a wasp that makes its cells in wood or hangs its nest to the branches of trees. **woodwind** (-wind), *n.* (*Mus.*) the wooden wind instruments in an orchestra etc. **wood-wool**, *n.* fine shavings, esp. of pine, used for dressing wounds, for packing etc. **woodwork**, *n.* things made of wood; the part of a building or other structure which is composed of wood. **wood-worker**, *n.* **woodworm**, *n.* any of various insect larvae that bore into furniture, wooden beams etc. **wood-wren**, *n.* a European warbler; a wood-warbler. **wooded**, *a.* (*usu. in comb.* as *well-wooded*). **wood-**

en, *a.* made of wood; (*fig.*) stiff, clumsy, ungainly, awkward; spiritless, expressionless. **wooden-head**, *n.* (*coll.*) a stupid person. **wooden-headed**, *a.* **woodenheadedness**, *n.* **wooden spoon**, *n.* a spoon made of wood, used in cooking; a booby prize, esp. in sports competitions. **woodenly**, *adv.* **woodenness**, *n.* **woodless**, *a.* **woodlessness**, *n.* **woody**, *a.* abounding in woods, well-wooded; of the nature of or consisting of wood. **woody fibre**, **tissue**, *n.* fibre or tissue consisting of wood-cells; the tissue of which wood is composed. **woody-nightshade** NIGHTSHADE. **woodiness**, *n.*
wooer WOO.
woof[1] (woof), *n.* the threads that cross the warp, the weft; cloth; texture.
woof[2] (wuf), *n.* the sound of a dog barking or growling. *v.i.* to produce this sound. **woofer**, *n.* a large loudspeaker used for low-frequency sounds.
wool (wul), *n.* the fine, soft, crisp or curly hair, forming the fleece of sheep, goats and some other animals, used as the raw material of cloth etc.; short, thick hair, underfur or down, resembling this; woollen yarn, worsted; fibrous or fleecy substance resembling wool. **dyed in the wool**, extremely committed to specified beliefs, opinions, politics etc. **to keep one's wool on**, (*coll.*) to keep one's temper. **to pull the wool over someone's eyes**, to deceive. **wool-ball**, *n.* a ball or mass of wool, esp. a lump of concreted wool frequently found in the stomach of sheep etc. **wool-bearing**, *a.* **wool-carding**, **-combing**, *n.* a process in the preparation of wool for spinning. **wool-classer**, *n.* (*Austral.*) a grader of wool. **wool-clip**, *n.* (*Austral.*) the annual amount of wool shorn. **wool-fat**, **-oil**, *n.* lanolin. **wool-fell**, *n.* a skin from which the wool has not been removed. **wool-gathering**, *a.* in a brown study, absent-minded. *n.* absent-mindedness, inattention. **wool-grower**, *n.* **wool-hall**, *n.* a market or exchange where wool-merchants do their business. **woolpack**, *n.* a pack or bale of wool; a fleecy cloud. **woolshed**, *n.* (*Austral.*) the building for shearing, packing and storing wool. **woolsorter**, *n.* a person who sorts wool according to quality etc. **woolsorter's disease**, *n.* pulmonary anthrax due to the inhalation of dust from infected wool. **wool-staple**, *n.* the fibre of wool. **wool-trade**, *n.* **woollen**, *a.* made or consisting of wool. *n.* cloth made of wool; (*pl.*) woollen goods. **woollen-draper**, *n.* one retailing woollens. **woolly**, *a.* consisting of or resembling, bearing or naturally covered with wool, or with a hair resembling wool; like wool in appearance, fleecy; (*Painting*) lacking clear definition, firmness or incisiveness; (*coll.*) with hazy ideas, muddled. *n.* a woollen pullover etc. **woolly-bear**, *n.* a hairy caterpillar, esp. of *Arctia virgo* or the tiger-moth. **woolly-haired**, **-headed**, *a.* **woolliness**, *n.*
woold (woold), *v.t.* (*Naut.*) to wind, esp. to wind (a rope etc.) round. **woolder**, *n.* a stick used for woolding.
woollen, **woolly**, etc. WOOL.
Woolsack (wul'sak), *n.* the seat of the Lord Chancellor in the House of Lords.
woolsey (wul'zi), LINSEY-WOOLSEY.
woomera (woo'mərə), WOMERAH.
woopie (woo'pi), *n.* (*coll.*, *facet.*) a well-off older person.
woozy (woo'zi), *a.* (*coll.*) suffering from giddiness, nausea etc.; dazed, fuddled. **woozily**, *adv.* **wooziness**, *n.*
wop (wop), *n.* (*offensive*) any person with a dark complexion, esp. an Italian.
Worcester, **Worcestershire sauce** (wus'tə, -shə), *n.* a dark sauce made by mixing soy sauce, vinegar, spices etc.
word (wœd), *n.* an articulate sound or combination of sounds uttered by the human voice or written, printed etc., expressing an idea or ideas and forming a constituent part, or the whole of or a substitute for a sentence; speech, discourse, talk; news, information, a message; a command, an injunction; a password, a watchword, a motto; one's assurance, promise or definite affirmation; (*Comput.*) a set of bits processed as one unit. *v.t.* to express in words, to select words to express. **a man** etc. **of his word**, one who can be relied upon to do what he says he will do. **as good as one's word**, reliable enough to do what one has promised. **big words**, boasting, exaggeration. **by word of mouth**, orally. **good word**, a commendation. **in a**, **one word**, briefly. **last word**, the latest improvement. **my word**, used to express surprise, indignation etc. **the**, **God's Word**, the Scriptures, or any part of them; Christ as the Logos. **to eat one's words**, to retract what one has said. **to have a word with**, to have a brief conversation with. **to have words with**, to dispute with. **to take someone at his**, **her word**, to assume that someone means what he, she says. **word for word**, in exactly the same words. **word-blind**, **-deaf**, *a.* unable to understand words owing to a cerebral lesion. **word-book**, *n.* a vocabulary. **word-break**, *n.* the place where a word is divided when it runs from one line to another in printing. **word order**, *n.* the order in which words are arranged in a sentence etc. **word-painter**, *n.* a writer who depicts scenes or events in a vivid and picturesque manner. **word-painting**, *n.* **word-perfect**, *a.* able to repeat something without a mistake. **word-picture**, *n.* a vivid description. **word-play**, *n.* a play upon words, a pun. **word processor**, *n.* an electronic device used for the automatic typing, editing and often printing of texts, usu. with a VDU. **word processing**, *n.* **wordsmith**, *n.* a person skilled in the use of words, esp. a writer. **word-square**, *n.* a series of words so arranged that the letters spell the same words when read across or downwards. **wordage**, *n.* number of words. **wording**, *n.* choice of words, phrasing etc.; contents of a document, advertisement etc. **wordless** (-lis), *a.* **wordy**, *a.* verbose, prolix, verbal. **wordily**, *adv.* **wordiness**, *n.*
Wordsworthian (wœdzwœ'thiən), *a.* of, pertaining to or after the manner or spirit of the poet William Wordsworth (1770–1850) or his poetry. *n.* a devotee of Wordsworth.
wore (waw), *past* WEAR[1].
work (wœk), *n.* exertion of energy (physical or mental), effort or activity directed to some purpose; labour, toil; that upon which labour is expended, an undertaking, a task; the materials used or to be used in this; employment as a means of livelihood; that which is done, an action, deed, performance or achievement; a thing made; a product of nature or art; a large engineering structure; a place of employment; a literary or musical production; (*Phys.*) the exertion of force in producing or maintaining motion against the action of a resisting force; (*pl.*, *often sing. in constr.*) an industrial establishment, a factory; building operations, esp. under the management of a public authority; the working part or mechanism (of a watch etc.); (*Theol.*) moral duties or the performance of meritorious acts. *v.i.* (*past*, *p.p.* **worked**, (*old-fashioned*) **wrought** (rawt)) to exert physical or mental energy for some purpose, to be engaged in labour, toil or effort, to be employed or occupied (at, in, on etc.); to be in continuous activity, to do the work or perform the motions appointed, to act, to operate; to take effect, to be effective, to exercise influence; to be in a state of motion or agitation, to ferment; to make way with effort or difficulty; to reach a certain condition gradually. *v.t.* to exert energy in or upon; to cause to do work, to keep in operation, to employ, to keep busy; to make or embroider with needlework; to carry on, to manage; to bring about, to effect, to produce as a result; to prepare or alter the condition, shape or consistency of by some process, to knead, to mould, to fashion; to earn through paid work; to treat, to investigate, to solve; to excite. **in**, **out of work**, employed or not employed. **the works**, (*coll.*) everything; the appropriate treatment; a violent beating. **to have one's work cut out**, to have a hard task. **to make short work of** SHORT. **to set to work**, to employ; to start working. **to**

work in, to introduce by manipulation; to intermix or admit of being introduced. **to work off,** to get rid of; to pay off (a debt etc.) by working; to find customers for. **to work out,** to compute, to solve, to find out; to exhaust; to accomplish, to effect; to expiate; to undertake a series of physical exercises. **to work over,** to examine carefully; to beat severely. **to work up,** to elaborate, to bring gradually into shape or efficiency; to stir up, to rouse; to study (a subject) perseveringly. **work to rule,** a form of industrial action in which workers follow regulations very strictly and thus slow down the rate of work. **workaday** (-ədā), *a.* everyday, ordinary, plain, practical. **work-bag, -basket, -box,** *n.* one holding materials etc. for work, esp. for sewing. **workbench,** *n.* one specially designed for woodworking, metalworking etc. **workbook,** *n.* an exercise book with spaces for answers to printed problems; a book for recording work done. **work-day,** *n.* a working-day. **work ethic,** *n.* a (positive) attitude towards work. **work-force,** *n.* the total number of workers employed or employable. **workhorse,** *n.* a person or thing that does or is capable of doing a great deal of work. **workhouse,** *n.* (*formerly*) a public establishment maintained for paupers. **work-in,** *n.* a form of protest in which workers occupy a factory and continue working. **workload,** *n.* the amount of work expected from a person, machine etc. **workman,** *n.* any man employed in manual labour. **workmanlike,** *a.* done in the manner of a good workman. **workmanship,** *n.* comparative skill, finish or execution shown in making something or in the thing made; the result of working or making. **workmate,** *n.* a person with whom one works. **workout,** *n.* a series of exercises for physical fitness. **work-people,** *n.pl.* workmen or workwomen. **workpiece,** *n.* any item on which work is being done. **workplace,** *n.* the place where one works. **workroom,** *n.* a room in which work is done. **workshop,** *n.* a room or building in which a handicraft or other work is carried on. **work-shy,** *a.* evading work. **work station,** *n.* an area where one person works; (*Comput.*) a unit consisting of a VDU and keyboard for use by one worker. **work study,** *n.* the investigation of the methods and practice of particular work with a view to increasing efficiency. **work-table,** *n.* one fitted for keeping sewing-materials etc. in. **worktop,** *n.* a flat surface fixed to the top of kitchen units, used to prepare food. **workwoman,** *n.fem.* **workable,** *a.* capable of being worked, practicable; that will work or operate; worth working or developing. **workaholic** (-əhol'ik), *n.* (*coll.*) a person addicted to working. **worked,** *a.* **worked up,** *a.* excited, worried. **worker,** *n.* a person who works, esp. a member of the working-class; a sterile female insect, esp. a bee which specializes in gathering food, caring for the young etc. **workless** (-lis), *a.*

working (wœ'king), *n.* engaged in work, esp manual labour; during which work is done or business discussed; able to function; taking an active part in a business. *n.* the act of labouring; operation, mode of operation; a mine or quarry or a portion of it which has been worked or in which work is going on; fermentation, movement. **working capital,** *n.* funds employed for the actual carrying on of a business. **working-class,** *n.* those who earn their living by manual labour. **working-day,** *n.* any day upon which work is ordinarily performed; the period daily devoted to work. **working drawing, plan,** *n.* a drawing or plan of a work prepared to guide a builder etc. in executing work. **working-out,** *n.* the act of working out. **working party,** *n.* a committee set up specifically to investigate a particular issue.

world (wœld), *n.* the whole system of things, the universe; a system of things, an orderly or organic whole; the earth with its lands and seas; a celestial body regarded as similar to this; a large natural or other division of the earth; the human inhabitants of the world; human society, the public; fashionable or prominent people; human affairs, social life and intercourse; a particular section or class of people, animals or things, a realm, a domain; a vast quantity, amount, number, degree etc. (of); all things external to oneself as related to the individual; man's inner life; any time, state or sphere of existence; the present state of existence as dist. from the future life; the ungodly or unregenerate portion of mankind. **all the world,** everybody. **for all the world,** exactly, precisely. **for the world,** on any account. **in the world,** at all, possibly. **out of this world,** (*coll.*) remarkable, striking; excellent. **to think the world of,** to love or respect greatly. **world without end,** to all eternity, everlastingly. **World Bank,** *n.* an agency of the United Nations set up in 1945 to lend money to poorer countries. **World Court,** *n.* popular name for the International Court of Justice at the Hague set up in 1921 to settle disputes between states. **World Health Organization,** *n.* a specialized agency of the United Nations which helps countries to develop their health administration. **world music,** *n.* folk music based on a variety of different national styles. **world power,** *n.* a sovereign state strong enough to affect the policy of other states throughout the world. **World Series,** *n.* a series of baseball games played in the US to decide the professional championship. **world-view** WELTANSCHAUUNG. **world war,** *n.* one involving most of the earth's major nations, esp. the 1914–18 or 1939–45 wars. **world-wearied, -weary,** *a.* tired of existence. **world-wide,** *a.* spread over the whole world; existing everywhere.

worldly (wœld'li), *a.* pertaining to the present, temporal or material world; earthly, secular, material, not spiritual. **worldly-minded,** *a.* devoted to worldly things. **worldly-mindedness,** *n.* **worldly-wise,** *a.* wise in the things of this world. **worldliness,** *n.* **worldling** (-ling), *n.* a worldly person.

WORM, (*acronym*) (*Comput.*) write once read many times.

worm (wœm), *n.* an invertebrate creeping animal with a long limbless segmented body; an intestinal parasite, a tapeworm, a fluke; any small creeping animal with very small or undeveloped feet, as larvae, grubs, caterpillars, maggots etc.; (*pl.*) any disease caused by parasitic worms, esp. in the intestine; a grovelling or despised person; a spiral part or thing; the spiral part of a screw; a spiral tool for boring rock; a spiral device for extracting cartridges etc.; the spiral condensing-pipe of a still; a ligament under a dog's tongue. *v.i.* to crawl, creep, wriggle or progress with a worm-like motion; to work stealthily or underhandedly. *v.t.* to insinuate (oneself), to make (one's way) in a worm-like manner; to draw (out) or extract by craft and perseverance; to free (a dog etc.) from worms; to cut the worm from under the tongue of (a dog). **worm's eye view,** a view from below or from a humble position. **worm-cast,** *n.* a cylindrical mass of earth voided by an earth-worm. **worm-eaten,** *a.* gnawed or bored by worms; old, shabby, decayed. **worm-fishing,** *n.* fishing with worms for bait. **worm-gear,** *n.* gear having a toothed or cogged wheel engaging with a revolving spiral. **wormhole,** *n.* a hole made by a worm in wood, fruit etc. **worm-holed,** *a.* **worm-powder,** *n.* one used as vermifuge. **worm-wheel,** *n.* the toothed wheel of worm-gear. **wormless** (-lis), *a.* **worm-like,** *a.* **wormy,** *a.*

wormwood (wœm'wud), *n.* a perennial herb having bitter and tonic properties, used in the manufacture of vermouth and absinthe and in medicine; bitterness, gall, mortification.

worn (wawn), *p.p.* WEAR[1]. **worn-out,** *a.* rendered useless by long wear; (*coll.*) exhausted, tired.

worry (wŭr'i), *v.t.* of dogs, to bite or keep on biting, to choke or pull about with the teeth; to tease, harass, bother, persecute or wear out with importunity etc.; to cause mental distress to. *v.i.* of dogs, to bite, pull about etc.; to be unduly anxious or troubled, to fret. *n.* the act

worse (wœs), *a.* (*comp. of* BAD) more bad; (*predicatively*) in a poorer state of health; in a less favourable state, position or circumstances. *adv.* more badly; into a poorer state of health etc.; less. *n.* a worse thing or things; loss, disadvantage, defeat. **the worse for,** damaged or harmed by. **the worse for drink,** (*coll.*) drunk. **the worse for wear,** shabby, worn; (*coll.*) tired, untidy etc. **worse off,** in a poorer condition or financial situation. **worsen,** *v.t., v.i.* to (cause to) grow worse.

worship (wœ'ship), *n.* the act of paying divine honour to God, esp. in religious services; an act or feeling of adoration or loving or admiring devotion or submissive respect (to a person, principle etc.). *v.t.* (*past, p.p.* **worshipped**) to pay divine honours to; to perform religious service to; to reverence with supreme respect and admiration; to treat as divine. *v.i.* to take part in a religious service. **place of worship,** church, chapel etc., where religious services are held. **Your Worship,** form of address to a mayor or magistrate. **worshipful,** *a.* deserving of worship (phrase applied to certain magistrates etc.). **worshipfully,** *adv.* **worshipfulness,** *n.* **worshipper,** *n.* one who worships; an attender at a place of worship.

worst (wœst), *a.* bad in the highest degree. *adv.* most badly. *n.* that which is most bad; the most bad, evil, severe or calamitous part, event, state etc. *v.t.* to get the better of in a contest etc., to defeat, to overthrow. **at worst,** in the worst circumstances; in the least favourable view. **if the worst comes to the worst,** if the worst of all possible things happens. **to come off worst, to get the worst of it,** to be defeated.

worsted (wus'tid), *n.* woollen yarn used for knitting stockings, carpets etc. *a.* made of worsted.

wort (wœt), *a.* a plant, a herb (*usu. in comb.,* as *moneywort, soapwort*); an infusion of malt for fermenting into beer.

worth (wœth), *a.* equal in value or price to; deserving, worthy of; having property to the value of, possessed of. *n.* that which a person or thing is worth, value, the equivalent of anything, esp. in money; merit, high character, excellence. **for all one is worth,** (*coll.*) with all one's strength, energy etc. **worthless** (-lis), *a.* **worthlessly,** *adv.* **worthlessness,** *n.*

worthwhile (wœthwīl'), *a.* worth the time, expense or effort involved.

worthy (wœ'dhi), *a.* having worth, estimable; deserving of respect, praise or honour, respectable; deserving (of, to be etc.); fit, suitable, adequate, appropriate, equivalent or adequate to the worth (of). *n.* a person of eminent worth; a person of some note or distinction in his or her time, locality etc. **worthily,** *adv.* **worthiness,** *n.*

-worthy (-wœdhi), *comb. form.* safe or suitable for, as in *seaworthy;* deserving of, as in *praiseworthy.*

wot (wot), *first and second sing.* WIT [1].

wotcher (woch'ə), *int.* (*sl.*) a Cockney form of greeting.

would (wud), *past, cond.* WILL [1].

wouldn't (wud'ənt), contr. form of *would not.*

Woulfe bottle (wulf), *n.* a bottle with three or more necks used in the handling and washing of gases. [British chemist P. *Woulfe,* 1727–1803]

wound [1] (woond), *n.* an injury caused by violence to the skin and flesh of an animal or the bark or substance of plants, esp. one involving disruption of the tissues; any damage, hurt or pain to feelings, reputation etc. *v.t.* to inflict a wound on. *v.i.* to cause a wound. **woundwort,** *n.* a plant formerly used to heal wounds. **woundable,** *a.* **wounder,** *n.* **woundless** (-lis), *a.*

wound [2] (wownd), *past, p.p.* WIND [1,2].

wourali (woorah'li), CURARE.

wove (wōv), *past, p.p.* **woven,** *p.p.* WEAVE.

wow [1] (wow), *int.* an exclamation of astonishment, wonder etc. *n.* a sensational or spectacular success. *v.t.* (*coll.*) to cause to feel great enthusiasm.

wow [2] (wow), *n.* a variation in pitch occurring at low frequencies in sound-reproducing systems.

wowser (wow'zə), *n.* (*Austral. sl.*) a spoil-sport, a puritan. **wowserism,** *n.*

wow-wow (wow'wow), *n.* the silvery gibbon of Java and Sumatra.

WP, (*abbr.*) word processor.

wpb, (*abbr.*) waste paper basket.

WPC, (*abbr.*) woman police constable.

wpm, (*abbr.*) words per minute.

WRAC (rak), (*abbr.*) Women's Royal Army Corps.

wrack (rak), *n.* seaweed thrown on the shore; cloud-rack; wreck, destruction, ruin.

WRAF (raf), (*abbr.*) Women's Royal Air Force.

wraith (rāth), *n.* the double or phantom of a living person; an apparition, a ghost appearing after death.

wrangle (rang'gl), *v.i.* to dispute, argue or quarrel angrily, peevishly or noisily, to brawl. *n.* an angry or noisy dispute or quarrel, an altercation, a brawl. **wrangler,** *n.* one who wrangles; (*esp. N Am.*) a cowboy; (*Camb. Univ.*) one of those placed in the first class in the mathematical tripos. **senior wrangler,** formerly the student who took the first place in this. **wranglership,** *n.*

wrap (rap), *v.t.* (*past, p.p.* **wrapped**) to fold or arrange so as to cover or enclose something; to enfold, envelop, muffle, pack, surround or conceal in some soft material; to hide, to disguise; (*in p.p.*) to absorb, to engross, to comprise (with *up*). *v.i.* to fold, to lap. *n.* something intended to wrap, as a cloak, shawl, rug etc. **to wrap up,** to dress warmly; (*coll.*) to bring to a conclusion; (*sl.*) to fall silent. **under wraps,** secret. **wraparound,** *a.* of a skirt etc., designed to be wrapped around the body; of a windscreen etc., curving round at the sides. **wrappage** (-ij), *n.* the act of wrapping; that which wraps or envelops, a wrapping or wrappings. **wrapper,** *n.* one who wraps; that in which anything is wrapped, esp. a dust jacket for a book; a woman's loose outer garment for indoor wear. **wrapping,** *n.* that which wraps; that in which something is wrapped or packaged; a wrapper, a cloak, a shawl, a rug.

wrasse (ras), *n.* an edible sea-fish of numerous species haunting coasts and rocks.

wrath (roth), *n.* deep or violent anger, indignation, rage. **wrathful,** *a.* **wrathfully,** *adv.* **wrathfulness,** *n.* **wrathless** (-lis), *a.* **wrathy,** *a.* **wrathily,** *adv.*

wreak (rēk), *v.t.* to carry out, to inflict, to execute. **wreaker,** *n.*

wreath (rēth, *pl.* rēdhz), *n.* a band or ring of flowers or leaves tied or woven together for wearing on the head, decorating statues, graves etc.; a representation of this in wood, stone etc.; a similar circlet of twisted silk etc.; a ring, a twist, a curl (of cloud, smoke etc.). **wreathless** (rēth'lis), *a.*

wreathe (rēdh), *v.t.* to form (flowers, leaves etc.) into a wreath; to surround, encircle, entwine (as if) with a wreath or with anything twisted; to form a wreath round; to cause (the face) to take on a certain expression, esp. a smiling one. *v.i.* to be curled, folded or entwined (round etc.). **wreather,** *n.*

wreck (rek), *n.* destruction, ruin, esp. of a ship at sea; the remains of anything irretrievably shattered or ruined; a dilapidated or worn-out person or thing; wreckage. *v.t.* to destroy (a vessel etc.) by collision, driving ashore etc.; to involve in shipwreck; to ruin or destroy. *v.i.* to suffer

shipwreck. **wreck-master**, *n*. an official appointed to take charge of goods etc. cast ashore after a shipwreck. **wreckage** (-ij), *n*. the debris, fragments or material from a wreck. **wrecker**, *n*. one who lures vessels to shipwreck with intent to plunder; a person or ship employed in recovering a wreck or a wrecked cargo; a person or thing that brings about ruin or destruction; (*esp. N Am.*) a recovery vehicle. **wrecking**, *n.*, *a*. **wrecking-car**, *n*. (*N Am.*) a railway-car carrying appliances for removing wreckage and obstructions from the line.
Wren (ren), (*abbr.*) a member of the Women's Royal Naval Service.
wren (ren), *n*. a small insessorial songbird with a short erect tail and short wings.
wrench (rench), *n*. a violent twist or sideways pull; an injury caused by twisting, a sprain; pain or distress caused by a parting, loss etc.; a tool for twisting or untwisting screws, bolts, nuts etc. *v.t.* to pull, wrest or twist with force or violence; to pull (off or away) thus; to strain, to sprain; to pervert, to distort.
wrest (rest), *v.t.* to twist, to turn aside by a violent effort; to pull, extort or wrench (away) forcibly; to pervert, to distort, to twist or deflect from its natural meaning. *n*. a violent wrench or twist; a turning instrument, esp. a tuning-key for a harp etc. **wrester**, *n*.
wrestle (res'l), *v.i.* to contend by grappling with and trying to throw one's opponent, esp. in a match under recognized rules; to struggle, to contend, to strive vehemently; *v.t.* to contend with in a wrestling-match; to move with difficulty, to manhandle. *n*. a bout at wrestling, a wrestling-match; a struggle. **wrestler**, *n*. **wrestling**, *n*.
wretch (rech), *n*. a miserable or unfortunate person; a despicable, mean, base or vile person. **wretched** (-id), *a*. miserable, unhappy, sunk in deep affliction or distress; calamitous, pitiable, afflictive; worthless, paltry, contemptible; (*coll.*) extremely unsatisfactory, uncomfortable or unpleasant. **wretchedly**, *adv*. **wretchedness**, *n*.
wrick (rik), *v.t.* to sprain or strain. *n*. a sprain or strain.
wriggle (rig'l), *v.i.* to turn, twist or move the body to and fro with short motions like an eel; to move or go (along, in, out etc.) with writhing contortions or twistings; to manoeuvre by clever or devious means. *v.t.* to move (one's body etc.) with a wriggling motion; to effect or make (one's way etc.) by wriggling. *n*. a wriggling motion. **wriggler**, *n*. **wriggly**, *a*.
wright (rit), *n*. one who makes or repairs (*esp. in comb.*, as *shipwright, playwright*).
wring (ring), *v.t.* (*past, p.p.* **wrung** (rŭng)) to twist and squeeze or compress; to turn, twist or strain forcibly; to press or squeeze (water etc. out) thus; to distort (a meaning etc.); to pain, to torture, to distress; to extract, to extort. *n*. a press, a squeeze, **to wring one's withers**, to appeal passionately to one's pity. **to wring the hands**, to press the hands together convulsively, as in great distress. **wringer**, *n*. one who or that which wrings; a wringing-machine. **wringing**, *n.*, *a*. **wringing-machine**, *n*. one for wringing water out of wet clothes etc. **wringing-wet**, *a*. so wet that moisture can be wrung out.
wrinkle[1] (ring'kl), *n*. a small ridge, crease or furrow caused by the folding or contraction of a flexible surface. *v.t., v.i.* to fold or contract into furrows, creases or ridges. **wrinkly**, *a*. **wrinklies**, *n.pl.* (*sl.*) old people.
wrinkle[2] (ring'kl), *n*. a useful bit of information or advice, a tip, a dodge.
wrist (rist), *n*. the joint uniting the hand to the forearm; the part of a sleeve over the wrist; a wrist-pin. **wristband**, *n*. a band or part of a sleeve covering the wrist, a cuff. **wrist-drop**, *n*. paralysis of the muscles of the forearm through lead-poisoning. **wrist-pin**, *n*. a pin or stud projecting from a crank for a connecting-rod to turn on. **wrist-watch**, *n*. a watch worn on a strap round the wrist.

wristlet (-lit), *n*. a band worn round the wrist; a bracelet; a handcuff.
writ (rit), *n*. that which is written; a written command or precept issued by a court in the name of the sovereign to a person commanding him to do or refrain from doing some specified act. **holy writ** HOLY.
write (rīt), *v.t.* (*past* **wrote** (rōt), *p.p.* **written** (rit'n)) to form or trace (esp. words, a sentence etc.) in letters or symbols, with a pen, pencil or the like on paper or other material; to trace (signs, characters etc.) thus, to set (down), to record, to describe, to state or convey by writing; to compose or produce as an author; to cover or fill with writing; to impress or stamp (disgrace etc.) on a person's face; to designate, to call, to put (oneself down as etc.) in writing; (*coll.*) to send a letter to; (*Comput.*) to record (data) in a storage device. *v.i.* to trace letters or symbols representing words on paper etc.; to write or send a letter; to compose or produce articles, books etc. as an author. **to write down**, to put in writing, to record; to criticize unfavourably; to write in such a way that one appeals to low standards of taste, intelligence etc. **to write in**, to insert in writing; to apply in writing (for). **to write off**, to record the cancelling of (a debt etc.); to compose rapidly and easily; to discard as useless, damaged etc.; to damage (a car) beyond repair; to write and send a letter etc. **to write oneself out**, to exhaust one's powers of literary production. **to write out**, to write the whole of; to remove (a character) from a drama series. **to write up**, to praise in writing, to puff; to post up (account-books etc.); to give full details in writing. **writer**, *n*. one who writes; an author, a journalist etc.; a clerk, an amanuensis; (*Sc.*) a solicitor. **writer to the signet**, (*Sc.*) a solicitor. **writer's cramp**, *n*. a spasmodic pain in the fingers or hand caused by prolonged writing. **writership**, *n*. writing, *n*. the act of one who writes; that which is written; an inscription; a book, article or other literary composition; a legal instrument. **writing on the wall**, a warning of coming disaster. **writing-case**, *n*. a portable case for writing materials. **writing desk**, *n*. a portable desk with space for papers etc. **writing-ink**, *n*. ink for writing, opp. to printer's ink. **writing-paper**, *n*. paper with a smooth surface for writing on. **writing-table**, *n*. a table used for writing on, usu. with a knee-hole, drawers etc.
writhe (rīdh). *v.i.* to twist, turn or roll the body about, as in pain; to shrink, to squirm (at, with shame etc.). *v.t.* to twist or distort (the limbs etc.). *n*. an act of writhing.
writher, *n*. **writhingly**, *adv*.
writing, written (rit'n), etc. WRITE.
WRNS (renz), (*abbr.*) Women's Royal Naval Service.
wrong (rong), *a*. not morally right, contrary to morality, conscience or law, wicked; not the right (one etc.), not that which is required, intended, proper, best etc.; not according to truth or reality; out of order, in bad condition, not suitable etc.; false, inaccurate, mistaken, erroneous. *adv*. wrongly, unjustly. *n*. that which is wrong; a wrong act, an injustice, a trespass, an injury, hurt or pain; deviation from what is right; wrongness, error. *v.t.* to treat unjustly, to do wrong to; to impute evil motives to unjustly. **in the wrong**, in a wrong position; in error. **to get on the wrong side of**, to fall into disfavour with. **to go wrong**, to fall into sin; to fail to operate correctly; to fall into error. **wrong side out**, inside out. **wrongdoer**, *n*. **wrongdoing**, *n*. **wrong-foot**, *v.t.* to cause to be off-balance; to gain an advantage over. **wrong fount**, *a*. of type, not of the right fount, size or pattern. **wrong-headed**, *a*. perverse, obstinate, crotchety. **wrong-headedness**, *n*. **wronger**, *n*. **wrongful**, *a*. injurious, unjust, wrong. **wrongfully**, *adv*. **wrongfulness**, *n*. **wrongly**, *adv*. **wrongness**, *n*.
wrote (rōt), *past* WRITE.
wroth (rŏth, roth), *a*. (*poet.*) angry, wrathful.
wrought (rawt), *a*. worked; (*often in comb.*) formed or fash-

ioned; decorated or ornamented. **wrought iron,** *n.* iron made malleable by having non-metallic impurities burned out of it; iron made malleable by forging or rolling. **wrought-up,** *a.* very tense or excited.
wrung (rŭng), *past*, *p.p.* WRING.
WRVS, (*abbr.*) Women's Royal Voluntary Service.
wry (rī), *a.* twisted, distorted, crooked, skew; showing distaste, disgust etc.; wrong, false, perverted. **wrybill,** *n.* a species of plover. **wry-mouth,** *n.* an eel-like sea-fish with a vertical mouth. **wry-mouthed,** *a.* having a distorted mouth or a cynical or distorted expression. **wry-neck,** *n.* a bird allied to the woodpeckers, with a habit of twisting its head round as on a pivot; stiff-neck. **wrynecked,** *a.* **wryly,** *adv.* **wryness,** *n.*
WSW, (*abbr.*) west south-west.
wt., (*abbr.*) weight.
wyandotte (wī'əndot), *n.* a breed of domestic fowl.

wych-elm (wich), *n.* the Scotch elm. **wych-hazel,** *n.* a N American shrub from whose bark an astringent lotion is prepared.
Wycliffite (wik'lifīt), *a.* pertaining to Wycliffe, his tenets or his followers. *n.* a follower of Wycliffe. [John *Wycliffe, c.* 1330–84, English ecclesiastical reformer and Lollard]
wye (wī), *n.* a Y-shaped thing.
Wykehamist (wik'əmist), *n.* a member (past or present) of Winchester College founded by William of Wykeham (1324–1404) Bishop of Winchester. *a.* of or pertaining to this.
wynd (wīnd), *n.* (*Sc.*) an alley.
WYSIWYG (wiz'iwig), (*acronym*) *what you see is what you get.*
wyvern (wī'vən), *n.* (*Her.*) a two-legged dragon with erect wings and barbed tail.

X

X, x, the 24th letter, and the 18th consonant, of the English alphabet (*pl.* **Xs, X's, Exes**), as a medial letter has the sound of *ks*, as in *axis, taxes,* or of *gz*, as in *exhaust, exult*; as an initial letter (chiefly in words of Greek origin) it has the sound of *z*; (Roman numeral) 10 (xx, 20; xxx, 30; xc, 90); (*Alg.*, *x*) the first unknown quantity or variable; (*usu. cap.*) an unknown thing or person; Christ, Christian (first letter of Christ in Greek); before 1982, a film for over 18-year-olds only; a kiss; a choice, esp. in voting; an error; a location. **x-axis,** *n.* the horizontal axis, along which **x-coordinates** are plotted in the Cartesian coordinate system. **X chromosome,** *n.* a sex chromosome which is found paired in women, and paired with the Y chromosome in men. **X-ray,** *n.* a röntgen ray, used in producing a photographic image of internal parts of the body, such as organs or bones, and in medical diagnosis; a picture thus produced. *v.t.* to produce such an image of part of the body; to treat with X-rays. **X-ray tube,** *n.* an evacuated tube in which electrons are beamed onto a metal target to produce X-rays. **xx** or **double-x, xxx** or **triple-x,** marks indicating the strength of ale etc. placed on brewers' casks.
xanthate, xanthein XANTH(O)-.
xanthic (zan'thik), *a.* of a yellowish colour. **xanthic acid,** *n.* a colourless oily liquid, prepared by decomposing xanthate of potassium with sulphuric or hydrochloric acid. **xanthin** (-thin), **-thine** (-thēn, -thin), *n.* the part of the yellow colouring-matter of flowers that is insoluble in water; a yellow colouring-matter obtained from madder; a crystalline compound found in blood, urine, the liver etc. **xanth(o)-,** *comb. form.* yellow. **xanthate** (zan'thāt), *n.* a salt of xanthic acid. **xanthein** (-thiin), *n.* the part of the yellow colouring-matter of flowers that is soluble in water.
Xanthochroi (zanthok'rōi), *n.pl.* (of peoples) fair whites or blonds, those having yellow or red hair, blue eyes and fair complexion. **xanthochroic** (-krō'-), **-chrous** (-rōəs), *a.* **xanthochroism,** *n.* a condition where all skin pigments apart from yellow disappear, as in some goldfish.
xanthoma (zanthō'mə), *n.* a skin disease characterized by a yellowish warty growth, usu. in flat patches, on the eyelids. **xanthomatous,** *a.*
xanthomelanous (zanthōmel'ənəs), *a.* having black hair and yellow or brownish skin.
xanthophyll (zan'thəfil), *n.* the yellow colouring-matter of withered leaves. **xanthophyllous,** *a.*
xanthopsia (zanthop'siə), *n.* (*Path.*) an affection of the sight in which objects appear yellowish.
Xantippe (zantip'i), *n.* a shrewish wife, a scold.
Xe, (*chem. symbol*) xenon.
xebec (zē'bek), *n.* a small three-masted vessel with lateen and square sails, used in the Mediterranean.
xen(o)-, *comb. form.* strange, foreign.
xenarthral (zenah'thrəl), *a.* (*Anat.*) peculiarly jointed (of certain vertebrae).
xenial (zē'niəl), *a.* of or pertaining to hospitality or the relations between host and guest.
xenogamy (zenog'əmi), *n.* cross-fertilization. **xenogamous,** *a.*
xenogenesis (zenōjen'əsis), *n.* heterogenesis, the production of offspring totally unlike the parents.
xenoglossia (zenōglos'iə), *n.* in psychical research, the knowledge of a language one has not learned, claimed by some mediums.
xenograft (zen'ōgrahft), *n.* a tissue graft from a member of a different species.
xenolith (zen'əlith), *n.* a fragment of rock enclosed in a different type of rock.
xenomania (zenōmā'niə), *n.* inordinate liking for everything foreign.
xenomenia (zenōmē'niə), *n.* loss of menstrual blood elsewhere than from the uterus, e.g. from the nose; vicarious menstruation.
xenomorphic (zenōmaw'fik), *a.* (*Petrol.*) not having its own proper crystalline form but an irregular shape due to surrounding minerals.
xenon (zen'ən), *n.* an inert gaseous element, at. no. 54; chem. symbol Xe, found in very small quantities in the atmosphere and solidifying at the temperature of liquid air.
xenophile (zen'əfīl), *n.* someone who likes foreign people and things.
xenophobia (zenəfō'biə), *n.* fear of, or aversion to, strangers or foreigners. **xenophobe** (zen'-), *n.* a person having such fear or aversion. **xenophobic,** *a.*
xeranthemum (ziəran'thiməm), *n.* an annual plant of the thistle family with everlasting flowers.
xerasia (ziərā'siə), *n.* a disease of the hair in which it becomes dry and powdery.
xer(o)-, *comb. form.* dry. **xerophile** (ziə'rəfil), *n.* **xerophily** (-li), *n.* **xeransis** (ziəran'sis), *n.* the state of being dried up, desiccation. **xerantic,** *a.*
xerodermia (ziərədœ'miə), *n.* morbid dryness of the skin. **xerodermatic** (-mat'-), **xerodermatous,** *a.*
xerography (zirog'rəfi), *n.* a photographic process in which the plate is sensitized electrically, and the latent image developed by a resinous powder. **xerographic** (-graf'-), *a.*
xerophagy (zirof'əji), *n.* the Christian rule of fasting; the act or habit of living on dry food or a meagre diet.
xerophilous (zirof'iləs), *a.* of plants, adapted to living in a hot, dry climate. **xerophile** (ziə'rəfil), *n.* **xerophily** (-li), *n.*
xerophthalmia (ziərəfthal'miə), *n.* a dry inflammation of the lining membrane of the eye, caused by a deficiency of vitamin A. **xeropthalmic,** *a.*
xerophyte (ze'rəfīt), *n.* a plant adapted to living in a region of little moisture, such as a cactus. **xerophytic** (-fit'-), *a.* **xerophytism,** *n.*
xerostomia (ziərəstō'miə), *n.* abnormal dryness of the mouth.
Xerox®(ziə'roks), *n.* a xerographic copying process; a copy produced by this process; the machine used for this process. *v.t.* to produce a copy of (an original document) by this process.
Xhosa (khō'sə, khaw'-), *n.* (a member of) one of the Bantu-speaking peoples in the Cape district of S Africa; their language, which is characterized by a sound system involving a series of clicks. **Xhosan,** *a.*
xi (zī. ksī. sī, ksē), *n.* the 14th letter of the Greek alphabet.
Xian (*abbr.*) CHRISTIAN.
xiph-, xiphi-, xipho- *comb. form.* sword. **xiphisternum** (zifistœ'nəm), *n.* the lower segment or sword-shaped process of the sternum. **xiphoid** (zif'oid), *a.* sword-shaped. **xiphoid appendage, cartilage** or **process,** *n.* the

xiphisternum.
Xmas (eks'məs, kris'-), CHRISTMAS.
Xn, Xnty, (*abbr.*) CHRISTIAN, CHRISTIANITY.
X-ray X.
Xt. (*abbr.*) CHRIST.
xylem (zī'ləm), *n.* woody tissue, opp. to phloem.
xylene (zī'lēn), *n.* any one of three isomeric colourless, volatile, liquid hydrocarbons distilled from coal- or wood-tar.
xyl(o)-, *comb. form.* wood.
xylobalsamum (zīlōbawl'səməm), *n.* the wood of, or a balsam obtained by decoction of the twigs and leaves of, the Balm of Gilead tree.
xylocarp (zī'lōkahp), *n.* a hard, woody fruit, or a tree bearing this. **xylocarpous** (-kah'-), *a.*
xylograph (zī'ləgrahf), *n.* an engraving on wood, esp. in a primitive style, or an impression from such an engraving; an impression obtained from the grain of wood used for surface decoration. **xylographer** (-log'-), *n.* **xylographic** (-graf'-), *a.* **xylography** (-log'-), *n.*
xyloid (zī'loid), *a.* woody, ligneous.
xyloidine (zī'loidin), *n.* a high explosive prepared by the action of nitric acid on starch or wood-fibre.
Xylonite® (zī'lənit), *n.* celluloid.
xylophagous (zīlof'əgəs), *a.* boring into wood (of insects). **xylophagan,** *n.*, *a.*
xylophilous (zīlof'iləs), *a.* living or growing on wood; liking wood.
xylophone (zī'ləfōn), *n.* a musical instrument consisting of a graduated series of wooden bars vibrating when struck or rubbed. **xylophonic** (-fon'-), *a.* **xylophonist** (-lof'-), *n.*
xylose (zī'lōz), *n.* wood sugar.
xylotomous (zīlot'əməs), *a.* of insects, boring into wood.
xyster (zis'tə), *n.* (*Surg.*) an instrument for scraping bones.
xystus (zis'təs), *n.* (*pl.* **-ti** (-tī)) in antiquity, a long covered portico or colonnade used for athletic exercises; a garden walk or terrace.

Y

Y¹, y, the 25th letter of the English alphabet (*pl.* **Y's, Ys, wyes**) is both a vowel and a palatal semi-vowel; as a vowel it has the same values as *i*; at the beginning of syllables and followed by a vowel, it corresponds to the Latin *i* or *j*, as in *ye, you;* (*Alg., y*) the second unknown quantity or variable; a Y-shaped branch, pipe, fork, coupling, figure etc. **y-axis**, *n.* the vertical axis along which y-coordinates are plotted in the Cartesian coordinate system. **Y-chromosome**, *n.* a sex chromosome which is found paired with the X chromosome in men, and is not present at all in women. **Y-fronts**® (wī′frŭnts), *n.pl.* men's or boys' underpants with an inverted Y-shaped front opening. **Y-level**, *n.* a surveying level mounted on a pair of Y-shaped supports. **Y-moth**, *n.* the gamma moth, from the Y-shaped mark on its wings.
Y², (*chem. symbol*) yttrium.
Y, (*abbr.*) YMCA, YWCA.
y., (*abbr.*) year.
-y (-i), *comb. form.* forming abstract nouns etc., as in *memory, remedy*; forming adjectives, as in *mighty, trusty;* forming diminutives of proper names etc., as *Jimmy, sonny*; forming nouns, as *army, treaty.*
yabber (yab′ə), *v.i.* to talk, to chatter. *n.* Aboriginal talk.
yabbie, yabby (yab′i), *n.* (*Austral.*) a fresh-water crayfish.
yacca (yak′ə), *n.* either of two W Indian evergreen trees of the yew family yielding wood used for cabinet work.
yacht (yot), *n.* a light sailing-vessel specially designed for racing; a vessel smaller than a liner, propelled by steam, sails, electricity or other motive power, used for pleasure trips, cruising, travel or as a state vessel to convey royal personages or government officials. *v.i.* to sail or cruise about in a yacht. **yacht-club**, *n.* a club for yacht-racing etc. **yachtsman, -woman**, *n.* one who keeps or sails a yacht. **yachtsmanship**, *n.* **yachter**, *n.* **yachting**, *n.*
yack, yak (yak), *n.* (*coll.*) noisy, unceasing, trivial chatter. *v.t.* to talk in this way. **yackety-yak** (yakəti-), *n.* trivial, persistent chatter.
yaffle (yaf′l), **yaffingale** (-ingāl), *n.* the green woodpecker.
Yager (yä′gə), *n.* a member of certain German corps of light infantry, esp. of sharp-shooters.
yah (yah), *int.* an exclamation of derision; (*coll.*) yes.
yahoo (yah′hoo, yəhoo′), *n.* (**Yahoo**) one of a race of brutes in human shape; a coarse, brutish or vicious and degraded person.
Yahveh (yäh′vä), **Yahweh** (-wä), **Yahvist** etc. JEHOVAH.
yak¹ (yak), *n.* a long-haired ruminant from the mountainous regions of Central Asia, intermediate between the ox and the bison.
yak² YACK.
Yakut (yəkut′), *n.* (one of) a mixed Turkish people dwelling in the basin of the Lena, in E Siberia.
Yale® **lock** (yāl), *n.* protected trade name of a type of lock with a revolving barrel. [Linus *Yale*, 1821–68, inventor]
yam (yam), *n.* (the fleshy edible tuber of) various species of tropical climber orig. from India.
yamen (yah′men), YAMUN.
yammer (yam′ə), *v.i.* to cry out, to whine, to complain peevishly. *n.* a whining sound. **yammerer**, *n.* **yammering**, *n., a.*
yamun (yah′mən), *n.* the office or official residence of a Chinese mandarin.

yang (yang), *n.* the masculine, positive, bright principle in nature, according to Chinese philosophy, which interacts with its complement, *yin.*
yank (yangk), *v.t.* to pull sharply, to jerk (off, out of etc.). *v.i.* to jerk vigorously. *n.* a sharp jerk, a pull.
Yank YANKEE.
Yankee (yang′ki), *n.* an inhabitant of New England (applied by foreigners to all the inhabitants of the US), (*sometimes derog.*) an American; (*Hist.*) a Federal soldier or Northerner in the American Civil War (1861–5); in horse-racing, a complicated type of bet relying on several different horses winning their respective races. *a.* pertaining to America or the Yankees. **Yankee Doodle**, *n.* a tune (probably of English origin) and song regarded as a national air of the US; an American. **Yank**, *n.* (*sometimes derog.*) short form of YANKEE. **Yankeedom** (-dəm), **Yankeeism**, *n.* **Yankeefied** (-fid), *a.*
yap (yap), *v.i.* of dogs etc., to yelp or bark snappishly; of humans, to talk constantly in a shrill, foolish manner. *n.* such a bark; foolish chatter. **yapper**, *n.* **yappy**, *a.* **yapster** (-stə), *n.* (*sl.*) a dog.
yapok (yap′ək), *n.* a small opossum with webbed hind feet and aquatic habits.
yapon (yaw′pən), *n.* an evergreen shrub growing in the southern US, the leaves of which are used for tea and by the Indians for their 'black drink', an emetic and purgative medicine.
yapp (yap), *n.* a style of book-binding, usu. in leather, with flaps at the edges.
yarborough (yah′bərə), *n.* (*Whist etc.*) a hand containing no card higher than a nine. [2nd Earl of *Yarborough*, who bet £1000 that no one could hold such a hand]
yard¹ (yahd), *n.* a unit of length, 3 ft. or 36 in. (0·9144 m); a measuring rod of this length, or this length of material; a cylindrical spar tapering each way from the middle slung horizontally or slantwise on a mast to extend a sail. **to man the yards**, to place men, or (of sailors) to stand along the yards, as a salute at reviews etc. **yard of ale**, a tall, narrow drinking glass for beer or ale; the amount of beer or ale in such a glass. **yard-arm**, *n.* either half of a sail-yard from the centre to the end. **yard-measure, -stick, -wand**, *n.* a tape or stick, 3 ft. in length and usu. graduated in feet, inches, etc., used for measuring. **yardage** (-dij), *n.* total length in yards; the amount of excavation in cubic yards.
yard² (yahd), *n.* a small piece of enclosed ground, esp. enclosed by, enclosing or adjoining a house or other building; a garden; (*esp. in comb. form.*) such an enclosure used for some specified manufacture or other purpose, as a dockyard, graveyard, timber-yard etc.; a series of tracks near a railway used e.g. for freight or for the storage and maintenance of rolling-stock. *v.t.* to collect or pen (cattle etc.) in a yard. **Yardie**, *n.* a member of a drug-trafficking gang originating in the West Indies. **The Yard** SCOTLAND YARD. **stock-yard** STOCK¹. **yard-man**, *n.* a man employed in a farm- or railway-yard. **yard-master**, *n.* the manager of this. **yardage** (-dij), *n.* a railway yard used as a cattle enclosure; the charge levied for such a use.
yarmulka, yarmulke (yah′mulkə), *n.* a skullcap worn all the time by Orthodox male Jews, and during prayer by others.

yarn (yahn), *n.* any spun fibre prepared for weaving, knitting, rope-making etc.; (*coll.*) a story or tale told by a sailor, a long or rambling story, esp. one of doubtful truth or accuracy. *v.i.* to tell a yarn. **yarn-dye,** *v.t.* to dye the yarn before it is spun or woven. **yarn-dyed,** *a.*
yarrow (ya'rō), *n.* a perennial herb with white flowers, pungent odour and astringent properties, the milfoil.
yashmak (yash'mak), *n.* the veil worn by Muslim women in public.
yataghan (yat'əgan), *n.* a Turkish sword or scimitar with double-curved blade and without a guard or cross-piece.
yaupon (yaw'pən), YAPON.
yaw (yaw), *v.i.* to steer out of the direct course, to move unsteadily (of a ship); (of a ship, aircraft etc.) to turn about its vertical axis. *v.t.* to cause to deviate from its course. *n.* an unsteady motion or temporary deviation from a course; the motion of an aircraft about its vertical axis. **yawing,** *n.*
yawl[1] (yawl), *v.i.* to howl, to yell. *n.* a howl or yell.
yawl[2] (yawl), *n.* a small boat, esp. a ship's jolly-boat; a small sailing-vessel with two masts.
yawn (yawn), *v.i.* to gape, to be or stand wide open; to open the mouth wide or to have the mouth open involuntarily through drowsiness, boredom, bewilderment etc. *v.t.* to express or utter by or with a yawn. *n.* an act of yawning. **yawner,** *n.* **yawning,** *n.* **yawningly,** *adv.*
yaws (yawz), *n.pl.* (*Med.*) an infectious tropical disease whose symptoms include sores, framboesia.
Yb, (*chem. symbol*) ytterbium.
yd., (*abbr.*) yard, yards.
ye[1] (yē. yi), *pron.*, *2nd pers. pl.* properly the nominative of *you,* for which it is now used only in the Bible, dialect, poetry etc.
ye[2], **y**[e] (dhē. dhə), the old method of printing THE, (but never pron. yē), from a confusion between the letters þ (th) and y.
yea (yā), *adv.* (now current only in the Bible or in voting) yes; verily, truly, indeed; not only so but also. *n.* an affirmative; one who votes in the affirmative.
yeah (ye), *int.* (*coll.*) yes. **oh yeah!** an expression of sarcasm or incredulity.
year (yiə), *n.* the period of time occupied by the revolution of the earth round the sun (the astronomical, equinoctial, natural, solar or tropical year, the time taken by the sun in returning to the same equinox, in mean length, 365 days, 5 hr, 48 min., 46 sec.; the astral or sidereal year, in which the sun apparently returns to the same place in relation to the fixed stars, 365 days, 6 hr, 9 min., 9 sec.; the Platonic, great or perfect year, estimated by early Greek and Hindu astronomers at about 26,000 years, at the end of which all the heavenly bodies were imagined to return to the same places as they occupied at the Creation); the period of 365 days, from 1 Jan. to 31 Dec., divided into 12 months, adopted as the calendar, legal or civil year, one day being added every fourth year (with the exception of centuries not divisible by 400), called bissextile or leap year; any period of this length taken as a unit of time; a body of students who enter a school or university in the same year; (*pl.*) age, length or time of life; (*pl.*, *coll.*) a long time; old age. **financial year** FINANCE. **historical year,** 12 months beginning on 1 Jan. **regnal year** REGNAL. **since the year dot,** (*coll.*) since as long ago as can be remembered. **year by year,** as the years go by. **year of grace, of our Lord,** a year of the Christian era. **year in year out,** year after year, without cessation. **yearbook,** *n.* a book published annually giving up-to-date information on some subject liable to change. **year-end,** *n.* the end of the (financial) year. **year-long,** *a.* lasting a year, all year. **year-round,** *a.* open, operating, happening, all year. **yearly,** *a.* happening or recurring once a year or every year, annual; lasting a year. *adv.* annually; once a year.

yearling (yiə'ling), *n.*, *a.* (of) an animal more than one and less than two years old; (of) a colt a year old dating from 1 Jan. of the year of foaling.
yearn (yœn), *v.i.* to feel a longing or desire (for, after etc.). **yearner,** *n.* **yearning,** *n.*, *a.* **yearningly,** *adv.*
yeast (yēst), *n.* a yellowish, viscous substance consisting of a growth of fungous cells developed in contact with saccharine liquids and producing alcoholic fermentation by means of enzymes, used in brewing, distilling etc., or for raising dough for bread etc. **yeast-plant,** *n.* any of several single-celled fungi that produce alcoholic fermentation in saccharine liquids. **yeast-powder,** *n.* a baking-powder used as a substitute for yeast. **yeasty,** *a.* containing or resembling yeast, esp. in causing or being characterized by fermentation; (*fig.*) frothy, foamy; unsubstantial, empty, superficial. **yeastily,** *adv.* **yeastiness,** *n.*
yegg (yeg), *n.* (*N Am. sl.*) a safe-breaker or burglar.
yell (yel), *v.i.* to cry out with a loud, sharp, or inarticulate cry as in rage, agony, terror or uncontrollable laughter. *v.t.* to utter or express thus. *n.* such a cry or shout; (*N Am.*) a distinctive shout used by college students etc. for encouragement, applause etc. **yeller,** *n.* **yelling,** *a.*
yellow (yel'ō), *a.* of a colour like that between green and orange in the spectrum or like that of gold, brass, sulphur, lemon; (*coll.*) cowardly; (*offensive*) of one of the Asiatic peoples of mongoloid ancestry. *n.* this colour, a yellow pigment, dye etc.; a sulphur butterfly (and other yellow butterflies and moths); (*pl.*) jaundice; (*N Am., pl.*) a disease of unknown origin attacking peach-trees etc.; egg yolk; the yellow ball in snooker. *v.t.* to make yellow. *v.i.* to turn yellow. **yellow-back,** *n.* a type of cheap novel. **yellow-backed, -bellied, -billed, -headed, -legged** etc., *a.* having a yellow back, belly etc. as specified. **yellow-belly, -bill, -head, -legs, -poll, -rump, -seed,** *n.* used as a name for animals, birds, fish and plants. **yellow-bird,** *n.* the American goldfinch, the yellow warbler, the golden oriole and other yellow birds. **yellow-blossomed,** *a.* **yellow-boy,** *n.* (*sl.*) a gold coin. **yellow card,** *n.* a yellow card shown by a referee to a competitor who has violated a rule. **yellow-earth,** *n.* a yellow ochre, sometimes used as a pigment. **yellow fever,** *n.* a malignant tropical fever caused by the bite of the mosquito, having symptoms such as jaundice and black vomit. **yellow flag,** *n.* a flag hoisted by a ship in quarantine or with an infectious disease on board. **yellow-hammer,** *n.* a bird of the bunting family with yellow head, neck and breast. **yellow jack,** *n.* yellow fever. **yellow-jacket,** *n.* a species of social wasp. **yellow line,** *n.* a line on a road showing parking restrictions. **yellow men,** *n.* the Xanthochroi. **yellow metal,** *n.* an alloy of three parts of copper and two of zinc. **yellow pages**®, *n.pl.* (that part of) a telephone directory, printed on yellow paper, which lists subscribers according to business. **yellow peril,** *n.* the alleged danger that the Asiatic races may overwhelm the Western civilizations. **Yellow Press,** *n.* journalism, or the newspaper press, of sensational and jingoist tendencies. **yellow-rattle,** *n.* an annual herb with yellow flowers and winged seeds that rattle in the capsules when ripe. **yellow spot,** *n.* the area at the centre of the retina in vertebrates where vision is acutest in daylight. **yellow streak,** *n.* (a tendency toward) cowardice. **yellow-wort,** *n.* an annual of the gentian family, used for dyeing things yellow. **yellowish, yellowy,** *a.* **yellowly,** *adv.* **yellowness,** *n.*
yelp (yelp), *v.i.* to utter a sharp, quick cry, as a dog in pain, fear or anticipation. *n.* such a bark or cry. **yelper,** *n.* **yelping,** *n.*
yen[1] (yen), *n.* (*pl.* **yen**) the Japanese monetary unit.
yen[2] (yen), *n.* an ambition, a yearning, a desire, a longing. *v.i.* to yearn.
yeoman (yō'mən), *n.* (*pl.* **-men**) a freeholder not ranking as one of the gentry; (*formerly*) a man qualified to serve on

juries and to vote etc., as holding free land of £2 annual value; a farmer, esp. a freeholder; a small landowner; a member of the yeomanry force; a petty or non-commissioned officer who carries out signalling or clerical duties in the navy. **yeoman of the guard**, a Beefeater. **yeoman service**, *n*. good service, hearty support. **yeomanlike**, *a*. **yeomanly**, *a*. **yeomanry** (-ri), *n*. (*collect*.) yeomen; a British force of volunteer cavalry consisting largely of country gentlemen and farmers, now forming part of the Territorial Army.
yep (yep), (*coll*.) yes.
-yer (-yə), *comb. form*. denoting an agent, as in *lawyer*, *sawyer*.
yerba (yɑ'bə), *n*. Paraguay tea, maté.
yes (yes), *adv*. as you say, it is true, agreed (indicating affirmation or consent); I hear (in answer to a summons etc.). *n*. (*pl*. **yeses**) the word 'yes'; an affirmative reply. **yes-man**, *n*. (*coll*.) an unquestioning follower, a sycophant.
yeshiva(h) (yəshē'və), *n*. (*pl*. **-va(h)s** or **-voth**) a Jewish school devoted to the study of the Talmud; an Orthodox Jewish day school providing religious and secular instruction.
yester-, *pref*. of or pertaining to the day preceding today. **yesterday** (yes'tədi), *n*. the day immediately before today; (*fig*.) time in the immediate past. *adv*. on or during yesterday; in the recent past. **yesteryear**, *n*., *adv*. (*poet*.) last year; the recent past.
yet (yet), *adv*. still, up to this or that time; by this or that time, so soon or early as the present, so far; in addition, further, besides; eventually, at some future time; even (*with compar*.); nevertheless, in spite of that. *conj*. nevertheless, notwithstanding, but still. **as yet**, up to this or that time, so far. **just yet**, in the immediate future (*with neg*.). **not yet**, not up to the present time.
yeti (yet'i), *n*. the hypothetical creature whose tracks are alleged to have been found in the snows of the Himalayas, also called the Abominable Snowman.
yew (ū), *n*. a dark-leaved evergreen shrub or tree, esp. a large tree with spreading branches the wood of which has long been valued for making bows; its wood. **yew-tree**, *n*.
Yggdrasil(l) (ig'drəsil), *n*. (*Scand. Myth*.) the world-tree binding together heaven, earth and hell with its roots and branches.
YHA, (*abbr*.) Youth Hostels Association.
YHVH, **YHWH**, *n*. Yahweh, the tetragrammaton.
Yiddish (yid'ish), *n*. a language spoken by Jews of E Europe and N America, based on a Hebraicized Middle German, with an admixture of Polish, French and English, and usually written in Hebrew characters. *a*. pertaining to this language. **Yid**, *n*. (*offensive*) a Jew. **Yiddisher**, *n*. a Yiddish speaker, a Jew. *a*. pertaining to Yiddish; Jewish.
yield (yēld), *v.t*. to produce, to bear, to bring forth as fruit, reward or result; to give up, to surrender, to concede, to relinquish, to resign. *v.i*. to give a return, to repay one's labour in cultivation etc., to bear, produce or bring forth (well or ill); to give way, to assent, to submit, to comply, to surrender, to make submission (to); to give place or precedence, or admit inferiority (to). *n*. that which is yielded or produced, output, return; annual return from an investment. **yield point**, *n*. the stress point at which a material, under increasing stress, ceases to behave elastically. **yieldable**, *a*. **yielder**, *n*. **yielding**, *a*. compliant; pliable. **yieldingly**, *adv*. **yieldingness**, *n*.
yin (yin), *n*. the feminine, negative, dark principle in nature, according to Chinese philosophy, which interacts with its complement and opposite, *yang*.
yip (yip), *n*. a short, sudden cry, yell. *v.t*. (*past*, *p.p*. **yipped**) to give a short, sudden cry.
yippee (yipē'), *int*. an exclamation of delight, pleasure, exuberant anticipation.

-yl (-il, -īl), *comb. form*. (*chem*.) denoting a radical, as in *ethyl*, *methyl*.
ylang-ylang (ēlangē'lang), *n*. a Malayan tree of the custard-apple family; an oil from the flowers of this tree, used in perfumes.
YMCA, (*abbr*.) (a hostel run by the) Young Men's Christian Association.
-yne (-in), *comb. form*. (*Chem*.) denoting a triple bond, as in *alkyne*.
yob (yob), **yobbo** (-ō), *n*. an aggressive, loutish youth; a hooligan. **yobbish**, *a*.
yodel (yō'dl), *v.t*., *v.i*. to sing or shout in a musical fashion with alternation from the natural voice to the falsetto. *n*. such a shout or musical cry, peculiar to Swiss and Tyrolese mountaineers; a yodelling contest. **yodeller**, *n*.
yoga (yō'gə), *n*. a Hindu system of abstract meditation and rigid asceticism by which the soul is supposed to become united with the eternal spirit of the universe; any system of exercises and practices assisting this. **hatha yoga** (hath'ə), that form of yoga most common in the West, which emphasizes the importance of physical exercises and breathing control. **yogi** (-gi), *n*. a devotee or adept of yoga. **yogic**, *a*. **yogism**, *n*.
yogourt, **yoghourt**, **yogurt**, **yoghurt** (yog'ət, yō'-), *n*. a milk food fermented in a special way and often flavoured with fruit, sugar etc.
yo-heave-ho (yōhēvhō'), (*int. Naut*.) formerly, a sailor's cry while heaving the anchor etc.
yo-ho (yōhō'), *int*. an exclamation calling attention.
yoicks (yoiks), *n*., *int*. (*now often facet*.) a foxhunter's cry.
yojan (yō'jən), *n*. an E Indian measure of distance, usu. about 5 miles (8 km).
yoke (yōk), *n*. a frame or cross-bar fitting over the necks of oxen or other draught animals and attaching them to a plough or vehicle; a pair of draught animals, esp. oxen yoked together; a device resembling this; a frame fitting a person's shoulders for carrying a pair of buckets suspended from the ends; a frame or cross-bar on which a bell swings; the cross-bar of a rudder to which the steering-lines are fastened; a coupling for two pipes discharging into one; a coupling, guiding or controlling piece in a machine; a tie-beam, tie-rod etc.; a part of a garment made to support the rest, as at the shoulders or hips; (*fig*.) a bond, a link, a tie, esp. that of love or wedlock; (*Rom. Hist*.) two upright spears with a third resting across them at the top, under which vanquished enemies were made to pass; hence, servitude, slavery, submission. *v.t*. to put a yoke upon; to unite by a yoke, to couple, esp. in marriage; to enslave. *v.i*. to go or work (well or badly together etc.). **yoke-fellow**, **-mate**, *n*. a person associated with one in marriage, work etc., a companion, a partner. **yoke-line**, **-rope**, *n*. one of the pair of ropes by which a rudder-yoke is worked.
yokel (yō'kəl), *n*. a rustic, a country bumpkin. **yokelish**, *a*.
yolk (yōk), *n*. the yellow part of an egg, the contents of the ovum, esp. that nourishing the embryo; the unctuous secretion from the sebaceous glands of sheep, wool-oil. **yolk-sac**, *n*. the thin, membranous bag enclosing the yolk in an egg. **yolked**, **yolky**, *a*.
Yom Kippur (yom kip'ə, -puə'), *n*. the Day of Atonement, a Jewish day of fasting.
yomp (yomp), *v.i*. (*Mil. sl*.) to trek, often with heavy equipment, over heavy terrain.
yon (yon), *a*., *adv*. (*dial*.) yonder. *pron*. (*dial*.) yonder person, thing or place; that. **yond** (yond), *a*. the most distant. *adv*. yonder.
yonder (yon'də), *a*. that over there; being at a distance, but in the direction looked at or pointed out, and within view. *adv*. over there; at a distance but within view, or where one is looking or pointing.
yoni (yō'ni), *n*. the Hindu symbol of the fertility of nature under which the consort of a male deity is worshipped,

represented by an oval design (the female sexual organs).

yonks (yongks), *n.pl.* (*sl.*) a long time, ages.

yoo-hoo (yoo'hoo), *int.* a call to attract someone's attention.

YOP, *n.* a former training scheme in Britain. [acronym for Youth Opportunity Programme]

yore (yaw), *n.* long ago, old time. **of yore**, formerly, of old time, long ago.

yorker (yaw'kə), *n.* (*Cricket*) a ball bowled so as to pitch immediately in front of the bat. **york**, *v.t.* to bowl (at) with a yorker.

Yorkie, yorkie (yaw'ki), *n.* (*coll.*) a Yorkshire terrier.

Yorkist (yaw'kist), *a.* of or pertaining to the house descended from Edmund Duke of York, son of Edward III, or the White Rose party supporting this in the Wars of the Roses. *n.* an adherent of this house or party.

Yorks. (yawks), (*abbr.*) Yorkshire.

Yorkshire (yawk'shə), *a.* of or derived from Yorkshire. **Yorkshire flannel**, *n.* flannel of undyed wool. **Yorkshire grit**, *n.* a grit used for polishing. **Yorkshire pudding**, *n.* batter baked and served with meat, esp. roast beef. **Yorkshire terrier**, *n.* a small shaggy variety of toy terrier.

Yoruba (yo'rəbə), *n.* (*pl.* **Yoruba, Yorubas**) a member of the people living in the coastal regions of W Africa, especially SW Nigeria; the Kwa language of this people. **Yoruban**, *a.*

you (ū, yu), *2nd pers. pron., sing. and pl.* (*pl. v.*) the person, animal, thing, or persons, etc., addressed; (*reflex.*) yourself, yourselves; (*indef.*) one, anyone, people generally. **you-all**, (*N Am. pl.*; *coll.*) you. **you-know-what, -who**, someone unspecified but known to both the speaker and the hearer.

you'd (yud), *contr.* you had, you would.

you'll (yul), *contr.* you shall, you will.

young (yŭng), *a.* being in the early stage of life, growth or development; of recent birth or beginning, newly formed, produced, come into action or operation etc.; not infirm or decayed with age, vigorous, fresh; immature, raw, inexperienced; pertaining to or characteristic of youth. *n. pl.* offspring, esp. of animals; those who are young. **with young**, (*poet. or Bibl.*) pregnant. **young blood**, *n.* (an injection of) people with vigour or enterprise. **young England, Ireland, Italy, Turks**, etc. *n.* names of political parties striving to sweep away abuses and introduce radical reforms. **(someone's) young man, woman**, *n.* a sweetheart. **young person**, *n.* (*Law*) somebody aged between 14 and 17. **younger** (-gə), *a.* **youngish**, *a.* **youngness**, *n.* **youngster** (-stə), *n.* a young person, a child, a lad; a young animal, such as a young horse.

your (yaw, yə, ūə), *a.* (often used indefinitely instead of *a* or *the* with a suggestion of disparagement) pertaining or belonging to you. **yours** (-z), *pron.* that or those belonging or pertaining to you. **you and yours**, you and your family or belongings. **yours faithfully, obediently, sincerely, truly**, etc., formal expressions preceding the signature in a letter. **yours truly**, (*coll.*) I, this person. **yourself** (-self'), *pron.* (*pl.* **-selves** (-selvz')) you and not another or others, you alone; you in particular; you in your normal condition, health etc.; also used reflexively. **by yourself**, alone; unaided.

you're (yaw, yə, ūə), *contr.* you are.

youth (ūth), *n.* (*pl.* **youths** (ūdhz)) the state of being young; the period of life from infancy to manhood or womanhood, youthfulness; the vigour, freshness, inexperience etc. of this period; a young man; young men and women collectively. **youth club**, *n.* a club which provides leisure time and social activities for young people. **Youth Hostel**, *n.* an organized establishment where hikers (of any age) etc. may put up for the night. **youth hosteller**, *n.* **youth hostelling**, *n.* **youth leader**, *n.* a social worker who works with young people in a particular community. **Youth Training Scheme**, *n.* a government-funded scheme of first-hand job training for unemployed school-leavers. **youthful**, *a.* **youthfully**, *adv.* **youthfulness**, *n.*

you've (ūv, yuv), *contr.* you have.

yowl (yowl), YAWL [1].

yo-yo (yō'yō), *n.* a toy which consists of a spool winding up and down on a string, originally a trade-name. *v.i.* to move up and down rapidly, like a yo-yo.

ypsilon (ipsi'lon, ip'si-), *n.* the 20th letter of the Greek alphabet.

yr. (*abbr.*) year; younger; your.

yrs. (*abbr.*) years; yours.

YTS, (*abbr.*) Youth Training Scheme.

ytterbium (itœ'biəm), *n.* a rare metallic element, at. no. 70; chem. symbol Yb. **ytterbia** (-biə), *n.* ytterbium oxide. **ytterbic**, *a.*

yttrium (it'riəm), *n.* a rare metallic element, at. no. 39; chem. symbol Y. **yttria** (-riə), *n.* yttrium oxide. **yttric, yttrious, yttriferous** (-rif'-), *a.*

yttro-, *comb. form.* yttrium.

yttrocerite (itrōsiə'rit), *n.* a violet-blue fluoride of yttrium.

yttrotantalite (-tan'təlit), *n.* an orthorhombic tantalite of yttrium.

yuan (ywahn), *n.* (*pl.* **yuan**) the standard monetary unit of the People's Republic of China.

yucca (yŭk'ə), *n.* a liliaceous subtropical American flowering plant, with rigid lanceolate leaves and an erect main stalk of white flowers, many species of which are grown for ornament.

yuck (yŭk), *int.* (*coll.*) an exclamation of disgust. *n.* (*coll.*) something unpleasant. **yucky**, *a.* (*coll.*) disgusting, unpleasant.

yuga (yoo'gə), *n.* one of the Hindu ages or cycles of the world.

Yugoslav, Jugoslav (yoo'gəslahv), *a.* of or pertaining to the southern Slav peoples or countries, esp. Yugoslavia. *n.* a native or inhabitant of Yugoslavia.

yulan (yoo'lən), *n.* a Chinese variety of magnolia tree with large, brilliant, snow-white or rosy flowers.

Yule, yule (yool), *n.* Christmas time or the festival of Christmas. **yule log**, *n.* a log formerly burned on Christmas Eve; a chocolate and sponge-cake representation of this. **yuletide**, *n.*

yum-yum (yŭmyŭm'), *int.* (*coll.*) an expression of pleasure, especially anticipation of delicious food. **yummy** (yŭm'i), *int. a.* delicious.

Yuppie, yuppie (yŭp'i), *n.* (*often derog.*) a young, financially successful professional person who spends much money on his/her lifestyle. *a.* pertaining to, designed to appeal to such people. **Yuppie, yuppie flu**, *n.* (*often derog.*) myalgic encephalomyelitis. **yuppi(e)fy**, *v.t.* to make suitable for or typical of yuppies. **yuppification**, *n.* [acronym for young upwardly-mobile (or urban) professional]

yurt, yourt (yuət), *n.* a circular, collapsible tent made of skins and used by nomads in Central Asia.

YWCA, (*abbr.*) (a hostel run by the) Young Women's Christian Association.

Z

Z, z, the last letter of the English alphabet (*pl.* **Zs, Z's, zeds**) has the sound of a voiced or sonant *s*, as in *zeal, lazy, reason*, or of a voiced *sh*, as in *azure*; (*Alg., z*) the third unknown quantity or variable; (*Chem.*) atomic number; (*Phys.*) impedance; something in the shape of a z. **z-axis,** *n.* a reference axis in the Cartesian coordinate system.
z (*abbr.*) zero; zone.
zabaglione (zabalyō'ni), **zabaione** (-bayō'-), *n.* a warm whipped dessert of egg yolks, sugar and marsala.
zabra (zah'brə), *n.* a small sailing vessel formerly used on the coasts of the Iberian Peninsula.
zaffre, zaffer (zaf'ə), *n.* impure oxide of cobalt used for enamelling and as a blue pigment for painting on glass, porcelain etc.
zaibatsu (zībat'soo), *n.* an elite group of wealthy families dominating Japanese industry and commerce.
Zairean (ziiə'riən), *n.* a native or inhabitant of the central African republic of Zaire. *a.* of or pertaining to Zaire.
zakuska (zəkus'kə), *n.* (*pl.* **-ski** (-ki)) a snack, an hors d'oeuvre.
Zambian (zam'biən), *n.* a native or inhabitant of the central African republic of Zambia. *a.* of or pertaining to Zambia.
zambomba (thambom'bə), *n.* a toy musical instrument made by stretching a piece of parchment over a widemouthed jar and inserting a stick through it which is rubbed with the fingers.
zamindar ZEMINDAR.
zamouse (zəmoos'), *n.* the W African short-horned buffalo.
zampogna (tsampōn'yə), *n.* an Italian bagpipe.
zanje (than'hē), *n.* a canal for irrigation. **zanjero** (-heə'rō), *n.* a person employed in working this.
Zante, Zante-wood, (zan'ti), *n.* the wood of the smoke-tree, satinwood.
zany (zā'ni), *n.* a buffoon in old theatrical entertainments who mimicked the clown; a simpleton, a fool; one who acts the fool. *a.* outrageous, fantastical, comical, absurd (e.g. of a comedy show). **zanily,** *adv.* **zaniness,** *n.* **zanyism,** *n.*
zap (zap), *v.t.* (*past, p.p.* **zapped**) (*coll.*) to hit, smack, strike; to overwhelm; to kill or destroy; to cause to go quickly. *v.i.* to move suddenly or quickly; to switch rapidly between television channels using a remote-control device. *n.* energy, go, vitality. *int.* expressing a sudden action. **zapper,** *n.* (*coll.*) one who habitually switches rapidly between television channels. **zappy,** *a.* (*coll.*) energetic, fast-moving; punchy, snappy.
zapateado (thapatäah'dō), *n.* (*pl.* **-dos**) a Spanish dance characterized by much clicking of the heels, and stamping and tapping of the feet.
zapotilla (zapətil'ə), SAPODILLA.
Zarathustrian (zarəthus'triən) etc. ZOROASTRIAN.
zaratite (za'rətit), *n.* a hydrous carbonate of nickel, usu. occurring as an incrustation.
zareba, zareeba, zariba (zərē'bə), *n.* a stockade, hedge or other enclosure for a camp or village in the Sudan.
zarf (zahf), *n.* an ornamental cup-shaped holder for a hot coffee-cup.
zarzuela (zahzwä'lə), *n.* a traditional Spanish form of vaudeville or comic opera.
zastruga (zastroo'gə), **sastruga** (sas-), *n.* (*pl.* **-gi** (-gi)) a ridge of snow caused by the wind.
zax SAX [1].
zeal (zēl), *n.* ardour, earnestness, enthusiasm, intense and eager pursuit or endeavour to attain or accomplish some object. **zealot** (zel'ət), *n.* one full of zeal, esp. one carried away by excess of zeal; a fanatical (esp. religious) partisan. **zealotism, zealotry** (-tri), *n.* **zealous** (zel'-), *a.* **zealously,** *adv.* **zealousness,** *n.*
zebec, zebeck (zē'bek), XEBEC.
zebra (zeb'rə, zē'-), *n.* (*pl.* **-bra, -bras**) a black and white striped, ass-like mammal of the horse family, esp. the true or mountain zebra from the mountainous regions of S Africa (in combination); any of various kinds or species of plant, bird, fish or mammal with similar markings, including the **zebra-antelope, -caterpillar, -finch, -fish, -mouse, -wolf, -wood, -woodpecker. zebra crossing,** *n.* a street-crossing marked by stripes where pedestrians have precedence over all other traffic. **zebrine** (-brin), **zebroid** (-broid), *a.*
zebu (zē'boo), *n.* the humped Indian ox.
zebub (zē'bŭb), *n.* an E African fly similar to the tsetse, a zimb.
zecchino, zechino (zekē'nō), *n.* a sequin (coin).
Zech., (*abbr.*) Zechariah.
Zechariah (zekəri'ə), *n.* an Old Testament book named after the Hebrew prophet to whom it is attributed.
zechstein (zekh'stīn), *n.* (*Geol.*) a German magnesian limestone.
zed (zed), *n.* the letter Z.
zedoary (zed'ōəri), *n.* a substance made from the rootstock of some species of the ginger family, used in medicine, dyeing, perfumery etc.
zee (zē), *n.* (*N Am.*) zed.
zein (zē'in), *n.* a protein found in Indian corn, used in the manufacture of inks, coatings, adhesives etc.
zeitgeist (tsīt'gīst), *n.* the spirit, or moral and intellectual tendency, of a period.
Zelanian (zilā'niən), *a.* (*Zool.*) of or pertaining to New Zealand.
zeloso (zelō'sō), *adv.* (*Mus.*) with energy.
zemindar (zemindah'), **zamindar** (zam-), *n.* one of a class of Bengali landowners formerly paying a certain land-tax to the British government; orig. a local governor and farmer of the revenue under the Mogul empire paying a fixed sum for his district. **zemindary** (zem'-), *n.*
zemstvo (zemst'vō), *n.* a pre-revolutionary Russian elective local assembly dealing with economic affairs.
Zen (zen), *n.* a Japanese Buddhist sect teaching that truth is in one's heart and can be learned only by meditation and self-mastery. **Zenic,** *a.* **Zenist,** *n.*
zenana (zinah'nə), *n.* in the East (esp. India or Iran), the portion of the house in a Hindu or Muslim household which is reserved for the women. **Zenana Mission,** *n.* a mission undertaken by women for spreading educational, medical and religious reforms among the inmates of zenanas.
Zend (zend), *n.* a former name for the ancient Iranian language, closely allied to Sanskrit, in which the sacred writings of the Zoroastrians are set down, now usu. called

Avestan; a name for the Zend-Avesta. **Zend-Avesta** (-əves'tə), *n.* a collection of the sacred scriptures of the Parsees or Zoroastrians together with a commentary.
zenith (zen'ith), *n.* the point in the heavens directly overhead to an observer, opp. to *nadir*; the highest or culminating point. **zenith-distance**, *n.* the angular distance of a heavenly body from the zenith. **zenithal**, *a.* **zenithal projection**, *n.*
zeolite (zē'əlīt), *n.* any one of a group of hydrous silicates found in cavities of eruptive rocks, which absorb radioactivity and which gelatinize in acid owing to the liberation of silica; any of various synthetic silicates resembling this. **zeolithiform** (-lith'ifawm), **zeolitic** (-lit'-), *a.*
Zeph., (*abbr.*) Zephaniah.
Zephaniah (zefəni'ə), *n.* an Old Testament book named after the Hebrew prophet to whom it is attributed.
zephyr (zef'ə), *n.* the west wind personified; any soft, gentle breeze; a light, gauzy fabric or worsted or woollen yarn, used for shawls, jerseys etc.; a jersey or other garment made of this.
Zeppelin, zeppelin (zep'əlin), *n.* a large dirigible airship. [Count *Zeppelin*, 1838–1918, German general and aeronaut]
zerda (zœ'də), *n.* the fennec fox.
zero (ziə'rō), *n.* (*pl.* **-oes**) the figure 0, a cipher, nothing, nought, nil; the point on a scale from which positive or negative quantities are reckoned, esp. on a thermometer (e.g. in Fahrenheit's thermometer 32° below the freezing-point of water; in the Centigrade and Réaumur's scales zero is the freezing-point); the lowest point in any scale or standard of comparison, the nadir, nullity. *a.* having no measurable quantity, size etc.; of a cloud ceiling, limiting visibility to 15 m (approx. 50 ft.) or less; of horizontal visibility, limited to 50 m (approx. 165 ft.) or less; (*coll.*) not any, nothing. *v.t.* to adjust or set to zero (of an instrument, scale, gauge etc.). **absolute zero** ABSOLUTE. **to zero in on**, (*coll.*) to focus attention on, to fix on; to aim for; to converge upon, to home in on. **zero hour**, *n.* the precise hour for the commencement of a prearranged military movement or other action, operation etc. **zero option**, *n.* a proposal that both sides in international nuclear arms negotiations agree to limit or remove shorter-range nuclear missiles. **zero-rated**, *a.* referring to goods on which the buyer need pay no value-added tax, but on which the seller can claim back any value-added tax he/she has already paid. **zeroth** (-th), *a.* (*Math.*) referring to a term in a series of terms which precedes what is usu. regarded as the first term.
zest (zest), *n.* a piece of lemon or orange peel, or the oil extracted from this, used to give a flavour to soups, wines etc.; hence, that which makes a thing enjoyable, piquancy, relish; keen enjoyment. **zestful**, *a.* **zestfully**, *adv.* **zestfulness**, *n.* **zesty**, *a.*
ZETA (zē'tə), *n.* a British apparatus used in investigating the controlled production of nuclear energy by the fusion of hydrogen. [acronym for *Zero Energy Thermonuclear Apparatus*]
zeta (zē'tə), *n.* the sixth letter of the Greek alphabet (Z, ζ).
zetetic (zitet'ik), *a.* proceeding by enquiry. *n.* a seeker; an enquiry, a search.
zeugma (zūg'mə), *n.* a figure of speech in which a verb or adjective governs or modifies two nouns to only one of which it is logically applicable. **zeugmatic** (-mat'-), *a.*
Zeus (zūs), *n.* the supreme deity in Greek mythology, corresponding to the Roman Jupiter.
zeuxite (zūk'sīt), *n.* a variety of tourmaline.
zho (zhō), **zo, dso, dzo** (zō), *n.* (*pl.* **-s**) a hybrid breed of Himalayan cattle developed from crossing the yak with common horned cattle.
zibel(l)ine (zib'əlin, -lin), *a.* of, pertaining to or resembling the sable. *n.* the sable or its fur; a thick woollen fabric with a soft nap.
zibet (zib'it), *n.* the Indian or Asiatic civet.
zidovudine (zīdov'ədēn), *n.* an alternative name for the AIDS drug AZT.
ziggurat (zig'ərat), *n.* an ancient Mesopotamian temple-tower of a rectangular or tiered design.
zigzag (zig'zag), *a.* having or taking sharp alternate turns or angles to left and right; of a sewing machine, capable of executing zigzag stitches. *n.* a zigzag line, road, path, pattern, moulding, series of trenches, stitches etc. *adv.* in a zigzag course or manner. *v.t.* (*past, p.p.* **zigzagged**) to form or do in a zigzag fashion. *v.i.* to move in a zigzag course. **zigzaggery**, *n.* **zigzaggy**, *a.*
zilch (zilch), *n.* (*esp. N Am sl.*) nothing, zero.
zilla(h), zila (zil'ah), *n.* an administrative district in India or Bangladesh.
zillion (zil'yən), *n.* (*coll.*) a huge unspecified amount, quantity or number.
zimb (zimb), *n.* a dipterous insect common in Ethiopia resembling the tsetse, and hurtful to cattle.
Zimbabwean (zimbahb'wiən), *n.* a native or inhabitant of the southern African republic of Zimbabwe. *a.* of or pertaining to Zimbabwe.
zimmer® (zim'ə), *n.* a metal walking-frame used as a means of support by those with walking difficulties.
zimocca (zimok'ə), *n.* a soft, fine, cup-shaped bath-sponge from the Mediterranean.
zinc (zingk), *n.* a bluish-white metallic element, at. no. 30; chem. symbol Zn, used in the manufacture of brass and nickel-silver, for coating sheet-iron, as roofing-material, for printing-blocks etc. *v.t.* to coat or cover with zinc. *a.* of or containing zinc. **flowers of zinc**, zinc oxide. **zinc-blende**, *n.* native sulphide of zinc, sphalerite. **zinc ointment**, *n.* a medical preparation of zinc oxide in an ointment base such as petroleum jelly. **zinc oxide**, *n.* white powdery oxide of zinc used as a white pigment, and in cements, ointments etc. **zinc sulphate**, *n.* white vitriol. **zinc-white**, *n.* oxide of zinc used as a pigment. **zincic, zinciferous, zinkiferous** (-kif'-), *a.* **zincite** (-īt), *n.* a native oxide of zinc. **zincograph** (-graf), *n.* a zinc plate on which a picture or design has been etched in relief for printing; an impression from this. **zincographer** (-kog'-), *n.* **zincographic, -ical** (-graf'-), *a.* **zincography** (-kog'-), *n.* **zincoid** (-koid), **zincous**, *a.* **zincy, zin(c)ky**, *a.*
zing (zing), *n.* (*coll.*) a shrill buzzing noise as of a bullet or a vibrating string; energy, go, zest. *v.i.* (*coll.*) to move very quickly esp. (as) with a high-pitched humming sound. **zinger**, *n.* (*esp. N Am. coll.*) a sharp, sarcastic or witty retort. **zingy** (-i), *a.* (*coll.*).
zingaro (zing'gərō), *n.* (*pl.* **-ri** (rē)) a gypsy.
Zinjanthropus (zinjan'thrəpəs), *n.* name given to the manlike fossil found in Tanzania in 1959 now known as *Australopithecus boisei*.
zinke (tsing'kə), *n.* (*pl.* **-ken** (-kən)) an old musical instrument consisting of a leather-covered tube with seven finger-holes, the precursor of the cornet.
zinky ZINC.
zinnia (zin'iə), *n.* a plant of the aster family with showy-rayed flowers in single terminal heads. [J G *Zinn*, 1727-59, German botanist]
Zion (zī'ən), **Sion** (sī'-), *n.* a hill in ancient Jerusalem, the royal residence of David and his successors; the ancient Hebrew theocracy, the Church of Christ, the heavenly Jerusalem, heaven; the Jewish homeland or people; the modern Jewish nation of Israel; used as a name for a Nonconformist chapel. **Zionism**, *n.* orig. a movement for establishing the resettlement of Palestine as the Jewish homeland, and the development of the state of Israel. **Zionist**, *n.*, *a.* **Zionistic** (-nis'-), *a.*
zip (zip), *n.* a zip-fastener, a zipper; the sharp sound made by a bullet or other missile striking an object or flying through the air; (*coll.*) energy, zest. *v.i.* (*past, p.p.*

zipped) to move or fly (as) with such a sound. **to zip along, through,** to move swiftly; to rush, to finish quickly. **to zip up,** to fasten by means of a zip. **zipfastener, zipper,** *n.* a fastening device with interlocking teeth, which opens or closes with a single motion. **zippy,** *a.* (*coll.*) energetic, speedy.

zip code (zip), *n.* (*N Am.*) postal code. [acronym for zone improvement plan]

zircon (zœ'kon), *n.* a translucent, variously-coloured silicate of zirconium, some varieties of which are cut into gems. **zircalloy** (-al'oi), *n.* an alloy of zirconium and small quantities of nickel, chromium and tin used in nuclear reactors. **zirconate** (-āt), *n.* a salt of zirconic acid. **zirconic** (-kon'-), *a.* **zirconium** (-kō'niəm), *n.* an earthy metallic element, at. no. 40; chem. symbol Zr, found chiefly in zircon.

zit (zit), *n.* (*sl.*) a spot, a pimple. **zitty,** *a.*

zither (zidh'ə), *n.* a simple stringed instrument consisting of a flat sounding-board and strings plucked by the fingers. **zitherist,** *n.*

zizz (ziz), *n.* (*coll.*) a nap. *v.i.* to doze.

zloty (zlot'i), *n.* (*pl.* **zloty, -s**) a coin and monetary unit of Poland.

Zn, (*chem. symbol*) zinc.

zo ZHO.

zo-, *comb. form.* zoo; as in *zoology.*

zoa (zō'ə), *pl.* -ZOON.

zoa, *comb. form.* animals, as in *Metazoa, Protozoa.*

zoanthropy (zōan'thrəpi), *n.* a mental disorder in which the patient believes himself transformed into one of the lower animals. **zoanthropic** (-throp'-), *a.*

zocco (zok'ō), **zoccolo** (-lō), *n.* (*pl.* **-ccos, -ccolos**) a plinth for a statue, a socle.

zodiac (zō'diak), *n.* the zone or broad belt of the heavens, extending about 8° to each side of the ecliptic, which the sun traverses during the year, anciently divided into 12 equal parts called the **signs of the zodiac**, which orig. corresponded to the **zodiacal constellations** bearing the same names, but now coinciding with the constellations bearing the names next in order. **zodiacal** (-di'ə-), *a.* pertaining to the zodiac. **zodiacal light,** *n.* a triangular tract or pillar of light sometimes seen, esp. in the tropics, rising from the point at which the sun is just about to rise or has just set.

zoetic (zōet'ik), *a.* pertaining to or of the nature of life, vital.

zoic (zō'ik), *a.* of or pertaining to animals or animal life; of rocks, containing fossils or other evidences of plant or animal life.

-zoic, *comb. form.* indicating a geological era, as in *Mesozoic, Palaeozoic.*

zoisite (zoi'sit), *n.* a translucent silicate of calcium and aluminium, first found in Carinthia. [Baron von *Zois*, 1747–1819, Slovenian geologist]

zoism (zō'izm), *n.* the doctrine that life originates from a specific principle, and is not merely the result of various forces. **zoist,** *n.* **zoistic** (-is'-), *a.*

Zolaism (zō'ləizm), *n.* excessive naturalism, unshrinking realism dealing with the sordid and repulsive aspects of life. **Zolaesque** (-esk'), **Zolaistic** (-is'-), *a.* **Zolaist,** *n.* [Emile *Zola*, 1840–1902, French novelist]

Zollverein (tsol'fərin), *n.* a customs union among states maintaining a tariff against imports and usu. having free trade with each other; a customs union among German states in the early 1830s led by Prussia.

zombi(e) (zom'bi), *n.* orig. an African snake god; in W Indian voodooism, a reanimated dead person capable of slow automatic movements; the supernatural spirit animating the dead person's body; a stupid, apathetic or slow-moving person. **zombielike,** *a.* **zombiism,** *n.*

zonal, zonary etc. ZONE.

zonda (zon'də), *n.* a hot dry west wind blowing from the Andes, usu. during July and August, in the Argentine.

zone (zōn), *n.* a well-marked band or stripe encircling an object; any one of the five great divisions of the earth bounded by circles parallel to the equator (the **torrid zone** between the tropics extending 23½° on each side of the equator, the **temperate zones** between the tropics and the polar circles and the **frigid zones** situated within the polar circles); any well-defined belt or tract of land distinguished by climate, the character of its organisms etc.; the part of the surface of a sphere or of a cone or cylinder enclosed between two parallel planes perpendicular to the axis; a stratum or area of rock distinguished by particular fossil remains; an area sectioned off for a particular function or restriction (e.g. a smoke-free zone); an area characterized by a particular form of government, business practice etc. (e.g. a duty-free zone). *v.t.* to encircle with or as with a zone; to allocate to certain districts or zones; to divide into zones. **zonal, zonary,** *a.* **zonally,** *adv.* **zonate** (-āt), **-ated,** *a.* marked with zones or concentric bands of colour. **zonation,** *n.* arrangement or division into zones. **zoned,** *a.* **zoner,** *n.* **zoning,** *n.* division into, or allocation to, zones; the marking off in town-planning of certain areas for specific purposes, e.g. residence, shopping etc.

zonked (zongkt), *a.* (*sl.*) intoxicated by drugs or alcohol, extremely drunk or stoned; (*sl.*) completely tired out, exhausted.

zoo (zoo), *n.* a zoological garden or a collection of living wild animals, orig. the Zoological Gardens in London.

zoo-, *comb. form.* pertaining to animals or to animal life.

zooblast (zō'əblast), *n.* an animal cell.

zoochemistry (zōkəm'istri), *n.* the chemistry of the substances occurring in the animal body. **zoochemical,** *a.*

zooecium (zōē'shiəm), *n.* (*pl.* **-cia,** -shiə), one of the cells forming the enclosing structure of polyzoans.

zoogamy (zōog'əmi), *n.* sexual reproduction of animals. **zoogamous,** *a.*

zoogeny (zōoj'əni), *n.* the original of living animals. **zoogenic** (-jen'-), **-genous** (-oj'-), *a.* produced from animals.

zoogony (zōəog'əni), *n.* the formation of animal organs.

zoogeography (zōəjiog'rəfi), *n.* the study of the distribution of animals, faunal geography. **zoogeographer,** *n.* **zoogeographic, -ical** (-graf'-), *a.* **zoogeographically,** *adv.*

zoography (-og'-), *n.* descriptive zoology. **zoographer,** *n.* **zoographic, -ical** (-graf'-), *a.*

zooid (zō'oid), *a.* having the nature of an animal; having organic life and motion. *n.* a more or less independent organism developed by fission or gemmation; a member of a compound organism; an organic body or cell capable of independent motion. **zooidal** (-oid'-), *a.*

zool., (*abbr.*) zoological; zoology.

zoolatry (zōol'ətri), *n.* animal-worship. **zoolater** (-tə), *n.* animal-worshipper. **zoolatrous,** *a.*

zoolite (zō'əlīt), **zoolith** (-lith), *n.* a fossil animal or animal substance. **zoolitic** (-lit'-), **-litic** (-lit'-), *a.*

zoology (zoo·ol'əji, zōol'-), *n.* the natural history of animals, the branch of biology dealing with the structure, physiology, classification, habits and distribution of animals. **zoological** (-loj'-), *a.* **zoological garden,** *n.* a public garden or park in which a collection of wild and other animals is kept. **zoologically,** *adv.* **zoologist,** *n.*

zoom (zoom), *v.i.* to turn an aircraft upwards suddenly at a very sharp angle; to make a continuous, deep, loud buzzing noise; to move quickly (as) with this noise; to rise rapidly, to soar (e.g. of prices). *n.* an act, instance or sound of zooming; a zoom lens; in cinematography, a shot taken with a lens whose focal length is adjusted during the shot. **to zoom in** or **out,** to increase or decrease rapidly the focal length of a zoom lens when taking a photograph, a film shot etc. in order to change the size of the image. **zoom lens,** *n.* a lens in a camera or microscope which has a variable focal length and can increase or de-

crease the size of an image continuously without moving nearer the object.
zoomagnetism (zōəmag'nitizm), *n.* animal magnetism.
zoomancy (zō'əmansi), *n.* divination by means of observation of the movements and behaviour of animals.
zoometry (zōom'itri), *n.* comparative measurement of the parts of animals. **zoometric, -ical** (-met'-), *a.*
zoomorphic (zōəmaw'fik), *a.* pertaining to or exhibiting animal forms; of religious symbolism representing animals; of gods represented under the form of animals. **zoomorphism, -morphy** (zō'-), *n.*
-zoon, *comb. form.* animal, as in *spermatozoon*.
zoon (zō'on), *n.* (*pl.* **zoa** (-ə)) the total product of a fertilized ovum; a developed individual of a compound organism. **zoonal,** *a.* **zoonic** (-on'-), *a.* **zoonomy** (-on'əmi), *n.* the science of the laws of animal life. **zoonomic** (-nom'-), *a.* **zoonomist,** *n.*
zoonosis (zōon'əsis), *n.* a disease which can be transmitted to humans by animals, e.g. rabies. **zoonotic** (-not'-), *a.*
zoopathology (zōəpəthol'əji), *n.* animal pathology.
zoophagous (zōof'əgəs), *a.* feeding on animals, carnivorous. **zoophagan,** *n.,* *a.*
zoophile (zō'əfil), *n.* an animal lover; a defender of animal rights and welfare. **zoophilia** (-fil'iə), **zoophilism** (-of'-), **zoophily** (-of'-), *n.* love of animals; sexual attraction towards animals. **zoophilist,** *n.* **zoophilous,** *a.*
zoophobia (zōəfō'biə), *n.* morbid fear or hatred of animals. **zoophobous** (-of'-), *a.*
zoophorus (zōof'ərəs), *n.* a continuous frieze carved with figures of men and animals in relief. **zoophoric** (-fo'rik), *a.*
zoophyte (zō'əfit), *n.* an invertebrate animal presenting many external resemblances to a plant, as a coral seaanemone, holothurian, sponge etc. **zoophytic, -ical** (-fit'-), *a.* **zoophytoid** (-of'itoid), *a.* **zoophytology** (-tol'-), *n.* the natural history of zoophytes. **zoophytological** (-loj'-), *a.* **zoophytologist** (-tol'-), *n.*
zooplankton (zō'əplangktən), *n.* the minute floating animal life of a body of water. **zooplanktonic** (-ton'-), *a.*
zoospore (zō'əspaw), *n.* a spore having the power of independent motion, usu. by means of cilia. **zoosporic** (-spo'-), **-sporous** (-os'-), *a.*
zootaxy (zō'ətaksi), *n.* the science of the classification of animals.
zootechnics (zōətek'niks), *n.* the science of breeding and domestication of animals.
zootomy (zōot'əmi), *n.* the dissection or anatomy of animals. **zootomic, -ical** (-tom'-), *a.* **zootomically,** *adv.* **zootomist,** *n.*
zootoxin (zōətok'sin), *n.* a toxin produced by an animal, e.g. snake venom.
zootrophic (zōətrof'ik), *a.* pertaining to the nourishment of animals.
zoot suit (zoot), *n.* a man's baggy suit popular in the late 1940s, consisting of a long jacket with fitted waist and padded shoulders, and wide trousers tapering into narrow cuffs.
zoril (zo'ril), **zorilla** (-ril'ə), **zorille** (-ril'), *n.* a small carnivorous quadruped allied to the skunks and polecats, found in Africa and Asia Minor. **zorra** (zo'rə), **zorillo,** *n.* a S American skunk. **zorro** (zo'rō), *n.* a S American fox-wolf.
Zoroastrian (zorōas'triən), *a.* pertaining to Zoroaster or the religious system set forth by him and his followers in the Zend-Avesta, based on the ancient Persian religion of the Magi and still held by the Parsees, sometimes called fire-worshippers. *n.* a follower of Zoroaster; an adherent of Zoroastrianism. **Zoroastrianism,** *n.*
zorra, zorro ZORIL.
zoster (zos'tə), *n.* an ancient Greek girdle or belt, worn esp. by men; shingles; herpes zoster.
Zouave (zooahv'), *n.* a soldier belonging to a French light infantry corps, orig. composed of Algerian recruits and still wearing an Oriental uniform; a zouave jacket. **zouave jacket,** *n.* a short, round-fronted jacket, usu. sleeveless, worn by women.
ZOUK (zook), *n.* a kind of lively music originating in the French Caribbean.
zounds (zowndz, zoondz), *int.* an exclamation of anger, surprise etc. [contr. from *God's wounds*, an obsolete oath]
Zr, (*chem. symbol*) zirconium.
zucchini (zukē'ni), *n.* (*chiefly N Am. and Austral.*) (*pl.* **-ni, -nis**) a courgette.
zuchetto (tsuket'ō), **zuchetta** (-ta), *n.* (*pl.* **-os**) the skullcap of a Roman Catholic ecclesiastic, black for priest, purple for bishop, red for cardinal, white for pope.
zugzwang (tsook'tsvang), *n.* a blocking position in chess making any move by an opponent disadvantageous. *v.t.* to place an opponent in this position.
Zulu (zoo'loo), *n.* a member of a branch of the Bantu people of SE Africa; the language of the Zulus, of or pertaining to this people or their language.
Zuñi (zoon'yē), *n.* a member of an American Indian people of New Mexico; the language of this people. **Zuñian,** *n.,* *a.*
Zwinglian (zwing'gliən), *a.* of or pertaining to Ulrich Zwingli (1485–1531), Swiss leader of the Reformation, or his doctrines. *n.* a believer in Zwinglian doctrine; a follower of Zwingli.
zwitterion (tsvit'əriən), *n.* an ion that carries both a positive and negative electric charge.
zydeco (zi'dikō), *n.* a blues-orientated Cajun music of the Louisiana Creole community, characterised by an individual style of accordion playing.
zyg-, zygo-, *comb. form.* yoke, union. **zygal** (zī'gəl), *a.* H-shaped, of the nature of a zygon.
zygobranchiate (zīgōbrang'kiət), *a.* having the right and the left gills alike (of certain gasteropods). **zygobranch** (zī'-), *n.* one of the Zygobranchia, an order of zygobranchiate gasteropods.
zygodactyl (zīgōdak'til), *a.* of birds, having the toes disposed in pairs, two in front and two behind; belonging to the Zygodactylae, a group of birds with two toes pointed forwards and two backwards, as in the parrots. *n.* one of the Zygodactylae. **zygodactylic** (-til'-), **zygodactylous,** *a.*
zygoma (zīgō'mə, zi-), *n.* (*pl.* **-mata** (-tə)) the arch joining the malar and temporal bones of the skull. **zygomatic** (-mat'-), *a.* **zygomatic arch,** *n.* the zygoma. **zygomatic bone,** *n.* the cheekbone.
zygomorphic (zīgōmaw'fik), **zygomorphous,** *a.* of flowers, divisible into similar halves only in one plane. **zygomorphism, -morphy** (zi'-), *n.*
zygon (zi'gon), *n.* a connecting bar, as the cross-bar of an H-shaped fissure of the brain.
zygophyte (zi'gəfit), *n.* a plant reproduced by means of zygospores.
zygopleural (zīgōploo'rəl), *a.* bilaterally symmetrical.
zygosis (zīgō'sis), *n.* (*Biol.*) conjugation. **zygose** (zī'-), *a.* relating to zygosis.
zygospore (zī'gəspaw), *n.* a spore formed by conjugation of two similar gametes. **zygosporic** (-spo'-), *a.*
zygote (zī'gōt), *n.* the product of the fusion between the oocyte and the spermatozoon; the fertilized ovum. **zygotic** (-got'-), *a.*
zyme (zim), *n.* a ferment, a disease-germ; the supposed cause of a zymotic disease. **zymase** (-mās), *n.* an enzyme causing fermentation in carbohydrates. **zymic,** *a.* relating to fermentation.
zym(o)-, *comb. form.* indicating fermentation.
zymogen (zī'məjən), *n.* a substance developing by internal change into a ferment or enzyme.
zymology (zimol'əji), *n.* the science of fermentation. **zymologic, -ical** (-loj'-), *a.* **zymologist,** *n.*
zymolysis (zimol'isis), *n.* zymosis.

zymometer (zīmom'itə), **zymosimeter** (-mōsim'-), *n.* an instrument for measuring the degree of fermentation.
zymosis (zīmō'sis), *n.* the process of fermentation, esp. that by which disease is introduced into the system; any zymotic disease. **zymotic** (-mot'-), *a.* of, pertaining to or produced by fermentation. **zymotic disease**, *n.* an epidemic, endemic or contagious disease produced by the multiplication of germs introduced from without. **zymotically**, *adv.*
zymurgy (zī'mœji), *n.* the department of technological chemistry treating of processes in which fermentation plays the principal part.
zythum (zī'thəm), *n.* a malt beverage of the ancient Egyptians.

Appendix I
Register of New Words

INTRODUCTION

Since the *Cassell Concise Dictionary* was published in 1989, the English language has changed and expanded quite noticeably. There have been advances in the scientific and medical fields; the worlds of entertainment, politics and communications have produced many new terms; and of course the vocabulary of slang and colloquial English has shifted as rapidly as it has ever done. New objects, ideas and lifestyles need new words or new uses of words to describe them.

Some of our 'new' words are in fact several years old, but have only very recently come to be widely used. Others are genuinely new. Of these, some will remain as a permanent part of the language, and others will disappear again as they become less useful or attractive. But they are all very much alive now, and we believe that our Register of New Words will be interesting and useful to all users of this dictionary.

Quotations from newspapers or magazines are provided with some of the entries where they can illustrate the use or the context of a word, or contribute some interesting information beyond the basic definition.

A

abled (ā'bəld), *a.* having full (physical) ability, able-bodied, used esp. as opp. of *disabled*. **differently abled,** *a.* disabled.
ace (ās), *a.* (*coll.*) excellent, of high quality.
acid (as'id), *a.* of colours, of a hard, very bright shade; relating to acid house music.
ADC, (*abbr.*) analogue–digital converter.
additive (ad'itiv), *n.* something which is added, esp. a chemical added to a processed substance such as food, in order to improve its appearance, longevity etc.
additive-free, *a.* of food etc., containing no chemical additives.
ADP, (*abbr.*) automatic data-processing.
affinity card (əfin'iti), *n.* a credit card linked to a charity, which receives a donation from the sponsoring company on the issue of the card, and subsequent payments in proportion to the money spent using the card.
agent (ā'jənt) *n.* (*Comput.*) a kind of computer language which enables a computer to program itself, and process data semi-autonomously.
AIDS-related complex (ādz), *n.* the stage in the development of AIDS when physical symptoms (such as loss of weight, herpes infection) start to appear, before the onset of full-blown AIDS.
Alar ® (ā'lah), *n.* a chemical, daminozide, used by fruit (esp. apple) growers, sprayed on to the growing fruit to produce larger and better-looking fruit.

aliterate (ālit'ərət), *n.*, *a.* (one who is) able to read, not illiterate, but who prefers to get information and entertainment by other means. [Gr. *a-*, not, LITERATE]
all (awl), *a.* **all-seater,** *a.* referring to a sports ground at which all the spectators are seated, for safety reasons. **all-singing all-dancing,** *a.* (*coll.*) having every available feature; performing any function required. **all-terrain bicycle** MOUNTAIN BICYCLE (see Dict.). **all-terrain vehicle,** *n.* a small, sturdily-built four-wheeled motor vehicle, with a wide wheelbase and very large cushion-like tyres which make the vehicle suitable for use even on very rough ground.
American pit bull terrier PIT BULL TERRIER.
animatronics (animətron'iks), *n. sing.* the technique of making lifelike robots programmed electronically to move and/or make noises. [*anima*te elec*tronics*]
antivirus (an'tīvīrəs), *n.* (*Comput.*) software designed to protect a computer system from a virus. **antiviral,** *a.* (*Comput.* and *Med.*).
antsy (ant'si), *a.* (*sl.*) excited, eager.
aqualeather (akwə'ledh'ə), *n.* tanned fish-skin.
arb (ahb), *n.* short for ARBITRAGEUR (see Dict.).
ARC (ahk), (*acronym*) AIDS-related complex.
architectural salvage (ahkitek'tūrəl), *n.* the practice of removing architectural features (such as fireplaces, doorframes,

969

mouldings) from buildings about to be demolished, so that they can be reused elsewhere.

atrium (ā'triəm), *n. (pl.* **atria, atriums**) a central hall in a large building, glass-roofed and stretching the whole height of the building, often overlooked by galleries on the upper floors.

awesome (aw'səm), *a. (sl.)* strikingly good, great, marvellous.

B

bad-mouth (bad'mowth), *v.t. (sl.)* to criticize, speak maliciously of.

bail bandit (bāl), *n.* an offender who breaks the law while on bail.

The ... police forces said last year that offences by 'bail bandits' had helped to inflate the record crime rate. The statistics led to calls for more people to be remanded in custody. *Independent*, 7 February 1992

Baker day (bā'ke), *n.* any one of the days during the normal school year which are set aside for the in-service training of teachers, and when pupils do not attend school.

ballistic (bəlis'tik), *a.* **to go ballistic,** *(sl.)* to become explosively angry, lose one's temper.

ballpark (bawl'pahk), *n. (N Am.)* a baseball field; *(coll.)* a sphere of activity. **ballpark figure,** *n. (coll., orig. N Am.)* an approximate amount.

bang (bang), *n. (sl.)* an injection of heroin. *v.t.* to inject (heroin). **bang on,** exactly right. **to bang on,** to talk loudly and at great length. **to bang up,** to imprison; to lock up in a cell.

bankroll (bangk'rōl), *n. (orig. N Am.)* a supply of money; a sum of money used to buy or invest in something, *v.t.* to supply the funding for (a purchase or investment).

basho (bash'ō), *n.* a tournament in sumo wrestling. [Jap.]

bells and whistles (belz, wis'əlz), *n. pl. (coll.)* showy but inessential accessories or additional features.

belt-bag BUM-BAG.

best-before date (bestbifaw'), *n.* the date marked on the packaging of a perishable product (usu. food or drink) indicating the date before which it is safe to use it; *(coll.)* the optimum age of a person, machine etc., before decline sets in.

bhangra (ban'grə), *n.* music based on a fusion of Asian and contemporary pop music. [from *bhangra*, traditional music of the Punjab]

bias crime (bī'əs), *n. (N Am.)* a racially motivated crime.

biosensor (bī'ōsensə), *n.* an electronic chip covered with a layer of e.g. an enzyme which can sense the presence of a particular substance, used in diagnosing illness, detecting the presence of drugs etc.

bioturbation (bīōtœbā'shən), *n.* the disturbance of layers in soil by the action of living creatures.

One factor not yet fully evaluated is the extent to which the pottery may have sunk through the silts to a lower and earlier level than its age suggests: archaeologists are well aware of the amount of bioturbation that can be caused by earthworms and other burrowing creatures, which enables heavy objects to move downwards. *The Times*, 30 August 1993

birth mother (bœth), *n.* the woman who has given birth to a child, not the adoptive mother.

blag (blag), *n. (sl.)* robbery, esp. violent robbery. *v.t., v.i.* to rob; to cadge or scrounge. **blagger,** *n.*

... the old-style blagger who might keep a sawn-off shotgun behind the chimney-piece, bringing it out every few months to hit a building society. *The Times*, 14 March 1992

blip (blip), *n.* a temporary movement in the performance of something especially in an unexpected or unwelcome direction.

blue flag (bloo), *n.* the European Community flag, allowed to be flown at beach areas which meet EC standards of cleanness of the beach and pollution-free water.

blush wine (blŭsh), *n. (orig. N Am.)* rosé wine.

bobbitt (bob'it), *v. (N Am.)* to sever the penis of one's husband or lover. [after the action of the American Lorena Bobbitt who attacked her husband with a carving knife]

bodacious (bodā'shəs), *a. (sl., esp. N Am.)* excellent, wonderful.

body (bod'i), *n.* (also **bodysuit,** -soot, -sūt) a woman's close-fitting, one-piece garment, made of stretch material and fastened by press-studs at the crotch. **body-bag,** *n.* a strong plastic bag in which a dead body (of e.g. a soldier killed in battle) is transported home. **body count,** *n.* a calculation or estimate of the number of violent deaths, in battle or in a film etc. **body scanner,** *n. (Med.)* an X-ray or ultrasound machine that uses a computer to produce cross-sectional pictures of the body.

bogus (bō'gəs), *a. (sl., esp. N Am.)* awful, bad, disappointing.

bouncy castle (bown'si), *n.* a children's play area consisting of an inflatable base (for bouncing on) and high inflatable sides decorated to look like (usu.) a castle.

BSE, *(abbr.)* bovine spongiform encephalopathy.

The chances of a human catching bovine spongiform encephalopathy – a disease that affects the brains of cattle – from contaminated meat are extremely remote.
New Scientist, 4 March 1989

bubble beat (bŭb'l), *n.* a form of techno music toned down for a mass audience, with a slower beat and fewer electronic effects.

Most chart Techno is derisively known as bubble beat to true ravers who prefer the head-on attack and breakneck speeds (over 130 bpm, or beats per minute) of Hardcore. *Guardian*, 29 February 1992

buddy (bŭd'i), *a.* of a film or story, dealing with the adventures of and relationship between usu. two male partners. *v.i.* to act as BUDDY (see Dict.) to a person with AIDS.

Buddy movies, celebrating the comradeship between men who work or fight or play together, are almost as old as Hollywood itself. *Listener*, 30 March 1989

bum-bag (bŭm'bag), *n.* (*coll.*) (also **belt-bag**) a pouch, usu. fastened with a zip, for holding money and other small personal belongings, worn on a belt round the waist or hips. Usu. worn at the front of the body, but orig. at the back, hence the name.

bungee jumping (bŭn'jē), *n.* the sport of jumping off high places with a rubber rope tied round one's ankles or attached to a body harness, the rope stretching to break one's fall only a few feet above the ground. [etym. uncertain]

Bungee jumping originated in Pentecost Island, off the north-east coast of Australia, as an initiation ceremony for boys ... The natives would build a very high platform in a tree, attach vines to their ankles and jump off. *Independent*, 20 February 1991

C

Canto-pop (kan'tō), *n.* popular music originating in Hong Kong and sung in Cantonese.

carbon tax (kah'bən), *n.* a suggested range of taxes on all forms of fossil fuel, but esp. on petrol, which would reduce the use of such fuel and so protect the environment.

Whitehall sources yesterday ducked the politically sensitive question of a 'carbon tax' – although ... the Environment Secretary has urged an agreed reduction in the emissions of carbon dioxide. *Guardian*, 5 March 1992

cardboard city (kahd'bawd), *n.* an area of a city where the homeless set up makeshift shelters constructed from cardboard boxes etc.

carphone (kah'fōn), *n.* a CELLULAR TELEPHONE (see Dict.) suitable for operating in a car.

cash (kash), *n.* **cash cow,** *n.* (*coll.*) a business, product etc. which consistently produces large profits; (*loosely*) a source of wealth or profit for relatively little outlay or work.

to cash out (*sl., esp. N Am.*) to give up one's job in a large company to work at home for less money but better quality of life.

cathart (kəthaht), *v.* (*N Am.*) to reveal one's innermost secrets and thoughts in public.

CD-i, (*abbr.*) compact disc interactive (a compact disc which integrates audio, visual and game-playing, often for educational information, and allows the user to interact with the images on the screen rather than just being a passive viewer).

centimillionaire (sen'timilyənea), *n.* someone who possesses a hundred million pounds, dollars etc.

Centralia (sentrāl'yə), *n.* Central Australia.

channel surfing (chan'əl), *n.* moving swiftly between television channels, using a remote control device.

chaos theory (kā'os), *n.* the theory that the universe or any dynamic system does not wholly work according to the laws of Newtonian physics, but in an apparently random and unpredictable way.

charge (chahj), *v.t.* **charge-cap,** *v.t.* of central government, to put an upper limit on the amount that a local council can levy in community charge. **charge-capping,** *n.* **charge card,** *n.* a credit card issued by a shop or a retail chain for use in its own outlets. **charge-payer,** *n.* one who pays the community charge.

chronic (kron'ik), *n.* (*sl.*) marijuana.

CIS, (*abbr.*) Commonwealth of Independent States (a federation made up of those republics of the old USSR, such as Russia and Ukraine, who wished to retain some voluntary links with each other).

CNN, (*abbr.*) Cable News Network.

cold (kōld), *a.* **cold call,** *v.t.* to make an unsolicited telephone call or visit to a prospective customer, in order to try to sell goods or services. *n.* such a telephone call or visit. **cold calling,** *n.* **cold fusion,** *n.* nuclear fusion at room temperature, a possible energy source cheaper and less polluting than existing sources.

The Utah claim was the starting gun in what has become a race for 'cold fusion' – so called because it would operate at room temperatures. *Newsweek*, 8 May 1989

comping (kom'ping), *n.* the hobby or habit of entering competitions that offer prizes to members of the public for solving puzzles, writing slogans etc.

comper, *n.* [*competition*]

complexity theory (kompleks'iti), *n.* the theory that certain dynamic systems are never in equilibrium, and have many interlocking parts that are not easily described by mathematics.

The essence of chaos theory is that certain phenomena involve so many factors that they are inherently unpredictable ... Complexity theory examines the systems that lie in the middle ground between the predictable and the chaotic. *Time*, 22 February 1993

consumer terrorism (kənsū'mə), *n.* the contamination of food products with poisonous substances, sharp objects etc., in order to blackmail the manufacturers or retailers of the foods.

consumer terrorist, *n.*

corn circle CROP CIRCLE.

corporate advertising (kaw'pərət), *n.* the advertising of the name and policies of a company or group of companies, as opposed to the promotion of a particular product.

council tax (kown'sl), *n.* a tax to finance local government (the successor to the COMMUNITY CHARGE), levied on the household and calculated on the capital value of the property.

crippleware (krip'l'weə), *n.* (*Comput.*) demonstration software which has been deliberately disabled in such a way that it can be evaluated by a prospective purchaser but not used to its full capacity.

crop circle (krop), *n.* a circle or a pattern

971

based on circles appearing as flattened stalks in the middle of a field of standing corn, of uncertain origin (perh. meteorological or man-made).
cruelty-free (kroo'əltifrē), *a.* of household, pharmaceutical etc. products, produced without being tested on animals.
cryonics (krīon'iks), *n. sing.* (also **cryonic suspension**) the practice of preserving a dead body by deep-freezing it, to await discovery of a cure for the condition which caused its death.
cryosurgery (krīōscœ'jəri), *n.* surgery using the application of very low temperatures to specific tissues in order to cut or remove them.
cryosurgeon, *n.*
CSCE, (*abbr.*) Conference for Security and Cooperation in Europe.

D

DAT, (*abbr.*) digital audio tape.
date rape (dāt), *n.* rape of a woman, after a social outing, committed by the man she has just been out with.
decaf (dē'kaf), *n.* (*coll.*) short for DE-CAFFEINATED (coffee) (see Dict.).
demographic timebomb (deməgraf'ik), *n.* a fall in the size of the national workforce caused by an earlier decline in the birthrate leading to a drop in the number of school leavers.
derivative (deriv'itiv), *n.* one of a range of financial products which allows an investor to buy or sell something at an agreed price at a fixed date in the future.
deskill (dēskil'), *v.t.* to reduce the level of skill required for (a job), e.g. by automation; to cause (workers) to perform tasks that do not use their skills.
detox (dē'toks), *n.* (*coll.*) short for DETOXIFICATION (removal of drugs or other toxins from the body by fasting or special diet) (see Dict.); (*esp. N Am.*) a treatment centre for drug addicts and alcoholics. *v.i.* to clear one's body by fasting etc.
devore, dévoré (devaw'rā), *a.* (of velvet and other materials) with designs etched by acid. [Fr. *dévoré*, eaten]
The velvet scarf is this season's ultimate accessory and a stocking-filler guaranteed to win hearts. Choose from panné, bouclé, dévoré, crushed or printed in jet black, caramel, gold or fruits of the forest.
Evening Standard, 17 December 1993
differently abled ABLED.
diss (diss), *v.t.* (*sl.*) to fail to show enough respect to; to treat with rudeness.
They're saying: 'We don't care who you are, we don't care how much rep [reputation] you have, we're gonna diss yo ... 'cause we're out for us.' *The Times*, 14 March 1992
dongle (dong'gl), *n.* (*Comput.*) an electronic device which is plugged into a computer, and without which a particular program will not work.
[etym. uncertain]
dope (dōp), *adj.* (*sl.*) very good; excellent.
[from *sl. dope*, drugs, and *sl. dope out*, to understand, to work out]
dork (dawk), *n.* (*sl.*) a stupid or contemptible person.
... and one turns to the photo section for verification to find a corkscrew-haired dork who looks like the bass player in The Groundhogs.
dorkish, dorky, *a.* [etym. uncertain]
double whammy (dŭb'l), *n.* (*sl.*, *esp. N Am.*) a particularly severe, or a twofold, curse or misfortune.
Among the characters produced from the fertile mind of Mr Capp was that of 'Evil-Eye Fleegle', master of the whammy. His single whammy, it was said, could stop a rampaging elephant in its tracks, his double whammy could melt a locomotive in full flight, while his triple whammy could 'turn Lake Erie into a mudflat or steam up Teddy Roosevelt's glasses on Mount Rushmore'.
Independent, 18 March 1992
[orig. *whammy*, evil eye, using one eye, *double whammy*, using both eyes, from the comic strip *L'il Abner*, by Al Capp]
doughnutting (dō'nŭting), *n.* (*coll.*) the practice among MPs of sitting immediately behind and around a speaker during the televising of parliament, to give the impression of a crowded house. **doughnut,** *v.t.*
down (down), *a.*, *adv.* **downlighter,** *n.* a lamp or spotlight in a room whose light is directed downwards. **download,** *v.t.* (*Comput.*) to transfer (programs or data) directly from one computer to another.
drice (drīs), *n.* frozen carbon dioxide in granular form. [*dry ice*]
drive-by (drīv'bī), *a.* of a crime, committed from a moving vehicle, applied esp. to a murder committed by shooting from a passing car.
DSS, (*abbr.*) Department of Social Security.
DTP, (*abbr.*) desk-top publishing.
dweeb (dwēb), *n.* (*sl.*, *esp. N Am.*) fool, idiot, stupid or contemptibly weak person.
The summer phenomenon *Wayne's World* countered with the more basic 'blowing chunks' (vomiting), 'sphincter boy' (creep) and 'dweeb' (wimp).
Sunday Times, 28 March 1993
DWEM, (*abbr.*) dead white European male.
It [political correctness] has led to the junking of the classics of western culture, written by 'DWEMs' ... in favour of works of dubious merit that have been elevated to cult status. *Independent*, 17 February 1992

E

E, (*abbr.*) ecstasy (drug).
E-mail (ē'māl), *n.* (also **email**) short for ELECTRONIC MAIL (see Dict.).
ecocide (iko'sīd), *n.* the destruction of an ecology. **ecocidal,** *a.*
ecofriendly (ek'ōfrendli), *a.* not damaging to the environment, environment-friendly.
ecolabel (ek'ōlābl), *n.* a label on a food or household product asserting that the product contains nothing damaging to the environment. **ecolabelling,** *n.*
ecstasy (ek'stəsi), *n.* methylenedioxymethamphetamine (MDMA), a synthetic stimulant and hallucinogenic drug, based on amphetamine.

ECU, ecu (ek'ū, ā'kū), *n.* a currency unit used as a unit of account in the European Community, based on the values of several different European currencies. [acronym of *E*uropean *C*urrency *U*nit]

egg, wobbly egg (eg), *n.* a sleeping tablet that can induce a sense of euphoria when taken with alcohol.

EMS, (*abbr.*) European Monetary System.

EMU, (*abbr.*) economic and monetary union.

energy tax (en'əji), *n.* a range of possible taxes on the use of non-renewable energy sources, to reduce their use and so protect the environment.

ENG, (*abbr.*) electronic news gathering.

environment-friendly ECOFRIENDLY.

EPOS, (*abbr.*) electronic point of sale.

ERM, (*abbr.*) exchange rate mechanism.

ethnic cleansing (klenz'ing), *n.* the systematic persecution of a distinct ethnic group within a community, members of which are killed, imprisoned, forced to move to another area, etc.

Eurosceptic (ū'rōskeptik), *n.* one who is sceptical about the benefits to the UK of membership of the European Community, and who is not in favour of further integration into the Community.

Eurotunnel (ū'rōtŭnl), *n.* the Channel tunnel.

F

---

fanny-pack, (*N Am.*) BUM-BAG.

fantasy (fan'təsi), *n.* a synthetic hallucinogenic drug, a mixture of ecstasy and LSD or mescalin.

fashion victim (fash'ən), *n.* a (usu. young) person who follows the extremes of fashion in clothes, hair etc., regardless of suitability or expense.
   Sartorial standards are inevitably a little lower than in go-ahead Glasgow, but it has its share of serious fashion victims among the Levis and Next separates.
     Scotland on Sunday, 4 June 1989

fast track (fahst), *n., a.* (being or relating to) a faster than normal route to advancement or promotion. *v.t.* to promote by such a route.

fast tracker, *n.*

fatwa, Fatwa, fatwah (fat'wa), *n.* a legal decision or ruling given by an Islamic religious leader. [Arab.]

feel-bad factor (fēl'bad), *n.* (*coll.*) a set of circumstances that together contributes to a feeling of dissatisfaction and/or pessimism.

feel-good factor (fēl'gud), *n.* (*coll.*) a set of circumstances that together contribute to a feeling of satisfaction and/or optimism.

fitness walker (fit'nis), *n.* one who walks regularly not only for pleasure but also as a form of keep-fit exercise.
   Fashionable fitness walkers ... in America now take along an electronic mileage counter to tell them how far they have travelled and, what is more, how many steps they have taken.
     Scotland on Sunday, 18 March 1990

flat-screen (flat'skrēn), *a.* of or relating to a television system using a flat rather than a curved screen, avoiding distortion of the picture.

fly-by-wire (flībīwīə), *n.* a system of aircraft control which uses electronic rather than mechanical connections between the control and the aircraft parts.

food court (food), *n.* an area in a shopping mall that contains several fast-food outlets with seating accommodation.
   Just as the food court is *de rigueur* for the shopping mall, the restaurant remains indispensable for any proper department store.
     QC, February 1990

footprint (fut'print), *n.* the area of the earth's surface within which the signal of a geostationary communications satellite can be received; the area within which a spacecraft is to land; the space on a desktop occupied by a computer.

Footsie (fut'si), *n.* (*coll.*) the Financial Times Stock Exchange 100 Index. [the initial letters FTSE]

fractal (frak'təl), *n.* a computer-generated figure produced from a simpler figure by applying a fixed set of rules, and itself giving rise to a more complex figure by repeated application of the same rules, and so on. **fractality,** *n.*

freestyle (frē'stīl), *v.* of a rapper, to make up rhymes spontaneously in front of an audience, rather than to prepare them beforehand.

friendly fire (frend'li), *n.* (*Mil.*) fire from one's own side during a military engagement which accidentally causes casualties or damage.

fullerene (ful'ərēn), *n.* (*Chem.*) a molecule consisting of 60 carbon atoms arranged in a spherical shape, with possible uses as a lubricant, superconductor etc.
   The fullerene molecule is constructed of carbon atoms that fit together mostly in hexagonal tiles, forming a sphere like chicken wire wrapped round a balloon. In each molecule there are 12 pentagons among the hexagons. This is just like a geodesic dome.
     Independent, 27 May 1991
[Buckminster *Fuller*, 1895–1983, US inventor of the geodesic dome]

fun-run (fŭn'rŭn), *n.* a long-distance run organized not as a competition but for enjoyment, to raise money for charity etc.

fundage (fŭnd'ij), *n.* (*sl.*, *esp. N Am.*) money.

FX, (*abbr.*) (*coll.*) (visual) effects.
   Most raves combine the spectacular visual FX of a Jean-Michel Jarre concert with the aural explosion of heavy metal.
     Time Out, 11 December 1991

G

---

galleria (galərē'ə), *n.* a number of small independent shops in one building, on one floor or arranged in galleries on several floors. [It., arcade]

gangbanger (gang'bang'ə), *n.* (*sl.*, *esp. N Am.*) a member of an urban teenage gang.

gender person (jen'də), *n.* (*esp. N Am.*) a deliberately non-sexist term for a human being of either sex.

gender reassignment (jen'də), *n.* sex change by surgery.

glass ceiling (glahs), *n.* the obstacle that many believe women face in the business world, sex discrimination ensuring that they rarely achieve the positions they merit.

global warming (glō'bəl), *n.* a slow warming of the earth's surface, caused by the greenhouse effect, and leading to climatic change, melting of the polar icecaps, a rise in sea-level etc.

granny-dumping (gran'i), *n.* (*esp. N Am.*) the practice of abandoning a confused aged relative, whose care has become too burdensome or expensive, outside a hospital, so that responsibility passes to the social services.

grant-maintained (grahnt), *a.* of a school, having opted out of local authority control, and funded directly by central government.

graphic (graf'ik), **-ical**, *a.* **graphical user interface**, (*Comput.*) a user interface which displays icons rather than written words. **graphic novel**, *n.* a full-length novel for adults in the form of a comic strip.

But the graphic novel is only remotely born of high art. Its immediate forerunner is the comic-strip tradition that reached its heyday in America in the 1940s, and the superhero strips that came over here with the GIs.
Sunday Times, 15 July 1990

graphics pad, tablet, *n.* (*Comput.*) a flat board on which one draws with a stylus, the pattern traced being transferred directly on to the computer screen.

green (grēn), *a.* **green card**, *n.* an international insurance certification for motorists.

greenfield, *a.* applied to a development site which has not been built on previously. **greenhouse gas**, *n.* any of the gases (principally carbon dioxide and methane) which, when they build up in the atmosphere, trap the sun's heat, contributing to the GREENHOUSE EFFECT (see Dict.). **green label, labelling** ECOLABEL. **green pound**, *n.* an adjustable financial unit of account used in agricultural dealings between the UK and the EC.

greige (grāzh), *a.* a colour midway between grey and beige.

gridlock (grid'lok), *n.* (*orig. N Am.*) a prolonged traffic jam covering large areas of a city, caused by simultaneous blocking of several important road intersections by stationary traffic.

grift (grift), *n.* (*esp. N Am.*) making money dishonestly by confidence tricks, crooked gambling etc.; the money so made. *v.i.* to make money thus. **grifter**, *n.* a swindler, confidence trickster or crooked gambler.

But the most remorselessly survivalist grifter is Roy's mother Lily, employed by the Mob to place bets at racetracks to bring down the price of nags.
Time Out, 31 July 1991
[perh. from GRAFT[3]]

grockle (grok'l), *n.* (*coll.*) a tourist, used orig. in SW England to refer to tourists from the Midlands and North of England.

grunge (grŭ'nj), *n.* (*sl., orig. N Am.*) of rock music, fashion etc., a style which emphasizes discordant, even deliberately ugly elements and rejects the artificiality of more sophisticated styles. In rock music, grunge depends upon the use of loud, deliberately distorted sequences; in fashion, it is the antithesis of conventional high fashion, being characterized by loose-fitting, multi-layered and uncoordinated outfits, which typically incorporate recycled fibres.

We are not grunge. I think grunge is a terrible term because it conjures up images of sloppy, badly-played music.
Guardian, 30 March 1993

GUI (goo'i), (*acronym*) graphical user interface.

H

halon (hā'lon), *n.* (*Chem.*) any of a class of compounds, made by combining carbon with a halogen, used in fire extinguishers etc, and ozone-depleting.

This plays a key role in the chemical reactions sparked off by chlorofluorocarbons and halons (used in aerosols, refrigerators and fire extinguishers) and ending in ozone loss.
Independent, 8 April 1992

hand (hand), *n.* **handbag**, *v.t.* (*coll.*) to hit (a person) with a handbag; of a woman, to scold vigorously or attack verbally. **hands off**, *a.* not requiring manual control; allowing things or people to follow their own course without intervention. **hands on**, *a.* requiring manual operation; exercising personal control, being personally involved.

Hanta virus (han'tə), *n.* an often-fatal virus whose symptoms are similar to those of flu. [from the Hantaan river in Korea where a number of related viruses were first isolated]

HDTV (*abbr.*) HIGH-DEFINITION TELEVISION.

herstory (hœ'stawri), *n.* in feminist jargon, history emphasizing the role of women or told from a woman's point of view.

hidden agenda (hid'n), *n.* a set of social, moral etc. assumptions, or an undetected bias underlying a discussion, a programme of action etc.

high-definition television (hīdefinish'ən), *n.* a television system using a larger number of scanning lines to give a much clearer picture, and a greater width-to-height ratio for the screen.

Television will be transformed in stages. The first step is high-definition television, a vast improvement over current broadcasting that will match the quality of motion pictures.
Newsweek, 1 January 1990

highlighter (hī'lītə), *n.* a fibre-tipped pen with a very wide nib, containing brightly-coloured transparent ink, used for drawing over words in a text to focus attention on them. **highlight**, *v.t.* to draw over (words in a text) with a highlighter.

himbo (him'bō), *n.* a man who has good looks, but who lacks depth and intelligence; the male equivalent of a bimbo.

hit skins (hit), *v.i.* (*sl., esp. N Am.*) to have sexual intercourse.

HIV-positive (āch'īvē), *a.* having a positive result from a blood test for antibodies to

974

human immunodeficiency virus; infected with HIV but not yet showing any symptoms of AIDS.
homophobia (homōfō'biə), *n*. hatred and fear of lesbians and gay men or homosexuality; persecution of lesbians and gay men.

The scale of homophobic discrimination in employment is probably as great as the bias against women and black people. However, it is much less visible because, lacking even the minimal legal protection enjoyed by women and black people, lesbian and gay victims are often afraid to fight the discrimination they've suffered.
OutRage, 2 April 1992
homophobe (hom'-), *n*. one who hates and fears homosexuality; one who persecutes lesbians and gay men. **homophobic** (-fō'-), *a*.
human bowling (hū'mən), *n*. a game in which a player strapped into a rolling globular cage attempts to knock down large soft skittles.
hypercolor (hī'pəkŭlə), *n*. a process for dyeing fabric so that it changes colour when exposed to even slightly raised temperatures.

You can put on what seems like a perfectly unremarkable grey cotton T-shirt, step out into the hot summer sun and watch it turn a startling shade of candyfloss pink. Hand prints will appear within instants of the cloth being touched. *Esquire*, 9 February 1992

I

---

ice (īs), *n*. (*sl*.) a very addictive synthetic form of crystallized methamphetamine, which can be smoked and gives a long-lasting high.

The smokable variant of the drug, which looks like ice cubes, is expected to be more popular. *Independent*, 23 January 1990
ice diving (īs), *n*. scuba diving beneath the surface of a frozen lake.
ident (ī'dent), *n*. (*sl*.) a distinctive image or logo that is regularly shown before television programmes to reinforce viewers' awareness of which channel they are watching.

Idents are 'channel identities', the little films that tell the viewers which channel they're watching. They are the prime agents of channel branding, establishing 'image' and a sense of continuity between the various programmes.
Independent on Sunday, 11 April 1993
immunocompromised (imūnōkom'prəmīzd), *a*. (*Med*.) having a defective immune system.
immunodeficiency (imūnōdifish'ənsi), *n*. (*Med*.) a deficiency in, or breakdown of, a person's immune system.
incentivize, -ise (insen'tivīz), *v.t.* to stimulate to further activity by offering incentives.
inclusive (inkloo'siv), *a*. of or pertaining to forms of language chosen so as not to put down or exclude a group of people, used esp. of non-sexist language in Christian liturgies.
indie (in'di), *a*. (*coll*.) short for INDE-PENDENT (used esp. of popular music).
INF, (*abbr*.) intermediate-range nuclear forces.
infotainment (infōtān'mənt), *n*. (*TV*) the presentation of news and current affairs in a bright and superficial way, as entertainment.
Inkatha (inkah'ta), *n*. a black political organization in South Africa, revived as a black liberation movement in 1975 under the Zulu Chief Mangosuthu Buthelezi. [Zulu, grass coil that Zulu women use when carrying loads on their heads]
Internet (in'tənet) *n*. (*Comput*.) a worldwide series of networks interconnecting numerous computers which range enormously in size and type.
ITC, (*abbr*.) Independent Television Commission (replaced the Independent Broadcasting Authority (IBA) in 1990).
Item (ī'təm), *n*. (*coll*.) two people having a stable, long-term relationship.

The two met on the set of *Mosquito Coast*, but didn't exactly become an item. 'I've always had trouble with definitions of boyfriend and girlfriend,' he protests.
Scotland on Sunday, 25 June 1989
IVF, (*abbr*.) in vitro fertilization.

J

---

jobsworth (jobz'wœth), *n*. (*coll*.) a minor functionary who is quite inflexible in his/her interpretation of the law, and is very unhelpful to the public. [from the expression 'It's more than my job's worth to ...']
jock¹ (jok), *n*. (*sl*.) a jockey; a disk jockey. Hence the freedom given to its jocks and producers to programme their music from a wider base than the Top 40.
Listener, 29 June 1989
[short for JOCKEY]
jock² (jok), *n*. (*sl*., *orig*. *N Am*.) an athlete. His aim, he says, is to prove that US football players are not just mindless jocks.
Mail on Sunday, 6 August 1989
[short for JOCKSTRAP]
joined-up (joind'ŭp), *a*. referring to cursive handwriting, with the letters joined up rather than printed separately as by a small child; (*coll*.) adult, sophisticated, characterized by intelligent and coherent thinking.
junk call, fax (jŭngk), *n*. an unsolicited telephone call or fax, usu. trying to sell some commodity or service.

K

K, (*abbr*.) one thousand. *n*. (kā) (*sl*.) one thousand pounds (£1000). [KILO-]
karaoke (kariō'ki), *n*. the technique or entertainment of singing in conjunction with a machine which provides a prerecorded backing track and mixes the voice to blend with the accompaniment.

Up until now, karaoke singers have been singing along to backing tracks with the aid of a songsheet. (The machines were designed originally to help solo cabaret performers; *karaoke* means 'empty orchestra'.)
Independent on Sunday, 10 June 1990

ketamine (ket′əmēn), *n.* an anaesthetic drug which does not send the patient to sleep but causes hallucinations and insensitivity to pain; of limited use in medicine because of the unpleasant side effects which include nightmares; sometimes also used as a recreational drug. What appears to have happened in recent months ... is that 'serious ravers' who were sold ketamine as a substitute for ecstasy have begun to ask for 'special K' in its own right, instead of being angry with the dealer.
Independent, 2 April 1992
key (kē), *n.* **keyboard,** *n.* an electronic musical instrument which is played by means of a keyboard and which is programmed to produce rhythmic backing and a wide range of different sounds. **keyhole surgery,** *n.* a surgical technique in which a very small incision is made and minute fibre-optic and surgical instruments are passed through it into the tissues beneath. The technique has been refined using laparoscopy. A surgeon need make only two tiny incisions, one for the cauterizing instrument, one for the tiny fibre-optic light and fine telescope. Every movement in the patient's chest cavity is magnified and monitored on a television screen.
Independent, 24 March 1992

L

---

lambada (lambah′də), *n.* a fast and erotic dance, orig. Brazilian, in which the couple press their bodies together while they gyrate their hips, the music for this dance combining Latin-American and Caribbean styles. [Port., a whipping]
LAN, *(abbr.)* local area network.
laser card (lā′zə), *n.* a plastic card, like a credit card, with information stored on it in the form of microscopic pits in the surface, which can be read by scanning with laser beam.
lead-free (led′frē), *a.* of petrol, UNLEADED (see Dict.).
lemon law (lem′ən), *n.* in the US, a law designed to protect buyers of faulty or substandard cars.
level playing field (lev′əl), *n.* a situation that is fair, because the various participants in it have equal advantages.
leverage (lē′vərij), *v.t.* of a small company, to finance (a business deal, e.g. a takeover) by borrowing capital in the expectation of being able to repay if the deal is successful. **leveraged buy-out,** *n.* a takeover bid financed in this way.
linkage (lingk′ij), *n.* the linking of two separate political problems in a negotiation so that agreement on one is unacceptable without agreement on the other.
liposuction (lī′pōsŭkshən), *n. (Surg.)* the removal of excess subcutaneous fat from the body by suction through a small incision in the skin.
LMS, *(abbr.)* LOCAL MANAGEMENT OF SCHOOLS.
local (lō′kəl), *a.* **local area network,** *n.* a group of computer systems close enough to each other (e.g. in the same building) to be linked directly by cable so as to share data storage facilities etc. The use of LANs (Local Area Networks) to connect up business PCs is growing, but it has yet to reach the dizzy heights predicted by the industry.
Practical Computing, November 1989
local management of schools, a system in which the governors and head teacher of a state school are responsible for managing a budget allocated to it by the local authority.
logic bomb (loj′ik), *n. (Comput.)* an instruction programmed into a computer that will trigger a breakdown if in the future a specified set of circumstances occurs.
lookism (luk′iz′m), *n.* prejudice in favour of people of attractive appearance.

M

---

mad cow disease (mad kow), *n. (coll.)* bovine spongiform encephalopathy.
magalog (mag′əlog), *n.* a mail-order catalogue designed to look like a magazine by the use of e.g. glossy paper, editorial material etc. [*maga*zine, cata*log*(ue)]
mailshot (māl′shot), *n.* an item of unsolicited mail, usu. an advertisement; a batch of such items sent out at one time by a company.
management buy-out (man′ijmənt), *n.* a takeover bid by the managers of a company, often in order to protect the company from closure or absorption into a rival company.
market maker (mah′kit), *n.* a dealer in securities on the Stock Exchange who combines the roles of STOCKBROKER and STOCKJOBBER (see Dict.).
MDMA, *(abbr.)* methylenedioxymethamphetamine (the drug ecstasy).
meat-free (mēt′frē), *a.* vegetarian (of food); containing, or prepared using, no animal products.
megabuck (meg′əbŭk), *n. (N Am. sl.)* a million dollars; *(pl., sl., orig. N Am.)* a very large sum of money.
micro (mī′krō), *n. (pl.* **-cros**) *(coll.)* short for MICROWAVE (oven) (see Dict.). **micro, microwave,** *v.t.* to cook in a microwave oven. **microwave, microwav(e)able,** *a,* suitable, or specially prepared, for cooking in a microwave oven.
MIDI (mi′dī), *(acronym) m*usical *i*nstrument *d*igital *i*nterface (a computer interface which allows electronic musical instruments and synthesizers to be controlled from a computer).
minke whale (ming′kə), *n.* a small whale, the lesser rorqual, *Balaenoptera acutorostrata,* with a pointed snout. [after a Norw. whaler called *Meincke*]
morphing (mawf′ing), *n. (Comput.)* a technique of transforming one image into another through a series of gradually changing images which appear on the computer screen in rapid succession.
movers and shakers (moo′vəz, shā′kəz), *n. pl. (coll., orig. N Am.)* people with power and influence who can introduce

976

new trends or policies in politics, business etc.
MRE, (*abbr.*) (*Mil.*) meals ready to eat.
MRM, (*abbr.*) mechanically recovered meat.
MSM, (*abbr.*) men having sex with men.
multi-access (mŭltiak′ses), *a.* (*Comput.*) referring to a computer system to which several users can have simultaneous access from different terminals.
multimedia (mŭltimē′diə), *a.* (*Comput.*) referring to the provision of information as graphics, animation, sound and text, from a single compact disc.
multipack (mŭl′tipak), *n.* a packet containing several items of e.g. a food product, for sale at less than the price of the same number of individual items.
multiskill (mŭltiskil′), *v.t.* to train (employees) in several different skills in different parts or aspects of the manufacturing process. **multi-skilling,** *n.*
multivitamin (mŭltivit′əmin), *n.* (a tablet etc. containing) a measured amount of several vitamins in one dose.

N

naff (naf), *a.* (*coll.*) unfashionable, lacking in style or credibility.
nanotechnology (nanōteknol′əji), *n.* the technology which deals with measuring, making or manipulating objects of extremely small size.
 Nanotechnology is the ability to shape and use materials at scales as small as individual atoms. The 'nano' prefix refers to the unit of size known as the nanometre or one thousandth of a millionth of a metre. This is about three times the distance between atoms in everyday materials – and, with the help of new devices, some scientists have begun to manipulate substances, atom by atom. *The Times*, 10 May 1990
national curriculum (nash′ənəl), *n.* a school curriculum standardized over all state schools in the country and specifying the subjects considered necessary for all pupils, and levels of attainment for various ages in each subject.
neural network (nū′ral), *n.* (*Comput.*) a computing system that is modelled on the human brain, with interconnected processors working simultaneously and sharing information, and capable of learning from experience.
 Neural networks may some day give robots enough artificial sense to, say, vacuum the carpet in a simple office without knocking over the water cooler, at least not more than once. *Newsweek*, 1 January 1990
neurocomputer, *n.* a computer using a neural network.
New Age (nū), *n.* a cultural movement which emphasizes the spiritual and mystical at the expense of the materialistic and egocentric aspect of modern western society, and which encourages interest in eastern religion, astrology, alternative medicine, ecology etc. *a.* referring to this cultural movement, esp. to a characteristic style of music, melodic and gentle, played mainly on piano, synthesizer etc.
 A massive shift in the collective consciousness, from an aggressive, destructive masculine principle to a nurturing feminine one, is upon us. We call it the New Age.
Arena, Spring 1990
Nicam (nī′kam), *n.* (*TV*) a sound system in which audio signals are converted into digital form and transmitted along with the standard TV signal.
 Nicam ...is coming, and is about to revolutionize the way you view your television. You will soon be able to watch your favourite programmes with a sound quality similar to CD stereo. *CD Review*, March 1990 [acronym for *n*ear-*i*nstantaneous *c*ompanded *a*udio *m*ultiplexing]
Nikkei index (nik′ā), *n.* an index of the relative prices of stocks and shares on the Tokyo Stock Exchange. [*Nihon Keizai Shimbun*, a Jap. financial newspaper which publishes the index]
Nirex (nī′reks), (*acronym*) *N*uclear *I*ndustry *R*adioactive Waste *Ex*ecutive (a government body which oversees the disposal of nuclear waste in the UK).
no-fault compensation (nō′fawlt), *n.* a system of compensation for medical etc. accidental injury which does not depend on proving another person guilty of misconduct, negligence etc.
 Victims too often have to start litigation to prove liability – and risk heavy legal costs. A 'no-fault' scheme would cut waiting time and ensure that the money would go directly to the victim and not to teams of lawyers.
The Times, 7 December 1988
no frills (nō′frilz), *a.* basic, unadorned, with no unnecessary extra features.
 ... wants to put the fun back into budget shopping. 'Your no-frills fun store with fantastic savings ... 'No frills' means no baskets or free carrier bags.
Independent, 1 February 1992
no-no (nō′nō), *n.* (*coll.*) something to be avoided at all costs.
 I ask if there is anything he cannot stock because of international sensibilities? 'Air disaster books are a bit of a no-no. Apart from that, there isn't. If we are in doubt we check with the Airports Authority.'
Bookseller, 17 April 1992
no-show (nōshō), *n.* a person who does not take up something reserved (e.g. an aeroplane seat, a restaurant table) but does not bother to cancel, or to cancel far enough in advance.
notebook (nōt′buk), *n.*, *a.* (a computer) smaller than a lap-top, about the size of an exercise book, battery-operated and portable.
 The promise of a notebook computer is larger and more simple [than that of a lap-top]. It will be a virtual Filofax, as flexible as the real thing, but more powerful.
Independent, 10 February 1992
nuclear-free (nū′kliəfrē), *a.* containing no nuclear weapons or installations, and across which nuclear waste etc. may not be transported, esp. in **nuclear-free zone,** an administrative area, usu. a borough or county declared by its elected authority to be nuclear-free.
NVQ, (*abbr.*) *N*ational *V*ocational *Q*ualifi-

cation (an industrial training qualification based on practical competence rather than theoretical knowledge).

O

orimulsion (o'rimulshən), *n.* a mixture of bitumen, water and detergents, used in fossil-fuelled power stations.
OTT, (*abbr.*) over the top (see Dict.).
out (owt), *a.* (*esp. N Am.*) of a gay man or lesbian, having declared publicly his/her sexual orientation. *v.t.* to make public, without permission, a claim that another person is lesbian or gay.
Why should a politician enjoy the Parliamentary privilege of staying in the closet while men are forcibly outed following cottaging prosecutions ... Outing is a tool of accountability, not a punishment.
Capital Gay, 3 April 1992
outer, *n.* **outing**, *n.*
outplacement (owtplās'mənt), *n.* professional relocation of redundant employees arranged by their former employer.
Outplacement agencies teach companies how to sack staff – and then they try to find the staff new jobs. *GQ*, June 1989
outplace (owt'-), *v.t.* **outplacer**, *n.* a person or agency providing this service.
outsource (owt'saws), *v.* to subcontract work to another company, especially as a means of reducing costs. **outsourcing**, *n.*
... the TSB, which has outsourced four of its computer and communications systems, including the network that links the bank with its subsidiaries.
The Times, 12 November 1993
ozone (ō'zōn), *n.* **ozone depletion**, *n.* reduction in the amount of ozone in the upper atmosphere caused by atmospheric pollution, esp. by chlorofluorocarbons. **ozone-depleting**, *a.* referring to atmospheric pollutants which contribute to ozone depletion. **ozone hole**, *n.* a hole in the OZONE LAYER (see Dict.), allowing ultraviolet radiation on to the earth's surface beneath it.
Every spring 40 to 50 per cent of the atmospheric ozone over the Antarctic disappears to form the best-known ozone hole ... Ozone depletion is the result of a complicated interplay of artificial and natural factors. *Independent*, 8 April 1992

P

paintball (pānt'bawl), *n.* a form of wargame in which the opposing teams shoot paintballs at each other; a pellet filled with coloured paint, used in the paintball game, which bursts on impact, covering clothing with paint.
There are two teams, each with a flag. The object is to capture the opposition's one and return it to your own base, before they do the same to yours. To make this more than a fancy dress version of hide-and-seek, the players are armed with pistols which fire pellets of coloured paint. If a pellet explodes on your body, you're excluded from the rest of the bout – or, in common parlance – dead. *Scotland on Sunday*, 18 March 1990
palm-top (pahm'top), *n.* (*Comput.*) a portable computer smaller than a notebook, but still with a full keyboard, that can be held in one hand while being used.
PC, (*abbr.*) political correctness; politically correct.
peace dividend (pēs), *n.* money which, since the signing of disarmament treaties and the end of the Cold War, no longer needs to be put towards the government's defence budget, and which is in theory available for civilian projects.
President Bush, however, has rejected the notion of the 'peace dividend' – diverting resources from the military to the domestic budget. *Glasgow Herald*, 20 January 1990
people (pē'pl), *n.* **people carrier, mover**, *n.* any of various methods of moving large numbers of people, e.g. moving pavements or driverless shuttles; a large van-like estate car with an extra row of seats to carry more people. **people power**, *n.* the power to influence the course of events exerted by the mass of the people, not through conventional political channels but through mass demonstrations etc.
At a time when people power seems to be at its most effective in changing what were once political certainties – the most striking example being the downfall of Ceaucescu ...
Economist, 17 March 1990
perfume dynamics (pœ'fūm), *n. sing.* the use of fresh, pleasant aromas diffused in the air in a building to make the environment more congenial to workers, customers etc., and so to improve performance.
photo-opportunity (fōtōopətū'niti), *n.* an event arranged for or by a politician, pressure group etc. primarily in order to get favourable media coverage through good pictures; a PHOTOCALL (see Dict.).
physically challenged (fiz'ikəli), *a.* a euphemistic term meaning disabled.
pink (pingk), *a.* referring to lesbians or gay men or to lesbian and gay concerns, as *pink vote, pink pounds.*
Now many of his friends are gay and he even had a pink birthday party in October at which pink potatoes were served.
Time Out, 25 January 1989
pink-collar, *a.* of jobs, traditionally associated with women (cp. WHITE- or BLUE-COLLAR.)
There are still few women working in the railways, mining or the fire service; too many are still confined to pink-collar secretarial jobs. *Sunday Times*, 23 April 1989
pink noise, *n.* recordings of mid- to low-frequency background noise, designed to be soothing to small babies.
A cassette version, renamed 'pink noise' in baby-friendly speak, has been available in the United Kingdom for some time. Satisfied customers attest that a few minutes' blast of the irregular Baby Soother tape puts a crying baby to sleep.
Independent, 24 February 1992
pit bull terrier (pit bul), *n.* a large variety of bull terrier bred in America specifically as a fighting dog.
political correctness (pəlit'ikəl), *n.* a political and moral movement, orig. American,

which aims to discourage discrimination, whether in language or action, against minorities such as women, people of colour, lesbians and gays or people with disabilities, and to promote equality of opportunity for these minorities. **politically correct,** *a.*

posse (pos′i), *n.* a criminal gang; (*sl.*) a group of people from the same area, or sharing a background or interest.

... threatened by violence from the newer menaces of Chinese mobs and drug-dealing Jamaican posses ... *Sky*, December 1989

Post-it Note ® (pōs′tit), *n.* a small sheet of paper for writing messages with a strip of adhesive along one edge, which can be stuck to a surface and removed again without causing damage to the surface.

post-traumatic stress disorder (or **syndrome**) (pōst′trawmat′ik), *n.* (*Psych.*) a psychologically disturbed condition suffered by a person who has been through a very traumatic experience (e.g. a shipwreck or aeroplane crash), the symptoms including anxiety, guilt and depression.

post-viral (fatigue) syndrome (pōstvī′rəl), *n.* a condition occurring after a viral infection, characterized by prolonged fatigue, depression etc.

power (pow′ə), *a.* designed to demonstrate and reinforce one's social or political strength, or dominant position in an organization (as *power dressing, power lunch*).

prion (prē′on), *n.* a self-replicating protein, thought to be responsible for debilitating brain disease.

Unlike viruses, the gene does not have to be present in an infected dose in order for the prion protein to cause disease: it is, at least in part, an enemy from within the body, and that explains why there is no immune response to prions when they do manage to enter a new host.
The Times, 2 September 1993

pro-choice (prō′choise), *a.* in favour of a woman's right to choose whether or not to have an abortion.

Still, the very day after the huge pro-choice rally, the Administration renewed its call for the US Supreme Court to overturn *Roe v. Wade*, the 1973 case that established a constitutional right to abortion.
Time, 20 April 1992

pro-choicer, *n.*

PSDR, (*abbr.*) public sector debt repayment.

psychobabble (sī′kōbabl), *n.* (excessive or unnecessary use of) the jargon of psychology.

Even *EastEnders'* Michelle speaks in psychobabble at times: 'I've got to find myself, Lofty! I've got to grow!'
Arena, Summer 1988

public domain (pŭb′lik), *a.* (*Comput.*) referring to a software which is not copyrighted or sold commercially, but is available for a very small fee to cover the cost of the disk and the copying. [see PUBLIC DOMAIN in Dict.]

puffin crossing (pŭf′in), *n.* a type of PELICAN CROSSING (see Dict.) incorporating sensors which monitor the movements of pedestrians and keep the traffic lights red until the crossing is clear.

Q

quality time (kwol′iti), *n.* (leisure) time spent constructively in improving one's family relationships, lifestyle etc.

queer (kwiə), *a.* (*sl.*) applied derogatorily to lesbians and gay men; (*esp. N Am.*) also applied by lesbians and gay men to themselves as a means of confirming their identity as a group in/out of society.

Queer denotes a very radical break from the old order. It sticks two fingers up at the Sanitisation of lesbian and gay identities by authority figures within the community ... Queer emerged out of the AIDS epidemic in the US but seems to reflect many of the energies that powered gay liberation and subsequently got respectabilised ...
Pink Paper, 12 January 1992

query language (kwiə′ri), *n.* (*Comput.*) a set of instructions and commands used for retrieving information from a database.

R

rainbow (rān′bō), *a.* consisting of, or bringing together (esp. minority) groups from different races or of different political views (as *rainbow alliance, rainbow coalition*).

ram-raid (ram′rād), *n.* a robbery from a shop carried out by using a car or van to break through the windows, security grilles etc.

Police are to ask the government for new powers to tackle the phenomenon of 'ram-raids', smash-and-grab attacks in which stolen high-performance cars are used as battering rams against store fronts.
The Times, 31 July 1991

rapper (rap′ə), *n.* a performer of RAP MUSIC (see Dict.).

rave (rāv), *n.* (*sl.*) a wild party, esp. a very large party with rock music held e.g. outdoors or in a tent or warehouse, and for which tickets may be sold; a session at a club where various types of popular music are played. **rave dancing, raving,** *n.* a very fast and athletic form of dance associated orig. with House and then with techno music.

Rave dancing is in sync with the video age: full of abrupt changes in tempo, jerky movements and high-energy turns. Just as house music borrows from a host of musical sources, raving steals from hip-hop, disco, vogueing and aerobics.
Independent, 6 March 1992

raver, *n.* a person who attends raves.

recycle (risik′lāt), *n.* material produced by recycling.

red (red), *a.* **red-eye,** *n.*, *a.* (*sl.*) (referring to) an overnight air journey from west to east, the traveller crossing several time zones and missing a large part of a night's sleep. *v.i.* to travel on such a flight.

At San Diego, last stop before the dreaded red-eye home for breakfast ...
Sunday Times, 25 March 1990

red route, *n*. one of a proposed network of roads in London where parking, unloading etc. are forbidden and heavily penalized, such roads being marked by red instead of yellow lines.

The 'red route' scheme ... introduced last January, was designed to test whether imposing special parking and stopping regulations on stretches of urban trunk roads would ease congestion.
Independent, 13 January 1992

renewable (rinū'əbl), *a*. of a form of energy derived from the harnessing of a natural force, e.g. the sun, wind, tides etc., either directly (as a windmill) or to generate electricity (as a wind turbine). *n*. such an energy source.

Methane from landfills is one aspect of energy from waste materials, an important strand in Britain's development of renewable energy. It is much nearer to commercial viability than most renewables, which is why it is growing so rapidly and requires the lowest subsidy. *Independent*, 13 April 1992

rescuer (res'kūə), *n*. an activist who opposes the practice of abortion (usually belonging to one of a number of US and British anti-abortionist groups).

The activists, who intend to step up action, say their numbers are swelling as more supporters cross the Atlantic this weekend. Clinics have increased security and are preparing to repel the militant 'rescuers'.
Sunday Times, 28 March 1993

right-on (rīton'), *a*. (*sl*.) up-to-date, modern, fashionable.

RISC, (*abbr*.) reduced instruction set computer (one whose central processing unit has a much smaller set of instructions than usual, and which consequently can work faster).

road pricing (rōd), *n*. a method of reducing traffic on a particular road or set of roads (e.g. in a city centre) by making drivers pay to use the roads.

road rage (rōd), *n*. (*coll*.) angry behaviour by motorists, usually caused by frustration at road conditions.

rock (rok), *n*. (*sl*.) crack (form of cocaine).

Crack's convenient packaging – it comes in neat little lumps that resemble rock salt – makes it especially appealing to the young and poor. A single rock, about a quarter the size of a sugar lump ...
Scotsman, 6 February 1989

Rocket ® (rok'it), *n*. a form of ready-mixed cocktail served in a syringe and drunk by squirting the liquid directly to the back of the throat.

Ben and his friends hold weekly 'Rocket races ...' 'The secret is to push the syringe as fast as possible, it makes you feel instantly pissed and it's a great way to begin the weekend.' *Independent*, 23 March 1992

RSI, (*abbr*.) REPETITIVE STRAIN INJURY (see Dict.).

rush (rŭsh), *n*. (*sl*.) a surge of euphoria induced by, or as if by, a drug.

'The music doesn't work if it's slow. You need a loud bass, a kicking track; you get such a rush you want to explode.' His friends say he dances like 'a rottweiler on speed'.
Independent, 6 March 1992

S

sab (sab), *n*. (*coll*.) short for (hunt) SABOTEUR (see Dict.).

SAD, (*abbr*.) SEASONAL AFFECTIVE DISORDER.

safe sex (sāf), *n*. sexual activity avoiding penetration or using physical protection such as condoms, to prevent the transmission of disease, esp. AIDS.

scratch card (skrach), *n*. a card (usu. sold in a lottery) whose surface has an opaque coating which is scratched off with a fingernail to reveal the figures, letters etc., printed underneath.

seasonal affective disorder (sē'zənəl), *n*. (*Psych*.) a state of fatigue and depression which occurs only during the winter months, thought to be caused by lack of sunlight.

second-guess (sekəndges'), *v.t.* (*coll.*, *orig. N Am*.) to forestall (a person) by guessing his/her actions, reactions etc, in advance; to re-evaluate (a situation etc.) with hindsight.

sedentism (sed'əntizm), *n*. the abandonment of a nomadic way of life in favour of living in permanent settlements.

sell-by date (sel'bī), *n*. a date marked on the packaging of a perishable product, by which day the product should be withdrawn from sale if not already sold; (*coll*.) a date after which decay or decline begins.

set-aside (set'əsīd), *a*. referring to an agricultural policy under which farmers are paid to leave land uncultivated or turn it over to a non-agricultural use (e.g. forestry or recreation) in order to reduce agricultural surpluses; of or pertaining to such land. *n*. set-aside land.

From next year, farmers will be paid to keep part of their land out of production under the Government's 'set-aside' scheme.
Environment Now, August 1988

sex (seks), *n*. **sex-worker**, *n*. (*euphem*.) a prostitute. **sexy**, *a*. (*sl*.) exciting, fashionable, desirable.

The scientists picked the insulin gene not because they were particularly interested in medical solutions but because they considered insulin a 'sexier' protein than any other they could think of.
Spare Rib, 22 May 1989

shareware (sheə'weə), *n*. (*Comput*.) software which, although in copyright and commercially produced, is available free on approval, with a registration fee payable after a trial period.

shell suit (shel), *n*. a type of tracksuit consisting of a usu. brightly-coloured waterproof nylon outer layer with a cotton-knit lining.

shiitake (shēitah'ki), *n*. a dark-brown mushroom with pale-beige gills used esp. in oriental cookery. [Jap. *shii*, a tree, *take*, mushroom]

shooting gallery (shoo'ting), *n*. (*sl*.) a place where drug-users congregate to inject their drugs.

In Switzerland they call this a *Fixerraum*, a

'fix room'. American drug addicts would call it a 'shooting gallery'.
Newsweek, 1 May 1989

SIDS, (*abbr.*) sudden infant death syndrome (same as *cot death*, see COT² in Dict.).

single market (sing'gl), *n.* (also **single European market**) the free movement of goods, services and labour between the various countries of the European Community after the removal of customs barriers and trade and labour restrictions.

sky park (skī), *n.* a residential development in which homes are serviced by their own airstrip.

SLD, (*abbr.*) Social and Liberal Democratic Party.

Slim (disease) (slim), *n.* an African name for AIDS (from the marked weight-loss which occurs at the onset of symptoms).

smart (smaht), *a.* **smart bomb,** *n.* (*coll.*) a missile with a computerized internal guidance system which enables it to steer itself very precisely on to its target. **smart card,** *n.* (*coll.*) a credit card incorporating a microchip which can store information on all transactions made with the card.

smokehood (smōk'hud), *n.* a transparent plastic hood, with an airtight seal round the neck and an independent air supply, which aircraft passengers can put over their heads if the cabin fills with smoke.

social (sō'shəl), *a.* **Social Charter,** *n.* a declaration of workers' rights (minimum wages, job security, training programmes etc.) within the European Community. **Social Fund,** *n.* a fund provided by the government and administered by the DSS which provides grants or loans for specific purposes to people in need.

sorted (sawt'ed), (*past, p.p.* SORT) (see Dict.). *a.* (*sl.*) arranged, fixed, okay. [orig. in a drugged state]

sound bite (sownd), *n.* a short extract from a recorded interview, speech, etc. used as part of a news bulletin or party political broadcast, and often spoken to be used in this way.

Special K (spesh'əl), *n.* (*sl.*) ketamine, used as a recreational drug. [from the name of the breakfast cereal]

speedball (spēd'bawl), *n.* (*sl.*) cocaine mixed with heroin or morphine, usu. injected.

spell-checker (spel'chekə), *n.* (also **spelling-checker**) (*Comput.*) a program used in a word-processor which checks all the words in a text, and draws the user's attention to any unrecognized spellings.

spin-doctor (spindok'tə), *n.* a person working for e.g. a political party, who tries to influence the interpretation given to news stories, statements etc., by journalists, broadcasters and ultimately by the public.

News broadcasts are so important ... that the so-called spin-doctors will be trying to persuade the broadcasters to accept their interpretation and their agenda.
Guardian, 12 March 1992

splatter movie (splat'ə), *n.* (*sl.*) a very violent film with a strong visual emphasis on the many bloody deaths and mutilations it contains.

SSSI, (*abbr.*) site of special scientific interest.

stagedive (stāj'dīv), *v.t.* to jump from a stage into the audience, usually during a rock concert.

The band ... does a dazzling version of 'Screemager' to a regular accompaniment of stagediving, which involves punters clambering on stage then throwing themselves back into the audience. *Guardian*, 30 March 1993

standalone (stan'dəlōn), *a.* (*Comput.*) referring to a system or device that is able to function independently, not connected to any other system or device; (*generally*) separate, self-contained, not dependent on another organization etc.

[The] chief executive of the Telegraph says: 'A stand-alone Sunday newspaper selling 600,000 is not a viable proposition except during an advertising boom.'
Guardian, 11 September 1989

standee (stand'ē), *n.* a standing passenger in a bus or railway carriage.

stealth bomber (stelth), *n.* a military aircraft designed using special technology which renders it undetectable to radar or infrared sensors.

Whatever its merits as a weapon, the new American 'stealth' bomber is easily the most stylish and camera-loving piece of military hardware since the boomerang.
Sunday Times, 23 July 1989

steaming (stē'ming), *n.* (*sl.*) a method of robbery practised in crowded public places, esp. fairgrounds and trains, in which a gang of youths runs very fast through the crowds stealing money, jewellery etc., as they go. **steam,** *v.i.* to steal in this way. **steamer,** *n.*

... letter recommends that train passengers carry cameras as a deterrent to the violent crime of 'steaming'. If I were a steamer rampaging through a train in search of small, portable valuable objects to steal, a carriage full of photographers would be a veritable Aladdin's cave.
letter in the *Independent*, 24 February 1989

stepping (step'ing), *n.* a form of keep-fit exercise in which one steps up on to and then down from a step while swinging one's arms in time to music.

Stepping itself is as simple as the equipment. The gym supplies the plastic step and all you need are exercise clothes and a pair of trainers. Then you're ready to begin the basic step. *Independent*, 13 February 1992

stinger (sting'ə), *n.* a device of metal spikes used by police to puncture the tyres of speeding stolen cars.

store card CHARGE CARD.

storyboard (staw'ribawd), *n.* a series of drawings of the sequence of shots in a film showing the camera angles, relative positions of the actors etc.

The more complete drawings are particularly important for commercials, where the storyboards are shown to the client. A 30-second commercial ... can have about 40 frames.
Independent, 18 February 1992

subsidiarity (sŭbsidiar'iti), *n.* the principle in government that decisions should wherever possible be taken at the lowest, or

981

most local, level.
   The doctrine of 'subsidiarity' – leaving decisions to the most appropriate level of government in the new Europe – should not be taken for a willingness to halt, let alone reverse, the continued movement to full European Union. *Guardian*, 21 March 1992

subtext (sŭb'tekst), *n*. an unstated message or theme in a speech or piece of writing, conveyed in the tone of voice, choice and arrangement of words etc.

supercomputer (soo'pəkəmpūtə), *n*. a very powerful computer capable of over 100 million arithmetic operations per second.
   The other boon has been supercomputers. Cosmologists use these silicon behemoths to model the universe's structure.
Newsweek, 13 June 1988

supercomputing, *n*.

supermodel (soo'pə'mod'l), *n*. a highly paid fashion model.

superwaif (soo'pə'wāf), *n*. a very thin and young fashion model, usu. an adolescent, who can often appear as though she has barely achieved puberty.

surfing (scef'ing), *n*. the sport of riding on the roof or clinging to the outside of an urban train.
   Surfing is what they [youths] do in their spare time, riding between the carriages on the Underground.
Independent, 23 December 1988
surfer, *n*.

sustainable (səstān'əbl), *a*. of sources of raw materials, e.g. forests, capable of being replaced at the same rate as they are used, so causing minimal damage to the environment.

swipe (swīp), *v.t.* to pass (a credit or debit card) along a narrow groove in a machine which can read the information contained in its magnetic strip and enable a transaction to be completed.

T

---

tag (tag), *n*. (*sl*.) a signature, a stylized name, set of initials or logo written on a wall etc., by a graffiti-writer. *v.i.* to sign one's name in this way.
   Until recently, no film about New York was complete without a creaking subway car sprayed with indecipherable 'tags'. Seen by some as a form of street art.
Sunday Times, 14 May 1989
   Last month [he] was sentenced to 200 hours community service for 'tagging' his name in paint across London Underground property.
Independent, 2 January 1992
tagger, *n*. **tagging,** *n*.

tamper-evident (tam'pə), *a*. of packaging for food, medicament etc., designed so that any attempt to open or remove the packaging before purchase by a consumer will be obvious, a device to discourage consumer terrorism.

tapas (tap'əs), *n.pl*. light savoury snacks or appetizers. **tapas bar,** *n*. a bar where such food is served along with the drinks.
   Tapas, as those in the know will tell you, are not really about eating at all. They are small, tasty items of food found in bars throughout Spain, where up to 40 different dishes can be ranged along the counter.
'You' Magazine, 13 August 1989
[Sp., pl. of *tapa*, a cover, a dish of hors d'oeuvres]

TEC (tek), (*acronym*) *T*raining and *E*nterprise *C*ouncil.

Teflon ® (tef'lon), *a*. (*coll*.) of a politician, criminal etc., able to ensure that his/her reputation or career is never harmed by successful prosecutions or allegations of malpractice.
   [He] is known to millions of fascinated Americans as the 'Teflon Don', for the government's failure to convict him in earlier trials. *Independent*, 3 April 1992

telecommuting (telikəmūt'ing), *n*. working at home, while keeping in touch with central office, with clients etc. by telephone, fax, computer links etc. **telecommute,** *v.i.* **telecommuter,** *n*.

teleconference (telikon'fərəns), *n*. a conference held between people physically separated but communicating by telephone, video and computer links. **teleconferencing,** *n*. the holding of such conferences.

telecottage (tel'ikotij), *n*. a building, or a public room, in a rural area containing electronic equipment that might be needed by telecommuters living nearby.
   Today the telecottage functions as a training centre, library, electronic post office, data processing service bureau ... It is a meeting place where neighbours get together for informal discussions or drop in to use the equipment. *Sunday Times*, 20 January 1990
telecottaging, *n*. telecommuting, using not one's own equipment but that in a telecottage.

telegery (telij'eri), *n*. collecting phonecards. **telegerist,** *n*.

teleprocessing (teliprō'sesing), *n*. (*Comput.*) the processing of data which is carried out from a remote terminal.

telesoftware (telisoft'weə), *n*. (*Comput.*) software which is transmitted to its users by a teletext system.

teleworking TELECOMMUTING.

terrestrial (tərəs'triəl), *a*. (*TV*) referring to television signals sent from a ground-based transmitter (as opposed to a satellite).

Tessa (tes'ə), *n*. a bank or building-society savings account which offers tax-free interest provided that the capital is left untouched for at least five years. [acronym for *t*ax-*e*xempt *s*pecial *s*avings *a*ccount]

thirtysomething (thœ'tisŭmthing), *n.*, *a*. (*coll*.) (a person) aged between 30 and 39, esp. referring to middle-class couples and their social, emotional etc. worries.
   Its thirtysomething characters discuss life's eternally fascinating imponderables – men, women, sex and friendship – against a dreamlike backdrop of sophisticated Manhattan.
Marie Claire, November 1989

tidal (tī'dəl), *a*. **tidal power,** *n*. electricity generated by the harnessing of the rise and fall of the tides, using a *tidal barrage* (a dam built across e.g. an estuary to trap water at high tide).
   In essence barrages are simple things. As

the tide rises sluices are opened and the water level inside the device rises. At high tide the sluices shut and the entire structure is made watertight, keeping a high water level inside while the tide falls outside. Once there is sufficient difference in water levels – about 10 feet – water is allowed to flow gradually out to sea through huge turbines that drive generators.
Independent, 3 February 1992

toon (toon), *n.* (*coll.*) a character from an animated cartoon film.

Tourette syndrome (turet'), *n.* (*Med.*) a neurological disorder which causes sufferers to make involuntary jerky movements, and also sudden involuntary utterances, sometimes obscene. [G. Gilles de la *Tourette*, 1857–1904, F neurologist]

traffic calming (traf'ik), *n.* the deliberate slowing down of traffic on a road by installing road humps, narrowing the road, lowering the speed limit etc.
The council has decided to build a series of little roundabouts to slow the traffic down enough for the existing roads to be used. They call it 'traffic calming'.
Guardian, 3 November 1989

train surfing SURFING.

Trojan (horse) (trō'jən), *n.* (also **trojan**) (*Comput.*) a program which is designed to harm a computer system, and which gains access to the system by being inserted into a legitimate program, and differs from a computer virus in that it cannot replicate itself.
There are certain programs which will destroy your data, but they are not viruses. They are what have come to be known as Trojans.
Guardian, 19 January 1989
[see TROJAN in Dict.]

tutee (tū'tē), *n.* a person being tutored.

twocking (twok'ing), *n.* (*sl.*) (also **twoccing**) taking a car without the owner's permission. **twocker, twoccer,** *n.* [acronym for *t*aking *w*ithout the *o*wner's *c*onsent]

U

unbundle (ŭnbŭn'dl), *v.t.* to divide up a package of goods or services and sell the constituent parts separately; in business, to break up a conglomerate after taking it over, to sell the subsidiaries and keep only the core business. **unbundler,** *n.* **unbundling,** *n.*

upfront (ŭpfrŭnt'), *a.* (*coll.*) out in front, foremost; of money, paid in advance; of a person's manner, open, direct, positive. **upfrontness,** *n.*

use-by date BEST-BEFORE DATE.

user (ū'zə), *n.* **user-definable, user-defined,** *a.* (*Comput.*) referring to a key on the keyboard which can be allocated a frequently-needed routine by the user, to avoid typing out in full each time. **user interface,** *n.* (*Comput.*) communication between a computer system and its user by means of software which enables the computer to request and accept instructions.

V

valet parking (val'it, val'ā), *n.* a service provided at some hotels, restaurants etc., in which an attendant takes the patron's car at the door, parks it and brings it back when required. **valet-park,** *v.t.*

vertically challenged (vœ'tikəli), *a.* below the ordinary size, short, stunted.
The unpleasant trend of political correctness which attempts to sanitize terms such as dwarf into 'vertically challenged' and human beings into 'gender persons' is creeping daily into these islands.
Sunday Times, 9 February 1992

vibewatch (vīb'woch), *v.i.* at a meeting, discussion forum etc., to keep the tone of the discussion calm, to prevent angry exchanges etc., by intervening as soon as tempers begin to rise, or to censor offensive language or behaviour. *n.* an instance of such intervention. **vibewatcher,** *n.* one of the participants in a meeting, appointed to regulate the atmosphere of the discussion. [*vibes, watch*]

video nasty (vid'iō), *n.* an extremely violent or horrific film on video, thought by some people to be responsible for 'copycat' violent behaviour.

videophone (vid'iōfōn), *n.* (also **video telephone**) a telephone with a small screen and a video camera built in, allowing callers to see as well as hear each other.
Within the next 10 years a business executive on an aircraft would be able to chat to groups of colleagues at locations around the world and make sure at the same time – via a videophone – that they were wearing the new corporate tie.
Independent, 19 March 1992

virino (virī'nō), *n.* a very small molecule of DNA or RNA, that behaves like a virus without a permanent protein coat.

virtual (vœ'chuəl), *a.* **virtual memory,** *n.* (*Comput.*) a facility which allows more data to be stored than the capacity of the main memory by transferring data to and from disks. **virtual reality,** *n.* a computer system which allows a person wearing a special helmet with eyepieces, and special gloves, to enter a computer-generated environment, and move about and experience it.
The VR user, he explains, is surrounded by a three-dimensional computer-generated representation of reality. He or she sees a complete environment, from fields to rooms, from a variety of different angles, and can move through that world, entering buildings, touching surfaces, just as in the 'real' world. *Bookcase*, Issue 38 (March 1992)

VR, (*abbr.*) virtual reality.

W

WAN, (*abbr.*) WIDE AREA NETWORK.
whammy DOUBLE WHAMMY.
wheelie bin (wēl'i) *n.* (also **wheely bin**) a large plastic dustbin with two wheels so that it can be tilted and wheeled to the

983

dustcart, and a bar at the back and a hinged lid so that it can be lifted and emptied automatically by a mechanism on the dustcart.

whirlpooling (wœl'pool'ing), (*sl., esp. N Am.*) sexual harassment by a group of young males at a swimming pool.

wide area network (wīd), *n.* a group of computer systems widely separated geographically, and linked to each other by telephone, satellite etc., for the exchange of data etc.

wigger (wi'gə), *n.* a white teenager who imitates black culture.

The wigger uniform of baggy shorts and T-shirts, usually accompanied by a cap worn backwards or at a cocky tilt and basketball shoes, can be found modelled by kids in shopping malls across the country – especially those in well-to-do, white neighbourhoods.
Independent on Sunday, 22 August 1993

wilding (wīl'ding), *n.* (*sl., esp. N Am.*) a rampage through the streets by a gang of youths who rob or assault passers-by as they go.

to go wilding, to go on such a rampage.

... her attackers who, on that terrible night, went 'wilding', a new, cold police expression to describe a rampage against individuals who are harassed and physically assaulted by a mob. *Glasgow Herald*, 4 May 1989
[perh. orig. *to go wild-thing*, a reference to the title of a rap song]

wind farm (wind), *n.* (also **wind park**) a number of windmills or wind turbines grouped on open, windy ground, e.g. a hillside, for generating electric power.

These projects have been guaranteed a premium price for their electricity, to help lure the private sector into building wind farms.
Independent, 6 January 1992

wipe SWIPE.

world music (wœld), *n.* rock or pop music incorporating elements from a variety of national (esp. Third World) styles.

While the current world music explosion is certainly a welcome phenomenon, I must confess that I have my reservations about the future of Latin music. When will it be taken as seriously as African?
20/20, July 1989

WORM (wœm) (*acronym*) (*Comput.*) write once read many times (referring to an optical disk on which a user can store data, but cannot alter it later).

wysiwyg (wiz'iwig), (*acronym*) (*Comput.*) what you see is what you get (referring to a computer or word processor which can print out exactly what is shown on the screen).

Y

Yardie (yah'di), *n.* (*sl.*) a member of a group of orig. West Indian criminal gangs trading in illegal drugs.

... links between the growing cocaine market and the Yardies. They are a loose collection of black gangs originating in Jamaica which established a stranglehold on the cocaine market in many American cities and have strongholds in the black areas of London, Birmingham, Manchester and Liverpool. *Sunday Times*, 15 January 1989
[*Yard*, Jamaican dial. word for home, used by expatriates to refer to Jamaica]

Z

zap (zap), *n.* (*sl.*) a public meeting or demonstration designed to target one particular issue.

Over the past two and a half years ACT-UP (London) has been noteworthy in having repeatedly planned and implemented a series of original actions and zaps which have for the most part been unique as street demonstrations in Britain.
Pink Paper, 19 January 1992

Zike ® (zīk), *n.* a type of battery-operated bicycle.

Sir Clive Sinclair launched the Zike yesterday, the electric bicycle that is his first journey into the consumer market since the launch of his C5 electric tricycle.
The Times, 5 March 1992

zinfandel (zin'fəndel), *n.* a variety of grape grown esp. in California for making red wine. [etym. uncertain]

Appendix II
Gazetteer and Index to Maps

A

Aachen (Aix-la-Chapelle) a historic university city and spa town in W Germany (pop. 250,000) 28B2
Äänekoski Finland 30F3
Abadan a major oil-refining port on an island in the Shatt al Arab waterway, Iran (pop. 296,000) 38C2
Abadla Algeria 42C1
Abakan Russian Federation 33L4
Abbeville France 24C1
Abéché Chad 43F3
Abeokuta an industrial town in West Nigeria (pop. 301,100) 42D4
Aberdeen a major North Sea oil city and fishing port in NE Scotland (pop. 214,000) 22F3
Aberfeldy Scotland 22E4
Abergavenney Wales 21D6
Abersoch Wales 21C5
Aberystwyth Wales 21C5
Abidjan a major port and the chief city of Côte d'Ivoire (pop. 1,850,000) 42C4
Åbo see **Turku** 30E3
Aboyne Scotland 22F3
Abrantes Portugal 25A2
Abu Zabi (Abu Dhabi) the largest sheikhdom of the United Arab Emirates, of which the city of Abu Dhabi is the capital (67,350 sq km/26,000 sq miles; pop. emirate 535,700/ city 244,000) 38D3
Abuja the new capital of Nigeria, in the centre of the country, inaugurated in 1992. 42D4
Acapulco a large port and beach resort on the Pacific coast of Mexico (pop. 800,000) 6D4
Accra the capital and main port of Ghana (pop. 1,045,400) 42C4
Achill Island Ireland 23A3
Acklins Island The Bahamas 14E2
Aconcagua the highest mountain of the Andes, in Argentina (6960 m/22,835 ft) 17C6
Adan (Aden) a major port in S Yemen, formerly the capital of South Yemen (pop. 264,350) 38C4
Adana a city and province in S Turkey (pop. city 776,000) 38B2
Ad Dakhla Morocco 42B2
Ad Dammam Saudi Arabia 38D3
Ad Dawhah (Doha) the capital of Qatar (pop. 180,000) 38D3
Addis Ababa see **Adis Abeba** 43G4
Adelaide the state capital of South Australia (pop. 969,000) 45C4
Aden see **Adan** 38C4
Adirondack Mountains a range in New York State, USA. The highest peak is Mt Marcy (1629 m/5344 ft) 8E2
Adis Abeba (Addis Ababa) the capital of Ethiopia, in the centre of the country (pop. 1,500,000) 43G4
Admiralty Islands Pacific Ocean 37G6
Adriatic Sea a branch of the Mediterranean Sea, between Italy, Slovenia and Croatia 26A2
Aegean Sea a branch of the Mediterranean Sea between Greece and Turkey 27F3
Afghanistan [*area* 652,090 sq km (251,772 sq miles); *pop.* 15,810,000; *cap.* Kabul; *major cities* Herat, Kandahar, Mazar-i-Sharif; *govt* people's republic; *religions* Sunni Islam, Shia Islam; *currency* Afghani] a landlocked country in S Asia 39E2
Africa the second largest continent in the world, with the Mediterranean Sea to the N, the Atlantic to the W and the Indian Ocean to the E. It has 52 nations, excluding Western Sahara (30,300,000 sq km/11,700,000 sq miles; pop. 537,000,000) 41-44
Agadez Niger 42D3
Agadir a port and popular tourist resort in Morocco (pop. 111,000) 42C1
Agen France 24C3
Agra a city in central India, and site of the Taj Mahal (pop. 747,000) 39F3
Agrigento Italy 26C3
Agropoli Italy 26C2
Aguascalientes Mexico 6C3
Ahmadabad (Ahmedabad) India 39F3
Ahvaz Iran 38C2
Aïn Sefra Algeria 42C1
Ajaccio France 26B2
Ajmer India 39F3
Aketi Zaïre 43F4
Akita Japan 35P7
Akron a city in NE Ohio, USA (pop. city 226,900/metropolitan area 650,100) 7E1
Aksu China 39G1
Akureyri Iceland 30B1
Alabama a state in S USA; state capital Montgomery (133,667 sq km/51,606 sq miles; pop. 4,021,000) 7E2
Alagoinhas Brazil 16F4
Alajuela Costa Rica 14C5
Åland Island Finland 30E3
Alaska the largest and most northerly state of the USA; state capital Juneau (1,518,800 sq km/586,400 sq miles; pop. 521,000) 6J
Alaska Peninsula USA 5C4
Albacete a town and province of SE Spain (pop. town 177,100) 25B2
Albania [*area* 28,748 sq km (11,100 sq miles); *pop.* 3,200,000; *cap.* Tiranë; *major cities* Durrës, Shkodër, Elbasan; *govt* socialist republic; *religion* constitutionally atheist but mainly Sunni Islam; *currency* lek] a small mountainous country in the E Mediterranean. Its immediate neighbours are Greece and the former Yugoslavian republics of Serbia and Macedonia 27D2
Albany Australia 45A4
Albany the capital city of New York State, USA (pop. 99,500) 7F1
Al Basrah (Basra) the second city of Iraq, and its main port (pop. 1,200,000) 38B2
Al Bayda Libya 43F1
Albert, Lake in the Great Rift Valley, is shared between Uganda and Zaïre (also known as Lake Mobuto Sese Seko) (5180 sq km/2000 sq miles) 43G4
Alberta a province of W Canada; capital Edmonton (661,190 sq km/255,285 sq miles; pop. 2,238,000) 5G4
Ålborg a city port in N Denmark (pop. 160,000) 16C4
Alborz (Elburz) Mountains a range in N Iraq between Tehran and the Caspian Sea. The highest peak is the extinct volcano Damavand (5760m/18,600 ft) 38D2
Albuquerque a university city on the Rio Grande in New

Mexico, USA (pop. 350,600) 6C2
Albury Australia 45D4
Alcalá de Henares a historic town in central Spain, the birthplace of Miguel de Cervantes (pop. 142,900) 25B1
Alcañiz Spain 25B1
Alcázar de San Juan Spain 25B2
Alcoy Spain 25B2
Aldan Russian Federation 33O4
Alderney one of the Channel Islands 21E8
Aldershot England 21G6
Aleppo see **Halab** 38B2
Alessandria Italy 26B2
Ålesund Norway 30B3
Aleutian Islands USA 6J
Alexander Archipelago USA 6J
Alexander Island Antarctica 48
Alexandria see **El Iskandarîya** 43F1
Alexandroúpolis Greece 27F2
Al Furat (Euphrates, River) a great river of the Middle East, flowing from E Turkey, across Syria and central Iraq to The Gulf (length 2720 km/1690 miles) 38C2
Algeciras Spain 25A2
Alger (El Djazair, Algiers) the capital of Algeria, on the Mediterranean coast (pop. 1,800,000) 42D1
Algeria [area 2,381,741 sq km (919,590 sq miles); pop. 25,360,000; cap. Alger (Algiers); major cities Oran, Constantine, Annaba; govt republic; religion Sunni Islam; currency Algerian dinar] a huge country in N Africa, which fringes the Mediterranean Sea 42D1
Al Hudaydah Yemen 38C4
Al Hufuf Saudi Arabia 38C3
Alicante a port and popular beach resort on the Mediterranean coast of Spain (pop. town 251,400) 25B2
Alice Springs a desert settlement in the Northern Territory of Australia (pop. 18,400) 45C3
Al Jawf Saudi Arabia 38B3
Al Khaburah Oman 38D3
Al Kuwayt Kuwait 38C3
Al Ladhiqiyah (Latakia) a city on the Mediterranean coast of Syria, founded by the Romans and now that country's main port 38B2
Allahabad a holy city in India on the confluence of the rivers Ganges and Yamuna (pop. 650,000) 39G3
Alma-Ata (Almaty) a trading and industrial city and capital of Kazakhstan (pop. 1,046,100) 32J5
Almada Portugal 25A2
Al Madinah (Medina) the second holiest city of Islam after Mecca (pop. 210,000) 38B3
Al Manamah the capital and main port of Bahrain (pop. 115,054) 38D3
Al Mawsil (Mosul) a historic trading city on the River Tigris, NW Iraq; an important centre for the surrounding oil-producing region (pop. 1,500,000) 38B2
Almería Spain 25B2
Al Mukalla Yemen 38C4
Alnwick England 20F2
Alor Island Indonesia 37E4
Alotau Papua New Guinea 37H5
Alps, The a mountain range in S central Europe that spans the borders of Switzerland, France, Germany, Austria, Slovenia and Italy 26B1
Alsace a region in the NE of France 24D2
Alta Norway 30E2
Altay (Altai) an area of high mountain ranges in central Asia on the borders of China and the Russian Federation at the W end of Mongolia 34D2
Amami O-shima Island Japan 35G4
Amarillo Texas, USA 6C2
Amazonas (Amazon) River the world's second longest river flowing E from the Andes of Peru through Brazil to the Atlantic (length 6440 km/4000 miles) 16B4
Ambon (Amboina) an island and the capital of the Spice Islands in the Maluku group in E central Indonesia (813 sq km/314 sq miles; pop. 73,000) 37E4
America the continent between the Atlantic and Pacific Oceans, in three zones: North America (USA, Canada, Mexico, Greenland: 23,500,000 sq km/9,000,000 sq miles, pop. 354,000,000); Central America (between S Mexico border and Panama-Colombia border with the Caribbean: 1,849,000 sq km/714,000 sq miles, pop. 63,000,000); South America (S of Panama-Colombia border: 17,600,000 sq km/6,800,000 sq miles, pop. 284,000,000) 4-17
American Samoa an American territory, comprising a group of five islands, in the central South Pacific 46
Amfilokhía Greece 27E3
Amiens France 24C2
Amindivi Islands see **Lakshadweep** 39F4
Amirante Islands Indian Ocean 44F1
Amman the capital of Jordan, in the NE of the country (pop. 1,232,600) 38B2
Amorgós Island Greece 27F3
Ampia Western Samoa 46
Amposta Spain 25C1
Amritsar an industrial city in N India, home of the Golden Temple, the most sacred shrine of the Sikhs (pop. 595,000) 39F2
Amsterdam the capital of the Netherlands, a historic port on the IJsselmeer (pop. 712,300) 28A2
Amudar'ya, River a central Asian river (ancient name Oxus) forming much of the border between Tajikistan and Afghanistan before flowing through Uzbekistan into the Aral Sea (length 2620 km/1630 miles) 32H6
Amundsen Sea Antarctica 48
Amur, River (Heilong Jiang) a river along the border between China and the Russian Federation, flowing E into the Pacific (length 4510 km/2800 miles) 35G1
Anáfi Island Greece 27F3
An Najaf Iraq 38B2
Anápolis Brazil 16E4
Anchorage the largest city and port in Alaska, USA, on its S coast (pop. 226,700) 5D3
Ancona Italy 26C2
Andalucía (Andalusia) a region of SW Spain (pop. 6,441,800) 25A2
Andaman and Nicobar Islands two groups of islands in the Bay of Bengal, administered by India 39H4
Andaman Sea a branch of the Bay of Bengal, lying between the Andaman Islands and Burma 36B2
Andes a high mountain range that runs down the entire length of the W coast of South America. The highest peak is Aconcagua 16-17
Andizhan Uzbekistan 32J5
Andorra [area 453 sq km (175 sq miles); pop. 51,400; cap. Andorra-la-Vella; govt co-principality; religion RC; currency franc, peseta] a tiny state, high in the E Pyrénées, between France and Spain 25C1
Andros Greece 27E3
Andros the largest of the islands of the Bahamas (4144 sq km/1600 sq miles; pop. 8900) 14D2
Angara, River a river in the Russian Federation flowing from Lake Baikal into the Yenisey River (length 1825 km/1135 miles) 33L4
Angarsk Russian Federation 33M4
Angers France 24B2
Anglesey an island off the NW tip of Wales (715 sq km/276 sq miles; pop. 69,000) 20C4
Angola [area 1,246,700 sq km (481,351 sq miles); pop. 10,020,000; cap. Luanda; major cities Huambo, Lobito, Benguela; govt people's republic; religions RC, Animism; currency kwanza] a country on the Atlantic coast of W central Africa, 10°S of the Equator 44B2
Angoulême France 24C2
Anguilla an island in the Leeward Islands group of the

Anjou | Arua

Caribbean, now a self-governing British dependency (91 sq km/35 sq miles; pop. 6500) 14G3
Anjou a former province of W France, in the valley of the River Loire 24B2
Ankara the capital of Turkey, in the E central part of Asian Turkey (pop. 2,252,000) 38B2
Annaba (Bone) a historic town and seaport on the Mediterranean coast of Algeria (pop. 245,000) 42D1
Annapolis the capital of the state of Maryland, USA (pop. 31,900) 7F2
Annapurna a mountain of the Himalayas, in Nepal (8172 m/26,810 ft) 34C4
Annecy France 24D2
Annobón Island Equatorial Guinea 42D5
Anqing China 35F3
Anshan China 35G2
Anshun China 34E4
Antalya Turkey 38B2
Antananarivo the capital of Madagascar (pop. 663,000) 44E2
Antarctica an ice-covered continent around the South Pole consisting of a plateau and mountain ranges reaching a height of 4500 m (15,000 ft). It is uninhabited apart from temporary staff at research stations (14,000,000 sq km/5,100,000 sq miles) 48
Antarctic Ocean (Southern Ocean) the waters that surround Antarctica made up of the S waters of the Atlantic, Indian and Pacific Oceans 48
Antequera Spain 25B2
Anticosti Island Canada 5M5
Antigua and Barbuda [area 440 sq km (170 sq miles); pop. 85,000; cap. St John's; govt constitutional monarchy; religion Christianity (mainly Anglicanism); currency East Caribbean dollar] a tiny state comprising three islands—Antigua, Barbuda and uninhabited Redonda—on the E side of the Leeward Islands 14G3
Antipodes Islands New Zealand 46
Antofagasta Chile 17B5
Antrim a county in Northern Ireland (2831 sq km/1093 sq miles; pop. county 642,267) 23E2
Antrim a town in Northern Ireland (pop. 22,242) 23E2
Antsirabe Madagascar 44E2
Antsiranana Madagascar 44E2
An Uaimh Ireland 23E3
Antwerpen (Antwerp, Anvers) the main port of Belgium (pop. 488,000) 24C1
Anyang an ancient city in Henan province, China (pop. 500,000) 35F3
Aomori Japan 35J2
Aosta Italy 26B1
Aparri Philippines 37E2
Apatity Russian Federation 30G2
Apeldoorn Netherlands 28B2
Apennines (Appennino) the mountain range which forms the "backbone" of Italy. The highest peak is Monte Corno (2912 m/9554 ft) 26C2
Apia the capital of Western Samoa (pop. 34,000) 46
Appalachian Mountains a chain stretching 2570 km (1600 miles) down E North America from Canada to Alabama in the USA. The highest peak is Mount Mitchell (2037 m/6684 ft) 7E2
Appleby-in-Westmorland England 20E3
Aqaba the only port in Jordan, on the Gulf of Aqaba (pop. 35,000) 38B3
Arabian Gulf see **Gulf, The** 38D3
Arabian Sea a branch of the Indian Ocean between India and Arabia 38D5
Aracaju Brazil 16F4
Aracena Spain 25A2
Arad Romania 29E3
Arafura Sea a stretch of the Pacific Ocean between New Guinea and Australia 45C1

Aral Sea a large, salty lake, to the E of the Caspian Sea, on the border between Uzbekistan and Kazakhstan (64,750 sq km/25,000 sq miles) 32G5
Aranda de Duero Spain 25B1
Aran Island Ireland 23C1
Aran Islands (Oileáin Arann) Ireland 23B3
Ararat, Mount (Büjük Agri Dagi) the mountain peak in E Turkey where Noah's Ark is said to have come to rest (5165 m/17,000 ft) 38C2
Arbatax Italy 26B3
Arbroath Scotland 22F4
Arcachon France 24B3
Archangel see **Arkhangel'sk** 32F3
Arch de los Chonos Chile 17B7
Arctic Ocean the ice-laden sea to the N of the Arctic Circle (14,100,000 sq km/5,440,000 sq miles) 48
Ardabàl (Ardabil) a town in Iran, famous for its knotted carpets (pop. 222,000) 38C2
Ardara Ireland 23C2
Ardennes a hilly and forested region straddling the borders of Belgium, Luxembourg and France 28B2
Arendil Norway 30C4
Arequipa Peru 16B4
Arezzo Italy 26C2
Argentina [area 2,766,889 sq km (1,068,296 sq miles); pop. 32,690,000; cap. Buenos Aires; major cities Cordoba, Rosaria, Mendoza, La Plata; govt federal republic; religion RC; currency austral] the world's eighth largest country, stretching from the Tropic of Capricorn to Cape Horn on the S tip of the South America. To the W the Andes, form the border with Chile 17C6
Argenton-S-Creuse France 24C2
Argos Greece 27E3
Århus a port and the second largest city in Denmark (pop. 246,700) 30C4
Arica Chile 16B4
Arisaig Scotland 22C4
Arizona a state in the SW of the USA; capital Phoenix (295,024 sq km/113,902 sq miles; pop. 3,137,000) 6B2
Arkansas a state in S USA; capital Little Rock (137,539 sq km/53,104 sq miles; pop. 2,359,000) 7D2
Arkansas, River a tributary of the Mississippi, flowing from the Rocky Mountains through Kansas, Oklahoma and Arkansas (length 2335 km/1450 miles) 7D2
Arkhangel'sk (Archangel) a port on the White Sea, Russian Federation (pop. 403,000) 32F3
Arklow Ireland 23E4
Arles France 24C3
Armagh a county in Northern Ireland (1254 sq km/484 sq miles; pop. county 118,820) 23E2
Armagh a city in Northern Ireland (pop. 12,700) 23E2
Armenia [area 29,800 sq km (11,500 sq miles); pop. 3,267,000; cap. Yerevan; major city Kumayri (Leninkan); govt republic; religion Armenian Orthodox; currency rouble] the smallest republic of the former USSR and part of the former republic of Armenia which was divided between Turkey, Iran and the USSR. It declared independence from the USSR in 1991 32F5
Arnhem a town in the Netherlands, scene of a battle in 1944 between British (and Polish) paratroops and the German army (pop. 128,600) 28B2
Arnhem Land an Aboriginal reserve in the Northern Territory of Australia 45C2
Arno, River a river of Italy, flowing W through Florence to Pisa (length 245 km/152 miles) 26C2
Arran Island Scotland 22C4
Ar Riyad (Riyadh) the capital and commercial centre of Saudi Arabia (pop. 300,000) 38C3
Arrochar Scotland 22D4
Arta Greece 27E3
Artois (province) France 24C1
Arua Uganda 43G4

Aruba   988   Bahrain

Aruba a Caribbean island off the coast of Venezuela, formerly one of the Netherlands Antilles   14F4
Arusha Tanzania   43G5
Arvide Canada   5L5
Arvika Sweden   16C4
Asahikawa Japan   35J2
Asansol India   39G3
Ascension Island a tiny volcanic island in the South Atlantic Ocean   41
Aseb Ethiopia   43H3
Ashford England   21H6
Ashikaga Japan   35N8
Ashington England   20F2
Ashkhabad capital of Turkmenistan (pop. 347,000)   32G6
Asia the largest continent, bounded by the Arctic, Pacific and Indian Oceans, plus the Mediterranean and Red Seas (43,600,000 sq km/16,800,000 sq miles; pop. 3,075,000,000). East Asia is taken to mean those countries to the NE of Bangladesh; South Asia refers to the countries on the Indian subcontinent; and SE Asia includes those countries SE of China, including the islands to the W of New Guinea   31-39
Asmera (Asmara) the main city of Eritrea in Ethiopia (pop. 430,000)   43G3
Astípalaia Island Greece   27F3
Astrakhan' a port of the Russian Federation near the Caspian Sea, situated on the delta of the River Volga (pop. 487,000)   32F5
Asturias a region of N Spain. The capital is Oviedo (pop. 1,227,000)   25A1
Asunción the capital and the only major city of Paraguay (pop. 456,000)   17D5
Aswân a city in S Egypt by the River Nile. Aswan High Dam is 13 km (8 miles) to the S (pop. 200,000)   43G2
Asyût Egypt   43G2
Atacama Desert an extremely dry desert lying mainly in N Chile   17C5
Atar Mauritania   42B2
Atbara Sudan   43G3
Athabasca, River a river in Canada which flows N from the Rocky Mountains to Lake Athabasca (length 1231 km/765 miles)   5G4
Athenry Ireland   23C3
Athens see **Athínai**
Athínai (Athens) the historic capital, and the principal city, of Greece (pop. city 885,700/metropolitan area 3,027,300)   27E3
Athlone Ireland   23D3
Atlanta the capital and largest city of Georgia, USA (pop. 426,100)   7E2
Atlantic Ocean the second largest ocean, lying between North and South America, Europe and Africa (82,200,000 sq km/31,700,000 sq miles)   2
Atlas Mountains a series of mountain chains stretching across North Africa from Morocco to Tunisia   42C1
At Ta'if Saudi Arabia   38C3
Auckland the largest city and chief port of New Zealand, on North Island (pop. 769,600)   45G4
Auckland Island New Zealand   46
Augsburg Germany   28B3
Augusta Australia   45A4
Augusta a city and river port on the Savannah River in Georgia, USA (pop. city 46,000/metropolitan area 368,300)   7E2
Augusta the capital of Maine, USA (pop. 22,000)   7G1
Augustow Poland   29E2
Austin the capital of Texas, USA (pop. city 397,000/metropolitan area 645,400)   6D2
Australia [area 7,686,848 sq km (2,967,892 sq miles); pop. 17,100,000; cap. Canberra; major cities Adelaide, Brisbane, Melbourne, Perth, Sydney; govt federal parliamentary state; religion Christianity; currency Australian dollar] the world's smallest continent, a vast and sparsely populated island state in the S hemisphere   45C3
Australian Capital Territory (ACT) the small region which surrounds Canberra, the capital of Australia (2432 sq km/939 sq miles; pop. 240,000)   45D4
Austria [area 83,853 sq km (32,376 sq miles); pop. 7,600,000; cap. Vienna; major cities Graz, Linz, Salzburg; govt federal republic; religion RC; currency schilling] a landlocked country in central Europe   28C3
Auvergne a mountainous region of central France   24C2
Auxerre France   24C2
Aveiro Portugal   25A1
Avellino Italy   26C2
Avesta Sweden   30D3
Avezzano Italy   26C2
Aviemore Scotland   22E3
Avignon a historic city on the River Rhone in S France, the seat of the Pope, 1309-77 (pop. 177,500)   24C3
Avila a town and province in the mountainous central region of Spain, famous as the birthplace of St Teresa (pop. town 41,800)   25B1
Avilés Spain   25A1
Avon a county in the W of England; county town Bristol (1338 sq km/517 sq miles; pop. 936,000)   21E6
Axel Heiberg Island Canada   5J2
Axminster England   21E7
Ayers Rock a huge rock, sacred to the Aborigines, rising sharply out of the plains in the Northern Territory of Australia (348 m/1142 ft)   45C3
Ayios Evstrátios Island Greece   27F3
Aylesbury England   21G6
Ayr Scotland   22D5
Azerbaijan [area 87,000 sq km (33,600 sq miles); pop. 6,506,000; cap. Baku; major cities Kirovabad, Sumgait; govt republic; religions Shia Islam, Russian Orthodox; currency rouble] a republic of the former USSR, which declared itself independent in 1991, on the SW coast of the Caspian Sea   32F5
Azores three groups of small islands in the North Atlantic Ocean, belonging to Portugal   2
Az Zawiyah Libya   42E1

B

Babar Island Indonesia   37E4
Babuyan Island Philippines   37E2
Bacau Romania   29F3
Bacolod Philippines   37E2
Badajoz Spain   25A2
Badalona Spain   25C1
Bafatá Guinea-Bissau   42B3
Baffin Bay a huge bay within the Arctic Circle between Baffin Island in Canada and Greenland   5M2
Baffin Island a large, mainly ice-bound island in NE Canada (507,451 sq km/195,927 sq miles)   5K2
Bagé Brazil   17D5
Baghdad the capital of Iraq, in the centre of the country, on the River Tigris (pop. 3,300,000)   38B2
Baghlan Afghanistan   39E2
Baguio Philippines   37E2
Bahamas, The [area 13,878 sq km (5358 sq miles); pop. 256,000; cap. Nassau; major city Freeport; govt constitutional monarchy; religion Christianity; currency Bahamian dollar] an archipelago of 700 islands in the Atlantic Ocean off the SE coast of Florida   14E1
Bahawalpur Pakistan   39F3
Bahía Blanca Argentina   17C6
Bahrain [area 678 sq km (262 sq miles); pop. 486,000; cap. Manama; govt monarchy (emirate); religions Shia Islam, Sunni Islam; currency Bahraini dollar] a Gulf state comprising 33 low-lying islands between the

Qatar peninsula and Saudi Arabia	38D3	**Bangor** Wales	20C4
Baia Mare Romania	29E3	**Bangui** the capital of the Central African Republic (pop. 387,000)	43E4
Baicheng China	34K2		
Baikal, Lake *see* **Baykal, Ozero**		**Banja Luka** a city of ancient origins on the Vrbas River in NW Bosnia Herzegovina (pop. 183,000)	26D2
Baja California a huge 1300-km (800-mile) long peninsula belonging to Mexico which stretches S from California, USA, into the Pacific (pop. 1,400,000)	6B3	**Banjarmasin** Indonesia	36D4
		Banjul the capital of the Gambia, formerly called Bathurst (pop. 42,000)	42B3
Baker Island Pacific Ocean	46		
Bakhtaran Iran	38C2	**Bank Islands** Vanuatu	45F2
Baku (Baky) a port on the Caspian Sea and the capital of the republic of Azerbaijan (pop. 1,661,000)	32F5	**Banks Island** Canada	5F2
		Bantry Ireland	23B5
Bala Wales	21D5	**Banyuwangi** Indonesia	36D4
Balaklava *see* **Sevastopol**	32E5	**Baoding** China	35F3
Balaton, Lake Hungary	27D1	**Baoji** China	34E3
Balbriggan Ireland	23E3	**Baotou** China	34E2
Bâle *see* **Basel**	24D2	**Baracaldo** Spain	25B1
Balearic Islands *see* **Islas Baleares**	25C2	**Barbados** [*area* 430 sq km (166 sq miles); *pop.* 260,000; *cap.* Bridgetown; *govt* constitutional monarchy; *religions* Anglicanism, Methodism; *currency* Barbados dollar] the most easterly island of the West Indies	14H4
Bali a small island off the E tip of Java, the only island in Indonesia to have preserved a predominantly Hindu culture intact	36D4		
Balikpapan Indonesia	36D4	**Barbuda** *see* **Antigua and Barbuda**	14G3
Balkans the SE corner of Europe, a broad, mountainous peninsula bordered by the Adriatic, Ionian, Aegean and Black Seas. Albania, Bulgaria, Greece, Romania, Slovenia, Croatia, Bosnia, the rest of former Yugoslavia and European Turkey are in the Balkans	26-27	**Barcelona** the second largest city in Spain, and the name of the surrounding province. It is a major port on the Mediterranean (pop. city 1,754,900)	25C1
		Barcelona Venezuela	16C1
		Bareilly India	39F3
		Barents Sea part of the Arctic Ocean, N of Norway	48
Balkhash Kazakhstan	32J5	**Bari** a major port on the Adriatic coast of Italy (pop. 370,000)	26D2
Balkhash, Ozero (Lake Balkhash) a massive lake in Kazakhstan, near the border with China (22,000 sq km/8500 sq miles)	32J5		
		Barletta Italy	26D2
Ballachulish Scotland	22C4	**Barnaul** Russian Federation	32K4
Ballantrae Scotland	22D5	**Barnstaple** England	21C6
Ballarat a historic gold-mining town in Victoria, Australia (pop. 62,600)	45D4	**Baroda** *see* **Vadodara**	39F3
		Barquisimeto Venezuela	16C1
Ballina Ireland	23B2	**Barra** Scotland	22A3
Ballinasloe Ireland	23C3	**Barranquilla** Colombia	16B1
Ballygawley Northern Ireland	23D2	**Barrow** USA	5C2
Ballymena Northern Ireland	23E2	**Barrow-in-Furness** England	20D3
Baltic Sea a shallow sea in N Europe, completely surrounded by land masses except for the narrow straits that connect it to the North Sea	29D1	**Barry** Wales	21D6
		Basel (Basle, Bâle) a city in N Switzerland and the name of the surrounding canton (pop. city 200,000)	24D2
Baltimore the largest city in the state of Maryland, USA (pop. city 763,000/metropolitan area 2,244,700)	7F2	**Basilan Island** Philippines	37E3
		Basildon England	21H6
Bamako the capital of Mali (pop. 405,000)	42C3	**Basingstoke** England	21F6
Bambari Central African Republic	43F4	**Basra** *see* **Al Basrah**	38B2
Bamberg Germany	28B3	**Bassein** Myanmar	36B2
Banbury England	21F5	**Bass Strait** the stretch of water (290 km/180 miles) separating mainland Australia from Tasmania	45D4
Banda Aceh Indonesia	36B3		
Bandar 'Abbas a port in S Iran on the Strait of Hormuz, at the neck of The Gulf (pop. 89,200)	38D3	**Bastia** France	26B2
		Bata Equatorial Guinea	42D4
Bandar Seri Begawan the capital of Brunei (Pop. 50,000)	36D3	**Batan Island** Philippines	37E1
		Batang China	34D3
Banda Sea Indonesia	37E4	**Batangas** Philippines	37E2
Bandon Ireland	23C5	**Bath** a beautifully preserved spa town SW England (pop. 85,000)	21E6
Bandundu Zaïre	43E5		
Bandung a large inland city in W Java, Indonesia (pop. 1,462,700)	36C4	**Bathurst** *see* **Banjul**	42B3
		Bathurst Island Australia	45B2
Banff Scotland	22F3	**Bathurst Island** Canada	5H2
Bangalore a large industrial city in central S India (pop. 2,921,800)	39F4	**Batley** England	20F4
		Baton Rouge the state capital of Louisiana, USA, on the Mississippi River (pop. city 238,900/metropolitan area 538,000)	11D4
Bangassou Central African Republic	43F4		
Banghazi (Benghazi) a major port on the Gulf of Sirte in Libya (pop. 485,000)	43F1		
		Battambang Cambodia	36C2
Bangka Island Indonesia	36C4	**Batumi** Georgia	32F5
Bangkok (Krung Thep) the capital of Thailand, on the River Chao Phraya (pop. 5,900,000)	36C2	**Bayeux** a market town in Normandy, France, home of an 11th-century tapestry depicting the Norman conquest of England (pop. 15,300)	24B2
Bangladesh [*area* 143,998 sq km (55,598 sq miles); *pop.* 113,340,000; *cap.* Dacca (Dhaka); *major cities* Chittagong, Khulna; *govt* republic; *religion* Sunni Islam; *currency* taka] a country bounded almost entirely by India and to the S by the Bay of Bengal	39G/H3		
		Baykal, Ozero (Lake Baikal) the world's deepest freshwater lake and largest by volume, in SE Siberia, Russian Federation (31,500 sq km/12,150 sq miles)	33M4
		Bayonne the capital of the French Basque region (pop. 129,730)	24B3
Bangor Northern Ireland	23F2		

Bayreuth | Birmingham

Bayreuth Germany — 28B3
Baza Spain — 25B2
Beaufort Sea a part of the Arctic Ocean to the N of North America — 48
Beaufort West South Africa — 44C4
Beauly Scotland — 22D3
Beauvais France — 23C2
Béchar Algeria — 42C1
Bedford England — 21G5
Bedfordshire a county in central S England; county town Bedford (1235 sq km/477 sq miles; pop. 75,000) — 21G5
Beijing (Peking) the capital of China, in the NE of the country (pop. 9,231,000) — 35F3
Beira Mozambique — 44D2
Beirut see **Beyrouth** — 38B2
Beja Portugal — 25A2
Béjar Spain — 25A1
Belarus (Belorussia, Byelorussia) [area : 207,600 sq km (80,150 sq miles); pop. 9,878,000; cap. Minsk; major cities Gomel, Vitebsk, Mogilev; govt republic; religions Russian Orthodox, RC; currency rouble] a republic of the former USSR, which declared itself independent in 1991 — 29F2
Belau a republic consisting of a group of islands in the W Pacific formerly known as Palau (494 sq km/191 sq miles; pop. 14,000; currency US dollar) — 37F3
Belcher Islands Canada — 5K4
Belcoo Northern Ireland — 23D2
Belém a major port of Brazil situated to the N of the mouth of the River Amazon (pop. 934,000) — 16E2
Belfast the capital and largest city of Northern Ireland (pop. 360,000) — 23F2
Belfort France — 24D2
Belgium [area 30,519 sq km (11,783 sq miles); pop. 9,930,000; cap. Brussels; major cities Antwerp, Ghent, Charleroi, Liege; govt constitutional monarchy; religion RC; currency Belgian franc] a small country in NW Europe with a short coastline on the North Sea — 28A2
Belgrade see **Beograd** — 27E2
Beihai China — 34E4
Belitung Indonesia — 36C4
Belize [area 22,965 sq km (8867 sq miles); pop. 193,000; cap. Belmopan; major city Belize City; govt constitutional monarchy; religion RC; currency Belize dollar] a small Central American country on the SE of the Yucatan Peninsula — 14B3
Bellary India — 39F4
Belle Ile France — 24B2
Bellinghausen Sea a part of the Pacific Ocean off Antarctica, due S of South America — 48
Bello Colombia — 16B2
Belmopan the capital of Belize (pop. 5000) — 14B3
Belmullet Ireland — 23A2
Belogorsk Russian Federation — 33O4
Belo Horizonte Brazil — 16E4
Belopan Guatemala — 14B3
Beloye More (White Sea) an arm of the Barents Sea off the NW of the Russian Federation — 32E3
Belorussia see **Belarus** — 29F2
Belostock see **Bialystok** — 29E2
Benares see **Varanasi** — 39G3
Benavente Spain — 25A1
Benbecula Scotland — 22A3
Bendigo Australia — 45D4
Benevento Italy — 26C2
Bengal, Bay of a massive bay of the Indian Ocean between India and Myanmar, S of Bangladesh — 39G4
Bengbu China — 35F3
Benghazi see **Banghazi** — 43F1
Benguela Angola — 44B2
Benicarló Spain — 25C1
Benidorm a popular Mediterranean seaside resort of Spain (pop. 25,600) — 25B2

Benin [area 112,622 sq km (43,483 sq miles); pop. 4,760,000; cap. Porto-Novo; major city Contonou; govt republic; religions Animism, RC, Sunni Islam; currency franc CFA] a cone-shaped country in West Africa with a short coastline on the Bight of Benin — 42D4
Benin City Nigeria — 42D4
Ben Nevis see **Grampian Mountains** — 22C4
Benue, River a river of Cameroon and Nigeria flowing to the Gulf of Guinea (length 1390 km/865 miles) — 42D4
Benxi an industrial city in Liaoning, China (pop. 1,200,000) — 35G2
Beograd (Belgrade) the capital of Serbia, on the confluence of the Danube and Sava (pop. 1,407,100) — 27E2
Berat Albania — 27D2
Berbera Somalia — 43H3
Bergamo Italy — 26B1
Bergen an old port in SW Norway, and now that country's second largest city (pop. 181,000) — 30B3
Bering Sea a part of the Pacific Ocean between Alaska and E Russian Federation — 5B3
Bering Strait a stretch of sea, 88 km (55 miles) wide, separating the Russian Federation from Alaska — 5B3
Berkner Island Antarctica — 48
Berkshire a county of central S England; county town Reading (1256 sq km/485 sq miles; pop. 708,000) — 21F6
Berlin the capital of Germany, in the N of the country on the River Spree (pop. 3,097,000) — 28C2
Bermuda [area 53 sq km (21 sq miles); pop. 59,066; cap. Hamilton; govt colony under British administration; religion Protestantism; currency Bermuda dollar] a group of 150 small islands in the W Atlantic Ocean — 7G2
Bern (Berne) the historic capital of Switzerland, and also the name of the surrounding canton (pop. city 150,000) — 24D2
Berwick-upon-Tweed England — 20F2
Besançon a town of ancient origins in the Jura region of E France (pop. 120,800) — 24D2
Betanzos Spain — 25A1
Bethlehem a town in the West Bank area of Israel, celebrated by Christians as the birthplace of Jesus Christ (pop. 30,000) — 40C3
Beuten see **Bytom** — 29D2
Beverley England — 20G4
Beyla Guinea — 42C4
Beyrouth (Beirut) the capital and main port of Lebanon (pop. 938,000) — 38B2
Béziers France — 24C3
Bhagalpur India — 39G3
Bhamo Myanmar — 36B1
Bhopal an industrial city in central India (pop. 671,000) — 39F3
Bhutan [area 47,000 sq km (18,147 sq miles); pop. 1,400,000; cap. Thimphu; govt constitutional monarchy; religion Buddhism; currency ngultrum] a small country surrounded by India to the S and China to the N — 39G3
Biak Island Indonesia — 37F4
Bialystok (Belostock) Poland — 29E2
Bideford England — 21C6
Biel Switzerland — 24D2
Bikaner India — 39F3
Bilaspur India — 39G3
Bilbao a port and industrial city in the Basque region of N Spain (pop. 433,000) — 25B1
Billings USA — 6C1
Bioko an island in the Gulf of Guinea (formerly Fernando Póo) now governed by Equatorial Guinea (2017 sq km/780 sq miles; pop. 57,000) — 42D4
Birkenhead England — 20E4
Birlad Romania — 29F3
Birmingham an industrial city and second largest city in the UK (pop. 976,000) — 21F3
Birmingham the largest city in Alabama, USA (pop. city 279,800/metropolitan area 895,200) — 7E2

Bir Moghrein   991   Bratsk

Bir Moghrein Mauritania 42B2
Birobidzhan Russian Federation 33P5
Birr Ireland 23D3
Biscay, Bay of a broad bay of the Atlantic Ocean between N Spain and Brittany in NW France 24B2
Bishkek, formerly Frunze, the capital of Kyrgyzstan (pop. 577,000) 32J5
Bishop Auckland England 20F3
Biskra Algeria 42D1
Bismarck the state capital of North Dakota, USA (pop. city 47,600/metropolitan area 86,100) 6C1
Bismarck Archipelago Pacific Ocean 37G4
Bismarck Sea a branch of the Pacific Ocean to the N of Papua New Guinea 37G4
Bissau (Bissão) a port and the capital of Guinea-Bissau (pop. 109,000) 42B3
Bitola Yugoslavia 27E2
Bizerte Tunisia 42D1
Blace Croatia 27D2
Blackburn England 20E4
Black Forest *see* **Schwarzwald** 28B3
Black Hills a range rising to 2207 m (7242 ft) on the border between South Dakota and Wyoming, USA 10B2
Blackpool the largest seaside holiday resort in the UK, in Lancashire (pop. 147,000) 20D4
Black Sea a sea lying between SE Europe and W Asia; it is surrounded by land except for the Bosphorus channel, leading to the Mediterranean Sea 32E5
Blagævgrad Bulgaria 27E2
Blagoveshchensk Russian Federation 33O4
Blair Atholl Scotland 22E4
Blanc, Mont (Monte Bianco) the highest mountain in W Europe, on the border between France and Italy (4808 m/15,770 ft) 24D2
Blantyre the largest city in Malawi (pop. 333,800) 44D2
Blida Algeria 42D1
Bloemfontein the judicial capital of South Africa, and capital of the Orange Free State (pop. 256,000) 44C3
Blönduós Iceland 30A1
Bluefields Nicaragua 14C4
Blyth England 20F2
Bo Sierra Leone 42B4
Boa Vista Brazil 16C2
Bobo Dioulasso Burkina Faso 42C3
Bocas del Toro Panama 14C5
Boden Sweden 30E2
Bodensee (Lake Constance) a lake surrounded by Germany to the N, Switzerland to the S and Austria to the E (536 sq km/207 sq miles) 28B3
Bodmin England 21C7
Bodö Norway 30C2
Bodrum (Halicarnassus) a port on the SE coast of Turkey (pop. 13,090) 27F3
Bognor Regis England 21G7
Bogor Indonesia 36C4
Bogotá capital of Colombia, on a plateau of the E Andes in the centre of the country (pop. 5,789,000) 16B2
Bohol Island Philippines 37E3
Boise the state capital of Idaho, USA (pop. city 107,200/metropolitan area 189,300) 6B1
Boké Guinea 42B3
Bolivia [*area* 1,098,581 sq km (424,162 sq miles); *pop.* 6,410,000; *cap.* La Paz (administrative), Sucre (legal); *major city* Cochabamba; *govt* republic; *religion* RC; *currency* boliviano] a landlocked republic of central South America through which the Andes run 16C4
Bollnäs Sweden 30D3
Bologna Italy 26C2
Bolton England 20E4
Boma Zaïre 42E5
Bombay a major port in central W India, and now its most important industrial city (pop. 8,234,400) 39F4

Bonaire a Caribbean island off the coast of Venezuela and part of the Netherland Antilles (288 sq km/111 sq miles; pop. 9700) 14F4
Bone *see* **Annaba** 42D1
Bonn the capital of former West Germany, it is the administrative centre of Germany until the government moves to Berlin (pop. 300,000) 28B2
Boothia Peninsula Canada 5J2
Borås Sweden 30C4
Bordeaux a major port on the Gironde estuary in SW France (pop. 650,125) 24B3
Borders an administrative region of S Scotland, created out of the old counties of Berwickshire, Peeblesshire, Roxburghshire, Selkirkshire and part of Midlothian (4662 sq km/1800 sq miles; pop. 101,000) 22E5
Borneo one of the largest islands in the world, divided between three countries. Most of it is known as Kalimantan, part of Indonesia. The N coast is divided into the states of Sarawak and Sabah, part of Malaysia, and the small independent sultanate of Brunei (751,900 sq km/290,320 sq miles) 36D3
Bornholm Island Denmark 30C4
Borzya Russian Federation 33N4
Bosanski Brod Croatia 27D1
Bosnia Herzegovina [*area* 51,129 sq km (19,736 sq miles); *pop.* 4,124,000; *cap.* Sarajevo; *major cities* Banja Luka, Tuzla; *govt* republic; *religions* Eastern Orthodox, Sunni Islam, RC; *currency* dinar] a republic of former Yugoslavia, which was recognized as an independent state in 1992 27D2
Boston England 21G5
Boston an Atlantic port and state capital of Massachusetts, USA (pop. city 570,700/metropolitan area 2,820,700) 7F1
Bothnia, Gulf of the most northerly arm of the Baltic Sea, bordered by Finland and Sweden 30E3
Botswana [*area* 581,730 sq km (224,606 sq miles); *pop.* 1,260,000; *cap.* Gaborone; *major cities* Mahalapye, Serowe, Francistown; *govt* republic; *religions* Animism, Anglicanism; *currency* pula] a landlocked republic in S Africa which straddles the Tropic of Capricorn 44C3
Bouaké the second largest city of the Côte d'Ivoire (pop. 640,000) 42C4
Bouar Central African Republic 42E4
Bouârfa Morocco 42C1
Boulogne France 24C1
Bounty Islands New Zealand 46
Bourg-en-Bresse France 24D2
Bourges France 24C2
Bourgogne (Burgundy) a region of central France, famous for its wine 24C2
Bournemouth England 21F7
Boyle Ireland 23C3
Boyne, River a river flowing into the Irish Sea on the E coast of the Republic of Ireland, site of a famous battle (1690) (length 115 km/70 miles) 23E3
Bräcke Sweden 30D3
Bradford a city in West Yorkshire, England, centre of the 19th-century woollen industry (pop. 281,000) 20F4
Braga Portugal 25A1
Brahmaputra a major river of South Asia, flowing from the Himalayas in Tibet through Assam in N India to join the River Ganges in Bangladesh (length 2900 km/1802 miles) 39H3
Braila a port in Romania on the Danube, 140 km (87 miles) inland from the Black Sea (pop. 214,000) 39F3
Brasília the capital, since 1960, of Brazil (pop. 412,000) 16E4
Brasov Romania 29F3
Bratislava (Pressburg) the second largest city in former Czechoslovakia, and capital of Slovakia (pop. 402,000) 28D3
Bratsk Russian Federation 33M4

Braunschweig 992 Bytom

Braunschweig (Brunswick) a historic town in N Germany (pop. 255,000) 28C2
Bray Ireland 23E3
Brazil [area 8,511,965 sq km (3,285,470 sq miles); pop. 115,600,000; cap. Brasília; major cities Belo Horizonte, Porto Alegre, Recife, Rio de Janeiro, Salvador, São Paulo; govt federal republic; religion RC; currency cruzeirois] the fifth largest country in the world, which covers nearly half of South America 16B/F4
Brazzaville capital of the Congo, on the River Zaïre (pop. 425,000) 42E5
Breda Netherlands 28A2
Bremen a major port on the River Weser, near to the North Sea coast of Germany, and also the name of the surrounding state (pop. 550,000) 28B2
Bremerhaven a port on the North Sea coast of Germany (pop. 135,000) 28B2
Brescia a city in N Italy (pop. 206,000) 26C1
Breslau see **Wrocław** 28D2
Bressay Scotland 22J7
Bressuire France 24B2
Brest (Brzesc) an inland port in Belarus on the River Bug on the Polish border (pop. 214,000) 29E2
Brest a naval port in NW France (pop. 205,000) 24B2
Bretagne (Brittany) the region of France which occupies the extreme NW peninsula 24B2
Briançon France 24D3
Briare France 24C2
Bridgetown the capital of Barbados (pop. 97,000) 14H4
Bridgwater England 21D6
Bridlington England 20G3
Brighton a famous seaside resort on the S coast of England in East Sussex (pop. 149,000) 21G7
Brindisi a port on the E coast of Italy at the S end of the Adriatic Sea (pop. 92,000) 27D2
Brisbane a port on the E coast of Australia, and the state capital of Queensland (pop. 942,400) 45E3
Bristol a major city and port in SW England; administrative centre of Avon (pop. 399,000) 21E6
British Columbia the W seaboard province of Canada; capital Victoria (929,730 sq km/358,968 sq miles; pop. 2,744,000) 5F4
British Isles the name given to the group of islands in NW Europe formed by the main islands of Great Britain and Ireland and the many surrounding islands 19
Brittany see **Bretagne** 24B2
Brive-la-Gaillarde France 24C2
Brno (Brünn) Czech Republic 28D3
Broken Hill Australia 45E4
Broome Australia 45B2
Brora Scotland 22E2
Brownsville USA 6D3
Bruck an der Mur Austria 28D3
Brunei [area 5,765 sq km (2,226 sq miles); pop. 267,000; cap. Bandar Seri Begawan; major cities Kuala Belait, Seria; govt monarchy (sultanate); religion Sunni Islam; currency Brunei dollar] a sultanate on the NW coast of Borneo, bounded on all sides by Sarawak (Malaysia), which splits it into two separate parts 36D3
Brünn see **Brno** 28D3
Brunswick see **Braunschweig** 28C2
Bruxelles (Brussels, Brussel) a historic city and capital of Belgium. It is the centre for the administration of the European Community (pop. 1,000,000) 24C1
Bryansk Russian Federation 32E4
Bucaramanga Colombia 16B2
Buchanan Liberia 42B4
Bucharest see **Bucuresti** 27F2
Buckie Scotland 22F3
Buckingham England 21F6
Buckinghamshire a county in central S England; county

town Aylesbury (1883 sq km/727 sq miles; pop. 609,000) 21G6
Bucuresti (Bucharest) the capital of Romania, in the SE of the country (pop. 1,861,000) 27F2
Budapest the capital of Hungary, comprising Buda and Pest, on opposite sides of the River Danube (pop. 2,064,400) 29D3
Bude England 21C7
Budweiss see **České Budějovice** 28C3
Buenaventura Colombia 16B2
Buenos Aires the capital of Argentina (pop. city 3,325,000/metropolitan area 9,948,000) 17D6
Buffalo a city and port in New York state at the E end of Lake Erie (pop. city 339,000/metropolitan area 1,205,000) 7F1
Buffalo Wyoming, USA 6C1
Bug, River a river which flows NW from the Ukraine, forming the border with Poland before turning W into Poland and joining the Narew and Vistula rivers (length 813 km/480 miles) 29E2
Builth Wells Wales 21D5
Bujumbura the capital of Burundi, at the N end of Lake Tanganyika (pop. 180,000) 43F5
Bukavu Zaïre 43F5
Bukhara an old trading city in Uzbekistan (pop. 204,000) 32H6
Bukittinggi Indonesia 36C4
Bulawayo the second city of Zimbabwe, in the SW of the country (pop. 414,000) 44C3
Bulgaria [area 110,912 sq km (42,823 sq miles); pop. 8,970,000; cap. Sofia (Sofiya); major cities Burgas, Plovdiv, Ruse, Varna; govt republic; religion Eastern Orthodox; currency lev] a country on the E Balkan peninsula with a coast on the Black Sea 27E2
Bumba Zaïre 43F4
Bunbury Australia 45A4
Bundaberg Australia 45E3
Bundoran Ireland 23C2
Buraydah Saudi Arabia 38C3
Burco Somalia 43H4
Burgas Bulgaria 27F2
Burgos Spain 25B1
Burgundy see **Bourgogne** 24C2
Burkina (Burkina Faso) [area 274,200 sq km (105,869 sq miles); pop. 8,760,000; cap. Ouagadougou; govt republic; religions Animist, Sunni Islam; currency franc CFA] a landlocked state in West Africa, on the S fringe of the Sahara Desert 42C3
Burley USA 6B1
Burma see **Myanmar** 36B1
Bursa a city in NW Turkey, and the name of the surrounding province, of which it is the capital (pop. city 614,100) 38A1
Buru Island Indonesia 37E4
Burundi [area 27,834 sq km (10,747 sq miles); pop. 5,540,000; cap. Bujumbura; govt republic; religion RC; currency Burundi franc] a small densely populated country in central E Africa, bounded by Rwanda to the N, Tanzania to the E and S, and Zaïre to the W 43F5
Bury St Edmunds England 21H5
Buta Zaïre 43F4
Bute Scotland 22C5
Butte USA 6B1
Butuan Philippines 37E3
Butung Island Indonesia 37E4
Buzau Romania 29F3
Byala Bulgaria 27F2
Bydgoszcz (Bromberg) Poland 29D2
Byelorussia see **Belarus** 29F2
Bylot Island Canada 5L2
Bytom (Beuthen) Poland 29D2

C

Cabanatuan Philippines 37E2
Cabimas Venezuela 16B1
Cabinda Angola 42E5
Cacak Yugoslavia 27E2
Cáceres Spain 25A2
Cádiz a port of Phoenician origins on the Atlantic coast of S Spain; also the surrounding province (pop. town 157,800) 25A2
Caen France 24B2
Caernarfon Wales 20C4
Cagayan de Oro Philippines 37E3
Cagliari the capital of the Italian island of Sardinia (pop. 232,800) 26B3
Caher Ireland 23D4
Cahors France 24C3
Caicos Islands Caribbean Sea 14E2
Cairngorms a range forming part of the Grampian Mountains in Scotland 22E3
Cairns a port on the NE coast of Queensland, Australia, and a tourist resort visitors to the Great Barrier Reef (pop. 48,000) 45D2
Cairo *see* **El Qâhira** 43G1
Calahorra Spain 25B1
Calais an old port in N France on the narrowest part of the English Channel, opposite Dover in England (pop. 101,500) 24C1
Calama Chile 16B5
Calamian Group Philippines 36D2
Calatayud Spain 25B1
Calcutta the largest city in India, a major port and industrial centre situated in the NE of the country, on the Hugli River (pop. 9,194,000) 39G3
Calgary the second largest city of Alberta, Canada (pop. 593,000) 5G4
Cali Colombia 16B2
California the most populous state of the USA on the Pacific coast; state capital Sacramento. Los Angeles is the biggest city (411,015 sq km/158,693 sq km; pop. 26,365,000) 6A2
California, Golfo de (Sea of Cortes) the narrow inlet separating mainland Mexico from the peninsula of Baja California 6B3
Callao the port serving Lima, the capital of Peru (pop. 440,500) 16B4
Calvi France 26B2
Camagüey Cuba 14D2
Cambodia [*area* 181,035 sq km (69,898 sq miles); *pop.* 8,300,000; *cap.* Phnom Penh; *major cities* Kampong Cham, Battambang; *govt* people's republic; *religion* Buddhism; *currency* riel] a SE Asian state on the Gulf of Thailand 36C2
Cambrian Mountains a range which forms the "backbone" of Wales 21D5
Cambridge a famous university city in E England (pop. 95,300) 21H5
Cambridge a city in Massachusetts, USA, home of Harvard University (pop. 103,000) 8E2
Cambridgeshire a county in E England (3409 sq km/1316 sq miles; pop. 578,700) 21G5
Camden USA 7F2
Cameroon [*area* 475,442 sq km (183,568 sq miles); *pop.* 11,540,000; *cap.* Yaoundé; *major city* Douala; *govt* republic; *religions* Animism, RC, Sunni Islam; *currency* franc CFA] a country in W central Africa, stretching from Lake Chad at its apex to the N borders of Equatorial Guinea, Gabon and the Congo in the S 42E4
Cameroun, Mount an active volcano in W Cameroon (4095 m/13,435 ft) 42D4

Campbell Island New Zealand 46
Campbeltown Scotland 22C5
Campinas Brazil 16E5
Campo Grande Brazil 17D5
Campos Brazil 16E5
Canada [*area* 9,976,139 sq km (3,851,787 sq miles); *pop.* 26,600,000; *cap.* Ottawa; *major cities* Toronto, Montréal, Vancouver, Québec; *govt* federal parliamentary state; *religions* RC, United Church of Canada, Anglicanism; *currency* Canadian dollar] the second largest country in the world, lying N of the USA 5
Canary Islands *see* **Islas Canarias** 42B2
Canaveral, Cape a long spit of land on the E coast of Florida, USA, the USA's main launch site for space missions 9C5
Canberra the capital of Australia, lying about halfway between Sydney and Melbourne in the SE of the country (pop. 255,900) 45D4
Cangzjou China 37F3
Cannes a famous beach resort on the Côte d'Azur in S France (pop. 72,800) 24D3
Canterbury a small cathedral city in Kent in S England (pop. 36,000) 21J6
Can Tho Vietnam 36C2
Canton *see* **Guangzhou** 35F4
Cape Breton Island part of the province of Nova Scotia lying off the E coast of Canada 5M5
Cape Town a major port on the SW tip of South Africa, the country's legislative capital (pop. 1,912,000) 44B4
Cape Verde [*area* 4033 sq km (1575 sq miles); *pop.* 369,000; *cap.* Praia; *govt* republic; *religion* RC; *currency* Cape Verde escudo] one of the world's smallest nations, situated in the Atlantic Ocean, about 640 km (400 miles) NW of Senegal 2
Cap-Haïtien Haiti 14E3
Capri a rocky island at the S end of the Bay of Naples on the W coast of Italy 26C2
Caprivi Strip a narrow corridor of land, 450 km (280 miles) long, which belongs to Namibia and gives it access to the Zambezi River 44C2
Caracal Romania 27E2
Caracaraí Brazil 16C2
Caracas the capital of Venezuala, in the NE of the country (pop. 3,500,000) 16C1
Caransebes Romania 29E3
Carbonia Italy 26B3
Carcassonne France 24C3
Cardiff (Caerdydd) the capital of Wales, in the SE of the principality, in South Glamorgan (pop. 281,000) 21D6
Cardigan Wales 21C5
Cardigan Bay the long, curving bay which, as part of the Irish Sea, forms much of the W coast of Wales 21C5
Caribbean Sea a part of the W Atlantic Ocean, bounded by the E coast of Central America, the N coast of South America and the West Indies 14
Carlisle England 20E3
Carlow a landlocked county in SE Republic of Ireland; county town Carlow (pop. county 39,000) 23E4
Carmarthen Wales 21C6
Carnarvon Australia 45A3
Carolina Brazil 16E3
Caroline Island Kiribati 46
Caroline Islands a scattered group of islands in the W Pacific Ocean which make up the Federated States of Micronesia and the separate state of Belau 37G3
Carpathian Mountains a broad range stretching for nearly 1000 km (625 miles) down the border between Slovakia and Poland and into central Romania. They rise to 2663 m (8737 ft) at their highest point 29F3
Carpentaria, Gulf of the broad gulf of shallow sea between the two hornlike peninsulas of N Australia 45C2
Carrickmacross Ireland 23E3
Carrick on Shannon Ireland 23C3

Carson City   994   Chicago

Carson City the state capital of Nevada, USA (pop. 35,900)   6B2
Cartagena a major port on the Caribbean coast of Colombia (pop. 548,000)   16B1
Cartagena a port of ancient origins on the Mediterranean coast of Spain (pop. 172,800)   25B2
Cartago Costa Rica   14C5
Casablanca *see* **Dar el Beida**   42C1
Cascade Range a range of mountains stretching 1125 km (700 miles) parallel to the coast of N California and into S Canada. The highest point is Mount Rainier (4392 m/ 14,410 ft) in Washington State   12B1
Caserta Italy   26C2
Cashel Ireland   23D4
Casper USA   6C1
Caspian Sea the largest inland (salt) sea in the world, supplied mainly by the River Volga. It lies to the N of Iran, which shares its coasts with Azerbaijan, Georgia, Kazakhstan and Turkmenistan   32G5
Cassino Italy   26C2
Castellón de la Plana Spain   25B2
Castlebar Ireland   23B3
Castleford England   20F4
Castries the capital of St Lucia (pop. 45,000)   14G4
Castrovillari Italy   26D3
Cataluña (Catalonia) an autonomous region of Spain, in the NE, centring on Barcelona (pop. 5,958,000)   25C1
Catania a major port and the second largest city in Sicily (pop. 378,500)   26D3
Catanzaro Italy   26D3
Cateraggio France   26B2
Cat Island The Bahamas   14D1
Catskill Mountains a range in New York State, USA. The highest peak is Slide Mountain (1281 m/4204 ft)   8E2
Caucasus (Kavkaz) the mountainous region between the Black and Caspian Seas, bounded by the Russian Federation, Georgia, Armenia and Azerbaijan. It contains Europe's highest point, Mount Elbrus   32F5
Cavan a county in N Republic of Ireland, part of the ancient province of Ulster; county town Cavan (1890 sq km/ 730 sq miles; pop. county 53,900)   23D3
Cawnpore *see* **Kanpur**   39G3
Cayenne capital of Guiana (French) (pop. 38,000)   16D2
Cayman Islands a group of three islands in the Caribbean Sea, 240 km (150 miles) NW of Jamaica, which form a British Crown colony   14C3
Cebu Island Philippines   37E2
Cecina Italy   26C2
Cefalù Italy   26C3
Celebes *see* **Sulawesi**   37E4
Celebes Sea a sea between the islands of E Indonesia and the Philippines   37E3
Central African Republic [*area* 622,984 sq km (240,534 sq miles); *pop.* 2,900,000; *cap.* Bangui; *govt* republic; *religions* Animism, RC; *currency* franc CFA] a landlocked country in central Africa   43E4
Central Region a local government area of Scotland formed in 1975 out of the old counties of Clackmannanshire and parts of Perthshire and Stirlingshire (2590 sq km/1000 sq miles; pop. 273,000)   22D4
Cerignola Italy   26D2
České Budějovice (Budweiss) a historic town in S Bohemia, Czech Republic, famous for its Budvar beer (pop. 92,800)   28C3
Ceuta a Spanish-administered enclave in N Morocco (pop. 80,000)   25A2
Ceylon *see* **Sri Lanka**   39G5
Chad [*area* 1,284,000 sq km (495,752 sq miles); *cap.* N'Djamena; *major cities* Sarh, Moundou; *govt* republic; *religions* Sunni Islam, Animism; *currency* franc CFA] a landlocked country in central N Africa   43E3
Chad, Lake a large lake in W Chad, on the border with Niger and Nigeria (26,000 sq km/10,000 sq miles)   42E3
Châlons-sur-Marne France   24C2
Chalon-sur-Saône France   24D2
Chambéry France   24D2
Champagne a region of NE France   24C2
Chañaral Chile   17B5
Chandigarh India   39F2
Changchun China   35G2
Changde China   35F4
Chang-hua Taiwan   37E1
Chang Jiang (Yangtze) the world's third longest river, rising in Tibet and flowing across China to the East China Sea (length 6380 km/3965 miles)   34D
Changsha China   35F4
Changzhi China   35F3
Channel Islands a group of islands in the English Channel, close to the coast of France, which are British Crown dependencies   21E8
Channel Port-aux-Basques Canada   5N5
Chardzhou Turkmenistan   32H6
Charleroi Belgium   24C1
Charleston the state capital of West Virginia, USA. (pop. city 59,400/metropolitan area 267,000)   7D2
Charleston an old port on the Atlantic coast of South Carolina, USA (pop. city 67,100/metropolitan area 472,500)   7F2
Charleville Australia   45D3
Charlotte USA   7E2
Charlottetown a port and the provincial capital of Prince Edward Island, Canada (pop. 45,000)   5M5
Chartres France   24C2
Châteauroux France   24C2
Chatham Islands New Zealand   46
Chattanooga an industrial city and railway town in Tennessee, USA (pop. city 164,400/metropolitan area 422,500)   7E2
Chaumont France   24D2
Chaves Portugal   25A1
Cheboksary Russian Federation   32F4
Cheju do an island belonging to South Korea, lying 90 km (56 miles) off its S tip, and dominated by the sacred volcano, Mount Halla (1950 m/6398 ft)   35G3
Chelmsford England   21H6
Cheltenham England   21E6
Chelyabinsk Russian Federation   32H4
Chemnitz an industrial city in SE Germany, named Karl-Marx-Stadt in former Communist East Germany (until 1990) (pop. 319,000)   28C2
Chengdu the capital of Sichuan province, China (pop. 2,470,000)   34E3
Chen Xian China   35F4
Cherbourg France   24B2
Cheremkhovo Russian Federation   33M4
Cherepovets Russian Federation   32E4
Chernigov Ukraine   32E4
Chernovtsy Ukraine   32D5
Chesapeake Bay an inlet, 314 km (195 miles) long, on the E coast of the USA   8D3
Chesham England   21G6
Cheshire a county in NW England; county town Chester (2322 sq km/897 sq miles; pop. 933,000)   20E4
Chester England   20E4
Chesterfield Inlet Canada   4J3
Chetumal Mexico   14B3
Cheviot Hills a range, 60 km (37 miles) long, which line the border between Scotland and England   20E2
Cheyenne the state capital of Wyoming, USA (pop. 50,900)   6C1
Chiang Mai (Chiengmai) the second largest city in Thailand, in the NW of the country (pop. 200,700)   36B2
Chiba Japan   35P9
Chicago the largest city in Illinois, and the third largest

Chichester 995 Compiègne

city in the USA (after New York and Los Angeles) (pop. city 2,992,500/metropolitan area 8,035,000) 7E1
Chichester England 21G7
Chiclayo Peru 16B3
Chifeng China 35F2
Chihuahua a city in N central Mexico, and the name of the surrounding province, of which it is the capital (pop. city 410,000) 6C3
Chile [*area* 756,945 sq km (292,256 sq miles); *pop.* 12,960,000; *cap.* Santiago; *major cities* Arica, Talcahuano, Viña del Mar; *govt* republic; *religion* RC; *currency* Chilean peso] a country in South America with a Pacific coastline 4200 km (2610 miles) long 17B6
Chiltern Hills a range to the NW of London, England, rising to 260 m (850 ft) 21F/G6
Chi-lung Taiwan 37E1
Chimbote Peru 16B3
Chimkent Kazakhstan 32H5
China [*area* 9,596,961 sq km (3,705,387 sq miles) ; *pop.* 1,114,000,000; *cap.* Beijing (Peking); *major cities* Chengdu, Guangzhou, Shanghai, Tianjin, Wuhan; *govt* people's republic; *religions* Buddhism, Confucianism, Taoism; *currency* yuan] the third largest country in the world, covering a large area of E Asia 34/35
China Sea *see* **East China Sea, South China Sea** 35/36
Chindwin, River a river in Myanmar, flowing parallel to the NW border before joining the Irrawaddy in the centre of the country (length 1130 km/700 miles) 34D4
Chingola Zambia 44C2
Chios *see* **Khíos** 27F3
Chippenham England 21E6
Chisinau *see* **Kishinev** 32D5
Chita Russian Federation 33N4
Chittagong the main port of Bangladesh and its second largest city (pop. 1,392,000) 39H3
Chojnice Poland 28D2
Ch'ongjin North Korea 35G2
Chongqing (Chungking) China 34E4
Chonnam *see* **Kwangju** 35G3
Chorley England 20E4
Choybalsan Mongolia 35F2
Christchurch the largest city on South Island, New Zealand (pop. 300,000) 45G5
Christmas Island an island in the Indian Ocean, 400 km (250 miles) S of Java, administered by Australia 36C5
Christmas Island *see* **Kiritimati** 46
Chumphon Thailand 36B2
Chungking *see* **Chongqing** 34E4
Churchill Canada 5J4
Churchill, River a river flowing into the Hudson Bay at Churchill after a journey through Saskatchewan and Manitoba (length 1600 km/1000 miles) 5H/J4
Church Stretton England 21E5
Cienfuegos Cuba 14C2
Cieza Spain 25B2
Cîmpina Romania 29F3
Cincinnati Ohio, USA 7E2
Cirebon Indonesia 36C4
Citaltépetl a volcanic peak to the SE of Mexico City; at 5747 m (18,855 ft) the highest point in Mexico 6D4
Citta di Castello Italy 26C2
Civitanova Marche Italy 26C2
Civitavecchia Italy 26C2
Clacton-on-Sea England 21J6
Clare a county on the W coast of Ireland; county town Ennis (3188 sq km/1230 sq miles; pop. 87,500) 23C4
Clare Island Ireland 23A3
Claremorris Ireland 23B3
Clermont Ferrand France 24C2
Cleveland a county of NE England created in 1974 out of Durham and Yorkshire to administer the industrial region along the River Tees, known as Teeside (583 sq km/225 sq miles; pop. 565,000) 20F3
Cleveland a port and industrial city on the S side of Lake Erie, in Ohio, USA (pop. city 546,500/metropolitan area 1,867,000) 7E1
Clifden Ireland 23A3
Cloghan Ireland 23D3
Cloncurry Australia 45D2
Clonmel Ireland 23D4
Cluj-Napoca Romania 29E3
Clwyd a county in NE Wales created in 1974 out of the county of Flintshire and parts of Merionethshire and Denbighshire (2425 sq km/936 sq miles; pop. 395,000) 20D4
Clyde, River a river in SW Scotland flowing NW to form an estuary 100 km (60 miles) long, with Glasgow at its head (length 170 km/105 miles) 22D/E5
Coast Ranges USA 6A1
Coatbridge Scotland 22D5
Cobán Guatemala 14A3
Cobh a town and port in Cork Harbour on the S coast of Ireland, 10 km (6 miles) from the city of Cork (pop. 6600) 23C5
Cochabamba Bolivia 16C4
Cochin a port on the SW tip of India (pop. 551,600) 39F5
Cocos Islands (Keeling Islands) a cluster of 28 small coral islands in the E Indian Ocean, equidistant from Sumatra and Australia 36B5
Cod, Cape a narrow, low-lying peninsula on the coast of Massachusetts, USA, where the Pilgrim Fathers landed in 1620 8E2
Codó Brazil 16E3
Coimbatore India 39F4
Coimbra Portugal 25A1
Colchester England 21H6
Coleraine Northern Ireland 23E1
Coll Island Scotland 22B4
Collooney Ireland 23C2
Colmar France 24D2
Cologne *see* **Köln** 28B2
Colombia [*area* 1,138,914 sq km (439,735 sq miles); *pop.* 33,000,000; *cap.* Bogotá; *major cities* Barranquilla, Cali, Cartagena, Medellin, *govt* republic; *religion* RC; *currency* peso] a country in NW South America 16B2
Colombo a major port and the capital of Sri Lanka (pop. 600,000) 39F5
Colón Panama 14D5
Colonsay Island Scotland 22B4
Colorado an inland state of central W USA; state capital Denver (270,000 sq km/104,247 sq miles; pop. 3,231,000) 6C2
Colorado, River a river rising in the Rocky Mountains in Colorado, USA, and flowing SW to the Gulf of California, forming the Grand Canyon on its way (length 2330 km/1450 miles) 12C3
Colorada Springs Colorado, USA 6C2
Columbia the state capital of South Carolina (pop. city 98,600/metropolitan area 433,200) 7E2
Columbia, River flows N from its source in British Columbia, Canada, before turning S into Washington State, USA and entering the Pacific Ocean at Portland, Oregon (length 1950 km/1210 miles) 5F5
Columbus Georgia USA 7E2
Columbus the state capital of Ohio, USA (pop. city 566,100/metropolitan area 1,279,600) 7E2
Como Italy 26B1
Comodoro Rivadavia Argentina 17C7
Comoros [*area* 2235 sq km (863 sq miles); *pop.* 503,000; *cap.* Moroni; *govt* federal Islamic republic; *religion* Sunni Islam; *currency* Comorian franc] a country consisting of three volcanic islands in the Indian Ocean between mainland Africa and Madagascar 44E2
Compiègne France 24C2

Conakry

Conakry the capital of Guinea, a port partly located on the island of Tumbo (pop. 763,000) — 42B4
Concepción Chile — 17B6
Concepción Paraguay — 17D5
Concord the state capital of New Hampshire, USA (pop. 30,900) — 7F1
Concordia Argentina — 17D6
Congo [area 342,000 sq km (132,046 sq miles); pop. 2,260,000; cap. Brazzaville; major city Pointe-Noire; govt republic; religion RC; currency franc CFA] a country in W-central Africa, straddling the Equator — 42E5
Congo, River see **Zaïre, River** — 43E4
Connecticut a state on the NE seaboard of the USA, in New England; the capital is Hartford (12,973 sq km/5009 sq miles; pop. 3,174,000) — 7F1
Consett England — 20F3
Constance, Lake see **Bodensee** — 28B3
Constanta a major port on the Black Sea coast of Romania (pop. 283,600) — 27F2
Constantine (Qacentina) an ancient walled city in the NE corner of Algeria (pop. 430,500) — 42D1
Constantinople see **Istanbul** — 38A1
Contonou Benin — 42D4
Cook, Mount the highest mountain in New Zealand, on South Island (3753 m/12,316 ft) — 45G5
Cook Islands a group of islands in the South Pacific, independent since 1965 but associated with New Zealand; capital Avarua (240 sq km/93 sq miles; pop. 17,700; currency Cook Islands/New Zealand dollar) — 46
Cook Strait the strait that separates North Island and South Island of New Zealand, 26 km (16 miles) across at its widest point — 45G5
Cooktown Australia — 45D2
Cooper Creek a river flowing into Lake Eyre in South Australia from central Queensland. The upper stretch is known as Barcoo (length 1420 km/800 miles) — 45C3
Copenhagen see **København** — 30C4
Coral Sea a part of the Pacific Ocean, off the NE coast of Australia — 45E2
Corby England — 21G5
Corcubíon Spain — 25A1
Córdoba the second city of Argentina, and the name of the surrounding province (pop. city 969,000) — 17C6
Córdoba (Cordova) a city in S Spain, famous for its cathedral, built originally as a mosque; also the name of the surrounding province (pop. city 284,700) — 25B2
Corfu see **Kérkira** — 27D3
Corigliano Italy — 26D3
Corinth see **Kórinthos** — 27E3
Corinto Nicaragua — 14B4
Cork a county in SW Republic of Ireland; county town Cork (7459 sq km/2880 sq miles; pop. 402,300) — 23C4
Cork the second largest city in the Republic of Ireland, at the head of a large natural harbour which cuts into the S coast (pop. city 136,300) — 23C5
Cornwall a county at the SW tip of England; county town Truro (3546 sq km/1369 sq miles; pop. 432,000) — 21C7
Coro Venezuela — 16C1
Corpus Christi a port in Texas on the Gulf of Mexico (pop. city 258,100/metropolitan area 361,300) — 6D3
Corrientes Argentina — 17D5
Corse (Corsica) a large island in the Mediterranean Sea N of Sardinia, governed by France (8680 sq km/3350 sq miles; pop. 240,000) — 26B2
Corumbá Brazil — 16D4
Corunna see **La Coruña** — 25A1
Cosenza Italy — 26D3
Costa Brava Spain — 25C1
Costa Rica [area 51,100 sq km (19,730 sq miles); pop. 2,910,000; cap. San José; major city Limón; govt republic; religion RC; currency Costa Rican colon] a country with the Pacific Ocean to the S and W and the Caribbean Sea

Cyprus

to the E, between Nicaragua and Panama — 14C5
Cotabato Philippines — 37E3
Côte d'Azur the coast of SE France, famed for its beaches and resorts such as St Tropez — 24D3
Côte d'Ivoire [area 322,463 sq km (124,503 sq miles); pop. 12,100,000; cap. Yamoussoukro; major cities Abidjan, Bouaké, Daloa; govt republic; religions Animism, Sunni Islam, RC; currency franc CFA] a former French colony in W Africa, on the Gulf of Guinea — 42C4
Cotonou a port and the main business centre of Benin (pop. 488,000) — 42D4
Cotswold Hills a range in W central England, lying E of the Severn — 21E5
Cottbus Germany — 28C2
Coventry an industrial city in the West Midlands of England (pop. 315,900) — 21F5
Cowes Isle of Wight — 21F7
Cracow see **Kraków** — 29D2
Craiova Romania — 27E2
Crawley England — 21G6
Cres Island Croatia — 26C2
Crete see **Kriti** — 27E3
Crewe England — 20E4
Crianlarich Scotland — 22D4
Crieff Scotland — 22E4
Croatia (Hrvatska) [area 56,538 sq km (21,824 sq miles); pop. 4,601,500; cap. Zagreb; major cities Rijeka, Split; govt republic; religions RC, Eastern Orthodox; currency dinar] a republic of former Yugoslavia, which declared itself independent in 1991 and was formally recognized in 1992 — 26D1
Cromer England — 21J5
Crotone Italy — 26D3
Cruzeiro do Sul Brazil — 16B3
Cuango, River see **Kwango, River** — 44A1
Cuba [area 110,861 sq km (42,803 sq miles); pop. 10,580,000; cap. Havana (La Habana); major cities Camaguey, Holguin, Santiago de Cuba; govt socialist republic; religion RC; currency Cuban peso] an island state in the Caribbean, home to a third of the West Indian population — 14D2
Cubango, River (Kavango River) a river flowing SE from central Angola to form the border with Namibia before petering out in the Okavango Delta, N Botswana (length 1600 km/1000 miles) — 44B2
Cucuí Brazil — 16C2
Cúcuta Colombia — 16B2
Cuddalore India — 39F4
Cuenca a city in S Ecuador, the site of a number of important Inca ruins (pop. 272,500) — 16B3
Cuenca Spain — 25B1
Cuiabá Brazil — 16D4
Culiacán Mexico — 6C3
Cumbernauld Scotland — 22D5
Cumbria a county in NW England, created in 1974 from the old counties of Cumberland, Westmorland and a part of Lancashire (6809 sq km/2629 sq miles; pop. 483,000) — 20E3
Cumnock Scotland — 22D5
Cuneo Italy — 26C2
Cupar Scotland — 22E4
Curaçao an island in the Caribbean lying just off the coast of Venezuela — 14F4
Curitiba Brazil — 16E5
Cuttack India — 39G3
Cuxhaven Germany — 28B2
Cuzco a city in the Andes in Peru, and the name of the surrounding province with numerous Inca remains, including Machu Picchu (pop. city 184,600) — 16B4
Cwmbran Wales — 21D6
Cyclades see **Kikládhes** — 27E3
Cyprus [area 9251 sq km (3572 sq miles); pop. 698,800; cap. Nicosia; major cities Limassol, Larnaca; govt

republic; *religions* Greek Orthodox, Sunni Islam; *currency* Cyprus pound] an island state in the E Mediterranean 40B1

Czech Republic [*area* 78,864 sq km (30,449 sq miles); *pop.* 10,291,900; *cap.* Prague (Praha); *major cities* Brno, Ostrava, Plzen; *govt* republic; *religions* RC, Protestantism; *currency* koruna] a landlocked country in central Europe, constituted on January 1,1993, with the dissolution of the federal republic of Czechoslovakia that it had previously formed with Slovakia 28D3

Czestochowa (Chenstokhov) Poland 29D2

D

Dacca *see* **Dhaka** 39H3
Dagupan Philippines 37E2
Dakar the main port and capital of Senegal (pop. 1,000,000) 42B3
Da Lat Vietnam 36C2
Dali China 34E4
Dallas a city in NE Texas, USA (pop. city 974,200/metropolitan area 2,203,700) 6D2
Dalmally Scotland 22C4
Daloa Côte d'Ivoire 42C4
Damascus *see* **Dimashq** 38B2
Damavand, Mount an extinct volcano, highest peak in the Alborz Mountains, Iran (5670 m/18,600 ft) 38D2
Da Nang Vietnam 36C2
Dandong China 35G2
Danube, River *see* **Donau** 28/29B3/D
Danzig *see* **Gdansk** 29D2
Dar'a Syria 38B2
Dar el Beida (Casablanca) the main and largest city of Morocco (pop.2,140,000) 42C1
Dar es Salaam the largest town and main port of Tanzania and national capital until 1974 (pop. 757,346) 43G5
Darling, River a river flowing from S Queensland to New South Wales in Australia before converging with the Murray (length 3057 km/1900 miles) 45D3/4
Darlington England 20F3
Darmstadt Germany 28B3
Daroca Spain 25B1
Daru Papua New Guinea 37G4
Darwin the capital of the Northern Territory, Australia (pop. 50,000) 45C2
Datong (Tatung) China 35F2
Daugavpils Latvia 32D4
Dauphine (province) France 24D3
Davao a city in S Mindanao, Philippines, and the country's second largest city (pop. city 540,000) 37E3
David Panama 14C5
Davis Strait the broad strait, 290 km (180 miles) across at its narrowest, separating Baffin Island in Canada and Greenland 5N3
Dawson Creek Canada 5F4
Dax France 24B3
Dayton USA 7E2
Daytona Beach USA 7E3
De Aar South Africa 44C4
Dead Sea a small sea on the Israel-Jordan border into which the River Jordan flows. It is one of the lowest places on earth (396 m/1299 ft below sea level) and the body of water with the highest salt content (1049 sq km/395 sq miles) 40C3
Death Valley a low-lying desert area in E California, USA 13C3
Debrecen Hungary 29E3
Debre Mark'os Ethiopia 43G3
Deccan the broad, triangular plateau which forms much of the S part of India 39F/G4

Dehra Dun a town in N India, in the foothills of the Himalayas, famed as the supposed home of the Hindu god Shiva (pop. 293,000) 39F2
Delaware a state on the E coast of the USA, the second smallest in the USA after Rhode Island; capital Dover (5328 sq km/2057 sq miles; pop. 622,000) 7F1
Delhi, including New Delhi, the capital of India, in the N of the country (pop. 5,729,300) 39F3
Denbigh Wales 20E4
Den Helder Netherlands 28A2
Denizli Turkey 38A2
Denmark [*area* 43,077 sq km (16,632 sq miles); *pop.* 5,140,000; *cap.* Copenhagen (København); *major cities* Ålborg, Århus, Odense; *govt* constitutional monarchy; *religion* Lutheranism; *currency* Danish krone] a small country between the North Sea and the entrance to the Baltic. It consists of a peninsula and an archipelago of 406 islands, only 89 of which are populated 30C4
Denmark Strait the arm of the North Atlantic Ocean separating Iceland and Greenland, 290 km (180 miles) apart 48
D'Entrecasteaux Island Papua New Guinea 37H4
Denver the state capital of Colorado, USA (pop. city 504,600/metropolitan area 1,582,500) 6C2
Derby England 21F5
Derbyshire a county in N central England; county town Matlock (2631 sq km/1016 sq miles; pop. 911,000) 20F4
Derry *see* **Londonderry** 23D2
Dese Ethiopia 43G3
Des Moines the state capital of Iowa, USA (pop. city 190,800/metropolitan area 377,100) 7D1
Dessau Germany 28C2
Detroit a major industrial city and Great Lakes port in Michigan, USA (pop. city 1,089,000/metropolitan area 4,315,800) 7E1
Deva Romania 29E3
Devon a county in SW England; county town Exeter (6715 sq km/2593 sq miles; pop. 980,000) 21D7
Devon Island Canada 5K2
Dezful Iran 38C2
Dezhou China 35F3
Dhaka (Dacca) the capital of Bangladesh, on the delta of the Ganges and Brahmaputra (pop. 3,458,600) 39H3
Dhodhekánisos (Dodecanese) a group of twelve islands belonging to Greece in the E Aegean Sea 27F3
Dibrugarh India 39H3
Didcot England 21F6
Dieppe France 24C2
Dijon the historic capital of the Bourgogne region in W central France (pop. 221,900) 24C2
Dili Indonesia 37E4
Dimashq (Damascus) the capital of Syria, an oasis town (pop. 1,042,500) 38B2
Dimitrovgrad Bulgaria 27F2
Dingle Ireland 23A4
Dingwall Scotland 22D3
Dire Dawa Ethiopia 43H4
Disko Island Greenland 5N3
Diyarbakir Turkey 38B2
Djelfa Algeria 42D1
Djibouti [*area* 23,200 sq km (8958 sq miles); *pop.* 484,000; *cap.* Djibouti; *govt* republic; *religion* Sunni Islam; *currency* Djibouti franc] a country in NE Africa bounded almost entirely by Ethiopia, except in the SE where it shares a border with Somalia 43H3
Dnepr (Dnieper), River the third longest river in Europe after the Volga and the Danube, flowing S through the Russian Federation and Ukraine to the Black Sea via Kiev (length 2285 km/1420 miles) 32E4
Dnepropetrovsk an industrial and agricultural city on the River Dnieper in the Ukraine. A former name was Ekaterinoslav (pop. 1,140,000) 32E5
Dnestr (Dniester), River a river flowing through the Ukraine

and Moldova to the Black Sea (length 1411 km/877 miles) **32D5**
Dobreta-Turnu-Severin Romania **27E2**
Dodecanese *see* **Dhodhekánisos** **27F3**
Dodoma the capital (since 1974) of Tanzania, in its centre (pop. 45,700) **43G5**
Doha *see* **Ad Dawhah** **38D3**
Dôle France **24D2**
Dolgellau Wales **21D5**
Dolomiti (Dolomites) a range of mountains in NE Italy, near the Austrian border. The highest point is Mount Marmolada (3342 m/10,964 ft) **26C1**
Dombås Norway **30B3**
Dominica [*area* 751 sq km (290 sq miles); *pop.* 81,200; *cap.* Roseau; *govt* republic; *religion* RC; *currency* franc] the most northerly of the Windward Islands in the West Indies **14G3**
Dominican Republic [*area* 48,734 sq km (18,816 sq miles); *pop.* 7,200,000; *cap.* Santo Domingo; *major city* Santiago de los Caballeros; *govt* republic; *currency* Dominican peso] a republic forming the E two-thirds of the island of Hispaniola in the West Indies **14F3**
Domodossola Italy **26B1**
Don, River a river flowing S into the Sea of Azov from S of Moscow (length 1870 km/1165 miles) **32F4/5**
Donau (Danube) River the longest river in W Europe, rising in the Black Forest in Germany, and passing through Austria, Slovakia, Hungary and Serbia, forming much of the border between Bulgaria and Romania before turning N and forming a delta on the Black Sea (length 2850 km/1770 miles) **28B3**
Doncaster England **20F4**
Donegal the northernmost county of the Republic of Ireland, on the W coast, and the name of a town in it; county town Lifford (pop. county 125,100) **23C2**
Donetsk Ukraine **32E5**
Dorchester England **21E7**
Dordogne, River a river of SW France which rises in the Massif Central and flows W to the Gironde estuary (length 475 km/295 miles) **24B/C3**
Dornie Scotland **22C3**
Dorset a county of SW England; county town Dorchester (2654 sq km/1025 sq miles; pop. 618,000) **21E7**
Dortmund an industrial city in the Ruhr region of Germany (pop. 620,000) **28B2**
Douala the main port of Cameroon, on the Gulf of Guinea (pop. 800,000) **42D4**
Douglas Isle of Man **20C3**
Dourados Brazil **17D5**
Douro (Duero), River a river flowing W from N central Spain across N Portugal to the Atlantic Ocean (length 895 km/555 miles) **25A/B1**
Dover a port in Kent, England, overlooking the English Channel at its narrowest point **21J6**
Dover state capital of Delaware, USA (pop. 22,500) **7F2**
Dover, Strait of the stretch of water separating England and France, where the English Channel meets the North Sea. Dover and Calais are situated on either side of its narrowest point, 34 km (21 miles) **21J6**
Down a county of Northern Ireland, on the E coast; county town Downpatrick (2448 sq km/945 sq miles; pop. 362,100) **23F2**
Drakensberg Mountains a range stretching 1125 km (700 miles) across Lesotho and neighbouring regions. The highest point is Thabana Ntlenyana (3482 m/11,424 ft) **44C4/D3**
Dráma Greece **27E2**
Drammen Norway **30C4**
Drava (Drau), River a river flowing from E Austria to Croatia and Serbia, where it forms much of the border with Hungary before joining the Danube (length 718 km/447 miles) **28D3**

Dresden a historic city on the River Elbe in the S of E Germany. Formerly the capital of Saxony, it was noted for its fine porcelain (pop. 522,500) **28C2**
Drogheda Ireland **23E3**
Dubayy (Dubai) the second largest of the United Arab Emirates, at the E end of The Gulf. Most of the population live in the capital, Dubai (3900 sq km/1506 sq miles; pop. emirate 296,000/city 265,700) **38D3**
Dublin (Baile Atha Cliath) the capital of the Republic of Ireland, on the River Liffey. Its main port area is at Dun Laoghaire (pop. 525,400) **23E3**
Dublin the county surrounding Dublin, the capital of the Republic of Ireland (pop. 1,002,000) **23E3**
Dubrovnik (Ragusa) a medieval port on the Adriatic coast of Croatia (pop. 31,200) **27D2**
Ducie Island Pacific Ocean **46**
Dudley England **21E5**
Dugi Island Croatia **26C2**
Duisburg a major inland port at the confluence of the Rhine and Ruhr in Germany (pop. 541,800) **28B2**
Dukou China **34E4**
Duluth Michigan, USA **7D1**
Dumbarton Scotland **22D5**
Dumfries Scotland **22E5**
Dumfries and Galloway a region of SW Scotland created out of the old counties of Dumfriesshire, Kirkcudbrightshire and Wigtownshire; regional capital Dumfries (6370 sq km/2459 sq miles; pop. 145,200) **22D5**
Dunbar Scotland **22F4**
Dundalk Ireland **23E2**
Dundee a port on the E coast of Scotland, on the N side of the Firth of Tay, and administrative centre of Tayside region (pop. 180,000) **22F4**
Dundrum Northern Ireland **23F2**
Dunedin New Zealand **45G5**
Dunfermline Scotland **22E4**
Dungarvan Ireland **23D4**
Dungiven Northern Ireland **23E2**
Dunkeld Scotland **22E4**
Dunkerque (Dunkirk) a port and industrial town in NE France, close to the Belgian border (pop. 196,600) **24C1**
Dun Laoghaire Ireland **23E3**
Dunleer Ireland **23E3**
Durban a port on the E coast of South Africa, and the largest city of Natal (pop. 960,800) **44D3**
Durham a cathedral city of NE England, and county town of County Durham (pop. 26,000) **20F3**
Durham a county of NE England; county town Durham (2436 sq km/940 sq miles; pop. 607,000) **20F3**
Durham USA **7F2**
Durness Scotland **22D2**
Durrës Albania **27D2**
Durrow Ireland **23D4**
Dushanbe an industrial city and the capital of Tajikistan (pop. 539,000) **32H6**
Düsseldorf a major commercial and industrial centre in the Ruhr region of Germany (pop. 579,800) **28B2**
Duyun China **34E4**
Dyfed a county in SW Wales, created in 1974 out of the old counties of Cardiganshire, Carmarthenshire and Pembrokeshire; county town Carmarthen (5765 sq km/2226 sq miles; pop. 377,000) **21C5**
Dzhambul Kazakhstan **32J5**

E

Eastbourne England **21H7**
East China Sea a part of the Pacific Ocean, off the E coast of China **35G4**
East Falkland Island South Atlantic Ocean **17D8**

East Kilbride — Estonia

East Kilbride Scotland 22D5
East London South Africa 44C4
East Sussex a county in SE England; county town Lewes (1795 sq km/693 sq miles; pop. 655,000) 21H7
Eboli Italy 26D2
Ebro, River a river flowing across NE Spain to the Mediterranean (length 909 km/565 miles) 25B/C1
Ecija Spain 25A2
Ecuador [*area* 283,561 sq km (109,483 sq miles); *pop.* 10,490,000; *cap.* Quito; *major cities* Guayaquil, Cuenca; *govt* republic; *religion* RC; *currency* sucre] an Andean country in the NW of South American 16B3
Eday Island Scotland 22F1
Edgeworthstown Ireland 23D3
Edinburgh capital of Scotland, a commercial centre, on the estuary of the River Forth (pop. 439,000) 22E5
Edmonton the capital of Alberta, Canada (pop. 657,000) 5G4
Edo *see* **Tokyo** 35H9
Edward (Rutanzige), Lake a lake in the Great Rift Valley, on the border between Uganda and Zaïre (2135 sq km/820 sq miles) 43F5
Efate Island Vanuatu 45F2
Egadi Island Italy 26C3
Egersund Norway 30B4
Egypt [*area* 1,001,449 sq km (386,659 sq miles); *pop.* 50,740,000; *cap.* Cairo (El Qahira); *major cities* Alexandria, El Gîza; *govt* republic; *religions* Sunni Islam, Christianity; *currency* Egyptian pound] a country in NE Africa, straddling the River Nile 43G2
Eigg Island Scotland 22B4
Eilat *see* **Elat** 40C4
Eindhoven an industrial city in S central Netherlands (pop. 194,600) 28B2
Eire *see* **Ireland, Republic of** 23
Eisenach Germany 28C2
Ekaterinburg *see* **Yeketerinburg** 32H4
Elat (Eilat) a port and tourist resort in the very S of Israel at the tip of the Gulf of Aqaba, an arm of the Red Sea (pop. 18,800) 40C4
Elba an island lying about 10 km (6 miles) off the coast of Italy (223 sq km/86 sq miles; pop. 28,400) 26C2
Elbe, River a largely navigable river flowing N from the Czech Republic through Germany and into the North Sea (length 1160 km/720 miles) 28C2
Elblag Poland 29D2
Elbrus, Mount the highest mountain in Europe, in W Caucasus, Russian Federation (5642 m/18,510 ft) 32F5
Elburz Mountains *see* **Alborz Mountains** 38D2
Elche Spain 25B2
El Djazair *see* **Alger** 42D1
El Dorado Venezuela 16C2
Eldoret Kenya 43G4
Eleuthera Island The Bahamas 14D1
El Faiyûm (Fayum) a large and fertile oasis W of the Nile in Egypt (pop. 167,080) 43G2
El Fasher Sudan 43F3
El Ferrol Spain 25A1
Elgin Scotland 22E3
El Gîza a suburb of Cairo, Egypt, at the edge of which stand three famous pyramids (pop. 1,230,500) 43G1
El Golea Algeria 42D1
El Iskandarîya (Alexandria) the main port of Egypt, on the Nile delta (pop. 2,320,000) 43F1
El Khartum (Khartoum) the capital of Sudan, at the confluence of the Blue and White Niles (pop. 561,000) 43G3
Ellesmere Island Canada 5K2
Ellesmere Port England 20E4
Ellon Scotland 22F3
El Minya Egypt 43G2
El Obeid Sudan 43G3
El Paso Texas, USA 6C2

El Qâhira (Cairo) the capital of Egypt, in the N of the country on Nile; it is the largest city in Africa (pop. 8,540,000) 43G1
El Salvador [*area* 21,041 sq km (8123 sq miles); *pop.* 5,220,000; *cap.* San Salvador; *major cities* Santa Ana, San Miguel; *govt* republic; *religion* RC; *currency* colón] the smallest and most densely populated state in Central America 14B4
Elvas Portugal 25A2
Ely England 21H5
Enarración Paraguay 17D5
Ende Indonesia 37E4
Enggano Island Indonesia 36C4
England the country occupying the greater part of Great Britain, and the largest of the countries that make up the United Kingdom. Scotland lies to the N and Wales to the W. The capital is London (130,357 sq km/50,331 sq miles; pop. 46,795,000) 20-21
English Channel the arm of the E Atlantic Ocean separating the S coast of England from France 21D7
Enna Italy 26C3
Ennis Ireland 23B4
Enniscorthy Ireland 23E4
Enniskillen Northern Ireland 23D2
Ennistymon Ireland 23B4
Enschede an industrial town in E Netherlands, close to the border with Germany (pop. 144,900) 28B2
Entebbe a town with an international airport on Lake Victoria, Uganda. It was the capital until 1962 (pop. 30,000) 43G4
Enugu a coal-mining centre in S central Nigeria, the capital of Biafra during the Civil War (1967-70) (pop. 222,600) 42D4
Epi Island Vanuatu 45F2
Episkopi Cyprus 40B1
Eptanisos *see* **Iónioi Nísoi** 27D3
Equatorial Guinea [*area* 28,051 sq km (10,830 sq miles); *pop.* 417,000; *cap.* Malabo; *major city* Bata; *govt* republic; *religion* RC; *currency* franc CFA] a W African country, consisting of a mainland area, a few offshore islets, and the islands of Bioko and Pagalu 42D4
Erenhot China 35F2
Erfurt a historic town and tourist centre in central Germany (pop. 215,000) 28C2
Erie, Lake one of the five Great Lakes (second smallest after Lake Ontario), on the border between Canada and the USA (25,670 sq km/9910 sq miles) 7E1
Eriskay Island Scotland 22A3
Eritrea an autonomous province of N Ethiopia, bordering the Red Sea. The capital is Asmara (117,400 sq km/45,316 sq miles; pop. 3,000,000) 43G3
Erlangen Germany 28B3
Erromanga Island Vanuatu 45F2
Erzurum Turkey 38B2
Esbjerg Denmark 30B4
Esfahan (Isfahan) a city in central Iran noted for its magnificent Islamic buildings (pop. 926,700) 38D2
Eskisehir a spa town in W Turkey and the name of the surrounding province (pop. town 367,300) 38B2
Esperance Australia 45B4
Espírito Santo Brazil 16E4
Espiritu Santo Island Vanuatu 45F2
Espoo Finland 30E3
Essaouira Morocco 42C1
Esseg *see* **Osijek** 27D1
Essen an industrial city in W Germany, and the largest in the Ruhr region (pop. 635,200) 28B2
Essex a county in SE England; county town Chelmsford (3674 sq km/1419 sq miles; pop. 1,492,000) 21H6
Estonia [*area* 45,100 sq km (17,413 sq miles); *pop.* 1,573,000; *cap.* Tallinn; *major cities* Tartu, Narva; *govt* republic; *religion* Eastern Orthodox, Lutheranism; *cur-*

Estremoz

rency rouble] the smallest of the three previous Soviet Baltic Republics, NW of the Russian Federation 30F4
Estremoz Portugal 25A2
Ethiopia [*area* 1,221,900 sq km (471,776 sq miles); *pop.* 50,000,000; *cap.* Adis Abeba; *major cities* Asmara, Dire Dawa; *govt* people's republic; *religion* Ethiopian Orthodox, Sunni Islam; *currency* Ethiopian birr] one of Africa's largest countries, stretching from the Red Sea to N of Kenya 43G4
Etna, Mount the largest volcano in Europe, near the E coast of Sicily, Italy, and still highly active (3323 m/10,902 ft) 26C3
Euphrates, River *see* **Al Furat** 38C2
Europe a continent divided from Asia by a border that runs down the Urals to the Caspian Sea and then W to the Black Sea. It has into two areas: Eastern Europe (countries that have or had Communist governments since World War II) and Western Europe (10,498,000 sq km/4,053,300 sq miles; pop. 682,000,000) 18
Evansville USA 7E2
Everest, Mount the highest mountain in the world, on the border between Nepal and China in the E Himalayas (8848 m/29,028 ft) 34C4
Evvoia (Euboea) a large island in the Aegean close to the E coast of Greece and joined to it by a bridge 27E3
Exeter England 21D7
Eyre, Lake a large salt lake in South Australia (8900 sq km/3400 sq miles) 45C3

F

Faeroe (Faroe) Islands *see* **Føroyar** 30A2
Fairbanks USA 5D3, 6J
Fair Isle a small island between the Orkney and Shetland Islands, famous for its knitwear (pop. 75) 22J8
Faisalabad (Lyallpur) an industrial city and agricultural centre in NE Pakistan (pop. 1,092,000) 39F2
Faiyum *see* **El Faiyûm** 43G2
Fakfak Indonesia 37F4
Falcarragh Ireland 23C1
Falkirk Scotland 22F4
Falkland Islands (Islas Malvinas) a British Crown colony consisting of two large islands and 200 smaller ones, 650 km (410 miles) E of S Argentina; capital Port Stanley (12,173 sq km/4700 sq miles; pop. 1800) 16D8
Falmouth England 20B7
Falster Island Denmark 30C5
Falun Sweden 30D3
Famagusta Cyprus 40B1
Fano Italy 26C2
Farah Afghanistan 38E2
Fareham England 21F7
Fargo USA 6D1
Faro the capital of the Algarve province of Portugal (pop. 28,200) 25A2
Faroe Islands *see* **Føroyar** 30A2
Farquhar Islands Indian Ocean 44F1
Fauske Norway 30D2
Faya-Largeau Chad 43E3
Fayum *see* **El Faiyûm** 43G2
Fdérik Mauritania 42B2
Felixstowe England 21J6
Fergana Uzbekistan 32J5
Ferkessédougou Côte d'Ivoire 42C4
Fermanagh a lakeland county in SW Northern Ireland; county town Enniskillen (1676 sq km/647 sq miles; pop. 51,400) 23D2
Fermoy Ireland 23C4
Fernando Póo *see* **Bioko** 42D4
Ferrara Italy 26C2

Fort William

Fès (Fez) a city in N Morocco, the oldest of that country's four imperial cities (pop. 448,823) 42C1
Fetlar Island Scotland 22K7
Feyzabad Afghanistan 39F2
Fianarantsoa Madagascar 44E3
Fife a region of E Scotland; administrative centre Glenrothes (1308 sq km/505 sq miles; pop. 344,000) 22E4
Figueras Spain 25C1
Fiji [*area* 18,274 sq km (7056 sq miles); *pop.* 727,104; *cap.* Suva; *govt* republic; *religion* Christianity, Hinduism; *currency* Fiji dollar] a country consisting of some 320 islands and atolls, situated around the 180° International Date Line and about 17° S of the Equator 46
Filiasi Romania 27E2
Finisterre, Cape the NW corner of Spain 25A1
Finland [*area* 338,127 sq km (130,551 sq miles); *pop.* 4,970,000; *cap.* Helsinki (Helsingfors); *major cities* Turku, Tampere; *govt* republic; *religion* Lutheranism; *currency* markka] a country in N Europe, W of the Russian Federation and E of the Gulf of Bothnia 30E3
Finland, Gulf of the E arm of the Baltic Sea, with Finland to the N, St Petersburg at its E end, and Estonia to the S 30F3/4
Firenze (Florence) one of the great Renaissance cities of Italy, straddling the River Arno, and capital of the region of Tuscany (pop. 453,300) 26C2
Fishguard Wales 21C6
Fitzroy Crossing Australia 45B2
Fiume *see* **Rijeka** 26C1
Fleetwood England 20D4
Flensburg Germany 28B2
Flinders Range mountains in the E part of South Australia, stretching over 400 km (250 miles). St Mary Peak is the highest (1188 m/3898 ft) 45C4
Flint Island Kiribati 46
Florence *see* **Firenze** 26C2
Flores Guatemala 14B3
Flores Island a volcanic island in the Sunda group in Indonesia, lying in the chain due E of Java (17,150 sq km/6622 sq miles; pop. 803,000) 37E4
Flores Sea a stretch of the Pacific Ocean between Flores Island and Sulawesi 36D4
Florianópolis Brazil 16E5
Florida a state occupying the peninsula in the SE corner of the USA; state capital Tallahassee (151,670 sq km/58,560 sq miles; pop. 11,366,000) 7E3
Florida, Straits of the waterway separating the S tip of Florida from Cuba, 145 km (90 miles) to the S 7E3
Fly, River a largely navigable river flowing from the central mountains in W Papua New Guinea to the Gulf of Papua (length 1200 km/750 miles) 37G4
Focsani Romania 29F3
Foggia Italy 26D2
Foligno Italy 26C2
Follonica Italy 26C2
Forfar Scotland 22F4
Forli Italy 26C2
Formentera Spain 25C2
Formia Italy 26C2
Føroyar (Faeroe (Faroe) Islands) a group of 18 islands in the North Atlantic belonging to Denmark, (1399 sq km/540 sq miles; pop. 44,500) 30A2
Fortaleza Brazil 16F3
Fort Augustus Scotland 22D3
Fort-de-France a port and the capital of Martinique (pop. 100,000) 14G4
Fort Lamy *see* **Ndjamena** 42E3
Fort Lauderdale a city and resort on the E coast of Florida, USA, 40 km (25 miles) N of Miami (pop. city 149,900/metropolitan area 1,093,300) 7E3
Fort Simpson Canada 5F3
Fort William Scotland 22C4

Fort Wayne   1001   Georgia

Fort Wayne USA 7E1
Fort Worth a city in NE Texas, USA, just W of Dallas and with it part of the Southwest Metroplex conurbation (pop. city 414,600/metropolitan area 1,144,400) 6D2
Fort Yukon USA 5D3
Foshan China 35F4
Fougères France 24B2
Foula Island Scotland 22H7
Foz do Iguacu Brazil 17D5
Fraga Spain 25C1
Franca Brazil 16E5
France [area 551,500 sq km (212,934 sq miles); pop. 56,180,000; cap. Paris; major cities Bordeaux, Lyon, Marseille, Toulouse; govt republic; religion RC; currency franc] the largest country in W Europe, with coasts on the English Channel, Mediterranean and Atlantic 24
Franceville Gabon 42E5
Franche-Comte (province) France 24D2
Francistown Botswana 44C3
Frankfort the state capital of Kentucky, USA (pop. 26,800) 7E2
Frankfurt (Frankfurt am Main) a major financial, trade and communications centre in central W Germany, on the River Main (pop. 614,700) 28B2
Fraser, River a river flowing through S British Columbia, Canada, from the Rockies to the Strait of Georgia by Vancouver (length 1370 km/850 miles) 5F4
Fraserburgh Scotland 22F3
Fraser Island Australia 45E3
Frederikshåb Greenland 5O3
Frederikshavn Denmark 30C4
Frederikstad Norway 30C4
Freetown the main port and capital of Sierra Leone (pop. 316,300) 42B4
Freiburg (Freiburg im Breisgau) the largest city in the Black Forest in SW Germany, close to the border with France (pop. 175,000) 28B3
Fremantle Australia 45A4
French Guiana (Guyane) see **Guiana (French)** 16D2
French Polynesia about 130 islands in the South Pacific administered as overseas territories by France 47
Fresno California USA 48B2
Frosinone Italy 26C2
Fuerteventura Island Canary Islands 42B2
Fuji-san (Mount Fuji, Fujiyama) the highest peak in Japan, a distinctive volcanic cone 100 km (62 miles) SW of Tokyo (3776 m/12,389 ft) 35N9
Fukui Japan 35M8
Fukuoka Japan 35H3
Fukushima Japan 35P8
Funchal the capital of Madeira (pop. 45,600) 42B1
Fundy, Bay of a bay between Nova Scotia and New Brunswick, Canada. It has the world's largest tidal range—15 m (50 ft) between low and high tide 7G1
Furneaux Group Australia 45D5
Fürth Germany 28B3
Furukawa Japan 35P7
Fushun a mining city in NE China, on one of the largest coalfields in the world (pop. 1,800,000) 35G2
Fuxin China 35G2
Fuzhou China 35F4
Fyn (Fünen) the second largest of the islands of Denmark, in the centre of the country (2976 sq km/1048 sq miles; pop. 433,800) 30C4

G

Gabès *Tunisia* 42D1
Gabon [area 267,667 sq km (103,346 sq miles); pop. 1,220,000; cap. Libreville; major city Port Gentile; govt republic; religion RC, Animism; currency franc CFA] a small country in W Africa straddling the Equator 42E5
Gaborone the capital of Botswana, in the SE of the country (pop. 79,000) 44C3
Gainsborough England 20G4
Gairloch Scotland 22C3
Galashiels Scotland 22F5
Galati an inland port on the Danube in E Romania, close to the Moldovan border (pop. 261,000) 29F3
Galicia a region in the very NW corner of Spain (pop. 2,754,000) 25A1
Galle Sri Lanka 39G5
Gallipoli Italy 27D2
Gällivare Sweden 30E2
Galveston Texas, USA 7D3
Galway a county in the central part of the W coast of Ireland; county town Galway (5940 sq km/2293 sq miles; pop. 171,800) 23B2
Galway (Galway City) the county town of Galway county in the Republic of Ireland (pop. 37,700) 23B3
Gambia [area 11,295 sq km (4361 sq miles); pop. 875,000; cap. Banjul; govt republic; religion Sunni Islam; currency dalasi] the smallest country in Africa, in W Africa, divided along its length by the River Gambia 42B3
Gambia, River a major river of W Africa, flowing into the Atlantic Ocean from its source in Guinea, through Senegal and Gambia (length 483 km/300 miles) 42B3
Gambier Islands Pacific Ocean 46
Gamboma Congo 42E5
Gandia Spain 25B2
Ganga, River (Ganges) the holy river of the Hindus, flowing from the Himalayas, across N India and forming a delta in Bangladesh as it flows into the Bay of Bengal (length 2525 km/1568 miles) 39D
Ganzhou China 35F4
Gao Mali 42C3
Garonne, River a major river of SW France, flowing N from the central Pyrénées to the Gironde estuary (length 575 km/355 miles) 24C/D3
Garve Scotland 22D3
Gascogne (Gascony) an historic province in the SW corner of France bordering Spain. 24B3
Gateshead England 20F3
Gauhati India 39H3
Gävle Sweden 30D3
Gaziantep Turkey 38B2
Gdansk (Danzig) the main port of Poland, on the Baltic Sea (pop. 464,500) 29D2
Gdynia (Gdingen) a port on the Baltic coast of Poland 16 km (10 miles) NW of Gdansk (pop. 240,200) 29D2
Gedaref Sudan 43G3
Geelong a port and second city of Victoria, Australia (pop. 142,000) 45D4
Gejiu China 34E4
Gela Italy 26C3
Gelati Romania 29F3
General Santos Philippines 37E3
Genève (Geneva; Genf) a city in SW Switzerland, at the W end of Lake Geneva; also the name of the surrounding canton (pop. city 165,000) 24D2
Genova (Genoa) the major seaport of NW Italy, and the capital of Liguria (pop. 760,300) 26B2
Gent (Ghent; Gand) a medieval city spanning the Rivers Lys and Schelde and the capital of the province of East Flanders, Belgium (pop. city 235,000/metropolitan area 490,000) 24C1
George Town a port and the main city of Penang Island, Malaysia (pop. 250,600) 36C3
Georgetown the main port and capital of Guyana (pop. 200,000) 16C2
Georgia [area 69,700 sq km (26,900 sq miles); pop. 5,976,000; cap. Tbilisi; major cities Kutaisi, Rustavi, Batumi; govt re-

public; *religion* Russian Orthodox; *currency* rouble] a republic in the SW of the former USSR which declared itself independent in 1991 32F5
Georgia a state in SE USA, named after George II by English colonists; state capital Atlanta (152,490 sq km/58,876 sq miles; pop. 5,837,000) 7E2
Gera Germany 28C2
Geraldton Australia 45A3
Germany [*area* 356,910 sq km (137,803 sq miles); population 79,070,000; *cap.* Berlin, Bonn (seat of government); *major cities* Cologne, Frankfurt, Hamburg, Leipzig, Munich, Stuttgart; *govt* federal republic; *religions* Lutheranism, RC; *currency* Deutsche Mark] a large industrialized country in N central Europe, which comprises the former East and West German Republics, reunified in 1990 28
Gerona Spain 25C1
Getafe Spain 25B1
Gevgelija Yugososlavia 27E2
Ghadamis Libya 42D1
Ghana [*area* 238,533 sq km (92,098 sq miles); *pop.* 14,900,000; *cap.* Accra; *major cities* Kumasi, Tamale, Sekondi-Takoradi; *govt* republic; *religion* Protestant, Animism, RC; *currency* cedi] a country in West Africa between Côte d'Ivoire and Togo 42C4
Ghat Libya 42E2
Ghats two ranges of mountains lining the coasts of the Deccan peninsula, India: the Eastern Ghats (rising to about 600 m/2000 ft) and the Western Ghats (1500 m/5000 ft) 39F/G4
Ghent *see* **Gent** 24C1
Gibraltar a self-governing British Crown Colony on the SW tip of Spain, where a limestone hill called the Rock of Gibraltar rises to 425 m (1394 ft) 25A2
Gibraltar, Strait of a waterway, 13 km (8 miles) at its narrowest, connecting the Mediterranean to the Atlantic, with Spain to the N and Morocco to the S 25A2
Gibson Desert a desert of sand and salt marshes in central W Australia 45B3
Gifu Japan 35M9
Giglio Island Italy 26C2
Gijón a port and industrial town in Asturias, in the centre of the N coast of Spain (pop.256,000) 25A1
Gilbert Islands *see* **Kiribati** 46
Gilgit a mountain district in N Pakistan, noted for its great beauty. Also the name of a small town 39F2
Gironde the long, thin estuary stretching some 80 km (50 miles) that connects the Rivers Dordogne and Garonne to the Atlantic coast of SW France 24B2
Girvan Scotland 22D5
Gisborne New Zealand 45G4
Giurgiu Romania 27F2
Giza *see* **El Gîza** 43G1
Glasgow a major industrial city and cultural centre on the River Clyde; the largest city in Scotland (pop. 751,000) 22D5
Glenrothes Scotland 22E4
Gliwice (Gleiwitz) Poland 29D2
Gloucester England 21E6
Gloucestershire a county in W England (2638 sq km/1019 sq miles; pop. 508,000) 21E6
Gniezno Poland 28D2
Gobabis Namibia 44B3
Godavari, River a river running E across the Deccan peninsula, India (length 1465 km/910 miles) 39F4
Godthåb (Nuuk) capital of Greenland (pop. 10,500) 5N3
Godwin Austen *see* **K2** 39F2
Goiânia Brazil 16E4
Gol Norway 30B3
Golan Heights an area of high ground in SW Syria on the border with Israel. They were captured by Israel in the Arab-Israeli War and annexed in 1981 (2225 m/ 7300 ft) 40C2

Golmund China 34D3, 39H2
Gomel' Belarus 32E4
Gomera Canary Islands 42B2
Gonaïves Haiti 14E3
Gonder Ethiopia 43G3
Good Hope, Cape of the tip of Cape Peninsula extending from the SW corner of South Africa 44B4
Goole England 20G4
Gor'kiy (Gorky) *see* **Nizhniy Novgorod** 32F4
Gorontalo Indonesia 37E3
Gort Ireland 23C3
Gorzów Wielkopolski Poland 28D2
Gospic Croatia 26D2
Göteborg (Gothenburg) a major port on the Kattegat and second largest city in Sweden (pop. 425,500) 30C4
Gotland an island in the Baltic which forms a county of Sweden (3140 sq km/1210 sq miles; pop. 56,100) 30D4
Göttingen a university town and medieval trading centre in central Germany (pop. 138,000) 28B2
Goulburn Australia 45D4
Gozo *see* **Malta** 26C3
Grampian an administrative region of NE Scotland created in 1975 out of the former counties of Aberdeenshire, Kincardineshire, Banffshire and part of Morayshire; capital Aberdeen (8550 sq km/3301 sq miles; pop. 497,000) 22E3
Grampian Mountains a range stretching across N Scotland. The highest peak is Ben Nevis (1344 m/4409 ft), the highest point in the UK 22D4/E3
Granada Nicaragua 14B4
Granada a city in the Sierra Nevada of central S Spain, an administrative centre in the Moorish occupation, when its Alhambra Palace was built; also the name of the surrounding province (pop. city 262,200) 25B2
Gran Canaria Canary Islands 42B2
Grand Bahama Island The Bahamas 14D1
Grand Canyon a dramatic gorge of the Colorado river, in parts over 1.5km (1 mile) deep, in NW Arizona 13D2
Grandola Portugal 25A2
Grand Rapids Michigan USA 7E1
Graz the second largest city in Austria, in the SE (pop. 243,000) 28D3
Great Abaco Island The Bahamas 14D1
Great Australian Bight the arm of the Southern Ocean which forms the deep indentation in the centre of the S coastline of Australia 45B4
Great Barrier Reef the world's most extensive coral reef which lines the coast of Queensland, Australia, stretching some 2000 km (1250 miles) 45D2
Great Bear Lake the fourth largest lake in North America, in NW Canada. It drains into the Mackenzie (31,153 sq km/12,028 sq miles) 5F3
Great Britain the island shared by England, Scotland and Wales, which forms the principal part of the United Kingdom of Great Britain and Northern Ireland.
Great Dividing Range a range of mountains which runs down the E coast of Australia, some 3600 km (2250 miles). The highest point is Mount Kosciusko (2230 m/7316 ft) 45D2/3
Greater Antilles Islands Caribbean Sea 14D3
Great Exuma Island The Bahamas 14D2
Great Inagua Island The Bahamas 14E2
Great Nicobar Island India 39H5
Great Salt Lake a salt lake in NW Utah, USA, NW of Salt Lake City (5200 sq km/2000 sq miles) 12D2
Great Sandy Desert the desert region in N Western Australia 45B3
Great Slave Lake a lake drained by the Mackenzie River in the S part of the Northwest Territories of Canada (28,570 sq km/11,030 sq miles) 5G3
Great Smoky Mountains part of the Appalachian Mountains, running along the border between Tennessee and North

Carolina. The highest point is Clingmans Dome (2025 m/ 6643 ft) 9C3
Great Victoria Desert a vast area of sand dunes straddling the border between Western Australia and South Australia 45B3
Great Yarmouth England 21J5
Greece [*area* 131,990 sq km (50,961 sq miles); *pop.* 10,140,000; *cap.* Athens (Athinai); *major cities* Patras, Piraeus, Thessaloníki; *govt* republic; *religion* Greek Orthodox; *currency* drachma] a peninsular-shaped country, the most SE extension of Western Europe 27E3
Greenland a huge island NE of North America, most of which lies within the Arctic Circle. A province of Denmark, it was granted home rule in 1979 5O2
Greenock Scotland 22D5
Greensboro USA 7F2
Grenada [*area* 344 sq km (133 sq miles); *pop.* 110,000; *cap.* St. Georges; *govt* constitutional monarchy; *religion* RC, Anglicanism, Methodism; *currency* East Caribbean dollar] the most southerly of the Windward Island chain in the Caribbean 14G4
Grenadines a string of 600 small islands that lie between St Vincent to the N and Grenada to the S 14G4
Grenoble a manufacturing city in SE France (pop. 396,800) 24D2
Gretna Scotland 22E5
Grimsby England 20G4
Grong Norway 30C3
Groningen the largest city in NE Netherlands (pop. city 205,700) 28B2
Groote Eylandt Island Australia 45C2
Grootfontein Namibia 44B2
Grosseto Italy 26C2
Groznyy Russian Federation 32F5
Grudziadz Poland 29D2
Guadalajara a major city of central W Mexico (pop. 2,300,000) 26C3
Guadalajara Spain 25B1
Guadalcanal the largest of the Solomon Islands in the SW Pacific and site of the capital, Honiara 45E2
Guadalupe Island Mexico 6B3
Guadeloupe a group of islands in the Leeward Islands, E Caribbean, an overseas department of France 14G3
Guam the largest of the Mariana Islands in the W Pacific Ocean 37G2
Guangzhou (Canton) a major port in SE China, the country's sixth largest city (pop. 5,350,000) 35F4
Guantánamo a city in SE Cuba. The USA has a naval base at nearby Guantánamo Bay (pop. 205,000) 14D3
Guarda Portugal 25A1
Guatemala [*area* 108,889 sq km (42,042 sq miles); *pop.* 9,000,000; *cap.* Guatemala City; *major cities* Puerto Barrios, Quezaltenango; *govt* republic; *religion* RC; *currency* quetzal] a country lying where North America meets Central America 14A4
Guatemala City the capital of Guatemala, in the SE of the country (pop. 1,329,600) 14A4
Guayaquil the main port and largest city of Ecuador (pop. 1,223,500) 16B3
Guernsey one of the Channel Islands, 50 km (30 miles) off the coast of France 21E8
Guiana (French) *or* **Guyane** [*area* 90,000 sq km (34,749 sq miles); *pop.* 73,800; *cap.* Cayenne; *govt* French overseas department; *religion* RC; *currency* franc] a country on the NE coast of South America 16D2
Guildford England 21G6
Guilin China 35F4
Guinea [*area* 245,857 sq km (94,925 sq miles); *pop.* 6,710,000; *cap.* Conakry; *major cities* Kankan, Labé; *govt* republic; *religion* Sunni Islam; *currency* Guinea franc] a country on the coast at the "bulge" in Africa 42B3
Guinea, Gulf of an arm of the Atlantic Ocean creating the deep, indent in the W coast of Africa 42D4
Guinea Bissau [*area* 36,125 sq km (13,948 sq miles); *pop.* 966,000; *cap.* Bissau; *govt* republic; *religion* Animism, Sunni Islam; *currency* peso] a country S of Senegal on the Atlantic coast of West Africa 42B3
Güiria Venezuela 16C1
Guiyang an industrial city in central S China, and capital of Guizhou province (pop. 1,260,000) 34E4
Gulf, The a huge inlet to the S of Iran which is connected to the Arabian Sea by the Strait of Hormuz; also called the Persian Gulf, or the Arabian Gulf 38C3
Gulu Uganda 43G4
Gur'yev Kazakhstan 32G5
Guyana [*area* 214,969 sq km (83,000 sq miles); *pop.* 990,000; *cap.* Georgetown; *major cities* New Amsterdam; *govt* cooperative republic; *religion* Hinduism, protestantism, RC; *currency* Guyana dollar] a South American country on the NE coast of the continent 16D2
Guyane *see* **Guiana (French)** 16D2
Guyenne (province) France 24B3
Gwalior a large city in central India, SE of Delhi (pop. 555,900) 39F3
Gwent a county in SE Wales, bordering the Severn estuary. It was created in 1974 and coincides with the old county of Monmouthshire; county town Cwmbran (1376 sq km/ 532 sq miles; pop. 440,000) 21E6
Gweru Zimbabwe 44C2
Gwynedd a county in NW Wales which includes Anglesey. It was created in 1974 out of the former county of Caernarfonshire, and parts of Denbighshire and Merionethshire; administrative centre Caernarfon (3868 sq km/1493 sq miles; pop. 232,000) 20C4
Gyandzha Azerbaijan 32F5
Györ Hungary 29D3

H

Haarlem Netherlands 28A2
Hachinohe *Japan* 35P6
Hagen Germany 28B2
Hague, The *see* **'s-Gravenhage** 28A2
Haifa the main port of Israel (pop. 224,700) 38B2
Haikou China 35F4
Ha'il Saudi Arabia 38C3
Hailar China 35F2
Hainan Dao Island a large tropical island in the South China Sea belonging to China and its S extremity (33,670 sq km/13,000 sq miles; pop. 5,400,000) 35F5
Haiphong a port in the N of Vietnam and its third largest city (pop. 1,379,000) 36C1
Haiti [*area* 27,750 sq km (10,714 sq miles); *pop.* 5,700,000; *cap.* Port-au-Prince; *major cities* Les Cayes, Gonaïves, Jérémie; *govt* republic; *religion* RC, Voodooism; *currency* gourde] a country occupying the W third of the island of Hispaniola 14E3
Hakodate a port at the S tip of Hokkaido Island, Japan (pop. 319,200) 35J2
Halab (Aleppo) Syria 38B2
Halden Norway 30C4
Halifax the capital of Nova Scotia, Canada (pop. city 114,595/metropolitan area 278,000) 5M5
Halifax a town in West Yorkshire, England (pop. 88,000) 20F4
Halle an industrial town and inland port served by the Saale River in central Germany (pop. 236,500) 28C2
Halmahera Island Indonesia 37E3
Halmstad Sweden 30C4
Hamadan Iran 38C2
Hamamatsu Japan 35M9
Hamar Norway 30C3

Hamburg

Hamburg the main port of Germany, situated on the River Elbe (pop. 1,617,800) 28C2
Hamersley Range part of the Pilbara Range in Western Australia. The highest peak is Mount Bruce (1235 m/4052 ft) 45A3
Hamhung (Hamheung) a port and industrial city on the E coast of North Korea (pop. 420,000) 35G2
Hami China 34D2
Hamilton a port and industrial city at the W end of Lake Ontario, Canada (pop. city 306,430/metropolitan area 542,090) 5K5
Hamilton a town in the NW North Island, New Zealand (pop. 97,900) 45C4
Hamm Germany 28B2
Hammerfest a town in the far N of Norway, one of the world's most northerly settlements (pop. 7400) 30E1
Hampshire a county of S England; county town Winchester (3773 sq km/1456 sq miles; pop. 1,500,000) 21F6
Hanamaki Japan 35P7
Handan China 35F3
Hangzhou (Hangchow) a port and industrial city on the E coast of central China, at the S end of the Grand Canal, linking it to Beijing 1100 km (690 miles) to the N (pop. 1,105,000) 35G3
Hankow *see* **Wuhan** 35F3
Hannover (Hanover) a historic city in central N Germany (pop. 514,000) 28B2
Hanoi the capital of Vietnam, in the N of the country (pop. 2,570,900) 36C1
Hanzhong China 34E3
Haora (Howrah) an industrial city in West Bengal, India, on the Hugli River (pop. 744,400) 39G3
Harare the capital of Zimbabwe; it was formerly called Salisbury (until 1982) (pop. 656,000) 44D2
Harbin the largest city of N China, situated in central Dongbei (Manchuria) (pop. 2,100,000) 35G2
Harer Ethiopia 43H4
Hargeysa Somalia 43H4
Harlow England 21H6
Harrisburg the state capital of Pennsylvania, USA (pop. city 52,100/metropolitan area 570,200) 7F1
Harris Island Scotland 22B3
Harrogate England 20F4
Hartford the state capital of Connecticut, USA (pop. city 136,400/metropolitan area 1,030,400) 7F1
Hartlepool England 20F3
Harwich England 21J6
Hässleholm Sweden 30C4
Hastings a historic port and resort on the S coast of England, in East Sussex (pop. 77,000) 21H7
Hastings New Zealand 45G4
Hatteras, Cape the tip of a chain of islands off the coast of North Carolina, USA 9D3
Haugesund Norway 30B4
Havana *see* **La Habana** 14C2
Havant England 21G7
Hawaii (Hawaiian Islands) a group of 122 islands just S of the Tropic of Cancer, 3700 km (2300 miles) from the coast of California. Since 1959 they have formed a state of the USA; state capital Honolulu (16,705 sq km/6450 sq miles; pop. 1,054,000) 6H
Hawick Scotland 22F5
Hay River Canada 5G3
Heanor England 20F4
Heard Islands Indian Ocean 46
Hebrides some 500 islands off the W coast of Scotland, consisting of the Inner Hebrides to the SE (main islands Tiree, Jura, Coll, Mull, Eigg, Skye) and the Outer Hebrides to the NW (including Lewis and Harris, the Uists, Benbecula, Barra) 22A1
Hefei an industrial city in central E China, capital of Anhui province (pop. 1,484,000) 35F3

1004

Hohhot

Hegang China 35H2
Heidelberg a university town in SW Germany (pop. 130,000) 28B3
Heilong Jiang, River *see* **Amur, River** 35G1
Helena the capital of Montana, USA (pop. 24,600) 6B1
Hella Iceland 30A2
Hellín Spain 25B2
Helmsdale Scotland 22E2
Helsinborg Sweden 30C4
Helsinki (Helsingfors) the capital and chief industrial centre and port of Finland (pop. 482,900) 30F3
Hengyang China 35F4
Henzada Myanmar 36B2
Heraklion *see* **Iráklion** 27F3
Herat a city in W Afghanistan on the Hari Rud River (pop. 150,500) 38E2
Hercegovina *see* **Bosnia Herzegovina** 27D2
Hereford England 21E5
Hereford and Worcester a county in W England, on the Welsh border, created 1974 from the counties of Herefordshire and Worcestershire; county town Worcester (3927 sq km/1516 sq miles; pop. 648,000) 21E5
Hermosillo USA 6B3
Hertfordshire a county in SE England; county town Hertford (1634 sq km/631 sq miles; pop. 980,000) 21G6
Herzegovina *see* **Bosnia Herzegovina** 27D2
Hexham England 20E3
Hierro Canary Islands 42B2
Highland Region an administrative region in N Scotland created 1975 from the counties of Caithness, Nairnshire, Sutherland, most of Inverness-shire, Ross and Cromarty and parts of Argyll and Morayshire; capital Inverness (26,136 sq km/10,091 sq miles; pop. 196,000) 22C3
Hiiumaa Island Estonia 30E4
Himalayas a massive mountain range stretching some 2400 km (1500 miles) from the N tip of India, across Nepal, Bhutan and S Tibet to Assam in NE India. Their average height is 6100 m (20,000 ft); the tallest peak is Mount Everest (8848 m/29,028 ft) 39G3
Himeji Japan 35L9
Hims (Homs) an industrial city of ancient origins in Syria (pop. 414,401) 38B2
Hinckley England 21F5
Hindu Kush a range of mountains stretching 600 km (370 miles) at the W end of the Himalayas, straddling the borders where Afghanistan, Tajikistan, China, India and Pakistan meet. The highest peak is Tirich Mir (7690 m/25,229 ft) in Pakistan 39E2
Hinnøy Island Norway 30D2
Hinnöy Island Sweden 30D2
Hiroshima an industrial city in SW Honshu Island, Japan. Three-quarters of it was destroyed on August 6, 1945 by the world's first atomic bomb, which killed 78,000 people (pop. 899,400) 35H3
Hirsova Romania 27F2
Hispaniola the large Caribbean island that is shared by Haiti and the Dominican Republic (76,200 sq km/29,400 sq miles) 14E3
Hitachi an industrial city on E Honshu Island, Japan (pop. 206,100) 35P8
Hobart a port and capital of Tasmania, Australia (pop. 173,700) 45D5
Ho Chi Minh City (Saigon) the largest city in Vietnam, and capital of former independent South Vietnam (pop. 3,500,000) 36C2
Höfn Iceland 30B2
Hoggar (Ahaggar) a remote mountain range of S Algeria noted for its rock formations. The highest peak is Tahat (2918 m/9573 ft) 42D2
Hohhot an industrial city and capital of the Nei Mongol Autonomous Region (Inner Mongolia), China (pop. 1,130,000) 35F2

Hokitika / Imperatriz

Hokitika New Zealand 45G5
Hokkaido Island the farthest N of the main islands of Japan and second largest after Honshu; capital Sapporo (78,509 sq km/30,312 sq miles; pop. 5,679,400) 35J2
Holguín Cuba 14D2
Holyhead Wales 20C4
Holy Island a small island, also called Lindisfarne, off the coast of NE England, with an 11th-century priory 20F2
Homs *see* **Hims** 38B2
Honduras [area 112,088 sq km (43,277 sq miles); pop. 4,440,000; cap. Tegucigalpa; govt republic; religion RC; currency lempira] a fan-shaped country in Central America 14B4
Hong (Song Hong; Yuan Jiang; Red River) a river that rises in SW China and flows SE across N Vietnam to the Gulf of Tongking (length 800 km/500 miles) 34E4
Hong Kong [area 1045 sq km (403 sq miles); pop. 5,760,000; govt colony under British administration until 1997 when China will take over; religion Buddhism, Taoism, Christianity; currency Hong Kong dollar] a colony in the South China Sea, consisting of Hong Kong Island, the peninsula of Kowloon and about 1000 sq km (386 sq miles) of adjacent land known as the New Territories 35F5
Honiara the capital of the Solomon Islands, situated on Guadalcanal (pop. 23,500) 45E1
Honolulu the state capital of Hawaii, USA, on the S coast of the island of Oahu (pop. city 373,000/metropolitan area 805,300) 6H
Honshu the central and largest of the islands of Japan 35J2
Hormuz (Ormuz), Strait of the narrow strait at the mouth of The Gulf between Oman and Iran 38D3
Horn, Cape (Cabo de Hornos) the S tip of South America 17C8
Horsham England 21G6
Hotan China 34C3
Hoting Sweden 30D3
Houghton-le-Spring England 20F3
Houston the largest city in Texas, USA (pop. city 1,705,700/ metropolitan area 3,164,400) 7D3
Hovd Mongolia 34D2
Howrah *see* **Haora** 39G3
Hoy Island Scotland 22F2
Hradec-Králové Czech Republic 28D2
Hrvatska *see* **Croatia** 26D1
Huainan China 35F3
Huambo Angola 44B2
Huancayo Peru 16B4
Huang He (Hwang Ho; Yellow River) the second longest river in China, flowing from the Qinghai mountains across N central China to the Yellow Sea, S of Beijing (length 5464 km/3395 miles) 35F3
Huangshi China 35F3
Huascaran a peak in the Andes in central Peru, and that country's highest mountain (6768 m/22,205 ft) 16B3
Hubei (province) China 35F3
Hubli India 39F4
Huddersfield England 20F4
Hudiksvall Sweden 30D3
Hudson Bay a huge bay in NE Canada, hemmed in to the N by Baffin Island, and connected to the Atlantic Ocean by the Hudson Strait 5K4
Hudson River a river flowing from the Adirondack Mountains in New York State, USA, to the Atlantic Ocean at New York City. The Erie Canal joins it to link New York to the Great Lakes (length 492 km/306 miles) 8E2
Hué the capital of the rulers of Vietnam from 200BC to the 19th century, located in the central coastal region of the country (pop. 190,100) 36C2
Huelva Spain 25A2
Hughenden Australia 45D3

Hull *see* **Kingston upon Hull** 20G4
Humaitá Brazil 16C3
Humber the estuary of the Rivers Ouse and Trent which cuts deep into the E coast of England to the N of the Wash (length 60 km/35 miles) 20G4
Humberside a county on the NE coast of England, centring on the Humber. It was created in 1974 out of parts of the East and West Ridings of Yorkshire and Lincolnshire; county town Beverley (3512 sq km/1356 sq miles; pop. 854,000) 20G4
Hungary [area 93,032 sq km (35,920 sq miles); pop. 10,590,000; cap. Budapest; major cities Debrecen, Miskolc, Pécs, Szeged; govt republic; religion RC, Calvinism, Lutheranism; currency forint] a landlocked country in the heart of Europe, dominated by the great plain to the E of the river Danube 29D3
Hunstanton England 21H5
Huntly Scotland 22F3
Huntsville USA 7E2
Huron, Lake one of the Great Lakes, at the centre of the group on the border between Canada and Michigan, USA (59,570 sq km/23,000 sq miles) 7E1
Húsavík Iceland 30B1
Hvar Island Croatia 26D2
Hwang Ho *see* **Huang He** 35F3
Hwange Zimbabwe 44C2
Hyderabad the capital of the state of Andhra Pradesh in E S India (pop. 2,093,500) 39F4
Hyderabad a city on the Indus delta 160 km (100 miles) NE of Karachi, Pakistan (pop. 795,000) 39E3
Hydra *see* **Idhra** 27E3
Hythe England 21J6

I

Ibadan the second largest city in Nigeria, some 120 km (75 miles) N of Lagos (pop. 1,009,000) 42D4
Ibiza Spain 25C2
Ibiza Island Spain 25C2
Ica Peru 16B4
Iceland [area 103,000 sq km (39,768 sq miles); pop. 253,500; capital Reykjavík; govt republic; religion Lutheranism; currency Icelandic krónais] a large island state in the North Atlantic, S of the Arctic Circle 5R3
Idaho an inland state in the NW of the USA; state capital Boise (216,413 sq km/83,557 sq miles; pop. 1,005,000) 6B1
Idhra (Hydra) a small island in the Aegean Sea, off the E coast of the Peloponnese, Greece 27E3
Igarka Russian Federation 32K3
Igoumenítsa Greece 27E3
IJsselmeer formerly a large inlet of the North Sea on the NE coast of the Netherlands, which was dammed and filled with water from the River IJssel and is now a freshwater lake, bordered by polders 28B2
Ikaría Island Greece 27F3
Ilebo Zaïre 43F5
Ile de Noirmoutier France 24B2
Ile de Ré France 24B2
Ile d'Oléron France 24B2
Iles d'Hyères France 24D3
Ilfracombe England 21C6
Iliodhrómia Island Greece 27E3
Illinois a state in the Midwest of the USA, bordering Lake Michigan to the N; capital Springfield. Chicago is its main city (146,075 sq km/56,400 sq miles; pop. 11,535,000) 7E1
Iloilo Philippines 37E2
Ilorin Nigeria 42D4
Imperatriz Brazil 16E3

Impfondo Congo 43E4
Imphal India 39H3
India [*area* 3,287,590 sq km (1,269,338 sq miles); *pop.* 843,930,000; *cap.* New Delhi; *major cities* Bangalore, Bombay, Calcutta, Delhi, Hyderabad, Madras; *govt* federal republic; *religion* Hinduism, Sunni Islam, Christianity; *currency* rupee] a vast country in South Asia, dominated by the N by the Himalayas 39F3
Indiana a state in the Midwest of the USA, SE of Lake Michigan; state capital Indianapolis (93,994 sq km/36,291 sq miles; pop. 5,499,000) 7E1
Indianapolis the state capital of Indiana (pop. city 710,300/metropolitan area 1,194,600) 7E2
Indian Desert *see* **Thar Desert** 39F3
Indian Ocean the third largest ocean, bounded by Asia to the N, Africa to the W and Australia to the E. The S waters merge with the Antarctic Ocean (73,481,000 sq km/28,364,000 sq miles) 46
Indonesia [*area* 1,904,569 sq km (735,354 sq miles); *pop.* 179,100,000; *cap.* Jakarta; *major cities* Badung, Medan, Semarang, Surabaya; *govt* republic; *religion* Sunni Islam, Christianity, Hinduism; *currency* rupiah] a country made up of 13,667 islands scattered across the Indian and Pacific Oceans in a huge crescent 36/74
Indore India 39F3
Indus, River a great river of Asia, flowing from Tibet, across the N tip of India before turning S through Pakistan to its estuary on the Arabian Sea, S of Karachi (length 3059 km/1900 miles) 39E3
Inhambane Mozambique 44D3
Inner Hebrides *see* **Hebrides** 22B4
Innsbruck Austria 28C3
In Salah Algeria 42D2
Inuvik Canada 5E3
Inveraray Scotland 22C4
Invercargill New Zealand 45F5
Inverness a town in NE Scotland at the head of the Moray Firth (pop. 40,000) 22D3
Inverurie Scotland 22F3
Ioannina Greece 27E3
Ionian Sea that part of the Mediterranean Sea between S Italy and Greece 27D3
Ióniol Nísoi (Ionian Islands; Eptanisos) the seven largest of the islands which lie scattered along the W coast of Greece in the Ionian Sea 27D3
Ios Island (Nios) 27F3
Iowa a state in the Midwest of the USA bounded E and W by the Mississippi and Missouri; capital Des Moines (145,791 sq km/56,290 sq miles; pop. 2,884,000) 7D1
Ipoh Malaysia 36C3
Ipswich England 21J5
Iquique Chile 16B5
Iquitos Peru 16B3
Iráklion (Heraklion) the capital and main port of Crete (pop. 111,000) 27F3
Iran [*area* 1,648,000 sq km (636,293,sq miles); *pop.* 53,920,000; *cap.* Tehran; *major cities* Esfahan, Mashhad, Tabriz; *govt* Islamic republic; *religion* Shia Islam; *currency* rial] a country lying across The Gulf from the Arabian peninsula 38D2
Iraq [*area* 438,317 sq km (169,234 sq miles); *pop.* 17,060,000; *cap.* Baghdad; *major cities* Al-Basrah, Al Mawsil; *govt* republic; *religion* Islam; *currency* Iraqi dinar] a country in SW Asia, wedged between The Gulf and Syria 38C2
Ireland an island off the W coast of Great Britain, four-fifths of which is the Republic of Ireland, while the remainder is Northern Ireland, a province of the UK (80,400 sq km/32,588 sq miles, pop. 5,202,000) 23
Ireland, Republic of [*area* 70,284 sq km (27,137 sq miles); *pop.* 3,540,000; *cap.* Dublin (Baile Atha Cliath); *major cities* Cork, Galway, Limerick, Waterford; *govt* republic; *religion* RC; *currency* punt] one of Europe's most W coun-

tries, situated in the Atlantic Ocean and separated from Great Britain by the Irish Sea 23
Irian Jaya the W half of the island of New Guinea, part of Indonesia since 1963 (410,660 sq km/158,556 sq miles; pop. 2,584,000) 37F4
Iringa Tanzania 44D1
Irish Sea the arm of the Atlantic that separates Ireland and Great Britain 19B3
Irkutsk an industrial city on the Trans-Siberian Railway lying near the S end of Lake Baikal in the Russian Federation (pop. 590,000) 33M4
Irrawaddy, River the central focus of Myanmar (Burma), flowing from its two primary sources in the N to Mandalay and then S to its delta in the Bay of Bengal (length 2000 km/1250 miles) 37B1
Irtysh, River a largely frozen river flowing N from near the border between NW China and Mongolia across Kazakhstan and through Omsk to join the River Ob' (length 4440 km/2760 miles) 32J4
Irvine Scotland 22D5
Isafördhur Iceland 30A1
Ischia Island Italy 26C2
Ise Japan 35M9
Isfahan *see* **Esfahan** 38D2
Ishinomaki Japan 35P7
Isiro Zaïre 43F4
Isla Blanquilla Island Venezuela 14G4
Isla Coiba Panama 14C5
Isla de Chiloé Island Chile 17B7
Isla de la Bahia Honduras 14B3
Isla de la Juventud Cuba 14C2
Isla del Rey Panama 14D5
Isla Los Roques Island Venezuela 14F4
Islamabad the capital of Pakistan since 1967, in the N of the country (pop. 201,000) 39F2
Isla Margarita Island Venezuela 14G4
Isla Santa Inés Island Chile 17B8
Islas Baleares (Balearic Islands) a group of islands in the W Mediterranean Sea belonging to Spain and famous as tourist resorts 25C2
Islas Canarias (Canary Islands) a group of islands belonging to Spain, situated some 95 km (60 miles) off the coast of Western Sahara 42B2
Islay Island Scotland 22B5
Isles Marquises (Marquesas Islands) a group of a dozen fertile, volcanic islands in the NE sector of French Polynesia, 1400 km (875 miles) NE of Tahiti 46
Isles of Scilly *see* **Scilly, Isles of** 20A8
Israel [*area* 20,770 sq km (8019 sq miles); *pop.* 4,820,000; *cap.* Jerusalem; *major cities* Tel Aviv-Jaffa, Haifa; *govt* republic; *religion* Judaism, Sunni Islam, Christianity; *currency* shekel] a country occupying a long narrow stretch of land in the SE of the Mediterranean 40C2
Issyk-Kul' a lake in S central Kazakhstan in the high mountains that line the border with China (6280 sq km/2424 sq miles) 34B2
Istanbul the largest city in Turkey, mainly on the W bank of the Bosphorus. It was founded by the Greeks in 660BC as Byzantium; between AD330 and 1930 it was called Constantinople (pop. 5,858,600) 38A1
Itabuna Brazil 16F4
Itaituba Brazil 16D3
Italy [*area* 301,268 sq km (116,320 sq miles); *pop.* 57,600,000; *cap.* Rome; *major cities* Milan, Naples, Turin, Genoa, Palermo; *govt* republic; *religion* RC; *currency* lira] a republic in S Europe, comprising a large, boot-shaped peninsula and the two main islands of Sicily and Sardinia 26
Iturup Island Russsian Federation 35J2
Ivalo Finland 30F2
Ivangrad Yugoslavia 26D2
Ivanovo Russian Federation 32F4

Ivory Coast — Jyväskylä

Ivory Coast see **Côte d'Ivoire** 42C4
Iwaki Japan 35P8
Iwo Jima the largest in the Volcano Islands group belonging to Japan, which lie some 1200 km (745 miles) S of Tokyo in the Pacific Ocean 37G1
Izhevsk Russian Federation 32G4
Izmir (Smyrna) a port of ancient Greek origin on the Aegean coast of Turkey (pop. 1,489,800) 38A2

J

Jabalpur India 39F3
Jackson the state capital of Mississippi (pop. city 208,800/metropolitan area 382,400) 7E2
Jacksonville a port on the NE coast of Florida, USA (pop. city 578,000/metropolitan area 795,300) 7E2
Jacmel Haiti 14E3
Jaén Spain 25B2
Jaffna a port on the tip of the N peninsula of Sri Lanka, and main centre for the Tamil population (pop. 118,200) 39G5
Jaipur the capital of the state of Rajasthan, India (pop. 1,015,200) 39F3
Jajce Bosnia 26D2
Jakarta the capital of Indonesia, a port on the NW tip of Java (pop. 6,503,000) 38C4
Jalgaon India 39F3
Jamaica [area 10,990 sq km (4243 sq miles); pop. 2,400,000; cap. Kingston; major cities Montego Bay, Spanish Town; govt constitutional monarchy; religion Anglicanism, RC, other Protestantism; currency Jamaican dollar] an island state in the Caribbean Sea 150 km (93 miles) S of Cuba 14D3
Jambi Indonesia 36C4
James Bay the S arm of Hudson Bay, Canada 5K4
Jammu India 39F2
Jamnagar India 39E2
Jämsänkoski Finland 30F3
Jamshedpur India 39G3
Japan [area 377,801 sq km (145,869 sq miles); pop. 123,260,000; capital Tokyo; major cities Osaka, Nagoya, Sapporo, Kobe, Kyoto, Yokohama; govt constitutional monarchy; religion Shintoism, Buddhism, Christianity; currency yen] a country on the E margin of Asia, consisting of four major islands, Honshu, Hokkaido, Kyushu and Shikoku, and many small islands 35H3
Japan, Sea of a part of the Pacific Ocean between Japan and Korea 35H3
Jardines de la Reina Cuba 14D2
Jarvis Island Pacific Ocean 46
Java Sea an arm of the Pacific Ocean that separates Java and Borneo 37D4
Jawa (Java) the central island in the S chain of islands of Indonesia. The capital is Jakarta (130,987 sq km/50,574 sq miles; pop. 91,269,600) 36C4
Jayapura Indonesia 37G4
Jedburgh Scotland 22F5
Jedda see **Jiddah** 38B3
Jefferson City the state capital of Missouri, USA (pop. 35,000) 7D2
Jelenia Góra Poland 28D2
Jena Germany 28C2
Jequié Brazil 16E4
Jerez de la Frontera (Jerez) a town in SW Spain, inland from Cadiz, famous for the wine to which it has given its name, sherry (pop. 176,200) 25A2
Jericho a town in the West Bank area occupied by Israel since 1967, on the site of a city that dates back to about 7000 BC (pop. 15,000) 40C3
Jersey largest of the British Channel Islands; capital St Helier (117 sq km/45 sq miles; pop. 77,000) 21E8
Jerusalem capital of Israel, and a city considered to be holy by Muslims, Christians and Jews (pop. 446,500) 38B2
Jhansi India 39F3
Jiamusi China 35H2
Ji'an China 35F4
Jiddah (Jedda) a port on the Red Sea coast of Saudi Arabia, and a centre of population (pop. 750,000) 38B3
Jihlava Czech Republic 28D3
Jima Ethiopia 43G4
Jinan China 35F3
Jingdezhen China 35F4
Jinhua China 35F4
Jining China 34/35F3
Jinja Uganda 43G4
Jinzhou China 35G2
Jiujiang China 35F4
João Pessoa Brazil 16F3
Jodhpur a city in India, on the perimeter of the Thar Desert. It has given its name to the riding breeches that first became popular here (pop. 506,300) 39F3
Jogjakarta see **Yogyakarta** 36D4
Johannesburg the centre of the Rand goldmining area of South Africa and now that country's largest town (pop. 1,536,500) 44C3
John o'Groats the village held to be the most northerly point of mainland Scotland and Great Britain 22E2
Johnson Island Pacific Ocean 46
Johor Baharu (Johore) a major port and growing city in Malaysia, on the S tip of the Malay Peninsula opposite Singapore, to which it is connected by a causeway; also capital of the state of Johor (pop. city 246,400) 36C3
Jokkmokk Sweden 30D2
Jólo Island Philippines 37E3
Jönköping Sweden 30C4
Jordan [area 97,740 sq km (37,737 sq miles); pop. 3,170,000; cap. Amman; major cities Irbid, Zarga; govt constitutional monarchy; religion Sunni Islam; currency Jordan dinar] a country, almost landlocked except for a short coastline on the Gulf of Aqaba, bounded by Saudi Arabia, Syria, Iraq and Israel 38B2
Jordan, River a river flowing S from Lebanon through Israel and Jordan to the Dead Sea, where it evaporates. Its West Bank, N of the Dead Sea, is disputed territory, occupied by Israel since 1967 (length 256 km/159 miles) 40C2
Jörn Sweden 30E2
Jos Nigeria 42D4
Jotunheimen Norway 30B3
Juan de Fuca, Strait of a channel S of Vancouver Island on the border between Canada and the USA, connecting with the Pacific Ocean 12B1
Juázeiro Brazil 16E3
Juba Sudan 43G4
Julianehåb Greenland 5O3
Jumna, River see **Yamuna, River** 39F3
Juneau the state capital of Alaska (pop. 23,800) 6J
Jura a large upland band in E central France lining the border with Switzerland, giving its name to a department in France and a canton in Switzerland 24D2
Jura Island Scotland 22C5
Jutland see **Jylland** 30B4
Jylland (Jutland) a large peninsula stretching 400 km (250 miles) N from Germany and separating the North and Baltic Seas. Mostly occupied by mainland Denmark, the S part belongs to Germany 30B4
Jyväskylä Finland 30F3

K

K2 (Godwin Austen) the second highest mountain in the world after Mount Everest, in the Karakoram mountain range on the disputed border between Pakistan and China (8611 m/28,250 ft)   39F2
Kabul the capital and main city of Afghanistan, in the NE of the country on the Kabul River (pop. 1,036,400)   39E2
Kaduna Nigeria   42D3
Kaédi Mauritania   42B3
Kaesong North Korea   35G3
Kagoshima a port on the S coast of Kyushu Island, Japan (pop. 530,500)   35H3
Kaifeng a city of ancient origins in Henan province, China (pop. 500,000)   35F3
Kailua Hawaii USA   6H
Kairouan a city in N Tunisia, to Muslims the most holy city of the Maghreb (pop. 72,300)   42D1
Kajaani Finland   30F3
Kakinada India   39G4
Kalabáka Greece   27E3
Kalahari a region of semi-desert occupying much of S Botswana and straddling the border with South Africa and Namibia   44C3
Kalajoki Finland   30E3
Kalámai Greece   27E3
Kalaupapa Hawaii USA   6H
Kalémié Zaïre   43F5
Kalgoorlie a small town in S Western Australia that has grown up around gold and nickel reserves (pop. 19,800)   45B4
Kalimantan the greater part of Borneo, which is governed by Indonesia (538,718 sq km/208,000 sq miles; pop. 6,724,000)   36D3
Kálimnos Island Greece   27F3
Kalinin *see* **Tver'**   32E4
Kaliningrad (Königsberg) a port and city on the Baltic coast belonging to the Russian Federation, an enclave between Lithuania and Poland (pop. 380,000)   32D4
Kalisz Poland   29D2
Kalmar Sweden   30D4
Kamaishi Japan   35P7
Kamchatka a peninsula, 1200 km (750 miles) long, dropping S from E Siberia into the N Pacific   33R4
Kamina Zaïre   44C1
Kamloops Canada   5F4
Kampala the capital and main city of Uganda, situated on Lake Victoria (pop. 500,000)   43G4
Kampuchea *see* **Cambodia**   36C2
Kananga a city in central S Zaïre, founded in 1894 as Luluabourg (pop. 704,000)   43F5
Kanazawa a historic port on the central N coast of Honshu Island, Japan (pop. 430,500)   35M8
Kandahar the second largest city in Afghanistan, in the SE, near the border with Pakistan (pop. 191,400)   39E2
Kandalaksha Russian Federation   32E3
Kandangan Indonesia   36D4
Kandy a town in the central mountains of Sri Lanka, which is sacred to Buddhists (pop. 101,300)   39G5
Kaneohe Hawaii USA   6H
Kangaroo Island Australia   45C4
Kangchenjunga the world's third highest mountain (after Mount Everest and K2), in the E Himalayas, on the borders between Nepal, China and India (8585 m/28,165 ft)   39G3
Kankan Guinea   42C3
Kano a historic trading city of the Hausa people of N Nigeria, the third largest city in Nigeria after Lagos and Ibadan (pop. 475,000)   42D3
Kanpur (Cawnpore) an industrial city in N central India (pop. 1,639,100)   39G3
Kansas a state in the Great Plains of the USA; state capital Topeka (213,064 sq km/82,264 sq miles; pop. 2,450,000)   6D2
Kansas City an industrial city on the Missouri river, straddling the border between Missouri and Kansas (pop. city 603,600/metropolitan area 1,476,700)   7D2
Kao-hsiung the second largest city in Taiwan and a major port (pop. 1,269,000)   37E1
Kaolack Senegal   42B3
Karachi a port and industrial centre, and the largest city in Pakistan (pop. 5,103,000)   39E3
Karaganda an industrial city in the mining region of Kazakhstan (pop. 608,000)   32J5
Karakoram a range of mountains at the W end of the Himalayas   39F2
Karakumy (Kara Kum) a sand desert in S Turkmenistan, E of the Caspian Sea   32G5
Kara Sea a branch of the Arctic Ocean off the central N coast of the Russian Federation   32J2
Karbala a town in central Iraq, 90 km (55 miles) S of Baghdad, sacred to Shia Muslims (pop. 107,500)   38B2
Karcag Hungary   29E3
Kariba Dam a hydroelectric dam on the River Zambezi on the border between Zambia and Zimbabwe   44C2
Karl-Marx-Stadt *see* **Chemnitz**   28C2
Karlobag Croatia   26D2
Karlovac Croatia   26D1
Karlshamn Sweden   30C4
Karlskoga Sweden   30C4
Karlskrona Sweden   30D4
Karlsruhe (Carlsruhe) an industrial city in the valley of the River Rhine, SW Germany (pop. 275,000)   28B3
Karlstad Sweden   30C4
Karoo (Karroo) two regions of semi-desert, the Great Karoo and the Little Karoo, between the mountain ranges of S Cape Province, South Africa   44C4
Kárpathos Island Greece   27F3
Karshi Uzbekistan   32H6
Kasai (Cassai), River a major river of Zaïre (length 2150 km/1350 miles)   44C1
Kasama Zimbabwe   44D2
Kasese Uganda   43G4
Kashi China   34B3
Kásos Island Greece   27F3
Kassala Sudan   43G3
Kassel (Cassel) an industrial city in central Germany (pop. 190,400)   28B2
Kastoria Greece   27E2
Kateríni Greece   27E2
Katherine Australia   45C2
Kathmandu (Katmandu) the capital and principal city of Nepal (pop. 195,260)   39G3
Katowice (Kattowitz) an industrial city in central S Poland (pop. 361,300)   29D2
Katsina Nigeria   42D3
Kattegat (Cattegat) the strait, 34 km (21 miles) at its narrowest, at the entrance to the Baltic Sea   29C1
Kauai Island Hawaii USA   6H
Kaunas (Kovno) an industrial city and former capital of Lithuania (pop. 400,000)   30E4
Kaválla Greece   27E2
Kavango, River *see* **Cubango, River**   44B2
Kavkaz *see* **Caucasus**   32F5
Kawaihae Hawaii USA   6H
Kawasaki an industrial city on the E coast of Honshu Island, Japan (pop. 1,088,600)   35N9
Kayes Mali   42B3
Kayseri Turkey   38B2
Kazakhstan [*area* 2,717,000 sq km (1,050,000 sq miles); *pop.* 15,654,000; *cap.* Alma Ata; *major city* Karaganda; *govt*

Kazan' 1009 **Kirov**

republic; *religion* Sunni Islam; *currency* rouble]the second largest republic of the former USSR, which declared itself independent in 1991 and which extends from the Caspian Sea to Mongolia 32H5
Kazan' an industrial city and capital of Tatar Republic in central Russian Federation (pop. 1,039,000) 32F4
Kazanlŭk Bulgaria 27F2
Kazan-rettó Japan 37G1
Kéa Greece 27E3
Kecskemét Hungary 29D3
Kediri Indonesia 36D4
Keeling Islands *see* **Cocos Islands** 36B5
Keetmanshoop Namibia 44B3
Kefallinía Island Greece 27E3
Keflavik Iceland 30A2
Keighley England 20F4
Keith Scotland 22F3
Kelang Malaysia 36C3
Kells a market town in County Meath, Ireland. It was the site of a monastery founded in the 6th century, the source of the illuminated Book of Kells 23E3
Kemerovo Russian Federation 32K4
Kemi Finland 30E2
Kemijärvi Finland 30F2
Kendal England 20E3
Kendari Indonesia 37E4
Kengtung Myanmar 36B1
Kenitra Morocco 42C1
Kenmare Ireland 23B5
Kenora Canada 5J5
Kent a county in SE England; county town Maidstone (3732 sq km/1441 sq miles; pop. 1,494,000) 21H6
Kentucky a state in E central USA; capital Frankfort (104,623 sq km/40,395 sq miles; pop. 3,726,000) 7E2
Kenya [*area* 580,367 sq km (224,080 sq miles); *pop.* 24,080,000; *cap.* Nairobi; *major cities* Mombasa, Kisumu; *govt* republic; *religions* RC, Protestantism, other Christianity, Animism; *currency* Kenya shilling] a country in E Africa, straddling the Equator and extending from Lake Victoria in the W to the Indian Ocean in the E 43G4
Kenya, Mount a towering extinct volcano in central Kenya, the second highest mountain in Africa after Mount Kilimanjaro (5200 m/17,058 ft) 43G5
Kepno Poland 29D2
Kepulauan Tanimbar Island Indonesia 37F4
Kerch Ukraine 32E5
Kerguelen the largest in a group of 300 islands in the S Indian Ocean forming part of the French Southern and Antarctic Territories, now occupied only by scientists (3414 sq km/1318 sq miles) 46
Kérkira (Corfu) the most northerly of the Ionian Islands, in W Greece; the capital is also called Kérkira (592 sq km/229 sq miles; pop. 97,100) 27D3
Kermadec Islands Pacific Ocean 46
Kerman Iran 38D2
Kerry a county in SW Republic of Ireland, noted for its rugged beauty and dairy pastures; county town Tralee (4701 sq km/1815 sq miles; pop. 122,800) 23B4
Keswick England 20E3
Key West a port and resort at the S end of Florida Keys, a chain of coral islands off the S tip of Florida, USA (pop. 24,900) 7E3
Khabarovsk a major industrial city in SE Siberia in the Russian Federation, 35 km (22 miles) N of the border with China (pop. 569,000) 33P5
Khalkis Greece 27E3
Khaniá Greece 27E3
Kharagpur India 39G4
Khar'kov a major industrial and commercial centre of the Ukraine (pop. 1,536,000) 32E4
Khartoum *see* **El Khartum** 43G3
Khartoum North Sudan 43G3

Khíos (Chios) a Greek island in the Aegean Sea, 8 km (5 miles) from the coast of Turkey (904 sq km/349 sq miles; pop. 49,900) 27F3
Khulna Bangladesh 39G3
Khyber Pass a strategic route (1072 m/3518 ft) over the Safed Koh mountains connecting Peshawar in Pakistan with Kabul in Afghanistan 39F2
Kiel a port and shipbuilding city on the Baltic coast of N Germany, at the mouth of the Kiel Ship Canal crossing Jutland from the Baltic to the North Sea (pop. 248,400) 28C2
Kielce (Kelsty) an industrial city in central S Poland (pop. 197,000) 29E2
Kiev *see* **Kiyev** 32E4
Kigali the capital of Rwanda (pop. 170,000) 43G5
Kigoma Tanzania 43F5
Kikládhes (Cyclades) a group of some 220 islands in the middle of the Aegean belonging to Greece (pop. 88,400) 27E3
Kikwit Zaïre 43E5
Kildare a county in SE Republic of Ireland; county town Naas (1694 sq km/654 sq miles; pop. 104,100) 23E3
Kilimanjaro, Mount Africa's highest mountain, in NE Tanzania (5895 m/19,340 ft) 43G5
Kilkenny a county in SE Republic of Ireland (2062 sq km/769 sq miles; pop. 70,800) 23D4
Kilkenny the county town of Kilkenny in the SE of the Republic of Ireland (pop. 10,100) 23D4
Killarney a market town in county Kerry, Republic of Ireland, at the centre of a much admired landscape of lakes and mountains (pop. 7700) 23B4
Kilmarnock Scotland 22D5
Kilrush Ireland 23B4
Kimberley a town in the N of Cape Province, South Africa, at the centre of South Africa's diamond mining industry (pop. 153,900) 44C3
Kimberley Plateau a vast plateau in the N of Western Australia (420,000 sq km/162,000 sq miles) 45B2
Kindia Guinea 42B3
Kindu Zaïre 43F5
King Island Australia 45D4
Kings Lynn England 21H5
Kingston the capital and main port of Jamaica (pop. 700,000) 14D3
Kingston upon Hull (Hull) a port in E England, on the N side of the Humber estuary (pop. 270,000) 20G4
Kingstown the capital of St Vincent and a port (pop. 22,800) 14G4
Kingswood England 21E6
Kingussie Scotland 22D3
Kinnegad Ireland 23D3
Kinshasa the capital of Zaïre, formerly called Léopold-ville, on the banks of the River Zaïre. It is the largest city in Central Africa (pop. 2,444,000) 42E5
Kintyre Scotland 22C5
Kinvarra Ireland 23C3
Kiribati [*area* 726 sq km (280 sq miles); *pop.* 66,250; *cap.* Tarawa; *govt* republic; *religions* RC, Protestantism; *currency* Australian dollar] a country comprising three groups of coral atolls and one isolated volcanic island spread over a large expanse of the central Pacific 46
Kirin *see* **Jilin** 35G2
Kiritimati (Christmas Island) the Pacific's largest coral atoll, at the NE end of the Kiribati group 46
Kirkby Stephen England 20E3
Kirkcaldy Scotland 22E4
Kirkenes Norway 30G2
Kirkuk an industrial city and regional capital in the Kurdish N of Iraq (pop. 650,000) 38B2
Kirkwall Scotland 22F2
Kirov an industrial city in E central Russian Federation, founded in the 12th century (pop. 407,000) 32F4

Kiruna 1010 Kunming

Kiruna Sweden 30E2
Kisangani a commercial centre and regional capital in N Zaïre, on the River Zaïre, originally called Stanleyville (pop. 339,000) 43F4
Kishinev (Chisinau) the capital of Moldova (pop. 605,000) 32D5
Kiskunfélegyháza Hungary 29D3
Kismaayo Somalia 43H4
Kistna, River *see* **Krishna, River** 39F4
Kisumu Kenya 43G5
Kita-Kyushu a major industrial city in the N of Kyushu Island, Japan (pop. 1,056,400) 35H3
Kithira Island Greece 27E3
Kíthnos Greece 27E3
Kitwe Zambia 44C2
Kivu, Lake a lake in the Great Rift Valley on the Rwanda-Zaïre border (2850 sq km/1100 sq miles) 43F5
Kiyev (Kiev) the capital of Ukraine, on the Dnieper, and a major industrial city (pop. 2,411,000) 32E4
Kizil Irmak the longest river in Turkey, flowing W and N to the Black Sea (length 1130 km/700 miles) 38B1
Kladno Czech Republic 28C2
Klagenfurt Austria 28C3
Klaipeda (Memel) a major port and shipbuilding centre on the Baltic coast of Lithuania (pop. 181,000) 32D4
Klerksdorp South Africa 44C3
Knoxville an industrial city in E Tennessee, USA, and a port on the Tennessee River (pop. city 174,000/metropolitan area 589,400) 7E2
Kobe a major port and shipbuilding centre at the S end of Honshu Island, Japan (pop. 1,410,800) 35L9
København (Copenhagen) a port and the capital of Denmark, located on the islands of Zealand and Amager (pop. 641,900) 30C4
Koblenz (Coblenz) a city at the confluence of the Rhine and Moselle in W Germany, and a centre for the winemaking industry (pop. 113,000) 28B2
Kochi Japan 35H3
Kodiak Island USA 6J
Kofu Japan 35N9
Kokkola Finland 30E3
Kolding Denmark 30B4
Kolhapur India 39F4
Köln (Cologne) a city and industrial centre on the River Rhine, Germany (pop. 932,400) 28B2
Koloma Russian Federation 32E4
Kol'skiy Poluostrov (Kola Peninsula) a bulging peninsula in the Barents Sea in the extreme NW of the Russian Federation, E of Murmansk 32E3
Kolwezi Zaïre 44C2
Komatsu Japan 35M8
Komotiní Greece 27F2
Kompong Cham Cambodia 36C2
Kompong Som Cambodia 36C2
Komsomol'sk na-Amure Russian Federation 33P4
Königsberg *see* **Kaliningrad** 32D4
Konin Poland 29D2
Konjic Bosnia 27D2
Konya a carpet-making town and capital of the province of the same name in central S Turkey (pop. town 438,900) 38B2
Kópavogur Iceland 30A2
Korcë Albania 27E2
Korcula Island Croatia 26D2
Korea, North [*area* 120,538 sq km (46,540 sq miles); *pop.* 22,420,000; *cap.* Pyongyang; *major cities* Chongjin, Nampo; *govt* socialist republic; *religions* Chondoism, Buddhism; *currency* North Korean won] a country occupying just over half of the Korean peninsula in E Asia 35G3
Korea, South [*area* 99,016 sq km (38,230 sq miles); *pop.* 42,800,000; *cap.* Seoul (Soul); *major cities* Pusan, Taegu,

Inch'on; *govt* republic; *religions* Buddhism, Christianity; *currency* South Korean won] a country occupying the S half of the Korean peninsula 35G3
Korea Strait a stretch of water, 64 km (40 miles) at its narrowest, separating South Korea from Japan 35G3
Kórinthos (Corinth) a town in W Greece, near the Corinth Ship Canal (pop. 22,700) 27E3
Koriyama Japan 35P8
Kornat Island Croatia 26D2
Korsör Denmark 30C4
Kos (Cos) one of the Dodecanese Islands, belonging to Greece, in the Aegean Sea (290 sq km/112 sq miles; pop. 20,300) 27F3
Kosciusko, Mount the highest mountain in Australia, a peak in the Snowy Mountains range in S New South Wales (2230 m/7316 ft) 45D4
Kosice a rapidly growing industrial city, and regional capital of E Slovakia (pop. 214,300) 29E3
Kosovo an autonomous province in SW Serbia. About 75% of the population are ethnic Albanians 27E2
Kosovska-Mitrovica Yugoslavia 27E2
Koszalin Poland 28D2
Kota India 39F3
Kota Baharu Malaysia 36C3
Kota Kinabalu Malaysia 36D3
Kotka Finland 30F3
Kotlas Russian Federation 32F3
Kotor Yugoslavia 27D2
Kouvola Finland 30E4
Kovno *see* **Kaunas** 30E4
Kowloon Hong Kong 36D1
Kra, Isthmus of a narrow neck of land, 50 km (30 miles) wide, shared by Myanmar and Thailand, which joins Peninsular Malaysia to mainland SE Asia 36B3
Kragujevac Yugoslavia 27E2
Kraków (Cracow) the third largest city in Poland, and capital during medieval times (pop. 520,700) 29D2
Kramsfors Sweden 30D3
Kranj Slovenia 26C1
Krasnodar Russian Federation 32E5
Krasnovodsk Turkmenistan 32G5
Krasnoyarsk a mining city on the Trans-Siberian Railway in central S Siberia (pop. 860,000) 33L4
Krefeld a textile town in W Germany, near the border with the Netherlands (pop. 224,000) 28B2
Krishna (Kistna) a river that flows through S India from the Western Ghats to the Bay of Bengal (length 1401 km/871 miles) 39F4
Kristiania *see* **Oslo** 39C4
Kristiansand Norway 30B4
Kristianstad Sweden 30C4
Kríti (Crete) the largest and most southerly of the islands of Greece, with important ruins of the Minoan civilization at Knossos; capital Heraklion. (8366 sq km/3229 sq miles; pop. 502,100) 27E3
Krivoy Rog Ukraine 32E5
Krk (Veglia) a richly fertile island belonging to Croatia, in the N Adriatic Sea 26C1
Krung Thep *see* **Bangkok** 36C2
Kruscevac Yugoslavia 27E2
Kuala Lumpur the capital of Malaysia, on the banks of the Kelang and Gombak Rivers (pop. 937,900) 36C3
Kuala Terengganu Malaysia 36C3
Kuching Malaysia 36D3
Kulata Bulgaria 27E2
Kumamoto a city in the W of Kyushu Island, Japan, noted for its electronics industries (pop. 555,700) 35H3
Kumanovo Yugoslavia 27E2
Kumasi a town in central S Ghana, and capital of the Ashanti people (pop. 415,300) 42C4
Kunashir Island Russian Federation 35J2
Kunming an industrial and trading city and capital of

Kuopio   Lastovo Island

Yunnan province in SW China (pop. 1,930,000)	34E4	**Lairg** Scotland	22D2
Kuopio Finland	30F3	**Lajes** Brazil	17D5
Kupang Indonesia	37E5	**Lake District** a region of lakes and mountains in NW England, which includes England's highest peak, Scafell Pike (978 m/3208 ft)	20D3
Kuril'Skiye Ostrova (Kuril Islands) a long chain of some 56 volcanic islands between the S coast of the Kamchatka peninsula, E Russian Federation and Hokkaido Island, Japan. It was taken from Japan by the former USSR in 1945 (15,600 sq km/6020 sq miles)	33R5	**Lake of the Woods** a lake spattered with some 17,000 islands in SW Ontario, Canada, on the USA border (4390 sq km/1695 sq miles)	10D1
Kurnool India	39F4	**Lakshadweep Islands** a territory of India consisting of 27 small islands (the Amindivi Islands, Laccadive Islands and Minicoy Islands) lying 300 km (186 miles) off the SW coast of mainland India	39F4
Kursk a major industrial city in the Russian Federation, 450 km (280 miles) S of Moscow (pop. 423,000)	32E4	**La Mancha** a high, arid plateau in central Spain, 160 km (100 miles) S of Madrid	25B2
Kushiro Japan	35J2	**Lambaréné** a provincial capital in E central Gabon, the site of a hospital founded by Albert Schweitzer (pop. 28,000)	42E5
Kutch (Kachchh) an inhospitable coastal region on the border between Pakistan and India, which floods in the monsoon season and then dries out into a baking, salty desert (44,185 sq km/17,060 sq miles)	39E3	**Lamía** Greece	27E3
Kuwait [*area* 17,818 sq km (6880 sq miles); *pop.* 2,040,000; *cap.* Kuwait (Al Kuwayt); *govt* constitutional monarchy; *religion* Sunni Islam, Shia Islam; *currency* Kuwait dinar] a tiny state on The Gulf, wedged between Iraq and Saudi Arabia	38C3	**Lampang** Thailand	36B2
		Lampedusa Island Italy	26C3
		Lampione Island Italy	26C3
		Lanai Island Hawaii USA	6H
Kuznatsk *see* **Novokuznatsk**	32K4	**Lanark** Scotland	22E5
Kwangju (Gwangju; Chonnam) an industrial city and regional capital in the SW corner of South Korea (pop. 727,600)	35G3	**Lancashire** a county of NW England, once the heart of industrial Britain; county town Preston (3043 sq km/1175 sq miles; pop. 1,378,000)	20E4
Kwango (Cuango), River a river which rises in N Angola and flows N to join the River Kasai in Zaïre (length 110 km/68 miles)	44B1	**Lancaster** England	20E3
		Land's End the tip of the peninsula formed by Cornwall, the most W point of mainland England	21B7
Kyle of Lochalsh Scotland	22C3	**Langres** France	24D2
Kyoto a city in central S Honshu Island, a former imperial capital of Japan (pop. 1,479,100)	35L9	**Lang Son** Vietnam	36C1
Kyrgyzstan [*area* 198,500 sq km (76,600 sq miles); *pop.* 3,886,000; *cap.* Bishkek; *govt* republic; *religion* Sunni Islam; *currency* rouble] a central Asian republic of the former USSR, on the border with NW China, which declared itself independent in 1991	32J5	**Lansing** the state capital of Michigan, USA (pop. city 128,000/metropolitan area 416,200)	7E1
		Lanzarote Island Canary Islands	42B2
		Lanzhou a major industrial city and capital of Gansu province, central China (pop. 2,260,000)	34E3
		Laoag Philippines	37E2
Kyushu the most southerly of Japan's main islands, and the third largest after Honshu and Hokkaido (43,065 sq km/ 16,627 sq miles; pop. 13,276,000)	35H3	**Lao Cai** Vietnam	36C1
		Laois a county in the centre of the Republic of Ireland; county town Portlaoise (1718 sq km/664 sq miles; pop. 51,200)	23D4
Kyzyl Russian Federation	33L4	**Laos** [*area* 236,800 sq km (91,428 sq miles); *pop.* 4,050,000; *cap.* Vientiane; *govt* people's republic; *religion* Buddhism; *currency* kip] a landlocked country in SE Asia	36C2
Kzyl Orda Kazakhstan	32H5	**La Palma** Canary Islands	42B2
		La Paz a city set high in the Andes of Bolivia, and the capital and seat of government (pop. 900,000)	16C4

L

		La Plata a port on the estuary of the River Plate (Rio de la Plata) in NE Argentina (pop. 455,000)	17D6
Laâyoune Western Sahara	42B2	**Lappland (Lapland)** a region of N Scandinavia and the adjoining territory of the Russian Federation, traditionally inhabited by the nomadic Lapp people	30E2
Labé Guinea	42B3	
Labrador the mainland part of Newfoundland, on the E coast of Canada (295,800 sq km/112,826 sq miles)	5M4	
Lábrea Brazil	16C3	**Laptev Sea** part of the Arctic Ocean bordering central N Siberia	33O2
Labytnangi Russian Federation	32H3	**Laredo** USA	6D3
Laccadive Islands *see* **Lakshadweep Islands**	39F4	**Largs** Scotland	22D5
La Ceiba Honduras	14B4	**La Rioja** an autonomous area in the S of the Basque region of Spain, famous for its wines (pop. 254,000)	25B1
La Coruña (Corunna) a port and manufacturing town in NW Spain, and also the name of the surrounding province (pop. town 232,400)	25A1	**Lárisa** Greece	27E3
		Larnaca a port, with an international airport, on the SE coast of Cyprus (pop. 48,400)	40B1
Ladozhskoye Ozero (Lake Ladoga; Lake Laatokka) Europe's largest lake, in the Russian Federation, NE of St Petersburg (18,390 sq km/7100 sq miles)	32E3	**Larne** Northern Ireland	23F2
Lae Papua New Guinea	39G4	**La Rochelle** France	24B2
La Flèche France	24B2	**La Roda** Spain	25B2
Lagos the principal port and former capital of Nigeria, on the Bight of Benin (pop. 1,477,000)	42D4	**La Romana** Dominican Republic	14F3
		La Serena Chile	17B5
Lagos Portugal	25A2	**Lashio** Myanmar	36B1
La Habana (Havana) the capital of Cuba, a port on the NW coast of the island. It is also the name of the surrounding province (pop. 1,925,000)	14C2	**Las Palmas de Gran Canaria** the main port and largest city of the Canary Islands, on the island of Gran Canaria (pop. 366,500)	42B2
Lahore Pakistan	39F2	**La Spezia** Italy	26B2
Lahti Finland	30F3	**Lastovo Island** Croatia	26D2
Lai Chau Vietnam	36C1		

Las Vegas   Lincoln

Las Vegas a city in SE Nevada, USA. The state's gaming laws have allowed it to develop as a gambling centre (pop. city 183,200/metropolitan area 536,500) 6B2
Latakia see **Al Ladhiqiyah** 38B2
Latina Italy 26C2
La Tortuga Island Venezuela 14F4
Latvia [area 63,700 sq km (24,595 sq miles); pop. 2,681,000; cap. Riga; major cities Daugavpils, Jurmala, Liepaja; govt republic; religion Lutheranism; currency rouble] a Baltic state that regained its independence in 1991 with the break-up of the USSR, sandwiched between Estonia and Lithuania 30E4
Launceston Tasmania 45D5
Laurencekirk Scotland 22F4
Lausanne a city on the N shore of Lake Geneva, Switzerland, and capital of the French-speaking canton of Vaud (pop. 140,000) 24D2
Laut Island Indonesia 36D4
Laval France 24B2
Lebanon [area 10,400 sq km (4015 sq miles); pop. 2,800,000; cap. Beirut (Beyrouth); major cities Tripoli, Zahle; govt republic; religions Shia Islam, Sunni Islam, Christianity; currency Lebanese pound] a mountainous country in the E Mediterranean 40C1
Leeds an important industrial town on the River Aire in West Yorkshire, in N England (pop. 450,000) 20F4
Leeuwarden Netherlands 28B2
Leeward and Windward Islands the Lesser Antilles in the S Caribbean are divided into two groups: the N islands, from the Virgin Islands to Guadeloupe are the Leeward Islands; the islands further S, from Dominica to Grenada, are the Windward Islands 14G3
Legazpi Philippines 37E2
Leghorn see **Livorno** 26C2
Le Havre the largest port on the N coast of France (pop. 255,900) 24C2
Leicester a historic cathedral city, and county town of Leicestershire (pop. 280,000) 21F5
Leicestershire a county in central England, which since 1974 has included the former county of Rutland (2553 sq km/ 986 sq miles; pop. 864,000) 21F5
Leipzig an industrial city and important cultural centre in SE Germany (pop. 559,000) 28C2
Leiria Portugal 25A2
Leitrim a county in NW Republic of Ireland, with a small strip of coast and a border with Northern Ireland; county town Carrick-on-Shannon (1525 sq km/589 sq miles; pop. 27,600) 23C2
Le Mans a university city in NW France, famous for a 24-hour annual car race (pop. 194,000) 24C2
Lemberg see **L'vov** 32D5
Lena a river which flows N across E Siberia, from close to Lake Baikal to the Laptev Sea (length 4270 km/2650 miles) 33O3
Leningrad see **Sankt Peterburg** 32E4
Lens France 24C1
León a major manufacturing city in central Mexico (pop. 675,000) 6C3
León Nicaragua 14B4
León a historic city, founded by the Romans, in NW Spain, and capital of a province of the same name (pop. city 131,200) 25A1
Leonarisso Cyprus 40C1
Lérida Spain 25C1
Léros Island Greece 27F3
Lerwick Scotland 22J7
Lesbos see **Lésvós** 27F3
Les Cayes Haiti 14E3
Leshan China 34E4
Leskovac Yugoslavia 27E2
Lesotho [area 30,355 sq km (11,720 sq miles); pop. 1,720,000; cap. Maseru; govt monarchy; religions RC, other Christianity; currency loti] a small landlocked kingdom surrounded by the Republic of South Africa 44C3
Lesser Antilles Islands Caribbean 14F4
Lésvós (Lesbos) a large Greek island in the Aegean, only 10 km (6 miles) from Turkey 27F3
Letterkenny Ireland 23D2
Levkás Island Greece 27E3
Lewes England 21H7
Lewis Island Scotland 22B2
Lexington a city in central Kentucky, USA (pop. city 210,200/metropolitan area 327,200) 8C3
Leyte Island Philippines 37E2
Lhasa the capital of Tibet. It lies 3606 m (11,830 ft) above sea level (pop. 120,000) 34D4, 39H3
Lianoyang China 35G2
Lianyungang China 35F3
Liaoyuan China 35G2
Liberec Czech Republic 28C2
Liberia [area 111,369 sq km (43,000 sq miles); pop. 2,440,000; cap. Monrovia; govt republic; religion Animism, Sunni Islam, Christianity; currency Liberian dollar] a country in West Africa between Sierra Leone and Côte d'Ivoire 42E2
Libreville the capital and main port of Gabon, originally a settlement for freed slaves (pop. 308,000) 42D4
Libya [area 1,759,540 sq km (679,358 sq miles); pop. 4,000,000; cap. Tripoli (Tarabulus); major cities Benghazi, Misurata; govt Socialist people's republic; religion Sunni Islam; currency Libyan dinar] a large N African country stretching from the Mediterranean to, and in some parts beyond, the Tropic of Cancer 42E2
Lichinga Mozambique 44D2
Liechtenstein [area 160 sq km (62 sq miles); pop. 28,181; cap. Vaduz; govt constitutional monarchy; religion RC; currency Swiss franc] a tiny central European mountainous principality on the River Rhine between Austria and Switzerland 28B3
Liège (Luik) a historic city in E Belgium, at the confluence of the Meuse and Ourthe rivers (pop. city 203,000/metropolitan area 609,000) 24D1
Liepaja Latvia 30E4
Liffey, River the river on which Dublin, capital of the Republic of Ireland, is set (length 80 km/49 miles) 23E3
Ligurian Sea the N arm of the Mediterranean Sea to the W of Italy, which includes the Gulf of Genoa 26B2
Likasi Zaïre 44C2
Lille France 24C1
Lillehammer Norway 30C3
Lilongwe the capital of Malawi, and the second largest city after Blantyre (pop. 172,000) 44D2
Lima the capital of Peru, on the River Rimac, 13 km (8 miles) from the coast (pop. 5,500,000) 16B4
Limassol the main port of Cyprus, in the S (pop. 107,200) 40B1
Limerick a city and port on the River Shannon, and county town of Limerick county (pop. 60,700) 23C4
Limerick a county in SW Republic of Ireland (2686 sq km/ 1037 sq miles; pop. 161,700) 23B4
Límnos Greece 27F3
Limoges a city in E France, famous for its porcelain, capital of Limousin region (pop. 144,200) 24C2
Limón Costa Rica 14C5
Limousin a region of E France in the foothills of the Massif Central, famous for its cattle (pop. 737,000) 24C2
Limpopo a river in South Africa flowing N to form part of the Botswana border before crossing Mozambique to the Indian Ocean (length 1610 km/1000 miles) 44D3
Linares Spain 25B2
Lincang China 34E4
Lincoln a historic cathedral city and the county town of Lincolnshire, England (pop. 77,000) 20G4
Lincoln the state capital of Nebraska, USA (pop. city

Lincolnshire   Luoyang

180,400/metropolitan area 203,000) 6D1
Lincolnshire a county on the E coast of central England; county town Lincoln (5885 sq km/2272 sq miles; pop. 558,000) 20G4
Lindisfarne *see* **Holy Island** 20F2
Linfen China 35F3
Linköping Sweden 30D4
Linosa Island Italy 26C3
Linz Austria 28C3
Lion, Golfe du (Gulf of Lions) an arm of the Mediterranean which forms a deep indent in the S coast of France 24C3
Lipari Islands (Eoli, Eolian Islands) a group of small volcanic islands which lie between the N coast of Sicily and mainland Italy (pop. 12,500) 26C3
Lisboa (Lisbon) the capital and principal port of Portugal, on the River Tagus, near the Atlantic coast (pop. 817,600) 25A2
Lisburn Northern Ireland 23E2
Lisieux France 24C2
Lithuania [*area* 65,200 sq km (25,174 sq miles); *pop.* 3,690,000; *cap.* Vilnius; *major cities* Kaunas, Klaipeda, Siauliai; *govt* republic; *religion* RC; *currency* rouble] the largest of the three former Soviet Baltic Republics, NW of the Russian Federation and Belarus 30E4
Little Rock the state capital of Arkansas, USA (pop. city 170,100/metropolitan area 492,700) 7D2
Liuzhou China 34E4
Livanátais Greece 27E3
Liverpool a major port on the estuary of the River Mersey in NW England; it is the administrative centre of Merseyside (pop. 497,000) 20E4
Livingston Scotland 22E5
Livingstone Zambia 44C2
Livno Bosnia 26D2
Livorno (Leghorn) a port and industrial city on the coast of Tuscany, N Italy (pop. 175,300) 26C2
Ljubljana an industrial city on the River Sava, and the capital of Slovenia (pop. 305,200) 26C1
Llandrindod Wells Wales 21D5
Lobito Angola 44B2
Lochboisdale Scotland 22A3
Lochgilphead Scotland 22C4
Lochinver Scotland 22C2
Lochmaddy Scotland 22A3
Locri Italy 26D3
Lódz an industrial city, second largest in Poland, in the centre of the country (pop. 848,500) 29D2
Logan, Mount the highest mountain in Canada, and second highest in North America, in SW Yukon, on the border with Alaska (5951 m/19,524 ft) 5E3
Logroño Spain 25B1
Loire, River the longest river in France, flowing N from the SE Massif Central, then W to the Atlantic Ocean W of Nantes. Its middle reaches are famous for their châteaux (length: 1020 km/635 miles) 24B2
Loja Ecuador 16B3
Loja Spain 25B2
Lolland Island Denmark 30C5
Lom Bulgaria 27E2
Lombok an island of the Lesser Sunda group, E of Bali (5435 sq km/2098 sq miles; pop. 1,300,200) 36D4
Lomé the capital and main port of Togo, situated close to the border with Ghana (pop. 283,000) 42D4
London the capital of the United Kingdom, in SE England, straddling both banks of the River Thames near its estuary (pop. 6,755,000) 21G6
Londonderry a county in Northern Ireland; county town also Londonderry (2076 sq km/801 sq miles; pop. 84,000) 23D2
Londonderry (Derry) the second largest city in Northern Ireland after Belfast, and county town of the county of Londonderry (pop. 62,000) 23D1

Longford a county in the centre of the Republic of Ireland, with a county town of the same name (1044 sq km/403 sq miles; pop. 31,100) 23D3
Long Island an island off the coast of New York State, stretching 190 km (118 miles) NE away from New York City. Its W end forms part of the city, but the rest is a mixture of residential suburbs, farmland and resort beaches (3685 sq km/1423 sq miles) 7F1
Long Island The Bahamas 14E2
Lorca Spain 25B2
Lord Howe Island Australia 45F4
Lorient France 24B2
Lorraine a region of NE France, with a border shared by Belgium, Luxembourg and Germany. The regional capital is Metz (pop. 2,320,000) 24D2
Los Angeles a vast, sprawling city on the Pacific Ocean in S California, USA, the second largest city in the USA after New York (pop. city 3,096,700/conurbation 12,372,600) 6B2
Losinj Island Croatia 26C2
Los Mochis Mexico 6C3
Lothian a local government region in SE central Scotland, with Edinburgh as its administrative centre. It was created in 1975 out of the former counties of Midlothian, and East and West Lothian (1756 sq km/678 sq miles; pop. 745,000) 22E5
Louisiana a state in central S USA, on the lower reaches of the Mississippi River, with a coastline on the Gulf of Mexico; state capital Baton Rouge (125,675 sq km/48,523 sq miles; pop. 4,481,000) 7D2
Louisville Kentucky USA 7E2
Loukhi Russian Federation 30G2
Lourenço Marques *see* **Maputo** 44D3
Louth a county on the E coast of the Republic of Ireland, bordering Northern Ireland; county town Dundalk (823 sq km/318 sq miles; pop. 88,500) 23E3
Louth England 20G4
Loznica Yugoslavia 27D2
Lualaba, River a river that flows N across E Zaïre from the border with Zambia before joining the Lomami to form the River Zaïre (length 1800 km/1120 miles) 43F5
Lu'an China 35F3
Luanda the capital of Angola, and a major port on the Atlantic Ocean (pop. 700,000) 44B1
Luang Prabang Laos 36C2
Lubango Angola 44B2
Lubbock USA 6C2
Lübeck a Baltic port in N Germany, 20 km (12 miles) from the coast on the River Trave (pop. 80,000) 28C2
Lublin (Lyublin) Poland 29E2
Lubumbashi the principal mining town of Zaïre, and capital of Shaba region in the SE. It was known as Elisabethville until 1966 (pop. 600,000) 44C2
Lucca a town in NW Tuscany, Italy (pop. 89,100) 26C2
Lucerne *see* **Luzern** 24D2, 26B1
Lucknow the capital of the state of Uttar Pradesh in central N India (pop. 1,007,600) 39G3
Lüda (Dalian) an industrial city and port in Liaoning province, NE China (pop. 4,000,000) 35G3
Lüderitz Namibia 44B3
Ludhiana India 39F2
Ludvika Sweden 30D3
Luga Russian Federation 30F4
Lugansk a major industrial city of the E Ukraine in the Donets Basin (pop. 491,000) 32F5
Lugo Spain 25A1
Luik *see* **Liège** 24D1
Luleå Sweden 30E2
Luluabourg *see* **Kananga** 43F5
Lundy Island England 21C6
Luohe China 35F3
Luoyang a city of ancient origins in E central China. As a

Lurgan | Malatya

principal centre of the Shang dynasty, the area is rich in archaeological remains (pop. 500,000) 35F3
Lurgan Northern Ireland 23E2
Lusaka the capital of Zambia, situated in the SE of the country (pop. 538,500) 44C2
Luton England 21G6
Luxembourg [*area* 2586 sq km (998 sq miles); *pop.* 378,400; *cap.* Luxembourg; *govt* constitutional monarchy; *religion* RC; *currency* Luxembourg franc] a small country bounded by Belgium on the W, France on the S and Germany on the E 24D2
Luxor a town that has grown up at one of the great archaeological sites of ancient Egypt, on the E bank of the Nile in the centre of the country (pop. 78,000) 43G2
Luzern (Lucerne) a small city set on Lake Lucerne in central Switzerland, retaining much of its medieval past; also the name of the surrounding canton (pop. city 67,500) 24D2, 26B1
Luzhou China 34E4
Luzon the largest island of the Philippines, with the nation's capital, Manila, at its centre (104,688 sq km/40,420 sq miles; pop. 29,400,000) 37E2
L'vov (Lemberg) a major industial city of medieval origins in the W Ukraine (pop. 688,000) 32D5
Lyallpur *see* **Faisalabad** 39F2
Lybster Scotland 22E2
Lycksele Sweden 30D3
Lyon (Lyons) the second largest city in France, at the confluence of the Rhône and Saone in the SE of the country (pop. 1,236,100) 24C2
Lyublin *see* **Lublin** 29E2

M

Ma'an Jordan 38B2
Maas, River *see* **Meuse** 24D2
Macapá Brazil 16D2
Macáu (Macao) a tiny Portuguese province on the S coast of China, due to be returned to China in 1999 36D1
Macclesfield England 20E4
Macdonnell Ranges parallel ranges of mountains of central Australia, near Alice Springs. The highest peak is Mount Ziel (1510 m/4954 ft) 45C3
Macedonia the most S republic of former Yugoslavia, with Skopje as its capital (25,713 sq km/9928 sq miles; pop. 1,912,200) 27E2
Maceió Brazil 16F3
Mackay Australia 45D3
Mackenzie, River a river flowing N through W Northwest Territories, Canada, from the Great Slave Lake to the Arctic (length 4250 km/2640 miles) 5F3
MacMurdo Sound Antarctica 48
Macomer Italy 26B2
Macon USA 7E2
Mâcon France 24C2
Macquarie Island New Zealand 46
Madagascar [*area* 587,041 sq km (226,657 sq miles); *pop.* 11,440,000; *cap.* Antananarivo; *major cities* Fianarantsoa, Mahajanga, Toamasina; *govt* republic; *religions* Animism, RC, Protestantism; *currency* Malagasy franc] an island state off the SE coast of Africa, the fourth largest island in the world 44E3
Madang Papua New Guinea 37G4
Madeira the main island in a small group in the E Atlantic Ocean belonging to Portugal; capital Funchal (740 sq km/ 286 sq miles; pop. 248,500) 42B1
Madison the state capital of Wisconsin, USA (pop. 170,700/metropolitan area 333,000) 7E1
Madras the main port on the E coast of India, and capital of the state of Tamil Nadu (pop. 4,289,300) 39G4

Madrid the capital of Spain, situated in the middle of the country, and also the name of the surrounding province (pop. city 3,188,300/province 4,727,000) 25B1
Madura an island off the NE coast of Java in Indonesia (5290 sq km/2042 sq miles; pop. 1,860,000) 36D4
Madurai a textile city in Tamil Nadu, in the S tip of India (pop. 907,700) 39F5
Mae Nam Khong, River *see* **Salween, River** 36B2
Mafia Island Tanzania 44D1
Mafikeng (Mafeking) a town in the state of Bophuthatswana, South Africa (pop. 29,400) 44C3
Magaden Russian Federation 33R4
Magdalena, River a river which flows N through W Colombia and into the Caribbean at Barranquila (length 1550 km/965 miles) 16B2
Magdeburg a city and inland port on the River Elbe in E Germany (pop. 289,000) 28C2
Magellanes, Estrecho de (Magellan, Strait of) the waterway, 3 km (2 miles) across at its narrowest point, separating Tierra Del Fuego from mainland South America, discovered by the Portuguese explorer Ferdinand Magellan in 1520 17C8
Magnitogorsk a steel-making town, in the S Urals, Russian Federation (pop. 421,000) 32G4
Mahajanga Madagascar 44E2
Mahalapye Botswana 44C3
Mahón Spain 25C2
Mahore *see* **Mayotte** 44E2
Maidstone England 21H6
Maiduguri Nigeria 42E3
Maine (province) France 24B2
Maine a state in the NE corner of the USA, bordering Canada; state capital Augusta (86,027 sq km/33,215 sq miles; pop. 1,164,000) 7F1
Mainz a city and inland port on the confluence of the Rivers Rhine and Main in W central Germany (pop. 185,000) 28B3
Maitland Australia 45E4
Maizuru Japan 35L9
Majene Indonesia 36D4
Majorca (Mallorca) the largest of the Balearic Islands, in the W Mediterranean; capital Palma (3639 sq km/1405 sq miles; pop. 460,000) 25C2
Makarska Croatia 26D2
Makassar *see* **Ujung Padang** 36D4
Makassar Strait the broad stretch of water, 130 km (81 miles) across at its narrowest, which separates Borneo and Sulawesi in Indonesia 36D4
Makhachkala a port and industrial city on the W coast of the Caspian Sea, and capital of the republic of Dagestan in the Russian Federation (pop. 269,000) 32F5
Makkah (Mecca) a city in central W Saudi Arabia, 64 km (40 miles) E of the Red Sea port of Jiddah. It was the birthplace of the Prophet Mohammed, and is the holiest city of Islam (pop. 375,000) 38B3
Makó Hungary 29E3
Makurdi Nigeria 42D4
Malabo a port and capital of Equatorial Guinea, on the N coast of Bioko Island (pop. 37,200) 42D4
Malacca *see* **Melaka** 36C3
Malacca, Strait of a busy waterway, 50 km (31 miles) wide at its narrowest, separating Sumatra in Indonesia from Malaysia, with Singapore at its E end 36C3
Malaga a port, manufacturing city, and tourist resort on the Mediterranean coast of Andalucia, S Spain. Also the name of the province of which it is the capital (pop. city 503,300) 25B2
Malagasy Republic *see* **Madagascar** 44E3
Malakal Sudan 43G4
Malang Indonesia 36D4
Malanje Angola 44B2
Malatya Turkey 38B2

Malawi [area 118,484 sq km (45,747 sq miles); pop. 7,980,000; cap. Lilongwe; major cities Blantyre, Mzuzu, Zomba; govt republic; religions Animism, RC, Presbyterianism; currency kwacha] a country in Africa, along the S and W shores of Lake Malawi 44D2
Malawi (Nyasa), Lake a long, narrow lake, the third largest in Africa, running down most of the E side of Malawi and forming its border with Tanzania and Mozambique (23,300 sq km/9000 sq miles) 44D2
Malaysia [area 329,749 sq km (127,316 sq miles); pop. 17,810,000; cap. Kuala Lumpur; major cities Ipoh, Georgetown, Johor Baharu; govt federal constitutional monarchy; religion Sunni Islam; currency Malaysian ringgit] a federation in SE Asia comprising Peninsular Malaysia and the states of Sabah and Sarawak on the island of Borneo 36C3
Malden Island Kiribati 46
Maldives [area 298 sq km (115 sq miles); pop. 214,139; cap. Malé; govt republic; religion Sunni Islam; currency rufiyaa] a republic in the Indian Ocean, comprising 1200 lowlying coral islands grouped into 12 atolls 39F5
Malekula Island Vanuatu 45F2
Mali [area 1,240,192 sq km (478,838 sq miles); pop. 9,090,000; cap. Bamako; other major cities Segou, Mopti; govt republic; religions Sunni Islam, Animism; currency franc CFA] a landlocked state in West Africa 42C3
Mallaig Scotland 22C3
Mallorca see **Majorca** 25C2
Mallow Ireland 23C4
Malmö a port in SW Sweden, on the channel separating Sweden from Denmark (pop. 229,900) 30C4
Malta [area 316 sq km (122 sq miles); pop. 354,900; cap. Valletta; govt republic; religion RC; currency Maltese pound] a state in the Mediterranean, S of Sicily, comprising the islands of Malta, Gozo and Comino 26C3
Malton England 20G3
Maluku (Moluccas) a group of some 1000 islands in E Indonesia, known as the Spice Islands. The capital is Ambon (74,505 sq km/28,766 sq miles; pop. 1,411,000) 37E4
Mamou Guinea 42B3
Man Côte d'Ivoire 42C4
Man, Isle of an island of the British Isles, in the Irish Sea. It is a British Crown possession, not a part of the UK, and has its own parliament; capital Douglas (585 sq km/226 sq miles; pop. 66,000) 20C3
Mana Hawaii USA 6H
Manacor Spain 25C2
Manado Indonesia 37E3
Managua the capital of Nicaragua, situated on the edge of Lake Managua (pop. 630,000) 14B4
Manakara Madagascar 44E3
Manaus a major port on the River Amazon in Brazil, 1600 km (1000 miles) from the sea (pop. 635,000) 16C3
Manchester a major industrial and commercial city in NW England, connected to the Mersey estuary by the Manchester Ship Canal (pop. 448,000) 20E4
Mandal Norway 30B4
Mandalay the principal city of central Myanmar and a port on the River Irrawaddy (pop. 417,300) 36B1
Manfredonia Italy 26D2
Mangalia Romania 27F2
Mangalore India 39F4
Manila the capital of the Philippines, an important port and commercial centre, sited on Luzon island (pop. 6,000,000) 37E2
Manitoba the most E of the prairie provinces of Canada; capital Winnipeg (650,087 sq km/250,998 sq miles; pop. 1,026,000) 5H4
Manizales Colombia 16B2
Mannheim an inland port and industrial city of Germany, on the confluence of the Rivers Rhine and Neckar (pop. 300,000) 28B3

Manokwari Indonesia 37F4
Mansfield England 20F4
Manta Ecuador 16B3
Mantes France 24C2
Manzanillo Cuba 14D2
Manzhouli China 35F2
Maoming China 35F4
Maputo capital and main port of Mozambique, formerly known as Lourenço Marques (pop. 785,500) 44D3
Maracaibo the second largest city in Venezuela, in the NW (pop. 1,100,000) 16B1
Maradi Niger 42D3
Marbella a popular resort on the Mediterranean coast of S Spain (pop. 67,900) 25B2
Marburg Germany 28B2
Mardan Pakistan 39F2
Mar del Plata a coastal city and beach resort on the NE coast of Argentina (pop. 424,000) 17D6
Margate England 21J6
Maribor (Marburg) an industrial city in Slovenia (pop. 185,700) 26D1
Marie-Galante Island Caribbean Sea 14G3
Mariestad Sweden 30C4
Marília Brazil 16E5
Mariupol Ukraine 32E5
Marmara, Sea of a small sea between the Dardanelles and the Bosphorus, in the link between the Mediterranean and the Black Sea. The surrounding coasts all belong to Turkey 27F2
Marmaris Turkey 27F3
Maroua Cameroon 42E3
Marquesas Islands see **Isles Marquises** 46
Marrakech (Marrakesh) a historic oasis city in central W Morocco, founded in the 11th century and formerly the country's capital (pop. 440,000) 42C1
Marseille (Marseilles) the largest port in France, on the Mediterranean coast, and France's third largest city after Paris and Lyons (pop. 1,110,500) 24D3
Marshall Islands a scattered group of some 1250 islands in Micronesia, in the W Pacific Ocean. They form a self-governing republic which remains in free association with the USA. 46
Martinique one of the larger of the islands in the Windward Islands group in the S Caribbean, administered as a department of France 14G4
Mary Turkmenistan 32H6
Maryland a state on the central E coast of the USA, almost divided by Chesapeake Bay; capital Annapolis (27,394 sq km/10,577 sq miles; pop. 4,392,000) 7F1
Masaya Nicaragua 14B4
Masbate Island Philippines 37E2
Maseru the capital of Lesotho (pop. 45,000) 44C3
Mashhad (Meshed) a major trading centre and capital of Khorasan province, NE Iran (pop. 1,120,000) 38D2
Masírah Island Oman 38D3
Masqat (Muscat) the historic capital of Oman. Its neighbouring port, Muttrah, has developed rapidly to form the commercial centre (pop. with Muttrah 80,000) 38D3
Massa Italy 26C2
Massachusetts one of the New England states on the NE coast of the USA; capital Boston (21,386 sq km/8257 sq miles; pop. 5,822,000) 7F1
Massif Central the rugged upland region occupying much of S central France W of the Rhone. The highest point is at Puy de Sancy (1885 m/6184 ft) 24C2
Matadi Zaïre 42E5
Matagalpa Nicaragua 14B4
Matamoros Mexico 6D3
Matanzas Cuba 14C2
Mataram Indonesia 36D4
Matlock England 20F4
Matsue Japan 35H3

Matsumoto Japan 35M8
Matsusaka Japan 35M9
Matsuyama Japan 35H3
Matterhorn (Monte Cervino) a distinctive, pyramid-shaped peak on the border between Italy and Switzerland, S of Zermatt (4477 m/14,688 ft) 24D2
Maui the second largest island of Hawaii, USA (1885 sq km/727 sq miles; pop. 63,000) 6H
Mauritania [area 1,025,520 sq km (395,953 sq miles); pop. 1,970,000; cap. Nouakchott; govt republic; religion Sunni Islam; currency ouguiya] a country on the W coast of Africa, about 47% of which is desert 42B2
Mauritius [area 2040 sq km (788 sq miles); pop. 1,081,669; cap. Port Louis; govt constitutional monarchy; religions Hinduism, RC, Sunni Islam; currency Mauritius rupee] an island state E of Madagascar 44F4
Mayaguana Island The Bahamas 14E2
Maybole Scotland 22D5
Mayo a county on the W coast of the Republic of Ireland; county town Castlebar (4831 sq km/1865 sq miles; pop. 114,700) 23B3
Mayotte (Mahore) France 44E2
Mazar-e Sharif Afghanistan 39E2
Mazatlán Mexico 6C3
Mbabane the capital of Swaziland (pop. 36,000) 44D3
Mbandaka Zaïre 43F4
Mbarara Uganda 43G5
Mbeya Tanzania 44D1
Mbuji-Mayi Zaïre 43F5
McKinley, Mount the highest mountain in North America, in S Alaska, USA (6194 m/20,320 ft) 5C3
Meath a county on the E coast of the Republic of Ireland, N of Dublin; county town Navan (2336 sq km/902 sq miles; pop. 95,400) 23D3
Meaux France 24C2
Mecca see **Makkah** 38B3
Medan Indonesia 36B3
Medellín the second largest city in Colombia after Bogota, in the centre of the country (pop. 1,998,000) 16B2
Medgidia Romania 27F2
Medicine Hat Canada 5G4
Medina see **Al Madinah** 38B3
Mediterranean Sea a large sea bounded by S Europe, N Africa and SW Asia. It is connected to the Atlantic Ocean by the Strait of Gibraltar 25, 26-7
Meerut an industrial town of N India, 60 km (40 miles) NE of Delhi. The Indian Mutiny began here in 1857 (pop. 536,600) 39F3
Meiktila Myanmar 36B1
Meknès a former capital, with a fine 17th-century royal palace, in N Morocco (pop. 320,000) 42C1
Mekong, River a great river of SE Asia, flowing from Tibet, through S China, Laos and Cambodia before forming a massive and highly fertile delta in S Vietnam and flowing into the South China Sea (length 4184 km/2562 miles) 36C1
Melaka (Malacca) a port on the SW coast of Malaysia, overlooking the Straits of Malacca (pop. 87,500) 36C3
Melbourne the second largest city in Australia, and capital of the state of Victoria (pop. 2,700,000) 45D4
Melilla Spain 25B2
Melitopol' Ukraine 32E5
Melo Uruguay 17D6
Melos see **Mílos** 27E3
Melun France 24C2
Melvich Scotland 22E2
Melville Island Australia 45C2
Melville Island Canada 5G2
Melville Peninsula Canada 5D3
Memel see **Klaipeda** 32D4
Memphis a city on the River Mississippi in SW Tennessee, USA, on the border with and extending into Arkansas (pop. city 648,000/metropolitan area 934,600) 7E2
Menai Strait a narrow strait, 180 m (590 ft) across at its narrowest, separating mainland Wales from Anglesey, spanned by road and rail bridges 20C4
Mende France 24C3
Mendoza Argentina 17C6
Menongue Angola 44B2
Menorca see **Minorca** 25C2
Menzanares Spain 25B2
Merauke Indonesia 37G4
Mercedes Argentina 17C6
Mergui Archipelago Myanmar 36B2
Mérida the historic capital of the Yucatan province of E Mexico (pop. 424,500) 7E3
Mérida Spain 25A2
Mersey, River a river in NW England, forming an estuary S of Liverpool deep and wide enough to permit access for ships to Liverpool and Manchester via the Manchester Ship Canal (length 110 km/70 miles) 20E4
Merseyside a metropolitan county created in 1974 out of parts of Lancashire and Cheshire, with Liverpool as its administrative centre (652 sq km/252 sq miles; pop. 1,501,000) 20E4
Merthyr Tydfil Wales 21D6
Meshed see **Mashhad** 38D2
Mesolóngian (Missolonghi) Greece 27E3
Messina a port, founded in the 8th century BC, in NE Sicily, overlooking the narrow Strait of Messina (6 km/4 miles wide at its narrowest) which separates Sicily from mainland Italy (pop. 266,300) 26D3
Metz capital of the Lorraine region in E France, situated on the River Moselle (pop. 194,800) 24D2
Meuse (Maas), River a river which flows NW from the Lorraine region of France, across central Belgium and into the Netherlands, where it joins the delta of the Rhine before entering the North Sea (length 935 km/580 miles) 24D2
Mexicali USA 6B2
Mexico [area 1,958,201 sq km (756,061 sq miles); pop. 81,140,000; cap. México City; major cities Guadalajara, Monterrey, Puebla de Zaragoza; govt federal republic; religion RC; currency : Mexican peso] the most southerly country in North America 6D3
Mexico, Gulf of an arm of the Atlantic Ocean, bounded by the Florida peninsula in the SE USA and the Yucatan peninsula in Mexico, with the island of Cuba in the middle of its entrance 7D3
México City the capital of Mexico, the most populous city in the world, in the S of the country on a plateau 2200 m (7350 ft) above sea level (pop. 17,000,000) 6D4
Meymaneh Afghanistan 39E2
Miami a major city and resort on the Atlantic coast of SE Florida, USA (pop. city 372,600/metropolitan area 1,706,000) 7E3
Mianyany China 34E3
Michigan a state in N central USA, formed out of two peninsulas between the Great Lakes, with Lake Michigan in the middle; capital Lansing (150,780 sq km/58,216 sq miles; pop. 9,088,000) 7E1
Michigan, Lake one of the Great Lakes, and the only one to lie entirely within the USA (57,750 sq km/22,300 sq miles) 7E1
Michurin Bulgaria 27F2
Middlesbrough the county town of Cleveland, England (pop. 149,800) 20F3
Mid Glamorgan a county in central S Wales, which was formed in 1974 out of part of the former counties of Breconshire, Glamorgan and Monmouthshire; administrative centre Cardiff (1000 sq km/393 sq miles; pop. 538,000) 21D6
Midway Islands two atolls belonging to the USA, in the N Pacific Ocean, 2000 km (1242 miles) NW of Hawaii

Mikkeli 1017 Monza

(3 sq km/2 sq miles; pop. 2200) 46
Mikkeli Finland 30F3
Mikonos Island Greece 27F3
Milano (Milan) the major industrial and commercial centre of N Italy, and the country's second largest city after Rome (pop. 1,605,000) 26B1
Mildura Australia 45E4
Milford Haven Wales 20B6
Millau France 24C3
Mílos Greece 27E3
Milton Keynes England 21G5
Milwaukee a port on the W side of Lake Michigan, and the main industrial centre of Wisconsin, USA (pop. city 620,800/metropolitan area 1,393,800) 7E1
Minatinán Mexico 7D4
Minch, The broad channel separating NW Scotland from the Western Isles 19B2
Mindanao the second largest island of the Philippines (94,631 sq km/36,537 sq miles; pop. 11,100,000) 37E3
Mindoro an island in W central Philippines (9736 sq km/ 3759 sq miles) 37E2
Minna Nigeria 42D4
Minneapolis a major agricultural and commercial centre in SE Minnesota, USA, on the River Mississippi, and adjoining St Paul (pop. city 258,300/ metropolitan area 2,230,900) 7D1
Minnesota a state in N central USA; capital St Paul (217,736 sq km/84,068 sq miles; pop. 4,193,000) 7D1
Minorca (Menorca) the second largest of the Balearic Islands (after Majorca); capital Mahon (702 sq km/271 sq miles; pop. 50,200) 25C2
Minsk a major industrial city, and the capital of Belarus (pop. 1,442,000) 32D4
Miranda de Ebro Spain 25B1
Miri Malaysia 36D3
Mirzapur India 39G3
Miskolc a city in NE Hungary, and the country's second largest city after Budapest (pop. 210,000) 29E3
Misoöl Indonesia 37F4
Misratah Libya 42E1
Mississippi a state in central S USA with a small coastline on the Gulf of Mexico. The state capital is Jackson (123,585 sq km/47,716 sq miles; pop. 2,613,000) 7D2
Mississippi, River the second longest river in the USA. It rises in Minnesota and runs S to the Gulf of Mexico (length 3779 km/2348 miles) 7D2
Missolonghi see **Mesolóngion**
Missouri a state in the Midwest of the USA; state capital Jefferson City (180,487 sq km/69,686 sq miles; pop. 5,029,000) 7D2
Missouri, River the main tributary of the Mississippi from which it is the longest river in N America, rising in Montana and flowing N, E and SE to join the Mississippi at St Louis (length 3969 km/2466 miles) 10D3
Mito Japan 35P8
Mits'iwa Ethiopia 43G3
Miyako Japan 35P7
Miyazaki Japan 35H3
Mizusawa Japan 35P7
Mjölby Sweden 30D4
Mlawa Poland 29E2
Mljet Island Croatia 26D2
Mobile Alabama USA 7E2
Mobutu Sese Seko, Lake see **Albert, Lake** 43G4
Moçambique Mozambique 44E2
Modena Italy 26C2
Moffat Scotland 22E5
Mogadishu see **Muqdisho** 43H4
Mogilev Belarus 32E4
Mohave Desert see **Mojave Desert** 13C4
Mo-i-Rana Norway 30C2
Mojave (Mohave) Desert a desert in S California, stretching

from Death Valley to Los Angeles (38,850 sq km/15,000 sq miles) 13C4
Mokp'o South Korea 35G3
Moldavia see **Moldova** 29F3
Molde Norway 30B3
Moldova (Moldavia) [area 33,700 sq km (13,000 sq miles); pop. 4,052,000; cap. Kishinev; major cities Tiraspol, Bendery; govt republic; religion Russian Orthodox; currency rouble] a Soviet socialist republic from 1940 until 1991 when it became independent. It is bounded by Romania and Ukraine 29F3
Mollendo Peru 16B4
Molokai Island Hawaii USA 6H
Molotov see **Perm'** 32G4
Moluccas see **Maluku** 37E4
Mombasa the second city of Kenya and an important port on the Indian Ocean (pop. 500,000) 43G5
Monaco [area 195 hectares (48 acres); pop. 29,876; cap. Monaco-Ville; govt constitutional monarchy; religion RC; currency: franc] a tiny principality on the Mediterranean, surrounded landwards by France 24D3
Monaghan a county in the central N of the Republic of Ireland, with a county town of the same name (1291 sq km/ 498 sq miles; pop. county 51,200) 23D2
Mondovi Italy 26B2
Mongolia [area 1,566,500 sq km (604,826 sq miles); pop. 2,095,000; cap. Ulan Bator (Ulaanbaatar); major cities Darhan, Erdenet; govt republic; religion Buddhism; but this religion is now suppressed; currency tugrik] a landlocked country in NE Asia, bounded to the N by the Russian Federation and by China to the S, W and E 33L5
Mongu Zambia 44C2
Monopoli Italy 27D2
Monreal del Campo Spain 25B1
Monrovia the capital and principal port of Liberia (pop. 425,000) 42B4
Montana a state in NW USA, on the border with Canada; state capital Helena (381,087 sq km/147,138 sq miles; pop. 826,000) 6B1
Montargis France 24C2
Montauban France 24C3
Montbéliard France 24D2
Monte Cristi Haiti 14E3
Montego Bay Jamaica 14D3
Montenegro (Crna Gora) the smallest of the republics of former Yugoslavia, in the SW on the Adriatic Sea and bordering Albania. The capital is Titograd (13,812 sq km/ 5331 sq miles; pop. 584,300) 27D2
Montería Colombia 16B2
Monterrey an industrial city in NE Mexico, the country's third largest city after Mexico City and Guadaljara (pop. 1,916,500) 6C3
Montes Claros Brazil 16E4
Montevideo the capital of Uruguay, and an important port on the River Plate estuary (pop. 1,500,000) 17D6
Montgomery the state capital of Alabama, USA (pop. city 185,000/metropolitan area 284,800) 7E2
Montluçon France 24C2
Montpelier the state capital of Vermont, USA (pop. 8200) 7F1
Montpellier a university and trading city in central S France (pop. 225,300) 24C3
Montréal the second largest city in Canada after Toronto, on the St Lawrence River, in the S of the province of Quebec. Two-thirds of the population are French-speaking Québecois (pop. 2,828,250) 5L5
Montrose Scotland 22F4
Montserrat a British Crown colony in the Leeward Islands in the S Caribbean; capital Plymouth (102 sq km/39 sq miles) 14G3
Monza Italy 26B1

Mopti Mali 42C3
Mora Sweden 30C3
Moradabad India 39F3
Moray Firth an inlet of the North Sea cutting some 56 km (35 miles) into the E coast of NE Scotland, with Inverness at its head 19C2
Morioka Japan 35P7
Morocco [*area* 446,550 sq km (172,413 sq miles); *pop.* 24,500,000; *cap.* Rabat; *major cities* Casablanca, Fez, Marrakech; *govt* constitutional monarchy; *religion* Sunni Islam; *currency* dirham] a country in NW Africa, with the Atlas mountains in the N, the Sahara in the S, and Atlantic and Mediterranean coasts 42C1
Moroni the capital of the Comoros islands (pop. 20,000) 44E2
Morotai Island Indonesia 37E3
Morphou Cyprus 40B1
Morwell Australia 45D4
Moscow *see* **Moskva** 32E4
Moselle (Mosel), River a river which flows N from SE Lorraine in France to form part of the border between Luxembourg and Germany before flowing E to the Rhine at Koblenz (length 550 km/340 miles) 28B3
Moshi Tanzania 43G5
Mosjöen Norway 30C2
Moskva (Moscow) the capital of the Russian Federation, on the Moskva River. It is an ancient city with a rich heritage, and is the political, industrial and cultural focus of the country (pop. 8,600,000) 32E4
Moss Norway 30C4
Mossoró Brazil 16F3
Mostaganem Algeria 42D1
Mostar Bosnia 27D2
Mosul *see* **Al Mawsil** 38B2
Motherwell Scotland 22E5
Motril Spain 25B2
Moulins France 24C2
Moulmein Myanmar 36B2
Moundou Chad 42E4
Mount Gambier Australia 45D4
Mount Isa Australia 45C3
Mourne Mountains a range of noted beauty in the S of County Down, Northern Ireland. The highest point is Slieve Donard (852 m/2795 ft) 23E2
Mozambique [*area* 801,590 sq km (309,494 sq miles); *pop.* 14,900,000; *cap.* Maputo; *major cities* Beira, Nampula; *govt* republic; *religions* Animism, RC, Sunni Islam; *currency* metical] a republic located in SE Africa 44D3
Mozambique Channel a broad strait, some 400 km (250 miles) across at its narrowest, separating Madagascar from mainland Africa 44D3
Mt Magnet Australia 45A3
Mtwara Tanzania 44E2
Mudanjiang China 35G2
Mufulira Zambia 44C2
Muhos Finland 30F3
Mulhouse an industrial city in Alsace, E France (pop. 222,700) 24D2
Mull an island off the central W coast of Scotland (925 sq km/357 sq miles) 22C4
Mullingar Ireland 23D3
Multan Pakistan 39F2
Muna Island Indonesia 37E4
München (Munich) a historic and industrial city in S Germany, capital of Bavaria (pop. 1,300,000) 28B3
Mungbere Zaïre 43F4
Munich *see* **München** 28B3
Münster Germany 28B2
Muonio Finland 30E2
Muqdisho (Mogadishu) the capital and main port of Somalia (pop. 400,000) 43H4
Murcia a trading and manufacturing city in SE Spain,

and capital of the province of the same name (pop. city 288,600) 25B2
Murmansk the largest city N of the Arctic Circle, a port and industrial centre on the Kola Peninsula in the NW corner of the Russian Federation (pop. 412,000) 32E3
Murray, River a major river of SE Australia, flowing W from the Snowy Mountains to form much of the New South Wales-Victoria boundary 45D4
Muscat *see* **Masqat** 38D3
Musselburgh Scotland 22E5
Mutare Zimbabwe 44D2
Mwanza Tanzania 43G5
Mwene Ditu Zaïre 43F5
My Tho Vietnam 36C2
Myanmar, Union of (formerly **Burma**) [*area* 676,578 sq km (261,227 sq miles); *pop.* 39,300,000; *cap.* Yangon (formerly Rangoon); *major cities* Mandalay, Moulmein, Pegu; *govt* republic; *religion* Buddhism; *currency* kyat] the second largest country in SE Asia, on the Bay of Bengal and bordering Thailand to the E 36B1
Myingyan Myanmar 36B1
Myitkyina Myanmar 36B1
Mymensingh Bangladesh 39H3
Mysore India 39F4

N

Naas Ireland 23E3
Naga Philippines 37E2
Nagano Japan 35N8
Nagaoka Japan 35N8
Nagasaki a port and industrial city on the W coast of Kyushu Island, Japan. The second atomic bomb was dropped here (August 9, 1945), killing 40,000 people (pop. 446,300) 35G3
Nagercoil India 39F5
Nagoya Japan 35M9
Nagpur an important commercial centre and textile manufacturing city on the Deccan plateau, India (pop. 1,302,100) 39F3
Nagykanizsa Hungary 38D3
Nain Canada 5M4
Nairn Scotland 22E3
Nairobi the capital of Kenya and a commercial centre, in the SW highland region (pop. 1,250,000) 43G5
Nakhodka Russian Federation 33P5
Nakhon Ratchasima Thailand 36C2
Nakhon Sawan Thailand 36C2
Nakhon Si Thammarat Thailand 36B3
Nakuru Kenya 43G5
Namangan Kyrgyzstan 32J5
Nam Dinh Vietnam 36C1
Namib Desert a desert lining the coast of Namibia 44B3
Namibe Angola 44B2
Namibia [*area* 824,292 sq km (318,259 sq miles); *pop.* 1,290,000; *cap.* Windhoek; *govt* republic; *religions* Lutheranism, RC, other Christianity; *currency* rand] a country on the Atlantic coast of SW Africa 44B3
Nampula Mozambique 44D2
Nanchang China 35F4
Nanchong China 34E3
Nancy a manufacturing city in NE France, and former capital of Lorraine (pop. 314,200) 24D2
Nanjing (Nanking) a major industrial and trading city on the lower reaches of the Chang Jiang river in central E China (pop. 3,551,000) 35F3
Nanning the capital of the Guangxi-Zhuang autonomous region in SE China (pop. 607,500) 34E4
Nanping China 35F4
Nansei-shoto (Ryukyu Islands) a chain of islands belonging

Nantes 1019 Nicaragua

to Japan stretching 1200 km (750 miles) between Japan and Taiwan (pop. 1,366,600) 35G4
Nantes France 24B2
Nantong China 35G3
Nanyang China 35F3
Napoli (Naples) the third largest city in Italy after Rome and Milan, and a port situated on the spectacular Bay of Naples (pop. 1,203,900) 26C2
Narbonne France 24C3
Narva Estonia 30F4
Narvik Norway 30D2
Nar'yan Mar Russian Federation 32G3
Nashville the state capital of Tennessee, USA, an industrial city famed for its Country and Western music (pop. city 462,500/ metropolitan area 890,300) 7E2
Nassau the capital of the Bahamas, on the N side of New Providence Island (pop. 120,000) 14D1
Nasser, Lake a massive artificial lake on the River Nile in S Egypt, created when the Aswan High Dam was completed in 1971 (5000 sq km/1930 sq miles) 43G2
Natal a port city on the NE tip of Brazil, capital of the state of Rio Grande do Norte (pop. 417,000) 16F3
Natuna Besar Island Indonesia 36C3
Nauru [area 21 sq km (8 sq miles); pop. 8100; cap. Yaren; govt republic; religions Protestantism, RC; currency Australian dollar] the world's smallest republic, a coral island just S of the Equator, halfway between Australia and Hawaii 46
Navarra (Navarre) a province in the mountainous NE part of Spain. The capital is Pamplona (10,420 sq km/4023 sq miles; pop. 507,400) 25B1
Náxos a fertile island in the S Aegean, the largest of the Cyclades (428 sq km/165 sq miles) 27F3
Ndjamena (N'Djamena) the capital of Chad (formerly Fort Lamy) in the SE of the country (pop. 303,000) 42E3
Ndola Zambia 44C2
Neagh, Lough the largest freshwater lake in the British Isles, in the E of Northern Ireland (381 sq km/147 sq miles) 19B3
Neápolos Greece 27E3
Near Islands USA 6J
Nebraska a state in the Midwest of the USA; capital Lincoln (200,018 sq km/77,227 sq miles; pop. 1,606,000) 6C1
Negev a desert in S Israel 40C3
Negros the fourth largest island of the Philippines (12,704 sq km/4905 sq miles; pop. 2,750,000) 37E3
Neiva Colombia 16B2
Nellore India 39G4
Nelson England 20E4
Nelson New Zealand 45G5
Nenagh Ireland 23C4
Nepal [area 140,797 sq km (54,362 sq miles); pop. 18,000,000; cap. Kathmandu; govt constitutional monarchy; religion Hinduism, Buddhism; currency Nepalese rupee] a long narrow country, landlocked between China and India on the flanks of the E Himalayas 39G3
Netherlands, The (area 40,844 sq km (15,770 sq miles); pop. 14,890,000; cap. Amsterdam; seat of government The Hague (Den Haag, 's-Gravenhage); major cities Eindhoven, Rotterdam; govt constitutional monarchy; religions RC, Dutch reformed, Calvinism; currency guilder] a country in NW Europe, bounded by the N and W by the North Sea 28A2
Neubrandenburg Germany 28C2
Neumünster Germany 28B2
Neuquén Argentina 17C6
Nevada a state in the W of the USA, consisting mostly of desert; state capital Carson City (286,298 sq km/110,540 sq miles; pop. 936,000) 6B2
Nevers France 24C2
Nevis see **St Kitts and Nevis** 14G3

Newark a major port city in New Jersey (pop. 314,400/ metropolitan area 1,875,300) 7F1
Newark-on-Trent England 20G4
New Britain the largest offshore island belonging to Papua New Guinea, in the Bismarck Archipelago (36,500 sq km/ 14,100 sq miles; pop. 237,000) 37G4
New Brunswick a province on the coast in SE Canada, bordering the USA; capital Fredericton (73,436 sq km/28,354 sq miles; pop. 696,000) 5M5
New Caledonia see **Nouvelle Calédonie** 45E1
Newcastle a port and industrial city in New South Wales, Australia (pop. 259,000) 45E4
Newcastle upon Tyne a historic and industrial city in Tyne and Wear, NE England (pop. 280,000) 20F3
New Delhi the official capital of India as of 1931 (pop. 273,000) 39F3
Newfoundland a province in the extreme E of Canada; capital St John's (372,000 sq km/143,634 sq miles; pop. 568,000) 5N4
New Georgia Solomon Islands 45E1
New Guinea one of the world's largest islands, divided into two: independent Papua New Guinea in the E and Irian Jaya, a state of Indonesia, in the W 37F4
New Hampshire a state of New England, in the NW of the USA; state capital Concord (24,097 sq km/9304 sq miles; pop. 998,000) 7F1
New Hebrides see **Vanuatu** 45F2
New Jersey a state on the Atlantic coast in the NE of the USA; state capital Trenton (20,295 sq km/7836 sq miles; pop. 7,562,000) 7F1
New Mexico a state in the SW of the USA, bordering Mexico; state capital Santa Fe (315,115 sq km/121,666 sq miles; pop. 1,450,000) 6C2
New Orleans an important and historic port in S Louisiana, on the Mississippi delta (pop. city 559,100/metropolitan area 1,318,800) 7E3
Newport Isle of Wight 21F7
Newport Wales 21E6
Newquay England 20B7
New Ross Ireland 23E4
Newry Northern Ireland 23E2
New South Wales the most populous state of Australia, in the SE of the country; capital Sydney (801,430 sq km/ 309,433 sq miles; pop. 5,379,000) 45D4
Newtown-Abbey Northern Ireland 23F2
Newton Aycliffe England 20F3
Newton Stewart Scotland 22D6
New York (City) the most populous city in the USA, its most important port, and a major financial centre. It is sited at the mouth of the Hudson River (pop. city 7,322,600/metropolitan area 8,376,900) 7F1
New York (State) a populous state in NE USA, on the Atlantic coast; state capital Albany (128,402 sq km/49,576 sq miles; pop. 17,783,000) 7F1
New Zealand [area 270,986 sq km (104,629 sq miles); pop. 3,390,000; cap. Wellington; major cities Auckland, Christchurch, Dunedin, Hamilton; govt constitutional monarchy; religions Anglicanism, RC, Presbyterianism; currency New Zealand dollar] a country, lying SE of Australia in the South Pacific, comprising two large main islands 45G5
Ngaliema, Mount see **Ruwenzori** 43G4
Ngaoundéré Cameroon 42E4
Nguru Nigeria 42E3
Nha Trang Vietnam 36C2
Niagara Falls spectacular waterfalls on the Niagara River, on the Canada-USA border between Lakes Erie and Ontario 7F1
Niamey the capital of Niger (pop. 400,000) 42D3
Nias Island Indonesia 36B4
Nicaragua [area 130,000 sq km (50,193 sq miles); pop. 3,750,000; cap. Managua; govt republic; religion RC; cur-

rency córdoba] the largest of the countries on the isthmus of Central America, between Honduras and Costa Rica 14B4
Nicaragua, Lago de (Lake Nicaragua) a large lake in SW Nicaragua (8264 sq km/3191 sq miles) 14B4
Nice a city, port and famous resort town of the Côte d'Azur, SE France (pop. 451,500) 24D3
Nicobar Island India 39H5
Nicosia the capital of Cyprus, situated in the centre of the island (pop. 161,100) 38B2
Niger [*area* 1,267,000 sq km (489,189 sq miles); *pop.* 7,450,000; *cap.* Niamey; *govt* republic; *religion* Sunni Islam; *currency* franc CFA] a landlocked republic in W Africa 42D3
Niger, River a river in W Africa flowing through Guinea, Mali, Niger and Nigeria to the Gulf of Guinea (length 4170 km/2590 miles) 42D3
Nigeria (*area* 923,768 sq km (356,667 sq miles); *pop.* 118,700,000; *cap.* Abuja; *major cities* Ibadan, Kano, Ogbomsho; *govt* federal republic; *religions* Sunni Islam, Christianity; *currency* Nairais] a large and populous country in W Africa 42D4
Niigata Japan 35N8
Nijmegen a city of E central Netherlands, close to the border with Germany (pop. 234,000) 28B2
Nikel Russian Federation 30G2
Nikolayev a port and industrial city on the N coast of the Black Sea, in the Ukraine (pop. 480,000) 32E5
Nile, River (An Nil) a major river of Africa and the longest in the world, rising in Burundi, flowing into Lake Victoria and then N through Uganda, Sudan and Egypt to its delta on the Mediterranean. It is called the White Nile (Bahr el Abiad) until it reaches Khartoum, where it is joined by the Blue Nile (Bahr el Azraq), which rises in Ethiopia (length 6695 km/4160 miles) 43G2
Nîmes France 24C3
Ningbo China 35G4
Nioro du Sahel Mali 42C3
Niort France 24B2
Nis (Nish) a historic city in the E of Serbia (pop. 230,000) 27E2
Nitra Slovakia 29D3
Niue Island Pacific Ocean 46
Nivernais (province) France 24C2
Nizamabad India 39F4
Nizhniy Novgorod an industrial city in the Russian Federation on the River Volga, formerly known as Gor'kiy (Gorky) (pop. 1,392,000) 32F4
Nizhniy Tagil Russian Federation 32H4
Nkongsamba Cameroon 42D4
Nong Khai Thailand 36C2
Nordkapp (North Cape) one of Europe's most northerly points—500 km (310 miles) N of the Arctic Circle in Norway 30F1
Norfolk a county of E England; county town Norwich (5355 sq km/2068 sq miles; pop. 714,000) 21H5
Norfolk a port and naval base in Virginia, USA (pop. city 279,700/metropolitan area (with Newport News) 1,261,200) 7F2
Norfolk Island Australia 45F3
Noril'sk Russian Federation 32K3
Normandie (Normandy) an area of central N France, now divided into two regions, Haute Normandie and Basse Normandie (pop. 3,006,000) 24B2
Norrköping Sweden 30D4
Norseman Australia 45B4
Northampton the county town of Northamptonshire (pop. 164,000) 21G5
Northamptonshire a county in central England; county town Northampton (2367 sq km/914 sq miles; pop. 547,000) 21G5
North Bay Canada 5L5
North Cape *see* **Nordkapp** 30F1
North Carolina a state on the SE coast of the USA; state capital Raleigh (136,198 sq km/52,586 sq miles; pop. 6,255,000) 7E2
North Dakota a state in W USA; state capital Bismarck (183,022 sq km/70,665 sq miles; pop. 685,000) 6C1
Northern Ireland a province of the UK, occupying most of the N of the island of Ireland, divided into six counties; capital Belfast (14,121 sq km/5452 sq miles; pop. 1,572,000) 19B3
Northern Marianas a group of 14 islands in the W Pacific which in 1978 became a commonwealth of the USA; capital Susepe, on the island of Saipan 37G2, 46
Northern Territory a territory of N Australia; capital Darwin (1,346,200 sq km/519,770 sq miles; pop. 136,800) 45C2
North Island New Zealand 45G4
North Korea *see* **Korea** 35G3
North Pole the most N point on the earth's axis 48
North Ronaldsay Island Scotland 22F1
North Sea a comparatively shallow branch of the Atlantic Ocean that separates the British Isles from the European mainland 19D2
North Uist Island Scotland 22A3
Northumberland a county in NE England; county town Morpeth (5033 sq km/1943 sq miles; pop. 302,000) 20E2
Northwest Territories a vast area of N Canada, occupying almost a third of the country's whole land area; capital Yellowknife (3,246,000 sq km/1,253,400 sq miles; pop. 45,740) 5G3
North Yorkshire a county in NE England created in 1974 from the division of the former county of Yorkshire; administrative centre Northallerton (8309 sq km/3207 sq miles; pop. 666,000) 20F3
Norway [*area* 323,895 sq km (125,056 sq miles); *pop.* 4,200,000; *cap.* Oslo; *major cities* Bergen, Trondheim, Stavanger; *govt* constitutional monarchy; *religion* Lutheranism; *currency* Norwegian krone] a country in the W half of the Scandinavian peninsula, N Europe 30B3
Norwegian Sea a sea lying between Norway, Greenland and Iceland 30B3
Norwich the county town of Norfolk, in E England (pop. 122,000) 21J5
Notodden Norway 30B4
Nottingham the historic county town of Nottinghamshire, situated on the River Trent (pop. 277,000) 21F5
Nottinghamshire a county in the Midlands of England; county town Nottingham, (2164 sq km/836 sq miles; pop. 1,000,000) 20F4
Nouadhibou Mauritania. 42B2
Nouakchott the capital city of Mauritania, near the Atlantic coast (pop. 135,000) 42B3
Nouméa the capital and chief port of Nouvelle Calédonie (pop. 85,000) 45E1
Nouvelle Calédonie (New Caledonia) the main island of a group of the same name in the South Pacific that forms an overseas territory of France (19,103 sq km/7376 sq miles; pop. 155,000) 45E1
Novara Italy 26B1
Nova Scotia a province on the E coast of Canada; capital Halifax (52,841 sq km/20,401 sq miles; pop. 847,000) 5M5
Novaya Zemlya Russian Federation 32G2
Novi Pazar Yugoslavia 27E2
Novi Sad (Ujvidek; Neusatz) a city on the River Danube and capital of Vojvodina, an autonomous province of Serbia (pop. 257,700) 27D1
Novokuznetsk an industrial city in central S Siberia (pop. 572,000) 32K4
Novorossiysk Russian Federation 32E5
Novosibirsk a major industrial city in central Russian

Federation (pop. 1,386,000) 32K4
Novosibirskiye Ostrova Island 33Q2
Nuku'alofa the capital and main port of Tonga (pop. 21,000) 46
Nukus Uzbekistan 32G5
Nullarbor Plain a huge, dry and treeless plain bordering the Great Australian Bight, in Western and South Australia 45B4
Numazu Japan 35N9
Nunivak Island USA 5B3
Nürnberg (Nuremberg) a city in Bavaria, central S Germany (pop. 486,000) 28B3
Nuuk see Godthåb 5N3
Nyala Sudan 43F3
Nyíregyháza Hungary 29E3
Nyköping Sweden 30D4
Nyngan Australia 45D4
Nzérekoré Guinea 42C4

O

Oahu the third largest of the islands of Hawaii, where the state capital, Honolulu, and Pearl Harbor are located (1549 sq km/598 sq miles; pop. 797,400) 6H
Ob' a river in the Russian Federation which rises near the border with Mongolia and flows N to the Kara Sea (length 5570 km/3460 miles) 32J3
Oban Scotland 22C4
Obi Island Indonesia 37E4
Odawara Japan 35N9
Odda Norway 30B3
Odemira Portugal 25A2
Odense Denmark 30C4
Oder, River a river in central Europe rising in the Czech Republic and flowing N and W to the Baltic Sea, forming part of the border between Germany and Poland (length 912 km/567 miles) 28C2
Odessa a major Black Sea port in the Ukraine (pop. 1,113,000) 32D5
Odessa USA 6C2
Offaly a county in the centre of the Republic of Ireland; county town Tullamore (1998 sq km/771 sq miles; pop. 58,300) 23D3
Offenbach Germany 28B2
Ogaki Japan 35M9
Ogasawara-Shotó Japan 37G1
Ogbomosho Nigeria 42D4
Ogden USA 6B1
Ohio a Midwest state of the USA, with a shore on Lake Erie; capital Columbus (106,765 sq km/41,220 sq miles; pop. 10,744,00) 7E1
Ohio River a river in E USA, formed at the confluence of the Allegheny and Monongahela rivers. It flows W and S and joins the Mississippi at Cairo, Illinois (length 1575 km/980 miles) 8C3
Ohrid Yugoslavia 27E2
Okaya Japan 35N8
Okayama Japan 35H3
Okazaki Japan 35M9
Okehampton England 21C7
Okhotsk Russian Federation 33Q4
Okhotsk, Sea of a part of the NW Pacific bounded by the Kamchatka peninsula, the Kuril islands, and the E coast of Siberia 33Q4
Okinawa Island Japan 35G4
Oklahoma a state in the SW of the USA; state capital Oklahoma City (173,320 sq km/66,919 sq miles; pop. 3,301,000) 6D2
Oklahoma City the state capital of Oklahoma (pop. 443,200/metropolitan area 962,600) 6D2

Öland Island Sweden 30D4
Olbia Italy 26B2
Oldenburg Germany 28B2
Olimbos, Mount (Mount Olympus) in central mainland Greece, the home of the gods of ancient Greek myth (2917 m/9570 ft) 27E2
Olomouc Czech Republic 28D3
Olsztyn Poland 29E2
Olympia a port and state capital of Washington, on the W coast of the USA (pop. city 29,200/metropolitan area 138,300) 6A1
Olympus, Mount see **Olimbos** 27E2
Omagh Northern Ireland 23D2
Omaha Nebraska USA 6D1
Oman [area 212,457 sq km (82,030 sq miles); pop. 2,000,000; cap. Masqat; govt monarchy (sultanate); religion Ibadi Islam, Sunni Islam; currency rial Omani] a small country in two parts on the Gulf of Oman 38D3
Oman, Gulf of a branch of the Arabian Sea leading to the Strait of Hormuz 38D3
Omdurman a city situated across the River Nile from Khartoum, the capital of Sudan (pop. 526,300) 43G3
Omsk an industrial city in central W Siberia, on the Trans-Siberian Railway (pop. 1,094,000) 32J4
Onitsha Nigeria 42D4
Ontario a province of central Canada; capital Toronto (1,068,582 sq km/412,580 sq miles; pop. 8,625,000) 5J4
Ontario, Lake the smallest and most easterly of the Great Lakes; it drains into the St Lawrence River (19,550 sq km/7550 sq miles) 6L5
Oostende (Ostend) Belgium 24C1
Opole Poland 29D2
Oporto see **Porto** 25A1
Oppdal Norway 30B3
Oradea Romania 29E3
Oran (Wahran) a Mediterranean port and second largest city of Algeria (pop. 670,000) 42C1
Orange Australia 45D4
Orange France 24C3
Orange, River the longest river in S Africa, rising in Lesotho and flowing W to the Atlantic (length 2090 km/1299 miles) 44B3
Orbetello Italy 26C2
Örebro Sweden 30D4
Oregon a state in the NW of the USA, on the Pacific; state capital Salem (251,180 sq km/96,981 sq miles; pop. 2,687,000) 6A1
Orel Russian Federation 32E4
Orenburg Russian Federation 32G4
Orëse Spain 25A1
Orinoco, River a river in N South America, rising in S Venezuela and flowing W, then N and E to the Atlantic. It forms part of the border between Colombia and Venezuela (length 2200 km/1370 miles) 16C2
Oristano Italy 26B3
Orkney Islands a group of some 90 islands off the NE coast of Scotland; capital Kirkwall (976 sq km/377 sq miles; pop. 19,000) 22E1
Orlando a city in central Florida, and focus for visitors to Disney World and Cape Canaveral (pop. city 137,100/metropolitan area 824,100) 7E3
Orléans France 24C2
Örnsköldsvik Sweden 30D3
Orsk Russian Federation 32G4
Oruro Bolivia 16C4
Osaka a port on S Honshu Island, and the third largest city in Japan (pop. 2,636,300) 35L9
Oshogbo Nigeria 42D4
Osijek a city in E Croatia, on the Drava River. It was formerly called Esseg (pop. 158,800) 27D1
Oskarshamn Sweden 30D4
Oslo the capital of Norway, and its main port, in the SE

Osnabrück — Peking

of the country. From 1624 to 1925 it was called Christiania (or Kristiania) (pop. city 448,800/metropolitan area 566,500) 39C4
Osnabrück Germany 28B2
Osorno Chile 17B7
Ostend see **Oostende** 24C1
Östersund Sweden 30C3
Ostia Italy 26C2
Ostrava Czech Republic 29D3
Osumi-shoto Japan 35H3
Oswestry England 21D5
Otaru Japan 35J2
Otranto, Strait of the waterway separating the heel of Italy from Albania 27D2
Ottawa the capital of Canada, in E Ontario, on the Ottawa River (pop. 718,000) 5L5
Ottawa, River a river of central Canada which flows into the St Lawrence at Montreal (length 1271 km/790 miles) 7F1
Ouagadougou the capital of Burkina, situated in the centre of the country (pop. 286,500) 42C3
Ouahigouya Burkina Faso 42C3
Ouargla Algeria 42D1
Oudtshoorn South Africa 44C4
Oujda Morocco 42C1
Oulu Finland 30F2
Outer Hebrides Scotland 22A3
Oviedo Spain 25A1
Oxford an old university city, and county town of Oxfordshire, England (pop. 117,000) 21F6
Oxfordshire a county in S central England (2611 sq km/1008 sq miles; pop. 558,000) 21F6
Oxus, River see **Amudar'ya, River** 32H6
Oyem Gabon 42E4

P

Pacific Ocean the largest and deepest ocean on Earth, between Asia and Australia to the W and the Americas to the E (165,384,000 sq km/63,838,000 sq miles) 46
Padang a port and the capital of West Sumatra, Indonesia (pop. 480,900) 36C4
Paderborn Germany 28B2
Pagai Selatan Island Indonesia 36C4
Pagai Utara Island Indonesia 36B4
Pag Island Croatia 26D2
Pahala Hawaii USA 6H
Painted Desert a desert of colourful rocks in N Arizona, USA (19,400 sq km/7500 sq miles) 13D3
Paisley Scotland 22D5
Pais Vasco (region) Spain 25B1
Pakistan [area 796,095 sq km (307,372 sq miles); pop. 105,400,000; cap. Islamabad; major cities Faisalabad, Hyderabad, Karachi, Lahore; govt federal Islamic republic; religion Sunni Islam, Shia Islam; currency Pakistan rupee] a country lying just N of the Tropic of Cancer and the Arabian Sea 39F2
Palan Island Pacific Ocean 46
Palangkaraya Indonesia 36D4
Palau see **Belau** 37F3
Palawan Island Philippines 36D2
Palembang a port and the capital of South Sumatra, on the SE coast (pop. 787,200) 36C4
Palencia Spain 25B1
Palermo the capital of Sicily, Italy, on the NW coast (pop. 718,900) 26C3
Palma (Palma de Mallorca) the capital of Majorca and of the Balearic Islands (pop. 304,400) 25C2
Palma, La Canary Islands 42B2
Palmerston North New Zealand 45G5

Palmi Italy 26D3
Palmyra Pacific Ocean 46
Palu Indonesia 36D4
Pamplona a city in NE Spain, famous for its bull-running festival in July (pop. 183,100) 25B1
Panamá [area 77,082 sq km (29,761 sq miles); pop. 2,320,000; cap. Panama City; major cities San Miguelito, Colón; govt republic; religion RC; currency balboa] a country at the narrowest point in Central America, where only 58 km (36 miles) of land separates the Caribbean from the Pacific, and the Panama Canal provides a major shipping route 14D5
Panama City the capital of Panama, situated at the Pacific end of the Panama Canal (pop. 502,000) 14D5
Panay Island Philippines 37E2
Pangkalpinang Indonesia 36D4
Pantelleria Island Italy 26C3
Papa Westray Island Scotland 22F1
Paphos Cyprus 40B1
Papua New Guinea [area 462,840 sq km (178,703 sq miles); population 3,800,000; cap. Port Moresby; govt constitutional monarchy; religion Protestantism, RC; currency kina] a country in the SW Pacific comprising the E half of the island of New Guinea and hundreds of other islands 37G4
Paracel Islands a group of islands lying some 300 km (185 miles) E of Vietnam, owned by China 36D2
Paracin Yugoslavia 27E2
Paraguay [area 406,752 sq km (157,047 sq miles); pop. 4,160,000; cap. Asunción; major city Ciudad Alfredo Stroessner; govt republic; religion RC; currency guaraní] a landlocked country in central South America, bordered by Bolivia, Brazil and Argentina 17D5
Paraguay, River a major river of South America, flowing S from Brazil through Paraguay to join the Parana (length 1920 km/1190 miles) 17D5
Parakou Benin 42D4
Paramaribo the capital and main port of Suriname (pop. 180,000) 16D2
Paraná Argentina 17C6
Parana, River the second longest river in South America, rising in Brazil and flowing S to join the River Plate (length 4200 km/2610 miles) 17D5
Parepare Indonesia 36D4
Paris the capital of France, in the N of the country, on the River Seine (pop. city 2,188,900/Greater Paris 8,761,700) 24C2
Parkano Finland 30E3
Parma Italy 26C2
Parnaíba Brazil 16E3
Pärnu Estonia 30E4
Páros an island in the Cyclades, Greece (194 sq km/75 sq miles; pop. 7400) 27F3
Pasadena California USA 6B2
Passo Fundo Brazil 17D5
Pasto Colombia 16B2
Patagonia a cold desert in S Argentina and Chile 17B7
Patna the capital of the state of Bihar, in NE India, on the River Ganges (pop. 918,900) 39G3
Pátrai (Patras) a port and the main city of the Peloponnese, Greece (pop. 154,000) 27E3
Pau France 24B3
Pavlodar Kazakhstan 32J4
Peace River a river in W Canada, a tributary of the Slave/Mackenzie River, rising in British Columbia (length 1923 km/1195 miles) 5G4
Pec Yugoslavia 27E2
Pécs the main city of SW Hungary (pop. 174,500) 29D3
Peebles Scotland 22E5
Pegu Myanmar 36B2
Pekanbaru Indonesia 36C3
Peking see **Beijing** 35F3

Pello   1023   Pôrto Alegre

Pello Finland 30E2
Pelopónnisos (Peloponnese) a broad peninsula of S Greece, joined to the N part of the country by the isthmus of Corinth 27E3
Pemba Mozambique 44E2
Pemba Island Tanzania 43G5
Pembroke Wales 21C6
Pennines a range of hills that runs down the middle of N England from the Scottish border to the Midlands, rising to 894 m (2087 ft) at Cross Fell 19C3
Pennsylvania a state of the NE USA situated mainly in the Appalachians; capital Harrisburg (117,412 sq km/45,333 sq miles; pop. 11,853,000) 7F1
Penrith England 20E3
Penza Russian Federation 32F4
Penzance England 20B7
Pereira Colombia 16B2
Périgueux France 24C2
Perm' an industrial port on the Kama River in the W Urals of the Russian Federation. It was known as Molotov 1940-57 (pop. 1,049,000) 32G4
Perpignan France 24C3
Persia *see* **Iran** 38D2
Persian Gulf *see* **Gulf, The** 38C/D3
Perth the state capital of Western Australia, which includes the port of Fremantle (pop. 969,000) 45A4
Perth a city and former capital of Scotland, 55 km (35 miles) N of Edinburgh (pop. 42,000) 22E4
Peru [*area* 1,285,216 sq km (496,235 sq miles); *pop.* 22,330,000; *cap.* Lima; *major cities* Arequipa, Callao, Cuzco, Trujillo; *govt* republic; *religion* RC; *currency* sol] a country just S of the Equator, on the Pacific coast of South America 16B4
Perugia Italy 26C2
Pescara Italy 26C2
Peshawar a historic town in NW Pakistan at the foot of the Khyber Pass (pop. 555,000) 39F2
Peterborough England 21G5
Peterhead Scotland 22G3
Petrograd *see* **Sankt Peterburg** 32E4
Petropavlovsk Kamchatskiy Russian Federation 33R4
Petropavlovsk Kazakhstan 32H4
Petrozavodsk Russian Federation 32E3
Philadelphia a port and city in SE Pennsylvania, the fourth largest city in the USA (pop. city 1,688,700/metropolitan area 4,768,400) 7F1
Philippines [*area* 300,000 sq km (115,830 sq miles); *pop.* 60,500,000; *cap.* Manila; *major cities* Cebu, Davao, Quezon City; *govt* republic; *religions* RC, Aglipayan, Sunni Islam; *currency* : Philippine peso] a country comprising a group of widely scattered mountainous islands in the W Pacific 37E2
Phitsanulok Thailand 36C2
Phnom Penh the capital of Cambodia, in the S of the country (pop. 500,000) 36C2
Phoenix the state capital of Arizona, USA (pop. city 853,300/metropolitan area 1,714,800) 6B2
Phoenix Islands Kiribati 46
Phuket Island Thailand 36B3
Piacenza Italy 26B1
Pierre the capital of South Dakota, USA (pop. 121,400) 6C1
Pietermaritzburg a city in E South Africa and capital of Natal (pop. 180,000) 44D3
Pietersburg South Africa 44C3
Pila Poland 28D2
Pílos Greece 27E3
Pilsen *see* **Plzen**
Pinar del Río Cuba 14C2
Pingliang China 34E3
Piombino Italy 26C2
Piraiévs (Piraeus) the main port of Greece, close to Athens, on the Aegean Sea (pop. 196,400) 27E3
Pírgos Greece 27E3
Pirot Yugoslavia 27E2
Pisa a city in NW Italy on the River Arno, famous for its leaning bell tower (pop. 104,300) 26C2
Pitcairn Island an island and British colony in the S Pacific, where mutineers from *HMS Bounty* settled 47
Piteå Sweden 30E2
Pitesti Romania 27E2
Pitlochry Scotland 22E4
Pittsburgh Pennsylvania USA 7F1
Piura Peru 16A3
Plasencia Spain 25A1
Plate, River *see* **Rio de la Plata** 17D6
Plenty, Bay of the broad inlet on the N coast of the North Island of New Zealand 45G4
Pleven Bulgaria 27E2
Ploiesti Romania 27F2
Plovdiv a major market town in Bulgaria (pop. 373,000) 27E2
Plymouth a port and naval base in SW England and the place from which the Pilgrim Fathers set sail in the *Mayflower* in 1620 (pop. 255,000) 21C7
Plzen (Pilsen) an industrial city in W Bohemia, Czech Republic. Pilsner lager beer was first produced here in 1842 (pop. 174,100) 28C3
Po, River the longest river in Italy, flowing E from the Alps across a fertile plain to the Adriatic Sea (length 642 km/405 miles) 27B1
Pointe-à-Pitre the main port of Guadeloupe (pop. 23,000) 14G3
Pointe Noire Congo 42E5
Poitiers France 24C2
Poitou (province) France 24B2
Poland [*area* 312,677 sq km (120,725 sq miles); *pop.* 37,930,000; *cap.* Warsaw (Warszawa); *major cities* Gdansk, Kraków, Lódz, Wroclow; *govt* republic; *religion* RC; *currency* zloty] a low-lying country on the North European Plain 29D2
Polis Cyprus 40B1
Polla Italy 26D2
Poltava Ukraine 32E5
Ponce Puerto Rico 14F3
Ponferrada Spain 25A1
Pontevedra Spain 25A1
Pontianak Indonesia 36C4
Poole England 21F7
Poona *see* **Pune** 39F4
Pori Finland 30E3
Port Augusta Australia 45C4
Port-au-Prince the main port and capital of Haiti (pop. 888,000) 14E3
Port Elizabeth South Africa 44C4
Port Gentil Gabon 42D5
Port Harcourt the second port of Nigeria after Lagos (pop. 288,900) 42D4
Port Headland Australia 45A3
Portland Maine 6A1
Port Laoise Ireland 23D3
Port Louis the capital and main port of Mauritius, on the E coast of the island (pop. 160,000) 44F4
Port Moresby the capital and main port of Papua New Guinea, in the SE (pop. 126,000) 37G4
Porto a port in NW Portugal, and the country's second largest city after Lisbon (pop. 330,200) 25A1
Port of Spain the capital and chief port of Trinidad and Tobago (pop. city 62,700/metropolitan area 443,000) 14G4
Porto Novo the administrative capital of Benin (pop. 209,000) 42D4
Porto Torres Italy 26B2
Porto Vecchio France 26B2
Pôrto Alegre Brazil 17D5

Pôrto Velho Brazil 16C3
Port Pirie Australia 45C4
Portrush Northern Ireland 23E1
Port Said the port at the Mediterranean end of the Suez Canal, Egypt (pop. 342,000) 43G1
Portsmouth a port and major naval base in S England (pop. 192,000) 21F7
Port Stanley *see* **Stanley**
Port Sudan Sudan 17D8
Port Talbot Wales 43G3
Portugal [*area* 92,389 sq km (35,671 sq miles); *pop.* 10,300,000; *cap.* Lisbon (Lisboa); *major cities* Braga, Coimbra, Oporto, Setúbal; *govt* republic; *religion* : RC; *currency* escudo] a country in the SW corner of Europe which borders Spain, making up about 15% of the Iberian peninsula 25A2
Posen *see* **Poznan**
Poso Indonesia 37E4
Potchefstroom South Africa 44C3
Potenza Italy 26D2
Potosí Bolivia 16C4
Potsdam Germany 28C2
Powys a county in mid-Wales created in 1974 out of Breconshire, Montgomeryshire, and Radnorshire; administrative centre Llandrindod Wells (5077 sq km/1960 sq miles; pop. 111,000) 21D5
Poznan (Posen) Poland 28D2
Praha (Prague) the capital and main city of the Czech Republic, on the Vltava River (pop. 1,235,000) 28C2
Prato Italy 26C2
Pressburg *see* **Bratislava**
Preston England 20E4
Preswick Scotland 22D5
Pretoria the administrative capital of South Africa, 48 km (30 miles) N of Johannesburg (pop. 739,000) 44C3
Prince Edward Island the smallest province of Canada, an island in the Gulf of St Lawrence; capital Charlottetown (5660 sq km/2185 sq miles; pop. 123,000) 5M5
Prince George Canada 5F4
Prince of Wales Island Canada 5H2
Prince Rupert Canada 5F4
Príncipe Island West Africa 42D4
Pristina the capital of the autonomous province of Kosovo in Serbia (pop. 216,000) 27E2
Prokop'yevsk Russian Federation 32K4
Prome Myanmar 36B2
Provence a historical region of coastal SE France 24D3
Providence a port and capital of Rhode Island, USA (pop. city 154,100/metropolitan area 1,095,000) 7F1
Prudhoe Bay USA 6J
Przemys'l Poland 29E3
Pskov Russian Federation 32D4
Pucallpa Peru 16B3
Puebla a major city 120 km (75 miles) SE of Mexico City, and the capital of a state of the same name (pop. city 835,000) 6D4
Puerto Armuelles Panama 14C5
Puerto Ayacucho Venezuela 16C2
Puerto Barrios Guatemala 14B3
Puerto Cabezas Nicaragua 14C4
Puerto Cortés Honduras 14B4
Puerto Juárez Mexico 7E3
Puertollano Spain 25B2
Puerto Montt Chile 17B7
Puerto Plata Dominican Republic 14E3
Puerto Rico (*area* 8897 sq km (3435 sq miles); *pop.* 3,196,520; *cap.* San Juan; *govt* self-governing commonwealth (USA); *religion* RC, Protestantism; *currency* US dollar] a self-governing commonwealth in association with the USA, in the Caribbean 14F3
Pula Croatia 26C2
Pune (Poona) a historical and industrial city E of Bombay in W India (pop. 1,203,400) 39F4
Puntarenas Costa Rica 14C5
Punta Arenas Chile 16B8
Pusan a major port, and the second largest city in South Korea after Seoul (pop. 3,160,000) 35G3
Puttgarden Germany 28C2
Putumayo, River a river of NW South America, rising in the Andes and flowing SE to the Amazon (length 1900 km/ 1180 miles) 16B3
Pyongyang (Pyeongyang) an industrial city and the capital of North Korea (pop. 1,700,000) 35G3
Pyrénées a range of mountains that runs from the Bay of Biscay to the Mediterranean. The highest point is Pico d'Aneto (3404 m/11,170 ft) 25B1

Q

Qamdo China 34D3
Qatar [*area* 11,000 sq km (4247 sq miles); *pop.* 371,863; *cap.* Doha (Ad Dawhah); *govt* monarchy; *religions* Wahhabi Sunni Islam; *currency* Qatari riyal] a small emirate halfway along the coast of The Gulf 38D3
Qazvin (Kasvin) a historic town in NW Iran (pop. 244,300) 38C2
Qena Egypt 43G2
Qingdao China 35G3
Qingjiang China 35F3
Qinhuangdao China 35F2
Qiqihar China 35G2
Qom (Qum) a holy city in central N Iran (pop. 424,100) 38D2
Quanzhou China 35F4
Québec the largest province of Canada, in the E of the country; capital Québec. The majority of the population are French-speaking. (1,358,000 sq km/524,300 sq miles; pop. 6,438,000) 5L4
Québec the capital of the province of Québec in Canada (pop. 164,580) 5L5
Queen Elizabeth Islands Canada 5G2
Queensland the NE state of Australia; capital Brisbane (1,272,200 sq km/491,200 sq miles; pop. 2,488,000) 45D3
Quelimane Mozambique 44D2
Quetta the capital of the province of Baluchistan, Pakistan (pop. 285,000) 39E3
Quezaltenango Guatemala 14A4
Quezon City a major city and university town, now part of Metro Manila, and administrative capital of the Philippines 1948-1976 (pop. 1,165,000) 37E2
Quilon India 39F5
Quimper France 24B2
Qui Nhon Vietnam 36C2
Quito the capital of Ecuador, just S of the Equator, 2850 m (9350 ft) high in the Andes (pop. 1,110,000) 16B3
Qum *see* **Qom** 38D2
Qu Xian China 35F4

R

Raasay Island Scotland 22B3
Raba Indonesia 36D4
Rab Island Croatia 26C2
Rabat the capital of Morocco, in the NW, on the Atlantic coast (pop. 520,000) 42C1
Radom Poland 29E2
Ragusa *see* **Dubrovnik** 27D2
Rainier, Mount *see* **Cascade Range** 12B1
Raipur India 39G3
Rajkot India 39F3

Raleigh   1025   Rügen Island

Raleigh the state capital of North Carolina, USA (pop. city 169,300/metropolitan area 609,300)   7F2
Ramsey Isle of Man   20C3
Rancagua Chile   17B6
Ranchi India   39G3
Randers Denmark   30C4
Rangoon see **Yangon**   36B2
Rangpur Bangladesh   39G3
Rapid City USA   6C1
Rasht Iran   38C2
Rathlin Island Northern Ireland   23E1
Ráth Luirc Ireland   23C4
Ratlam India   39F3
Rauma Finland   30E3
Ravenna a city in NE Italy, noted for its Byzantine churches (pop. 136,500)   26C2
Rawalpindi a military town of ancient origins in N Pakistan (pop. 928,000)   39F2
Razgrad Bulgaria   27F2
Reading England   21G6
Recife a regional capital of E Brazil (pop. 1,205,000) 16F3
Redon France   24B2
Red River a river of S USA, rising in Texas and flowing E to the Mississippi (length 1639km/1018miles)   7D2
Red River see **Hong**   34E4
Red Sea a long, narrow sea lying between the Arabian Peninsula and the coast of NE Africa   38B3
Regensburg Germany   28B3
Reggane Algeria   42D2
Reggio di Calabria a port on the toe of S Italy (pop. 177,700)   26D3
Reggio nell'Emilia a town of Roman origins in NE Italy (pop. 130,300)   26C2
Regina the capital of the province of Saskatchewan, Canada (pop. 164,000)   5H4
Reims (Rheims) a historic city in France, the centre of the production of champagne (pop. 204,000)   24C2
Renell Island Solomon Islands   45F2
Rennes France   24B2
Reno a gambling centre in Nevada, USA (pop. city 105,600/metropolitan area 211,500)   6B2
Resistencia Argentina   17D5
Resolution Island Canada   5M3
Réunion an island to the E of Madagascar, an overseas department of France. The capital is Saint-Denis (2515 sq km/970 sq miles; pop. 530,000)   44F4
Reykjavik the capital and main port of Iceland, on the SW coast (pop. 87,300)   30A2
Rheims see **Reims**   24C2
Rhein (Rhine, Rhin, Rijn), River an important river of Europe, rising in the Swiss Alps, flowing N through Germany and then W through the Netherlands to the North Sea (length 1320km/ 825miles)   28B3
Rhode Island the smallest state in the USA; capital Providence (3144 sq km/1214 sq miles; pop. 968,000)   7F1
Rhodes see **Ródhos**   27F3
Rhône, River a major river of Europe, rising in the Swiss Alps and flowing W into France, and then S to the Golfe de Lion (length 812 km/505 miles)   24C3
Rhum Island Scotland   22B4
Rhyl Wales   20E4
Richmond the state capital of Virginia (pop. city 219,100/metropolitan area 796,100)   7F2
Riga a Baltic port, and the capital of Latvia (pop. 875,000)   32D4
Rijeka (Fiume) a port on the Adriatic, in Croatia (pop. 193,000)   26C1
Rijn, River see **Rhein, River**   28B3
Rimini Italy   26C2
Rîmnicu Vîlcea Romania   29E3
Ringwood England   21F7
Rio Branco Brazil   16C3

Rio Bravo see **Rio Grande**   7C3
Rio de Janeiro a major port and former capital (1763-1960) of Brazil, in the SE (pop. 5,094,000)   16E5
Rio de la Plata (Plate, River) the estuary of the Parana and Uruguay Rivers in SE South America   17D6
Rio Grande (Rio Bravo) a river of North America, rising in Colorado, USA, and flowing SE to the Gulf of Mexico, for much of its length forming the USA-Mexico border (length 3078 km/1885 miles)   7C3
Rio Grande Brazil   17D5
Río Gallegos Argentina   16C8
Rioja, La see **La Rioja**   25B1
Ripon England   20F3
Riyadh see **Ar Riyad**   38C3
Roanne France   24C2
Rochdale England   20E4
Rochester England   21H6
Rochester USA   7D1
Rockford USA   7E1
Rockhampton Australia   45E3
Rocky Mountains (Rockies) a huge range in W North America, extending 4800 km (3000 miles) from British Columbia to New Mexico   5F4
Rødbyhavn Denmark   30C5
Ródhos (Rhodes) the largest of the Dodecanese group of islands belonging to Greece (1399 sq km/540 sq miles; pop. 88,500)   27F3
Roma (Rome) the historic capital of Italy, on the River Tiber, in the centre of the country near the W coast (pop. 2,831,300)   26C2
Roman Romania   29F3
Romania [area 237,500 sq km (91,699 sq miles); pop. 23,000,000; cap. Bucharest (Bucuresti); major cities Brasov, Constanta, Timisoara; govt republic; religions Romanian Orthodox, RC; currency leu] an almost circular country, apart from a small extension towards the Black Sea, located in SE Europe   27E1
Rome see **Roma**   26C2
Ronda Spain   25A2
Rosario an industrial and commercial city on the River Parana in Argentina (pop. 935,500)   17C6
Roscoff France   24B2
Roscommon a county in NW Republic of Ireland, with a county town of the same name. (2462 sq km/950 sq miles; pop. county 54,500)   23C3
Roscrea Ireland   23D4
Roseau the capital of Dominica (pop. 17,000)   14G3
Rosslare Ireland   23E4
Ross Sea a large branch of the Antarctic Ocean, S of New Zealand   48
Rostock a major port on the Baltic coast of Germany (pop. 242,000)   28C2
Rostov-na-Donu (Rostov-on-Don) a major industrial city on the River Don, near the NW end of the Sea of Azov in SE Russian Federation (pop. 983,000)   32E5
Rotherham England   20F4
Roti Island Indonesia   37E5
Rotterdam the largest city in the Netherlands and the busiest port in the world (pop. city 558,800/Greater Rotterdam 1,024,700)   28A2
Rouen France   24C2
Round Island Mauritius   44F4
Rousay Island Scotland   22E1
Roussillon (province) France   24C3
Rovaniemi Finland   30F2
Royal Tunbridge Wells England   21H6
Rub al-Khali the "Empty Quarter", a vast area of desert straddling the borders of Saudi Arabia, Oman and Yemen (650,000 sq km/251,000 sq miles)   38C4
Ruffec France   24C2
Rugby England   21F5
Rügen Island Germany   28C2

Ruhr, River

Ruhr, River a river in NW Germany whose valley forms the industrial heartland of W Germany. It joins the Rhine at Duisburg (length 235 km/146 miles) 28B2
Ruma Yugoslavia 27D1
Runcorn England 20E4
Ruoqiang China 34C3
Ruse Bulgaria 27F2
Russian Federation, The, [*area* 17,075,400 sq km (6,592,800 sq miles) ; *pop.* 142,117,000 ; *cap.* Moscow (Moskva) ; *major cities* St Petersburg, Nizhniy Novgorod, Novosibirsk; *govt* republic; *religions* Russian Orthodox, Sunni Islam, Shia Islam, RC; *currency* rouble] the largest country in the world, it extends from Eastern Europe through the Urals E to the Pacific Ocean. It became independent in 1991. 32/33
Rutanzige, Lake *see* **Edward, Lake** 43F5
Ruteng Indonesia 37E4
Ruwenzori a mountain range on the border between Zaïre and Uganda, also called the Mountains of the Moon. The highest peak is Mount Ngaliema (Mount Stanley) (5109 m/16,763 ft) 43G4
Rwanda [*area* 26,338 sq km (10,169 sq miles); *pop.* 6,710,000; *cap.* Kigali; *govt* republic; *religions* RC, Animism; *currency* Rwanda franc] a small republic in the heart of central Africa, 2° S of the Equator 43F5
Ryazan Russian Federation 32E4
Rybinsk Russian Federation 32E4
Rybnik Poland 29D2
Ryukyu Islands *see* **Nansei-shoto** 35G4
Rzeszów Poland 29E2

S

Saarbrücken Germany 28B3
Saaremaa Island Estonia 30E4
Sabac Yugoslavia 27D2
Sabadell Spain 25C1
Sabah the more E of the two states of Malaysia on the N coast of the island of Borneo (73,700 sq km/28,450 sq miles; pop. 1,034,000) 36D3
Sabha Libya 42E2
Sacramento the state capital of California (pop. city 304,100/metropolitan area 1,219,600) 6A2
Sadiya India 39H3
Safi Morocco 42C1
Sagunto Spain 25B2
Sahara Desert the world's largest desert, spanning N Africa, from the Atlantic to the Red Sea, and the Mediterranean to Mali, Niger, Chad and Sudan 42
Sahel a semi-arid belt crossing Africa from Senegal to Sudan, separating the Sahara from tropical Africa to the S 42D3
Saigon *see* **Ho Chi Minh City** 36C2
Saintes France 24B2
Saint John *see* **St John** 5M5
Sakai Japan 35L9
Sakata Japan 35N7
Sakhalin a large island N of Japan, belonging to the Russian Federation (76,400 sq km/29,500 sq miles; pop. 660,000) 33Q4
Sakishima gunto Japan 34K6
Salalah Oman 38D4
Salamanca Spain 25A1
Salangen Norway 30D2
Salayar Island Indonesia 37E4
Salbris France 24C2
Salem India 39F4
Salem the state capital of Oregon, USA (pop. city 90,300/metropolitan area 255,200) 6A1
Salerno Italy 26C2

Sanmenxia

Salford England 20E4
Salisbury England 21F6
Salo Finland 30E3
Saloniki *see* **Thessaloníki** 27E2
Salonta Romania 29E3
Salta Argentina 16C5
Saltillo Mexico 6C3
Salt Lake City the state capital of Utah (pop. city 164,800/metropolitan area 1,025,300) 6B1
Salto Uruguay 17D6
Salvador a port on the central E coast of Brazil and capital of the state of Bahia (pop. 1,507,000) 16F4
Salvador, El *see* **El Salvador** 14B4
Salween, River a river rising in Tibet and flowing S to Burma, forming a large part of the border with Thailand, to the Andaman Sea (length 2900 km/1800 miles) 36B2
Salzburg a city in central N Austria, and the name of the surrounding state, of which it is the capital (pop. city 140,000) 28C3
Salzgitter-Bad Germany 28C2
Samar the third largest island of the Philippines (13,080 sq km/5050 sq miles; pop. 1,100,000) 37E2
Samara Russian Federation 32G4
Samarinda Indonesia 36D4
Samarkand an ancient city in Uzbekistan (pop. 515,000) 32H6
Samoa *see* **American Samoa; Western Samoa** 46
Sámos a Greek island 2 km (1 mile) off the coast of Turkey (pop. 31,600) 27F3
Samothráki Island Greece 27F2
Samsun Turkey 38B1
San Mali 42C3
San'a the capital of Yemen, in the middle of the country (pop. 210,000) 38C4
San Antonio Texas 6D3
San Benedetto del Tronto Italy 26C2
San Cristobal Island Solomon Islands 45F2
San Cristóbal Venezuela 16B2
Sancti Spíritus Cuba 14D2
Sandakan Malaysia 36D3
Sanday Island Scotland 22F1
San Diego California 6B2
Sandoy Island Denmark 30A2
San Fernando Philippines 37E2
San Francisco a Pacific port and commercial centre in California (pop. city 712,800/metropolitan area 5,684,600) 6A2
Sanjo Japan 35N8
San Jose a city in California, USA, and centre of the electronics industry (pop. city 686,200/metropolitan area 1,371,500) 6A2
San José the capital of Costa Rica, in its centre (pop. 249,000) 14C5
San Juan Argentina 17C6
San Juan del Norte Nicaragua 14C4
San Juan del Sur Nicaragua 14B4
San Juan the capital of Puerto Rico, and a major port (pop. 435,000) 14F3
San Julián Argentina 17C7
Sankt Peterburg (St Petersburg) a former capital of Russia and the current Russian Federation's second-largest city. It is an important industrial city, cultural centre and port on the Baltic Sea. It was called Petrograd 1914-24 and, until 1991, Leningrad 32E4
San Luis Potosi an elegant colonial city and provincial capital in N-central Mexico (pop. city 407,000) 6C3
San Marino [*area* 61 sq km (24 sq miles); *pop.* 22,746; *cap.* San Marino; *govt* republic; *religion* RC; *currency* lira] a tiny landlocked state in central Italy, in the E foothills of the Apennines 26D2
Sanmenxia China 35F3

San Miguel de Tucumán   1027   Serbia

San Miguel de Tucumán a regional capital in NW Argentina (pop. 497,000) 16C5
San Miguel El Salvador 14B4
San Pedro Sula the second largest city in Honduras (pop. 398,000) 14B3
San Remo Italy 26B2
San Salvador the capital and major city of El Salvador (pop. 884,100) 14B4
San Salvador Island a small island in the centre of the Bahamas, the first place in the New World reached by Columbus (1492) (pop. 850) 14E1
San Sebastián Spain 25B1
San Severo Italy 26D2
Santa Ana El Salvador 14B4
Santa Clara Cuba 14C2
Santa Cruz Bolivia 16C4
Santa Cruz de Tenerife Canary Islands 42B2
Santa Cruz Islands Solomon Islands 45F2
Santa Fe the state capital of New Mexico, USA (pop. city 52,300/metropolitan area 100,500) 6C2
Santa Fé Argentina 17C6
Santa Isabel Island Solomon Islands 45E1
Santa Marta Colombia 16B1
Santander Spain 25B1
Santarém Brazil 16D3
Santarém Portugal 25A2
Santa Rosa Argentina 17C6
Santiago the capital and principal city of Chile (pop. 4,132,000) 17B6
Santiago Dominican Republic 14E3
Santiago Panama 14C5
Santiago de Compostela Spain 25A1
Santiago de Cuba Cuba 14D3
Santo Domingo the capital and main port of the Dominican Republic (pop. 1,313,000) 14F3
Santorini *see* **Thíra** 27F3
São Carlos Brazil 16E5
São Francisco, River a river of E Brazil, with important hydroelectric dams (length 2900 km/1800 miles) 16E4
São Luis Brazil 16E3
Saône, River a river of E France which merges with the Rhône at Lyons (length 480 km/300 miles) 24C2
São Paulo a major industrial city in SE Brazil, and capital of the state of São Paulo (pop. city 8,500,000/metropolitan area 16,000,000) 16E5
São Tomé and Príncipe [*area* 964 sq km (372 sq miles); *pop.* 115,600; *cap.* São Tomé; *govt* republic; *religion* RC; *currency* dobra] a country consisting of two volcanic islands off the W coast of Africa 42D4
São Vicente, Cabo de (Cape St Vincent) the SW corner of Portugal 25B2
Sapporo a modern city, founded as the capital of Hokkaido Island, Japan (pop. 1,543,000) 35J2
Sapri Italy 26D2
Saragossa *see* **Zaragoza** 25B1
Sarajevo the capital of Bosnia Herzegovina (pop. 448,500) 27D2
Saratov an industrial city and river port on the Volga, Russian Federation (pop. 894,000) 32F4
Sarawak a state of Malaysia occupying much of the NW coast of Borneo (125,204 sq km/48,342 sq miles; pop. 1,323,000) 36D3
Sardegna (Sardinia) the second largest island of the Mediterranean after Sicily, also belonging to Italy, lying just S of Corsica. The capital is Cagliari 26B2
Sarh Chad 43E4
Sark Island UK 21E8
Sarrion Spain 25B1
Sasebo Japan 35G3
Saskatchewan a province of W Canada, in the Great Plains; capital Regina (651,900 sq km/251,000 sq miles; pop. 968,000) 5H4

Saskatchewan, River a river of Canada, rising in the Rocky Mountains and flowing W into Lake Winnipeg (length 1930 km/1200 miles) 5H4
Saskatoon Saskatchewan Canada 5H4
Sassandra Côte d'Ivoire 42C4
Sassari Sardegna 26B2
Sassnitz Germany 28C2
Satu Mare Romania 29E3
Saudi Arabia [*area* 2,149,690 sq km (829,995 sq miles); *pop.* 12,000,000; *cap.* Riyadh (Ar Riyah); *major cities* Mecca, Jeddah, Medina, Ta'if; *govt* monarchy; *religions* Sunni Islam, Shia Islam; *currency* rial] a country occupying over 70% of the Arabian Peninsula 38C3
Saul Ste Marie Canada 5K5
Savannah the main port of Georgia, USA (pop. city 145,400/metropolitan area 323,900) 7E2
Savannakhet Laos 36C2
Savona Italy 26B2
Savonlinna Finland 30F3
Saxmundham England 21J5
Saynshand Mongolia 35F2
Scafell Pike *see* **Lake District** 20D3
Scapa Flow an anchorage surrounded by the Orkney Islands, famous as a wartime naval base 22E2
Scarborough England 20G3
Schwarzwald (Black Forest) an extensive area of mountainous pine forests in SW Germany 28B3
Schwerin Germany 28C2
Scilly, Isles of a group of islands off the SW tip of England. The main islands are St Mary's, St Martin's and Tresco (pop. 2000) 20A8
Scotland a country of the UK, occupying the N part of Great Britain; capital Edinburgh (78,762 sq km/30,410 sq miles; pop. 5,035,000) 22
Scourie Scotland 22C2
Scunthorpe England 20G4
Seattle a port in Washington State, USA (pop. city 490,000/metropolitan area 1,677,000) 6A1
Seaward Peninsula USA 5B3
Sebes Romania 29E3
Ségou Mali 42C3
Segovia Spain 25B1
Seinäjoki Finland 30E3
Seine, River a river of N France, flowing through Paris to the English Channel (length 775 km/482 miles) 24C2
Sekondi Ghana 42C4
Selat Sunda (Sunda Strait) a strait, 26 km (16 miles) at its narrowest, separating Java and Sumatra 36C4
Selby England 20F4
Semarang Indonesia 36D4
Semipalatinsk Kazakhstan 32K4
Sendai Japan 35P7
Senegal [*area* 196,722 sq km (75,954 sq miles); *pop.* 7,170,000; *cap.* Dakar; *major cities* Kaolack, Thies,St Louis; *govt* republic; *religions* Sunni Islam, RC; *currency* franc CFA] a former French colony in West Africa extending from the most W point in Africa, Cape Verde, to the border with Mali 42B3
Sénégal River a West African river that flows through Guinea, Mali, Mauritania and Senegal to the Atlantic (length 1790 km/1110 miles) 42B3
Senlis France 24C2
Sennen England 20B7
Sens France 24C2
Seoul *see* **Soul** 35G3
Sepik a major river of Papua New Guinea (length 1200 km/750 miles) 37G4
Seram (Ceram) an island in the Maluku group, Indonesia (17,148 sq km/6621 sq miles) 37E4
Serbia (Srbija) the largest republic of former Yugoslavia. The capital is Belgrade (88,361 sq km/34,107 sq miles; pop. 9,314,000) 27E2

Sergino Russian Federation 32H3
Sérifos Greece 27E3
Serov Russian Federation 32H4
Serpukhov Russian Federation 32E4
Sérrai Greece 27E2
Sétif Algeria 42D1
Setúbal Portugal 25A2
Sevastopol' a Black Sea port of the Ukraine (pop. 335,000) 32E5
Severn, River the longest river in the UK, flowing through Wales and W England (length 350 km/220 miles) 21E6
Severnaya Zemlya Russian Federation 33L2
Severodvinsk Russian Federation 32E3
Sevilla (Seville) a historic, now industrial city in S Spain, and also the name of the surrounding province (pop. 653,800) 25A2
Seychelles [*area* 280 sq km (108 sq miles); *pop.* 67,378; *cap.* Victoria; *govt* republic; *religion* RC; *currency* Seychelles rupee] a group of volcanic islands in the Indian Ocean 1200 km (746 miles) from E Africa 44F1
Seydhisfödhur Iceland 30C1
Sézanne France 24C2
Sfax Tunisia 42E1
's-Gravenhage (The Hague; Den Haag) the administrative centre of the Netherlands, on the W coast (pop. 449,300) 28A2
Shado Shima Island Japan 35N7
Shahjahanpur India 39G3
Shakhty Russian Federation 32F5
Shanghai the largest city in China. An important port, it is situated on the delta of the Chang Jiang river (pop. 11,860,000) 35G3
Shangrao China 35F4
Shannon, River a river of the Republic of Ireland, the longest in the British Isles, flowing SW into the Atlantic near Limerick (length 386 km/240 miles) 23C3
Shantou China 35F4
Shanxi (province) China 35F3
Shaoguan China 35F4
Shaoxing China 35G4
Shaoyang China 35F4
Shapinsay Island Scotland 22F1
Shashi China 35F3
Sheffield a major industrial city in South Yorkshire, England (pop. 545,000) 20F4
Shenyang the capital of Liaoning province, China (pop. 4,000,000) 35G2
Shetland Islands a group of some 100 islands, 160 km (100 miles) NE of mainland Scotland; capital Lerwick (1426 sq km/550 sq miles; pop. 28,000) 22J7
Shijiazhuang the capital of Hebei province, China (pop. 973,000) 35F3
Shikoku the smallest of the four main islands of Japan (pop. 4,227,200) 35H3
Shillong India 39H3
Shimizu Japan 35N9
Shingu Japan 35L10
Shiraz Iran 38D3
Shizuoka Japan 35N9
Shkodër Albania 27D2
Shreveport USA 7D2
Shrewsbury England 21E5
Shropshire a county of W central England; county town Shrewsbury (3490 sq km/1347 sq miles; pop. 390,000) 21E5
Shuangyashan China 35H2
Sialkot Pakistan 39F2
Šiauliai Lithuania 30E5
Sibenik Croatia 26D2
Siberia *see* **Sibirskoye** 33L3
Siberut Island Indonesia 36B4
Sibiu Romania 29E3
Sibolga Indonesia 36B3
Sibirskoye (Siberia) a huge tract of land, mostly in N Russian Federation, that extends from the Urals to the Pacific. It is renowned for its inhospitable climate, but many parts are fertile, and it is rich in valuable minerals 33L3
Sibu Malaysia 36D3
Sicilia (Sicily) an island off the "toe" of Italy and largest in the Mediterranean. The capital is Palermo (25,708 sq km/9926 sq miles; pop. 5,065,000) 26C3
Sidi Bel Abbès Algeria 42C1
Siedlce Poland 29E2
Siegen Germany 28B2
Siena (Sienna) Italy 26C2
Sierra Leone [*area* 71,740 sq km (27,699 sq miles); *pop.* 4,140,000; *cap.* Freetown; *govt* republic; *religion* Animism, Sunni Islam, Christianity; *currency* leone] a country on the Atlantic coast of W Africa 42B4
Sierra Madre Occidental Mexico 6C3
Sierra Madre Oriental Mexico 6C3
Sierra Nevada a mountain range in E California, USA 12B3
Sifnos Greece 27E3
Sigüenza Spain 25B1
Siguiri Guinea 42C3
Sikasso Mali 42C3
Sikinos Greece 27E3
Simeulue Island Indonesia 36B4
Simpson Desert an arid, uninhabited region in the centre of Australia 45C3
Sinai a mountainous peninsula in NE Egypt, bordering Israel 40B4
Singapore [*area* 618 sq km (239 sq miles); *pop.* 2,690,000; *cap.* Singapore; *govt* republic; *religions* Buddhism, Sunni Islam, Christianity; *currency* Singapore dollar] one of the world's smallest countries, comprising 60 islands at the foot of the Malay peninsula in SE Asia 36C3
Singkawang Indonesia 36C3
Sintra Portugal 25A2
Sioux Falls USA 6D1
Siping China 35G2
Sipora Island Indonesia 36B4
Siracusa (Syracuse) an ancient seaport on the E coast of Sicily, Italy (pop. 119,200) 26D3
Síros Greece 27E3
Sirte, Gulf of a huge indent of the Mediterranean Sea in the coastline of Libya 43E1
Sisak Croatia 26D1
Sittwe Myanmar 36B1
Sivas Turkey 38B2
Sjaelland (Zealand) the largest island of Denmark, on which the capital, Copenhagen, is sited (7014 sq km/2708 sq miles; pop. 1,855,500) 30C4
Skagerrak the channel, some 130 km (80 miles) wide, separating Denmark and Norway 30B4
Skara Sweden 30C4
Skegness England 20H4
Skellefteå Sweden 30E3
Skiathos the westernmost of the Greek Sporades (Dodecanese) Islands (pop. 4200) 27E3
Skien Norway 30B4
Skikda Algeria 42D1
Skiros Greece 27E3
Skopélos Greece 27E3
Skopje the capital of Macedonia, a republic in the S of former Yugoslavia (pop. 506,500) 27E2
Skovorodino Russian Federation 33O4
Skye an island off the NW coast of Scotland; the largest of the Inner Hebrides. The main town is Portree (1417 sq km/547 sq miles; pop. 8000) 22B3
Slatina Romania 27E2
Sligo a county on the NW coast of the Republic of Ireland,

Sliven with a county town of the same name (1796 sq km/693 sq miles; pop. county 55,400) 23C2
Sliven Bulgaria 27F2
Slovakia [*area* 49,032 sq km (18,931 sq miles); *pop.* 5,013,000; *cap.* Bratislava; *major city* Kovice; *govt* republic; *religion* RC; *currency* koruna] a landlocked country in central Europe, became a republic on January 1, 1993 with the dissolution of Czechoslovakia 29D3
Slovenia [*area* 20,251 sq km (7817 sq miles); *pop.* 1,998,912; *cap.* Ljubljana; *major cities* Maribor, Celje; *govt* republic; *religion* RC; *currency* Slovene tolar] a country that made a unilateral declaration of independence from former Yugoslavia on June 25, 1991 26C1
Smolensk an industrial city in the Russian Federation, on the River Dnieper (pop. 326,000) 32E4
Smyrna *see* **Izmir** 38A2
Snake, River a river of the NW USA, which flows into the Columbia River in Washington State (length 1670 km/1038 miles) 6B1
Snowdonia a mountainous region in N Wales. The highest peak is Mount Snowdon (1085 m/3560 ft) 20C4
Snowy Mountains a range in SE Australia, where the River Snowy has been dammed. The highest peak is Mount Kosciusko (2230 m/7316 ft) 45D4
Sobral Brazil 16E3
Socotra an island in the NW Indian Ocean, belonging to Yemen 38D4
Sodankylä Finland 30F2
Söderhamn Sweden 30D3
Södertälje Sweden 30D4
Sofiya (Sofia) the capital of Bulgaria, in the W of the country (pop. 1,093,800) 27E2
Sokodé Togo 42D4
Sokoto Nigeria 42D3
Solapur India 39F4
Solent, The a strait in the English Channel that separates the Isle of Wight from mainland England 21F7
Sollefteå Sweden 30D3
Solomon Islands [*area* 28,896 sq km (11,157 sq miles); *pop.* 308,796; *cap.* Honiara; *govt* constitutional monarchy; *religions* Anglicanism, RC, other Christianity; *currency* Solomon Island dollar] a nation consisting of six large islands and numerous smaller ones between 5 and 12° S of the Equator, to the E of Papua New Guinea 45F2
Somalia [*area* 637,657 sq km (246,199 sq miles); *pop.* 6,260,000; *cap.* Mogadishu; *major cities* Hargeisa, Baidoa, Burao, Kismaayo; *govt* republic; *religion* Sunni Islam; *currency* Somali shilling] a on the "horn" of Africa's E coast 43H4
Somerset a county in SW England; county town Taunton (3458 sq km/1335 sq miles; pop. 440,000) 21D6
Somerset Island Canada 5J2
Sondrio Italy 26B1
Song Hong *see* **Hong** 34E4
Songkhla Thailand 36C3
Sorocaba Brazil 16E5
Sorong Indonesia 37F4
Soroti Uganda 43G4
Sorrento Italy 26C2
Sorsele Sweden 30D2
Sosnowiec Poland 29D2
Souillac France 24C3
Soul (Seoul) the capital of South Korea, in its NW (pop. 8,364,000) 35G3
South Africa [*area* 1,221,037 sq km (471,442 sq miles); *pop.* 30,190,000; *cap.* Pretoria (administrative), Cape Town (legislative); *major cities.* Johannesburg, Durban, Port Elizabeth, Bloemfontein; *govt* republic; *religions* Dutch Reform, Independent African, other Christianity, Hinduism; *currency* rand] a republic at the S tip of Africa with a huge coastline on the Atlantic and Indian Oceans 44C4
Southampton a major port in S England (pop. 206,000) 21F7
Southampton Island Canada 5K3
South Australia a state in central S Australia, on the Great Australian Bight; capital Adelaide (984,380 sq km/380,069 sq miles; pop. 1,347,000) 45C3
South Carolina a state in SE USA, with a coast on the Atlantic Ocean; capital Columbia (80,432 sq km/31,055 sq miles; pop. 3,347,000) 7E2
South China Sea an arm of the Pacific Ocean between SE China, Malaysia and the Philippines 36D2
South Dakota a state in the W USA; capital Pierre (199,552 sq km/77,047 sq miles; pop. 708,000) 6C1
Southend-on-Sea England 21H6
Southern Alps a range of mountains on the South Island of New Zealand 45F5
South Georgia an island in the South Atlantic, and a dependency of the Falkland Islands (3755 sq km/1450 sq km) 16F8
South Glamorgan a county in S Wales; administrative centre Cardiff (416 sq km/161 sq miles; pop. 384,700) 21D6
South Island New Zealand 45F5
South Korea *see* **Korea** 35G3
South Pole the most southerly point of the Earth's axis, in Antarctica 48
Southport England 20D4
South Ronaldsay Scotland 22F2
South Shields England 20F3
South Uist Island Scotland 22A3
South Yorkshire a county of NE England created in 1974 by the division of the former county of Yorkshire; administrative centre Barnsley (1560 sq km/602 sq miles; pop. 1,302,000) 20F4
Soweto a group of black townships to the S of Johannesburg, South Africa (pop. 829,400) 44C3
Spain [*area* 504,782 sq km (194,896 sq miles); *pop.* 39,540,000; *cap.* Madrid; *major cities* Barcelona, Seville, Zaragosa, Malaga, Bilbao; *govt* constitutional monarchy; *religion* RC; *currency* peseta] a country occupying the greater part of the Iberian peninsula 25
Spalding England 21G5
Spitsbergen a large island group in the Svalbard archipelago, 580 km (360 miles) N of Norway (39,000 sq km/15,060 sq miles; pop. 2000) 32C2
Split the largest city on the coast of Dalmatia, Croatia (pop. 236,000) 26D2
Spokane USA 6B1
Spratly Islands 36D2
Springfield the state capital of Illinois, USA (pop. 101,600/metropolitan area 190,100) 7E2
Springfield a manufacturing city in Massachusetts, USA (pop. city 150,300/metropolitan area 515,900) 8E2
Springfield Missouri USA 7D2
Springs South Africa 44C3
Srbija *see* **Serbia** 27E2
Sri Lanka [*area* 65,610 sq km (25,332 sq miles); *pop.* 16,810,000; *cap.* Colombo; *major cities* Dehiwela-Mt Lavinia, Moratuwa, Jaffna; *govt* republic; *religions* Buddhism, Hinduism, Christianity, Sunni Islam; *currency* Sri Lankan rupee] a large island in the Indian Ocean, S of India 39G5
Srinagar the capital of the state of Jammu and Kashmir, N India (pop. 606,000) 39F2
Stafford England 21E5
Staffordshire a Midlands county of England; county town Stafford (2716 sq km/1049 sq miles; pop. 1,018,000) 21E5
St Albans England 21G6
St Andrews Scotland 22F4
Stanley (Port Stanley) the capital of the Falkland Islands (pop. 1000) 17D8

Stanley, Mount *see* **Ruwenzori**	43G4
Stanleyville *see* **Kisangani**	43F4
Stara Zagora Bulgaria	27F2
Starbuck Island Kiribati	46
St Austell England	21C7
Stavanger Norway	30B4
Stavropol' Russian Federation	32F5
St Brieuc France	24B2
St Christopher and Nevis *see* **St Kitts-Nevis**	14G3
St Croix the largest of the US Virgin Islands. The main town is Christiansted (218 sq km/84 sq miles; pop. 50,000)	14F3
St David's Wales	20B6
St Denis Réunion	44F4
St Dizier France	24C2
Steinkjer Norway	30C3
St-Etienne France	24C2
Stettin *see* **Szczecin**	28C2
Stewart Island New Zealand	45F5
St Gaudens France	24C3
St George's the capital of Grenada, and the island's main port (pop. 30,800)	14G4
St Helens England	20E4
St Helier Jersey	21E8
Stirling Scotland	22E4
St Ives England	20B7
St John (Saint John) a port at the mouth of the Saint John River, on the Atlantic coast of New Brunswick, Canada (pop. 114,000)	5M5
St John's the capital and main port of Antigua (pop. 30,000)	14G3
St John's a port and the capital of Newfoundland, Canada (pop. 155,000)	5N5
St Kitts-Nevis (St Christopher and Nevis) [*area* 261 sq km (101 sq miles); *pop.* 43,410; *cap.* Basseterre; *govt* constitutional monarchy; *religions* Anglicanism, Methodism; *currency* East Caribbean dollar] the islands of St Christopher and Nevis lying in the Leeward group in the E Caribbean	14G3
St Lawrence, Gulf of an arm of the Atlantic Ocean in NE Canada, into which the St Lawrence River flows	5M5
St Lawrence, River a commercially important river of SE Canada, flowing NE from Lake Ontario to the Gulf of St Lawrence, forming part of the border with the USA (length 1197 km/744 miles)	7F1
St Lawrence Island USA	5A3, 6J, 33U3
St Louis Senegal	42B3
St Louis a city in E Missouri, USA, on the Mississippi (pop. city 429,300/metropolitan area 2,398,400)	7D2
St Lucia [*area* 622 sq km (240 sq miles); *pop.* 146,600; *cap.* Castries; *govt* constitutional monarchy; *religion* RC; *currency* East Caribbean dollar] one of the Windward Islands in the E Caribbean, formed of extinct volcanoes	14G4
St Malo France	24B2
St Martin one of the Leeward Islands in the SE Caribbean, divided politically between Guadeloupe (France) and the Netherlands Antilles	14G3
St Martin's England	20A8
St Mary's England	20A8
St Nazaire France	24B2
Stockholm the capital of Sweden, and an important port on the Baltic Sea (pop. 1,435,500)	30D4
Stockport England	20E4
Stoke-on-Trent England	20E4
Stonehaven Scotland	22F4
Stören Norway	30C3
Storlien Sweden	30C3
Stornoway Scotland	22B2
Storuman Sweden	30D2
St Paul the state capital of Minnesota, twinned with the adjoining city of Minneapolis (pop. city 265,900/ metropolitan area 2,230,900)	7D1
St Peter Port Guernsey	21E8
St Petersburg *see* **Sankt Peterburg**	32E4
St Petersburg USA	7E3
St Quentin France	24C2
Strabane Northern Ireland	23D2
Stralsund Germany	28C2
Stranraer Scotland	22D6
Strasbourg an industrial city and river port in E France, the capital of Alsace region, and seat of the European Parliament (pop. 378,500)	24D2
Stratford-on-Avon (Stratford-upon-Avon) a town in Warwickshire, England, the birthplace of William Shakespeare (pop. 22,000)	21F5
Strathclyde an administrative region in W Scotland; administrative centre Glasgow. It was created in 1975 out of the former counties of Ayrshire, Lanarkshire, Renfrewshire, Bute, Dunbartonshire and parts of Stirlingshire and Argyll (13,856 sq km/5350 sq miles; pop. 2,373,000)	22D5
Streymoy Island Denmark	30A2
Strömsund Sweden	30D3
Stronsay Island Scotland	22F1
Stroud England	21E6
St Tropez France	24D3
Stuttgart a major industrial centre and river port of the Neckar river in SW Germany (pop. 600,000)	28B3
St Vincent and the Grenadines [*area* 388 sq km (150 sq miles); *pop.* 113,950; *cap.* Kingstown; *govt* constitutional monarchy; *religions* Anglicanism, Methodism, RC; *currency* East Caribbean dollar] an island in the E Caribbean, separated from Grenada by the Grenadines, the N islands of are part of the country	14G4
St Vincent, Cape *see* **São Vicente, Cabo de**	25B2
Subotica Yugoslavia	27D1
Suceava Romania	29F3
Sucre the legal capital of Bolivia (pop. 70,000)	16C4
Sudan [*area* 2,505,813 sq km (967,494 sq miles); *pop.* 25,560,000; *cap.* Khartoum (El Khartum); *major cities* Omdurman, Khartoum North, Port Sudan; *govt* republic; *religions* Sunni Islam, Animism, Christianity; *currency* Sudanese pound] the largest country in Africa, covering much of the upper Nile basin	43G3
Suduroy Island Denmark	30A2
Suez (El Suweis) a town at the S end of the Suez Canal (pop. 195,000)	43G2
Suez, Gulf of a N arm of the Red Sea that leads to the Suez Canal	43G2
Suez Canal a canal in NE Egypt, linking the Mediterranean to the Red Sea, completed in 1869	43G2
Suffolk a county in East Anglia, England; county town Ipswich (3800 sq km/1467 sq miles; pop. 619,000)	21H5
Sukhumi Georgia	32F5
Sukkur Pakistan	39E3
Sulawesi (Celebes) a large island in the centre of Indonesia (179,370 sq km/69,255 sq miles; pop. 10,409,600)	37E4
Sulu Archipelago a chain of over 400 islands off the SW Philippines, between there and Borneo	37E3
Sulu Sea a part of the Pacific Ocean which lies between the Philippines and Borneo	37E3
Sumatra the main island of W Indonesia (473,607 sq km/ 182,860 sq miles; pop. 28,016,200)	36B3
Sumba one of the Lesser Sunda Islands, Indonesia (11,153 sq km/4306 sq miles; pop. 251,100)	36D4
Sumbawa one of the Lesser Sunda Islands, Indonesia (15,448 sq km/5965 sq miles; pop. 195,000)	36D4
Sumen Bulgaria	27F2
Sumy Ukraine	32E4
Sunda Strait *see* **Selat Sunda**	36C4
Sunderland an industrial town in Tyne and Wear, England (pop. 200,000)	20F3
Sundsvall Sweden	30D3
Suntar Russian Federation	33N3

Superior, Lake the largest and most westerly of the Great Lakes (82,400 sq km/31,800 sq miles) 7E1
Surabaya the second largest city of Indonesia after Jakarta, on the NE coast of Java (pop. 2,470,000) 36D4
Surakarta Indonesia 36D4
Surat India 39F3
Surgut Russian Federation 32J3
Suriname [*area* 163,265 sq km (63,037 sq miles); *pop.* 416,839; *cap.* Paramaribo; *govt* republic; *religions* Hinduism, RC, Sunni Islam; *currency* Suriname guilder] a country in NE South America 16D2
Surrey a county of central S England; county town Guildford (1655 sq km/639 sq miles; pop. 1,012,000) 21G6
Surtsey Island Iceland 30A2
Suva the capital of Fiji (pop. city 74,000/metropolitan area 133,000) 46
Suzhou China 35G3
Svalbard an archipelago in the Arctic Ocean to the N of Norway, which has sovereignty (62,049 sq km/23,958 sq miles; pop. 3500) 32C2
Sveg Sweden 30C3
Sverdrup Islands Canada 5H2
Svetozarevo Yugoslavia 27E2
Swakopmund South Africa 44B3
Swan Island Honduras 14C3
Swansea a port in S Wales (pop. 168,000) 21D6
Swaziland [*area* 1736 sq km (670 sq miles); *pop.* 681,059; capital Mbabane; *major cities* Big Bend, Manzini, Mhlume; *govt* monarchy; *religion* Christianity, Animism; *currency* emalangeni] a landlocked hilly enclave almost entirely within the Republic of South Africa 44D3
Sweden [*area* 449,964 sq km (173,731 sq miles); *pop.* 8,500,000; *cap.* Stockholm; *major cities* Göteborg, Malmö, Uppsala, Orebro; *govt* constitutional monarchy; *religion* Lutheranism; *currency* krona] a large country in N Europe, making up half the Scandinavian peninsula 30
Swindon England 21F6
Switzerland [*area* 41,293 sq km (15,943 sq miles); *pop.* 6,700,000; Berne (Bern); *major cities* Zürich, Basle, Geneva, Lausanne; *govt* federal republic; *religions* RC, Protestantism; *currency* Swiss franc] a landlocked country in central Europe 28B3
Sydney the largest city and port in Australia, and the capital of New South Wales (pop. 3,332,600) 45E4
Syktyvkar Russian Federation 32G3
Syracuse *see* **Siracusa** 26D3
Syracuse a city in the centre of New York State (pop. city 164,200/metropolitan area 650,000) 7F1
Syrdar'ya, River a river of central Asia, flowing through Kazakhstan to the Aral Sea (length 2860 km/1780 miles) 32H5
Syria [*area* 185,180 sq km (71,498 sq miles); *pop.* 11,300,000; *cap.* Damascus (Dimashq); *major cities* Aleppo, Homs,Lattakia, Hama; *govt* republic; *religion* Sunni Islam; *currency* Syrian pound] a country in SW Asia bordering on the Mediterranean Sea in the W 38B2
Syzran' Russian Federation 32F4
Szczecin (Stettin) Poland 28C2
Szczecinek Poland 28D2
Szeged Hungary 29E3
Székesfehérvár Hungary 29D3
Szekszárd Hungary 29D3
Szolnok Hungary 29E3

T

Tabora Tanzania 43G5
Tabriz a city in NW Iran (pop. 599,000) 38C2
Tabuaeran Island Kiribati 46
Tabuk Saudi Arabia 38B3
Taegu South Korea 35G3
Taejon South Korea 35G3
Tagus *see* **Tajo** 25B2
Tahiti the largest of the islands of French Polynesia in the South Pacific. The capital is Papeete (1005 sq km/ 388 sq miles; pop. 96,000) 47
Tahoua Niger 42D3
T'ai-nan a city in SW Taiwan (pop. 595,000) 37E1
T'aipei the capital and largest city of Taiwan, in the N (pop. 2,272,000) 37E1
Taiwan [*area* 36,179 sq km (13,969 sq miles); *pop.* 20,300,000; *cap.* Taipei; *major cities* Kaohsiung, Tai-chung, Tainan; *govt* republic; *religions* Taoism, Buddhism, Christianity; *currency* New Taiwan dollar] a large mountainous island about 160 km (99 miles) off the SE coast of mainland China 37E1
Taiwan Strait the stretch of water between Taiwan and China 36D1
Taiyuan China 35F3
Ta'izz Yemen 38C4
Tajikistan [*area* 143,100 sq km (55,250 sq miles); *pop.* 5,100,000; *cap.* Dushanbe; *govt* republic; *religion* Shia Islam; *currency* rouble] a republic of S central former USSR, which declared its independence in 1991 32H6
Tajo (Tagus; Tejo), River a major river of SW Europe, rising in E Spain and flowing W and SW through Portugal to the Atlantic (iength 1007 km/626 miles) 25B2
Takada Japan 35N8
Takaoka Japan 35M8
Takasaki Japan 35N8
Taklimakan Desert the largest desert in China, consisting mainly of sand, in the W of the country 39G2
Takoradi Ghana 42C4
Talavera de la Reina Spain 25B2
Talca Chile 17B6
Talcahuano Chile 17B6
Tallahassee the state capital of Florida, USA (pop. city 112,000/metropolitan area 207,600) 7E2
Tallinn a port on the Baltic Sea, and capital of Estonia (pop. 458,000) 32D4
Tamale Ghana 42C4
Tamanrasset Algeria 42D2
Tambacounda Senegal 42B3
Tampa Florida USA 7E3
Tampere (Tammerfors) the second largest city in Finland, in the SW of the country (pop. 167,000) 30E3
Tampico Mexico 6D3
Tamworth Australia 45E4
Tana (Tsana), Lake a lake in the mountains of NW Ethiopia, source of the Blue Nile (3673 sq km/1418 sq miles) 43G3
Tanga Tanzania 43G5
Tanganyika, Lake the second largest lake in Africa after Lake Victoria, in the Great Rift Valley, between Tanzania and Zaïre, although Burundi and Zambia also share the shoreline (32,893 sq km/12,700 sq miles) 43G5
Tanger (Tangier) a port on the N coast of Morocco, on the Strait of Gibraltar (pop. 188,000) 42C1
Tangshan China 35F3
Tanna Island Vanuatu 45F2
Tanta Egypt 43G1
Tanzania [*area* 945,087 sq km (364,898 sq miles); *pop.* 24,800,000; *cap.* Dodoma; *major cities* Dar es Salaam, Zanzibar, Mwanza, Tanga; *govt* republic; *religions* Sunni Islam, RC, Anglicanism, Hinduism; *currency* Tanzanian shilling] a country comprising a mainland in E Africa and the islands of Pemba and Zanzibar 43G5
Taolanaro Madagascar 44E3
Tarabulus (Tripoli) the capital and main port of Libya, in the NW (pop. 620,000) 42E1
Tarakan Indonesia/Malaysia 36D3
Tarancón Spain 25B1

Taranto | 1032 | Titicaca, Lake

Taranto Italy 27D2
Taranto, Golfo de an inlet of the Mediterranean Sea between the "toe" and "heel" of Italy 27D3
Tarbert Ireland 23B4
Tarbert Strathclyde Scotland 22C5
Tarbert Western Isles Scotland 22B3
Tarbes France 24C3
Tarcoola Australia 45C4
Tarfaya Morocco 42B2
Tarnów Poland 29E2
Tarragona Spain 25C1
Tarrasa Spain 25C1
Tarutung Indonesia 36B3
Tashkent the capital of Uzbekistan, in the NE (pop. 1,987,000) 32H5
Tasmania an island state to the S of Australia, separated from it by the Bass Strait; capital Hobart (68,332 sq km/ 26,383 sq miles; pop. 435,000) 45D5
Tasman Sea a branch of the Pacific Ocean that separates Australia and New Zealand 45E4
Tatung *see* **Datong** 35F2
Taunton England 21D6
Tavira Portugal 25A2
Tavoy Myanmar 36B2
Tawau Malaysia 36D3
Tawitawi Philippines 37E3
Tayside an administrative region of Scotland formed in 1975 out of the former counties of Angus, Kinrossshire and part of Perthshire; administrative centre Dundee (7511 sq km/2900 sq miles; pop. 392,000) 22E4
Tbilisi the capital of Georgia, in the centre of the republic (pop. 1,140,000) 32F5
Tecuci Romania 29F3
Tegucigalpa the capital of Honduras, in its S (pop. 473,700) 14B4
Tehran the capital of Iran, in the central N of the country (pop. 6,000,000) 38D2
Tehuantepec Mexico 7D4
Tel Aviv-Jaffa the largest city of Israel, former capital and main financial centre (pop. 324,000) 38B2
Telford England 21E5
Telukbetung Indonesia 36C4
Temuco Chile 17B6
Tenerife the largest of the Canary Islands; capital Santa Cruz (2058 sq km/795 sq miles; pop. 558,000) 42B2
Tennant Creek Australia 45C2
Tennessee a state in S central USA; capital Nashville (109,412 sq km/42,244 sq miles; pop. 4,762,000) 7E2
Tennessee, River a river which flows SW from the Appalachian Mountains of North Carolina, then to Alabama, Tennessee and Kentucky to join the Ohio River (length 1049km/ 652miles) 11E3
Teófilo Otôni Brazil 16E4
Teresina Brazil 16E3
Termez Uzbekistan 32H6
Termoli Italy 26C2
Terni Italy 26C2
Teruel Spain 25B1
Tessalit Mali 42D2
Tete Mozambique 44D2
Tétouan Morocco 42C1
Tevere, River a river of central Italy, rising E of Florence and flowing S to Rome and into the Mediterranean (length 405 km/252 miles) 26C2
Teviothead Scotland 22F5
Texas a state in SW USA, bordering Mexico, the nation's second largest state; capital Austin (678,927 sq km/262,134 sq miles; pop. 16,370,000) 6C2
Thailand [*area* 513,115 sq km (198,114 sq miles); *pop.* 55,900,00; *cap.* Bangkok (Krung Thep); *major cities* Chiengmai, Hat Yai, Songkhla; *govt* constitutional monarchy; *religions* Buddhism, Sunni Islam; *currency* baht] a tropical country in SE Asia 36C2
Thailand, Gulf of a branch of the South China Sea lying between the Malay peninsula and the coasts of Thailand, Cambodia and Vietnam 36C2
Thames, River a major river of S England flowing E from the Cotswolds, through London to its estuary on the North Sea (length 338 km/210 miles) 21F6
Thar Desert (Indian Desert) a desert in NW India, covering the border with Pakistan 29F3
Thásos Island Greece 27E2
Thessaloníki (Saloniki) the second largest city in Greece (pop. 706,200) 27E2
Thetford England 21H5
Thiès Senegal 42B3
Thimphu (Thimbu) the capital of Bhutan, in its W (pop. 8922) 39G3
Thionville France 24D2
Thíra (Santorini) a volcanic island in the Cyclades, Greece (84 sq km/32 sq miles; pop. 7100) 27F3
Thival Greece 27E3
Thiviers France 24C2
Thon Buri Thailand 36C2
Thule Greenland 5M2
Thunder Bay Canada 5K5
Thurles Ireland 23D4
Thurso Scotland 22E2
Thurston Island Antarctica P170
Tianjin (Tientsin) a major industrial city in Hebei province, the third largest city in China after Shanghai and Beijing (pop. 7,390,000) 35F3
Tianshui China 34E3
Tiber, River *see* **Tevere** 26C2
Tibet (Xizang Autonomous Region) a region of SW China, consisting of a huge plateau in the Himalayas. Formerly a Buddhist kingdom, it was invaded by China in 1950 and has been desecrated (1,221,600 sq km/471,660 sq miles; pop. 1,893,000) 34C3
Tidjikdja Mauritania 42B3
Tientsin *see* **Tianjin** 35F3
Tierp Sweden 30D3
Tierra del Fuego an archipelago at the S tip of South America, belonging to Argentina and Chile, separated from the mainland by the Strait of Magellan 16C8
Tigris a major river of the Middle East, rising in E Turkey, flowing through Syria and Iraq and joining the Euphrates to form a delta at the Shatt al Arab as it enters The Gulf (length 1900 km/1180 miles) 43H1
Tijuana Mexico 6B2
Tilburg Netherlands 28A2
Tílos Island Greece 27F3
Timbákion Greece 27E3
Timbuktu *see* **Tombouctou** 42C3
Timimoun Algeria 42D2
Timisoara Romania 29E3
Timor an island at the E end of the Lesser Sunda Islands, Indonesia. The E half was a possession of Portugal, but was annexed in 1975 by Indonesia (30,775 sq km/11,883 sq miles; pop. 3,085,000) 37E4
Timor Sea the arm of the Indian Ocean between the NW coast of Australia and the island of Timor 45B2
Tindouf Algeria 42C2
Tipperary a county in S Republic of Ireland. It includes the town of Tipperary; county town Clonmel (4255 sq km/ 1643 sq miles; pop. 135,000) 23C4
Tiranë (Tirana) the largest city and the capital of Albania, in the centre of the country (pop. 220,000) 27D2
Tiree Island Scotland 22B4
Tirgu Mures Romania 29E3
Tiruchirāppalli (Trichinopoly) India 39F4
Titicaca, Lake the largest lake in South America, in the Andes, on the border between Bolivia and Peru (8135 sq km/3141 sq miles) 16B4

Titograd   Tyne and Wear

Titograd Yugoslavia 27D2
Titov Veles Yugoslavia 27E2
Tlemcen Algeria 42C1
Toamasina Madagascar 44E2
Tobago *see* **Trinidad and Tobago** 14G4
Tobermory Scotland 22B4
Tobol'sk Russian Federation 32H4
Togo [*area* 56,785 sq km (21,925 sq miles); *pop.* 3,400,000; *cap.* Lomé; *govt* republic; *religions* Animism, RC, Sunni Islam; *currency* franc CFA] a very small W African country with a narrow coastal plain on the Gulf of Guinea 42D4
Tokelau Islands Pacific Ocean 46
Tokushima Japan 35H3, 35L9
Tokyo the capital of Japan, a port on the E coast of Honshu Island. Its original name was Edo (until 1868) (pop. city 8,353,700/Greater Tokyo 11,680,000) 35H9
Toledo a historic city of central Spain. (pop. 60,100) 25B2
Toledo a city and Great Lake port in Ohio, USA (pop. city 343,900/metropolitan area 610,800) 7E1
Toliara Madagascar 44E3
Tombouctou (Timbuktu) a town in central Mali at the edge of the Sahara (pop. 20,000) 42C3
Tomsk Russian Federation 32K4
Tonbridge England 21H6
Tonga [*area* 750 sq km (290 sq miles); *pop.* 95,200; *cap.* Nuku'alofa; *govt* constitutional monarchy; *religions* Methodism, RC; *currency* pa'anga] a country comprising over 170 islands, about one-fifth of which are inhabited, situated 20° S of the Equator, W of the International Date Line in the Pacific 46
Tonghua China 14G2
Tongling China 14F3
Tongue Scotland 22D2
Tonle Sap a lake in central Cambodia which swells and quadruples in size when the River Mekong floods (In flood 10,400 sq km/4000 sq miles) 36C2
Toowoomba Australia 45E3
Topeka the state capital of Kansas, USA (pop. city 119,000/metropolitan area 159,000) 6D2
Tordesillas Spain 25A1
Torino (Turin) Italy 26B1
Tornio Finland 30E2
Toronto the largest city of Canada and capital of Ontario, situated on Lake Ontario (pop. 2,999,000) 5L5
Torquay England 21D7
Torreón Mexico 6C3
Torres Strait between the NE tip of Australia and New Guinea 45D2
Torridon Scotland 22C3
Tortosa Spain 25C1
Torun Poland 29D2
Tóshavn Faroes 30A2
Tottori Japan 35L9
Touggourt Algeria 42D1
Toulon a major naval base and port in SE France (pop. 418,000) 24D3
Toulouse a city of SW France, on the Garonne River (pop. 551,000) 24C3
Tourcoing France 24C1
Tours France 24C2
Townsville Australia 45D2
Toyama Japan 35M8
Toyohashi Japan 35M9
Tozeur Tunisia 42D1
Trâblous a port in N Lebanon (pop. 175,000) 40C1
Trabzon (Trebizond) a port on the Black Sea in NE Turkey (pop. 156,000) 38B1
Tralee Ireland 23B4
Trangan Island Indonesia 37F4
Trapani Italy 26C3
Trent, River the main river of the Midlands of England, flowing NE from Staffordshire to the Humber (length 270 km/170 miles) 20G4
Trenton a city in E USA on the Delaware River in W New Jersey, of which it is the capital (pop. 92,124) 7F1
Trichinopoly *see* **Tiruchiràppalli** 39F4
Trieste Italy 26C1
Trincomalee Sri Lanka 39G5
Trinidad and Tobago [*area* 5130 sq km (1981 sq miles); *pop.* 1,240,000; *cap.* Port-of-Spain; *govt* republic; *religions* RC, Hinduism, Anglicanism, Sunni Islam; *currency* Trinidad and Tobago dollar] a country consisting of the two most southerly of the Lesser Antilles, situated off NE Venezuela 14G4
Tripoli *see* **Tarabulus** 42E1
Tripolis Greece 27E3
Tromsö Norway 30D2
Trondheim Norway 30C3
Trowbridge England 21E6
Troyes France 24C2
Trujillo Peru 16B3
Truro England 20B7
Tselinograd Kazakhstan 32J4
Tsetserleg Mongolia 34E2
Tsuchiura Japan 35P8
Tsumeb Namibia 44B2
Tsuruga Japan 35M9
Tsuruoka Japan 35N7
Tsushima Strait *see* **Korea Strait** 35G3
Tuamotu Islands Pacific Ocean 46
Tubruq Libya 43F1
Tubuai Islands Pacific Ocean 46
Tucson Arizona USA 6B2
Tudela Spain 25B1
Tula Russian Federation 32E4
Tullamore Ireland 23D3
Tulsa Oklahoma 6D2
Tunis the capital and main port of Tunisia (pop. 550,000) 42E1
Tunisia [*area* 163,610 sq km (63,170 sq miles); *pop.* 7,750,000; *cap.* Tunis; *major cities* Sfax, Bizerta, Djerba; *govt* republic; *religion* Sunni Islam; *currency* Tunisian dinar] a N African country N of the Sahara 42D1
Turda Romania 29E3
Turin *see* **Torino** 26B1
Turkey [*area* 779,452 sq km (300,946 sq miles); *pop.* 50,670,000; *cap.* Ankara; *major cities* Istanbul, Izmir, Adana, Bursa; *govt* republic; *religion* Sunni Islam; *currency* Turkish lira] a country with land on the continents of Europe and Asia 27F3, 32E5
Turkmenistan [*area* 488,100 sq km (186,400 sq miles); *pop.* 3,600,000; *cap.* Ashkhabad; *govt* republic; *religion* Sunni Islam; *currency* rouble] a central Asian republic of the former USSR, which became independent in 1991 32H6
Turks and Caicos Islands a British colony in the NE West Indies consisting of some 14 main islands. The capital is Cockburn Town on Grand Turk (430 sq km/166 sq miles; pop. 7400) 14E2
Turku (Åbo) Finland 30E3
Turneff Islands Belize 14B3
Tuticorin India 39F5
Tuvalu [*area* 26 sq km (10 sq miles); *pop.* 8229; *cap.* Funafuti; *govt* constitutional monarchy; *religion* Protestantism; *currency* Australian dollar] a small country consisting of nine coral atolls N of Fiji in the S Pacific 46
Tuzla Bosnia 27D2
Tver Russian Federation 32E4
Tyne and Wear a metropolitan county in NE England, created in 1974 out of parts of Durham and Northumberland; administrative centre Sunderland (540 sq km/208 sq miles; pop. 1,145,000) 20F3

Tyrone   Vatican City State

Tyrone a county in W Northern Ireland; county town Omagh (3266 sq km/1260 sq miles; pop. 160,000)   23D2
Tyrrhenian Sea a part of the Mediterranean Sea between Sicily, Sardinia, and mainland Italy   26C2
Tyumen' Russian Federation   32H4
Tyy Spain   25A1

U

Uberaba Brazil   37E4
Udaipur India   39F3
Uddevalla Sweden   30C4
Ufa Russian Federation   32G4
Uganda [*area* 235,880 sq km (91,073 sq miles); *pop.* 17,000,000; *cap.* Kampala; *major cities* Jinja, Masaka, Mbale; *govt* republic; *religions* RC, Protestantism, Animism, Sunni Islam; *currency* Uganda shilling] a landlocked country in E central Africa   43G4
Uig Scotland   22B3
Ujung Padang a major port in SW Sulawesi, Indonesia, formerly known as Makassar (pop. 709,000)   36D4
Ukhta Russian Federation   32G3
Ukraine [*area* 603,700 sq km (233,100 sq miles); *pop.* 51,700,000; *cap.* Kiev; *major cities* Dnepropetrovsk, Donetsk, Kharkov, Odessa; *govt* republic; *religions* Russian Orthodox, RC; *currency* rouble] formerly a Soviet socialist republic, which became independent in 1991   32E5
Ulaanbaatar (Ulan Bator) the capital of Mongolia, in the central N of the country (pop. 440,000)   34E2
Ulan Ude Russian Federation   33M4
Ullapool Scotland   22C3
Ulm Germany   28B3
Ul'yanovsk Russian Federation   32F4
Umeå Sweden   30E3
Umtata South Africa   44C4
Ungava Peninsula Canada   5L3
Unimark Island USA   6J
United Arab Emirates (UAE) [*area* 83,600 sq km (32,278 sq miles); *pop.* 1,600,000; *cap.* Abu Dhabi; *major cities* Dubai, Sharjh, Ras al Khaymah; *govt:* monarchy (emirates); *religion* Sunni Islam; *currency* dirham] a federation of seven oil-rich sheikdoms in The Gulf   38D3
United Kingdom (UK) [*area* 244,100 sq km (94,247 sq miles); *pop.* 57,240,000; *cap.* London; *major cities* Birmingham, Manchester, Glasgow, Liverpool; *govt* constitutional monarchy; *religion* Anglicanism, RC, Presbyterianism, Methodism; *currency* pound sterling] a country in NW Europe, comprising the island of Great Britain and the NE of Ireland, plus many smaller islands   19C3
United States of America (USA) [*area* 9,372,614 sq km (3,618,766 sq miles); *pop.* 249,630,000; *cap.* Washington DC; *major cities* New York, Chicago, Detroit, Houston, Los Angeles, Philadelphia, San Diego, San Francisco; *govt* federal republic; *religion* Protestantism, RC, Judaism, Eastern Orthodox; *currency* US dollar] a country stretching across central North America, from the Atlantic to the Pacific, with fifty states, including outlying Alaska and Hawaii   6/7
Unst Island Scotland   22K7
Upington South Africa   44C3
Uppsala Sweden   30D4
Uralskiy Khrebet (Ural Mountains; Urals) a mountain range in W Russian Federation. Running N to S from the Arctic to the Aral Sea, it is the traditional dividing line between Europe and Asia. The highest point is Mount Narodnaya (1894 m/6214 ft)   32G4
Ural'sk Russian Federation   32G4
Urgench Uzbekistan   32H5

Urosevac Yugoslavia   27E2
Uruguay [*area* 177,414 sq km (68,500 sq miles); *pop.* 3,100,000; *cap.* Montevideo; *govt* republic; *religions* RC, Protestantism; *currency* Uruguayan nuevo peso] a small country on the E coast of South America   17D6
Ürümqi (Urumchi) China   34C2
Urziceni Romania   27F2
Üsküdar Turkey   38A1
Ussuriysk Russian Federation   33P5
Usti Czech Republic   28C2
Ust' Kamchatsk Russian Federation   33S4
Ust' Nera Russian Federation   33Q3
Utah a state in W USA; capital Salt Lake City. (212,628 sq km/82,096 sq miles; pop. 1,645,000)   6B2
Utiel Spain   25B2
Utrecht Netherlands   28A2
Utsjoki Finland   30F2
Utsunomiya Japan   35N8
Uzbekistan [*area* 449,500 sq km (173,546 sq miles); *pop.* 20,300,000; *cap.* Tashkent; *major city* Samarkand; *govt* republic; *religion* Sunni Islam; *currency* rouble] a central Asian republic of the former USSR, which declared itself independent in 1991   32H5

V

Vaasa Finland   30E3
Vadodara (Baroda) India   39F3
Vaduz Liechtenstein   24D2
Valdepeñas Spain   25B2
Valence France   24C3
Valencia Spain   25B2
Valencia Venezuela   16C1
Valencia de Alcántara Spain   25A2
Valenciennes France   24C1
Valladolid Spain   25B1
Valledupar Colombia   16B1
Valletta the capital of Malta (pop. 14,000)   42E1
Valls Spain   25C1
Valparaíso the main port of Chile, in the centre of the country (pop. 267,000)   17B6
Vancouver a major port and industrial centre in SE British Columbia, Canada, on the mainland opposite Vancouver Island (pop. 1,268,000)   5F5
Vancouver Island the largest island off the Pacific coast of North America, in SW Canada; capital Victoria (32,137 sq km/12,408 sq miles; pop. 390,000)   5F5
Vang Norway   30B3
Vannas Sweden   30D3
Vannes France   24B2
Vantaa Finland   30F3
Vanuatu [*area* 12,189 sq km (4706 sq miles); *pop.* 142,630; *cap.* Vila; *govt* republic; *religion* Protestantism, Animism; *currency* vatu] a country in the W Pacific comprising about eighty islands   45F2
Varanasi (Benares) a holy Hindu city on the banks of the River Ganges, NE India (pop. 798,000)   39G3
Varazdin Croatia   26D1
Vardö Norway   30G1
Varna Bulgaria   27F2
Varnamo Sweden   30C4
Västerås Sweden   30D4
Västervik Sweden   30D4
Vasto Italy   26C2
Vatican City State [*area* 44 hectares (108.7 acres); *pop.* 1000; *cap.* Vatican City (Citta del Vaticano); *govt* Papal Commission; *religion* RC; *currency* Vatican City lira] the world's smallest independent state and headquarters of the Roman Catholic Church in the heart of Rome.   262C

Växjö   Waterford

Växjö Sweden	30C4
Veglia *see* **Krk**	26C1
Veliko Turnovo Bulgaria	27F2
Vellore India	39F4
Venezia (Venice) a historic port built on islands at the head of the Adriatic Sea in NE Italy. The principal thoroughfares are canals (pop. 346,000)	26C1
Venezuela [area 912,050 sq km (352,143 sq miles); pop. 9,250,000; cap. Caracas; major cities Maracaibo, Valencia; govt federal republic; religion RC; currency bolívar] a country forming the N crest of South America	16C2
Venice *see* **Venezia**	26C1
Veracruz Mexico	6D3
Vereeniging South Africa	44C3
Verín Spain	25A1
Vermont a state in the NE of the USA; state capital Montpelier (24,887 sq km/9609 sq miles; pop. 535,000)	7F1
Versailles a town just W of Paris, France, which has a palace built by Louis XIV (pop. 96,000)	24C2
Vestmannaeyjar Island Iceland	30A2
Vesuvio (Vesuvius) an active volcano SE of Naples, in SW Italy (1281 m/4203 ft)	26C2
Viana do Castelo Portugal	25A1
Viangchan *see* **Vientiane**	36C2
Vibo Valentia Italy	26D3
Victoria a state in SE Australia; state capital Melbourne (227,620 sq km/87,884 sq miles; pop. 4,054,000)	45D4
Victoria, Lake the largest lake in Africa, and the second largest freshwater lake in the world after Lake Superior. Its shoreline is shared by Uganda, Kenya and Tanzania (69,485 sq km/26,828 sq miles)	43G5
Victoria Falls one of the world's greatest waterfalls on the River Zambezi (108 m/355 ft)	44C2
Victoria Island Canada	5G2
Videle Romania	27F2
Vidin Bulgaria	27E2
Viella Spain	25C1
Vienna *see* **Wien**	28D3
Vientiane (Viangchan) the capital of Laos, on the River Mekong in the NE of the country (pop. 177,000)	36C2
Vierzon France	24C2
Vietnam [area 331,689 sq km (128,065 sq miles); pop. 65,000,000; cap. Hanoi; major cities Ho Chi Minh City, Haiphong; govt socialist republic; religion Buddhism, Taoism, RC; currency dong] a long narrow country in SE Asia, running down the coast of the South China Sea	36C2
Vigo Spain	25A1
Vijayawada India	39G4
Vila Nova de Gaia Portugal	25A1
Vila Real Portugal	25A1
Vila Vanuatu	45F2
Villach Austria	28C3
Vilnius the capital of Lithuania (pop. 536,000)	32D4
Viña del Mar Chile	17B6
Vinh Vietnam	36C2
Vinnitsa Ukraine	32D5
Virgin Islands, British a British Crown colony in the E Caribbean, E of Puerto Rico	14G3
Virgin Islands, US a territory of the USA in the E Caribbean, to the E of Puerto Rico	14G3
Virginia a coastal state in the E of the USA; capital Richmond (103,03sq km/39,780 sq miles; pop. 5,706,000)	7F2
Visby Sweden	30D4
Vishakhapatnam India	39G4
Vis Island Yugoslavia	26D2
Vistula *see* **Wisła**	29D2
Vitebsk Belarus	32E4
Viterbo Italy	26C2
Vitoria Spain	25B1
Vitória Brazil	16E5
Vitry-le-Francois France	24C2
Vladimir Russian Federation	32F4
Vladivostok a major port on the Pacific coast in E Russian Federation (pop. 590,000)	33P5
Vlorë Albania	27D2
Vojvodina (province) Serbia	27D2
Volga, River a largely navigable river of W Russian Federation, flowing S from Moscow to the Caspian Sea. It is the longest river in Europe (lenth 3690 km/2293 miles)	32F4
Volgograd a port and major industrial city on the Volga (pop. 990,000)	32F5
Vologda Russian Federation	32F4
Vólos Greece	27E3
Volta a river in Ghana which flows S to the Bight of Benin (length 480 km/298 miles)	42C4
Volta, Lake a major artificial lake that occupies much of E Ghana, formed by the damming of the Volta River (8480 sq km/3251 sq miles)	42C4
Vorkuta Russian Federation	32H3
Voronezh Russian Federation	32E4
Voss Norway	30B3
Vranje Yugoslavia	27E2
Vratsa Bulgaria	27E2
Vrbovsko Croatia	26D1
Vrsac Yugoslavia	27E1

W

Wadi Halfa Sudan	43G2
Wad Medani Sudan	43G3
Wagga Wagga Australia	45D4
Wahiawa Hawaii USA	6H
Waigeo Island Indonesia	37F4
Wailuka Hawaii USA	6H
Wakayama Japan	35L9
Wake Island Pacific Ocean	5
Wakkanai Japan	35J2
Wales a principality in the SW of Great Britain, forming a part of the UK; capital Cardiff (20,768 sq km/8017 sq miles; pop. 2,749,600)	21D5
Wallis and Futuna Islands three small islands forming an overseas territory of France	46
Walsall England	21F5
Walvis Bay South Africa	44B3
Wanganui New Zealand	45G4
Warangal India	39F4
Warley England	21F5
Warminster England	21E6
Warrnambool Australia	45E4
Warszawa (Warsaw) the capital of Poland, on the River Vistula, in the central E (pop. 1,641,000)	29E2
Warwick England	21F5
Warwickshire a county of central England; county town Warwick (1981 sq km/765 sq miles; pop. 475,000)	21F5
Wash, The a shallow inlet formed by the North Sea in the E coast of England	20H5
Washington a state in NW USA, with a coast on the Pacific; capital Olympia (172,416 sq km/66,570 sq miles; pop. 4,409,000)	6A1
Washington DC capital of the USA, standing in its own territory, District of Columbia (DC), between Virginia and Maryland (179 sq km/69 sq miles; pop. city 622,800/metropolitan area 3,429,400)	7F2
Waterford a county in S Republic of Ireland; county town Waterford (1838 sq km/710 sq miles; pop. county 89,000)	23D3

Watford   1036   Xingtai

Watford England 21G6
Wau Papua New Guinea 37G4
Wau Sudan 43F4
Weifang China 35F3
Welkom South Africa 44C3
Wellesley Islands Australia 45C2
Wellington the capital of New Zealand and a port in the SW of North Island (pop. 342,000) 45G5
Welshpool Wales 21D5
Wenzhou China 35G4
Wesel Germany 28B2
Weser a river in the NW of Germany, flowing to the North Sea (477 sq km/196 sq miles) 28B2
Wessel Islands Australia 45C2
West Bromwich England 21F5
Western Australia a state occupying much of the W half of Australia; capital Perth (2,527,636 sq km/975,920 sq miles; pop. 1,300,000) 45B3
Western Isles the regional island authority covering the Outer Hebrides of W Scotland; administrative centre Stornaway, on the Isle of Lewis (2900 sq km/1120 sq miles; pop. 32,000) 22A3
Western Sahara a disputed territory of W Africa, with a coastline on the Atlantic. Consisting mainly of desert, it is rich in phosphates. The main town is Laâyoune (El Aaiún) (266,770 sq km/103,000 sq miles; pop. 200,000) 42B2
Western Samoa [*area* 2831 sq km (1093 sq miles); *pop.* 163,000; *cap.* Apia; *govt* constitutional monarchy; *religion* Protestantism; *currency* tala] a small country, in the Polynesian sector of the Pacific Ocean, consisting of seven small islands and two larger volcanic islands 46
West Falkland Island South Atlantic Ocean 17C8
West Glamorgan a county in South Wales, created in 1974 from part of Glamorgan and the borough of Swansea; administrative centre Swansea (817 sq km/315 sq miles; pop. 368,000) 21D6
West Indies a general term for the islands of the Caribbean Sea 14F2
Westmeath a county in the central N of the Republic of Ireland; county town Mullingar (1764 sq km/681 sq miles; pop. 61,500) 23D3
West Midlands a metropolitan county of central England, created in 1974; administrative centre Birmingham (889 sq km/347 sq miles; pop. 2,658,000) 21F5
Weston-super-Mare England 21E6
Westport Ireland 23B3
Westray Island Scotland 22E1
West Sussex a county in S England; county town Chichester (1989 sq km/768 sq miles; pop. 660,000) 21G6
West Virginia a state of E USA; capital Charleston (62,341 sq km/24,070 sq miles; pop. 1,936,000) 7E2
West Yorkshire a county in NE England created in 1974 from the division of the former county of Yorkshire; administrative centre Wakefield (2039 sq km/787 sq miles; pop. 2,038,000) 20F4
Wetar Island Indonesia 37E4
Wewak Papua New Guinea 37G4
Wexford a county in SE Republic of Ireland; county town Wexford (2352 sq km/908 sq miles; pop. county 13,293) 23E4
Weymouth England 21E7
Whalsay Island Scotland 22K7
Whangarei New Zealand 45G4
Whitby England 20G3
Whitchurch England 21E5
Whitehaven England 20D3
Whitehorse Canada 5E3
White Sea *see* **Beloye More** 32E3
Whitney, Mount in E California, the highest peak in the USA outside Alaska (4418 m/14,495 ft) 13C3

Whyalla Australia 45C4
Wichita a city in S Kansas, USA, on the Arkansas River (pop. city 283,500/metropolitan area 428,600) 6D2
Wick Scotland 22E2
Wicklow a county in SW Republic of Ireland; county town Wicklow (2025 sq km/782 sq miles; pop. county 87,000) 23E3
Widnes England 20E4
Wien (Vienna) the capital of Austria, on the River Danube, in the NE of the country (pop. 1,531,000) 28D3
Wigan England 20E4
Wight, Isle of an island county off the S coast of England, separated from it by the Solent; county town Newport (380 sq km/147 sq miles; pop. 120,000) 21F7
Wilhelmshaven Germany 28B2
Wilmington USA 7F2
Wilmslow England 20E4
Wiltshire a county in S England; county town Trowbridge (3481 sq km/1344 sq miles; pop. 510,000) 21F6
Winchester a historic city in S England, and the county town of Hampshire (pop. 31,000) 21F6
Windhoek Namibia 44B3
Windsor England 21G6
Windward Islands *see* **Leeward and Windward Islands** 14G3
Winnemucca USA 6A1
Winnipeg the capital of Manitoba, Canada, in the S of the state (pop. 585,000) 5J5
Winnipeg, Lake a lake in the S of Manitoba, Canada, which drains into Hudson Bay via the Nelson River (23,553 sq km/9094 sq miles) 5J5
Wisconsin a state in N central USA, bordering Lake Superior and Lake Michigan; capital Madison (141,061 sq km/54,464 sq miles; pop. 4,775,000) 7D1
Wista (Vistula) a river of central and N Poland, flowing N to the Baltic Sea (length 1090 km/677 miles) 29D2
Wloclawek Poland 29D2
Wokam Island Indonesia 37F4
Wolfsburg Germany 28C2
Wollongong New South Wales, Australia 45E4
Wolverhampton England 21E5
Wonsan North Korea 35G3
Worcester England 21E5
Worcester South Africa 44B4
Workington England 20D3
Worksop England 20F4
Worthing England 21G7
Wrexham Wales 20E4
Wroclaw (Breslau) Poland 28D2
Wuhan (Hankow) China 35F3
Wuhu China 35F3
Würzburg Germany 28B3
Wuxi China 35G3
Wuzhou China 35F4
Wyndham Australia 45B2
Wyoming a state in W USA; capital Cheyenne (253,597 sq km/97,914 sq miles; pop. 509,000) 6C1

X

Xai Xai Mozambique 44D3
Xi (Xi Jiang; Si Kiang) the third longest river in China, flowing across the SW to the South China Sea near Guangzhou (length 2300 km/1437 miles) 35F4
Xiamen China 35F4
Xi'an the capital of Shaanxi province and an industrial centre and former capital of China. (Pop. 2,330,000) 34E3
Xiangfan China 35F3
Xiangtan China 35F4
Xingtai China 35F3

Xining   1037   Zimbabwe

Xining China (pop. 860,000)	34E3
Xinxiang China	35F3
Xinyang China	35F3
Xizang Autonomous Region *see* **Tibet**	34C3
Xuchang China	35F3
Xuzhou China	35F3

Y

Ya'an China	34E3
Yakutsk Russian Federation	33O3
Yamagata Japan	35P7
Yamdena Island Indonesia	37F4
Yamoussoukro the new capital of Côte d'Ivoire, in its centre (pop. 45,000)	42C4
Yamuna (Jumna), River a river of N India, a tributary of the Ganges (length 1376 km/855 miles)	39F3
Yangon (Rangoon) the capital of Myanmar (Burma) and an important port on the mouth of the Rangoon River (pop. 2,549,000)	36B2
Yangquan China	35F3
Yangtze Kiang *see* **Chang Jiang**	34/35D/G4
Yangzhou China	35F3
Yantai China	35G3
Yaoundé the capital of Cameroon, in its SW (pop. 500,000)	42E4
Yapen Island Indonesia	37F4
Yap Island Pacific Ocean	37F3
Yaroslavl' Russian Federation	32E4
Yazd Iran	38D2
Ye Myanmar	36B2
Yekaterinburg Russian Federation	32H4
Yell Island Scotland	22J7
Yellowknife a city on the Great Slave Lake, Canada, capital of the Northwest Territories (pop. 11,077)	5G3
Yellow River *see* **Huang He**	35F3
Yellow Sea a branch of the Pacific Ocean between the NE coast of China and Korea	35G3
Yemen [*area* 195,000 sq km (75,290 sq miles); *pop.* 12,000,000; *cap.* Sana'a (administrative) *cap.* Aden (commercial); *govt* republic; *religion* Zaidism, Shia Islam, Sunni Islam; *currency* riyal and dinar] a country bounded by Saudi Arabia in the N, Oman in the E, the Gulf of Aden in the S, and the Red Sea in the W	38C4
Yeniseysk Russian Federation	33L4
Yeovil England	21E7
Yerevan an industrial city and the capital of Armenia (pop. 1,114,000)	32F5
Yibin China	34E4
Yichang China	35F3
Yinchuan China	34E3
Yingkou China	35G2
Yiyang China	35F4
Yogyakarta (Jogjakarta) Indonesia	36D4
Yokkaichu Japan	35M9
Yokohama the main port of Japan, and its second largest city after neighbouring Tokyo, on the SE coast of Honshu (pop. 3,012,900)	35N9
Yokosuka Japan	35N9
York England	20F4
Youghal Ireland	23D4
Ystad Sweden	30C4
Yuan Jiang *see* **Hong**	34E4
Yucatán a state on a broad peninsula of SE Mexico (pop. 1,100,000)	7E4
Yugoslavia [*area* 127,886 sq km (49,377 sq miles); *pop.* 11,807,098; *cap.* Belgrade (Beograd); *major cities* Nis, Skopje, Titograd; *govt* federal republic; *religion* Eastern Orthodox; *currency* dinar] the name Yugoslavia now refers to the republics of Serbia, Montenegro and Macedonia, the others having gained independence in 1991/92	27E2
Yukon Territory a mountainous territory in NW Canada centred on the River Yukon and including the River Klondike (536,372 sq km/207,076 sq miles; pop. 23,500)	5E3
Yumen China	34D2
Yuzhno-Sakhalinsk Russian Federation	33Q5

Z

Zaandam Netherlands	28A2
Zadar Yugoslavia	26D2
Zafra Spain	25A2
Zagreb the capital of Croatia (pop. 1,180,000)	26D1
Zagros Mountains (Kuhha-ye Zagros) a range in SW Iran, running parallel to the border with Iraq. The highest point is Zard Kuh (4548 m/14,918 ft)	38D2
Zahedan Iran	38E3
Zaïre [*area* 2,345,409 sq km (905,562 sq miles); *pop.* 34,140,000; *cap.* Kinshasa; *major cities* Lubumbashi, Mbuji-Mayi, Kananga; *govt* republic; *religion* RC, Protestantism, Animism; *currency* zaïre] a vast country in W central Africa	43F5
Zaïre (Congo), River the second longest river in Africa after the Nile, rising as the Lualaba in the S of Zaïre, then flowing N, NW and SW, forming the border between Zaïre and the Congo before entering the Atlantic Ocean (length 4800 km/3000 miles)	43E4
Zákinthos Island Greece	27E3
Zambezi, River a river of S Africa, rising in Zambia, flowing S to form the border with Zimbabwe, and then SE across Mozambique to the Indian Ocean (length 2740 km/1700 miles)	44C2
Zambia [*area* 752,614 sq km (290,584 sq miles); *pop.* 8,500,000; *cap.* Lusaka; *major cities* Kitwe, Ndola, Mufulira; *govt* republic; *religion* Christianity, Animism; *currency* kwacha] a country in central Africa	44C2
Zamboanga Philippines	37E3
Zamora Spain	25A1
Zanzibar an island lying just off the east coast of Tanzania, in the Indian Ocean. The main town is the port also called Zanzibar (2461 sq km/950 sq miles; pop. 556,000)	43G5
Zaporozh'ye a major industrial city on the River Dnieper in the Ukraine (pop. 844,000)	32E5
Zaragón (region) Spain	25B1
Zaragoza, (Saragossa) a historic and industrial city in NE Spain, on the River Ebro (pop. 590,000)	25B1
Zaysan Kazakhstan	32K5
Zealand *see* **Sjaelland**	30C4
Zenica Bosnia	27D2
Zhangjiakou China	35F2
Zhangzhou China	35F4
Zhanjiang China	35F4
Zhengzhou the capital of Henan province, in east central China (pop. 1,271,000)	35F3
Zhenjiang China	35F3
Zhitomir Ukraine	32D4
Zhuzhou China	35F4
Zibo an industrial city in Shangdong province, northeastern China (pop. 2,000,000)	35F3
Zielona Gora Poland	28D2
Ziguinchor Senegal	42B3
Zilina Slovakia	29D3
Zimbabwe [*area* 390,580 sq km (150,803 sq miles); *pop.* 9,370,000; *cap.* Harare; *major cities* Bulawayo, Mutare, Gweru; *govt* republic; *religion* Animism, Anglicanism, RC; *currency* Zimbabwe dollar] a landlocked country in S Africa	44C2

Zinder Niger	42D3
Zlatoust Russian Federation	32G4
Znojmo Czech Republic	28D3
Zomba Malawi	44D2
Zouerate Mauritania	42B2
Zunyi China	34E4
Zürich the largest city in Switzerland, in the NE of the country, and a major industrial and financial centre (pop. 422,000)	24D2
Zwickau Germany	28C2

CONTENTS AND LEGEND 1

	2-3	The World *Political*

NORTH & CENTRAL AMERICA

4	North America *Political*
5	Canada
6-7	United States of America & Mexico
8-9	U.S.A. Eastern States
10-11	U.S.A. Central States
12-13	U.S.A. Western States
14	Central America & The Caribbean

SOUTH AMERICA

15	South America *Political*
16-17	South America

EUROPE

18	Europe *Political*
19	British Isles
20-21	England & Wales
22	Scotland
23	Ireland
24	France
25	Spain & Portugal
26-27	Italy & The Balkans
28-29	Central Europe
30	Scandinavia & The Baltic

ASIA

31	Asia *Political*
32-33	Russian Federation Transcaucasia & Central Asia
34-35	East Asia
36-37	South-East Asia
38-39	South Asia & The Middle East
40	Israel & Lebanon

AFRICA

41	Africa *Political*
42-43	Northern Africa
44	Southern Africa

AUSTRALASIA & OCEANIA

45	Australasia
46-47	Oceania *Political*

48	The Polar Regions

Roads – *at scales larger than 1:3 million*
— Motorway/Highway
— Other Main Road
– at scales smaller than 1:3 million
— Principal Road: Motorway/Highway
— Other Main Road
— Main Railway

Towns & Cities

☐	Population > 5,000,000
☐	1-5,000,000
○	500,000-1,000,000
○	< 500,000
☐ **Paris**	National Capital
✈	Airport
——	International Boundary
– – –	International Boundary – not defined or in dispute
——	Internal Boundary
——	River
....	Canal
	Marsh or Swamp

Relief

▲ 1510 Peak (in meters)

5000 meters (16405 feet)
4000 (13124)
3000 (9843)
2000 (6562)
1000 (3281)
500 (1641)
200 (656)
100 (328)
0
Land below sea level

Note – The 0-100 contour layer appears only at scales larger than 1:3 million

© Geddes & Grosset

2 THE WORLD Political

CANADA 5

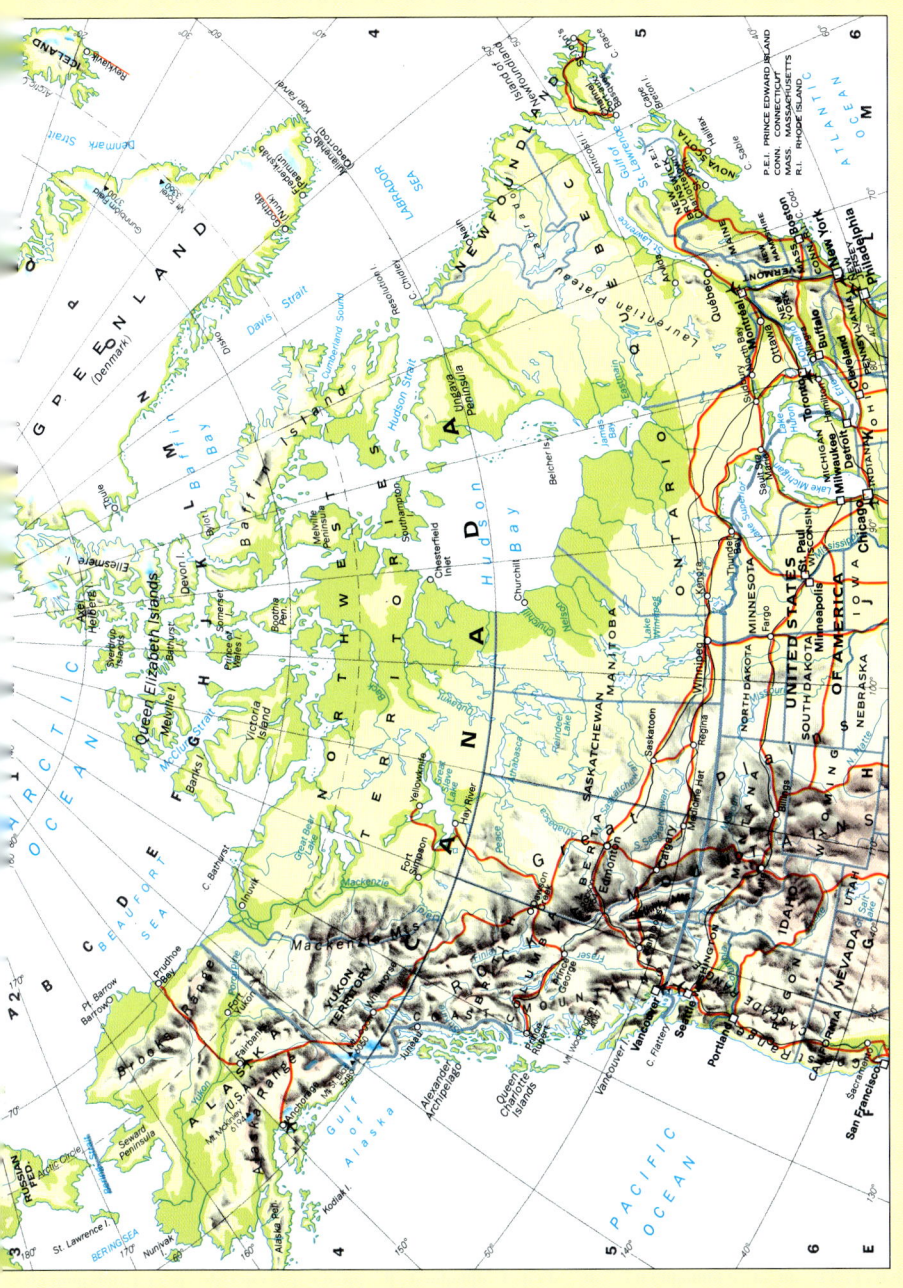

UNITED STATES OF AMERICA & MEXICO

UNITED STATES OF AMERICA & MEXICO 7

8 U.S.A. EASTERN STATES

U. S. A. EASTERN STATES 9

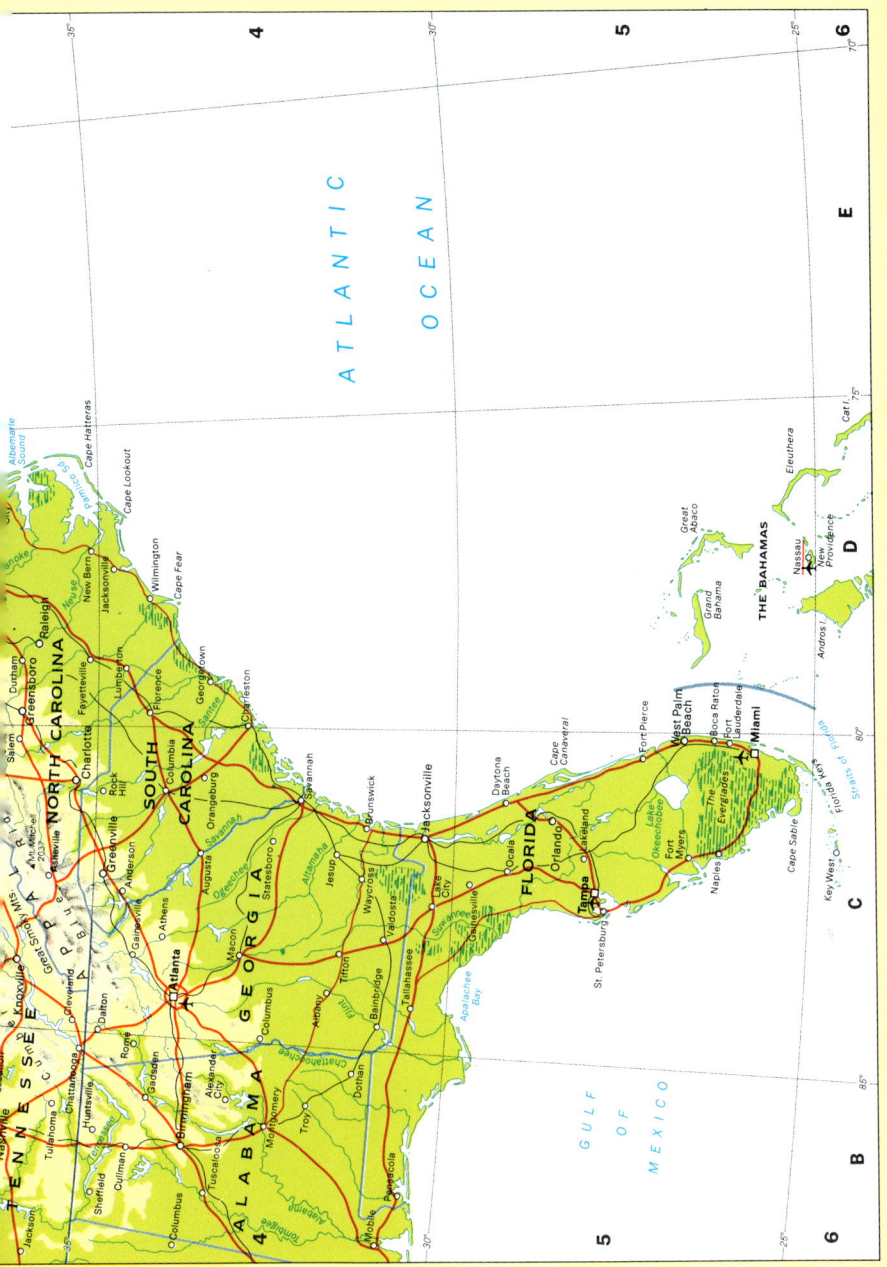

10 U.S.A. CENTRAL STATES

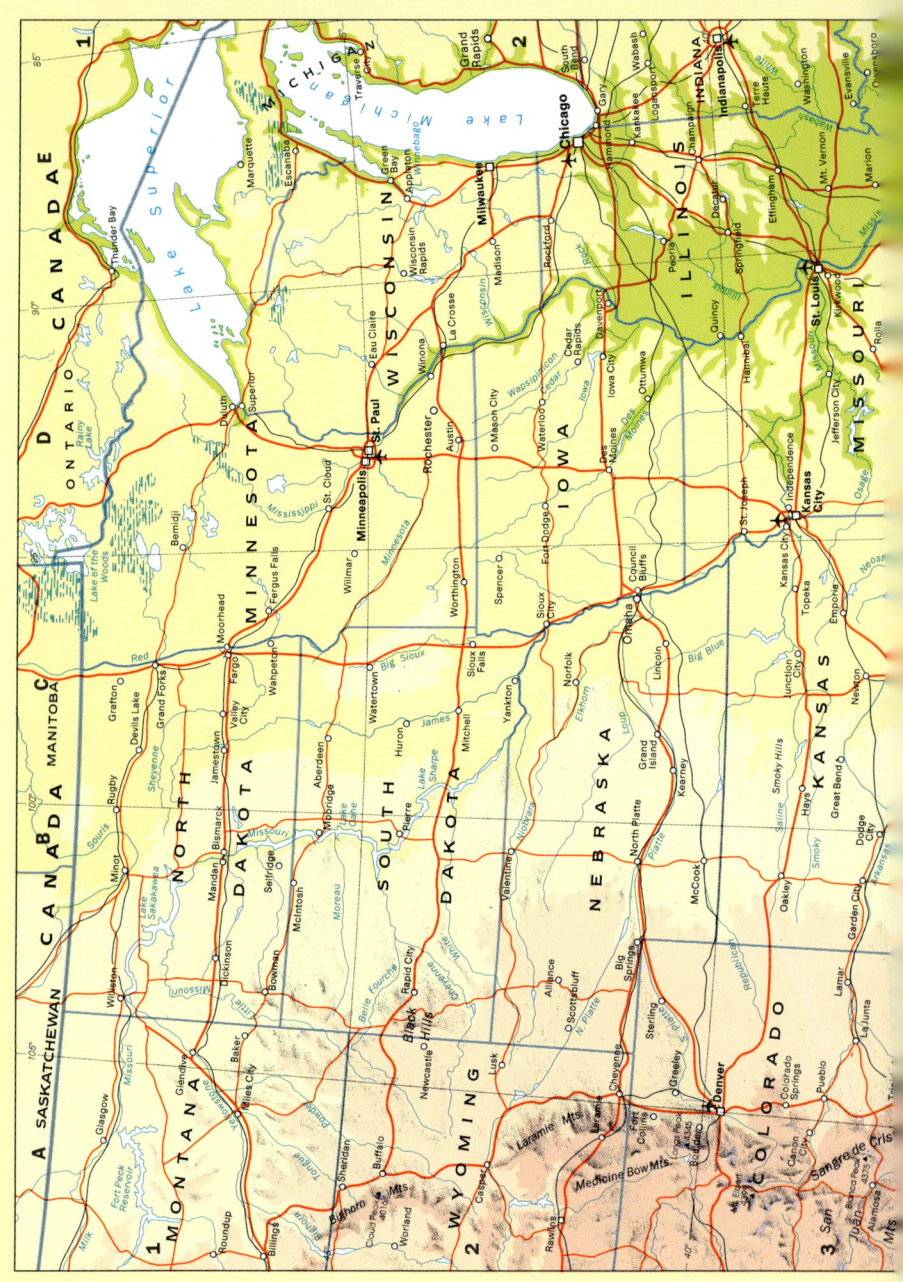

U.S.A. CENTRAL STATES 11

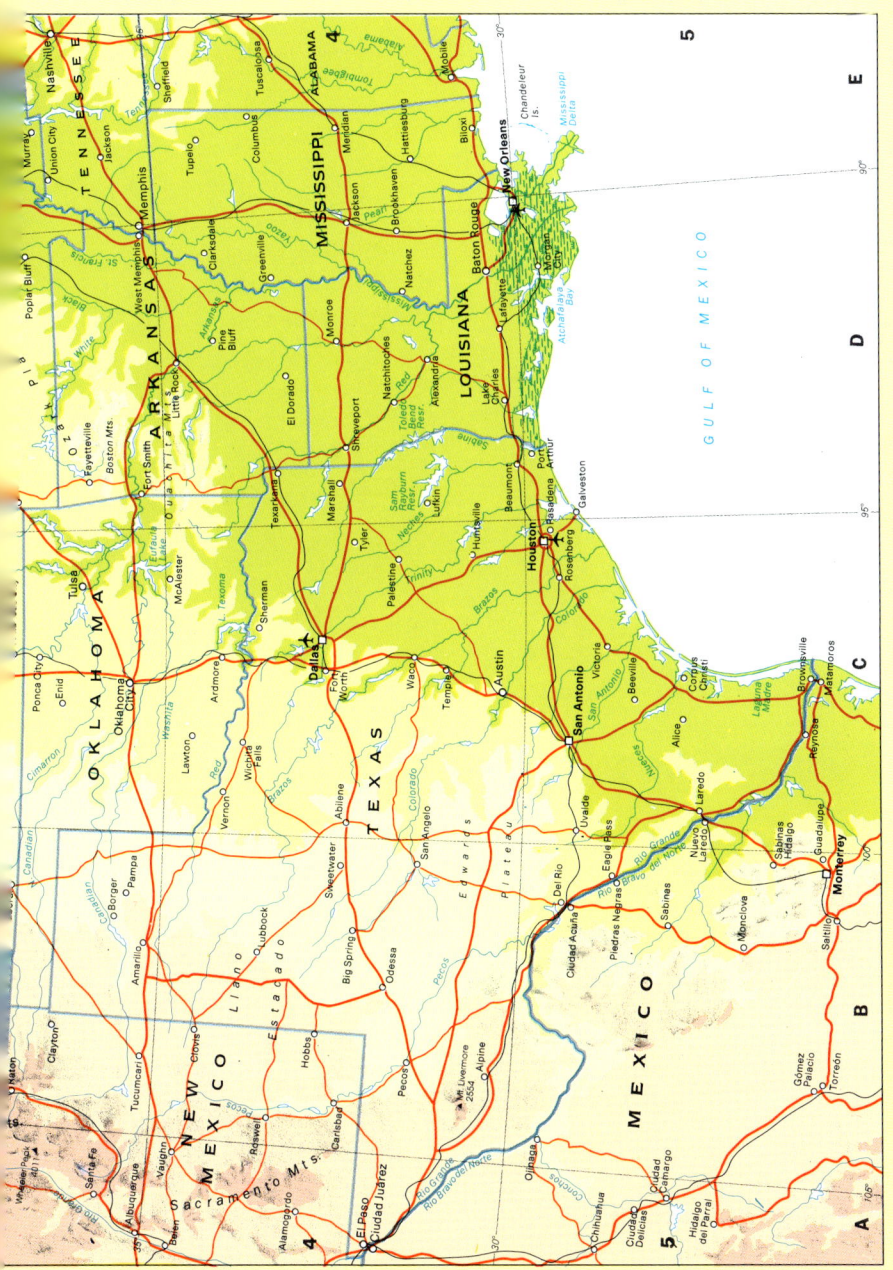

© Geddes & Grosset

12 U.S.A. WESTERN STATES

U. S. A. WESTERN STATES 13

14 CENTRAL AMERICA & THE CARIBBEAN

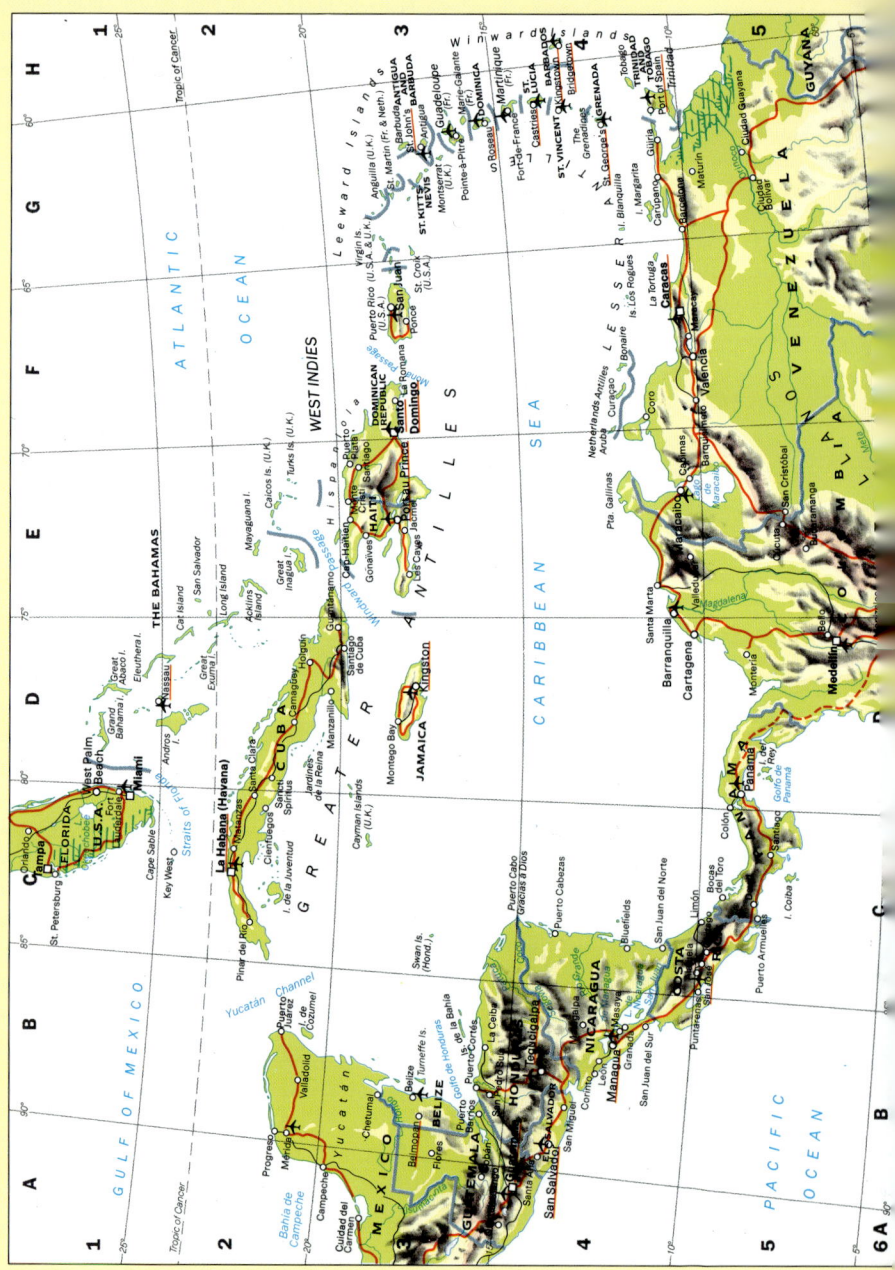

SOUTH AMERICA Political 15

16 SOUTH AMERICA

SOUTH AMERICA 17

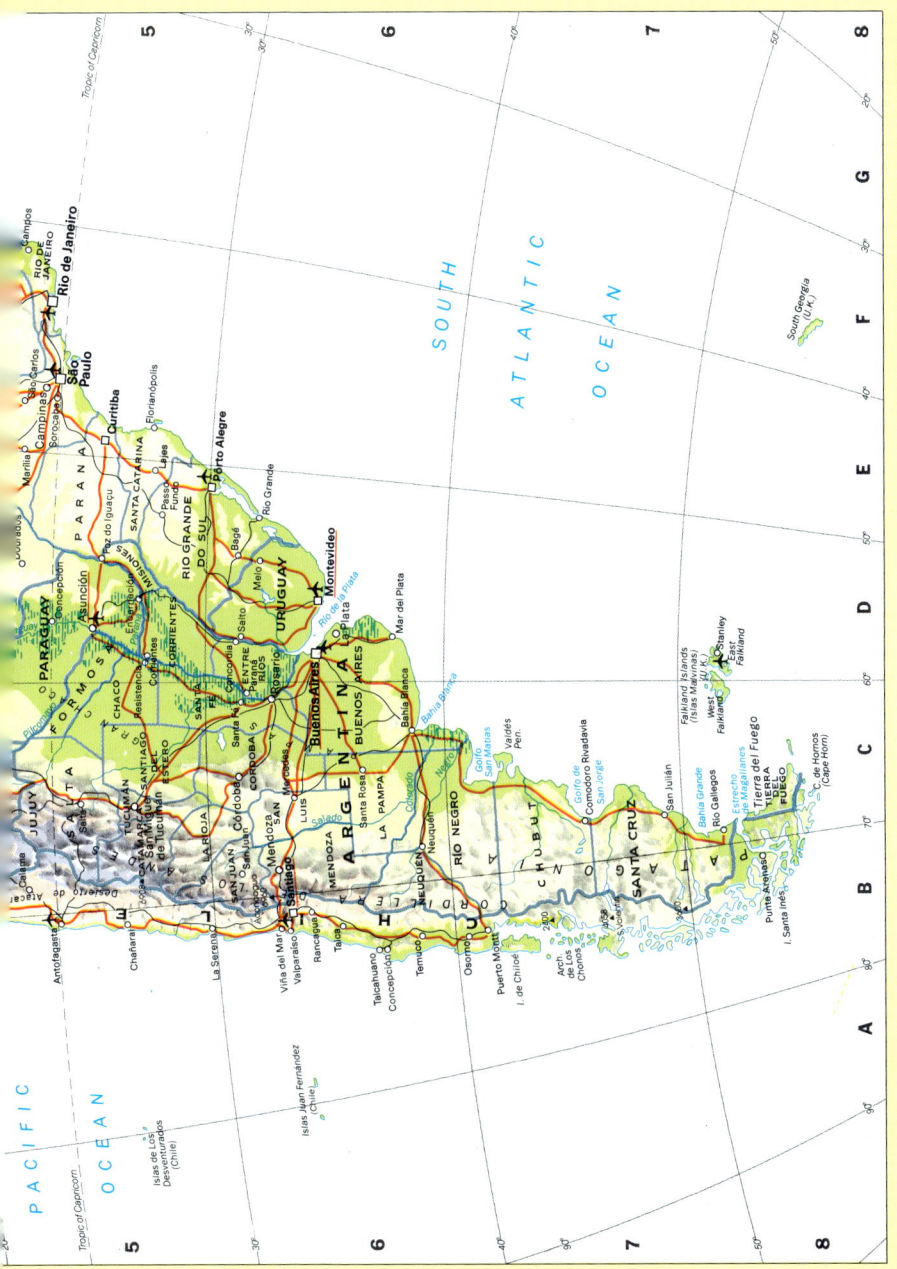

© Geddes & Grosset

18 EUROPE Political

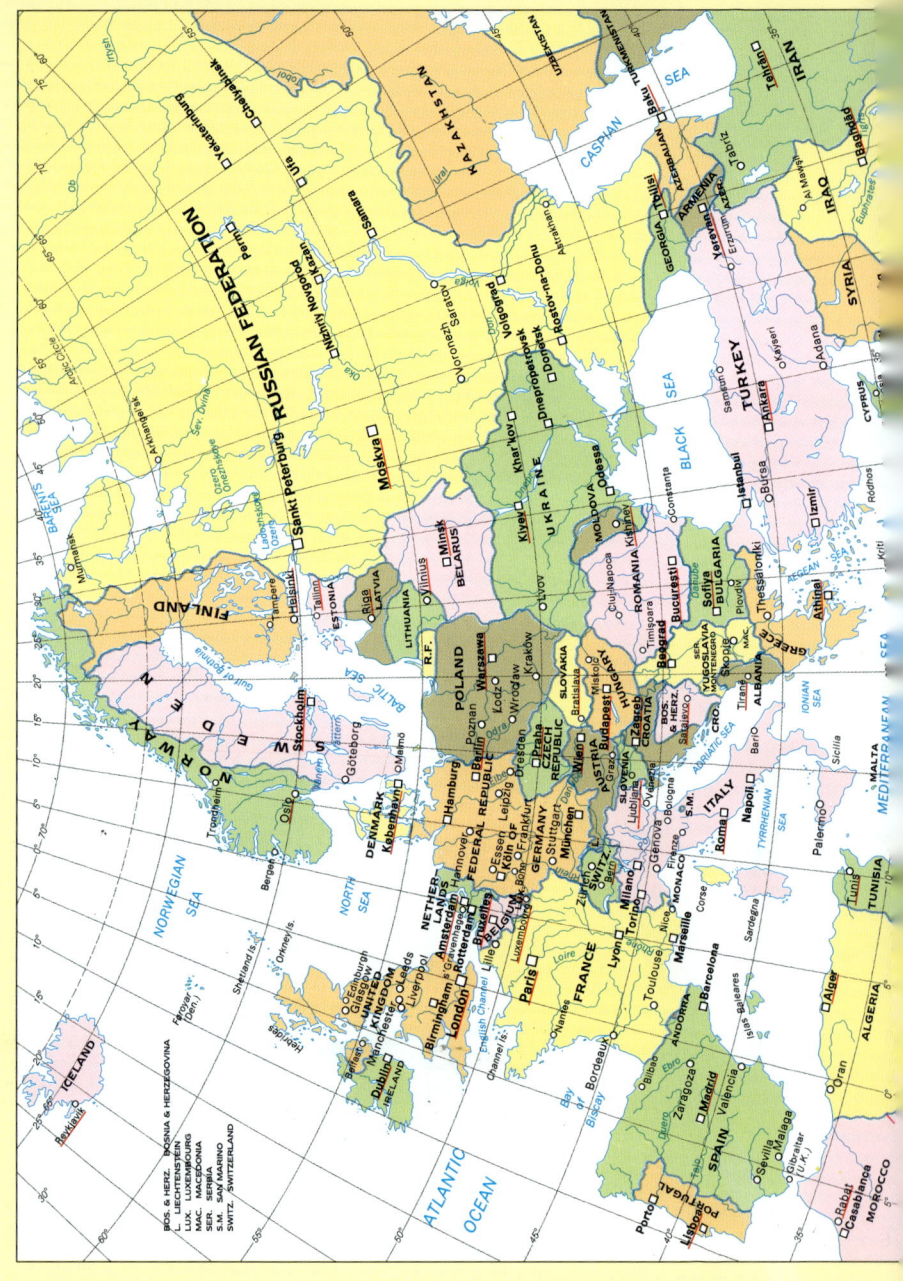

BRITISH ISLES 19

20 ENGLAND AND WALES

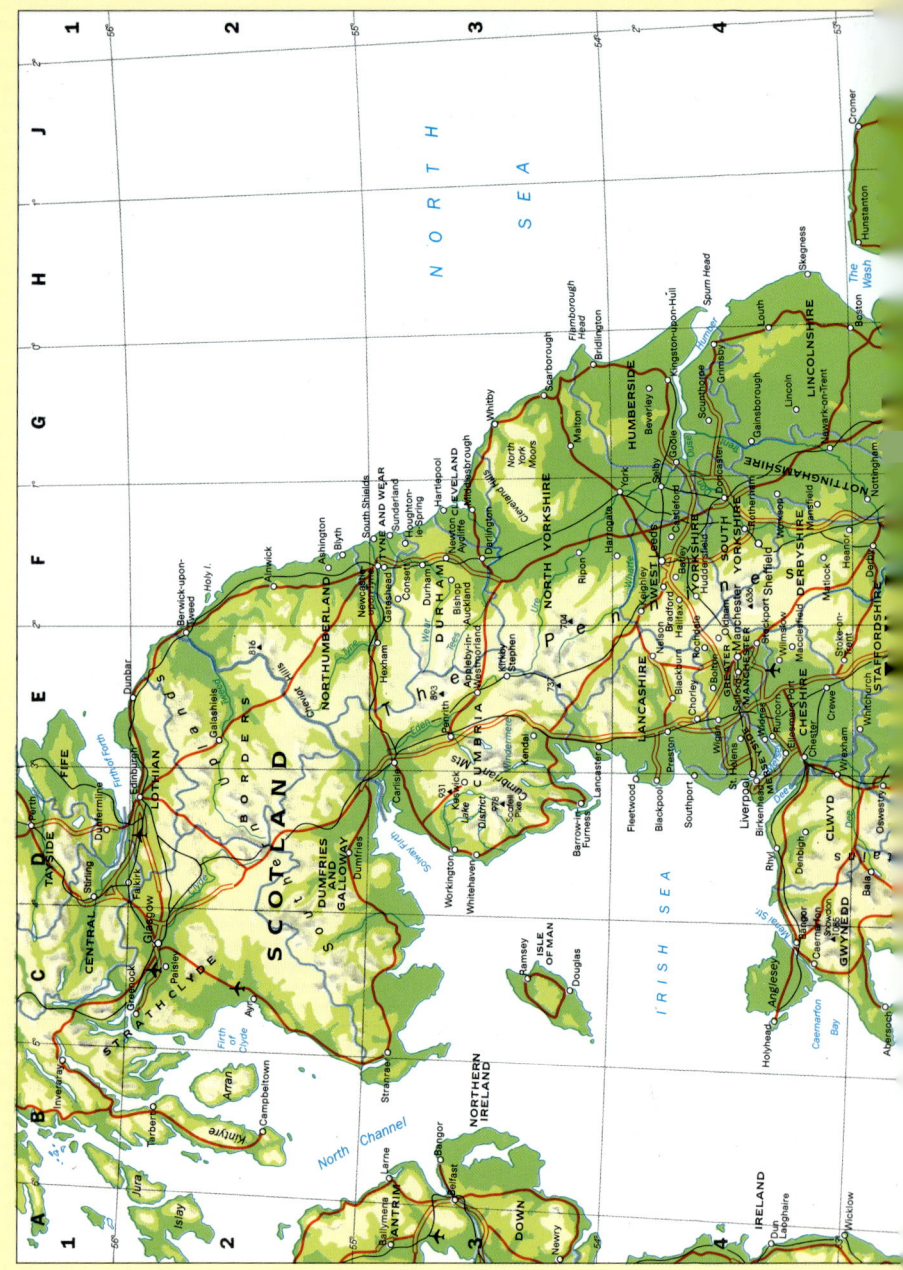

ENGLAND AND WALES 21

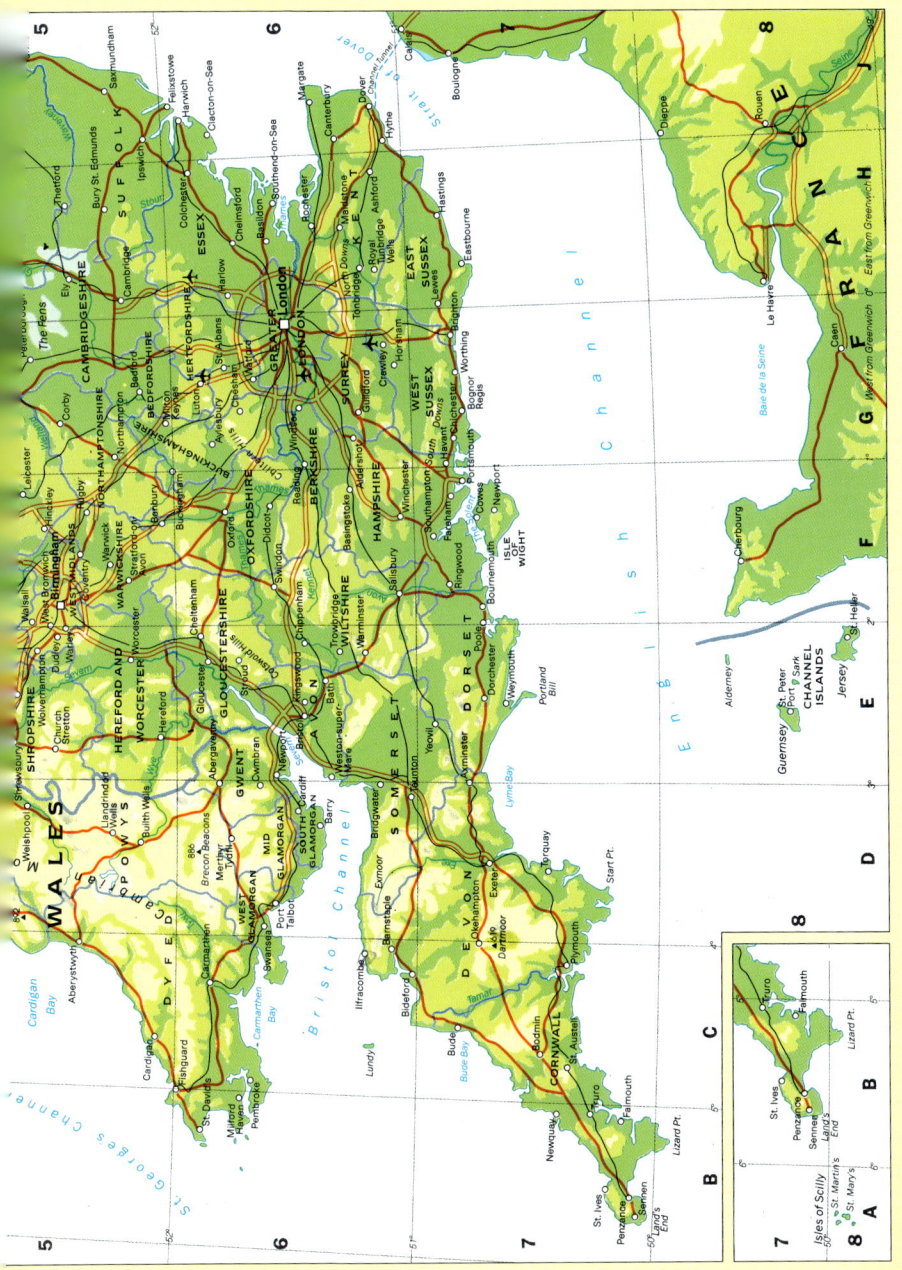

© Geddes & Grosset

22 SCOTLAND

IRELAND 23

24 FRANCE

SPAIN AND PORTUGAL 25

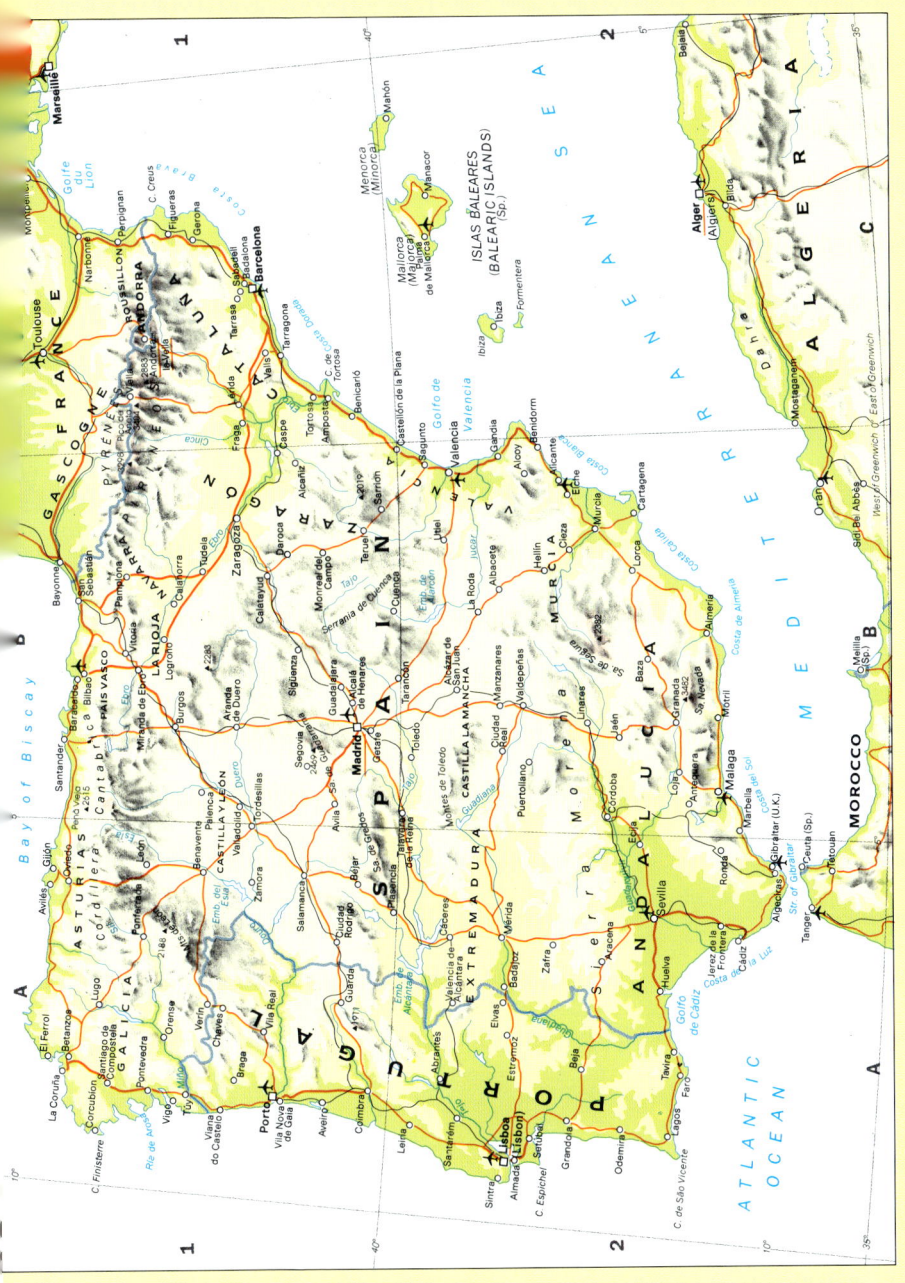

26 ITALY AND THE BALKANS

ITALY AND THE BALKANS 27

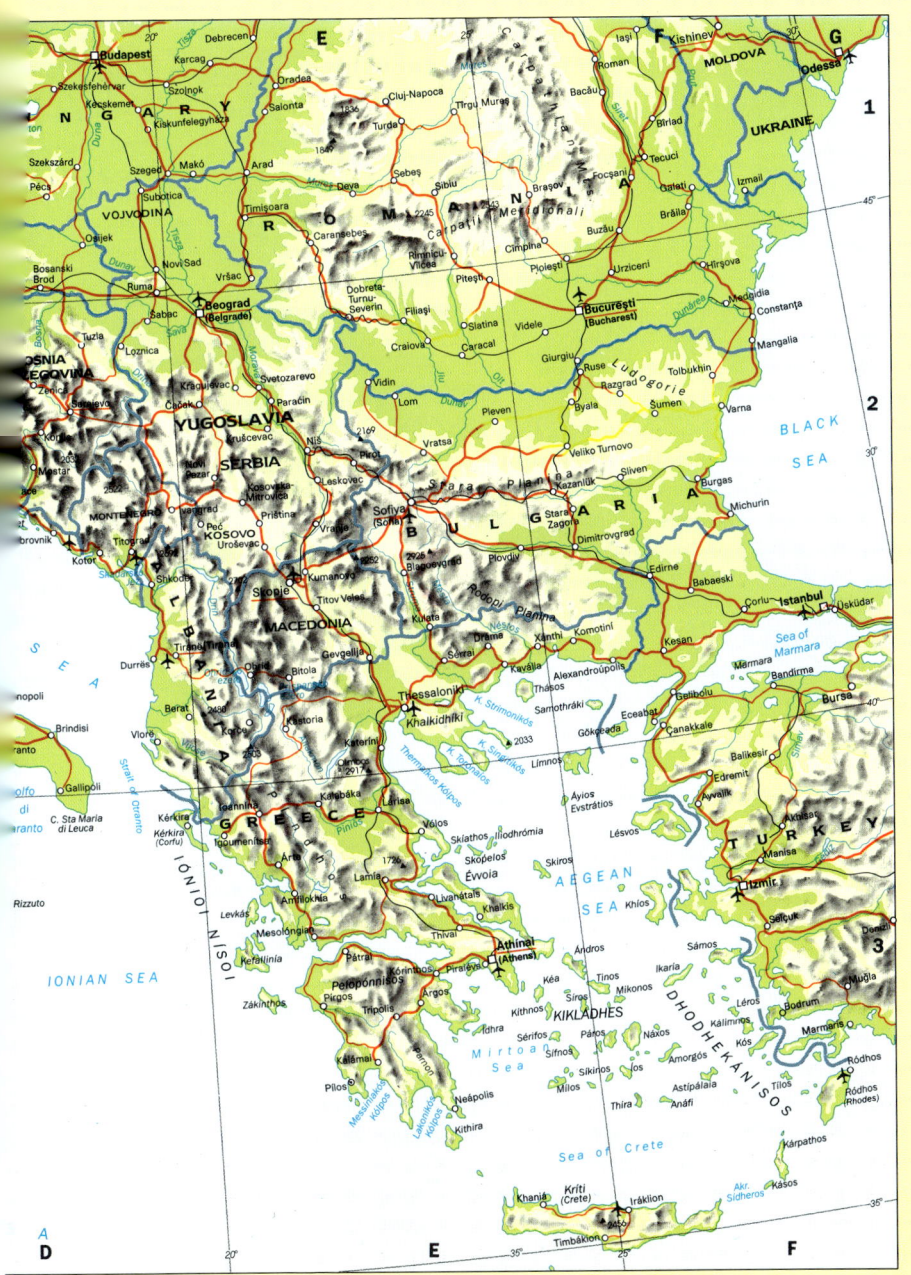

28 CENTRAL EUROPE

CENTRAL EUROPE

30 SCANDINAVIA AND THE BALTIC

ASIA Political 31

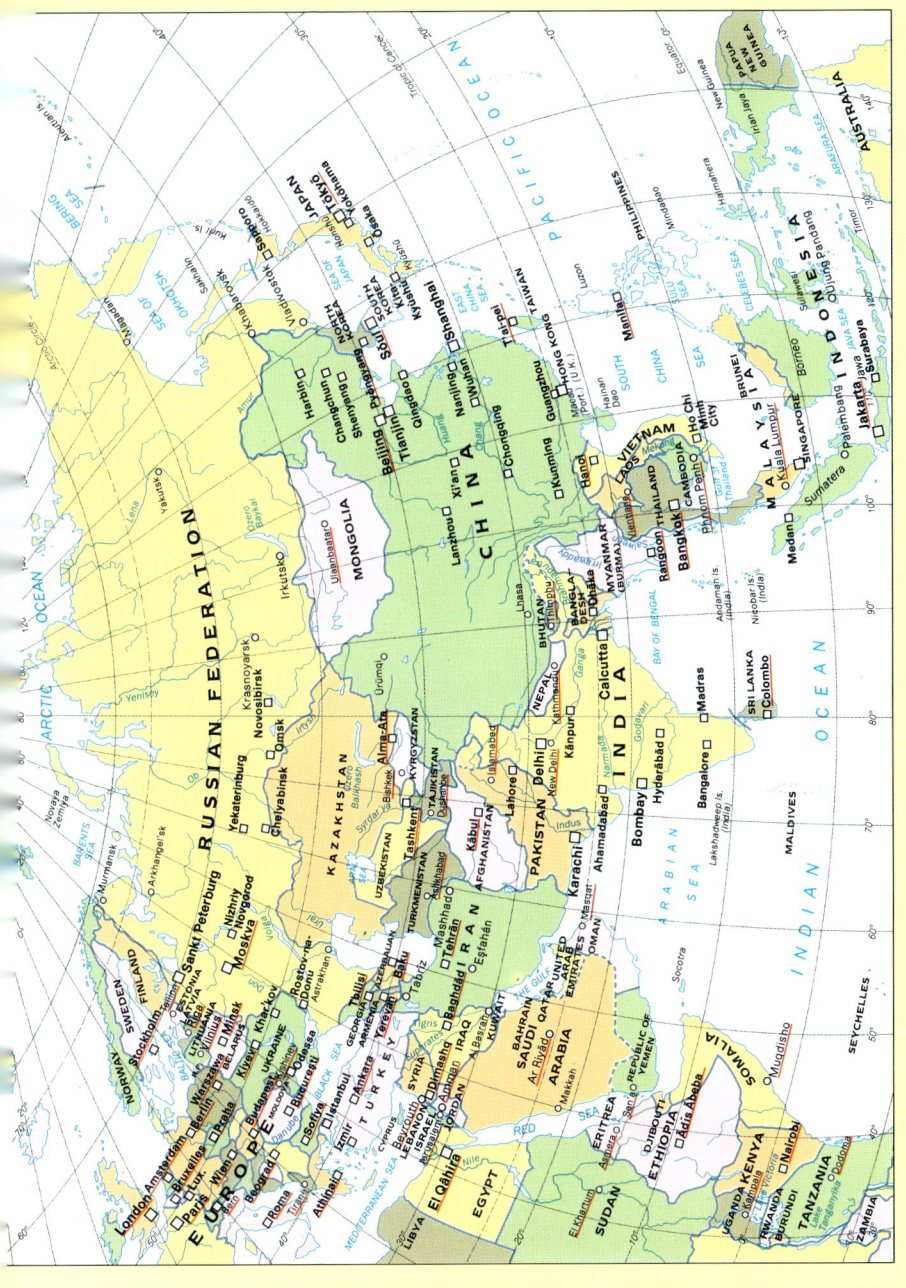

32 RUSSIAN FEDERATION, TRANSCAUCASIA AND CENTRAL ASI

RUSSIAN FEDERATION, TRANSCAUCASIA AND CENTRAL ASIA 33

34 EAST ASIA

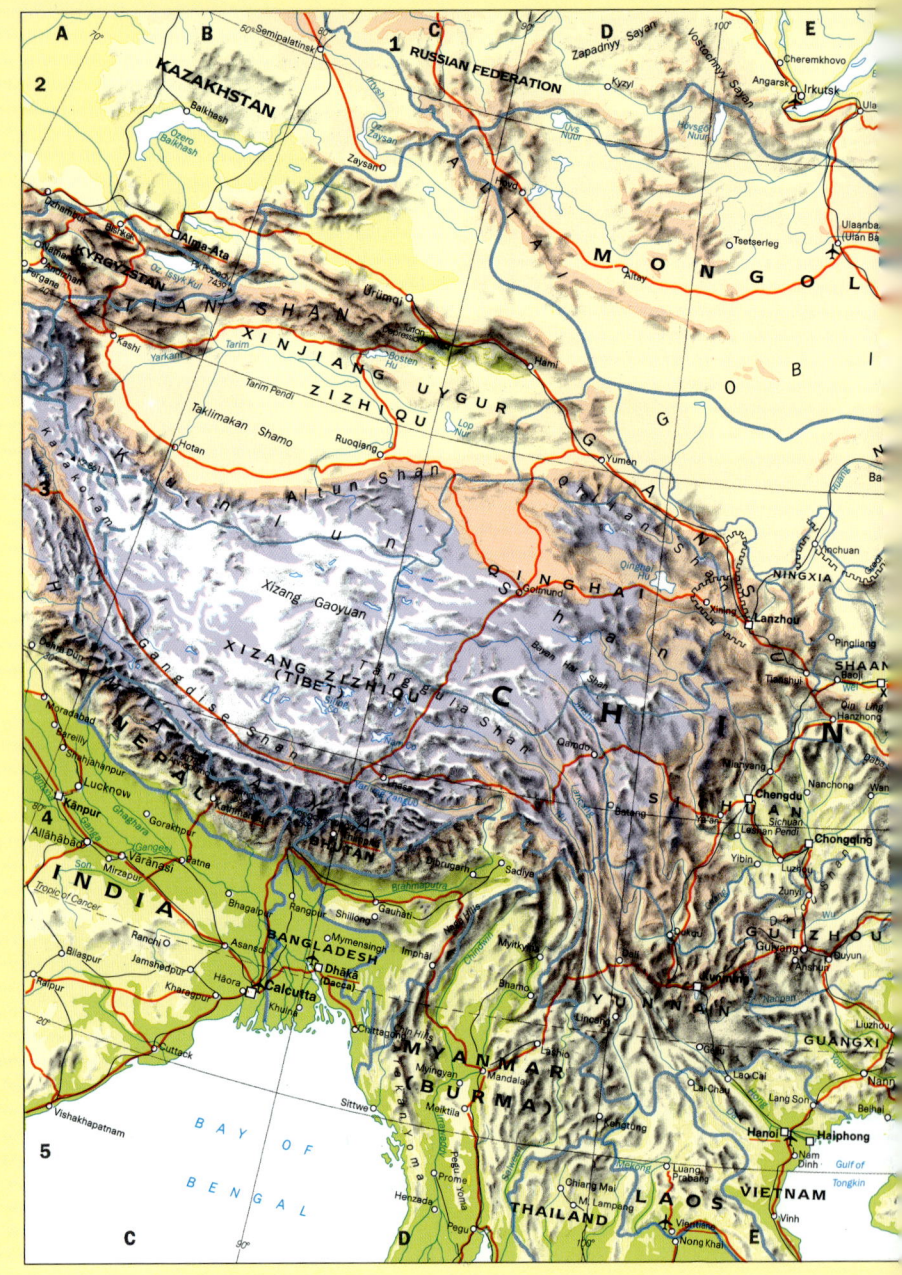

EAST ASIA 35

SOUTH-EAST ASIA

SOUTH-EAST ASIA 37

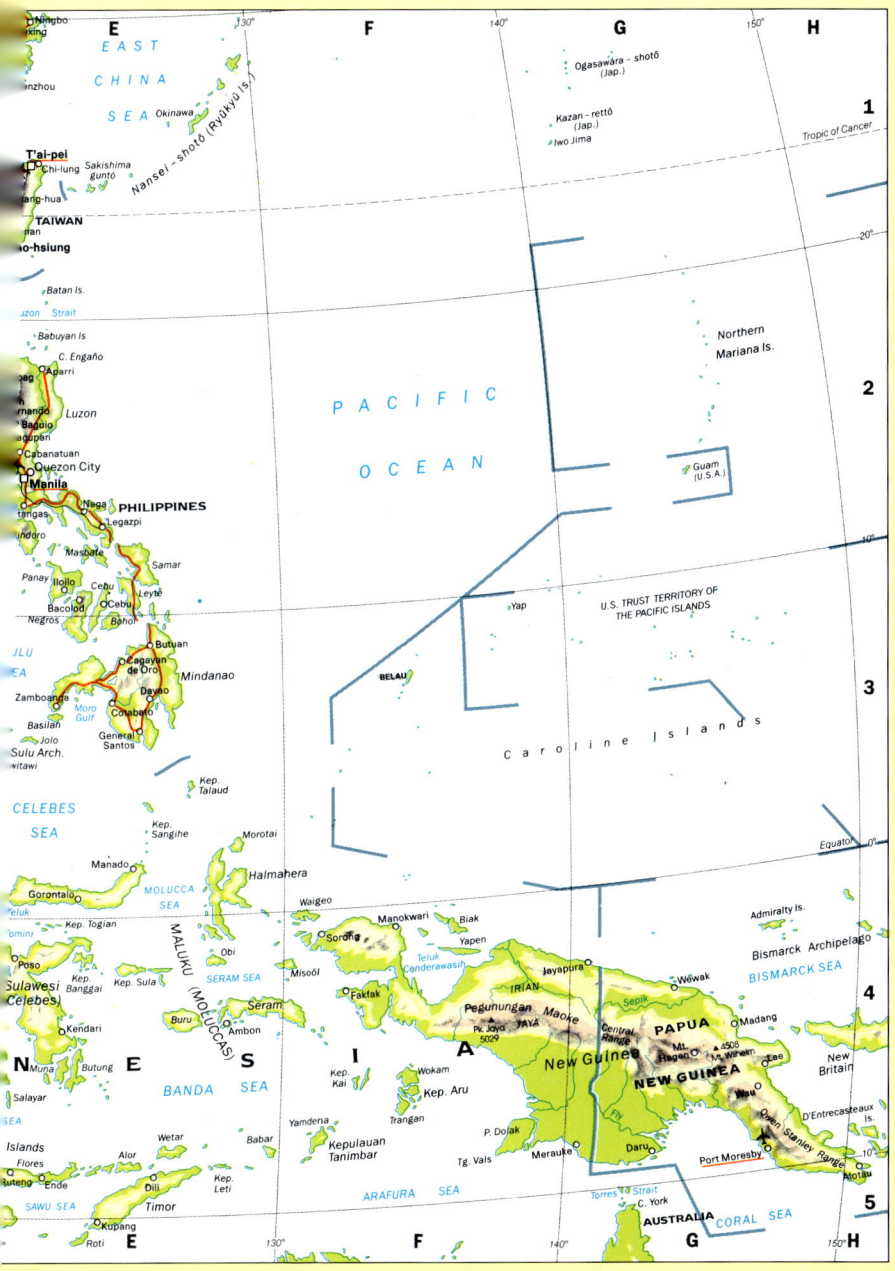

38 SOUTH ASIA AND THE MIDDLE EAST

SOUTH ASIA AND THE MIDDLE EAST 39

40 ISRAEL AND LEBANON

AFRICA Political 41

© Geddes & Grosset

NORTHERN AFRICA

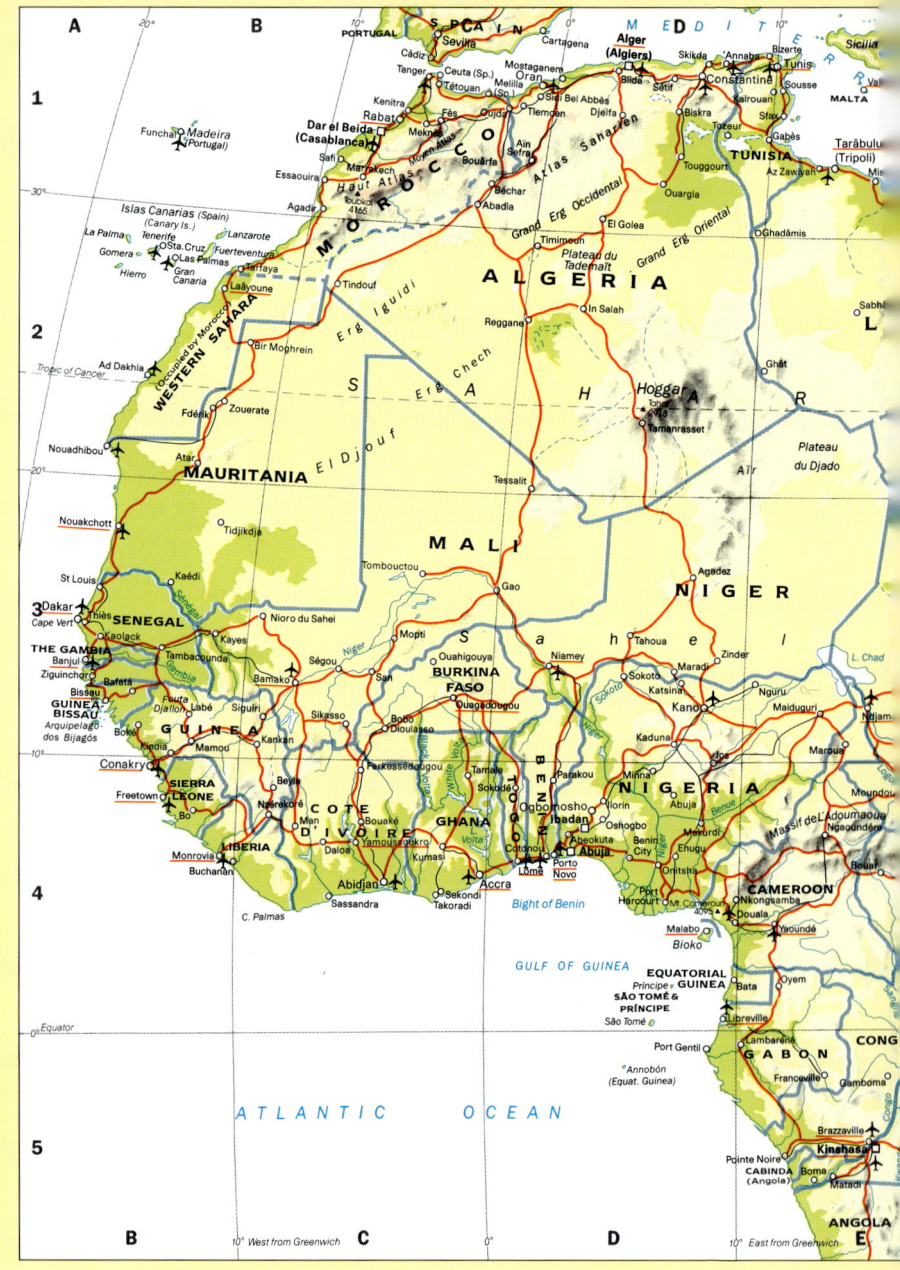

NORTHERN AFRICA

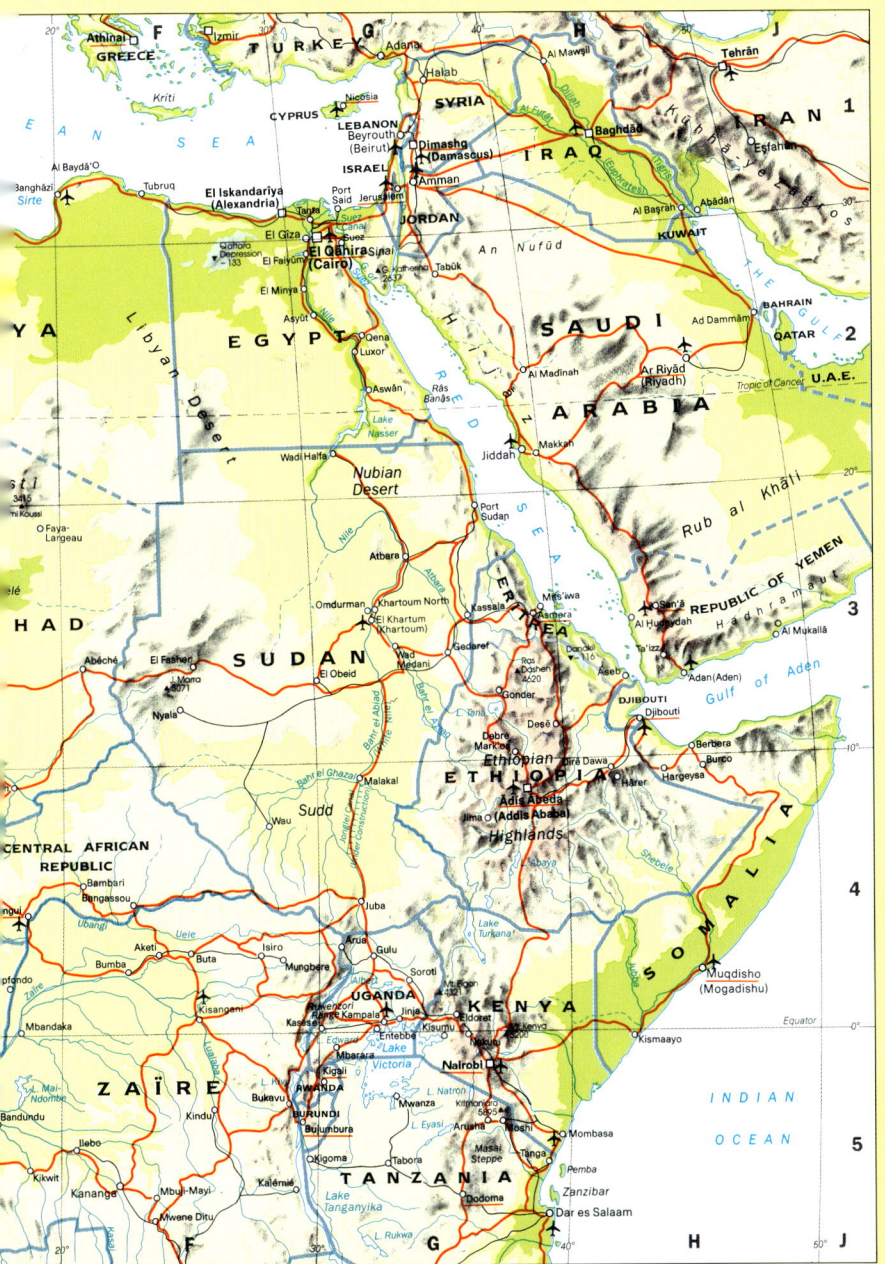

44 SOUTHERN AFRICA

AUSTRALASIA 45

46 OCEANIA Political

OCEANIA Political 47

POLAR REGIONS

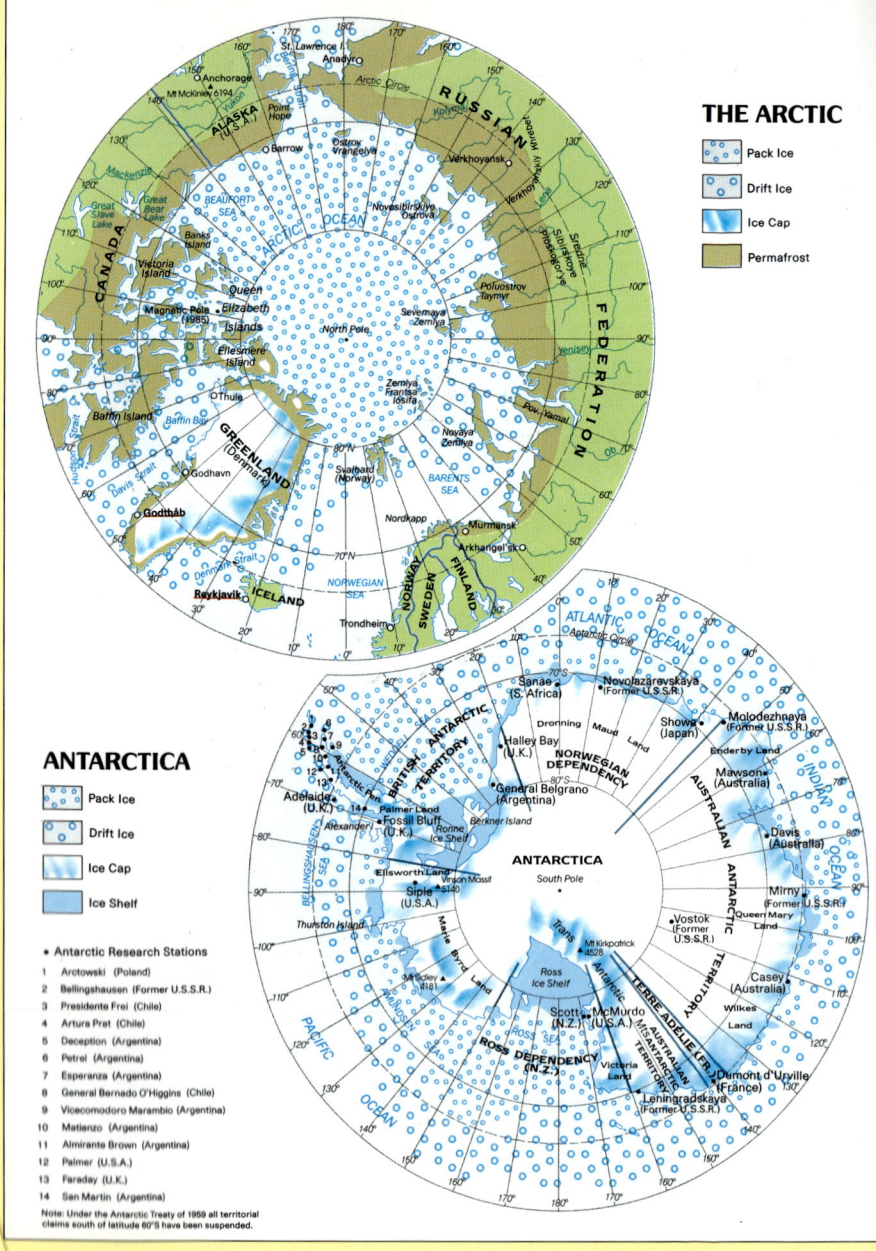